CHILTON'S

FUEL INJECTION & ELECTRONIC ENGINE CONTROLS–1988-90

NISSAN • SUBARU • SUZUKI • TOYOTA

CHILTON BOOK COMPANY

ONE OF THE ***ABC PUBLISHING COMPANIES,***
A PART OF ***CAPITAL CITIES/ABC, INC.***

Manufactured in USA

Chilton Way, Radnor, PA 19089
ISBN 0-8019-8044-5
1234567890 9876543210

HOW TO USE THIS MANUAL

For ease of use, this manual is divided into sections as follows:

SECTION 1 Basic Electricity
SECTION 2 Troubleshooting and Diagnosis
SECTION 3 Self-Diagnostic Systems
SECTION 4 Fuel Injection Systems

The **CONTENTS** summarize the subjects covered in each section.

To quickly locate the proper service section, use the application chart on the following pages. It references applicable **CAR AND TRUCK MODELS** and **SERVICE SECTIONS** for major electronic engine control systems.

It is recommended that the user be familiar with the applicable **GENERAL INFORMATION, SERVICE PRECAUTIONS** and **TROUBLESHOOTING AND DIAGNOSIS TECHNIQUES** before testing or servicing any engine control system.

Major service sections are grouped by vehicle manufacturer, with each engine control system subsection containing:

- **GENERAL INFORMATION** pertaining to the operation of the system, individual components and the overall logic by which components work together.

- **SERVICE PRECAUTIONS** (if any) of which the user should be aware to prevent injury or damage to the vehicle or components.

- **FAULT DIAGNOSIS** in the form of diagnostic charts or test procedures which lead the user through the various system circuit tests and explain the trouble codes stored in the computer memory.

SAFETY NOTICE

Proper service and repair procedures are vital to the safe, reliable operation of all motor vehicles, as well as the personal safety of those performing service or repairs. This manual outlines procedures for servicing and repairing vehicles using safe, effective methods. The procedures contain many NOTES and CAUTIONS which should be followed along with standard safety procedures to eliminate the possibility of personal injury or improper service which could damage the vehicle or compromise its safety.

It is important to note that repair procedures and techniques, tools and parts for servicing motor vehicles, as well as the skill and experience of the individual performing the work vary widely. It is not possible to anticipate all of the hazards that may result. Standard and accepted safety precautions and equipment should be used when handling toxic or flammable fluids and safety goggles or other protection should be used during cutting, grinding, chiseling, prying or any other process that that can cause material removal or projectiles. Similar protection against the high voltages generated in all electronic ignition systems should be employed during service procedures.

Some procedures require the use of tools or test equipment specially designed for a specific purpose. Before substituting another tool or procedure, you must be completely satisfied that neither your personal safety, nor the performance of the vehicle will be endangered.

PART NUMBERS

Part numbers listed in this reference are not recommendations by Chilton for any product by brand name. They are references that can be used with interchange manuals and aftermarket supplier catalogs to locate each brand supplier's discrete part number.

Although information in this manual is based on industry sources and is complete as possible at the time of publication, the possibilty exists that some car manufacturers made later changes which could not be included here. While striving for total accuracy, Chilton Book Company cannot assume responsibity for any errors, changes or omissions that may occur in the compilation of this data.

Contents

ENGINE CONTROL SYSTEM APPLICATION CHART
NISSAN

NISSAN					
Year	**Model**	**Engine cc liter**	**Engine code**	**Fuel system**	**Ignition system**
1988	200SX	1809 (1.8)	CA18ET ④	EFI ②	ECCS ①
		1974 (2.0)	CA20E ④	EFI	ECCS ①
		2960 (3.0)	VG30E ④	EFI	ECCS ①
	Pulsar	1597 (1.6)	E16I ③	EFI	ECCS ①
	Stanza	1809 (1.8)	CA18DE ④	EFI	ECCS ①
		1974 (2.0)	CA20A ④	EFI	ECCS ①
	300ZX	2960 (3.0)	VG30E ④	EFI	ECCS ①
		2960 (3.0)	VG30ET ④	EFI ②	ECCS ①
	Maxima	2960 (3.0)	VG30E ④	EFI	ECCS ①
	Sentra	1597 (1.6)	E16I ③	EFI	ECCS ①
	Pathfinder & Truck	2960 (3.0)	VG30I ③	EFI	ECCS ①
		2389 (2.4)	Z24I ③	EFI	ECCS ①
	Van	2389 (2.4)	Z24I ③	EFI	ECCS ①
1989	240SX	2389 (2.4)	KA24E ④	EFI	ECCS ①
	Pulsar	1597 (1.6)	GA16I ③	EFI	ECCS ①
	Stanza	1809 (1.8)	GA18DE ④	EFI	ECCS ①
		1974 (2.0)	CA20E ④	EFI	ECCS ①
	3002X	2960 (3.0)	VG30E ④	EFI	ECCS ①
		2960 (3.0)	VG30ET ④	EFI ②	ECCS ①
	Maxima	2960 (3.0)	VG30E ④	EFI	ECCS ①
	Sentra	1597 (1.6)	GA16T ③	EFI	ECCS ①
	Pathfinder & Truck	2960 (3.0)	VG30I ③	EFI	ECCS ①
		2389 (2.4)	Z24I ③	EFI	ECCS ①
1990	240SX	2389 (2.4)	KA24E ④	EFI	ECCS ①
	Pulsar	1597 (1.6)	GA16I ③	EFI	ECCS ①
	Stanza	2389 (2.4)	KA24E ④	EFI	ECCS ①
	300ZX	2960 (3.0)	VG30E ④	EFI	ECCS ①
		2960 (3.0)	VG30ET ②	EFI	ECCS ①
	Maxima	2960 (3.0)	VG30E ④	EFI	ECCS ①
	Sentra	1597 (1.6)	GA16I ③	EFI	ECCS ①
	Truck & Pathfinder	2960 (3.0)	VG30E ④	EFI	ECCS ①
		2389 (2.4)	KA24E ④	EFI	ECCS ①
	Axxess	2389 (2.4)	KA24E ④	EFI	ECCS ①

① Electronic concentrated control system
② Turbocharged engine
③ Throttle body injection
④ Port fuel injection

SUBARU

Year	Model	Engine cc liter	Engine code	Fuel system	Ignition system
1988	Justy	1189 (1.2)	7	carb	EEM
	Justy 4WD	1189 (1.2)	8	carb	EEM
	XT	1781 (1.8)	4	MPI	EEM
	XT6	2672 (2.7)	8	MPI	EEM
	XT6 4WD	2672 (2.7)	9	MPI	EEM
	Sedan	1781 (1.8)	4	SPI	EEM
	Station wagon	1781 (1.8)	5	SPI	EEM
	3 door	1781 (1.8)	4	SPI	EEM
	Sedan Turbo	1781 (1.8)	5	MPI	EEM
1989	Justy	1189 (1.2)	7	carb	EEM
	Justy 4WD	1189 (1.2)	8	carb	EEM
	XT	1781 (1.8)	4	MPI	EEM
	XT6	2672 (2.7)	8	MPI	EEM
	XT6 4WD	2672 (2.7)	9	MPI	EEM
	Sedan	1781 (1.8)	4	SPI	EEM
	Station wagon	1781 (1.8)	5	SPI	EEM
	3 door	1781 (1.8)	4	SPI	EEM
	Sedan Turbo	1781 (1.8)	5	MPI	EEM
1990	Justy	1189 (1.2)	7	2bbl-MPI	EEM
	Justy 4WD	1189 (1.2)	8	2bbl-MPI	EEM
	Justy 5 door	1189 (1.2)	8	MPI	EEM
	XT	1781 (1.8)	4	MPI	EEM
	XT6	2672 (2.7)	8	MPI	EEM
	Loyale sedan	1781 (1.8)	4	SPI	EEM
	Loyale wagon	1781 (1.8)	5	SPI	EEM
	Loyale coupe	1781 (1.8)	5	SPI	EEM
	Loyale Turbo	1781 (1.8)	7	MPI	EEM
	Legacy wagon	2200 (2.2)	6	MPI	DIS
	Legacy sedan	2200 (2.2)	6	MPI	DIS

MPI —multi port injection
SPI —single port injection
EEM—electronic engine management
DIS —distributorless ignition system

ENGINE CONTROL SYSTEM APPLICATION CHART
SUZUKI AND TOYOTA

SUZUKI

Year	Model	Engine cc liter	Engine code	Fuel system	Ignition system
1988	Samurai	1324 (1.3)	5	carb	Electronic
1989	Samurai	1324 (1.3)	5	carb	Electronic
	Swift GLX	1324 (1.3)	5	TBI	Electronic
	Swift GTI	1324 (1.3)	—	MPI	Electronic
	Sidekick	1298 (1.3)	5	carb	Electronic
	Sidekick	1590 (1.6)	0	TBI	Electronic
1990	Samurai	1324 (1.3)	5	TBI	Electronic
	Swift GL	1324 (1.3)	5	TBI	Electronic
	Swift GT	1324 (1.3)	—	MPI	Electronic
	Sidekick	1590 (1.6)	0	TBI	Electronic

TBI —Throttle body injection
MPI—Multi port injection

TOYOTA

Year	Model	Engine cc (liter)	Engine Code	Fuel System	Ignition System
1988	Camry	1998 (2.0)	3S-FE	EFI	ESA
	Celica	1998 (2.0)	3S-GE	EFI	ESA
		1998 (2.0)	3S-FE	EFI	ESA
	Corolla	1587 (1.6)	4A-GE	EFI	ESA
		1587 (1.6)	4A-C	Carb	IIA
		1587 (1.6)	4A-F	Carb	IIA
	Cressida	2759 (2.8)	5M-GE	EFI	ESA
	MR-2	1587 (1.6)	4A-GE	EFI	ESA
		1587 (1.6)	4A-GZE	EFI	ESA
	Supra	2954 (3.0)	7M-GE	EFI	ESA
		2954 (3.0)	7M-GTE	EFI	ESA
	Tercel	1452 (1.4)	3A-C	Carb	IIA
		1456 (1.4)	3E	Carb	IIA
	Van	2237 (2.2)	4Y-EC	EFI	ESA
	Truck & 4-Runner	2366 (2.4)	22R	Carb	ESA
		2366 (2.4)	22R-E	EFI	ESA
		2366 (2.4)	22R-TE	EFI	ESA
1989	Camry	1998 (2.0)	3S-FE	EFI	ESA
		2507 (2.5)	2VZ-FE	EFI	ESA
	Celica	1998 (2.0)	3S-FE	EFI	ESA
		1998 (2.0)	3S-GE	EFI	ESA
		1998 (2.0)	3S-GTE	EFI	ESA
	Corolla	1587 (1.6)	4A-F	Carb	IIA
		1587 (1.6)	4A-FE	EFI	IIA
		1587 (1.6)	4A-FE	EFI	ESA
		1587 (1.6)	4A-GE	EFI	ESA
	Cressida	2954 (3.0)	7M-GE	EFI	ESA

TOYOTA

Year	Model	Engine cc liter	Engine code	Fuel system	Ignition system
1989	MR-2	1587 (1.6)	4A-GE	EFI	ESA
		1587 (1.6)	4A-GZE	EFI	ESA
	Supra	2954 (3.0)	7M-GE	EFI	ESA
		2954 (3.0)	7M-GTE	EFI	ESA
	Tercel	1456 (1.4)	3E	Carb	IIA
	Van	2237 (2.2)	4Y-EC	EFI	ESA
	Truck & 4-Runner	2366 (2.4)	22R	Carb	ESA
		2366 (2.4)	22R-E	EFI	ESA
		2958 (3.0)	3VZ-E	EFI	ESA
1990	Camry	1998 (2.0)	3S-FE	EFI	ESA
		2507 (2.5)	2VZ-FE	EFI	ESA
	Celica	1587 (1.6)	4A-FE	EFI	ESA
		2164 (2.2)	5S-FE	EFI	ESA
		1998 (2.0)	3S-GTE	EFI	ESA
	Corolla	1587 (1.6)	4A-FE	EFI	ESA
		1587 (1.6)	4A-GE	EFI	ESA
	Cressida	2954 (3.0)	7M-GE	EFI	ESA
	MR-2	2954 (3.0)	7M-GE	EFI	ESA
		2954 (3.0)	7M-GTE	EFI	ESA
	Supra	2954 (3.0)	7M-GE	EFI	ESA
		2954 (3.0)	7M-GTE	EFI	ESA
	Tercel	1456 (1.4)	3E	Carb	IIA
		1456 (1.4)	3E-E	EFI	ESA
	Truck & 4-Runner	2366 (2.4)	22R	Carb	ESA
		2366 (2.4)	22R-E	EFI	ESA
		2958 (3.0)	3VZ-E	EFI	ESA

Basic Electricity

SECTION 1

INDEX

FUNDAMENTALS OF ELECTRICITY

A good understanding of basic electrical theory and how circuits work is necessary to successfully perform the service and testing outlined in this manual. Therefore, this section should be read before attempting any diagnosis and repair.

All matter is made up of tiny particles called molecules. Each molecule is made up of two or more atoms. Atoms may be divided into even smaller particles called protons, neutrons and electrons. These particles are the same in all matter and differences in materials (hard or soft, conductive or non-conductive) occur only because of the number and arrangement of these particles. In other words, the protons, neutrons and electrons in a drop of water are the same as those in an ounce of lead, there are just more of them (arranged differently) in a lead molecule than in a water molecule. Protons and neutrons packed together form the nucleus of the atom, while electrons orbit around the nucleus much the same way as the planets of the solar system orbit around the sun.

The proton is a small positive natural charge of electricity, while the neutron has no electrical charge. The electron carries a negative charge equal to the positive charge of the proton. Every electrically neutral atom contains the same number of protons and electrons, the exact number of which determines the element. The only difference between a conductor and an insulator is that a conductor possesses free electrons in large quantities, while an insulator has only a few. An element must have very few free electrons to be a good insulator and vice-versa. When we speak of electricity, we're talking about these free electrons.

In a conductor, the movement of the free electrons is hindered by collisions with the adjoining atoms of the element (matter). This hindrance to movement is called **RESISTANCE** and it varies with different materials and temperatures. As temperature increases, the movement of the free electrons increases, causing more frequent collisions and therefore increasing resistance to the movement of the electrons. The number of collisions (resistance) also increases with the number of electrons flowing (current). Current is defined as the movement of electrons through a conductor such as a wire. In a conductor (such as copper) electrons can be caused to leave their atoms and move to other atoms. This flow is continuous in that every time an atom gives up an electron, it collects another one to take its place. This movement of electrons is called electric current and is measured in amperes. When 6.28 billion, billion electrons pass a certain point in the circuit in one second, the amount of current flow is called 1 ampere.

The force or pressure which causes electrons to flow in any conductor (such as a wire) is called **VOLTAGE.** It is measured in volts and is similar to the pressure that causes water to flow in a pipe. Voltage is the difference in electrical pressure measured between 2 different points in a circuit. In a 12 volt system, for example, the force measured between the two battery posts is 12 volts. Two important concepts are voltage potential and polarity. Voltage potential is the amount of voltage or electrical pressure at a certain point in the circuit with respect to another point. For example, if the voltage potential at one post of the 12 volt battery is 0, the voltage potential at the other post is 12 volts with respect to the first post. One post of the battery is said to be positive (+); the other post is negative (−) and the conventional direction of current flow is from positive to negative in an electrical circuit. It should be noted that the electron flow in the wire is opposite the current flow. In other words, when the circuit is energized, the current flows from positive to negative, but the electrons actually move from negative to positive. The voltage or pressure needed to produce a current flow in a circuit must be greater than the resistance present in the circuit. In other words, if the voltage drop across the resistance is greater than or equal to the voltage input, the voltage potential will be zero—no voltage will flow through the circuit. Resistance to the flow of electrons is measured in ohms. One volt will cause 1 ampere to flow through a resistance of 1 ohm.

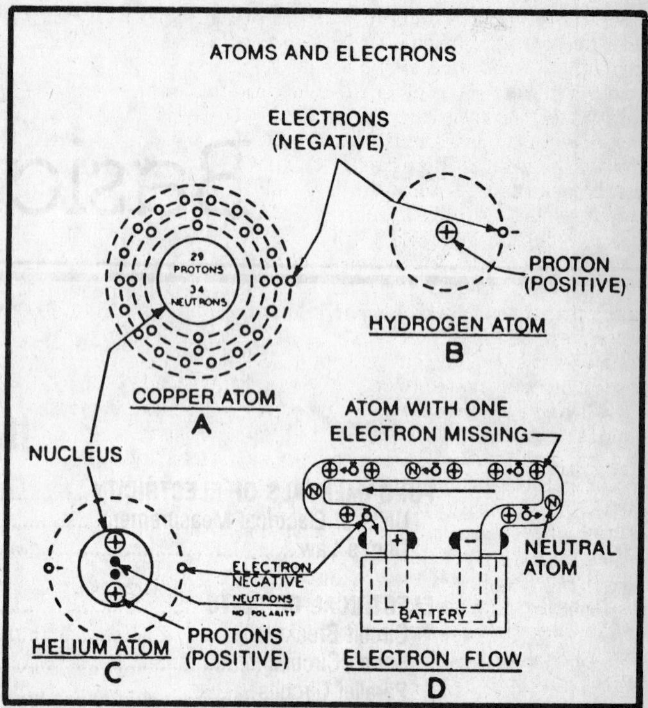
ATOMS AND ELECTRONS
ELECTRONS (NEGATIVE)
29 PROTONS
34 NEUTRONS
PROTON (POSITIVE)
HYDROGEN ATOM
B
COPPER ATOM
A
ATOM WITH ONE ELECTRON MISSING
NUCLEUS
NEUTRAL ATOM
ELECTRON NEGATIVE
NEUTRONS NO POLARITY
BATTERY
PROTONS (POSITIVE)
HELIUM ATOM
C
ELECTRON FLOW
D

Typical atoms of copper (A), hydrogen (B) and helium (C). Electron flow in battery circuit (D)

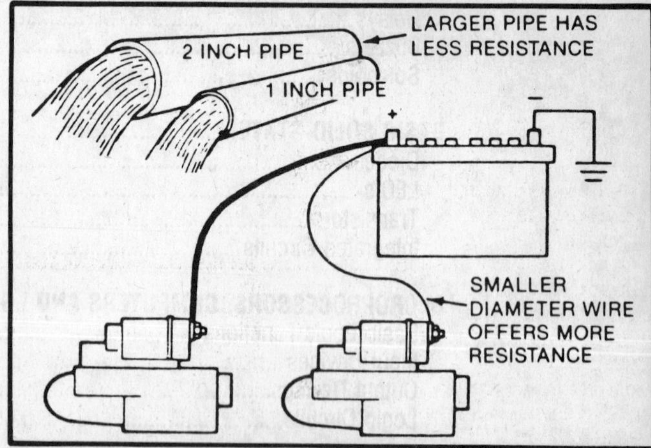
2 INCH PIPE
LARGER PIPE HAS LESS RESISTANCE
1 INCH PIPE
SMALLER DIAMETER WIRE OFFERS MORE RESISTANCE

Electrical resistance can be compared to water flow through a pipe. The smaller the wire (pipe), the more resistance to the flow of electrons (water)

Units Of Electrical Measurement

There are 3 fundamental characteristics of a direct-current electrical circuit: volts, amperes and ohms.

VOLTAGE in a circuit controls the intensity with which the loads in the circuit operate. The brightness of a lamp, the heat of an electrical defroster, the speed of a motor are all directly proportional to the voltage, if the resistance in the circuit and/or mechanical load on electric motors remains constant. Voltage available from the battery is constant (normally 12 volts), but as it operates the various loads in the circuit, voltage decreases (drops).

AMPERE is the unit of measurement of current in an electrical circuit. One ampere is the quantity of current that will flow through a resistance of 1 ohm at a pressure of 1 volt. The amount of current that flows in a circuit is controlled by the voltage and the resistance in the circuit. Current flow is directly proportional to resistance. Thus, as voltage is increased or decreased, current is increased or decreased accordingly. Current is decreased as resistance is increased. However, current is also increased as resistance is decreased. With little or no resistance in a circuit, current is high.

OHM is the unit of measurement of resistance, represented by the Greek letter Omega (Ω). One ohm is the resistance of a conductor through which a current of one ampere will flow at a pressure of one volt. Electrical resistance can be measured on an instrument called an ohmmeter. The loads (electrical devices) are the primary resistances in a circuit. Loads such as lamps, solenoids and electric heaters have a resistance that is essentially fixed; at a normal fixed voltage, they will draw a fixed current. Motors, on the other hand, do not have a fixed resistance. Increasing the mechanical load on a motor (such as might be caused by a misadjusted track in a power window system) will decrease the motor speed. The drop in motor rpm has the effect of reducing the internal resistance of the motor because the current draw of the motor varies directly with the mechanical load on the motor, although its actual resistance is unchanged. Thus, as the motor load increases, the current draw of the motor increases, and may increase up to the point where the motor stalls (cannot move the mechanical load).

Circuits are designed with the total resistance of the circuit taken into account. Troubles can arise when unwanted resistances enter into a circuit. If corrosion, dirt, grease, or any other contaminant occurs in places like switches, connectors and grounds, or if loose connections occur, resistances will develop in these areas. These resistances act like additional loads in the circuit and cause problems.

OHM'S LAW

Ohm's law is a statement of the relationship between the 3 fundamental characteristics of an electrical circuit. These rules apply to direct current (DC) only.

Ohm's law provides a means to make an accurate circuit analysis without actually seeing the circuit. If, for example, one wanted to check the condition of the rotor winding in a alternator whose specifications indicate that the field (rotor) current draw is normally 2.5 amperes at 12 volts, simply connect the rotor to a 12 volt battery and measure the current with an ammeter. If it measures about 2.5 amperes, the rotor winding can be assumed good.

An ohmmeter can be used to test components that have been removed from the vehicle in much the same manner as an ammeter. Since the voltage and the current of the rotor windings used as an earlier example are known, the resistance can be calculated using Ohms law. The formula would be ohms equals volts divided by amperes.

If the rotor resistance measures about 4.8 ohms when checked with an ohmmeter, the winding can be assumed good. By plugging in different specifications, additional circuit information can be determined such as current draw, etc.

Electrical Circuits

An electrical circuit must start from a source of electrical supply and return to that source through a continuous path. Circuits are designed to handle a certain maximum current flow. The

E / I | R

$$I = \frac{E}{R} \quad \text{or} \quad \text{AMPERES} = \frac{\text{VOLTS}}{\text{OHMS}}$$

$$R = \frac{E}{I} \quad \text{or} \quad \text{OHMS} = \frac{\text{VOLTS}}{\text{AMPERES}}$$

$$E = I \times R \quad \text{or} \quad \text{VOLTS} = \text{AMPERES} \times \text{OHMS}$$

Ohms Law is the basis for all electrical measurements. By simply plugging in two values, the third can be calculated using the illustrated formula.

$$R = \frac{E}{I} \qquad \text{Where: } E = 12 \text{ volts}, \quad I = 2.5 \text{ amperes}$$

$$R = \frac{12 \text{ volts}}{2.5 \text{ amps}} = 4.8 \text{ ohms}$$

An example of calculating resistance (R) when the voltage (E) and amperage (I) is known.

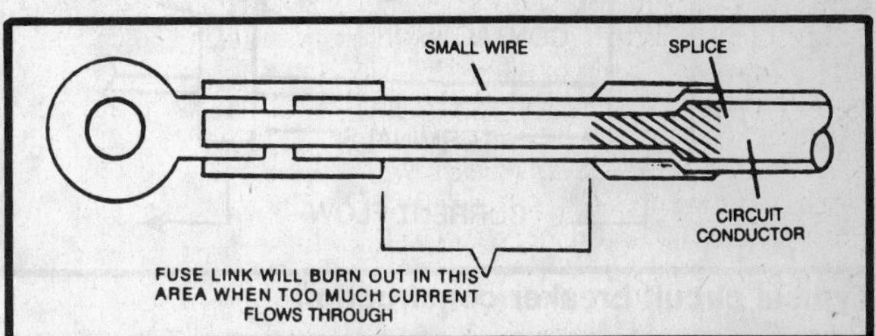
SMALL WIRE
SPLICE
CIRCUIT CONDUCTOR
FUSE LINK WILL BURN OUT IN THIS AREA WHEN TOO MUCH CURRENT FLOWS THROUGH

Typical fusible link wire

maximum allowable current flow is designed higher than the normal current requirements of all the loads in the circuit. Wire size, connections, insulation, etc., are designed to prevent undesirable voltage drop, overheating of conductors, arcing of contacts and other adverse effects. If the safe maximum current flow level is exceeded, damage to the circuit components will result; it is this condition that circuit protection devices are designed to prevent.

Protection devices are fuses, fusible links or circuit breakers designed to open or break the circuit quickly whenever an overload, such as a short circuit, occurs. By opening the circuit quickly, the circuit protection device prevents damage to the wiring, battery and other circuit components. Fuses and fusible links are designed to carry a preset maximum amount of current and to melt when that maximum is exceeded, while circuit breakers merely break the connection and may be manually reset. The maximum amperage rating of each fuse is marked on the fuse body and all contain a see-through portion that shows the break in the fuse element when blown. Fusible link maximum amperage rating is indicated by gauge or thickness of the wire. Never replace a blown fuse or fusible link with one of a higher amperage rating.

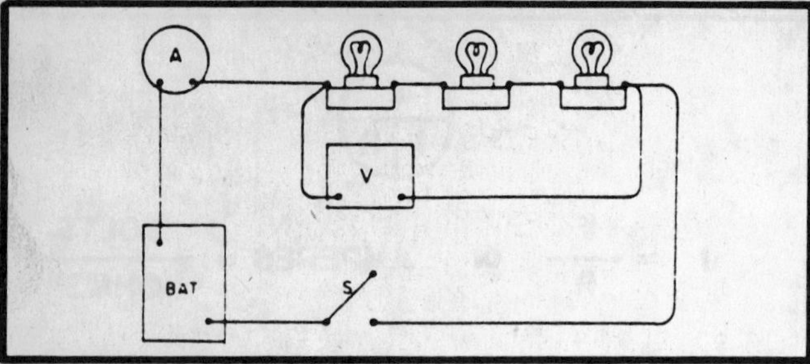
A
V
BAT
S
Example of a series circuit

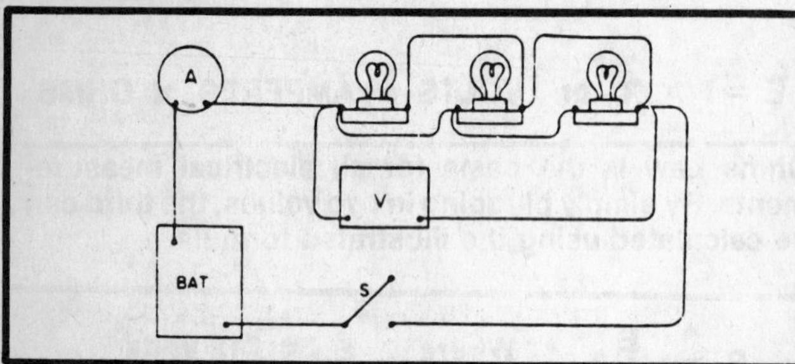
A
V
BAT
S
Example of a parallel circuit

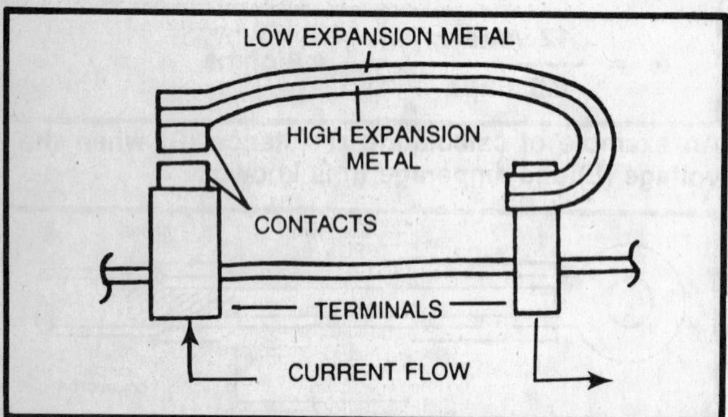
LOW EXPANSION METAL
HIGH EXPANSION METAL
CONTACTS
TERMINALS
CURRENT FLOW
Typical circuit breaker construction

CAUTION

Resistance wires, like fusible links, are also spliced into conductors in some areas. Do not make the mistake of replacing a fusible link with a resistance wire. Resistance wires are longer than fusible links and are stamped "RESISTOR-DO NOT CUT OR SPLICE."

Circuit breakers consist of 2 strips of metal which have different coefficients of expansion. As an overload or current flows through the bimetallic strip, the high-expansion metal will elongate due to heat and break the contact. With the circuit open, the bimetal strip cools and shrinks, drawing the strip down until contact is re-established and current flows once again. In actual operation, the contact is broken very quickly if the overload is continuous and the circuit will be repeatedly broken and remade until the source of the overload is corrected.

The self-resetting type of circuit breaker is the one most generally used in automotive electrical systems. On manually reset circuit breakers, a button will pop up on the circuit breaker case. This button must be pushed in to reset the circuit breaker and restore power to the circuit. Always repair the source of the overload before resetting a circuit breaker or replacing a fuse or fusible link. When searching for overloads, keep in mind that the circuit protection devices protect only against overloads between the protection device and ground.

There are 2 basic types of circuit; Series and Parallel. In a se-

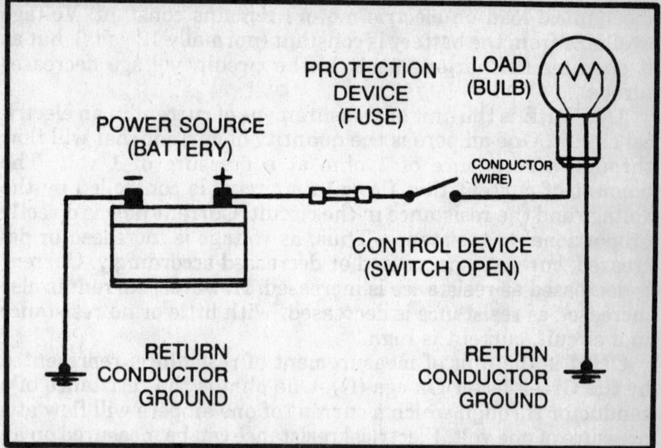
PROTECTION DEVICE (FUSE)
LOAD (BULB)
POWER SOURCE (BATTERY)
CONDUCTOR (WIRE)
CONTROL DEVICE (SWITCH OPEN)
RETURN CONDUCTOR GROUND
RETURN CONDUCTOR GROUND
Typical circuit with all essential components

ries circuit, all of the elements are connected in chain fashion with the same amount of current passing through each element or load. No matter where an ammeter is connected in a series circuit, it will always read the same. The most important fact to remember about a series circuit is that the sum of the voltages across each element equals the source voltage. The total resistance of a series circuit is equal to the sum of the individual resistances within each element of the circuit. Using ohms law, one can determine the voltage drop across each element in the circuit. If the total resistance and source voltage is known, the amount of current can be calculated. Once the amount of current (amperes) is known, values can be substituted in the Ohms law formula to calculate the voltage drop across each individual element in the series circuit. The individual voltage drops must add up to the same value as the source voltage.

A parallel circuit, unlike a series circuit, contains 2 or more branches, each branch a separate path independent of the others. The total current draw from the voltage source is the sum of all the currents drawn by each branch. Each branch of a parallel circuit can be analyzed separately. The individual branches can be either simple circuits, series circuits or combinations of series-parallel circuits. Ohms law applies to parallel circuits just as it applies to series circuits, by considering each branch independently of the others. The most important thing to remember is that the voltage across each branch is the same as the source voltage. The current in any branch is that voltage divided by the resistance of the branch. A practical method of determining the resistance of a parallel circuit is to divide the product of the 2 resistances by the sum of 2 resistances at a time. Amperes through a parallel circuit is the sum of the amperes through the separate branches. Voltage across a parallel circuit is the same as the voltage across each branch.

By measuring the voltage drops the resistance of each element within the circuit is being measured. The greater the voltage drop, the greater the resistance. Voltage drop measurements are a common way of checking circuit resistances in automotive electrical systems. When part of a circuit develops excessive resistance (due to a bad connection) the element will show a higher than normal voltage drop. Normally, automotive wiring is selected to limit voltage drops to a few tenths of a volt. In parallel circuits, the total resistance is less than the sum of the individual resistances; because the current has 2 paths to take, the total resistance is lower.

Magnetism and Electromagnets

Electricity and magnetism are very closely associated because when electric current passes through a wire, a magnetic field is created around the wire. When a wire carrying electric current

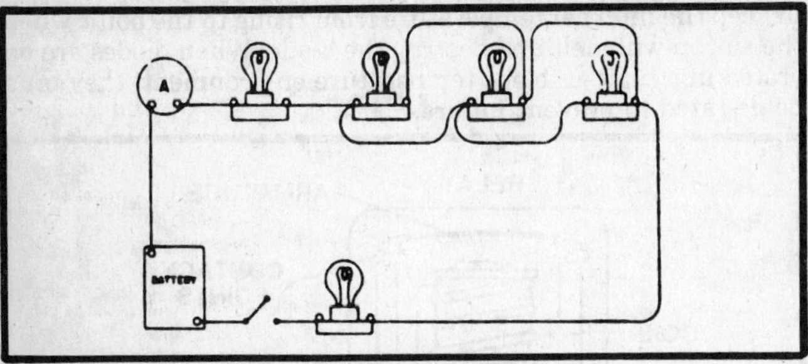
A
BATTERY
Example of a series-parallel circuit

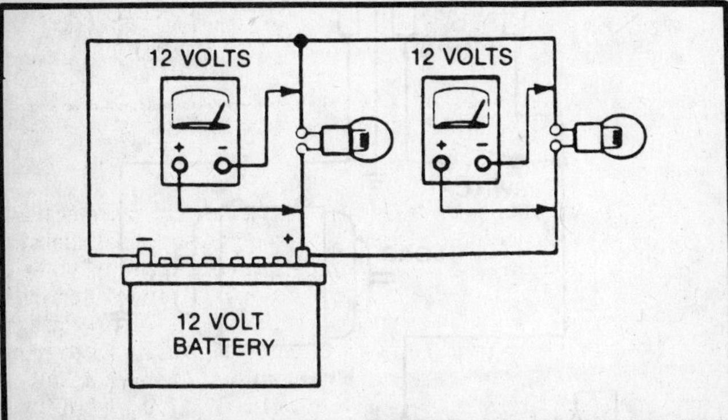
12 VOLTS
12 VOLTS
12 VOLT
BATTERY
Voltage drop in a parallel circuit. Voltage drop across each lamp is 12 volts

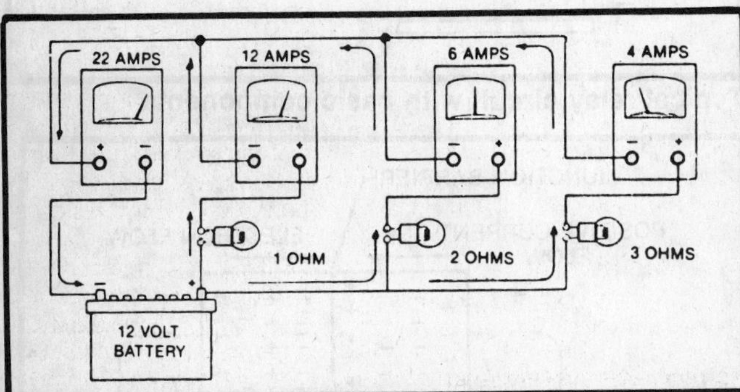
22 AMPS
12 AMPS
6 AMPS
4 AMPS
1 OHM
2 OHMS
3 OHMS
12 VOLT
BATTERY
Total current in parallel circuit: 4 + 6 + 12 = 22 amps

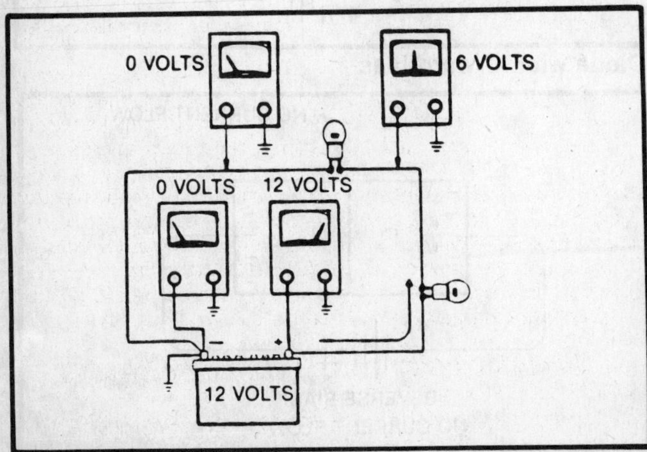
0 VOLTS
6 VOLTS
0 VOLTS
12 VOLTS
12 VOLTS
Voltage drop in a series circuit

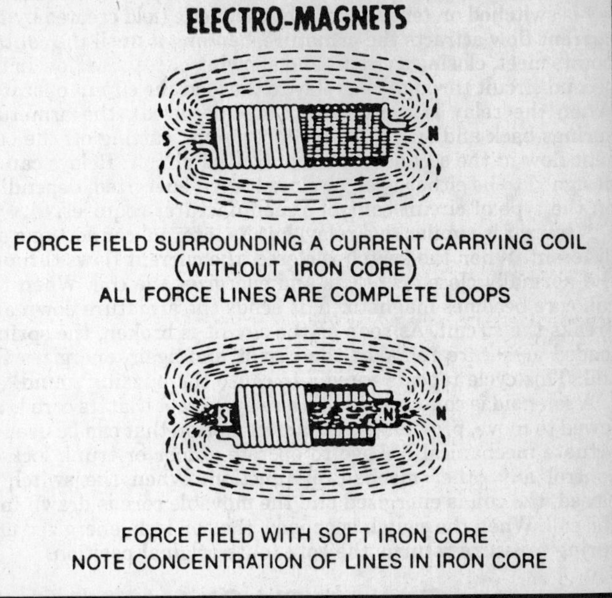
ELECTRO-MAGNETS
FORCE FIELD SURROUNDING A CURRENT CARRYING COIL
(WITHOUT IRON CORE)
ALL FORCE LINES ARE COMPLETE LOOPS
FORCE FIELD WITH SOFT IRON CORE
NOTE CONCENTRATION OF LINES IN IRON CORE
Magnetic field surrounding an electromagnet

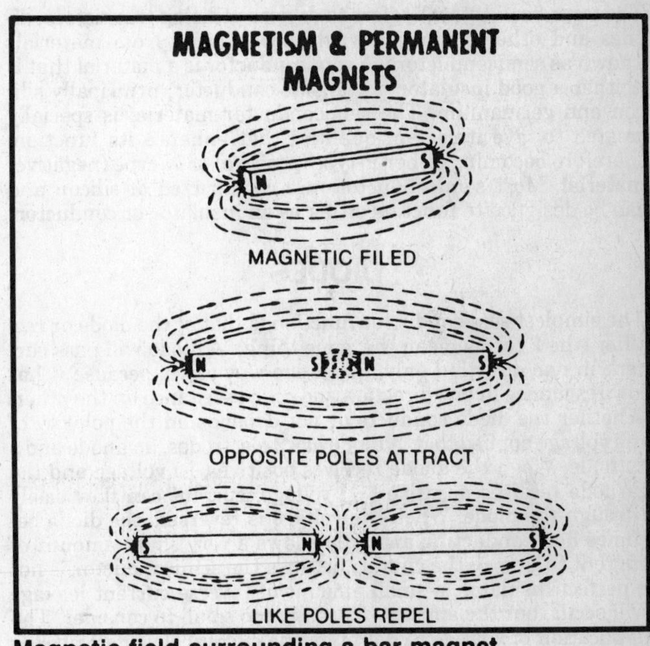
MAGNETISM & PERMANENT MAGNETS
MAGNETIC FILED
OPPOSITE POLES ATTRACT
LIKE POLES REPEL
Magnetic field surrounding a bar magnet

is wound into a coil, a magnetic field with North and South poles is created just like in a bar magnet. If an iron core is placed within the coil, the magnetic field becomes stronger because iron conducts magnetic lines much easier than air. This arrangement is called an electromagnet and is the basic principle behind the operation of such components as relays, buzzers and solenoids.

A relay is basically just a remote-controlled switch that uses a small amount of current to control the flow of a large amount of current. The simplest relay contains an electromagnetic coil in series with a voltage source (battery) and a switch. A movable armature made of some magnetic material pivots at one end and is held a small distance away from the electromagnet by a spring or the spring steel of the armature itself. A contact point, made of a good conductor, is attached to the free end of the armature with another contact point a small distance away. When the relay is switched on (energized), the magnetic field created by the current flow attracts the armature, bending it until the contact points meet, closing a circuit and allowing current to flow in the second circuit through the relay to the load the circuit operates. When the relay is switched off (de-energized), the armature springs back and opens the contact points, cutting off the current flow in the secondary, or controlled, circuit. Relays can be designed to be either open or closed when energized, depending on the type of circuit control a manufacturer requires.

A buzzer is similar to a relay, but its internal connections are different. When the switch is closed, the current flows through the normally closed contacts and energizes the coil. When the coil core becomes magnetized, it bends the armature down and breaks the circuit. As soon as the circuit is broken, the spring-loaded armature remakes the circuit and again energizes the coil. This cycle repeats rapidly to cause the buzzing sound.

A solenoid is constructed like a relay, except that its core is allowed to move, providing mechanical motion that can be used to actuate mechanical linkage to operate a door or trunk lock or control any other mechanical function. When the switch is closed, the coil is energized and the movable core is drawn into the coil. When the switch is opened, the coil is de-energized and spring pressure returns the core to its original position.

Basic Solid State

The term "solid state" refers to devices utilizing transistors, diodes and other components which are made from materials known as semiconductors. A semiconductor is a material that is neither a good insulator nor a good conductor; principally silicon and germanium. The semiconductor material is specially treated to give it certain qualities that enhance its function, therefore becoming either P-type (positive) or N-type (negative) material. Most semiconductors are constructed of silicon and can be designed to function either as an insulator or conductor.

DIODES

The simplest semiconductor function is that of the diode or rectifier (the 2 terms mean the same thing). A diode will pass current in one direction only, like a one-way valve, because it has low resistance in one direction and high resistance on the other. Whether the diode conducts or not depends on the polarity of the voltage applied to it. A diode has 2 electrodes, an anode and a cathode. When the anode receives positive (+) voltage and the cathode receives negative (−) voltage, current can flow easily through the diode. When the voltage is reversed, the diode becomes non-conducting and only allows a very slight amount of current to flow in the circuit. Because the semiconductor is not a perfect insulator, a small amount of reverse current leakage will occur, but the amount is usually too small to consider. The application of voltage to maintain the current flow described is called "forward bias."

A light-emitting diode (LED) is made of a particular type of crystal that glows when current is passed through it. LED's are used in display faces of many digital or electronic instrument clusters. LED's are usually arranged to display numbers (digital readout), but can be used to illuminate a variety of electronic graphic displays.

Like any other electrical device, diodes have certain ratings that must be observed and should not be exceeded. The forward current rating (or bias) indicates how much current can safely pass through the diode without causing damage or destroying it. Forward current rating is usually given in either amperes or milliamperes. The voltage drop across a diode remains constant regardless of the current flowing through it. Small diodes designed to carry low amounts of current need no special provision for dissipating the heat generated in any electrical device, but large current carrying diodes are usually mounted on heat sinks to keep the internal temperature from rising to the point where the silicon will melt and destroy the diode. When diodes are operated in a high ambient temperature environment, they must be de-rated to prevent failure.

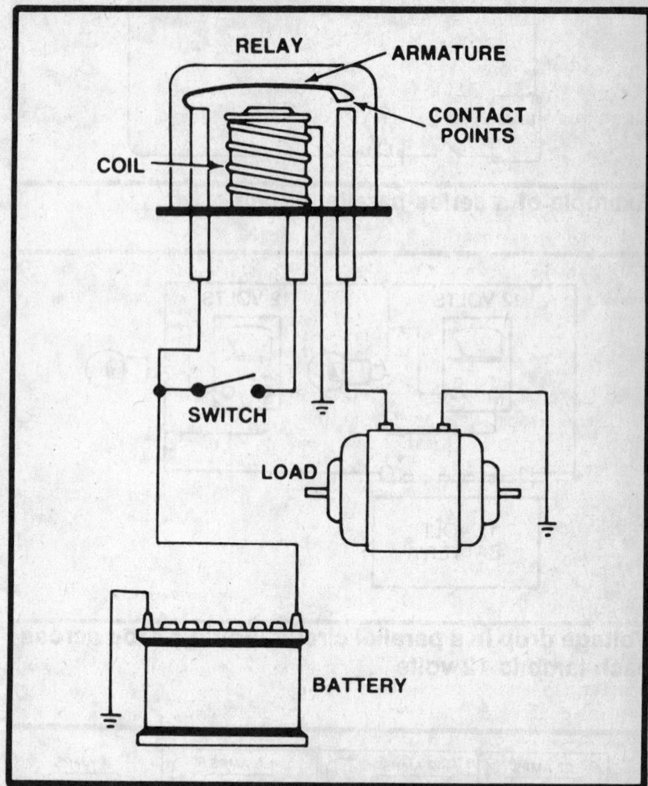
RELAY
ARMATURE
CONTACT POINTS
COIL
SWITCH
LOAD
BATTERY

Typical relay circuit with basic components

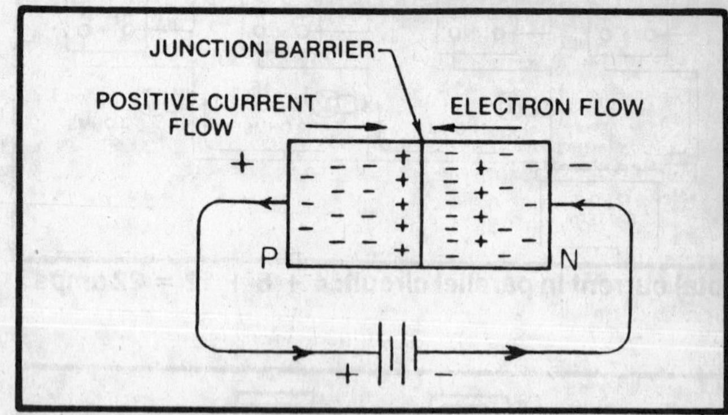
JUNCTION BARRIER
POSITIVE CURRENT FLOW
ELECTRON FLOW
+
−
P
N

Diode with forward bias

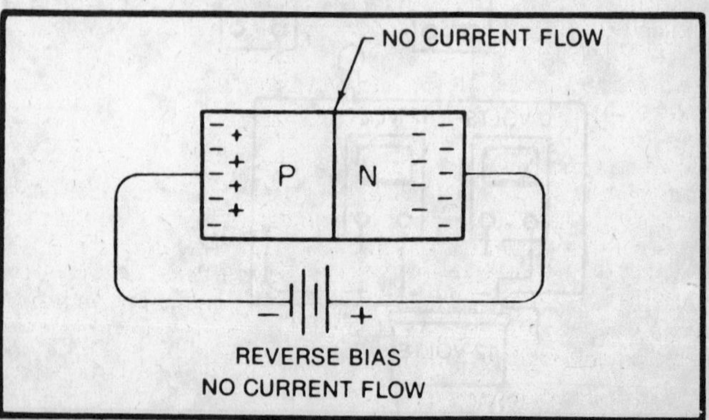
NO CURRENT FLOW
P
N
−
+
REVERSE BIAS
NO CURRENT FLOW

Diode with reverse bias

Another diode specification is its peak inverse voltage rating. This value is the maximum amount of voltage the diode can safely handle when operating in the blocking mode. This value can be anywhere from 50–1000 volts, depending on the diode. If voltage amount is exceeded, it will damage the diode just as too much forward current will. Most semiconductor failures are caused by excessive voltage or internal heat.

One can test a diode with a small battery and a lamp with the same voltage rating. With this arrangement one can find a bad diode and determine the polarity of a good one. A diode can fail and cause either a short or open circuit, but in either case it fails to function as a diode. Testing is simply a matter of connecting the test bulb first in one direction and then the other and making sure that current flows in one direction only. If the diode is shorted, the test bulb will remain on no matter how the light is connected.

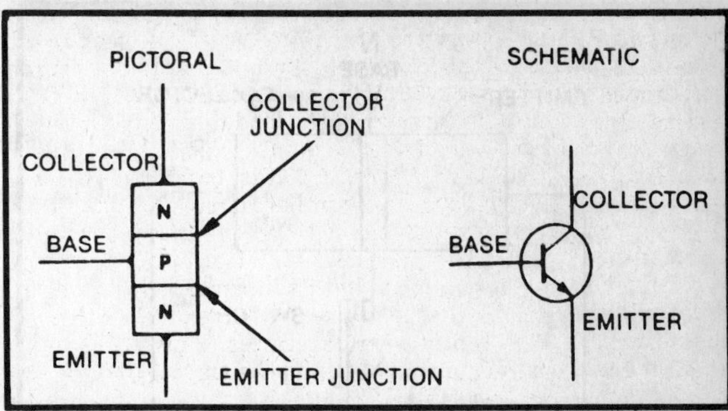
PICTORAL
SCHEMATIC
COLLECTOR JUNCTION
COLLECTOR
N
BASE
P
N
EMITTER
EMITTER JUNCTION
COLLECTOR
BASE
EMITTER
NPN transistor illustrations (pictorial and schematic)

TRANSISTORS

The transistor is an electrical device used to control voltage within a circuit. A transistor can be considered a "controllable diode" in that, in addition to passing or blocking current, the transistor can control the amount of current passing through it. Simple transistors are composed of 3 pieces of semiconductor material, P and N type, joined together and enclosed in a container. If 2 sections of P material and 1 section of N material are used, it is known as a PNP transistor; if the reverse is true, then it is known as an NPN transistor. The 2 types cannot be interchanged.

Most modern transistors are made from silicon (earlier transistors were made from germanium) and contain 3 elements; the emitter, the collector and the base. In addition to passing or blocking current, the transistor can control the amount of current passing through it and because of this can function as an amplifier or a switch. The collector and emitter form the main current-carrying circuit of the transistor. The amount of current that flows through the collector-emitter junction is controlled by the amount of current in the base circuit. Only a small amount of base-emitter current is necessary to control a large amount of collector-emitter current (the amplifier effect). In automotive applications, however, the transistor is used primarily as a switch.

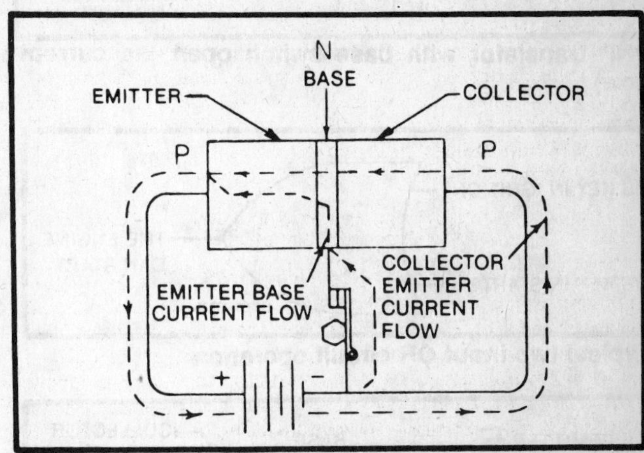
N
BASE
EMITTER
COLLECTOR
P
P
COLLECTOR EMITTER CURRENT FLOW
EMITTER BASE CURRENT FLOW
+
−
PNP transistor with base switch closed (base emitter and collector emitter current flow)

When no current flows in the base-emitter junction, the collector-emitter circuit has a high resistance, like to open contacts of a relay. Almost no current flows through the circuit and transistor is considered OFF. By bypassing a small amount of current into the base circuit, the resistance is low, allowing current to flow through the circuit and turning the transistor ON. This condition is known as "saturation" and is reached when the base current reaches the maximum value designed into the transistor that allows current to flow. Depending on various factors, the transistor can turn on and off (go from cutoff to saturation) in less than one millionth of a second.

Much of what was said about ratings for diodes applies to transistors, since they are constructed of the same materials. When transistors are required to handle relatively high currents, such as in voltage regulators or ignition systems, they are generally mounted on heat sinks in the same manner as diodes. They can be damaged or destroyed in the same manner if their voltage ratings are exceeded. A transistor can be checked for proper operation by measuring the resistance with an ohmmeter between the base-emitter terminals and then between the base-collector terminals. The forward resistance should be small, while the reverse resistance should be large. Compare the readings with those from a known good transistor. As a final check, measure the forward and reverse resistance between the collector and emitter terminals.

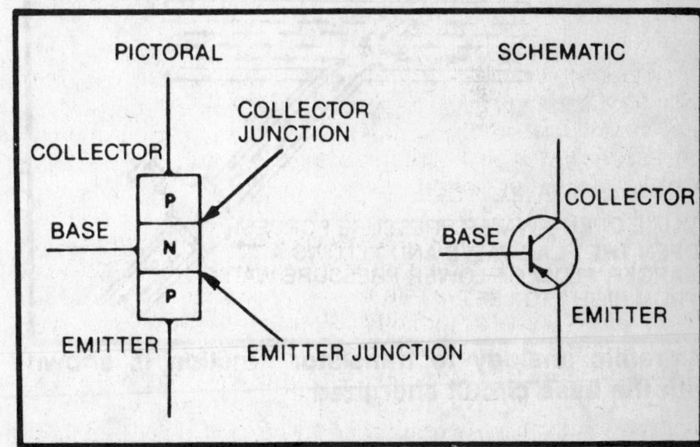
PICTORAL
SCHEMATIC
COLLECTOR JUNCTION
COLLECTOR
P
BASE
N
P
EMITTER
EMITTER JUNCTION
COLLECTOR
BASE
EMITTER
PNP transistor illustrations (pictorial and schematic)

INTEGRATED CIRCUITS

The integrated circuit (IC) is an extremely sophisticated solid state device that consists of a silicone wafer (or chip) which has been doped, insulated and etched many times so that it contains an entire electrical circuit with transistors, diodes, conductors and capacitors miniaturized within each tiny chip. Integrated circuits are often referred to as "computers on a chip" and are largely responsible for the current boom in electronic control technology.

Microprocessors, Computers and Logic Systems

Mechanical or electromechanical control devices lack the precision necessary to meet the requirements of modern control standards. They do not have the ability to respond to a variety of

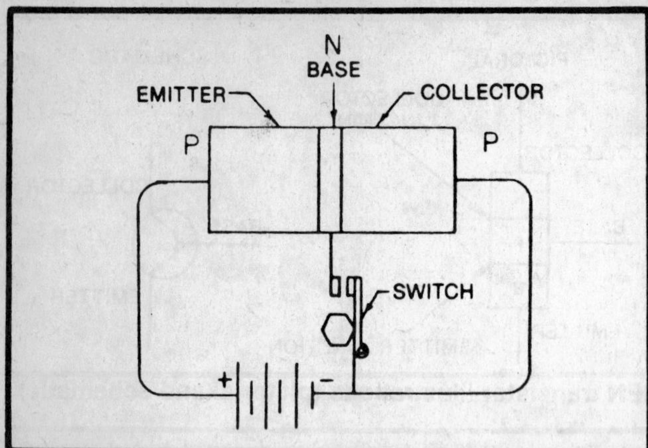
N
BASE
EMITTER
COLLECTOR
P
P
SWITCH
+
−

PNP transistor with base switch open (no current flow)

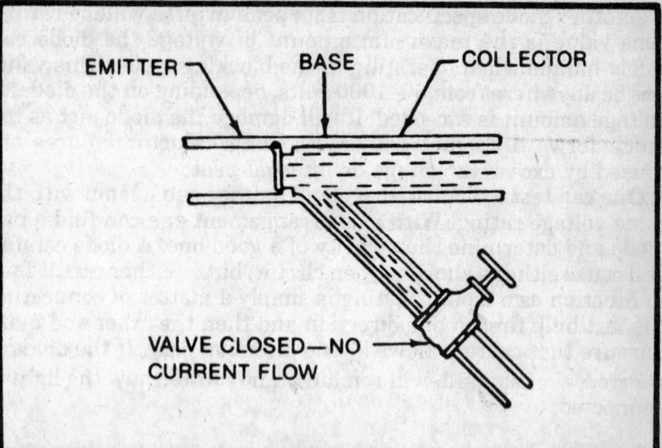
EMITTER
BASE
COLLECTOR
VALVE CLOSED—NO
CURRENT FLOW

Hydraulic analogy to transistor function is shown with the base circuit shut off

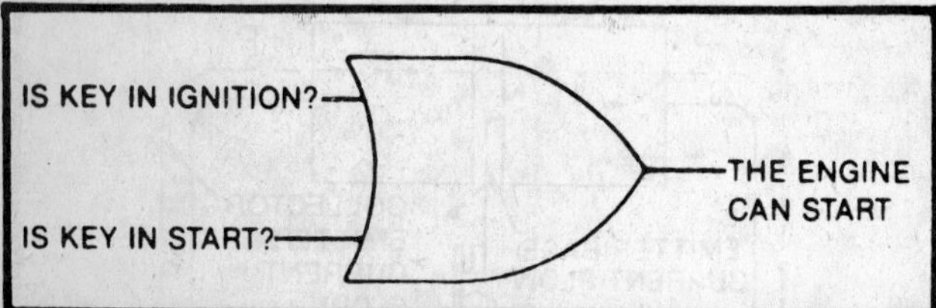
IS KEY IN IGNITION?
IS KEY IN START?
THE ENGINE
CAN START

Typical two-input OR circuit operation

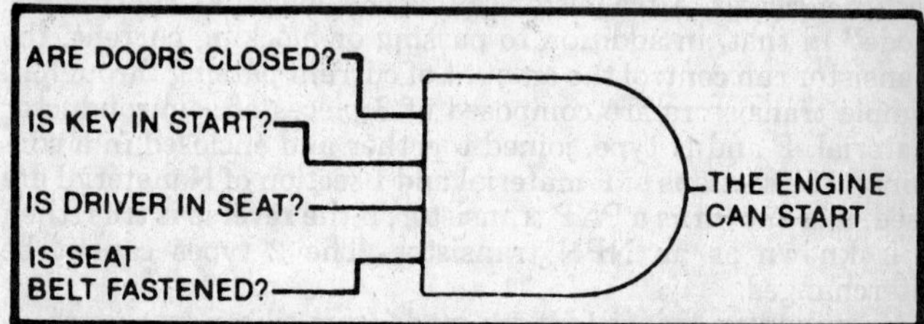
ARE DOORS CLOSED?
IS KEY IN START?
IS DRIVER IN SEAT?
IS SEAT
BELT FASTENED?
THE ENGINE
CAN START

Multiple input AND operation in a typical automotive starting circuit

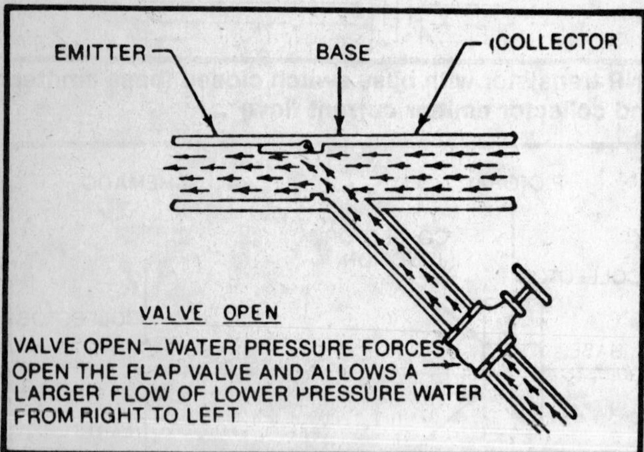
EMITTER
BASE
COLLECTOR
VALVE OPEN
VALVE OPEN—WATER PRESSURE FORCES
OPEN THE FLAP VALVE AND ALLOWS A
LARGER FLOW OF LOWER PRESSURE WATER
FROM RIGHT TO LEFT

Hydraulic analogy to transistor function is shown with the base circuit energized

input conditions common to antilock brakes, climate control and electronic suspension operation. To meet these requirements, manufacturers have gone to solid state logic systems and microprocessors to control the basic functions of suspension, brake and temperature control, as well as other systems and accessories.

One of the more vital roles of microprocessor-based systems is their ability to perform logic functions and make decisions. Logic designers use a shorthand notation to indicate whether a voltage is present in a circuit (the number 1) or not present (the number 0). Their systems are designed to respond in different ways depending on the output signal (or the lack of it) from various control devices.

There are 3 basic logic functions or "gates" used to construct a microprocessor control system: the AND gate, the OR gate or the NOT gate. Stated simply, the AND gate works when voltage is present in 2 or more circuits which then energize a third (A and B energize C). The OR gate works when voltage is present at either circuit A or circuit B which then energizes circuit C. The NOT function is performed by a solid state device called an "inverter" which reverses the input from a circuit so that, if voltage is going in, no voltage comes out and vice versa. With these three basic building blocks, a logic designer can create complex systems easily. In actual use, a logic or decision making system may employ many logic gates and receive inputs from a number of sources (sensors), but for the most part, all utilize the basic logic gates discussed above.

Stripped to its bare essentials, a computerized decision-making system is made up of three subsystems:

a. Input devices (sensors or switches)
b. Logic circuits (computer control unit)
c. Output devices (actuators or controls)

The input devices are usually nothing more than switches or sensors that provide a voltage signal to the control unit logic circuits that is read as a 1 or 0 (on or off) by the logic circuits. The output devices are anything from a warning light to solenoid-operated valves, motors, linkage, etc. In most cases, the logic circuits themselves lack sufficient output power to operate these devices directly. Instead, they operate some intermediate device such as a relay or power transistor which in turn operates the appropriate device or control. Many problems diagnosed as computer failures are really the result of a malfunctioning intermediate device like a relay. This must be kept in mind whenever troubleshooting any microprocessor-based control system.

The logic systems discussed above are called "hardware" systems, because they consist only of the physical electronic components (gates, resistors, transistors, etc.). Hardware systems do not contain a program and are designed to perform specific or "dedicated" functions which cannot readily be changed. For many simple automotive control requirements, such dedicated logic systems are perfectly adequate. When more complex logic functions are required, or where it may be desirable to alter these functions (e.g. from one model vehicle to another) a true computer system is used. A computer can be programmed through its software to perform many different functions and, if that program is stored on a separate integrated circuit chip called a ROM (Read Only Memory), it can be easily changed simply by plugging in a different ROM with the desired program. Most on-board automotive computers are designed with this capability. The on-board computer method of engine control offers the manufacturer a flexible method of responding to data from a variety of input devices and of controlling an equally large variety of output controls. The computer response can be changed quickly and easily by simply modifying its software program.

The microprocessor is the heart of the microcomputer. It is the thinking part of the computer system through which all the data from the various sensors passes. Within the microprocessor, data is acted upon, compared, manipulated or stored for future use. A microprocessor is not necessarily a microcomputer, but the differences between the 2 are becoming very minor. Originally, a microprocessor was a major part of a microcomputer, but nowadays microprocessors are being called "single-chip microcomputers". They contain all the essential elements to make them behave as a computer, including the most important ingredient–the program.

All computers require a program. In a general purpose computer, the program can be easily changed to allow different tasks to be performed. In a "dedicated" computer, such as most on-board automotive computers, the program isn't quite so easily altered. These automotive computers are designed to perform one or several specific tasks, such as maintaining the passenger compartment temperature at a specific, predetermined level. A program is what makes a computer smart; without a program a computer can do absolutely nothing. The term "software" refers to the computer's program that makes the hardware preform the function needed.

The software program is simply a listing in sequential order of the steps or commands necessary to make a computer perform the desired task. Before the computer can do anything at all, the program must be fed into it by one of several possible methods. A computer can never be "smarter" than the person programming it, but it is a lot faster. Although it cannot perform any calculation or operation that the programmer himself cannot perform, its processing time is measured in millionths of a second.

Because a computer is limited to performing only those operations (instructions) programmed into its memory, the program must be broken down into a large number of very simple steps. Two different programmers can come up with 2 different programs, since there is usually more than one way to perform any task or solve a problem. In any computer, however, there is only so much memory space available, so an overly long or inefficient program may not fit into the memory. In addition to performing arithmetic functions (such as with a trip computer), a computer can also store data, look up data in a table and perform the logic functions previously discussed. A Random Access Memory (RAM) allows the computer to store bits of data temporarily while waiting to be acted upon by the program. It may also be used to store output data that is to be sent to an output device. Whatever data is stored in a RAM is lost when power is removed from the system by turning **OFF** the ignition key, for example.

Computers have another type of memory called a Read Only Memory (ROM) which is permanent. This memory is not lost when the power is removed from the system. Most programs for automotive computers are stored on a ROM memory chip. Data is usually in the form of a look-up table that saves computing time and program steps. For example, a computer designed to control the amount of distributor advance can have this information stored in a table. The information that determines distributor advance (engine rpm, manifold vacuum and temperature) is coded to produce the correct amount of distributor advance over a wide range of engine operating conditions. Instead of the computer computing the required advance, it simply looks it up in a pre-programmed table. However, not all electronic control functions can be handled in this manner; some must be computed. On an antilock brake system, for example, the computer must measure the rotation of each separate wheel and then calculate how much brake pressure to apply in order to prevent one wheel from locking up and causing a loss of control.

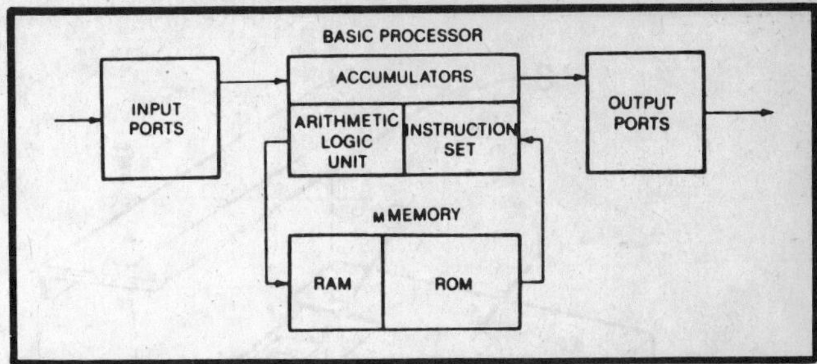
BASIC PROCESSOR
INPUT PORTS
ACCUMULATORS
ARITHMETIC LOGIC UNIT
INSTRUCTION SET
OUTPUT PORTS
MEMORY
RAM
ROM

Schematic of typical microprocessor based on-board computer showing essential components

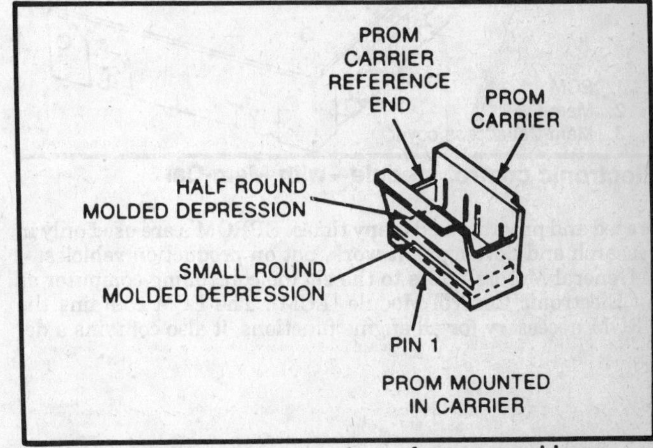
PROM CARRIER REFERENCE END
PROM CARRIER
HALF ROUND MOLDED DEPRESSION
SMALL ROUND MOLDED DEPRESSION
PIN 1
PROM MOUNTED IN CARRIER

Typical PROM showing carrier refernce markings

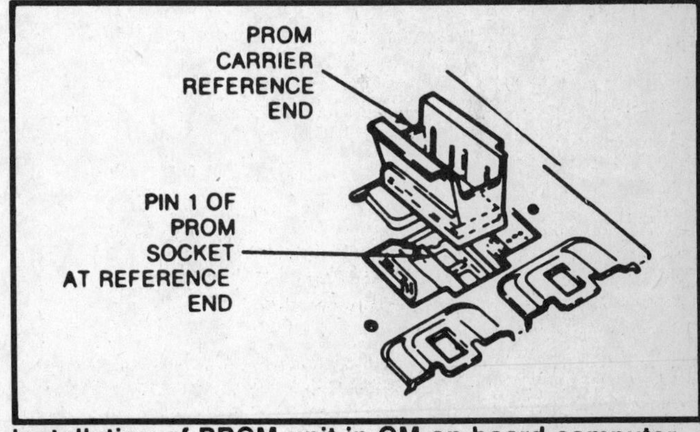
PROM CARRIER REFERENCE END
PIN 1 OF PROM SOCKET AT REFERENCE END

Installation of PROM unit in GM on-board computer

There are several ways of programming a ROM, but once programmed the ROM cannot be changed. If the ROM is made on the same chip that contains the microprocessor, the whole computer must be altered if a program change is needed. For this reason, a ROM is usually placed on a separate chip. Another type of memory is the Programmable Read Only Memory (PROM) that has the program "burned in" with the appropriate programming machine. Like the ROM, once a PROM has been programmed, it cannot be changed. The advantage of the PROM is that it can be produced in small quantities economically, since it is manufactured with a blank memory. Program changes for various vehicles can be made readily. There is still another type of memory called an EPROM (Erasable PROM) which can be

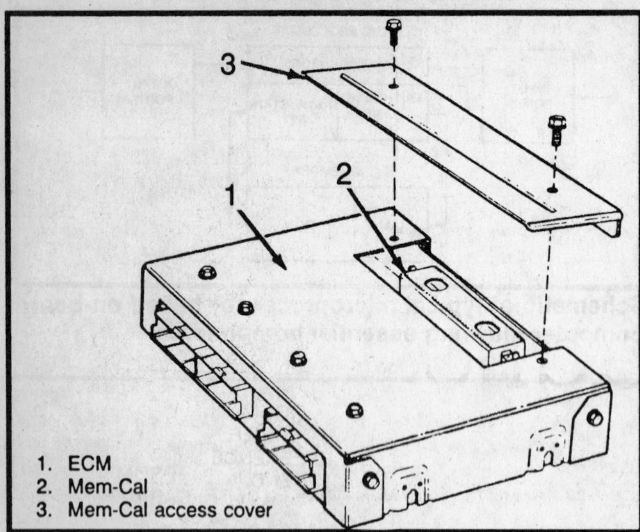
3
2
1
1. ECM
2. Mem-Cal
3. Mem-Cal access cover

Electronic control module—with Mem-Cal

erased and programmed many times. EPROM's are used only in research and development work, not on production vehicles.

General Motors refers to the engine controlling computer as an Electronic Control Module (ECM). The ECM contains the PROM necessary for all engine functions, it also contains a de-

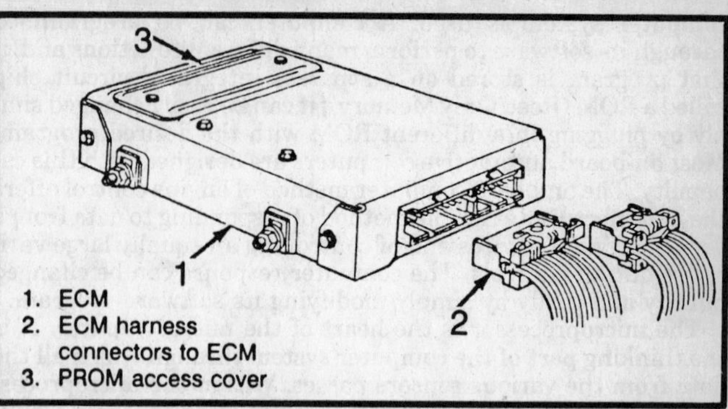
3
1
2
1. ECM
2. ECM harness connectors to ECM
3. PROM access cover

Electronic control module—with PROM and CalPak

vice called a CalPak. This allows the fuel delivery function should other parts of the ECM become damaged. It has an access door in the ECM, like the PROM has. There is a third type control module used in some ECMs called a Mem-Cal. The Mem-Cal contains the function of PROM, CalPak and Electronic Spark Control (EST) module. Like the PROM, it contains the calibrations needed for a specific vehicle, as well as the back-up fuel control circuitry required if the rest of the ECM should become damaged and the spark control. An ECM containing a PROM and CalPak can be identified by the 2 connector harnesses, while the ECM containing the Mem-Cal has 3 connector harnesses attached to it.

Troubleshooting and Diagnosis

SECTION 2

TROUBLESHOOTING AND DIAGNOSIS

Diagnostic Equipment and Special Tools

While we may think that with no moving parts, electronic components should never wear out, in the real world malfunctions do occur. The problem is that any computer-based system is extremely sensitive to electrical voltages and cannot tolerate careless or haphazard testing or service procedures. An inexperienced individual can literally do major damage looking for a minor problem by using the wrong kind of test equipment or connecting test leads or connectors with the ignition switch ON. Therefore, when selecting test equipment, make sure the manufacturers instructions state that the tester is compatible with whatever type of electronic control system is being serviced. Read all instructions carefully and double check all test points before installing probes or making any connections.

The following section outlines basic diagnosis techniques for dealing with computerized engine control systems. Along with a general explanation of the various types of test equipment available to aid in servicing modern electronic automotive systems, basic repair techniques for wiring harnesses and connectors is given. Read the basic information before attempting any repairs or testing on any computerized system, to provide the background of information necessary to avoid the most common and obvious mistakes that can cost both time and money. Likewise, the individual system sections for engine controls, fuel injection and feedback carburetors should be read from the beginning to the end before any repairs or diagnosis is attempted. Although the replacement and testing procedures are simple in themselves, the systems are not, and unless one has a thorough understanding of all components and their function within a particular fuel injection system (for example), the logical test sequence these systems demand cannot be followed. Minor malfunctions can make a big difference, so it is important to know how each component affects the operation of the overall electronic system to find the ultimate cause of a problem without replacing good components unnecessarily. It is not enough to use the correct test equipment; the test equipment must be used correctly.

Safety Precautions

CAUTION

Whenever working on or around any computer-based microprocessor control system, always observe these general precautions to prevent the possibility of personal injury or damage to electronic components:

- Never install or remove battery cables with the key ON or the engine running. Jumper cables should be connected with the key OFF to avoid power surges that can damage electronic control units. Engines equipped with computer controlled systems should avoid both giving and getting jump starts due to the possibility of serious damage to components from arcing in the engine compartment when connections are made with the ignition ON.
- Always remove the battery cables before charging the battery. Never use a high-output charger on an installed battery or attempt to use any type of "hot shot" (24 volt) starting aid.
- Exercise care when inserting test probes into connectors to insure good connections without damaging the connector or spreading the pins. Always probe connectors from the rear (wire) side, NOT the pin side, to avoid accidental shorting of terminals during test procedures.
- Never remove or attach wiring harness connectors with the ignition switch ON, especially to an electronic control unit.
- Do not drop any components during service procedures and never apply 12 volts directly to any component (like a solenoid or relay) unless instructed specifically to do so. Some component electrical windings are designed to safely handle only 4 or 5 volts and can be destroyed in seconds if 12 volts are applied directly to the connector.
- Remove the electronic control unit if the vehicle is to be placed in an environment where temperatures exceed approximately 176°F (80°C), such as a paint spray booth or when arc- or gas-welding near the control unit location in the car.

Organized Troubleshooting

When diagnosing a specific problem, organized troubleshooting is a must. The complexity of a modern automobile demands that you approach any problem in a logical, organized manner. There are certain troubleshooting techniques that are standard:

1. Establish when the problem occurs. Does the problem appear only under certain conditions? Were there any noises, odors, or other unusual symptoms? Make notes on any symptoms found, including warning lights and trouble codes, if applicable.
2. Isolate the problem area. To do this, make some simple tests and observations; then eliminate the systems that are working properly. Check for obvious problems such as broken wires or split or disconnected vacuum hoses. Always check the obvious before assuming something complicated is the cause.
3. Test for problems systematically to determine the cause once the problem area is isolated. Are all the components functioning properly? Is there power going to electrical switches and motors? Is there vacuum at vacuum switches and/or actuators? Is there a mechanical problem such as bent linkage or loose mounting screws? Doing careful, systematic checks will often turn up most causes on the first inspection without wasting time checking components that have little or no relationship to the problem.
4. Test all repairs after the work is done to make sure that the problem is fixed. Some causes can be traced to more than one component, so a careful verification of repair work is important to pick up additional malfunctions that may cause a problem to reappear or a different problem to arise. A blown fuse, for example, is a simple problem that may require more than just replacing a fuse. If you don't look for a problem that caused a fuse to blow, a shorted wire may go undetected.

The diagnostic tree charts are designed to help solve problems by leading the user through closely defined conditions and tests so that only the most likely components, vacuum and electrical circuits are checked for proper operation when troubleshooting a particular malfunction. By using the trouble trees to eliminate those systems and components which normally will not cause the condition described, a problem can be isolated within one or more systems or circuits without wasting time on unnecessary testing. Experience has shown that most problems tend to be the result of a fairly simple and obvious cause, such as loose or corroded connectors or air leaks in the intake system. A careful inspection of components during testing is essential to quick and accurate troubleshooting. Frequent references to special test equipment will be found in the text and in the diagnosis charts. These devices or compatible equivalents are necessary to perform some of the more complicated test procedures listed, but many components can be functionally tested with the quick checks outlined in the "On-Car Service" procedures. Aftermarket testers are available from a variety of sources, as well as from the vehicle manufacturer, but care should be taken that any test equipment being used is designed to diagnose that particular system accurately without damaging the control unit (ECU) or components being tested.

NOTE: Pinpointing the exact cause of trouble in an electrical system can sometimes only be done using special test equipment. The following describes commonly used test equipment and explains how to put it to best use in diagnosis. In addition to the information covered below, the manufacturer's instructions booklet provided with the tester should be read and clearly understood before attempting any test procedures.

Jumper Wires

Jumper wires are simple, yet extremely valuable pieces of test equipment. Jumper wires are merely wires that are used to bypass sections of a circuit. The simplest type of jumper wire is merely a length of multistrand wire with an alligator clip at each end. Jumper wires are usually fabricated from lengths of standard automotive wire and whatever type of connector (alligator clip, spade connector or pin connector) that is required for the particular vehicle being tested. The well-equipped tool box will have several different styles of jumper wires in several different lengths. Some jumper wires are made with three or more terminals coming from a common splice for special-purpose testing. In cramped, hard-to-reach areas it is advisable to have insulated boots over the jumper wire terminals in order to prevent accidental grounding, sparks, and possible fire, especially when testing fuel system components.

Jumper wires are used primarily to locate open electrical circuits, on either the ground (–) side of the circuit or on the hot (+) side. If an electrical component fails to operate, connect the jumper wire between the component and a good ground. If the component operates only with the jumper installed, the ground circuit is open. If the ground circuit is good, but the component does not operate, the circuit between the power feed and component is open. You can sometimes connect the jumper wire directly from the battery to the hot terminal of the component, but first make sure the component uses 12 volts in operation. Some electrical components, such as fuel injectors, are designed to operate on about 4 volts and running 12 volts directly to the injector terminals can burn out the wiring. By inserting an in-line fuseholder between a set of test leads, a fused jumper wire can be used for bypassing open circuits. Use a 5 amp fuse to provide protection against voltage spikes. When in doubt, use a voltmeter to check the voltage input to the component and measure how much voltage is being applied normally. By moving the jumper wire successively back from the lamp toward the power source, you can isolate the area of the circuit where the open is located. When the component stops functioning, or the power is cut off, the open is in the segment of wire between the jumper and the point previously tested.

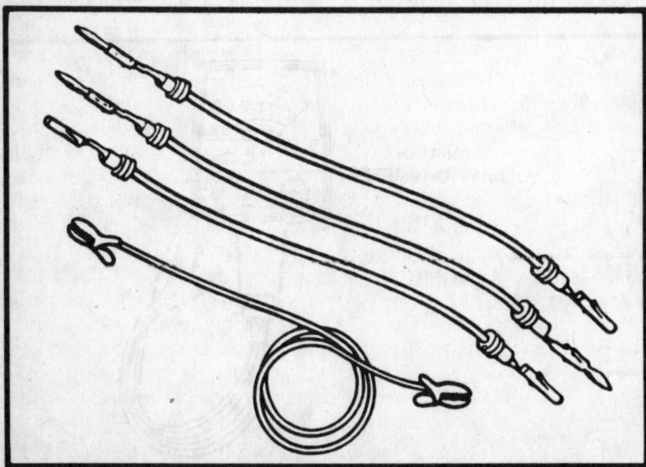

Typical jumper wires with various terminal ends

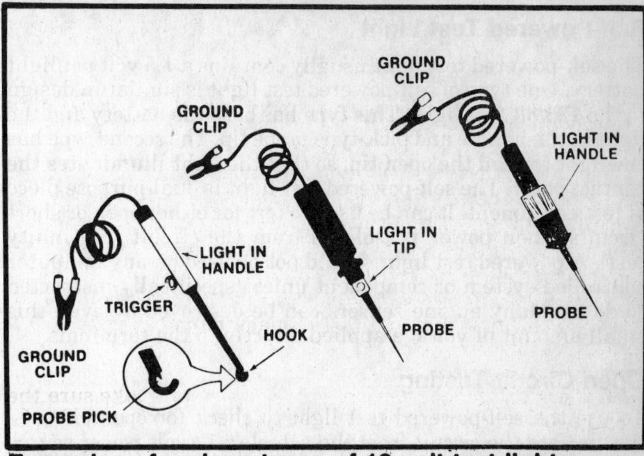
GROUND CLIP
GROUND CLIP
LIGHT IN HANDLE
LIGHT IN TIP
LIGHT IN HANDLE
TRIGGER
PROBE
HOOK
PROBE
GROUND CLIP
PROBE PICK

Examples of various types of 12 volt test lights

CAUTION

Never use jumpers made from wire that is of lighter gauge than used in the circuit under test. If the jumper wire is of too small gauge, it may overheat and possibly melt. Never use jumpers to bypass high-resistance loads (such as motors) in a circuit. Bypassing resistances, in effect, creates a short circuit which may, in turn, cause damage and fire. Never use a jumper for anything other than temporary bypassing of components in a circuit.

12 Volt Test Light

The 12 volt test light is used to check circuits and components while electrical current is flowing through them. It is used for voltage and ground tests. Twelve volt test lights come in different styles but all have three main parts; a ground clip, a probe, and a light. The most commonly used 12 volt test lights have pick-type probes. To use a 12 volt test light, connect the ground clip to a good ground and probe wherever necessary with the pick. The pick should be sharp so that it can penetrate wire insulation to make contact with the wire, without making a large hole in the insulation. The wrap-around light is handy in hard to reach areas or where it is difficult to support a wire to push a probe pick into it. To use the wrap around light, hook the wire to be probed with the hook and pull the trigger. A small pick will be forced through the wire insulation into the wire core.

CAUTION

Do not use a test light to probe electronic ignition spark plug or coil wires. Never use a pick-type test light to probe wiring on computer controlled systems unless specifically instructed to do so.

Like the jumper wire, the 12 volt test light is used to isolate opens in circuits. But, whereas the jumper wire is used to bypass the open to operate the load, the 12 volt test light is used to locate the presence of voltage in a circuit. If the test light glows, you know that there is power up to that point; if the 12 volt test light does not glow when its probe is inserted into the wire or connector, you know that there is an open circuit (no power). Move the test light in successive steps back toward the power source until the light in the handle does glow. When it does glow, the open is between the probe and point previously probed.

NOTE: The test light does not detect that 12 volts (or any particular amount of voltage) is present; it only detects that some voltage is present. It is advisable before using the test light to touch its terminals across the battery posts to make sure the light is operating properly.

Self-Powered Test Light

The self-powered test light usually contains a 1.5 volt penlight battery. One type of self-powered test light is similar in design to the 12 volt test light. This type has both the battery and the light in the handle and pick-type probe tip. The second type has the light toward the open tip, so that the light illuminates the contact point. The self-powered test light is dual-purpose piece of test equipment. It can be used to test for either open or short circuits when power is isolated from the circuit (continuity test). A powered test light should not be used on any computer controlled system or component unless specifically instructed to do so. Many engine sensors can be destroyed by even this small amount of voltage applied directly to the terminals.

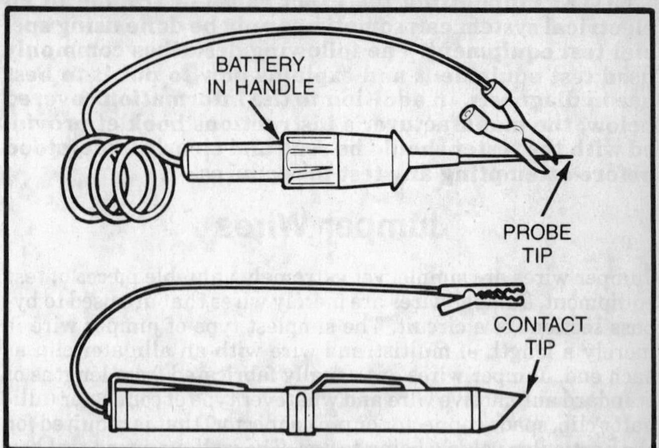
BATTERY
IN HANDLE
PROBE
TIP
CONTACT
TIP
Two types of self-powered test lights

Open Circuit Testing

To use the self-powered test light to check for open circuits, first isolate the circuit from the vehicle's 12 volt power source by disconnecting the battery or wiring harness connector. Connect the test light ground clip to a good ground and probe sections of the circuit sequentially with the test light. (start from either end of the circuit). If the light is out, the open is between the probe and the circuit ground. If the light is on, the open is between the probe and end of the circuit toward the power source.

Short Circuit Testing

By isolating the circuit both from power and from ground, and using a self-powered test light, you can check for shorts to ground in the circuit. Isolate the circuit from power and ground. Connect the test light ground clip to a good ground and probe any easy-to-reach test point in the circuit. If the light comes on, there is a short somewhere in the circuit. To isolate the short, probe a test point at either end of the isolated circuit (the light should be on). Leave the test light probe connected and open connectors, switches, remove parts, etc., sequentially, until the light goes out. When the light goes out, the short is between the last circuit component opened and the previous circuit opened.

NOTE: The 1.5 volt battery in the test light does not provide much current. A weak battery may not provide enough power to illuminate the test light even when a complete circuit is made (especially if there are high resistances in the circuit). Always make sure that the test battery is strong. To check the battery, briefly touch the ground clip to the probe; if the light glows brightly the battery is strong enough for testing. Never use a self-powered test light to perform checks for opens or shorts when power is applied to the electrical system under test. The 12-volt vehicle power will quickly burn out the 1.5 volt light bulb in the test light.

Voltmeter

A voltmeter is used to measure voltage at any point in a circuit, or to measure the voltage drop across any part of a circuit. It can also be used to check continuity in a wire or circuit by indicating current flow from one end to the other. Voltmeters usually have various scales on the meter dial and a selector switch to allow the selection of different voltages. The voltmeter has a positive and a negative lead. To avoid damage to the meter, always connect the negative lead to the negative (–) side of circuit (to ground or nearest the ground side of the circuit) and connect the positive lead to the positive (+) side of the circuit (to the power source or the nearest power source). Note that the negative voltmeter lead will always be black and that the positive voltmeter will always be some color other than black (usually red). Depending on how the voltmeter is connected into the circuit, it has several uses.

A voltmeter can be connected either in parallel or in series with a circuit and it has a very high resistance to current flow. When connected in parallel, only a small amount of current will flow through the voltmeter current path; the rest will flow through the normal circuit current path and the circuit will work normally. When the voltmeter is connected in series with a circuit, only a small amount of current can flow through the circuit. The circuit will not work properly, but the voltmeter reading will show if the circuit is complete or not.

Available Voltage Measurement

Set the voltmeter selector switch to the 20V position and connect the meter negative lead to the negative post of the battery. Connect the positive meter lead to the positive post of the battery and turn the ignition switch ON to provide a load. Read the voltage on the meter or digital display. A well-charged battery should register over 12 volts. If the meter reads below 11.5 volts, the battery power may be insufficient to operate the electrical system properly. This test determines voltage available from the battery and should be the first step in any electrical trouble diagnosis procedure. Many electrical problems, especially on computer controlled systems, can be caused by a low state of charge in the battery. Excessive corrosion at the battery cable terminals can cause a poor contact that will prevent proper charging and full battery current flow.

Normal battery voltage is 12 volts when fully charged. When the battery is supplying current to one or more circuits it is said to be "under load". When everything is off the electrical system is under a "no-load" condition. A fully charged battery

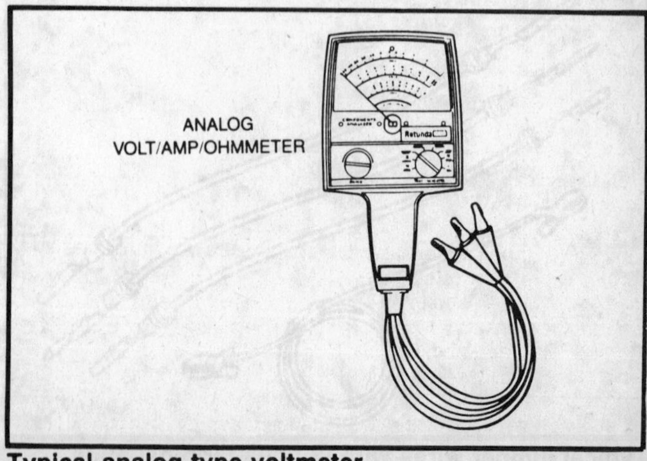
ANALOG
VOLT/AMP/OHMMETER
Typical analog-type voltmeter

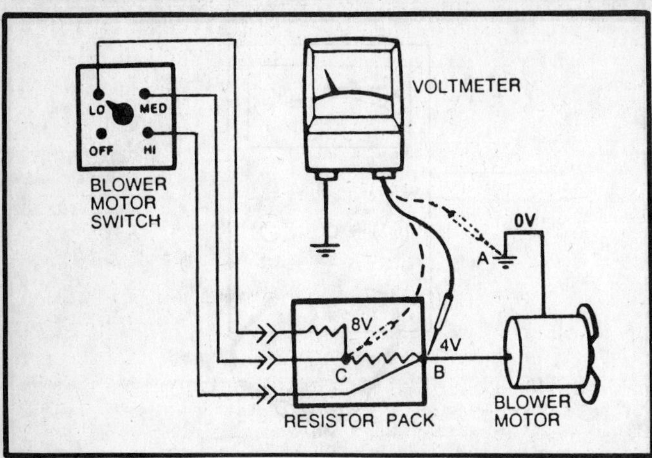
VOLTMETER
LO
MED
OFF
HI
BLOWER
MOTOR
SWITCH
0V
A
8V
4V
C
B
RESISTOR PACK
BLOWER
MOTOR

Measuring available voltage in a blower circuit

may show about 12.5 volts at no load; will drop to 12 volts under medium load; and will drop even lower under heavy load. If the battery is partially discharged the voltage decrease under heavy load may be excessive, even though the battery shows 12 volts or more at no load. When allowed to discharge further, the battery's available voltage under load will decrease more severely. For this reason, it is important that the battery be fully charged during all testing procedures to avoid errors in diagnosis and incorrect test results.

VOLTAGE DROP

When current flows through a resistance, the voltage beyond the resistance is reduced (the larger the current, the greater the reduction in voltage). When no current is flowing, there is no voltage drop because there is no current flow. All points in the circuit which are connected to the power source are at the same voltage as the power source. The total voltage drop always equals the total source voltage. In a long circuit with many connectors, a series of small, unwanted voltage drops due to corrosion at the connectors can add up to a total loss of voltage which impairs the operation of the normal loads in the circuit.

Indirect Computation of Voltage Drops

1. Set the voltmeter selector switch to the 20 volt position.
2. Connect the meter negative lead to a good ground.
3. Probe all resistances in the circuit with the positive meter lead.
4. Operate the circuit in all modes and observe the voltage readings.

Direct Measurement of Voltage Drops

1. Set the voltmeter switch to the 20 volt position.
2. Connect the voltmeter negative lead to the ground side of the resistance load to be measured.
3. Connect the positive lead to the positive side of the resistance or load to be measured.
4. Read the voltage drop directly on the 20 volt scale.

Too high a voltage indicates too high a resistance. If, for example, a blower motor runs too slowly, you can determine if there is too high a resistance in the resistor pack. By taking voltage drop readings in all parts of the circuit, you can isolate the problem. Too low a voltage drop indicates too low a resistance. If, for example, a blower motor runs too fast in the MED and/or LOW position, the problem can be isolated in the resistor pack by taking voltage drop readings in all parts of the circuit to locate a possibly shorted resistor. The maximum allowable voltage drop under load is critical, especially if there is more than one high resistance problem in a circuit because all voltage drops are cumulative. A small drop is normal due to the resistance of the conductors.

High Resistance Testing

1. Set the voltmeter selector switch to the 4 volt position.
2. Connect the voltmeter positive lead to the positive post of the battery.
3. Turn on the headlights and heater blower to provide a load.
4. Probe various points in the circuit with the negative voltmeter lead.
5. Read the voltage drop on the 4 volt scale. Some average maximum allowable voltage drops are:

FUSE PANEL – 7 volts
IGNITION SWITCH – 5volts
HEADLIGHT SWITCH – 7 volts
IGNITION COIL (+) – 5 volts
ANY OTHER LOAD – 1.3 volts

NOTE: Voltage drops are all measured while a load is operating; without current flow, there will be no voltage drop.

Ohmmeter

The ohmmeter is designed to read resistance (ohms) in a circuit or component. Although there are several different styles of ohmmeters, all will usually have a selector switch which permits the measurement of different ranges of resistance (usually the selector switch allows the multiplication of the meter reading by 10, 100, 1000, and 10,000). A calibration knob al-

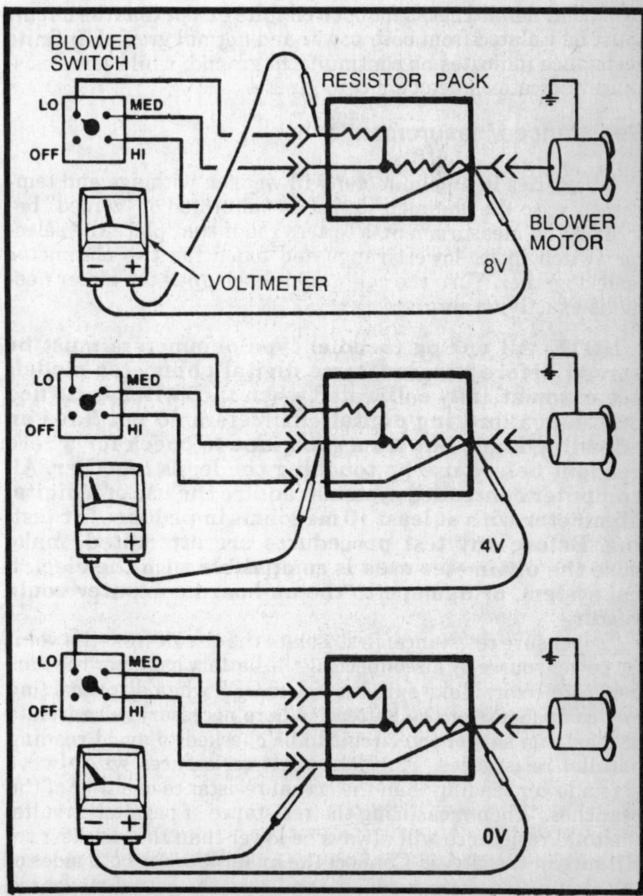
BLOWER
SWITCH
RESISTOR PACK
LO
MED
OFF
HI
BLOWER
MOTOR
8V
VOLTMETER
4V
0V

Direct measurement of voltage drops in a circuit

lows the meter to be set at zero for accurate measurement. Since all ohmmeters are powered by an internal battery (usually 9 volts), the ohmmeter can be used as a self-powered test light. When the ohmmeter is connected, current from the ohmmeter flows through the circuit or component being tested. Since the ohmmeter's internal resistance and voltage are known values, the amount of current flow through the meter depends on the resistance of the circuit or component being tested.

The ohmmeter can be used to perform continuity test for opens or shorts (either by observation of the meter needle or as a self-powered test light), and to read actual resistance in a circuit. It should be noted that the ohmmeter is used to check the resistance of a component or wire while there is no voltage applied to the circuit. Current flow from an outside voltage source (such as the vehicle battery) can damage the ohmmeter, so the circuit or component should be isolated from the vehicle electrical system before any testing is done. Since the ohmmeter uses its own voltage source, either lead can be connected to any test point.

NOTE: When checking diodes or other solid state components, the ohmmeter leads can only be connected one way in order to measure current flow in a single direction. Make sure the positive (+) and negative (-) terminal connections are as described in the test procedures to verify the one-way diode operation.

In using the meter for making continuity checks, do not be concerned with the actual resistance readings. Zero resistance, or any resistance readings, indicate continuity in the circuit. Infinite resistance indicates an open in the circuit. A high resistance reading where there should be none indicates a problem in the circuit. Checks for short circuits are made in the same manner as checks for open circuits except that the circuit must be isolated from both power and normal ground. Infinite resistance indicates no continuity to ground, while zero resistance indicates a dead short to ground.

Resistance Measurement

The batteries in an ohmmeter will weaken with age and temperature, so the ohmmeter must be calibrated or "zeroed" before taking measurements. To zero the meter, place the selector switch in its lowest range and touch the two ohmmeter leads together. Turn the calibration knob until the meter needle is exactly on zero.

NOTE: All analog (needle) type ohmmeters must be zeroed before use, but some digital ohmmeter models are automatically calibrated when the switch is turned on. Self-calibrating digital ohmmeters do not have an adjusting knob, but it's a good idea to check for a zero readout before use by touching the leads together. All computer controlled systems require the use of a digital ohmmeter with at least 10 megohms impedance for testing. Before any test procedures are attempted, make sure the ohmmeter used is compatible with the electrical system, or damage to the on-board computer could result.

To measure resistance, first isolate the circuit from the vehicle power source by disconnecting the battery cables or the harness connector. Make sure the key is OFF when disconnecting any components or the battery. Where necessary, also isolate at least one side of the circuit to be checked to avoid reading parallel resistances. Parallel circuit resistances will always give a lower reading than the actual resistance of either of the branches. When measuring the resistance of parallel circuits, the total resistance will always be lower than the smallest resistance in the circuit. Connect the meter leads to both sides of the circuit (wire or component) and read the actual measured ohms on the meter scale. Make sure the selector switch is set to

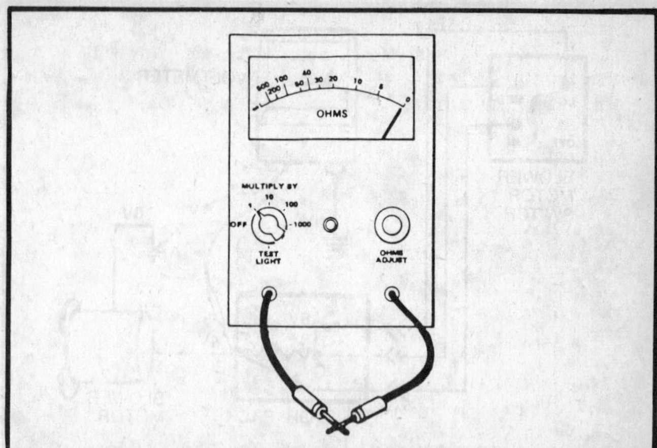
OHMS
MULTIPLY BY
OFF
TEST LIGHT
OHMS ADJUST

Analog ohmmeters must be calibrated before use by touching the probes together and adjusting the knob

the proper ohm scale for the circuit being tested to avoid misreading the ohmmeter test value.

CAUTION

Never use an ohmmeter with power applied to the circuit. Like the self-powered test light, the ohmmeter is designed to operate on its own power supply. The normal 12 volt automotive electrical system current could damage the meter.

Ammeters

An ammeter measures the amount of current flowing through a circuit in units called amperes or amps. Amperes are units of electron flow which indicate how fast the electrons are flowing through the circuit. Since Ohm's Law dictates that current flow in a circuit is equal to the circuit voltage divided by the total circuit resistance, increasing voltage also increases the current level (amps). Likewise, any decrease in resistance will increase the amount of amps in a circuit. At normal operating voltage, most circuits have a characteristic amount of amperes, called "current draw" which can be measured using an ammeter. By referring to a specified current draw rating, measuring the amperes, and comparing the two values, one can determine what is happening within the circuit to aid in diagnosis. An open circuit, for example, will not allow any current to flow so the ammeter reading will be zero. More current flows through a heavily loaded circuit or when the charging system is operating.

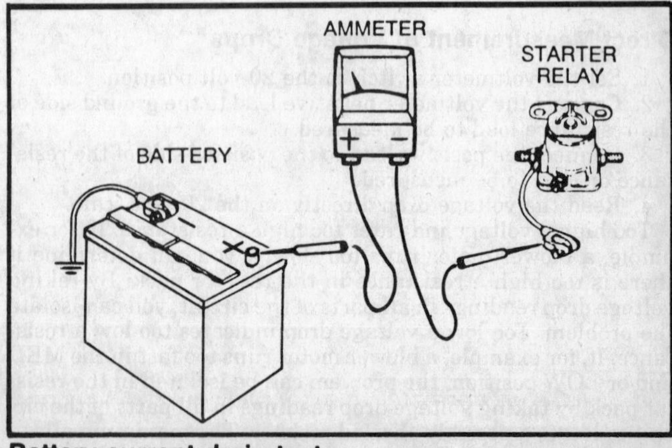
AMMETER
STARTER RELAY
BATTERY

Battery current drain test

An ammeter is always connected in series with the circuit being tested. All of the current that normally flows through the circuit must also flow through the ammeter; if there is any other path for the current to follow, the ammeter reading will not be accurate. The ammeter itself has very little resistance to current flow and therefore will not affect the circuit, but it will measure current draw only when the circuit is closed and electricity is flowing. Excessive current draw can blow fuses and drain the battery, while a reduced current draw can cause motors to run slowly, lights to dim and other components not to operate properly. The ammeter can help diagnose these conditions by locating the cause of the high or low reading.

Multimeters

Different combinations of test meters can be built into a single unit designed for specific tests. Some of the more common combination test devices are known as Volt-Amp testers, Tach-Dwell meters, or Digital Multimeters. The Volt-Amp tester is used for charging system, starting system or battery tests and consists of a voltmeter, an ammeter and a variable resistance carbon pile. The voltmeter will usually have at least two ranges for use with 6, 12 and 24 volt systems. The ammeter also has more than one range for testing various levels of battery loads and starter current draw and the carbon pile can be adjusted to offer different amounts of resistance. The Volt-Amp tester has heavy leads to carry large amounts of current and many later models have an inductive ammeter pickup that clamps around the wire to simplify test connections. On some models, the ammeter also has a zero-center scale to allow testing of charging and starting systems without switching leads or polarity. A digital multimeter is a voltmeter, ammeter and ohmmeter combined in an instrument which gives a digital readout. These are often used when testing solid state circuits because of their high input impedence (usually 10 megohms or more).

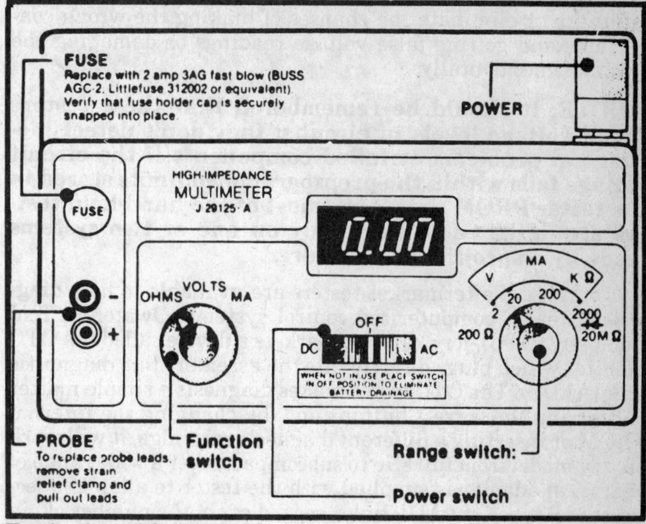
FUSE
Replace with 2 amp 3AG fast blow (BUSS AGC-2, Littlefuse 312002 or equivalent) Verify that fuse holder cap is securely snapped into place.
POWER
HIGH IMPEDANCE MULTIMETER J 29125-A
FUSE
0.00
OHMS VOLTS MA
MA
V
K Ω
20
200
2
2000
200 Ω
20M Ω
OFF
DC
AC
WHEN NOT IN USE PLACE SWITCH IN OFF POSITION TO ELIMINATE BATTERY DRAINAGE
PROBE
To replace probe leads, remove strain relief clamp and pull out leads
Function switch
Range switch:
Power switch
Typical multimeter

UF9

The tach-dwell meter combines a tachometer and a dwell (cam angle) meter and is a specialized kind of voltmeter. The tachometer scale is marked to show engine speed in rpm and the dwell scale is marked to show degrees of distributor shaft rotation. In most electronic ignition systems, dwell is determined by the control unit, but the dwell meter can also be used to check the duty cycle (operation) of some electronic engine control systems. Some tach-dwell meters are powered by an internal battery, while others take their power from the car battery in use. The battery powered testers usually require calibration much like an ohmmeter before testing.

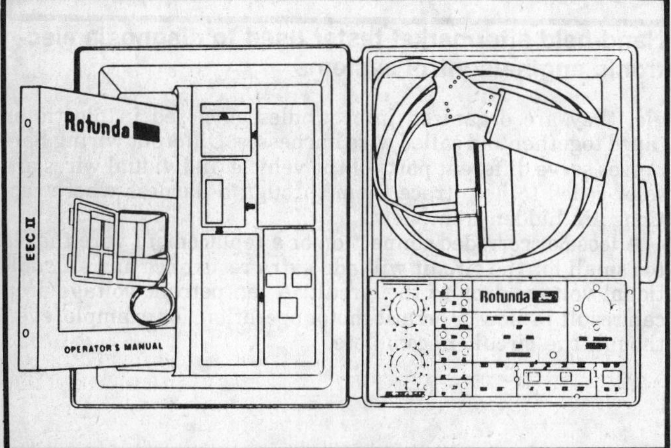
Rotunda
EEC II
OPERATOR'S MANUAL
Rotunda
Typical electronic engine control tester

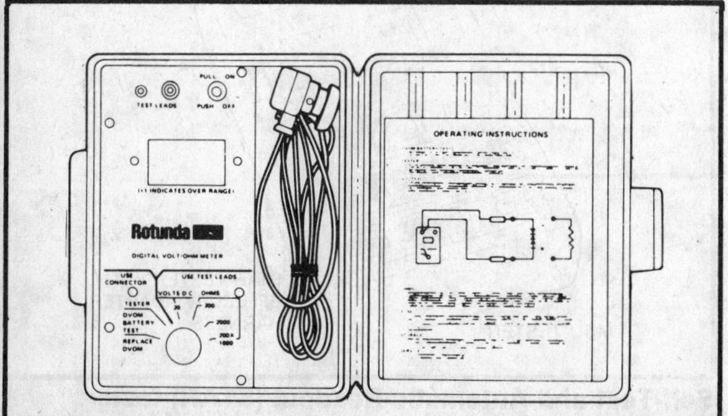
TEST LEADS
PULL ON
PUSH OFF
Rotunda
DIGITAL VOLT/OHM METER
OPERATING INSTRUCTIONS
Digital volt-ohmmeter

Special Test Equipment

A variety of diagnostic tools are available to help troubleshoot and repair computerized engine control systems. The most sophisticated of these devices are the console-type engine analyzers that usually occupy a garage service bay, but there are several types of aftermarket electronic testers available that will allow quick circuit tests of the engine control system by plugging directly into a special connector located in the engine compartment or under the dashboard. Several tool and equipment manufacturers offer simple, hand-held testers that measure various circuit voltage levels on command to check all system components for proper operation. Although these testers usually cost about $300–500, consider that the average computer control unit (or ECM) can cost just as much and the money saved by not replacing perfectly good sensors or components in an attempt to correct a problem could justify the purchase price of a special diagnostic tester the first time it's used.

These computerized testers can allow quick and easy test measurements while the engine is operating or while the car is being driven. In addition, the on-board computer memory can be read to access any stored trouble codes; in effect allowing the computer to tell you where it hurts and aid trouble diagnosis by pinpointing exactly which circuit or component is malfunctioning. In the same manner, repairs can be tested to make sure the problem has been corrected. The biggest advantage these special testers have is their relatively easy hookups that minimize or eliminate the chances of making the wrong connections and getting false voltage readings or damaging the computer accidentally.

NOTE: It should be remembered that these testers check voltage levels in circuits; they don't detect mechanical problems or failed components if the circuit voltage falls within the preprogrammed limits stored in the tester PROM unit. Also, most of the hand-held testers are designed to work only on one or two systems made by a specific manufacturer.

A variety of aftermarket testers are available to help diagnose different computerized control systems. Owatonna Tool Company (OTC), for example, markets a device called the OTC Monitor which plugs directly into the assembly line diagnostic link (ALDL). The OTC tester makes diagnosis a simple matter of pressing the correct buttons and, by changing the internal PROM or inserting a different diagnosis cartridge, it will work on any model from full size to subcompact, over a wide range of years. An adapter is supplied with the tester to allow connection to all types of ALDL links, regardless of the number of pin terminals used. By inserting an updated PROM into the OTC tester, it can be easily updated to diagnose any new modifications of computerized control systems.

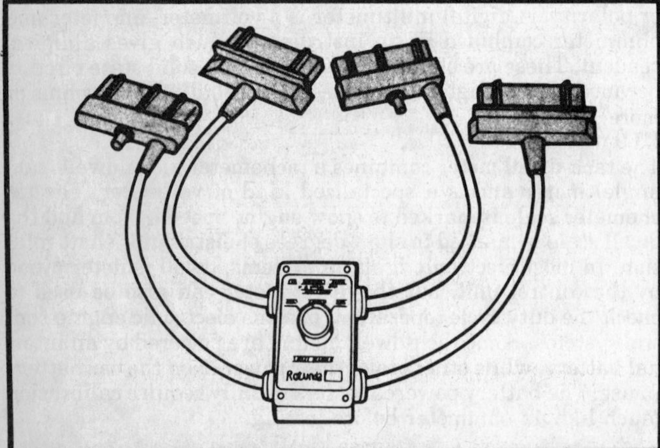
Rotunda

Typical adapter wiring harness for connecting tester to diagnostic terminal

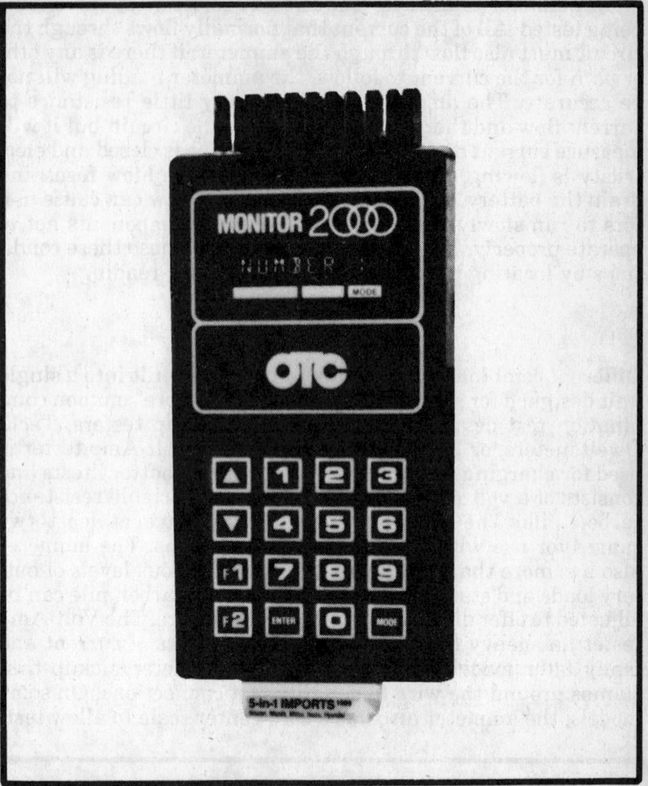
MONITOR 2000
NUMBER
MODE
OTC
1
2
3
4
5
6
F1
7
8
9
F2
ENTER
0
MODE
5-in-1 IMPORTS

Hand-held aftermarket tester used to diagnosis electronic engine control systems

Wiring Diagrams

The average automobile contains about ½ mile of wiring, with hundreds of individual connections. To protect the many wires from damage and to keep them from becoming a confusing tangle, they are organized into bundles, enclosed in plastic or taped together and called wire harnesses. Different wiring harnesses serve different parts of the vehicle. Individual wires are color-coded to help trace them through a harness where sections are hidden from view.

A loose or corroded connection or a replacement wire that is too small for the circuit will add extra resistance and an additional voltage drop to the circuit. A ten percent voltage drop can result in slow or erratic motor operation, for example, even though the circuit is complete.

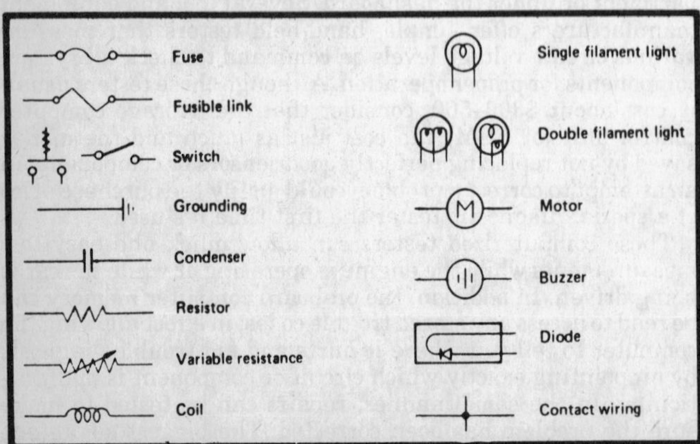
Fuse
Fusible link
Switch
Grounding
Condenser
Resistor
Variable resistance
Coil
Single filament light
Double filament light
Motor
M
Buzzer
Diode
Contact wiring

Typical electrical symbols found on wiring diagrams

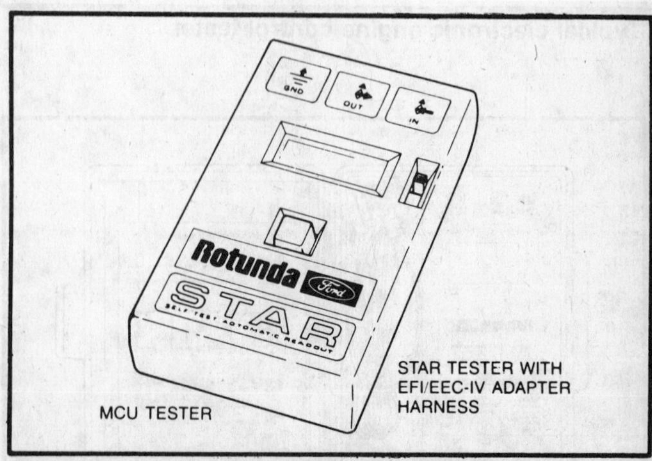
GND
OUT
IN
Rotunda
Ford
STAR
SELF TEST AUTOMATIC READOUT
MCU TESTER
STAR TESTER WITH EFI/EEC-IV ADAPTER HARNESS

Self-Test and Automatic Readout (STAR) tester

Self-Diagnostic Systems

3 SECTION

INDEX

NISSAN ELECTRONIC CONCENTRATED CONTROL SYSTEM (ECCS)

General Information

The Nissan Electro Injection System utilizes Throttle Body Fuel Injection system (TBI) on 1988–90 models equipped with the E16i, VG30i, GA16i or Z24i engines. The electronic control unit consists of a microcomputer, inspection lamps, a diagnostic mode selector and connectors for signal input and output and for power supply. The Electronic Concentrated Control System (ECCS) computer controls the amount of fuel injected, ignition timing, mixture ratio feedback, idle speed, fuel pump operation, mixture heating, Air Injection Valve (AIV) operation, Exhaust Gas Recirculation (EGR) and vapor canister purge operation.

The Nissan Electronic Concentrated Control System (ECCS) is an air flow controlled, port fuel injection and engine control system. It is used on 1988–90 models equipped with VG30E, VG30ET, VG30DE, CA20E, CA18ET, CA18DE and KA24E engines. The ECCS electronic control unit consists of a microcomputer, inspection lamps, a diagnostic mode selector and connectors for signal input and output and for power supply.

SYSTEM OPERATION

The self-diagnostic function is useful for diagnosing malfunctions in major sensors and actuators of the ECCS system. There are 5 modes in self-diagnostics, however only mode 2 and Mode 3 will be covered in this section.

In the self-diagnostic mode, the control unit constantly monitors the function of sensors and actuators regardless of ignition switch position. If a malfunction occurs, the information is stored in the control unit and can be retrieved from the memory by turning **ON** the diagnostic mode selector on the side of the control unit. When activated, the malfunction is indicated by flashing a red and green LED (also located on the control unit). Since all the self-diagnostic results are stored in the control unit memory, even intermittent malfunctions can be diagnosed. A malfunctioning part's group is indicated by the number of both red and green LED's flashing. First, the red LED flashes and the green flashes follow. The red LED refers to the number of tens, while the green refers to the number of units. If the red LED flashes twice and the green LED flashes once, a Code 21 is being displayed. All malfunctions are classified by their trouble code number.

The diagnostic result is retained in the control unit memory until the starter is operated 50 times after a diagnostic item is judged to be malfunctioning. The diagnostic result will then be canceled automatically. If a diagnostic item which has been judged malfunctioning and stored in memory is again judged to be malfunctioning before the starter is operated 50 times, the second result will replace the previous one and stored in the memory until the starter is operated 50 more times.

To switch the modes, turn the ignition switch **ON**, then turn the diagnostic mode selector on the control unit fully clockwise and wait for the inspection lamps to flash. Count the number of flashes until the inspection lamps have flashed the number of the desired mode, then immediately turn the diagnostic mode selector fully counterclockwise.

NOTE: When the ignition switch is turned OFF during diagnosis in each mode and then turned back on again after the power to the control unit has dropped off completely, the diagnosis will then automatically return to Mode 1.

The stored memory will be lost if the battery terminal is disconnected, or Mode 4 is selected after selecting Mode 3. However, if the diagnostic mode selector is kept turned fully clockwise, it will continue to change in the order of Mode 1, 2, 3, etc. and in this case, the stored memory will not be erased.

Self-Diagnostic System

ENTERING SELF-DIAGNOSIS AND DIAGNOSTIC CODE DISPLAY

Mode 2 Self Diagnosis Procedure

1990 STANZA AND 1990 300ZX

Both the 1990 Stanza and 1990 300ZX self diagnosis is preform in Mode 2. In this mode, a malfunction code is indicated by the number of flashes from the red LED or the CHECK ENGINE light.

The long (0.6 sec.) blinking indicates the number of 10 digits and the short (0.3 sec.) blinking indicates the number of single digits. For example, the red LED flashes 1 time for 0.6 sec. and then it flashes 2 times for 0.3 sec. This indicates the number 12 (code 12) and refers to a malfunction in the airflow meter. Mode 2 may be entered as follows:

1. Turn the diagnostic mode selector on the ECU fully clockwise and wait 2 seconds.

ACCESSING SELF-DIAGNOSTIC MODE TYPICAL

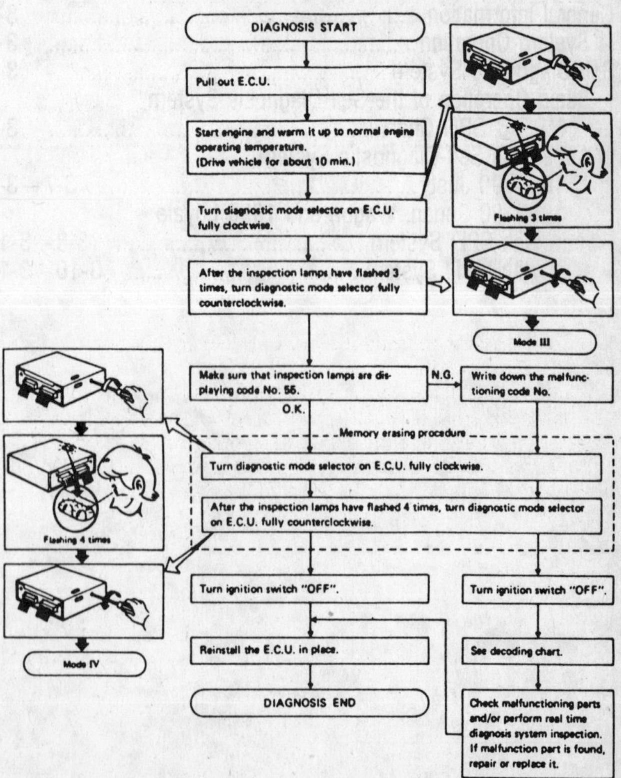
DIAGNOSIS START
Pull out E.C.U.
Start engine and warm it up to normal engine operating temperature. (Drive vehicle for about 10 min.)
Turn diagnostic mode selector on E.C.U. fully clockwise.
Flashing 3 times
After the inspection lamps have flashed 3 times, turn diagnostic mode selector fully counterclockwise.
Mode III
Make sure that inspection lamps are displaying code No. 55.
N.G.
Write down the malfunctioning code No.
O.K.
Memory erasing procedure
Turn diagnostic mode selector on E.C.U. fully clockwise.
Flashing 4 times
After the inspection lamps have flashed 4 times, turn diagnostic mode selector on E.C.U. fully counterclockwise.
Turn ignition switch "OFF".
Turn ignition switch "OFF".
Reinstall the E.C.U. in place.
See decoding chart.
Mode IV
DIAGNOSIS END
Check malfunctioning parts and/or perform real time diagnosis system inspection. If malfunction part is found, repair or replace it.

CAUTION:
During displaying code No. in self-diagnosis mode (mode III), if the other diagnostic mode should be done, make sure to write down the malfunctioning code No. before turning diagnostic mode selector on E.C.U. fully clockwise, or select the diagnostic mode after turning switch "OFF". Otherwise self-diagnosis information stored in E.C.U. memory until now would be lost.

1988-90 NISSAN DIAGNOSTIC TROUBLE CODES

Trouble Codes	200SX	240SX	300ZX	Axxess	Maxima	Pulsar	Stanza	Sentra	Truck Van Pathfinder
11—Crank angle sensor circuit	X	X	X	X	X	X	X	X	X
12—Air flow meter circuit	X	X	X	X	X	X	X	X	X
13—Engine temperature sensor circuit	X	X	X	X	X	X	X	X	X
14—Vehicle speed sensor circuit	X	X	X	X	X	X	X	X	X
15—Mixture ratio feedback control slips out								X	
21—Ignition signal missing in primary coil	X	X	X	X	X	X	X	X	X
22—Fuel pump circuit	X		②③	X	X		X		
23—Idle switch circuit	X		②③	X	②	X	X	X	
24—Full switch circuit	X			X			X		
25—A.A.C. valve circuit						X		X	
31—E.C.U.	X	X	X	X	X	X	X	X	X
32—E.G.R. function (California only)		X	X	X	③④	X	X	X	X
33—Exhaust gas sensor circuit	X	X	⑤	X	X	X	X	X	X
34—Detonation sensor	X		X		③④	X			X
35—Exhaust gas temperature sensor circuit (California only)		X	X	X	③④	X	X	X	X
41—Air temperature sensor circuit	X			X			X		
42—Fuel temperature sensor circuit	X		X		②				
43—Throttle sensor circuit		X	X	X	③④	X	X	X	X
44—No malfunctioning in circuits	X								
45—Injector leak (California only)		X	X	X	③④	X	X	X	X
51—Injector circuit (California only)			④		③④				X
53—Exhaust gas sensor circuit (right side)			④						
54—Signal circuit from A/T control unit to ECU			④		③④				
55—No malfunctioning circuits		X	X	X	X	X	X	X	X

① California only
② 1988
③ 1989
④ 1990
⑤ Left side

2. Turn the diagnostic mode selector fully counterclockwise, Mode 2 is in entered.

3. Observe trouble codes and repair system as needed.

4. When trouble is found and corrected, trouble codes can be cleared by turning the ignition switch off and then on again after power to the ECU has dropped off completely, the diagnosis will automatically return to Mode 1, thus clearing the ECU.

NOTE: Some vehicle computers have a learning ability. If a change is noted in vehicle performance after clearing codes, it may be due to the computers learning ability. To restore performance, warm vehicle to normal operating temperature and drive at part throttle with moderate acceleration, before performing any additional diagnosis.

SUBARU FUEL INJECTION AND CARBURETOR CONTROL SYSTEMS

General Information

The fuel injection systems covered in this section will be and Multi-Point Fuel Injection (MPFI) and Single-Point Fuel Injection (SPFI). In that self-diagnosis is virtually the same as in the SPFI and MPFI, the Electronic Fuel-Controlled Carburetor (EFC) self-diagnosis have been included in this section.

The MPFI system supplies the optimum air/fuel mixture to the engine under all various operating conditions. System fuel, which is pressurized at a constant pressure, is injected into the intake air passage of the cylinder head. The amount of fuel injected is controlled by the intermittent injection system where the electro-magnetic injection valve (fuel injector) opens only for a short period of time, depending on the amount of air required for 1 cycle of operation. During system operation, the amount injection is determined by the duration of an electric pulse sent to the fuel injector, which permits precise metering of the fuel.

All the operating conditions of the engine are converted into electric signals, resulting in additional features of the system, such as improved adaptability and easier addition of compensating element.

The SPFI system electronically controls the amount of injection from the fuel injector, and supplies the optimum air/fuel mixture under all operating conditions of the engine. Features of the SPFI system are as follows:

- Precise control of of the air/fuel mixture is accomplished by an increased number of input signals transmitting engine operating conditions to the control unit.
- The use of hot wire type air flow meter not only eliminates the need for high altitude compensation, but improves driving performance at high altitudes.
- The air control valve automatically regulates the idle speed to the set value under all engine operating conditions.
- Ignition timing is electrically controlled, thereby allowing the use of complicated spark advances characteristics.
- Wear of the air flow meter and fuel injector is automatically corrected so that they maintain their original performance.

The EFC system controls the exhaust gas emissions before entering the catalyst (in the catalytic converter) by regulation of the air/fuel ratio which enables the 3-way catalyst to reduce hydrocarbons (HC), carbon monoxide (CO) and oxides of nitrogen.

When there is little oxygen content in the exhaust gas and the output voltage of the oxygen sensor is larger than the "Slice Level" (averaged voltage which has been calculated from the air/fuel ratio becomes lean. The amount of controlled fuel is determined by the duty ratio which is defined as a percent of the opening time of the solenoid valve to 1 pulse cycle.

When the oxygen content in the exhaust gas is large and the output voltage of the oxygen sensor is smaller than the Slice Level, the air/fuel ratio becomes rich.

Outside the feedback control zone or under specific operating conditions, such as cold starting, acceleration and the warning-up of the oxygen sensor, the control unit fixes control signals at a constant flow rate to control the air/fuel, regardless of the output voltage from the oxygen sensor, to maintain good driveability under all operating conditions. The speed of increasing or decreasing the amount of air/fuel vary with the operating conditions of the vehicle, such as idling, acceleration and cruising.

If a predetermined level is not satisfied, or a fault is found, the warning lamp signals the driver. When this occurs, the self-diagnosis function is performed.

SYSTEM OPERATION

The self-diagnosis system detects and indicates faults, in vari-

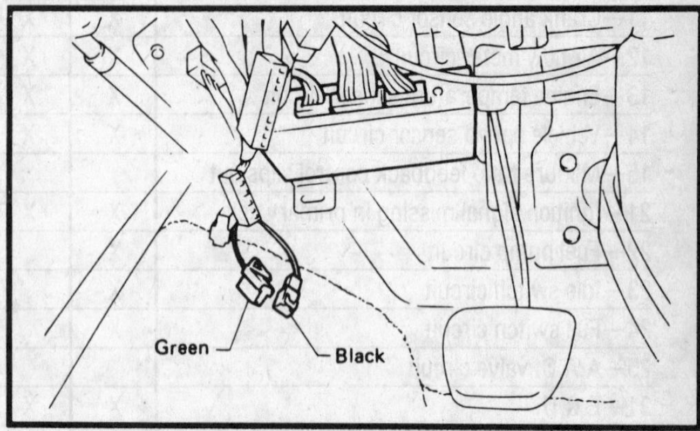
Green
Black
Location of the ECU test connectors – Justy

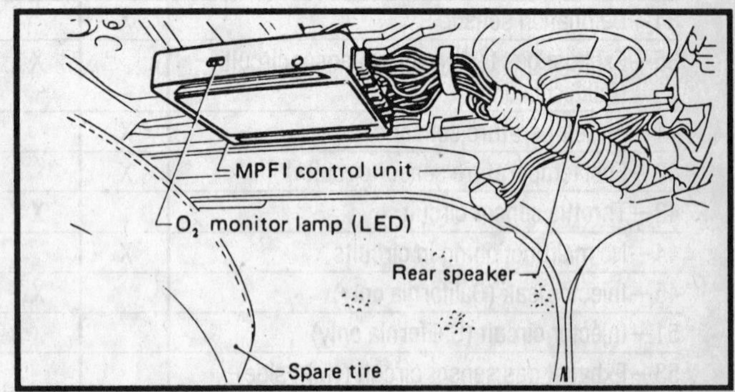
MPFI control unit
O₂ monitor lamp (LED)
Rear speaker
Spare tire
Location of the light emitting diode (LED) – self diagnosis system

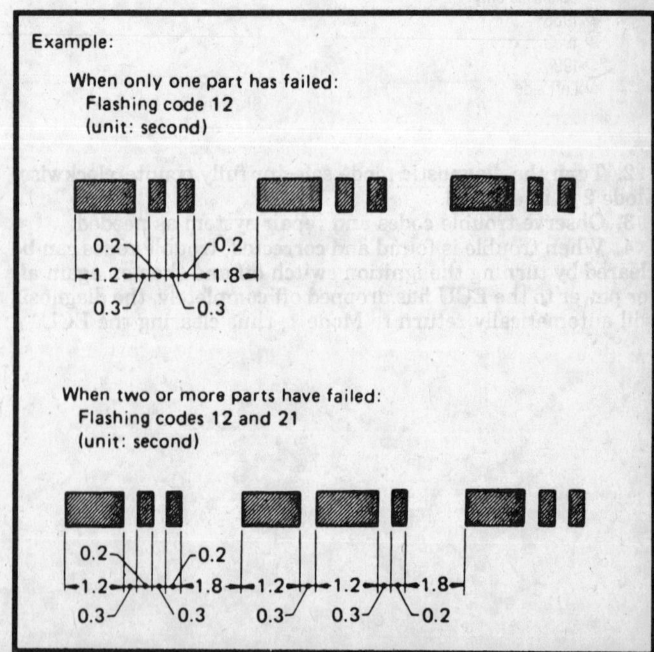
Example:
When only one part has failed:
Flashing code 12
(unit: second)
0.2
0.2
1.2
1.8
0.3
0.3
When two or more parts have failed:
Flashing codes 12 and 21
(unit: second)
0.2
0.2
1.2
1.8
1.2
1.2
1.8
0.3
0.3
0.3
0.3
0.2
How to read trouble codes – flashing

Mode	Engine	Read memory connector	Test mode connector
U-check	Ignition ON	DISCONNECT	DISCONNECT
Read memory	Ignition ON	CONNECT	DISCONNECT
D-check	Ignition ON	DISCONNECT	CONNECT
Clear memory	Ignition ON (engine on)	CONNECT	CONNECT

READ MEMORY CONNECTOR
TEST MODE CONNECTOR
MPFI CONTROL UNIT
REAR TRUNK AREA

Preparing for self diagnosis in each mode position–XT model

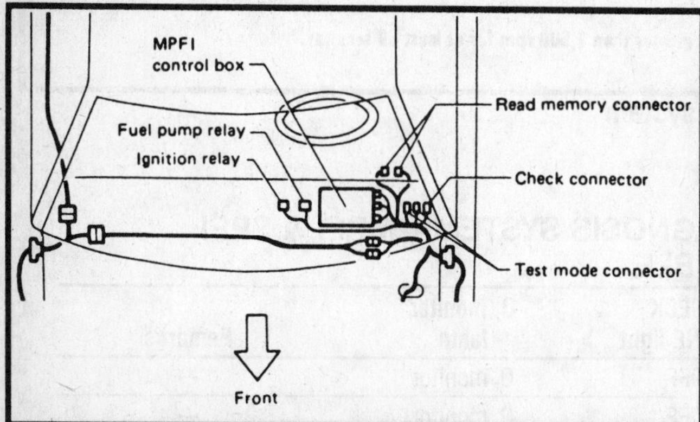
MPFI
control box
Read memory connector
Fuel pump relay
Ignition relay
Check connector
Test mode connector
Front

Location of the ECU and related connectors–Sedan, Wagon, 3 door, Loyale and Legacy

ous inputs and outputs of the electronic control unit. The warning lamp (CHECK ENGINE light), located on the instrument panel indicates a fault or trouble and the LED (light emitting diode) in the control unit indicates a trouble code. A fail-safe function is incorporated into the system to ensure minimal driveability if a failure of a sensor occurs.

The self-diagnosis function consists of 4 modes: U-check mode, Read memory mode, D-check mode and Clear code mode. Two connectors (Read memory and Test mode) and 2 lamps (CHECK ENGINE and O_2 monitor) are used. The connectors are used for mode selection and the lamps monitor the type of problem.

U-CHECK MODE

The U-Check mode diagnostics only the MPFI components necessary for start-up and driveabilty. When a fault occurs, the warning lamp (CHECK ENGINE) lights to indicate to the user that inspection is necessary. The diagnosis of other components, which do not effect start-up and driveabilty, are excluded from this mode.

READ MEMORY CODE

This mode is used to read past problems (even when the vehicle monitor lamps are off). It is most effective in detecting poor contacts or loose connections of the connectors, harness, etc.

D-CHECK MODE

This mode is used to check the entire MPFI system and to detect faulty components.

CLEAR MEMORY MODE

This mode is used to clear the trouble code from the memory after all faults have been corrected.

Self-Diagnostic System

NOTE: Some vehicle computers have a learning ability. If a change is noted in vehicle performance after clearing codes, it may be due to the computers learning ability. To restore performance, warm vehicle to normal operating temperature and drive at part throttle with moderate acceleration, before performing any additional diagnosis.

NO TROUBLE

Engine	Read memory connector	Test mode connector	CHECK ENGINE light	O_2 monitor lamp	Remarks
ON	X	X	OFF	O_2 monitor	
ON	O	X	Blinking	OFF	
ON	X	O	Blinking	Vehicle specification code	
OFF (Ignition switch ON)	X	X	ON	Vehicle specification code	Before starting the engine, the self-diagnosis system assumes the engine to be in NO TROUBLE condition.
OFF (Ignition switch ON)	O	X	Blinking	OFF	
OFF (Ignition switch ON)	X	O	• ON → Blinking	Vehicle specification code	

Basic operation of the self diagnosis system–EFC system

TROUBLE

Engine	Read memory connector	Test mode connector	CHECK ENGINE light	O_2 monitor lamp	Remarks
ON	X	X	ON	Trouble code	
ON	O	X	ON	Trouble code (memory)	
ON	X	O	* OFF → ON	Trouble code	Vehicle specification code is outputted when CHECK ENGINE light is OFF.
OFF (Ignition switch ON)	X	X	ON	Trouble code	
OFF (Ignition switch ON)	O	X	ON	Trouble code (memory)	
OFF (Ignition switch ON)	X	O	ON	* Vehicle specification code → Trouble code	

* : The indication is not changed until engine is operated at speed greater than 1,500 rpm for at least 39 seconds.

Basic operation of the self diagnosis system (cont.) – EFC system

BASIC OPERATION OF THE SELF DIAGNOSIS SYSTEM—MPFI & SPFI

NO TROUBLE

Engine	Read memory connector	Test mode connector	CHECK ENGINE light	O_2 monitor lamp	Remarks
ON	X	X	OFF	O_2 monitor	—
ON	O	X	OFF	O_2 monitor	—
ON ①	X	O	OFF→Blink ②	OFF	Vehicle specification code is outputted when CHECK ENGINE light is OFF.
ON ①	O	O	OFF→Blink	OFF	All memory stored in control unit is cleared after CHECK ENGINE light blinks.
OFF (Ignition switch ON)	O	X	ON	Vehicle specification code	Before starting the engine, the self-diagnosis system assumes the engine to be in a NO TROUBLE condition.
OFF (Ignition switch ON)	X	X	ON	Vehicle specification code	
OFF (Ignition switch ON)	X	O	ON	Vehicle specification code	
OFF (Ignition switch ON)	O	O	ON	Vehicle specification code	

TROUBLE

Engine	Read memory connector	Test mode connector	CHECK ENGINE light	O_2 monitor lamp	Remarks
ON	X	X	ON	Trouble code	
ON	O	X	ON	Trouble code (memory)	
ON ①	X	O	OFF→ON ②	Trouble code	Vehicle specification code is outputted when CHECK ENGINE light is OFF.
ON ①	O	O	OFF→ON	Trouble code	

TROUBLE

Engine	Read memory connector	Test mode connector	CHECK ENGINE light	O_2 monitor lamp	Remarks
OFF (Ignition switch ON)	○	X	ON	Trouble code (memory)	
STALL (Ignition switch ON)	X	X	ON	Trouble code	
STALL (Ignition switch ON)	X	○	ON	Trouble code	
STALL (Ignition switch ON)	○	○	ON	Trouble code	

① Ignition timing is set to 20° BTDC (when the engine is on, test mode connector is connected, and idle switch is ON).
② CHECK ENGINE light remains off until engine is operated at speed greater than 2,000 rpm for at least 40 seconds.

SELF-DIAGNOSIS SYSTEM – 1989–90 JUSTY

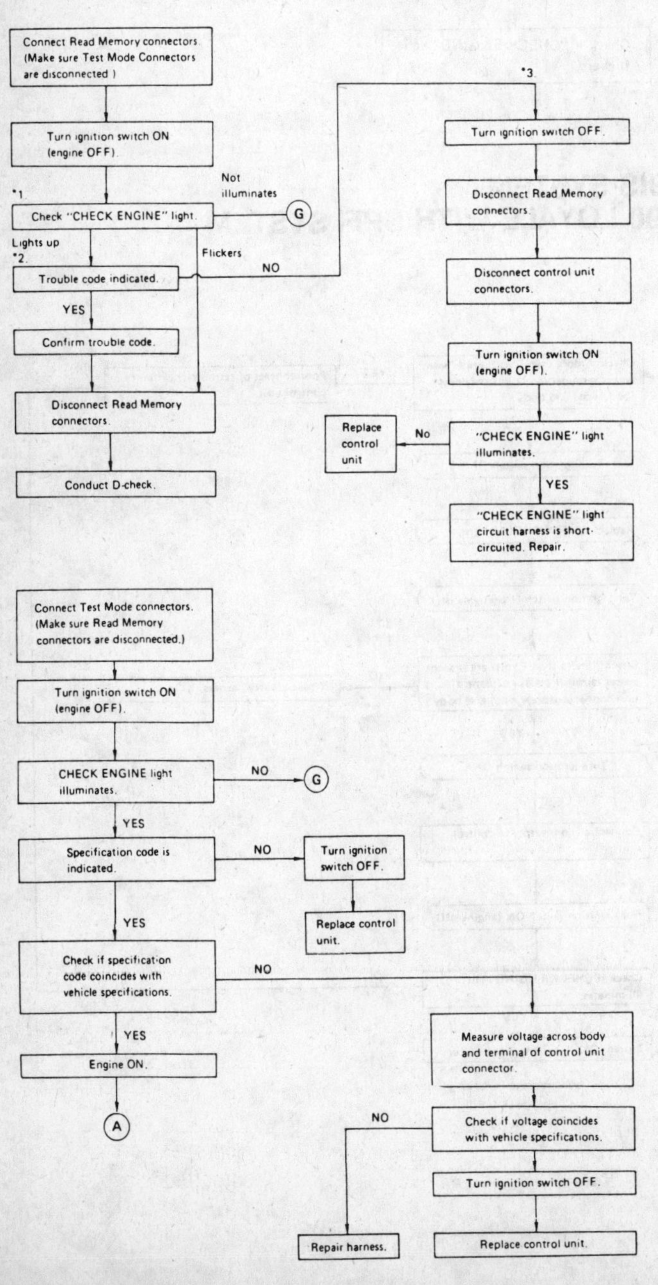
Connect Read Memory connectors. (Make sure Test Mode Connectors are disconnected.)
Turn ignition switch ON (engine OFF).
*1
Check "CHECK ENGINE" light.
Not illuminates
G
Lights up
*2
Flickers
Trouble code indicated.
NO
YES
Confirm trouble code.
Disconnect Read Memory connectors.
Conduct D-check.
*3
Turn ignition switch OFF.
Disconnect Read Memory connectors.
Disconnect control unit connectors.
Turn ignition switch ON (engine OFF).
"CHECK ENGINE" light illuminates.
No
Replace control unit
YES
"CHECK ENGINE" light circuit harness is short-circuited. Repair.
Connect Test Mode connectors. (Make sure Read Memory connectors are disconnected.)
Turn ignition switch ON (engine OFF).
CHECK ENGINE light illuminates.
NO
G
YES
Specification code is indicated.
NO
Turn ignition switch OFF.
Replace control unit.
YES
Check if specification code coincides with vehicle specifications.
NO
YES
Engine ON.
A
Measure voltage across body and terminal of control unit connector.
Check if voltage coincides with vehicle specifications.
NO
Turn ignition switch OFF.
Repair harness.
Replace control unit.

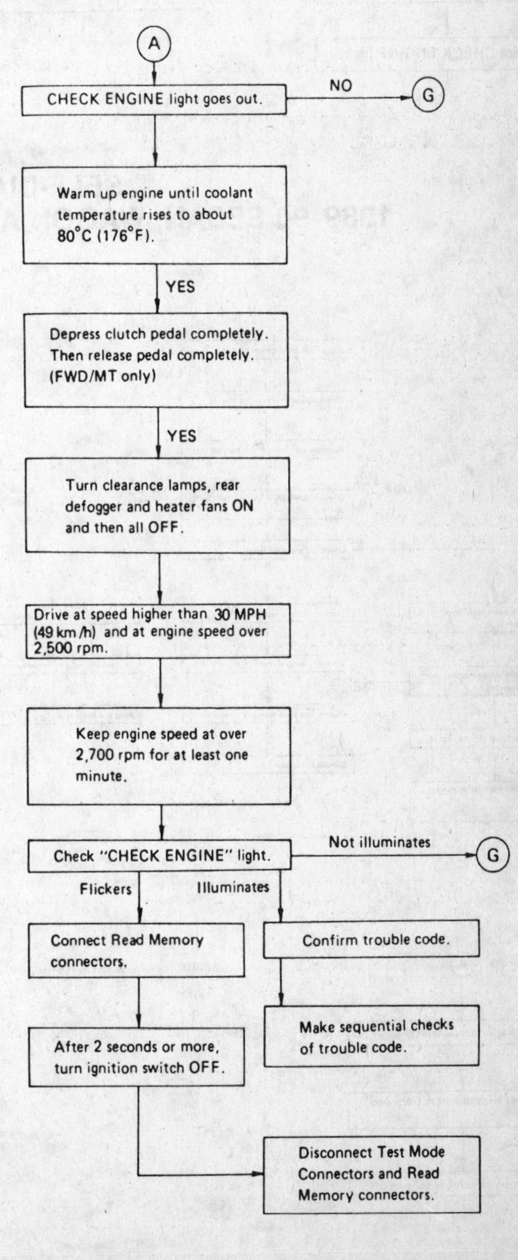
A
CHECK ENGINE light goes out.
NO
G
Warm up engine until coolant temperature rises to about 80°C (176°F).
YES
Depress clutch pedal completely. Then release pedal completely. (FWD/MT only)
YES
Turn clearance lamps, rear defogger and heater fans ON and then all OFF.
Drive at speed higher than 30 MPH (49 km/h) and at engine speed over 2,500 rpm.
Keep engine speed at over 2,700 rpm for at least one minute.
Check "CHECK ENGINE" light.
Not illuminates
G
Flickers
Illuminates
Connect Read Memory connectors.
Confirm trouble code.
After 2 seconds or more, turn ignition switch OFF.
Make sequential checks of trouble code.
Disconnect Test Mode Connectors and Read Memory connectors.

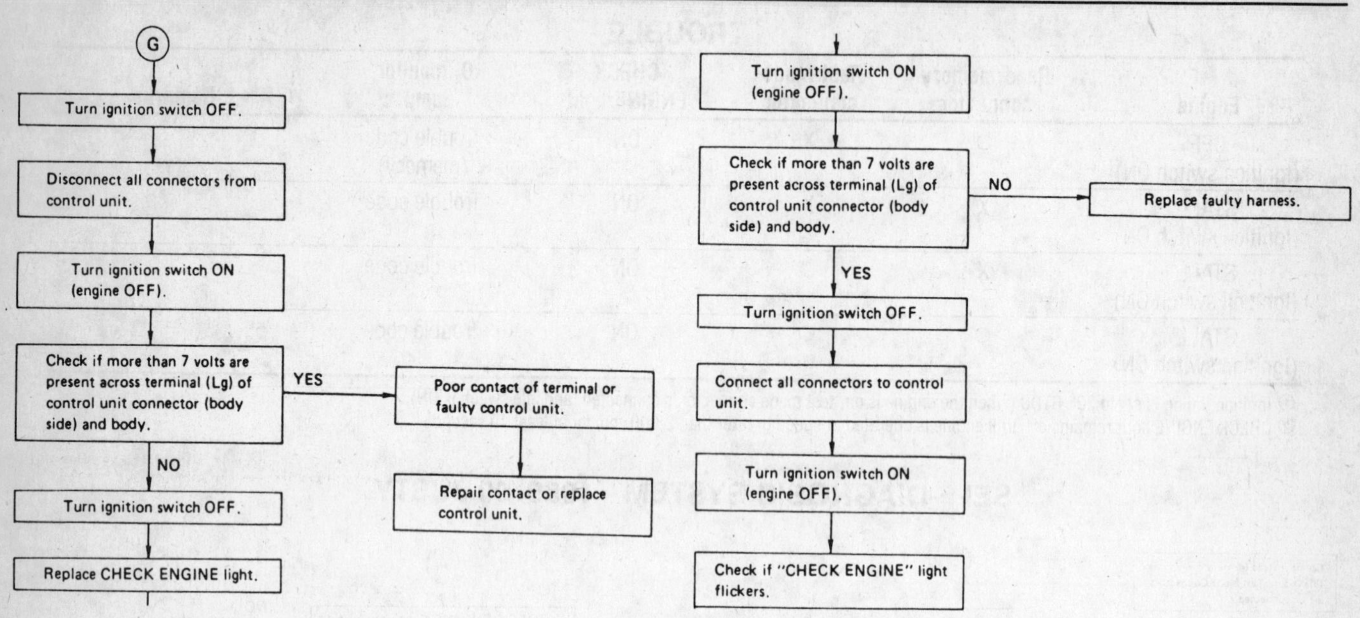
G
Turn ignition switch OFF.
Disconnect all connectors from control unit.
Turn ignition switch ON (engine OFF).
Check if more than 7 volts are present across terminal (Lg) of control unit connector (body side) and body.
YES
Poor contact of terminal or faulty control unit.
Repair contact or replace control unit.
NO
Turn ignition switch OFF.
Replace CHECK ENGINE light.
Turn ignition switch ON (engine OFF).
Check if more than 7 volts are present across terminal (Lg) of control unit connector (body side) and body.
NO
Replace faulty harness.
YES
Turn ignition switch OFF.
Connect all connectors to control unit.
Turn ignition switch ON (engine OFF).
Check if "CHECK ENGINE" light flickers.

SELF-DIAGNOSIS SYSTEM 1989–90 SEDAN, WAGON AND 1990 LOYALE WITH SPFI SYSTEM

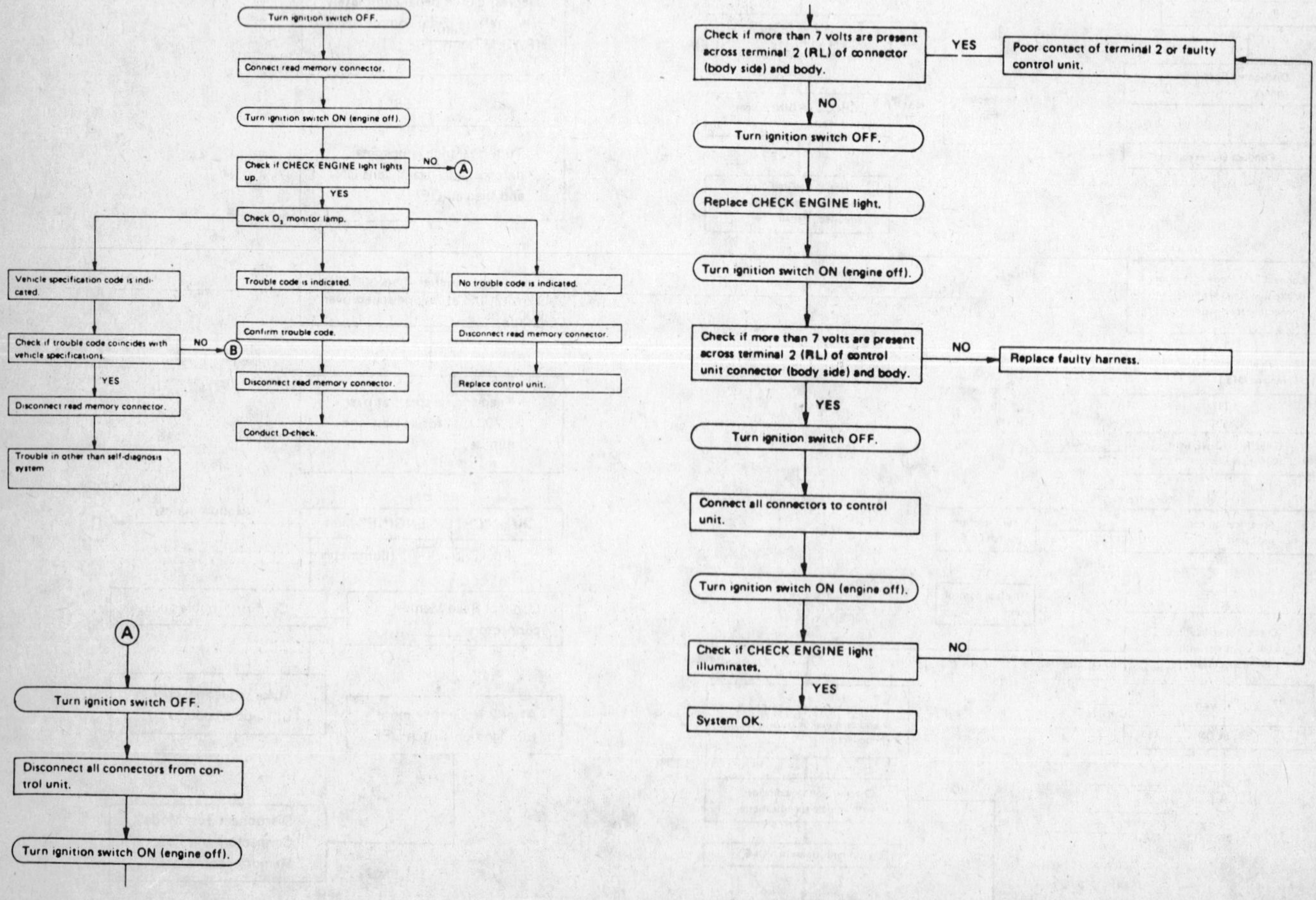

Turn ignition switch OFF.

Connect read memory connector.

Turn ignition switch ON (engine off).

Check if CHECK ENGINE light lights up. — NO → A

YES

Check O_2 monitor lamp.

Vehicle specification code is indicated.

Check if trouble code coincides with vehicle specifications. — NO → B

YES

Disconnect read memory connector.

Trouble in other than self-diagnosis system

Trouble code is indicated.

Confirm trouble code.

Disconnect read memory connector.

Conduct D-check.

No trouble code is indicated.

Disconnect read memory connector.

Replace control unit.

A

Turn ignition switch OFF.

Disconnect all connectors from control unit.

Turn ignition switch ON (engine off).

Check if more than 7 volts are present across terminal 2 (RL) of connector (body side) and body. — YES → Poor contact of terminal 2 or faulty control unit.

NO

Turn ignition switch OFF.

Replace CHECK ENGINE light.

Turn ignition switch ON (engine off).

Check if more than 7 volts are present across terminal 2 (RL) of control unit connector (body side) and body. — NO → Replace faulty harness.

YES

Turn ignition switch OFF.

Connect all connectors to control unit.

Turn ignition switch ON (engine off).

Check if CHECK ENGINE light illuminates. — NO → Poor contact of terminal 2 or faulty control unit.

YES

System OK.

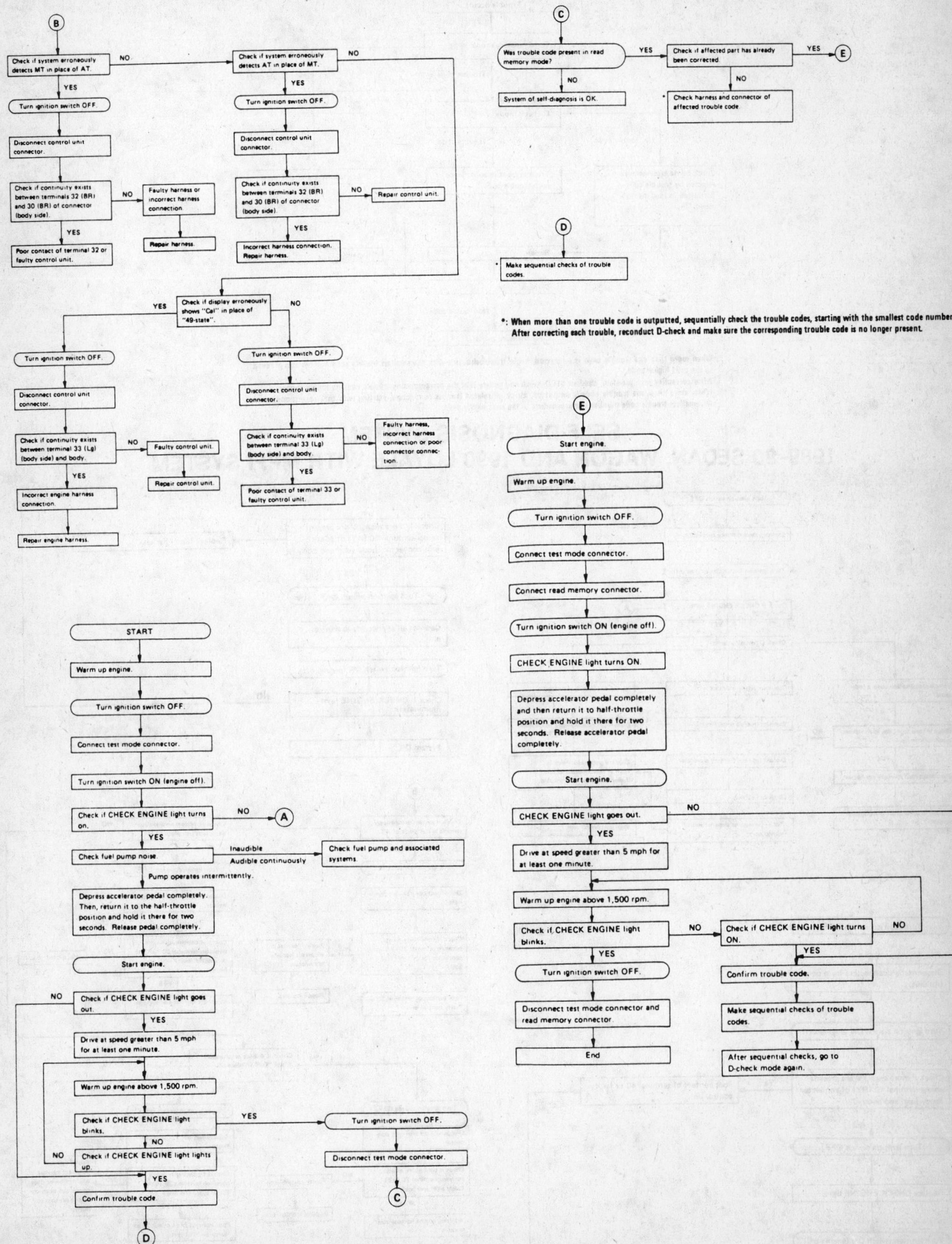
B
Check if system erroneously detects MT in place of AT.
NO
YES
Turn ignition switch OFF.
Disconnect control unit connector.
Check if continuity exists between terminals 32 (BR) and 30 (BR) of connector (body side).
NO
Faulty harness or incorrect harness connection.
Repair harness.
YES
Poor contact of terminal 32 or faulty control unit.
Check if system erroneously detects AT in place of MT.
NO
YES
Turn ignition switch OFF.
Disconnect control unit connector.
Check if continuity exists between terminals 32 (BR) and 30 (BR) of connector (body side).
NO
Repair control unit.
YES
Incorrect harness connection. Repair harness.
Check if display erroneously shows "Cal" in place of "49-state".
YES
NO
Turn ignition switch OFF.
Disconnect control unit connector.
Check if continuity exists between terminal 33 (Lg) (body side) and body.
NO
Faulty control unit.
Repair control unit.
YES
Incorrect engine harness connection.
Repair engine harness.
Turn ignition switch OFF.
Disconnect control unit connector.
Check if continuity exists between terminal 33 (Lg) (body side) and body.
NO
Faulty harness, incorrect harness connection or poor connector connection.
YES
Poor contact of terminal 33 or faulty control unit.
C
Was trouble code present in read memory mode?
YES
Check if affected part has already been corrected.
YES
E
NO
System of self-diagnosis is OK.
NO
* Check harness and connector of affected trouble code.
D
* Make sequential checks of trouble codes.
START
Warm up engine.
Turn ignition switch OFF.
Connect test mode connector.
Turn ignition switch ON (engine off).
Check if CHECK ENGINE light turns on.
NO
A
YES
Check fuel pump noise.
Inaudible
Audible continuously
Check fuel pump and associated systems.
Pump operates intermittently.
Depress accelerator pedal completely. Then, return it to the half-throttle position and hold it there for two seconds. Release pedal completely.
Start engine.
NO
Check if CHECK ENGINE light goes out.
YES
Drive at speed greater than 5 mph for at least one minute.
Warm up engine above 1,500 rpm.
Check if CHECK ENGINE light blinks.
YES
Turn ignition switch OFF.
Disconnect test mode connector.
C
NO
NO
Check if CHECK ENGINE light lights up.
YES
Confirm trouble code.
D
E
Start engine.
Warm up engine.
Turn ignition switch OFF.
Connect test mode connector.
Connect read memory connector.
Turn ignition switch ON (engine off).
CHECK ENGINE light turns ON.
Depress accelerator pedal completely and then return it to half-throttle position and hold it there for two seconds. Release accelerator pedal completely.
Start engine.
CHECK ENGINE light goes out.
NO
YES
Drive at speed greater than 5 mph for at least one minute.
Warm up engine above 1,500 rpm.
Check if CHECK ENGINE light blinks.
NO
Check if CHECK ENGINE light turns ON.
NO
YES
Confirm trouble code.
Make sequential checks of trouble codes.
After sequential checks, go to D-check mode again.
YES
Turn ignition switch OFF.
Disconnect test mode connector and read memory connector.
End

***: When more than one trouble code is outputted, sequentially check the trouble codes, starting with the smallest code number. After correcting each trouble, reconduct D-check and make sure the corresponding trouble code is no longer present.**

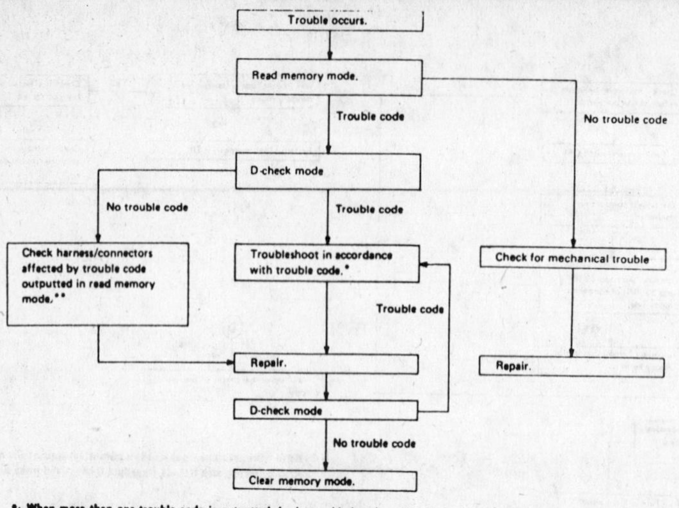
Trouble occurs.
Read memory mode.
Trouble code
No trouble code
D-check mode
No trouble code
Trouble code
Check harness/connectors affected by trouble code outputted in read memory mode.**
Troubleshoot in accordance with trouble code.*
Check for mechanical trouble
Trouble code
Repair.
Repair.
D-check mode
No trouble code
Clear memory mode.

*: When more than one trouble code is outputted, begin troubleshooting with the smallest trouble code number and proceed to the next higher code.
After correcting each problem, conduct the D-check and ensure that the corresponding trouble code no longer appears.
**: When more than one trouble code is outputted, check all related harness connectors, starting with that corresponding to the smallest trouble code number and proceeding to the next higher code.

SELF-DIAGNOSIS SYSTEM
1989–90 SEDAN, WAGON AND 1990 LOYALE WITH MPFI SYSTEM

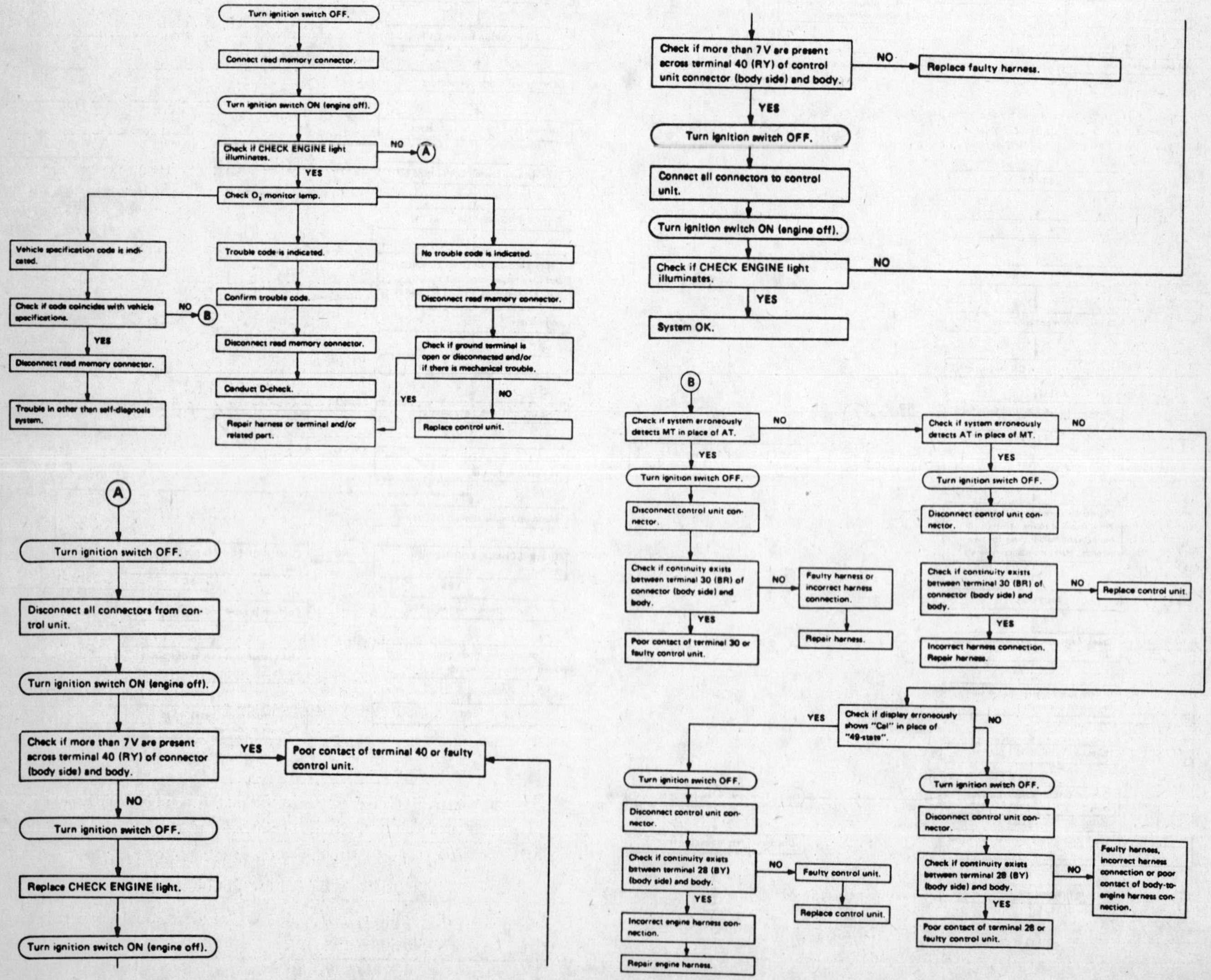

Turn ignition switch OFF.
Connect read memory connector.
Turn ignition switch ON (engine off).
Check if CHECK ENGINE light illuminates. NO → (A)
YES
Check O_2 monitor lamp.

Vehicle specification code is indicated.
Check if code coincides with vehicle specifications. NO → (B)
YES
Disconnect read memory connector.
Trouble in other than self-diagnosis system.

Trouble code is indicated.
Confirm trouble code.
Disconnect read memory connector.
Conduct D-check.
Repair harness or terminal and/or related part.

No trouble code is indicated.
Disconnect read memory connector.
Check if ground terminal is open or disconnected and/or if there is mechanical trouble.
YES → Repair harness or terminal and/or related part.
NO → Replace control unit.

(A)
Turn ignition switch OFF.
Disconnect all connectors from control unit.
Turn ignition switch ON (engine off).
Check if more than 7 V are present across terminal 40 (RY) of connector (body side) and body. YES → Poor contact of terminal 40 or faulty control unit.
NO
Turn ignition switch OFF.
Replace CHECK ENGINE light.
Turn ignition switch ON (engine off).

Check if more than 7 V are present across terminal 40 (RY) of control unit connector (body side) and body. NO → Replace faulty harness.
YES
Turn ignition switch OFF.
Connect all connectors to control unit.
Turn ignition switch ON (engine off).
Check if CHECK ENGINE light illuminates. NO
YES
System OK.

(B)
Check if system erroneously detects MT in place of AT. NO → Check if system erroneously detects AT in place of MT.
YES
Turn ignition switch OFF.
Disconnect control unit connector.
Check if continuity exists between terminal 30 (BR) of connector (body side) and body. NO → Faulty harness or incorrect harness connection. → Repair harness.
YES
Poor contact of terminal 30 or faulty control unit.

Check if system erroneously detects AT in place of MT. NO
YES
Turn ignition switch OFF.
Disconnect control unit connector.
Check if continuity exists between terminal 30 (BR) of connector (body side) and body. NO → Replace control unit.
YES
Incorrect harness connection. Repair harness.

Check if display erroneously shows "Cal" in place of "49-state".

YES
Turn ignition switch OFF.
Disconnect control unit connector.
Check if continuity exists between terminal 28 (BY) (body side) and body. NO → Faulty control unit. → Replace control unit.
YES
Incorrect engine harness connection.
Repair engine harness.

NO
Turn ignition switch OFF.
Disconnect control unit connector.
Check if continuity exists between terminal 28 (BY) (body side) and body. NO → Faulty harness, incorrect harness connection or poor contact of body-to-engine harness connection.
YES
Poor contact of terminal 28 or faulty control unit.

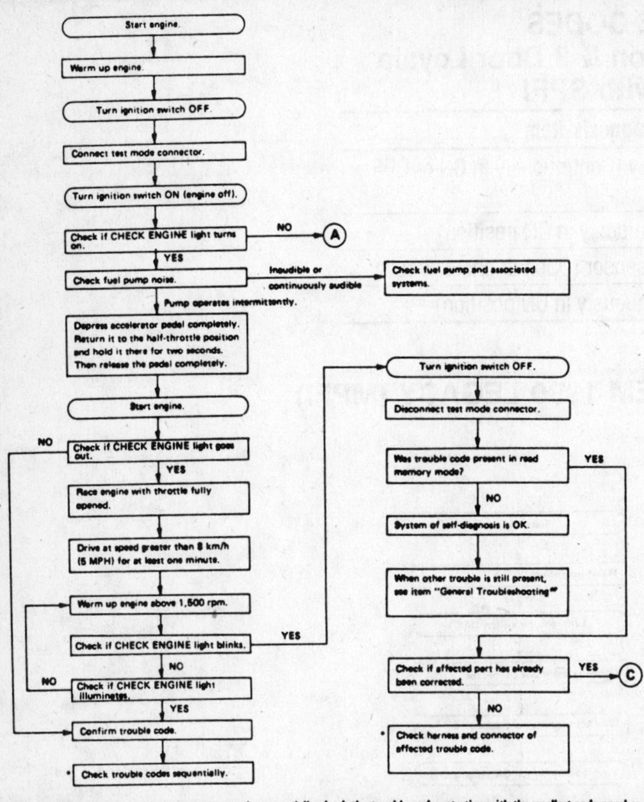
Start engine.
Warm up engine.
Turn ignition switch OFF.
Connect test mode connector.
Turn ignition switch ON (engine off).
Check if CHECK ENGINE light turns on.
NO
A
YES
Check fuel pump noise.
Inaudible or continuously audible
Check fuel pump and associated systems.
Pump operates intermittently.
Depress accelerator pedal completely. Return it to the half-throttle position and hold it there for two seconds. Then release the pedal completely.
Start engine.
NO
Check if CHECK ENGINE light goes out.
YES
Race engine with throttle fully opened.
Drive at speed greater than 8 km/h (5 MPH) for at least one minute.
Warm up engine above 1,500 rpm.
Check if CHECK ENGINE light blinks.
YES
NO
NO
Check if CHECK ENGINE light illuminates.
YES
Confirm trouble code.
* Check trouble codes sequentially.
Turn ignition switch OFF.
Disconnect test mode connector.
Was trouble code present in read memory mode?
YES
NO
System of self-diagnosis is OK.
When other trouble is still present, see item "General Troubleshooting"
Check if affected part has already been corrected.
YES
C
NO
* Check harness and connector of affected trouble code.

*: When more than one trouble code is outputted, sequentially check the trouble codes, starting with the smallest code number. After correcting each trouble, reconduct D-check and make sure the corresponding trouble code is no longer present. If another trouble code is outputted, carry out troubleshooting again.

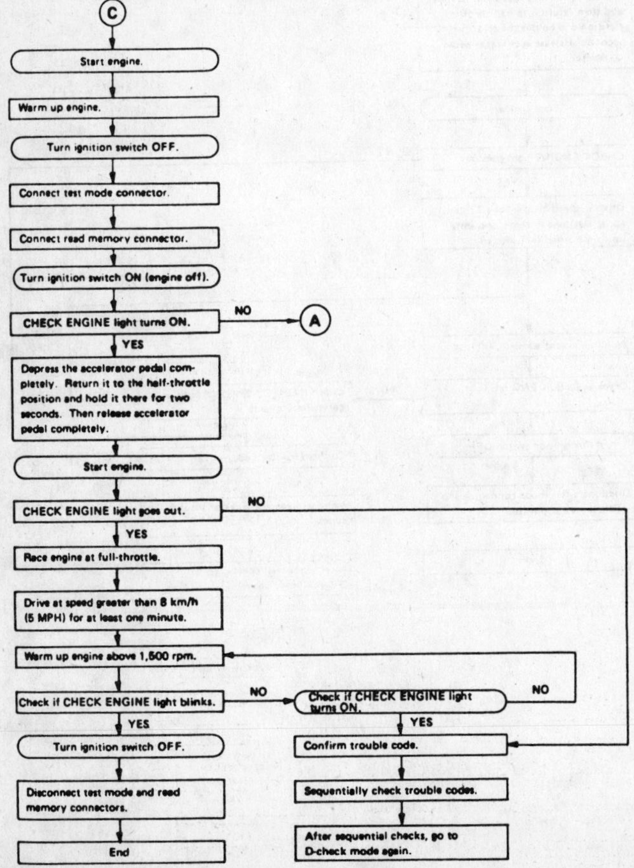
C
Start engine.
Warm up engine.
Turn ignition switch OFF.
Connect test mode connector.
Connect read memory connector.
Turn ignition switch ON (engine off).
CHECK ENGINE light turns ON.
NO
A
YES
Depress the accelerator pedal completely. Return it to the half-throttle position and hold it there for two seconds. Then release accelerator pedal completely.
Start engine.
CHECK ENGINE light goes out.
NO
YES
Race engine at full-throttle.
Drive at speed greater than 8 km/h (5 MPH) for at least one minute.
Warm up engine above 1,500 rpm.
Check if CHECK ENGINE light blinks.
NO
Check if CHECK ENGINE light turns ON.
NO
YES
YES
Turn ignition switch OFF.
Confirm trouble code.
Disconnect test mode and read memory connectors.
Sequentially check trouble codes.
End
After sequential checks, go to D-check mode again.

TROUBLE CODES 1989–90 JUSTY WITH EFC-1200 cc ENGINE

Code No.	U	D	Diagnosis Item
14	○	○	Duty solenoid valve control system
15	○	○	CFC system ①
16		○	Feed back system
17	○	○	Fuel pump & Auto choke
21	○	○	Thermosensor
22	○	○	VLC solenoid valve
23	○	○	Pressure sensor
24	○	○	Idle-up solenoid valve
25	○	○	FCV solenoid valve
32	○	○	O_2 sensor
33	○	○	Car speed sensor
34	○	○	EGR solenoid valve
35	○	○	CPC solenoid valve
41	○	○	Feed back system ②
46	○	○	Radiator fan control system
52		○	Clutch switch ③
53	○	○	HAC solenoid valve
55	○	○	EGR sensor ②
56	○	○	EGR system ②
62		○	Idle-up system (1)
63		○	Idle-up system (2)

① MT only
② California only
③ FWD/MT only

TROUBLE CODES 1989–90 Sedan, Wagon & 3 Door Loyale Vehicles with SPFI

Code No.	Diagnosis Item
11	Crank angle sensor (No reference pulse)
12	Starter switch (Continuously in ON or OFF position while cranking)
13	Crank angle sensor (No position pulse)
14	Fuel injector (Abnormal injector output)
21	Water temperature sensor (Open or shorted circuit)
23	Air flow meter (Open or shorted circuit)
24	Air control valve (Open or shorted circuit)
31	Throttle sensor (Open or shorted circuit)
32	O_2 sensor (Abnormal sensor signal)
33	Car-speed sensor (No signal is present during operation)
34	EGR solenoid valve (Solenoid switch continuously in ON or OFF position, or clogged EGR line) ①
35	Purge control solenoid valve (Solenoid switch continuously in ON or OFF position)
42	Idle switch (Abnormal idle switch signal in relation to throttle sensor output)

TROUBLE CODES
1989–90 Sedan, Wagon & 3 Door Loyale Vehicles with SPFI

Code No.	Diagnosis Item
45	Kick-down control relay (Continuously in ON or OFF position)
51	Neutral switch (Continuously in ON position)
55 ①	EGR gas temperature sensor (Open or short circuit)
61	Parking switch (Continuously in ON position)

① California model only

SELF-DIAGNOSIS SYSTEM 1990 LEGACY (MPFI)

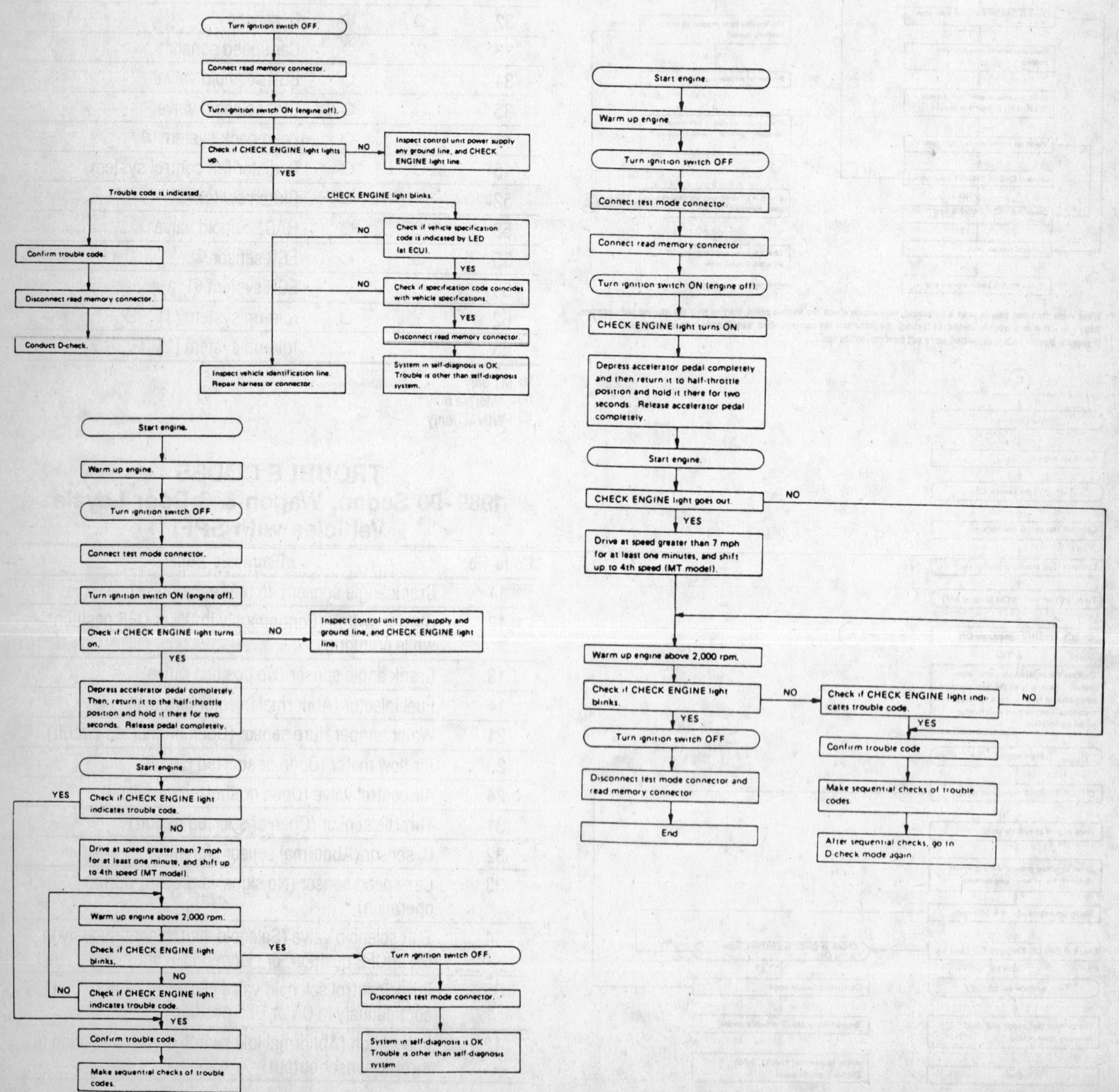
Turn ignition switch OFF.
Connect read memory connector.
Turn ignition switch ON (engine off).
Check if CHECK ENGINE light lights up.
NO
Inspect control unit power supply any ground line, and CHECK ENGINE light line.
YES
Trouble code is indicated.
CHECK ENGINE light blinks.
Confirm trouble code.
Disconnect read memory connector.
Conduct D-check.
Check if vehicle specification code is indicated by LED (at ECU).
Check if specification code coincides with vehicle specifications.
Disconnect read memory connector.
System in self-diagnosis is OK. Trouble is other than self-diagnosis system.
Inspect vehicle identification line. Repair harness or connector.
Start engine.
Warm up engine.
Turn ignition switch OFF.
Connect test mode connector.
Turn ignition switch ON (engine off).
Check if CHECK ENGINE light turns on.
Inspect control unit power supply and ground line, and CHECK ENGINE light line.
Depress accelerator pedal completely. Then, return it to the half-throttle position and hold it there for two seconds. Release pedal completely.
Start engine.
Check if CHECK ENGINE light indicates trouble code.
Drive at speed greater than 7 mph for at least one minute, and shift up to 4th speed (MT model).
Warm up engine above 2,000 rpm.
Check if CHECK ENGINE light blinks.
Turn ignition switch OFF.
Check if CHECK ENGINE light indicates trouble code.
Disconnect test mode connector.
Confirm trouble code.
System in self-diagnosis is OK Trouble is other than self-diagnosis system
Make sequential checks of trouble codes.
Start engine.
Warm up engine.
Turn ignition switch OFF.
Connect test mode connector.
Connect read memory connector.
Turn ignition switch ON (engine off).
CHECK ENGINE light turns ON.
Depress accelerator pedal completely and then return it to half-throttle position and hold it there for two seconds. Release accelerator pedal completely.
Start engine.
CHECK ENGINE light goes out.
Drive at speed greater than 7 mph for at least one minutes, and shift up to 4th speed (MT model).
Warm up engine above 2,000 rpm.
Check if CHECK ENGINE light blinks.
Check if CHECK ENGINE light indicates trouble code.
Turn ignition switch OFF.
Disconnect test mode connector and read memory connector.
End
Confirm trouble code.
Make sequential checks of trouble codes.
After sequential checks, go to D-check mode again.

TROUBLE CODES
1989-90 Sedan, Wagon & 3 Door Vehicles with 1800 cc Engine (MPFI)

Code No.	Diagnosis Item
11	Crank angle sensor (No reference pulse)
12	Starter switch (Continuously in ON position or continuously in OFF position while cranking)
13	Crank angle sensor (No position pulse)
14	Fuel injectors #1 and #2 (Abnormal injector output)
15	Fuel injectors #3 and #4 (Abnormal injector output)
21	Water temperature sensor (Open or shorted circuit)
22	Knock sensor (Open or shorted circuit)
23	Air flow meter (Open or shorted circuit)
31	Throttle sensor (Open or shorted circuit)
32	O_2 sensor (Abnormal sensor signal)
33	Car-speed sensor (No signal is present during operation)
34 ①	EGR solenoid valve (Solenoid switch continuously in ON or OFF position)
35	Purge control solenoid valve (Solenoid switch continuously in ON or OFF position)
41	System too lean
42	Idle switch (Abnormal idle switch signal in relation to throttle sensor output)
44	Duty solenoid valve (Waste gate control)
51	Neutral switch (Continuously in ON position)

① Except California model

TROUBLE CODES
1989-90 XT CPE (MPFI) with 1800 cc and 2700 cc Engine

Code No.	Diagnosis Item	U-check	D-check
11	Crank angle sensor (No reference pulse)	○	○
12	Starter switch (Continuously in ON position or continuously in OFF position while cranking)	○	○
13	Crank angle sensor (No position pulse)	○	○
14	Fuel injectors ① #1 and #2, ② #5 and #6 (Abnormal injector output)	○	○
15	Fuel injectors ① #3 and #4, ② #1 and #2 (Abnormal injector output)	○	○
21	Water temperature sensor (Open or shorted circuit)	○	○
22 ②	Knock sensor (Open or shorted circuit)	○	○
23	Air flow meter (Open or shorted circuit)	○	○
24 ②	By-pass air control valve (Open or shorted circuit)	○	○
25 ②	Fuel injectors #3 and #4 (Abnormal injector output)	○	○
31	Throttle sensor (Open or shorted circuit)	○	○
32	O_2 sensor (Abnormal sensor signal)	○	○
33	Car-speed sensor (No signal is present during operation)	○	○
35	Purge control solenoid valve (Solenoid switch continuously in ON or OFF position)	–	○
41	System too lean	○	○
42	Idle switch (Abnormal idle switch signal in relation to throttle sensor output)	–	○
51	Neutral switch (No signal is present)	–	○

① 1800 cc model
② 2700 cc model

TROUBLE CODES
1990 Legacy with 2200 Engine (MPFI)

Code No.		Diagnosis Item
11.	Crank angle sensor	No signal entered from crank angle sensor, but signal entered from cam angle sensor.
12.	Starter switch	Abnormal signal emitted from ignition switch.
13.	Cam angle sensor	No signal entered from cam angle sensor, but signal entered from crank angle sensor.
14.	Injector #1	Fuel injector inoperative. (Abnormal signal emitted from monitor circuit.)
15.	Injector #2	Fuel injector inoperative. (Abnormal signal emitted from monitor circuit.)
16.	Injector #3	Fuel injector inoperative. (Abnormal signal emitted from monitor circuit.)
17.	Injector #4	Fuel injector inoperative. (Abnormal signal emitted from monitor circuit.)

TROUBLE CODES
1990 Legacy with 2200 Engine (MPFI)

Code No.		Diagnosis Item
21.	Water temperature sensor	Abnormal signal emitted from water temperature sensor.
22.	Knock sensor	Abnormal voltage produced in knock sensor monitor circuit.
23.	Air flow sensor	Abnormal voltage input entered from air flow sensor.
24.	Air control valve	Air control valve inoperative. (Abnormal signal emitted from monitor circuit.)
31.	Throttle position sensor	Abnormal voltage input entered from throttle sensor.
32.	O_2 sensor	O_2 sensor inoperative.
33.	Vehicle speed sensor	Abnormal voltage input entered from speed sensor.
35.	Canister purge solenoid valve	Solenoid valve inoperative.
41.	AF (Air/fuel) learning control	Faulty learning control function.
42.	Idle switch	Abnormal voltage input entered from idle switch.
45.	Atmospheric sensor	Faulty sensor.
49.	Air flow sensor	Use of improper air flow sensor.
51.	Neutral switch (MT)	Abnormal signal entered from neutral switch.
51.	Inhibitor switch (AT)	Abnormal signal entered from inhibitor switch.
52.	Parking switch	Abnormal signal entered from parking switch.

SPECIFICATION CODES
1989-90

		Engines			
		1800cc FWD	1800cc 4WD	2200cc	2700cc
MT	49 state and Canada	05	05	01	01
	California	06	06	02	02
AT	49 state and Canada	07	03	03	03
	California	08	04	04	04

Specification		Spec. code
49-S and Canada	FWD/MT	01
	FWD/ECVT	02
	4WD/MT	03
California	FWD/MT	10
	FWD/FCVT	20
	4WD/MT	30

Specification codes – 1988-90 Justy

SUZUKI ELECTRONIC FUEL INJECTION (EFI)

General Information

The EFI system supplies the vehicle's combustion chambers with air/fuel mixture of optimized ratio under varying driving conditions.

The Throttle Body Injection (TBI) system system consists of a single injector which injects fuel into a throttle body bore. This system consists of 2 major sub-systems: An air/fuel delivery system and the Electronic control system. The main components of the air/fuel delivery system consists of the fuel tank, fuel pump, fuel filter, throttle body assembly, fuel feed and return lines, air cleaner and the Idle Speed Control (ISC) solenoid valve.

The electronic control system consists of the ECM, which controls various devices according to signals received from sensors. Functionally, the air/fuel control system is divided into 5 subsystems. These sub-systems are as follows:

- Fuel injection control system
- ISC solenoid valve control system
- Fuel pump control system
- EGR control system (California model only)
- Shift-up indicator light control system (if equipped)

Also, vehicles equipped with automatic transmission, the ECM sends a throttle valve opening signal to an automatic transmission control module to control the transmission.

The Multi-Point Fuel Injection (MPFI) system consists of a single injector per cylinder, which injects fuel into each intake port of the cylinder head. This system consists of 3 major subsystem: An air intake system, fuel delivery system and an electronic control system.

The main components of the air intake system consists of the air cleaner, air flow meter, throttle body, air valve, ISC solenoid valve and the intake manifold. The main components of the fuel delivery system consists of the fuel tank, fuel pump, fuel filter, fuel feed and return lines and the pressure regulator.

The electronic control system consists of the ECM, which controls various devices according to signals received from sensors. Other controlling functions of the ECM are as follows:

- EGR control system, equipped in only California spec. vehicles
- Evaporative emission control system
- Throttle valve opening signal output for automatic transmission
- Electronic Spark Advance (ESA) system

Both the MPFI and TBI system includes a self-diagnosis function which is controlled by the ECM. If a fault is detected when the ignition switch is **ON** and the engine is running, the ECM will response by turning ON or flashing the CHECK ENGINE light. The self-diagnosis system includes the following components; however, not all vehicles used every components listed below:

- Oxygen Sensor
- Water Temperature Sensor (WTS)
- Throttle switch (TS) – manual transmission only
- Throttle Position Sensor (TPS) – automatic transmission
- Air temperature sensor (ATS)
- Pressure sensor (PS)
- Ignition signal
- Speed sensor
- EGR system
- Air Flow Meter (AFM)
- Crank Angle Sensor
- Idle Switch Circuit
- Lock-up Circuit – automatic transmission
- 5th Switch Circuit – manual transmission only
- Central Processing Unit (CPU) of the ECM

When the ignition switch is turned **ON** and the engine is stop, the CHECK ENGINE light will light. This is only to check the CHECK ENGINE light bulb and circuit. However, if the self-diagnosis system detects trouble in the EFI system, the ECM will turn ON the CHECK ENGINE light with the engine running to warn the driver of such trouble and at the same time it stores the trouble area in the ECM backup memory. The CHECK ENGINE light will remain ON as long as the trouble exists but will turn OFF when the normal condition is restored.

The EFI system also includes a fail-safe function. Should a malfunction occur in the EFI system, the ECM will control such functions as the injector, ISC solenoid valve and others on the basic of a standard data programmed in the ECM. During a fail-safe condition, the ECM will ignore the failure signal and/or the Central Processing Unit (CPU) and provide the vehicle with a diminished level of engine performance.

SYSTEM OPERATION

When the ignition switch is turn **ON**, power is supply to the fuel pump via the fuel pump relay. The fuel pump is activated and the fuel system is pressurized. Simultaneously, an ignition signal is sent from the ignition coil primary circuit. An engine start signal is also sent to the ECM via the engine starter circuit. The ECM uses the engine start signal to determine if the engine is cranking or not and thus control the fuel injector and fuel pump relay.

While the engine is cranking, the ECM keeps the Idle Speed Control (ISC) solenoid valve ON. This provides the engine with a better start. After the engine has started, the ECM gradually reduce the ISC solenoid valve ON time to maintain the specified idle speed.

When the injector (solenoid coil) is energized by the ECM, the needle valve which is incorporated with the plunger opens and the injector, which is under pressure, injects fuel in a conic dispersion into the throttle body bore. The injected fuel is mixed wit the air which has been filtered through the air cleaner in the throttle body. The air/fuel mixture is drawn through clearance between the throttle valve, throttle bore and an idle by-pass passage into the intake manifold. The intake manifold then distributes the air/fuel mixture to each combustion chamber. Should the fuel system pressure exceed a preset level, a valve, located in the fuel pressure regulator opens and excess fuel returns to the fuel tank via the return line.

An air valve, located in the throttle body supplies bypass air into the intake manifold without letting the air pass through the throttle valve when the engine is cold. This condition causes the engine speed to increase (fast idle state) and thus provides engine warm-up. As the engine is warmed up, a piston inside the air valve gradually blocks the amount of air passing through the air valve and simultaneously the engine speed is reduced. As the engine coolant temperature reaches approximately 176°F (80°C), the valve is fully open and the engine speed returns to normal idle speed.

The ECM also uses the following signals to compensates for engine speeds and/or fuel injection ON time:

- Air Conditioning Signal – The ECM uses this signal to determine whether the air conditioner is operating or not and uses it as 1 of the signals for controlling the ISC valve operation.
- Battery Voltage – The fuel injector is driven by its solenoid coil based upon the ECM output. However, there is some delay called "ineffective injection time", which doesn't provide fuel, between the ECM signal and the valve action. The ineffective injection time depends on the battery voltage signal. The ECM takes this information to compensate for fuel injection time.
- **R**, **D**, **2** or **L** Range Signal – When in these range, the automatic transmission (automatic transmission only) module sends a battery voltage signal to the ECM. The ECM uses this signal as 1 of the signals to control the fuel injector and the ISC solenoid valve.
- Illumination Light Signal – If equipped, this signal is sent from the illumination light circuit. It is used to reduce intensity of the shift-up indicator light when the illumination light is ON.

Self-Diagnostic System

ENTERING SELF-DIAGNOSTIC SYSTEM

The system has a diagnosis switch in the fuse box under the instrument panel. When the ignition switch is turned **ON** with the diagnosis switch is grounded (spare fuse connected) the CHECK ENGINE light begins flashing diagnostic codes stored in the ECU.

On some models there is a diagnosic monitor connector under the hood. Diagnostic codes will also be obtained by grounding this terminal.

OBTAINING AND CLEARING DIAGNOSTIC CODES – SWIFT

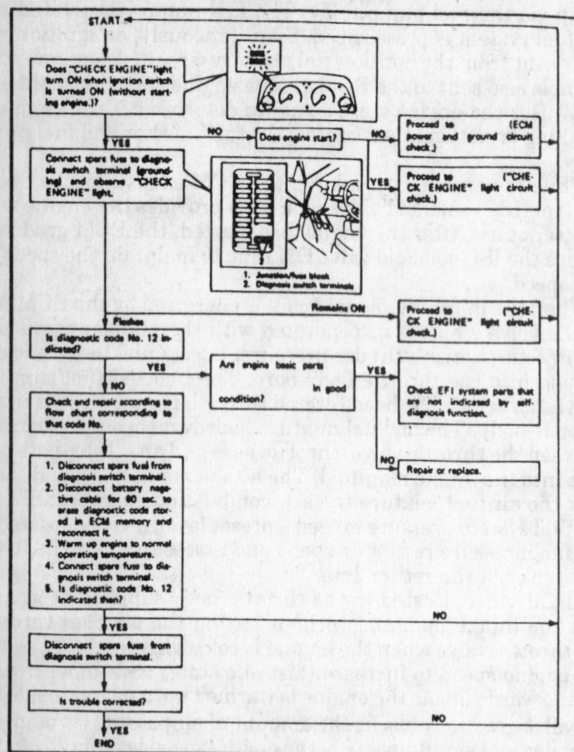
START
Does "CHECK ENGINE" light turn ON when ignition switch is turned ON (without starting engine.)?
YES
NO
Does engine start?
NO
Proceed to (ECM power and ground circuit check.)
YES
Proceed to ("CHECK ENGINE" light circuit check.)
Connect spare fuse to diagnosis switch terminal (grounding) and observe "CHECK ENGINE" light.
1. Junction/fuse block
2. Diagnosis switch terminals
Remains ON
Proceed to ("CHECK ENGINE" light circuit check.)
Flashes
Is diagnostic code No. 12 indicated?
YES
Are engine basic parts in good condition?
YES
Check EFI system parts that are not indicated by self-diagnosis function.
NO
Repair or replace.
NO
Check and repair according to flow chart corresponding to that code No.
1. Disconnect spare fuse from diagnosis switch terminal.
2. Disconnect battery negative cable for 60 sec. to erase diagnostic code stored in ECM memory and reconnect it.
3. Warm up engine to normal operating temperature.
4. Connect spare fuse to diagnosis switch terminal.
5. Is diagnostic code No. 12 indicated then?
NO
YES
Disconnect spare fuse from diagnosis switch terminal.
Is trouble corrected?
NO
YES
END

OBTAINING AND CLEARING DIAGNOSTIC CODES – SIDEKICK 1600

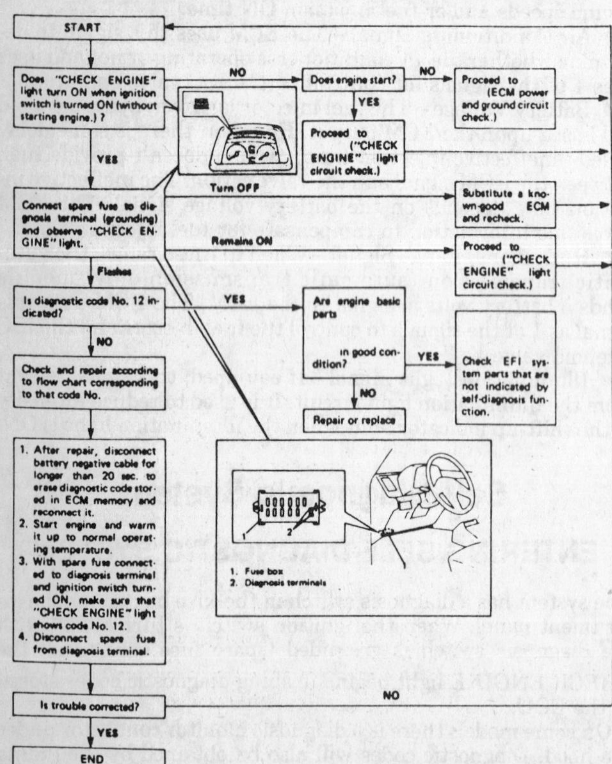
START
Does "CHECK ENGINE" light turn ON when ignition switch is turned ON (without starting engine.)?
NO
Does engine start?
NO
Proceed to (ECM power and ground circuit check.)
YES
Proceed to ("CHECK ENGINE" light circuit check.)
YES
Connect spare fuse to diagnosis terminal (grounding) and observe "CHECK ENGINE" light.
Turn OFF
Substitute a known-good ECM and recheck.
Remains ON
Proceed to ("CHECK ENGINE" light circuit check.)
Flashes
Is diagnostic code No. 12 indicated?
YES
Are engine basic parts in good condition?
YES
Check EFI system parts that are not indicated by self-diagnosis function.
NO
Repair or replace
NO
Check and repair according to flow chart corresponding to that code No.
1. After repair, disconnect battery negative cable for longer than 20 sec. to erase diagnostic code stored in ECM memory and reconnect it.
2. Start engine and warm it up to normal operating temperature.
3. With spare fuse connected to diagnosis terminal and ignition switch turned ON, make sure that "CHECK ENGINE" light shows code No. 12.
4. Disconnect spare fuse from diagnosis terminal.
Is trouble corrected?
NO
YES
END
1. Fuse box
2. Diagnosis terminals

If more than a single code is stored, the CHECK ENGINE light indicates the codes 3 times each, for as long as the ignition switch is **ON** and the diagnostic terminal is grounded.

CLEARING TROUBLE CODES

1. After repair, disconnect the negative battery cable for 60 seconds or more.
2. Start engine and allow to reach normal operating temperature.
3. Ground diagnostic switch and check for a Code 12.
4. Disconnect diagnostic switch ground.

NOTE: Some vehicle computers have a learning ability. If a change is noted in vehicle performance after clearing codes, it may be due to the computers learning ability. To restore performance, warm vehicle to normal operating temperature and drive at part throttle with moderate acceleration, before performing any additional diagnosis.

OBTAINING AND CLEARING DIAGNOSTIC CODES – SIDEKICK 1300

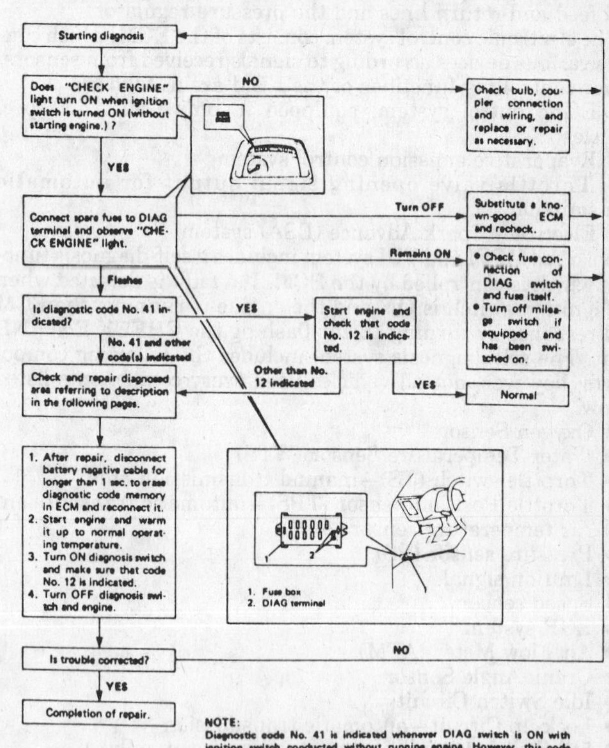
Starting diagnosis
Does "CHECK ENGINE" light turn ON when ignition switch is turned ON (without starting engine.)?
NO
Check bulb, coupler connection and wiring, and replace or repair as necessary.
YES
Connect spare fuses to DIAG terminal and observe "CHECK ENGINE" light.
Turn OFF
Substitute a known-good ECM and recheck.
Remains ON
• Check fuse connection of DIAG switch and fuse itself.
• Turn off mileage switch if equipped and has been switched on.
Flashes
Is diagnostic code No. 41 indicated?
YES
Start engine and check that code No. 12 is indicated.
No. 41 and other code(s) indicated
Other than No. 12 indicated
YES
Normal
Check and repair diagnosed area referring to description in the following pages.
1. After repair, disconnect battery negative cable for longer than 1 sec. to erase diagnostic code memory in ECM and reconnect it.
2. Start engine and warm it up to normal operating temperature.
3. Turn ON diagnosis switch and make sure that code No. 12 is indicated.
4. Turn OFF diagnosis switch and engine.
1. Fuse box
2. DIAG terminal
NO
Is trouble corrected?
YES
Completion of repair.
NOTE: Diagnostic code No. 41 is indicated whenever DIAG switch is ON with ignition switch conducted without running engine. However, this code disappears when engine is started and is replaced with code No. 12 if the system is normal.

ENGINE DIAGNOSTIC CODES SWIFT EXCEPT GTI

DIAGNOSTIC CODE NO.	DIAGNOSTIC CODE MODE	DIAGNOSTIC AREA
12	1 2	Normal
13	1 3	Oxygen sensor
14	1 4	Water temperature sensor
15	1 5	
21	2 1	Throttle switch (M/T model only)
21	2 1	Throttle position sensor (A/T model only)
22	2 2	
23	2 3	Air temperature sensor
25	2 5	
24	2 4	Speed sensor
31	3 1	Pressure sensor
32	3 2	
41	4 1	Ignition signal
51	5 1	EGR system (California spec. model only)

ENGINE DIAGNOSTIC CODES—SWIFT GTI

DIAGNOSTIC CODE NO.	"CHECK ENGINE" LIGHT FLASHING PATTERN	DIAGNOSTIC ITEM
13	1 3	Oxygen sensor
14	1 4	Water temperature sensor
15	1 5	
21	2 1	Throttle position sensor
22	2 2	
23	2 3	Air temperature sensor
25	2 5	
31	3 1	Pressure sensor
32	3 2	
41	4 1	Ignition signal
42	4 2	Lock-up signal (For AT vehicle) 5th switch (For MT vehicle)
44	4 4	Idle switch of throttle position sensor
45	4 5	
51	5 1	EGR system (For California spec. vehicle)
53	5 3	Ground circuit (For california spec. vehicle)

ENGINE DIAGNOSTIC CODES SIDEKICK 1600

DIAGNOSTIC CODE NO.	DIAGNOSTIC CODE MODE	DIAGNOSTIC AREA
12	1 2	Normal
13	1 3	Oxygen sensor
14	1 4	Water temperature sensor
15	1 5	
21	2 1	Throttle position sensor
22	2 2	
24	2 4	Speed sensor
33	3 3	Air flow sensor
34	3 4	
41	4 1	Ignition signal
42	4 2	Crank angle sensor
51	5 1	EGR system (California spec. model only)
52	5 2	Fuel leakage from fuel injector (California spec. model only)
ON		ECM

ENGINE DIAGNOSTIC CODES SIDEKICK 1300

Diagnostic Code No.	Check Engine Light Flashing Parttern	Diagnostic Area
ON		ECM
12	1 2	Normal
12		Normal
13		Oxygen sensor
14		Water temperature sensor (WTS)
21		Wide open (WO) micro switch and idle micro switch
33		Idle up vacuum switching valve (VSV)
41		Ignition signal circuit
51		ECM
52		Fuel cut solenoid valve
53		EGR vacuum switching valve (VSV)
54		Mixture control solenoid valve
55		Vent solenoid valve
ON		ECM

TOYOTA ELECTRONIC FUEL INJECTION (EFI) SYSTEM

General Information

This system is broken down into 3 major systems; the Fuel System, Air Induction System and the Electronic Control System. Most of the engines are equipped with a Toyota Computer Control System (TCCS) which centrally controls the electronic fuel injection, electronic spark advance and the exhaust gas recirculation valve. The systems can be diagnosed by means of an Electronic Control Unit (ECU) which employs a microcomputer. The ECU and the TCCS control the following functions:

ELECTRONIC FUEL INJECTION (EFI)

The ECU receives signals from the various sensors indicating changing engine operations conditions such as:

1. Intake air volume
2. Intake air temperature
3. Coolant temperature sensor
4. Engine rpm
5. Acceleration/deceleration
6. Exhaust oxygen content

These signals are utilized by the ECU to determine the injection duration necessary for an optimum air-fuel ratio.

The Electronic Spark Advance (ESA)

The ECU is programmed with data for optimum ignition timing during any and all operating conditions. Using the data provided by sensors which monitor various engine functions (rpm, intake air volume, coolant temperature, etc.), the microcomputer (ECU) triggers the spark at precisely the right moment.

Idle Speed Control (ISC)

The ECU is programmed with specific engine speed values to respond to different engine conditions (coolant temperature, air conditioner on/off, etc.). Sensors transmit signals to the ECU which controls the flow of air through the by-pass of the throttle valve and adjusts the idle speed to the specified value. Some vehicles use a ISC valve while others use an air valve to control throttle body by-pass air flow.

Exhaust Gas Recirculation (EGR)

The ECU detects the coolant temperature and controls the EGR operations accordingly.

Fail-Safe Function

In the event of a computer malfunction, a backup circuit will take over to provide minimal driveability. Simultaneously, the CHECK ENGINE warning light is activated.

Turbo Indicator

The ECU detects turbocharger pressure, which is determined by the intake volume and the engine rpm, and lights a green colored turbocharger indicator light located in the combination meter. Moreover, if the turbocharger pressure increases abnormally, the ECU will light the CHECK ENGINE warning light on the instrument panel.

Self-Diagnostic System

ENTERING SELF-DIAGNOSTIC AND DIAGNOSTIC CODE DISPLAY

The ECU contains a built-in self diagnosis system by which troubles with the engine signal the engine signal network are detected and a "CHECK ENGINE" warning light on the instrument panel flashes Code No. 12–71 (these code numbers vary from model to model). The CHECK ENGINE light on the instrument panel informs the driver that a malfunction has been detected. The light goes out automatically when the malfunction has been cleared.

The diagnostic code can be read by the number of blinks of the CHECK ENGINE warning light when the proper terminals of the check connector are short-circuited. If the vehicle is equipped with a super monitor display, the diagnostic code is indicated on the display screen.

1. The battery voltage should be above 11 volts. Throttle valve fully closed (throttle position sensor IDL points closed).
2. Place the transmission in **P** or **N** range. Turn the A/C switch **OFF**. Start the engine and let it run to reach its normal operating temperature.

Without Super Monitor Display

NORMAL MODE

1. Turn the ignition switch to the **ON** position. Do not start the engine. Use a suitable jumper wire and short the terminals of the check connector.
2. Read the diagnostic code as indicated by the number of flashes of the CHECK ENGINE warning light.
3. If the system is operating normally (no malfunction), the light will blink once every ¼ second.
4. In the event of a malfunction, the light will blink once every ½ second. The 1st number of blinks will equal the 1st digit of a 2 digit diagnostic code. After a 1.5 second pause, the 2nd number of blinks will equal the 2nd number of a 2 digit diagnostic code. If there are 2 or more codes, there will be a 2.5 second pause between each.
5. After all the codes have been output, there will be a 4.5 second pause and they will be repeated as long as the terminals of the check connector are shorted.

NOTE: In event of multiple trouble codes, indication will begin from the smaller value and continue to the larger in order.

6. After the diagnosis check, remove the jumper wire from the check connector.

TEST MODE

The Cressida provides a diagnostic test mode.

1. Using a jumper wire, connect the TE2 and E1 terminals of the Toyota Diagnostic Communication Link (TDCL), then turn the ignition switch **ON** to begin the diagnostic test mode.
2. Start the engine and drive the vehicle at a speed of 10 mph or more. Simulate the conditions where the malfunction has be reported to happen.
3. Using a jumper wire, connect the TE1 and E1 terminals of the TDCL connector.
4. Read the diagnosis code as indicated by the number of engine flashes.
5. After diagnosis check remove the jumper wires.

With Super Monitor Display

The Super Monitor was offered as an option on the 1988 Cressida and 1988 Supra vehicles.

1. Turn the ignition switch to the **ON** position. Do not start the engine.
2. Simultaneously push and hold in the SELECT and INPUT M keys for at least 3 seconds. The letters DIAG will appear on the screen.
3. After a short pause, hold the SET key in for at least 3 sec-

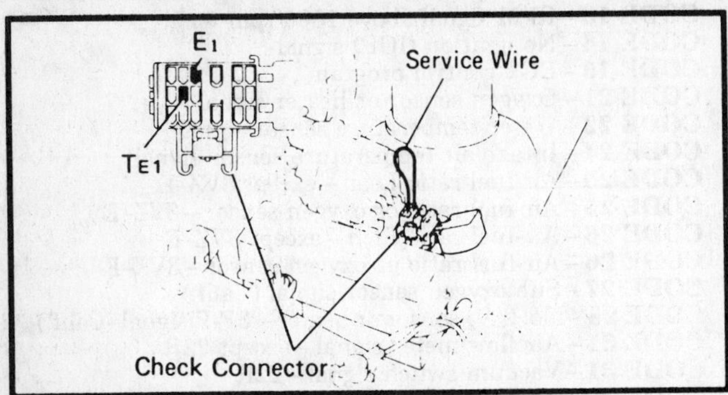
E1
Service Wire
TE1
Check Connector
Diagnostic service connector – except Van

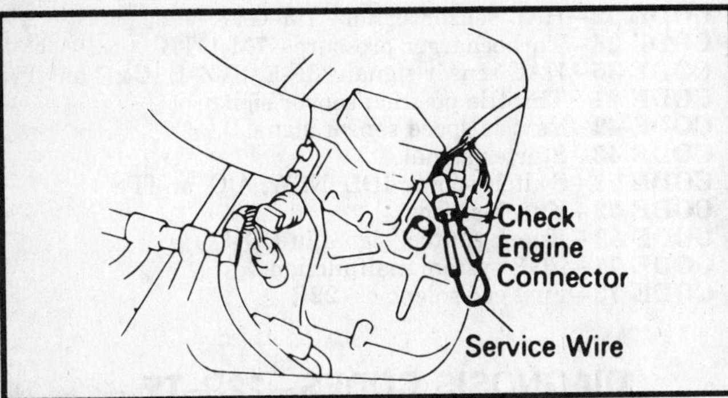
Check
Engine
Connector
Service Wire
Diagnostic service connector – Van

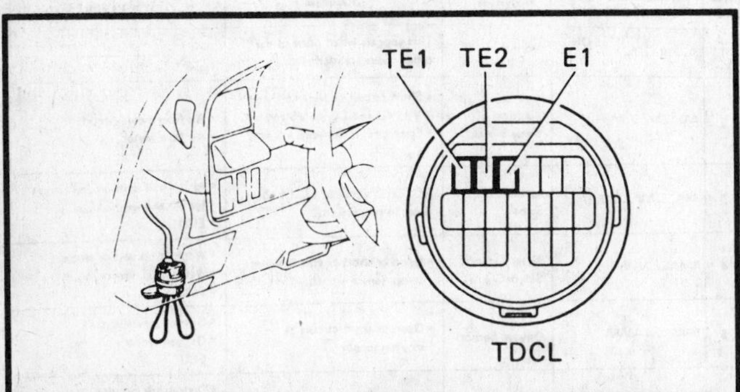
TE1
TE2
E1
TDCL
Toyota Diagnostic Communication Link (TDCL) – Cressida

onds. If the system is normal (no malfunctions), ENG-OK will appear on the screen.

4. If there is a malfunction, the code number for it will appear on the screen. In the event of 2 or more numbers, there will be a 3 second pause between each (Example ENG-42).

5. After confirmation of the diagnostic code, either turn **OFF** the ignition switch or push the super monitor display key **ON** so the time appears.

DIAGNOSIS INDICATION

1. Including "Normal" or "ENG–OK", the ECU is programmed several diagnostic codes.

2. When more than a single code is indicated, the lowest number code will appear first. However, no other code will appear along with Code 11.

3. All detected diagnostic codes, except Code 51 and 53, will be retained in memory by the ECU from the time of detection until canceled out.

4. Once the malfunction is cleared, the CHECK ENGINE warning light on the instrument panel will go out but the diagnostic code(s) remain stored in the ECU memory (except for Code 51 and 53).

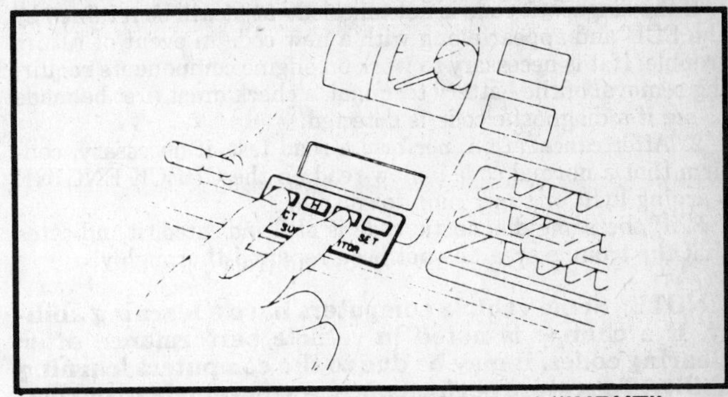
SET
Simultaneously pushing "SELECT" and "INPUT"

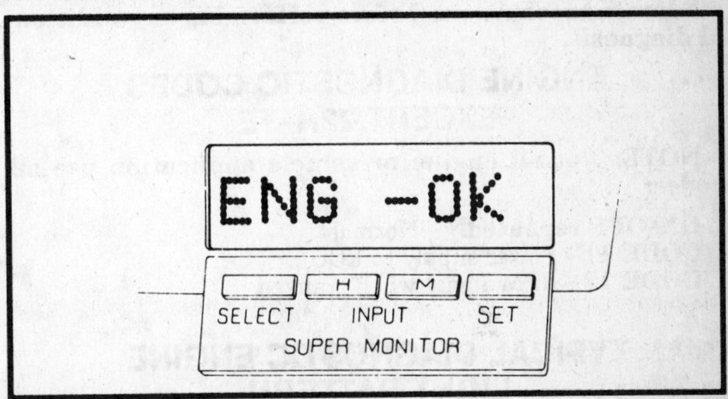
ENG -OK
H
M
SELECT
INPUT
SET
SUPER MONITOR
No malfunction message

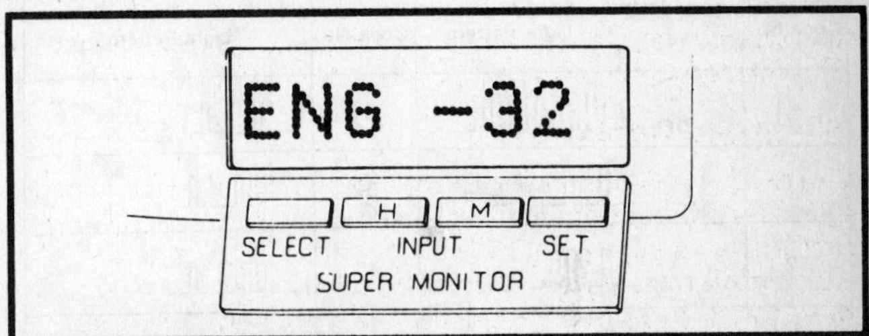
ENG -32
H
M
SELECT
INPUT
SET
SUPER MONITOR
Engine Code 32 message

CLEARING TROUBLE CODES

1. After repairing the trouble area, the diagnostic code that is retained in the ECU memory must be canceled out by removing the ECU memory fuse fuse for 30 seconds or more, depending on the ambient temperature (the lower temperature, the longer the fuse must be left out with the ignition switch **OFF**). To clear ECU memory remove the following fuse:

Camry – 15A EFI fuse
Celica – 15A EFI fuse
Corolla – 15A Stop fuse
1988 Cressida – 15A EFI fuse
1989–90 Cressida – 20A EFI fuse
MR-2 – 7.5 AM2 fuse
Supra – 15A EFI fuse
Tercel – 20A Stop fuse
Van – 15A EFI fuse
Pick-up – 15A EFI fuse
4Runner – 15A EFI fuse
Land Cruiser – 15A EFI fuse

NOTE: Cancellation can also be done by removing the battery negative terminal, but keep in mind, when removing the negative battery cable, the other memory systems (radio, ETR, clock, etc.) will also be canceled out.

If the diagnostic code is not canceled out, it will be retained by the ECU and appear along with a new code in event of future trouble. If it is necessary to work on engine components requiring removal of the battery terminal, a check must first be made to see if a diagnostic code is detected.

2. After cancellation, perform a road test, if necessary, confirm that a normal code is now read on the CHECK ENGINE warning light or super monitor display.

3. If the same diagnostic code is still indicated, it indicates that the trouble area has not been repaired thoroughly.

NOTE: Some vehicle computers have a learning ability. If a change is noted in vehicle performance after clearing codes, it may be due to the computers learning ability. To restore performance, warm vehicle to normal operating temperature and drive at part throttle with moderate acceleration, before performing any additional diagnosis.

ENGINE DIAGNOSTIC CODES EXCEPT 22R–TE

NOTE: Not all engine or vehicle application use all codes:

ON–OFF repeatedly – Normal
CODE 11 – Power supply to ECU
CODE 12 – RPM signal when cranked
CODE 13 – RPM signal above 1500 rpm
CODE 14 – No ignition (IGF) signal
CODE 16 – ECT control program
CODE 21 – Oxygen sensor or heater signal
CODE 22 – Water temperature sensor signal
CODE 24 – Intake air temperature sensor signal
CODE 25 – Air-fuel ratio Lean – except 3VZ–E
CODE 25 – Air-fuel ratio or oxygen sensor – 3VZ–E
CODE 26 – Air-fuel ratio Rich – except 3VZ–E
CODE 26 – Air-fuel ratio or oxygen sensor – 3VZ–E
CODE 27 – Sub oxygen sensor signal (Calif.)
CODE 28 – No. 2 O_2 sensor or heater – 3F-E signal (Calif.)
CODE 31 – Air flow meter signal – except 22R
CODE 31 – Vacuum switch signal – 22R
CODE 32 – Air flow meter signal – except 7M-GTE & 3VZ- E
CODE 32 – HAC sensor signal – 7M–GTE
CODE 34 – Turbocharger pressure – 7M–GTE
CODE 35 – HAC sensor signal – 3F-E, 3VZ–E (C&C only)
CODE 41 – Throttle position sensor signal
CODE 42 – Vehicle speed sensor signal
CODE 43 – Starter signal
CODE 51 – Switch signal; IDL, NSW, A/C, or TPS
CODE 52 – Knock sensor
CODE 53 – Knock control signal in ECU
CODE 71 – EGR system malfunction
CODE 72 – Fuel cut solenoid – 22R

TYPICAL DIAGNOSTIC ENGINE LIGHT PATTERN

Code No.	Light Pattern	Code No.	Light Pattern
–	ON / OFF	31	
11		32	
12		41	
13		42	
14		43	
21		51	
22		52	
23		53	

DIAGNOSIS CODES – 22R–TE

Code No.	Number of blinks "CHECK ENGINE"	System	Diagnosis	Trouble area
1	ON OFF	Normal	This appears when none of the other codes are identified.	————
2		Air flow. Meter Signal	• Short circuit between VC and VB, VC and E_2, or VS and VC. • Open circuit between VC and E_2.	• Air flow meter circuit • Air flow meter
3		Ignition Signal	NO "IGf" signal to ECU 4–5 times in succession.	• Igniter and ignition coil circuit • Igniter and ignition coil • ECU
4		Water Temp. Sensor Signal	• Open or short circuit in water temp. sensor signal.	• Water temp. sensor circuit • Water temp. sensor • ECU
5		Oxygen Sensor	• Open or short circuit in oxygen sensor	• Oxygen sensor circuit • Oxygen sensor • ECU
6		RPM Signal	NO "Ne signal to ECU when engine speed is above 1,500 rpm.	• Distributor circuit • Distributor • Igniter circuit • Igniter • ECU
7		Throttle Position Sensor Signal	• Open or short circuit in throttle position sensor signal.	• Throttle position sensor circuit • Throttle position sensor
8		Intake Air Temp. Sensor Signal	• Open or short circuit in intake air temp. sensor signal.	• Intake air temp. sensor circuit • Intake air temp. sensor ECU
10		Starter Signal	• NO "STA" signal to ECU until engine speed reaches 800 rpm with vehicle not moving.	• IG switch circuit • IG switch • ECU
11		Switch Signal	• Air conditioner switch ON, neutral start switch OFF, idle switch OFF during diagnosis check.	• A/C switch circuit • A/C switch • A/C Amplifire • Throttle position sensor circuit • Throttle position sensor • ECU
12		Knock Sensor Signal	• Open or short circuit in knock sensor signal.	• Knock sensor circuit • Knock sensor • ECU
13		Knock Control Signal	• Knock control ECU faulty.	• ECU
14		Turbocharger Pressure	• When the fuel cut-off due to high boost is occured.	• Turbocharger • Air flow meter circuit • Air flow meter • ECU

Fuel Injection Systems

SECTION 4

INDEX

NISSAN ELECTRONIC CONCENTRATED CONTROL SYSTEM (ECCS) THROTTLE BODY INJECTION SYSTEM

NISSAN ELECTRONIC CONCENTRATED CONTROL SYSTEM (ECCS) PORT FUEL INJECTION SYSTEM

SECTION 4 FUEL INJECTION SYSTEMS

SUBARU MULTI-POINT FUEL INJECTION (MPFI) SYSTEM

SUBARU SINGLE-POINT FUEL INJECTION (SPFI) SYSTEM

1990 SUBARU LEGACY MULTI-POINT FUEL INJECTION (MPFI) SYSTEM

SUZUKI ELECTRONIC FUEL INJECTION (EFI) SYSTEMS

TOYOTA ELECTRONIC FUEL INJECTION (EFI) SYSTEM

NISSAN ELECTRONIC CONCENTRATED CONTROL (THROTTLE BODY INJECTION) SYSTEM (ECCS)

General Information

The Nissan Electro Injection System is a throttle body fuel injection system used on 1988–90 models equipped with the E16i, VG30i, GA16i or Z24i engines. The electronic control unit consists of a microcomputer, inspection lamps, a diagnostic mode selector and connectors for signal input and output and for power supply. The Electronic Concentrated Control System (ECCS) computer controls the amount of fuel injected, ignition timing, mixture ratio feedback, idle speed, fuel pump operation, mixture heating, Air Injection Valve (AIV) operation, Exhaust Gas Recirculation (EGR) and vapor canister purge operation.

SYSTEM COMPONENTS

Crank Angle Sensor

The crank angle sensor is a basic component of the entire system. It monitors engine speed and piston position and sends other signals which the control unit uses to calculate ignition timing and other functions.

The crank angle sensor has a rotor plate and a wave forming circuit. On all models, the signal rotor plate has 360 slits for 1 degree signals (crank angle). On models equipped with the VG30I engine, the rotor plate also consists of 6 slits for 120 degrees signal (engine speed). On models equipped with Z24I, E16I and GA16I engines, the rotor plate also consists of 4 slits for 180 degrees signal (engine speed). Light emitting diodes (LED's) and photo diodes are built in the wave forming circuit. When the rotor plate passes the space between the LED and the photo diode, the slits of the rotor plate continually cut the light which is sent to the photo diode from the LED, causing rough shaped pulses. These pulses are converted into on-off signals by the wave forming circuit and sent to the control unit as input signals.

Air Flow Meter

The air flow meter measures the intake air flow rate by taking a part of the entire flow. Measurement are made in such a manner that the control unit receives electrical output signals varied by the amount of heat emitted from a hot wire placed in the stream of intake air. When intake air flows into the intake manifold through a route around the hot wire, the heat generated by the wire is taken away by the passing air. The amount of heat removed depends on the air flow, but the maximum temperature of the hot wire is automatically controlled, requiring more electrical current to maintain the controlled temperature in the wire as the amount of intake air increases. By measuring the amount of current necessary to maintain the hot wire temperature, the control unit measures the amount of intake air passing the wire and therefore knows the volume of air entering the engine.

Water Temperature Sensor

The water temperature sensor, located on the front side of the intake manifold, detects engine coolant temperature and sends signals to the control unit. The air temperature sensor is installed in the air cleaner and senses the temperature of the intake air. The water and air temperature sensors employs a thermistor which is sensitive to changes in temperature. The electrical resistance of a thermistor decreases as temperature rises.

Exhaust Gas Sensor

The exhaust gas sensor, which is placed in the exhaust pipe, monitors the amount of oxygen in the exhaust gas. The sensor is made of ceramic titania which changes electrical resistance at the ideal air/fuel ratio (14.7:1). The control unit supplies the sensor with approximately 1 volt and takes the output voltage of the sensor depending on its resistance. The oxygen sensor is equipped with a heater to bring it to operating temperature quickly.

Throttle Sensor/Idle Switch

The throttle sensor/idle switch is attached to the throttle body and operates in response to accelerator pedal movement. This sensor has 2 functions: it contains an idle switch and throttle position sensor. The idle switch closes when the throttle valve is positioned at idle and opens when it is in any other position. The throttle sensor is a potentiometer which transforms the throttle valve position into output voltage and feeds the voltage signal to the control unit. In addition, the throttle sensor detects the opening or closing speed of the throttle valve and feeds the rate of voltage change to the control unit.

Power Steering Oil Pressure Switch

A power steering oil pressure switch is attached to the power steering high pressure line and detects the power steering load, sending a load signal to the control unit which then sends the idle-up signal to the idle speed control (ISC) valve.

Fuel Pressure Regulator

A fuel pressure regulator is built into the side of the throttle body. It maintains fuel pressure at a constant 14 psi. Since the injected fuel amount depends on injection pulse duration, it is necessary to keep the fuel pressure constant. The fuel pump with a fuel damper is located in the fuel tank. The pump is an electric, vane roller type.

Fuel Injector

The fuel injector is basically a small solenoid valve. As the control unit sends injection signals to the injector, high pressure fuel, which is supplied to the coil built into the injector, pulls the ball valve back and the fuel is injected onto the throttle valve through the nozzle. The amount of injected fuel is controlled by the computer by means of longer or shorter signals (pulse duration) to the injector. A mixture heater is located between the throttle valve and the intake manifold. This is designed and operated for atomizing fuel in the cold engine start condition. The heater is also controlled by the computer.

Mixture Heater

The mixture heater is located between the throttle valve and the intake manifold. This is designed and operated for atomizing fuel in the cold engine start condition. The ECU controls the heater.

Idle Speed Control (ISC) Valve

The Idle Speed Control (ISC) valve is a rotary solenoid valve that receives a pulse signal from the control unit. This pulse signal determines the position of the slider, thereby varying bypass air quantity which raises or lowers the idle speed. The ISC valve

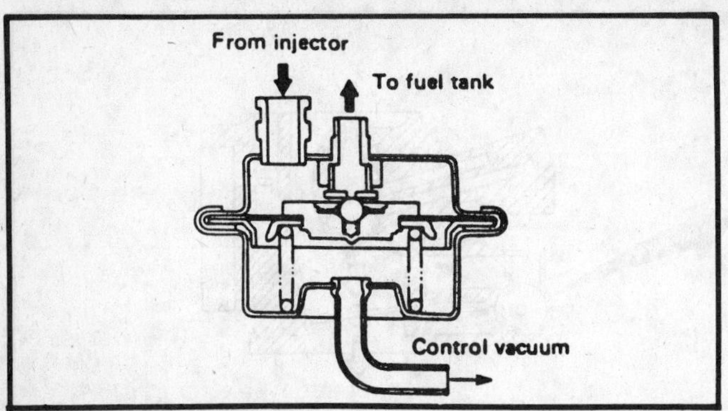
From injector
To fuel tank
Control vacuum

Fuel pressure regulator

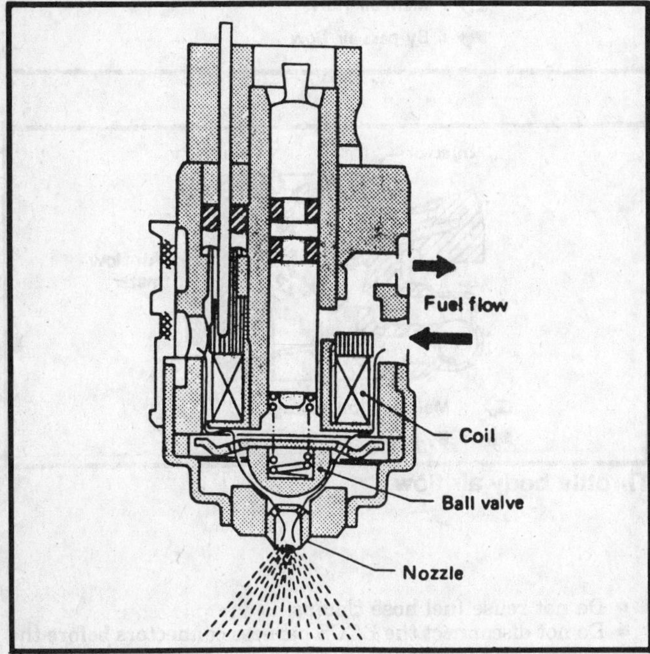
Fuel flow
Coil
Ball valve
Nozzle

Fuel injector

Mixture heater

has additional functions which include idle-up after cold start (fast idle), idle speed feedback control, idle-up for air conditioner and power steering (fast idle control device) and deceleration vacuum control.

Power Transistor and Ignition Coil

The ignition signal from the ECU is amplified by the power transistor, which turns the ignition coil primary circuit on and off, inducing the proper high voltage in the secondary circuit. The ignition coil is a small, molded type.

Electronic Control Unit (ECU)

The ECU consists of a microcomputer, inspection lamps, a diagnostic mode selector and connectors for signal input and output, and for power supply. The unit has control of the injected fuel amount, ignition timing, mixture ratio feedback, idle speed, fuel pump operation, mixture heating, AIV operation, and EGR and canister purge operation.

SYSTEM OPERATION

In operation, the on-board computer (control unit) calculates the basic injection pulse width by processing signals from the crank angle sensor and air flow meter. Receiving signals from each sensor which detects various engine operating conditions, the computer adds various enrichments (which are preprogrammed) to the basic injection amount. In this manner, the optimum amount of fuel is delivered through the injectors. The fuel is enriched when starting, during warm-up, when accelerating and when operating under a heavy load. The fuel is leaned during deceleration according to the closing rate of the throttle valve.

The mixture ratio feedback system (closed loop control) is designed to control the air/fuel mixture precisely to the stoichiometric or optimum point so that the 3-way catalytic converter can minimize CO, HC and NOx emissions simultaneously. The optimum air/fuel fuel mixture is 14.7:1. This system uses an exhaust gas (oxygen) sensor located in the exhaust manifold to give an indication of whether the fuel mixture is richer or leaner than the stoichiometric point. The control unit adjusts the injection pulse width according to the sensor voltage so the mixture ratio will be within the narrow window around the stoichiometric fuel ratio. The system goes into closed loop as soon as the oxygen sensor heats up enough to register. The system will operate under open loop when starting the engine, when the engine temperature is cold, when exhaust gas sensor temperature is cold, when driving at high speeds or under heavy load, at idle (after mixture ratio learning is completed), when the exhaust gas sensor monitors a rich condition for more than 10 seconds and during deceleration.

Ignition timing is controlled in response to engine operating conditions. The optimum ignition timing in each driving condition is preprogrammed in the computer. The signal from the control unit is transmitted to the power transistor and controls ignition timing. The idle speed is also controlled according to engine operating conditions, temperature and gear position. On manual transmission models, if battery voltage is less than 12 volts for a few seconds, a higher idle speed will be maintained by the control unit to improve charging function.

The control unit energizes the mixture heating relay when the engine is running and the water temperature is below 122°F (50°C). The mixture heating relay will be shut off when several minutes have passed after the water temperature exceeds 122°F (50°C). In addition, the Air Injection Valve (AIV), which supplies secondary air to the exhaust manifold, is controlled by the computer according to engine temperature. When the engine is cold, the AIV system operates to reduce HC and CO emissions. In extremely cold conditions, the AIV control system does not operate to reduce afterburning.

A signal from the control unit is also sent to the EGR and fuel vapor canister purge cut solenoid valve, which cuts the vacuum for the EGR and canister control valve. The EGR and canister purge activates when the vehicle speed is above 6 mph, the water temperature is above 140°F (60°C) and the engine is under light load at low rpm. The vacuum will be interrupted unless all of the conditions are met.

Finally, the control unit operates the air flow meter self-cleaning system. After the engine is stopped, the control unit heats up the hot wire to approximately 1832°F (1000°C) to burn off dust adhering to the hot wire. The self-cleaning function will

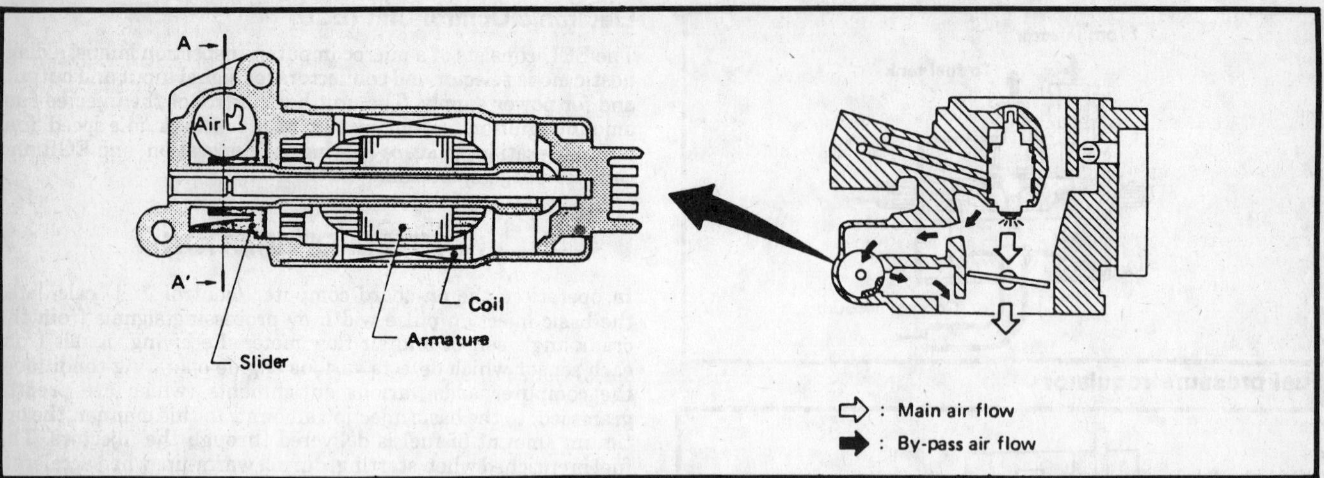
A
Air
A'
Coil
Armature
Slider
: Main air flow
: By-pass air flow

Idle speed control

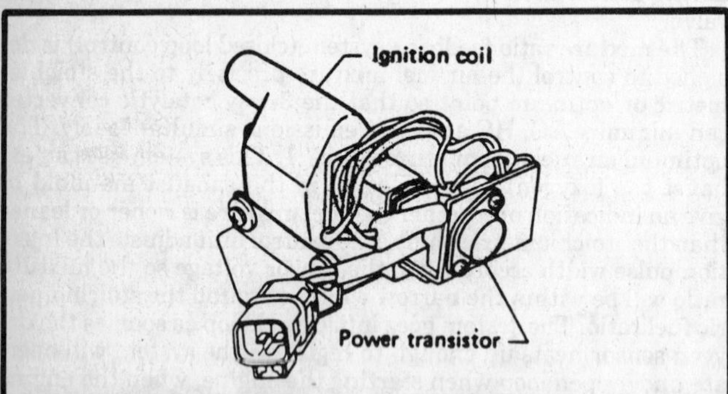
Ignition coil
Power transistor

Power transistor and ignition coil—typical

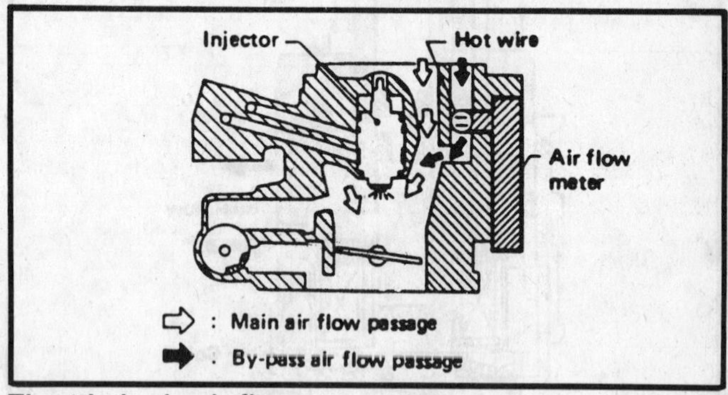
Injector
Hot wire
Air flow meter
: Main air flow passage
: By-pass air flow passage

Throttle body air flow

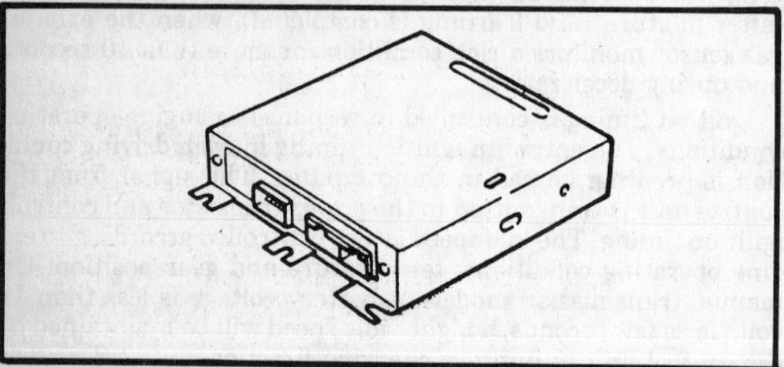

Electronic control unit (ECU)

activate if the engine speed has exceeded 2000 rpm before the key is turned **OFF**, vehicle speed has exceeded 12 mph before the key is turned **OFF**, the water temperature is between 140–203°F (60–95°C), or the engine has been stopped by turning the ignition key **OFF**. Self-cleaning will be activated only if all of the above conditions are met. The hot wire will be heated for 0.3 seconds, 5 seconds after the ignition is switched **OFF**.

There is a fail-safe system built into the control unit should the air flow meter malfunction. If the air flow meter output voltage is higher or lower than the specified value, the control unit senses an air flow meter malfunction and substitutes the throttle sensor signal for the air flow meter input. It is possible to drive the vehicle and start the engine, but the engine speed will not rise more than 2400 rpm in order to inform the driver of fail-safe system operation while driving.

SERVICE PRECAUTIONS

- Do not operate the fuel pump when the fuel lines are empty.
- Do not reuse fuel hose clamps.
- Do not disconnect the ECCS harness connectors before the battery ground cable has been disconnected.
- Make sure all ECCS connectors are fastened securely. A poor connection can cause an extremely high surge voltage in the coil and condenser and result in damage to integrated circuits.
- Keep the ECCS harness at least 4 in. away from adjacent harnesses to prevent an ECCS system malfunction due to external electronic "noise."
- Keep all parts and harnesses dry during service.
- Before attempting to remove any parts, turn off the ignition switch and disconnect the battery ground cable.
- Always use a 12 volt battery as a power source.
- Do not attempt to disconnect the battery cables with the engine running.
- Do not depress the accelerator pedal when starting.
- Do not rev up the engine immediately after starting or just prior to shutdown.
- Do not attempt to disassemble the ECCS control unit under any circumstances.
- If a battery cable is disconnected, the memory will return to the ROM (programmed) values. Engine operation may vary slightly, but this is not an indication of a problem. Do not replace parts because of a slight variation.
- If installing a 2-way or CB radio, keep the antenna as far as possible away from the electronic control unit. Keep the antenna feeder line at least 8 in. away from the ECCS harness and do not let the 2 run parallel for a long distance. Be sure to ground the radio to the vehicle body.

Diagnosis and Testing

SELF-DIAGNOSTIC SYSTEM

The self-diagnostic function is useful for diagnosing malfunctions in major sensors and actuators of the ECCS system. There are 5 modes in self-diagnostics:

MODE 1

During closed loop operation, the green inspection lamp turns ON when a lean condition is detected and OFF when a rich condition is detected. During open loop operation, the red inspection lamp stays OFF.

MODE 2

The green inspection lamp function is the same as in Mode 1. During closed loop operation, the red inspection lamp turns ON and OFF simultaneously with the green inspection lamp when the mixture ratio is controlled within the specified value. During open loop operation, the red inspection lamp stays OFF.

MODE 3

This mode is the same as the former self-diagnosis mode.

MODE 4

During this mode, the inspection lamps monitor the ON/OFF condition of the idle switch, starter switch and vehicle speed sensor.

In switches ON/OFF diagnosis system, ON/OFF operation of the following switches can be detected continuously:

- Idle switch
- Starter switch
- Vehicle speed sensor (if equipped)

1. Idle Switch and Starter Switch—The switches ON/OFF status at the point when Mode IV is selected is stored in ECU memory. When either switch is turned from **ON** to **OFF** or **OFF** to **ON**, the red LED on ECU alternately comes on and goes off each time switching is detected.

2. Vehicle Speed Sensor—The switches ON/OFF status at the point when Mode IV is selected is stored in ECU memory. When vehicle speed is 12 mph (20 km/h) or slower, the green LED on ECU is off. When vehicle speed exceeds 12 mph (20 km/h), the green LED on ECU comes ON.

MODE 5

The moment a malfunction is detected, the display will be presented immediately by flashing the inspection lamps during the driving test.

In real time diagnosis, if any of the following items are judged to be faulty, a malfunction is indicated immediately:

- Crank angle sensor (180 degrees signal and 1 degree signal)
- Ignition signal
- Air flow meter output signal
- Fuel pump

Consequently, this diagnosis is a very effective measure to diagnose whether the above systems cause the malfunction or not, during driving test. Compared with self-diagnosis, real time diagnosis is very sensitive, and can detect malfunctioning conditions in a moment. Further, items regarded to be malfunctions in this diagnosis are not stored in ECU memory.

To switch the modes, turn the ignition switch **ON**, then turn the diagnostic mode selector on the control unit fully clockwise and wait for the inspection lamps to flash. Count the number of flashes until the inspection lamps have flashed the number of the desired mode, then immediately turn the diagnostic mode selector fully counterclockwise.

NOTE: When the ignition switch is turned OFF during diagnosis in each mode, and then turned back on again after the power to the control unit has dropped off completely, the diagnosis will automatically return to Mode

The stored memory will be lost if the battery terminal is disconnected, or Mode 4 is selected after selecting Mode 3. However, if the diagnostic mode selector is kept turned fully clockwise, it will continue to change in the order of Mode 1, 2, 3, etc., and in this case, the stored memory will not be erased.

In Mode 3, the self-diagnostic mode, the control unit constantly monitors the function of sensors and actuators regardless of ignition key position. If a malfunction occurs, the information is stored in the control unit and can be retrieved from the memory by turning **ON** the diagnostic mode selector on the side of the control unit. When activated, the malfunction is indicated by flashing a red and green LED (also located on the control unit). Since all the self-diagnostic results are stored in the control unit memory, even intermittent malfunctions can be diagnosed. A malfunctioning part's group is indicated by the number of both red and green LED's flashing. First, the red LED flashes and the green flashes follow. The red LED refers to the number of tens, while the green refers to the number of units. If the red LED flashes twice and the green LED flashes once, a Code 21 is being displayed. All malfunctions are classified by their trouble code number.

The diagnostic result is retained in the control unit memory until the starter is operated 50 times after a diagnostic item is judged to be malfunctioning. The diagnostic result will then be canceled automatically. If a diagnostic item which has been judged malfunctioning and stored in memory is again judged to be malfunctioning before the starter is operated 50 times, the second result will replace the previous one and stored in the memory until the starter is operated 50 more times.

In Mode 5 (real time diagnosis), if the crank angle sensor, ignition signal or air flow meter output signal are judged to be malfunctioning, the malfunction will be indicated immediately. This diagnosis is very effective for determining whether these systems are causing a malfunction during the driving test. Compared with self-diagnosis, real time diagnosis is very sensitive and can detect malfunctioning conditions immediately. However, malfunctioning items in this diagnosis mode are not stored in memory.

TESTING PRECAUTIONS

- Before connecting or disconnecting control unit ECU harness connectors, make sure the ignition switch is **OFF** and the negative battery cable is disconnected to avoid the possibility of damage to the control unit.
- When performing ECU input/output signal diagnosis, remove the pin terminal retainer from the 20 and 16-pin connectors to make it easier to insert tester probes into the connector.
- When connecting or disconnecting pin connectors from the ECU, take care not to bend or break any pin terminals. Check that there are no bends or breaks on ECU pin terminals before attempting any connections.
- Before replacing any ECU, perform the ECU input/output signal diagnosis to make sure the ECU is functioning properly or not.

Keys to symbols

DISCONNECT : Check after disconnecting the connector to be measured.

CONNECT : Check after connecting the connector to be measured.

When measuring voltage or resistance at connector with tester probes, there are two methods of measurement; one is done from terminal side and the other from harness side. Before measuring, confirm symbol mark again.

HS : Inspection should be done from harness side.

TS : Inspection should be done from terminal side.

ECCS test symbols and identification chart

- After performing the Electronic Control System Inspection, perform the ECCS self-diagnosis and driving test.
- When measuring supply voltage of ECU controlled components with a circuit tester, separate one tester probe from another. If the two tester probes accidentally make contact with each other during measurement, a short circuit will result and damage the power transistor in the ECU.

ELECTRONIC CONTROL UNIT (ECU)

Location

1988–89 Truck/Pathfinder–Under right (passenger) seat
1988 Van–Behind left (driver) trim panel behind driver's seat
1988–90 Pulsar–Under right (passenger) seat
1988–90 Sentra–Under right (passenger) seat
1990 Sentra With 4WD–Under left (driver) seat

FUEL SYSTEM

Pressure Releasing

CAUTION

Fuel system pressure must be relieved before disconnecting any fuel lines or attempting to remove any fuel system components.

1. Remove the fuel pump fuse from the fuse box.
2. Start the engine.
3. After the engine stalls, crank it over for a few seconds to make sure all fuel is exhausted from the lines.
4. Turn the ignition switch **OFF**. Install the fuel pump fuse.

Pressure Testing

1. Relieve fuel system pressure as previously described.
2. Disconnect the fuel inlet hose at the electro injection unit.
3. Install a fuel pressure gauge.
4. Start the engine and check the fuel line for leakage.
5. Read the pressure on the pressure gauge. Pressure gauge should read as follows:
   1988 Pulsar and 2WD Sentra–14 psi
   1988 Van and 4WD Sentra–36.3 psi
   1989 Pulsar and Sentra–43.4 psi
   1988–89 Truck and Pathfinder–36.3 psi
   1990 Pulsar and Sentra
6. Relieve fuel system pressure again.
7. Remove the pressure gauge from the fuel line and reconnect the fuel inlet hose.

NOTE: When reconnecting a fuel line, always use a new clamp. Make sure that the screw of the clamp does not contact any adjacent parts and tighten the hose clamp to 1 ft. lb.

AIR FLOW METER

Testing

Before removing the air flow meter, remove the throttle valve switch. When removing the air flow meter, pull it out vertically, taking care not to bend or damage the plug portion. Never touch the sensor portion with your finger and apply silicone grease to the mating surface between the air flow meter and throttle body when installing to allow heat to escape.

NOTE: Failure to use silicone grease for heat dissipation will result in air flow meter failure.

1988 PULSAR AND SENTRA

Apply battery voltage between terminals E (+) and C (–), then measure the voltage between terminals A (+) and D (–). Voltage should be 1.5–2.0 volts without air flow at the sensor, and 2.5–4.0 volts when air is blown through the sensor. Use oral air pressure only when blowing through the sensor. If any other results are obtained during testing, replace the air flow meter.

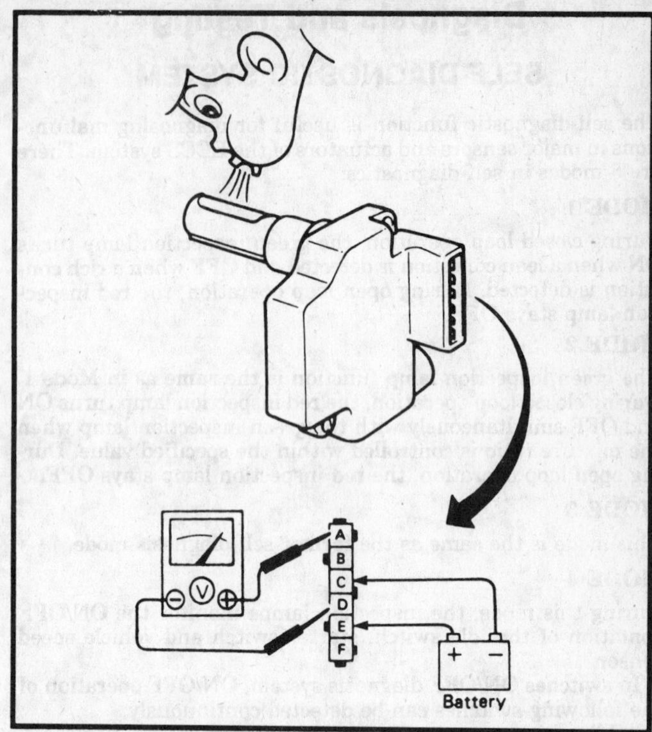
A
B
C
D
E
F
V
Battery
Air flow meter testing–1988 Pulsar and Sentra

ALL MODELS EXCEPT 1988 PULSAR AND SENTRA

1. Remove the air duct assembly.
2. Turn the ignition switch to the **ON** position and disconnect the air flow meter electrical connector.
3. Using a voltmeter, check between terminal B and ground. Voltmeter should read approximately 12 volts.
4. Using a voltmeter, check between terminal D and ground. Voltmeter should read approximately 0.0–0.5 volts.
5. If above specifications are not as indicated, replace the air flow meter.

IDLE SPEED CONTROL (ISC) VALVE

Testing

1988 PULSAR AND SENTRA

1. Use an ohmmeter to check the resistance between the terminals. Resistance between terminals A and B should be 9.5–10.5 ohms. Resistance between terminals B and C should be 8.5–9.5 ohms.
2. Check the insulation between each terminal and the ISC valve body. No continuity should exist when a probes are touched to the terminal and valve body. If continuity exists, replace the ISC valve.
3. Apply battery voltage between terminals B (+) and C (–), then check that the ISC valve is fully closed.
4. Apply battery voltage between terminals B (+) and A (–), then check that the ISC valve is fully open.
5. Check the opening clearance of the ISC valve without applying voltage to terminals. The opening clearance should be 0–0.08 in. (0–2mm).
6. If any test results are different than described, replace the ISC valve. When installing the ISC valve to the throttle body, tighten the bolts in the sequence illustrated to 3–4 ft. lbs. (4–5 Nm).

Air flow meter testing—Except 1988 Pulsar and Sentra

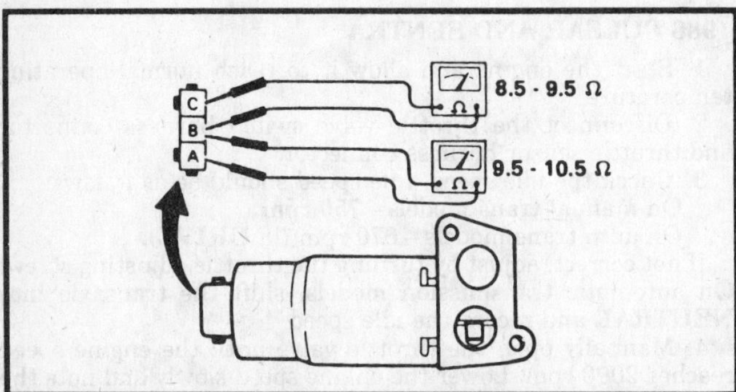
C
B
A
8.5 - 9.5 Ω
9.5 - 10.5 Ω

Idle speed control (ISC) valve testing

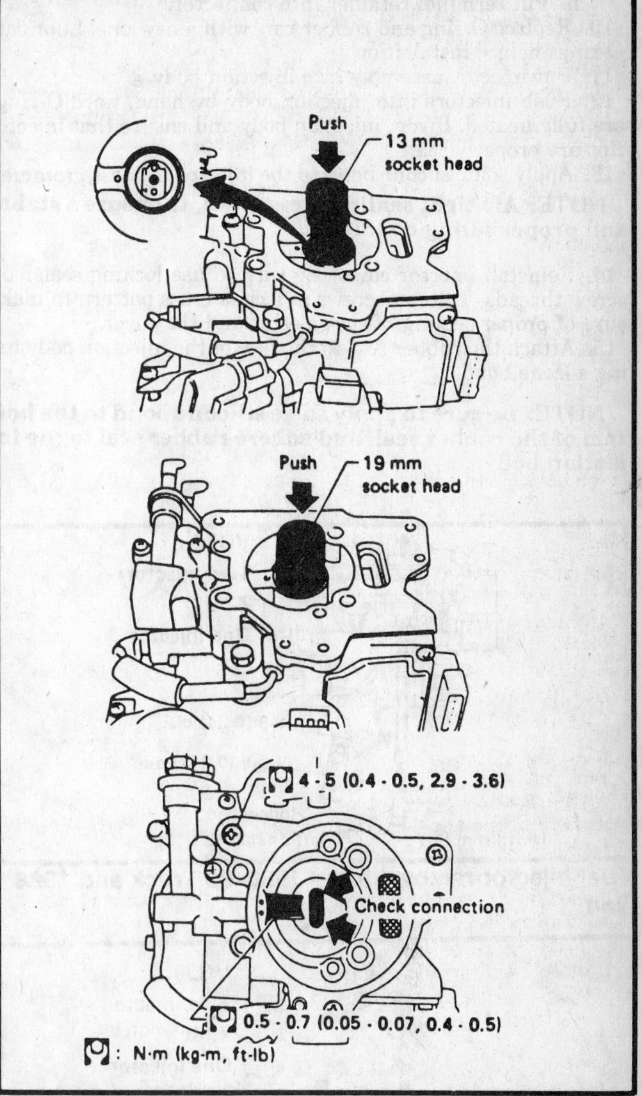
Push
13 mm socket head
Push
19 mm socket head
4 - 5 (0.4 - 0.5, 2.9 - 3.6)
Check connection
0.5 - 0.7 (0.05 - 0.07, 0.4 - 0.5)
: N·m (kg-m, ft-lb)

Fuel injector installation—All except 1988–89 Truck and 1988 Van

Component Replacement

INJECTOR

Removal and Installation

ALL EXCEPT 1988 VAN AND 1988–89 TRUCK

1. Relieve fuel system pressure.
2. Remove the injector cover and pull the injector straight up to remove it from the throttle body. Be careful not to break or bend the injector terminal.
3. Remove the O-rings from the injector.
4. Coat the lower O-ring with a small amount of clean engine oil and install it on the injector.
5. Install the injector and push it into place using a suitable tool or 13mm socket. Align the direction of the injector terminals and be careful not to bend or break any terminals.
6. Install the upper O-ring by pushing it into place with a suitable tool or 19mm socket.
7. Install the upper (white) plate.
8. Install the injector cover with the rubber plug removed. Make sure that the two O-rings are installed in the injector cover.
9. Check for proper connection between the injector terminal and injector cover terminal, then install the rubber plug. After installation, make sure there is no fuel leakage and that the engine is idling properly.

1988 VAN AND 1988–89 TRUCK

1. Relieve the fuel pressure from the fuel system.
2. Drain approximately 1⅛ quart (1 liter) of engine coolant.
3. Remove or disconnect the following parts:
   a. Air cleaner
   b. Harness connectors for throttle sensor, idle switch, injectors and air flow meter
   c. Accelerator wire
   d. Fuel hose
   e. Coolant hose
   f. ASCD wire (if equipped)
4. Remove injection body from intake manifold.
5. Remove rubber seal and injector harness grommet from the injection body.
6. Remove injector body.
7. With the throttle valve kept in the fully opened position, tap bottom of fuel injector using a suitable tool.

NOTE: If nozzle tip is damaged or deformed by the tool, replace the injector.

8. Disconnect harness of of a malfunctioning injector from the harness connector.
9. Put harness of a new injector into injector harness grommet and harness tube.

NOTE: Harness grommet should be replaced with a new one every time it is removed.

   a. Fix boots and terminals in harness with terminal pliers and then put harness in connector.

b. Put terminal retainer into connector.

10. Replace O-ring and rubber ring with a new one. Lubricate O-rings before installation.

11. Put injector assembly into injection body.

12. Push injectors into injection body by hand, until O-rings are fully heated. Invert injection body and ensure that injector tips are properly seated.

13. Apply some silcone bond to the injector harness grommet.

NOTE: Air-tight sealing is essential, to ensure a stable and proper idling condition.

14. Reinstall injector cover. Be sure to use locking sealer on screw threads. Tighten screws in a criss-cross pattern to make sure of proper seating of the injector and the cover.

15. Attach the rubber seal to the face of the injection body using silcone bond.

NOTE: Be sure to apply some silicone bond to the bottom of the rubber seal, and adhere rubber seal to the injection body.

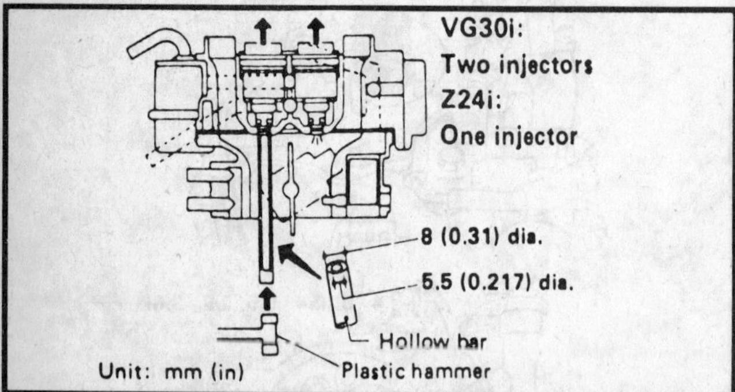
VG30i:
Two injectors
Z24i:
One injector
8 (0.31) dia.
5.5 (0.217) dia.
Hollow bar
Unit: mm (in)
Plastic hammer

Fuel injector removal tool – 1988–89 Truck and 1988 Van

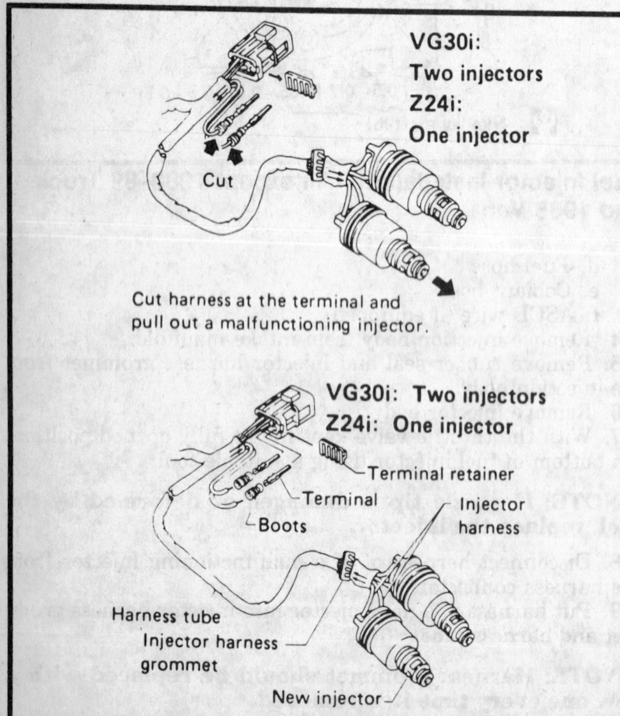
VG30i:
Two injectors
Z24i:
One injector
Cut
Cut harness at the terminal and
pull out a malfunctioning injector.
VG30i: Two injectors
Z24i: One injector
Terminal retainer
Terminal
Boots
Injector
harness
Harness tube
Injector harness
grommet
New injector

Fuel injector-from-harness removal and installation – 1988–89 Truck and 1988 Van

16. Reinstall or connect the following parts:
   a. Harness connectors
   b. Accelerator wire
   c. Coolant hose
   d. Fuel hose
   e. ASCD wire (if equipped)

17. Add approximately 1⅛ quart (1 liter) of engine coolant to the radiator.

18. Start the engine and make sure that no fuel leaks from between the injector cover and the injection body.

19. Perform mixture ratio feedback system inspection to make sure that there is no fuel leakage at injector top seal.

THROTTLE VALVE SWITCH

Adjustment

1988 PULSAR AND SENTRA

1. Start the engine and allow it to reach normal operating temperature.

2. Disconnect the throttle valve switch harness connector and throttle sensor harness connector.

3. Check the idle speed. Idle speed should be as follows:
   On manual trans models – 750 rpm
   On auto trans models – 670 rpm (in **DRIVE**).

If not correct, adjust by turning the throttle adjusting screw. On automatic transmission models, shift the transaxle into **NEUTRAL** and record the idle speed.

4. Manually open the throttle valve until the engine speed reaches 2000 rpm. Lower the engine speed slowly and note the rpm at which the idle contact turns from OFF to ON. It should be 900–1200 rpm on manual transmission models. On automatic transmission models, it should be approximately 300 rpm over whatever engine speed was recorded in NEUTRAL (Step 3) ± 150 rpm. If not correct, adjust by loosening the the throttle valve switch securing screws and turning the throttle valve switch.

5. Reconnect the throttle valve switch harness connector and throttle sensor harness connector.

DASH POT

Inspection and Adjustment

1. Start the engine and allow it to reach normal operating temperature. Make sure the idle speed is adjusted properly.

2. Turn the throttle by hand and read the engine speed when the dashpot just touches the adjusting screw.
   1988 Pulsar 2200–3000 rpm
   1988 Sentra – 2000–2800 rpm
   1988–89 Truck with VG30i engine and manual trans – 1600– 1800 rpm
   1988–89 Truck with VG30i engine and auto trans – 2000–2200 rpm
   1988 Van and 1988–89 Truck with Z24i engine – 2700–3300 rpm
   1989–90 Pulsar and Sentra with GA16i engine and manual trans – 1800–2200 rpm
   1989–90 Pulsar and Sentra with GA16i engine and auto trans – 2000–2800 rpm (in **NEUTRAL**)

3. If out of specification, adjust it by turning the adjusting screw.

FAST IDLE

Inspection and Adjustment

1. Warm up engine sufficiently.

2. Make sure that aligning mark stamped on fast idle cam meets center of roller installed on cam follow lever. If not, cor-

rect location of fast idle cam by turning adjusting screw. If not adjustable, replace thermo element.

3. Check clearance (G) between roller and fast idle cam. Clearance should be as folows:

Z24i engine – 0.028–0.118 in. (0.7–3.0mm)
VG30i engine – 0.020–0.118 in. (0.5–3.0mm)
GA16i engine with manual trans – 0.071–0.193 in. (1.8–4.9mm)
GA16i engine with auto trans – 0.075–0.213 in. (1.9–5.4mm)

4. If above specifications are not as indicated, adjust clearance (G) by turning adjusting screw (S2). Adjusting clearance should be as follows:

Z24i engine – 0.047–0.063 in. (1.2–1.6mm)
VG30i engine–0.031–0.047 in. (0.8–1.2mm)
GA16i engine with manual trans – 0.083–0.161 in. (2.1–4.1mm)
GA16i engine with auto trans – 0.094–0.193 in. (2.4–4.9mm)

NOTE: Make sure that the engine has sufficiently been warmed up when adjusting clearance.

F.I.C.D. SOLENOID

Inspection and Adjustment

1. Warm up engine sufficiently.
2. Check idle speeds:

1988 Van – Z24i engine with manual trans – 800 ± 50 rpm
1988 Van – Z24i engine with auto trans – 700 ± 50 rpm in **DRIVE**

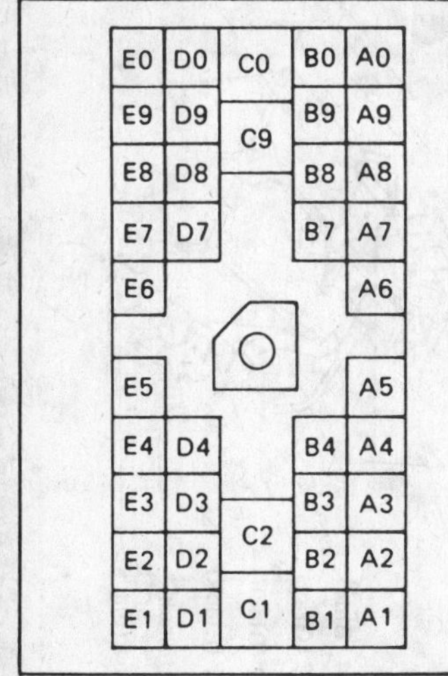
SUPER MULTIPLE JUNCTION (S.M.J.)
MAIN HARNESS
INSTRUMENT HARNESS

Super Multiple Junction (SMJ) terminal identification – All models except 1988–89 Truck and 1988 Van

1988 – 89 Truck – Z24i engine with manual trans – 800 ± 50 rpm
1988 – 89 Truck – Z24i engine with auto trans – 650 ± 50 rpm in **DRIVE**
1988 – 89 Truck – VG30i engine with manual trans – 800 ± 50 rpm
1988 – 89 Truck – VG30i engine with auto trans – 700 ± 50 rpm in **DRIVE**

3. Turn air conditioner switch **ON** and check idle speed. Idle speed should be 900 ± 50 rpm in **NEUTRAL**. If out of specification, adjust idle speed by turning adjusting screw.
4. If F.I.C.D. solenoid valve does not work, check harness and solenoid valve as follows:
   a. Disconnect 6 pin (8 pin on VG30i) and check battery voltage with ignition **ON** and A/C **ON**.
   b. Check continuity of solenoid valve.
   c. Repair and/or replace as necessary.

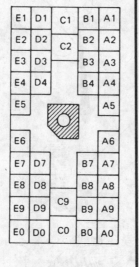

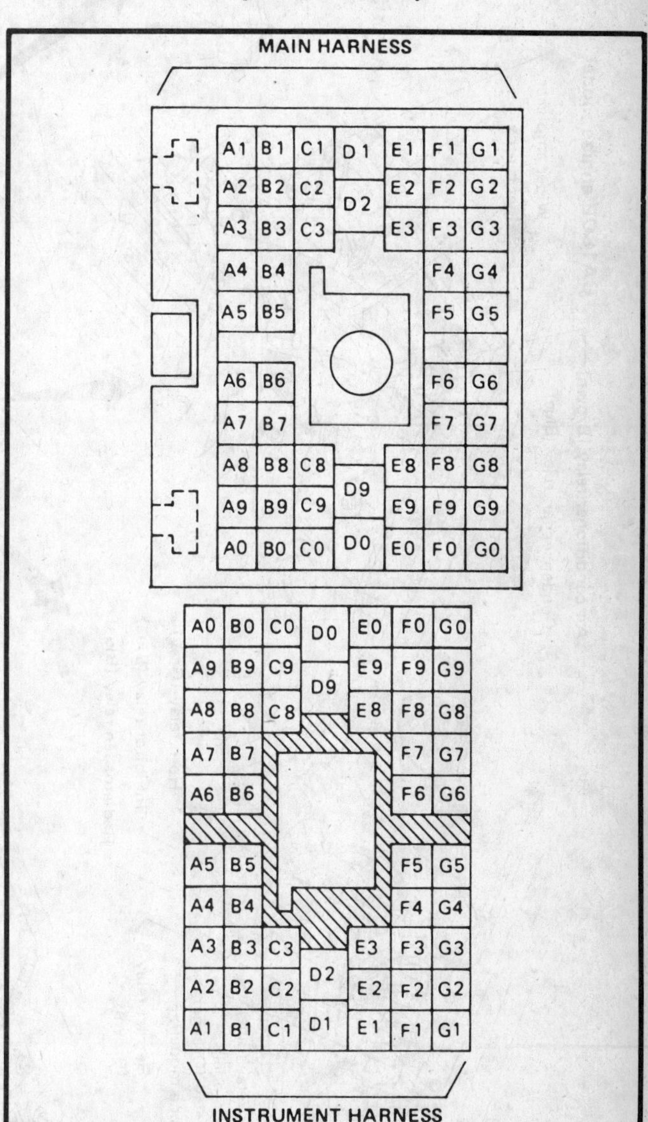
MAIN HARNESS
INSTRUMENT HARNESS

Super Multiple Junction (SMJ) terminal

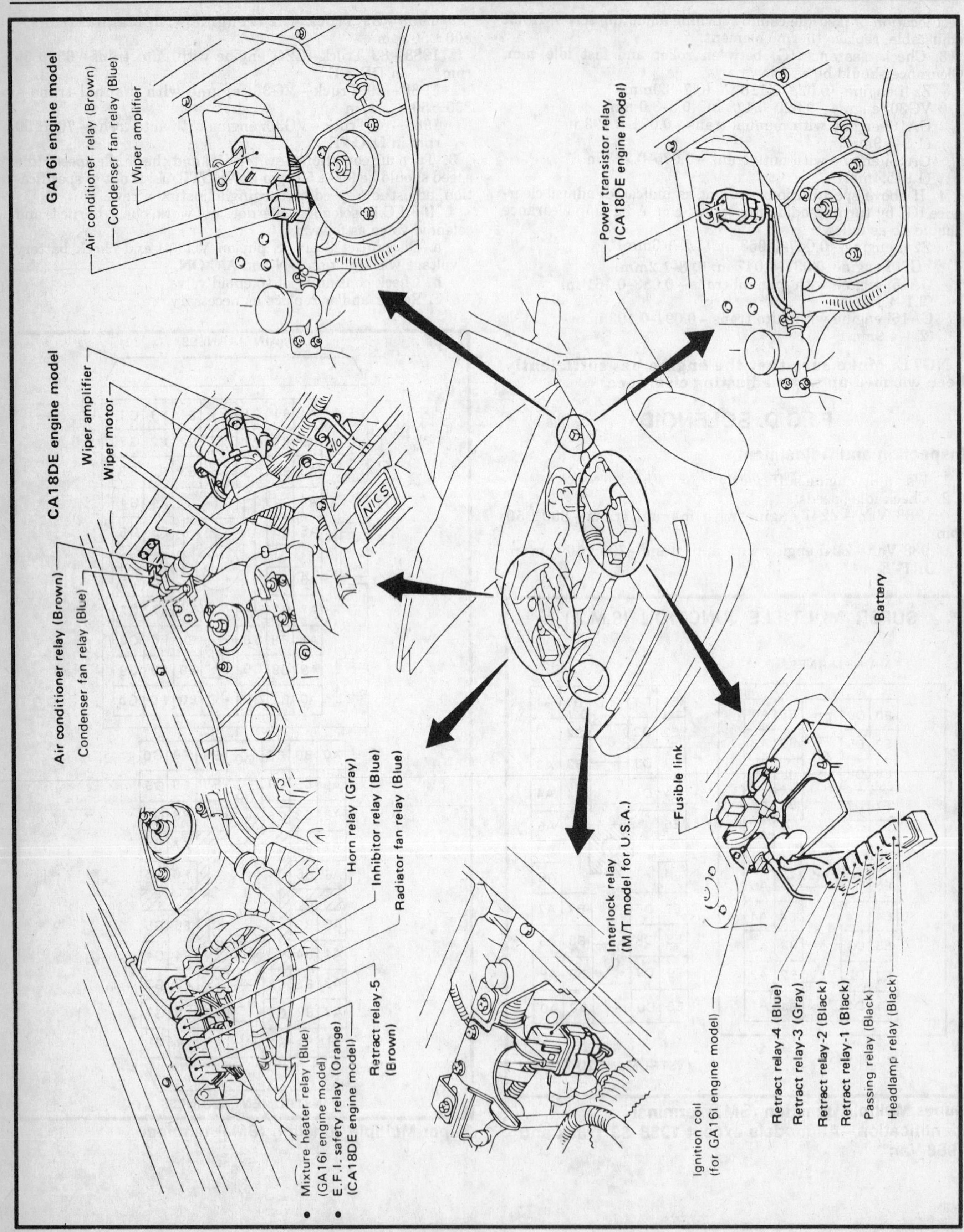
GA16i engine model
Air conditioner relay (Brown)
Condenser fan relay (Blue)
Wiper amplifier
Power transistor relay
(CA18DE engine model)
CA18DE engine model
Wiper amplifier
Wiper motor
NICS
Air conditioner relay (Brown)
Condenser fan relay (Blue)
Horn relay (Gray)
Inhibitor relay (Blue)
Radiator fan relay (Blue)
Battery
Fusible link
Interlock relay
(M/T model for U.S.A.)
• Mixture heater relay (Blue)
(GA16i engine model)
• E.F.I. safety relay (Orange)
(CA18DE engine model)
Retract relay-5
(Brown)
Ignition coil
(for GA16i engine model)
Retract relay-4 (Blue)
Retract relay-3 (Gray)
Retract relay-2 (Black)
Retract relay-1 (Black)
Passing relay (Black)
Headlamp relay (Black)

Location of engine compartment electrical components

ECCS COMPONENTS LOCATION—1988 PULSAR

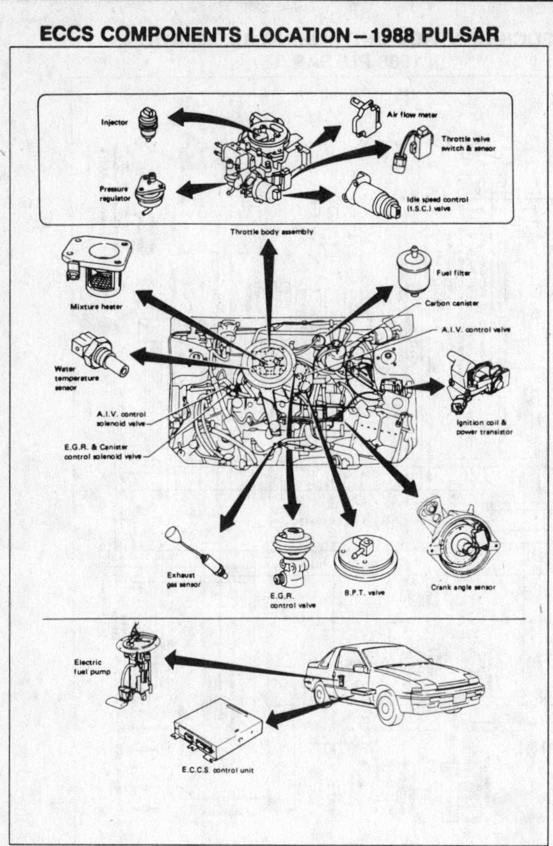
Injector
Air flow meter
Throttle valve switch & sensor
Pressure regulator
Idle speed control (I.S.C.) valve
Throttle body assembly
Fuel filter
Mixture heater
Carbon canister
A.I.V. control valve
Water temperature sensor
A.I.V. control solenoid valve
Ignition coil & power transistor
E.G.R. & Canister control solenoid valve
Exhaust gas sensor
E.G.R. control valve
B.P.T. valve
Crank angle sensor
Electric fuel pump
E.C.C.S. control unit

ECCS SYSTEM SCHEMATIC—1988 PULSAR

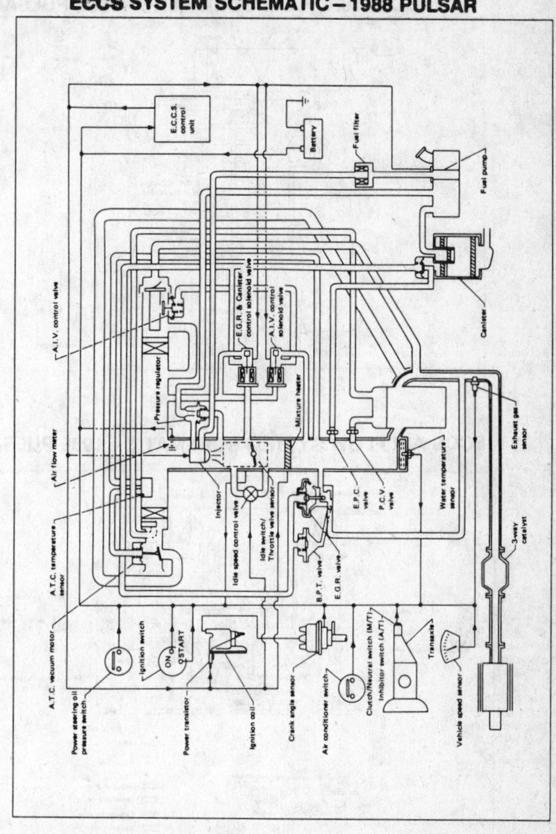
E.C.C.S. control unit
Battery
Fuel filter
Fuel pump
Canister
A.I.V. control valve
E.G.R. & Canister control solenoid valve
A.I.V. control solenoid valve
Pressure regulator
Mixture heater
Exhaust gas sensor
Air flow meter
Injector
Idle speed control valve
Idle switch/ Throttle valve sensor
E.P.C. valve
P.C.V. valve
Water temperature sensor
3-way catalyst
A.T.C. temperature sensor
B.P.T. valve
E.G.R. valve
A.T.C. vacuum motor
Ignition switch
ON
START
Power steering oil pressure switch
Power transistor
Ignition coil
Crank angle sensor
Air conditioner switch
Clutch/Neutral switch (M/T)
Inhibitor switch (A/T)
Transaxle
Vehicle speed sensor

ECCS VACUUM CONTROL SCHEMATIC—1988 PULSAR

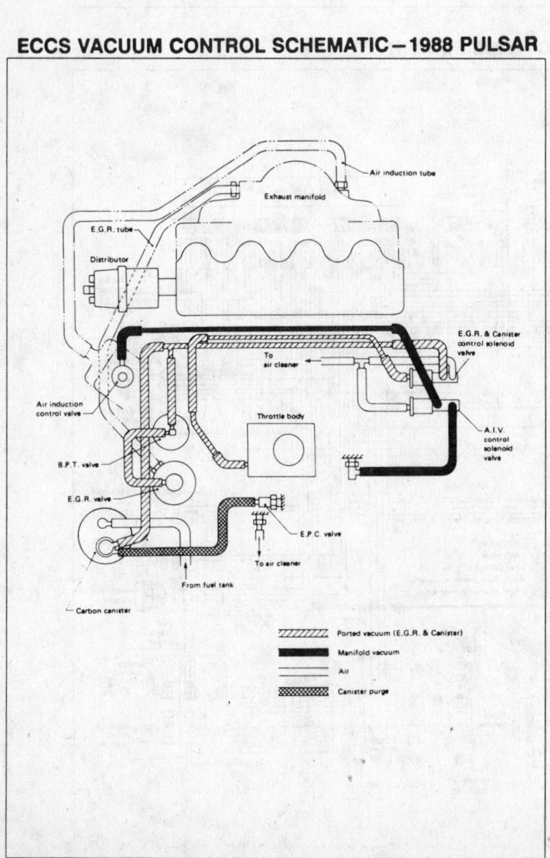
Air induction tube
Exhaust manifold
E.G.R. tube
Distributor
E.G.R. & Canister control solenoid valve
To air cleaner
Air induction control valve
Throttle body
A.I.V. control solenoid valve
B.P.T. valve
E.G.R. valve
E.P.C. valve
To air cleaner
From fuel tank
Carbon canister
Ported vacuum (E.G.R. & Canister)
Manifold vacuum
Air
Canister purge

ECCS CONTROL SYSTEM CHART—1988 PULSAR

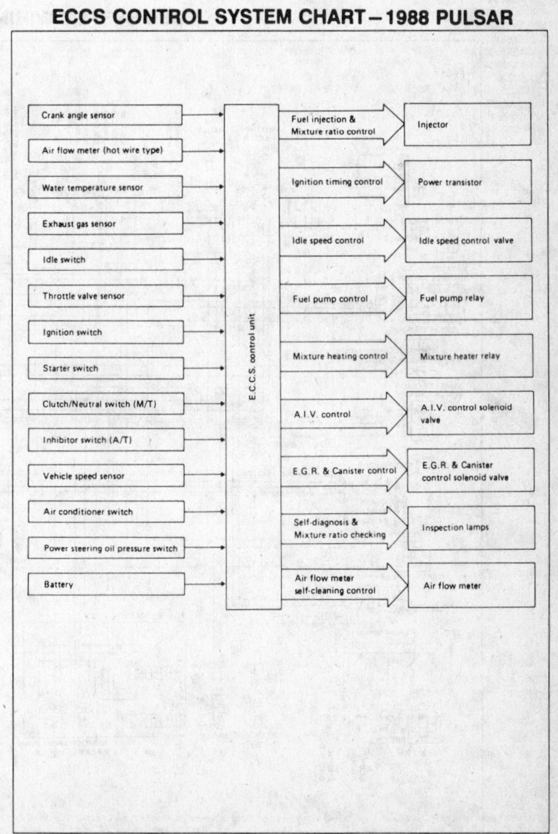
Crank angle sensor
Air flow meter (hot wire type)
Water temperature sensor
Exhaust gas sensor
Idle switch
Throttle valve sensor
Ignition switch
Starter switch
Clutch/Neutral switch (M/T)
Inhibitor switch (A/T)
Vehicle speed sensor
Air conditioner switch
Power steering oil pressure switch
Battery
E.C.C.S. control unit
Fuel injection & Mixture ratio control
Ignition timing control
Idle speed control
Fuel pump control
Mixture heating control
A.I.V. control
E.G.R. & Canister control
Self-diagnosis & Mixture ratio checking
Air flow meter self-cleaning control
Injector
Power transistor
Idle speed control valve
Fuel pump relay
Mixture heater relay
A.I.V. control solenoid valve
E.G.R. & Canister control solenoid valve
Inspection lamps
Air flow meter

ECCS FUEL FLOW SCHEMATIC—1988 PULSAR

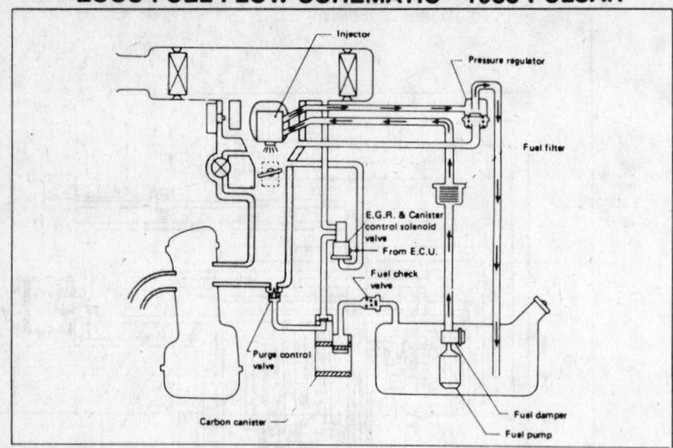
Injector
Pressure regulator
Fuel filter
E.G.R. & Canister control solenoid valve
From E.C.U.
Fuel check valve
Purge control valve
Carbon canister
Fuel damper
Fuel pump

ECCS AIR FLOW SYSTEM SCHEMATIC—1988 PULSAR

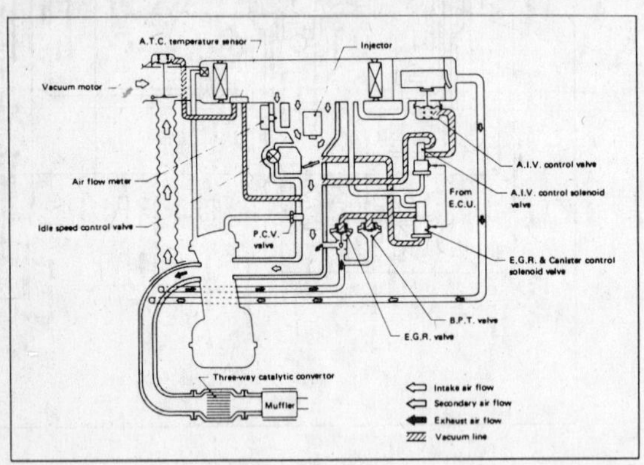
A.T.C. temperature sensor
Injector
Vacuum motor
Air flow meter
Idle speed control valve
A.I.V. control valve
A.I.V. control solenoid valve
From E.C.U.
P.C.V. valve
E.G.R. & Canister control solenoid valve
B.P.T. valve
E.G.R. valve
Three-way catalytic convertor
Muffler
Intake air flow
Secondary air flow
Exhaust air flow
Vacuum line

ECCS CONTROL UNIT WIRING SCHEMATIC—1988 PULSAR

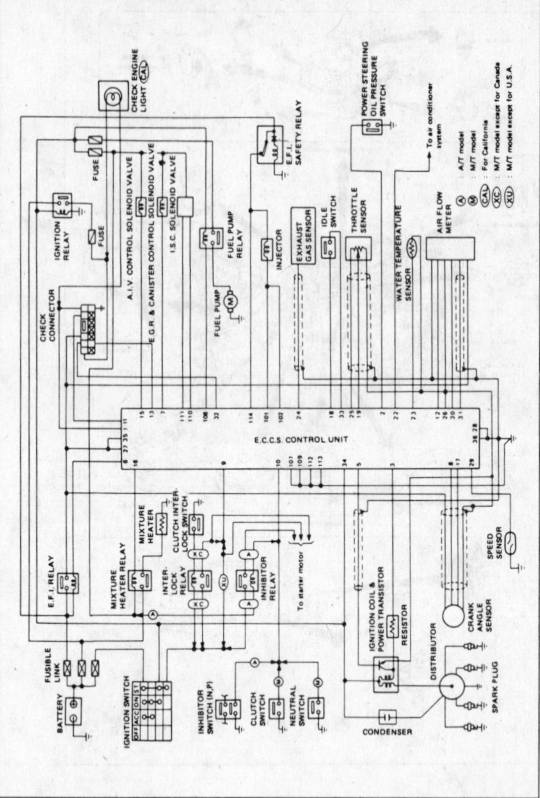
CHECK ENGINE LIGHT
E.F.I. SAFETY RELAY
POWER STEERING OIL PRESSURE SWITCH
To air conditioner system
A/T model
M/T model
For California
M/T model except for Canada
M/T model except for U.S.A.
FUSE
IGNITION RELAY
A.I.V. CONTROL SOLENOID VALVE
E.G.R. & CANISTER CONTROL SOLENOID VALVE
I.S.C. SOLENOID VALVE
FUEL PUMP RELAY
INJECTOR
EXHAUST GAS SENSOR
IDLE SWITCH
THROTTLE SENSOR
WATER TEMPERATURE SENSOR
AIR FLOW METER
CHECK CONNECTOR
FUEL PUMP
E.C.C.S. CONTROL UNIT
E.F.I. RELAY
MIXTURE HEATER RELAY
MIXTURE HEATER
INTER LOCK RELAY
CLUTCH INTER LOCK SWITCH
INHIBITOR RELAY
To starter motor
SPEED SENSOR
IGNITION COIL & POWER TRANSISTOR
RESISTOR
CRANK ANGLE SENSOR
DISTRIBUTOR
SPARK PLUG
FUSIBLE LINK
BATTERY
IGNITION SWITCH
INHIBITOR SWITCH
CLUTCH SWITCH
NEUTRAL SWITCH
CONDENSER

ECCS SYSTEM WIRING SCHEMATIC—1988 PULSAR

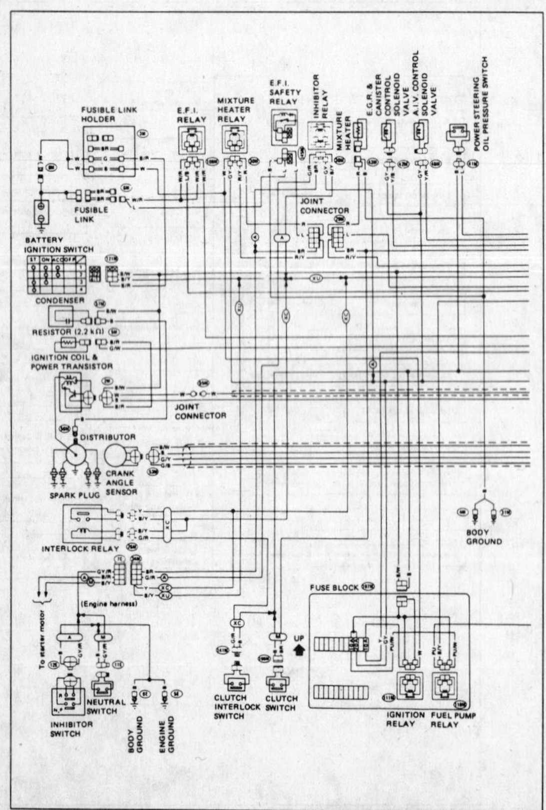
FUSIBLE LINK HOLDER
E.F.I. RELAY
MIXTURE HEATER RELAY
E.F.I. SAFETY RELAY
INHIBITOR RELAY
MIXTURE HEATER
E.G.R. & CANISTER CONTROL SOLENOID VALVE
A.I.V. CONTROL SOLENOID VALVE
POWER STEERING OIL PRESSURE SWITCH
FUSIBLE LINK
BATTERY
IGNITION SWITCH
JOINT CONNECTOR
CONDENSER
RESISTOR (2.2 kΩ)
IGNITION COIL & POWER TRANSISTOR
DISTRIBUTOR
CRANK ANGLE SENSOR
SPARK PLUG
INTERLOCK RELAY
BODY GROUND
(Engine harness)
To starter motor
FUSE BLOCK
UP
NEUTRAL SWITCH
INHIBITOR SWITCH
ENGINE GROUND
CLUTCH INTERLOCK SWITCH
CLUTCH SWITCH
IGNITION RELAY
FUEL PUMP RELAY

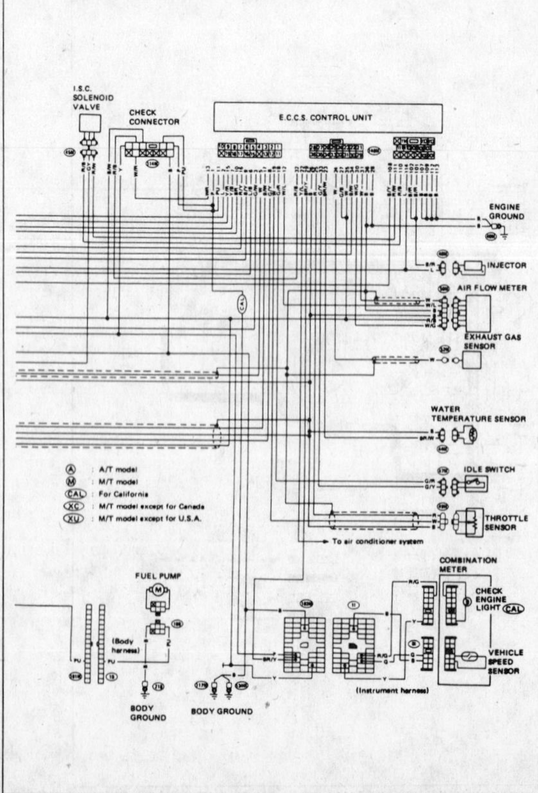
I.S.C. SOLENOID VALVE
CHECK CONNECTOR
E.C.C.S. CONTROL UNIT
ENGINE GROUND
INJECTOR
AIR FLOW METER
EXHAUST GAS SENSOR
WATER TEMPERATURE SENSOR
A/T model
M/T model
For California
M/T model except for Canada
M/T model except for U.S.A.
IDLE SWITCH
THROTTLE SENSOR
To air conditioner system
COMBINATION METER
FUEL PUMP
CHECK ENGINE LIGHT
VEHICLE SPEED SENSOR
(Body harness)
(Instrument harness)
BODY GROUND

ECCS DIAGNOSTIC PROCEDURE—1988 PULSAR

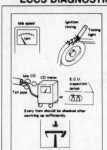

Driveability

1. Make sure that the following items are in the proper condition.

CHECK DATA

1) Idle speed
   M/T: 800±100 rpm
   A/T: 700±100 rpm (in "D" position)
2) Ignition timing
   7°±5° B.T.D.C.
3) Idle CO
   - Self-diagnosis mode: mode II
   - Race engine, and immediately after the engine returns to the idle condition, confirm that E.C.U. red and green inspection lamps flash.
4) Mixture ratio at approximately 2,000 rpm of engine speed.
   - Self-diagnosis mode: mode II
   - Check the simultaneous flashes of E.C.U. green and red lamps.
5) Idle switch operation
   "ON" at idle position and "OFF" when accelerator pedal is depressed.

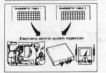

2. Perform driving test.
   Evaluate effectiveness of adjustments by driving vehicle.
   During driving test, perform real time diagnostic test.

3. Perform E.C.C.S. self-diagnosis.

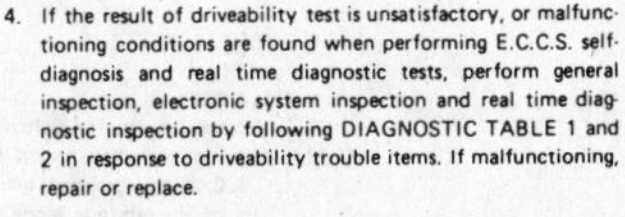

4. If the result of driveability test is unsatisfactory, or malfunctioning conditions are found when performing E.C.C.S. self-diagnosis and real time diagnostic tests, perform general inspection, electronic system inspection and real time diagnostic inspection by following DIAGNOSTIC TABLE 1 and 2 in response to driveability trouble items. If malfunctioning, repair or replace.

5. Perform ON/OFF diagnostic test on switches.

6. Perform driving test.
   Re-evaluate vehicle performance after all the inspections are performed.

ECCS DIAGNOSTIC TABLE 1—1988 PULSAR

Sensor & actuator \ System	Fuel injection & mixture ratio feedback control	Ignition timing control	Idle speed control	Mixture heater control	A.I.V. control	E.G.R. & Canister control
Crank angle sensor	O	O	O	O		O
Air flow meter	O	O				O
Water temperature sensor	O	O	O	O	O	O
Exhaust gas sensor	O					
Idle switch	O	O	O		O	
Throttle sensor	O					
Ignition switch	O	O	O			
Neutral & clutch switch	O		O			
Vehicle speed sensor	O		O			O
Air conditioner switch			O			
Power steering switch			O			
Battery voltage	O					
Injector	O					
Power transistor		O				
I.S.C. valve			O			
Mixture heater relay				O		
A.I.V. control solenoid valve					O	
E.G.R. & Canister control solenoid valve						O

This table indicates the inspection items for the E.C.C.S. control system. For each system, it is necessary to check sensors or actuators marked "O".

ECCS DIAGNOSTIC TABLE 2—1988 PULSAR

DRIVEABILITY INSPECTION TABLE

Trouble items concerned with driveability \ Inspection item		General inspection: Fuel flow system: Fuel pump	Fuel filter	Fuel line	Electric system: Battery	Spark plug	Ignition wire	Alternator	Starter	Air flow system: Air cleaner	Air flow line	E.G.R. B.P.T. valve	I.S.C. valve	P.C.V.	E.C.C.S. system inspection: Crank angle sensor: 180° signal noise	180° signal faults	1° signal faults	Air flow meter: Poor connection	Water temperature sensor: Short	Open	Poor connection
SURGE — Road/load driving	Heavy load		O	O		◎	O			O	O	O		O	O	O	O	◎	O	O	O
SURGE — Road/load driving	Middle load		O	O		◎	O			O	O	◎		O	O	O	O	◎	O	O	O
SURGE — Road/load driving	Light load		O	O		◎	O			O	O	◎		O	O	O	O	◎	O	O	O
SURGE — Accelerating driving	Slow acceleration		O	O		◎	O			O	O	◎		O	O	O	O	◎	O	O	O
SURGE — Decelerating driving	Rapid acceleration		O	O		◎	O			O	O	O		O	O	O	O	◎	O	O	O
SURGE — Decelerating driving	Slow deceleration		O	O		◎	O			O	O	O		O	O	O	O	◎	O	O	O
HESITATION	Rapid acceleration		O	O		◎	O			O	O	O		O	O	O	O	◎	O	O	O
HESITATION	Slow acceleration		O	O		◎	O			O	O	O		O	O	O	O	◎	O	O	O
STUMBLE	Rapid acceleration		O	O		◎	O			O	O	O		O	O	O	O	◎	O	O	O
STUMBLE	Slow acceleration		O	O		◎	O			O	O	O		O	O	O	O	◎	O	O	O
BACKFIRE			O	O			O				O			O	O	O	O	◎	O		O
AFTER FIRE							O			O			O	O	O	O	O	◎		O	O
IDLE STABILITY						◎	O	O		O	O	◎	O	O	O	O	O	O	O	O	O
ENGINE STALL		O	O	O	O	◎	O	O		O	O	O	O	O	◎	◎	◎	◎	O	O	O
STARTABILITY		O	[illegible]		O	◎	O	O	O	O	O		O	O	◎	◎	◎	O	O	O	O

Trouble items concerned with driveability \ E.C.C.S. system inspection		Idle switch: Short	Open	Throttle sensor: Short	Open	E.G.R. & canister control solenoid valve: Short	Open	Exhaust gas sensor: Short	Open	Injector: Clogging	Neutral/clutch switch: Short	Starter signal: Short	Open	Ignition signal: Short	Open	Battery voltage: Low voltage	Fuel pump circuit: Short	Open	Poor connection	Mixture heater: Open (for only cold condition)	A.I.V. control solenoid valve: Short	Open
SURGE — Road/load driving	Heavy load						◎	O	O	◎									O	O	O	
SURGE — Road/load driving	Middle load						◎	O	O	◎									O	O	O	
SURGE — Road/load driving	Light load						◎	O	O	◎									O	O	O	
SURGE — Accelerating driving	Slow acceleration						◎	O	O	◎									O	O	O	
SURGE — Decelerating driving	Rapid acceleration						O	O	O	◎									O	O		
SURGE — Decelerating driving	Slow deceleration						O	O	O	◎									O	O		
HESITATION	Rapid acceleration			◎	◎		O	O	O	◎	O								O	O	O	
HESITATION	Slow acceleration			◎	◎		O	O	O	◎	O								O	O	O	
STUMBLE	Rapid acceleration			◎	◎		O			◎	O								O	O		
STUMBLE	Slow acceleration			◎	◎		O			◎	O								O	O		
BACKFIRE		O	O	O	O			O	O	◎	O								O	O		
AFTER FIRE		O	O				◎	O	O											O		O
IDLE STABILITY			◎				O	O	O	◎									O	O		
ENGINE STALL			◎					O	O	◎				O	O	O	O	O	O	O		
STARTABILITY										◎		O	O	O	O	O	O	O	O			

This table indicates the inspection items for each type of symptom. It is necessary for each symptom to check sensors or actuators marked "◎" or "O".
Items marked "◎" have a significant influence on driveability. Prior to items marked "O", check items marked "◎".
Improper mixture ratio, improper ignition condition, and an excess of E.G.R. volume can cause any symptom.
* If air flow meter circuit is short or open, the fail-safe system operates and engine revolution does not rise to 2,400 rpm or more.

ECCS SELF-DIAGNOSIS DESCRIPTION—1988 PULSAR

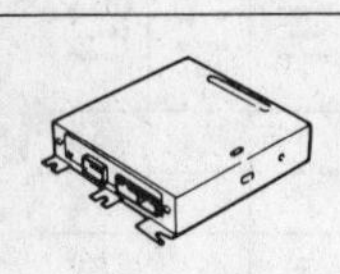

Description

The self-diagnosis is useful to diagnose malfunctions in major sensors and actuators of the E.C.C.S. system. There are 5 modes in the self-diagnosis system.

1. **Mode I – Mixture ratio feedback control monitor A**
   - During closed loop condition:
     The green inspection lamp turns ON when lean condition is detected and goes OFF by rich condition.
   - During open loop condition:
     The green inspection lamp remains ON or OFF.
2. **Mode II – Mixture ratio feedback control monitor B**
   The green inspection lamp function is the same as Mode I.
   - During closed loop condition:
     The red inspection lamp turns ON and OFF simultaneously with the green inspection lamp when the mixture ratio is controlled within the specified value.
   - During open loop condition:
     The red inspection lamp remains ON or OFF.
3. **Mode III – Self-diagnosis**
   This mode is the same as the former self-diagnosis in self-diagnosis mode.
4. **Mode IV – Switches ON/OFF diagnosis**
   During this mode, the inspection lamps monitor the switch ON-OFF condition.
   - Idle switch
   - Starter switch
   - Vehicle speed sensor
5. **Mode V – Real time diagnosis**
   The moment the malfunction is detected, the display will be presented immediately. That is, the condition at which the malfunction occurs can be found by observing the inspection lamps during driving test.

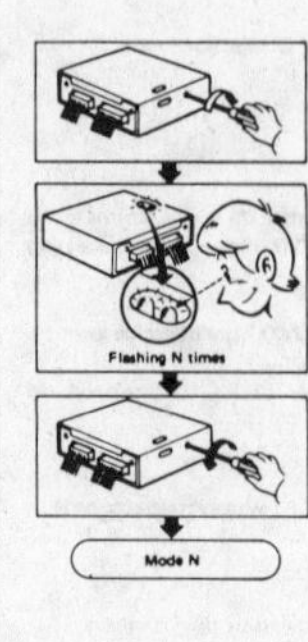
Flashing N times
Mode N

Description (Cont'd)

SWITCHING THE MODES

1. Turn ignition switch "ON".
2. Turn diagnostic mode selector on E.C.U. fully clockwise and wait the inspection lamps flash.
3. Count the number of the flashing time, and after the inspection lamps have flashed the number of the required mode, turn diagnostic mode selector fully counterclockwise immediately.

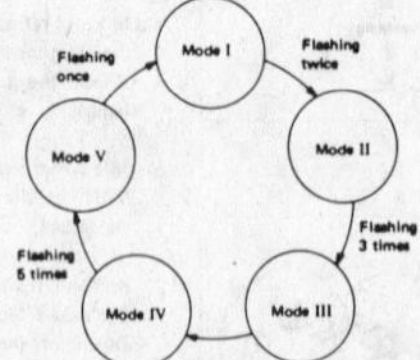
Mode I
Flashing twice
Mode II
Flashing 3 times
Mode III
Flashing 4 times
Mode IV
Flashing 5 times
Mode V
Flashing once

NOTE:
When the ignition switch is turned off during diagnosis, in each mode, and then turned back on again after the power to the E.C.U. has dropped off completely, the diagnosis will automatically return to Mode I.
The stored memory would be lost if:

1. Battery terminal is disconnected.
2. After selecting Mode III, Mode IV is selected.
   However, if the diagnostic mode selector is kept turned fully clockwise, it will continue to change in the order of Mode I → II → III → IV → V → I ... etc., and in this state the stored memory will not be erased.

ECCS SELF-DIAGNOSIS DESCRIPTION (CONT.)—PULSAR

CHECK
CHECK ENGINE LIGHT

Description (Cont'd)

CHECK ENGINE LIGHT (For California only)

This vehicle has a check engine light on instrument panel. This light comes ON under the following conditions:

1) When ignition switch is turned "ON" (for bulb check).
2) When systems related to emission performance malfunction in Mode I (with engine running).

- **This check engine light always illuminates and is synchronous with red L.E.D.**
- **Malfunction systems related to emission performance can be detected by self-diagnosis, and they are clarified as self-diagnostic codes in Mode III.**

Code No.	Malfunction
12	Air flow meter circuit
14	Vehicle speed sensor circuit
23	Idle switch circuit
31	E.C.U. (E.C.C.S. control unit)
33	Exhaust gas sensor circuit

Use the following diagnostic flowchart to check and repair a malfunctioning system.

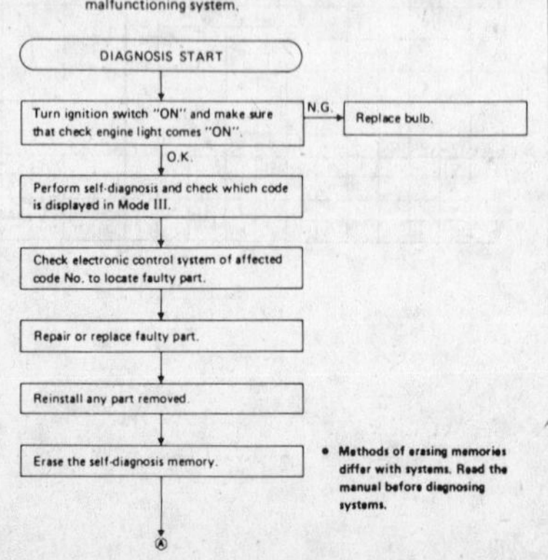
DIAGNOSIS START
Turn ignition switch "ON" and make sure that check engine light comes "ON".
N.G.
Replace bulb.
O.K.
Perform self-diagnosis and check which code is displayed in Mode III.
Check electronic control system of affected code No. to locate faulty part.
Repair or replace faulty part.
Reinstall any part removed.
Erase the self-diagnosis memory.
• Methods of erasing memories differ with systems. Read the manual before diagnosing systems.
Ⓐ

Description (Cont'd)

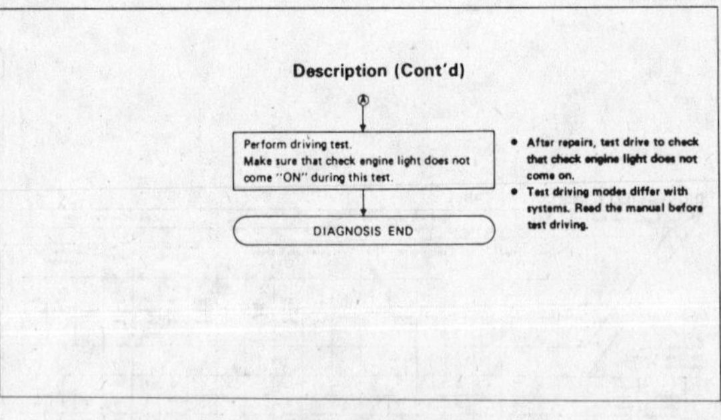
Ⓐ
Perform driving test.
Make sure that check engine light does not come "ON" during this test.
• After repairs, test drive to check that check engine light does not come on.
• Test driving modes differ with systems. Read the manual before test driving.
DIAGNOSIS END

ECCS SELF-DIAGNOSTIC MODES 1 AND II—1988 PULSAR

Modes I & II —Mixture Ratio Feedback Control Monitors A & B

In these modes, the control unit provides the Air-fuel ratio monitor presentation and the Air-fuel ratio feedback coefficient monitor presentation.

Mode	LED	Engine stopped (Ignition switch "ON")	Engine running: Open loop condition	Engine running: Closed loop condition		
Mode I (Monitor A)	Green	ON	*Remains ON or OFF	Blinks		
	Red	ON	Except for California model • OFF	For California model • ON: when CHECK ENGINE LIGHT ITEMS are stored in the E.C.U. • OFF: except for the above condition		
Mode II (Monitor B)	Green	ON	*Remains ON or OFF	Blinks		
	Red	OFF	*Remains ON or OFF (synchronous with green LED)	Compensating mixture ratio		
				More than 5% rich	Between 5% lean and 5% rich	More
				OFF	Synchronized with green LED	Remains ON

*: Maintains conditions just before switching to open loop

ECCS SELF-DIAGNOSTIC MODE III—1988 PULSAR

Mode III — Self-Diagnostic System

The E.C.U. constantly monitors the function of these sensors and actuators, regardless of ignition key position. If a malfunction occurs, the information is stored in the E.C.U. and can be retrieved from the memory by turning on the diagnostic mode selector, located on the side of the E.C.U. When activated, the malfunction is indicated by flashing a red and a green L.E.D. (Light Emitting Diode), also located on the E.C.U. Since all the self-diagnostic results are stored in the E.C.U.'s memory even intermittent malfunctions can be diagnosed.

A malfunctioning part's group is indicated by the number of both the red and the green L.E.D.s flashing. First, the red L.E.D. flashes and the green flashes follow. The red L.E.D. refers to the number of tens while the green one refers to the number of units. For example, when the red L.E.D. flashes once and then the green one flashes twice, this means the number "12" showing the air flow meter signal is malfunctioning. In this way, all the problems are classified by the code numbers.

- **When engine fails to start, crank engine more than two seconds before starting self-diagnosis.**
- **Before starting self-diagnosis, do not erase stored memory. If doing so, self-diagnosis function for intermittent malfunctions would be lost.**

The stored memory would be lost if:

1. **Battery terminal is disconnected.**
2. **After selecting Mode III, Mode IV is selected.**

DISPLAY CODE TABLE

Code No.	Detected items
11	Crank angle sensor circuit
12	Air flow meter circuit
13	Water temperature sensor circuit
14	Vehicle speed sensor circuit
15	Mixture ratio feedback control slips out
21	Ignition signal missing in primary coil
23	Idle switch circuit
25	Idle speed control slips out
31	E.C.U. (E.C.C.S. control unit)
33	Exhaust gas sensor circuit
43	Throttle sensor circuit
55	No malfunctioning in the above circuit

Mode III — Self-Diagnostic System (Cont'd)

RETENTION OF DIAGNOSTIC RESULTS

The diagnostic result is retained in E.C.U. memory until the starter is operated fifty times after a diagnostic item is judged to be malfunctioning. The diagnostic result will then be cancelled automatically. If a diagnostic item which has been judged to be malfunctioning and stored in memory is again judged to be malfunctioning before the starter is operated fifty times, the second result will replace the previous one. It will be stored in E.C.U. memory until the starter is operated fifty times more.

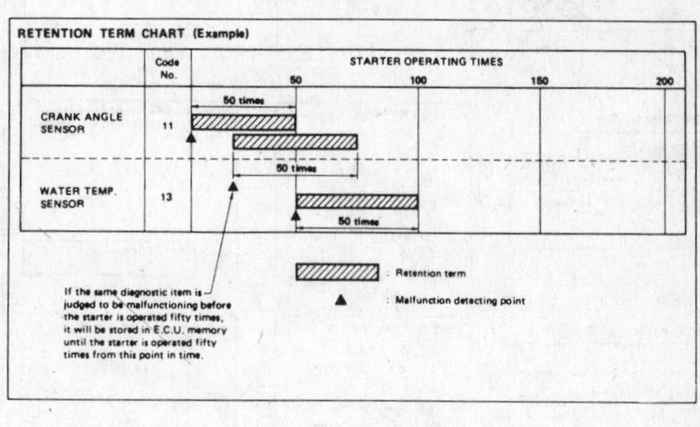
RETENTION TERM CHART (Example)
Code No.
STARTER OPERATING TIMES
50
100
150
200
CRANK ANGLE SENSOR
11
50 times
WATER TEMP. SENSOR
13
50 times
50 times
: Retention term
: Malfunction detecting point
If the same diagnostic item is judged to be malfunctioning before the starter is operated fifty times, it will be stored in E.C.U. memory until the starter is operated fifty times from this point in time.

ECCS SELF-DIAGNOSTIC MODE III (CONT.)—1988 PULSAR

Mode III — Self-Diagnostic System (Cont'd)

SELF-DIAGNOSTIC PROCEDURE

DIAGNOSIS START

Pull out E.C.U. from under the passenger seat.

Start engine and warm it up to normal engine operating temperature. (Drive vehicle for about 10 min.)

Turn diagnostic mode selector on E.C.U. fully clockwise.

Flashing 3 times

After the inspection lamps have flashed 3 times, turn diagnostic mode selector fully counterclockwise.

Mode III

Make sure that inspection lamps are displaying code No. 55.

N.G. → Write down the malfunctioning code No.

O.K.

Memory erasing procedure

Turn diagnostic mode selector on E.C.U. fully clockwise.

After the inspection lamps have flashed 4 times, turn diagnostic mode selector on E.C.U. fully counterclockwise.

Flashing 4 times

Mode IV

Turn ignition switch "OFF".

Reinstall the E.C.U. in place.

DIAGNOSIS END

(N.G. branch:) Turn ignition switch "OFF".

See decoding chart.

Check malfunctioning parts and/or perform real time diagnosis system inspection. If malfunction part is found, repair or replace it.

CAUTION:

During displaying code No. in self-diagnosis mode (mode III), if the other diagnostic mode should be done, make sure to write down the malfunctioning code No. before turning diagnostic mode selector on E.C.U. fully clockwise, or select the diagnostic mode after turning switch "OFF". Otherwise self-diagnosis information stored in E.C.U. memory until now would be lost.

ECCS SELF-DIAGNOSTIC MODE III DISPLAY CODES—1988 PULSAR

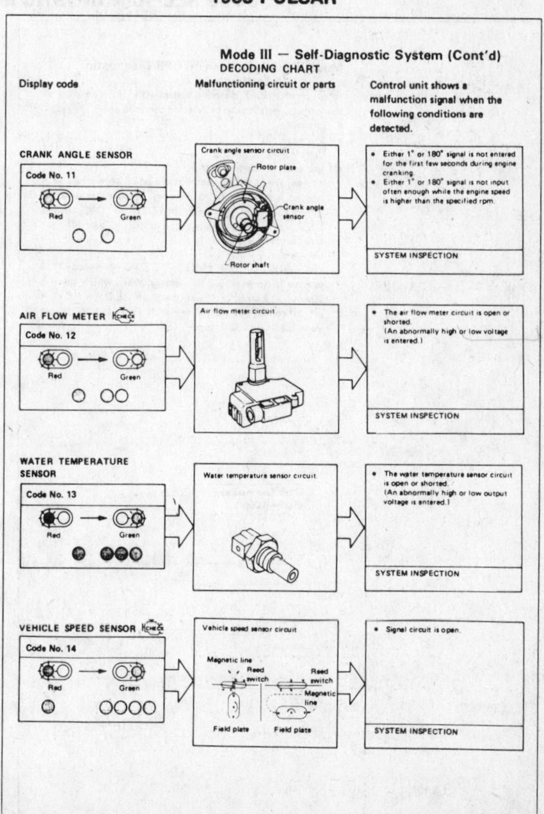
Mode III — Self-Diagnostic System (Cont'd)
DECODING CHART
Display code
Malfunctioning circuit or parts
Control unit shows a malfunction signal when the following conditions are detected.
CRANK ANGLE SENSOR
Code No. 11
Red
Green
Crank angle sensor circuit
Rotor plate
Crank angle sensor
Rotor shaft
Either 1° or 180° signal is not entered for the first few seconds during engine cranking.
Either 1° or 180° signal is not input often enough while the engine speed is higher than the specified rpm.
SYSTEM INSPECTION
AIR FLOW METER
Code No. 12
Red
Green
Air flow meter circuit
The air flow meter circuit is open or shorted.
(An abnormally high or low voltage is entered.)
SYSTEM INSPECTION
WATER TEMPERATURE SENSOR
Code No. 13
Red
Green
Water temperature sensor circuit
The water temperature sensor circuit is open or shorted.
(An abnormally high or low output voltage is entered.)
SYSTEM INSPECTION
VEHICLE SPEED SENSOR
Code No. 14
Red
Green
Vehicle speed sensor circuit
Magnetic line
Reed switch
Reed switch
Magnetic line
Field plate
Field plate
Signal circuit is open.
SYSTEM INSPECTION

ECCS SELF-DIAGNOSTIC MODE III DISPLAY CODES (CONT.)—1988 PULSAR

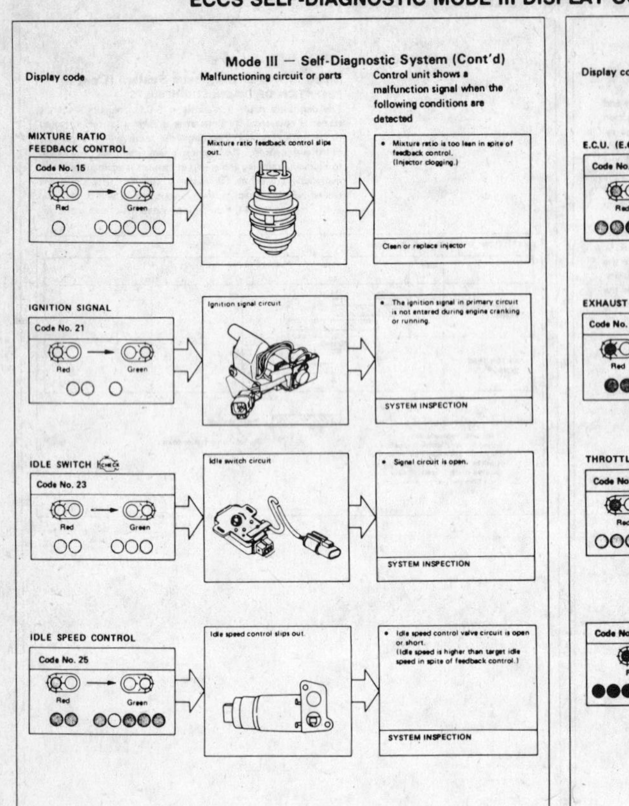
Mode III — Self-Diagnostic System (Cont'd)
Display code
Malfunctioning circuit or parts
Control unit shows a malfunction signal when the following conditions are detected
MIXTURE RATIO FEEDBACK CONTROL
Code No. 15
Red
Green
Mixture ratio feedback control slips out.
• Mixture ratio is too lean in spite of feedback control. (Injector clogging.)
Clean or replace injector
IGNITION SIGNAL
Code No. 21
Red
Green
Ignition signal circuit
• The ignition signal in primary circuit is not entered during engine cranking or running.
SYSTEM INSPECTION
IDLE SWITCH
Code No. 23
Red
Green
Idle switch circuit
• Signal circuit is open.
SYSTEM INSPECTION
IDLE SPEED CONTROL
Code No. 25
Red
Green
Idle speed control slips out.
• Idle speed control valve circuit is open or short. (Idle speed is higher than target idle speed in spite of feedback control.)
SYSTEM INSPECTION

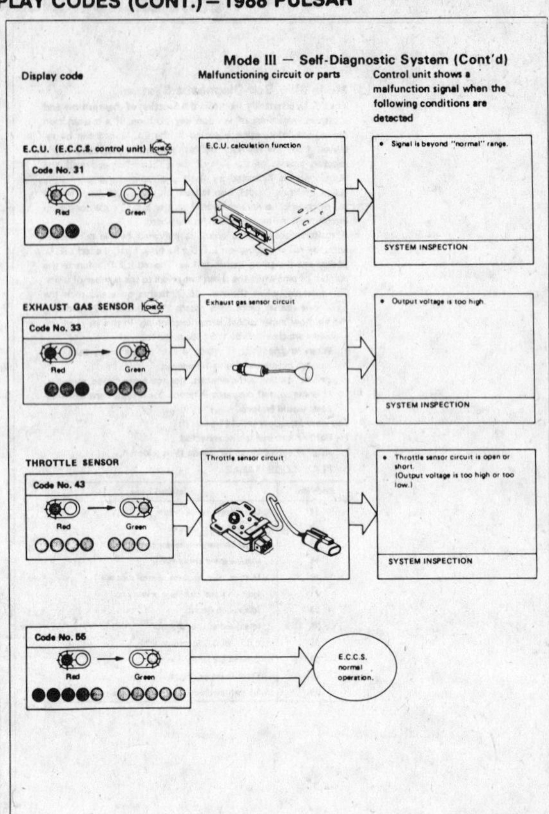
Mode III — Self-Diagnostic System (Cont'd)
Display code
Malfunctioning circuit or parts
Control unit shows a malfunction signal when the following conditions are detected
E.C.U. (E.C.C.S. control unit)
Code No. 31
Red
Green
E.C.U. calculation function
• Signal is beyond "normal" range.
SYSTEM INSPECTION
EXHAUST GAS SENSOR
Code No. 33
Red
Green
Exhaust gas sensor circuit
• Output voltage is too high.
SYSTEM INSPECTION
THROTTLE SENSOR
Code No. 43
Red
Green
Throttle sensor circuit
• Throttle sensor circuit is open or short. (Output voltage is too high or too low.)
SYSTEM INSPECTION
Code No. 55
Red
Green
E.C.C.S. normal operation.

ECCS SELF-DIAGNOSTIC MODE IV—1988 PULSAR

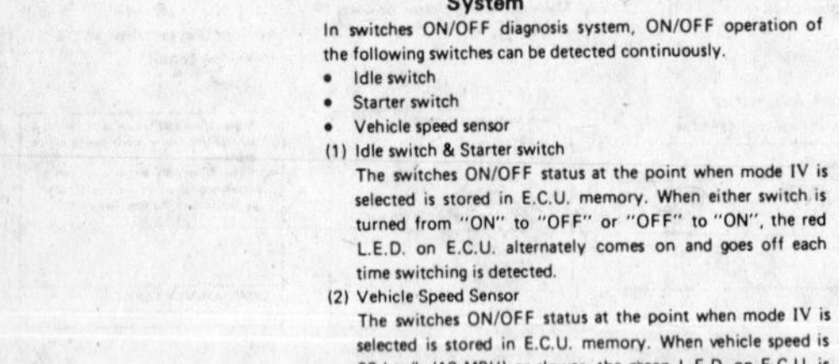

Mode IV — Switches ON/OFF Diagnostic System

In switches ON/OFF diagnosis system, ON/OFF operation of the following switches can be detected continuously.

- Idle switch
- Starter switch
- Vehicle speed sensor

(1) Idle switch & Starter switch
The switches ON/OFF status at the point when mode IV is selected is stored in E.C.U. memory. When either switch is turned from "ON" to "OFF" or "OFF" to "ON", the red L.E.D. on E.C.U. alternately comes on and goes off each time switching is detected.

(2) Vehicle Speed Sensor
The switches ON/OFF status at the point when mode IV is selected is stored in E.C.U. memory. When vehicle speed is 20 km/h (12 MPH) or slower, the green L.E.D. on E.C.U. is off. When vehicle speed exceeds 20 km/h (12 MPH), the green L.E.D. on E.C.U. comes "ON".

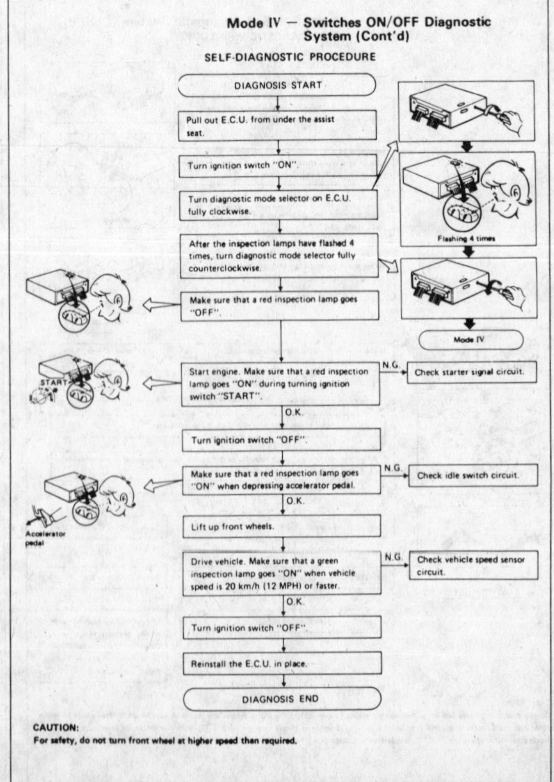
Mode IV — Switches ON/OFF Diagnostic System (Cont'd)
SELF-DIAGNOSTIC PROCEDURE
DIAGNOSIS START
Pull out E.C.U. from under the assist seat.
Turn ignition switch "ON".
Turn diagnostic mode selector on E.C.U. fully clockwise.
After the inspection lamps have flashed 4 times, turn diagnostic mode selector fully counterclockwise.
Flashing 4 times
Mode IV
Make sure that a red inspection lamp goes "OFF".
Start engine. Make sure that a red inspection lamp goes "ON" during turning ignition switch "START".
START
N.G.
Check starter signal circuit.
O.K.
Turn ignition switch "OFF".
Make sure that a red inspection lamp goes "ON" when depressing accelerator pedal.
N.G.
Check idle switch circuit.
Accelerator pedal
O.K.
Lift up front wheels.
Drive vehicle. Make sure that a green inspection lamp goes "ON" when vehicle speed is 20 km/h (12 MPH) or faster.
N.G.
Check vehicle speed sensor circuit.
O.K.
Turn ignition switch "OFF".
Reinstall the E.C.U. in place.
DIAGNOSIS END
CAUTION:
For safety, do not turn front wheel at higher speed than required.

ECCS SELF-DIAGNOSTIC MODE V—1988 PULSAR

Mode V — Real Time Diagnostic System

In real time diagnosis, if any of the following items are judged to be faulty, a malfunction is indicated immediately.

- Crank angle sensor (180° signal & 1° signal)
- Ignition signal
- Air flow meter output signal

Consequently, this diagnosis is a very effective measure to diagnose whether the above systems cause the malfunction or not, during driving test. Compared with self-diagnosis, real time diagnosis is very sensitive, and can detect malfunctioning conditions in a moment. Further, items regarded to be malfunctions in this diagnosis are not stored in E.C.U. memory.

SELF-DIAGNOSTIC PROCEDURE

DIAGNOSIS START

Pull out E.C.U. from under the passenger seat.

Start engine.

Turn diagnostic mode selector on E.C.U. fully clockwise.

Flashing 5 times

After the inspection lamps have flashed 5 times, turn diagnostic mode selector fully counterclockwise.

Mode V

Make sure that inspection lamps are not flashing for 5 min. when idling or racing.

O.K. → Turn ignition switch "OFF". → Reinstall the E.C.U. in place. → DIAGNOSIS END

N.G. → If flashing, count no. of flashes. → Turn ignition switch "OFF". → See decoding chart. → Perform real time-diagnosis system inspection. If malfunction part is found, repair or replace it.

CAUTION:
In real time diagnosis, pay attention to inspection lamp flashing. E.C.U. displays the malfunction code only once, and does not memorize the inspection.

ECCS SELF-DIAGNOSTIC MODE V DECODING CHART—1988 PULSAR

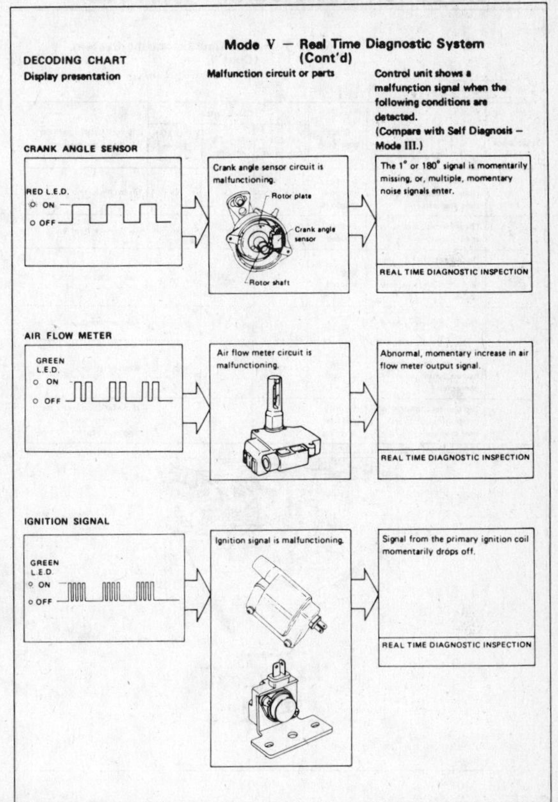
Mode V — Real Time Diagnostic System (Cont'd)
DECODING CHART
Display presentation
Malfunction circuit or parts
Control unit shows a malfunction signal when the following conditions are detected. (Compare with Self Diagnosis — Mode III.)
CRANK ANGLE SENSOR
RED L.E.D.
ON
OFF
Crank angle sensor circuit is malfunctioning.
Rotor plate
Crank angle sensor
Rotor shaft
The 1° or 180° signal is momentarily missing, or, multiple, momentary noise signals enter.
REAL TIME DIAGNOSTIC INSPECTION
AIR FLOW METER
GREEN L.E.D.
ON
OFF
Air flow meter circuit is malfunctioning.
Abnormal, momentary increase in air flow meter output signal.
REAL TIME DIAGNOSTIC INSPECTION
IGNITION SIGNAL
GREEN L.E.D.
ON
OFF
Ignition signal is malfunctioning.
Signal from the primary ignition coil momentarily drops off.
REAL TIME DIAGNOSTIC INSPECTION

ECCS SELF-DIAGNOSTIC MODE V INSPECTION CHART—1988 PULSAR

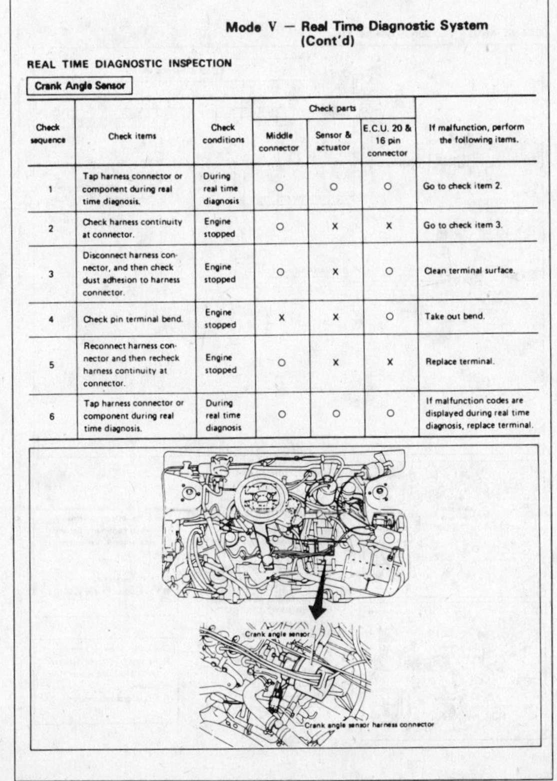

Mode V — Real Time Diagnostic System (Cont'd)

REAL TIME DIAGNOSTIC INSPECTION

Crank Angle Sensor

Check sequence	Check items	Check conditions	Check parts: Middle connector	Sensor & actuator	E.C.U. 20 & 16 pin connector	If malfunction, perform the following items.
1	Tap harness connector or component during real time diagnosis.	During real time diagnosis	O	O	O	Go to check item 2.
2	Check harness continuity at connector.	Engine stopped	O	X	X	Go to check item 3.
3	Disconnect harness connector, and then check dust adhesion to harness connector.	Engine stopped	O	X	O	Clean terminal surface.
4	Check pin terminal bend.	Engine stopped	X	X	O	Take out bend.
5	Reconnect harness connector and then recheck harness continuity at connector.	Engine stopped	O	X	X	Replace terminal.
6	Tap harness connector or component during real time diagnosis.	During real time diagnosis	O	O	O	If malfunction codes are displayed during real time diagnosis, replace terminal.

Crank angle sensor
Crank angle sensor harness connector

Mode V — Real Time Diagnostic System (Cont'd)

Air Flow Meter

Check sequence	Check items	Check conditions	Check parts: Middle connector	Sensor & actuator	E.C.U. 20 & 16 pin connector	If malfunction, perform the following items.
1	Tap harness connector or component during real time diagnosis.	During real time diagnosis	O	O	O	Go to check item 2.
2	Check harness continuity at connector.	Engine stopped	O	X	X	Go to check item 3.
3	Disconnect harness connector, and then check dust adhesion to harness connector.	Engine stopped	O	X	O	Clean terminal surface.
4	Check pin terminal bend.	Engine stopped	X	X	O	Take out bend.
5	Reconnect harness connector and then recheck harness continuity at connector.	Engine stopped	O	X	X	Replace terminal.
6	Tap harness connector or component during real time diagnosis.	During real time diagnosis	O	O	O	If malfunction codes are displayed during real time diagnosis, replace terminal.

Air flow meter

Air flow meter harness connector

ECCS SELF-DIAGNOSTIC MODE V INSPECTION CHART (CONT.)—1988 PULSAR

Mode V — Real Time Diagnostic System (Cont'd)

Ignition Signal

Check sequence	Check items	Check conditions	Check parts: Middle connector	Check parts: Sensor & actuator	Check parts: E.C.U. 20 & 16 pin connector	If malfunction, perform the following items.
1	Tap harness connector or component during real time diagnosis.	During real time diagnosis	O	O	O	Go to check item 2.
2	Check harness continuity at connector.	Engine stopped	O	X	X	Go to check item 3.
3	Disconnect harness connector, and then check dust adhesion to harness connector.	Engine stopped	O	X	O	Clean terminal surface.
4	Check pin terminal bend.	Engine stopped	X	X	O	Take out bend.
5	Reconnect harness connector and then recheck harness continuity at connector.	Engine stopped	O	X	X	Replace terminal.
6	Tap harness connector or component during real time diagnosis.	During real time diagnosis	O	O	O	If malfunction codes are displayed during real time diagnosis, replace terminal.

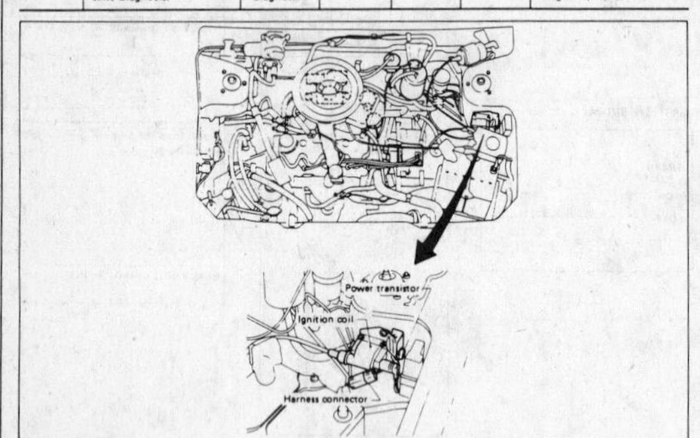
Power transistor
Ignition coil
Harness connector

ECCS SYSTEM INSPECTION CAUTION—1988 PULSAR

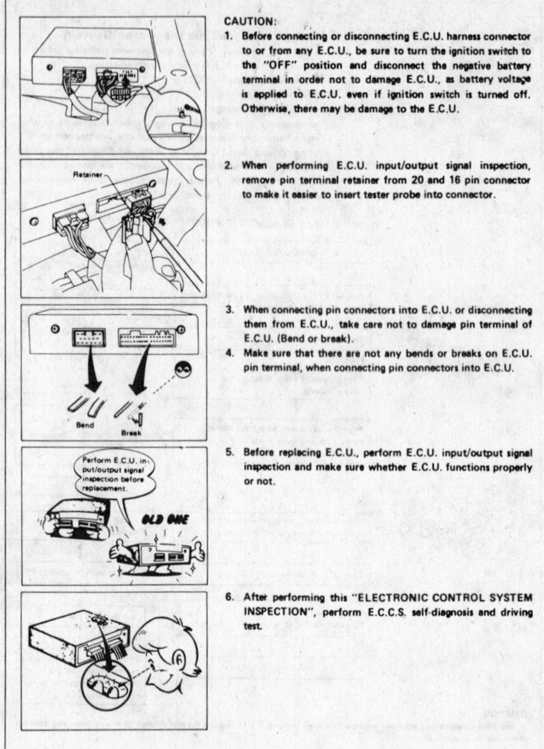
Retainer
Bend
Break
Perform E.C.U. input/output signal inspection before replacement.
OLD ONE

CAUTION:

1. Before connecting or disconnecting E.C.U. harness connector to or from any E.C.U., be sure to turn the ignition switch to the "OFF" position and disconnect the negative battery terminal in order not to damage E.C.U., as battery voltage is applied to E.C.U. even if ignition switch is turned off. Otherwise, there may be damage to the E.C.U.
2. When performing E.C.U. input/output signal inspection, remove pin terminal retainer from 20 and 16 pin connector to make it easier to insert tester probe into connector.
3. When connecting pin connectors into E.C.U. or disconnecting them from E.C.U., take care not to damage pin terminal of E.C.U. (Bend or break).
4. Make sure that there are not any bends or breaks on E.C.U. pin terminal, when connecting pin connectors into E.C.U.
5. Before replacing E.C.U., perform E.C.U. input/output signal inspection and make sure whether E.C.U. functions properly or not.
6. After performing this "ELECTRONIC CONTROL SYSTEM INSPECTION", perform E.C.C.S. self-diagnosis and driving test.

ECCS SYSTEM INSPECTION CAUTION—1988 PULSAR

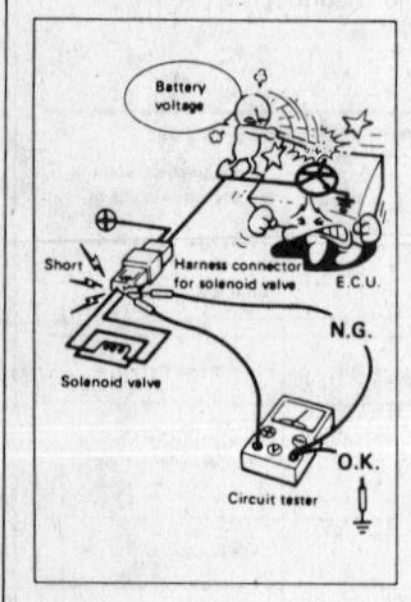
Battery voltage
Short
Harness connector for solenoid valve
E.C.U.
N.G.
Solenoid valve
O.K.
Circuit tester

7. When measuring supply voltage of E.C.U. controlled components with a circuit tester, separate one tester probe from the other.
   If the two tester probes accidentally make contact with each other during measurement, the circuit will be shorted, resulting in damage to the power transistor of the control unit.
8. Keys to symbols
   - DISCONNECT : Check after disconnecting the connector to be measured.
   - CONNECT : Check after connecting the connector to be measured.
9. When measuring voltage or resistance at connector with tester probes, there are two methods of measurement; one is done from terminal side and the other from harness side. Before measuring, confirm symbol mark again.
   - : Inspection should be done from harness side.
   - : Inspection should be done from terminal side.
10. As for continuity check of joint connector.

ECCS SELF-DIAGNOSTIC CHART—1988 PULSAR

CRANK ANGLE SENSOR (Code No. 11)

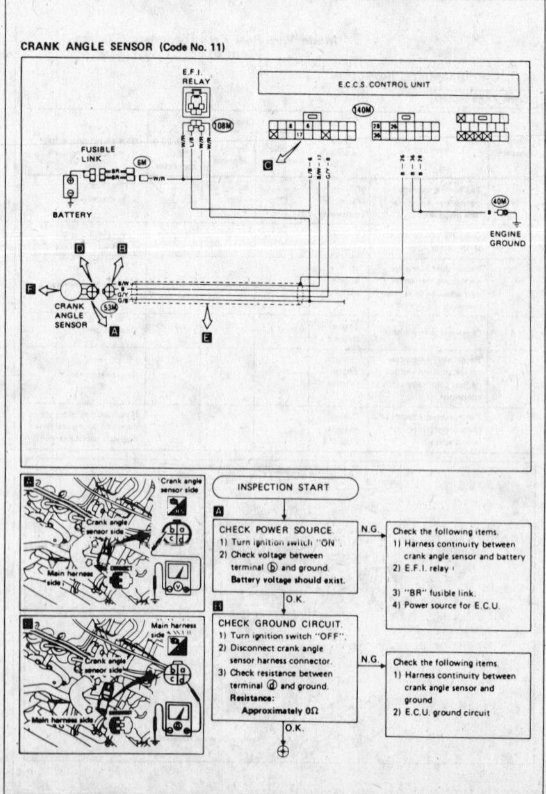
E.F.I. RELAY
E.C.C.S. CONTROL UNIT
FUSIBLE LINK
BATTERY
ENGINE GROUND
CRANK ANGLE SENSOR
Crank angle sensor side
Main harness side
INSPECTION START
CHECK POWER SOURCE.
1) Turn ignition switch "ON".
2) Check voltage between terminal ⓑ and ground.
Battery voltage should exist.
N.G.
Check the following items.
1) Harness continuity between crank angle sensor and battery
2) E.F.I. relay
3) "BR" fusible link
4) Power source for E.C.U.
O.K.
CHECK GROUND CIRCUIT.
1) Turn ignition switch "OFF".
2) Disconnect crank angle sensor harness connector.
3) Check resistance between terminal ⓓ and ground.
Resistance:
Approximately 0Ω
N.G.
Check the following items.
1) Harness continuity between crank angle sensor and ground
2) E.C.U. ground circuit
O.K.

ECCS SELF-DIAGNOSTIC CHART—1988 PULSAR

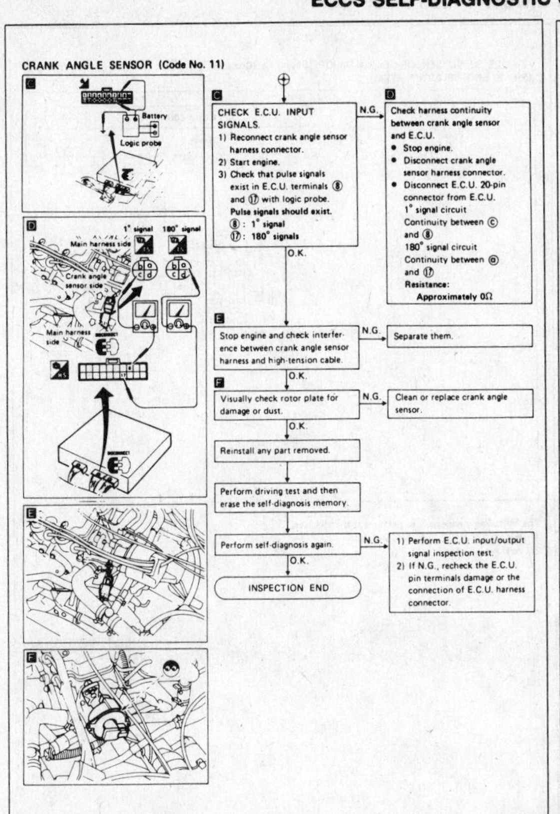
CRANK ANGLE SENSOR (Code No. 11)
Battery
Logic probe
1° signal
180° signal
Main harness side
Crank angle sensor side
CHECK E.C.U. INPUT SIGNALS.
1) Reconnect crank angle sensor harness connector.
2) Start engine.
3) Check that pulse signals exist in E.C.U. terminals ⑧ and ⑰ with logic probe.
Pulse signals should exist.
⑧: 1° signal
⑰: 180° signals
N.G.
Check harness continuity between crank angle sensor and E.C.U.
• Stop engine.
• Disconnect crank angle sensor harness connector.
• Disconnect E.C.U. 20-pin connector from E.C.U.
1° signal circuit
Continuity between ⓒ and ⑧
180° signal circuit
Continuity between ⓓ and ⑰
Resistance:
Approximately 0Ω
O.K.
Stop engine and check interference between crank angle sensor harness and high-tension cable.
N.G.
Separate them.
O.K.
Visually check rotor plate for damage or dust.
N.G.
Clean or replace crank angle sensor.
O.K.
Reinstall any part removed.
Perform driving test and then erase the self-diagnosis memory.
Perform self-diagnosis again.
N.G.
1) Perform E.C.U. input/output signal inspection test.
2) If N.G., recheck the E.C.U. pin terminals damage or the connection of E.C.U. harness connector.
O.K.
INSPECTION END

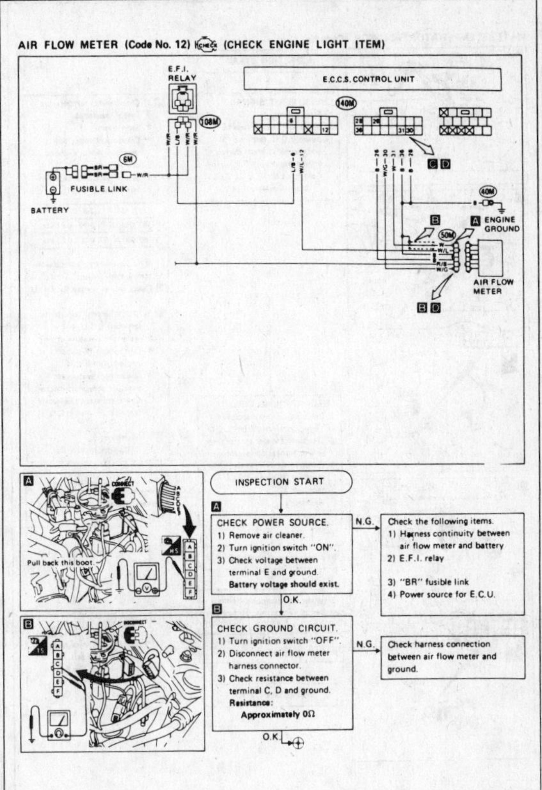
AIR FLOW METER (Code No. 12) (CHECK ENGINE LIGHT ITEM)
E.F.I. RELAY
E.C.C.S. CONTROL UNIT
BATTERY
FUSIBLE LINK
ENGINE GROUND
AIR FLOW METER
Pull back this boot.
INSPECTION START
CHECK POWER SOURCE.
1) Remove air cleaner.
2) Turn ignition switch "ON".
3) Check voltage between terminal E and ground.
Battery voltage should exist.
N.G.
Check the following items.
1) Harness continuity between air flow meter and battery
2) E.F.I. relay
3) "BR" fusible link
4) Power source for E.C.U.
O.K.
CHECK GROUND CIRCUIT.
1) Turn ignition switch "OFF".
2) Disconnect air flow meter harness connector.
3) Check resistance between terminal C, D and ground.
Resistance:
Approximately 0Ω
N.G.
Check harness connection between air flow meter and ground.
O.K.

ECCS SELF-DIAGNOSTIC CHART—1988 PULSAR

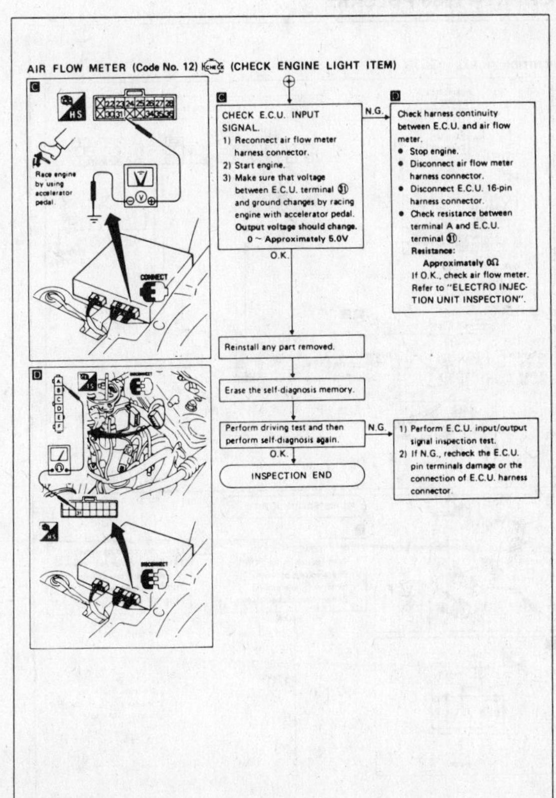
AIR FLOW METER (Code No. 12) (CHECK ENGINE LIGHT ITEM)
Race engine by using accelerator pedal
CHECK E.C.U. INPUT SIGNAL.
1) Reconnect air flow meter harness connector.
2) Start engine.
3) Make sure that voltage between E.C.U. terminal ㉛ and ground changes by racing engine with accelerator pedal.
Output voltage should change.
0 ~ Approximately 5.0V
N.G.
Check harness continuity between E.C.U. and air flow meter.
• Stop engine.
• Disconnect air flow meter harness connector.
• Disconnect E.C.U. 16-pin harness connector.
• Check resistance between terminal A and E.C.U. terminal ㉛.
Resistance:
Approximately 0Ω
If O.K., check air flow meter. Refer to "ELECTRO INJECTION UNIT INSPECTION".
O.K.
Reinstall any part removed.
Erase the self-diagnosis memory.
Perform driving test and then perform self-diagnosis again.
N.G.
1) Perform E.C.U. input/output signal inspection test.
2) If N.G., recheck the E.C.U. pin terminals damage or the connection of E.C.U. harness connector.
O.K.
INSPECTION END

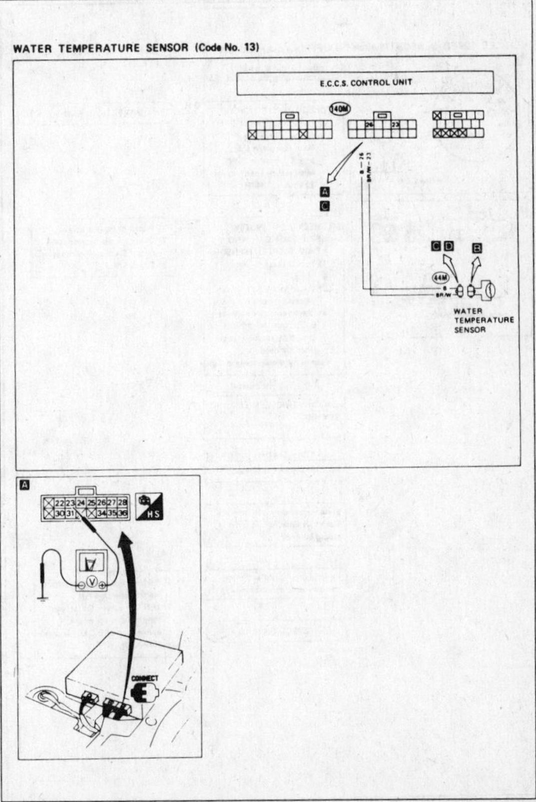
WATER TEMPERATURE SENSOR (Code No. 13)
E.C.C.S. CONTROL UNIT
WATER TEMPERATURE SENSOR
CONNECT

ECCS SELF-DIAGNOSTIC CHART—1988 PULSAR

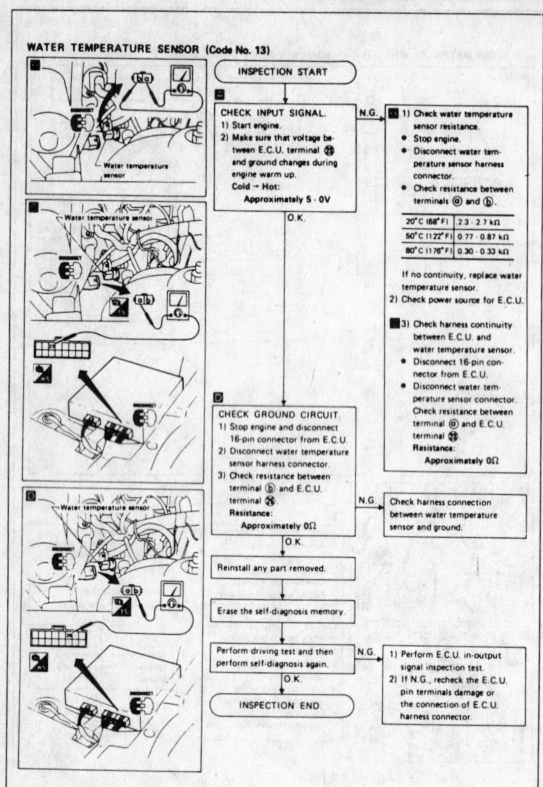
WATER TEMPERATURE SENSOR (Code No. 13)
INSPECTION START
CHECK INPUT SIGNAL.
1) Start engine.
2) Make sure that voltage between E.C.U. terminal ⑳ and ground changes during engine warm up.
Cold → Hot:
Approximately 5 - 0V
N.G.
1) Check water temperature sensor resistance.
• Stop engine.
• Disconnect water temperature sensor harness connector.
• Check resistance between terminals ⓐ and ⓑ.
20°C (68°F) 2.3 - 2.7 kΩ
50°C (122°F) 0.77 - 0.87 kΩ
80°C (176°F) 0.30 - 0.33 kΩ
If no continuity, replace water temperature sensor.
2) Check power source for E.C.U.
3) Check harness continuity between E.C.U. and water temperature sensor.
• Disconnect 16-pin connector from E.C.U.
• Disconnect water temperature sensor connector.
Check resistance between terminal ⓐ and E.C.U. terminal ⑳.
Resistance:
Approximately 0Ω
O.K.
CHECK GROUND CIRCUIT.
1) Stop engine and disconnect 16-pin connector from E.C.U.
2) Disconnect water temperature sensor harness connector.
3) Check resistance between terminal ⓑ and E.C.U. terminal ㉖.
Resistance:
Approximately 0Ω
N.G.
Check harness connection between water temperature sensor and ground.
O.K.
Reinstall any part removed.
Erase the self-diagnosis memory.
Perform driving test and then perform self-diagnosis again.
N.G.
1) Perform E.C.U. in-output signal inspection test.
2) If N.G., recheck the E.C.U. pin terminals damage or the connection of E.C.U. harness connector.
O.K.
INSPECTION END
Water temperature sensor
Water temperature sensor

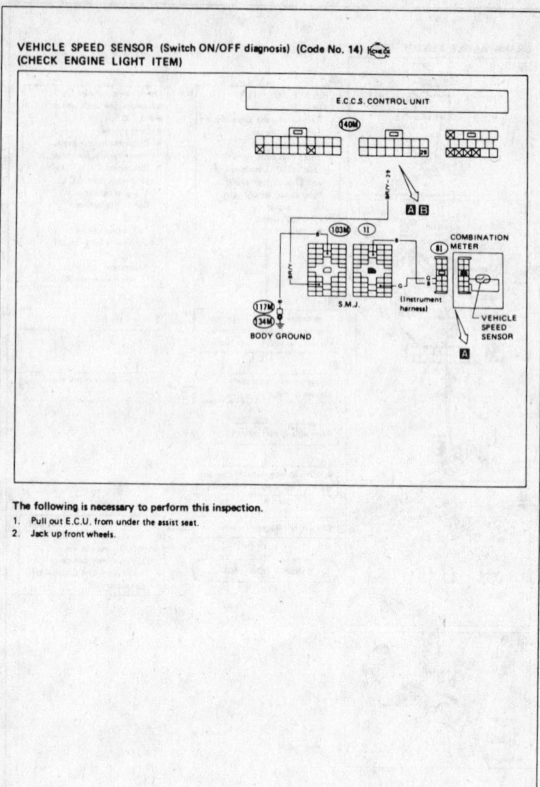
VEHICLE SPEED SENSOR (Switch ON/OFF diagnosis) (Code No. 14) (CHECK ENGINE LIGHT ITEM)
E.C.C.S. CONTROL UNIT
140M
103M
11
81
COMBINATION METER
117M
134M
BODY GROUND
S.M.J.
(Instrument harness)
VEHICLE SPEED SENSOR
The following is necessary to perform this inspection.
1. Pull out E.C.U. from under the assist seat.
2. Jack up front wheels.

ECCS SELF-DIAGNOSTIC CHART—1988 PULSAR

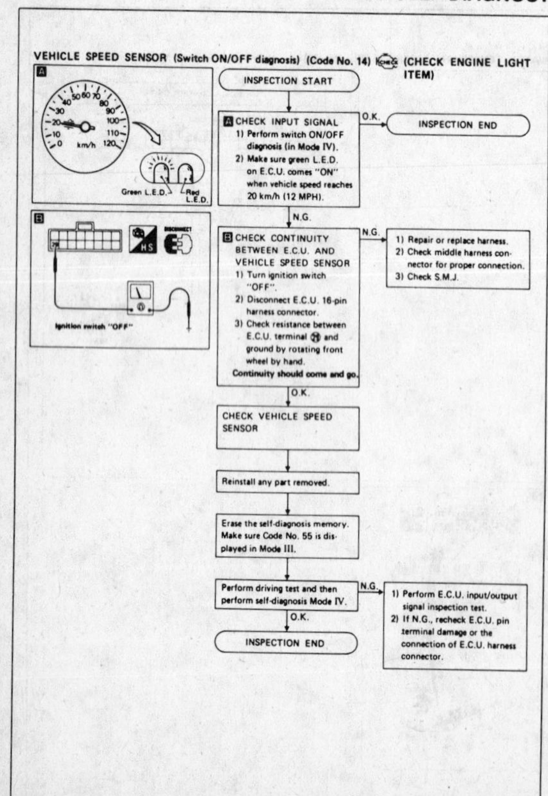
VEHICLE SPEED SENSOR (Switch ON/OFF diagnosis) (Code No. 14) (CHECK ENGINE LIGHT ITEM)
INSPECTION START
CHECK INPUT SIGNAL
1) Perform switch ON/OFF diagnosis (in Mode IV).
2) Make sure green L.E.D. on E.C.U. comes "ON" when vehicle speed reaches 20 km/h (12 MPH).
O.K.
INSPECTION END
N.G.
CHECK CONTINUITY BETWEEN E.C.U. AND VEHICLE SPEED SENSOR
1) Turn ignition switch "OFF".
2) Disconnect E.C.U. 16-pin harness connector.
3) Check resistance between E.C.U. terminal ㉙ and ground by rotating front wheel by hand.
Continuity should come and go.
N.G.
1) Repair or replace harness.
2) Check middle harness connector for proper connection.
3) Check S.M.J.
O.K.
CHECK VEHICLE SPEED SENSOR
Reinstall any part removed.
Erase the self-diagnosis memory. Make sure Code No. 55 is displayed in Mode III.
Perform driving test and then perform self-diagnosis Mode IV.
N.G.
1) Perform E.C.U. input/output signal inspection test.
2) If N.G., recheck E.C.U. pin terminal damage or the connection of E.C.U. harness connector.
O.K.
INSPECTION END
km/h
Green L.E.D.
Red L.E.D.
Ignition switch "OFF"

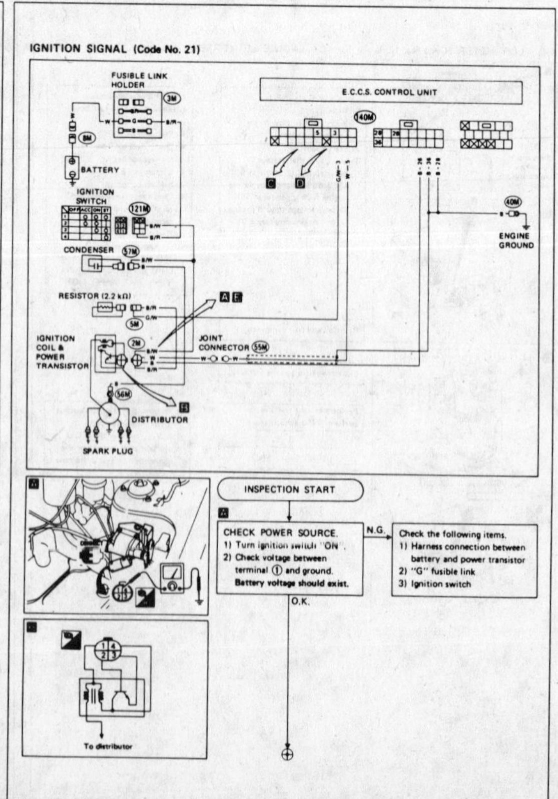
IGNITION SIGNAL (Code No. 21)
FUSIBLE LINK HOLDER
3M
8M
BATTERY
IGNITION SWITCH
121M
CONDENSER
37M
RESISTOR (2.2 kΩ)
5M
IGNITION COIL & POWER TRANSISTOR
2M
JOINT CONNECTOR
55M
56M
DISTRIBUTOR
SPARK PLUG
E.C.C.S. CONTROL UNIT
140M
40M
ENGINE GROUND
INSPECTION START
CHECK POWER SOURCE.
1) Turn ignition switch "ON".
2) Check voltage between terminal ① and ground.
Battery voltage should exist.
N.G.
Check the following items.
1) Harness connection between battery and power transistor
2) "G" fusible link
3) Ignition switch
O.K.
To distributor

ECCS SELF-DIAGNOSTIC CHART—1988 PULSAR

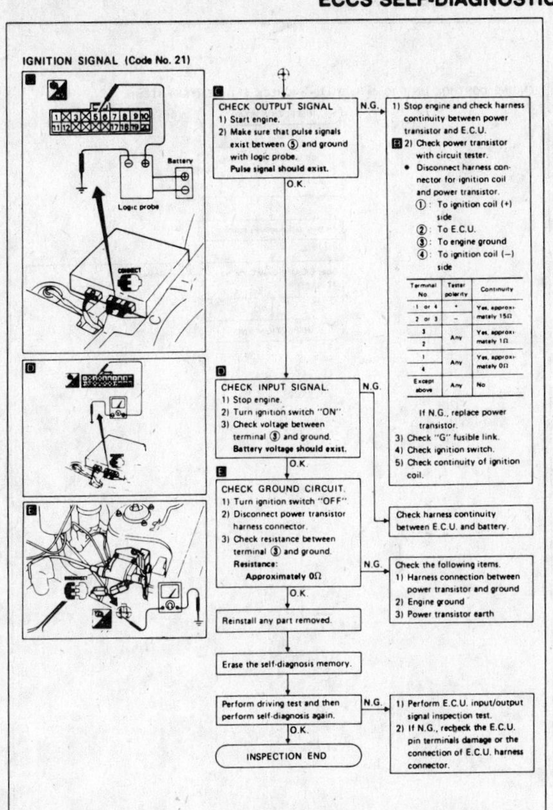
IGNITION SIGNAL (Code No. 21)
CHECK OUTPUT SIGNAL
1) Start engine.
2) Make sure that pulse signals exist between ⑤ and ground with logic probe.
Pulse signal should exist.
N.G.
O.K.
1) Stop engine and check harness continuity between power transistor and E.C.U.
2) Check power transistor with circuit tester.
• Disconnect harness connector for ignition coil and power transistor.
①: To ignition coil (+) side
②: To E.C.U.
③: To engine ground
④: To ignition coil (−) side
If N.G., replace power transistor.
3) Check "G" fusible link.
4) Check ignition switch.
5) Check continuity of ignition coil.
CHECK INPUT SIGNAL.
1) Stop engine.
2) Turn ignition switch "ON".
3) Check voltage between terminal ③ and ground.
Battery voltage should exist.
CHECK GROUND CIRCUIT.
1) Turn ignition switch "OFF".
2) Disconnect power transistor harness connector.
3) Check resistance between terminal ③ and ground.
Resistance: Approximately 0Ω
Check harness continuity between E.C.U. and battery.
Check the following items.
1) Harness connection between power transistor and ground
2) Engine ground
3) Power transistor earth
Reinstall any part removed.
Erase the self-diagnosis memory.
Perform driving test and then perform self-diagnosis again.
1) Perform E.C.U. input/output signal inspection test.
2) If N.G., recheck the E.C.U. pin terminals damage or the connection of E.C.U. harness connector.
INSPECTION END
Battery
Logic probe

Terminal No.	Tester polarity	Continuity
1 or 4	+	Yes, approximately 15Ω
2 or 3	−	
3 / 2	Any	Yes, approximately 1Ω
1 / 4	Any	Yes, approximately 0Ω
Except above	Any	No

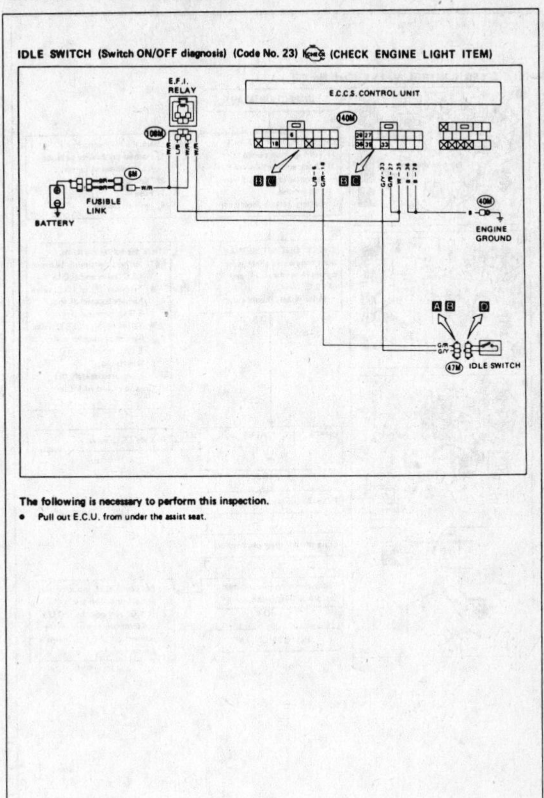
IDLE SWITCH (Switch ON/OFF diagnosis) (Code No. 23) (CHECK ENGINE LIGHT ITEM)
E.F.I. RELAY
E.C.C.S. CONTROL UNIT
FUSIBLE LINK
BATTERY
ENGINE GROUND
IDLE SWITCH
The following is necessary to perform this inspection.
• Pull out E.C.U. from under the assist seat.

ECCS SELF-DIAGNOSTIC CHART—1988 PULSAR

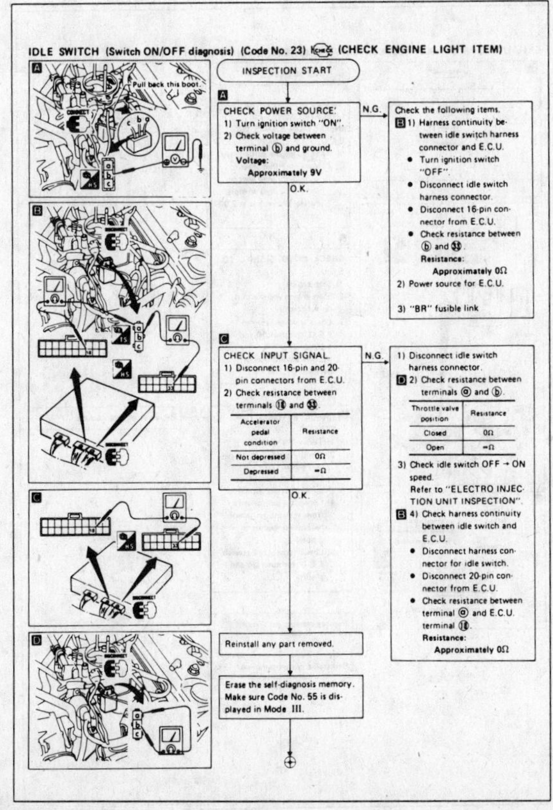
IDLE SWITCH (Switch ON/OFF diagnosis) (Code No. 23) (CHECK ENGINE LIGHT ITEM)
Pull back this boot.
INSPECTION START
CHECK POWER SOURCE:
1) Turn ignition switch "ON".
2) Check voltage between terminal ⓑ and ground.
Voltage: Approximately 9V
N.G.
O.K.
Check the following items.
1) Harness continuity between idle switch harness connector and E.C.U.
• Turn ignition switch "OFF".
• Disconnect idle switch harness connector.
• Disconnect 16-pin connector from E.C.U.
• Check resistance between ⓑ and ㊸.
Resistance: Approximately 0Ω
2) Power source for E.C.U.
3) "BR" fusible link
CHECK INPUT SIGNAL.
1) Disconnect 16-pin and 20-pin connectors from E.C.U.
2) Check resistance between terminals ⑯ and ㊸.
1) Disconnect idle switch harness connector.
2) Check resistance between terminals ⓒ and ⓑ.
3) Check idle switch OFF → ON speed.
Refer to "ELECTRO INJECTION UNIT INSPECTION".
4) Check harness continuity between idle switch and E.C.U.
• Disconnect harness connector for idle switch.
• Disconnect 20-pin connector from E.C.U.
• Check resistance between terminal ⓒ and E.C.U. terminal ⑯.
Resistance: Approximately 0Ω
Reinstall any part removed.
Erase the self-diagnosis memory. Make sure Code No. 55 is displayed in Mode III.

Accelerator pedal condition	Resistance
Not depressed	0Ω
Depressed	∞Ω

Throttle valve position	Resistance
Closed	0Ω
Open	∞Ω

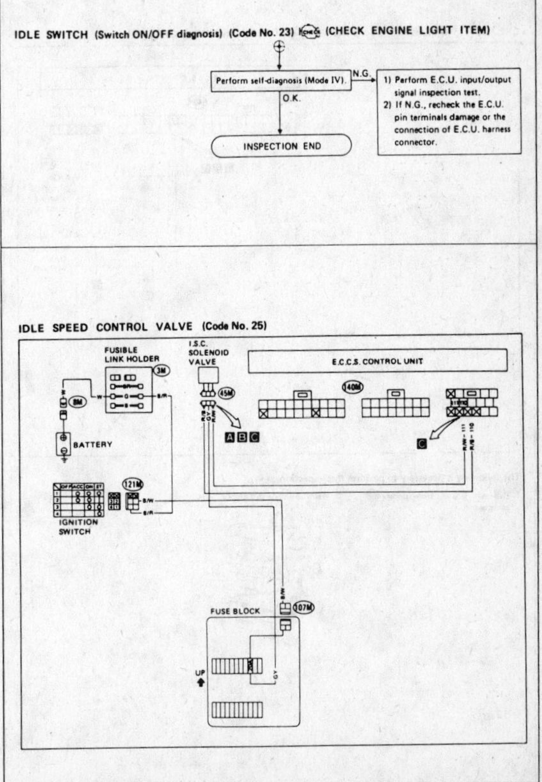
IDLE SWITCH (Switch ON/OFF diagnosis) (Code No. 23) (CHECK ENGINE LIGHT ITEM)
Perform self-diagnosis (Mode IV).
N.G.
O.K.
1) Perform E.C.U. input/output signal inspection test.
2) If N.G., recheck the E.C.U. pin terminals damage or the connection of E.C.U. harness connector.
INSPECTION END
IDLE SPEED CONTROL VALVE (Code No. 25)
FUSIBLE LINK HOLDER
I.S.C. SOLENOID VALVE
E.C.C.S. CONTROL UNIT
BATTERY
IGNITION SWITCH
FUSE BLOCK
UP

ECCS SELF-DIAGNOSTIC CHART—1988 PULSAR

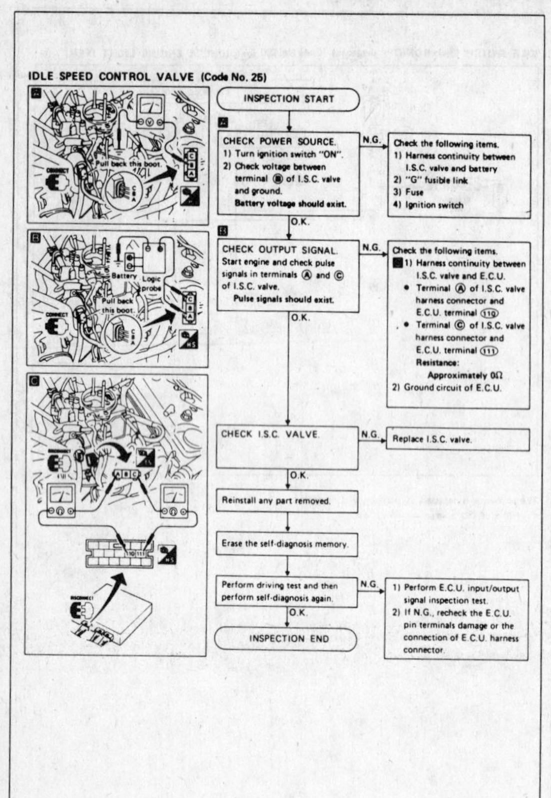
IDLE SPEED CONTROL VALVE (Code No. 25)
INSPECTION START
CHECK POWER SOURCE.
1) Turn ignition switch "ON".
2) Check voltage between terminal Ⓑ of I.S.C. valve and ground.
Battery voltage should exist.
N.G.
Check the following items.
1) Harness continuity between I.S.C. valve and battery
2) "G" fusible link
3) Fuse
4) Ignition switch
O.K.
CHECK OUTPUT SIGNAL.
Start engine and check pulse signals in terminals Ⓐ and Ⓒ of I.S.C. valve.
Pulse signals should exist.
N.G.
Check the following items.
1) Harness continuity between I.S.C. valve and E.C.U.
• Terminal Ⓐ of I.S.C. valve harness connector and E.C.U. terminal 110
• Terminal Ⓒ of I.S.C. valve harness connector and E.C.U. terminal 111
Resistance:
Approximately 0Ω
2) Ground circuit of E.C.U.
O.K.
CHECK I.S.C. VALVE.
N.G.
Replace I.S.C. valve.
O.K.
Reinstall any part removed.
Erase the self-diagnosis memory.
Perform driving test and then perform self-diagnosis again.
N.G.
1) Perform E.C.U. input/output signal inspection test.
2) If N.G., recheck the E.C.U. pin terminals damage or the connection of E.C.U. harness connector.
O.K.
INSPECTION END
Pull back this boot.
Battery
Logic probe
Pull back this boot.

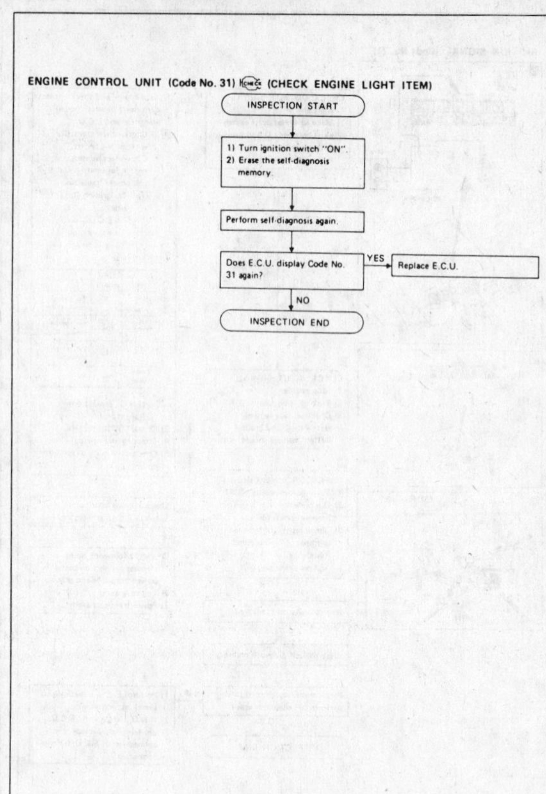
ENGINE CONTROL UNIT (Code No. 31) (CHECK ENGINE LIGHT ITEM)
INSPECTION START
1) Turn ignition switch "ON".
2) Erase the self-diagnosis memory.
Perform self-diagnosis again.
Does E.C.U. display Code No. 31 again?
YES
Replace E.C.U.
NO
INSPECTION END

ECCS SELF-DIAGNOSTIC CHART—1988 PULSAR

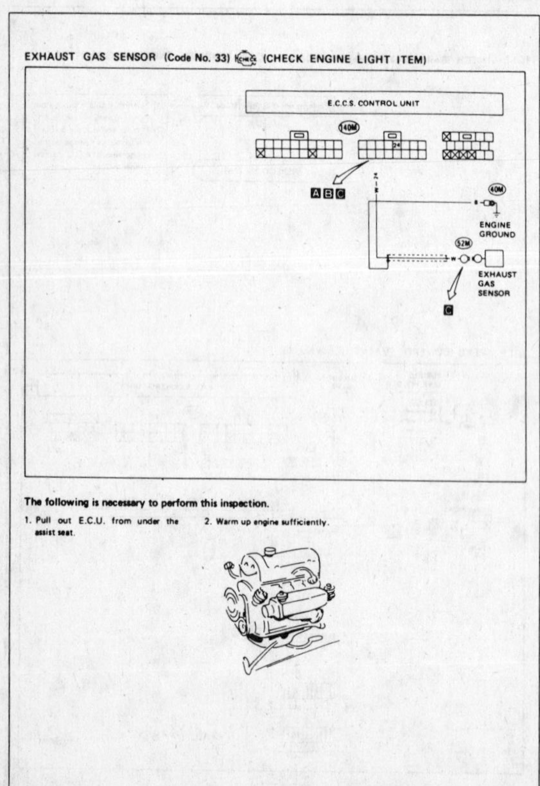
EXHAUST GAS SENSOR (Code No. 33) (CHECK ENGINE LIGHT ITEM)
E.C.C.S. CONTROL UNIT
40M
ENGINE GROUND
52M
EXHAUST GAS SENSOR
The following is necessary to perform this inspection.
1. Pull out E.C.U. from under the assist seat.
2. Warm up engine sufficiently.

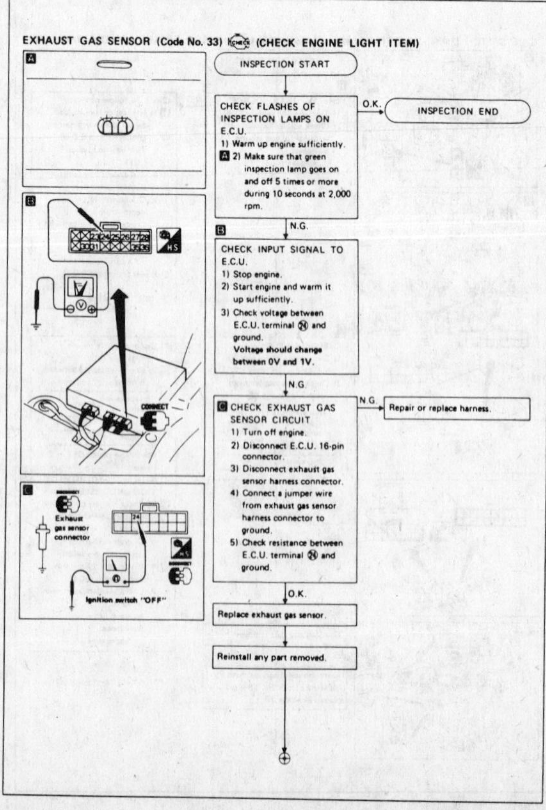
EXHAUST GAS SENSOR (Code No. 33) (CHECK ENGINE LIGHT ITEM)
INSPECTION START
CHECK FLASHES OF INSPECTION LAMPS ON E.C.U.
1) Warm up engine sufficiently.
2) Make sure that green inspection lamp goes on and off 5 times or more during 10 seconds at 2,000 rpm.
O.K.
INSPECTION END
N.G.
CHECK INPUT SIGNAL TO E.C.U.
1) Stop engine.
2) Start engine and warm it up sufficiently.
3) Check voltage between E.C.U. terminal 24 and ground.
Voltage should change between 0V and 1V.
N.G.
CHECK EXHAUST GAS SENSOR CIRCUIT
1) Turn off engine.
2) Disconnect E.C.U. 16-pin connector.
3) Disconnect exhaust gas sensor harness connector.
4) Connect a jumper wire from exhaust gas sensor harness connector to ground.
5) Check resistance between E.C.U. terminal 24 and ground.
N.G.
Repair or replace harness.
O.K.
Replace exhaust gas sensor.
Reinstall any part removed.
Exhaust gas sensor connector
Ignition switch "OFF"

ECCS SELF-DIAGNOSTIC CHART—1988 PULSAR

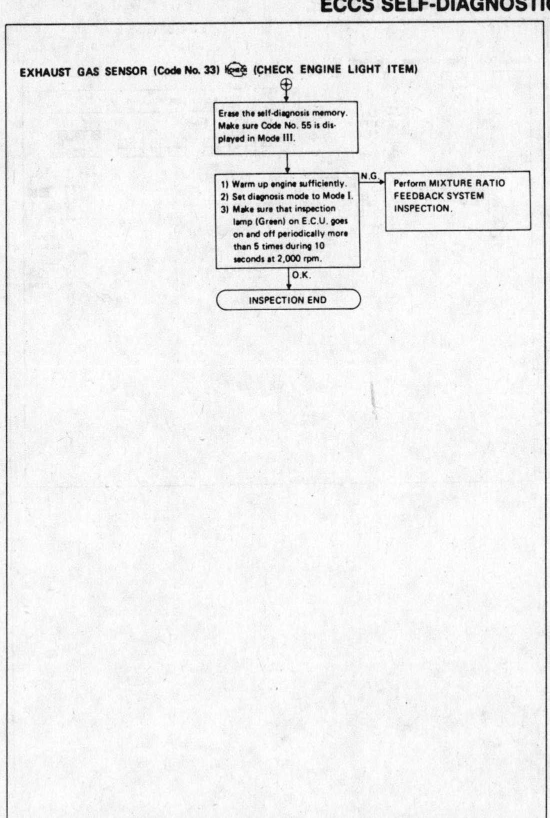
EXHAUST GAS SENSOR (Code No. 33) (CHECK ENGINE LIGHT ITEM)
Erase the self-diagnosis memory. Make sure Code No. 55 is displayed in Mode III.
1) Warm up engine sufficiently.
2) Set diagnosis mode to Mode I.
3) Make sure that inspection lamp (Green) on E.C.U. goes on and off periodically more than 5 times during 10 seconds at 2,000 rpm.
N.G.
Perform MIXTURE RATIO FEEDBACK SYSTEM INSPECTION.
O.K.
INSPECTION END

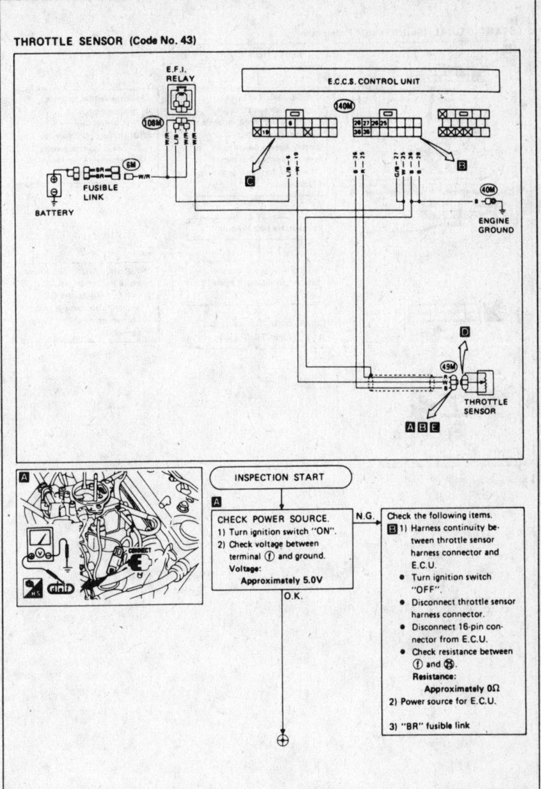
THROTTLE SENSOR (Code No. 43)
E.F.I. RELAY
E.C.C.S. CONTROL UNIT
FUSIBLE LINK
BATTERY
ENGINE GROUND
THROTTLE SENSOR
INSPECTION START
CHECK POWER SOURCE.
1) Turn ignition switch "ON".
2) Check voltage between terminal ⓕ and ground.
Voltage:
Approximately 5.0V
N.G.
Check the following items.
1) Harness continuity between throttle sensor harness connector and E.C.U.
• Turn ignition switch "OFF".
• Disconnect throttle sensor harness connector.
• Disconnect 16-pin connector from E.C.U.
• Check resistance between ⓕ and ㉕.
Resistance:
Approximately 0Ω
2) Power source for E.C.U.
3) "BR" fusible link
O.K.

ECCS SELF-DIAGNOSTIC CHART—1988 PULSAR

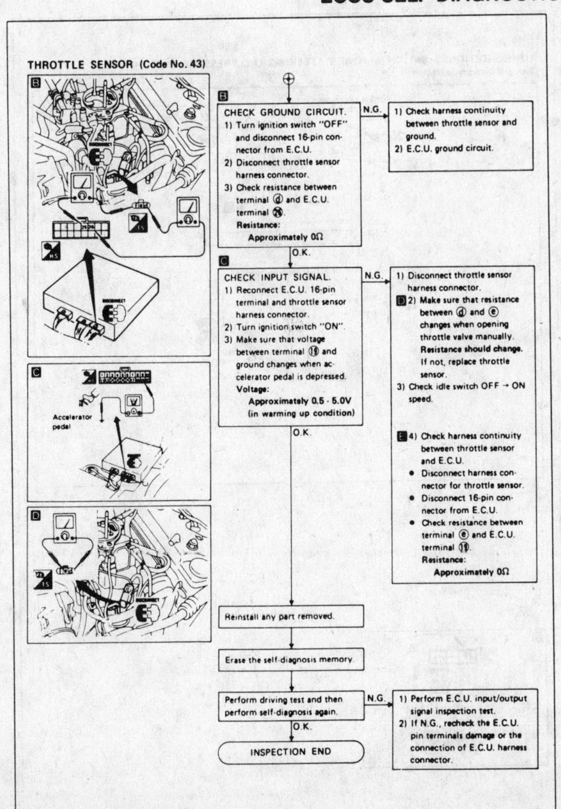
THROTTLE SENSOR (Code No. 43)
CHECK GROUND CIRCUIT.
1) Turn ignition switch "OFF" and disconnect 16-pin connector from E.C.U.
2) Disconnect throttle sensor harness connector.
3) Check resistance between terminal ⓓ and E.C.U. terminal ㉖.
Resistance:
Approximately 0Ω
N.G.
1) Check harness continuity between throttle sensor and ground.
2) E.C.U. ground circuit.
O.K.
CHECK INPUT SIGNAL.
1) Reconnect E.C.U. 16-pin terminal and throttle sensor harness connector.
2) Turn ignition switch "ON".
3) Make sure that voltage between terminal ⑲ and ground changes when accelerator pedal is depressed.
Voltage:
Approximately 0.5 - 5.0V
(in warming up condition)
N.G.
1) Disconnect throttle sensor harness connector.
2) Make sure that resistance between ⓓ and ⓔ changes when opening throttle valve manually.
Resistance should change.
If not, replace throttle sensor.
3) Check idle switch OFF → ON speed.
4) Check harness continuity between throttle sensor and E.C.U.
• Disconnect harness connector for throttle sensor.
• Disconnect 16-pin connector from E.C.U.
• Check resistance between terminal ⓔ and E.C.U. terminal ⑲.
Resistance:
Approximately 0Ω
O.K.
Accelerator pedal
Reinstall any part removed.
Erase the self-diagnosis memory.
Perform driving test and then perform self-diagnosis again.
N.G.
1) Perform E.C.U. input/output signal inspection test.
2) If N.G., recheck the E.C.U. pin terminals damage or the connection of E.C.U. harness connector.
O.K.
INSPECTION END

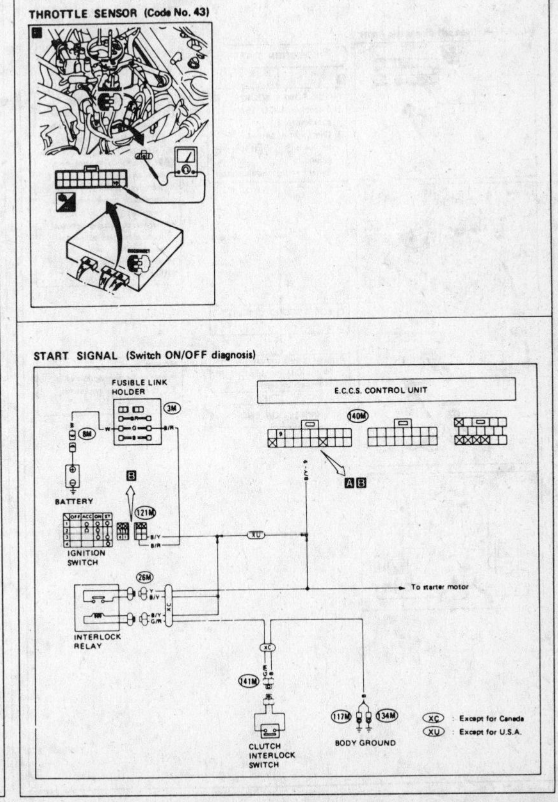
THROTTLE SENSOR (Code No. 43)
START SIGNAL (Switch ON/OFF diagnosis)
FUSIBLE LINK HOLDER
E.C.C.S. CONTROL UNIT
BATTERY
IGNITION SWITCH
To starter motor
INTERLOCK RELAY
CLUTCH INTERLOCK SWITCH
BODY GROUND
XC : Except for Canada
XU : Except for U.S.A.

ECCS SELF-DIAGNOSTIC CHART—1988 PULSAR

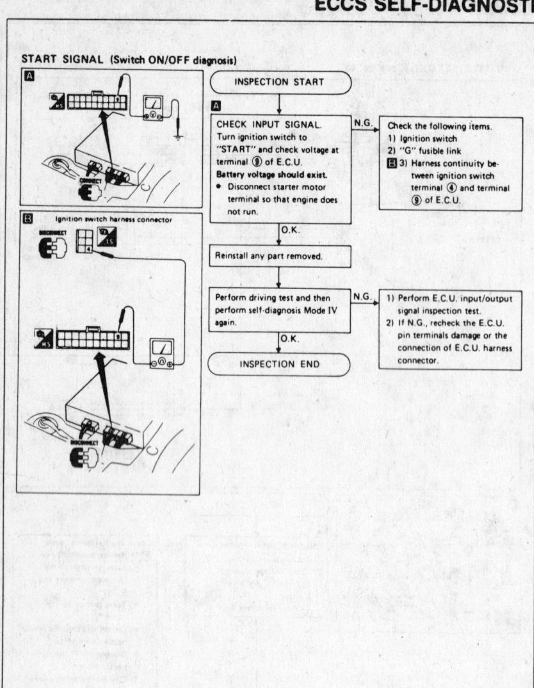
START SIGNAL (Switch ON/OFF diagnosis)
A
Ignition switch harness connector
INSPECTION START
A
CHECK INPUT SIGNAL.
Turn ignition switch to "START" and check voltage at terminal ⑨ of E.C.U.
Battery voltage should exist.
• Disconnect starter motor terminal so that engine does not run.
N.G.
Check the following items.
1) Ignition switch
2) "G" fusible link
B 3) Harness continuity between ignition switch terminal ④ and terminal ⑨ of E.C.U.
O.K.
Reinstall any part removed.
Perform driving test and then perform self-diagnosis Mode IV again.
N.G.
1) Perform E.C.U. input/output signal inspection test.
2) If N.G., recheck the E.C.U. pin terminals damage or the connection of E.C.U. harness connector.
O.K.
INSPECTION END

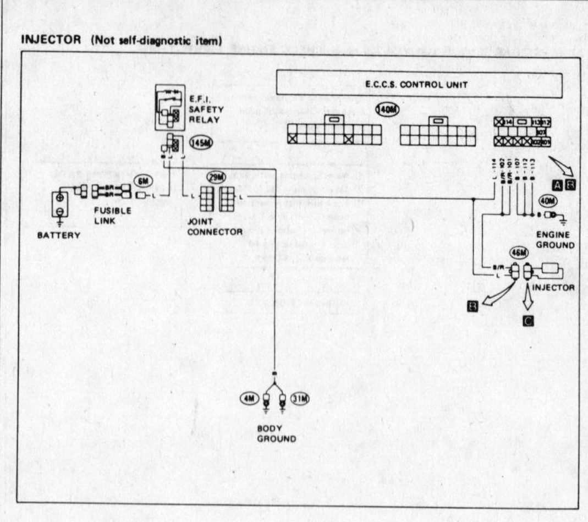
INJECTOR (Not self-diagnostic item)
E.C.C.S. CONTROL UNIT
E.F.I. SAFETY RELAY
BATTERY
FUSIBLE LINK
JOINT CONNECTOR
ENGINE GROUND
INJECTOR
BODY GROUND

ECCS SELF-DIAGNOSTIC CHART—1988 PULSAR

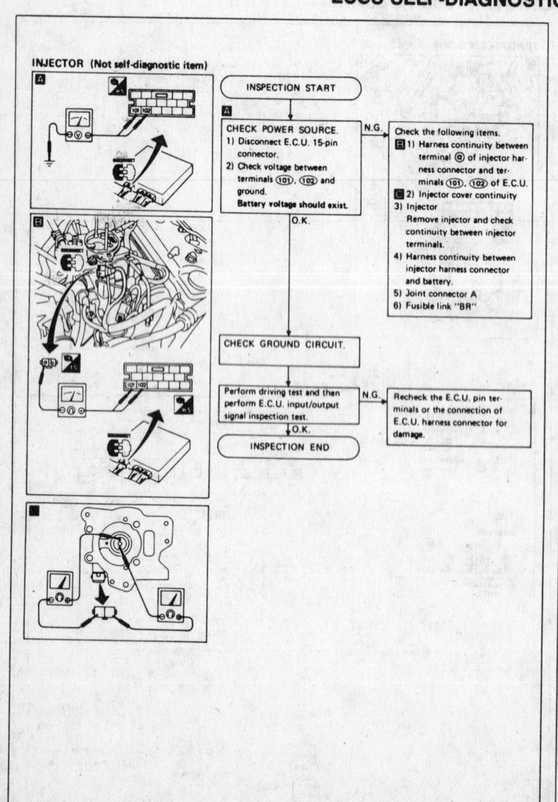
INJECTOR (Not self-diagnostic item)
INSPECTION START
A
CHECK POWER SOURCE.
1) Disconnect E.C.U. 15-pin connector.
2) Check voltage between terminals (101), (102) and ground.
Battery voltage should exist.
N.G.
Check the following items.
B 1) Harness continuity between terminal ⓞ of injector harness connector and terminals (101), (102) of E.C.U.
C 2) Injector cover continuity
3) Injector
Remove injector and check continuity between injector terminals.
4) Harness continuity between injector harness connector and battery.
5) Joint connector A
6) Fusible link "BR"
O.K.
CHECK GROUND CIRCUIT.
Perform driving test and then perform E.C.U. input/output signal inspection test.
N.G.
Recheck the E.C.U. pin terminals or the connection of E.C.U. harness connector for damage.
O.K.
INSPECTION END

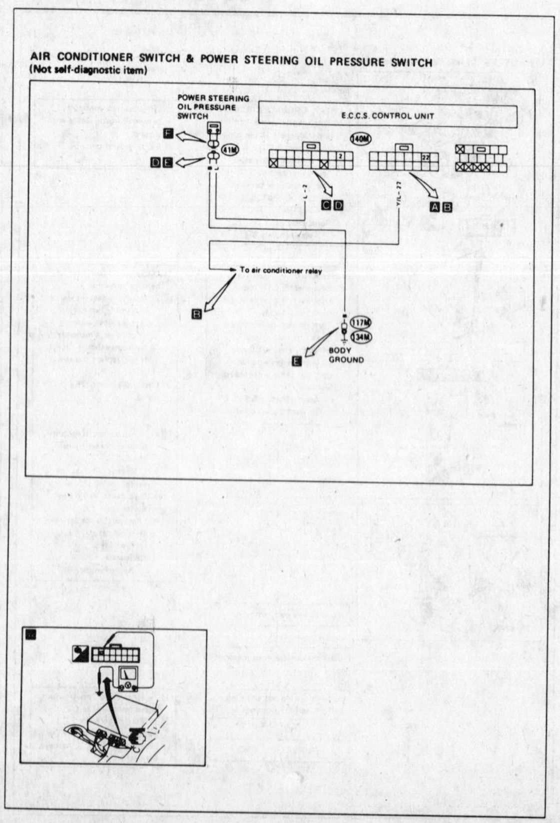
AIR CONDITIONER SWITCH & POWER STEERING OIL PRESSURE SWITCH
(Not self-diagnostic item)
POWER STEERING OIL PRESSURE SWITCH
E.C.C.S. CONTROL UNIT
To air conditioner relay
BODY GROUND

ECCS SELF-DIAGNOSTIC CHART—1988 PULSAR

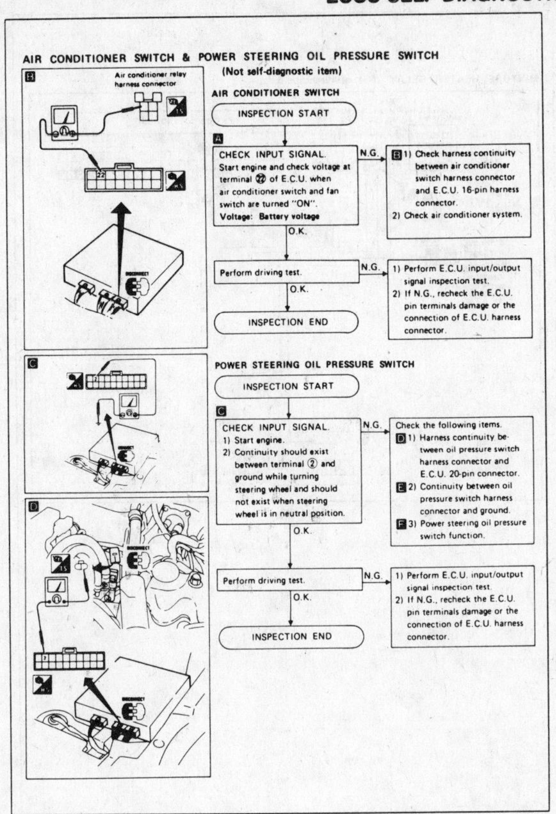
AIR CONDITIONER SWITCH & POWER STEERING OIL PRESSURE SWITCH
(Not self-diagnostic item)
Air conditioner relay harness connector
AIR CONDITIONER SWITCH
INSPECTION START
CHECK INPUT SIGNAL.
Start engine and check voltage at terminal 22 of E.C.U. when air conditioner switch and fan switch are turned "ON".
Voltage: Battery voltage
N.G.
1) Check harness continuity between air conditioner switch harness connector and E.C.U. 16-pin harness connector.
2) Check air conditioner system.
O.K.
Perform driving test.
N.G.
1) Perform E.C.U. input/output signal inspection test.
2) If N.G., recheck the E.C.U. pin terminals damage or the connection of E.C.U. harness connector.
O.K.
INSPECTION END
POWER STEERING OIL PRESSURE SWITCH
INSPECTION START
CHECK INPUT SIGNAL.
1) Start engine.
2) Continuity should exist between terminal ② and ground while turning steering wheel and should not exist when steering wheel is in neutral position.
N.G.
Check the following items.
1) Harness continuity between oil pressure switch harness connector and E.C.U. 20-pin connector.
2) Continuity between oil pressure switch harness connector and ground.
3) Power steering oil pressure switch function.
O.K.
Perform driving test.
N.G.
1) Perform E.C.U. input/output signal inspection test.
2) If N.G., recheck the E.C.U. pin terminals damage or the connection of E.C.U. harness connector.
O.K.
INSPECTION END

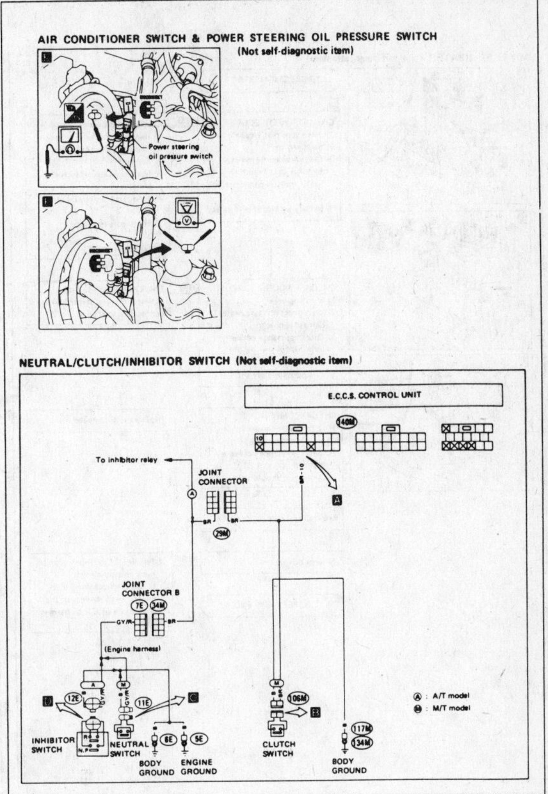
AIR CONDITIONER SWITCH & POWER STEERING OIL PRESSURE SWITCH
(Not self-diagnostic item)
Power steering oil pressure switch
NEUTRAL/CLUTCH/INHIBITOR SWITCH (Not self-diagnostic item)
E.C.C.S. CONTROL UNIT
To inhibitor relay
JOINT CONNECTOR
JOINT CONNECTOR B
(Engine harness)
INHIBITOR SWITCH
NEUTRAL SWITCH
BODY GROUND
ENGINE GROUND
CLUTCH SWITCH
BODY GROUND
Ⓐ : A/T model
Ⓜ : M/T model

ECCS SELF-DIAGNOSTIC CHART—1988 PULSAR

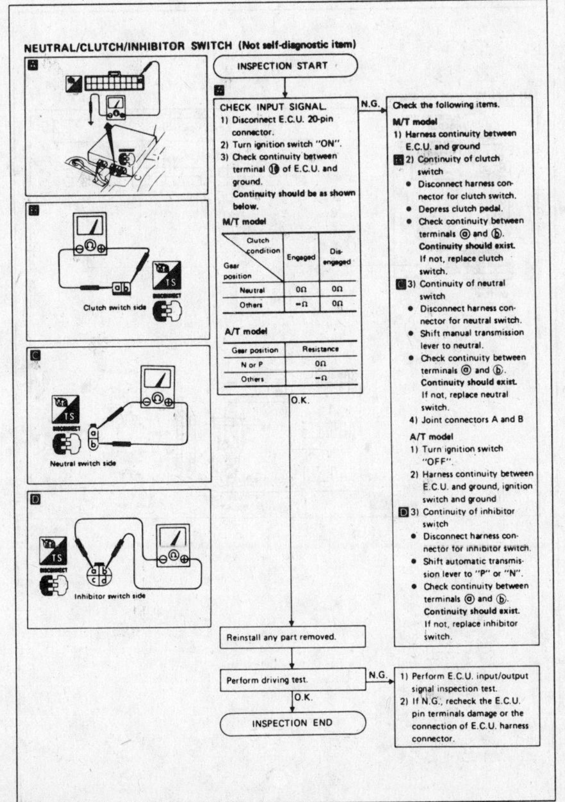
NEUTRAL/CLUTCH/INHIBITOR SWITCH (Not self-diagnostic item)
INSPECTION START
CHECK INPUT SIGNAL.
1) Disconnect E.C.U. 20-pin connector.
2) Turn ignition switch "ON".
3) Check continuity between terminal ⑪ of E.C.U. and ground.
Continuity should be as shown below.
M/T model
Clutch condition / Gear position | Engaged | Dis-engaged
Neutral | 0Ω | 0Ω
Others | ∞Ω | 0Ω
A/T model
Gear position | Resistance
N or P | 0Ω
Others | ∞Ω
N.G.
Check the following items.
M/T model
1) Harness continuity between E.C.U. and ground
2) Continuity of clutch switch
• Disconnect harness connector for clutch switch.
• Depress clutch pedal.
• Check continuity between terminals ⓐ and ⓑ.
Continuity should exist.
If not, replace clutch switch.
3) Continuity of neutral switch
• Disconnect harness connector for neutral switch.
• Shift manual transmission lever to neutral.
• Check continuity between terminals ⓐ and ⓑ.
Continuity should exist.
If not, replace neutral switch.
4) Joint connectors A and B
A/T model
1) Turn ignition switch "OFF".
2) Harness continuity between E.C.U. and ground, ignition switch and ground
3) Continuity of inhibitor switch
• Disconnect harness connector for inhibitor switch.
• Shift automatic transmission lever to "P" or "N".
• Check continuity between terminals ⓐ and ⓑ.
Continuity should exist.
If not, replace inhibitor switch.
O.K.
Reinstall any part removed.
Perform driving test.
N.G.
1) Perform E.C.U. input/output signal inspection test.
2) If N.G., recheck the E.C.U. pin terminals damage or the connection of E.C.U. harness connector.
O.K.
INSPECTION END
Clutch switch side
Neutral switch side
Inhibitor switch side

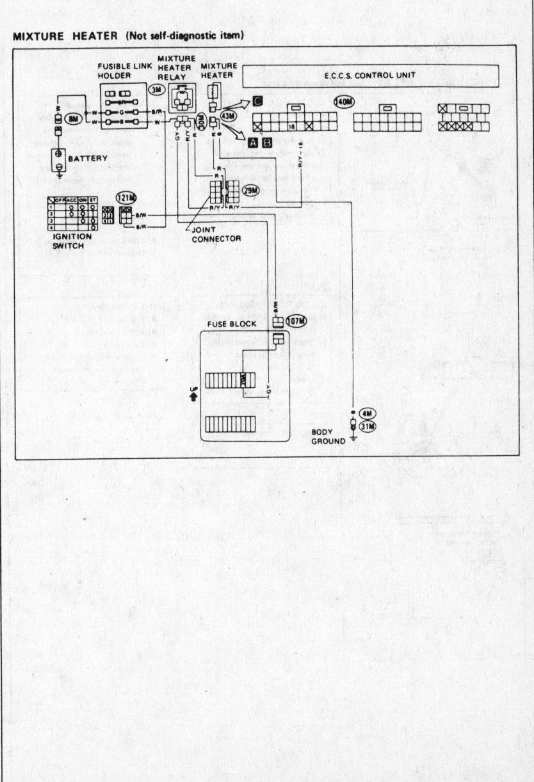
MIXTURE HEATER (Not self-diagnostic item)
FUSIBLE LINK HOLDER
MIXTURE HEATER RELAY
MIXTURE HEATER
E.C.C.S. CONTROL UNIT
BATTERY
IGNITION SWITCH
JOINT CONNECTOR
FUSE BLOCK
BODY GROUND

ECCS SELF-DIAGNOSTIC CHART—1988 PULSAR

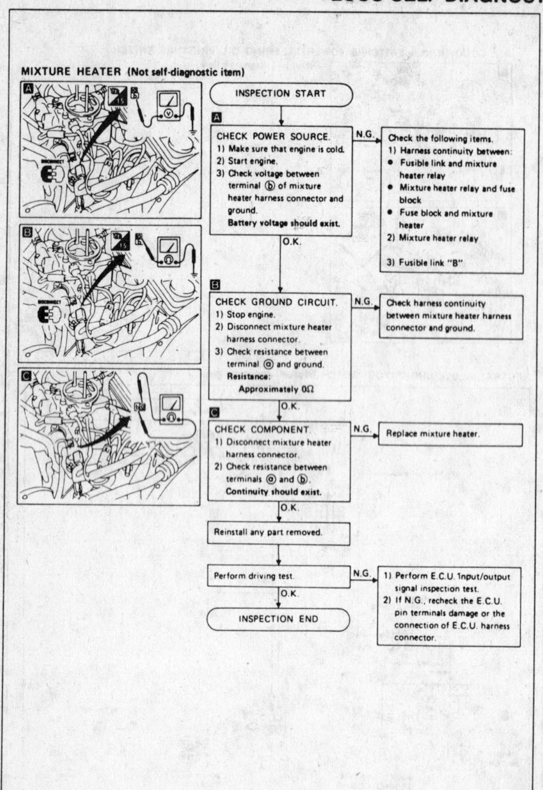
MIXTURE HEATER (Not self-diagnostic item)
INSPECTION START
A
CHECK POWER SOURCE.
1) Make sure that engine is cold.
2) Start engine.
3) Check voltage between terminal ⓑ of mixture heater harness connector and ground.
Battery voltage should exist.
N.G.
Check the following items.
1) Harness continuity between:
• Fusible link and mixture heater relay
• Mixture heater relay and fuse block
• Fuse block and mixture heater
2) Mixture heater relay
3) Fusible link "B"
O.K.
B
CHECK GROUND CIRCUIT.
1) Stop engine.
2) Disconnect mixture heater harness connector.
3) Check resistance between terminal ⓐ and ground.
Resistance:
Approximately 0Ω
N.G.
Check harness continuity between mixture heater harness connector and ground.
O.K.
C
CHECK COMPONENT.
1) Disconnect mixture heater harness connector.
2) Check resistance between terminals ⓐ and ⓑ.
Continuity should exist.
N.G.
Replace mixture heater.
O.K.
Reinstall any part removed.
Perform driving test.
N.G.
1) Perform E.C.U. Input/output signal inspection test.
2) If N.G., recheck the E.C.U. pin terminals damage or the connection of E.C.U. harness connector.
O.K.
INSPECTION END

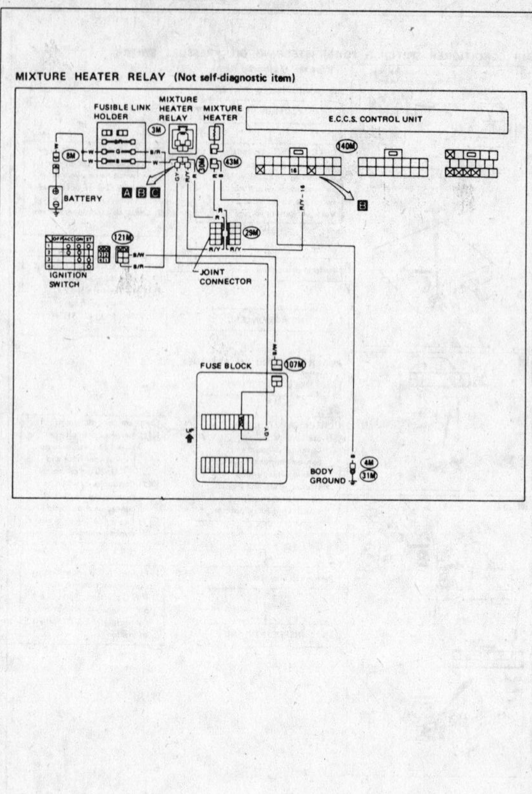
MIXTURE HEATER RELAY (Not self-diagnostic item)
FUSIBLE LINK HOLDER
MIXTURE HEATER RELAY
MIXTURE HEATER
E.C.C.S. CONTROL UNIT
BATTERY
IGNITION SWITCH
JOINT CONNECTOR
FUSE BLOCK
BODY GROUND

ECCS SELF-DIAGNOSTIC CHART—1988 PULSAR

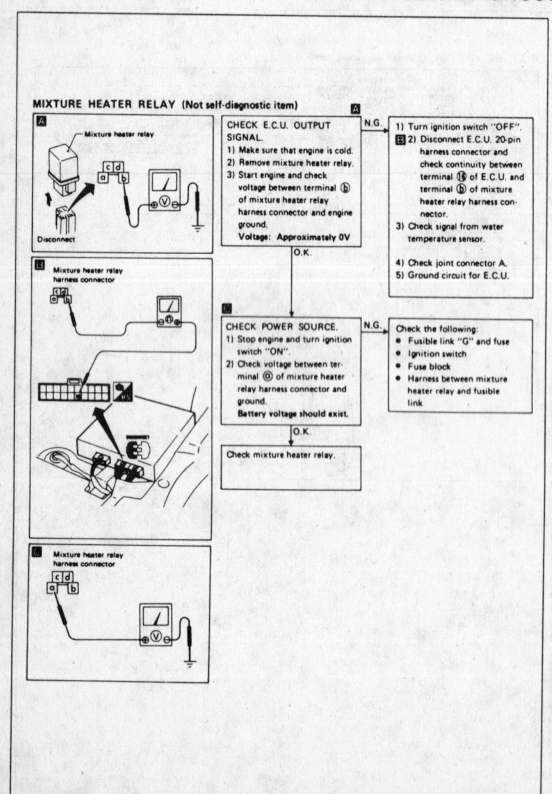
MIXTURE HEATER RELAY (Not self-diagnostic item)
Mixture heater relay
Disconnect
Mixture heater relay harness connector
A
CHECK E.C.U. OUTPUT SIGNAL.
1) Make sure that engine is cold.
2) Remove mixture heater relay.
3) Start engine and check voltage between terminal ⓑ of mixture heater relay harness connector and engine ground.
Voltage: Approximately 0V
N.G.
1) Turn ignition switch "OFF".
B 2) Disconnect E.C.U. 20-pin harness connector and check continuity between terminal ⑯ of E.C.U. and terminal ⓑ of mixture heater relay harness connector.
3) Check signal from water temperature sensor.
4) Check joint connector A.
5) Ground circuit for E.C.U.
O.K.
C
CHECK POWER SOURCE.
1) Stop engine and turn ignition switch "ON".
2) Check voltage between terminal ⓐ of mixture heater relay harness connector and ground.
Battery voltage should exist.
N.G.
Check the following:
• Fusible link "G" and fuse
• Ignition switch
• Fuse block
• Harness between mixture heater relay and fusible link
O.K.
Check mixture heater relay.

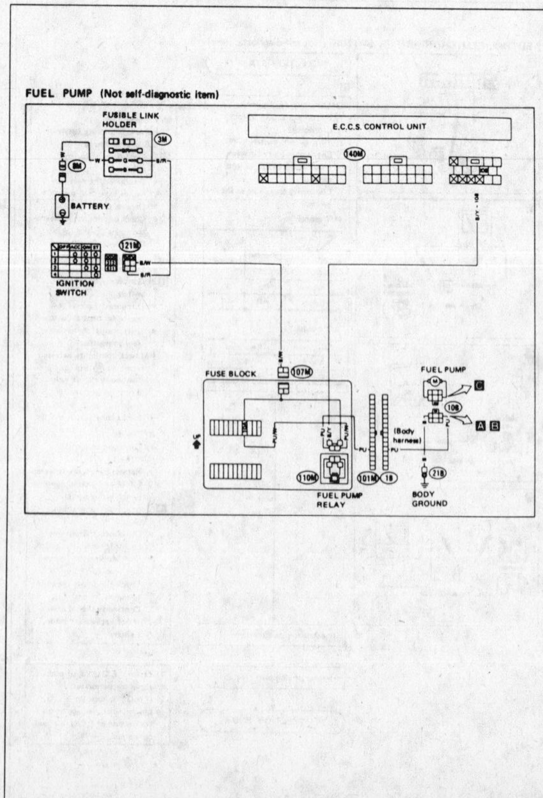
FUEL PUMP (Not self-diagnostic item)
FUSIBLE LINK HOLDER
E.C.C.S. CONTROL UNIT
BATTERY
IGNITION SWITCH
FUSE BLOCK
FUEL PUMP
FUEL PUMP RELAY
(Body harness)
BODY GROUND

ECCS SELF-DIAGNOSTIC CHART–1988 PULSAR

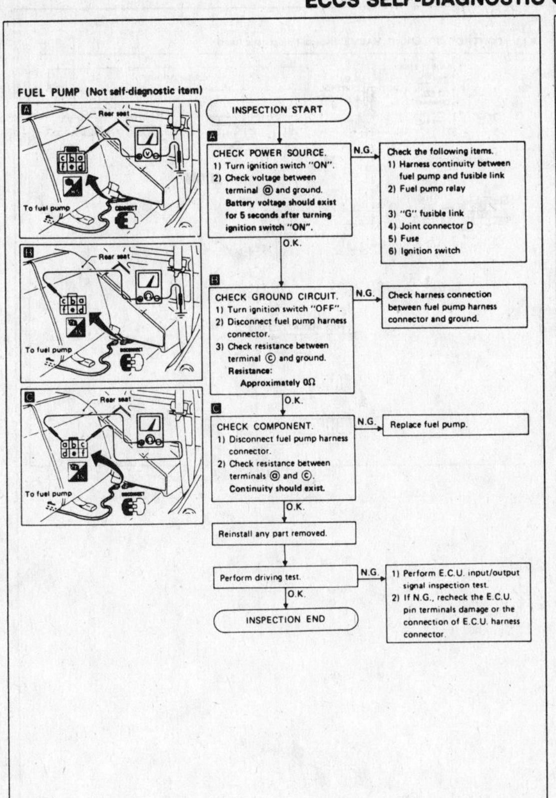
FUEL PUMP (Not self-diagnostic item)
INSPECTION START
A
CHECK POWER SOURCE.
1) Turn ignition switch "ON".
2) Check voltage between terminal ⓐ and ground.
Battery voltage should exist for 5 seconds after turning ignition switch "ON".
N.G.
Check the following items.
1) Harness continuity between fuel pump and fusible link
2) Fuel pump relay
3) "G" fusible link
4) Joint connector D
5) Fuse
6) Ignition switch
O.K.
B
CHECK GROUND CIRCUIT.
1) Turn ignition switch "OFF".
2) Disconnect fuel pump harness connector.
3) Check resistance between terminal ⓒ and ground.
Resistance:
Approximately 0Ω
N.G.
Check harness connection between fuel pump harness connector and ground.
O.K.
C
CHECK COMPONENT.
1) Disconnect fuel pump harness connector.
2) Check resistance between terminals ⓐ and ⓒ.
Continuity should exist.
N.G.
Replace fuel pump.
O.K.
Reinstall any part removed.
Perform driving test.
N.G.
1) Perform E.C.U. input/output signal inspection test.
2) If N.G., recheck the E.C.U. pin terminals damage or the connection of E.C.U. harness connector.
O.K.
INSPECTION END
Rear seat
To fuel pump
CONNECT
DISCONNECT

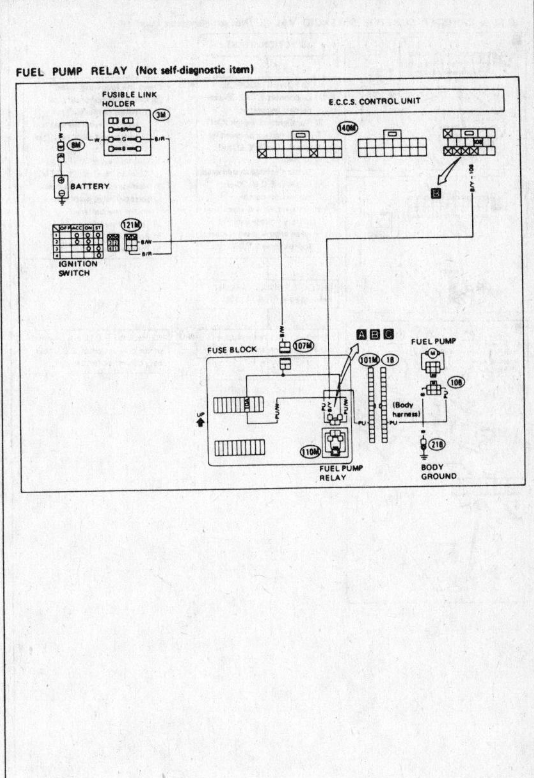
FUEL PUMP RELAY (Not self-diagnostic item)
FUSIBLE LINK HOLDER
E.C.C.S. CONTROL UNIT
BATTERY
IGNITION SWITCH
FUSE BLOCK
FUEL PUMP
FUEL PUMP RELAY
BODY GROUND
(Body harness)

ECCS SELF-DIAGNOSTIC CHART–1988 PULSAR

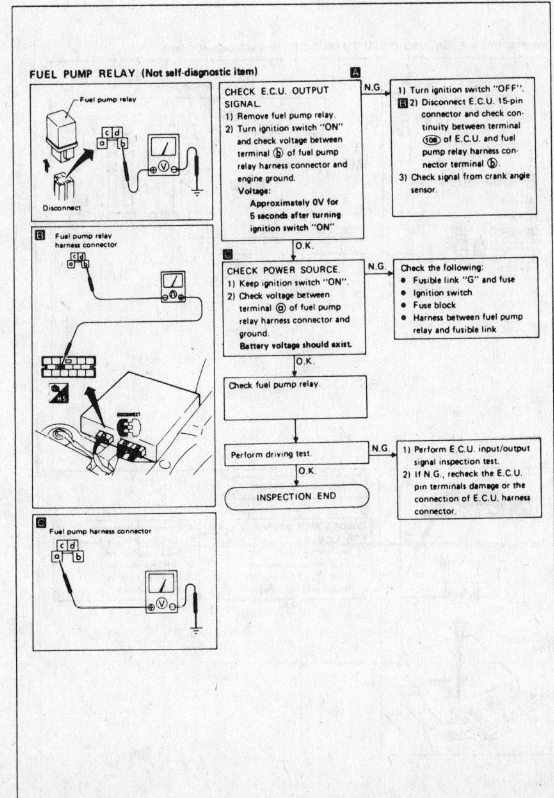
FUEL PUMP RELAY (Not self-diagnostic item)
A
CHECK E.C.U. OUTPUT SIGNAL.
1) Remove fuel pump relay.
2) Turn ignition switch "ON" and check voltage between terminal ⓑ of fuel pump relay harness connector and engine ground.
Voltage:
Approximately 0V for 5 seconds after turning ignition switch "ON"
N.G.
1) Turn ignition switch "OFF".
2) Disconnect E.C.U. 15-pin connector and check continuity between terminal ⑩⑧ of E.C.U. and fuel pump relay harness connector terminal ⓑ.
3) Check signal from crank angle sensor.
O.K.
B
CHECK POWER SOURCE.
1) Keep ignition switch "ON".
2) Check voltage between terminal ⓐ of fuel pump relay harness connector and ground.
Battery voltage should exist.
N.G.
Check the following:
• Fusible link "G" and fuse
• Ignition switch
• Fuse block
• Harness between fuel pump relay and fusible link
O.K.
Check fuel pump relay.
Perform driving test.
N.G.
1) Perform E.C.U. input/output signal inspection test.
2) If N.G., recheck the E.C.U. pin terminals damage or the connection of E.C.U. harness connector.
O.K.
INSPECTION END
Fuel pump relay
Disconnect
Fuel pump relay harness connector
Fuel pump harness connector

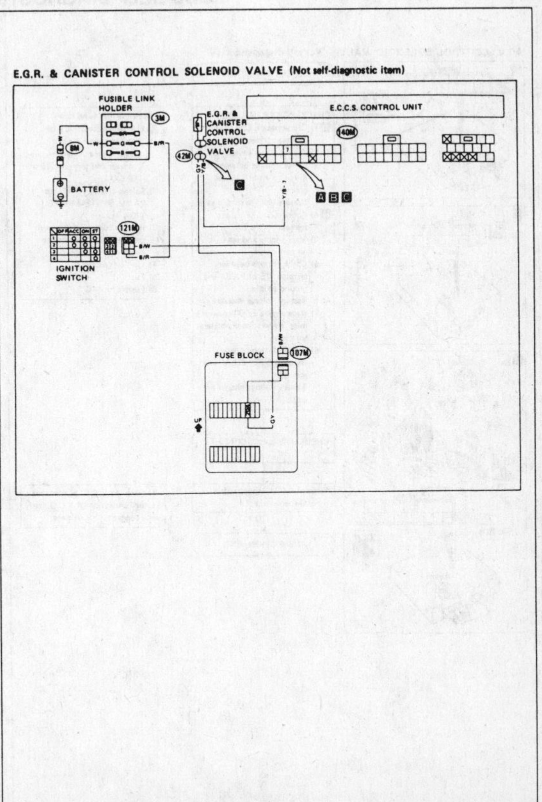
E.G.R. & CANISTER CONTROL SOLENOID VALVE (Not self-diagnostic item)
FUSIBLE LINK HOLDER
E.G.R. & CANISTER CONTROL SOLENOID VALVE
E.C.C.S. CONTROL UNIT
BATTERY
IGNITION SWITCH
FUSE BLOCK

ECCS SELF-DIAGNOSTIC CHART—1988 PULSAR

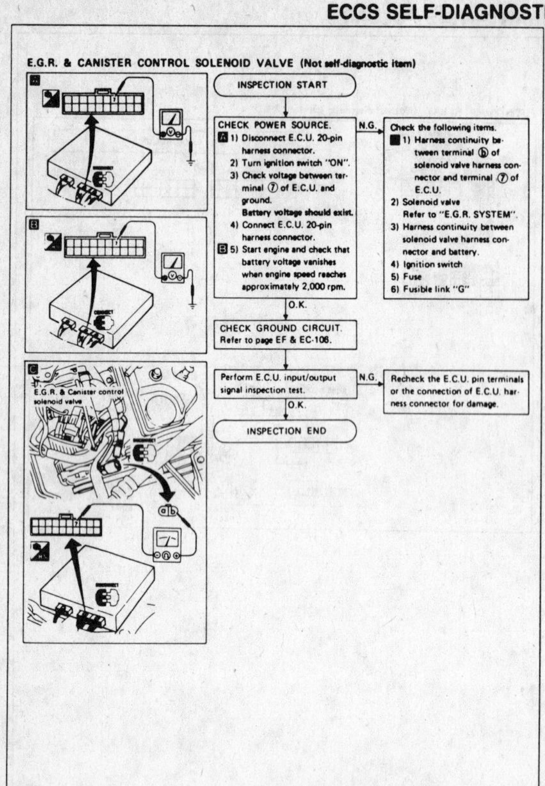
E.G.R. & CANISTER CONTROL SOLENOID VALVE (Not self-diagnostic item)
INSPECTION START
CHECK POWER SOURCE.
1) Disconnect E.C.U. 20-pin harness connector.
2) Turn ignition switch "ON".
3) Check voltage between terminal ⑦ of E.C.U. and ground.
Battery voltage should exist.
4) Connect E.C.U. 20-pin harness connector.
5) Start engine and check that battery voltage vanishes when engine speed reaches approximately 2,000 rpm.
N.G.
Check the following items.
1) Harness continuity between terminal ⓑ of solenoid valve harness connector and terminal ⑦ of E.C.U.
2) Solenoid valve
Refer to "E.G.R. SYSTEM".
3) Harness continuity between solenoid valve harness connector and battery.
4) Ignition switch
5) Fuse
6) Fusible link "G"
O.K.
CHECK GROUND CIRCUIT.
Refer to page EF & EC-108.
Perform E.C.U. input/output signal inspection test.
N.G.
Recheck the E.C.U. pin terminals or the connection of E.C.U. harness connector for damage.
O.K.
INSPECTION END
E.G.R. & Canister control solenoid valve

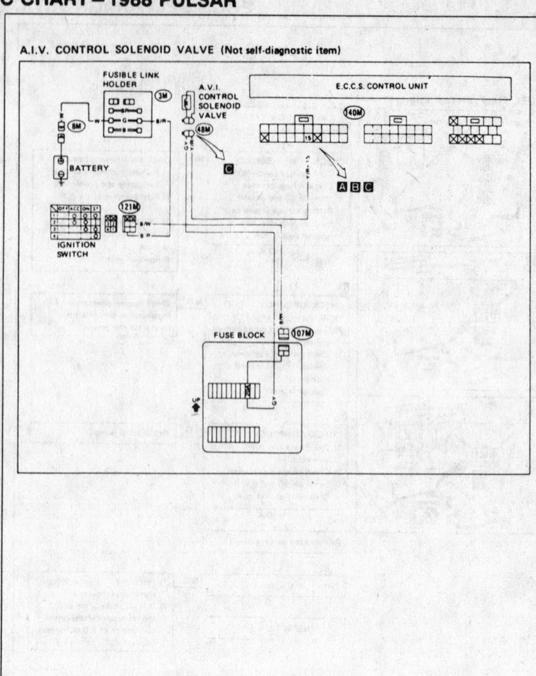
A.I.V. CONTROL SOLENOID VALVE (Not self-diagnostic item)
FUSIBLE LINK HOLDER
3M
A.I.V. CONTROL SOLENOID VALVE
48M
E.C.C.S. CONTROL UNIT
140M
8M
BATTERY
121M
IGNITION SWITCH
FUSE BLOCK
107M

ECCS SELF-DIAGNOSTIC CHART—1988 PULSAR

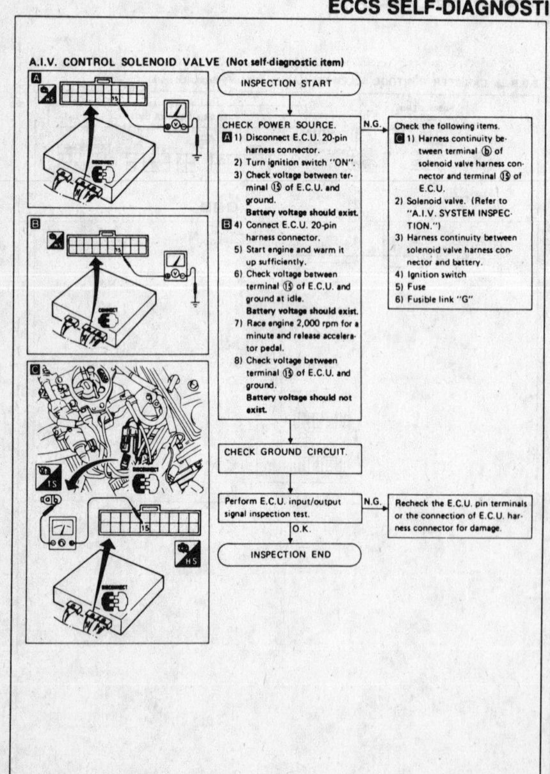
A.I.V. CONTROL SOLENOID VALVE (Not self-diagnostic item)
INSPECTION START
CHECK POWER SOURCE.
1) Disconnect E.C.U. 20-pin harness connector.
2) Turn ignition switch "ON".
3) Check voltage between terminal ⑮ of E.C.U. and ground.
Battery voltage should exist.
4) Connect E.C.U. 20-pin harness connector.
5) Start engine and warm it up sufficiently.
6) Check voltage between terminal ⑮ of E.C.U. and ground at idle.
Battery voltage should exist.
7) Race engine 2,000 rpm for a minute and release accelerator pedal.
8) Check voltage between terminal ⑮ of E.C.U. and ground.
Battery voltage should not exist.
N.G.
Check the following items.
1) Harness continuity between terminal ⓑ of solenoid valve harness connector and terminal ⑮ of E.C.U.
2) Solenoid valve. (Refer to "A.I.V. SYSTEM INSPECTION.")
3) Harness continuity between solenoid valve harness connector and battery.
4) Ignition switch
5) Fuse
6) Fusible link "G"
CHECK GROUND CIRCUIT.
Perform E.C.U. input/output signal inspection test.
N.G.
Recheck the E.C.U. pin terminals or the connection of E.C.U. harness connector for damage.
O.K.
INSPECTION END

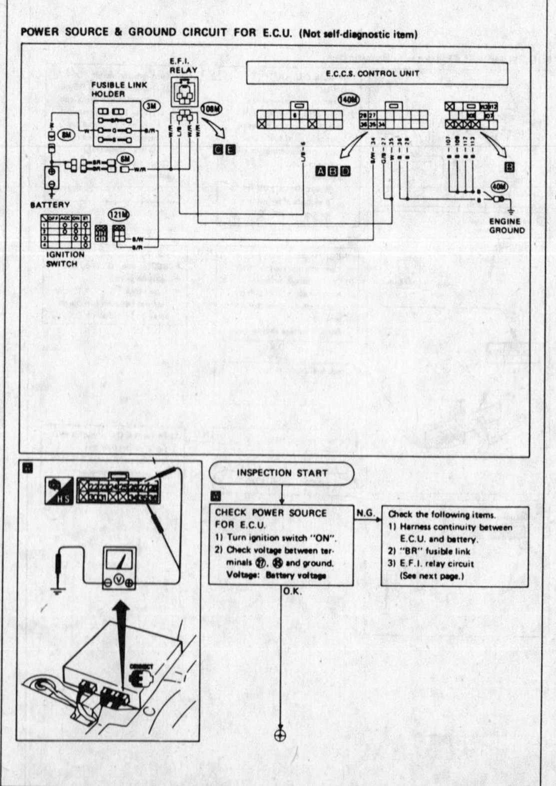
POWER SOURCE & GROUND CIRCUIT FOR E.C.U. (Not self-diagnostic item)
E.F.I. RELAY
FUSIBLE LINK HOLDER
3M
108M
E.C.C.S. CONTROL UNIT
140M
8M
40M
BATTERY
121M
IGNITION SWITCH
ENGINE GROUND
INSPECTION START
CHECK POWER SOURCE FOR E.C.U.
1) Turn ignition switch "ON".
2) Check voltage between terminals ㉗, ㊱ and ground.
Voltage: Battery voltage
N.G.
Check the following items.
1) Harness continuity between E.C.U. and battery.
2) "BR" fusible link
3) E.F.I. relay circuit (See next page.)
O.K.

ECCS SELF-DIAGNOSTIC CHART—1988 PULSAR

ECU SIGNAL INSPECTION—1988 PULSAR

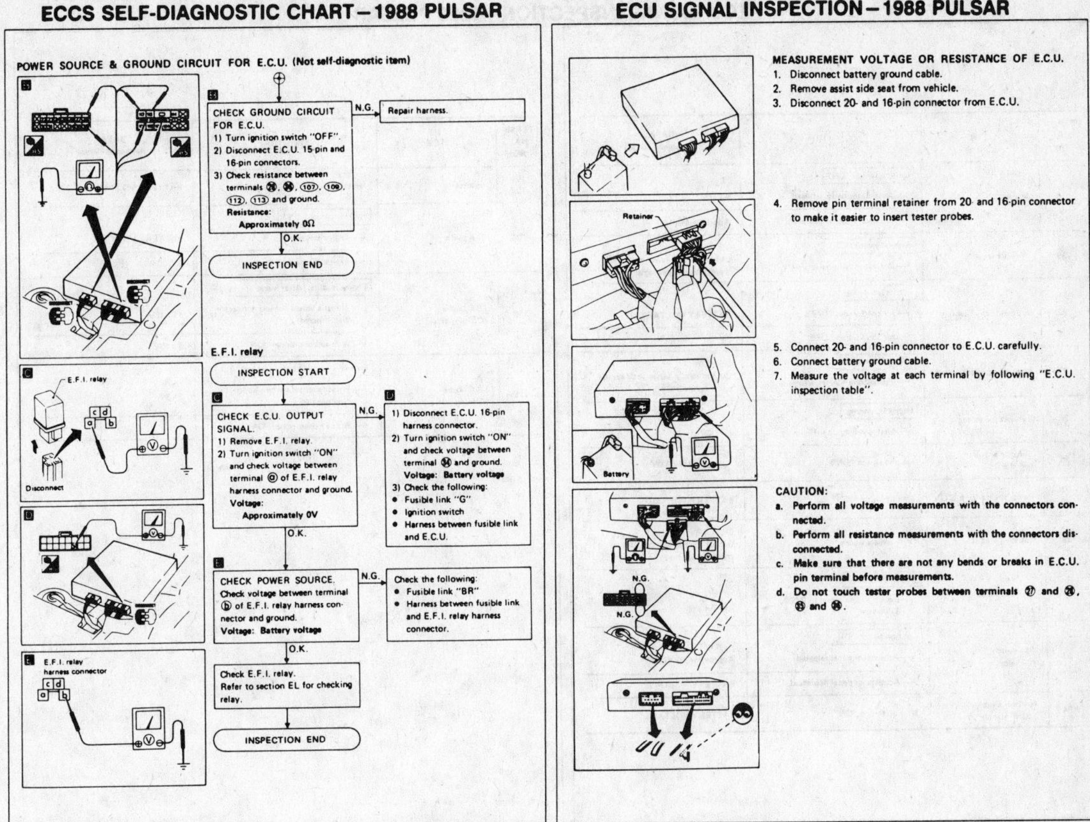
POWER SOURCE & GROUND CIRCUIT FOR E.C.U. (Not self-diagnostic item)
CHECK GROUND CIRCUIT FOR E.C.U.
1) Turn ignition switch "OFF".
2) Disconnect E.C.U. 15-pin and 16-pin connectors.
3) Check resistance between terminals (28), (36), (107), (109), (112), (113) and ground.
Resistance: Approximately 0Ω
N.G.
Repair harness.
O.K.
INSPECTION END
E.F.I. relay
INSPECTION START
CHECK E.C.U. OUTPUT SIGNAL.
1) Remove E.F.I. relay.
2) Turn ignition switch "ON" and check voltage between terminal ⓐ of E.F.I. relay harness connector and ground.
Voltage: Approximately 0V
N.G.
1) Disconnect E.C.U. 16-pin harness connector.
2) Turn ignition switch "ON" and check voltage between terminal (34) and ground.
Voltage: Battery voltage
3) Check the following:
• Fusible link "G"
• Ignition switch
• Harness between fusible link and E.C.U.
O.K.
CHECK POWER SOURCE.
Check voltage between terminal ⓑ of E.F.I. relay harness connector and ground.
Voltage: Battery voltage
N.G.
Check the following:
• Fusible link "BR"
• Harness between fusible link and E.F.I. relay harness connector.
O.K.
Check E.F.I. relay.
Refer to section EL for checking relay.
INSPECTION END
E.F.I. relay
Disconnect
E.F.I. relay harness connector
MEASUREMENT VOLTAGE OR RESISTANCE OF E.C.U.
1. Disconnect battery ground cable.
2. Remove assist side seat from vehicle.
3. Disconnect 20- and 16-pin connector from E.C.U.
Retainer
4. Remove pin terminal retainer from 20- and 16-pin connector to make it easier to insert tester probes.
Battery
5. Connect 20- and 16-pin connector to E.C.U. carefully.
6. Connect battery ground cable.
7. Measure the voltage at each terminal by following "E.C.U. inspection table".
N.G.
N.G.
CAUTION:
a. Perform all voltage measurements with the connectors connected.
b. Perform all resistance measurements with the connectors disconnected.
c. Make sure that there are not any bends or breaks in E.C.U. pin terminal before measurements.
d. Do not touch tester probes between terminals (27) and (28), (35) and (36).

ECU SIGNAL INSPECTION—1988 PULSAR

E16i PIN CONNECTOR TERMINAL LAYOUT

20-pin connector: 1 2 3 ⊠ 5 6 7 8 9 10 / 11 12 ⊠ ⊠ 15 16 17 18 19 ⊠

16-pin connector: 21 22 23 24 25 26 27 28 / 29 30 31 ⊠ 33 34 35 36

15-pin connector: 112 113 114 ⊠ / 107 108 109 110 111 / 101 102 ⊠ ⊠ ⊠ ⊠

H.S. H.S. H.S.

E.C.U. INSPECTION TABLE

TERMINAL NO.	ITEM	CONDITION	DATA (Reference values)
2	Power steering oil pressure switch	Engine is idling. — When turning steering wheel.	Approximately 0V
		— With wheels straight ahead.	8 - 9V
3	Ignition signal (from resistor)	Ignition switch "ON"	11 - 12V
5	Ignition signal (from power transistor)	Ignition switch "ON"	Approximately 0.1V
		Engine is running. — Rev up engine speed to 2,000 - 3,000 rpm.	0.5 - 2.0V
6	Main relay	Ignition switch "ON"	0.8 - 1.0V
7	E.G.R. & canister control solenoid valve	Vehicle speed is below 10 km/h (6 MPH).	0.7 - 1.0V
		Vehicle speed is above 10 km/h (6 MPH) and load is applied. — Engine is cold or during warm-up. [Water temperature is below 60°C (140°F).]	0.7 - 1.0V
		— After warming up. [Water temperature is above 80°C (176°F)]	BATTERY VOLTAGE
8	Crank angle sensor (1° signal)	Engine is running. — Rev up engine speed to 2,000 - 3,000 rpm.	2.3 - 2.7V
9	Start signal	Cranking (Starter motor "S" terminal disconnected)	BATTERY VOLTAGE

TERMINAL NO.	ITEM	CONDITION	DATA (Reference values)
10	Neutral/Clutch/Inhibitor switch	Ignition switch "ON" — • M/T model Clutch pedal depressed or transaxle in "N" position — • A/T model Transaxle in "N" or "P" position	0V
		— • M/T model Clutch pedal released and transaxle except "N" position — • A/T model Transaxle except "N" and "P" position	BATTERY VOLTAGE
12	Hot wire self-cleaning (Air flow meter)	Drive vehicle • Rev up engine speed to more than 2,000 rpm and return to idle. • Drive vehicle at more than 20 km/h (12 MPH). • Water temperature is between 60°C (140°F) and 95°C (203°F). — Stop engine.	Voltage should appear. Key "OFF" Approximately 0V Several sec.
15	A.I.V. control solenoid valve	After warm-up — After idling has continued for several seconds.	0.7 - 0.9V
		— Except above	BATTERY VOLTAGE
16	Mixture heater	Start engine when engine is cold [Water temperature is below 50°C (122°F)]. — Until several minutes pass after water temperature exceeds 50°C (122°F).	0.7 - 0.9V
		— After warm-up	BATTERY VOLTAGE
17	Crank angle sensor (180° signal)	Engine is running. — Rev up engine speed to 2,000 - 3,000 rpm.	0.1 - 0.4V

ECU SIGNAL INSPECTION—1988 PULSAR

TERMINAL NO.	ITEM	CONDITION	DATA (Reference values)
18	Idle switch	Ignition switch "ON" — Accelerator pedal released.	9 - 10V
		Ignition switch "ON" — Accelerator pedal depressed.	Approximately 0V
19	Throttle sensor	Ignition switch "ON" — Depress throttle valve slowly.	0.5 - 5.0V Output voltage varies with the throttle valve opening angle.
22	Air conditioner switch	Engine is running. — Air conditioner "ON".	BATTERY VOLTAGE
		Engine is running. — Air conditioner "OFF".	0V
23	Water temperature sensor	Engine is running. — Engine cold → After warm-up.	0.5 - 5.0V Output voltage varies with engine water temperature.
24	Exhaust gas sensor	Engine is running. — After warming up sufficiently.	0 - 1.0V
25	Throttle sensor (⊕ side)	Ignition switch "ON"	Approximately 5.0V
27 35	Power source for E.C.U.	Ignition switch "ON"	BATTERY VOLTAGE
29	Vehicle speed sensor	Drive vehicle slowly.	0V or 4.8V Output voltage changes, one to another, repeatedly.
31	Air flow meter	Engine is running. — Rev up engine speed to 2,000 - 3,000 rpm.	2.4 - 4.0V
33	Idle switch (⊕ side)	Ignition switch "ON" — Accelerator pedal released.	Approximately 10V
		Ignition switch "ON" — Accelerator pedal depressed.	BATTERY VOLTAGE
34	Ignition switch ("ON" signal)	Ignition switch "ON"	BATTERY VOLTAGE

TERMINAL NO.	ITEM	CONDITION	DATA (Reference values)
101 102	Injector	Ignition switch "OFF"	BATTERY VOLTAGE
108	Fuel pump	Ignition switch "ON" — Engine stopped.	BATTERY VOLTAGE
		Ignition switch "ON" — Engine is running.	0.7 - 0.9V
110	I.S.C. valve (Opening side)	Engine is idling (After warm-up). — Throttle sensor harness connector disconnected.	11 - 12V
		Engine is idling (After warm-up). — With air conditioner "ON"	8 - 10V
111	I.S.C. valve (Closing side)	Engine is idling (After warm-up). — Throttle sensor harness connector disconnected.	2 - 3V
		Engine is idling (After warm-up). — With air conditioner "ON"	4 - 6V

ECCS MIXTURE RATIO FEEDBACK SYSTEM INSPECTION CHART—1988 PULSAR

PREPARATION

1. **Make sure that the following parts are in good order.**
   - **Battery**
   - **Ignition system**
   - **Engine oil and coolant levels**
   - **Fuses**
   - **E.C.C.S. harness connectors**
   - **Vacuum hoses**
   - **Air intake system (oil filler cap, oil level gauge, etc.)**
   - **Fuel pressure**
   - **A.I.V. hose**
   - **Engine compression**
   - **E.G.R. valve operation**
   - **Throttle valve**
2. **On air conditioner equipped models, checks should be carried out while the air conditioner is "OFF".**
3. **When measuring "CO" percentage, insert probe more than 40 cm (15.7 in) into tail pipe.**

Overall inspection sequence

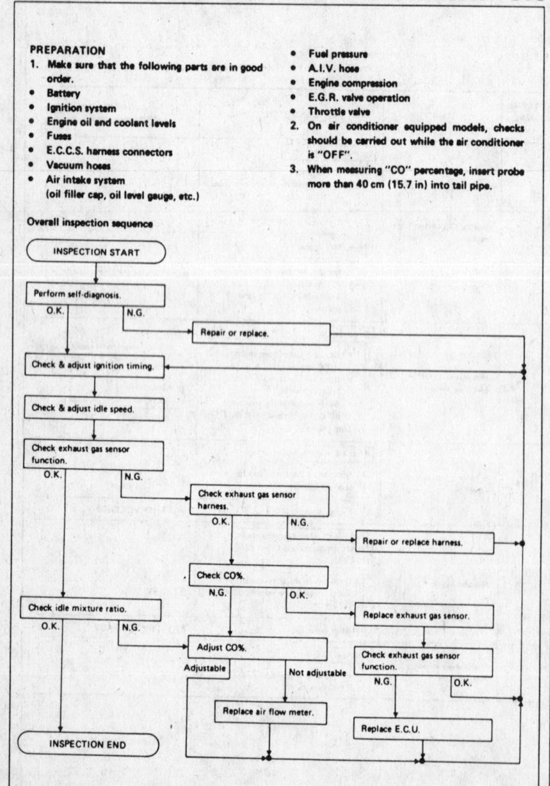
INSPECTION START
Perform self-diagnosis.
O.K.
N.G.
Repair or replace.
Check & adjust ignition timing.
Check & adjust idle speed.
Check exhaust gas sensor function.
O.K.
N.G.
Check exhaust gas sensor harness.
O.K.
N.G.
Repair or replace harness.
Check CO%.
N.G.
O.K.
Replace exhaust gas sensor.
Check idle mixture ratio.
O.K.
N.G.
Adjust CO%.
Adjustable
Not adjustable
Check exhaust gas sensor function.
N.G.
O.K.
Replace air flow meter.
Replace E.C.U.
INSPECTION END

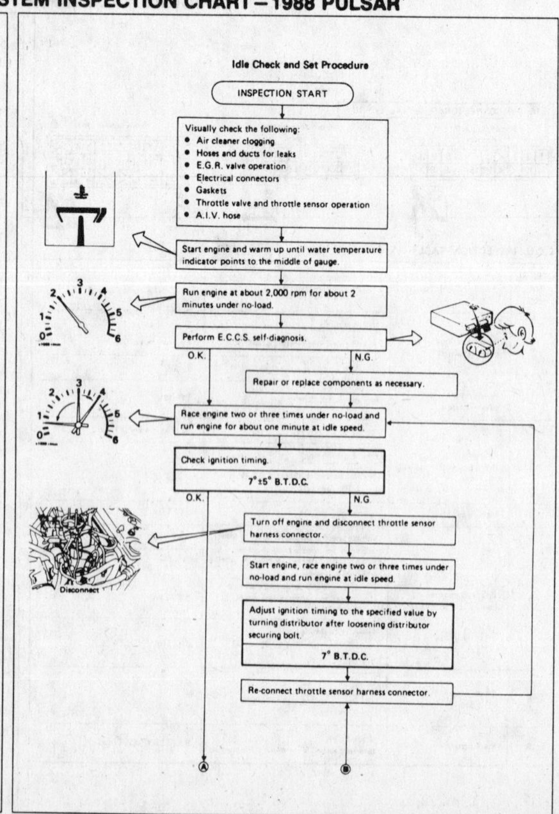
Idle Check and Set Procedure
INSPECTION START
Visually check the following:
Air cleaner clogging
Hoses and ducts for leaks
E.G.R. valve operation
Electrical connectors
Gaskets
Throttle valve and throttle sensor operation
A.I.V. hose
Start engine and warm up until water temperature indicator points to the middle of gauge.
Run engine at about 2,000 rpm for about 2 minutes under no-load.
Perform E.C.C.S. self-diagnosis.
O.K.
N.G.
Repair or replace components as necessary.
Race engine two or three times under no-load and run engine for about one minute at idle speed.
Check ignition timing.
7°±5° B.T.D.C.
O.K.
N.G.
Turn off engine and disconnect throttle sensor harness connector.
Disconnect
Start engine, race engine two or three times under no-load and run engine at idle speed.
Adjust ignition timing to the specified value by turning distributor after loosening distributor securing bolt.
7° B.T.D.C.
Re-connect throttle sensor harness connector.
A
B

4
SECTION

ECCS MIXTURE RATIO FEEDBACK SYSTEM INSPECTION CHART – 1988 PULSAR

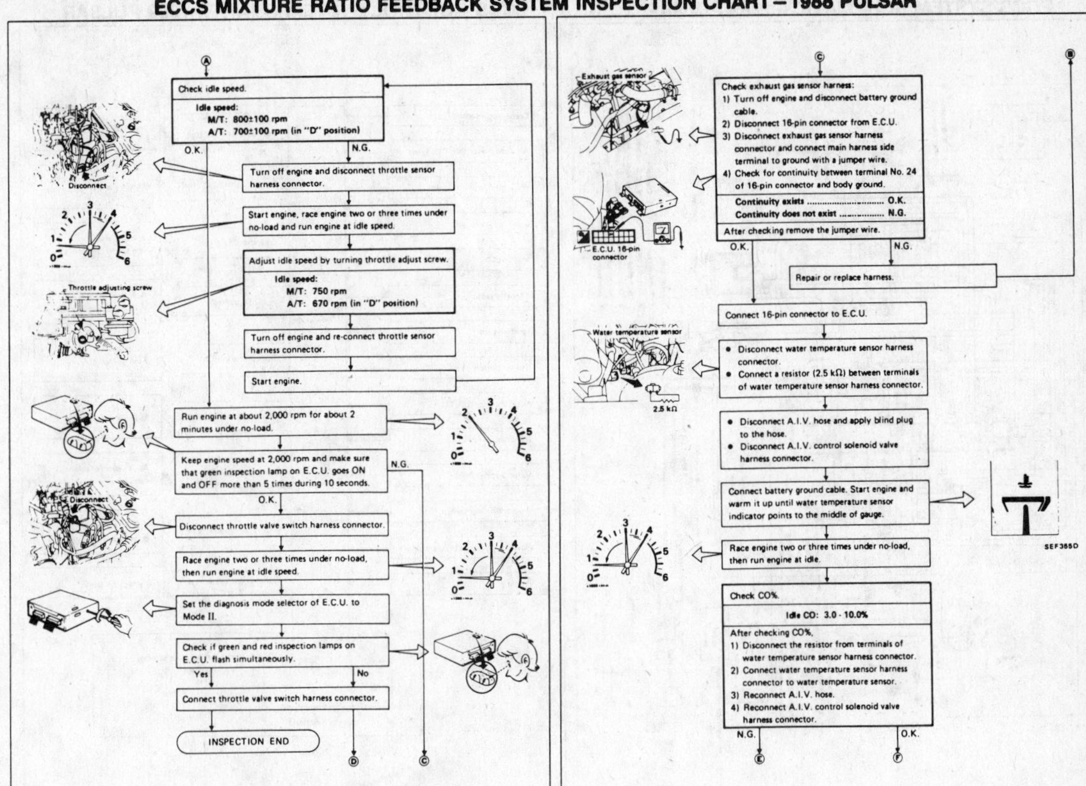

A
Check idle speed.
Idle speed:
M/T: 800±100 rpm
A/T: 700±100 rpm (in "D" position)
O.K.
N.G.
Disconnect
Turn off engine and disconnect throttle sensor harness connector.
Start engine, race engine two or three times under no-load and run engine at idle speed.
Throttle adjusting screw
Adjust idle speed by turning throttle adjust screw.
Idle speed:
M/T: 750 rpm
A/T: 670 rpm (in "D" position)
Turn off engine and re-connect throttle sensor harness connector.
Start engine.
Run engine at about 2,000 rpm for about 2 minutes under no-load.
Keep engine speed at 2,000 rpm and make sure that green inspection lamp on E.C.U. goes ON and OFF more than 5 times during 10 seconds.
N.G.
O.K.
Disconnect
Disconnect throttle valve switch harness connector.
Race engine two or three times under no-load, then run engine at idle speed.
Set the diagnosis mode selector of E.C.U. to Mode II.
Check if green and red inspection lamps on E.C.U. flash simultaneously.
Yes
No
Connect throttle valve switch harness connector.
INSPECTION END
D
C
C
B
Exhaust gas sensor
E.C.U. 16-pin connector
Check exhaust gas sensor harness:
1) Turn off engine and disconnect battery ground cable.
2) Disconnect 16-pin connector from E.C.U.
3) Disconnect exhaust gas sensor harness connector and connect main harness side terminal to ground with a jumper wire.
4) Check for continuity between terminal No. 24 of 16-pin connector and body ground.
Continuity exists .... O.K.
Continuity does not exist .... N.G.
After checking remove the jumper wire.
O.K.
N.G.
Repair or replace harness.
Connect 16-pin connector to E.C.U.
Water temperature sensor
2.5 kΩ
• Disconnect water temperature sensor harness connector.
• Connect a resistor (2.5 kΩ) between terminals of water temperature sensor harness connector.
• Disconnect A.I.V. hose and apply blind plug to the hose.
• Disconnect A.I.V. control solenoid valve harness connector.
Connect battery ground cable. Start engine and warm it up until water temperature sensor indicator points to the middle of gauge.
SEF355D
Race engine two or three times under no-load, then run engine at idle.
Check CO%.
Idle CO: 3.0 - 10.0%
After checking CO%,
1) Disconnect the resistor from terminals of water temperature sensor harness connector.
2) Connect water temperature sensor harness connector to water temperature sensor.
3) Reconnect A.I.V. hose.
4) Reconnect A.I.V. control solenoid valve harness connector.
N.G.
O.K.
E
F

ECCS MIXTURE RATIO FEEDBACK SYSTEM INSPECTION CHART – 1988 PULSAR

D E F B

- Replace exhaust gas sensor.
- Run engine at 2,000 rpm and make sure that green inspection lamp on E.C.U. goes ON and OFF more than 5 times during 10 seconds.
  - O.K.
  - N.G. → Replace E.C.U.
- Connect exhaust gas sensor harness connector to exhaust gas sensor.
- Turn off engine and remove throttle body from vehicle. (Air flow meter)
- Drill a hole in seal plug which seals variable resistor of air flow meter and remove seal plug. (Drill)
- Install throttle body on vehicle.
- Start engine and warm it up until water temperature sensor indicator points to the middle of gauge. (SEF355D)
- Set the diagnosis mode selector of E.C.U. to Mode II.
- Adjust idle mixture ratio by turning variable resistor of air flow meter so that green and red inspection lamps on E.C.U. flash simultaneously at 2,000 rpm under no-load.
  - Adjustable
  - Not adjustable → Replace air flow meter.
- Turn off engine and remove throttle body from vehicle. (Air flow meter)
- Install new seal plug into variable resistor hole of air flow meter. (Push; New seal plug)
- Install throttle body on vehicle.

ECCS COMPONENTS LOCATION – 1989 PULSAR

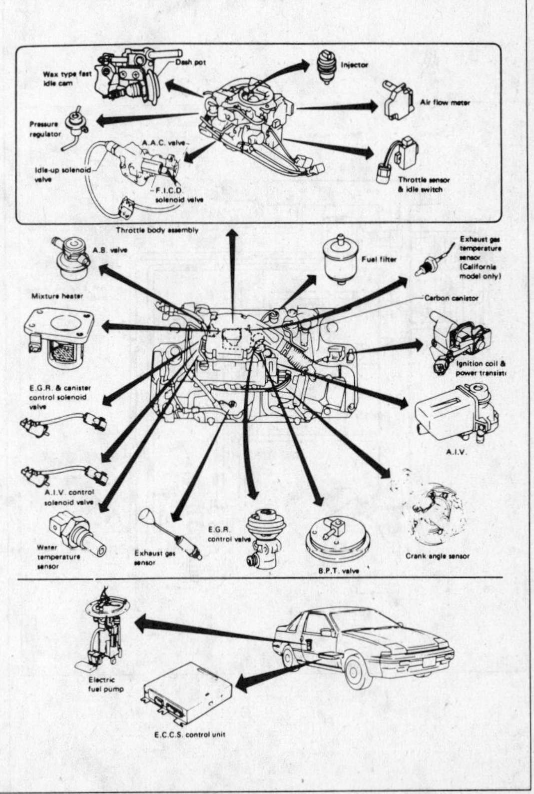

Wax type fast idle cam
Dash pot
Injector
Air flow meter
Pressure regulator
A.A.C. valve
Idle-up solenoid valve
F.I.C.D. solenoid valve
Throttle sensor & idle switch
Throttle body assembly
A.B. valve
Fuel filter
Exhaust gas temperature sensor (California model only)
Mixture heater
Carbon canistor
Ignition coil & power transist
E.G.R. & canister control solenoid valve
A.I.V.
A.I.V. control solenoid valve
Water temperature sensor
Exhaust gas sensor
E.G.R. control valve
B.P.T. valve
Crank angle sensor
Electric fuel pump
E.C.C.S. control unit

ECCS SYSTEM SCHEMATIC—1989 PULSAR

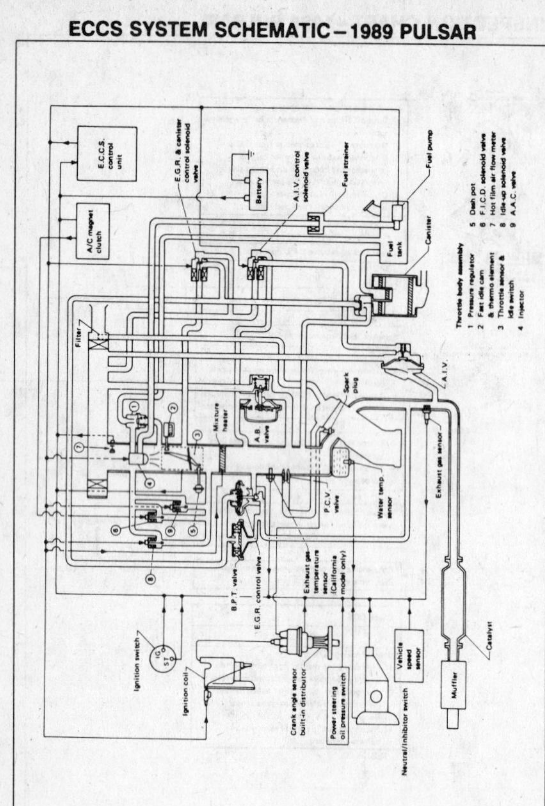
E.C.C.S. control unit
A/C magnet clutch
Battery
Fuel pump
Fuel strainer
Fuel tank
Canister
Filter
A.I.V.
Exhaust gas sensor
Ignition switch
Ignition coil
Muffler
Catalyst
Vehicle speed sensor
Power steering oil pressure switch
Neutral/Inhibitor switch
Throttle body assembly
1 Pressure regulator
2 Fast idle cam & thermo element
3 Throttle sensor & idle switch
4 Injector
5 Dash pot
6 F.I.C.D. solenoid valve
7 Hot film air flow meter
8 Idle-up solenoid valve
9 A.A.C. valve

ECCS CONTROL SYSTEM CHART—1989 PULSAR

Input (sensors)		Control (E.C.C.S. control unit)	Output
Crank angle sensor		Fuel injection & mixture ratio control	Injector
Air flow meter (hot film type)		Ignition timing control	Power transistor
Water temperature sensor		Idle speed control	A.A.C. valve & idle speed control valve
Exhaust gas sensor		F.I.C.D. control	A.A.C. valve & F.I.C.D. solenoid valve
Idle switch		Fuel pump control	Fuel pump relay
Throttle sensor		Mixture heating control	Mixture heater relay
Ignition switch		A.I.V. control	A.I.V. control solenoid valve
Starter switch		E.G.R. & canister control	E.G.R. & canister control solenoid valve
Neutral switch (M/T)		Exhaust gas sensor monitor & self-diagnosis	Inspection lamps (on the control unit)
Inhibitor switch (A/T)			
Vehicle speed sensor			
Air conditioner switch			
Power steering oil pressure switch			
Battery			

Fail-safe function:
- Air flow meter
- Water temperature sensor
- Throttle sensor

ECCS FUEL FLOW SYSTEM SCHEMATIC—1989 PULSAR

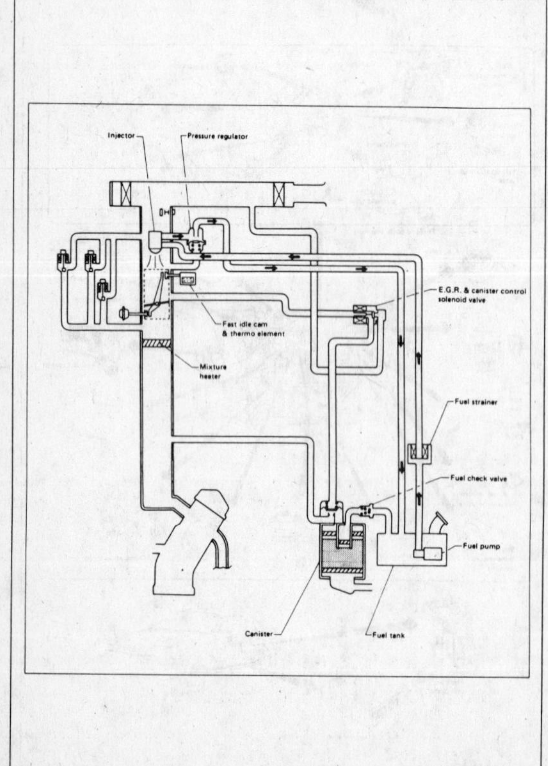
Injector
Pressure regulator
E.G.R. & canister control solenoid valve
Fast idle cam & thermo element
Mixture heater
Fuel strainer
Fuel check valve
Fuel pump
Canister
Fuel tank

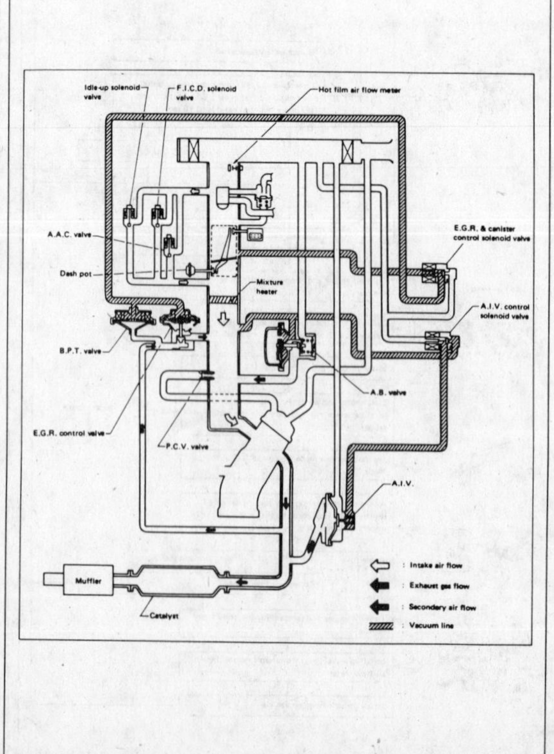
Idle-up solenoid valve
F.I.C.D. solenoid valve
Hot film air flow meter
A.A.C. valve
Dash pot
Mixture heater
E.G.R. & canister control solenoid valve
A.I.V. control solenoid valve
B.P.T. valve
A.B. valve
E.G.R. control valve
P.C.V. valve
A.I.V.
Muffler
Catalyst
: Intake air flow
: Exhaust gas flow
: Secondary air flow
: Vacuum line

ECCS CONTROL UNIT SCHEMATIC—1989 PULSAR

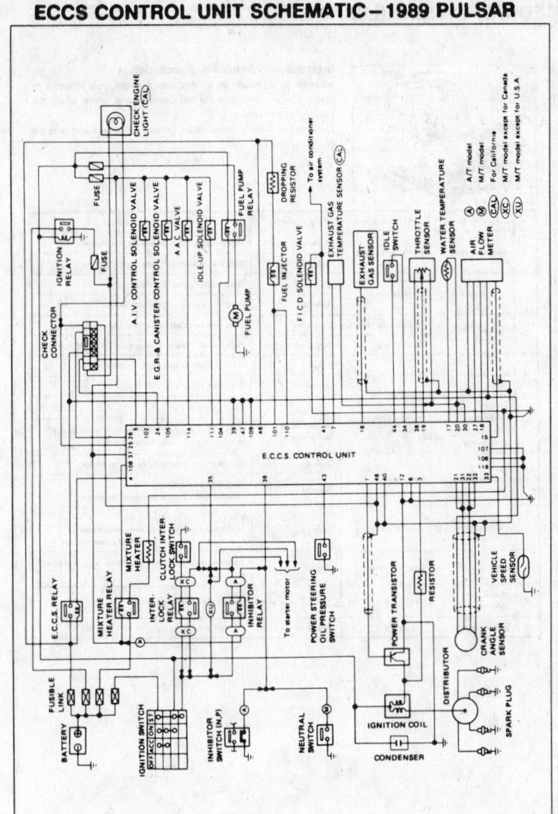
E.C.C.S. CONTROL UNIT
CHECK ENGINE LIGHT (CAL)
IGNITION RELAY
FUSE
CHECK CONNECTOR
A.I.V. CONTROL SOLENOID VALVE
E.G.R. & CANISTER CONTROL SOLENOID VALVE
A.A.C. VALVE
IDLE UP SOLENOID VALVE
FUEL PUMP
FUEL PUMP RELAY
DROPPING RESISTOR
FUEL INJECTOR
F.I.C.D. SOLENOID VALVE
EXHAUST GAS TEMPERATURE SENSOR (CAL)
EXHAUST GAS SENSOR
IDLE SWITCH
THROTTLE SENSOR
WATER TEMPERATURE SENSOR
AIR FLOW METER
E.C.C.S. RELAY
MIXTURE HEATER RELAY
MIXTURE HEATER
INTER. LOCK RELAY
CLUTCH INTER LOCK SWITCH
INHIBITOR RELAY
To starter motor
POWER STEERING OIL PRESSURE SWITCH
POWER TRANSISTOR
RESISTOR
VEHICLE SPEED SENSOR
CRANK ANGLE SENSOR
DISTRIBUTOR
SPARK PLUG
IGNITION COIL
CONDENSER
FUSIBLE LINK
BATTERY
IGNITION SWITCH
INHIBITOR SWITCH (N,P)
NEUTRAL SWITCH

ECCS WIRING SCHEMATIC—1989 PULSAR

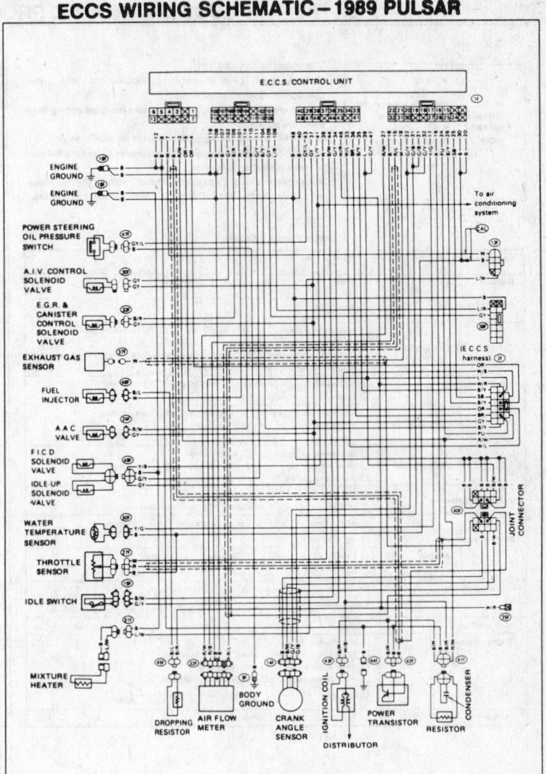
E.C.C.S. CONTROL UNIT
ENGINE GROUND
POWER STEERING OIL PRESSURE SWITCH
A.I.V. CONTROL SOLENOID VALVE
E.G.R. & CANISTER CONTROL SOLENOID VALVE
EXHAUST GAS SENSOR
FUEL INJECTOR
A.A.C. VALVE
F.I.C.D. SOLENOID VALVE
IDLE-UP SOLENOID VALVE
WATER TEMPERATURE SENSOR
THROTTLE SENSOR
IDLE SWITCH
MIXTURE HEATER
To air conditioning system
(E.C.C.S. harness)
JOINT CONNECTOR
DROPPING RESISTOR
AIR FLOW METER
BODY GROUND
CRANK ANGLE SENSOR
IGNITION COIL
DISTRIBUTOR
POWER TRANSISTOR
RESISTOR
CONDENSER

ECCS WIRING SCHEMATIC (CONT.)—1989 PULSAR

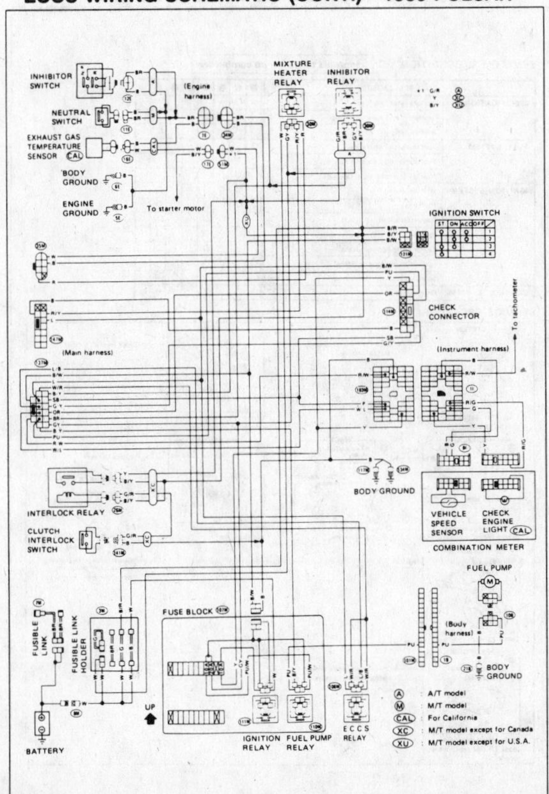
INHIBITOR SWITCH
NEUTRAL SWITCH
EXHAUST GAS TEMPERATURE SENSOR (CAL)
BODY GROUND
ENGINE GROUND
To starter motor
MIXTURE HEATER RELAY
INHIBITOR RELAY
(Engine harness)
IGNITION SWITCH
CHECK CONNECTOR
To tachometer
(Main harness)
(Instrument harness)
INTERLOCK RELAY
CLUTCH INTERLOCK SWITCH
BODY GROUND
VEHICLE SPEED SENSOR
CHECK ENGINE LIGHT (CAL)
COMBINATION METER
FUEL PUMP
(Body harness)
FUSE BLOCK
FUSIBLE LINK
FUSIBLE LINK HOLDER
BATTERY
UP
IGNITION RELAY
FUEL PUMP RELAY
E.C.C.S. RELAY
Ⓐ : A/T model
Ⓜ : M/T model
(CAL) : For California
(XC) : M/T model except for Canada
(XU) : M/T model except for U.S.A.

ECCS DIAGNOSTIC PROCEDURE—1989 PULSAR

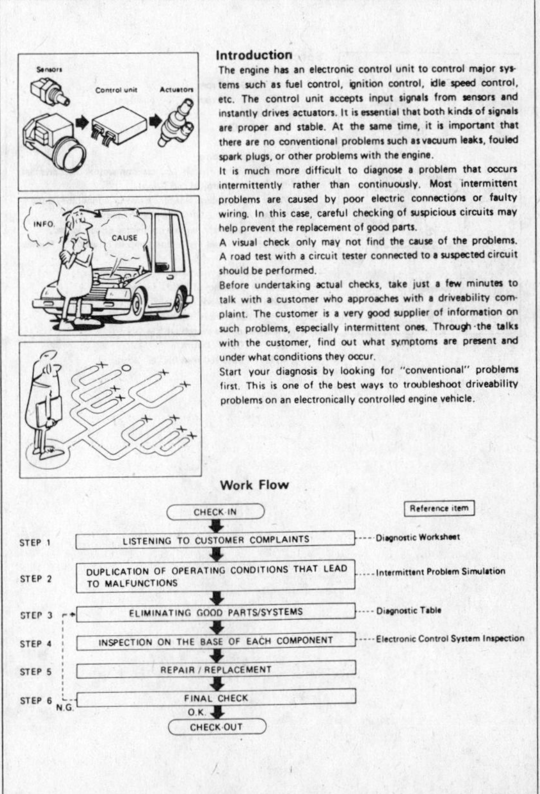
Sensors
Control unit
Actuators
INFO.
CAUSE

Introduction

The engine has an electronic control unit to control major systems such as fuel control, ignition control, idle speed control, etc. The control unit accepts input signals from sensors and instantly drives actuators. It is essential that both kinds of signals are proper and stable. At the same time, it is important that there are no conventional problems such as vacuum leaks, fouled spark plugs, or other problems with the engine.

It is much more difficult to diagnose a problem that occurs intermittently rather than continuously. Most intermittent problems are caused by poor electric connections or faulty wiring. In this case, careful checking of suspicious circuits may help prevent the replacement of good parts.

A visual check only may not find the cause of the problems. A road test with a circuit tester connected to a suspected circuit should be performed.

Before undertaking actual checks, take just a few minutes to talk with a customer who approaches with a driveability complaint. The customer is a very good supplier of information on such problems, especially intermittent ones. Through the talks with the customer, find out what symptoms are present and under what conditions they occur.

Start your diagnosis by looking for "conventional" problems first. This is one of the best ways to troubleshoot driveability problems on an electronically controlled engine vehicle.

Work Flow

		Reference item
	CHECK IN	
STEP 1	LISTENING TO CUSTOMER COMPLAINTS	Diagnostic Worksheet
STEP 2	DUPLICATION OF OPERATING CONDITIONS THAT LEAD TO MALFUNCTIONS	Intermittent Problem Simulation
STEP 3	ELIMINATING GOOD PARTS/SYSTEMS	Diagnostic Table
STEP 4	INSPECTION ON THE BASE OF EACH COMPONENT	Electronic Control System Inspection
STEP 5	REPAIR / REPLACEMENT	
STEP 6	FINAL CHECK (N.G. → STEP 3; O.K. → CHECK OUT)	
	CHECK OUT	

ECCS DIAGNOSTIC PROCEDURE—1989 PULSAR

KEY POINTS

WHAT ..... Vehicle & engine model
WHEN ..... Date, Frequencies
WHERE ..... Road conditions
HOW ..... Operating conditions, Weather conditions, Symptoms

Diagnostic Worksheet

There are many kinds of operating conditions that lead to malfunctions on engine components.
A good grasp of such conditions can make troubleshooting faster and more accurate.
In general, feelings for a problem depend on each customer. It is important to fully understand the symptoms or under what conditions a customer complains..
Make good use of a diagnostic worksheet such as the one shown below in order to utilize all the complaints for troubleshooting.

WORKSHEET SAMPLE

Customer name MR/MS	Model & Year	VIN
Engine #	Trans.	Mileage
Incident Date	Manuf. Date	In Service Date

Symptoms	☐ Startability	☐ Impossible to start ☐ No combustion ☐ Partial combustion ☐ Partial combustion affected by throttle position ☐ Partial combustion NOT affected by throttle position ☐ Possible but hard to start ☐ Others [ ]
	☐ Idling	☐ No fast idle ☐ Unstable ☐ High idle ☐ Low idle ☐ Others [ ]
	☐ Driveability	☐ Stumble ☐ Surge ☐ Detonation ☐ Lack of power ☐ Intake backfire ☐ Exhaust backfire ☐ Others [ ]
	☐ Engine stall	☐ At the time of start ☐ While idling ☐ While accelerating ☐ While decelerating ☐ Just after stopping ☐ While loading
Incident occurrence		☐ Just after delivery ☐ Recently ☐ In the morning ☐ At night ☐ In the daytime
Frequency		☐ All the time ☐ Under certain conditions ☐ Sometimes
Weather conditions		☐ Not effected
	Weather	☐ Fine ☐ Raining ☐ Snowing ☐ Others [ ]
	Temperature	☐ Hot ☐ Warm ☐ Cool ☐ Cold ☐ Humid °F
Engine conditions		☐ Cold ☐ During warm-up ☐ After warm-up Engine speed 0 2,000 4,000 6,000 8,000 rpm
Road conditions		☐ In town ☐ In suburbs ☐ Highway ☐ Off road (up/down)
Driving conditions		☐ Not affected ☐ At starting ☐ While idling ☐ At racing ☐ While accelerating ☐ While cruising ☐ While decelerating ☐ While turning (RH/LH) Vehicle speed 0 10 20 30 40 50 60 MPH
Check engine light		☐ Turned on ☐ Not turned on

GRRRRR GRRRR

THIS is the symptom, isn't it ?

Intermittent Problem Simulation

In order to duplicate an intermittent problem, it is effective to create similar conditions for component parts, under which the problem might occur.
Perform the activity listed under Service procedure and note the result.

	Variable factor	Influential part	Target condition	Service procedure
1	Mixture ratio	Pressure regulator	Made lean	Remove vacuum hose and apply vacuum.
			Made rich	Remove vacuum hose and apply pressure.
2	Ignition timing	Distributor	Advanced	Rotate distributor clockwise.
			Retarded	Rotate distributor counterclockwise.
3	Mixture ratio feedback control	Exhaust gas sensor	Suspended	Disconnect exhaust gas sensor harness connector.
		Control unit	Operation check	Perform self-diagnosis (Mode I/II) at 2,000 rpm.
4	Idle speed	Throttle body	Raised	Turn idle adjust screw counterclockwise.
			Lowered	Turn idle adjust screw clockwise.
5	Electric connection (Electric continuity)	Harness connectors and wires	Poor electric connection or faulty wiring	Tap or wiggle.
				Race engine rapidly. See if the torque reaction of the engine unit causes electric breaks.
6	Temperature	Control unit	Cooled	Cool with an icing spray or similar device.
			Warmed	Heat with a hair drier. [WARNING: Do not overheat the unit.]
7	Moisture	Electric parts	Damp	Wet [WARNING: Do not directly pour water on components. Use a mist sprayer.]
8	Electric loads	Load switches	Loaded	Turn on head lights, air conditioner, rear defogger, etc.
9	Idle switch condition	Control unit	ON-OFF switching	Perform self-diagnosis (Mode IV).
10	Ignition spark	Timing light	Spark power check	Try to flash timing light for each cylinder.

ECCS DIAGNOSTIC PROCEDURE SPECIFICATIONS—1989 PULSAR

Idle rpm
Ignition timing
Timing light
Idle CO
CO meter
E.C.U. inspection lamps
Tail pipe
or
Every item should be checked after warming up sufficiently.

Specifications

1) Idle speed
   M/T: 800±100 rpm
   A/T: 750±100 rpm (in "D" position)
2) Ignition timing
   7°±5° B.T.D.C.
3) Idle CO
   - 3.0 - 10.0% (in tail pipe)
     - Throttle valve switch harness connector disconnected (No A.I.V. controlled condition).
     - Water temperature sensor harness connector disconnected and then 2.5 kΩ resistor connected.
     - Exhaust gas sensor harness connector disconnected.
   - Flashes of E.C.U. red inspection lamp in Mode II (If flashes, O.K.)
4) Mixture ratio at approximately 2,000 rpm of engine speed.
   Number of flashes of E.C.U. inspection green lamp in Mode I:
   5 times or more/10 seconds
5) Engine speed of idle switch OFF → ON
   M/T: 1,050±150 rpm
   A/T: 1,150±150 rpm (in "N" position)

ECCS DIAGNOSTIC CHART—1989 PULSAR

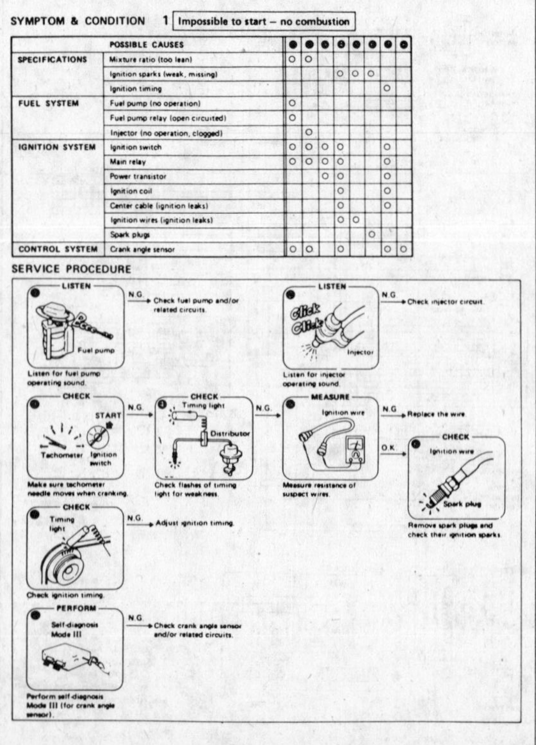

SYMPTOM & CONDITION 1 Impossible to start – no combustion

POSSIBLE CAUSES		●	●	●	●	●	●	●	●
SPECIFICATIONS	Mixture ratio (too lean)	○	○						
	Ignition sparks (weak, missing)				○	○	○		
	Ignition timing							○	
FUEL SYSTEM	Fuel pump (no operation)	○							
	Fuel pump relay (open circuited)	○							
	Injector (no operation, clogged)		○						
IGNITION SYSTEM	Ignition switch	○	○	○	○			○	
	Main relay	○	○	○	○			○	
	Power transistor			○	○			○	
	Ignition coil				○			○	
	Center cable (ignition leaks)				○			○	
	Ignition wires (ignition leaks)				○	○			
	Spark plugs						○		
CONTROL SYSTEM	Crank angle sensor	○	○		○			○	○

SERVICE PROCEDURE

LISTEN
Fuel pump
N.G.
Check fuel pump and/or related circuits.
Listen for fuel pump operating sound.
LISTEN
Click
Click
Injector
N.G
Check injector circuit.
Listen for injector operating sound.
CHECK
START
Tachometer
Ignition switch
N.G.
Make sure tachometer needle moves when cranking.
CHECK
Timing light
Distributor
N.G.
Check flashes of timing light for weakness.
MEASURE
Ignition wire
N.G
Replace the wire.
O.K.
Measure resistance of suspect wires.
CHECK
Ignition wire
Spark plug
Remove spark plugs and check their ignition sparks.
CHECK
Timing light
N.G.
Adjust ignition timing.
Check ignition timing.
PERFORM
Self-diagnosis Mode III
N.G.
Check crank angle sensor and/or related circuits.
Perform self-diagnosis Mode III (for crank angle sensor).

ECCS DIAGNOSTIC CHART—1989 PULSAR

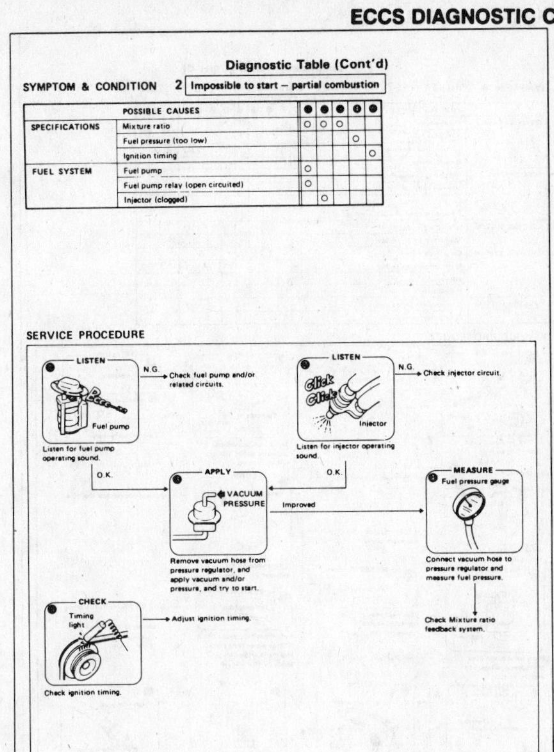

Diagnostic Table (Cont'd)

SYMPTOM & CONDITION 2 Impossible to start – partial combustion

POSSIBLE CAUSES		❶	❷	❸	❹	❺
SPECIFICATIONS	Mixture ratio	○	○	○		
	Fuel pressure (too low)				○	
	Ignition timing					○
FUEL SYSTEM	Fuel pump	○				
	Fuel pump relay (open circuited)	○				
	Injector (clogged)		○			

SERVICE PROCEDURE

LISTEN
Fuel pump
Listen for fuel pump operating sound.
N.G.
Check fuel pump and/or related circuits.
O.K.
LISTEN
Click Click
Injector
Listen for injector operating sound.
N.G.
Check injector circuit.
O.K.
APPLY
VACUUM PRESSURE
Remove vacuum hose from pressure regulator, and apply vacuum and/or pressure, and try to start.
Improved
MEASURE
Fuel pressure gauge
Connect vacuum hose to pressure regulator and measure fuel pressure.
Check Mixture ratio feedback system.
CHECK
Timing light
Check ignition timing.
Adjust ignition timing.

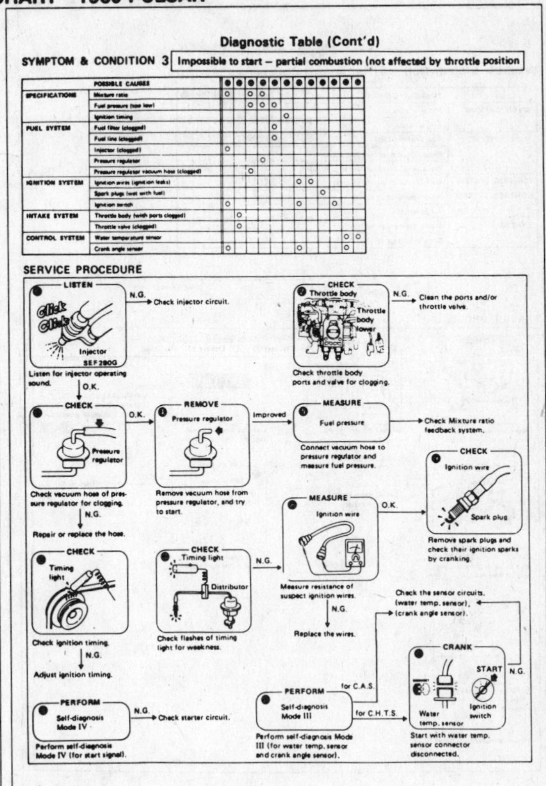

Diagnostic Table (Cont'd)

SYMPTOM & CONDITION 3 Impossible to start – partial combustion (not affected by throttle position

POSSIBLE CAUSES		❶	❷	❸	❹	❺	❻	❼	❽	❾	❿	⓫	⓬
SPECIFICATIONS	Mixture ratio	○		○	○								
	Fuel pressure (too low)			○	○	○							
	Ignition timing						○						
FUEL SYSTEM	Fuel filter (clogged)					○							
	Fuel line (clogged)					○							
	Injector (clogged)	○											
	Pressure regulator				○								
	Pressure regulator vacuum hose (clogged)			○									
IGNITION SYSTEM	Ignition wires (ignition leaks)							○	○				
	Spark plugs (wet with fuel)									○			
	Ignition switch	○						○			○		
INTAKE SYSTEM	Throttle body (with ports clogged)		○										
	Throttle valve (clogged)		○										
CONTROL SYSTEM	Water temperature sensor											○	○
	Crank angle sensor	○						○				○	

SERVICE PROCEDURE

LISTEN
Click Click
Injector
SEF280G
Listen for injector operating sound.
N.G.
Check injector circuit.
O.K.
CHECK
Throttle body
Throttle body lower
Check throttle body ports and valve for clogging.
N.G.
Clean the ports and/or throttle valve.
CHECK
Pressure regulator
Check vacuum hose of pressure regulator for clogging.
N.G.
Repair or replace the hose.
O.K.
REMOVE
Pressure regulator
Remove vacuum hose from pressure regulator, and try to start.
Improved
MEASURE
Fuel pressure
Connect vacuum hose to pressure regulator and measure fuel pressure.
Check Mixture ratio feedback system.
CHECK
Ignition wire
Spark plug
Remove spark plugs and check their ignition sparks by cranking.
O.K.
MEASURE
Ignition wire
Measure resistance of suspect ignition wires.
N.G.
Replace the wires.
CHECK
Timing light
Check ignition timing.
N.G.
Adjust ignition timing.
CHECK
Timing light
Distributor
Check flashes of timing light for weakness.
N.G.
Check the sensor circuits. (water temp. sensor), (crank angle sensor).
PERFORM
Self-diagnosis Mode IV
Perform self-diagnosis Mode IV (for start signal).
N.G.
Check starter circuit.
PERFORM
Self-diagnosis Mode III
Perform self-diagnosis Mode III (for water temp. sensor and crank angle sensor).
for C.A.S.
for C.H.T.S.
CRANK
START
Water temp. sensor
Ignition switch
Start with water temp. sensor connector disconnected.
N.G.

ECCS DIAGNOSTIC CHART—1989 PULSAR

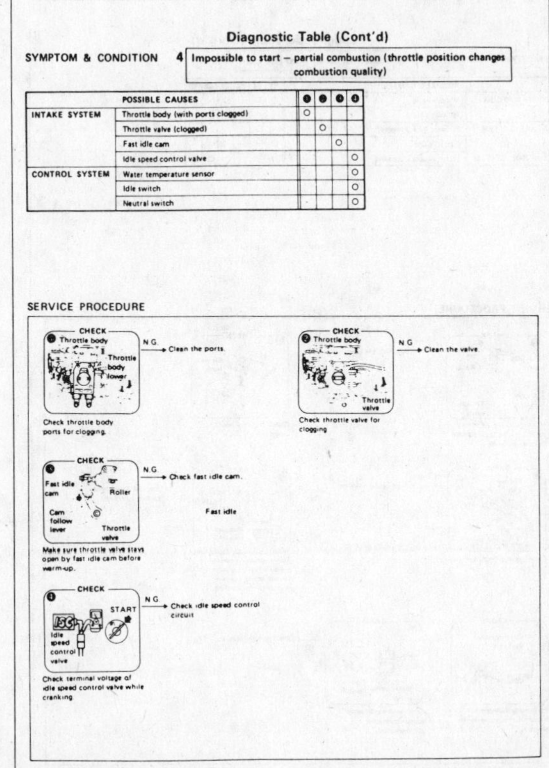

Diagnostic Table (Cont'd)

SYMPTOM & CONDITION 4 Impossible to start – partial combustion (throttle position changes combustion quality)

POSSIBLE CAUSES		❶	❷	❸	❹
INTAKE SYSTEM	Throttle body (with ports clogged)	○			
	Throttle valve (clogged)		○		
	Fast idle cam			○	
	Idle speed control valve				○
CONTROL SYSTEM	Water temperature sensor				○
	Idle switch				○
	Neutral switch				○

SERVICE PROCEDURE

CHECK
Throttle body
Throttle body lower
Check throttle body ports for clogging.
N.G.
Clean the ports.
CHECK
Throttle body
Throttle valve
Check throttle valve for clogging
N.G.
Clean the valve.
CHECK
Fast idle cam
Roller
Cam follow lever
Throttle valve
Make sure throttle valve stays open by fast idle cam before warm-up.
N.G.
Check fast idle cam.
Fast idle
CHECK
ISC
START
Idle speed control valve
Check terminal voltage of idle speed control valve while cranking
N.G.
Check idle speed control circuit

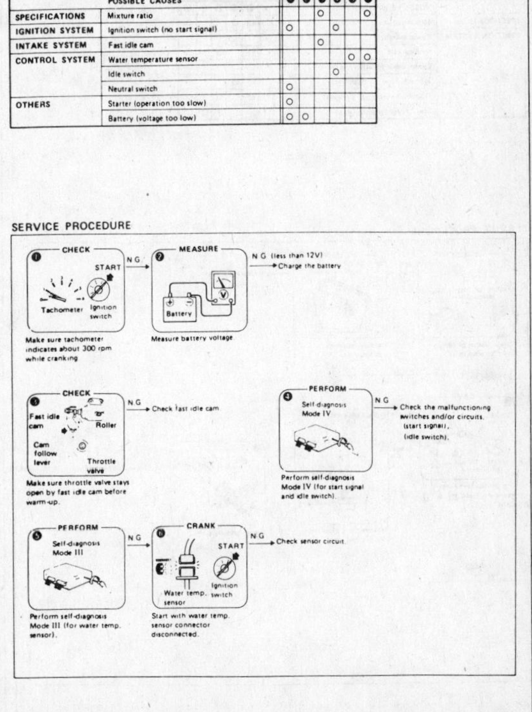

Diagnostic Table (Cont'd)

SYMPTOM & CONDITION 5 Hard to start – before warm-up

POSSIBLE CAUSES		❶	❷	❸	❹	❺	❻
SPECIFICATIONS	Mixture ratio			○			○
IGNITION SYSTEM	Ignition switch (no start signal)	○			○		
INTAKE SYSTEM	Fast idle cam			○			
CONTROL SYSTEM	Water temperature sensor					○	○
	Idle switch				○		
	Neutral switch	○					
OTHERS	Starter (operation too slow)	○					
	Battery (voltage too low)	○	○				

SERVICE PROCEDURE

CHECK
START
Tachometer
Ignition switch
Make sure tachometer indicates about 300 rpm while cranking
N.G.
MEASURE
Battery
Measure battery voltage
N.G (less than 12V)
Charge the battery
CHECK
Fast idle cam
Roller
Cam follow lever
Throttle valve
Make sure throttle valve stays open by fast idle cam before warm-up.
N.G
Check fast idle cam.
PERFORM
Self-diagnosis Mode IV
Perform self-diagnosis Mode IV (for start signal and idle switch).
N.G
Check the malfunctioning switches and/or circuits. (start signal), (idle switch).
PERFORM
Self-diagnosis Mode III
Perform self-diagnosis Mode III (for water temp. sensor).
N.G
CRANK
START
Water temp. sensor
Ignition switch
Start with water temp. sensor connector disconnected.
N.G
Check sensor circuit.

ECCS DIAGNOSTIC CHART—1989 PULSAR

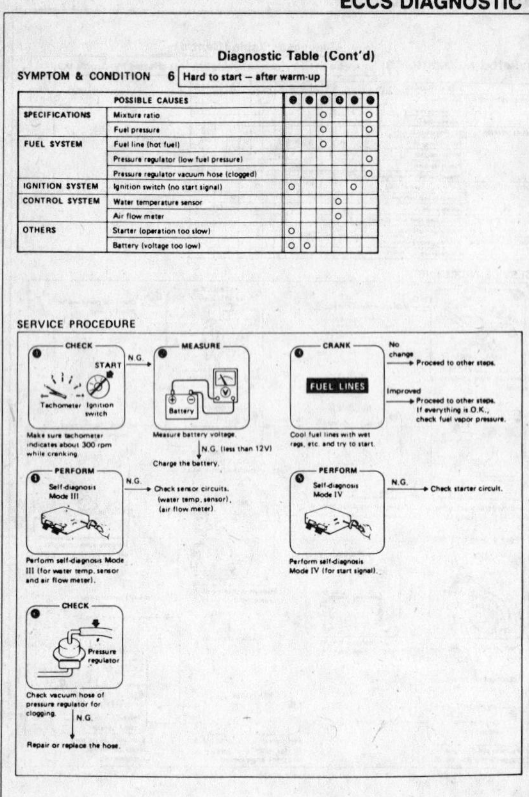

Diagnostic Table (Cont'd)

SYMPTOM & CONDITION 6 Hard to start – after warm-up

POSSIBLE CAUSES		1	2	3	4	5	6
SPECIFICATIONS	Mixture ratio			O			O
	Fuel pressure			O			O
FUEL SYSTEM	Fuel line (hot fuel)			O			
	Pressure regulator (low fuel pressure)						O
	Pressure regulator vacuum hose (clogged)						O
IGNITION SYSTEM	Ignition switch (no start signal)	O				O	
CONTROL SYSTEM	Water temperature sensor				O		
	Air flow meter				O		
OTHERS	Starter (operation too slow)	O					
	Battery (voltage too low)	O	O				

SERVICE PROCEDURE

CHECK
START
Tachometer
Ignition switch
Make sure tachometer indicates about 300 rpm while cranking.
N.G.
MEASURE
Battery
Measure battery voltage.
N.G. (less than 12V)
Charge the battery.
CRANK
FUEL LINES
Cool fuel lines with wet rags, etc. and try to start.
No change
Proceed to other steps.
Improved
Proceed to other steps. If everything is O.K., check fuel vapor pressure.
PERFORM
Self-diagnosis Mode III
Perform self-diagnosis Mode III (for water temp. sensor and air flow meter).
N.G.
Check sensor circuits. (water temp. sensor), (air flow meter).
PERFORM
Self-diagnosis Mode IV
Perform self-diagnosis Mode IV (for start signal).
N.G.
Check starter circuit.
CHECK
Pressure regulator
Check vacuum hose of pressure regulator for clogging.
N.G.
Repair or replace the hose.

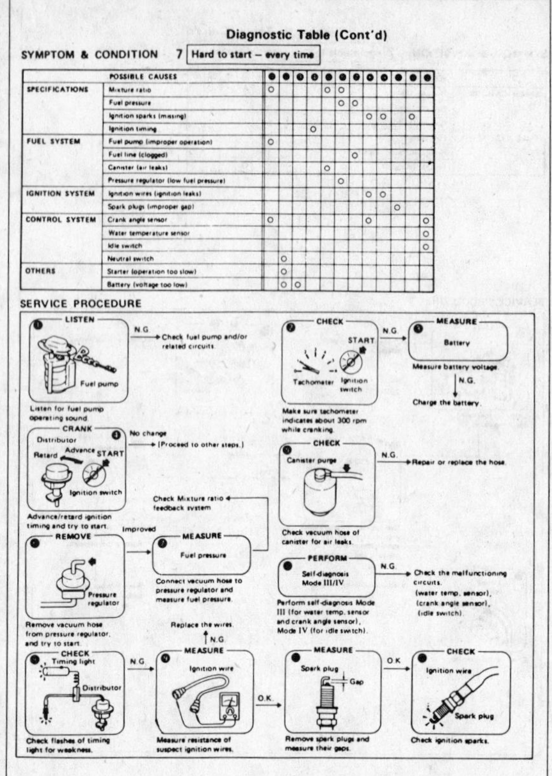

Diagnostic Table (Cont'd)

SYMPTOM & CONDITION 7 Hard to start – every time

POSSIBLE CAUSES		1	2	3	4	5	6	7	8	9	10	11	12
SPECIFICATIONS	Mixture ratio	O				O	O						
	Fuel pressure						O	O					
	Ignition sparks (missing)								O	O		O	
	Ignition timing				O								
FUEL SYSTEM	Fuel pump (improper operation)	O											
	Fuel line (clogged)							O					
	Canister (air leaks)					O							
	Pressure regulator (low fuel pressure)						O						
IGNITION SYSTEM	Ignition wires (ignition leaks)								O	O			
	Spark plugs (improper gap)										O		
CONTROL SYSTEM	Crank angle sensor	O							O				O
	Water temperature sensor												O
	Idle switch												O
	Neutral switch		O										
OTHERS	Starter (operation too slow)		O										
	Battery (voltage too low)		O	O									

SERVICE PROCEDURE

LISTEN
Fuel pump
Listen for fuel pump operating sound.
N.G.
Check fuel pump and/or related circuits.
CHECK
START
Tachometer
Ignition switch
Make sure tachometer indicates about 300 rpm while cranking.
N.G.
MEASURE
Battery
Measure battery voltage.
N.G.
Charge the battery.
CRANK
Distributor
Advance
Retard
START
Ignition switch
Advance/retard ignition timing and try to start.
No change
(Proceed to other steps.)
CHECK
Canister purge
Check vacuum hose of canister for air leaks.
N.G.
Repair or replace the hose.
Check Mixture ratio feedback system
Improved
REMOVE
Pressure regulator
Remove vacuum hose from pressure regulator, and try to start.
MEASURE
Fuel pressure
Connect vacuum hose to pressure regulator and measure fuel pressure.
PERFORM
Self-diagnosis Mode III/IV
Perform self-diagnosis Mode III (for water temp. sensor and crank angle sensor), Mode IV (for idle switch).
N.G.
Check the malfunctioning circuits. (water temp. sensor), (crank angle sensor), (idle switch).
Replace the wires.
N.G.
CHECK
Timing light
Distributor
Check flashes of timing light for weakness.
N.G.
MEASURE
Ignition wire
Measure resistance of suspect ignition wires.
O.K.
MEASURE
Spark plug
Gap
Remove spark plugs and measure their gaps.
O.K.
CHECK
Ignition wire
Spark plug
Check ignition sparks.

ECCS DIAGNOSTIC CHART—1989 PULSAR

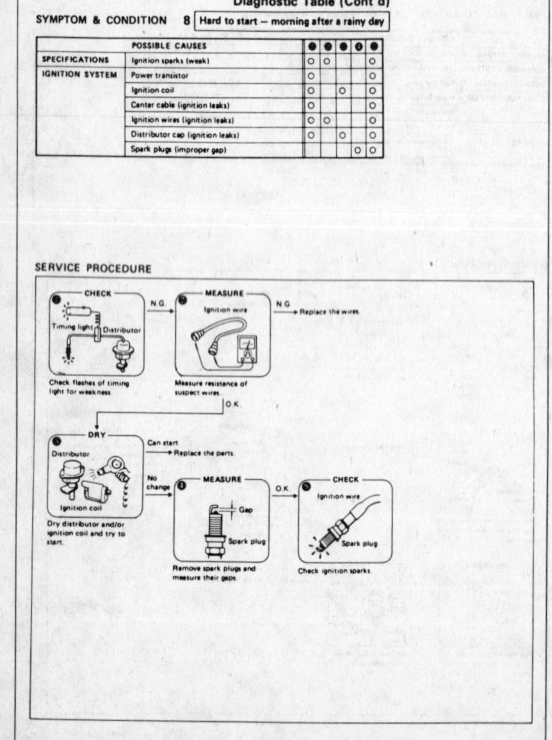

Diagnostic Table (Cont'd)

SYMPTOM & CONDITION 8 Hard to start – morning after a rainy day

POSSIBLE CAUSES		1	2	3	4	5
SPECIFICATIONS	Ignition sparks (weak)	O	O			O
IGNITION SYSTEM	Power transistor	O				O
	Ignition coil	O		O		O
	Center cable (ignition leaks)	O				O
	Ignition wires (ignition leaks)	O	O			O
	Distributor cap (ignition leaks)	O		O		O
	Spark plugs (improper gap)				O	O

SERVICE PROCEDURE

CHECK
Timing light
Distributor
Check flashes of timing light for weakness.
N.G.
MEASURE
Ignition wire
Measure resistance of suspect wires.
N.G.
Replace the wires.
O.K.
DRY
Distributor
Ignition coil
Dry distributor and/or ignition coil and try to start.
Can start
Replace the parts.
No change
MEASURE
Gap
Spark plug
Remove spark plugs and measure their gaps.
O.K.
CHECK
Ignition wire
Spark plug
Check ignition sparks.

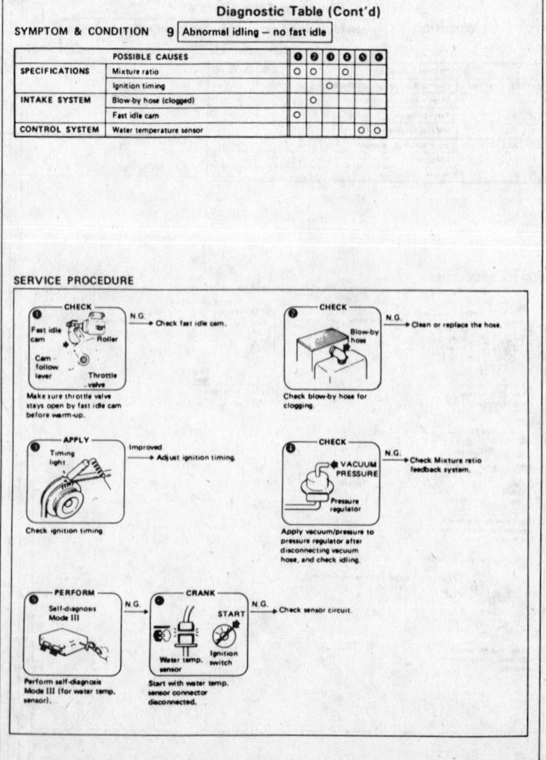

Diagnostic Table (Cont'd)

SYMPTOM & CONDITION 9 Abnormal idling – no fast idle

POSSIBLE CAUSES		1	2	3	4	5	6
SPECIFICATIONS	Mixture ratio	O	O		O		
	Ignition timing			O			
INTAKE SYSTEM	Blow-by hose (clogged)		O				
	Fast idle cam	O					
CONTROL SYSTEM	Water temperature sensor					O	O

SERVICE PROCEDURE

CHECK
Fast idle cam
Roller
Cam follow lever
Throttle valve
Make sure throttle valve stays open by fast idle cam before warm-up.
N.G.
Check fast idle cam.
CHECK
Blow-by hose
Check blow-by hose for clogging.
N.G.
Clean or replace the hose.
APPLY
Timing light
Check ignition timing.
Improved
Adjust ignition timing.
CHECK
VACUUM PRESSURE
Pressure regulator
Apply vacuum/pressure to pressure regulator after disconnecting vacuum hose, and check idling.
N.G.
Check Mixture ratio feedback system.
PERFORM
Self-diagnosis Mode III
Perform self-diagnosis Mode III (for water temp. sensor).
N.G.
CRANK
START
Water temp. sensor
Ignition switch
Start with water temp. sensor connector disconnected.
N.G.
Check sensor circuit.

ECCS DIAGNOSTIC CHART – 1989 PULSAR

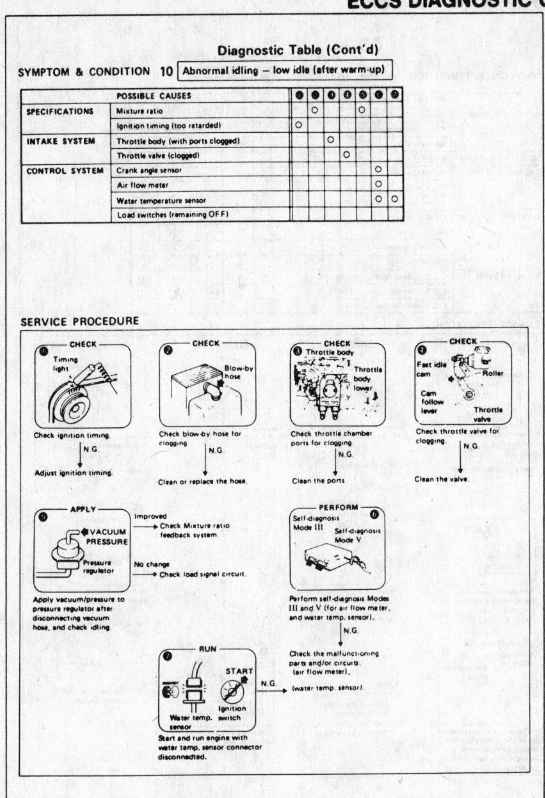

Diagnostic Table (Cont'd)

SYMPTOM & CONDITION 10 Abnormal idling – low idle (after warm-up)

	POSSIBLE CAUSES	1	2	3	4	5	6	7
SPECIFICATIONS	Mixture ratio		O			O		
	Ignition timing (too retarded)	O						
INTAKE SYSTEM	Throttle body (with ports clogged)			O				
	Throttle valve (clogged)				O			
CONTROL SYSTEM	Crank angle sensor						O	
	Air flow meter						O	
	Water temperature sensor						O	O
	Load switches (remaining OFF)							

SERVICE PROCEDURE
CHECK
Timing light
Check ignition timing.
N.G.
Adjust ignition timing.
CHECK
Blow-by hose
Check blow-by hose for clogging.
N.G.
Clean or replace the hose.
CHECK
Throttle body
Throttle body lower
Check throttle chamber ports for clogging.
N.G.
Clean the ports
CHECK
Fast idle cam
Roller
Cam follow lever
Throttle valve
Check throttle valve for clogging.
N.G.
Clean the valve.
APPLY
VACUUM PRESSURE
Pressure regulator
Improved
Check Mixture ratio feedback system.
No change
Check load signal circuit.
Apply vacuum/pressure to pressure regulator after disconnecting vacuum hose, and check idling.
PERFORM
Self-diagnosis Mode III
Self-diagnosis Mode V
Perform self-diagnosis Modes III and V (for air flow meter, and water temp. sensor).
N.G.
Check the malfunctioning parts and/or circuits. (air flow meter), (water temp. sensor).
RUN
START
Water temp. sensor
Ignition switch
N.G.
Start and run engine with water temp. sensor connector disconnected.

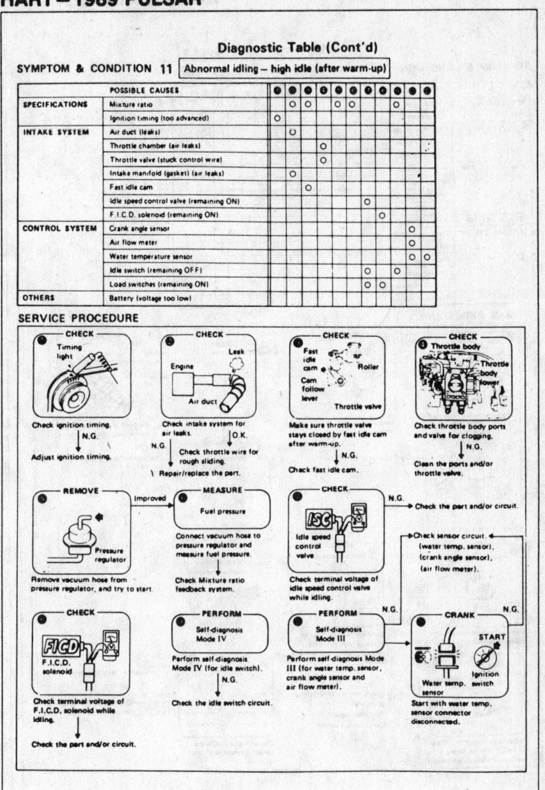

Diagnostic Table (Cont'd)

SYMPTOM & CONDITION 11 Abnormal idling – high idle (after warm-up)

	POSSIBLE CAUSES	1	2	3	4	5	6	7	8	9	10	11
SPECIFICATIONS	Mixture ratio		O	O		O	O			O		
	Ignition timing (too advanced)	O										
INTAKE SYSTEM	Air duct (leaks)		O									
	Throttle chamber (air leaks)				O							
	Throttle valve (stuck control wire)				O							
	Intake manifold (gasket) (air leaks)		O									
	Fast idle cam			O								
	Idle speed control valve (remaining ON)							O				
	F.I.C.D. solenoid (remaining ON)								O			
CONTROL SYSTEM	Crank angle sensor										O	
	Air flow meter										O	
	Water temperature sensor										O	O
	Idle switch (remaining OFF)							O		O		
	Load switches (remaining ON)							O	O			
OTHERS	Battery (voltage too low)											

SERVICE PROCEDURE
CHECK
Timing light
Check ignition timing.
N.G.
Adjust ignition timing.
CHECK
Leak
Engine
Air duct
Check intake system for air leaks.
O.K.
N.G.
Check throttle wire for rough sliding.
Repair/replace the part.
CHECK
Fast idle cam
Roller
Cam follow lever
Throttle valve
Make sure throttle valve stays closed by fast idle cam after warm-up.
N.G.
Check fast idle cam.
CHECK
Throttle body
Throttle body lower
Check throttle body ports and valve for clogging.
N.G.
Clean the ports and/or throttle valve.
REMOVE
Pressure regulator
Remove vacuum hose from pressure regulator, and try to start.
Improved
MEASURE
Fuel pressure
Connect vacuum hose to pressure regulator and measure fuel pressure.
Check Mixture ratio feedback system.
CHECK
ISC
Idle speed control valve
N.G.
Check the part and/or circuit.
Check terminal voltage of idle speed control valve while idling.
Check sensor circuit.
(water temp. sensor),
(crank angle sensor),
(air flow meter).
CHECK
FICD
F.I.C.D. solenoid
Check terminal voltage of F.I.C.D. solenoid while idling.
Check the part and/or circuit.
PERFORM
Self-diagnosis Mode IV
Perform self-diagnosis Mode IV (for idle switch).
N.G.
Check the idle switch circuit.
PERFORM
Self-diagnosis Mode III
N.G.
Perform self-diagnosis Mode III (for water temp. sensor, crank angle sensor and air flow meter).
CRANK
N.G.
START
Water temp. sensor
Ignition switch
Start with water temp. sensor connector disconnected.

ECCS DIAGNOSTIC CHART – 1989 PULSAR

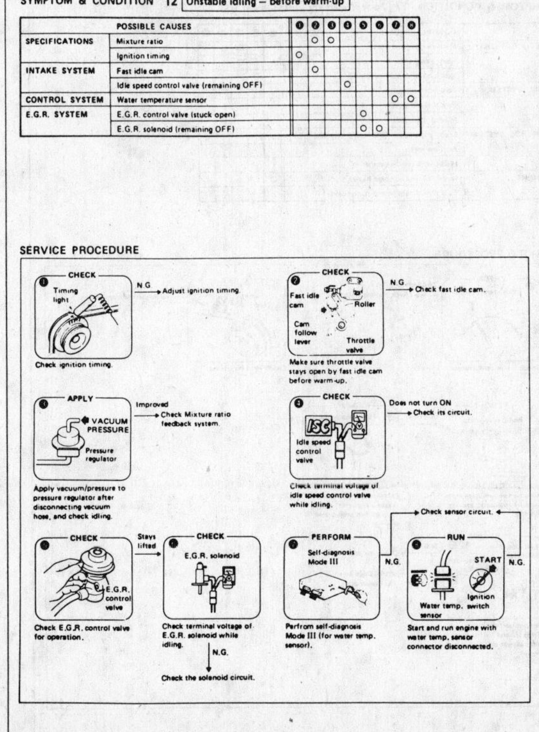

Diagnostic Table (Cont'd)

SYMPTOM & CONDITION 12 Unstable idling – before warm-up

	POSSIBLE CAUSES	1	2	3	4	5	6	7	8
SPECIFICATIONS	Mixture ratio		O	O					
	Ignition timing	O							
INTAKE SYSTEM	Fast idle cam		O						
	Idle speed control valve (remaining OFF)				O				
CONTROL SYSTEM	Water temperature sensor							O	O
E.G.R. SYSTEM	E.G.R. control valve (stuck open)					O			
	E.G.R. solenoid (remaining OFF)					O	O		

SERVICE PROCEDURE
CHECK
Timing light
N.G.
Adjust ignition timing.
Check ignition timing.
CHECK
Fast idle cam
Roller
Cam follow lever
Throttle valve
N.G.
Check fast idle cam.
Make sure throttle valve stays open by fast idle cam before warm-up.
APPLY
VACUUM PRESSURE
Pressure regulator
Improved
Check Mixture ratio feedback system.
Apply vacuum/pressure to pressure regulator after disconnecting vacuum hose, and check idling.
CHECK
ISC
Idle speed control valve
Does not turn ON
Check its circuit.
Check terminal voltage of idle speed control valve while idling.
CHECK
E.G.R. control valve
Stays lifted
Check E.G.R. control valve for operation.
CHECK
E.G.R. solenoid
Check terminal voltage of E.G.R. solenoid while idling.
N.G.
Check the solenoid circuit.
PERFORM
Self-diagnosis Mode III
N.G.
Perfrom self-diagnosis Mode III (for water temp. sensor).
Check sensor circuit.
RUN
START
Water temp. sensor
Ignition switch
N.G.
Start and run engine with water temp. sensor connector disconnected.

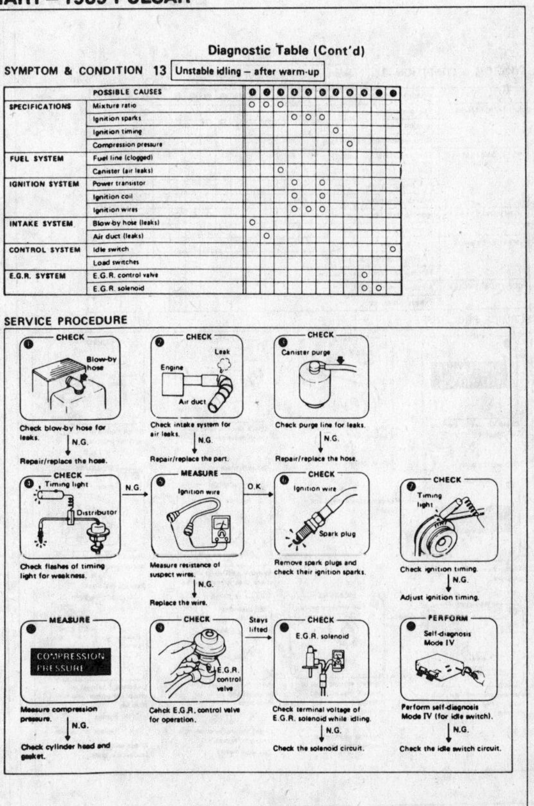

Diagnostic Table (Cont'd)

SYMPTOM & CONDITION 13 Unstable idling – after warm-up

	POSSIBLE CAUSES	1	2	3	4	5	6	7	8	9	10	11
SPECIFICATIONS	Mixture ratio	O	O	O								
	Ignition sparks				O	O	O					
	Ignition timing							O				
	Compression pressure								O			
FUEL SYSTEM	Fuel line (clogged)											
	Canister (air leaks)			O								
IGNITION SYSTEM	Power transistor				O		O					
	Ignition coil				O		O					
	Ignition wires				O	O	O					
INTAKE SYSTEM	Blow-by hose (leaks)	O										
	Air duct (leaks)		O									
CONTROL SYSTEM	Idle switch											O
	Load switches											
E.G.R. SYSTEM	E.G.R. control valve									O		
	E.G.R. solenoid									O	O	

SERVICE PROCEDURE
CHECK
Blow-by hose
Check blow-by hose for leaks.
N.G.
Repair/replace the hose.
CHECK
Leak
Engine
Air duct
Check intake system for air leaks.
N.G.
Repair/replace the part.
CHECK
Canister purge
Check purge line for leaks.
N.G.
Repair/replace the hose.
CHECK
Timing light
Distributor
N.G.
Check flashes of timing light for weakness.
MEASURE
Ignition wire
O.K.
Measure resistance of suspect wires.
N.G.
Replace the wire.
CHECK
Ignition wire
Spark plug
Remove spark plugs and check their ignition sparks.
CHECK
Timing light
Check ignition timing.
N.G.
Adjust ignition timing.
MEASURE
COMPRESSION PRESSURE
Measure compression pressure.
N.G.
Check cylinder head and gasket.
CHECK
E.G.R. control valve
Stays lifted
Cehck E.G.R. control valve for operation.
CHECK
E.G.R. solenoid
Check terminal voltage of E.G.R. solenoid while idling.
N.G.
Check the solenoid circuit.
PERFORM
Self-diagnosis Mode IV
Perform self-diagnosis Mode IV (for idle switch).
N.G.
Check the idle switch circuit.

ECCS DIAGNOSTIC CHART—1989 PULSAR

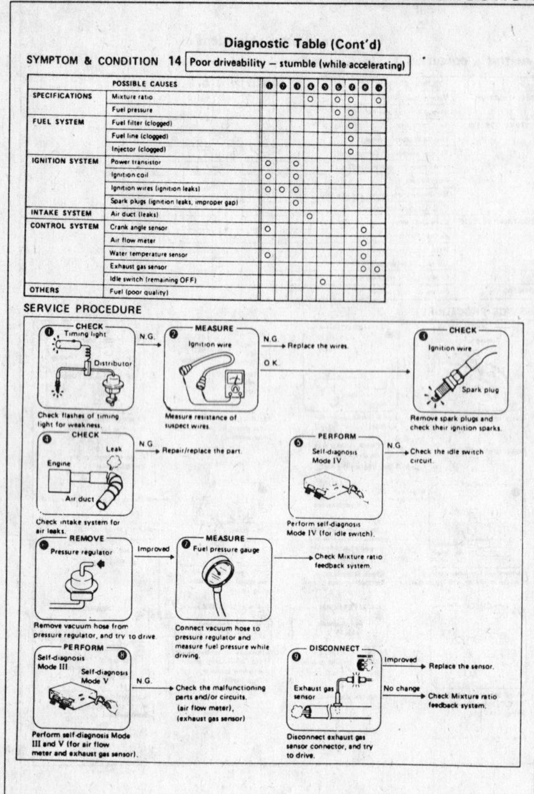

Diagnostic Table (Cont'd)

SYMPTOM & CONDITION 14 Poor driveability – stumble (while accelerating)

	POSSIBLE CAUSES	1	2	3	4	5	6	7	8	9
SPECIFICATIONS	Mixture ratio				O		O	O		O
	Fuel pressure						O	O		
FUEL SYSTEM	Fuel filter (clogged)							O		
	Fuel line (clogged)							O		
	Injector (clogged)							O		
IGNITION SYSTEM	Power transistor	O		O						
	Ignition coil	O		O						
	Ignition wires (ignition leaks)	O	O	O						
	Spark plugs (ignition leaks, improper gap)			O						
INTAKE SYSTEM	Air duct (leaks)				O					
CONTROL SYSTEM	Crank angle sensor	O							O	
	Air flow meter								O	
	Water temperature sensor	O							O	
	Exhaust gas sensor								O	O
	Idle switch (remaining OFF)					O				
OTHERS	Fuel (poor quality)									

SERVICE PROCEDURE

1 CHECK Timing light Distributor
Check flashes of timing light for weakness.
N.G.
2 MEASURE Ignition wire
Measure resistance of suspect wires.
N.G. Replace the wires.
O.K.
3 CHECK Ignition wire Spark plug
Remove spark plugs and check their ignition sparks.
4 CHECK Engine Leak Air duct
Check intake system for air leaks.
N.G. Repair/replace the part.
5 PERFORM Self-diagnosis Mode IV
Perform self-diagnosis Mode IV (for idle switch).
N.G. Check the idle switch circuit.
6 REMOVE Pressure regulator
Remove vacuum hose from pressure regulator, and try to drive.
Improved
7 MEASURE Fuel pressure gauge
Connect vacuum hose to pressure regulator and measure fuel pressure while driving.
Check Mixture ratio feedback system.
8 PERFORM Self-diagnosis Mode III Self-diagnosis Mode V
Perform self-diagnosis Mode III and V (for air flow meter and exhaust gas sensor).
N.G. Check the malfunctioning parts and/or circuits. (air flow meter), (exhaust gas sensor)
9 DISCONNECT Exhaust gas sensor
Disconnect exhaust gas sensor connector, and try to drive.
Improved Replace the sensor.
No change Check Mixture ratio feedback system.

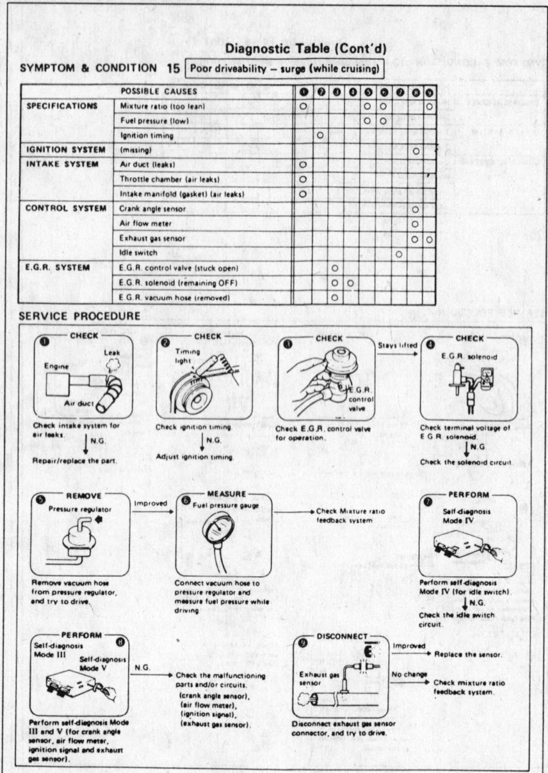

Diagnostic Table (Cont'd)

SYMPTOM & CONDITION 15 Poor driveability – surge (while cruising)

	POSSIBLE CAUSES	1	2	3	4	5	6	7	8	9
SPECIFICATIONS	Mixture ratio (too lean)	O				O	O			O
	Fuel pressure (low)					O	O			
	Ignition timing		O							
IGNITION SYSTEM	(missing)								O	
INTAKE SYSTEM	Air duct (leaks)	O								
	Throttle chamber (air leaks)	O								
	Intake manifold (gasket) (air leaks)	O								
CONTROL SYSTEM	Crank angle sensor								O	
	Air flow meter								O	
	Exhaust gas sensor								O	O
	Idle switch							O		
E.G.R. SYSTEM	E.G.R. control valve (stuck open)			O						
	E.G.R. solenoid (remaining OFF)			O	O					
	E.G.R. vacuum hose (removed)			O						

SERVICE PROCEDURE

1 CHECK Engine Leak Air duct
Check intake system for air leaks.
N.G.
Repair/replace the part.
2 CHECK Timing light
Check ignition timing.
N.G.
Adjust ignition timing.
3 CHECK E.G.R. control valve
Check E.G.R. control valve for operation.
Stays lifted
4 CHECK E.G.R. solenoid
Check terminal voltage of E.G.R. solenoid.
N.G.
Check the solenoid circuit.
5 REMOVE Pressure regulator
Remove vacuum hose from pressure regulator, and try to drive.
Improved
6 MEASURE Fuel pressure gauge
Connect vacuum hose to pressure regulator and measure fuel pressure while driving.
Check Mixture ratio feedback system.
7 PERFORM Self-diagnosis Mode IV
Perform self-diagnosis Mode IV (for idle switch).
N.G.
Check the idle switch circuit.
8 PERFORM Self-diagnosis Mode III Self-diagnosis Mode V
Perform self-diagnosis Mode III and V (for crank angle sensor, air flow meter, ignition signal and exhaust gas sensor).
N.G. Check the malfunctioning parts and/or circuits. (crank angle sensor), (air flow meter), (ignition signal), (exhaust gas sensor).
9 DISCONNECT Exhaust gas sensor
Disconnect exhaust gas sensor connector, and try to drive.
Improved Replace the sensor.
No change Check mixture ratio feedback system.

ECCS DIAGNOSTIC CHART—1989 PULSAR

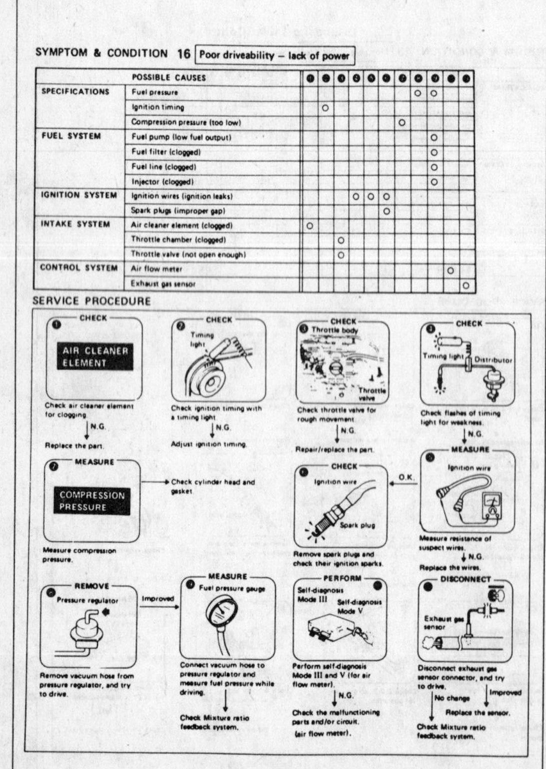

SYMPTOM & CONDITION 16 Poor driveability – lack of power

	POSSIBLE CAUSES	1	2	3	4	5	6	7	8	9	10	11
SPECIFICATIONS	Fuel pressure								O	O		
	Ignition timing		O									
	Compression pressure (too low)							O				
FUEL SYSTEM	Fuel pump (low fuel output)									O		
	Fuel filter (clogged)									O		
	Fuel line (clogged)									O		
	Injector (clogged)									O		
IGNITION SYSTEM	Ignition wires (ignition leaks)				O	O	O					
	Spark plugs (improper gap)						O					
INTAKE SYSTEM	Air cleaner element (clogged)	O										
	Throttle chamber (clogged)			O								
	Throttle valve (not open enough)			O								
CONTROL SYSTEM	Air flow meter										O	
	Exhaust gas sensor											O

SERVICE PROCEDURE

1 CHECK AIR CLEANER ELEMENT
Check air cleaner element for clogging.
N.G.
Replace the part.
2 CHECK Timing light
Check ignition timing with a timing light.
N.G.
Adjust ignition timing.
3 CHECK Throttle body Throttle valve
Check throttle valve for rough movement.
N.G.
Repair/replace the part.
4 CHECK Timing light Distributor
Check flashes of timing light for weakness.
N.G.
7 MEASURE COMPRESSION PRESSURE
Measure compression pressure.
Check cylinder head and gasket.
6 CHECK Ignition wire Spark plug
Remove spark plugs and check their ignition sparks.
O.K.
5 MEASURE Ignition wire
Measure resistance of suspect wires.
N.G.
Replace the wires.
8 REMOVE Pressure regulator
Remove vacuum hose from pressure regulator, and try to drive.
Improved
9 MEASURE Fuel pressure gauge
Connect vacuum hose to pressure regulator and measure fuel pressure while driving.
Check Mixture ratio feedback system.
10 PERFORM Self-diagnosis Mode III Self-diagnosis Mode V
Perform self-diagnosis Mode III and V (for air flow meter).
N.G.
Check the malfunctioning parts and/or circuit. (air flow meter).
11 DISCONNECT Exhaust gas sensor
Disconnect exhaust gas sensor connector, and try to drive.
N.G.
No change
Improved
Replace the sensor.
Check Mixture ratio feedback system.

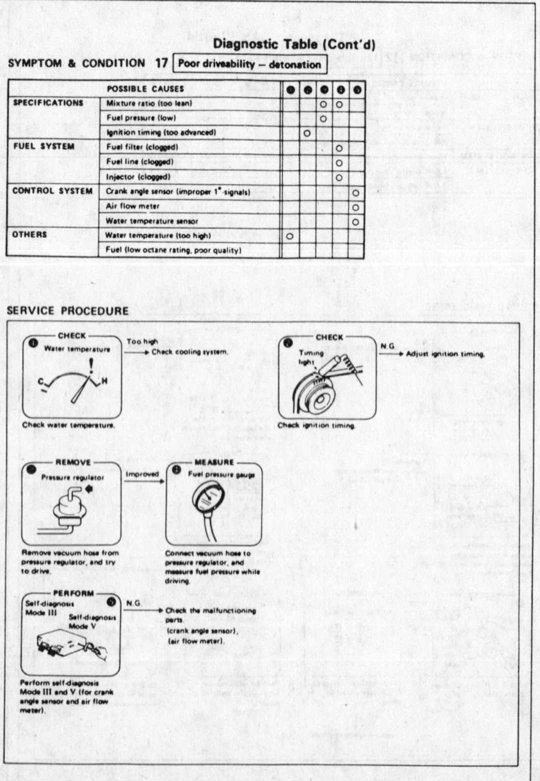

Diagnostic Table (Cont'd)

SYMPTOM & CONDITION 17 Poor driveability – detonation

	POSSIBLE CAUSES	1	2	3	4	5
SPECIFICATIONS	Mixture ratio (too lean)			O	O	
	Fuel pressure (low)			O		
	Ignition timing (too advanced)		O			
FUEL SYSTEM	Fuel filter (clogged)				O	
	Fuel line (clogged)				O	
	Injector (clogged)				O	
CONTROL SYSTEM	Crank angle sensor (improper 1° signals)					O
	Air flow meter					O
	Water temperature sensor					O
OTHERS	Water temperature (too high)	O				
	Fuel (low octane rating, poor quality)					

SERVICE PROCEDURE

1 CHECK Water temperature
Too high
Check cooling system.
Check water temperature.
2 CHECK Timing light
N.G. Adjust ignition timing.
Check ignition timing.
3 REMOVE Pressure regulator
Remove vacuum hose from pressure regulator, and try to drive.
Improved
4 MEASURE Fuel pressure gauge
Connect vacuum hose to pressure regulator, and measure fuel pressure while driving.
5 PERFORM Self-diagnosis Mode III Self-diagnosis Mode V
N.G. Check the malfunctioning parts. (crank angle sensor), (air flow meter).
Perform self-diagnosis Mode III and V (for crank angle sensor and air flow meter).

ECCS DIAGNOSTIC CHART—1989 PULSAR

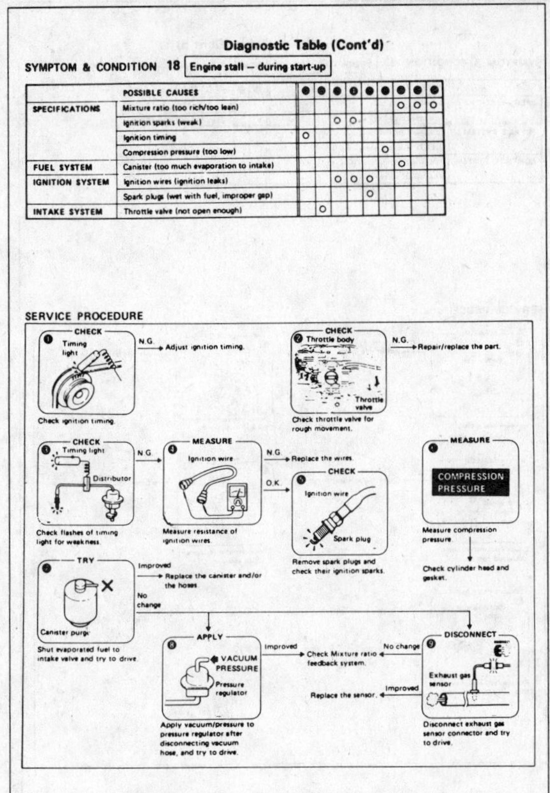

Diagnostic Table (Cont'd)

SYMPTOM & CONDITION 18 Engine stall – during start-up

	POSSIBLE CAUSES	1	2	3	4	5	6	7	8	9
SPECIFICATIONS	Mixture ratio (too rich/too lean)							O	O	O
	Ignition sparks (weak)			O	O					
	Ignition timing	O								
	Compression pressure (too low)						O			
FUEL SYSTEM	Canister (too much evaporation to intake)							O		
IGNITION SYSTEM	Ignition wires (ignition leaks)			O	O	O				
	Spark plugs (wet with fuel, improper gap)					O				
INTAKE SYSTEM	Throttle valve (not open enough)		O							

SERVICE PROCEDURE

CHECK
1 Timing light
N.G. → Adjust ignition timing.
Check ignition timing.
CHECK
2 Throttle body
Throttle valve
N.G. → Repair/replace the part.
Check throttle valve for rough movement.
CHECK
3 Timing light
Distributor
Check flashes of timing light for weakness.
N.G.
MEASURE
4 Ignition wire
Measure resistance of ignition wires.
N.G. → Replace the wires.
O.K.
CHECK
5 Ignition wire
Spark plug
Remove spark plugs and check their ignition sparks.
MEASURE
6 COMPRESSION PRESSURE
Measure compression pressure.
Check cylinder head and gasket.
TRY
7 Canister purge
Shut evaporated fuel to intake valve and try to drive.
Improved → Replace the canister and/or the hoses.
No change
APPLY
8 VACUUM PRESSURE
Pressure regulator
Apply vacuum/pressure to pressure regulator after disconnecting vacuum hose, and try to drive.
Improved → Check Mixture ratio feedback system.
No change
DISCONNECT
9 Exhaust gas sensor
Disconnect exhaust gas sensor connector and try to drive.
Improved
Replace the sensor.

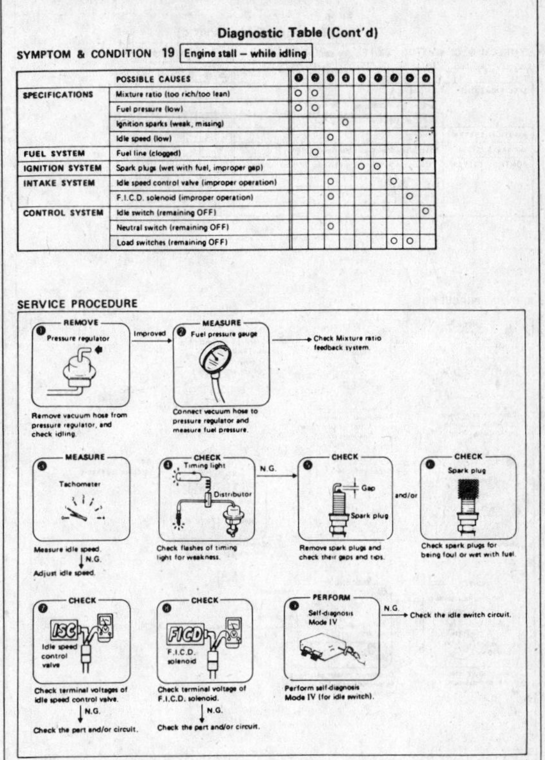

Diagnostic Table (Cont'd)

SYMPTOM & CONDITION 19 Engine stall – while idling

	POSSIBLE CAUSES	1	2	3	4	5	6	7	8	9
SPECIFICATIONS	Mixture ratio (too rich/too lean)	O	O							
	Fuel pressure (low)	O	O							
	Ignition sparks (weak, missing)				O					
	Idle speed (low)			O						
FUEL SYSTEM	Fuel line (clogged)		O							
IGNITION SYSTEM	Spark plugs (wet with fuel, improper gap)					O	O			
INTAKE SYSTEM	Idle speed control valve (improper operation)			O				O		
	F.I.C.D. solenoid (improper operation)			O					O	
CONTROL SYSTEM	Idle switch (remaining OFF)									O
	Neutral switch (remaining OFF)			O						
	Load switches (remaining OFF)							O	O	

SERVICE PROCEDURE

REMOVE
1 Pressure regulator
Remove vacuum hose from pressure regulator, and check idling.
Improved
MEASURE
2 Fuel pressure gauge
Connect vacuum hose to pressure regulator and measure fuel pressure.
→ Check Mixture ratio feedback system.
MEASURE
3 Tachometer
Measure idle speed.
N.G.
Adjust idle speed.
CHECK
4 Timing light
Distributor
Check flashes of timing light for weakness.
N.G.
CHECK
5 Gap
Spark plug
Remove spark plugs and check their gaps and tips.
and/or
CHECK
6 Spark plug
Check spark plugs for being foul or wet with fuel.
CHECK
7 ISC
Idle speed control valve
Check terminal voltages of idle speed control valve.
N.G.
Check the part and/or circuit.
CHECK
8 FICD
F.I.C.D. solenoid
Check terminal voltage of F.I.C.D. solenoid.
N.G.
Check the part and/or circuit.
PERFORM
9 Self-diagnosis Mode IV
Perform self-diagnosis Mode IV (for idle switch).
N.G. → Check the idle switch circuit.

ECCS DIAGNOSTIC CHART—1989 PULSAR

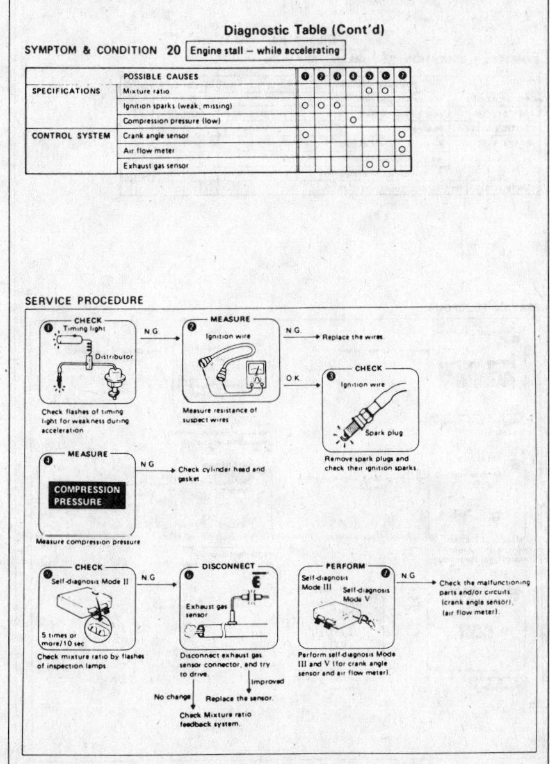

Diagnostic Table (Cont'd)

SYMPTOM & CONDITION 20 Engine stall – while accelerating

	POSSIBLE CAUSES	1	2	3	4	5	6	7
SPECIFICATIONS	Mixture ratio					O	O	
	Ignition sparks (weak, missing)	O	O	O				
	Compression pressure (low)				O			
CONTROL SYSTEM	Crank angle sensor	O						O
	Air flow meter							O
	Exhaust gas sensor					O	O	

SERVICE PROCEDURE

CHECK
1 Timing light
Distributor
Check flashes of timing light for weakness during acceleration
N.G.
MEASURE
2 Ignition wire
Measure resistance of suspect wires
N.G. → Replace the wires.
O.K.
CHECK
3 Ignition wire
Spark plug
Remove spark plugs and check their ignition sparks
MEASURE
4 COMPRESSION PRESSURE
Measure compression pressure
N.G. → Check cylinder head and gasket
CHECK
5 Self-diagnosis Mode II
5 times or more/10 sec.
Check mixture ratio by flashes of inspection lamps.
N.G.
DISCONNECT
6 Exhaust gas sensor
Disconnect exhaust gas sensor connector, and try to drive.
Improved
Replace the sensor.
No change
Check Mixture ratio feedback system.
PERFORM
7 Self-diagnosis Mode III
Self-diagnosis Mode V
Perform self-diagnosis Mode III and V (for crank angle sensor and air flow meter).
N.G. → Check the malfunctioning parts and/or circuits (crank angle sensor), (air flow meter).

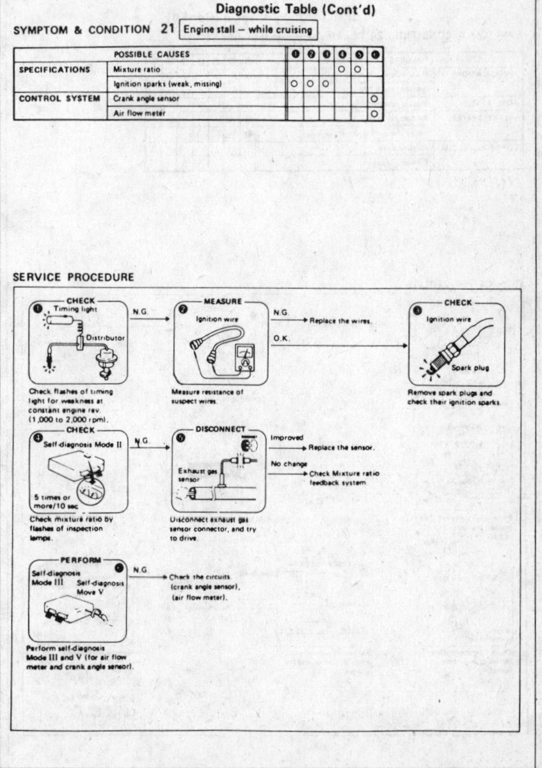

Diagnostic Table (Cont'd)

SYMPTOM & CONDITION 21 Engine stall – while cruising

	POSSIBLE CAUSES	1	2	3	4	5	6
SPECIFICATIONS	Mixture ratio				O	O	
	Ignition sparks (weak, missing)	O	O	O			
CONTROL SYSTEM	Crank angle sensor						O
	Air flow meter						O

SERVICE PROCEDURE

CHECK
1 Timing light
Distributor
Check flashes of timing light for weakness at constant engine rev. (1,000 to 2,000 rpm).
N.G.
MEASURE
2 Ignition wire
Measure resistance of suspect wires.
N.G. → Replace the wires.
O.K.
CHECK
3 Ignition wire
Spark plug
Remove spark plugs and check their ignition sparks.
CHECK
4 Self-diagnosis Mode II
5 times or more/10 sec
Check mixture ratio by flashes of inspection lamps.
N.G.
DISCONNECT
5 Exhaust gas sensor
Disconnect exhaust gas sensor connector, and try to drive.
Improved → Replace the sensor.
No change → Check Mixture ratio feedback system.
PERFORM
6 Self-diagnosis Mode III
Self-diagnosis Move V
Perform self-diagnosis Mode III and V (for air flow meter and crank angle sensor).
N.G. → Check the circuits (crank angle sensor), (air flow meter).

ECCS DIAGNOSTIC CHART—1989 PULSAR

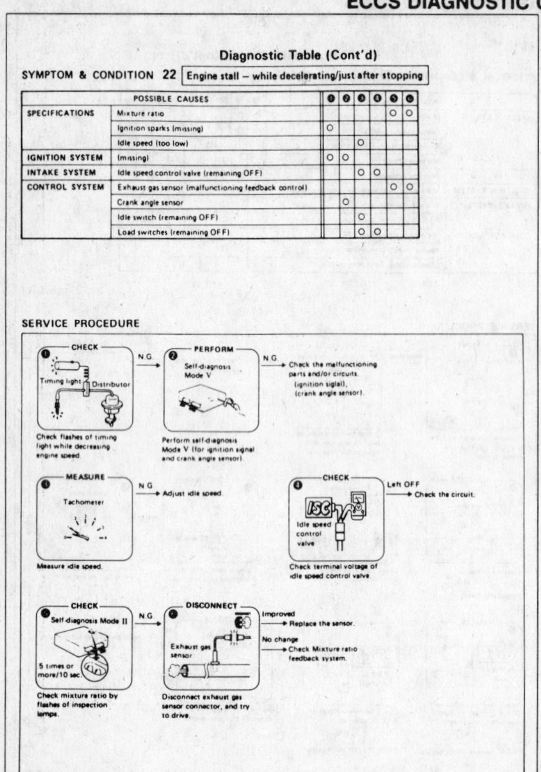

Diagnostic Table (Cont'd)

SYMPTOM & CONDITION 22 Engine stall – while decelerating/just after stopping

POSSIBLE CAUSES		❶	❷	❸	❹	❺	❻
SPECIFICATIONS	Mixture ratio					○	○
	Ignition sparks (missing)	○					
	Idle speed (too low)			○			
IGNITION SYSTEM	(missing)	○	○				
INTAKE SYSTEM	Idle speed control valve (remaining OFF)			○	○		
CONTROL SYSTEM	Exhaust gas sensor (malfunctioning feedback control)					○	○
	Crank angle sensor		○				
	Idle switch (remaining OFF)			○			
	Load switches (remaining OFF)			○	○		

SERVICE PROCEDURE
❶ CHECK
Timing light
Distributor
N.G.
Check flashes of timing light while decreasing engine speed.
❷ PERFORM
Self-diagnosis Mode V
N.G.
Check the malfunctioning parts and/or circuits. (ignition siglal), (crank angle sensor).
Perform self-diagnosis Mode V (for ignition signal and crank angle sensor).
❸ MEASURE
Tachometer
N.G.
Adjust idle speed.
Measure idle speed.
❹ CHECK
Idle speed control valve
Left OFF
Check the circuit.
Check terminal voltage of idle speed control valve.
❺ CHECK
Self-diagnosis Mode II
5 times or more/10 sec.
N.G.
Check mixture ratio by flashes of inspection lamps.
❻ DISCONNECT
Exhaust gas sensor
Improved
Replace the sensor.
No change
Check Mixture ratio feedback system.
Disconnect exhaust gas sensor connector, and try to drive.

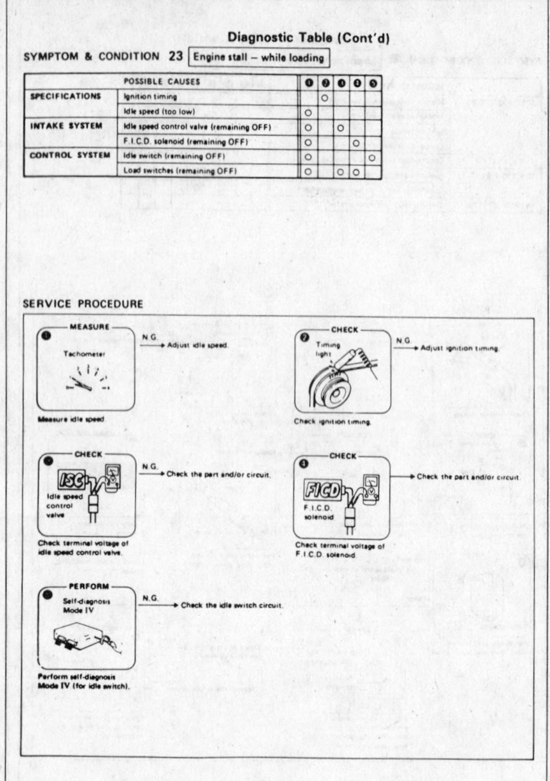

Diagnostic Table (Cont'd)

SYMPTOM & CONDITION 23 Engine stall – while loading

POSSIBLE CAUSES		❶	❷	❸	❹	❺
SPECIFICATIONS	Ignition timing		○			
	Idle speed (too low)	○				
INTAKE SYSTEM	Idle speed control valve (remaining OFF)	○		○		
	F.I.C.D. solenoid (remaining OFF)	○			○	
CONTROL SYSTEM	Idle switch (remaining OFF)	○				○
	Load switches (remaining OFF)	○		○	○	

SERVICE PROCEDURE
❶ MEASURE
Tachometer
N.G.
Adjust idle speed.
Measure idle speed.
❷ CHECK
Timing light
N.G.
Adjust ignition timing.
Check ignition timing.
❸ CHECK
Idle speed control valve
N.G.
Check the part and/or circuit.
Check terminal voltage of idle speed control valve.
❹ CHECK
F.I.C.D. solenoid
Check the part and/or circuit.
Check terminal voltage of F.I.C.D. solenoid.
❺ PERFORM
Self-diagnosis Mode IV
N.G.
Check the idle switch circuit.
Perform self-diagnosis Mode IV (for idle switch).

ECCS DIAGNOSTIC CHART—1989 PULSAR

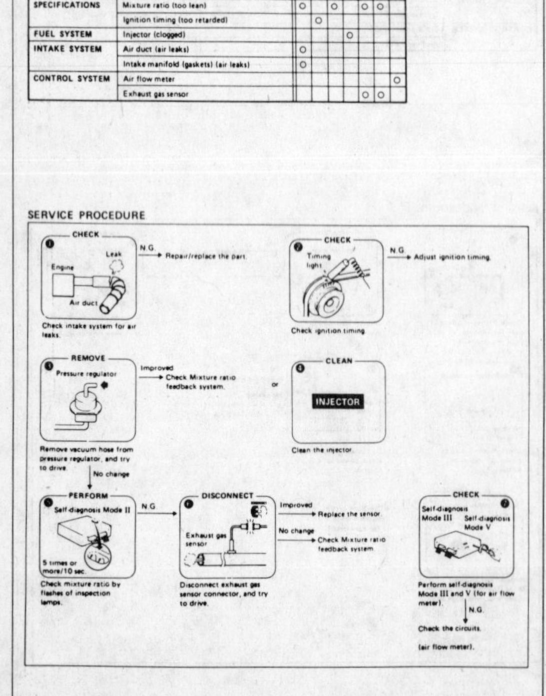

Diagnostic Table (Cont'd)

SYMPTOM & CONDITION 24 Backfire – through the intake

POSSIBLE CAUSES		❶	❷	❸	❹	❺	❻	❼
SPECIFICATIONS	Mixture ratio (too lean)	○		○		○	○	
	Ignition timing (too retarded)		○					
FUEL SYSTEM	Injector (clogged)				○			
INTAKE SYSTEM	Air duct (air leaks)	○						
	Intake manifold (gaskets) (air leaks)	○						
CONTROL SYSTEM	Air flow meter							○
	Exhaust gas sensor					○	○	

SERVICE PROCEDURE
❶ CHECK
Leak
Engine
Air duct
N.G.
Repair/replace the part.
Check intake system for air leaks.
❷ CHECK
Timing light
N.G.
Adjust ignition timing.
Check ignition timing.
❸ REMOVE
Pressure regulator
Improved
Check Mixture ratio feedback system.
or
Remove vacuum hose from pressure regulator, and try to drive.
No change
❹ CLEAN
INJECTOR
Clean the injector.
❺ PERFORM
Self-diagnosis Mode II
5 times or more/10 sec.
N.G.
Check mixture ratio by flashes of inspection lamps.
❻ DISCONNECT
Exhaust gas sensor
Improved
Replace the sensor.
No change
Check Mixture ratio feedback system.
Disconnect exhaust gas sensor connector, and try to drive.
❼ CHECK
Self-diagnosis Mode III
Self-diagnosis Mode V
Perform self-diagnosis Mode III and V (for air flow meter).
N.G.
Check the circuits. (air flow meter).

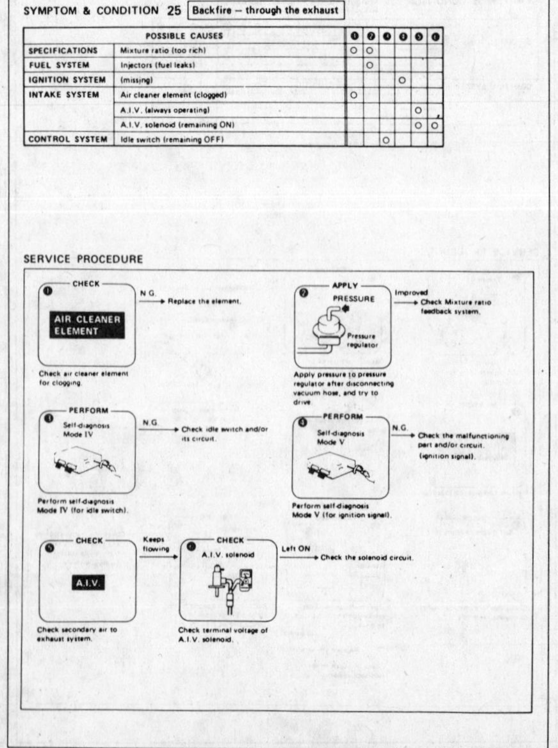

Diagnostic Table (Cont'd)

SYMPTOM & CONDITION 25 Backfire – through the exhaust

POSSIBLE CAUSES		❶	❷	❸	❹	❺	❻
SPECIFICATIONS	Mixture ratio (too rich)	○	○				
FUEL SYSTEM	Injectors (fuel leaks)		○				
IGNITION SYSTEM	(missing)				○		
INTAKE SYSTEM	Air cleaner element (clogged)	○					
	A.I.V. (always operating)					○	
	A.I.V. solenoid (remaining ON)					○	○
CONTROL SYSTEM	Idle switch (remaining OFF)			○			

SERVICE PROCEDURE
❶ CHECK
AIR CLEANER ELEMENT
N.G.
Replace the element.
Check air cleaner element for clogging.
❷ APPLY
PRESSURE
Pressure regulator
Improved
Check Mixture ratio feedback system.
Apply pressure to pressure regulator after disconnecting vacuum hose, and try to drive.
❸ PERFORM
Self-diagnosis Mode IV
N.G.
Check idle switch and/or its circuit.
Perform self-diagnosis Mode IV (for idle switch).
❹ PERFORM
Self-diagnosis Mode V
N.G.
Check the malfunctioning part and/or circuit. (ignition signal).
Perform self-diagnosis Mode V (for ignition signal).
❺ CHECK
A.I.V.
Keeps flowing
Check secondary air to exhaust system.
❻ CHECK
A.I.V. solenoid
Left ON
Check the solenoid circuit.
Check terminal voltage of A.I.V. solenoid.

ECCS DIAGNOSTIC CHART—1989 PULSAR

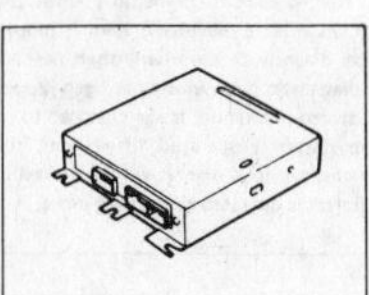

Description

The self-diagnosis is useful to diagnose malfunctions in major sensors and actuators of the E.C.C.S. system. There are 5 modes in the self-diagnosis system.

1. **Mode I – Mixture ratio feedback control monitor A**
   - During closed loop condition:
     The green inspection lamp turns ON when lean condition is detected and goes OFF by rich condition.
   - During open loop condition:
     The green inspection lamp remains ON or OFF.
2. **Mode II – Mixture ratio feedback control monitor B**
   The green inspection lamp function is the same as Mode I.
   - During closed loop condition:
     The red inspection lamp turns ON and OFF simultaneously with the green inspection lamp when the mixture ratio is controlled within the specified value.
   - During open loop condition:
     The red inspection lamp remains ON or OFF.
3. **Mode III – Self-diagnosis**
   This mode is the same as the former self-diagnosis in self-diagnosis mode.
4. **Mode IV – Switches ON/OFF diagnosis**
   During this mode, the inspection lamps monitor the switch ON-OFF condition.
   - Idle switch
   - Starter switch
   - Vehicle speed sensor
5. **Mode V – Real time diagnosis**
   The moment the malfunction is detected, the display will be presented immediately. That is, the condition at which the malfunction occurs can be found by observing the inspection lamps during driving test.

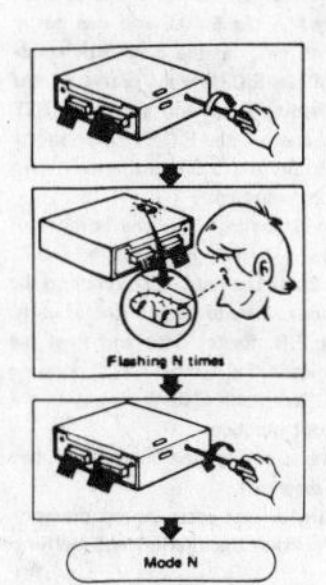
Flashing N times
Mode N

Description (Cont'd)

SWITCHING THE MODES

1. Turn ignition switch "ON".
2. Turn diagnostic mode selector on E.C.U. fully clockwise and wait the inspection lamps flash.
3. Count the number of the flashing time, and after the inspection lamps have flashed the number of the required mode, turn diagnostic mode selector fully counterclockwise immediately.

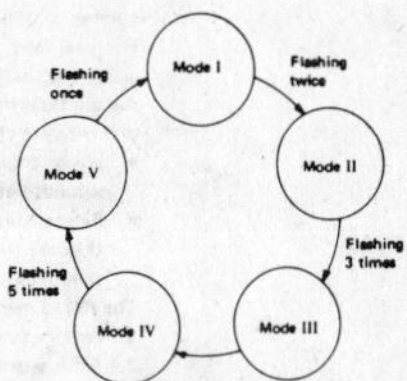
Mode I
Flashing twice
Mode II
Flashing 3 times
Mode III
Flashing 4 times
Mode IV
Flashing 5 times
Mode V
Flashing once

NOTE:
When the ignition switch is turned off during diagnosis, in each mode, and then turned back on again after the power to the E.C.U. has dropped off completely, the diagnosis will automatically return to Mode I.
The stored memory would be lost if:

1. Battery terminal is disconnected.
2. After selecting Mode III, Mode IV is selected.
   However, if the diagnostic mode selector is kept turned fully clockwise, it will continue to change in the order of Mode I → II → III → IV → V → I ... etc., and in this state the stored memory will not be erased.

ECCS DIAGNOSTIC CHART—1989 PULSAR

CHECK
CHECK ENGINE LIGHT

Description (Cont'd)

CHECK ENGINE LIGHT (For California only)

This vehicle has a check engine light on instrument panel. This light comes ON under the following conditions:

1) When ignition switch is turned "ON" (for bulb check).
2) When systems related to emission performance malfunction in Mode I (with engine running).
- **This check engine light always illuminates and is synchronous with red L.E.D.**
- **Malfunction systems related to emission performance can be detected by self-diagnosis, and they are clarified as self-diagnostic codes in Mode III.**

3) Check engine light will come "ON" only when malfunction is sensed.
   The check engine light will turn off when normal operation is resumed. Mode III memory must be cleared as the contents remain stored.

Code No.	Malfunction
12	Air flow meter circuit
13	Water temperature sensor circuit
14	Vehicle speed sensor circuit
23	Idle switch circuit
31	E.C.U. (E.C.C.S. control unit)
32	E.G.R. function
33	Exhaust gas sensor circuit
35	Exhaust gas temperature sensor circuit
43	Throttle sensor circuit
45	Injector leak

Use the following diagnostic flowchart to check and repair a malfunctioning system.

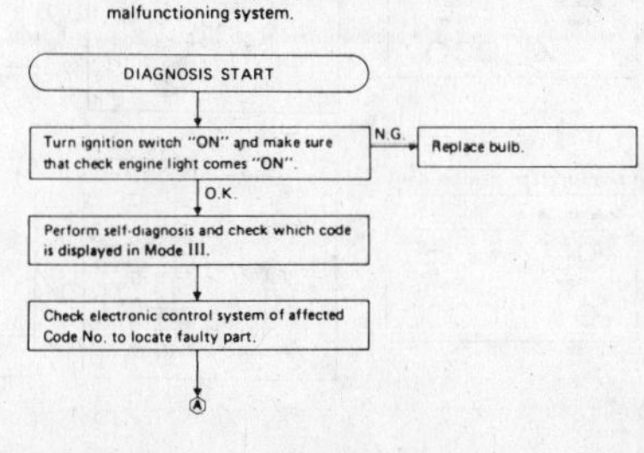
DIAGNOSIS START
Turn ignition switch "ON" and make sure that check engine light comes "ON".
N.G.
Replace bulb.
O.K.
Perform self-diagnosis and check which code is displayed in Mode III.
Check electronic control system of affected Code No. to locate faulty part.
Ⓐ

Description (Cont'd)

Ⓐ

Repair or replace faulty part.

Reinstall any part removed.

Erase the self-diagnosis memory.
- **Methods of erasing memories differ with systems. Read the manual before diagnosing systems.**

Perform driving test.
Make sure that check engine light does not come "ON" during this test.
- **After repairs, test drive to check that check engine light does not come on.**
- **Test driving modes differ with systems. Read the manual before test driving.**

DIAGNOSIS END

ECCS SELF-DIAGNOSTIC MODES I AND II—1989 PULSAR

Modes I & II —Mixture Ratio Feedback Control Monitors A & B

In these modes, the control unit provides the Air-fuel ratio monitor presentation and the Air-fuel ratio feedback coefficient monitor presentation.

Mode	LED	Engine stopped (Ignition switch "ON")	Engine running			
			Open loop condition	Closed loop condition		
Mode I (Monitor A)	Green	ON	*Remains ON or OFF	Blinks		
	Red	ON	Except for California model • OFF	For California model • ON: when CHECK ENGINE LIGHT ITEMS are stored in the E.C.U. • OFF: except for the above condition		
Mode II (Monitor B)	Green	ON	*Remains ON or OFF	Blinks		
	Red	OFF	*Remains ON or OFF (synchronous with green LED)	Compensating mixture ratio		
				More than 5% rich	Between 5% lean and 5% rich	More
				OFF	Synchronized with green LED	Remains ON

*: Maintains conditions just before switching to open loop

ECCS SELF-DIAGNOSTIC MODE III—1989 PULSAR

Mode III — Self-diagnostic System

The E.C.U. constantly monitors the function of these sensors and actuators, regardless of ignition key position. If a malfunction occurs, the information is stored in the E.C.U. and can be retrieved from the memory by turning on the diagnostic mode selector, located on the side of the E.C.U. When activated, the malfunction is indicated by flashing a red and a green L.E.D. (Light Emitting Diode), also located on the E.C.U. Since all the self-diagnostic results are stored in the E.C.U.'s memory even intermittent malfunctions can be diagnosed.

A malfunctioning part's group is indicated by the number of both the red and the green L.E.D.s flashing. First, the red L.E.D. flashes and the green flashes follow. The red L.E.D. refers to the number of tens while the green one refers to the number of units. For example, when the red L.E.D. flashes once and then the green one flashes twice, this means the number "12" showing the air flow meter signal is malfunctioning. In this way, all the problems are classified by the code numbers.

- **When engine fails to start, crank engine more than two seconds before starting self-diagnosis.**
- **Before starting self-diagnosis, do not erase stored memory. If doing so, self-diagnosis function for intermittent malfunctions would be lost.**

The stored memory would be lost if:

1. **Battery terminal is disconnected.**
2. **After selecting Mode III, Mode IV is selected.**

DISPLAY CODE TABLE

Code No.	Detected items	California	Non-California
11	Crank angle sensor circuit	X	X
12	Air flow meter circuit	X*	X
13	Water temperature sensor circuit	X*	X
14	Vehicle speed sensor circuit	X*	X
21	Ignition signal missing in primary coil	X	X
23	Idle switch circuit	X*	X
25	A.A.C. valve circuit	X	X
31	E.C.U. (E.C.C.S. control unit)	X*	X
32	E.G.R. function	X*	–
33	Exhaust gas sensor circuit	X*	X
35	Exhaust gas temperature sensor circuit	X*	–
43	Throttle sensor circuit	X*	X
45	Injector leak	X*	–
55	No malfunctioning in the above circuit	X	X

X: Available
–: Not available
*: Check engine light item

Mode III — Self-diagnostic System (Cont'd)

RETENTION OF DIAGNOSTIC RESULTS

The diagnostic result is retained in E.C.U. memory until the starter is operated fifty times after a diagnostic item is judged to be malfunctioning. The diagnostic result will then be cancelled automatically. If a diagnostic item which has been judged to be malfunctioning and stored in memory is again judged to be malfunctioning before the starter is operated fifty times, the second result will replace the previous one. It will be stored in E.C.U. memory until the starter is operated fifty times more.

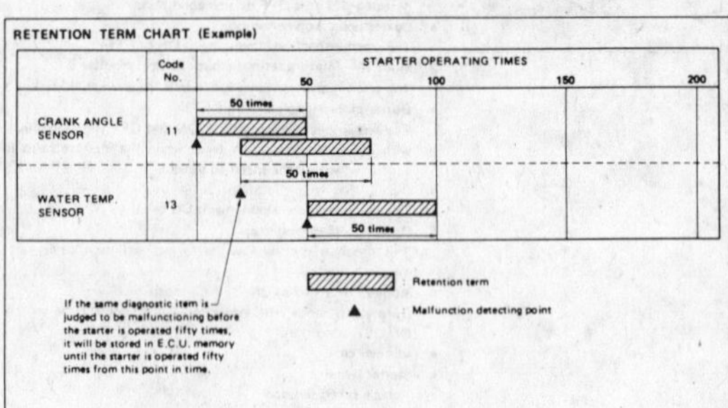
RETENTION TERM CHART (Example)
Code No.
STARTER OPERATING TIMES
50
100
150
200
CRANK ANGLE SENSOR
11
50 times
WATER TEMP. SENSOR
13
50 times
50 times
If the same diagnostic item is judged to be malfunctioning before the starter is operated fifty times, it will be stored in E.C.U. memory until the starter is operated fifty times from this point in time.
Retention term
Malfunction detecting point

ECCS SELF-DIAGNOSTIC MODE III PROCEDURE—1989 PULSAR

ECCS SELF-DIAGNOSTIC MODE III DISPLAY CODES—1989 PULSAR

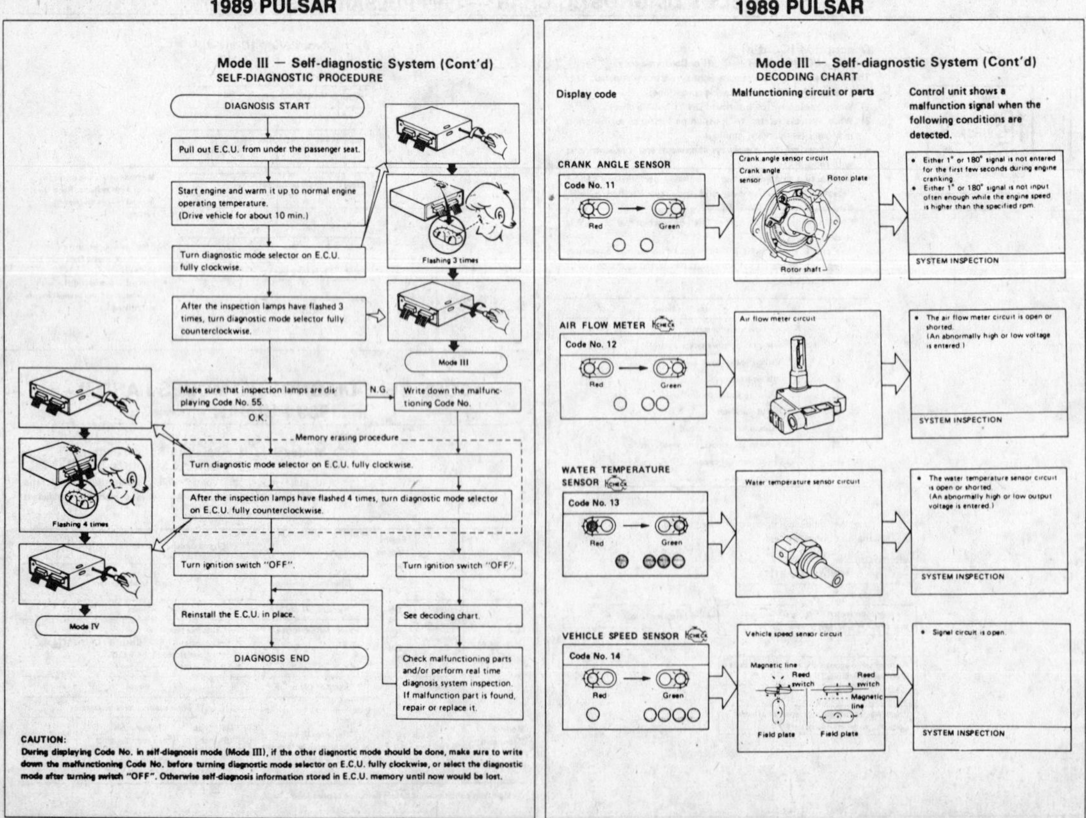
Mode III — Self-diagnostic System (Cont'd)
SELF-DIAGNOSTIC PROCEDURE
DIAGNOSIS START
Pull out E.C.U. from under the passenger seat.
Start engine and warm it up to normal engine operating temperature. (Drive vehicle for about 10 min.)
Turn diagnostic mode selector on E.C.U. fully clockwise.
Flashing 3 times
After the inspection lamps have flashed 3 times, turn diagnostic mode selector fully counterclockwise.
Mode III
Make sure that inspection lamps are displaying Code No. 55.
N.G.
Write down the malfunctioning Code No.
O.K.
Memory erasing procedure
Turn diagnostic mode selector on E.C.U. fully clockwise.
After the inspection lamps have flashed 4 times, turn diagnostic mode selector on E.C.U. fully counterclockwise.
Flashing 4 times
Turn ignition switch "OFF".
Turn ignition switch "OFF".
Reinstall the E.C.U. in place.
See decoding chart.
Mode IV
DIAGNOSIS END
Check malfunctioning parts and/or perform real time diagnosis system inspection. If malfunction part is found, repair or replace it.
CAUTION:
During displaying Code No. in self-diagnosis mode (Mode III), if the other diagnostic mode should be done, make sure to write down the malfunctioning Code No. before turning diagnostic mode selector on E.C.U. fully clockwise, or select the diagnostic mode after turning switch "OFF". Otherwise self-diagnosis information stored in E.C.U. memory until now would be lost.
Mode III — Self-diagnostic System (Cont'd)
DECODING CHART
Display code
Malfunctioning circuit or parts
Control unit shows a malfunction signal when the following conditions are detected.
CRANK ANGLE SENSOR
Code No. 11
Red
Green
Crank angle sensor circuit
Crank angle sensor
Rotor plate
Rotor shaft
Either 1° or 180° signal is not entered for the first few seconds during engine cranking.
Either 1° or 180° signal is not input often enough while the engine speed is higher than the specified rpm.
SYSTEM INSPECTION
AIR FLOW METER
Code No. 12
Red
Green
Air flow meter circuit
The air flow meter circuit is open or shorted. (An abnormally high or low voltage is entered.)
SYSTEM INSPECTION
WATER TEMPERATURE SENSOR
Code No. 13
Red
Green
Water temperature sensor circuit
The water temperature sensor circuit is open or shorted. (An abnormally high or low output voltage is entered.)
SYSTEM INSPECTION
VEHICLE SPEED SENSOR
Code No. 14
Red
Green
Vehicle speed sensor circuit
Magnetic line
Reed switch
Reed switch
Magnetic line
Field plate
Field plate
Signal circuit is open.
SYSTEM INSPECTION

ECCS SELF-DIAGNOSTIC MODE III DISPLAY CODES—1989 PULSAR

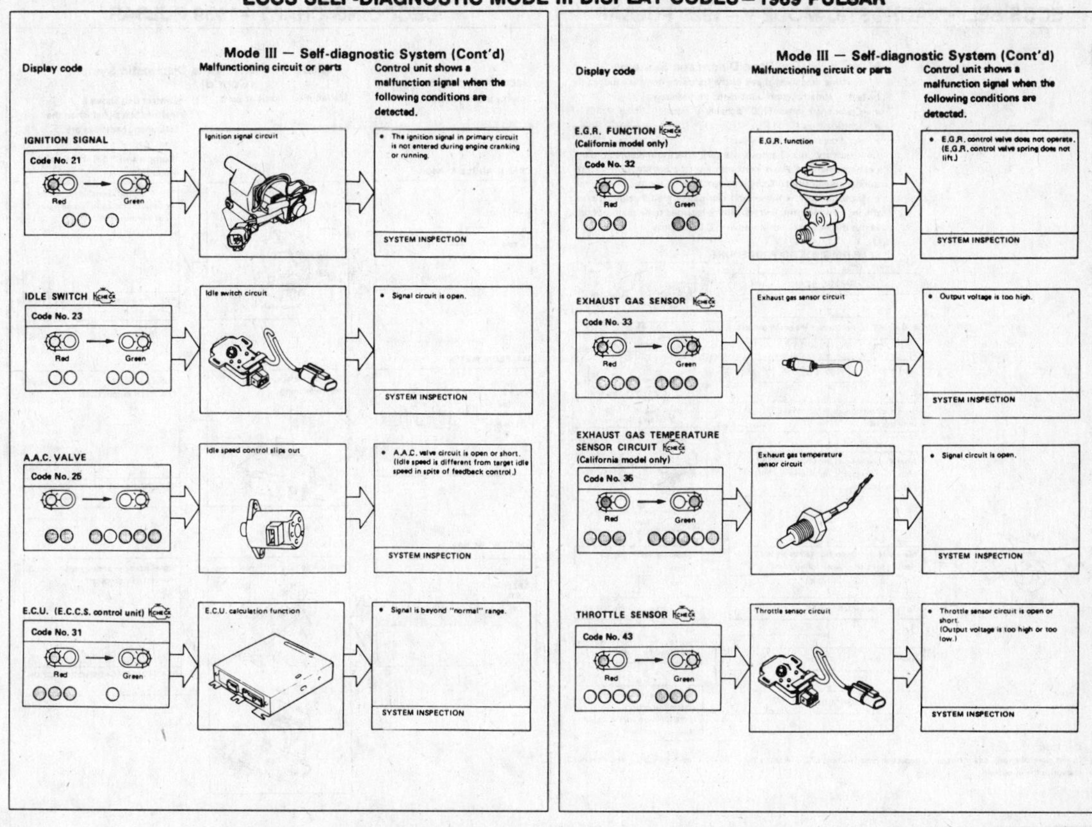
Mode III — Self-diagnostic System (Cont'd)
Display code
Malfunctioning circuit or parts
Control unit shows a malfunction signal when the following conditions are detected.
IGNITION SIGNAL
Code No. 21
Red
Green
Ignition signal circuit
• The ignition signal in primary circuit is not entered during engine cranking or running.
SYSTEM INSPECTION
IDLE SWITCH
Code No. 23
Idle switch circuit
• Signal circuit is open.
A.A.C. VALVE
Code No. 25
Idle speed control slips out
• A.A.C. valve circuit is open or short. (Idle speed is different from target idle speed in spite of feedback control.)
E.C.U. (E.C.C.S. control unit)
Code No. 31
E.C.U. calculation function
• Signal is beyond "normal" range.
E.G.R. FUNCTION
(California model only)
Code No. 32
E.G.R. function
• E.G.R. control valve does not operate. (E.G.R. control valve spring does not lift.)
EXHAUST GAS SENSOR
Code No. 33
Exhaust gas sensor circuit
• Output voltage is too high.
EXHAUST GAS TEMPERATURE SENSOR CIRCUIT
(California model only)
Code No. 35
Exhaust gas temperature sensor circuit
• Signal circuit is open.
THROTTLE SENSOR
Code No. 43
Throttle sensor circuit
• Throttle sensor circuit is open or short.
(Output voltage is too high or too low.)

ECCS SELF-DIAGNOSTIC MODE III DISPLAY CODES—1989 PULSAR

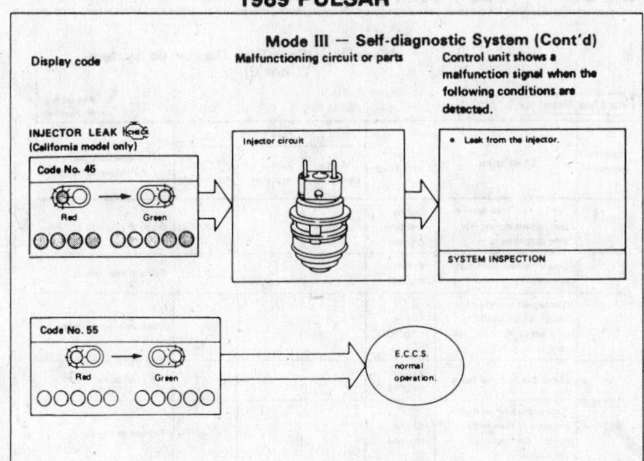
Mode III — Self-diagnostic System (Cont'd)
Display code
Malfunctioning circuit or parts
Control unit shows a malfunction signal when the following conditions are detected.
INJECTOR LEAK
(California model only)
Code No. 45
Red
Green
Injector circuit
• Leak from the injector.
SYSTEM INSPECTION
Code No. 55
E.C.C.S. normal operation.

ECCS SELF-DIAGNOSTIC MODE IV—1989 PULSAR

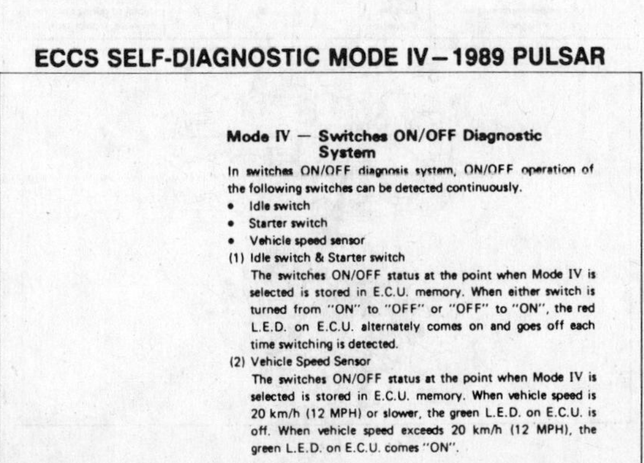

Mode IV — Switches ON/OFF Diagnostic System

In switches ON/OFF diagnosis system, ON/OFF operation of the following switches can be detected continuously.

- Idle switch
- Starter switch
- Vehicle speed sensor

(1) Idle switch & Starter switch

The switches ON/OFF status at the point when Mode IV is selected is stored in E.C.U. memory. When either switch is turned from "ON" to "OFF" or "OFF" to "ON", the red L.E.D. on E.C.U. alternately comes on and goes off each time switching is detected.

(2) Vehicle Speed Sensor

The switches ON/OFF status at the point when Mode IV is selected is stored in E.C.U. memory. When vehicle speed is 20 km/h (12 MPH) or slower, the green L.E.D. on E.C.U. is off. When vehicle speed exceeds 20 km/h (12 MPH), the green L.E.D. on E.C.U. comes "ON".

ECCS SELF-DIAGNOSTIC MODE IV—1989 PULSAR

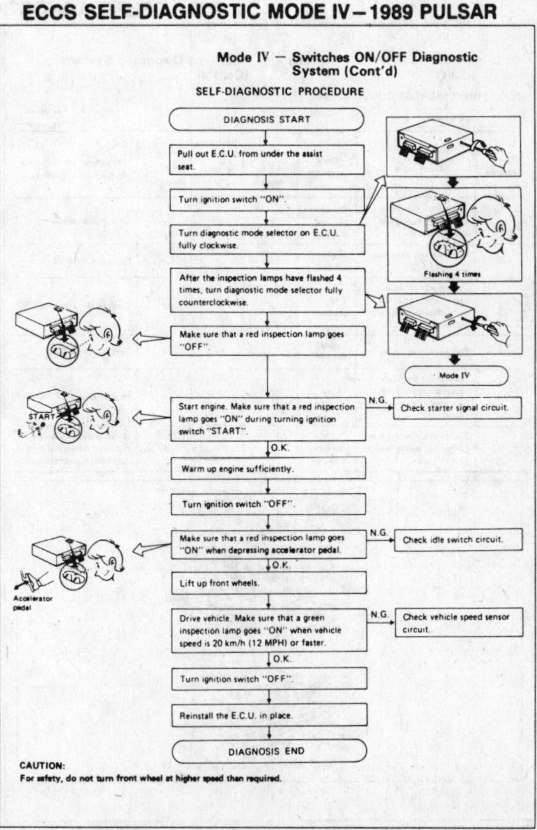
Mode IV — Switches ON/OFF Diagnostic System (Cont'd)
SELF-DIAGNOSTIC PROCEDURE
DIAGNOSIS START
Pull out E.C.U. from under the assist seat.
Turn ignition switch "ON".
Turn diagnostic mode selector on E.C.U. fully clockwise.
After the inspection lamps have flashed 4 times, turn diagnostic mode selector fully counterclockwise.
Flashing 4 times
Make sure that a red inspection lamp goes "OFF".
Mode IV
START
Start engine. Make sure that a red inspection lamp goes "ON" during turning ignition switch "START".
N.G.
Check starter signal circuit.
O.K.
Warm up engine sufficiently.
Turn ignition switch "OFF".
Make sure that a red inspection lamp goes "ON" when depressing accelerator pedal.
Check idle switch circuit.
Accelerator pedal
Lift up front wheels.
Drive vehicle. Make sure that a green inspection lamp goes "ON" when vehicle speed is 20 km/h (12 MPH) or faster.
Check vehicle speed sensor circuit.
Turn ignition switch "OFF".
Reinstall the E.C.U. in place.
DIAGNOSIS END
CAUTION:
For safety, do not turn front wheel at higher speed than required.

ECCS SELF-DIAGNOSTIC MODE V—1989 PULSAR

Mode V — Real Time Diagnostic System

In real time diagnosis, if any of the following items are judged to be faulty, a malfunction is indicated immediately.

- Crank angle sensor (180° signal & 1° signal)
- Ignition signal
- Air flow meter output signal

Consequently, this diagnosis is a very effective measure to diagnose whether the above systems cause the malfunction or not, during driving test. Compared with self-diagnosis, real time diagnosis is very sensitive, and can detect malfunctioning conditions in a moment. Further, items regarded to be malfunctions in this diagnosis are not stored in E.C.U. memory.

SELF-DIAGNOSTIC PROCEDURE

DIAGNOSIS START

Pull out E.C.U. from under the passenger seat.

Start engine.

Turn diagnostic mode selector on E.C.U. fully clockwise.

Flashing 5 times

After the inspection lamps have flashed 5 times, turn diagnostic mode selector fully counterclockwise.

Mode V

Make sure that inspection lamps are not flashing for 5 min. when idling or racing.

N.G.: If flashing, count no. of flashes. → Turn ignition switch "OFF". → See decoding chart. → Perform real time-diagnosis system inspection. If malfunction part is found, repair or replace it.

O.K.: Turn ignition switch "OFF".

Reinstall the E.C.U. in place.

DIAGNOSIS END

CAUTION:
In real time diagnosis, pay attention to inspection lamp flashing. E.C.U. displays the malfunction code only once, and does not memorize the inspection.

ECCS SELF-DIAGNOSTIC MODE V DECODING CHART—1989 PULSAR

Mode V — Real Time Diagnostic System (Cont'd)

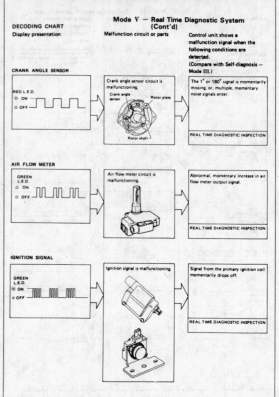

DECODING CHART

Display presentation	Malfunction circuit or parts	Control unit shows a malfunction signal when the following conditions are detected. (Compare with Self-diagnosis — Mode III.)
CRANK ANGLE SENSOR RED L.E.D. ON OFF	Crank angle sensor circuit is malfunctioning. Crank angle sensor Rotor plate Rotor shaft	The 1° or 180° signal is momentarily missing, or, multiple, momentary noise signals enter. REAL TIME DIAGNOSTIC INSPECTION
AIR FLOW METER GREEN L.E.D. ON OFF	Air flow meter circuit is malfunctioning.	Abnormal, momentary increase in air flow meter output signal. REAL TIME DIAGNOSTIC INSPECTION
IGNITION SIGNAL GREEN L.E.D. ON OFF	Ignition signal is malfunctioning.	Signal from the primary ignition coil momentarily drops off. REAL TIME DIAGNOSTIC INSPECTION

ECCS SELF-DIAGNOSTIC MODE V—1989 PULSAR

Mode V — Real Time Diagnostic System (Cont'd)

REAL TIME DIAGNOSTIC INSPECTION

Crank Angle Sensor

X: Available
–: Not available

Check sequence	Check items	Check conditions	Check parts: Middle connector	Check parts: Sensor & actuator	Check parts: E.C.U. connectors	If malfunction, perform the following items.
1	Tap harness connector or component during real time diagnosis.	During real time diagnosis	X	X	X	Go to check item 2.
2	Check harness continuity at connector.	Engine stopped	X	–	–	Go to check item 3.
3	Disconnect harness connector, and then check dust adhesion to harness connector.	Engine stopped	X	–	X	Clean terminal surface.
4	Check pin terminal bend.	Engine stopped	–	–	X	Take out bend.
5	Reconnect harness connector and then recheck harness continuity at connector.	Engine stopped	X	–	–	Replace terminal.
6	Tap harness connector or component during real time diagnosis.	During real time diagnosis	X	X	X	If malfunction codes are displayed during real time diagnosis, replace terminal.

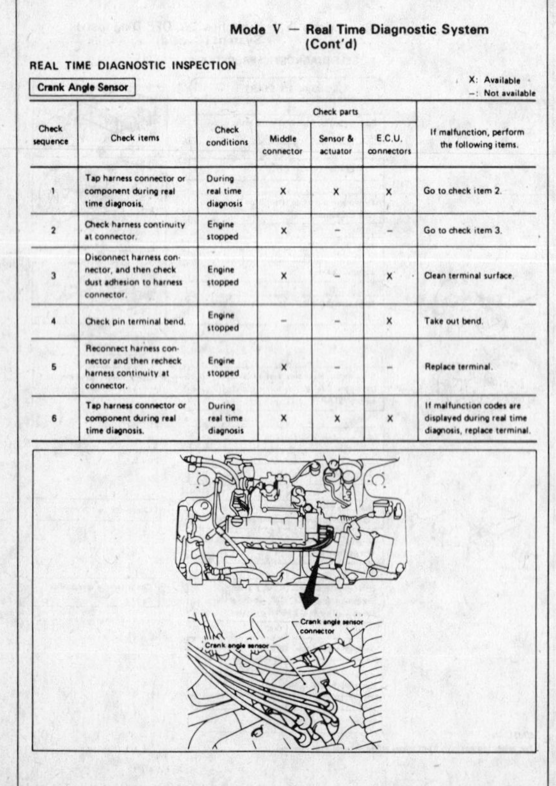
Crank angle sensor connector
Crank angle sensor

Mode V — Real Time Diagnostic System (Cont'd)

Air Flow Meter

X: Available
–: Not available

Check sequence	Check items	Check conditions	Check parts: Middle connector	Check parts: Sensor & actuator	Check parts: E.C.U. connectors	If malfunction, perform the following items.
1	Tap harness connector or component during real time diagnosis.	During real time diagnosis	X	X	X	Go to check item 2.
2	Check harness continuity at connector.	Engine stopped	X	–	–	Go to check item 3.
3	Disconnect harness connector, and then check dust adhesion to harness connector.	Engine stopped	X	–	X	Clean terminal surface.
4	Check pin terminal bend.	Engine stopped	–	–	X	Take out bend.
5	Reconnect harness connector and then recheck harness continuity at connector.	Engine stopped	X	–	–	Replace terminal.
6	Tap harness connector or component during real time diagnosis.	During real time diagnosis	X	X	X	If malfunction codes are displayed during real time diagnosis, replace terminal.

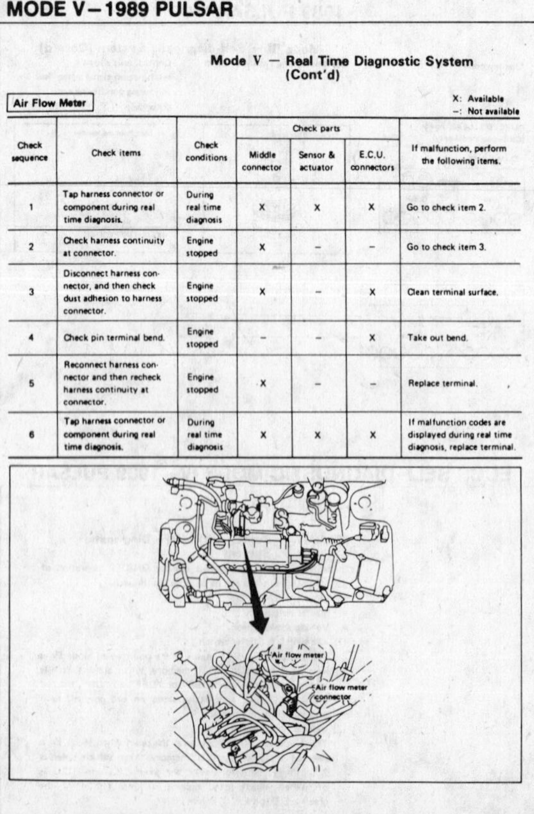
Air flow meter
Air flow meter connector

ECCS SELF-DIAGNOSTIC MODE V—1989 PULSAR

Mode V — Real Time Diagnostic System (Cont'd)

Ignition Signal

X: Available
–: Not available

Check sequence	Check items	Check conditions	Check parts: Middle connector	Check parts: Sensor & actuator	Check parts: E.C.U. connectors	If malfunction, perform the following items.
1	Tap harness connector or component during real time diagnosis.	During real time diagnosis	X	X	X	Go to check item 2.
2	Check harness continuity at connector.	Engine stopped	X	–	–	Go to check item 3.
3	Disconnect harness connector, and then check dust adhesion to harness connector.	Engine stopped	X	–	X	Clean terminal surface.
4	Check pin terminal bend.	Engine stopped	–	–	X	Take out bend.
5	Reconnect harness connector and then recheck harness continuity at connector.	Engine stopped	X	–	–	Replace terminal.
6	Tap harness connector or component during real time diagnosis.	During real time diagnosis	X	X	X	If malfunction codes are displayed during real time diagnosis, replace terminal.

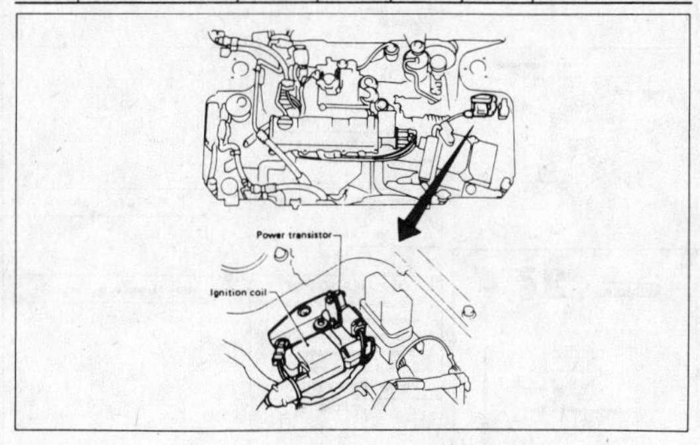
Power transistor
Ignition coil

ECCS SYSTEM INSPECTION CAUTION—1989 PULSAR

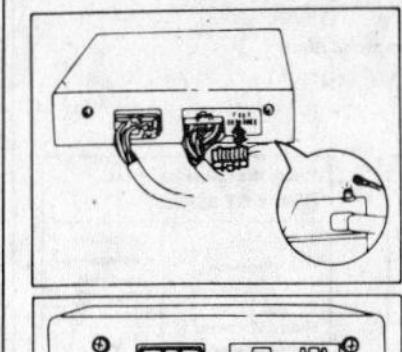

CAUTION:

1. Before connecting or disconnecting E.C.U. harness connector to or from any E.C.U., be sure to turn the ignition switch to the "OFF" position and disconnect the negative battery terminal in order not to damage E.C.U. as battery voltage is applied to E.C.U. even if ignition switch is turned off. Otherwise, there may be damage to the E.C.U.

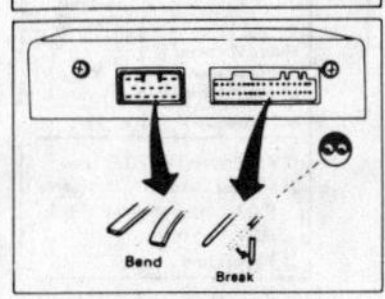
Bend
Break

2. When connecting pin connectors into E.C.U. or disconnecting them from E.C.U., take care not to damage pin terminal of E.C.U. (Bend or break).
3. Make sure that there are not any bends or breaks on E.C.U. pin terminal, when connecting pin connectors into E.C.U.

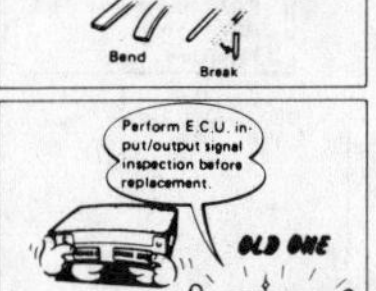
Perform E.C.U. input/output signal inspection before replacement.
OLD ONE

4. Before replacing E.C.U., perform E.C.U. input/output signal inspection and make sure whether E.C.U. functions properly or not.

5. After performing this "ELECTRONIC CONTROL SYSTEM INSPECTION", perform E.C.C.S. self-diagnosis and driving test.

ECCS SYSTEM INSPECTION CAUTION (CONT.)—1989 PULSAR

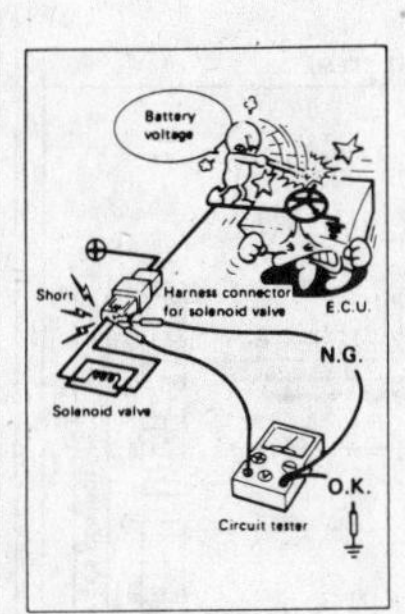
Battery voltage
Short
Harness connector for solenoid valve
E.C.U.
N.G.
Solenoid valve
O.K.
Circuit tester

6. When measuring supply voltage of E.C.U. controlled components with a circuit tester, separate one tester probe from the other.
   If the two tester probes accidentally make contact with each other during measurement, the circuit will be shorted, resulting in damage to the power transistor of the control unit.

7. When measuring voltage or resistance at connector with tester probes, there are two methods of measurement; one is done from terminal side and the other from harness side. Before measuring, confirm symbol mark again.
   HS : Inspection should be done from harness side.
   TS : Inspection should be done from terminal side.

Refer to GI section.

8. As for continuity check of joint connector, refer to EL section.
9. Key to symbols
   DISCONNECT : Check after disconnecting the connector to be measured.
   CONNECT : Check after connecting the connector to be measured.

Thin wire
Tester probe

10. Improve tester probe as shown to perform test easily.
11. For the first trouble-shooting procedure, perform POWER SOURCE & GROUND CIRCUIT FOR E.C.U. check.

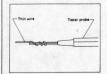

ECCS SELF-DIAGNOSTIC CHART—1989 PULSAR

POWER SOURCE & GROUND CIRCUIT FOR E.C.U. (Not self-diagnosis item)

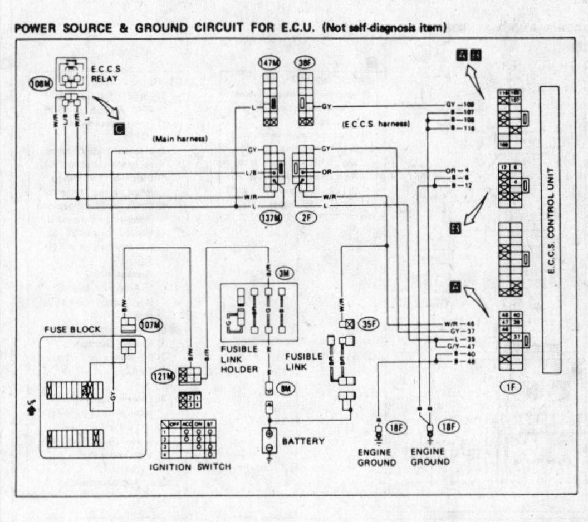
E.C.C.S RELAY
(Main harness)
(E.C.C.S. harness)
E.C.C.S. CONTROL UNIT
FUSE BLOCK
FUSIBLE LINK HOLDER
FUSIBLE LINK
BATTERY
IGNITION SWITCH
ENGINE GROUND
ENGINE GROUND

ECCS SELF-DIAGNOSTIC CHART—1989 PULSAR

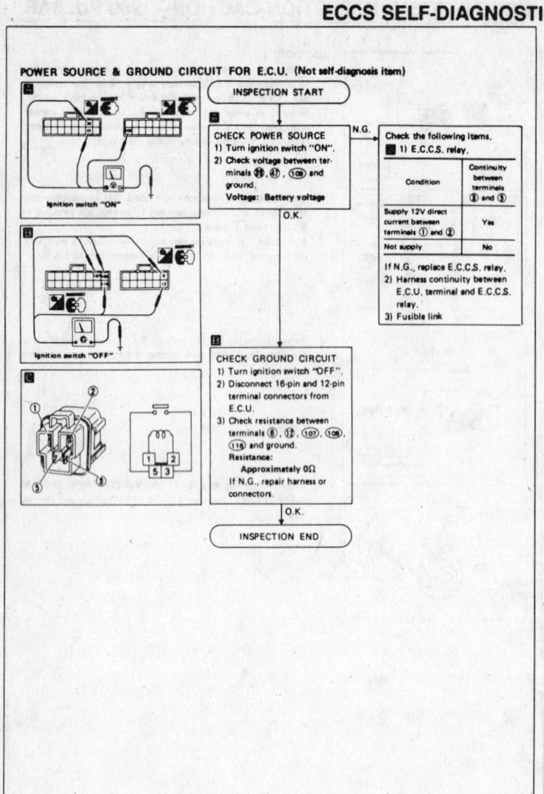
POWER SOURCE & GROUND CIRCUIT FOR E.C.U. (Not self-diagnosis item)
INSPECTION START
CHECK POWER SOURCE
1) Turn ignition switch "ON".
2) Check voltage between terminals ㊴, ㊼, ⑩⑧ and ground.
Voltage: Battery voltage
N.G.
Check the following items.
1) E.C.C.S. relay.
Condition
Continuity between terminals ③ and ⑤
Supply 12V direct current between terminals ① and ②
Yes
Not supply
No
If N.G., replace E.C.C.S. relay.
2) Harness continuity between E.C.U. terminal and E.C.C.S. relay.
3) Fusible link
O.K.
CHECK GROUND CIRCUIT
1) Turn ignition switch "OFF".
2) Disconnect 16-pin and 12-pin terminal connectors from E.C.U.
3) Check resistance between terminals ⑥, ⑫, ⑩⑦, ⑩⑧, ⑪⑥ and ground.
Resistance: Approximately 0Ω
If N.G., repair harness or connectors.
O.K.
INSPECTION END
Ignition switch "ON"
Ignition switch "OFF"

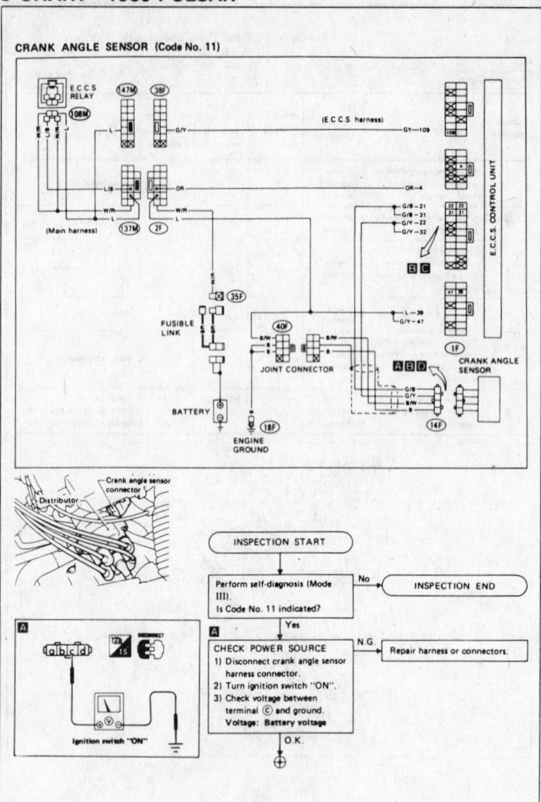
CRANK ANGLE SENSOR (Code No. 11)
ECCS RELAY
(E.C.C.S. harness)
(Main harness)
FUSIBLE LINK
JOINT CONNECTOR
BATTERY
ENGINE GROUND
CRANK ANGLE SENSOR
E.C.C.S. CONTROL UNIT
Crank angle sensor connector
Distributor
INSPECTION START
Perform self-diagnosis (Mode III).
Is Code No. 11 indicated?
No
INSPECTION END
Yes
CHECK POWER SOURCE
1) Disconnect crank angle sensor harness connector.
2) Turn ignition switch "ON".
3) Check voltage between terminal © and ground.
Voltage: Battery voltage
N.G.
Repair harness or connectors.
O.K.
Ignition switch "ON"

ECCS SELF-DIAGNOSTIC CHART—1989 PULSAR

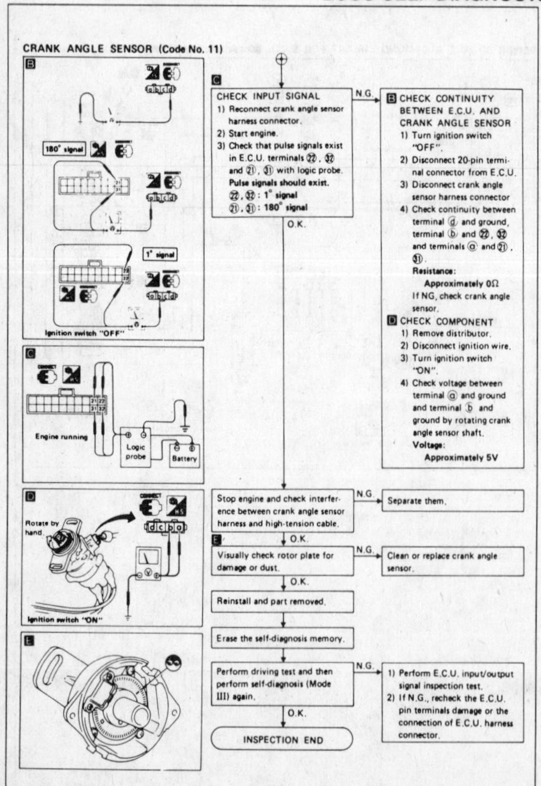
CRANK ANGLE SENSOR (Code No. 11)
CHECK INPUT SIGNAL
1) Reconnect crank angle sensor harness connector.
2) Start engine.
3) Check that pulse signals exist in E.C.U. terminals ㉒, ㉜ and ㉑, ㉛ with logic probe.
Pulse signals should exist.
㉒, ㉜: 1° signal
㉑, ㉛: 180° signal
N.G.
CHECK CONTINUITY BETWEEN E.C.U. AND CRANK ANGLE SENSOR
1) Turn ignition switch "OFF".
2) Disconnect 20-pin terminal connector from E.C.U.
3) Disconnect crank angle sensor harness connector
4) Check continuity between terminal ⓓ and ground, terminal ⓑ and ㉒, ㉜ and terminals ⓐ and ㉑, ㉛.
Resistance: Approximately 0Ω
If NG, check crank angle sensor.
CHECK COMPONENT
1) Remove distributor.
2) Disconnect ignition wire.
3) Turn ignition switch "ON".
4) Check voltage between terminal ⓐ and ground and terminal ⓑ and ground by rotating crank angle sensor shaft.
Voltage: Approximately 5V
O.K.
Stop engine and check interference between crank angle sensor harness and high-tension cable.
Separate them.
Visually check rotor plate for damage or dust.
Clean or replace crank angle sensor.
Reinstall and part removed.
Erase the self-diagnosis memory.
Perform driving test and then perform self-diagnosis (Mode III) again.
1) Perform E.C.U. input/output signal inspection test.
2) If N.G., recheck the E.C.U. pin terminals damage or the connection of E.C.U. harness connector.
INSPECTION END
180° signal
1° signal
Ignition switch "OFF"
Engine running
Logic probe
Battery
Rotate by hand
Ignition switch "ON"

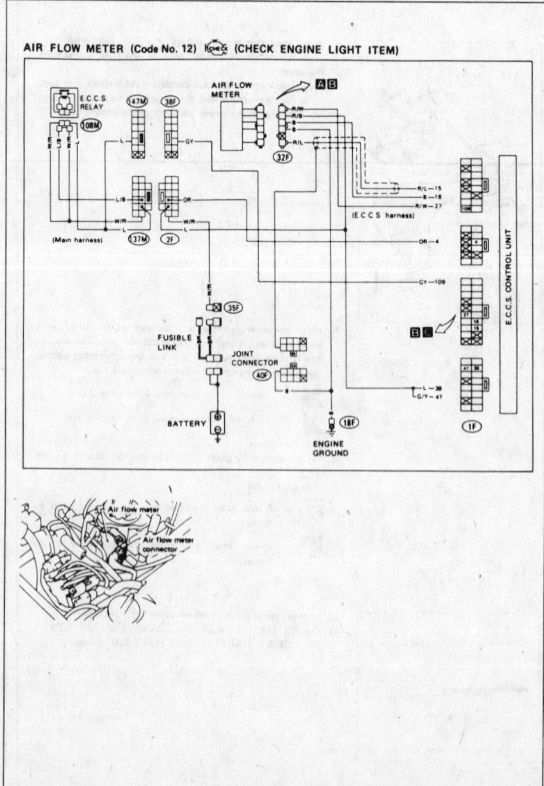
AIR FLOW METER (Code No. 12) (CHECK ENGINE LIGHT ITEM)
ECCS RELAY
AIR FLOW METER
(E.C.C.S. harness)
(Main harness)
FUSIBLE LINK
JOINT CONNECTOR
BATTERY
ENGINE GROUND
E.C.C.S. CONTROL UNIT
Air flow meter
Air flow meter connector

ECCS SELF-DIAGNOSTIC CHART—1989 PULSAR

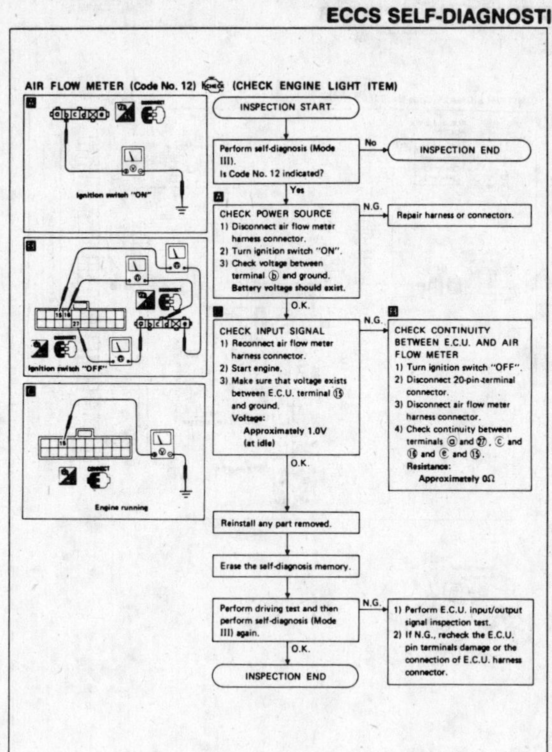
AIR FLOW METER (Code No. 12) (CHECK ENGINE LIGHT ITEM)
Ignition switch "ON"
Ignition switch "OFF"
Engine running
INSPECTION START
Perform self-diagnosis (Mode III).
Is Code No. 12 indicated?
No
INSPECTION END
Yes
CHECK POWER SOURCE
1) Disconnect air flow meter harness connector.
2) Turn ignition switch "ON".
3) Check voltage between terminal ⓑ and ground.
Battery voltage should exist.
N.G.
Repair harness or connectors.
O.K.
CHECK INPUT SIGNAL
1) Reconnect air flow meter harness connector.
2) Start engine.
3) Make sure that voltage exists between E.C.U. terminal ⑮ and ground.
Voltage:
Approximately 1.0V (at idle)
N.G.
CHECK CONTINUITY BETWEEN E.C.U. AND AIR FLOW METER
1) Turn ignition switch "OFF".
2) Disconnect 20-pin terminal connector.
3) Disconnect air flow meter harness connector.
4) Check continuity between terminals ⓐ and ㉗, ⓒ and ⑯ and ⓔ and ⑮.
Resistance:
Approximately 0Ω
O.K.
Reinstall any part removed.
Erase the self-diagnosis memory.
Perform driving test and then perform self-diagnosis (Mode III) again.
N.G.
1) Perform E.C.U. input/output signal inspection test.
2) If N.G., recheck the E.C.U. pin terminals damage or the connection of E.C.U. harness connector.
O.K.
INSPECTION END

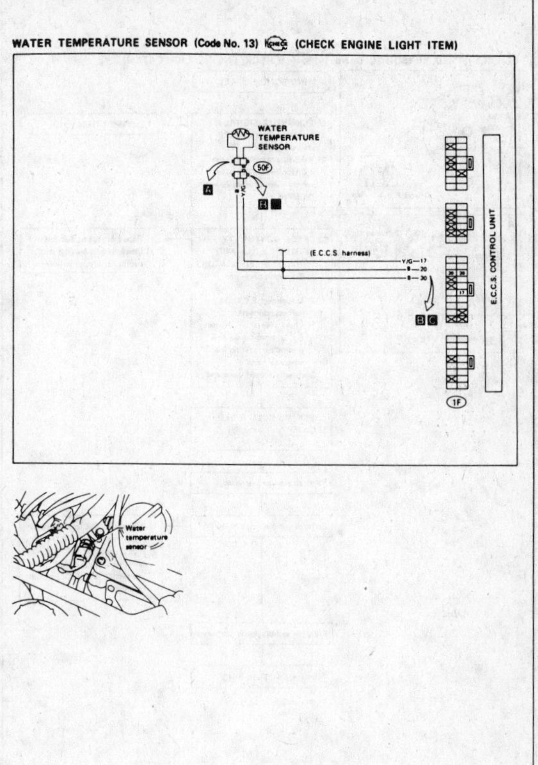
WATER TEMPERATURE SENSOR (Code No. 13) (CHECK ENGINE LIGHT ITEM)
WATER TEMPERATURE SENSOR
50F
(E.C.C.S. harness)
1F
E.C.C.S. CONTROL UNIT
Water temperature sensor

ECCS SELF-DIAGNOSTIC CHART—1989 PULSAR

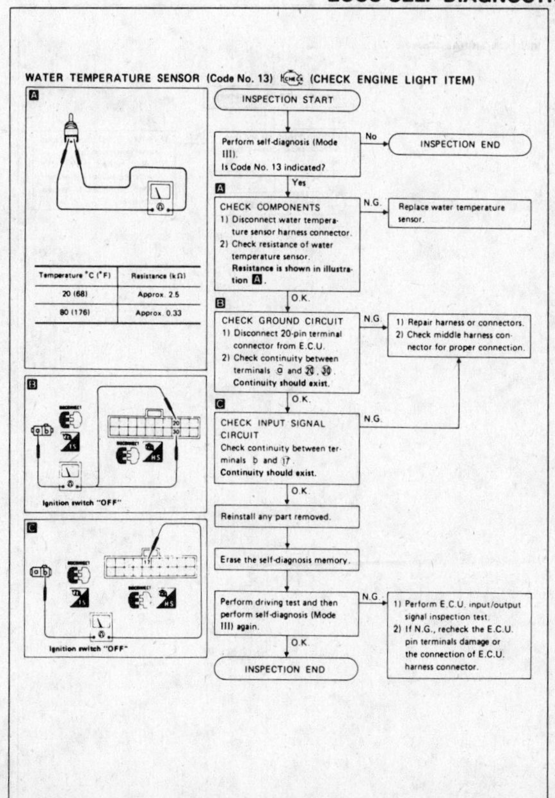
WATER TEMPERATURE SENSOR (Code No. 13) (CHECK ENGINE LIGHT ITEM)
Ignition switch "OFF"
Ignition switch "OFF"
INSPECTION START
Perform self-diagnosis (Mode III).
Is Code No. 13 indicated?
No
INSPECTION END
Yes
CHECK COMPONENTS
1) Disconnect water temperature sensor harness connector.
2) Check resistance of water temperature sensor.
Resistance is shown in illustration A.
N.G.
Replace water temperature sensor.
O.K.
CHECK GROUND CIRCUIT
1) Disconnect 20-pin terminal connector from E.C.U.
2) Check continuity between terminals ⓐ and ⑳, ㉚.
Continuity should exist.
N.G.
1) Repair harness or connectors.
2) Check middle harness connector for proper connection.
O.K.
CHECK INPUT SIGNAL CIRCUIT
Check continuity between terminals ⓑ and ⑰.
Continuity should exist.
N.G.
O.K.
Reinstall any part removed.
Erase the self-diagnosis memory.
Perform driving test and then perform self-diagnosis (Mode III) again.
N.G.
1) Perform E.C.U. input/output signal inspection test.
2) If N.G., recheck the E.C.U. pin terminals damage or the connection of E.C.U. harness connector.
O.K.
INSPECTION END

Temperature °C (°F)	Resistance (kΩ)
20 (68)	Approx. 2.5
80 (176)	Approx. 0.33

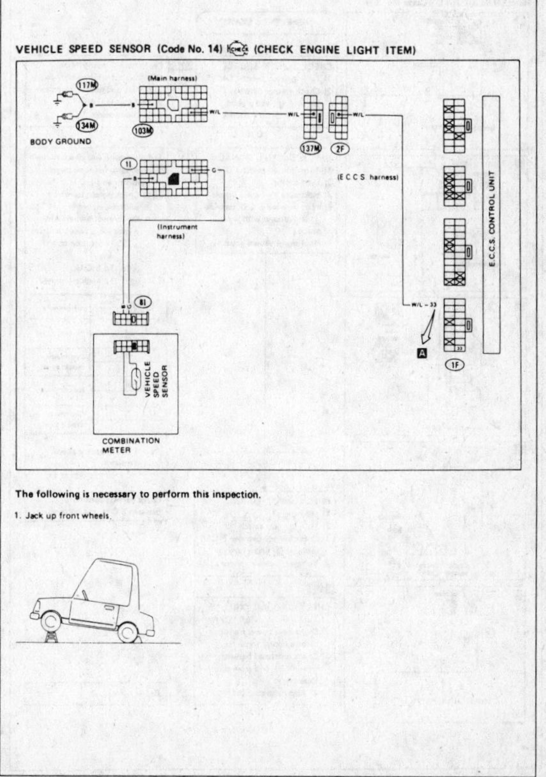
VEHICLE SPEED SENSOR (Code No. 14) (CHECK ENGINE LIGHT ITEM)
(Main harness)
117M
134M
103M
BODY GROUND
137M
2F
(E.C.C.S. harness)
(Instrument harness)
81
VEHICLE SPEED SENSOR
COMBINATION METER
E.C.C.S. CONTROL UNIT
1F

The following is necessary to perform this inspection.

1. Jack up front wheels.

ECCS SELF-DIAGNOSTIC CHART—1989 PULSAR

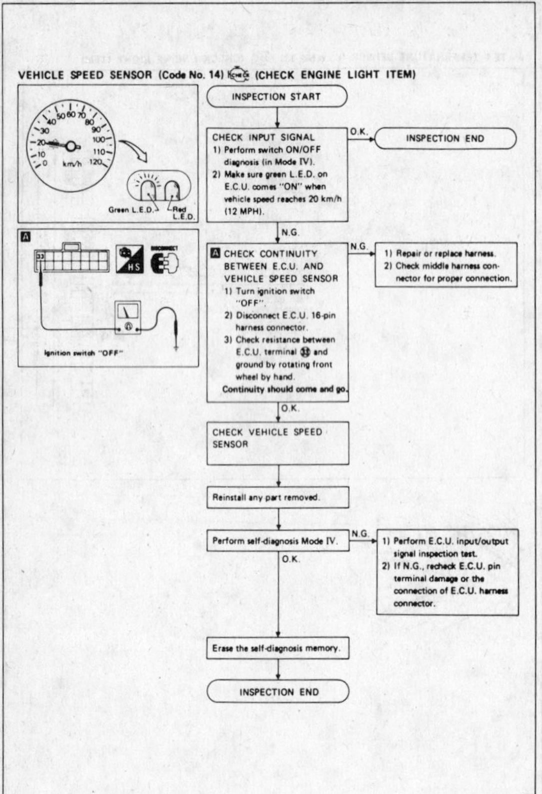
VEHICLE SPEED SENSOR (Code No. 14) (CHECK ENGINE LIGHT ITEM)
km/h
Green L.E.D.
Red L.E.D.
Ignition switch "OFF"
INSPECTION START
CHECK INPUT SIGNAL
1) Perform switch ON/OFF diagnosis (in Mode IV).
2) Make sure green L.E.D. on E.C.U. comes "ON" when vehicle speed reaches 20 km/h (12 MPH).
O.K.
INSPECTION END
N.G.
A CHECK CONTINUITY BETWEEN E.C.U. AND VEHICLE SPEED SENSOR
1) Turn ignition switch "OFF".
2) Disconnect E.C.U. 16-pin harness connector.
3) Check resistance between E.C.U. terminal 33 and ground by rotating front wheel by hand.
Continuity should come and go.
N.G.
1) Repair or replace harness.
2) Check middle harness connector for proper connection.
O.K.
CHECK VEHICLE SPEED SENSOR
Reinstall any part removed.
Perform self-diagnosis Mode IV.
N.G.
1) Perform E.C.U. input/output signal inspection test.
2) If N.G., recheck E.C.U. pin terminal damage or the connection of E.C.U. harness connector.
O.K.
Erase the self-diagnosis memory.
INSPECTION END

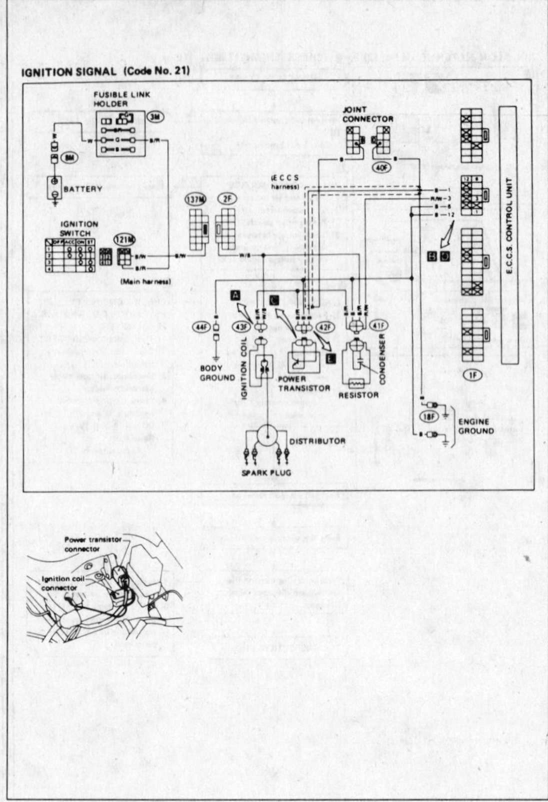
IGNITION SIGNAL (Code No. 21)
FUSIBLE LINK HOLDER
JOINT CONNECTOR
BATTERY
ECCS harness
IGNITION SWITCH
(Main harness)
E.C.C.S CONTROL UNIT
BODY GROUND
IGNITION COIL
POWER TRANSISTOR
CONDENSER
RESISTOR
DISTRIBUTOR
SPARK PLUG
ENGINE GROUND
Power transistor connector
Ignition coil connector

ECCS SELF-DIAGNOSTIC CHART—1989 PULSAR

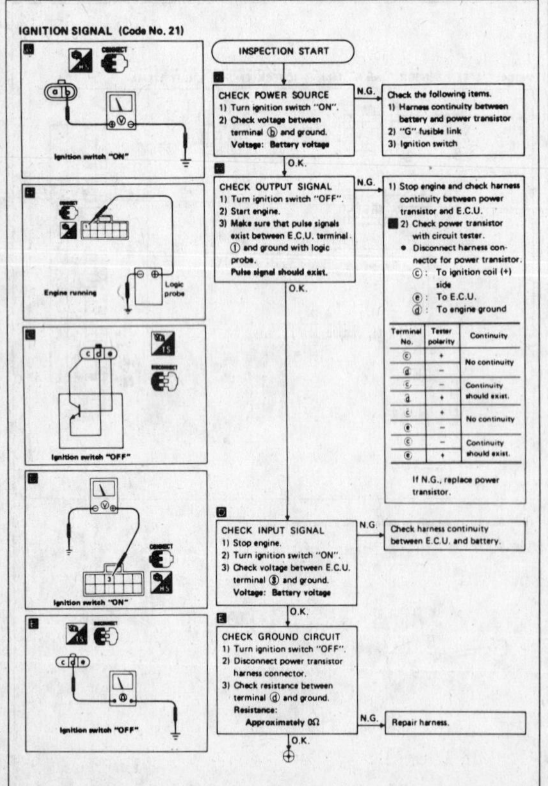
IGNITION SIGNAL (Code No. 21)
Ignition switch "ON"
Engine running
Logic probe
Ignition switch "OFF"
Ignition switch "ON"
Ignition switch "OFF"
INSPECTION START
CHECK POWER SOURCE
1) Turn ignition switch "ON".
2) Check voltage between terminal ⓑ and ground.
Voltage: Battery voltage
N.G.
Check the following items.
1) Harness continuity between battery and power transistor
2) "G" fusible link
3) Ignition switch
O.K.
CHECK OUTPUT SIGNAL
1) Turn ignition switch "OFF".
2) Start engine.
3) Make sure that pulse signals exist between E.C.U. terminal ① and ground with logic probe.
Pulse signal should exist.
N.G.
1) Stop engine and check harness continuity between power transistor and E.C.U.
2) Check power transistor with circuit tester.
• Disconnect harness connector for power transistor.
ⓒ: To ignition coil (+) side
ⓔ: To E.C.U.
ⓓ: To engine ground
Terminal No. | Tester polarity | Continuity
ⓒ | + | No continuity
ⓓ | − |
ⓒ | − | Continuity should exist.
ⓓ | + |
ⓒ | + | No continuity
ⓔ | − |
ⓒ | − | Continuity should exist.
ⓔ | + |
If N.G., replace power transistor.
O.K.
CHECK INPUT SIGNAL
1) Stop engine.
2) Turn ignition switch "ON".
3) Check voltage between E.C.U. terminal ③ and ground.
Voltage: Battery voltage
N.G.
Check harness continuity between E.C.U. and battery.
O.K.
CHECK GROUND CIRCUIT
1) Turn ignition switch "OFF".
2) Disconnect power transistor harness connector.
3) Check resistance between terminal ⓓ and ground.
Resistance: Approximately 0Ω
N.G.
Repair harness.
O.K.

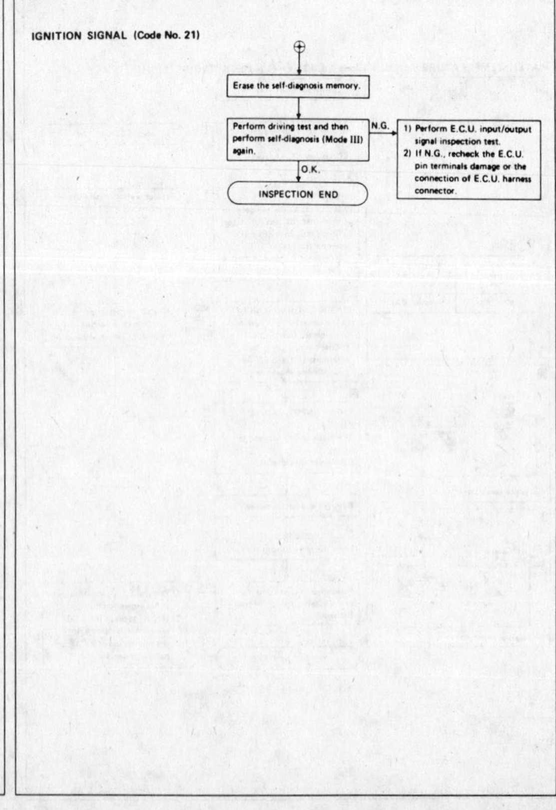
IGNITION SIGNAL (Code No. 21)
Erase the self-diagnosis memory.
Perform driving test and then perform self-diagnosis (Mode III) again.
N.G.
1) Perform E.C.U. input/output signal inspection test.
2) If N.G., recheck the E.C.U. pin terminals damage or the connection of E.C.U. harness connector.
O.K.
INSPECTION END

ECCS SELF-DIAGNOSTIC CHART—1989 PULSAR

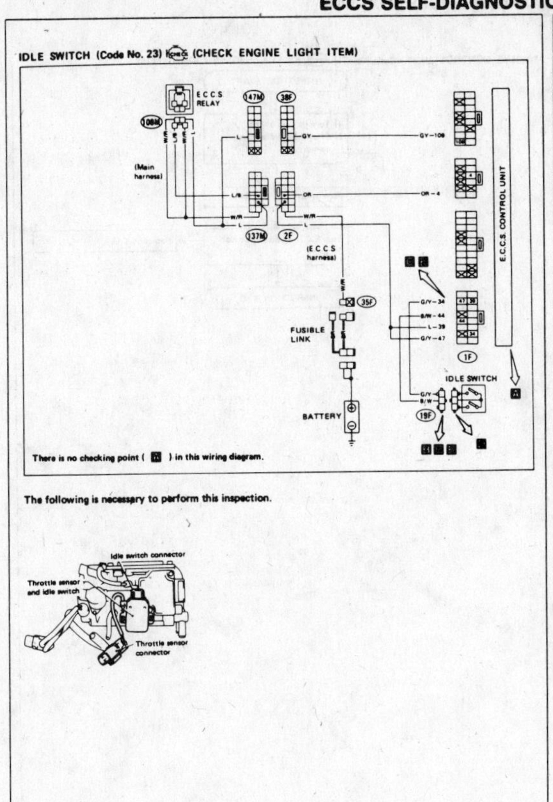
IDLE SWITCH (Code No. 23) (CHECK ENGINE LIGHT ITEM)
ECCS RELAY
(Main harness)
(ECCS harness)
FUSIBLE LINK
BATTERY
IDLE SWITCH
E.C.C.S. CONTROL UNIT
There is no checking point ( ) in this wiring diagram.
The following is necessary to perform this inspection.
Idle switch connector
Throttle sensor and idle switch
Throttle sensor connector

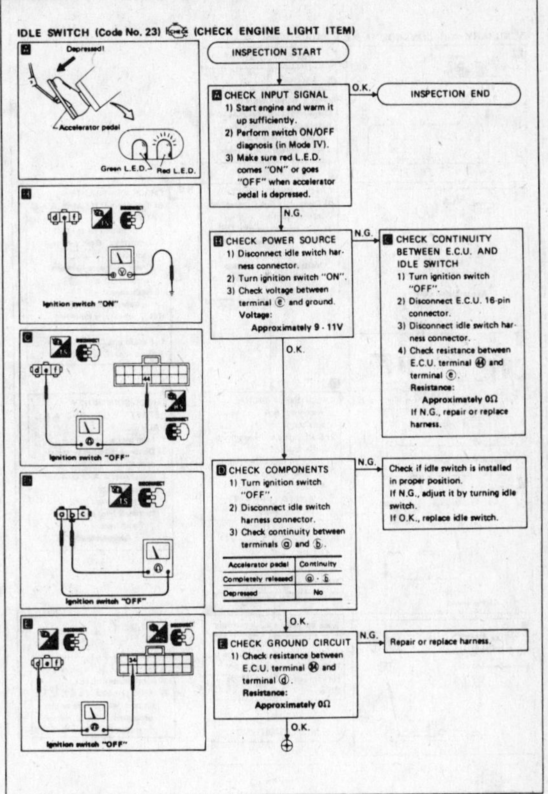
IDLE SWITCH (Code No. 23) (CHECK ENGINE LIGHT ITEM)
Depressed!
Accelerator pedal
Green L.E.D.
Red L.E.D.
Ignition switch "ON"
Ignition switch "OFF"
INSPECTION START
A CHECK INPUT SIGNAL
1) Start engine and warm it up sufficiently.
2) Perform switch ON/OFF diagnosis (in Mode IV).
3) Make sure red L.E.D. comes "ON" or goes "OFF" when accelerator pedal is depressed.
O.K.
INSPECTION END
N.G.
B CHECK POWER SOURCE
1) Disconnect idle switch harness connector.
2) Turn ignition switch "ON".
3) Check voltage between terminal ⓔ and ground.
Voltage:
Approximately 9 - 11V
N.G.
C CHECK CONTINUITY BETWEEN E.C.U. AND IDLE SWITCH
1) Turn ignition switch "OFF".
2) Disconnect E.C.U. 16-pin connector.
3) Disconnect idle switch harness connector.
4) Check resistance between E.C.U. terminal ㊹ and terminal ⓔ.
Resistance:
Approximately 0Ω
If N.G., repair or replace harness.
O.K.
D CHECK COMPONENTS
1) Turn ignition switch "OFF".
2) Disconnect idle switch harness connector.
3) Check continuity between terminals ⓐ and ⓑ.
Accelerator pedal | Continuity
Completely released | ⓐ - ⓑ
Depressed | No
N.G.
Check if idle switch is installed in proper position.
If N.G., adjust it by turning idle switch.
If O.K., replace idle switch.
O.K.
E CHECK GROUND CIRCUIT
1) Check resistance between E.C.U. terminal ㉞ and terminal ⓓ.
Resistance:
Approximately 0Ω
N.G.
Repair or replace harness.
O.K.

ECCS SELF-DIAGNOSTIC CHART—1989 PULSAR

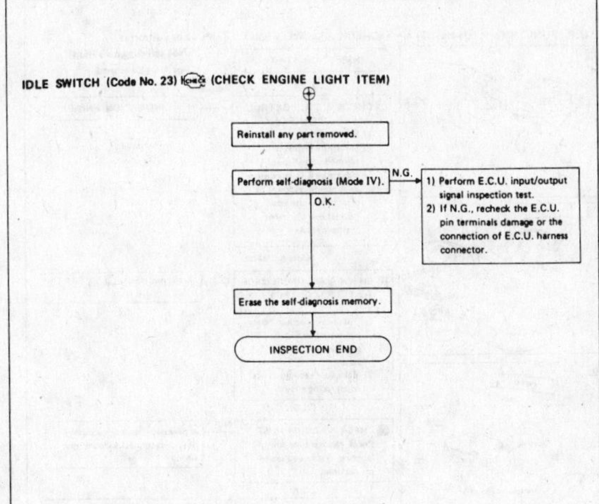
IDLE SWITCH (Code No. 23) (CHECK ENGINE LIGHT ITEM)
Reinstall any part removed.
Perform self-diagnosis (Mode IV).
N.G.
1) Perform E.C.U. input/output signal inspection test.
2) If N.G., recheck the E.C.U. pin terminals damage or the connection of E.C.U. harness connector.
O.K.
Erase the self-diagnosis memory.
INSPECTION END

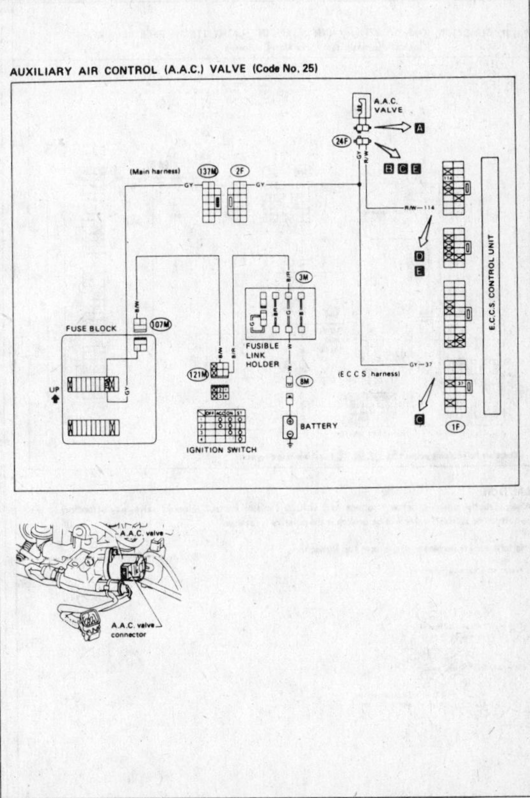
AUXILIARY AIR CONTROL (A.A.C.) VALVE (Code No. 25)
A.A.C. VALVE
(Main harness)
FUSE BLOCK
UP
FUSIBLE LINK HOLDER
(ECCS harness)
BATTERY
IGNITION SWITCH
E.C.C.S. CONTROL UNIT
A.A.C. valve
A.A.C. valve connector

ECCS SELF-DIAGNOSTIC CHART—1989 PULSAR

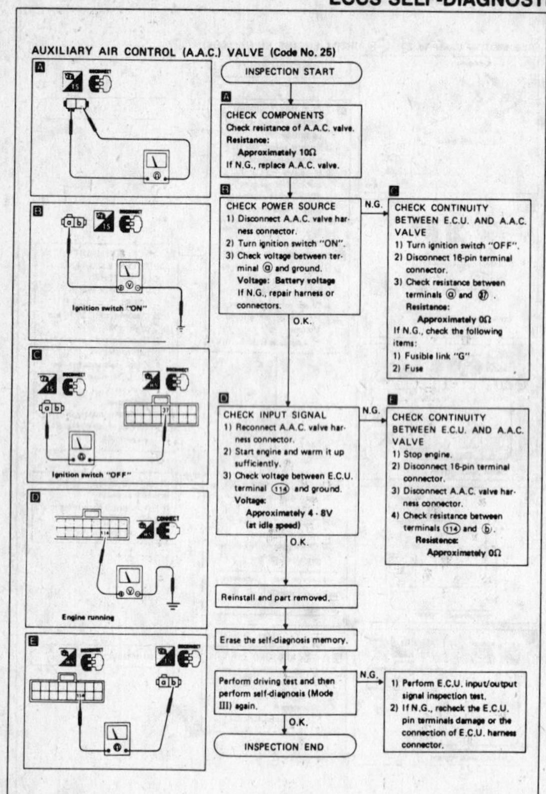
AUXILIARY AIR CONTROL (A.A.C.) VALVE (Code No. 25)
INSPECTION START
A CHECK COMPONENTS
Check resistance of A.A.C. valve.
Resistance: Approximately 10Ω
If N.G., replace A.A.C. valve.
B CHECK POWER SOURCE
1) Disconnect A.A.C. valve harness connector.
2) Turn ignition switch "ON".
3) Check voltage between terminal ⓐ and ground.
Voltage: Battery voltage
If N.G., repair harness or connectors.
N.G.
C CHECK CONTINUITY BETWEEN E.C.U. AND A.A.C. VALVE
1) Turn ignition switch "OFF".
2) Disconnect 16-pin terminal connector.
3) Check resistance between terminals ⓐ and 37.
Resistance: Approximately 0Ω
If N.G., check the following items:
1) Fusible link "G"
2) Fuse
O.K.
D CHECK INPUT SIGNAL
1) Reconnect A.A.C. valve harness connector.
2) Start engine and warm it up sufficiently.
3) Check voltage between E.C.U. terminal 114 and ground.
Voltage: Approximately 4 - 8V (at idle speed)
N.G.
E CHECK CONTINUITY BETWEEN E.C.U. AND A.A.C. VALVE
1) Stop engine.
2) Disconnect 16-pin terminal connector.
3) Disconnect A.A.C. valve harness connector.
4) Check resistance between terminals 114 and ⓑ.
Resistance: Approximately 0Ω
O.K.
Reinstall and part removed.
Erase the self-diagnosis memory.
Perform driving test and then perform self-diagnosis (Mode III) again.
N.G.
1) Perform E.C.U. input/output signal inspection test.
2) If N.G., recheck the E.C.U. pin terminals damage or the connection of E.C.U. harness connector.
O.K.
INSPECTION END
Ignition switch "ON"
Ignition switch "OFF"
Engine running

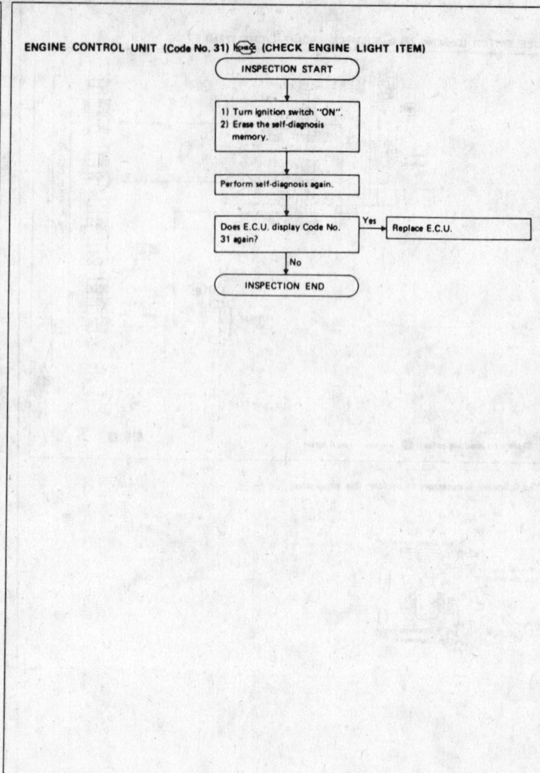
ENGINE CONTROL UNIT (Code No. 31) (CHECK ENGINE LIGHT ITEM)
INSPECTION START
1) Turn ignition switch "ON".
2) Erase the self-diagnosis memory.
Perform self-diagnosis again.
Does E.C.U. display Code No. 31 again?
Yes
Replace E.C.U.
No
INSPECTION END

ECCS SELF-DIAGNOSTIC CHART—1989 PULSAR

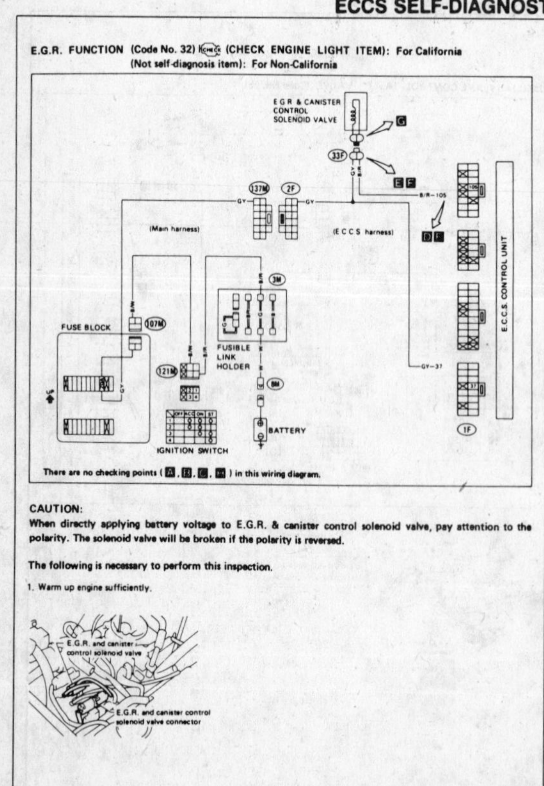
E.G.R. FUNCTION (Code No. 32) (CHECK ENGINE LIGHT ITEM): For California
(Not self-diagnosis item): For Non-California
E G R & CANISTER CONTROL SOLENOID VALVE
(Main harness)
(E C C S harness)
FUSE BLOCK
FUSIBLE LINK HOLDER
BATTERY
IGNITION SWITCH
E.C.C.S. CONTROL UNIT
There are no checking points (A, B, C, H) in this wiring diagram.
CAUTION:
When directly applying battery voltage to E.G.R. & canister control solenoid valve, pay attention to the polarity. The solenoid valve will be broken if the polarity is reversed.
The following is necessary to perform this inspection.
1. Warm up engine sufficiently.
E.G.R. and canister control solenoid valve
E.G.R. and canister control solenoid valve connector

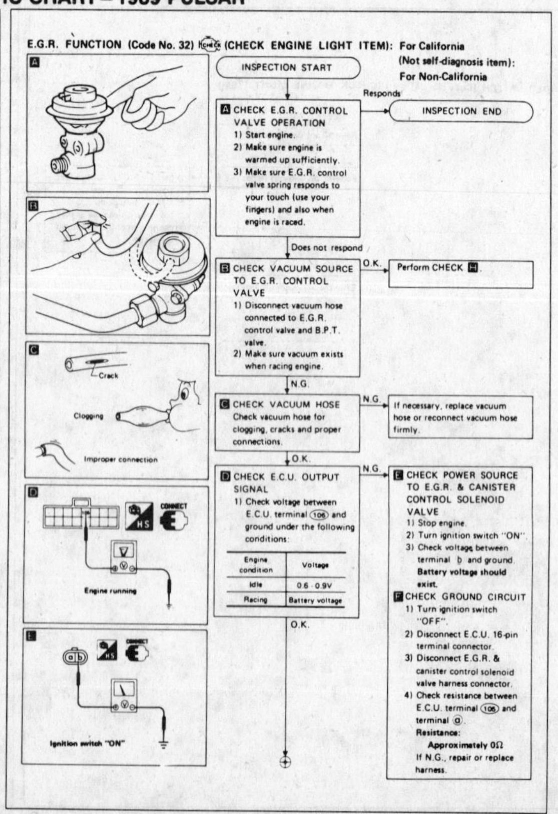
E.G.R. FUNCTION (Code No. 32) (CHECK ENGINE LIGHT ITEM): For California
(Not self-diagnosis item): For Non-California
INSPECTION START
A CHECK E.G.R. CONTROL VALVE OPERATION
1) Start engine.
2) Make sure engine is warmed up sufficiently.
3) Make sure E.G.R. control valve spring responds to your touch (use your fingers) and also when engine is raced.
Responds
INSPECTION END
Does not respond
B CHECK VACUUM SOURCE TO E.G.R. CONTROL VALVE
1) Disconnect vacuum hose connected to E.G.R. control valve and B.P.T. valve.
2) Make sure vacuum exists when racing engine.
O.K.
Perform CHECK H.
N.G.
C CHECK VACUUM HOSE
Check vacuum hose for clogging, cracks and proper connections.
N.G.
If necessary, replace vacuum hose or reconnect vacuum hose firmly.
O.K.
D CHECK E.C.U. OUTPUT SIGNAL
1) Check voltage between E.C.U. terminal 106 and ground under the following conditions:
Engine condition | Voltage
Idle | 0.6 - 0.9V
Racing | Battery voltage
N.G.
E CHECK POWER SOURCE TO E.G.R. & CANISTER CONTROL SOLENOID VALVE
1) Stop engine.
2) Turn ignition switch "ON".
3) Check voltage between terminal ⓑ and ground.
Battery voltage should exist.
F CHECK GROUND CIRCUIT
1) Turn ignition switch "OFF".
2) Disconnect E.C.U. 16-pin terminal connector.
3) Disconnect E.G.R. & canister control solenoid valve harness connector.
4) Check resistance between E.C.U. terminal 106 and terminal ⓐ.
Resistance: Approximately 0Ω
If N.G., repair or replace harness.
O.K.
Crack
Clogging
Improper connection
Engine running
Ignition switch "ON"

ECCS SELF-DIAGNOSTIC CHART – 1989 PULSAR

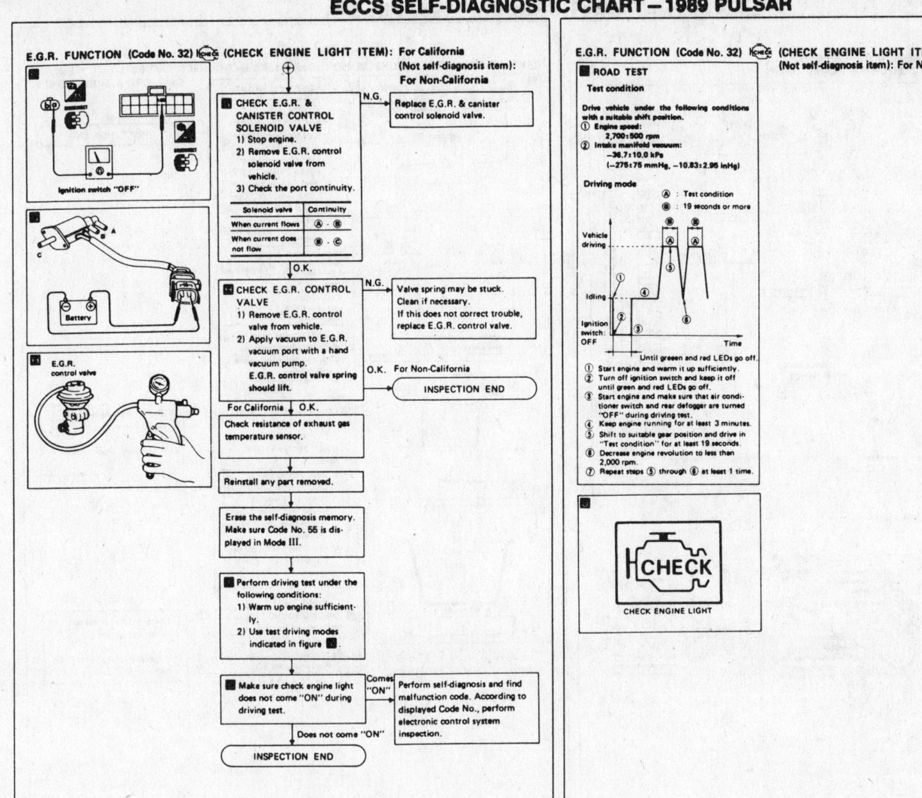
E.G.R. FUNCTION (Code No. 32) CHECK (CHECK ENGINE LIGHT ITEM): For California
(Not self-diagnosis item): For Non-California
Ignition switch "OFF"
Battery
E.G.R. control valve
CHECK E.G.R. & CANISTER CONTROL SOLENOID VALVE
1) Stop engine.
2) Remove E.G.R. control solenoid valve from vehicle.
3) Check the port continuity.
Solenoid valve | Continuity
When current flows | A - B
When current does not flow | B - C
N.G.
Replace E.G.R. & canister control solenoid valve.
O.K.
CHECK E.G.R. CONTROL VALVE
1) Remove E.G.R. control valve from vehicle.
2) Apply vacuum to E.G.R. vacuum port with a hand vacuum pump.
E.G.R. control valve spring should lift.
N.G.
Valve spring may be stuck. Clean if necessary.
If this does not correct trouble, replace E.G.R. control valve.
O.K. For Non-California
INSPECTION END
For California O.K.
Check resistance of exhaust gas temperature sensor.
Reinstall any part removed.
Erase the self-diagnosis memory. Make sure Code No. 55 is displayed in Mode III.
Perform driving test under the following conditions:
1) Warm up engine sufficiently.
2) Use test driving modes indicated in figure.
Make sure check engine light does not come "ON" during driving test.
Comes "ON"
Perform self-diagnosis and find malfunction code. According to displayed Code No., perform electronic control system inspection.
Does not come "ON"
INSPECTION END

E.G.R. FUNCTION (Code No. 32) (CHECK ENGINE LIGHT ITEM): For California (Not self-diagnosis item): For Non-California

ROAD TEST

Test condition

Drive vehicle under the following conditions with a suitable shift position.

① Engine speed: 2,700±500 rpm

② Intake manifold vacuum: −36.7±10.0 kPa (−275±75 mmHg, −10.83±2.95 inHg)

Driving mode

Ⓐ : Test condition

Ⓑ : 19 seconds or more

Vehicle driving — Idling — Ignition switch: OFF — Time

Until green and red LEDs go off.

① Start engine and warm it up sufficiently.

② Turn off ignition switch and keep it off until green and red LEDs go off.

③ Start engine and make sure that air conditioner switch and rear defogger are turned "OFF" during driving test.

④ Keep engine running for at least 3 minutes.

⑤ Shift to suitable gear position and drive in "Test condition" for at least 19 seconds.

⑥ Decrease engine revolution to less than 2,000 rpm.

⑦ Repeat steps ⑤ through ⑥ at least 1 time.

CHECK

CHECK ENGINE LIGHT

ECCS SELF-DIAGNOSTIC CHART – 1989 PULSAR

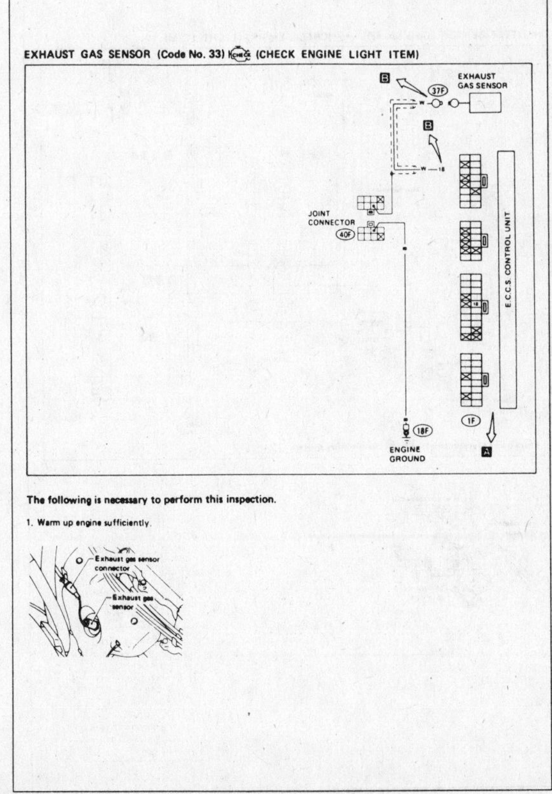
EXHAUST GAS SENSOR (Code No. 33) (CHECK ENGINE LIGHT ITEM)
EXHAUST GAS SENSOR
37F
JOINT CONNECTOR
40F
E.C.C.S CONTROL UNIT
18F
1F
ENGINE GROUND
The following is necessary to perform this inspection.
1. Warm up engine sufficiently.
Exhaust gas sensor connector
Exhaust gas sensor

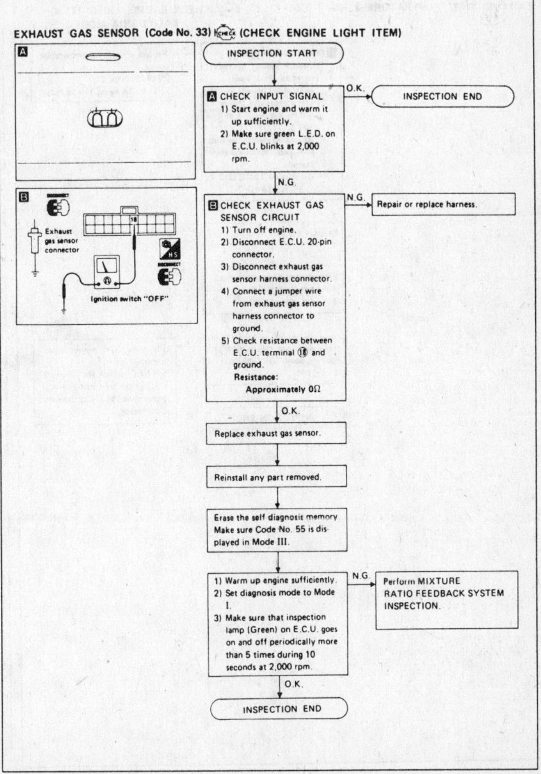
EXHAUST GAS SENSOR (Code No. 33) (CHECK ENGINE LIGHT ITEM)
INSPECTION START
CHECK INPUT SIGNAL
1) Start engine and warm it up sufficiently.
2) Make sure green L.E.D. on E.C.U. blinks at 2,000 rpm.
O.K.
INSPECTION END
N.G.
Exhaust gas sensor connector
Ignition switch "OFF"
CHECK EXHAUST GAS SENSOR CIRCUIT
1) Turn off engine.
2) Disconnect E.C.U. 20-pin connector.
3) Disconnect exhaust gas sensor harness connector.
4) Connect a jumper wire from exhaust gas sensor harness connector to ground.
5) Check resistance between E.C.U. terminal ⑱ and ground.
Resistance: Approximately 0Ω
N.G.
Repair or replace harness.
O.K.
Replace exhaust gas sensor.
Reinstall any part removed.
Erase the self diagnosis memory. Make sure Code No. 55 is displayed in Mode III.
1) Warm up engine sufficiently.
2) Set diagnosis mode to Mode I.
3) Make sure that inspection lamp (Green) on E.C.U. goes on and off periodically more than 5 times during 10 seconds at 2,000 rpm.
N.G.
Perform MIXTURE RATIO FEEDBACK SYSTEM INSPECTION.
O.K.
INSPECTION END

ECCS SELF-DIAGNOSTIC CHART—1989 PULSAR

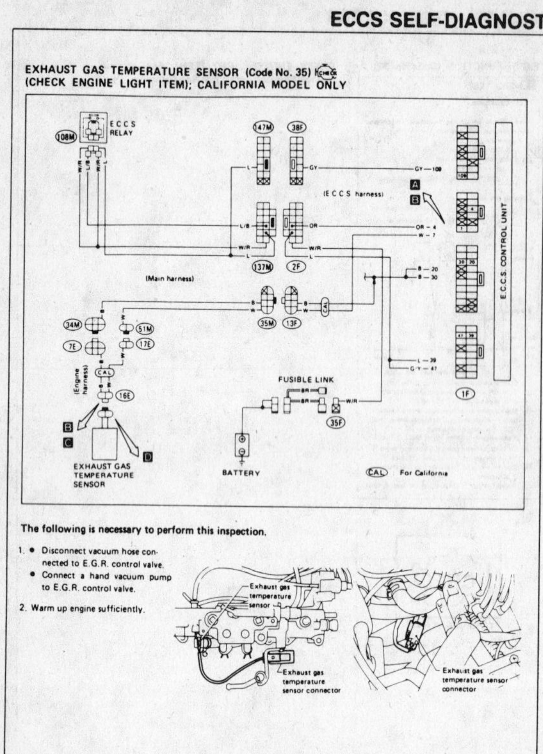
EXHAUST GAS TEMPERATURE SENSOR (Code No. 35) (CHECK ENGINE LIGHT ITEM); CALIFORNIA MODEL ONLY
ECCS RELAY
(Main harness)
(ECCS harness)
E.C.C.S. CONTROL UNIT
FUSIBLE LINK
EXHAUST GAS TEMPERATURE SENSOR
BATTERY
CAL For California
The following is necessary to perform this inspection.
1. • Disconnect vacuum hose connected to E.G.R. control valve.
• Connect a hand vacuum pump to E.G.R. control valve.
2. Warm up engine sufficiently.
Exhaust gas temperature sensor
Exhaust gas temperature sensor connector
Exhaust gas temperature sensor connector

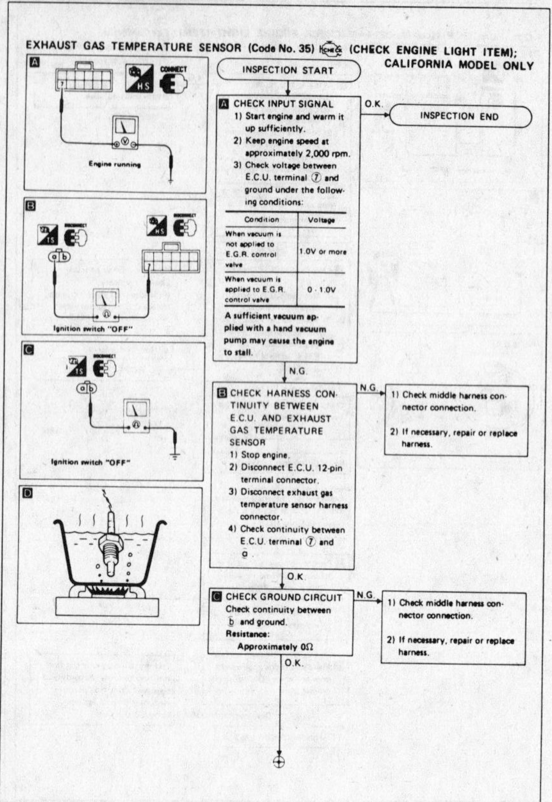
EXHAUST GAS TEMPERATURE SENSOR (Code No. 35) (CHECK ENGINE LIGHT ITEM); CALIFORNIA MODEL ONLY
Engine running
Ignition switch "OFF"
Ignition switch "OFF"
INSPECTION START
A CHECK INPUT SIGNAL
1) Start engine and warm it up sufficiently.
2) Keep engine speed at approximately 2,000 rpm.
3) Check voltage between E.C.U. terminal ⑦ and ground under the following conditions:
Condition | Voltage
When vacuum is not applied to E.G.R. control valve | 1.0V or more
When vacuum is applied to E.G.R. control valve | 0 - 1.0V
A sufficient vacuum applied with a hand vacuum pump may cause the engine to stall.
O.K.
INSPECTION END
N.G.
B CHECK HARNESS CONTINUITY BETWEEN E.C.U. AND EXHAUST GAS TEMPERATURE SENSOR
1) Stop engine.
2) Disconnect E.C.U. 12-pin terminal connector.
3) Disconnect exhaust gas temperature sensor harness connector.
4) Check continuity between E.C.U. terminal ⑦ and a.
N.G.
1) Check middle harness connector connection.
2) If necessary, repair or replace harness.
O.K.
C CHECK GROUND CIRCUIT
Check continuity between b and ground.
Resistance: Approximately 0Ω
N.G.
1) Check middle harness connector connection.
2) If necessary, repair or replace harness.
O.K.

ECCS SELF-DIAGNOSTIC CHART—1989 PULSAR

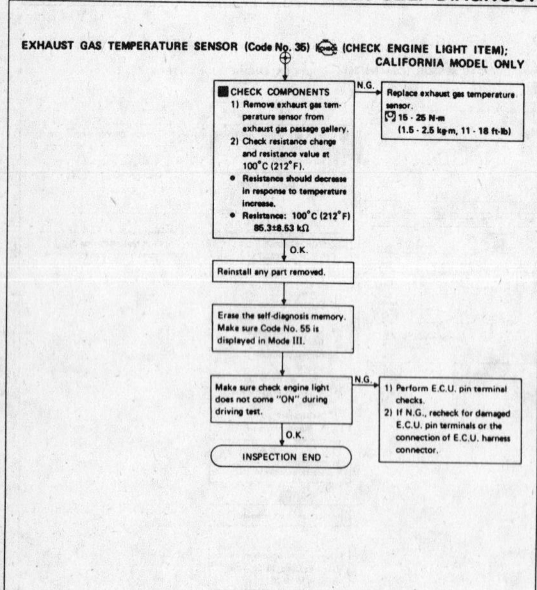
EXHAUST GAS TEMPERATURE SENSOR (Code No. 35) (CHECK ENGINE LIGHT ITEM); CALIFORNIA MODEL ONLY
CHECK COMPONENTS
1) Remove exhaust gas temperature sensor from exhaust gas passage gallery.
2) Check resistance change and resistance value at 100°C (212°F).
• Resistance should decrease in response to temperature increase.
• Resistance: 100°C (212°F) 85.3±8.53 kΩ
N.G.
Replace exhaust gas temperature sensor.
15 - 25 N·m (1.5 - 2.5 kg-m, 11 - 18 ft-lb)
O.K.
Reinstall any part removed.
Erase the self-diagnosis memory. Make sure Code No. 55 is displayed in Mode III.
Make sure check engine light does not come "ON" during driving test.
N.G.
1) Perform E.C.U. pin terminal checks.
2) If N.G., recheck for damaged E.C.U. pin terminals or the connection of E.C.U. harness connector.
O.K.
INSPECTION END

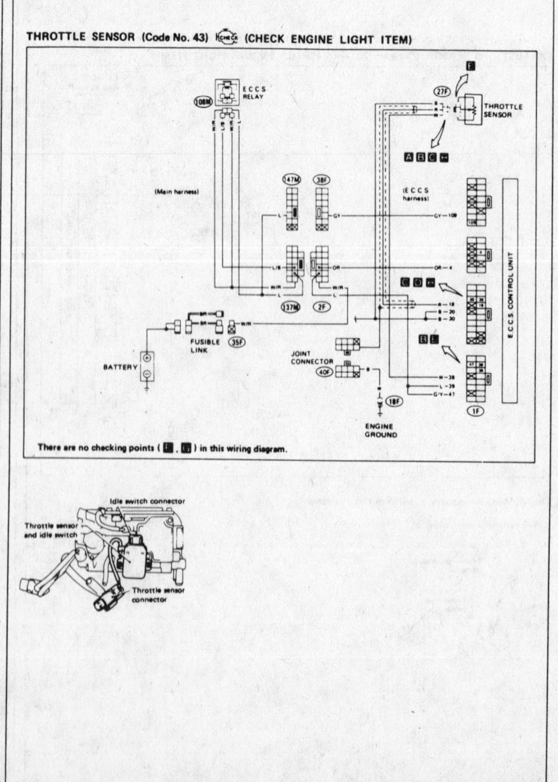
THROTTLE SENSOR (Code No. 43) (CHECK ENGINE LIGHT ITEM)
ECCS RELAY
THROTTLE SENSOR
(Main harness)
(ECCS harness)
E.C.C.S. CONTROL UNIT
FUSIBLE LINK
BATTERY
JOINT CONNECTOR
ENGINE GROUND
There are no checking points (F , G) in this wiring diagram.
Idle switch connector
Throttle sensor and idle switch
Throttle sensor connector

ECCS SELF-DIAGNOSTIC CHART—1989 PULSAR

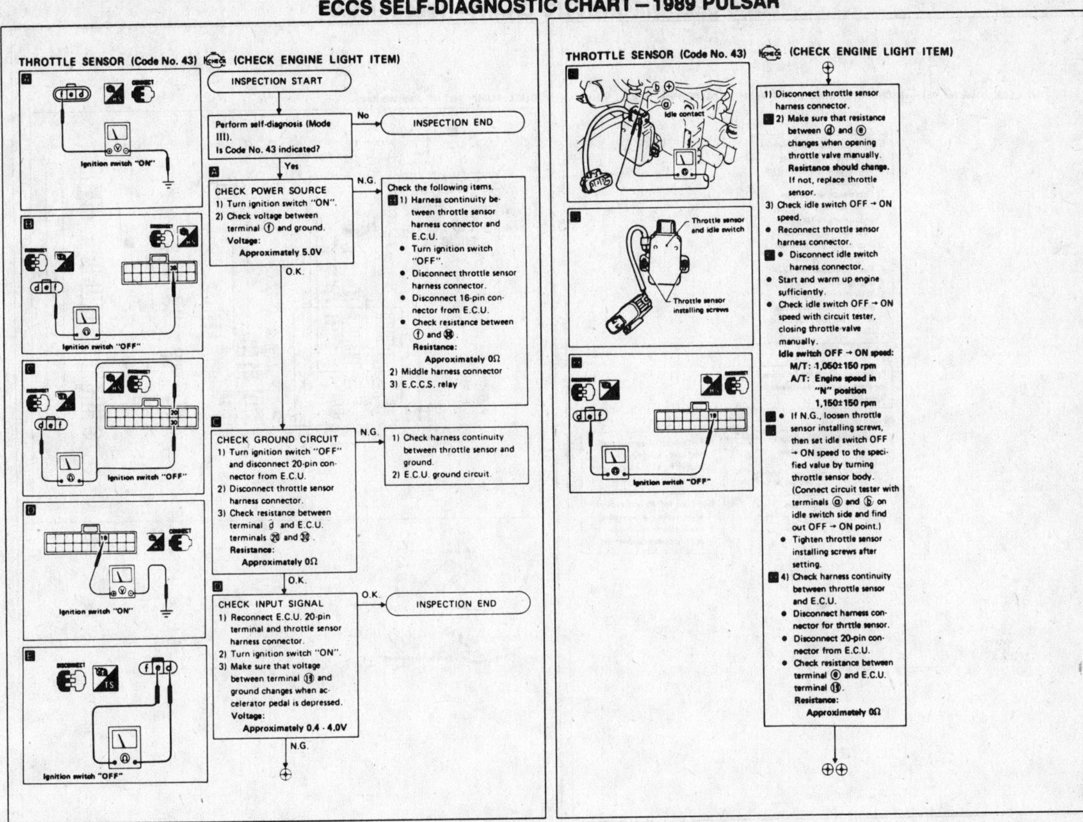
THROTTLE SENSOR (Code No. 43) (CHECK ENGINE LIGHT ITEM)
INSPECTION START
Perform self-diagnosis (Mode III). Is Code No. 43 indicated?
No
INSPECTION END
Yes
CHECK POWER SOURCE
1) Turn ignition switch "ON".
2) Check voltage between terminal ⓕ and ground.
Voltage: Approximately 5.0V
N.G.
Check the following items.
1) Harness continuity between throttle sensor harness connector and E.C.U.
• Turn ignition switch "OFF".
• Disconnect throttle sensor harness connector.
• Disconnect 16-pin connector from E.C.U.
• Check resistance between ⓕ and 38.
Resistance: Approximately 0Ω
2) Middle harness connector
3) E.C.C.S. relay
O.K.
CHECK GROUND CIRCUIT
1) Turn ignition switch "OFF" and disconnect 20-pin connector from E.C.U.
2) Disconnect throttle sensor harness connector.
3) Check resistance between terminal ⓓ and E.C.U. terminals 20 and 30.
Resistance: Approximately 0Ω
N.G.
1) Check harness continuity between throttle sensor and ground.
2) E.C.U. ground circuit.
O.K.
CHECK INPUT SIGNAL
1) Reconnect E.C.U. 20-pin terminal and throttle sensor harness connector.
2) Turn ignition switch "ON".
3) Make sure that voltage between terminal 19 and ground changes when accelerator pedal is depressed.
Voltage: Approximately 0.4 - 4.0V
O.K.
INSPECTION END
N.G.
Ignition switch "ON"
Ignition switch "OFF"
Ignition switch "OFF"
Ignition switch "ON"
Ignition switch "OFF"
THROTTLE SENSOR (Code No. 43) (CHECK ENGINE LIGHT ITEM)
Idle contact
Throttle sensor and idle switch
Throttle sensor installing screws
Ignition switch "OFF"
1) Disconnect throttle sensor harness connector.
2) Make sure that resistance between ⓓ and ⓔ changes when opening throttle valve manually. Resistance should change. If not, replace throttle sensor.
3) Check idle switch OFF → ON speed.
• Reconnect throttle sensor harness connector.
• Disconnect idle switch harness connector.
• Start and warm up engine sufficiently.
• Check idle switch OFF → ON speed with circuit tester, closing throttle valve manually.
Idle switch OFF → ON speed:
M/T: 1,050±150 rpm
A/T: Engine speed in "N" position 1,150±150 rpm
• If N.G., loosen throttle sensor installing screws, then set idle switch OFF → ON speed to the specified value by turning throttle sensor body. (Connect circuit tester with terminals ⓐ and ⓑ on idle switch side and find out OFF → ON point.)
• Tighten throttle sensor installing screws after setting.
4) Check harness continuity between throttle sensor and E.C.U.
• Disconnect harness connector for thrttle sensor.
• Disconnect 20-pin connector from E.C.U.
• Check resistance between terminal ⓔ and E.C.U. terminal 19.
Resistance: Approximately 0Ω

ECCS SELF-DIAGNOSTIC CHART—1989 PULSAR

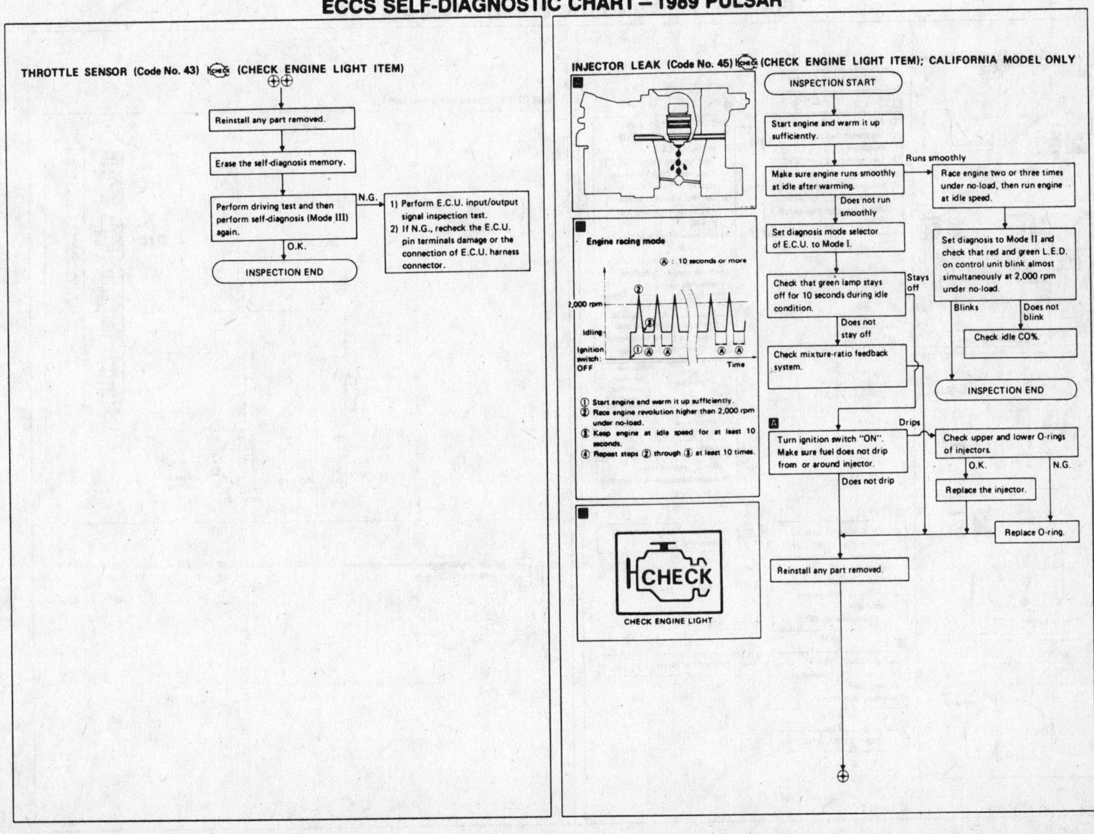
THROTTLE SENSOR (Code No. 43) (CHECK ENGINE LIGHT ITEM)
Reinstall any part removed.
Erase the self-diagnosis memory.
Perform driving test and then perform self-diagnosis (Mode III) again.
N.G.
1) Perform E.C.U. input/output signal inspection test.
2) If N.G., recheck the E.C.U. pin terminals damage or the connection of E.C.U. harness connector.
O.K.
INSPECTION END
INJECTOR LEAK (Code No. 45) (CHECK ENGINE LIGHT ITEM); CALIFORNIA MODEL ONLY
INSPECTION START
Start engine and warm it up sufficiently.
Make sure engine runs smoothly at idle after warming.
Runs smoothly
Race engine two or three times under no-load, then run engine at idle speed.
Does not run smoothly
Set diagnosis mode selector of E.C.U. to Mode I.
Set diagnosis to Mode II and check that red and green L.E.D. on control unit blink almost simultaneously at 2,000 rpm under no-load.
Blinks
Does not blink
Check idle CO%.
Check that green lamp stays off for 10 seconds during idle condition.
Stays off
Does not stay off
Check mixture-ratio feedback system.
INSPECTION END
Turn ignition switch "ON". Make sure fuel does not drip from or around injector.
Drips
Check upper and lower O-rings of injectors.
O.K.
N.G.
Replace the injector.
Replace O-ring.
Does not drip
Reinstall any part removed.
Engine racing mode
Ⓐ: 10 seconds or more
2,000 rpm
Idling
Ignition switch OFF
Time
① Start engine and warm it up sufficiently.
② Race engine revolution higher than 2,000 rpm under no-load.
③ Keep engine at idle speed for at least 10 seconds.
④ Repeat steps ② through ③ at least 10 times.
CHECK
CHECK ENGINE LIGHT

ECCS SELF-DIAGNOSTIC CHART—1989 PULSAR

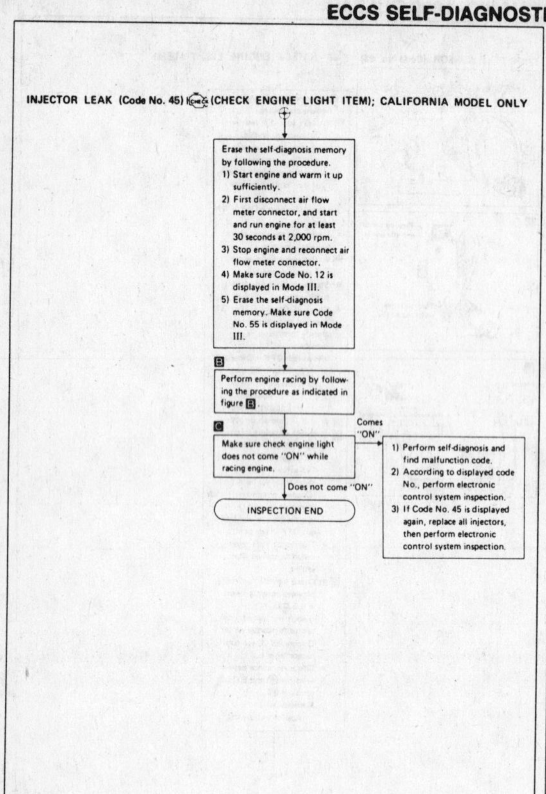
INJECTOR LEAK (Code No. 45) (CHECK ENGINE LIGHT ITEM); CALIFORNIA MODEL ONLY
Erase the self-diagnosis memory by following the procedure.
1) Start engine and warm it up sufficiently.
2) First disconnect air flow meter connector, and start and run engine for at least 30 seconds at 2,000 rpm.
3) Stop engine and reconnect air flow meter connector.
4) Make sure Code No. 12 is displayed in Mode III.
5) Erase the self-diagnosis memory. Make sure Code No. 55 is displayed in Mode III.
B
Perform engine racing by following the procedure as indicated in figure B.
C
Make sure check engine light does not come "ON" while racing engine.
Comes "ON"
1) Perform self-diagnosis and find malfunction code.
2) According to displayed code No., perform electronic control system inspection.
3) If Code No. 45 is displayed again, replace all injectors, then perform electronic control system inspection.
Does not come "ON"
INSPECTION END

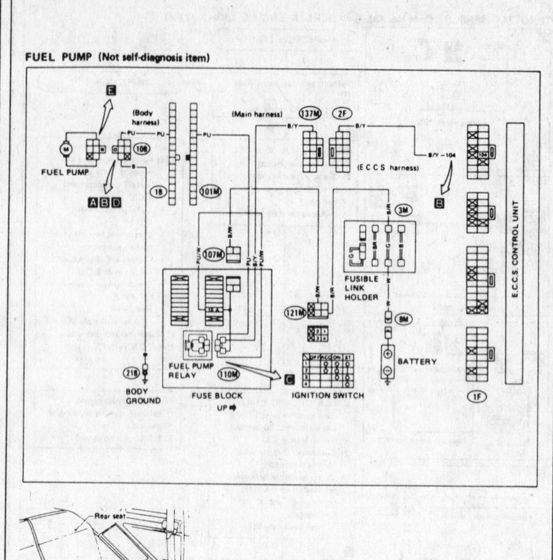
FUEL PUMP (Not self-diagnosis item)
FUEL PUMP
(Body harness)
(Main harness)
(E C C S harness)
FUSIBLE LINK HOLDER
BATTERY
FUEL PUMP RELAY
BODY GROUND
FUSE BLOCK
IGNITION SWITCH
ECCS CONTROL UNIT

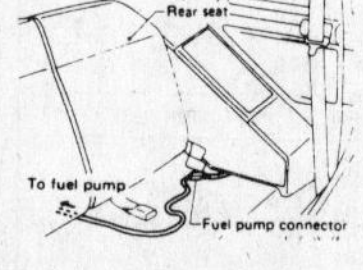
Rear seat
To fuel pump
Fuel pump connector

ECCS SELF-DIAGNOSTIC CHART—1989 PULSAR

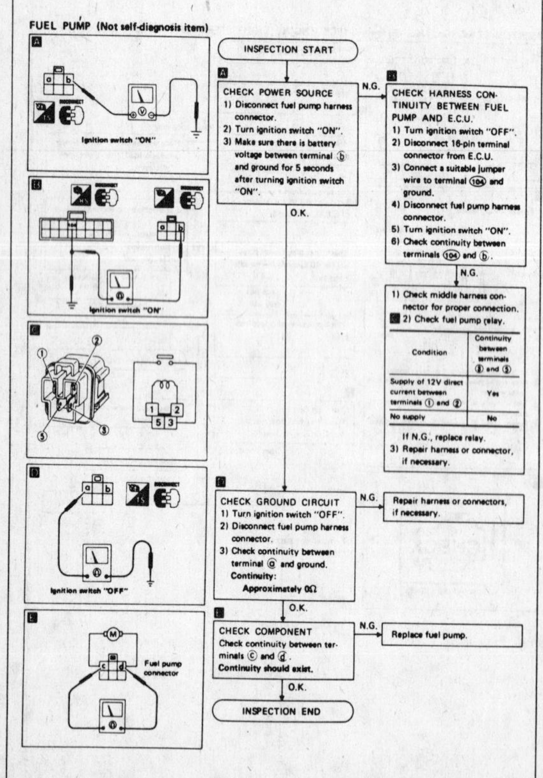
FUEL PUMP (Not self-diagnosis item)
Ignition switch "ON"
Ignition switch "ON"
Ignition switch "OFF"
Fuel pump connector
INSPECTION START
A
CHECK POWER SOURCE
1) Disconnect fuel pump harness connector.
2) Turn ignition switch "ON".
3) Make sure there is battery voltage between terminal ⓑ and ground for 5 seconds after turning ignition switch "ON".
N.G.
B
CHECK HARNESS CONTINUITY BETWEEN FUEL PUMP AND E.C.U.
1) Turn ignition switch "OFF".
2) Disconnect 16-pin terminal connector from E.C.U.
3) Connect a suitable jumper wire to terminal (104) and ground.
4) Disconnect fuel pump harness connector.
5) Turn ignition switch "ON".
6) Check continuity between terminals (104) and ⓑ.
N.G.
1) Check middle harness connector for proper connection.
C 2) Check fuel pump relay.
Condition | Continuity between terminals ③ and ⑤
Supply of 12V direct current between terminals ① and ② | Yes
No supply | No
If N.G., replace relay.
3) Repair harness or connector, if necessary.
O.K.
D
CHECK GROUND CIRCUIT
1) Turn ignition switch "OFF".
2) Disconnect fuel pump harness connector.
3) Check continuity between terminal ⓐ and ground.
Continuity: Approximately 0Ω
N.G.
Repair harness or connectors, if necessary.
O.K.
E
CHECK COMPONENT
Check continuity between terminals ⓒ and ⓓ.
Continuity should exist.
N.G.
Replace fuel pump.
O.K.
INSPECTION END

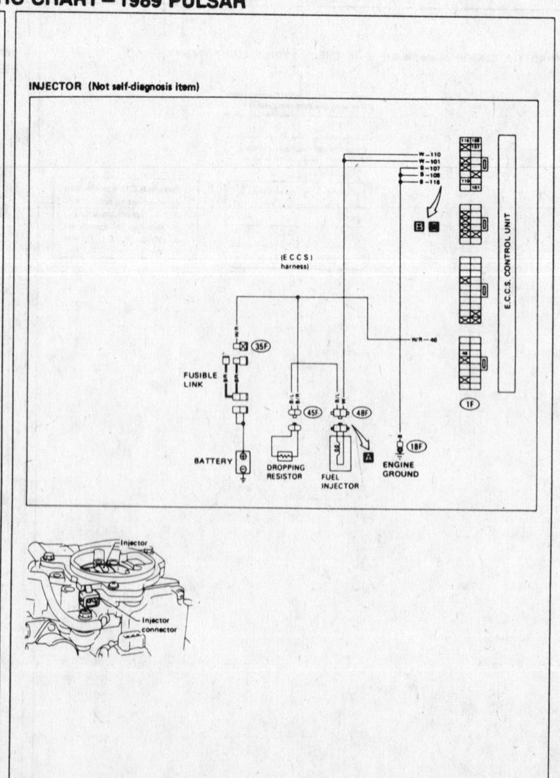
INJECTOR (Not self-diagnosis item)
(E C C S) harness
FUSIBLE LINK
BATTERY
DROPPING RESISTOR
FUEL INJECTOR
ENGINE GROUND
ECCS CONTROL UNIT
Injector
Injector connector

ECCS SELF-DIAGNOSTIC CHART—1989 PULSAR

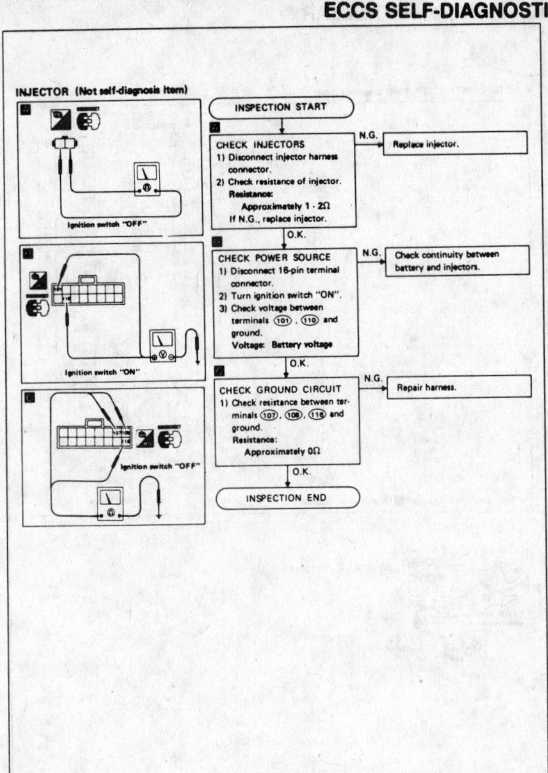
INJECTOR (Not self-diagnosis item)
Ignition switch "OFF"
Ignition switch "ON"
Ignition switch "OFF"
INSPECTION START
CHECK INJECTORS
1) Disconnect injector harness connector.
2) Check resistance of injector.
Resistance:
Approximately 1 - 2Ω
If N.G., replace injector.
N.G.
Replace injector.
O.K.
CHECK POWER SOURCE
1) Disconnect 16-pin terminal connector.
2) Turn ignition switch "ON".
3) Check voltage between terminals 101, 110 and ground.
Voltage: Battery voltage
N.G.
Check continuity between battery and injectors.
O.K.
CHECK GROUND CIRCUIT
1) Check resistance between terminals 107, 108, 116 and ground.
Resistance:
Approximately 0Ω
N.G.
Repair harness.
O.K.
INSPECTION END

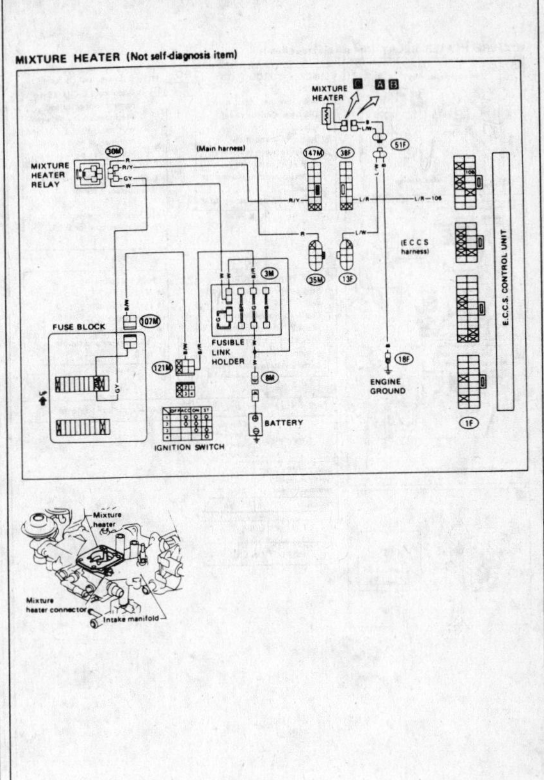
MIXTURE HEATER (Not self-diagnosis item)
MIXTURE HEATER
(Main harness)
MIXTURE HEATER RELAY
FUSE BLOCK
FUSIBLE LINK HOLDER
IGNITION SWITCH
BATTERY
ENGINE GROUND
(ECCS harness)
E.C.C.S. CONTROL UNIT
Mixture heater
Mixture heater connector
Intake manifold

ECCS SELF-DIAGNOSTIC CHART—1989 PULSAR

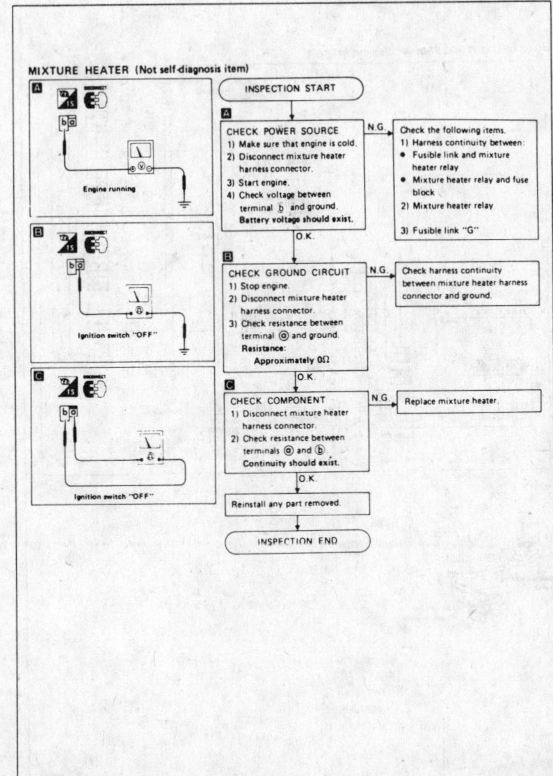
MIXTURE HEATER (Not self-diagnosis item)
Engine running
Ignition switch "OFF"
Ignition switch "OFF"
INSPECTION START
CHECK POWER SOURCE
1) Make sure that engine is cold.
2) Disconnect mixture heater harness connector.
3) Start engine.
4) Check voltage between terminal ⓑ and ground.
Battery voltage should exist.
N.G.
Check the following items.
1) Harness continuity between:
• Fusible link and mixture heater relay
• Mixture heater relay and fuse block
2) Mixture heater relay
3) Fusible link "G"
O.K.
CHECK GROUND CIRCUIT
1) Stop engine.
2) Disconnect mixture heater harness connector.
3) Check resistance between terminal ⓐ and ground.
Resistance:
Approximately 0Ω
N.G.
Check harness continuity between mixture heater harness connector and ground.
O.K.
CHECK COMPONENT
1) Disconnect mixture heater harness connector.
2) Check resistance between terminals ⓐ and ⓑ.
Continuity should exist.
N.G.
Replace mixture heater.
O.K.
Reinstall any part removed.
INSPECTION END

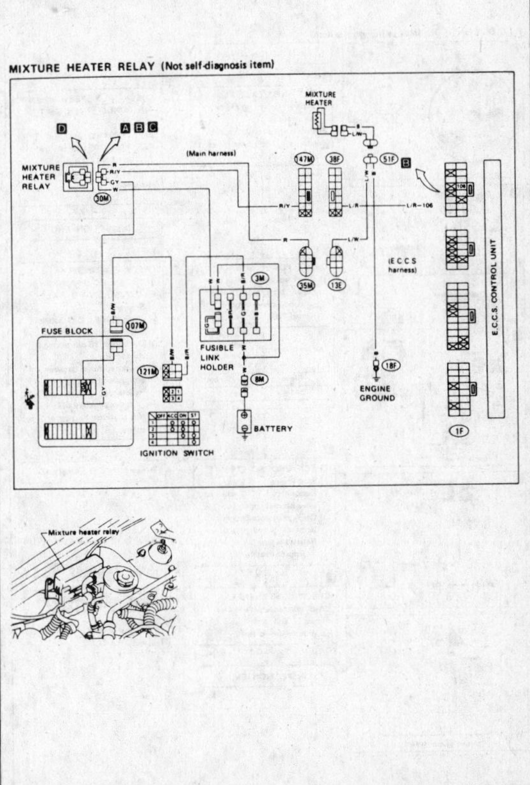
MIXTURE HEATER RELAY (Not self-diagnosis item)
MIXTURE HEATER
(Main harness)
MIXTURE HEATER RELAY
FUSE BLOCK
FUSIBLE LINK HOLDER
IGNITION SWITCH
BATTERY
ENGINE GROUND
(ECCS harness)
E.C.C.S. CONTROL UNIT
Mixture heater relay

ECCS SELF-DIAGNOSTIC CHART—1989 PULSAR

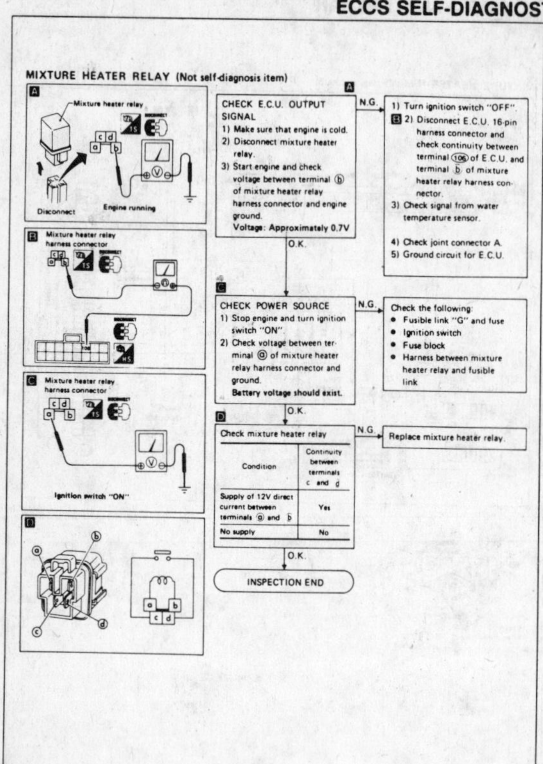
MIXTURE HEATER RELAY (Not self-diagnosis item)
Mixture heater relay
Disconnect
Engine running
CHECK E.C.U. OUTPUT SIGNAL
1) Make sure that engine is cold.
2) Disconnect mixture heater relay.
3) Start engine and check voltage between terminal ⓑ of mixture heater relay harness connector and engine ground.
Voltage: Approximately 0.7V
N.G.
1) Turn ignition switch "OFF".
2) Disconnect E.C.U. 16-pin harness connector and check continuity between terminal 106 of E.C.U. and terminal ⓑ of mixture heater relay harness connector.
3) Check signal from water temperature sensor.
4) Check joint connector A.
5) Ground circuit for E.C.U.
O.K.
Mixture heater relay harness connector
CHECK POWER SOURCE
1) Stop engine and turn ignition switch "ON".
2) Check voltage between terminal ⓐ of mixture heater relay harness connector and ground.
Battery voltage should exist.
Check the following:
Fusible link "G" and fuse
Ignition switch
Fuse block
Harness between mixture heater relay and fusible link
Ignition switch "ON"
Check mixture heater relay
Replace mixture heater relay.
Condition
Continuity between terminals c and d
Supply of 12V direct current between terminals ⓐ and ⓑ
Yes
No supply
No
INSPECTION END

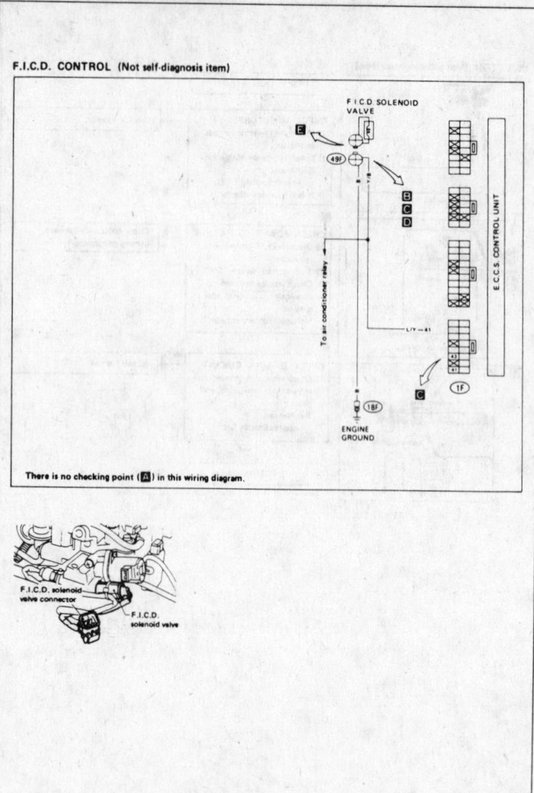
F.I.C.D. CONTROL (Not self-diagnosis item)
F.I.C.D. SOLENOID VALVE
To air conditioner relay
E.C.C.S. CONTROL UNIT
ENGINE GROUND
There is no checking point (A) in this wiring diagram.
F.I.C.D. solenoid valve connector
F.I.C.D. solenoid valve

ECCS SELF-DIAGNOSTIC CHART—1989 PULSAR

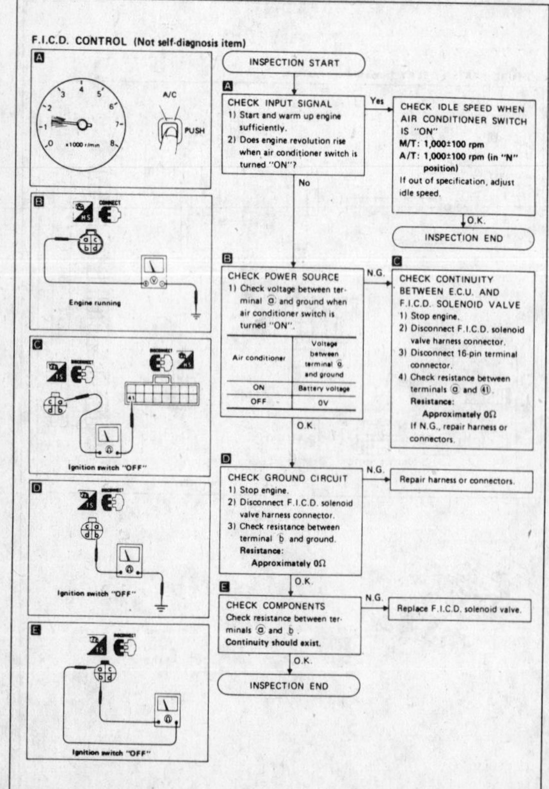
F.I.C.D. CONTROL (Not self-diagnosis item)
INSPECTION START
A/C
PUSH
CHECK INPUT SIGNAL
1) Start and warm up engine sufficiently.
2) Does engine revolution rise when air conditioner switch is turned "ON"?
Yes
CHECK IDLE SPEED WHEN AIR CONDITIONER SWITCH IS "ON"
M/T: 1,000±100 rpm
A/T: 1,000±100 rpm (in "N" position)
If out of specification, adjust idle speed.
O.K.
INSPECTION END
No
Engine running
CHECK POWER SOURCE
1) Check voltage between terminal ⓐ and ground when air conditioner switch is turned "ON".
Air conditioner
Voltage between terminal ⓐ and ground
ON
Battery voltage
OFF
0V
N.G.
CHECK CONTINUITY BETWEEN E.C.U. AND F.I.C.D. SOLENOID VALVE
1) Stop engine.
2) Disconnect F.I.C.D. solenoid valve harness connector.
3) Disconnect 16-pin terminal connector.
4) Check resistance between terminals ⓐ and 41.
Resistance: Approximately 0Ω
If N.G., repair harness or connectors.
Ignition switch "OFF"
CHECK GROUND CIRCUIT
1) Stop engine.
2) Disconnect F.I.C.D. solenoid valve harness connector.
3) Check resistance between terminal ⓑ and ground.
Resistance: Approximately 0Ω
Repair harness or connectors.
CHECK COMPONENTS
Check resistance between terminals ⓐ and ⓑ.
Continuity should exist.
Replace F.I.C.D. solenoid valve.
INSPECTION END

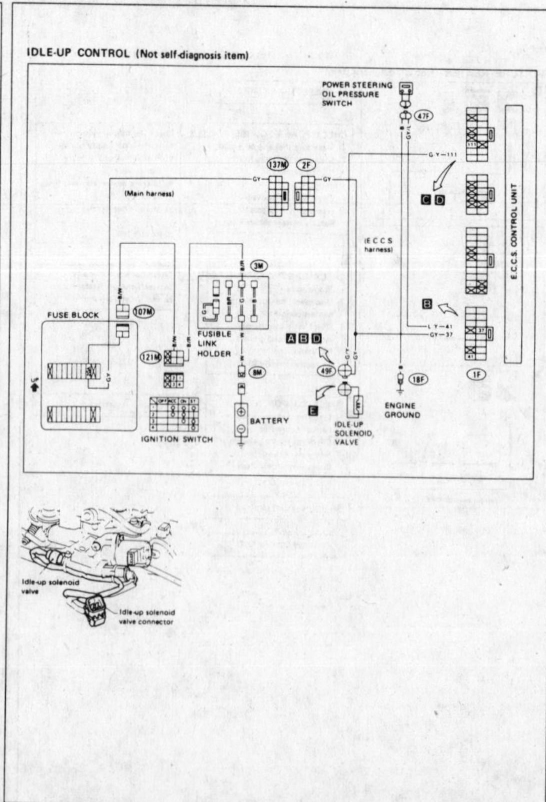
IDLE-UP CONTROL (Not self-diagnosis item)
POWER STEERING OIL PRESSURE SWITCH
(Main harness)
(ECCS harness)
E.C.C.S. CONTROL UNIT
FUSE BLOCK
FUSIBLE LINK HOLDER
BATTERY
IGNITION SWITCH
IDLE-UP SOLENOID VALVE
ENGINE GROUND
Idle-up solenoid valve
Idle-up solenoid valve connector

ECCS SELF-DIAGNOSTIC CHART—1989 PULSAR

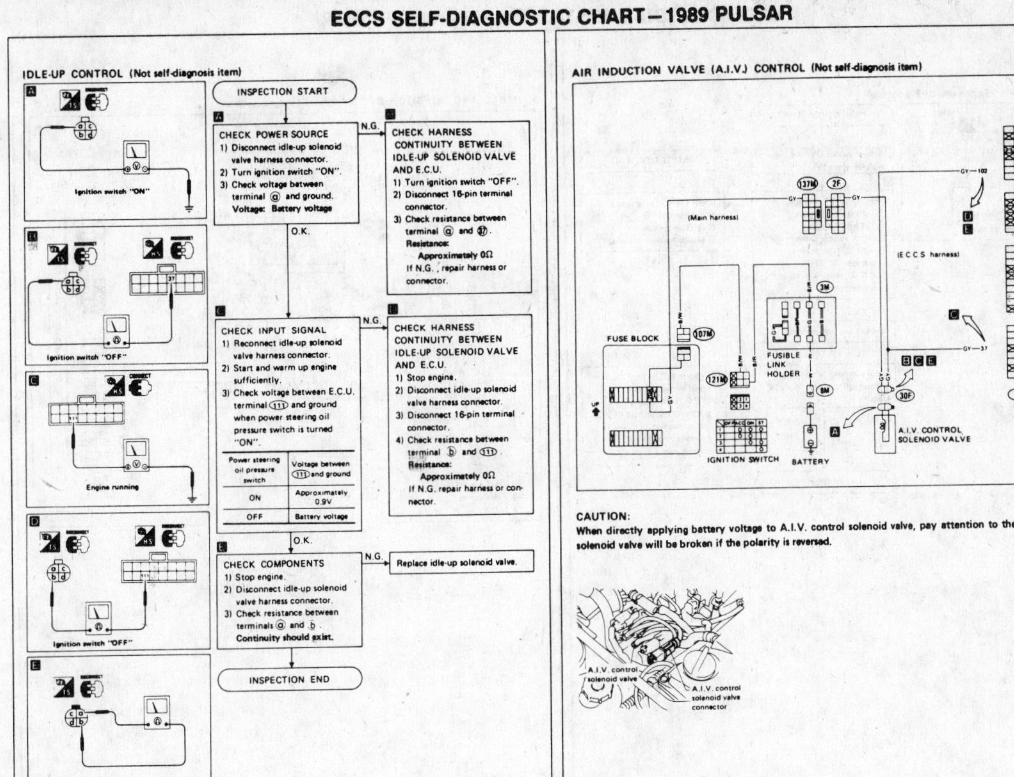
IDLE-UP CONTROL (Not self-diagnosis item)
INSPECTION START
CHECK POWER SOURCE
1) Disconnect idle-up solenoid valve harness connector.
2) Turn ignition switch "ON".
3) Check voltage between terminal ⓐ and ground.
Voltage: Battery voltage
CHECK HARNESS CONTINUITY BETWEEN IDLE-UP SOLENOID VALVE AND E.C.U.
1) Turn ignition switch "OFF".
2) Disconnect 16-pin terminal connector.
3) Check resistance between terminal ⓐ and 37.
Resistance: Approximately 0Ω
If N.G., repair harness or connector.
CHECK INPUT SIGNAL
1) Reconnect idle-up solenoid valve harness connector.
2) Start and warm up engine sufficiently.
3) Check voltage between E.C.U. terminal 111 and ground when power steering oil pressure switch is turned "ON".
Power steering oil pressure switch | Voltage between 111 and ground
ON | Approximately 0.9V
OFF | Battery voltage
CHECK HARNESS CONTINUITY BETWEEN IDLE-UP SOLENOID VALVE AND E.C.U.
1) Stop engine.
2) Disconnect idle-up solenoid valve harness connector.
3) Disconnect 16-pin terminal connector.
4) Check resistance between terminal ⓑ and 111.
Resistance: Approximately 0Ω
If N.G. repair harness or connector.
CHECK COMPONENTS
1) Stop engine.
2) Disconnect idle-up solenoid valve harness connector.
3) Check resistance between terminals ⓐ and ⓑ.
Continuity should exist.
Replace idle-up solenoid valve.
INSPECTION END
N.G.
O.K.
Ignition switch "ON"
Ignition switch "OFF"
Engine running
AIR INDUCTION VALVE (A.I.V.) CONTROL (Not self-diagnosis item)
(Main harness)
(E C C S harness)
E.C.C.S. CONTROL UNIT
FUSE BLOCK
FUSIBLE LINK HOLDER
IGNITION SWITCH
BATTERY
A.I.V. CONTROL SOLENOID VALVE
CAUTION:
When directly applying battery voltage to A.I.V. control solenoid valve, pay attention to the polarity. The solenoid valve will be broken if the polarity is reversed.
A.I.V. control solenoid valve
A.I.V. control solenoid valve connector

ECCS SELF-DIAGNOSTIC CHART—1989 PULSAR

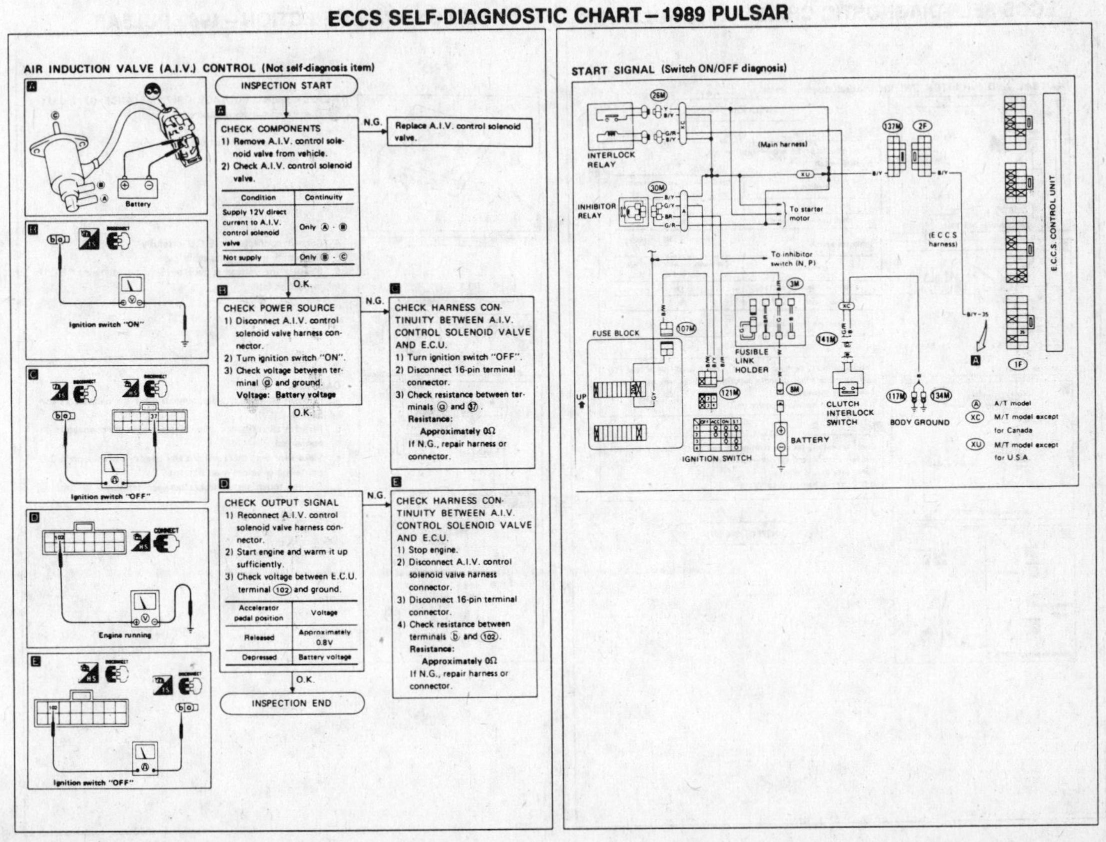
AIR INDUCTION VALVE (A.I.V.) CONTROL (Not self-diagnosis item)
INSPECTION START
CHECK COMPONENTS
1) Remove A.I.V. control solenoid valve from vehicle.
2) Check A.I.V. control solenoid valve.
Condition | Continuity
Supply 12V direct current to A.I.V. control solenoid valve | Only Ⓐ - Ⓑ
Not supply | Only Ⓑ - Ⓒ
Replace A.I.V. control solenoid valve.
Battery
CHECK POWER SOURCE
1) Disconnect A.I.V. control solenoid valve harness connector.
2) Turn ignition switch "ON".
3) Check voltage between terminal ⓐ and ground.
Voltage: Battery voltage
CHECK HARNESS CONTINUITY BETWEEN A.I.V. CONTROL SOLENOID VALVE AND E.C.U.
1) Turn ignition switch "OFF".
2) Disconnect 16-pin terminal connector.
3) Check resistance between terminals ⓐ and 37.
Resistance: Approximately 0Ω
If N.G., repair harness or connector.
CHECK OUTPUT SIGNAL
1) Reconnect A.I.V. control solenoid valve harness connector.
2) Start engine and warm it up sufficiently.
3) Check voltage between E.C.U. terminal 102 and ground.
Accelerator pedal position | Voltage
Released | Approximately 0.8V
Depressed | Battery voltage
CHECK HARNESS CONTINUITY BETWEEN A.I.V. CONTROL SOLENOID VALVE AND E.C.U.
1) Stop engine.
2) Disconnect A.I.V. control solenoid valve harness connector.
3) Disconnect 16-pin terminal connector.
4) Check resistance between terminals ⓑ and 102.
Resistance: Approximately 0Ω
If N.G., repair harness or connector.
INSPECTION END
N.G.
O.K.
Ignition switch "ON"
Ignition switch "OFF"
Engine running
START SIGNAL (Switch ON/OFF diagnosis)
INTERLOCK RELAY
INHIBITOR RELAY
(Main harness)
To starter motor
To inhibitor switch (N, P)
(E C C S harness)
E.C.C.S. CONTROL UNIT
FUSE BLOCK
FUSIBLE LINK HOLDER
IGNITION SWITCH
BATTERY
CLUTCH INTERLOCK SWITCH
BODY GROUND
Ⓐ A/T model
XC M/T model except for Canada
XU M/T model except for U.S.A.

ECCS SELF-DIAGNOSTIC CHART—1989 PULSAR

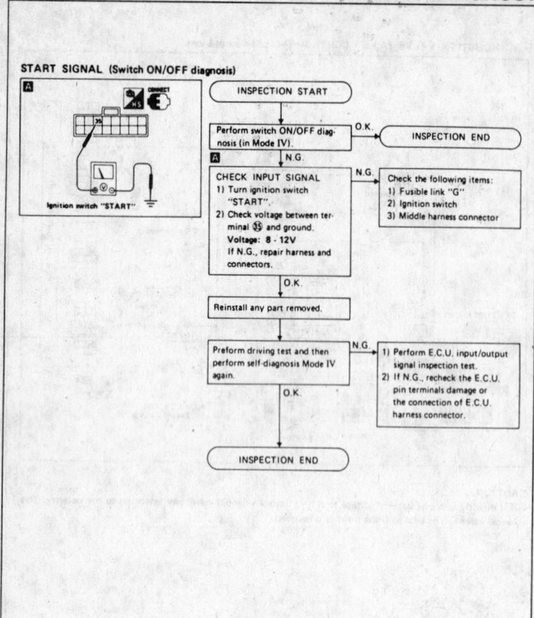
START SIGNAL (Switch ON/OFF diagnosis)
A
CONNECT
Ignition switch "START"
INSPECTION START
Perform switch ON/OFF diagnosis (in Mode IV).
O.K.
INSPECTION END
A
N.G.
CHECK INPUT SIGNAL
1) Turn ignition switch "START".
2) Check voltage between terminal (35) and ground.
Voltage: 8 - 12V
If N.G., repair harness and connectors.
N.G.
Check the following items:
1) Fusible link "G"
2) Ignition switch
3) Middle harness connector
O.K.
Reinstall any part removed.
Preform driving test and then perform self-diagnosis Mode IV again.
N.G.
1) Perform E.C.U. input/output signal inspection test.
2) If N.G., recheck the E.C.U. pin terminals damage or the connection of E.C.U. harness connector.
O.K.
INSPECTION END

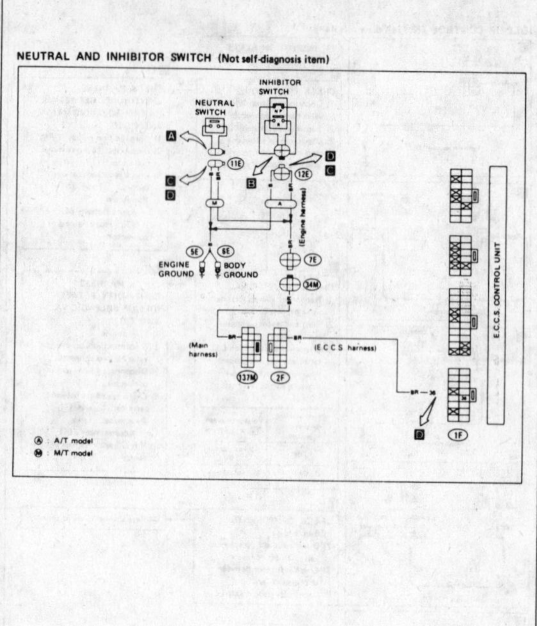
NEUTRAL AND INHIBITOR SWITCH (Not self-diagnosis item)
INHIBITOR SWITCH
NEUTRAL SWITCH
11E
12E
ENGINE GROUND
BODY GROUND
5E
6E
7E
34M
(Main harness)
(E C C S harness)
137M
2F
1F
E.C.C.S. CONTROL UNIT
Ⓐ A/T model
Ⓜ M/T model

ECCS SELF-DIAGNOSTIC CHART—1989 PULSAR

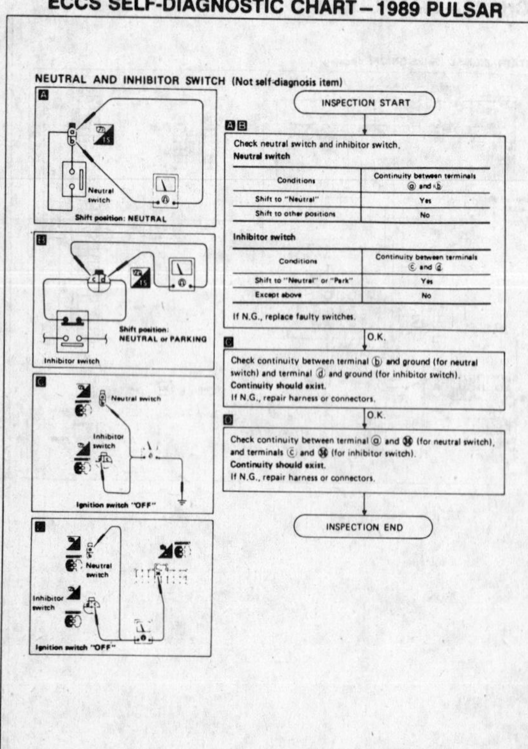
NEUTRAL AND INHIBITOR SWITCH (Not self-diagnosis item)
A
Neutral switch
Shift position: NEUTRAL
B
Shift position: NEUTRAL or PARKING
Inhibitor switch
C
Neutral switch
Inhibitor switch
Ignition switch "OFF"
D
Neutral switch
Inhibitor switch
Ignition switch "OFF"
INSPECTION START
A B
Check neutral switch and inhibitor switch.
Neutral switch
Conditions | Continuity between terminals ⓐ and ⓑ
Shift to "Neutral" | Yes
Shift to other positions | No
Inhibitor switch
Conditions | Continuity between terminals ⓒ and ⓓ
Shift to "Neutral" or "Park" | Yes
Except above | No
If N.G., replace faulty switches.
O.K.
C
Check continuity between terminal ⓑ and ground (for neutral switch) and terminal ⓓ and ground (for inhibitor switch).
Continuity should exist.
If N.G., repair harness or connectors.
O.K.
D
Check continuity between terminal ⓐ and (36) (for neutral switch), and terminals ⓒ and (36) (for inhibitor switch).
Continuity should exist.
If N.G., repair harness or connectors.
INSPECTION END

ECU SIGNAL INSPECTION—1989 PULSAR

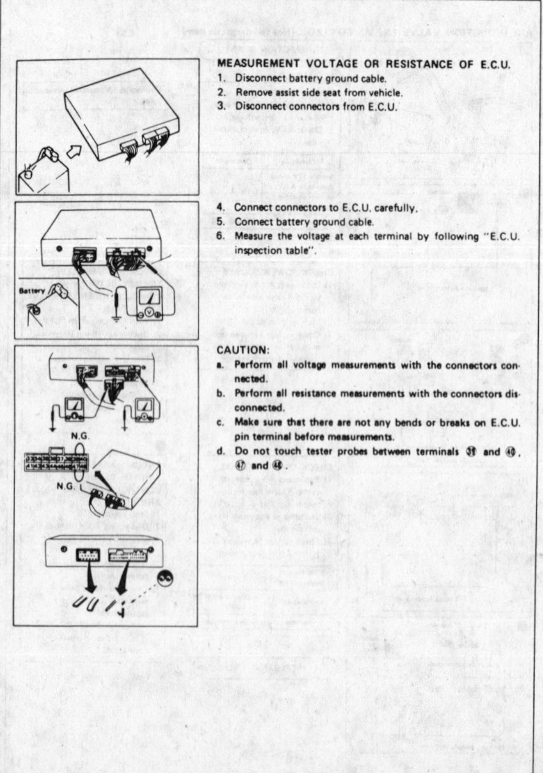
MEASUREMENT VOLTAGE OR RESISTANCE OF E.C.U.
1. Disconnect battery ground cable.
2. Remove assist side seat from vehicle.
3. Disconnect connectors from E.C.U.
4. Connect connectors to E.C.U. carefully.
5. Connect battery ground cable.
6. Measure the voltage at each terminal by following "E.C.U. inspection table".
Battery
N.G.
N.G.
CAUTION:
a. Perform all voltage measurements with the connectors connected.
b. Perform all resistance measurements with the connectors disconnected.
c. Make sure that there are not any bends or breaks on E.C.U. pin terminal before measurements.
d. Do not touch tester probes between terminals (39) and (40), (47) and (48).

ECU SIGNAL INSPECTION—1989 PULSAR

E.C.U. inspection table

*Data are reference values.

TERMINAL NO.	ITEM	CONDITION	*DATA
1	Ignition signal	Engine is running.	0 - 1.0V
7	Exhaust gas temperature sensor	Engine is running. — Idle speed	1.0V or more
		Engine is running. — E.G.R. system is operating.	0 - 1.0V
15	Air flow meter	Engine is running. Do not run engine at high speed under no-load.	0.5 - 4.0V Output voltage varies with engine revolution.
17	Water temperature sensor	Engine is running.	0.5 - 5.0V Output voltage varies with engine water temperature.
18	Exhaust gas sensor	Engine is running. — After warming up sufficiently	0 - Approximately 1.0V
19	Throttle sensor	Ignition switch "ON"	0.4 - 4.0V Output voltage varies with the throttle valve opening angle.
21 31	Crank angle sensor (Reference signal)	Engine is running. Do not run engine at high speed under no-load.	0.3 - 0.5V
22 32	Crank angle sensor (Position signal)	Engine is running. Do not run engine at high speed under no-load.	2.0 - 2.7V
34	Idle switch (− side)	Ignition switch "ON" — Throttle valve: Idle position	Approximately 10V
		Ignition switch "ON" — Throttle valve: Any position except idle position	Approximately 0V
35	Start signal	Cranking	8 - 12V

*Data are reference values.

TERMINAL NO.	ITEM	CONDITION	*DATA
36	Neutral switch & Inhibitor switch	Ignition switch "ON" — Neutral/Parking	0V
		Ignition switch "ON" — Except the above gear position	Approximately 6.5V
37	Ignition switch	Ignition switch "OFF"	0V
		Ignition switch "ON"	BATTERY VOLTAGE (11 - 14V)
41	Air conditioner	Engine is running. — Both air conditioner switch and blower switch are "ON".	BATTERY VOLTAGE (11 - 14V)
43	Power steering switch	Engine is running. — Steering wheel is turned.	Approximately 0V
		Engine is running. — Except above operation	Approximately 7.0 - 9.0V
44	Idle switch (⊕ side)	Ignition switch "ON" — Throttle valve: Idle position	Approximately 9 - 11V
		Ignition switch "ON" — Throttle valve: Except idle position	BATTERY VOLTAGE (11 - 14V)
46	Power source	Ignition switch "OFF"	BATTERY VOLTAGE (11 - 14V)
39 47	Power source for E.C.U.	Ignition switch "ON"	BATTERY VOLTAGE (11 - 14V)
101 110	Injector	Engine is running.	BATTERY VOLTAGE (11 - 14V)

ECU SIGNAL INSPECTION—1989 PULSAR

*Data are reference values.

TERMINAL NO.	ITEM	CONDITION	*DATA
102	A.I.V. control solenoid valve	Engine is running. — Idle speed	Approximately 0.8V
		Engine is running. — Accelerator pedal is depressed. — After warming up	BATTERY VOLTAGE (11 - 14V)
104	Fuel pump relay	Ignition switch "ON" — For 5 seconds after turning ignition switch "ON" / Engine is running.	0.7 - 0.9V
		Ignition switch "ON" — In 5 seconds after turning ignition switch "ON"	BATTERY VOLTAGE (11 - 14V)
105	E.G.R. & canister control solenoid valve	Engine is running. — Engine is cold. [Water temperature is below 60°C (140°F).]	0.6 - 0.9V
		Engine is running. — After warming up [Water temperature is above 60°C (140°F).]	BATTERY VOLTAGE (11 - 14V)
106	Mixture heater relay	Engine is running. — Engine is cold. [Water temperature is below 50°C (122°F).]	Approximately 0.7V
		Engine is running. — After warming up [Water temperature is above 50° (122°F).]	BATTERY VOLTAGE (11 - 14V)

*Data are reference values.

TERMINAL NO.	ITEM	CONDITION	*DATA
111	Idle-up solenoid valve	Engine is running. — After warming up — Idle speed	BATTERY VOLTAGE (11 - 14V)
		Engine is running. — After warming up — Idle speed — Power steering switch is "ON"	Approximately 0.9V
114	A.A.C. valve	Engine is running. — Idle speed	7 - 8V
		Engine is running. — Power steering switch is "ON" — Air conditioner is operating. — Rear defogger is "ON". — Headlights are in high position.	4 - 7V

E.C.U. PIN CONNECTOR TERMINAL LAYOUT

16-pin connector

101	102	103	104	105	106	107	108
109	110	111	112	113	114	115	116

12-pin connector

1	2	3	4	5	6
7	8	9	10	11	12

20-pin connector

13	14	15	16	17	18	19	20	21	22
23	24	25	26	27	28	29	30	31	32

16-pin connector

33	34	35	36	37	38	39	40
41	42	43	44	45	46	47	48

ECCS MIXTURE RATIO FEEDBACK SYSTEM INSPECTION—1989 PULSAR

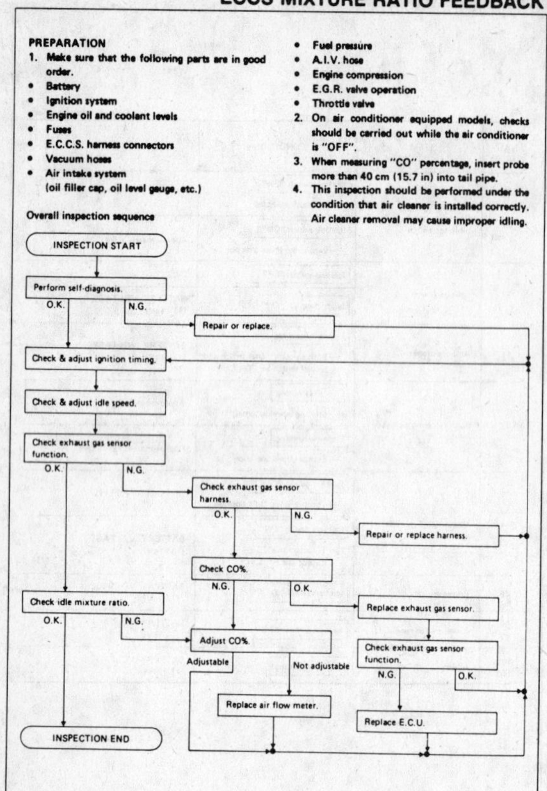
PREPARATION
1. Make sure that the following parts are in good order.
• Battery
• Ignition system
• Engine oil and coolant levels
• Fuses
• E.C.C.S. harness connectors
• Vacuum hoses
• Air intake system (oil filler cap, oil level gauge, etc.)
• Fuel pressure
• A.I.V. hose
• Engine compression
• E.G.R. valve operation
• Throttle valve
2. On air conditioner equipped models, checks should be carried out while the air conditioner is "OFF".
3. When measuring "CO" percentage, insert probe more than 40 cm (15.7 in) into tail pipe.
4. This inspection should be performed under the condition that air cleaner is installed correctly. Air cleaner removal may cause improper idling.
Overall inspection sequence
INSPECTION START
Perform self-diagnosis.
O.K.
N.G.
Repair or replace.
Check & adjust ignition timing.
Check & adjust idle speed.
Check exhaust gas sensor function.
O.K.
N.G.
Check exhaust gas sensor harness.
O.K.
N.G.
Repair or replace harness.
Check CO%.
N.G.
O.K.
Check idle mixture ratio.
O.K.
N.G.
Replace exhaust gas sensor.
Adjust CO%.
Adjustable
Not adjustable
Check exhaust gas sensor function.
N.G.
O.K.
Replace air flow meter.
Replace E.C.U.
INSPECTION END

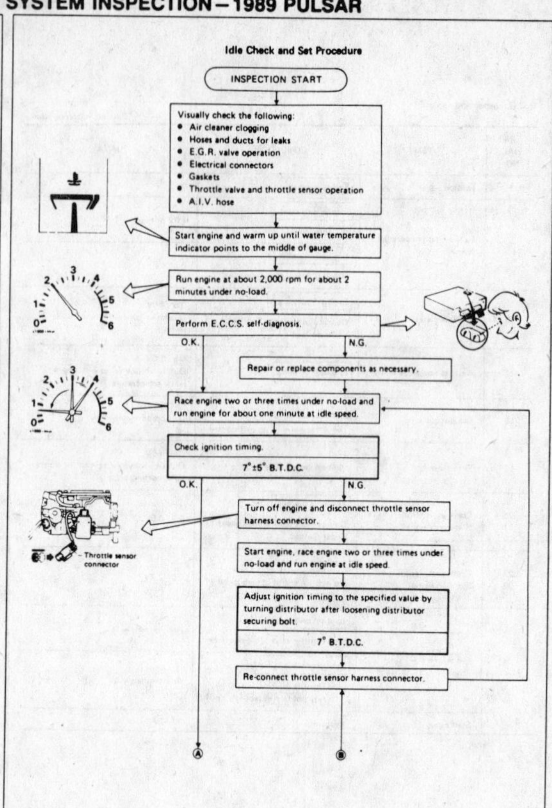
Idle Check and Set Procedure
INSPECTION START
Visually check the following:
• Air cleaner clogging
• Hoses and ducts for leaks
• E.G.R. valve operation
• Electrical connectors
• Gaskets
• Throttle valve and throttle sensor operation
• A.I.V. hose
Start engine and warm up until water temperature indicator points to the middle of gauge.
Run engine at about 2,000 rpm for about 2 minutes under no-load.
Perform E.C.C.S. self-diagnosis.
O.K.
N.G.
Repair or replace components as necessary.
Race engine two or three times under no-load and run engine for about one minute at idle speed.
Check ignition timing.
7°±5° B.T.D.C.
O.K.
N.G.
Turn off engine and disconnect throttle sensor harness connector.
Throttle sensor connector
Start engine, race engine two or three times under no-load and run engine at idle speed.
Adjust ignition timing to the specified value by turning distributor after loosening distributor securing bolt.
7° B.T.D.C.
Re-connect throttle sensor harness connector.
A
B

ECCS MIXTURE RATIO FEEDBACK SYSTEM INSPECTION—1989 PULSAR

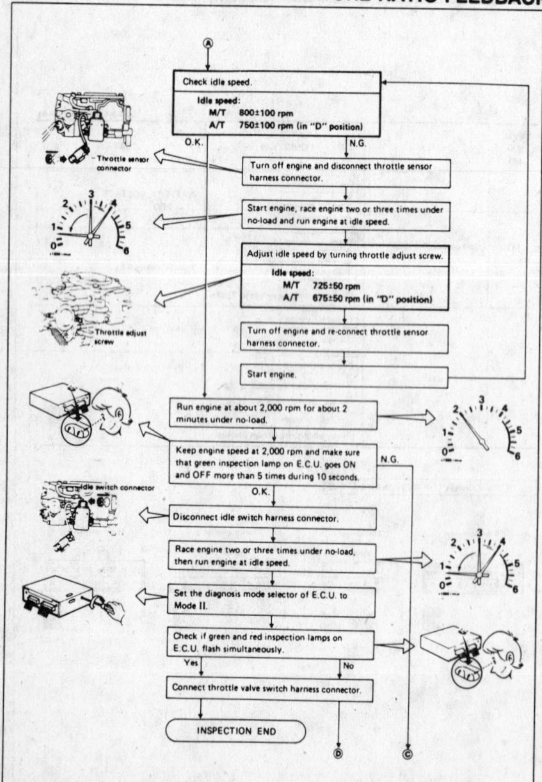
A
Check idle speed.
Idle speed:
M/T 800±100 rpm
A/T 750±100 rpm (in "D" position)
O.K.
N.G.
Throttle sensor connector
Turn off engine and disconnect throttle sensor harness connector.
Start engine, race engine two or three times under no-load and run engine at idle speed.
Adjust idle speed by turning throttle adjust screw.
Idle speed:
M/T 725±50 rpm
A/T 675±50 rpm (in "D" position)
Throttle adjust screw
Turn off engine and re-connect throttle sensor harness connector.
Start engine.
Run engine at about 2,000 rpm for about 2 minutes under no-load.
Keep engine speed at 2,000 rpm and make sure that green inspection lamp on E.C.U. goes ON and OFF more than 5 times during 10 seconds.
N.G.
O.K.
Idle switch connector
Disconnect idle switch harness connector.
Race engine two or three times under no-load, then run engine at idle speed.
Set the diagnosis mode selector of E.C.U. to Mode II.
Check if green and red inspection lamps on E.C.U. flash simultaneously.
Yes
No
Connect throttle valve switch harness connector.
INSPECTION END
D
C

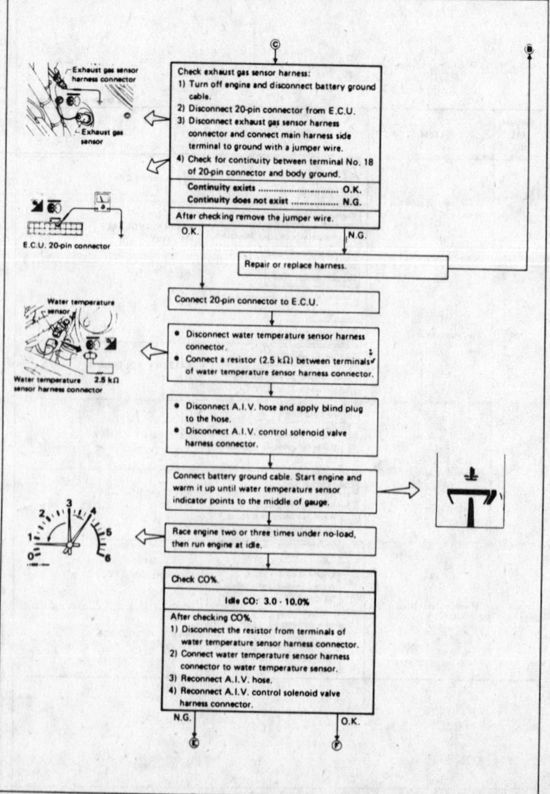
C
B
Exhaust gas sensor harness connector
Exhaust gas sensor
Check exhaust gas sensor harness:
1) Turn off engine and disconnect battery ground cable.
2) Disconnect 20-pin connector from E.C.U.
3) Disconnect exhaust gas sensor harness connector and connect main harness side terminal to ground with a jumper wire.
4) Check for continuity between terminal No. 18 of 20-pin connector and body ground.
Continuity exists .... O.K.
Continuity does not exist .... N.G.
After checking remove the jumper wire.
E.C.U. 20-pin connector
O.K.
N.G.
Repair or replace harness.
Connect 20-pin connector to E.C.U.
Water temperature sensor
• Disconnect water temperature sensor harness connector.
• Connect a resistor (2.5 kΩ) between terminals of water temperature sensor harness connector.
Water temperature sensor harness connector
2.5 kΩ
• Disconnect A.I.V. hose and apply blind plug to the hose.
• Disconnect A.I.V. control solenoid valve harness connector.
Connect battery ground cable. Start engine and warm it up until water temperature sensor indicator points to the middle of gauge.
Race engine two or three times under no-load, then run engine at idle.
Check CO%.
Idle CO: 3.0 - 10.0%
After checking CO%,
1) Disconnect the resistor from terminals of water temperature sensor harness connector.
2) Connect water temperature sensor harness connector to water temperature sensor.
3) Reconnect A.I.V. hose.
4) Reconnect A.I.V. control solenoid valve harness connector.
N.G.
O.K.
E
F

ECCS MIXTURE RATIO FEEDBACK SYSTEM INSPECTION—1989 PULSAR

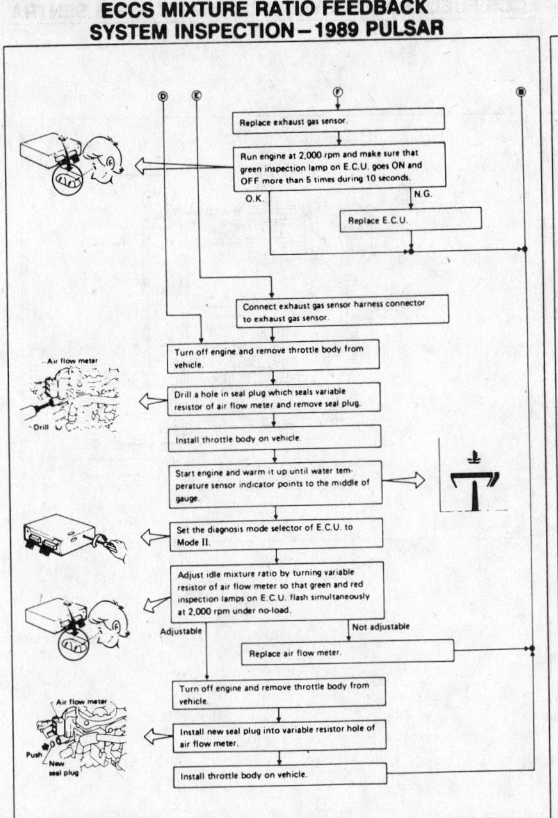
D
E
F
B
Replace exhaust gas sensor.
Run engine at 2,000 rpm and make sure that green inspection lamp on E.C.U. goes ON and OFF more than 5 times during 10 seconds.
O.K.
N.G.
Replace E.C.U.
Connect exhaust gas sensor harness connector to exhaust gas sensor.
Turn off engine and remove throttle body from vehicle.
Air flow meter
Drill
Drill a hole in seal plug which seals variable resistor of air flow meter and remove seal plug.
Install throttle body on vehicle.
Start engine and warm it up until water temperature sensor indicator points to the middle of gauge.
Set the diagnosis mode selector of E.C.U. to Mode II.
Adjust idle mixture ratio by turning variable resistor of air flow meter so that green and red inspection lamps on E.C.U. flash simultaneously at 2,000 rpm under no-load.
Adjustable
Not adjustable
Replace air flow meter.
Turn off engine and remove throttle body from vehicle.
Air flow meter
Push
New seal plug
Install new seal plug into variable resistor hole of air flow meter.
Install throttle body on vehicle.

ECCS COMPONENTS LOCATION—1988 SENTRA

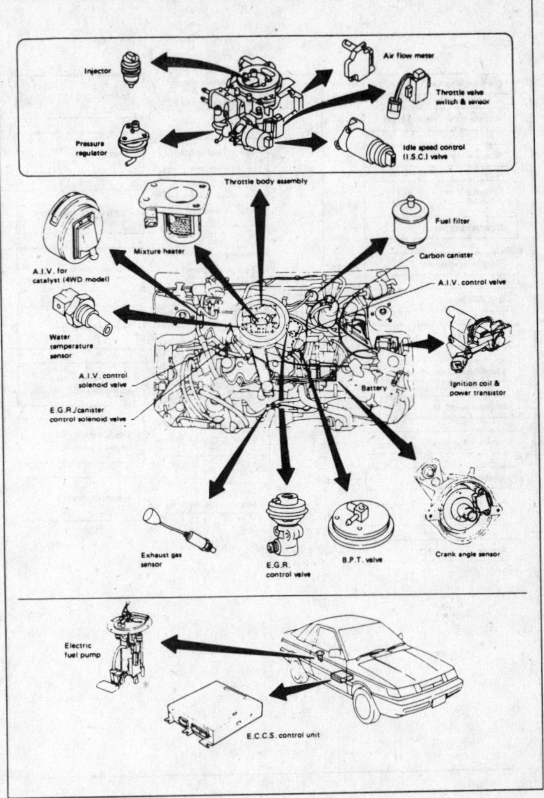
Injector
Air flow meter
Throttle valve switch & sensor
Pressure regulator
Idle speed control (I.S.C.) valve
Throttle body assembly
Mixture heater
Fuel filter
Carbon canister
A.I.V. for catalyst (4WD model)
A.I.V. control valve
Water temperature sensor
A.I.V. control solenoid valve
E.G.R./canister control solenoid valve
Battery
Ignition coil & power transistor
Exhaust gas sensor
E.G.R. control valve
B.P.T. valve
Crank angle sensor
Electric fuel pump
E.C.C.S. control unit

ECCS SYSTEM SCHEMATIC—1988 SENTRA

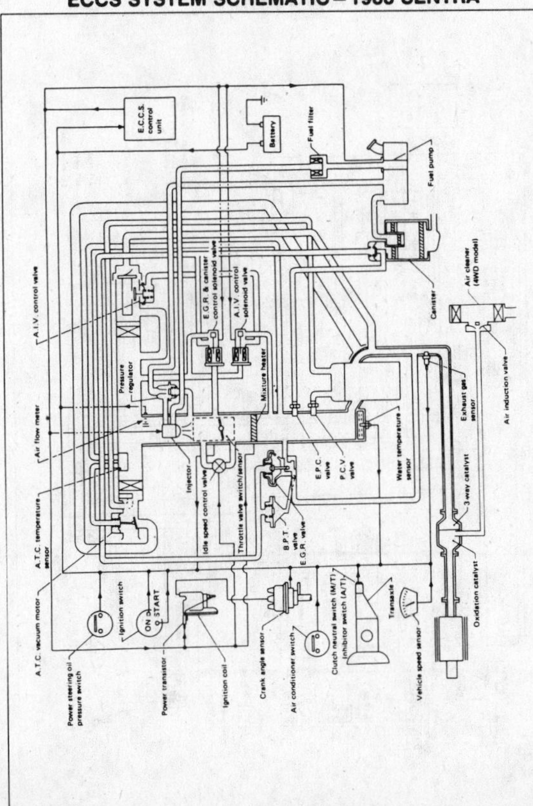
E.C.C.S. control unit
Battery
Fuel filter
Fuel pump
Canister
Air cleaner (4WD model)
A.I.V. control valve
E.G.R. & canister control solenoid valve
A.I.V. control solenoid valve
Pressure regulator
Mixture heater
Air flow meter
Injector
Idle speed control valve
Throttle valve switch/sensor
E.P.C. valve
P.C.V. valve
Water temperature sensor
Exhaust gas sensor
Air induction valve
3-way catalyst
B.P.T. valve
E.G.R. valve
A.T.C. temperature sensor
A.T.C. vacuum motor
Power steering oil pressure switch
Ignition switch
ON START
Power transistor
Ignition coil
Crank angle sensor
Air conditioner switch
Clutch neutral switch (M/T) Inhibitor switch (A/T)
Transaxle
Vehicle speed sensor
Oxidation catalyst

ECCS VACUUM CONTROL SCHEMATIC—1988 SENTRA

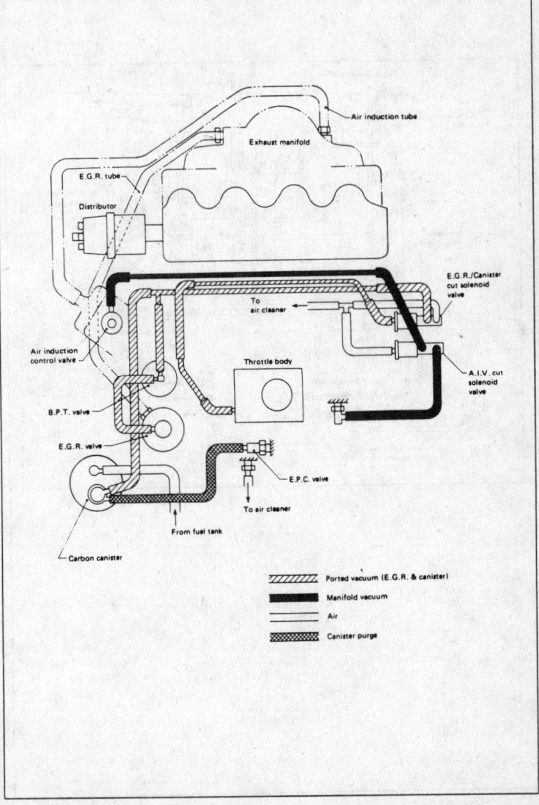
Air induction tube
Exhaust manifold
E.G.R. tube
Distributor
E.G.R./Canister cut solenoid valve
To air cleaner
Air induction control valve
Throttle body
A.I.V. cut solenoid valve
B.P.T. valve
E.G.R. valve
E.P.C. valve
To air cleaner
From fuel tank
Carbon canister
Ported vacuum (E.G.R. & canister)
Manifold vacuum
Air
Canister purge

ECCS CONTROL SYSTEM CHART—1988 SENTRA

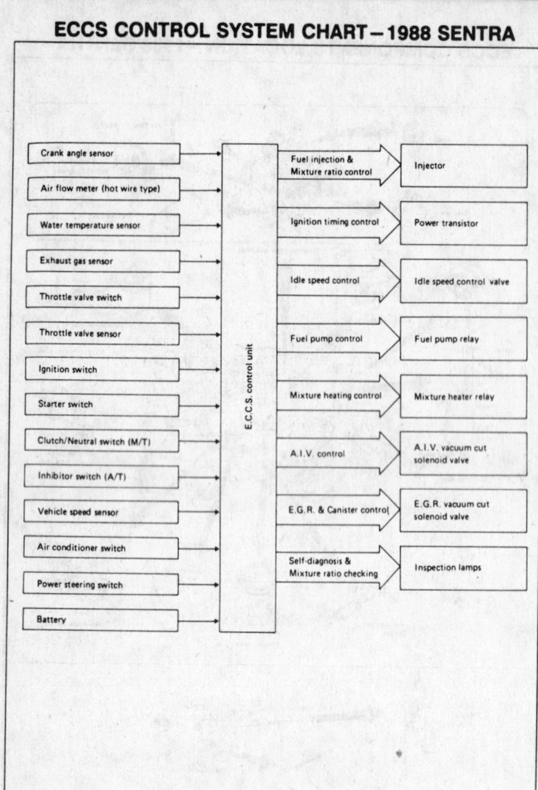
Crank angle sensor
Air flow meter (hot wire type)
Water temperature sensor
Exhaust gas sensor
Throttle valve switch
Throttle valve sensor
Ignition switch
Starter switch
Clutch/Neutral switch (M/T)
Inhibitor switch (A/T)
Vehicle speed sensor
Air conditioner switch
Power steering switch
Battery
E.C.C.S. control unit
Fuel injection & Mixture ratio control
Ignition timing control
Idle speed control
Fuel pump control
Mixture heating control
A.I.V. control
E.G.R. & Canister control
Self-diagnosis & Mixture ratio checking
Injector
Power transistor
Idle speed control valve
Fuel pump relay
Mixture heater relay
A.I.V. vacuum cut solenoid valve
E.G.R. vacuum cut solenoid valve
Inspection lamps

ECCS FUEL FLOW SYSTEM SCHEMATIC—1988 SENTRA

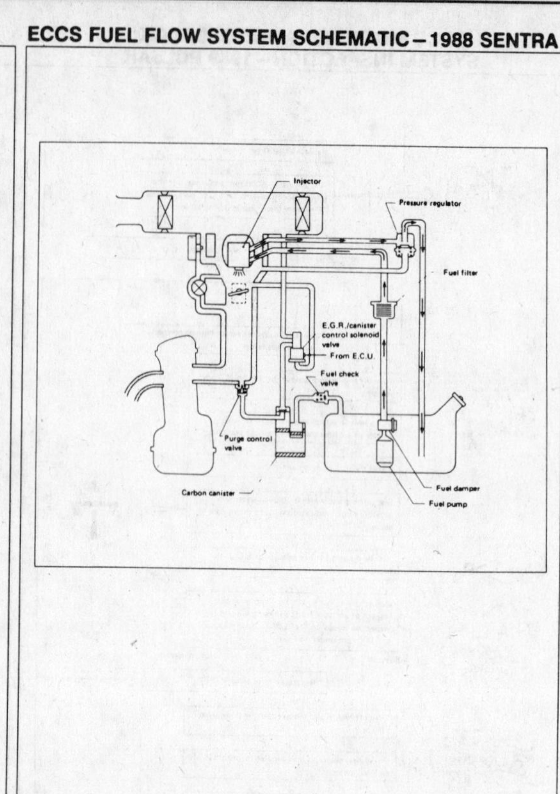
Injector
Pressure regulator
Fuel filter
E.G.R./canister control solenoid valve
From E.C.U.
Fuel check valve
Purge control valve
Carbon canister
Fuel damper
Fuel pump

ECCS AIR FLOW SYSTEM SCHEMATIC—1988 SENTRA

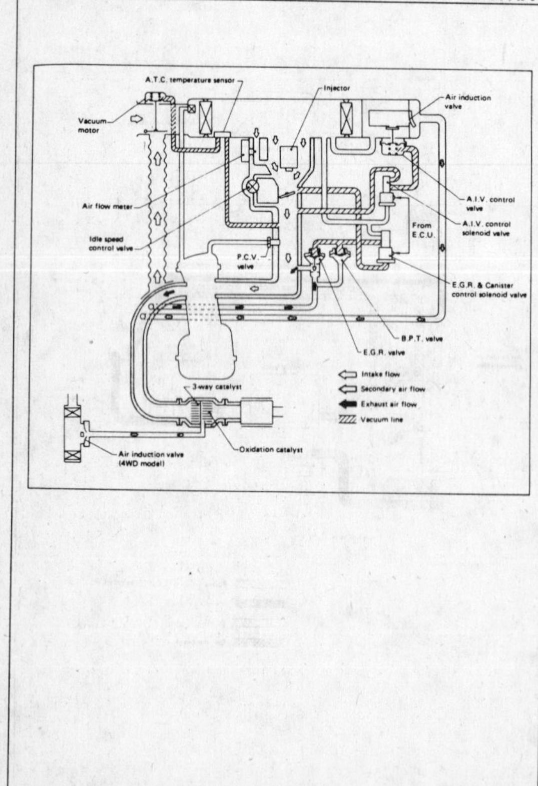
A.T.C. temperature sensor
Injector
Air induction valve
Vacuum motor
Air flow meter
Idle speed control valve
P.C.V. valve
A.I.V. control valve
A.I.V. control solenoid valve
From E.C.U.
E.G.R. & Canister control solenoid valve
B.P.T. valve
E.G.R. valve
3-way catalyst
Oxidation catalyst
Air induction valve (4WD model)
Intake flow
Secondary air flow
Exhaust air flow
Vacuum line

ECCS CONTROL UNIT WIRING SCHEMATIC—1988 SENTRA

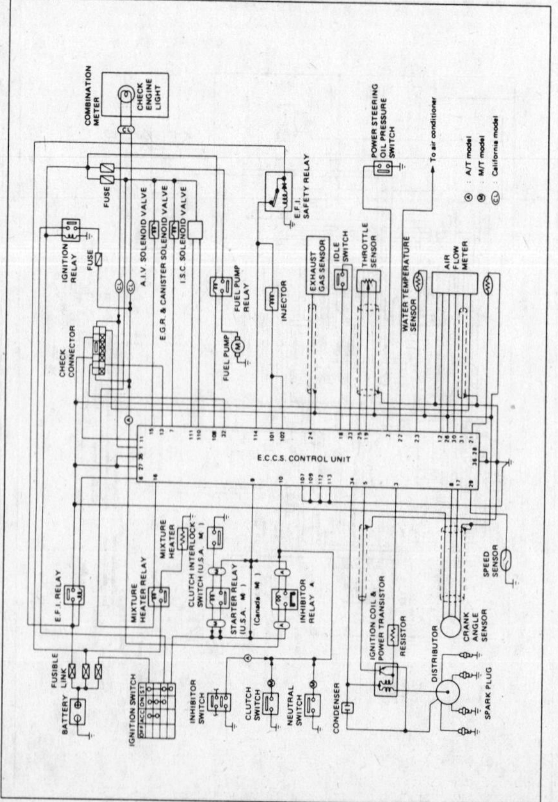
COMBINATION METER
CHECK ENGINE LIGHT
POWER STEERING OIL PRESSURE SWITCH
To air conditioner
A/T model
M/T model
California model
E.F.I. SAFETY RELAY
FUSE
A.I.V. SOLENOID VALVE
E.G.R. & CANISTER SOLENOID VALVE
I.S.C. SOLENOID VALVE
IGNITION RELAY
FUEL PUMP RELAY
INJECTOR
EXHAUST GAS SENSOR
IDLE SWITCH
THROTTLE SENSOR
WATER TEMPERATURE SENSOR
AIR FLOW METER
CHECK CONNECTOR
FUEL PUMP
E.C.C.S. CONTROL UNIT
E.F.I. RELAY
MIXTURE HEATER RELAY
MIXTURE HEATER
CLUTCH INTERLOCK SWITCH
STARTER RELAY
INHIBITOR RELAY
IGNITION COIL & POWER TRANSISTOR
RESISTOR
SPEED SENSOR
CRANK ANGLE SENSOR
DISTRIBUTOR
SPARK PLUG
CONDENSER
NEUTRAL SWITCH
CLUTCH SWITCH
INHIBITOR SWITCH
IGNITION SWITCH
FUSIBLE LINK
BATTERY

ECCS SYSTEM WIRING SCHEMATIC—1988 SENTRA

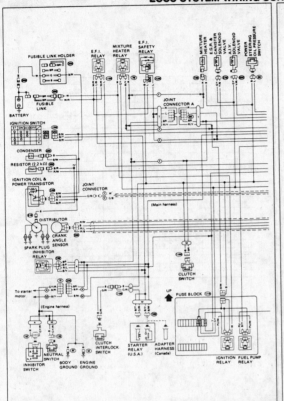

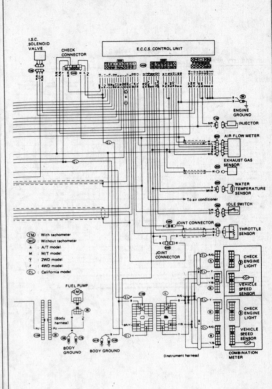

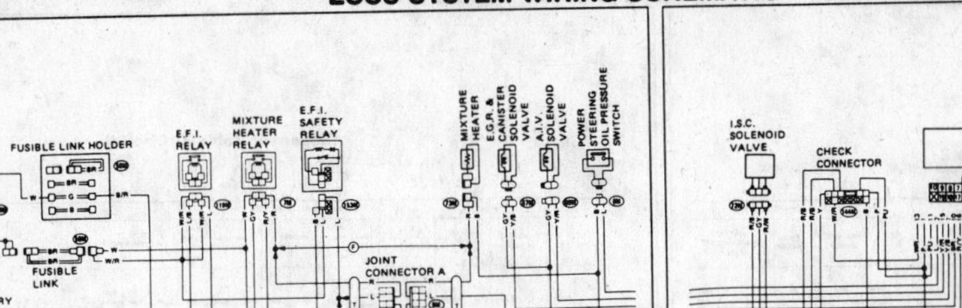
FUSIBLE LINK HOLDER
E.F.I. RELAY
MIXTURE HEATER RELAY
E.F.I. SAFETY RELAY
MIXTURE HEATER
E.G.R. & CANISTER SOLENOID VALVE
A.I.V. SOLENOID VALVE
POWER STEERING OIL PRESSURE SWITCH
I.S.C. SOLENOID VALVE
CHECK CONNECTOR
FUSIBLE LINK
BATTERY
JOINT CONNECTOR A

ECCS DIAGNOSTIC PROCEDURE—1988 SENTRA

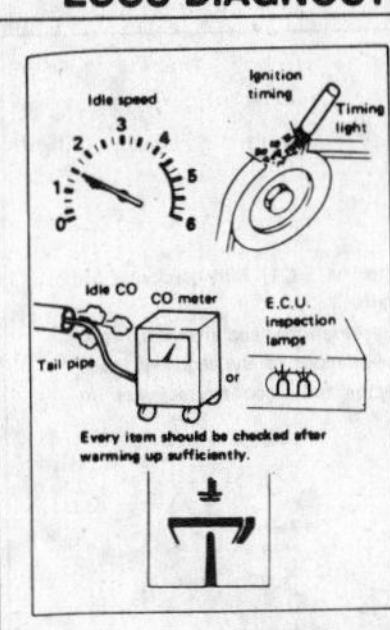
Idle speed
Ignition timing
Timing light
Idle CO
CO meter
E.C.U. inspection lamps
Tail pipe
or
Every item should be checked after warming up sufficiently.

Driveability

1. Make sure that the following items are in the proper condition.

CHECK DATA

1) Idle speed
   M/T: 800±100 rpm
   A/T: 700±100 rpm (in "D" position)
2) Ignition timing
   7°±5° B.T.D.C.
3) Idle CO
   - Self-diagnosis mode: mode II
   - Race engine, and immediately after the engine returns to the idle condition, confirm that E.C.U. red and green inspection lamps flash.
4) Mixture ratio at approximately 2,000 rpm of engine speed.
   - Self-diagnosis mode: mode II
   - Check the simultaneous flashes of E.C.U. green and red lamps.
5) Idle switch operation
   "ON" at idle position and "OFF" when accelerator pedal is depressed.

2. Perform driving test.
   Evaluate effectiveness of adjustments by driving vehicle. During driving test, perform real time diagnostic test.

3. Perform E.C.C.S. self-diagnosis.

4. If the result of driveability test is unsatisfactory, or malfunctioning conditions are found when performing E.C.C.S. self-diagnosis and real time diagnostic tests, perform general inspection, electronic system inspection and real time diagnostic inspection by following DIAGNOSTIC TABLE 1 and 2 in response to driveability trouble items. If malfunctioning, repair or replace.

Driveability (Cont'd)

5. Perform ON/OFF diagnostic test on switches.

6. Perform driving test.
   Re-evaluate vehicle performance after all the inspections are performed.

ECCS DIAGNOSTIC PROCEDURE—1988 SENTRA

Diagnostic Table 1

Sensor & actuator \ System	Fuel injection & mixture ratio feedback control	Ignition timing control	Idle speed control	Mixture heater control	A.I.V. control	E.G.R. & Canister control
Crank angle sensor	○	○	○	○		○
Air flow meter	○	○				○
Water temperature sensor	○	○	○	○	○	○
Exhaust gas sensor	○					
Idle switch	○	○	○		○	
Throttle sensor	○					
Ignition switch	○	○	○			
Neutral & clutch switch			○			
Vehicle speed sensor	○		○			○
Air conditioner switch			○			
Power steering switch			○			
Battery voltage	○					
Injector	○					
Power transistor		○				
I.S.C. valve			○			
Mixture heater relay				○		
A.I.V. control solenoid valve					○	
E.G.R. & Canister control solenoid valve						○

This table indicates the inspection items for the E.C.C.S. control system. For each system, it is necessary to check sensors or actuators marked "O".

ECCS DIAGNOSTIC PROCEDURE—1988 SENTRA

Diagnostic Table 2

DRIVEABILITY INSPECTION TABLE

TROUBLE ITEMS CONCERNED WITH DRIVEABILITY \ INSPECTION ITEM			GENERAL INSPECTION: FUEL FLOW SYSTEM			ELECTRIC SYSTEM				AIR FLOW SYSTEM					E C C S SYSTEM INSPECTION: CRANK ANGLE SENSOR			AIR FLOW METER	WATER TEMPERATURE SENSOR		
			Fuel pump	Fuel filter	Fuel line	Battery	Spark plug	Ignition wire	Alternator	Starter	Air cleaner	Air flow line	E G R B P T valve	I S C valve	P C V	180° signal noise	180° signal faults	1° signal faults	Poor connection	Short	Open	Poor connection
SURGE	ROAD LOAD DRIVING	Heavy load		○	○		◎	○			○	○	○		○	○	○	○	◎	○	○	○
		Middle load		○	○		◎	○			○	○	◎		○	○	○	○	◎	○	○	○
		Light load		○	○		◎	○			○	○	◎		○	○	○	○	◎	○	○	○
	ACCELERATING DRIVING	Slow acceleration		○	○		◎	○			○	○	◎		○	○	○	○	◎	○	○	○
	DECELERATING DRIVING	Rapid acceleration		○	○		◎	○			○	○	○		○	○	○	○	◎	○	○	○
		Slow deceleration		○	○		◎	○			○	○	○		○	○	○	○	◎	○	○	○
HESITATION		Rapid acceleration		○	○		◎	○			○	○	○		○	○	○	○	◎	○	○	○
		Slow acceleration		○	○		◎	○			○	○	○		○	○	○	○	◎	○	○	○
STUMBLE		Rapid acceleration		○	○		◎	○			○	○	○		○	○	○	○	◎	○	○	○
		Slow acceleration		○	○		◎	○			○	○	○		○	○	○	○	◎	○	○	○
BACKFIRE				○	○			○				○			○	○	○	○	◎	○		○
AFTER FIRE								○			○			○	○	○	○	○	◎		○	○
IDLE STABILITY							◎	○	○		○	○	◎	○	○	○	○	○	○	○	○	○
ENGINE STALL			○	○	○	○	◎	○	○		○	○	○	○	○	◎	◎	◎	◎	○	○	○
STARTABILITY			○			○	◎	○	○	○	○	○		○	○	◎	◎	◎	○	○	○	○

This table indicates the inspection items for each type of symptom. It is necessary for each symptom to check sensors or actuators marked "◎" or "○".
Items marked "◎" have a significant influence on driveability. Prior to items marked "○", check items marked "◎".
Improper mixture ratio, improper ignition condition, and an excess of E.G.R. volume can cause any symptom.
* If air flow meter circuit is short or open, the fail-safe system operates and engine revolution does not rise to 2,400 rpm or more.

Diagnostic Table 2 (Cont'd)

TROUBLE ITEMS			E C C S SYSTEM INSPECTION: IDLE SWITCH		THROTTLE SENSOR		E G R CUT SOLENOID VALVE		EXHAUST GAS SENSOR		INJECTOR	NEUTRAL/CLUTCH SWITCH	STARTER SIGNAL		IGNITION SIGNAL		BATTERY VOLTAGE	FUEL PUMP CIRCUIT			MIXTURE HEATER	A.I.V CONTROL SOLENOID VALVE	
			Short	Open	Short	Open	Short	Open	Short	Open	Clogging	Short	Short	Open	Short	Open	Low voltage	Short	Open	Poor connection	Open (for only cold condition)	Short	Open
SURGE	ROAD LOAD DRIVING	Heavy load						◎	○	○	◎									○	○	○	
		Middle load						◎	○	○	◎									○	○	○	
		Light load						◎	○	○	◎									○	○	○	
	ACCELERATING DRIVING	Slow acceleration						◎	○	○	◎									○	○	○	
	DECELERATING DRIVING	Rapid acceleration						○	○	○	◎									○	○		
		Slow deceleration						○	○	○	◎									○	○		
HESITATION		Rapid acceleration			◎	◎		○	○	○	◎	○								○	○	○	
		Slow acceleration			◎	◎		○	○	○	◎	○								○	○	○	
STUMBLE		Rapid acceleration			◎	◎		○			◎	○								○	○		
		Slow acceleration			◎	◎		○			◎	○								○	○		
BACKFIRE			○	○	○	○			○	○	◎	○								○	○		
AFTER FIRE			○	○				◎	○	○											○		○
IDLE STABILITY				◎				○	○	○	◎									○	○		
ENGINE STALL				◎					○	○	◎				○	○	○	○	○	○	○		
STARTABILITY											◎		○	○	○	○	○	○	○	○			

ECCS SELF-DIAGNOSTIC DESCRIPTION—1988 SENTRA

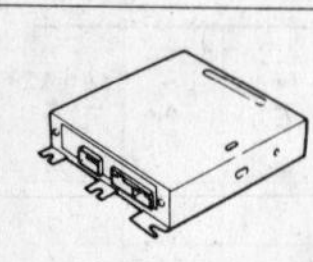

Description

The self-diagnosis is useful to diagnose malfunctions in major sensors and actuators of the E.C.C.S. system. There are 5 modes in the self-diagnosis system.

1. **Mode I – Mixture ratio feedback control monitor A**
   - During closed loop condition:
     The green inspection lamp turns ON when lean condition is detected and goes OFF by rich condition.
   - During open loop condition:
     The green inspection lamp remains ON or OFF.
2. **Mode II – Mixture ratio feedback control monitor B**
   The green inspection lamp function is the same as Mode I.
   - During closed loop condition:
     The red inspection lamp turns ON and OFF simultaneously with the green inspection lamp when the mixture ratio is controlled within the specified value.
   - During open loop condition:
     The red inspection lamp remains ON or OFF.
3. **Mode III – Self-diagnosis**
   This mode is the same as the former self-diagnosis in self-diagnosis mode.
4. **Mode IV – Switches ON/OFF diagnosis**
   During this mode, the inspection lamps monitor the switch ON/OFF condition.
   - Idle switch
   - Starter switch
   - Vehicle speed sensor
5. **Mode V – Real time diagnosis**
   The moment the malfunction is detected, the display will be presented immediately. That is, the condition at which the malfunction occurs can be found by observing the inspection lamps during driving test.

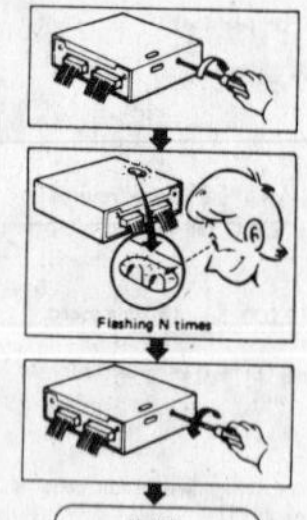
Flashing N times
Mode N

Description (Cont'd)

SWITCHING THE MODES

1. Turn ignition switch "ON".
2. Turn diagnostic mode selector on E.C.U. fully clockwise and wait the inspection lamps flash.
3. Count the number of the flashing time, and after the inspection lamps have flashed the number of the required mode, turn diagnostic mode selector fully counterclockwise immediately.

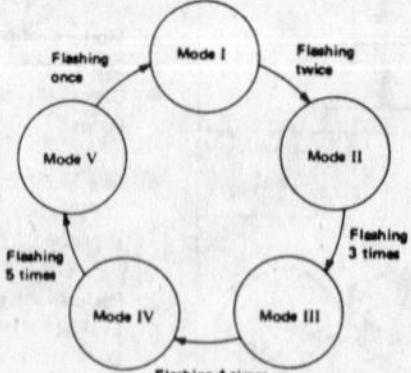
Mode I
Flashing twice
Mode II
Flashing 3 times
Mode III
Flashing 4 times
Mode IV
Flashing 5 times
Mode V
Flashing once

NOTE:
When the ignition switch is turned off during diagnosis, in each mode, and then turned back on again after the power to the E.C.U. has dropped off completely, the diagnosis will automatically return to Mode I.
The stored memory would be lost if:

1. Battery terminal is disconnected.
2. After selecting Mode III, Mode IV is selected.
   However, if the diagnostic mode selector is kept turned fully clockwise, it will continue to change in the order of Mode I → II → III → IV → V → I ... etc., and in this state the stored memory will not be erased.

ECCS SELF-DIAGNOSTIC DESCRIPTION—1988 SENTRA

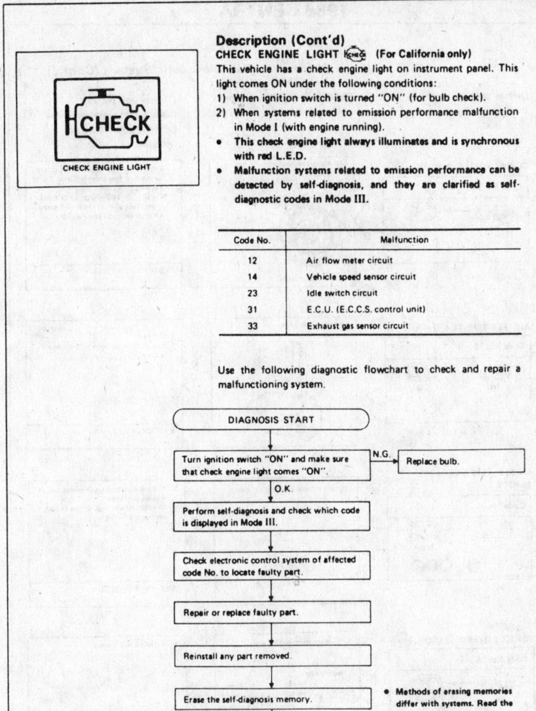
CHECK
CHECK ENGINE LIGHT

Description (Cont'd)

CHECK ENGINE LIGHT (For California only)

This vehicle has a check engine light on instrument panel. This light comes ON under the following conditions:

1) When ignition switch is turned "ON" (for bulb check).
2) When systems related to emission performance malfunction in Mode I (with engine running).

- **This check engine light always illuminates and is synchronous with red L.E.D.**
- **Malfunction systems related to emission performance can be detected by self-diagnosis, and they are clarified as self-diagnostic codes in Mode III.**

Code No.	Malfunction
12	Air flow meter circuit
14	Vehicle speed sensor circuit
23	Idle switch circuit
31	E.C.U. (E.C.C.S. control unit)
33	Exhaust gas sensor circuit

Use the following diagnostic flowchart to check and repair a malfunctioning system.

DIAGNOSIS START
Turn ignition switch "ON" and make sure that check engine light comes "ON".
N.G.
Replace bulb.
O.K.
Perform self-diagnosis and check which code is displayed in Mode III.
Check electronic control system of affected code No. to locate faulty part.
Repair or replace faulty part.
Reinstall any part removed.
Erase the self-diagnosis memory.
• Methods of erasing memories differ with systems. Read the manual before diagnosing systems.
Ⓐ

Description (Cont'd)

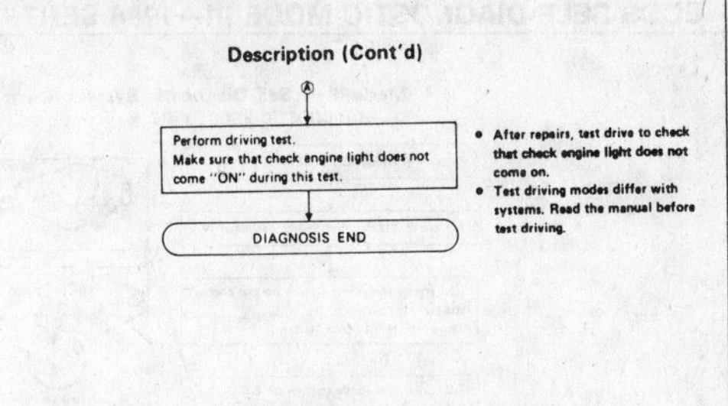
Ⓐ
Perform driving test.
Make sure that check engine light does not come "ON" during this test.
DIAGNOSIS END
• After repairs, test drive to check that check engine light does not come on.
• Test driving modes differ with systems. Read the manual before test driving.

ECCS SELF-DIAGNOSTIC MODES I AND II—1988 SENTRA

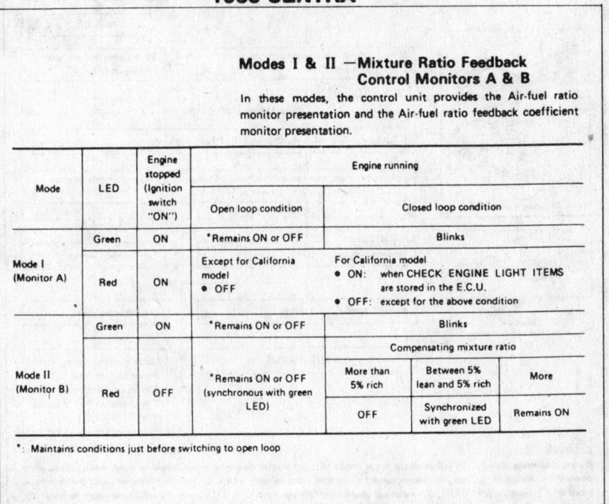

Modes I & II —Mixture Ratio Feedback Control Monitors A & B

In these modes, the control unit provides the Air-fuel ratio monitor presentation and the Air-fuel ratio feedback coefficient monitor presentation.

Mode	LED	Engine stopped (Ignition switch "ON")	Engine running			
			Open loop condition	Closed loop condition		
Mode I (Monitor A)	Green	ON	*Remains ON or OFF	Blinks		
	Red	ON	Except for California model • OFF	For California model • ON: when CHECK ENGINE LIGHT ITEMS are stored in the E.C.U. • OFF: except for the above condition		
Mode II (Monitor B)	Green	ON	*Remains ON or OFF	Blinks		
	Red	OFF	*Remains ON or OFF (synchronous with green LED)	Compensating mixture ratio		
				More than 5% rich	Between 5% lean and 5% rich	More
				OFF	Synchronized with green LED	Remains ON

*: Maintains conditions just before switching to open loop

ECCS SELF-DIAGNOSTIC MODE III—1988 SENTRA

Mode III — Self-Diagnostic System

The E.C.U. constantly monitors the function of these sensors and actuators, regardless of ignition key position. If a malfunction occurs, the information is stored in the E.C.U. and can be retrieved from the memory by turning on the diagnostic mode selector, located on the side of the E.C.U. When activated, the malfunction is indicated by flashing a red and a green L.E.D. (Light Emitting Diode), also located on the E.C.U. Since all the self-diagnostic results are stored in the E.C.U.'s memory even intermittent malfunctions can be diagnosed.

A malfunctioning part's group is indicated by the number of both the red and the green L.E.D.s flashing. First, the red L.E.D. flashes and the green flashes follow. The red L.E.D. refers to the number of tens while the green one refers to the number of units. For example, when the red L.E.D. flashes once and then the green one flashes twice, this means the number "12" showing the air flow meter signal is malfunctioning. In this way, all the problems are classified by the code numbers.

- **When engine fails to start, crank engine more than two seconds before starting self-diagnosis.**
- **Before starting self-diagnosis, do not erase stored memory. If doing so, self-diagnosis function for intermittent malfunctions would be lost.**

The stored memory would be lost if:

1. Battery terminal is disconnected.
2. After selecting Mode III, Mode IV is selected.

DISPLAY CODE TABLE

Code No.	Detected items
11	Crank angle sensor circuit
12	Air flow meter circuit
13	Water temperature sensor circuit
14	Vehicle speed sensor circuit
15	Mixture ratio feedback control slips out
21	Ignition signal missing in primary coil
23	Idle switch circuit
25	Idle speed control slips out
31	E.C.U. (E.C.C.S. control unit)
33	Exhaust gas sensor circuit
43	Throttle sensor circuit
55	No malfunctioning in the above circuit

Mode III — Self-Diagnostic System (Cont'd)

RETENTION OF DIAGNOSTIC RESULTS

The diagnostic result is retained in E.C.U. memory until the starter is operated fifty times after a diagnostic item is judged to be malfunctioning. The diagnostic result will then be cancelled automatically. If a diagnostic item which has been judged to be malfunctioning and stored in memory is again judged to be malfunctioning before the starter is operated fifty times, the second result will replace the previous one. It will be stored in E.C.U. memory until the starter is operated fifty times more.

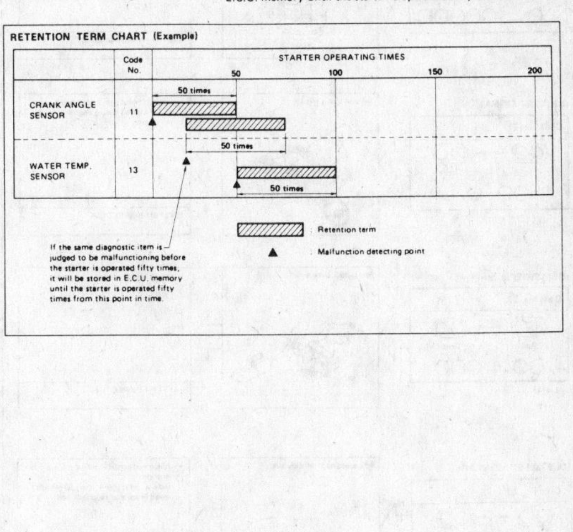
RETENTION TERM CHART (Example)
Code No.
STARTER OPERATING TIMES
50
100
150
200
50 times
CRANK ANGLE SENSOR
11
50 times
WATER TEMP. SENSOR
13
50 times
If the same diagnostic item is judged to be malfunctioning before the starter is operated fifty times, it will be stored in E.C.U. memory until the starter is operated fifty times from this point in time.
: Retention term
: Malfunction detecting point

ECCS SELF-DIAGNOSTIC MODE III—1988 SENTRA

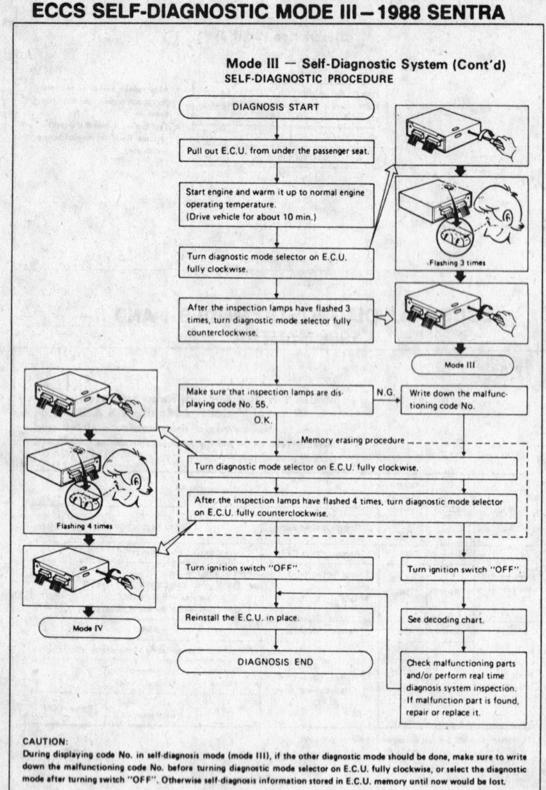
Mode III — Self-Diagnostic System (Cont'd)
SELF-DIAGNOSTIC PROCEDURE
DIAGNOSIS START
Pull out E.C.U. from under the passenger seat.
Start engine and warm it up to normal engine operating temperature. (Drive vehicle for about 10 min.)
Turn diagnostic mode selector on E.C.U. fully clockwise.
Flashing 3 times
After the inspection lamps have flashed 3 times, turn diagnostic mode selector fully counterclockwise.
Mode III
Make sure that inspection lamps are displaying code No. 55.
N.G.
Write down the malfunctioning code No.
O.K.
Memory erasing procedure
Turn diagnostic mode selector on E.C.U. fully clockwise.
After the inspection lamps have flashed 4 times, turn diagnostic mode selector on E.C.U. fully counterclockwise.
Flashing 4 times
Mode IV
Turn ignition switch "OFF".
Turn ignition switch "OFF".
Reinstall the E.C.U. in place.
See decoding chart.
DIAGNOSIS END
Check malfunctioning parts and/or perform real time diagnosis system inspection. If malfunction part is found, repair or replace it.
CAUTION:
During displaying code No. in self-diagnosis mode (mode III), if the other diagnostic mode should be done, make sure to write down the malfunctioning code No. before turning diagnostic mode selector on E.C.U. fully clockwise, or select the diagnostic mode after turning switch "OFF". Otherwise self-diagnosis information stored in E.C.U. memory until now would be lost.

ECCS SELF-DIAGNOSTIC MODE III DISPLAY CODES—1988 SENTRA

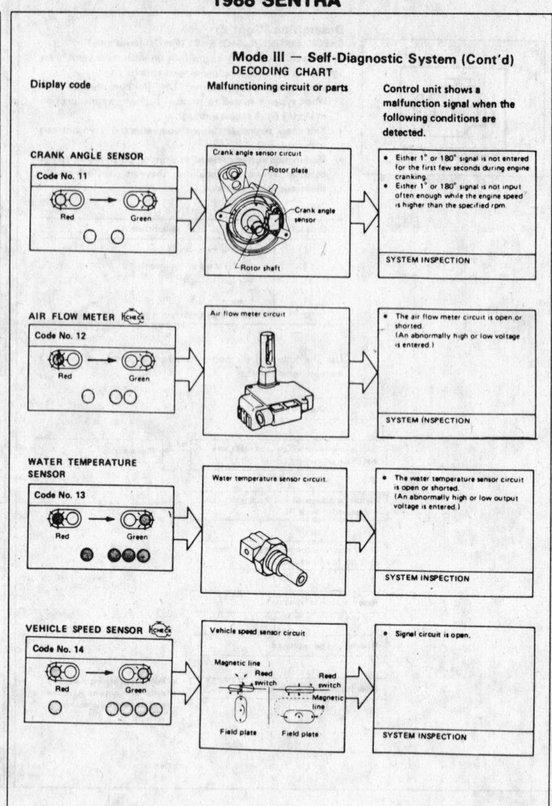
Mode III — Self-Diagnostic System (Cont'd)
DECODING CHART
Display code
Malfunctioning circuit or parts
Control unit shows a malfunction signal when the following conditions are detected.
CRANK ANGLE SENSOR
Code No. 11
Red
Green
Crank angle sensor circuit
Rotor plate
Crank angle sensor
Rotor shaft
Either 1° or 180° signal is not entered for the first few seconds during engine cranking.
Either 1° or 180° signal is not input often enough while the engine speed is higher than the specified rpm.
SYSTEM INSPECTION
AIR FLOW METER
Code No. 12
Air flow meter circuit
The air flow meter circuit is open or shorted. (An abnormally high or low voltage is entered.)
SYSTEM INSPECTION
WATER TEMPERATURE SENSOR
Code No. 13
Water temperature sensor circuit.
The water temperature sensor circuit is open or shorted. (An abnormally high or low output voltage is entered.)
SYSTEM INSPECTION
VEHICLE SPEED SENSOR
Code No. 14
Vehicle speed sensor circuit
Magnetic line
Reed switch
Reed switch
Magnetic line
Field plate
Field plate
Signal circuit is open.
SYSTEM INSPECTION

ECCS SELF-DIAGNOSTIC MODE III DISPLAY CODES—1988 SENTRA

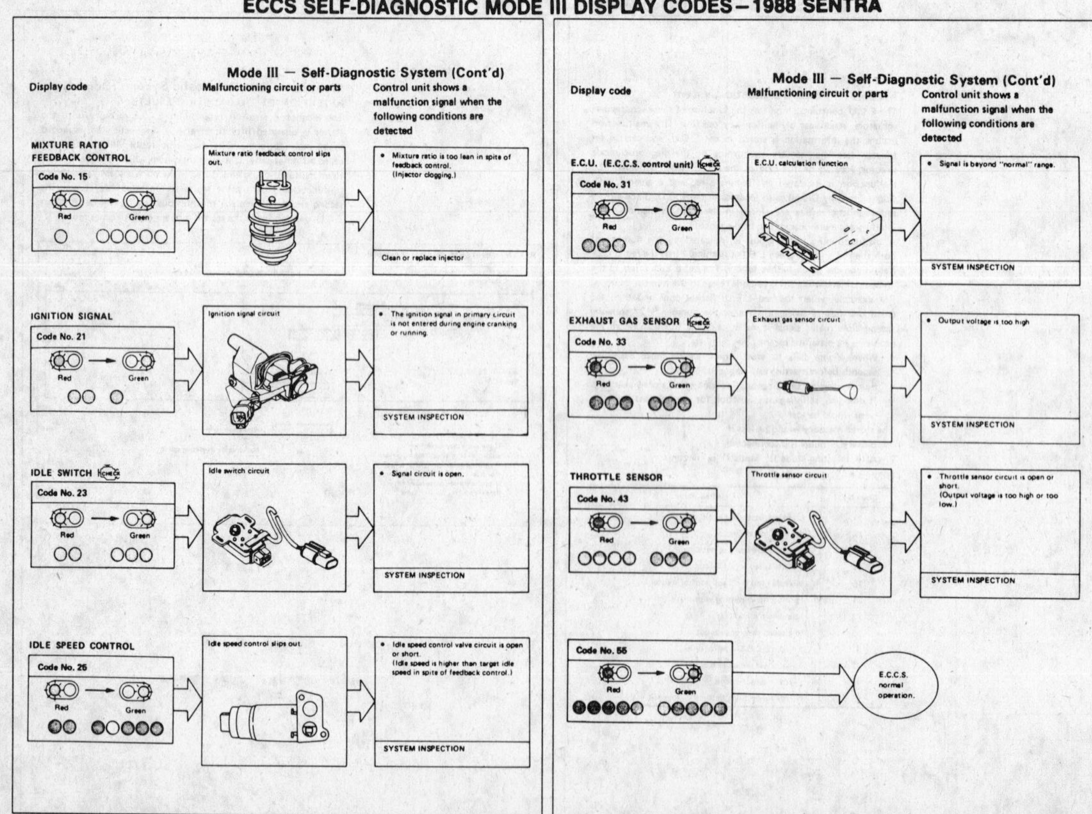
Mode III — Self-Diagnostic System (Cont'd)
Display code
Malfunctioning circuit or parts
Control unit shows a malfunction signal when the following conditions are detected
MIXTURE RATIO FEEDBACK CONTROL
Code No. 15
Red
Green
Mixture ratio feedback control slips out.
Mixture ratio is too lean in spite of feedback control. (Injector clogging.)
Clean or replace injector
IGNITION SIGNAL
Code No. 21
Ignition signal circuit
The ignition signal in primary circuit is not entered during engine cranking or running.
SYSTEM INSPECTION
IDLE SWITCH
Code No. 23
Idle switch circuit
Signal circuit is open.
SYSTEM INSPECTION
IDLE SPEED CONTROL
Code No. 25
Idle speed control slips out.
Idle speed control valve circuit is open or short. (Idle speed is higher than target idle speed in spite of feedback control.)
SYSTEM INSPECTION
Mode III — Self-Diagnostic System (Cont'd)
Display code
Malfunctioning circuit or parts
Control unit shows a malfunction signal when the following conditions are detected
E.C.U. (E.C.C.S. control unit)
Code No. 31
E.C.U. calculation function
Signal is beyond "normal" range.
SYSTEM INSPECTION
EXHAUST GAS SENSOR
Code No. 33
Exhaust gas sensor circuit
Output voltage is too high
SYSTEM INSPECTION
THROTTLE SENSOR
Code No. 43
Throttle sensor circuit
Throttle sensor circuit is open or short. (Output voltage is too high or too low.)
SYSTEM INSPECTION
Code No. 55
E.C.C.S. normal operation.

ECCS SELF-DIAGNOSTIC MODE IV—1988 SENTRA

Mode IV — Switches ON/OFF Diagnostic System

In switches ON/OFF diagnosis system, ON/OFF operation of the following switches can be detected continuously.

- Idle switch
- Starter switch
- Vehicle speed sensor

(1) Idle switch & Starter switch
The switches ON/OFF status at the point when mode IV is selected is stored in E.C.U. memory. When either switch is turned from "ON" to "OFF" or "OFF" to "ON", the red L.E.D. on E.C.U. alternately comes on and goes off each time switching is detected.

(2) Vehicle Speed Sensor
The switches ON/OFF status at the point when mode IV is selected is stored in E.C.U. memory. When vehicle speed is 20 km/h (12 MPH) or slower, the green L.E.D. on E.C.U. is off. When vehicle speed exceeds 20 km/h (12 MPH), the green L.E.D. on E.C.U. comes "ON".

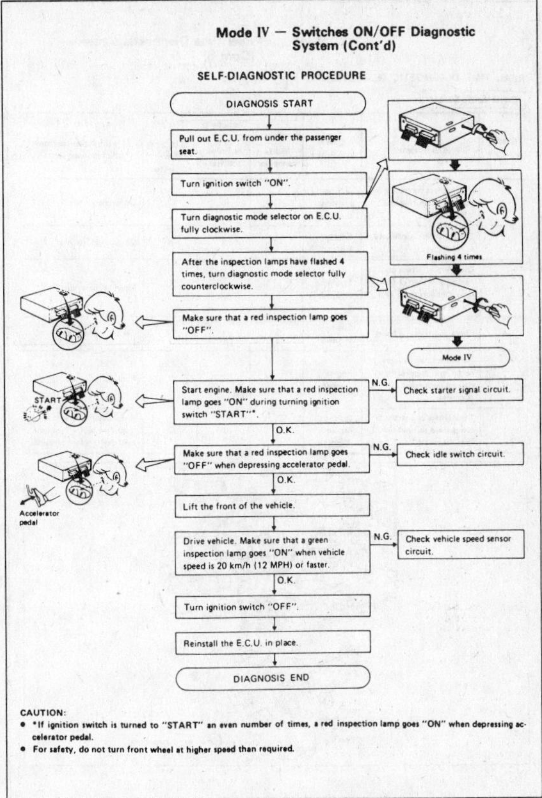
Mode IV — Switches ON/OFF Diagnostic System (Cont'd)
SELF-DIAGNOSTIC PROCEDURE
DIAGNOSIS START
Pull out E.C.U. from under the passenger seat.
Turn ignition switch "ON".
Turn diagnostic mode selector on E.C.U. fully clockwise.
After the inspection lamps have flashed 4 times, turn diagnostic mode selector fully counterclockwise.
Flashing 4 times
Mode IV
Make sure that a red inspection lamp goes "OFF".
Start engine. Make sure that a red inspection lamp goes "ON" during turning ignition switch "START"*.
START
N.G.
Check starter signal circuit.
O.K.
Make sure that a red inspection lamp goes "OFF" when depressing accelerator pedal.
N.G.
Check idle switch circuit.
O.K.
Accelerator pedal
Lift the front of the vehicle.
Drive vehicle. Make sure that a green inspection lamp goes "ON" when vehicle speed is 20 km/h (12 MPH) or faster.
N.G.
Check vehicle speed sensor circuit.
O.K.
Turn ignition switch "OFF".
Reinstall the E.C.U. in place.
DIAGNOSIS END
CAUTION:
• *If ignition switch is turned to "START" an even number of times, a red inspection lamp goes "ON" when depressing accelerator pedal.
• For safety, do not turn front wheel at higher speed than required.

ECCS SELF-DIAGNOSTIC MODE V—1988 SENTRA

Mode V — Real Time Diagnostic System

In real time diagnosis, if any of the following items are judged to be faulty, a malfunction is indicated immediately.

- Crank angle sensor (180° signal & 1° signal)
- Ignition signal
- Air flow meter output signal

Consequently, this diagnosis is a very effective measure to diagnose whether the above systems cause the malfunction or not, during driving test. Compared with self-diagnosis, real time diagnosis is very sensitive, and can detect malfunctioning conditions in a moment. Further, items regarded to be malfunctions in this diagnosis are not stored in E.C.U. memory.

SELF-DIAGNOSTIC PROCEDURE

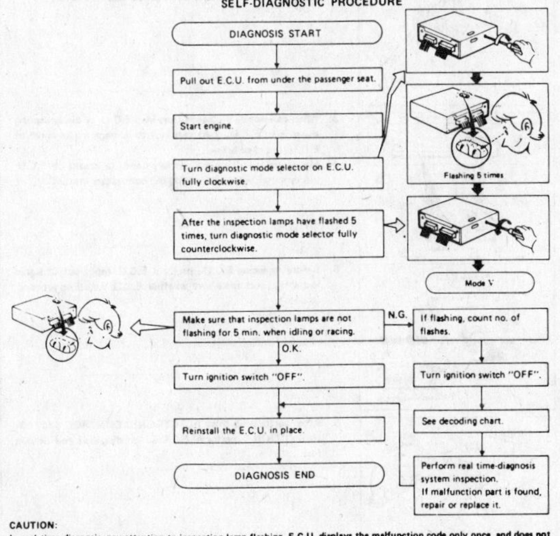
DIAGNOSIS START
Pull out E.C.U. from under the passenger seat.
Start engine.
Turn diagnostic mode selector on E.C.U. fully clockwise.
Flashing 5 times
After the inspection lamps have flashed 5 times, turn diagnostic mode selector fully counterclockwise.
Mode V
Make sure that inspection lamps are not flashing for 5 min. when idling or racing.
N.G.
If flashing, count no. of flashes.
O.K.
Turn ignition switch "OFF".
Turn ignition switch "OFF".
See decoding chart.
Reinstall the E.C.U. in place.
DIAGNOSIS END
Perform real time-diagnosis system inspection.
If malfunction part is found, repair or replace it.
CAUTION:
In real time diagnosis, pay attention to inspection lamp flashing. E.C.U. displays the malfunction code only once, and does not memorize the inspection.

ECCS SELF DIAGNOSTIC MODE V DECODING CHART—1988 SENTRA

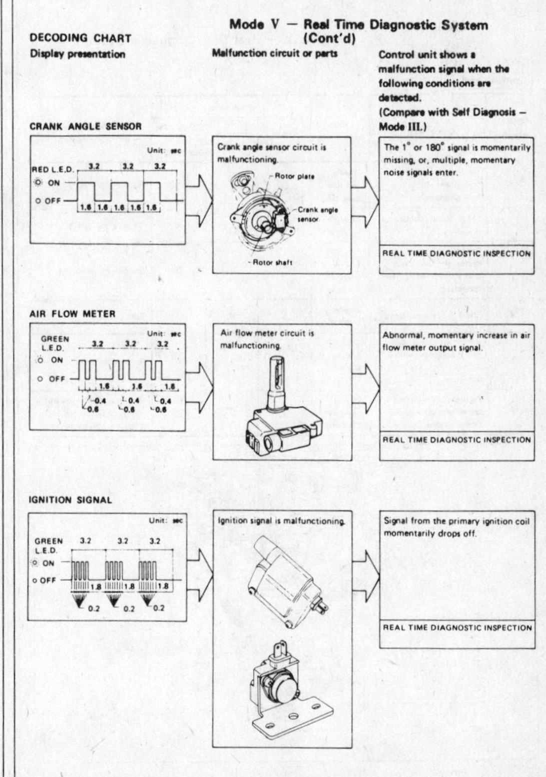
Mode V — Real Time Diagnostic System (Cont'd)
DECODING CHART
Display presentation
Malfunction circuit or parts
Control unit shows a malfunction signal when the following conditions are detected.
(Compare with Self Diagnosis – Mode III.)
CRANK ANGLE SENSOR
Unit: sec
RED L.E.D.
ON
OFF
3.2
3.2
3.2
1.6
Crank angle sensor circuit is malfunctioning.
Rotor plate
Crank angle sensor
Rotor shaft
The 1° or 180° signal is momentarily missing, or, multiple, momentary noise signals enter.
REAL TIME DIAGNOSTIC INSPECTION
AIR FLOW METER
Unit: sec
GREEN L.E.D.
ON
OFF
3.2
3.2
3.2
1.6
0.4
0.8
Air flow meter circuit is malfunctioning.
Abnormal, momentary increase in air flow meter output signal.
REAL TIME DIAGNOSTIC INSPECTION
IGNITION SIGNAL
Unit: sec
GREEN L.E.D.
ON
OFF
3.2
3.2
3.2
1.8
0.2
Ignition signal is malfunctioning.
Signal from the primary ignition coil momentarily drops off.
REAL TIME DIAGNOSTIC INSPECTION

ECCS SELF-DIAGNOSTIC MODE V INSPECTION—1988 SENTRA

Mode V — Real Time Diagnostic System (Cont'd)

REAL TIME DIAGNOSTIC INSPECTION

Crank Angle Sensor

Check sequence	Check items	Check conditions	Check parts: Middle connector	Sensor & actuator	E.C.U. 20 & 16 pin connector	If malfunction, perform the following items.
1	Tap harness connector or component during real time diagnosis.	During real time diagnosis	O	O	O	Go to check item 2.
2	Check harness continuity at connector.	Engine stopped	O	X	X	Go to check item 3.
3	Disconnect harness connector, and then check dust adhesion to harness connector.	Engine stopped	O	X	O	Clean terminal surface.
4	Check pin terminal bend.	Engine stopped	X	X	O	Take out bend.
5	Reconnect harness connector and then recheck harness continuity at connector.	Engine stopped	O	X	X	Replace terminal.
6	Tap harness connector or component during real time diagnosis.	During real time diagnosis	O	O	O	If malfunction codes are displayed during real time diagnosis, replace terminal.

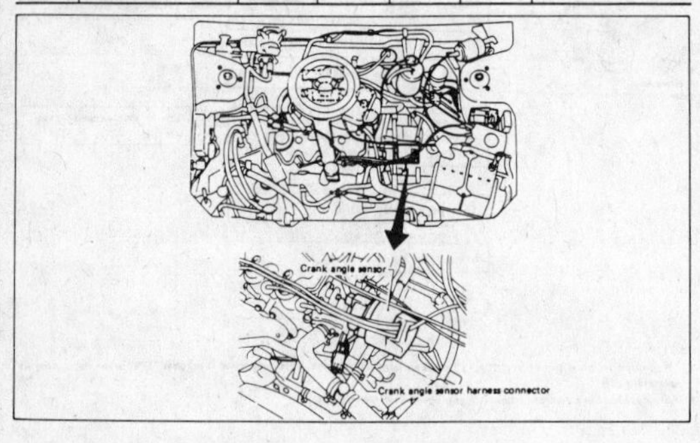
Crank angle sensor
Crank angle sensor harness connector

Mode V — Real Time Diagnostic System (Cont'd)

Air Flow Meter

Check sequence	Check items	Check conditions	Check parts: Middle connector	Sensor & actuator	E.C.U. 20 & 16 pin connector	If malfunction, perform the following items.
1	Tap harness connector or component during real time diagnosis.	During real time diagnosis	O	O	O	Go to check item 2.
2	Check harness continuity at connector.	Engine stopped	O	X	X	Go to check item 3.
3	Disconnect harness connector, and then check dust adhesion to harness connector.	Engine stopped	O	X	O	Clean terminal surface.
4	Check pin terminal bend.	Engine stopped	X	X	O	Take out bend.
5	Reconnect harness connector and then recheck harness continuity at connector.	Engine stopped	O	X	X	Replace terminal.
6	Tap harness connector or component during real time diagnosis.	During real time diagnosis	O	O	O	If malfunction codes are displayed during real time diagnosis, replace terminal.

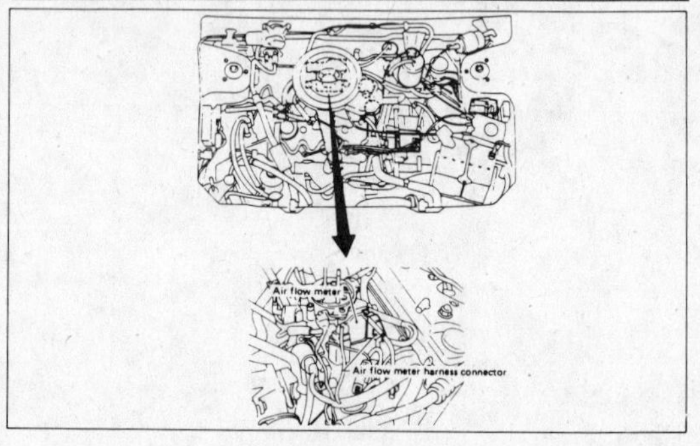
Air flow meter
Air flow meter harness connector

ECCS SELF-DIAGNOSTIC MODE V INSPECTION—1988 SENTRA

Mode V — Real Time Diagnostic System (Cont'd)

Ignition Signal

Check sequence	Check items	Check conditions	Check parts: Middle connector	Sensor & actuator	E.C.U. 20 & 16 pin connector	If malfunction, perform the following items.
1	Tap harness connector or component during real time diagnosis.	During real time diagnosis	O	O	O	Go to check item 2.
2	Check harness continuity at connector.	Engine stopped	O	X	X	Go to check item 3.
3	Disconnect harness connector, and then check dust adhesion to harness connector.	Engine stopped	O	X	O	Clean terminal surface.
4	Check pin terminal bend.	Engine stopped	X	X	O	Take out bend.
5	Reconnect harness connector and then recheck harness continuity at connector.	Engine stopped	O	X	X	Replace terminal.
6	Tap harness connector or component during real time diagnosis.	During real time diagnosis	O	O	O	If malfunction codes are displayed during real time diagnosis, replace terminal.

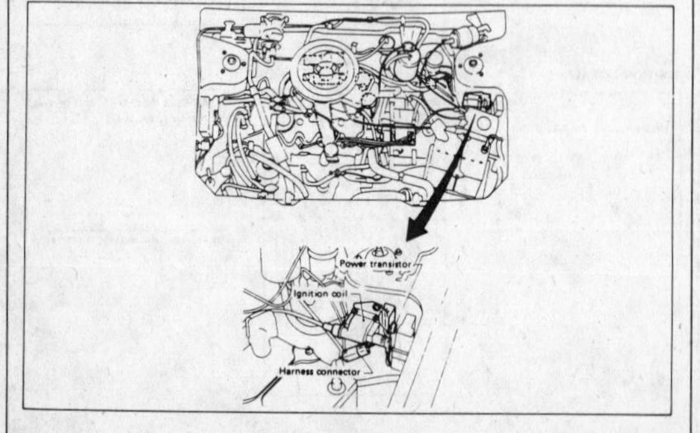
Power transistor
Ignition coil
Harness connector

ECCS SYSTEM INSPECTION CAUTION—1988 SENTRA

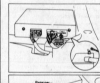

CAUTION:

1. Before connecting or disconnecting E.C.U. harness connector to or from any E.C.U., be sure to turn the ignition switch to the "OFF" position and disconnect the negative battery terminal in order not to damage E.C.U., as battery voltage is applied to E.C.U. even if ignition switch is turned off. Otherwise, there may be damage to the E.C.U.

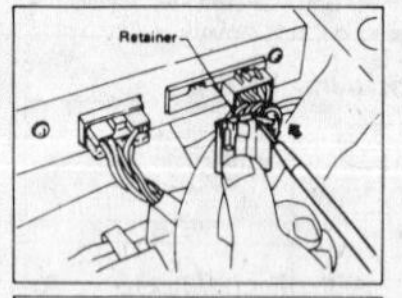
Retainer

2. When performing E.C.U. input/output signal inspection, remove pin terminal retainer from 20 and 16 pin connector to make it easier to insert tester probe into connector.

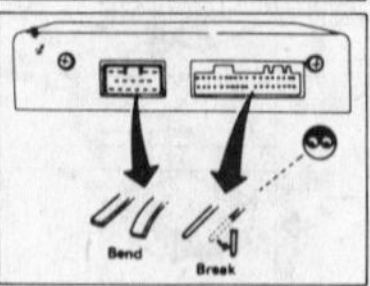
Bend
Break

3. When connecting pin connectors into E.C.U. or disconnecting them from E.C.U., take care not to damage pin terminal of E.C.U. (Bend or break).
4. Make sure that there are not any bends or breaks on E.C.U. pin terminal, when connecting pin connectors into E.C.U.

Perform E.C.U. input/output signal inspection before replacement.
OLD ONE

5. Before replacing E.C.U., perform E.C.U. input/output signal inspection and make sure whether E.C.U. functions properly or not.

6. After performing this "ELECTRONIC CONTROL SYSTEM INSPECTION", perform E.C.C.S. self-diagnosis and driving test.

ECCS SYSTEM INSPECTION CAUTION—1988 SENTRA

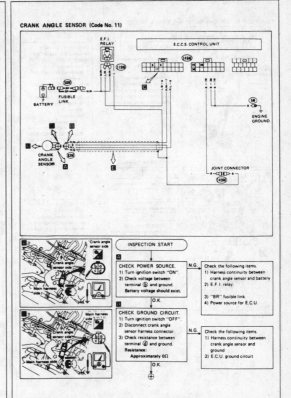

7. When measuring supply voltage of E.C.U. controlled components with a circuit tester, separate one tester probe from the other.
   If the two tester probes accidentally make contact with each other during measurement, the circuit will be shorted, resulting in damage to the power transistor of the control unit.

8. Keys to symbols
   - (DISCONNECT) : Check after disconnecting the connector to be measured.
   - (CONNECT) : Check after connecting the connector to be measured.

9. When measuring voltage or resistance at connector with tester probes, there are two methods of measurement; one is done from terminal side and the other from harness side. Before measuring, confirm symbol mark again.
   - (HS) : Inspection should be done from harness side.
   - (TS) : Inspection should be done from terminal side.

ECCS SELF-DIAGNOSTIC CHART—1988 SENTRA

CRANK ANGLE SENSOR (Code No. 11)

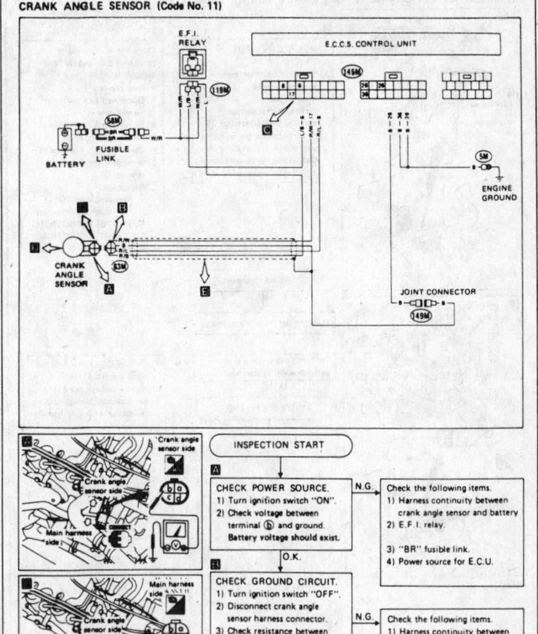
E.F.I. RELAY
E.C.C.S. CONTROL UNIT
145M
119M
58M
FUSIBLE LINK
BATTERY
5M
ENGINE GROUND
CRANK ANGLE SENSOR
53M
JOINT CONNECTOR
149M

INSPECTION START

A — CHECK POWER SOURCE.
1) Turn ignition switch "ON".
2) Check voltage between terminal ⓑ and ground.
Battery voltage should exist.

N.G. → Check the following items.
1) Harness continuity between crank angle sensor and battery
2) E.F.I. relay.
3) "BR" fusible link.
4) Power source for E.C.U.

O.K. ↓

B — CHECK GROUND CIRCUIT.
1) Turn ignition switch "OFF".
2) Disconnect crank angle sensor harness connector.
3) Check resistance between terminal ⓓ and ground.
Resistance: Approximately 0Ω

N.G. → Check the following items.
1) Harness continuity between crank angle sensor and ground
2) E.C.U. ground circuit

O.K. ↓ ⊕

ECCS SELF-DIAGNOSTIC CHART—1988 SENTRA

CRANK ANGLE SENSOR (Code No. 11)

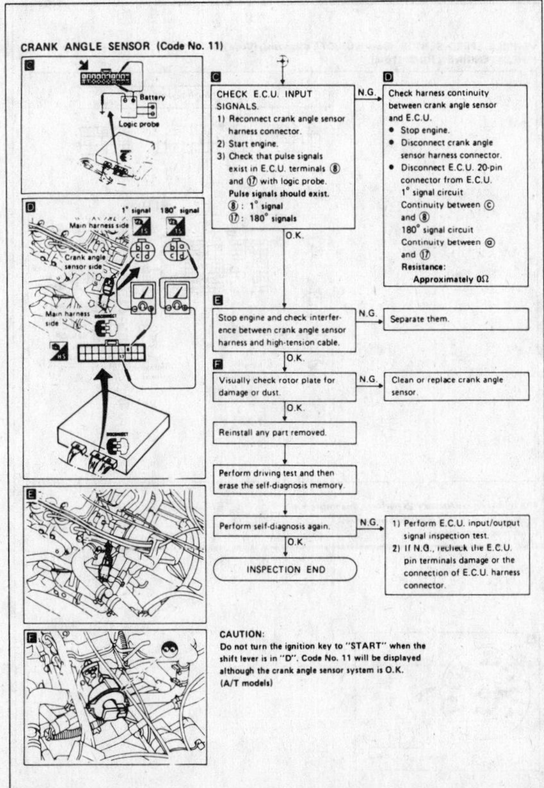
Battery
Logic probe
1° signal
180° signal
Main harness side
Crank angle sensor side
Main harness side

C — CHECK E.C.U. INPUT SIGNALS.
1) Reconnect crank angle sensor harness connector.
2) Start engine.
3) Check that pulse signals exist in E.C.U. terminals ⑧ and ⑰ with logic probe.
Pulse signals should exist.
⑧ : 1° signal
⑰ : 180° signals

N.G. → D — Check harness continuity between crank angle sensor and E.C.U.
- Stop engine.
- Disconnect crank angle sensor harness connector.
- Disconnect E.C.U. 20-pin connector from E.C.U.

1° signal circuit
Continuity between ⓒ and ⑧
180° signal circuit
Continuity between ⓔ and ⑰
Resistance: Approximately 0Ω

O.K. ↓

E — Stop engine and check interference between crank angle sensor harness and high-tension cable.
N.G. → Separate them.

O.K. ↓

F — Visually check rotor plate for damage or dust.
N.G. → Clean or replace crank angle sensor.

O.K. ↓

Reinstall any part removed.

↓

Perform driving test and then erase the self-diagnosis memory.

↓

Perform self-diagnosis again.
N.G. → 1) Perform E.C.U. input/output signal inspection test.
2) If N.G., recheck the E.C.U. pin terminals damage or the connection of E.C.U. harness connector.

O.K. ↓

INSPECTION END

CAUTION:
Do not turn the ignition key to "START" when the shift lever is in "D". Code No. 11 will be displayed although the crank angle sensor system is O.K. (A/T models)

AIR FLOW METER (Code No. 12) (CHECK ENGINE LIGHT ITEM)

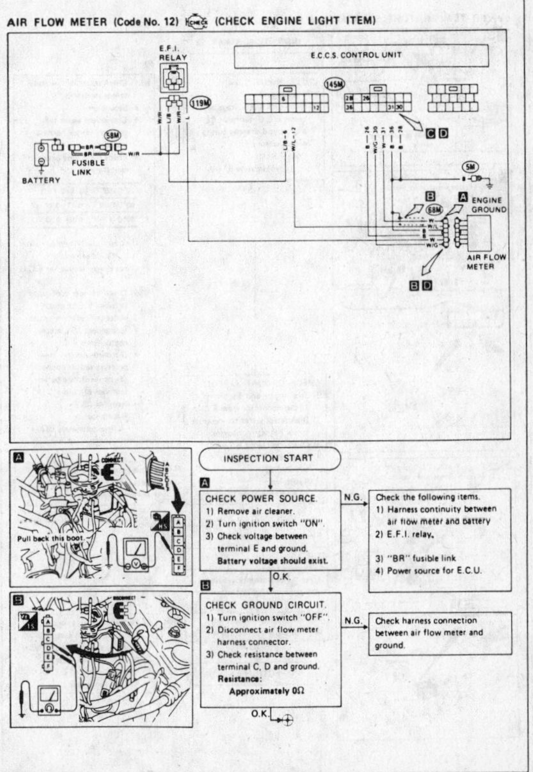
E.F.I. RELAY
E.C.C.S. CONTROL UNIT
145M
119M
58M
FUSIBLE LINK
BATTERY
5M
ENGINE GROUND
AIR FLOW METER
Pull back this boot

INSPECTION START

A — CHECK POWER SOURCE.
1) Remove air cleaner.
2) Turn ignition switch "ON".
3) Check voltage between terminal E and ground.
Battery voltage should exist.

N.G. → Check the following items.
1) Harness continuity between air flow meter and battery
2) E.F.I. relay.
3) "BR" fusible link
4) Power source for E.C.U.

O.K. ↓

B — CHECK GROUND CIRCUIT.
1) Turn ignition switch "OFF".
2) Disconnect air flow meter harness connector.
3) Check resistance between terminal C, D and ground.
Resistance: Approximately 0Ω

N.G. → Check harness connection between air flow meter and ground.

O.K. → ⊕

ECCS SELF-DIAGNOSTIC CHART—1988 SENTRA

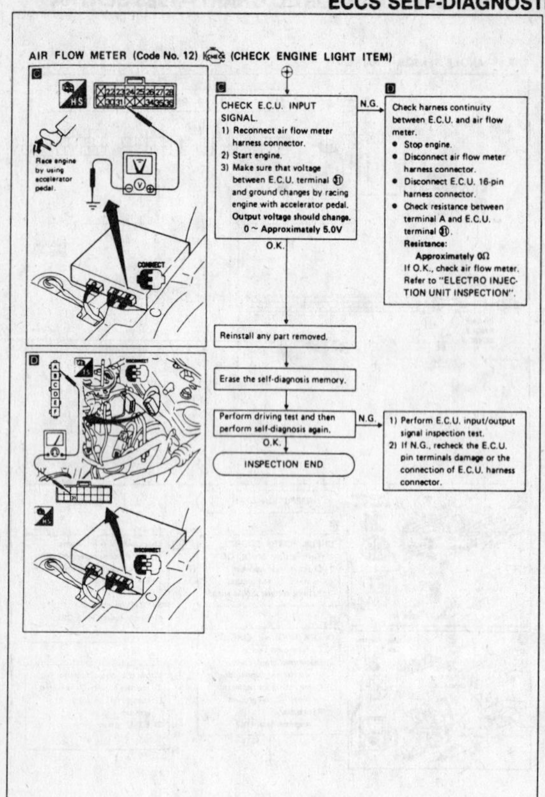
AIR FLOW METER (Code No. 12) (CHECK ENGINE LIGHT ITEM)
Race engine by using accelerator pedal.
CONNECT
CHECK E.C.U. INPUT SIGNAL.
1) Reconnect air flow meter harness connector.
2) Start engine.
3) Make sure that voltage between E.C.U. terminal ㉛ and ground changes by racing engine with accelerator pedal.
Output voltage should change.
0 ~ Approximately 5.0V
N.G.
Check harness continuity between E.C.U. and air flow meter.
• Stop engine.
• Disconnect air flow meter harness connector.
• Disconnect E.C.U. 16-pin harness connector.
• Check resistance between terminal A and E.C.U. terminal ㉛.
Resistance:
Approximately 0Ω
If O.K., check air flow meter. Refer to "ELECTRO INJECTION UNIT INSPECTION".
O.K.
Reinstall any part removed.
Erase the self-diagnosis memory.
Perform driving test and then perform self-diagnosis again.
N.G.
1) Perform E.C.U. input/output signal inspection test.
2) If N.G., recheck the E.C.U. pin terminals damage or the connection of E.C.U. harness connector.
O.K.
INSPECTION END

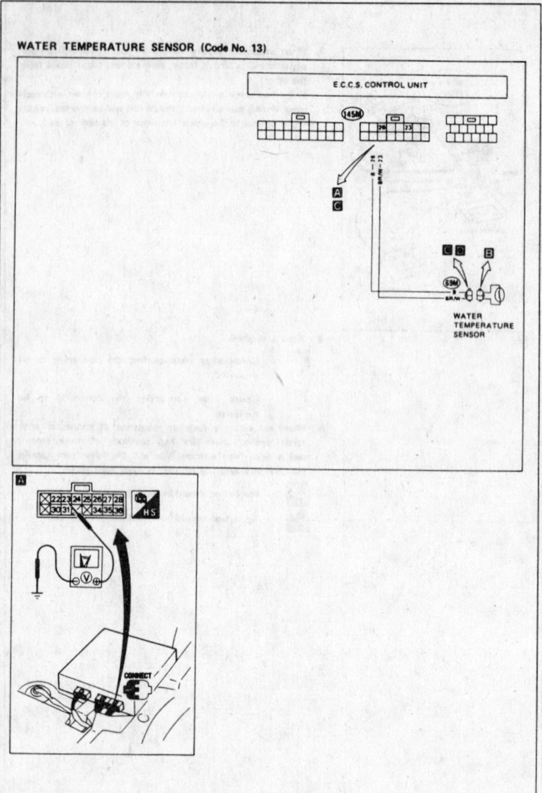
WATER TEMPERATURE SENSOR (Code No. 13)
E.C.C.S. CONTROL UNIT
145M
WATER TEMPERATURE SENSOR
CONNECT

ECCS SELF-DIAGNOSTIC CHART—1988 SENTRA

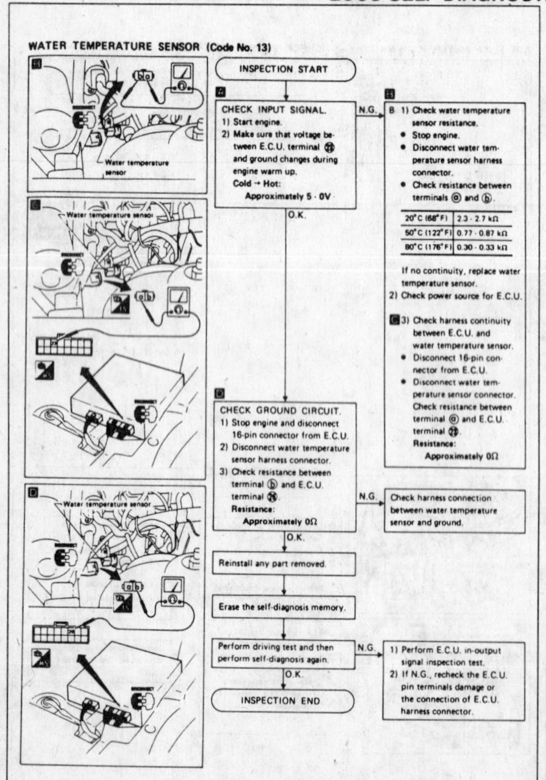
WATER TEMPERATURE SENSOR (Code No. 13)
Water temperature sensor
INSPECTION START
CHECK INPUT SIGNAL.
1) Start engine.
2) Make sure that voltage between E.C.U. terminal ㉓ and ground changes during engine warm up.
Cold → Hot:
Approximately 5 - 0V
N.G.
B 1) Check water temperature sensor resistance.
• Stop engine.
• Disconnect water temperature sensor harness connector.
• Check resistance between terminals ⓐ and ⓑ.
20°C (68°F) 2.3 - 2.7 kΩ
50°C (122°F) 0.77 - 0.87 kΩ
80°C (176°F) 0.30 - 0.33 kΩ
If no continuity, replace water temperature sensor.
2) Check power source for E.C.U.
C 3) Check harness continuity between E.C.U. and water temperature sensor.
• Disconnect 16-pin connector from E.C.U.
• Disconnect water temperature sensor connector.
Check resistance between terminal ⓐ and E.C.U. terminal ㉓.
Resistance:
Approximately 0Ω
O.K.
CHECK GROUND CIRCUIT.
1) Stop engine and disconnect 16-pin connector from E.C.U.
2) Disconnect water temperature sensor harness connector.
3) Check resistance between terminal ⓑ and E.C.U. terminal ㉖.
Resistance:
Approximately 0Ω
N.G.
Check harness connection between water temperature sensor and ground.
O.K.
Reinstall any part removed.
Erase the self-diagnosis memory.
Perform driving test and then perform self-diagnosis again.
N.G.
1) Perform E.C.U. in-output signal inspection test.
2) If N.G., recheck the E.C.U. pin terminals damage or the connection of E.C.U. harness connector.
O.K.
INSPECTION END

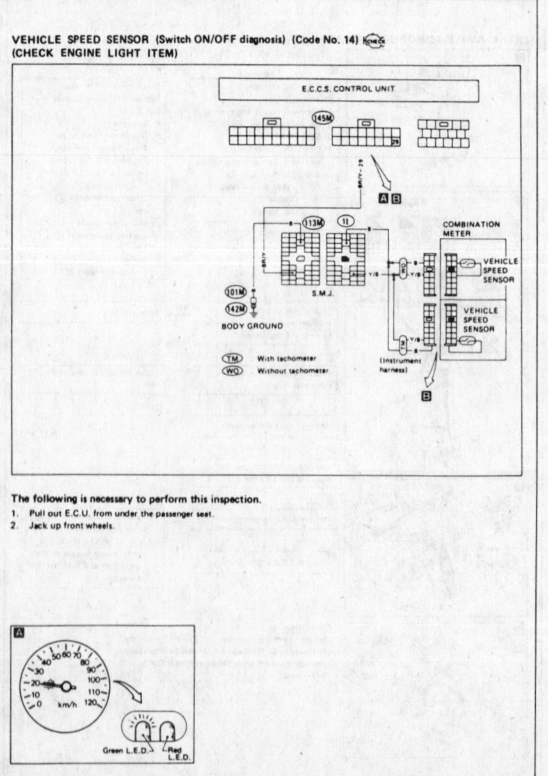
VEHICLE SPEED SENSOR (Switch ON/OFF diagnosis) (Code No. 14) (CHECK ENGINE LIGHT ITEM)
E.C.C.S. CONTROL UNIT
145M
113M
11
COMBINATION METER
VEHICLE SPEED SENSOR
VEHICLE SPEED SENSOR
S.M.J.
101M
142M
BODY GROUND
TM With tachometer
WO Without tachometer
(Instrument harness)
The following is necessary to perform this inspection.
1. Pull out E.C.U. from under the passenger seat.
2. Jack up front wheels.
km/h
Green L.E.D.
Red L.E.D.

ECCS SELF-DIAGNOSTIC CHART—1988 SENTRA

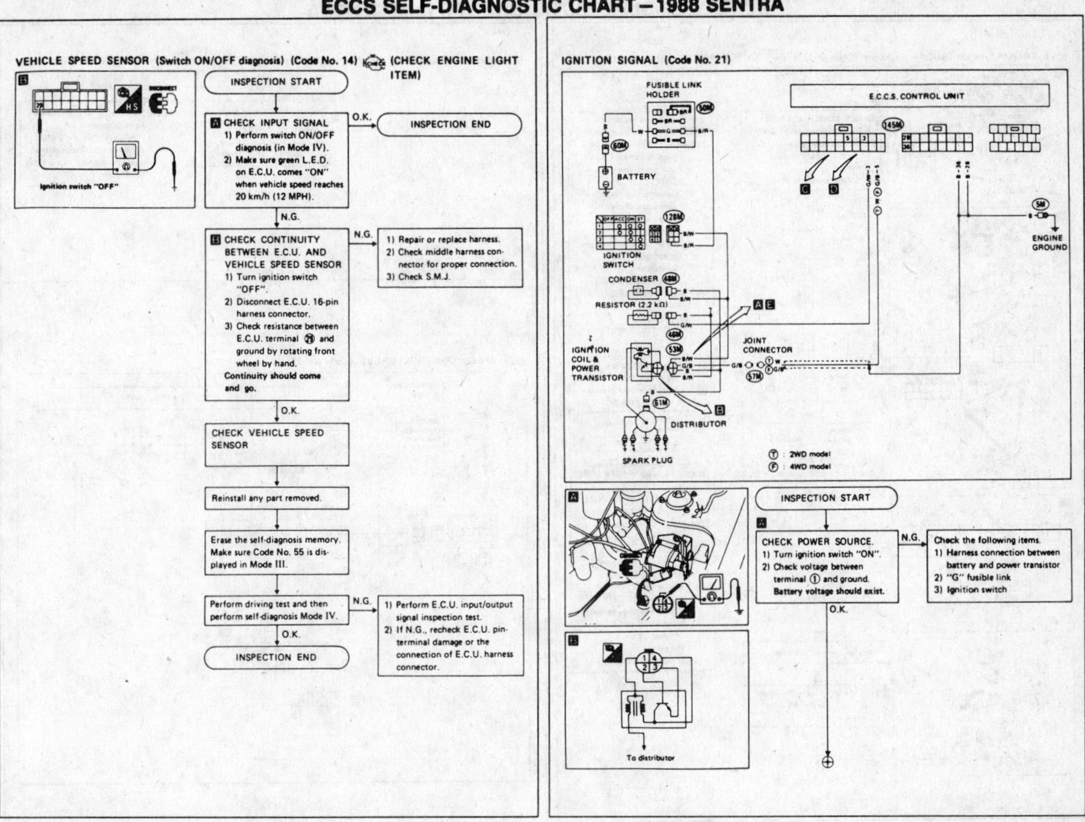
VEHICLE SPEED SENSOR (Switch ON/OFF diagnosis) (Code No. 14) (CHECK ENGINE LIGHT ITEM)
Ignition switch "OFF"
INSPECTION START
A CHECK INPUT SIGNAL
1) Perform switch ON/OFF diagnosis (in Mode IV).
2) Make sure green L.E.D. on E.C.U. comes "ON" when vehicle speed reaches 20 km/h (12 MPH).
O.K.
INSPECTION END
N.G.
B CHECK CONTINUITY BETWEEN E.C.U. AND VEHICLE SPEED SENSOR
1) Turn ignition switch "OFF".
2) Disconnect E.C.U. 16-pin harness connector.
3) Check resistance between E.C.U. terminal 29 and ground by rotating front wheel by hand.
Continuity should come and go.
N.G.
1) Repair or replace harness.
2) Check middle harness connector for proper connection.
3) Check S.M.J.
O.K.
CHECK VEHICLE SPEED SENSOR
Reinstall any part removed.
Erase the self-diagnosis memory. Make sure Code No. 55 is displayed in Mode III.
Perform driving test and then perform self-diagnosis Mode IV.
N.G.
1) Perform E.C.U. input/output signal inspection test.
2) If N.G., recheck E.C.U. pin-terminal damage or the connection of E.C.U. harness connector.
O.K.
INSPECTION END
IGNITION SIGNAL (Code No. 21)
FUSIBLE LINK HOLDER
E.C.C.S. CONTROL UNIT
BATTERY
IGNITION SWITCH
CONDENSER
RESISTOR (2.2 kΩ)
IGNITION COIL & POWER TRANSISTOR
JOINT CONNECTOR
DISTRIBUTOR
SPARK PLUG
ENGINE GROUND
T : 2WD model
F : 4WD model
INSPECTION START
A CHECK POWER SOURCE.
1) Turn ignition switch "ON".
2) Check voltage between terminal 1 and ground.
Battery voltage should exist.
N.G.
Check the following items.
1) Harness connection between battery and power transistor
2) "G" fusible link
3) Ignition switch
O.K.
To distributor

ECCS SELF-DIAGNOSTIC CHART—1988 SENTRA

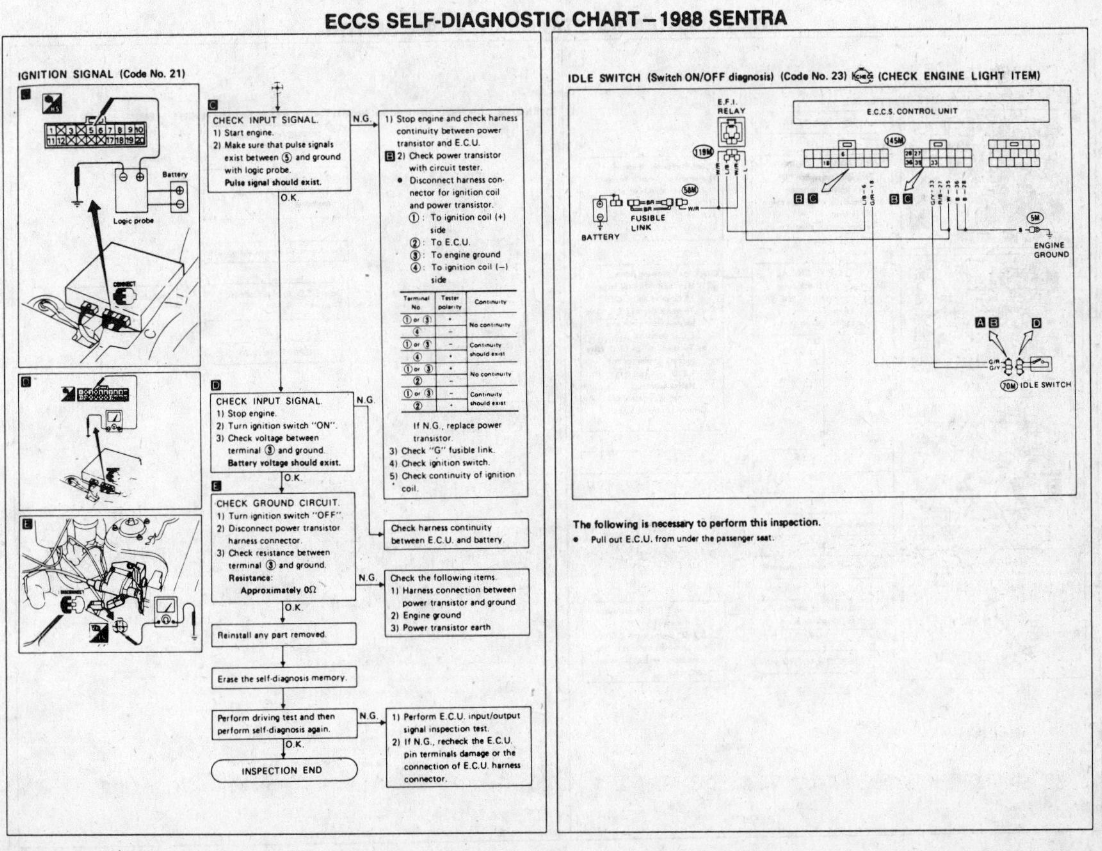
IGNITION SIGNAL (Code No. 21)
Battery
Logic probe
C CHECK INPUT SIGNAL.
1) Start engine.
2) Make sure that pulse signals exist between 5 and ground with logic probe.
Pulse signal should exist.
N.G.
1) Stop engine and check harness continuity between power transistor and E.C.U.
B 2) Check power transistor with circuit tester.
• Disconnect harness connector for ignition coil and power transistor.
1 : To ignition coil (+) side
2 : To E.C.U.
3 : To engine ground
4 : To ignition coil (−) side
Terminal No.
Tester polarity
Continuity
No continuity
Continuity should exist
If N.G., replace power transistor.
3) Check "G" fusible link.
4) Check ignition switch.
5) Check continuity of ignition coil.
O.K.
D CHECK INPUT SIGNAL.
1) Stop engine.
2) Turn ignition switch "ON".
3) Check voltage between terminal 3 and ground.
Battery voltage should exist.
N.G.
O.K.
E CHECK GROUND CIRCUIT.
1) Turn ignition switch "OFF".
2) Disconnect power transistor harness connector.
3) Check resistance between terminal 3 and ground.
Resistance: Approximately 0Ω
Check harness continuity between E.C.U. and battery.
N.G.
Check the following items.
1) Harness connection between power transistor and ground
2) Engine ground
3) Power transistor earth
O.K.
Reinstall any part removed.
Erase the self-diagnosis memory.
Perform driving test and then perform self-diagnosis again.
N.G.
1) Perform E.C.U. input/output signal inspection test.
2) If N.G., recheck the E.C.U. pin terminals damage or the connection of E.C.U. harness connector.
O.K.
INSPECTION END
IDLE SWITCH (Switch ON/OFF diagnosis) (Code No. 23) (CHECK ENGINE LIGHT ITEM)
E.F.I. RELAY
E.C.C.S. CONTROL UNIT
FUSIBLE LINK
BATTERY
ENGINE GROUND
IDLE SWITCH
The following is necessary to perform this inspection.
• Pull out E.C.U. from under the passenger seat.

ECCS SELF-DIAGNOSTIC CHART—1988 SENTRA

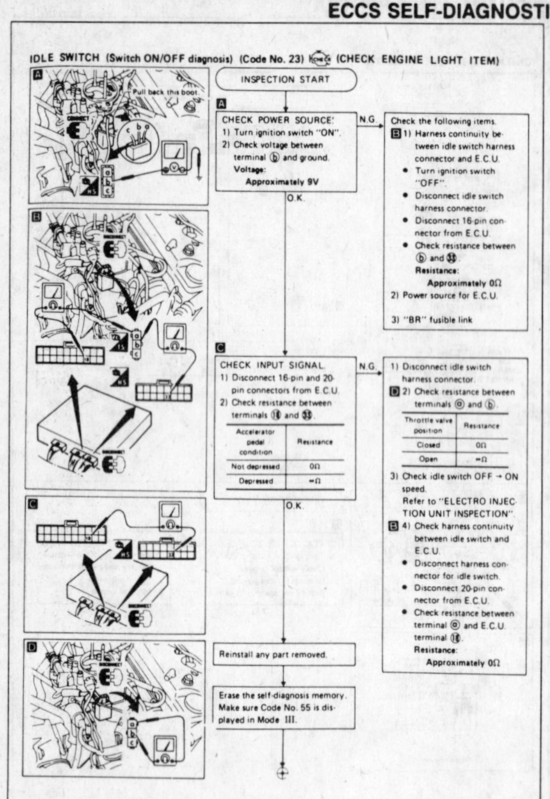
IDLE SWITCH (Switch ON/OFF diagnosis) (Code No. 23) (CHECK ENGINE LIGHT ITEM)
Pull back this boot.
INSPECTION START
A
CHECK POWER SOURCE.
1) Turn ignition switch "ON".
2) Check voltage between terminal ⓑ and ground.
Voltage:
Approximately 9V
N.G.
Check the following items.
B 1) Harness continuity between idle switch harness connector and E.C.U.
• Turn ignition switch "OFF".
• Disconnect idle switch harness connector.
• Disconnect 16-pin connector from E.C.U.
• Check resistance between ⓑ and ㉝.
Resistance:
Approximately 0Ω
2) Power source for E.C.U.
3) "BR" fusible link
O.K.
C
CHECK INPUT SIGNAL.
1) Disconnect 16-pin and 20-pin connectors from E.C.U.
2) Check resistance between terminals ⑱ and ㉝.
Accelerator pedal condition | Resistance
Not depressed | 0Ω
Depressed | ∞Ω
N.G.
1) Disconnect idle switch harness connector.
D 2) Check resistance between terminals ⓐ and ⓑ.
Throttle valve position | Resistance
Closed | 0Ω
Open | ∞Ω
3) Check idle switch OFF → ON speed.
Refer to "ELECTRO INJECTION UNIT INSPECTION".
B 4) Check harness continuity between idle switch and E.C.U.
• Disconnect harness connector for idle switch.
• Disconnect 20-pin connector from E.C.U.
• Check resistance between terminal ⓐ and E.C.U. terminal ⑱.
Resistance:
Approximately 0Ω
O.K.
Reinstall any part removed.
Erase the self-diagnosis memory. Make sure Code No. 55 is displayed in Mode III.

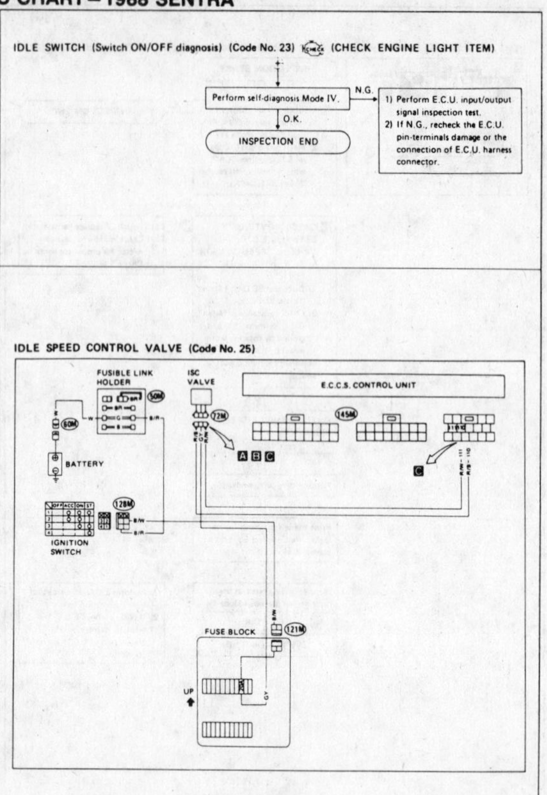
IDLE SWITCH (Switch ON/OFF diagnosis) (Code No. 23) (CHECK ENGINE LIGHT ITEM)
Perform self-diagnosis Mode IV.
N.G.
1) Perform E.C.U. input/output signal inspection test.
2) If N.G., recheck the E.C.U. pin-terminals damage or the connection of E.C.U. harness connector.
O.K.
INSPECTION END
IDLE SPEED CONTROL VALVE (Code No. 25)
FUSIBLE LINK HOLDER
ISC VALVE
E.C.C.S. CONTROL UNIT
BATTERY
IGNITION SWITCH
FUSE BLOCK
UP

ECCS SELF-DIAGNOSTIC CHART—1988 SENTRA

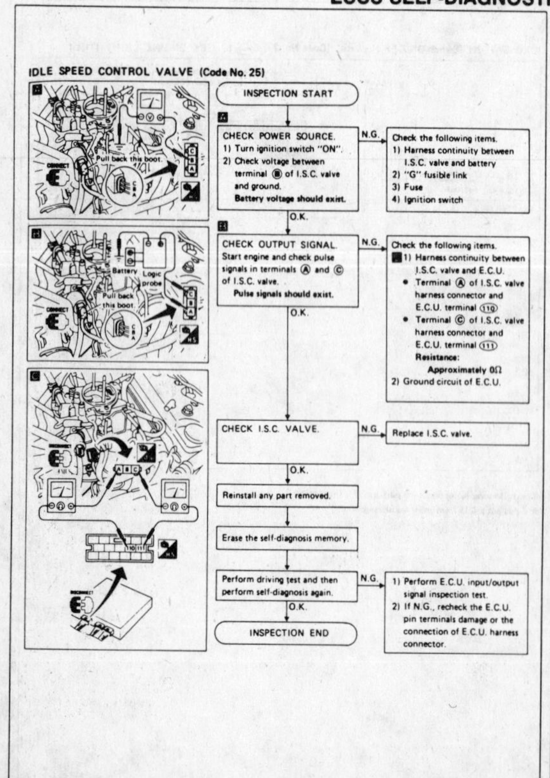
IDLE SPEED CONTROL VALVE (Code No. 25)
Pull back this boot.
Battery
Logic probe
Pull back this boot
INSPECTION START
A
CHECK POWER SOURCE.
1) Turn ignition switch "ON".
2) Check voltage between terminal Ⓑ of I.S.C. valve and ground.
Battery voltage should exist.
N.G.
Check the following items.
1) Harness continuity between I.S.C. valve and battery
2) "G" fusible link
3) Fuse
4) Ignition switch
O.K.
B
CHECK OUTPUT SIGNAL.
Start engine and check pulse signals in terminals Ⓐ and Ⓒ of I.S.C. valve.
Pulse signals should exist.
N.G.
Check the following items.
1) Harness continuity between I.S.C. valve and E.C.U.
• Terminal Ⓐ of I.S.C. valve harness connector and E.C.U. terminal ⑪⓪
• Terminal Ⓒ of I.S.C. valve harness connector and E.C.U. terminal ⑪①
Resistance:
Approximately 0Ω
2) Ground circuit of E.C.U.
O.K.
CHECK I.S.C. VALVE.
N.G.
Replace I.S.C. valve.
O.K.
Reinstall any part removed.
Erase the self-diagnosis memory.
Perform driving test and then perform self-diagnosis again.
N.G.
1) Perform E.C.U. input/output signal inspection test.
2) If N.G., recheck the E.C.U. pin terminals damage or the connection of E.C.U. harness connector.
O.K.
INSPECTION END

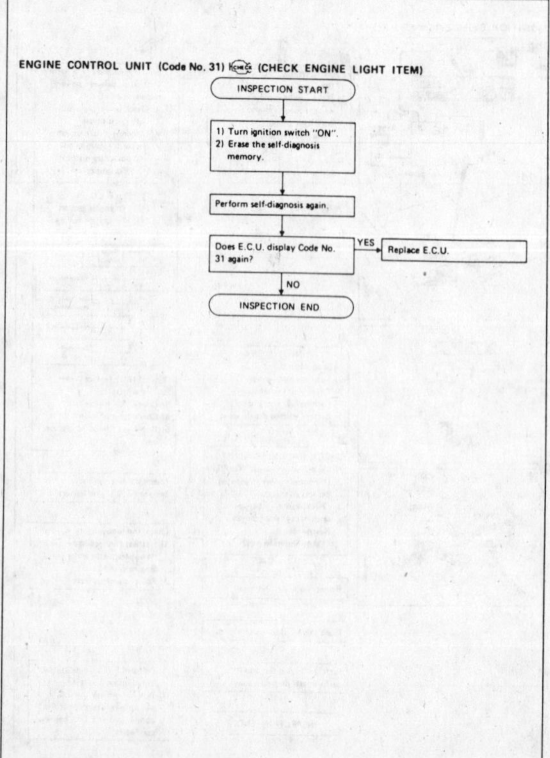
ENGINE CONTROL UNIT (Code No. 31) (CHECK ENGINE LIGHT ITEM)
INSPECTION START
1) Turn ignition switch "ON".
2) Erase the self-diagnosis memory.
Perform self-diagnosis again.
Does E.C.U. display Code No. 31 again?
YES
Replace E.C.U.
NO
INSPECTION END

ECCS SELF-DIAGNOSTIC CHART—1988 SENTRA

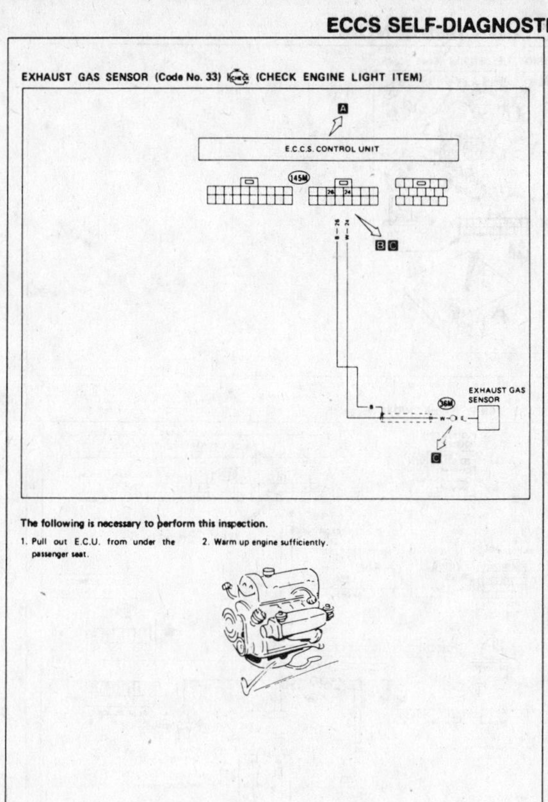
EXHAUST GAS SENSOR (Code No. 33) (CHECK ENGINE LIGHT ITEM)
A
E.C.C.S. CONTROL UNIT
145M
26
24
B C
36M
EXHAUST GAS SENSOR
C
The following is necessary to perform this inspection.
1. Pull out E.C.U. from under the passenger seat.
2. Warm up engine sufficiently.

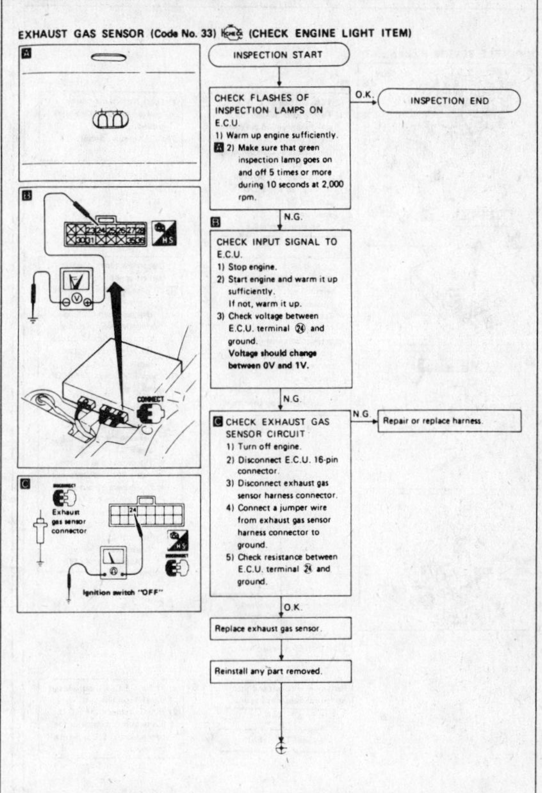
EXHAUST GAS SENSOR (Code No. 33) (CHECK ENGINE LIGHT ITEM)
A
B
CONNECT
C
DISCONNECT
Exhaust gas sensor connector
Ignition switch "OFF"
INSPECTION START
CHECK FLASHES OF INSPECTION LAMPS ON E.C.U.
1) Warm up engine sufficiently.
A 2) Make sure that green inspection lamp goes on and off 5 times or more during 10 seconds at 2,000 rpm.
O.K.
INSPECTION END
N.G.
B
CHECK INPUT SIGNAL TO E.C.U.
1) Stop engine.
2) Start engine and warm it up sufficiently.
If not, warm it up.
3) Check voltage between E.C.U. terminal ㉔ and ground.
Voltage should change between 0V and 1V.
N.G.
C CHECK EXHAUST GAS SENSOR CIRCUIT
1) Turn off engine.
2) Disconnect E.C.U. 16-pin connector.
3) Disconnect exhaust gas sensor harness connector.
4) Connect a jumper wire from exhaust gas sensor harness connector to ground.
5) Check resistance between E.C.U. terminal ㉔ and ground.
N.G.
Repair or replace harness.
O.K.
Replace exhaust gas sensor.
Reinstall any part removed.

ECCS SELF-DIAGNOSTIC CHART—1988 SENTRA

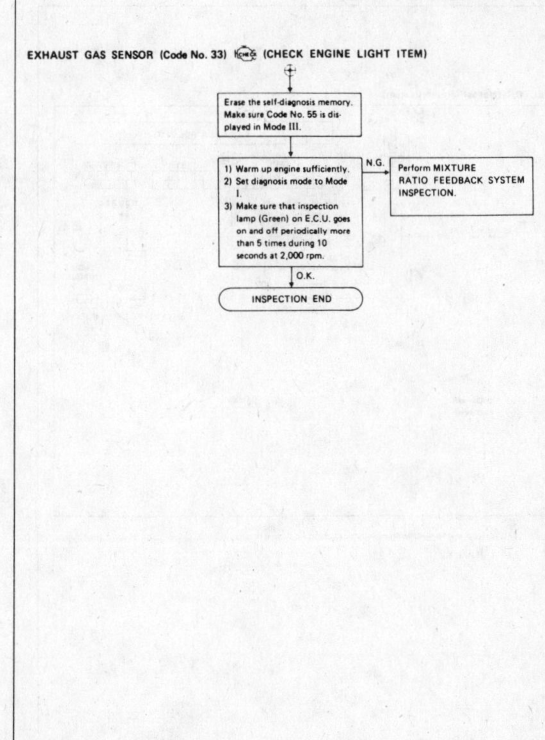
EXHAUST GAS SENSOR (Code No. 33) (CHECK ENGINE LIGHT ITEM)
Erase the self-diagnosis memory. Make sure Code No. 55 is displayed in Mode III.
1) Warm up engine sufficiently.
2) Set diagnosis mode to Mode I.
3) Make sure that inspection lamp (Green) on E.C.U. goes on and off periodically more than 5 times during 10 seconds at 2,000 rpm.
N.G.
Perform MIXTURE RATIO FEEDBACK SYSTEM INSPECTION.
O.K.
INSPECTION END

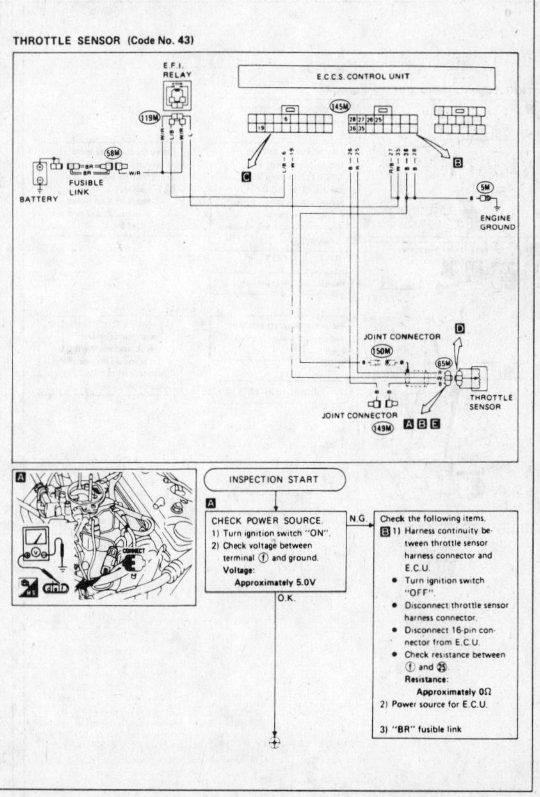
THROTTLE SENSOR (Code No. 43)
E.F.I. RELAY
E.C.C.S. CONTROL UNIT
119M
145M
58M
BATTERY
FUSIBLE LINK
C
B
5M
ENGINE GROUND
JOINT CONNECTOR
150M
D
65M
THROTTLE SENSOR
JOINT CONNECTOR
149M
A B E
A
CONNECT
INSPECTION START
A
CHECK POWER SOURCE.
1) Turn ignition switch "ON".
2) Check voltage between terminal ⓕ and ground.
Voltage:
Approximately 5.0V
O.K.
N.G.
Check the following items.
B 1) Harness continuity between throttle sensor harness connector and E.C.U.
• Turn ignition switch "OFF".
• Disconnect throttle sensor harness connector.
• Disconnect 16-pin connector from E.C.U.
• Check resistance between ⓕ and ㉕.
Resistance:
Approximately 0Ω
2) Power source for E.C.U.
3) "BR" fusible link

ECCS SELF-DIAGNOSTIC CHART—1988 SENTRA

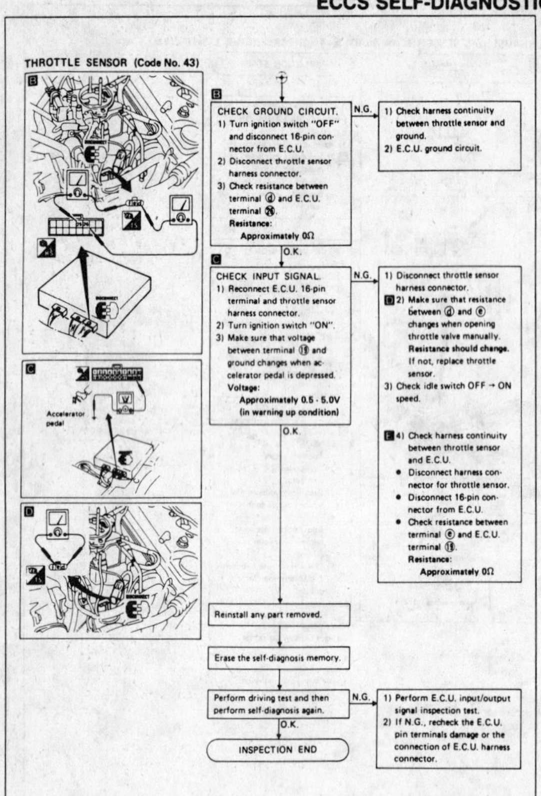
THROTTLE SENSOR (Code No. 43)
CHECK GROUND CIRCUIT.
1) Turn ignition switch "OFF" and disconnect 16-pin connector from E.C.U.
2) Disconnect throttle sensor harness connector.
3) Check resistance between terminal ⓓ and E.C.U. terminal ⓐ.
Resistance:
Approximately 0Ω
N.G.
1) Check harness continuity between throttle sensor and ground.
2) E.C.U. ground circuit.
O.K.
CHECK INPUT SIGNAL.
1) Reconnect E.C.U. 16-pin terminal and throttle sensor harness connector.
2) Turn ignition switch "ON".
3) Make sure that voltage between terminal ⑪ and ground changes when accelerator pedal is depressed.
Voltage:
Approximately 0.5 - 5.0V
(in warming up condition)
N.G.
1) Disconnect throttle sensor harness connector.
2) Make sure that resistance between ⓓ and ⓔ changes when opening throttle valve manually.
Resistance should change.
If not, replace throttle sensor.
3) Check idle switch OFF → ON speed.
4) Check harness continuity between throttle sensor and E.C.U.
• Disconnect harness connector for throttle sensor.
• Disconnect 16-pin connector from E.C.U.
• Check resistance between terminal ⓔ and E.C.U. terminal ⑪.
Resistance:
Approximately 0Ω
O.K.
Accelerator pedal
Reinstall any part removed.
Erase the self-diagnosis memory.
Perform driving test and then perform self-diagnosis again.
N.G.
1) Perform E.C.U. input/output signal inspection test.
2) If N.G., recheck the E.C.U. pin terminals damage or the connection of E.C.U. harness connector.
O.K.
INSPECTION END

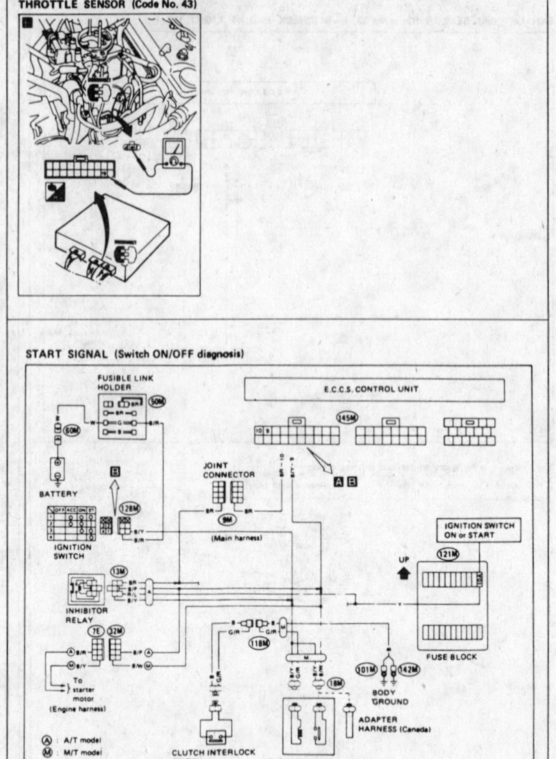
THROTTLE SENSOR (Code No. 43)
START SIGNAL (Switch ON/OFF diagnosis)
FUSIBLE LINK HOLDER
E.C.C.S. CONTROL UNIT
BATTERY
JOINT CONNECTOR
(Main harness)
IGNITION SWITCH
IGNITION SWITCH ON or START
UP
INHIBITOR RELAY
FUSE BLOCK
To starter motor
(Engine harness)
BODY GROUND
ADAPTER HARNESS (Canada)
Ⓐ : A/T model
Ⓜ : M/T model
CLUTCH INTERLOCK SWITCH
STARTER RELAY (U.S.A.)

ECCS SELF-DIAGNOSTIC CHART—1988 SENTRA

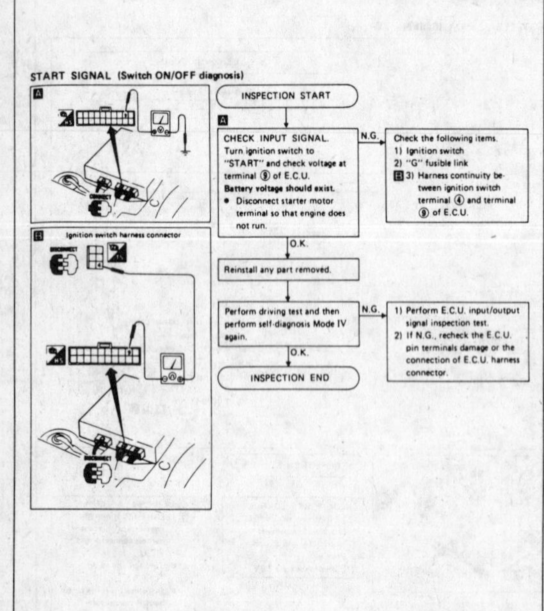
START SIGNAL (Switch ON/OFF diagnosis)
Ignition switch harness connector
INSPECTION START
CHECK INPUT SIGNAL.
Turn ignition switch to "START" and check voltage at terminal ⑨ of E.C.U.
Battery voltage should exist.
• Disconnect starter motor terminal so that engine does not run.
N.G.
Check the following items.
1) Ignition switch
2) "G" fusible link
3) Harness continuity between ignition switch terminal ④ and terminal ⑨ of E.C.U.
O.K.
Reinstall any part removed.
Perform driving test and then perform self-diagnosis Mode IV again.
N.G.
1) Perform E.C.U. input/output signal inspection test.
2) If N.G., recheck the E.C.U. pin terminals damage or the connection of E.C.U. harness connector.
O.K.
INSPECTION END

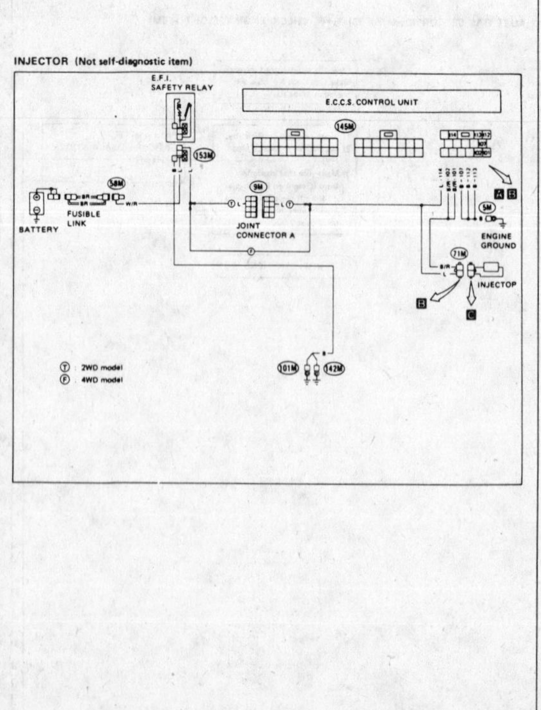
INJECTOR (Not self-diagnostic item)
E.F.I. SAFETY RELAY
E.C.C.S. CONTROL UNIT
BATTERY
FUSIBLE LINK
JOINT CONNECTOR A
ENGINE GROUND
INJECTOR
Ⓣ : 2WD model
Ⓕ : 4WD model

ECCS SELF-DIAGNOSTIC CHART—1988 SENTRA

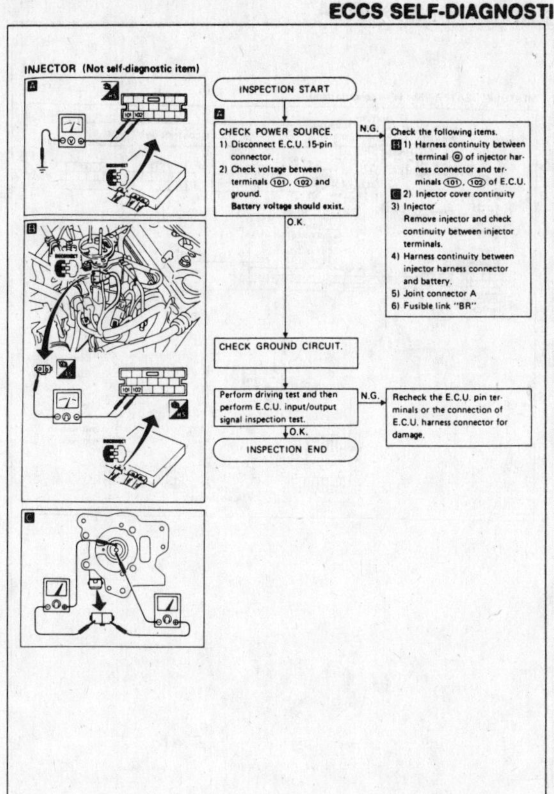
INJECTOR (Not self-diagnostic item)
INSPECTION START
CHECK POWER SOURCE.
1) Disconnect E.C.U. 15-pin connector.
2) Check voltage between terminals (101), (102) and ground.
Battery voltage should exist.
N.G.
Check the following items.
1) Harness continuity between terminal Ⓖ of injector harness connector and terminals (101), (102) of E.C.U.
2) Injector cover continuity
3) Injector
Remove injector and check continuity between injector terminals.
4) Harness continuity between injector harness connector and battery.
5) Joint connector A
6) Fusible link "BR"
O.K.
CHECK GROUND CIRCUIT.
Perform driving test and then perform E.C.U. input/output signal inspection test.
N.G.
Recheck the E.C.U. pin terminals or the connection of E.C.U. harness connector for damage.
O.K.
INSPECTION END

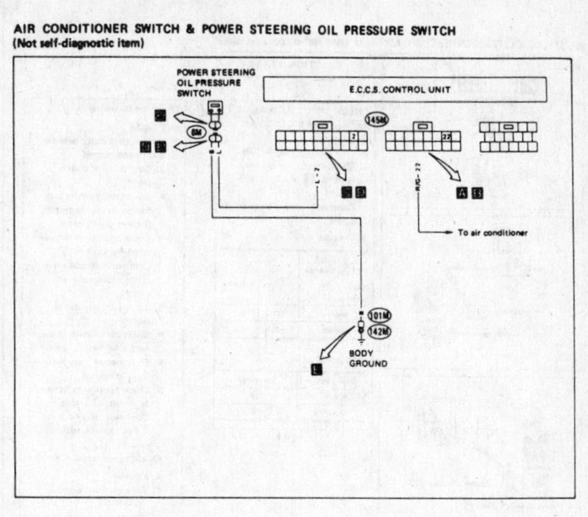
AIR CONDITIONER SWITCH & POWER STEERING OIL PRESSURE SWITCH
(Not self-diagnostic item)
POWER STEERING OIL PRESSURE SWITCH
E.C.C.S. CONTROL UNIT
To air conditioner
BODY GROUND

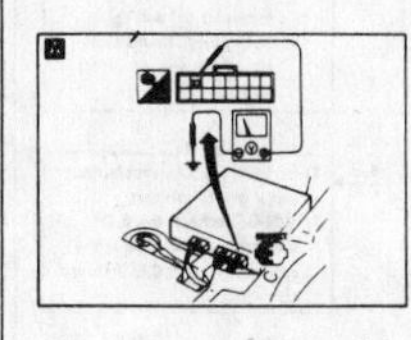

ECCS SELF-DIAGNOSTIC CHART—1988 SENTRA

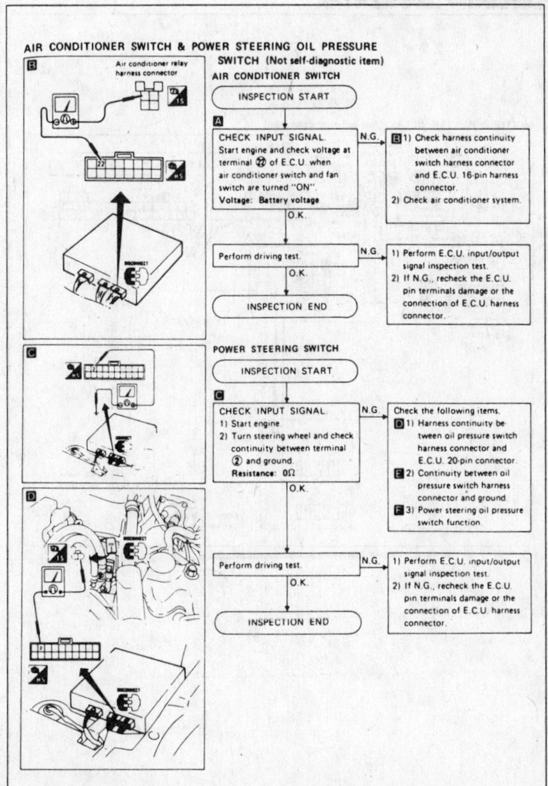
AIR CONDITIONER SWITCH & POWER STEERING OIL PRESSURE SWITCH (Not self-diagnostic item)
Air conditioner relay harness connector
AIR CONDITIONER SWITCH
INSPECTION START
CHECK INPUT SIGNAL.
Start engine and check voltage at terminal ㉒ of E.C.U. when air conditioner switch and fan switch are turned "ON".
Voltage: Battery voltage
N.G.
1) Check harness continuity between air conditioner switch harness connector and E.C.U. 16-pin harness connector.
2) Check air conditioner system.
O.K.
Perform driving test.
N.G.
1) Perform E.C.U. input/output signal inspection test.
2) If N.G., recheck the E.C.U. pin terminals damage or the connection of E.C.U. harness connector.
O.K.
INSPECTION END
POWER STEERING SWITCH
INSPECTION START
CHECK INPUT SIGNAL.
1) Start engine
2) Turn steering wheel and check continuity between terminal ② and ground.
Resistance: 0Ω
N.G.
Check the following items.
1) Harness continuity between oil pressure switch harness connector and E.C.U. 20-pin connector.
2) Continuity between oil pressure switch harness connector and ground.
3) Power steering oil pressure switch function.
O.K.
Perform driving test.
N.G.
1) Perform E.C.U. input/output signal inspection test.
2) If N.G., recheck the E.C.U. pin terminals damage or the connection of E.C.U. harness connector.
O.K.
INSPECTION END

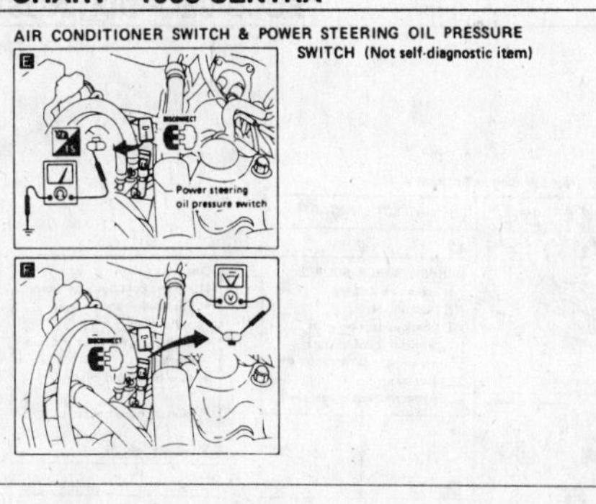
AIR CONDITIONER SWITCH & POWER STEERING OIL PRESSURE SWITCH (Not self-diagnostic item)
Power steering oil pressure switch

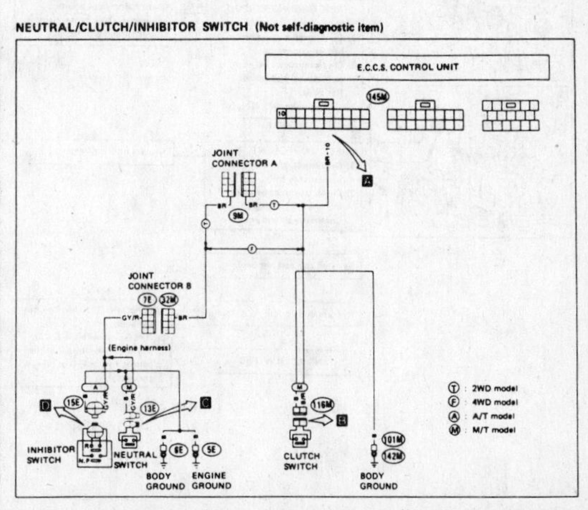
NEUTRAL/CLUTCH/INHIBITOR SWITCH (Not self-diagnostic item)
E.C.C.S. CONTROL UNIT
JOINT CONNECTOR A
JOINT CONNECTOR B
(Engine harness)
INHIBITOR SWITCH
NEUTRAL SWITCH
BODY GROUND
ENGINE GROUND
CLUTCH SWITCH
BODY GROUND
Ⓣ: 2WD model
Ⓕ: 4WD model
Ⓐ: A/T model
Ⓜ: M/T model

ECCS SELF-DIAGNOSTIC CHART—1988 SENTRA

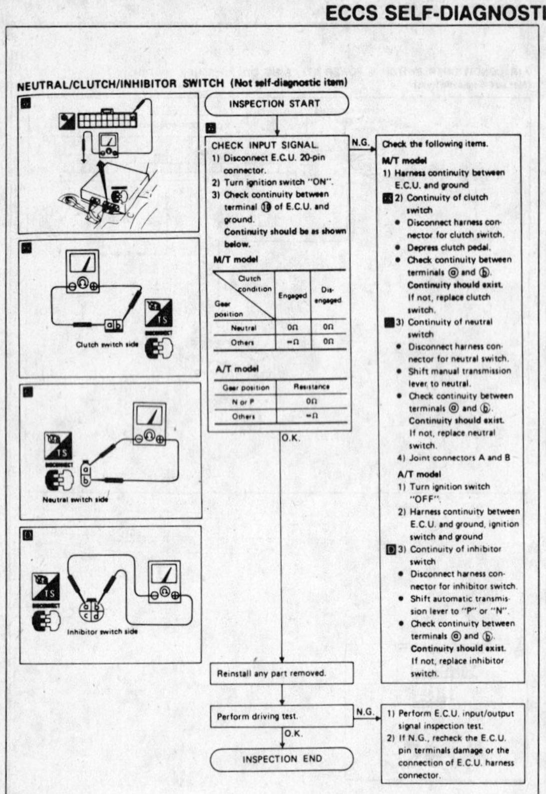
NEUTRAL/CLUTCH/INHIBITOR SWITCH (Not self-diagnostic item)
INSPECTION START
CHECK INPUT SIGNAL.
1) Disconnect E.C.U. 20-pin connector.
2) Turn ignition switch "ON".
3) Check continuity between terminal ⑲ of E.C.U. and ground.
Continuity should be as shown below.
M/T model
Clutch condition / Gear position | Engaged | Dis-engaged
Neutral | 0Ω | 0Ω
Others | ∞Ω | 0Ω
A/T model
Gear position | Resistance
N or P | 0Ω
Others | ∞Ω
N.G.
Check the following items.
M/T model
1) Harness continuity between E.C.U. and ground
2) Continuity of clutch switch
• Disconnect harness connector for clutch switch.
• Depress clutch pedal.
• Check continuity between terminals ⓐ and ⓑ.
Continuity should exist.
If not, replace clutch switch.
3) Continuity of neutral switch
• Disconnect harness connector for neutral switch.
• Shift manual transmission lever to neutral.
• Check continuity between terminals ⓐ and ⓑ.
Continuity should exist.
If not, replace neutral switch.
4) Joint connectors A and B
A/T model
1) Turn ignition switch "OFF".
2) Harness continuity between E.C.U. and ground, ignition switch and ground
D 3) Continuity of inhibitor switch
• Disconnect harness connector for inhibitor switch.
• Shift automatic transmission lever to "P" or "N".
• Check continuity between terminals ⓐ and ⓑ.
Continuity should exist.
If not, replace inhibitor switch.
O.K.
Reinstall any part removed.
Perform driving test.
N.G.
1) Perform E.C.U. input/output signal inspection test.
2) If N.G., recheck the E.C.U. pin terminals damage or the connection of E.C.U. harness connector.
O.K.
INSPECTION END
Clutch switch side
Neutral switch side
Inhibitor switch side
DISCONNECT

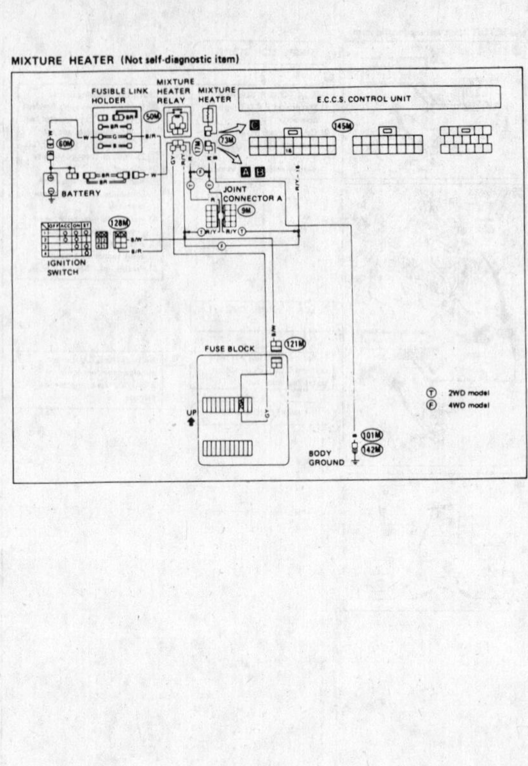
MIXTURE HEATER (Not self-diagnostic item)
FUSIBLE LINK HOLDER
MIXTURE HEATER RELAY
MIXTURE HEATER
E.C.C.S. CONTROL UNIT
BATTERY
JOINT CONNECTOR A
IGNITION SWITCH
FUSE BLOCK
UP
BODY GROUND
Ⓣ : 2WD model
Ⓕ : 4WD model

ECCS SELF-DIAGNOSTIC CHART—1988 SENTRA

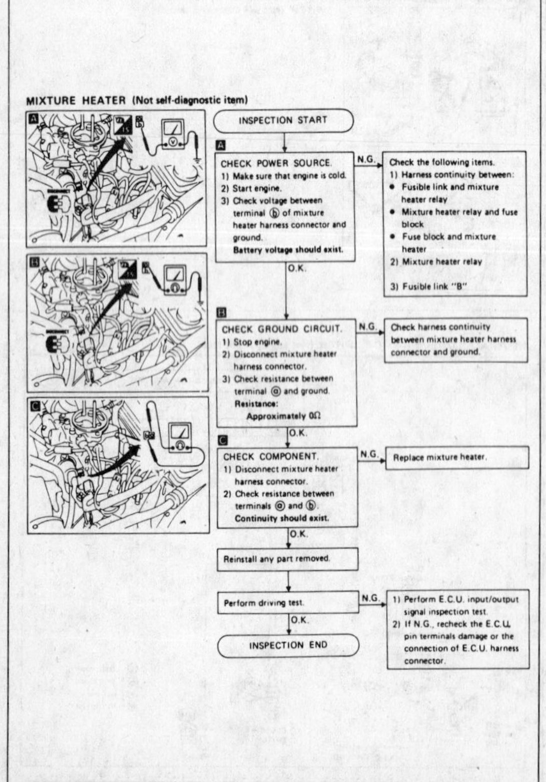
MIXTURE HEATER (Not self-diagnostic item)
INSPECTION START
A
CHECK POWER SOURCE.
1) Make sure that engine is cold.
2) Start engine.
3) Check voltage between terminal ⓑ of mixture heater harness connector and ground.
Battery voltage should exist.
N.G.
Check the following items.
1) Harness continuity between:
• Fusible link and mixture heater relay
• Mixture heater relay and fuse block
• Fuse block and mixture heater
2) Mixture heater relay
3) Fusible link "B"
O.K.
B
CHECK GROUND CIRCUIT.
1) Stop engine.
2) Disconnect mixture heater harness connector.
3) Check resistance between terminal ⓐ and ground.
Resistance:
Approximately 0Ω
N.G.
Check harness continuity between mixture heater harness connector and ground.
O.K.
C
CHECK COMPONENT.
1) Disconnect mixture heater harness connector.
2) Check resistance between terminals ⓐ and ⓑ.
Continuity should exist.
N.G.
Replace mixture heater.
O.K.
Reinstall any part removed.
Perform driving test.
N.G.
1) Perform E.C.U. input/output signal inspection test.
2) If N.G., recheck the E.C.U. pin terminals damage or the connection of E.C.U. harness connector.
O.K.
INSPECTION END

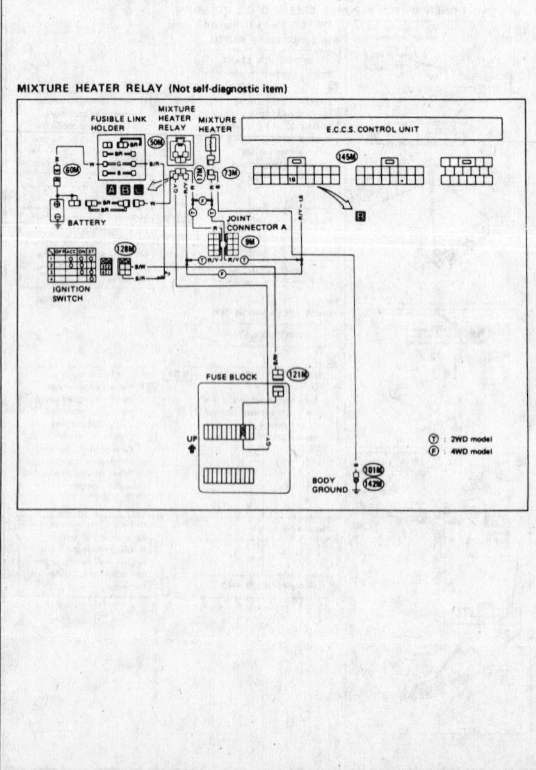
MIXTURE HEATER RELAY (Not self-diagnostic item)
FUSIBLE LINK HOLDER
MIXTURE HEATER RELAY
MIXTURE HEATER
E.C.C.S. CONTROL UNIT
BATTERY
JOINT CONNECTOR A
IGNITION SWITCH
FUSE BLOCK
UP
BODY GROUND
Ⓣ : 2WD model
Ⓕ : 4WD model

ECCS SELF-DIAGNOSTIC CHART—1988 SENTRA

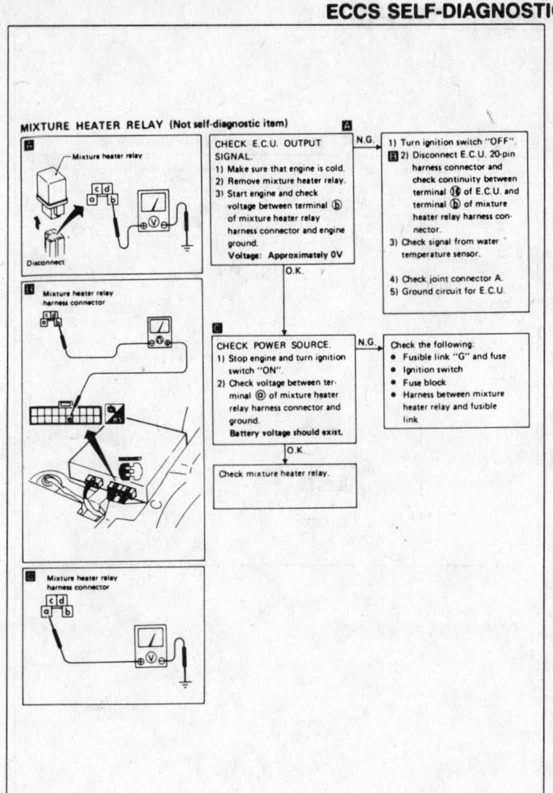
MIXTURE HEATER RELAY (Not self-diagnostic item)
Mixture heater relay
Disconnect
Mixture heater relay harness connector
Mixture heater relay harness connector
A
CHECK E.C.U. OUTPUT SIGNAL.
1) Make sure that engine is cold.
2) Remove mixture heater relay.
3) Start engine and check voltage between terminal ⓑ of mixture heater relay harness connector and engine ground.
Voltage: Approximately 0V
N.G.
1) Turn ignition switch "OFF".
B 2) Disconnect E.C.U. 20-pin harness connector and check continuity between terminal ⑯ of E.C.U. and terminal ⓑ of mixture heater relay harness connector.
3) Check signal from water temperature sensor.
4) Check joint connector A.
5) Ground circuit for E.C.U.
O.K.
C
CHECK POWER SOURCE.
1) Stop engine and turn ignition switch "ON".
2) Check voltage between terminal ⓒ of mixture heater relay harness connector and ground.
Battery voltage should exist.
N.G.
Check the following:
• Fusible link "G" and fuse
• Ignition switch
• Fuse block
• Harness between mixture heater relay and fusible link
O.K.
Check mixture heater relay.

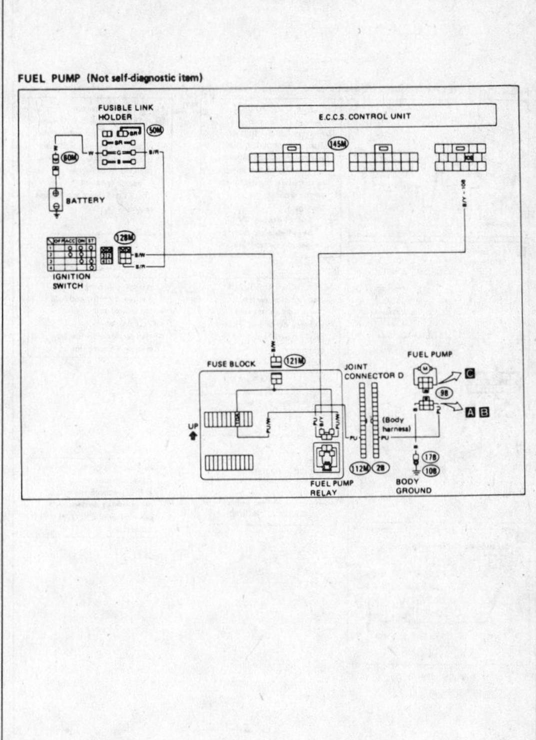
FUEL PUMP (Not self-diagnostic item)
FUSIBLE LINK HOLDER
E.C.C.S. CONTROL UNIT
BATTERY
IGNITION SWITCH
FUSE BLOCK
JOINT CONNECTOR D
FUEL PUMP
(Body harness)
UP
FUEL PUMP RELAY
BODY GROUND

ECCS SELF-DIAGNOSTIC CHART—1988 SENTRA

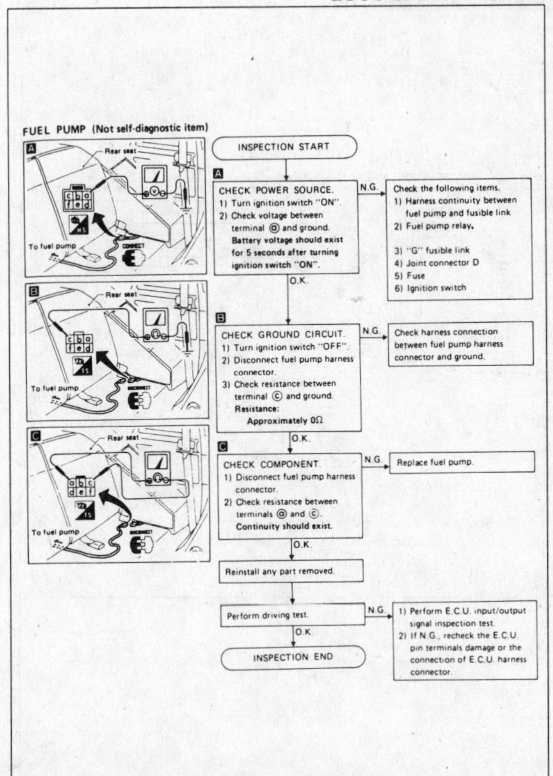
FUEL PUMP (Not self-diagnostic item)
Rear seat
To fuel pump
INSPECTION START
A
CHECK POWER SOURCE.
1) Turn ignition switch "ON".
2) Check voltage between terminal ⓐ and ground.
Battery voltage should exist for 5 seconds after turning ignition switch "ON".
N.G.
Check the following items.
1) Harness continuity between fuel pump and fusible link
2) Fuel pump relay.
3) "G" fusible link
4) Joint connector D
5) Fuse
6) Ignition switch
O.K.
B
CHECK GROUND CIRCUIT.
1) Turn ignition switch "OFF".
2) Disconnect fuel pump harness connector.
3) Check resistance between terminal ⓒ and ground.
Resistance:
Approximately 0Ω
N.G.
Check harness connection between fuel pump harness connector and ground.
O.K.
C
CHECK COMPONENT.
1) Disconnect fuel pump harness connector.
2) Check resistance between terminals ⓐ and ⓒ.
Continuity should exist.
N.G.
Replace fuel pump.
O.K.
Reinstall any part removed.
Perform driving test.
N.G.
1) Perform E.C.U. input/output signal inspection test.
2) If N.G., recheck the E.C.U. pin terminals damage or the connection of E.C.U. harness connector.
O.K.
INSPECTION END

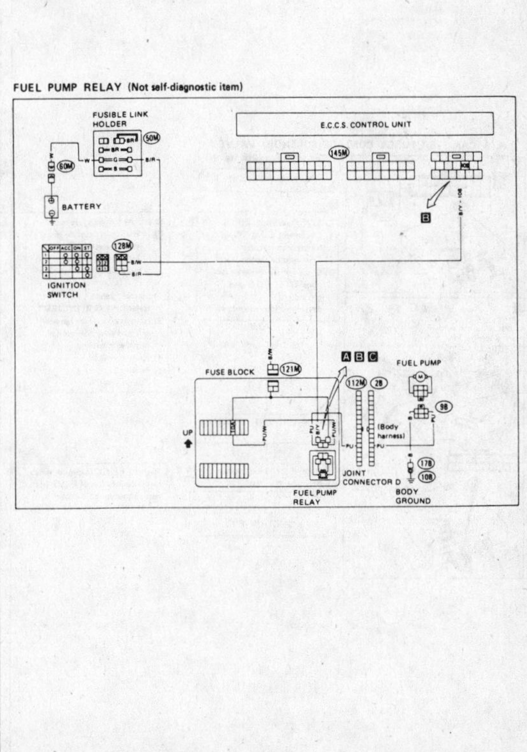
FUEL PUMP RELAY (Not self-diagnostic item)
FUSIBLE LINK HOLDER
E.C.C.S. CONTROL UNIT
BATTERY
IGNITION SWITCH
FUSE BLOCK
FUEL PUMP
(Body harness)
UP
JOINT CONNECTOR D
FUEL PUMP RELAY
BODY GROUND

ECCS SELF-DIAGNOSTIC CHART—1988 SENTRA

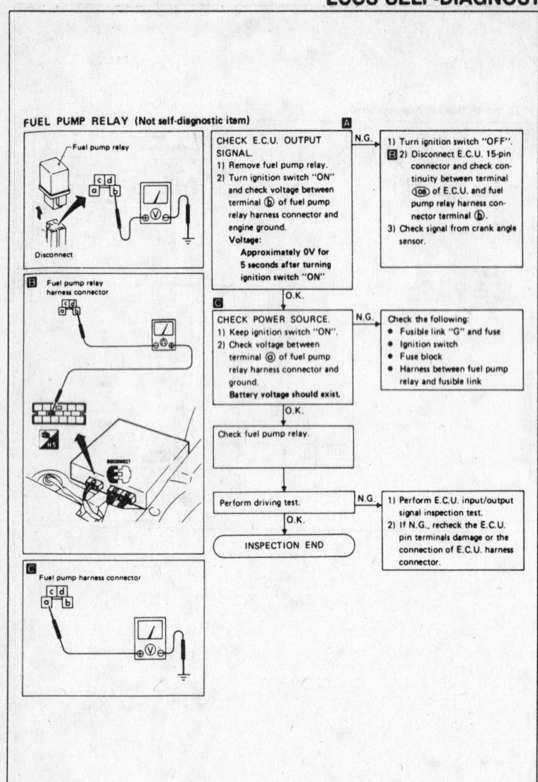
FUEL PUMP RELAY (Not self-diagnostic item)
Fuel pump relay
Disconnect
A
CHECK E.C.U. OUTPUT SIGNAL.
1) Remove fuel pump relay.
2) Turn ignition switch "ON" and check voltage between terminal ⓑ of fuel pump relay harness connector and engine ground.
Voltage:
Approximately 0V for 5 seconds after turning ignition switch "ON"
N.G.
1) Turn ignition switch "OFF".
B 2) Disconnect E.C.U. 15-pin connector and check continuity between terminal (108) of E.C.U. and fuel pump relay harness connector terminal ⓑ.
3) Check signal from crank angle sensor.
O.K.
B Fuel pump relay harness connector
C
CHECK POWER SOURCE.
1) Keep ignition switch "ON".
2) Check voltage between terminal ⓐ of fuel pump relay harness connector and ground.
Battery voltage should exist.
N.G.
Check the following:
• Fusible link "G" and fuse
• Ignition switch
• Fuse block
• Harness between fuel pump relay and fusible link
O.K.
Check fuel pump relay.
Perform driving test.
N.G.
1) Perform E.C.U. input/output signal inspection test.
2) If N.G., recheck the E.C.U. pin terminals damage or the connection of E.C.U. harness connector.
O.K.
INSPECTION END
C Fuel pump harness connector

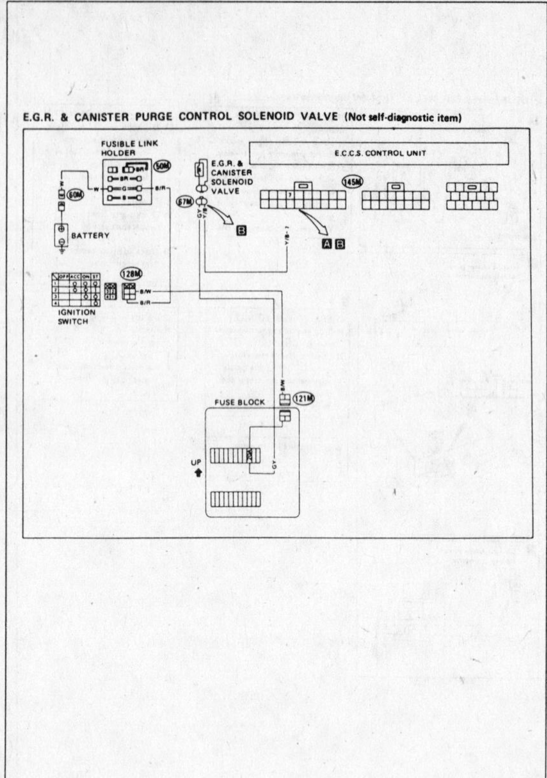
E.G.R. & CANISTER PURGE CONTROL SOLENOID VALVE (Not self-diagnostic item)
FUSIBLE LINK HOLDER
E.G.R. & CANISTER SOLENOID VALVE
E.C.C.S. CONTROL UNIT
BATTERY
IGNITION SWITCH
FUSE BLOCK
UP

ECCS SELF-DIAGNOSTIC CHART—1988 SENTRA

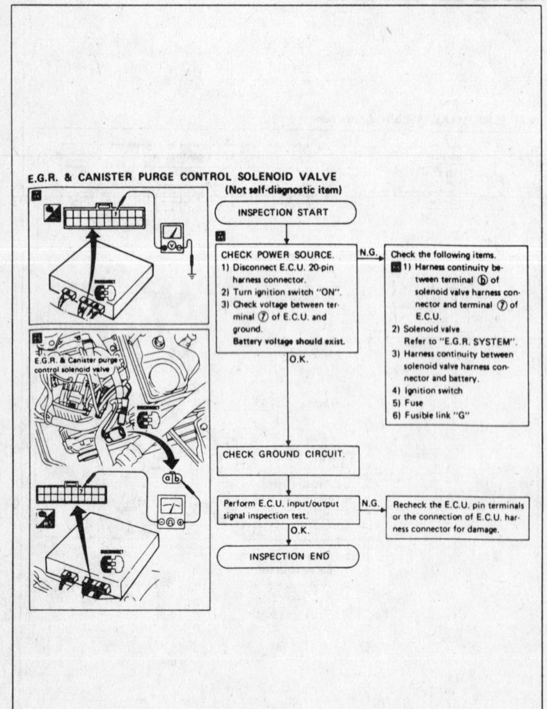
E.G.R. & CANISTER PURGE CONTROL SOLENOID VALVE
(Not self-diagnostic item)
INSPECTION START
CHECK POWER SOURCE.
1) Disconnect E.C.U. 20-pin harness connector.
2) Turn ignition switch "ON".
3) Check voltage between terminal ⑦ of E.C.U. and ground.
Battery voltage should exist.
N.G.
Check the following items.
1) Harness continuity between terminal ⓑ of solenoid valve harness connector and terminal ⑦ of E.C.U.
2) Solenoid valve
Refer to "E.G.R. SYSTEM".
3) Harness continuity between solenoid valve harness connector and battery.
4) Ignition switch
5) Fuse
6) Fusible link "G"
O.K.
E.G.R. & Canister purge control solenoid valve
CHECK GROUND CIRCUIT.
Perform E.C.U. input/output signal inspection test.
N.G.
Recheck the E.C.U. pin terminals or the connection of E.C.U. harness connector for damage.
O.K.
INSPECTION END

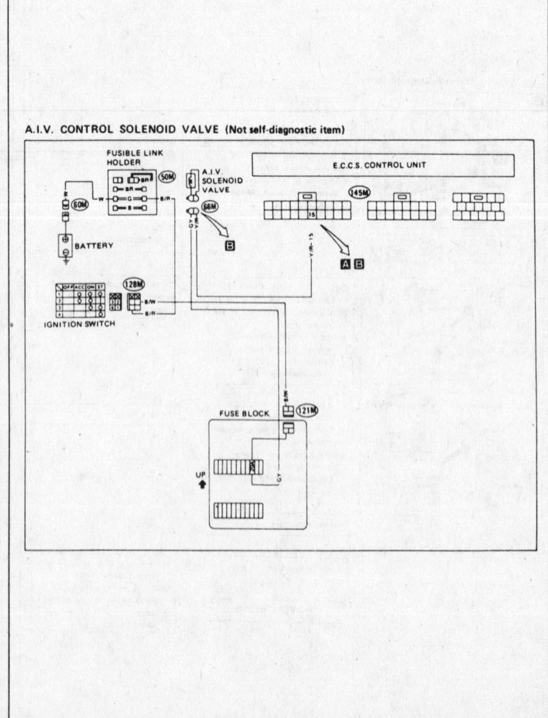
A.I.V. CONTROL SOLENOID VALVE (Not self-diagnostic item)
FUSIBLE LINK HOLDER
A.I.V. SOLENOID VALVE
E.C.C.S. CONTROL UNIT
BATTERY
IGNITION SWITCH
FUSE BLOCK
UP

ECCS SELF-DIAGNOSTIC CHART—1988 SENTRA

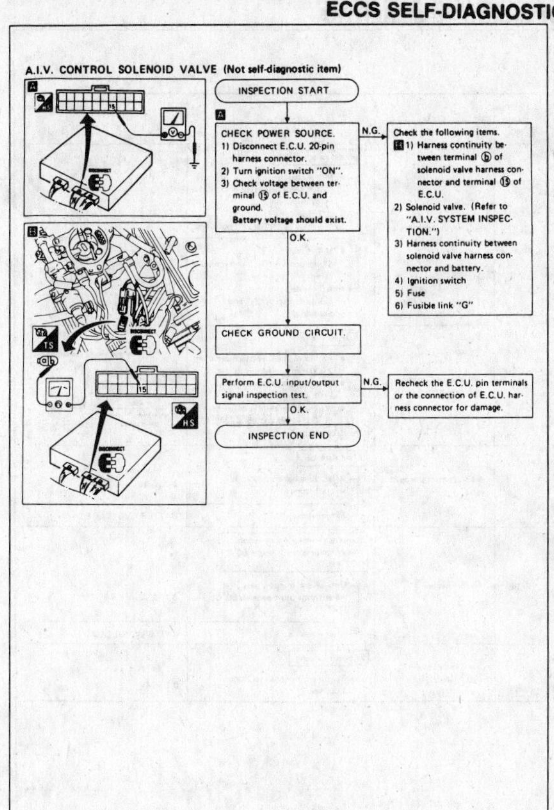
A.I.V. CONTROL SOLENOID VALVE (Not self-diagnostic item)
INSPECTION START
CHECK POWER SOURCE.
1) Disconnect E.C.U. 20-pin harness connector.
2) Turn ignition switch "ON".
3) Check voltage between terminal ⑮ of E.C.U. and ground.
Battery voltage should exist.
N.G.
Check the following items.
1) Harness continuity between terminal ⓑ of solenoid valve harness connector and terminal ⑮ of E.C.U.
2) Solenoid valve. (Refer to "A.I.V. SYSTEM INSPECTION.")
3) Harness continuity between solenoid valve harness connector and battery.
4) Ignition switch
5) Fuse
6) Fusible link "G"
O.K.
CHECK GROUND CIRCUIT.
Perform E.C.U. input/output signal inspection test.
N.G.
Recheck the E.C.U. pin terminals or the connection of E.C.U. harness connector for damage.
O.K.
INSPECTION END

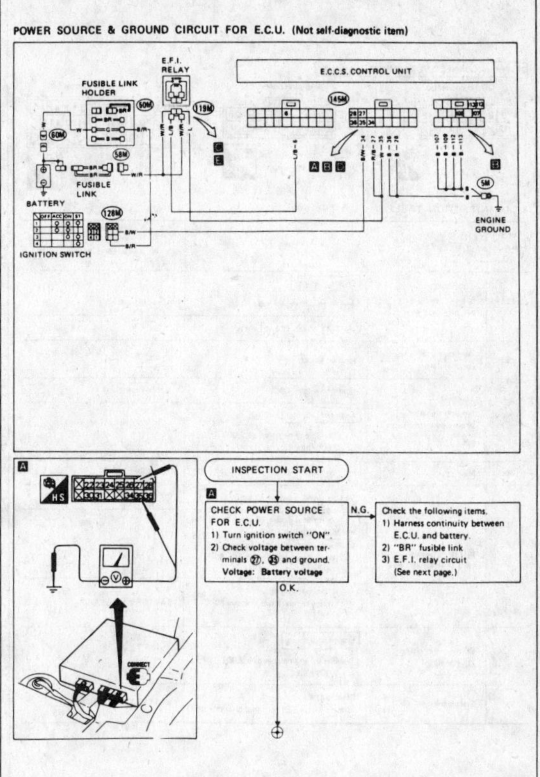
POWER SOURCE & GROUND CIRCUIT FOR E.C.U. (Not self-diagnostic item)
E.F.I. RELAY
FUSIBLE LINK HOLDER
E.C.C.S. CONTROL UNIT
BATTERY
FUSIBLE LINK
IGNITION SWITCH
ENGINE GROUND
INSPECTION START
CHECK POWER SOURCE FOR E.C.U.
1) Turn ignition switch "ON".
2) Check voltage between terminals ㉗, ㉟ and ground.
Voltage: Battery voltage
N.G.
Check the following items.
1) Harness continuity between E.C.U. and battery.
2) "BR" fusible link
3) E.F.I. relay circuit (See next page.)
O.K.

ECCS SELF-DIAGNOSTIC CHART—1988 SENTRA

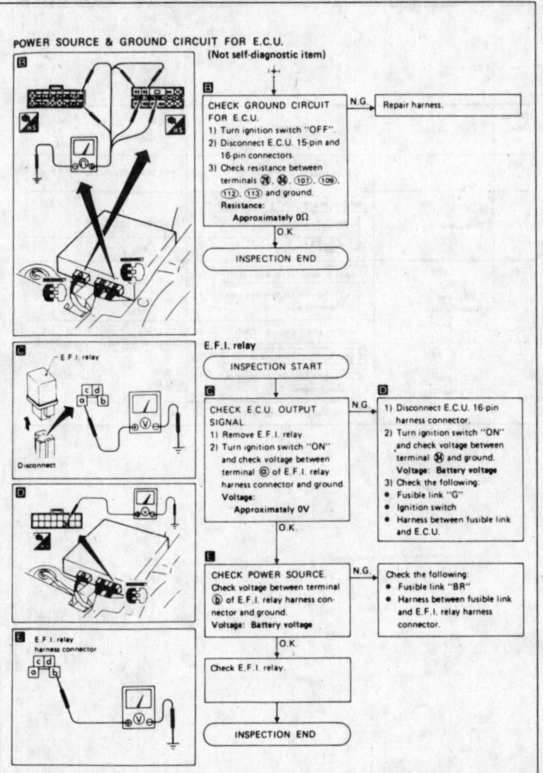
POWER SOURCE & GROUND CIRCUIT FOR E.C.U. (Not self-diagnostic item)
CHECK GROUND CIRCUIT FOR E.C.U.
1) Turn ignition switch "OFF".
2) Disconnect E.C.U. 15-pin and 16-pin connectors.
3) Check resistance between terminals ㉖, ㉞, 107, 109, 112, 113 and ground.
Resistance: Approximately 0Ω
N.G.
Repair harness.
O.K.
INSPECTION END
E.F.I. relay
INSPECTION START
CHECK E.C.U. OUTPUT SIGNAL.
1) Remove E.F.I. relay.
2) Turn ignition switch "ON" and check voltage between terminal ⓒ of E.F.I. relay harness connector and ground.
Voltage: Approximately 0V
N.G.
1) Disconnect E.C.U. 16-pin harness connector.
2) Turn ignition switch "ON" and check voltage between terminal ㉞ and ground.
Voltage: Battery voltage
3) Check the following.
• Fusible link "G"
• Ignition switch
• Harness between fusible link and E.C.U.
O.K.
CHECK POWER SOURCE.
Check voltage between terminal ⓑ of E.F.I. relay harness connector and ground.
Voltage: Battery voltage
N.G.
Check the following:
• Fusible link "BR"
• Harness between fusible link and E.F.I. relay harness connector.
O.K.
Check E.F.I. relay.
INSPECTION END
E.F.I. relay
Disconnect
E.F.I. relay harness connector

ECU SIGNAL INSPECTION CHART—1988 SENTRA

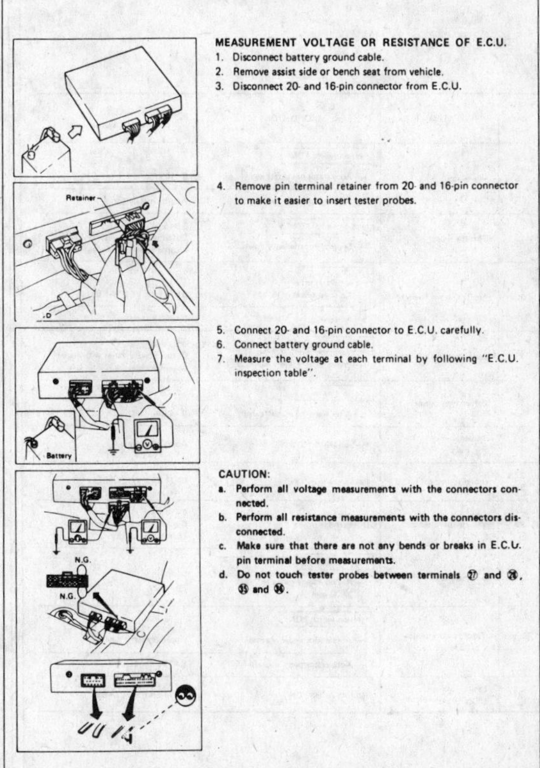
MEASUREMENT VOLTAGE OR RESISTANCE OF E.C.U.
1. Disconnect battery ground cable.
2. Remove assist side or bench seat from vehicle.
3. Disconnect 20- and 16-pin connector from E.C.U.
Retainer
4. Remove pin terminal retainer from 20- and 16-pin connector to make it easier to insert tester probes.
5. Connect 20- and 16-pin connector to E.C.U. carefully.
6. Connect battery ground cable.
7. Measure the voltage at each terminal by following "E.C.U. inspection table".
Battery
CAUTION:
a. Perform all voltage measurements with the connectors connected.
b. Perform all resistance measurements with the connectors disconnected.
c. Make sure that there are not any bends or breaks in E.C.U. pin terminal before measurements.
d. Do not touch tester probes between terminals ㉗ and ㉘, ㉟ and ㊱.
N.G.

ECU SIGNAL INSPECTION CHART—1988 SENTRA

E16i PIN CONNECTOR TERMINAL LAYOUT

20-pin connector: 1 2 3 ⊠ 5 6 7 8 9 10 / 11 12 ⊠ ⊠ 15 16 17 18 19 ⊠ (H.S.)

16-pin connector: 21 22 23 24 25 26 27 28 / 29 30 31 ⊠ 33 34 35 36 (H.S.)

15-pin connector: 112 113 114 ⊠ / 107 108 109 110 111 / 101 102 ⊠ ⊠ ⊠ ⊠ (H.S.)

E.C.U. INSPECTION TABLE

TERMINAL NO.	ITEM	CONDITION	DATA (Reference values)
2	Power steering oil pressure switch	[Engine is idling.] — When turning steering wheel.	Approximately 0V
		— With wheels straight ahead.	8 - 9V
3	Ignition signal (from resistor)	Ignition switch "ON"	11 - 12V
5	Ignition signal (from power transistor)	Ignition switch "ON"	Approximately 0.1V
		[Engine is running.] — Rev up engine speed to 2,000 - 3,000 rpm.	0.5 - 2.0V
6	Main relay	Ignition switch "ON"	0.8 - 1.0V
7	E.G.R. & canister purge control solenoid valve	Vehicle speed is below 10 km/h (6 MPH).	0.7 - 1.0V
		[Vehicle speed is above 10 km/h (6 MPH) and load is applied.] — Engine is cold or during warm-up. [Water temperature is below 60°C (140°F).]	0.7 - 1.0V
		— After warming up. [Water temperature is above 80°C (176°F)]	BATTERY VOLTAGE
8	Crank angle sensor (1° signal)	[Engine is running.] — Rev up engine speed to 2,000 - 3,000 rpm.	2.3 - 2.7V
9	Start signal	Cranking (Starter motor "S" terminal disconnected)	BATTERY VOLTAGE

TERMINAL NO.	ITEM	CONDITION	DATA (Reference values)
10	Neutral/Clutch/Inhibitor switch	[Ignition switch "ON"] — • M/T model Clutch pedal depressed or transaxle in "N" position • A/T model Transaxle in "N" or "P" position	0V
		— • M/T model Clutch pedal released and transaxle except "N" position • A/T model, Transaxle except "N" and "P" position	BATTERY VOLTAGE
12	Hot wire self-cleaning (Air flow meter)	[Drive vehicle] • Rev up engine speed to more than 2,000 rpm and return to idle. • Drive vehicle at more than 20 km/h (12 MPH). • Water temperature is between 60°C (140°F) and 95°C (203°F). — Stop engine.	Voltage should appear. Key "OFF" Approximately 0V Several sec.
15	A.I.V. control solenoid valve	[After warm-up] — After idling has continued for several seconds.	0.7 - 0.9V
		— Except above	BATTERY VOLTAGE
16	Mixture heater	[Start engine when engine is cold [Water temperature is below 50°C (122°F)].] — Until several minutes pass after water temperature exceeds 50°C (122°F).	0.7 - 0.9V
		— After warm-up	BATTERY VOLTAGE
17	Crank angle sensor (180° signal)	[Engine is running.] — Rev up engine speed to 2,000 - 3,000 rpm.	0.1 - 0.4V

ECU SIGNAL INSPECTION CHART—1988 SENTRA

TERMINAL NO.	ITEM	CONDITION	DATA (Reference values)
18	Idle switch	[Ignition switch "ON"] — Accelerator pedal released.	9 - 10V
		— Accelerator pedal depressed.	Approximately 0V
19	Throttle sensor	[Ignition switch "ON"] — Depress throttle valve slowly.	0.5 - 5.0V Output voltage varies with the throttle valve opening angle.
22	Air conditioner switch	[Engine is running.] — Air conditioner "ON".	BATTERY VOLTAGE
		— Air conditioner "OFF".	0V
23	Water temperature sensor	[Engine is running.] — Engine cold → After warm-up.	0.5 - 5.0V Output voltage varies with engine water temperature.
24	Exhaust gas sensor	[Engine is running.] — After warming up sufficiently.	0 - 1.0V
25	Throttle sensor (⊕ side)	Ignition switch "ON"	Approximately 5.0V
27 35	Power source for E.C.U.	Ignition switch "ON"	BATTERY VOLTAGE
29	Vehicle speed sensor	Drive vehicle slowly.	0V or 4.8V Output voltage changes, one to another, repeatedly.
31	Air flow meter	[Engine is running.] — Rev up engine speed to 2,000 - 3,000 rpm.	2.4 - 4.0V
33	Throttle valve switch (⊕ side)	[Ignition switch "ON"] — Accelerator pedal released.	Approximately 10V
		— Accelerator pedal depressed.	BATTERY VOLTAGE
34	Ignition switch ("ON" signal)	Ignition switch "ON"	BATTERY VOLTAGE

TERMINAL NO.	ITEM	CONDITION	DATA (Reference values)
101 102	Injector	Ignition switch "OFF"	BATTERY VOLTAGE
108	Fuel pump	[Ignition switch "ON"] — Engine stopped.	BATTERY VOLTAGE
		— Engine is running.	0.7 - 0.9V
110	I.S.C. valve (Opening side)	[Engine is idling (After warm-up).] — Throttle sensor harness connector disconnected.	11 - 12V
		— With air conditioner "ON"	8 - 10V
111	I.S.C. valve (Closing side)	[Engine is idling (After warm-up).] — Throttle sensor harness connector disconnected.	2 - 3V
		— With air conditioner "ON"	4 - 6V

ECCS MIXTURE RATIO FEEDBACK SYSTEM INPSECTION CHART—1988 SENTRA

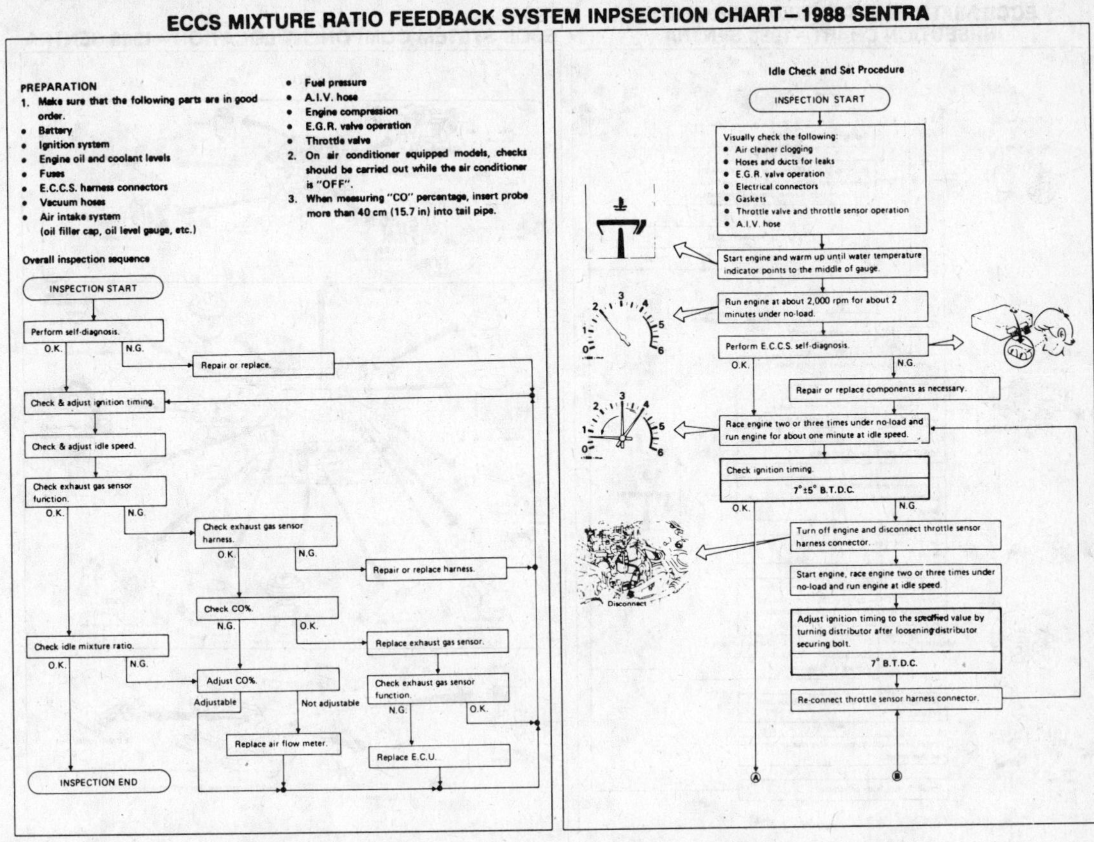
PREPARATION
1. Make sure that the following parts are in good order.
• Battery
• Ignition system
• Engine oil and coolant levels
• Fuses
• E.C.C.S. harness connectors
• Vacuum hoses
• Air intake system (oil filler cap, oil level gauge, etc.)
• Fuel pressure
• A.I.V. hose
• Engine compression
• E.G.R. valve operation
• Throttle valve
2. On air conditioner equipped models, checks should be carried out while the air conditioner is "OFF".
3. When measuring "CO" percentage, insert probe more than 40 cm (15.7 in) into tail pipe.
Overall inspection sequence
INSPECTION START
Perform self-diagnosis.
O.K.
N.G.
Repair or replace.
Check & adjust ignition timing.
Check & adjust idle speed.
Check exhaust gas sensor function.
O.K.
N.G.
Check exhaust gas sensor harness.
O.K.
N.G.
Repair or replace harness.
Check CO%.
N.G.
O.K.
Replace exhaust gas sensor.
Check idle mixture ratio.
O.K.
N.G.
Adjust CO%.
Adjustable
Not adjustable
Check exhaust gas sensor function.
N.G.
O.K.
Replace air flow meter.
Replace E.C.U.
INSPECTION END
Idle Check and Set Procedure
INSPECTION START
Visually check the following:
• Air cleaner clogging
• Hoses and ducts for leaks
• E.G.R. valve operation
• Electrical connectors
• Gaskets
• Throttle valve and throttle sensor operation
• A.I.V. hose
Start engine and warm up until water temperature indicator points to the middle of gauge.
Run engine at about 2,000 rpm for about 2 minutes under no-load.
Perform E.C.C.S. self-diagnosis.
O.K.
N.G.
Repair or replace components as necessary.
Race engine two or three times under no-load and run engine for about one minute at idle speed.
Check ignition timing.
7°±5° B.T.D.C.
O.K.
N.G.
Turn off engine and disconnect throttle sensor harness connector.
Disconnect
Start engine, race engine two or three times under no-load and run engine at idle speed.
Adjust ignition timing to the specified value by turning distributor after loosening distributor securing bolt.
7° B.T.D.C.
Re-connect throttle sensor harness connector.
Ⓐ
Ⓑ

ECCS MIXTURE RATIO FEEDBACK SYSTEM INPSECTION CHART—1988 SENTRA

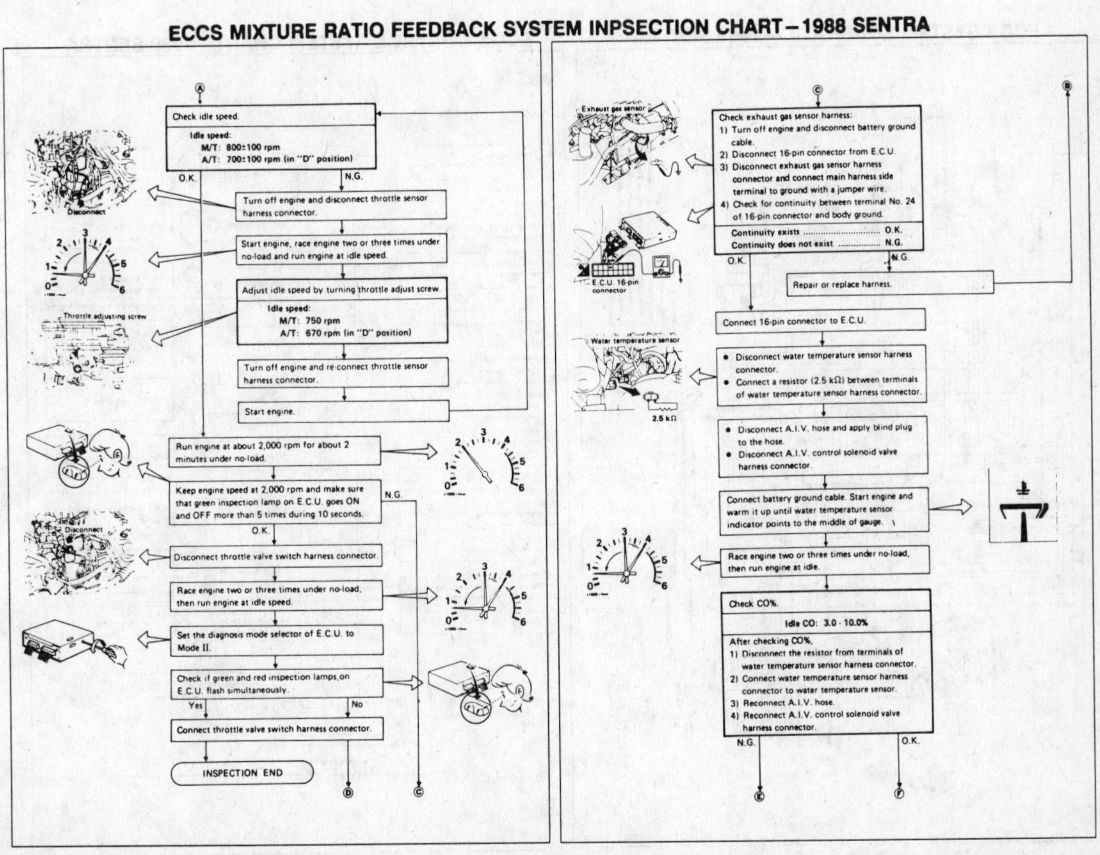
Ⓐ
Check idle speed.
Idle speed:
M/T: 800±100 rpm
A/T: 700±100 rpm (in "D" position)
O.K.
N.G.
Turn off engine and disconnect throttle sensor harness connector.
Disconnect
Start engine, race engine two or three times under no-load and run engine at idle speed.
Adjust idle speed by turning throttle adjust screw.
Idle speed:
M/T: 750 rpm
A/T: 670 rpm (in "D" position)
Throttle adjusting screw
Turn off engine and re-connect throttle sensor harness connector.
Start engine.
Run engine at about 2,000 rpm for about 2 minutes under no-load.
Keep engine speed at 2,000 rpm and make sure that green inspection lamp on E.C.U. goes ON and OFF more than 5 times during 10 seconds.
N.G.
O.K.
Disconnect
Disconnect throttle valve switch harness connector.
Race engine two or three times under no-load, then run engine at idle speed.
Set the diagnosis mode selector of E.C.U. to Mode II.
Check if green and red inspection lamps on E.C.U. flash simultaneously.
Yes
No
Connect throttle valve switch harness connector.
INSPECTION END
Ⓓ
Ⓒ
Ⓒ
Ⓑ
Exhaust gas sensor
Check exhaust gas sensor harness:
1) Turn off engine and disconnect battery ground cable.
2) Disconnect 16-pin connector from E.C.U.
3) Disconnect exhaust gas sensor harness connector and connect main harness side terminal to ground with a jumper wire.
4) Check for continuity between terminal No. 24 of 16-pin connector and body ground.
Continuity exists .... O.K.
Continuity does not exist .... N.G.
E.C.U. 16-pin connector
O.K.
N.G.
Repair or replace harness.
Connect 16-pin connector to E.C.U.
Water temperature sensor
• Disconnect water temperature sensor harness connector.
• Connect a resistor (2.5 kΩ) between terminals of water temperature sensor harness connector.
2.5 kΩ
• Disconnect A.I.V. hose and apply blind plug to the hose.
• Disconnect A.I.V. control solenoid valve harness connector.
Connect battery ground cable. Start engine and warm it up until water temperature sensor indicator points to the middle of gauge.
Race engine two or three times under no-load, then run engine at idle.
Check CO%.
Idle CO: 3.0 - 10.0%
After checking CO%,
1) Disconnect the resistor from terminals of water temperature sensor harness connector.
2) Connect water temperature sensor harness connector to water temperature sensor.
3) Reconnect A.I.V. hose.
4) Reconnect A.I.V. control solenoid valve harness connector.
N.G.
O.K.
Ⓔ
Ⓕ

ECCS MIXTURE RATIO FEEDBACK SYSTEM INPSECTION CHART – 1988 SENTRA

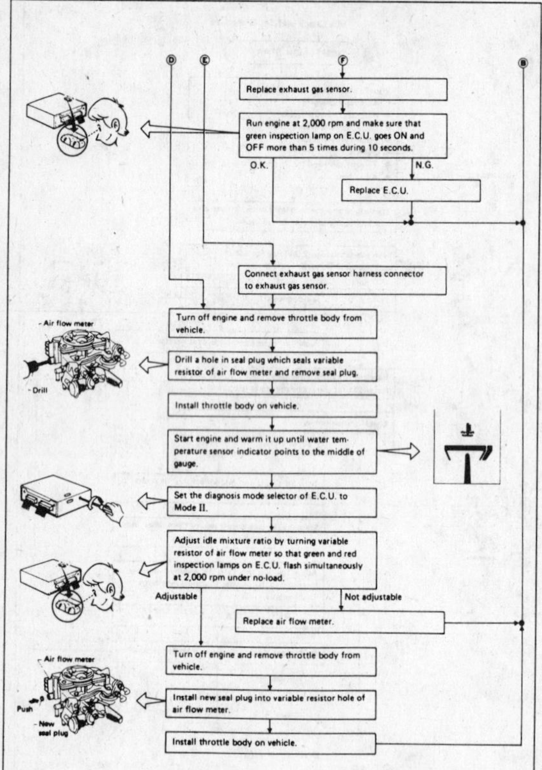
D
E
F
B
Replace exhaust gas sensor.
Run engine at 2,000 rpm and make sure that green inspection lamp on E.C.U. goes ON and OFF more than 5 times during 10 seconds.
O.K.
N.G.
Replace E.C.U.
Connect exhaust gas sensor harness connector to exhaust gas sensor.
Turn off engine and remove throttle body from vehicle.
Air flow meter
Drill
Drill a hole in seal plug which seals variable resistor of air flow meter and remove seal plug.
Install throttle body on vehicle.
Start engine and warm it up until water temperature sensor indicator points to the middle of gauge.
Set the diagnosis mode selector of E.C.U. to Mode II.
Adjust idle mixture ratio by turning variable resistor of air flow meter so that green and red inspection lamps on E.C.U. flash simultaneously at 2,000 rpm under no-load.
Adjustable
Not adjustable
Replace air flow meter.
Turn off engine and remove throttle body from vehicle.
Air flow meter
Push
New seal plug
Install new seal plug into variable resistor hole of air flow meter.
Install throttle body on vehicle.

ECCS SYSTEM COMPONENT LOCATION – 1989 SENTRA

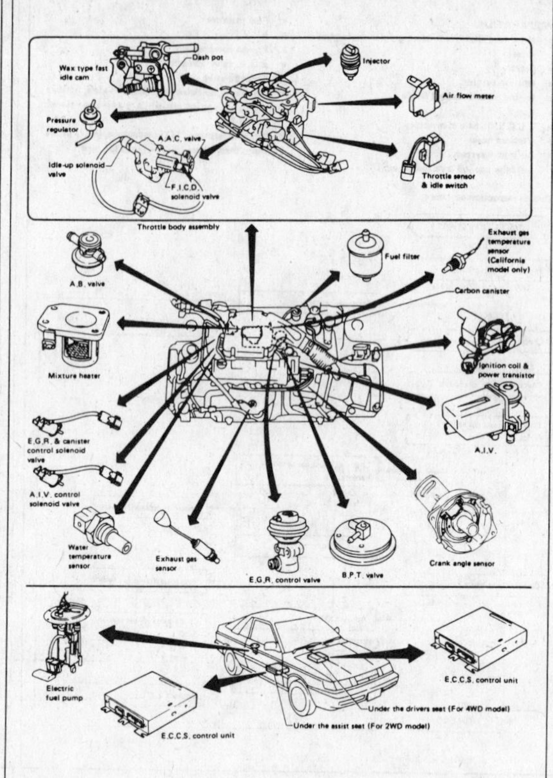
Dash pot
Wax type fast idle cam
Injector
Air flow meter
Pressure regulator
A.A.C. valve
Idle-up solenoid valve
F.I.C.D. solenoid valve
Throttle sensor & idle switch
Throttle body assembly
A.B. valve
Fuel filter
Exhaust gas temperature sensor (California model only)
Carbon canister
Mixture heater
Ignition coil & power transistor
A.I.V.
E.G.R. & canister control solenoid valve
A.I.V. control solenoid valve
Water temperature sensor
Exhaust gas sensor
E.G.R. control valve
B.P.T. valve
Crank angle sensor
Electric fuel pump
E.C.C.S. control unit
Under the drivers seat (For 4WD model)
Under the assist seat (For 2WD model)
E.C.C.S. control unit

ECCS SYSTEM SCHEMATIC – 1989 SENTRA

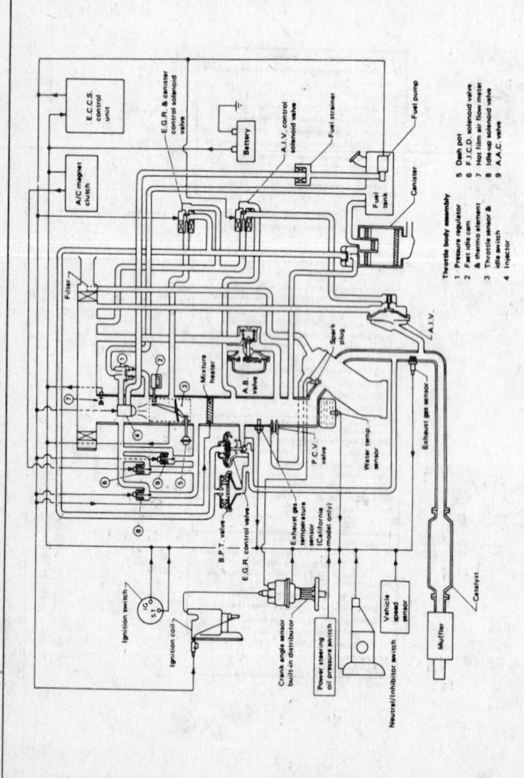
E.C.C.S. control unit
A/C magnet clutch
E.G.R. & canister control solenoid valve
Battery
A.I.V. control solenoid valve
Fuel strainer
Fuel pump
Fuel tank
Canister
Filter
Mixture heater
A.B. valve
Spark plug
A.I.V.
P.C.V. valve
Water temp. sensor
Exhaust gas sensor
B.P.T. valve
E.G.R. control valve
Exhaust gas temperature sensor (California model only)
Catalyst
Muffler
Ignition switch
Ignition coil
Crank angle sensor built-in distributor
Power steering oil pressure switch
Neutral/Inhibitor switch
Vehicle speed sensor
Throttle body assembly
1 Pressure regulator
2 Fast idle cam & thermo element
3 Throttle sensor & idle switch
4 Injector
5 Dash pot
6 F.I.C.D. solenoid valve
7 Hot film air flow meter
8 Idle-up solenoid valve
9 A.A.C. valve

ECCS CONTROL SYSTEM CHART – 1989 SENTRA

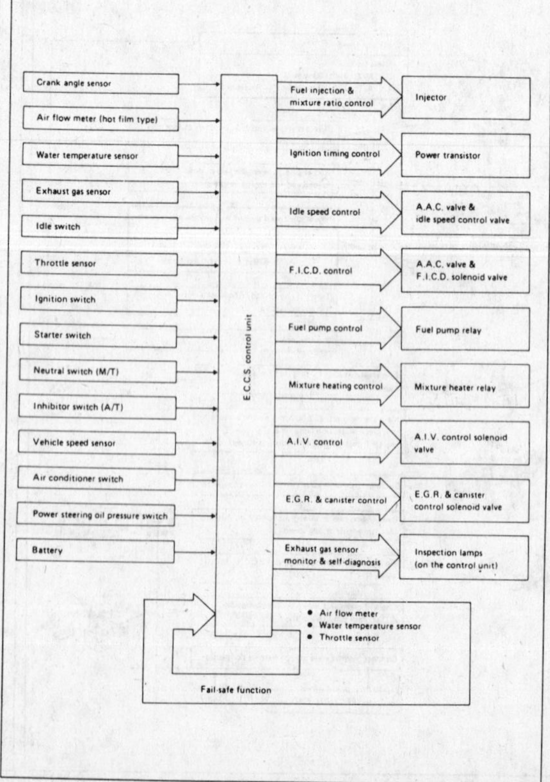
Crank angle sensor
Air flow meter (hot film type)
Water temperature sensor
Exhaust gas sensor
Idle switch
Throttle sensor
Ignition switch
Starter switch
Neutral switch (M/T)
Inhibitor switch (A/T)
Vehicle speed sensor
Air conditioner switch
Power steering oil pressure switch
Battery
E.C.C.S. control unit
Fuel injection & mixture ratio control
Ignition timing control
Idle speed control
F.I.C.D. control
Fuel pump control
Mixture heating control
A.I.V. control
E.G.R. & canister control
Exhaust gas sensor monitor & self-diagnosis
Injector
Power transistor
A.A.C. valve & idle speed control valve
A.A.C. valve & F.I.C.D. solenoid valve
Fuel pump relay
Mixture heater relay
A.I.V. control solenoid valve
E.G.R. & canister control solenoid valve
Inspection lamps (on the control unit)
• Air flow meter
• Water temperature sensor
• Throttle sensor
Fail-safe function

ECCS FUEL FLOW SYSTEM SCHEMATIC—1989 SENTRA

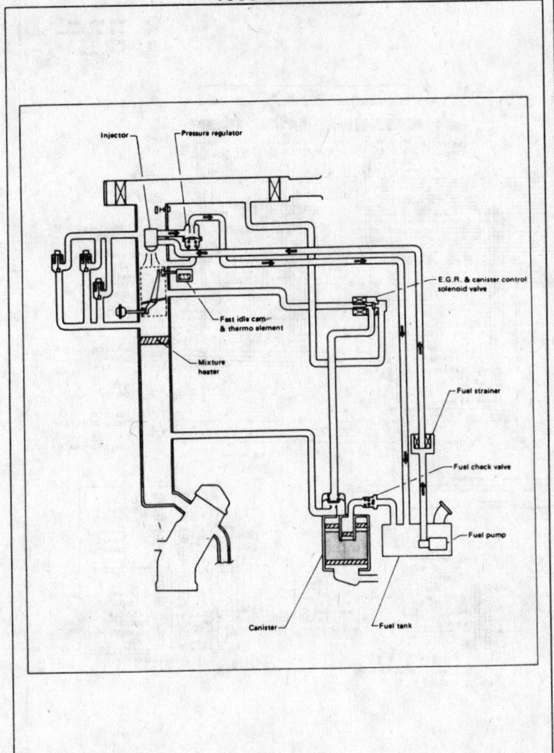
Injector
Pressure regulator
E.G.R. & canister control solenoid valve
Fast idle cam & thermo element
Mixture heater
Fuel strainer
Fuel check valve
Fuel pump
Canister
Fuel tank

ECCS AIR FLOW SYSTEM SCHEMATIC—1989 SENTRA

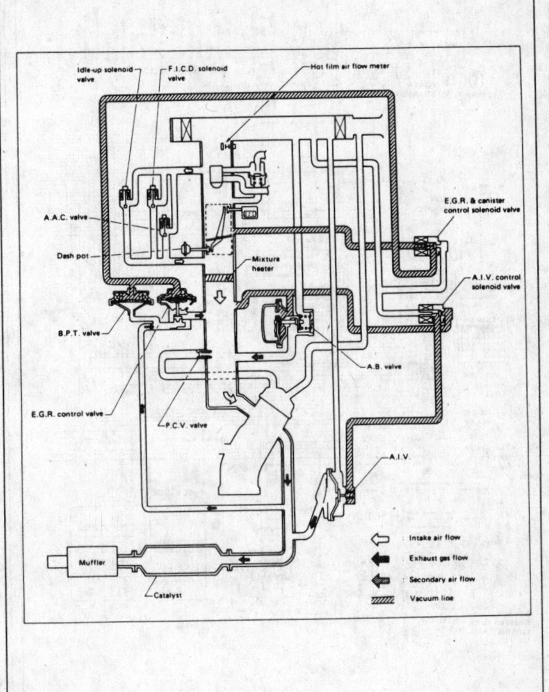
Idle-up solenoid valve
F.I.C.D. solenoid valve
Hot film air flow meter
E.G.R. & canister control solenoid valve
A.A.C. valve
Dash pot
Mixture heater
A.I.V. control solenoid valve
B.P.T. valve
A.B. valve
E.G.R. control valve
P.C.V. valve
A.I.V.
Muffler
Catalyst
Intake air flow
Exhaust gas flow
Secondary air flow
Vacuum line

ECCS CONTROL UNIT WIRING SCHEMATIC—1989 SENTRA

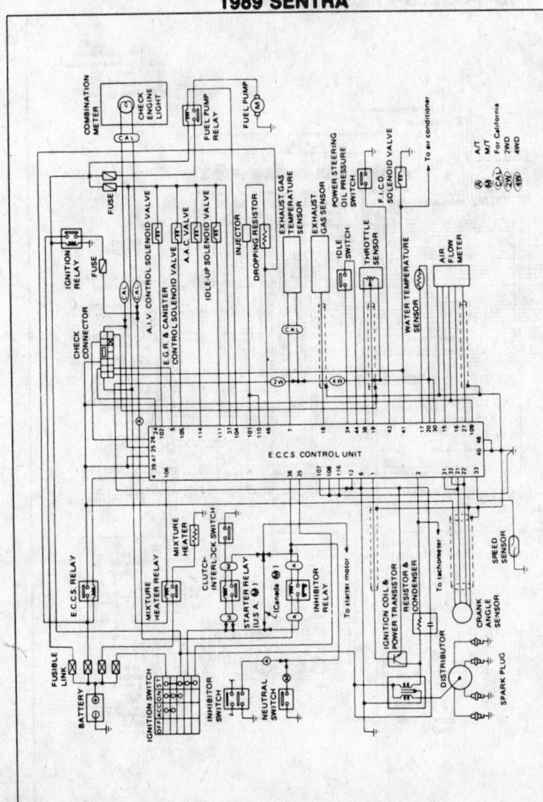
COMBINATION METER
CHECK ENGINE LIGHT
FUEL PUMP RELAY
FUEL PUMP
E.C.C.S. CONTROL UNIT
E.C.C.S. RELAY
MIXTURE HEATER RELAY
MIXTURE HEATER
IGNITION COIL & POWER TRANSISTOR
DISTRIBUTOR
SPARK PLUG
SPEED SENSOR
FUSIBLE LINK
BATTERY
IGNITION SWITCH

ECCS DIAGNOSTIC PROCEDURE—1989 SENTRA

Sensors
Control unit
Actuators
INFO
CAUSE

Introduction

The engine has an electronic control unit to control major systems such as fuel control, ignition control, idle speed control, etc. The control unit accepts input signals from sensors and instantly drives actuators. It is essential that both kinds of signals are proper and stable. At the same time, it is important that there are no conventional problems such as vacuum leaks, fouled spark plugs, or other problems with the engine.

It is much more difficult to diagnose a problem that occurs intermittently rather than continuously. Most intermittent problems are caused by poor electric connections or faulty wiring. In this case, careful checking of suspicious circuits may help prevent the replacement of good parts.

A visual check only may not find the cause of the problems. A road test with a circuit tester connected to a suspected circuit should be performed.

Before undertaking actual checks, take just a few minutes to talk with a customer who approaches with a driveability complaint. The customer is a very good supplier of information on such problems, especially intermittent ones. Through the talks with the customer, find out what symptoms are present and under what conditions they occur.

Start your diagnosis by looking for "conventional" problems first. This is one of the best ways to troubleshoot driveability problems on an electronically controlled engine vehicle.

Work Flow

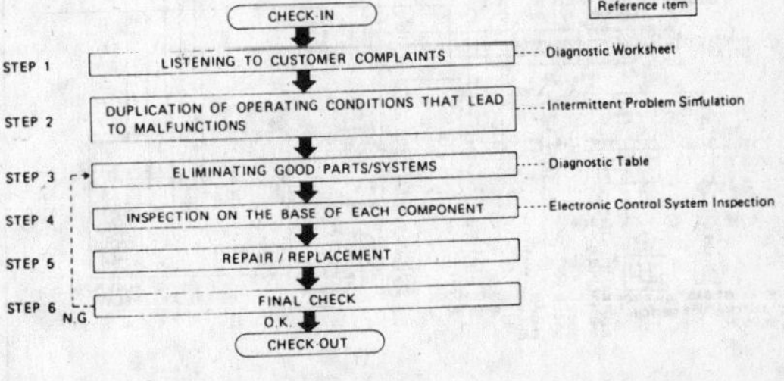
Reference item
CHECK-IN
STEP 1
LISTENING TO CUSTOMER COMPLAINTS
Diagnostic Worksheet
STEP 2
DUPLICATION OF OPERATING CONDITIONS THAT LEAD TO MALFUNCTIONS
Intermittent Problem Simulation
STEP 3
ELIMINATING GOOD PARTS/SYSTEMS
Diagnostic Table
STEP 4
INSPECTION ON THE BASE OF EACH COMPONENT
Electronic Control System Inspection
STEP 5
REPAIR / REPLACEMENT
STEP 6
FINAL CHECK
N.G.
O.K.
CHECK-OUT

ECCS SYSTEM WIRING SCHEMATIC 2WD MODELS–1989 SENTRA

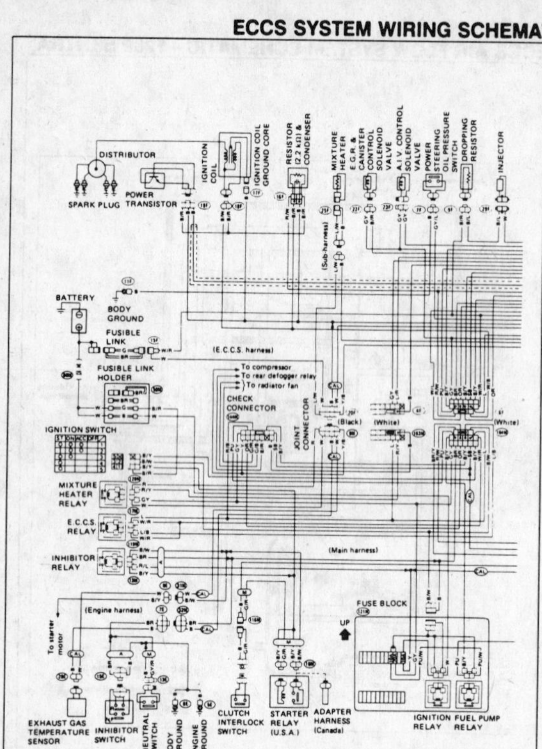
DISTRIBUTOR
SPARK PLUG
POWER TRANSISTOR
IGNITION COIL
IGNITION COIL GROUND CORE
RESISTOR (2.2 kΩ) & CONDENSER
MIXTURE HEATER
E.G.R. & CANISTER CONTROL SOLENOID VALVE
A.I.V. CONTROL SOLENOID VALVE
POWER STEERING OIL PRESSURE SWITCH
DROPPING RESISTOR
INJECTOR
BATTERY
BODY GROUND
FUSIBLE LINK
FUSIBLE LINK HOLDER
(E.C.C.S. harness)
To compressor
To rear defogger relay
To radiator fan
CHECK CONNECTOR
JOINT CONNECTOR
IGNITION SWITCH
MIXTURE HEATER RELAY
E.C.C.S. RELAY
INHIBITOR RELAY
(Main harness)
(Engine harness)
FUSE BLOCK
To starter motor
EXHAUST GAS TEMPERATURE SENSOR
INHIBITOR SWITCH
NEUTRAL SWITCH
BODY GROUND
ENGINE GROUND
CLUTCH INTERLOCK SWITCH
STARTER RELAY (U.S.A.)
ADAPTER HARNESS (Canada)
IGNITION RELAY
FUEL PUMP RELAY

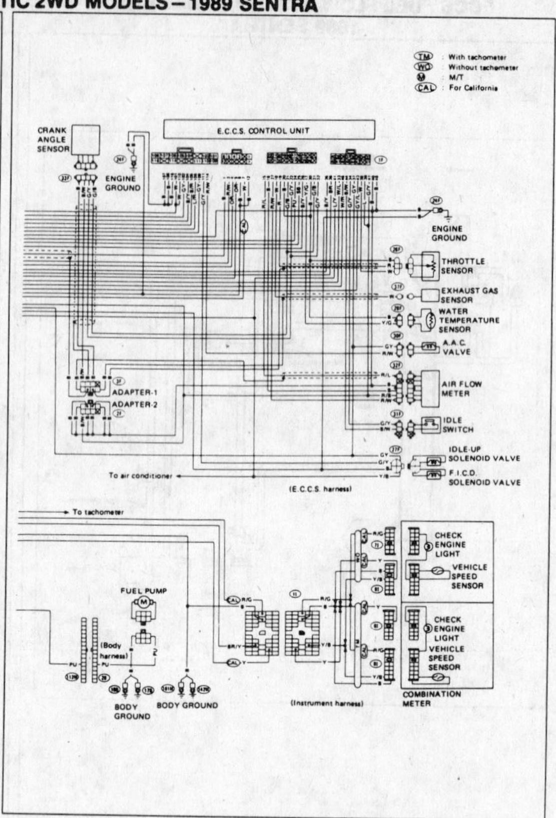
TM : With tachometer
WO : Without tachometer
M : M/T
CAL : For California
CRANK ANGLE SENSOR
E.C.C.S. CONTROL UNIT
ENGINE GROUND
THROTTLE SENSOR
EXHAUST GAS SENSOR
WATER TEMPERATURE SENSOR
A.A.C. VALVE
AIR FLOW METER
ADAPTER-1
ADAPTER-2
IDLE SWITCH
IDLE-UP SOLENOID VALVE
F.I.C.D. SOLENOID VALVE
To air conditioner
(E.C.C.S. harness)
To tachometer
CHECK ENGINE LIGHT
VEHICLE SPEED SENSOR
FUEL PUMP
(Body harness)
BODY GROUND
(Instrument harness)
COMBINATION METER

ECCS SYSTEM WIRING SCHEMATIC 4WD MODELS–1989 SENTRA

4WD MODEL

To compressor
FUSIBLE LINK HOLDER
E.C.C.S. RELAY
MIXTURE HEATER RELAY
MIXTURE HEATER
(Sub-harness)
E.G.R. & CANISTER CONTROL SOLENOID VALVE
A.I.V. CONTROL SOLENOID VALVE
POWER STEERING OIL PRESSURE SWITCH
DROPPING RESISTOR
INJECTOR
JOINT CONNECTOR
BATTERY
FUSIBLE LINK
IGNITION SWITCH
RESISTOR (2.2 kΩ) & CONDENSER
IGNITION COIL GROUND CORE
(Main harness)
IGNITION COIL
To tachometer
POWER TRANSISTOR
DISTRIBUTOR
SPARK PLUG
CRANK ANGLE SENSOR
INHIBITOR RELAY
(Engine harness)
FUSE BLOCK
To starter motor
EXHAUST GAS TEMPERATURE SENSOR
INHIBITOR SWITCH
NEUTRAL SWITCH
BODY GROUND
ENGINE GROUND
CLUTCH INTERLOCK SWITCH
STARTER RELAY (U.S.A.)
ADAPTER HARNESS (Canada)
IGNITION RELAY
FUEL PUMP RELAY

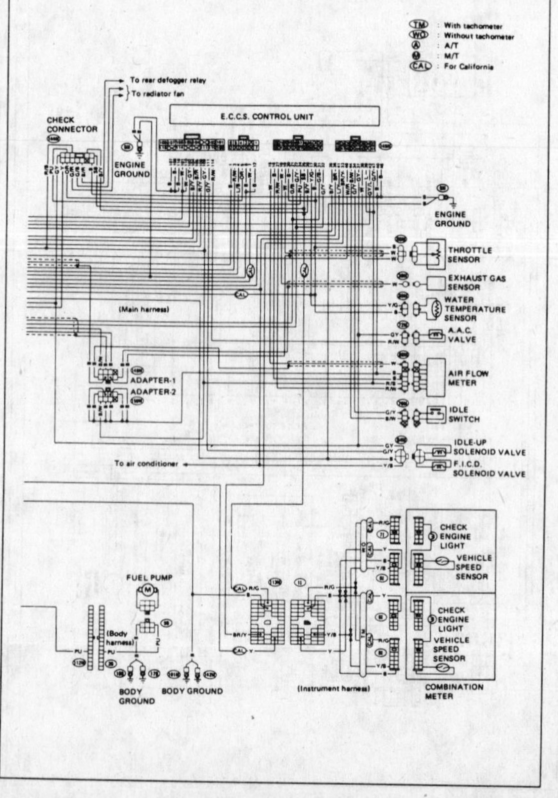
TM : With tachometer
WO : Without tachometer
A : A/T
M : M/T
CAL : For California
To rear defogger relay
To radiator fan
CHECK CONNECTOR
E.C.C.S. CONTROL UNIT
ENGINE GROUND
THROTTLE SENSOR
EXHAUST GAS SENSOR
WATER TEMPERATURE SENSOR
(Main harness)
A.A.C. VALVE
AIR FLOW METER
ADAPTER-1
ADAPTER-2
IDLE SWITCH
IDLE-UP SOLENOID VALVE
F.I.C.D. SOLENOID VALVE
To air conditioner
CHECK ENGINE LIGHT
VEHICLE SPEED SENSOR
FUEL PUMP
(Body harness)
BODY GROUND
(Instrument harness)
COMBINATION METER

ECCS DIAGNOSTIC PROCEDURE—1989 SENTRA

KEY POINTS

- WHAT ..... Vehicle & engine model
- WHEN ..... Date, Frequencies
- WHERE ..... Road conditions
- HOW ..... Operating conditions, Weather conditions, Symptoms

Diagnostic Worksheet

There are many kinds of operating conditions that lead to malfunctions on engine components.

A good grasp of such conditions can make troubleshooting faster and more accurate.

In general, feelings for a problem depend on each customer. It is important to fully understand the symptoms or under what conditions a customer complains.

Make good use of a diagnostic worksheet such as the one shown below in order to utilize all the complaints for troubleshooting.

WORKSHEET SAMPLE

Customer name MR/MS		Model & Year	VIN
Engine #		Trans.	Mileage
Incident Date		Manuf. Date	In Service Date
Symptoms	☐ Startability	☐ Impossible to start ☐ No combustion ☐ Partial combustion ☐ Partial combustion affected by throttle position ☐ Partial combustion NOT affected by throttle position ☐ Possible but hard to start ☐ Others [ ]	
	☐ Idling	☐ No fast idle ☐ Unstable ☐ High idle ☐ Low idle ☐ Others [ ]	
	☐ Driveability	☐ Stumble ☐ Surge ☐ Detonation ☐ Lack of power ☐ Intake backfire ☐ Exhaust backfire ☐ Others [ ]	
	☐ Engine stall	☐ At the time of start ☐ While idling ☐ While accelerating ☐ While decelerating ☐ Just after stopping ☐ While loading	
Incident occurrence		☐ Just after delivery ☐ Recently ☐ In the morning ☐ At night ☐ In the daytime	
Frequency		☐ All the time ☐ Under certain conditions ☐ Sometimes	
Weather conditions		☐ Not effected	
	Weather	☐ Fine ☐ Raining ☐ Snowing ☐ Others [ ]	
	Temperature	☐ Hot ☐ Warm ☐ Cool ☐ Cold ☐ Humid °F	
Engine conditions		☐ Cold ☐ During warm-up ☐ After warm-up Engine speed 0 2,000 4,000 6,000 8,000 rpm	
Road conditions		☐ In town ☐ In suburbs ☐ Highway ☐ Off road (up/down)	
Driving conditions		☐ Not affected ☐ At starting ☐ While idling ☐ At racing ☐ While accelerating ☐ While cruising ☐ While decelerating ☐ While turning (RH/LH) Vehicle speed 0 10 20 30 40 50 60 MPH	
Check engine light		☐ Turned on ☐ Not turned on	

Intermittent Problem Simulation

In order to duplicate an intermittent problem, it is effective to create similar conditions for component parts, under which the problem might occur.

Perform the activity listed under Service procedure and note the result.

	Variable factor	Influential part	Target condition	Service procedure
1	Mixture ratio	Pressure regulator	Made lean	Remove vacuum hose and apply vacuum.
			Made rich	Remove vacuum hose and apply pressure.
2	Ignition timing	Distributor	Advanced	Rotate distributor clockwise.
			Retarded	Rotate distributor counterclockwise.
3	Mixture ratio feedback control	Exhaust gas sensor	Suspended	Disconnect exhaust gas sensor harness connector.
		Control unit	Operation check	Perform self-diagnosis (Mode I/II) at 2,000 rpm.
4	Idle speed	Throttle body	Raised	Turn idle adjust screw counterclockwise.
			Lowered	Turn idle adjust screw clockwise.
5	Electric connection (Electric continuity)	Harness connectors and wires	Poor electric connection or faulty wiring	Tap or wiggle.
				Race engine rapidly. See if the torque reaction of the engine unit causes electric breaks.
6	Temperature	Control unit	Cooled	Cool with an icing spray or similar device.
			Warmed	Heat with a hair drier. [WARNING: Do not overheat the unit.]
7	Moisture	Electric parts	Damp	Wet [WARNING: Do not directly pour water on components. Use a mist sprayer.]
8	Electric loads	Load switches	Loaded	Turn on head lights, air conditioner, rear defogger, etc.
9	Idle switch condition	Control unit	ON-OFF switching	Perform self-diagnosis (Mode IV).
10	Ignition spark	Timing light	Spark power check	Try to flash timing light for each cylinder.

ECCS DIAGNOSTIC PROCEDURE SPECIFICATIONS—1989 SENTRA

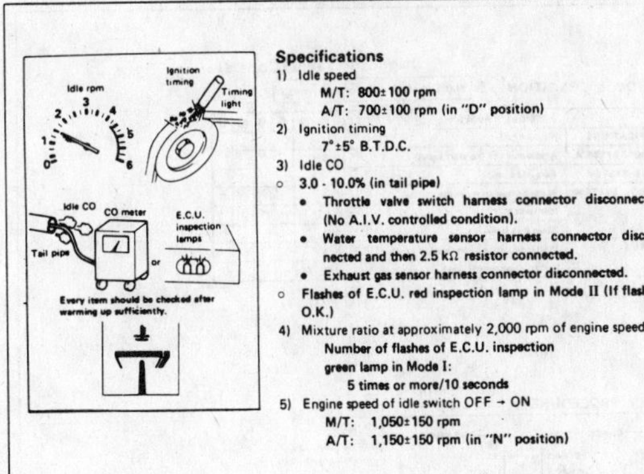
Idle rpm
Ignition timing
Timing light
Idle CO
CO meter
E.C.U. inspection lamps
Tail pipe
or
Every item should be checked after warming up sufficiently.

Specifications

1) Idle speed
   - M/T: 800±100 rpm
   - A/T: 700±100 rpm (in "D" position)
2) Ignition timing
   - 7°±5° B.T.D.C.
3) Idle CO
   - ○ 3.0 - 10.0% (in tail pipe)
     - • Throttle valve switch harness connector disconnected (No A.I.V. controlled condition).
     - • Water temperature sensor harness connector disconnected and then 2.5 kΩ resistor connected.
     - • Exhaust gas sensor harness connector disconnected.
   - ○ Flashes of E.C.U. red inspection lamp in Mode II (If flashes, O.K.)
4) Mixture ratio at approximately 2,000 rpm of engine speed.
   - Number of flashes of E.C.U. inspection green lamp in Mode I:
     - 5 times or more/10 seconds
5) Engine speed of idle switch OFF → ON
   - M/T: 1,050±150 rpm
   - A/T: 1,150±150 rpm (in "N" position)

ECCS DIAGNOSTIC CHART—1989 SENTRA

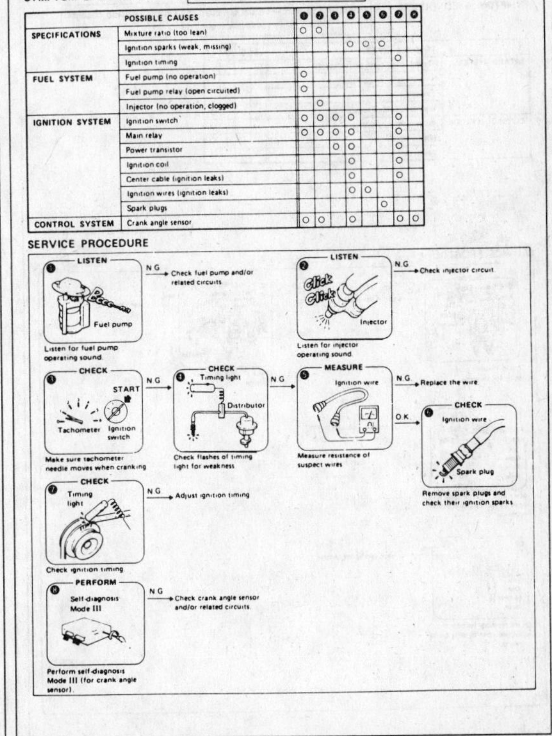

Diagnostic Table (Cont'd)

SYMPTOM & CONDITION 1 Impossible to start – no combustion

	POSSIBLE CAUSES	❶	❷	❸	❹	❺	❻	❼	❽
SPECIFICATIONS	Mixture ratio (too lean)	○	○						
	Ignition sparks (weak, missing)				○	○	○		
	Ignition timing							○	
FUEL SYSTEM	Fuel pump (no operation)	○							
	Fuel pump relay (open circuited)	○							
	Injector (no operation, clogged)		○						
IGNITION SYSTEM	Ignition switch	○	○	○	○			○	
	Main relay	○	○	○	○			○	
	Power transistor			○	○			○	
	Ignition coil				○			○	
	Center cable (ignition leaks)				○			○	
	Ignition wires (ignition leaks)				○	○			
	Spark plugs						○		
CONTROL SYSTEM	Crank angle sensor	○	○		○			○	○

SERVICE PROCEDURE

❶ LISTEN
Fuel pump
N.G → Check fuel pump and/or related circuits
Listen for fuel pump operating sound.
❷ LISTEN
Click Click
Injector
N.G → Check injector circuit
Listen for injector operating sound.
❸ CHECK
START
Tachometer
Ignition switch
N.G →
Make sure tachometer needle moves when cranking
❹ CHECK
Timing light
Distributor
N.G →
Check flashes of timing light for weakness
❺ MEASURE
Ignition wire
N.G → Replace the wire
O.K →
Measure resistance of suspect wires
❻ CHECK
Ignition wire
Spark plug
Remove spark plugs and check their ignition sparks
❼ CHECK
Timing light
N.G → Adjust ignition timing
Check ignition timing
❽ PERFORM
Self-diagnosis Mode III
N.G → Check crank angle sensor and/or related circuits
Perform self-diagnosis Mode III (for crank angle sensor).

ECCS DIAGNOSTIC CHART—1989 SENTRA

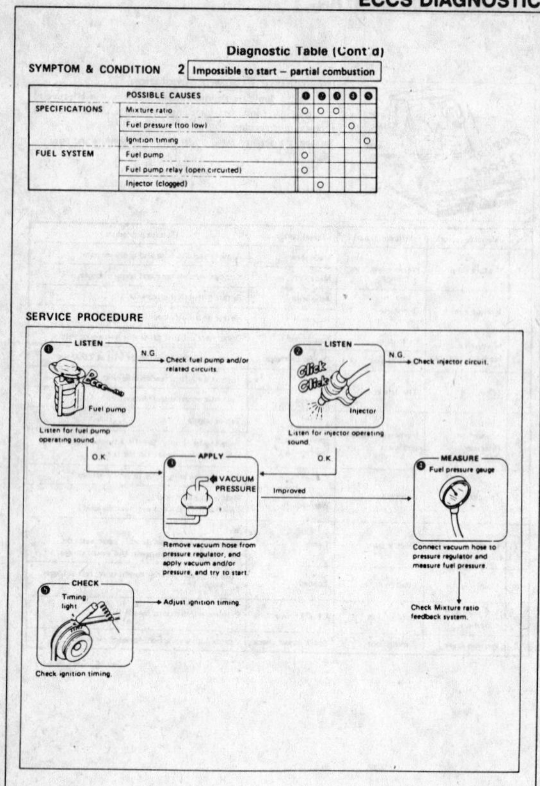

Diagnostic Table (Cont'd)

SYMPTOM & CONDITION 2 Impossible to start – partial combustion

POSSIBLE CAUSES		1	2	3	4	5
SPECIFICATIONS	Mixture ratio	O	O	O		
	Fuel pressure (too low)				O	
	Ignition timing					O
FUEL SYSTEM	Fuel pump	O				
	Fuel pump relay (open circuited)	O				
	Injector (clogged)		O			

SERVICE PROCEDURE

LISTEN
Fuel pump
N.G.
Check fuel pump and/or related circuits.
Listen for fuel pump operating sound.
LISTEN
Injector
N.G.
Check injector circuit.
Listen for injector operating sound.
O.K.
O.K.
APPLY
VACUUM PRESSURE
Improved
Remove vacuum hose from pressure regulator, and apply vacuum and/or pressure, and try to start.
MEASURE
Fuel pressure gauge
Connect vacuum hose to pressure regulator and measure fuel pressure.
Check Mixture ratio feedback system.
CHECK
Timing light
Adjust ignition timing
Check ignition timing.

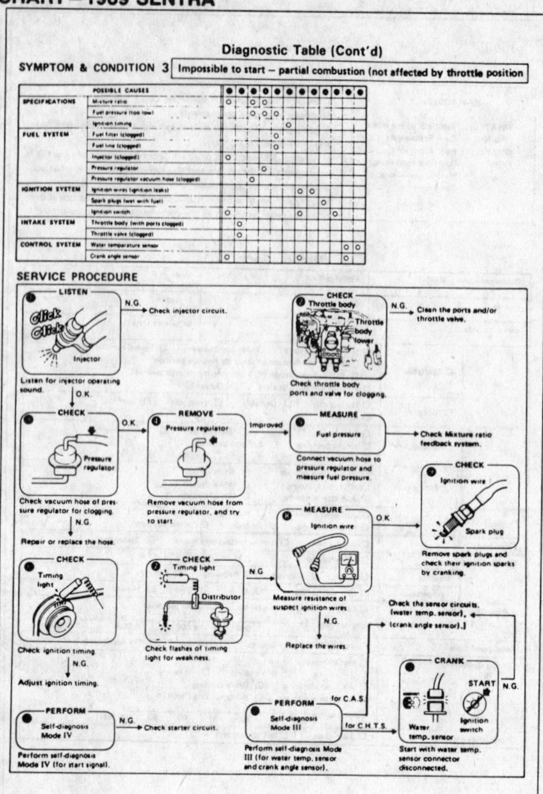

Diagnostic Table (Cont'd)

SYMPTOM & CONDITION 3 Impossible to start – partial combustion (not affected by throttle position

POSSIBLE CAUSES		1	2	3	4	5	6	7	8	9	10	11	12
SPECIFICATIONS	Mixture ratio	O		O	O								
	Fuel pressure (too low)			O	O	O							
	Ignition timing						O						
FUEL SYSTEM	Fuel filter (clogged)					O							
	Fuel line (clogged)					O							
	Injector (clogged)	O											
	Pressure regulator				O								
	Pressure regulator vacuum hose (clogged)			O									
IGNITION SYSTEM	Ignition wires (ignition leaks)							O	O				
	Spark plugs (wet with fuel)									O			
	Ignition switch	O						O			O		
INTAKE SYSTEM	Throttle body (with ports clogged)		O										
	Throttle valve (clogged)		O										
CONTROL SYSTEM	Water temperature sensor											O	O
	Crank angle sensor	O						O				O	

SERVICE PROCEDURE

LISTEN
Injector
N.G.
Check injector circuit.
Listen for injector operating sound.
O.K.
CHECK
Throttle body
Throttle body lower
N.G.
Clean the ports and/or throttle valve.
Check throttle body ports and valve for clogging.
CHECK
Pressure regulator
O.K.
REMOVE
Pressure regulator
Improved
MEASURE
Fuel pressure
Check Mixture ratio feedback system.
Check vacuum hose of pressure regulator for clogging.
N.G.
Repair or replace the hose.
Remove vacuum hose from pressure regulator, and try to start.
Connect vacuum hose to pressure regulator and measure fuel pressure.
MEASURE
Ignition wire
O.K.
CHECK
Ignition wire
Spark plug
Remove spark plugs and check their ignition sparks by cranking.
CHECK
Timing light
Check ignition timing.
N.G.
Adjust ignition timing.
CHECK
Timing light
Distributor
N.G.
Check flashes of timing light for weakness.
Measure resistance of suspect ignition wires.
N.G.
Replace the wires.
Check the sensor circuits. [water temp. sensor], [crank angle sensor].]
CRANK
START
N.G.
Water temp. sensor
Ignition switch
Start with water temp. sensor connector disconnected.
PERFORM
Self-diagnosis Mode IV
N.G.
Check starter circuit.
Perform self-diagnosis Mode IV (for start signal).
PERFORM
Self-diagnosis Mode III
for C.A.S.
for C.H.T.S.
Perform self-diagnosis Mode III (for water temp. sensor and crank angle sensor).

ECCS DIAGNOSTIC CHART—1989 SENTRA

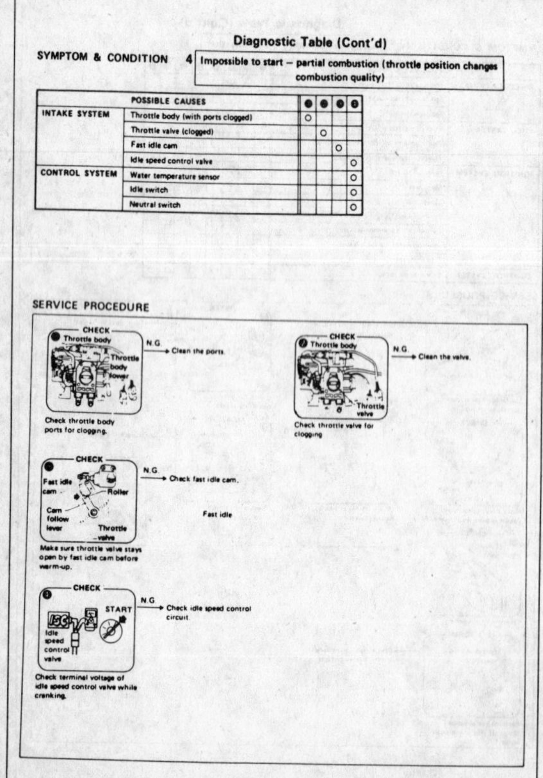

Diagnostic Table (Cont'd)

SYMPTOM & CONDITION 4 Impossible to start – partial combustion (throttle position changes combustion quality)

POSSIBLE CAUSES		1	2	3	4
INTAKE SYSTEM	Throttle body (with ports clogged)	O			
	Throttle valve (clogged)		O		
	Fast idle cam			O	
	Idle speed control valve				O
CONTROL SYSTEM	Water temperature sensor				O
	Idle switch				O
	Neutral switch				O

SERVICE PROCEDURE

CHECK
Throttle body
Throttle body lower
N.G.
Clean the ports.
Check throttle body ports for clogging.
CHECK
Throttle body
Throttle valve
N.G.
Clean the valve.
Check throttle valve for clogging.
CHECK
Fast idle cam
Roller
Cam follow lever
Throttle valve
N.G.
Check fast idle cam.
Fast idle
Make sure throttle valve stays open by fast idle cam before warm-up.
CHECK
START
Idle speed control valve
N.G.
Check idle speed control circuit.
Check terminal voltage of idle speed control valve while cranking.

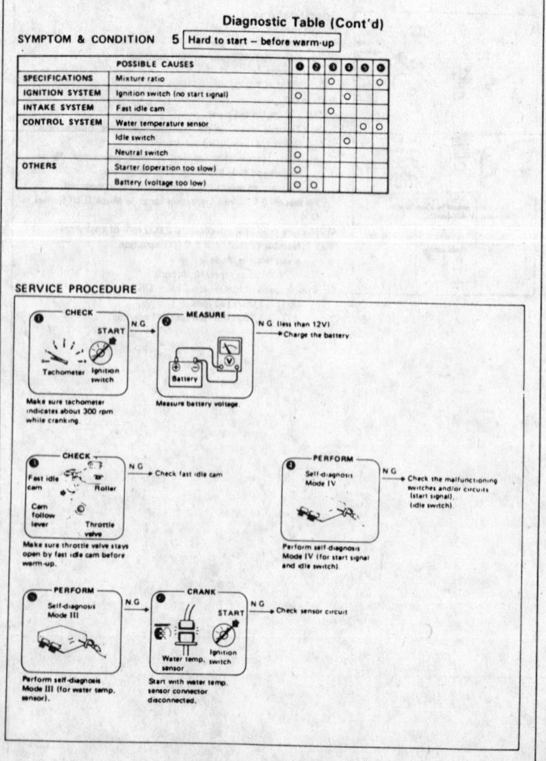

Diagnostic Table (Cont'd)

SYMPTOM & CONDITION 5 Hard to start – before warm-up

POSSIBLE CAUSES		1	2	3	4	5	6
SPECIFICATIONS	Mixture ratio			O			O
IGNITION SYSTEM	Ignition switch (no start signal)	O			O		
INTAKE SYSTEM	Fast idle cam			O			
CONTROL SYSTEM	Water temperature sensor					O	O
	Idle switch				O		
	Neutral switch	O					
OTHERS	Starter (operation too slow)	O					
	Battery (voltage too low)	O	O				

SERVICE PROCEDURE

CHECK
START
Tachometer
Ignition switch
N.G.
Make sure tachometer indicates about 300 rpm while cranking.
MEASURE
Battery
N.G. (less than 12V)
Charge the battery
Measure battery voltage.
CHECK
Fast idle cam
Roller
Cam follow lever
Throttle valve
N.G.
Check fast idle cam
Make sure throttle valve stays open by fast idle cam before warm-up.
PERFORM
Self-diagnosis Mode IV
N.G.
Check the malfunctioning switches and/or circuits (start signal) (idle switch).
Perform self-diagnosis Mode IV (for start signal and idle switch).
PERFORM
Self-diagnosis Mode III
N.G.
CRANK
START
Water temp. sensor
Ignition switch
N.G.
Check sensor circuit
Perform self-diagnosis Mode III (for water temp. sensor).
Start with water temp. sensor connector disconnected.

ECCS DIAGNOSTIC CHART—1989 SENTRA

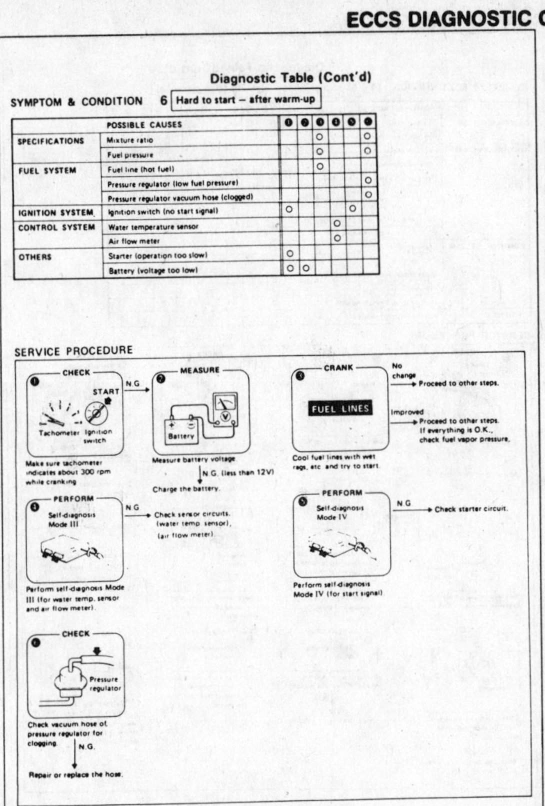

Diagnostic Table (Cont'd)

SYMPTOM & CONDITION 6 Hard to start – after warm-up

	POSSIBLE CAUSES	1	2	3	4	5	6
SPECIFICATIONS	Mixture ratio			○			○
	Fuel pressure			○			○
FUEL SYSTEM	Fuel line (hot fuel)			○			
	Pressure regulator (low fuel pressure)						○
	Pressure regulator vacuum hose (clogged)						○
IGNITION SYSTEM	Ignition switch (no start signal)	○				○	
CONTROL SYSTEM	Water temperature sensor				○		
	Air flow meter				○		
OTHERS	Starter (operation too slow)	○					
	Battery (voltage too low)	○	○				

SERVICE PROCEDURE

1 CHECK
START
Tachometer
Ignition switch
N.G.
Make sure tachometer indicates about 300 rpm while cranking.
2 MEASURE
Battery
Measure battery voltage.
N.G. (less than 12V)
Charge the battery.
3 CRANK
FUEL LINES
No change
Proceed to other steps.
Improved
Proceed to other steps. If everything is O.K., check fuel vapor pressure.
Cool fuel lines with wet rags, etc. and try to start.
4 PERFORM
Self-diagnosis Mode III
N.G.
Check sensor circuits. (water temp. sensor), (air flow meter).
Perform self-diagnosis Mode III (for water temp. sensor and air flow meter).
5 PERFORM
Self-diagnosis Mode IV
N.G.
Check starter circuit.
Perform self-diagnosis Mode IV (for start signal).
6 CHECK
Pressure regulator
Check vacuum hose of pressure regulator for clogging.
N.G.
Repair or replace the hose.

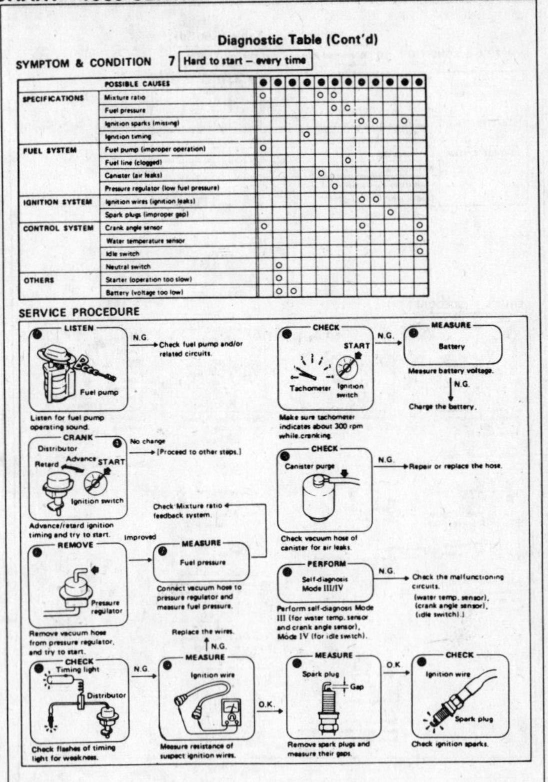

Diagnostic Table (Cont'd)

SYMPTOM & CONDITION 7 Hard to start – every time

	POSSIBLE CAUSES	1	2	3	4	5	6	7	8	9	10	11	12
SPECIFICATIONS	Mixture ratio	○				○	○						
	Fuel pressure						○	○					
	Ignition sparks (missing)								○	○		○	
	Ignition timing				○								
FUEL SYSTEM	Fuel pump (improper operation)	○											
	Fuel line (clogged)							○					
	Canister (air leaks)					○							
	Pressure regulator (low fuel pressure)						○						
IGNITION SYSTEM	Ignition wires (ignition leaks)								○	○			
	Spark plugs (improper gap)										○		
CONTROL SYSTEM	Crank angle sensor	○							○				○
	Water temperature sensor												○
	Idle switch												○
	Neutral switch		○										
OTHERS	Starter (operation too slow)		○										
	Battery (voltage too low)		○	○									

SERVICE PROCEDURE

LISTEN
Fuel pump
N.G.
Check fuel pump and/or related circuits.
Listen for fuel pump operating sound.
CHECK
START
Tachometer
Ignition switch
N.G.
Make sure tachometer indicates about 300 rpm while cranking.
MEASURE
Battery
Measure battery voltage.
N.G.
Charge the battery.
CRANK
Distributor
Retard
Advance
START
Ignition switch
No change
(Proceed to other steps.)
Advance/retard ignition timing and try to start.
CHECK
Canister purge
N.G.
Repair or replace the hose.
Check vacuum hose of canister for air leaks.
Check Mixture ratio feedback system.
Improved
REMOVE
Pressure regulator
Remove vacuum hose from pressure regulator, and try to start.
MEASURE
Fuel pressure
Connect vacuum hose to pressure regulator and measure fuel pressure.
PERFORM
Self-diagnosis Mode III/IV
N.G.
Check the malfunctioning circuits. (water temp. sensor), (crank angle sensor), (idle switch).
Perform self-diagnosis Mode III (for water temp. sensor and crank angle sensor), Mode IV (for idle switch).
Replace the wires.
N.G.
CHECK
Timing light
Distributor
N.G.
Check flashes of timing light for weakness.
MEASURE
Ignition wire
O.K.
Measure resistance of suspect ignition wires.
MEASURE
Spark plug
Gap
O.K.
Remove spark plugs and measure their gaps.
CHECK
Ignition wire
Spark plug
Check ignition sparks.

ECCS DIAGNOSTIC CHART—1989 SENTRA

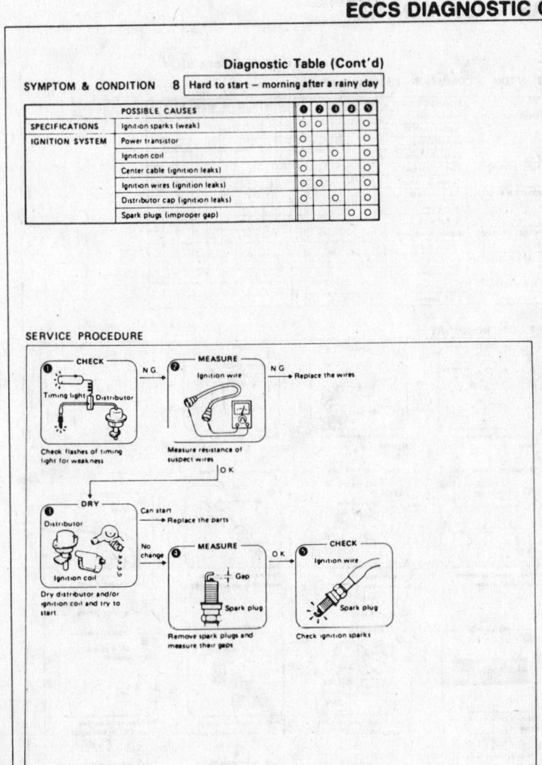

Diagnostic Table (Cont'd)

SYMPTOM & CONDITION 8 Hard to start – morning after a rainy day

	POSSIBLE CAUSES	1	2	3	4	5
SPECIFICATIONS	Ignition sparks (weak)	○	○			○
IGNITION SYSTEM	Power transistor	○				○
	Ignition coil	○		○		○
	Center cable (ignition leaks)	○				○
	Ignition wires (ignition leaks)	○	○			○
	Distributor cap (ignition leaks)	○		○		○
	Spark plugs (improper gap)				○	○

SERVICE PROCEDURE

1 CHECK
Timing light
Distributor
N.G.
Check flashes of timing light for weakness.
2 MEASURE
Ignition wire
N.G.
Replace the wires.
Measure resistance of suspect wires.
O.K.
3 DRY
Distributor
Ignition coil
Can start
Replace the parts.
No change
Dry distributor and/or ignition coil and try to start.
4 MEASURE
Gap
Spark plug
O.K.
Remove spark plugs and measure their gaps.
5 CHECK
Ignition wire
Spark plug
Check ignition sparks.

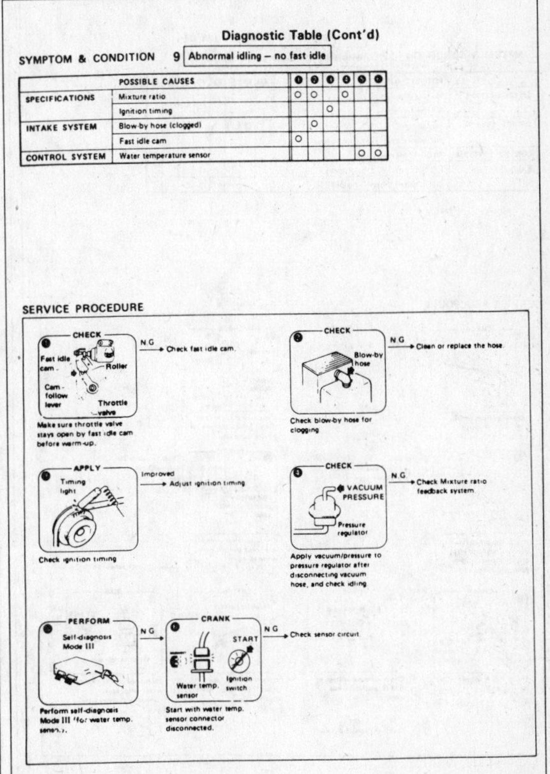

Diagnostic Table (Cont'd)

SYMPTOM & CONDITION 9 Abnormal idling – no fast idle

	POSSIBLE CAUSES	1	2	3	4	5	6
SPECIFICATIONS	Mixture ratio	○	○		○		
	Ignition timing			○			
INTAKE SYSTEM	Blow-by hose (clogged)		○				
	Fast idle cam	○					
CONTROL SYSTEM	Water temperature sensor					○	○

SERVICE PROCEDURE

1 CHECK
Fast idle cam
Roller
Cam-follow lever
Throttle valve
N.G.
Check fast idle cam.
Make sure throttle valve stays open by fast idle cam before warm-up.
2 CHECK
Blow-by hose
N.G.
Clean or replace the hose.
Check blow-by hose for clogging.
3 APPLY
Timing light
Improved
Adjust ignition timing.
Check ignition timing.
4 CHECK
VACUUM PRESSURE
Pressure regulator
N.G.
Check Mixture ratio feedback system.
Apply vacuum/pressure to pressure regulator after disconnecting vacuum hose, and check idling.
5 PERFORM
Self-diagnosis Mode III
N.G.
Perform self-diagnosis Mode III (for water temp. sensor).
6 CRANK
START
Water temp. sensor
Ignition switch
N.G.
Check sensor circuit.
Start with water temp. sensor connector disconnected.

ECCS DIAGNOSTIC CHART – 1989 SENTRA

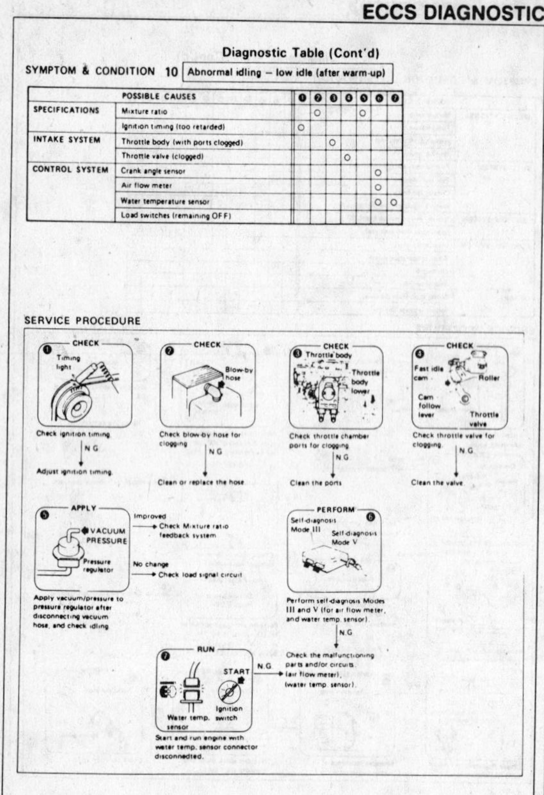

Diagnostic Table (Cont'd)

SYMPTOM & CONDITION 10 Abnormal idling – low idle (after warm-up)

	POSSIBLE CAUSES	1	2	3	4	5	6	7
SPECIFICATIONS	Mixture ratio		○			○		
	Ignition timing (too retarded)	○						
INTAKE SYSTEM	Throttle body (with ports clogged)			○				
	Throttle valve (clogged)				○			
CONTROL SYSTEM	Crank angle sensor						○	
	Air flow meter						○	
	Water temperature sensor						○	○
	Load switches (remaining OFF)							

SERVICE PROCEDURE
1 CHECK
Timing light
Check ignition timing.
N.G
Adjust ignition timing.
2 CHECK
Blow-by hose
Check blow-by hose for clogging.
N.G
Clean or replace the hose.
3 CHECK
Throttle body
Throttle body lower
Check throttle chamber ports for clogging.
N.G
Clean the ports.
4 CHECK
Fast idle cam
Roller
Cam follow lever
Throttle valve
Check throttle valve for clogging.
N.G
Clean the valve.
5 APPLY
VACUUM PRESSURE
Pressure regulator
Improved
Check Mixture ratio feedback system
No change
Check load signal circuit
Apply vacuum/pressure to pressure regulator after disconnecting vacuum hose, and check idling.
6 PERFORM
Self-diagnosis Mode III
Self-diagnosis Mode V
Perform self-diagnosis Modes III and V (for air flow meter, and water temp. sensor).
N.G
Check the malfunctioning parts and/or circuits. (air flow meter), (water temp. sensor).
7 RUN
START
Water temp. sensor
Ignition switch
N.G
Start and run engine with water temp. sensor connector disconnedted.

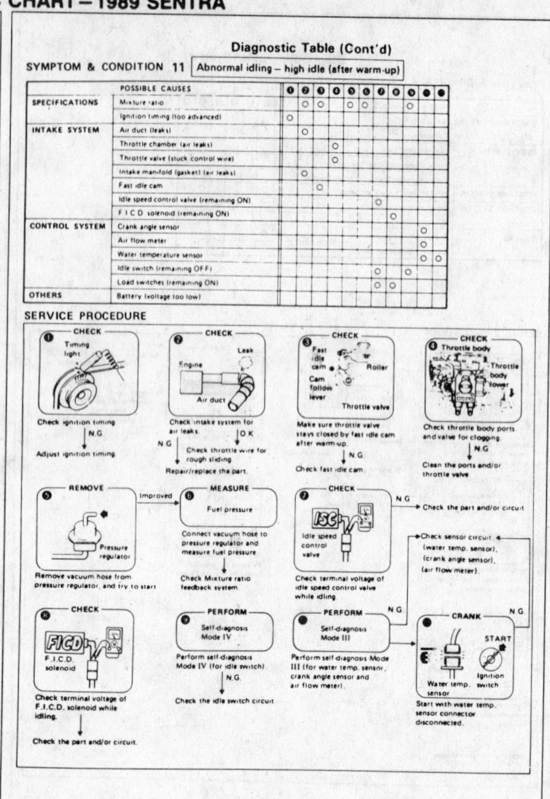

Diagnostic Table (Cont'd)

SYMPTOM & CONDITION 11 Abnormal idling – high idle (after warm-up)

	POSSIBLE CAUSES	1	2	3	4	5	6	7	8	9	10	11
SPECIFICATIONS	Mixture ratio		○	○		○	○			○		
	Ignition timing (too advanced)	○										
INTAKE SYSTEM	Air duct (leaks)		○									
	Throttle chamber (air leaks)				○							
	Throttle valve (stuck control wire)				○							
	Intake manifold (gasket) (air leaks)		○									
	Fast idle cam			○								
	Idle speed control valve (remaining ON)							○				
	F.I.C.D. solenoid (remaining ON)								○			
CONTROL SYSTEM	Crank angle sensor										○	
	Air flow meter										○	
	Water temperature sensor										○	○
	Idle switch (remaining OFF)							○		○		
	Load switches (remaining ON)							○	○			
OTHERS	Battery (voltage too low)											

SERVICE PROCEDURE
1 CHECK
Timing light
Check ignition timing.
N.G
Adjust ignition timing.
2 CHECK
Engine
Leak
Air duct
Check intake system for air leaks.
O.K
Check throttle wire for rough sliding.
N.G
Repair/replace the part.
3 CHECK
Fast idle cam
Roller
Cam follow lever
Throttle valve
Make sure throttle valve stays closed by fast idle cam after warm-up.
N.G
Check fast idle cam.
4 CHECK
Throttle body
Throttle body lower
Check throttle body ports and valve for clogging.
N.G
Clean the ports and/or throttle valve.
5 REMOVE
Pressure regulator
Remove vacuum hose from pressure regulator, and try to start.
Improved
6 MEASURE
Fuel pressure
Connect vacuum hose to pressure regulator and measure fuel pressure.
Check Mixture ratio feedback system.
7 CHECK
ISC
Idle speed control valve
Check terminal voltage of idle speed control valve while idling.
N.G
Check the part and/or circuit.
Check sensor circuit. (water temp. sensor), (crank angle sensor), (air flow meter).
8 CHECK
FICD
F.I.C.D. solenoid
Check terminal voltage of F.I.C.D. solenoid while idling.
Check the part and/or circuit.
9 PERFORM
Self-diagnosis Mode IV
Perform self-diagnosis Mode IV (for idle switch).
N.G.
Check the idle switch circuit.
10 PERFORM
Self-diagnosis Mode III
Perform self-diagnosis Mode III (for water temp. sensor, crank angle sensor and air flow meter).
N.G.
11 CRANK
START
Water temp. sensor
Ignition switch
N.G.
Start with water temp. sensor connector disconnected.

ECCS DIAGNOSTIC CHART – 1989 SENTRA

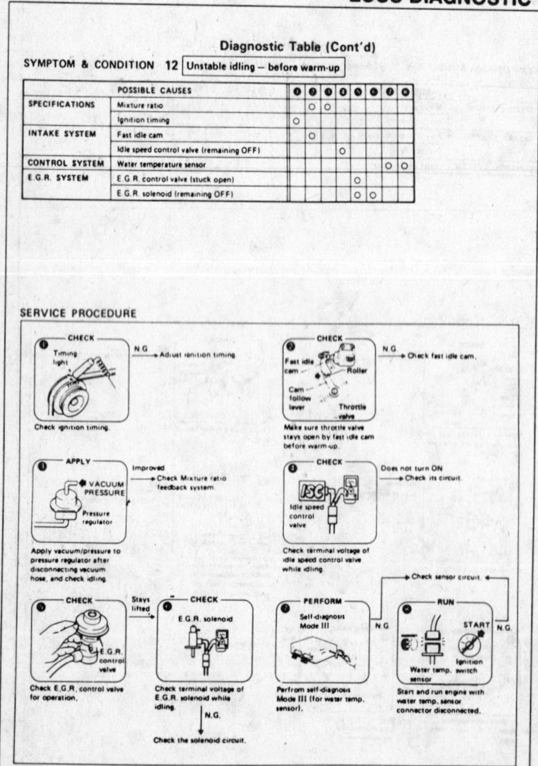

Diagnostic Table (Cont'd)

SYMPTOM & CONDITION 12 Unstable idling – before warm-up

	POSSIBLE CAUSES	1	2	3	4	5	6	7	8
SPECIFICATIONS	Mixture ratio		○	○					
	Ignition timing	○							
INTAKE SYSTEM	Fast idle cam		○						
	Idle speed control valve (remaining OFF)				○				
CONTROL SYSTEM	Water temperature sensor							○	○
E.G.R. SYSTEM	E.G.R. control valve (stuck open)					○			
	E.G.R. solenoid (remaining OFF)					○	○		

SERVICE PROCEDURE
1 CHECK
Timing light
N.G
Adjust ignition timing
Check ignition timing.
2 CHECK
Fast idle cam
Roller
Cam follow lever
Throttle valve
N.G.
Check fast idle cam.
Make sure throttle valve stays open by fast idle cam before warm-up.
3 APPLY
VACUUM PRESSURE
Pressure regulator
Improved
Check Mixture ratio feedback system
Apply vacuum/pressure to pressure regulator after disconnecting vacuum hose, and check idling
4 CHECK
ISC
Idle speed control valve
Does not turn ON
Check its circuit.
Check terminal voltage of idle speed control valve while idling
5 CHECK
E.G.R. control valve
Stays lifted
Check E.G.R. control valve for operation.
6 CHECK
E.G.R. solenoid
Check terminal voltage of E.G.R. solenoid while idling.
N.G.
Check the solenoid circuit.
7 PERFORM
Self-diagnosis Mode III
N.G
Perfrom self-diagnosis Mode III (for water temp. sensor).
Check sensor circuit.
8 RUN
START
Water temp. sensor
Ignition switch
N.G.
Start and run engine with water temp. sensor connector disconnected.

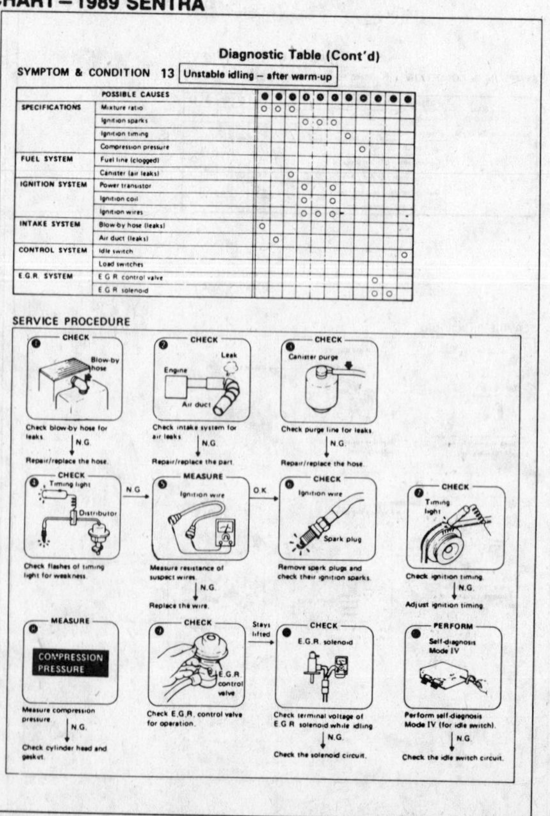

Diagnostic Table (Cont'd)

SYMPTOM & CONDITION 13 Unstable idling – after warm-up

	POSSIBLE CAUSES	1	2	3	4	5	6	7	8	9	10	11
SPECIFICATIONS	Mixture ratio	○	○	○								
	Ignition sparks				○	○	○					
	Ignition timing							○				
	Compression pressure								○			
FUEL SYSTEM	Fuel line (clogged)											
	Canister (air leaks)			○								
IGNITION SYSTEM	Power transistor				○		○					
	Ignition coil				○		○					
	Ignition wires				○	○	○					
INTAKE SYSTEM	Blow-by hose (leaks)	○										
	Air duct (leaks)		○									
CONTROL SYSTEM	Idle switch											○
	Load switches											
E.G.R. SYSTEM	E.G.R. control valve									○		
	E.G.R. solenoid									○	○	

SERVICE PROCEDURE
1 CHECK
Blow-by hose
Check blow-by hose for leaks.
N.G.
Repair/replace the hose.
2 CHECK
Engine
Leak
Air duct
Check intake system for air leaks.
N.G.
Repair/replace the part.
3 CHECK
Canister purge
Check purge line for leaks.
N.G.
Repair/replace the hose.
4 CHECK
Timing light
Distributor
Check flashes of timing light for weakness.
N.G
5 MEASURE
Ignition wire
Measure resistance of suspect wires.
N.G.
Replace the wire.
O.K.
6 CHECK
Ignition wire
Spark plug
Remove spark plugs and check their ignition sparks.
7 CHECK
Timing light
Check ignition timing.
N.G.
Adjust ignition timing.
8 MEASURE
COMPRESSION PRESSURE
Measure compression pressure.
N.G.
Check cylinder head and gasket.
9 CHECK
E.G.R. control valve
Check E.G.R. control valve for operation.
Stays lifted
10 CHECK
E.G.R. solenoid
Check terminal voltage of E.G.R. solenoid while idling.
N.G.
Check the solenoid circuit.
11 PERFORM
Self-diagnosis Mode IV
Perform self-diagnosis Mode IV (for idle switch).
N.G.
Check the idle switch circuit.

ECCS DIAGNOSTIC CHART—1989 SENTRA

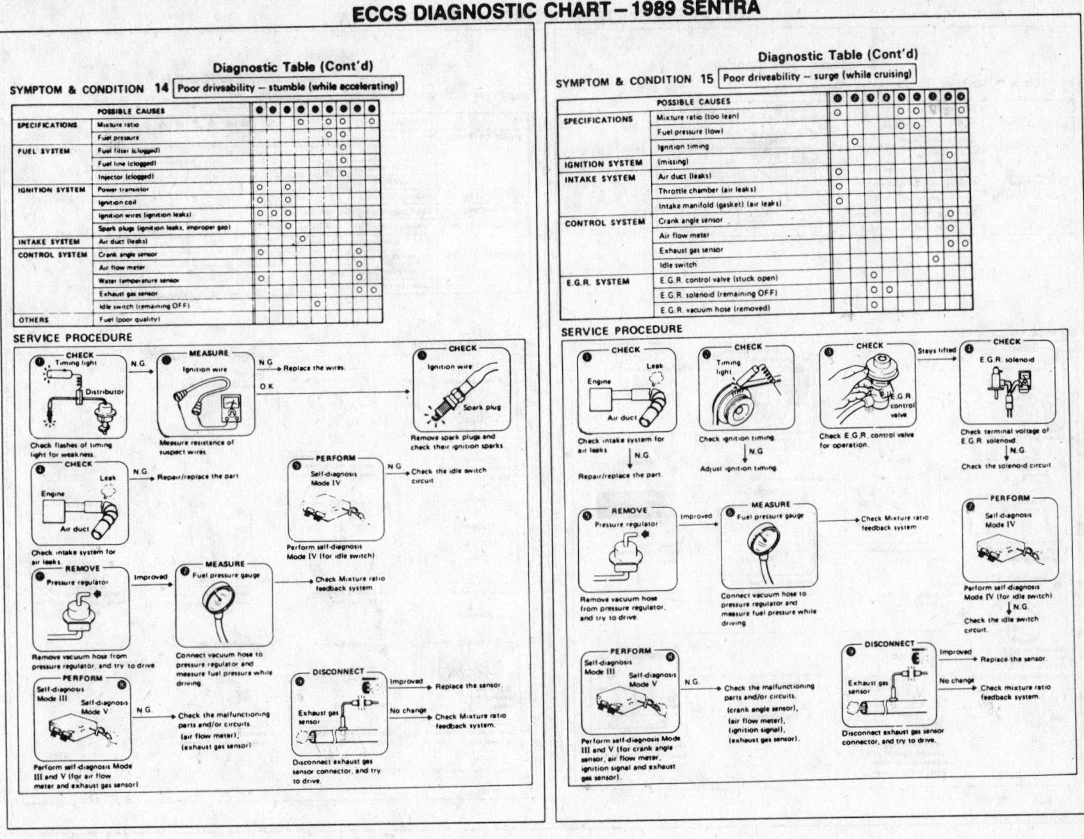

Diagnostic Table (Cont'd)
SYMPTOM & CONDITION 14 Poor driveability – stumble (while accelerating)
POSSIBLE CAUSES
SPECIFICATIONS Mixture ratio
Fuel pressure
FUEL SYSTEM Fuel filter (clogged)
Fuel line (clogged)
Injector (clogged)
IGNITION SYSTEM Power transistor
Ignition coil
Ignition wires (ignition leaks)
Spark plugs (ignition leaks, improper gap)
INTAKE SYSTEM Air duct (leaks)
CONTROL SYSTEM Crank angle sensor
Air flow meter
Water temperature sensor
Exhaust gas sensor
Idle switch (remaining OFF)
OTHERS Fuel (poor quality)
SERVICE PROCEDURE
CHECK Timing light Distributor
Check flashes of timing light for weakness.
MEASURE Ignition wire
Measure resistance of suspect wires.
N.G. Replace the wires.
O.K.
CHECK Ignition wire Spark plug
Remove spark plugs and check their ignition sparks.
CHECK Leak Engine Air duct
Check intake system for air leaks.
N.G. Repair/replace the part
PERFORM Self-diagnosis Mode IV
Perform self-diagnosis Mode IV (for idle switch)
N.G. Check the idle switch circuit
REMOVE Pressure regulator
Remove vacuum hose from pressure regulator, and try to drive.
Improved
MEASURE Fuel pressure gauge
Connect vacuum hose to pressure regulator and measure fuel pressure while driving.
Check Mixture ratio feedback system.
PERFORM Self-diagnosis Mode III Self-diagnosis Mode V
Perform self-diagnosis Mode III and V (for air flow meter and exhaust gas sensor)
N.G. Check the malfunctioning parts and/or circuits. (air flow meter), (exhaust gas sensor)
DISCONNECT Exhaust gas sensor
Disconnect exhaust gas sensor connector, and try to drive.
Improved Replace the sensor.
No change Check Mixture ratio feedback system.
Diagnostic Table (Cont'd)
SYMPTOM & CONDITION 15 Poor driveability – surge (while cruising)
POSSIBLE CAUSES
SPECIFICATIONS Mixture ratio (too lean)
Fuel pressure (low)
Ignition timing
IGNITION SYSTEM (missing)
INTAKE SYSTEM Air duct (leaks)
Throttle chamber (air leaks)
Intake manifold (gasket) (air leaks)
CONTROL SYSTEM Crank angle sensor
Air flow meter
Exhaust gas sensor
Idle switch
E.G.R. SYSTEM E.G.R. control valve (stuck open)
E.G.R. solenoid (remaining OFF)
E.G.R. vacuum hose (removed)
SERVICE PROCEDURE
CHECK Engine Leak Air duct
Check intake system for air leaks.
N.G. Repair/replace the part.
CHECK Timing light
Check ignition timing.
N.G. Adjust ignition timing.
CHECK E.G.R. control valve
Check E.G.R. control valve for operation.
Stays lifted
CHECK E.G.R. solenoid
Check terminal voltage of E.G.R. solenoid.
N.G. Check the solenoid circuit.
REMOVE Pressure regulator
Remove vacuum hose from pressure regulator, and try to drive.
Improved
MEASURE Fuel pressure gauge
Connect vacuum hose to pressure regulator and measure fuel pressure while driving.
Check Mixture ratio feedback system
PERFORM Self-diagnosis Mode IV
Perform self-diagnosis Mode IV (for idle switch).
N.G. Check the idle switch circuit.
PERFORM Self-diagnosis Mode III Self-diagnosis Mode V
Perform self-diagnosis Mode III and V (for crank angle sensor, air flow meter, ignition signal and exhaust gas sensor).
N.G. Check the malfunctioning parts and/or circuits. (crank angle sensor), (air flow meter), (ignition signal), (exhaust gas sensor).
DISCONNECT Exhaust gas sensor
Disconnect exhaust gas sensor connector, and try to drive.
Improved Replace the sensor.
No change Check mixture ratio feedback system.

ECCS DIAGNOSTIC CHART—1989 SENTRA

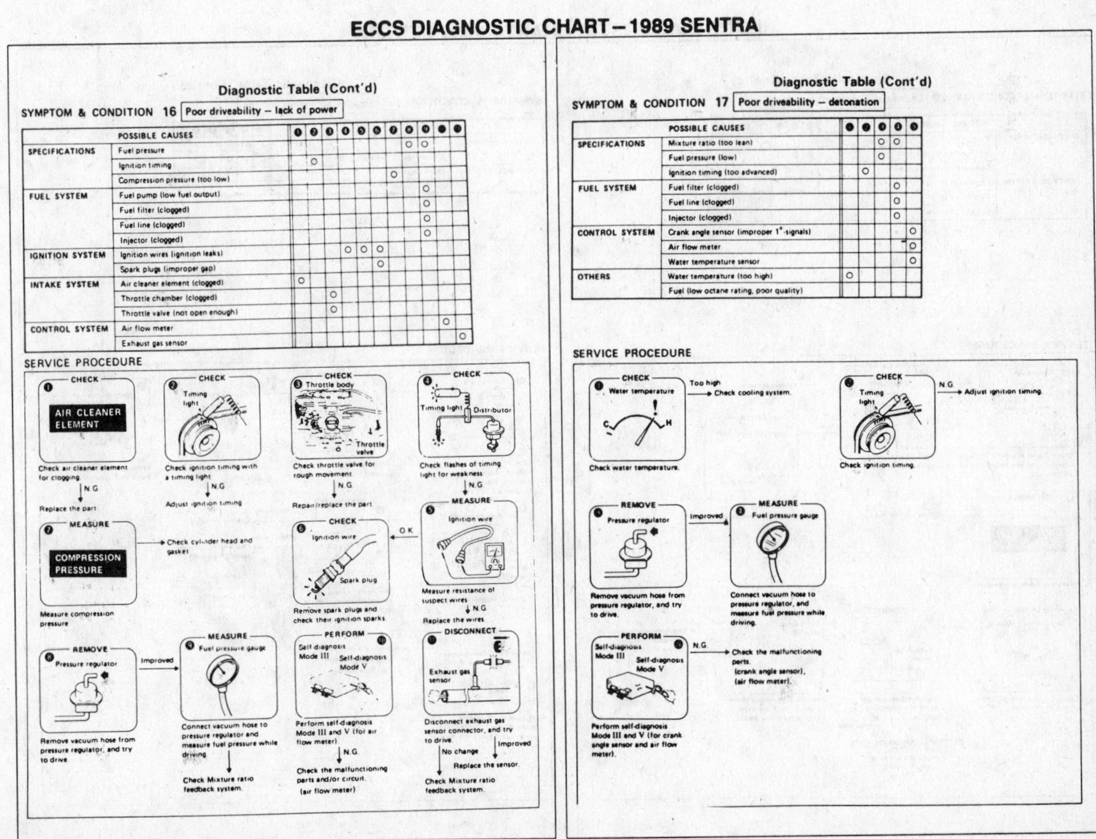

Diagnostic Table (Cont'd)
SYMPTOM & CONDITION 16 Poor driveability – lack of power
POSSIBLE CAUSES
SPECIFICATIONS Fuel pressure
Ignition timing
Compression pressure (too low)
FUEL SYSTEM Fuel pump (low fuel output)
Fuel filter (clogged)
Fuel line (clogged)
Injector (clogged)
IGNITION SYSTEM Ignition wires (ignition leaks)
Spark plugs (improper gap)
INTAKE SYSTEM Air cleaner element (clogged)
Throttle chamber (clogged)
Throttle valve (not open enough)
CONTROL SYSTEM Air flow meter
Exhaust gas sensor
SERVICE PROCEDURE
CHECK AIR CLEANER ELEMENT
Check air cleaner element for clogging.
N.G. Replace the part.
CHECK Timing light
Check ignition timing with a timing light.
N.G. Adjust ignition timing.
CHECK Throttle body Throttle valve
Check throttle valve for rough movement.
N.G. Repair/replace the part.
CHECK Timing light Distributor
Check flashes of timing light for weakness.
N.G.
MEASURE COMPRESSION PRESSURE
Measure compression pressure.
Check cylinder head and gasket
CHECK Ignition wire Spark plug
Remove spark plugs and check their ignition sparks.
O.K.
MEASURE Ignition wire
Measure resistance of suspect wires.
N.G. Replace the wires.
REMOVE Pressure regulator
Remove vacuum hose from pressure regulator, and try to drive.
Improved
MEASURE Fuel pressure gauge
Connect vacuum hose to pressure regulator and measure fuel pressure while driving.
Check Mixture ratio feedback system.
PERFORM Self-diagnosis Mode III Self-diagnosis Mode V
Perform self-diagnosis Mode III and V (for air flow meter).
N.G. Check the malfunctioning parts and/or circuit. (air flow meter)
DISCONNECT Exhaust gas sensor
Disconnect exhaust gas sensor connector, and try to drive.
No change Improved
Replace the sensor.
Check Mixture ratio feedback system.
Diagnostic Table (Cont'd)
SYMPTOM & CONDITION 17 Poor driveability – detonation
POSSIBLE CAUSES
SPECIFICATIONS Mixture ratio (too lean)
Fuel pressure (low)
Ignition timing (too advanced)
FUEL SYSTEM Fuel filter (clogged)
Fuel line (clogged)
Injector (clogged)
CONTROL SYSTEM Crank angle sensor (improper 1° signals)
Air flow meter
Water temperature sensor
OTHERS Water temperature (too high)
Fuel (low octane rating, poor quality)
SERVICE PROCEDURE
CHECK Water temperature
Check water temperature.
Too high Check cooling system.
CHECK Timing light
Check ignition timing.
N.G. Adjust ignition timing.
REMOVE Pressure regulator
Remove vacuum hose from pressure regulator, and try to drive.
Improved
MEASURE Fuel pressure gauge
Connect vacuum hose to pressure regulator, and measure fuel pressure while driving.
PERFORM Self-diagnosis Mode III Self-diagnosis Mode V
Perform self-diagnosis Mode III and V (for crank angle sensor and air flow meter).
N.G. Check the malfunctioning parts. (crank angle sensor), (air flow meter).

ECCS DIAGNOSTIC CHART—1989 SENTRA

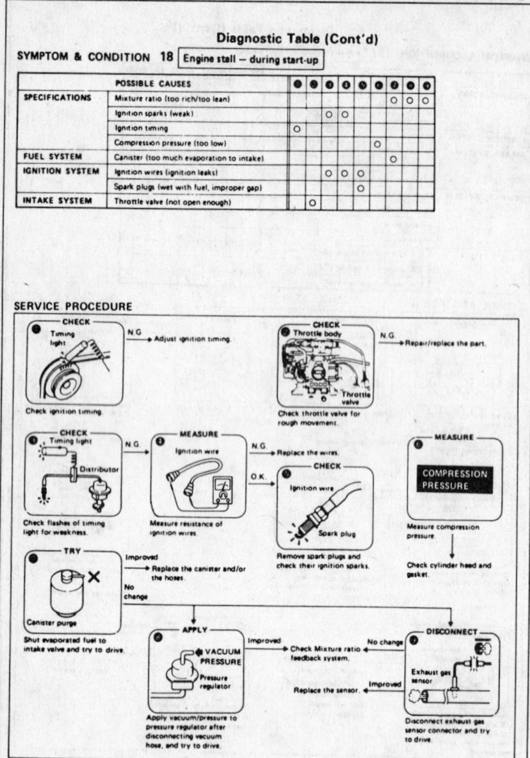

Diagnostic Table (Cont'd)

SYMPTOM & CONDITION 18 Engine stall — during start-up

	POSSIBLE CAUSES	1	2	3	4	5	6	7	8	9
SPECIFICATIONS	Mixture ratio (too rich/too lean)							○	○	○
	Ignition sparks (weak)			○	○					
	Ignition timing	○								
	Compression pressure (too low)						○			
FUEL SYSTEM	Canister (too much evaporation to intake)							○		
IGNITION SYSTEM	Ignition wires (ignition leaks)			○	○	○				
	Spark plugs (wet with fuel, improper gap)					○				
INTAKE SYSTEM	Throttle valve (not open enough)		○							

SERVICE PROCEDURE
CHECK
1
Timing light
N.G
Adjust ignition timing.
Check ignition timing.
CHECK
2
Throttle body
Throttle valve
N.G.
Repair/replace the part.
Check throttle valve for rough movement.
CHECK
3
Timing light
Distributor
Check flashes of timing light for weakness.
N.G.
MEASURE
4
Ignition wire
Measure resistance of ignition wires.
N.G.
Replace the wires.
O.K.
CHECK
5
Ignition wire
Spark plug
Remove spark plugs and check their ignition sparks.
MEASURE
6
COMPRESSION PRESSURE
Measure compression pressure.
Check cylinder head and gasket.
TRY
7
Canister purge
Shut evaporated fuel to intake valve and try to drive.
Improved
Replace the canister and/or the hoses.
No change
APPLY
8
VACUUM PRESSURE
Pressure regulator
Apply vacuum/pressure to pressure regulator after disconnecting vacuum hose, and try to drive.
Improved
Check Mixture ratio feedback system.
No change
DISCONNECT
9
Exhaust gas sensor
Disconnect exhaust gas sensor connector and try to drive.
Improved
Replace the sensor.

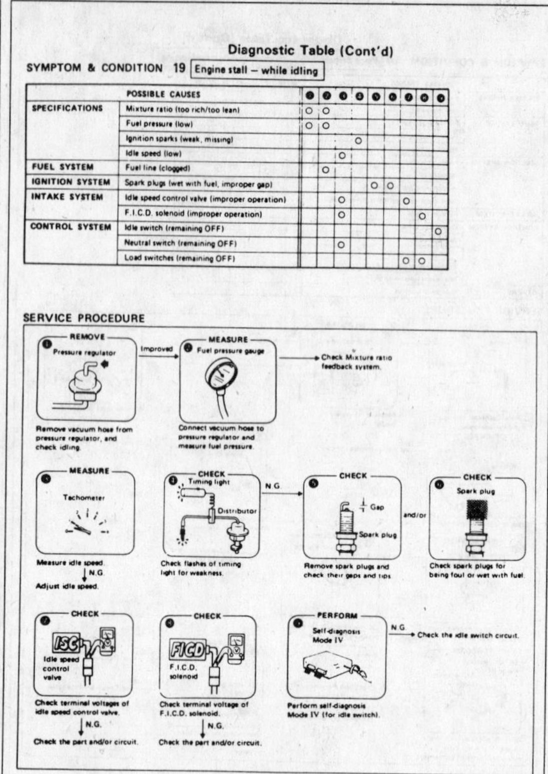

Diagnostic Table (Cont'd)

SYMPTOM & CONDITION 19 Engine stall — while idling

	POSSIBLE CAUSES	1	2	3	4	5	6	7	8	9
SPECIFICATIONS	Mixture ratio (too rich/too lean)	○	○							
	Fuel pressure (low)	○	○							
	Ignition sparks (weak, missing)				○					
	Idle speed (low)			○						
FUEL SYSTEM	Fuel line (clogged)		○							
IGNITION SYSTEM	Spark plugs (wet with fuel, improper gap)					○	○			
INTAKE SYSTEM	Idle speed control valve (improper operation)			○				○		
	F.I.C.D. solenoid (improper operation)			○					○	
CONTROL SYSTEM	Idle switch (remaining OFF)									○
	Neutral switch (remaining OFF)			○						
	Load switches (remaining OFF)							○	○	

SERVICE PROCEDURE
REMOVE
1
Pressure regulator
Remove vacuum hose from pressure regulator, and check idling.
Improved
MEASURE
2
Fuel pressure gauge
Connect vacuum hose to pressure regulator and measure fuel pressure.
Check Mixture ratio feedback system.
MEASURE
3
Tachometer
Measure idle speed.
N.G.
Adjust idle speed.
CHECK
4
Timing light
Distributor
Check flashes of timing light for weakness.
N.G.
CHECK
5
Gap
Spark plug
Remove spark plugs and check their gaps and tips.
and/or
CHECK
6
Spark plug
Check spark plugs for being foul or wet with fuel.
CHECK
7
ISC
Idle speed control valve
Check terminal voltages of idle speed control valve.
N.G.
Check the part and/or circuit.
CHECK
8
FICD
F.I.C.D. solenoid
Check terminal voltage of F.I.C.D. solenoid.
N.G.
Check the part and/or circuit.
PERFORM
9
Self-diagnosis Mode IV
Perform self-diagnosis Mode IV (for idle switch).
N.G.
Check the idle switch circuit.

ECCS DIAGNOSTIC CHART—1989 SENTRA

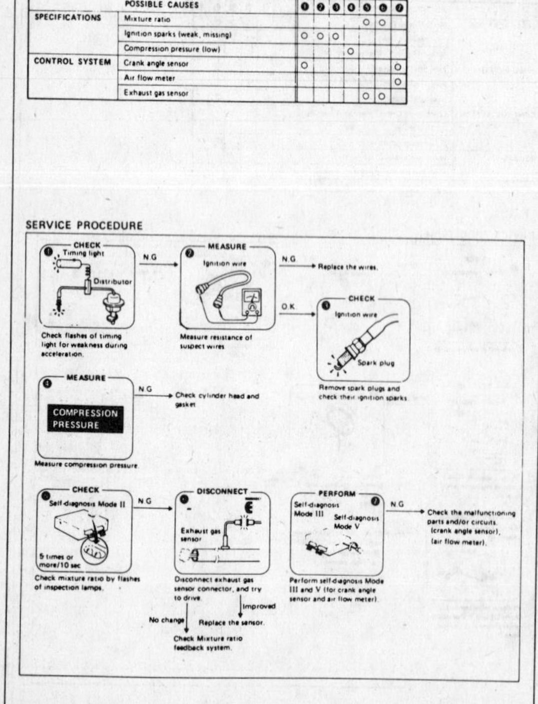

Diagnostic Table (Cont'd)

SYMPTOM & CONDITION 20 Engine stall — while accelerating

	POSSIBLE CAUSES	1	2	3	4	5	6	7
SPECIFICATIONS	Mixture ratio					○	○	
	Ignition sparks (weak, missing)	○	○	○				
	Compression pressure (low)				○			
CONTROL SYSTEM	Crank angle sensor	○						○
	Air flow meter							○
	Exhaust gas sensor					○	○	

SERVICE PROCEDURE
CHECK
1
Timing light
Distributor
Check flashes of timing light for weakness during acceleration.
N.G
MEASURE
2
Ignition wire
Measure resistance of suspect wires.
N.G
Replace the wires.
O.K.
CHECK
3
Ignition wire
Spark plug
Remove spark plugs and check their ignition sparks.
MEASURE
4
COMPRESSION PRESSURE
Measure compression pressure.
N.G
Check cylinder head and gasket.
CHECK
5
Self-diagnosis Mode II
5 times or more/10 sec
Check mixture ratio by flashes of inspection lamps.
N.G
DISCONNECT
6
Exhaust gas sensor
Disconnect exhaust gas sensor connector, and try to drive.
Improved
Replace the sensor.
No change
Check Mixture ratio feedback system.
PERFORM
7
Self-diagnosis Mode III
Self-diagnosis Mode V
Perform self-diagnosis Mode III and V (for crank angle sensor and air flow meter).
N.G
Check the malfunctioning parts and/or circuits. (crank angle sensor), (air flow meter).

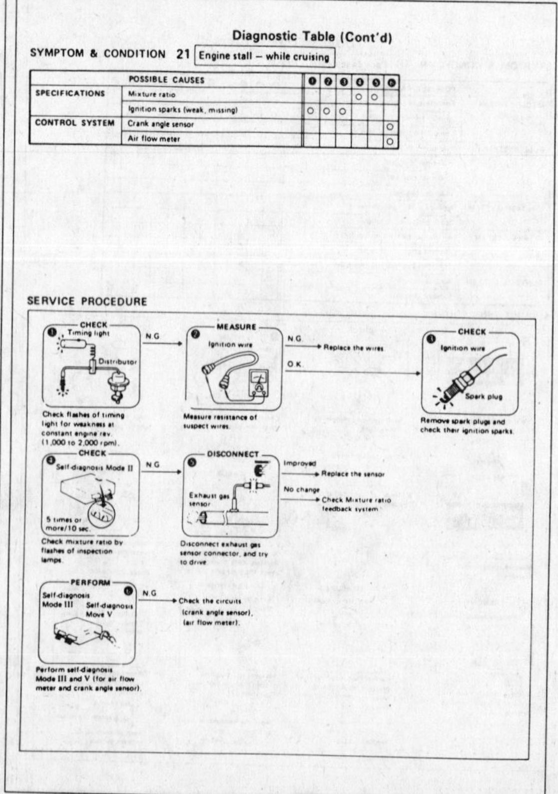

Diagnostic Table (Cont'd)

SYMPTOM & CONDITION 21 Engine stall — while cruising

	POSSIBLE CAUSES	1	2	3	4	5	6
SPECIFICATIONS	Mixture ratio				○	○	
	Ignition sparks (weak, missing)	○	○	○			
CONTROL SYSTEM	Crank angle sensor						○
	Air flow meter						○

SERVICE PROCEDURE
CHECK
1
Timing light
Distributor
Check flashes of timing light for weakness at constant engine rev. (1,000 to 2,000 rpm).
N.G
MEASURE
2
Ignition wire
Measure resistance of suspect wires.
N.G.
Replace the wires.
O.K.
CHECK
3
Ignition wire
Spark plug
Remove spark plugs and check their ignition sparks.
CHECK
4
Self-diagnosis Mode II
5 times or more/10 sec.
Check mixture ratio by flashes of inspection lamps.
N.G
DISCONNECT
5
Exhaust gas sensor
Disconnect exhaust gas sensor connector, and try to drive.
Improved
Replace the sensor
No change
Check Mixture ratio feedback system.
PERFORM
6
Self-diagnosis Mode III
Self-diagnosis Move V
Perform self-diagnosis Mode III and V (for air flow meter and crank angle sensor).
N.G
Check the circuits (crank angle sensor), (air flow meter).

ECCS DIAGNOSTIC CHART—1989 SENTRA

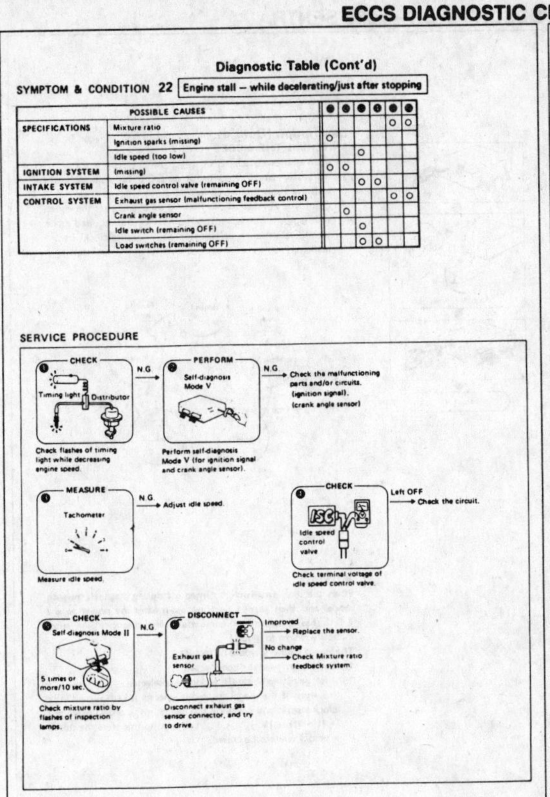

Diagnostic Table (Cont'd)

SYMPTOM & CONDITION 22 Engine stall – while decelerating/just after stopping

POSSIBLE CAUSES		❶	❷	❸	❹	❺	❻
SPECIFICATIONS	Mixture ratio					○	○
	Ignition sparks (missing)	○					
	Idle speed (too low)			○			
IGNITION SYSTEM	(missing)	○	○				
INTAKE SYSTEM	Idle speed control valve (remaining OFF)			○	○		
CONTROL SYSTEM	Exhaust gas sensor (malfunctioning feedback control)					○	○
	Crank angle sensor		○				
	Idle switch (remaining OFF)			○			
	Load switches (remaining OFF)			○	○		

SERVICE PROCEDURE

CHECK
Timing light
Distributor
N.G.
Check flashes of timing light while decreasing engine speed.
PERFORM
Self-diagnosis Mode V
N.G.
Check the malfunctioning parts and/or circuits. (ignition signal). (crank angle sensor)
Perform self-diagnosis Mode V (for ignition signal and crank angle sensor).
MEASURE
Tachometer
N.G.
Adjust idle speed.
Measure idle speed.
CHECK
ISC
Idle speed control valve
Left OFF
Check the circuit.
Check terminal voltage of idle speed control valve.
CHECK
Self-diagnosis Mode II
5 times or more/10 sec
N.G.
Check mixture ratio by flashes of inspection lamps.
DISCONNECT
Exhaust gas sensor
Improved
Replace the sensor.
No change
Check Mixture ratio feedback system.
Disconnect exhaust gas sensor connector, and try to drive.

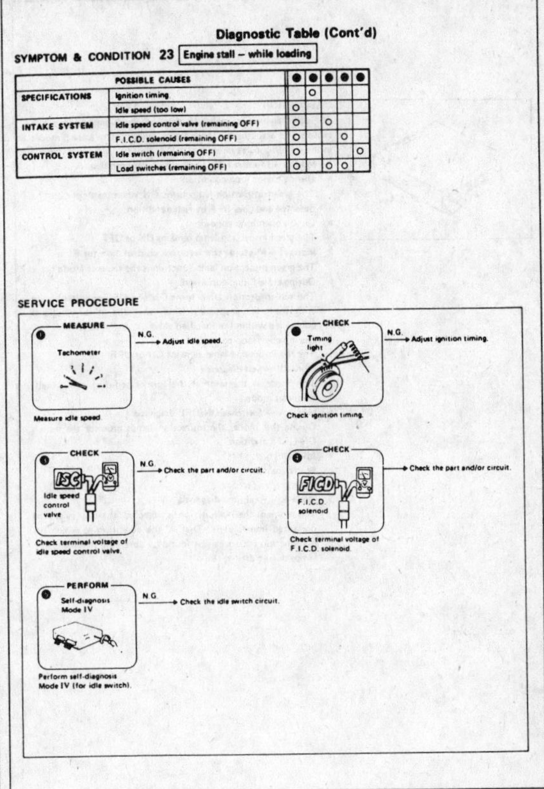

Diagnostic Table (Cont'd)

SYMPTOM & CONDITION 23 Engine stall – while loading

POSSIBLE CAUSES		❶	❷	❸	❹	❺
SPECIFICATIONS	Ignition timing		○			
	Idle speed (too low)	○				
INTAKE SYSTEM	Idle speed control valve (remaining OFF)	○		○		
	F.I.C.D. solenoid (remaining OFF)	○			○	
CONTROL SYSTEM	Idle switch (remaining OFF)	○				○
	Load switches (remaining OFF)	○		○	○	

SERVICE PROCEDURE

MEASURE
Tachometer
N.G.
Adjust idle speed.
Measure idle speed
CHECK
Timing light
N.G.
Adjust ignition timing.
Check ignition timing.
CHECK
ISC
Idle speed control valve
N.G.
Check the part and/or circuit.
Check terminal voltage of idle speed control valve.
CHECK
FICD
F.I.C.D. solenoid
Check the part and/or circuit.
Check terminal voltage of F.I.C.D. solenoid.
PERFORM
Self-diagnosis Mode IV
N.G.
Check the idle switch circuit.
Perform self-diagnosis Mode IV (for idle switch).

ECCS DIAGNOSTIC CHART—1989 SENTRA

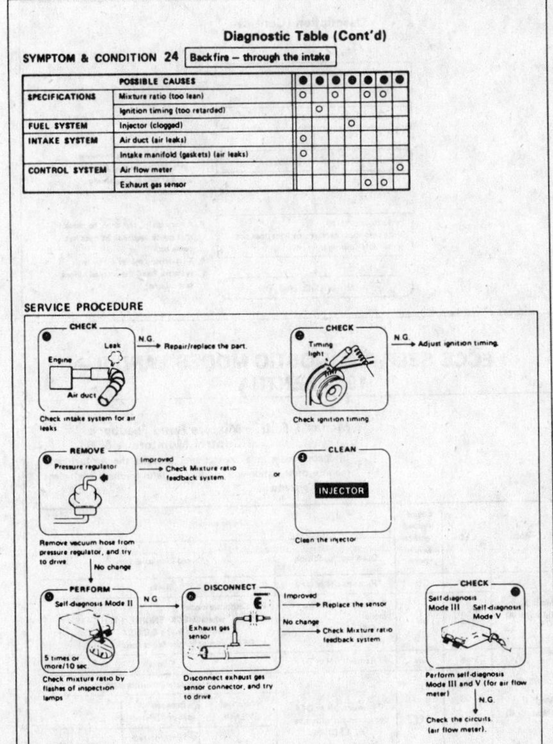

Diagnostic Table (Cont'd)

SYMPTOM & CONDITION 24 Backfire – through the intake

POSSIBLE CAUSES		❶	❷	❸	❹	❺	❻	❼
SPECIFICATIONS	Mixture ratio (too lean)	○		○		○	○	
	Ignition timing (too retarded)		○					
FUEL SYSTEM	Injector (clogged)				○			
INTAKE SYSTEM	Air duct (air leaks)	○						
	Intake manifold (gaskets) (air leaks)	○						
CONTROL SYSTEM	Air flow meter							○
	Exhaust gas sensor					○	○	

SERVICE PROCEDURE

CHECK
Leak
Engine
Air duct
N.G.
Repair/replace the part.
Check intake system for air leaks
CHECK
Timing light
N.G.
Adjust ignition timing.
Check ignition timing
REMOVE
Pressure regulator
Improved
Check Mixture ratio feedback system.
or
Remove vacuum hose from pressure regulator, and try to drive.
No change
CLEAN
INJECTOR
Clean the injector
PERFORM
Self-diagnosis Mode II
5 times or more/10 sec
N.G.
Check mixture ratio by flashes of inspection lamps
DISCONNECT
Exhaust gas sensor
Improved
Replace the sensor.
No change
Check Mixture ratio feedback system
Disconnect exhaust gas sensor connector, and try to drive
CHECK
Self-diagnosis Mode III
Self-diagnosis Mode V
Perform self-diagnosis Mode III and V (for air flow meter)
N.G.
Check the circuits. (air flow meter).

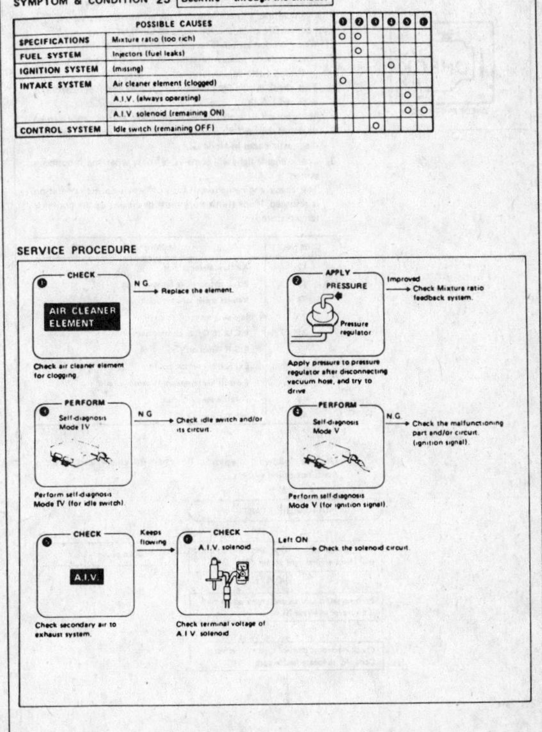

Diagnostic Table (Cont'd)

SYMPTOM & CONDITION 25 Backfire – through the exhaust

POSSIBLE CAUSES		❶	❷	❸	❹	❺	❻
SPECIFICATIONS	Mixture ratio (too rich)	○	○				
FUEL SYSTEM	Injectors (fuel leaks)		○				
IGNITION SYSTEM	(missing)				○		
INTAKE SYSTEM	Air cleaner element (clogged)	○					
	A.I.V. (always operating)					○	
	A.I.V. solenoid (remaining ON)					○	○
CONTROL SYSTEM	Idle switch (remaining OFF)			○			

SERVICE PROCEDURE

CHECK
AIR CLEANER ELEMENT
N.G.
Replace the element.
Check air cleaner element for clogging
APPLY PRESSURE
Pressure regulator
Improved
Check Mixture ratio feedback system.
Apply pressure to pressure regulator after disconnecting vacuum hose, and try to drive
PERFORM
Self-diagnosis Mode IV
N.G.
Check idle switch and/or its circuit.
Perform self-diagnosis Mode IV (for idle switch)
PERFORM
Self-diagnosis Mode V
N.G.
Check the malfunctioning part and/or circuit. (ignition signal).
Perform self-diagnosis Mode V (for ignition signal).
CHECK
A.I.V.
Keeps flowing
Check secondary air to exhaust system.
CHECK
A.I.V. solenoid
Left ON
Check the solenoid circuit.
Check terminal voltage of A.I.V. solenoid.

ECCS SELF-DIAGNOSTIC DESCRIPTION—1989 SENTRA

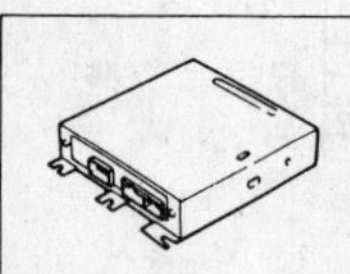

Description

The self-diagnosis is useful to diagnose malfunctions in major sensors and actuators of the E.C.C.S. system. There are 5 modes in the self-diagnosis system.

1. **Mode I – Mixture ratio feedback control monitor A**
   - During closed loop condition:
     The green inspection lamp turns ON when lean condition is detected and goes OFF by rich condition.
   - During open loop condition:
     The green inspection lamp remains ON or OFF.
2. **Mode II – Mixture ratio feedback control monitor B**
   The green inspection lamp function is the same as Mode I.
   - During closed loop condition:
     The red inspection lamp turns ON and OFF simultaneously with the green inspection lamp when the mixture ratio is controlled within the specified value.
   - During open loop condition:
     The red inspection lamp remains ON or OFF.
3. **Mode III – Self-diagnosis**
   This mode is the same as the former self-diagnosis in self-diagnosis mode.
4. **Mode IV – Switches ON/OFF diagnosis**
   During this mode, the inspection lamps monitor the switch ON-OFF condition.
   - Idle switch
   - Starter switch
   - Vehicle speed sensor
5. **Mode V – Real time diagnosis**
   The moment the malfunction is detected, the display will be presented immediately. That is, the condition at which the malfunction occurs can be found by observing the inspection lamps during driving test.

Flashing N times
Mode N

Description (Cont'd)

SWITCHING THE MODES

1. Turn ignition switch "ON".
2. Turn diagnostic mode selector on E.C.U. fully clockwise and wait the inspection lamps flash.
3. Count the number of the flashing time, and after the inspection lamps have flashed the number of the required mode, turn diagnostic mode selector fully counterclockwise immediately.

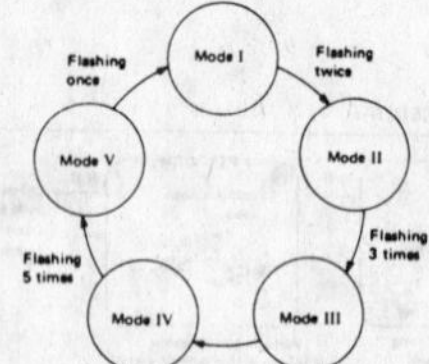
Mode I
Flashing twice
Mode II
Flashing 3 times
Mode III
Flashing 4 times
Mode IV
Flashing 5 times
Mode V
Flashing once

NOTE:
When the ignition switch is turned off during diagnosis, in each mode, and then turned back on again after the power to the E.C.U. has dropped off completely, the diagnosis will automatically return to Mode I.
The stored memory would be lost if:

1. Battery terminal is disconnected.
2. After selecting Mode III, Mode IV is selected.
   However, if the diagnostic mode selector is kept turned fully clockwise, it will continue to change in the order of Mode I → II → III → IV → V → I ... etc., and in this state the stored memory will not be erased.

ECCS SELF-DIAGNOSTIC DESCRIPTION—1989 SENTRA

CHECK
CHECK ENGINE LIGHT

Description (Cont'd)

CHECK ENGINE LIGHT (For California only)

This vehicle has a check engine light on instrument panel. This light comes ON under the following conditions:

1) When ignition switch is turned "ON" (for bulb check).
2) When systems related to emission performance malfunction in Mode I (with engine running).
   - **This check engine light always illuminates and is synchronous with red L.E.D.**
   - **Malfunction systems related to emission performance can be detected by self-diagnosis, and they are clarified as self-diagnostic codes in Mode III.**
3) Check engine light will come "ON" only when malfunction is sensed.
   The check engine light will turn off when normal operation is resumed. Mode III memory must be cleared as the contents remain stored.

Code No.	Malfunction
12	Air flow meter circuit
13	Water temperature sensor circuit
14	Vehicle speed sensor circuit
23	Idle switch circuit
31	E.C.U. (E.C.C.S. control unit)
32	E.G.R. function
33	Exhaust gas sensor circuit
35	Exhaust gas temperature sensor circuit
43	Throttle sensor circuit
45	Injector leak

Use the following diagnostic flowchart to check and repair a malfunctioning system.

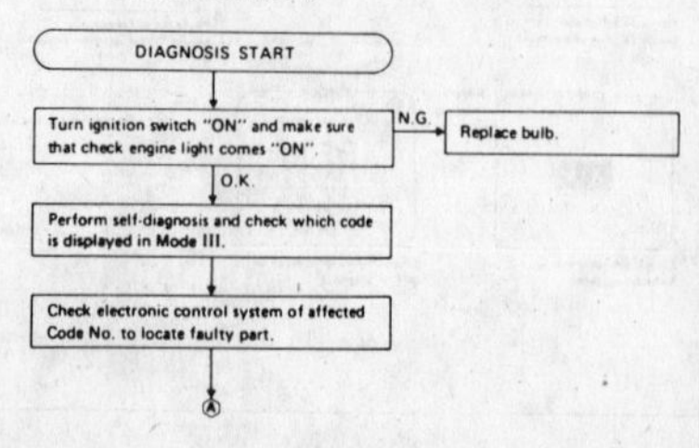
DIAGNOSIS START
Turn ignition switch "ON" and make sure that check engine light comes "ON".
N.G.
Replace bulb.
O.K.
Perform self-diagnosis and check which code is displayed in Mode III.
Check electronic control system of affected Code No. to locate faulty part.
A

Description (Cont'd)

Ⓐ

Repair or replace faulty part.
↓
Reinstall any part removed.
↓
Erase the self-diagnosis memory.
- **Methods of erasing memories differ with systems. Read the manual before diagnosing systems.**

↓
Perform driving test.
Make sure that check engine light does not come "ON" during this test.
- **After repairs, test drive to check that check engine light does not come on.**
- **Test driving modes differ with systems. Read the manual before test driving.**

↓
DIAGNOSIS END

ECCS SELF-DIAGNOSTIC MODES I AND II—1989 SENTRA

Modes I & II —Mixture Ratio Feedback Control Monitors A & B

In these modes, the control unit provides the Air-fuel ratio monitor presentation and the Air-fuel ratio feedback coefficient monitor presentation.

Mode	LED	Engine stopped (Ignition switch "ON")	Engine running			
			Open loop condition	Closed loop condition		
Mode I (Monitor A)	Green	ON	*Remains ON or OFF	Blinks		
	Red	ON	Except for California model • OFF	For California model • ON: when CHECK ENGINE LIGHT ITEMS are stored in the E.C.U. • OFF: except for the above condition		
Mode II (Monitor B)	Green	ON	*Remains ON or OFF	Blinks		
	Red	OFF	*Remains ON or OFF (synchronous with green LED)	Compensating mixture ratio		
				More than 5% rich	Between 5% lean and 5% rich	More
				OFF	Synchronized with green LED	Remains ON

*: Maintains conditions just before switching to open loop

ECCS SELF-DIAGNOSTIC MODE III—1989 SENTRA

Mode III — Self-diagnostic System

The E.C.U. constantly monitors the function of these sensors and actuators, regardless of ignition key position. If a malfunction occurs, the information is stored in the E.C.U. and can be retrieved from the memory by turning on the diagnostic mode selector, located on the side of the E.C.U. When activated, the malfunction is indicated by flashing a red and a green L.E.D. (Light Emitting Diode), also located on the E.C.U. Since all the self-diagnostic results are stored in the E.C.U.'s memory even intermittent malfunctions can be diagnosed.

A malfunctioning part's group is indicated by the number of both the red and the green L.E.D.s flashing. First, the red L.E.D. flashes and the green flashes follow. The red L.E.D. refers to the number of tens while the green one refers to the number of units. For example, when the red L.E.D. flashes once and then the green one flashes twice, this means the number "12" showing the air flow meter signal is malfunctioning. In this way, all the problems are classified by the code numbers.

- **When engine fails to start, crank engine more than two seconds before starting self-diagnosis.**
- **Before starting self-diagnosis, do not erase stored memory. If doing so, self-diagnosis function for intermittent malfunctions would be lost.**

The stored memory would be lost if:

1. **Battery terminal is disconnected.**
2. **After selecting Mode III, Mode IV is selected.**

DISPLAY CODE TABLE

Code No.	Detected items	California	Non-California
11	Crank angle sensor circuit	X	X
12	Air flow meter circuit	X*	X
13	Water temperature sensor circuit	X*	X
14	Vehicle speed sensor circuit	X*	X
21	Ignition signal missing in primary coil	X	X
23	Idle switch circuit	X*	X
25	A.A.C. valve circuit	X	X
31	E.C.U. (E.C.C.S. control unit)	X*	X
32	E.G.R. function	X*	–
33	Exhaust gas sensor circuit	X*	X
35	Exhaust gas temperature sensor circuit	X*	–
43	Throttle sensor circuit	X*	X
45	Injector leak	X*	–
55	No malfunctioning in the above circuit	X	X

X: Available
–: Not available
*: Check engine light item

Mode III — Self-diagnostic System (Cont'd)

RETENTION OF DIAGNOSTIC RESULTS

The diagnostic result is retained in E.C.U. memory until the starter is operated fifty times after a diagnostic item is judged to be malfunctioning. The diagnostic result will then be cancelled automatically. If a diagnostic item which has been judged to be malfunctioning and stored in memory is again judged to be malfunctioning before the starter is operated fifty times, the second result will replace the previous one. It will be stored in E.C.U. memory until the starter is operated fifty times more.

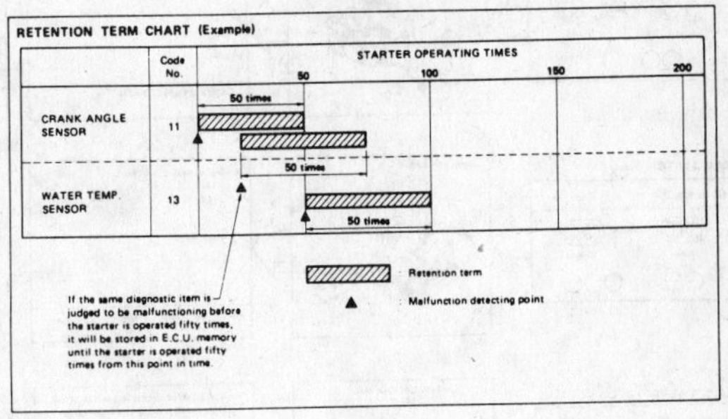
RETENTION TERM CHART (Example)
Code No.
STARTER OPERATING TIMES
50
100
150
200
CRANK ANGLE SENSOR
11
50 times
50 times
WATER TEMP. SENSOR
13
50 times
: Retention term
: Malfunction detecting point
If the same diagnostic item is judged to be malfunctioning before the starter is operated fifty times, it will be stored in E.C.U. memory until the starter is operated fifty times from this point in time.

ECCS SELF-DIAGNOSTIC MODE III—1989 SENTRA

Mode III — Self-diagnostic System (Cont'd)

SELF-DIAGNOSTIC PROCEDURE

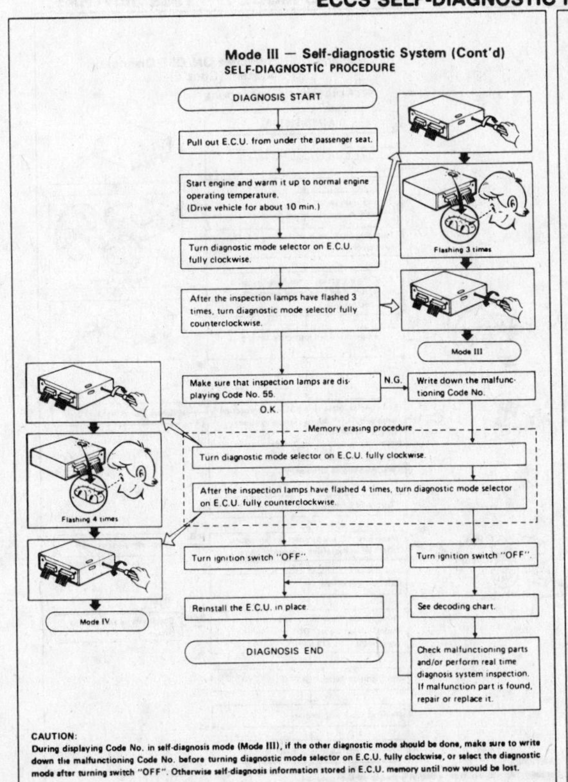
DIAGNOSIS START
Pull out E.C.U. from under the passenger seat.
Start engine and warm it up to normal engine operating temperature. (Drive vehicle for about 10 min.)
Turn diagnostic mode selector on E.C.U. fully clockwise.
Flashing 3 times
After the inspection lamps have flashed 3 times, turn diagnostic mode selector fully counterclockwise.
Mode III
Make sure that inspection lamps are displaying Code No. 55.
N.G.
Write down the malfunctioning Code No.
O.K.
Memory erasing procedure
Turn diagnostic mode selector on E.C.U. fully clockwise.
After the inspection lamps have flashed 4 times, turn diagnostic mode selector on E.C.U. fully counterclockwise.
Flashing 4 times
Turn ignition switch "OFF".
Turn ignition switch "OFF".
Reinstall the E.C.U. in place.
See decoding chart.
Mode IV
DIAGNOSIS END
Check malfunctioning parts and/or perform real time diagnosis system inspection. If malfunction part is found, repair or replace it.

CAUTION:
During displaying Code No. in self-diagnosis mode (Mode III), if the other diagnostic mode should be done, make sure to write down the malfunctioning Code No. before turning diagnostic mode selector on E.C.U. fully clockwise, or select the diagnostic mode after turning switch "OFF". Otherwise self-diagnosis information stored in E.C.U. memory until now would be lost.

Mode III — Self-diagnostic System (Cont'd)

DECODING CHART

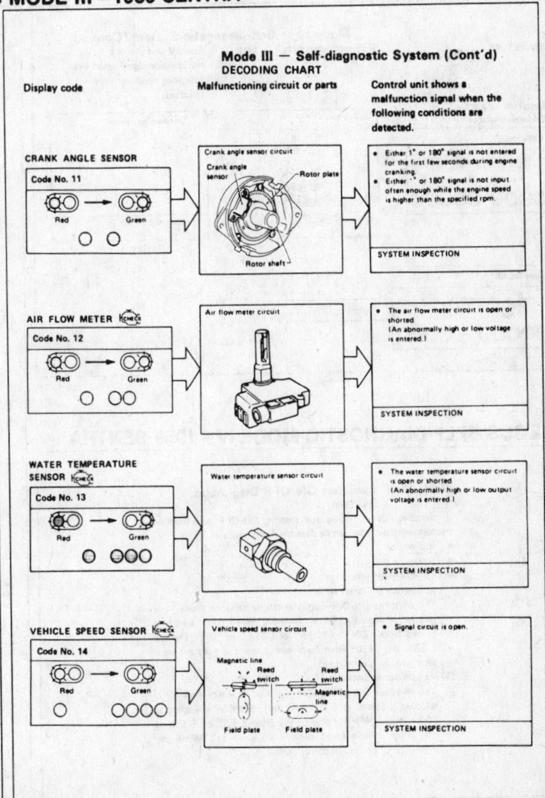
Display code
Malfunctioning circuit or parts
Control unit shows a malfunction signal when the following conditions are detected.
CRANK ANGLE SENSOR
Code No. 11
Red
Green
Crank angle sensor circuit
Crank angle sensor
Rotor plate
Rotor shaft
• Either 1° or 180° signal is not entered for the first few seconds during engine cranking.
• Either 1° or 180° signal is not input often enough while the engine speed is higher than the specified rpm.
SYSTEM INSPECTION
AIR FLOW METER CHECK
Code No. 12
Red
Green
Air flow meter circuit
• The air flow meter circuit is open or shorted. (An abnormally high or low voltage is entered.)
SYSTEM INSPECTION
WATER TEMPERATURE SENSOR CHECK
Code No. 13
Red
Green
Water temperature sensor circuit
• The water temperature sensor circuit is open or shorted. (An abnormally high or low output voltage is entered.)
SYSTEM INSPECTION
VEHICLE SPEED SENSOR CHECK
Code No. 14
Red
Green
Vehicle speed sensor circuit
Magnetic line
Reed switch
Reed switch
Magnetic line
Field plate
Field plate
• Signal circuit is open.
SYSTEM INSPECTION

ECCS SELF-DIAGNOSTIC MODE III DISPLAY CODES—1989 SENTRA

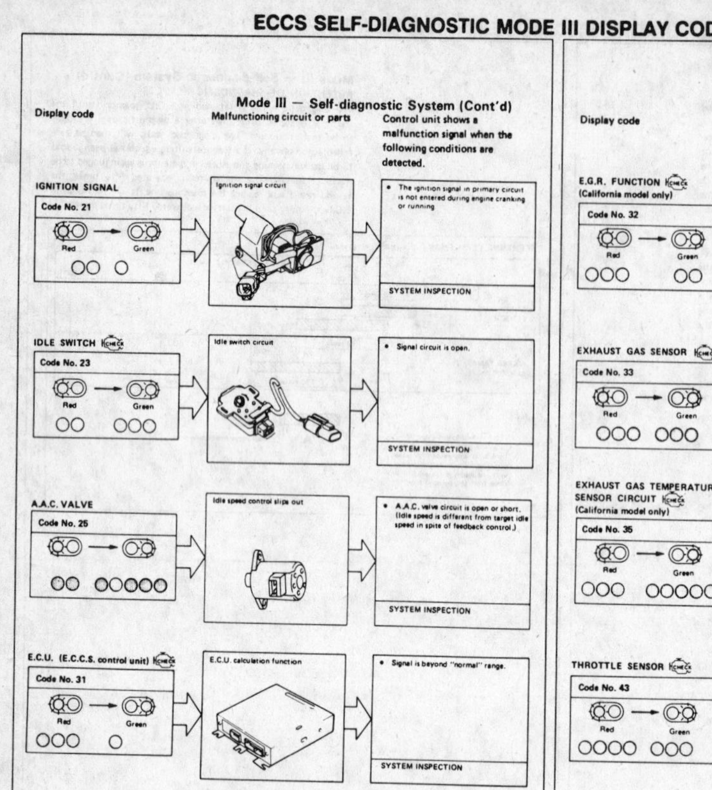
Mode III — Self-diagnostic System (Cont'd)
Display code
Malfunctioning circuit or parts
Control unit shows a malfunction signal when the following conditions are detected.
IGNITION SIGNAL
Code No. 21
Red
Green
Ignition signal circuit
• The ignition signal in primary circuit is not entered during engine cranking or running
SYSTEM INSPECTION
IDLE SWITCH
Code No. 23
Red
Green
Idle switch circuit
• Signal circuit is open.
SYSTEM INSPECTION
A.A.C. VALVE
Code No. 25
Idle speed control slips out
• A.A.C. valve circuit is open or short. (Idle speed is different from target idle speed in spite of feedback control.)
SYSTEM INSPECTION
E.C.U. (E.C.C.S. control unit)
Code No. 31
Red
Green
E.C.U. calculation function
• Signal is beyond "normal" range.
SYSTEM INSPECTION

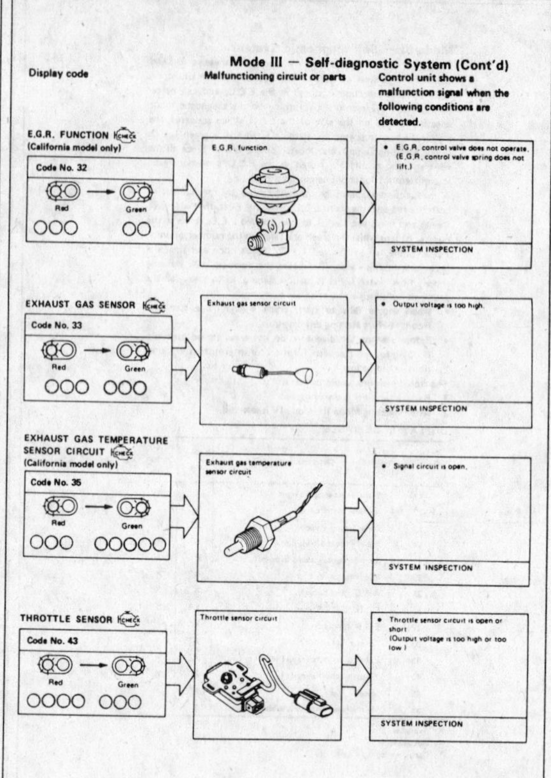
Mode III — Self-diagnostic System (Cont'd)
Display code
Malfunctioning circuit or parts
Control unit shows a malfunction signal when the following conditions are detected.
E.G.R. FUNCTION
(California model only)
Code No. 32
Red
Green
E.G.R. function
• E.G.R. control valve does not operate. (E.G.R. control valve spring does not lift.)
SYSTEM INSPECTION
EXHAUST GAS SENSOR
Code No. 33
Red
Green
Exhaust gas sensor circuit
• Output voltage is too high.
SYSTEM INSPECTION
EXHAUST GAS TEMPERATURE SENSOR CIRCUIT
(California model only)
Code No. 35
Red
Green
Exhaust gas temperature sensor circuit
• Signal circuit is open.
SYSTEM INSPECTION
THROTTLE SENSOR
Code No. 43
Red
Green
Throttle sensor circuit
• Throttle sensor circuit is open or short
(Output voltage is too high or too low.)
SYSTEM INSPECTION

ECCS SELF-DIAGNOSTIC MODE III DISPLAY CODES —1989 SENTRA

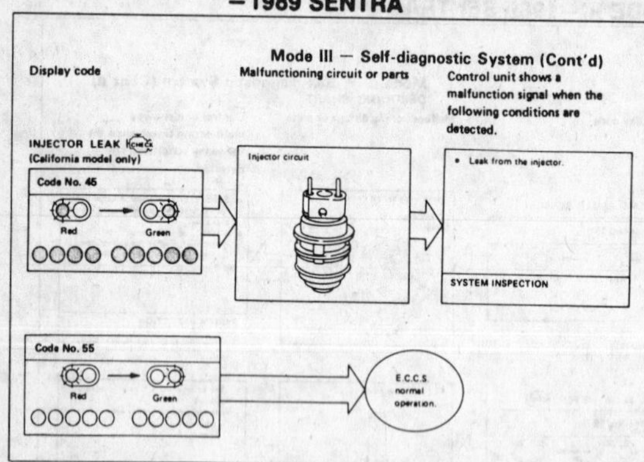
Mode III — Self-diagnostic System (Cont'd)
Display code
Malfunctioning circuit or parts
Control unit shows a malfunction signal when the following conditions are detected.
INJECTOR LEAK
(California model only)
Code No. 45
Red
Green
Injector circuit
• Leak from the injector.
SYSTEM INSPECTION
Code No. 55
Red
Green
E.C.C.S. normal operation.

ECCS SELF-DIAGNOSTIC MODE IV—1989 SENTRA

Mode IV — Switches ON/OFF Diagnostic System

In switches ON/OFF diagnosis system, ON/OFF operation of the following switches can be detected continuously.

- Idle switch
- Starter switch
- Vehicle speed sensor

(1) Idle switch & Starter switch

The switches ON/OFF status at the point when Mode IV is selected is stored in E.C.U. memory. When either switch is turned from "ON" to "OFF" or "OFF" to "ON", the red L.E.D. on E.C.U. alternately comes on and goes off each time switching is detected.

(2) Vehicle Speed Sensor

The switches ON/OFF status at the point when Mode IV is selected is stored in E.C.U. memory. When vehicle speed is 20 km/h (12 MPH) or slower, the green L.E.D. on E.C.U. is off. When vehicle speed exceeds 20 km/h (12 MPH), the green L.E.D. on E.C.U. comes "ON".

ECCS SELF-DIAGNOSTIC MODE IV—1989 SENTRA

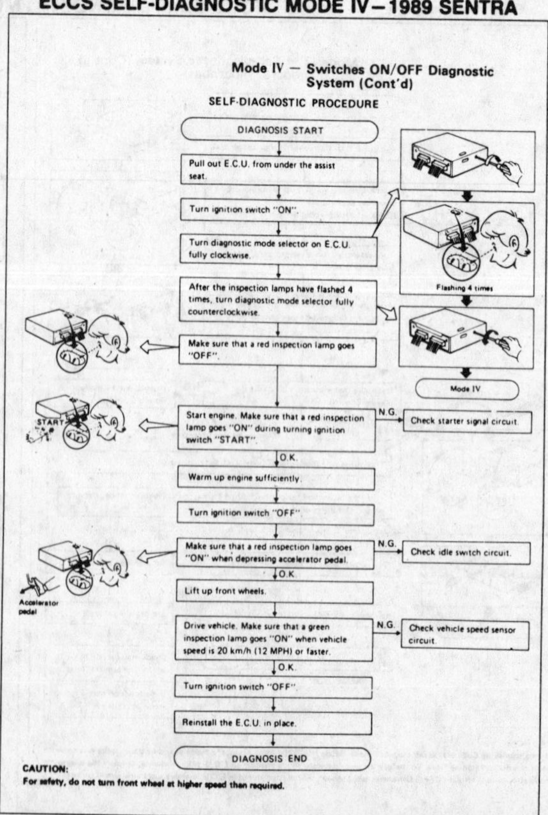
Mode IV — Switches ON/OFF Diagnostic System (Cont'd)
SELF-DIAGNOSTIC PROCEDURE
DIAGNOSIS START
Pull out E.C.U. from under the assist seat.
Turn ignition switch "ON".
Turn diagnostic mode selector on E.C.U. fully clockwise.
After the inspection lamps have flashed 4 times, turn diagnostic mode selector fully counterclockwise.
Flashing 4 times
Mode IV
Make sure that a red inspection lamp goes "OFF".
Start engine. Make sure that a red inspection lamp goes "ON" during turning ignition switch "START".
START
N.G.
Check starter signal circuit.
O.K.
Warm up engine sufficiently.
Turn ignition switch "OFF".
Make sure that a red inspection lamp goes "ON" when depressing accelerator pedal.
Accelerator pedal
N.G.
Check idle switch circuit.
O.K.
Lift up front wheels.
Drive vehicle. Make sure that a green inspection lamp goes "ON" when vehicle speed is 20 km/h (12 MPH) or faster.
N.G.
Check vehicle speed sensor circuit.
O.K.
Turn ignition switch "OFF".
Reinstall the E.C.U. in place.
DIAGNOSIS END
CAUTION:
For safety, do not turn front wheel at higher speed than required.

ECCS SELF-DIAGNOSTIC MODE V—1989 SENTRA

ECCS SELF-DIAGNOSTIC MODE V DECODING CHART—1989 SENTRA

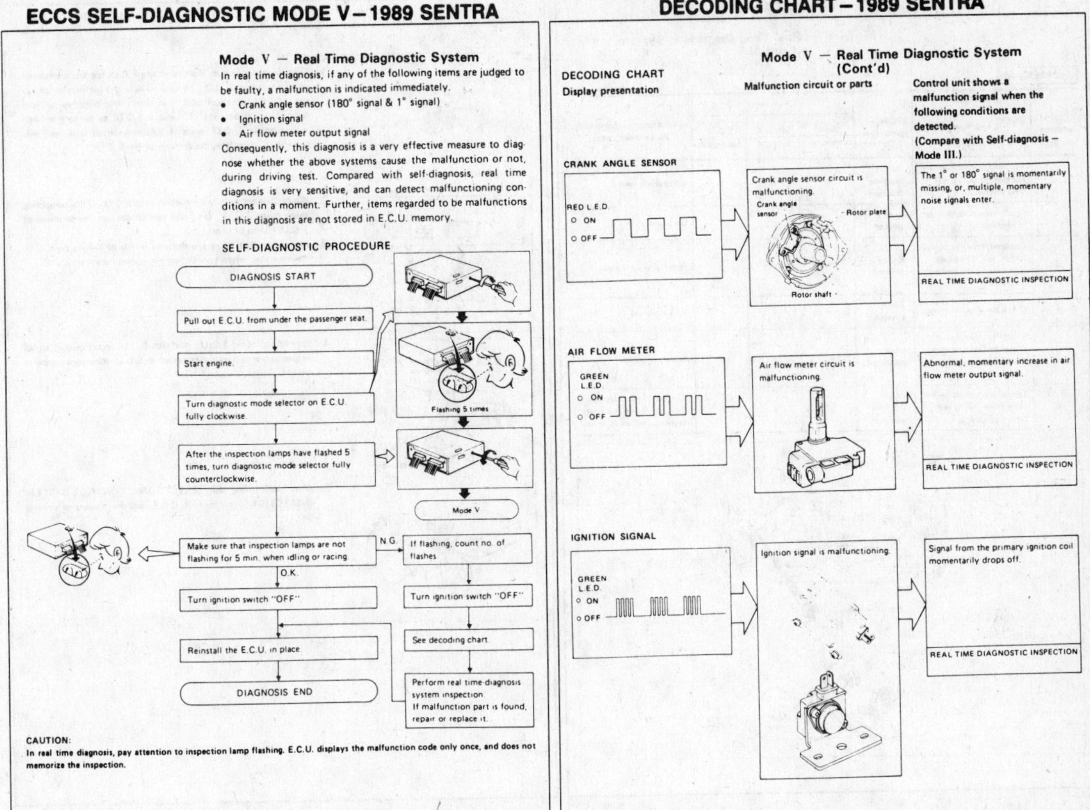

Mode V — Real Time Diagnostic System

In real time diagnosis, if any of the following items are judged to be faulty, a malfunction is indicated immediately.

- Crank angle sensor (180° signal & 1° signal)
- Ignition signal
- Air flow meter output signal

Consequently, this diagnosis is a very effective measure to diagnose whether the above systems cause the malfunction or not, during driving test. Compared with self-diagnosis, real time diagnosis is very sensitive, and can detect malfunctioning conditions in a moment. Further, items regarded to be malfunctions in this diagnosis are not stored in E.C.U. memory.

SELF-DIAGNOSTIC PROCEDURE

DIAGNOSIS START
Pull out E.C.U. from under the passenger seat.
Start engine.
Turn diagnostic mode selector on E.C.U. fully clockwise.
Flashing 5 times
After the inspection lamps have flashed 5 times, turn diagnostic mode selector fully counterclockwise.
Mode V
Make sure that inspection lamps are not flashing for 5 min. when idling or racing.
O.K.
Turn ignition switch "OFF".
Reinstall the E.C.U. in place.
DIAGNOSIS END
N.G.
If flashing, count no. of flashes.
Turn ignition switch "OFF".
See decoding chart.
Perform real time-diagnosis system inspection. If malfunction part is found, repair or replace it.

CAUTION:
In real time diagnosis, pay attention to inspection lamp flashing. E.C.U. displays the malfunction code only once, and does not memorize the inspection.

Mode V — Real Time Diagnostic System (Cont'd)

DECODING CHART Display presentation	Malfunction circuit or parts	Control unit shows a malfunction signal when the following conditions are detected. (Compare with Self-diagnosis — Mode III.)
CRANK ANGLE SENSOR RED L.E.D. ○ ON ○ OFF	Crank angle sensor circuit is malfunctioning. Crank angle sensor Rotor plate Rotor shaft	The 1° or 180° signal is momentarily missing, or, multiple, momentary noise signals enter. REAL TIME DIAGNOSTIC INSPECTION
AIR FLOW METER GREEN L.E.D. ○ ON ○ OFF	Air flow meter circuit is malfunctioning.	Abnormal, momentary increase in air flow meter output signal. REAL TIME DIAGNOSTIC INSPECTION
IGNITION SIGNAL GREEN L.E.D. ○ ON ○ OFF	Ignition signal is malfunctioning.	Signal from the primary ignition coil momentarily drops off. REAL TIME DIAGNOSTIC INSPECTION

ECCS SELF-DIAGNOSTIC MODE V—1989 SENTRA

Mode V — Real Time Diagnostic System (Cont'd)

REAL TIME DIAGNOSTIC INSPECTION

Crank Angle Sensor

X: Available
–: Not available

Check sequence	Check items	Check conditions	Check parts: Middle connector	Check parts: Sensor & actuator	Check parts: E.C.U. connectors	If malfunction, perform the following items.
1	Tap harness connector or component during real time diagnosis.	During real time diagnosis	X	X	X	Go to check item 2.
2	Check harness continuity at connector.	Engine stopped	X	–	–	Go to check item 3.
3	Disconnect harness connector, and then check dust adhesion to harness connector.	Engine stopped	X	–	X	Clean terminal surface.
4	Check pin terminal bend.	Engine stopped	–	–	X	Take out bend.
5	Reconnect harness connector and then recheck harness continuity at connector.	Engine stopped	X	–	–	Replace terminal.
6	Tap harness connector or component during real time diagnosis.	During real time diagnosis	X	X	X	If malfunction codes are displayed during real time diagnosis, replace terminal.

Crank angle sensor connector
Crank angle sensor

Mode V — Real Time Diagnostic System (Cont'd)

Air Flow Meter

X: Available
–: Not available

Check sequence	Check items	Check conditions	Check parts: Middle connector	Check parts: Sensor & actuator	Check parts: E.C.U. connectors	If malfunction, perform the following items.
1	Tap harness connector or component during real time diagnosis.	During real time diagnosis	X	X	X	Go to check item 2.
2	Check harness continuity at connector.	Engine stopped	X	–	–	Go to check item 3.
3	Disconnect harness connector, and then check dust adhesion to harness connector.	Engine stopped	X	–	X	Clean terminal surface.
4	Check pin terminal bend.	Engine stopped	–	–	X	Take out bend.
5	Reconnect harness connector and then recheck harness continuity at connector.	Engine stopped	X	–	–	Replace terminal.
6	Tap harness connector or component during real time diagnosis.	During real time diagnosis	X	X	X	If malfunction codes are displayed during real time diagnosis, replace terminal.

Air flow meter
Air flow meter connector

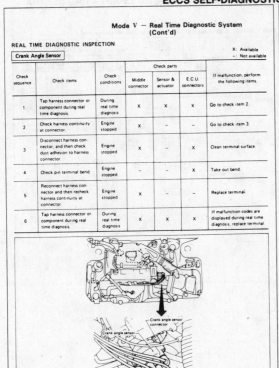

ECCS SELF-DIAGNOSTIC MODE V—1989 SENTRA

Mode V — Real Time Diagnostic System (Cont'd)

Ignition Signal

X: Available
–: Not available

Check sequence	Check items	Check conditions	Check parts			If malfunction, perform the following items.
			Middle connector	Sensor & actuator	E.C.U. connectors	
1	Tap harness connector or component during real time diagnosis.	During real time diagnosis	X	X	X	Go to check item 2.
2	Check harness continuity at connector.	Engine stopped	X	–	–	Go to check item 3.
3	Disconnect harness connector, and then check dust adhesion to harness connector.	Engine stopped	X	–	X	Clean terminal surface.
4	Check pin terminal bend.	Engine stopped	–	–	X	Take out bend.
5	Reconnect harness connector and then recheck harness continuity at connector.	Engine stopped	X	–	–	Replace terminal.
6	Tap harness connector or component during real time diagnosis.	During real time diagnosis	X	X	X	If malfunction codes are displayed during real time diagnosis, replace terminal.

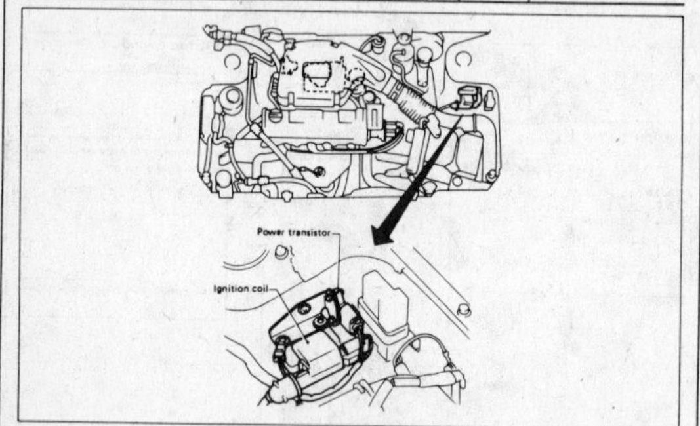
Power transistor
Ignition coil

ECCS SYSTEM INPSECTION CAUTION—1989 SENTRA

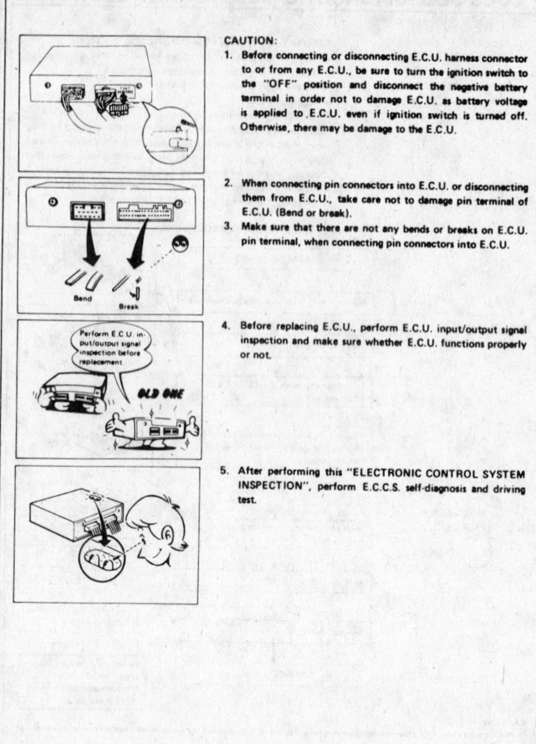
Bend
Break
Perform E.C.U. input/output signal inspection before replacement
OLD ONE

CAUTION:

1. Before connecting or disconnecting E.C.U. harness connector to or from any E.C.U., be sure to turn the ignition switch to the "OFF" position and disconnect the negative battery terminal in order not to damage E.C.U. as battery voltage is applied to E.C.U. even if ignition switch is turned off. Otherwise, there may be damage to the E.C.U.
2. When connecting pin connectors into E.C.U. or disconnecting them from E.C.U., take care not to damage pin terminal of E.C.U. (Bend or break).
3. Make sure that there are not any bends or breaks on E.C.U. pin terminal, when connecting pin connectors into E.C.U.
4. Before replacing E.C.U., perform E.C.U. input/output signal inspection and make sure whether E.C.U. functions properly or not.
5. After performing this "ELECTRONIC CONTROL SYSTEM INSPECTION", perform E.C.C.S. self-diagnosis and driving test.

ECCS SYSTEM INPSECTION CAUTION—1989 SENTRA

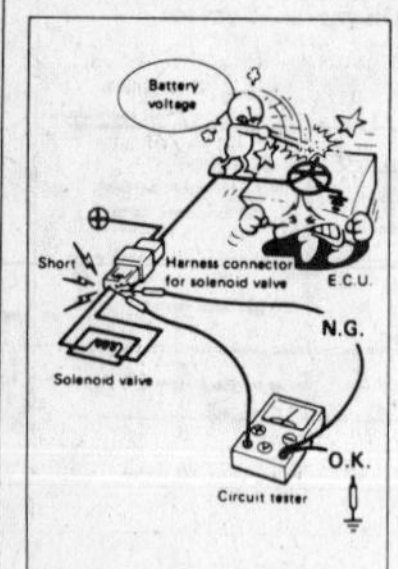
Battery voltage
Short
Harness connector for solenoid valve
E.C.U.
N.G.
Solenoid valve
O.K.
Circuit tester

6. When measuring supply voltage of E.C.U. controlled components with a circuit tester, separate one tester probe from the other.
   If the two tester probes accidentally make contact with each other during measurement, the circuit will be shorted, resulting in damage to the power transistor of the control unit.
7. When measuring voltage or resistance at connector with tester probes, there are two methods of measurement; one is done from terminal side and the other from harness side. Before measuring, confirm symbol mark again.
   - HS : Inspection should be done from harness side.
   - TS : Inspection should be done from terminal side.

   Refer to GI section.
8. As for continuity check of joint connector, refer to EL section.
9. Key to symbols
   - DISCONNECT : Check after disconnecting the connector to be measured.
   - CONNECT : Check after connecting the connector to be measured.
10. Improve tester probe as shown to perform test easily.
11. For the first trouble-shooting procedure, perform POWER SOURCE & GROUND CIRCUIT FOR E.C.U. check.

Thin wire
Tester probe

ECCS SELF—DIAGNOSTIC CHART—1989 SENTRA

POWER SOURCE & GROUND CIRCUIT FOR E.C.U. (Not self-diagnosis item)

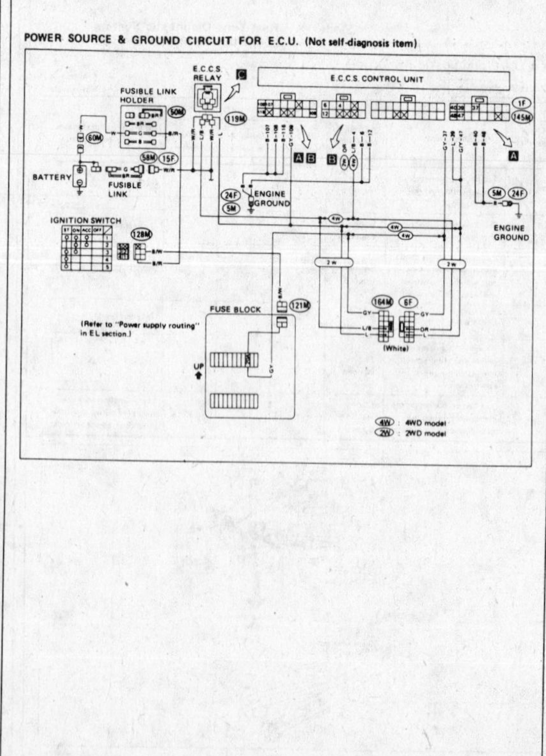
E.C.C.S. RELAY
E.C.C.S. CONTROL UNIT
FUSIBLE LINK HOLDER
BATTERY
FUSIBLE LINK
ENGINE GROUND
IGNITION SWITCH
FUSE BLOCK
(Refer to "Power supply routing" in EL section.)
UP
(White)
ENGINE GROUND
4WD model
2WD model

ECCS SELF-DIAGNOSTIC CHART—1989 SENTRA

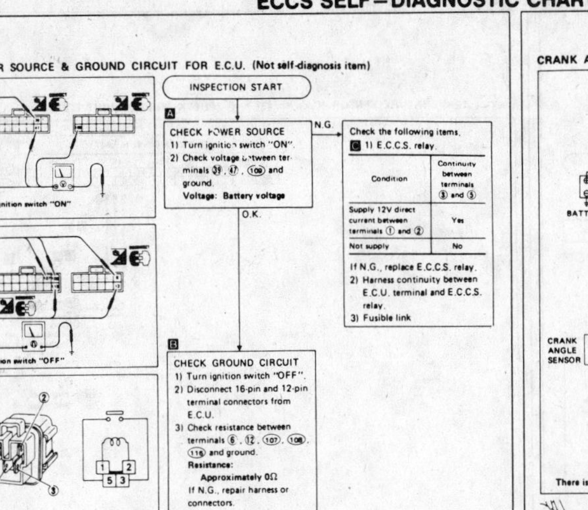
POWER SOURCE & GROUND CIRCUIT FOR E.C.U. (Not self-diagnosis item)
INSPECTION START
CHECK POWER SOURCE
1) Turn ignition switch "ON".
2) Check voltage between terminals ㊴, ㊼, ⑩⑨ and ground.
Voltage: Battery voltage
N.G.
Check the following items.
1) E.C.C.S. relay.
Condition
Continuity between terminals ③ and ⑤
Supply 12V direct current between terminals ① and ②
Yes
Not supply
No
If N.G., replace E.C.C.S. relay.
2) Harness continuity between E.C.U. terminal and E.C.C.S. relay.
3) Fusible link
O.K.
CHECK GROUND CIRCUIT
1) Turn ignition switch "OFF".
2) Disconnect 16-pin and 12-pin terminal connectors from E.C.U.
3) Check resistance between terminals ⑥, ⑫, ⑩⑦, ⑩⑧, ⑪⑥ and ground.
Resistance: Approximately 0Ω
If N.G., repair harness or connectors.
O.K.
INSPECTION END
Ignition switch "ON"
Ignition switch "OFF"

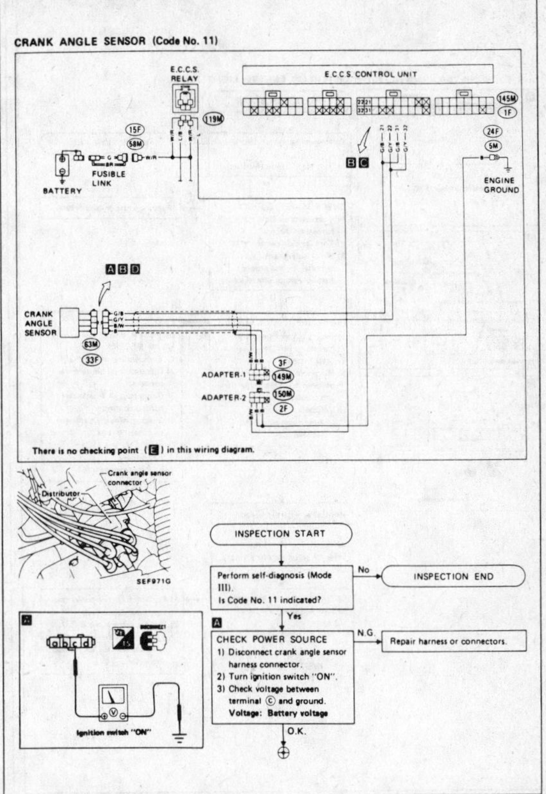
CRANK ANGLE SENSOR (Code No. 11)
E.C.C.S. RELAY
E.C.C.S. CONTROL UNIT
BATTERY
FUSIBLE LINK
ENGINE GROUND
CRANK ANGLE SENSOR
ADAPTER-1
ADAPTER-2
There is no checking point in this wiring diagram.
Crank angle sensor connector
Distributor
SEF971G
INSPECTION START
Perform self-diagnosis (Mode III).
Is Code No. 11 indicated?
No
INSPECTION END
Yes
CHECK POWER SOURCE
1) Disconnect crank angle sensor harness connector.
2) Turn ignition switch "ON".
3) Check voltage between terminal ⓒ and ground.
Voltage: Battery voltage
N.G.
Repair harness or connectors.
O.K.
Ignition switch "ON"

ECCS SELF-DIAGNOSTIC CHART—1989 SENTRA

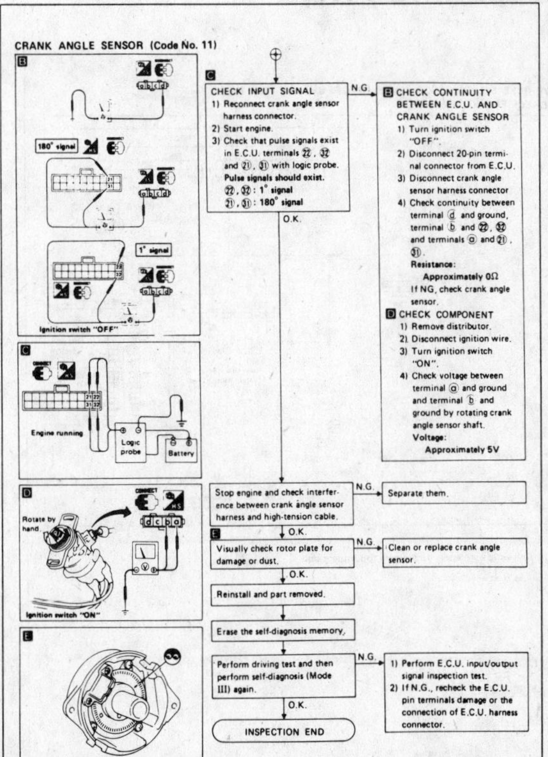
CRANK ANGLE SENSOR (Code No. 11)
CHECK INPUT SIGNAL
1) Reconnect crank angle sensor harness connector.
2) Start engine.
3) Check that pulse signals exist in E.C.U. terminals ㉒, ㉜ and ㉑, ㉛ with logic probe.
Pulse signals should exist.
㉒, ㉜ : 1° signal
㉑, ㉛ : 180° signal
N.G.
CHECK CONTINUITY BETWEEN E.C.U. AND CRANK ANGLE SENSOR
1) Turn ignition switch "OFF".
2) Disconnect 20-pin terminal connector from E.C.U.
3) Disconnect crank angle sensor harness connector
4) Check continuity between terminal ⓓ and ground, terminal ⓑ and ㉒, ㉜ and terminals ⓐ and ㉑, ㉛.
Resistance: Approximately 0Ω
If N.G., check crank angle sensor.
CHECK COMPONENT
1) Remove distributor.
2) Disconnect ignition wire.
3) Turn ignition switch "ON".
4) Check voltage between terminal ⓐ and ground and terminal ⓑ and ground by rotating crank angle sensor shaft.
Voltage: Approximately 5V
O.K.
Stop engine and check interference between crank angle sensor harness and high-tension cable.
N.G.
Separate them.
O.K.
Visually check rotor plate for damage or dust.
N.G.
Clean or replace crank angle sensor.
O.K.
Reinstall and part removed.
Erase the self-diagnosis memory.
Perform driving test and then perform self-diagnosis (Mode III) again.
N.G.
1) Perform E.C.U. input/output signal inspection test.
2) If N.G., recheck the E.C.U. pin terminals damage or the connection of E.C.U. harness connector.
O.K.
INSPECTION END
180° signal
1° signal
Ignition switch "OFF"
Engine running
Logic probe
Battery
Rotate by hand
Ignition switch "ON"

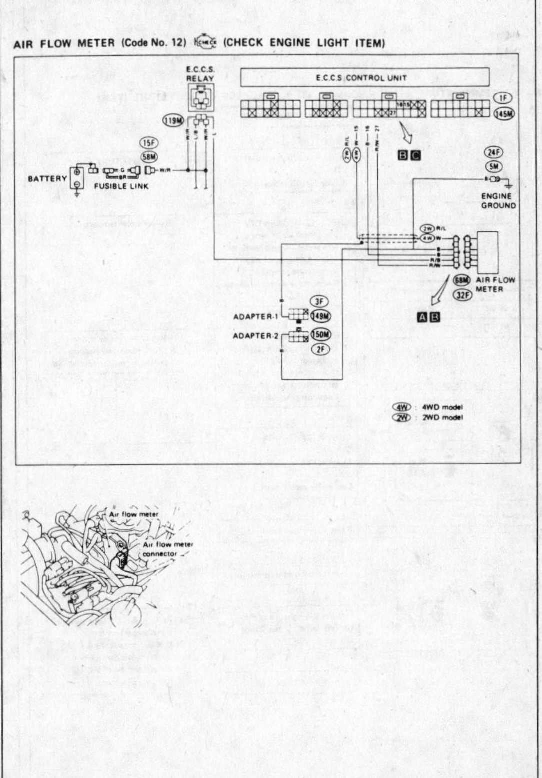
AIR FLOW METER (Code No. 12) (CHECK ENGINE LIGHT ITEM)
E.C.C.S. RELAY
E.C.C.S. CONTROL UNIT
BATTERY
FUSIBLE LINK
ENGINE GROUND
AIR FLOW METER
ADAPTER-1
ADAPTER-2
4W : 4WD model
2W : 2WD model
Air flow meter
Air flow meter connector

ECCS SELF—DIAGNOSTIC CHART—1989 SENTRA

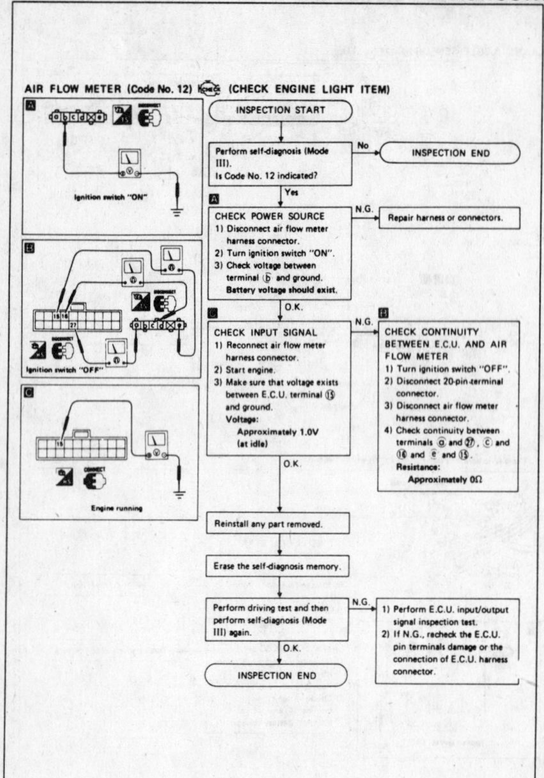
AIR FLOW METER (Code No. 12) (CHECK ENGINE LIGHT ITEM)
INSPECTION START
Perform self-diagnosis (Mode III). Is Code No. 12 indicated?
No
INSPECTION END
Yes
CHECK POWER SOURCE
1) Disconnect air flow meter harness connector.
2) Turn ignition switch "ON".
3) Check voltage between terminal ⓑ and ground. Battery voltage should exist.
N.G.
Repair harness or connectors.
O.K.
CHECK INPUT SIGNAL
1) Reconnect air flow meter harness connector.
2) Start engine.
3) Make sure that voltage exists between E.C.U. terminal ⑮ and ground.
Voltage: Approximately 1.0V (at idle)
N.G.
CHECK CONTINUITY BETWEEN E.C.U. AND AIR FLOW METER
1) Turn ignition switch "OFF".
2) Disconnect 20-pin terminal connector.
3) Disconnect air flow meter harness connector.
4) Check continuity between terminals ⓐ and ㉗, ⓒ and ⑯ and ⓔ and ⑮.
Resistance: Approximately 0Ω
O.K.
Reinstall any part removed.
Erase the self-diagnosis memory.
Perform driving test and then perform self-diagnosis (Mode III) again.
N.G.
1) Perform E.C.U. input/output signal inspection test.
2) If N.G., recheck the E.C.U. pin terminals damage or the connection of E.C.U. harness connector.
O.K.
INSPECTION END
Ignition switch "ON"
Ignition switch "OFF"
Engine running

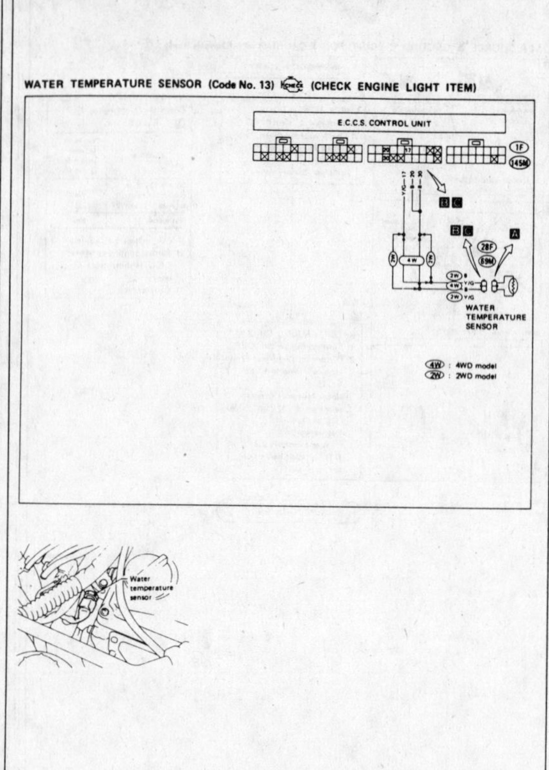
WATER TEMPERATURE SENSOR (Code No. 13) (CHECK ENGINE LIGHT ITEM)
E.C.C.S. CONTROL UNIT
1F
145M
28F
59M
WATER TEMPERATURE SENSOR
4W : 4WD model
2W : 2WD model
Water temperature sensor

ECCS SELF—DIAGNOSTIC CHART—1989 SENTRA

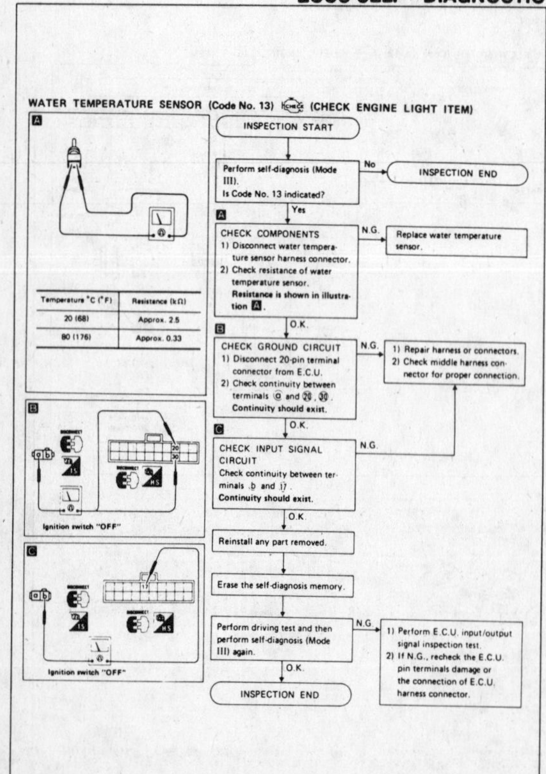
WATER TEMPERATURE SENSOR (Code No. 13) (CHECK ENGINE LIGHT ITEM)
INSPECTION START
Perform self-diagnosis (Mode III). Is Code No. 13 indicated?
No
INSPECTION END
Yes
CHECK COMPONENTS
1) Disconnect water temperature sensor harness connector.
2) Check resistance of water temperature sensor.
Resistance is shown in illustration A.
N.G.
Replace water temperature sensor.
Temperature °C (°F) | Resistance (kΩ)
20 (68) | Approx. 2.5
80 (176) | Approx. 0.33
O.K.
CHECK GROUND CIRCUIT
1) Disconnect 20-pin terminal connector from E.C.U.
2) Check continuity between terminals ⓐ and ⑳, ㉚.
Continuity should exist.
N.G.
1) Repair harness or connectors.
2) Check middle harness connector for proper connection.
O.K.
CHECK INPUT SIGNAL CIRCUIT
Check continuity between terminals ⓑ and ⑰.
Continuity should exist.
N.G.
O.K.
Reinstall any part removed.
Erase the self-diagnosis memory.
Perform driving test and then perform self-diagnosis (Mode III) again.
N.G.
1) Perform E.C.U. input/output signal inspection test.
2) If N.G., recheck the E.C.U. pin terminals damage or the connection of E.C.U. harness connector.
O.K.
INSPECTION END
Ignition switch "OFF"
Ignition switch "OFF"

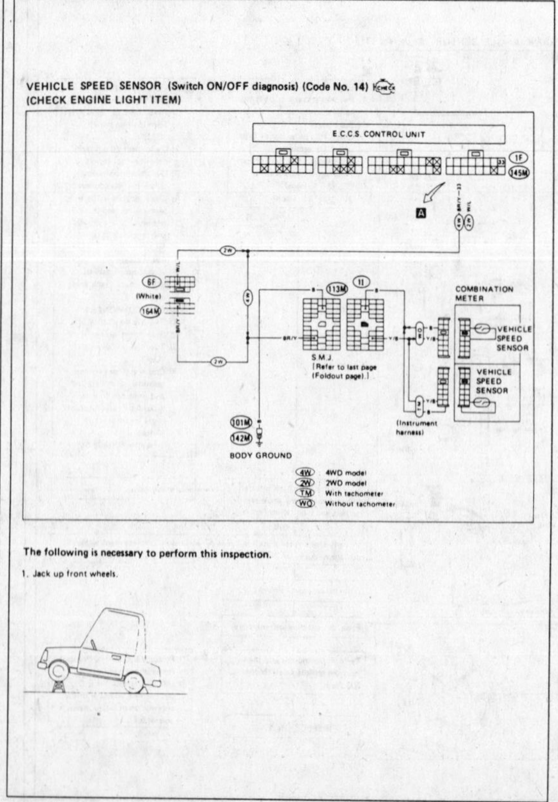
VEHICLE SPEED SENSOR (Switch ON/OFF diagnosis) (Code No. 14) (CHECK ENGINE LIGHT ITEM)
E.C.C.S. CONTROL UNIT
1F
145M
5F (White)
164M
113M
11
COMBINATION METER
VEHICLE SPEED SENSOR
VEHICLE SPEED SENSOR
S.M.J. [Refer to last page (Foldout page).]
(Instrument harness)
101M
142M
BODY GROUND
4W 4WD model
2W 2WD model
TM With tachometer
WO Without tachometer
The following is necessary to perform this inspection.
1. Jack up front wheels.

ECCS SELF—DIAGNOSTIC CHART—1989 SENTRA

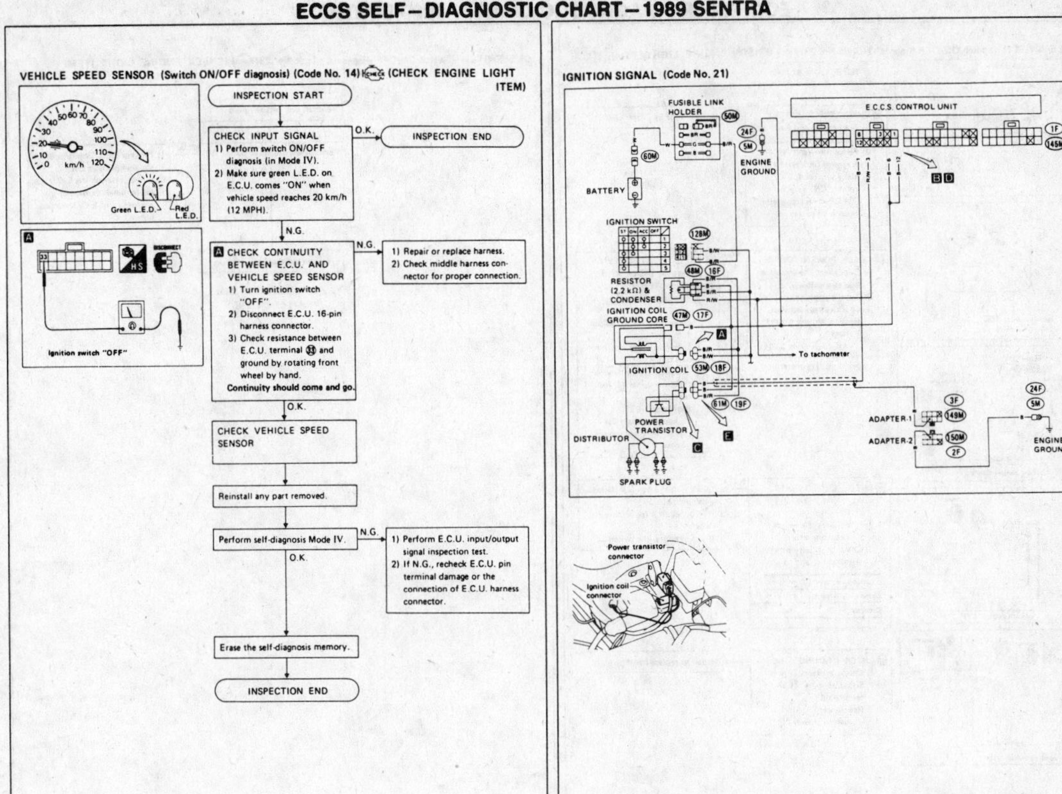
VEHICLE SPEED SENSOR (Switch ON/OFF diagnosis) (Code No. 14) (CHECK ENGINE LIGHT ITEM)
INSPECTION START
CHECK INPUT SIGNAL
1) Perform switch ON/OFF diagnosis (in Mode IV).
2) Make sure green L.E.D. on E.C.U. comes "ON" when vehicle speed reaches 20 km/h (12 MPH).
O.K.
INSPECTION END
N.G.
Green L.E.D.
Red L.E.D.
km/h
CHECK CONTINUITY BETWEEN E.C.U. AND VEHICLE SPEED SENSOR
1) Turn ignition switch "OFF".
2) Disconnect E.C.U. 16-pin harness connector.
3) Check resistance between E.C.U. terminal ㉝ and ground by rotating front wheel by hand.
Continuity should come and go.
1) Repair or replace harness.
2) Check middle harness connector for proper connection.
Ignition switch "OFF"
DISCONNECT
CHECK VEHICLE SPEED SENSOR
Reinstall any part removed.
Perform self-diagnosis Mode IV.
1) Perform E.C.U. input/output signal inspection test.
2) If N.G., recheck E.C.U. pin terminal damage or the connection of E.C.U. harness connector.
Erase the self-diagnosis memory.
INSPECTION END
IGNITION SIGNAL (Code No. 21)
E.C.C.S. CONTROL UNIT
FUSIBLE LINK HOLDER
BATTERY
ENGINE GROUND
IGNITION SWITCH
RESISTOR (2.2 kΩ) & CONDENSER
IGNITION COIL GROUND CORE
IGNITION COIL
POWER TRANSISTOR
DISTRIBUTOR
SPARK PLUG
To tachometer
ADAPTER-1
ADAPTER-2
ENGINE GROUND
Power transistor connector
Ignition coil connector

ECCS SELF—DIAGNOSTIC CHART—1989 SENTRA

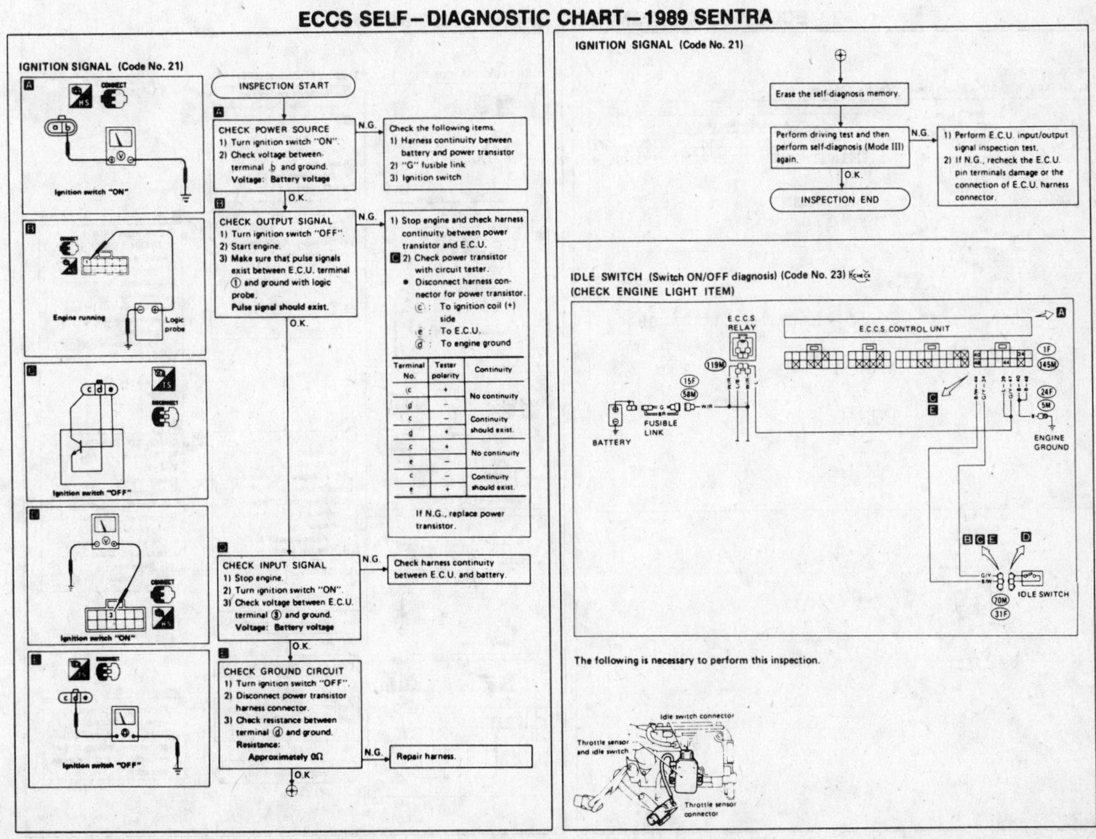
IGNITION SIGNAL (Code No. 21)
INSPECTION START
CHECK POWER SOURCE
1) Turn ignition switch "ON".
2) Check voltage between terminal ⓑ and ground.
Voltage: Battery voltage
Check the following items.
1) Harness continuity between battery and power transistor
2) "G" fusible link
3) Ignition switch
Ignition switch "ON"
CHECK OUTPUT SIGNAL
1) Turn ignition switch "OFF".
2) Start engine.
3) Make sure that pulse signals exist between E.C.U. terminal ① and ground with logic probe.
Pulse signal should exist.
1) Stop engine and check harness continuity between power transistor and E.C.U.
2) Check power transistor with circuit tester.
• Disconnect harness connector for power transistor.
ⓒ : To ignition coil (+) side
ⓔ : To E.C.U.
ⓓ : To engine ground
Engine running
Logic probe
Terminal No. | Tester polarity | Continuity
c | + | No continuity
d | − |
c | − | Continuity should exist.
d | + |
c | + | No continuity
e | − |
c | − | Continuity should exist.
e | + |
If N.G., replace power transistor.
Ignition switch "OFF"
CHECK INPUT SIGNAL
1) Stop engine.
2) Turn ignition switch "ON".
3) Check voltage between E.C.U. terminal ③ and ground.
Voltage: Battery voltage
Check harness continuity between E.C.U. and battery.
Ignition switch "ON"
CHECK GROUND CIRCUIT
1) Turn ignition switch "OFF".
2) Disconnect power transistor harness connector.
3) Check resistance between terminal ⓓ and ground.
Resistance: Approximately 0Ω
Repair harness.
Ignition switch "OFF"
IGNITION SIGNAL (Code No. 21)
Erase the self-diagnosis memory.
Perform driving test and then perform self-diagnosis (Mode III) again.
1) Perform E.C.U. input/output signal inspection test.
2) If N.G., recheck the E.C.U. pin terminals damage or the connection of E.C.U. harness connector.
INSPECTION END
IDLE SWITCH (Switch ON/OFF diagnosis) (Code No. 23) (CHECK ENGINE LIGHT ITEM)
E.C.C.S. RELAY
E.C.C.S. CONTROL UNIT
BATTERY
FUSIBLE LINK
ENGINE GROUND
IDLE SWITCH
The following is necessary to perform this inspection.
Idle switch connector
Throttle sensor and idle switch
Throttle sensor connector

ECCS SELF—DIAGNOSTIC CHART—1989 SENTRA

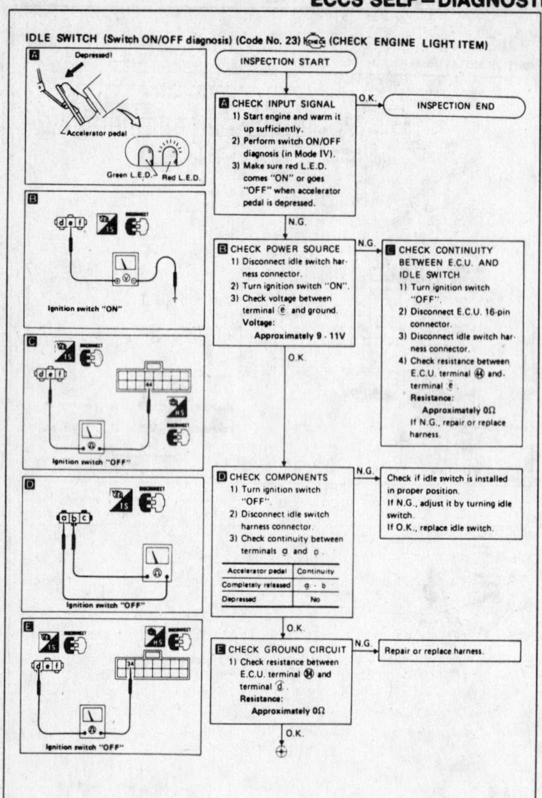
IDLE SWITCH (Switch ON/OFF diagnosis) (Code No. 23) (CHECK ENGINE LIGHT ITEM)
Depressed!
Accelerator pedal
Green L.E.D.
Red L.E.D.
Ignition switch "ON"
Ignition switch "OFF"
INSPECTION START
A CHECK INPUT SIGNAL
1) Start engine and warm it up sufficiently.
2) Perform switch ON/OFF diagnosis (in Mode IV).
3) Make sure red L.E.D. comes "ON" or goes "OFF" when accelerator pedal is depressed.
O.K.
INSPECTION END
N.G.
B CHECK POWER SOURCE
1) Disconnect idle switch harness connector.
2) Turn ignition switch "ON".
3) Check voltage between terminal e and ground.
Voltage:
Approximately 9 - 11V
C CHECK CONTINUITY BETWEEN E.C.U. AND IDLE SWITCH
1) Turn ignition switch "OFF".
2) Disconnect E.C.U. 16-pin connector.
3) Disconnect idle switch harness connector.
4) Check resistance between E.C.U. terminal 44 and terminal e.
Resistance:
Approximately 0Ω
If N.G., repair or replace harness.
D CHECK COMPONENTS
1) Turn ignition switch "OFF".
2) Disconnect idle switch harness connector.
3) Check continuity between terminals a and b.
Accelerator pedal | Continuity
Completely released | a - b
Depressed | No
Check if idle switch is installed in proper position.
If N.G., adjust it by turning idle switch.
If O.K., replace idle switch.
E CHECK GROUND CIRCUIT
1) Check resistance between E.C.U. terminal 34 and terminal d.
Resistance:
Approximately 0Ω
Repair or replace harness.

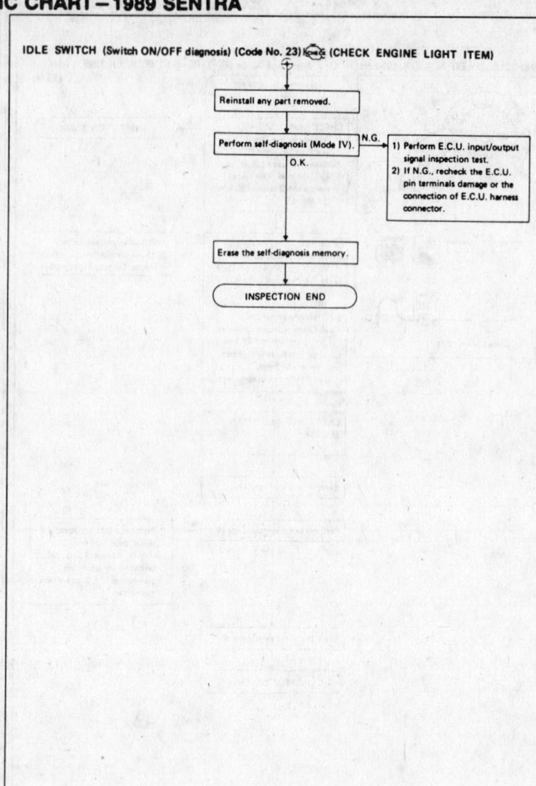
IDLE SWITCH (Switch ON/OFF diagnosis) (Code No. 23) (CHECK ENGINE LIGHT ITEM)
Reinstall any part removed.
Perform self-diagnosis (Mode IV).
N.G.
1) Perform E.C.U. input/output signal inspection test.
2) If N.G., recheck the E.C.U. pin terminals damage or the connection of E.C.U. harness connector.
O.K.
Erase the self-diagnosis memory.
INSPECTION END

ECCS SELF—DIAGNOSTIC CHART—1989 SENTRA

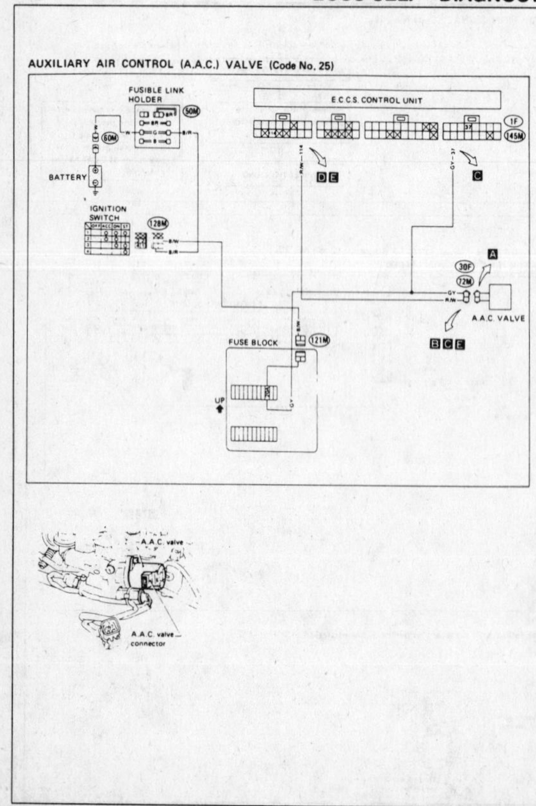
AUXILIARY AIR CONTROL (A.A.C.) VALVE (Code No. 25)
FUSIBLE LINK HOLDER
BATTERY
IGNITION SWITCH
E.C.C.S. CONTROL UNIT
A.A.C. VALVE
FUSE BLOCK
UP
A.A.C. valve
A.A.C. valve connector

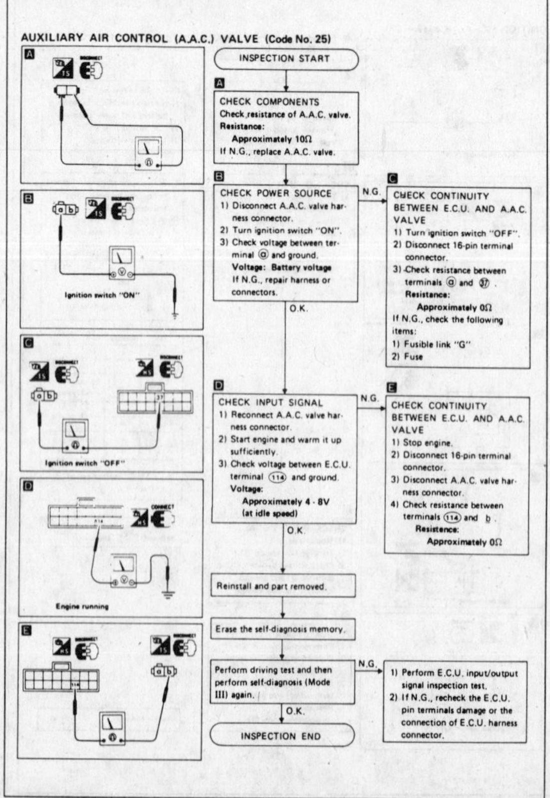
AUXILIARY AIR CONTROL (A.A.C.) VALVE (Code No. 25)
INSPECTION START
A CHECK COMPONENTS
Check resistance of A.A.C. valve.
Resistance:
Approximately 10Ω
If N.G., replace A.A.C. valve.
B CHECK POWER SOURCE
1) Disconnect A.A.C. valve harness connector.
2) Turn ignition switch "ON".
3) Check voltage between terminal a and ground.
Voltage: Battery voltage
If N.G., repair harness or connectors.
C CHECK CONTINUITY BETWEEN E.C.U. AND A.A.C. VALVE
1) Turn ignition switch "OFF".
2) Disconnect 16-pin terminal connector.
3) Check resistance between terminals a and 37.
Resistance:
Approximately 0Ω
If N.G., check the following items:
1) Fusible link "G"
2) Fuse
D CHECK INPUT SIGNAL
1) Reconnect A.A.C. valve harness connector.
2) Start engine and warm it up sufficiently.
3) Check voltage between E.C.U. terminal 114 and ground.
Voltage:
Approximately 4 - 8V (at idle speed)
E CHECK CONTINUITY BETWEEN E.C.U. AND A.A.C. VALVE
1) Stop engine.
2) Disconnect 16-pin terminal connector.
3) Disconnect A.A.C. valve harness connector.
4) Check resistance between terminals 114 and b.
Resistance:
Approximately 0Ω
Reinstall and part removed.
Erase the self-diagnosis memory.
Perform driving test and then perform self-diagnosis (Mode III) again.
1) Perform E.C.U. input/output signal inspection test.
2) If N.G., recheck the E.C.U. pin terminals damage or the connection of E.C.U. harness connector.
INSPECTION END
Ignition switch "ON"
Ignition switch "OFF"
Engine running

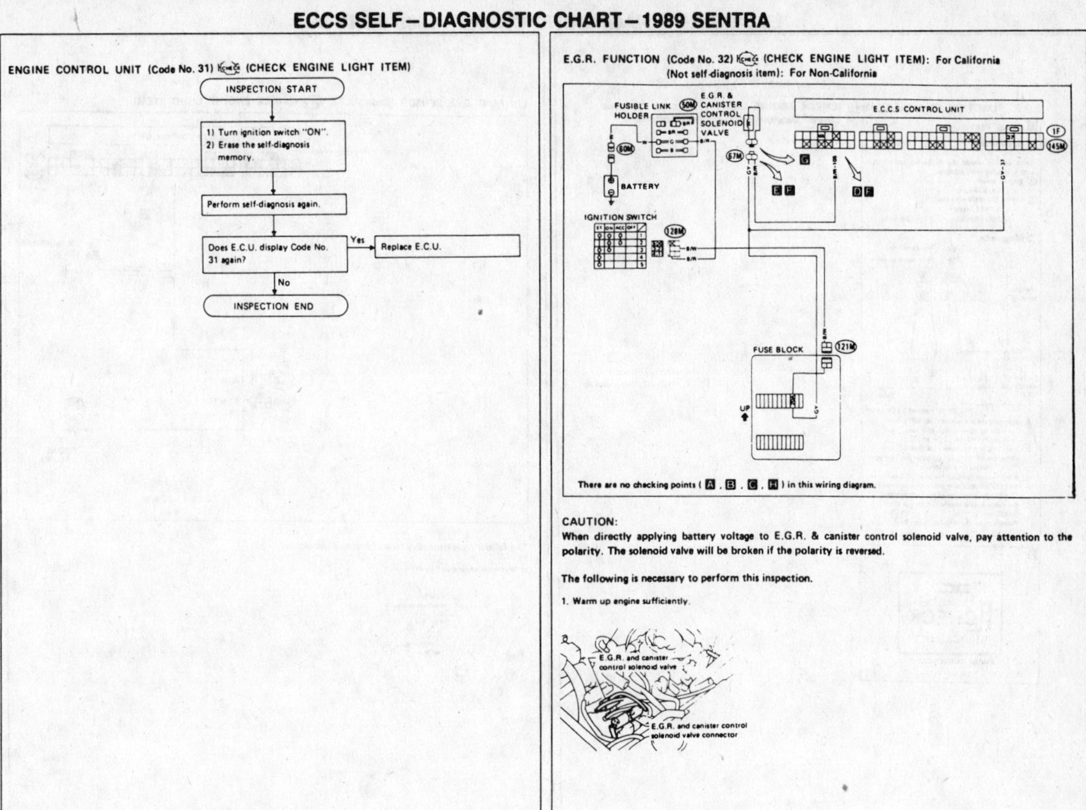

ECCS SELF—DIAGNOSTIC CHART—1989 SENTRA

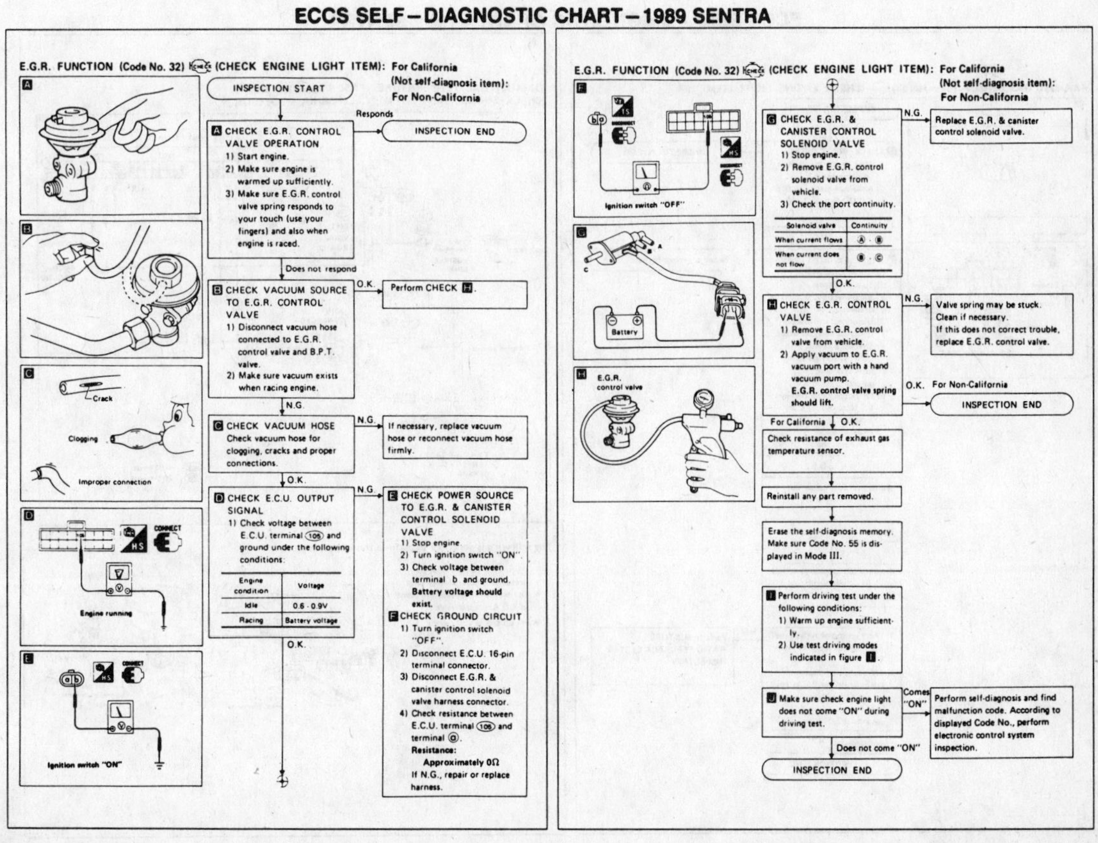

ECCS SELF—DIAGNOSTIC CHART—1989 SENTRA

ECCS SELF—DIAGNOSTIC CHART—1989 SENTRA

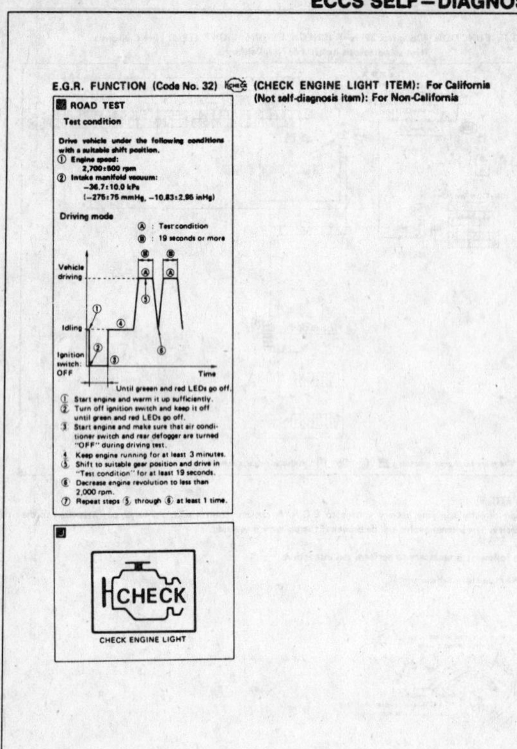
E.G.R. FUNCTION (Code No. 32) (CHECK ENGINE LIGHT ITEM): For California
(Not self-diagnosis item): For Non-California
ROAD TEST
Test condition
Drive vehicle under the following conditions with a suitable shift position.
① Engine speed: 2,700±500 rpm
② Intake manifold vacuum: −36.7±10.0 kPa (−275±75 mmHg, −10.83±2.95 inHg)
Driving mode
Ⓐ : Test condition
Ⓑ : 19 seconds or more
Vehicle driving
Idling
Ignition switch: OFF
Time
Until green and red LEDs go off.
① Start engine and warm it up sufficiently.
② Turn off ignition switch and keep it off until green and red LEDs go off.
③ Start engine and make sure that air conditioner switch and rear defogger are turned "OFF" during driving test.
④ Keep engine running for at least 3 minutes.
⑤ Shift to suitable gear position and drive in "Test condition" for at least 19 seconds.
⑥ Decrease engine revolution to less than 2,000 rpm.
⑦ Repeat steps ⑤ through ⑥ at least 1 time.
CHECK
CHECK ENGINE LIGHT

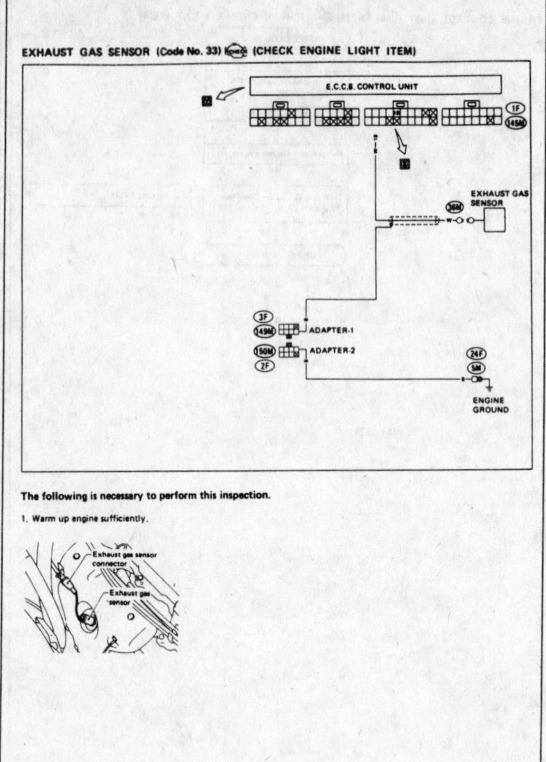
EXHAUST GAS SENSOR (Code No. 33) (CHECK ENGINE LIGHT ITEM)
E.C.C.S. CONTROL UNIT
EXHAUST GAS SENSOR
ADAPTER-1
ADAPTER-2
ENGINE GROUND
The following is necessary to perform this inspection.
1. Warm up engine sufficiently.
Exhaust gas sensor connector
Exhaust gas sensor

ECCS SELF—DIAGNOSTIC CHART—1989 SENTRA

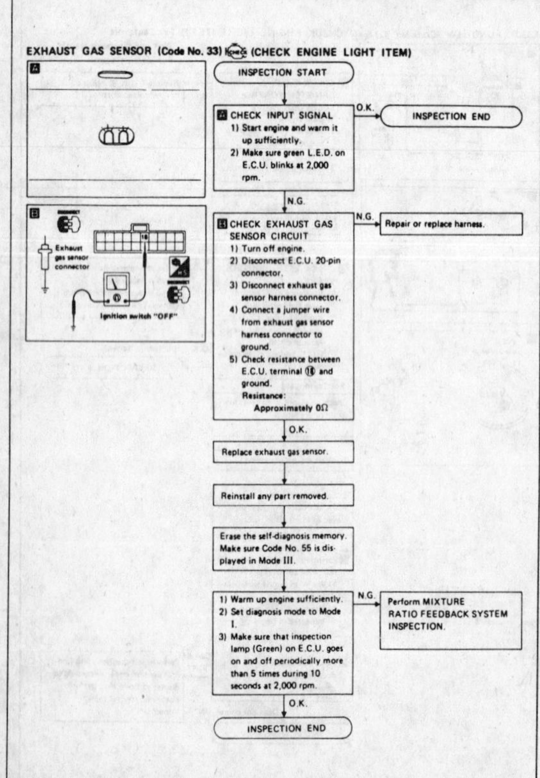
EXHAUST GAS SENSOR (Code No. 33) (CHECK ENGINE LIGHT ITEM)
Exhaust gas sensor connector
Ignition switch "OFF"
INSPECTION START
A CHECK INPUT SIGNAL
1) Start engine and warm it up sufficiently.
2) Make sure green L.E.D. on E.C.U. blinks at 2,000 rpm.
O.K.
INSPECTION END
N.G.
B CHECK EXHAUST GAS SENSOR CIRCUIT
1) Turn off engine.
2) Disconnect E.C.U. 20-pin connector.
3) Disconnect exhaust gas sensor harness connector.
4) Connect a jumper wire from exhaust gas sensor harness connector to ground.
5) Check resistance between E.C.U. terminal ⑱ and ground.
Resistance: Approximately 0Ω
N.G.
Repair or replace harness.
O.K.
Replace exhaust gas sensor.
Reinstall any part removed.
Erase the self-diagnosis memory. Make sure Code No. 55 is displayed in Mode III.
1) Warm up engine sufficiently.
2) Set diagnosis mode to Mode I.
3) Make sure that inspection lamp (Green) on E.C.U. goes on and off periodically more than 5 times during 10 seconds at 2,000 rpm.
N.G.
Perform MIXTURE RATIO FEEDBACK SYSTEM INSPECTION.
O.K.
INSPECTION END

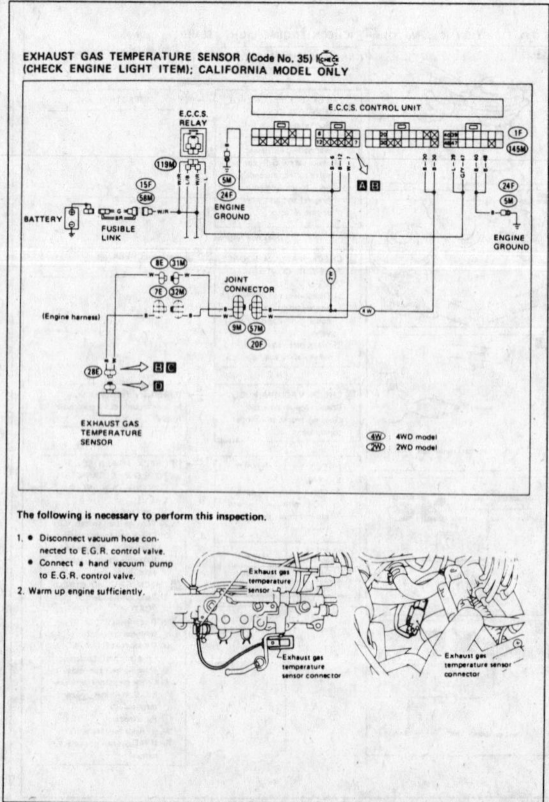
EXHAUST GAS TEMPERATURE SENSOR (Code No. 35) (CHECK ENGINE LIGHT ITEM); CALIFORNIA MODEL ONLY
E.C.C.S. CONTROL UNIT
E.C.C.S. RELAY
BATTERY
FUSIBLE LINK
ENGINE GROUND
JOINT CONNECTOR
(Engine harness)
EXHAUST GAS TEMPERATURE SENSOR
4WD model
2WD model
The following is necessary to perform this inspection.
1. • Disconnect vacuum hose connected to E.G.R. control valve.
• Connect a hand vacuum pump to E.G.R. control valve.
2. Warm up engine sufficiently.
Exhaust gas temperature sensor
Exhaust gas temperature sensor connector

ECCS SELF—DIAGNOSTIC CHART—1989 SENTRA

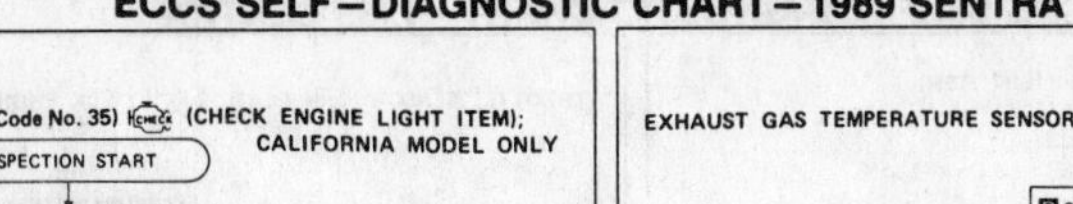

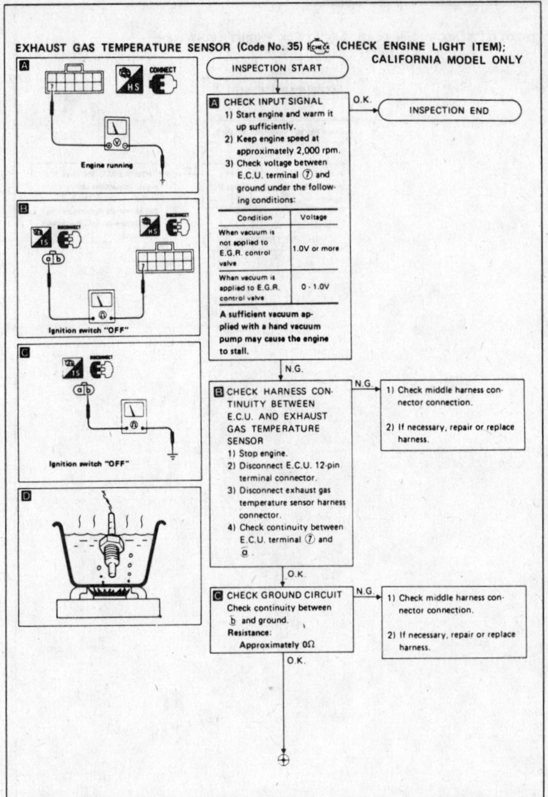
EXHAUST GAS TEMPERATURE SENSOR (Code No. 35) (CHECK ENGINE LIGHT ITEM); CALIFORNIA MODEL ONLY
Engine running
Ignition switch "OFF"
Ignition switch "OFF"
INSPECTION START
A CHECK INPUT SIGNAL
1) Start engine and warm it up sufficiently.
2) Keep engine speed at approximately 2,000 rpm.
3) Check voltage between E.C.U. terminal ⑦ and ground under the following conditions:
Condition | Voltage
When vacuum is not applied to E.G.R. control valve | 1.0V or more
When vacuum is applied to E.G.R. control valve | 0 - 1.0V
A sufficient vacuum applied with a hand vacuum pump may cause the engine to stall.
O.K.
INSPECTION END
N.G.
B CHECK HARNESS CONTINUITY BETWEEN E.C.U. AND EXHAUST GAS TEMPERATURE SENSOR
1) Stop engine.
2) Disconnect E.C.U. 12-pin terminal connector.
3) Disconnect exhaust gas temperature sensor harness connector.
4) Check continuity between E.C.U. terminal ⑦ and a.
N.G.
1) Check middle harness connector connection.
2) If necessary, repair or replace harness.
O.K.
C CHECK GROUND CIRCUIT
Check continuity between b and ground.
Resistance:
Approximately 0Ω
N.G.
1) Check middle harness connector connection.
2) If necessary, repair or replace harness.
O.K.

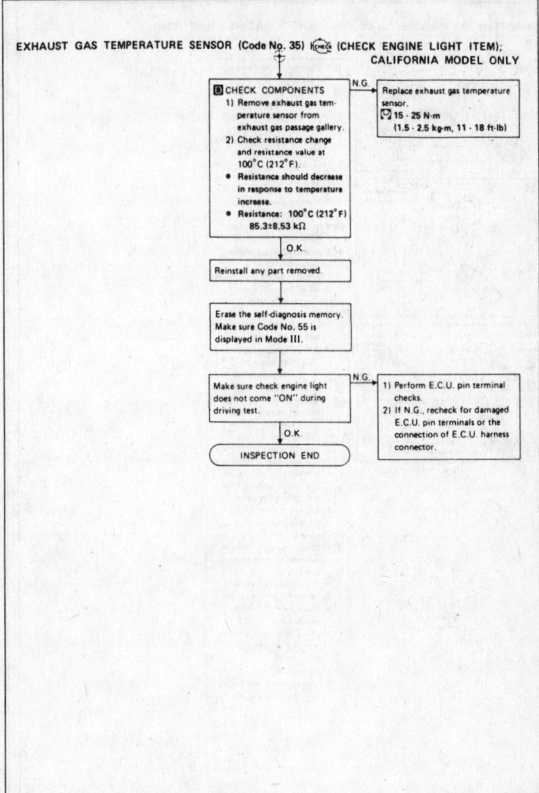
EXHAUST GAS TEMPERATURE SENSOR (Code No. 35) (CHECK ENGINE LIGHT ITEM); CALIFORNIA MODEL ONLY
D CHECK COMPONENTS
1) Remove exhaust gas temperature sensor from exhaust gas passage gallery.
2) Check resistance change and resistance value at 100°C (212°F).
• Resistance should decrease in response to temperature increase.
• Resistance: 100°C (212°F) 85.3±8.53 kΩ
N.G.
Replace exhaust gas temperature sensor.
15 - 25 N·m (1.5 - 2.5 kg-m, 11 - 18 ft-lb)
O.K.
Reinstall any part removed.
Erase the self-diagnosis memory. Make sure Code No. 55 is displayed in Mode III.
Make sure check engine light does not come "ON" during driving test.
N.G.
1) Perform E.C.U. pin terminal checks.
2) If N.G., recheck for damaged E.C.U. pin terminals or the connection of E.C.U. harness connector.
O.K.
INSPECTION END

ECCS SELF—DIAGNOSTIC CHART—1989 SENTRA

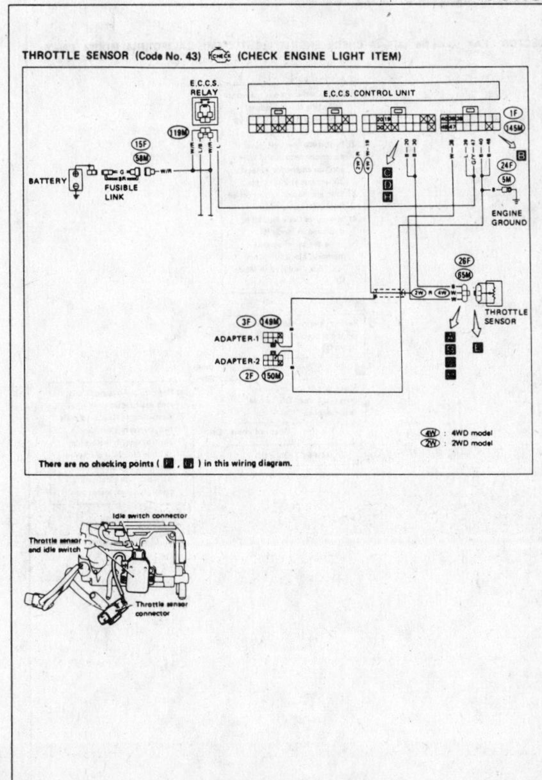
THROTTLE SENSOR (Code No. 43) (CHECK ENGINE LIGHT ITEM)
E.C.C.S. RELAY
E.C.C.S. CONTROL UNIT
BATTERY
FUSIBLE LINK
ENGINE GROUND
THROTTLE SENSOR
ADAPTER-1
ADAPTER-2
4WD model
2WD model
There are no checking points (,) in this wiring diagram.
Idle switch connector
Throttle sensor and idle switch
Throttle sensor connector

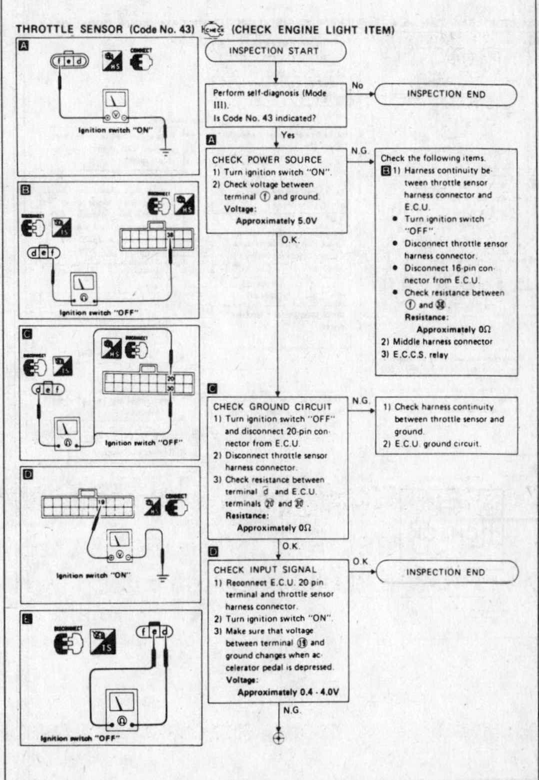
THROTTLE SENSOR (Code No. 43) (CHECK ENGINE LIGHT ITEM)
Ignition switch "ON"
Ignition switch "OFF"
Ignition switch "OFF"
Ignition switch "ON"
Ignition switch "OFF"
INSPECTION START
Perform self-diagnosis (Mode III).
Is Code No. 43 indicated?
No
INSPECTION END
Yes
A CHECK POWER SOURCE
1) Turn ignition switch "ON".
2) Check voltage between terminal ⓕ and ground.
Voltage:
Approximately 5.0V
N.G.
Check the following items.
B 1) Harness continuity between throttle sensor harness connector and E.C.U.
• Turn ignition switch "OFF".
• Disconnect throttle sensor harness connector.
• Disconnect 16-pin connector from E.C.U.
• Check resistance between ⓕ and ㊳.
Resistance:
Approximately 0Ω
2) Middle harness connector
3) E.C.C.S. relay
O.K.
C CHECK GROUND CIRCUIT
1) Turn ignition switch "OFF" and disconnect 20-pin connector from E.C.U.
2) Disconnect throttle sensor harness connector.
3) Check resistance between terminal ⓓ and E.C.U. terminals ⑳ and ㉚.
Resistance:
Approximately 0Ω
N.G.
1) Check harness continuity between throttle sensor and ground.
2) E.C.U. ground circuit.
O.K.
D CHECK INPUT SIGNAL
1) Reconnect E.C.U. 20-pin terminal and throttle sensor harness connector.
2) Turn ignition switch "ON".
3) Make sure that voltage between terminal ⑲ and ground changes when accelerator pedal is depressed.
Voltage:
Approximately 0.4 - 4.0V
O.K.
INSPECTION END
N.G.

ECCS SELF—DIAGNOSTIC CHART—1989 SENTRA

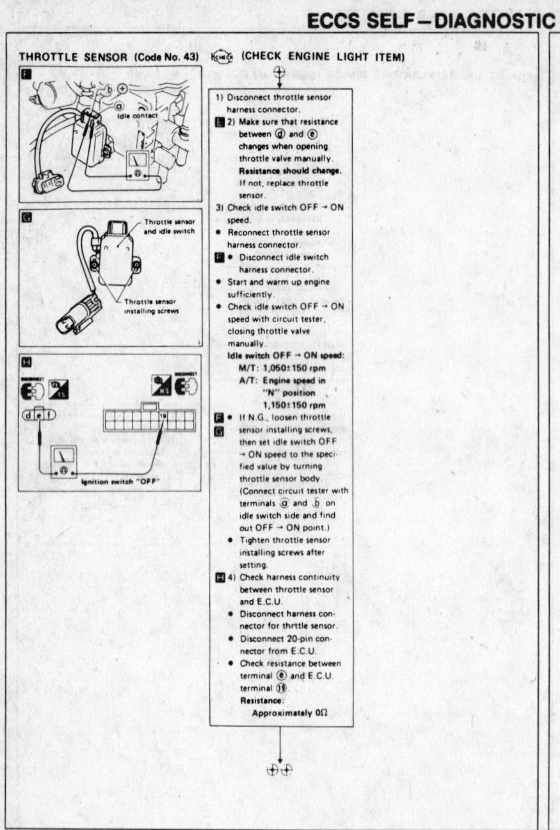
THROTTLE SENSOR (Code No. 43) (CHECK ENGINE LIGHT ITEM)
F
Idle contact
G
Throttle sensor and idle switch
Throttle sensor installing screws
H
Ignition switch "OFF"
1) Disconnect throttle sensor harness connector.
F 2) Make sure that resistance between ⓓ and ⓔ changes when opening throttle valve manually. Resistance should change. If not, replace throttle sensor.
3) Check idle switch OFF → ON speed.
• Reconnect throttle sensor harness connector.
F • Disconnect idle switch harness connector.
• Start and warm up engine sufficiently.
• Check idle switch OFF → ON speed with circuit tester, closing throttle valve manually.
Idle switch OFF → ON speed:
M/T: 1,050±150 rpm
A/T: Engine speed in "N" position 1,150±150 rpm
F G • If N.G., loosen throttle sensor installing screws, then set idle switch OFF → ON speed to the specified value by turning throttle sensor body. (Connect circuit tester with terminals ⓐ and ⓑ on idle switch side and find out OFF → ON point.)
• Tighten throttle sensor installing screws after setting.
H 4) Check harness continuity between throttle sensor and E.C.U.
• Disconnect harness connector for thrttle sensor.
• Disconnect 20-pin connector from E.C.U.
• Check resistance between terminal ⓔ and E.C.U. terminal ⑲.
Resistance: Approximately 0Ω

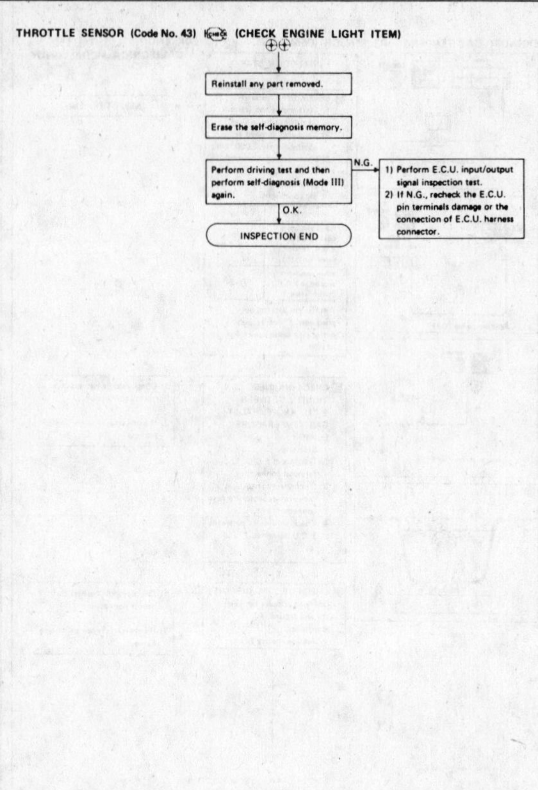
THROTTLE SENSOR (Code No. 43) (CHECK ENGINE LIGHT ITEM)
Reinstall any part removed.
Erase the self-diagnosis memory.
Perform driving test and then perform self-diagnosis (Mode III) again.
N.G.
1) Perform E.C.U. input/output signal inspection test.
2) If N.G., recheck the E.C.U. pin terminals damage or the connection of E.C.U. harness connector.
O.K.
INSPECTION END

ECCS SELF—DIAGNOSTIC CHART—1989 SENTRA

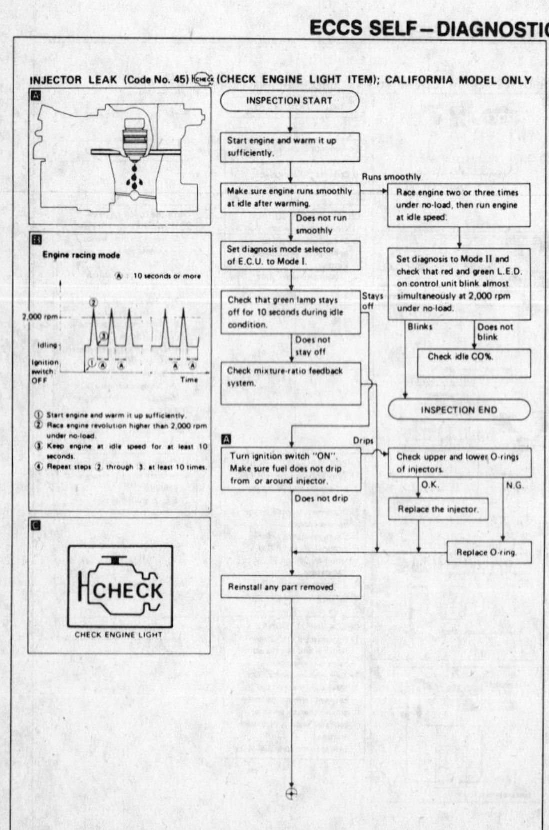
INJECTOR LEAK (Code No. 45) (CHECK ENGINE LIGHT ITEM); CALIFORNIA MODEL ONLY
A
B
Engine racing mode
Ⓐ : 10 seconds or more
2,000 rpm
Idling
Ignition switch OFF
Time
① Start engine and warm it up sufficiently.
② Race engine revolution higher than 2,000 rpm under no-load.
③ Keep engine at idle speed for at least 10 seconds.
④ Repeat steps ② through ③ at least 10 times.
C
CHECK
CHECK ENGINE LIGHT
INSPECTION START
Start engine and warm it up sufficiently.
Runs smoothly
Make sure engine runs smoothly at idle after warming.
Race engine two or three times under no-load, then run engine at idle speed.
Does not run smoothly
Set diagnosis mode selector of E.C.U. to Mode I.
Set diagnosis to Mode II and check that red and green L.E.D. on control unit blink almost simultaneously at 2,000 rpm under no-load.
Check that green lamp stays off for 10 seconds during idle condition.
Stays off
Blinks
Does not blink
Does not stay off
Check idle CO%.
Check mixture-ratio feedback system.
INSPECTION END
A
Drips
Turn ignition switch "ON". Make sure fuel does not drip from or around injector.
Check upper and lower O-rings of injectors.
O.K.
N.G.
Does not drip
Replace the injector.
Replace O-ring.
Reinstall any part removed

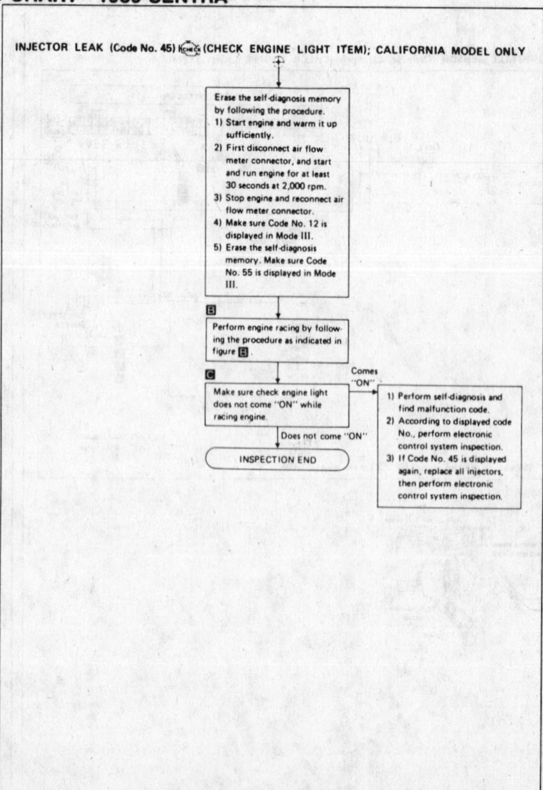
INJECTOR LEAK (Code No. 45) (CHECK ENGINE LIGHT ITEM); CALIFORNIA MODEL ONLY
Erase the self-diagnosis memory by following the procedure.
1) Start engine and warm it up sufficiently.
2) First disconnect air flow meter connector, and start and run engine for at least 30 seconds at 2,000 rpm.
3) Stop engine and reconnect air flow meter connector.
4) Make sure Code No. 12 is displayed in Mode III.
5) Erase the self-diagnosis memory. Make sure Code No. 55 is displayed in Mode III.
B
Perform engine racing by following the procedure as indicated in figure B.
C
Make sure check engine light does not come "ON" while racing engine.
Comes "ON"
1) Perform self-diagnosis and find malfunction code.
2) According to displayed code No., perform electronic control system inspection.
3) If Code No. 45 is displayed again, replace all injectors, then perform electronic control system inspection.
Does not come "ON"
INSPECTION END

ECCS SELF-DIAGNOSTIC CHART—1989 SENTRA

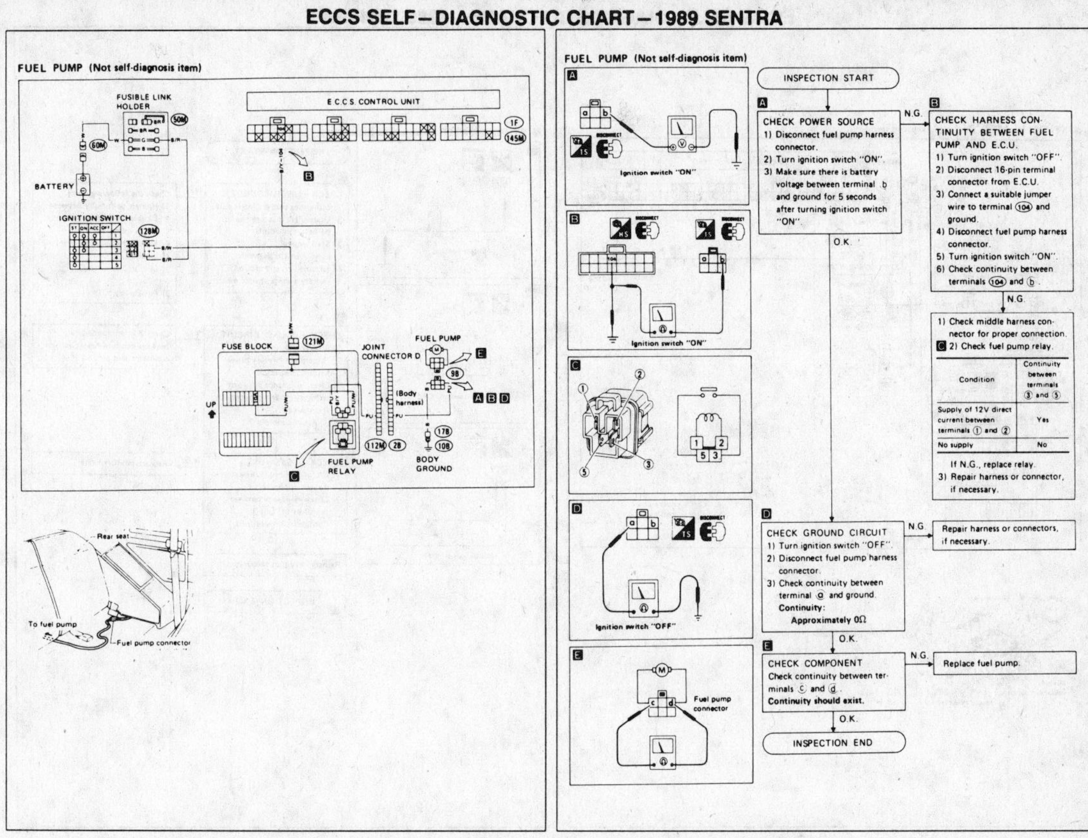
FUEL PUMP (Not self-diagnosis item)
INSPECTION START
CHECK POWER SOURCE
1) Disconnect fuel pump harness connector.
2) Turn ignition switch "ON".
3) Make sure there is battery voltage between terminal ⓑ and ground for 5 seconds after turning ignition switch "ON".
CHECK HARNESS CONTINUITY BETWEEN FUEL PUMP AND E.C.U.
1) Turn ignition switch "OFF".
2) Disconnect 16-pin terminal connector from E.C.U.
3) Connect a suitable jumper wire to terminal (104) and ground.
4) Disconnect fuel pump harness connector.
5) Turn ignition switch "ON".
6) Check continuity between terminals (104) and ⓑ.
1) Check middle harness connector for proper connection.
2) Check fuel pump relay.
If N.G., replace relay.
3) Repair harness or connector, if necessary.
CHECK GROUND CIRCUIT
1) Turn ignition switch "OFF".
2) Disconnect fuel pump harness connector.
3) Check continuity between terminal ⓐ and ground.
Continuity: Approximately 0Ω
Repair harness or connectors, if necessary.
CHECK COMPONENT
Check continuity between terminals ⓒ and ⓓ.
Continuity should exist.
Replace fuel pump.
INSPECTION END
Ignition switch "ON"
Ignition switch "OFF"
E.C.C.S. CONTROL UNIT
FUSIBLE LINK HOLDER
BATTERY
IGNITION SWITCH
FUSE BLOCK
JOINT CONNECTOR D
FUEL PUMP
FUEL PUMP RELAY
BODY GROUND
Rear seat
To fuel pump
Fuel pump connector

Condition	Continuity between terminals ③ and ⑤
Supply of 12V direct current between terminals ① and ②	Yes
No supply	No

ECCS SELF-DIAGNOSTIC CHART—1989 SENTRA

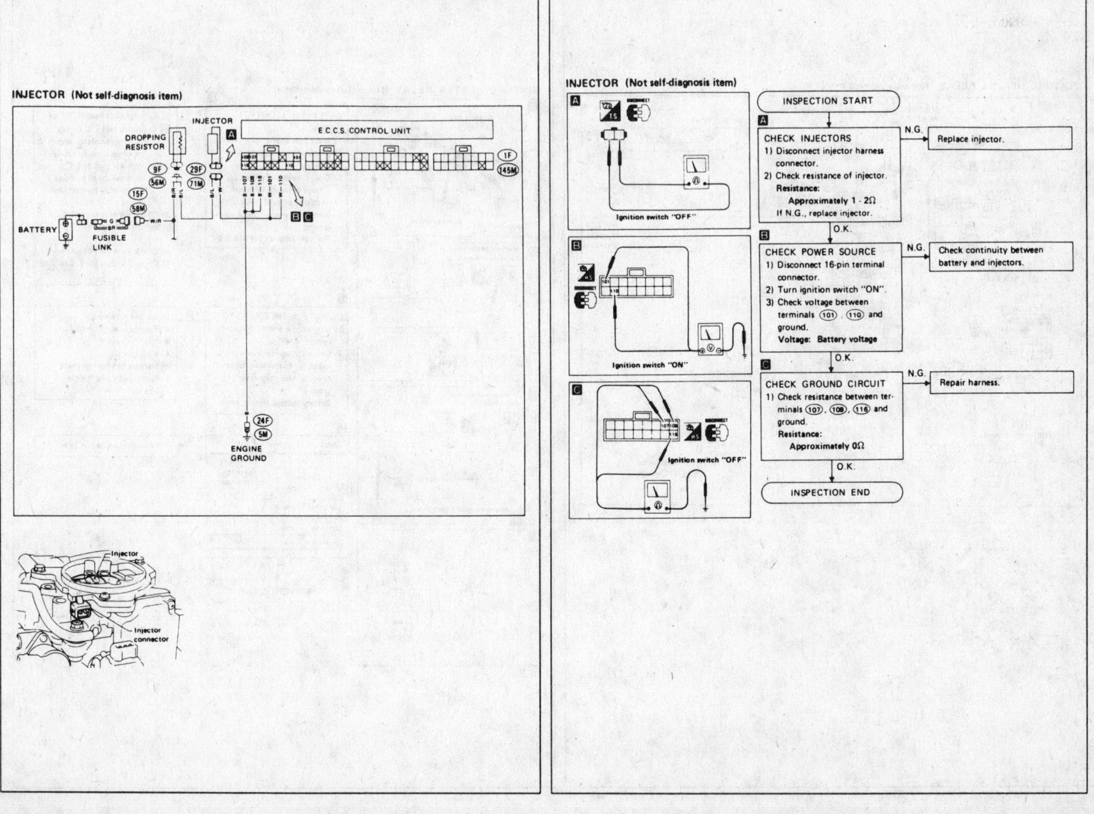
INJECTOR (Not self-diagnosis item)
INSPECTION START
CHECK INJECTORS
1) Disconnect injector harness connector.
2) Check resistance of injector.
Resistance: Approximately 1 - 2Ω
If N.G., replace injector.
Replace injector.
CHECK POWER SOURCE
1) Disconnect 16-pin terminal connector.
2) Turn ignition switch "ON".
3) Check voltage between terminals (101), (110) and ground.
Voltage: Battery voltage
Check continuity between battery and injectors.
CHECK GROUND CIRCUIT
1) Check resistance between terminals (107), (108), (116) and ground.
Resistance: Approximately 0Ω
Repair harness.
INSPECTION END
Ignition switch "OFF"
Ignition switch "ON"
E.C.C.S. CONTROL UNIT
DROPPING RESISTOR
INJECTOR
BATTERY
FUSIBLE LINK
ENGINE GROUND
Injector
Injector connector

ECCS SELF—DIAGNOSTIC CHART—1989 SENTRA

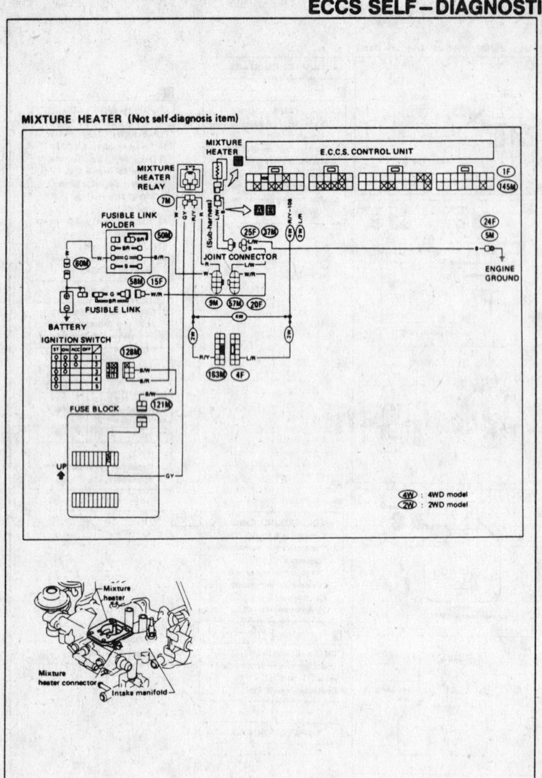
MIXTURE HEATER (Not self-diagnosis item)
MIXTURE HEATER
MIXTURE HEATER RELAY
E.C.C.S. CONTROL UNIT
FUSIBLE LINK HOLDER
JOINT CONNECTOR
ENGINE GROUND
FUSIBLE LINK
BATTERY
IGNITION SWITCH
FUSE BLOCK
UP
4WD model
2WD model
Mixture heater
Mixture heater connector
Intake manifold

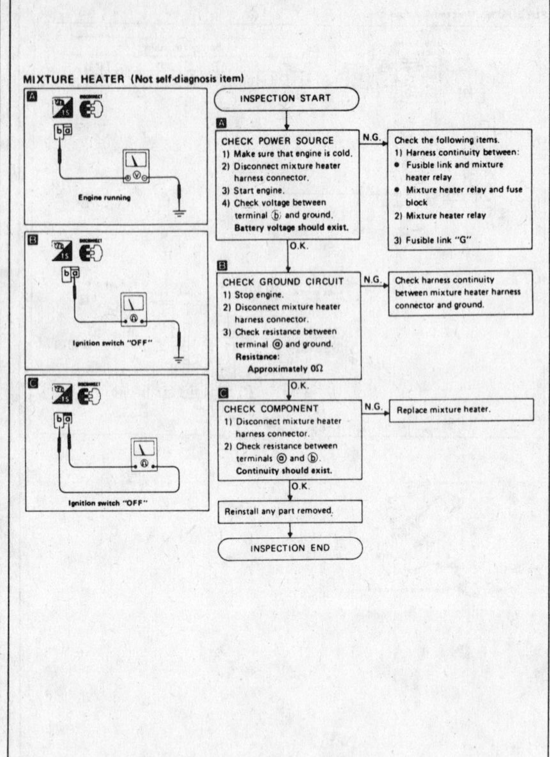
MIXTURE HEATER (Not self-diagnosis item)
Engine running
Ignition switch "OFF"
Ignition switch "OFF"
INSPECTION START
A
CHECK POWER SOURCE
1) Make sure that engine is cold.
2) Disconnect mixture heater harness connector.
3) Start engine.
4) Check voltage between terminal ⓑ and ground.
Battery voltage should exist.
N.G.
Check the following items.
1) Harness continuity between:
• Fusible link and mixture heater relay
• Mixture heater relay and fuse block
2) Mixture heater relay
3) Fusible link "G"
O.K.
B
CHECK GROUND CIRCUIT
1) Stop engine.
2) Disconnect mixture heater harness connector.
3) Check resistance between terminal ⓐ and ground.
Resistance:
Approximately 0Ω
N.G.
Check harness continuity between mixture heater harness connector and ground.
O.K.
C
CHECK COMPONENT
1) Disconnect mixture heater harness connector.
2) Check resistance between terminals ⓐ and ⓑ.
Continuity should exist.
N.G.
Replace mixture heater.
O.K.
Reinstall any part removed.
INSPECTION END

ECCS SELF—DIAGNOSTIC CHART—1989 SENTRA

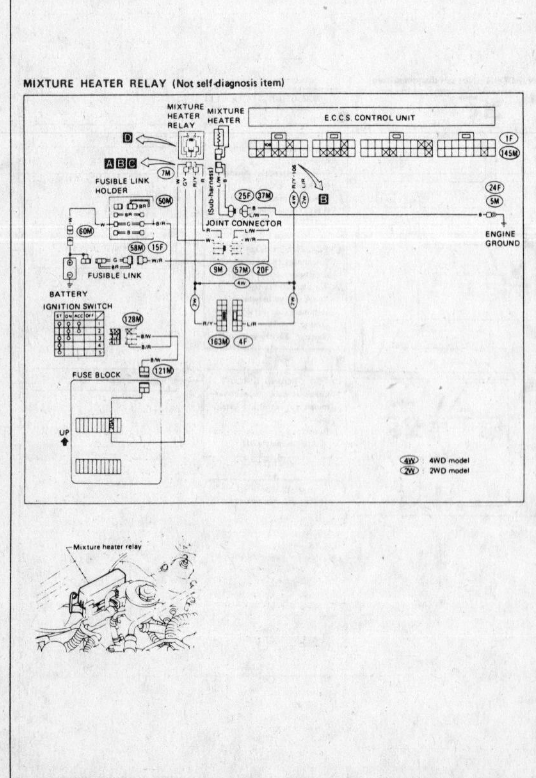
MIXTURE HEATER RELAY (Not self-diagnosis item)
MIXTURE HEATER RELAY
MIXTURE HEATER
E.C.C.S. CONTROL UNIT
FUSIBLE LINK HOLDER
JOINT CONNECTOR
ENGINE GROUND
FUSIBLE LINK
BATTERY
IGNITION SWITCH
FUSE BLOCK
UP
4WD model
2WD model
Mixture heater relay

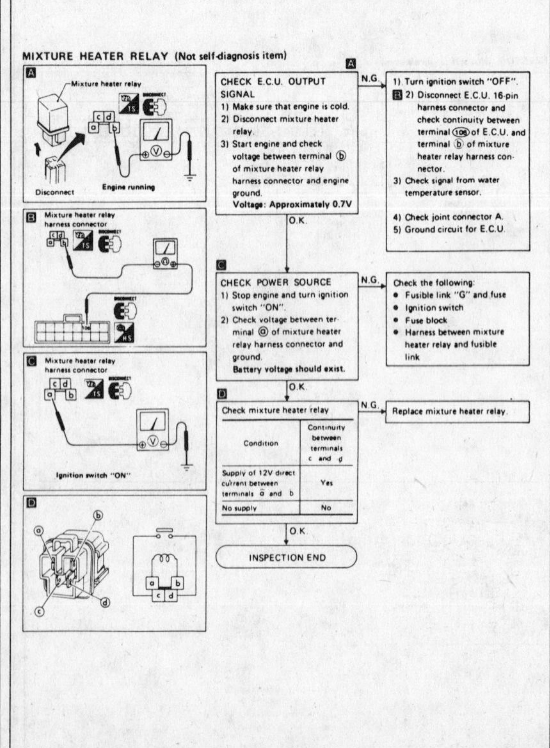
MIXTURE HEATER RELAY (Not self-diagnosis item)
Mixture heater relay
Disconnect
Engine running
Mixture heater relay harness connector
Mixture heater relay harness connector
Ignition switch "ON"
A
CHECK E.C.U. OUTPUT SIGNAL
1) Make sure that engine is cold.
2) Disconnect mixture heater relay.
3) Start engine and check voltage between terminal ⓑ of mixture heater relay harness connector and engine ground.
Voltage: Approximately 0.7V
N.G.
1) Turn ignition switch "OFF".
B 2) Disconnect E.C.U. 16-pin harness connector and check continuity between terminal (108) of E.C.U. and terminal ⓑ of mixture heater relay harness connector.
3) Check signal from water temperature sensor.
4) Check joint connector A.
5) Ground circuit for E.C.U.
O.K.
C
CHECK POWER SOURCE
1) Stop engine and turn ignition switch "ON".
2) Check voltage between terminal ⓐ of mixture heater relay harness connector and ground.
Battery voltage should exist.
N.G.
Check the following:
• Fusible link "G" and fuse
• Ignition switch
• Fuse block
• Harness between mixture heater relay and fusible link
O.K.
D
Check mixture heater relay
Condition | Continuity between terminals c and d
Supply of 12V direct current between terminals a and b | Yes
No supply | No
N.G.
Replace mixture heater relay.
O.K.
INSPECTION END

ECCS SELF-DIAGNOSTIC CHART—1989 SENTRA

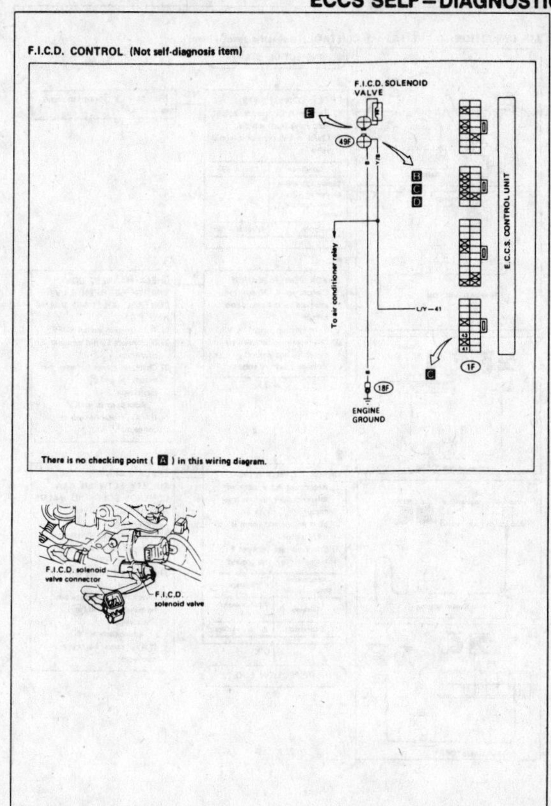
F.I.C.D. CONTROL (Not self-diagnosis item)
F.I.C.D. SOLENOID VALVE
E.C.C.S. CONTROL UNIT
To air conditioner relay
L/Y—41
ENGINE GROUND
There is no checking point (A) in this wiring diagram.
F.I.C.D. solenoid valve connector
F.I.C.D. solenoid valve

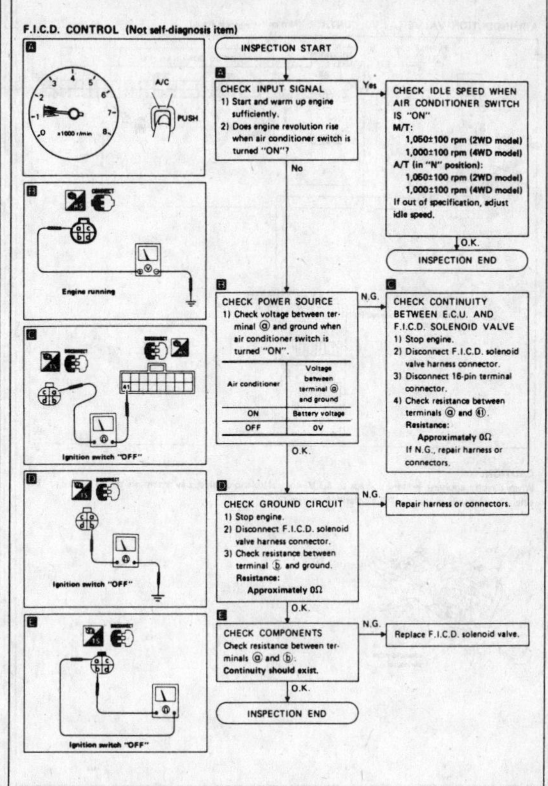
F.I.C.D. CONTROL (Not self-diagnosis item)
A/C
PUSH
Engine running
Ignition switch "OFF"
INSPECTION START
A CHECK INPUT SIGNAL
1) Start and warm up engine sufficiently.
2) Does engine revolution rise when air conditioner switch is turned "ON"?
Yes
CHECK IDLE SPEED WHEN AIR CONDITIONER SWITCH IS "ON"
M/T:
1,050±100 rpm (2WD model)
1,000±100 rpm (4WD model)
A/T (in "N" position):
1,050±100 rpm (2WD model)
1,000±100 rpm (4WD model)
If out of specification, adjust idle speed.
O.K.
INSPECTION END
No
B CHECK POWER SOURCE
1) Check voltage between terminal ⓐ and ground when air conditioner switch is turned "ON".
Air conditioner | Voltage between terminal ⓐ and ground
ON | Battery voltage
OFF | 0V
N.G.
C CHECK CONTINUITY BETWEEN E.C.U. AND F.I.C.D. SOLENOID VALVE
1) Stop engine.
2) Disconnect F.I.C.D. solenoid valve harness connector.
3) Disconnect 16-pin terminal connector.
4) Check resistance between terminals ⓐ and ㊶.
Resistance:
Approximately 0Ω
If N.G., repair harness or connectors.
O.K.
D CHECK GROUND CIRCUIT
1) Stop engine.
2) Disconnect F.I.C.D. solenoid valve harness connector.
3) Check resistance between terminal ⓑ and ground.
Resistance:
Approximately 0Ω
N.G.
Repair harness or connectors.
O.K.
E CHECK COMPONENTS
Check resistance between terminals ⓐ and ⓑ.
Continuity should exist.
N.G.
Replace F.I.C.D. solenoid valve.
O.K.
INSPECTION END

ECCS SELF-DIAGNOSTIC CHART—1989 SENTRA

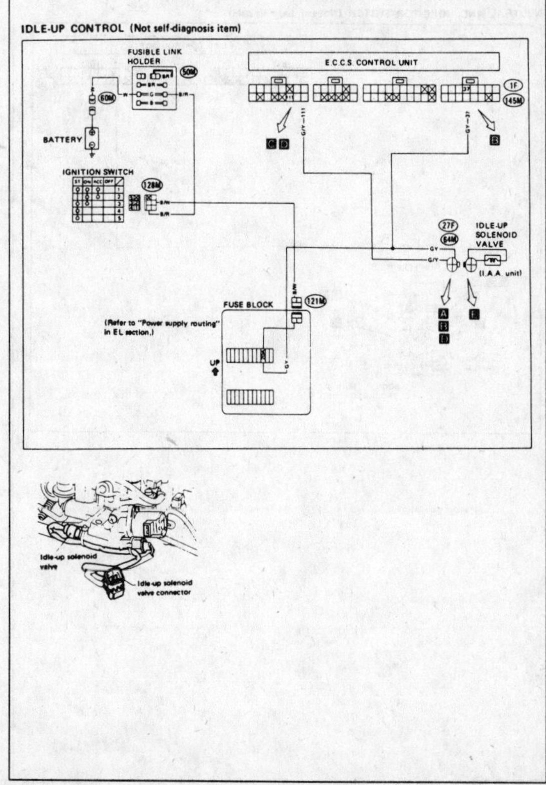
IDLE-UP CONTROL (Not self-diagnosis item)
FUSIBLE LINK HOLDER
BATTERY
IGNITION SWITCH
E.C.C.S. CONTROL UNIT
IDLE-UP SOLENOID VALVE
(I.A.A. unit)
FUSE BLOCK
(Refer to "Power supply routing" in EL section.)
UP
Idle-up solenoid valve
Idle-up solenoid valve connector

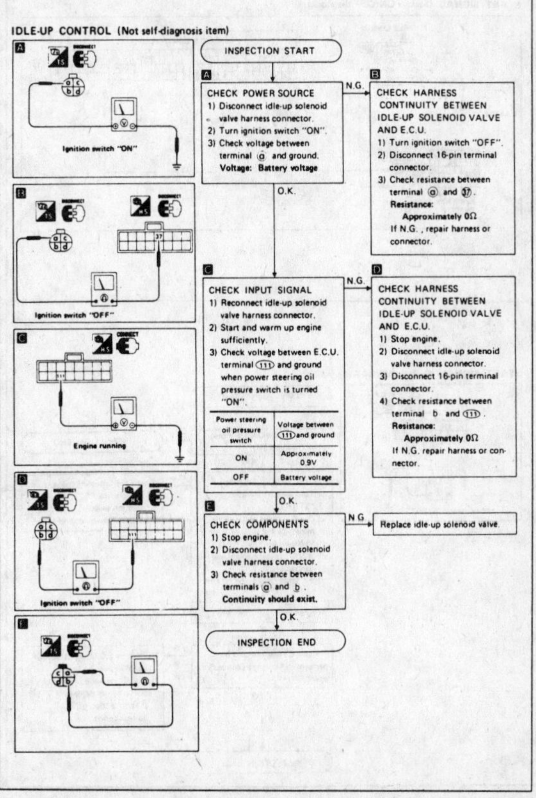
IDLE-UP CONTROL (Not self-diagnosis item)
Ignition switch "ON"
Ignition switch "OFF"
Engine running
INSPECTION START
A CHECK POWER SOURCE
1) Disconnect idle-up solenoid valve harness connector.
2) Turn ignition switch "ON".
3) Check voltage between terminal ⓐ and ground.
Voltage: Battery voltage
N.G.
B CHECK HARNESS CONTINUITY BETWEEN IDLE-UP SOLENOID VALVE AND E.C.U.
1) Turn ignition switch "OFF".
2) Disconnect 16-pin terminal connector.
3) Check resistance between terminal ⓐ and ㊲.
Resistance:
Approximately 0Ω
If N.G., repair harness or connector.
O.K.
C CHECK INPUT SIGNAL
1) Reconnect idle-up solenoid valve harness connector.
2) Start and warm up engine sufficiently.
3) Check voltage between E.C.U. terminal (111) and ground when power steering oil pressure switch is turned "ON".
Power steering oil pressure switch | Voltage between (111) and ground
ON | Approximately 0.9V
OFF | Battery voltage
N.G.
D CHECK HARNESS CONTINUITY BETWEEN IDLE-UP SOLENOID VALVE AND E.C.U.
1) Stop engine.
2) Disconnect idle-up solenoid valve harness connector.
3) Disconnect 16-pin terminal connector.
4) Check resistance between terminal b and (111).
Resistance:
Approximately 0Ω
If N.G. repair harness or connector.
O.K.
E CHECK COMPONENTS
1) Stop engine.
2) Disconnect idle-up solenoid valve harness connector.
3) Check resistance between terminals ⓐ and b.
Continuity should exist.
N G
Replace idle-up solenoid valve.
O.K.
INSPECTION END

ECCS SELF-DIAGNOSTIC CHART—1989 SENTRA

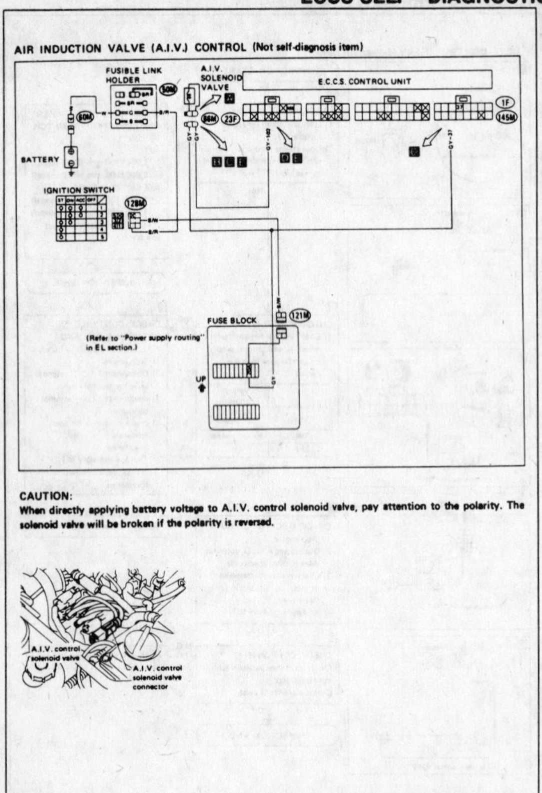
AIR INDUCTION VALVE (A.I.V.) CONTROL (Not self-diagnosis item)
FUSIBLE LINK HOLDER
A.I.V. SOLENOID VALVE
E.C.C.S. CONTROL UNIT
BATTERY
IGNITION SWITCH
FUSE BLOCK
(Refer to "Power supply routing" in EL section.)
UP
CAUTION:
When directly applying battery voltage to A.I.V. control solenoid valve, pay attention to the polarity. The solenoid valve will be broken if the polarity is reversed.
A.I.V. control solenoid valve
A.I.V. control solenoid valve connector

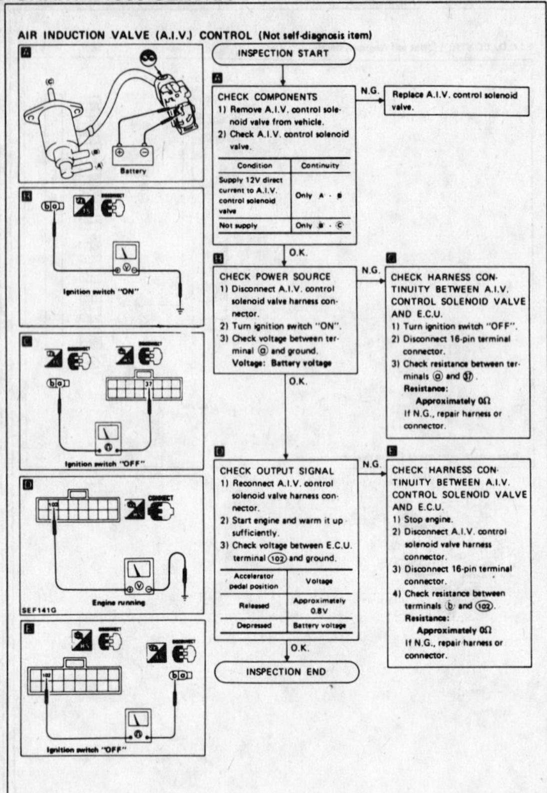
AIR INDUCTION VALVE (A.I.V.) CONTROL (Not self-diagnosis item)
Battery
Ignition switch "ON"
Ignition switch "OFF"
Engine running
SEF141G
INSPECTION START
CHECK COMPONENTS
1) Remove A.I.V. control solenoid valve from vehicle.
2) Check A.I.V. control solenoid valve.
Condition | Continuity
Supply 12V direct current to A.I.V. control solenoid valve | Only A - B
Not supply | Only B - C
N.G.
Replace A.I.V. control solenoid valve.
O.K.
CHECK POWER SOURCE
1) Disconnect A.I.V. control solenoid valve harness connector.
2) Turn ignition switch "ON".
3) Check voltage between terminal ⓐ and ground.
Voltage: Battery voltage
N.G.
CHECK HARNESS CONTINUITY BETWEEN A.I.V. CONTROL SOLENOID VALVE AND E.C.U.
1) Turn ignition switch "OFF".
2) Disconnect 16-pin terminal connector.
3) Check resistance between terminals ⓐ and ㊲.
Resistance:
Approximately 0Ω
If N.G., repair harness or connector.
O.K.
CHECK OUTPUT SIGNAL
1) Reconnect A.I.V. control solenoid valve harness connector.
2) Start engine and warm it up sufficiently.
3) Check voltage between E.C.U. terminal (102) and ground.
Accelerator pedal position | Voltage
Released | Approximately 0.8V
Depressed | Battery voltage
N.G.
CHECK HARNESS CONTINUITY BETWEEN A.I.V. CONTROL SOLENOID VALVE AND E.C.U.
1) Stop engine.
2) Disconnect A.I.V. control solenoid valve harness connector.
3) Disconnect 16-pin terminal connector.
4) Check resistance between terminals ⓑ and (102).
Resistance:
Approximately 0Ω
If N.G., repair harness or connector.
O.K.
INSPECTION END

ECCS SELF-DIAGNOSTIC CHART—1989 SENTRA

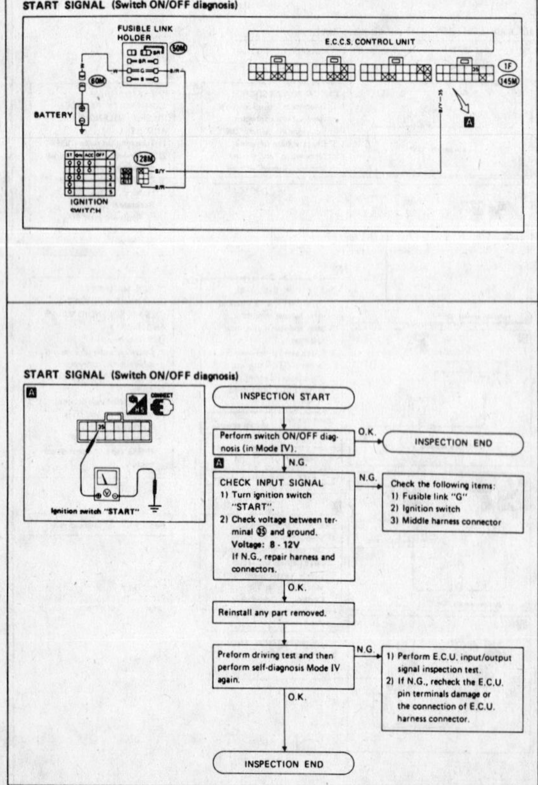
START SIGNAL (Switch ON/OFF diagnosis)
FUSIBLE LINK HOLDER
E.C.C.S. CONTROL UNIT
BATTERY
IGNITION SWITCH
START SIGNAL (Switch ON/OFF diagnosis)
Ignition switch "START"
INSPECTION START
Perform switch ON/OFF diagnosis (in Mode IV).
O.K.
INSPECTION END
N.G.
CHECK INPUT SIGNAL
1) Turn ignition switch "START".
2) Check voltage between terminal ㉟ and ground.
Voltage: 8 - 12V
If N.G., repair harness and connectors.
N.G.
Check the following items:
1) Fusible link "G"
2) Ignition switch
3) Middle harness connector
O.K.
Reinstall any part removed.
Preform driving test and then perform self-diagnosis Mode IV again.
N.G.
1) Perform E.C.U. input/output signal inspection test.
2) If N.G., recheck the E.C.U. pin terminals damage or the connection of E.C.U. harness connector.
O.K.
INSPECTION END

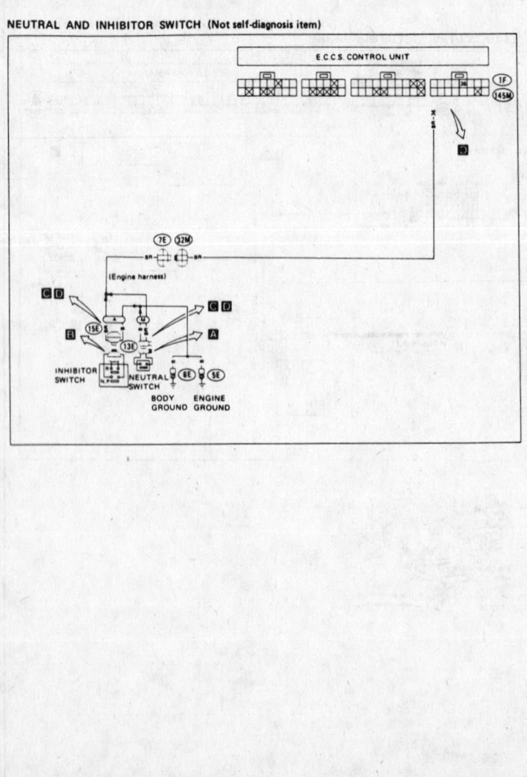
NEUTRAL AND INHIBITOR SWITCH (Not self-diagnosis item)
E.C.C.S. CONTROL UNIT
(Engine harness)
INHIBITOR SWITCH
NEUTRAL SWITCH
BODY GROUND
ENGINE GROUND

ECCS SELF—DIAGNOSTIC CHART—1989 SENTRA

ECU SIGNAL INSPECTION—1989 SENTRA

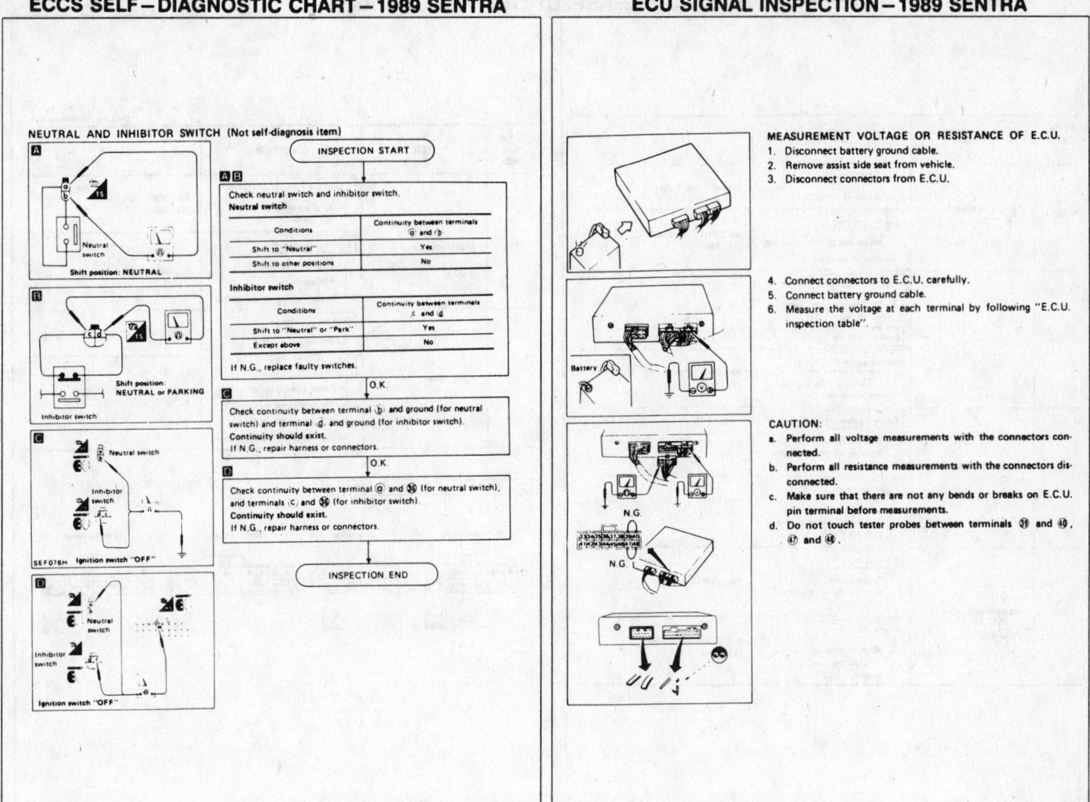
NEUTRAL AND INHIBITOR SWITCH (Not self-diagnosis item)
A
Neutral switch
Shift position: NEUTRAL
B
Shift position: NEUTRAL or PARKING
Inhibitor switch
C
Neutral switch
Inhibitor switch
SEF076H
Ignition switch "OFF"
D
Neutral switch
Inhibitor switch
Ignition switch "OFF"
INSPECTION START
A B
Check neutral switch and inhibitor switch.
Neutral switch
Conditions
Continuity between terminals ⓐ and ⓑ
Shift to "Neutral"
Yes
Shift to other positions
No
Inhibitor switch
Conditions
Continuity between terminals ⓒ and ⓓ
Shift to "Neutral" or "Park"
Yes
Except above
No
If N.G., replace faulty switches.
O.K.
C
Check continuity between terminal ⓑ and ground (for neutral switch) and terminal ⓓ and ground (for inhibitor switch).
Continuity should exist.
If N.G., repair harness or connectors.
O.K.
D
Check continuity between terminal ⓐ and ㊱ (for neutral switch), and terminals ⓒ and ㊱ (for inhibitor switch).
Continuity should exist.
If N.G., repair harness or connectors.
INSPECTION END
MEASUREMENT VOLTAGE OR RESISTANCE OF E.C.U.
1. Disconnect battery ground cable.
2. Remove assist side seat from vehicle.
3. Disconnect connectors from E.C.U.
4. Connect connectors to E.C.U. carefully.
5. Connect battery ground cable.
6. Measure the voltage at each terminal by following "E.C.U. inspection table".
Battery
N.G.
N.G.
CAUTION:
a. Perform all voltage measurements with the connectors connected.
b. Perform all resistance measurements with the connectors disconnected.
c. Make sure that there are not any bends or breaks on E.C.U. pin terminal before measurements.
d. Do not touch tester probes between terminals ㊴ and ㊵, ㊼ and ㊽.

ECU SIGNAL INSPECTION—1989 SENTRA

E.C.U. inspection table

*Data are reference values.

TERMINAL NO.	ITEM	CONDITION	*DATA
1	Ignition signal	Engine is running.	0 - 1.0V
7	Exhaust gas temperature sensor	Engine is running. — Idle speed	1.0V or more
		Engine is running. — E.G.R. system is operating.	0 - 1.0V
15	Air flow meter	Engine is running. Do not run engine at high speed under no-load.	0.5 - 4.0V Output voltage varies with engine revolution.
17	Water temperature sensor	Engine is running.	0.5 - 5.0V Output voltage varies with engine water temperature.
18	Exhaust gas sensor	Engine is running. — After warming up sufficiently	0 - Approximately 1.0V
19	Throttle sensor	Ignition switch "ON"	0.4 - 4.0V Output voltage varies with the throttle valve opening angle.
21 31	Crank angle sensor (Reference signal)	Engine is running. Do not run engine at high speed under no-load.	0.3 - 0.5V
22 32	Crank angle sensor (Position signal)	Engine is running. Do not run engine at high speed under no-load.	2.0 - 2.7V
34	Idle switch (⊖ side)	Ignition switch "ON" — Throttle valve: Idle position	Approximately 10V
		Ignition switch "ON" — Throttle valve: Any position except idle position	Approximately 0V
35	Start signal	Cranking	8 - 12V

*Data are reference values.

TERMINAL NO.	ITEM	CONDITION	*DATA
36	Neutral switch & Inhibitor switch	Ignition switch "ON" — Neutral/Parking	0V
		Ignition switch "ON" — Except the above gear position	Approximately 6.5V
37	Ignition switch	Ignition switch "OFF"	0V
		Ignition switch "ON"	BATTERY VOLTAGE (11 - 14V)
41	Air conditioner	Engine is running. — Both air conditioner switch and blower switch are "ON".	BATTERY VOLTAGE (11 - 14V)
43	Power steering switch	Engine is running. — Steering wheel is turned.	Approximately 0V
		Engine is running. — Except above operation	Approximately 7.0 - 9.0V
44	Idle switch (⊕ side)	Ignition switch "ON" — Throttle valve: Idle position	Approximately 9 - 11V
		Ignition switch "ON" — Throttle valve. Except idle position	BATTERY VOLTAGE (11 - 14V)
46	Power source	Ignition switch "OFF"	BATTERY VOLTAGE (11 - 14V)
39 47	Power source for E.C.U.	Ignition switch "ON"	BATTERY VOLTAGE (11 - 14V)
101 110	Injector	Engine is running.	BATTERY VOLTAGE (11 - 14V)

ECU SIGNAL INSPECTION—1989 SENTRA

TERMINAL NO.	ITEM	CONDITION	*DATA
102	A.I.V. control solenoid valve	Engine is running. — Idle speed	Approximately 0.8V
		Engine is running. — Accelerator pedal is depressed. — After warming up	BATTERY VOLTAGE (11 - 14V)
104	Fuel pump relay	Ignition switch "ON" — For 5 seconds after turning ignition switch "ON"; Engine is running.	0.7 - 0.9V
		Ignition switch "ON" — In 5 seconds after turning ignition switch "ON"	BATTERY VOLTAGE (11 - 14V)
105	E.G.R. & canister control solenoid valve	Engine is running. — Engine is cold. [Water temperature is below 60°C (140°F).]	0.6 - 0.9V
		Engine is running. — After warming up [Water temperature is above 60°C (140°F).]	BATTERY VOLTAGE (11 - 14V)
106	Mixture heater relay	Engine is running. — Engine is cold. [Water temperature is below 50°C (122°F).]	Approximately 0.7V
		Engine is running. — After warming up [Water temperature is above 50° (122°F).]	BATTERY VOLTAGE (11 - 14V)

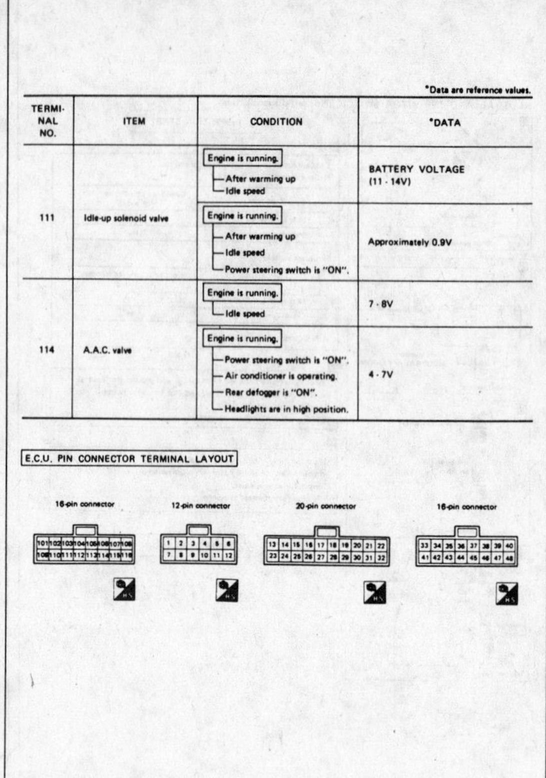

*Data are reference values.

TERMINAL NO.	ITEM	CONDITION	*DATA
111	Idle-up solenoid valve	Engine is running. — After warming up — Idle speed	BATTERY VOLTAGE (11 - 14V)
		Engine is running. — After warming up — Idle speed — Power steering switch is "ON".	Approximately 0.9V
114	A.A.C. valve	Engine is running. — Idle speed	7 - 8V
		Engine is running. — Power steering switch is "ON". — Air conditioner is operating. — Rear defogger is "ON". — Headlights are in high position.	4 - 7V

E.C.U. PIN CONNECTOR TERMINAL LAYOUT
16-pin connector
101 102 103 104 105 106 107 108
109 110 111 112 113 114 115 116
12-pin connector
1 2 3 4 5 6
7 8 9 10 11 12
20-pin connector
13 14 15 16 17 18 19 20 21 22
23 24 25 26 27 28 29 30 31 32
16-pin connector
33 34 35 36 37 38 39 40
41 42 43 44 45 46 47 48

ECCS MIXTURE RATIO FEEDBACK SYSTEM INSPECTION CHART—1989 SENTRA

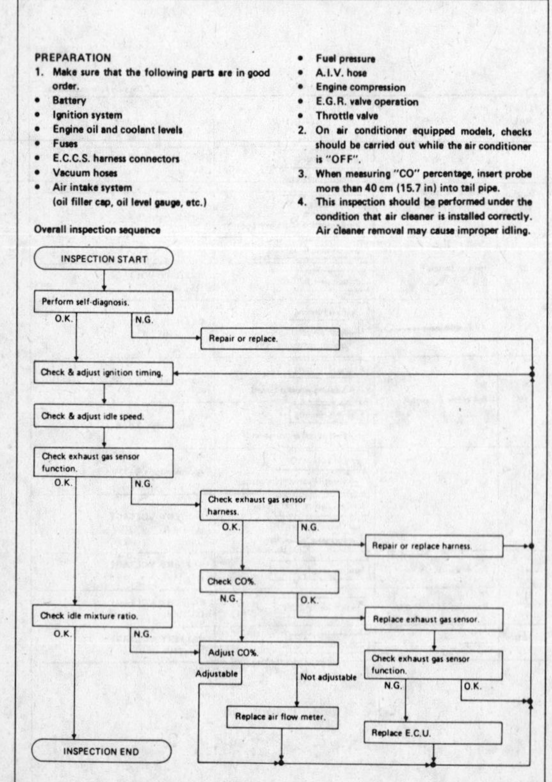
PREPARATION
1. Make sure that the following parts are in good order.
• Battery
• Ignition system
• Engine oil and coolant levels
• Fuses
• E.C.C.S. harness connectors
• Vacuum hoses
• Air intake system (oil filler cap, oil level gauge, etc.)
• Fuel pressure
• A.I.V. hose
• Engine compression
• E.G.R. valve operation
• Throttle valve
2. On air conditioner equipped models, checks should be carried out while the air conditioner is "OFF".
3. When measuring "CO" percentage, insert probe more than 40 cm (15.7 in) into tail pipe.
4. This inspection should be performed under the condition that air cleaner is installed correctly. Air cleaner removal may cause improper idling.
Overall inspection sequence
INSPECTION START
Perform self-diagnosis.
O.K.
N.G.
Repair or replace.
Check & adjust ignition timing.
Check & adjust idle speed.
Check exhaust gas sensor function.
O.K.
N.G.
Check exhaust gas sensor harness.
O.K.
N.G.
Repair or replace harness.
Check CO%.
N.G.
O.K.
Replace exhaust gas sensor.
Check idle mixture ratio.
O.K.
N.G.
Adjust CO%.
Adjustable
Not adjustable
Check exhaust gas sensor function.
N.G.
O.K.
Replace air flow meter.
Replace E.C.U.
INSPECTION END

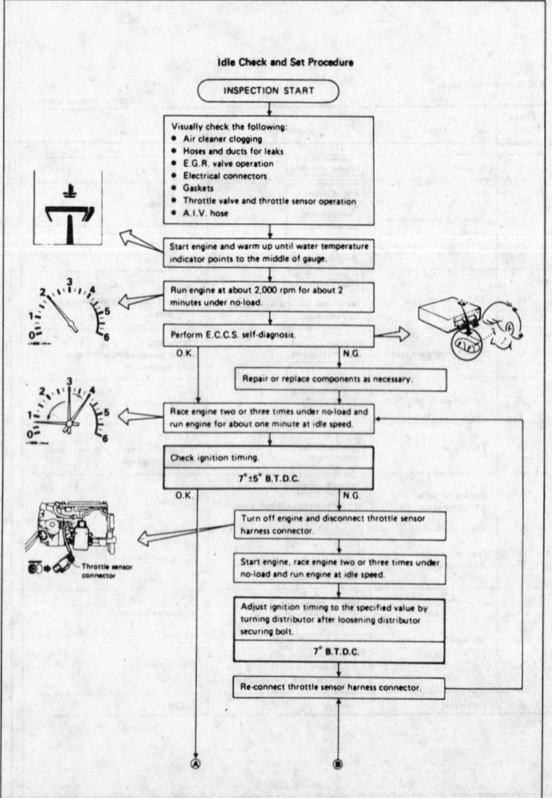
Idle Check and Set Procedure
INSPECTION START
Visually check the following:
• Air cleaner clogging
• Hoses and ducts for leaks
• E.G.R. valve operation
• Electrical connectors
• Gaskets
• Throttle valve and throttle sensor operation
• A.I.V. hose
Start engine and warm up until water temperature indicator points to the middle of gauge.
Run engine at about 2,000 rpm for about 2 minutes under no-load.
Perform E.C.C.S. self-diagnosis.
O.K.
N.G.
Repair or replace components as necessary.
Race engine two or three times under no-load and run engine for about one minute at idle speed.
Check ignition timing.
7° ±5° B.T.D.C.
O.K.
N.G.
Turn off engine and disconnect throttle sensor harness connector.
Start engine, race engine two or three times under no-load and run engine at idle speed.
Adjust ignition timing to the specified value by turning distributor after loosening distributor securing bolt.
7° B.T.D.C.
Re-connect throttle sensor harness connector.
Throttle sensor connector
0 1 2 3 4 5 6
Ⓐ
Ⓑ

ECCS MIXTURE RATIO FEEDBACK SYSTEM INSPECTION CHART—1989 SENTRA

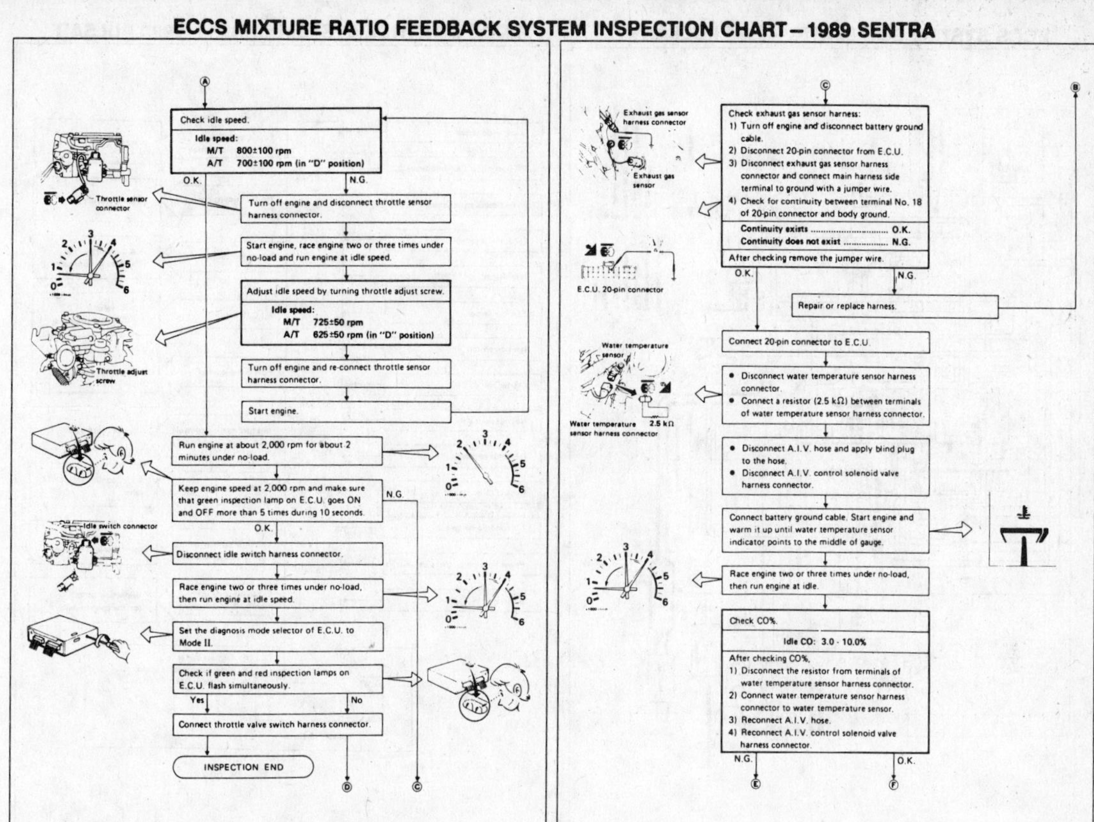
A
Check idle speed.
Idle speed:
M/T 800±100 rpm
A/T 700±100 rpm (in "D" position)
O.K.
N.G.
Throttle sensor connector
Turn off engine and disconnect throttle sensor harness connector.
Start engine, race engine two or three times under no-load and run engine at idle speed.
Adjust idle speed by turning throttle adjust screw.
Idle speed:
M/T 725±50 rpm
A/T 625±50 rpm (in "D" position)
Throttle adjust screw
Turn off engine and re-connect throttle sensor harness connector.
Start engine.
Run engine at about 2,000 rpm for about 2 minutes under no-load.
Keep engine speed at 2,000 rpm and make sure that green inspection lamp on E.C.U. goes ON and OFF more than 5 times during 10 seconds.
N.G.
O.K.
Idle switch connector
Disconnect idle switch harness connector.
Race engine two or three times under no-load, then run engine at idle speed.
Set the diagnosis mode selector of E.C.U. to Mode II.
Check if green and red inspection lamps on E.C.U. flash simultaneously.
Yes
No
Connect throttle valve switch harness connector.
INSPECTION END
D
C
C
B
Exhaust gas sensor harness connector
Exhaust gas sensor
Check exhaust gas sensor harness:
1) Turn off engine and disconnect battery ground cable.
2) Disconnect 20-pin connector from E.C.U.
3) Disconnect exhaust gas sensor harness connector and connect main harness side terminal to ground with a jumper wire.
4) Check for continuity between terminal No. 18 of 20-pin connector and body ground.
Continuity exists .... O.K.
Continuity does not exist .... N.G.
After checking remove the jumper wire.
E.C.U. 20-pin connector
O.K.
N.G.
Repair or replace harness.
Connect 20-pin connector to E.C.U.
Water temperature sensor
Water temperature sensor harness connector
2.5 kΩ
Disconnect water temperature sensor harness connector.
Connect a resistor (2.5 kΩ) between terminals of water temperature sensor harness connector.
Disconnect A.I.V. hose and apply blind plug to the hose.
Disconnect A.I.V. control solenoid valve harness connector.
Connect battery ground cable. Start engine and warm it up until water temperature sensor indicator points to the middle of gauge.
Race engine two or three times under no-load, then run engine at idle.
Check CO%.
Idle CO: 3.0 - 10.0%
After checking CO%,
1) Disconnect the resistor from terminals of water temperature sensor harness connector.
2) Connect water temperature sensor harness connector to water temperature sensor.
3) Reconnect A.I.V. hose.
4) Reconnect A.I.V. control solenoid valve harness connector.
N.G.
O.K.
E
F

ECCS MIXTURE RATIO FEEDBACK SYSTEM INSPECTION CHART—1989 SENTRA

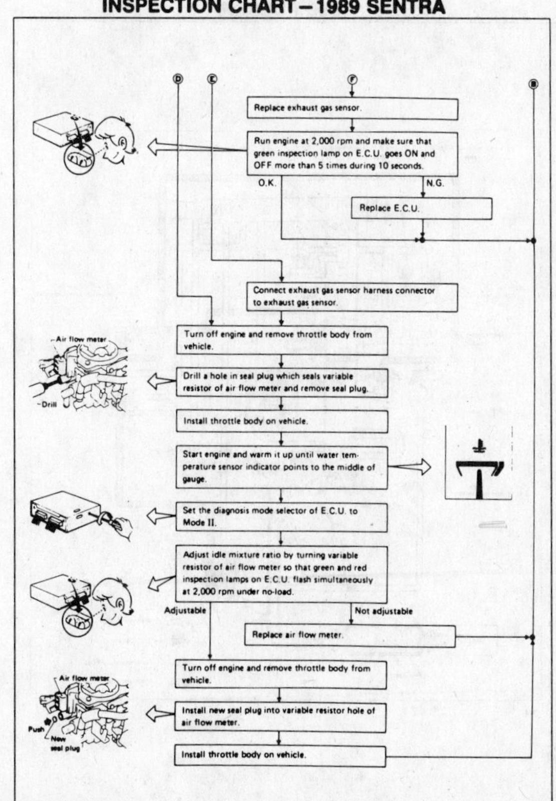
D
E
F
B
Replace exhaust gas sensor.
Run engine at 2,000 rpm and make sure that green inspection lamp on E.C.U. goes ON and OFF more than 5 times during 10 seconds.
O.K.
N.G.
Replace E.C.U.
Connect exhaust gas sensor harness connector to exhaust gas sensor.
Air flow meter
Drill
Turn off engine and remove throttle body from vehicle.
Drill a hole in seal plug which seals variable resistor of air flow meter and remove seal plug.
Install throttle body on vehicle.
Start engine and warm it up until water temperature sensor indicator points to the middle of gauge.
Set the diagnosis mode selector of E.C.U. to Mode II.
Adjust idle mixture ratio by turning variable resistor of air flow meter so that green and red inspection lamps on E.C.U. flash simultaneously at 2,000 rpm under no-load.
Adjustable
Not adjustable
Replace air flow meter.
Turn off engine and remove throttle body from vehicle.
Install new seal plug into variable resistor hole of air flow meter.
Install throttle body on vehicle.
Push
New seal plug

ECCS COMPONENTS LOCATION—1990 PULSAR

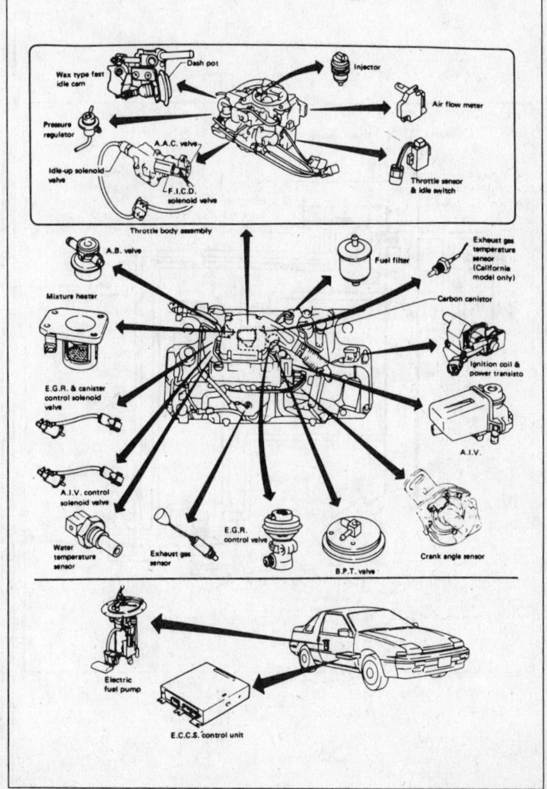
Dash pot
Injector
Wax type fast idle cam
Air flow meter
Pressure regulator
A.A.C. valve
Idle-up solenoid valve
F.I.C.D. solenoid valve
Throttle sensor & idle switch
Throttle body assembly
A.B. valve
Fuel filter
Exhaust gas temperature sensor (California model only)
Mixture heater
Carbon canister
Ignition coil & power transisto
E.G.R. & canister control solenoid valve
A.I.V.
A.I.V. control solenoid valve
E.G.R. control valve
Water temperature sensor
Exhaust gas sensor
B.P.T. valve
Crank angle sensor
Electric fuel pump
E.C.C.S. control unit

ECCS SYSTEM SCHEMATIC—1990 PULSAR

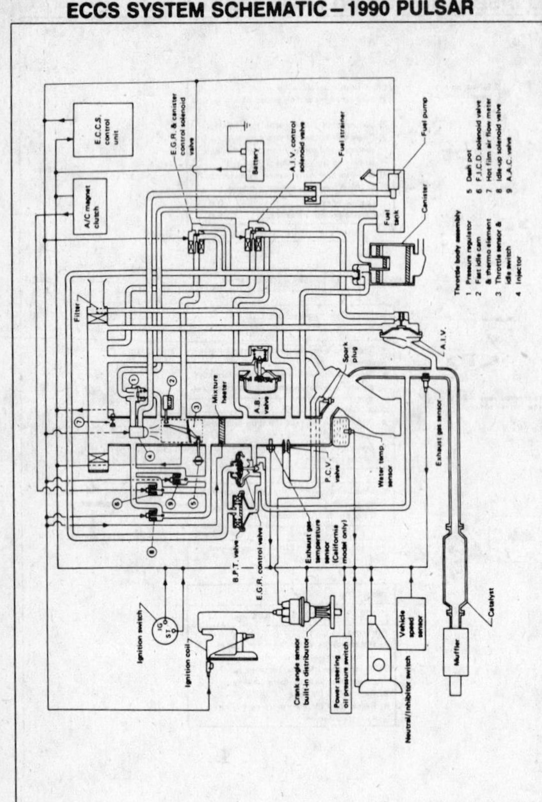
E.C.C.S. control unit
A/C magnet clutch
E.G.R. & canister control solenoid valve
Battery
A.I.V. control solenoid valve
Fuel strainer
Fuel pump
Fuel tank
Canister
Throttle body assembly
1 Pressure regulator
2 Fast idle cam & thermo element
3 Throttle sensor & idle switch
4 Injector
5 Dash pot
6 F.I.C.D. solenoid valve
7 Hot film air flow meter
8 Idle-up solenoid valve
9 A.A.C. valve
Filter
Mixture heater
A.B. valve
Spark plug
A.I.V.
P.C.V. valve
Water temp. sensor
Exhaust gas sensor
B.P.T. valve
E.G.R. control valve
Exhaust gas temperature sensor (California models only)
Ignition switch
Ignition coil
Crank angle sensor built-in distributor
Power steering oil pressure switch
Vehicle speed sensor
Neutral/inhibitor switch
Catalyst
Muffler

ECCS CONTROL SYSTEM CHART—1990 PULSAR

Inputs to the E.C.C.S. control unit:

- Crank angle sensor
- Air flow meter (hot film type)
- Water temperature sensor
- Exhaust gas sensor
- Idle switch
- Throttle sensor
- Ignition switch
- Starter switch
- Neutral switch (M/T)
- Inhibitor switch (A/T)
- Vehicle speed sensor
- Air conditioner switch
- Power steering oil pressure switch
- Battery

E.C.C.S. control unit

Control	Output
Fuel injection & mixture ratio control	Injector
Ignition timing control	Power transistor
Idle speed control	A.A.C. valve & idle speed control valve
F.I.C.D. control	A.A.C. valve & F.I.C.D. solenoid valve
Fuel pump control	Fuel pump relay
Mixture heating control	Mixture heater relay
A.I.V. control	A.I.V. control solenoid valve
E.G.R. & canister control	E.G.R. & canister control solenoid valve
Exhaust gas sensor monitor & self-diagnosis	Inspection lamps (on the control unit)

Fail-safe function

- Air flow meter
- Water temperature sensor
- Throttle sensor

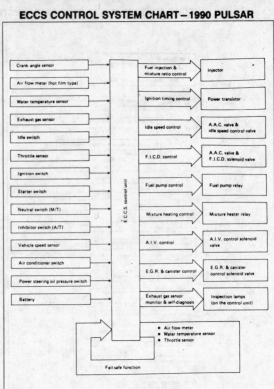

ECCS FUEL FLOW SYSTEM SCHEMATIC—1990 PULSAR

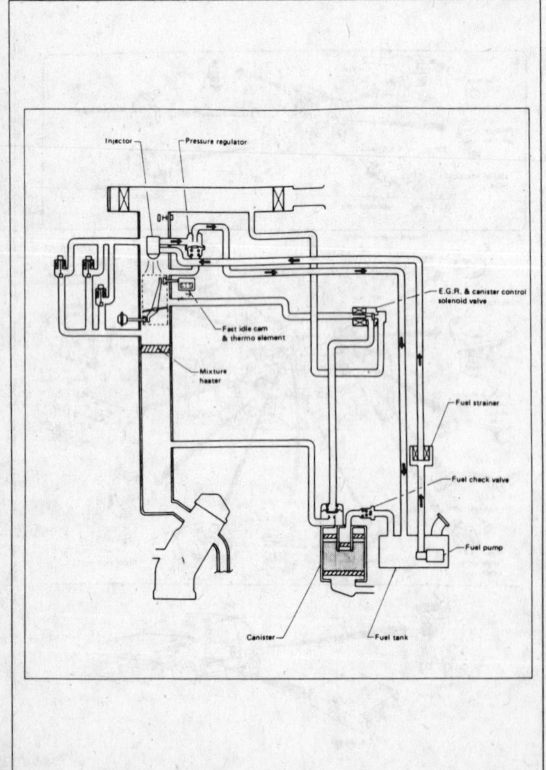
Injector
Pressure regulator
E.G.R. & canister control solenoid valve
Fast idle cam & thermo element
Mixture heater
Fuel strainer
Fuel check valve
Fuel pump
Canister
Fuel tank

ECCS AIR FLOW SYSTEM SCHEMATIC—1990 PULSAR

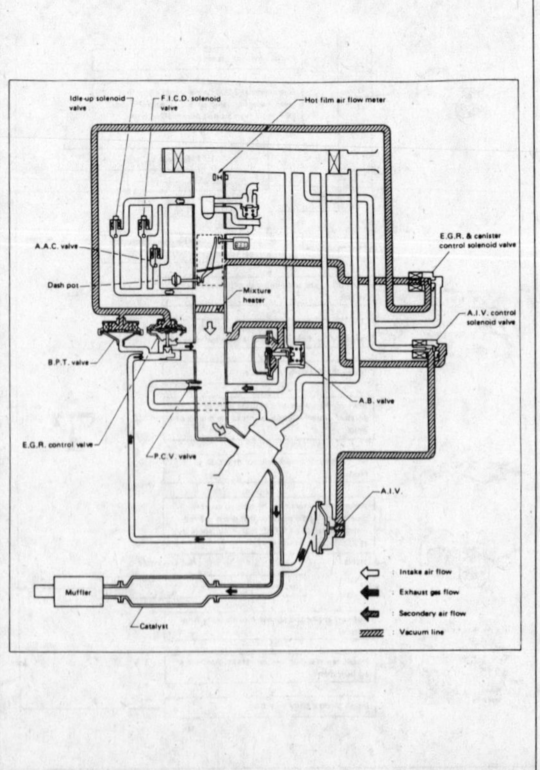
Idle-up solenoid valve
F.I.C.D. solenoid valve
Hot film air flow meter
E.G.R. & canister control solenoid valve
A.A.C. valve
Dash pot
Mixture heater
A.I.V. control solenoid valve
B.P.T. valve
A.B. valve
E.G.R. control valve
P.C.V. valve
A.I.V.
Muffler
Catalyst
: Intake air flow
: Exhaust gas flow
: Secondary air flow
: Vacuum line

ECCS CONTROL UNIT WIRING SCHEMATIC – 1990 PULSAR

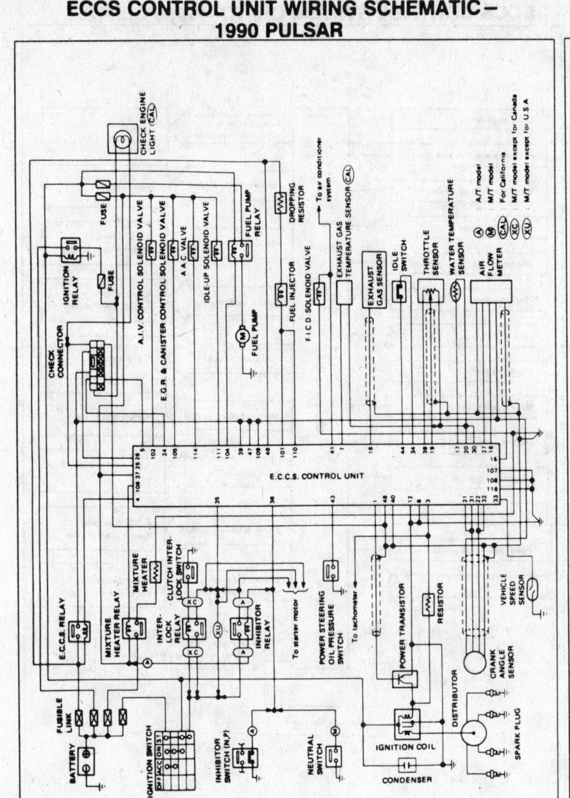

ECCS WIRING SCHEMATIC – 1990 PULSAR

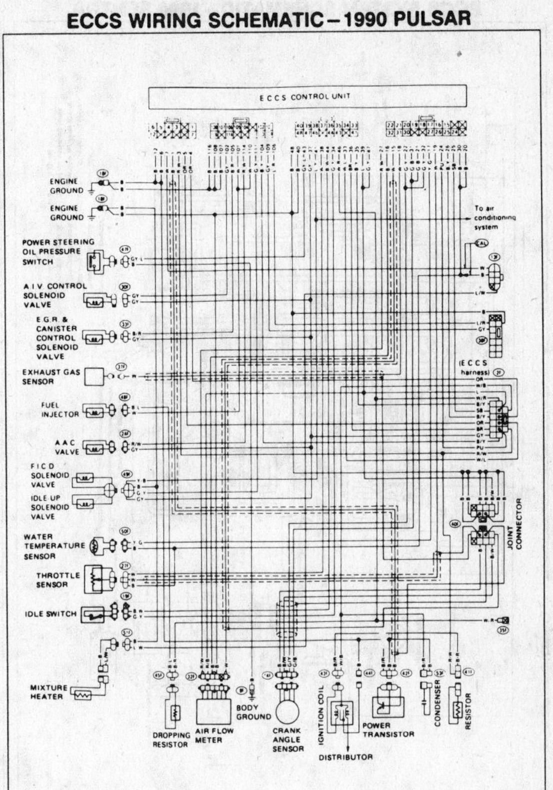

ECCS WIRING SCHEMATIC (CONT.) – 1990 PULSAR

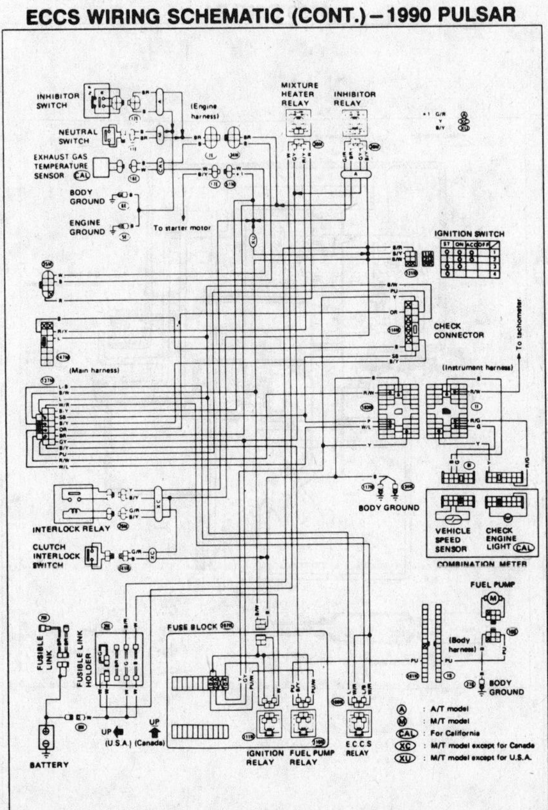

ECCS COMPONENTS LOCATION – 1990 SENTRA

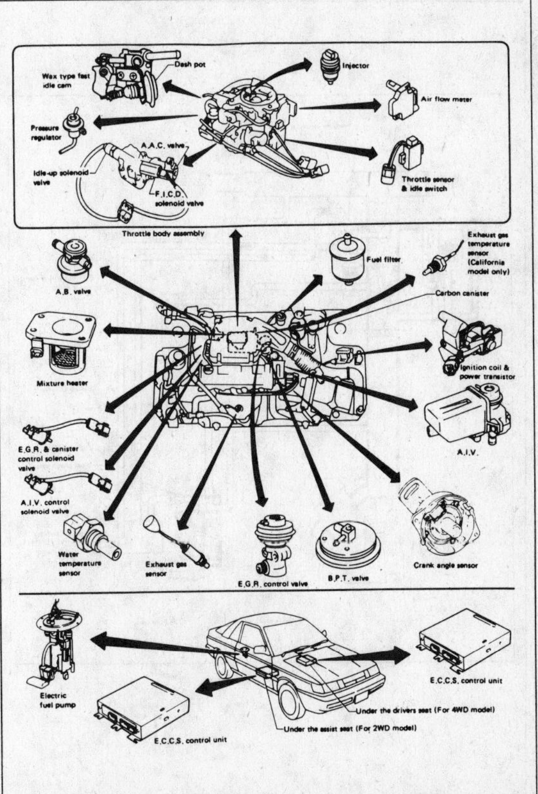

Dash pot
Injector
Wax type fast idle cam
Air flow meter
Pressure regulator
A.A.C. valve
Idle-up solenoid valve
F.I.C.D. solenoid valve
Throttle sensor & idle switch
Throttle body assembly
A.B. valve
Fuel filter
Exhaust gas temperature sensor (California model only)
Carbon canister
Mixture heater
Ignition coil & power transistor
A.I.V.
E.G.R. & canister control solenoid valve
A.I.V. control solenoid valve
Water temperature sensor
Exhaust gas sensor
E.G.R. control valve
B.P.T. valve
Crank angle sensor
Electric fuel pump
E.C.C.S. control unit
Under the drivers seat (For 4WD model)
Under the assist seat (For 2WD model)
E.C.C.S. control unit

ECCS SYSTEM SCHEMATIC—1990 SENTRA

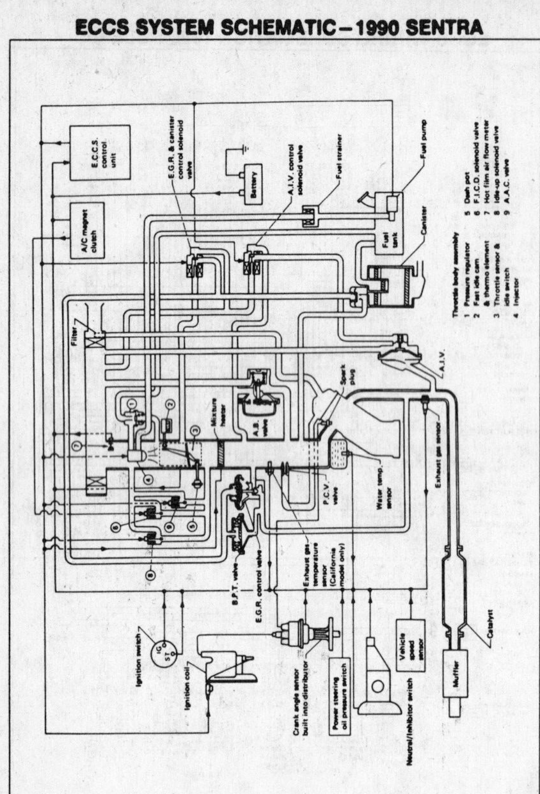
E.C.C.S. control unit
A/C magnet clutch
E.G.R. & canister control solenoid valve
Battery
A.I.V. control solenoid valve
Fuel strainer
Fuel pump
Fuel tank
Canister
Filter
Mixture heater
A.B. valve
Spark plug
A.I.V.
Throttle body assembly
1 Pressure regulator
2 Fast idle cam & thermo element
3 Idle switch
4 Injector
5 Dash pot
6 F.I.C.D. solenoid valve
7 Hot film air flow meter
8 Idle-up solenoid valve
9 A.A.C. valve
Exhaust gas sensor
P.C.V. valve
Water temp. sensor
B.P.T. valve
E.G.R. control valve
Exhaust gas temperature sensor (California models only)
Catalyst
Muffler
Vehicle speed sensor
Ignition switch
Ignition coil
Crank angle sensor built into distributor
Power steering oil pressure switch
Neutral/inhibitor switch

ECCS CONTROL SYSTEM CHART—1990 SENTRA

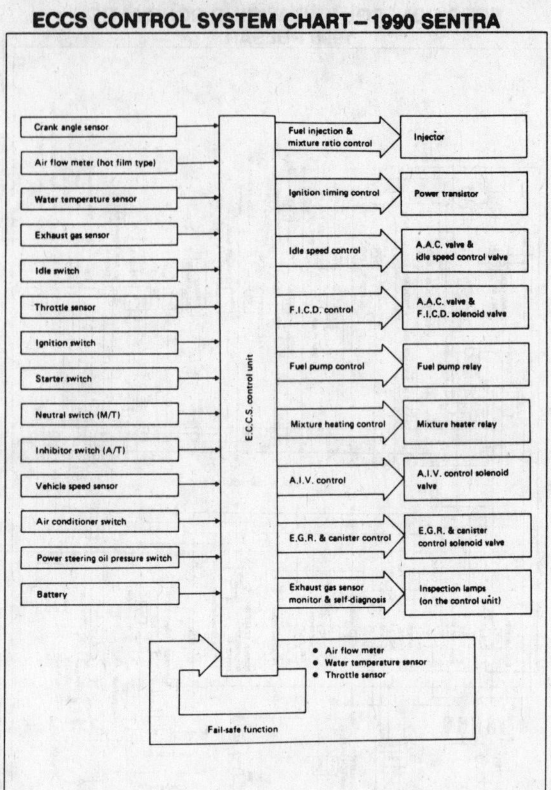
Crank angle sensor
Air flow meter (hot film type)
Water temperature sensor
Exhaust gas sensor
Idle switch
Throttle sensor
Ignition switch
Starter switch
Neutral switch (M/T)
Inhibitor switch (A/T)
Vehicle speed sensor
Air conditioner switch
Power steering oil pressure switch
Battery
E.C.C.S. control unit
Fuel injection & mixture ratio control
Injector
Ignition timing control
Power transistor
Idle speed control
A.A.C. valve & idle speed control valve
F.I.C.D. control
A.A.C. valve & F.I.C.D. solenoid valve
Fuel pump control
Fuel pump relay
Mixture heating control
Mixture heater relay
A.I.V. control
A.I.V. control solenoid valve
E.G.R. & canister control
E.G.R. & canister control solenoid valve
Exhaust gas sensor monitor & self-diagnosis
Inspection lamps (on the control unit)
Air flow meter
Water temperature sensor
Throttle sensor
Fail-safe function

ECCS FUEL FLOW SYSTEM SCHEMATIC—1990 SENTRA

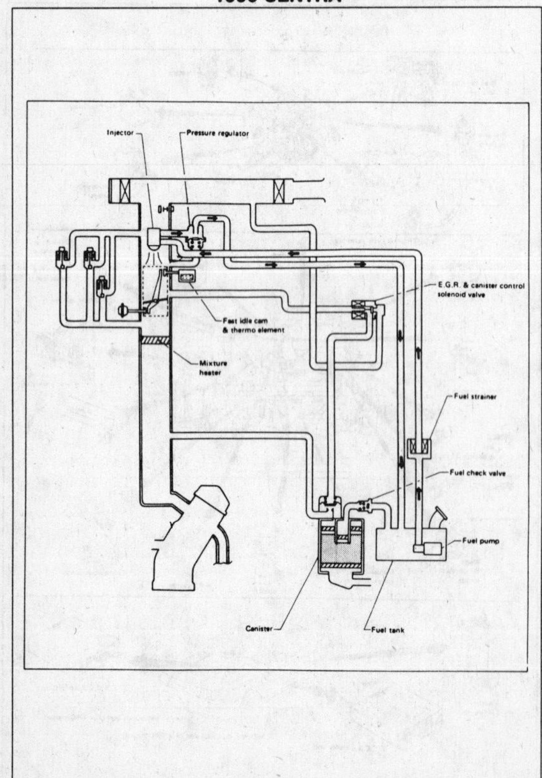
Injector
Pressure regulator
E.G.R. & canister control solenoid valve
Fast idle cam & thermo element
Mixture heater
Fuel strainer
Fuel check valve
Fuel pump
Canister
Fuel tank

ECCS AIR FLOW SYSTEM SCHEMATIC—1990 SENTRA

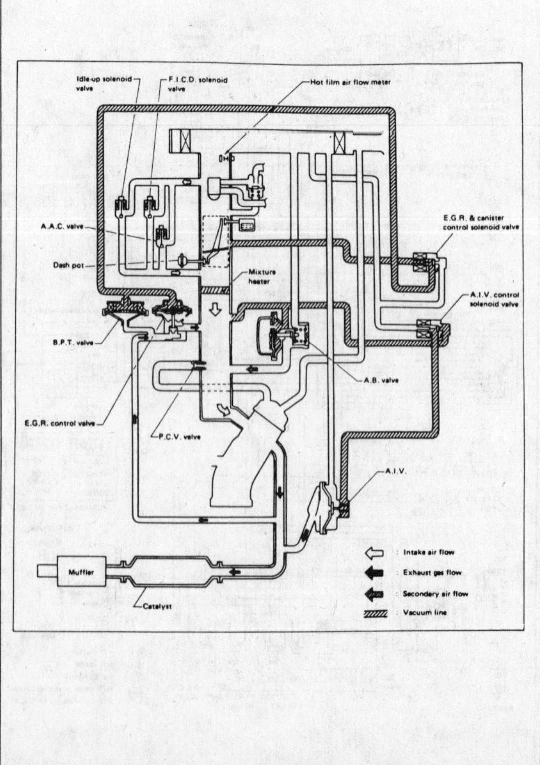
Idle-up solenoid valve
F.I.C.D. solenoid valve
Hot film air flow meter
A.A.C. valve
Dash pot
Mixture heater
E.G.R. & canister control solenoid valve
A.I.V. control solenoid valve
B.P.T. valve
A.B. valve
E.G.R. control valve
P.C.V. valve
A.I.V.
Muffler
Catalyst
Intake air flow
Exhaust gas flow
Secondary air flow
Vacuum line

ECCS WIRING SCHEMATIC—1990 SENTRA (2WD)

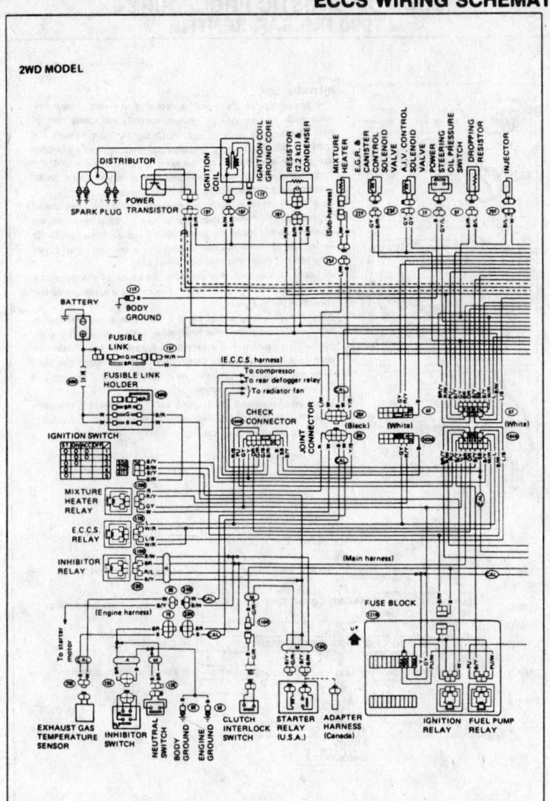

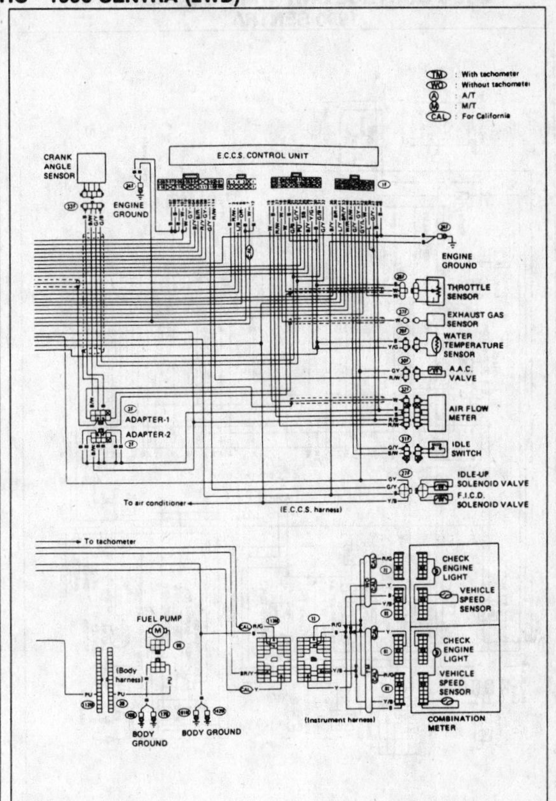

ECCS WIRING SCHEMATIC—1990 SENTRA (4WD)

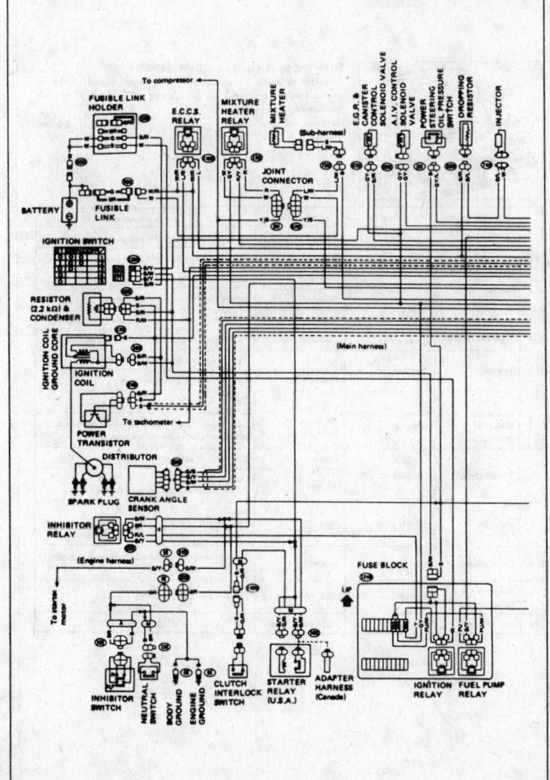

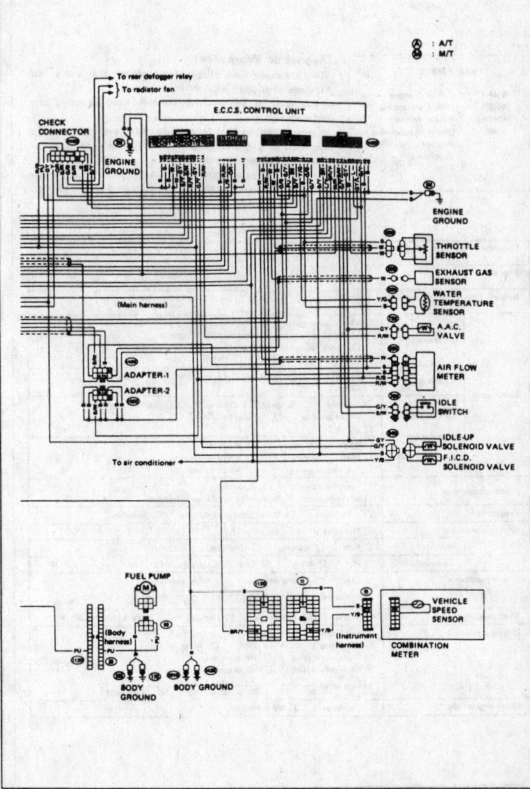

ECCS CONTROL UNIT WIRING SCHEMATIC—1990 SENTRA

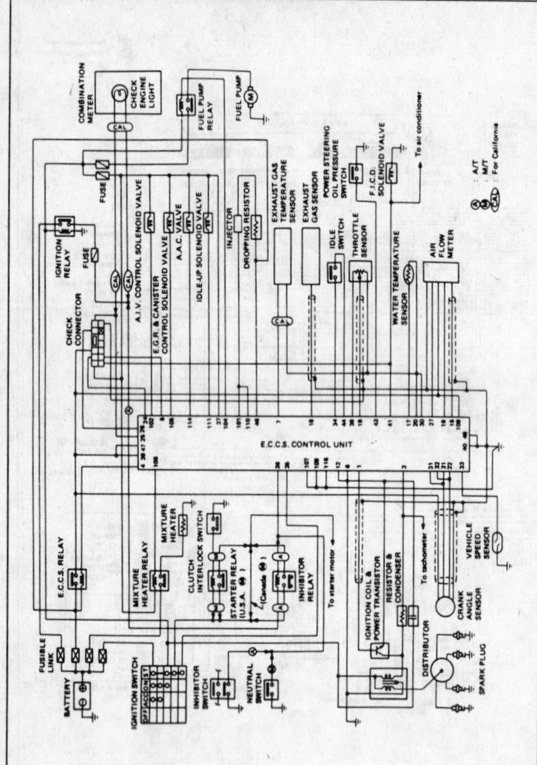
COMBINATION METER
CHECK ENGINE LIGHT
FUEL PUMP RELAY
FUEL PUMP
FUSE
IGNITION RELAY
FUSE
CHECK CONNECTOR
A.I.V. CONTROL SOLENOID VALVE
E.G.R. & CANISTER CONTROL SOLENOID VALVE
A.A.C. VALVE
IDLE-UP SOLENOID VALVE
INJECTOR
DROPPING RESISTOR
EXHAUST GAS TEMPERATURE SENSOR
EXHAUST GAS SENSOR
IDLE SWITCH
THROTTLE SENSOR
POWER STEERING OIL PRESSURE SWITCH
F.I.C.D. SOLENOID VALVE
To air conditioner
WATER TEMPERATURE SENSOR
AIR FLOW METER
A/T
M/T
For California
E.C.C.S. CONTROL UNIT
E.C.C.S. RELAY
MIXTURE HEATER RELAY
MIXTURE HEATER
CLUTCH INTERLOCK SWITCH
STARTER RELAY
INHIBITOR RELAY
To starter motor
IGNITION COIL & POWER TRANSISTOR
RESISTOR & CONDENSER
To tachometer
VEHICLE SPEED SENSOR
CRANK ANGLE SENSOR
DISTRIBUTOR
SPARK PLUG
FUSIBLE LINK
BATTERY
IGNITION SWITCH
INHIBITOR SWITCH
NEUTRAL SWITCH

ECCS DIAGNOSTIC PROCEDURE—1990 PULSAR/SENTRA

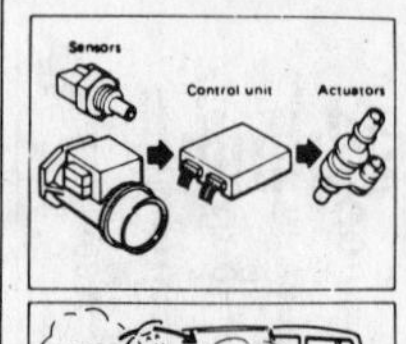
Sensors
Control unit
Actuators

INFO.
CAUSE

Introduction

The engine has an electronic control unit to control major systems such as fuel control, ignition control, idle speed control, etc. The control unit accepts input signals from sensors and instantly drives actuators. It is essential that both kinds of signals are proper and stable. At the same time, it is important that there are no conventional problems such as vacuum leaks, fouled spark plugs, or other problems with the engine.

It is much more difficult to diagnose a problem that occurs intermittently rather than continuously. Most intermittent problems are caused by poor electric connections or faulty wiring. In this case, careful checking of suspicious circuits may help prevent the replacement of good parts.

A visual check only may not find the cause of the problems. A road test with a circuit tester connected to a suspected circuit should be performed.

Before undertaking actual checks, take just a few minutes to talk with a customer who approaches with a driveability complaint. The customer is a very good supplier of information on such problems, especially intermittent ones. Through the talks with the customer, find out what symptoms are present and under what conditions they occur.

Start your diagnosis by looking for "conventional" problems first. This is one of the best ways to troubleshoot driveability problems on an electronically controlled engine vehicle.

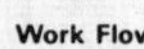

Work Flow

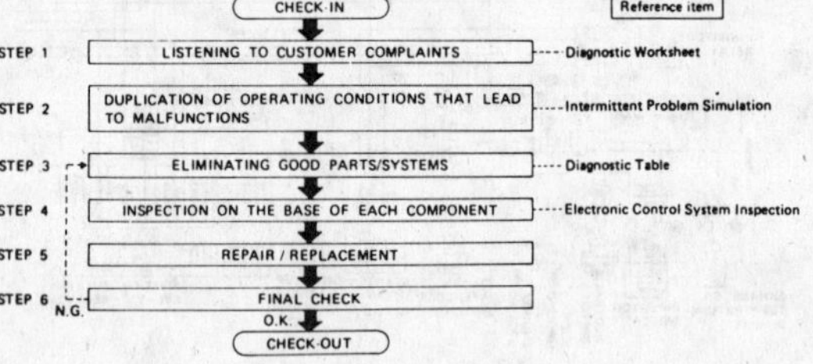
CHECK-IN
Reference item
STEP 1
LISTENING TO CUSTOMER COMPLAINTS
Diagnostic Worksheet
STEP 2
DUPLICATION OF OPERATING CONDITIONS THAT LEAD TO MALFUNCTIONS
Intermittent Problem Simulation
STEP 3
ELIMINATING GOOD PARTS/SYSTEMS
Diagnostic Table
STEP 4
INSPECTION ON THE BASE OF EACH COMPONENT
Electronic Control System Inspection
STEP 5
REPAIR / REPLACEMENT
STEP 6
FINAL CHECK
N.G.
O.K.
CHECK-OUT

ECCS DIAGNOSTIC PROCEDURE—1990 PULSAR/SENTRA

KEY POINTS

WHAT ..... Vehicle & engine model
WHEN ..... Date, Frequencies
WHERE ..... Road conditions
HOW ..... Operating conditions, Weather conditions, Symptoms

Diagnostic Worksheet

There are many kinds of operating conditions that lead to malfunctions on engine components.

A good grasp of such conditions can make troubleshooting faster and more accurate.

In general, feelings for a problem depend on each customer. It is important to fully understand the symptoms or under what conditions a customer complains.

Make good use of a diagnostic worksheet such as the one shown below in order to utilize all the complaints for troubleshooting.

WORKSHEET SAMPLE

Customer name MR/MS	Model & Year	VIN
Engine #	Trans.	Mileage
Incident Date	Manuf. Date	In Service Date
Symptoms	☐ Startability	☐ Impossible to start ☐ No combustion ☐ Partial combustion ☐ Partial combustion affected by throttle position ☐ Partial combustion NOT affected by throttle position ☐ Possible but hard to start ☐ Others [ ]
	☐ Idling	☐ No fast idle ☐ Unstable ☐ High idle ☐ Low idle ☐ Others [ ]
	☐ Driveability	☐ Stumble ☐ Surge ☐ Detonation ☐ Lack of power ☐ Intake backfire ☐ Exhaust backfire ☐ Others [ ]
	☐ Engine stall	☐ At the time of start ☐ While idling ☐ While accelerating ☐ While decelerating ☐ Just after stopping ☐ While loading
Incident occurrence		☐ Just after delivery ☐ Recently ☐ In the morning ☐ At night ☐ In the daytime
Frequency		☐ All the time ☐ Under certain conditions ☐ Sometimes
Weather conditions		☐ Not effected
	Weather	☐ Fine ☐ Raining ☐ Snowing ☐ Others [ ]
	Temperature	☐ Hot ☐ Warm ☐ Cool ☐ Cold ☐ Humid °F
Engine conditions		☐ Cold ☐ During warm-up ☐ After warm-up Engine speed 0 2,000 4,000 6,000 8,000 rpm
Road conditions		☐ In town ☐ In suburbs ☐ Highway ☐ Off road (up/down)
Driving conditions		☐ Not affected ☐ At starting ☐ While idling ☐ At racing ☐ While accelerating ☐ While cruising ☐ While decelerating ☐ While turning (RH/LH) Vehicle speed 0 10 20 30 40 50 60 MPH
Check engine light		☐ Turned on ☐ Not turned on

GRRRRR
GRRRR
THIS is the symptom, isn't it ?

Intermittent Problem Simulation

In order to duplicate an intermittent problem, it is effective to create similar conditions for component parts, under which the problem might occur.

Perform the activity listed under Service procedure and note the result.

	Variable factor	Influential part	Target condition	Service procedure
1	Mixture ratio	Pressure regulator	Made lean	Remove vacuum hose and apply vacuum.
			Made rich	Remove vacuum hose and apply pressure.
2	Ignition timing	Distributor	Advanced	Rotate distributor clockwise.
			Retarded	Rotate distributor counterclockwise.
3	Mixture ratio feedback control	Exhaust gas sensor	Suspended	Disconnect exhaust gas sensor harness connector.
		Control unit	Operation check	Perform self-diagnosis (Mode I/II) at 2,000 rpm.
4	Idle speed	Throttle body	Raised	Turn idle adjust screw counterclockwise.
			Lowered	Turn idle adjust screw clockwise.
5	Electric connection (Electric continuity)	Harness connectors and wires	Poor electric connection or faulty wiring	Tap or wiggle.
				Race engine rapidly. See if the torque reaction of the engine unit causes electric breaks.
6	Temperature	Control unit	Cooled	Cool with an icing spray or similar device.
			Warmed	Heat with a hair drier. [WARNING: Do not overheat the unit.]
7	Moisture	Electric parts	Damp	Wet [WARNING: Do not directly pour water on components. Use a mist sprayer.]
8	Electric loads	Load switches	Loaded	Turn on head lights, air conditioner, rear defogger, etc.
9	Idle switch condition	Control unit	ON-OFF switching	Perform self-diagnosis (Mode IV).
10	Ignition spark	Timing light	Spark power check	Try to flash timing light for each cylinder.

ECCS SPECIFICATIONS CHART—1990 PULSAR/SENTRA

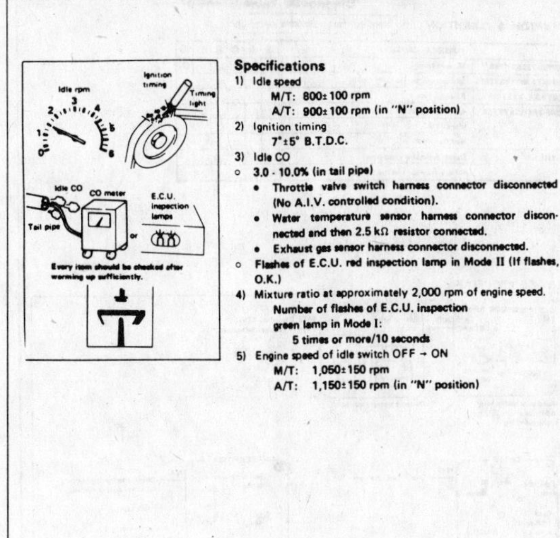
Idle rpm
Ignition timing
Timing light
Idle CO
CO meter
Tail pipe
E.C.U. inspection lamps
or
Every item should be checked after warming up sufficiently.

Specifications

1) Idle speed
   M/T: 800±100 rpm
   A/T: 900±100 rpm (in "N" position)
2) Ignition timing
   7°±5° B.T.D.C.
3) Idle CO
   ○ 3.0 - 10.0% (in tail pipe)
   - Throttle valve switch harness connector disconnected (No A.I.V. controlled condition).
   - Water temperature sensor harness connector disconnected and then 2.5 kΩ resistor connected.
   - Exhaust gas sensor harness connector disconnected.

   ○ Flashes of E.C.U. red inspection lamp in Mode II (If flashes, O.K.)
4) Mixture ratio at approximately 2,000 rpm of engine speed.
   Number of flashes of E.C.U. inspection green lamp in Mode I:
   5 times or more/10 seconds
5) Engine speed of idle switch OFF → ON
   M/T: 1,050±150 rpm
   A/T: 1,150±150 rpm (in "N" position)

ECCS DIAGNOSTIC CHART—1990 PULSAR/SENTRA

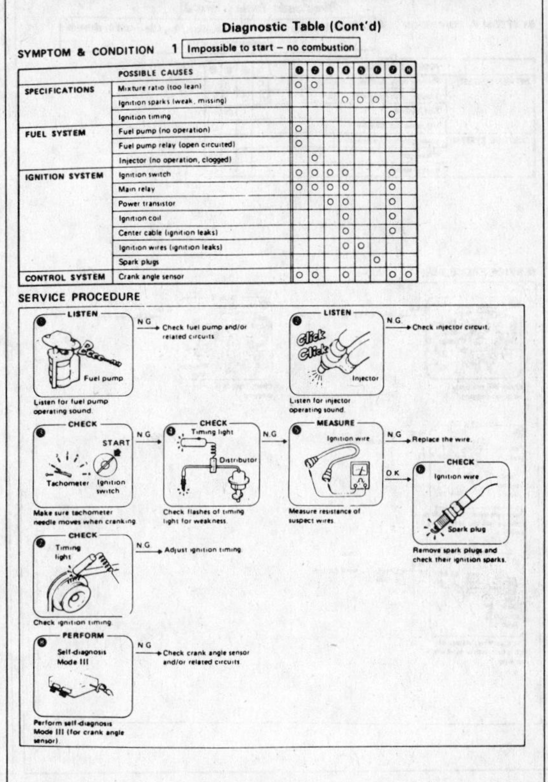

Diagnostic Table (Cont'd)

SYMPTOM & CONDITION 1 Impossible to start – no combustion

	POSSIBLE CAUSES	1	2	3	4	5	6	7	8
SPECIFICATIONS	Mixture ratio (too lean)	○	○						
	Ignition sparks (weak, missing)				○	○	○		
	Ignition timing							○	
FUEL SYSTEM	Fuel pump (no operation)	○							
	Fuel pump relay (open circuited)	○							
	Injector (no operation, clogged)		○						
IGNITION SYSTEM	Ignition switch	○	○	○	○			○	
	Main relay	○	○	○	○			○	
	Power transistor			○	○			○	
	Ignition coil				○			○	
	Center cable (ignition leaks)				○			○	
	Ignition wires (ignition leaks)				○	○			
	Spark plugs						○		
CONTROL SYSTEM	Crank angle sensor	○	○		○			○	○

SERVICE PROCEDURE

LISTEN
Fuel pump
N.G. → Check fuel pump and/or related circuits.
Listen for fuel pump operating sound.
LISTEN
Click Click
Injector
N.G. → Check injector circuit.
Listen for injector operating sound.
CHECK
START
Tachometer
Ignition switch
N.G.
Make sure tachometer needle moves when cranking.
CHECK
Timing light
Distributor
N.G.
Check flashes of timing light for weakness.
MEASURE
Ignition wire
N.G. → Replace the wire.
Measure resistance of suspect wires.
O.K.
CHECK
Ignition wire
Spark plug
Remove spark plugs and check their ignition sparks.
CHECK
Timing light
N.G. → Adjust ignition timing.
Check ignition timing.
PERFORM
Self-diagnosis Mode III
N.G. → Check crank angle sensor and/or related circuits.
Perform self-diagnosis Mode III (for crank angle sensor).

ECCS DIAGNOSTIC CHART—1990 PULSAR/SENTRA

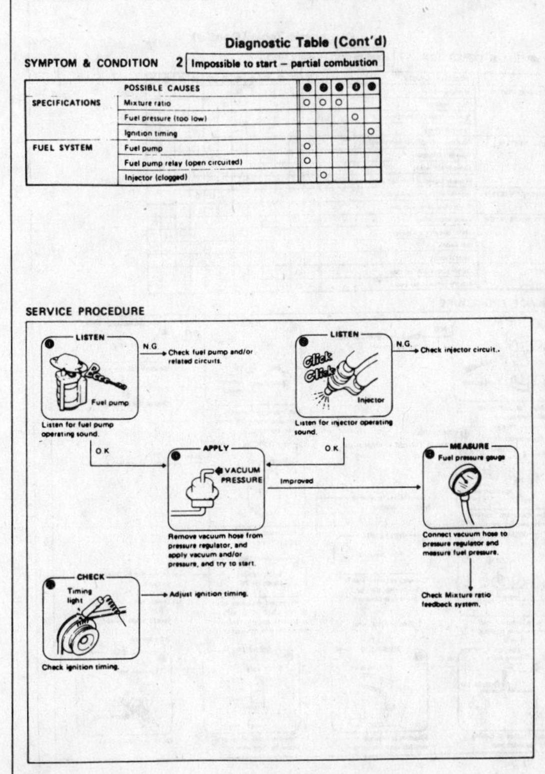

Diagnostic Table (Cont'd)

SYMPTOM & CONDITION 2 Impossible to start – partial combustion

	POSSIBLE CAUSES	1	2	3	4	5
SPECIFICATIONS	Mixture ratio	○	○	○		
	Fuel pressure (too low)				○	
	Ignition timing					○
FUEL SYSTEM	Fuel pump	○				
	Fuel pump relay (open circuited)	○				
	Injector (clogged)		○			

SERVICE PROCEDURE

LISTEN
Fuel pump
N.G. → Check fuel pump and/or related circuits.
Listen for fuel pump operating sound.
LISTEN
Click Click
Injector
N.G. → Check injector circuit.
Listen for injector operating sound.
O.K.
O.K.
APPLY
VACUUM PRESSURE
Improved
Remove vacuum hose from pressure regulator, and apply vacuum and/or pressure, and try to start.
MEASURE
Fuel pressure gauge
Connect vacuum hose to pressure regulator and measure fuel pressure.
Check Mixture ratio feedback system.
CHECK
Timing light
→ Adjust ignition timing.
Check ignition timing.

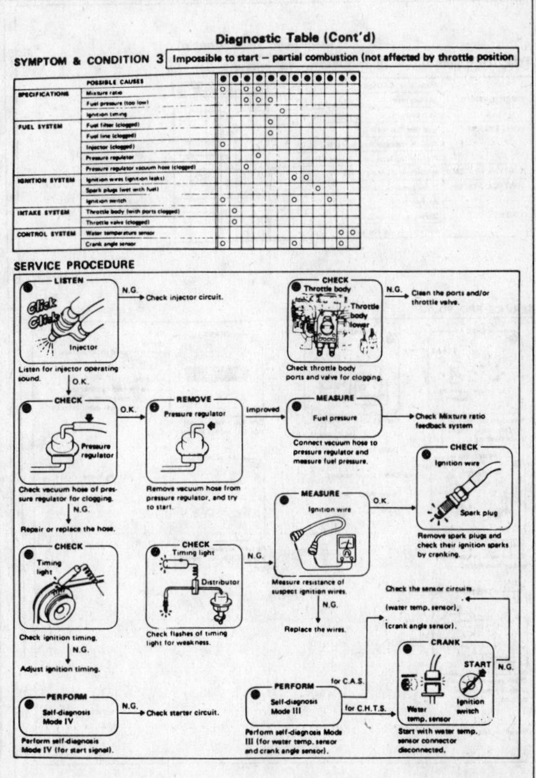

Diagnostic Table (Cont'd)

SYMPTOM & CONDITION 3 Impossible to start – partial combustion (not affected by throttle position)

	POSSIBLE CAUSES	1	2	3	4	5	6	7	8	9	10	11	12
SPECIFICATIONS	Mixture ratio	○		○	○								
	Fuel pressure (too low)			○	○	○							
	Ignition timing						○						
FUEL SYSTEM	Fuel filter (clogged)					○							
	Fuel line (clogged)					○							
	Injector (clogged)	○											
	Pressure regulator				○								
	Pressure regulator vacuum hose (clogged)			○									
IGNITION SYSTEM	Ignition wires (ignition leaks)							○	○				
	Spark plugs (wet with fuel)									○			
	Ignition switch	○						○			○		
INTAKE SYSTEM	Throttle body (with ports clogged)		○										
	Throttle valve (clogged)		○										
CONTROL SYSTEM	Water temperature sensor											○	○
	Crank angle sensor	○						○				○	

SERVICE PROCEDURE

LISTEN
Click Click
Injector
N.G. → Check injector circuit.
Listen for injector operating sound.
O.K.
CHECK
Throttle body
Throttle body lower
N.G. → Clean the ports and/or throttle valve.
Check throttle body ports and valve for clogging.
CHECK
Pressure regulator
O.K.
Check vacuum hose of pressure regulator for clogging.
N.G.
Repair or replace the hose.
REMOVE
Pressure regulator
Improved
Remove vacuum hose from pressure regulator, and try to start.
MEASURE
Fuel pressure
→ Check Mixture ratio feedback system
Connect vacuum hose to pressure regulator and measure fuel pressure.
CHECK
Ignition wire
Spark plug
Remove spark plugs and check their ignition sparks by cranking.
MEASURE
Ignition wire
O.K.
Measure resistance of suspect ignition wires.
N.G.
Replace the wires.
CHECK
Timing light
Check ignition timing.
N.G.
Adjust ignition timing.
CHECK
Timing light
Distributor
N.G.
Check flashes of timing light for weakness.
Check the sensor circuits. (water temp. sensor) (crank angle sensor)
CRANK
START
Water temp. sensor
Ignition switch
N.G.
Start with water temp. sensor connector disconnected.
PERFORM
Self-diagnosis Mode IV
N.G. → Check starter circuit.
Perform self-diagnosis Mode IV (for start signal).
PERFORM
Self-diagnosis Mode III
for C.A.S.
for C.H.T.S.
Perform self-diagnosis Mode III (for water temp. sensor and crank angle sensor).

ECCS DIAGNOSTIC CHART—1990 PULSAR/SENTRA

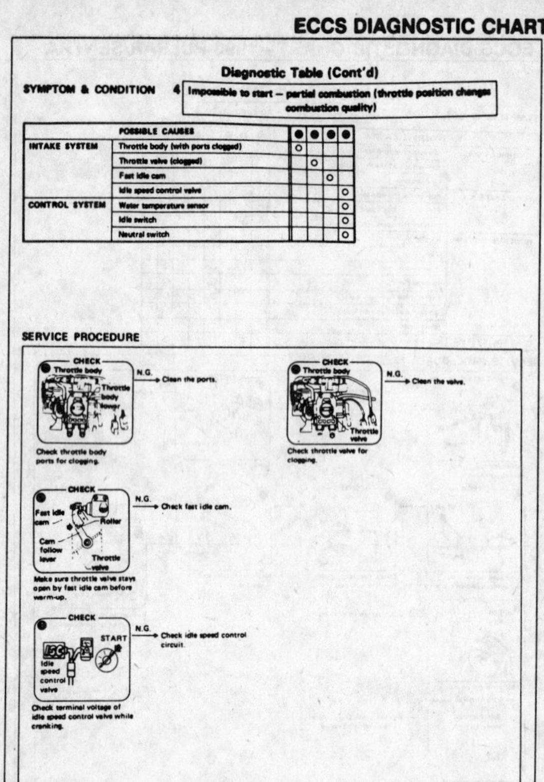

Diagnostic Table (Cont'd)

SYMPTOM & CONDITION 4 Impossible to start – partial combustion (throttle position changes combustion quality)

	POSSIBLE CAUSES	●	●	●	●
INTAKE SYSTEM	Throttle body (with ports clogged)	O			
	Throttle valve (clogged)		O		
	Fast idle cam			O	
	Idle speed control valve				O
CONTROL SYSTEM	Water temperature sensor				O
	Idle switch				O
	Neutral switch				O

SERVICE PROCEDURE

CHECK Throttle body
Throttle body
Throttle body
N.G. Clean the ports.
Check throttle body ports for clogging.
CHECK Throttle body
Throttle valve
N.G. Clean the valve.
Check throttle valve for clogging.
CHECK
Fast idle cam
Roller
Cam follow lever
Throttle valve
N.G. Check fast idle cam.
Make sure throttle valve stays open by fast idle cam before warm-up.
CHECK
START
Idle speed control valve
N.G. Check idle speed control circuit.
Check terminal voltage of idle speed control valve while cranking.

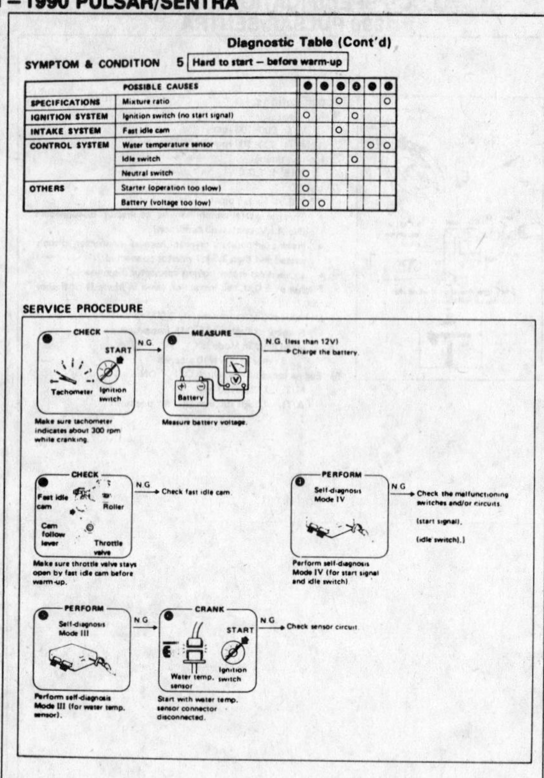

Diagnostic Table (Cont'd)

SYMPTOM & CONDITION 5 Hard to start – before warm-up

	POSSIBLE CAUSES	●	●	●	●	●	●
SPECIFICATIONS	Mixture ratio			O			O
IGNITION SYSTEM	Ignition switch (no start signal)	O			O		
INTAKE SYSTEM	Fast idle cam			O			
CONTROL SYSTEM	Water temperature sensor					O	O
	Idle switch				O		
	Neutral switch	O					
OTHERS	Starter (operation too slow)	O					
	Battery (voltage too low)	O	O				

SERVICE PROCEDURE

CHECK
START
Tachometer
Ignition switch
Make sure tachometer indicates about 300 rpm while cranking.
N.G.
MEASURE
Battery
Measure battery voltage.
N.G. (less than 12V) Charge the battery.
CHECK
Fast idle cam
Roller
Cam follow lever
Throttle valve
N.G. Check fast idle cam.
Make sure throttle valve stays open by fast idle cam before warm-up.
PERFORM
Self-diagnosis Mode IV
N.G. Check the malfunctioning switches and/or circuits.
(start signal),
(idle switch).)
Perform self-diagnosis Mode IV (for start signal and idle switch).
PERFORM
Self-diagnosis Mode III
Perform self-diagnosis Mode III (for water temp. sensor).
N.G.
CRANK
START
Water temp. sensor
Ignition switch
N.G. Check sensor circuit.
Start with water temp. sensor connector disconnected.

ECCS DIAGNOSTIC CHART—1990 PULSAR/SENTRA

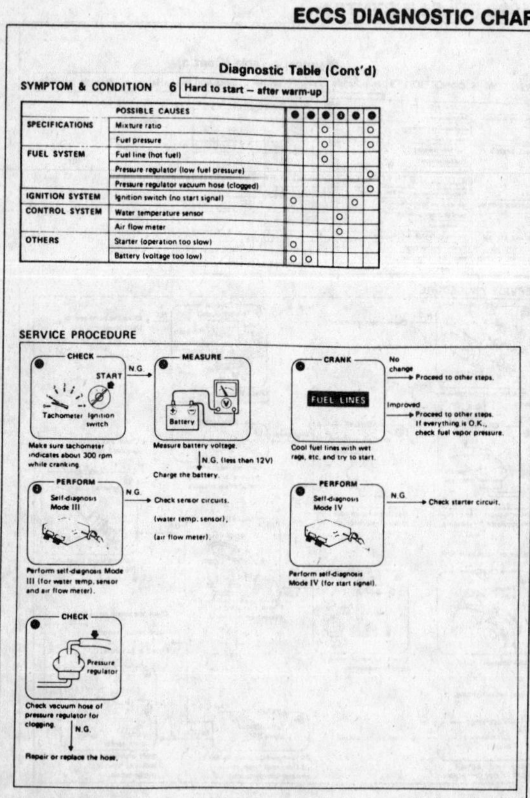

Diagnostic Table (Cont'd)

SYMPTOM & CONDITION 6 Hard to start – after warm-up

	POSSIBLE CAUSES	●	●	●	●	●	●
SPECIFICATIONS	Mixture ratio			O			O
	Fuel pressure			O			O
FUEL SYSTEM	Fuel line (hot fuel)			O			
	Pressure regulator (low fuel pressure)						O
	Pressure regulator vacuum hose (clogged)						O
IGNITION SYSTEM	Ignition switch (no start signal)	O				O	
CONTROL SYSTEM	Water temperature sensor				O		
	Air flow meter				O		
OTHERS	Starter (operation too slow)	O					
	Battery (voltage too low)	O	O				

SERVICE PROCEDURE

CHECK
START
Tachometer
Ignition switch
Make sure tachometer indicates about 300 rpm while cranking.
N.G.
MEASURE
Battery
Measure battery voltage.
N.G. (less than 12V)
Charge the battery.
CRANK
FUEL LINES
No change
Proceed to other steps.
Improved
Proceed to other steps. If everything is O.K., check fuel vapor pressure.
Cool fuel lines with wet rags, etc. and try to start.
PERFORM
Self-diagnosis Mode III
N.G. Check sensor circuits.
(water temp. sensor),
(air flow meter).
Perform self-diagnosis Mode III (for water temp. sensor and air flow meter).
PERFORM
Self-diagnosis Mode IV
N.G. Check starter circuit.
Perform self-diagnosis Mode IV (for start signal).
CHECK
Pressure regulator
Check vacuum hose of pressure regulator for clogging.
N.G.
Repair or replace the hose.

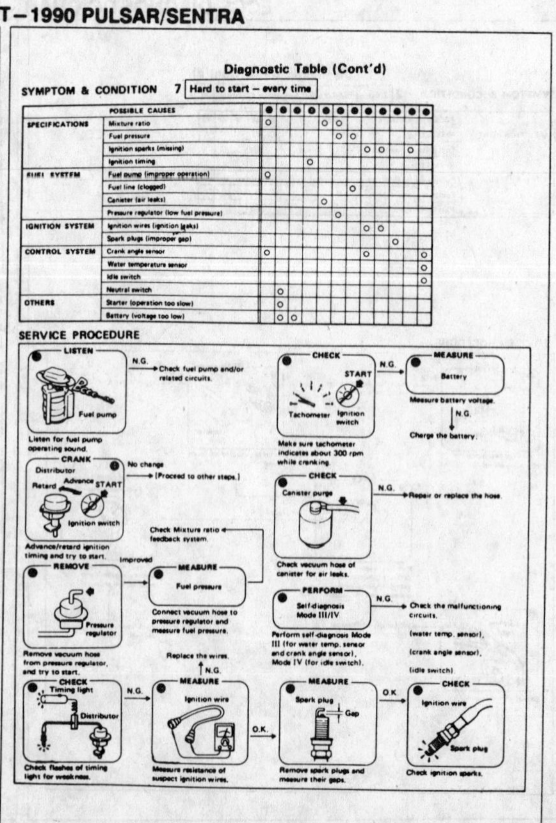

Diagnostic Table (Cont'd)

SYMPTOM & CONDITION 7 Hard to start – every time

	POSSIBLE CAUSES	●	●	●	●	●	●	●	●	●	●	●	●
SPECIFICATIONS	Mixture ratio	O				O	O						
	Fuel pressure						O	O					
	Ignition sparks (missing)								O	O		O	
	Ignition timing				O								
FUEL SYSTEM	Fuel pump (improper operation)	O											
	Fuel line (clogged)							O					
	Canister (air leaks)					O							
	Pressure regulator (low fuel pressure)						O						
IGNITION SYSTEM	Ignition wires (ignition leaks)								O	O			
	Spark plugs (improper gap)										O		
CONTROL SYSTEM	Crank angle sensor	O							O				O
	Water temperature sensor												O
	Idle switch												O
	Neutral switch		O										
OTHERS	Starter (operation too slow)		O										
	Battery (voltage too low)		O	O									

SERVICE PROCEDURE

LISTEN
Fuel pump
N.G. Check fuel pump and/or related circuits.
Listen for fuel pump operating sound.
CHECK
START
Tachometer
Ignition switch
Make sure tachometer indicates about 300 rpm while cranking.
N.G.
MEASURE
Battery
Measure battery voltage.
N.G.
Charge the battery.
CRANK
Distributor
Retard
Advance
START
Ignition switch
No change
(Proceed to other steps.)
Advance/retard ignition timing and try to start.
Improved
Check Mixture ratio feedback system.
CHECK
Canister purge
N.G. Repair or replace the hose.
Check vacuum hose of canister for air leaks.
REMOVE
Pressure regulator
Remove vacuum hose from pressure regulator, and try to start.
MEASURE
Fuel pressure
Connect vacuum hose to pressure regulator and measure fuel pressure.
PERFORM
Self-diagnosis Mode III/IV
N.G. Check the malfunctioning circuits.
(water temp. sensor),
(crank angle sensor),
(idle switch)
Perform self-diagnosis Mode III (for water temp. sensor and crank angle sensor), Mode IV (for idle switch).
Replace the wires.
N.G.
CHECK
Timing light
Distributor
Check flashes of timing light for weakness.
N.G.
MEASURE
Ignition wire
Measure resistance of suspect ignition wires.
O.K.
MEASURE
Spark plug
Gap
Remove spark plugs and measure their gaps.
O.K.
CHECK
Ignition wire
Spark plug
Check ignition sparks.

ECCS DIAGNOSTIC CHART—1990 PULSAR/SENTRA

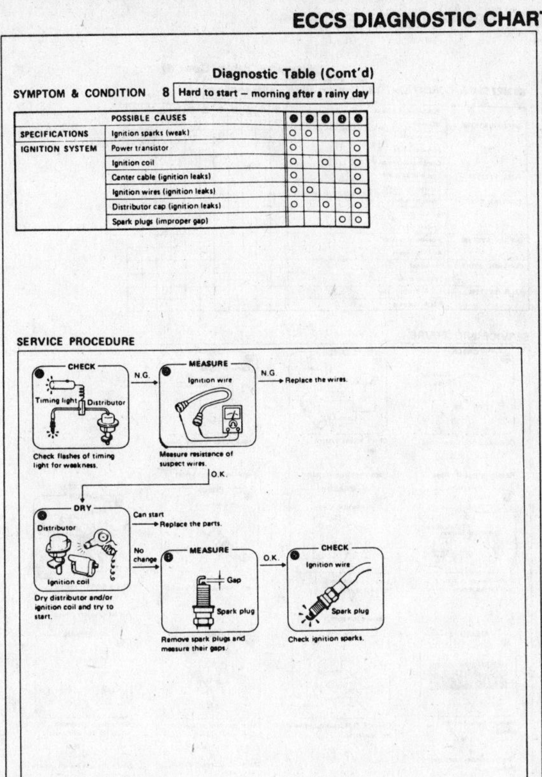

Diagnostic Table (Cont'd)

SYMPTOM & CONDITION 8 Hard to start – morning after a rainy day

	POSSIBLE CAUSES	1	2	3	4	5
SPECIFICATIONS	Ignition sparks (weak)	○	○			○
IGNITION SYSTEM	Power transistor	○				○
	Ignition coil	○		○		○
	Center cable (ignition leaks)	○				○
	Ignition wires (ignition leaks)	○	○			○
	Distributor cap (ignition leaks)	○		○		○
	Spark plugs (improper gap)				○	○

SERVICE PROCEDURE
CHECK
Timing light
Distributor
Check flashes of timing light for weakness.
N.G.
MEASURE
Ignition wire
Measure resistance of suspect wires.
N.G.
Replace the wires.
O.K.
DRY
Distributor
Ignition coil
Dry distributor and/or ignition coil and try to start.
Can start
Replace the parts.
No change
MEASURE
Gap
Spark plug
Remove spark plugs and measure their gaps.
O.K.
CHECK
Ignition wire
Spark plug
Check ignition sparks.

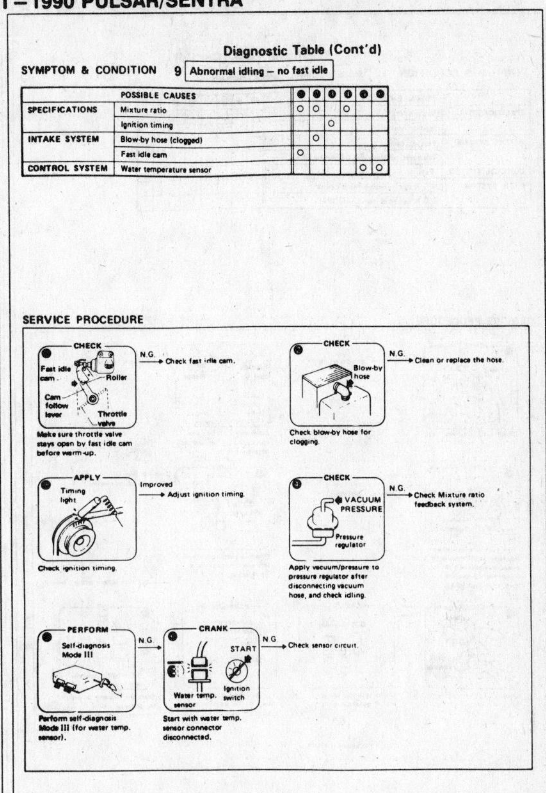

Diagnostic Table (Cont'd)

SYMPTOM & CONDITION 9 Abnormal idling – no fast idle

	POSSIBLE CAUSES	1	2	3	4	5	6
SPECIFICATIONS	Mixture ratio	○	○		○		
	Ignition timing			○			
INTAKE SYSTEM	Blow-by hose (clogged)		○				
	Fast idle cam	○					
CONTROL SYSTEM	Water temperature sensor					○	○

SERVICE PROCEDURE
CHECK
Fast idle cam
Roller
Cam follow lever
Throttle valve
Make sure throttle valve stays open by fast idle cam before warm-up.
N.G.
Check fast idle cam.
CHECK
Blow-by hose
Check blow-by hose for clogging.
N.G.
Clean or replace the hose.
APPLY
Timing light
Check ignition timing.
Improved
Adjust ignition timing.
CHECK
VACUUM PRESSURE
Pressure regulator
Apply vacuum/pressure to pressure regulator after disconnecting vacuum hose, and check idling.
N.G.
Check Mixture ratio feedback system.
PERFORM
Self-diagnosis Mode III
Perform self-diagnosis Mode III (for water temp. sensor).
N.G.
CRANK
START
Water temp. sensor
Ignition switch
Start with water temp. sensor connector disconnected.
N.G.
Check sensor circuit.

ECCS DIAGNOSTIC CHART—1990 PULSAR/SENTRA

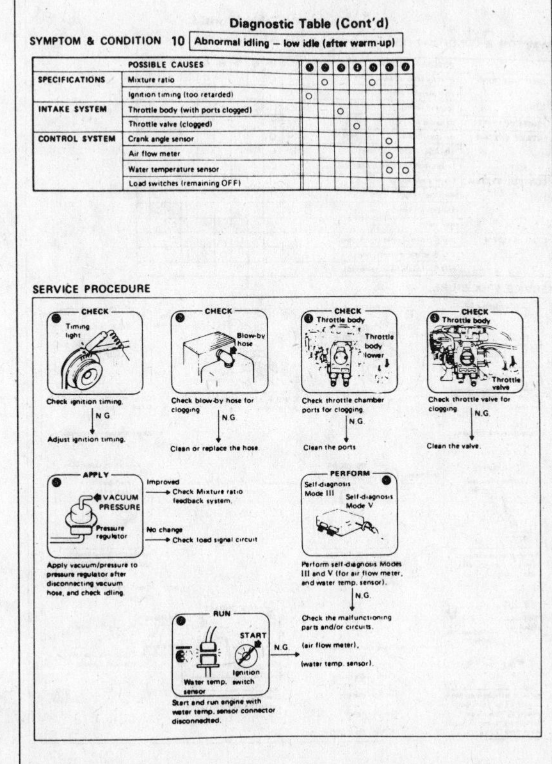

Diagnostic Table (Cont'd)

SYMPTOM & CONDITION 10 Abnormal idling – low idle (after warm-up)

	POSSIBLE CAUSES	1	2	3	4	5	6	7
SPECIFICATIONS	Mixture ratio		○			○		
	Ignition timing (too retarded)	○						
INTAKE SYSTEM	Throttle body (with ports clogged)			○				
	Throttle valve (clogged)				○			
CONTROL SYSTEM	Crank angle sensor						○	
	Air flow meter						○	
	Water temperature sensor						○	○
	Load switches (remaining OFF)							

SERVICE PROCEDURE
CHECK
Timing light
Check ignition timing.
N.G.
Adjust ignition timing.
CHECK
Blow-by hose
Check blow-by hose for clogging.
N.G.
Clean or replace the hose.
CHECK
Throttle body
Throttle body lower
Check throttle chamber ports for clogging.
N.G.
Clean the ports.
CHECK
Throttle body
Throttle valve
Check throttle valve for clogging.
N.G.
Clean the valve.
APPLY
VACUUM PRESSURE
Pressure regulator
Apply vacuum/pressure to pressure regulator after disconnecting vacuum hose, and check idling.
Improved
Check Mixture ratio feedback system.
No change
Check load signal circuit.
PERFORM
Self-diagnosis Mode III
Self-diagnosis Mode V
Perform self-diagnosis Modes III and V (for air flow meter, and water temp. sensor).
N.G.
Check the malfunctioning parts and/or circuits.
RUN
START
Water temp. sensor
Ignition switch
Start and run engine with water temp. sensor connector disconnected.
N.G.
(air flow meter),
(water temp. sensor).

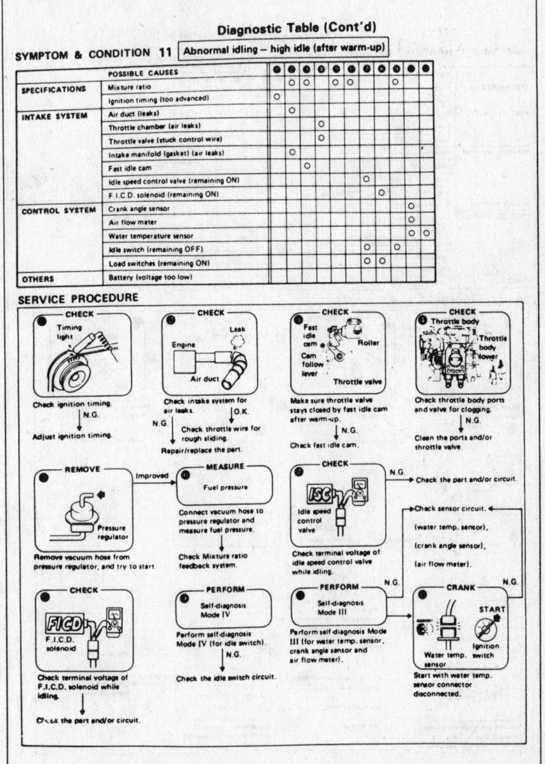

Diagnostic Table (Cont'd)

SYMPTOM & CONDITION 11 Abnormal idling – high idle (after warm-up)

	POSSIBLE CAUSES	1	2	3	4	5	6	7	8	9	10	11
SPECIFICATIONS	Mixture ratio		○	○		○	○			○		
	Ignition timing (too advanced)	○										
INTAKE SYSTEM	Air duct (leaks)		○									
	Throttle chamber (air leaks)				○							
	Throttle valve (stuck control wire)				○							
	Intake manifold (gasket) (air leaks)		○									
	Fast idle cam			○								
	Idle speed control valve (remaining ON)							○				
	F.I.C.D. solenoid (remaining ON)								○			
CONTROL SYSTEM	Crank angle sensor										○	
	Air flow meter										○	
	Water temperature sensor										○	○
	Idle switch (remaining OFF)							○		○		
	Load switches (remaining ON)							○	○			
OTHERS	Battery (voltage too low)											

SERVICE PROCEDURE
CHECK
Timing light
Check ignition timing.
N.G.
Adjust ignition timing.
CHECK
Engine
Leak
Air duct
Check intake system for air leaks.
N.G.
Repair/replace the part.
O.K.
Check throttle wire for rough sliding.
N.G.
CHECK
Fast idle cam
Roller
Cam follow lever
Throttle valve
Make sure throttle valve stays closed by fast idle cam after warm-up.
N.G.
Check fast idle cam.
CHECK
Throttle body
Throttle body lower
Check throttle body ports and valve for clogging.
N.G.
Clean the ports and/or throttle valve.
REMOVE
Pressure regulator
Remove vacuum hose from pressure regulator, and try to start.
Improved
MEASURE
Fuel pressure
Connect vacuum hose to pressure regulator and measure fuel pressure.
Check Mixture ratio feedback system.
CHECK
ISC
Idle speed control valve
Check terminal voltage of idle speed control valve while idling.
N.G.
Check the part and/or circuit.
Check sensor circuit.
(water temp. sensor),
(crank angle sensor),
(air flow meter).
CHECK
FICD
F.I.C.D. solenoid
Check terminal voltage of F.I.C.D. solenoid while idling.
Check the part and/or circuit.
PERFORM
Self-diagnosis Mode IV
Perform self-diagnosis Mode IV (for idle switch).
N.G.
Check the idle switch circuit.
PERFORM
Self-diagnosis Mode III
Perform self-diagnosis Mode III (for water temp. sensor, crank angle sensor and air flow meter).
N.G.
CRANK
START
Water temp. sensor
Ignition switch
Start with water temp. sensor connector disconnected.
N.G.

ECCS DIAGNOSTIC CHART—1990 PULSAR/SENTRA

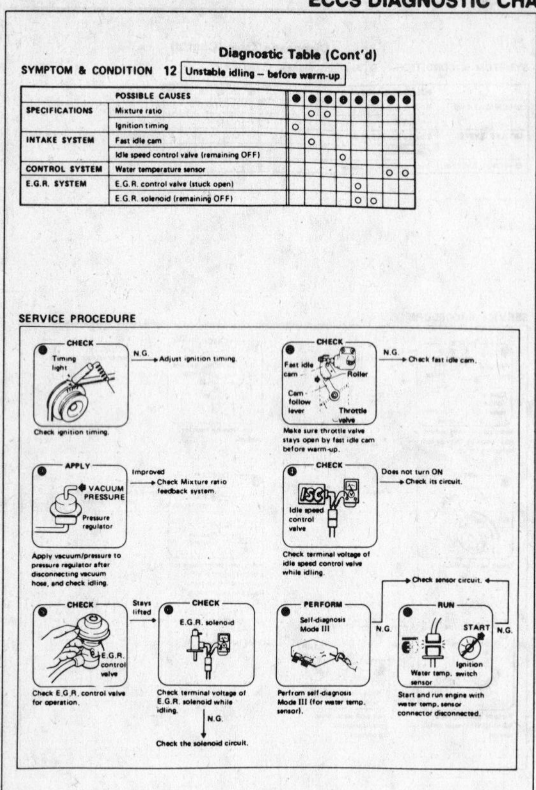

Diagnostic Table (Cont'd)

SYMPTOM & CONDITION 12 Unstable idling – before warm-up

	POSSIBLE CAUSES	●	●	●	●	●	●	●	●
SPECIFICATIONS	Mixture ratio		○	○					
	Ignition timing	○							
INTAKE SYSTEM	Fast idle cam		○						
	Idle speed control valve (remaining OFF)				○				
CONTROL SYSTEM	Water temperature sensor							○	○
E.G.R. SYSTEM	E.G.R. control valve (stuck open)					○			
	E.G.R. solenoid (remaining OFF)					○	○		

SERVICE PROCEDURE

CHECK – Timing light – N.G. → Adjust ignition timing.
Check ignition timing.

CHECK – Fast idle cam, Roller, Cam-follow lever, Throttle valve – N.G. → Check fast idle cam.
Make sure throttle valve stays open by fast idle cam before warm-up.

APPLY – VACUUM PRESSURE, Pressure regulator – Improved → Check Mixture ratio feedback system.
Apply vacuum/pressure to pressure regulator after disconnecting vacuum hose, and check idling.

CHECK – ISC, Idle speed control valve – Does not turn ON → Check its circuit.
Check terminal voltage of idle speed control valve while idling.

CHECK – E.G.R. control valve – Stays lifted →
Check E.G.R. control valve for operation.

CHECK – E.G.R. solenoid
Check terminal voltage of E.G.R. solenoid while idling.
N.G. → Check the solenoid circuit.

PERFORM – Self-diagnosis Mode III – N.G. → Check sensor circuit.
Perfrom self-diagnosis Mode III (for water temp. sensor).

RUN – START, Ignition switch, Water temp. sensor – N.G. → Check sensor circuit.
Start and run engine with water temp. sensor connector disconnected.

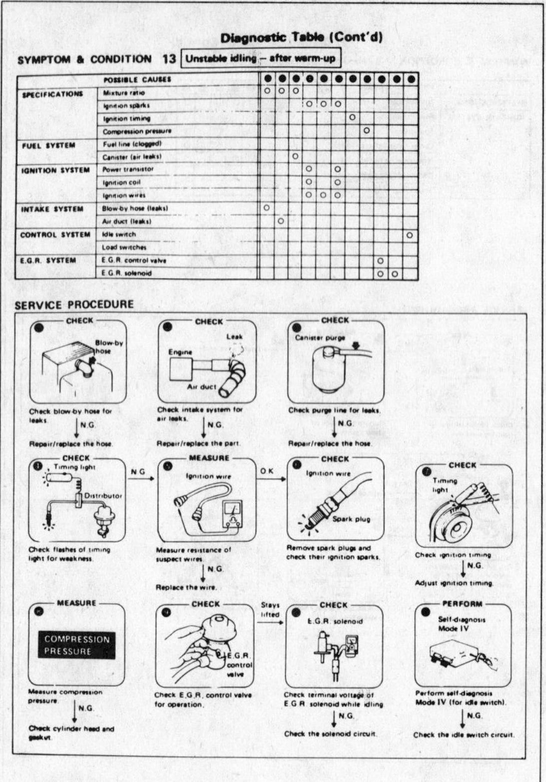

Diagnostic Table (Cont'd)

SYMPTOM & CONDITION 13 Unstable idling – after warm-up

	POSSIBLE CAUSES	●	●	●	●	●	●	●	●	●	●	●
SPECIFICATIONS	Mixture ratio	○	○	○								
	Ignition sparks				○	○	○					
	Ignition timing							○				
	Compression pressure								○			
FUEL SYSTEM	Fuel line (clogged)											
	Canister (air leaks)			○								
IGNITION SYSTEM	Power transistor				○		○					
	Ignition coil				○		○					
	Ignition wires				○	○	○					
INTAKE SYSTEM	Blow-by hose (leaks)	○										
	Air duct (leaks)		○									
CONTROL SYSTEM	Idle switch											○
	Load switches											
E.G.R. SYSTEM	E.G.R. control valve									○		
	E.G.R. solenoid									○	○	

SERVICE PROCEDURE

CHECK – Blow-by hose
Check blow-by hose for leaks.
N.G. → Repair/replace the hose.

CHECK – Engine, Leak, Air duct
Check intake system for air leaks.
N.G. → Repair/replace the part.

CHECK – Canister purge
Check purge line for leaks.
N.G. → Repair/replace the hose.

CHECK – Timing light, Distributor – N.G. →
Check flashes of timing light for weakness.

MEASURE – Ignition wire – O.K →
Measure resistance of suspect wires.
N.G. → Replace the wire.

CHECK – Ignition wire, Spark plug
Remove spark plugs and check their ignition sparks.

CHECK – Timing light
Check ignition timing.
N.G. → Adjust ignition timing.

MEASURE – COMPRESSION PRESSURE
Measure compression pressure.
N.G. → Check cylinder head and gaskut.

CHECK – E.G.R. control valve – Stays lifted →
Check E.G.R. control valve for operation.

CHECK – E.G.R. solenoid
Check terminal voltage of E.G.R. solenoid while idling.
N.G. → Check the solenoid circuit.

PERFORM – Self-diagnosis Mode IV
Perform self-diagnosis Mode IV (for idle switch).
N.G. → Check the idle switch circuit.

ECCS DIAGNOSTIC CHART—1990 PULSAR/SENTRA

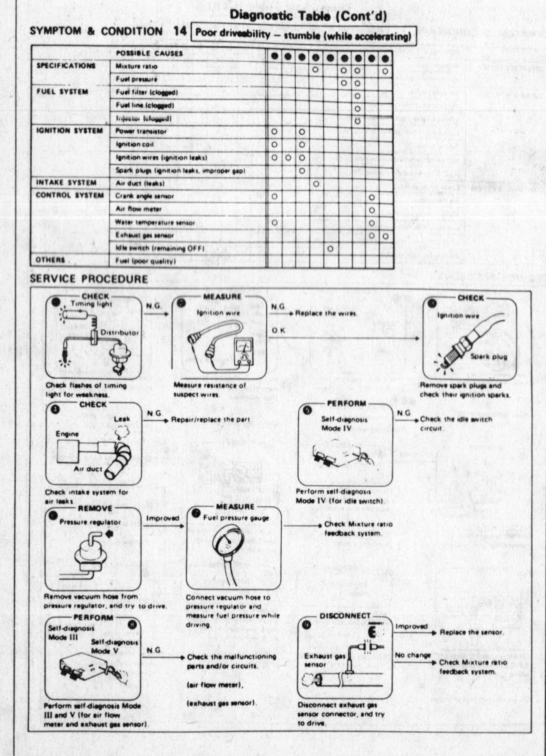

Diagnostic Table (Cont'd)

SYMPTOM & CONDITION 14 Poor driveability – stumble (while accelerating)

	POSSIBLE CAUSES	●	●	●	●	●	●	●	●	●
SPECIFICATIONS	Mixture ratio				○		○	○		○
	Fuel pressure						○	○		
FUEL SYSTEM	Fuel filter (clogged)							○		
	Fuel line (clogged)							○		
	Injector (clogged)							○		
IGNITION SYSTEM	Power transistor	○		○						
	Ignition coil	○		○						
	Ignition wires (ignition leaks)	○	○	○						
	Spark plugs (ignition leaks, improper gap)			○						
INTAKE SYSTEM	Air duct (leaks)				○					
CONTROL SYSTEM	Crank angle sensor	○							○	
	Air flow meter								○	
	Water temperature sensor	○							○	
	Exhaust gas sensor								○	○
	Idle switch (remaining OFF)					○				
OTHERS	Fuel (poor quality)									

SERVICE PROCEDURE

CHECK – Timing light, Distributor – N.G. →
Check flashes of timing light for weakness.

MEASURE – Ignition wire – N.G. → Replace the wires. O.K →
Measure resistance of suspect wires.

CHECK – Ignition wire, Spark plug
Remove spark plugs and check their ignition sparks.

CHECK – Engine, Leak, Air duct – N.G. → Repair/replace the part
Check intake system for air leaks.

PERFORM – Self-diagnosis Mode IV – N.G. → Check the idle switch circuit.
Perform self-diagnosis Mode IV (for idle switch).

REMOVE – Pressure regulator – Improved →
Remove vacuum hose from pressure regulator, and try to drive.

MEASURE – Fuel pressure gauge → Check Mixture ratio feedback system.
Connect vacuum hose to pressure regulator and measure fuel pressure while driving.

PERFORM – Self-diagnosis Mode III, Self-diagnosis Mode V – N.G. → Check the malfunctioning parts and/or circuits. (air flow meter), (exhaust gas sensor).
Perform self-diagnosis Mode III and V (for air flow meter and exhaust gas sensor).

DISCONNECT – Exhaust gas sensor – Improved → Replace the sensor. No change → Check Mixture ratio feedback system.
Disconnect exhaust gas sensor connector, and try to drive.

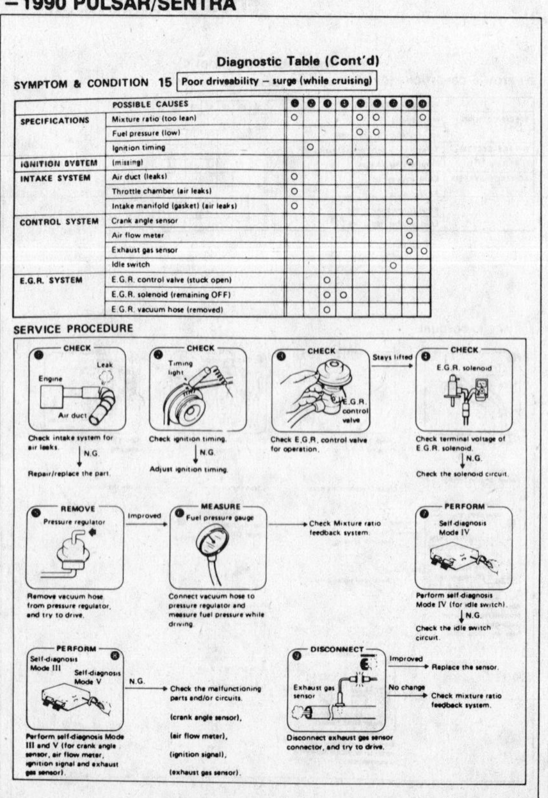

Diagnostic Table (Cont'd)

SYMPTOM & CONDITION 15 Poor driveability – surge (while cruising)

	POSSIBLE CAUSES	●	●	●	●	●	●	●	●	●
SPECIFICATIONS	Mixture ratio (too lean)	○				○	○			○
	Fuel pressure (low)					○	○			
	Ignition timing		○							
IGNITION SYSTEM	(missing)								○	
INTAKE SYSTEM	Air duct (leaks)	○								
	Throttle chamber (air leaks)	○								
	Intake manifold (gasket) (air leaks)	○								
CONTROL SYSTEM	Crank angle sensor								○	
	Air flow meter								○	
	Exhaust gas sensor								○	○
	Idle switch							○		
E.G.R. SYSTEM	E.G.R. control valve (stuck open)			○						
	E.G.R. solenoid (remaining OFF)			○	○					
	E.G.R. vacuum hose (removed)			○						

SERVICE PROCEDURE

CHECK – Engine, Leak, Air duct
Check intake system for air leaks.
N.G. → Repair/replace the part.

CHECK – Timing light
Check ignition timing.
N.G. → Adjust ignition timing.

CHECK – E.G.R. control valve – Stays lifted →
Check E.G.R. control valve for operation.

CHECK – E.G.R. solenoid
Check terminal voltage of E.G.R. solenoid.
N.G. → Check the solenoid circuit.

REMOVE – Pressure regulator – Improved →
Remove vacuum hose from pressure regulator, and try to drive.

MEASURE – Fuel pressure gauge → Check Mixture ratio feedback system.
Connect vacuum hose to pressure regulator and measure fuel pressure while driving.

PERFORM – Self-diagnosis Mode IV
Perform self-diagnosis Mode IV (for idle switch).
N.G. → Check the idle switch circuit.

PERFORM – Self-diagnosis Mode III, Self-diagnosis Mode V – N.G. → Check the malfunctioning parts and/or circuits. (crank angle sensor), (air flow meter), (ignition signal), (exhaust gas sensor).
Perform self-diagnosis Mode III and V (for crank angle sensor, air flow meter, ignition signal and exhaust gas sensor).

DISCONNECT – Exhaust gas sensor – Improved → Replace the sensor. No change → Check mixture ratio feedback system.
Disconnect exhaust gas sensor connector, and try to drive.

ECCS DIAGNOSTIC CHART—1990 PULSAR/SENTRA

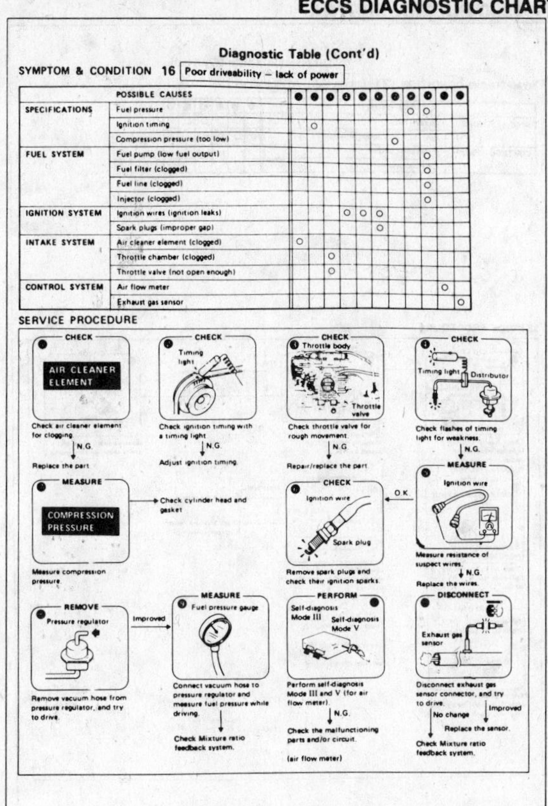

Diagnostic Table (Cont'd)

SYMPTOM & CONDITION 16 Poor driveability – lack of power

	POSSIBLE CAUSES	1	2	3	4	5	6	7	8	9	10	11
SPECIFICATIONS	Fuel pressure								○	○		
	Ignition timing		○									
	Compression pressure (too low)							○				
FUEL SYSTEM	Fuel pump (low fuel output)									○		
	Fuel filter (clogged)									○		
	Fuel line (clogged)									○		
	Injector (clogged)									○		
IGNITION SYSTEM	Ignition wires (ignition leaks)				○	○	○					
	Spark plugs (improper gap)						○					
INTAKE SYSTEM	Air cleaner element (clogged)	○										
	Throttle chamber (clogged)			○								
	Throttle valve (not open enough)			○								
CONTROL SYSTEM	Air flow meter										○	
	Exhaust gas sensor											○

SERVICE PROCEDURE
CHECK
AIR CLEANER ELEMENT
Check air cleaner element for clogging.
N.G.
Replace the part.
CHECK
Timing light
Check ignition timing with a timing light.
N.G.
Adjust ignition timing.
CHECK
Throttle body
Throttle valve
Check throttle valve for rough movement.
N.G.
Repair/replace the part.
CHECK
Timing light
Distributor
Check flashes of timing light for weakness.
N.G.
MEASURE
COMPRESSION PRESSURE
Measure compression pressure.
Check cylinder head and gasket.
CHECK
Ignition wire
Spark plug
Remove spark plugs and check their ignition sparks.
O.K.
MEASURE
Ignition wire
Measure resistance of suspect wires.
N.G.
Replace the wires.
REMOVE
Pressure regulator
Remove vacuum hose from pressure regulator, and try to drive.
Improved
MEASURE
Fuel pressure gauge
Connect vacuum hose to pressure regulator and measure fuel pressure while driving.
Check Mixture ratio feedback system.
PERFORM
Self-diagnosis Mode III
Self-diagnosis Mode V
Perform self-diagnosis Mode III and V (for air flow meter).
N.G.
Check the malfunctioning parts and/or circuit.
(air flow meter)
DISCONNECT
Exhaust gas sensor
Disconnect exhaust gas sensor connector, and try to drive.
Improved
No change
Replace the sensor.
Check Mixture ratio feedback system.

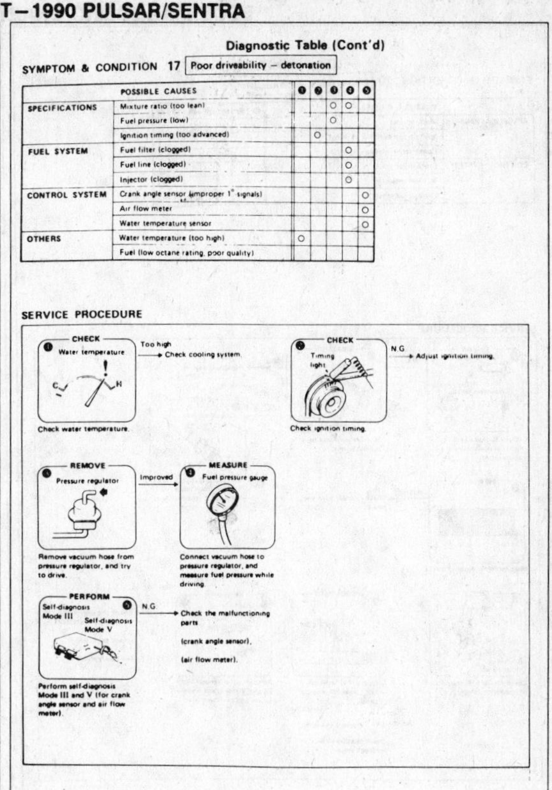

Diagnostic Table (Cont'd)

SYMPTOM & CONDITION 17 Poor driveability – detonation

	POSSIBLE CAUSES	1	2	3	4	5
SPECIFICATIONS	Mixture ratio (too lean)			○	○	
	Fuel pressure (low)			○		
	Ignition timing (too advanced)		○			
FUEL SYSTEM	Fuel filter (clogged)				○	
	Fuel line (clogged)				○	
	Injector (clogged)				○	
CONTROL SYSTEM	Crank angle sensor (improper 1° signals)					○
	Air flow meter					○
	Water temperature sensor					○
OTHERS	Water temperature (too high)	○				
	Fuel (low octane rating, poor quality)					

SERVICE PROCEDURE
CHECK
Water temperature
Too high
Check cooling system.
Check water temperature.
CHECK
Timing light
N.G.
Adjust ignition timing.
Check ignition timing.
REMOVE
Pressure regulator
Improved
MEASURE
Fuel pressure gauge
Remove vacuum hose from pressure regulator, and try to drive.
Connect vacuum hose to pressure regulator, and measure fuel pressure while driving.
PERFORM
Self-diagnosis Mode III
Self-diagnosis Mode V
N.G.
Check the malfunctioning parts
(crank angle sensor),
(air flow meter).
Perform self-diagnosis Mode III and V (for crank angle sensor and air flow meter).

ECCS DIAGNOSTIC CHART—1990 PULSAR/SENTRA

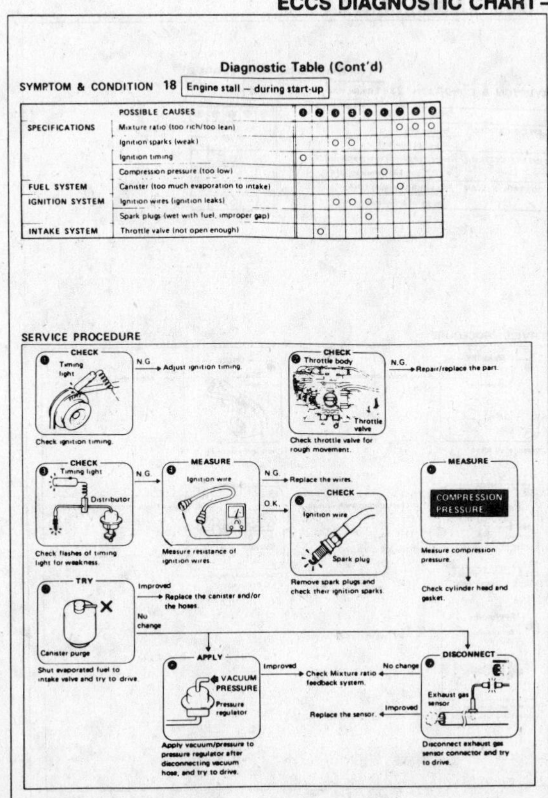

Diagnostic Table (Cont'd)

SYMPTOM & CONDITION 18 Engine stall – during start-up

	POSSIBLE CAUSES	1	2	3	4	5	6	7	8	9
SPECIFICATIONS	Mixture ratio (too rich/too lean)							○	○	○
	Ignition sparks (weak)			○	○					
	Ignition timing	○								
	Compression pressure (too low)						○			
FUEL SYSTEM	Canister (too much evaporation to intake)							○		
IGNITION SYSTEM	Ignition wires (ignition leaks)			○	○	○				
	Spark plugs (wet with fuel, improper gap)					○				
INTAKE SYSTEM	Throttle valve (not open enough)		○							

SERVICE PROCEDURE
CHECK
Timing light
N.G.
Adjust ignition timing.
Check ignition timing.
CHECK
Throttle body
Throttle valve
N.G.
Repair/replace the part.
Check throttle valve for rough movement.
CHECK
Timing light
Distributor
Check flashes of timing light for weakness.
N.G.
MEASURE
Ignition wire
N.G.
Replace the wires.
Measure resistance of ignition wires.
O.K.
CHECK
Ignition wire
Spark plug
Remove spark plugs and check their ignition sparks.
MEASURE
COMPRESSION PRESSURE
Measure compression pressure.
Check cylinder head and gasket.
TRY
Canister purge
Shut evaporated fuel to intake valve and try to drive.
Improved
Replace the canister and/or the hoses.
No change
APPLY
VACUUM PRESSURE
Pressure regulator
Apply vacuum/pressure to pressure regulator after disconnecting vacuum hose, and try to drive.
Improved
Check Mixture ratio feedback system.
No change
DISCONNECT
Exhaust gas sensor
Improved
Replace the sensor.
Disconnect exhaust gas sensor connector and try to drive.

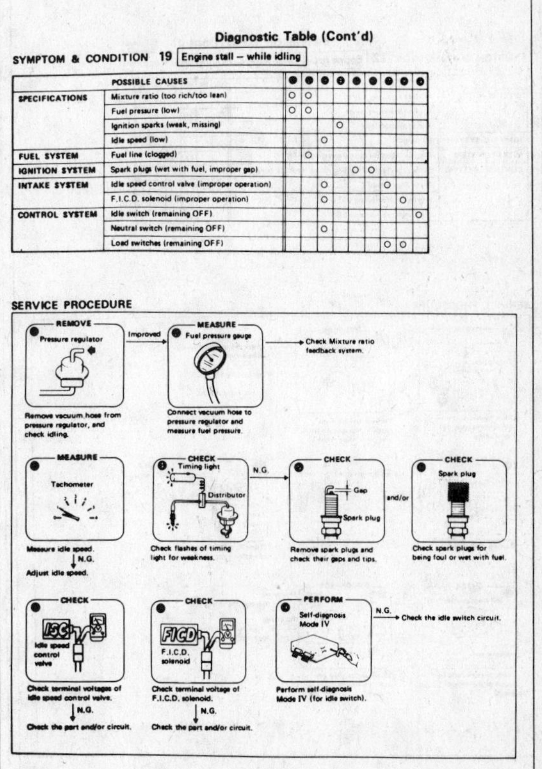

Diagnostic Table (Cont'd)

SYMPTOM & CONDITION 19 Engine stall – while idling

	POSSIBLE CAUSES	1	2	3	4	5	6	7	8	9
SPECIFICATIONS	Mixture ratio (too rich/too lean)	○	○							
	Fuel pressure (low)	○	○							
	Ignition sparks (weak, missing)				○					
	Idle speed (low)			○						
FUEL SYSTEM	Fuel line (clogged)		○							
IGNITION SYSTEM	Spark plugs (wet with fuel, improper gap)					○	○			
INTAKE SYSTEM	Idle speed control valve (improper operation)			○				○		
	F.I.C.D. solenoid (improper operation)			○					○	
CONTROL SYSTEM	Idle switch (remaining OFF)									○
	Neutral switch (remaining OFF)			○						
	Load switches (remaining OFF)							○	○	

SERVICE PROCEDURE
REMOVE
Pressure regulator
Improved
MEASURE
Fuel pressure gauge
Check Mixture ratio feedback system.
Remove vacuum hose from pressure regulator, and check idling.
Connect vacuum hose to pressure regulator and measure fuel pressure.
MEASURE
Tachometer
Measure idle speed.
N.G.
Adjust idle speed.
CHECK
Timing light
Distributor
N.G.
Check flashes of timing light for weakness.
CHECK
Gap
Spark plug
Remove spark plugs and check their gaps and tips.
and/or
CHECK
Spark plug
Check spark plugs for being foul or wet with fuel.
CHECK
ISC
Idle speed control valve
Check terminal voltages of idle speed control valve.
N.G.
Check the part and/or circuit.
CHECK
FICD
F.I.C.D. solenoid
Check terminal voltage of F.I.C.D. solenoid.
N.G.
Check the part and/or circuit.
PERFORM
Self-diagnosis Mode IV
N.G.
Check the idle switch circuit.
Perform self-diagnosis Mode IV (for idle switch).

ECCS DIAGNOSTIC CHART–1990 PULSAR/SENTRA

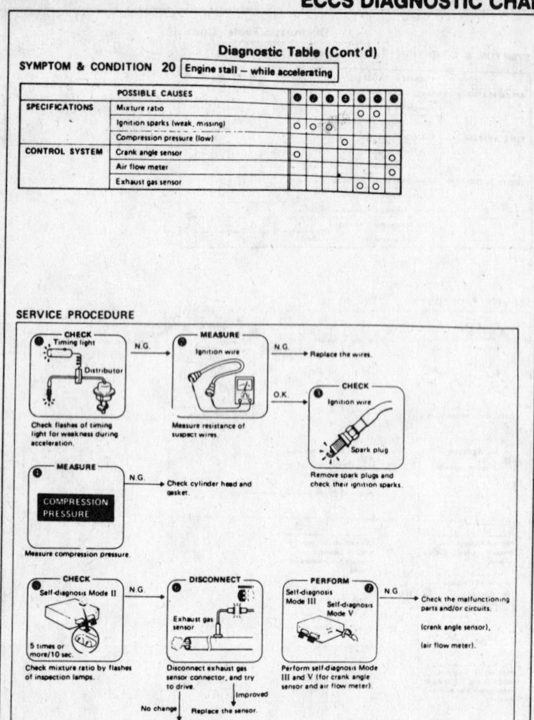

Diagnostic Table (Cont'd)

SYMPTOM & CONDITION 20 Engine stall – while accelerating

POSSIBLE CAUSES		1	2	3	4	5	6	7
SPECIFICATIONS	Mixture ratio					○	○	
	Ignition sparks (weak, missing)	○	○	○				
	Compression pressure (low)				○			
CONTROL SYSTEM	Crank angle sensor	○						○
	Air flow meter							○
	Exhaust gas sensor					○	○	

SERVICE PROCEDURE

CHECK
Timing light
Distributor
Check flashes of timing light for weakness during acceleration.
N.G.
MEASURE
Ignition wire
Measure resistance of suspect wires.
N.G.
Replace the wires.
O.K.
CHECK
Ignition wire
Spark plug
Remove spark plugs and check their ignition sparks.
MEASURE
COMPRESSION PRESSURE
Measure compression pressure.
N.G.
Check cylinder head and gasket.
CHECK
Self-diagnosis Mode II
5 times or more/10 sec.
Check mixture ratio by flashes of inspection lamps.
N.G.
DISCONNECT
Exhaust gas sensor
Disconnect exhaust gas sensor connector, and try to drive.
Improved
Replace the sensor.
No change
Check Mixture ratio feedback system.
PERFORM
Self-diagnosis Mode III
Self-diagnosis Mode V
Perform self-diagnosis Mode III and V (for crank angle sensor and air flow meter).
N.G.
Check the malfunctioning parts and/or circuits.
(crank angle sensor),
(air flow meter).

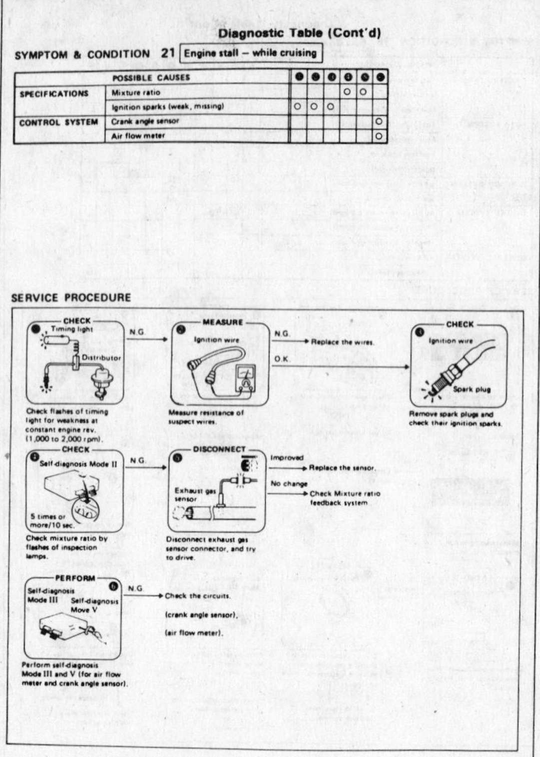

Diagnostic Table (Cont'd)

SYMPTOM & CONDITION 21 Engine stall – while cruising

POSSIBLE CAUSES		1	2	3	4	5	6
SPECIFICATIONS	Mixture ratio				○	○	
	Ignition sparks (weak, missing)	○	○	○			
CONTROL SYSTEM	Crank angle sensor						○
	Air flow meter						○

SERVICE PROCEDURE

CHECK
Timing light
Distributor
Check flashes of timing light for weakness at constant engine rev. (1,000 to 2,000 rpm).
N.G.
MEASURE
Ignition wire
Measure resistance of suspect wires.
N.G.
Replace the wires.
O.K.
CHECK
Ignition wire
Spark plug
Remove spark plugs and check their ignition sparks.
CHECK
Self-diagnosis Mode II
5 times or more/10 sec.
Check mixture ratio by flashes of inspection lamps.
N.G.
DISCONNECT
Exhaust gas sensor
Disconnect exhaust gas sensor connector, and try to drive.
Improved
Replace the sensor.
No change
Check Mixture ratio feedback system.
PERFORM
Self-diagnosis Mode III
Self-diagnosis Move V
Perform self-diagnosis Mode III and V (for air flow meter and crank angle sensor).
N.G.
Check the circuits.
(crank angle sensor),
(air flow meter).

ECCS DIAGNOSTIC CHART–1990 PULSAR/SENTRA

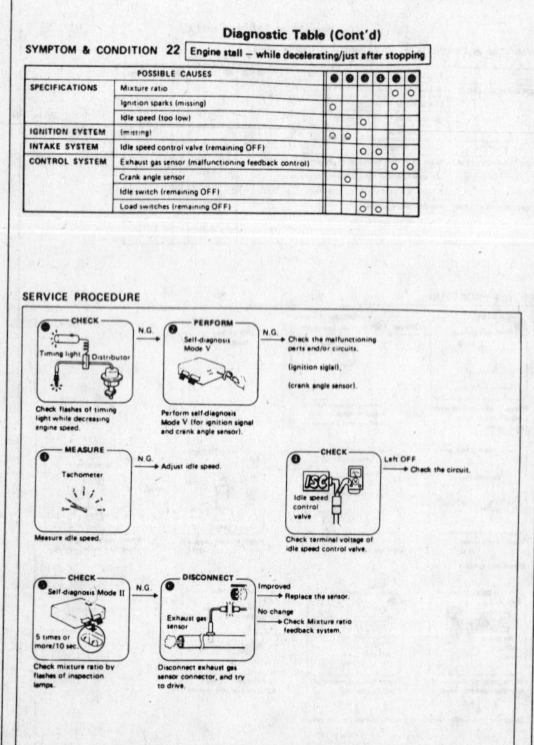

Diagnostic Table (Cont'd)

SYMPTOM & CONDITION 22 Engine stall – while decelerating/just after stopping

POSSIBLE CAUSES		1	2	3	4	5	6
SPECIFICATIONS	Mixture ratio					○	○
	Ignition sparks (missing)	○					
	Idle speed (too low)			○			
IGNITION SYSTEM	(missing)	○	○				
INTAKE SYSTEM	Idle speed control valve (remaining OFF)			○	○		
CONTROL SYSTEM	Exhaust gas sensor (malfunctioning feedback control)					○	○
	Crank angle sensor		○				
	Idle switch (remaining OFF)			○			
	Load switches (remaining OFF)			○	○		

SERVICE PROCEDURE

CHECK
Timing light
Distributor
Check flashes of timing light while decreasing engine speed.
N.G.
PERFORM
Self-diagnosis Mode V
Perform self-diagnosis Mode V (for ignition signal and crank angle sensor).
N.G.
Check the malfunctioning parts and/or circuits.
(ignition sigial),
(crank angle sensor).
MEASURE
Tachometer
Measure idle speed.
N.G.
Adjust idle speed.
CHECK
ISC
Idle speed control valve
Check terminal voltage of idle speed control valve.
Left OFF
Check the circuit.
CHECK
Self-diagnosis Mode II
5 times or more/10 sec.
Check mixture ratio by flashes of inspection lamps.
N.G.
DISCONNECT
Exhaust gas sensor
Disconnect exhaust gas sensor connector, and try to drive.
Improved
Replace the sensor.
No change
Check Mixture ratio feedback system.

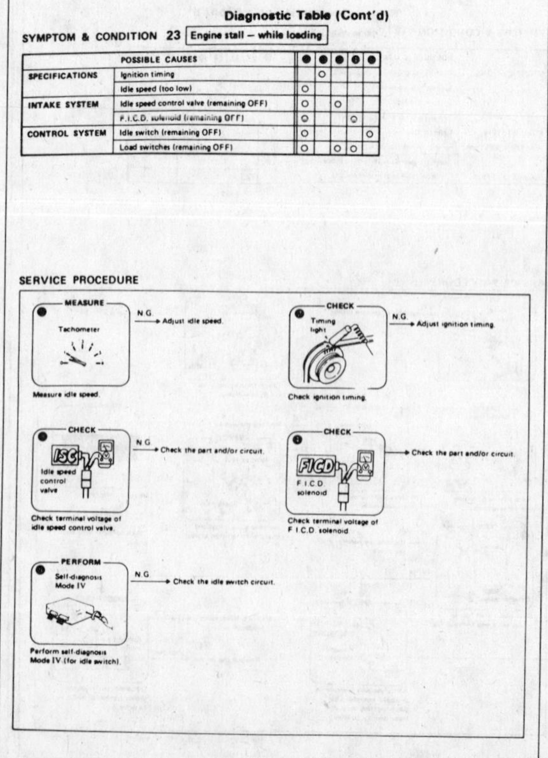

Diagnostic Table (Cont'd)

SYMPTOM & CONDITION 23 Engine stall – while loading

POSSIBLE CAUSES		1	2	3	4	5
SPECIFICATIONS	Ignition timing		○			
	Idle speed (too low)	○				
INTAKE SYSTEM	Idle speed control valve (remaining OFF)	○		○		
	F.I.C.D. solenoid (remaining OFF)	○			○	
CONTROL SYSTEM	Idle switch (remaining OFF)	○				○
	Load switches (remaining OFF)	○		○	○	

SERVICE PROCEDURE

MEASURE
Tachometer
Measure idle speed.
N.G.
Adjust idle speed.
CHECK
Timing light
Check ignition timing.
N.G.
Adjust ignition timing.
CHECK
ISC
Idle speed control valve
Check terminal voltage of idle speed control valve.
N.G.
Check the part and/or circuit.
CHECK
FICD
F.I.C.D. solenoid
Check terminal voltage of F.I.C.D. solenoid.
Check the part and/or circuit.
PERFORM
Self-diagnosis Mode IV
Perform self-diagnosis Mode IV (for idle switch).
N.G.
Check the idle switch circuit.

ECCS DIAGNOSTIC CHART—1990 PULSAR/SENTRA

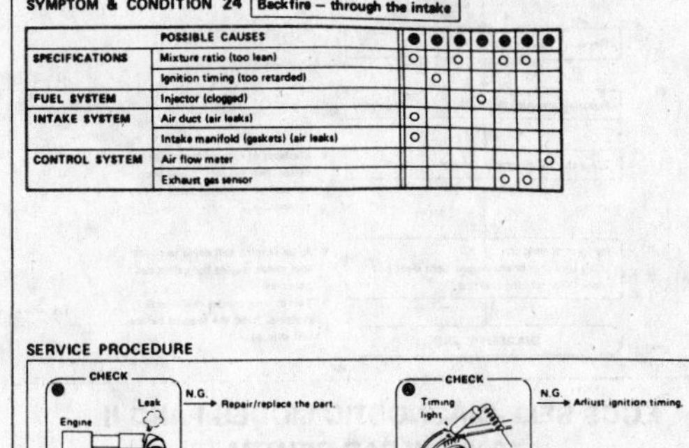

Diagnostic Table (Cont'd)

SYMPTOM & CONDITION 24 Backfire – through the intake

	POSSIBLE CAUSES	●	●	●	●	●	●	●
SPECIFICATIONS	Mixture ratio (too lean)	○		○		○	○	
	Ignition timing (too retarded)		○					
FUEL SYSTEM	Injector (clogged)				○			
INTAKE SYSTEM	Air duct (air leaks)	○						
	Intake manifold (gaskets) (air leaks)	○						
CONTROL SYSTEM	Air flow meter							○
	Exhaust gas sensor					○	○	

SERVICE PROCEDURE

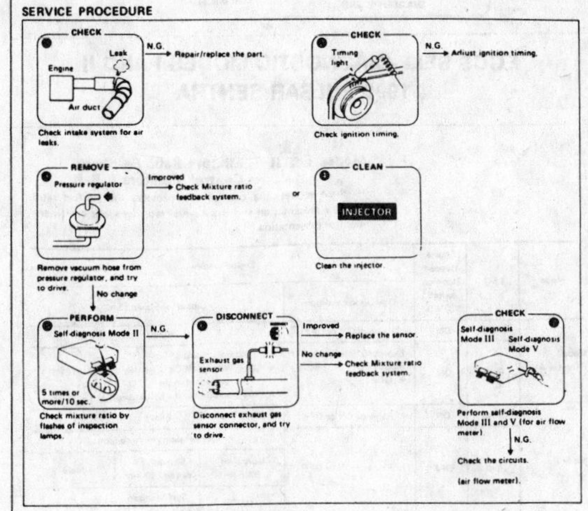
CHECK
Leak
Engine
Air duct
N.G.
Repair/replace the part.
Check intake system for air leaks.
CHECK
Timing light
N.G.
Adjust ignition timing.
Check ignition timing.
REMOVE
Pressure regulator
Improved
Check Mixture ratio feedback system.
or
CLEAN
INJECTOR
Clean the injector.
Remove vacuum hose from pressure regulator, and try to drive.
No change
PERFORM
Self-diagnosis Mode II
5 times or more/10 sec.
Check mixture ratio by flashes of inspection lamps.
N.G.
DISCONNECT
Exhaust gas sensor
Disconnect exhaust gas sensor connector, and try to drive.
Improved
Replace the sensor.
No change
Check Mixture ratio feedback system.
CHECK
Self-diagnosis Mode III
Self-diagnosis Mode V
Perform self-diagnosis Mode III and V (for air flow meter).
N.G.
Check the circuits.
(air flow meter).

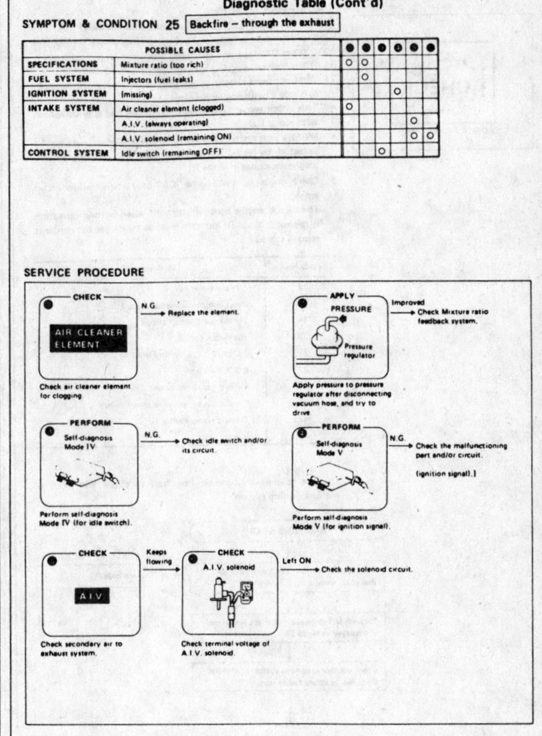

Diagnostic Table (Cont'd)

SYMPTOM & CONDITION 25 Backfire – through the exhaust

	POSSIBLE CAUSES	●	●	●	●	●	●
SPECIFICATIONS	Mixture ratio (too rich)	○	○				
FUEL SYSTEM	Injectors (fuel leaks)		○				
IGNITION SYSTEM	(missing)				○		
INTAKE SYSTEM	Air cleaner element (clogged)	○					
	A.I.V. (always operating)					○	
	A.I.V. solenoid (remaining ON)					○	○
CONTROL SYSTEM	Idle switch (remaining OFF)			○			

SERVICE PROCEDURE
CHECK
AIR CLEANER ELEMENT
N.G.
Replace the element.
Check air cleaner element for clogging.
APPLY
PRESSURE
Pressure regulator
Improved
Check Mixture ratio feedback system.
Apply pressure to pressure regulator after disconnecting vacuum hose, and try to drive.
PERFORM
Self-diagnosis Mode IV
N.G.
Check idle switch and/or its circuit.
Perform self-diagnosis Mode IV (for idle switch).
PERFORM
Self-diagnosis Mode V
N.G.
Check the malfunctioning part and/or circuit.
(ignition signal).]
Perform self-diagnosis Mode V (for ignition signal).
CHECK
AIV
Check secondary air to exhaust system.
Keeps flowing
CHECK
A.I.V. solenoid
Check terminal voltage of A.I.V. solenoid.
Left ON
Check the solenoid circuit.

ECCS SELF-DIAGNOSTIC DESCRIPTION—1990 PULSAR/SENTRA

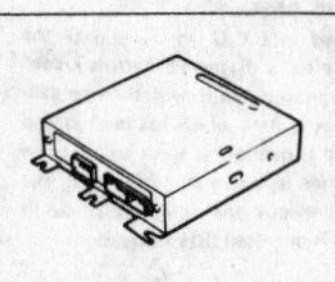

Description

The self-diagnosis is useful to diagnose malfunctions in major sensors and actuators of the E.C.C.S. system. There are 5 modes in the self-diagnosis system.

1. **Mode I – Mixture ratio feedback control monitor A**
- During closed loop condition:
  The green inspection lamp turns ON when lean condition is detected and goes OFF by rich condition.
- During open loop condition:
  The green inspection lamp remains ON or OFF.

2. **Mode II – Mixture ratio feedback control monitor B**
   The green inspection lamp function is the same as Mode I.
- During closed loop condition:
  The red inspection lamp turns ON and OFF simultaneously with the green inspection lamp when the mixture ratio is controlled within the specified value.
- During open loop condition:
  The red inspection lamp remains ON or OFF.

3. **Mode III – Self-diagnosis**
   This mode is the same as the former self-diagnosis in self-diagnosis mode.
4. **Mode IV – Switches ON/OFF diagnosis**
   During this mode, the inspection lamps monitor the switch ON-OFF condition.
- Idle switch
- Starter switch
- Vehicle speed sensor

5. **Mode V – Real time diagnosis**
   The moment the malfunction is detected, the display will be presented immediately. That is, the condition at which the malfunction occurs can be found by observing the inspection lamps during driving test.

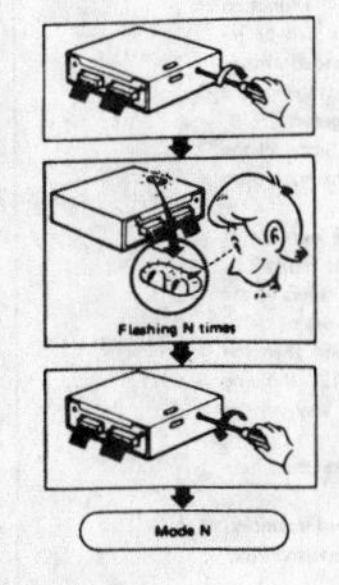
Flashing N times
Mode N

Description (Cont'd)

SWITCHING THE MODES

1. Turn ignition switch "ON".
2. Turn diagnostic mode selector on E.C.U. fully clockwise and wait the inspection lamps flash.
3. Count the number of the flashing time, and after the inspection lamps have flashed the number of the required mode, turn diagnostic mode selector fully counterclockwise immediately.

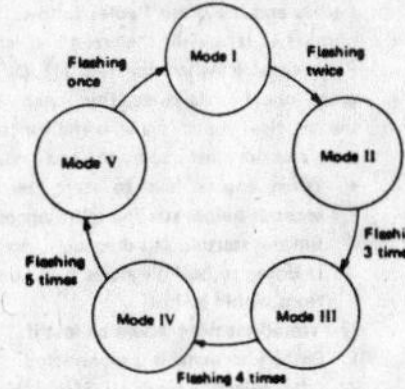
Flashing once
Mode I
Flashing twice
Mode V
Mode II
Flashing 3 times
Flashing 5 times
Mode IV
Mode III
Flashing 4 times

NOTE:
When the ignition switch is turned off during diagnosis, in each mode, and then turned back on again after the power to the E.C.U. has dropped off completely, the diagnosis will automatically return to Mode I.
The stored memory would be lost if:

1. Battery terminal is disconnected.
2. After selecting Mode III, Mode IV is selected.
   However, if the diagnostic mode selector is kept turned fully clockwise, it will continue to change in the order of Mode I → II → III → IV → V → I ... etc., and in this state the stored memory will not be erased.

ECCS SELF-DIAGNOSTIC DESCRIPTION – 1990 PULSAR/SENTRA

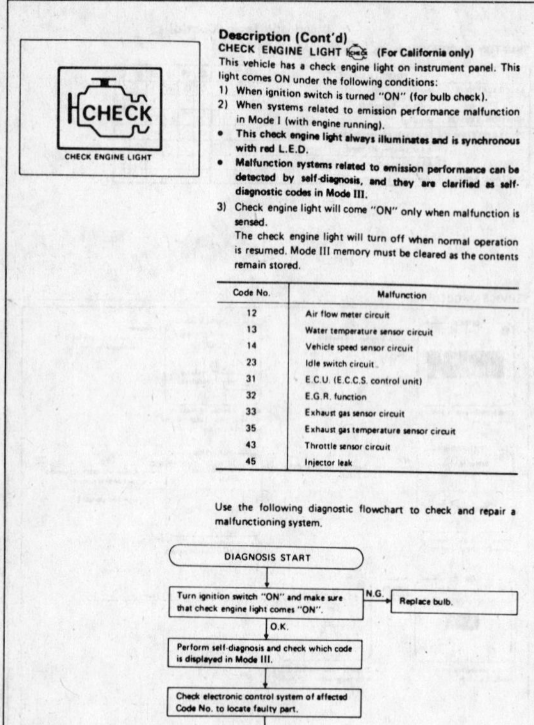
CHECK
CHECK ENGINE LIGHT

Description (Cont'd)

CHECK ENGINE LIGHT (For California only)

This vehicle has a check engine light on instrument panel. This light comes ON under the following conditions:

1) When ignition switch is turned "ON" (for bulb check).
2) When systems related to emission performance malfunction in Mode I (with engine running).

- **This check engine light always illuminates and is synchronous with red L.E.D.**
- **Malfunction systems related to emission performance can be detected by self-diagnosis, and they are clarified as self-diagnostic codes in Mode III.**

3) Check engine light will come "ON" only when malfunction is sensed.
   The check engine light will turn off when normal operation is resumed. Mode III memory must be cleared as the contents remain stored.

Code No.	Malfunction
12	Air flow meter circuit
13	Water temperature sensor circuit
14	Vehicle speed sensor circuit
23	Idle switch circuit
31	E.C.U. (E.C.C.S. control unit)
32	E.G.R. function
33	Exhaust gas sensor circuit
35	Exhaust gas temperature sensor circuit
43	Throttle sensor circuit
45	Injector leak

Use the following diagnostic flowchart to check and repair a malfunctioning system.

DIAGNOSIS START

↓

Turn ignition switch "ON" and make sure that check engine light comes "ON". — N.G. → Replace bulb.

↓ O.K.

Perform self-diagnosis and check which code is displayed in Mode III.

↓

Check electronic control system of affected Code No. to locate faulty part.

↓

Ⓐ

Description (Cont'd)

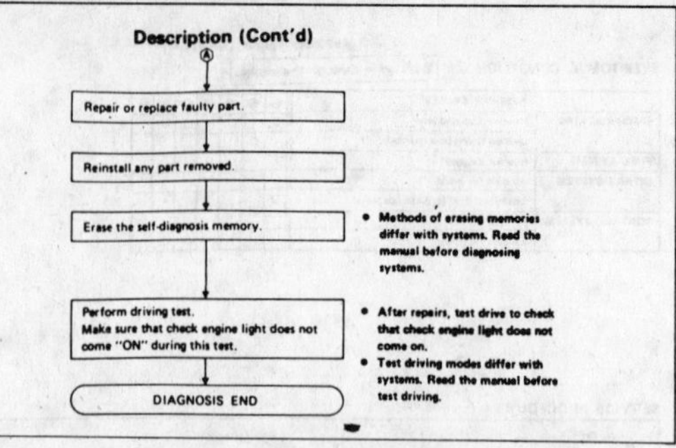
Ⓐ
Repair or replace faulty part.
Reinstall any part removed.
Erase the self-diagnosis memory.
Perform driving test.
Make sure that check engine light does not come "ON" during this test.
DIAGNOSIS END
• Methods of erasing memories differ with systems. Read the manual before diagnosing systems.
• After repairs, test drive to check that check engine light does not come on.
• Test driving modes differ with systems. Read the manual before test driving.

ECCS SELF-DIAGNOSTIC MODES I AND II 1990 PULSAR/SENTRA

Modes I & II — Mixture Ratio Feedback Control Monitors A & B

In these modes, the control unit provides the Air-fuel ratio monitor presentation and the Air-fuel ratio feedback coefficient monitor presentation.

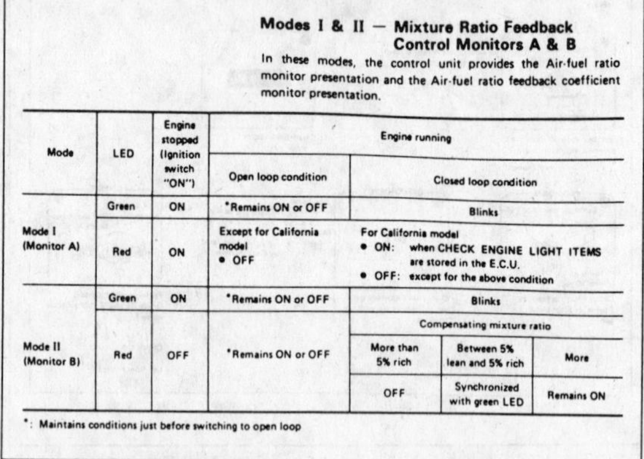

<table>
<tr><th rowspan="2">Mode</th><th rowspan="2">LED</th><th rowspan="2">Engine stopped (Ignition switch "ON")</th><th colspan="4">Engine running</th></tr>
<tr><th>Open loop condition</th><th colspan="3">Closed loop condition</th></tr>
<tr><td rowspan="2">Mode I (Monitor A)</td><td>Green</td><td>ON</td><td>*Remains ON or OFF</td><td colspan="3">Blinks</td></tr>
<tr><td>Red</td><td>ON</td><td>Except for California model
• OFF</td><td colspan="3">For California model
• ON: when CHECK ENGINE LIGHT ITEMS are stored in the E.C.U.
• OFF: except for the above condition</td></tr>
<tr><td rowspan="4">Mode II (Monitor B)</td><td>Green</td><td>ON</td><td>*Remains ON or OFF</td><td colspan="3">Blinks</td></tr>
<tr><td rowspan="3">Red</td><td rowspan="3">OFF</td><td rowspan="3">*Remains ON or OFF</td><td colspan="3">Compensating mixture ratio</td></tr>
<tr><td>More than 5% rich</td><td>Between 5% lean and 5% rich</td><td>More</td></tr>
<tr><td>OFF</td><td>Synchronized with green LED</td><td>Remains ON</td></tr>
</table>

*: Maintains conditions just before switching to open loop

ECCS SELF-DIAGNOSTIC MODE III – 1990 PULSAR/SENTRA

Mode III — Self-diagnostic System

The E.C.U. constantly monitors the function of these sensors and actuators, regardless of ignition key position. If a malfunction occurs, the information is stored in the E.C.U. and can be retrieved from the memory by turning on the diagnostic mode selector, located on the side of the E.C.U. When activated, the malfunction is indicated by flashing a red and a green L.E.D. (Light Emitting Diode), also located on the E.C.U. Since all the self-diagnostic results are stored in the E.C.U.'s memory even intermittent malfunctions can be diagnosed.

A malfunctioning part's group is indicated by the number of both the red and the green L.E.D.s flashing. First, the red L.E.D. flashes and the green flashes follow. The red L.E.D. refers to the number of tens while the green one refers to the number of units. For example, when the red L.E.D. flashes once and then the green one flashes twice, this means the number "12" showing the air flow meter signal is malfunctioning. In this way, all the problems are classified by the code numbers.

- **When engine fails to start, crank engine more than two seconds before starting self-diagnosis.**
- **Before starting self-diagnosis, do not erase stored memory. If doing so, self-diagnosis function for intermittent malfunctions would be lost.**

The stored memory would be lost if:

1. **Battery terminal is disconnected.**
2. **After selecting Mode III, Mode IV is selected.**

DISPLAY CODE TABLE

Code No.	Detected items	California	Non-California
11	Crank angle sensor circuit	X	X
12	Air flow meter circuit	X*	X
13	Water temperature sensor circuit	X*	X
14	Vehicle speed sensor circuit	X*	X
21	Ignition signal missing in primary coil	X	X
23	Idle switch circuit	X*	X
25	A.A.C. valve circuit	X	X
31	E.C.U. (E.C.C.S. control unit)	X*	X
32	E.G.R. function	X*	–
33	Exhaust gas sensor circuit	X*	X
35	Exhaust gas temperature sensor circuit	X*	–
43	Throttle sensor circuit	X*	X
45	Injector leak	X*	–
55	No malfunctioning in the above circuit	X	X

X: Available
–: Not available
*: Check engine light item

Mode III — Self-diagnostic System (Cont'd)

RETENTION OF DIAGNOSTIC RESULTS

The diagnostic result is retained in E.C.U. memory until the starter is operated fifty times after a diagnostic item is judged to be malfunctioning. The diagnostic result will then be cancelled automatically. If a diagnostic item which has been judged to be malfunctioning and stored in memory is again judged to be malfunctioning before the starter is operated fifty times, the second result will replace the previous one. It will be stored in E.C.U. memory until the starter is operated fifty times more.

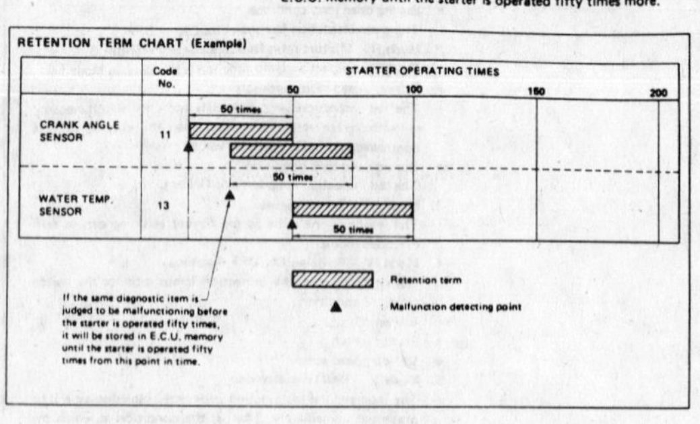
RETENTION TERM CHART (Example)
Code No.
STARTER OPERATING TIMES
50
100
150
200
CRANK ANGLE SENSOR
11
50 times
50 times
WATER TEMP. SENSOR
13
50 times
Retention term
Malfunction detecting point
If the same diagnostic item is judged to be malfunctioning before the starter is operated fifty times, it will be stored in E.C.U. memory until the starter is operated fifty times from this point in time.

ECCS SELF-DIAGNOSTIC MODE III—1990 PULSAR/SENTRA

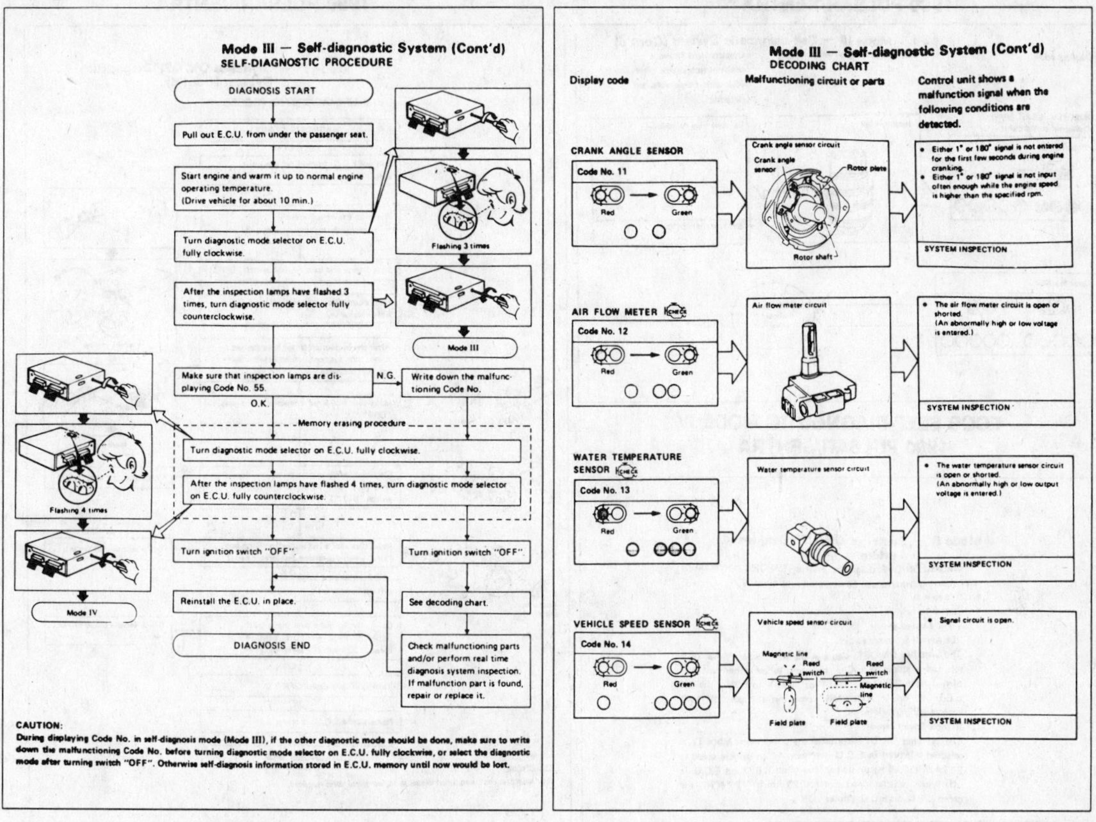

ECCS SELF-DIAGNOSTIC MODE III DISPLAY CODES—1990 PULSAR/SENTRA

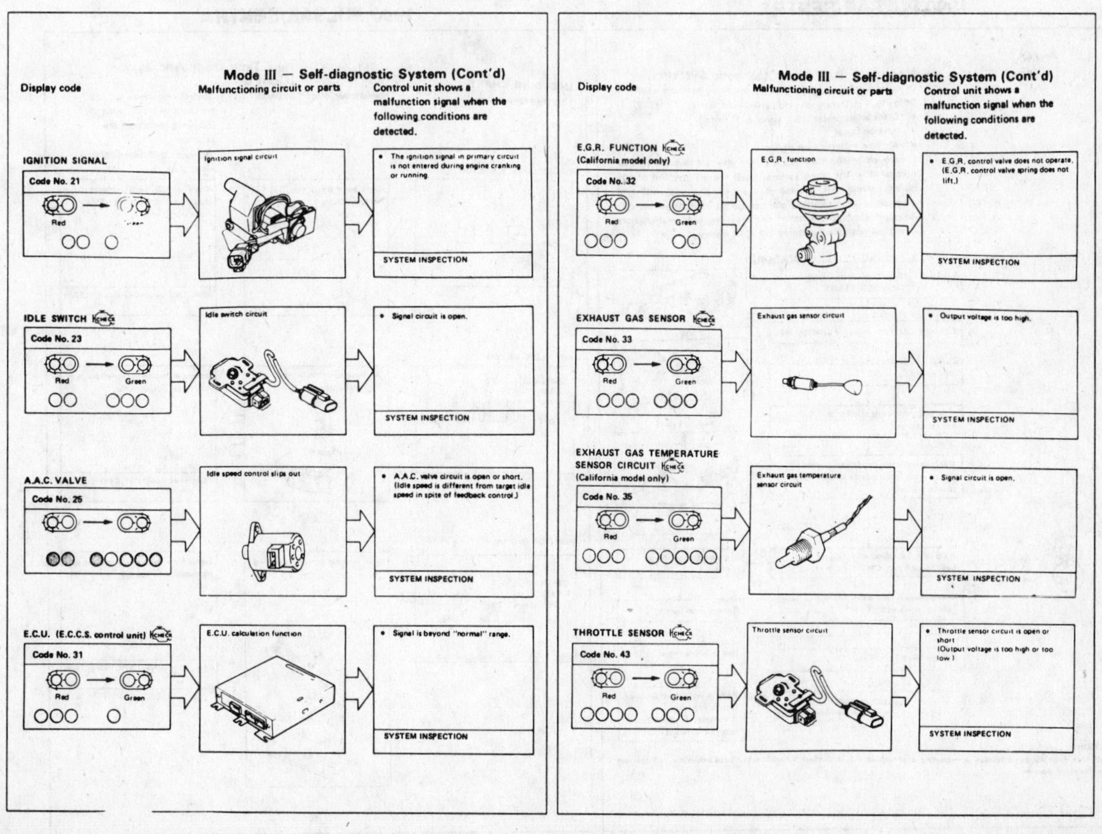

ECCS SELF-DIAGNOSTIC MODE III DISPLAY CODES 1990 PULSAR/SENTRA

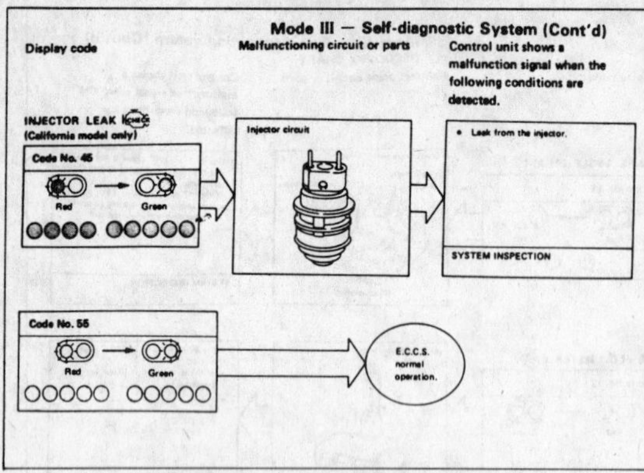
Mode III — Self-diagnostic System (Cont'd)
Display code
Malfunctioning circuit or parts
Control unit shows a malfunction signal when the following conditions are detected.
INJECTOR LEAK
(California model only)
Code No. 45
Red
Green
Injector circuit
• Leak from the injector.
SYSTEM INSPECTION
Code No. 55
Red
Green
E.C.C.S. normal operation.

ECCS SELF-DIAGNOSTIC MODE IV 1990 PULSAR/SENTRA

Mode IV — Switches ON/OFF Diagnostic System

In switches ON/OFF diagnosis system, ON/OFF operation of the following switches can be detected continuously.

- Idle switch
- Starter switch
- Vehicle speed sensor

(1) Idle switch & Starter switch
The switches ON/OFF status at the point when Mode IV is selected is stored in E.C.U. memory. When either switch is turned from "ON" to "OFF" or "OFF" to "ON", the red L.E.D. on E.C.U. alternately comes on and goes off each time switching is detected.

(2) Vehicle Speed Sensor
The switches ON/OFF status at the point when Mode IV is selected is stored in E.C.U. memory. When vehicle speed is 20 km/h (12 MPH) or slower, the green L.E.D. on E.C.U. is off. When vehicle speed exceeds 20 km/h (12 MPH), the green L.E.D. on E.C.U. comes "ON".

ECCS SELF-DIAGNOSTIC MODE IV 1990 PULSAR/SENTRA

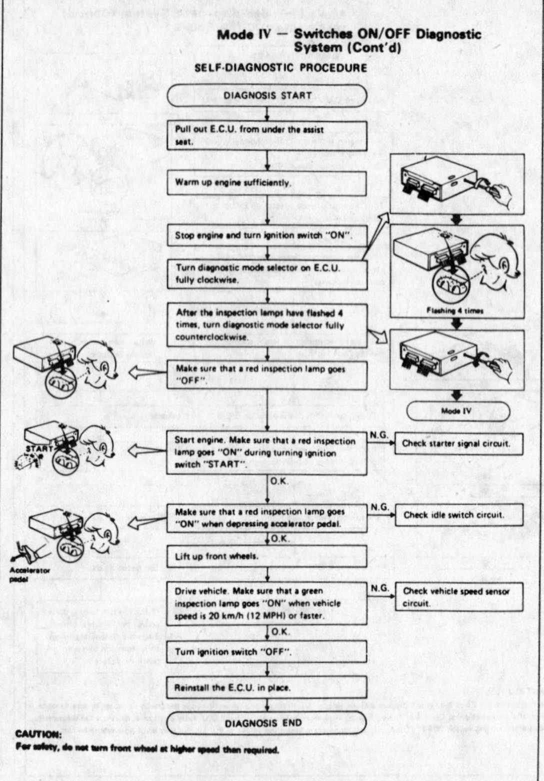
Mode IV — Switches ON/OFF Diagnostic System (Cont'd)
SELF-DIAGNOSTIC PROCEDURE
DIAGNOSIS START
Pull out E.C.U. from under the assist seat.
Warm up engine sufficiently.
Stop engine and turn ignition switch "ON".
Turn diagnostic mode selector on E.C.U. fully clockwise.
After the inspection lamps have flashed 4 times, turn diagnostic mode selector fully counterclockwise.
Flashing 4 times
Mode IV
Make sure that a red inspection lamp goes "OFF".
Start engine. Make sure that a red inspection lamp goes "ON" during turning ignition switch "START".
START
N.G.
Check starter signal circuit.
O.K.
Make sure that a red inspection lamp goes "ON" when depressing accelerator pedal.
N.G.
Check idle switch circuit.
O.K.
Accelerator pedal
Lift up front wheels.
Drive vehicle. Make sure that a green inspection lamp goes "ON" when vehicle speed is 20 km/h (12 MPH) or faster.
N.G.
Check vehicle speed sensor circuit.
O.K.
Turn ignition switch "OFF".
Reinstall the E.C.U. in place.
DIAGNOSIS END
CAUTION:
For safety, do not turn front wheel at higher speed than required.

ECCS SELF-DIAGNOSTIC MODE V 1990 PULSAR/SENTRA

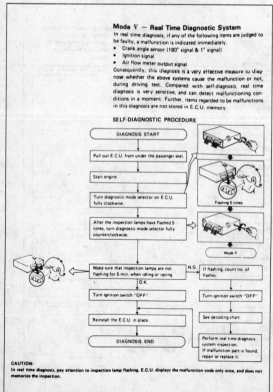

Mode V — Real Time Diagnostic System

In real time diagnosis, if any of the following items are judged to be faulty, a malfunction is indicated immediately.

- Crank angle sensor (180° signal & 1° signal)
- Ignition signal
- Air flow meter output signal

Consequently, this diagnosis is a very effective measure to diagnose whether the above systems cause the malfunction or not, during driving test. Compared with self-diagnosis, real time diagnosis is very sensitive, and can detect malfunctioning conditions in a moment. Further, items regarded to be malfunctions in this diagnosis are not stored in E.C.U. memory.

SELF-DIAGNOSTIC PROCEDURE

DIAGNOSIS START

Pull out E.C.U. from under the passenger seat.

Start engine.

Turn diagnostic mode selector on E.C.U. fully clockwise.

Flashing 5 times

After the inspection lamps have flashed 5 times, turn diagnostic mode selector fully counterclockwise.

Mode V

Make sure that inspection lamps are not flashing for 5 min. when idling or racing.

N.G. → If flashing, count no. of flashes. → Turn ignition switch "OFF". → See decoding chart. → Perform real time-diagnosis system inspection. If malfunction part is found, repair or replace it.

O.K.

Turn ignition switch "OFF".

Reinstall the E.C.U. in place.

DIAGNOSIS END

CAUTION:
In real time diagnosis, pay attention to inspection lamp flashing. E.C.U. displays the malfunction code only once, and does not memorize the inspection.

ECCS SELF-DIAGNOSTIC MODE V DECODING CHART 1990 PULSAR/SENTRA

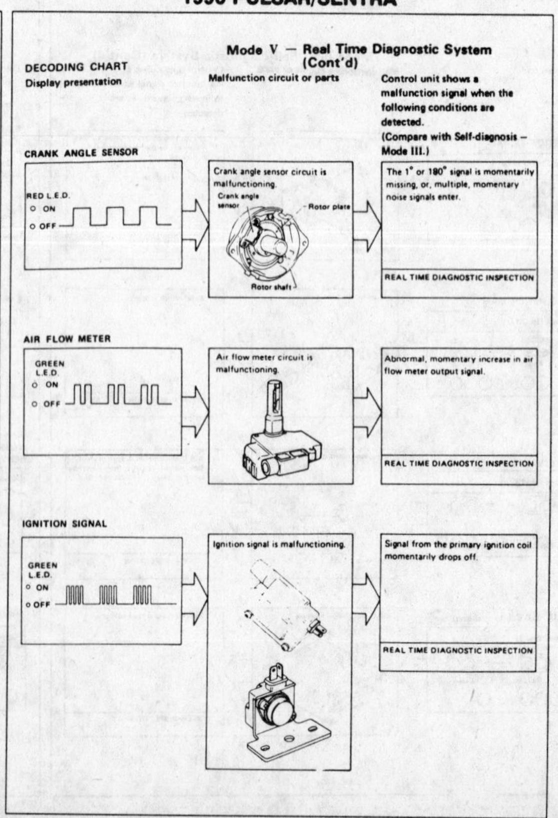
Mode V — Real Time Diagnostic System (Cont'd)
DECODING CHART
Display presentation
Malfunction circuit or parts
Control unit shows a malfunction signal when the following conditions are detected.
(Compare with Self-diagnosis — Mode III.)
CRANK ANGLE SENSOR
RED L.E.D.
ON
OFF
Crank angle sensor circuit is malfunctioning.
Crank angle sensor
Rotor plate
Rotor shaft
The 1° or 180° signal is momentarily missing, or, multiple, momentary noise signals enter.
REAL TIME DIAGNOSTIC INSPECTION
AIR FLOW METER
GREEN L.E.D.
ON
OFF
Air flow meter circuit is malfunctioning.
Abnormal, momentary increase in air flow meter output signal.
REAL TIME DIAGNOSTIC INSPECTION
IGNITION SIGNAL
GREEN L.E.D.
ON
OFF
Ignition signal is malfunctioning.
Signal from the primary ignition coil momentarily drops off.
REAL TIME DIAGNOSTIC INSPECTION

ECCS SELF-DIAGNOSTIC MODE V—1990 PULSAR/SENTRA

Mode V — Real Time Diagnostic System (Cont'd)

REAL TIME DIAGNOSTIC INSPECTION

Crank Angle Sensor

X: Available
—: Not available

Check sequence	Check items	Check conditions	Check parts			If malfunction, perform the following items.
			Middle connector	Sensor & actuator	E.C.U. connectors	
1	Tap harness connector or component during real time diagnosis.	During real time diagnosis	X	X	X	Go to check item 2.
2	Check harness continuity at connector.	Engine stopped	X	—	—	Go to check item 3.
3	Disconnect harness connector, and then check dust adhesion to harness connector.	Engine stopped	X	—	X	Clean terminal surface.
4	Check pin terminal bend.	Engine stopped	—	—	X	Take out bend.
5	Reconnect harness connector and then recheck harness continuity at connector.	Engine stopped	X	—	—	Replace terminal.
6	Tap harness connector or component during real time diagnosis.	During real time diagnosis	X	X	X	If malfunction codes are displayed during real time diagnosis, replace terminal.

Mode V — Real Time Diagnostic System (Cont'd)

Air Flow Meter

X: Available
—: Not available

Check sequence	Check items	Check conditions	Check parts			If malfunction, perform the following items.
			Middle connector	Sensor & actuator	E.C.U. connectors	
1	Tap harness connector or component during real time diagnosis.	During real time diagnosis	X	X	X	Go to check item 2.
2	Check harness continuity at connector.	Engine stopped	X	—	—	Go to check item 3.
3	Disconnect harness connector, and then check dust adhesion to harness connector.	Engine stopped	X	—	X	Clean terminal surface.
4	Check pin terminal bend.	Engine stopped	—	—	X	Take out bend.
5	Reconnect harness connector and then recheck harness continuity at connector.	Engine stopped	X	—	—	Replace terminal.
6	Tap harness connector or component during real time diagnosis.	During real time diagnosis	X	X	X	If malfunction codes are displayed during real time diagnosis, replace terminal.

ECCS SELF-DIAGNOSTIC MODE V 1990 PULSAR/SENTRA

Mode V — Real Time Diagnostic System (Cont'd)

Ignition Signal

X: Available
—: Not available

Check sequence	Check items	Check conditions	Check parts			If malfunction, perform the following items.
			Middle connector	Sensor & actuator	E.C.U. connectors	
1	Tap harness connector or component during real time diagnosis.	During real time diagnosis	X	X	X	Go to check item 2.
2	Check harness continuity at connector.	Engine stopped	X	—	—	Go to check item 3.
3	Disconnect harness connector, and then check dust adhesion to harness connector.	Engine stopped	X	—	X	Clean terminal surface.
4	Check pin terminal bend.	Engine stopped	—	—	X	Take out bend.
5	Reconnect harness connector and then recheck harness continuity at connector.	Engine stopped	X	—	—	Replace terminal.
6	Tap harness connector or component during real time diagnosis.	During real time diagnosis	X	X	X	If malfunction codes are displayed during real time diagnosis, replace terminal.

ECCS SYSTEM INSPECTION CAUTION 1990 PULSAR/SENTRA

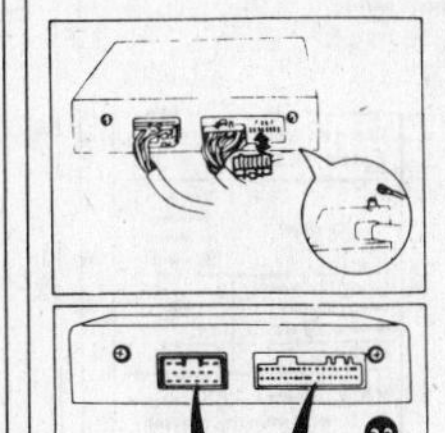

CAUTION:

1. Before connecting or disconnecting E.C.U. harness connector to or from any E.C.U., be sure to turn the ignition switch to the "OFF" position and disconnect the negative battery terminal in order not to damage E.C.U. as battery voltage is applied to E.C.U. even if ignition switch is turned off. Otherwise, there may be damage to the E.C.U.

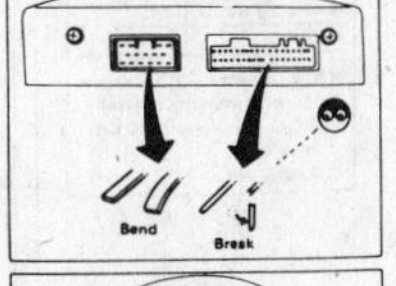
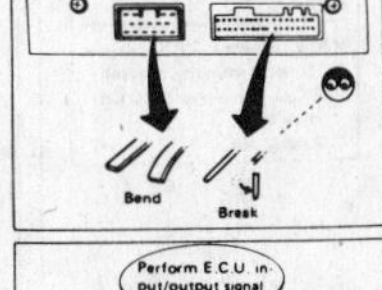
Bend
Break

2. When connecting pin connectors into E.C.U. or disconnecting them from E.C.U., take care not to damage pin terminal of E.C.U. (Bend or break).
3. Make sure that there are not any bends or breaks on E.C.U. pin terminal, when connecting pin connectors into E.C.U.

Perform E.C.U. input/output signal inspection before replacement.
OLD ONE

4. Before replacing E.C.U., perform E.C.U. input/output signal inspection and make sure whether E.C.U. functions properly or not.

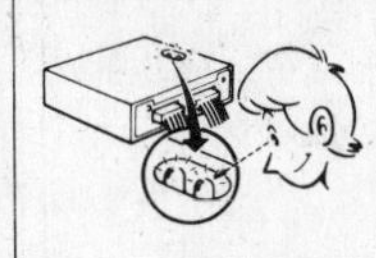

5. After performing this "ELECTRONIC CONTROL SYSTEM INSPECTION", perform E.C.C.S. self-diagnosis and driving test.

ECCS SYSTEM INSPECTION CAUTION 1990 PULSAR/SENTRA

6. When measuring supply voltage of E.C.U. controlled components with a circuit tester, separate one tester probe from the other.
   If the two tester probes accidentally make contact with each other during measurement, the circuit will be shorted, resulting in damage to the power transistor of the control unit.

7. When measuring voltage or resistance at connector with tester probes, there are two methods of measurement; one is done from terminal side and the other from harness side. Before measuring, confirm symbol mark again.
   - H.S. : Inspection should be done from harness side.
   - T.S. : Inspection should be done from terminal side.

8. As for continuity check of joint connector

9. Key to symbols
   - DISCONNECT : Check after disconnecting the connector to be measured.
   - CONNECT : Check after connecting the connector to be measured.

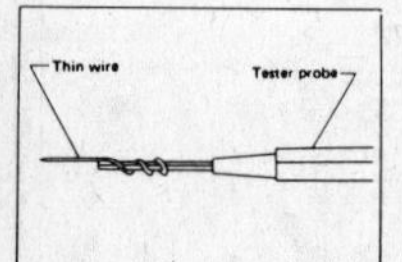
Thin wire
Tester probe

10. Improve tester probe as shown to perform test easily.
11. For the first trouble-shooting procedure, perform POWER SOURCE & GROUND CIRCUIT FOR E.C.U. check.

ECCS SELF-DIAGNOSTIC CHART 1990 PULSAR/SENTRA

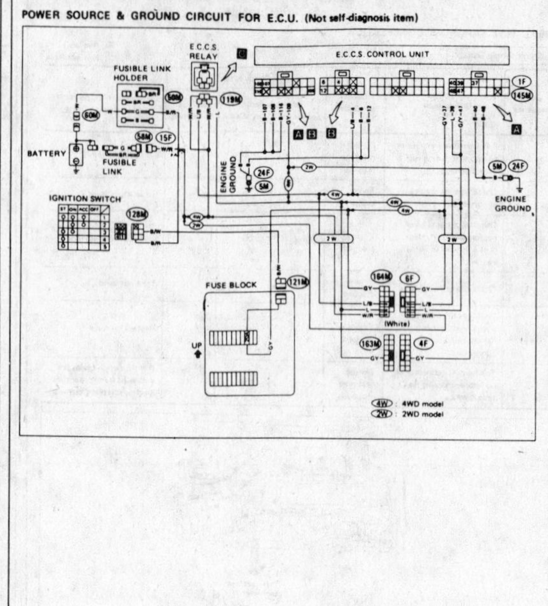
POWER SOURCE & GROUND CIRCUIT FOR E.C.U. (Not self-diagnosis item)
E.C.C.S. RELAY
E.C.C.S. CONTROL UNIT
FUSIBLE LINK HOLDER
BATTERY
FUSIBLE LINK
IGNITION SWITCH
ENGINE GROUND
ENGINE GROUND
FUSE BLOCK
(White)
UP
4W : 4WD model
2W : 2WD model

ECCS SELF-DIAGNOSTIC CHART—1990 PULSAR/SENTRA

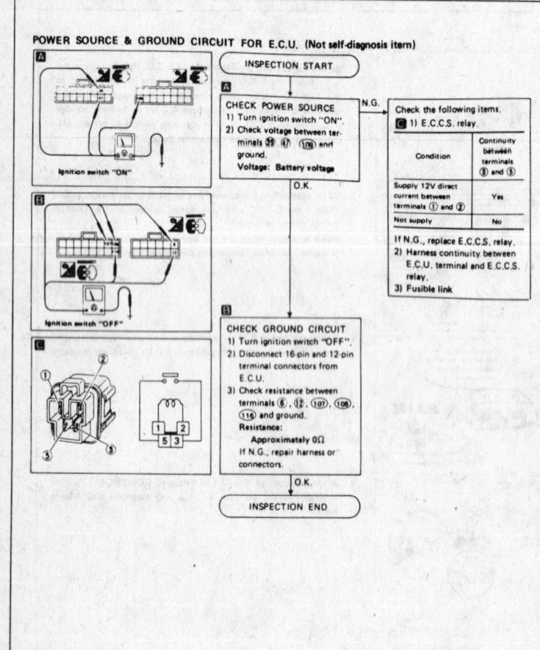
POWER SOURCE & GROUND CIRCUIT FOR E.C.U. (Not self-diagnosis item)
Ignition switch "ON"
Ignition switch "OFF"
INSPECTION START
A
CHECK POWER SOURCE
1) Turn ignition switch "ON".
2) Check voltage between terminals 38, 47, 109 and ground.
Voltage: Battery voltage
N.G.
Check the following items.
C 1) E.C.C.S. relay.
Condition | Continuity between terminals 3 and 5
Supply 12V direct current between terminals 1 and 2 | Yes
Not supply | No
If N.G., replace E.C.C.S. relay.
2) Harness continuity between E.C.U. terminal and E.C.C.S. relay.
3) Fusible link
O.K.
B
CHECK GROUND CIRCUIT
1) Turn ignition switch "OFF".
2) Disconnect 16-pin and 12-pin terminal connectors from E.C.U.
3) Check resistance between terminals 6, 12, 107, 108, 116 and ground.
Resistance:
Approximately 0Ω
If N.G., repair harness or connectors.
O.K.
INSPECTION END

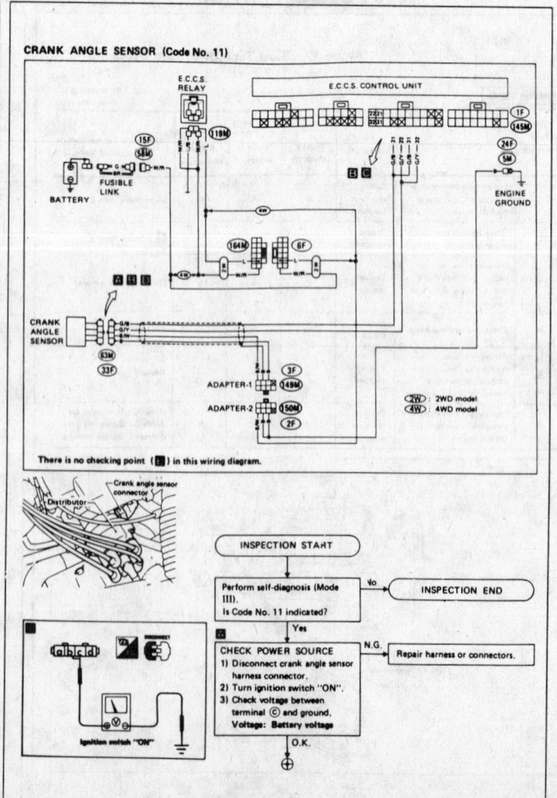
CRANK ANGLE SENSOR (Code No. 11)
E.C.C.S. RELAY
E.C.C.S. CONTROL UNIT
BATTERY
FUSIBLE LINK
ENGINE GROUND
CRANK ANGLE SENSOR
ADAPTER-1
ADAPTER-2
2W : 2WD model
4W : 4WD model
There is no checking point in this wiring diagram.
Distributor
Crank angle sensor connector
Ignition switch "ON"
INSPECTION START
Perform self-diagnosis (Mode III).
Is Code No. 11 indicated?
No
INSPECTION END
Yes
CHECK POWER SOURCE
1) Disconnect crank angle sensor harness connector.
2) Turn ignition switch "ON".
3) Check voltage between terminal © and ground.
Voltage: Battery voltage
N.G.
Repair harness or connectors.
O.K.

ECCS SELF-DIAGNOSTIC CHART—1990 PULSAR/SENTRA

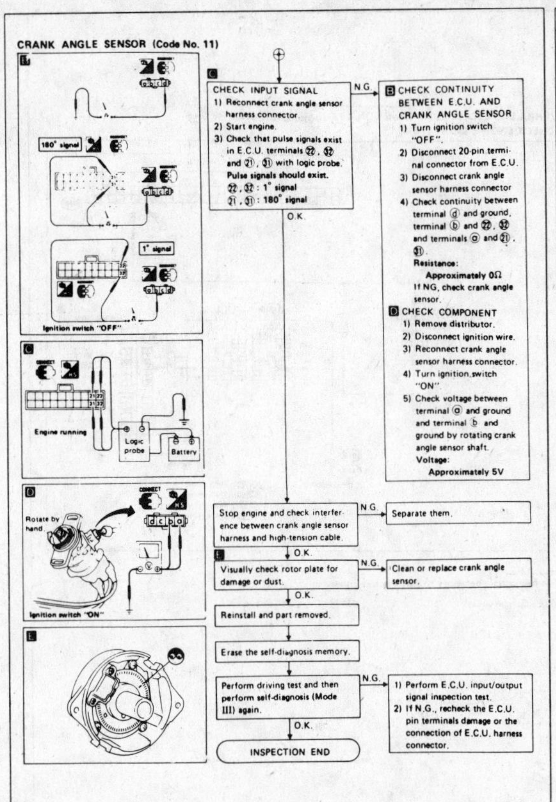
CRANK ANGLE SENSOR (Code No. 11)
180° signal
1° signal
Ignition switch "OFF"
Engine running
Logic probe
Battery
Rotate by hand
Ignition switch "ON"
C CHECK INPUT SIGNAL
1) Reconnect crank angle sensor harness connector.
2) Start engine.
3) Check that pulse signals exist in E.C.U. terminals (22), (32) and (21), (31) with logic probe.
Pulse signals should exist.
(22), (32) : 1° signal
(21), (31) : 180° signal
N.G.
B CHECK CONTINUITY BETWEEN E.C.U. AND CRANK ANGLE SENSOR
1) Turn ignition switch "OFF".
2) Disconnect 20-pin terminal connector from E.C.U.
3) Disconnect crank angle sensor harness connector
4) Check continuity between terminal (d) and ground, terminal (b) and (22), (32) and terminals (a) and (21), (31).
Resistance:
Approximately 0Ω
If NG, check crank angle sensor.
D CHECK COMPONENT
1) Remove distributor.
2) Disconnect ignition wire.
3) Reconnect crank angle sensor harness connector.
4) Turn ignition switch "ON".
5) Check voltage between terminal (a) and ground and terminal (b) and ground by rotating crank angle sensor shaft.
Voltage:
Approximately 5V
O.K.
Stop engine and check interference between crank angle sensor harness and high-tension cable.
N.G.
Separate them.
O.K.
E Visually check rotor plate for damage or dust.
N.G.
Clean or replace crank angle sensor.
O.K.
Reinstall and part removed.
Erase the self-diagnosis memory.
Perform driving test and then perform self-diagnosis (Mode III) again.
N.G.
1) Perform E.C.U. input/output signal inspection test.
2) If N.G., recheck the E.C.U. pin terminals damage or the connection of E.C.U. harness connector.
O.K.
INSPECTION END

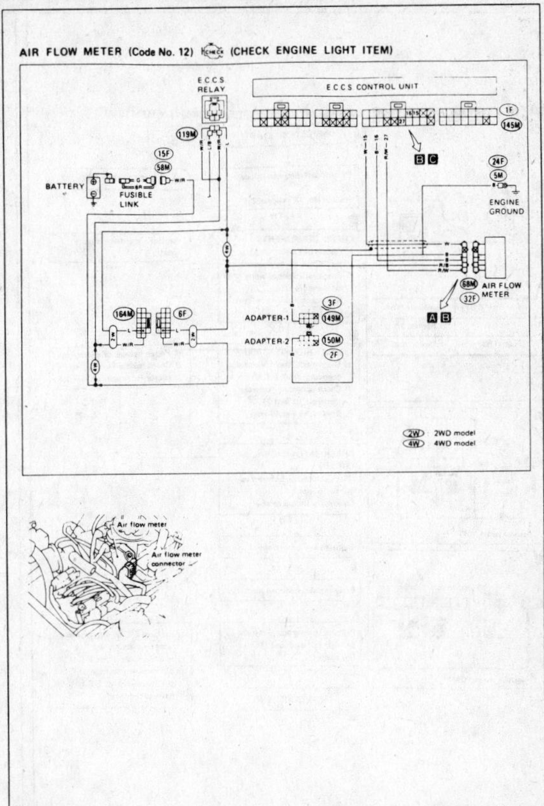
AIR FLOW METER (Code No. 12) (CHECK ENGINE LIGHT ITEM)
ECCS RELAY
ECCS CONTROL UNIT
BATTERY
FUSIBLE LINK
ENGINE GROUND
AIR FLOW METER
ADAPTER-1
ADAPTER-2
2WD model
4WD model
Air flow meter
Air flow meter connector

ECCS SELF-DIAGNOSTIC CHART—1990 PULSAR/SENTRA

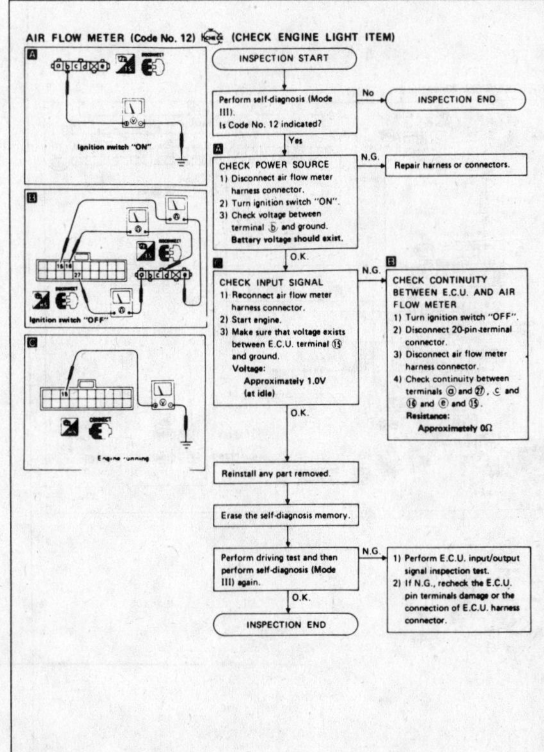
AIR FLOW METER (Code No. 12) (CHECK ENGINE LIGHT ITEM)
Ignition switch "ON"
Ignition switch "OFF"
Engine running
INSPECTION START
Perform self-diagnosis (Mode III).
Is Code No. 12 indicated?
No
INSPECTION END
Yes
A CHECK POWER SOURCE
1) Disconnect air flow meter harness connector.
2) Turn ignition switch "ON".
3) Check voltage between terminal (b) and ground.
Battery voltage should exist.
N.G.
Repair harness or connectors.
O.K.
C CHECK INPUT SIGNAL
1) Reconnect air flow meter harness connector.
2) Start engine.
3) Make sure that voltage exists between E.C.U. terminal (15) and ground.
Voltage:
Approximately 1.0V (at idle)
N.G.
B CHECK CONTINUITY BETWEEN E.C.U. AND AIR FLOW METER
1) Turn ignition switch "OFF".
2) Disconnect 20-pin terminal connector.
3) Disconnect air flow meter harness connector.
4) Check continuity between terminals (a) and (27), (c) and (16) and (e) and (15).
Resistance:
Approximately 0Ω
O.K.
Reinstall any part removed.
Erase the self-diagnosis memory.
Perform driving test and then perform self-diagnosis (Mode III) again.
N.G.
1) Perform E.C.U. input/output signal inspection test.
2) If N.G., recheck the E.C.U. pin terminals damage or the connection of E.C.U. harness connector.
O.K.
INSPECTION END

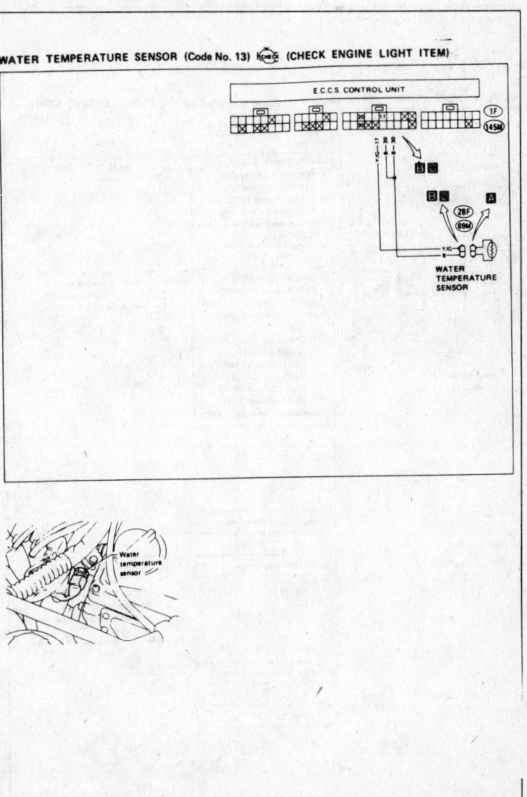
WATER TEMPERATURE SENSOR (Code No. 13) (CHECK ENGINE LIGHT ITEM)
ECCS CONTROL UNIT
WATER TEMPERATURE SENSOR
Water temperature sensor

ECCS SELF-DIAGNOSTIC CHART—1990 PULSAR/SENTRA

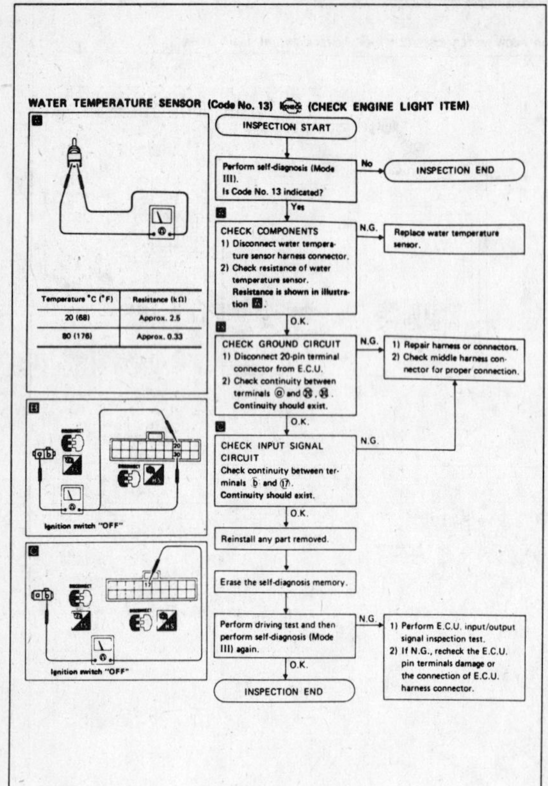
WATER TEMPERATURE SENSOR (Code No. 13) (CHECK ENGINE LIGHT ITEM)
INSPECTION START
Perform self-diagnosis (Mode III).
Is Code No. 13 indicated?
No
INSPECTION END
Yes
CHECK COMPONENTS
1) Disconnect water temperature sensor harness connector.
2) Check resistance of water temperature sensor.
Resistance is shown in illustration.
N.G.
Replace water temperature sensor.
Temperature °C (°F)
Resistance (kΩ)
20 (68)
Approx. 2.5
80 (176)
Approx. 0.33
O.K.
CHECK GROUND CIRCUIT
1) Disconnect 20-pin terminal connector from E.C.U.
2) Check continuity between terminals and.
Continuity should exist.
N.G.
1) Repair harness or connectors.
2) Check middle harness connector for proper connection.
O.K.
CHECK INPUT SIGNAL CIRCUIT
Check continuity between terminals and.
Continuity should exist.
N.G.
O.K.
Reinstall any part removed.
Erase the self-diagnosis memory.
Perform driving test and then perform self-diagnosis (Mode III) again.
N.G.
1) Perform E.C.U. input/output signal inspection test.
2) If N.G., recheck the E.C.U. pin terminals damage or the connection of E.C.U. harness connector.
O.K.
INSPECTION END
Ignition switch "OFF"
Ignition switch "OFF"

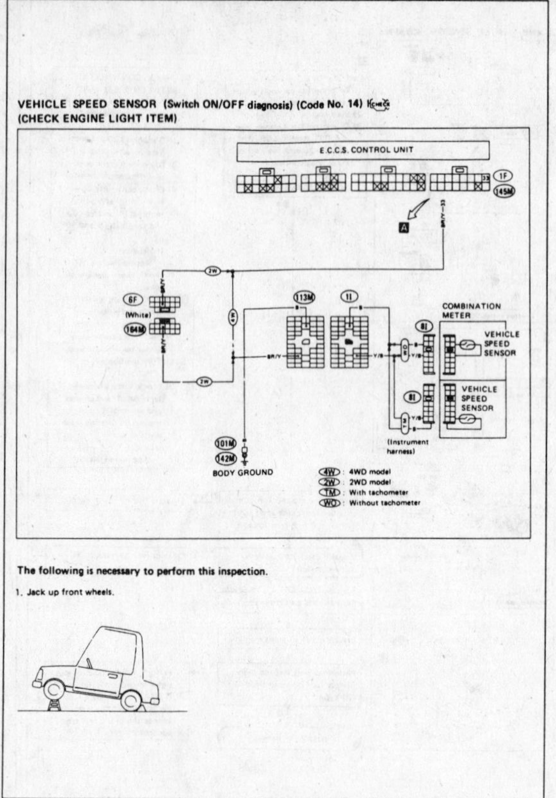
VEHICLE SPEED SENSOR (Switch ON/OFF diagnosis) (Code No. 14)
(CHECK ENGINE LIGHT ITEM)
E.C.C.S. CONTROL UNIT
1F
145M
2W
6F
(White)
164M
113M
11
COMBINATION METER
VEHICLE SPEED SENSOR
VEHICLE SPEED SENSOR
(Instrument harness)
101M
142M
BODY GROUND
4W : 4WD model
2W : 2WD model
TM : With tachometer
WO : Without tachometer
The following is necessary to perform this inspection.
1. Jack up front wheels.

ECCS SELF-DIAGNOSTIC CHART—1990 PULSAR/SENTRA

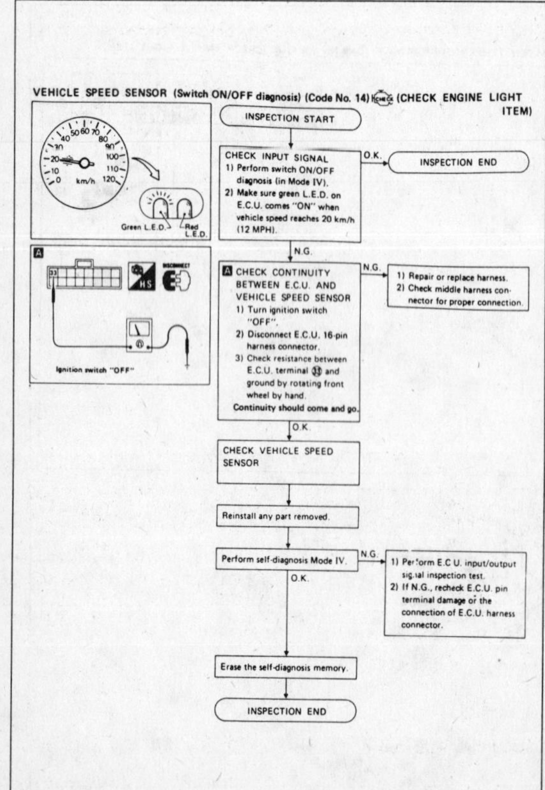
VEHICLE SPEED SENSOR (Switch ON/OFF diagnosis) (Code No. 14) (CHECK ENGINE LIGHT ITEM)
INSPECTION START
CHECK INPUT SIGNAL
1) Perform switch ON/OFF diagnosis (in Mode IV).
2) Make sure green L.E.D. on E.C.U. comes "ON" when vehicle speed reaches 20 km/h (12 MPH).
O.K.
INSPECTION END
N.G.
CHECK CONTINUITY BETWEEN E.C.U. AND VEHICLE SPEED SENSOR
1) Turn ignition switch "OFF".
2) Disconnect E.C.U. 16-pin harness connector.
3) Check resistance between E.C.U. terminal and ground by rotating front wheel by hand.
Continuity should come and go.
N.G.
1) Repair or replace harness.
2) Check middle harness connector for proper connection.
O.K.
CHECK VEHICLE SPEED SENSOR
Reinstall any part removed.
Perform self-diagnosis Mode IV.
N.G.
1) Perform E.C.U. input/output signal inspection test.
2) If N.G., recheck E.C.U. pin terminal damage or the connection of E.C.U. harness connector.
O.K.
Erase the self-diagnosis memory.
INSPECTION END
km/h
Green L.E.D.
Red L.E.D.
Ignition switch "OFF"

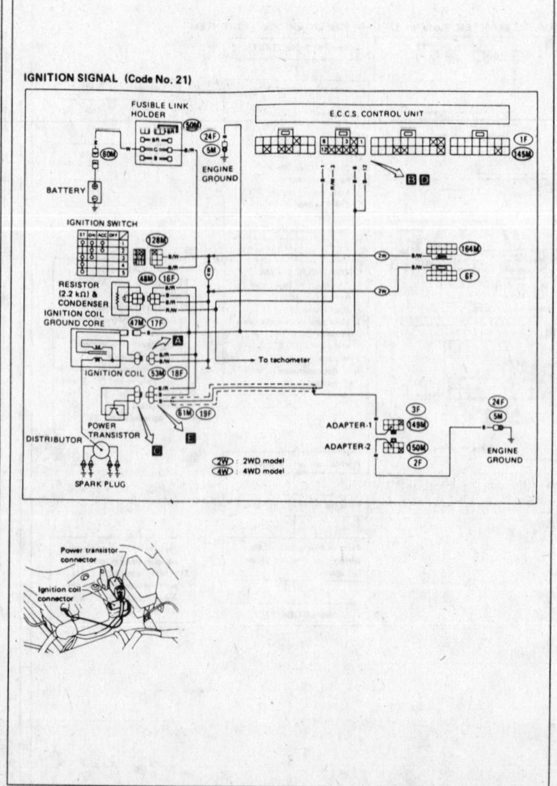
IGNITION SIGNAL (Code No. 21)
FUSIBLE LINK HOLDER
E.C.C.S. CONTROL UNIT
1F
145M
50M
24F
5M
ENGINE GROUND
80M
BATTERY
IGNITION SWITCH
128M
164M
6F
RESISTOR (2.2 kΩ) & CONDENSER
48M
16F
IGNITION COIL GROUND CORE
47M
17F
To tachometer
IGNITION COIL
53M
18F
51M
19F
POWER TRANSISTOR
DISTRIBUTOR
SPARK PLUG
ADAPTER-1
ADAPTER-2
3F
149M
150M
2F
24F
5M
ENGINE GROUND
2W : 2WD model
4W : 4WD model
Power transistor connector
Ignition coil connector

ECCS SELF-DIAGNOSTIC CHART–1990 PULSAR/SENTRA

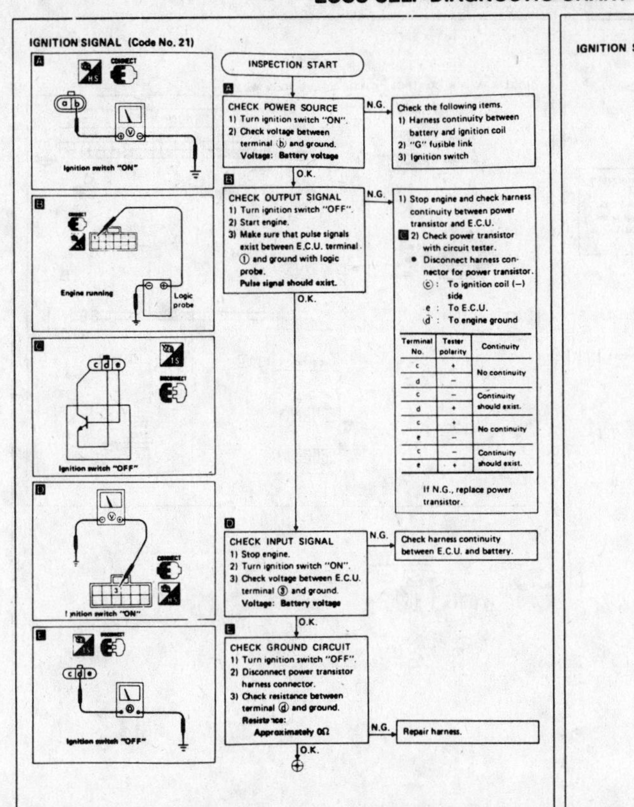
IGNITION SIGNAL (Code No. 21)
Ignition switch "ON"
Engine running
Logic probe
Ignition switch "OFF"
Ignition switch "ON"
Ignition switch "OFF"
INSPECTION START
A CHECK POWER SOURCE
1) Turn ignition switch "ON".
2) Check voltage between terminal ⓑ and ground.
Voltage: Battery voltage
N.G.
Check the following items.
1) Harness continuity between battery and ignition coil
2) "G" fusible link
3) Ignition switch
O.K.
B CHECK OUTPUT SIGNAL
1) Turn ignition switch "OFF".
2) Start engine.
3) Make sure that pulse signals exist between E.C.U. terminal ① and ground with logic probe.
Pulse signal should exist.
N.G.
1) Stop engine and check harness continuity between power transistor and E.C.U.
C 2) Check power transistor with circuit tester.
• Disconnect harness connector for power transistor.
ⓒ: To ignition coil (–) side
e: To E.C.U.
ⓓ: To engine ground
Terminal No. | Tester polarity | Continuity
c + / d – : No continuity
c – / d + : Continuity should exist.
c + / e – : No continuity
c – / e + : Continuity should exist.
If N.G., replace power transistor.
O.K.
D CHECK INPUT SIGNAL
1) Stop engine.
2) Turn ignition switch "ON".
3) Check voltage between E.C.U. terminal ③ and ground.
Voltage: Battery voltage
N.G.
Check harness continuity between E.C.U. and battery.
O.K.
E CHECK GROUND CIRCUIT
1) Turn ignition switch "OFF".
2) Disconnect power transistor harness connector.
3) Check resistance between terminal ⓓ and ground.
Resistance: Approximately 0Ω
N.G.
Repair harness.
O.K.

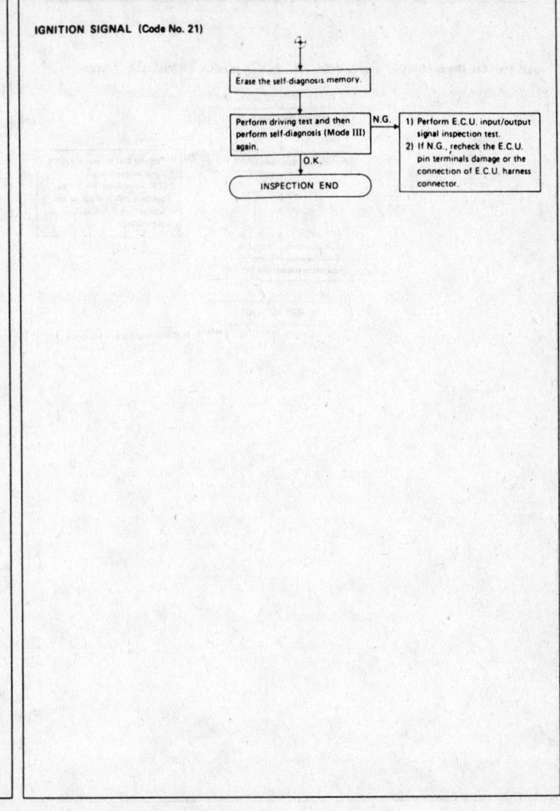
IGNITION SIGNAL (Code No. 21)
Erase the self-diagnosis memory.
Perform driving test and then perform self-diagnosis (Mode III) again.
N.G.
1) Perform E.C.U. input/output signal inspection test.
2) If N.G., recheck the E.C.U. pin terminals damage or the connection of E.C.U. harness connector.
O.K.
INSPECTION END

ECCS SELF-DIAGNOSTIC CHART–1990 PULSAR/SENTRA

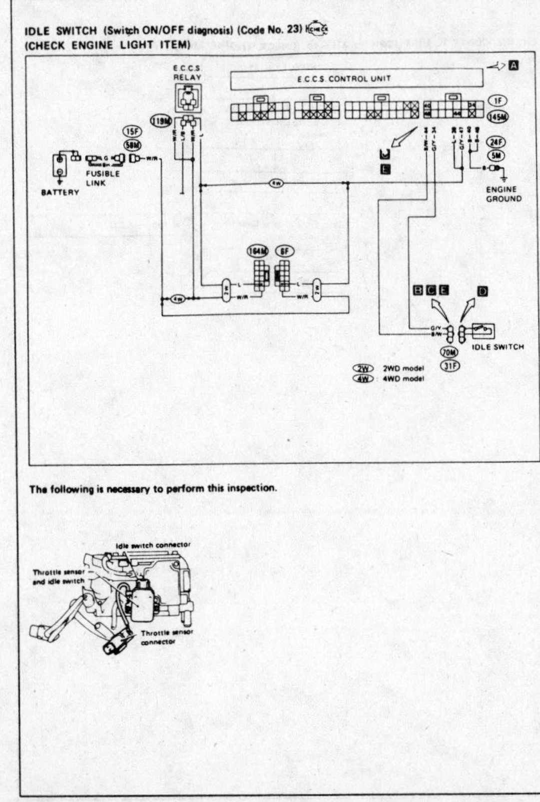
IDLE SWITCH (Switch ON/OFF diagnosis) (Code No. 23) (CHECK ENGINE LIGHT ITEM)
E.C.C.S. RELAY
E.C.C.S. CONTROL UNIT
FUSIBLE LINK
BATTERY
ENGINE GROUND
IDLE SWITCH
2W 2WD model
4W 4WD model
The following is necessary to perform this inspection.
Idle switch connector
Throttle sensor and idle switch
Throttle sensor connector

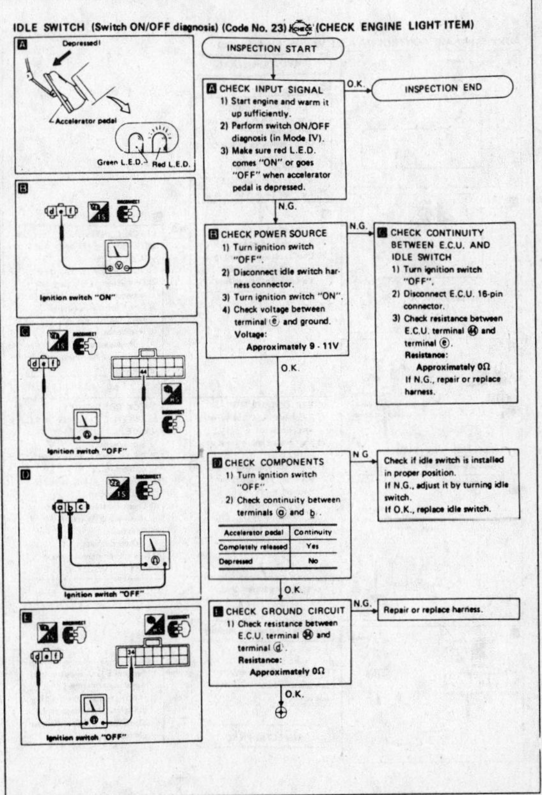
IDLE SWITCH (Switch ON/OFF diagnosis) (Code No. 23) (CHECK ENGINE LIGHT ITEM)
Depressed!
Accelerator pedal
Green L.E.D.
Red L.E.D.
Ignition switch "ON"
Ignition switch "OFF"
Ignition switch "OFF"
Ignition switch "OFF"
INSPECTION START
A CHECK INPUT SIGNAL
1) Start engine and warm it up sufficiently.
2) Perform switch ON/OFF diagnosis (in Mode IV).
3) Make sure red L.E.D. comes "ON" or goes "OFF" when accelerator pedal is depressed.
O.K.
INSPECTION END
N.G.
B CHECK POWER SOURCE
1) Turn ignition switch "OFF".
2) Disconnect idle switch harness connector.
3) Turn ignition switch "ON".
4) Check voltage between terminal ⓔ and ground.
Voltage: Approximately 9 - 11V
N.G.
C CHECK CONTINUITY BETWEEN E.C.U. AND IDLE SWITCH
1) Turn ignition switch "OFF".
2) Disconnect E.C.U. 16-pin connector.
3) Check resistance between E.C.U. terminal ⓐ and terminal ⓔ.
Resistance: Approximately 0Ω
If N.G., repair or replace harness.
O.K.
D CHECK COMPONENTS
1) Turn ignition switch "OFF".
2) Check continuity between terminals ⓐ and ⓑ.
Accelerator pedal | Continuity
Completely released | Yes
Depressed | No
N.G.
Check if idle switch is installed in proper position.
If N.G., adjust it by turning idle switch.
If O.K., replace idle switch.
O.K.
E CHECK GROUND CIRCUIT
1) Check resistance between E.C.U. terminal ⓐ and terminal ⓓ.
Resistance: Approximately 0Ω
N.G.
Repair or replace harness.
O.K.

ECCS SELF-DIAGNOSTIC CHART—1990 PULSAR/SENTRA

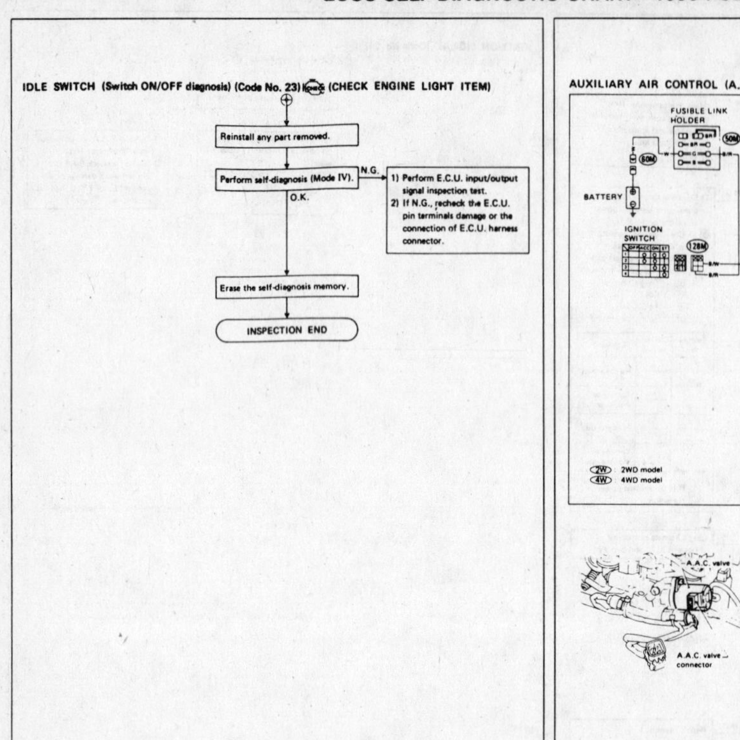
IDLE SWITCH (Switch ON/OFF diagnosis) (Code No. 23) (CHECK ENGINE LIGHT ITEM)
Reinstall any part removed.
Perform self-diagnosis (Mode IV).
N.G.
1) Perform E.C.U. input/output signal inspection test.
2) If N.G., recheck the E.C.U. pin terminals damage or the connection of E.C.U. harness connector.
O.K.
Erase the self-diagnosis memory.
INSPECTION END

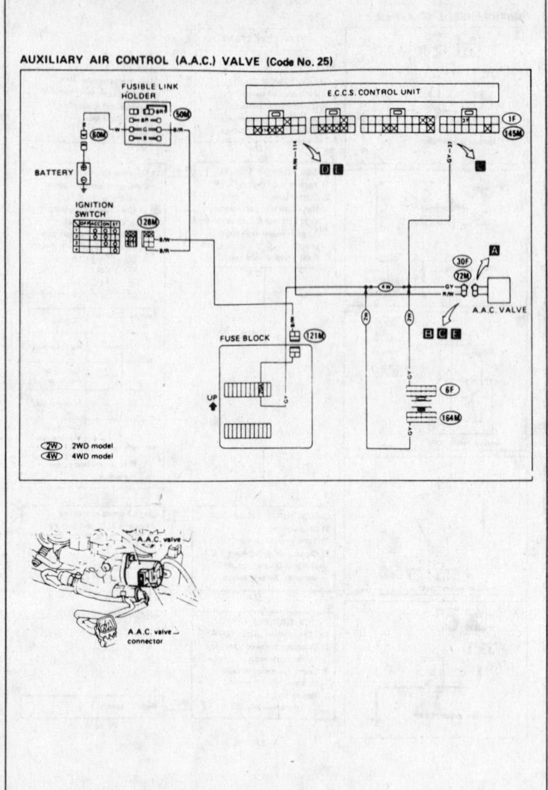
AUXILIARY AIR CONTROL (A.A.C.) VALVE (Code No. 25)
FUSIBLE LINK HOLDER
E.C.C.S. CONTROL UNIT
BATTERY
IGNITION SWITCH
FUSE BLOCK
UP
A.A.C. VALVE
2WD model
4WD model
A.A.C. valve
A.A.C. valve connector

ECCS SELF-DIAGNOSTIC CHART—1990 PULSAR/SENTRA

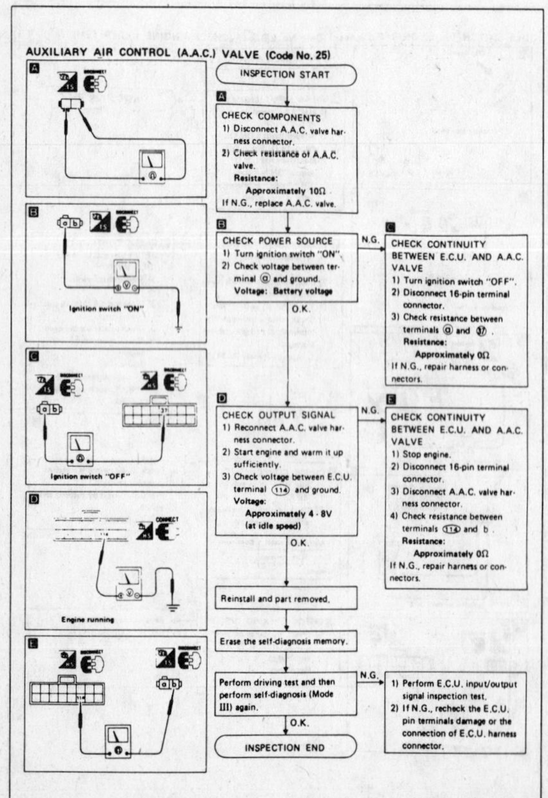
AUXILIARY AIR CONTROL (A.A.C.) VALVE (Code No. 25)
INSPECTION START
A
CHECK COMPONENTS
1) Disconnect A.A.C. valve harness connector.
2) Check resistance of A.A.C. valve.
Resistance:
Approximately 10Ω
If N.G., replace A.A.C. valve.
B
CHECK POWER SOURCE
1) Turn ignition switch "ON".
2) Check voltage between terminal ⓐ and ground.
Voltage: Battery voltage
N.G.
C
CHECK CONTINUITY BETWEEN E.C.U. AND A.A.C. VALVE
1) Turn ignition switch "OFF".
2) Disconnect 16-pin terminal connector.
3) Check resistance between terminals ⓐ and 37.
Resistance:
Approximately 0Ω
If N.G., repair harness or connectors.
O.K.
D
CHECK OUTPUT SIGNAL
1) Reconnect A.A.C. valve harness connector.
2) Start engine and warm it up sufficiently.
3) Check voltage between E.C.U. terminal 114 and ground.
Voltage:
Approximately 4 - 8V (at idle speed)
N.G.
E
CHECK CONTINUITY BETWEEN E.C.U. AND A.A.C. VALVE
1) Stop engine.
2) Disconnect 16-pin terminal connector.
3) Disconnect A.A.C. valve harness connector.
4) Check resistance between terminals 114 and b.
Resistance:
Approximately 0Ω
If N.G., repair harness or connectors.
O.K.
Reinstall and part removed.
Erase the self-diagnosis memory.
Perform driving test and then perform self-diagnosis (Mode III) again.
N.G.
1) Perform E.C.U. input/output signal inspection test.
2) If N.G., recheck the E.C.U. pin terminals damage or the connection of E.C.U. harness connector.
O.K.
INSPECTION END
Ignition switch "ON"
Ignition switch "OFF"
Engine running
DISCONNECT
CONNECT

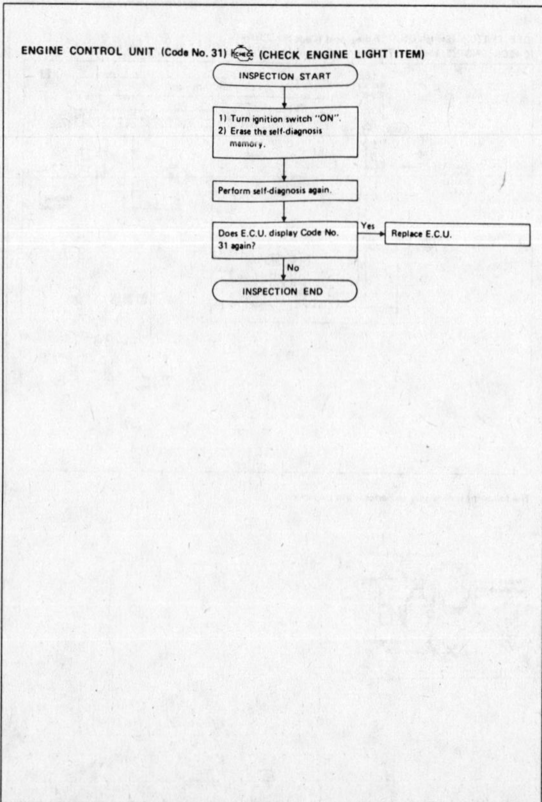
ENGINE CONTROL UNIT (Code No. 31) (CHECK ENGINE LIGHT ITEM)
INSPECTION START
1) Turn ignition switch "ON".
2) Erase the self-diagnosis memory.
Perform self-diagnosis again.
Does E.C.U. display Code No. 31 again?
Yes
Replace E.C.U.
No
INSPECTION END

ECCS DIAGNOSTIC CHART—1990 PULSAR/SENTRA

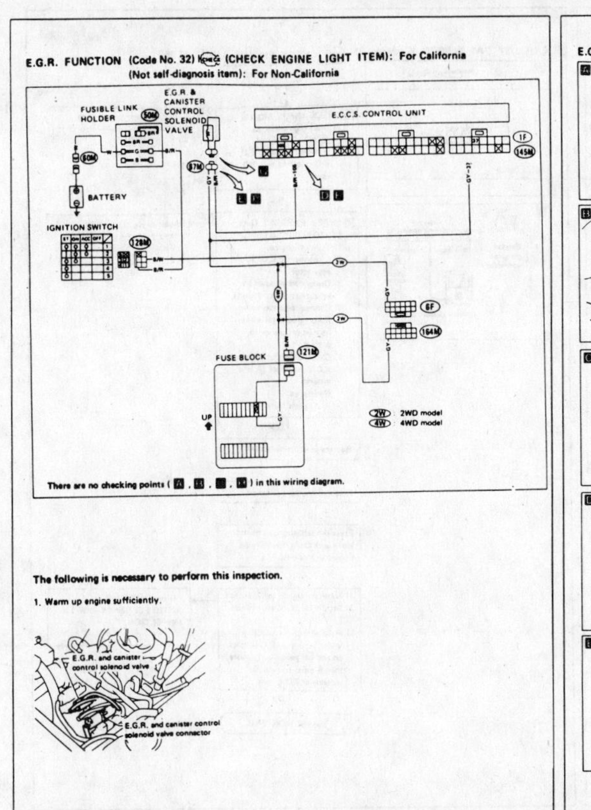
E.G.R. FUNCTION (Code No. 32) (CHECK ENGINE LIGHT ITEM): For California
(Not self-diagnosis item): For Non-California
FUSIBLE LINK HOLDER
E.G.R & CANISTER CONTROL SOLENOID VALVE
E.C.C.S. CONTROL UNIT
BATTERY
IGNITION SWITCH
FUSE BLOCK
UP
2WD model
4WD model
There are no checking points in this wiring diagram.
The following is necessary to perform this inspection.
1. Warm up engine sufficiently.
E.G.R. and canister control solenoid valve
E.G.R. and canister control solenoid valve connector

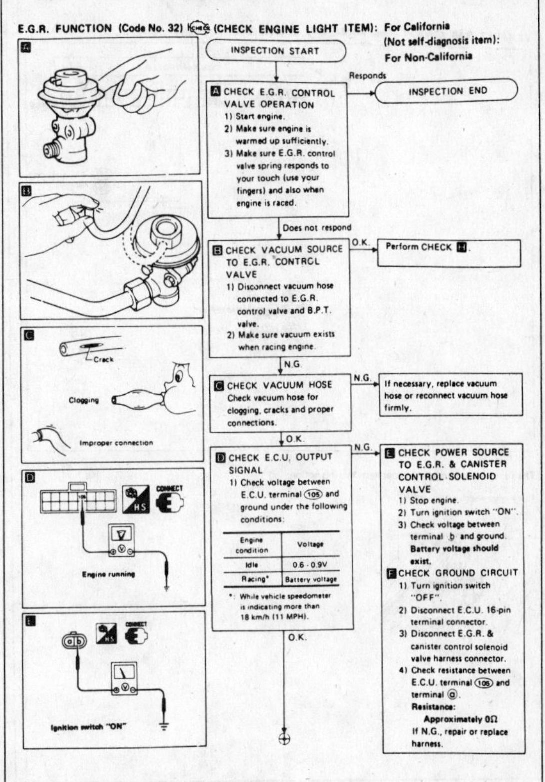
E.G.R. FUNCTION (Code No. 32) (CHECK ENGINE LIGHT ITEM): For California
(Not self-diagnosis item): For Non-California
INSPECTION START
A CHECK E.G.R. CONTROL VALVE OPERATION
1) Start engine.
2) Make sure engine is warmed up sufficiently.
3) Make sure E.G.R. control valve spring responds to your touch (use your fingers) and also when engine is raced.
Responds
INSPECTION END
Does not respond
B CHECK VACUUM SOURCE TO E.G.R. CONTROL VALVE
1) Disconnect vacuum hose connected to E.G.R. control valve and B.P.T. valve.
2) Make sure vacuum exists when racing engine.
O.K.
Perform CHECK H.
N.G.
C CHECK VACUUM HOSE
Check vacuum hose for clogging, cracks and proper connections.
N.G.
If necessary, replace vacuum hose or reconnect vacuum hose firmly.
O.K.
D CHECK E.C.U. OUTPUT SIGNAL
1) Check voltage between E.C.U. terminal 106 and ground under the following conditions:
Engine condition
Voltage
Idle
0.6 - 0.9V
Racing*
Battery voltage
*: While vehicle speedometer is indicating more than 18 km/h (11 MPH).
N.G.
E CHECK POWER SOURCE TO E.G.R. & CANISTER CONTROL SOLENOID VALVE
1) Stop engine.
2) Turn ignition switch "ON".
3) Check voltage between terminal b and ground.
Battery voltage should exist.
F CHECK GROUND CIRCUIT
1) Turn ignition switch "OFF".
2) Disconnect E.C.U. 16-pin terminal connector.
3) Disconnect E.G.R. & canister control solenoid valve harness connector.
4) Check resistance between E.C.U. terminal 106 and terminal a.
Resistance:
Approximately 0Ω
If N.G., repair or replace harness.
O.K.
Crack
Clogging
Improper connection
CONNECT
Engine running
Ignition switch "ON"

ECCS DIAGNOSTIC CHART—1990 PULSAR/SENTRA

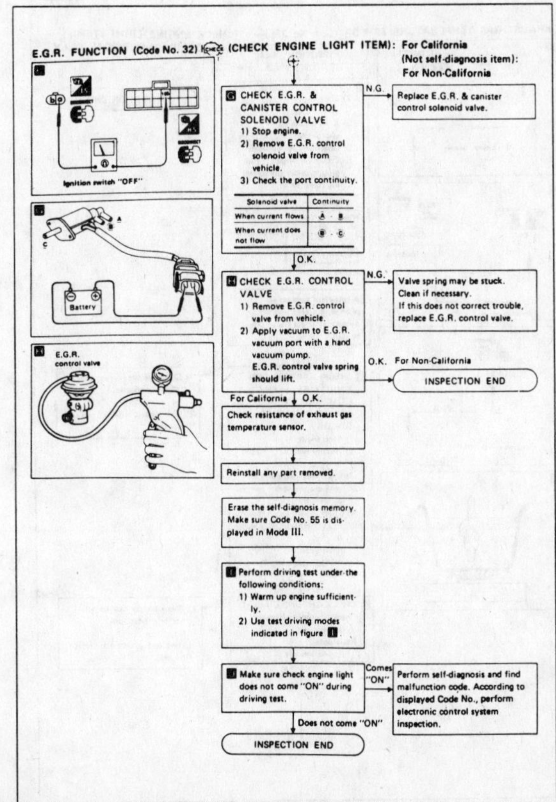
E.G.R. FUNCTION (Code No. 32) (CHECK ENGINE LIGHT ITEM): For California
(Not self-diagnosis item): For Non-California
G CHECK E.G.R. & CANISTER CONTROL SOLENOID VALVE
1) Stop engine.
2) Remove E.G.R. control solenoid valve from vehicle.
3) Check the port continuity.
Solenoid valve
Continuity
When current flows
A - B
When current does not flow
B - C
N.G.
Replace E.G.R. & canister control solenoid valve.
O.K.
H CHECK E.G.R. CONTROL VALVE
1) Remove E.G.R. control valve from vehicle.
2) Apply vacuum to E.G.R. vacuum port with a hand vacuum pump.
E.G.R. control valve spring should lift.
N.G.
Valve spring may be stuck. Clean if necessary.
If this does not correct trouble, replace E.G.R. control valve.
O.K. For Non-California
INSPECTION END
For California O.K.
Check resistance of exhaust gas temperature sensor.
Reinstall any part removed.
Erase the self-diagnosis memory. Make sure Code No. 55 is displayed in Mode III.
I Perform driving test under the following conditions:
1) Warm up engine sufficiently.
2) Use test driving modes indicated in figure I.
J Make sure check engine light does not come "ON" during driving test.
Comes "ON"
Perform self-diagnosis and find malfunction code. According to displayed Code No., perform electronic control system inspection.
Does not come "ON"
INSPECTION END
Ignition switch "OFF"
Battery
E.G.R. control valve

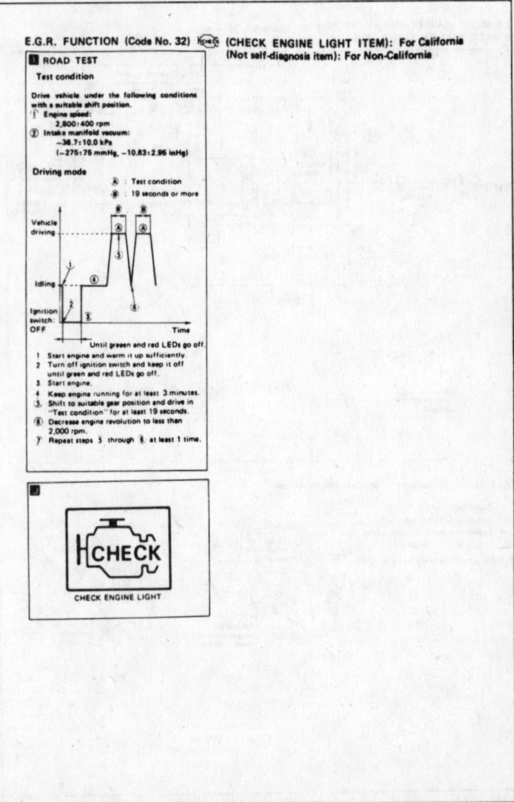
E.G.R. FUNCTION (Code No. 32) (CHECK ENGINE LIGHT ITEM): For California
(Not self-diagnosis item): For Non-California
I ROAD TEST
Test condition
Drive vehicle under the following conditions with a suitable shift position.
① Engine speed: 2,800±400 rpm
② Intake manifold vacuum: −36.7±10.0 kPa (−275±75 mmHg, −10.83±2.95 inHg)
Driving mode
A : Test condition
B : 19 seconds or more
Vehicle driving
Idling
Ignition switch: OFF
Time
Until green and red LEDs go off.
1 Start engine and warm it up sufficiently.
2 Turn off ignition switch and keep it off until green and red LEDs go off.
3 Start engine.
4 Keep engine running for at least 3 minutes.
5 Shift to suitable gear position and drive in "Test condition" for at least 19 seconds.
6 Decrease engine revolution to less than 2,000 rpm.
7 Repeat steps 5 through 6 at least 1 time.
CHECK
CHECK ENGINE LIGHT

ECCS SELF-DIAGNOSTIC CHART—1990 PULSAR/SENTRA

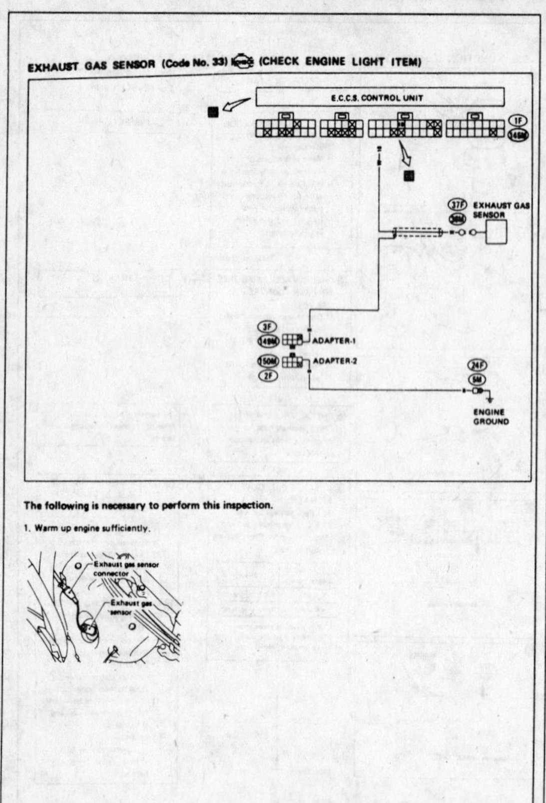
EXHAUST GAS SENSOR (Code No. 33) (CHECK ENGINE LIGHT ITEM)
E.C.C.S. CONTROL UNIT
EXHAUST GAS SENSOR
ADAPTER-1
ADAPTER-2
ENGINE GROUND
The following is necessary to perform this inspection.
1. Warm up engine sufficiently.
Exhaust gas sensor connector
Exhaust gas sensor

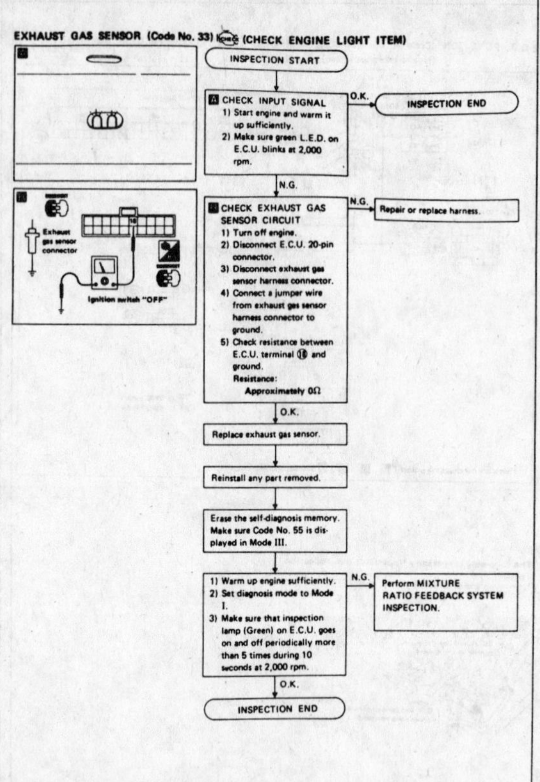
EXHAUST GAS SENSOR (Code No. 33) (CHECK ENGINE LIGHT ITEM)
Exhaust gas sensor connector
Ignition switch "OFF"
INSPECTION START
A CHECK INPUT SIGNAL
1) Start engine and warm it up sufficiently.
2) Make sure green L.E.D. on E.C.U. blinks at 2,000 rpm.
O.K.
INSPECTION END
N.G.
B CHECK EXHAUST GAS SENSOR CIRCUIT
1) Turn off engine.
2) Disconnect E.C.U. 20-pin connector.
3) Disconnect exhaust gas sensor harness connector.
4) Connect a jumper wire from exhaust gas sensor harness connector to ground.
5) Check resistance between E.C.U. terminal ⑱ and ground.
Resistance:
Approximately 0Ω
N.G.
Repair or replace harness.
O.K.
Replace exhaust gas sensor.
Reinstall any part removed.
Erase the self-diagnosis memory. Make sure Code No. 55 is displayed in Mode III.
1) Warm up engine sufficiently.
2) Set diagnosis mode to Mode I.
3) Make sure that inspection lamp (Green) on E.C.U. goes on and off periodically more than 5 times during 10 seconds at 2,000 rpm.
N.G.
Perform MIXTURE RATIO FEEDBACK SYSTEM INSPECTION.
O.K.
INSPECTION END

ECCS SELF-DIAGNOSTIC CHART—1990 PULSAR/SENTRA

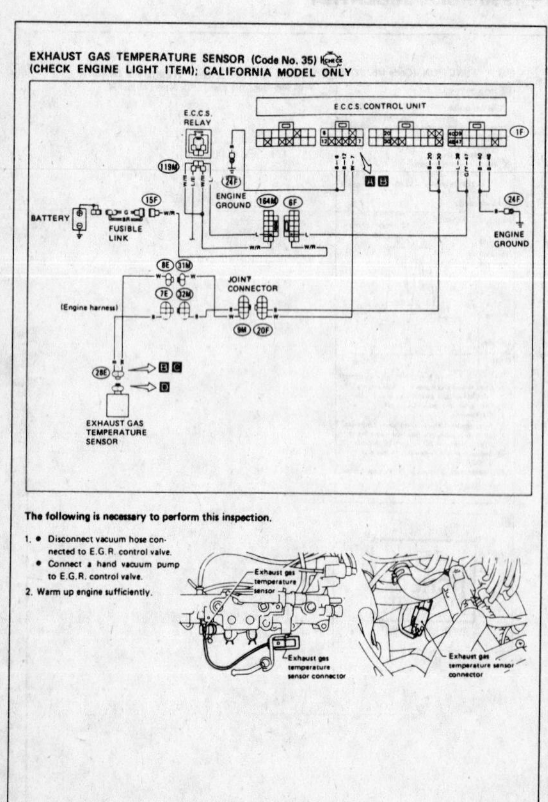
EXHAUST GAS TEMPERATURE SENSOR (Code No. 35) (CHECK ENGINE LIGHT ITEM); CALIFORNIA MODEL ONLY
E.C.C.S. CONTROL UNIT
E.C.C.S. RELAY
BATTERY
FUSIBLE LINK
ENGINE GROUND
JOINT CONNECTOR
(Engine harness)
EXHAUST GAS TEMPERATURE SENSOR
The following is necessary to perform this inspection.
1. • Disconnect vacuum hose connected to E.G.R. control valve.
• Connect a hand vacuum pump to E.G.R. control valve.
2. Warm up engine sufficiently.
Exhaust gas temperature sensor
Exhaust gas temperature sensor connector

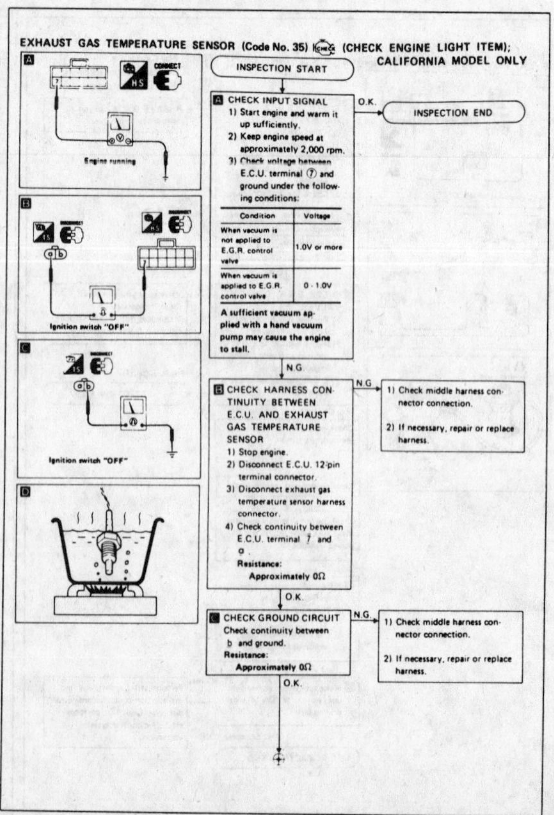
EXHAUST GAS TEMPERATURE SENSOR (Code No. 35) (CHECK ENGINE LIGHT ITEM); CALIFORNIA MODEL ONLY
Engine running
Ignition switch "OFF"
INSPECTION START
A CHECK INPUT SIGNAL
1) Start engine and warm it up sufficiently.
2) Keep engine speed at approximately 2,000 rpm.
3) Check voltage between E.C.U. terminal ⑦ and ground under the following conditions:
Condition | Voltage
When vacuum is not applied to E.G.R. control valve | 1.0V or more
When vacuum is applied to E.G.R. control valve | 0 - 1.0V
A sufficient vacuum applied with a hand vacuum pump may cause the engine to stall.
O.K.
INSPECTION END
N.G.
B CHECK HARNESS CONTINUITY BETWEEN E.C.U. AND EXHAUST GAS TEMPERATURE SENSOR
1) Stop engine.
2) Disconnect E.C.U. 12-pin terminal connector.
3) Disconnect exhaust gas temperature sensor harness connector.
4) Check continuity between E.C.U. terminal 7 and a.
Resistance:
Approximately 0Ω
N.G.
1) Check middle harness connector connection.
2) If necessary, repair or replace harness.
O.K.
C CHECK GROUND CIRCUIT
Check continuity between b and ground.
Resistance:
Approximately 0Ω
N.G.
1) Check middle harness connector connection.
2) If necessary, repair or replace harness.
O.K.

ECCS SELF-DIAGNOSTIC CHART—1990 PULSAR/SENTRA

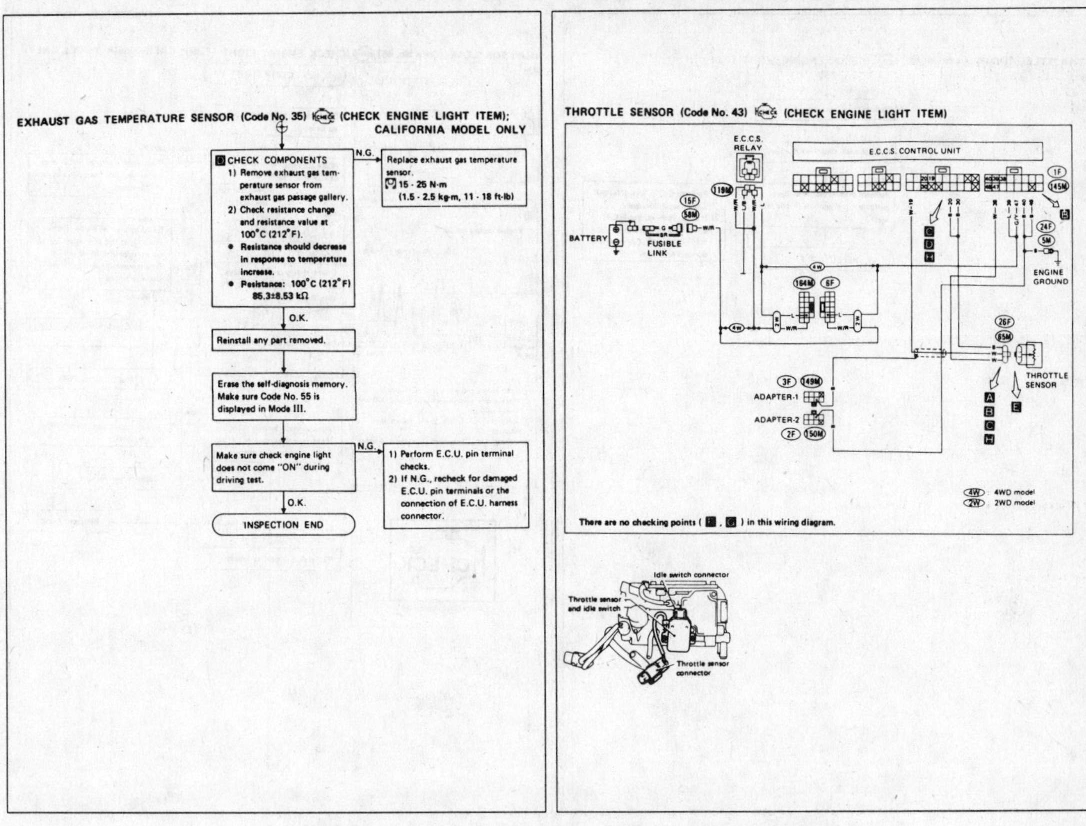
EXHAUST GAS TEMPERATURE SENSOR (Code No. 35) (CHECK ENGINE LIGHT ITEM); CALIFORNIA MODEL ONLY
D CHECK COMPONENTS
1) Remove exhaust gas temperature sensor from exhaust gas passage gallery.
2) Check resistance change and resistance value at 100°C (212°F).
• Resistance should decrease in response to temperature increase.
• Resistance: 100°C (212°F) 85.3±8.53 kΩ
N.G.
Replace exhaust gas temperature sensor.
15 - 25 N·m (1.5 - 2.5 kg-m, 11 - 18 ft-lb)
O.K.
Reinstall any part removed.
Erase the self-diagnosis memory. Make sure Code No. 55 is displayed in Mode III.
Make sure check engine light does not come "ON" during driving test.
N.G.
1) Perform E.C.U. pin terminal checks.
2) If N.G., recheck for damaged E.C.U. pin terminals or the connection of E.C.U. harness connector.
O.K.
INSPECTION END
THROTTLE SENSOR (Code No. 43) (CHECK ENGINE LIGHT ITEM)
E.C.C.S. RELAY
E.C.C.S. CONTROL UNIT
BATTERY
FUSIBLE LINK
ENGINE GROUND
THROTTLE SENSOR
ADAPTER-1
ADAPTER-2
4WD model
2WD model
There are no checking points (,) in this wiring diagram.
Idle switch connector
Throttle sensor and idle switch
Throttle sensor connector

ECCS SELF-DIAGNOSTIC CHART—1990 PULSAR/SENTRA

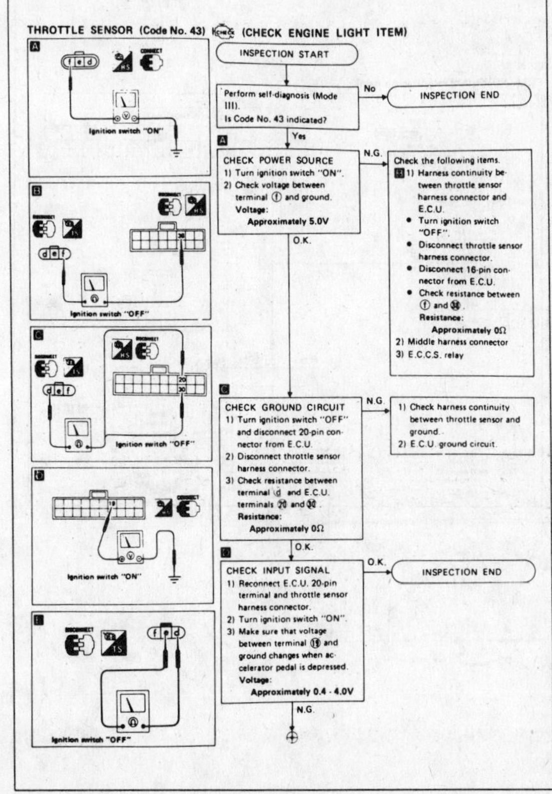
THROTTLE SENSOR (Code No. 43) (CHECK ENGINE LIGHT ITEM)
Ignition switch "ON"
Ignition switch "OFF"
INSPECTION START
Perform self-diagnosis (Mode III).
Is Code No. 43 indicated?
No
INSPECTION END
Yes
A CHECK POWER SOURCE
1) Turn ignition switch "ON".
2) Check voltage between terminal ⓕ and ground.
Voltage:
Approximately 5.0V
N.G.
Check the following items.
B 1) Harness continuity between throttle sensor harness connector and E.C.U.
• Turn ignition switch "OFF".
• Disconnect throttle sensor harness connector.
• Disconnect 16-pin connector from E.C.U.
• Check resistance between ⓕ and ㊳.
Resistance:
Approximately 0Ω
2) Middle harness connector
3) E.C.C.S. relay
O.K.
C CHECK GROUND CIRCUIT
1) Turn ignition switch "OFF" and disconnect 20-pin connector from E.C.U.
2) Disconnect throttle sensor harness connector.
3) Check resistance between terminal ⓓ and E.C.U. terminals ⑳ and ㉚.
Resistance:
Approximately 0Ω
N.G.
1) Check harness continuity between throttle sensor and ground.
2) E.C.U. ground circuit.
O.K.
D CHECK INPUT SIGNAL
1) Reconnect E.C.U. 20-pin terminal and throttle sensor harness connector.
2) Turn ignition switch "ON".
3) Make sure that voltage between terminal ⑲ and ground changes when accelerator pedal is depressed.
Voltage:
Approximately 0.4 - 4.0V
O.K.
INSPECTION END
N.G.

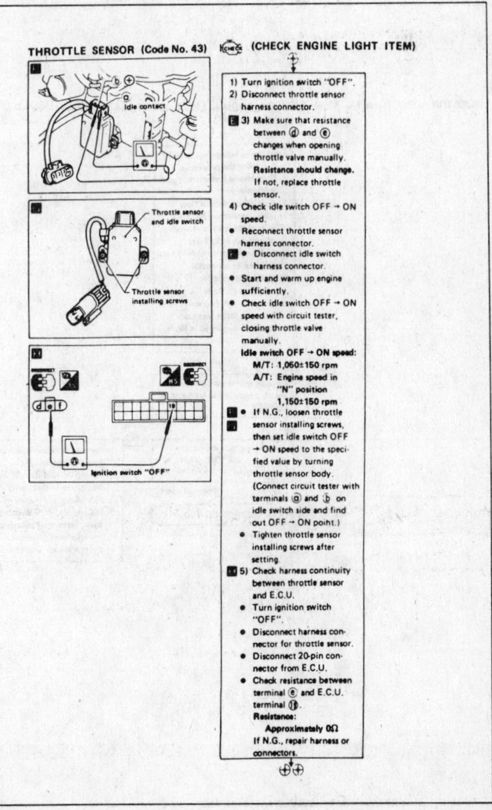
THROTTLE SENSOR (Code No. 43) (CHECK ENGINE LIGHT ITEM)
Idle contact
Throttle sensor and idle switch
Throttle sensor installing screws
Ignition switch "OFF"
1) Turn ignition switch "OFF".
2) Disconnect throttle sensor harness connector.
E 3) Make sure that resistance between ⓓ and ⓔ changes when opening throttle valve manually.
Resistance should change.
If not, replace throttle sensor.
4) Check idle switch OFF → ON speed.
• Reconnect throttle sensor harness connector.
• Disconnect idle switch harness connector.
• Start and warm up engine sufficiently.
• Check idle switch OFF → ON speed with circuit tester, closing throttle valve manually.
Idle switch OFF → ON speed:
M/T: 1,060±150 rpm
A/T: Engine speed in "N" position 1,150±150 rpm
• If N.G., loosen throttle sensor installing screws, then set idle switch OFF → ON speed to the specified value by turning throttle sensor body. (Connect circuit tester with terminals ⓐ and ⓑ on idle switch side and find out OFF → ON point.)
• Tighten throttle sensor installing screws after setting.
5) Check harness continuity between throttle sensor and E.C.U.
• Turn ignition switch "OFF".
• Disconnect harness connector for throttle sensor.
• Disconnect 20-pin connector from E.C.U.
• Check resistance between terminal ⓔ and E.C.U. terminal ⑲.
Resistance:
Approximately 0Ω
If N.G., repair harness or connectors.

ECCS SELF-DIAGNOSTIC CHART—1990 PULSAR/SENTRA

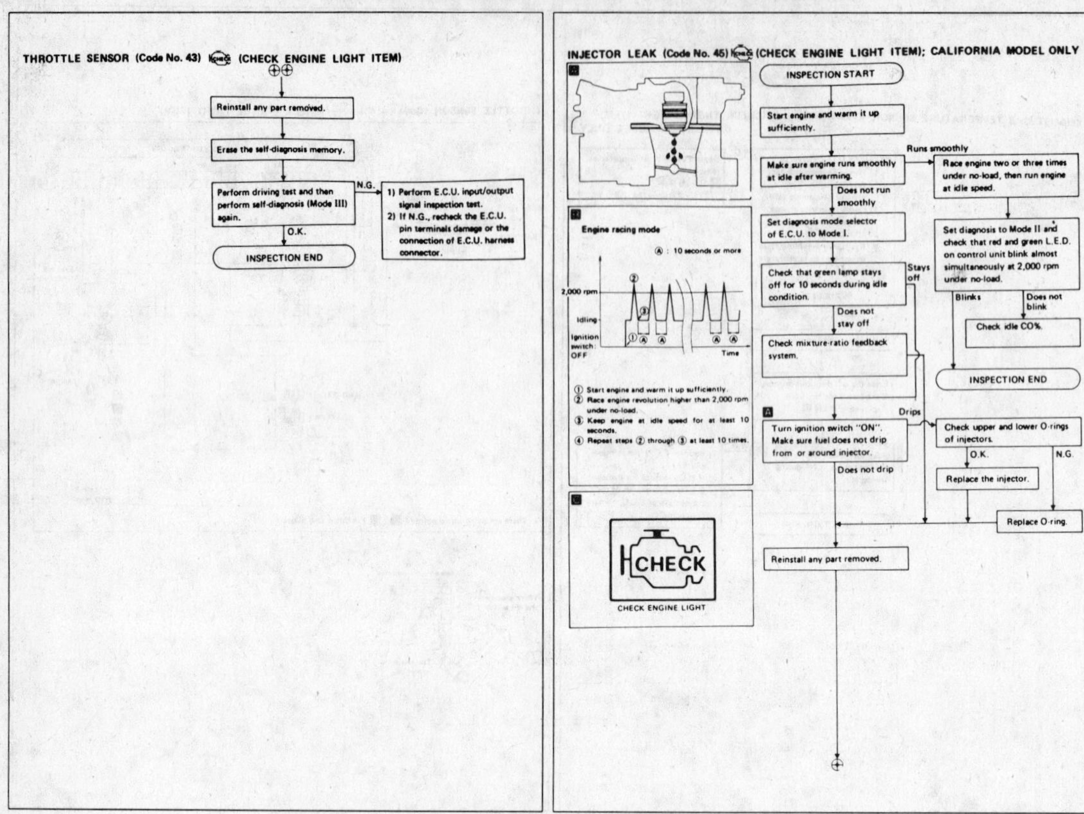
THROTTLE SENSOR (Code No. 43) (CHECK ENGINE LIGHT ITEM)
Reinstall any part removed.
Erase the self-diagnosis memory.
Perform driving test and then perform self-diagnosis (Mode III) again.
N.G.
1) Perform E.C.U. input/output signal inspection test.
2) If N.G., recheck the E.C.U. pin terminals damage or the connection of E.C.U. harness connector.
O.K.
INSPECTION END
INJECTOR LEAK (Code No. 45) (CHECK ENGINE LIGHT ITEM); CALIFORNIA MODEL ONLY
Engine racing mode
Ⓐ : 10 seconds or more
2,000 rpm
Idling
Ignition switch: OFF
Time
① Start engine and warm it up sufficiently.
② Race engine revolution higher than 2,000 rpm under no-load.
③ Keep engine at idle speed for at least 10 seconds.
④ Repeat steps ② through ③ at least 10 times.
CHECK
CHECK ENGINE LIGHT
INSPECTION START
Start engine and warm it up sufficiently.
Make sure engine runs smoothly at idle after warming.
Runs smoothly
Race engine two or three times under no-load, then run engine at idle speed.
Does not run smoothly
Set diagnosis mode selector of E.C.U. to Mode I.
Set diagnosis to Mode II and check that red and green L.E.D. on control unit blink almost simultaneously at 2,000 rpm under no-load.
Blinks
Does not blink
Check idle CO%.
Check that green lamp stays off for 10 seconds during idle condition.
Stays off
Does not stay off
Check mixture-ratio feedback system.
INSPECTION END
Turn ignition switch "ON". Make sure fuel does not drip from or around injector.
Drips
Check upper and lower O-rings of injectors.
O.K.
N.G.
Replace the injector.
Replace O-ring.
Does not drip
Reinstall any part removed.

ECCS SELF-DIAGNOSTIC CHART—1990 PULSAR/SENTRA

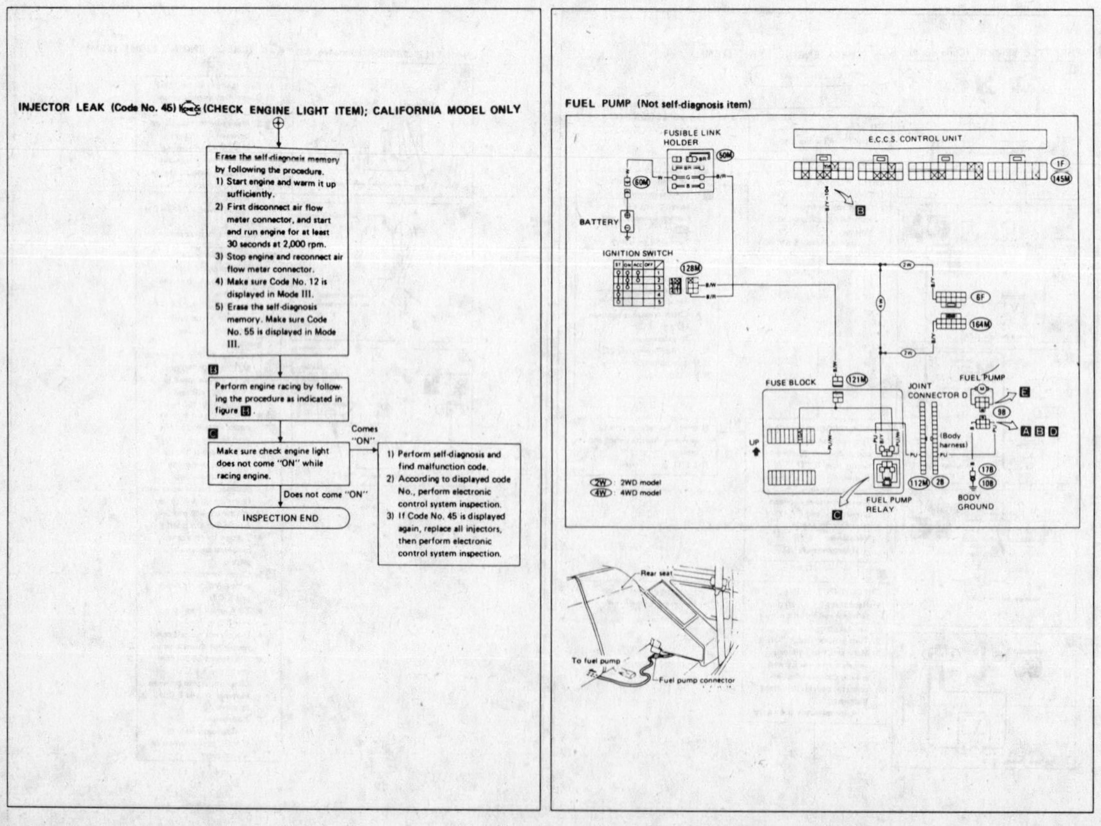
INJECTOR LEAK (Code No. 45) (CHECK ENGINE LIGHT ITEM); CALIFORNIA MODEL ONLY
Erase the self-diagnosis memory by following the procedure.
1) Start engine and warm it up sufficiently.
2) First disconnect air flow meter connector, and start and run engine for at least 30 seconds at 2,000 rpm.
3) Stop engine and reconnect air flow meter connector.
4) Make sure Code No. 12 is displayed in Mode III.
5) Erase the self-diagnosis memory. Make sure Code No. 55 is displayed in Mode III.
Perform engine racing by following the procedure as indicated in figure B.
Make sure check engine light does not come "ON" while racing engine.
Comes "ON"
1) Perform self-diagnosis and find malfunction code.
2) According to displayed code No., perform electronic control system inspection.
3) If Code No. 45 is displayed again, replace all injectors, then perform electronic control system inspection.
Does not come "ON"
INSPECTION END
FUEL PUMP (Not self-diagnosis item)
FUSIBLE LINK HOLDER
50M
60M
BATTERY
IGNITION SWITCH
ST ON ACC OFF
128M
B/W
B/R
E.C.C.S. CONTROL UNIT
1F
145M
2W
6F
164M
4W
FUSE BLOCK
121M
JOINT CONNECTOR D
(Body harness)
FUEL PUMP
9B
UP
112M
2B
17B
10B
FUEL PUMP RELAY
BODY GROUND
2W : 2WD model
4W : 4WD model
Rear seat
To fuel pump
Fuel pump connector

ECCS SELF-DIAGNOSTIC CHART—1990 PULSAR/SENTRA

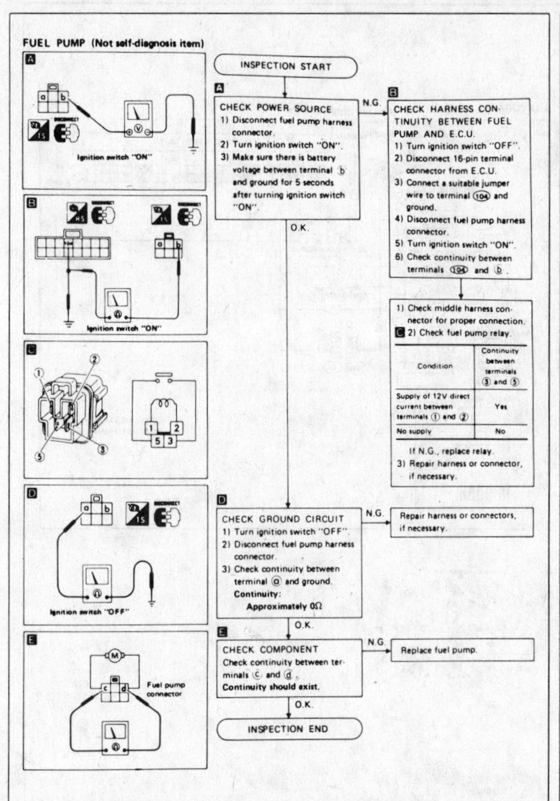
FUEL PUMP (Not self-diagnosis item)
Ignition switch "ON"
Ignition switch "ON"
Ignition switch "OFF"
Fuel pump connector
INSPECTION START
A
CHECK POWER SOURCE
1) Disconnect fuel pump harness connector.
2) Turn ignition switch "ON".
3) Make sure there is battery voltage between terminal ⓑ and ground for 5 seconds after turning ignition switch "ON".
N.G.
B
CHECK HARNESS CONTINUITY BETWEEN FUEL PUMP AND E.C.U.
1) Turn ignition switch "OFF".
2) Disconnect 16-pin terminal connector from E.C.U.
3) Connect a suitable jumper wire to terminal (104) and ground.
4) Disconnect fuel pump harness connector.
5) Turn ignition switch "ON".
6) Check continuity between terminals (104) and ⓑ.
1) Check middle harness connector for proper connection.
C 2) Check fuel pump relay.
Condition | Continuity between terminals ③ and ⑤
Supply of 12V direct current between terminals ① and ② | Yes
No supply | No
If N.G., replace relay.
3) Repair harness or connector, if necessary.
O.K.
D
CHECK GROUND CIRCUIT
1) Turn ignition switch "OFF".
2) Disconnect fuel pump harness connector.
3) Check continuity between terminal ⓐ and ground.
Continuity:
Approximately 0Ω
N.G.
Repair harness or connectors, if necessary.
O.K.
E
CHECK COMPONENT
Check continuity between terminals ⓒ and ⓓ.
Continuity should exist.
N.G.
Replace fuel pump.
O.K.
INSPECTION END

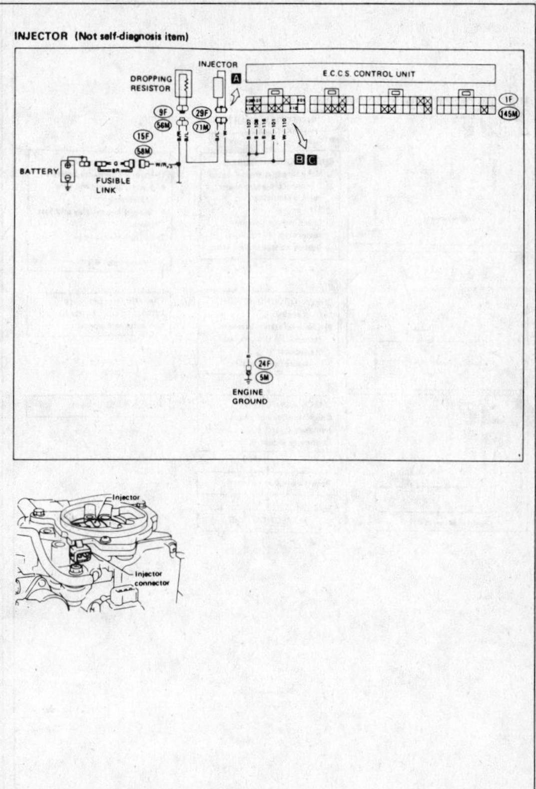
INJECTOR (Not self-diagnosis item)
INJECTOR
DROPPING RESISTOR
E.C.C.S. CONTROL UNIT
BATTERY
FUSIBLE LINK
ENGINE GROUND
Injector
Injector connector

ECCS SELF-DIAGNOSTIC CHART—1990 PULSAR/SENTRA

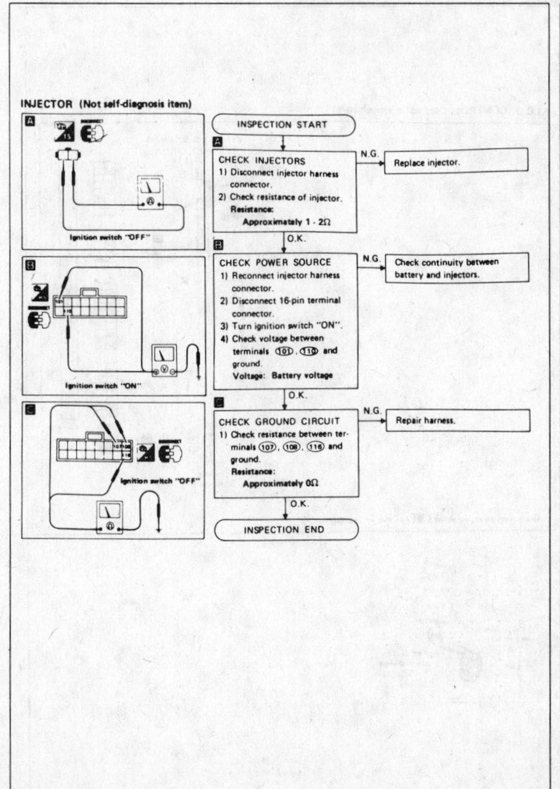
INJECTOR (Not self-diagnosis item)
Ignition switch "OFF"
Ignition switch "ON"
Ignition switch "OFF"
INSPECTION START
A
CHECK INJECTORS
1) Disconnect injector harness connector.
2) Check resistance of injector.
Resistance:
Approximately 1 - 2Ω
N.G.
Replace injector.
O.K.
B
CHECK POWER SOURCE
1) Reconnect injector harness connector.
2) Disconnect 16-pin terminal connector.
3) Turn ignition switch "ON".
4) Check voltage between terminals (101), (110) and ground.
Voltage: Battery voltage
N.G.
Check continuity between battery and injectors.
O.K.
C
CHECK GROUND CIRCUIT
1) Check resistance between terminals (107), (108), (116) and ground.
Resistance:
Approximately 0Ω
N.G.
Repair harness.
O.K.
INSPECTION END

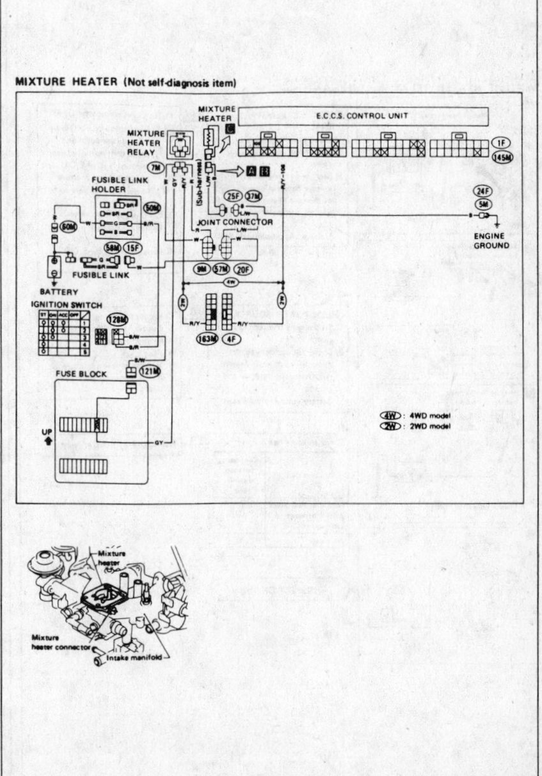
MIXTURE HEATER (Not self-diagnosis item)
MIXTURE HEATER
MIXTURE HEATER RELAY
E.C.C.S. CONTROL UNIT
FUSIBLE LINK HOLDER
JOINT CONNECTOR
ENGINE GROUND
FUSIBLE LINK
BATTERY
IGNITION SWITCH
FUSE BLOCK
UP
4WD: 4WD model
2WD: 2WD model
Mixture heater
Mixture heater connector
Intake manifold

ECCS SELF-DIAGNOSTIC CHART—1990 PULSAR/SENTRA

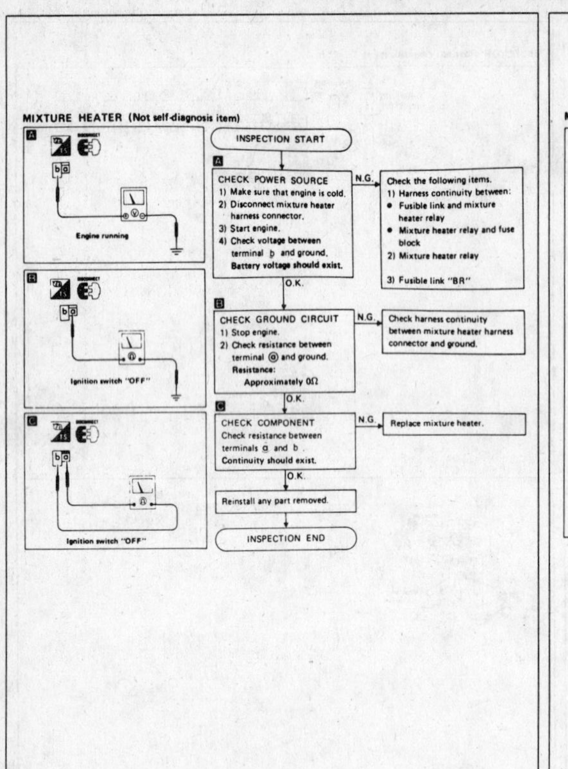
MIXTURE HEATER (Not self-diagnosis item)
Engine running
Ignition switch "OFF"
Ignition switch "OFF"
INSPECTION START
A CHECK POWER SOURCE
1) Make sure that engine is cold.
2) Disconnect mixture heater harness connector.
3) Start engine.
4) Check voltage between terminal b and ground.
Battery voltage should exist.
N.G.
Check the following items.
1) Harness continuity between:
• Fusible link and mixture heater relay
• Mixture heater relay and fuse block
2) Mixture heater relay
3) Fusible link "BR"
O.K.
B CHECK GROUND CIRCUIT
1) Stop engine.
2) Check resistance between terminal ⓐ and ground.
Resistance:
Approximately 0Ω
N.G.
Check harness continuity between mixture heater harness connector and ground.
O.K.
C CHECK COMPONENT
Check resistance between terminals a and b.
Continuity should exist.
N.G.
Replace mixture heater.
O.K.
Reinstall any part removed.
INSPECTION END

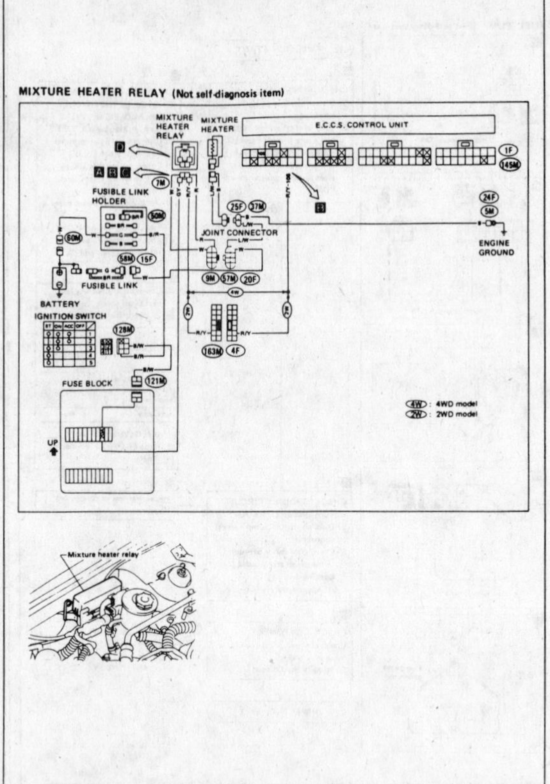
MIXTURE HEATER RELAY (Not self-diagnosis item)
MIXTURE HEATER RELAY
MIXTURE HEATER
E.C.C.S. CONTROL UNIT
FUSIBLE LINK HOLDER
JOINT CONNECTOR
ENGINE GROUND
FUSIBLE LINK
BATTERY
IGNITION SWITCH
FUSE BLOCK
UP
4W : 4WD model
2W : 2WD model
Mixture heater relay

ECCS SELF-DIAGNOSTIC CHART—1990 PULSAR/SENTRA

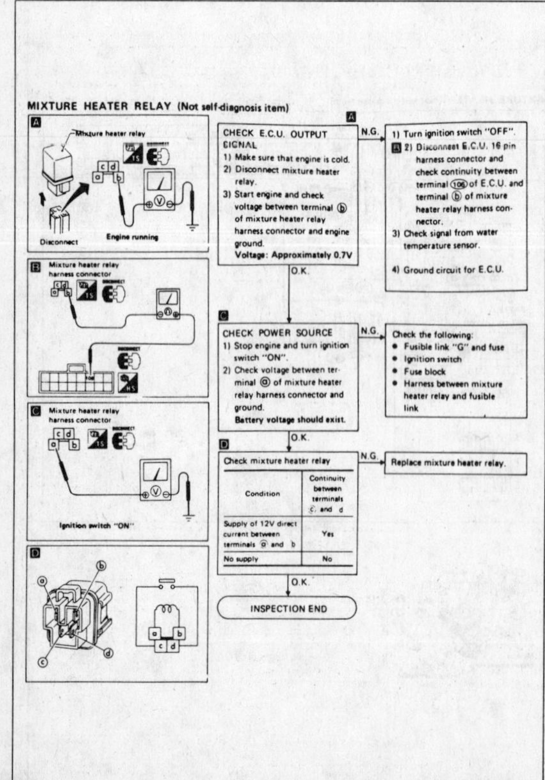
MIXTURE HEATER RELAY (Not self-diagnosis item)
Mixture heater relay
Disconnect
Engine running
Mixture heater relay harness connector
Mixture heater relay harness connector
Ignition switch "ON"
A CHECK E.C.U. OUTPUT SIGNAL
1) Make sure that engine is cold.
2) Disconnect mixture heater relay.
3) Start engine and check voltage between terminal ⓑ of mixture heater relay harness connector and engine ground.
Voltage: Approximately 0.7V
N.G.
1) Turn ignition switch "OFF".
B 2) Disconnect E.C.U. 16 pin harness connector and check continuity between terminal (106) of E.C.U. and terminal ⓑ of mixture heater relay harness connector.
3) Check signal from water temperature sensor.
4) Ground circuit for E.C.U.
O.K.
C CHECK POWER SOURCE
1) Stop engine and turn ignition switch "ON".
2) Check voltage between terminal ⓐ of mixture heater relay harness connector and ground.
Battery voltage should exist.
N.G.
Check the following:
• Fusible link "G" and fuse
• Ignition switch
• Fuse block
• Harness between mixture heater relay and fusible link
O.K.
D Check mixture heater relay
Condition | Continuity between terminals c and d
Supply of 12V direct current between terminals a and b | Yes
No supply | No
N.G.
Replace mixture heater relay.
O.K.
INSPECTION END

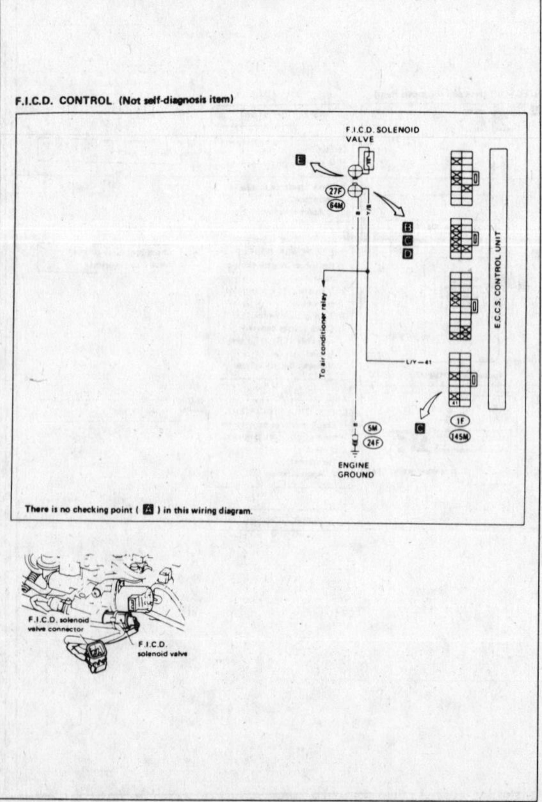
F.I.C.D. CONTROL (Not self-diagnosis item)
F.I.C.D. SOLENOID VALVE
E.C.C.S. CONTROL UNIT
To air conditioner relay
ENGINE GROUND
There is no checking point (A) in this wiring diagram.
F.I.C.D. solenoid valve connector
F.I.C.D. solenoid valve

ECCS SELF-DIAGNOSTIC CHART—1990 PULSAR/SENTRA

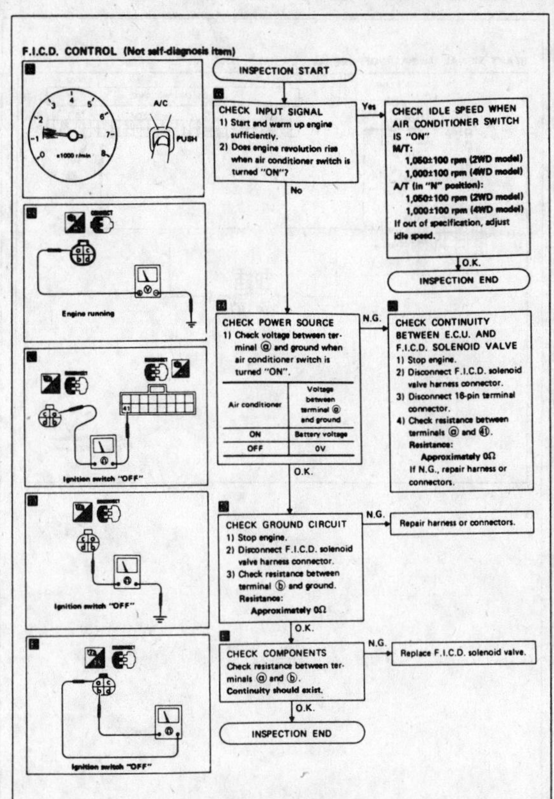
F.I.C.D. CONTROL (Not self-diagnosis item)
A/C
PUSH
x1000 r/min
Engine running
Ignition switch "OFF"
Ignition switch "OFF"
Ignition switch "OFF"
INSPECTION START
CHECK INPUT SIGNAL
1) Start and warm up engine sufficiently.
2) Does engine revolution rise when air conditioner switch is turned "ON"?
Yes
CHECK IDLE SPEED WHEN AIR CONDITIONER SWITCH IS "ON"
M/T:
1,050±100 rpm (2WD model)
1,000±100 rpm (4WD model)
A/T (in "N" position):
1,050±100 rpm (2WD model)
1,000±100 rpm (4WD model)
If out of specification, adjust idle speed.
O.K.
INSPECTION END
No
CHECK POWER SOURCE
1) Check voltage between terminal ⓐ and ground when air conditioner switch is turned "ON".
Air conditioner | Voltage between terminal ⓐ and ground
ON | Battery voltage
OFF | 0V
N.G.
CHECK CONTINUITY BETWEEN E.C.U. AND F.I.C.D. SOLENOID VALVE
1) Stop engine.
2) Disconnect F.I.C.D. solenoid valve harness connector.
3) Disconnect 16-pin terminal connector.
4) Check resistance between terminals ⓐ and ㊶.
Resistance:
Approximately 0Ω
If N.G., repair harness or connectors.
O.K.
CHECK GROUND CIRCUIT
1) Stop engine.
2) Disconnect F.I.C.D. solenoid valve harness connector.
3) Check resistance between terminal ⓑ and ground.
Resistance:
Approximately 0Ω
N.G.
Repair harness or connectors.
O.K.
CHECK COMPONENTS
Check resistance between terminals ⓐ and ⓑ.
Continuity should exist.
N.G.
Replace F.I.C.D. solenoid valve.
O.K.
INSPECTION END

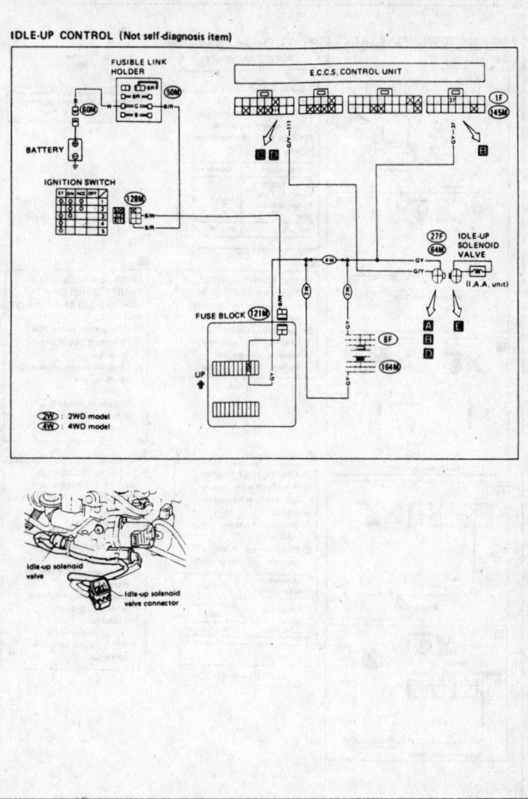
IDLE-UP CONTROL (Not self-diagnosis item)
FUSIBLE LINK HOLDER
E.C.C.S. CONTROL UNIT
BATTERY
IGNITION SWITCH
IDLE-UP SOLENOID VALVE
(I.A.A. unit)
FUSE BLOCK
UP
2W : 2WD model
4W : 4WD model
Idle-up solenoid valve
Idle-up solenoid valve connector

ECCS SELF-DIAGNOSTIC CHART—1990 PULSAR/SENTRA

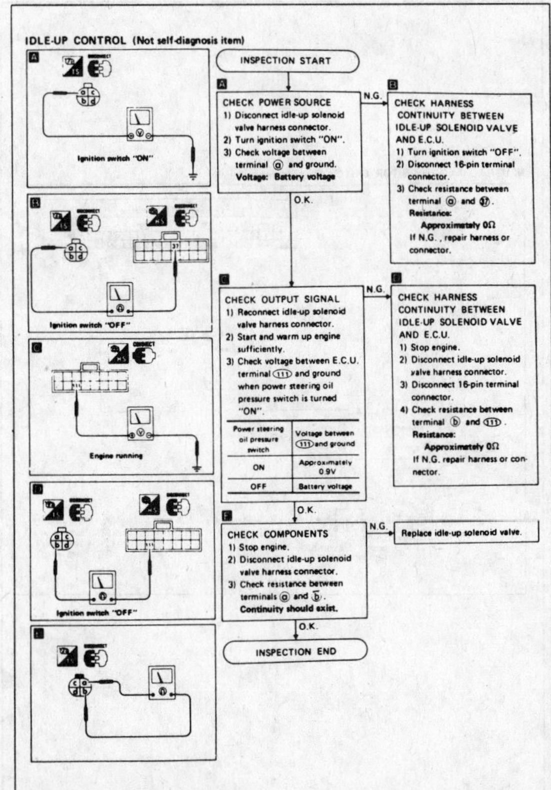
IDLE-UP CONTROL (Not self-diagnosis item)
Ignition switch "ON"
Ignition switch "OFF"
Engine running
Ignition switch "OFF"
INSPECTION START
A
CHECK POWER SOURCE
1) Disconnect idle-up solenoid valve harness connector.
2) Turn ignition switch "ON".
3) Check voltage between terminal ⓐ and ground.
Voltage: Battery voltage
N.G.
B
CHECK HARNESS CONTINUITY BETWEEN IDLE-UP SOLENOID VALVE AND E.C.U.
1) Turn ignition switch "OFF".
2) Disconnect 16-pin terminal connector.
3) Check resistance between terminal ⓐ and ㊲.
Resistance:
Approximately 0Ω
If N.G., repair harness or connector.
O.K.
C
CHECK OUTPUT SIGNAL
1) Reconnect idle-up solenoid valve harness connector.
2) Start and warm up engine sufficiently.
3) Check voltage between E.C.U. terminal (111) and ground when power steering oil pressure switch is turned "ON".
Power steering oil pressure switch | Voltage between (111) and ground
ON | Approximately 0.9V
OFF | Battery voltage
N.G.
D
CHECK HARNESS CONTINUITY BETWEEN IDLE-UP SOLENOID VALVE AND E.C.U.
1) Stop engine.
2) Disconnect idle-up solenoid valve harness connector.
3) Disconnect 16-pin terminal connector.
4) Check resistance between terminal ⓑ and (111).
Resistance:
Approximately 0Ω
If N.G. repair harness or connector.
O.K.
E
CHECK COMPONENTS
1) Stop engine.
2) Disconnect idle-up solenoid valve harness connector.
3) Check resistance between terminals ⓐ and ⓑ.
Continuity should exist.
N.G.
Replace idle-up solenoid valve.
O.K.
INSPECTION END

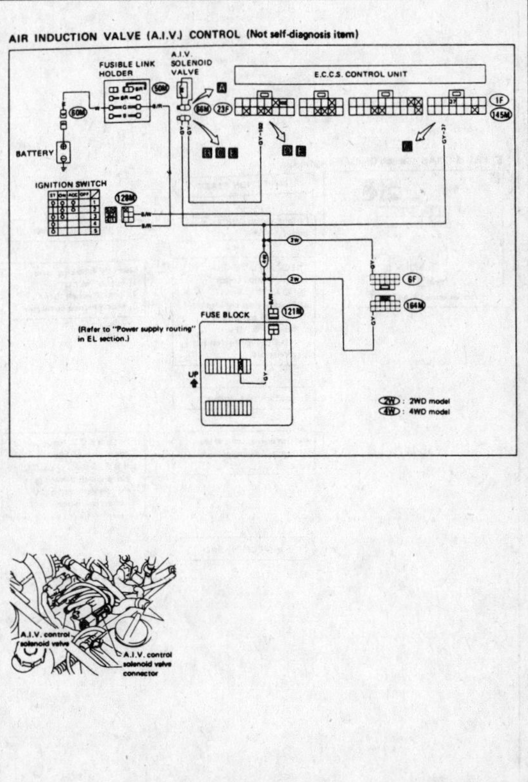
AIR INDUCTION VALVE (A.I.V.) CONTROL (Not self-diagnosis item)
FUSIBLE LINK HOLDER
A.I.V. SOLENOID VALVE
E.C.C.S. CONTROL UNIT
BATTERY
IGNITION SWITCH
FUSE BLOCK
(Refer to "Power supply routing" in EL section.)
UP
2W : 2WD model
4W : 4WD model
A.I.V. control solenoid valve
A.I.V. control solenoid valve connector

ECCS SELF-DIAGNOSTIC CHART—1990 PULSAR/SENTRA

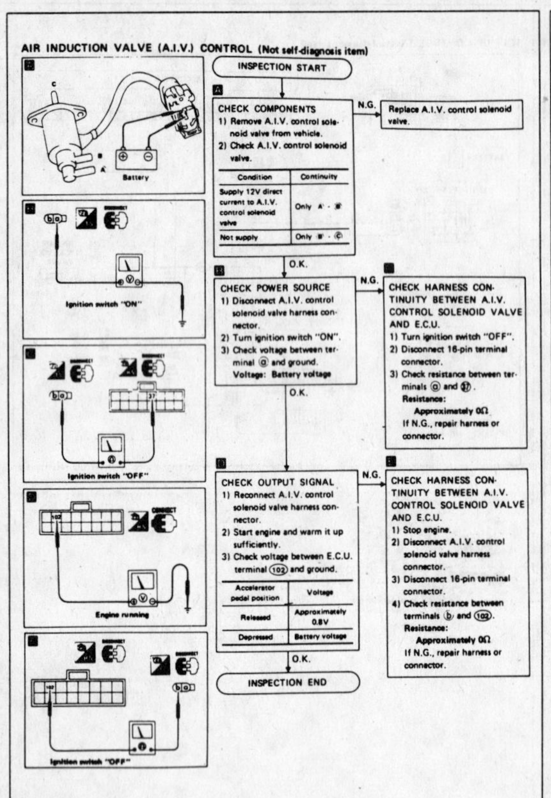
AIR INDUCTION VALVE (A.I.V.) CONTROL (Not self-diagnosis item)
INSPECTION START
CHECK COMPONENTS
1) Remove A.I.V. control solenoid valve from vehicle.
2) Check A.I.V. control solenoid valve.
Condition
Continuity
Supply 12V direct current to A.I.V. control solenoid valve
Not supply
N.G.
Replace A.I.V. control solenoid valve.
Battery
O.K.
CHECK POWER SOURCE
1) Disconnect A.I.V. control solenoid valve harness connector.
2) Turn ignition switch "ON".
3) Check voltage between terminal ⓐ and ground.
Voltage: Battery voltage
CHECK HARNESS CONTINUITY BETWEEN A.I.V. CONTROL SOLENOID VALVE AND E.C.U.
1) Turn ignition switch "OFF".
2) Disconnect 16-pin terminal connector.
3) Check resistance between terminals ⓐ and 37.
Resistance:
Approximately 0Ω
If N.G., repair harness or connector.
Ignition switch "ON"
Ignition switch "OFF"
CHECK OUTPUT SIGNAL
1) Reconnect A.I.V. control solenoid valve harness connector.
2) Start engine and warm it up sufficiently.
3) Check voltage between E.C.U. terminal 102 and ground.
Accelerator pedal position
Voltage
Released
Approximately 0.8V
Depressed
Battery voltage
CHECK HARNESS CONTINUITY BETWEEN A.I.V. CONTROL SOLENOID VALVE AND E.C.U.
1) Stop engine.
2) Disconnect A.I.V. control solenoid valve harness connector.
3) Disconnect 16-pin terminal connector.
4) Check resistance between terminals ⓑ and 102.
Resistance:
Approximately 0Ω
If N.G., repair harness or connector.
Engine running
INSPECTION END
Ignition switch "OFF"

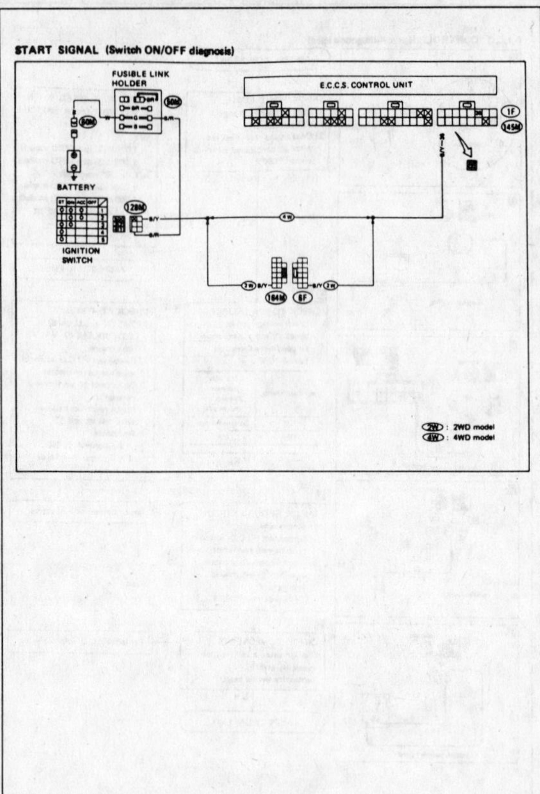
START SIGNAL (Switch ON/OFF diagnosis)
FUSIBLE LINK HOLDER
E.C.C.S. CONTROL UNIT
BATTERY
IGNITION SWITCH
2WD : 2WD model
4WD : 4WD model

ECCS SELF-DIAGNOSTIC CHART—1990 PULSAR/SENTRA

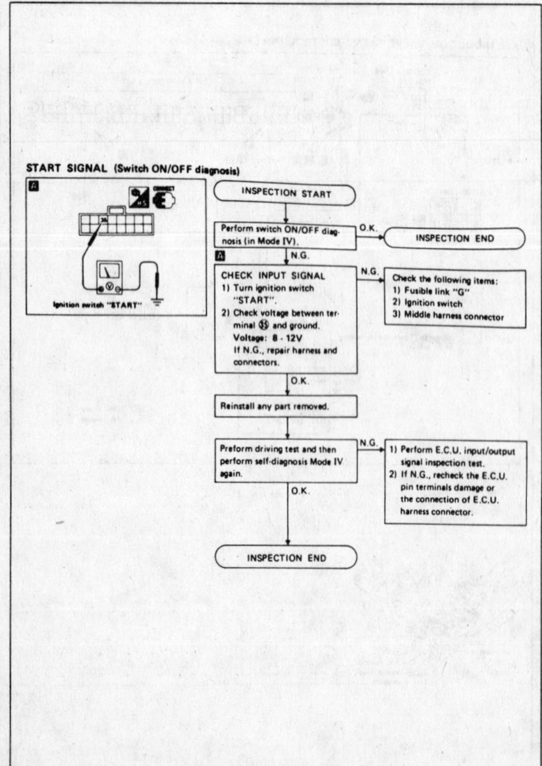
START SIGNAL (Switch ON/OFF diagnosis)
INSPECTION START
Perform switch ON/OFF diagnosis (in Mode IV).
O.K.
INSPECTION END
N.G.
CHECK INPUT SIGNAL
1) Turn ignition switch "START".
2) Check voltage between terminal 35 and ground.
Voltage: 8 - 12V
If N.G., repair harness and connectors.
Check the following items:
1) Fusible link "G"
2) Ignition switch
3) Middle harness connector
Ignition switch "START"
O.K.
Reinstall any part removed.
Preform driving test and then perform self-diagnosis Mode IV again.
1) Perform E.C.U. input/output signal inspection test.
2) If N.G., recheck the E.C.U. pin terminals damage or the connection of E.C.U. harness connector.
INSPECTION END

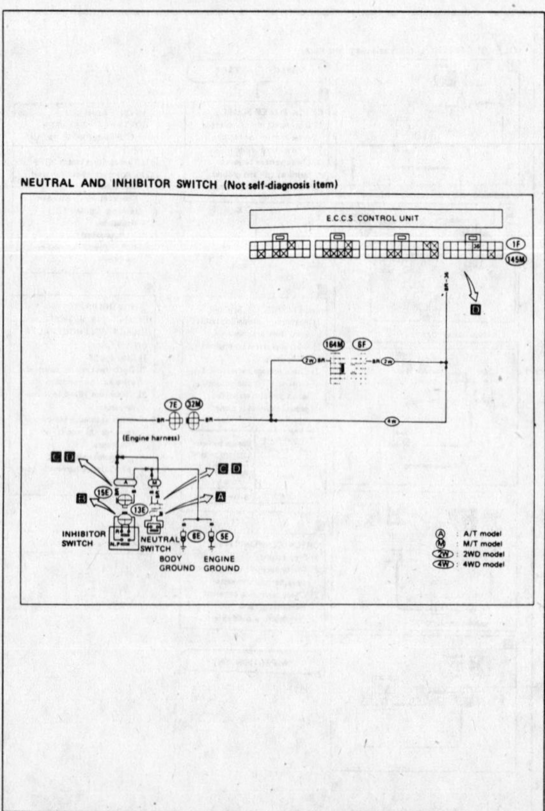
NEUTRAL AND INHIBITOR SWITCH (Not self-diagnosis item)
E.C.C.S. CONTROL UNIT
(Engine harness)
INHIBITOR SWITCH
NEUTRAL SWITCH
BODY GROUND
ENGINE GROUND
Ⓐ : A/T model
Ⓜ : M/T model
2WD : 2WD model
4WD : 4WD model

ECCS SELF-DIAGNOSTIC CHART 1990 PULSAR/SENTRA

ECU SIGNAL INSPECTION—1990 PULSAR/SENTRA

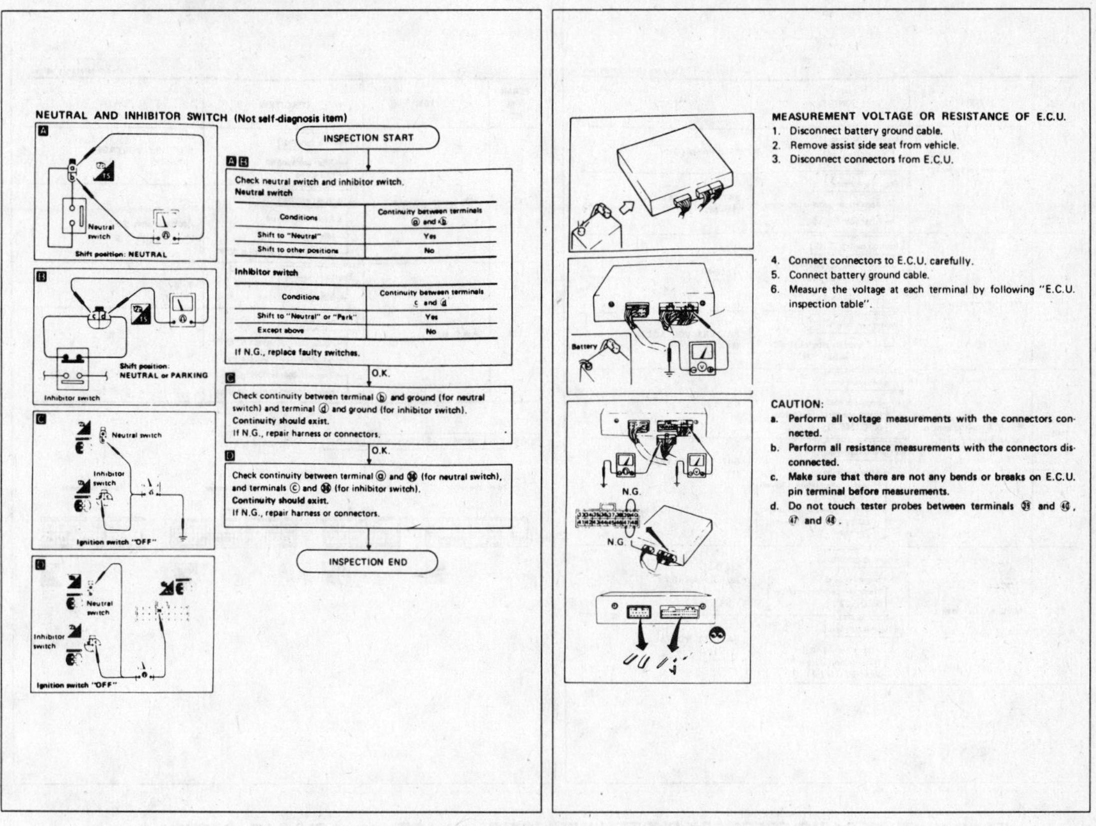
NEUTRAL AND INHIBITOR SWITCH (Not self-diagnosis item)
INSPECTION START
Check neutral switch and inhibitor switch.
Neutral switch
Conditions
Shift to "Neutral"
Shift to other positions
Yes
No
Inhibitor switch
Shift to "Neutral" or "Park"
Except above
If N.G., replace faulty switches.
O.K.
Check continuity between terminal ⓑ and ground (for neutral switch) and terminal ⓓ and ground (for inhibitor switch).
Continuity should exist.
If N.G., repair harness or connectors.
O.K.
Continuity should exist.
If N.G., repair harness or connectors.
INSPECTION END
Shift position: NEUTRAL
Shift position: NEUTRAL or PARKING
Inhibitor switch
Neutral switch
Ignition switch "OFF"
MEASUREMENT VOLTAGE OR RESISTANCE OF E.C.U.
1. Disconnect battery ground cable.
2. Remove assist side seat from vehicle.
3. Disconnect connectors from E.C.U.
4. Connect connectors to E.C.U. carefully.
5. Connect battery ground cable.
6. Measure the voltage at each terminal by following "E.C.U. inspection table".
Battery
CAUTION:
a. Perform all voltage measurements with the connectors connected.
b. Perform all resistance measurements with the connectors disconnected.
c. Make sure that there are not any bends or breaks on E.C.U. pin terminal before measurements.
N.G.
N.G.

ECU SIGNAL INSPECTION—1990 PULSAR/SENTRA

E.C.U. inspection table

*Data are reference values.

TERMINAL NO.	ITEM	CONDITION	*DATA
1	Ignition signal	Engine is running.	0 - 1.0V
7	Exhaust gas temperature sensor	Engine is running. — Idle speed	1.0V or more
		Engine is running. — E.G.R. system is operating.	0 - 1.0V
15	Air flow meter	Engine is running. Do not run engine at high speed under no-load.	0.5 - 4.0V Output voltage varies with engine revolution.
17	Water temperature sensor	Engine is running.	0.5 - 5.0V Output voltage varies with engine water temperature.
18	Exhaust gas sensor	Engine is running. — After warming up sufficiently	0 - Approximately 1.0V
19	Throttle sensor	Ignition switch "ON"	0.4 - 4.0V Output voltage varies with the throttle valve opening angle.
21 31	Crank angle sensor (Reference signal)	Engine is running. Do not run engine at high speed under no-load.	0.3 - 0.5V
22 32	Crank angle sensor (Position signal)	Engine is running. Do not run engine at high speed under no-load.	2.0 - 2.7V
34	Idle switch (− side)	Ignition switch "ON" — Throttle valve: Idle position	Approximately 10V
		Ignition switch "ON" — Throttle valve: Any position except idle position	Approximately 0V
35	Start signal	Cranking	8 - 12V

*Data are reference values.

TERMINAL NO.	ITEM	CONDITION	*DATA
36	Neutral switch & Inhibitor switch	Ignition switch "ON" — Neutral/Parking	0V
		Ignition switch "ON" — Except the above gear position	Approximately 6.5V
37	Ignition switch	Ignition switch "OFF"	0V
		Ignition switch "ON"	BATTERY VOLTAGE (11 - 14V)
41	Air conditioner	Engine is running. — Both air conditioner switch and blower switch are "ON".	BATTERY VOLTAGE (11 - 14V)
43	Power steering switch	Engine is running. — Steering wheel is turned.	Approximately 0V
		Engine is running. — Except above operation	Approximately 7.0 - 9.0V
44	Idle switch (+ side)	Ignition switch "ON" — Throttle valve: Idle position	Approximately 9 - 11V
		Ignition switch "ON" — Throttle valve: Except idle position	BATTERY VOLTAGE (11 - 14V)
46	Power source	Ignition switch "OFF"	BATTERY VOLTAGE (11 - 14V)
39 47	Power source for E.C.U.	Ignition switch "ON"	BATTERY VOLTAGE (11 - 14V)
101 110	Injector	Engine is running.	BATTERY VOLTAGE (11 - 14V)

ECU SIGNAL INSPECTION—1990 PULSAR/SENTRA

*Data are reference values.

TERMINAL NO.	ITEM	CONDITION	*DATA
102	A.I.V. control solenoid valve	Engine is running. — Idle speed	Approximately 0.8V
		Engine is running. — Accelerator pedal is depressed. — After warming up	BATTERY VOLTAGE (11 - 14V)
104	Fuel pump relay	Ignition switch "ON" — For 5 seconds after turning ignition switch "ON"; Engine is running.	0.7 - 0.9V
		Ignition switch "ON" — In 5 seconds after turning ignition switch "ON"	BATTERY VOLTAGE (11 - 14V)
105	E.G.R. & canister control solenoid valve	Engine is running. — Engine is cold. [Water temperature is below 60°C (140°F).]	0.6 - 0.9V
		Engine is running. — After warming up [• Water temperature is above 60°C (140°F). • Vehicle speedometer is indicating more than 18 km/h (11 MPH).]	BATTERY VOLTAGE (11 - 14V)
106	Mixture heater relay	Engine is running. — Engine is cold. [Water temperature is below 50°C (122°F).]	Approximately 0.7V
		Engine is running. — After warming up [Water temperature is above 50° (122°F).]	BATTERY VOLTAGE (11 - 14V)

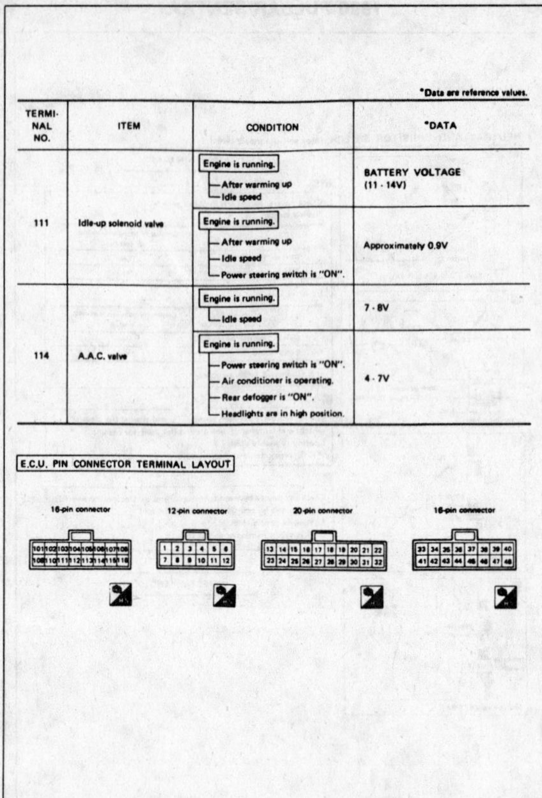

*Data are reference values.

TERMINAL NO.	ITEM	CONDITION	*DATA
111	Idle-up solenoid valve	Engine is running. — After warming up — Idle speed	BATTERY VOLTAGE (11 - 14V)
		Engine is running. — After warming up — Idle speed — Power steering switch is "ON".	Approximately 0.9V
114	A.A.C. valve	Engine is running. — Idle speed	7 - 8V
		Engine is running. — Power steering switch is "ON". — Air conditioner is operating. — Rear defogger is "ON". — Headlights are in high position.	4 - 7V

E.C.U. PIN CONNECTOR TERMINAL LAYOUT
16-pin connector
101 102 103 104 105 106 107 108
109 110 111 112 113 114 115 116
12-pin connector
1 2 3 4 5 6
7 8 9 10 11 12
20-pin connector
13 14 15 16 17 18 19 20 21 22
23 24 25 26 27 28 29 30 31 32
16-pin connector
33 34 35 36 37 38 39 40
41 42 43 44 45 46 47 48

ECCS MIXTURE RATIO FEEDBACK SYSTEM INSPECTION CHART—1990 PULSAR/ SENTRA

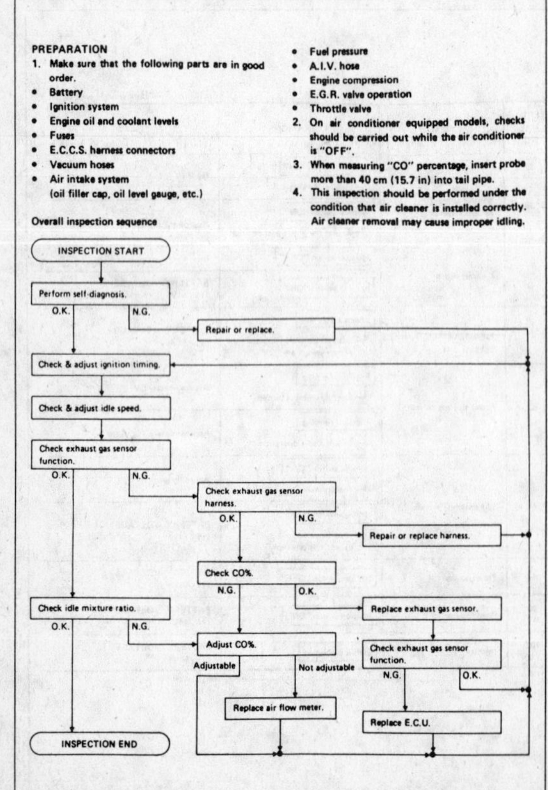

PREPARATION

1. Make sure that the following parts are in good order.
- Battery
- Ignition system
- Engine oil and coolant levels
- Fuses
- E.C.C.S. harness connectors
- Vacuum hoses
- Air intake system (oil filler cap, oil level gauge, etc.)
- Fuel pressure
- A.I.V. hose
- Engine compression
- E.G.R. valve operation
- Throttle valve

2. On air conditioner equipped models, checks should be carried out while the air conditioner is "OFF".
3. When measuring "CO" percentage, insert probe more than 40 cm (15.7 in) into tail pipe.
4. This inspection should be performed under the condition that air cleaner is installed correctly. Air cleaner removal may cause improper idling.

Overall inspection sequence

INSPECTION START
Perform self-diagnosis.
O.K.
N.G.
Repair or replace.
Check & adjust ignition timing.
Check & adjust idle speed.
Check exhaust gas sensor function.
O.K.
N.G.
Check exhaust gas sensor harness.
O.K.
N.G.
Repair or replace harness.
Check CO%.
N.G.
O.K.
Replace exhaust gas sensor.
Check idle mixture ratio.
O.K.
N.G.
Adjust CO%.
Adjustable
Not adjustable
Check exhaust gas sensor function.
N.G.
O.K.
Replace air flow meter.
Replace E.C.U.
INSPECTION END

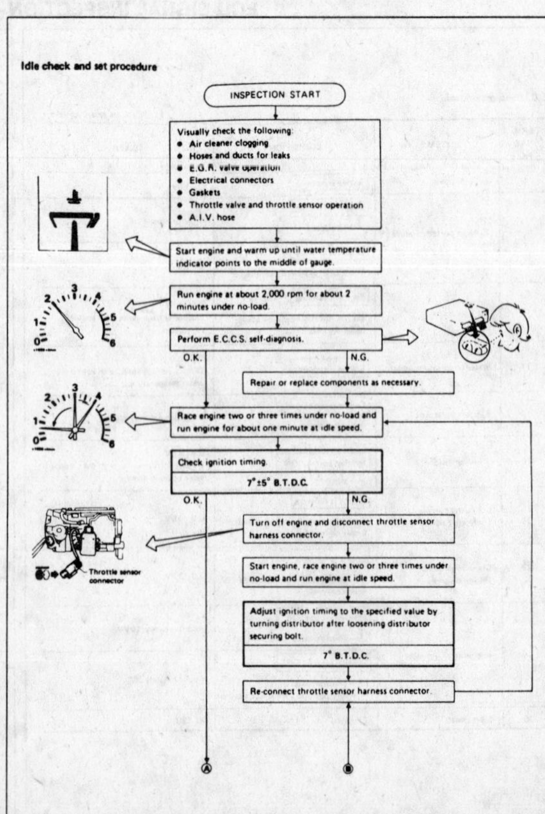

Idle check and set procedure
INSPECTION START
Visually check the following:
• Air cleaner clogging
• Hoses and ducts for leaks
• E.G.R. valve operation
• Electrical connectors
• Gaskets
• Throttle valve and throttle sensor operation
• A.I.V. hose
Start engine and warm up until water temperature indicator points to the middle of gauge.
Run engine at about 2,000 rpm for about 2 minutes under no-load.
Perform E.C.C.S. self-diagnosis.
O.K.
N.G.
Repair or replace components as necessary.
Race engine two or three times under no-load and run engine for about one minute at idle speed.
Check ignition timing.
7°±5° B.T.D.C.
O.K.
N.G.
Turn off engine and disconnect throttle sensor harness connector.
Start engine, race engine two or three times under no-load and run engine at idle speed.
Adjust ignition timing to the specified value by turning distributor after loosening distributor securing bolt.
7° B.T.D.C.
Re-connect throttle sensor harness connector.
Throttle sensor connector
A
B

ECCS MIXTURE RATIO FEEDBACK SYSTEM INSPECTION CHART—1990 PULSAR/ SENTRA

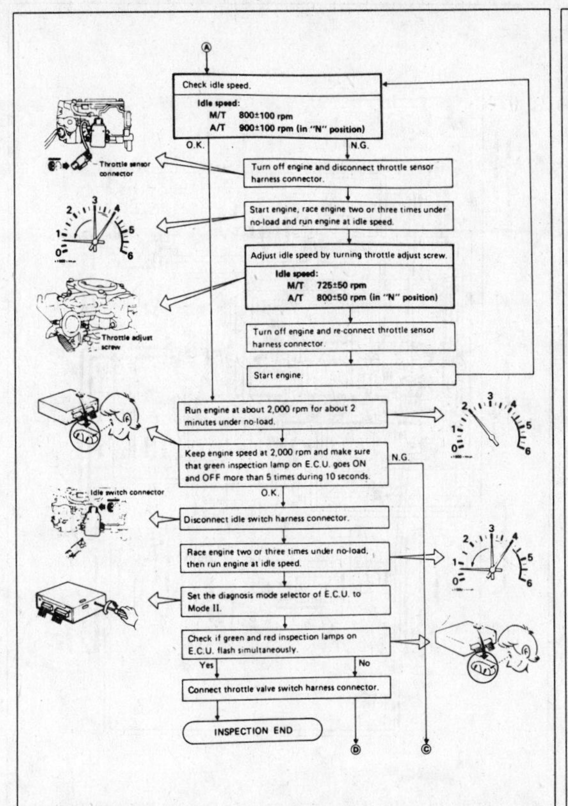
A
Check idle speed.
Idle speed:
M/T 800±100 rpm
A/T 900±100 rpm (in "N" position)
O.K.
N.G.
Throttle sensor connector
Turn off engine and disconnect throttle sensor harness connector.
Start engine, race engine two or three times under no-load and run engine at idle speed.
Adjust idle speed by turning throttle adjust screw.
Idle speed:
M/T 725±50 rpm
A/T 800±50 rpm (in "N" position)
Throttle adjust screw
Turn off engine and re-connect throttle sensor harness connector.
Start engine.
Run engine at about 2,000 rpm for about 2 minutes under no-load.
Keep engine speed at 2,000 rpm and make sure that green inspection lamp on E.C.U. goes ON and OFF more than 5 times during 10 seconds.
N.G.
O.K.
Idle switch connector
Disconnect idle switch harness connector.
Race engine two or three times under no-load, then run engine at idle speed.
Set the diagnosis mode selector of E.C.U. to Mode II.
Check if green and red inspection lamps on E.C.U. flash simultaneously.
Yes
No
Connect throttle valve switch harness connector.
INSPECTION END
D
C

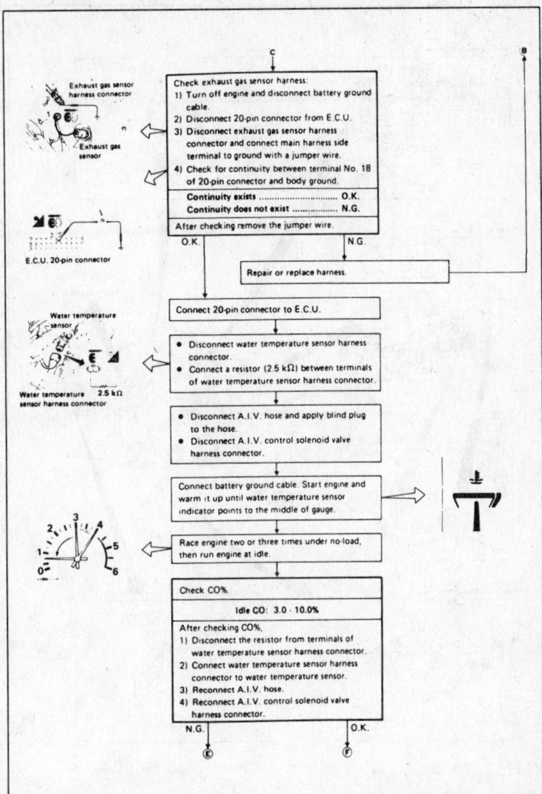
C
Exhaust gas sensor harness connector
Exhaust gas sensor
Check exhaust gas sensor harness:
1) Turn off engine and disconnect battery ground cable.
2) Disconnect 20-pin connector from E.C.U.
3) Disconnect exhaust gas sensor harness connector and connect main harness side terminal to ground with a jumper wire.
4) Check for continuity between terminal No. 18 of 20-pin connector and body ground.
Continuity exists .... O.K.
Continuity does not exist .... N.G.
After checking remove the jumper wire.
O.K.
N.G.
E.C.U. 20-pin connector
Repair or replace harness.
B
Connect 20-pin connector to E.C.U.
Water temperature sensor
Water temperature sensor harness connector
2.5 kΩ
• Disconnect water temperature sensor harness connector.
• Connect a resistor (2.5 kΩ) between terminals of water temperature sensor harness connector.
• Disconnect A.I.V. hose and apply blind plug to the hose.
• Disconnect A.I.V. control solenoid valve harness connector.
Connect battery ground cable. Start engine and warm it up until water temperature sensor indicator points to the middle of gauge.
Race engine two or three times under no-load, then run engine at idle.
Check CO%.
Idle CO: 3.0 - 10.0%
After checking CO%,
1) Disconnect the resistor from terminals of water temperature sensor harness connector.
2) Connect water temperature sensor harness connector to water temperature sensor.
3) Reconnect A.I.V. hose.
4) Reconnect A.I.V. control solenoid valve harness connector.
N.G.
O.K.
E
F

ECCS COMPONENTS LOCATION—1988 VAN

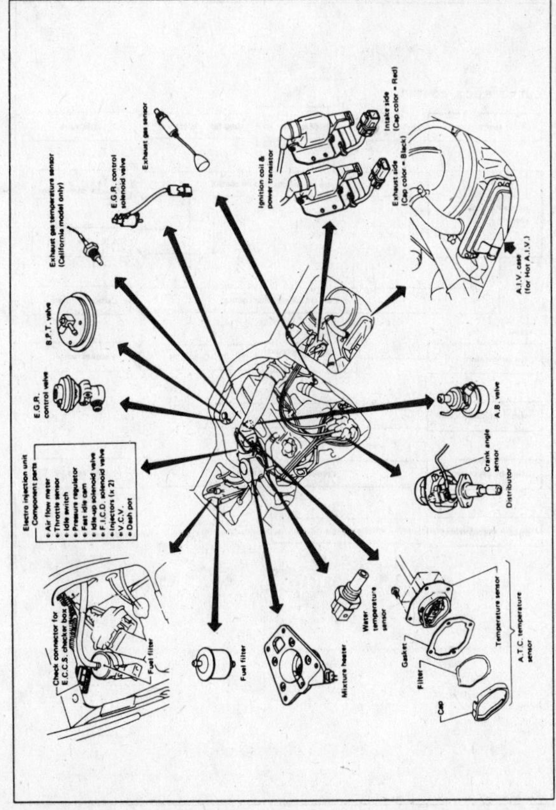
Exhaust gas sensor
E.G.R. control solenoid valve
Ignition coil & power transistor
A.I.V. case (for Hot A.I.V.)
B.P.T. valve
E.G.R. control valve
A.B. valve
Crank angle sensor
Distributor
Fuel filter
Mixture heater
Water temperature sensor
Temperature sensor
A.T.C. temperature sensor
Gasket
Filter
Cap

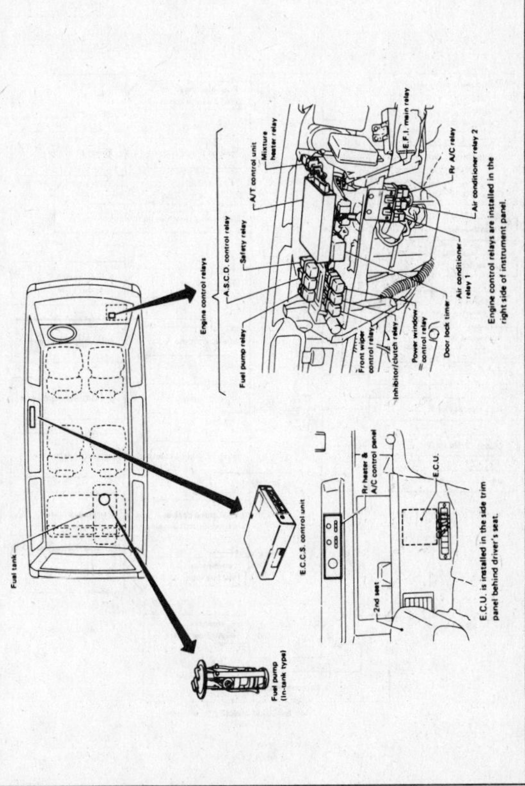
Engine control relays
Fuel tank
Fuel pump (In-tank type)
E.C.C.S. control unit
E.C.U.
Engine control relays are installed in the right side of instrument panel.
E.C.U. is installed in the side trim panel behind driver's seat.

ECCS COMPONENTS LOCATION (CONT.)—1988 VAN

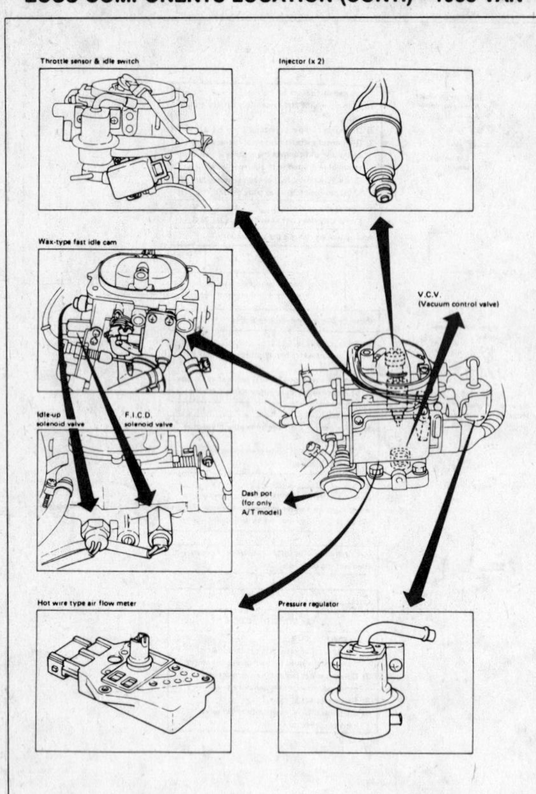
Throttle sensor & idle switch
Injector (x 2)
Wax-type fast idle cam
V.C.V.
(Vacuum control valve)
Idle-up solenoid valve
F.I.C.D. solenoid valve
Dash pot (for only A/T model)
Hot wire type air flow meter
Pressure regulator

ECCS SYSTEM SCHEMATIC—1988 VAN

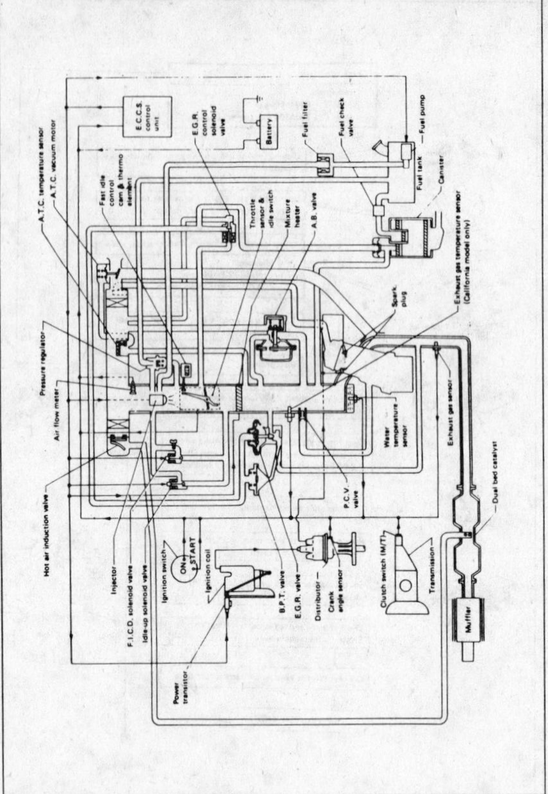
E.C.C.S. control unit
E.G.R. control solenoid valve
Battery
Fuel filter
Fuel check valve
Fuel pump
Fuel tank
Canister
A.T.C. temperature sensor
A.T.C. vacuum motor
Fast idle control cam & thermo element
Throttle sensor & idle switch
Mixture heater
A.B. valve
Exhaust gas temperature sensor (California model only)
Spark plug
Pressure regulator
Air flow meter
Hot air induction valve
Injector
F.I.C.D. solenoid valve
Idle-up solenoid valve
Ignition switch
ON
START
Ignition coil
Power transistor
B.P.T. valve
E.G.R. valve
Distributor—Crank angle sensor
Clutch switch (M/T)
Transmission
Water temperature sensor
P.C.V. valve
Exhaust gas sensor
Dual bed catalyst
Muffler

ECCS CONTROL SYSTEM CHART—1988 VAN

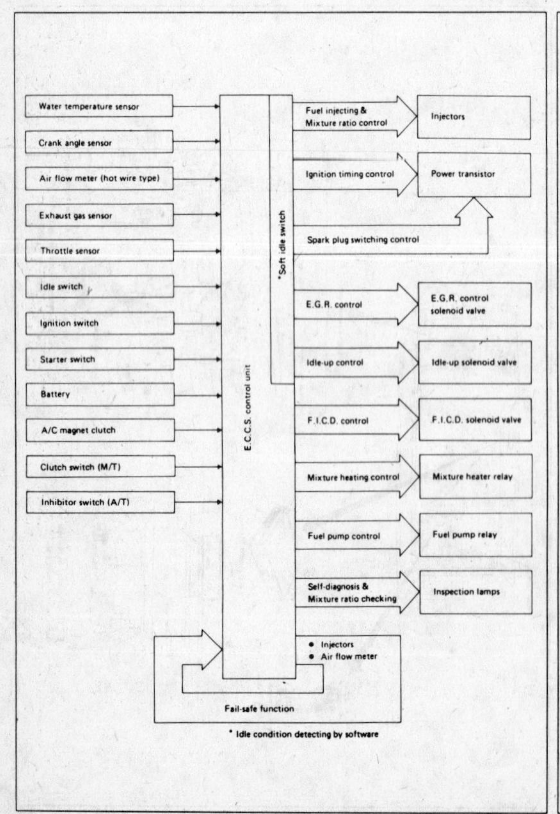
Water temperature sensor
Crank angle sensor
Air flow meter (hot wire type)
Exhaust gas sensor
Throttle sensor
Idle switch
Ignition switch
Starter switch
Battery
A/C magnet clutch
Clutch switch (M/T)
Inhibitor switch (A/T)
E.C.C.S. control unit
*Soft idle switch
Fuel injecting & Mixture ratio control
Injectors
Ignition timing control
Power transistor
Spark plug switching control
E.G.R. control
E.G.R. control solenoid valve
Idle-up control
Idle-up solenoid valve
F.I.C.D. control
F.I.C.D. solenoid valve
Mixture heating control
Mixture heater relay
Fuel pump control
Fuel pump relay
Self-diagnosis & Mixture ratio checking
Inspection lamps
• Injectors
• Air flow meter
Fail-safe function
* Idle condition detecting by software

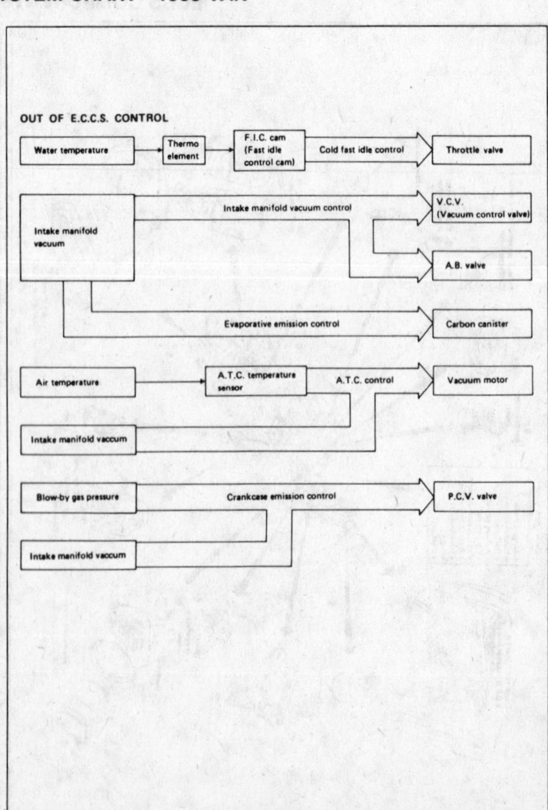
OUT OF E.C.C.S. CONTROL
Water temperature
Thermo element
F.I.C. cam (Fast idle control cam)
Cold fast idle control
Throttle valve
Intake manifold vacuum
Intake manifold vacuum control
V.C.V. (Vacuum control valve)
A.B. valve
Evaporative emission control
Carbon canister
Air temperature
A.T.C. temperature sensor
A.T.C. control
Vacuum motor
Intake manifold vaccum
Blow-by gas pressure
Crankcase emission control
P.C.V. valve
Intake manifold vaccum

ECCS FUEL FLOW SYSTEM SCHEMATIC—1988 VAN

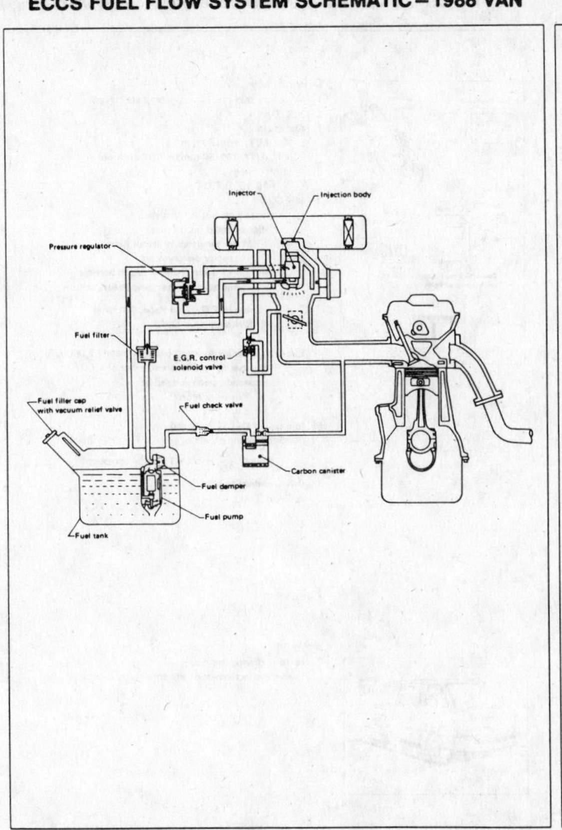
Injector
Injection body
Pressure regulator
Fuel filter
E.G.R. control solenoid valve
Fuel filler cap with vacuum relief valve
Fuel check valve
Carbon canister
Fuel damper
Fuel pump
Fuel tank

ECCS AIR FLOW SYSTEM SCHEMATIC—1988 VAN

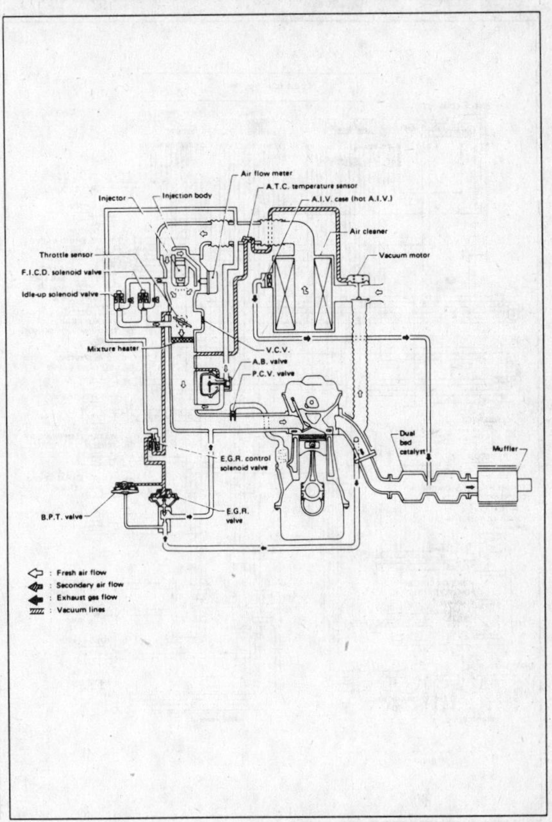
Air flow meter
A.T.C. temperature sensor
Injector
Injection body
A.I.V. case (hot A.I.V.)
Air cleaner
Throttle sensor
Vacuum motor
F.I.C.D. solenoid valve
Idle-up solenoid valve
Mixture heater
V.C.V.
A.B. valve
P.C.V. valve
Dual bed catalyst
Muffler
E.G.R. control solenoid valve
B.P.T. valve
E.G.R. valve
: Fresh air flow
: Secondary air flow
: Exhaust gas flow
: Vacuum lines

ECCS CONTROL UNIT WIRING SCHEMATIC—1988 VAN

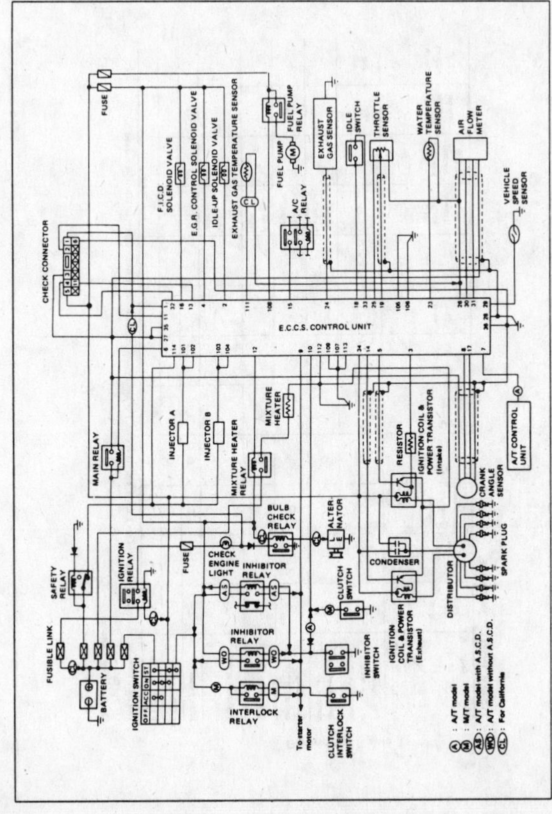
FUSE
F.I.C.D SOLENOID VALVE
E.G.R. CONTROL SOLENOID VALVE
IDLE-UP SOLENOID VALVE
EXHAUST GAS TEMPERATURE SENSOR
FUEL PUMP
FUEL PUMP RELAY
A/C RELAY
EXHAUST GAS SENSOR
IDLE SWITCH
THROTTLE SENSOR
WATER TEMPERATURE SENSOR
AIR FLOW METER
VEHICLE SPEED SENSOR
CHECK CONNECTOR
E.C.C.S. CONTROL UNIT
MAIN RELAY
INJECTOR A
INJECTOR B
MIXTURE HEATER RELAY
MIXTURE HEATER
RESISTOR
IGNITION COIL & POWER TRANSISTOR (Intake)
CRANK ANGLE SENSOR
A/T CONTROL UNIT
BULB CHECK RELAY
ALTERNATOR
SAFETY RELAY
IGNITION RELAY
CHECK ENGINE LIGHT
INHIBITOR RELAY
CONDENSER
CLUTCH SWITCH
INHIBITOR SWITCH
IGNITION COIL & POWER TRANSISTOR (Exhaust)
DISTRIBUTOR
SPARK PLUG
FUSIBLE LINK
BATTERY
IGNITION SWITCH
INTERLOCK RELAY
To starter motor
CLUTCH INTERLOCK SWITCH
A/T model
M/T model
A/T model with A.S.C.D.
A/T model without A.S.C.D.
For California

ECCS WIRING SCHEMATIC—1988 VAN

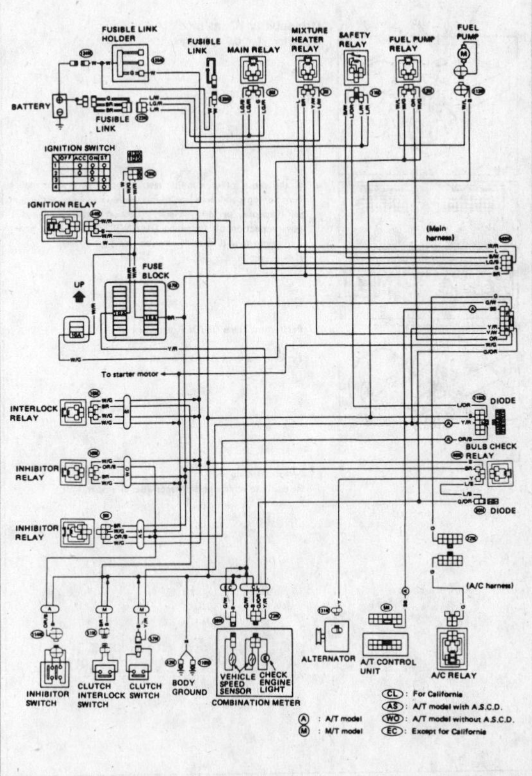
FUSIBLE LINK HOLDER
FUSIBLE LINK
MAIN RELAY
MIXTURE HEATER RELAY
SAFETY RELAY
FUEL PUMP RELAY
FUEL PUMP
BATTERY
IGNITION SWITCH
IGNITION RELAY
FUSE BLOCK
UP
(Main harness)
To starter motor
INTERLOCK RELAY
INHIBITOR RELAY
DIODE
BULB CHECK RELAY
(A/C harness)
INHIBITOR SWITCH
CLUTCH INTERLOCK SWITCH
CLUTCH SWITCH
BODY GROUND
VEHICLE SPEED SENSOR
CHECK ENGINE LIGHT
COMBINATION METER
ALTERNATOR
A/T CONTROL UNIT
A/C RELAY
A : A/T model
M : M/T model
CL : For California
AS : A/T model with A.S.C.D.
WO : A/T model without A.S.C.D.
EC : Except for California

ECCS WIRING SCHEMATIC (CONT.)—1988 VAN

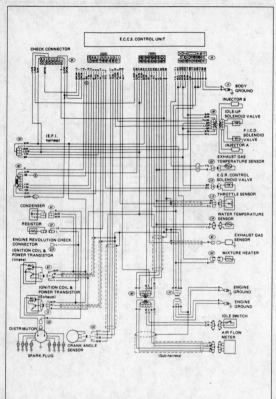

ECCS DIAGNOSTIC PROCEDURE—1988 VAN

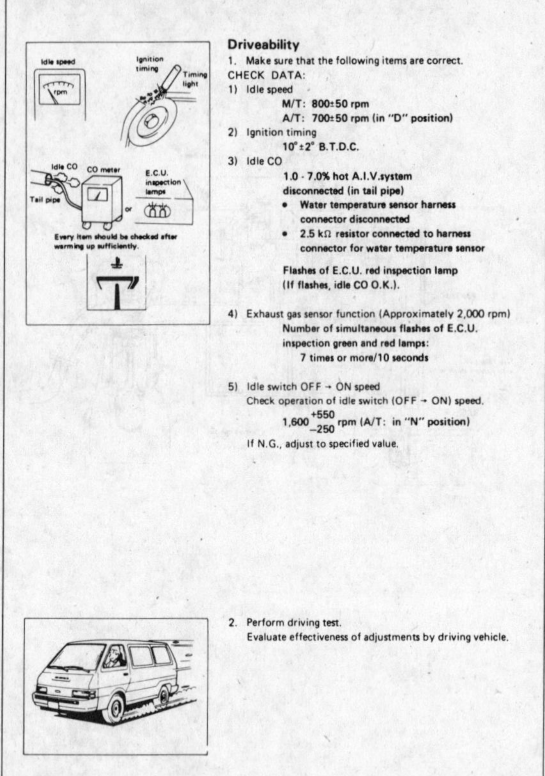

Driveability

1. Make sure that the following items are correct.

CHECK DATA:

1) Idle speed
   M/T: 800±50 rpm
   A/T: 700±50 rpm (in "D" position)
2) Ignition timing
   10°±2° B.T.D.C.
3) Idle CO
   1.0 - 7.0% hot A.I.V.system disconnected (in tail pipe)
   - Water temperature sensor harness connector disconnected
   - 2.5 kΩ resistor connected to harness connector for water temperature sensor

   Flashes of E.C.U. red inspection lamp (If flashes, idle CO O.K.).
4) Exhaust gas sensor function (Approximately 2,000 rpm)
   Number of simultaneous flashes of E.C.U. inspection green and red lamps:
   7 times or more/10 seconds
5) Idle switch OFF → ON speed
   Check operation of idle switch (OFF → ON) speed.
   $1{,}600^{+550}_{-250}$ rpm (A/T: in "N" position)
   If N.G., adjust to specified value.

2. Perform driving test.
   Evaluate effectiveness of adjustments by driving vehicle.

Idle speed
rpm
Ignition timing
Timing light
Idle CO
CO meter
Tail pipe
E.C.U. inspection lamps
or
Every item should be checked after warming up sufficiently.

ECCS DIAGNOSTIC PROCEDURE—1988 VAN

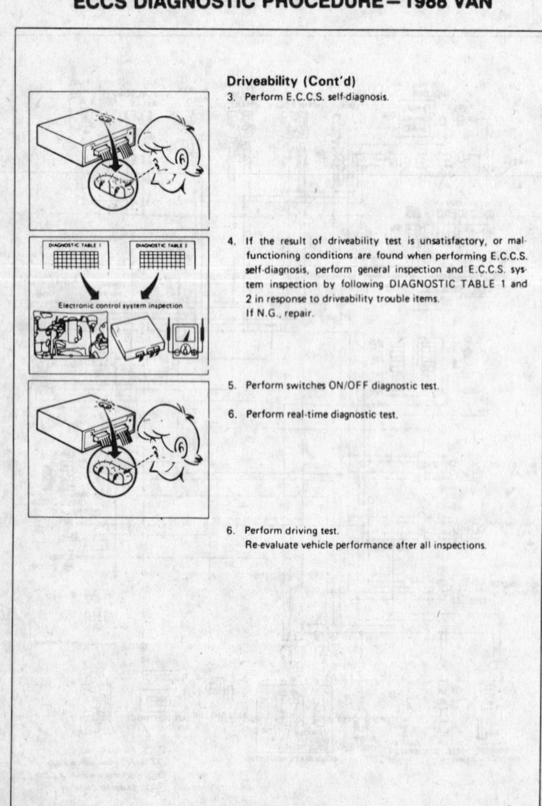

Driveability (Cont'd)

3. Perform E.C.C.S. self-diagnosis.
4. If the result of driveability test is unsatisfactory, or malfunctioning conditions are found when performing E.C.C.S. self-diagnosis, perform general inspection and E.C.C.S. system inspection by following DIAGNOSTIC TABLE 1 and 2 in response to driveability trouble items.
   If N.G., repair.
5. Perform switches ON/OFF diagnostic test.
6. Perform real-time diagnostic test.
6. Perform driving test.
   Re-evaluate vehicle performance after all inspections.

DIAGNOSTIC TABLE 1
DIAGNOSTIC TABLE 2
Electronic control system inspection

ECCS DIAGNOSTIC TABLE 1—1988 VAN

SYSTEM INSPECTION TABLE

System \ Sensor & actuator	Crank angle sensor	Air flow meter	Water temperature sensor	Ignition switch	Injector	Throttle sensor	Clutch switch	Exhaust gas sensor	Battery voltage	E.G.R. control solenoid valve	Idle-up solenoid valve	A/C magnet clutch
Fuel injection & mixture ratio feedback control	O	O	O	O	O	O	O	O	O			
Ignition timing control	O	O	O	O		O			Power transistor			
Spark plug switching control	O	O	O						Power transistor			
Fuel pump control	O			O					Fuel pump relay			
Mixture heater control	O		O	O	O				Mixture heater relay			
Idle-up control	O		O	O					O		O	
E.G.R. control	O		O	O		O				O		
F.I.C.D. control	O		O	O								O

This table indicates the inspection items for the E.C.C.S. control system. For each system, it is necessary to check sensors or actuators marked "O".

*1: For California

*2: Except for California

Diagnostic Table 1

ECCS DIAGNOSTIC TABLE 2—1988 VAN

Diagnostic Table 2

DRIVEABILITY INSPECTION TABLE

TROUBLE ITEMS CONCERNED WITH DRIVEABILITY \ INSPECTION ITEM			GENERAL INSPECTION															ECCS SYSTEM INSPECTION					
			FUEL FLOW SYSTEM				ELECTRIC SYSTEM					AIR FLOW SYSTEM						IDLE-UP SOLENOID VALVE		CRANK ANGLE SENSOR			AIR FLOW METER *
			Fuel level	Fuel pump	Fuel filter	Fuel line	Battery	Spark plug	Ignition wire	Alternator	Starter	Air cleaner	Air flow line	E.G.R. valve	A.B. valve	Fast idle cam	P.C.V.	Short	Open	180° signal noise	180° signal faults	1° signal faults	Poor connection
SURGE	ROAD/LOAD DRIVING	Heavy load			○	○		◎	○			○	○	○			○			○	○	○	◎
		Middle load			○	○		◎	○			○	○	◎			○			○	○	○	◎
		Light load			○	○		◎	○			○	○	◎			○			○	○	○	◎
	ACCELERATING DRIVING	Slow acceleration			○	○		◎	○			○	○	◎			○			○	○	○	◎
	DECELERATING DRIVING	Rapid deceleration			○	○		◎	○			○	○	○			○			○	○	○	◎
		Slow deceleration			○	○		◎	○			○	○	○			○			○	○	○	◎
HESITATION		Rapid acceleration			○	○		◎	○			○	○	○			○			○	○	○	◎
		Slow acceleration			○	○		◎	○			○	○	○			○			○	○	○	◎
STUMBLE		Rapid acceleration			○	○		◎	○			○	○	○			○			○	○	○	◎
		Slow acceleration			○	○		◎	○			○	○	○			○			○	○	○	◎
BACKFIRE					○	○			○				○		○		○			○	○	○	◎
AFTER FIRE									○			○					○			○	○	○	◎
IDLE STABILITY								◎	○	○		○	○	◎	○	◎	○		◎	○	○	○	○
ENGINE STALL			○	○	○	○	○	◎	○	○		○	○	○	○	◎	○			◎	◎	◎	◎
STARTABILITY			○	○			○	◎	○	○	○	○	○			○	○			◎	◎	◎	○

This table indicates the inspection items for each type of symptom. It is necessary for each symptom to check sensors or actuators marked "◎" or "○".
Items marked "◎" have a significant influence on driveability. Prior to items marked "○", check items marked "◎".
Improper mixture ratio, improper ignition condition, and an excess of E.G.R. volume can cause any symptom.

Diagnostic Table 2 (Cont'd)

ECCS SYSTEM INSPECTION																				
WATER TEMPERATURE SENSOR			THROTTLE SENSOR		E.G.R. CONTROL SOLENOID VALVE		EXHAUST GAS SENSOR		INJECTOR *1	CLUTCH SWITCH	STARTER SIGNAL		IGNITION SIGNAL		BATTERY VOLTAGE	FUEL PUMP CIRCUIT			MIXTURE HEATER	F.I.C.D. SOLENOID VALVE
Short	Open	Poor connection	Short	Open	Short	Open	Short	Open	Clogging	Short	Short	Open	Short	Open	Low voltage	Short	Open	Poor connection	Open (for only cold condition)	Open
○	○	○				◎	○	○	◎									○	○	
○	○	○				◎	○	○	◎									○	○	
○	○	○				◎	○	○	◎									○	○	
○	○	○				◎	○	○	◎									○	○	
○	○	○				○	○	○	◎									○	○	
○	○	○				○	○	○	◎									○	○	
○	○	○	◎	◎		○	○	○	◎	○								○	○	
○	○	○	◎	◎		○	○	○	◎	○								○	○	
○	○	○	◎	◎		○			◎	○								○	○	
○	○	○	◎	◎		○			◎	○								○	○	
○		○	○	○			○	○	◎	○								○	○	
	○	○	○	○		◎	○	○											○	
○	○	○				○	○	○	◎									○	○	○
○	○	○					○	○	◎				○	○	○	○	○	○	○	○
○	○	○							◎		○	○	○	○	○	○	○	○		

*1: If injector or air flow meter circuit is short or open, the fail-safe system operates and engine revolution does not rise to 2,800 rpm or more.
*2: For California
*3: Except for California

ECCS SELF-DIAGNOSIS DESCRIPTION—1988 VAN

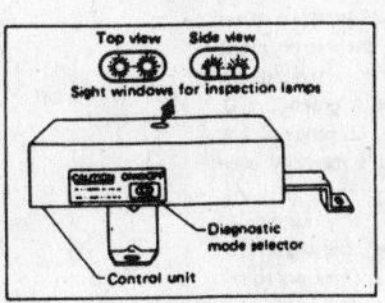
Top view
Side view
Sight windows for inspection lamps
Diagnostic mode selector
Control unit

Description

The self-diagnosis is useful to diagnose malfunctions in major sensors and actuators of the E.C.C.S. system. There are 5 modes in the self-diagnosis system.

1. **Mode I – Mixture ratio feedback control monitor A**
   - During closed loop condition:
     The green inspection lamp turns ON when lean condition is detected and goes OFF by rich condition.
   - During open loop condition:
     The green inspection lamp remains ON or OFF.
2. **Mode II – Mixture ratio feedback control monitor B**
   The green inspection lamp function is the same as Mode I.
   - During closed loop condition:
     The red inspection lamp turns ON and OFF simultaneously with the green inspection lamp when the mixture ratio is controlled within the specified value.
   - During open loop condition:
     The red inspection lamp remains ON or OFF.
3. **Mode III – Self-diagnosis**
   This mode is the same as the former self-diagnosis in self-diagnosis mode.
4. **Mode IV – Switches ON/OFF diagnosis**
   During this mode, the inspection lamps monitor the switch ON-OFF condition.
   - Idle switch
   - Starter switch
   - Vehicle speed sensor
5. **Mode V – Real time diagnosis**
   The moment the malfunction is detected, the display will be presented immediately. That is, the condition at which the malfunction occurs can be found by observing the inspection lamps during driving test.

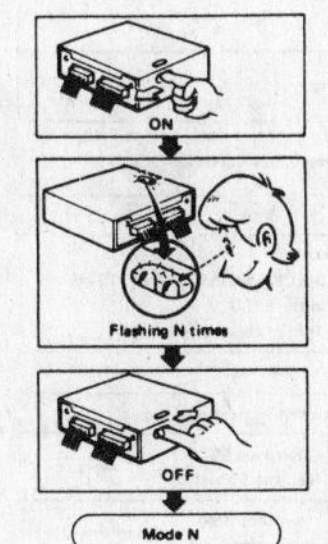
ON
Flashing N times
OFF
Mode N

Description (Cont'd)

1. Turn ignition switch "ON".
2. Turn diagnostic mode selector "ON" and wait until the inspection lamps flash.
3. After the inspection lamps have flashed the number of the required mode, turn diagnostic mode selector "OFF" immediately.

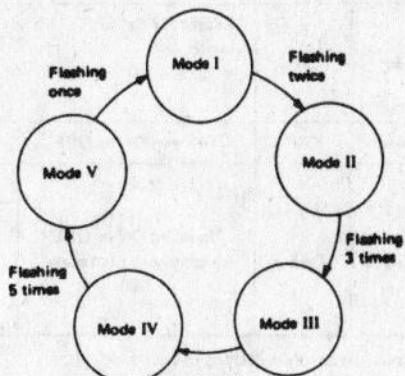
Mode I
Flashing twice
Mode II
Flashing 3 times
Mode III
Flashing 4 times
Mode IV
Flashing 5 times
Mode V
Flashing once

When the ignition switch is turned off during diagnosis, in any mode, and then turned back on again after the power to the E.C.U. has dropped off completely, the diagnosis will automatically return to Mode I.

The stored memory would be lost if:

1. Battery terminal is disconnected.
2. After selecting Mode III, Mode IV is selected.
   However, if the diagnostic mode selector is kept turned fully clockwise, it will continue to change in the order of Mode I → II → III → IV → V → I ... etc., and in this state the stored memory will not be erased.

ECCS SELF-DIAGNOSIS DESCRIPTION (CONT.)—1988 VAN

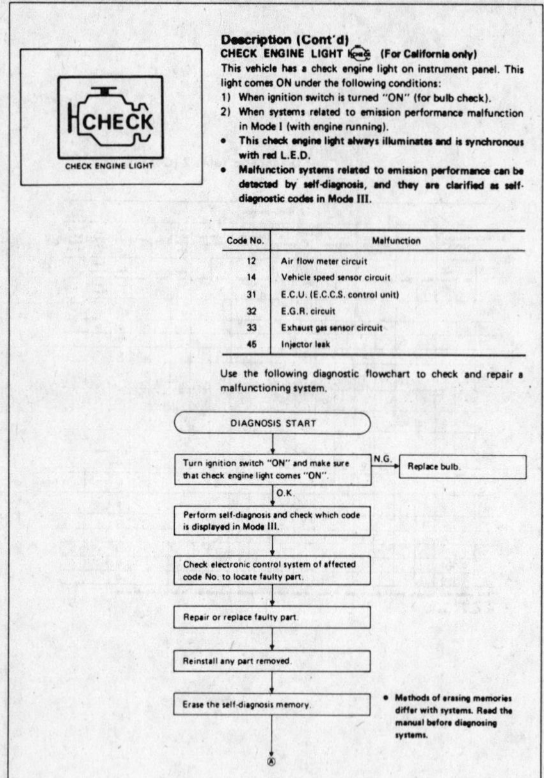

CHECK

CHECK ENGINE LIGHT

Description (Cont'd)

CHECK ENGINE LIGHT (For California only)

This vehicle has a check engine light on instrument panel. This light comes ON under the following conditions:

1) When ignition switch is turned "ON" (for bulb check).
2) When systems related to emission performance malfunction in Mode I (with engine running).

- This check engine light always illuminates and is synchronous with red L.E.D.
- Malfunction systems related to emission performance can be detected by self-diagnosis, and they are clarified as self-diagnostic codes in Mode III.

Code No.	Malfunction
12	Air flow meter circuit
14	Vehicle speed sensor circuit
31	E.C.U. (E.C.C.S. control unit)
32	E.G.R. circuit
33	Exhaust gas sensor circuit
45	Injector leak

Use the following diagnostic flowchart to check and repair a malfunctioning system.

DIAGNOSIS START

Turn ignition switch "ON" and make sure that check engine light comes "ON". — N.G. → Replace bulb.

O.K.

Perform self-diagnosis and check which code is displayed in Mode III.

Check electronic control system of affected code No. to locate faulty part.

Repair or replace faulty part.

Reinstall any part removed.

Erase the self-diagnosis memory.

- Methods of erasing memories differ with systems. Read the manual before diagnosing systems.

Ⓐ

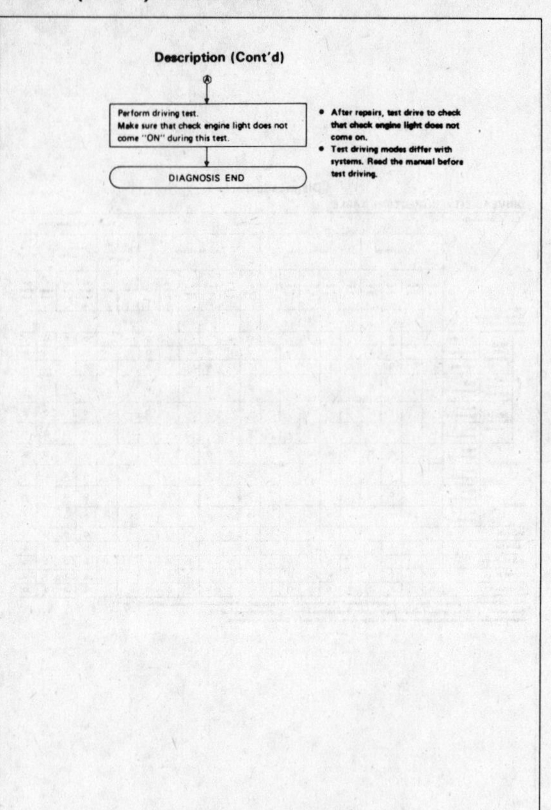

Description (Cont'd)

Ⓐ

Perform driving test.
Make sure that check engine light does not come "ON" during this test.

- After repairs, test drive to check that check engine light does not come on.
- Test driving modes differ with systems. Read the manual before test driving.

DIAGNOSIS END

ECCS SELF-DIAGNOSTIC MODES 1 AND II—1988 VAN

Modes I & II — Mixture Ratio Feedback Control Monitors A & B

In these modes, L.E.D. at the E.C.U. give an air-fuel ratio monitor presentation and the air-fuel ratio feedback coefficient monitor presentation.

Mode	LED	Engine stopped (Ignition switch "ON")	Engine running			
			Open loop condition	Closed loop condition		
Mode I (Monitor A)	Green	ON	*Remains ON or OFF	Blinks		
	Red	ON	Except for California model • OFF	For California model • ON: when CHECK ENGINE LIGHT ITEMS are stored in the E.C.U. • OFF: except for the above condition		
Mode II (Monitor B)	Green	ON	*Remains ON or OFF	Blinks		
	Red	OFF	*Remains ON or OFF (synchronous with green LED)	Compensating mixture ratio		
				More than 5% rich	Between 5% lean and 5% rich	More
				OFF	Synchronized with green LED	Remains ON

*: Maintains conditions just before switching to open loop

ECCS SELF-DIAGNOSTIC MODE III—1988 VAN

Mode III — Self-diagnostic System

The E.C.U. constantly monitors the function of many sensors and actuators, regardless of ignition key position. If a malfunction occurs, the information is stored in the E.C.U. and can be retrieved from the memory by turning on the diagnostic mode selector, located on the side of the E.C.U. When activated, the malfunction is indicated by flashing a red and a green L.E.D. (Light Emitting Diode), also located on the E.C.U. Since all the self-diagnostic results are stored in the E.C.U.'s memory even intermittent malfunctions can be diagnosed.

A malfunctioning part's group is indicated by the number of both the red and the green L.E.D.s flashing. First, the red L.E.D. flashes and the green flashes follow. The red L.E.D. refers to the number of tens while the green one refers to the number of units. For example, when the red L.E.D. flashes once and then the green one flashes twice, this means the number "12" showing the air flow meter signal is malfunctioning. In this way, all the problems are classified by the code numbers.

- When engine fails to start, crank engine more than two seconds before starting self-diagnosis.
- Before starting self-diagnosis, do not erase stored memory. If doing so, self-diagnosis function for intermittent malfunctions would be lost.

The stored memory would be lost if:

1. Battery terminal is disconnected.
2. After selecting Mode III, Mode IV is selected.

DISPLAY CODE TABLE

Code No.	Detected items	California	Except for California
11	Crank angle sensor circuit	X	X
12	Air flow meter circuit	X*	X
13	Water temperature sensor circuit	X	X
14	Vehicle speed sensor circuit	X*	X
21	Ignition signal missing in primary coil	X	X
31	E.C.U. (E.C.C.S. control unit)	X*	X
32	E.G.R. function	X*	–
33	Exhaust gas sensor circuit	X*	X
35	Exhaust gas temperature sensor circuit	X	–
43	Throttle sensor circuit	X	X
45	Injector leak	X*	–
51	Injector circuit	X	X
55	No malfunctioning in the above circuit	X	X

X: Available
–: Not available
* CHECK ENGINE LIGHT ITEM

ECCS SELF-DIAGNOSTIC MODE III (CONT.)—1988 VAN

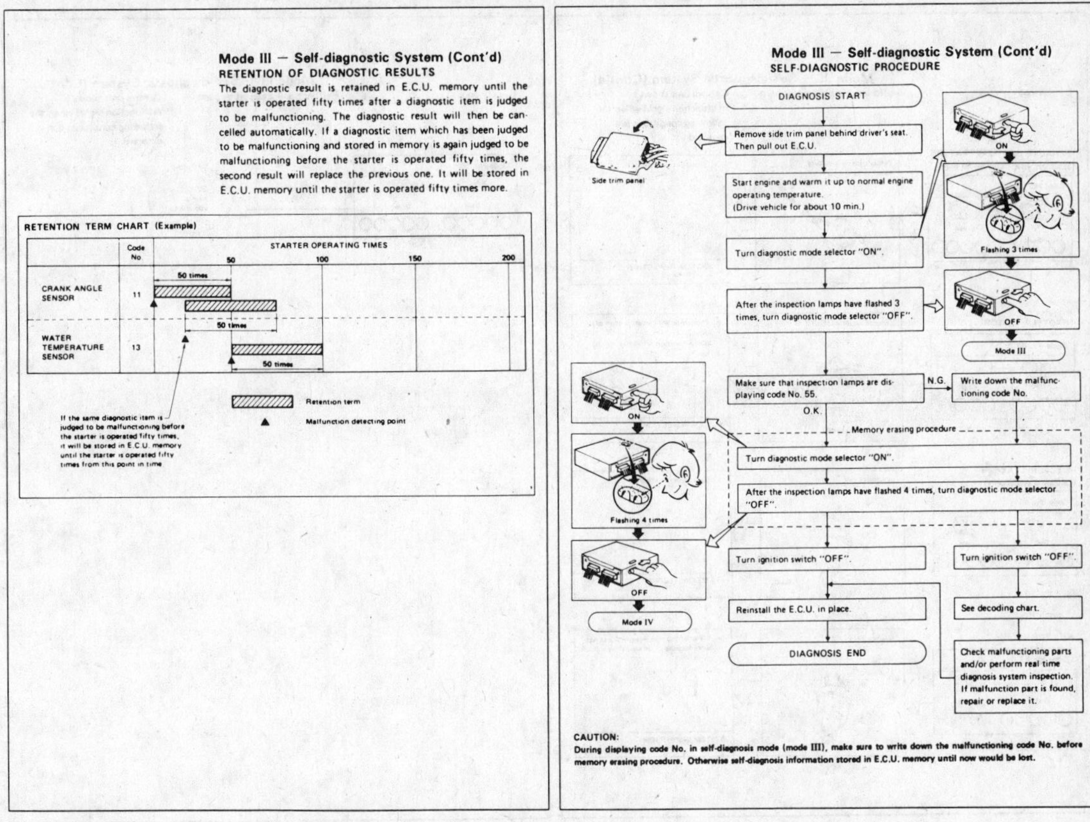
Mode III — Self-diagnostic System (Cont'd)
RETENTION OF DIAGNOSTIC RESULTS
The diagnostic result is retained in E.C.U. memory until the starter is operated fifty times after a diagnostic item is judged to be malfunctioning. The diagnostic result will then be cancelled automatically. If a diagnostic item which has been judged to be malfunctioning and stored in memory is again judged to be malfunctioning before the starter is operated fifty times, the second result will replace the previous one. It will be stored in E.C.U. memory until the starter is operated fifty times more.
RETENTION TERM CHART (Example)
Code No.
STARTER OPERATING TIMES
50
100
150
200
CRANK ANGLE SENSOR
11
WATER TEMPERATURE SENSOR
13
50 times
Retention term
Malfunction detecting point
If the same diagnostic item is judged to be malfunctioning before the starter is operated fifty times, it will be stored in E.C.U. memory until the starter is operated fifty times from this point in time.
Mode III — Self-diagnostic System (Cont'd)
SELF-DIAGNOSTIC PROCEDURE
DIAGNOSIS START
Remove side trim panel behind driver's seat. Then pull out E.C.U.
Side trim panel
Start engine and warm it up to normal engine operating temperature. (Drive vehicle for about 10 min.)
Turn diagnostic mode selector "ON".
ON
Flashing 3 times
After the inspection lamps have flashed 3 times, turn diagnostic mode selector "OFF".
OFF
Mode III
Make sure that inspection lamps are displaying code No. 55.
N.G.
Write down the malfunctioning code No.
O.K.
Memory erasing procedure
Turn diagnostic mode selector "ON".
After the inspection lamps have flashed 4 times, turn diagnostic mode selector "OFF".
Flashing 4 times
Turn ignition switch "OFF".
Turn ignition switch "OFF".
Reinstall the E.C.U. in place.
See decoding chart.
DIAGNOSIS END
Check malfunctioning parts and/or perform real time diagnosis system inspection. If malfunction part is found, repair or replace it.
Mode IV
CAUTION:
During displaying code No. in self-diagnosis mode (mode III), make sure to write down the malfunctioning code No. before memory erasing procedure. Otherwise self-diagnosis information stored in E.C.U. memory until now would be lost.

ECCS SELF-DIAGNOSTIC MODE III DISPLAY CODES—1988 VAN

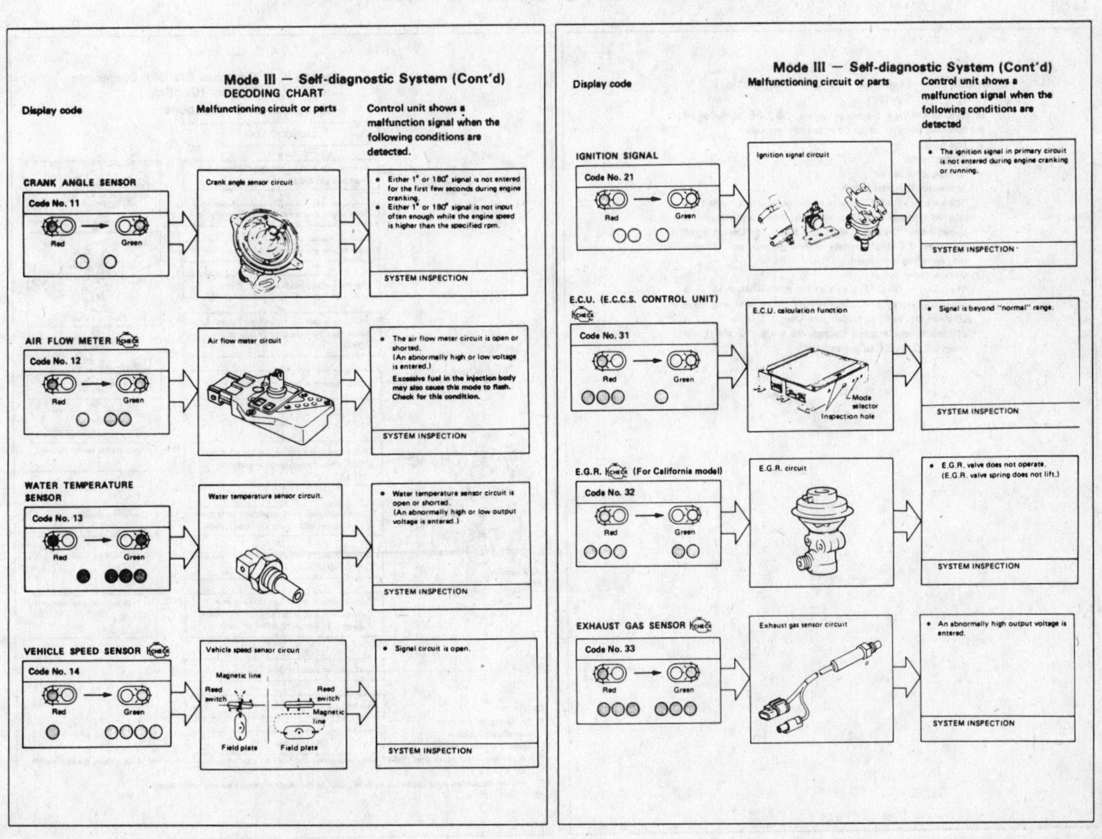
Mode III — Self-diagnostic System (Cont'd)
DECODING CHART
Display code
Malfunctioning circuit or parts
Control unit shows a malfunction signal when the following conditions are detected.
CRANK ANGLE SENSOR
Code No. 11
Red
Green
Crank angle sensor circuit
Either 1° or 180° signal is not entered for the first few seconds during engine cranking.
Either 1° or 180° signal is not input often enough while the engine speed is higher than the specified rpm.
SYSTEM INSPECTION
AIR FLOW METER
Code No. 12
Air flow meter circuit
The air flow meter circuit is open or shorted. (An abnormally high or low voltage is entered.)
Excessive fuel in the injection body may also cause this mode to flash. Check for this condition.
SYSTEM INSPECTION
WATER TEMPERATURE SENSOR
Code No. 13
Water temperature sensor circuit.
Water temperature sensor circuit is open or shorted. (An abnormally high or low output voltage is entered.)
SYSTEM INSPECTION
VEHICLE SPEED SENSOR
Code No. 14
Vehicle speed sensor circuit
Magnetic line
Reed switch
Field plate
Signal circuit is open.
SYSTEM INSPECTION
Mode III — Self-diagnostic System (Cont'd)
IGNITION SIGNAL
Code No. 21
Ignition signal circuit
The ignition signal in primary circuit is not entered during engine cranking or running.
SYSTEM INSPECTION
E.C.U. (E.C.C.S. CONTROL UNIT)
Code No. 31
E.C.U. calculation function
Mode selector
Inspection hole
Signal is beyond "normal" range.
SYSTEM INSPECTION
E.G.R. (For California model)
Code No. 32
E.G.R. circuit
E.G.R. valve does not operate. (E.G.R. valve spring does not lift.)
SYSTEM INSPECTION
EXHAUST GAS SENSOR
Code No. 33
Exhaust gas sensor circuit
An abnormally high output voltage is entered.
SYSTEM INSPECTION

ECCS SELF-DIAGNOSTIC MODE III DISPLAY CODES (CONT.)—1988 VAN

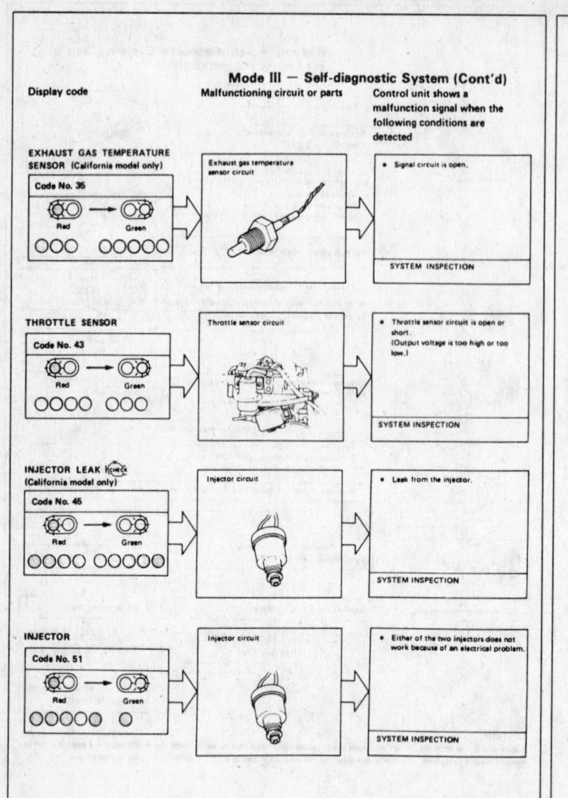
Mode III — Self-diagnostic System (Cont'd)
Display code
Malfunctioning circuit or parts
Control unit shows a malfunction signal when the following conditions are detected
EXHAUST GAS TEMPERATURE SENSOR (California model only)
Code No. 35
Red
Green
Exhaust gas temperature sensor circuit
• Signal circuit is open.
SYSTEM INSPECTION
THROTTLE SENSOR
Code No. 43
Throttle sensor circuit
• Throttle sensor circuit is open or short. (Output voltage is too high or too low.)
SYSTEM INSPECTION
INJECTOR LEAK
(California model only)
Code No. 45
Injector circuit
• Leak from the injector.
SYSTEM INSPECTION
INJECTOR
Code No. 51
Injector circuit
• Either of the two injectors does not work because of an electrical problem.
SYSTEM INSPECTION

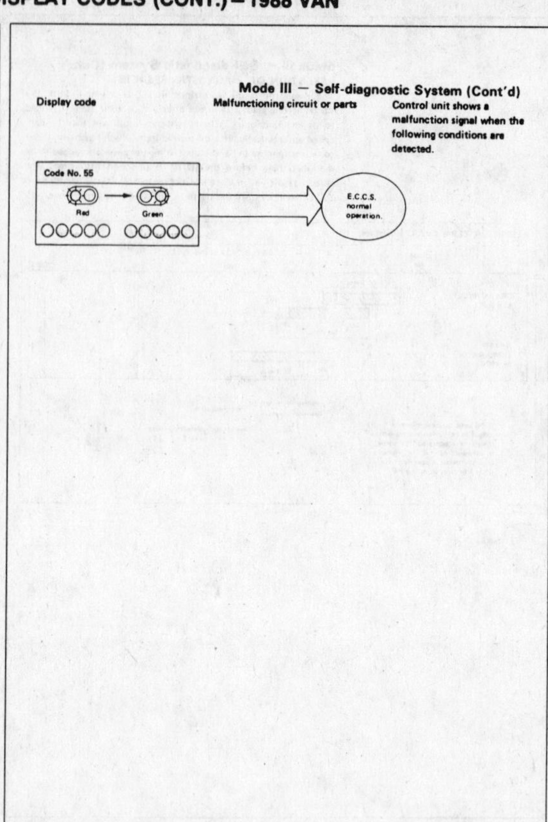
Mode III — Self-diagnostic System (Cont'd)
Display code
Malfunctioning circuit or parts
Control unit shows a malfunction signal when the following conditions are detected.
Code No. 55
Red
Green
E.C.C.S. normal operation.

ECCS SELF-DIAGNOSTIC MODE IV—1988 VAN

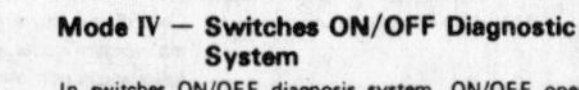
Mode IV — Switches ON/OFF Diagnostic System

In switches ON/OFF diagnosis system, ON/OFF operation of the following switches can be detected continuously.

- Idle switch
- Starter switch
- Vehicle speed sensor

(1) Idle switch & Starter switch

The switches ON/OFF status at the point when mode IV is selected is stored in E.C.U. memory. When either switch is turned from "ON" to "OFF" or "OFF" to "ON", the red L.E.D. on E.C.U. alternately comes on and goes off each time switching is detected.

(2) Vehicle Speed Sensor

The switches ON/OFF status at the point when mode IV is selected is stored in E.C.U. memory. When vehicle speed is 20 km/h (12 MPH) or slower, the green L.E.D. on E.C.U. is off. When vehicle speed exceeds 20 km/h (12 MPH), the green L.E.D. on E.C.U. comes "ON".

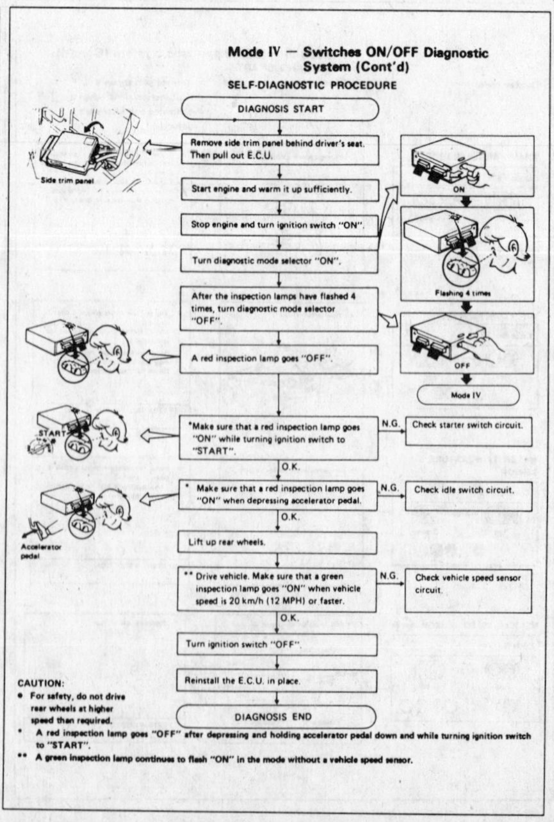
Mode IV — Switches ON/OFF Diagnostic System (Cont'd)
SELF-DIAGNOSTIC PROCEDURE
DIAGNOSIS START
Remove side trim panel behind driver's seat. Then pull out E.C.U.
Side trim panel
ON
Start engine and warm it up sufficiently.
Stop engine and turn ignition switch "ON".
Turn diagnostic mode selector "ON".
Flashing 4 times
After the inspection lamps have flashed 4 times, turn diagnostic mode selector "OFF".
OFF
A red inspection lamp goes "OFF".
Mode IV
START
*Make sure that a red inspection lamp goes "ON" while turning ignition switch to "START".
N.G.
Check starter switch circuit.
O.K.
* Make sure that a red inspection lamp goes "ON" when depressing accelerator pedal.
N.G.
Check idle switch circuit.
O.K.
Accelerator pedal
Lift up rear wheels.
** Drive vehicle. Make sure that a green inspection lamp goes "ON" when vehicle speed is 20 km/h (12 MPH) or faster.
N.G.
Check vehicle speed sensor circuit.
O.K.
Turn ignition switch "OFF".
Reinstall the E.C.U. in place.
DIAGNOSIS END
CAUTION:
• For safety, do not drive rear wheels at higher speed than required.
* A red inspection lamp goes "OFF" after depressing and holding accelerator pedal down and while turning ignition switch to "START".
** A green inspection lamp continues to flash "ON" in the mode without a vehicle speed sensor.

ECCS SELF-DIAGNOSTIC MODE V—1988 VAN

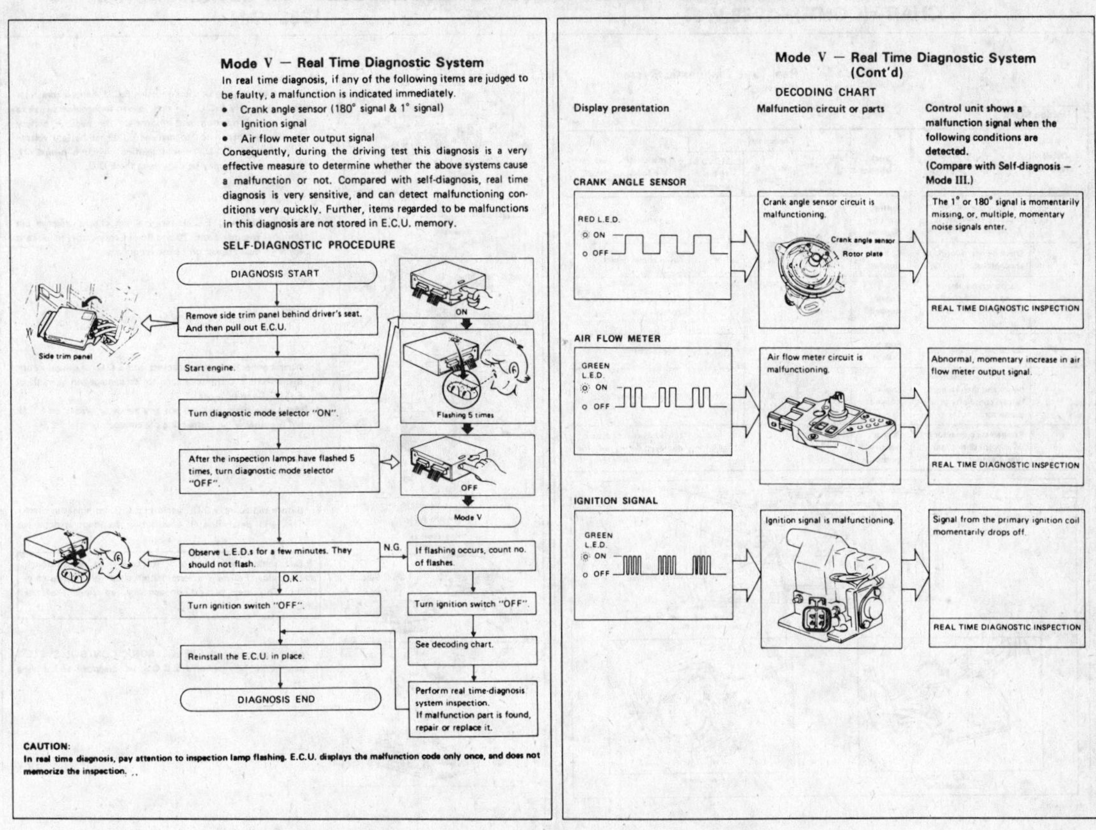

Mode V — Real Time Diagnostic System

In real time diagnosis, if any of the following items are judged to be faulty, a malfunction is indicated immediately.

- Crank angle sensor (180° signal & 1° signal)
- Ignition signal
- Air flow meter output signal

Consequently, during the driving test this diagnosis is a very effective measure to determine whether the above systems cause a malfunction or not. Compared with self-diagnosis, real time diagnosis is very sensitive, and can detect malfunctioning conditions very quickly. Further, items regarded to be malfunctions in this diagnosis are not stored in E.C.U. memory.

SELF-DIAGNOSTIC PROCEDURE

- DIAGNOSIS START
- Remove side trim panel behind driver's seat. And then pull out E.C.U.
- Start engine.
- Turn diagnostic mode selector "ON".
- After the inspection lamps have flashed 5 times, turn diagnostic mode selector "OFF".
- Observe L.E.D.s for a few minutes. They should not flash.
  - O.K.: Turn ignition switch "OFF". → Reinstall the E.C.U. in place. → DIAGNOSIS END
  - N.G.: If flashing occurs, count no. of flashes. → Turn ignition switch "OFF". → See decoding chart. → Perform real time-diagnosis system inspection. If malfunction part is found, repair or replace it.

Side trim panel
ON
Flashing 5 times
OFF
Mode V

CAUTION:
In real time diagnosis, pay attention to inspection lamp flashing. E.C.U. displays the malfunction code only once, and does not memorize the inspection.

Mode V — Real Time Diagnostic System (Cont'd)

DECODING CHART

Display presentation	Malfunction circuit or parts	Control unit shows a malfunction signal when the following conditions are detected. (Compare with Self-diagnosis — Mode III.)
CRANK ANGLE SENSOR — RED L.E.D. ON / OFF	Crank angle sensor circuit is malfunctioning.	The 1° or 180° signal is momentarily missing, or, multiple, momentary noise signals enter. — REAL TIME DIAGNOSTIC INSPECTION
AIR FLOW METER — GREEN L.E.D. ON / OFF	Air flow meter circuit is malfunctioning.	Abnormal, momentary increase in air flow meter output signal. — REAL TIME DIAGNOSTIC INSPECTION
IGNITION SIGNAL — GREEN L.E.D. ON / OFF	Ignition signal is malfunctioning.	Signal from the primary ignition coil momentarily drops off. — REAL TIME DIAGNOSTIC INSPECTION

Crank angle sensor
Rotor plate

ECCS SELF-DIAGNOSTIC MODE V INSPECTION CHART—1988 VAN

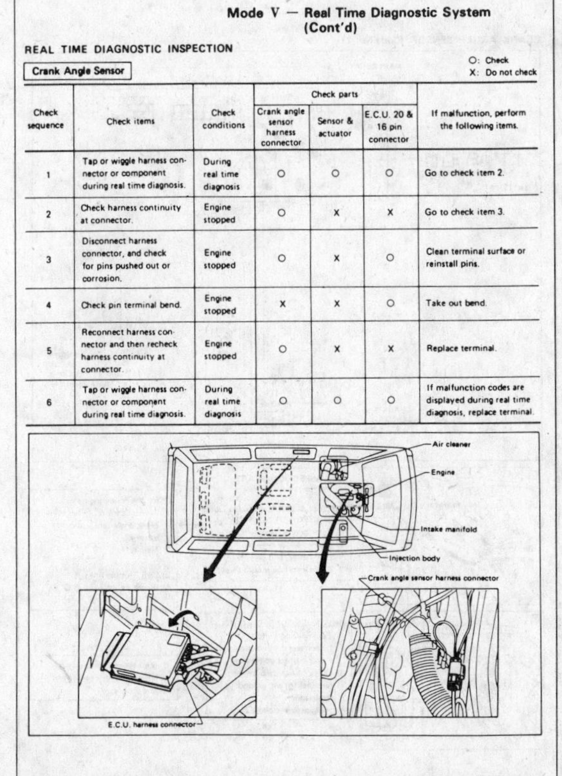

Mode V — Real Time Diagnostic System (Cont'd)

REAL TIME DIAGNOSTIC INSPECTION

Crank Angle Sensor

O: Check
X: Do not check

Check sequence	Check items	Check conditions	Check parts: Crank angle sensor harness connector	Check parts: Sensor & actuator	Check parts: E.C.U. 20 & 16 pin connector	If malfunction, perform the following items.
1	Tap or wiggle harness connector or component during real time diagnosis.	During real time diagnosis	O	O	O	Go to check item 2.
2	Check harness continuity at connector.	Engine stopped	O	X	X	Go to check item 3.
3	Disconnect harness connector, and check for pins pushed out or corrosion.	Engine stopped	O	X	O	Clean terminal surface or reinstall pins.
4	Check pin terminal bend.	Engine stopped	X	X	O	Take out bend.
5	Reconnect harness connector and then recheck harness continuity at connector.	Engine stopped	O	X	X	Replace terminal.
6	Tap or wiggle harness connector or component during real time diagnosis.	During real time diagnosis	O	O	O	If malfunction codes are displayed during real time diagnosis, replace terminal.

Air cleaner
Engine
Intake manifold
Injection body
Crank angle sensor harness connector
E.C.U. harness connector

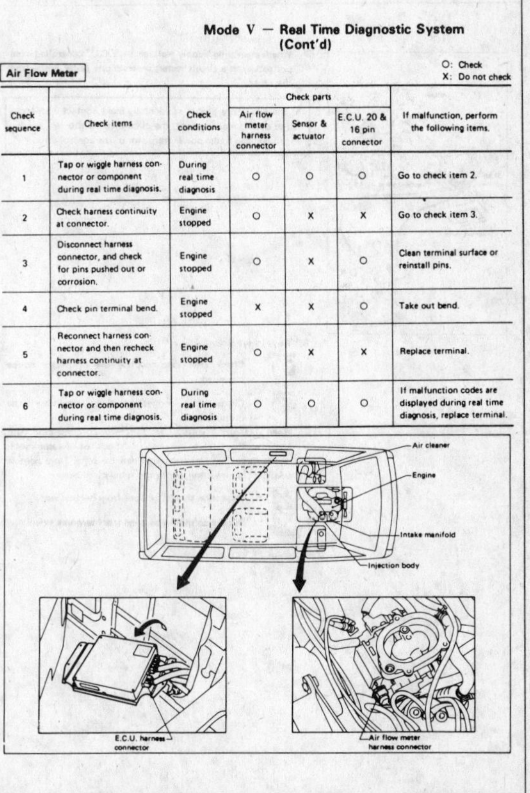

Mode V — Real Time Diagnostic System (Cont'd)

Air Flow Meter

O: Check
X: Do not check

Check sequence	Check items	Check conditions	Check parts: Air flow meter harness connector	Check parts: Sensor & actuator	Check parts: E.C.U. 20 & 16 pin connector	If malfunction, perform the following items.
1	Tap or wiggle harness connector or component during real time diagnosis.	During real time diagnosis	O	O	O	Go to check item 2.
2	Check harness continuity at connector.	Engine stopped	O	X	X	Go to check item 3.
3	Disconnect harness connector, and check for pins pushed out or corrosion.	Engine stopped	O	X	O	Clean terminal surface or reinstall pins.
4	Check pin terminal bend.	Engine stopped	X	X	O	Take out bend.
5	Reconnect harness connector and then recheck harness continuity at connector.	Engine stopped	O	X	X	Replace terminal.
6	Tap or wiggle harness connector or component during real time diagnosis.	During real time diagnosis	O	O	O	If malfunction codes are displayed during real time diagnosis, replace terminal.

Air cleaner
Engine
Intake manifold
Injection body
E.C.U. harness connector
Air flow meter harness connector

ECCS SELF-DIAGNOSTIC MODE V INSPECTION CHART (CONT.)—1988 VAN

Mode V — Real Time Diagnostic System (Cont'd)

Ignition Signal

O: Check
X: Do not check

Check sequence	Check items	Check conditions	Check parts: Ignition coil & power transistor harness connector	Check parts: Sensor & actuator	Check parts: E.C.U. 20 & 16 pin connector	If malfunction, perform the following items.
1	Tap or wiggle harness connector or component during real time diagnosis.	During real time diagnosis	O	O	O	Go to check item 2.
2	Check harness continuity at connector.	Engine stopped	O	X	X	Go to check item 3.
3	Disconnect harness connector, and check for pins pushed out or corrosion.	Engine stopped	O	X	O	Clean terminal surface or reinstall pins.
4	Check pin terminal bend.	Engine stopped	X	X	O	Take out bend.
5	Reconnect harness connector and then recheck harness continuity at connector.	Engine stopped	O	X	X	Replace terminal.
6	Tap harness connector or component during real time diagnosis.	During real time diagnosis	O	O	O	If malfunction codes are displayed during real time diagnosis, replace terminal.

Air cleaner
Engine
Intake manifold
Injection body
Ignition coil & power transistor harness connector
E.C.U. harness connector
Air cleaner

ECCS SYSTEM INSPECTION CAUTION 1988 VAN

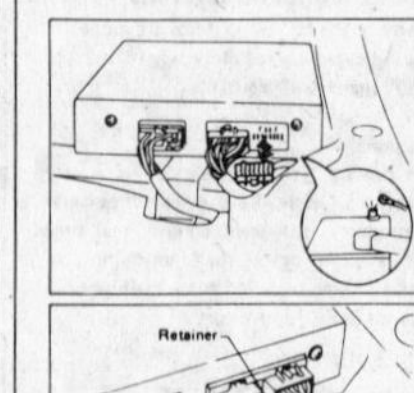

CAUTION:

1. Before connecting or disconnecting E.C.U. harness connector to or from any E.C.U., be sure to turn the ignition switch to the "OFF" position and disconnect the negative battery terminal in order not to damage E.C.U. as battery voltage is applied to E.C.U. even if ignition switch is turned off. Otherwise, there may be damage to the E.C.U.

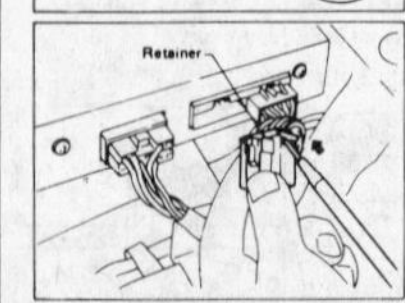
Retainer

2. When performing E.C.U. terminal pin checks, remove pin terminal retainer from 20-and 16-pin connectors to make it easier to insert tester probe into connector.

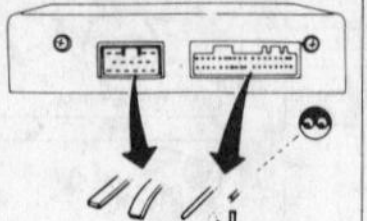
Bend
Break

3. When connecting pin connectors into E.C.U. or disconnecting them from E.C.U., take care not to damage pin terminal of E.C.U. (Bend or break).
4. Make sure that there are not any bends or breaks on E.C.U. pin terminal when connecting pin connectors into E.C.U.

Perform E.C.U. terminal pin checks before replacement.
OLD ONE

5. Before replacing E.C.U., perform E.C.U. terminal pin checks. This will determine if a sensor or an input is defective.

> E.C.U. seldom fails on its own. It is usually damaged through carelessness. Follow the terminal pin check procedures carefully. This may prevent unnecessary replacement of the E.C.U.

6. After performing this "ELECTRONIC CONTROL SYSTEM INSPECTION", perform E.C.C.S. self-diagnosis and driving test.

ECCS SYSTEM INPSECTION CAUTION (CONT.) 1988 VAN

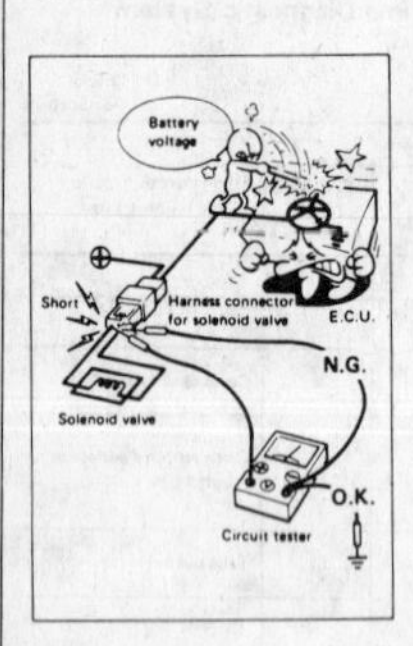
Battery voltage
Short
Harness connector for solenoid valve
E.C.U.
N.G.
Solenoid valve
O.K.
Circuit tester

7. When measuring supply voltage of E.C.U. controlled components with a circuit tester, separate one tester probe from the other.

> If the two tester probes accidentally make contact with each other during measurement, the circuit will be shorted, resulting in damage to the power transistor of the control unit.

8. Keys to symbols
   - (DISCONNECT symbol) : Check after disconnecting the connector to be measured.
   - (CONNECT symbol) : Check after connecting the connector to be measured.
9. When measuring voltage or resistance at connector with tester probes, there are two methods of measurement; one is done from terminal side and the other from harness side. Before measuring, confirm symbol mark again.
   - (HS symbol) : Inspection should be done from harness side.
   - (TS symbol) : Inspection should be done from terminal side.

ECCS SELF-DIAGNOSTIC CHART—1988 VAN

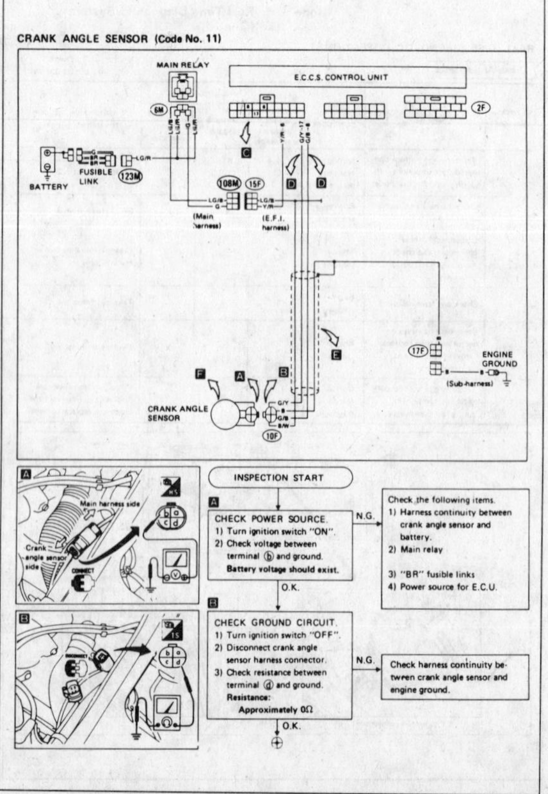
CRANK ANGLE SENSOR (Code No. 11)
MAIN RELAY
E.C.C.S. CONTROL UNIT
BATTERY
FUSIBLE LINK
(Main harness)
(E.F.I. harness)
ENGINE GROUND
(Sub-harness)
CRANK ANGLE SENSOR
Main harness side
Crank angle sensor side
INSPECTION START
CHECK POWER SOURCE.
1) Turn ignition switch "ON".
2) Check voltage between terminal ⓑ and ground.
Battery voltage should exist.
N.G.
Check the following items.
1) Harness continuity between crank angle sensor and battery.
2) Main relay
3) "BR" fusible links
4) Power source for E.C.U.
O.K.
CHECK GROUND CIRCUIT.
1) Turn ignition switch "OFF".
2) Disconnect crank angle sensor harness connector.
3) Check resistance between terminal ⓓ and ground.
Resistance:
Approximately 0Ω
N.G.
Check harness continuity between crank angle sensor and engine ground.
O.K.

ECCS SELF-DIAGNOSTIC CHART—1988 VAN

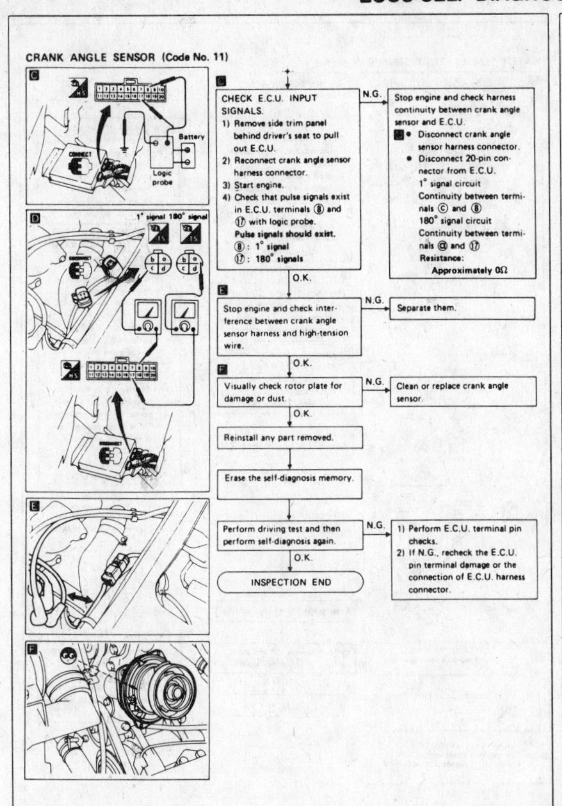
CRANK ANGLE SENSOR (Code No. 11)
Battery
Logic probe
1° signal 180° signal
C
CHECK E.C.U. INPUT SIGNALS.
1) Remove side trim panel behind driver's seat to pull out E.C.U.
2) Reconnect crank angle sensor harness connector.
3) Start engine.
4) Check that pulse signals exist in E.C.U. terminals ⑧ and ⑰ with logic probe.
Pulse signals should exist.
⑧ : 1° signal
⑰ : 180° signals
N.G.
Stop engine and check harness continuity between crank angle sensor and E.C.U.
D • Disconnect crank angle sensor harness connector.
• Disconnect 20-pin connector from E.C.U.
1° signal circuit
Continuity between terminals Ⓒ and ⑧
180° signal circuit
Continuity between terminals Ⓓ and ⑰
Resistance:
Approximately 0Ω
O.K.
E
Stop engine and check interference between crank angle sensor harness and high-tension wire.
N.G.
Separate them.
O.K.
F
Visually check rotor plate for damage or dust.
N.G.
Clean or replace crank angle sensor.
O.K.
Reinstall any part removed.
Erase the self-diagnosis memory.
Perform driving test and then perform self-diagnosis again.
N.G.
1) Perform E.C.U. terminal pin checks.
2) If N.G., recheck the E.C.U. pin terminal damage or the connection of E.C.U. harness connector.
O.K.
INSPECTION END

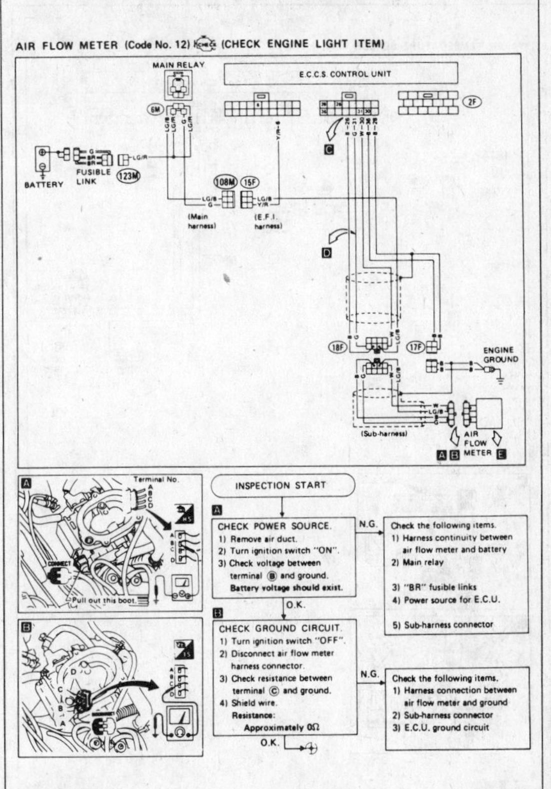
AIR FLOW METER (Code No. 12) (CHECK ENGINE LIGHT ITEM)
MAIN RELAY
E.C.C.S. CONTROL UNIT
BATTERY
FUSIBLE LINK
(Main harness)
(E.F.I. harness)
(Sub-harness)
ENGINE GROUND
AIR FLOW METER
Terminal No.
Pull out this boot.
INSPECTION START
A
CHECK POWER SOURCE.
1) Remove air duct.
2) Turn ignition switch "ON".
3) Check voltage between terminal Ⓑ and ground.
Battery voltage should exist.
N.G.
Check the following items.
1) Harness continuity between air flow meter and battery
2) Main relay
3) "BR" fusible links
4) Power source for E.C.U.
5) Sub-harness connector
O.K.
B
CHECK GROUND CIRCUIT.
1) Turn ignition switch "OFF".
2) Disconnect air flow meter harness connector.
3) Check resistance between terminal Ⓒ and ground.
4) Shield wire.
Resistance:
Approximately 0Ω
N.G.
Check the following items.
1) Harness connection between air flow meter and ground
2) Sub-harness connector
3) E.C.U. ground circuit
O.K.

ECCS SELF-DIAGNOSTIC CHART—1988 VAN

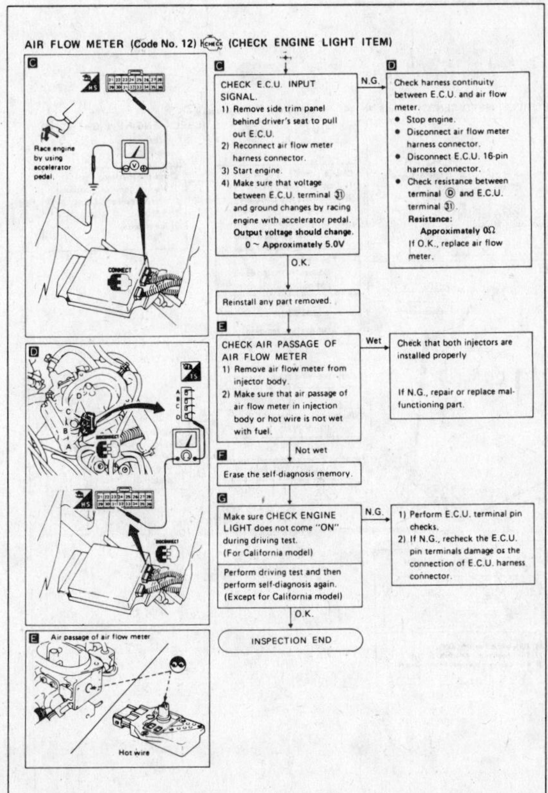
AIR FLOW METER (Code No. 12) (CHECK ENGINE LIGHT ITEM)
Race engine by using accelerator pedal.
C
CHECK E.C.U. INPUT SIGNAL.
1) Remove side trim panel behind driver's seat to pull out E.C.U.
2) Reconnect air flow meter harness connector.
3) Start engine.
4) Make sure that voltage between E.C.U. terminal ㉛ and ground changes by racing engine with accelerator pedal.
Output voltage should change.
0 ~ Approximately 5.0V
N.G.
D
Check harness continuity between E.C.U. and air flow meter.
• Stop engine.
• Disconnect air flow meter harness connector.
• Disconnect E.C.U. 16-pin harness connector.
• Check resistance between terminal Ⓓ and E.C.U. terminal ㉛.
Resistance:
Approximately 0Ω
If O.K., replace air flow meter.
O.K.
Reinstall any part removed.
E
CHECK AIR PASSAGE OF AIR FLOW METER
1) Remove air flow meter from injector body.
2) Make sure that air passage of air flow meter in injection body or hot wire is not wet with fuel.
Wet
Check that both injectors are installed properly
If N.G., repair or replace malfunctioning part.
Not wet
F
Erase the self-diagnosis memory.
G
Make sure CHECK ENGINE LIGHT does not come "ON" during driving test.
(For California model)
Perform driving test and then perform self-diagnosis again.
(Except for California model)
N.G.
1) Perform E.C.U. terminal pin checks.
2) If N.G., recheck the E.C.U. pin terminals damage or the connection of E.C.U. harness connector.
O.K.
INSPECTION END
Air passage of air flow meter
Hot wire

AIR FLOW METER (Code No. 12) (CHECK ENGINE LIGHT ITEM)
F
Diagnostic mode selector
1) Turn ignition switch "ON".
2) Turn diagnostic mode selector "ON".
3) And then turn diagnostic mode selector "OFF".
G
CHECK
CHECK ENGINE LIGHT

ECCS SELF-DIAGNOSTIC CHART—1988 VAN

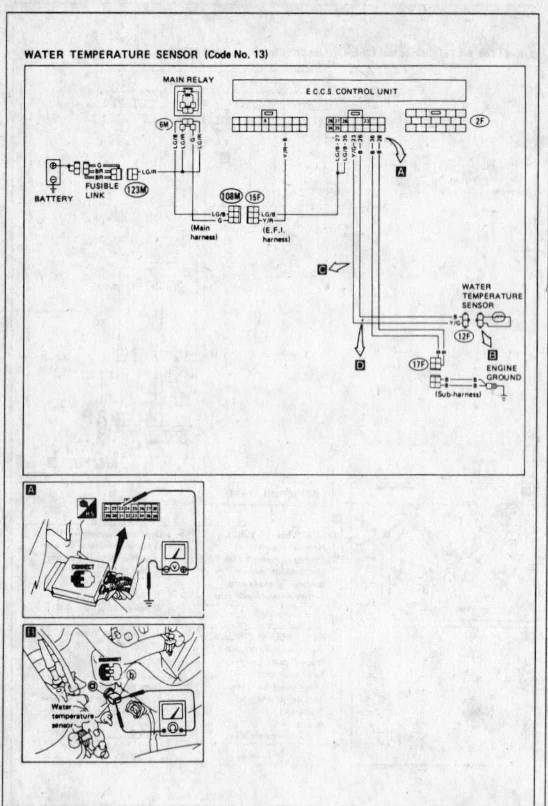
WATER TEMPERATURE SENSOR (Code No. 13)
MAIN RELAY
E.C.C.S. CONTROL UNIT
BATTERY
FUSIBLE LINK
(Main harness)
(E.F.I. harness)
WATER TEMPERATURE SENSOR
ENGINE GROUND
(Sub-harness)
Water temperature sensor

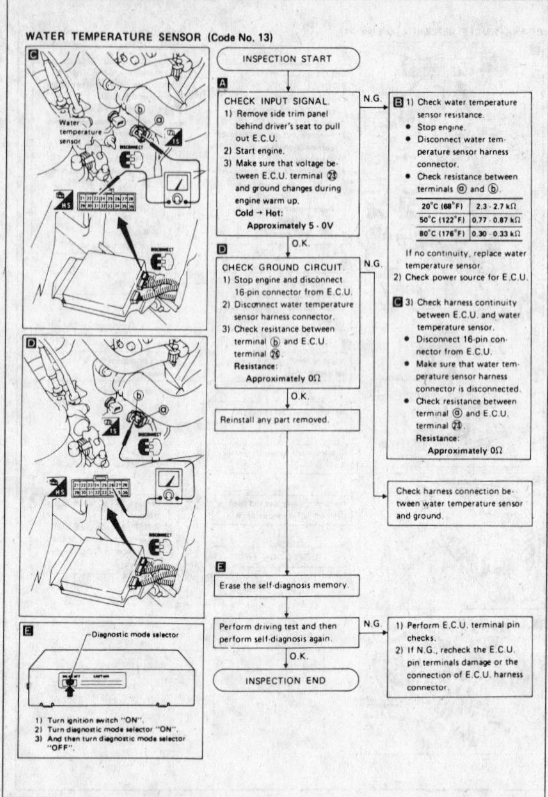
WATER TEMPERATURE SENSOR (Code No. 13)
INSPECTION START
A
CHECK INPUT SIGNAL.
1) Remove side trim panel behind driver's seat to pull out E.C.U.
2) Start engine.
3) Make sure that voltage between E.C.U. terminal 23 and ground changes during engine warm up.
Cold → Hot:
Approximately 5 - 0V
N.G.
B 1) Check water temperature sensor resistance.
• Stop engine.
• Disconnect water temperature sensor harness connector.
• Check resistance between terminals ⓐ and ⓑ.
20°C (68°F) 2.3 - 2.7 kΩ
50°C (122°F) 0.77 - 0.87 kΩ
80°C (176°F) 0.30 - 0.33 kΩ
If no continuity, replace water temperature sensor.
2) Check power source for E.C.U.
C 3) Check harness continuity between E.C.U. and water temperature sensor.
• Disconnect 16-pin connector from E.C.U.
• Make sure that water temperature sensor harness connector is disconnected.
• Check resistance between terminal ⓐ and E.C.U. terminal 23.
Resistance:
Approximately 0Ω
O.K.
D
CHECK GROUND CIRCUIT.
1) Stop engine and disconnect 16-pin connector from E.C.U.
2) Disconnect water temperature sensor harness connector.
3) Check resistance between terminal ⓑ and E.C.U. terminal 26.
Resistance:
Approximately 0Ω
N.G.
Check harness connection between water temperature sensor and ground.
O.K.
Reinstall any part removed.
E
Erase the self-diagnosis memory.
Perform driving test and then perform self-diagnosis again.
N.G.
1) Perform E.C.U. terminal pin checks.
2) If N.G., recheck the E.C.U. pin terminals damage or the connection of E.C.U. harness connector.
O.K.
INSPECTION END
Water temperature sensor
Diagnostic mode selector
1) Turn ignition switch "ON".
2) Turn diagnostic mode selector "ON".
3) And then turn diagnostic mode selector "OFF".

ECCS SELF-DIAGNOSTIC CHART—1988 VAN

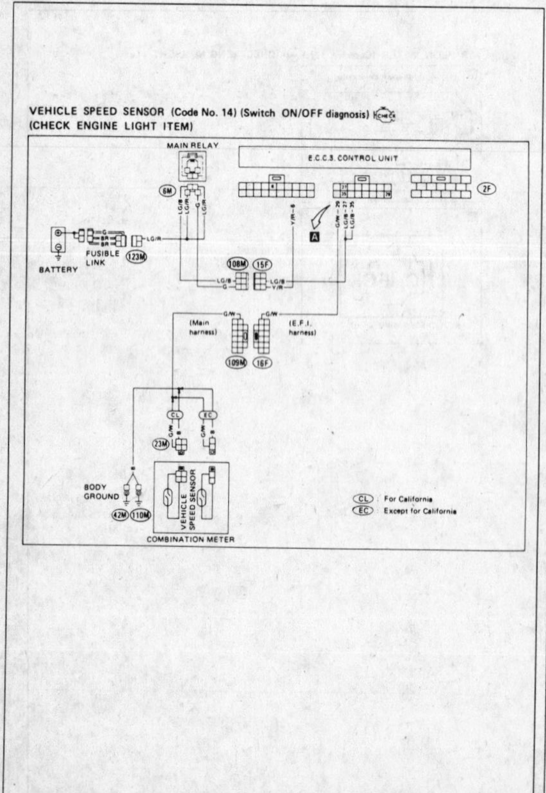
VEHICLE SPEED SENSOR (Code No. 14) (Switch ON/OFF diagnosis)
(CHECK ENGINE LIGHT ITEM)
MAIN RELAY
E.C.C.S. CONTROL UNIT
BATTERY
FUSIBLE LINK
(Main harness)
(E.F.I. harness)
BODY GROUND
VEHICLE SPEED SENSOR
COMBINATION METER
CL For California
EC Except for California

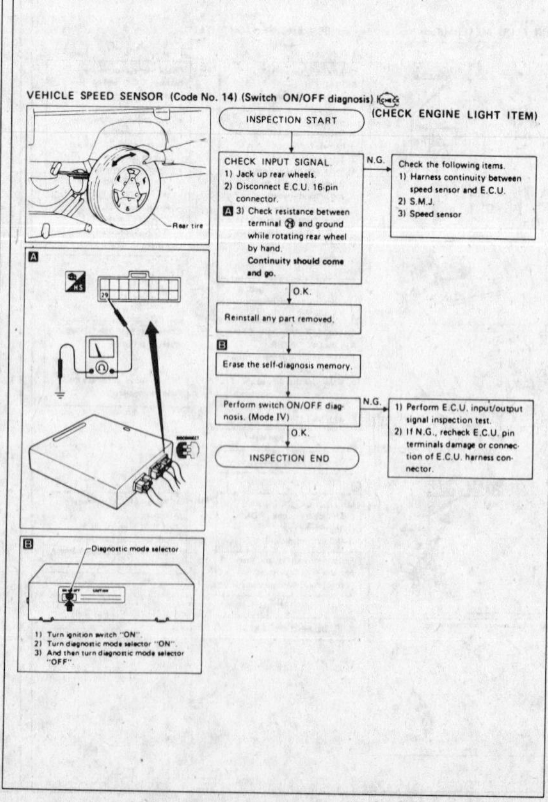
VEHICLE SPEED SENSOR (Code No. 14) (Switch ON/OFF diagnosis)
(CHECK ENGINE LIGHT ITEM)
INSPECTION START
CHECK INPUT SIGNAL.
1) Jack up rear wheels.
2) Disconnect E.C.U. 16-pin connector.
A 3) Check resistance between terminal 29 and ground while rotating rear wheel by hand.
Continuity should come and go.
N.G.
Check the following items.
1) Harness continuity between speed sensor and E.C.U.
2) S.M.J.
3) Speed sensor
O.K.
Reinstall any part removed.
B
Erase the self-diagnosis memory.
Perform switch ON/OFF diagnosis. (Mode IV)
N.G.
1) Perform E.C.U. input/output signal inspection test.
2) If N.G., recheck E.C.U. pin terminals damage or connection of E.C.U. harness connector.
O.K.
INSPECTION END
Rear tire
Diagnostic mode selector
1) Turn ignition switch "ON".
2) Turn diagnostic mode selector "ON".
3) And then turn diagnostic mode selector "OFF".

ECCS SELF-DIAGNOSTIC CHART—1988 VAN

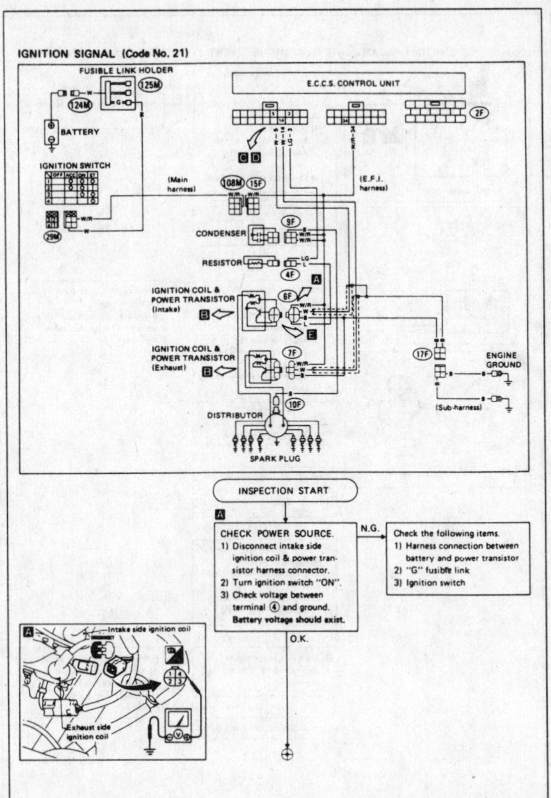

IGNITION SIGNAL (Code No. 21)
FUSIBLE LINK HOLDER
BATTERY
IGNITION SWITCH
E.C.C.S. CONTROL UNIT
(Main harness)
(E.F.I. harness)
CONDENSER
RESISTOR
IGNITION COIL & POWER TRANSISTOR (Intake)
IGNITION COIL & POWER TRANSISTOR (Exhaust)
DISTRIBUTOR
SPARK PLUG
ENGINE GROUND
(Sub-harness)
INSPECTION START
CHECK POWER SOURCE.
1) Disconnect intake side ignition coil & power transistor harness connector.
2) Turn ignition switch "ON".
3) Check voltage between terminal ④ and ground.
Battery voltage should exist.
N.G.
Check the following items.
1) Harness connection between battery and power transistor
2) "G" fusible link
3) Ignition switch
O.K.
Intake side ignition coil
Exhaust side ignition coil

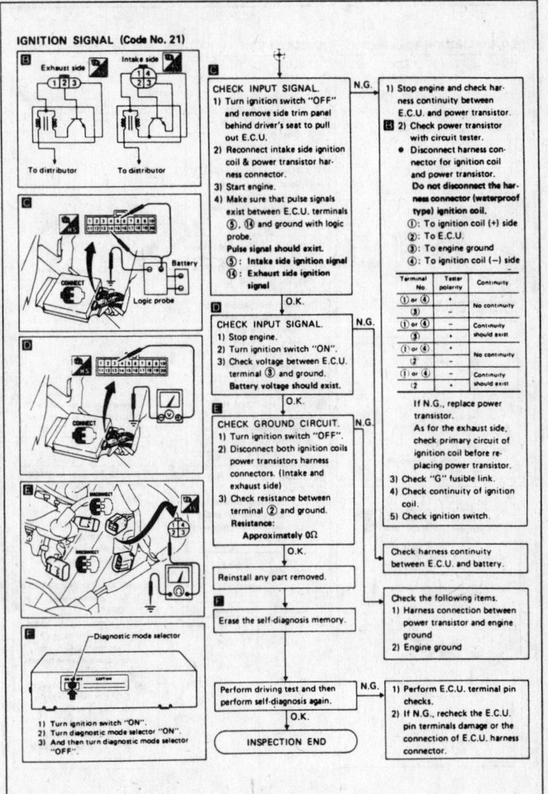

IGNITION SIGNAL (Code No. 21)
Exhaust side
Intake side
To distributor
To distributor
Battery
Logic probe
CHECK INPUT SIGNAL.
1) Turn ignition switch "OFF" and remove side trim panel behind driver's seat to pull out E.C.U.
2) Reconnect intake side ignition coil & power transistor harness connector.
3) Start engine.
4) Make sure that pulse signals exist between E.C.U. terminals ⑤, ⑭ and ground with logic probe.
Pulse signal should exist.
⑤: Intake side ignition signal
⑭: Exhaust side ignition signal
N.G.
1) Stop engine and check harness continuity between E.C.U. and power transistor.
2) Check power transistor with circuit tester.
• Disconnect harness connector for ignition coil and power transistor.
Do not disconnect the harness connector (waterproof type) ignition coil.
①: To ignition coil (+) side
②: To E.C.U.
③: To engine ground
④: To ignition coil (−) side
If N.G., replace power transistor.
As for the exhaust side, check primary circuit of ignition coil before replacing power transistor.
3) Check "G" fusible link.
4) Check continuity of ignition coil.
5) Check ignition switch.
O.K.
CHECK INPUT SIGNAL.
1) Stop engine.
2) Turn ignition switch "ON".
3) Check voltage between E.C.U. terminal ③ and ground.
Battery voltage should exist.
N.G.
Check harness continuity between E.C.U. and battery.
O.K.
CHECK GROUND CIRCUIT.
1) Turn ignition switch "OFF".
2) Disconnect both ignition coils power transistors harness connectors. (Intake and exhaust side)
3) Check resistance between terminal ② and ground.
Resistance:
Approximately 0Ω
N.G.
Check the following items.
1) Harness connection between power transistor and engine ground
2) Engine ground
O.K.
Reinstall any part removed.
Erase the self-diagnosis memory.
Perform driving test and then perform self-diagnosis again.
N.G.
1) Perform E.C.U. terminal pin checks.
2) If N.G., recheck the E.C.U. pin terminals damage or the connection of E.C.U. harness connector.
O.K.
INSPECTION END
Diagnostic mode selector
1) Turn ignition switch "ON".
2) Turn diagnostic mode selector "ON".
3) And then turn diagnostic mode selector "OFF".

Terminal No.	Tester polarity	Continuity
① or ④	+	No continuity
③	−	
① or ④	−	Continuity should exist
③	+	
① or ④	+	No continuity
②	−	
① or ④	−	Continuity should exist
②	+	

ECCS SELF-DIAGNOSTIC CHART—1988 VAN

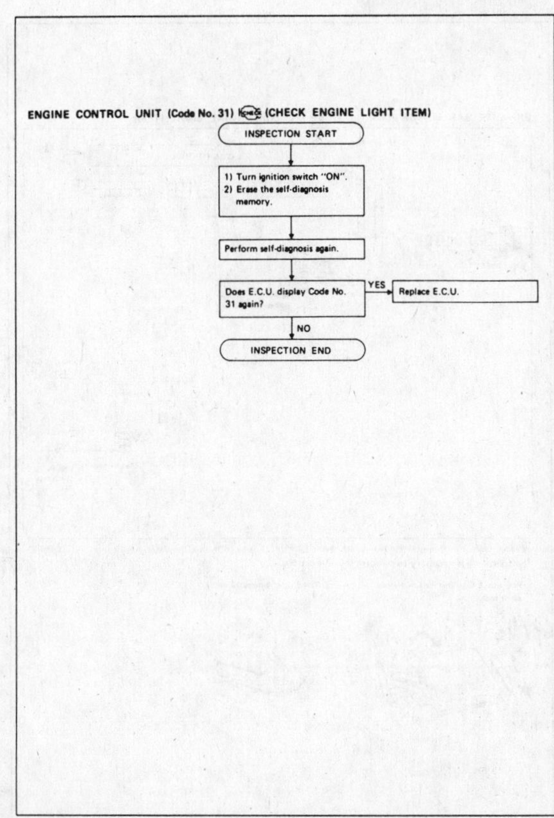

ENGINE CONTROL UNIT (Code No. 31) (CHECK ENGINE LIGHT ITEM)
INSPECTION START
1) Turn ignition switch "ON".
2) Erase the self-diagnosis memory.
Perform self-diagnosis again.
Does E.C.U. display Code No. 31 again?
YES
Replace E.C.U.
NO
INSPECTION END

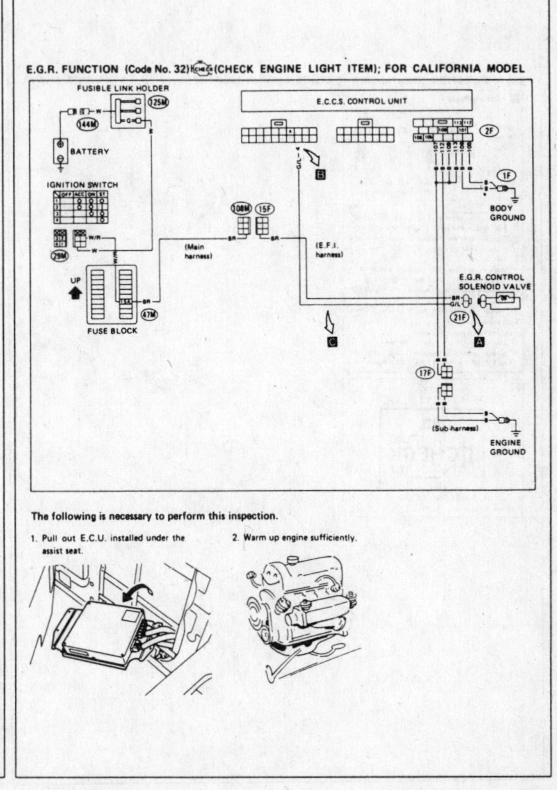

E.G.R. FUNCTION (Code No. 32) (CHECK ENGINE LIGHT ITEM); FOR CALIFORNIA MODEL
FUSIBLE LINK HOLDER
BATTERY
IGNITION SWITCH
E.C.C.S. CONTROL UNIT
BODY GROUND
(Main harness)
(E.F.I. harness)
FUSE BLOCK
E.G.R. CONTROL SOLENOID VALVE
(Sub-harness)
ENGINE GROUND
The following is necessary to perform this inspection.
1. Pull out E.C.U. installed under the assist seat.
2. Warm up engine sufficiently.

ECCS SELF-DIAGNOSTIC CHART—1988 VAN

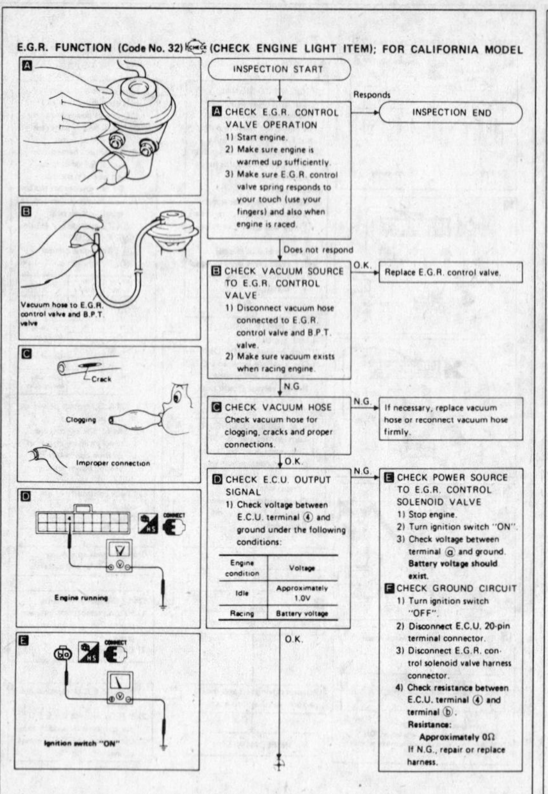
E.G.R. FUNCTION (Code No. 32) (CHECK ENGINE LIGHT ITEM); FOR CALIFORNIA MODEL
A
B
Vacuum hose to E.G.R. control valve and B.P.T. valve
C
Crack
Clogging
Improper connection
D
CONNECT
Engine running
E
CONNECT
Ignition switch "ON"
INSPECTION START
Responds
INSPECTION END
A CHECK E.G.R. CONTROL VALVE OPERATION
1) Start engine.
2) Make sure engine is warmed up sufficiently.
3) Make sure E.G.R. control valve spring responds to your touch (use your fingers) and also when engine is raced.
Does not respond
B CHECK VACUUM SOURCE TO E.G.R. CONTROL VALVE
1) Disconnect vacuum hose connected to E.G.R. control valve and B.P.T. valve.
2) Make sure vacuum exists when racing engine.
O.K.
Replace E.G.R. control valve.
N.G.
C CHECK VACUUM HOSE
Check vacuum hose for clogging, cracks and proper connections.
N.G.
If necessary, replace vacuum hose or reconnect vacuum hose firmly.
O.K.
D CHECK E.C.U. OUTPUT SIGNAL
1) Check voltage between E.C.U. terminal ④ and ground under the following conditions:
Engine condition | Voltage
Idle | Approximately 1.0V
Racing | Battery voltage
N.G.
E CHECK POWER SOURCE TO E.G.R. CONTROL SOLENOID VALVE
1) Stop engine.
2) Turn ignition switch "ON".
3) Check voltage between terminal ⓐ and ground. Battery voltage should exist.
F CHECK GROUND CIRCUIT
1) Turn ignition switch "OFF".
2) Disconnect E.C.U. 20-pin terminal connector.
3) Disconnect E.G.R. control solenoid valve harness connector.
4) Check resistance between E.C.U. terminal ④ and terminal ⓑ.
Resistance:
Approximately 0Ω
If N.G., repair or replace harness.
O.K.

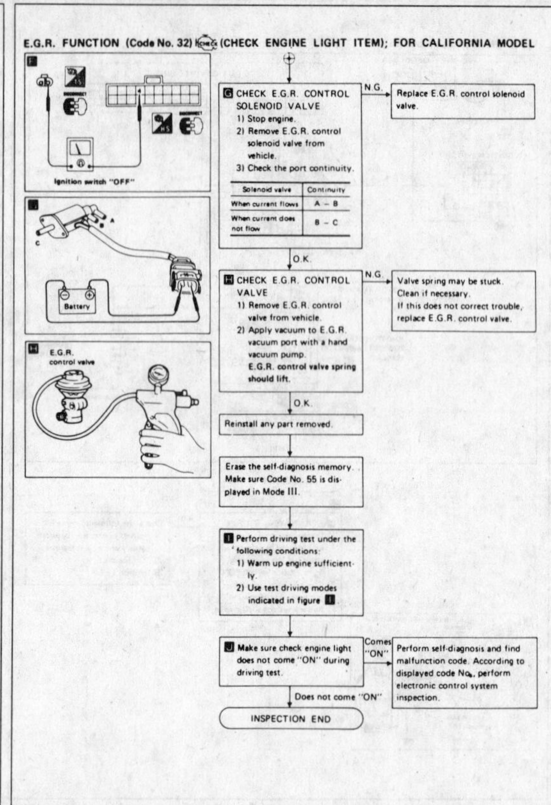
E.G.R. FUNCTION (Code No. 32) (CHECK ENGINE LIGHT ITEM); FOR CALIFORNIA MODEL
F
Ignition switch "OFF"
G
A
B
C
Battery
H
E.G.R. control valve
G CHECK E.G.R. CONTROL SOLENOID VALVE
1) Stop engine.
2) Remove E.G.R. control solenoid valve from vehicle.
3) Check the port continuity.
Solenoid valve | Continuity
When current flows | A – B
When current does not flow | B – C
N.G.
Replace E.G.R. control solenoid valve.
O.K.
H CHECK E.G.R. CONTROL VALVE
1) Remove E.G.R. control valve from vehicle.
2) Apply vacuum to E.G.R. vacuum port with a hand vacuum pump.
E.G.R. control valve spring should lift.
N.G.
Valve spring may be stuck. Clean if necessary. If this does not correct trouble, replace E.G.R. control valve.
O.K.
Reinstall any part removed.
Erase the self-diagnosis memory. Make sure Code No. 55 is displayed in Mode III.
I Perform driving test under the following conditions:
1) Warm up engine sufficiently.
2) Use test driving modes indicated in figure I
J Make sure check engine light does not come "ON" during driving test.
Comes "ON"
Perform self-diagnosis and find malfunction code. According to displayed code No., perform electronic control system inspection.
Does not come "ON"
INSPECTION END

ECCS SELF-DIAGNOSTIC CHART—1988 VAN

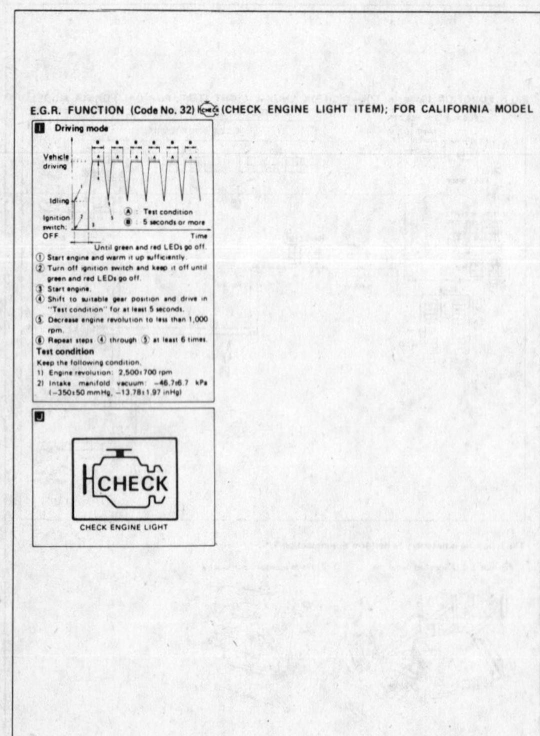
E.G.R. FUNCTION (Code No. 32) (CHECK ENGINE LIGHT ITEM); FOR CALIFORNIA MODEL
I Driving mode
Vehicle driving
Idling
Ignition switch: OFF
Ⓐ: Test condition
Ⓑ: 5 seconds or more
Time
Until green and red LEDs go off.
① Start engine and warm it up sufficiently.
② Turn off ignition switch and keep it off until green and red LEDs go off.
③ Start engine.
④ Shift to suitable gear position and drive in "Test condition" for at least 5 seconds.
⑤ Decrease engine revolution to less than 1,000 rpm.
⑥ Repeat steps ④ through ⑤ at least 6 times.
Test condition
Keep the following condition.
1) Engine revolution: 2,500±700 rpm
2) Intake manifold vacuum: −46.7±6.7 kPa (−350±50 mmHg, −13.78±1.97 inHg)
J
CHECK
CHECK ENGINE LIGHT

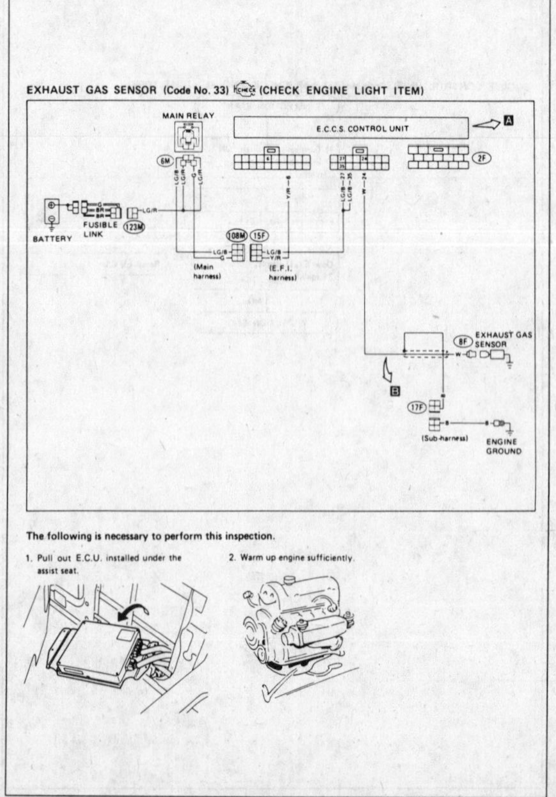
EXHAUST GAS SENSOR (Code No. 33) (CHECK ENGINE LIGHT ITEM)
MAIN RELAY
E.C.C.S. CONTROL UNIT
A
6M
2F
BATTERY
FUSIBLE LINK
123M
108M
15F
(Main harness)
(E.F.I. harness)
EXHAUST GAS SENSOR
8F
B
17F
(Sub-harness)
ENGINE GROUND
The following is necessary to perform this inspection.
1. Pull out E.C.U. installed under the assist seat.
2. Warm up engine sufficiently.

ECCS SELF-DIAGNOSTIC CHART—1988 VAN

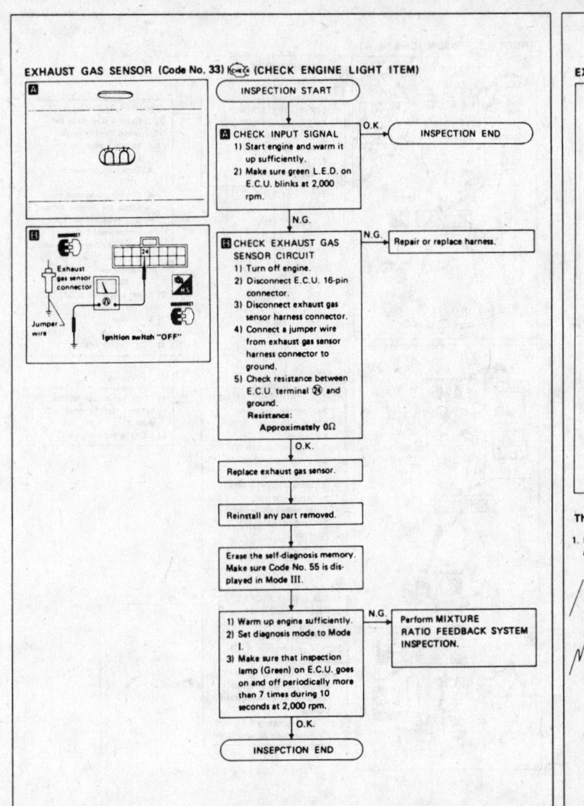
EXHAUST GAS SENSOR (Code No. 33) (CHECK ENGINE LIGHT ITEM)
INSPECTION START
A CHECK INPUT SIGNAL
1) Start engine and warm it up sufficiently.
2) Make sure green L.E.D. on E.C.U. blinks at 2,000 rpm.
O.K.
INSPECTION END
N.G.
B CHECK EXHAUST GAS SENSOR CIRCUIT
1) Turn off engine.
2) Disconnect E.C.U. 16-pin connector.
3) Disconnect exhaust gas sensor harness connector.
4) Connect a jumper wire from exhaust gas sensor harness connector to ground.
5) Check resistance between E.C.U. terminal 24 and ground.
Resistance:
Approximately 0Ω
N.G.
Repair or replace harness.
O.K.
Replace exhaust gas sensor.
Reinstall any part removed.
Erase the self-diagnosis memory. Make sure Code No. 55 is displayed in Mode III.
1) Warm up engine sufficiently.
2) Set diagnosis mode to Mode I.
3) Make sure that inspection lamp (Green) on E.C.U. goes on and off periodically more than 7 times during 10 seconds at 2,000 rpm.
N.G.
Perform MIXTURE RATIO FEEDBACK SYSTEM INSPECTION.
O.K.
INSPECTION END
Exhaust gas sensor connector
Jumper wire
Ignition switch "OFF"

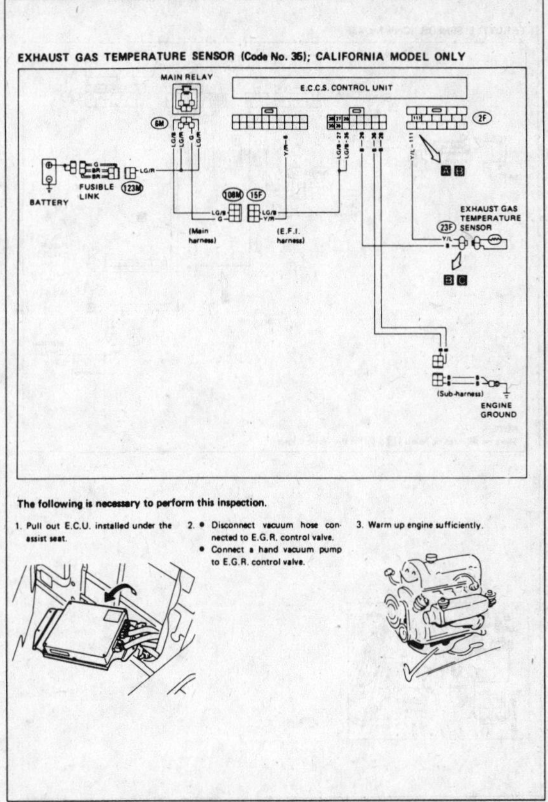
EXHAUST GAS TEMPERATURE SENSOR (Code No. 35); CALIFORNIA MODEL ONLY
MAIN RELAY
E.C.C.S. CONTROL UNIT
BATTERY
FUSIBLE LINK
(Main harness)
(E.F.I. harness)
EXHAUST GAS TEMPERATURE SENSOR
(Sub-harness)
ENGINE GROUND
The following is necessary to perform this inspection.
1. Pull out E.C.U. installed under the assist seat.
2. • Disconnect vacuum hose connected to E.G.R. control valve.
• Connect a hand vacuum pump to E.G.R. control valve.
3. Warm up engine sufficiently.

ECCS SELF-DIAGNOSTIC CHART—1988 VAN

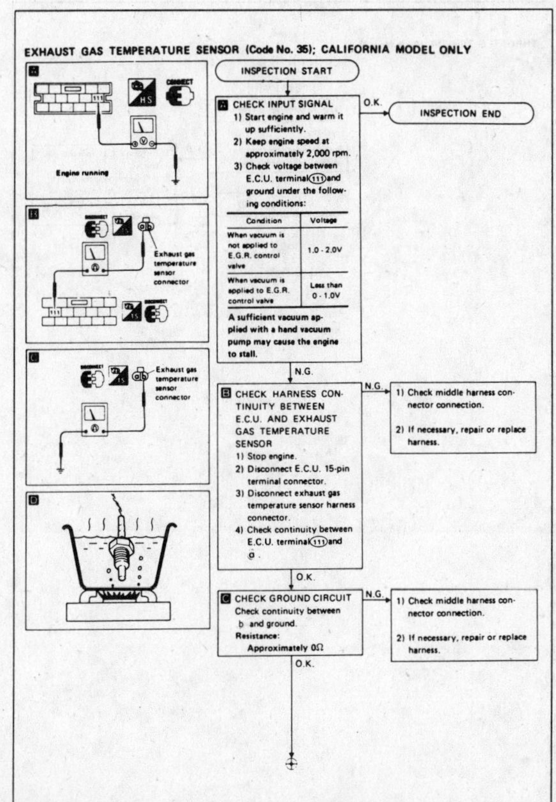
EXHAUST GAS TEMPERATURE SENSOR (Code No. 35); CALIFORNIA MODEL ONLY
INSPECTION START
A CHECK INPUT SIGNAL
1) Start engine and warm it up sufficiently.
2) Keep engine speed at approximately 2,000 rpm.
3) Check voltage between E.C.U. terminal 111 and ground under the following conditions:
Condition | Voltage
When vacuum is not applied to E.G.R. control valve | 1.0 - 2.0V
When vacuum is applied to E.G.R. control valve | Less than 0 - 1.0V
A sufficient vacuum applied with a hand vacuum pump may cause the engine to stall.
O.K.
INSPECTION END
N.G.
B CHECK HARNESS CONTINUITY BETWEEN E.C.U. AND EXHAUST GAS TEMPERATURE SENSOR
1) Stop engine.
2) Disconnect E.C.U. 15-pin terminal connector.
3) Disconnect exhaust gas temperature sensor harness connector.
4) Check continuity between E.C.U. terminal 111 and a.
N.G.
1) Check middle harness connector connection.
2) If necessary, repair or replace harness.
O.K.
C CHECK GROUND CIRCUIT
Check continuity between b and ground.
Resistance:
Approximately 0Ω
N.G.
1) Check middle harness connector connection.
2) If necessary, repair or replace harness.
O.K.
Engine running
Exhaust gas temperature sensor connector

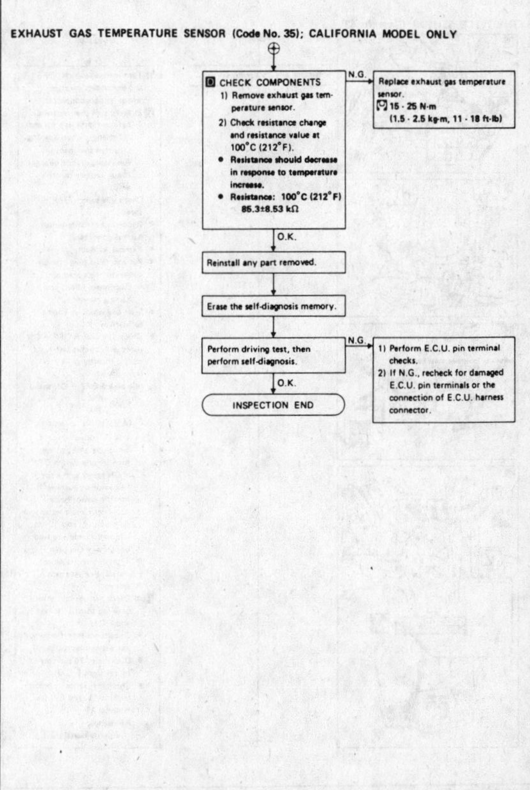
EXHAUST GAS TEMPERATURE SENSOR (Code No. 35); CALIFORNIA MODEL ONLY
D CHECK COMPONENTS
1) Remove exhaust gas temperature sensor.
2) Check resistance change and resistance value at 100°C (212°F).
• Resistance should decrease in response to temperature increase.
• Resistance: 100°C (212°F) 85.3±8.53 kΩ
N.G.
Replace exhaust gas temperature sensor.
15 - 25 N·m (1.5 - 2.5 kg-m, 11 - 18 ft-lb)
O.K.
Reinstall any part removed.
Erase the self-diagnosis memory.
Perform driving test, then perform self-diagnosis.
N.G.
1) Perform E.C.U. pin terminal checks.
2) If N.G., recheck for damaged E.C.U. pin terminals or the connection of E.C.U. harness connector.
O.K.
INSPECTION END

ECCS SELF-DIAGNOSTIC CHART—1988 VAN

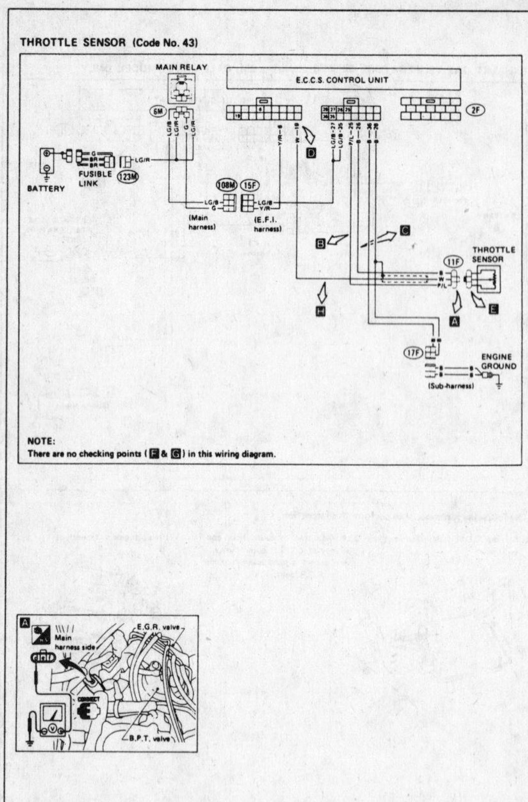
THROTTLE SENSOR (Code No. 43)
MAIN RELAY
E.C.C.S. CONTROL UNIT
BATTERY
FUSIBLE LINK
(Main harness)
(E.F.I. harness)
THROTTLE SENSOR
ENGINE GROUND
(Sub-harness)
NOTE:
There are no checking points (F & G) in this wiring diagram.
E.G.R. valve
Main harness side
B.P.T. valve

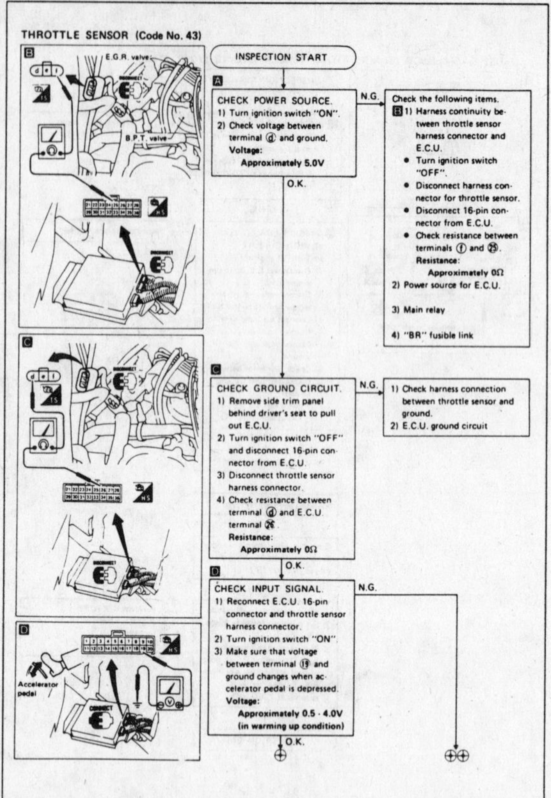
THROTTLE SENSOR (Code No. 43)
INSPECTION START
CHECK POWER SOURCE.
1) Turn ignition switch "ON".
2) Check voltage between terminal ⓓ and ground.
Voltage:
Approximately 5.0V
N.G.
Check the following items.
1) Harness continuity between throttle sensor harness connector and E.C.U.
• Turn ignition switch "OFF".
• Disconnect harness connector for throttle sensor.
• Disconnect 16-pin connector from E.C.U.
• Check resistance between terminals ⓕ and ㉕.
Resistance:
Approximately 0Ω
2) Power source for E.C.U.
3) Main relay
4) "BR" fusible link
O.K.
CHECK GROUND CIRCUIT.
1) Remove side trim panel behind driver's seat to pull out E.C.U.
2) Turn ignition switch "OFF" and disconnect 16-pin connector from E.C.U.
3) Disconnect throttle sensor harness connector.
4) Check resistance between terminal ⓓ and E.C.U. terminal ㉖.
Resistance:
Approximately 0Ω
N.G.
1) Check harness connection between throttle sensor and ground.
2) E.C.U. ground circuit
O.K.
CHECK INPUT SIGNAL.
1) Reconnect E.C.U. 16-pin connector and throttle sensor harness connector.
2) Turn ignition switch "ON".
3) Make sure that voltage between terminal ⑲ and ground changes when accelerator pedal is depressed.
Voltage:
Approximately 0.5 - 4.0V
(in warming up condition)
N.G.
O.K.
E.G.R. valve
B.P.T. valve
Accelerator pedal

ECCS SELF-DIAGNOSTIC CHART—1988 VAN

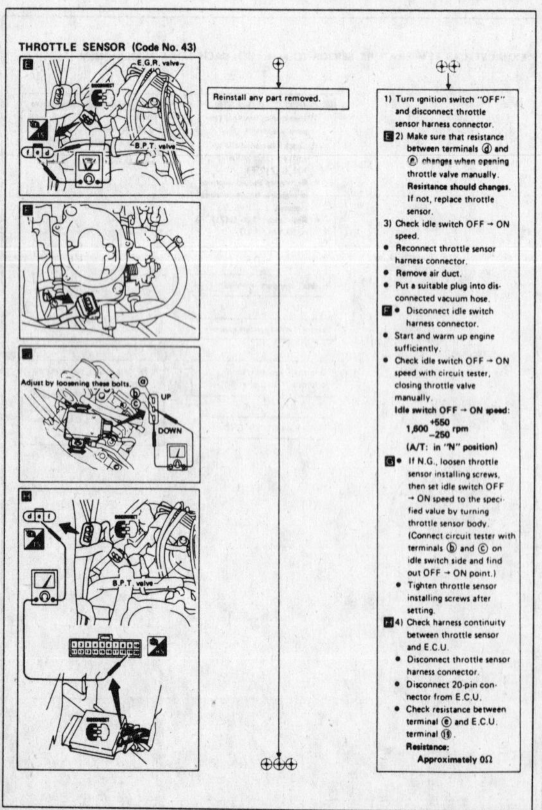
THROTTLE SENSOR (Code No. 43)
Reinstall any part removed.
1) Turn ignition switch "OFF" and disconnect throttle sensor harness connector.
2) Make sure that resistance between terminals ⓓ and ⓔ changes when opening throttle valve manually.
Resistance should changes.
If not, replace throttle sensor.
3) Check idle switch OFF → ON speed.
• Reconnect throttle sensor harness connector.
• Remove air duct.
• Put a suitable plug into disconnected vacuum hose.
• Disconnect idle switch harness connector.
• Start and warm up engine sufficiently.
• Check idle switch OFF → ON speed with circuit tester, closing throttle valve manually.
Idle switch OFF → ON speed:
1,800 +550 -250 rpm
(A/T: in "N" position)
• If N.G., loosen throttle sensor installing screws, then set idle switch OFF → ON speed to the specified value by turning throttle sensor body. (Connect circuit tester with terminals ⓑ and ⓒ on idle switch side and find out OFF → ON point.)
• Tighten throttle sensor installing screws after setting.
4) Check harness continuity between throttle sensor and E.C.U.
• Disconnect throttle sensor harness connector.
• Disconnect 20-pin connector from E.C.U.
• Check resistance between terminal ⓔ and E.C.U. terminal ⑲.
Resistance:
Approximately 0Ω
E.G.R. valve
B.P.T. valve
Adjust by loosening these bolts
UP
DOWN

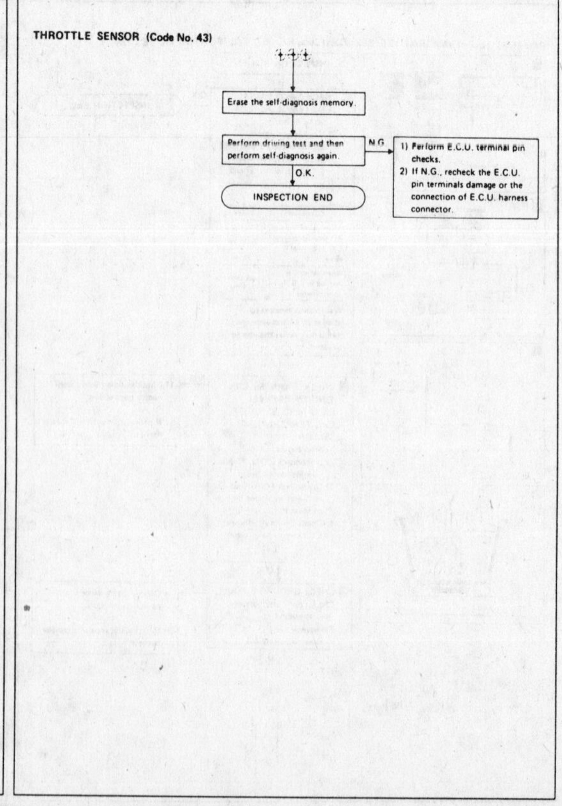
THROTTLE SENSOR (Code No. 43)
Erase the self-diagnosis memory.
Perform driving test and then perform self-diagnosis again.
N.G.
1) Perform E.C.U. terminal pin checks.
2) If N.G., recheck the E.C.U. pin terminals damage or the connection of E.C.U. harness connector.
O.K.
INSPECTION END

ECCS SELF-DIAGNOSTIC CHART—1988 VAN

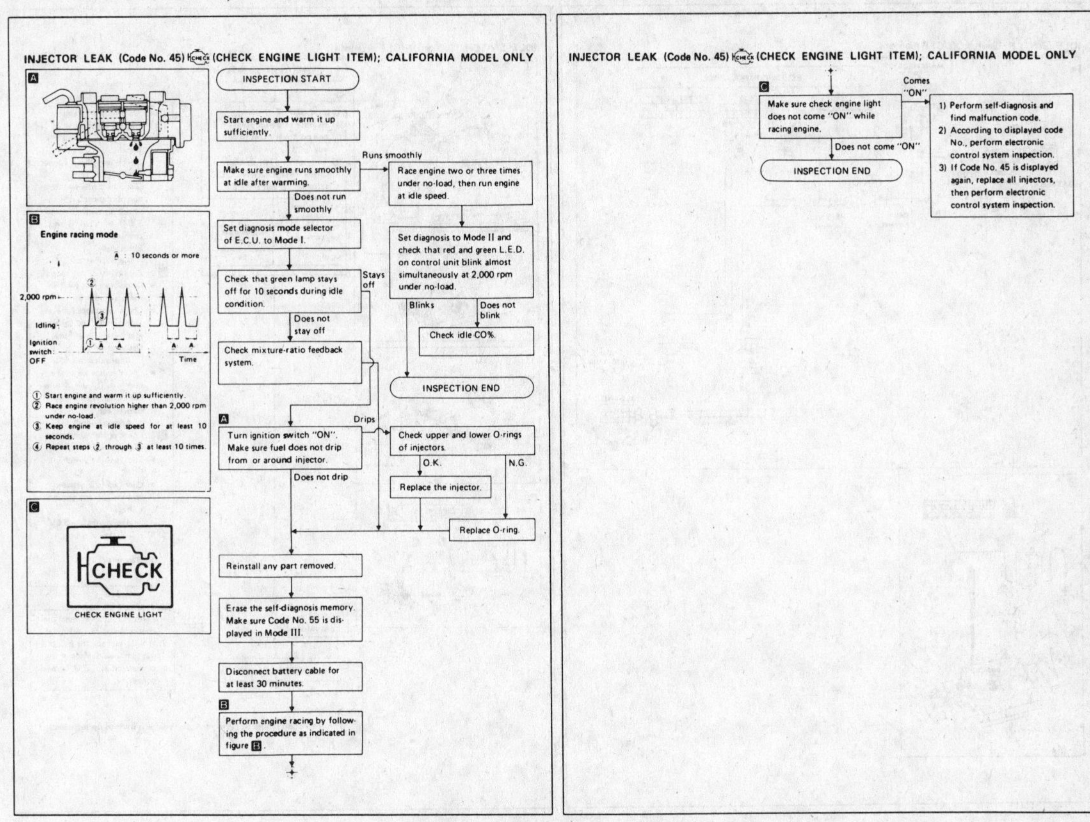
INJECTOR LEAK (Code No. 45) (CHECK ENGINE LIGHT ITEM); CALIFORNIA MODEL ONLY
INSPECTION START
Start engine and warm it up sufficiently.
Make sure engine runs smoothly at idle after warming.
Runs smoothly
Race engine two or three times under no-load, then run engine at idle speed.
Does not run smoothly
Set diagnosis mode selector of E.C.U. to Mode I.
Set diagnosis to Mode II and check that red and green L.E.D. on control unit blink almost simultaneously at 2,000 rpm under no-load.
Check that green lamp stays off for 10 seconds during idle condition.
Stays off
Blinks
Does not blink
Check idle CO%.
Does not stay off
Check mixture-ratio feedback system.
INSPECTION END
Drips
Turn ignition switch "ON". Make sure fuel does not drip from or around injector.
Check upper and lower O-rings of injectors.
O.K.
N.G.
Replace the injector.
Does not drip
Replace O-ring.
Reinstall any part removed.
Erase the self-diagnosis memory. Make sure Code No. 55 is displayed in Mode III.
Disconnect battery cable for at least 30 minutes.
Perform engine racing by following the procedure as indicated in figure B.
Engine racing mode
A: 10 seconds or more
2,000 rpm
Idling
Ignition switch: OFF
Time
① Start engine and warm it up sufficiently.
② Race engine revolution higher than 2,000 rpm under no-load.
③ Keep engine at idle speed for at least 10 seconds.
④ Repeat steps ② through ③ at least 10 times.
CHECK
CHECK ENGINE LIGHT
INJECTOR LEAK (Code No. 45) (CHECK ENGINE LIGHT ITEM); CALIFORNIA MODEL ONLY
Make sure check engine light does not come "ON" while racing engine.
Comes "ON"
1) Perform self-diagnosis and find malfunction code.
2) According to displayed code No., perform electronic control system inspection.
3) If Code No. 45 is displayed again, replace all injectors, then perform electronic control system inspection.
Does not come "ON"
INSPECTION END

ECCS SELF-DIAGNOSTIC CHART—1988 VAN

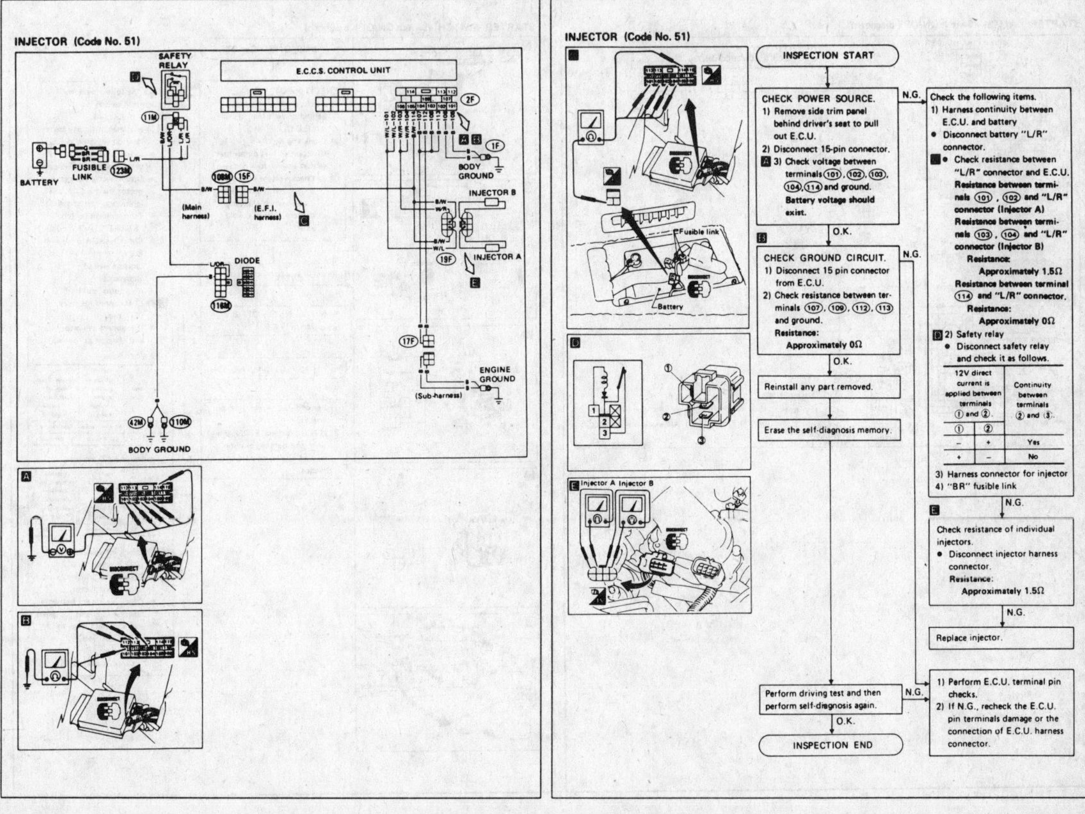
INJECTOR (Code No. 51)
SAFETY RELAY
E.C.C.S. CONTROL UNIT
BATTERY
FUSIBLE LINK
(Main harness)
(E.F.I. harness)
BODY GROUND
INJECTOR B
INJECTOR A
DIODE
ENGINE GROUND
(Sub-harness)
BODY GROUND
INJECTOR (Code No. 51)
INSPECTION START
CHECK POWER SOURCE.
1) Remove side trim panel behind driver's seat to pull out E.C.U.
2) Disconnect 15-pin connector.
3) Check voltage between terminals 101, 102, 103, 104, 114 and ground.
Battery voltage should exist.
N.G.
Check the following items.
1) Harness continuity between E.C.U. and battery
• Disconnect battery "L/R" connector.
• Check resistance between "L/R" connector and E.C.U.
Resistance between terminals 101, 102 and "L/R" connector (Injector A)
Resistance between terminals 103, 104 and "L/R" connector (Injector B)
Resistance: Approximately 1.5Ω
Resistance between terminal 114 and "L/R" connector.
Resistance: Approximately 0Ω
2) Safety relay
• Disconnect safety relay and check it as follows.
12V direct current is applied between terminals ① and ②.
Continuity between terminals ② and ③.
Yes
No
3) Harness connector for injector
4) "BR" fusible link
O.K.
CHECK GROUND CIRCUIT.
1) Disconnect 15 pin connector from E.C.U.
2) Check resistance between terminals 107, 109, 112, 113 and ground.
Resistance: Approximately 0Ω
Fusible link
Battery
Reinstall any part removed.
Erase the self-diagnosis memory.
Injector A Injector B
Check resistance of individual injectors.
• Disconnect injector harness connector.
Resistance: Approximately 1.5Ω
Replace injector.
Perform driving test and then perform self-diagnosis again.
1) Perform E.C.U. terminal pin checks.
2) If N.G., recheck the E.C.U. pin terminals damage or the connection of E.C.U. harness connector.
INSPECTION END

ECCS SELF-DIAGNOSTIC CHART—1988 VAN

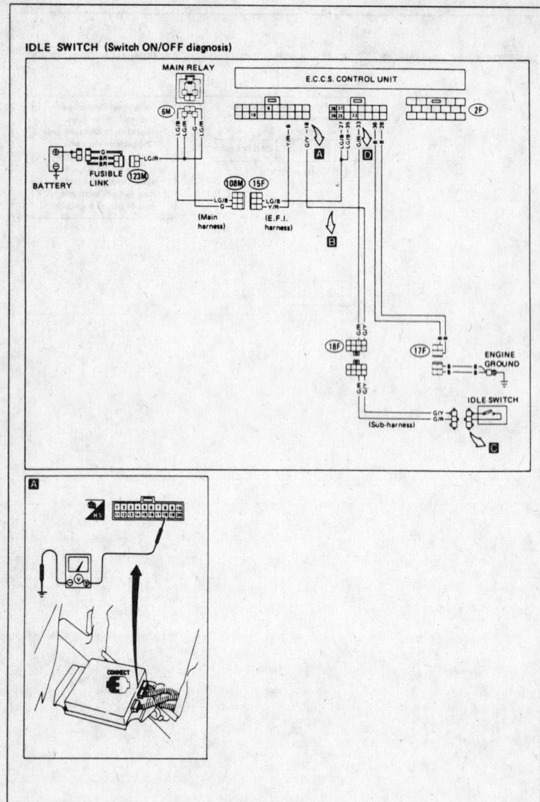
IDLE SWITCH (Switch ON/OFF diagnosis)
MAIN RELAY
E.C.C.S. CONTROL UNIT
BATTERY
FUSIBLE LINK
(Main harness)
(E.F.I. harness)
ENGINE GROUND
IDLE SWITCH
(Sub-harness)

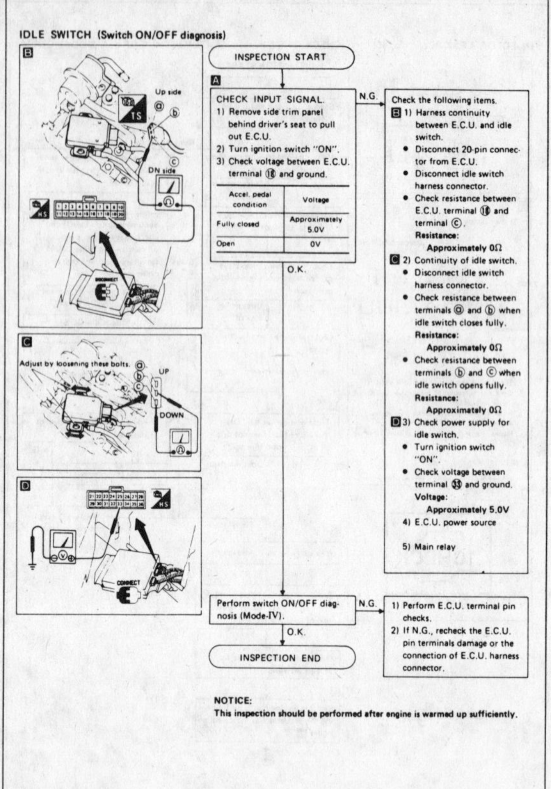
IDLE SWITCH (Switch ON/OFF diagnosis)
INSPECTION START
CHECK INPUT SIGNAL.
1) Remove side trim panel behind driver's seat to pull out E.C.U.
2) Turn ignition switch "ON".
3) Check voltage between E.C.U. terminal ⑱ and ground.
Accel. pedal condition | Voltage
Fully closed | Approximately 5.0V
Open | 0V
N.G.
Check the following items.
B 1) Harness continuity between E.C.U. and idle switch.
• Disconnect 20-pin connector from E.C.U.
• Disconnect idle switch harness connector.
• Check resistance between E.C.U. terminal ⑱ and terminal ⓒ.
Resistance: Approximately 0Ω
C 2) Continuity of idle switch.
• Disconnect idle switch harness connector.
• Check resistance between terminals ⓐ and ⓑ when idle switch closes fully.
Resistance: Approximately 0Ω
• Check resistance between terminals ⓑ and ⓒ when idle switch opens fully.
Resistance: Approximately 0Ω
D 3) Check power supply for idle switch.
• Turn ignition switch "ON".
• Check voltage between terminal ㉝ and ground.
Voltage: Approximately 5.0V
4) E.C.U. power source
5) Main relay
O.K.
Perform switch ON/OFF diagnosis (Mode-IV).
N.G.
1) Perform E.C.U. terminal pin checks.
2) If N.G., recheck the E.C.U. pin terminals damage or the connection of E.C.U. harness connector.
O.K.
INSPECTION END
Up side
DN side
Adjust by loosening these bolts.
UP
DOWN
NOTICE:
This inspection should be performed after engine is warmed up sufficiently.

ECCS SELF-DIAGNOSTIC CHART—1988 VAN

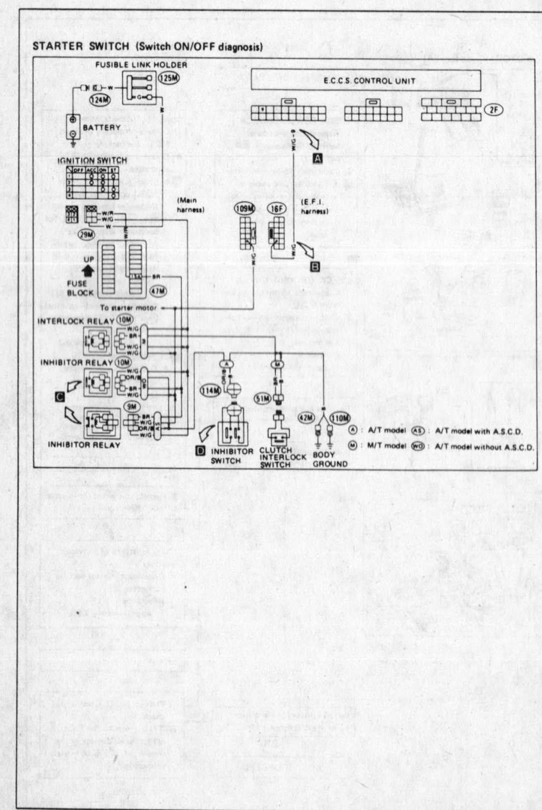
STARTER SWITCH (Switch ON/OFF diagnosis)
FUSIBLE LINK HOLDER
E.C.C.S. CONTROL UNIT
BATTERY
IGNITION SWITCH
(Main harness)
(E.F.I. harness)
UP
FUSE BLOCK
To starter motor
INTERLOCK RELAY
INHIBITOR RELAY
INHIBITOR RELAY
INHIBITOR SWITCH
CLUTCH INTERLOCK SWITCH
BODY GROUND
Ⓐ: A/T model
Ⓜ: M/T model
A/T model with A.S.C.D.
A/T model without A.S.C.D.

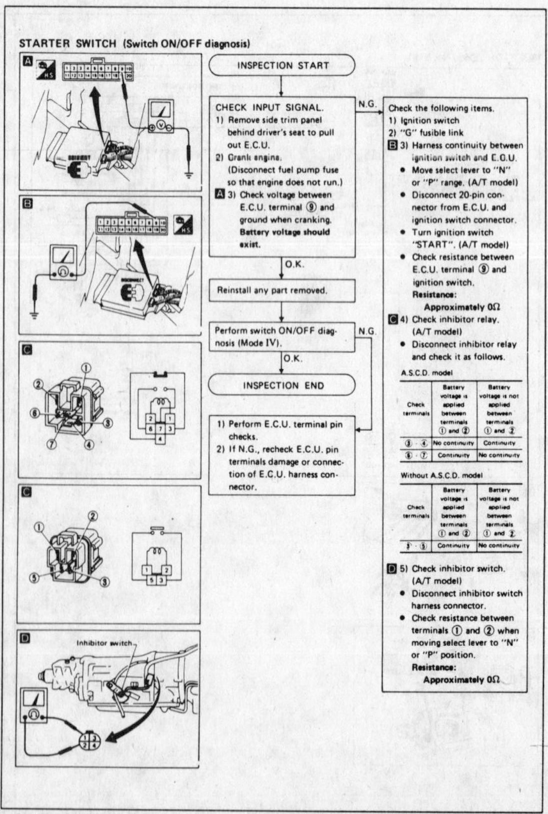
STARTER SWITCH (Switch ON/OFF diagnosis)
INSPECTION START
CHECK INPUT SIGNAL.
1) Remove side trim panel behind driver's seat to pull out E.C.U.
2) Crank engine. (Disconnect fuel pump fuse so that engine does not run.)
A 3) Check voltage between E.C.U. terminal ⑨ and ground when cranking. Battery voltage should exist.
N.G.
Check the following items.
1) Ignition switch
2) "G" fusible link
B 3) Harness continuity between ignition switch and E.C.U.
• Move select lever to "N" or "P" range. (A/T model)
• Disconnect 20-pin connector from E.C.U. and ignition switch connector.
• Turn ignition switch "START". (A/T model)
• Check resistance between E.C.U. terminal ⑨ and ignition switch.
Resistance: Approximately 0Ω
C 4) Check inhibitor relay. (A/T model)
• Disconnect inhibitor relay and check it as follows.
A.S.C.D. model
Check terminals | Battery voltage is applied between terminals ① and ② | Battery voltage is not applied between terminals ① and ②
③-④ | No continuity | Continuity
⑥-⑦ | Continuity | No continuity
Without A.S.C.D. model
Check terminals | Battery voltage is applied between terminals ① and ② | Battery voltage is not applied between terminals ① and ②
③-⑤ | Continuity | No continuity
D 5) Check inhibitor switch. (A/T model)
• Disconnect inhibitor switch harness connector.
• Check resistance between terminals ① and ② when moving select lever to "N" or "P" position.
Resistance: Approximately 0Ω
O.K.
Reinstall any part removed.
Perform switch ON/OFF diagnosis (Mode IV).
N.G.
1) Perform E.C.U. terminal pin checks.
2) If N.G., recheck E.C.U. pin terminals damage or connection of E.C.U. harness connector.
O.K.
INSPECTION END
Inhibitor switch

ECCS SELF-DIAGNOSTIC CHART—1988 VAN

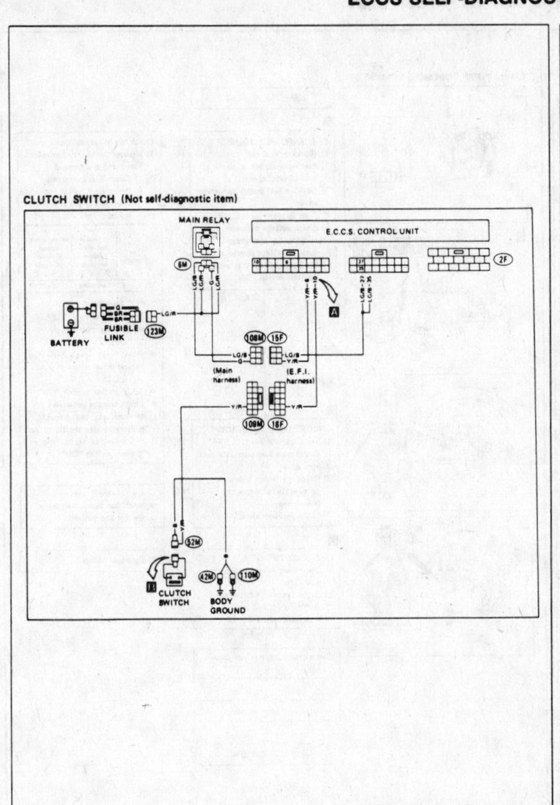
CLUTCH SWITCH (Not self-diagnostic item)
MAIN RELAY
E.C.C.S. CONTROL UNIT
BATTERY
FUSIBLE LINK
(Main harness)
(E.F.I. harness)
CLUTCH SWITCH
BODY GROUND

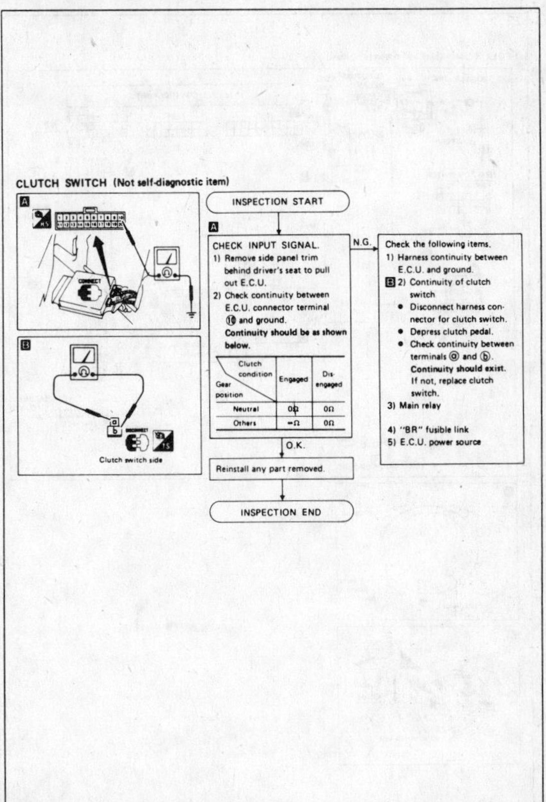
CLUTCH SWITCH (Not self-diagnostic item)
INSPECTION START
A
CHECK INPUT SIGNAL.
1) Remove side panel trim behind driver's seat to pull out E.C.U.
2) Check continuity between E.C.U. connector terminal ⑩ and ground.
Continuity should be as shown below.
Clutch condition / Gear position | Engaged | Dis-engaged
Neutral | 0Ω | 0Ω
Others | ∞Ω | 0Ω
N.G.
Check the following items.
1) Harness continuity between E.C.U. and ground.
B 2) Continuity of clutch switch
• Disconnect harness connector for clutch switch.
• Depress clutch pedal.
• Check continuity between terminals ⓐ and ⓑ.
Continuity should exist.
If not, replace clutch switch.
3) Main relay
4) "BR" fusible link
5) E.C.U. power source
O.K.
Reinstall any part removed.
INSPECTION END
Clutch switch side

ECCS SELF-DIAGNOSTIC CHART—1988 VAN

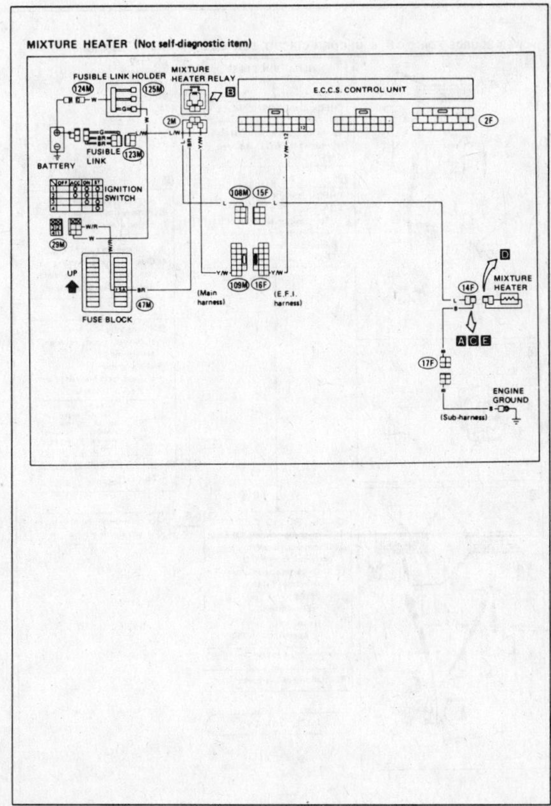
MIXTURE HEATER (Not self-diagnostic item)
FUSIBLE LINK HOLDER
MIXTURE HEATER RELAY
E.C.C.S. CONTROL UNIT
BATTERY
FUSIBLE LINK
IGNITION SWITCH
FUSE BLOCK
UP
(Main harness)
(E.F.I. harness)
MIXTURE HEATER
ENGINE GROUND
(Sub-harness)

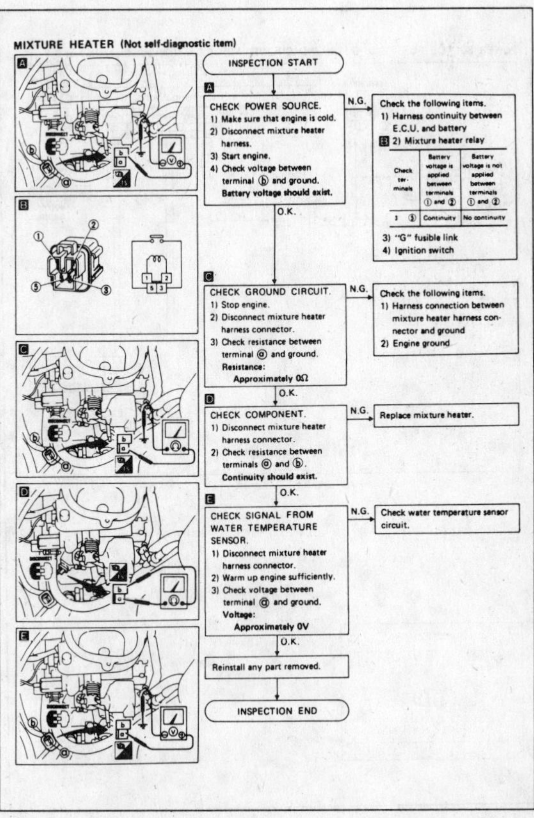
MIXTURE HEATER (Not self-diagnostic item)
INSPECTION START
A
CHECK POWER SOURCE.
1) Make sure that engine is cold.
2) Disconnect mixture heater harness.
3) Start engine.
4) Check voltage between terminal ⓑ and ground.
Battery voltage should exist.
N.G.
Check the following items.
1) Harness continuity between E.C.U. and battery
B 2) Mixture heater relay
Check terminals | Battery voltage is applied between terminals ① and ② | Battery voltage is not applied between terminals ① and ②
3 ⑤ | Continuity | No continuity
3) "G" fusible link
4) Ignition switch
O.K.
C
CHECK GROUND CIRCUIT.
1) Stop engine.
2) Disconnect mixture heater harness connector.
3) Check resistance between terminal ⓐ and ground.
Resistance:
Approximately 0Ω
N.G.
Check the following items.
1) Harness connection between mixture heater harness connector and ground
2) Engine ground
O.K.
D
CHECK COMPONENT.
1) Disconnect mixture heater harness connector.
2) Check resistance between terminals ⓐ and ⓑ.
Continuity should exist.
N.G.
Replace mixture heater.
O.K.
E
CHECK SIGNAL FROM WATER TEMPERATURE SENSOR.
1) Disconnect mixture heater harness connector.
2) Warm up engine sufficiently.
3) Check voltage between terminal ⓐ and ground.
Voltage:
Approximately 0V
N.G.
Check water temperature sensor circuit.
O.K.
Reinstall any part removed.
INSPECTION END

ECCS SELF-DIAGNOSTIC CHART—1988 VAN

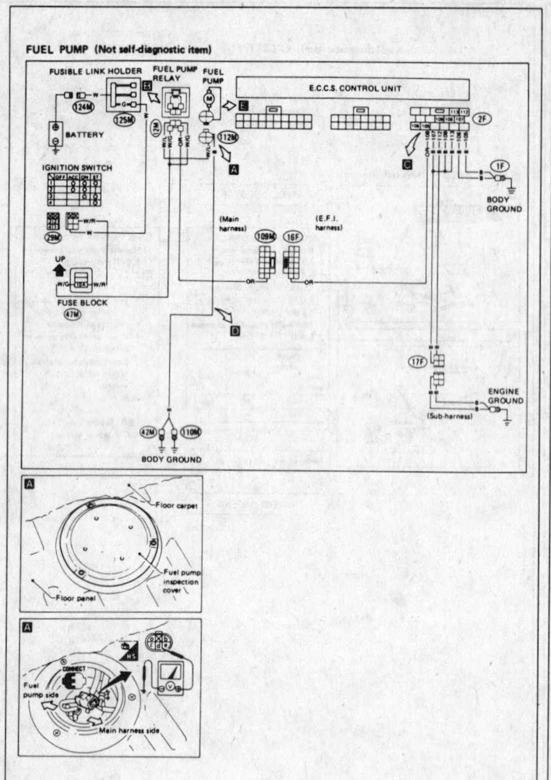
FUEL PUMP (Not self-diagnostic item)
FUSIBLE LINK HOLDER
FUEL PUMP RELAY
FUEL PUMP
E.C.C.S. CONTROL UNIT
BATTERY
IGNITION SWITCH
BODY GROUND
(Main harness)
(E.F.I. harness)
UP
FUSE BLOCK
ENGINE GROUND
(Sub-harness)
BODY GROUND
Floor carpet
Fuel pump inspection cover
Floor panel
Fuel pump side
Main harness side

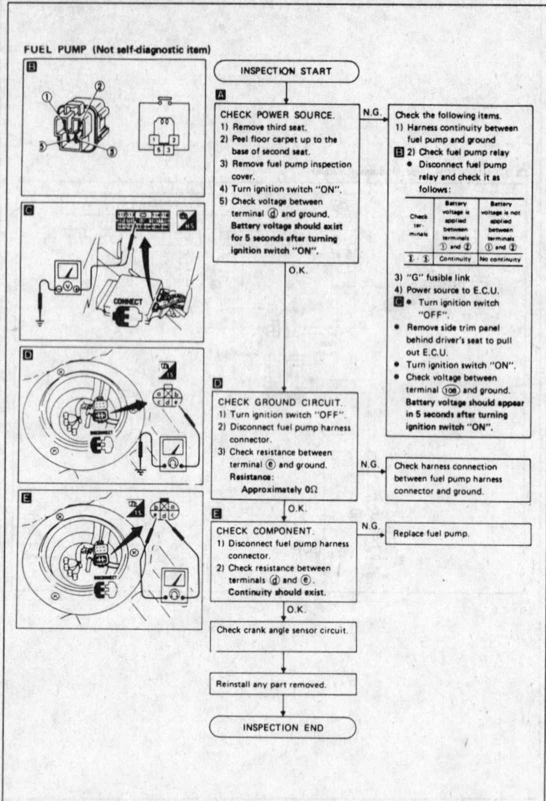
FUEL PUMP (Not self-diagnostic item)
INSPECTION START
A
CHECK POWER SOURCE.
1) Remove third seat.
2) Peel floor carpet up to the base of second seat.
3) Remove fuel pump inspection cover.
4) Turn ignition switch "ON".
5) Check voltage between terminal ⓓ and ground.
Battery voltage should exist for 5 seconds after turning ignition switch "ON".
N.G.
Check the following items.
1) Harness continuity between fuel pump and ground
B 2) Check fuel pump relay
• Disconnect fuel pump relay and check it as follows:
Check terminals
Battery voltage is applied between terminals ① and ②
Battery voltage is not applied between terminals ① and ②
③ · ⑤
Continuity
No continuity
3) "G" fusible link
4) Power source to E.C.U.
C • Turn ignition switch "OFF".
• Remove side trim panel behind driver's seat to pull out E.C.U.
• Turn ignition switch "ON".
• Check voltage between terminal (108) and ground.
Battery voltage should appear in 5 seconds after turning ignition switch "ON".
O.K.
D
CHECK GROUND CIRCUIT.
1) Turn ignition switch "OFF".
2) Disconnect fuel pump harness connector.
3) Check resistance between terminal ⓔ and ground.
Resistance:
Approximately 0Ω
N.G.
Check harness connection between fuel pump harness connector and ground.
O.K.
E
CHECK COMPONENT.
1) Disconnect fuel pump harness connector.
2) Check resistance between terminals ⓓ and ⓔ.
Continuity should exist.
N.G.
Replace fuel pump.
O.K.
Check crank angle sensor circuit.
Reinstall any part removed.
INSPECTION END

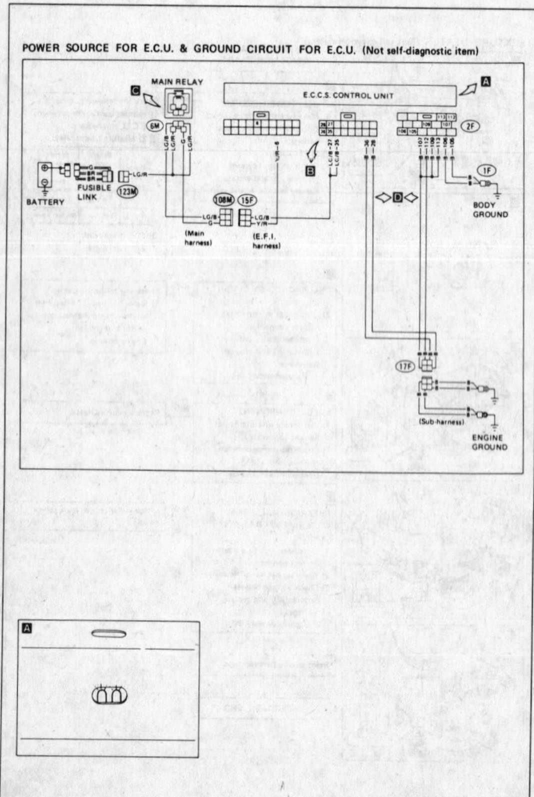
POWER SOURCE FOR E.C.U. & GROUND CIRCUIT FOR E.C.U. (Not self-diagnostic item)
MAIN RELAY
E.C.C.S. CONTROL UNIT
BATTERY
FUSIBLE LINK
(Main harness)
(E.F.I. harness)
BODY GROUND
(Sub-harness)
ENGINE GROUND

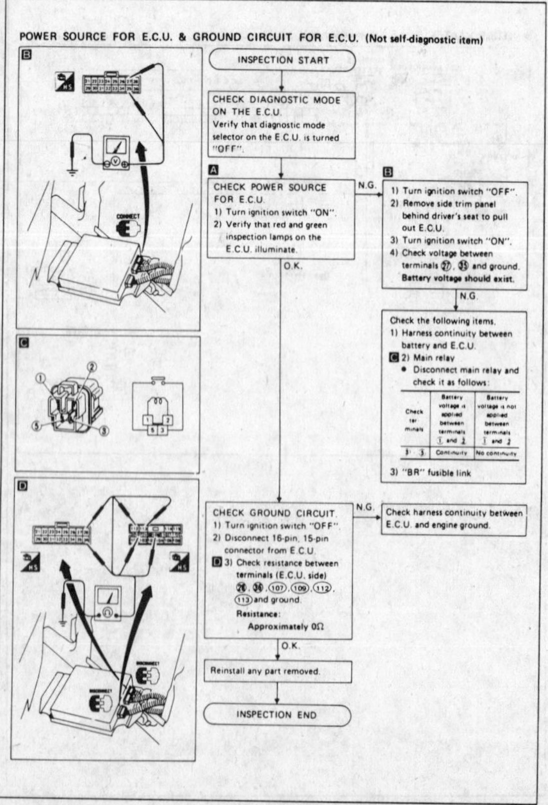
POWER SOURCE FOR E.C.U. & GROUND CIRCUIT FOR E.C.U. (Not self-diagnostic item)
INSPECTION START
CHECK DIAGNOSTIC MODE ON THE E.C.U.
Verify that diagnostic mode selector on the E.C.U. is turned "OFF".
A
CHECK POWER SOURCE FOR E.C.U.
1) Turn ignition switch "ON".
2) Verify that red and green inspection lamps on the E.C.U. illuminate.
N.G.
B
1) Turn ignition switch "OFF".
2) Remove side trim panel behind driver's seat to pull out E.C.U.
3) Turn ignition switch "ON".
4) Check voltage between terminals ㉗, ㉟ and ground.
Battery voltage should exist.
N.G.
Check the following items.
1) Harness continuity between battery and E.C.U.
C 2) Main relay
• Disconnect main relay and check it as follows:
Check terminals
Battery voltage is applied between terminals ① and ②
Battery voltage is not applied between terminals ① and ②
③ · ⑤
Continuity
No continuity
3) "BR" fusible link
O.K.
CHECK GROUND CIRCUIT.
1) Turn ignition switch "OFF".
2) Disconnect 16-pin, 15-pin connector from E.C.U.
D 3) Check resistance between terminals (E.C.U. side) ㉘, ㊱, (107), (109), (112), (113) and ground.
Resistance:
Approximately 0Ω
N.G.
Check harness continuity between E.C.U. and engine ground.
O.K.
Reinstall any part removed.
INSPECTION END

ECCS SELF-DIAGNOSTIC CHART—1988 VAN

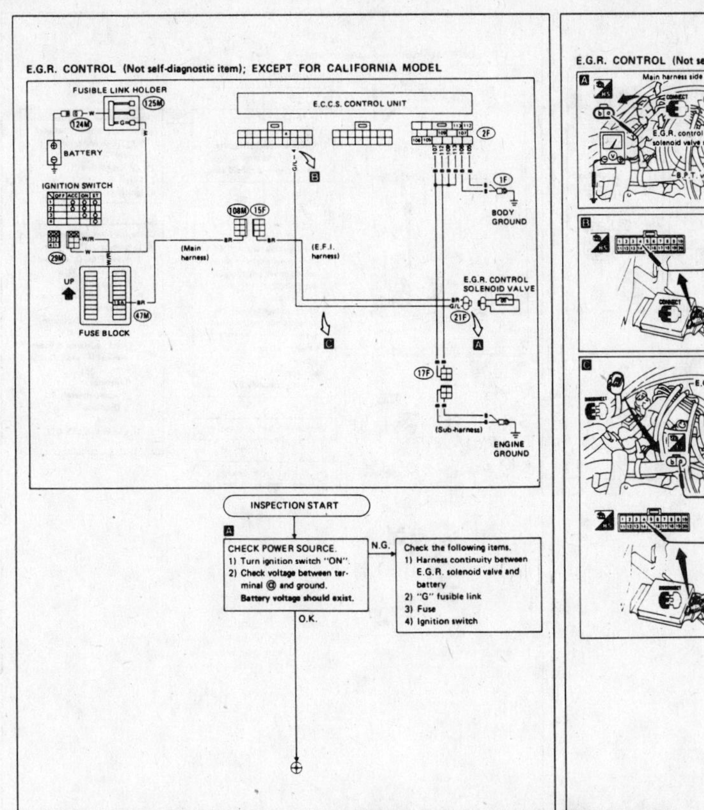
E.G.R. CONTROL (Not self-diagnostic item); EXCEPT FOR CALIFORNIA MODEL
FUSIBLE LINK HOLDER
BATTERY
IGNITION SWITCH
E.C.C.S. CONTROL UNIT
BODY GROUND
(Main harness)
(E.F.I. harness)
E.G.R. CONTROL SOLENOID VALVE
FUSE BLOCK
(Sub-harness)
ENGINE GROUND
INSPECTION START
CHECK POWER SOURCE.
1) Turn ignition switch "ON".
2) Check voltage between terminal ⓐ and ground.
Battery voltage should exist.
N.G.
Check the following items.
1) Harness continuity between E.G.R. solenoid valve and battery
2) "G" fusible link
3) Fuse
4) Ignition switch
O.K.

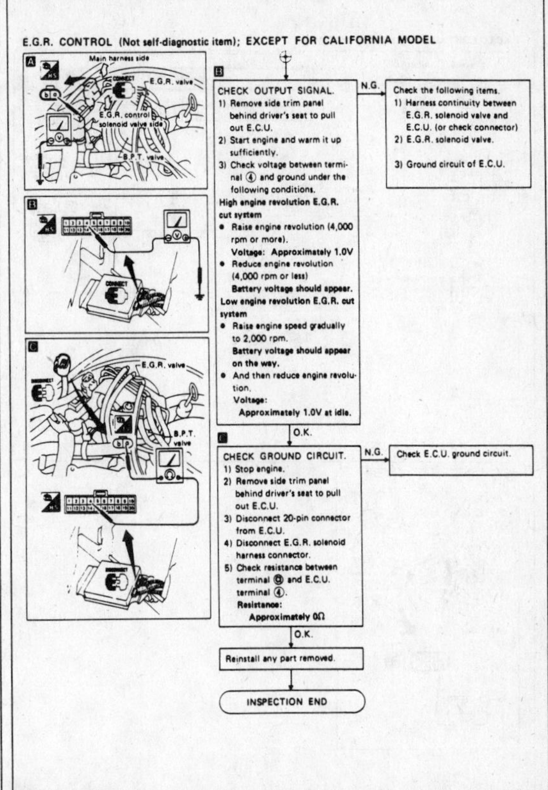
E.G.R. CONTROL (Not self-diagnostic item); EXCEPT FOR CALIFORNIA MODEL
Main harness side
E.G.R. valve
E.G.R. control solenoid valve side
B.P.T. valve
CHECK OUTPUT SIGNAL.
1) Remove side trim panel behind driver's seat to pull out E.C.U.
2) Start engine and warm it up sufficiently.
3) Check voltage between terminal ④ and ground under the following conditions.
High engine revolution E.G.R. cut system
• Raise engine revolution (4,000 rpm or more).
Voltage: Approximately 1.0V
• Reduce engine revolution (4,000 rpm or less)
Battery voltage should appear.
Low engine revolution E.G.R. cut system
• Raise engine speed gradually to 2,000 rpm.
Battery voltage should appear on the way.
• And then reduce engine revolution.
Voltage:
Approximately 1.0V at idle.
N.G.
Check the following items.
1) Harness continuity between E.G.R. solenoid valve and E.C.U. (or check connector)
2) E.G.R. solenoid valve.
3) Ground circuit of E.C.U.
O.K.
CHECK GROUND CIRCUIT.
1) Stop engine.
2) Remove side trim panel behind driver's seat to pull out E.C.U.
3) Disconnect 20-pin connector from E.C.U.
4) Disconnect E.G.R. solenoid harness connector.
5) Check resistance between terminal ⓑ and E.C.U. terminal ④.
Resistance:
Approximately 0Ω
N.G.
Check E.C.U. ground circuit.
O.K.
Reinstall any part removed.
INSPECTION END

ECCS SELF-DIAGNOSTIC CHART—1988 VAN

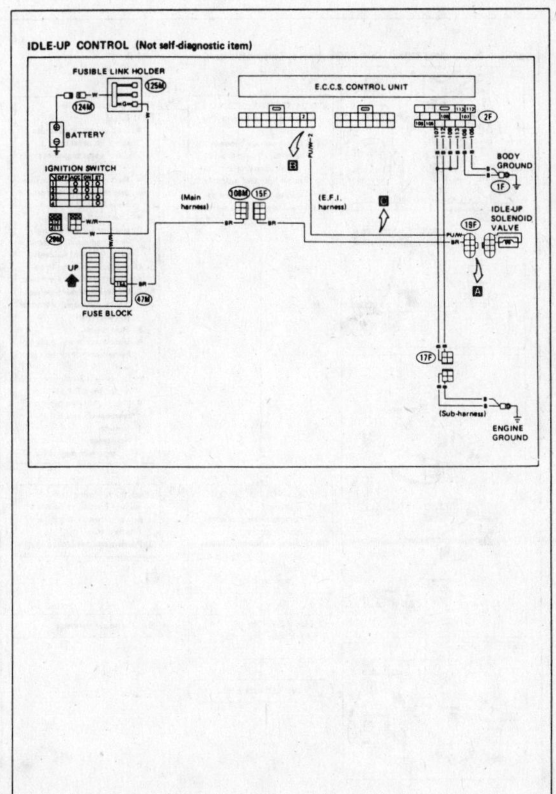
IDLE-UP CONTROL (Not self-diagnostic item)
FUSIBLE LINK HOLDER
BATTERY
IGNITION SWITCH
E.C.C.S. CONTROL UNIT
BODY GROUND
(Main harness)
(E.F.I. harness)
IDLE-UP SOLENOID VALVE
FUSE BLOCK
(Sub-harness)
ENGINE GROUND

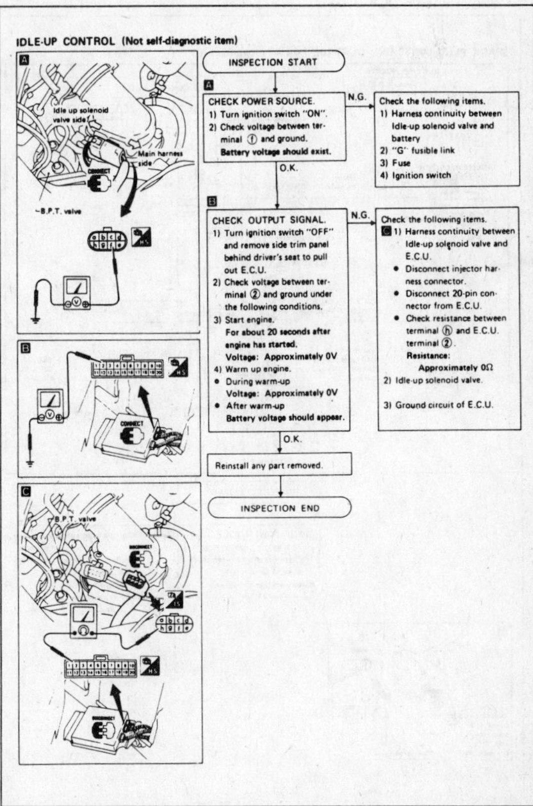
IDLE-UP CONTROL (Not self-diagnostic item)
Idle up solenoid valve side
Main harness side
B.P.T. valve
INSPECTION START
CHECK POWER SOURCE.
1) Turn ignition switch "ON".
2) Check voltage between terminal ① and ground.
Battery voltage should exist.
N.G.
Check the following items.
1) Harness continuity between Idle-up solenoid valve and battery
2) "G" fusible link
3) Fuse
4) Ignition switch
O.K.
CHECK OUTPUT SIGNAL.
1) Turn ignition switch "OFF" and remove side trim panel behind driver's seat to pull out E.C.U.
2) Check voltage between terminal ② and ground under the following conditions.
3) Start engine.
For about 20 seconds after engine has started.
Voltage: Approximately 0V
4) Warm up engine.
• During warm-up
Voltage: Approximately 0V
• After warm-up
Battery voltage should appear.
N.G.
Check the following items.
1) Harness continuity between Idle-up solenoid valve and E.C.U.
• Disconnect injector harness connector.
• Disconnect 20-pin connector from E.C.U.
• Check resistance between terminal ⓗ and E.C.U. terminal ②.
Resistance:
Approximately 0Ω
2) Idle-up solenoid valve.
3) Ground circuit of E.C.U.
O.K.
Reinstall any part removed.
INSPECTION END
B.P.T. valve

ECCS SELF-DIAGNOSTIC CHART—1988 VAN

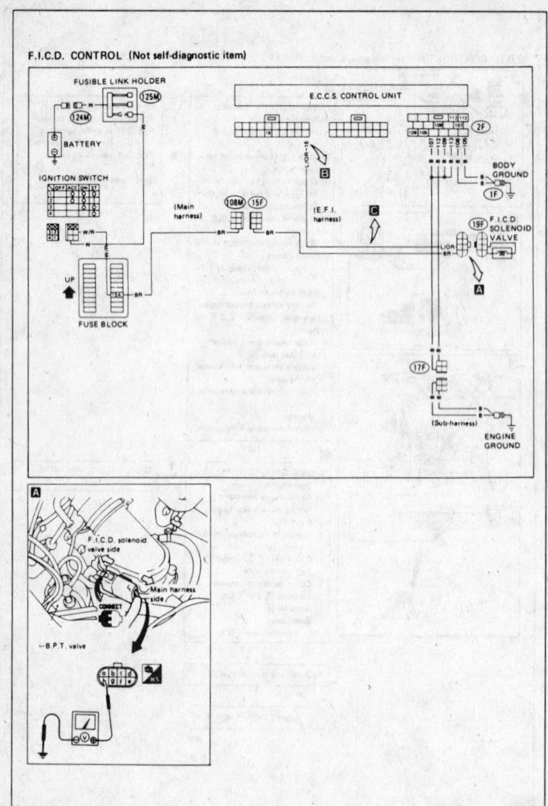
F.I.C.D. CONTROL (Not self-diagnostic item)
FUSIBLE LINK HOLDER
125M
124M
BATTERY
IGNITION SWITCH
E.C.C.S. CONTROL UNIT
2F
BODY GROUND
1F
(Main harness)
108M
15F
(E.F.I. harness)
19F
F.I.C.D. SOLENOID VALVE
UP
FUSE BLOCK
17F
(Sub-harness)
ENGINE GROUND
F.I.C.D. solenoid valve side
Main harness side
CONNECT
B.P.T. valve

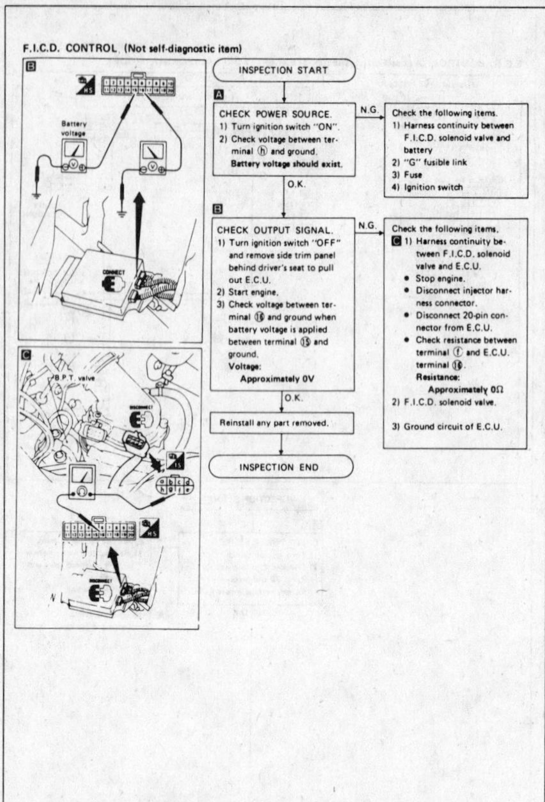
F.I.C.D. CONTROL (Not self-diagnostic item)
Battery voltage
CONNECT
B.P.T. valve
INSPECTION START
A
CHECK POWER SOURCE.
1) Turn ignition switch "ON".
2) Check voltage between terminal ⓗ and ground.
Battery voltage should exist.
N.G.
Check the following items.
1) Harness continuity between F.I.C.D. solenoid valve and battery
2) "G" fusible link
3) Fuse
4) Ignition switch
O.K.
B
CHECK OUTPUT SIGNAL.
1) Turn ignition switch "OFF" and remove side trim panel behind driver's seat to pull out E.C.U.
2) Start engine.
3) Check voltage between terminal ⑯ and ground when battery voltage is applied between terminal ⑮ and ground.
Voltage:
Approximately 0V
N.G.
Check the following items.
C 1) Harness continuity between F.I.C.D. solenoid valve and E.C.U.
• Stop engine.
• Disconnect injector harness connector.
• Disconnect 20-pin connector from E.C.U.
• Check resistance between terminal ⓕ and E.C.U. terminal ⑯.
Resistance:
Approximately 0Ω
2) F.I.C.D. solenoid valve.
3) Ground circuit of E.C.U.
O.K.
Reinstall any part removed.
INSPECTION END

ECCS SELF-DIAGNOSTIC CHART—1988 VAN

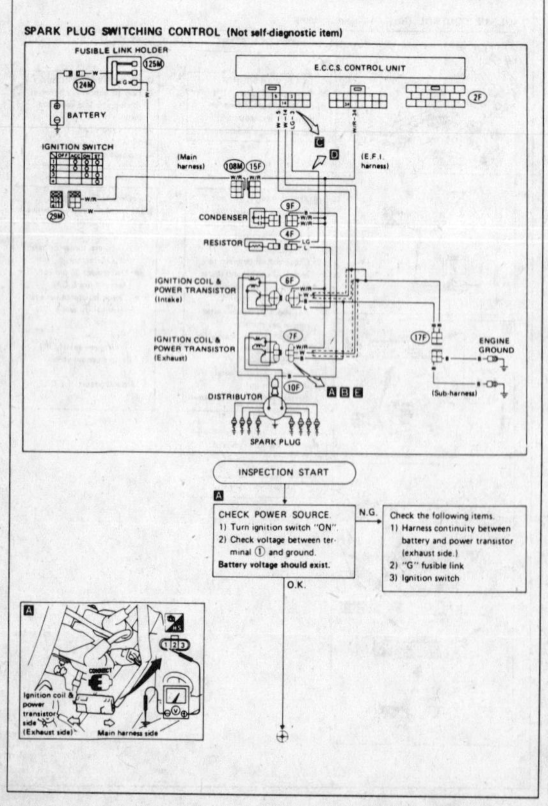
SPARK PLUG SWITCHING CONTROL (Not self-diagnostic item)
FUSIBLE LINK HOLDER
125M
124M
BATTERY
IGNITION SWITCH
E.C.C.S. CONTROL UNIT
2F
(Main harness)
108M
15F
(E.F.I. harness)
29M
CONDENSER
9F
RESISTOR
4F
IGNITION COIL & POWER TRANSISTOR (Intake)
6F
IGNITION COIL & POWER TRANSISTOR (Exhaust)
7F
17F
ENGINE GROUND
(Sub-harness)
DISTRIBUTOR
10F
SPARK PLUG
INSPECTION START
A
CHECK POWER SOURCE.
1) Turn ignition switch "ON".
2) Check voltage between terminal ① and ground.
Battery voltage should exist.
N.G.
Check the following items.
1) Harness continuity between battery and power transistor (exhaust side.)
2) "G" fusible link
3) Ignition switch
O.K.
CONNECT
Ignition coil & power transistor side (Exhaust side)
Main harness side

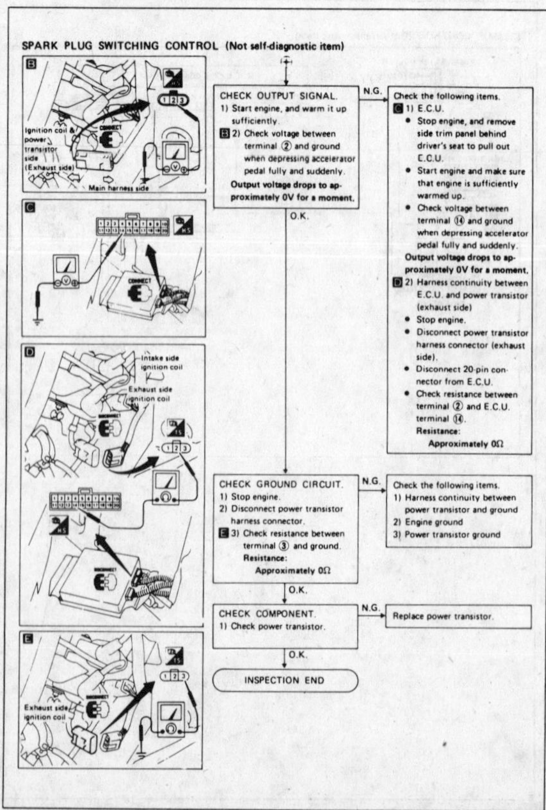
SPARK PLUG SWITCHING CONTROL (Not self-diagnostic item)
Ignition coil & power transistor side (Exhaust side)
Main harness side
CHECK OUTPUT SIGNAL.
1) Start engine, and warm it up sufficiently.
B 2) Check voltage between terminal ② and ground when depressing accelerator pedal fully and suddenly.
Output voltage drops to approximately 0V for a moment.
N.G.
Check the following items.
C 1) E.C.U.
• Stop engine, and remove side trim panel behind driver's seat to pull out E.C.U.
• Start engine and make sure that engine is sufficiently warmed up.
• Check voltage between terminal ⑭ and ground when depressing accelerator pedal fully and suddenly.
Output voltage drops to approximately 0V for a moment.
D 2) Harness continuity between E.C.U. and power transistor (exhaust side)
• Stop engine.
• Disconnect power transistor harness connector (exhaust side).
• Disconnect 20-pin connector from E.C.U.
• Check resistance between terminal ② and E.C.U. terminal ⑭.
Resistance:
Approximately 0Ω
O.K.
Intake side ignition coil
Exhaust side ignition coil
CHECK GROUND CIRCUIT.
1) Stop engine.
2) Disconnect power transistor harness connector.
E 3) Check resistance between terminal ③ and ground.
Resistance:
Approximately 0Ω
N.G.
Check the following items.
1) Harness continuity between power transistor and ground
2) Engine ground
3) Power transistor ground
O.K.
CHECK COMPONENT.
1) Check power transistor.
N.G.
Replace power transistor.
O.K.
INSPECTION END
Exhaust side ignition coil

ECU TERMINAL INSPECTION—1988 VAN

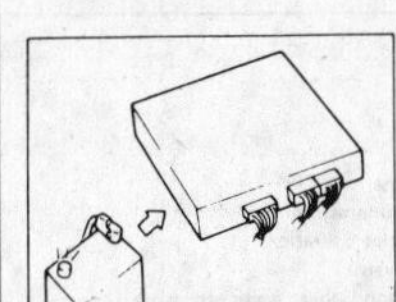

MEASUREMENT VOLTAGE OR RESISTANCE OF E.C.U.

1. Disconnect battery ground cable.
2. Remove assist side or bench seat from vehicle.
3. Disconnect 20- and 16-pin connectors from E.C.U.
4. Remove pin terminal retainer from 20- and 16-pin connectors to make it easier to insert tester probes.

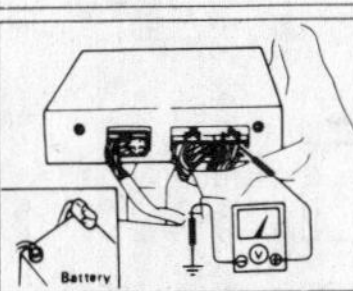

5. Connect 20- and 16-pin connectors to E.C.U. carefully.
6. Connect battery ground cable.
7. Measure the voltage at each terminal by following "E.C.U. inspection table".

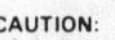

CAUTION:

a. **Perform all voltage measurements with the connectors connected.**
b. **Perform all resistance measurements with the connectors disconnected.**
c. **Make sure that there is not any bends or breaks on E.C.U. pin terminal before measurements.**
d. Do not touch tester probes between terminals ㉗ and ㉘, ㉟ and ㊱.

E.C.U. inspection table

*Data are reference values.

TERMINAL NO.	ITEM	CONDITION	*DATA
2	Idle-up solenoid	Engine is running. —For about 20 seconds after starting engine. —During warming up	Approximately 1.0V
		Engine is running. —Except the conditions shown above	BATTERY VOLTAGE (11 - 14V)
3	Ignition check	Engine is running. —Idle speed	9 - 12V (Decreases as engine revs up.)
4	E.G.R. control solenoid valve	Engine is running. —High engine revolutions —Idle speed	Approximately 1.0V
		Engine is running. —Except the above	BATTERY VOLTAGE (11 - 14V)
5	Ignition signal (Intake side)	Engine is running. —Idle speed	Approximately 1.0V
		Engine is running. —Engine speed is 2,000 rpm.	Approximately 0.8V
6	Main relay	Engine is running. ↓ Ignition switch "OFF" —For 16 seconds after turning off ignition switch.	0.8 - 1.0V
		Ignition switch "OFF" —In 16 seconds after turning off ignition switch.	BATTERY VOLTAGE (11 - 14V)
8	Crank angle sensor (Position signal)	Engine is running. **Do not run engine at high speed under no-load.**	2.5 - 2.7V
9	Start signal	Cranking	8 - 12V

*Data are reference values.

TERMINAL NO.	ITEM	CONDITION	*DATA
10	Neutral/clutch switch or Inhibitor switch	Ignition switch "ON" —Neutral/Parking	0V
		Ignition switch "ON" —Except the above gear position	Approximately 5V (0V: with clutch disengaged)
12	Mixture heater relay	Engine is running. —Engine is cold or during warm up. [Water temperature is below 70°C (158°F).]	0.7 - 0.9V
		Engine is running. —After warming up. [Water temperature is above 70°C (158°F).]	BATTERY VOLTAGE (11 - 14V)
14	Ignition signal (Exhaust side)	Engine is running. —Idle speed	0.2 - 0.4V
		Engine is running. —Engine speed is 2,000 rpm.	Approximately 0.8V
15	Air conditioner relay	Engine is running. —Air conditioner magnet clutch switch is "OFF".	0V
		Engine is running. —Air conditioner magnet clutch switch is "ON".	BATTERY VOLTAGE (11 - 14V)
16	F.I.C.D. solenoid valve	Engine is running. —Air conditioner magnet clutch switch is "OFF".	BATTERY VOLTAGE (11 - 14V)
		Engine is running. —Air conditioner magnet clutch switch is "ON".	0.7 - 0.9V
17	Crank angle sensor (Reference signal)	Engine is running. **Do not run engine at high speed under no-load.**	0.2 - 0.4V

*Data are reference values.

TERMINAL NO.	ITEM	CONDITION	*DATA
18	Idle switch (⊖ side)	Ignition switch "ON" —Throttle valve: idle position **Inspection should be done after warming up engine sufficiently.**	Approximately 5.0V
		Ignition switch "ON" —Throttle valve: except idle position	0V
19	Throttle sensor (Output line)	Ignition switch "ON" **Inspection should be done after warming up engine sufficiently.**	0.5 - 4.0V Output voltage varies with the throttle valve opening angle.
20	A.I.V. control solenoid valve	Engine is running. —Engine is cold. [Water temperature is below 15°C (59°F).]	0.8 - 0.9V
		Engine is running. —During warming up [Water temperature is between 15°C (59°F) and 40°C (104°F).]	BATTERY VOLTAGE (11 - 14V)
		Engine is running. —After warming up. [Water temperature is above 40°C (104°F).] Idle condition after 3,000 rpm no-load driving for 10 seconds.	0.8 - 0.9V
		When depressing accelerator pedal at the condition above.	BATTERY VOLTAGE (11 - 14V)
23	Water temperature sensor	Engine is running	1.0 - 5.0V Output voltage varies with water temperature.
24	Exhaust gas sensor	Engine is running. —After warming up sufficiently.	0 - Approximately 1.0V
25	Throttle sensor (⊕ side)	Engine is running.	Approximately 5.0V
33	Idle switch (⊕ side)	Ignition switch "ON"	Approximately 5.0V

ECU TERMINAL INSPECTION—1988 VAN

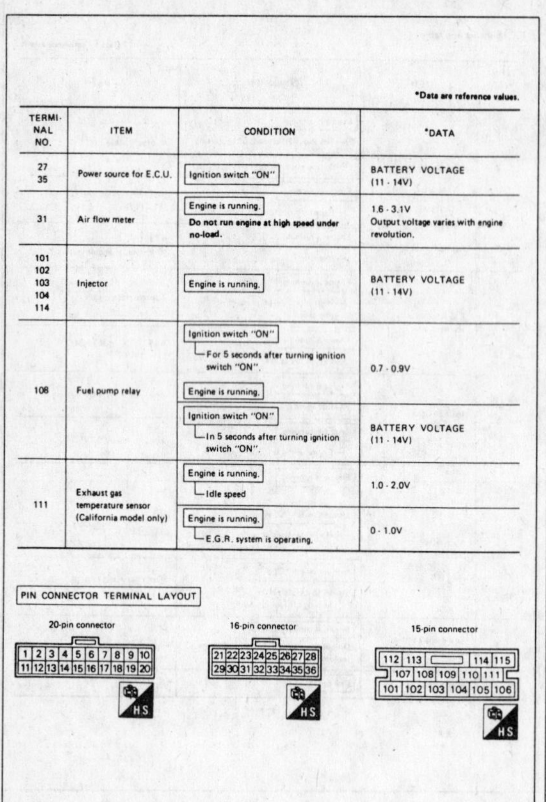

*Data are reference values.

TERMINAL NO.	ITEM	CONDITION	*DATA
27 35	Power source for E.C.U.	Ignition switch "ON"	BATTERY VOLTAGE (11 - 14V)
31	Air flow meter	Engine is running. Do not run engine at high speed under no-load.	1.6 - 3.1V Output voltage varies with engine revolution.
101 102 103 104 114	Injector	Engine is running.	BATTERY VOLTAGE (11 - 14V)
108	Fuel pump relay	Ignition switch "ON" — For 5 seconds after turning ignition switch "ON". Engine is running.	0.7 - 0.9V
		Ignition switch "ON" — In 5 seconds after turning ignition switch "ON".	BATTERY VOLTAGE (11 - 14V)
111	Exhaust gas temperature sensor (California model only)	Engine is running. — Idle speed	1.0 - 2.0V
		Engine is running. — E.G.R. system is operating.	0 - 1.0V

PIN CONNECTOR TERMINAL LAYOUT

20-pin connector
1 2 3 4 5 6 7 8 9 10
11 12 13 14 15 16 17 18 19 20
16-pin connector
21 22 23 24 25 26 27 28
29 30 31 32 33 34 35 36
15-pin connector
112 113 114 115
107 108 109 110 111
101 102 103 104 105 106
H.S.

ECCS MIXTURE RATIO FEEDBACK SYSTEM INSPECTION—1988 VAN

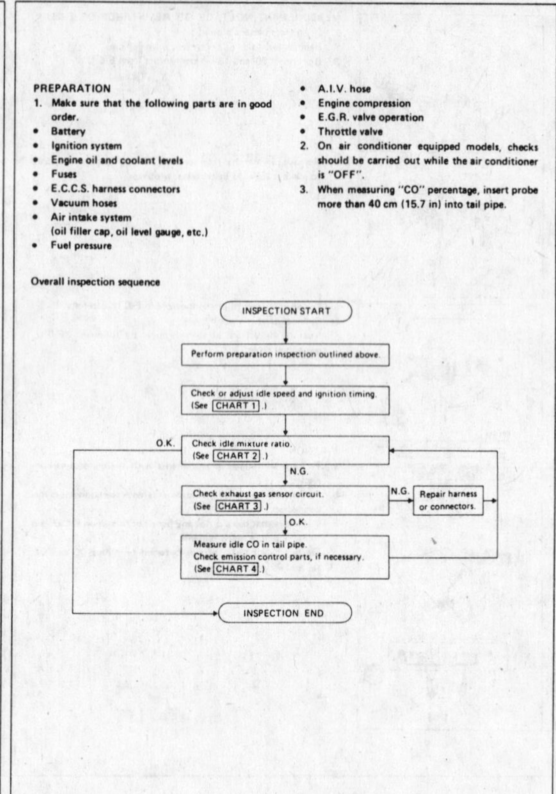

PREPARATION

1. Make sure that the following parts are in good order.
   - Battery
   - Ignition system
   - Engine oil and coolant levels
   - Fuses
   - E.C.C.S. harness connectors
   - Vacuum hoses
   - Air intake system (oil filler cap, oil level gauge, etc.)
   - Fuel pressure
   - A.I.V. hose
   - Engine compression
   - E.G.R. valve operation
   - Throttle valve
2. On air conditioner equipped models, checks should be carried out while the air conditioner is "OFF".
3. When measuring "CO" percentage, insert probe more than 40 cm (15.7 in) into tail pipe.

Overall inspection sequence

INSPECTION START
Perform preparation inspection outlined above.
Check or adjust idle speed and ignition timing. (See CHART 1.)
O.K.
Check idle mixture ratio. (See CHART 2.)
N.G.
Check exhaust gas sensor circuit. (See CHART 3.)
N.G.
Repair harness or connectors.
O.K.
Measure idle CO in tail pipe. Check emission control parts, if necessary. (See CHART 4.)
INSPECTION END

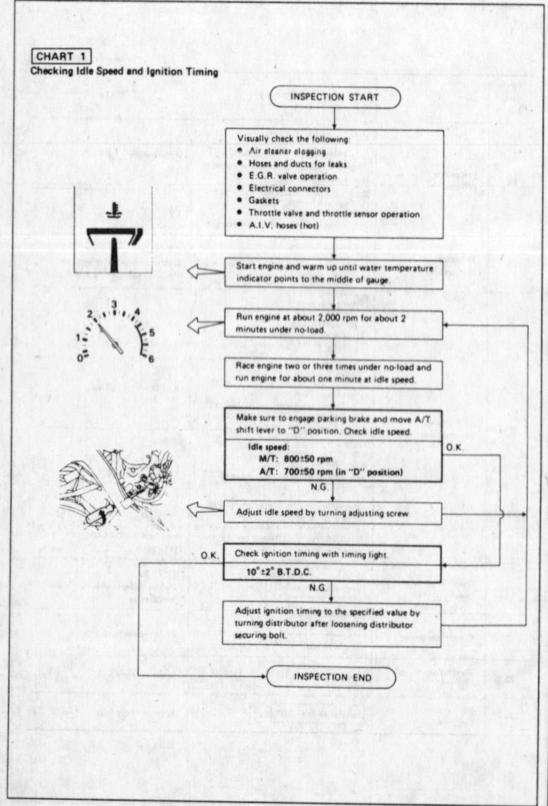

CHART 1

Checking Idle Speed and Ignition Timing

INSPECTION START
Visually check the following:
• Air cleaner clogging
• Hoses and ducts for leaks
• E.G.R. valve operation
• Electrical connectors
• Gaskets
• Throttle valve and throttle sensor operation
• A.I.V. hoses (hot)
Start engine and warm up until water temperature indicator points to the middle of gauge.
Run engine at about 2,000 rpm for about 2 minutes under no-load.
0 1 2 3 4 5 6
Race engine two or three times under no-load and run engine for about one minute at idle speed.
Make sure to engage parking brake and move A/T shift lever to "D" position. Check idle speed.
Idle speed:
M/T: 800±50 rpm
A/T: 700±50 rpm (in "D" position)
O.K.
N.G.
Adjust idle speed by turning adjusting screw.
O.K.
Check ignition timing with timing light.
10°±2° B.T.D.C.
N.G.
Adjust ignition timing to the specified value by turning distributor after loosening distributor securing bolt.
INSPECTION END

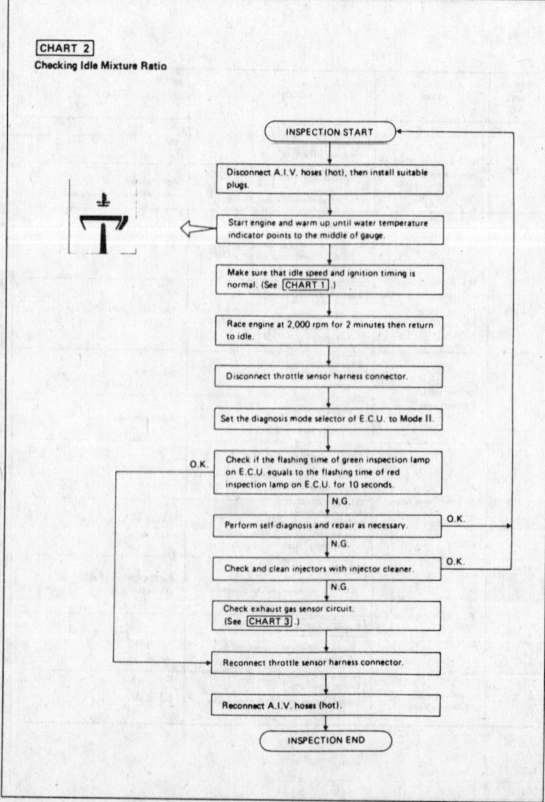

CHART 2

Checking Idle Mixture Ratio

INSPECTION START
Disconnect A.I.V. hoses (hot), then install suitable plugs.
Start engine and warm up until water temperature indicator points to the middle of gauge.
Make sure that idle speed and ignition timing is normal. (See CHART 1.)
Race engine at 2,000 rpm for 2 minutes then return to idle.
Disconnect throttle sensor harness connector.
Set the diagnosis mode selector of E.C.U. to Mode II.
O.K.
Check if the flashing time of green inspection lamp on E.C.U. equals to the flashing time of red inspection lamp on E.C.U. for 10 seconds.
N.G.
Perform self-diagnosis and repair as necessary.
O.K.
N.G.
Check and clean injectors with injector cleaner.
O.K.
N.G.
Check exhaust gas sensor circuit. (See CHART 3.)
Reconnect throttle sensor harness connector.
Reconnect A.I.V. hoses (hot).
INSPECTION END

ECCS MIXTURE RATIO FEEDBACK SYSTEM INSPECTION—1988 VAN

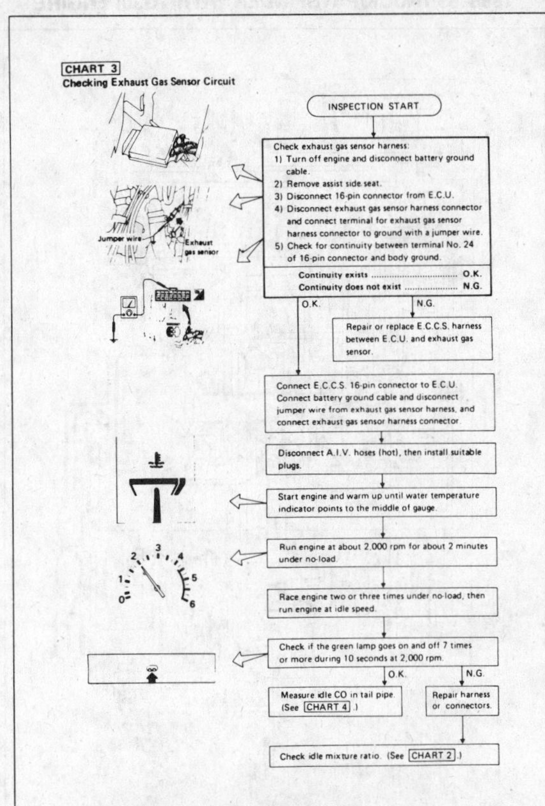
CHART 3
Checking Exhaust Gas Sensor Circuit
INSPECTION START
Check exhaust gas sensor harness:
1) Turn off engine and disconnect battery ground cable.
2) Remove assist side seat.
3) Disconnect 16-pin connector from E.C.U.
4) Disconnect exhaust gas sensor harness connector and connect terminal for exhaust gas sensor harness connector to ground with a jumper wire.
5) Check for continuity between terminal No. 24 of 16-pin connector and body ground.
Continuity exists .... O.K.
Continuity does not exist .... N.G.
Jumper wire
Exhaust gas sensor
O.K.
N.G.
Repair or replace E.C.C.S. harness between E.C.U. and exhaust gas sensor.
Connect E.C.C.S. 16-pin connector to E.C.U. Connect battery ground cable and disconnect jumper wire from exhaust gas sensor harness, and connect exhaust gas sensor harness connector.
Disconnect A.I.V. hoses (hot), then install suitable plugs.
Start engine and warm up until water temperature indicator points to the middle of gauge.
Run engine at about 2,000 rpm for about 2 minutes under no-load.
0
1
2
3
4
5
6
Race engine two or three times under no-load, then run engine at idle speed.
Check if the green lamp goes on and off 7 times or more during 10 seconds at 2,000 rpm.
O.K.
N.G.
Measure idle CO in tail pipe. (See CHART 4.)
Repair harness or connectors.
Check idle mixture ratio. (See CHART 2.)

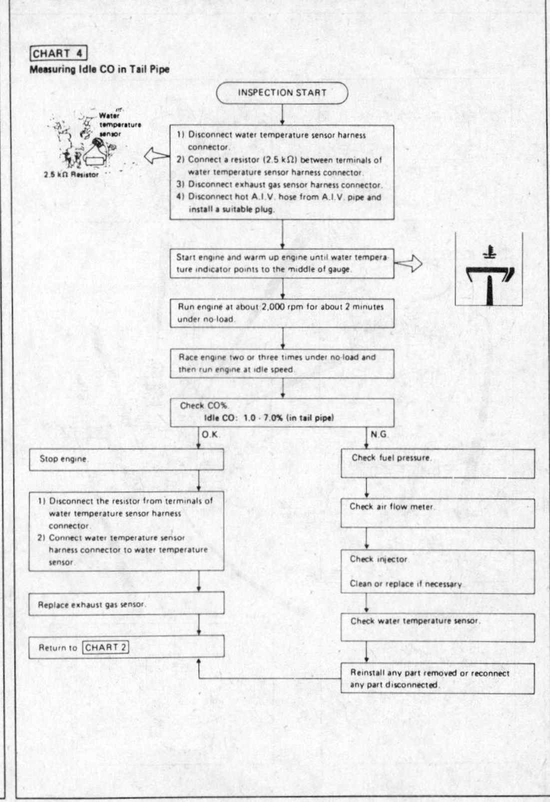
CHART 4
Measuring Idle CO in Tail Pipe
INSPECTION START
Water temperature sensor
2.5 kΩ Resistor
1) Disconnect water temperature sensor harness connector.
2) Connect a resistor (2.5 kΩ) between terminals of water temperature sensor harness connector.
3) Disconnect exhaust gas sensor harness connector.
4) Disconnect hot A.I.V. hose from A.I.V. pipe and install a suitable plug.
Start engine and warm up engine until water temperature indicator points to the middle of gauge.
Run engine at about 2,000 rpm for about 2 minutes under no-load.
Race engine two or three times under no-load and then run engine at idle speed.
Check CO%.
Idle CO: 1.0 - 7.0% (in tail pipe)
O.K.
N.G.
Stop engine.
1) Disconnect the resistor from terminals of water temperature sensor harness connector.
2) Connect water temperature sensor harness connector to water temperature sensor.
Replace exhaust gas sensor.
Return to CHART 2
Check fuel pressure.
Check air flow meter.
Check injector.
Clean or replace if necessary.
Check water temperature sensor.
Reinstall any part removed or reconnect any part disconnected.

ECCS COMPONENTS LOCATION 1988–89 TRUCK/PATHFINDER WITH VG30I ENGINE

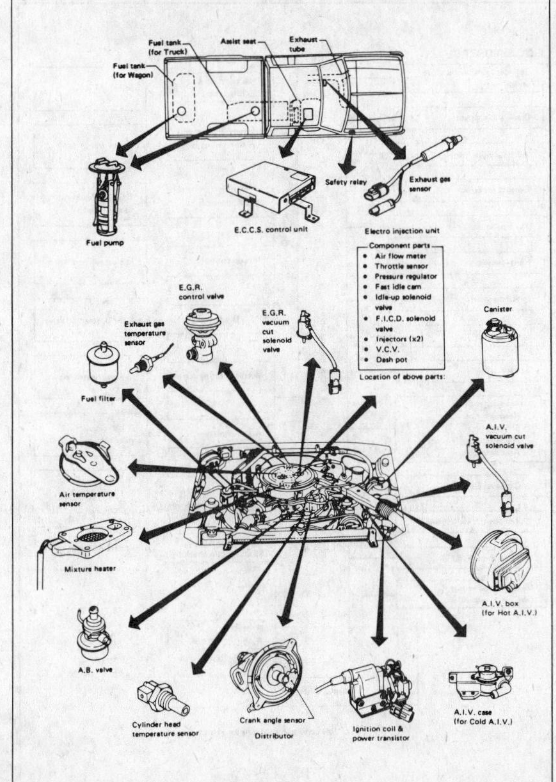
Fuel tank (for Truck)
Fuel tank (for Wagon)
Assist seat
Exhaust tube
Fuel pump
E.C.C.S. control unit
Safety relay
Exhaust gas sensor
Electro injection unit
Component parts
• Air flow meter
• Throttle sensor
• Pressure regulator
• Fast idle cam
• Idle-up solenoid valve
• F.I.C.D. solenoid valve
• Injectors (x2)
• V.C.V.
• Dash pot
Location of above parts:
E.G.R. control valve
Exhaust gas temperature sensor
E.G.R. vacuum cut solenoid valve
Canister
Fuel filter
A.I.V. vacuum cut solenoid valve
Air temperature sensor
Mixture heater
A.I.V. box (for Hot A.I.V.)
A.B. valve
Cylinder head temperature sensor
Crank angle sensor
Distributor
Ignition coil & power transistor
A.I.V. case (for Cold A.I.V.)

ECCS COMPONENTS LOCATION 1988–89 TRUCK/PATHFINDER WITH Z24I ENGINE

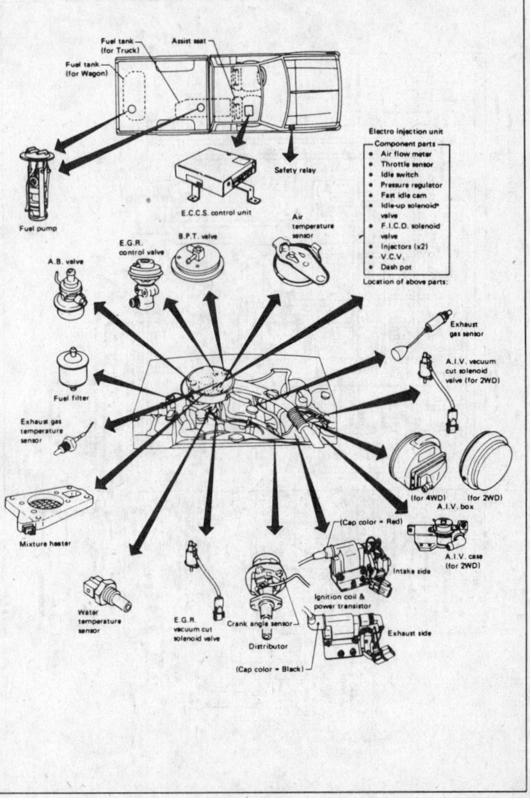
Fuel tank (for Truck)
Fuel tank (for Wagon)
Assist seat
Fuel pump
E.C.C.S. control unit
Safety relay
Electro injection unit
Component parts
• Air flow meter
• Throttle sensor
• Idle switch
• Pressure regulator
• Fast idle cam
• Idle-up solenoid valve
• F.I.C.D. solenoid valve
• Injectors (x2)
• V.C.V.
• Dash pot
Location of above parts:
Air temperature sensor
B.P.T. valve
E.G.R. control valve
A.B. valve
Exhaust gas sensor
A.I.V. vacuum cut solenoid valve (for 2WD)
Fuel filter
Exhaust gas temperature sensor
(for 4WD)
(for 2WD)
A.I.V. box
Mixture heater
(Cap color = Red)
A.I.V. case (for 2WD)
Intake side
Ignition coil & power transistor
Exhaust side
Water temperature sensor
E.G.R. vacuum cut solenoid valve
Crank angle sensor
Distributor
(Cap color = Black)

ECCS ELECTRO INJECTION UNIT PARTS LOCATION—1988–89 TRUCK/PATHFINDER

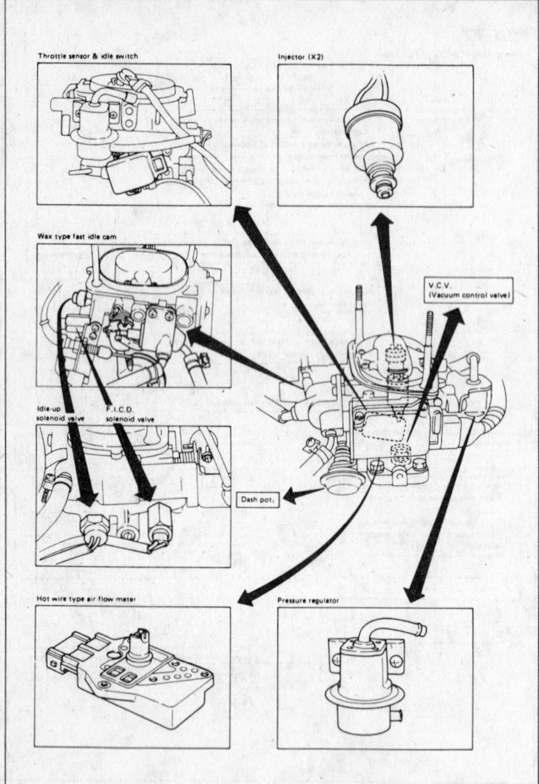
Throttle sensor & idle switch
Injector (X2)
Wax type fast idle cam
V.C.V.
(Vacuum control valve)
Idle-up solenoid valve
F.I.C.D. solenoid valve
Dash pot
Hot wire type air flow meter
Pressure regulator

ECCS SYSTEM SCHEMATIC 1988–89 TRUCK/PATHFINDER WITH VG30I ENGINE

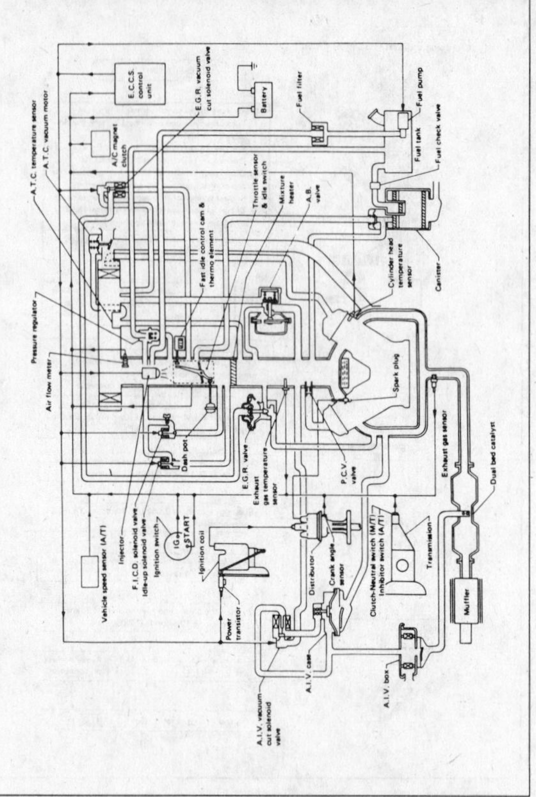

ECCS SYSTEM SCHEMATIC 1988–89 TRUCK/PATHFINDER WITH Z24I ENGINE

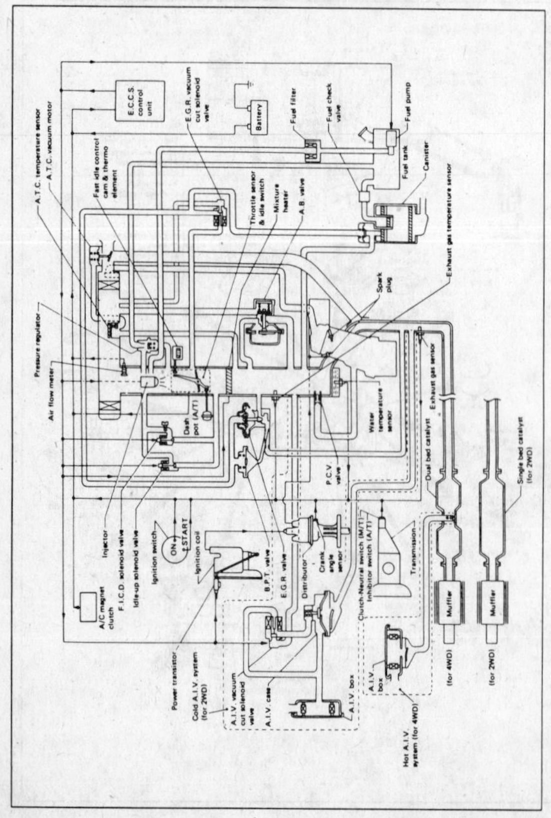

ECCS CONTROL SYSTEM CHART 1988–89 TRUCK/PATHFINDER WITH VG30I ENGINE

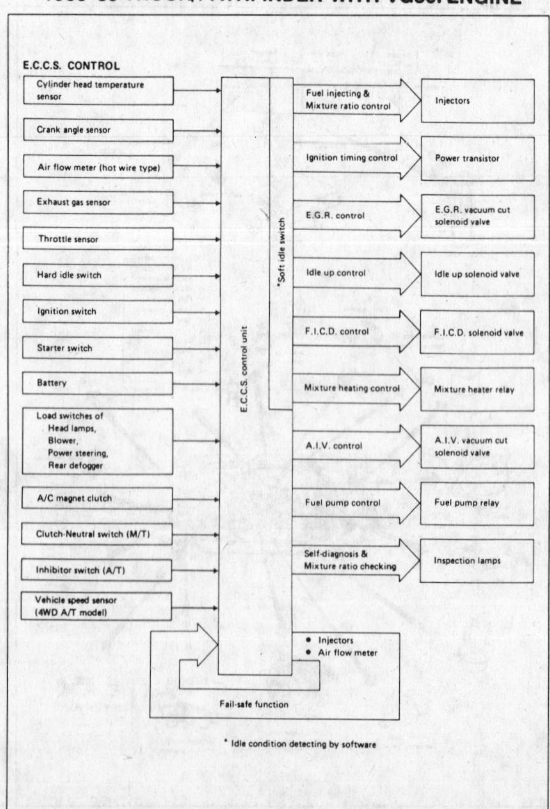
E.C.C.S. CONTROL
Cylinder head temperature sensor
Crank angle sensor
Air flow meter (hot wire type)
Exhaust gas sensor
Throttle sensor
Hard idle switch
Ignition switch
Starter switch
Battery
Load switches of
Head lamps,
Blower,
Power steering,
Rear defogger
A/C magnet clutch
Clutch-Neutral switch (M/T)
Inhibitor switch (A/T)
Vehicle speed sensor
(4WD A/T model)
E.C.C.S. control unit
*Soft idle switch
Fuel injecting &
Mixture ratio control
Ignition timing control
E.G.R. control
Idle up control
F.I.C.D. control
Mixture heating control
A.I.V. control
Fuel pump control
Self-diagnosis &
Mixture ratio checking
Injectors
Power transistor
E.G.R. vacuum cut solenoid valve
Idle up solenoid valve
F.I.C.D. solenoid valve
Mixture heater relay
A.I.V. vacuum cut solenoid valve
Fuel pump relay
Inspection lamps
• Injectors
• Air flow meter
Fail-safe function
* Idle condition detecting by software

ECCS CONTROL SYSTEM CHART (CONT.) 1988–89 TRUCK/PATHFINDER WITH VG30I ENGINE

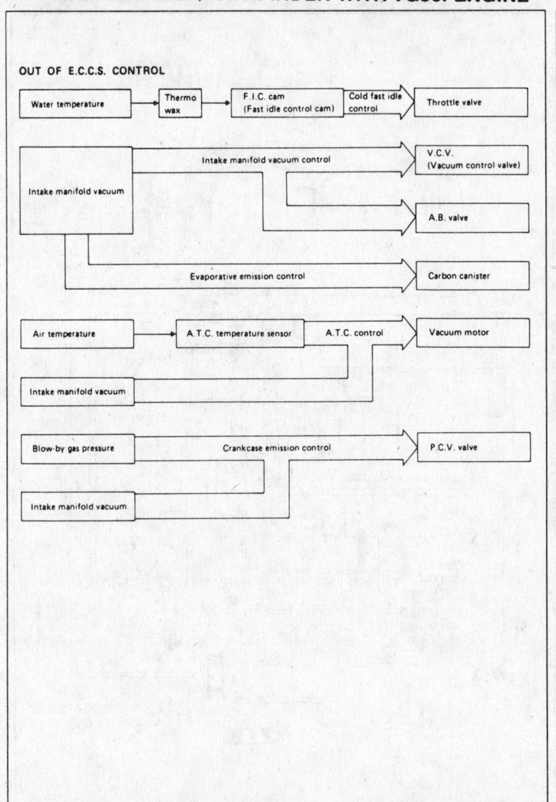
OUT OF E.C.C.S. CONTROL
Water temperature
Thermo wax
F.I.C. cam (Fast idle control cam)
Cold fast idle control
Throttle valve
Intake manifold vacuum
Intake manifold vacuum control
V.C.V. (Vacuum control valve)
A.B. valve
Evaporative emission control
Carbon canister
Air temperature
A.T.C. temperature sensor
A.T.C. control
Vacuum motor
Intake manifold vacuum
Blow-by gas pressure
Crankcase emission control
P.C.V. valve
Intake manifold vacuum

ECCS CONTROL SYSTEM CHART 1988–89 TRUCK/PATHFINDER WITH Z24I ENGINE

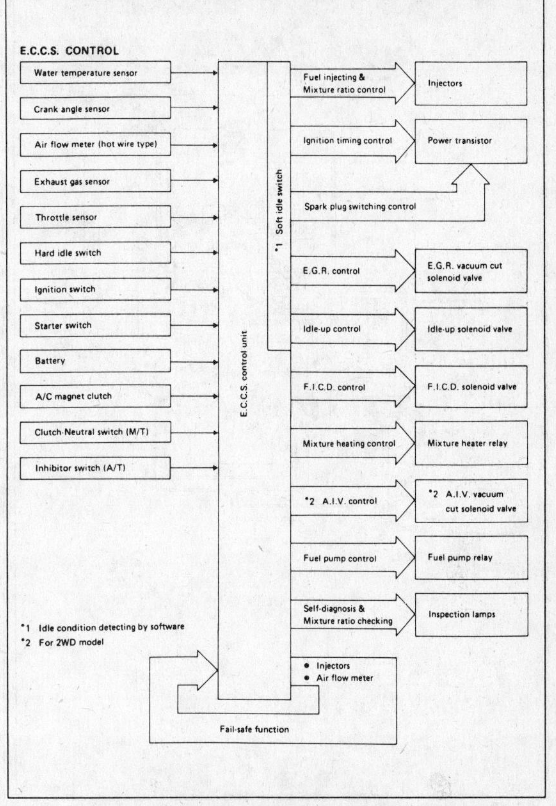
E.C.C.S. CONTROL
Water temperature sensor
Crank angle sensor
Air flow meter (hot wire type)
Exhaust gas sensor
Throttle sensor
Hard idle switch
Ignition switch
Starter switch
Battery
A/C magnet clutch
Clutch-Neutral switch (M/T)
Inhibitor switch (A/T)
E.C.C.S. control unit
*1 Soft idle switch
Fuel injecting & Mixture ratio control
Injectors
Ignition timing control
Power transistor
Spark plug switching control
E.G.R. control
E.G.R. vacuum cut solenoid valve
Idle-up control
Idle-up solenoid valve
F.I.C.D. control
F.I.C.D. solenoid valve
Mixture heating control
Mixture heater relay
*2 A.I.V. control
*2 A.I.V. vacuum cut solenoid valve
Fuel pump control
Fuel pump relay
Self-diagnosis & Mixture ratio checking
Inspection lamps
*1 Idle condition detecting by software
*2 For 2WD model
Injectors
Air flow meter
Fail-safe function

ECCS CONTROL SYSTEM CHART (CONT.) 1988–89 TRUCK/PATHFINDER WITH Z24I ENGINE

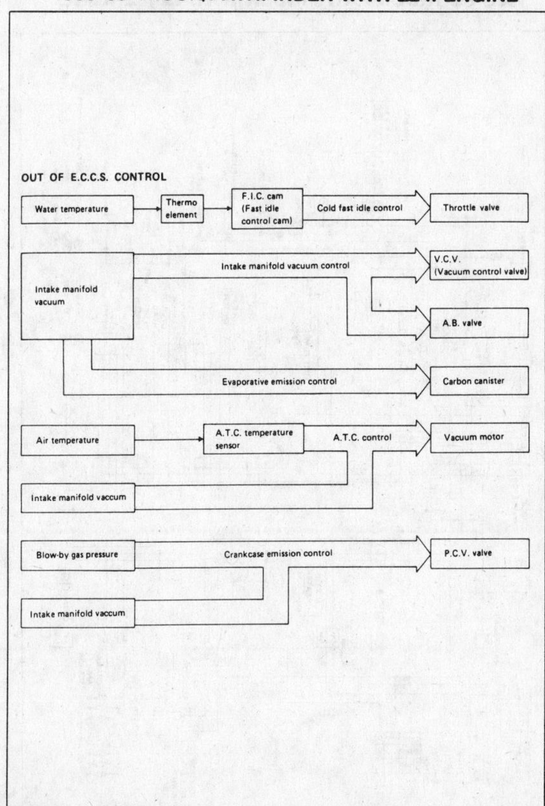
OUT OF E.C.C.S. CONTROL
Water temperature
Thermo element
F.I.C. cam (Fast idle control cam)
Cold fast idle control
Throttle valve
Intake manifold vacuum
Intake manifold vacuum control
V.C.V. (Vacuum control valve)
A.B. valve
Evaporative emission control
Carbon canister
Air temperature
A.T.C. temperature sensor
A.T.C. control
Vacuum motor
Intake manifold vaccum
Blow-by gas pressure
Crankcase emission control
P.C.V. valve
Intake manifold vaccum

ECCS FUEL FLOW SYSTEM SCHEMATIC 1988–89 TRUCK/PATHFINDER WITH VG30I ENGINE

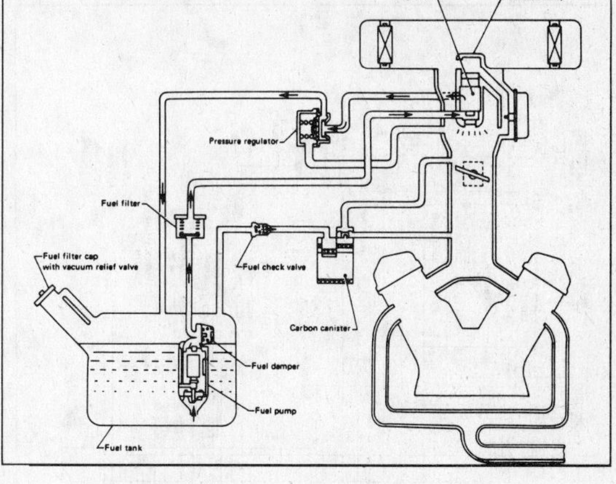
Injector
Throttle body
Pressure regulator
Fuel filter
Fuel filter cap with vacuum relief valve
Fuel check valve
Carbon canister
Fuel damper
Fuel pump
Fuel tank

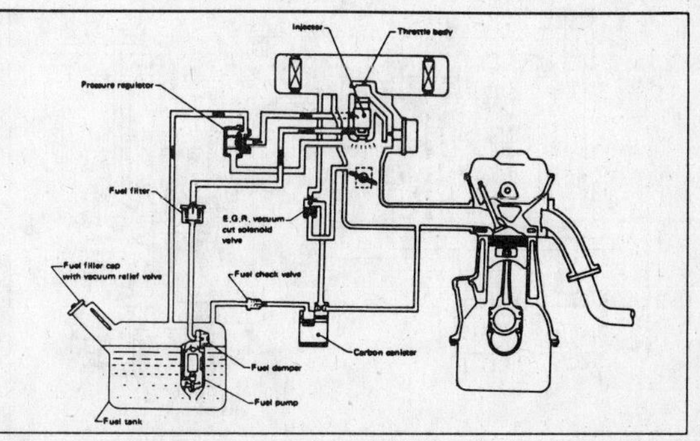
Injector
Throttle body
Pressure regulator
Fuel filter
E.G.R. vacuum cut solenoid valve
Fuel filler cap with vacuum relief valve
Fuel check valve
Carbon canister
Fuel damper
Fuel pump
Fuel tank

ECCS AIR FLOW SYSTEM SCHEMATIC 1988–89 TRUCK/PATHFINDER WITH VG30I ENGINE

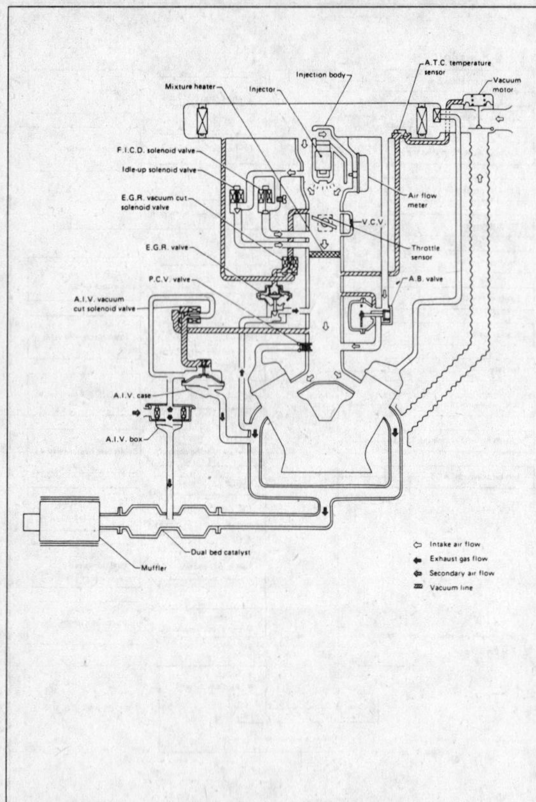
A.T.C. temperature sensor
Vacuum motor
Injection body
Mixture heater
Injector
F.I.C.D. solenoid valve
Idle-up solenoid valve
E.G.R. vacuum cut solenoid valve
Air flow meter
V.C.V.
Throttle sensor
E.G.R. valve
A.B. valve
P.C.V. valve
A.I.V. vacuum cut solenoid valve
A.I.V. case
A.I.V. box
Dual bed catalyst
Muffler
Intake air flow
Exhaust gas flow
Secondary air flow
Vacuum line

ECCS AIR FLOW SYSTEM SCHEMATIC 1988–89 TRUCK/PATHFINDER WITH Z24I ENGINE

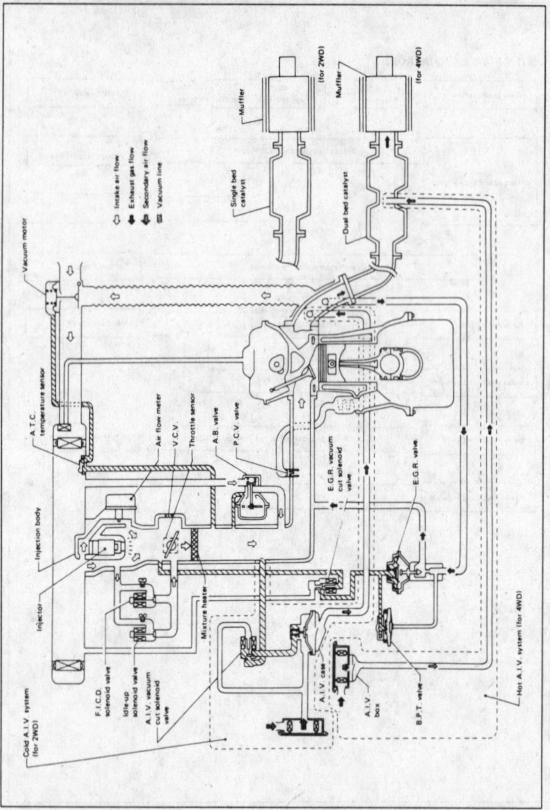

ECCS CONTROL UNIT WIRING SCHEMATIC 1988–89 TRUCK/PATHFINDER WITH VG30I ENGINE

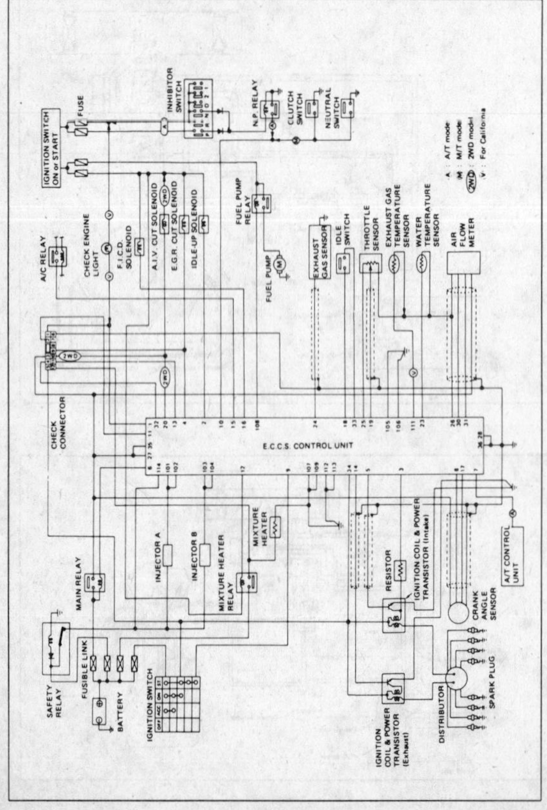

ECCS CONTROL UNIT WIRING SCHEMATIC 1988–89 TRUCK/PATHFINDER WITH Z24I ENGINE

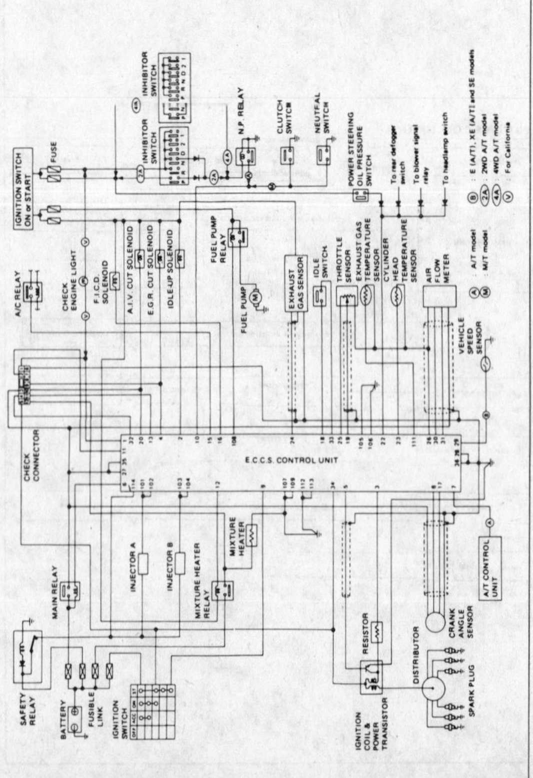

ECCS WIRING SCHEMATIC
1988–89 TRUCK/PATHFINDER WITH VG30I ENGINE

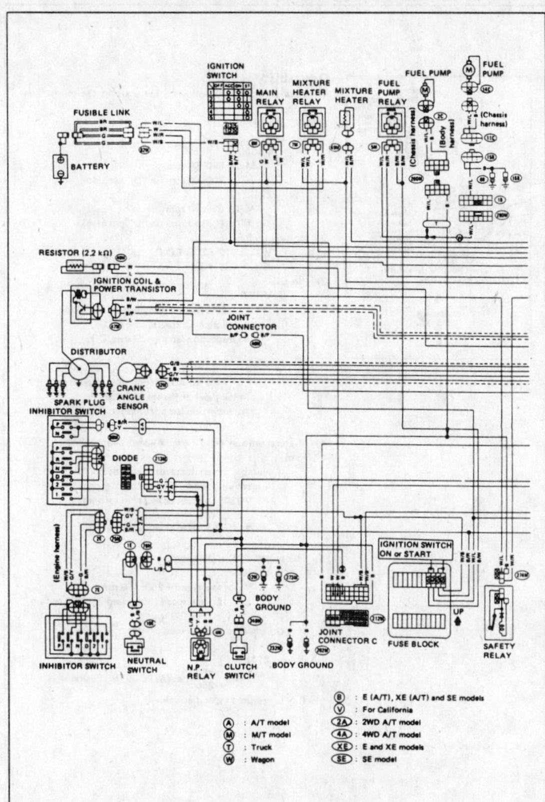

ECCS WIRING SCHEMATIC (CONT.)
1988–89 TRUCK/PATHFINDER WITH VG30I ENGINE

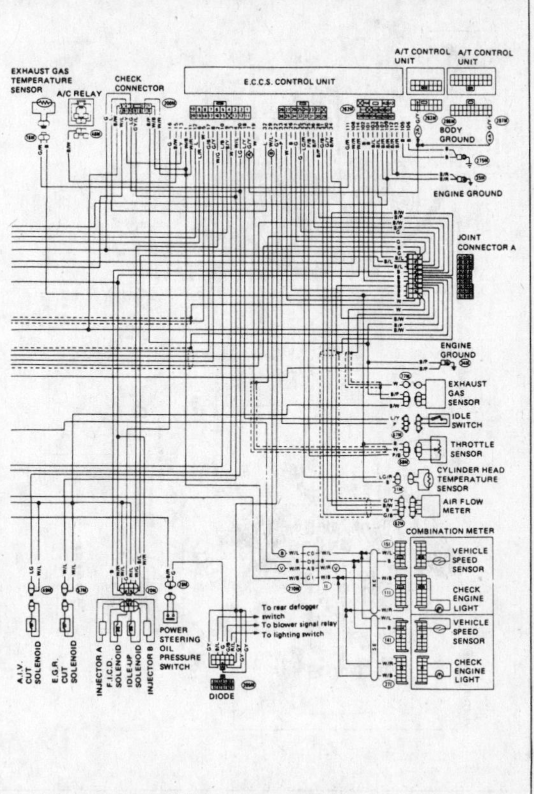

ECCS WIRING SCHEMATIC
1988–89 TRUCK/PATHFINDER With Z24I ENGINE

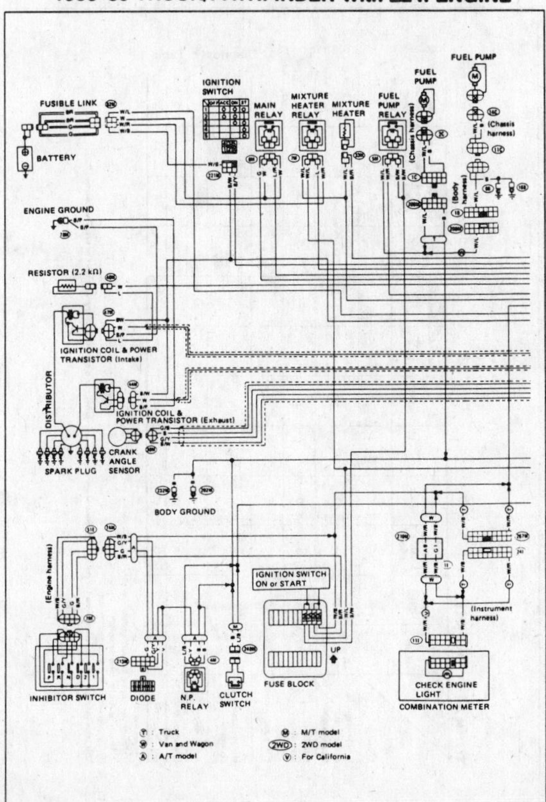

ECCS WIRING SCHEMATIC (CONT.)
1988–89 TRUCK/PATHFINDER WITH Z24I ENGINE

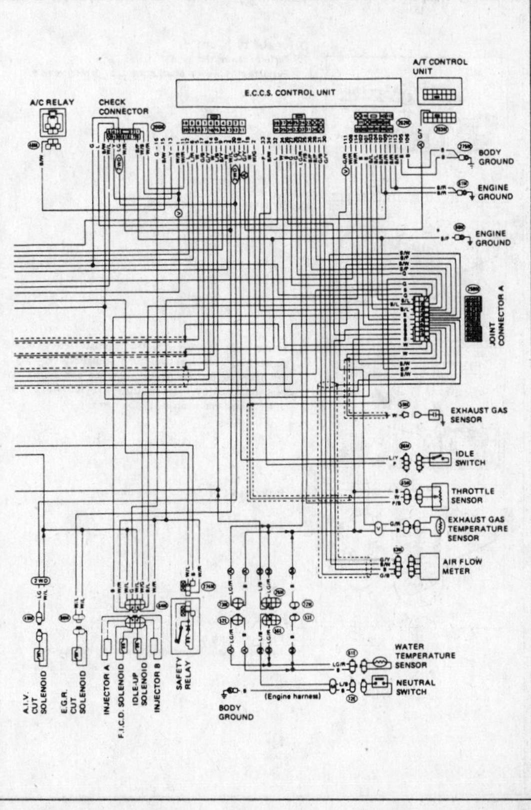

THROTTLE BODY ASSEMBLY EXPLODED
1988–89 TRUCK/PATHFINDER

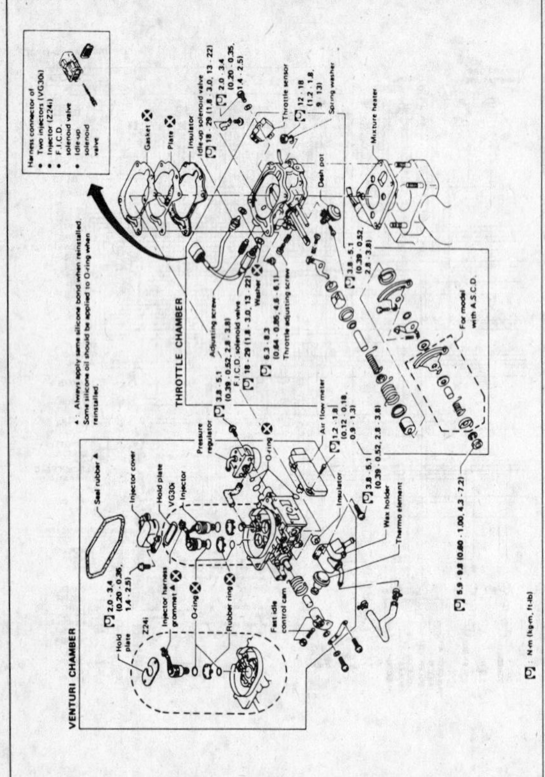
VENTURI CHAMBER
THROTTLE CHAMBER
Harness connector of
• Two injectors (VG30i)
• Injector (Z24i)
• F.I.C.D. solenoid valve
• Idle-up solenoid valve
Gasket
Plate
Insulator
Idle-up solenoid valve
Throttle sensor
Spring washer
Mixture heater
Dash pot
For model with A.S.C.D.
Adjusting screw
F.I.C.D. solenoid valve
Washer
Throttle adjusting screw
Air flow meter
O-ring
Pressure regulator
Insulator
Wax holder
Thermo element
Fast idle control cam
Rubber ring
O-ring
Injector harness grommet
Z24i
Hold plate
Injector
VG30i
Injector cover
Seal rubber
N·m (kg-m, ft-lb)

ECCS DIAGNOSTIC PROCEDURE
1988–89 TRUCK/PATHFINDER

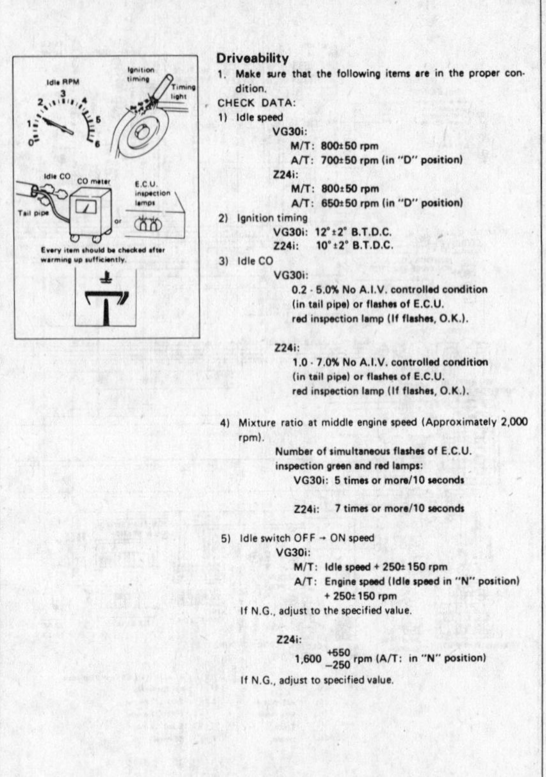
Idle RPM
Ignition timing
Timing light
Idle CO
CO meter
E.C.U. inspection lamps
Tail pipe
Every item should be checked after warming up sufficiently.

Driveability

1. Make sure that the following items are in the proper condition.

CHECK DATA:

1) Idle speed
   - VG30i:
     - M/T: 800±50 rpm
     - A/T: 700±50 rpm (in "D" position)
   - Z24i:
     - M/T: 800±50 rpm
     - A/T: 650±50 rpm (in "D" position)
2) Ignition timing
   - VG30i: 12°±2° B.T.D.C.
   - Z24i: 10°±2° B.T.D.C.
3) Idle CO
   - VG30i:
     0.2 - 5.0% No A.I.V. controlled condition (in tail pipe) or flashes of E.C.U. red inspection lamp (If flashes, O.K.).
   - Z24i:
     1.0 - 7.0% No A.I.V. controlled condition (in tail pipe) or flashes of E.C.U. red inspection lamp (If flashes, O.K.).
4) Mixture ratio at middle engine speed (Approximately 2,000 rpm).
   Number of simultaneous flashes of E.C.U. inspection green and red lamps:
   - VG30i: 5 times or more/10 seconds
   - Z24i: 7 times or more/10 seconds
5) Idle switch OFF → ON speed
   - VG30i:
     - M/T: Idle speed + 250±150 rpm
     - A/T: Engine speed (Idle speed in "N" position) + 250±150 rpm

     If N.G., adjust to the specified value.
   - Z24i:
     $1{,}600^{+550}_{-250}$ rpm (A/T: in "N" position)

     If N.G., adjust to specified value.

ECCS DIAGNOSTIC PROCEDURE (CONT.)
1988–89 TRUCK/PATHFINDER

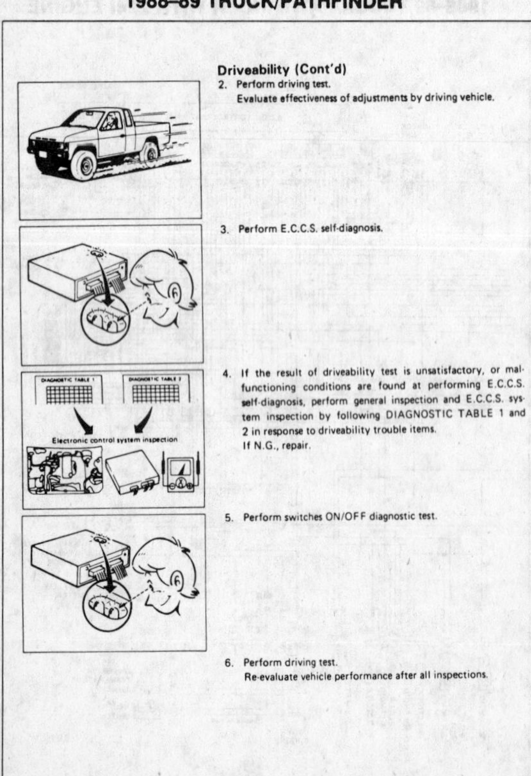
DIAGNOSTIC TABLE 1
DIAGNOSTIC TABLE 2
Electronic control system inspection

Driveability (Cont'd)

2. Perform driving test.
   Evaluate effectiveness of adjustments by driving vehicle.
3. Perform E.C.C.S. self-diagnosis.
4. If the result of driveability test is unsatisfactory, or malfunctioning conditions are found at performing E.C.C.S. self-diagnosis, perform general inspection and E.C.C.S. system inspection by following DIAGNOSTIC TABLE 1 and 2 in response to driveability trouble items.
   If N.G., repair.
5. Perform switches ON/OFF diagnostic test.
6. Perform driving test.
   Re-evaluate vehicle performance after all inspections.

ECCS DIAGNOSTIC TABLE 1
1988–89 TRUCK/PATHFINDER WITH VG30I ENGINE

Diagnostic Table 1

SYSTEM INSPECTION TABLE

System \ Sensor & actuator	Crank angle sensor	Air flow meter	Cylinder head temperature sensor	Ignition switch	Injector	Throttle sensor	Neutral/ clutch switch	Exhaust gas sensor	Battery voltage	A.I.V. cut solenoid valve	E.G.R. cut solenoid valve	Idle-up solenoid valve
Fuel injection & mixture ratio feedback control	O	O	O	O	O	O	O	O	O			
Ignition timing control	O	O	O	O		O			Power transistor			
Fuel pump control	O			O					Fuel pump relay			
Mixture heater control	O		O	O	O				Mixture heater relay			
A.I.V. control	O		O			O		O		O		
Idle-up control	O		O	O					O			O*
E.G.R. control	O		O	O		O					O	

* Input switch
① Power steering oil pressure switch
② Heater or Air conditioner switch
③ Lighting switch & rear defogger switch

This table indicates the inspection items for the E.C.C.S. control system. For each system, it is necessary to check sensors or actuators marked "O".

ECCS DIAGNOSTIC TABLE 1
1988–89 TRUCK/PATHFINDER WITH Z24I ENIGNE

Diagnostic Table 1 (Cont'd)

SYSTEM INSPECTION TABLE

System \ Sensor & actuator	Crank angle sensor	Air flow meter	Water temperature sensor	Ignition switch	Injector	Throttle sensor	Neutral/clutch switch	Exhaust gas sensor	Battery voltage	A.I.V. cut solenoid valve	E.G.R. cut solenoid valve	Idle-up solenoid valve
Fuel injection & mixture ratio feedback control	○	○	○	○	○	○	○	○	○			
Ignition timing control	○	○	○	○		○			Power transistor			
Spark plug switching control	○	○	○						Power transistor			
Fuel pump control	○			○					Fuel pump relay			
Mixture heater control	○		○	○	○				Mixture heater relay			
A.I.V. control	○		○			○				○		
Idle-up control	○		○	○					○			○
E.G.R. control	○		○	○		○					○	

This table indicates the inspection items for the E.C.C.S. control system. For each system, it is necessary to check sensors or actuators marked "○".

ECCS DIAGNOSTIC TABLE 2
1988–89 TRUCK/PATHFINDER

Diagnostic Table 2

DRIVEABILITY INSPECTION TABLE

Trouble items concerned with driveability \ Inspection item			General inspection: Fuel flow system				Electric system					Air flow system							Idle up solenoid valve		E.C.C.S. system inspection: Crank angle sensor		
			Fuel level	Fuel pump	Fuel filter	Fuel line	Battery	Spark plug	Ignition wire	Alternator	Starter	Air cleaner	Air flow line	E.G.R. valve	A.B. valve	F.I.C.D. solenoid valve	Fast idle cam	P.C.V.	Short	Open	120° (VG30i) 180° (Z24i) signal noise	120° (VG30i) 180° (Z24i) signal faults	1° signal faults
SURGE	ROAD/LOAD DRIVING	Heavy load			○	○		◎	○			○	○	○				○			○	○	○
		Middle load			○	○		◎	○			○	○	◎				○			○	○	○
		Light load			○	○		◎	○			○	○	◎				○			○	○	○
	ACCELERATING DRIVING	Slow acceleration			○	○		◎	○			○	○	◎				○			○	○	○
	DECELERATING DRIVING	Rapid deceleration			○	○		◎	○			○	○	○				○			○	○	○
		Slow deceleration			○	○		◎	○			○	○	○				○			○	○	○
HESITATION		Rapid acceleration			○	○		◎	○			○	○	○				○			○	○	○
		Slow acceleration			○	○		◎	○			○	○	○				○			○	○	○
STUMBLE		Rapid acceleration			○	○		◎	○			○	○	○				○			○	○	○
		Slow acceleration			○	○		◎	○			○	○	○				○			○	○	○
BACKFIRE					○	○			○				○		○			○			○	○	○
AFTER FIRE									○			○						○			○	○	○
IDLE STABILITY								◎	○	○		○	○	◎	○	○	◎	○		◎	○	○	○
ENGINE STALL			○	○	○	○	○	◎	○	○		○	○	○	○	○	◎	○			◎	◎	◎
STARTABILITY			○	○			○	◎	○	○	○	○	○				○	○			◎	◎	◎

This table indicates the inspection items for each type of symptom. It is necessary for each symptom to check sensors or actuators marked "◎" or "○". Items marked "◎" have a significant influence on driveability. Prior to items marked "○", check items marked "◎".
Improper mixture ratio, improper ignition condition, and an excess of E.G.R. volume can cause any symptom.

ECCS DIAGNOSTIC TABLE 2 (CONT.)
1988–89 TRUCK/PATHFINDER

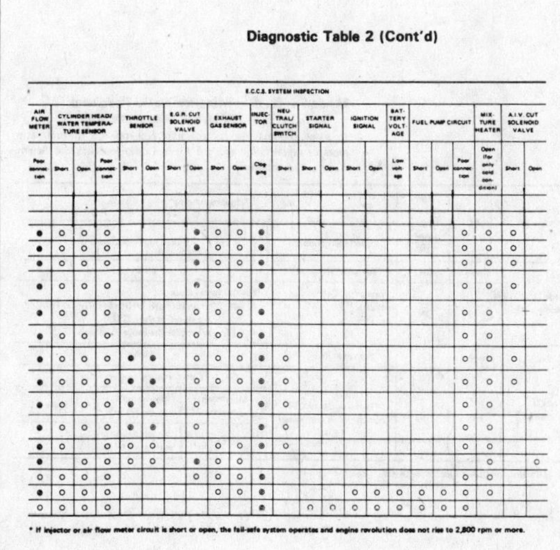

Diagnostic Table 2 (Cont'd)

E.C.C.S. SYSTEM INSPECTION

AIR FLOW METER*	CYLINDER HEAD/WATER TEMPERATURE SENSOR			THROTTLE SENSOR		E.G.R. CUT SOLENOID VALVE		EXHAUST GAS SENSOR		INJECTOR	NEUTRAL/CLUTCH SWITCH	STARTER SIGNAL		IGNITION SIGNAL		BATTERY VOLTAGE	FUEL PUMP CIRCUIT			MIXTURE HEATER	A.I.V. CUT SOLENOID VALVE	
Poor connection	Short	Open	Poor connection	Short	Open	Short	Open	Short	Open	Clogging	Short	Short	Open	Short	Open	Low voltage	Short	Open	Poor connection	Open (for only cold condition)	Short	Open
●	○	○	○				◎	○	○	◎									○	○	○	
●	○	○	○				◎	○	○	◎									○	○	○	
●	○	○	○				◎	○	○	◎									○	○	○	
●	○	○	○				◎	○	○	◎									○	○	○	
●	○	○	○				○	○	○	◎									○	○		
●	○	○	○				○	○	○	◎									○	○		
●	○	○	○	●	◎		○	○	○	◎	○								○	○	○	
●	○	○	○	●	◎		○	○	○	◎	○								○	○	○	
●	○	○	○	◎	◎		○			◎	○								○	○		
●	○	○	○	◎	◎		○			◎	○								○	○		
●	○		○	○	○			○	○	◎	○								○	○		
●		○	○	○	○		◎	○	○											○		○
○	○	○	○				○	○	○	◎									○	○		
●	○	○	○					○	○	◎				○	○	○	○	○	○	○		
○	○	○	○							◎		○	○	○	○	○	○	○	○			

* If injector or air flow meter circuit is short or open, the fail-safe system operates and engine revolution does not rise to 2,800 rpm or more.

ECCS SELF-DIAGNOSIS DESCRIPTION
1988–89 TRUCK/PATHFINDER

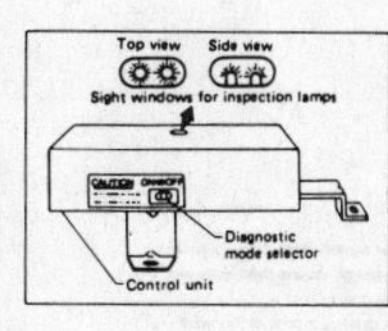
Top view
Side view
Sight windows for inspection lamps
Diagnostic mode selector
Control unit

Description

The self-diagnosis is useful to diagnose malfunctions in major sensors and actuators of the E.C.C.S. system. There are 5 modes in the self-diagnosis system.

1. **Mode I – Mixture ratio feedback control monitor A**
   - During closed loop condition:
     The green inspection lamp turns ON when lean condition is detected and goes OFF by rich condition.
   - During open loop condition:
     The green inspection lamp remains ON or OFF.
2. **Mode II – Mixture ratio feedback control monitor B**
   The green inspection lamp function is the same as Mode I.
   - During closed loop condition:
     The red inspection lamp turns ON and OFF simultaneously with the green inspection lamp when the mixture ratio is controlled within the specified value.
   - During open loop condition:
     The red inspection lamp remains ON or OFF.
3. **Mode III – Self-diagnosis**
   This mode is the same as the former self-diagnosis in self-diagnosis mode.
4. **Mode IV – Switches ON/OFF diagnosis**
   During this mode, the inspection lamps monitor the switch ON-OFF condition.
   - Idle switch
   - Starter switch
   - Vehicle speed sensor
5. **Mode V – Real time diagnosis**
   The moment the malfunction is detected, the display will be presented immediately. That is, the condition at which the malfunction occurs can be found by observing the inspection lamps during driving test.

ECCS SELF-DIAGNOSIS DESCRIPTION (CONT.)—1988–89 TRUCK/PATHFINDER

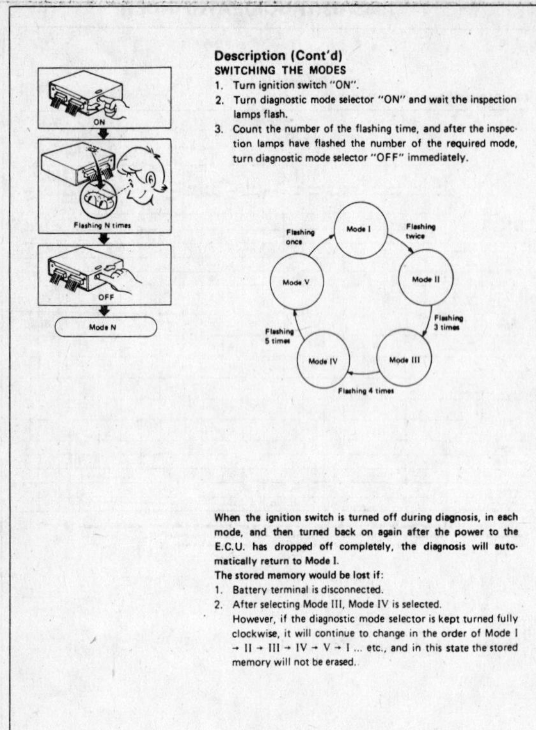

Description (Cont'd)

SWITCHING THE MODES

1. Turn ignition switch "ON".
2. Turn diagnostic mode selector "ON" and wait the inspection lamps flash.
3. Count the number of the flashing time, and after the inspection lamps have flashed the number of the required mode, turn diagnostic mode selector "OFF" immediately.

ON
Flashing N times
OFF
Mode N
Mode I
Flashing twice
Mode II
Flashing 3 times
Mode III
Flashing 4 times
Mode IV
Flashing 5 times
Mode V
Flashing once

When the ignition switch is turned off during diagnosis, in each mode, and then turned back on again after the power to the E.C.U. has dropped off completely, the diagnosis will automatically return to Mode I.

The stored memory would be lost if:

1. Battery terminal is disconnected.
2. After selecting Mode III, Mode IV is selected.
   However, if the diagnostic mode selector is kept turned fully clockwise, it will continue to change in the order of Mode I → II → III → IV → V → I ... etc., and in this state the stored memory will not be erased.

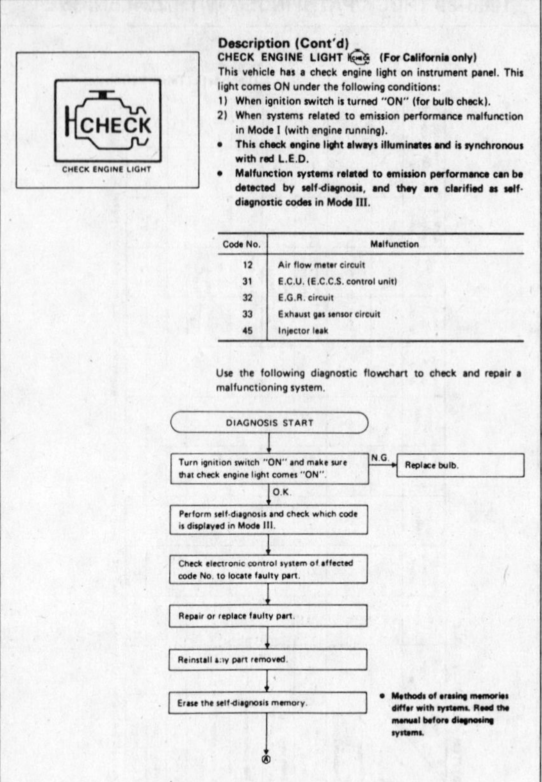

Description (Cont'd)

CHECK ENGINE LIGHT (For California only)

This vehicle has a check engine light on instrument panel. This light comes ON under the following conditions:

1) When ignition switch is turned "ON" (for bulb check).
2) When systems related to emission performance malfunction in Mode I (with engine running).

- **This check engine light always illuminates and is synchronous with red L.E.D.**
- **Malfunction systems related to emission performance can be detected by self-diagnosis, and they are clarified as self-diagnostic codes in Mode III.**

CHECK
CHECK ENGINE LIGHT

Code No.	Malfunction
12	Air flow meter circuit
31	E.C.U. (E.C.C.S. control unit)
32	E.G.R. circuit
33	Exhaust gas sensor circuit
45	Injector leak

Use the following diagnostic flowchart to check and repair a malfunctioning system.

DIAGNOSIS START
Turn ignition switch "ON" and make sure that check engine light comes "ON".
N.G.
Replace bulb.
O.K.
Perform self-diagnosis and check which code is displayed in Mode III.
Check electronic control system of affected code No. to locate faulty part.
Repair or replace faulty part.
Reinstall any part removed.
Erase the self-diagnosis memory.
• Methods of erasing memories differ with systems. Read the manual before diagnosing systems.
A

ECCS SELF-DIAGNOSIS DESCRIPTION (CONT.) 1988–89 TRUCK/PATHFINDER

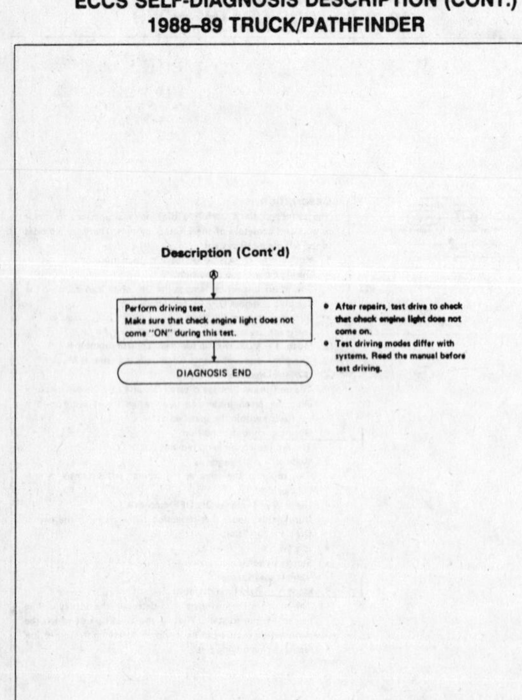

Description (Cont'd)

A
Perform driving test.
Make sure that check engine light does not come "ON" during this test.
DIAGNOSIS END
• After repairs, test drive to check that check engine light does not come on.
• Test driving modes differ with systems. Read the manual before test driving.

ECCS SELF-DIAGNOSTIC MODES 1 AND II 1988–89 TRUCK/PATHFINDER

Modes I & II —Mixture Ratio Feedback Control Monitors A & B

In these modes, the control unit provides the Air-fuel ratio monitor presentation and the Air-fuel ratio feedback coefficient monitor presentation.

Mode	LED	Engine stopped (Ignition switch "ON")	Engine running			
			Open loop condition	Closed loop condition		
Mode I (Monitor A)	Green	ON	*Remains ON or OFF	Blinks		
	Red	ON	Except for California model • OFF	For California model • ON: when CHECK ENGINE LIGHT ITEMS are stored in the E.C.U. • OFF: except for the above condition		
Mode II (Monitor B)	Green	ON	*Remains ON or OFF	Blinks		
	Red	OFF	*Remains ON or OFF (synchronous with green LED)	Compensating mixture ratio		
				More than 5% rich	Between 5% lean and 5% rich	More
				OFF	Synchronized with green LED	Remains ON

*: Maintains conditions just before switching to open loop

ECCS SELF-DIAGNOSTIC MODE III—1988–89 TRUCK/PATHFINDER

Mode III — Self-Diagnostic System

The E.C.U. constantly monitors the function of these sensors and actuators, regardless of ignition key position. If a malfunction occurs, the information is stored in the E.C.U. and can be retrieved from the memory by turning on the diagnostic mode selector, located on the side of the E.C.U. When activated, the malfunction is indicated by flashing a red and a green L.E.D. (Light Emitting Diode), also located on the E.C.U. Since all the self-diagnostic results are stored in the E.C.U.'s memory even intermittent malfunctions can be diagnosed.

A malfunctioning part's group is indicated by the number of both the red and the green L.E.D.s flashing. First, the red L.E.D. flashes and the green flashes follow. The red L.E.D. refers to the number of tens while the green one refers to the number of units. For example, when the red L.E.D. flashes once and then the green one flashes twice, this means the number "12" showing the air flow meter signal is malfunctioning. In this way, all the problems are classified by the code numbers.

- **When engine fails to start, crank engine more than two seconds before starting self-diagnosis.**
- **Before starting self-diagnosis, do not erase stored memory. If doing so, self-diagnosis function for intermittent malfunctions would be lost.**

The stored memory would be lost if:

1. **Battery terminal is disconnected.**
2. **After selecting Mode III, Mode IV is selected.**

DISPLAY CODE TABLE

X: Available
—: Not available

Code No.	Detected items	California	Non-California
11	Crank angle sensor circuit	X	X
12	Air flow meter circuit	X	X
13	Cylinder head/Water temperature sensor circuit	X	X
14	Vehicle speed sensor circuit (VG30i: 4WD A/T model only)	X	X
21	Ignition signal missing in primary coil	X	X
31	E.C.U. (E.C.C.S. control unit)	X	X
32	E.G.R. circuit	X	—
33	Exhaust gas sensor circuit	X	X
35	Exhaust gas temperature sensor circuit	X	—
43	Throttle sensor circuit	X	X
45	Injector leak	X	—
51	Injector	X	X
55	No malfunction in the above circuit	X	X

Mode III — Self-Diagnostic System (Cont'd)

RETENTION OF DIAGNOSTIC RESULTS

The diagnostic result is retained in E.C.U. memory until the starter is operated fifty times after a diagnostic item is judged to be malfunctioning. The diagnostic result will then be cancelled automatically. If a diagnostic item which has been judged to be malfunctioning and stored in memory is again judged to be malfunctioning before the starter is operated fifty times, the second result will replace the previous one. It will be stored in E.C.U. memory until the starter is operated fifty times more.

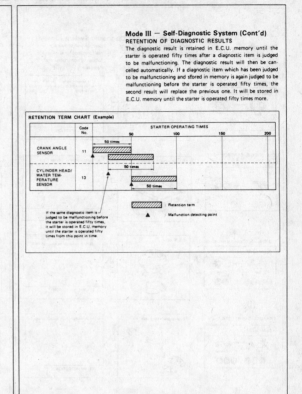

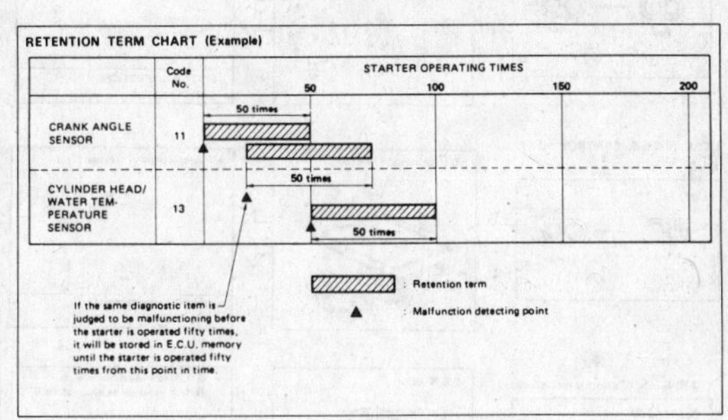

RETENTION TERM CHART (Example)
Code No.
STARTER OPERATING TIMES
50
100
150
200
CRANK ANGLE SENSOR
11
50 times
CYLINDER HEAD/ WATER TEMPERATURE SENSOR
13
50 times
50 times
If the same diagnostic item is judged to be malfunctioning before the starter is operated fifty times, it will be stored in E.C.U. memory until the starter is operated fifty times from this point in time.
: Retention term
: Malfunction detecting point

ECCS SELF-DIAGNOSTIC MODE III (CONT.)—1988–89 TRUCK/PATHFINDER

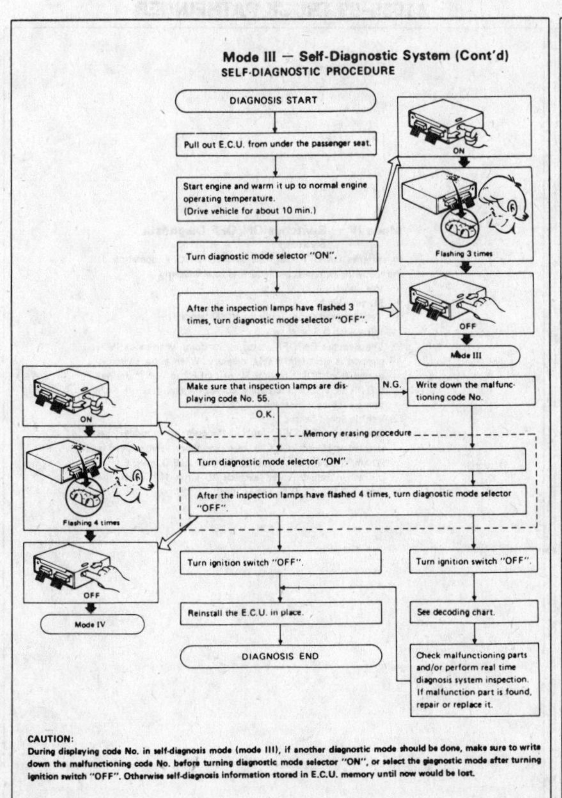

Mode III — Self-Diagnostic System (Cont'd)
SELF-DIAGNOSTIC PROCEDURE
DIAGNOSIS START
Pull out E.C.U. from under the passenger seat.
Start engine and warm it up to normal engine operating temperature. (Drive vehicle for about 10 min.)
Turn diagnostic mode selector "ON".
After the inspection lamps have flashed 3 times, turn diagnostic mode selector "OFF".
ON
Flashing 3 times
OFF
Mode III
Make sure that inspection lamps are displaying code No. 55.
N.G.
Write down the malfunctioning code No.
O.K.
Memory erasing procedure
Turn diagnostic mode selector "ON".
After the inspection lamps have flashed 4 times, turn diagnostic mode selector "OFF".
ON
Flashing 4 times
OFF
Mode IV
Turn ignition switch "OFF".
Turn ignition switch "OFF".
Reinstall the E.C.U. in place.
See decoding chart.
DIAGNOSIS END
Check malfunctioning parts and/or perform real time diagnosis system inspection. If malfunction part is found, repair or replace it.

CAUTION:
During displaying code No. in self-diagnosis mode (mode III), if another diagnostic mode should be done, make sure to write down the malfunctioning code No. before turning diagnostic mode selector "ON", or select the giagnostic mode after turning ignition switch "OFF". Otherwise self-diagnosis information stored in E.C.U. memory until now would be lost.

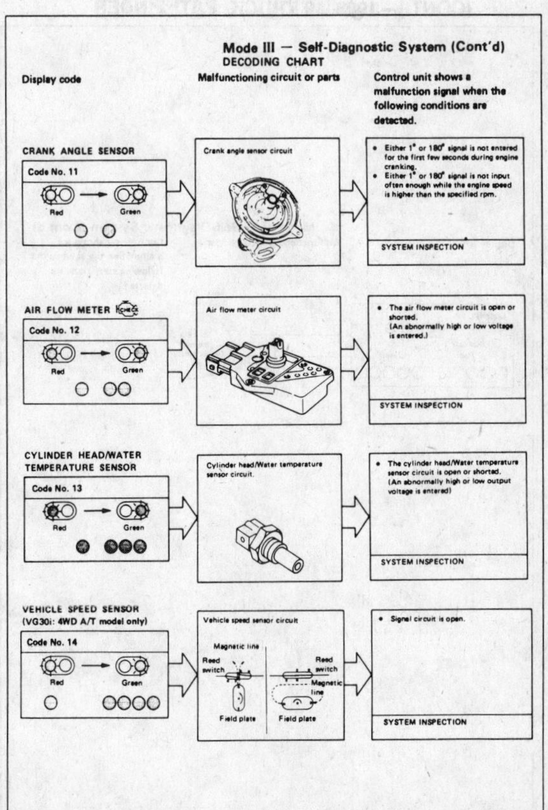

Mode III — Self-Diagnostic System (Cont'd)
DECODING CHART
Display code
Malfunctioning circuit or parts
Control unit shows a malfunction signal when the following conditions are detected.
CRANK ANGLE SENSOR
Code No. 11
Red
Green
Crank angle sensor circuit
• Either 1° or 180° signal is not entered for the first few seconds during engine cranking.
• Either 1° or 180° signal is not input often enough while the engine speed is higher than the specified rpm.
SYSTEM INSPECTION
AIR FLOW METER
Code No. 12
Red
Green
Air flow meter circuit
• The air flow meter circuit is open or shorted. (An abnormally high or low voltage is entered.)
SYSTEM INSPECTION
CYLINDER HEAD/WATER TEMPERATURE SENSOR
Code No. 13
Red
Green
Cylinder head/Water temperature sensor circuit.
• The cylinder head/Water temperature sensor circuit is open or shorted. (An abnormally high or low output voltage is entered)
SYSTEM INSPECTION
VEHICLE SPEED SENSOR (VG30i: 4WD A/T model only)
Code No. 14
Red
Green
Vehicle speed sensor circuit
Magnetic line
Reed switch
Reed switch
Magnetic line
Field plate
Field plate
• Signal circuit is open.
SYSTEM INSPECTION

ECCS SELF-DIAGNOSTIC MODE III DISPLAY CODES (CONT.)—1988–89 TRUCK/ PATHFINDER

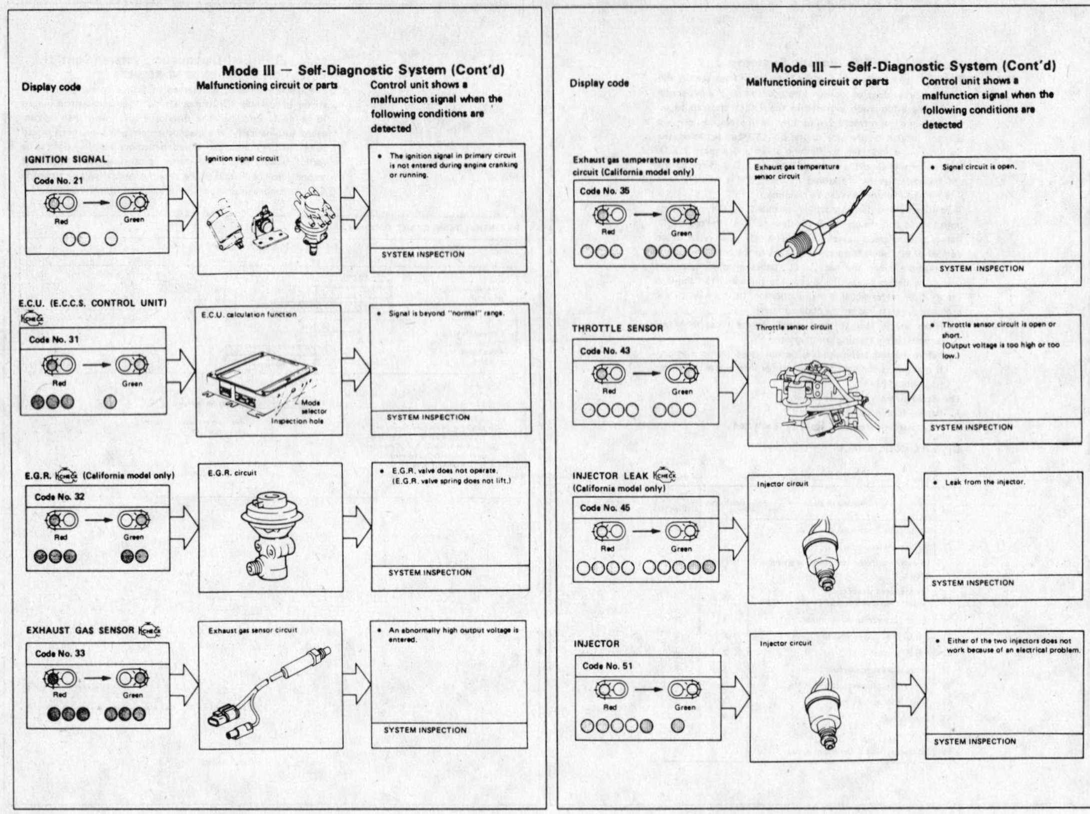
Mode III — Self-Diagnostic System (Cont'd)
Display code
Malfunctioning circuit or parts
Control unit shows a malfunction signal when the following conditions are detected
IGNITION SIGNAL
Code No. 21
Red
Green
Ignition signal circuit
• The ignition signal in primary circuit is not entered during engine cranking or running.
SYSTEM INSPECTION
E.C.U. (E.C.C.S. CONTROL UNIT)
Code No. 31
Red
Green
E.C.U. calculation function
Mode selector
Inspection hole
• Signal is beyond "normal" range.
SYSTEM INSPECTION
E.G.R. (California model only)
Code No. 32
Red
Green
E.G.R. circuit
• E.G.R. valve does not operate. (E.G.R. valve spring does not lift.)
SYSTEM INSPECTION
EXHAUST GAS SENSOR
Code No. 33
Red
Green
Exhaust gas sensor circuit
• An abnormally high output voltage is entered.
SYSTEM INSPECTION
Mode III — Self-Diagnostic System (Cont'd)
Display code
Malfunctioning circuit or parts
Control unit shows a malfunction signal when the following conditions are detected
Exhaust gas temperature sensor circuit (California model only)
Code No. 35
Red
Green
Exhaust gas temperature sensor circuit
• Signal circuit is open.
SYSTEM INSPECTION
THROTTLE SENSOR
Code No. 43
Red
Green
Throttle sensor circuit
• Throttle sensor circuit is open or short. (Output voltage is too high or too low.)
SYSTEM INSPECTION
INJECTOR LEAK (California model only)
Code No. 45
Red
Green
Injector circuit
• Leak from the injector.
SYSTEM INSPECTION
INJECTOR
Code No. 51
Red
Green
Injector circuit
• Either of the two injectors does not work because of an electrical problem.
SYSTEM INSPECTION

ECCS SELF-DIAGNOSTIC MODE III DISPLAY CODES (CONT.)—1988–89 TRUCK/ PATHFINDER

ECCS SELF-DIAGNOSTIC MODE IV 1988–89 TRUCK/PATHFINDER

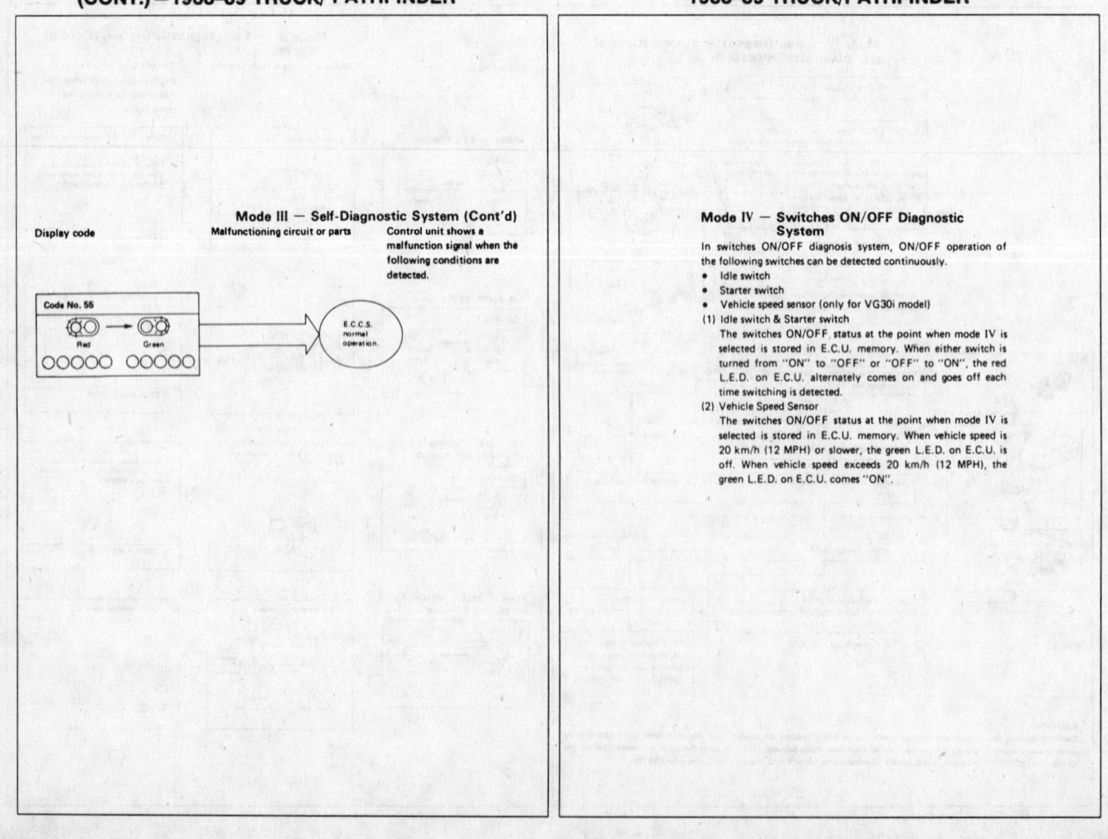
Mode III — Self-Diagnostic System (Cont'd)
Display code
Malfunctioning circuit or parts
Control unit shows a malfunction signal when the following conditions are detected.
Code No. 55
Red
Green
E.C.C.S. normal operation.

Mode IV — Switches ON/OFF Diagnostic System

In switches ON/OFF diagnosis system, ON/OFF operation of the following switches can be detected continuously.

- Idle switch
- Starter switch
- Vehicle speed sensor (only for VG30i model)

(1) Idle switch & Starter switch
The switches ON/OFF status at the point when mode IV is selected is stored in E.C.U. memory. When either switch is turned from "ON" to "OFF" or "OFF" to "ON", the red L.E.D. on E.C.U. alternately comes on and goes off each time switching is detected.

(2) Vehicle Speed Sensor
The switches ON/OFF status at the point when mode IV is selected is stored in E.C.U. memory. When vehicle speed is 20 km/h (12 MPH) or slower, the green L.E.D. on E.C.U. is off. When vehicle speed exceeds 20 km/h (12 MPH), the green L.E.D. on E.C.U. comes "ON".

ECCS SELF-DIAGNOSTIC MODE IV (CONT.) 1988–89 TRUCK/PATHFINDER

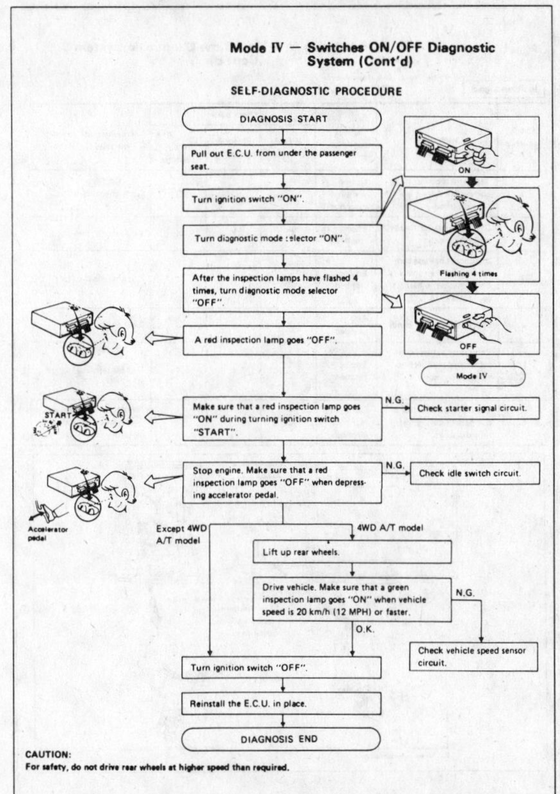
Mode IV — Switches ON/OFF Diagnostic System (Cont'd)
SELF-DIAGNOSTIC PROCEDURE
DIAGNOSIS START
Pull out E.C.U. from under the passenger seat.
Turn ignition switch "ON".
Turn diagnostic mode selector "ON".
After the inspection lamps have flashed 4 times, turn diagnostic mode selector "OFF".
A red inspection lamp goes "OFF".
Make sure that a red inspection lamp goes "ON" during turning ignition switch "START".
N.G.
Check starter signal circuit.
Stop engine. Make sure that a red inspection lamp goes "OFF" when depressing accelerator pedal.
N.G.
Check idle switch circuit.
Except 4WD A/T model
4WD A/T model
Lift up rear wheels.
Drive vehicle. Make sure that a green inspection lamp goes "ON" when vehicle speed is 20 km/h (12 MPH) or faster.
N.G.
O.K.
Check vehicle speed sensor circuit.
Turn ignition switch "OFF".
Reinstall the E.C.U. in place.
DIAGNOSIS END
ON
Flashing 4 times
OFF
Mode IV
START
Accelerator pedal
CAUTION:
For safety, do not drive rear wheels at higher speed than required.

ECCS SELF-DIAGNOSTIC MODE V 1988–89 TRUCK/PATHFINDER

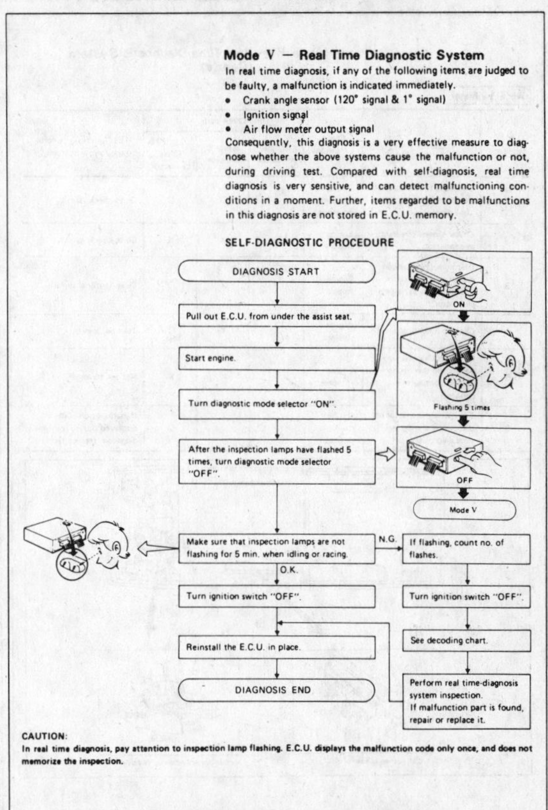
Mode V — Real Time Diagnostic System
In real time diagnosis, if any of the following items are judged to be faulty, a malfunction is indicated immediately.
• Crank angle sensor (120° signal & 1° signal)
• Ignition signal
• Air flow meter output signal
Consequently, this diagnosis is a very effective measure to diagnose whether the above systems cause the malfunction or not, during driving test. Compared with self-diagnosis, real time diagnosis is very sensitive, and can detect malfunctioning conditions in a moment. Further, items regarded to be malfunctions in this diagnosis are not stored in E.C.U. memory.
SELF-DIAGNOSTIC PROCEDURE
DIAGNOSIS START
Pull out E.C.U. from under the assist seat.
Start engine.
Turn diagnostic mode selector "ON".
After the inspection lamps have flashed 5 times, turn diagnostic mode selector "OFF".
Make sure that inspection lamps are not flashing for 5 min. when idling or racing.
N.G.
If flashing, count no. of flashes.
O.K.
Turn ignition switch "OFF".
Turn ignition switch "OFF".
See decoding chart.
Reinstall the E.C.U. in place.
Perform real time-diagnosis system inspection.
If malfunction part is found, repair or replace it.
DIAGNOSIS END
ON
Flashing 5 times
OFF
Mode V
CAUTION:
In real time diagnosis, pay attention to inspection lamp flashing. E.C.U. displays the malfunction code only once, and does not memorize the inspection.

ECCS SELF-DIAGNOSTIC MODE V DECODING CHART—1988–89 TRUCK/PATHFINDER

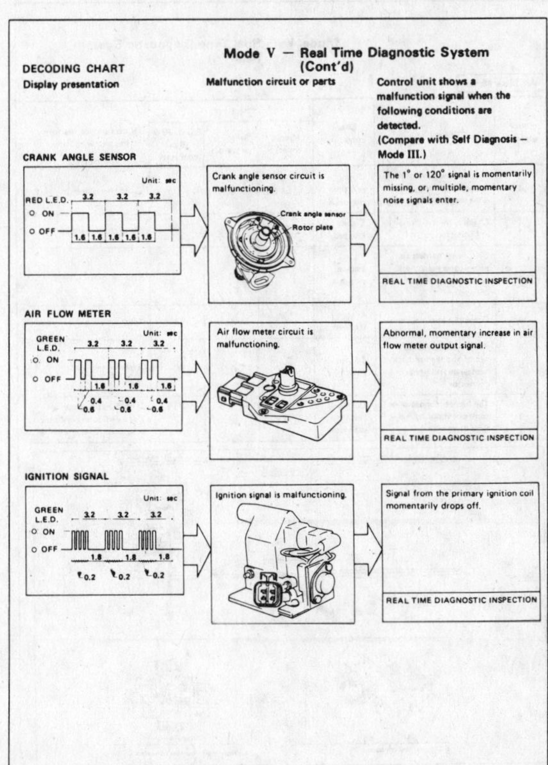
Mode V — Real Time Diagnostic System (Cont'd)
DECODING CHART
Display presentation
Malfunction circuit or parts
Control unit shows a malfunction signal when the following conditions are detected.
(Compare with Self Diagnosis — Mode III.)
CRANK ANGLE SENSOR
Unit: sec
RED L.E.D.
ON
OFF
3.2
1.6
Crank angle sensor circuit is malfunctioning.
Crank angle sensor
Rotor plate
The 1° or 120° signal is momentarily missing, or, multiple, momentary noise signals enter.
REAL TIME DIAGNOSTIC INSPECTION
AIR FLOW METER
GREEN L.E.D.
ON
OFF
3.2
1.6
0.4
0.6
Air flow meter circuit is malfunctioning.
Abnormal, momentary increase in air flow meter output signal.
REAL TIME DIAGNOSTIC INSPECTION
IGNITION SIGNAL
GREEN L.E.D.
ON
OFF
3.2
1.8
0.2
Ignition signal is malfunctioning.
Signal from the primary ignition coil momentarily drops off.
REAL TIME DIAGNOSTIC INSPECTION

ECCS SELF-DIAGNOSTIC MODE V INSPECTION CHART—1988–89 TRUCK/PATHFINDER WITH VG30I ENGINE

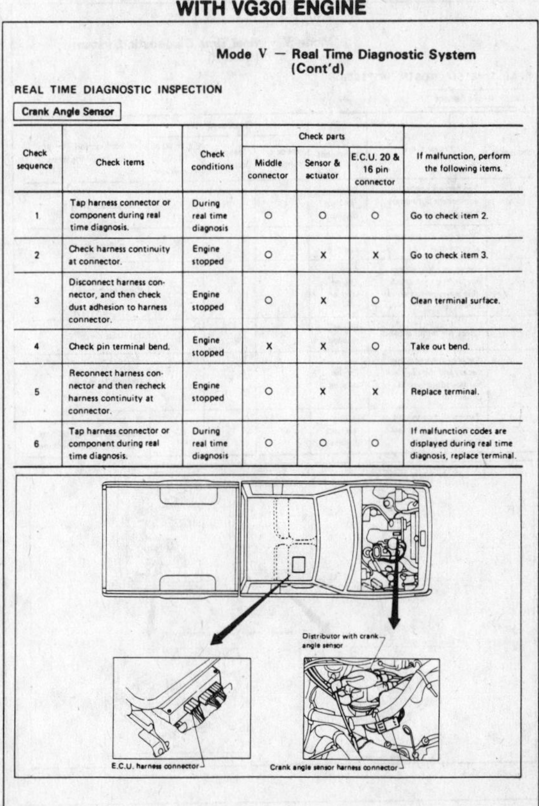

Mode V — Real Time Diagnostic System (Cont'd)

REAL TIME DIAGNOSTIC INSPECTION

Crank Angle Sensor

Check sequence	Check items	Check conditions	Check parts: Middle connector	Sensor & actuator	E.C.U. 20 & 16 pin connector	If malfunction, perform the following items.
1	Tap harness connector or component during real time diagnosis.	During real time diagnosis	O	O	O	Go to check item 2.
2	Check harness continuity at connector.	Engine stopped	O	X	X	Go to check item 3.
3	Disconnect harness connector, and then check dust adhesion to harness connector.	Engine stopped	O	X	O	Clean terminal surface.
4	Check pin terminal bend.	Engine stopped	X	X	O	Take out bend.
5	Reconnect harness connector and then recheck harness continuity at connector.	Engine stopped	O	X	X	Replace terminal.
6	Tap harness connector or component during real time diagnosis.	During real time diagnosis	O	O	O	If malfunction codes are displayed during real time diagnosis, replace terminal.

Distributor with crank-angle sensor
E.C.U. harness connector
Crank angle sensor harness connector

ECCS SELF-DIAGNOSTIC MODE V INSPECTION CHART (CONT.) 1988–89 TRUCK/ PATHFINDER WITH VG30I ENGINE

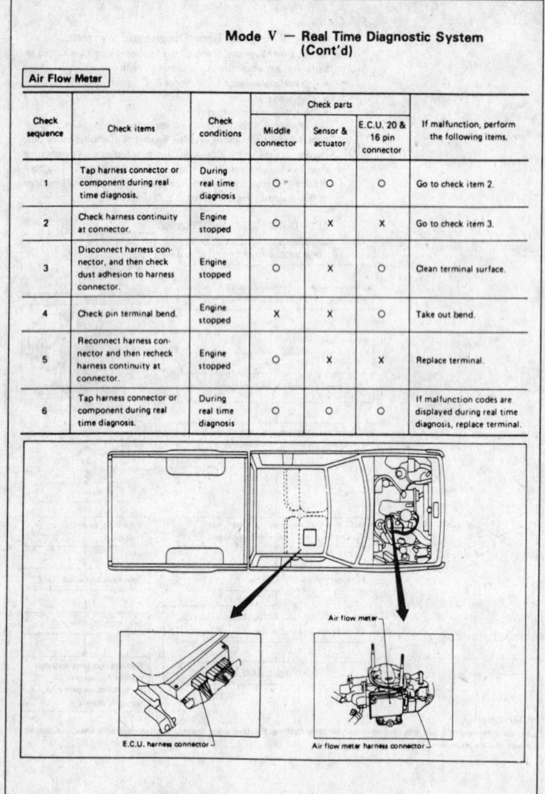

Mode V — Real Time Diagnostic System (Cont'd)

Air Flow Meter

Check sequence	Check items	Check conditions	Check parts: Middle connector	Sensor & actuator	E.C.U. 20 & 16 pin connector	If malfunction, perform the following items.
1	Tap harness connector or component during real time diagnosis.	During real time diagnosis	O	O	O	Go to check item 2.
2	Check harness continuity at connector.	Engine stopped	O	X	X	Go to check item 3.
3	Disconnect harness connector, and then check dust adhesion to harness connector.	Engine stopped	O	X	O	Clean terminal surface.
4	Check pin terminal bend.	Engine stopped	X	X	O	Take out bend.
5	Reconnect harness connector and then recheck harness continuity at connector.	Engine stopped	O	X	X	Replace terminal.
6	Tap harness connector or component during real time diagnosis.	During real time diagnosis	O	O	O	If malfunction codes are displayed during real time diagnosis, replace terminal.

Air flow meter
E.C.U. harness connector
Air flow meter harness connector

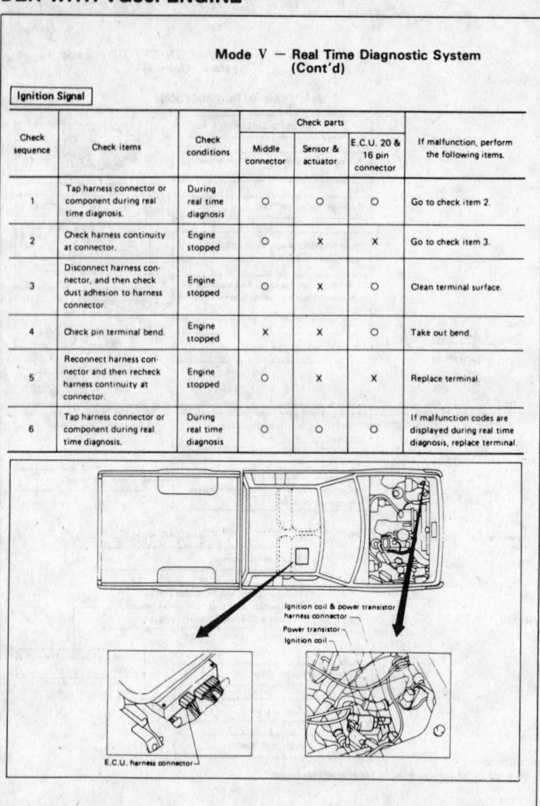

Mode V — Real Time Diagnostic System (Cont'd)

Ignition Signal

Check sequence	Check items	Check conditions	Check parts: Middle connector	Sensor & actuator	E.C.U. 20 & 16 pin connector	If malfunction, perform the following items.
1	Tap harness connector or component during real time diagnosis.	During real time diagnosis	O	O	O	Go to check item 2.
2	Check harness continuity at connector.	Engine stopped	O	X	X	Go to check item 3.
3	Disconnect harness connector, and then check dust adhesion to harness connector.	Engine stopped	O	X	O	Clean terminal surface.
4	Check pin terminal bend.	Engine stopped	X	X	O	Take out bend.
5	Reconnect harness connector and then recheck harness continuity at connector.	Engine stopped	O	X	X	Replace terminal.
6	Tap harness connector or component during real time diagnosis.	During real time diagnosis	O	O	O	If malfunction codes are displayed during real time diagnosis, replace terminal.

Ignition coil & power transistor harness connector
Power transistor
Ignition coil
E.C.U. harness connector

ECCS SELF-DIAGNOSTIC MODE V INSPECTION CHART — 1988–89 TRUCK/ PATHFINDER WITH Z24I ENGINE

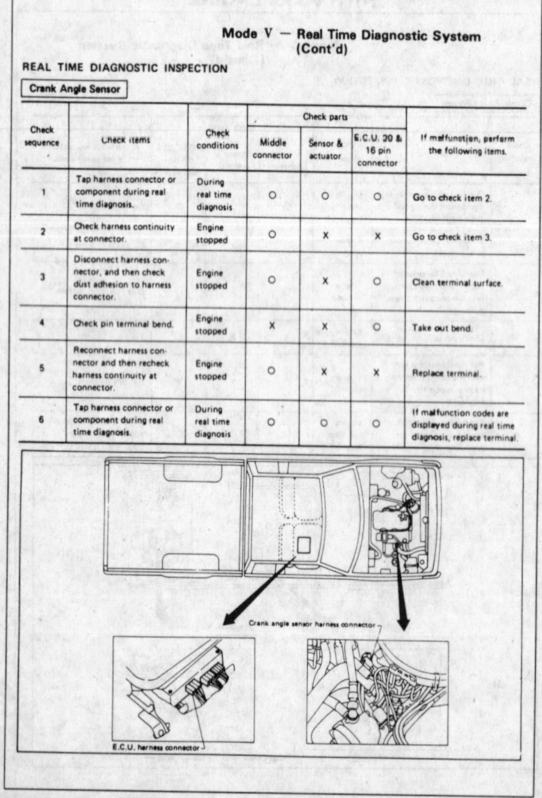

Mode V — Real Time Diagnostic System (Cont'd)

REAL TIME DIAGNOSTIC INSPECTION

Crank Angle Sensor

Check sequence	Check items	Check conditions	Check parts: Middle connector	Sensor & actuator	E.C.U. 20 & 16 pin connector	If malfunction, perform the following items.
1	Tap harness connector or component during real time diagnosis.	During real time diagnosis	O	O	O	Go to check item 2.
2	Check harness continuity at connector.	Engine stopped	O	X	X	Go to check item 3.
3	Disconnect harness connector, and then check dust adhesion to harness connector.	Engine stopped	O	X	O	Clean terminal surface.
4	Check pin terminal bend.	Engine stopped	X	X	O	Take out bend.
5	Reconnect harness connector and then recheck harness continuity at connector.	Engine stopped	O	X	X	Replace terminal.
6	Tap harness connector or component during real time diagnosis.	During real time diagnosis	O	O	O	If malfunction codes are displayed during real time diagnosis, replace terminal.

Crank angle sensor harness connector
E.C.U. harness connector

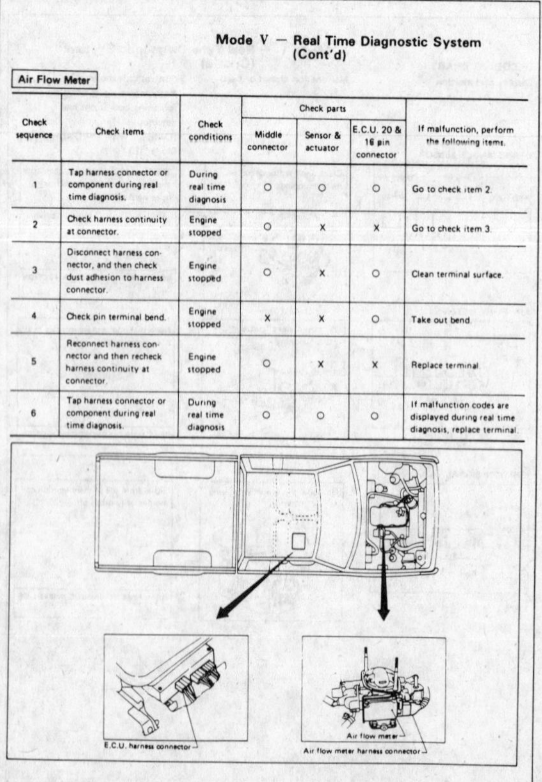

Mode V — Real Time Diagnostic System (Cont'd)

Air Flow Meter

Check sequence	Check items	Check conditions	Check parts: Middle connector	Sensor & actuator	E.C.U. 20 & 16 pin connector	If malfunction, perform the following items.
1	Tap harness connector or component during real time diagnosis.	During real time diagnosis	O	O	O	Go to check item 2.
2	Check harness continuity at connector.	Engine stopped	O	X	X	Go to check item 3.
3	Disconnect harness connector, and then check dust adhesion to harness connector.	Engine stopped	O	X	O	Clean terminal surface.
4	Check pin terminal bend.	Engine stopped	X	X	O	Take out bend.
5	Reconnect harness connector and then recheck harness continuity at connector.	Engine stopped	O	X	X	Replace terminal.
6	Tap harness connector or component during real time diagnosis.	During real time diagnosis	O	O	O	If malfunction codes are displayed during real time diagnosis, replace terminal.

Air flow meter
E.C.U. harness connector
Air flow meter harness connector

ECCS SELF-DIAGNOSTIC MODE V INSPECTION CHART (CONT.)—1988–89 TRUCK/PATHFINDER WITH Z24I ENGINE

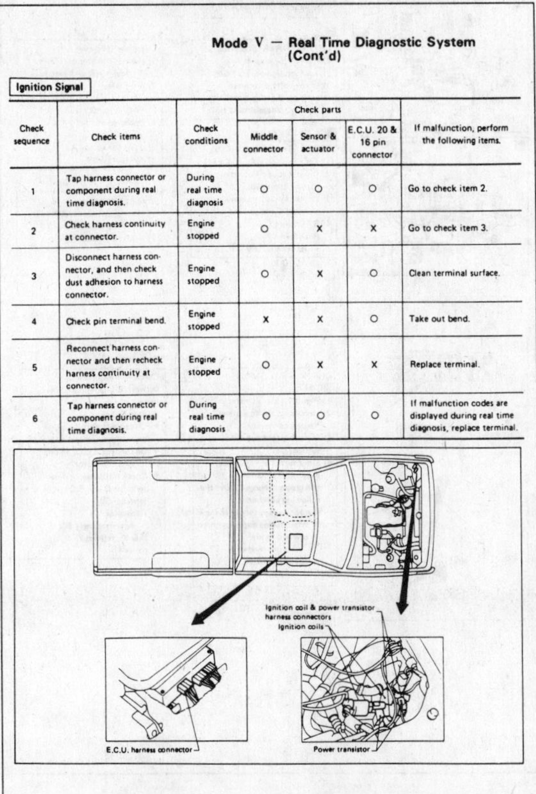

Mode V — Real Time Diagnostic System (Cont'd)

Ignition Signal

Check sequence	Check items	Check conditions	Check parts: Middle connector	Check parts: Sensor & actuator	Check parts: E.C.U. 20 & 16 pin connector	If malfunction, perform the following items.
1	Tap harness connector or component during real time diagnosis.	During real time diagnosis	O	O	O	Go to check item 2.
2	Check harness continuity at connector.	Engine stopped	O	X	X	Go to check item 3.
3	Disconnect harness connector, and then check dust adhesion to harness connector.	Engine stopped	O	X	O	Clean terminal surface.
4	Check pin terminal bend.	Engine stopped	X	X	O	Take out bend.
5	Reconnect harness connector and then recheck harness continuity at connector.	Engine stopped	O	X	X	Replace terminal.
6	Tap harness connector or component during real time diagnosis.	During real time diagnosis	O	O	O	If malfunction codes are displayed during real time diagnosis, replace terminal.

Ignition coil & power transistor harness connectors
Ignition coils
E.C.U. harness connector
Power transistor

ECCS SYSTEM INSPECTION CAUTION 1988–89 TRUCK/PATHFINDER

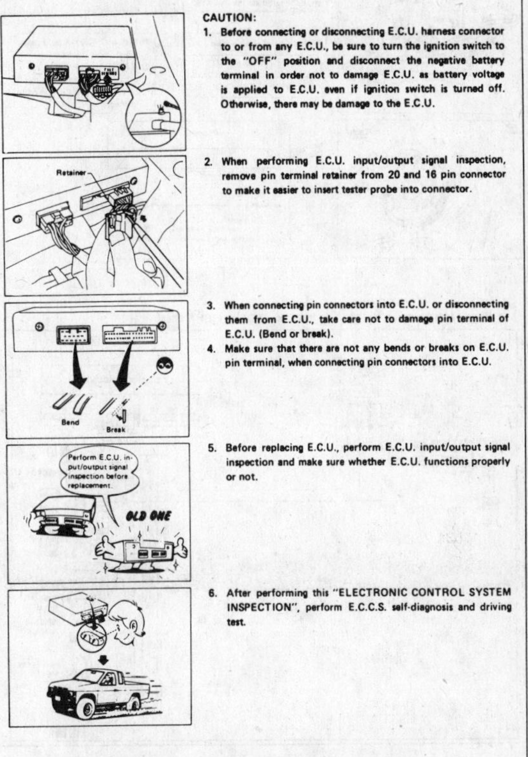

Retainer
Bend
Break
Perform E.C.U. input/output signal inspection before replacement
OLD ONE

CAUTION:

1. Before connecting or disconnecting E.C.U. harness connector to or from any E.C.U., be sure to turn the ignition switch to the "OFF" position and disconnect the negative battery terminal in order not to damage E.C.U. as battery voltage is applied to E.C.U. even if ignition switch is turned off. Otherwise, there may be damage to the E.C.U.
2. When performing E.C.U. input/output signal inspection, remove pin terminal retainer from 20 and 16 pin connector to make it easier to insert tester probe into connector.
3. When connecting pin connectors into E.C.U. or disconnecting them from E.C.U., take care not to damage pin terminal of E.C.U. (Bend or break).
4. Make sure that there are not any bends or breaks on E.C.U. pin terminal, when connecting pin connectors into E.C.U.
5. Before replacing E.C.U., perform E.C.U. input/output signal inspection and make sure whether E.C.U. functions properly or not.
6. After performing this "ELECTRONIC CONTROL SYSTEM INSPECTION", perform E.C.C.S. self-diagnosis and driving test.

ECCS SYSTEM INSPECTION CAUTION (CONT.) 1988–89 TRUCK/PATHFINDER

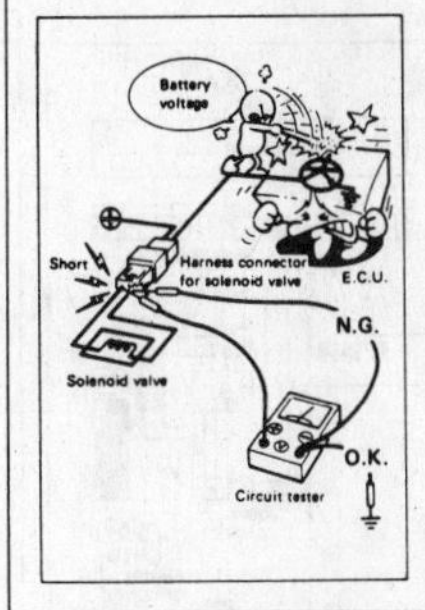

Battery voltage
Short
Harness connector for solenoid valve
E.C.U.
N.G.
Solenoid valve
O.K.
Circuit tester

7. When measuring supply voltage of E.C.U. controlled components with a circuit tester, separate one tester probe from the other.
   If the two tester probes accidentally make contact with each other during measurement, the circuit will be shorted, resulting in damage to the power transistor of the control unit.
8. Keys to symbols
   DISCONNECT : Check after disconnecting the connector to be measured.
   CONNECT : Check after connecting the connector to be measured.
9. When measuring voltage or resistance at connector with tester probes, there are two methods of measurement; one is done from terminal side and the other from harness side. Before measuring, confirm symbol mark again.
   HS : Inspection should be done from harness side.
   TS : Inspection should be done from terminal side.

ECCS SELF-DIAGNOSTIC CHART 1988–89 TRUCK/PATHFINDER WITH VG30I ENGINE

CRANK ANGLE SENSOR (Code No. 11)

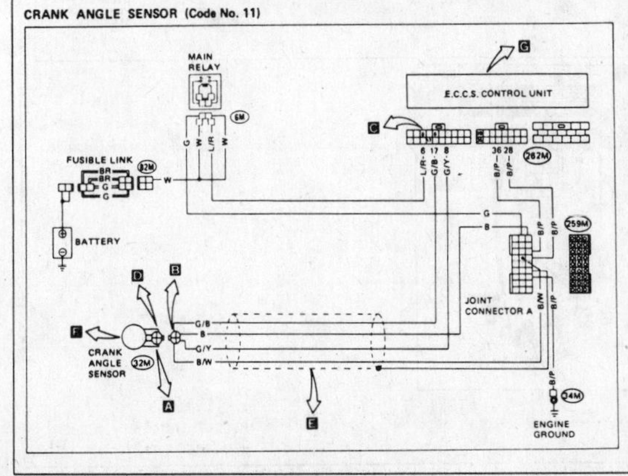

MAIN RELAY
FUSIBLE LINK
BATTERY
E.C.C.S. CONTROL UNIT
JOINT CONNECTOR A
CRANK ANGLE SENSOR
ENGINE GROUND

ECCS SELF-DIAGNOSTIC CHART 1988–89 TRUCK/PATHFINDER WITH VG30I ENGINE

CRANK ANGLE SENSOR (Code No. 11)

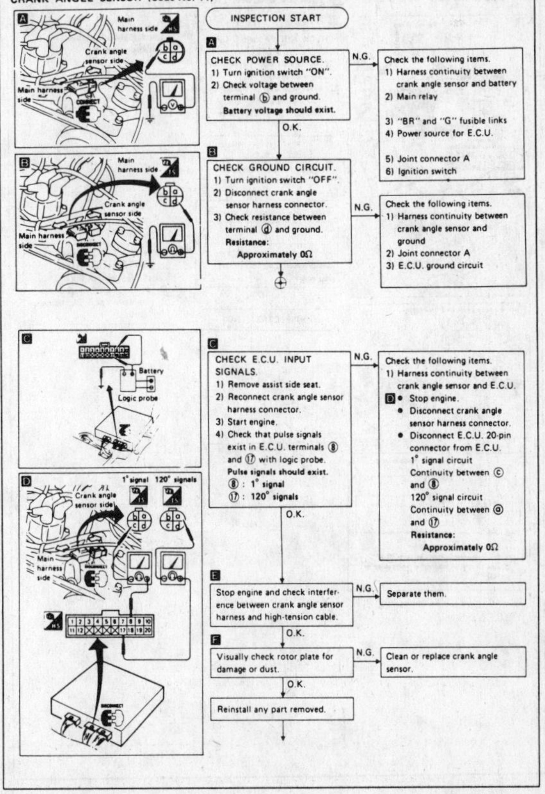

INSPECTION START
A
CHECK POWER SOURCE.
1) Turn ignition switch "ON".
2) Check voltage between terminal ⓑ and ground.
Battery voltage should exist.
N.G.
Check the following items.
1) Harness continuity between crank angle sensor and battery
2) Main relay
3) "BR" and "G" fusible links
4) Power source for E.C.U.
5) Joint connector A
6) Ignition switch
O.K.
B
CHECK GROUND CIRCUIT.
1) Turn ignition switch "OFF".
2) Disconnect crank angle sensor harness connector.
3) Check resistance between terminal ⓓ and ground.
Resistance:
Approximately 0Ω
N.G.
Check the following items.
1) Harness continuity between crank angle sensor and ground
2) Joint connector A
3) E.C.U. ground circuit
C
CHECK E.C.U. INPUT SIGNALS.
1) Remove assist side seat.
2) Reconnect crank angle sensor harness connector.
3) Start engine.
4) Check that pulse signals exist in E.C.U. terminals ⑧ and ⑰ with logic probe.
Pulse signals should exist.
⑧ : 1° signal
⑰ : 120° signals
N.G.
Check the following items.
1) Harness continuity between crank angle sensor and E.C.U.
D • Stop engine.
• Disconnect crank angle sensor harness connector.
• Disconnect E.C.U. 20-pin connector from E.C.U.
1° signal circuit
Continuity between ⓒ and ⑧
120° signal circuit
Continuity between ⓐ and ⑰
Resistance:
Approximately 0Ω
O.K.
E
Stop engine and check interference between crank angle sensor harness and high-tension cable.
N.G.
Separate them.
O.K.
F
Visually check rotor plate for damage or dust.
N.G.
Clean or replace crank angle sensor.
O.K.
Reinstall any part removed.
Main harness side
Crank angle sensor side
Battery
Logic probe
1° signal
120° signals

ECCS SELF-DIAGNOSTIC CHART—1988-89 TRUCK/PATHFINDER WITH VG30I ENGINE

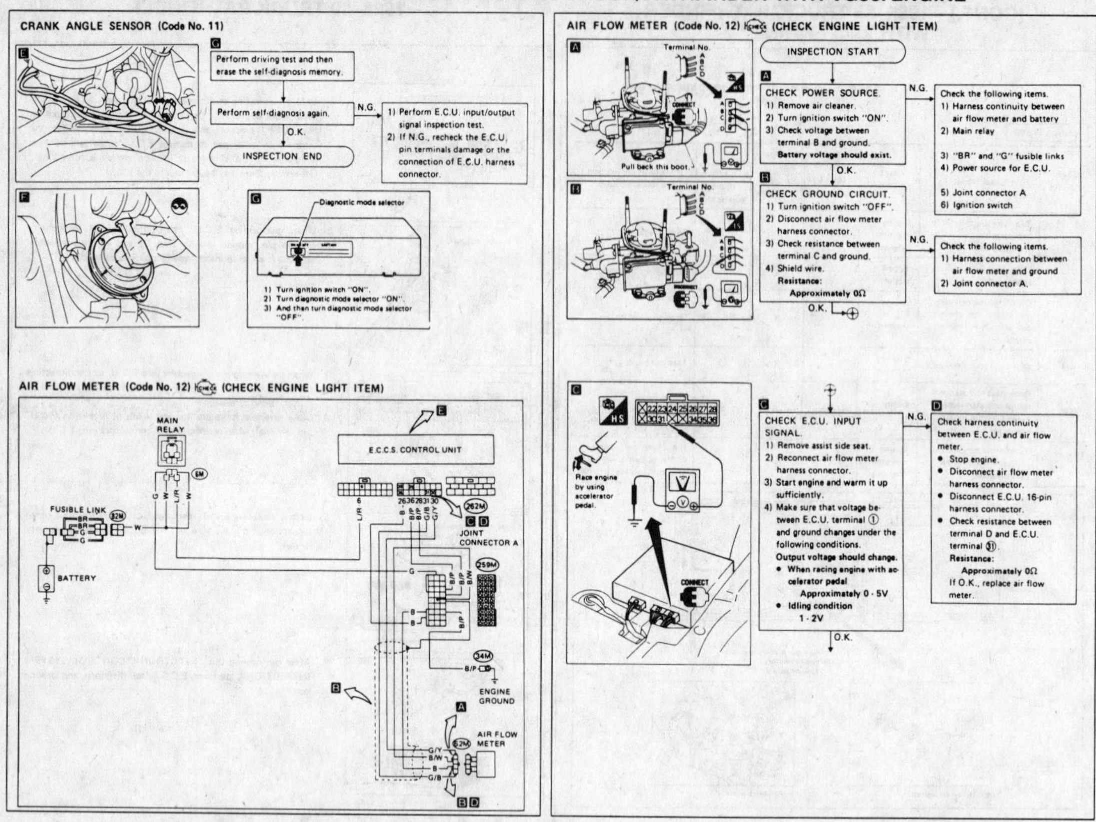
CRANK ANGLE SENSOR (Code No. 11)
Perform driving test and then erase the self-diagnosis memory.
Perform self-diagnosis again.
N.G.
1) Perform E.C.U. input/output signal inspection test.
2) If N.G., recheck the E.C.U. pin terminals damage or the connection of E.C.U. harness connector.
O.K.
INSPECTION END
Diagnostic mode selector
1) Turn ignition switch "ON".
2) Turn diagnostic mode selector "ON".
3) And then turn diagnostic mode selector "OFF".
AIR FLOW METER (Code No. 12) (CHECK ENGINE LIGHT ITEM)
MAIN RELAY
E.C.C.S. CONTROL UNIT
FUSIBLE LINK
BATTERY
JOINT CONNECTOR A
ENGINE GROUND
AIR FLOW METER
AIR FLOW METER (Code No. 12) (CHECK ENGINE LIGHT ITEM)
Terminal No.
Pull back this boot.
INSPECTION START
CHECK POWER SOURCE.
1) Remove air cleaner.
2) Turn ignition switch "ON".
3) Check voltage between terminal B and ground.
Battery voltage should exist.
N.G.
Check the following items.
1) Harness continuity between air flow meter and battery
2) Main relay
3) "BR" and "G" fusible links
4) Power source for E.C.U.
5) Joint connector A
6) Ignition switch
O.K.
CHECK GROUND CIRCUIT.
1) Turn ignition switch "OFF".
2) Disconnect air flow meter harness connector.
3) Check resistance between terminal C and ground.
4) Shield wire.
Resistance:
Approximately 0Ω
N.G.
Check the following items.
1) Harness connection between air flow meter and ground
2) Joint connector A.
O.K.
Race engine by using accelerator pedal.
CHECK E.C.U. INPUT SIGNAL.
1) Remove assist side seat.
2) Reconnect air flow meter harness connector.
3) Start engine and warm it up sufficiently.
4) Make sure that voltage between E.C.U. terminal ① and ground changes under the following conditions.
Output voltage should change.
• When racing engine with accelerator pedal
Approximately 0 - 5V
• Idling condition
1 - 2V
O.K.
N.G.
Check harness continuity between E.C.U. and air flow meter.
• Stop engine.
• Disconnect air flow meter harness connector.
• Disconnect E.C.U. 16-pin harness connector.
• Check resistance between terminal D and E.C.U. terminal ㉛.
Resistance:
Approximately 0Ω
If O.K., replace air flow meter.

ECCS SELF-DIAGNOSTIC CHART—1988-89 TRUCK/PATHFINDER WITH VG30I ENGINE

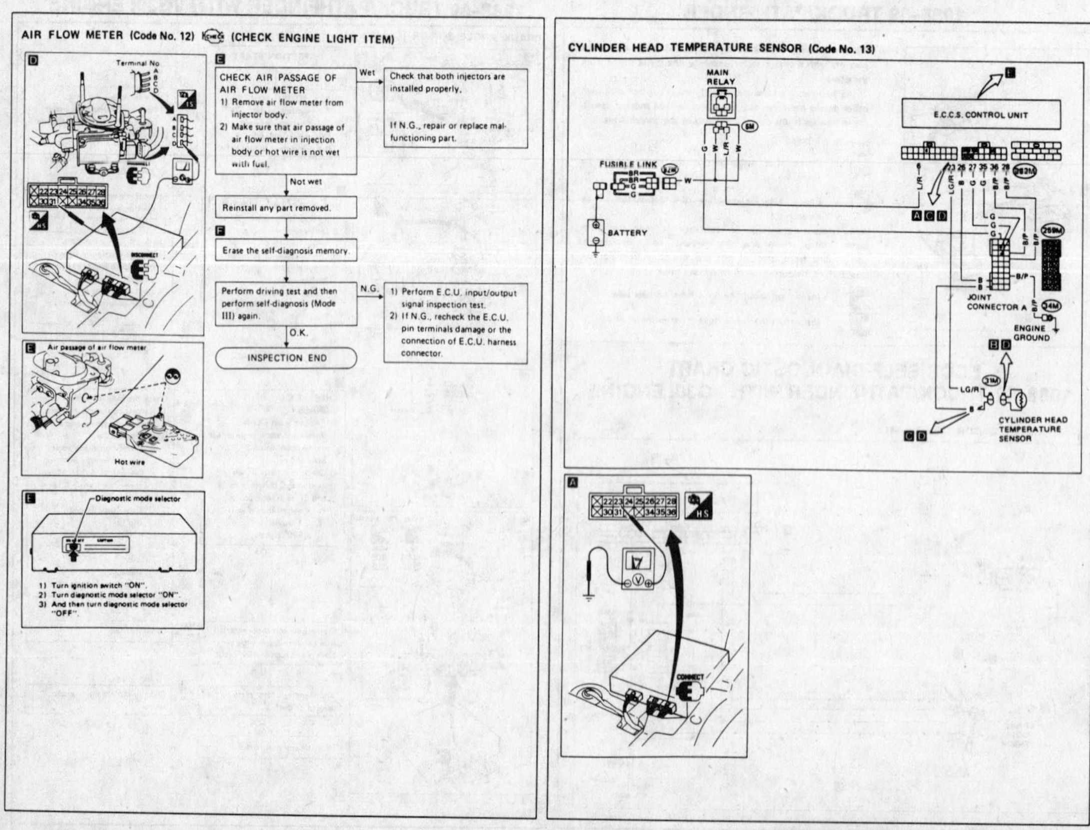
AIR FLOW METER (Code No. 12) (CHECK ENGINE LIGHT ITEM)
Terminal No.
CHECK AIR PASSAGE OF AIR FLOW METER
1) Remove air flow meter from injector body.
2) Make sure that air passage of air flow meter in injection body or hot wire is not wet with fuel.
Wet
Check that both injectors are installed properly.
If N.G., repair or replace malfunctioning part.
Not wet
Reinstall any part removed.
Erase the self-diagnosis memory.
Perform driving test and then perform self-diagnosis (Mode III) again.
N.G.
1) Perform E.C.U. input/output signal inspection test.
2) If N.G., recheck the E.C.U. pin terminals damage or the connection of E.C.U. harness connector.
O.K.
INSPECTION END
Air passage of air flow meter
Hot wire
Diagnostic mode selector
1) Turn ignition switch "ON".
2) Turn diagnostic mode selector "ON".
3) And then turn diagnostic mode selector "OFF".
CYLINDER HEAD TEMPERATURE SENSOR (Code No. 13)
MAIN RELAY
E.C.C.S. CONTROL UNIT
FUSIBLE LINK
BATTERY
JOINT CONNECTOR A
ENGINE GROUND
CYLINDER HEAD TEMPERATURE SENSOR

ECCS SELF-DIAGNOSTIC CHART—1988–89 TRUCK/PATHFINDER WITH VG30I ENGINE

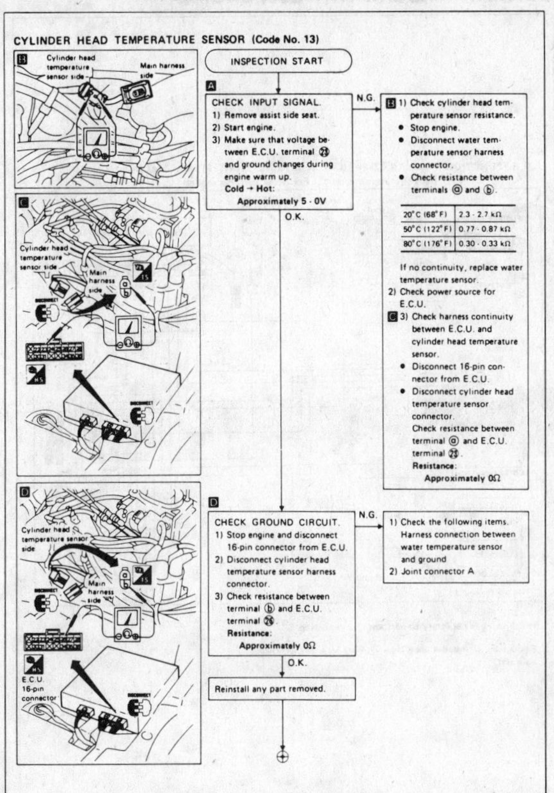
CYLINDER HEAD TEMPERATURE SENSOR (Code No. 13)
Cylinder head temperature sensor side
Main harness side
INSPECTION START
A
CHECK INPUT SIGNAL.
1) Remove assist side seat.
2) Start engine.
3) Make sure that voltage between E.C.U. terminal ㉓ and ground changes during engine warm up.
Cold → Hot:
Approximately 5 - 0V
N.G.
B 1) Check cylinder head temperature sensor resistance.
• Stop engine.
• Disconnect water temperature sensor harness connector.
• Check resistance between terminals ⓐ and ⓑ.
20°C (68°F) 2.3 - 2.7 kΩ
50°C (122°F) 0.77 - 0.87 kΩ
80°C (176°F) 0.30 - 0.33 kΩ
If no continuity, replace water temperature sensor.
2) Check power source for E.C.U.
C 3) Check harness continuity between E.C.U. and cylinder head temperature sensor.
• Disconnect 16-pin connector from E.C.U.
• Disconnect cylinder head temperature sensor connector.
Check resistance between terminal ⓐ and E.C.U. terminal ㉓.
Resistance:
Approximately 0Ω
O.K.
D
CHECK GROUND CIRCUIT.
1) Stop engine and disconnect 16-pin connector from E.C.U.
2) Disconnect cylinder head temperature sensor harness connector.
3) Check resistance between terminal ⓑ and E.C.U. terminal ㉖.
Resistance:
Approximately 0Ω
N.G.
1) Check the following items.
Harness connection between water temperature sensor and ground
2) Joint connector A
O.K.
Reinstall any part removed.
E.C.U. 16-pin connector

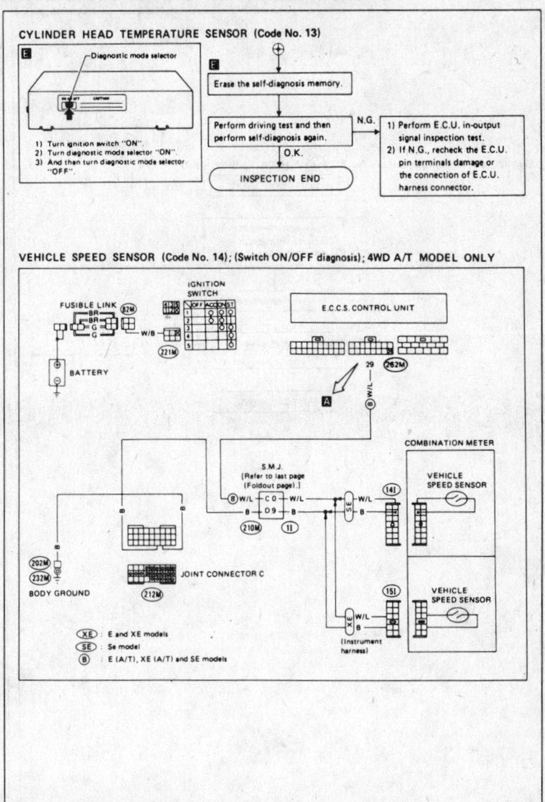
CYLINDER HEAD TEMPERATURE SENSOR (Code No. 13)
E
Diagnostic mode selector
1) Turn ignition switch "ON".
2) Turn diagnostic mode selector "ON".
3) And then turn diagnostic mode selector "OFF".
Erase the self-diagnosis memory.
Perform driving test and then perform self-diagnosis again.
N.G.
1) Perform E.C.U. in-output signal inspection test.
2) If N.G., recheck the E.C.U. pin terminals damage or the connection of E.C.U. harness connector.
O.K.
INSPECTION END
VEHICLE SPEED SENSOR (Code No. 14); (Switch ON/OFF diagnosis); 4WD A/T MODEL ONLY
FUSIBLE LINK
IGNITION SWITCH
E.C.C.S. CONTROL UNIT
BATTERY
COMBINATION METER
VEHICLE SPEED SENSOR
S.M.J. (Refer to last page (Foldout page).)
JOINT CONNECTOR C
BODY GROUND
E and XE models
Se model
E (A/T), XE (A/T) and SE models
(Instrument harness)

ECCS SELF-DIAGNOSTIC CHART—1988–89 TRUCK/PATHFINDER WITH VG30I ENGINE

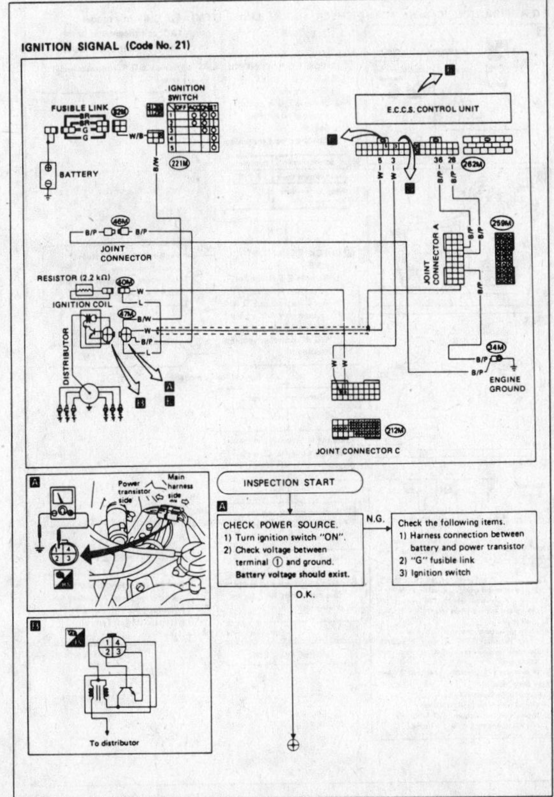
IGNITION SIGNAL (Code No. 21)
FUSIBLE LINK
IGNITION SWITCH
BATTERY
E.C.C.S. CONTROL UNIT
JOINT CONNECTOR
RESISTOR (2.2 kΩ)
IGNITION COIL
DISTRIBUTOR
JOINT CONNECTOR A
ENGINE GROUND
JOINT CONNECTOR C
Power transistor
Main harness side
INSPECTION START
A
CHECK POWER SOURCE.
1) Turn ignition switch "ON".
2) Check voltage between terminal ① and ground.
Battery voltage should exist.
N.G.
Check the following items.
1) Harness connection between battery and power transistor
2) "G" fusible link
3) Ignition switch
O.K.
To distributor

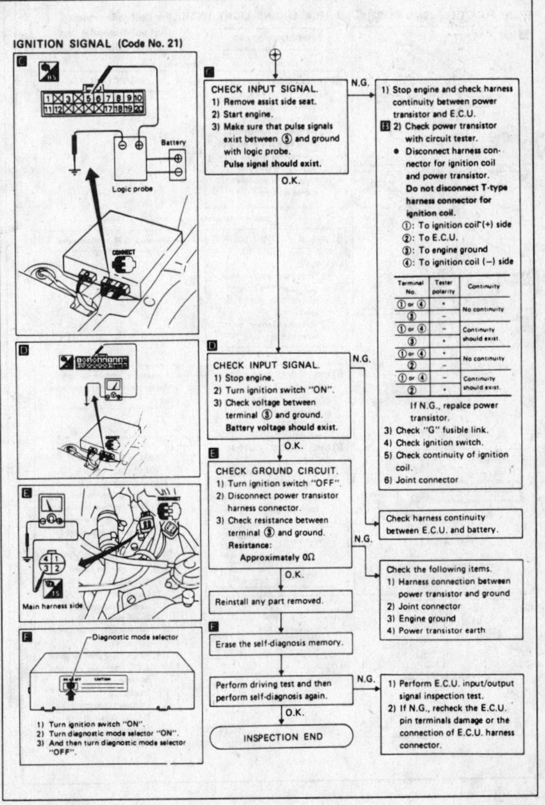
IGNITION SIGNAL (Code No. 21)
Battery
Logic probe
C
CHECK INPUT SIGNAL.
1) Remove assist side seat.
2) Start engine.
3) Make sure that pulse signals exist between ⑤ and ground with logic probe.
Pulse signal should exist.
N.G.
1) Stop engine and check harness continuity between power transistor and E.C.U.
B 2) Check power transistor with circuit tester.
• Disconnect harness connector for ignition coil and power transistor.
Do not disconnect T-type harness connector for ignition coil.
①: To ignition coil (+) side
②: To E.C.U.
③: To engine ground
④: To ignition coil (–) side
Terminal No. | Tester polarity | Continuity
① or ④ | + | No continuity
③ | – |
① or ④ | – | Continuity should exist.
③ | + |
① or ④ | + | No continuity
② | – |
① or ④ | – | Continuity should exist.
② | + |
If N.G., repalce power transistor.
3) Check "G" fusible link.
4) Check ignition switch.
5) Check continuity of ignition coil.
6) Joint connector
O.K.
D
CHECK INPUT SIGNAL.
1) Stop engine.
2) Turn ignition switch "ON".
3) Check voltage between terminal ③ and ground.
Battery voltage should exist.
N.G.
Check harness continuity between E.C.U. and battery.
O.K.
E
CHECK GROUND CIRCUIT.
1) Turn ignition switch "OFF".
2) Disconnect power transistor harness connector.
3) Check resistance between terminal ③ and ground.
Resistance:
Approximately 0Ω
N.G.
Check the following items.
1) Harness connection between power transistor and ground
2) Joint connector
3) Engine ground
4) Power transistor earth
O.K.
Reinstall any part removed.
Main harness side
F
Diagnostic mode selector
1) Turn ignition switch "ON".
2) Turn diagnostic mode selector "ON".
3) And then turn diagnostic mode selector "OFF".
Erase the self-diagnosis memory.
Perform driving test and then perform self-diagnosis again.
N.G.
1) Perform E.C.U. input/output signal inspection test.
2) If N.G., recheck the E.C.U. pin terminals damage or the connection of E.C.U. harness connector.
O.K.
INSPECTION END

ECCS SELF-DIAGNOSTIC CHART—1988–89 TRUCK/PATHFINDER WITH VG30I ENGINE

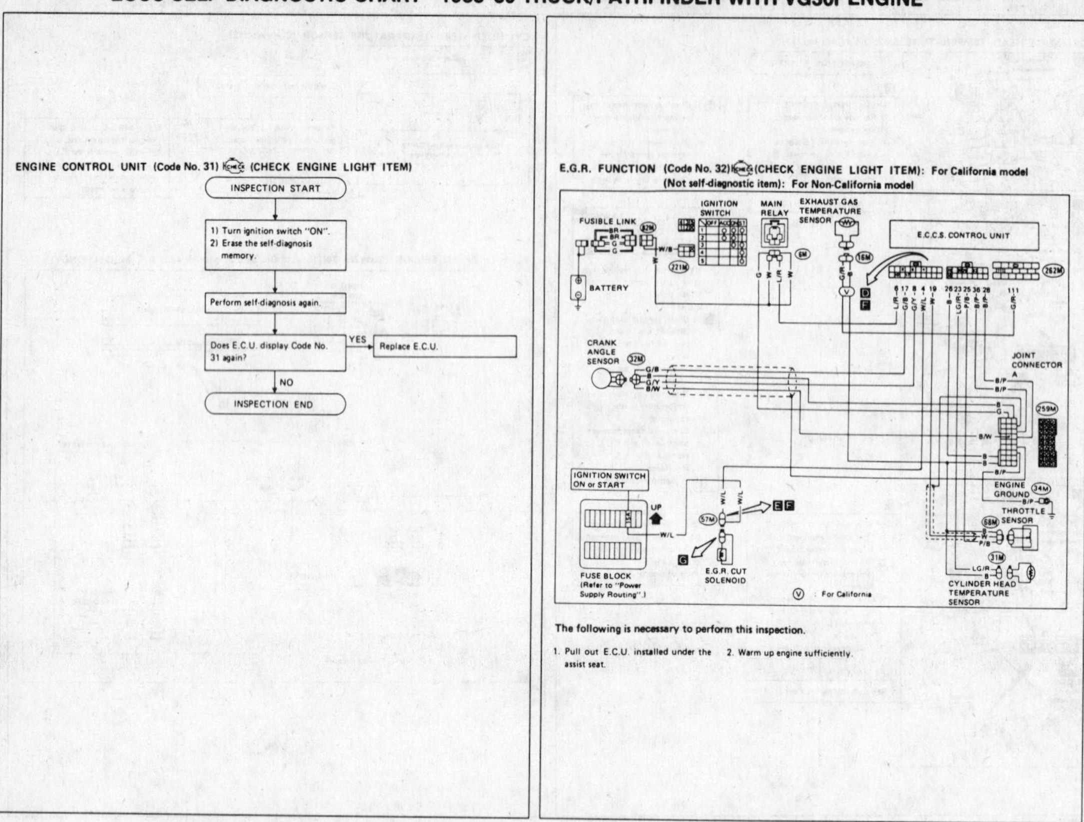
ENGINE CONTROL UNIT (Code No. 31) (CHECK ENGINE LIGHT ITEM)
INSPECTION START
1) Turn ignition switch "ON".
2) Erase the self-diagnosis memory.
Perform self-diagnosis again.
Does E.C.U. display Code No. 31 again?
YES
Replace E.C.U.
NO
INSPECTION END
E.G.R. FUNCTION (Code No. 32) (CHECK ENGINE LIGHT ITEM): For California model
(Not self-diagnostic item): For Non-California model
FUSIBLE LINK
IGNITION SWITCH
MAIN RELAY
EXHAUST GAS TEMPERATURE SENSOR
E.C.C.S. CONTROL UNIT
BATTERY
CRANK ANGLE SENSOR
JOINT CONNECTOR A
IGNITION SWITCH ON or START
FUSE BLOCK (Refer to "Power Supply Routing".)
E.G.R. CUT SOLENOID
ENGINE GROUND
THROTTLE SENSOR
CYLINDER HEAD TEMPERATURE SENSOR
Ⓥ : For California
The following is necessary to perform this inspection.
1. Pull out E.C.U. installed under the assist seat.
2. Warm up engine sufficiently.

ECCS SELF-DIAGNOSTIC CHART—1988–89 TRUCK/PATHFINDER WITH VG30I ENGINE

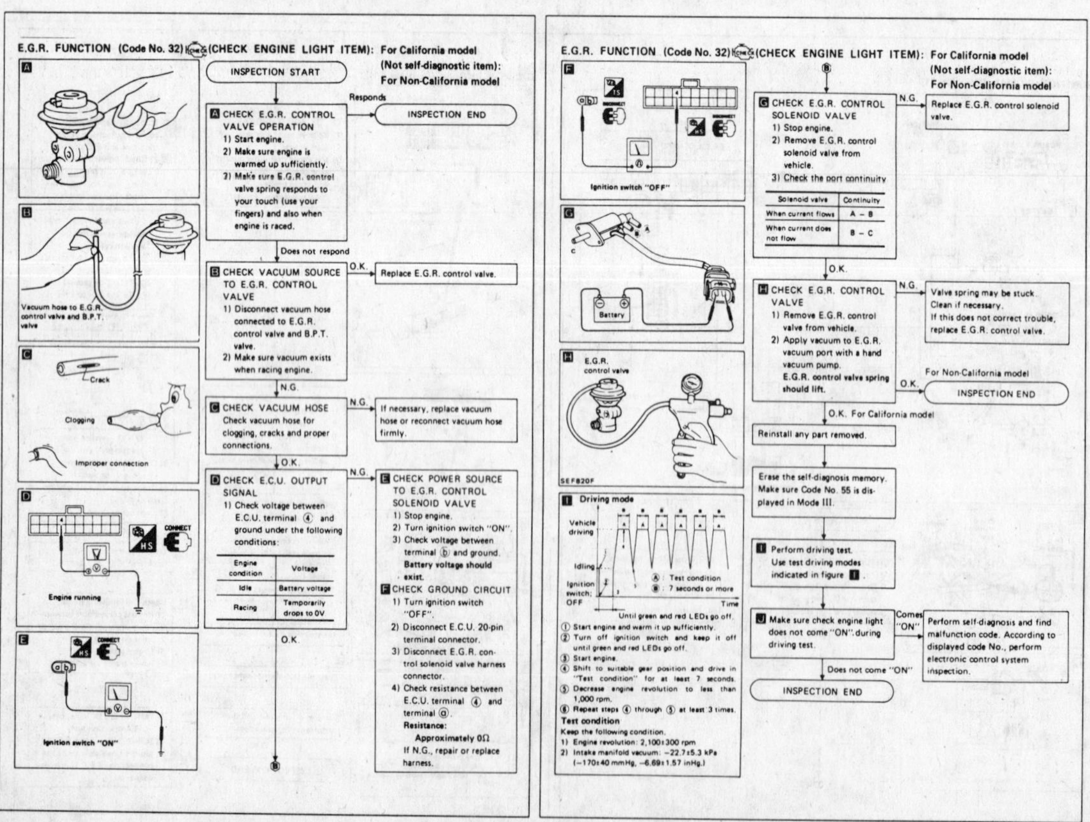
E.G.R. FUNCTION (Code No. 32) (CHECK ENGINE LIGHT ITEM): For California model
(Not self-diagnostic item):
For Non-California model
INSPECTION START
A CHECK E.G.R. CONTROL VALVE OPERATION
1) Start engine.
2) Make sure engine is warmed up sufficiently.
3) Make sure E.G.R. control valve spring responds to your touch (use your fingers) and also when engine is raced.
Responds
INSPECTION END
Does not respond
B CHECK VACUUM SOURCE TO E.G.R. CONTROL VALVE
1) Disconnect vacuum hose connected to E.G.R. control valve and B.P.T. valve.
2) Make sure vacuum exists when racing engine.
O.K.
Replace E.G.R. control valve.
N.G.
C CHECK VACUUM HOSE
Check vacuum hose for clogging, cracks and proper connections.
N.G.
If necessary, replace vacuum hose or reconnect vacuum hose firmly.
O.K.
D CHECK E.C.U. OUTPUT SIGNAL
1) Check voltage between E.C.U. terminal ④ and ground under the following conditions:
Engine condition | Voltage
Idle | Battery voltage
Racing | Temporarily drops to 0V
N.G.
E CHECK POWER SOURCE TO E.G.R. CONTROL SOLENOID VALVE
1) Stop engine.
2) Turn ignition switch "ON".
3) Check voltage between terminal ⓑ and ground.
Battery voltage should exist.
F CHECK GROUND CIRCUIT
1) Turn ignition switch "OFF".
2) Disconnect E.C.U. 20-pin terminal connector.
3) Disconnect E.G.R. control solenoid valve harness connector.
4) Check resistance between E.C.U. terminal ④ and terminal ⓐ.
Resistance:
Approximately 0Ω
If N.G., repair or replace harness.
O.K.
Vacuum hose to E.G.R. control valve and B.P.T. valve
Crack
Clogging
Improper connection
Engine running
Ignition switch "ON"
Ignition switch "OFF"
Battery
E.G.R. control valve
SEF820F
G CHECK E.G.R. CONTROL SOLENOID VALVE
1) Stop engine.
2) Remove E.G.R. control solenoid valve from vehicle.
3) Check the port continuity.
Solenoid valve | Continuity
When current flows | A – B
When current does not flow | B – C
N.G.
Replace E.G.R. control solenoid valve.
O.K.
H CHECK E.G.R. CONTROL VALVE
1) Remove E.G.R. control valve from vehicle.
2) Apply vacuum to E.G.R. vacuum port with a hand vacuum pump.
E.G.R. control valve spring should lift.
N.G.
Valve spring may be stuck. Clean if necessary.
If this does not correct trouble, replace E.G.R. control valve.
For Non-California model
O.K.
INSPECTION END
O.K. For California model
Reinstall any part removed.
Erase the self-diagnosis memory. Make sure Code No. 55 is displayed in Mode III.
I Perform driving test.
Use test driving modes indicated in figure I.
J Make sure check engine light does not come "ON" during driving test.
Comes "ON"
Perform self-diagnosis and find malfunction code. According to displayed code No., perform electronic control system inspection.
Does not come "ON"
INSPECTION END
I Driving mode
Vehicle driving
Idling
Ignition switch: OFF
Time
Ⓐ : Test condition
Ⓑ : 7 seconds or more
Until green and red LEDs go off.
① Start engine and warm it up sufficiently.
② Turn off ignition switch and keep it off until green and red LEDs go off.
③ Start engine.
④ Shift to suitable gear position and drive in "Test condition" for at least 7 seconds.
⑤ Decrease engine revolution to less than 1,000 rpm.
⑥ Repeat steps ④ through ⑤ at least 3 times.
Test condition
Keep the following condition.
1) Engine revolution: 2,100±300 rpm
2) Intake manifold vacuum: −22.7±5.3 kPa (−170±40 mmHg, −6.69±1.57 inHg)

ECCS SELF-DIAGNOSTIC CHART—1988–89 TRUCK/PATHFINDER WITH VG30I ENGINE

E.G.R. FUNCTION (Code No. 32) (CHECK ENGINE LIGHT ITEM): For California model
(Not self-diagnostic item): For Non-California model
J
CHECK
CHECK ENGINE LIGHT

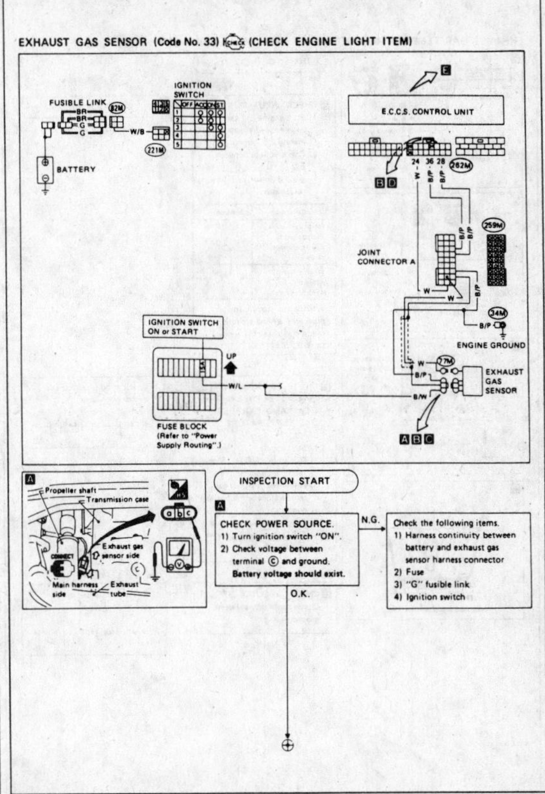
EXHAUST GAS SENSOR (Code No. 33) (CHECK ENGINE LIGHT ITEM)
FUSIBLE LINK
IGNITION SWITCH
BATTERY
E.C.C.S. CONTROL UNIT
24 36 28
262M
259M
JOINT CONNECTOR A
34M
ENGINE GROUND
IGNITION SWITCH ON or START
UP
W/L
FUSE BLOCK (Refer to "Power Supply Routing".)
EXHAUST GAS SENSOR
A
Propeller shaft
Transmission case
Exhaust gas sensor side
Main harness side
Exhaust tube
INSPECTION START
A
CHECK POWER SOURCE.
1) Turn ignition switch "ON".
2) Check voltage between terminal ⓒ and ground.
Battery voltage should exist.
N.G.
Check the following items.
1) Harness continuity between battery and exhaust gas sensor harness connector
2) Fuse
3) "G" fusible link
4) Ignition switch
O.K.

ECCS SELF-DIAGNOSTIC CHART—1988–89 TRUCK/PATHFINDER WITH VG30I ENGINE

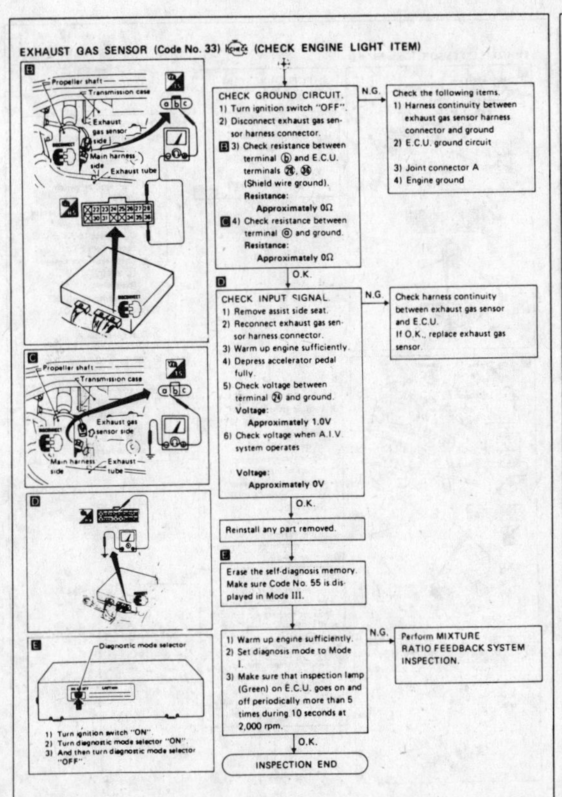
EXHAUST GAS SENSOR (Code No. 33) (CHECK ENGINE LIGHT ITEM)
B
Propeller shaft
Transmission case
Exhaust gas sensor side
Main harness side
Exhaust tube
CHECK GROUND CIRCUIT.
1) Turn ignition switch "OFF".
2) Disconnect exhaust gas sensor harness connector.
B 3) Check resistance between terminal ⓑ and E.C.U. terminals ㉖, ㊱ (Shield wire ground).
Resistance: Approximately 0Ω
C 4) Check resistance between terminal ⓐ and ground.
Resistance: Approximately 0Ω
N.G.
Check the following items.
1) Harness continuity between exhaust gas sensor harness connector and ground
2) E.C.U. ground circuit
3) Joint connector A
4) Engine ground
O.K.
C
Propeller shaft
Transmission case
Exhaust gas sensor side
Main harness side
Exhaust tube
D
CHECK INPUT SIGNAL.
1) Remove assist side seat.
2) Reconnect exhaust gas sensor harness connector.
3) Warm up engine sufficiently.
4) Depress accelerator pedal fully.
5) Check voltage between terminal ㉔ and ground.
Voltage: Approximately 1.0V
6) Check voltage when A.I.V. system operates
Voltage: Approximately 0V
N.G.
Check harness continuity between exhaust gas sensor and E.C.U.
If O.K., replace exhaust gas sensor.
O.K.
D
Reinstall any part removed.
E
Erase the self-diagnosis memory. Make sure Code No. 55 is displayed in Mode III.
E
Diagnostic mode selector
1) Turn ignition switch "ON".
2) Turn diagnostic mode selector "ON".
3) And then turn diagnostic mode selector "OFF".
1) Warm up engine sufficiently.
2) Set diagnosis mode to Mode I.
3) Make sure that inspection lamp (Green) on E.C.U. goes on and off periodically more than 5 times during 10 seconds at 2,000 rpm.
N.G.
Perform MIXTURE RATIO FEEDBACK SYSTEM INSPECTION.
O.K.
INSPECTION END

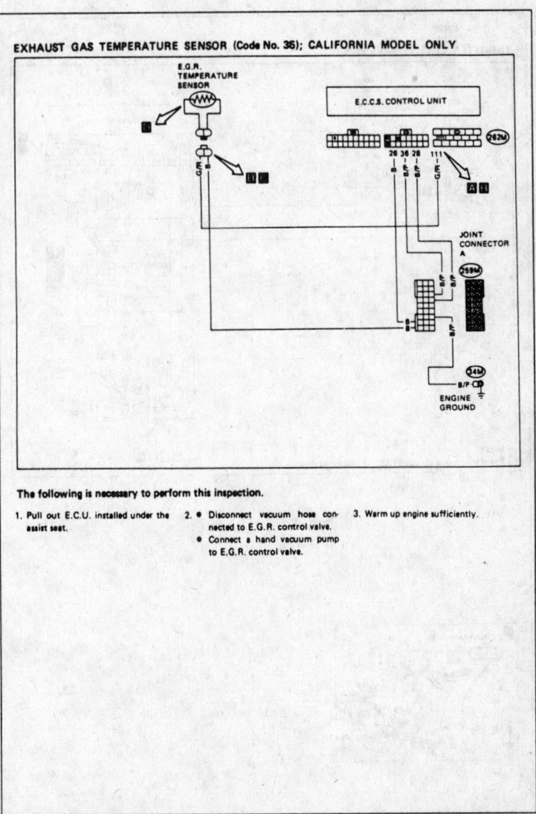
EXHAUST GAS TEMPERATURE SENSOR (Code No. 35); CALIFORNIA MODEL ONLY
E.G.R. TEMPERATURE SENSOR
E.C.C.S. CONTROL UNIT
262M
26 36 28 111
JOINT CONNECTOR A
259M
34M
ENGINE GROUND
The following is necessary to perform this inspection.
1. Pull out E.C.U. installed under the assist seat.
2. • Disconnect vacuum hose connected to E.G.R. control valve.
• Connect a hand vacuum pump to E.G.R. control valve.
3. Warm up engine sufficiently.

ECCS SELF-DIAGNOSTIC CHART—1988–89 TRUCK/PATHFINDER WITH VG30I ENGINE

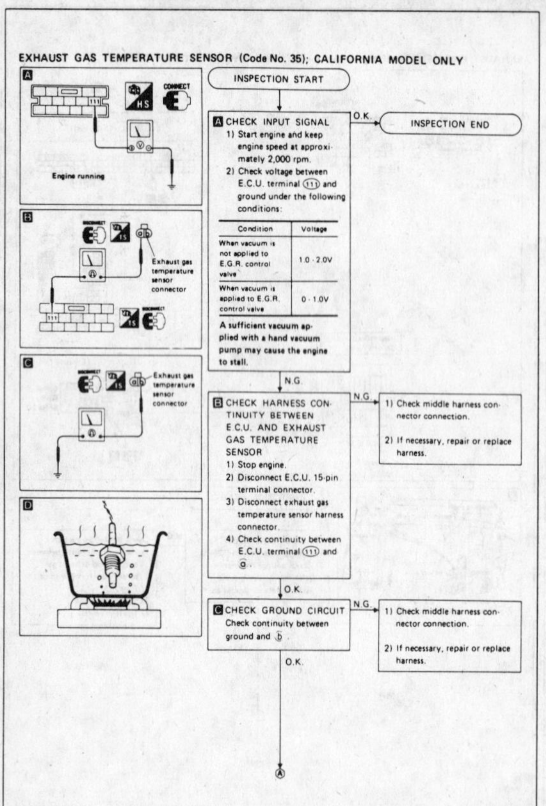
EXHAUST GAS TEMPERATURE SENSOR (Code No. 35); CALIFORNIA MODEL ONLY
Engine running
Exhaust gas temperature sensor connector
INSPECTION START
A CHECK INPUT SIGNAL
1) Start engine and keep engine speed at approximately 2,000 rpm.
2) Check voltage between E.C.U. terminal (111) and ground under the following conditions:
Condition | Voltage
When vacuum is not applied to E.G.R. control valve | 1.0 - 2.0V
When vacuum is applied to E.G.R. control valve | 0 - 1.0V
A sufficient vacuum applied with a hand vacuum pump may cause the engine to stall.
O.K.
INSPECTION END
N.G.
B CHECK HARNESS CONTINUITY BETWEEN E.C.U. AND EXHAUST GAS TEMPERATURE SENSOR
1) Stop engine.
2) Disconnect E.C.U. 15-pin terminal connector.
3) Disconnect exhaust gas temperature sensor harness connector.
4) Check continuity between E.C.U. terminal (111) and (a).
1) Check middle harness connector connection.
2) If necessary, repair or replace harness.
C CHECK GROUND CIRCUIT
Check continuity between ground and (b).
1) Check middle harness connector connection.
2) If necessary, repair or replace harness.

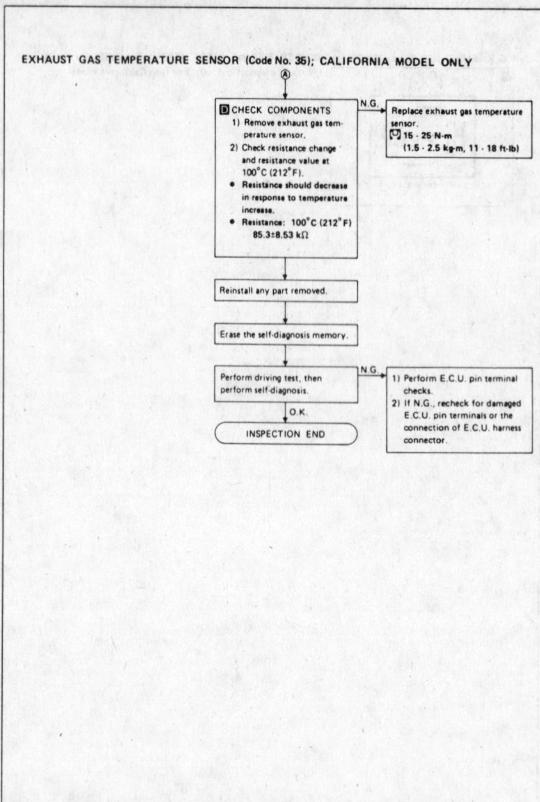
EXHAUST GAS TEMPERATURE SENSOR (Code No. 35); CALIFORNIA MODEL ONLY
D CHECK COMPONENTS
1) Remove exhaust gas temperature sensor.
2) Check resistance change and resistance value at 100°C (212°F).
• Resistance should decrease in response to temperature increase.
• Resistance: 100°C (212°F) 85.3±8.53 kΩ
N.G.
Replace exhaust gas temperature sensor.
15 - 25 N·m (1.5 - 2.5 kg-m, 11 - 18 ft-lb)
Reinstall any part removed.
Erase the self-diagnosis memory.
Perform driving test, then perform self-diagnosis.
1) Perform E.C.U. pin terminal checks.
2) If N.G., recheck for damaged E.C.U. pin terminals or the connection of E.C.U. harness connector.
O.K.
INSPECTION END

ECCS SELF-DIAGNOSTIC CHART—1988–89 TRUCK/PATHFINDER WITH VG30I ENGINE

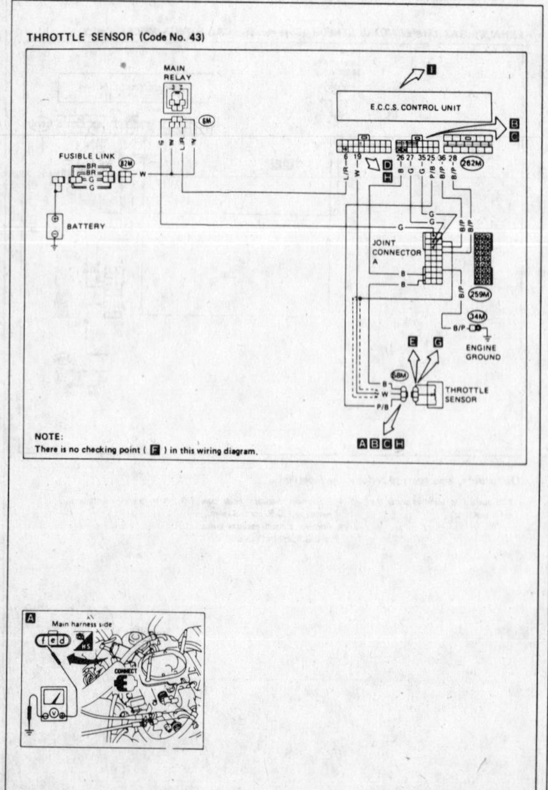
THROTTLE SENSOR (Code No. 43)
MAIN RELAY
E.C.C.S. CONTROL UNIT
FUSIBLE LINK
BATTERY
JOINT CONNECTOR A
ENGINE GROUND
THROTTLE SENSOR
NOTE:
There is no checking point (F) in this wiring diagram.
Main harness side

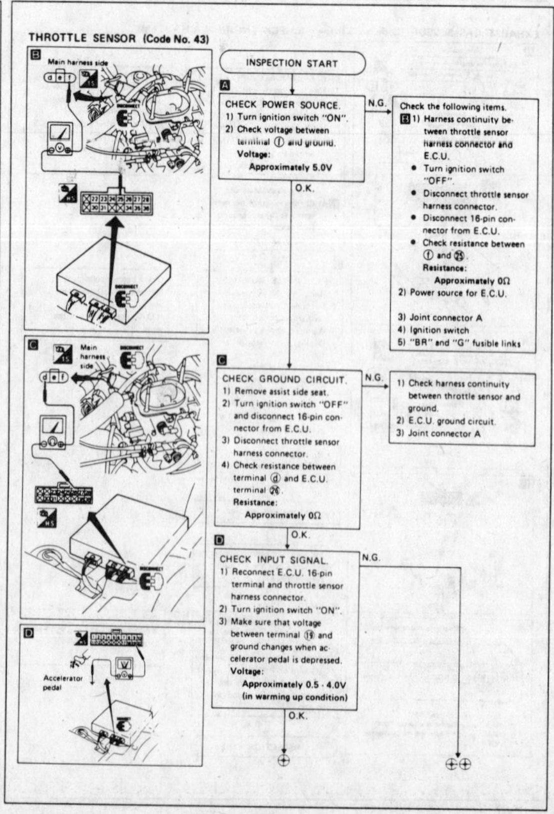
THROTTLE SENSOR (Code No. 43)
Main harness side
Accelerator pedal
INSPECTION START
A CHECK POWER SOURCE.
1) Turn ignition switch "ON".
2) Check voltage between terminal (f) and ground.
Voltage: Approximately 5.0V
N.G.
Check the following items.
B 1) Harness continuity between throttle sensor harness connector and E.C.U.
• Turn ignition switch "OFF".
• Disconnect throttle sensor harness connector.
• Disconnect 16-pin connector from E.C.U.
• Check resistance between (f) and (25).
Resistance: Approximately 0Ω
2) Power source for E.C.U.
3) Joint connector A
4) Ignition switch
5) "BR" and "G" fusible links
O.K.
C CHECK GROUND CIRCUIT.
1) Remove assist side seat.
2) Turn ignition switch "OFF" and disconnect 16-pin connector from E.C.U.
3) Disconnect throttle sensor harness connector.
4) Check resistance between terminal (d) and E.C.U. terminal (26).
Resistance: Approximately 0Ω
1) Check harness continuity between throttle sensor and ground.
2) E.C.U. ground circuit.
3) Joint connector A
D CHECK INPUT SIGNAL.
1) Reconnect E.C.U. 16-pin terminal and throttle sensor harness connector.
2) Turn ignition switch "ON".
3) Make sure that voltage between terminal (19) and ground changes when accelerator pedal is depressed.
Voltage: Approximately 0.5 - 4.0V (in warming up condition)

ECCS SELF-DIAGNOSTIC CHART—1988–89 TRUCK/PATHFINDER WITH VG30I ENGINE

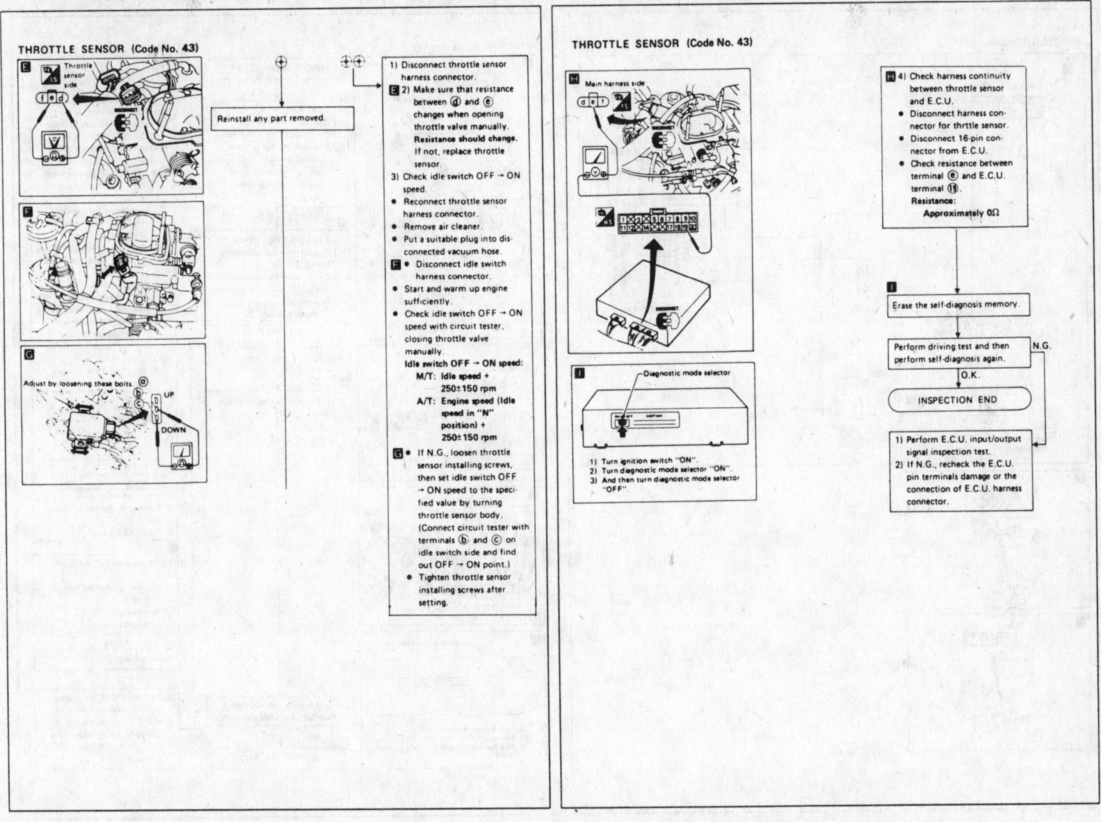
THROTTLE SENSOR (Code No. 43)
Reinstall any part removed.
1) Disconnect throttle sensor harness connector.
2) Make sure that resistance between ⓓ and ⓔ changes when opening throttle valve manually. Resistance should change. If not, replace throttle sensor.
3) Check idle switch OFF → ON speed.
• Reconnect throttle sensor harness connector.
• Remove air cleaner.
• Put a suitable plug into disconnected vacuum hose.
• Disconnect idle switch harness connector.
• Start and warm up engine sufficiently.
• Check idle switch OFF → ON speed with circuit tester, closing throttle valve manually.
Idle switch OFF → ON speed:
M/T: Idle speed + 250±150 rpm
A/T: Engine speed (Idle speed in "N" position) + 250±150 rpm
• If N.G., loosen throttle sensor installing screws, then set idle switch OFF → ON speed to the specified value by turning throttle sensor body. (Connect circuit tester with terminals ⓑ and ⓒ on idle switch side and find out OFF → ON point.)
• Tighten throttle sensor installing screws after setting.
Adjust by loosening these bolts.
UP
DOWN
THROTTLE SENSOR (Code No. 43)
Main harness side
Diagnostic mode selector
1) Turn ignition switch "ON".
2) Turn diagnostic mode selector "ON".
3) And then turn diagnostic mode selector "OFF".
4) Check harness continuity between throttle sensor and E.C.U.
• Disconnect harness connector for thrttle sensor.
• Disconnect 16-pin connector from E.C.U.
• Check resistance between terminal ⓔ and E.C.U. terminal ⑪.
Resistance: Approximately 0Ω
Erase the self-diagnosis memory.
Perform driving test and then perform self-diagnosis again.
N.G.
O.K.
INSPECTION END
1) Perform E.C.U. input/output signal inspection test.
2) If N.G., recheck the E.C.U. pin terminals damage or the connection of E.C.U. harness connector.

ECCS SELF-DIAGNOSTIC CHART—1988–89 TRUCK/PATHFINDER WITH VG30I ENGINE

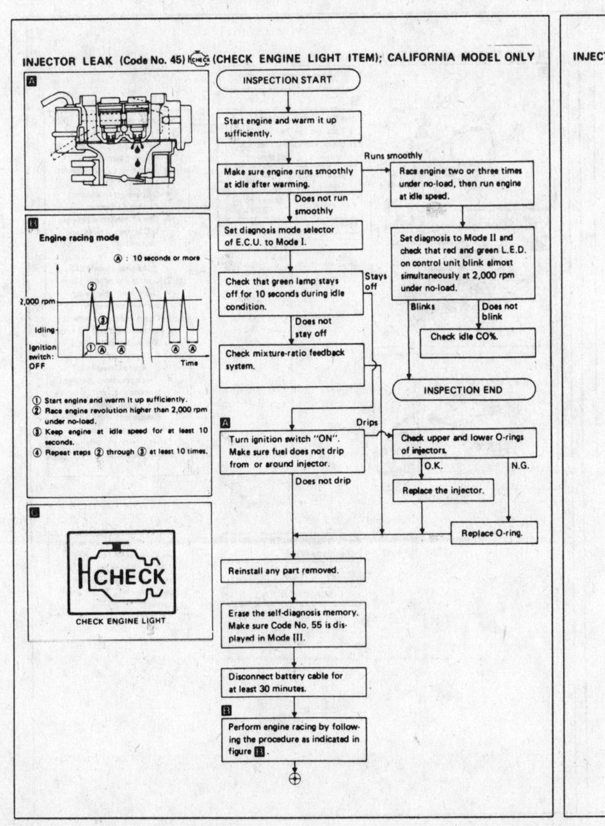
INJECTOR LEAK (Code No. 45) (CHECK ENGINE LIGHT ITEM); CALIFORNIA MODEL ONLY
INSPECTION START
Start engine and warm it up sufficiently.
Make sure engine runs smoothly at idle after warming.
Runs smoothly
Race engine two or three times under no-load, then run engine at idle speed.
Does not run smoothly
Set diagnosis mode selector of E.C.U. to Mode I.
Set diagnosis to Mode II and check that red and green L.E.D. on control unit blink almost simultaneously at 2,000 rpm under no-load.
Check that green lamp stays off for 10 seconds during idle condition.
Stays off
Blinks
Does not blink
Check idle CO%.
Does not stay off
Check mixture-ratio feedback system.
INSPECTION END
Engine racing mode
Ⓐ : 10 seconds or more
2,000 rpm
Idling
Ignition switch: OFF
Time
① Start engine and warm it up sufficiently.
② Race engine revolution higher than 2,000 rpm under no-load.
③ Keep engine at idle speed for at least 10 seconds.
④ Repeat steps ② through ③ at least 10 times.
Turn ignition switch "ON". Make sure fuel does not drip from or around injector.
Drips
Check upper and lower O-rings of injectors.
O.K.
N.G.
Does not drip
Replace the injector.
Replace O-ring.
CHECK
CHECK ENGINE LIGHT
Reinstall any part removed.
Erase the self-diagnosis memory. Make sure Code No. 55 is displayed in Mode III.
Disconnect battery cable for at least 30 minutes.
Perform engine racing by following the procedure as indicated in figure B.

INJECTOR LEAK (Code No. 45) (CHECK ENGINE LIGHT ITEM); CALIFORNIA MODEL ONLY
Make sure check engine light does not come "ON" while racing engine.
Comes "ON"
1) Perform self-diagnosis and find malfunction code.
2) According to displayed code No., perform electronic control system inspection.
3) If Code No. 45 is displayed again, replace all injectors, then perform electronic control system inspection.
Does not come "ON"
INSPECTION END

ECCS SELF-DIAGNOSTIC CHART—1988–89 TRUCK/PATHFINDER WITH VG30I ENGINE

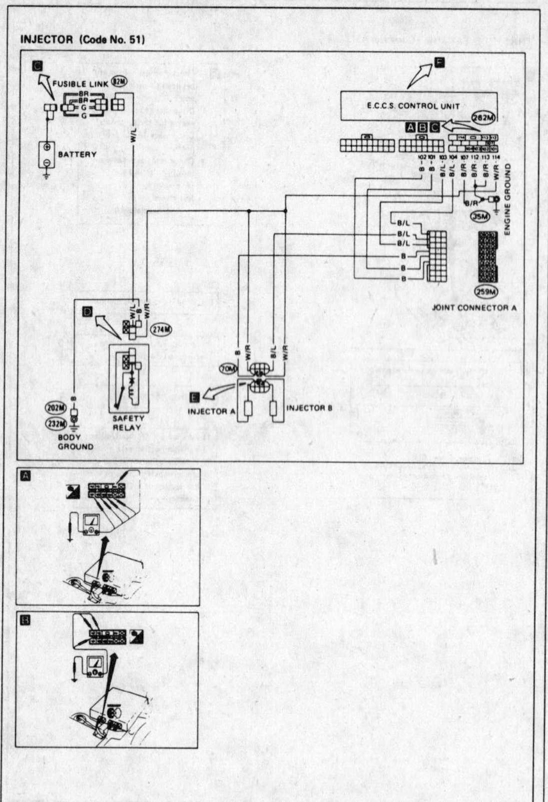
INJECTOR (Code No. 51)
FUSIBLE LINK
BATTERY
E.C.C.S. CONTROL UNIT
JOINT CONNECTOR A
ENGINE GROUND
SAFETY RELAY
BODY GROUND
INJECTOR A
INJECTOR B

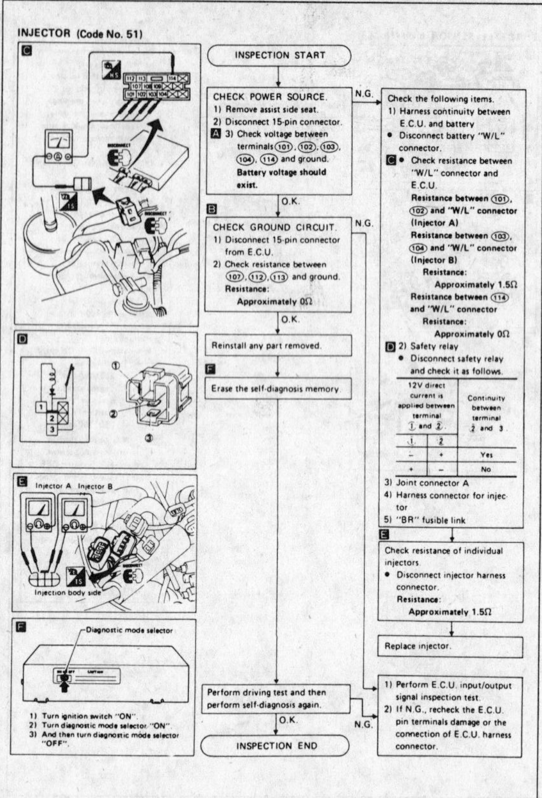
INJECTOR (Code No. 51)
INSPECTION START
CHECK POWER SOURCE.
1) Remove assist side seat.
2) Disconnect 15-pin connector.
3) Check voltage between terminals 101, 102, 103, 104, 114 and ground.
Battery voltage should exist.
N.G.
Check the following items.
1) Harness continuity between E.C.U. and battery
• Disconnect battery "W/L" connector.
• Check resistance between "W/L" connector and E.C.U.
Resistance between 101, 102 and "W/L" connector (Injector A)
Resistance between 103, 104 and "W/L" connector (Injector B)
Resistance: Approximately 1.5Ω
Resistance between 114 and "W/L" connector
Resistance: Approximately 0Ω
2) Safety relay
• Disconnect safety relay and check it as follows.
12V direct current is applied between terminal 1 and 2 | Continuity between terminal 2 and 3
— + Yes
+ — No
3) Joint connector A
4) Harness connector for injector
5) "BR" fusible link
O.K.
CHECK GROUND CIRCUIT.
1) Disconnect 15-pin connector from E.C.U.
2) Check resistance between 107, 112, 113 and ground.
Resistance: Approximately 0Ω
O.K.
Reinstall any part removed.
Erase the self-diagnosis memory
Check resistance of individual injectors.
• Disconnect injector harness connector.
Resistance: Approximately 1.5Ω
Replace injector.
Injector A Injector B
Injection body side
Diagnostic mode selector
1) Turn ignition switch "ON".
2) Turn diagnostic mode selector "ON".
3) And then turn diagnostic mode selector "OFF".
Perform driving test and then perform self-diagnosis again.
O.K.
N.G.
1) Perform E.C.U. input/output signal inspection test.
2) If N.G., recheck the E.C.U. pin terminals damage or the connection of E.C.U. harness connector.
INSPECTION END

ECCS SELF-DIAGNOSTIC CHART—1988–89 TRUCK/PATHFINDER WITH VG30I ENGINE

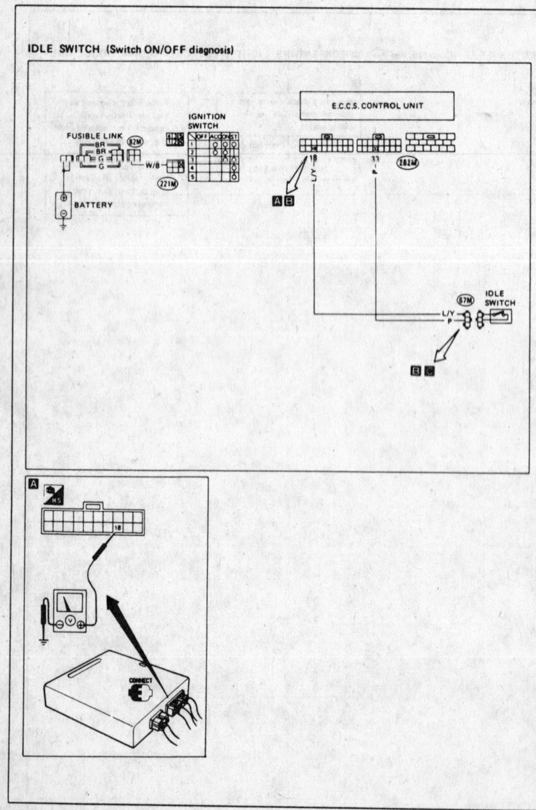
IDLE SWITCH (Switch ON/OFF diagnosis)
FUSIBLE LINK
BATTERY
IGNITION SWITCH
E.C.C.S. CONTROL UNIT
IDLE SWITCH

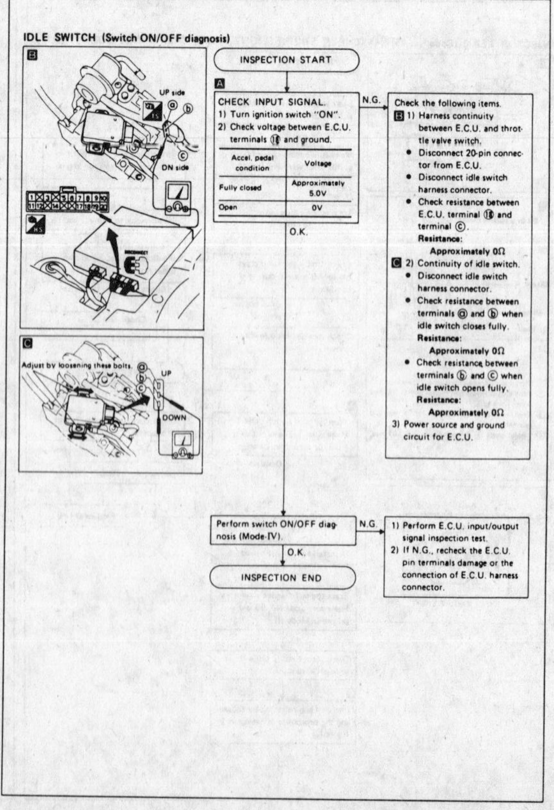
IDLE SWITCH (Switch ON/OFF diagnosis)
INSPECTION START
CHECK INPUT SIGNAL.
1) Turn ignition switch "ON".
2) Check voltage between E.C.U. terminals 18 and ground.
Accel. pedal condition | Voltage
Fully closed | Approximately 5.0V
Open | 0V
N.G.
Check the following items.
1) Harness continuity between E.C.U. and throttle valve switch.
• Disconnect 20-pin connector from E.C.U.
• Disconnect idle switch harness connector.
• Check resistance between E.C.U. terminal 18 and terminal ©.
Resistance: Approximately 0Ω
2) Continuity of idle switch.
• Disconnect idle switch harness connector.
• Check resistance between terminals ⓐ and ⓑ when idle switch closes fully.
Resistance: Approximately 0Ω
• Check resistance between terminals ⓑ and © when idle switch opens fully.
Resistance: Approximately 0Ω
3) Power source and ground circuit for E.C.U.
UP side
DN side
Adjust by loosening these bolts.
UP
DOWN
O.K.
Perform switch ON/OFF diagnosis (Mode-IV).
N.G.
1) Perform E.C.U. input/output signal inspection test.
2) If N.G., recheck the E.C.U. pin terminals damage or the connection of E.C.U. harness connector.
O.K.
INSPECTION END

ECCS SELF-DIAGNOSTIC CHART—1988–89 TRUCK/PATHFINDER WITH VG30I ENGINE

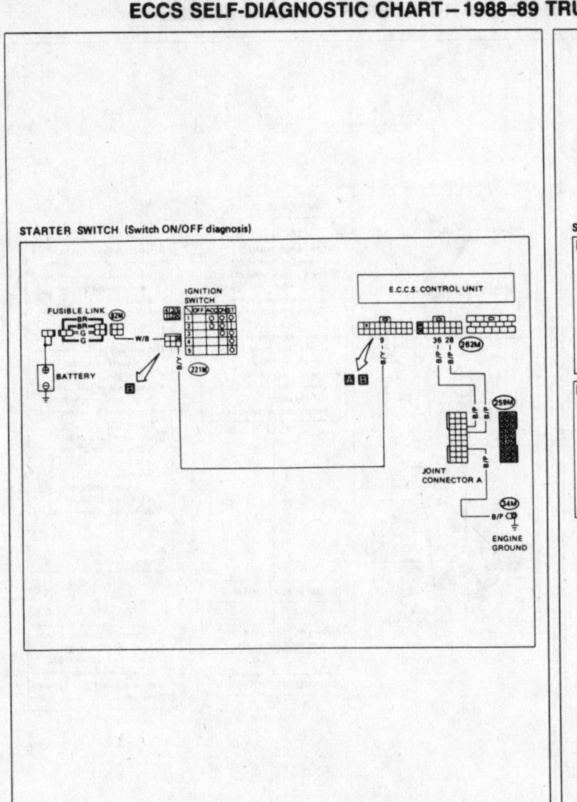
STARTER SWITCH (Switch ON/OFF diagnosis)
E.C.C.S. CONTROL UNIT
FUSIBLE LINK
IGNITION SWITCH
BATTERY
JOINT CONNECTOR A
ENGINE GROUND

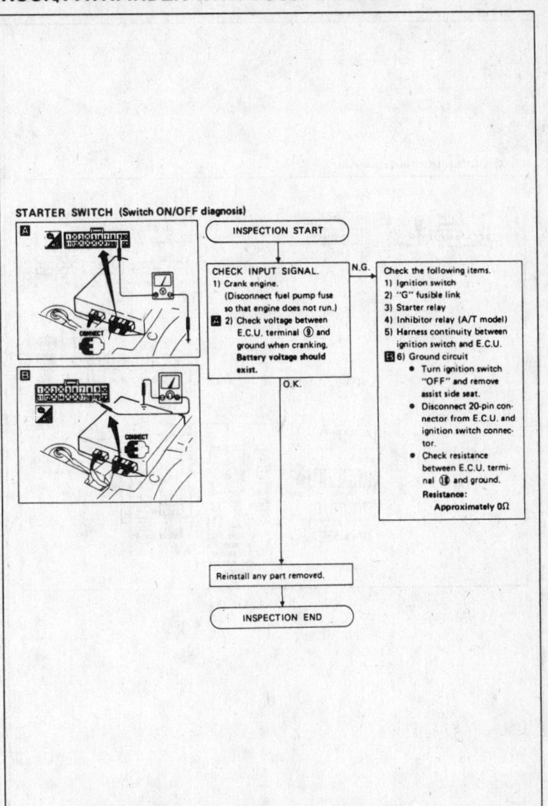
STARTER SWITCH (Switch ON/OFF diagnosis)
INSPECTION START
CHECK INPUT SIGNAL.
1) Crank engine.
(Disconnect fuel pump fuse so that engine does not run.)
A 2) Check voltage between E.C.U. terminal ⑨ and ground when cranking.
Battery voltage should exist.
N.G.
Check the following items.
1) Ignition switch
2) "G" fusible link
3) Starter relay
4) Inhibitor relay (A/T model)
5) Harness continuity between ignition switch and E.C.U.
B 6) Ground circuit
• Turn ignition switch "OFF" and remove assist side seat.
• Disconnect 20-pin connector from E.C.U. and ignition switch connector.
• Check resistance between E.C.U. terminal ⑱ and ground.
Resistance: Approximately 0Ω
O.K.
Reinstall any part removed.
INSPECTION END

ECCS SELF-DIAGNOSTIC CHART—1988–89 TRUCK/PATHFINDER WITH VG30I ENGINE

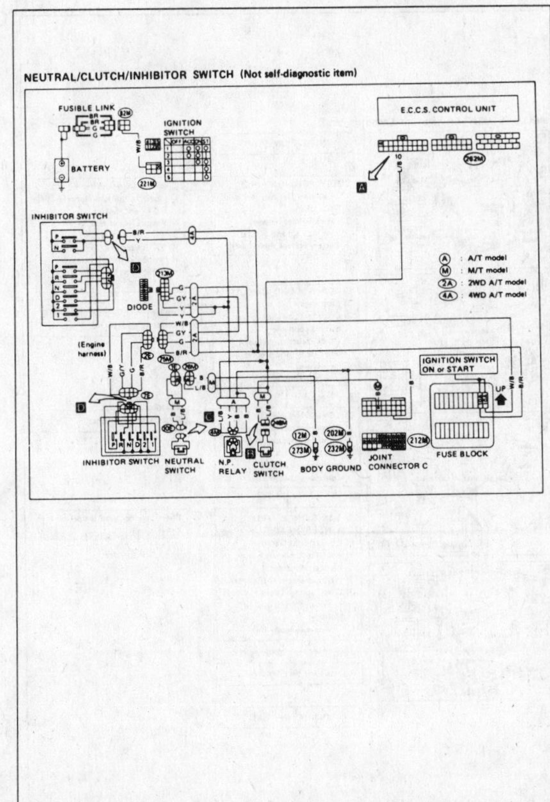
NEUTRAL/CLUTCH/INHIBITOR SWITCH (Not self-diagnostic item)
FUSIBLE LINK
E.C.C.S. CONTROL UNIT
IGNITION SWITCH
BATTERY
INHIBITOR SWITCH
DIODE
(Engine harness)
Ⓐ : A/T model
Ⓜ : M/T model
2A : 2WD A/T model
4A : 4WD A/T model
IGNITION SWITCH ON or START
INHIBITOR SWITCH
NEUTRAL SWITCH
N.P. RELAY
CLUTCH SWITCH
BODY GROUND
JOINT CONNECTOR C
FUSE BLOCK

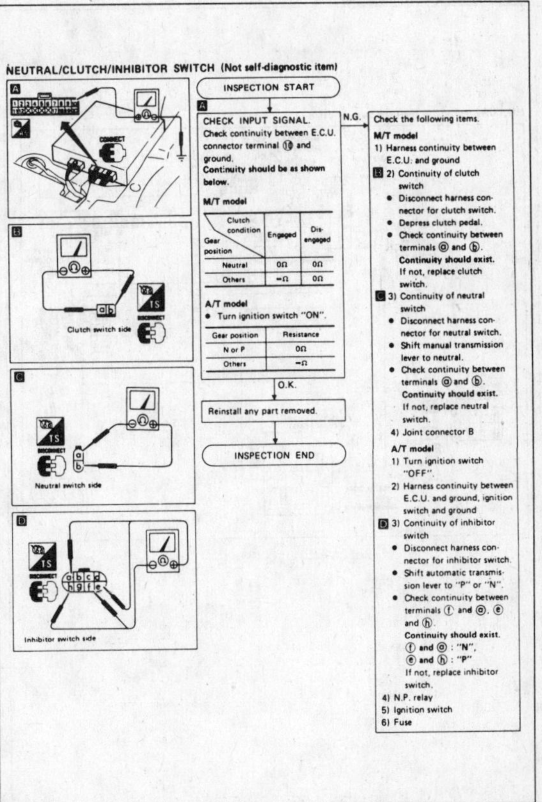
NEUTRAL/CLUTCH/INHIBITOR SWITCH (Not self-diagnostic item)
INSPECTION START
A
CHECK INPUT SIGNAL.
Check continuity between E.C.U. connector terminal ⑩ and ground.
Continuity should be as shown below.
M/T model
Gear position / Clutch condition | Engaged | Dis-engaged
Neutral | 0Ω | 0Ω
Others | ∞Ω | 0Ω
A/T model
• Turn ignition switch "ON".
Gear position | Resistance
N or P | 0Ω
Others | ∞Ω
N.G.
Check the following items.
M/T model
1) Harness continuity between E.C.U. and ground
B 2) Continuity of clutch switch
• Disconnect harness connector for clutch switch.
• Depress clutch pedal.
• Check continuity between terminals ⓐ and ⓑ.
Continuity should exist.
If not, replace clutch switch.
C 3) Continuity of neutral switch
• Disconnect harness connector for neutral switch.
• Shift manual transmission lever to neutral.
• Check continuity between terminals ⓐ and ⓑ.
Continuity should exist.
If not, replace neutral switch.
4) Joint connector B
A/T model
1) Turn ignition switch "OFF".
2) Harness continuity between E.C.U. and ground, ignition switch and ground
D 3) Continuity of inhibitor switch
• Disconnect harness connector for inhibitor switch.
• Shift automatic transmission lever to "P" or "N".
• Check continuity between terminals ⓕ and ⓖ, ⓔ and ⓗ.
Continuity should exist.
ⓕ and ⓖ : "N",
ⓔ and ⓗ : "P"
If not, replace inhibitor switch.
4) N.P. relay
5) Ignition switch
6) Fuse
O.K.
Reinstall any part removed.
INSPECTION END
Clutch switch side
Neutral switch side
Inhibitor switch side

ECCS SELF-DIAGNOSTIC CHART—1988–89 TRUCK/PATHFINDER WITH VG30I ENGINE

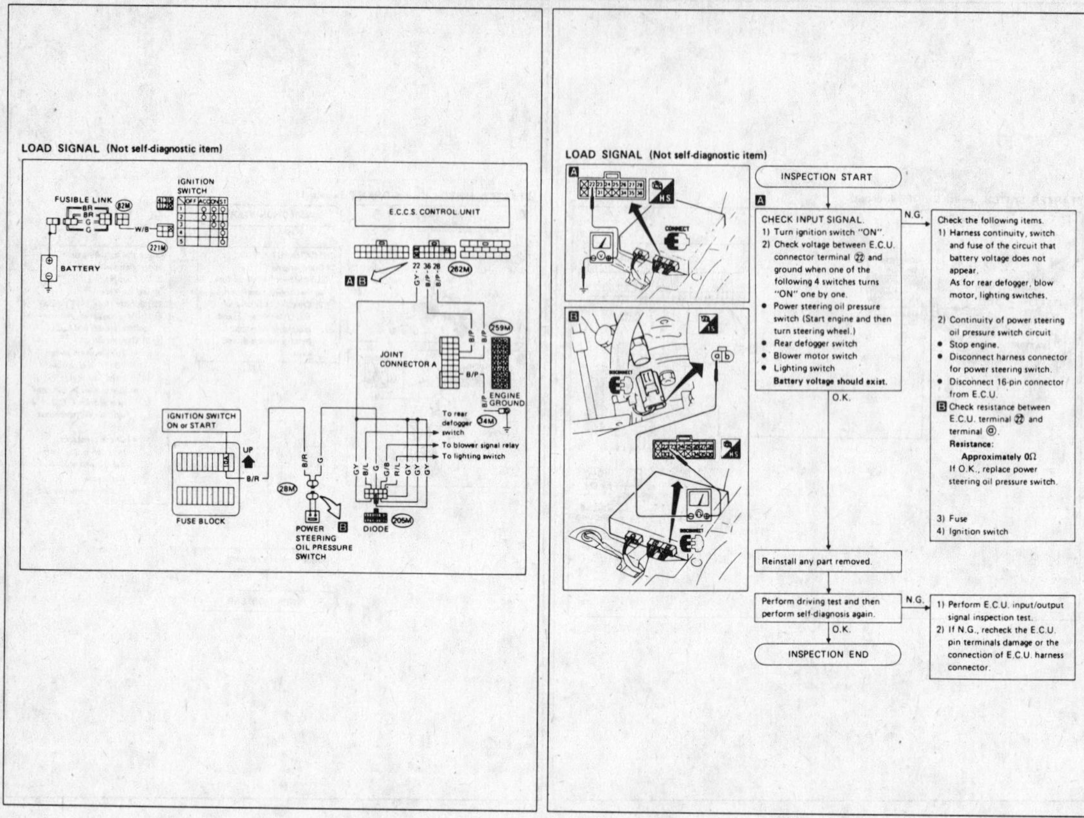
LOAD SIGNAL (Not self-diagnostic item)
FUSIBLE LINK
IGNITION SWITCH
BATTERY
E.C.C.S. CONTROL UNIT
JOINT CONNECTOR A
ENGINE GROUND
IGNITION SWITCH ON or START
FUSE BLOCK
To rear defogger switch
To blower signal relay
To lighting switch
POWER STEERING OIL PRESSURE SWITCH
DIODE
LOAD SIGNAL (Not self-diagnostic item)
INSPECTION START
CHECK INPUT SIGNAL.
1) Turn ignition switch "ON".
2) Check voltage between E.C.U. connector terminal ㉒ and ground when one of the following 4 switches turns "ON" one by one.
• Power steering oil pressure switch (Start engine and then turn steering wheel.)
• Rear defogger switch
• Blower motor switch
• Lighting switch
Battery voltage should exist.
N.G.
Check the following items.
1) Harness continuity, switch and fuse of the circuit that battery voltage does not appear.
As for rear defogger, blow motor, lighting switches.
2) Continuity of power steering oil pressure switch circuit
• Stop engine.
• Disconnect harness connector for power steering switch.
• Disconnect 16-pin connector from E.C.U.
B Check resistance between E.C.U. terminal ㉒ and terminal ⓐ.
Resistance:
Approximately 0Ω
If O.K., replace power steering oil pressure switch.
3) Fuse
4) Ignition switch
O.K.
Reinstall any part removed.
Perform driving test and then perform self-diagnosis again.
N.G.
1) Perform E.C.U. input/output signal inspection test.
2) If N.G., recheck the E.C.U. pin terminals damage or the connection of E.C.U. harness connector.
O.K.
INSPECTION END

ECCS SELF-DIAGNOSTIC CHART—1988–89 TRUCK/PATHFINDER WITH VG30I ENGINE

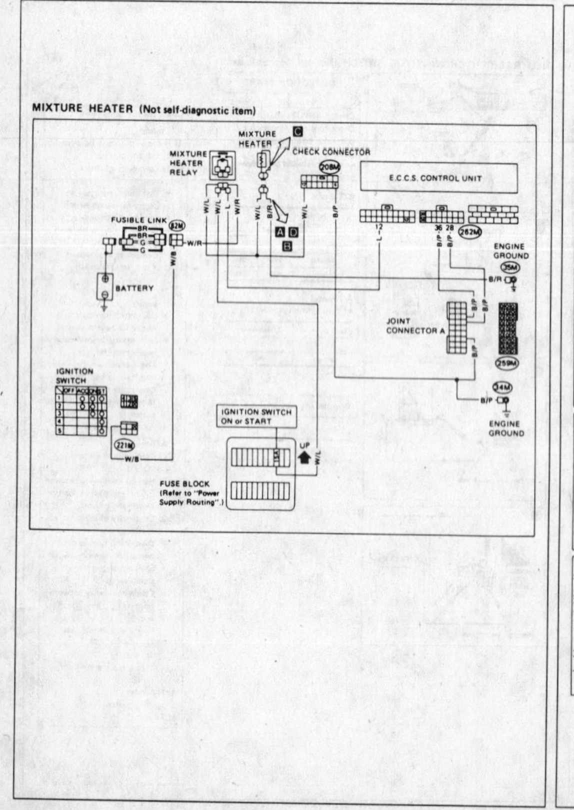
MIXTURE HEATER (Not self-diagnostic item)
MIXTURE HEATER
CHECK CONNECTOR
MIXTURE HEATER RELAY
E.C.C.S. CONTROL UNIT
FUSIBLE LINK
BATTERY
ENGINE GROUND
JOINT CONNECTOR A
IGNITION SWITCH
IGNITION SWITCH ON or START
FUSE BLOCK (Refer to "Power Supply Routing".)

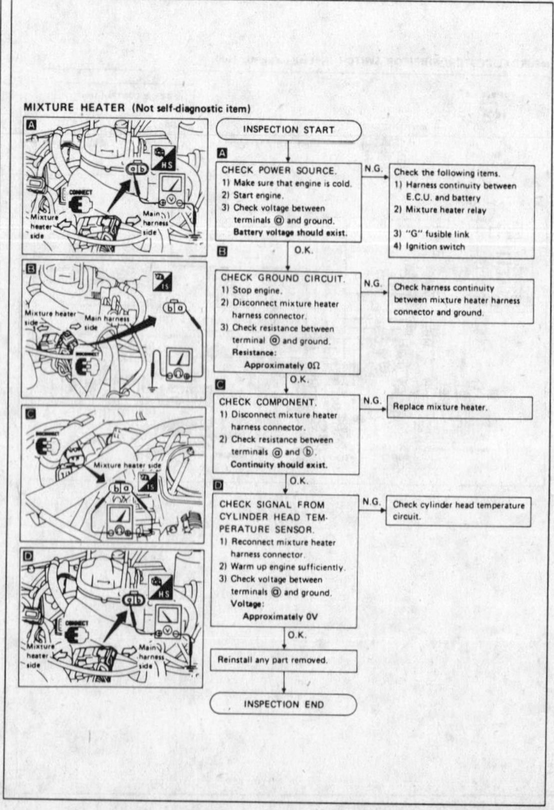
MIXTURE HEATER (Not self-diagnostic item)
Mixture heater side
Main harness side
INSPECTION START
CHECK POWER SOURCE.
1) Make sure that engine is cold.
2) Start engine.
3) Check voltage between terminals ⓐ and ground.
Battery voltage should exist.
N.G.
Check the following items.
1) Harness continuity between E.C.U. and battery
2) Mixture heater relay
3) "G" fusible link
4) Ignition switch
O.K.
CHECK GROUND CIRCUIT.
1) Stop engine.
2) Disconnect mixture heater harness connector.
3) Check resistance between terminal ⓐ and ground.
Resistance:
Approximately 0Ω
N.G.
Check harness continuity between mixture heater harness connector and ground.
O.K.
CHECK COMPONENT.
1) Disconnect mixture heater harness connector.
2) Check resistance between terminals ⓐ and ⓑ.
Continuity should exist.
N.G.
Replace mixture heater.
O.K.
CHECK SIGNAL FROM CYLINDER HEAD TEMPERATURE SENSOR.
1) Reconnect mixture heater harness connector.
2) Warm up engine sufficiently.
3) Check voltage between terminals ⓐ and ground.
Voltage:
Approximately 0V
N.G.
Check cylinder head temperature circuit.
O.K.
Reinstall any part removed.
INSPECTION END

ECCS SELF-DIAGNOSTIC CHART—1988–89 TRUCK/PATHFINDER WITH VG30I ENGINE

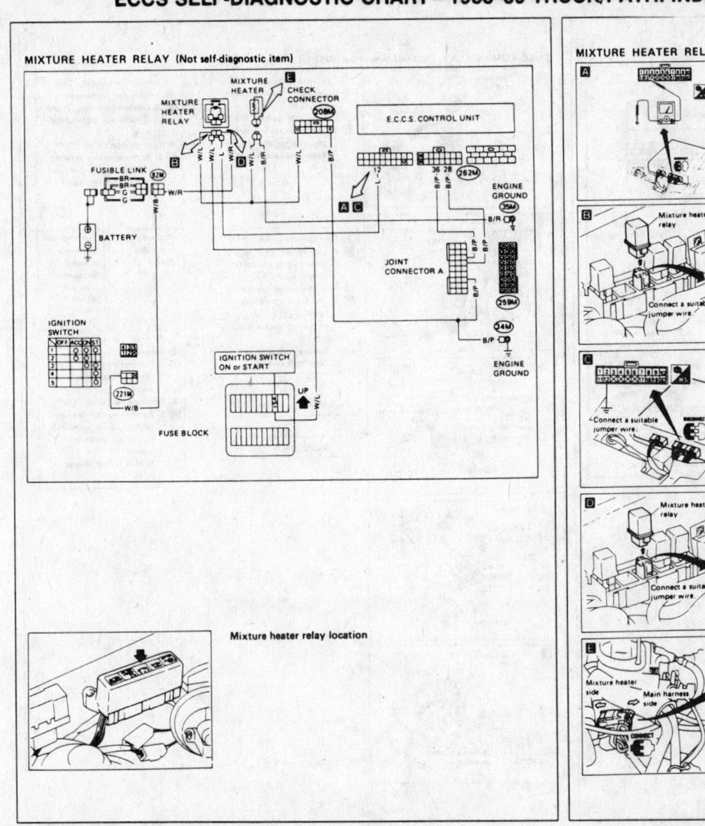
MIXTURE HEATER RELAY (Not self-diagnostic item)
MIXTURE HEATER
CHECK CONNECTOR
MIXTURE HEATER RELAY
E.C.C.S. CONTROL UNIT
FUSIBLE LINK
BATTERY
ENGINE GROUND
JOINT CONNECTOR A
IGNITION SWITCH
IGNITION SWITCH ON or START
FUSE BLOCK
Mixture heater relay location

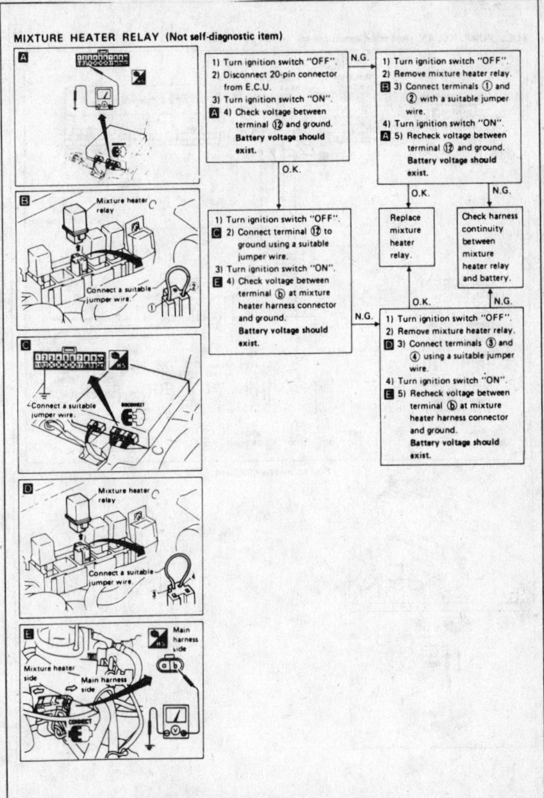
MIXTURE HEATER RELAY (Not self-diagnostic item)
1) Turn ignition switch "OFF".
2) Disconnect 20-pin connector from E.C.U.
3) Turn ignition switch "ON".
A 4) Check voltage between terminal ⑫ and ground.
Battery voltage should exist.
N.G.
1) Turn ignition switch "OFF".
2) Remove mixture heater relay.
B 3) Connect terminals ① and ② with a suitable jumper wire.
4) Turn ignition switch "ON".
A 5) Recheck voltage between terminal ⑫ and ground.
Battery voltage should exist.
O.K.
Replace mixture heater relay.
N.G.
Check harness continuity between mixture heater relay and battery.
O.K.
1) Turn ignition switch "OFF".
C 2) Connect terminal ⑫ to ground using a suitable jumper wire.
3) Turn ignition switch "ON".
E 4) Check voltage between terminal ⓑ at mixture heater harness connector and ground.
Battery voltage should exist.
N.G.
1) Turn ignition switch "OFF".
2) Remove mixture heater relay.
D 3) Connect terminals ③ and ④ using a suitable jumper wire.
4) Turn ignition switch "ON".
E 5) Recheck voltage between terminal ⓑ at mixture heater harness connector and ground.
Battery voltage should exist.
Mixture heater relay
Connect a suitable jumper wire.
Main harness side
Mixture heater side

ECCS SELF-DIAGNOSTIC CHART—1988–89 TRUCK/PATHFINDER WITH VG30I ENGINE

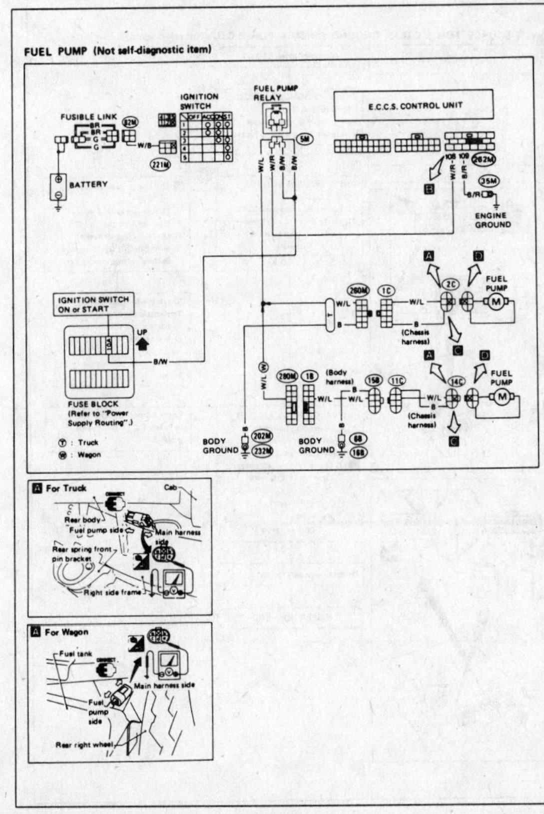
FUEL PUMP (Not self-diagnostic item)
FUSIBLE LINK
IGNITION SWITCH
FUEL PUMP RELAY
E.C.C.S. CONTROL UNIT
BATTERY
ENGINE GROUND
FUEL PUMP
IGNITION SWITCH ON or START
FUSE BLOCK (Refer to "Power Supply Routing".)
Ⓣ: Truck
Ⓦ: Wagon
BODY GROUND
For Truck
Cab
Rear body
Fuel pump side
Main harness side
Rear spring front pin bracket
Right side frame
For Wagon
Fuel tank
Fuel pump side
Rear right wheel

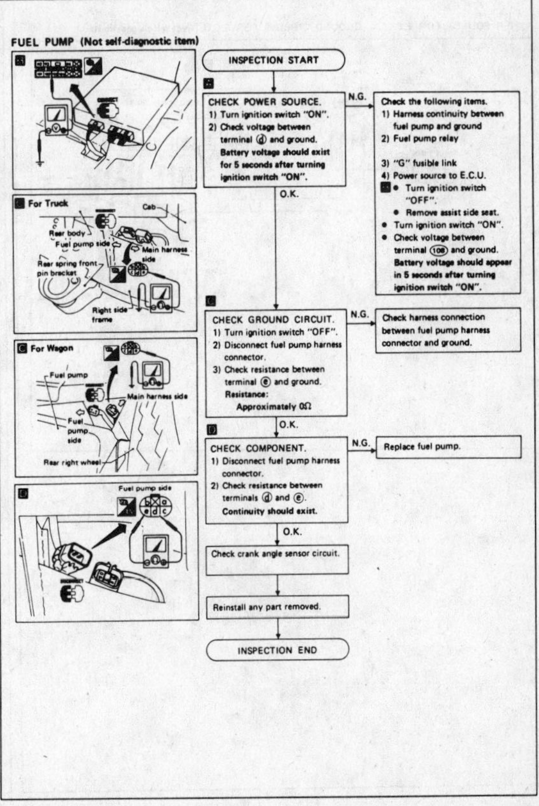
FUEL PUMP (Not self-diagnostic item)
INSPECTION START
CHECK POWER SOURCE.
1) Turn ignition switch "ON".
2) Check voltage between terminal ⓓ and ground.
Battery voltage should exist for 5 seconds after turning ignition switch "ON".
N.G.
Check the following items.
1) Harness continuity between fuel pump and ground
2) Fuel pump relay
3) "G" fusible link
4) Power source to E.C.U.
• Turn ignition switch "OFF".
• Remove assist side seat.
• Turn ignition switch "ON".
• Check voltage between terminal ⑩⑧ and ground.
Battery voltage should appear in 5 seconds after turning ignition switch "ON".
O.K.
CHECK GROUND CIRCUIT.
1) Turn ignition switch "OFF".
2) Disconnect fuel pump harness connector.
3) Check resistance between terminal ⓔ and ground.
Resistance: Approximately 0Ω
N.G.
Check harness connection between fuel pump harness connector and ground.
O.K.
CHECK COMPONENT.
1) Disconnect fuel pump harness connector.
2) Check resistance between terminals ⓓ and ⓔ.
Continuity should exist.
N.G.
Replace fuel pump.
O.K.
Check crank angle sensor circuit.
Reinstall any part removed.
INSPECTION END
For Truck
For Wagon

ECCS SELF-DIAGNOSTIC CHART—1988-89 TRUCK/PATHFINDER WITH VG30I ENGINE

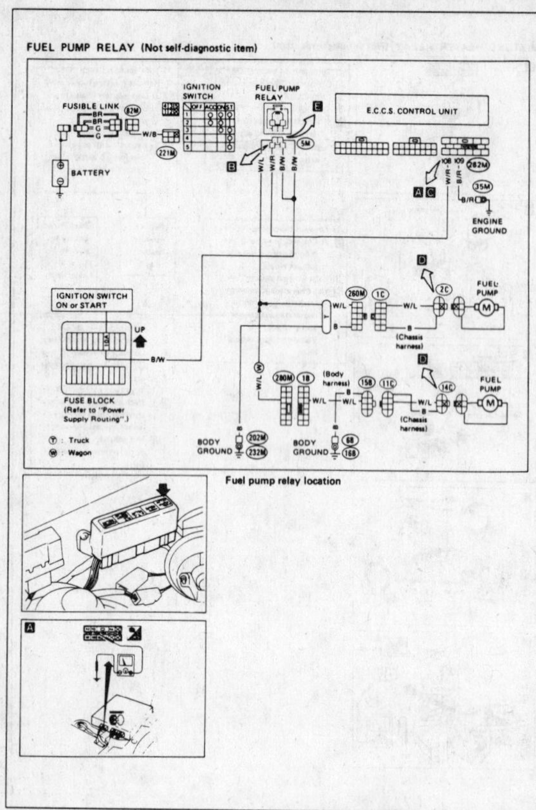
FUEL PUMP RELAY (Not self-diagnostic item)
FUSIBLE LINK
BATTERY
IGNITION SWITCH
FUEL PUMP RELAY
E.C.C.S. CONTROL UNIT
ENGINE GROUND
FUEL PUMP
(Chassis harness)
(Body harness)
IGNITION SWITCH ON or START
FUSE BLOCK (Refer to "Power Supply Routing".)
Ⓣ: Truck
Ⓦ: Wagon
BODY GROUND
Fuel pump relay location

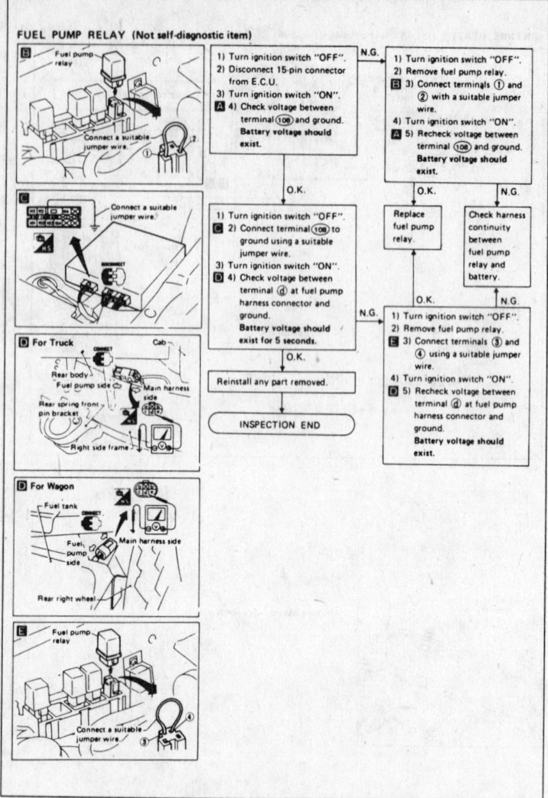
FUEL PUMP RELAY (Not self-diagnostic item)
Fuel pump relay
Connect a suitable jumper wire.
1) Turn ignition switch "OFF".
2) Disconnect 15-pin connector from E.C.U.
3) Turn ignition switch "ON".
A 4) Check voltage between terminal (108) and ground. Battery voltage should exist.
N.G.
1) Turn ignition switch "OFF".
2) Remove fuel pump relay.
B 3) Connect terminals ① and ② with a suitable jumper wire.
4) Turn ignition switch "ON".
A 5) Recheck voltage between terminal (108) and ground. Battery voltage should exist.
O.K.
Replace fuel pump relay.
N.G.
Check harness continuity between fuel pump relay and battery.
O.K.
1) Turn ignition switch "OFF".
C 2) Connect terminal (108) to ground using a suitable jumper wire.
3) Turn ignition switch "ON".
D 4) Check voltage between terminal ⓓ at fuel pump harness connector and ground. Battery voltage should exist for 5 seconds.
N.G.
1) Turn ignition switch "OFF".
2) Remove fuel pump relay.
E 3) Connect terminals ③ and ④ using a suitable jumper wire.
4) Turn ignition switch "ON".
D 5) Recheck voltage between terminal ⓓ at fuel pump harness connector and ground. Battery voltage should exist.
O.K.
Reinstall any part removed.
INSPECTION END
D For Truck
Cab
Rear body
Fuel pump side
Main harness side
Rear spring front pin bracket
Right side frame
D For Wagon
Fuel tank
Fuel pump side
Main harness side
Rear right wheel
E Fuel pump relay
Connect a suitable jumper wire.

ECCS SELF-DIAGNOSTIC CHART—1988-89 TRUCK/PATHFINDER WITH VG30I ENGINE

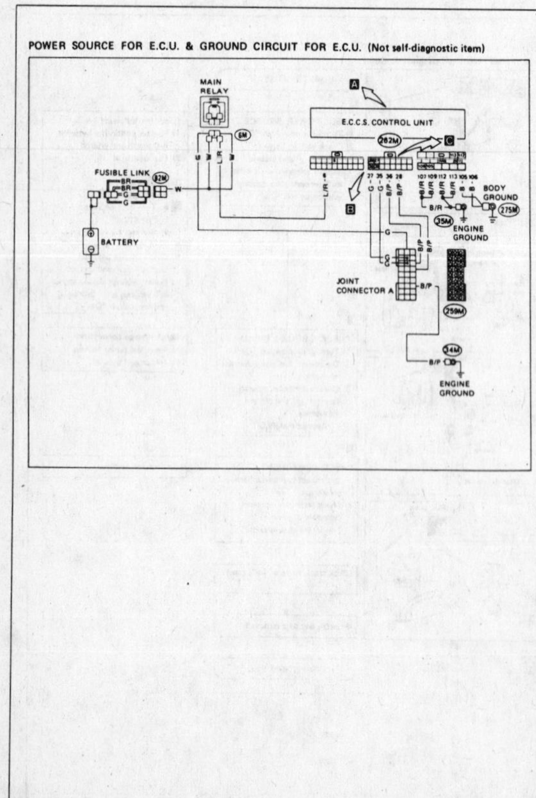
POWER SOURCE FOR E.C.U. & GROUND CIRCUIT FOR E.C.U. (Not self-diagnostic item)
MAIN RELAY
FUSIBLE LINK
BATTERY
E.C.C.S. CONTROL UNIT
BODY GROUND
ENGINE GROUND
JOINT CONNECTOR A

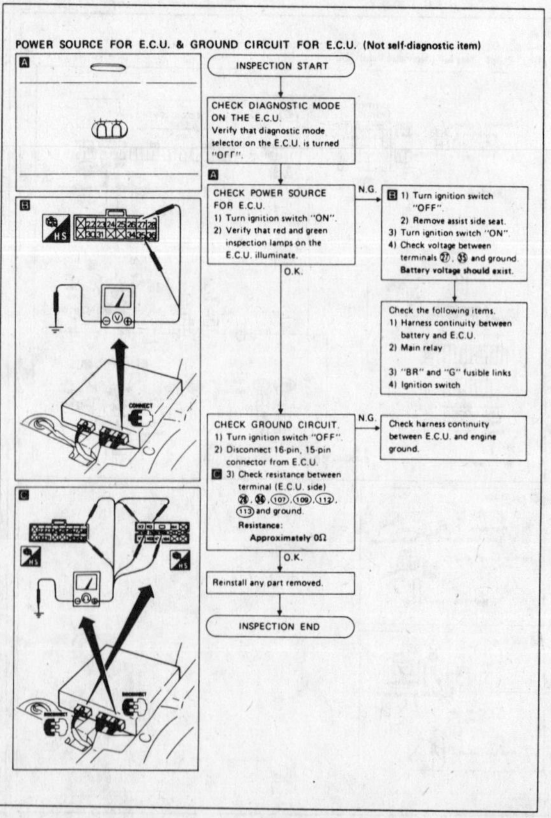
POWER SOURCE FOR E.C.U. & GROUND CIRCUIT FOR E.C.U. (Not self-diagnostic item)
INSPECTION START
CHECK DIAGNOSTIC MODE ON THE E.C.U.
Verify that diagnostic mode selector on the E.C.U. is turned "OFF".
A
CHECK POWER SOURCE FOR E.C.U.
1) Turn ignition switch "ON".
2) Verify that red and green inspection lamps on the E.C.U. illuminate.
N.G.
B 1) Turn ignition switch "OFF".
2) Remove assist side seat.
3) Turn ignition switch "ON".
4) Check voltage between terminals ㉗, ㉟ and ground. Battery voltage should exist.
Check the following items.
1) Harness continuity between battery and E.C.U.
2) Main relay
3) "BR" and "G" fusible links
4) Ignition switch
O.K.
CHECK GROUND CIRCUIT.
1) Turn ignition switch "OFF".
2) Disconnect 16-pin, 15-pin connector from E.C.U.
C 3) Check resistance between terminal (E.C.U. side) ㉘, ㊱, (107), (109), (112), (113) and ground.
Resistance: Approximately 0Ω
N.G.
Check harness continuity between E.C.U. and engine ground.
O.K.
Reinstall any part removed.
INSPECTION END

ECCS SELF-DIAGNOSTIC CHART—1988–89 TRUCK/PATHFINDER WITH VG30I ENGINE

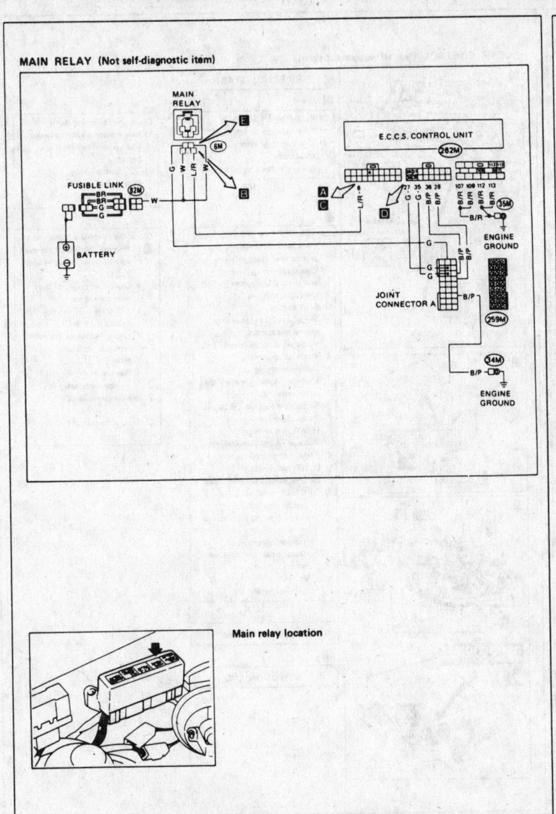
MAIN RELAY (Not self-diagnostic item)
MAIN RELAY
E.C.C.S. CONTROL UNIT
FUSIBLE LINK
BATTERY
ENGINE GROUND
JOINT CONNECTOR A
ENGINE GROUND
Main relay location

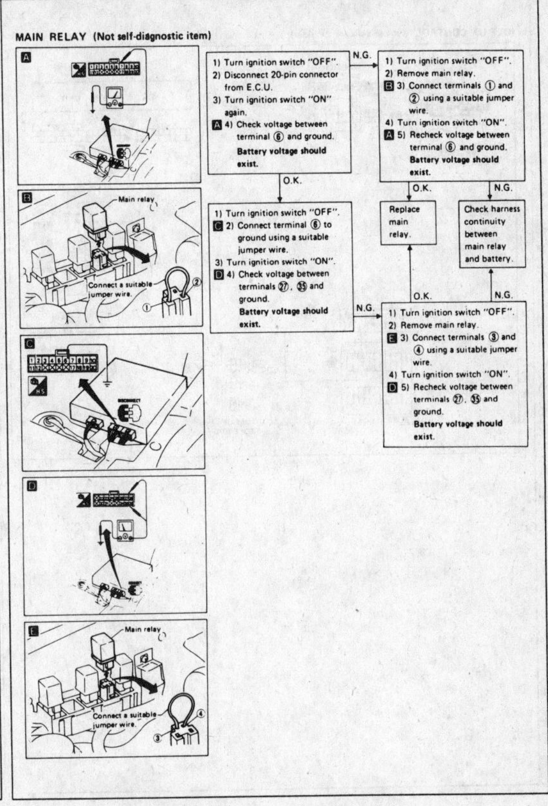
MAIN RELAY (Not self-diagnostic item)
1) Turn ignition switch "OFF".
2) Disconnect 20-pin connector from E.C.U.
3) Turn ignition switch "ON" again.
A 4) Check voltage between terminal ⑥ and ground.
Battery voltage should exist.
N.G.
1) Turn ignition switch "OFF".
2) Remove main relay.
B 3) Connect terminals ① and ② using a suitable jumper wire.
4) Turn ignition switch "ON".
A 5) Recheck voltage between terminal ⑥ and ground.
Battery voltage should exist.
O.K.
Replace main relay.
N.G.
Check harness continuity between main relay and battery.
O.K.
1) Turn ignition switch "OFF".
C 2) Connect terminal ⑥ to ground using a suitable jumper wire.
3) Turn ignition switch "ON".
D 4) Check voltage between terminals ㉗, ㉟ and ground.
Battery voltage should exist.
N.G.
1) Turn ignition switch "OFF".
2) Remove main relay.
E 3) Connect terminals ③ and ④ using a suitable jumper wire.
4) Turn ignition switch "ON".
D 5) Recheck voltage between terminals ㉗, ㉟ and ground.
Battery voltage should exist.
Main relay
Connect a suitable jumper wire.

ECCS SELF-DIAGNOSTIC CHART—1988–89 TRUCK/PATHFINDER WITH VG30I ENGINE

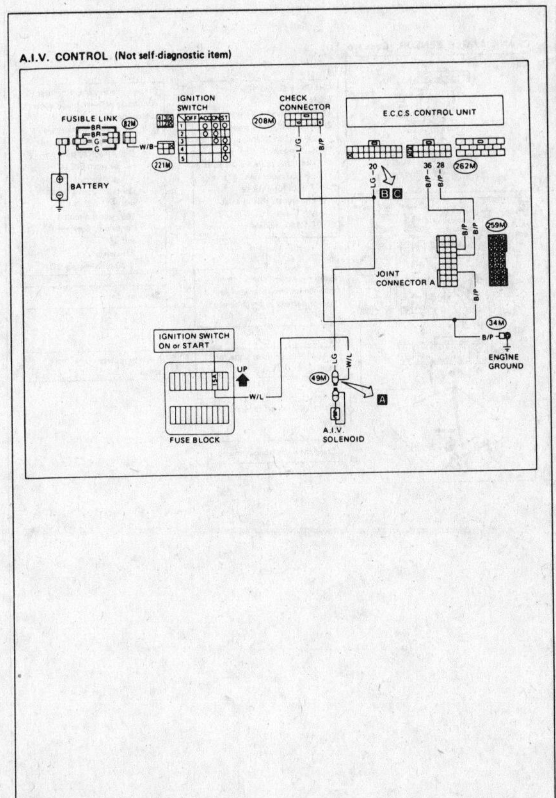
A.I.V. CONTROL (Not self-diagnostic item)
IGNITION SWITCH
CHECK CONNECTOR
E.C.C.S. CONTROL UNIT
FUSIBLE LINK
BATTERY
JOINT CONNECTOR A
IGNITION SWITCH ON or START
FUSE BLOCK
ENGINE GROUND
A.I.V. SOLENOID

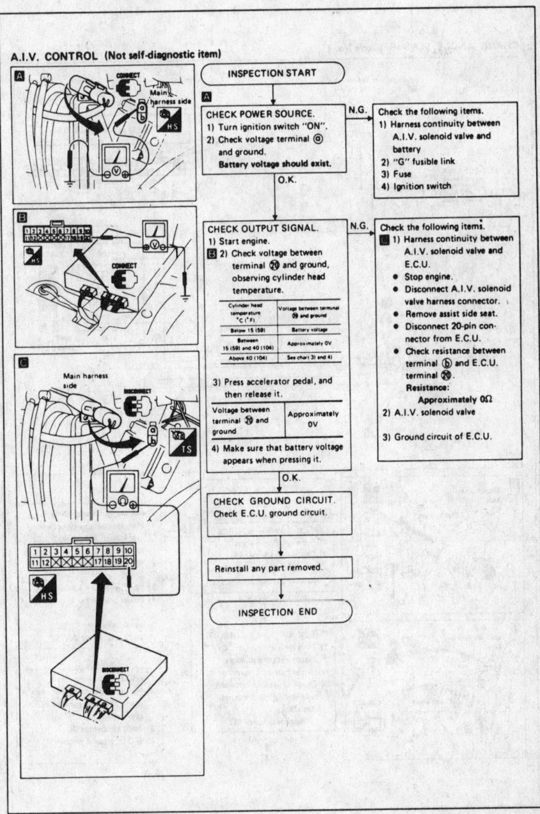
A.I.V. CONTROL (Not self-diagnostic item)
INSPECTION START
A
CHECK POWER SOURCE.
1) Turn ignition switch "ON".
2) Check voltage terminal ⓐ and ground.
Battery voltage should exist.
N.G.
Check the following items.
1) Harness continuity between A.I.V. solenoid valve and battery
2) "G" fusible link
3) Fuse
4) Ignition switch
O.K.
CHECK OUTPUT SIGNAL.
1) Start engine.
B 2) Check voltage between terminal ⑳ and ground, observing cylinder head temperature.
Cylinder head temperature °C (°F) | Voltage between terminal ⑳ and ground
Below 15 (59) | Battery voltage
Between 15 (59) and 40 (104) | Approximately 0V
Above 40 (104) | See chart 3) and 4)
3) Press accelerator pedal, and then release it.
Voltage between terminal ⑳ and ground | Approximately 0V
4) Make sure that battery voltage appears when pressing it.
N.G.
Check the following items.
C 1) Harness continuity between A.I.V. solenoid valve and E.C.U.
• Stop engine.
• Disconnect A.I.V. solenoid valve harness connector.
• Remove assist side seat.
• Disconnect 20-pin connector from E.C.U.
• Check resistance between terminal ⓑ and E.C.U. terminal ⑳.
Resistance: Approximately 0Ω
2) A.I.V. solenoid valve
3) Ground circuit of E.C.U.
O.K.
CHECK GROUND CIRCUIT.
Check E.C.U. ground circuit.
Reinstall any part removed.
INSPECTION END
Main harness side

ECCS SELF-DIAGNOSTIC CHART—1988-89 TRUCK/PATHFINDER WITH VG30I ENGINE

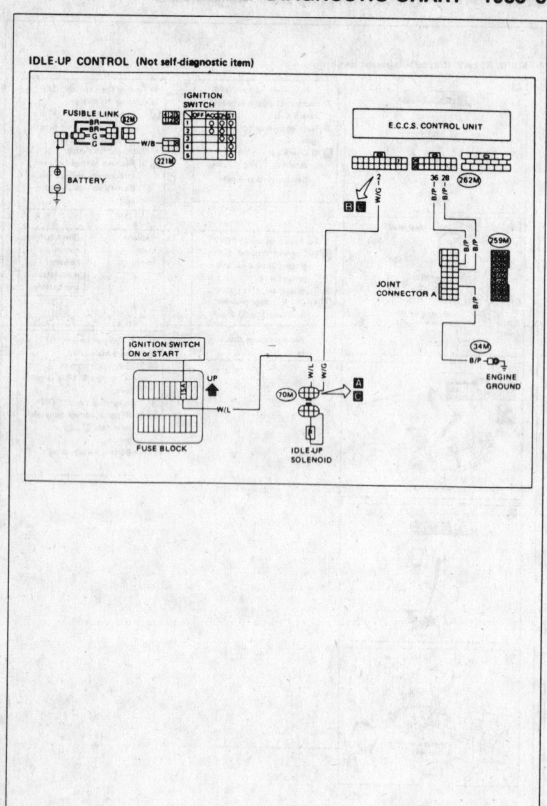
IDLE-UP CONTROL (Not self-diagnostic item)
FUSIBLE LINK
IGNITION SWITCH
BATTERY
E.C.C.S. CONTROL UNIT
JOINT CONNECTOR A
IGNITION SWITCH ON or START
UP
FUSE BLOCK
IDLE-UP SOLENOID
ENGINE GROUND

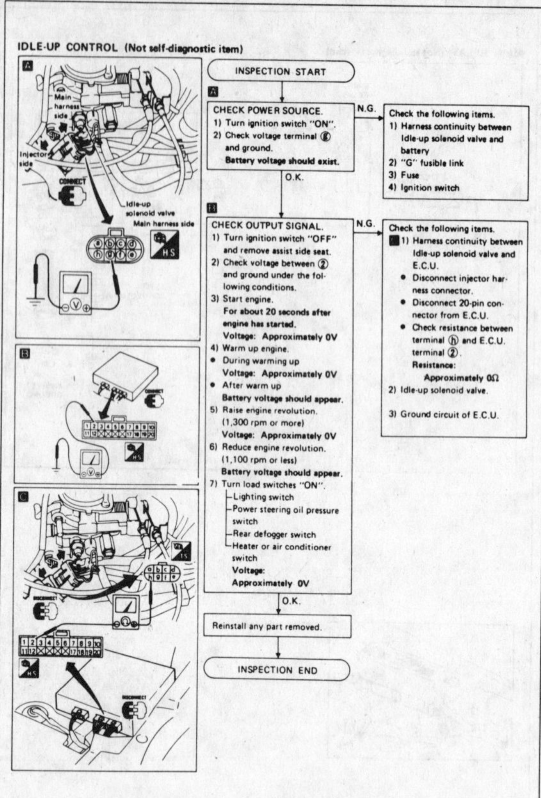
IDLE-UP CONTROL (Not self-diagnostic item)
INSPECTION START
CHECK POWER SOURCE.
1) Turn ignition switch "ON".
2) Check voltage terminal ⓗ and ground.
Battery voltage should exist.
N.G.
Check the following items.
1) Harness continuity between Idle-up solenoid valve and battery
2) "G" fusible link
3) Fuse
4) Ignition switch
O.K.
CHECK OUTPUT SIGNAL.
1) Turn ignition switch "OFF" and remove assist side seat.
2) Check voltage between ② and ground under the following conditions.
3) Start engine.
For about 20 seconds after engine has started.
Voltage: Approximately 0V
4) Warm up engine.
• During warming up
Voltage: Approximately 0V
• After warm up
Battery voltage should appear.
5) Raise engine revolution. (1,300 rpm or more)
Voltage: Approximately 0V
6) Reduce engine revolution. (1,100 rpm or less)
Battery voltage should appear.
7) Turn load switches "ON".
Lighting switch
Power steering oil pressure switch
Rear defogger switch
Heater or air conditioner switch
Voltage: Approximately 0V
N.G.
Check the following items.
1) Harness continuity between Idle-up solenoid valve and E.C.U.
• Disconnect injector harness connector.
• Disconnect 20-pin connector from E.C.U.
• Check resistance between terminal ⓗ and E.C.U. terminal ②.
Resistance: Approximately 0Ω
2) Idle-up solenoid valve.
3) Ground circuit of E.C.U.
O.K.
Reinstall any part removed.
INSPECTION END
Main harness side
Injector side
Idle-up solenoid valve
Main harness side

ECCS SELF-DIAGNOSTIC CHART—1988-89 TRUCK/PATHFINDER WITH Z24I ENGINE

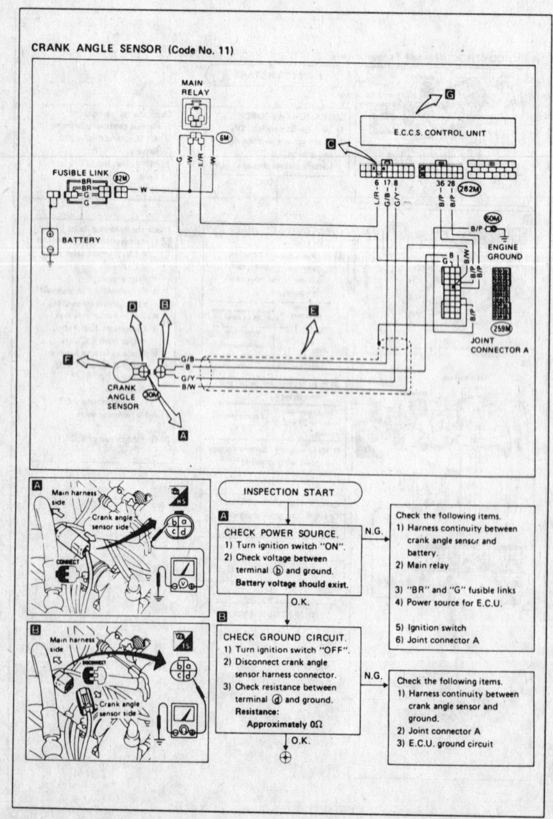
CRANK ANGLE SENSOR (Code No. 11)
MAIN RELAY
FUSIBLE LINK
BATTERY
E.C.C.S. CONTROL UNIT
ENGINE GROUND
JOINT CONNECTOR A
CRANK ANGLE SENSOR
INSPECTION START
CHECK POWER SOURCE.
1) Turn ignition switch "ON".
2) Check voltage between terminal ⓑ and ground.
Battery voltage should exist.
N.G.
Check the following items.
1) Harness continuity between crank angle sensor and battery.
2) Main relay
3) "BR" and "G" fusible links
4) Power source for E.C.U.
5) Ignition switch
6) Joint connector A
O.K.
CHECK GROUND CIRCUIT.
1) Turn ignition switch "OFF".
2) Disconnect crank angle sensor harness connector.
3) Check resistance between terminal ⓓ and ground.
Resistance: Approximately 0Ω
N.G.
Check the following items.
1) Harness continuity between crank angle sensor and ground.
2) Joint connector A
3) E.C.U. ground circuit
O.K.
Main harness side
Crank angle sensor side

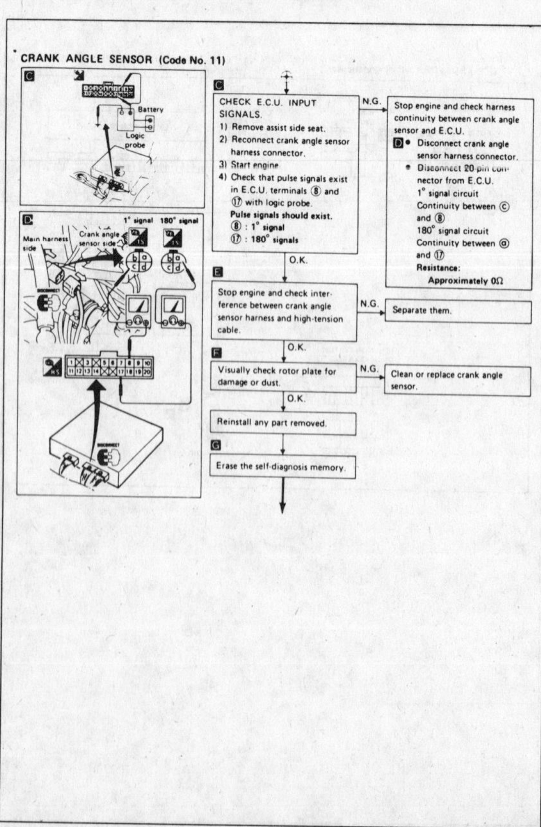
CRANK ANGLE SENSOR (Code No. 11)
CHECK E.C.U. INPUT SIGNALS.
1) Remove assist side seat.
2) Reconnect crank angle sensor harness connector.
3) Start engine.
4) Check that pulse signals exist in E.C.U. terminals ⑧ and ⑰ with logic probe.
Pulse signals should exist.
⑧ : 1° signal
⑰ : 180° signals
N.G.
Stop engine and check harness continuity between crank angle sensor and E.C.U.
• Disconnect crank angle sensor harness connector.
• Disconnect 20-pin connector from E.C.U.
1° signal circuit
Continuity between ⓒ and ⑧
180° signal circuit
Continuity between ⓐ and ⑰
Resistance: Approximately 0Ω
O.K.
Stop engine and check interference between crank angle sensor harness and high-tension cable.
N.G.
Separate them.
O.K.
Visually check rotor plate for damage or dust.
N.G.
Clean or replace crank angle sensor.
O.K.
Reinstall any part removed.
Erase the self-diagnosis memory.
Battery
Logic probe
Main harness side
Crank angle sensor side
1° signal
180° signal

ECCS SELF-DIAGNOSTIC CHART—1988–89 TRUCK/PATHFINDER WITH Z24I ENGINE

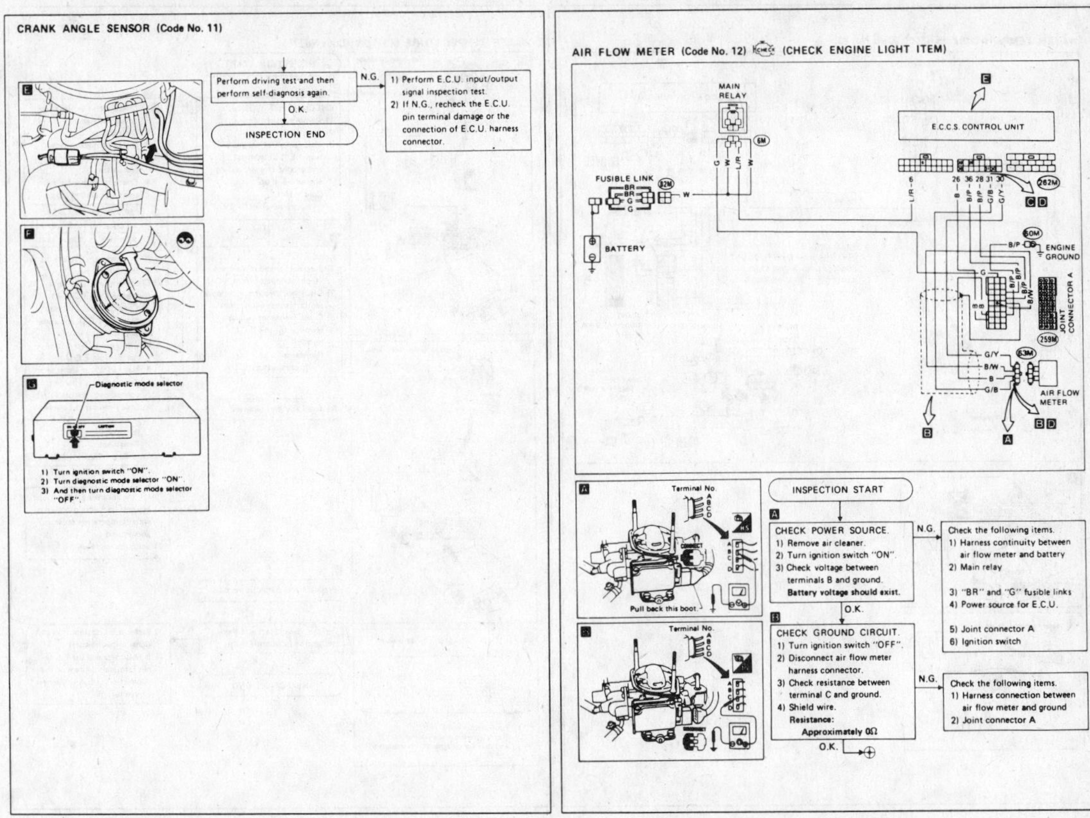
CRANK ANGLE SENSOR (Code No. 11)
Perform driving test and then perform self-diagnosis again.
N.G.
1) Perform E.C.U. input/output signal inspection test.
2) If N.G., recheck the E.C.U. pin terminal damage or the connection of E.C.U. harness connector.
O.K.
INSPECTION END
Diagnostic mode selector
1) Turn ignition switch "ON".
2) Turn diagnostic mode selector "ON".
3) And then turn diagnostic mode selector "OFF".
AIR FLOW METER (Code No. 12) (CHECK ENGINE LIGHT ITEM)
MAIN RELAY
E.C.C.S. CONTROL UNIT
FUSIBLE LINK
BATTERY
ENGINE GROUND
JOINT CONNECTOR A
AIR FLOW METER
Terminal No.
Pull back this boot.
INSPECTION START
CHECK POWER SOURCE.
1) Remove air cleaner.
2) Turn ignition switch "ON".
3) Check voltage between terminals B and ground.
Battery voltage should exist.
N.G.
Check the following items.
1) Harness continuity between air flow meter and battery
2) Main relay
3) "BR" and "G" fusible links
4) Power source for E.C.U.
5) Joint connector A
6) Ignition switch
O.K.
CHECK GROUND CIRCUIT.
1) Turn ignition switch "OFF".
2) Disconnect air flow meter harness connector.
3) Check resistance between terminal C and ground.
4) Shield wire.
Resistance:
Approximately 0Ω
N.G.
Check the following items.
1) Harness connection between air flow meter and ground
2) Joint connector A
O.K.

ECCS SELF-DIAGNOSTIC CHART—1988–89 TRUCK/PATHFINDER WITH Z24I ENGINE

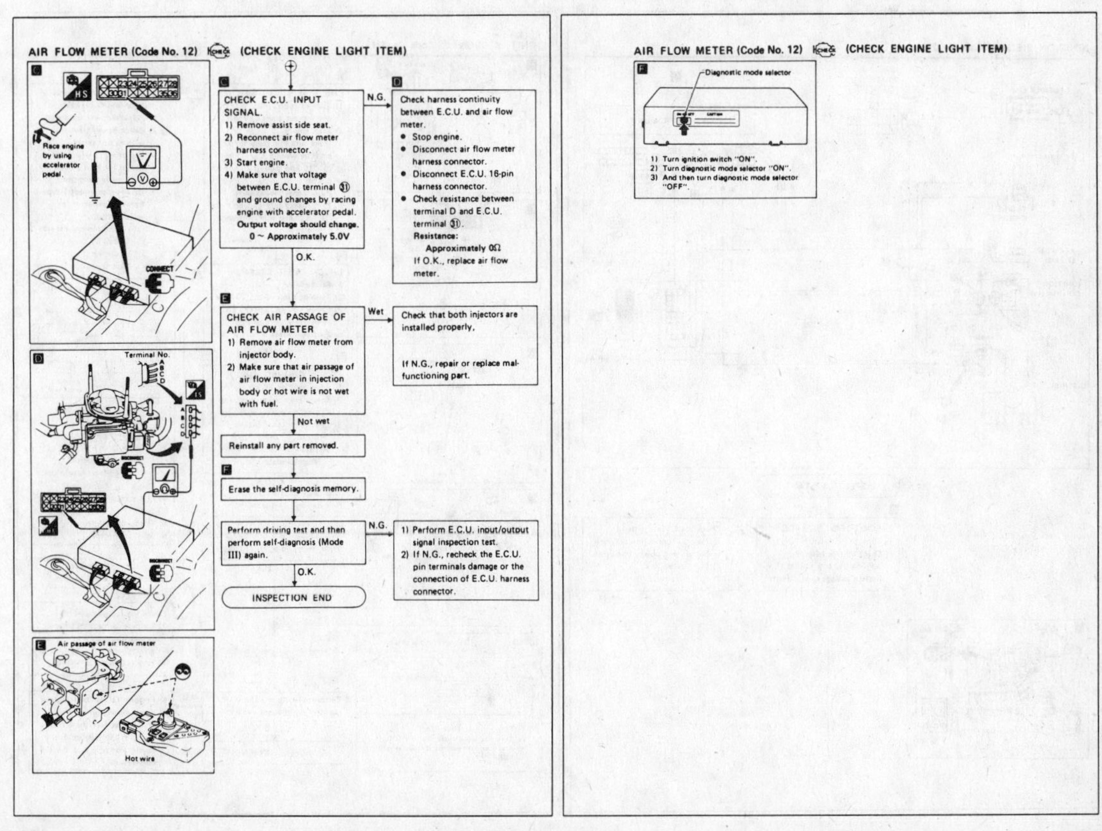
AIR FLOW METER (Code No. 12) (CHECK ENGINE LIGHT ITEM)
Race engine by using accelerator pedal.
CONNECT
CHECK E.C.U. INPUT SIGNAL.
1) Remove assist side seat.
2) Reconnect air flow meter harness connector.
3) Start engine.
4) Make sure that voltage between E.C.U. terminal 31 and ground changes by racing engine with accelerator pedal.
Output voltage should change.
0 ~ Approximately 5.0V
N.G.
Check harness continuity between E.C.U. and air flow meter.
• Stop engine.
• Disconnect air flow meter harness connector.
• Disconnect E.C.U. 16-pin harness connector.
• Check resistance between terminal D and E.C.U. terminal 31.
Resistance:
Approximately 0Ω
If O.K., replace air flow meter.
O.K.
Terminal No.
DISCONNECT
CHECK AIR PASSAGE OF AIR FLOW METER
1) Remove air flow meter from injector body.
2) Make sure that air passage of air flow meter in injection body or hot wire is not wet with fuel.
Wet
Check that both injectors are installed properly.
If N.G., repair or replace malfunctioning part.
Not wet
Reinstall any part removed.
Erase the self-diagnosis memory.
Perform driving test and then perform self-diagnosis (Mode III) again.
N.G.
1) Perform E.C.U. input/output signal inspection test.
2) If N.G., recheck the E.C.U. pin terminals damage or the connection of E.C.U. harness connector.
O.K.
INSPECTION END
Air passage of air flow meter
Hot wire
AIR FLOW METER (Code No. 12) (CHECK ENGINE LIGHT ITEM)
Diagnostic mode selector
1) Turn ignition switch "ON".
2) Turn diagnostic mode selector "ON".
3) And then turn diagnostic mode selector "OFF".

ECCS SELF-DIAGNOSTIC CHART—1988–89 TRUCK/PATHFINDER WITH Z24I ENGINE

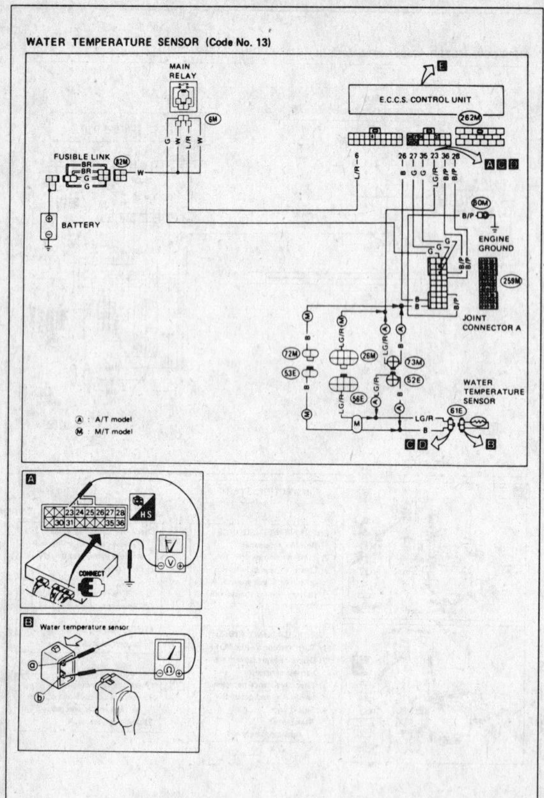
WATER TEMPERATURE SENSOR (Code No. 13)
MAIN RELAY
E.C.C.S. CONTROL UNIT
FUSIBLE LINK
BATTERY
ENGINE GROUND
JOINT CONNECTOR A
WATER TEMPERATURE SENSOR
Ⓐ: A/T model
Ⓜ: M/T model
Water temperature sensor

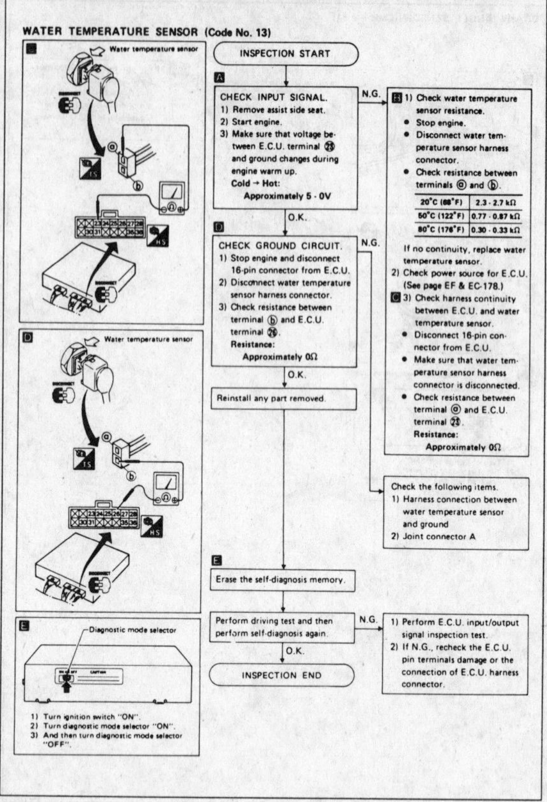
WATER TEMPERATURE SENSOR (Code No. 13)
Water temperature sensor
Diagnostic mode selector
1) Turn ignition switch "ON".
2) Turn diagnostic mode selector "ON".
3) And then turn diagnostic mode selector "OFF".
INSPECTION START
CHECK INPUT SIGNAL.
1) Remove assist side seat.
2) Start engine.
3) Make sure that voltage between E.C.U. terminal ㉓ and ground changes during engine warm up.
Cold → Hot:
Approximately 5 - 0V
N.G.
1) Check water temperature sensor resistance.
• Stop engine.
• Disconnect water temperature sensor harness connector.
• Check resistance between terminals ⓒ and ⓑ.
20°C (68°F) 2.3 - 2.7 kΩ
50°C (122°F) 0.77 - 0.87 kΩ
80°C (176°F) 0.30 - 0.33 kΩ
If no continuity, replace water temperature sensor.
2) Check power source for E.C.U. (See page EF & EC-178.)
3) Check harness continuity between E.C.U. and water temperature sensor.
• Disconnect 16-pin connector from E.C.U.
• Make sure that water temperature sensor harness connector is disconnected.
• Check resistance between terminal ⓒ and E.C.U. terminal ㉓.
Resistance:
Approximately 0Ω
O.K.
CHECK GROUND CIRCUIT.
1) Stop engine and disconnect 16-pin connector from E.C.U.
2) Disconnect water temperature sensor harness connector.
3) Check resistance between terminal ⓑ and E.C.U. terminal ㉖.
Resistance:
Approximately 0Ω
N.G.
Check the following items.
1) Harness connection between water temperature sensor and ground
2) Joint connector A
O.K.
Reinstall any part removed.
Erase the self-diagnosis memory.
Perform driving test and then perform self-diagnosis again.
N.G.
1) Perform E.C.U. input/output signal inspection test.
2) If N.G., recheck the E.C.U. pin terminals damage or the connection of E.C.U. harness connector.
O.K.
INSPECTION END

ECCS SELF-DIAGNOSTIC CHART—1988–89 TRUCK/PATHFINDER WITH Z24I ENGINE

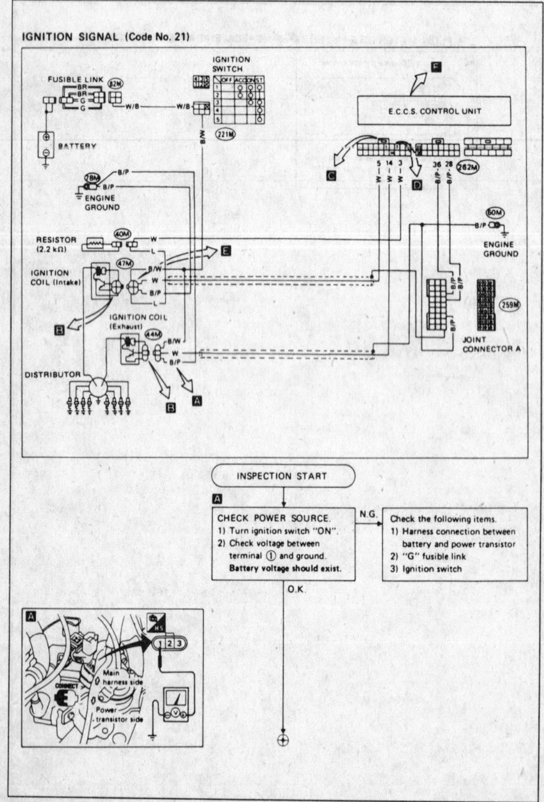
IGNITION SIGNAL (Code No. 21)
FUSIBLE LINK
IGNITION SWITCH
BATTERY
ENGINE GROUND
E.C.C.S. CONTROL UNIT
RESISTOR (2.2 kΩ)
IGNITION COIL (Intake)
IGNITION COIL (Exhaust)
DISTRIBUTOR
JOINT CONNECTOR A
INSPECTION START
CHECK POWER SOURCE.
1) Turn ignition switch "ON".
2) Check voltage between terminal ① and ground.
Battery voltage should exist.
N.G.
Check the following items.
1) Harness connection between battery and power transistor
2) "G" fusible link
3) Ignition switch
O.K.
Main harness side
Power transistor side

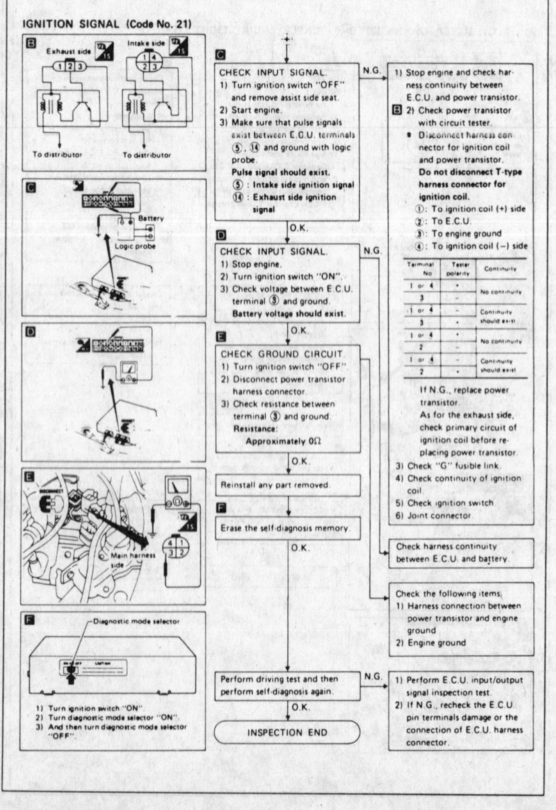
IGNITION SIGNAL (Code No. 21)
Exhaust side
Intake side
To distributor
Battery
Logic probe
Main harness side
Diagnostic mode selector
1) Turn ignition switch "ON".
2) Turn diagnostic mode selector "ON".
3) And then turn diagnostic mode selector "OFF".
CHECK INPUT SIGNAL.
1) Turn ignition switch "OFF" and remove assist side seat.
2) Start engine.
3) Make sure that pulse signals exist between E.C.U. terminals ⑤, ⑭ and ground with logic probe.
Pulse signal should exist.
⑤ : Intake side ignition signal
⑭ : Exhaust side ignition signal
N.G.
1) Stop engine and check harness continuity between E.C.U. and power transistor.
2) Check power transistor with circuit tester.
• Disconnect harness connector for ignition coil and power transistor.
Do not disconnect T-type harness connector for ignition coil.
①: To ignition coil (+) side
②: To E.C.U.
③: To engine ground
④: To ignition coil (−) side
Terminal No | Tester polarity | Continuity
1 or 4 | + | No continuity
3 | − |
1 or 4 | − | Continuity should exist
3 | + |
1 or 4 | + | No continuity
2 | − |
1 or 4 | − | Continuity should exist
2 | + |
If N.G., replace power transistor.
As for the exhaust side, check primary circuit of ignition coil before replacing power transistor.
3) Check "G" fusible link.
4) Check continuity of ignition coil.
5) Check ignition switch.
6) Joint connector.
O.K.
CHECK INPUT SIGNAL.
1) Stop engine.
2) Turn ignition switch "ON".
3) Check voltage between E.C.U. terminal ③ and ground.
Battery voltage should exist.
N.G.
Check harness continuity between E.C.U. and battery.
O.K.
CHECK GROUND CIRCUIT.
1) Turn ignition switch "OFF".
2) Disconnect power transistor harness connector.
3) Check resistance between terminal ③ and ground.
Resistance:
Approximately 0Ω
Check the following items.
1) Harness connection between power transistor and engine ground
2) Engine ground
O.K.
Reinstall any part removed.
Erase the self-diagnosis memory.
O.K.
Perform driving test and then perform self-diagnosis again.
N.G.
1) Perform E.C.U. input/output signal inspection test.
2) If N.G., recheck the E.C.U. pin terminals damage or the connection of E.C.U. harness connector.
O.K.
INSPECTION END

ECCS SELF-DIAGNOSTIC CHART—1988-89 TRUCK/PATHFINDER WITH Z24I ENGINE

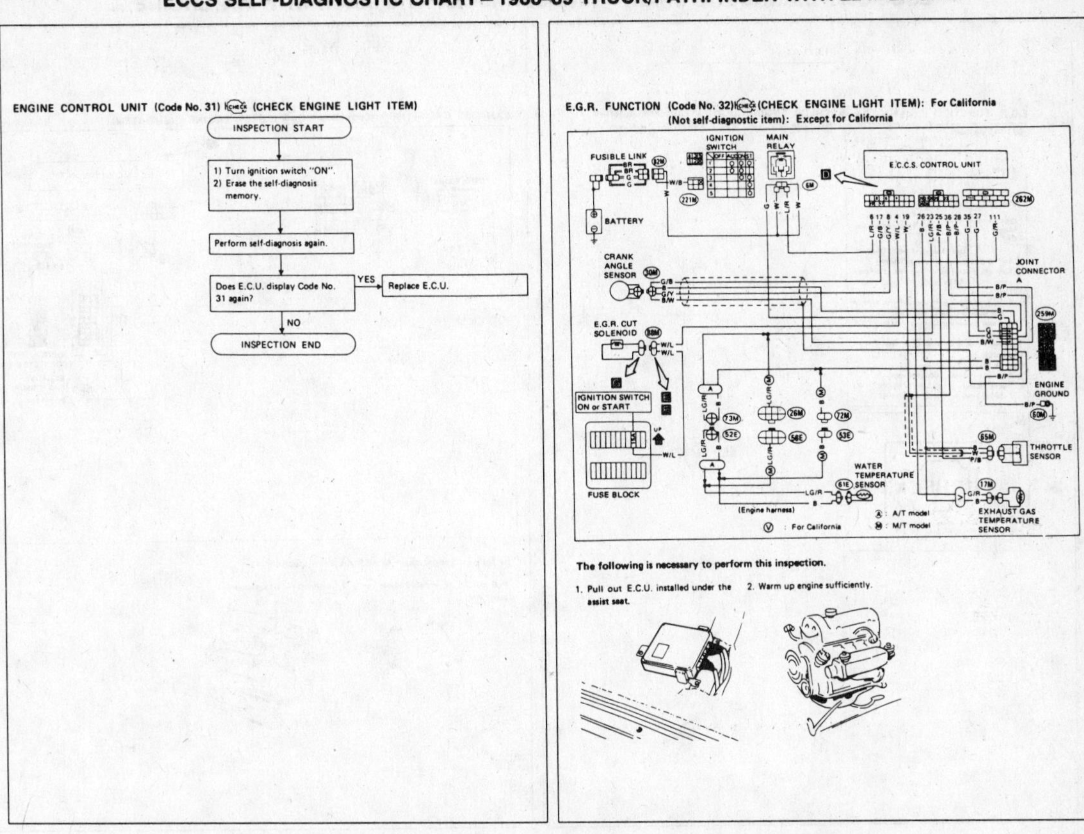

ECCS SELF-DIAGNOSTIC CHART—1988-89 TRUCK/PATHFINDER WITH Z24I ENGINE

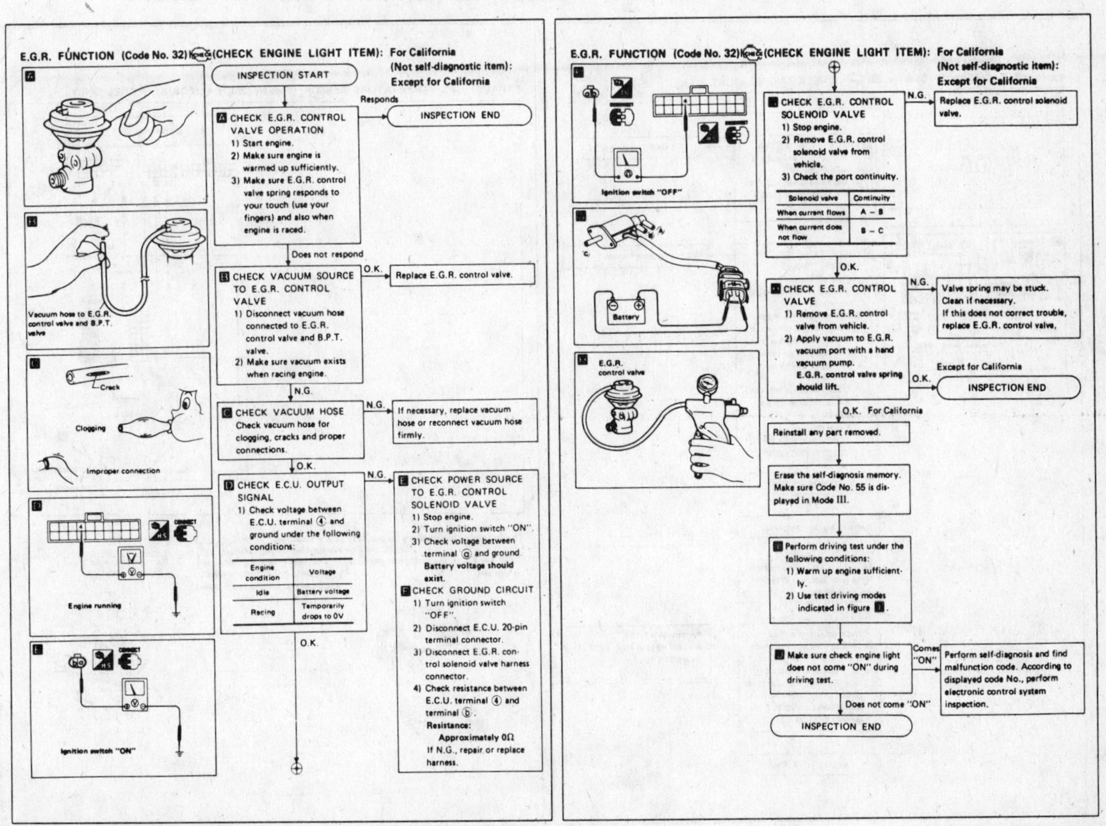

ECCS SELF-DIAGNOSTIC CHART—1988-89 TRUCK/PATHFINDER WITH Z24I ENGINE

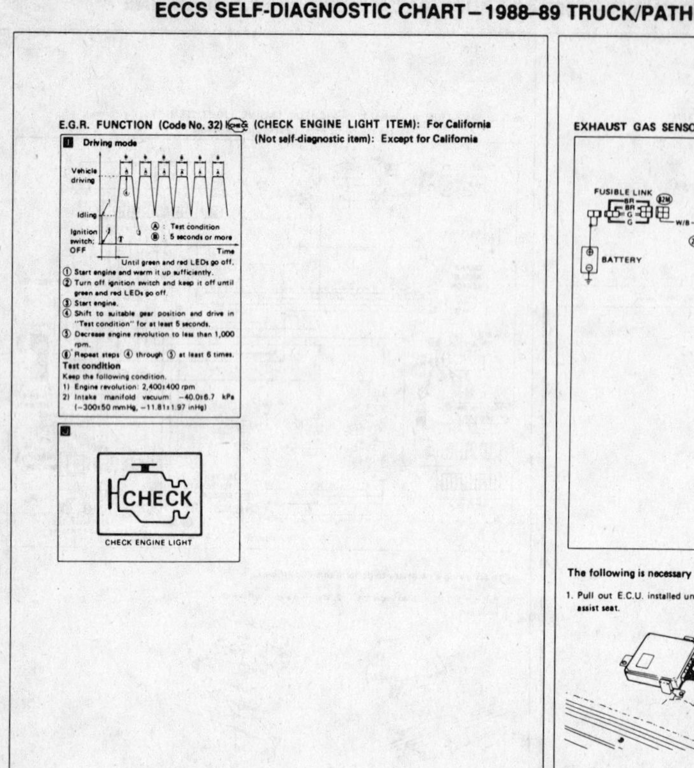
E.G.R. FUNCTION (Code No. 32) (CHECK ENGINE LIGHT ITEM): For California
(Not self-diagnostic item): Except for California
I Driving mode
Vehicle driving
Idling
Ignition switch; OFF
Ⓐ : Test condition
Ⓑ : 5 seconds or more
Time
Until green and red LEDs go off.
① Start engine and warm it up sufficiently.
② Turn off ignition switch and keep it off until green and red LEDs go off.
③ Start engine.
④ Shift to suitable gear position and drive in "Test condition" for at least 5 seconds.
⑤ Decrease engine revolution to less than 1,000 rpm.
⑥ Repeat steps ④ through ⑤ at least 6 times.
Test condition
Keep the following condition.
1) Engine revolution: 2,400±400 rpm
2) Intake manifold vacuum: −40.0±6.7 kPa (−300±50 mmHg, −11.81±1.97 inHg)
J
CHECK
CHECK ENGINE LIGHT

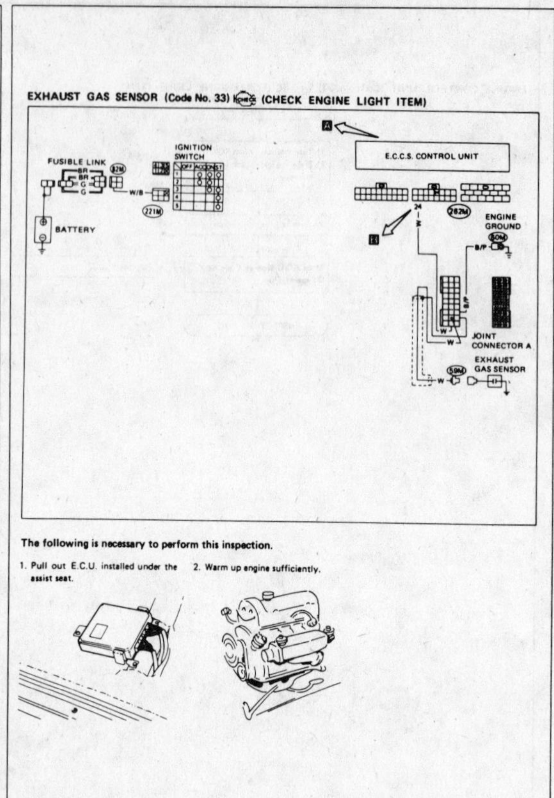
EXHAUST GAS SENSOR (Code No. 33) (CHECK ENGINE LIGHT ITEM)
A
E.C.C.S. CONTROL UNIT
FUSIBLE LINK
IGNITION SWITCH
BATTERY
24
B
ENGINE GROUND
JOINT CONNECTOR A
EXHAUST GAS SENSOR
The following is necessary to perform this inspection.
1. Pull out E.C.U. installed under the assist seat.
2. Warm up engine sufficiently.

ECCS SELF-DIAGNOSTIC CHART—1988-89 TRUCK/PATHFINDER WITH Z24I ENGINE

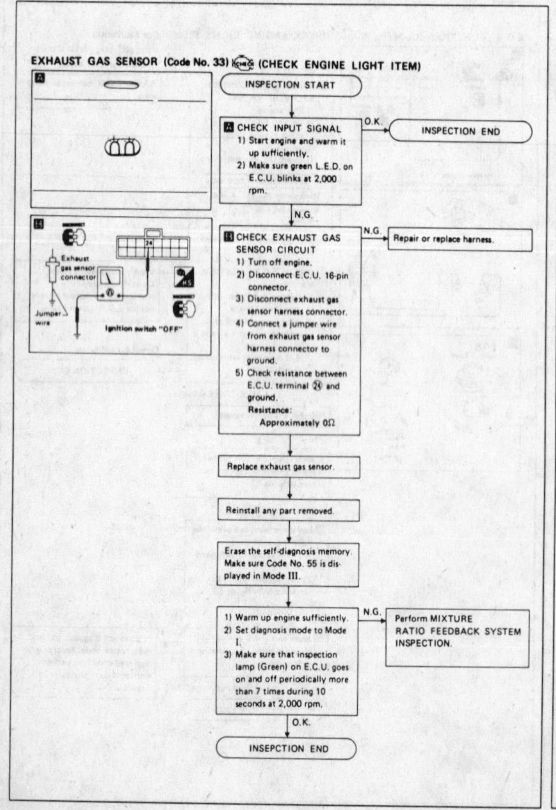
EXHAUST GAS SENSOR (Code No. 33) (CHECK ENGINE LIGHT ITEM)
A
B
Exhaust gas sensor connector
Jumper wire
24
Ignition switch "OFF"
INSPECTION START
A CHECK INPUT SIGNAL
1) Start engine and warm it up sufficiently.
2) Make sure green L.E.D. on E.C.U. blinks at 2,000 rpm.
O.K.
INSPECTION END
N.G.
B CHECK EXHAUST GAS SENSOR CIRCUIT
1) Turn off engine.
2) Disconnect E.C.U. 16-pin connector.
3) Disconnect exhaust gas sensor harness connector.
4) Connect a jumper wire from exhaust gas sensor harness connector to ground.
5) Check resistance between E.C.U. terminal ㉔ and ground.
Resistance:
Approximately 0Ω
N.G.
Repair or replace harness.
Replace exhaust gas sensor.
Reinstall any part removed.
Erase the self-diagnosis memory. Make sure Code No. 55 is displayed in Mode III.
1) Warm up engine sufficiently.
2) Set diagnosis mode to Mode I.
3) Make sure that inspection lamp (Green) on E.C.U. goes on and off periodically more than 7 times during 10 seconds at 2,000 rpm.
N.G.
Perform MIXTURE RATIO FEEDBACK SYSTEM INSPECTION.
O.K.
INSEPCTION END

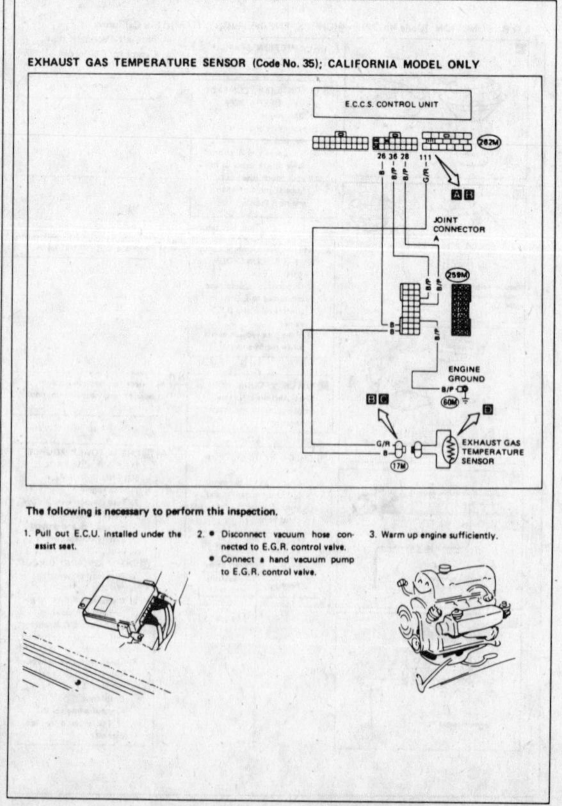
EXHAUST GAS TEMPERATURE SENSOR (Code No. 35); CALIFORNIA MODEL ONLY
E.C.C.S. CONTROL UNIT
26 36 28 111
A B
JOINT CONNECTOR
ENGINE GROUND
B C
D
EXHAUST GAS TEMPERATURE SENSOR
The following is necessary to perform this inspection.
1. Pull out E.C.U. installed under the assist seat.
2. • Disconnect vacuum hose connected to E.G.R. control valve.
• Connect a hand vacuum pump to E.G.R. control valve.
3. Warm up engine sufficiently.

ECCS SELF-DIAGNOSTIC CHART—1988–89 TRUCK/PATHFINDER WITH Z24I ENGINE

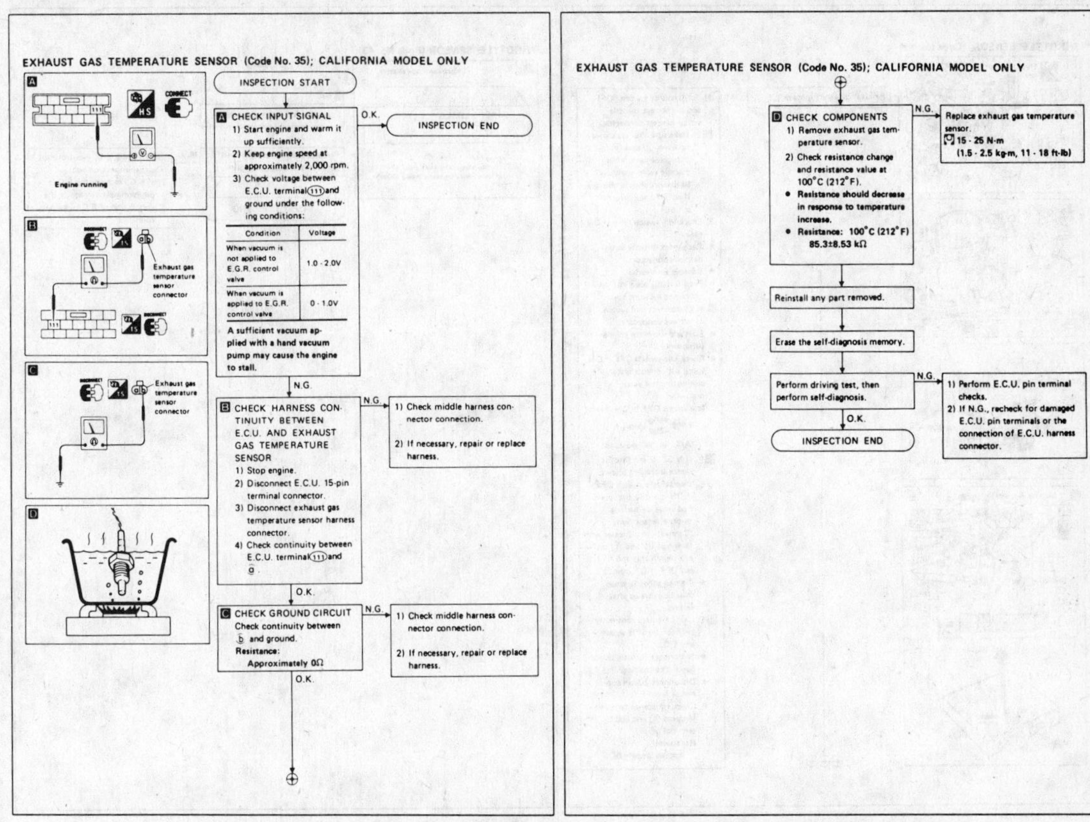
EXHAUST GAS TEMPERATURE SENSOR (Code No. 35); CALIFORNIA MODEL ONLY
INSPECTION START
A CHECK INPUT SIGNAL
1) Start engine and warm it up sufficiently.
2) Keep engine speed at approximately 2,000 rpm.
3) Check voltage between E.C.U. terminal (111) and ground under the following conditions:
Condition
Voltage
When vacuum is not applied to E.G.R. control valve
1.0 - 2.0V
When vacuum is applied to E.G.R. control valve
0 - 1.0V
A sufficient vacuum applied with a hand vacuum pump may cause the engine to stall.
O.K.
INSPECTION END
N.G.
B CHECK HARNESS CONTINUITY BETWEEN E.C.U. AND EXHAUST GAS TEMPERATURE SENSOR
1) Stop engine.
2) Disconnect E.C.U. 15-pin terminal connector.
3) Disconnect exhaust gas temperature sensor harness connector.
4) Check continuity between E.C.U. terminal (111) and a.
1) Check middle harness connector connection.
2) If necessary, repair or replace harness.
C CHECK GROUND CIRCUIT
Check continuity between b and ground.
Resistance:
Approximately 0Ω
1) Check middle harness connector connection.
2) If necessary, repair or replace harness.
Engine running
Exhaust gas temperature sensor connector
D CHECK COMPONENTS
1) Remove exhaust gas temperature sensor.
2) Check resistance change and resistance value at 100°C (212°F).
• Resistance should decrease in response to temperature increase.
• Resistance: 100°C (212°F) 85.3±8.53 kΩ
Replace exhaust gas temperature sensor.
15 - 25 N·m (1.5 - 2.5 kg-m, 11 - 18 ft-lb)
Reinstall any part removed.
Erase the self-diagnosis memory.
Perform driving test, then perform self-diagnosis.
1) Perform E.C.U. pin terminal checks.
2) If N.G., recheck for damaged E.C.U. pin terminals or the connection of E.C.U. harness connector.
INSPECTION END

ECCS SELF-DIAGNOSTIC CHART—1988–89 TRUCK/PATHFINDER WITH Z24I ENGINE

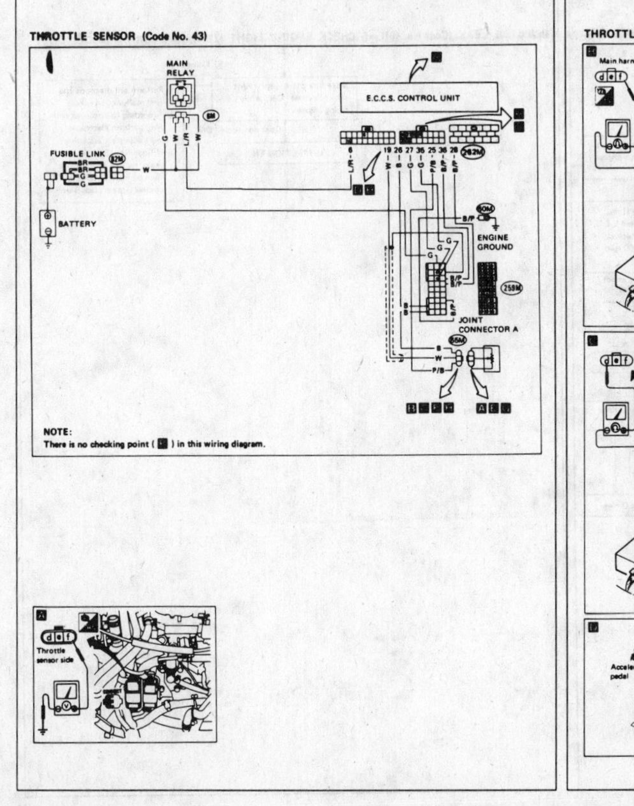
THROTTLE SENSOR (Code No. 43)
MAIN RELAY
FUSIBLE LINK
BATTERY
E.C.C.S. CONTROL UNIT
ENGINE GROUND
JOINT CONNECTOR A
NOTE:
There is no checking point in this wiring diagram.
Throttle sensor side

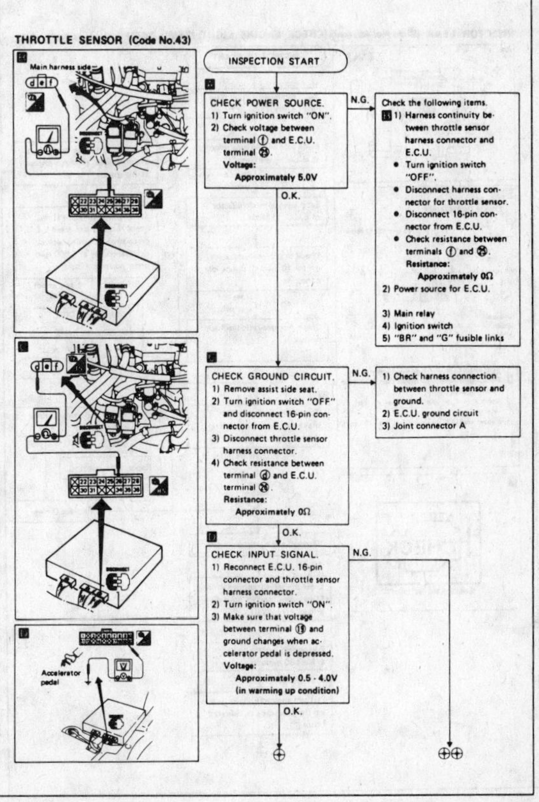
THROTTLE SENSOR (Code No.43)
Main harness side
INSPECTION START
CHECK POWER SOURCE.
1) Turn ignition switch "ON".
2) Check voltage between terminal f and E.C.U. terminal 26.
Voltage:
Approximately 5.0V
Check the following items.
1) Harness continuity between throttle sensor harness connector and E.C.U.
• Turn ignition switch "OFF".
• Disconnect harness connector for throttle sensor.
• Disconnect 16-pin connector from E.C.U.
• Check resistance between terminals f and 26.
Resistance:
Approximately 0Ω
2) Power source for E.C.U.
3) Main relay
4) Ignition switch
5) "BR" and "G" fusible links
CHECK GROUND CIRCUIT.
1) Remove assist side seat.
2) Turn ignition switch "OFF" and disconnect 16-pin connector from E.C.U.
3) Disconnect throttle sensor harness connector.
4) Check resistance between terminal d and E.C.U. terminal 26.
Resistance:
Approximately 0Ω
1) Check harness connection between throttle sensor and ground.
2) E.C.U. ground circuit
3) Joint connector A
CHECK INPUT SIGNAL.
1) Reconnect E.C.U. 16-pin connector and throttle sensor harness connector.
2) Turn ignition switch "ON".
3) Make sure that voltage between terminal 19 and ground changes when accelerator pedal is depressed.
Voltage:
Approximately 0.5 - 4.0V
(in warming up condition)
Accelerator pedal

ECCS SELF-DIAGNOSTIC CHART—1988–89 TRUCK/PATHFINDER WITH Z24I ENGINE

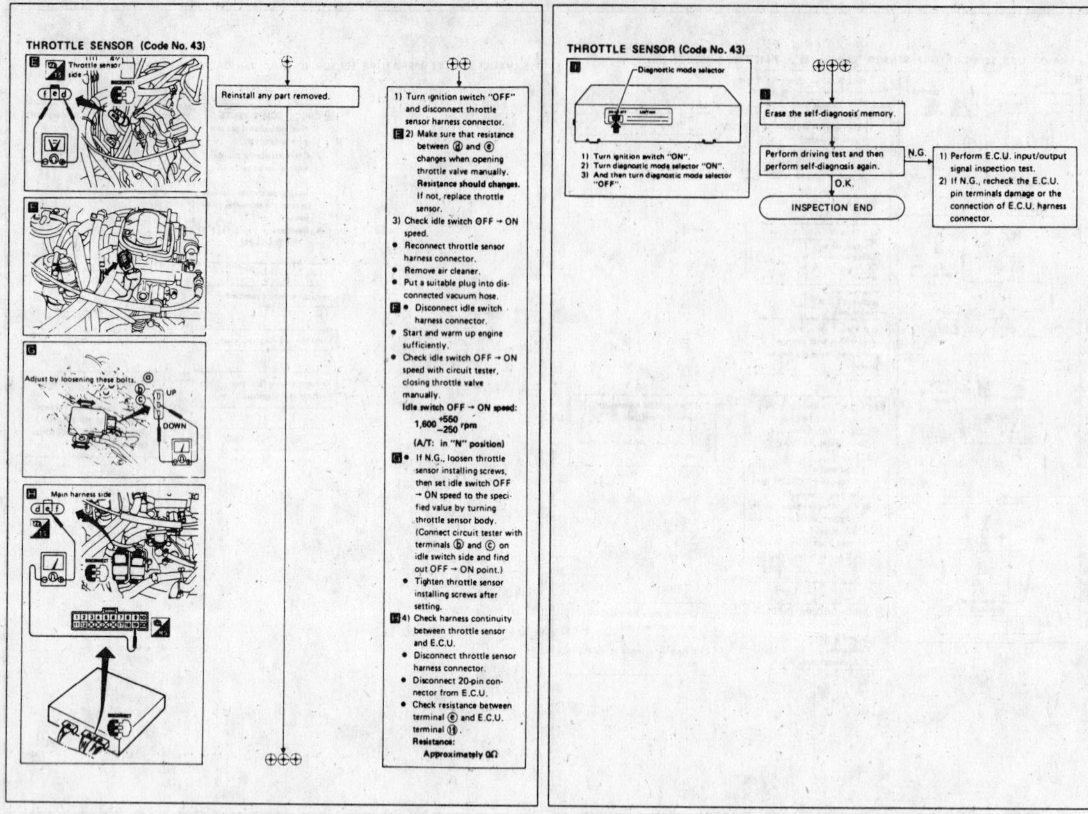
THROTTLE SENSOR (Code No. 43)
Throttle sensor side
Adjust by loosening these bolts.
UP
DOWN
Main harness side
Reinstall any part removed.
1) Turn ignition switch "OFF" and disconnect throttle sensor harness connector.
2) Make sure that resistance between ⓓ and ⓔ changes when opening throttle valve manually. Resistance should change. If not, replace throttle sensor.
3) Check idle switch OFF → ON speed.
• Reconnect throttle sensor harness connector.
• Remove air cleaner.
• Put a suitable plug into disconnected vacuum hose.
• Disconnect idle switch harness connector.
• Start and warm up engine sufficiently.
• Check idle switch OFF → ON speed with circuit tester, closing throttle valve manually.
Idle switch OFF → ON speed: 1,600 +550 −250 rpm
(A/T: in "N" position)
• If N.G., loosen throttle sensor installing screws, then set idle switch OFF → ON speed to the specified value by turning throttle sensor body. (Connect circuit tester with terminals ⓑ and ⓒ on idle switch side and find out OFF → ON point.)
• Tighten throttle sensor installing screws after setting.
4) Check harness continuity between throttle sensor and E.C.U.
• Disconnect throttle sensor harness connector.
• Disconnect 20-pin connector from E.C.U.
• Check resistance between terminal ⓔ and E.C.U. terminal ⑪.
Resistance: Approximately 0Ω
THROTTLE SENSOR (Code No. 43)
Diagnostic mode selector
1) Turn ignition switch "ON".
2) Turn diagnostic mode selector "ON".
3) And then turn diagnostic mode selector "OFF".
Erase the self-diagnosis memory.
Perform driving test and then perform self-diagnosis again.
N.G.
1) Perform E.C.U. input/output signal inspection test.
2) If N.G., recheck the E.C.U. pin terminals damage or the connection of E.C.U. harness connector.
O.K.
INSPECTION END

ECCS SELF-DIAGNOSTIC CHART—1988–89 TRUCK/PATHFINDER WITH Z24I ENGINE

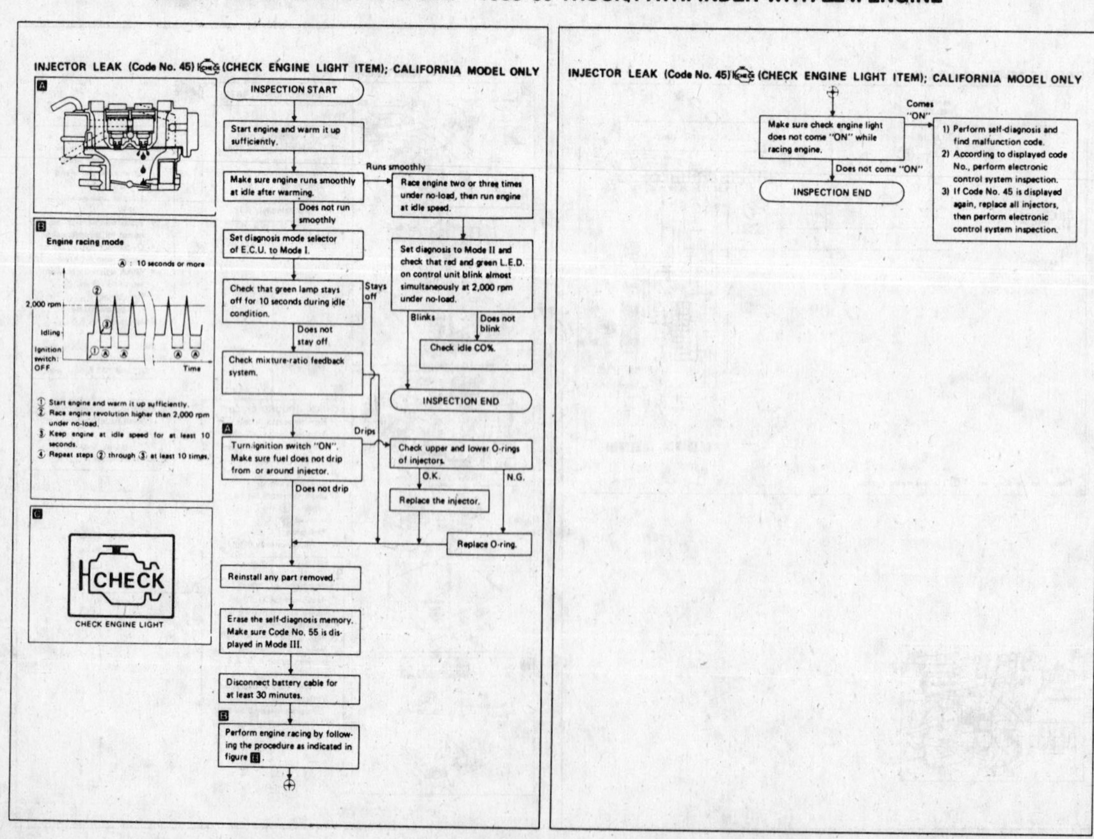
INJECTOR LEAK (Code No. 45) (CHECK ENGINE LIGHT ITEM); CALIFORNIA MODEL ONLY
INSPECTION START
Start engine and warm it up sufficiently.
Make sure engine runs smoothly at idle after warming.
Runs smoothly
Race engine two or three times under no-load, then run engine at idle speed.
Does not run smoothly
Set diagnosis mode selector of E.C.U. to Mode I.
Set diagnosis to Mode II and check that red and green L.E.D. on control unit blink almost simultaneously at 2,000 rpm under no-load.
Check that green lamp stays off for 10 seconds during idle condition.
Stays off
Blinks
Does not blink
Does not stay off
Check idle CO%.
Check mixture-ratio feedback system.
INSPECTION END
Drips
Turn ignition switch "ON". Make sure fuel does not drip from or around injector.
Check upper and lower O-rings of injectors.
O.K.
N.G.
Does not drip
Replace the injector.
Replace O-ring.
Reinstall any part removed.
Erase the self-diagnosis memory. Make sure Code No. 55 is displayed in Mode III.
Disconnect battery cable for at least 30 minutes.
Perform engine racing by following the procedure as indicated in figure B.
Engine racing mode
Ⓐ: 10 seconds or more
2,000 rpm
Idling
Ignition switch OFF
Time
① Start engine and warm it up sufficiently.
② Race engine revolution higher than 2,000 rpm under no-load.
③ Keep engine at idle speed for at least 10 seconds.
④ Repeat steps ② through ③ at least 10 times.
CHECK
CHECK ENGINE LIGHT
INJECTOR LEAK (Code No. 45) (CHECK ENGINE LIGHT ITEM); CALIFORNIA MODEL ONLY
Make sure check engine light does not come "ON" while racing engine.
Comes "ON"
1) Perform self-diagnosis and find malfunction code.
2) According to displayed code No., perform electronic control system inspection.
3) If Code No. 45 is displayed again, replace all injectors, then perform electronic control system inspection.
Does not come "ON"
INSPECTION END

ECCS SELF-DIAGNOSTIC CHART—1988–89 TRUCK/PATHFINDER WITH Z24I ENGINE

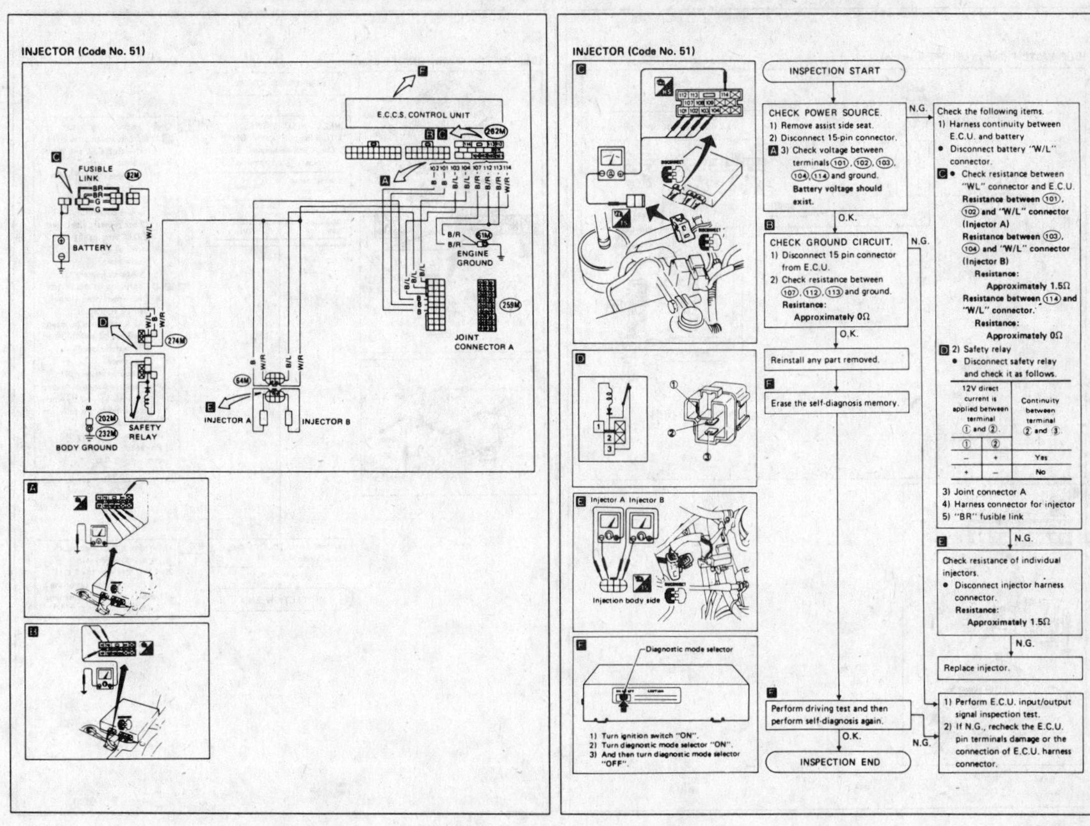
INJECTOR (Code No. 51)
E.C.C.S. CONTROL UNIT
FUSIBLE LINK
BATTERY
ENGINE GROUND
JOINT CONNECTOR A
SAFETY RELAY
BODY GROUND
INJECTOR A
INJECTOR B
INSPECTION START
CHECK POWER SOURCE.
1) Remove assist side seat.
2) Disconnect 15-pin connector.
3) Check voltage between terminals (101), (102), (103), (104), (114) and ground.
Battery voltage should exist.
O.K.
N.G.
Check the following items.
1) Harness continuity between E.C.U. and battery
• Disconnect battery "W/L" connector.
• Check resistance between "WL" connector and E.C.U.
Resistance between (101), (102) and "W/L" connector (Injector A)
Resistance between (103), (104) and "W/L" connector (Injector B)
Resistance: Approximately 1.5Ω
Resistance between (114) and "W/L" connector.
Resistance: Approximately 0Ω
2) Safety relay
• Disconnect safety relay and check it as follows.
12V direct current is applied between terminal ① and ②
Continuity between terminal ② and ③
CHECK GROUND CIRCUIT.
1) Disconnect 15 pin connector from E.C.U.
2) Check resistance between (107), (112), (113) and ground.
Resistance: Approximately 0Ω
Reinstall any part removed.
Erase the self-diagnosis memory.
3) Joint connector A
4) Harness connector for injector
5) "BR" fusible link
Check resistance of individual injectors.
• Disconnect injector harness connector.
Resistance: Approximately 1.5Ω
Replace injector.
Injector A Injector B
Injection body side
Diagnostic mode selector
1) Turn ignition switch "ON".
2) Turn diagnostic mode selector "ON".
3) And then turn diagnostic mode selector "OFF".
Perform driving test and then perform self-diagnosis again.
1) Perform E.C.U. input/output signal inspection test.
2) If N.G., recheck the E.C.U. pin terminals damage or the connection of E.C.U. harness connector.
INSPECTION END

①	②	
–	+	Yes
+	–	No

ECCS SELF-DIAGNOSTIC CHART—1988–89 TRUCK/PATHFINDER WITH Z24I ENGINE

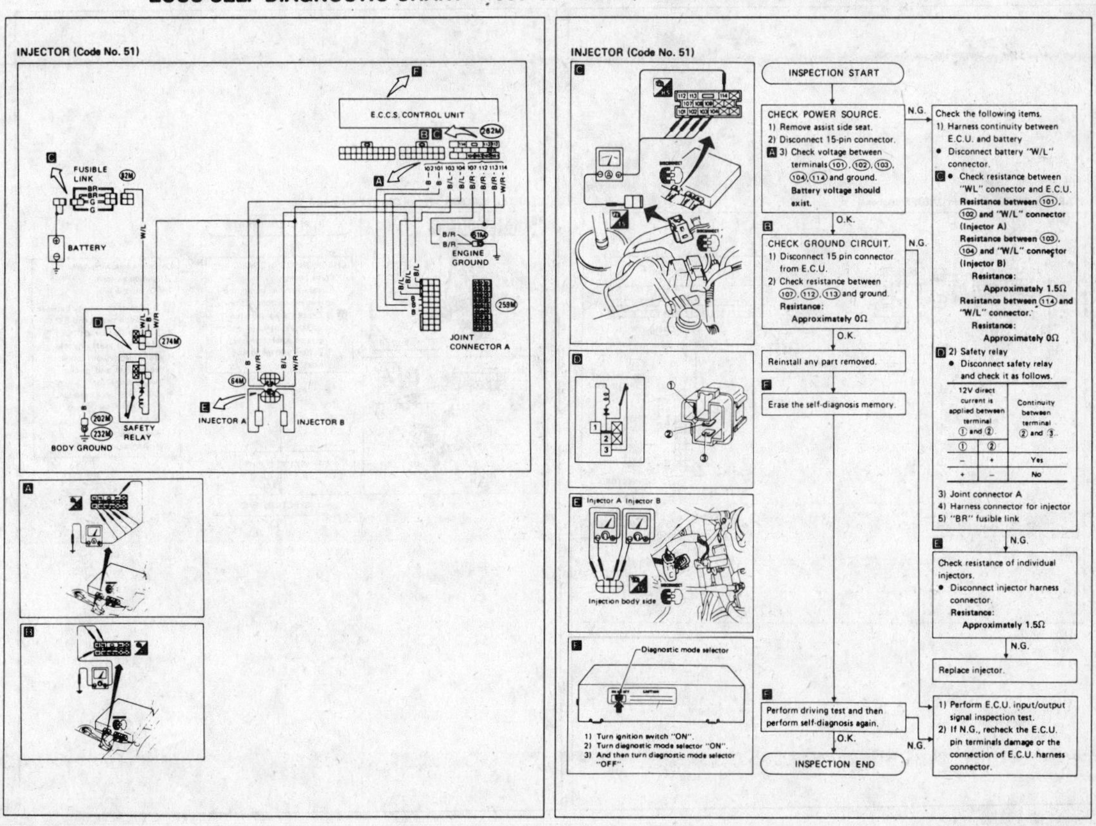

ECCS SELF-DIAGNOSTIC CHART—1988–89 TRUCK/PATHFINDER WITH Z24I ENGINE

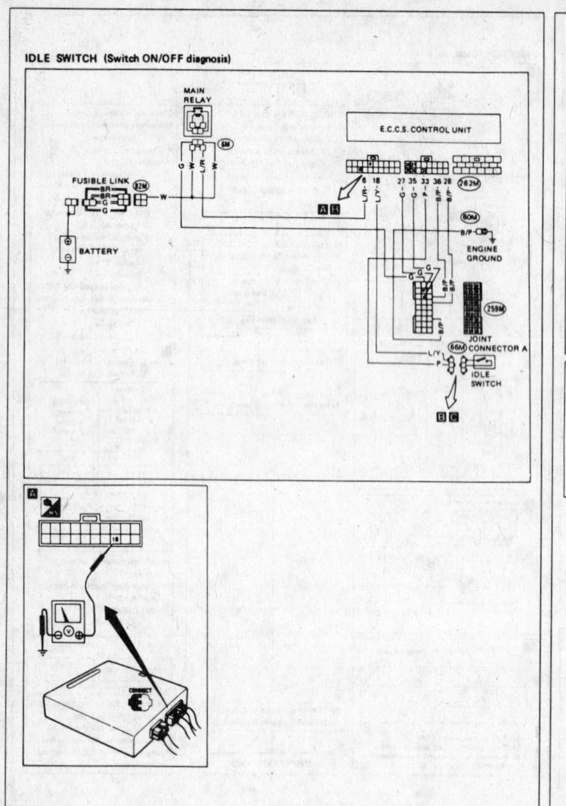
IDLE SWITCH (Switch ON/OFF diagnosis)
MAIN RELAY
E.C.C.S. CONTROL UNIT
FUSIBLE LINK
BATTERY
ENGINE GROUND
JOINT CONNECTOR A
IDLE SWITCH

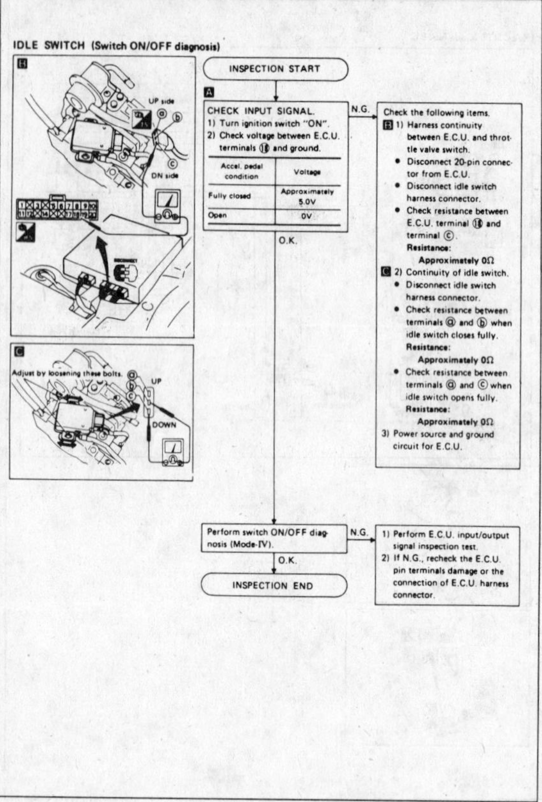
IDLE SWITCH (Switch ON/OFF diagnosis)
UP side
DN side
Adjust by loosening these bolts.
UP
DOWN
INSPECTION START
CHECK INPUT SIGNAL.
1) Turn ignition switch "ON".
2) Check voltage between E.C.U. terminals ⑱ and ground.
Accel. pedal condition | Voltage
Fully closed | Approximately 5.0V
Open | 0V
N.G.
Check the following items.
B 1) Harness continuity between E.C.U. and throttle valve switch.
• Disconnect 20-pin connector from E.C.U.
• Disconnect idle switch harness connector.
• Check resistance between E.C.U. terminal ⑱ and terminal ⓒ.
Resistance: Approximately 0Ω
C 2) Continuity of idle switch.
• Disconnect idle switch harness connector.
• Check resistance between terminals ⓐ and ⓑ when idle switch closes fully.
Resistance: Approximately 0Ω
• Check resistance between terminals ⓐ and ⓒ when idle switch opens fully.
Resistance: Approximately 0Ω
3) Power source and ground circuit for E.C.U.
O.K.
Perform switch ON/OFF diagnosis (Mode-IV).
N.G.
1) Perform E.C.U. input/output signal inspection test.
2) If N.G., recheck the E.C.U. pin terminals damage or the connection of E.C.U. harness connector.
O.K.
INSPECTION END

ECCS SELF-DIAGNOSTIC CHART—1988–89 TRUCK/PATHFINDER WITH Z24I ENGINE

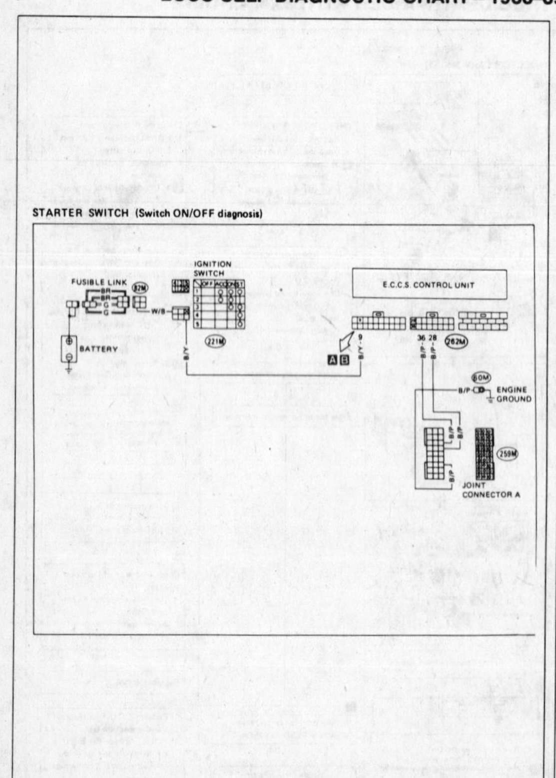
STARTER SWITCH (Switch ON/OFF diagnosis)
IGNITION SWITCH
FUSIBLE LINK
E.C.C.S. CONTROL UNIT
BATTERY
ENGINE GROUND
JOINT CONNECTOR A

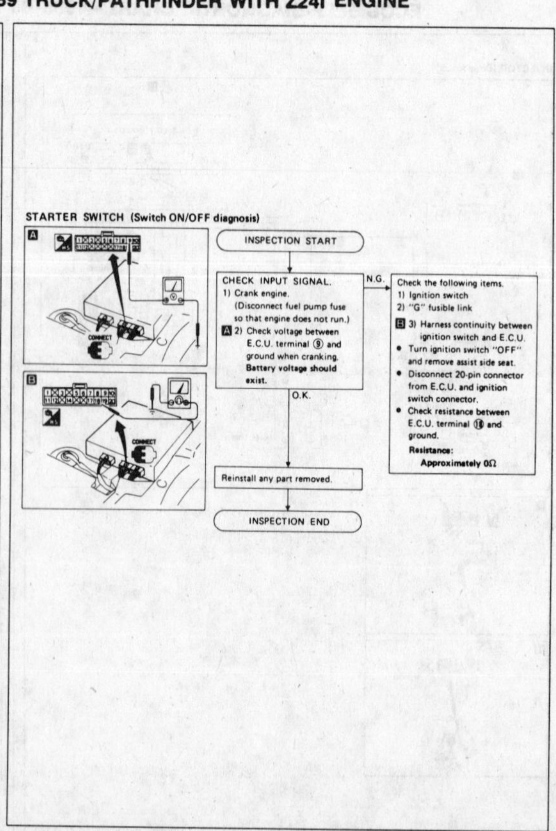
STARTER SWITCH (Switch ON/OFF diagnosis)
INSPECTION START
CHECK INPUT SIGNAL.
1) Crank engine.
(Disconnect fuel pump fuse so that engine does not run.)
A 2) Check voltage between E.C.U. terminal ⑨ and ground when cranking.
Battery voltage should exist.
N.G.
Check the following items.
1) Ignition switch
2) "G" fusible link
B 3) Harness continuity between ignition switch and E.C.U.
• Turn ignition switch "OFF" and remove assist side seat.
• Disconnect 20-pin connector from E.C.U. and ignition switch connector.
• Check resistance between E.C.U. terminal ⑱ and ground.
Resistance: Approximately 0Ω
O.K.
Reinstall any part removed.
INSPECTION END

ECCS SELF-DIAGNOSTIC CHART—1988–89 TRUCK/PATHFINDER WITH Z24I ENGINE

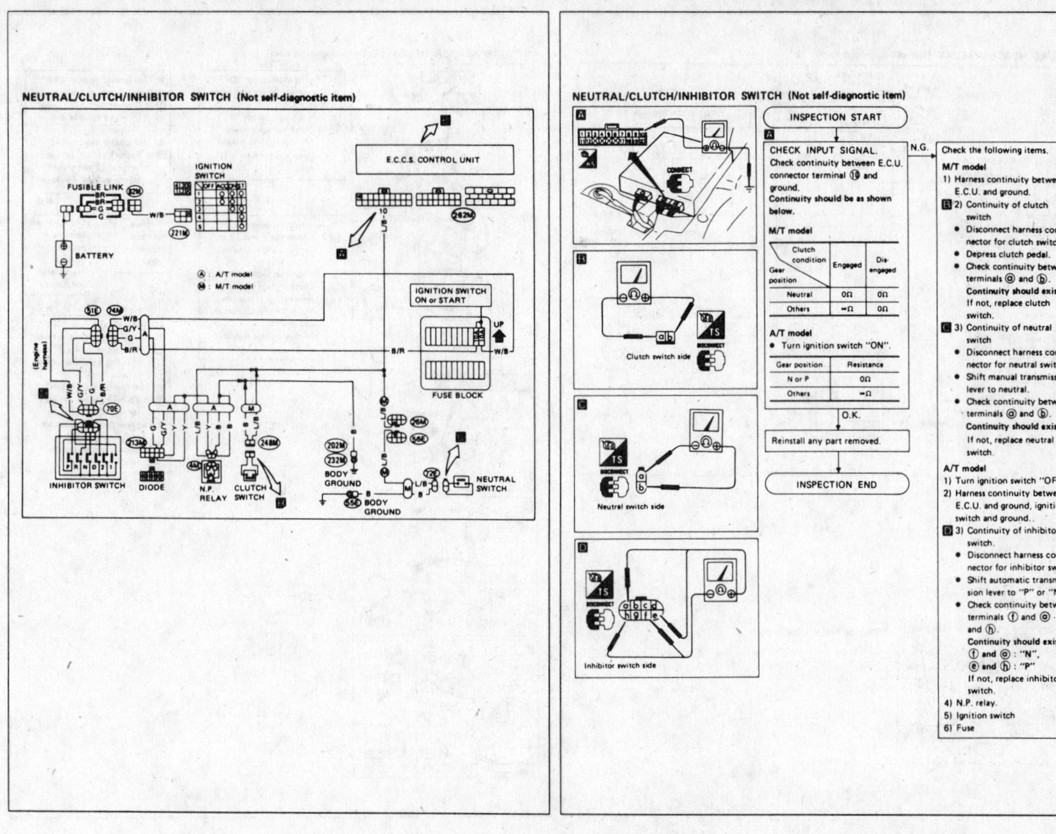
NEUTRAL/CLUTCH/INHIBITOR SWITCH (Not self-diagnostic item)
E.C.C.S. CONTROL UNIT
IGNITION SWITCH
FUSIBLE LINK
BATTERY
Ⓐ : A/T model
Ⓜ : M/T model
IGNITION SWITCH ON or START
FUSE BLOCK
INHIBITOR SWITCH
DIODE
N.P. RELAY
CLUTCH SWITCH
BODY GROUND
NEUTRAL SWITCH
NEUTRAL/CLUTCH/INHIBITOR SWITCH (Not self-diagnostic item)
INSPECTION START
CHECK INPUT SIGNAL.
Check continuity between E.C.U. connector terminal ⑲ and ground.
Continuity should be as shown below.
M/T model
Clutch condition / Gear position | Engaged | Dis-engaged
Neutral | 0Ω | 0Ω
Others | ∞Ω | 0Ω
A/T model
• Turn ignition switch "ON".
Gear position | Resistance
N or P | 0Ω
Others | ∞Ω
N.G.
Check the following items.
M/T model
1) Harness continuity between E.C.U. and ground.
2) Continuity of clutch switch
• Disconnect harness connector for clutch switch.
• Depress clutch pedal.
• Check continuity between terminals ⓐ and ⓑ.
Continuity should exist.
If not, replace clutch switch.
3) Continuity of neutral switch
• Disconnect harness connector for neutral switch.
• Shift manual transmission lever to neutral.
• Check continuity between terminals ⓐ and ⓑ.
Continuity should exist.
If not, replace neutral switch.
A/T model
1) Turn ignition switch "OFF".
2) Harness continuity between E.C.U. and ground, ignition switch and ground.
3) Continuity of inhibitor switch.
• Disconnect harness connector for inhibitor switch.
• Shift automatic transmission lever to "P" or "N".
• Check continuity between terminals ⓕ and ⓖ - ⓔ and ⓗ.
Continuity should exist.
ⓕ and ⓖ : "N".
ⓔ and ⓗ : "P"
If not, replace inhibitor switch.
4) N.P. relay.
5) Ignition switch
6) Fuse
O.K.
Reinstall any part removed.
INSPECTION END
CONNECT
DISCONNECT
Clutch switch side
Neutral switch side
Inhibitor switch side

ECCS SELF-DIAGNOSTIC CHART—1988–89 TRUCK/PATHFINDER WITH Z24I ENGINE

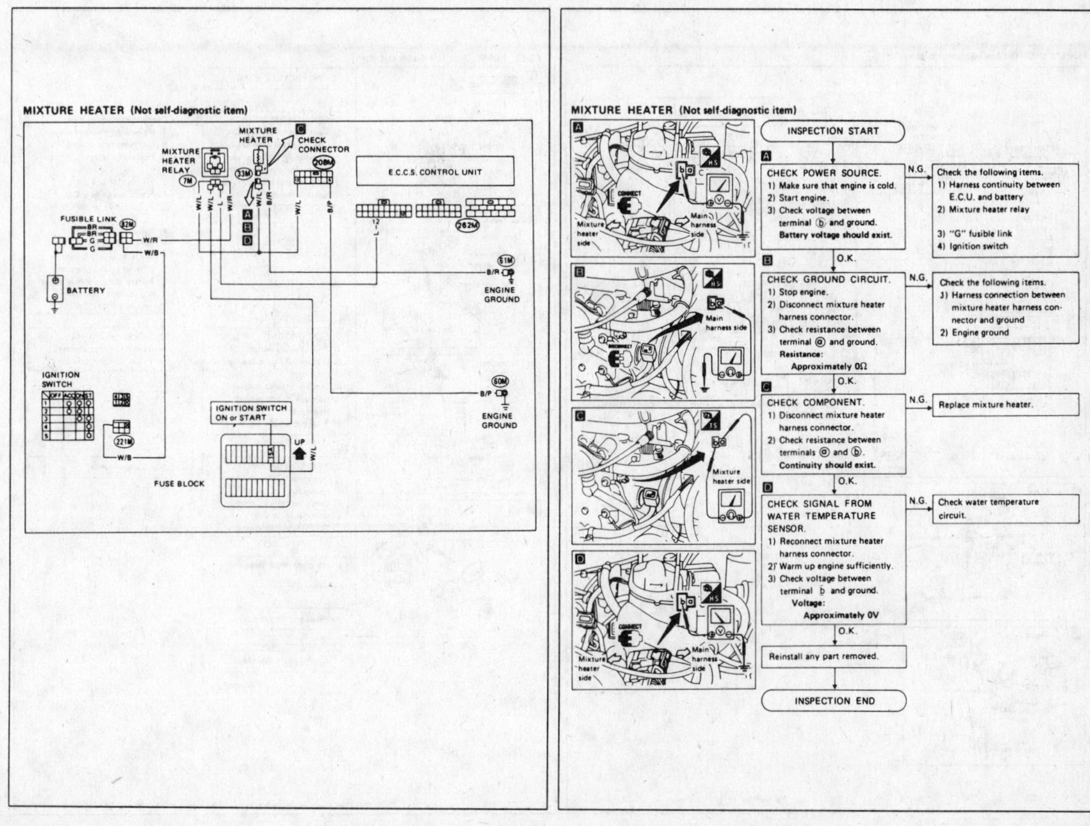
MIXTURE HEATER (Not self-diagnostic item)
MIXTURE HEATER RELAY
MIXTURE HEATER
CHECK CONNECTOR
E.C.C.S. CONTROL UNIT
FUSIBLE LINK
BATTERY
ENGINE GROUND
IGNITION SWITCH
IGNITION SWITCH ON or START
FUSE BLOCK
MIXTURE HEATER (Not self-diagnostic item)
INSPECTION START
CHECK POWER SOURCE.
1) Make sure that engine is cold.
2) Start engine.
3) Check voltage between terminal ⓑ and ground.
Battery voltage should exist.
N.G.
Check the following items.
1) Harness continuity between E.C.U. and battery
2) Mixture heater relay
3) "G" fusible link
4) Ignition switch
O.K.
CHECK GROUND CIRCUIT.
1) Stop engine.
2) Disconnect mixture heater harness connector.
3) Check resistance between terminal ⓐ and ground.
Resistance:
Approximately 0Ω
N.G.
Check the following items.
1) Harness connection between mixture heater harness connector and ground
2) Engine ground
O.K.
CHECK COMPONENT.
1) Disconnect mixture heater harness connector.
2) Check resistance between terminals ⓐ and ⓑ.
Continuity should exist.
N.G.
Replace mixture heater.
O.K.
CHECK SIGNAL FROM WATER TEMPERATURE SENSOR.
1) Reconnect mixture heater harness connector.
2) Warm up engine sufficiently.
3) Check voltage between terminal b and ground.
Voltage:
Approximately 0V
N.G.
Check water temperature circuit.
O.K.
Reinstall any part removed.
INSPECTION END
Mixture heater side
Main harness side
Mixture heater side

ECCS SELF-DIAGNOSTIC CHART—1988–89 TRUCK/PATHFINDER WITH Z24I ENGINE

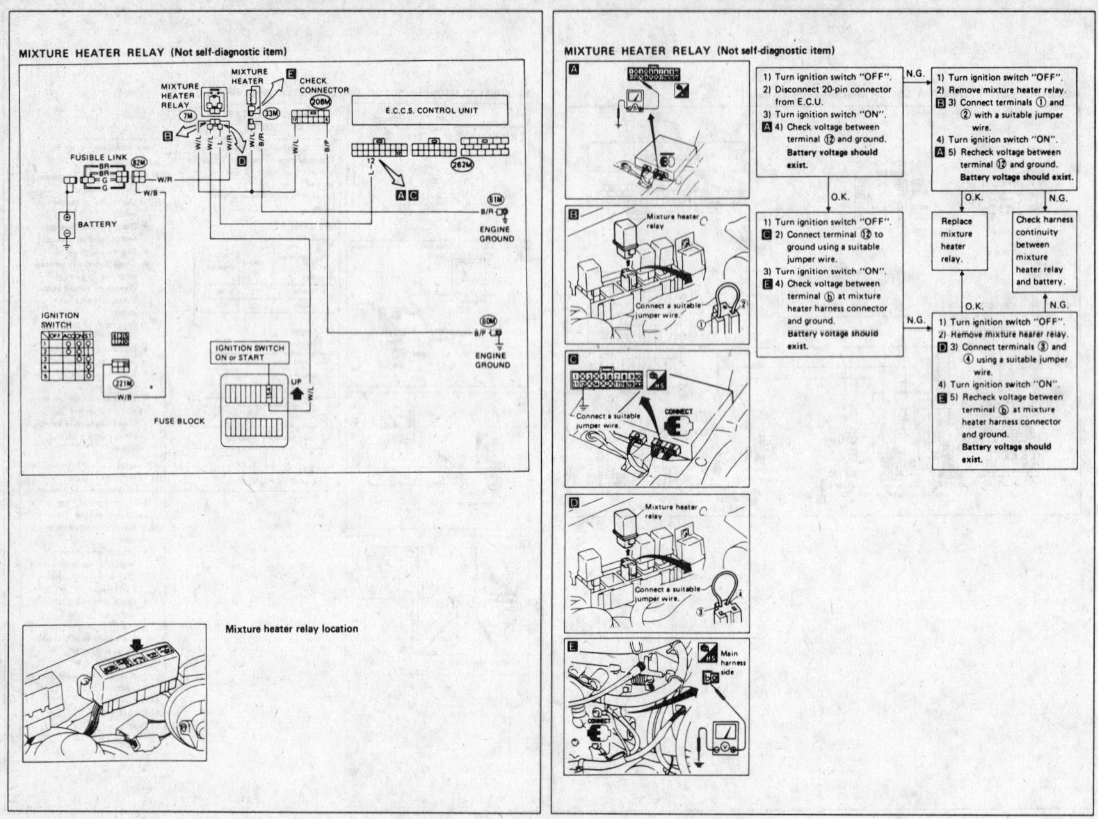
MIXTURE HEATER RELAY (Not self-diagnostic item)
MIXTURE HEATER RELAY
MIXTURE HEATER
CHECK CONNECTOR
E.C.C.S. CONTROL UNIT
FUSIBLE LINK
BATTERY
ENGINE GROUND
IGNITION SWITCH
IGNITION SWITCH ON or START
FUSE BLOCK
Mixture heater relay location
MIXTURE HEATER RELAY (Not self-diagnostic item)
1) Turn ignition switch "OFF".
2) Disconnect 20-pin connector from E.C.U.
3) Turn ignition switch "ON".
A 4) Check voltage between terminal ⑫ and ground. Battery voltage should exist.
N.G.
1) Turn ignition switch "OFF".
2) Remove mixture heater relay.
B 3) Connect terminals ① and ② with a suitable jumper wire.
4) Turn ignition switch "ON".
A 5) Recheck voltage between terminal ⑫ and ground. Battery voltage should exist.
O.K.
Replace mixture heater relay.
N.G.
Check harness continuity between mixture heater relay and battery.
O.K.
1) Turn ignition switch "OFF".
C 2) Connect terminal ⑫ to ground using a suitable jumper wire.
3) Turn ignition switch "ON".
E 4) Check voltage between terminal ⓑ at mixture heater harness connector and ground. Battery voltage should exist.
N.G.
1) Turn ignition switch "OFF".
2) Remove mixture heater relay.
D 3) Connect terminals ③ and ④ using a suitable jumper wire.
4) Turn ignition switch "ON".
E 5) Recheck voltage between terminal ⓑ at mixture heater harness connector and ground. Battery voltage should exist.
O.K.
N.G.
Mixture heater relay
Connect a suitable jumper wire
CONNECT
Main harness side

ECCS SELF-DIAGNOSTIC CHART—1988–89 TRUCK/PATHFINDER WITH Z24I ENGINE

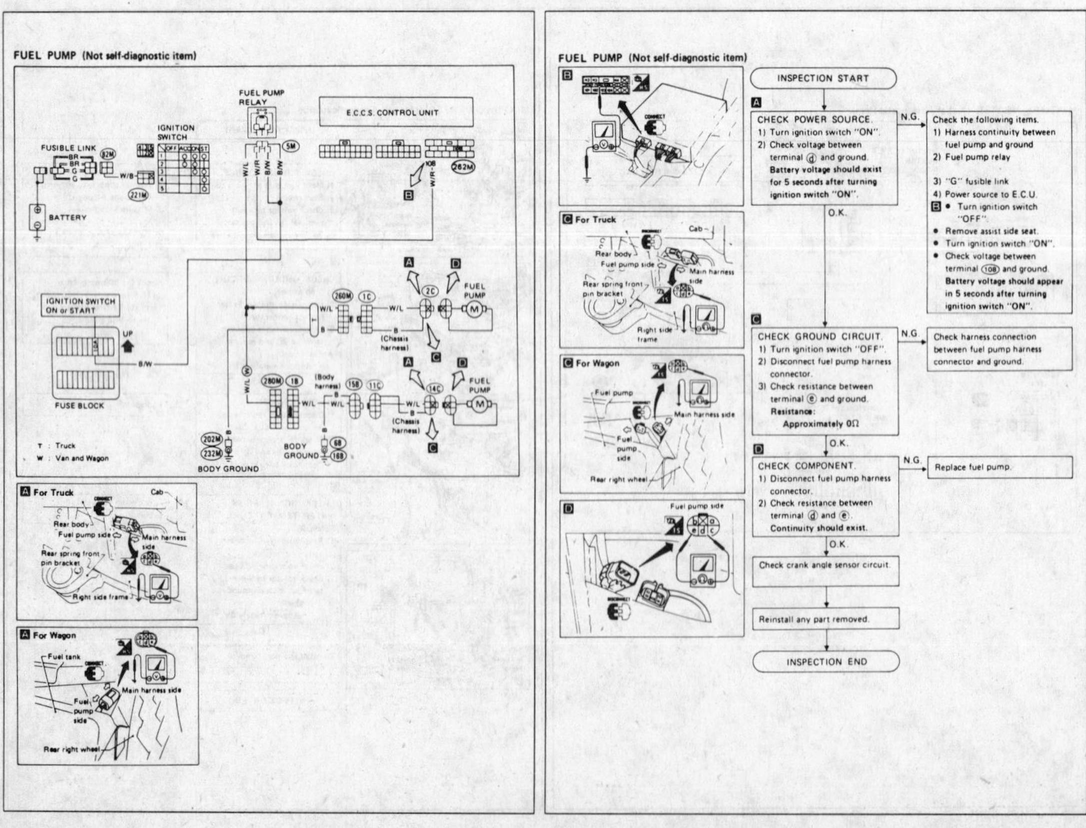
FUEL PUMP (Not self-diagnostic item)
FUEL PUMP RELAY
E.C.C.S. CONTROL UNIT
IGNITION SWITCH
FUSIBLE LINK
BATTERY
IGNITION SWITCH ON or START
FUSE BLOCK
FUEL PUMP
(Chassis harness)
(Body harness)
BODY GROUND
T : Truck
W : Van and Wagon
For Truck
For Wagon
Cab
Rear body
Fuel pump side
Main harness side
Rear spring front pin bracket
Right side frame
Fuel tank
Rear right wheel
FUEL PUMP (Not self-diagnostic item)
INSPECTION START
A
CHECK POWER SOURCE.
1) Turn ignition switch "ON".
2) Check voltage between terminal ⓓ and ground. Battery voltage should exist for 5 seconds after turning ignition switch "ON".
N.G.
Check the following items.
1) Harness continuity between fuel pump and ground
2) Fuel pump relay
3) "G" fusible link
4) Power source to E.C.U.
B • Turn ignition switch "OFF".
• Remove assist side seat.
• Turn ignition switch "ON".
• Check voltage between terminal (108) and ground. Battery voltage should appear in 5 seconds after turning ignition switch "ON".
O.K.
C
CHECK GROUND CIRCUIT.
1) Turn ignition switch "OFF".
2) Disconnect fuel pump harness connector.
3) Check resistance between terminal ⓔ and ground. Resistance: Approximately 0Ω
N.G.
Check harness connection between fuel pump harness connector and ground.
O.K.
D
CHECK COMPONENT.
1) Disconnect fuel pump harness connector.
2) Check resistance between terminal ⓓ and ⓔ. Continuity should exist.
N.G.
Replace fuel pump.
O.K.
Check crank angle sensor circuit.
Reinstall any part removed.
INSPECTION END

ECCS SELF-DIAGNOSTIC CHART—1988–89 TRUCK/PATHFINDER WITH Z24I ENGINE

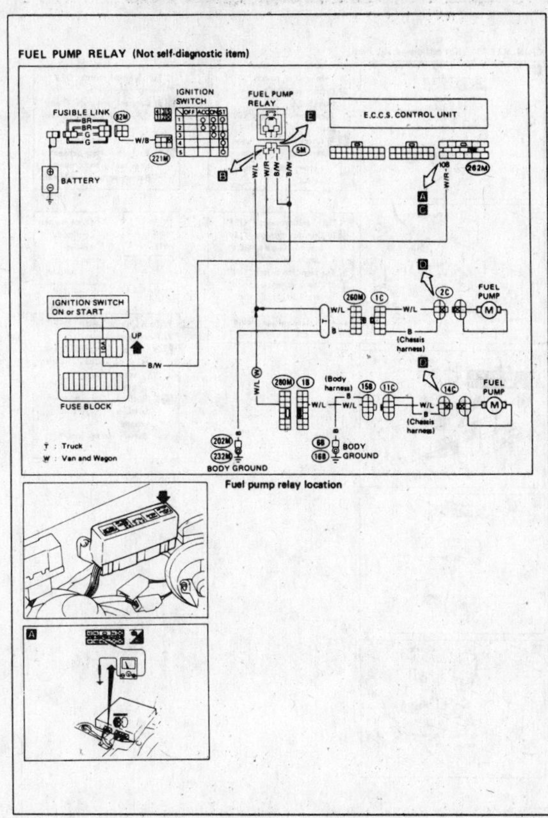
FUEL PUMP RELAY (Not self-diagnostic item)
FUSIBLE LINK
IGNITION SWITCH
FUEL PUMP RELAY
E.C.C.S. CONTROL UNIT
BATTERY
IGNITION SWITCH ON or START
FUSE BLOCK
FUEL PUMP
BODY GROUND
Truck
Van and Wagon
Fuel pump relay location

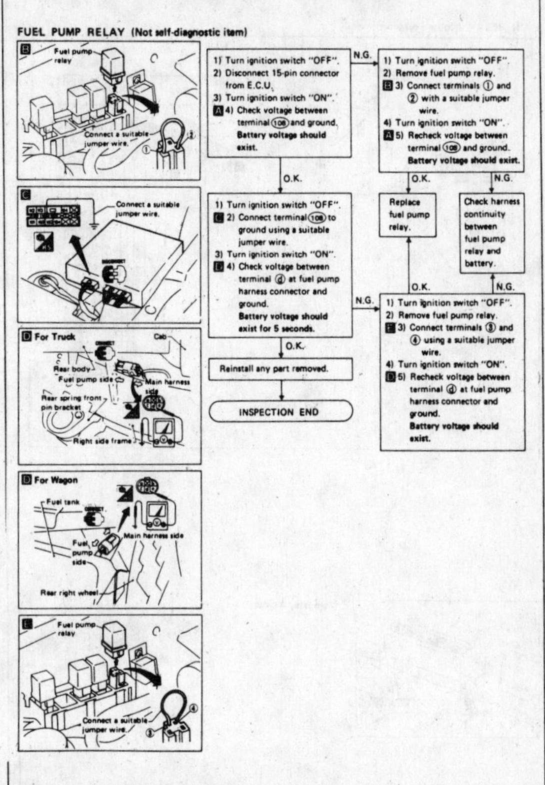
FUEL PUMP RELAY (Not self-diagnostic item)
Fuel pump relay
Connect a suitable jumper wire.
For Truck
For Wagon
Fuel tank
Main harness side
Fuel pump side
Rear right wheel
Right side frame
1) Turn ignition switch "OFF".
2) Disconnect 15-pin connector from E.C.U.
3) Turn ignition switch "ON".
A 4) Check voltage between terminal (108) and ground. Battery voltage should exist.
N.G.
1) Turn ignition switch "OFF".
2) Remove fuel pump relay.
B 3) Connect terminals ① and ② with a suitable jumper wire.
4) Turn ignition switch "ON".
A 5) Recheck voltage between terminal (108) and ground. Battery voltage should exist.
O.K.
Replace fuel pump relay.
N.G.
Check harness continuity between fuel pump relay and battery.
O.K.
1) Turn ignition switch "OFF".
C 2) Connect terminal (108) to ground using a suitable jumper wire.
3) Turn ignition switch "ON".
D 4) Check voltage between terminal ⓓ at fuel pump harness connector and ground. Battery voltage should exist for 5 seconds.
N.G.
1) Turn ignition switch "OFF".
2) Remove fuel pump relay.
E 3) Connect terminals ③ and ④ using a suitable jumper wire.
4) Turn ignition switch "ON".
D 5) Recheck voltage between terminal ⓓ at fuel pump harness connector and ground. Battery voltage should exist.
O.K.
Reinstall any part removed.
INSPECTION END

ECCS SELF-DIAGNOSTIC CHART—1988–89 TRUCK/PATHFINDER WITH Z24I ENGINE

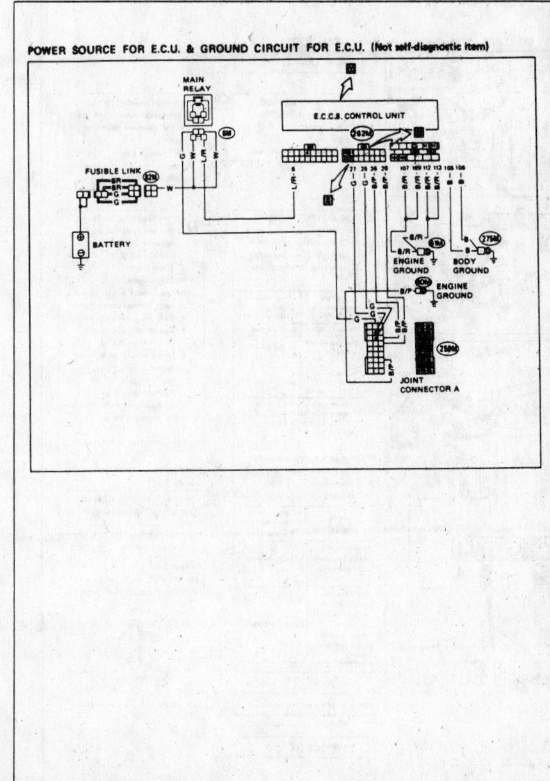
POWER SOURCE FOR E.C.U. & GROUND CIRCUIT FOR E.C.U. (Not self-diagnostic item)
MAIN RELAY
E.C.C.S. CONTROL UNIT
FUSIBLE LINK
BATTERY
ENGINE GROUND
BODY GROUND
JOINT CONNECTOR A

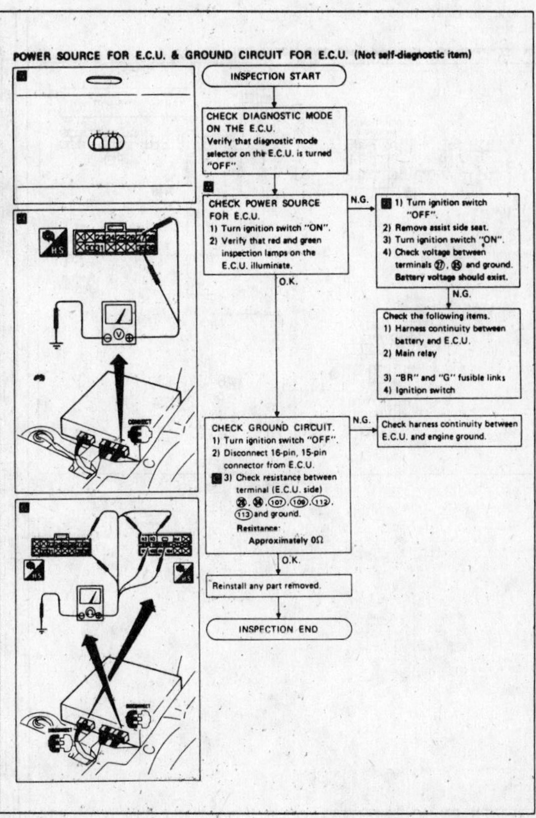
POWER SOURCE FOR E.C.U. & GROUND CIRCUIT FOR E.C.U. (Not self-diagnostic item)
INSPECTION START
CHECK DIAGNOSTIC MODE ON THE E.C.U.
Verify that diagnostic mode selector on the E.C.U. is turned "OFF".
CHECK POWER SOURCE FOR E.C.U.
1) Turn ignition switch "ON".
2) Verify that red and green inspection lamps on the E.C.U. illuminate.
N.G.
1) Turn ignition switch "OFF".
2) Remove assist side seat.
3) Turn ignition switch "ON".
4) Check voltage between terminals 37, 38 and ground. Battery voltage should exist.
N.G.
Check the following items.
1) Harness continuity between battery and E.C.U.
2) Main relay
3) "BR" and "G" fusible links
4) Ignition switch
O.K.
CHECK GROUND CIRCUIT.
1) Turn ignition switch "OFF".
2) Disconnect 16-pin, 15-pin connector from E.C.U.
3) Check resistance between terminal (E.C.U. side) 28, 34, 107, 108, 112, 113 and ground.
Resistance: Approximately 0Ω
N.G.
Check harness continuity between E.C.U. and engine ground.
O.K.
Reinstall any part removed.
INSPECTION END

ECCS SELF-DIAGNOSTIC CHART—1988–89 TRUCK/PATHFINDER WITH Z24I ENGINE

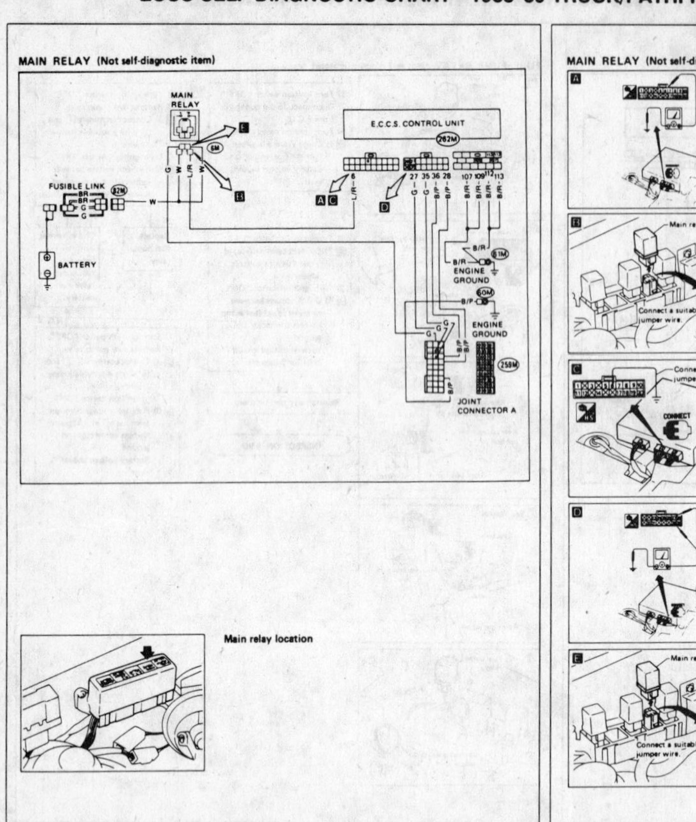
MAIN RELAY (Not self-diagnostic item)
MAIN RELAY
E.C.C.S. CONTROL UNIT
FUSIBLE LINK
BATTERY
ENGINE GROUND
JOINT CONNECTOR A
Main relay location

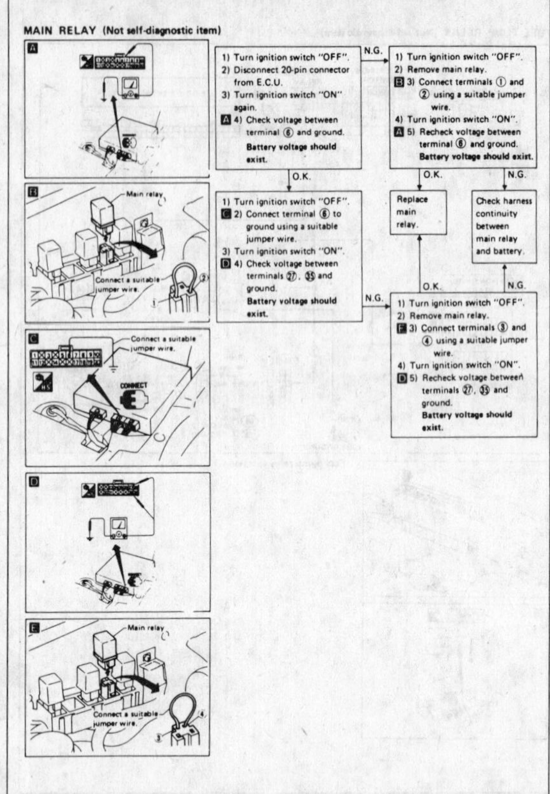
MAIN RELAY (Not self-diagnostic item)
1) Turn ignition switch "OFF".
2) Disconnect 20-pin connector from E.C.U.
3) Turn ignition switch "ON" again.
A 4) Check voltage between terminal ⑥ and ground.
Battery voltage should exist.
N.G.
1) Turn ignition switch "OFF".
2) Remove main relay.
B 3) Connect terminals ① and ② using a suitable jumper wire.
4) Turn ignition switch "ON".
A 5) Recheck voltage between terminal ⑥ and ground.
Battery voltage should exist.
O.K.
N.G.
Replace main relay.
Check harness continuity between main relay and battery.
1) Turn ignition switch "OFF".
C 2) Connect terminal ⑥ to ground using a suitable jumper wire.
3) Turn ignition switch "ON".
D 4) Check voltage between terminals ㉗, ㉟ and ground.
Battery voltage should exist.
1) Turn ignition switch "OFF".
2) Remove main relay.
E 3) Connect terminals ③ and ④ using a suitable jumper wire.
4) Turn ignition switch "ON".
D 5) Recheck voltage between terminals ㉗, ㉟ and ground.
Battery voltage should exist.
Main relay
Connect a suitable jumper wire.
CONNECT

ECCS SELF-DIAGNOSTIC CHART—1988–89 TRUCK/PATHFINDER WITH Z24I ENGINE

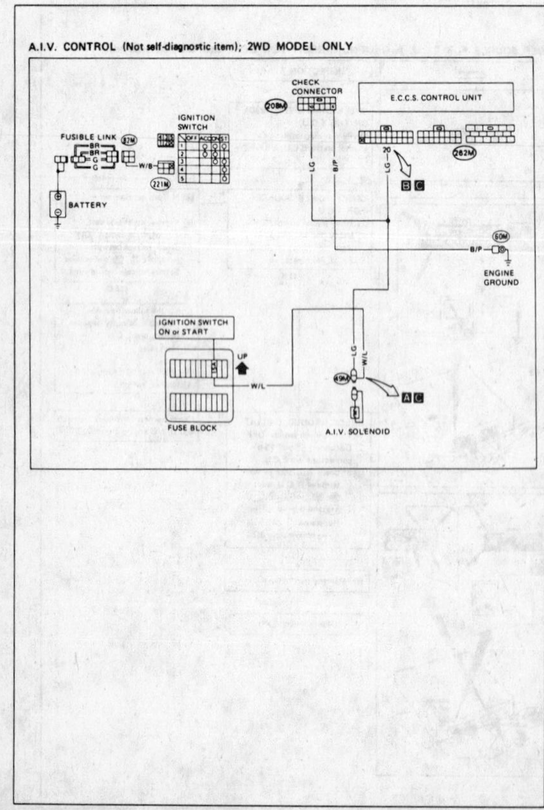
A.I.V. CONTROL (Not self-diagnostic item); 2WD MODEL ONLY
CHECK CONNECTOR
E.C.C.S. CONTROL UNIT
IGNITION SWITCH
FUSIBLE LINK
BATTERY
ENGINE GROUND
IGNITION SWITCH ON or START
UP
FUSE BLOCK
A.I.V. SOLENOID

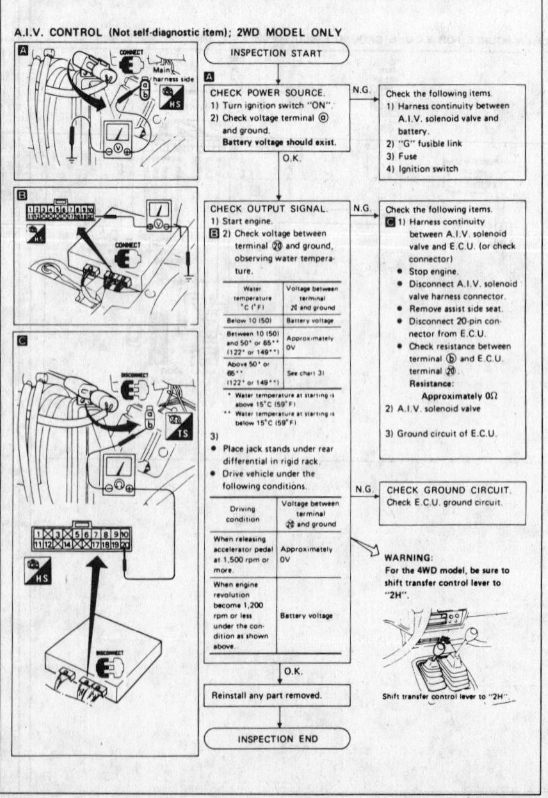
A.I.V. CONTROL (Not self-diagnostic item); 2WD MODEL ONLY
INSPECTION START
CHECK POWER SOURCE.
1) Turn ignition switch "ON".
2) Check voltage terminal ⓐ and ground.
Battery voltage should exist.
N.G.
Check the following items.
1) Harness continuity between A.I.V. solenoid valve and battery.
2) "G" fusible link
3) Fuse
4) Ignition switch
O.K.
CHECK OUTPUT SIGNAL.
1) Start engine.
B 2) Check voltage between terminal ⑳ and ground, observing water temperature.
Water temperature °C (°F)
Voltage between terminal ⑳ and ground
Below 10 (50)
Battery voltage
Between 10 (50) and 50* or 65** (122* or 149**)
Approximately 0V
Above 50* or 65** (122* or 149**)
See chart 3)
* Water temperature at starting is above 15°C (59°F)
** Water temperature at starting is below 15°C (59°F)
3)
Place jack stands under rear differential in rigid rack.
Drive vehicle under the following conditions.
Driving condition
Voltage between terminal ⑳ and ground
When releasing accelerator pedal at 1,500 rpm or more.
Approximately 0V
When engine revolution become 1,200 rpm or less under the condition as shown above.
Battery voltage
N.G.
Check the following items.
C 1) Harness continuity between A.I.V. solenoid valve and E.C.U. (or check connector)
Stop engine.
Disconnect A.I.V. solenoid valve harness connector.
Remove assist side seat.
Disconnect 20-pin connector from E.C.U.
Check resistance between terminal ⓑ and E.C.U. terminal ⑳.
Resistance: Approximately 0Ω
2) A.I.V. solenoid valve
3) Ground circuit of E.C.U.
N.G.
CHECK GROUND CIRCUIT.
Check E.C.U. ground circuit.
WARNING:
For the 4WD model, be sure to shift transfer control lever to "2H".
O.K.
Reinstall any part removed.
INSPECTION END
Shift transfer control lever to "2H".
CONNECT
DISCONNECT
Main harness side

ECCS SELF-DIAGNOSTIC CHART—1988–89 TRUCK/PATHFINDER WITH Z24I ENGINE

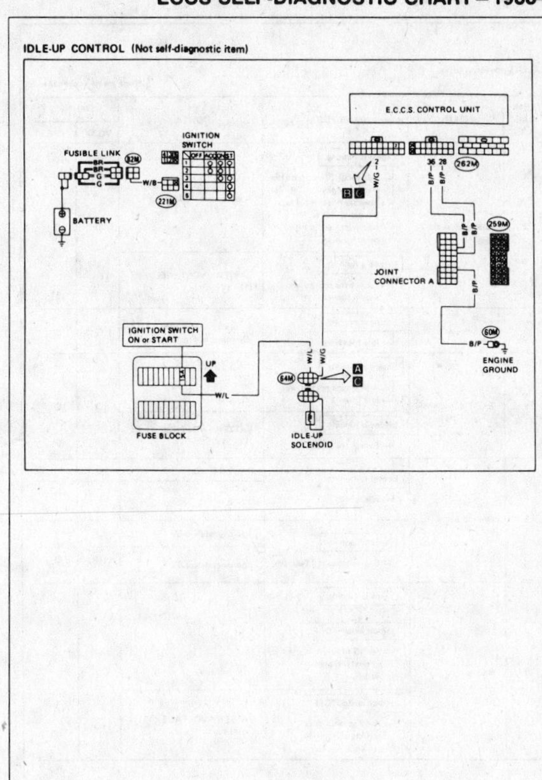
IDLE-UP CONTROL (Not self-diagnostic item)
E.C.C.S. CONTROL UNIT
IGNITION SWITCH
FUSIBLE LINK
BATTERY
JOINT CONNECTOR A
IGNITION SWITCH ON or START
UP
FUSE BLOCK
IDLE-UP SOLENOID
ENGINE GROUND

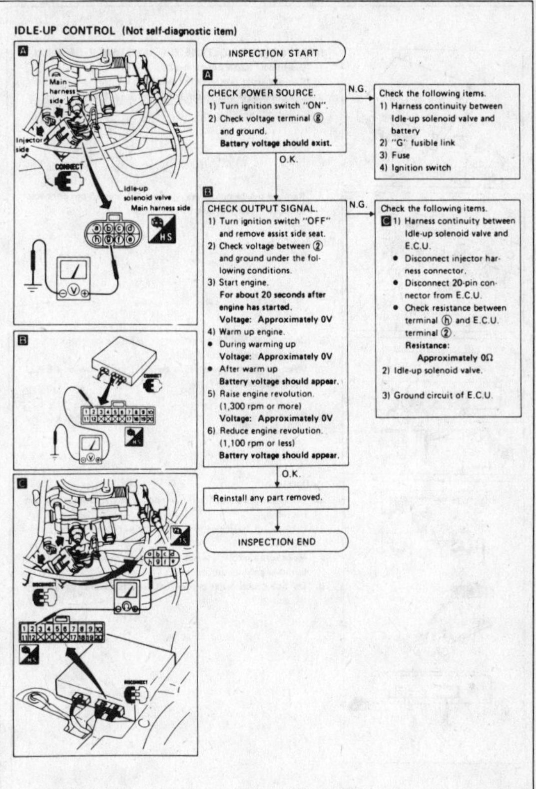
IDLE-UP CONTROL (Not self-diagnostic item)
Main harness side
Injector side
Idle-up solenoid valve
Main harness side
INSPECTION START
A
CHECK POWER SOURCE.
1) Turn ignition switch "ON".
2) Check voltage terminal ⑧ and ground.
Battery voltage should exist.
N.G.
Check the following items.
1) Harness continuity between Idle-up solenoid valve and battery
2) "G" fusible link
3) Fuse
4) Ignition switch
O.K.
B
CHECK OUTPUT SIGNAL.
1) Turn ignition switch "OFF" and remove assist side seat.
2) Check voltage between ② and ground under the following conditions.
3) Start engine.
For about 20 seconds after engine has started.
Voltage: Approximately 0V
4) Warm up engine.
• During warming up
Voltage: Approximately 0V
• After warm up
Battery voltage should appear.
5) Raise engine revolution.
(1,300 rpm or more)
Voltage: Approximately 0V
6) Reduce engine revolution.
(1,100 rpm or less)
Battery voltage should appear.
N.G.
Check the following items.
C 1) Harness continuity between Idle-up solenoid valve and E.C.U.
• Disconnect injector harness connector.
• Disconnect 20-pin connector from E.C.U.
• Check resistance between terminal ⓗ and E.C.U. terminal ②.
Resistance:
Approximately 0Ω
2) Idle-up solenoid valve.
3) Ground circuit of E.C.U.
O.K.
Reinstall any part removed.
INSPECTION END

ECCS SELF-DIAGNOSTIC CHART—1988–89 TRUCK/PATHFINDER WITH Z24I ENGINE

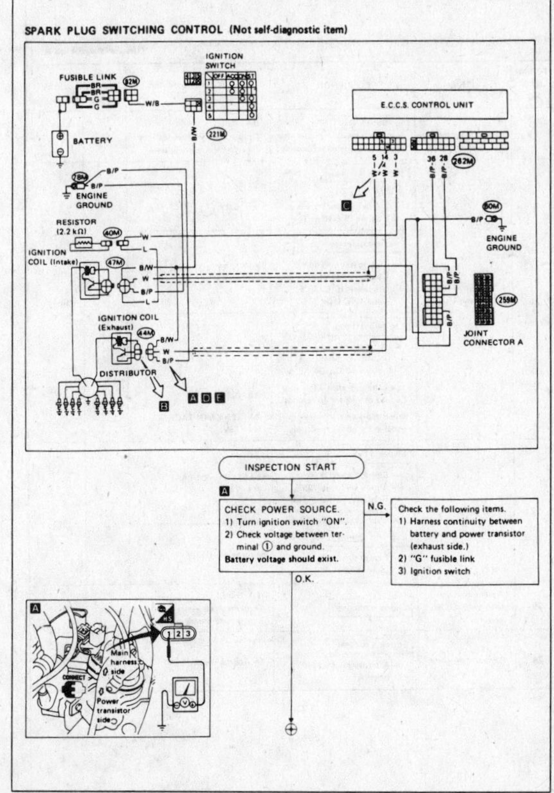
SPARK PLUG SWITCHING CONTROL (Not self-diagnostic item)
IGNITION SWITCH
FUSIBLE LINK
BATTERY
E.C.C.S. CONTROL UNIT
ENGINE GROUND
RESISTOR (2.2 kΩ)
IGNITION COIL (Intake)
IGNITION COIL (Exhaust)
DISTRIBUTOR
JOINT CONNECTOR A
INSPECTION START
A
CHECK POWER SOURCE.
1) Turn ignition switch "ON".
2) Check voltage between terminal ① and ground.
Battery voltage should exist.
N.G.
Check the following items.
1) Harness continuity between battery and power transistor (exhaust side.)
2) "G" fusible link
3) Ignition switch
O.K.
Main harness side
Power transistor side

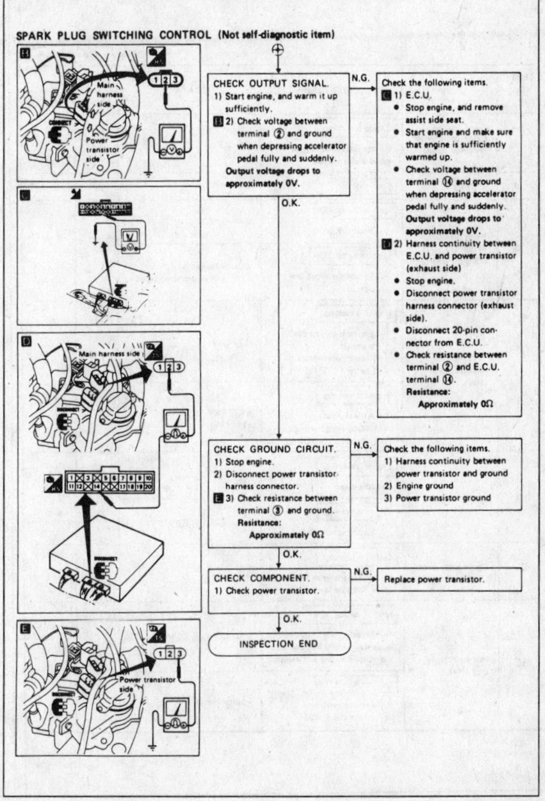
SPARK PLUG SWITCHING CONTROL (Not self-diagnostic item)
Main harness side
Power transistor side
CHECK OUTPUT SIGNAL.
1) Start engine, and warm it up sufficiently.
B 2) Check voltage between terminal ② and ground when depressing accelerator pedal fully and suddenly.
Output voltage drops to approximately 0V.
N.G.
Check the following items.
C 1) E.C.U.
• Stop engine, and remove assist side seat.
• Start engine and make sure that engine is sufficiently warmed up.
• Check voltage between terminal ⑭ and ground when depressing accelerator pedal fully and suddenly.
Output voltage drops to approximately 0V.
D 2) Harness continuity between E.C.U. and power transistor (exhaust side)
• Stop engine.
• Disconnect power transistor harness connector (exhaust side).
• Disconnect 20-pin connector from E.C.U.
• Check resistance between terminal ② and E.C.U. terminal ⑭.
Resistance:
Approximately 0Ω
O.K.
CHECK GROUND CIRCUIT.
1) Stop engine.
2) Disconnect power transistor harness connector.
E 3) Check resistance between terminal ③ and ground.
Resistance:
Approximately 0Ω
N.G.
Check the following items.
1) Harness continuity between power transistor and ground
2) Engine ground
3) Power transistor ground
O.K.
CHECK COMPONENT.
1) Check power transistor.
N.G.
Replace power transistor.
O.K.
INSPECTION END
Main harness side
Power transistor side

ECU SIGNAL INSPECTION—1988–89 TRUCK/PATHFINDER

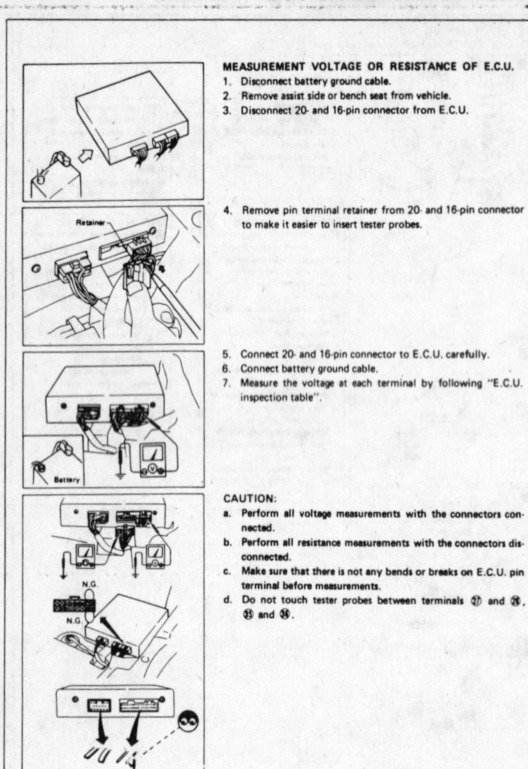
Retainer
Battery
N.G.
N.G.

MEASUREMENT VOLTAGE OR RESISTANCE OF E.C.U.

1. Disconnect battery ground cable.
2. Remove assist side or bench seat from vehicle.
3. Disconnect 20- and 16-pin connector from E.C.U.
4. Remove pin terminal retainer from 20- and 16-pin connector to make it easier to insert tester probes.
5. Connect 20- and 16-pin connector to E.C.U. carefully.
6. Connect battery ground cable.
7. Measure the voltage at each terminal by following "E.C.U. inspection table".

CAUTION:

a. Perform all voltage measurements with the connectors connected.
b. Perform all resistance measurements with the connectors disconnected.
c. Make sure that there is not any bends or breaks on E.C.U. pin terminal before measurements.
d. Do not touch tester probes between terminals ㉗ and ㉘, ㉟ and ㊱.

E.C.U. inspection table

*Data are reference values.

TERMINAL NO.	ITEM	CONDITION	*DATA	ENGINE	
				VG30i	Z24i
2	Idle-up solenoid	[Engine is running.] — For about 20 seconds after starting engine. — Steering wheel is turned. — Blower switch is "ON". — Headlamps are in high beam position.	Approximately 1.0V	○	○
		[Engine is running.] — Except the conditions shown above	BATTERY VOLTAGE (11 - 14V)		
3	Ignition check	[Engine is running.] — Idle speed	9 - 12V (Decreases as engine revs up.)	○	○
4	E.G.R. cut solenoid	[Engine is running.] — High engine revolution — Idle speed	0.8 - 1.0V	○	○
		[Engine is running.] — Except the above	BATTERY VOLTAGE (11 - 14V)		
5	Ignition signal	[Engine is running.] — Idle speed	0.4 - 0.6V	○	
			0.2 - 0.4V		○ (Intake)
		[Engine is running.] — Engine speed is 2,000 rpm.	Approximately 1.0V	○	
			Approximately 0.8V		○ (Intake)
6	Main relay	[Engine is running.] → [Ignition switch "OFF"] — For 15 seconds after turning off ignition switch.	0.8 - 1.0V	○	○
		[Ignition switch "OFF"] — In 15 seconds after turning off ignition switch.	BATTERY VOLTAGE (11 - 14V)		

ECU SIGNAL INSPECTION—1988–89 TRUCK/PATHFINDER

*Data are reference values.

TERMINAL NO.	ITEM	CONDITION	*DATA	ENGINE	
				VG30i	Z24i
8	Crank angle sensor (Position signal)	[Engine is running.] Do not run engine at high speed under no load.	2.5 - 2.7V	○	○
9	Start signal	[Cranking]	8 - 12V	○	○
10	Neutral/clutch switch or Inhibitor switch	[Ignition switch "ON"] — Neutral/Parking	0V	○	○
		[Ignition switch "ON"] — Except the above gear position	Approximately 5V (0V: with clutch disengaged)		
12	Mixture heater relay	[Engine is running.] — Engine is cold or during warm up. [Water temperature is below 70°C (158°F).]	0.7 - 0.9V	○	○
		[Engine is running.] — After warming up. [Water temperature is above 70°C (158°F).]	BATTERY VOLTAGE (11 - 14V)		
14	Ignition signal	[Engine is running.] — Idle speed	0.2 - 0.4V		○ (Exhaust)
		[Engine is running.] — Engine speed is 2,000 rpm	Approximately 0.8V		
17	Crank angle sensor (Reference signal)	[Engine is running.] Do not run engine at high speed under no load.	0.2 - 0.4V	○	○
18	Idle switch (⊖ side)	[Ignition switch "ON"] — Throttle valve: idle position. Inspection should be done after warming up engine sufficiently.	Approximately 8 - 10V	○	○
		[Ignition switch "ON"] — Throttle valve: except idle position	Approximately 0V		

*Data are reference values.

TERMINAL NO.	ITEM	CONDITION	*DATA	ENGINE	
				VG30i	Z24i
19	Throttle sensor	[Ignition switch "ON"] Inspection should be done after warming up engine sufficiently.	0.4 - 4.0V Output voltage varies with the throttle valve opening angle.	○	○
20	A.I.V. cut solenoid	[Engine is running.] — Engine is cold. [Water temperature is below 15°C (59°F).]	0.8 - 0.9V	○	○ (Only 2WD)
		[Engine is running.] — During warming up [Water temperature is between 15°C (59°F) and 40°C (104°F).]	BATTERY VOLTAGE (11 - 14V)		
		[Engine is running.] — After warming up. [Water temperature is above 40°C (104°F).] Idle condition after 3,000 rpm no load driving for 10 seconds	0.8 - 0.9V		
		When depressing accelerator pedal at the above condition.	BATTERY VOLTAGE (11 - 14V)		
22	Load signal	[Engine is running.] — Steering wheel is turned. — Blower switch is "ON". — Headlamps are in high beam.	BATTERY VOLTAGE (11 - 14V)	○	
		[Engine is running.] — Except the conditions shown above	Approximately 0V		
23	Water or cylinder head temperature sensor	[Engine is running.]	1.0 - 5.0V Output voltage varies with engine water temperature.	○	○
24	Exhaust gas sensor	[Engine is running.] — After warming up sufficiently.	0 - Approximately 1.0V	○	○

ECU SIGNAL INSPECTION 1988–89 TRUCK/PATHFINDER

*Data are reference values.

TERMINAL NO.	ITEM	CONDITION	*DATA	ENGINE VG30i	ENGINE Z24i
33	Idle switch (⊕ side)	Engine is running. — Idle speed	9 - 11V	○	○
		Engine is running. — Engine speed is 2,000 rpm.	BATTERY VOLTAGE (11 - 14V)		
27 35	Power source for E.C.U.	Ignition switch "ON"	BATTERY VOLTAGE (11 - 14V)	○	○
31	Air flow meter	Engine is running. Do not run engine at high speed under no load.	1.0 - 3.0V Output voltage varies with engine revolution and throttle valve movement.	○	○
101 102 103 104 114	Injector	Engine is running.	BATTERY VOLTAGE (11 - 14V)	○	○
108	Fuel pump relay	Ignition switch "ON" — For 5 seconds after turning ignition switch "ON". Engine is running.	0.7 - 0.9V	○	○
		Ignition switch "ON" — In 5 seconds after turning ignition switch "ON".	BATTERY VOLTAGE (11 - 14V)		
111	Exhaust gas temperature sensor	Engine is running. — Idle speed	1.0 - 2.0V	○ (Only California model)	○ (Only California model)
		Engine is running. — E.G.R. system is operating.	0 - 1.0V		

PIN CONNECTOR TERMINAL LAYOUT

20-pin connector: 1 2 3 4 5 6 7 8 9 10 / 11 12 13 14 15 16 17 18 19 20 — HS

16-pin connector: 21 22 23 24 25 26 27 28 / 29 30 31 32 33 34 35 36 — HS

15-pin connector: 112 113 114 115 / 107 108 109 110 111 / 101 102 103 104 105 106 — HS

ECCS MIXTURE RATIO FEEDBACK SYSTEM INSPECTION—1988–89 TRUCK/ PATHFINDER

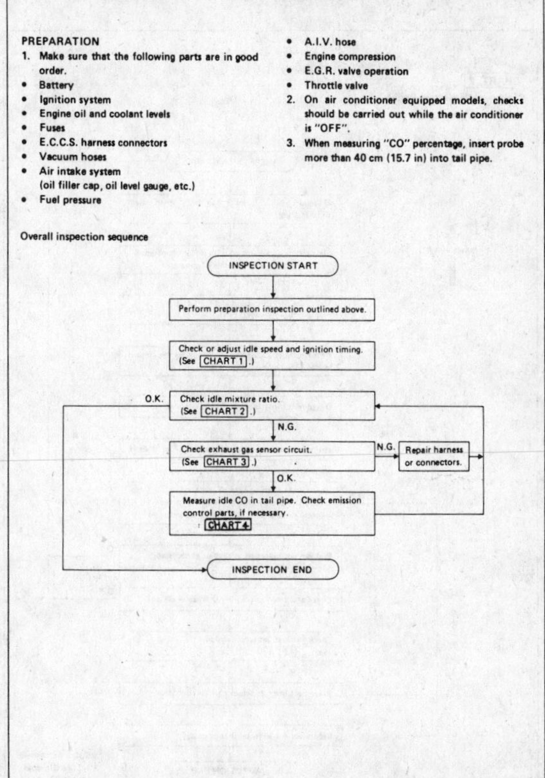
PREPARATION
1. Make sure that the following parts are in good order.
• Battery
• Ignition system
• Engine oil and coolant levels
• Fuses
• E.C.C.S. harness connectors
• Vacuum hoses
• Air intake system (oil filler cap, oil level gauge, etc.)
• Fuel pressure
• A.I.V. hose
• Engine compression
• E.G.R. valve operation
• Throttle valve
2. On air conditioner equipped models, checks should be carried out while the air conditioner is "OFF".
3. When measuring "CO" percentage, insert probe more than 40 cm (15.7 in) into tail pipe.
Overall inspection sequence
INSPECTION START
Perform preparation inspection outlined above.
Check or adjust idle speed and ignition timing. (See CHART 1.)
O.K.
Check idle mixture ratio. (See CHART 2.)
N.G.
Check exhaust gas sensor circuit. (See CHART 3.)
N.G.
Repair harness or connectors.
O.K.
Measure idle CO in tail pipe. Check emission control parts, if necessary. CHART 4
INSPECTION END

ECCS MIXTURE RATIO FEEDBACK SYSTEM INSPECTION (CONT.)—1988–89 TRUCK/PATHFINDER WITH VG30I ENGINE

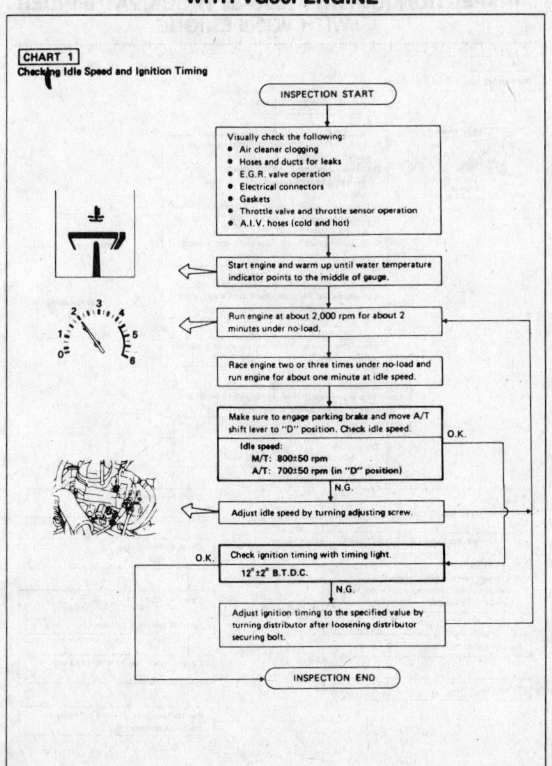
CHART 1
Checking Idle Speed and Ignition Timing
INSPECTION START
Visually check the following:
• Air cleaner clogging
• Hoses and ducts for leaks
• E.G.R. valve operation
• Electrical connectors
• Gaskets
• Throttle valve and throttle sensor operation
• A.I.V. hoses (cold and hot)
Start engine and warm up until water temperature indicator points to the middle of gauge.
Run engine at about 2,000 rpm for about 2 minutes under no-load.
Race engine two or three times under no-load and run engine for about one minute at idle speed.
Make sure to engage parking brake and move A/T shift lever to "D" position. Check idle speed.
Idle speed:
M/T: 800±50 rpm
A/T: 700±50 rpm (in "D" position)
O.K.
N.G.
Adjust idle speed by turning adjusting screw.
Check ignition timing with timing light.
12°±2° B.T.D.C.
O.K.
N.G.
Adjust ignition timing to the specified value by turning distributor after loosening distributor securing bolt.
INSPECTION END

ECCS MIXTURE RATIO FEEDBACK SYSTEM INSPECTION (CONT.)—1988–89 TRUCK/PATHFINDER WITH Z24I ENGINE

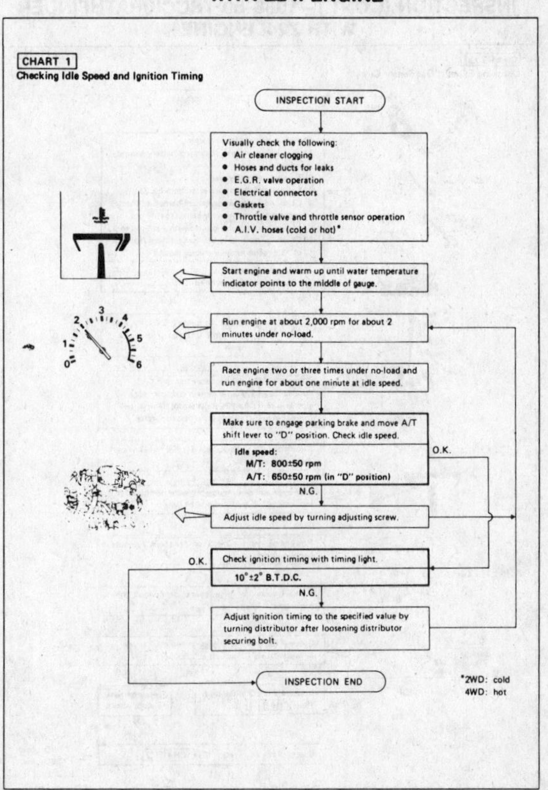
CHART 1
Checking Idle Speed and Ignition Timing
INSPECTION START
Visually check the following:
• Air cleaner clogging
• Hoses and ducts for leaks
• E.G.R. valve operation
• Electrical connectors
• Gaskets
• Throttle valve and throttle sensor operation
• A.I.V. hoses (cold or hot)*
Start engine and warm up until water temperature indicator points to the middle of gauge.
Run engine at about 2,000 rpm for about 2 minutes under no-load.
Race engine two or three times under no-load and run engine for about one minute at idle speed.
Make sure to engage parking brake and move A/T shift lever to "D" position. Check idle speed.
Idle speed:
M/T: 800±50 rpm
A/T: 650±50 rpm (in "D" position)
O.K.
N.G.
Adjust idle speed by turning adjusting screw.
Check ignition timing with timing light.
10°±2° B.T.D.C.
O.K.
N.G.
Adjust ignition timing to the specified value by turning distributor after loosening distributor securing bolt.
INSPECTION END
*2WD: cold
4WD: hot

INSPECTION (CONT.)—1988–89 TRUCK/PATHFINDER ECCS MIXTURE RATIO FEEDBACK SYSTEM

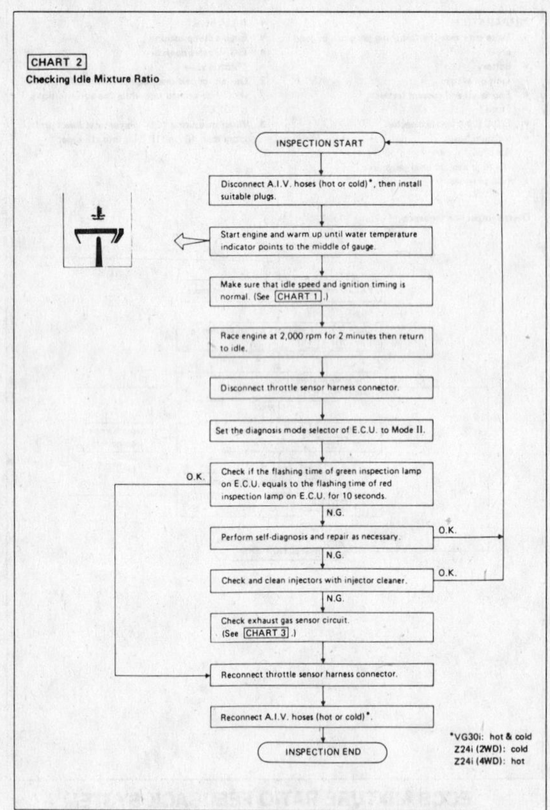
CHART 2
Checking Idle Mixture Ratio
INSPECTION START
Disconnect A.I.V. hoses (hot or cold)*, then install suitable plugs.
Start engine and warm up until water temperature indicator points to the middle of gauge.
Make sure that idle speed and ignition timing is normal. (See CHART 1.)
Race engine at 2,000 rpm for 2 minutes then return to idle.
Disconnect throttle sensor harness connector.
Set the diagnosis mode selector of E.C.U. to Mode II.
Check if the flashing time of green inspection lamp on E.C.U. equals to the flashing time of red inspection lamp on E.C.U. for 10 seconds.
O.K.
N.G.
Perform self-diagnosis and repair as necessary.
O.K.
N.G.
Check and clean injectors with injector cleaner.
O.K.
N.G.
Check exhaust gas sensor circuit. (See CHART 3.)
Reconnect throttle sensor harness connector.
Reconnect A.I.V. hoses (hot or cold)*.
INSPECTION END
*VG30i: hot & cold
Z24i (2WD): cold
Z24i (4WD): hot

ECCS MIXTURE RATIO FEEDBACK SYSTEM INSPECTION (CONT.)—1988–89 TRUCK/PATHFINDER WITH VG30I ENGINE

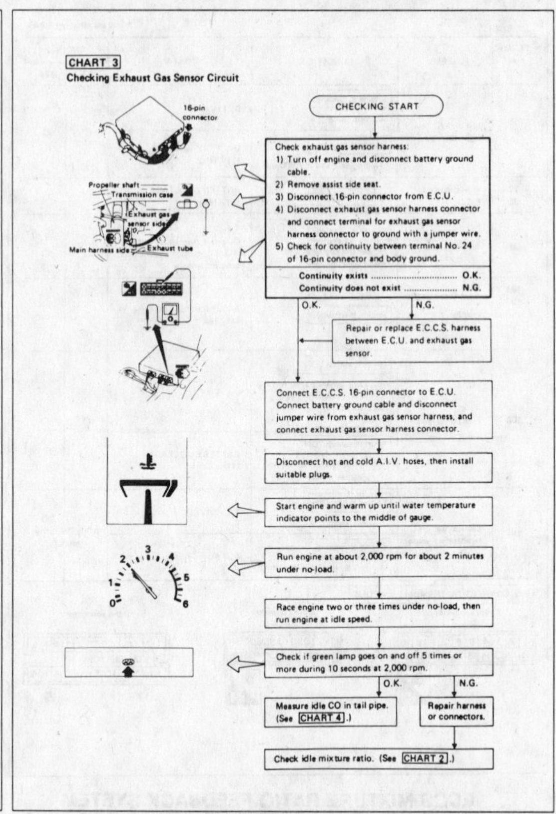
CHART 3
Checking Exhaust Gas Sensor Circuit
16-pin connector
Propeller shaft
Transmission case
Exhaust gas sensor side
Main harness side
Exhaust tube
CHECKING START
Check exhaust gas sensor harness:
1) Turn off engine and disconnect battery ground cable.
2) Remove assist side seat.
3) Disconnect 16-pin connector from E.C.U.
4) Disconnect exhaust gas sensor harness connector and connect terminal for exhaust gas sensor harness connector to ground with a jumper wire.
5) Check for continuity between terminal No. 24 of 16-pin connector and body ground.
Continuity exists .... O.K.
Continuity does not exist .... N.G.
O.K.
N.G.
Repair or replace E.C.C.S. harness between E.C.U. and exhaust gas sensor.
Connect E.C.C.S. 16-pin connector to E.C.U. Connect battery ground cable and disconnect jumper wire from exhaust gas sensor harness, and connect exhaust gas sensor harness connector.
Disconnect hot and cold A.I.V. hoses, then install suitable plugs.
Start engine and warm up until water temperature indicator points to the middle of gauge.
Run engine at about 2,000 rpm for about 2 minutes under no-load.
Race engine two or three times under no-load, then run engine at idle speed.
Check if green lamp goes on and off 5 times or more during 10 seconds at 2,000 rpm.
O.K.
N.G.
Measure idle CO in tail pipe. (See CHART 4.)
Repair harness or connectors.
Check idle mixture ratio. (See CHART 2.)

ECCS MIXTURE RATIO FEEDBACK SYSTEM INSPECTION (CONT.)—1988–89 TRUCK/PATHFINDER WITH Z24I ENGINE

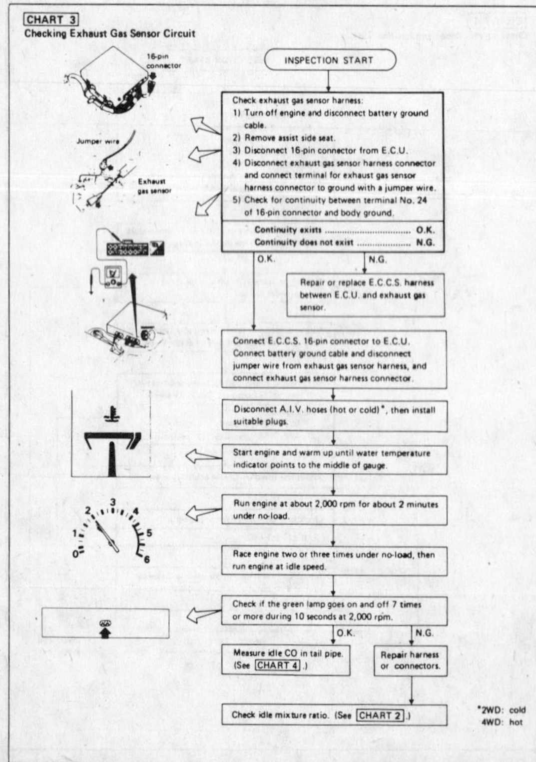
CHART 3
Checking Exhaust Gas Sensor Circuit
16-pin connector
Jumper wire
Exhaust gas sensor
INSPECTION START
Check exhaust gas sensor harness:
1) Turn off engine and disconnect battery ground cable.
2) Remove assist side seat.
3) Disconnect 16-pin connector from E.C.U.
4) Disconnect exhaust gas sensor harness connector and connect terminal for exhaust gas sensor harness connector to ground with a jumper wire.
5) Check for continuity between terminal No. 24 of 16-pin connector and body ground.
Continuity exists .... O.K.
Continuity does not exist .... N.G.
O.K.
N.G.
Repair or replace E.C.C.S. harness between E.C.U. and exhaust gas sensor.
Connect E.C.C.S. 16-pin connector to E.C.U. Connect battery ground cable and disconnect jumper wire from exhaust gas sensor harness, and connect exhaust gas sensor harness connector.
Disconnect A.I.V. hoses (hot or cold)*, then install suitable plugs.
Start engine and warm up until water temperature indicator points to the middle of gauge.
Run engine at about 2,000 rpm for about 2 minutes under no-load.
Race engine two or three times under no-load, then run engine at idle speed.
Check if the green lamp goes on and off 7 times or more during 10 seconds at 2,000 rpm.
O.K.
N.G.
Measure idle CO in tail pipe. (See CHART 4.)
Repair harness or connectors.
Check idle mixture ratio. (See CHART 2.)
*2WD: cold
4WD: hot

ECCS MIXTURE RATIO FEEDBACK SYSTEM INSPECTION (CONT.)—1988–89 TRUCK/PATHFINDER WITH VG30I ENIGNE

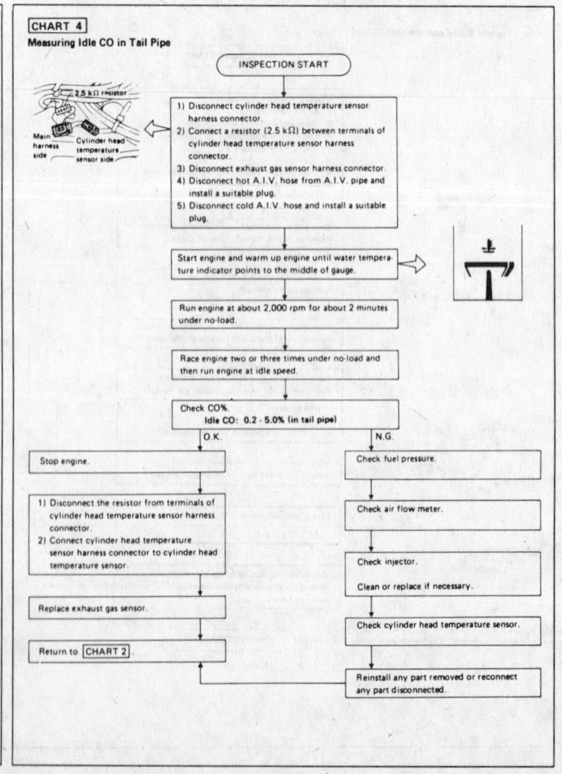
CHART 4
Measuring Idle CO in Tail Pipe
INSPECTION START
2.5 kΩ resistor
Main harness side
Cylinder head temperature sensor side
1) Disconnect cylinder head temperature sensor harness connector.
2) Connect a resistor (2.5 kΩ) between terminals of cylinder head temperature sensor harness connector.
3) Disconnect exhaust gas sensor harness connector.
4) Disconnect hot A.I.V. hose from A.I.V. pipe and install a suitable plug.
5) Disconnect cold A.I.V. hose and install a suitable plug.
Start engine and warm up engine until water temperature indicator points to the middle of gauge.
Run engine at about 2,000 rpm for about 2 minutes under no-load.
Race engine two or three times under no-load and then run engine at idle speed.
Check CO%.
Idle CO: 0.2 - 5.0% (in tail pipe)
O.K.
N.G.
Stop engine.
1) Disconnect the resistor from terminals of cylinder head temperature sensor harness connector.
2) Connect cylinder head temperature sensor harness connector to cylinder head temperature sensor.
Replace exhaust gas sensor.
Return to CHART 2.
Check fuel pressure.
Check air flow meter.
Check injector.
Clean or replace if necessary.
Check cylinder head temperature sensor.
Reinstall any part removed or reconnect any part disconnected.

ECCS MIXTURE RATIO FEEDBACK SYSTEM INSPECTION (CONT.) – 1988–89 TRUCK/PATHFINDER WITH Z24I ENIGNE

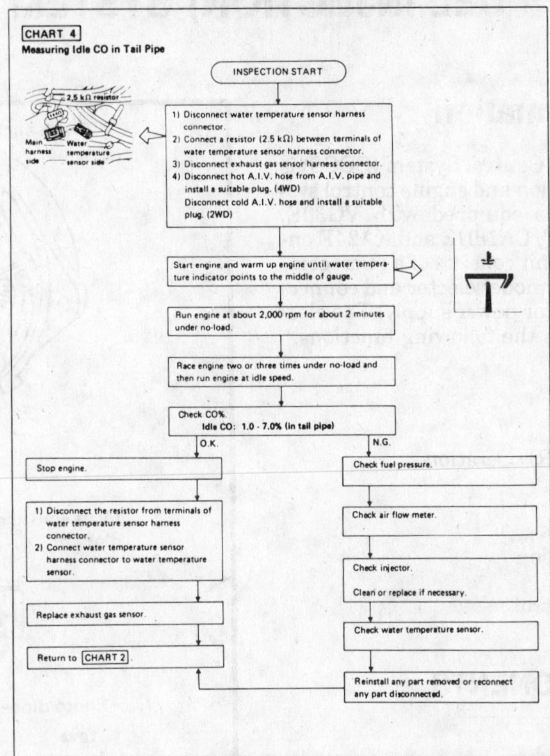
CHART 4
Measuring Idle CO in Tail Pipe
INSPECTION START
2.5 kΩ resistor
Main harness side
Water temperature sensor side
1) Disconnect water temperature sensor harness connector.
2) Connect a resistor (2.5 kΩ) between terminals of water temperature sensor harness connector.
3) Disconnect exhaust gas sensor harness connector.
4) Disconnect hot A.I.V. hose from A.I.V. pipe and install a suitable plug. (4WD)
Disconnect cold A.I.V. hose and install a suitable plug. (2WD)
Start engine and warm up engine until water temperature indicator points to the middle of gauge.
Run engine at about 2,000 rpm for about 2 minutes under no-load.
Race engine two or three times under no-load and then run engine at idle speed.
Check CO%.
Idle CO: 1.0 - 7.0% (in tail pipe)
O.K.
N.G.
Stop engine.
1) Disconnect the resistor from terminals of water temperature sensor harness connector.
2) Connect water temperature sensor harness connector to water temperature sensor.
Replace exhaust gas sensor.
Return to CHART 2.
Check fuel pressure.
Check air flow meter.
Check injector.
Clean or replace if necessary.
Check water temperature sensor.
Reinstall any part removed or reconnect any part disconnected.

EXPLODED VIEW OF ELECTRO INJECTION UNIT 1988–89 TRUCK/PATHFINDER

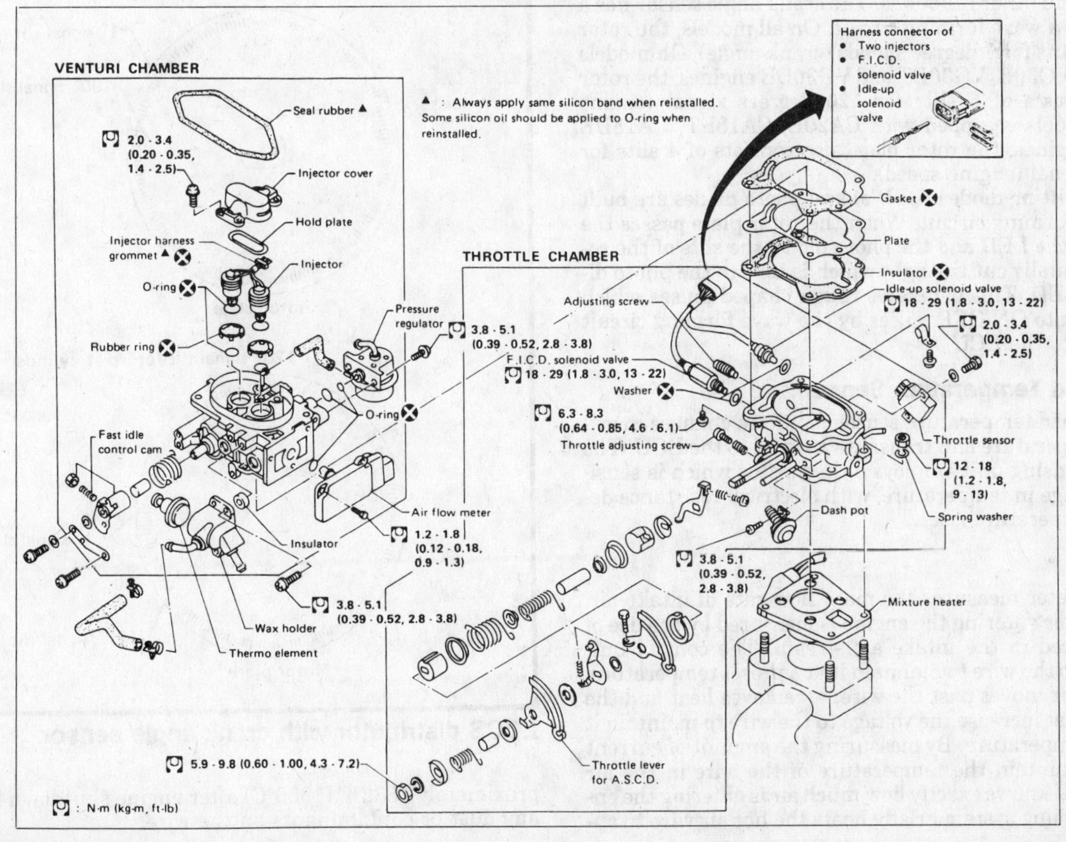
VENTURI CHAMBER
Seal rubber ▲
▲ : Always apply same silicon band when reinstalled. Some silicon oil should be applied to O-ring when reinstalled.
Harness connector of
• Two injectors
• F.I.C.D. solenoid valve
• Idle-up solenoid valve
2.0 - 3.4 (0.20 - 0.35, 1.4 - 2.5)
Injector cover
Hold plate
Injector harness grommet ▲
Injector
O-ring
Rubber ring
Pressure regulator
3.8 - 5.1 (0.39 - 0.52, 2.8 - 3.8)
O-ring
Fast idle control cam
Air flow meter
1.2 - 1.8 (0.12 - 0.18, 0.9 - 1.3)
Insulator
3.8 - 5.1 (0.39 - 0.52, 2.8 - 3.8)
Wax holder
Thermo element
THROTTLE CHAMBER
Adjusting screw
F.I.C.D. solenoid valve
18 - 29 (1.8 - 3.0, 13 - 22)
Washer
6.3 - 8.3 (0.64 - 0.85, 4.6 - 6.1)
Throttle adjusting screw
Gasket
Plate
Insulator
Idle-up solenoid valve
18 - 29 (1.8 - 3.0, 13 - 22)
2.0 - 3.4 (0.20 - 0.35, 1.4 - 2.5)
Throttle sensor
12 - 18 (1.2 - 1.8, 9 - 13)
Dash pot
Spring washer
3.8 - 5.1 (0.39 - 0.52, 2.8 - 3.8)
Mixture heater
5.9 - 9.8 (0.60 - 1.00, 4.3 - 7.2)
Throttle lever for A.S.C.D.
: N·m (kg-m, ft-lb)

NISSAN ELECTRONIC CONCENTRATED CONTROL (PORT FUEL INJECTION) SYSTEM (ECCS)

General Information

The Nissan Electronic Concentrated Control System (ECCS) is an air flow controlled, port fuel injection and engine control system. It is used on 1988–90 models equipped with VG30E, VG30ET, VG30DE, CA20E, CA18ET, CA18DE and KA24E engines. The ECCS electronic control unit consists of a microcomputer, inspection lamps, a diagnostic mode selector and connectors for signal input and output and for power supply. The electronic control unit, or ECU, controls the following functions:

- Amount of injected fuel
- Ignition timing
- Mixture ratio feedback
- Pressure regulator control
- Exhaust Gas Recirculation (EGR) operation
- Idle speed control
- Fuel pump operation
- Air regulator control
- Air Injection Valve (AIV) operation
- Self-diagnostics
- Air flow meter self-cleaning control
- Fail safe system

SYSTEM COMPONENTS

Crank Angle Sensor

The crank angle sensor is a basic component of the ECCS system. It monitors engine speed and piston position, as well as sending signals which the ECU uses to control fuel injection, ignition timing and other functions. The crank angle sensor has a rotor plate and a wave forming circuit. On all models, the rotor plate has 360 slits for 1 degree signals (crank angle). On models equipped with VG30E, VG30ET and VG30DE engines, the rotor plate also consists of 6 slits for 120 degrees signal (engine speed). On models equipped with CA20E, CA18ET, CA18DE and KA24E engines, the rotor plate also consists of 4 slits for 180 degrees signal (engine speed).

The light emitting diodes (LED's) and photo diodes are built into the wave forming circuit. When the rotor plate passes the space between the LED and the photo diode, the slits of the rotor plate continually cut the light which is sent to the photo diode from the LED. This generates rough shaped pulses which are converted into ON/OFF pulses by the wave forming circuit and then sent to the ECU.

Cylinder Head Temperature Sensor

The cylinder head temperature sensor monitors changes in cylinder head temperature and transmits a signal to the ECU. The temperature sensing unit employs a thermistor which is sensitive to the change in temperature, with electrical resistance decreasing as temperature rises.

Air Flow Meter

The air flow meter measures the mass flow rate of intake air. The volume of air entering the engine is measured by the use of a hot wire placed in the intake air stream. The control unit sends current to the wire to maintain it at a preset temperature. As the intake air moves past the wire, it removes heat and the control unit must increase the voltage to the wire to maintain it at the preset temperature. By measuring the amount of current necessary to maintain the temperature of the wire in the air stream, the ECU knows exactly how much air is entering the engine. A self-cleaning system briefly heats the hot air wire to approximately 1832°F (1000°C) after engine shutdown to burn off any dust or contaminants on the wire.

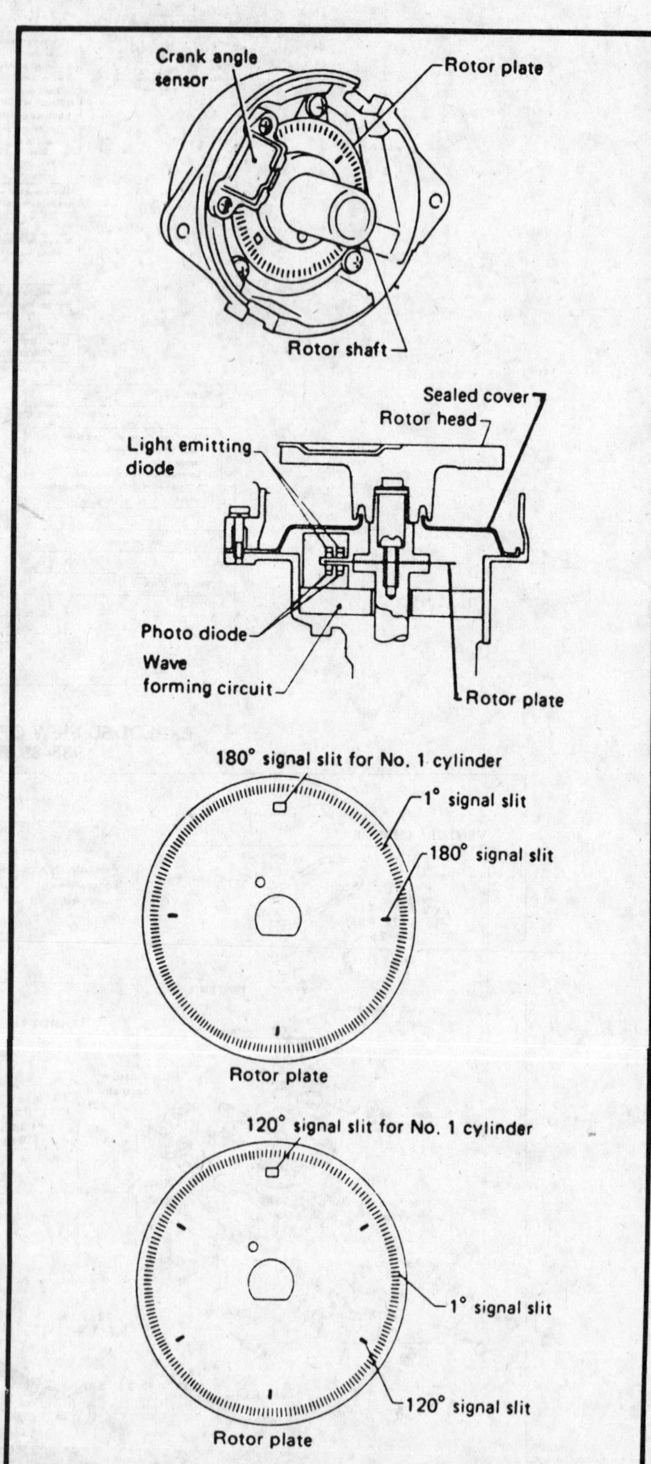
Crank angle sensor
Rotor plate
Rotor shaft
Sealed cover
Rotor head
Light emitting diode
Photo diode
Wave forming circuit
Rotor plate
180° signal slit for No. 1 cylinder
1° signal slit
180° signal slit
Rotor plate
120° signal slit for No. 1 cylinder
1° signal slit
120° signal slit
Rotor plate

ECCS distributor with crank angle sensor

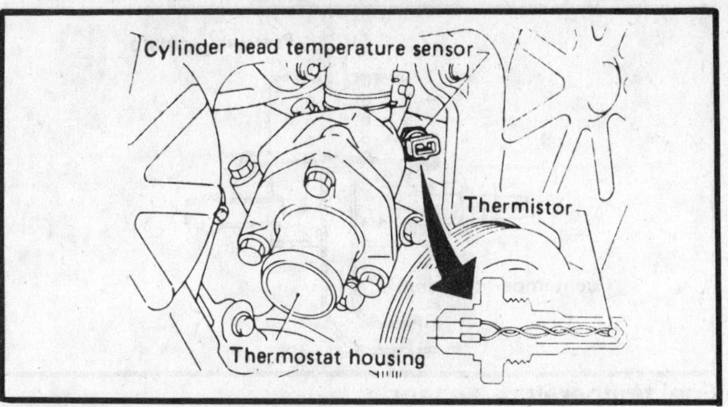
Cylinder head temperature sensor
Thermistor
Thermostat housing

Cylinder head temperature sensor

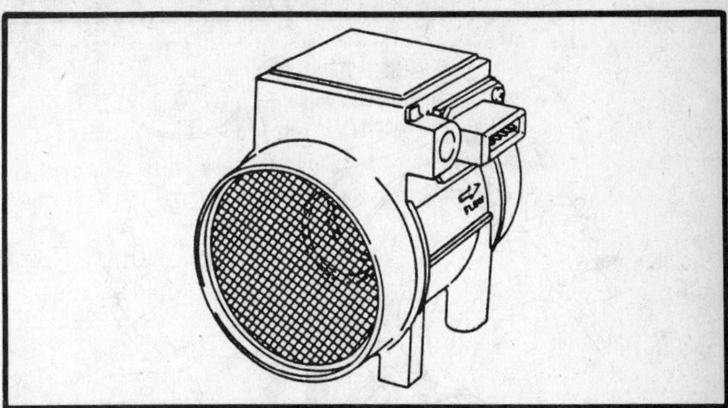

Air flow meter—except CA18ET engine

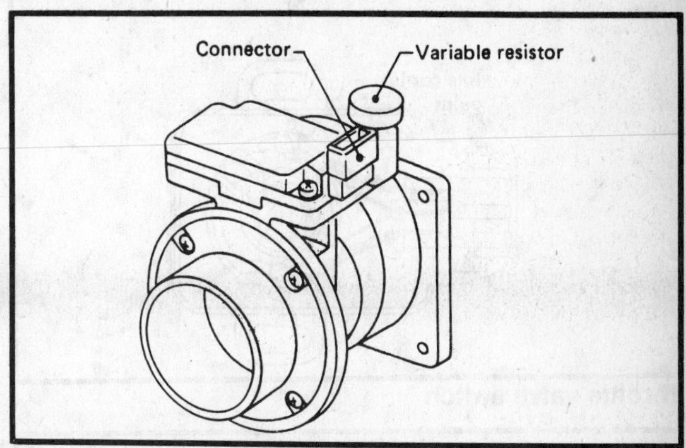
Connector
Variable resistor

Air flow meter—CA18ET engine

Exhaust Gas Sensor

The exhaust gas sensor, which is placed in the exhaust pipe, monitors the amount of oxygen in the exhaust gas. The sensor is made of ceramic titania which changes electrical resistance at the ideal air/fuel ratio (14.7:1). The control unit supplies the sensor with approximately 1 volt and takes the output voltage of the sensor depending on its resistance. The oxygen sensor is equipped with a heater to bring it to operating temperature quickly.

Throttle Valve Switch

A throttle valve switch is attached to the throttle chamber and operates in response to accelerator pedal movement. The switch has an idle contact and a full throttle contact. The idle contact closes when the throttle valve is positioned at idle and opens when it is in any other position.

Fuel Injector

The fuel injector is a small, precision solenoid valve. As the ECU sends an injection signal to each injector, the coil built into the injector pulls the needle valve back and fuel is injected through the nozzle and into the intake manifold. The amount of fuel injected is dependent on how long the signal is (pulse duration); the longer the signal, the more fuel delivered.

Detonation Sensor (Turbo Model)

The detonation sensor is attached to the cylinder block and senses engine knocking conditions. A knocking vibration from the cylinder block is applied as pressure to the piezoelectric element. This vibrational pressure is then converted into a voltage signal which is delivered as output.

Fuel Temperature Sensor

A fuel temperature sensor is built into the fuel pressure regulator. When the fuel temperature is higher than the preprogrammed level, the ECU will enrich the fuel injected to compensate for temperature expansion. The temperature sensor and pressure regulator should be replaced as an assembly if either malfunctions. The electric fuel pump with an integral damper is installed in the fuel tank. It is a vane roller type with the electric motor cooled by the fuel itself. The fuel filter is of metal construction in order to withstand the high fuel system pressure. The fuel pump develops 61–71 psi, but the pressure regulator keeps system pressure at 36 psi in operation.

Power Transistor

The ignition signal from the ECU is amplified by the power transistor, which turns the ignition coil primary circuit on and off, inducing the necessary high voltage in the secondary circuit to fire the spark plugs. Ignition timing is controlled according to engine operating conditions, with the optimum timing advance for each driving condition preprogrammed into the ECU memory.

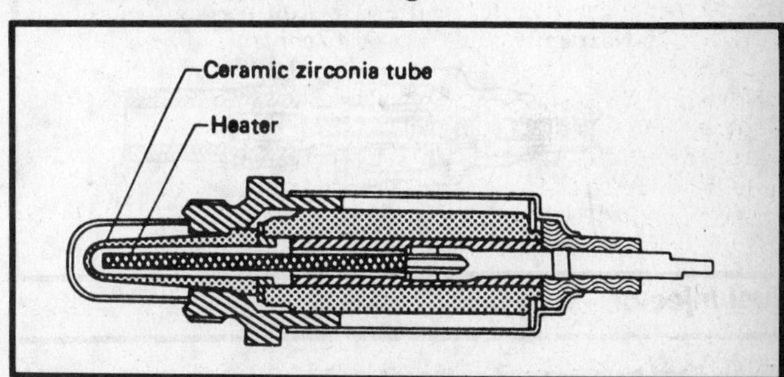
Ceramic zirconia tube
Heater

Exhaust gas senor—zirconia tube type

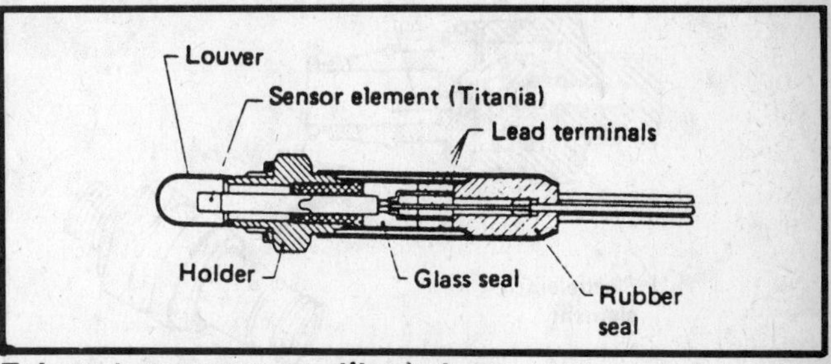
Louver
Sensor element (Titania)
Lead terminals
Holder
Glass seal
Rubber seal

Exhaust gas sensor—titania type

Vehicle Speed Sensor

The vehicle speed sensor provides a vehicle speed signal to the ECU. On conventional speedometers, the speed sensor consists of a reed switch which transforms vehicle speed into a pulse signal. On digital electronic speedometers, the speed sensor con-

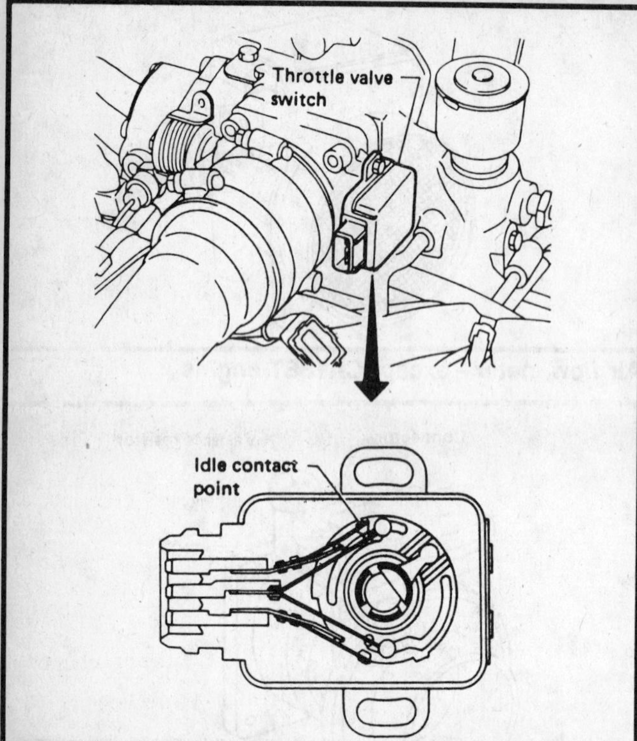
Throttle valve switch
Idle contact point
Throttle valve switch

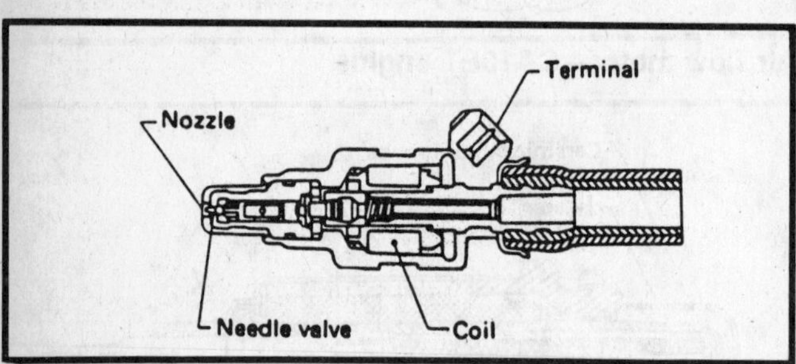
Terminal
Nozzle
Needle valve
Coil
Fuel injector

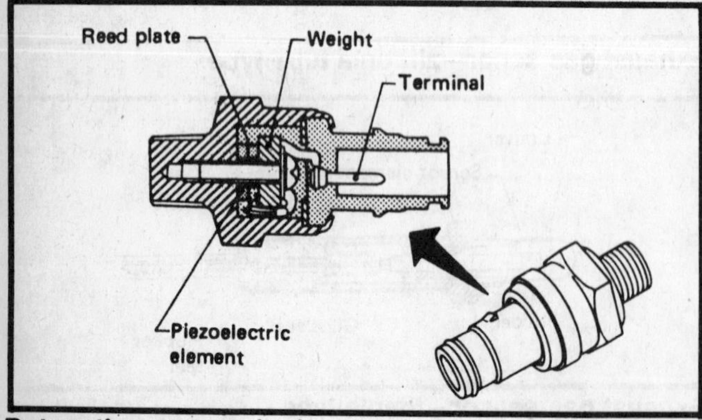
Reed plate
Weight
Terminal
Piezoelectric element
Detonation sensor (turbo sensor)

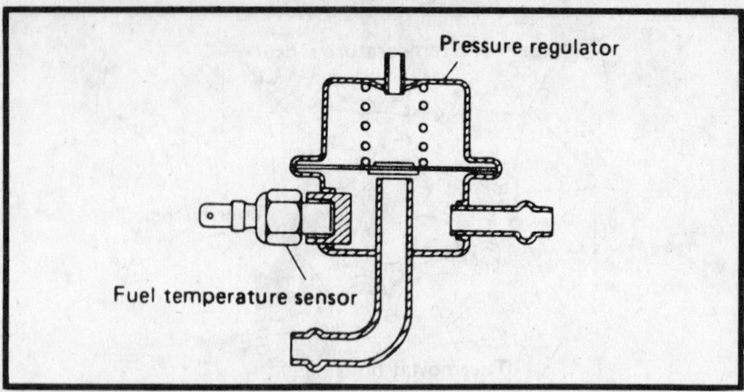
Pressure regulator
Fuel temperature sensor
Fuel temperature sensor

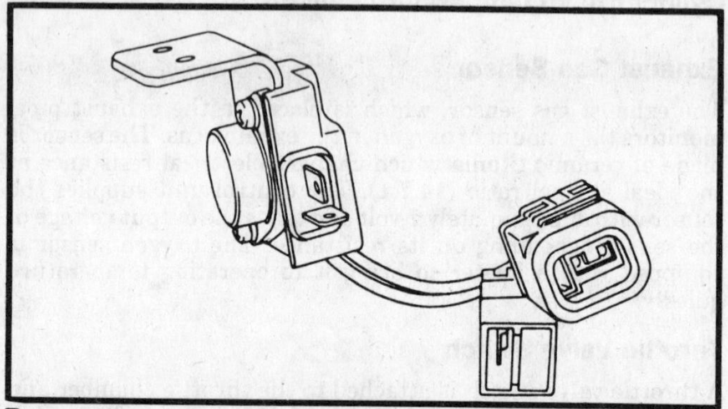
Power transistor – CA18ET engine

Power transistor and ignition coil – typical

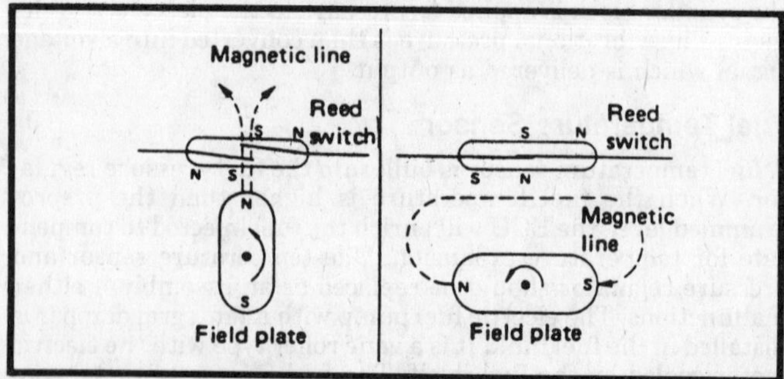
Magnetic line
Reed switch
Reed switch
Magnetic line
Field plate
Field plate
Vehicle speed sensor

sists of an LED, photo diode, shutter and wave forming circuit. It operates on the same principle as the crank angle sensor.

Swirl Control Valve (SCV) Control Solenoid Valve

The SCV control solenoid valve cuts the intake manifold vacuum signal for the swirl control valve. It responds to ON/OFF signal from the ECU. When the solenoid is off, the vacuum signal from the intake manifold is cut. When the control unit sends an ON signal, the coil pulls the plunger and feeds the vacuum signal to the swirl control valve actuator.

Idle-Up Solenoid Valve

An idle-up solenoid valve is attached to the intake collector to stabilize idle speed when the engine load is heavy because of electrical load, power steering load, etc. An air regulator pro-

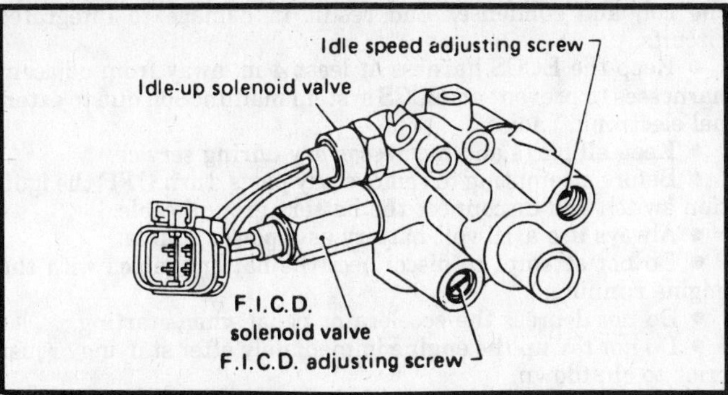
Idle speed adjusting screw
Idle-up solenoid valve
F.I.C.D.
solenoid valve
F.I.C.D. adjusting screw

Idle-up solenoid valve

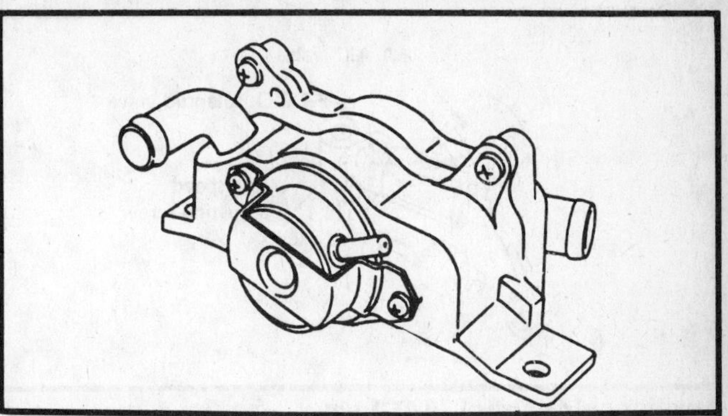

Air injection valve

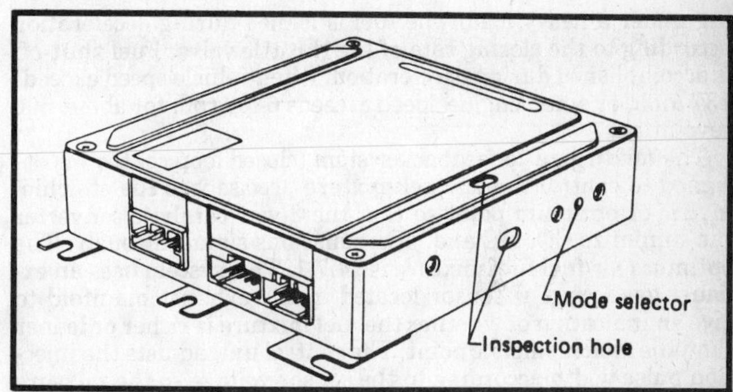
Mode selector
Inspection hole

Control unit

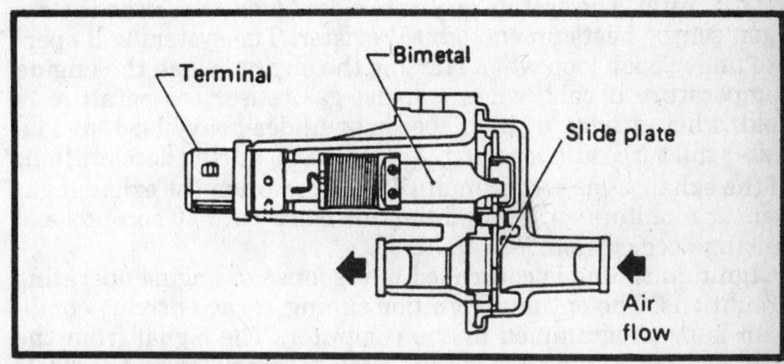
Terminal
Bimetal
Slide plate
Air
flow

Air regulator

vides an air bypass when the engine is cold in order to increase idle speed during warmup (fast idle). A bimetal, heater and rotary shutter are built into the air regulator. When bimetal temperature is low, the air bypass port is open. As the engine starts and electric current flows through a heater, the bimetal begins to rotate the shutter to close off the air bypass port. The air passage remains closed until the engine is stopped and the bimetal temperature drops.

Air Injection Valve (AIV)

The Air Injection Valve (AIV) sends secondary air to the exhaust manifold, utilizing a vacuum caused by exhaust pulsation in the exhaust manifold. When the exhaust pressure is below atmospheric pressure (negative pressure), secondary air is sent to the exhaust manifold. When the exhaust pressure is above atmospheric pressure, the reed valves prevent secondary air from being sent to the air cleaner. The AIV control solenoid valve cuts the intake manifold vacuum signal for AIV control. The solenoid valve actuates in response to the ON/OFF signal from the ECU. When the solenoid is off, the vacuum signal from the intake manifold is cut. As the control unit outputs an on signal, the coil pulls the plunger downward and feeds the vacuum signal to the AIV control valve.

Exhaust Gas Recirculation (EGR) Vacuum Cut Solenoid Valve

The EGR vacuum cut solenoid valve is the same type as that of the AIV. The EGR system is controlled by the ECU; at both low and high engine speed (rpm), the solenoid valve turns on and the EGR valve cuts the exhaust gas recirculation into the intake manifold. The pressure regulator control solenoid valve also actuates in response to the ON/OFF signal from the ECU. When it is off, a vacuum signal from the intake manifold is fed into the pressure regulator. As the control unit outputs an on signal, the coil pulls the plunger downward and cuts the vacuum signal.

Electronic Control Unit (ECU)

The ECU consists of a microcomputer, inspection lamps, a diagnostic mode selector, and connectors for signal input and output, and for power supply. The unit has control of the engine.

Air Regulator

The air regulator provides an air bypass when the engine is cold for the purpose of a fast idle during warm-up. A bimetal, heater and rotary shutter are built into the air regulator. When the bimetal temperature is low, the air bypass port is open. As the engine starts and electric current flows through a heater, the bimetal begins to rotate the shutter to close off the bypass port. The air passage remains closed until the engine is stopped and the bimetal temperature drops.

Idle Air Adjusting (IAA) Unit

The IAA consists of the AAC valve, FICD solenoid valve and an idle adjust screw. It receives signals from the ECU and controls the idle speed to the pre-set valve.

The FICD solenoid valve compensates for change in the idle speed caused by the operation of the air compressor. A vacuum control valve is installed in this unit to prevent an abnormal rise in the intake manifold vacuum pressure during deceleration

Auxiliary Air Control (AAC) Valve

The AAC valve is attached to the intake collector. The ECU actuates the AAC valve by an ON/OFF pulse of approximately 160 Hz. The longer that ON duty is left on, the larger the amount of air that will flow through the AAC valve.

SYSTEM OPERATION

In operation, the on-board computer (control unit) calculates the basic injection pulse width by processing signals from the crank angle sensor and air flow meter. Receiving signals from each sensor which detects various engine operating conditions, the computer adds various enrichments (which are preprogrammed) to the basic injection amount. In this manner, the optimum amount of fuel is delivered through the injectors. The fuel is enriched when starting, during warm-up, when accelerating, when cylinder head temperature is high and when operat-

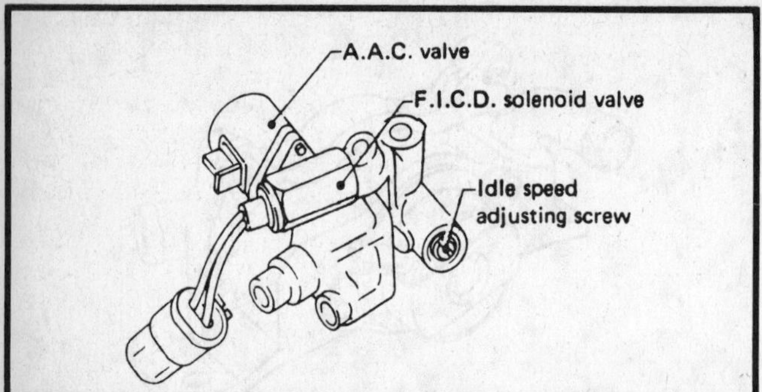
A.A.C. valve
F.I.C.D. solenoid valve
Idle speed
adjusting screw

Auxiliary air control (AAC) valve

ing under a heavy load. The fuel is leaned during deceleration according to the closing rate of the throttle valve. Fuel shut-off is accomplished during deceleration, when vehicle speed exceeds 137 mph, or when engine speed exceeds 6400 rpm for about 500 revolutions.

The mixture ratio feedback system (closed loop control) is designed to control the air/fuel mixture precisely to the stoichiometric or optimum point so that the 3-way catalytic converter can minimize CO, HC and NOx emissions simultaneously. The optimum air/fuel fuel mixture is 14.7:1. This system uses an exhaust gas (oxygen) sensor located in the exhaust manifold to give an indication of whether the fuel mixture is richer or leaner than the stoichiometric point. The control unit adjusts the injection pulse width according to the sensor voltage so the mixture ratio will be within the narrow window around the stoichiometric fuel ratio. The system goes into closed loop as soon as the oxygen sensor heats up enough to register. The system will operate under open loop when starting the engine, when the engine temperature is cold, when exhaust gas sensor temperature is cold, when driving at high speeds or under heavy load, at idle (after mixture ratio learning is completed), during deceleration, if the exhaust gas sensor malfunctions, or when the exhaust gas sensor monitors a rich condition for more than 10 seconds and during deceleration.

Ignition timing is controlled in response to engine operating conditions. The optimum ignition timing in each driving condition is preprogrammed in the computer. The signal from the control unit is transmitted to the power transistor and controls ignition timing. The idle speed is also controlled according to engine operating conditions, temperature and gear position. On manual transmission models, if battery voltage is less than 12 volts for a few seconds, a higher idle speed will be maintained by the control unit to improve charging function.

There is a fail-safe system built into the ECCS control unit. If the output voltage of the air flow meter is extremely low, the ECU will substitute a preprogrammed value for the air flow meter signal and allow the vehicle to be driven as long as the engine speed is kept below 2000 rpm. If the cylinder head temperature sensor circuit is open, the control unit clamps the warmup enrichment at a certain amount. This amount is almost the same as that when the cylinder head temperature is between 68–176°F (20–80°C). If the fuel pump circuit malfunctions, the fuel pump relay comes on until the engine stops. This allows the fuel pump to receive power from the relay.

SERVICE PRECAUTIONS

- Do not operate the fuel pump when the fuel lines are empty.
- Do not reuse fuel hose clamps.
- Do not disconnect the ECCS harness connectors before the battery ground cable has been disconnected.
- Make sure all ECCS connectors are fastened securely. A poor connection can cause an extremely high surge voltage in the coil and condenser and result in damage to integrated circuits.
- Keep the ECCS harness at least 4 in. away from adjacent harnesses to prevent an ECCS system malfunction due to external electronic "noise."
- Keep all parts and harnesses dry during service.
- Before attempting to remove any parts, turn **OFF** the ignition switch and disconnect the battery ground cable.
- Always use a 12 volt battery as a power source.
- Do not attempt to disconnect the battery cables with the engine running.
- Do not depress the accelerator pedal when starting.
- Do not rev up the engine immediately after starting or just prior to shutdown.
- Do not attempt to disassemble the ECCS control unit under any circumstances.
- If a battery cable is disconnected, the memory will return to the ROM (programmed) values. Engine operation may vary slightly, but this is not an indication of a problem. Do not replace parts because of a slight variation.
- If installing a 2-way or CB radio, keep the antenna as far as possible away from the electronic control unit. Keep the antenna feeder line at least 8 in. away from the ECCS harness and do not let the 2 run parallel for a long distance. Be sure to ground the radio to the vehicle body.

Diagnosis and Testing

SELF-DIAGNOSTIC SYSTEM

The self-diagnostic function is useful for diagnosing malfunctions in major sensors and actuators of the ECCS system. There are 5 modes in self-diagnostics on all models except 1990 300ZX and Stanza. On 1990 300ZX and Stanza, there are 2 modes in self-diagnostics

MODE 1

Except 1990 300ZX and Stanza

During closed loop operation, the green inspection lamp turns ON when a lean condition is detected and OFF when a rich condition is detected. During open loop operation, the red inspection lamp stays OFF.

1990 300ZX and Stanza

During this mode, the red LED in the ECU and the CHECK ENGINE LIGHT on the instrument panel stay ON. If either remain OFF, check the bulb in the CHECK ENGINE LIGHT or the red LED.

MODE 2

Except 1990 300ZX and Stanza

The green inspection lamp function is the same as in Mode 1. During closed loop operation, the red inspection lamp turns ON and OFF simultaneously with the green inspection lamp when the mixture ratio is controlled within the specified value. During open loop operation, the red inspection lamp stays OFF.

1990 300ZX and Stanza

These models use a Mode 2 for self-diagnostic results and exhaust gas sensor monitor.

When in Mode 2 (self-diagnostic results), a malfunction code is indicated by the number of flashes from the red LED or the CHECK ENGINE LIGHT.

When in Mode 2 (exhaust gas sensor monitor), the CHECK ENGINE LIGHT and red LED display the condition of the fuel mixture (rich/lean) which is monitored by the exhaust gas sensor. If 2 exhaust sensors are used (right side and left side), the left exhaust gas sensor monitor operates first, when selecting this mode.

MODE 3

This mode is the same as the former self-diagnosis mode.

MODE 4

During this mode, the inspection lamps monitor the ON/OFF condition of the idle switch, starter switch and vehicle speed sensor.

In switches ON/OFF diagnosis system, ON/OFF operation of the following switches can be detected continuously:

- Idle switch
- Starter switch
- Vehicle speed sensor

1. Idle switch and starter switch—the switches ON/OFF status at the point when Mode IV is selected is stored in ECU memory. When either switch is turned from **ON** to **OFF** or **OFF** to **ON**, the red LED on ECU alternately comes on and goes off each time switching is detected.
2. Vehicle speed sensor—The switches ON/OFF status at the point when Mode IV is selected is stored in ECU memory. When vehicle speed is 12 mph (20 km/h) or slower, the green LED on ECU is off. When vehicle speed exceeds 12 mph (20 km/h), the green LED on ECU comes ON.

MODE 5

The moment a malfunction is detected, the display will be presented immediately by flashing the inspection lamps during the driving test.

In real time diagnosis, if any of the following items are judged to be faulty, a malfunction is indicated immediately:

- Crank angle sensor
- Ignition signal
- Air flow meter output signal
- Fuel pump (some models)

Consequently, this diagnosis is a very effective measure to diagnose whether the above systems cause the malfunction or not, during driving test. Compared with self-diagnosis, real time diagnosis is very sensitive, and can detect malfunctioning conditions in a moment. Further, items regarded to be malfunctions in this diagnosis are not stored in ECU memory.

To switch the modes, turn the ignition switch **ON**, then turn the diagnostic mode selector on the control unit fully clockwise and wait for the inspection lamps to flash. Count the number of flashes until the inspection lamps have flashed the number of the desired mode, then immediately turn the diagnostic mode selector fully counterclockwise.

NOTE: When the ignition switch is turned OFF during diagnosis in each mode, and then turned back on again after the power to the control unit has dropped off completely, the diagnosis will automatically return to Mode 1.

The stored memory will be lost if the battery terminal is disconnected, or Mode 4 is selected after selecting Mode 3. However, if the diagnostic mode selector is kept turned fully clockwise, it will continue to change in the order of Mode 1, 2, 3, etc., and in this case, the stored memory will not be erased.

In Mode 3, the control unit constantly monitors the function of sensors and actuators regardless of ignition key position. If a malfunction occurs, the information is stored in the control unit and can be retrieved from the memory by turning **ON** the diagnostic mode selector on the side of the control unit. When activated, the malfunction is indicated by flashing a red and green LED (also located on the control unit). Since all the self-diagnostic results are stored in the control unit memory, even intermittent malfunctions can be diagnosed. A malfunctioning part's group is indicated by the number of both red and green LED's flashing. First, the red LED flashes and the green flashes follow. The red LED refers to the number of tens, while the green refers to the number of units. If the red LED flashes twice and the green LED flashes once, a Code 21 is being displayed. All malfunctions are classified by their trouble code number.

The diagnostic result is retained in the control unit memory until the starter is operated 50 times after a diagnostic item is judged to be malfunctioning. The diagnostic result will then be canceled automatically. If a diagnostic item which has been judged malfunctioning and stored in memory is again judged to be malfunctioning before the starter is operated 50 times, the second result will replace the previous one and stored in the memory until the starter is operated 50 more times.

In Mode 5 (real time diagnosis), if the crank angle sensor, ignition signal or air flow meter output signal are judged to be malfunctioning, the malfunction will be indicated immediately. This diagnosis is very effective for determining whether these systems are causing a malfunction during the driving test. Compared with self-diagnosis, real time diagnosis is very sensitive and can detect malfunctioning conditions immediately. However, malfunctioning items in this diagnosis mode are not stored in memory.

TESTING PRECAUTIONS

- Before connecting or disconnecting control unit ECU harness connectors, make sure the ignition switch is **OFF** and the negative battery cable is disconnected to avoid the possibility of damage to the control unit.
- When performing ECU input/output signal diagnosis, remove the pin terminal retainer from the 20 and 16-pin connectors to make it easier to insert tester probes into the connector.
- When connecting or disconnecting pin connectors from the ECU, take care not to bend or break any pin terminals. Check that there are no bends or breaks on ECU pin terminals before attempting any connections.
- Before replacing any ECU, perform the ECU input/output signal diagnosis to make sure the ECU is functioning properly or not.
- After performing the Electronic Control System Inspection, perform the ECCS self-diagnosis and driving test.
- When measuring supply voltage of ECU controlled components with a circuit tester, separate one tester probe from another. If the 2 tester probes accidentally make contact with each other during measurement, a short circuit will result and damage the power transistor in the ECU.

ELECTRONIC CONTROL UNIT (ECU)

Location

1988 200SX—Behind left (driver) kick panel
1989–90 240SX—Behind right (passenger) kick panel
1988 Maxima—Under right (passenger) seat
1989–90 Maxima—Under center console
1988–89 300ZX—Behind right (passenger) kick panel
1990 300ZX—behind glove box
1988–89 Stanza—Under right (passenger) seat
1988 Stanza Wagon—Under left (driver) seat
1990 Stanza—Under center console
1988–89 Pulsar NX—Under right (passenger) seat
1990 Axxess—Under center console
1990 Truck/Pathfinder—Under right (passenger) seat

FUEL SYSTEM

Releasing Pressure

CAUTION

Fuel system pressure must be relieved before disconnecting any fuel lines or attempting to remove any fuel system components.

EXCEPT 300ZX

1. Remove the fuel pump fuse from the fuse box.
2. Start the engine.

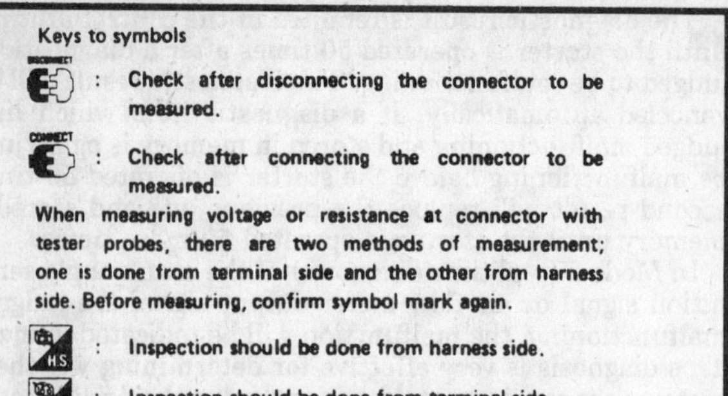
Keys to symbols
DISCONNECT : Check after disconnecting the connector to be measured.
CONNECT : Check after connecting the connector to be measured.
When measuring voltage or resistance at connector with tester probes, there are two methods of measurement; one is done from terminal side and the other from harness side. Before measuring, confirm symbol mark again.
HS : Inspection should be done from harness side.
TS : Inspection should be done from terminal side.

ECCS test symbols and identification chart

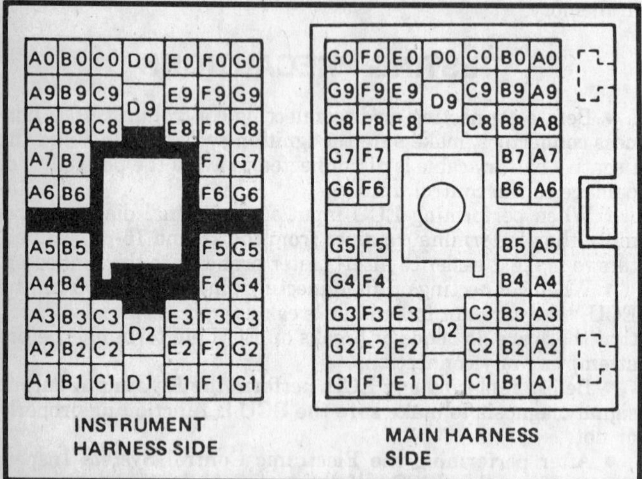
INSTRUMENT HARNESS SIDE
MAIN HARNESS SIDE

Super Multiple Junction (SMJ) terminal identification—1988 200SX

3. After the engine stalls, crank it over for a few seconds to make sure all fuel is exhausted from the lines.

4. Turn the ignition switch **OFF** and install the fuel pump fuse. Erase the trouble code memory of the ECCS control unit to eliminate false Code 22.

300ZX

1. On 1988–89 models, start the engine, then remove the fuse pump fuse from the fuse box.

2. On 1990 models, remove the fuel pump relay, then start the engine.

3. On all models, after the engine stalls, crank it over for a few seconds to make sure that all the fuel is exhausted from the lines.

4. Turn the ignition switch **OFF** and install the fuel pump fuse and/or relay.

FUEL SYSTEM

Pressure Testing

MODELS WITH VG30ET, CA18ET, CA20E AND 1988 VG30E ENGINES

1. Relieve fuel system pressure as previously described.
2. Disconnect the fuel inlet hose between the fuel filter and fuel line on the engine side.
3. Install a fuel pressure gauge.
4. Start the engine and check the fuel line for leakage.
5. Read the pressure on the pressure gauge. Fuel pressure gauge should read as follows:

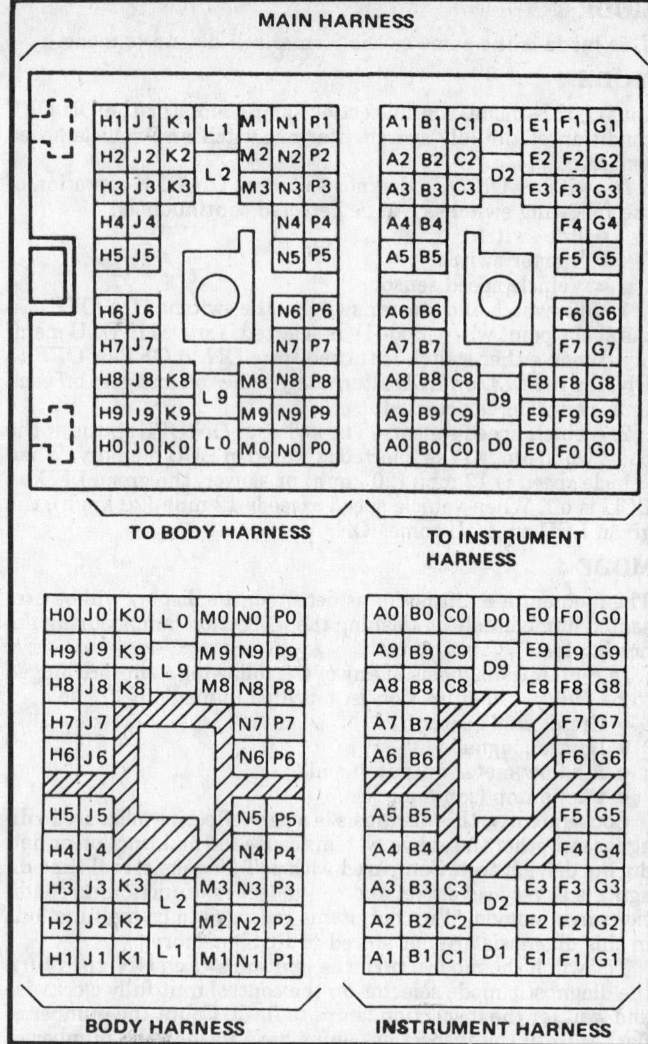
MAIN HARNESS
TO BODY HARNESS
TO INSTRUMENT HARNESS
BODY HARNESS
INSTRUMENT HARNESS

Super Multiple Junction (SMJ) terminal identification—1988–89 Stanza

AT IDLE:
VG30E engine—30 psi
VG30ET engine—30 psi
CA18ET engine—30 psi
CA20E engine—37 psi

ACCELERATOR PEDAL FULLY DEPRESSED:
VG30E engine—37 psi
CA18ET engine—37 psi
CA20E engine—44 psi
VG30ET engine—44 psi

6. Stop the engine and disconnect the fuel pressure regulator vacuum hose from the intake collector. Plug the intake collector with a rubber cap.
7. Connect a hand vacuum pump to the fuel pressure regulator.
8. Disconnect the ECCS harness connectors at the ECU.
9. Install a jumper wire to connect terminal No. 108 of the ECU to a body ground.
10. Turn the ignition switch **ON** and read the fuel pressure gauge as the vacuum is changed with the hand vacuum pump. Fuel pressure should decrease as vacuum increases. If not, replace the fuel pressure regulator.
11. Relieve fuel system pressure again.

4
SECTION

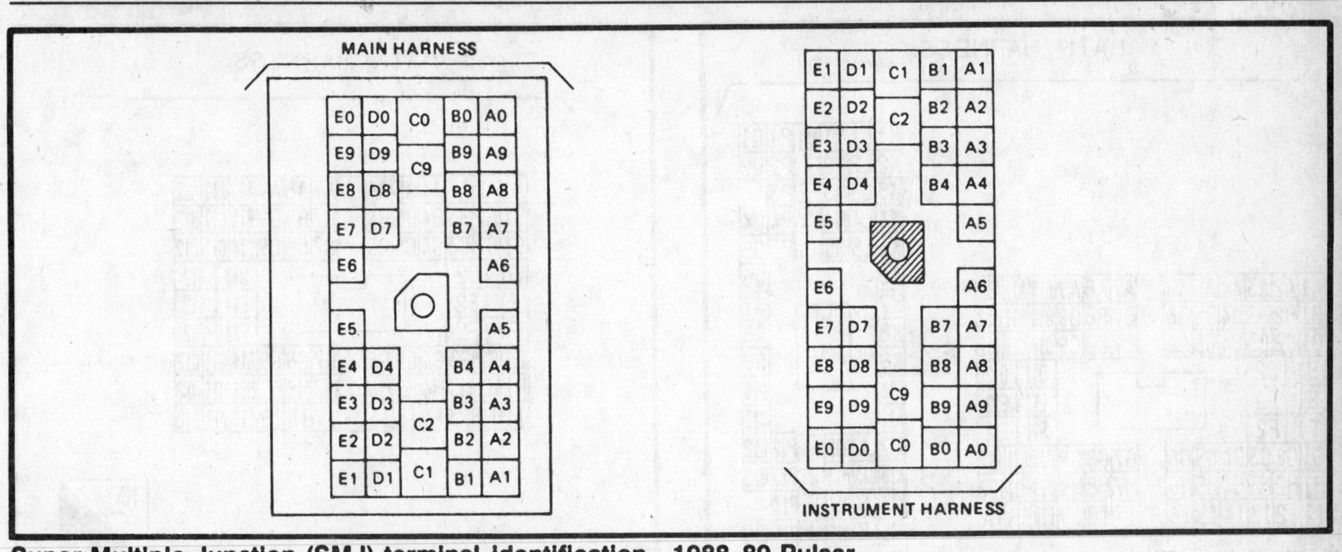
MAIN HARNESS
INSTRUMENT HARNESS

Super Multiple Junction (SMJ) terminal identification – 1988–89 Pulsar

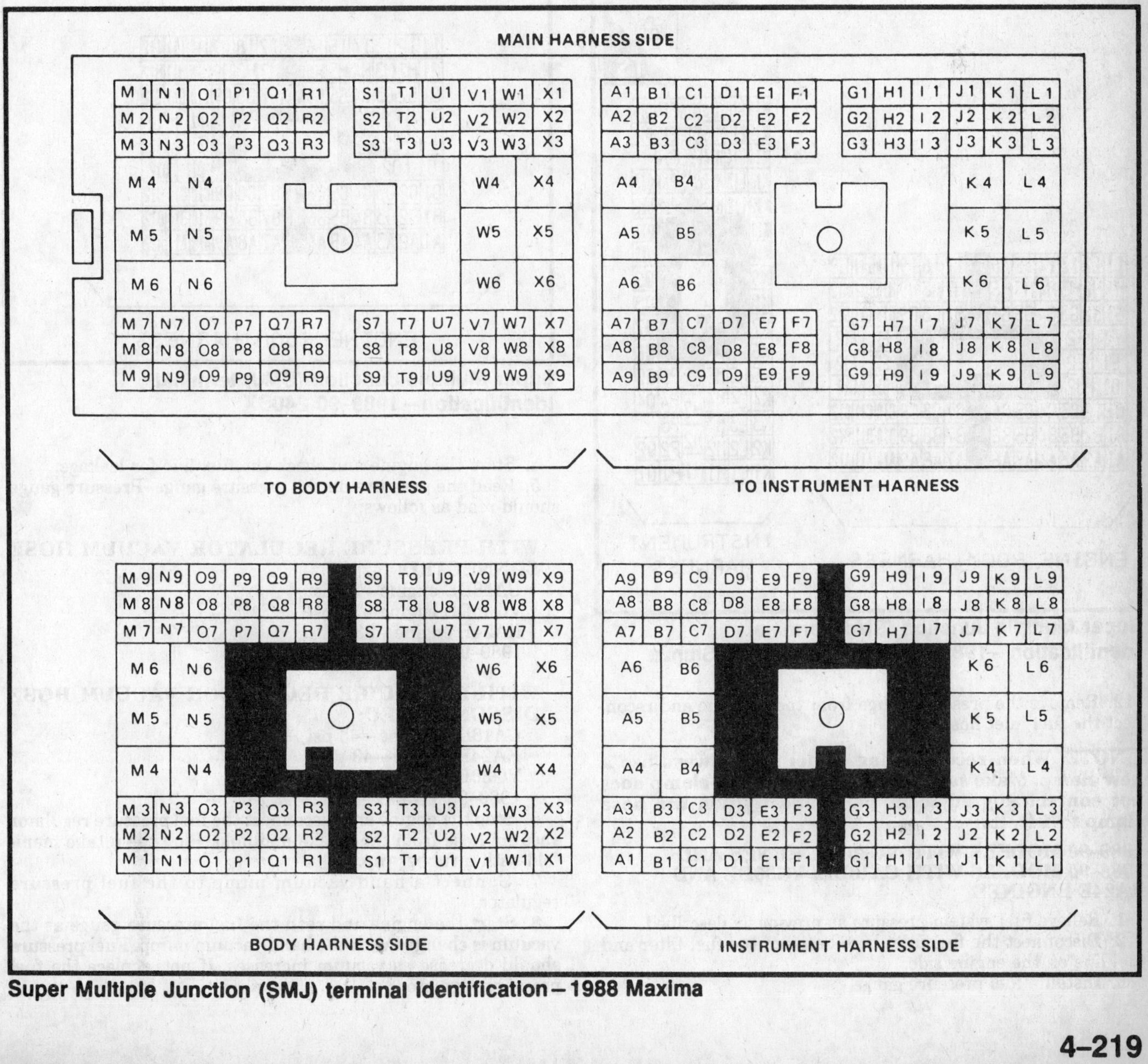

MAIN HARNESS SIDE

TO BODY HARNESS

TO INSTRUMENT HARNESS

BODY HARNESS SIDE

INSTRUMENT HARNESS SIDE

Super Multiple Junction (SMJ) terminal identification – 1988 Maxima

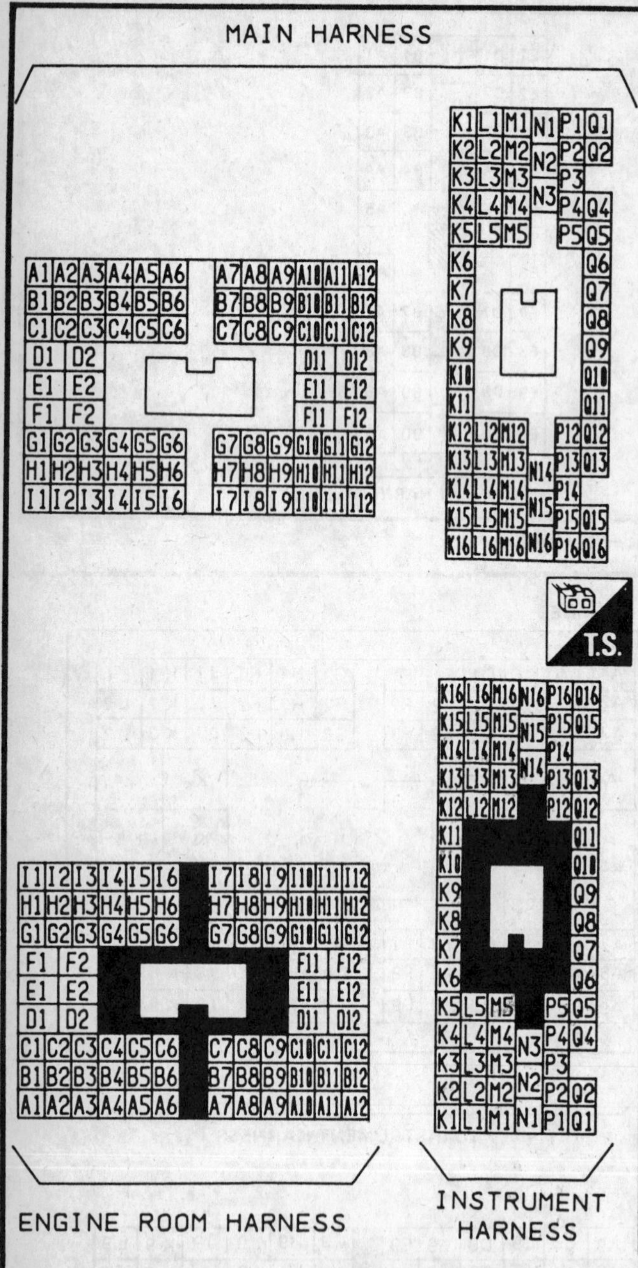
MAIN HARNESS
T.S.
ENGINE ROOM HARNESS
INSTRUMENT HARNESS

Super Multiple Junction (SMJ) terminal identification – 1989–90 Maxima and 1990 Stanza

12. Remove the pressure gauge from the fuel line and reconnect the fuel inlet hose.

NOTE: When reconnecting a fuel line, always use a new clamp. Make sure that the screw of the clamp does not contact any adjacent parts and tighten the hose clamp to 1 ft. lb.

1989–90 MODELS WITH VG30E ENGINES AND 1988–90 MODELS WITH CA18DE, VG30DE AND KA24E ENGINES

1. Relieve fuel system pressure as previously described.
2. Disconnect the fuel inlet hose between the fuel filter and fuel line on the engine side.
3. Install a fuel pressure gauge.

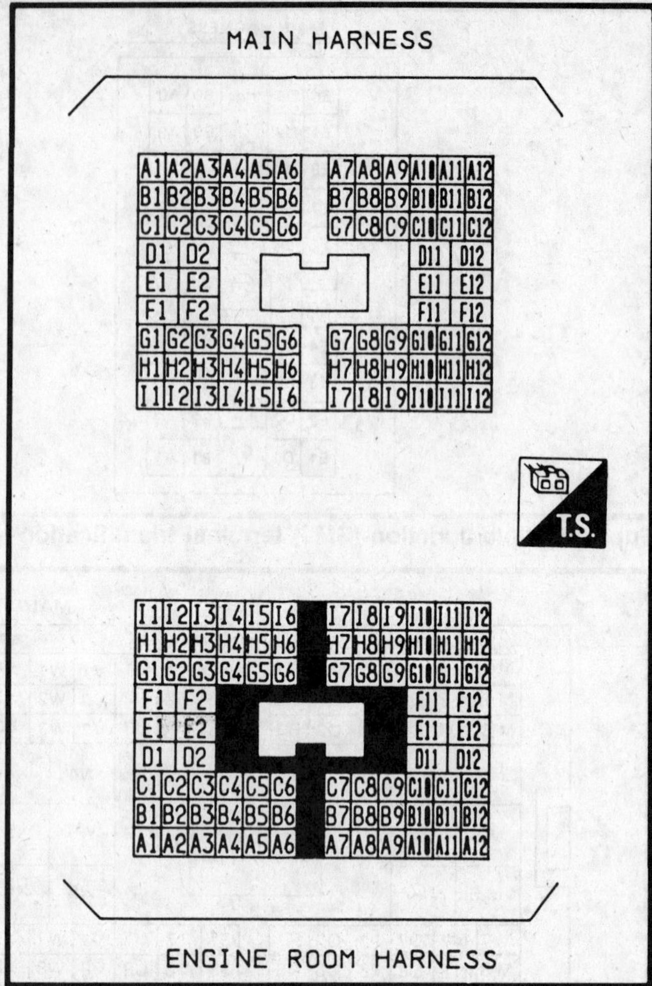
MAIN HARNESS
T.S.
ENGINE ROOM HARNESS

Super Multiple Junction (SMJ) terminal identification – 1989–90 240SX

4. Start the engine and check the fuel line for leakage.
5. Read the pressure on the pressure gauge. Pressure gauge should read as follows:

WITH PRESSURE REGULATOR VACUUM HOSE CONNECTED:

CA18DE engine – 36 psi
VG30DE engine – 36 psi
KA24E engine – 33 psi
1989–90 VG30E – 34 psi

WITH PRESSUIRE REGULATOR VACUUM HOSE DISCONNECTED:

CA18DE engine – 43 psi
KA24E engine – 43 psi
VG30DE engine – 43 psi
1989–90 VG30E engine – 43 psi

6. Stop the engine and disconnect the fuel pressure regulator vacuum hose from the intake manifold. Plug the intake manifold with a rubber cap.
7. Connect a hand vacuum pump to the fuel pressure regulator.
8. Start the engine and read the fuel pressure gauge as the vacuum is changed with the hand vacuum pump. Fuel pressure should decrease as vacuum increases. If not, replace the fuel pressure regulator.

MAIN HARNESS

INSTRUMENT HARNESS

ENGINE ROOM HARNESS

Super Multiple Junction (SMJ) terminal identification – 1988 Axxess

EGR VACUUM CUT AND AIR INJECTION CONTROL SOLENOID VALVE

Testing

1. Check the solenoid valve for continuity after disconnecting the harness connector. Continuity should exist and resistance should be 30–40 ohms. If not, replace the solenoid valve.
2. Check the solenoid valve for normal operation after disconnecting the harness connector and all vacuum hoses. Tag all hoses before removal.
3. Supply the solenoid valve with battery voltage and check whether there is continuity between ports A, B and C. With the solenoid OFF, there should be continuity between B and C. With the solenoid ON, there should be continuity between A and B.

Component Replacement

INJECTOR AND FUEL PIPE

Removal and Installation

1988 200SX CA18ET ENGINE

1. Relieve fuel pressure from system.
2. Disconnect air intake pipe.
3. Disconnect ECCS harness from injectors.
4. Disconnect ignition wire.
5. Disconnect accelerator wire.
6. Remove throttle chamber.
7. Disconnect fuel hoses and pressure regulator vacuum hose.
8. Remove injectors with fuel tube assembly.
9. Remove injector from fuel tube.
10. To install, reverse removal procedure.

1988–89 VG30E (EXCEPT 1989 MAXIMA) AND VG30ET ENGINE

1. Relieve fuel pressure from system.
2. Disconnect the following from intake collector:
   a. Air duct
   b. Accelerator wire
   c. Blow-by hoses
   d. Air regulator hose
   e. BCDD hose (Maxima)
   f. EGR tube
   g. Harness clamps
   h. Harness connectors
   i. Intake collector cover
   j. Water hoses (when necessary)
3. Disconnect fuel hoses.
4. Remove intake collector.
5. Remove bolts securing fuel tube.
6. Remove bolts securing injectors and remove injectors, fuel tubes and pressure regulator as an assembly.
7. To install, reverse removal procedure.

1988 200SX AND 1988–89 STANZA CA20E ENGINE

1. Relieve fuel pressure from system.
2. Disconnect ECCS harness and ignition wires.
3. Disconnect fuel hoses and pressure regulator vacuum hose.
4. Remove bolts securing fuel tube.
5. Remove bolts securing injectors; then take out fuel tube and injector as an assembly.
6. Remove injector from fuel tube.
7. Remove fuel hose.
8. To install, reverse removal procedure.

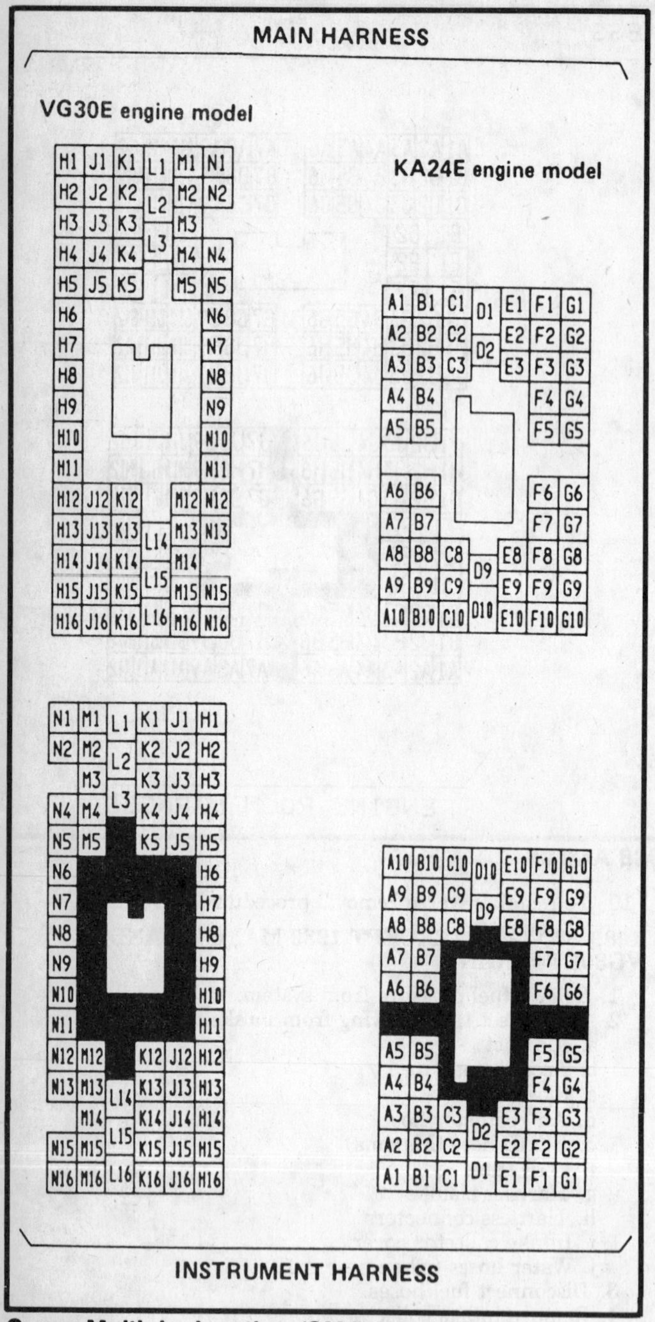

MAIN HARNESS

VG30E engine model

KA24E engine model

INSTRUMENT HARNESS

Super Multiple Junction (SMJ) terminal identification—1990 Tuck/Pathfinder

1988–89 PULSAR CA18DE ENGINE

1. Relieve fuel pressure from system.
2. Remove throttle chamber, intake manifold stay, IAA unit, intake side rocker cover and the PCU.
3. Disconnect fuel hoses and pressure regulator vacuum hose.
4. Remove injector assembly fixing bolts.
5. Remove injectors from fuel tube.
6. To install, reverse removal procedure.

1989–90 240SX AND 1990 TRUCK/PATHFINDER WITH KA24E ENGINE

1. Relieve fuel pressure from system.
2. Remove the BPT valve and the bolts securing the fuel tube.
3. Remove the bolts securing the injectors, then take out the fuel tube and injector as an assembly.
4. Remove injector from fuel tube.
5. Reverse procedure to install.

1990 STANZA AND AXXESS WITH KA24E ENGINE

1. Relieve fuel pressure from system.
2. Remove or disconnect the following:
   a. Air duct
   b. Fuel hoses
   c. Pressure regulator
   d. Accelerator wire bracket
   e. Injector harness connectors
3. Remove the bolts securing the fuel tube.
4. Remove bolts securing injectors and remove injectors and fuel tube as an assembly.
5. Reverse procedure to install.

1990 300ZX WITH VG30DE ENGINE

1. Relieve fuel pressure from system.
2. Open the drain cock on the radiator and drain the coolant into a suitable container.
3. Disconnect all electrical connectors, vacuum hoses and wires.
4. Remove intake manifold collector.
5. Remove injectors with fuel tube assembly.
6. Remove injectors from the fuel tube assembly.
7. Reverse procedure to install.

1990 TRACK/PATHFINDER WITH VG30E ENGINE

1. Relieve fuel pressure from system.
2. Remove the 2 drain plugs (from both sides of cylinder block) and drain coolant into a suitable container.
3. Disconnect the ASCD and accelerator control wire from the intake manifold collector.
4. Disconnect the following from the intake collector:
   a. AAC valve
   b. Throttle sensor and throttle valve switch
   c. Ignition coil
   d. EGR control solenoid valve
   e. Air regulator
   f. Exhaust gas temperature sensor (California models)
   g. Water and heater hoses
   h. PCV hose from RH rocker cover
   i. Air duct hose
   j. Ground wire
   k. EGR tube
   l. Purge hose from canister
5. Disconnect the master brake cylinder, pressure regulator and carbon canister vacuum hoses.
6. Remove the intake collector.
7. Remove the fuel hoses from the injector fuel tube assembly.
8. Disconnect the injector electrical connectors, then remove the injectors with fuel tube assembly.
9. Reverse procedure to install.

1989–90 MAXIMA WITH VG30E ENGINE

1. Relieve fuel pressure from system.
2. Disconnect the ASCD and accelerator control wire from the intake manifoild collector.
3. Disconnect the following connectors:
   a. AAC valve
   b. Throttle sensor
   c. Idle switch
4. Disconnect the water hose for the air cut valve and the PCV hose.
5. Disconnect the follwing vacuum hoses from:
   a. Vacuum gallery (carbon canister)
   b. Power valve actuator
   c. Master brake cylinder

d. EGR conrtol valve

6. Remove the lower intake manifold collector from the engine, then disconnect the ground wire.

7. Disconnect the fuel pressure regulator vacuum hose, then the fuel feed and return hose.

8. Disconnect injector connectors, then remove the injector fuel tube assembly.

9. Reverse procedure to install.

INJECTOR RUBBER HOSE

Removal and Installation

1. On injector rubber hose, measure off a point approximately 0.79 in. (20mm) from socket end.

2. Heat soldering iron (150 watt) for 15 minutes. Cut hose into braided reinforcement from mark to socket end.

NOTE: Do not feed soldering iron until it touches injector tail piece. Be careful not to damage socket, plastic connector, etc. with solder iron. Never place injector in a vise when disconnecting rubber hose.

3. Pull rubber hose out with hand.

4. To install, clean exterior of injector tail piece.

5. Wet inside of new rubber hose with fuel.

6. Push end of rubber hose with hose socket onto injector tail piece by hand as far as it will go. Clamp is not necessary at this connection.

NOTE: After properly connecting fuel hose to injector, check connection for fuel leakage.

FUEL PRESSURE REGULATOR

Removal and Installation

1. Relieve fuel pressure from system.

2. Disengage vacuum tube connecting regulator to intake manifold from pressure regulator.

3. Remove screws securing pressure regulator.

4. Unfasten hose clamps, and disconnect pressure regulator from fuel hose.

NOTE: Place a rag under fuel pipe to absorb any remaining fuel.

5. To install, reverse the removal procedure.

**ECCS COMPONENTS LOCATION
1988 200SX WITH VG30E ENGINE**

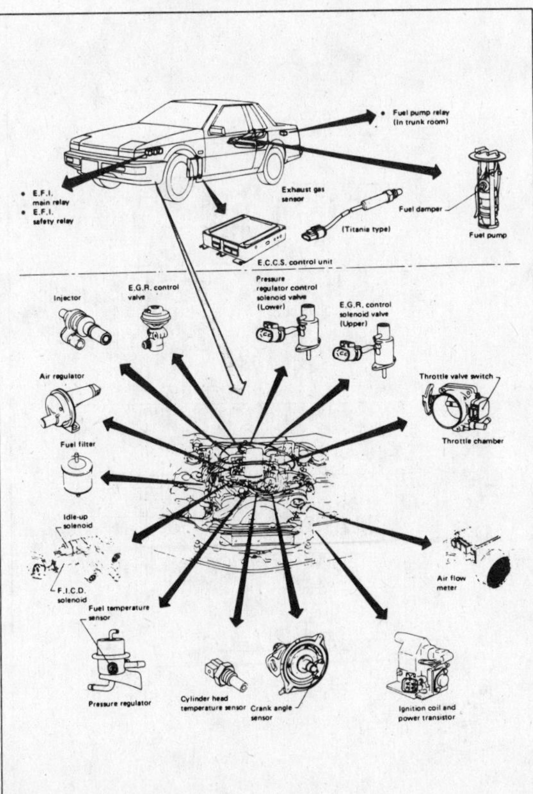
Fuel pump relay (In trunk room)
E.F.I. main relay
E.F.I. safety relay
Exhaust gas sensor
Fuel damper
(Titania type)
Fuel pump
E.C.C.S. control unit
Injector
E.G.R. control valve
Pressure regulator control solenoid valve (Lower)
E.G.R. control solenoid valve (Upper)
Air regulator
Throttle valve switch
Throttle chamber
Fuel filter
Idle-up solenoid
F.I.C.D. solenoid
Air flow meter
Fuel temperature sensor
Pressure regulator
Cylinder head temperature sensor
Crank angle sensor
Ignition coil and power transistor

**ECCS SYSTEM SCHEMATIC
1988 200SX WITH VG30E ENGINE**

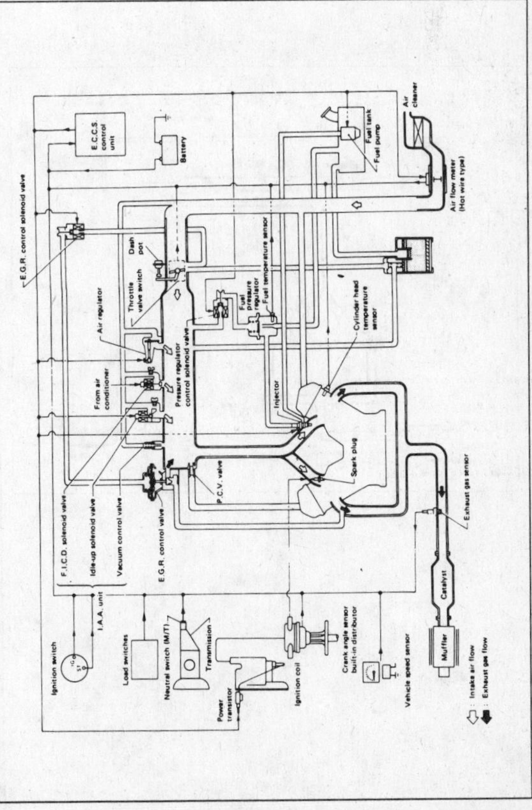
E.C.C.S. control unit
Battery
Fuel tank
Fuel pump
Air cleaner
Air flow meter (Hot wire type)
E.G.R. control solenoid valve
Dash pot
Throttle valve switch
Air regulator
Fuel pressure regulator
Fuel temperature sensor
Cylinder head temperature sensor
From air conditioner
Pressure regulator control solenoid valve
Injector
Spark plug
P.C.V. valve
Exhaust gas sensor
F.I.C.D. solenoid valve
Idle-up solenoid valve
Vacuum control valve
E.G.R. control valve
I.A.A. unit
Ignition switch
Load switches
Neutral switch (M/T)
Transmission
Power transistor
Ignition coil
Crank angle sensor built-in distributor
Vehicle speed sensor
Catalyst
Muffler
Intake air flow
Exhaust gas flow

**ECCS CONTROL SYSTEM CHART
1988 200SX WITH VG30 ENGINE**

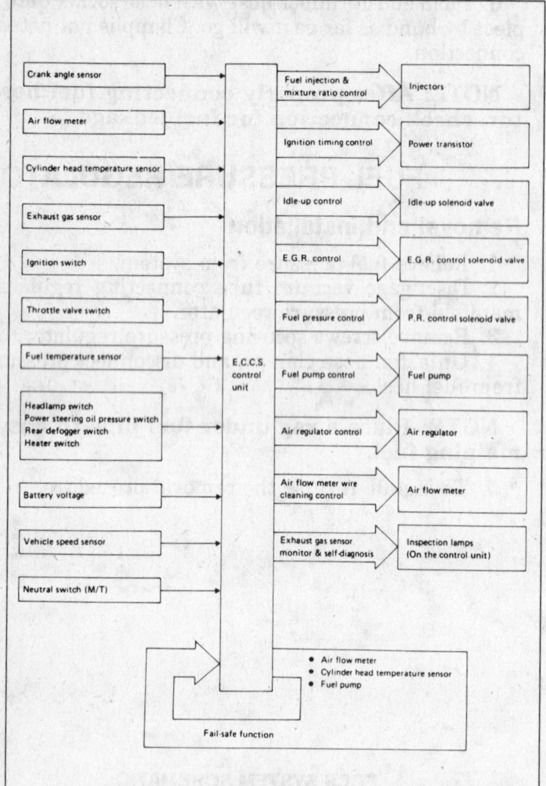
Crank angle sensor
Air flow meter
Cylinder head temperature sensor
Exhaust gas sensor
Ignition switch
Throttle valve switch
Fuel temperature sensor
Headlamp switch
Power steering oil pressure switch
Rear defogger switch
Heater switch
Battery voltage
Vehicle speed sensor
Neutral switch (M/T)
E.C.C.S. control unit
Fuel injection & mixture ratio control
Ignition timing control
Idle-up control
E.G.R. control
Fuel pressure control
Fuel pump control
Air regulator control
Air flow meter wire cleaning control
Exhaust gas sensor monitor & self-diagnosis
Injectors
Power transistor
Idle-up solenoid valve
E.G.R. control solenoid valve
P.R. control solenoid valve
Fuel pump
Air regulator
Air flow meter
Inspection lamps (On the control unit)
Air flow meter
Cylinder head temperature sensor
Fuel pump
Fail-safe function

**ECCS FUEL FLOW SYSTEM SCHEMATIC
1988 200SX WITH VG30E ENGINE**

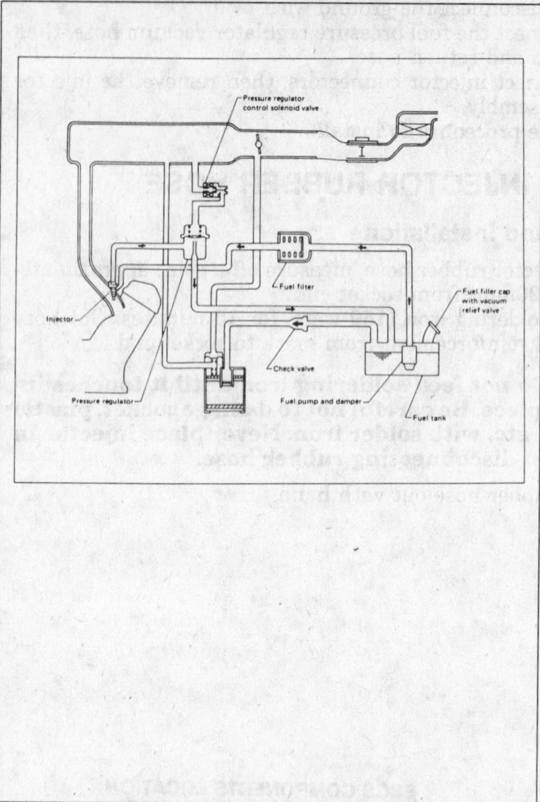
Pressure regulator control solenoid valve
Fuel filter
Fuel filler cap with vacuum relief valve
Injector
Check valve
Pressure regulator
Fuel pump and damper
Fuel tank

**ECCS AIR FLOW SYSTEM SCHEMATIC
1988 200SX WITH VG30E ENIGNE**

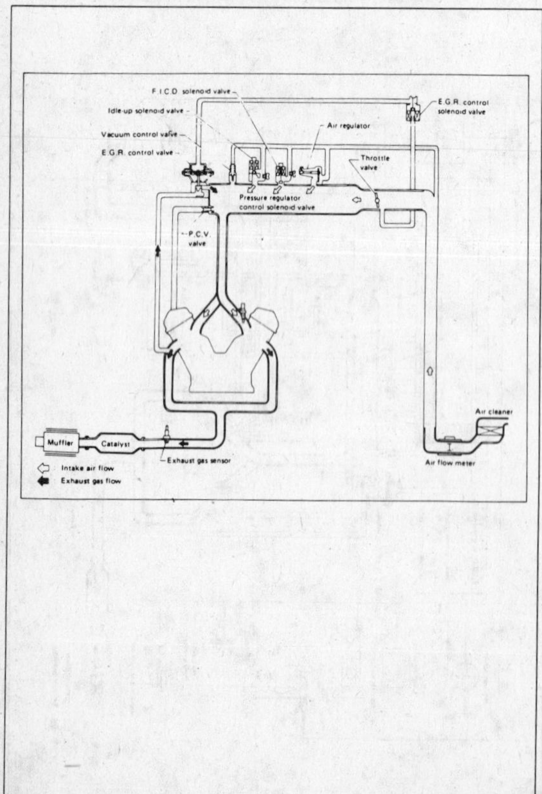
F.I.C.D. solenoid valve
Idle-up solenoid valve
Vacuum control valve
E.G.R. control valve
Air regulator
E.G.R. control solenoid valve
Throttle valve
Pressure regulator control solenoid valve
P.C.V. valve
Air cleaner
Muffler
Catalyst
Exhaust gas sensor
Air flow meter
Intake air flow
Exhaust gas flow

**ECCS CONTROL UNIT WIRING SCHEMATIC
1988 200SX WITH VG30E ENIGNE**

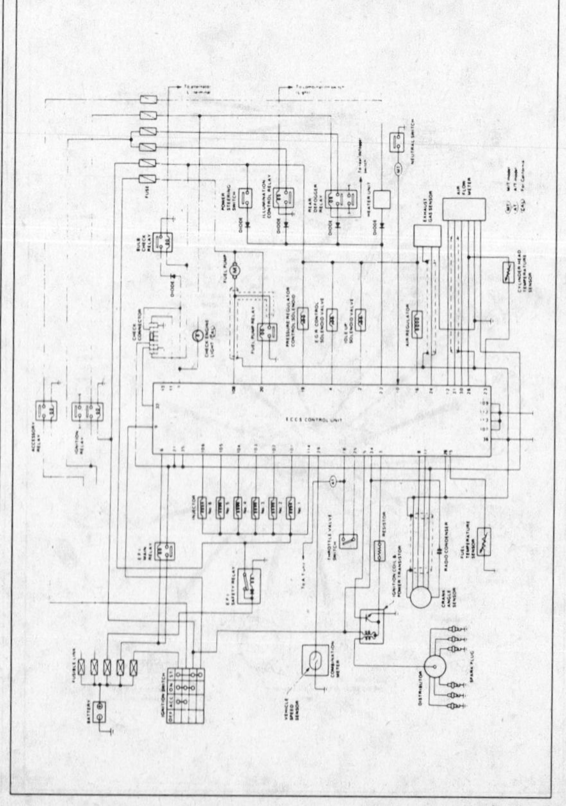

ECCS WIRING SCHEMATIC—1988 200SX WITH VG30E ENGINE

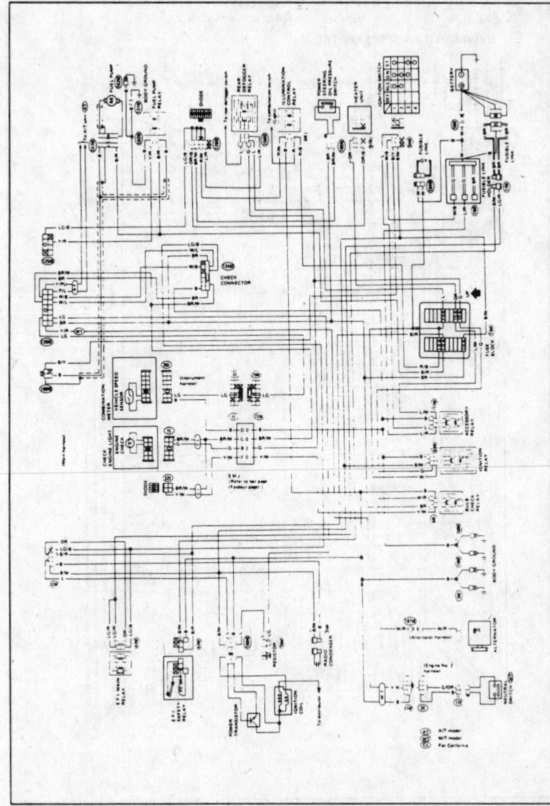

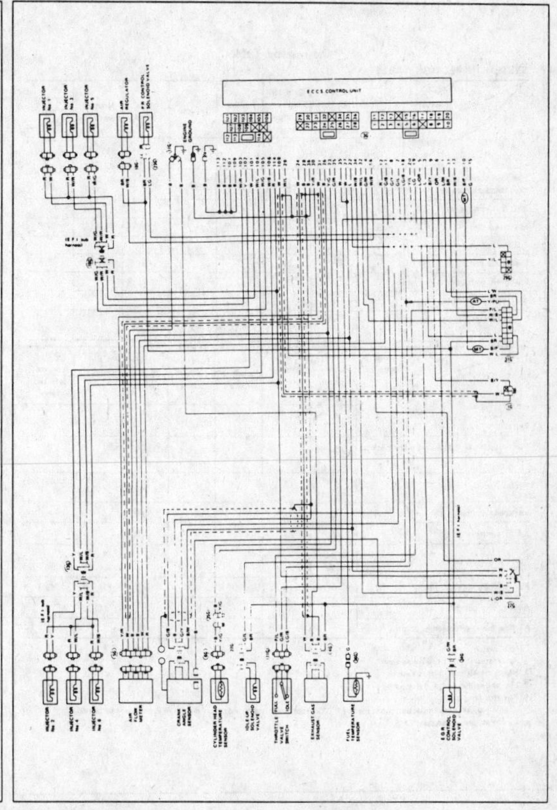

ECCS DIAGNOSTIC PROCEDURE—1988 200SX WITH VG30E ENGINE

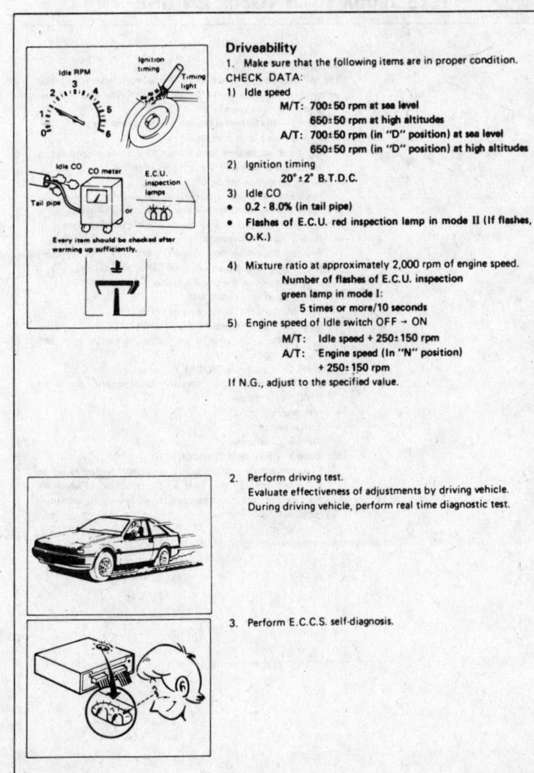
Idle RPM
Ignition timing
Timing light
Idle CO
CO meter
E.C.U. inspection lamps
Tail pipe
Every item should be checked after warming up sufficiently.

Driveability

1. Make sure that the following items are in proper condition.

CHECK DATA:

1) Idle speed
   - M/T: 700±50 rpm at sea level
     650±50 rpm at high altitudes
   - A/T: 700±50 rpm (in "D" position) at sea level
     650±50 rpm (in "D" position) at high altitudes
2) Ignition timing
   20°±2° B.T.D.C.
3) Idle CO
   - 0.2 - 8.0% (in tail pipe)
   - Flashes of E.C.U. red inspection lamp in mode II (If flashes, O.K.)
4) Mixture ratio at approximately 2,000 rpm of engine speed.
   Number of flashes of E.C.U. inspection green lamp in mode I:
   5 times or more/10 seconds
5) Engine speed of Idle switch OFF → ON
   - M/T: Idle speed + 250±150 rpm
   - A/T: Engine speed (In "N" position) + 250±150 rpm

If N.G., adjust to the specified value.

2. Perform driving test.
   Evaluate effectiveness of adjustments by driving vehicle.
   During driving vehicle, perform real time diagnostic test.

3. Perform E.C.C.S. self-diagnosis.

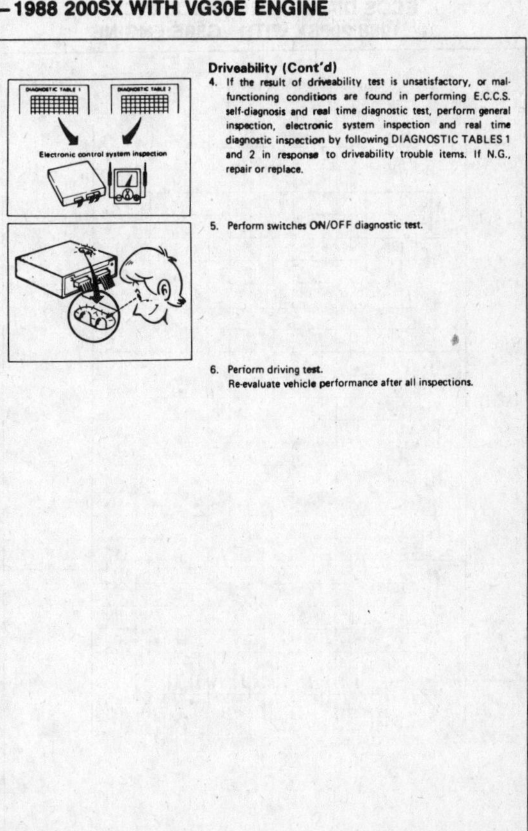
DIAGNOSTIC TABLE 1
DIAGNOSTIC TABLE 2
Electronic control system inspection

Driveability (Cont'd)

4. If the result of driveability test is unsatisfactory, or malfunctioning conditions are found in performing E.C.C.S. self-diagnosis and real time diagnostic test, perform general inspection, electronic system inspection and real time diagnostic inspection by following DIAGNOSTIC TABLES 1 and 2 in response to driveability trouble items. If N.G., repair or replace.

5. Perform switches ON/OFF diagnostic test.

6. Perform driving test.
   Re-evaluate vehicle performance after all inspections.

ECCS DIAGNOSTIC TABLE 1
1988 200SX WITH VG30E ENGINE

Diagnostic Table 1

SYSTEM INSPECTION TABLE

System \ Sensor & actuator	Crank angle sensor	Air flow meter	Cylinder head temperature sensor	Ignition switch	Injector	Idle switch	Neutral switch	Exhaust gas sensor
Fuel injection & mixture ratio feedback control	O	O	O	O	O	O	O	O
Ignition timing control	O	O	O	O		O		
Idle speed control	O		O	O		O	O	
E.G.R. control	O		O	O		O		
Fuel pump control	O		O	O		O		
Fuel pressure control				O				
Air regulator control	O			O				
Air flow meter self-cleaning control	O		O	O				

System \ Sensor & actuator	Battery voltage	E.G.R. control solenoid valve	Idle-up solenoid valve	Fuel temperature seosor	Vehicle speed sensor	Air regulator	P.R. control solenoid valve
Fuel injection & mixture ratio feedback control	O			O	O		
Ignition timing control	Power transistor				O		
Idle speed control	O		O*		O		
E.G.R. control		O					
Fuel pump control	Fuel pump relay						
Fuel pressure control				O			O
Air regulator control						O	
Air flow meter self-cleaning control					O		

* Input switch
① Power steering oil pressure switch
② Heater or Air conditioner switch
③ Lighting switch & rear defogger switch
④ Radiator fan switch

This table indicates the inspection items for the E.C.C.S. control system. For each system, it is necessary to check sensors or actuators marked "O".

ECCS DIAGNOSTIC TABLE 2
1988 200SX WITH VG30E ENGINE

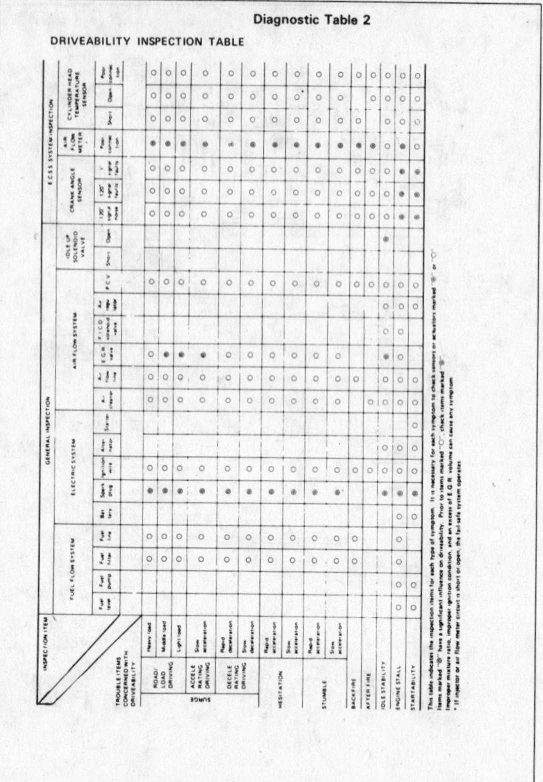

Diagnostic Table 2

DRIVEABILITY INSPECTION TABLE

Trouble items concerned with driveability \ Inspection item	General inspection – Fuel flow system: Fuel level	Fuel pump	Fuel filter	Fuel line	Electric system: Battery	Spark plug	Ignition wire	Alternator	Starter	Air flow system: Air cleaner	Air flow line	E.G.R. valve	F.I.C.D. solenoid valve	Air regulator	P.C.V.	Idle up solenoid valve: Short	Open	ECCS system inspection – Crank angle sensor: 120° signal noise	120° signal faults	1° signal faults	Air flow meter: Poor connection	Cylinder head temperature sensor: Short	Open	Poor connection
Surge – Road/load driving – Heavy load			O	O		◎	O			O	O	O			O			O	O	O	◎	O	O	O
Surge – Road/load driving – Middle load			O	O		◎	O			O	O	◎			O			O	O	O	◎	O	O	O
Surge – Road/load driving – Light load			O	O		◎	O			O	O	◎			O			O	O	O	◎	O	O	O
Surge – Accelerating driving – Slow acceleration			O	O		◎	O			O	O	◎			O			O	O	O	◎	O	O	O
Surge – Decelerating driving – Rapid deceleration			O	O		◎	O			O	O	O			O			O	O	O	◎	O	O	O
Surge – Decelerating driving – Slow deceleration			O	O		◎	O			O	O	O			O			O	O	O	◎	O	O	O
Hesitation – Rapid acceleration			O	O		◎	O			O	O	O			O			O	O	O	◎	O	O	O
Hesitation – Slow acceleration			O	O		◎	O			O	O	O			O			O	O	O	◎	O	O	O
Stumble – Rapid acceleration			O	O		◎	O			O	O	O			O			O	O	O	◎	O	O	O
Stumble – Slow acceleration			O	O		◎	O			O	O	O			O			O	O	O	◎	O	O	O
Backfire			O	O			O				O				O			O	O	O	◎	O		O
After fire							O			O					O			O	O	O	◎		O	O
Idle stability						◎	O	O		O	O	◎	O	O	O		◎	O	O	O	O	O	O	O
Engine stall	O	O	O	O	O	◎	O	O		O	O	O	O	O	O			◎	◎	◎	◎	O	O	O
Startability	O	O			O	◎	O	O	O	O	O			O	O			◎	◎	◎	O	O	O	O

This table indicates the inspection items for each type of symptom. It is necessary for each symptom to check sensors or actuators marked "◎" or "O".
Items marked "◎" have a significant influence on driveability. Prior to items marked "O", check items marked "◎".
Improper mixture ratio, improper ignition condition, and an excess of E.G.R. volume can cause any symptom.
* If injector or air flow meter circuit is short or open, the fail-safe system operates.

ECCS DIAGNOSTIC TABLE 2 (CONT.)
1988 200SX WITH VG30E ENGINE

Diagnostic Table 2 (Cont'd)

Trouble items concerned with driveability \ Inspection item (E.C.C.S. system inspection)	Idle switch: Short	Open	E.G.R. control solenoid valve: Short	Open	Exhaust gas sensor: Short	Open	Injector: Clogging	Neutral switch: Short	Starter signal: Short	Open	Ignition signal: Short	Open	Battery voltage: Low voltage	Fuel pump circuit: Short	Open	Poor connection	Fuel temp sensor: Short	Open	P.R. control solenoid valve: Short	Open	Vehicle speed sensor: Short	Open	Poor connection
Surge – Road/load driving – Heavy load	O			◎	O	O	◎									O							
Surge – Road/load driving – Middle load	O			◎	O	O	◎									O							
Surge – Road/load driving – Light load	O			◎	O	O	◎									O							
Surge – Accelerating driving – Slow acceleration	O			◎	O	O	◎									O							
Surge – Decelerating driving – Rapid deceleration				O	O	O	◎									O						O	O
Surge – Decelerating driving – Slow deceleration				O	O	O	◎									O						O	O
Hesitation – Rapid acceleration	O			O	O	O	◎	O								O							
Hesitation – Slow acceleration	O			O	O	O	◎	O								O							
Stumble – Rapid acceleration	O			O			◎	O								O			O	O			
Stumble – Slow acceleration	O			O			◎	O								O			O	O			
Backfire	O				O	O	◎	O								O							
After fire		O		◎	O	O																	
Idle stability		O		O	O	O	◎									O					O	O	O
Engine stall					O	O	◎				O	O	O	O	O	O			O	O			
Startability							◎		O	O	O	O	O	O	O	O	O	O	O	O			

ECCS SELF-DIAGNOSTIC DESCRIPTION
1988 200SX WITH VG30E ENGINE

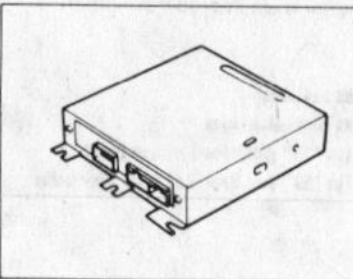

Description

The self-diagnosis is useful to diagnose malfunctions in major sensors and actuators of the E.C.C.S. system. There are 5 modes in the self-diagnosis system.

1. **Mode I – Mixture ratio feedback control monitor A**
   - During closed loop condition:
     The green inspection lamp turns ON when lean condition is detected and goes OFF by rich condition.
   - During open loop condition:
     The green inspection lamp remains ON or OFF.
2. **Mode II – Mixture ratio feedback control monitor B**
   The green inspection lamp function is the same as Mode I.
   - During closed loop condition:
     The red inspection lamp turns ON and OFF simultaneously with the green inspection lamp when the mixture ratio is controlled within the specified value.
   - During open loop condition:
     The red inspection lamp remains ON or OFF.
3. **Mode III – Self-diagnosis**
   This mode is the same as the former self-diagnosis in self-diagnosis mode.
4. **Mode IV – Switches ON/OFF diagnosis**
   During this mode, the inspection lamps monitor the switch ON-OFF condition.
   - Idle switch
   - Starter switch
   - Vehicle speed sensor
5. **Mode V – Real time diagnosis**
   The moment the malfunction is detected, the display will be presented immediately. That is, the condition at which the malfunction occurs can be found by observing the inspection lamps during driving test.

ECCS SELF-DIAGNOSTIC DESCRIPTION (CONT.)—1988 200SX WITH VG30E

Description (Cont'd)

SWITCHING THE MODES

1. Turn ignition switch "ON".
2. Turn diagnostic mode selector on E.C.U. fully clockwise and wait the inspection lamps flash.
3. Count the number of the flashing time, and after the inspection lamps have flashed the number of the required mode, turn diagnostic mode selector fully counterclockwise immediately.

Flashing N times

Mode N

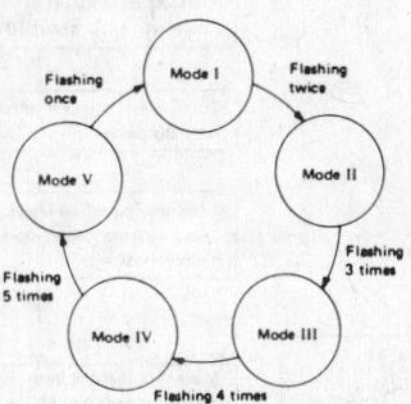
Mode I
Flashing twice
Mode II
Flashing 3 times
Mode III
Flashing 4 times
Mode IV
Flashing 5 times
Mode V
Flashing once

When the ignition switch is turned off during diagnosis, in each mode, and then turned back on again after the power to the E.C.U. has dropped off completely, the diagnosis will automatically return to Mode I.

The stored memory would be lost if:

1. Battery terminal is disconnected.
2. After selecting Mode III, Mode IV is selected.

   However, if the diagnostic mode selector is kept turned fully clockwise, it will continue to change in the order of Mode I → II → III → IV → V → I ... etc., and in this state the stored memory will not be erased.

Description (Cont'd)

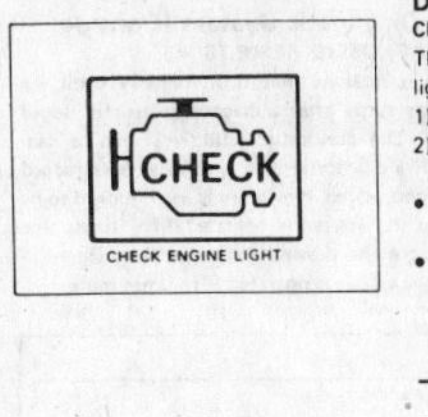
CHECK
CHECK ENGINE LIGHT

CHECK ENGINE LIGHT (For California only)

This vehicle has a check engine light on instrument panel. This light comes ON under the following conditions:

1) When ignition switch is turned "ON" (for bulb check).
2) When systems related to emission performance malfunction in Mode I (with engine running).

- **This check engine light always illuminates and is synchronous with red L.E.D.**
- **Malfunction systems related to emission performance can be detected by self-diagnosis, and they are clarified as self-diagnostic codes in Mode III.**

Code No.	Malfunction
12	Air flow meter circuit
14	Vehicle speed sensor circuit
23	Idle switch circuit
31	E.C.U. (E.C.C.S. control unit)
33	Exhaust gas sensor circuit

Use the following diagnostic flowchart to check and repair a malfunctioning system.

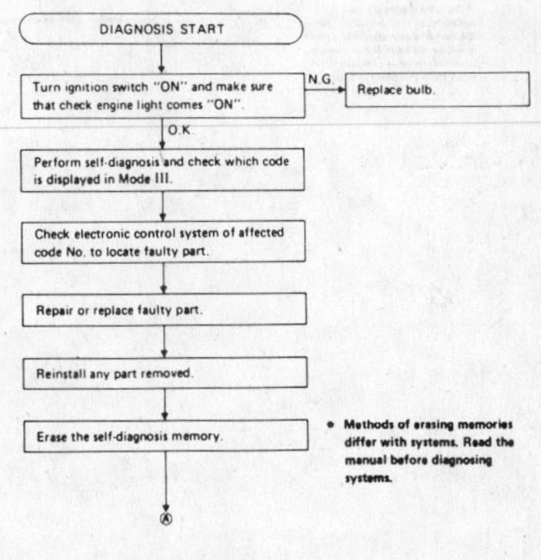
DIAGNOSIS START
Turn ignition switch "ON" and make sure that check engine light comes "ON".
N.G.
Replace bulb.
O.K.
Perform self-diagnosis and check which code is displayed in Mode III.
Check electronic control system of affected code No. to locate faulty part.
Repair or replace faulty part.
Reinstall any part removed.
Erase the self-diagnosis memory.
Methods of erasing memories differ with systems. Read the manual before diagnosing systems.
Ⓐ

ECCS SELF-DIAGNOSTIC MODES I AND II—1988 200SX WITH VG30E ENGINE

Description (Cont'd)

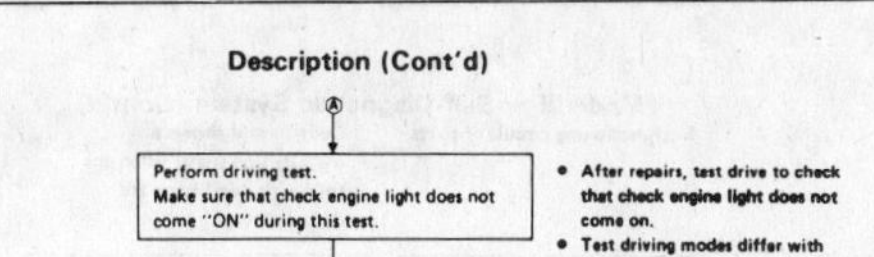
Ⓐ
Perform driving test.
Make sure that check engine light does not come "ON" during this test.
DIAGNOSIS END
After repairs, test drive to check that check engine light does not come on.
Test driving modes differ with systems. Read the manual before test driving.

Modes I & II —Mixture Ratio Feedback Control Monitors A & B

In these modes, the control unit provides the Air-fuel ratio monitor presentation and the Air-fuel ratio feedback coefficient monitor presentation.

Mode	LED	Engine stopped (Ignition switch "ON")	Engine running: Open loop condition	Engine running: Closed loop condition		
Mode I (Monitor A)	Green	ON	*Remains ON or OFF	Blinks		
	Red	ON	Except for California model • OFF	For California model • ON: when CHECK ENGINE LIGHT ITEMS are stored in the E.C.U. • OFF: except for the above condition		
Mode II (Monitor B)	Green	ON	*Remains ON or OFF	Blinks		
	Red	OFF	*Remains ON or OFF (synchronous with green LED)	Compensating mixture ratio		
				More than 5% rich	Between 5% lean and 5% rich	More
				OFF	Synchronized with green LED	Remains ON

*: Maintains conditions just before switching to open loop

ECCS SELF-DIAGNOSTIC MODE III—1988 200SX WITH VG30E ENGINE

Mode III — Self-Diagnostic System

The E.C.U. constantly monitors the function of these sensors and actuators, regardless of ignition key position. If a malfunction occurs, the information is stored in the E.C.U. and can be retrieved from the memory by turning on the diagnostic mode selector, located on the side of the E.C.U. When activated, the malfunction is indicated by flashing a red and a green L.E.D. (Light Emitting Diode), also located on the E.C.U. Since all the self-diagnostic results are stored in the E.C.U.'s memory even intermittent malfunctions can be diagnosed.

A malfunctioning part's group is indicated by the number of both the red and the green L.E.D.s flashing. First, the red L.E.D. flashes and the green flashes follow. The red L.E.D. refers to the number of tens while the green one refers to the number of units. For example, when the red L.E.D. flashes once and then the green one flashes twice, this means the number "12" showing the air flow meter signal is malfunctioning. In this way, all the problems are classified by the code numbers.

- **When engine fails to start, crank engine more than two seconds before starting self-diagnosis.**
- **Before starting self-diagnosis, do not erase stored memory. If doing so, self-diagnosis function for intermittent malfunctions would be lost.**

The stored memory would be lost if:

1. **Battery terminal is disconnected.**
2. **After selecting Mode III, Mode IV is selected.**

DISPLAY CODE TABLE

Code No.	Detected items
11	Crank angle sensor circuit
12	Air flow meter circuit
13	Cylinder head temperature sensor circuit
14	Vehicle speed sensor circuit
21	Ignition signal missing in primary coil
22	Fuel pump circuit
23	Idle switch circuit
31	E.C.U. (E.C.C.S. control unit)
33	Exhaust gas sensor circuit
42	Fuel temperature sensor circuit
55	No malfunction in the above circuit

ECCS SELF-DIAGNOSTIC MODE III (CONT.)—1988 200SX WITH VG30E ENGINE

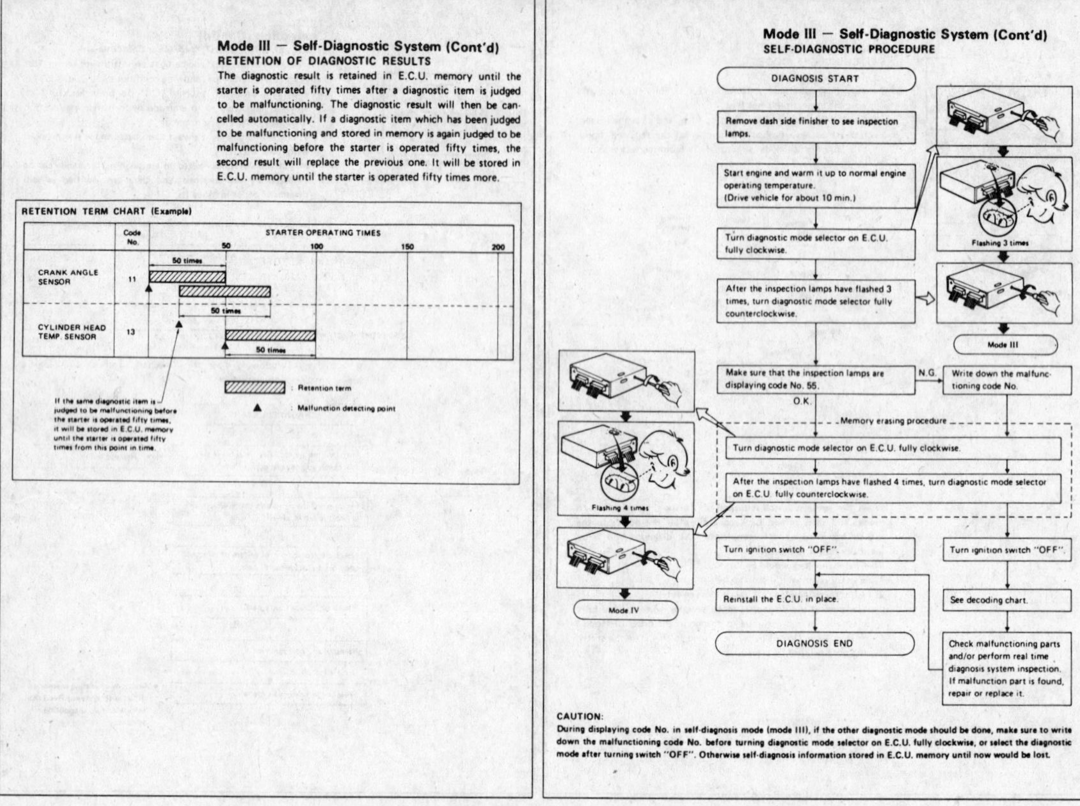

Mode III — Self-Diagnostic System (Cont'd)

RETENTION OF DIAGNOSTIC RESULTS

The diagnostic result is retained in E.C.U. memory until the starter is operated fifty times after a diagnostic item is judged to be malfunctioning. The diagnostic result will then be cancelled automatically. If a diagnostic item which has been judged to be malfunctioning and stored in memory is again judged to be malfunctioning before the starter is operated fifty times, the second result will replace the previous one. It will be stored in E.C.U. memory until the starter is operated fifty times more.

RETENTION TERM CHART (Example)
Code No.
STARTER OPERATING TIMES
50
100
150
200
50 times
CRANK ANGLE SENSOR
11
50 times
CYLINDER HEAD TEMP. SENSOR
13
50 times
If the same diagnostic item is judged to be malfunctioning before the starter is operated fifty times, it will be stored in E.C.U. memory until the starter is operated fifty times from this point in time.
: Retention term
: Malfunction detecting point

Mode III — Self-Diagnostic System (Cont'd)

SELF-DIAGNOSTIC PROCEDURE

DIAGNOSIS START
Remove dash side finisher to see inspection lamps.
Start engine and warm it up to normal engine operating temperature. (Drive vehicle for about 10 min.)
Turn diagnostic mode selector on E.C.U. fully clockwise.
Flashing 3 times
After the inspection lamps have flashed 3 times, turn diagnostic mode selector fully counterclockwise.
Mode III
Make sure that the inspection lamps are displaying code No. 55.
N.G.
Write down the malfunctioning code No.
O.K.
Memory erasing procedure
Turn diagnostic mode selector on E.C.U. fully clockwise.
Flashing 4 times
After the inspection lamps have flashed 4 times, turn diagnostic mode selector on E.C.U. fully counterclockwise.
Turn ignition switch "OFF".
Turn ignition switch "OFF".
Mode IV
Reinstall the E.C.U. in place.
See decoding chart.
DIAGNOSIS END
Check malfunctioning parts and/or perform real time diagnosis system inspection. If malfunction part is found, repair or replace it.

CAUTION:
During displaying code No. in self-diagnosis mode (mode III), if the other diagnostic mode should be done, make sure to write down the malfunctioning code No. before turning diagnostic mode selector on E.C.U. fully clockwise, or select the diagnostic mode after turning switch "OFF". Otherwise self-diagnosis information stored in E.C.U. memory until now would be lost.

ECCS SELF-DIAGNOSTIC MODE III DISPLAY CODES—1988 200SX WITH VG30E

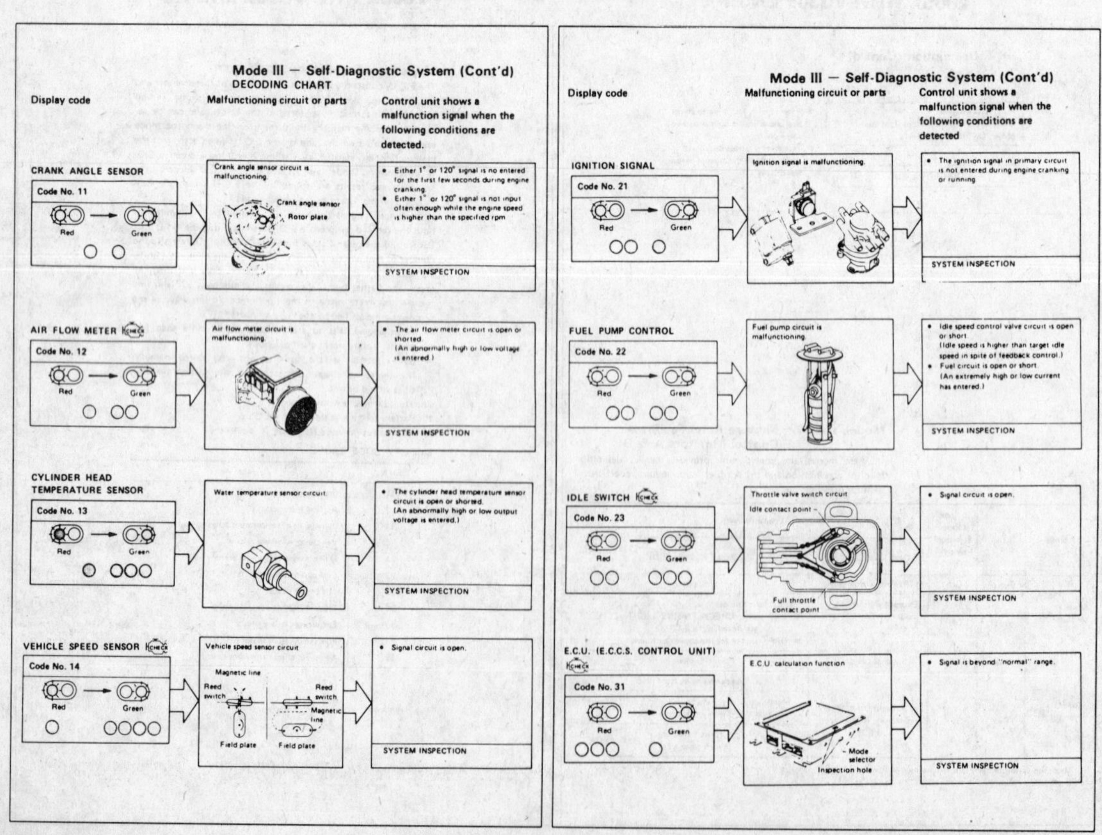

Mode III — Self-Diagnostic System (Cont'd)

DECODING CHART

Display code
Malfunctioning circuit or parts
Control unit shows a malfunction signal when the following conditions are detected.
CRANK ANGLE SENSOR
Code No. 11
Red
Green
Crank angle sensor circuit is malfunctioning.
Crank angle sensor
Rotor plate
• Either 1° or 120° signal is no entered for the first few seconds during engine cranking.
• Either 1° or 120° signal is not input often enough while the engine speed is higher than the specified rpm
SYSTEM INSPECTION
AIR FLOW METER
Code No. 12
Red
Green
Air flow meter circuit is malfunctioning.
• The air flow meter circuit is open or shorted. (An abnormally high or low voltage is entered.)
SYSTEM INSPECTION
CYLINDER HEAD TEMPERATURE SENSOR
Code No. 13
Red
Green
Water temperature sensor circuit.
• The cylinder head temperature sensor circuit is open or shorted. (An abnormally high or low output voltage is entered.)
SYSTEM INSPECTION
VEHICLE SPEED SENSOR
Code No. 14
Red
Green
Vehicle speed sensor circuit
Magnetic line
Reed switch
Reed switch
Magnetic line
Field plate
Field plate
• Signal circuit is open.
SYSTEM INSPECTION

Mode III — Self-Diagnostic System (Cont'd)

Display code
Malfunctioning circuit or parts
Control unit shows a malfunction signal when the following conditions are detected
IGNITION SIGNAL
Code No. 21
Red
Green
Ignition signal is malfunctioning.
• The ignition signal in primary circuit is not entered during engine cranking or running
SYSTEM INSPECTION
FUEL PUMP CONTROL
Code No. 22
Red
Green
Fuel pump circuit is malfunctioning.
• Idle speed control valve circuit is open or short. (Idle speed is higher than target idle speed in spite of feedback control.)
• Fuel circuit is open or short. (An extremely high or low current has entered.)
SYSTEM INSPECTION
IDLE SWITCH
Code No. 23
Red
Green
Throttle valve switch circuit
Idle contact point
Full throttle contact point
• Signal circuit is open.
SYSTEM INSPECTION
E.C.U. (E.C.C.S. CONTROL UNIT)
Code No. 31
Red
Green
E.C.U. calculation function
Mode selector
Inspection hole
• Signal is beyond "normal" range.
SYSTEM INSPECTION

ECCS SELF-DIAGNOSTIC MODE III DISPLAY CODES (CONT.)—1988 200SX WITH VG30E ENGINE

Mode III — Self-Diagnostic System (Cont'd)

Display code	Malfunctioning circuit or parts	Control unit shows a malfunction signal when the following conditions are detected.
EXHAUST GAS SENSOR (CHECK) Code No. 33 — Red ○○○ → Green ○○○	Exhaust gas sensor circuit (Louver; Sensor element (Titania); Lead terminals; Holder; Glass seal; Rubber seal)	• Signal circuit is open. SYSTEM INSPECTION
FUEL TEMPERATURE SENSOR Code No. 42 — Red ○○○○ → Green ○○	Fuel temperature sensor circuit is malfunctioning. (Fuel temperature sensor)	• Fuel temperature circuit is open or short. (An abnormally high or low voltage has entered.) SYSTEM INSPECTION
Code No. 55 — Red ○○○○○ → Green ○○○○○	E.C.C.S. normal operation.	

ECCS SELF-DIAGNOSTIC MODE IV 1988 200SX WITH VG30E ENGINE

Mode IV — Switches ON/OFF Diagnostic System

In switches ON/OFF diagnosis system, ON/OFF operation of the following switches can be detected continuously.

- Throttle valve switch
- Starter switch
- Vehicle speed sensor

(1) Throttle valve switch & Starter switch
The switches ON/OFF status at the point when mode IV is selected is stored in E.C.U. memory. When either switch is turned from "ON" to "OFF" or "OFF" to "ON", the red L.E.D. on E.C.U. alternately comes on and goes off each time switching is detected.

(2) Vehicle Speed Sensor
The switches ON/OFF status at the point when mode IV is selected is stored in E.C.U. memory. When vehicle speed is 20 km/h (12 MPH) or slower, the green L.E.D. on E.C.U. is off. When vehicle speed exceeds 20 km/h (12 MPH), the green L.E.D. on E.C.U. comes "ON".

ECCS SELF-DIAGNOSTIC MODE IV (CONT.) 200SX WITH VG30E ENGINE

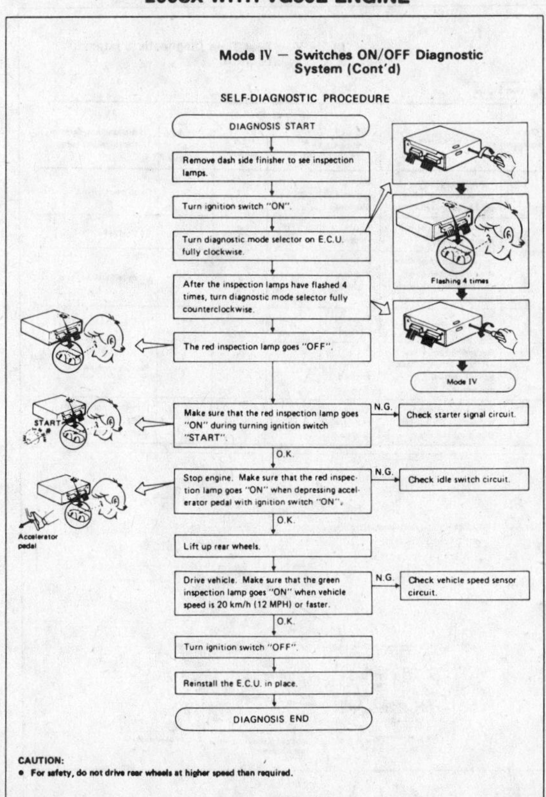
Mode IV — Switches ON/OFF Diagnostic System (Cont'd)
SELF-DIAGNOSTIC PROCEDURE
DIAGNOSIS START
Remove dash side finisher to see inspection lamps.
Turn ignition switch "ON".
Turn diagnostic mode selector on E.C.U. fully clockwise.
After the inspection lamps have flashed 4 times, turn diagnostic mode selector fully counterclockwise.
Flashing 4 times
Mode IV
The red inspection lamp goes "OFF".
Make sure that the red inspection lamp goes "ON" during turning ignition switch "START".
START
N.G.
Check starter signal circuit.
O.K.
Stop engine. Make sure that the red inspection lamp goes "ON" when depressing accelerator pedal with ignition switch "ON".
N.G.
Check idle switch circuit.
Accelerator pedal
O.K.
Lift up rear wheels.
Drive vehicle. Make sure that the green inspection lamp goes "ON" when vehicle speed is 20 km/h (12 MPH) or faster.
N.G.
Check vehicle speed sensor circuit.
O.K.
Turn ignition switch "OFF".
Reinstall the E.C.U. in place.
DIAGNOSIS END
CAUTION:
• For safety, do not drive rear wheels at higher speed than required.

ECCS SELF-DIAGNOSTIC MODE V 1988 200SX WITH VG30E ENGINE

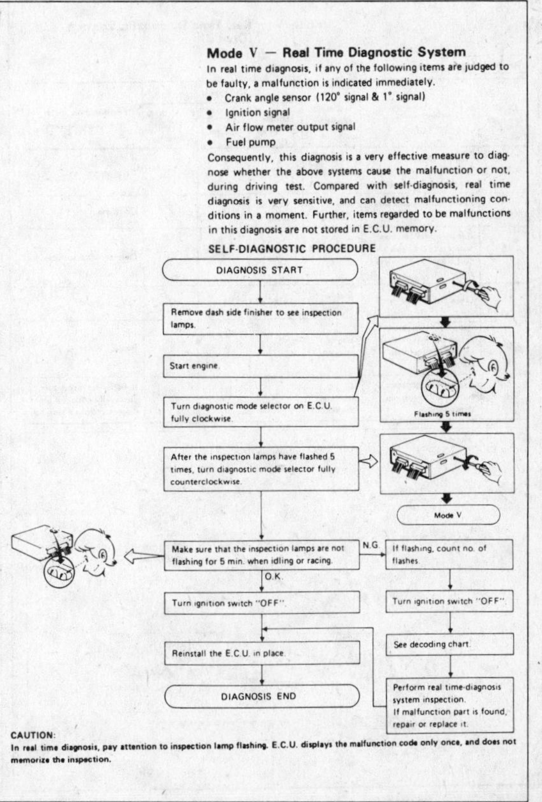
Mode V — Real Time Diagnostic System
In real time diagnosis, if any of the following items are judged to be faulty, a malfunction is indicated immediately.
• Crank angle sensor (120° signal & 1° signal)
• Ignition signal
• Air flow meter output signal
• Fuel pump
Consequently, this diagnosis is a very effective measure to diagnose whether the above systems cause the malfunction or not, during driving test. Compared with self-diagnosis, real time diagnosis is very sensitive, and can detect malfunctioning conditions in a moment. Further, items regarded to be malfunctions in this diagnosis are not stored in E.C.U. memory.
SELF-DIAGNOSTIC PROCEDURE
DIAGNOSIS START
Remove dash side finisher to see inspection lamps.
Start engine.
Turn diagnostic mode selector on E.C.U. fully clockwise.
After the inspection lamps have flashed 5 times, turn diagnostic mode selector fully counterclockwise.
Flashing 5 times
Mode V
Make sure that the inspection lamps are not flashing for 5 min. when idling or racing.
N.G.
If flashing, count no. of flashes.
O.K.
Turn ignition switch "OFF".
Turn ignition switch "OFF".
See decoding chart.
Reinstall the E.C.U. in place.
Perform real time-diagnosis system inspection. If malfunction part is found, repair or replace it.
DIAGNOSIS END
CAUTION:
In real time diagnosis, pay attention to inspection lamp flashing. E.C.U. displays the malfunction code only once, and does not memorize the inspection.

ECCS SELF-DIAGNOSTIC MODE V DECODING CHART—1988 200SX WITH VG30E

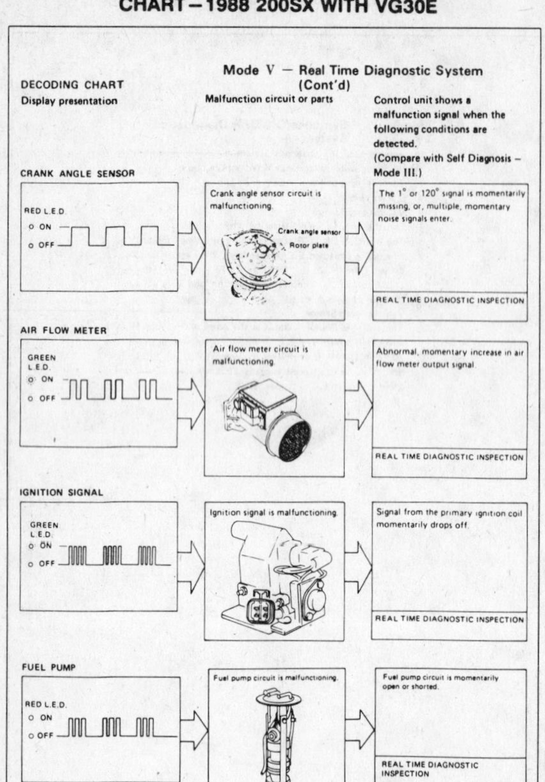
Mode V — Real Time Diagnostic System
(Cont'd)
DECODING CHART
Display presentation
Malfunction circuit or parts
Control unit shows a malfunction signal when the following conditions are detected.
(Compare with Self Diagnosis – Mode III.)
CRANK ANGLE SENSOR
RED L.E.D.
ON
OFF
Crank angle sensor circuit is malfunctioning.
Crank angle sensor
Rotor plate
The 1° or 120° signal is momentarily missing, or, multiple, momentary noise signals enter.
REAL TIME DIAGNOSTIC INSPECTION
AIR FLOW METER
GREEN L.E.D.
ON
OFF
Air flow meter circuit is malfunctioning.
Abnormal, momentary increase in air flow meter output signal.
REAL TIME DIAGNOSTIC INSPECTION
IGNITION SIGNAL
GREEN L.E.D.
ON
OFF
Ignition signal is malfunctioning.
Signal from the primary ignition coil momentarily drops off.
REAL TIME DIAGNOSTIC INSPECTION
FUEL PUMP
RED L.E.D.
ON
OFF
Fuel pump circuit is malfunctioning.
Fuel pump circuit is momentarily open or shorted.
REAL TIME DIAGNOSTIC INSPECTION

ECCS SELF-DIAGNOSTIC MODE V INSPECTION 1988 200SX WITH VG30E ENGINE

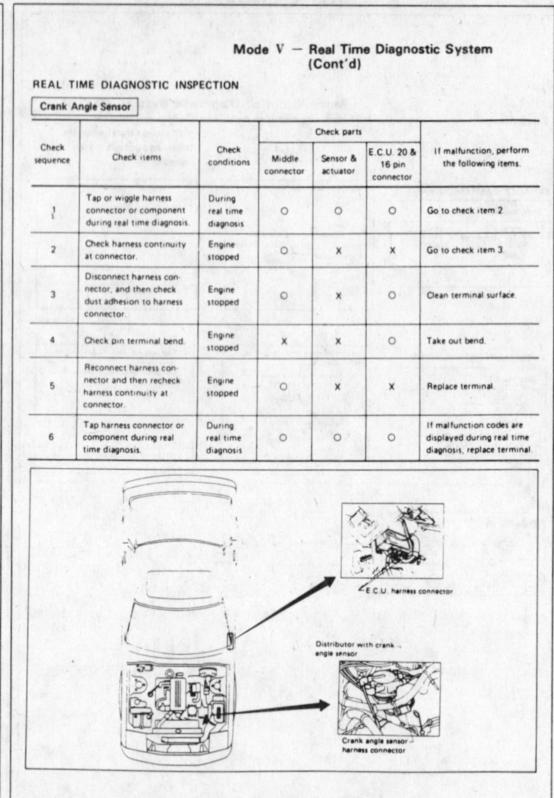

Mode V — Real Time Diagnostic System (Cont'd)

REAL TIME DIAGNOSTIC INSPECTION

Crank Angle Sensor

Check sequence	Check items	Check conditions	Check parts			If malfunction, perform the following items.
			Middle connector	Sensor & actuator	E.C.U. 20 & 16 pin connector	
1	Tap or wiggle harness connector or component during real time diagnosis.	During real time diagnosis	O	O	O	Go to check item 2.
2	Check harness continuity at connector.	Engine stopped	O	X	X	Go to check item 3.
3	Disconnect harness connector, and then check dust adhesion to harness connector.	Engine stopped	O	X	O	Clean terminal surface.
4	Check pin terminal bend.	Engine stopped	X	X	O	Take out bend.
5	Reconnect harness connector and then recheck harness continuity at connector.	Engine stopped	O	X	X	Replace terminal.
6	Tap harness connector or component during real time diagnosis.	During real time diagnosis	O	O	O	If malfunction codes are displayed during real time diagnosis, replace terminal.

E.C.U. harness connector
Distributor with crank angle sensor
Crank angle sensor harness connector

ECCS SELF-DIAGNOSTIC MODE V INPSECTION (CONT.)—1988 200SX WITH VG30E

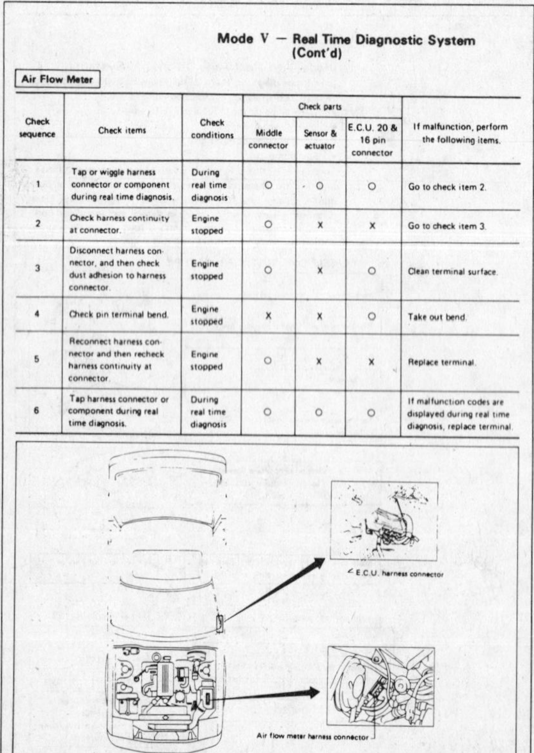

Mode V — Real Time Diagnostic System (Cont'd)

Air Flow Meter

Check sequence	Check items	Check conditions	Check parts			If malfunction, perform the following items.
			Middle connector	Sensor & actuator	E.C.U. 20 & 16 pin connector	
1	Tap or wiggle harness connector or component during real time diagnosis.	During real time diagnosis	O	O	O	Go to check item 2.
2	Check harness continuity at connector.	Engine stopped	O	X	X	Go to check item 3.
3	Disconnect harness connector, and then check dust adhesion to harness connector.	Engine stopped	O	X	O	Clean terminal surface.
4	Check pin terminal bend.	Engine stopped	X	X	O	Take out bend.
5	Reconnect harness connector and then recheck harness continuity at connector.	Engine stopped	O	X	X	Replace terminal.
6	Tap harness connector or component during real time diagnosis.	During real time diagnosis	O	O	O	If malfunction codes are displayed during real time diagnosis, replace terminal.

E.C.U. harness connector
Air flow meter harness connector

Mode V — Real Time Diagnostic System (Cont'd)

Ignition Signal

Check sequence	Check items	Check conditions	Check parts			If malfunction, perform the following items.
			Middle connector	Sensor & actuator	E.C.U. 20 & 16 pin connector	
1	Tap or wiggle harness connector or component during real time diagnosis.	During real time diagnosis	O	O	O	Go to check item 2.
2	Check harness continuity at connector.	Engine stopped	O	X	X	Go to check item 3.
3	Disconnect harness connector, and then check dust adhesion to harness connector.	Engine stopped	O	X	O	Clean terminal surface.
4	Check pin terminal bend.	Engine stopped	X	X	O	Take out bend.
5	Reconnect harness connector and then recheck harness continuity at connector.	Engine stopped	O	X	X	Replace terminal.
6	Tap harness connector or component during real time diagnosis.	During real time diagnosis	O	O	O	If malfunction codes are displayed during real time diagnosis, replace terminal.

E.C.U. harness connector

Ignition coil

Ignition coil & power transistor harness connector

Power transistor

ECCS SELF-DIAGNOSTIC MODE V INPSECTION (CONT.)—1988 200SX WITH VG30E

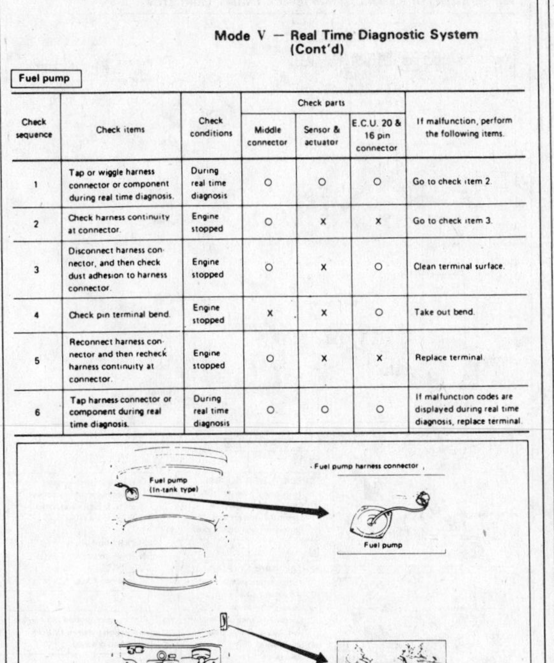

Mode V — Real Time Diagnostic System (Cont'd)

Fuel pump

Check sequence	Check items	Check conditions	Check parts: Middle connector	Check parts: Sensor & actuator	Check parts: E.C.U. 20 & 16 pin connector	If malfunction, perform the following items.
1	Tap or wiggle harness connector or component during real time diagnosis.	During real time diagnosis	O	O	O	Go to check item 2.
2	Check harness continuity at connector.	Engine stopped	O	X	X	Go to check item 3.
3	Disconnect harness connector, and then check dust adhesion to harness connector.	Engine stopped	O	X	O	Clean terminal surface.
4	Check pin terminal bend.	Engine stopped	X	X	O	Take out bend.
5	Reconnect harness connector and then recheck harness continuity at connector.	Engine stopped	O	X	X	Replace terminal.
6	Tap harness connector or component during real time diagnosis.	During real time diagnosis	O	O	O	If malfunction codes are displayed during real time diagnosis, replace terminal.

Fuel pump (In-tank type)
Fuel pump harness connector
Fuel pump
E.C.U. harness connector

ECCS SYSTEM INSPECTION CAUTION 1988 200SX WITH VG30E ENGINE

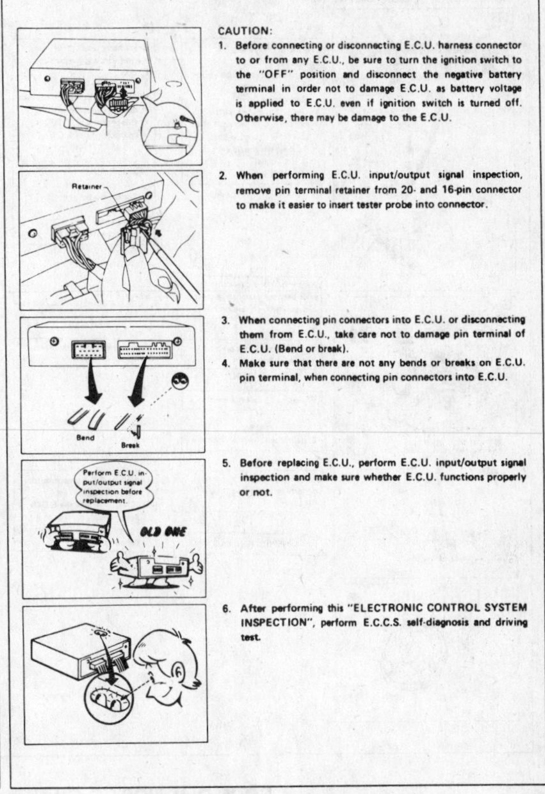

CAUTION:

1. Before connecting or disconnecting E.C.U. harness connector to or from any E.C.U., be sure to turn the ignition switch to the "OFF" position and disconnect the negative battery terminal in order not to damage E.C.U. as battery voltage is applied to E.C.U. even if ignition switch is turned off. Otherwise, there may be damage to the E.C.U.
2. When performing E.C.U. input/output signal inspection, remove pin terminal retainer from 20- and 16-pin connector to make it easier to insert tester probe into connector.
3. When connecting pin connectors into E.C.U. or disconnecting them from E.C.U., take care not to damage pin terminal of E.C.U. (Bend or break).
4. Make sure that there are not any bends or breaks on E.C.U. pin terminal, when connecting pin connectors into E.C.U.
5. Before replacing E.C.U., perform E.C.U. input/output signal inspection and make sure whether E.C.U. functions properly or not.
6. After performing this "ELECTRONIC CONTROL SYSTEM INSPECTION", perform E.C.C.S. self-diagnosis and driving test.

Retainer
Bend
Break
Perform E.C.U. input/output signal inspection before replacement.
OLD ONE

ECCS SYSTEM INSPECTION CAUTION 1988 200SX WITH VG30E ENGINE

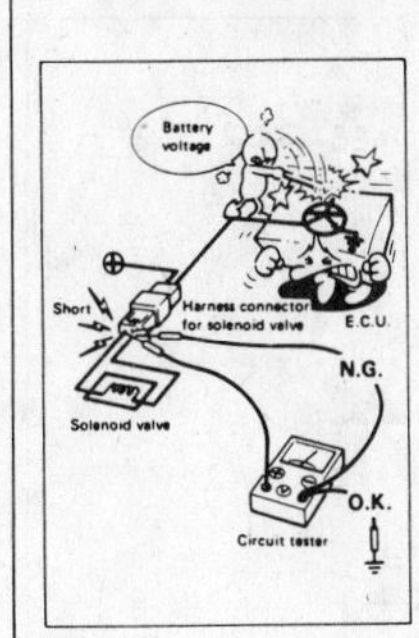

Battery voltage
Short
Harness connector for solenoid valve
E.C.U.
N.G.
Solenoid valve
O.K.
Circuit tester

7. When measuring supply voltage of E.C.U. controlled components with a circuit tester, separate one tester probe from the other.
   If the two tester probes accidentally make contact with each other during measurement, the circuit will be shorted, resulting in damage to the power transistor of the control unit.
8. Keys to symbols
   - DISCONNECT : Check after disconnecting the connector to be measured.
   - CONNECT : Check after connecting the connector to be measured.
9. When measuring voltage or resistance at connector with tester probes, there are two methods of measurement; one is done from terminal side and the other from harness side. Before measuring, confirm symbol mark again.
   - HS : Inspection should be done from harness side.
   - TS : Inspection should be done from terminal side.
10. As for continuity check of joint connector.

ECCS DIAGNOSTIC CHART 1988 200SX WITH VG30E ENGINE

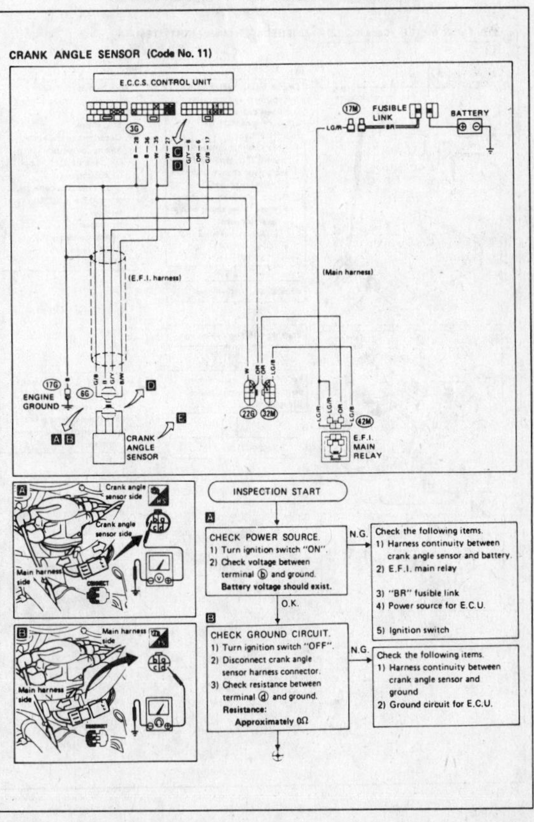

CRANK ANGLE SENSOR (Code No. 11)
E.C.C.S. CONTROL UNIT
FUSIBLE LINK
BATTERY
(E.F.I. harness)
(Main harness)
ENGINE GROUND
CRANK ANGLE SENSOR
E.F.I. MAIN RELAY
Crank angle sensor side
Main harness side
INSPECTION START
CHECK POWER SOURCE.
1) Turn ignition switch "ON".
2) Check voltage between terminal ⓑ and ground.
Battery voltage should exist.
N.G.
Check the following items.
1) Harness continuity between crank angle sensor and battery.
2) E.F.I. main relay
3) "BR" fusible link
4) Power source for E.C.U.
5) Ignition switch
O.K.
CHECK GROUND CIRCUIT.
1) Turn ignition switch "OFF".
2) Disconnect crank angle sensor harness connector.
3) Check resistance between terminal ⓓ and ground.
Resistance:
Approximately 0Ω
N.G.
Check the following items.
1) Harness continuity between crank angle sensor and ground
2) Ground circuit for E.C.U.

ECCS DIAGNOSTIC CHART—1988 200SX WITH VG30E ENGINE

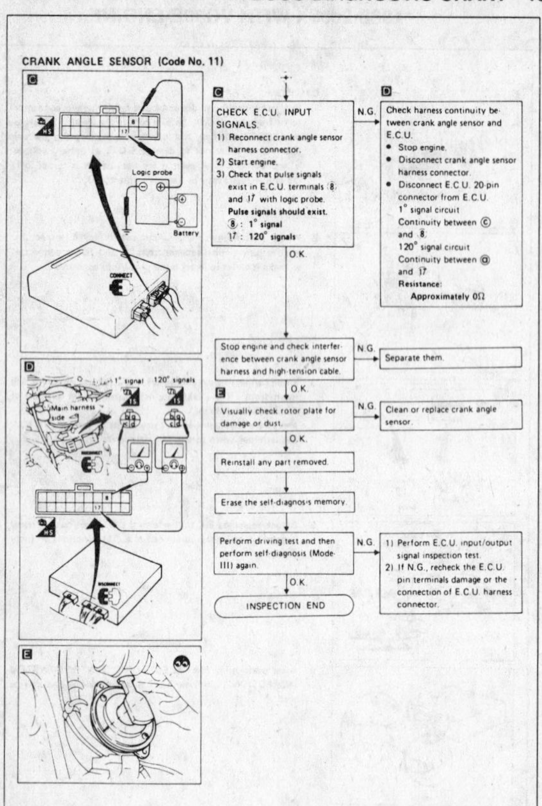
CRANK ANGLE SENSOR (Code No. 11)
C
Logic probe
Battery
CONNECT
C
CHECK E.C.U. INPUT SIGNALS.
1) Reconnect crank angle sensor harness connector.
2) Start engine.
3) Check that pulse signals exist in E.C.U. terminals 8 and 17 with logic probe.
Pulse signals should exist.
8 : 1° signal
17 : 120° signals
N.G.
D
Check harness continuity between crank angle sensor and E.C.U.
• Stop engine.
• Disconnect crank angle sensor harness connector.
• Disconnect E.C.U. 20-pin connector from E.C.U.
1° signal circuit
Continuity between © and 8
120° signal circuit
Continuity between ⓓ and 17
Resistance:
Approximately 0Ω
O.K.
Stop engine and check interference between crank angle sensor harness and high-tension cable.
N.G.
Separate them.
O.K.
E
Visually check rotor plate for damage or dust.
N.G.
Clean or replace crank angle sensor.
O.K.
Reinstall any part removed.
Erase the self-diagnosis memory.
Perform driving test and then perform self-diagnosis (Mode III) again.
N.G.
1) Perform E.C.U. input/output signal inspection test.
2) If N.G., recheck the E.C.U. pin terminals damage or the connection of E.C.U. harness connector.
O.K.
INSPECTION END
D
1° signal
120° signals
Main harness side
E

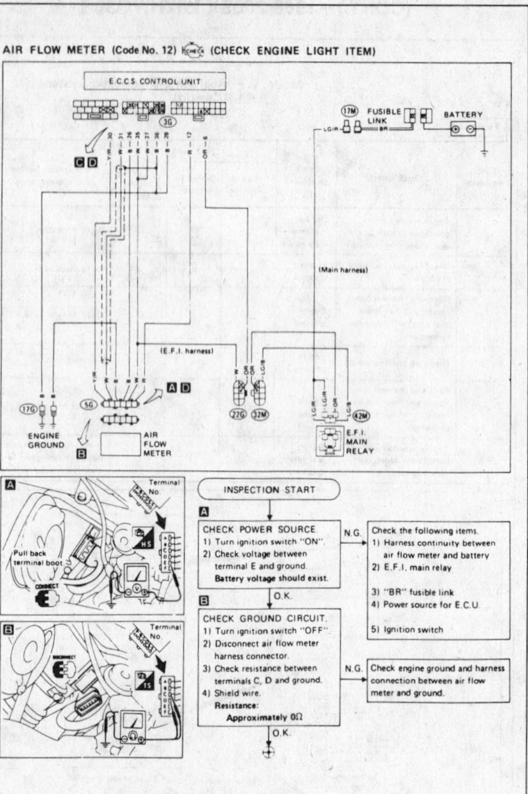
AIR FLOW METER (Code No. 12) (CHECK ENGINE LIGHT ITEM)
E.C.C.S. CONTROL UNIT
3G
17M
FUSIBLE LINK
BATTERY
C D
(Main harness)
(E.F.I. harness)
A D
17G
5G
22G
32M
42M
ENGINE GROUND
B
AIR FLOW METER
E.F.I. MAIN RELAY
A
Terminal No.
Pull back terminal boot
B
Terminal No.
INSPECTION START
A
CHECK POWER SOURCE.
1) Turn ignition switch "ON".
2) Check voltage between terminal E and ground.
Battery voltage should exist.
N.G.
Check the following items.
1) Harness continuity between air flow meter and battery
2) E.F.I. main relay
3) "BR" fusible link
4) Power source for E.C.U.
5) Ignition switch
O.K.
B
CHECK GROUND CIRCUIT.
1) Turn ignition switch "OFF".
2) Disconnect air flow meter harness connector.
3) Check resistance between terminals C, D and ground.
4) Shield wire.
Resistance:
Approximately 0Ω
N.G.
Check engine ground and harness connection between air flow meter and ground.
O.K.

ECCS DIAGNOSTIC CHART—1988 200SX WITH VG30E ENGINE

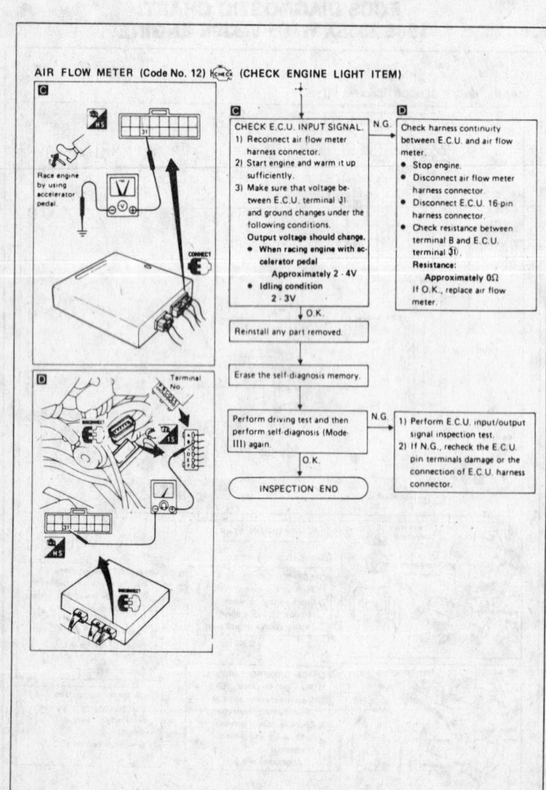
AIR FLOW METER (Code No. 12) (CHECK ENGINE LIGHT ITEM)
C
Race engine by using accelerator pedal
CONNECT
C
CHECK E.C.U. INPUT SIGNAL.
1) Reconnect air flow meter harness connector.
2) Start engine and warm it up sufficiently.
3) Make sure that voltage between E.C.U. terminal 31 and ground changes under the following conditions.
Output voltage should change.
• When racing engine with accelerator pedal
Approximately 2 - 4V
• Idling condition
2 - 3V
N.G.
D
Check harness continuity between E.C.U. and air flow meter.
• Stop engine.
• Disconnect air flow meter harness connector.
• Disconnect E.C.U. 16-pin harness connector.
• Check resistance between terminal B and E.C.U. terminal 31.
Resistance:
Approximately 0Ω
If O.K., replace air flow meter.
O.K.
Reinstall any part removed.
Erase the self-diagnosis memory.
Perform driving test and then perform self-diagnosis (Mode III) again.
N.G.
1) Perform E.C.U. input/output signal inspection test.
2) If N.G., recheck the E.C.U. pin terminals damage or the connection of E.C.U. harness connector.
O.K.
INSPECTION END
D
Terminal No.
31

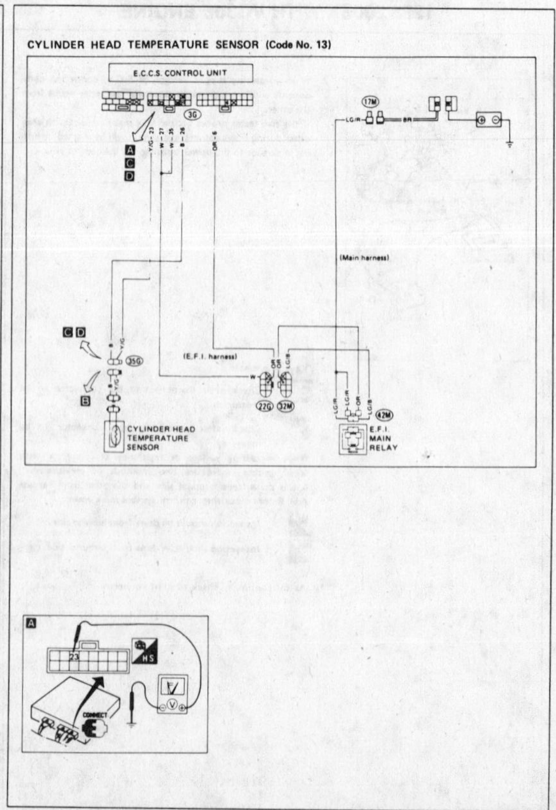
CYLINDER HEAD TEMPERATURE SENSOR (Code No. 13)
E.C.C.S. CONTROL UNIT
3G
17M
A
C
D
(Main harness)
C D
35G
B
(E.F.I. harness)
22G
32M
42M
CYLINDER HEAD TEMPERATURE SENSOR
E.F.I. MAIN RELAY
A
23
CONNECT

ECCS DIAGNOSTIC CHART—1988 200SX WITH VG30E ENGINE

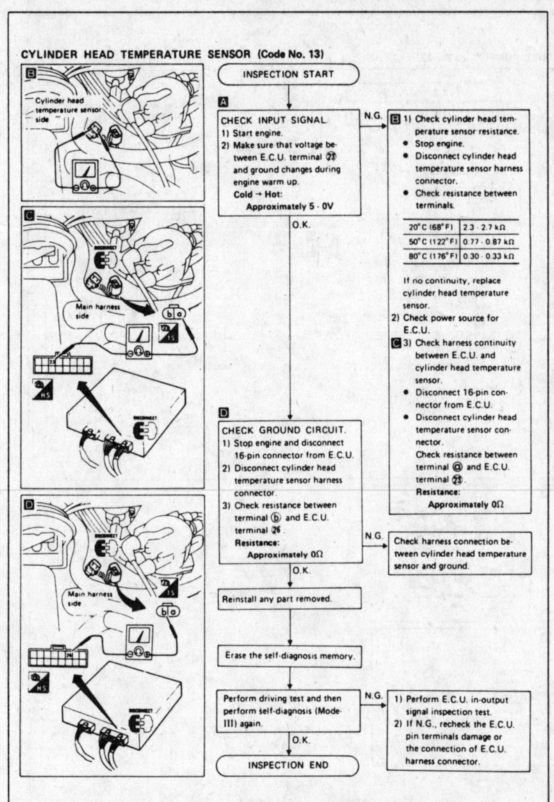
CYLINDER HEAD TEMPERATURE SENSOR (Code No. 13)
INSPECTION START
Cylinder head temperature sensor side
Main harness side
A CHECK INPUT SIGNAL.
1) Start engine.
2) Make sure that voltage between E.C.U. terminal 23 and ground changes during engine warm up.
Cold → Hot:
Approximately 5 - 0V
N.G.
O.K.
B 1) Check cylinder head temperature sensor resistance.
• Stop engine.
• Disconnect cylinder head temperature sensor harness connector.
• Check resistance between terminals.
20°C (68°F) 2.3 - 2.7 kΩ
50°C (122°F) 0.77 - 0.87 kΩ
80°C (176°F) 0.30 - 0.33 kΩ
If no continuity, replace cylinder head temperature sensor.
2) Check power source for E.C.U.
C 3) Check harness continuity between E.C.U. and cylinder head temperature sensor.
• Disconnect 16-pin connector from E.C.U.
• Disconnect cylinder head temperature sensor connector.
Check resistance between terminal ⓐ and E.C.U. terminal 23.
Resistance:
Approximately 0Ω
D CHECK GROUND CIRCUIT.
1) Stop engine and disconnect 16-pin connector from E.C.U.
2) Disconnect cylinder head temperature sensor harness connector.
3) Check resistance between terminal ⓑ and E.C.U. terminal 26.
Resistance:
Approximately 0Ω
Check harness connection between cylinder head temperature sensor and ground.
Reinstall any part removed.
Erase the self-diagnosis memory.
Perform driving test and then perform self-diagnosis (Mode III) again.
1) Perform E.C.U. in-output signal inspection test.
2) If N.G., recheck the E.C.U. pin terminals damage or the connection of E.C.U. harness connector.
INSPECTION END

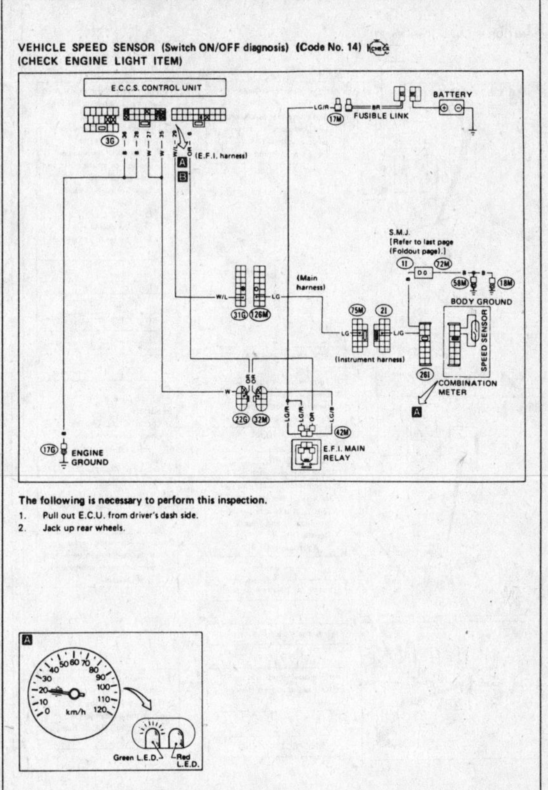
VEHICLE SPEED SENSOR (Switch ON/OFF diagnosis) (Code No. 14)
(CHECK ENGINE LIGHT ITEM)
E.C.C.S. CONTROL UNIT
BATTERY
FUSIBLE LINK
(E.F.I. harness)
S.M.J.
[Refer to last page (Foldout page).]
(Main harness)
BODY GROUND
(Instrument harness)
SPEED SENSOR
COMBINATION METER
ENGINE GROUND
E.F.I. MAIN RELAY
The following is necessary to perform this inspection.
1. Pull out E.C.U. from driver's dash side.
2. Jack up rear wheels.
km/h
Green L.E.D.
Red L.E.D.

ECCS DIAGNOSTIC CHART—1988 200SX WITH VG30E ENGINE

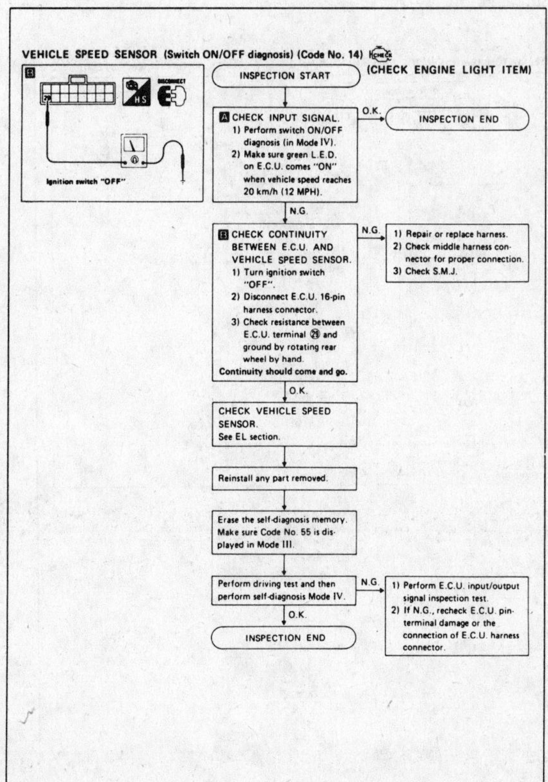
VEHICLE SPEED SENSOR (Switch ON/OFF diagnosis) (Code No. 14)
(CHECK ENGINE LIGHT ITEM)
Ignition switch "OFF"
INSPECTION START
A CHECK INPUT SIGNAL.
1) Perform switch ON/OFF diagnosis (in Mode IV).
2) Make sure green L.E.D. on E.C.U. comes "ON" when vehicle speed reaches 20 km/h (12 MPH).
O.K.
INSPECTION END
N.G.
B CHECK CONTINUITY BETWEEN E.C.U. AND VEHICLE SPEED SENSOR.
1) Turn ignition switch "OFF".
2) Disconnect E.C.U. 16-pin harness connector.
3) Check resistance between E.C.U. terminal 29 and ground by rotating rear wheel by hand.
Continuity should come and go.
1) Repair or replace harness.
2) Check middle harness connector for proper connection.
3) Check S.M.J.
CHECK VEHICLE SPEED SENSOR.
See EL section.
Reinstall any part removed.
Erase the self-diagnosis memory.
Make sure Code No. 55 is displayed in Mode III.
Perform driving test and then perform self-diagnosis Mode IV.
1) Perform E.C.U. input/output signal inspection test.
2) If N.G., recheck E.C.U. pin-terminal damage or the connection of E.C.U. harness connector.
INSPECTION END

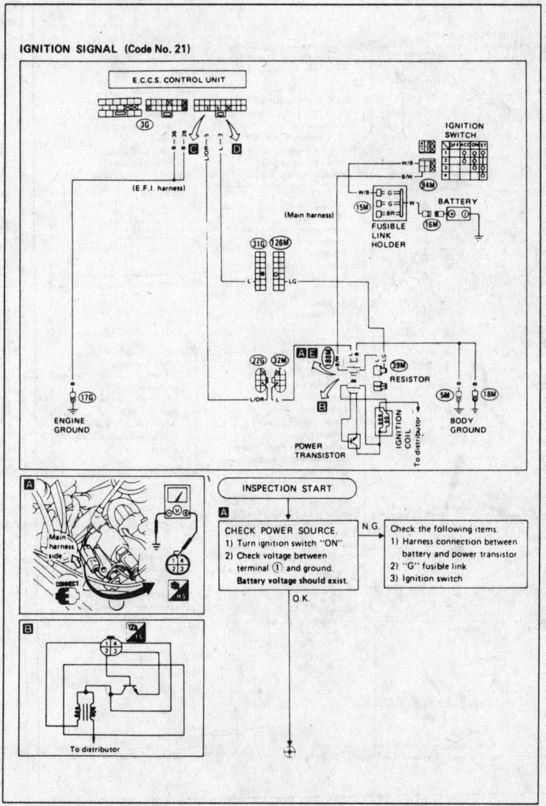
IGNITION SIGNAL (Code No. 21)
E.C.C.S. CONTROL UNIT
IGNITION SWITCH
BATTERY
FUSIBLE LINK HOLDER
(E.F.I. harness)
(Main harness)
RESISTOR
ENGINE GROUND
BODY GROUND
POWER TRANSISTOR
IGNITION COIL
To distributor
INSPECTION START
A CHECK POWER SOURCE.
1) Turn ignition switch "ON".
2) Check voltage between terminal ① and ground.
Battery voltage should exist.
N.G.
Check the following items.
1) Harness connection between battery and power transistor
2) "G" fusible link
3) Ignition switch
O.K.
Main harness side
To distributor

ECCS DIAGNOSTIC CHART—1988 200SX WITH VG30E ENGINE

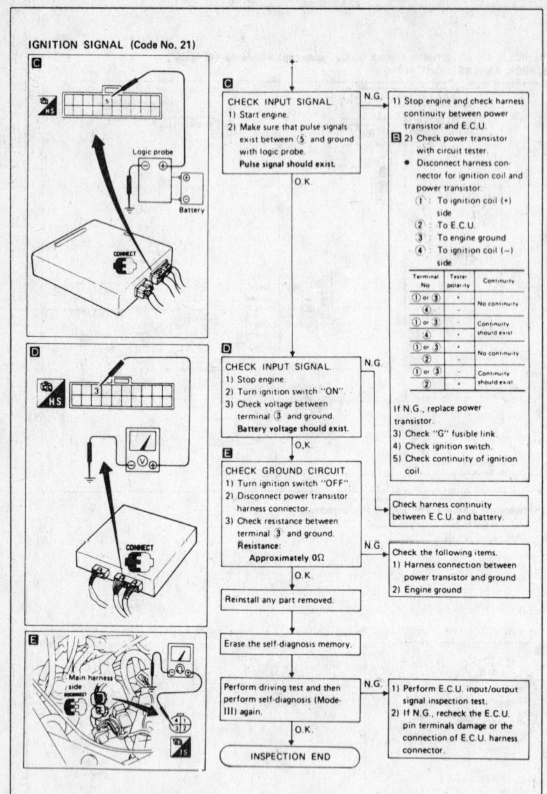
IGNITION SIGNAL (Code No. 21)
Logic probe
Battery
CONNECT
CHECK INPUT SIGNAL.
1) Start engine.
2) Make sure that pulse signals exist between ⑤ and ground with logic probe.
Pulse signal should exist.
N.G.
1) Stop engine and check harness continuity between power transistor and E.C.U.
2) Check power transistor with circuit tester.
● Disconnect harness connector for ignition coil and power transistor.
①: To ignition coil (+) side
②: To E.C.U.
③: To engine ground
④: To ignition coil (−) side
If N.G., replace power transistor.
3) Check "G" fusible link.
4) Check ignition switch.
5) Check continuity of ignition coil.
O.K.
CHECK INPUT SIGNAL.
1) Stop engine.
2) Turn ignition switch "ON".
3) Check voltage between terminal ③ and ground.
Battery voltage should exist.
Check harness continuity between E.C.U. and battery.
CHECK GROUND CIRCUIT.
1) Turn ignition switch "OFF".
2) Disconnect power transistor harness connector.
3) Check resistance between terminal ③ and ground.
Resistance: Approximately 0Ω
Check the following items.
1) Harness connection between power transistor and ground
2) Engine ground
Reinstall any part removed.
Erase the self-diagnosis memory.
Perform driving test and then perform self-diagnosis (Mode III) again.
1) Perform E.C.U. input/output signal inspection test.
2) If N.G., recheck the E.C.U. pin terminals damage or the connection of E.C.U. harness connector.
Main harness side
INSPECTION END

Terminal No.	Tester polarity	Continuity
① or ③	+	No continuity
④	−	
① or ③	−	Continuity should exist
④	+	
① or ③	+	No continuity
②	−	
① or ③	−	Continuity should exist
②	+	

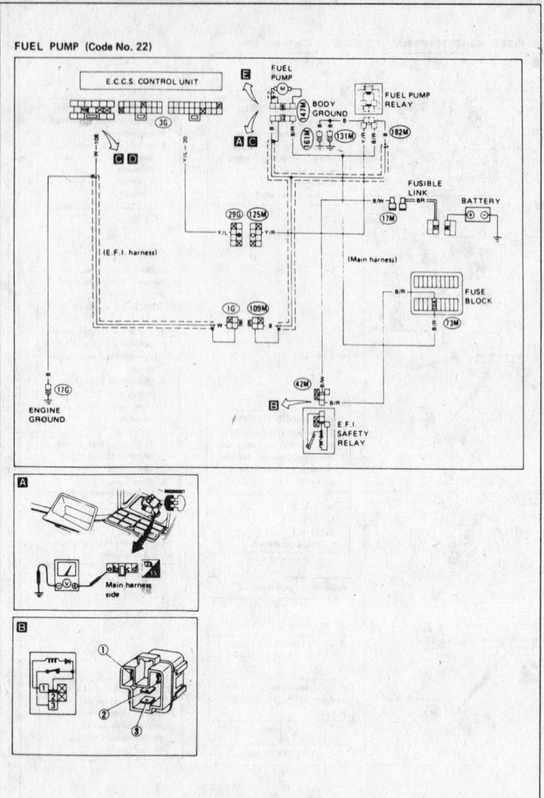
FUEL PUMP (Code No. 22)
E.C.C.S. CONTROL UNIT
FUEL PUMP
BODY GROUND
FUEL PUMP RELAY
FUSIBLE LINK
BATTERY
(E.F.I. harness)
(Main harness)
FUSE BLOCK
ENGINE GROUND
E.F.I. SAFETY RELAY
Main harness side

ECCS DIAGNOSTIC CHART—1988 200SX WITH VG30E ENGINE

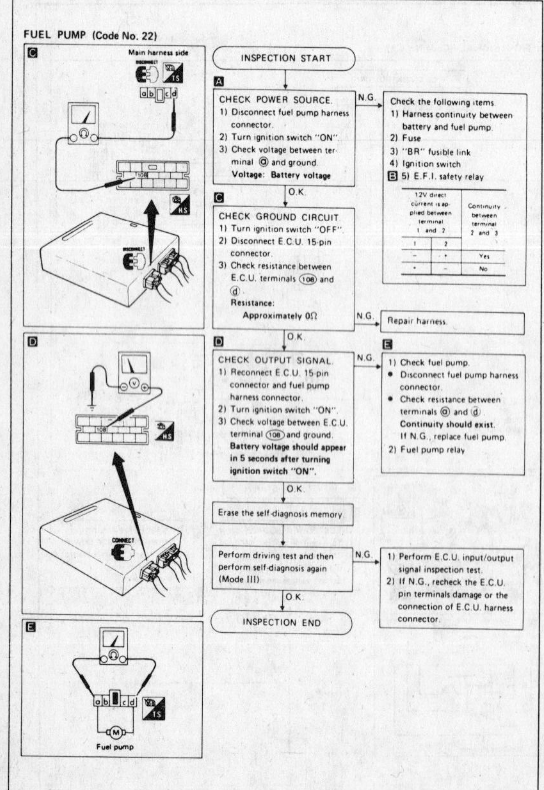
FUEL PUMP (Code No. 22)
Main harness side
INSPECTION START
CHECK POWER SOURCE.
1) Disconnect fuel pump harness connector.
2) Turn ignition switch "ON".
3) Check voltage between terminal ⓐ and ground.
Voltage: Battery voltage
N.G.
Check the following items.
1) Harness continuity between battery and fuel pump.
2) Fuse
3) "BR" fusible link
4) Ignition switch
5) E.F.I. safety relay
O.K.
CHECK GROUND CIRCUIT.
1) Turn ignition switch "OFF".
2) Disconnect E.C.U. 15-pin connector.
3) Check resistance between E.C.U. terminals (108) and ⓓ.
Resistance: Approximately 0Ω
Repair harness.
CHECK OUTPUT SIGNAL.
1) Reconnect E.C.U. 15-pin connector and fuel pump harness connector.
2) Turn ignition switch "ON".
3) Check voltage between E.C.U. terminal (108) and ground.
Battery voltage should appear in 5 seconds after turning ignition switch "ON".
1) Check fuel pump.
● Disconnect fuel pump harness connector.
● Check resistance between terminals ⓐ and ⓓ.
Continuity should exist.
If N.G., replace fuel pump.
2) Fuel pump relay
Erase the self-diagnosis memory.
Perform driving test and then perform self-diagnosis again (Mode III).
1) Perform E.C.U. input/output signal inspection test.
2) If N.G., recheck the E.C.U. pin terminals damage or the connection of E.C.U. harness connector.
INSPECTION END
CONNECT
Fuel pump

12V direct current is applied between terminal 1 and 2		Continuity between terminal 2 and 3
1	2	
−	+	Yes
+	−	No

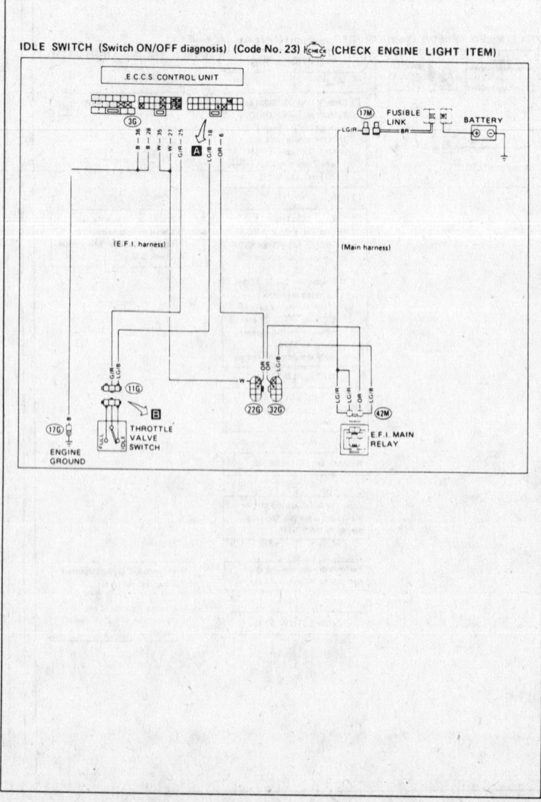
IDLE SWITCH (Switch ON/OFF diagnosis) (Code No. 23) (CHECK ENGINE LIGHT ITEM)
E.C.C.S. CONTROL UNIT
FUSIBLE LINK
BATTERY
(E.F.I. harness)
(Main harness)
ENGINE GROUND
THROTTLE VALVE SWITCH
E.F.I. MAIN RELAY

ECCS DIAGNOSTIC CHART—1988 200SX WITH VG30E ENGINE

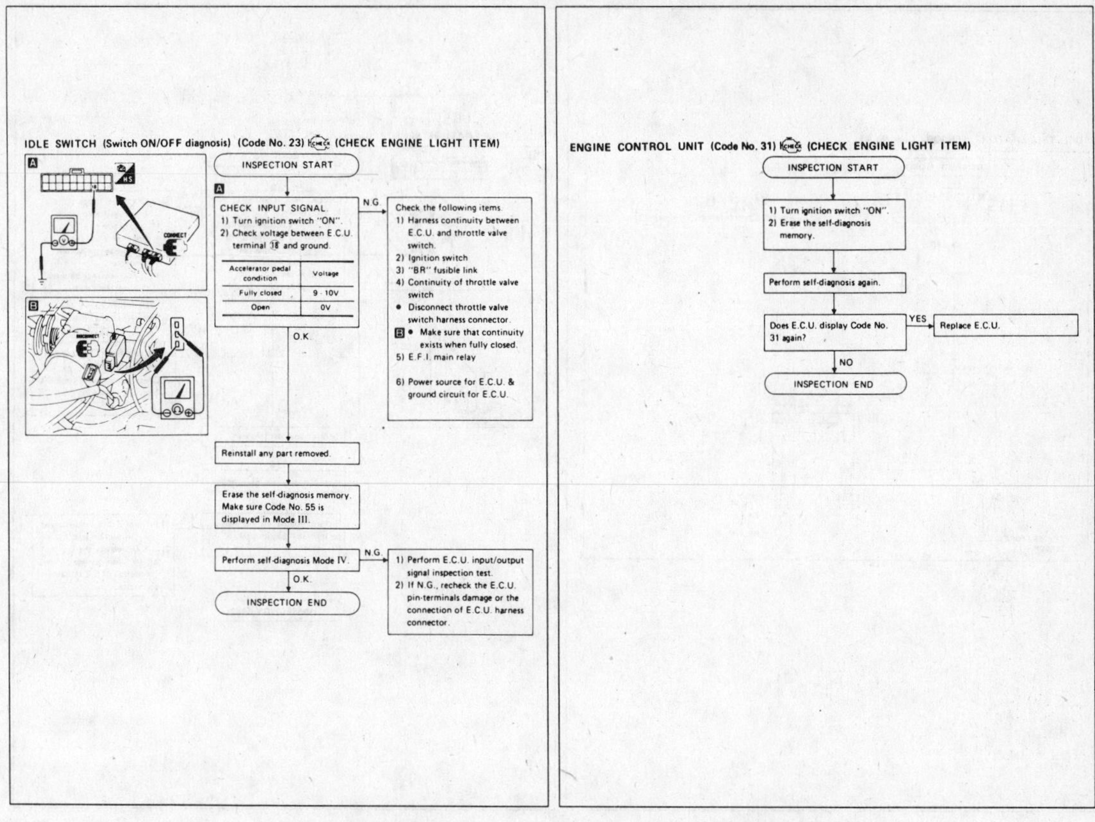
IDLE SWITCH (Switch ON/OFF diagnosis) (Code No. 23) (CHECK ENGINE LIGHT ITEM)
INSPECTION START
CHECK INPUT SIGNAL.
1) Turn ignition switch "ON".
2) Check voltage between E.C.U. terminal 18 and ground.
Accelerator pedal condition
Voltage
Fully closed
9 - 10V
Open
0V
N.G.
Check the following items.
1) Harness continuity between E.C.U. and throttle valve switch.
2) Ignition switch
3) "BR" fusible link
4) Continuity of throttle valve switch
• Disconnect throttle valve switch harness connector.
• Make sure that continuity exists when fully closed.
5) E.F.I. main relay
6) Power source for E.C.U. & ground circuit for E.C.U.
O.K.
Reinstall any part removed.
Erase the self-diagnosis memory. Make sure Code No. 55 is displayed in Mode III.
Perform self-diagnosis Mode IV.
1) Perform E.C.U. input/output signal inspection test.
2) If N.G., recheck the E.C.U. pin-terminals damage or the connection of E.C.U. harness connector.
INSPECTION END
ENGINE CONTROL UNIT (Code No. 31) (CHECK ENGINE LIGHT ITEM)
INSPECTION START
1) Turn ignition switch "ON".
2) Erase the self-diagnosis memory.
Perform self-diagnosis again.
Does E.C.U. display Code No. 31 again?
YES
Replace E.C.U.
NO
INSPECTION END

ECCS DIAGNOSTIC CHART—1988 200SX WITH VG30E ENGINE

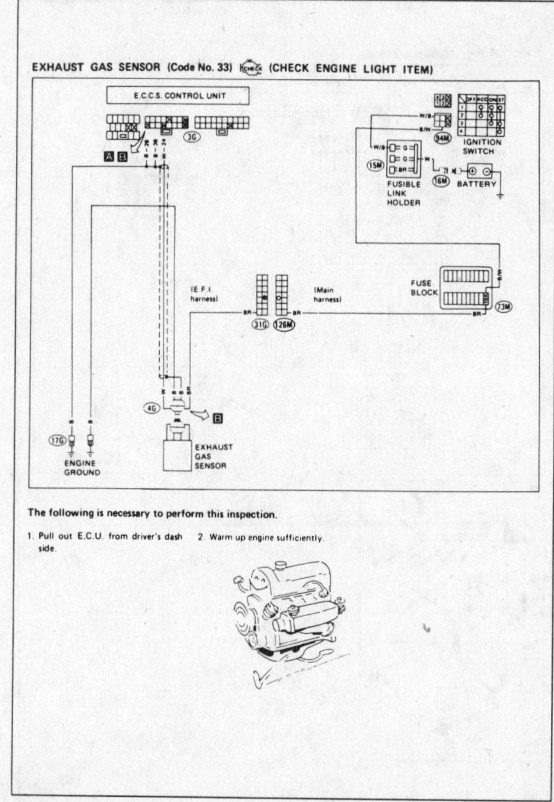
EXHAUST GAS SENSOR (Code No. 33) (CHECK ENGINE LIGHT ITEM)
E.C.C.S. CONTROL UNIT
IGNITION SWITCH
FUSIBLE LINK HOLDER
BATTERY
FUSE BLOCK
(E.F.I. harness)
(Main harness)
ENGINE GROUND
EXHAUST GAS SENSOR
The following is necessary to perform this inspection.
1. Pull out E.C.U. from driver's dash side.
2. Warm up engine sufficiently.

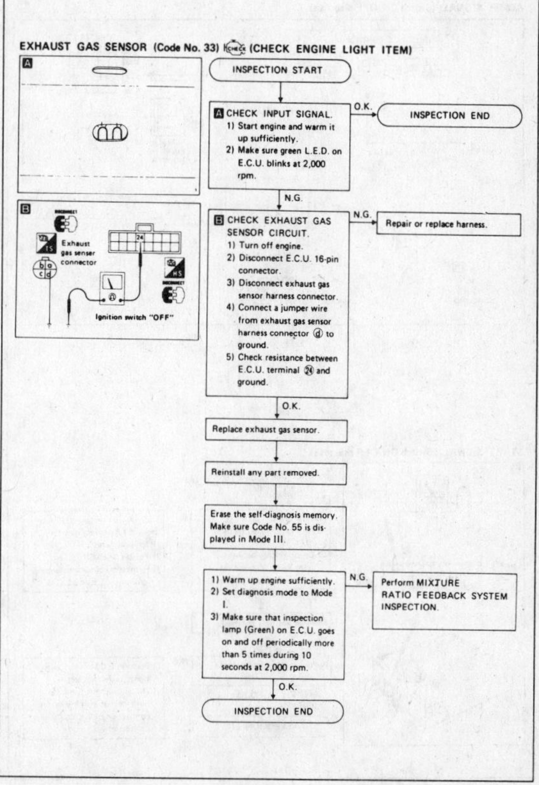
EXHAUST GAS SENSOR (Code No. 33) (CHECK ENGINE LIGHT ITEM)
INSPECTION START
CHECK INPUT SIGNAL.
1) Start engine and warm it up sufficiently.
2) Make sure green L.E.D. on E.C.U. blinks at 2,000 rpm.
O.K.
INSPECTION END
N.G.
CHECK EXHAUST GAS SENSOR CIRCUIT.
1) Turn off engine.
2) Disconnect E.C.U. 16-pin connector.
3) Disconnect exhaust gas sensor harness connector.
4) Connect a jumper wire from exhaust gas sensor harness connector ⓓ to ground.
5) Check resistance between E.C.U. terminal 24 and ground.
Repair or replace harness.
Exhaust gas sensor connector
Ignition switch "OFF"
Replace exhaust gas sensor.
Reinstall any part removed.
Erase the self-diagnosis memory. Make sure Code No. 55 is displayed in Mode III.
1) Warm up engine sufficiently.
2) Set diagnosis mode to Mode I.
3) Make sure that inspection lamp (Green) on E.C.U. goes on and off periodically more than 5 times during 10 seconds at 2,000 rpm.
Perform MIXTURE RATIO FEEDBACK SYSTEM INSPECTION.
INSPECTION END

ECCS DIAGNOSTIC CHART – 1988 200SX WITH VG30E ENGINE

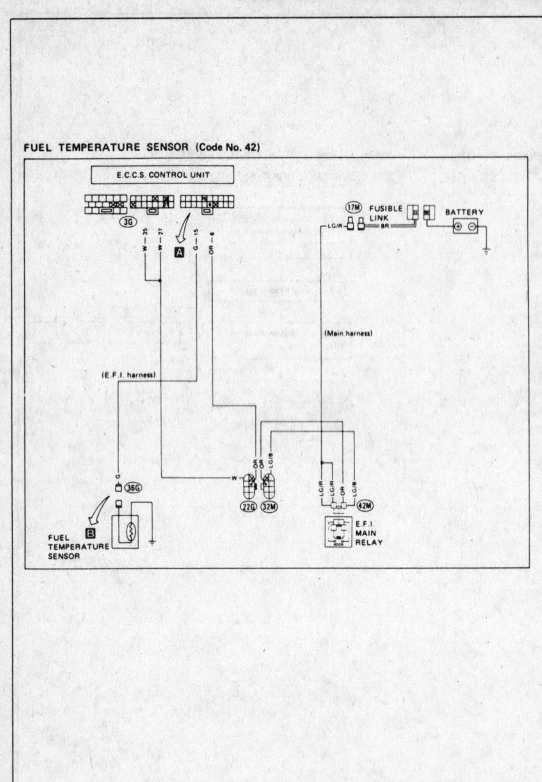
FUEL TEMPERATURE SENSOR (Code No. 42)
E.C.C.S. CONTROL UNIT
FUSIBLE LINK
BATTERY
(Main harness)
(E.F.I. harness)
FUEL TEMPERATURE SENSOR
E.F.I. MAIN RELAY

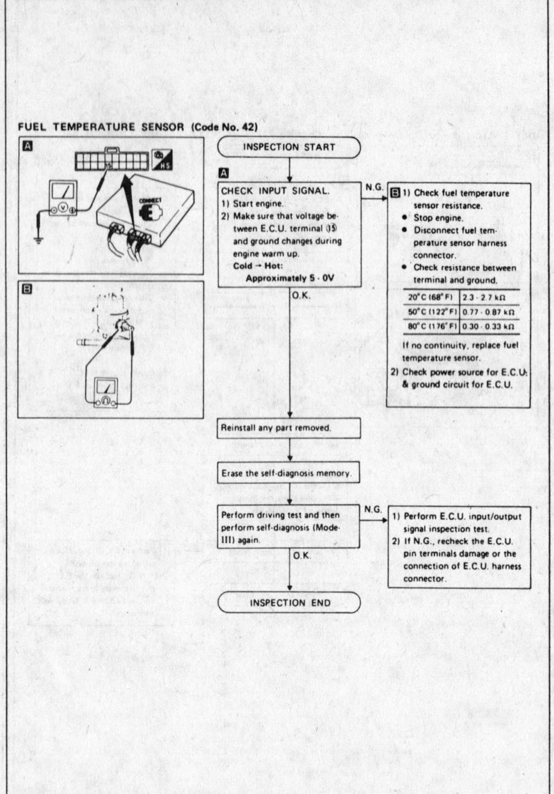
FUEL TEMPERATURE SENSOR (Code No. 42)
INSPECTION START
CHECK INPUT SIGNAL.
1) Start engine.
2) Make sure that voltage between E.C.U. terminal ⑮ and ground changes during engine warm up.
Cold → Hot:
Approximately 5 - 0V
N.G.
1) Check fuel temperature sensor resistance.
• Stop engine.
• Disconnect fuel temperature sensor harness connector.
• Check resistance between terminal and ground.
20°C (68°F) 2.3 - 2.7 kΩ
50°C (122°F) 0.77 - 0.87 kΩ
80°C (176°F) 0.30 - 0.33 kΩ
If no continuity, replace fuel temperature sensor.
2) Check power source for E.C.U. & ground circuit for E.C.U.
O.K.
Reinstall any part removed.
Erase the self-diagnosis memory.
Perform driving test and then perform self-diagnosis (Mode III) again.
N.G.
1) Perform E.C.U. input/output signal inspection test.
2) If N.G., recheck the E.C.U. pin terminals damage or the connection of E.C.U. harness connector.
O.K.
INSPECTION END

ECCS DIAGNOSTIC CHART – 1988 200SX WITH VG30E ENGINE

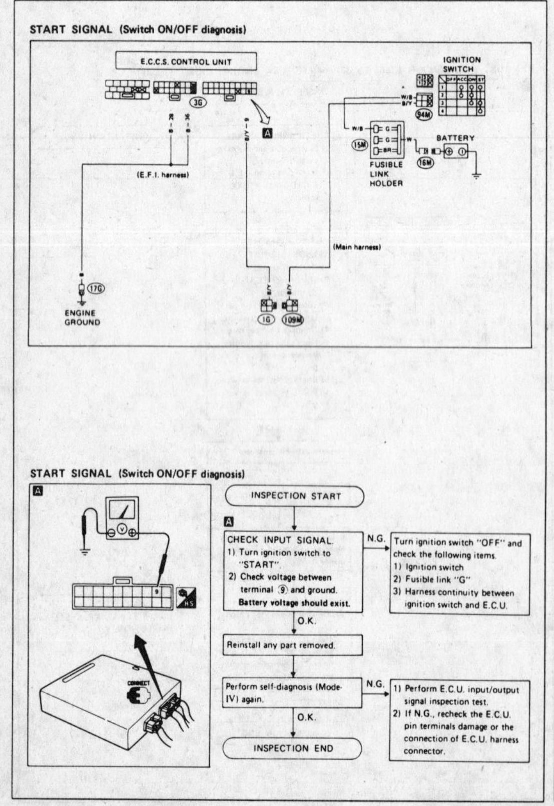
START SIGNAL (Switch ON/OFF diagnosis)
E.C.C.S. CONTROL UNIT
IGNITION SWITCH
BATTERY
FUSIBLE LINK HOLDER
(E.F.I. harness)
(Main harness)
ENGINE GROUND
START SIGNAL (Switch ON/OFF diagnosis)
INSPECTION START
CHECK INPUT SIGNAL.
1) Turn ignition switch to "START".
2) Check voltage between terminal ⑨ and ground.
Battery voltage should exist.
N.G.
Turn ignition switch "OFF" and check the following items.
1) Ignition switch
2) Fusible link "G"
3) Harness continuity between ignition switch and E.C.U.
O.K.
Reinstall any part removed.
Perform self-diagnosis (Mode IV) again.
N.G.
1) Perform E.C.U. input/output signal inspection test.
2) If N.G., recheck the E.C.U. pin terminals damage or the connection of E.C.U. harness connector.
O.K.
INSPECTION END

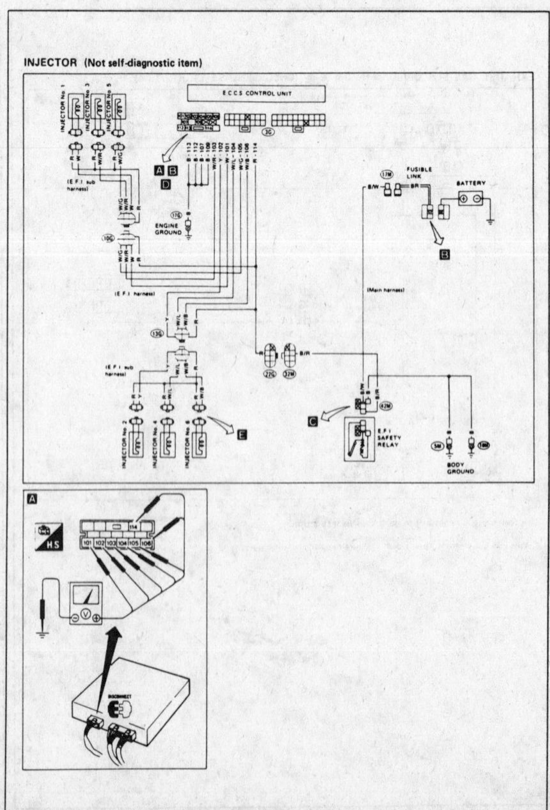
INJECTOR (Not self-diagnostic item)
E.C.C.S. CONTROL UNIT
FUSIBLE LINK
BATTERY
ENGINE GROUND
(E.F.I. harness)
(Main harness)
(E.F.I. sub harness)
E.F.I. SAFETY RELAY
BODY GROUND

ECCS DIAGNOSTIC CHART—1988 200SX WITH VG30E ENGINE

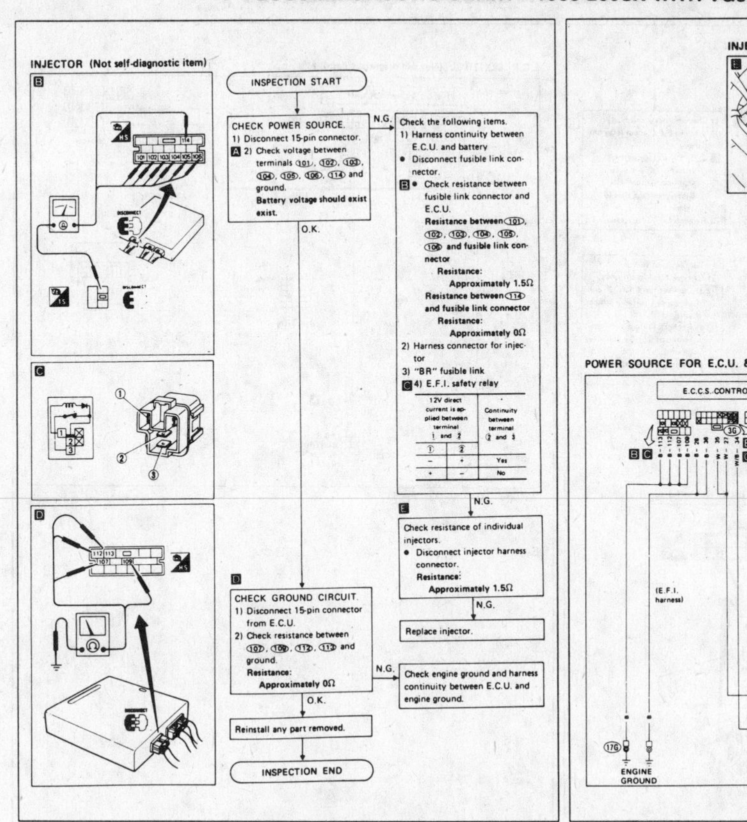
INJECTOR (Not self-diagnostic item)
INSPECTION START
CHECK POWER SOURCE.
1) Disconnect 15-pin connector.
A 2) Check voltage between terminals 101, 102, 103, 104, 105, 106, 114 and ground.
Battery voltage should exist.
N.G.
Check the following items.
1) Harness continuity between E.C.U. and battery
• Disconnect fusible link connector.
B • Check resistance between fusible link connector and E.C.U.
Resistance between 101, 102, 103, 104, 105, 106 and fusible link connector
Resistance: Approximately 1.5Ω
Resistance between 114 and fusible link connector
Resistance: Approximately 0Ω
2) Harness connector for injector
3) "BR" fusible link
C 4) E.F.I. safety relay
O.K.
N.G.
E
Check resistance of individual injectors.
• Disconnect injector harness connector.
Resistance: Approximately 1.5Ω
N.G.
Replace injector.
D
CHECK GROUND CIRCUIT.
1) Disconnect 15-pin connector from E.C.U.
2) Check resistance between 107, 109, 112, 113 and ground.
Resistance: Approximately 0Ω
N.G.
Check engine ground and harness continuity between E.C.U. and engine ground.
O.K.
Reinstall any part removed.
INSPECTION END
DISCONNECT

12V direct current is applied between terminal 1 and 2		Continuity between terminal 3 and 3
1	2	
−	+	Yes
+	−	No

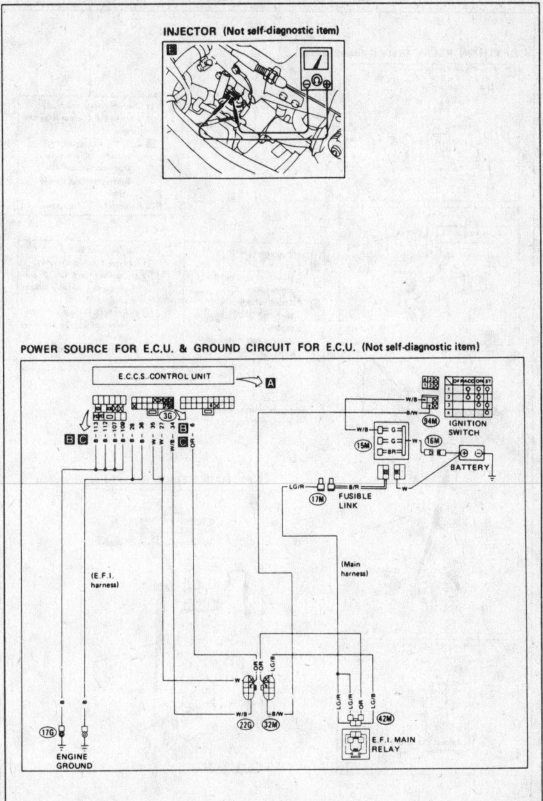
INJECTOR (Not self-diagnostic item)
POWER SOURCE FOR E.C.U. & GROUND CIRCUIT FOR E.C.U. (Not self-diagnostic item)
E.C.C.S. CONTROL UNIT
IGNITION SWITCH
BATTERY
FUSIBLE LINK
(E.F.I. harness)
(Main harness)
E.F.I. MAIN RELAY
ENGINE GROUND

ECCS DIAGNOSTIC CHART—1988 200SX WITH VG30E ENGINE

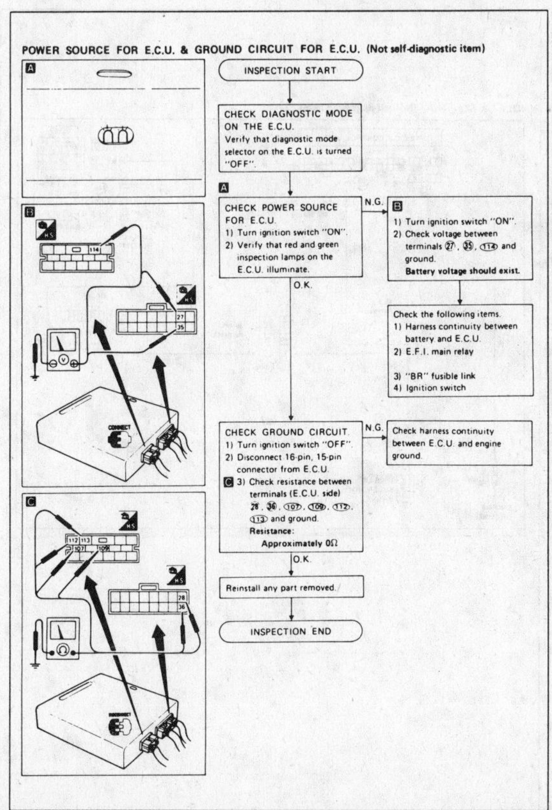
POWER SOURCE FOR E.C.U. & GROUND CIRCUIT FOR E.C.U. (Not self-diagnostic item)
INSPECTION START
CHECK DIAGNOSTIC MODE ON THE E.C.U.
Verify that diagnostic mode selector on the E.C.U. is turned "OFF".
A
CHECK POWER SOURCE FOR E.C.U.
1) Turn ignition switch "ON".
2) Verify that red and green inspection lamps on the E.C.U. illuminate.
N.G.
B
1) Turn ignition switch "ON".
2) Check voltage between terminals 27, 35, 114 and ground.
Battery voltage should exist.
Check the following items.
1) Harness continuity between battery and E.C.U.
2) E.F.I. main relay
3) "BR" fusible link
4) Ignition switch
O.K.
CHECK GROUND CIRCUIT.
1) Turn ignition switch "OFF".
2) Disconnect 16-pin, 15-pin connector from E.C.U.
C 3) Check resistance between terminals (E.C.U. side) 28, 36, 107, 109, 112, 113 and ground.
Resistance: Approximately 0Ω
N.G.
Check harness continuity between E.C.U. and engine ground.
O.K.
Reinstall any part removed.
INSPECTION END
CONNECT
DISCONNECT

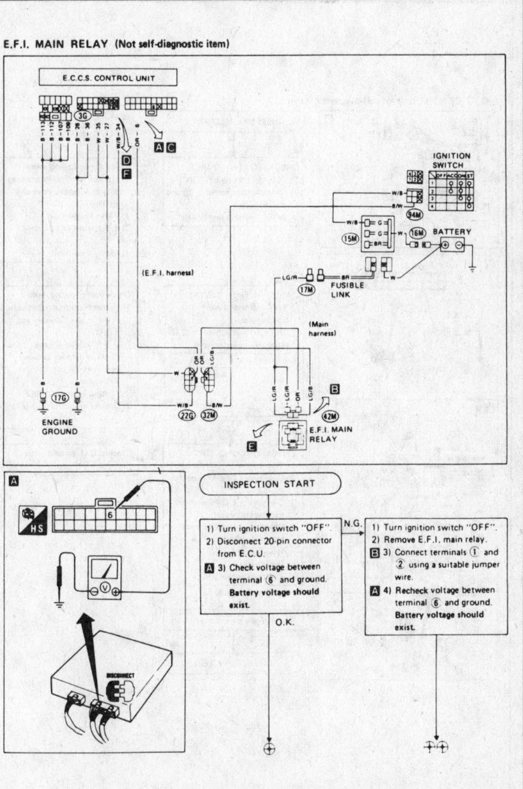
E.F.I. MAIN RELAY (Not self-diagnostic item)
E.C.C.S. CONTROL UNIT
IGNITION SWITCH
BATTERY
FUSIBLE LINK
(E.F.I. harness)
(Main harness)
ENGINE GROUND
E.F.I. MAIN RELAY
INSPECTION START
1) Turn ignition switch "OFF".
2) Disconnect 20-pin connector from E.C.U.
A 3) Check voltage between terminal 6 and ground.
Battery voltage should exist.
N.G.
1) Turn ignition switch "OFF".
2) Remove E.F.I. main relay.
B 3) Connect terminals 1 and 2 using a suitable jumper wire.
A 4) Recheck voltage between terminal 6 and ground.
Battery voltage should exist.
O.K.
DISCONNECT

ECCS DIAGNOSTIC CHART—1988 200SX WITH VG30E ENGINE

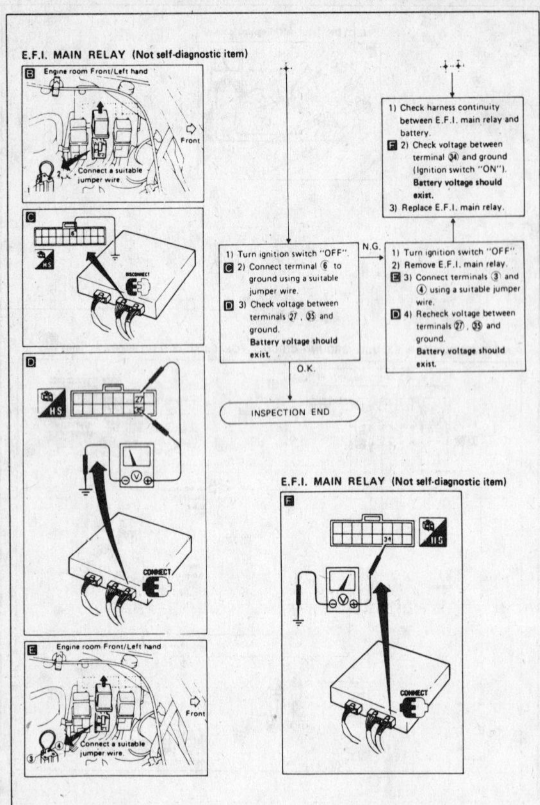
E.F.I. MAIN RELAY (Not self-diagnostic item)
B Engine room Front/Left hand
Front
Connect a suitable jumper wire.
C
D
E Engine room Front/Left hand
Front
Connect a suitable jumper wire.
1) Turn ignition switch "OFF".
C 2) Connect terminal ⑥ to ground using a suitable jumper wire.
D 3) Check voltage between terminals ㉗, ㉟ and ground.
Battery voltage should exist.
O.K.
INSPECTION END
N.G.
1) Turn ignition switch "OFF".
2) Remove E.F.I. main relay.
E 3) Connect terminals ③ and ④ using a suitable jumper wire.
D 4) Recheck voltage between terminals ㉗, ㉟ and ground.
Battery voltage should exist.
1) Check harness continuity between E.F.I. main relay and battery.
F 2) Check voltage between terminal ㉞ and ground (Ignition switch "ON").
Battery voltage should exist.
3) Replace E.F.I. main relay.
E.F.I. MAIN RELAY (Not self-diagnostic item)
F

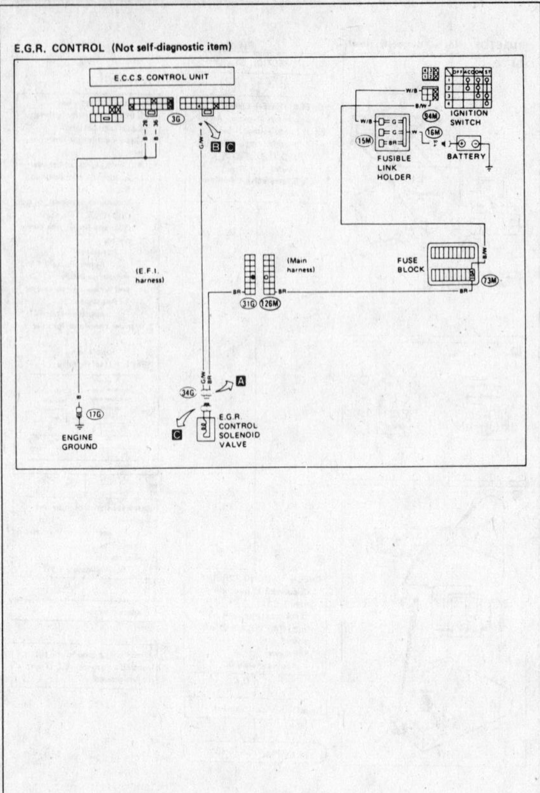
E.G.R. CONTROL (Not self-diagnostic item)
E.C.C.S. CONTROL UNIT
IGNITION SWITCH
FUSIBLE LINK HOLDER
BATTERY
(E.F.I. harness)
(Main harness)
FUSE BLOCK
ENGINE GROUND
E.G.R. CONTROL SOLENOID VALVE

ECCS DIAGNOSTIC CHART—1988 200SX WITH VG30E ENGINE

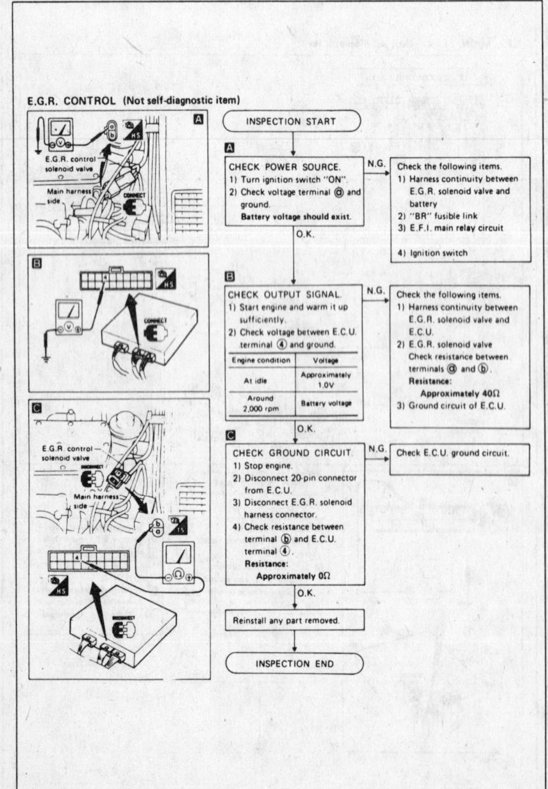
E.G.R. CONTROL (Not self-diagnostic item)
A
E.G.R. control solenoid valve
Main harness side
B
C
E.G.R. control solenoid valve
Main harness side
INSPECTION START
A
CHECK POWER SOURCE.
1) Turn ignition switch "ON".
2) Check voltage terminal ⓐ and ground.
Battery voltage should exist.
N.G.
Check the following items.
1) Harness continuity between E.G.R. solenoid valve and battery
2) "BR" fusible link
3) E.F.I. main relay circuit
4) Ignition switch
O.K.
B
CHECK OUTPUT SIGNAL.
1) Start engine and warm it up sufficiently.
2) Check voltage between E.C.U. terminal ④ and ground.
Engine condition | Voltage
At idle | Approximately 1.0V
Around 2,000 rpm | Battery voltage
N.G.
Check the following items.
1) Harness continuity between E.G.R. solenoid valve and E.C.U.
2) E.G.R. solenoid valve
Check resistance between terminals ⓐ and ⓑ.
Resistance:
Approximately 40Ω
3) Ground circuit of E.C.U.
O.K.
C
CHECK GROUND CIRCUIT.
1) Stop engine.
2) Disconnect 20-pin connector from E.C.U.
3) Disconnect E.G.R. solenoid harness connector.
4) Check resistance between terminal ⓑ and E.C.U. terminal ④.
Resistance:
Approximately 0Ω
N.G.
Check E.C.U. ground circuit.
O.K.
Reinstall any part removed.
INSPECTION END

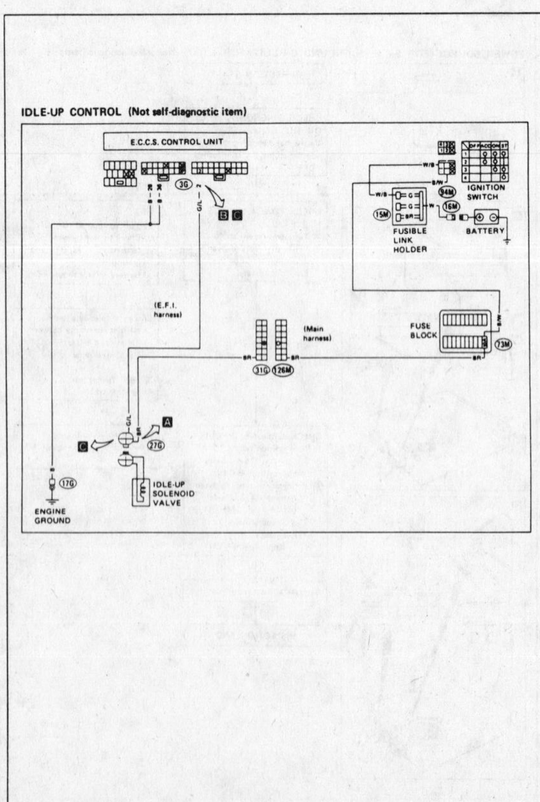
IDLE-UP CONTROL (Not self-diagnostic item)
E.C.C.S. CONTROL UNIT
IGNITION SWITCH
FUSIBLE LINK HOLDER
BATTERY
(E.F.I. harness)
(Main harness)
FUSE BLOCK
ENGINE GROUND
IDLE-UP SOLENOID VALVE

ECCS DIAGNOSTIC CHART—1988 200SX WITH VG30E ENGINE

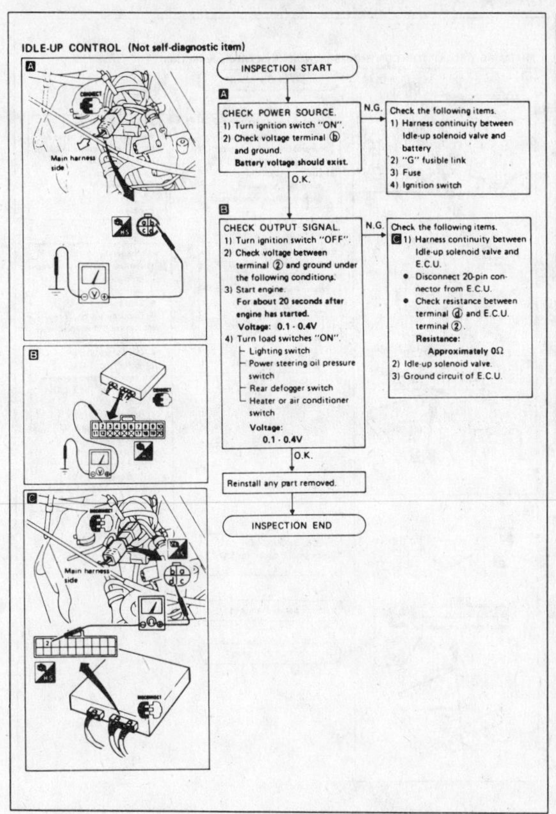
IDLE-UP CONTROL (Not self-diagnostic item)
Main harness side
INSPECTION START
A
CHECK POWER SOURCE.
1) Turn ignition switch "ON".
2) Check voltage terminal ⓑ and ground.
Battery voltage should exist.
N.G.
Check the following items.
1) Harness continuity between Idle-up solenoid valve and battery
2) "G" fusible link
3) Fuse
4) Ignition switch
O.K.
B
CHECK OUTPUT SIGNAL.
1) Turn ignition switch "OFF".
2) Check voltage between terminal ② and ground under the following conditions.
3) Start engine.
For about 20 seconds after engine has started.
Voltage: 0.1 - 0.4V
4) Turn load switches "ON".
Lighting switch
Power steering oil pressure switch
Rear defogger switch
Heater or air conditioner switch
Voltage:
0.1 - 0.4V
N.G.
Check the following items.
C 1) Harness continuity between Idle-up solenoid valve and E.C.U.
• Disconnect 20-pin connector from E.C.U.
• Check resistance between terminal ⓓ and E.C.U. terminal ②.
Resistance:
Approximately 0Ω
2) Idle-up solenoid valve.
3) Ground circuit of E.C.U.
O.K.
Reinstall any part removed.
INSPECTION END
Main harness side

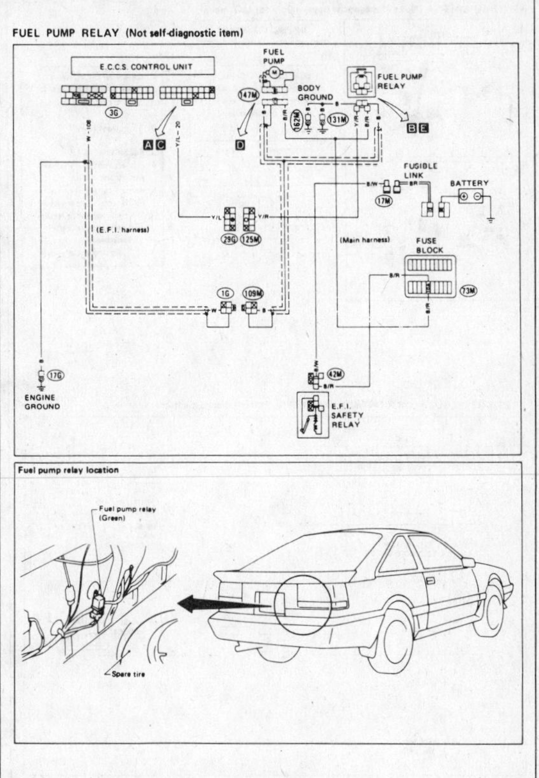
FUEL PUMP RELAY (Not self-diagnostic item)
E.C.C.S. CONTROL UNIT
FUEL PUMP
BODY GROUND
FUEL PUMP RELAY
FUSIBLE LINK
BATTERY
(E.F.I. harness)
(Main harness)
FUSE BLOCK
ENGINE GROUND
E.F.I. SAFETY RELAY
Fuel pump relay location
Fuel pump relay (Green)
Spare tire

ECCS DIAGNOSTIC CHART—1988 200SX WITH VG30E ENGINE

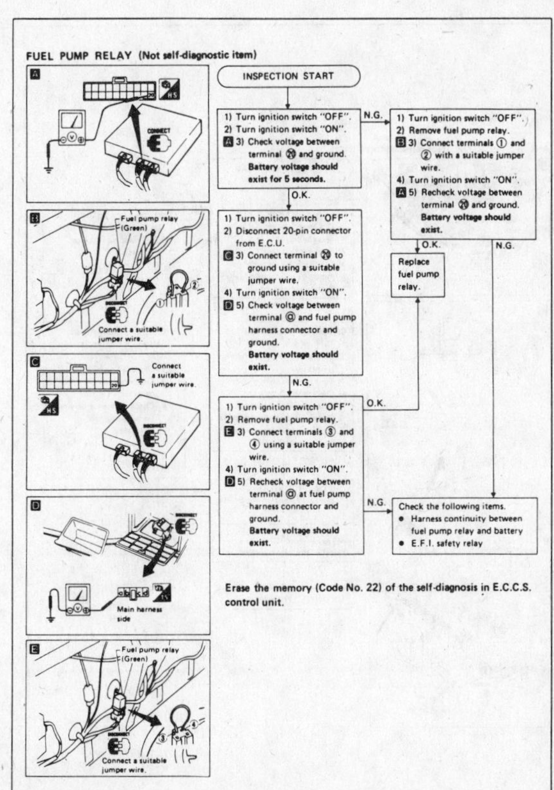
FUEL PUMP RELAY (Not self-diagnostic item)
INSPECTION START
1) Turn ignition switch "OFF".
2) Turn ignition switch "ON".
A 3) Check voltage between terminal ⑳ and ground.
Battery voltage should exist for 5 seconds.
N.G.
1) Turn ignition switch "OFF".
2) Remove fuel pump relay.
B 3) Connect terminals ① and ② with a suitable jumper wire.
4) Turn ignition switch "ON".
A 5) Recheck voltage between terminal ⑳ and ground.
Battery voltage should exist.
O.K.
Replace fuel pump relay.
N.G.
O.K.
1) Turn ignition switch "OFF".
2) Disconnect 20-pin connector from E.C.U.
C 3) Connect terminal ⑳ to ground using a suitable jumper wire.
4) Turn ignition switch "ON".
D 5) Check voltage between terminal ⓐ and fuel pump harness connector and ground.
Battery voltage should exist.
N.G.
1) Turn ignition switch "OFF".
2) Remove fuel pump relay.
E 3) Connect terminals ③ and ④ using a suitable jumper wire.
4) Turn ignition switch "ON".
D 5) Recheck voltage between terminal ⓐ at fuel pump harness connector and ground.
Battery voltage should exist.
O.K.
N.G.
Check the following items.
• Harness continuity between fuel pump relay and battery
• E.F.I. safety relay
Erase the memory (Code No. 22) of the self-diagnosis in E.C.C.S. control unit.
Fuel pump relay (Green)
Connect a suitable jumper wire.
Connect a suitable jumper wire.
Main harness side
Fuel pump relay (Green)
Connect a suitable jumper wire.

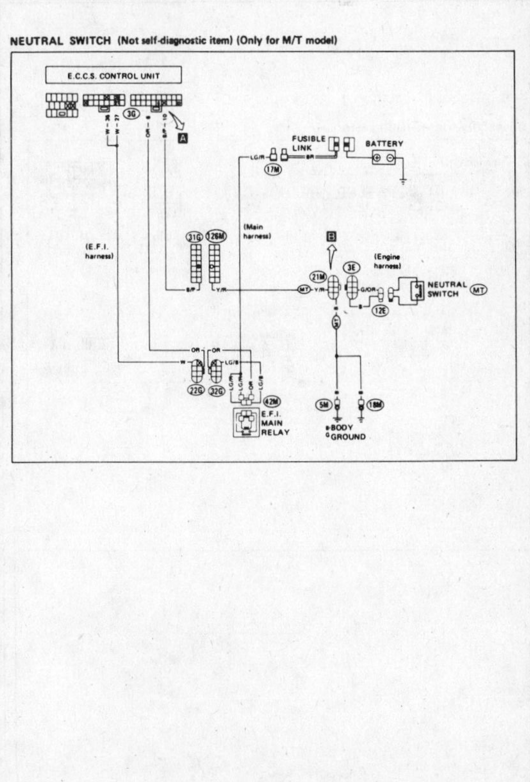
NEUTRAL SWITCH (Not self-diagnostic item) (Only for M/T model)
E.C.C.S. CONTROL UNIT
FUSIBLE LINK
BATTERY
(Main harness)
(E.F.I. harness)
(Engine harness)
NEUTRAL SWITCH
E.F.I. MAIN RELAY
BODY GROUND

ECCS DIAGNOSTIC CHART—1988 200SX WITH VG30E ENGINE

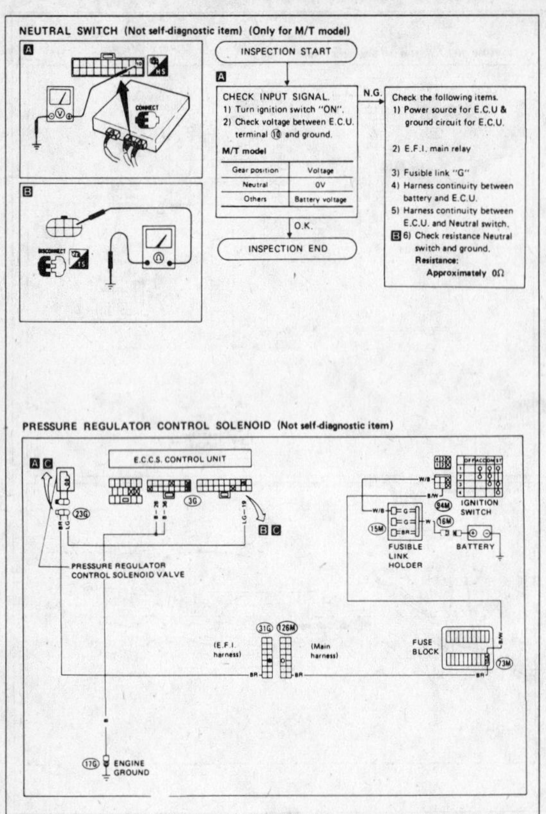
NEUTRAL SWITCH (Not self-diagnostic item) (Only for M/T model)
INSPECTION START
A
CHECK INPUT SIGNAL.
1) Turn ignition switch "ON".
2) Check voltage between E.C.U. terminal ⑩ and ground.
M/T model
Gear position | Voltage
Neutral | 0V
Others | Battery voltage
N.G.
Check the following items.
1) Power source for E.C.U & ground circuit for E.C.U.
2) E.F.I. main relay
3) Fusible link "G"
4) Harness continuity between battery and E.C.U.
5) Harness continuity between E.C.U. and Neutral switch.
B 6) Check resistance Neutral switch and ground.
Resistance:
Approximately 0Ω
O.K.
INSPECTION END
PRESSURE REGULATOR CONTROL SOLENOID (Not self-diagnostic item)
E.C.C.S. CONTROL UNIT
PRESSURE REGULATOR CONTROL SOLENOID VALVE
IGNITION SWITCH
FUSIBLE LINK HOLDER
BATTERY
(E.F.I. harness)
(Main harness)
FUSE BLOCK
ENGINE GROUND

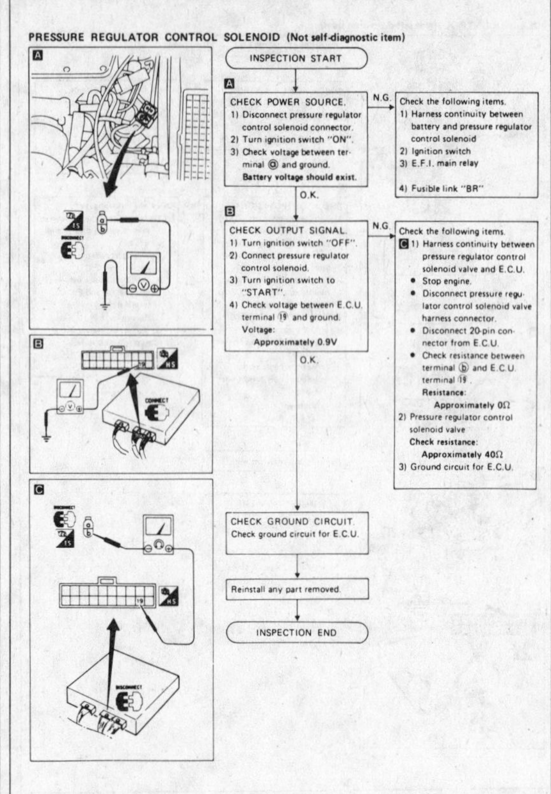
PRESSURE REGULATOR CONTROL SOLENOID (Not self-diagnostic item)
INSPECTION START
A
CHECK POWER SOURCE.
1) Disconnect pressure regulator control solenoid connector.
2) Turn ignition switch "ON".
3) Check voltage between terminal ⓐ and ground.
Battery voltage should exist.
N.G.
Check the following items.
1) Harness continuity between battery and pressure regulator control solenoid
2) Ignition switch
3) E.F.I. main relay
4) Fusible link "BR"
O.K.
B
CHECK OUTPUT SIGNAL.
1) Turn ignition switch "OFF".
2) Connect pressure regulator control solenoid.
3) Turn ignition switch to "START".
4) Check voltage between E.C.U. terminal ⑲ and ground.
Voltage:
Approximately 0.9V
N.G.
Check the following items.
C 1) Harness continuity between pressure regulator control solenoid valve and E.C.U.
• Stop engine.
• Disconnect pressure regulator control solenoid valve harness connector.
• Disconnect 20-pin connector from E.C.U.
• Check resistance between terminal ⓑ and E.C.U. terminal ⑲.
Resistance:
Approximately 0Ω
2) Pressure regulator control solenoid valve
Check resistance:
Approximately 40Ω
3) Ground circuit for E.C.U.
O.K.
CHECK GROUND CIRCUIT.
Check ground circuit for E.C.U.
Reinstall any part removed.
INSPECTION END

ECCS DIAGNOSTIC CHART—1988 200SX WITH VG30E ENGINE

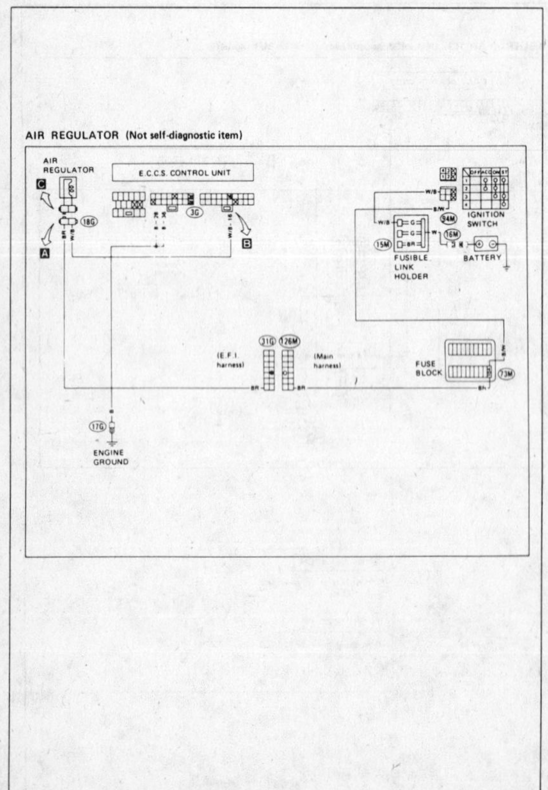
AIR REGULATOR (Not self-diagnostic item)
AIR REGULATOR
E.C.C.S. CONTROL UNIT
IGNITION SWITCH
FUSIBLE LINK HOLDER
BATTERY
(E.F.I. harness)
(Main harness)
FUSE BLOCK
ENGINE GROUND

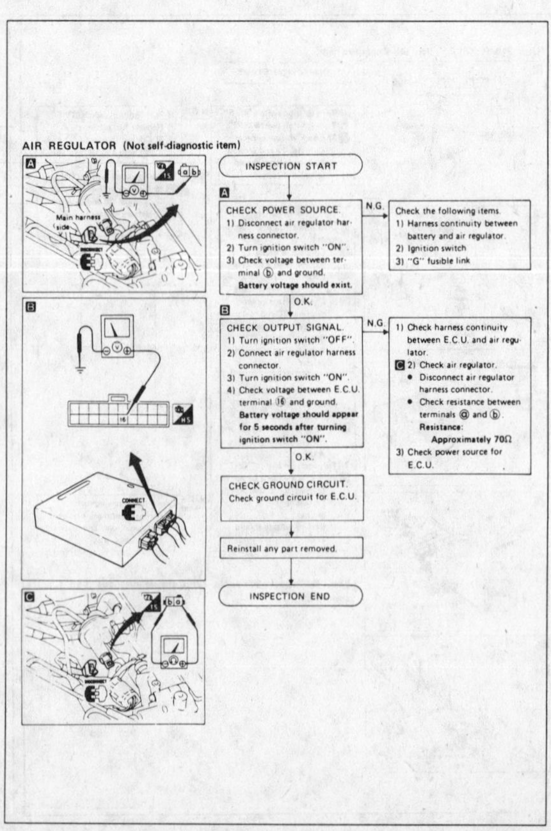
AIR REGULATOR (Not self-diagnostic item)
INSPECTION START
A
CHECK POWER SOURCE.
1) Disconnect air regulator harness connector.
2) Turn ignition switch "ON".
3) Check voltage between terminal ⓑ and ground.
Battery voltage should exist.
N.G.
Check the following items.
1) Harness continuity between battery and air regulator.
2) Ignition switch
3) "G" fusible link
O.K.
B
CHECK OUTPUT SIGNAL.
1) Turn ignition switch "OFF".
2) Connect air regulator harness connector.
3) Turn ignition switch "ON".
4) Check voltage between E.C.U. terminal ⑯ and ground.
Battery voltage should appear for 5 seconds after turning ignition switch "ON".
N.G.
1) Check harness continuity between E.C.U. and air regulator.
C 2) Check air regulator.
• Disconnect air regulator harness connector.
• Check resistance between terminals ⓐ and ⓑ.
Resistance:
Approximately 70Ω
3) Check power source for E.C.U.
O.K.
CHECK GROUND CIRCUIT.
Check ground circuit for E.C.U.
Reinstall any part removed.
INSPECTION END
Main harness side

ECU SIGNAL INPSECTION—1988 200SX WITH VG30E ENGINE

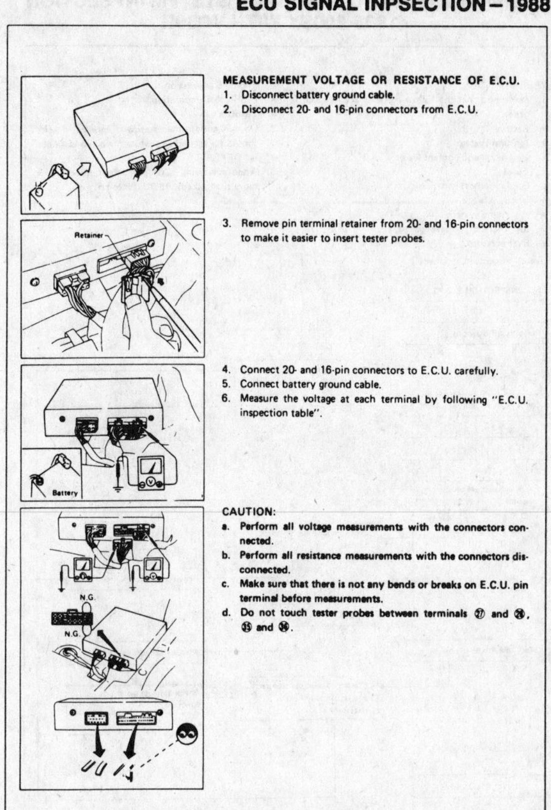
Retainer
Battery
N.G.
N.G.

MEASUREMENT VOLTAGE OR RESISTANCE OF E.C.U.

1. Disconnect battery ground cable.
2. Disconnect 20- and 16-pin connectors from E.C.U.
3. Remove pin terminal retainer from 20- and 16-pin connectors to make it easier to insert tester probes.
4. Connect 20- and 16-pin connectors to E.C.U. carefully.
5. Connect battery ground cable.
6. Measure the voltage at each terminal by following "E.C.U. inspection table".

CAUTION:

a. **Perform all voltage measurements with the connectors connected.**
b. **Perform all resistance measurements with the connectors disconnected.**
c. **Make sure that there is not any bends or breaks on E.C.U. pin terminal before measurements.**
d. **Do not touch tester probes between terminals ㉗ and ㉘, ㉟ and ㊱.**

E.C.U. inspection table

*Data are reference values.

TERMINAL NO.	ITEM	CONDITION	*DATA
2	Idle-up solenoid valve	Engine is running and gear position is in P or N (A/T). — For about 20 seconds after starting engine. — When turning steering wheel. — Heater or air conditioner switch is "ON". — Lighting switch position is "ON".	0.1 - 0.4V
		Engine is running. — Except the conditions shown above	BATTERY VOLTAGE (11 - 14V)
3	Ignition signal (from resistor)	Ignition switch "ON"	BATTERY VOLTAGE (11 - 14V)
4	E.G.R. control solenoid valve	Engine is running after being warmed up. — High engine revolution — Idle speed (Throttle valve switch "ON".)	Approximately 1.0V
		Engine is running. — Low engine revolution	BATTERY VOLTAGE (11 - 14V)
5	Ignition signal (from power transistor)	Engine is running. **Do not turn engine at high speed under no-load.**	0.5 - 2.0V
6	E.F.I. main relay	Ignition switch "ON"	0.7 - 0.9V
8	Crank angle sensor (position signal)	At idle	2.3 - 2.5V
9	Start signal	Cranking	BATTERY VOLTAGE (11 - 14V)
10	Neutral switch (M/T)	Ignition switch "ON" — Gear position is in Neutral	0V
		Ignition switch "ON" — Any gear position except Neutral	BATTERY VOLTAGE (11 - 14V)

ECU SIGNAL INPSECTION—1988 200SX WITH VG30E ENGINE

*Data are reference values.

TERMINAL NO.	ITEM	CONDITION	*DATA
12	Air flow meter self-cleaning signal	Race engine at more than 1,500 rpm and then turn ignition switch "OFF". — For 6 seconds	0V
		Race engine at more than 1,500 rpm and then turn ignition switch "OFF". — For one second after 6 seconds	9.0 - 10.0V
15	Fuel temperature sensor	At idle	0 - 5V Output voltage varies with engine temperature.
16	Air regulator	Engine is running.	0.6 - 0.8V
		1.5 seconds after ignition switch "OFF"	BATTERY VOLTAGE (11 - 14V)
17	Crank angle sensor (Reference signal)	Engine is running. **Do not turn engine at high speed under no-load.**	0.2 - 0.4V
18	Throttle valve switch (⊖ side)	Ignition switch "ON" — Release accelerator pedal. (Throttle valve switch "OFF")	9.0 - 10.0V
		Ignition switch "UN" — Depress accelerator pedal. (Throttle valve switch "ON")	0V
19	Pressure regulator control solenoid	Stop and restart engine after warming it up. — For 30 seconds	0.8 - 1.0V
		Stop and restart engine after warming it up. — After 3 minutes	BATTERY VOLTAGE (11 - 14V)

*Data are reference values.

TERMINAL NO.	ITEM	CONDITION	*DATA
20	Fuel pump relay	Engine is running.	BATTERY VOLTAGE (11 - 14V)
		1.5 seconds after ignition switch "OFF"	0V
22	Load signal	Engine is running and gear position is in P or N (A/T). — When turning steering wheel. — Heater or air conditioner switch is "ON". — Lighting switch position is "ON".	BATTERY VOLTAGE (11 - 14V)
		Engine is running. — Except the conditions shown above.	0V
23	Cylinder head temperature sensor	Engine is running.	0 - 5.0V Output voltage varies with engine temperature.
24	Exhaust gas sensor	Engine is running. — After warming up sufficiently	0 - Approximately 1.0V
25	Throttle valve switch (⊕ side)	Ignition switch "ON"	9.0 - 10.0V
27 35	Power source for E.C.U.	Ignition switch "ON"	BATTERY VOLTAGE (11 - 14V)
29	Vehicle speed sensor	Ignition switch "ON" — While rotating rear wheel slowly	0 or 7.4V
30	Air flow meter	Ignition switch "ON"	2.0 - 4.0V
31	Air quantity signal	Race engine from idle to 3,000 rpm	2.0 - 4.0V
34	Ignition switch signal	Ignition switch "ON"	BATTERY VOLTAGE (11 - 14V)

ECU SIGNAL INPSECTION 1988 200SX WITH VG30E ENGINE

ECCS MIXTURE RATIO FEEDBACK SYSTEM INPSECTION 1988 200SX WITH VG30E

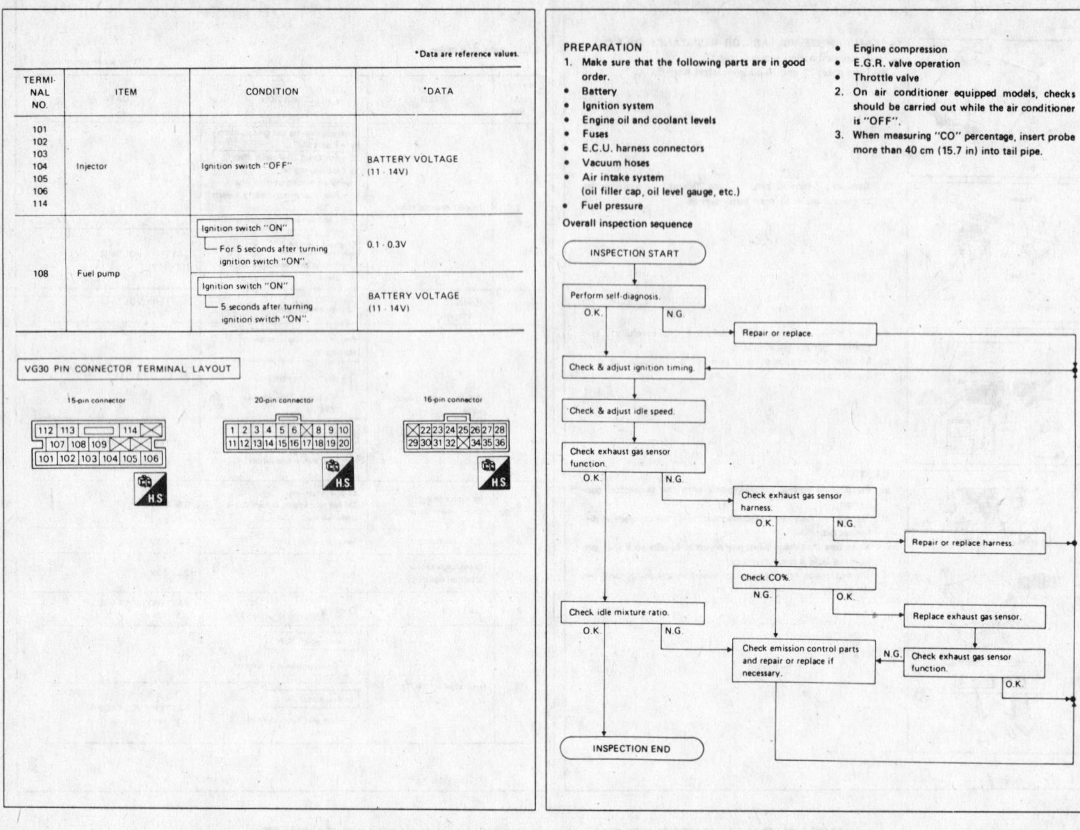

*Data are reference values.

TERMINAL NO.	ITEM	CONDITION	*DATA
101 102 103 104 105 106 114	Injector	Ignition switch "OFF"	BATTERY VOLTAGE (11-14V)
108	Fuel pump	Ignition switch "ON" — For 5 seconds after turning ignition switch "ON"	0.1-0.3V
		Ignition switch "ON" — 5 seconds after turning ignition switch "ON".	BATTERY VOLTAGE (11-14V)

VG30 PIN CONNECTOR TERMINAL LAYOUT
15-pin connector
112 113 114
107 108 109
101 102 103 104 105 106
20-pin connector
1 2 3 4 5 6 8 9 10
11 12 13 14 15 16 17 18 19 20
16-pin connector
22 23 24 25 26 27 28
29 30 31 32 34 35 36
HS
PREPARATION
1. Make sure that the following parts are in good order.
• Battery
• Ignition system
• Engine oil and coolant levels
• Fuses
• E.C.U. harness connectors
• Vacuum hoses
• Air intake system (oil filler cap, oil level gauge, etc.)
• Fuel pressure
• Engine compression
• E.G.R. valve operation
• Throttle valve
2. On air conditioner equipped models, checks should be carried out while the air conditioner is "OFF".
3. When measuring "CO" percentage, insert probe more than 40 cm (15.7 in) into tail pipe.
Overall inspection sequence
INSPECTION START
Perform self-diagnosis.
O.K.
N.G.
Repair or replace.
Check & adjust ignition timing.
Check & adjust idle speed.
Check exhaust gas sensor function.
O.K.
N.G.
Check exhaust gas sensor harness.
O.K.
N.G.
Repair or replace harness.
Check CO%.
N.G.
O.K.
Replace exhaust gas sensor.
Check idle mixture ratio.
O.K.
N.G.
Check emission control parts and repair or replace if necessary.
N.G.
Check exhaust gas sensor function.
O.K.
INSPECTION END

ECCS MIXTURE RATIO FEEDBACK SYSTEM INPSECTION (CONT.)—1988 200SX WITH VG30E ENGINE

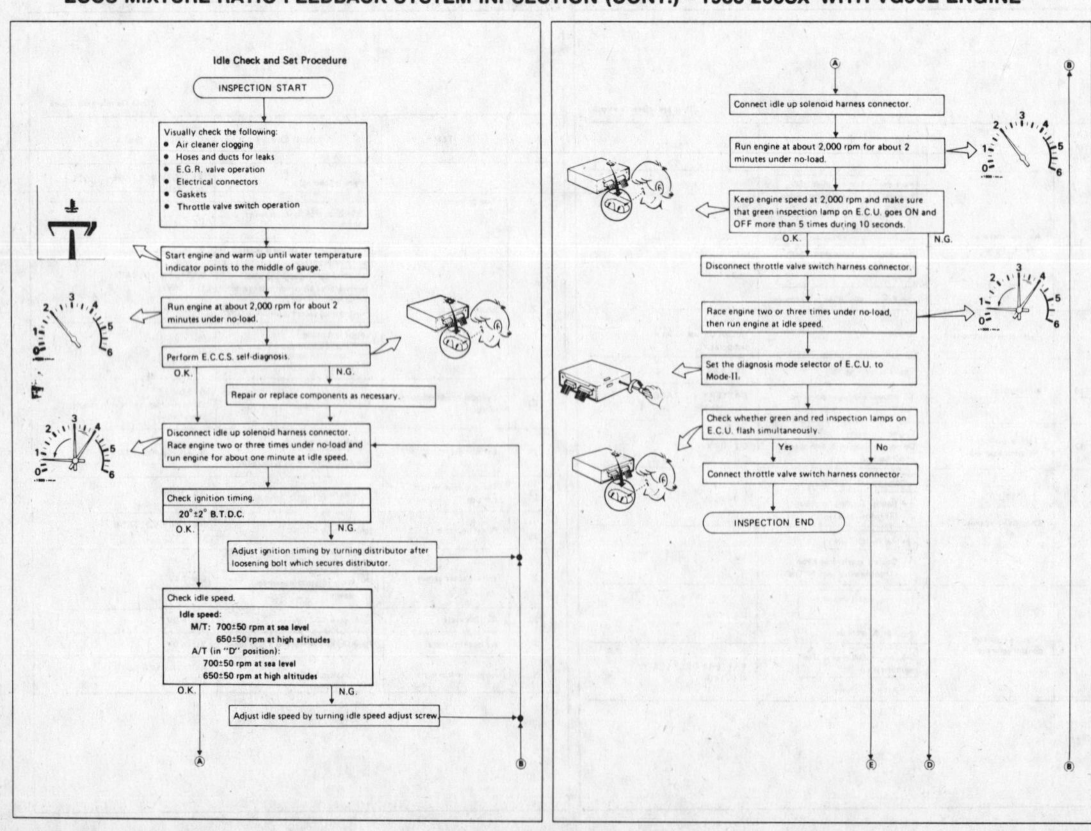

Idle Check and Set Procedure
INSPECTION START
Visually check the following:
• Air cleaner clogging
• Hoses and ducts for leaks
• E.G.R. valve operation
• Electrical connectors
• Gaskets
• Throttle valve switch operation
Start engine and warm up until water temperature indicator points to the middle of gauge.
Run engine at about 2,000 rpm for about 2 minutes under no-load.
Perform E.C.C.S. self-diagnosis.
O.K.
N.G.
Repair or replace components as necessary.
Disconnect idle up solenoid harness connector. Race engine two or three times under no-load and run engine for about one minute at idle speed.
Check ignition timing.
20°±2° B.T.D.C.
O.K.
N.G.
Adjust ignition timing by turning distributor after loosening bolt which secures distributor.
Check idle speed.
Idle speed:
M/T: 700±50 rpm at sea level
650±50 rpm at high altitudes
A/T (in "D" position):
700±50 rpm at sea level
650±50 rpm at high altitudes
O.K.
N.G.
Adjust idle speed by turning idle speed adjust screw.
A
B
Connect idle up solenoid harness connector.
Run engine at about 2,000 rpm for about 2 minutes under no-load.
Keep engine speed at 2,000 rpm and make sure that green inspection lamp on E.C.U. goes ON and OFF more than 5 times during 10 seconds.
O.K.
N.G.
Disconnect throttle valve switch harness connector.
Race engine two or three times under no-load, then run engine at idle speed.
Set the diagnosis mode selector of E.C.U. to Mode-II.
Check whether green and red inspection lamps on E.C.U. flash simultaneously.
Yes
No
Connect throttle valve switch harness connector.
INSPECTION END
E
D
0 1 2 3 4 5 6

ECCS MIXTURE RATIO FEEDBACK SYSTEM INPSECTION (CONT.)—1988 200SX WITH VG30E ENGINE

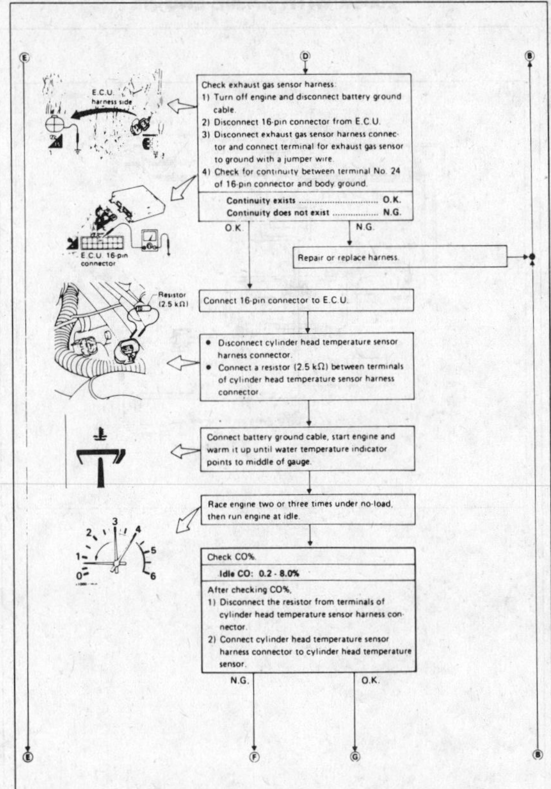

E
D
B
Check exhaust gas sensor harness:
1) Turn off engine and disconnect battery ground cable.
2) Disconnect 16-pin connector from E.C.U.
3) Disconnect exhaust gas sensor harness connector and connect terminal for exhaust gas sensor to ground with a jumper wire.
4) Check for continuity between terminal No. 24 of 16-pin connector and body ground.
Continuity exists .... O.K.
Continuity does not exist .... N.G.
E.C.U. harness side
E.C.U. 16-pin connector
O.K.
N.G.
Repair or replace harness.
Resistor (2.5 kΩ)
Connect 16-pin connector to E.C.U.
Disconnect cylinder head temperature sensor harness connector.
Connect a resistor (2.5 kΩ) between terminals of cylinder head temperature sensor harness connector.
Connect battery ground cable, start engine and warm it up until water temperature indicator points to middle of gauge.
Race engine two or three times under no-load, then run engine at idle.
Check CO%.
Idle CO: 0.2 - 8.0%
After checking CO%,
1) Disconnect the resistor from terminals of cylinder head temperature sensor harness connector.
2) Connect cylinder head temperature sensor harness connector to cylinder head temperature sensor.
N.G.
O.K.
E
F
G
B

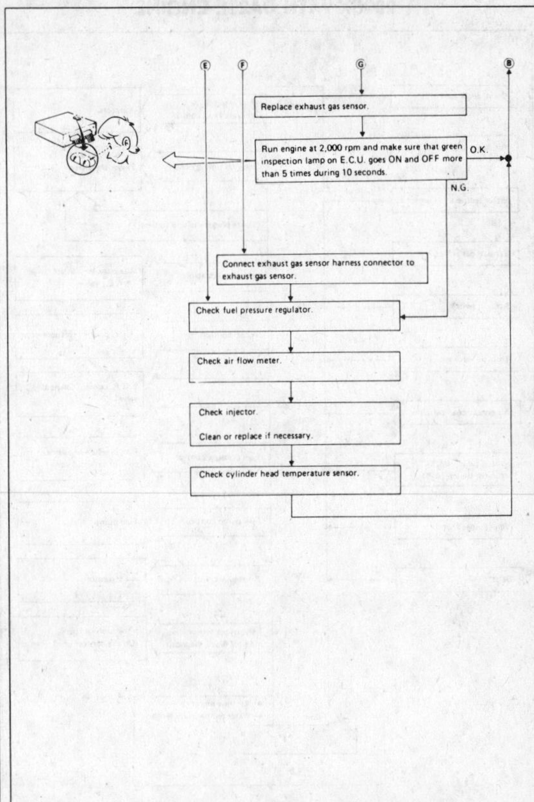

E
F
G
B
Replace exhaust gas sensor.
Run engine at 2,000 rpm and make sure that green inspection lamp on E.C.U. goes ON and OFF more than 5 times during 10 seconds.
O.K.
N.G.
Connect exhaust gas sensor harness connector to exhaust gas sensor.
Check fuel pressure regulator.
Check air flow meter.
Check injector.
Clean or replace if necessary.
Check cylinder head temperature sensor.

ECCS COMPONENTS LOCATION—1988 200SX WITH CA20E ENGINE

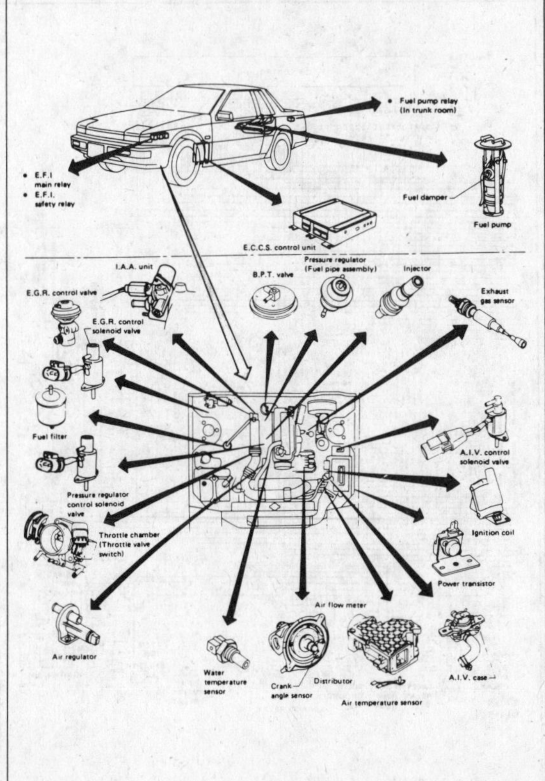

Fuel pump relay (In trunk room)
E.F.I main relay
E.F.I. safety relay
Fuel damper
Fuel pump
E.C.C.S. control unit
I.A.A. unit
B.P.T. valve
Pressure regulator (Fuel pipe assembly)
Injector
E.G.R. control valve
Exhaust gas sensor
E.G.R. control solenoid valve
Fuel filter
A.I.V. control solenoid valve
Pressure regulator control solenoid valve
Ignition coil
Throttle chamber (Throttle valve switch)
Power transistor
Air flow meter
Air regulator
Water temperature sensor
Crank angle sensor
Distributor
A.I.V. case
Air temperature sensor

ECCS SYSTEM SCHEMATIC—1988 200SX WITH CA20E ENGINE

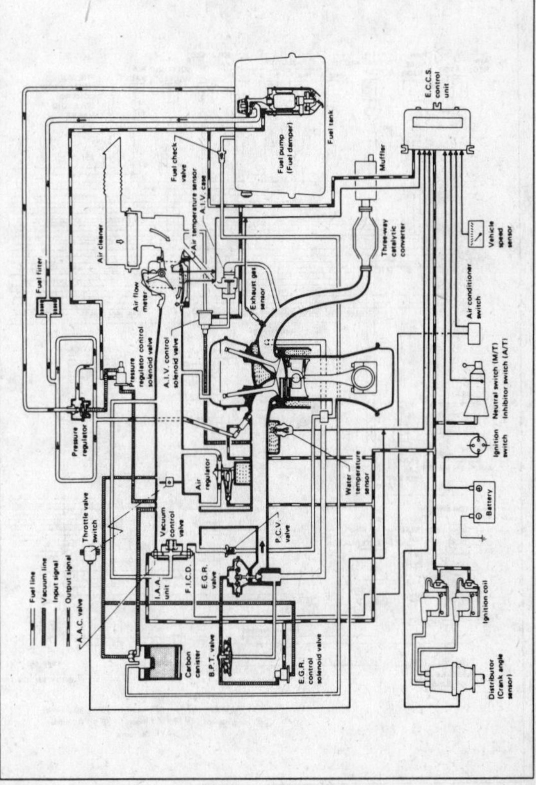

E.C.C.S. control unit
Fuel tank
Fuel pump (Fuel damper)
Fuel check valve
Muffler
Air temperature sensor
A.I.V. case
Air cleaner
Three-way catalytic converter
Vehicle speed sensor
Fuel filter
Air flow meter
Exhaust gas sensor
Air conditioner switch
Pressure regulator control solenoid valve
A.I.V. control solenoid valve
Neutral switch (M/T)
Inhibitor switch (A/T)
Pressure regulator
Ignition switch
Air regulator
Water temperature sensor
Battery
Throttle valve switch
Vacuum control valve
P.C.V. valve
Fuel line
Vacuum line
Input signal
Output signal
A.A.C. valve
I.A.A. unit
F.I.C.D.
E.G.R. valve
Ignition coil
Carbon canister
B.P.T. valve
E.G.R. control solenoid valve

ECCS CONTROL SYSTEM CHART—1988 200SX WITH CA20E ENGINE

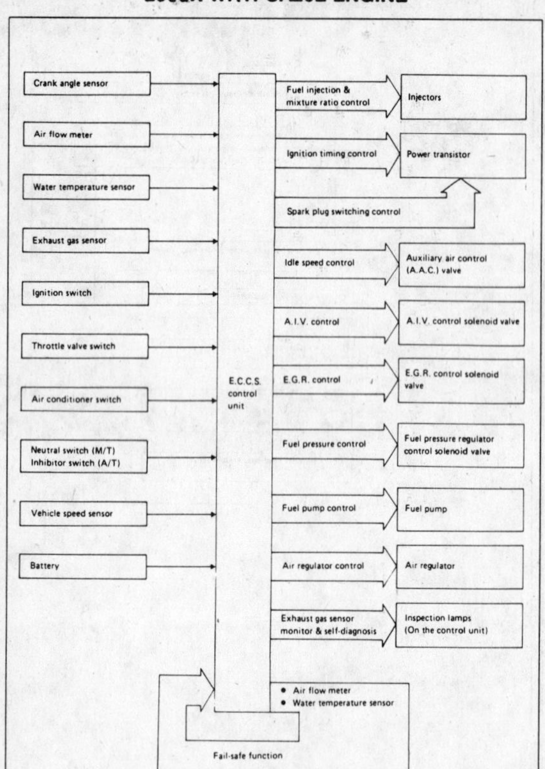
Crank angle sensor
Air flow meter
Water temperature sensor
Exhaust gas sensor
Ignition switch
Throttle valve switch
Air conditioner switch
Neutral switch (M/T)
Inhibitor switch (A/T)
Vehicle speed sensor
Battery
E.C.C.S. control unit
Fuel injection & mixture ratio control
Injectors
Ignition timing control
Power transistor
Spark plug switching control
Idle speed control
Auxiliary air control (A.A.C.) valve
A.I.V. control
A.I.V. control solenoid valve
E.G.R. control
E.G.R. control solenoid valve
Fuel pressure control
Fuel pressure regulator control solenoid valve
Fuel pump control
Fuel pump
Air regulator control
Air regulator
Exhaust gas sensor monitor & self-diagnosis
Inspection lamps (On the control unit)
Air flow meter
Water temperature sensor
Fail-safe function

ECCS FUEL FLOW SYSTEM SCHEMATIC—1988 200SX WITH CA20E ENGINE

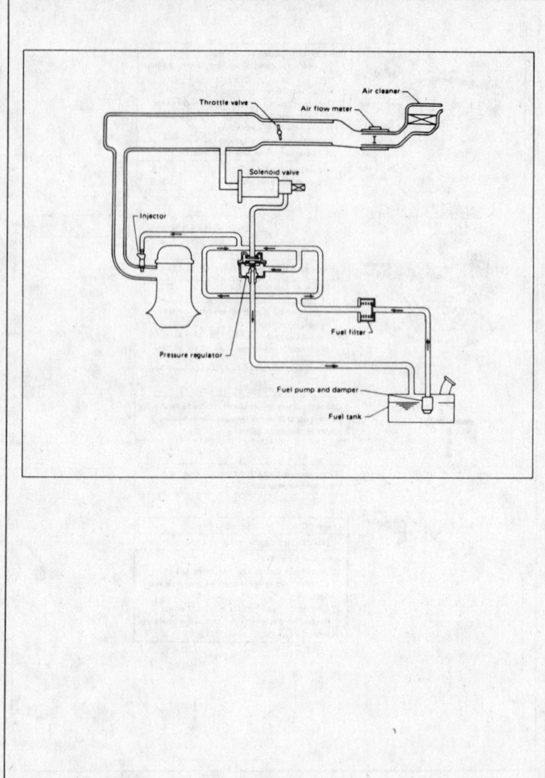
Air cleaner
Throttle valve
Air flow meter
Solenoid valve
Injector
Fuel filter
Pressure regulator
Fuel pump and damper
Fuel tank

ECCS WIRING SCHEMATIC—1988 200SX WITH CA20E ENGINE

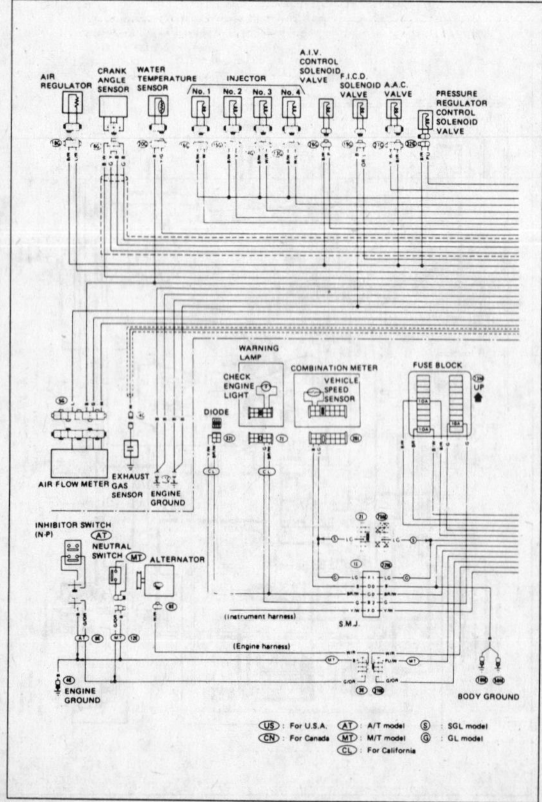
AIR REGULATOR
CRANK ANGLE SENSOR
WATER TEMPERATURE SENSOR
INJECTOR
No. 1
No. 2
No. 3
No. 4
A.I.V. CONTROL SOLENOID VALVE
F.I.C.D. SOLENOID VALVE
A.A.C. VALVE
PRESSURE REGULATOR CONTROL SOLENOID VALVE
WARNING LAMP
CHECK ENGINE LIGHT
COMBINATION METER
VEHICLE SPEED SENSOR
FUSE BLOCK
DIODE
AIR FLOW METER
EXHAUST GAS SENSOR
ENGINE GROUND
INHIBITOR SWITCH (N-P)
NEUTRAL SWITCH
ALTERNATOR
(Instrument harness)
S.M.J.
(Engine harness)
ENGINE GROUND
BODY GROUND
US : For U.S.A.
CN : For Canada
AT : A/T model
MT : M/T model
CL : For California
S : SGL model
G : GL model

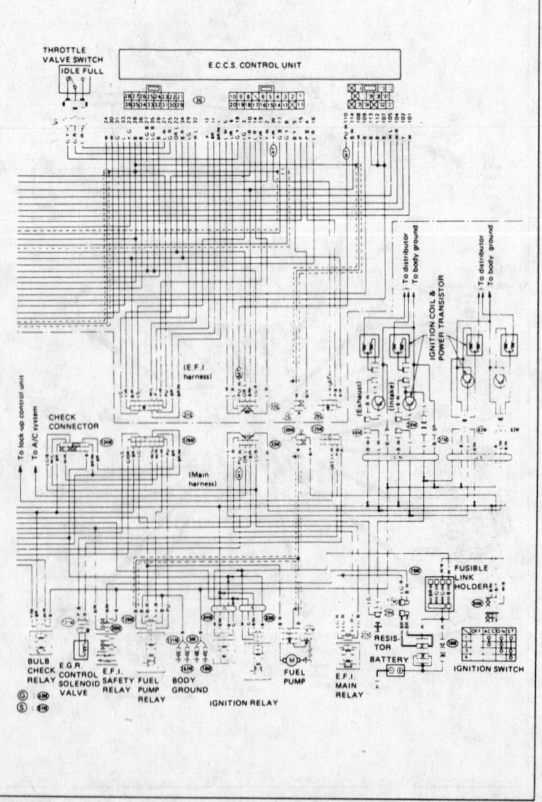
THROTTLE VALVE SWITCH
IDLE FULL
E.C.C.S. CONTROL UNIT
To distributor
To body ground
IGNITION COIL & POWER TRANSISTOR
(E.F.I. harness)
To lock-up control unit
To A/C system
CHECK CONNECTOR
(Main harness)
FUSIBLE LINK HOLDER
RESISTOR
BATTERY
IGNITION SWITCH
BULB CHECK RELAY
E.G.R. CONTROL SOLENOID VALVE
E.F.I. SAFETY RELAY
FUEL PUMP RELAY
BODY GROUND
IGNITION RELAY
FUEL PUMP
E.F.I. MAIN RELAY

ECCS CONTROL UNIT WIRING 1988 200SX WITH CA20E ENGINE

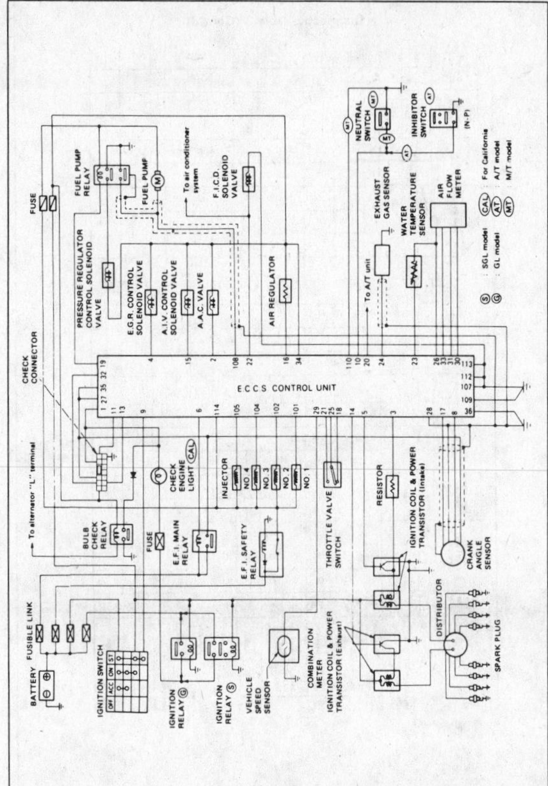
FUSE
FUEL PUMP RELAY
FUEL PUMP
To air conditioner system
F.I.C.D. SOLENOID VALVE
NEUTRAL SWITCH
INHIBITOR SWITCH
EXHAUST GAS SENSOR
WATER TEMPERATURE SENSOR
AIR FLOW METER
For California
A/T model
M/T model
SGL model
GL model
PRESSURE REGULATOR CONTROL SOLENOID VALVE
E.G.R. CONTROL SOLENOID VALVE
A.I.V. CONTROL SOLENOID VALVE
A.A.C. VALVE
AIR REGULATOR
To A/T unit
CHECK CONNECTOR
E.C.C.S. CONTROL UNIT
To alternator "L" terminal
CHECK ENGINE LIGHT
INJECTOR
RESISTOR
IGNITION COIL & POWER TRANSISTOR (Intake)
CRANK ANGLE SENSOR
BULB CHECK RELAY
FUSE
E.F.I. MAIN RELAY
E.F.I. SAFETY RELAY
THROTTLE VALVE SWITCH
DISTRIBUTOR
FUSIBLE LINK
BATTERY
IGNITION SWITCH
IGNITION RELAY
VEHICLE SPEED SENSOR
COMBINATION METER
IGNITION COIL & POWER TRANSISTOR (Exhaust)
SPARK PLUG

ECCS IGNITION TIMING CONTROL CHART—1988 200SX WITH CA20E ENGINE

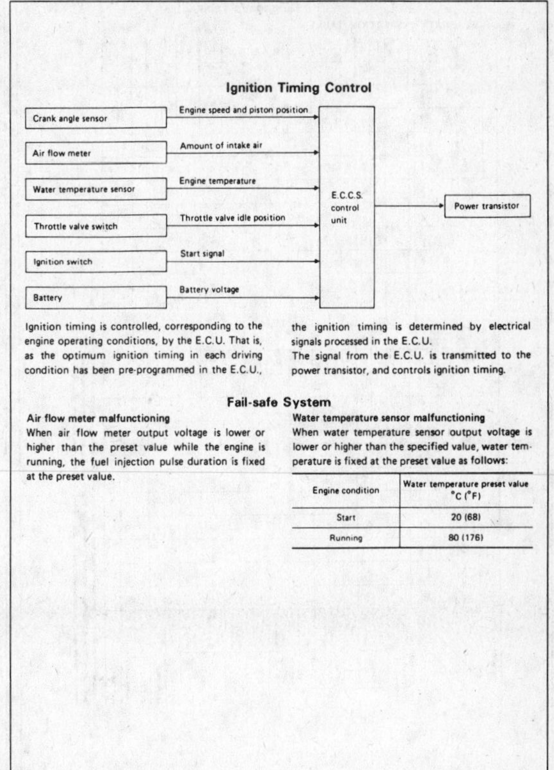
Ignition Timing Control
Crank angle sensor
Engine speed and piston position
Air flow meter
Amount of intake air
Water temperature sensor
Engine temperature
Throttle valve switch
Throttle valve idle position
Ignition switch
Start signal
Battery
Battery voltage
E.C.C.S. control unit
Power transistor

Ignition timing is controlled, corresponding to the engine operating conditions, by the E.C.U. That is, as the optimum ignition timing in each driving condition has been pre-programmed in the E.C.U., the ignition timing is determined by electrical signals processed in the E.C.U.

The signal from the E.C.U. is transmitted to the power transistor, and controls ignition timing.

Fail-safe System

Air flow meter malfunctioning

When air flow meter output voltage is lower or higher than the preset value while the engine is running, the fuel injection pulse duration is fixed at the preset value.

Water temperature sensor malfunctioning

When water temperature sensor output voltage is lower or higher than the specified value, water temperature is fixed at the preset value as follows:

Engine condition	Water temperature preset value °C (°F)
Start	20 (68)
Running	80 (176)

ECCS DIAGNOSTIC PROCEDURE—1988 200SX WITH CA20E ENGINE

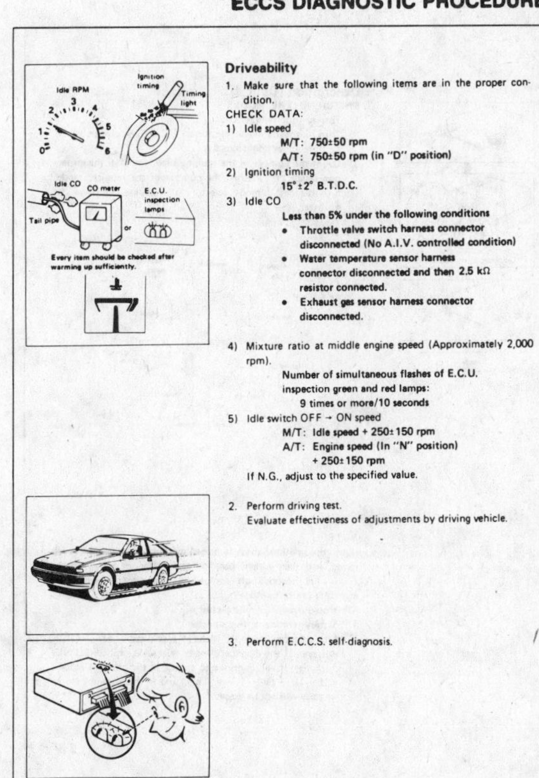
Idle RPM
Ignition timing
Timing light
Idle CO
CO meter
E.C.U. inspection lamps
Tail pipe
or
Every item should be checked after warming up sufficiently.

Driveability

1. Make sure that the following items are in the proper condition.

CHECK DATA:

1) Idle speed
   M/T: 750±50 rpm
   A/T: 750±50 rpm (in "D" position)
2) Ignition timing
   15°±2° B.T.D.C.
3) Idle CO
   Less than 5% under the following conditions
   - Throttle valve switch harness connector disconnected (No A.I.V. controlled condition)
   - Water temperature sensor harness connector disconnected and then 2.5 kΩ resistor connected.
   - Exhaust gas sensor harness connector disconnected.
4) Mixture ratio at middle engine speed (Approximately 2,000 rpm).
   Number of simultaneous flashes of E.C.U. inspection green and red lamps:
   9 times or more/10 seconds
5) Idle switch OFF → ON speed
   M/T: Idle speed + 250±150 rpm
   A/T: Engine speed (In "N" position) + 250±150 rpm

If N.G., adjust to the specified value.

2. Perform driving test.
   Evaluate effectiveness of adjustments by driving vehicle.

3. Perform E.C.C.S. self-diagnosis.

Driveability (Cont'd)

DIAGNOSTIC TABLE 1 DIAGNOSTIC TABLE 2

Electronic control system inspection

4. If the result of driveability test is unsatisfactory, or malfunctioning conditions are found in performing E.C.C.S. self-diagnosis, perform general inspection and E.C.C.S. system inspection by following DIAGNOSTIC TABLE 1 and 2 in response to driveability trouble items.
   If N.G., repair.

5. Perform driving test.
   Re-evaluate vehicle performance after the inspection.

Diagnostic Table 1

SYSTEM INSPECTION TABLE

System \ Sensor & actuator	Crank angle sensor	Air flow meter	Water temperature sensor	Ignition switch	Injector	Throttle valve switch	Neutral/ inhibitor switch	Exhaust gas sensor
Fuel injection & mixture ratio feedback control	O	O	O	O	O	O	O	O
Ignition timing control	O	O	O	O		O		
A.I.V. control	O		O			O		O
Fuel pump control	O			O				
Fuel pressure control	O		O	O	O			
Idle speed control	O		O	O		O	O	
E.G.R. control	O		O	O		O		

System \ Sensor & actuator	A.I.V. control solenoid valve	E.G.R. control solenoid valve	P.R. control solenoid valve/ Fuel pump relay	A.A.C. valve	Air regulator	Vehicle speed sensor	E.F.I. main relay
Fuel injection & mixture ratio feedback control						O	O
Ignition timing control							O
A.I.V. control	O						O
Fuel pump control							O
Fuel pressure control			O				O
Idle speed control				O	O	O	O
E.G.R. control		O					O

This table indicates the inspection items for the E.C.C.S. control system. For each system, it is necessary to check sensors or actuators marked "O".

ECCS DIAGNOSTIC TABLE 2—1988 200SX WITH CA20E ENGINE

Diagnostic Table 2

DRIVEABILITY INSPECTION TABLE

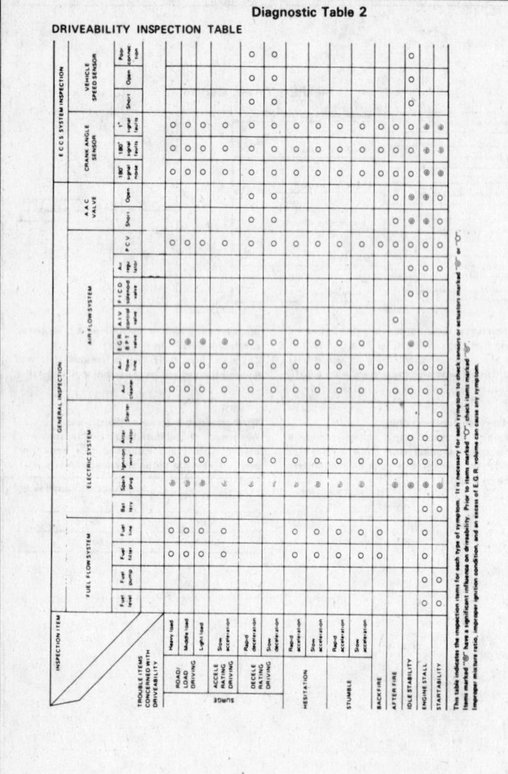

Diagnostic Table 2 (Cont'd)

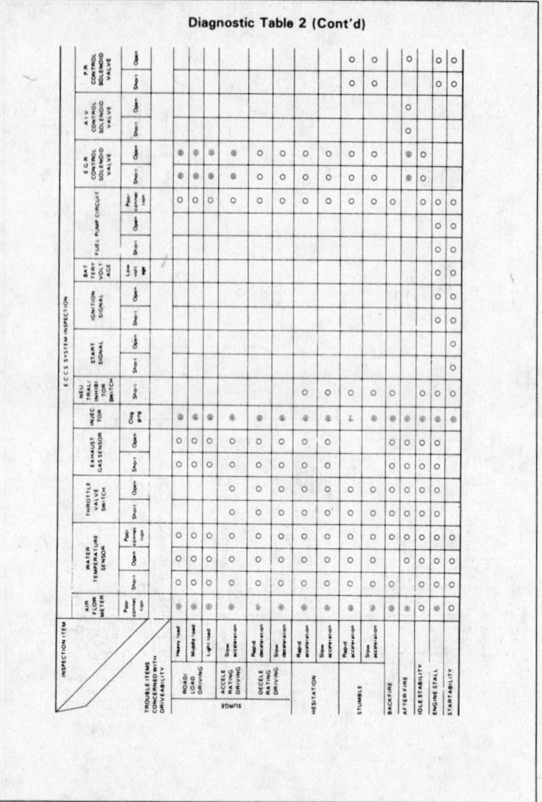

ECCS SELF-DIAGNOSTIC DESCRIPTION—1988 200SX WITH CA20E ENGINE

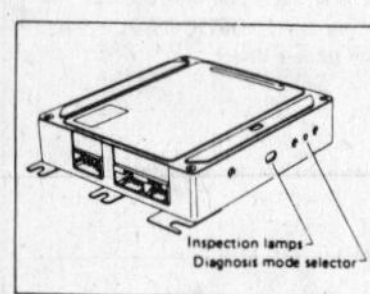
Inspection lamps
Diagnosis mode selector

Description

The self-diagnosis is useful to diagnose malfunctions in major sensors and actuators of the E.C.C.S. system. There are 5 modes in the self-diagnosis system.

1. **Mode I – Mixture ratio feedback control monitor A**
   - During closed loop condition:
     The green inspection lamp turns ON when lean condition is detected and goes OFF by rich condition.
   - During open loop condition:
     The green inspection lamp remains ON or OFF.
2. **Mode II – Mixture ratio feedback control monitor B**
   The green inspection lamp function is the same as Mode I.
   - During closed loop condition:
     The red inspection lamp turns ON and OFF simultaneously with the green inspection lamp when the mixture ratio is controlled within the specified value.
   - During open loop condition:
     The red inspection lamp remains ON or OFF.
3. **Mode III – Self-diagnosis**
   This mode is the same as the former self-diagnosis in self-diagnosis mode.
4. **Mode IV – Switches ON/OFF diagnosis**
   During this mode, the inspection lamps monitor the switch ON-OFF condition.
   - Idle switch
   - Starter switch
   - Vehicle speed sensor
5. **Mode V – Real time diagnosis**
   The moment the malfunction is detected, the display will be presented immediately. That is, the condition at which the malfunction occurs can be found by observing the inspection lamps during driving test.

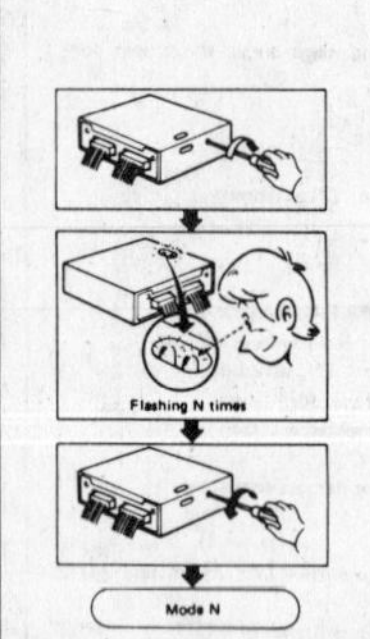
Flashing N times
Mode N

Description (Cont'd)

SWITCHING THE MODES

1. Turn ignition switch "ON".
2. Turn diagnostic mode selector on E.C.U. fully clockwise and wait the inspection lamps flash.
3. Count the number of the flashing time, and after the inspection lamps have flashed the number of the required mode, turn diagnostic mode selector fully counterclockwise immediately.

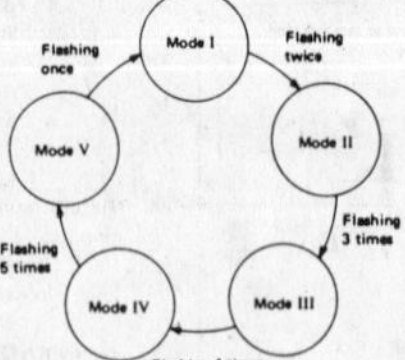
Mode I
Flashing twice
Mode II
Flashing 3 times
Mode III
Flashing 4 times
Mode IV
Flashing 5 times
Mode V
Flashing once

NOTE:
When the ignition switch is turned off during diagnosis, in each mode, and then turned back on again after the power to the E.C.U. has dropped off completely, the diagnosis will automatically return to Mode I.
The stored memory would be lost if:

1. Battery terminal is disconnected.
2. After selecting Mode III, Mode IV is selected.
   However, if the diagnostic mode selector is kept turned fully clockwise, it will continue to change in the order of Mode I → II → III → IV → V → I ... etc., and in this state the stored memory will not be erased.

ECCS SELF-DIAGNOSTIC DESCRIPTION (CONT.)—1988 200SX WITH CA20E

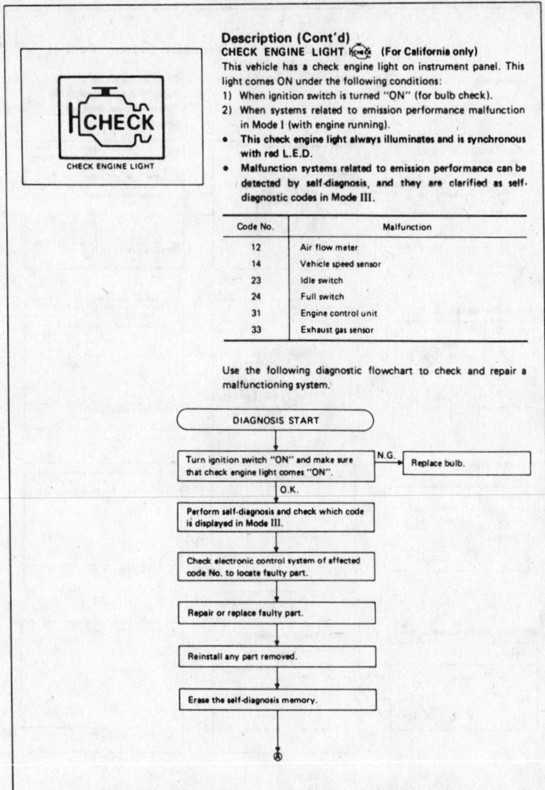

CHECK
CHECK ENGINE LIGHT

Description (Cont'd)

CHECK ENGINE LIGHT (For California only)

This vehicle has a check engine light on instrument panel. This light comes ON under the following conditions:

1) When ignition switch is turned "ON" (for bulb check).
2) When systems related to emission performance malfunction in Mode I (with engine running).

- **This check engine light always illuminates and is synchronous with red L.E.D.**
- **Malfunction systems related to emission performance can be detected by self-diagnosis, and they are clarified as self-diagnostic codes in Mode III.**

Code No.	Malfunction
12	Air flow meter
14	Vehicle speed sensor
23	Idle switch
24	Full switch
31	Engine control unit
33	Exhaust gas sensor

Use the following diagnostic flowchart to check and repair a malfunctioning system.

DIAGNOSIS START
Turn ignition switch "ON" and make sure that check engine light comes "ON".
N.G.
Replace bulb.
O.K.
Perform self-diagnosis and check which code is displayed in Mode III.
Check electronic control system of affected code No. to locate faulty part.
Repair or replace faulty part.
Reinstall any part removed.
Erase the self-diagnosis memory.
A

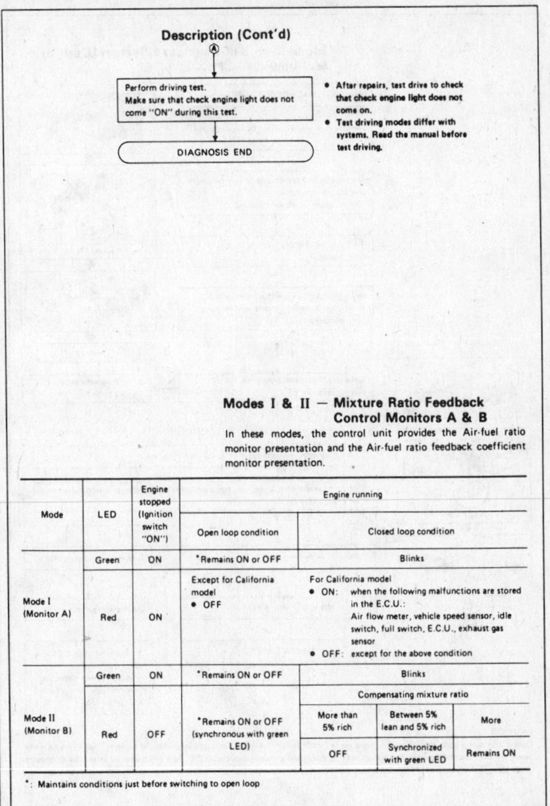

Description (Cont'd)

A
Perform driving test.
Make sure that check engine light does not come "ON" during this test.
DIAGNOSIS END

- **After repairs, test drive to check that check engine light does not come on.**
- **Test driving modes differ with systems. Read the manual before test driving.**

Modes I & II — Mixture Ratio Feedback Control Monitors A & B

In these modes, the control unit provides the Air-fuel ratio monitor presentation and the Air-fuel ratio feedback coefficient monitor presentation.

Mode	LED	Engine stopped (Ignition switch "ON")	Engine running: Open loop condition	Engine running: Closed loop condition		
Mode I (Monitor A)	Green	ON	*Remains ON or OFF	Blinks		
	Red	ON	Except for California model • OFF	For California model • ON: when the following malfunctions are stored in the E.C.U.: Air flow meter, vehicle speed sensor, idle switch, full switch, E.C.U., exhaust gas sensor • OFF: except for the above condition		
Mode II (Monitor B)	Green	ON	*Remains ON or OFF	Blinks		
	Red	OFF	*Remains ON or OFF (synchronous with green LED)	Compensating mixture ratio		
				More than 5% rich	Between 5% lean and 5% rich	More
				OFF	Synchronized with green LED	Remains ON

*: Maintains conditions just before switching to open loop

ECCS SELF-DIAGNOSTIC MODE III—1988 200SX WITH CA20E ENGINE

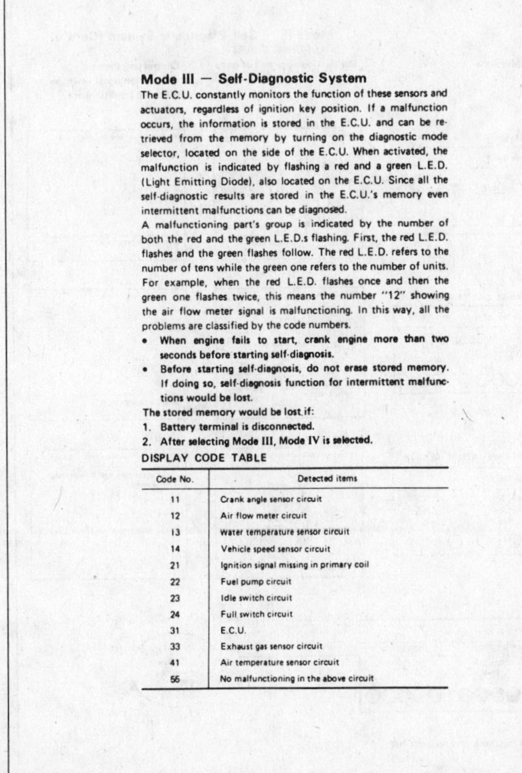

Mode III — Self-Diagnostic System

The E.C.U. constantly monitors the function of these sensors and actuators, regardless of ignition key position. If a malfunction occurs, the information is stored in the E.C.U. and can be retrieved from the memory by turning on the diagnostic mode selector, located on the side of the E.C.U. When activated, the malfunction is indicated by flashing a red and a green L.E.D. (Light Emitting Diode), also located on the E.C.U. Since all the self-diagnostic results are stored in the E.C.U.'s memory even intermittent malfunctions can be diagnosed.

A malfunctioning part's group is indicated by the number of both the red and the green L.E.D.s flashing. First, the red L.E.D. flashes and the green flashes follow. The red L.E.D. refers to the number of tens while the green one refers to the number of units. For example, when the red L.E.D. flashes once and then the green one flashes twice, this means the number "12" showing the air flow meter signal is malfunctioning. In this way, all the problems are classified by the code numbers.

- **When engine fails to start, crank engine more than two seconds before starting self-diagnosis.**
- **Before starting self-diagnosis, do not erase stored memory. If doing so, self-diagnosis function for intermittent malfunctions would be lost.**

The stored memory would be lost if:

1. **Battery terminal is disconnected.**
2. **After selecting Mode III, Mode IV is selected.**

DISPLAY CODE TABLE

Code No.	Detected items
11	Crank angle sensor circuit
12	Air flow meter circuit
13	Water temperature sensor circuit
14	Vehicle speed sensor circuit
21	Ignition signal missing in primary coil
22	Fuel pump circuit
23	Idle switch circuit
24	Full switch circuit
31	E.C.U.
33	Exhaust gas sensor circuit
41	Air temperature sensor circuit
55	No malfunctioning in the above circuit

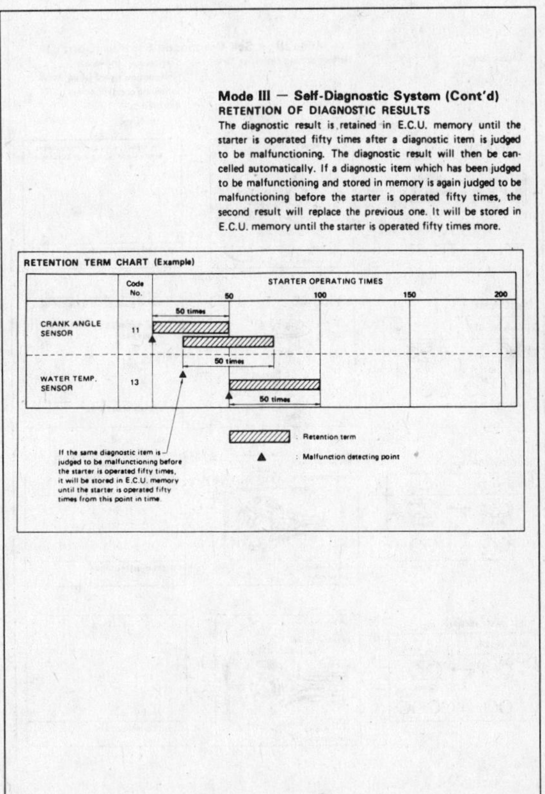

Mode III — Self-Diagnostic System (Cont'd)

RETENTION OF DIAGNOSTIC RESULTS

The diagnostic result is retained in E.C.U. memory until the starter is operated fifty times after a diagnostic item is judged to be malfunctioning. The diagnostic result will then be cancelled automatically. If a diagnostic item which has been judged to be malfunctioning and stored in memory is again judged to be malfunctioning before the starter is operated fifty times, the second result will replace the previous one. It will be stored in E.C.U. memory until the starter is operated fifty times more.

RETENTION TERM CHART (Example)
Code No.
STARTER OPERATING TIMES
50
100
150
200
CRANK ANGLE SENSOR
11
50 times
WATER TEMP. SENSOR
13
50 times
50 times
: Retention term
: Malfunction detecting point
If the same diagnostic item is judged to be malfunctioning before the starter is operated fifty times, it will be stored in E.C.U. memory until the starter is operated fifty times from this point in time.

ECCS SELF-DIAGNOSTIC MODE III DISPLAY CHARTS—1988 200SX WITH CA20E

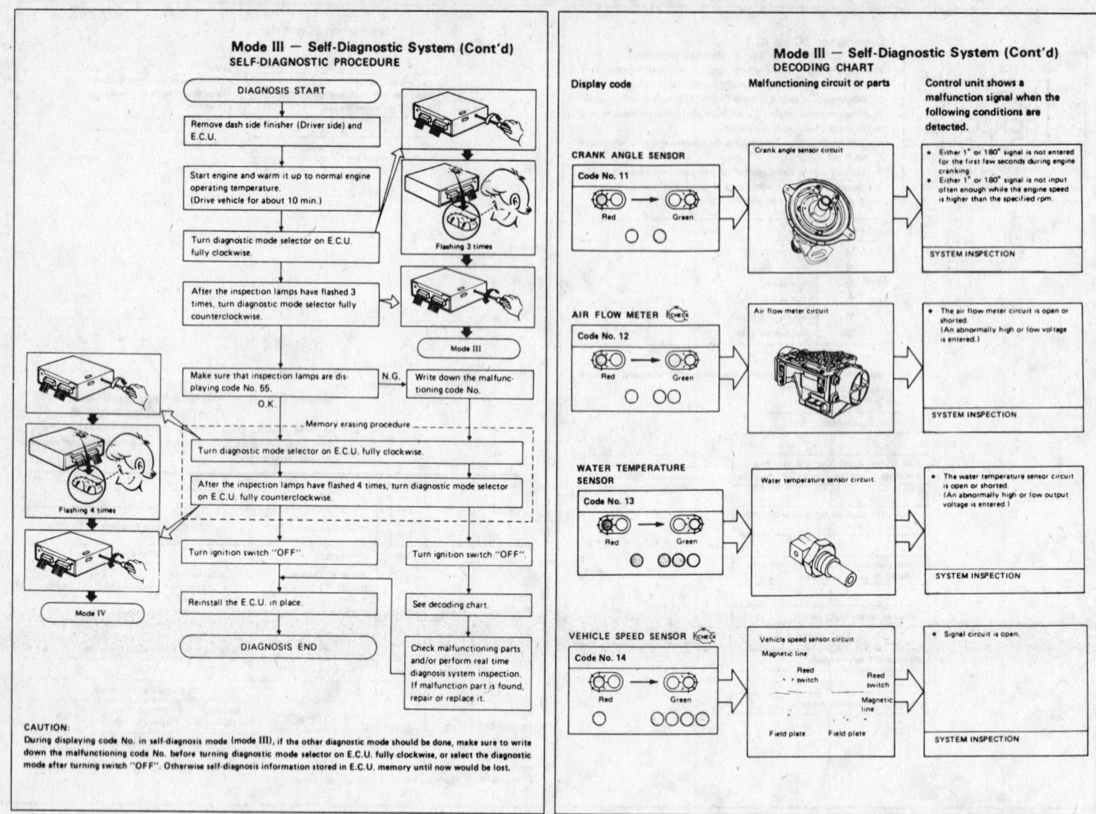
Mode III — Self-Diagnostic System (Cont'd)
SELF-DIAGNOSTIC PROCEDURE
DIAGNOSIS START
Remove dash side finisher (Driver side) and E.C.U.
Start engine and warm it up to normal engine operating temperature. (Drive vehicle for about 10 min.)
Turn diagnostic mode selector on E.C.U. fully clockwise.
Flashing 3 times
After the inspection lamps have flashed 3 times, turn diagnostic mode selector fully counterclockwise.
Mode III
Make sure that inspection lamps are displaying code No. 55.
N.G.
Write down the malfunctioning code No.
O.K.
Memory erasing procedure
Turn diagnostic mode selector on E.C.U. fully clockwise.
After the inspection lamps have flashed 4 times, turn diagnostic mode selector on E.C.U. fully counterclockwise.
Flashing 4 times
Turn ignition switch "OFF".
Turn ignition switch "OFF".
Mode IV
Reinstall the E.C.U. in place.
See decoding chart.
DIAGNOSIS END
Check malfunctioning parts and/or perform real time diagnosis system inspection. If malfunction part is found, repair or replace it.
CAUTION:
During displaying code No. in self-diagnosis mode (mode III), if the other diagnostic mode should be done, make sure to write down the malfunctioning code No. before turning diagnostic mode selector on E.C.U. fully clockwise, or select the diagnostic mode after turning switch "OFF". Otherwise self-diagnosis information stored in E.C.U. memory until now would be lost.
Mode III — Self-Diagnostic System (Cont'd)
DECODING CHART
Display code
Malfunctioning circuit or parts
Control unit shows a malfunction signal when the following conditions are detected.
CRANK ANGLE SENSOR
Code No. 11
Red
Green
Crank angle sensor circuit
Either 1° or 180° signal is not entered for the first few seconds during engine cranking.
Either 1° or 180° signal is not input often enough while the engine speed is higher than the specified rpm.
SYSTEM INSPECTION
AIR FLOW METER
Code No. 12
Air flow meter circuit
The air flow meter circuit is open or shorted. (An abnormally high or low voltage is entered.)
SYSTEM INSPECTION
WATER TEMPERATURE SENSOR
Code No. 13
Water temperature sensor circuit.
The water temperature sensor circuit is open or shorted. (An abnormally high or low output voltage is entered.)
SYSTEM INSPECTION
VEHICLE SPEED SENSOR
Code No. 14
Vehicle speed sensor circuit
Magnetic line
Reed switch
Field plate
Signal circuit is open.
SYSTEM INSPECTION

ECCS SELF-DIAGNOSTIC MODE III DISPLAY CHARTS (CONT.)—1988 200SX WITH CA20E ENGINE

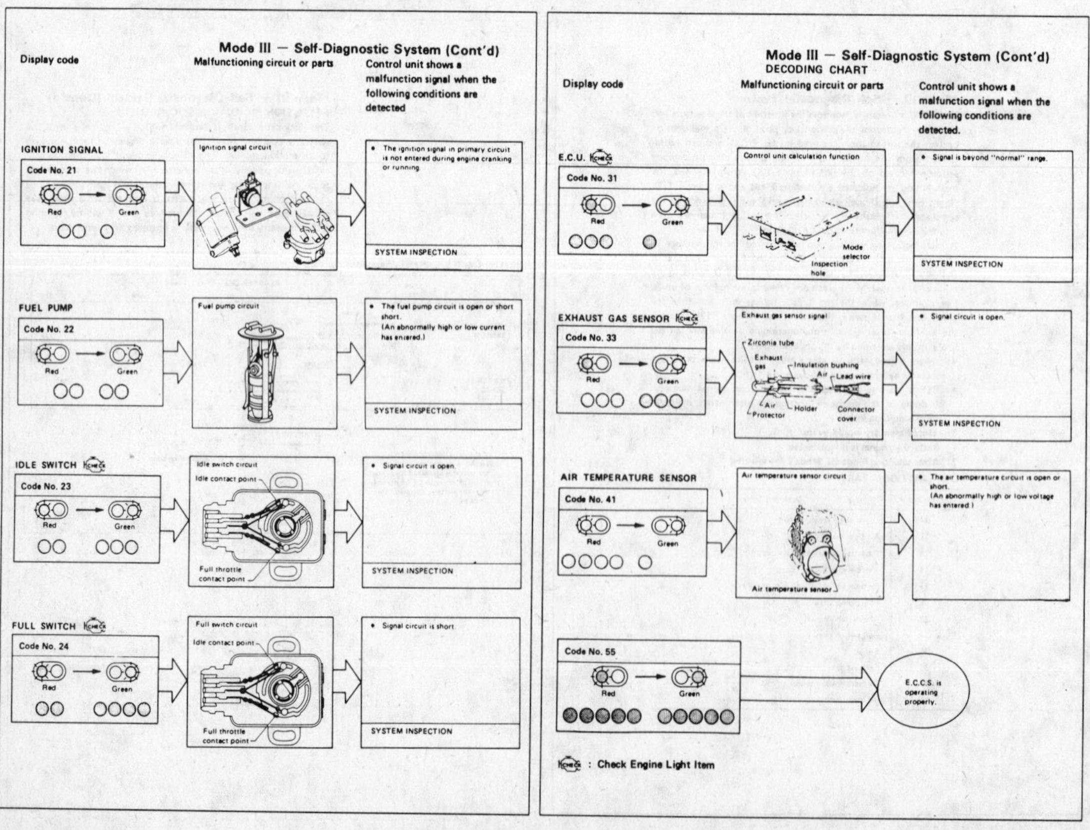
Mode III — Self-Diagnostic System (Cont'd)
Display code
Malfunctioning circuit or parts
Control unit shows a malfunction signal when the following conditions are detected
IGNITION SIGNAL
Code No. 21
Red
Green
Ignition signal circuit
The ignition signal in primary circuit is not entered during engine cranking or running.
SYSTEM INSPECTION
FUEL PUMP
Code No. 22
Fuel pump circuit
The fuel pump circuit is open or short short. (An abnormally high or low current has entered.)
SYSTEM INSPECTION
IDLE SWITCH
Code No. 23
Idle switch circuit
Idle contact point
Full throttle contact point
Signal circuit is open.
SYSTEM INSPECTION
FULL SWITCH
Code No. 24
Full switch circuit
Idle contact point
Full throttle contact point
Signal circuit is short.
SYSTEM INSPECTION
Mode III — Self-Diagnostic System (Cont'd)
DECODING CHART
E.C.U.
Code No. 31
Control unit calculation function
Mode selector
Inspection hole
Signal is beyond "normal" range.
SYSTEM INSPECTION
EXHAUST GAS SENSOR
Code No. 33
Exhaust gas sensor signal
Zirconia tube
Exhaust gas
Insulation bushing
Air
Lead wire
Holder
Connector cover
Protector
Signal circuit is open.
SYSTEM INSPECTION
AIR TEMPERATURE SENSOR
Code No. 41
Air temperature sensor circuit
Air temperature sensor
The air temperature circuit is open or short. (An abnormally high or low voltage has entered.)
Code No. 55
E.C.C.S. is operating properly.
: Check Engine Light Item

ECCS SELF-DIAGNOSTIC MODE IV—1988 200SX WITYH CA20E ENGINE

Mode IV — Switches ON/OFF Diagnostic System

In switches ON/OFF diagnosis system, ON/OFF operation of the following switches can be detected continuously.

- Idle switch
- Starter switch
- Vehicle speed sensor

(1) Idle switch & starter switch

The switches ON/OFF status at the point when mode IV is selected is stored in E.C.U. memory. When either switch is turned from "ON" to "OFF" or "OFF" to "ON", the red L.E.D. on E.C.U. alternately comes and goes off each time switching is detected.

(2) Vehicle speed sensor

The switches ON/OFF status at the point when mode IV is selected is stored in E.C.U. memory. When vehicle speed is 20 km/h (12 MPH) or slower, the green L.E.D. on E.C.U. is off. When vehicle speed exceeds 20 km/h (12 MPH), the green L.E.D. on E.C.U. turns "ON".

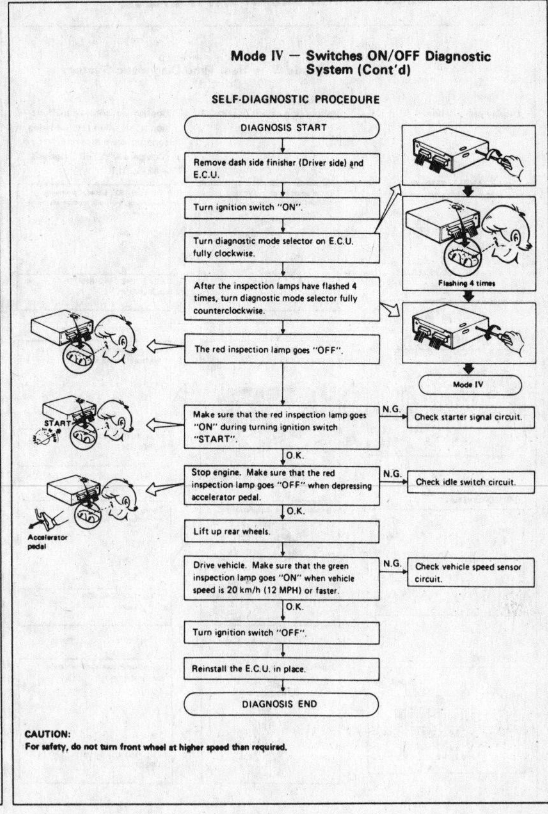
Mode IV — Switches ON/OFF Diagnostic System (Cont'd)
SELF-DIAGNOSTIC PROCEDURE
DIAGNOSIS START
Remove dash side finisher (Driver side) and E.C.U.
Turn ignition switch "ON".
Turn diagnostic mode selector on E.C.U. fully clockwise.
After the inspection lamps have flashed 4 times, turn diagnostic mode selector fully counterclockwise.
Flashing 4 times
Mode IV
The red inspection lamp goes "OFF".
Make sure that the red inspection lamp goes "ON" during turning ignition switch "START".
START
N.G.
Check starter signal circuit.
O.K.
Stop engine. Make sure that the red inspection lamp goes "OFF" when depressing accelerator pedal.
Accelerator pedal
N.G.
Check idle switch circuit.
O.K.
Lift up rear wheels.
Drive vehicle. Make sure that the green inspection lamp goes "ON" when vehicle speed is 20 km/h (12 MPH) or faster.
N.G.
Check vehicle speed sensor circuit.
O.K.
Turn ignition switch "OFF".
Reinstall the E.C.U. in place.
DIAGNOSIS END
CAUTION:
For safety, do not turn front wheel at higher speed than required.

ECCS SELF-DIAGNOSTIC MODE V—1988 200SX WITH CA20E ENGINE

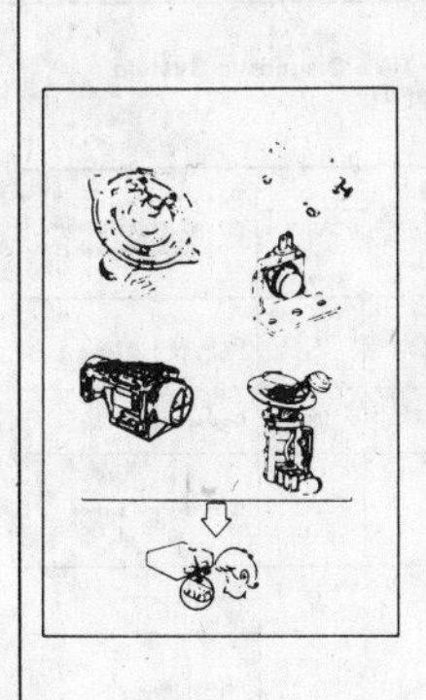

Mode V — Real Time Diagnostic System

In real time diagnosis, if any of the following items are judged to be faulty, a malfunction is indicated immediately.

- Crank angle sensor (180° signal & 1° signal)
- Ignition signal
- Air flow meter output signal
- Fuel pump

Consequently, this diagnosis is a very effective measure to diagnose whether the above systems cause the malfunction or not, during driving test. Compared with self-diagnosis, real time diagnosis is very sensitive, and can detect malfunctioning conditions in a moment. Further, items regarded to be malfunctions in this diagnosis are not stored in E.C.U. memory.

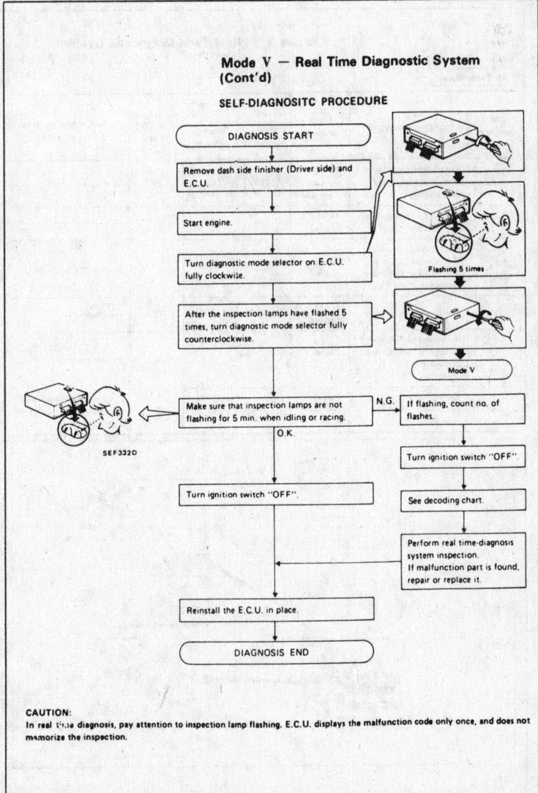
Mode V — Real Time Diagnostic System (Cont'd)
SELF-DIAGNOSITC PROCEDURE
DIAGNOSIS START
Remove dash side finisher (Driver side) and E.C.U.
Start engine.
Turn diagnostic mode selector on E.C.U. fully clockwise.
Flashing 5 times
After the inspection lamps have flashed 5 times, turn diagnostic mode selector fully counterclockwise.
Mode V
Make sure that inspection lamps are not flashing for 5 min. when idling or racing.
N.G.
If flashing, count no. of flashes.
O.K.
SEF332D
Turn ignition switch "OFF".
Turn ignition switch "OFF".
See decoding chart.
Perform real time-diagnosis system inspection.
If malfunction part is found, repair or replace it.
Reinstall the E.C.U. in place.
DIAGNOSIS END
CAUTION:
In real time diagnosis, pay attention to inspection lamp flashing. E.C.U. displays the malfunction code only once, and does not memorize the inspection.

ECCS SELF-DIAGNOSTIC MODE V DECODING CHART—1988 200SX WITH CA20E

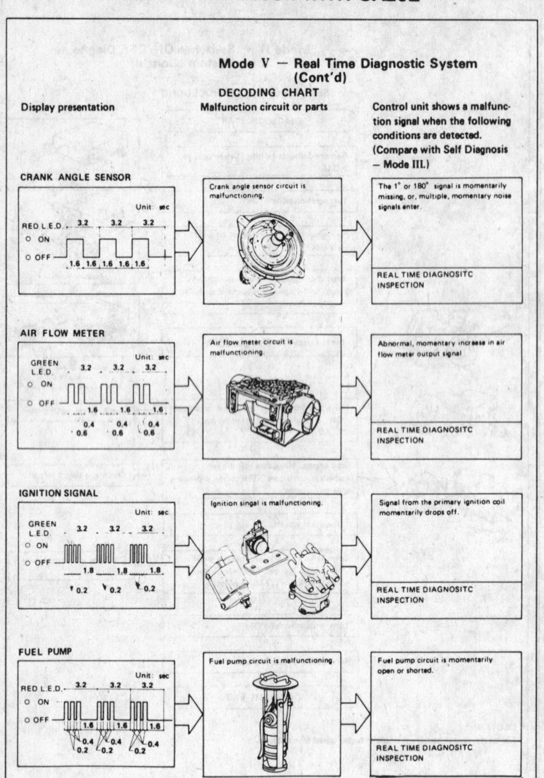
Mode V — Real Time Diagnostic System (Cont'd)
DECODING CHART
Display presentation
Malfunction circuit or parts
Control unit shows a malfunction signal when the following conditions are detected. (Compare with Self Diagnosis – Mode III.)
CRANK ANGLE SENSOR
Unit: sec
RED L.E.D.
ON
OFF
3.2
1.6
Crank angle sensor circuit is malfunctioning.
The 1° or 180° signal is momentarily missing, or, multiple, momentary noise signals enter.
REAL TIME DIAGNOSITC INSPECTION
AIR FLOW METER
GREEN L.E.D.
0.4
0.6
Air flow meter circuit is malfunctioning.
Abnormal, momentary increase in air flow meter output signal
REAL TIME DIAGNOSITC INSPECTION
IGNITION SIGNAL
1.8
0.2
Ignition signal is malfunctioning.
Signal from the primary ignition coil momentarily drops off.
REAL TIME DIAGNOSITC INSPECTION
FUEL PUMP
Fuel pump circuit is malfunctioning.
Fuel pump circuit is momentarily open or shorted.
REAL TIME DIAGNOSITC INSPECTION

ECCS SELF-DIAGNOSTIC MODE V INSPECTION CHART—1988 200SX WITH CA20E

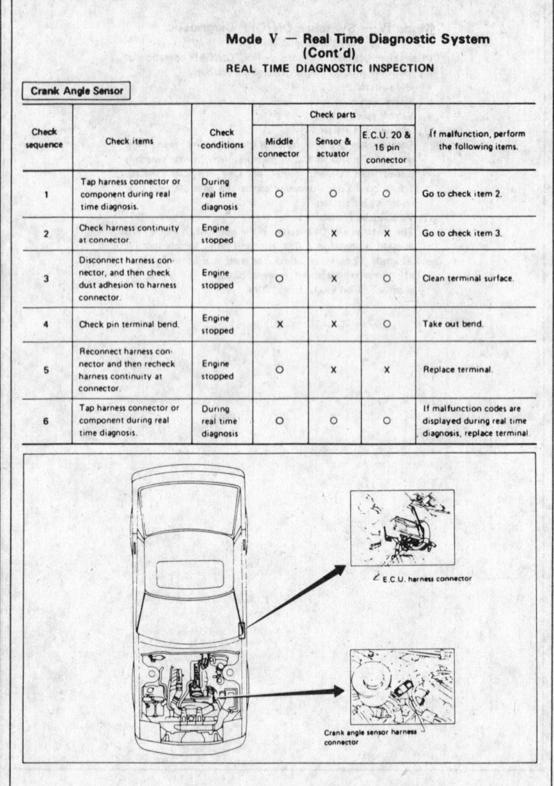

Mode V — Real Time Diagnostic System (Cont'd)
REAL TIME DIAGNOSTIC INSPECTION

Crank Angle Sensor

Check sequence	Check items	Check conditions	Check parts: Middle connector	Sensor & actuator	E.C.U. 20 & 16 pin connector	If malfunction, perform the following items.
1	Tap harness connector or component during real time diagnosis.	During real time diagnosis	O	O	O	Go to check item 2.
2	Check harness continuity at connector.	Engine stopped	O	X	X	Go to check item 3.
3	Disconnect harness connector, and then check dust adhesion to harness connector.	Engine stopped	O	X	O	Clean terminal surface.
4	Check pin terminal bend.	Engine stopped	X	X	O	Take out bend.
5	Reconnect harness connector and then recheck harness continuity at connector.	Engine stopped	O	X	X	Replace terminal.
6	Tap harness connector or component during real time diagnosis.	During real time diagnosis	O	O	O	If malfunction codes are displayed during real time diagnosis, replace terminal.

E.C.U. harness connector
Crank angle sensor harness connector

ECCS SELF-DIAGNOSTIC MODE V INSPECTION CHART (CONT.)—1988 200SX WITH CA20E ENGINE

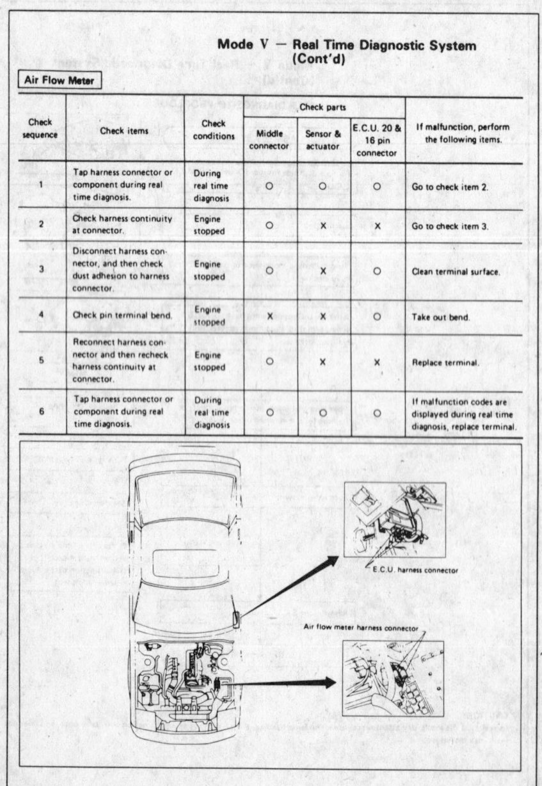

Mode V — Real Time Diagnostic System (Cont'd)

Air Flow Meter

Check sequence	Check items	Check conditions	Check parts: Middle connector	Sensor & actuator	E.C.U. 20 & 16 pin connector	If malfunction, perform the following items.
1	Tap harness connector or component during real time diagnosis.	During real time diagnosis	O	O	O	Go to check item 2.
2	Check harness continuity at connector.	Engine stopped	O	X	X	Go to check item 3.
3	Disconnect harness connector, and then check dust adhesion to harness connector.	Engine stopped	O	X	O	Clean terminal surface.
4	Check pin terminal bend.	Engine stopped	X	X	O	Take out bend.
5	Reconnect harness connector and then recheck harness continuity at connector.	Engine stopped	O	X	X	Replace terminal.
6	Tap harness connector or component during real time diagnosis.	During real time diagnosis	O	O	O	If malfunction codes are displayed during real time diagnosis, replace terminal.

E.C.U. harness connector
Air flow meter harness connector

Mode V — Real Time Diagnostic System (Cont'd)

Ignition Signal

Check sequence	Check items	Check conditions	Check parts: Middle connector	Sensor & actuator	E.C.U. 20 & 16 pin connector	If malfunction, perform the following items.
1	Tap harness connector or component during real time diagnosis.	During real time diagnosis	O	O	O	Go to check item 2.
2	Check harness continuity at connector.	Engine stopped	O	X	X	Go to check item 3.
3	Disconnect harness connector, and then check dust adhesion to harness connector.	Engine stopped	O	X	O	Clean terminal surface.
4	Check pin terminal bend.	Engine stopped	X	X	O	Take out bend.
5	Reconnect harness connector and then recheck harness continuity at connector.	Engine stopped	O	X	X	Replace terminal.
6	Tap harness connector or component during real time diagnosis.	During real time diagnosis	O	O	O	If malfunction codes are displayed during real time diagnosis, replace terminal.

E.C.U. harness connector
Exhaust side ignition coil
Check point
Intake side ignition coil
Power transistor
Power transistor
Ignition coil & power transistor harness connector

ECCS SELF-DIAGNOSTIC MODE V INPSECTION CHART (CONT.)—1988 CA20E

Mode V — Real Time Diagnostic System (Cont'd)

Fuel Pump

Check sequence	Check items	Check conditions	Check parts: Middle connector	Check parts: Sensor & actuator	Check parts: E.C.U. 20 & 16 pin connector	If malfunction, perform the following items.
1	Tap harness connector or component during real time diagnosis.	During real time diagnosis	O	O	O	Go to check item 2.
2	Check harness continuity at connector.	Engine stopped	O	X	X	Go to check item 3.
3	Disconnect harness connector, and then check dust adhesion to harness connector.	Engine stopped	O	X	O	Clean terminal surface.
4	Check pin terminal bend.	Engine stopped	X	X	O	Take out bend.
5	Reconnect harness connector and then recheck harness continuity at connector.	Engine stopped	O	X	X	Replace terminal.
6	Tap harness connector or component during real time diagnosis.	During real time diagnosis	O	O	O	If malfunction codes are displayed during real time diagnosis, replace terminal.

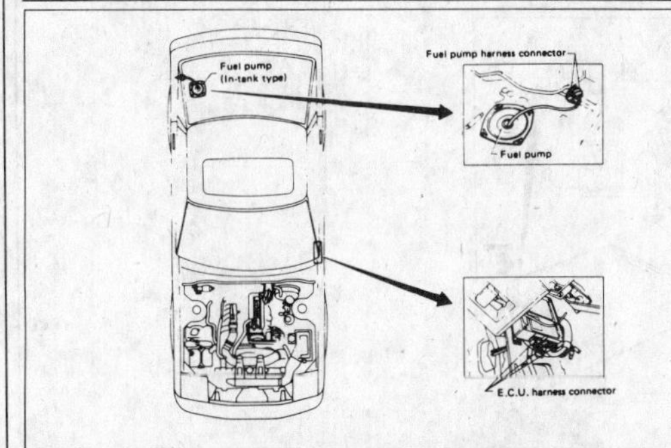
Fuel pump (In-tank type)
Fuel pump harness connector
Fuel pump
E.C.U. harness connector

ECCS SYSTEM CAUTION—1988 200SX WITH CA20E ENGINE

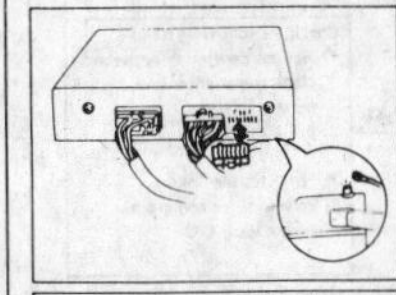

CAUTION:

1. Before connecting or disconnecting E.C.U. harness connector to or from any E.C.U., be sure to turn the ignition switch to the "OFF" position and disconnect the negative battery terminal in order not to damage E.C.U. as battery voltage is applied to E.C.U. even if ignition switch is turned off. Otherwise, there may be damage to the E.C.U.

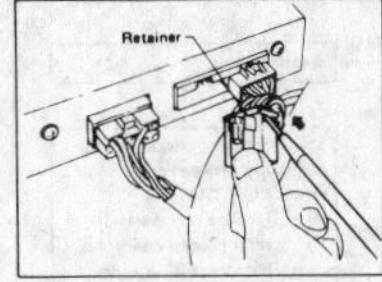
Retainer

2. When performing E.C.U. input/output signal inspection, remove pin terminal retainer from 20 and 16 pin connector to make it easier to insert tester probe into connector.

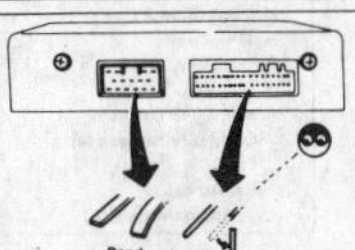
Bend
Break

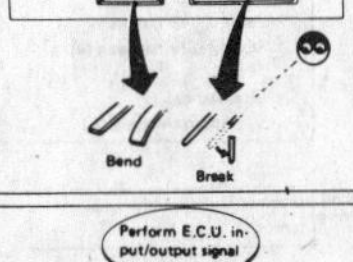

3. When connecting pin connectors into E.C.U. or disconnecting them from E.C.U., take care not to damage pin terminal of E.C.U. (Bend or break).
4. Make sure that there are not any bends or breaks on E.C.U. pin terminal, when connecting pin connectors into E.C.U.

Perform E.C.U. input/output signal inspection before replacement.
OLD ONE

5. Before replacing E.C.U., perform E.C.U. input/output signal inspection and make sure whether E.C.U. functions properly or not.

6. After performing this "ELECTRONIC CONTROL SYSTEM INSPECTION", perform E.C.C.S. self-diagnosis and driving test.

ECCS SELF-DIAGNOSTIC CHART—1988 200SX WITH CA20E ENGINE

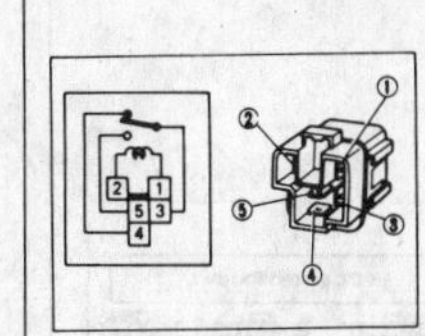

7. When measuring supply voltage of E.C.U. controlled components with a circuit tester, separate one tester probe from the other.
   If the two tester probes accidentally make contact with each other during measurement, the circuit will be shorted, resulting in damage to the power transistor of the control unit.

8. Keys to symbols
   - DISCONNECT : Check after disconnecting the connector to be measured.
   - CONNECT : Check after connecting the connector to be measured.
9. When measuring voltage or resistance at connector with tester probes, there are two methods of measurement; one is done from terminal side and the other from harness side. Before measuring, confirm symbol mark again.
   - HS : Inspection should be done from harness side.
   - TS : Inspection should be done from terminal side.

10. As for continuity check of joint connector.

CRANK ANGLE SENSOR (Code No. 11)

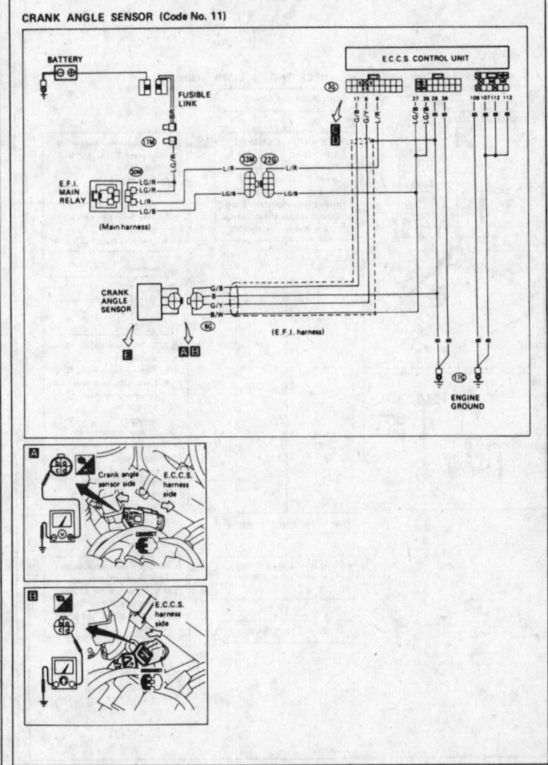
BATTERY
FUSIBLE LINK
E.C.C.S. CONTROL UNIT
E.F.I. MAIN RELAY
(Main harness)
CRANK ANGLE SENSOR
(E.F.I. harness)
ENGINE GROUND
Crank angle sensor side
E.C.C.S. harness side

ECCS SELF-DIAGNOSTIC CHART—1988 200SX WITH CA20E ENGINE

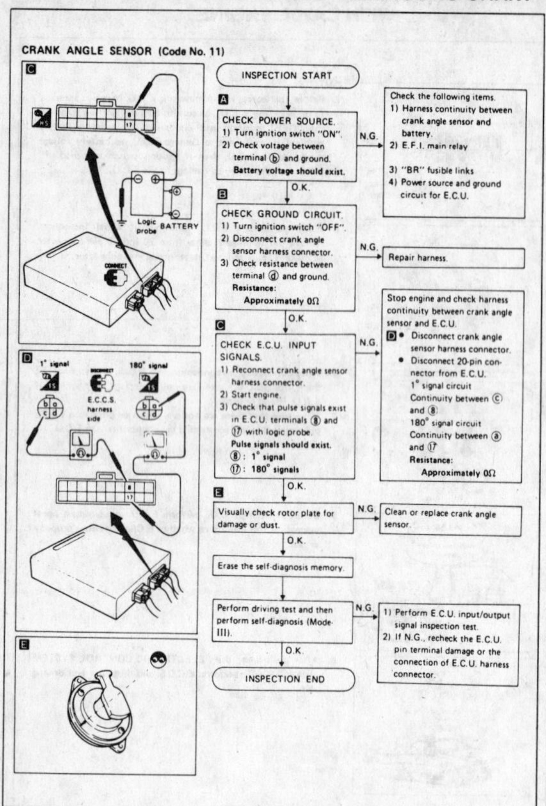
CRANK ANGLE SENSOR (Code No. 11)
INSPECTION START
A
CHECK POWER SOURCE.
1) Turn ignition switch "ON".
2) Check voltage between terminal ⓑ and ground.
Battery voltage should exist.
N.G.
Check the following items.
1) Harness continuity between crank angle sensor and battery.
2) E.F.I. main relay
3) "BR" fusible links
4) Power source and ground circuit for E.C.U.
O.K.
B
CHECK GROUND CIRCUIT.
1) Turn ignition switch "OFF".
2) Disconnect crank angle sensor harness connector.
3) Check resistance between terminal ⓓ and ground.
Resistance:
Approximately 0Ω
N.G.
Repair harness.
O.K.
C
CHECK E.C.U. INPUT SIGNALS.
1) Reconnect crank angle sensor harness connector.
2) Start engine.
3) Check that pulse signals exist in E.C.U. terminals ⑧ and ⑰ with logic probe.
Pulse signals should exist.
⑧ : 1° signal
⑰ : 180° signals
N.G.
Stop engine and check harness continuity between crank angle sensor and E.C.U.
D • Disconnect crank angle sensor harness connector.
• Disconnect 20-pin connector from E.C.U.
1° signal circuit
Continuity between ⓒ and ⑧
180° signal circuit
Continuity between ⓐ and ⑰
Resistance:
Approximately 0Ω
O.K.
E
Visually check rotor plate for damage or dust.
N.G.
Clean or replace crank angle sensor.
O.K.
Erase the self-diagnosis memory.
Perform driving test and then perform self-diagnosis (Mode III).
N.G.
1) Perform E.C.U. input/output signal inspection test.
2) If N.G., recheck the E.C.U. pin terminal damage or the connection of E.C.U. harness connector.
O.K.
INSPECTION END
C
Logic probe
BATTERY
CONNECT
D
1° signal
180° signal
DISCONNECT
E.C.C.S. harness side
E

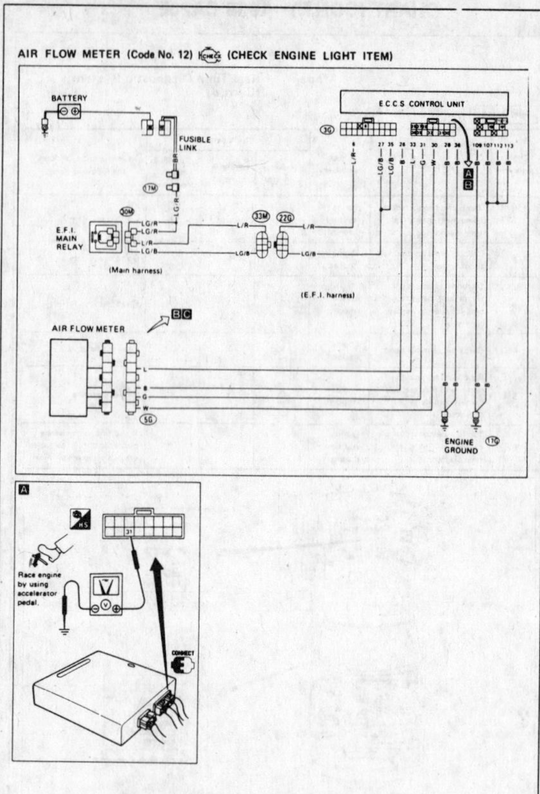
AIR FLOW METER (Code No. 12) (CHECK ENGINE LIGHT ITEM)
BATTERY
ECCS CONTROL UNIT
FUSIBLE LINK
E.F.I. MAIN RELAY
(Main harness)
(E.F.I. harness)
AIR FLOW METER
ENGINE GROUND
A
Race engine by using accelerator pedal.
CONNECT

ECCS SELF-DIAGNOSTIC CHART—1988 200SX WITH CA20E ENGINE

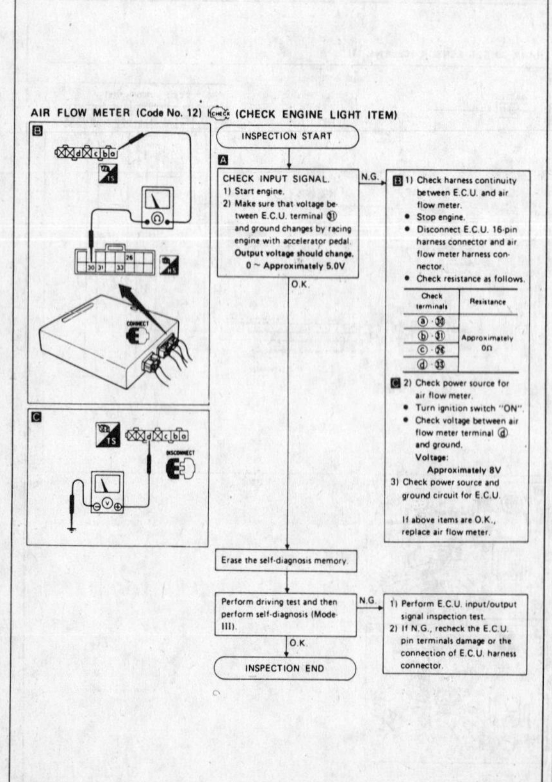
AIR FLOW METER (Code No. 12) (CHECK ENGINE LIGHT ITEM)
INSPECTION START
A
CHECK INPUT SIGNAL.
1) Start engine.
2) Make sure that voltage between E.C.U. terminal ㉛ and ground changes by racing engine with accelerator pedal.
Output voltage should change.
0 ~ Approximately 5.0V
N.G.
B 1) Check harness continuity between E.C.U. and air flow meter.
• Stop engine.
• Disconnect E.C.U. 16-pin harness connector and air flow meter harness connector.
• Check resistance as follows.
Check terminals
Resistance
ⓐ - ㉚
ⓑ - ㉛
ⓒ - ㉖
ⓓ - ㉝
Approximately 0Ω
C 2) Check power source for air flow meter.
• Turn ignition switch "ON".
• Check voltage between air flow meter terminal ⓓ and ground.
Voltage:
Approximately 8V
3) Check power source and ground circuit for E.C.U.
If above items are O.K., replace air flow meter.
O.K.
Erase the self-diagnosis memory.
Perform driving test and then perform self-diagnosis (Mode III).
N.G.
1) Perform E.C.U. input/output signal inspection test.
2) If N.G., recheck the E.C.U. pin terminals damage or the connection of E.C.U. harness connector.
O.K.
INSPECTION END
B
CONNECT
C
DISCONNECT

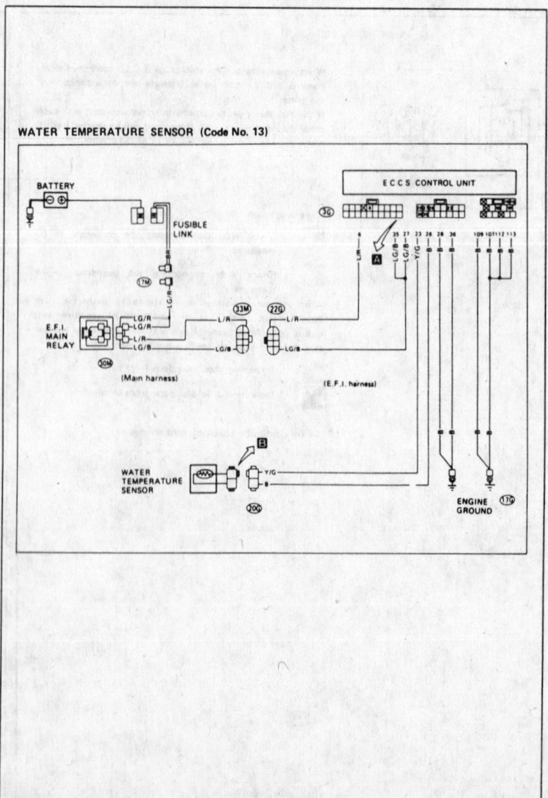
WATER TEMPERATURE SENSOR (Code No. 13)
BATTERY
ECCS CONTROL UNIT
FUSIBLE LINK
E.F.I. MAIN RELAY
(Main harness)
(E.F.I. harness)
WATER TEMPERATURE SENSOR
ENGINE GROUND

ECCS SELF-DIAGNOSTIC CHART—1988 200SX WITH CA20E ENGINE

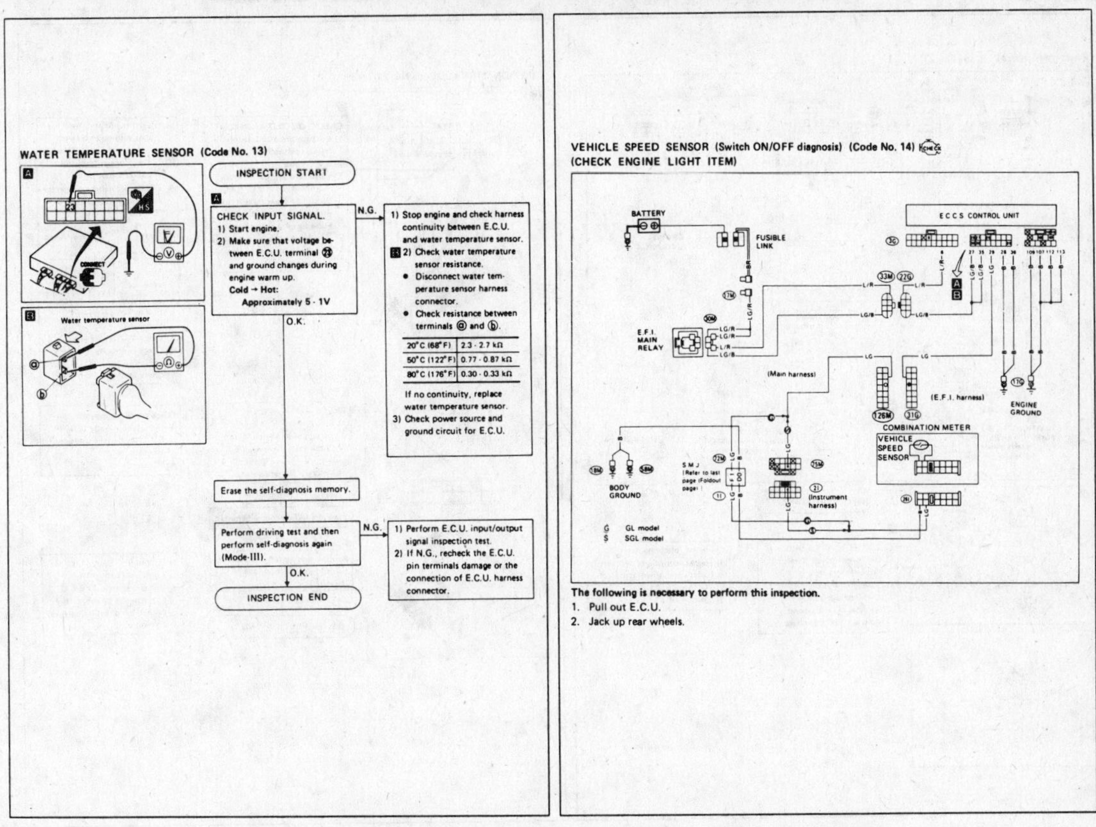
WATER TEMPERATURE SENSOR (Code No. 13)
INSPECTION START
CHECK INPUT SIGNAL.
1) Start engine.
2) Make sure that voltage between E.C.U. terminal ⑳ and ground changes during engine warm up.
Cold → Hot:
Approximately 5 - 1V
N.G.
1) Stop engine and check harness continuity between E.C.U. and water temperature sensor.
2) Check water temperature sensor resistance.
• Disconnect water temperature sensor harness connector.
• Check resistance between terminals ⓐ and ⓑ.
20°C (68°F) | 2.3 - 2.7 kΩ
50°C (122°F) | 0.77 - 0.87 kΩ
80°C (176°F) | 0.30 - 0.33 kΩ
If no continuity, replace water temperature sensor.
3) Check power source and ground circuit for E.C.U.
O.K.
Erase the self-diagnosis memory.
Perform driving test and then perform self-diagnosis again (Mode-III).
N.G.
1) Perform E.C.U. input/output signal inspection test.
2) If N.G., recheck the E.C.U. pin terminals damage or the connection of E.C.U. harness connector.
O.K.
INSPECTION END
Water temperature sensor
VEHICLE SPEED SENSOR (Switch ON/OFF diagnosis) (Code No. 14) (CHECK ENGINE LIGHT ITEM)
BATTERY
FUSIBLE LINK
ECCS CONTROL UNIT
E.F.I. MAIN RELAY
(Main harness)
(E.F.I. harness)
ENGINE GROUND
COMBINATION METER
VEHICLE SPEED SENSOR
BODY GROUND
S.M.J. (Refer to last page (Foldout page).)
(Instrument harness)
G GL model
S SGL model
The following is necessary to perform this inspection.
1. Pull out E.C.U.
2. Jack up rear wheels.

ECCS SELF-DIAGNOSTIC CHART—1988 200SX WITH CA20E ENGINE

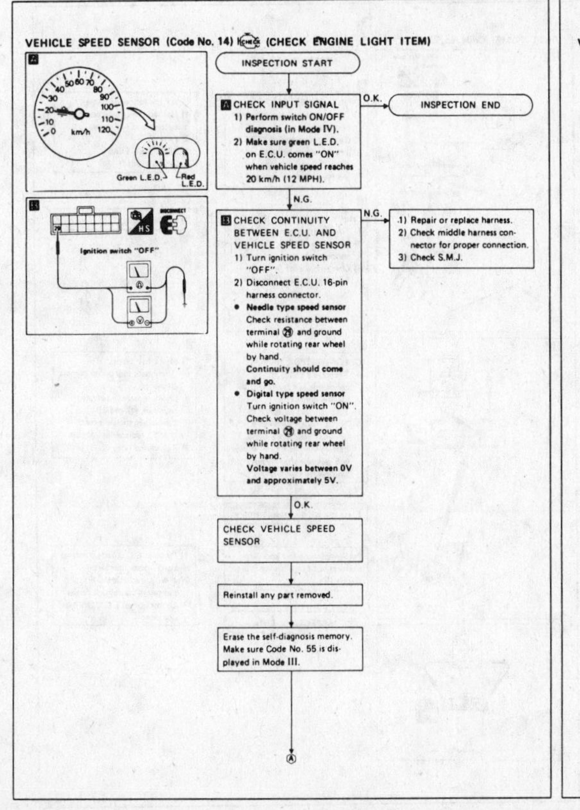
VEHICLE SPEED SENSOR (Code No. 14) (CHECK ENGINE LIGHT ITEM)
INSPECTION START
CHECK INPUT SIGNAL
1) Perform switch ON/OFF diagnosis (in Mode IV).
2) Make sure green L.E.D. on E.C.U. comes "ON" when vehicle speed reaches 20 km/h (12 MPH).
O.K.
INSPECTION END
km/h
Green L.E.D.
Red L.E.D.
N.G.
CHECK CONTINUITY BETWEEN E.C.U. AND VEHICLE SPEED SENSOR
1) Turn ignition switch "OFF".
2) Disconnect E.C.U. 16-pin harness connector.
• Needle type speed sensor
Check resistance between terminal ㉙ and ground while rotating rear wheel by hand.
Continuity should come and go.
• Digital type speed sensor
Turn ignition switch "ON".
Check voltage between terminal ㉙ and ground while rotating rear wheel by hand.
Voltage varies between 0V and approximately 5V.
N.G.
1) Repair or replace harness.
2) Check middle harness connector for proper connection.
3) Check S.M.J.
Ignition switch "OFF"
O.K.
CHECK VEHICLE SPEED SENSOR
Reinstall any part removed.
Erase the self-diagnosis memory. Make sure Code No. 55 is displayed in Mode III.
Ⓐ

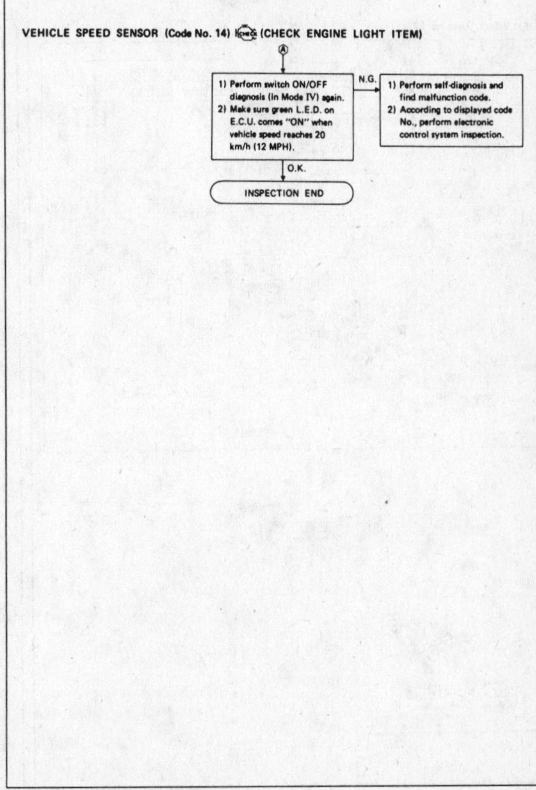
VEHICLE SPEED SENSOR (Code No. 14) (CHECK ENGINE LIGHT ITEM)
Ⓐ
1) Perform switch ON/OFF diagnosis (in Mode IV) again.
2) Make sure green L.E.D. on E.C.U. comes "ON" when vehicle speed reaches 20 km/h (12 MPH).
N.G.
1) Perform self-diagnosis and find malfunction code.
2) According to displayed code No., perform electronic control system inspection.
O.K.
INSPECTION END

ECCS SELF-DIAGNOSTIC CHART—1988 200SX WITH CA20E ENGINE

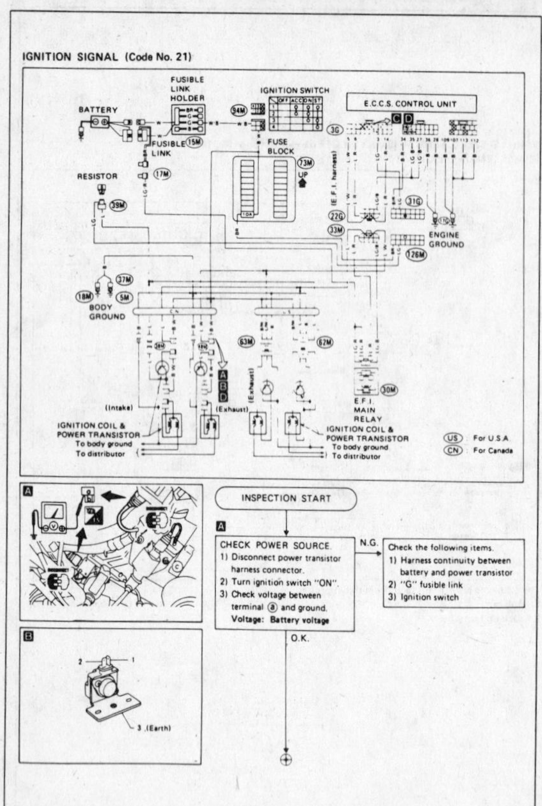
IGNITION SIGNAL (Code No. 21)
BATTERY
FUSIBLE LINK HOLDER
IGNITION SWITCH
E.C.C.S. CONTROL UNIT
FUSIBLE LINK
FUSE BLOCK
RESISTOR
ENGINE GROUND
BODY GROUND
(Intake)
(Exhaust)
E.F.I. MAIN RELAY
IGNITION COIL & POWER TRANSISTOR
To body ground
To distributor
US For U.S.A.
CN For Canada
INSPECTION START
A
CHECK POWER SOURCE.
1) Disconnect power transistor harness connector.
2) Turn ignition switch "ON".
3) Check voltage between terminal ⓐ and ground.
Voltage: Battery voltage
N.G.
Check the following items.
1) Harness continuity between battery and power transistor
2) "G" fusible link
3) Ignition switch
O.K.
3 (Earth)

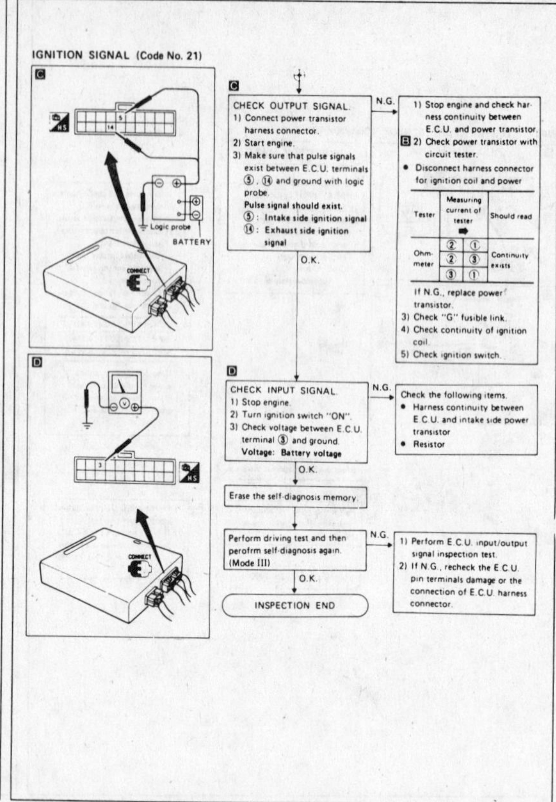
IGNITION SIGNAL (Code No. 21)
C
CHECK OUTPUT SIGNAL.
1) Connect power transistor harness connector.
2) Start engine.
3) Make sure that pulse signals exist between E.C.U. terminals ⑤, ⑭ and ground with logic probe.
Pulse signal should exist.
⑤: Intake side ignition signal
⑭: Exhaust side ignition signal
N.G.
1) Stop engine and check harness continuity between E.C.U. and power transistor.
B 2) Check power transistor with circuit tester.
• Disconnect harness connector for ignition coil and power transistor.
Tester
Measuring current of tester
Should read
Ohm-meter
Continuity exists
If N.G., replace power transistor.
3) Check "G" fusible link.
4) Check continuity of ignition coil.
5) Check ignition switch.
Logic probe
BATTERY
O.K.
D
CHECK INPUT SIGNAL.
1) Stop engine.
2) Turn ignition switch "ON".
3) Check voltage between E.C.U. terminal ③ and ground.
Voltage: Battery voltage
N.G.
Check the following items.
• Harness continuity between E.C.U. and intake side power transistor
• Resistor
O.K.
Erase the self-diagnosis memory.
Perform driving test and then perofrm self-diagnosis again. (Mode III)
N.G.
1) Perform E.C.U. input/output signal inspection test.
2) If N.G., recheck the E.C.U. pin terminals damage or the connection of E.C.U. harness connector.
O.K.
INSPECTION END

ECCS SELF-DIAGNOSTIC CHART—1988 200SX WITH CA20E ENGINE

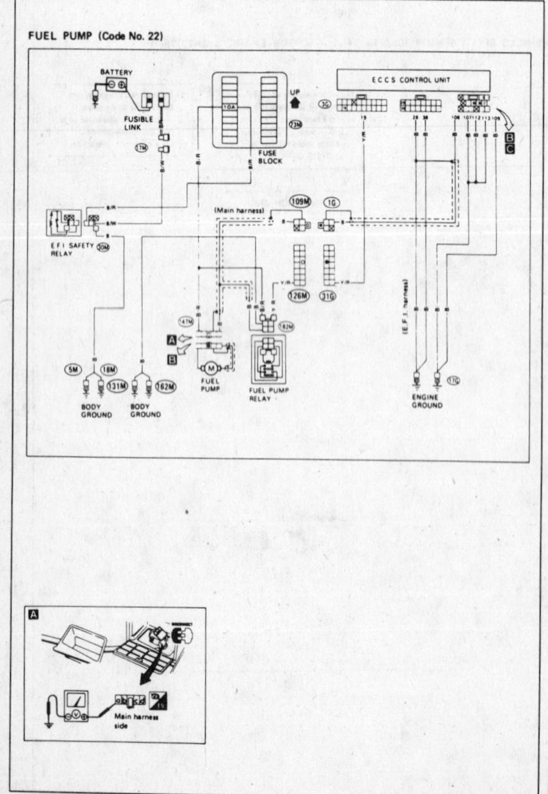
FUEL PUMP (Code No. 22)
BATTERY
FUSIBLE LINK
FUSE BLOCK
E.C.C.S. CONTROL UNIT
UP
(Main harness)
E.F.I. SAFETY RELAY
FUEL PUMP
FUEL PUMP RELAY
ENGINE GROUND
BODY GROUND
Main harness side

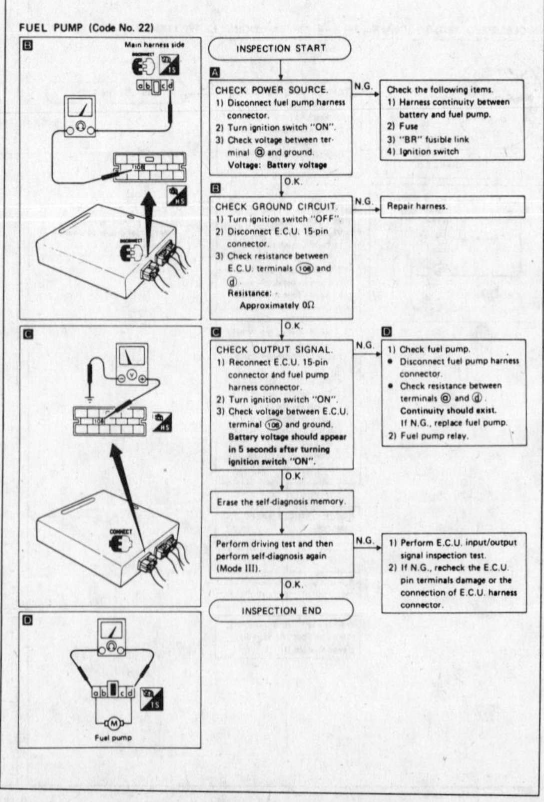
FUEL PUMP (Code No. 22)
Main harness side
INSPECTION START
A
CHECK POWER SOURCE.
1) Disconnect fuel pump harness connector.
2) Turn ignition switch "ON".
3) Check voltage between terminal ⓒ and ground.
Voltage: Battery voltage
N.G.
Check the following items.
1) Harness continuity between battery and fuel pump.
2) Fuse
3) "BR" fusible link
4) Ignition switch
O.K.
B
CHECK GROUND CIRCUIT.
1) Turn ignition switch "OFF".
2) Disconnect E.C.U. 15-pin connector.
3) Check resistance between E.C.U. terminals ⑩⑧ and ⓓ.
Resistance: Approximately 0Ω
N.G.
Repair harness.
O.K.
C
CHECK OUTPUT SIGNAL.
1) Reconnect E.C.U. 15-pin connector and fuel pump harness connector.
2) Turn ignition switch "ON".
3) Check voltage between E.C.U. terminal ⑩⑧ and ground.
Battery voltage should appear in 5 seconds after turning ignition switch "ON".
N.G.
D
1) Check fuel pump.
• Disconnect fuel pump harness connector.
• Check resistance between terminals ⓒ and ⓓ.
Continuity should exist.
If N.G., replace fuel pump.
2) Fuel pump relay.
O.K.
Erase the self-diagnosis memory.
Perform driving test and then perform self-diagnosis again (Mode III).
N.G.
1) Perform E.C.U. input/output signal inspection test.
2) If N.G., recheck the E.C.U. pin terminals damage or the connection of E.C.U. harness connector.
O.K.
INSPECTION END
Fuel pump

ECCS SELF-DIAGNOSTIC CHART—1988 200SX WITH CA20E ENGINE

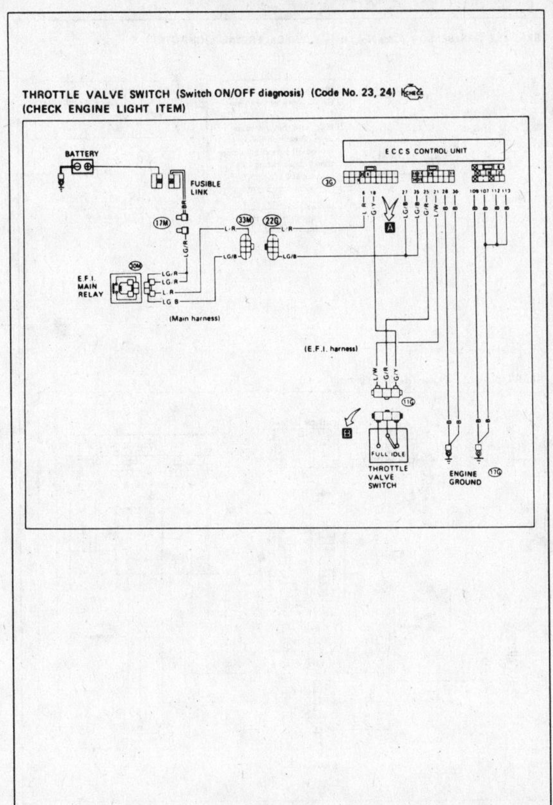
THROTTLE VALVE SWITCH (Switch ON/OFF diagnosis) (Code No. 23, 24)
(CHECK ENGINE LIGHT ITEM)
BATTERY
FUSIBLE LINK
E C C S CONTROL UNIT
E.F.I. MAIN RELAY
(Main harness)
(E.F.I. harness)
FULL IDLE
THROTTLE VALVE SWITCH
ENGINE GROUND

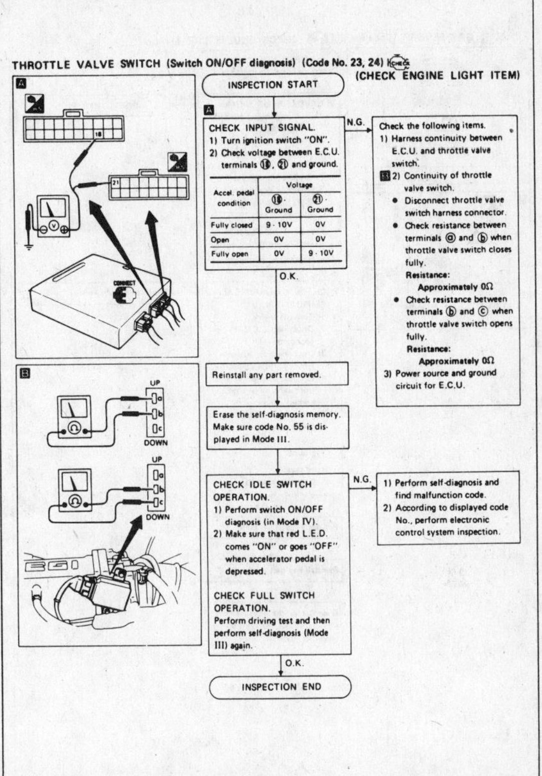
THROTTLE VALVE SWITCH (Switch ON/OFF diagnosis) (Code No. 23, 24)
(CHECK ENGINE LIGHT ITEM)
INSPECTION START
A
CHECK INPUT SIGNAL.
1) Turn ignition switch "ON".
2) Check voltage between E.C.U. terminals ⑱, ㉑ and ground.
Accel. pedal condition | Voltage ⑱ - Ground | ㉑ - Ground
Fully closed | 9 - 10V | 0V
Open | 0V | 0V
Fully open | 0V | 9 - 10V
N.G.
Check the following items.
1) Harness continuity between E.C.U. and throttle valve switch.
B 2) Continuity of throttle valve switch.
• Disconnect throttle valve switch harness connector.
• Check resistance between terminals ⓐ and ⓑ when throttle valve switch closes fully.
Resistance: Approximately 0Ω
• Check resistance between terminals ⓑ and ⓒ when throttle valve switch opens fully.
Resistance: Approximately 0Ω
3) Power source and ground circuit for E.C.U.
O.K.
Reinstall any part removed.
Erase the self-diagnosis memory. Make sure code No. 55 is displayed in Mode III.
CHECK IDLE SWITCH OPERATION.
1) Perform switch ON/OFF diagnosis (in Mode IV).
2) Make sure that red L.E.D. comes "ON" or goes "OFF" when accelerator pedal is depressed.
CHECK FULL SWITCH OPERATION.
Perform driving test and then perform self-diagnosis (Mode III) again.
N.G.
1) Perform self-diagnosis and find malfunction code.
2) According to displayed code No., perform electronic control system inspection.
O.K.
INSPECTION END
B
UP
DOWN
UP
DOWN

ECCS SELF-DIAGNOSTIC CHART—1988 200SX WITH CA20E ENGINE

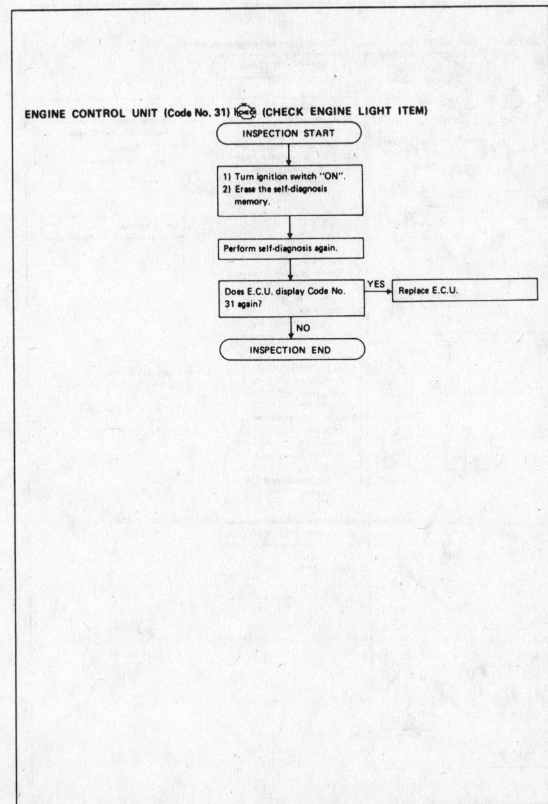
ENGINE CONTROL UNIT (Code No. 31) (CHECK ENGINE LIGHT ITEM)
INSPECTION START
1) Turn ignition switch "ON".
2) Erase the self-diagnosis memory.
Perform self-diagnosis again.
Does E.C.U. display Code No. 31 again?
YES
Replace E.C.U.
NO
INSPECTION END

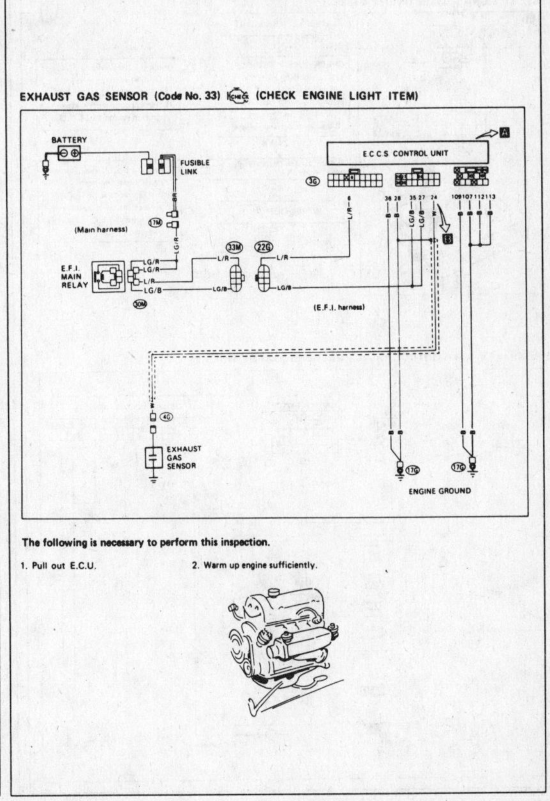
EXHAUST GAS SENSOR (Code No. 33) (CHECK ENGINE LIGHT ITEM)
BATTERY
FUSIBLE LINK
E C C S CONTROL UNIT
(Main harness)
E.F.I. MAIN RELAY
(E.F.I. harness)
EXHAUST GAS SENSOR
ENGINE GROUND
The following is necessary to perform this inspection.
1. Pull out E.C.U.
2. Warm up engine sufficiently.

ECCS SELF-DIAGNOSTIC CHART—1988 200SX WITH CA20E ENGINE

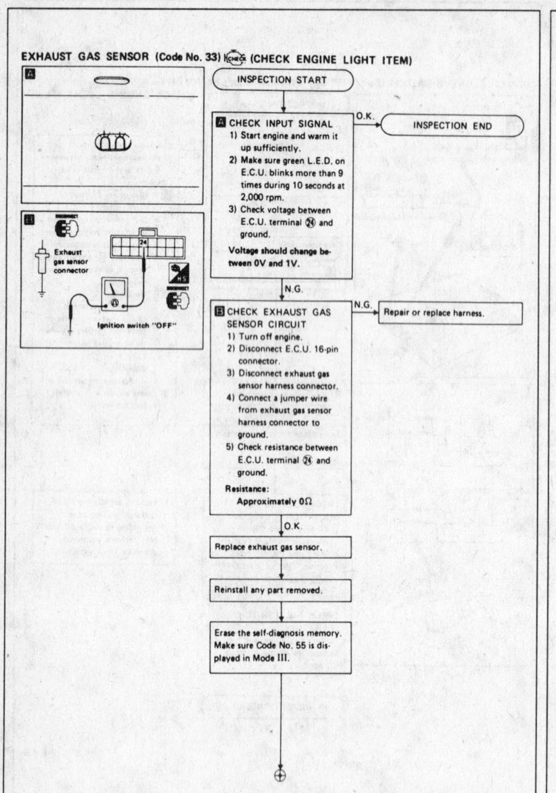
EXHAUST GAS SENSOR (Code No. 33) (CHECK ENGINE LIGHT ITEM)
INSPECTION START
A CHECK INPUT SIGNAL
1) Start engine and warm it up sufficiently.
2) Make sure green L.E.D. on E.C.U. blinks more than 9 times during 10 seconds at 2,000 rpm.
3) Check voltage between E.C.U. terminal 24 and ground.
Voltage should change between 0V and 1V.
O.K.
INSPECTION END
N.G.
B CHECK EXHAUST GAS SENSOR CIRCUIT
1) Turn off engine.
2) Disconnect E.C.U. 16-pin connector.
3) Disconnect exhaust gas sensor harness connector.
4) Connect a jumper wire from exhaust gas sensor harness connector to ground.
5) Check resistance between E.C.U. terminal 24 and ground.
Resistance: Approximately 0Ω
N.G.
Repair or replace harness.
O.K.
Replace exhaust gas sensor.
Reinstall any part removed.
Erase the self-diagnosis memory. Make sure Code No. 55 is displayed in Mode III.
Exhaust gas sensor connector
Ignition switch "OFF"

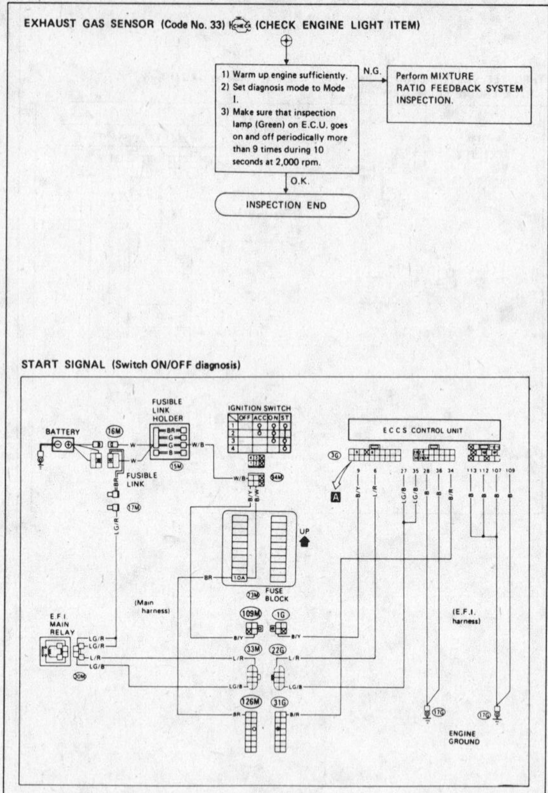
EXHAUST GAS SENSOR (Code No. 33) (CHECK ENGINE LIGHT ITEM)
1) Warm up engine sufficiently.
2) Set diagnosis mode to Mode I.
3) Make sure that inspection lamp (Green) on E.C.U. goes on and off periodically more than 9 times during 10 seconds at 2,000 rpm.
N.G.
Perform MIXTURE RATIO FEEDBACK SYSTEM INSPECTION.
O.K.
INSPECTION END
START SIGNAL (Switch ON/OFF diagnosis)
BATTERY
FUSIBLE LINK HOLDER
IGNITION SWITCH
ECCS CONTROL UNIT
FUSIBLE LINK
FUSE BLOCK
E.F.I. MAIN RELAY
(Main harness)
(E.F.I. harness)
ENGINE GROUND

ECCS SELF-DIAGNOSTIC CHART—1988 200SX WITH CA20E ENGINE

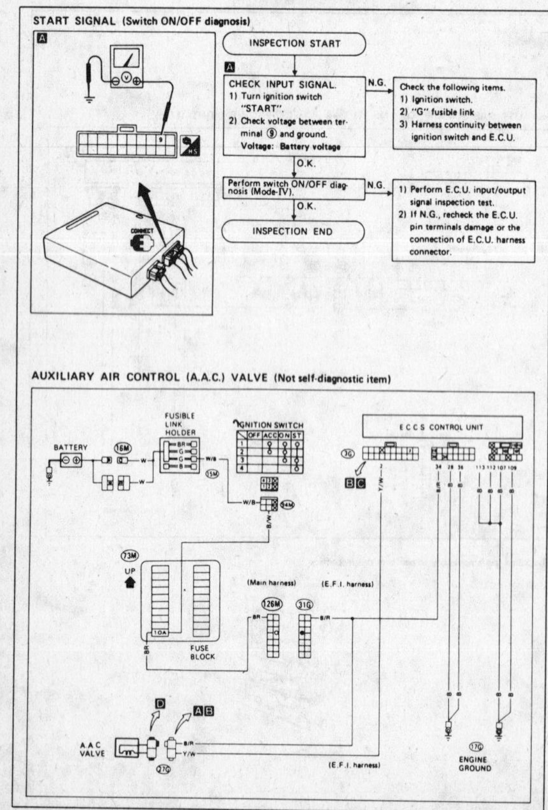
START SIGNAL (Switch ON/OFF diagnosis)
INSPECTION START
A CHECK INPUT SIGNAL.
1) Turn ignition switch "START".
2) Check voltage between terminal 9 and ground.
Voltage: Battery voltage
N.G.
Check the following items.
1) Ignition switch.
2) "G" fusible link
3) Harness continuity between ignition switch and E.C.U.
O.K.
Perform switch ON/OFF diagnosis (Mode-IV).
N.G.
1) Perform E.C.U. input/output signal inspection test.
2) If N.G., recheck the E.C.U. pin terminals damage or the connection of E.C.U. harness connector.
O.K.
INSPECTION END
AUXILIARY AIR CONTROL (A.A.C.) VALVE (Not self-diagnostic item)
BATTERY
FUSIBLE LINK HOLDER
IGNITION SWITCH
ECCS CONTROL UNIT
FUSE BLOCK
(Main harness)
(E.F.I. harness)
A.A.C. VALVE
ENGINE GROUND

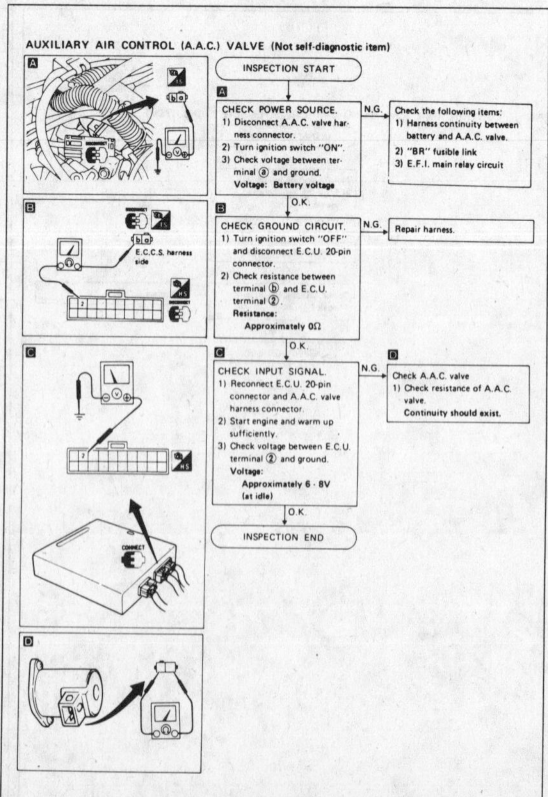
AUXILIARY AIR CONTROL (A.A.C.) VALVE (Not self-diagnostic item)
INSPECTION START
A CHECK POWER SOURCE.
1) Disconnect A.A.C. valve harness connector.
2) Turn ignition switch "ON".
3) Check voltage between terminal ⓐ and ground.
Voltage: Battery voltage
N.G.
Check the following items:
1) Harness continuity between battery and A.A.C. valve.
2) "BR" fusible link
3) E.F.I. main relay circuit
O.K.
B CHECK GROUND CIRCUIT.
1) Turn ignition switch "OFF" and disconnect E.C.U. 20-pin connector.
2) Check resistance between terminal ⓑ and E.C.U. terminal ②.
Resistance: Approximately 0Ω
N.G.
Repair harness.
O.K.
C CHECK INPUT SIGNAL.
1) Reconnect E.C.U. 20-pin connector and A.A.C. valve harness connector.
2) Start engine and warm up sufficiently.
3) Check voltage between E.C.U. terminal ② and ground.
Voltage: Approximately 6 - 8V (at idle)
N.G.
D Check A.A.C. valve
1) Check resistance of A.A.C. valve.
Continuity should exist.
O.K.
INSPECTION END
E.C.C.S. harness side

ECCS SELF-DIAGNOSTIC CHART—1988 200SX WITH CA20E ENGINE

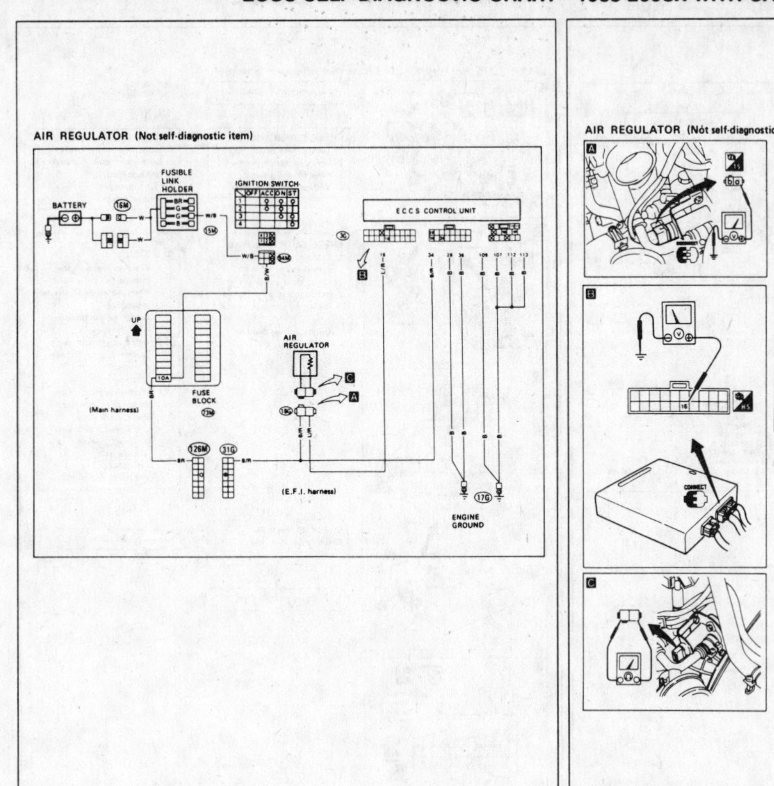
AIR REGULATOR (Not self-diagnostic item)
BATTERY
FUSIBLE LINK HOLDER
IGNITION SWITCH
E C C S CONTROL UNIT
UP
AIR REGULATOR
FUSE BLOCK
(Main harness)
(E.F.I. harness)
ENGINE GROUND

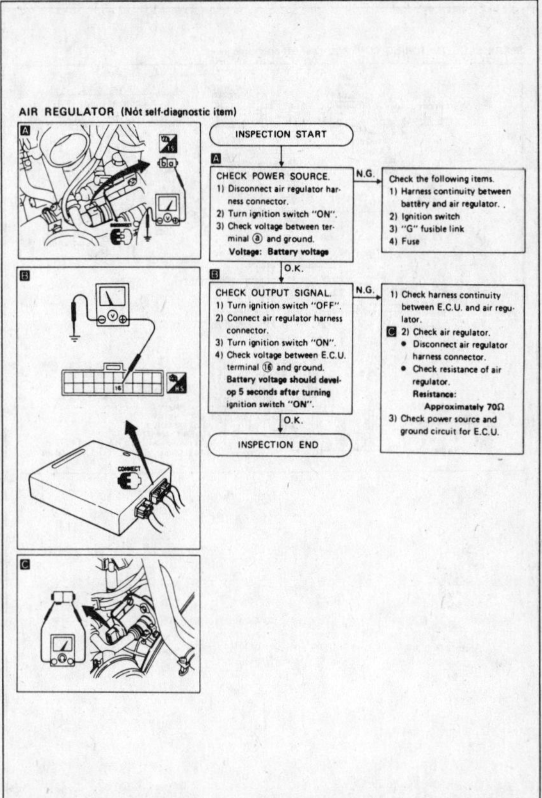
AIR REGULATOR (Not self-diagnostic item)
INSPECTION START
A
CHECK POWER SOURCE.
1) Disconnect air regulator harness connector.
2) Turn ignition switch "ON".
3) Check voltage between terminal ⓐ and ground.
Voltage: Battery voltage
N.G.
Check the following items.
1) Harness continuity between battery and air regulator.
2) Ignition switch
3) "G" fusible link
4) Fuse
O.K.
B
CHECK OUTPUT SIGNAL.
1) Turn ignition switch "OFF".
2) Connect air regulator harness connector.
3) Turn ignition switch "ON".
4) Check voltage between E.C.U. terminal ⑯ and ground.
Battery voltage should develop 5 seconds after turning ignition switch "ON".
N.G.
1) Check harness continuity between E.C.U. and air regulator.
C 2) Check air regulator.
• Disconnect air regulator harness connector.
• Check resistance of air regulator.
Resistance:
Approximately 70Ω
3) Check power source and ground circuit for E.C.U.
O.K.
INSPECTION END

ECCS SELF-DIAGNOSTIC CHART—1988 200SX WITH CA20E ENGINE

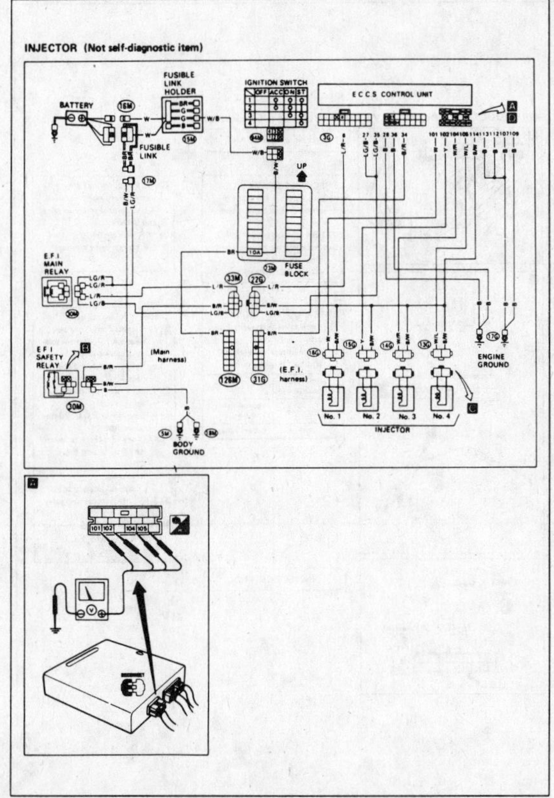
INJECTOR (Not self-diagnostic item)
BATTERY
FUSIBLE LINK HOLDER
FUSIBLE LINK
IGNITION SWITCH
E C C S CONTROL UNIT
UP
FUSE BLOCK
E.F.I. MAIN RELAY
E.F.I. SAFETY RELAY
(Main harness)
(E.F.I. harness)
ENGINE GROUND
BODY GROUND
No. 1
No. 2
No. 3
No. 4
INJECTOR

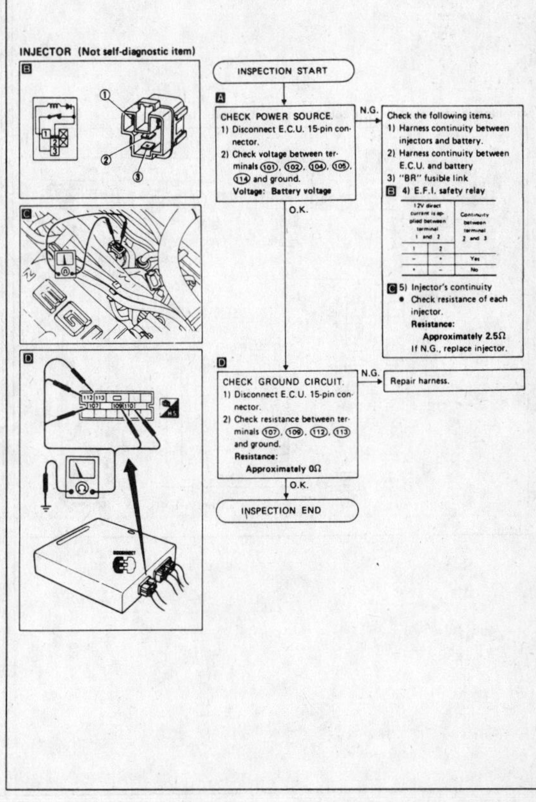
INJECTOR (Not self-diagnostic item)
INSPECTION START
A
CHECK POWER SOURCE.
1) Disconnect E.C.U. 15-pin connector.
2) Check voltage between terminals (101), (102), (104), (105), (114) and ground.
Voltage: Battery voltage
N.G.
Check the following items.
1) Harness continuity between injectors and battery.
2) Harness continuity between E.C.U. and battery
3) "BR" fusible link
B 4) E.F.I. safety relay
12V direct current is applied between terminal 1 and 2
Continuity between terminal 2 and 3
1
2
–
+
Yes
+
–
No
C 5) Injector's continuity
• Check resistance of each injector.
Resistance:
Approximately 2.5Ω
If N.G., replace injector.
O.K.
D
CHECK GROUND CIRCUIT.
1) Disconnect E.C.U. 15-pin connector.
2) Check resistance between terminals (107), (109), (112), (113) and ground.
Resistance:
Approximately 0Ω
N.G.
Repair harness.
O.K.
INSPECTION END

ECCS SELF-DIAGNOSTIC CHART—1988 200SX WITH CA20E ENGINE

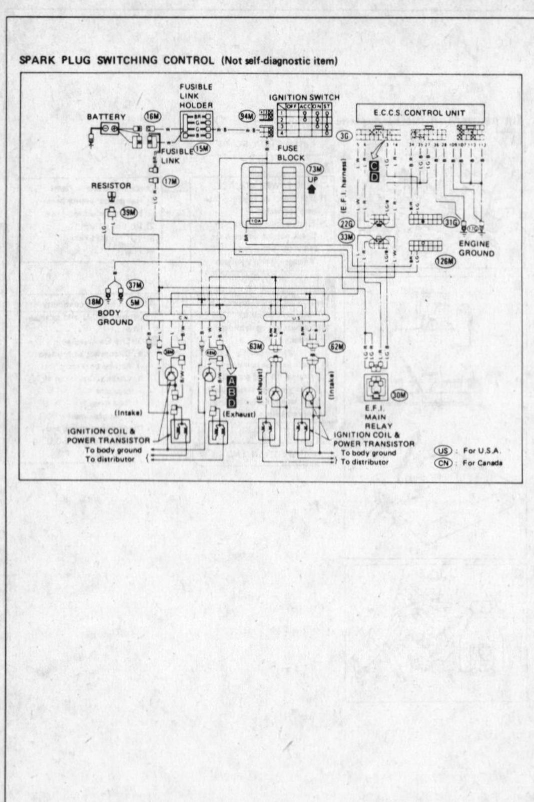
SPARK PLUG SWITCHING CONTROL (Not self-diagnostic item)
BATTERY
FUSIBLE LINK HOLDER
FUSIBLE LINK
IGNITION SWITCH
E.C.C.S. CONTROL UNIT
FUSE BLOCK
RESISTOR
ENGINE GROUND
BODY GROUND
(Intake)
(Exhaust)
E.F.I. MAIN RELAY
IGNITION COIL & POWER TRANSISTOR
To body ground
To distributor
US : For U.S.A.
CN : For Canada

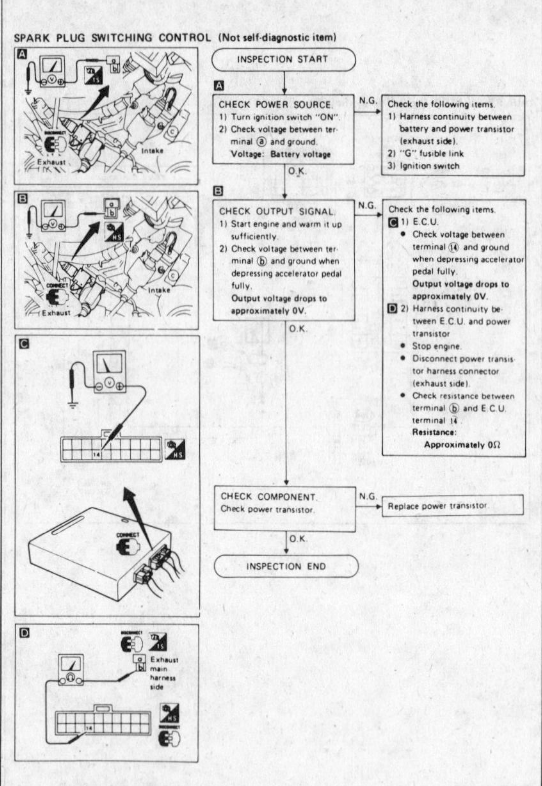
SPARK PLUG SWITCHING CONTROL (Not self-diagnostic item)
INSPECTION START
A
CHECK POWER SOURCE.
1) Turn ignition switch "ON".
2) Check voltage between terminal ⓐ and ground.
Voltage: Battery voltage
N.G.
Check the following items.
1) Harness continuity between battery and power transistor (exhaust side).
2) "G" fusible link
3) Ignition switch
O.K.
B
CHECK OUTPUT SIGNAL.
1) Start engine and warm it up sufficiently.
2) Check voltage between terminal ⓑ and ground when depressing accelerator pedal fully.
Output voltage drops to approximately 0V.
N.G.
Check the following items.
C 1) E.C.U.
• Check voltage between terminal ⑭ and ground when depressing accelerator pedal fully.
Output voltage drops to approximately 0V.
D 2) Harness continuity between E.C.U. and power transistor
• Stop engine.
• Disconnect power transistor harness connector (exhaust side).
• Check resistance between terminal ⓑ and E.C.U. terminal 14.
Resistance:
Approximately 0Ω
O.K.
CHECK COMPONENT.
Check power transistor.
N.G.
Replace power transistor.
O.K.
INSPECTION END
Exhaust
Intake
Exhaust main harness side

ECCS SELF-DIAGNOSTIC CHART—1988 200SX WITH CA20E ENGINE

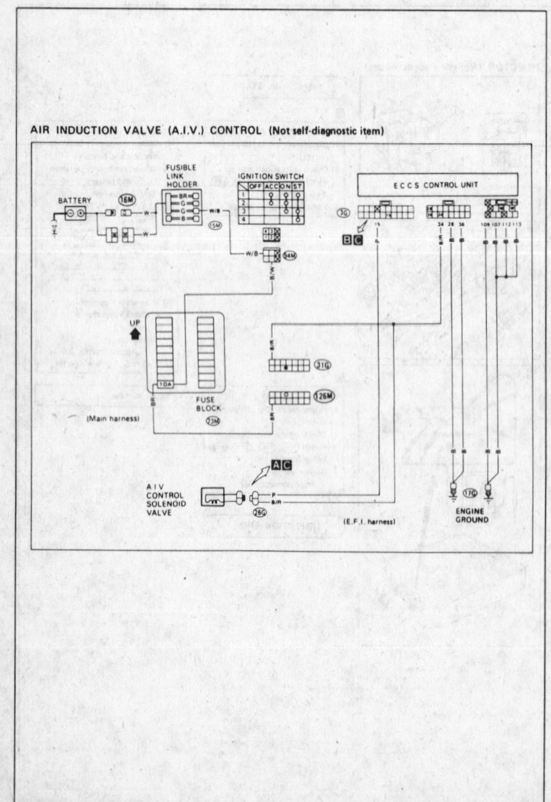
AIR INDUCTION VALVE (A.I.V.) CONTROL (Not self-diagnostic item)
BATTERY
FUSIBLE LINK HOLDER
IGNITION SWITCH
E.C.C.S. CONTROL UNIT
UP
FUSE BLOCK
(Main harness)
A.I.V. CONTROL SOLENOID VALVE
(E.F.I. harness)
ENGINE GROUND

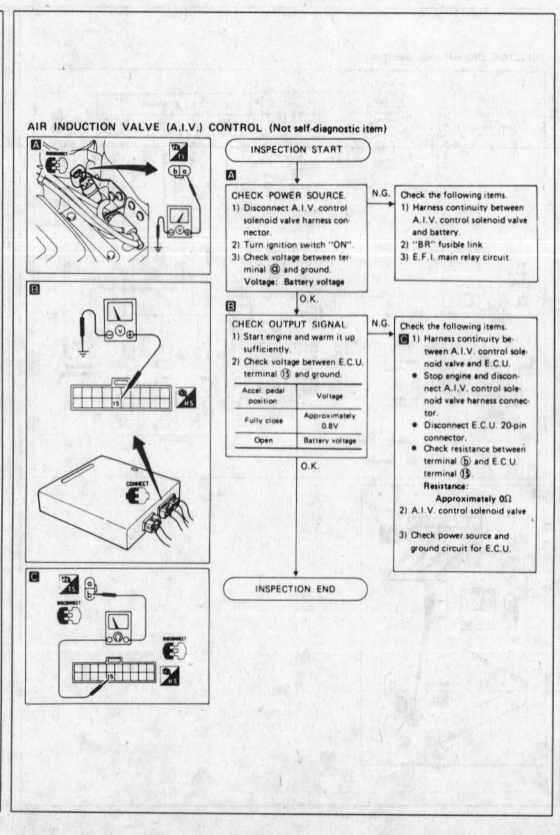
AIR INDUCTION VALVE (A.I.V.) CONTROL (Not self-diagnostic item)
INSPECTION START
A
CHECK POWER SOURCE.
1) Disconnect A.I.V. control solenoid valve harness connector.
2) Turn ignition switch "ON".
3) Check voltage between terminal ⓐ and ground.
Voltage: Battery voltage
N.G.
Check the following items.
1) Harness continuity between A.I.V. control solenoid valve and battery.
2) "BR" fusible link
3) E.F.I. main relay circuit
O.K.
B
CHECK OUTPUT SIGNAL.
1) Start engine and warm it up sufficiently.
2) Check voltage between E.C.U. terminal ⑮ and ground.
Accel. pedal position | Voltage
Fully close | Approximately 0.8V
Open | Battery voltage
N.G.
Check the following items.
C 1) Harness continuity between A.I.V. control solenoid valve and E.C.U.
• Stop engine and disconnect A.I.V. control solenoid valve harness connector.
• Disconnect E.C.U. 20-pin connector.
• Check resistance between terminal ⓑ and E.C.U. terminal ⑮.
Resistance:
Approximately 0Ω
2) A.I.V. control solenoid valve
3) Check power source and ground circuit for E.C.U.
O.K.
INSPECTION END

ECCS SELF-DIAGNOSTIC CHART—1988 200SX WITH CA20E ENGINE

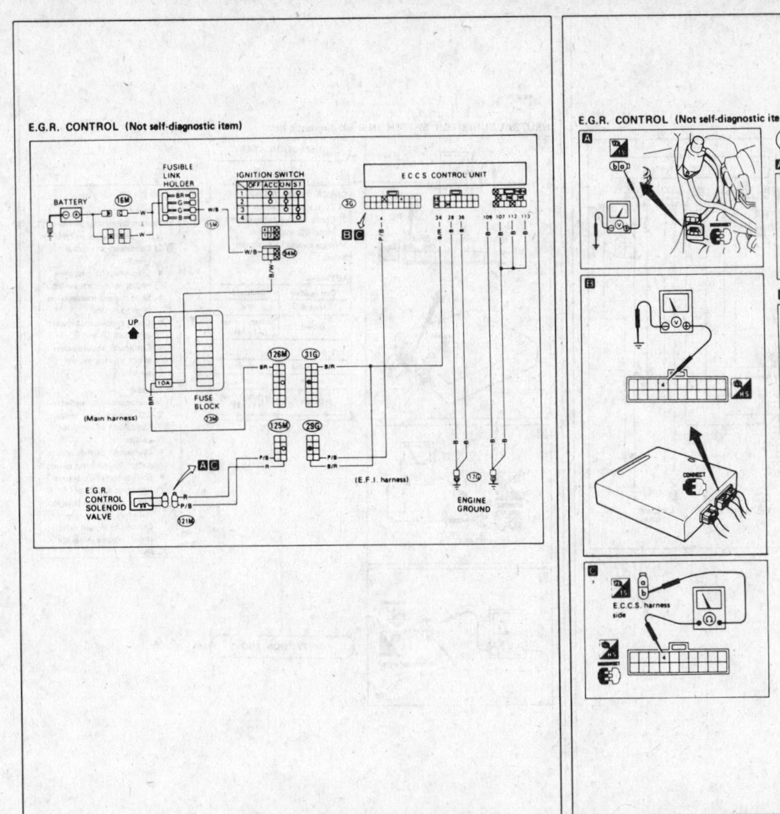
E.G.R. CONTROL (Not self-diagnostic item)
BATTERY
FUSIBLE LINK HOLDER
IGNITION SWITCH
OFF ACC ON ST
ECCS CONTROL UNIT
UP
FUSE BLOCK
(Main harness)
E.G.R. CONTROL SOLENOID VALVE
(E.F.I. harness)
ENGINE GROUND

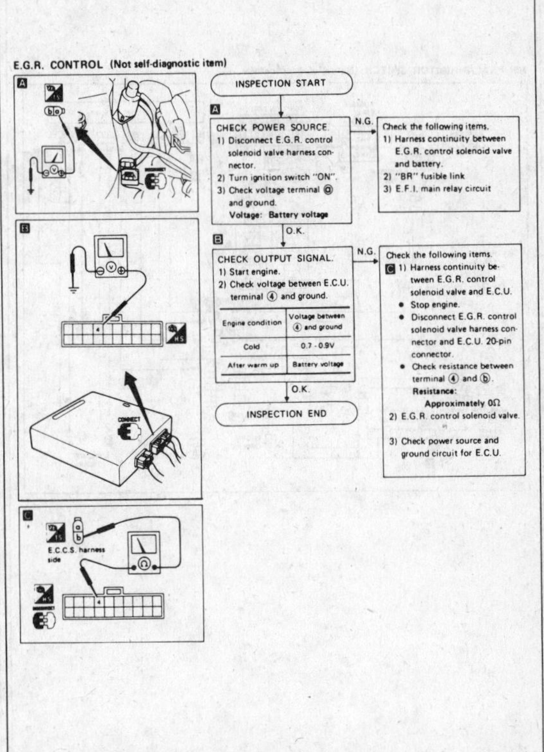
E.G.R. CONTROL (Not self-diagnostic item)
INSPECTION START
A
CHECK POWER SOURCE.
1) Disconnect E.G.R. control solenoid valve harness connector.
2) Turn ignition switch "ON".
3) Check voltage terminal ⓐ and ground.
Voltage: Battery voltage
N.G.
Check the following items.
1) Harness continuity between E.G.R. control solenoid valve and battery.
2) "BR" fusible link
3) E.F.I. main relay circuit
O.K.
B
CHECK OUTPUT SIGNAL.
1) Start engine.
2) Check voltage between E.C.U. terminal ④ and ground.
Engine condition | Voltage between ④ and ground
Cold | 0.7 - 0.9V
After warm up | Battery voltage
N.G.
Check the following items.
C 1) Harness continuity between E.G.R. control solenoid valve and E.C.U.
• Stop engine.
• Disconnect E.G.R. control solenoid valve harness connector and E.C.U. 20-pin connector.
• Check resistance between terminal ④ and ⓑ.
Resistance: Approximately 0Ω
2) E.G.R. control solenoid valve.
3) Check power source and ground circuit for E.C.U.
O.K.
INSPECTION END
E.C.C.S. harness side

ECCS SELF-DIAGNOSTIC CHART—1988 200SX WITH CA20E ENGINE

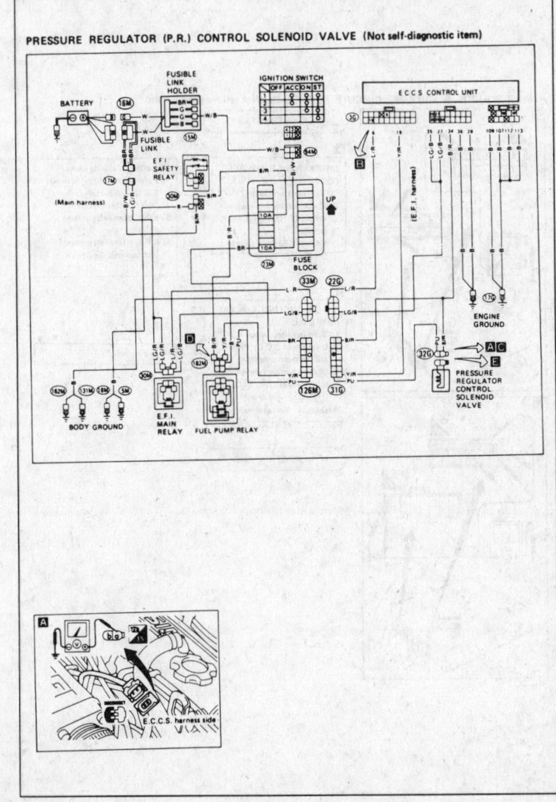
PRESSURE REGULATOR (P.R.) CONTROL SOLENOID VALVE (Not self-diagnostic item)
BATTERY
FUSIBLE LINK HOLDER
FUSIBLE LINK
IGNITION SWITCH
ECCS CONTROL UNIT
E.F.I. SAFETY RELAY
(Main harness)
UP
FUSE BLOCK
(E.F.I. harness)
ENGINE GROUND
PRESSURE REGULATOR CONTROL SOLENOID VALVE
BODY GROUND
E.F.I. MAIN RELAY
FUEL PUMP RELAY
E.C.C.S. harness side

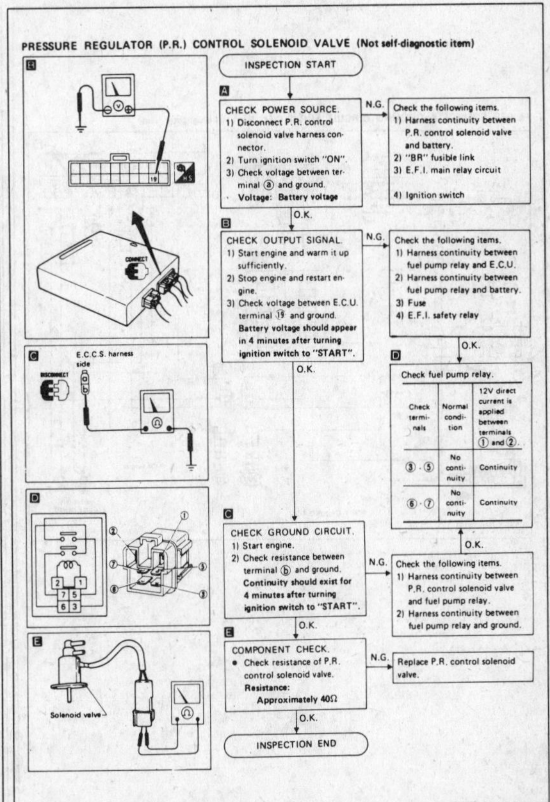
PRESSURE REGULATOR (P.R.) CONTROL SOLENOID VALVE (Not self-diagnostic item)
INSPECTION START
A
CHECK POWER SOURCE.
1) Disconnect P.R. control solenoid valve harness connector.
2) Turn ignition switch "ON".
3) Check voltage between terminal ⓐ and ground.
Voltage: Battery voltage
N.G.
Check the following items.
1) Harness continuity between P.R. control solenoid valve and battery.
2) "BR" fusible link
3) E.F.I. main relay circuit
4) Ignition switch
O.K.
B
CHECK OUTPUT SIGNAL.
1) Start engine and warm it up sufficiently.
2) Stop engine and restart engine.
3) Check voltage between E.C.U. terminal ⑲ and ground.
Battery voltage should appear in 4 minutes after turning ignition switch to "START".
N.G.
Check the following items.
1) Harness continuity between fuel pump relay and E.C.U.
2) Harness continuity between fuel pump relay and battery.
3) Fuse
4) E.F.I. safety relay
O.K.
D
Check fuel pump relay.
Check terminals | Normal condition | 12V direct current is applied between terminals ① and ②.
③ - ⑤ | No continuity | Continuity
⑥ - ⑦ | No continuity | Continuity
O.K.
C
CHECK GROUND CIRCUIT.
1) Start engine.
2) Check resistance between terminal ⓑ and ground.
Continuity should exist for 4 minutes after turning ignition switch to "START".
N.G.
Check the following items.
1) Harness continuity between P.R. control solenoid valve and fuel pump relay.
2) Harness continuity between fuel pump relay and ground.
O.K.
E
COMPONENT CHECK.
• Check resistance of P.R. control solenoid valve.
Resistance: Approximately 40Ω
N.G.
Replace P.R. control solenoid valve.
O.K.
INSPECTION END
E.C.C.S. harness side
Solenoid valve

ECCS SELF-DIAGNOSTIC CHART—1988 200SX WITH CA20E ENGINE

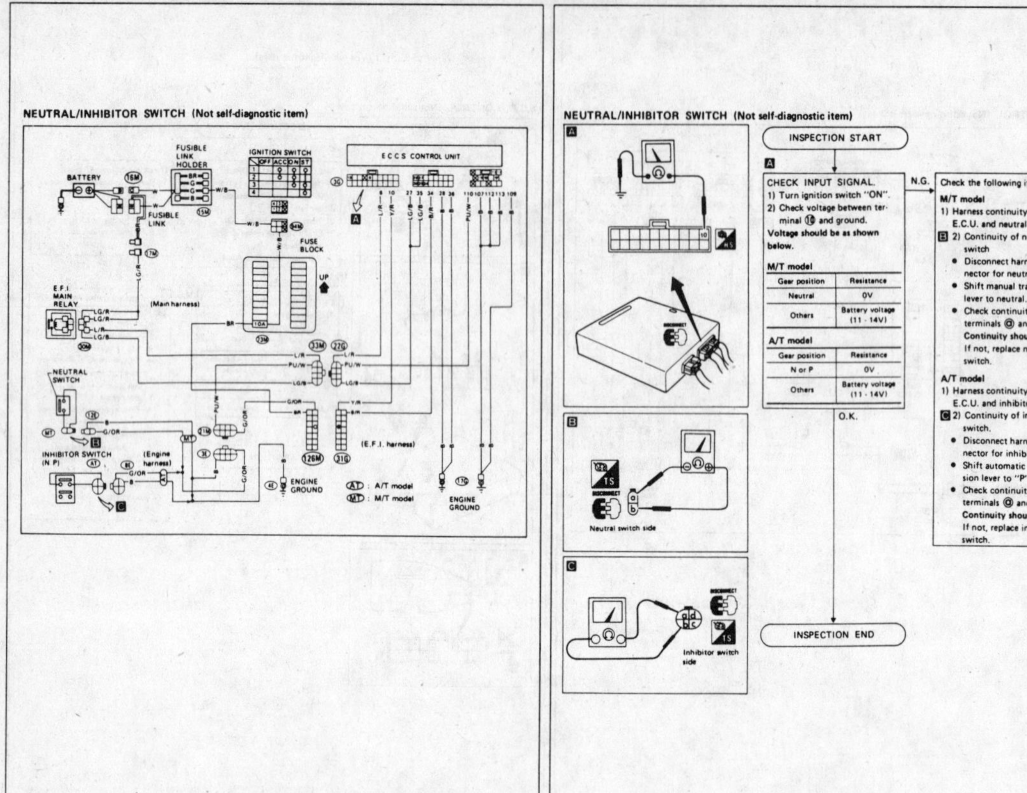
NEUTRAL/INHIBITOR SWITCH (Not self-diagnostic item)
BATTERY
FUSIBLE LINK HOLDER
FUSIBLE LINK
IGNITION SWITCH
E C C S CONTROL UNIT
FUSE BLOCK
UP
E.F.I. MAIN RELAY
(Main harness)
NEUTRAL SWITCH
INHIBITOR SWITCH (N P)
(Engine harness)
(E.F.I. harness)
ENGINE GROUND
AT : A/T model
MT : M/T model
NEUTRAL/INHIBITOR SWITCH (Not self-diagnostic item)
Neutral switch side
Inhibitor switch side
INSPECTION START
A
CHECK INPUT SIGNAL.
1) Turn ignition switch "ON".
2) Check voltage between terminal ⑩ and ground.
Voltage should be as shown below.
M/T model
Gear position | Resistance
Neutral | 0V
Others | Battery voltage (11 - 14V)
A/T model
Gear position | Resistance
N or P | 0V
Others | Battery voltage (11 - 14V)
N.G.
Check the following items.
M/T model
1) Harness continuity between E.C.U. and neutral switch.
B 2) Continuity of neutral switch
• Disconnect harness connector for neutral switch.
• Shift manual transmission lever to neutral.
• Check continuity between terminals ⓐ and ⓑ. Continuity should exist. If not, replace neutral switch.
A/T model
1) Harness continuity between E.C.U. and inhibitor switch.
C 2) Continuity of inhibitor switch.
• Disconnect harness connector for inhibitor switch.
• Shift automatic transmission lever to "P" or "N".
• Check continuity between terminals ⓐ and ⓑ. Continuity should exist. If not, replace inhibitor switch.
O.K.
INSPECTION END

ECCS SELF-DIAGNOSTIC CHART—1988 200SX WITH CA20E ENGINE

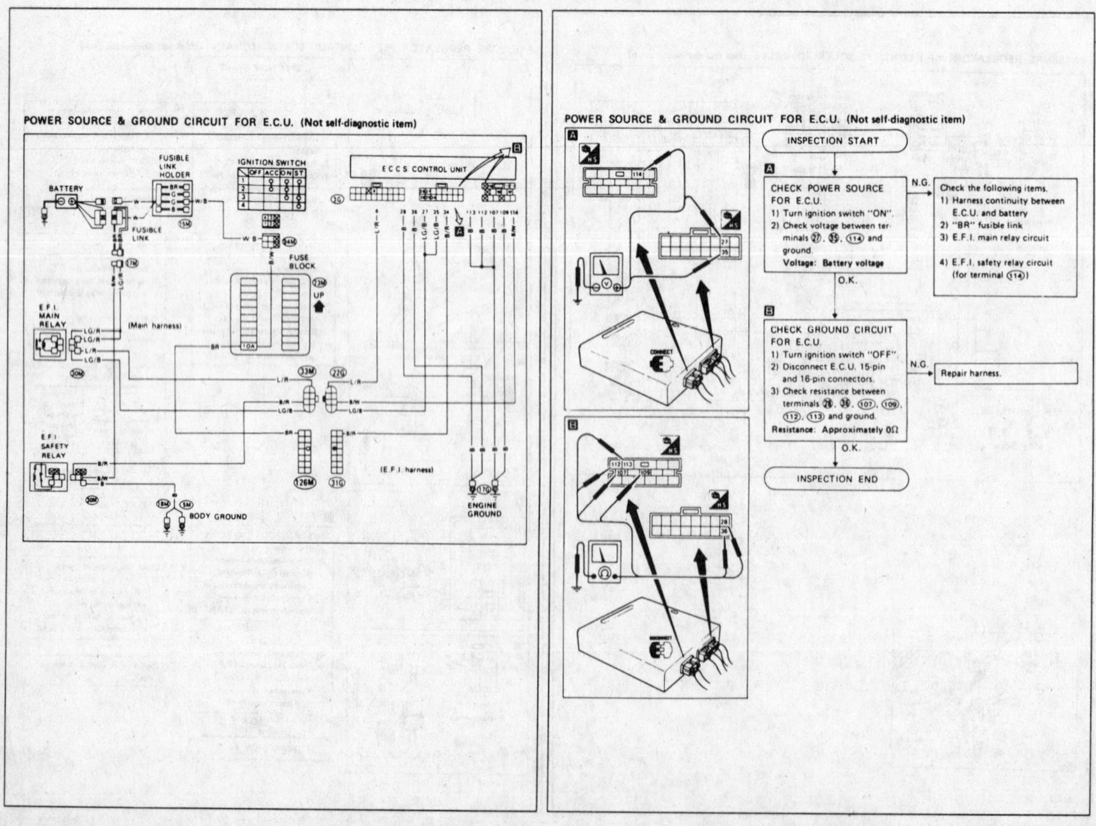
POWER SOURCE & GROUND CIRCUIT FOR E.C.U. (Not self-diagnostic item)
BATTERY
FUSIBLE LINK HOLDER
FUSIBLE LINK
IGNITION SWITCH
E C C S CONTROL UNIT
FUSE BLOCK
UP
E.F.I. MAIN RELAY
(Main harness)
E.F.I. SAFETY RELAY
BODY GROUND
(E.F.I. harness)
ENGINE GROUND
POWER SOURCE & GROUND CIRCUIT FOR E.C.U. (Not self-diagnostic item)
INSPECTION START
A
CHECK POWER SOURCE FOR E.C.U.
1) Turn ignition switch "ON".
2) Check voltage between terminals ㉗, ㉟, (114) and ground.
Voltage: Battery voltage
N.G.
Check the following items.
1) Harness continuity between E.C.U. and battery
2) "BR" fusible link
3) E.F.I. main relay circuit
4) E.F.I. safety relay circuit (for terminal (114))
O.K.
B
CHECK GROUND CIRCUIT FOR E.C.U.
1) Turn ignition switch "OFF".
2) Disconnect E.C.U. 15-pin and 16-pin connectors.
3) Check resistance between terminals ㉘, ㊱, (107), (109), (112), (113) and ground.
Resistance: Approximately 0Ω
N.G.
Repair harness.
O.K.
INSPECTION END

ECCS SELF-DIAGNOSTIC CHART—1988 200SX WITH CA20E ENGINE

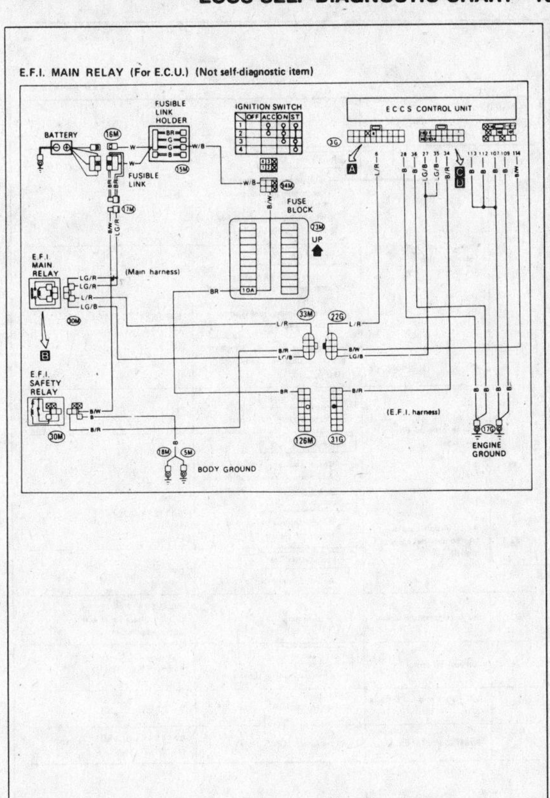
E.F.I. MAIN RELAY (For E.C.U.) (Not self-diagnostic item)
BATTERY
FUSIBLE LINK HOLDER
IGNITION SWITCH
OFF ACC ON ST
E C C S CONTROL UNIT
FUSIBLE LINK
FUSE BLOCK
UP
E.F.I. MAIN RELAY
(Main harness)
E.F.I. SAFETY RELAY
(E.F.I. harness)
ENGINE GROUND
BODY GROUND

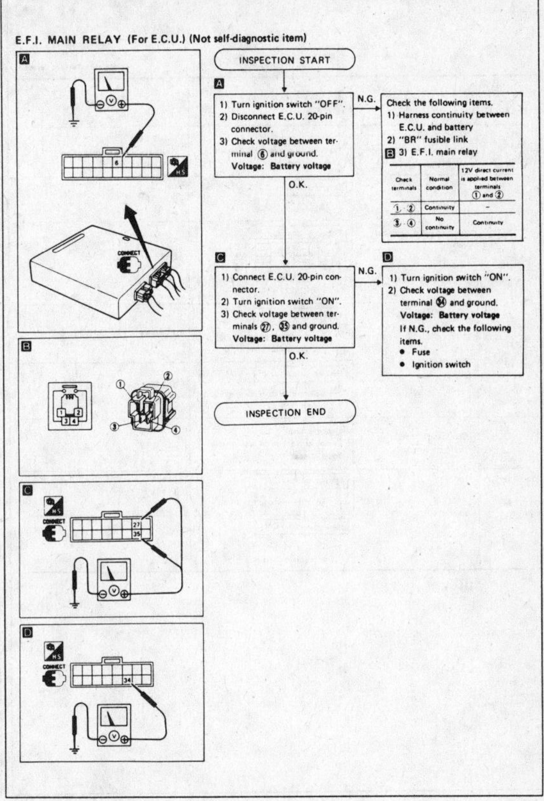
E.F.I. MAIN RELAY (For E.C.U.) (Not self-diagnostic item)
INSPECTION START
A
1) Turn ignition switch "OFF".
2) Disconnect E.C.U. 20-pin connector.
3) Check voltage between terminal ⑥ and ground.
Voltage: Battery voltage
N.G.
Check the following items.
1) Harness continuity between E.C.U. and battery
2) "BR" fusible link
B 3) E.F.I. main relay
Check terminals | Normal condition | 12V direct current is applied between terminals ① and ②
①-② | Continuity | –
③-④ | No continuity | Continuity
O.K.
C
1) Connect E.C.U. 20-pin connector.
2) Turn ignition switch "ON".
3) Check voltage between terminals ㉗, ㉟ and ground.
Voltage: Battery voltage
N.G.
D
1) Turn ignition switch "ON".
2) Check voltage between terminal ㉞ and ground.
Voltage: Battery voltage
If N.G., check the following items.
• Fuse
• Ignition switch
O.K.
INSPECTION END
CONNECT

ECU SIGNAL INSPECTION—1988 200SX WITH CA20E ENGINE

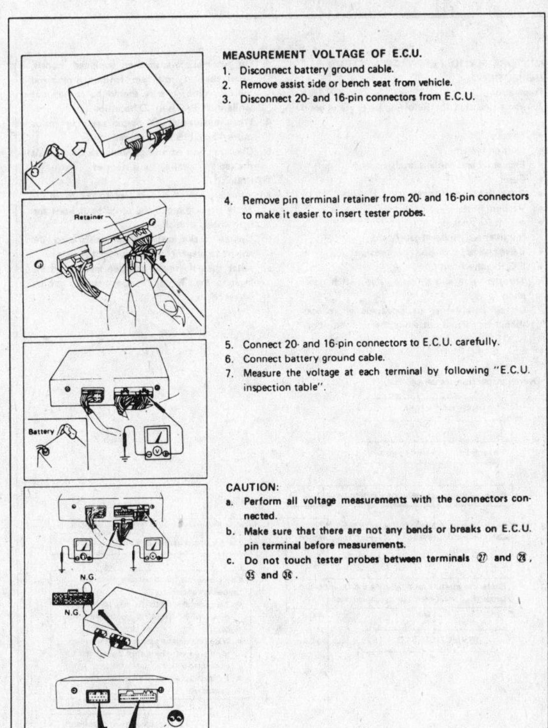
Retainer
Battery
N.G.

MEASUREMENT VOLTAGE OF E.C.U.

1. Disconnect battery ground cable.
2. Remove assist side or bench seat from vehicle.
3. Disconnect 20- and 16-pin connectors from E.C.U.
4. Remove pin terminal retainer from 20- and 16-pin connectors to make it easier to insert tester probes.
5. Connect 20- and 16-pin connectors to E.C.U. carefully.
6. Connect battery ground cable.
7. Measure the voltage at each terminal by following "E.C.U. inspection table".

CAUTION:

a. Perform all voltage measurements with the connectors connected.
b. Make sure that there are not any bends or breaks on E.C.U. pin terminal before measurements.
c. Do not touch tester probes between terminals ㉗ and ㉘, ㉟ and ㊱.

E.C.U. inspection table

*Data are reference values.

TERMINAL NO.	ITEM	CONDITION	*DATA
2	A.A.C. valve	Engine is running. — At idle (after warming up)	6.0 - 8.0V
3	Ignition signal (from resistor)	Ignition switch "ON"	BATTERY VOLTAGE (11 - 14V)
4	E.G.R. control solenoid valve	Engine is running. — Engine is cold. [Water temperature is below 60°C (140°F).]	0.7 - 0.9V
		Engine is running. — After warming up [Water temperature is between 60°C (140°F) and 105°C (221°F).]	BATTERY VOLTAGE (11 - 14V)
5	Ignition signal (from intake side power transistor)	Engine is running.	0.4 - 2.2V Output voltage varies with engine speed.
6	E.F.I. relay	Ignition switch "ON"	0.8 - 1.0V
8	Crank angle sensor (position signal)	Engine is running. **Do not turn engine at high speed under no-load.**	2.2 - 2.8V
9	Start signal	Ignition switch "START"	BATTERY VOLTAGE (11 - 14V)
10	Neutral signal	Ignition switch "ON" — Gear position: Neutral (M/T) : N or P range (A/T)	0V
		Ignition switch "ON" — Gear position: [Except neutral (M/T) Except N or P range (A/T)]	BATTERY VOLTAGE (11 - 14V)

ECU SIGNAL INSPECTION (CONT.)—1988 200SX WITH CA20E ENGINE

*Data are reference values.

TERMINAL NO.	ITEM	CONDITION	*DATA
14	Ignition signal (from exhaust side power transistor)	Engine is running. **Do not turn engine at high speed under no-load.**	0.4 - 2.2V (Output voltage varies with engine revolution.)
15	A.I.V. control solenoid valve	[Engine is running.] — At idle	0.7 - 0.9V
		[Engine is running.] — When depressing accelerator pedal [Water temperature is above 50°C (122°F).]	BATTERY VOLTAGE (11 - 14V)
16	Air regulator	[Ignition switch "ON"] — For 5 seconds after turning ignition switch "ON"	0.7 - 0.9V
		[Ignition switch "ON"] — 5 seconds after turning ignition switch "ON"	BATTERY VOLTAGE (11 - 14V)
17	Crank angle sensor (Reference signal)	Engine is running. **Do not turn engine at high speed under no-load.**	0.2 - 0.4V
18	Idle switch (⊖ side)	[Ignition switch "ON"] — Throttle valve: idle position	9 - 10V
		[Ignition switch "ON"] — Throttle valve: except idle position	0V

*Data are reference values.

TERMINAL NO.	ITEM	CONDITION	*DATA
19	Pressure regulator control solenoid valve (Fuel pump relay)	[Ignition switch "ON"] — For approximately 4 minutes after turning ignition switch to "START". [Water temperature is above 60°C (140°F).]	0.8 - 1.0V
		[Ignition switch "ON"] — Approximately 4 minutes after turning ignition switch to "START". [Water temperature is above 60°C (140°F).]	BATTERY VOLTAGE (11 - 14V)
		Ignition switch "ON" or "START". [Water temperature is below 60°C (140°F).]	
21	Full throttle switch (⊖ side)	[Ignition switch "ON"] — Throttle valve: fully open	9 - 10V
		[Ignition switch "ON"] — Throttle valve: Any position except full throttle	0V
22	Air conditioner signal (Air conditioner equipped model)	[Ignition switch "ON"] — Air conditioner switch and heater fan switch "ON"	BATTERY VOLTAGE (11 - 14V)
		[Ignition switch "ON"] — Air conditioner switch "OFF"	0V
23	Water temperature sensor	Engine is running.	1.0 - 5.0V Output voltage varies with engine water temperature.
24	Exhaust gas sensor	[Engine is running.] — After warming up sufficiently.	0 - Approximately 1.0V
25	Idle switch and full throttle switch (⊕ side)	Ignition switch "ON"	9 - 10V
27 35	Power source for E.C.U.	Ignition switch "ON"	BATTERY VOLTAGE (11 - 14V)

ECU SIGNAL INSPECTION (CONT.) 1988 200SX WITH CA20E ENGINE

*Data are reference values.

TERMINAL NO.	ITEM	CONDITION	*DATA
29	Vehicle speed sensor	[Ignition switch "ON"] — When rotating rear wheel slowly	Voltage varies between 0V and approximately 5V.
30	Air temperature sensor	Ignition switch "ON"	Approximately 3V [Air temperature is 20°C (68°F).] Output voltage varies with air temperature.
31	Air flow meter	Engine is running. **Do not turn engine at high speed under no-load.**	0 - Approximately 5V Output voltage varies with engine revolution.
33	Power source for air flow meter	Ignition switch "ON"	8V
34	Ignition switch signal	Ignition switch "ON"	BATTERY VOLTAGE (11 - 14V)
101 102 104 105 114	Injector	Ignition switch "OFF"	BATTERY VOLTAGE (11 - 14V)
108	Fuel pump [Water temperature is below 60°C (140°F).]	[Ignition switch "ON"] — For 5 seconds after turning ignition switch "ON".	0.7 - 0.9V
		[Ignition switch "ON"] — 5 seconds after turning ignition switch "ON".	BATTERY VOLTAGE (9 - 14V)

PIN CONNECTOR TERMINAL LAYOUT

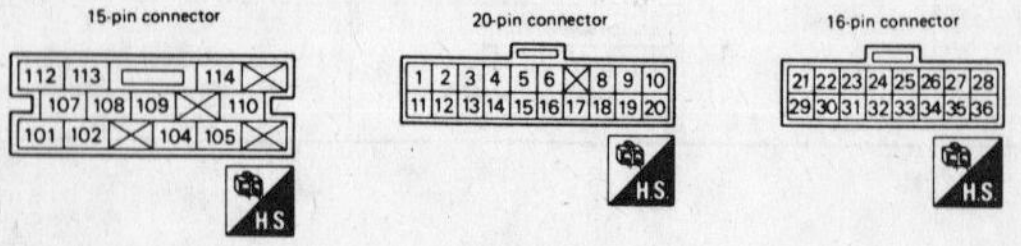

15-pin connector

112	113		114	✕
107	108	109	✕	110
101	102	✕	104	105	✕

H.S

20-pin connector

1	2	3	4	5	6	✕	8	9	10
11	12	13	14	15	16	17	18	19	20

H.S

16-pin connector

21	22	23	24	25	26	27	28
29	30	31	32	33	34	35	36

H.S

ECCS MIXTURE RATIO FEEDBACK SYSTEM INSPECTION—1988 200SX WITH CA20E

MIXTURE RATIO FEEDBACK SYSTEM INSPECTION

Preparation

1. Make sure that the following parts are in good order.
   - Battery
   - Ignition system
   - Engine oil and coolant levels
   - Fuses
   - E.C.C.S. harness connectors
   - Vacuum hoses
   - Air intake system (oil filler cap, oil level gauge, etc.)
   - Valve clearance, engine compression
   - E.G.R. valve operation
   - Throttle valve and throttle valve switch operation
2. On air conditioner equipped models, checks should be carried out while the air conditioner is "OFF".
3. On automatic transmission equipped models, when checking idle rpm, ignition timing and mixture ratio, checks should be carried out while shift lever is in "D" position.
4. When measuring "CO" percentage, insert probe more 40 cm (15.7 in) into tail pipe.
5. Checking and adjusting should be done while the radiator cooling fan is stopped.

WARNING:

a. When selector lever is shifted to "D" position, apply parking brake and block both front and rear wheels with chocks.
b. Depress brake pedal while accelerating the engine to prevent forward surge of vehicle.
c. After the adjustment has been made, shift the lever to the "N" or "P" position and remove wheel chocks.

Overall inspection sequence

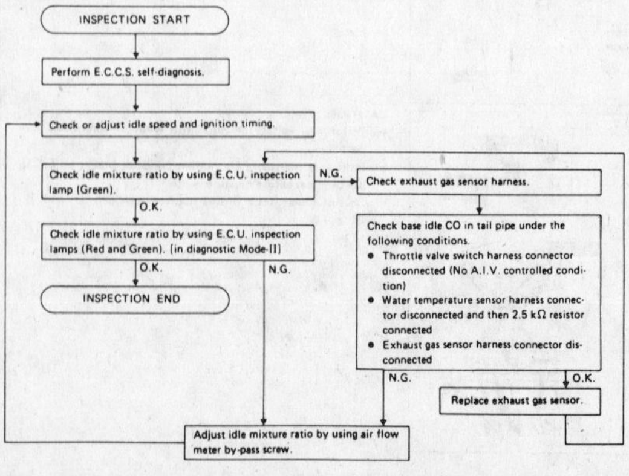

INSPECTION START
Perform E.C.C.S. self-diagnosis.
Check or adjust idle speed and ignition timing.
Check idle mixture ratio by using E.C.U. inspection lamp (Green).
N.G.
Check exhaust gas sensor harness.
O.K.
Check idle mixture ratio by using E.C.U. inspection lamps (Red and Green). [in diagnostic Mode-II]
O.K.
N.G.
INSPECTION END
Check base idle CO in tail pipe under the following conditions.
- Throttle valve switch harness connector disconnected (No A.I.V. controlled condition)
- Water temperature sensor harness connector disconnected and then 2.5 kΩ resistor connected
- Exhaust gas sensor harness connector disconnected
N.G.
O.K.
Replace exhaust gas sensor.
Adjust idle mixture ratio by using air flow meter by-pass screw.

ECCS MIXTURE RATIO FEEDBACK SYSTEM INSPECTION (CONT.)—1988 200SX WITH CA20E ENGINE

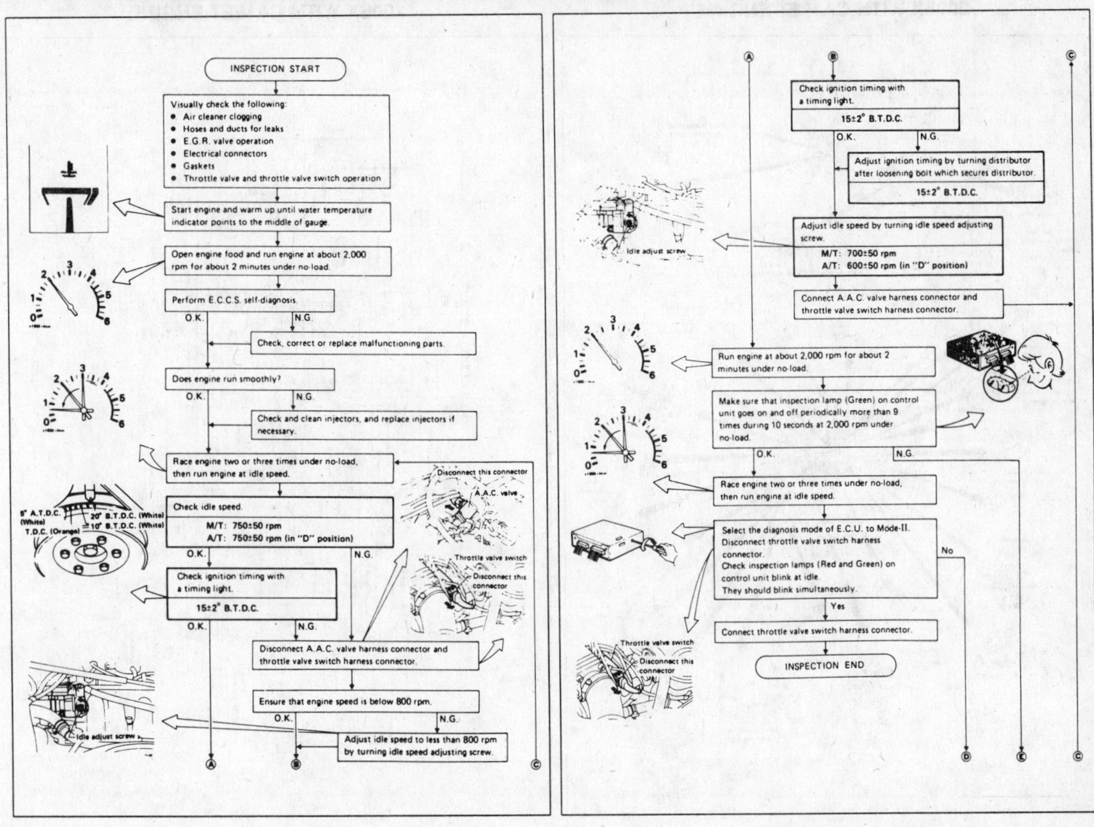
INSPECTION START
Visually check the following:
• Air cleaner clogging
• Hoses and ducts for leaks
• E.G.R. valve operation
• Electrical connectors
• Gaskets
• Throttle valve and throttle valve switch operation
Start engine and warm up until water temperature indicator points to the middle of gauge.
Open engine food and run engine at about 2,000 rpm for about 2 minutes under no-load.
Perform E.C.C.S. self-diagnosis.
O.K.
N.G.
Check, correct or replace malfunctioning parts.
Does engine run smoothly?
Check and clean injectors, and replace injectors if necessary.
Race engine two or three times under no-load, then run engine at idle speed.
Disconnect this connector
A.A.C. valve
5° A.T.D.C. (White)
20° B.T.D.C. (White)
10° B.T.D.C. (White)
T.D.C. (Orange)
Check idle speed.
M/T: 750±50 rpm
A/T: 750±50 rpm (in "D" position)
Throttle valve switch
Disconnect this connector
Check ignition timing with a timing light.
15±2° B.T.D.C.
Disconnect A.A.C. valve harness connector and throttle valve switch harness connector.
Ensure that engine speed is below 800 rpm.
Idle adjust screw
Adjust idle speed to less than 800 rpm by turning idle speed adjusting screw.
Adjust ignition timing by turning distributor after loosening bolt which secures distributor.
15±2° B.T.D.C.
Idle adjust screw
Adjust idle speed by turning idle speed adjusting screw.
M/T: 700±50 rpm
A/T: 600±50 rpm (in "D" position)
Connect A.A.C. valve harness connector and throttle valve switch harness connector.
Run engine at about 2,000 rpm for about 2 minutes under no-load.
Make sure that inspection lamp (Green) on control unit goes on and off periodically more than 9 times during 10 seconds at 2,000 rpm under no-load.
Race engine two or three times under no-load, then run engine at idle speed.
Select the diagnosis mode of E.C.U. to Mode-II.
Disconnect throttle valve switch harness connector.
Check inspection lamps (Red and Green) on control unit blink at idle.
They should blink simultaneously.
Yes
No
Connect throttle valve switch harness connector.
Throttle valve switch
Disconnect this connector
INSPECTION END

ECCS MIXTURE RATIO FEEDBACK SYSTEM INSPECTION (CONT.)—1988 200SX WITH CA20E ENGINE

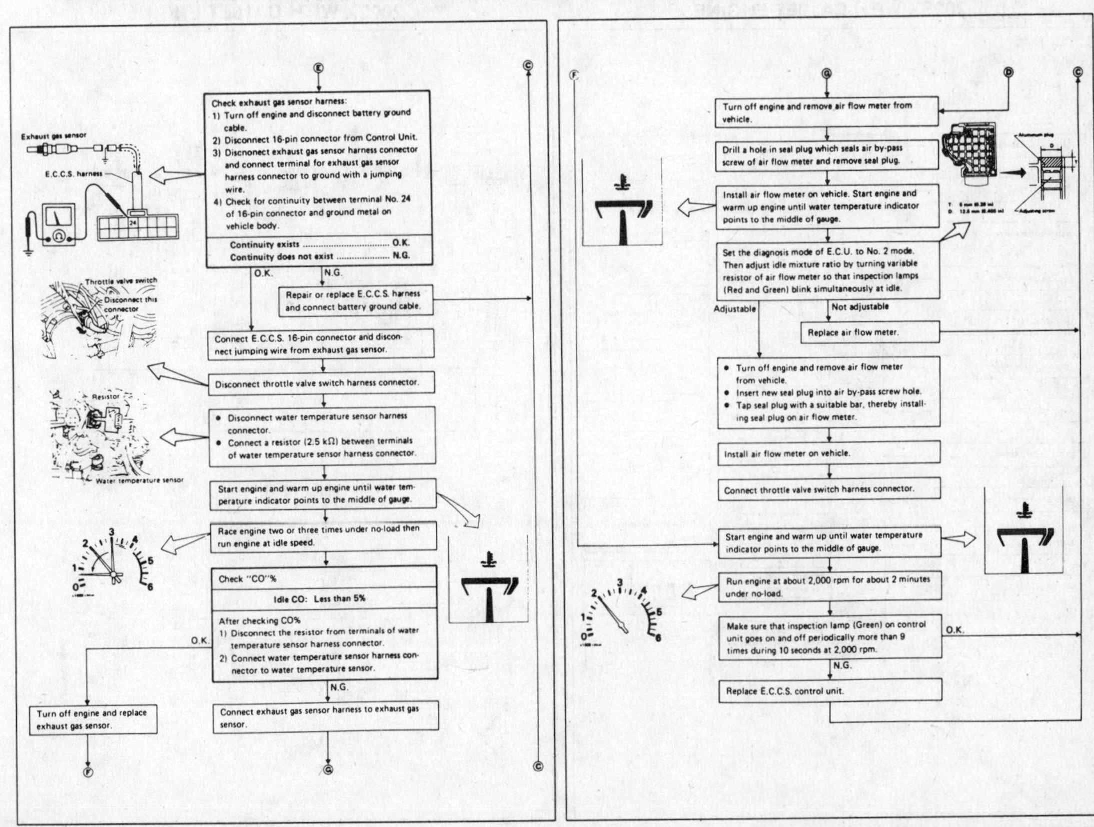
Check exhaust gas sensor harness:
1) Turn off engine and disconnect battery ground cable.
2) Disconnect 16-pin connector from Control Unit.
3) Disconnect exhaust gas sensor harness connector and connect terminal for exhaust gas sensor harness connector to ground with a jumping wire.
4) Check for continuity between terminal No. 24 of 16-pin connector and ground metal on vehicle body.
Continuity exists ..... O.K.
Continuity does not exist ..... N.G.
Exhaust gas sensor
E.C.C.S. harness
Repair or replace E.C.C.S. harness and connect battery ground cable.
Throttle valve switch
Disconnect this connector
Connect E.C.C.S. 16-pin connector and disconnect jumping wire from exhaust gas sensor.
Disconnect throttle valve switch harness connector.
Resistor
Water temperature sensor
• Disconnect water temperature sensor harness connector.
• Connect a resistor (2.5 kΩ) between terminals of water temperature sensor harness connector.
Start engine and warm up engine until water temperature indicator points to the middle of gauge.
Race engine two or three times under no-load then run engine at idle speed.
Check "CO"%
Idle CO: Less than 5%
After checking CO%
1) Disconnect the resistor from terminals of water temperature sensor harness connector.
2) Connect water temperature sensor harness connector to water temperature sensor.
Turn off engine and replace exhaust gas sensor.
Connect exhaust gas sensor harness to exhaust gas sensor.
Turn off engine and remove air flow meter from vehicle.
Drill a hole in seal plug which seals air by-pass screw of air flow meter and remove seal plug.
Install air flow meter on vehicle. Start engine and warm up engine until water temperature indicator points to the middle of gauge.
Set the diagnosis mode of E.C.U. to No. 2 mode. Then adjust idle mixture ratio by turning variable resistor of air flow meter so that inspection lamps (Red and Green) blink simultaneously at idle.
Adjustable
Not adjustable
Replace air flow meter.
• Turn off engine and remove air flow meter from vehicle.
• Insert new seal plug into air by-pass screw hole.
• Tap seal plug with a suitable bar, thereby installing seal plug on air flow meter.
Install air flow meter on vehicle.
Connect throttle valve switch harness connector.
Start engine and warm up until water temperature indicator points to the middle of gauge.
Run engine at about 2,000 rpm for about 2 minutes under no-load.
Make sure that inspection lamp (Green) on control unit goes on and off periodically more than 9 times during 10 seconds at 2,000 rpm.
Replace E.C.C.S. control unit.

ECCS COMPONENTS LOCATION—1988 200SX WITH CA18ET ENGINE

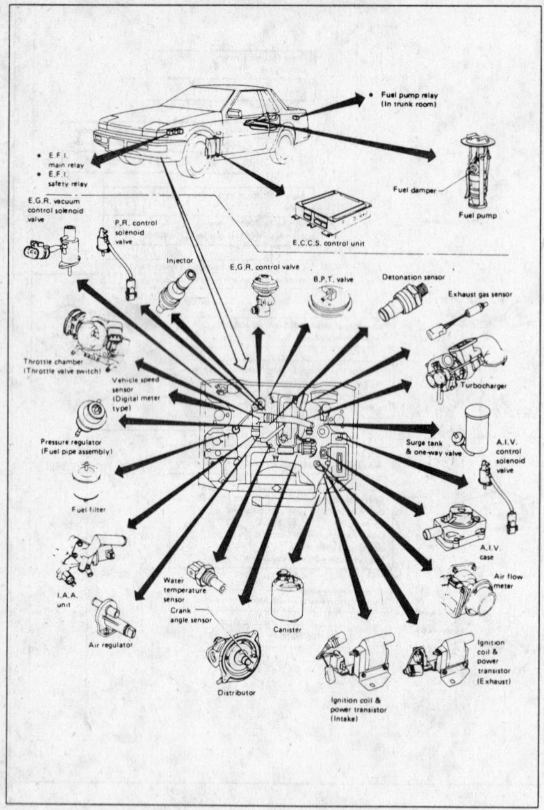
Fuel pump relay (In trunk room)
E.F.I. main relay
E.F.I. safety relay
E.G.R. vacuum control solenoid valve
P.R. control solenoid valve
Fuel damper
Fuel pump
E.C.C.S. control unit
Injector
E.G.R. control valve
B.P.T. valve
Detonation sensor
Exhaust gas sensor
Throttle chamber (Throttle valve switch)
Vehicle speed sensor (Digital meter type)
Turbocharger
Pressure regulator (Fuel pipe assembly)
Surge tank & one-way valve
A.I.V. control solenoid valve
Fuel filter
A.I.V. case
Air flow meter
I.A.A. unit
Water temperature sensor
Crank angle sensor
Canister
Air regulator
Ignition coil & power transistor (Exhaust)
Distributor
Ignition coil & power transistor (Intake)

ECCS SYSTEM SCHEMATIC—1988 200SX WITH CA18ET ENGINE

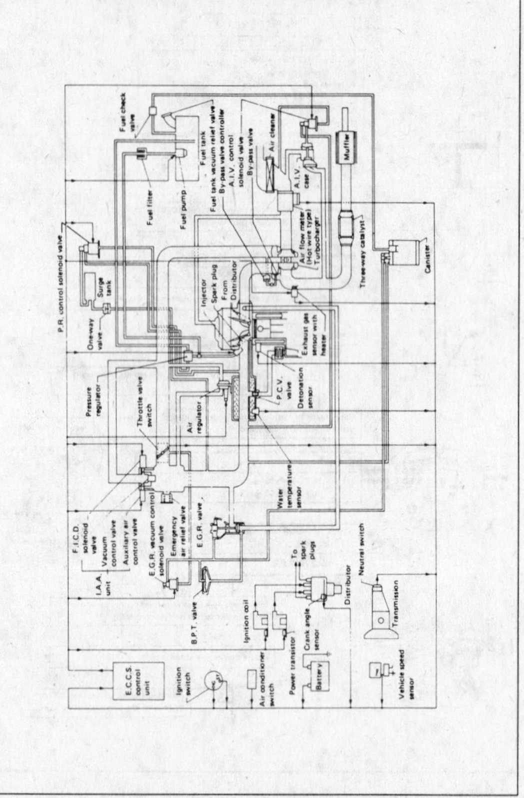

ECCS CONTROL SYSTEM CHART—1988 200SX WITH CA18ET ENGINE

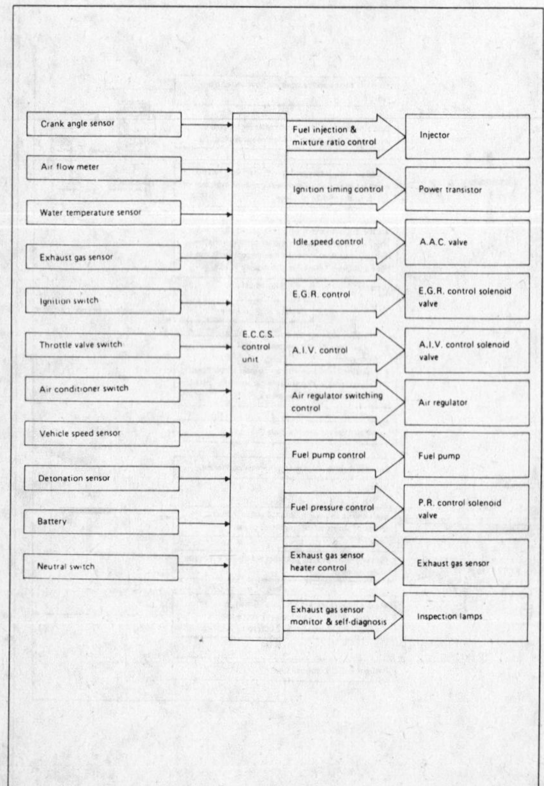
Crank angle sensor
Air flow meter
Water temperature sensor
Exhaust gas sensor
Ignition switch
Throttle valve switch
Air conditioner switch
Vehicle speed sensor
Detonation sensor
Battery
Neutral switch
E.C.C.S. control unit
Fuel injection & mixture ratio control
Injector
Ignition timing control
Power transistor
Idle speed control
A.A.C. valve
E.G.R. control
E.G.R. control solenoid valve
A.I.V. control
A.I.V. control solenoid valve
Air regulator switching control
Air regulator
Fuel pump control
Fuel pump
Fuel pressure control
P.R. control solenoid valve
Exhaust gas sensor heater control
Exhaust gas sensor
Exhaust gas sensor monitor & self-diagnosis
Inspection lamps

ECCS CONTROL UNIT WIRING SCHEMATIC—1988 200SX WITH CA18ET ENGINE

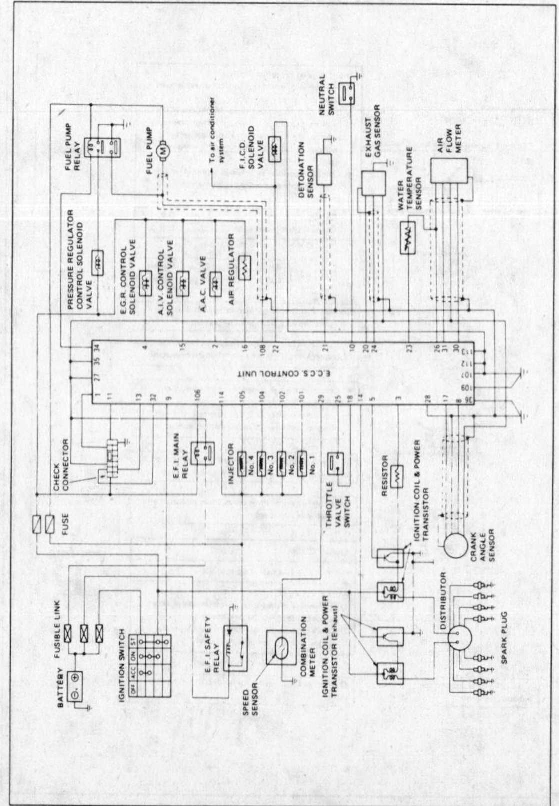

ECCS WIRING SCHEMATIC—1988 200SX WITH CA18ET ENGINE

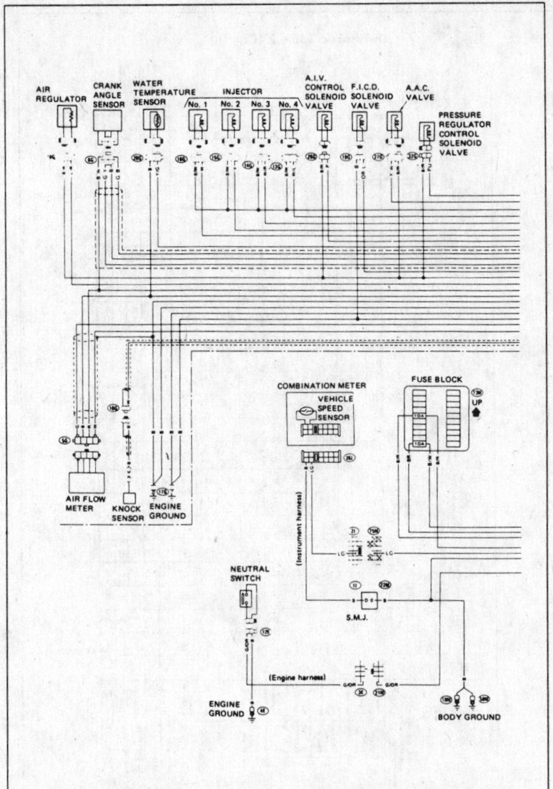
AIR REGULATOR
CRANK ANGLE SENSOR
WATER TEMPERATURE SENSOR
INJECTOR
No. 1
No. 2
No. 3
No. 4
A.I.V. CONTROL SOLENOID VALVE
F.I.C.D. SOLENOID VALVE
A.A.C. VALVE
PRESSURE REGULATOR CONTROL SOLENOID VALVE
COMBINATION METER
VEHICLE SPEED SENSOR
FUSE BLOCK
AIR FLOW METER
KNOCK SENSOR
ENGINE GROUND
NEUTRAL SWITCH
S.M.J.
(Engine harness)
ENGINE GROUND
BODY GROUND

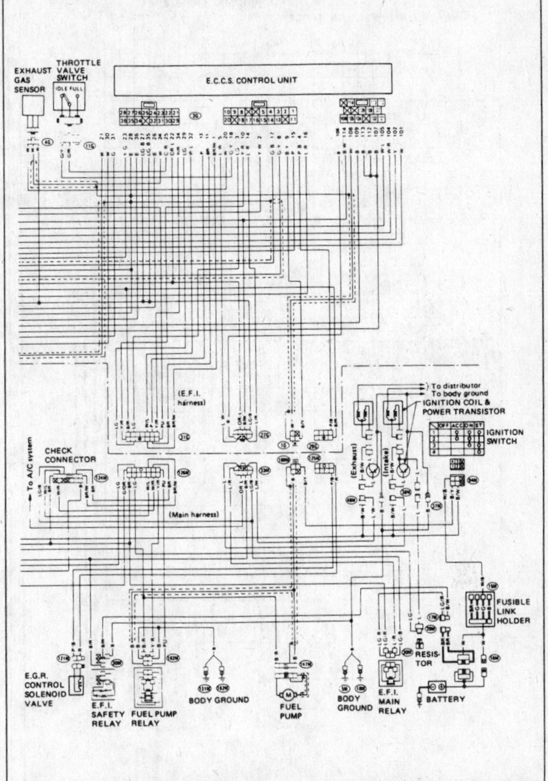
EXHAUST GAS SENSOR
THROTTLE VALVE SWITCH
E.C.C.S. CONTROL UNIT
To distributor
To body ground
IGNITION COIL & POWER TRANSISTOR
IGNITION SWITCH
CHECK CONNECTOR
To A/C system
(E.F.I. harness)
(Main harness)
FUSIBLE LINK HOLDER
RESISTOR
E.G.R. CONTROL SOLENOID VALVE
E.F.I. SAFETY RELAY
FUEL PUMP RELAY
BODY GROUND
FUEL PUMP
BODY GROUND
E.F.I. MAIN RELAY
BATTERY

ECCS DIAGNOSTIC PROCEDURE—1988 200SX WITH CA18ET ENGINE

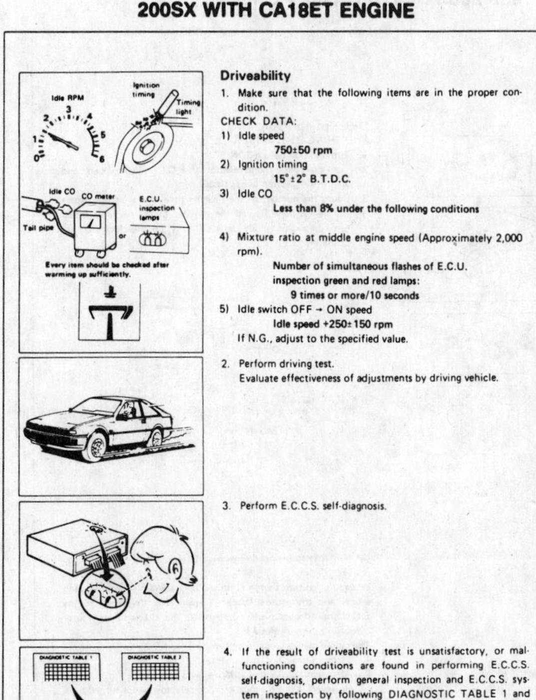
Idle RPM
Ignition timing
Timing light
Idle CO
CO meter
E.C.U. inspection lamps
Tail pipe
or
Every item should be checked after warming up sufficiently.
DIAGNOSTIC TABLE 1
DIAGNOSTIC TABLE 2
Electronic control system inspection

Driveability

1. Make sure that the following items are in the proper condition.

CHECK DATA:

1) Idle speed
   750±50 rpm
2) Ignition timing
   15°±2° B.T.D.C.
3) Idle CO
   Less than 8% under the following conditions
4) Mixture ratio at middle engine speed (Approximately 2,000 rpm).
   Number of simultaneous flashes of E.C.U. inspection green and red lamps:
   9 times or more/10 seconds
5) Idle switch OFF → ON speed
   Idle speed +250±150 rpm

If N.G., adjust to the specified value.

2. Perform driving test.
   Evaluate effectiveness of adjustments by driving vehicle.
3. Perform E.C.C.S. self-diagnosis.
4. If the result of driveability test is unsatisfactory, or malfunctioning conditions are found in performing E.C.C.S. self-diagnosis, perform general inspection and E.C.C.S. system inspection by following DIAGNOSTIC TABLE 1 and 2 in response to driveability trouble items.
   If N.G., repair.
5. Perform driving test.
   Re-evaluate vehicle performance after the inspection.

ECCS DIAGNOSTIC TABLE I—1988 200SX WITH CA18ET ENGINE

Diagnostic Table 1

SYSTEM INSPECTION TABLE

System \ Sensor & actuator	Crank angle sensor	Air flow meter	Water temperature sensor	Ignition switch	Injector	Throttle valve switch	Neutral switch	Exhaust gas sensor
Fuel injection & mixture ratio feedback control	○	○	○	○	○	○	○	○
Ignition timing control	○	○	○	○		○		
A.I.V. control	○		○			○		○
Fuel pump control	○			○				
Fuel pressure control	○		○	○	○			
Idle speed control	○		○	○		○	○	
E.G.R. control	○		○	○		○		

System \ Sensor & actuator	A.I.V. control solenoid valve	E.G.R. control solenoid valve	P.R. control solenoid valve/Fuel pump relay	A.A.C. valve	Air regulator	Vehicle speed sensor	E.F.I. main relay	Datonation sensor
Fuel injection & mixture ratio feedback control						○	○	
Ignition timing control							○	○
A.I.V. control	○						○	
Fuel pump control							○	
Fuel pressure control			○				○	
Idle speed control				○	○	○	○	
E.G.R. control		○					○	

This table indicates the inspection items for the E.C.C.S. control system. For each system, it is necessary to check sensors or actuators marked "○".

ECCS DIAGNOSTIC TABLE II—1988 200SX WITH CA18ET ENGINE

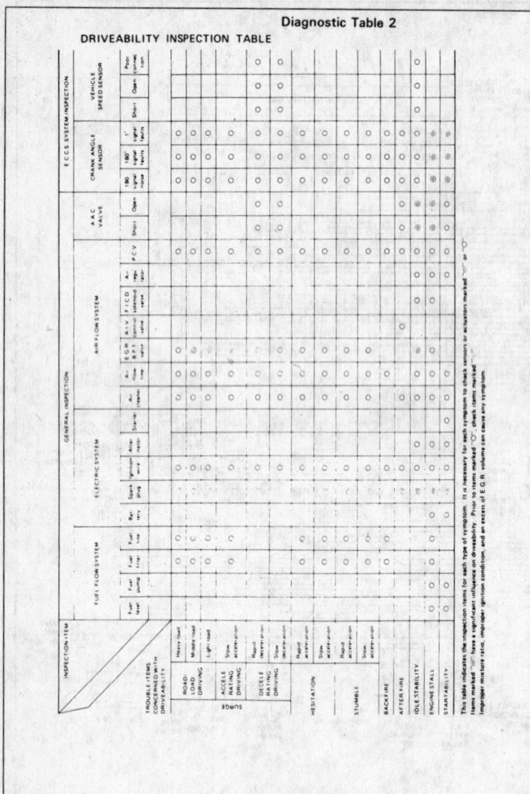

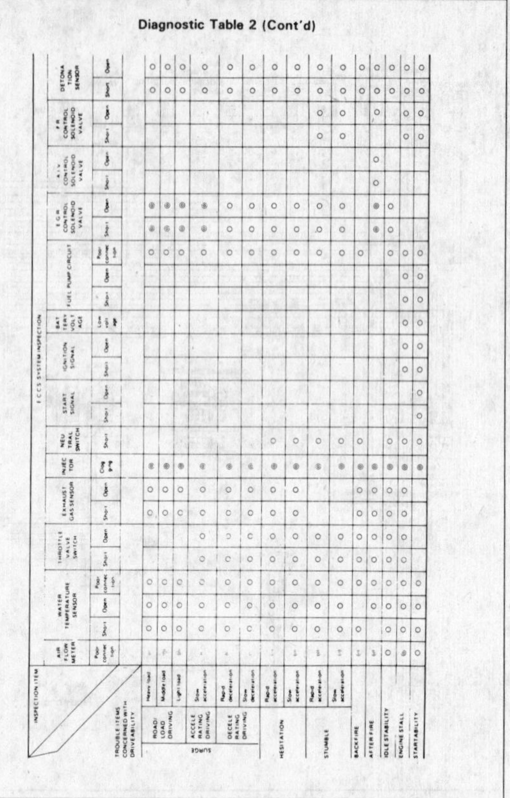

ECCS SELF-DIAGNOSTIC DESCRIPTION—1988 200SX WITH CA18ET ENGINE

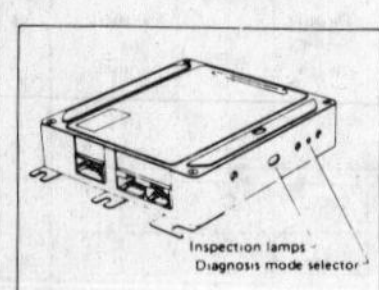
Inspection lamps
Diagnosis mode selector

Description

The self-diagnosis is useful to diagnose malfunctions in major sensors and actuators of the E.C.C.S. system. There are 5 modes in the self-diagnosis system.

1. **Mode I – Mixture ratio feedback control monitor A**
   - During closed loop condition:
     The green inspection lamp turns ON when lean condition is detected and goes OFF by rich condition.
   - During open loop condition:
     The green inspection lamp remains ON or OFF.
2. **Mode II – Mixture ratio feedback control monitor B**
   The green inspection lamp function is the same as Mode I.
   - During closed loop condition:
     The red inspection lamp turns ON and OFF simultaneously with the green inspection lamp when the mixture ratio is controlled within the specified value.
   - During open loop condition:
     The red inspection lamp remains ON or OFF.
3. **Mode III – Self-diagnosis**
   This mode is the same as the former self-diagnosis in self-diagnosis mode.
4. **Mode IV – Switches ON/OFF diagnosis**
   During this mode, the inspection lamps monitor the switch ON-OFF condition.
   - Idle switch
   - Starter switch
   - Vehicle speed sensor
5. **Mode V – Real time diagnosis**
   The moment the malfunction is detected, the display will be presented immediately. That is, the condition at which the malfunction occurs can be found by observing the inspection lamps during driving test.

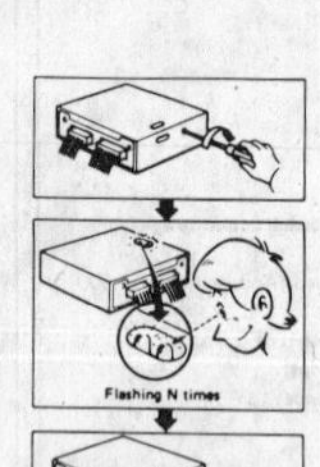
Flashing N times
Mode N

Description (Cont'd)

SWITCHING THE MODES

1. Turn ignition switch "ON".
2. Turn diagnostic mode selector on E.C.U. fully clockwise and wait the inspection lamps flash.
3. Count the number of the flashing time, and after the inspection lamps have flashed the number of the required mode, turn diagnostic mode selector fully counterclockwise immediately.

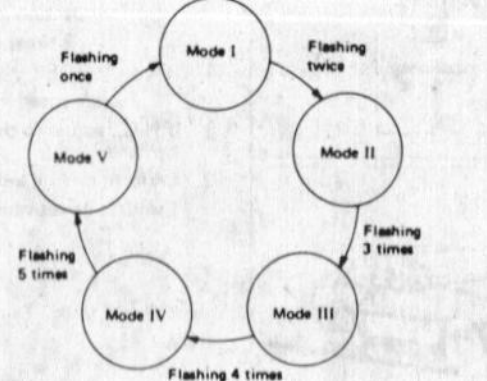
Mode I
Flashing twice
Mode II
Flashing 3 times
Mode III
Flashing 4 times
Mode IV
Flashing 5 times
Mode V
Flashing once

NOTE:
When the ignition switch is turned off during diagnosis, in each mode, and then turned back on again after the power to the E.C.U. has dropped off completely, the diagnosis will automatically return to Mode I.

The stored memory would be lost if:

1. Battery terminal is disconnected.
2. After selecting Mode III, Mode IV is selected.
   However, if the diagnostic mode selector is kept turned fully clockwise, it will continue to change in the order of Mode I → II → III → IV → V → I ... etc., and in this state the stored memory will not be erased.

ECCS SELF-DIAGNOSTIC MODES I AND II 1988 200SX WITH CA18ET ENGINE

Modes I & II — Mixture Ratio Feedback Control Monitors A & B

In these modes, the control unit provides the Air-fuel ratio monitor presentation and the Air-fuel ratio feedback coefficient monitor presentation.

Mode	LED	Engine stopped (Ignition switch "ON")	Engine running			
			Open loop condition	Closed loop condition		
Mode I (Monitor A)	Green	ON	* Remains ON or OFF	Blinks		
	Red	ON	OFF			
Mode II (Monitor B)	Green	ON	* Remains ON or OFF	Blinks		
	Red	OFF	* Remains ON or OFF (synchronous with green LED)	Compensating mixture ratio		
				More than 5% rich	Between 5% lean and 5% rich	More
				OFF	Synchronized with green LED	Remains ON

*: Maintains conditions just before switching to open loop

ECCS SELF-DIAGNOSTIC MODE III—1988 200SX WITH CA18ET ENGINE

Mode III — Self-Diagnostic System

The E.C.U. constantly monitors the function of these sensors and actuators, regardless of ignition key position. If a malfunction occurs, the information is stored in the E.C.U. and can be retrieved from the memory by turning on the diagnostic mode selector, located on the side of the E.C.U. When activated, the malfunction is indicated by flashing a red and a green L.E.D. (Light Emitting Diode), also located on the E.C.U. Since all the self-diagnostic results are stored in the E.C.U.'s memory even intermittent malfunctions can be diagnosed.

A malfunctioning part's group is indicated by the number of both the red and the green L.E.D.s flashing. First, the red L.E.D. flashes and the green flashes follow. The red L.E.D. refers to the number of tens while the green one refers to the number of units. For example, when the red L.E.D. flashes once and then the green one flashes twice, this means the number "12" showing the air flow meter signal is malfunctioning. In this way, all the problems are classified by the code numbers.

- **When engine fails to start, crank engine more than two seconds before starting self-diagnosis.**
- **Before starting self-diagnosis, do not erase stored memory. If doing so, self-diagnosis function for intermittent malfunctions would be lost.**

The stored memory would be lost if:

1. **Battery terminal is disconnected.**
2. **After selecting Mode III, Mode IV is selected.**

DISPLAY CODE TABLE

Code No.	Detected items
11	Crank angle sensor circuit
12	Air flow meter circuit
13	Water temperature sensor circuit
21	Ignition signal missing in primary coil
22	Fuel pump circuit
34	Detonation sensor
44	No malfunctioning in the above circuit

ECCS SELF-DIAGNOSTIC MODE III (CONT.)—1988 200SX WITH CA18ET ENGINE

Mode III — Self-Diagnostic System (Cont'd)

RETENTION OF DIAGNOSTIC RESULTS

The diagnostic result is retained in E.C.U. memory until the starter is operated fifty times after a diagnostic item is judged to be malfunctioning. The diagnostic result will then be cancelled automatically. If a diagnostic item which has been judged to be malfunctioning and stored in memory is again judged to be malfunctioning before the starter is operated fifty times, the second result will replace the previous one. It will be stored in E.C.U. memory until the starter is operated fifty times more.

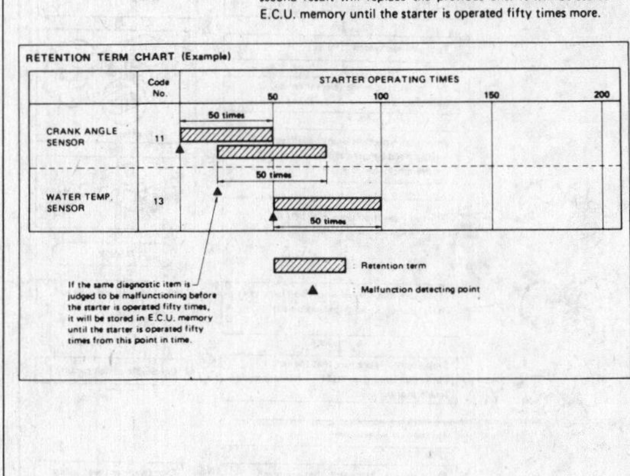
RETENTION TERM CHART (Example)
Code No.
STARTER OPERATING TIMES
50
100
150
200
CRANK ANGLE SENSOR
11
50 times
50 times
WATER TEMP. SENSOR
13
50 times
If the same diagnostic item is judged to be malfunctioning before the starter is operated fifty times, it will be stored in E.C.U. memory until the starter is operated fifty times from this point in time.
: Retention term
: Malfunction detecting point

Mode III — Self-Diagnostic System (Cont'd)

SELF-DIAGNOSTIC PROCEDURE

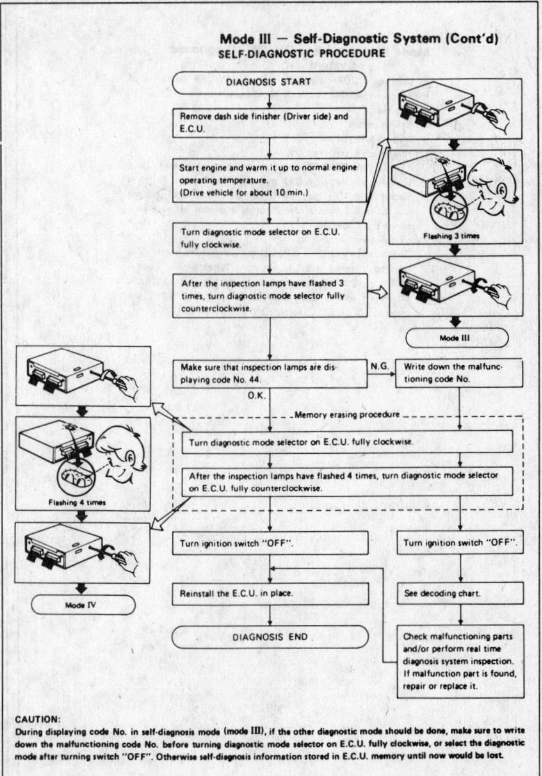
DIAGNOSIS START
Remove dash side finisher (Driver side) and E.C.U.
Start engine and warm it up to normal engine operating temperature. (Drive vehicle for about 10 min.)
Turn diagnostic mode selector on E.C.U. fully clockwise.
Flashing 3 times
After the inspection lamps have flashed 3 times, turn diagnostic mode selector fully counterclockwise.
Mode III
Make sure that inspection lamps are displaying code No. 44.
N.G.
Write down the malfunctioning code No.
O.K.
Memory erasing procedure
Turn diagnostic mode selector on E.C.U. fully clockwise.
After the inspection lamps have flashed 4 times, turn diagnostic mode selector on E.C.U. fully counterclockwise.
Flashing 4 times
Turn ignition switch "OFF".
Turn ignition switch "OFF".
Mode IV
Reinstall the E.C.U. in place.
See decoding chart.
DIAGNOSIS END
Check malfunctioning parts and/or perform real time diagnosis system inspection. If malfunction part is found, repair or replace it.

CAUTION:
During displaying code No. in self-diagnosis mode (mode III), if the other diagnostic mode should be done, make sure to write down the malfunctioning code No. before turning diagnostic mode selector on E.C.U. fully clockwise, or select the diagnostic mode after turning switch "OFF". Otherwise self-diagnosis information stored in E.C.U. memory until now would be lost.

ECCS SELF-DIAGNOSTIC MODE III DISPLAY CODES—1988 200SX WITH CA18ET

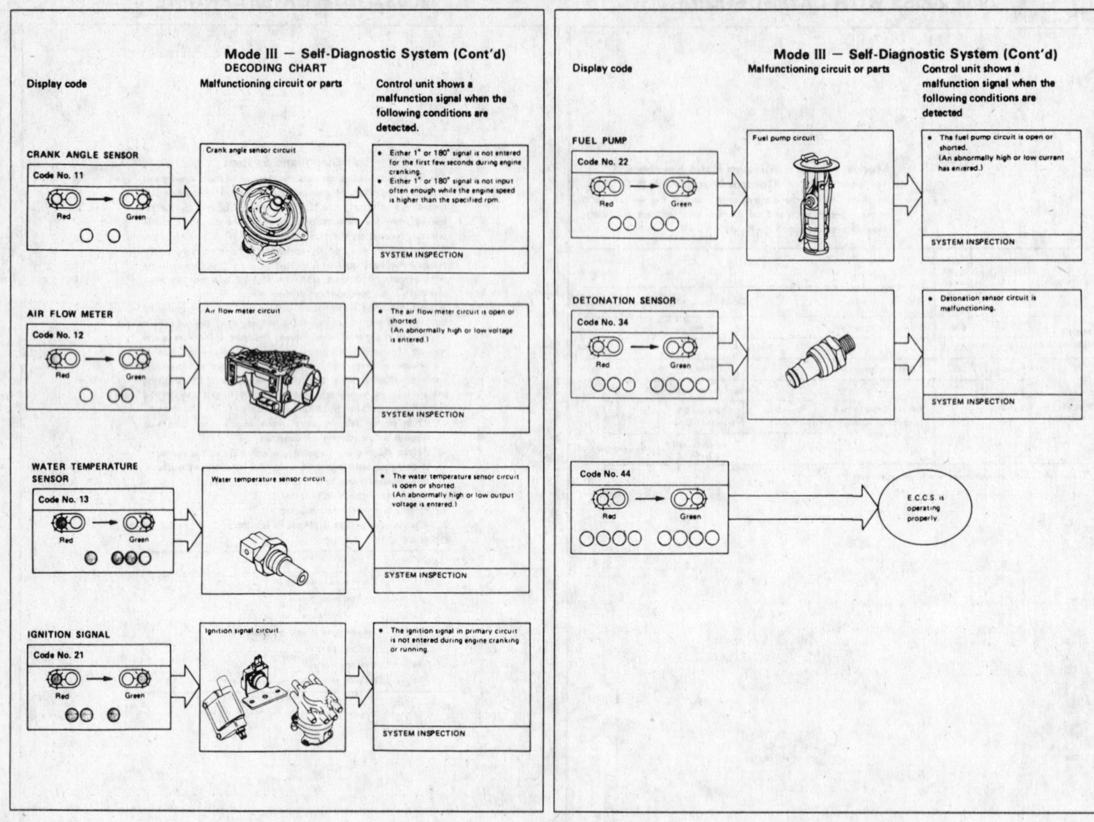
Mode III — Self-Diagnostic System (Cont'd)
DECODING CHART
Display code
Malfunctioning circuit or parts
Control unit shows a malfunction signal when the following conditions are detected.
CRANK ANGLE SENSOR
Code No. 11
Red
Green
Crank angle sensor circuit
• Either 1° or 180° signal is not entered for the first few seconds during engine cranking.
• Either 1° or 180° signal is not input often enough while the engine speed is higher than the specified rpm.
SYSTEM INSPECTION
AIR FLOW METER
Code No. 12
Red
Green
Air flow meter circuit
• The air flow meter circuit is open or shorted.
(An abnormally high or low voltage is entered.)
SYSTEM INSPECTION
WATER TEMPERATURE SENSOR
Code No. 13
Red
Green
Water temperature sensor circuit
• The water temperature sensor circuit is open or shorted.
(An abnormally high or low output voltage is entered.)
SYSTEM INSPECTION
IGNITION SIGNAL
Code No. 21
Red
Green
Ignition signal circuit
• The ignition signal in primary circuit is not entered during engine cranking or running.
SYSTEM INSPECTION
Mode III — Self-Diagnostic System (Cont'd)
Display code
Malfunctioning circuit or parts
Control unit shows a malfunction signal when the following conditions are detected
FUEL PUMP
Code No. 22
Red
Green
Fuel pump circuit
• The fuel pump circuit is open or shorted.
(An abnormally high or low current has entered.)
SYSTEM INSPECTION
DETONATION SENSOR
Code No. 34
Red
Green
• Detonation sensor circuit is malfunctioning.
SYSTEM INSPECTION
Code No. 44
Red
Green
E.C.C.S. is operating properly.

ECCS SELF-DIAGNOSTIC MODE IV—1988 200SX WITH CA18ET ENGINE

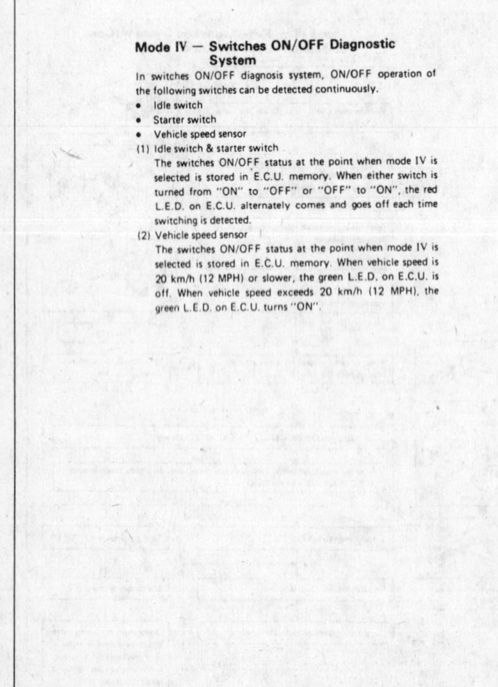

Mode IV — Switches ON/OFF Diagnostic System

In switches ON/OFF diagnosis system, ON/OFF operation of the following switches can be detected continuously.

- Idle switch
- Starter switch
- Vehicle speed sensor

(1) Idle switch & starter switch
The switches ON/OFF status at the point when mode IV is selected is stored in E.C.U. memory. When either switch is turned from "ON" to "OFF" or "OFF" to "ON", the red L.E.D. on E.C.U. alternately comes and goes off each time switching is detected.

(2) Vehicle speed sensor
The switches ON/OFF status at the point when mode IV is selected is stored in E.C.U. memory. When vehicle speed is 20 km/h (12 MPH) or slower, the green L.E.D. on E.C.U. is off. When vehicle speed exceeds 20 km/h (12 MPH), the green L.E.D. on E.C.U. turns "ON".

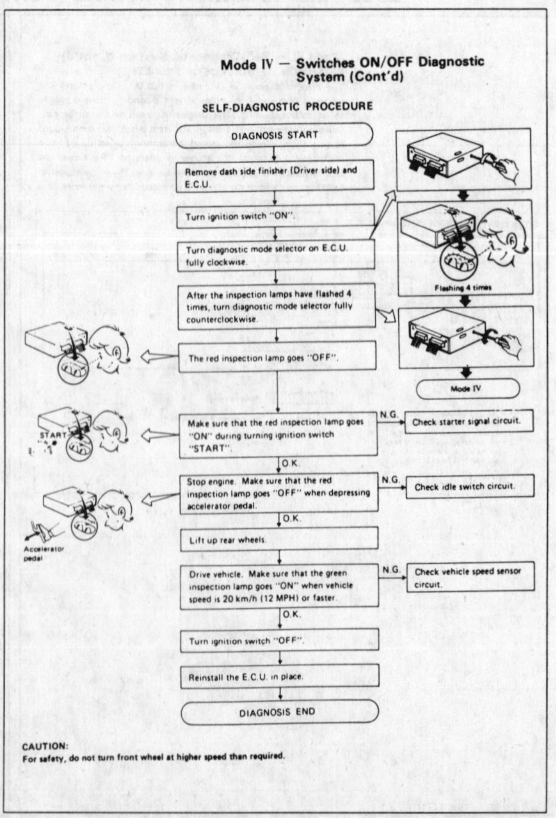
Mode IV — Switches ON/OFF Diagnostic System (Cont'd)
SELF-DIAGNOSTIC PROCEDURE
DIAGNOSIS START
Remove dash side finisher (Driver side) and E.C.U.
Turn ignition switch "ON".
Turn diagnostic mode selector on E.C.U. fully clockwise.
After the inspection lamps have flashed 4 times, turn diagnostic mode selector fully counterclockwise.
Flashing 4 times
Mode IV
The red inspection lamp goes "OFF".
Make sure that the red inspection lamp goes "ON" during turning ignition switch "START".
N.G.
Check starter signal circuit.
START
O.K.
Stop engine. Make sure that the red inspection lamp goes "OFF" when depressing accelerator pedal.
N.G.
Check idle switch circuit.
Accelerator pedal
O.K.
Lift up rear wheels.
Drive vehicle. Make sure that the green inspection lamp goes "ON" when vehicle speed is 20 km/h (12 MPH) or faster.
N.G.
Check vehicle speed sensor circuit.
O.K.
Turn ignition switch "OFF".
Reinstall the E.C.U. in place.
DIAGNOSIS END
CAUTION:
For safety, do not turn front wheel at higher speed than required.

ECCS SELF-DIAGNOSTIC MODE V-1988 200SX WITH CA18ET ENGINE

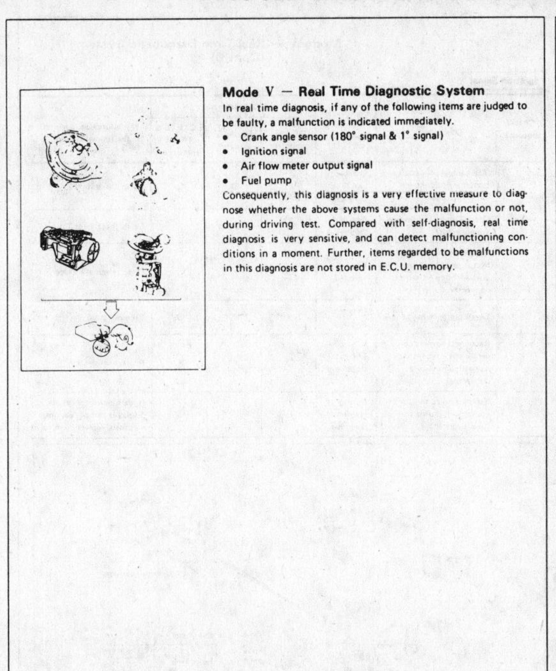

Mode V — Real Time Diagnostic System

In real time diagnosis, if any of the following items are judged to be faulty, a malfunction is indicated immediately.

- Crank angle sensor (180° signal & 1° signal)
- Ignition signal
- Air flow meter output signal
- Fuel pump

Consequently, this diagnosis is a very effective measure to diagnose whether the above systems cause the malfunction or not, during driving test. Compared with self-diagnosis, real time diagnosis is very sensitive, and can detect malfunctioning conditions in a moment. Further, items regarded to be malfunctions in this diagnosis are not stored in E.C.U. memory.

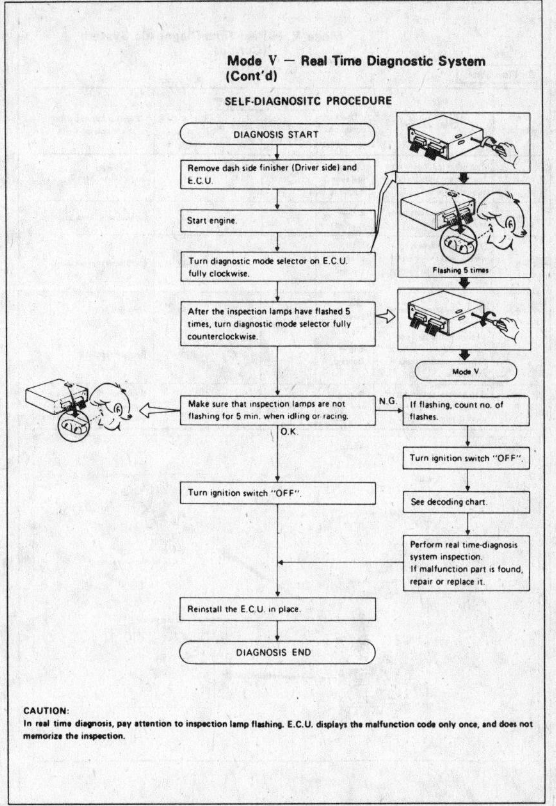
Mode V — Real Time Diagnostic System (Cont'd)
SELF-DIAGNOSITC PROCEDURE
DIAGNOSIS START
Remove dash side finisher (Driver side) and E.C.U.
Start engine.
Turn diagnostic mode selector on E.C.U. fully clockwise.
Flashing 5 times
After the inspection lamps have flashed 5 times, turn diagnostic mode selector fully counterclockwise.
Mode V
Make sure that inspection lamps are not flashing for 5 min. when idling or racing.
N.G.
If flashing, count no. of flashes.
O.K.
Turn ignition switch "OFF".
See decoding chart.
Perform real time-diagnosis system inspection. If malfunction part is found, repair or replace it.
Turn ignition switch "OFF".
Reinstall the E.C.U. in place.
DIAGNOSIS END

CAUTION:
In real time diagnosis, pay attention to inspection lamp flashing. E.C.U. displays the malfunction code only once, and does not memorize the inspection.

ECCS SELF-DIAGNOSTIC MODE V DECODING CHART—1988 200SX WITH CA18ET

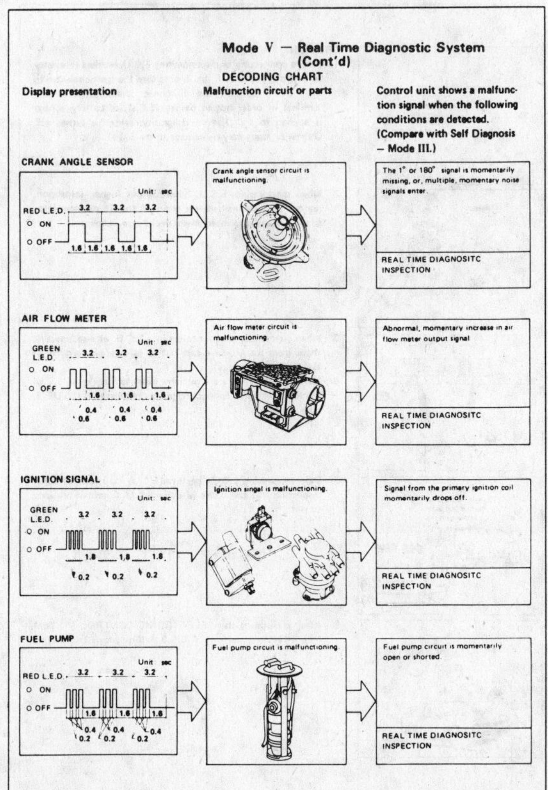

Mode V — Real Time Diagnostic System (Cont'd)

DECODING CHART

Display presentation	Malfunction circuit or parts	Control unit shows a malfunction signal when the following conditions are detected. (Compare with Self Diagnosis – Mode III.)
CRANK ANGLE SENSOR (RED L.E.D.; Unit: sec; 3.2, 3.2, 3.2; ON/OFF 1.6, 1.6, 1.6, 1.6, 1.6)	Crank angle sensor circuit is malfunctioning.	The 1° or 180° signal is momentarily missing, or, multiple, momentary noise signals enter. REAL TIME DIAGNOSITC INSPECTION
AIR FLOW METER (GREEN L.E.D.; Unit: sec; 3.2, 3.2, 3.2; ON/OFF 1.6, 1.6, 1.6; 0.4, 0.4, 0.4; 0.6, 0.6, 0.6)	Air flow meter circuit is malfunctioning.	Abnormal, momentary increase in air flow meter output signal REAL TIME DIAGNOSITC INSPECTION
IGNITION SIGNAL (GREEN L.E.D.; Unit: sec; 3.2, 3.2, 3.2; ON/OFF 1.8, 1.8, 1.8; 0.2, 0.2, 0.2)	Ignition singal is malfunctioning.	Signal from the primary ignition coil momentarily drops off. REAL TIME DIAGNOSITC INSPECTION
FUEL PUMP (RED L.E.D.; Unit: sec; 3.2, 3.2, 3.2; ON/OFF 1.6, 1.6, 1.6; 0.4, 0.4, 0.4; 0.2, 0.2, 0.2)	Fuel pump circuit is malfunctioning.	Fuel pump circuit is momentarily open or shorted. REAL TIME DIAGNOSITC INSPECTION

ECCS SELF-DIAGNOSTIC MODE V INSPECTION 1988 200SX WITH CA18ET

Mode V — Real Time Diagnostic System (Cont'd)

REAL TIME DIAGNOSTIC INSPECTION

Crank Angle Sensor

Check sequence	Check items	Check conditions	Check parts: Middle connector	Check parts: Sensor & actuator	Check parts: E.C.U. 20 & 16 pin connector	If malfunction, perform the following items.
1	Tap harness connector or component during real time diagnosis.	During real time diagnosis	O	O	O	Go to check item 2.
2	Check harness continuity at connector.	Engine stopped	O	X	X	Go to check item 3.
3	Disconnect harness connector, and then check dust adhesion to harness connector.	Engine stopped	O	X	O	Clean terminal surface.
4	Check pin terminal bend.	Engine stopped	X	X	O	Take out bend.
5	Reconnect harness connector and then recheck harness continuity at connector.	Engine stopped	O	X	X	Replace terminal.
6	Tap harness connector or component during real time diagnosis.	During real time diagnosis	O	O	O	If malfunction codes are displayed during real time diagnosis, replace terminal.

E.C.U. harness connector

Crank angle sensor harness connector

ECCS SELF-DIAGNOSTIC MODE V INSPECTION (CONT.)—1988 200SX WITH CA18ET ENGINE

Mode V — Real Time Diagnostic System (Cont'd)

Air Flow Meter

Check sequence	Check items	Check conditions	Check parts: Middle connector	Check parts: Sensor & actuator	Check parts: E.C.U. 20 & 16 pin connector	If malfunction, perform the following items.
1	Tap harness connector or component during real time diagnosis.	During real time diagnosis	O	O	O	Go to check item 2.
2	Check harness continuity at connector.	Engine stopped	O	X	X	Go to check item 3.
3	Disconnect harness connector, and then check dust adhesion to harness connector.	Engine stopped	O	X	O	Clean terminal surface.
4	Check pin terminal bend.	Engine stopped	X	X	O	Take out bend.
5	Reconnect harness connector and then recheck harness continuity at connector.	Engine stopped	O	X	X	Replace terminal.
6	Tap harness connector or component during real time diagnosis.	During real time diagnosis	O	O	O	If malfunction codes are displayed during real time diagnosis, replace terminal.

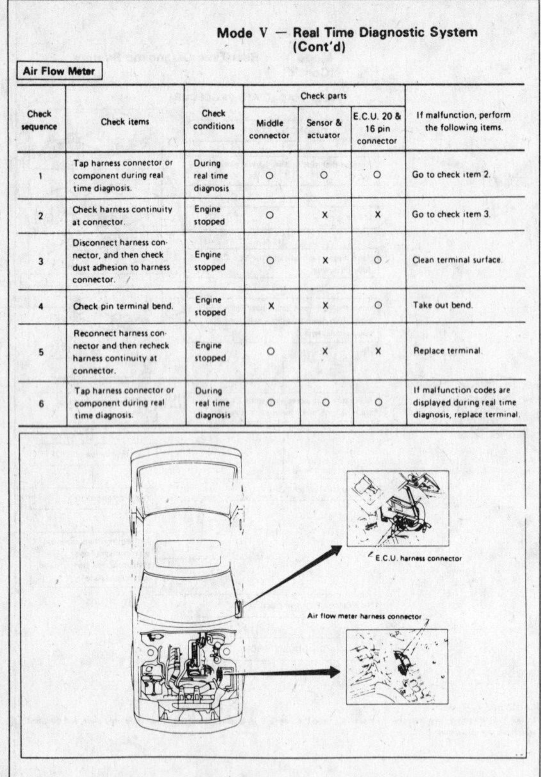
E.C.U. harness connector
Air flow meter harness connector

Mode V — Real Time Diagnostic System (Cont'd)

Ignition Signal

Check sequence	Check items	Check conditions	Check parts: Middle connector	Check parts: Sensor & actuator	Check parts: E.C.U. 20 & 16 pin connector	If malfunction, perform the following items.
1	Tap harness connector or component during real time diagnosis.	During real time diagnosis	O	O	O	Go to check item 2.
2	Check harness continuity at connector.	Engine stopped	O	X	X	Go to check item 3.
3	Disconnect harness connector, and then check dust adhesion to harness connector.	Engine stopped	O	X	O	Clean terminal surface.
4	Check pin terminal bend.	Engine stopped	X	X	O	Take out bend.
5	Reconnect harness connector and then recheck harness continuity at connector.	Engine stopped	O	X	X	Replace terminal.
6	Tap harness connector or component during real time diagnosis.	During real time diagnosis	O	O	O	If malfunction codes are displayed during real time diagnosis, replace terminal.

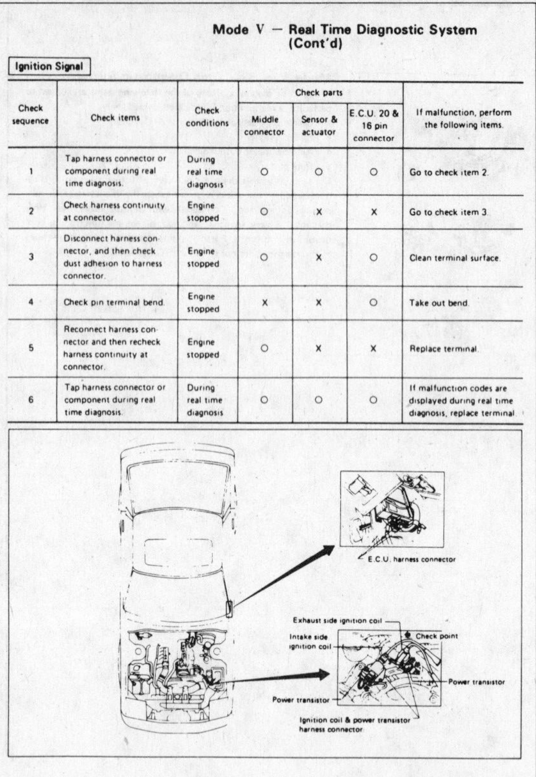
E.C.U. harness connector
Exhaust side ignition coil
Intake side ignition coil
Check point
Power transistor
Power transistor
Ignition coil & power transistor harness connector

ECCS SELF-DIAGNOSTIC MODE V INSPECTION (CONT.)—1988 200SX WITH CA18ET ENGINE

Mode V — Real Time Diagnostic System (Cont'd)

Fuel Pump

Check sequence	Check items	Check conditions	Check parts: Middle connector	Check parts: Sensor & actuator	Check parts: E.C.U. 20 & 16 pin connector	If malfunction, perform the following items.
1	Tap harness connector or component during real time diagnosis.	During real time diagnosis	O	O	O	Go to check item 2.
2	Check harness continuity at connector.	Engine stopped	O	X	X	Go to check item 3.
3	Disconnect harness connector, and then check dust adhesion to harness connector.	Engine stopped	O	X	O	Clean terminal surface.
4	Check pin terminal bend.	Engine stopped	X	X	O	Take out bend.
5	Reconnect harness connector and then recheck harness continuity at connector.	Engine stopped	O	X	X	Replace terminal.
6	Tap harness connector or component during real time diagnosis.	During real time diagnosis	O	O	O	If malfunction codes are displayed during real time diagnosis, replace terminal.

Fuel pump (In-tank type)
Fuel pump harness connector
Fuel pump
E.C.U. harness connector

ECCS SYSTEM CAUTION 1988 200SX WITH CA18ET ENGINE

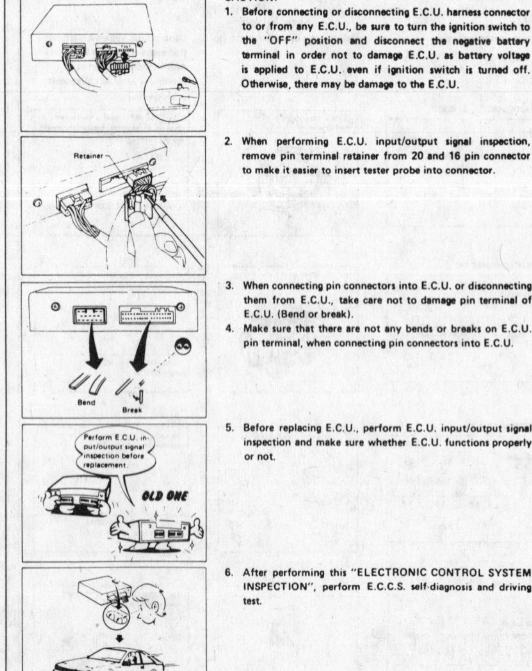
Retainer
Bend
Break
Perform E.C.U. input/output signal inspection before replacement.
OLD ONE

CAUTION:

1. Before connecting or disconnecting E.C.U. harness connector to or from any E.C.U., be sure to turn the ignition switch to the "OFF" position and disconnect the negative battery terminal in order not to damage E.C.U. as battery voltage is applied to E.C.U. even if ignition switch is turned off. Otherwise, there may be damage to the E.C.U.
2. When performing E.C.U. input/output signal inspection, remove pin terminal retainer from 20 and 16 pin connector to make it easier to insert tester probe into connector.
3. When connecting pin connectors into E.C.U. or disconnecting them from E.C.U., take care not to damage pin terminal of E.C.U. (Bend or break).
4. Make sure that there are not any bends or breaks on E.C.U. pin terminal, when connecting pin connectors into E.C.U.
5. Before replacing E.C.U., perform E.C.U. input/output signal inspection and make sure whether E.C.U. functions properly or not.
6. After performing this "ELECTRONIC CONTROL SYSTEM INSPECTION", perform E.C.C.S. self-diagnosis and driving test.

ECCS DIAGNOSTIC CHART—1988 200SX WITH CA18ET ENGINE

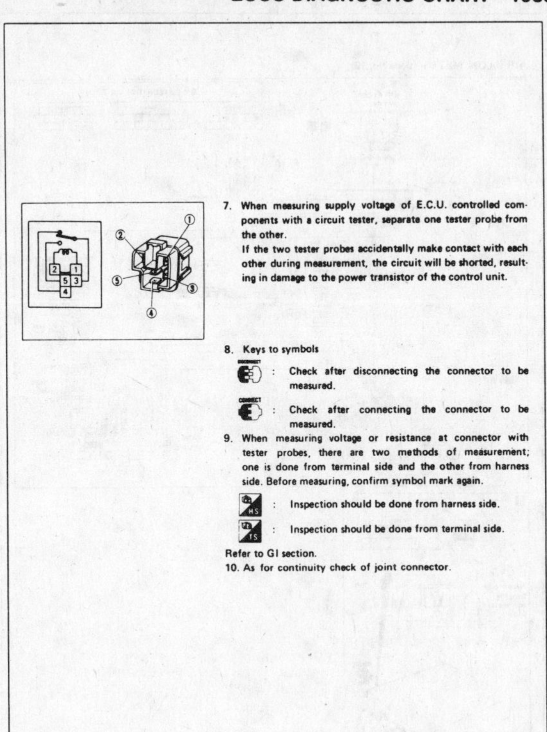

7. When measuring supply voltage of E.C.U. controlled components with a circuit tester, separate one tester probe from the other.
   If the two tester probes accidentally make contact with each other during measurement, the circuit will be shorted, resulting in damage to the power transistor of the control unit.
8. Keys to symbols
   - DISCONNECT : Check after disconnecting the connector to be measured.
   - CONNECT : Check after connecting the connector to be measured.
9. When measuring voltage or resistance at connector with tester probes, there are two methods of measurement; one is done from terminal side and the other from harness side. Before measuring, confirm symbol mark again.
   - HS : Inspection should be done from harness side.
   - TS : Inspection should be done from terminal side.

Refer to GI section.

10. As for continuity check of joint connector.

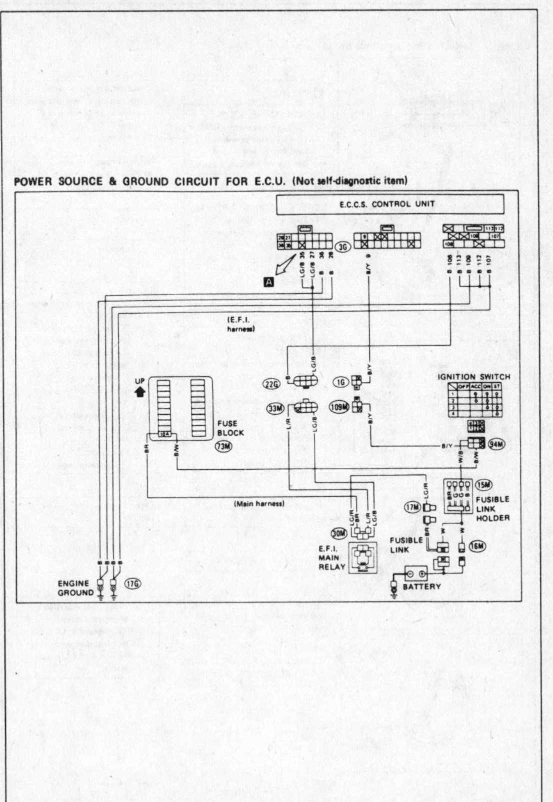
POWER SOURCE & GROUND CIRCUIT FOR E.C.U. (Not self-diagnostic item)
E.C.C.S. CONTROL UNIT
(E.F.I. harness)
UP
FUSE BLOCK
IGNITION SWITCH
(Main harness)
FUSIBLE LINK HOLDER
FUSIBLE LINK
E.F.I. MAIN RELAY
BATTERY
ENGINE GROUND

ECCS DIAGNOSTIC CHART—1988 200SX WITH CA18ET ENGINE

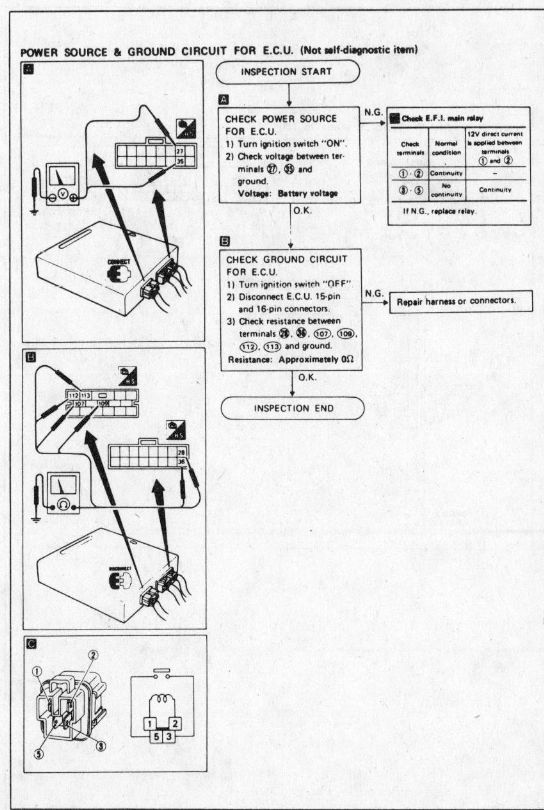
POWER SOURCE & GROUND CIRCUIT FOR E.C.U. (Not self-diagnostic item)
INSPECTION START
A
CHECK POWER SOURCE FOR E.C.U.
1) Turn ignition switch "ON".
2) Check voltage between terminals ㉗, ㉟ and ground.
Voltage: Battery voltage
N.G.
Check E.F.I. main relay
O.K.
B
CHECK GROUND CIRCUIT FOR E.C.U.
1) Turn ignition switch "OFF".
2) Disconnect E.C.U. 15-pin and 16-pin connectors.
3) Check resistance between terminals ㉘, ㊱, (107), (109), (112), (113) and ground.
Resistance: Approximately 0Ω
N.G.
Repair harness or connectors.
O.K.
INSPECTION END

Check terminals	Normal condition	12V direct current is applied between terminals ① and ②
①·②	Continuity	–
③·⑤	No continuity	Continuity

If N.G., replace relay.

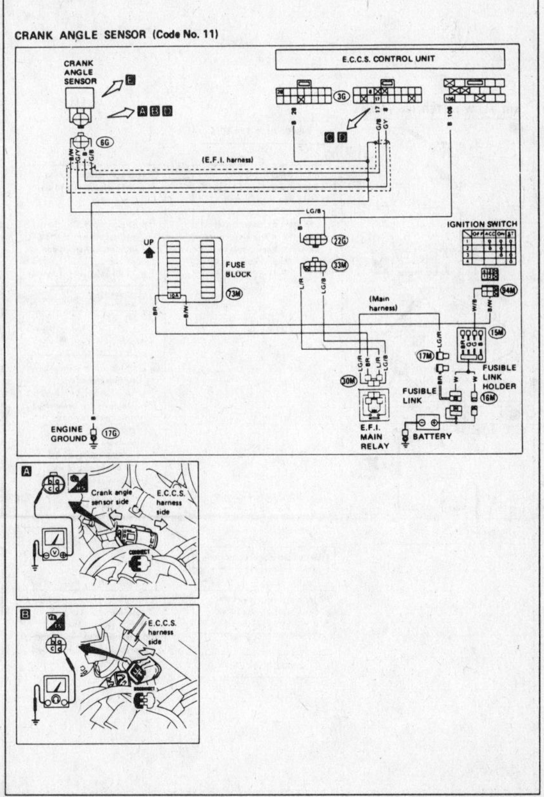
CRANK ANGLE SENSOR (Code No. 11)
CRANK ANGLE SENSOR
E.C.C.S. CONTROL UNIT
(E.F.I. harness)
IGNITION SWITCH
UP
FUSE BLOCK
(Main harness)
FUSIBLE LINK HOLDER
FUSIBLE LINK
E.F.I. MAIN RELAY
BATTERY
ENGINE GROUND
Crank angle sensor side
E.C.C.S. harness side

ECCS DIAGNOSTIC CHART—1988 200SX WITH CA18ET ENGINE

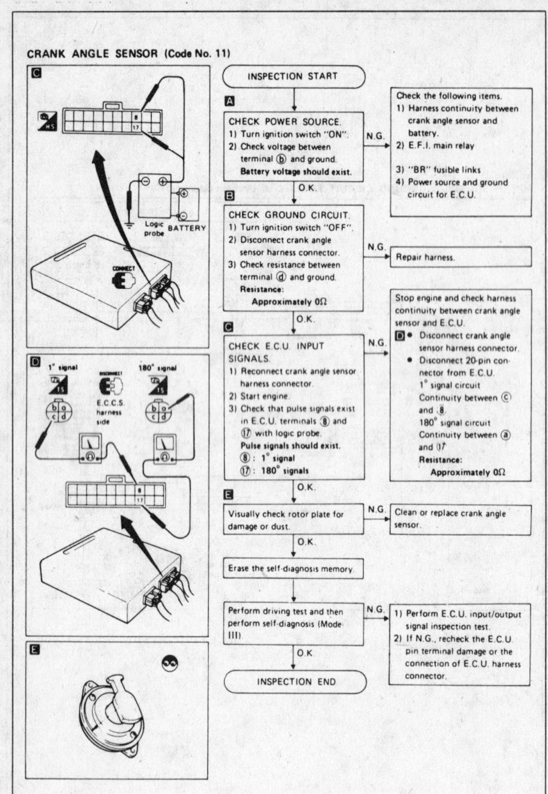
CRANK ANGLE SENSOR (Code No. 11)
INSPECTION START
CHECK POWER SOURCE.
1) Turn ignition switch "ON".
2) Check voltage between terminal ⓑ and ground.
Battery voltage should exist.
N.G.
Check the following items.
1) Harness continuity between crank angle sensor and battery.
2) E.F.I. main relay
3) "BR" fusible links
4) Power source and ground circuit for E.C.U.
O.K.
CHECK GROUND CIRCUIT.
1) Turn ignition switch "OFF".
2) Disconnect crank angle sensor harness connector.
3) Check resistance between terminal ⓓ and ground.
Resistance:
Approximately 0Ω
Repair harness.
CHECK E.C.U. INPUT SIGNALS.
1) Reconnect crank angle sensor harness connector.
2) Start engine.
3) Check that pulse signals exist in E.C.U. terminals ⑧ and ⑰ with logic probe.
Pulse signals should exist.
⑧: 1° signal
⑰: 180° signals
Stop engine and check harness continuity between crank angle sensor and E.C.U.
• Disconnect crank angle sensor harness connector.
• Disconnect 20-pin connector from E.C.U.
1° signal circuit
Continuity between ⓒ and ⑧
180° signal circuit
Continuity between ⓐ and ⑰
Resistance:
Approximately 0Ω
Visually check rotor plate for damage or dust.
Clean or replace crank angle sensor.
Erase the self-diagnosis memory.
Perform driving test and then perform self-diagnosis (Mode III).
1) Perform E.C.U. input/output signal inspection test.
2) If N.G., recheck the E.C.U. pin terminal damage or the connection of E.C.U. harness connector.
INSPECTION END
Logic probe
BATTERY
CONNECT
1° signal
180° signal
E.C.C.S. harness side

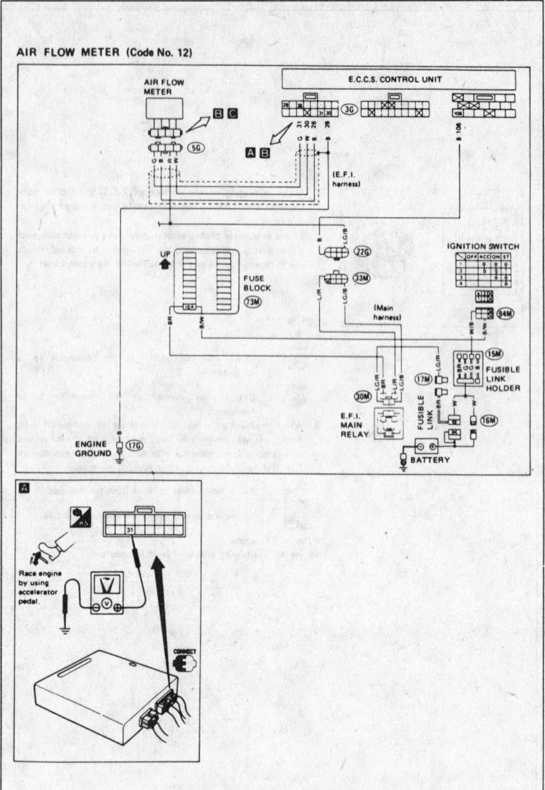
AIR FLOW METER (Code No. 12)
AIR FLOW METER
E.C.C.S. CONTROL UNIT
(E.F.I. harness)
IGNITION SWITCH
FUSE BLOCK
(Main harness)
FUSIBLE LINK HOLDER
E.F.I. MAIN RELAY
FUSIBLE LINK
BATTERY
ENGINE GROUND
Race engine by using accelerator pedal.
CONNECT

ECCS DIAGNOSTIC CHART—1988 200SX WITH CA18ET ENGINE

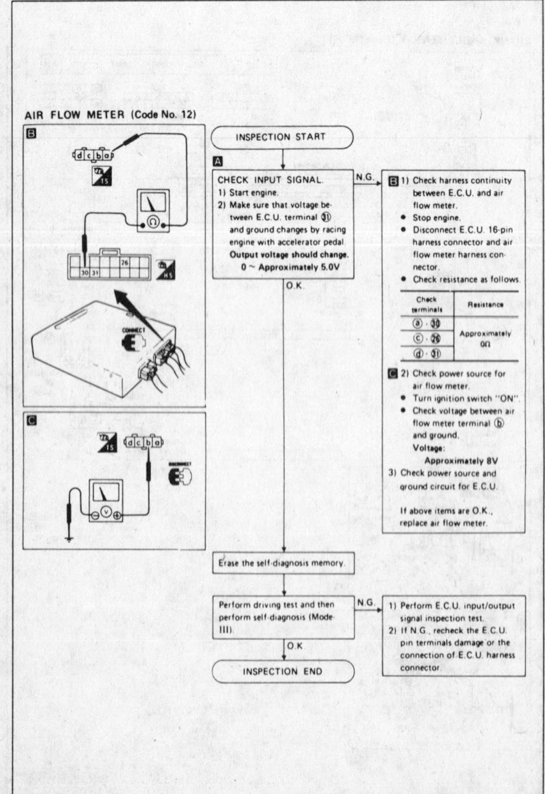
AIR FLOW METER (Code No. 12)
INSPECTION START
CHECK INPUT SIGNAL.
1) Start engine.
2) Make sure that voltage between E.C.U. terminal ㉛ and ground changes by racing engine with accelerator pedal.
Output voltage should change.
0 ~ Approximately 5.0V
N.G.
1) Check harness continuity between E.C.U. and air flow meter.
• Stop engine.
• Disconnect E.C.U. 16-pin harness connector and air flow meter harness connector.
• Check resistance as follows.
Check terminals
Resistance
ⓐ-㉚
ⓒ-㉖
ⓓ-㉛
Approximately 0Ω
2) Check power source for air flow meter.
• Turn ignition switch "ON".
• Check voltage between air flow meter terminal ⓑ and ground.
Voltage:
Approximately 8V
3) Check power source and ground circuit for E.C.U.
If above items are O.K., replace air flow meter.
O.K.
Erase the self-diagnosis memory.
Perform driving test and then perform self-diagnosis (Mode III).
1) Perform E.C.U. input/output signal inspection test.
2) If N.G., recheck the E.C.U. pin terminals damage or the connection of E.C.U. harness connector.
INSPECTION END
CONNECT
DISCONNECT

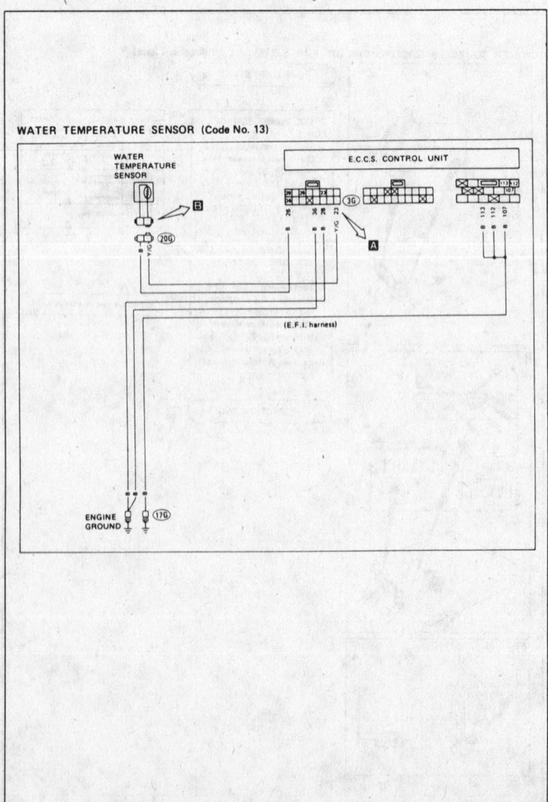
WATER TEMPERATURE SENSOR (Code No. 13)
WATER TEMPERATURE SENSOR
E.C.C.S. CONTROL UNIT
(E.F.I. harness)
ENGINE GROUND

ECCS DIAGNOSTIC CHART—1988 200SX WITH CA18ET ENGINE

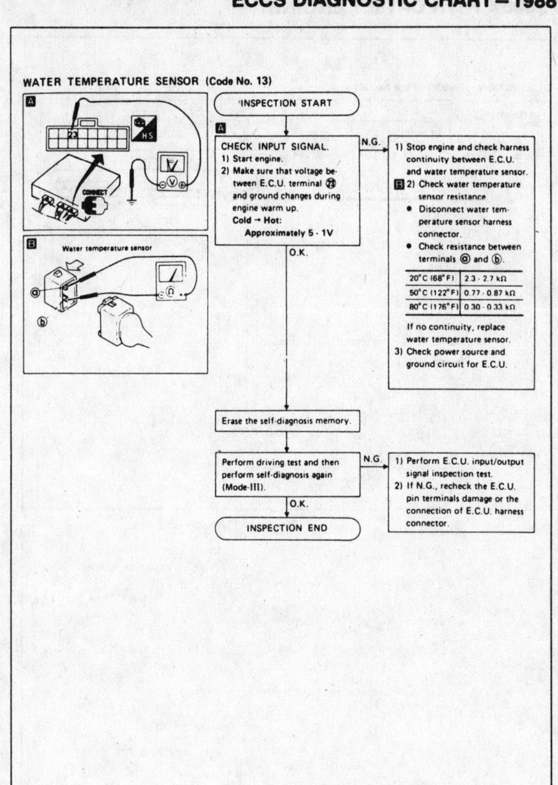
WATER TEMPERATURE SENSOR (Code No. 13)
INSPECTION START
CHECK INPUT SIGNAL.
1) Start engine.
2) Make sure that voltage between E.C.U. terminal ㉓ and ground changes during engine warm up.
Cold → Hot:
Approximately 5 - 1V
N.G.
1) Stop engine and check harness continuity between E.C.U. and water temperature sensor.
2) Check water temperature sensor resistance.
• Disconnect water temperature sensor harness connector.
• Check resistance between terminals ⓐ and ⓑ.
20°C (68°F) 2.3 - 2.7 kΩ
50°C (122°F) 0.77 - 0.87 kΩ
80°C (176°F) 0.30 - 0.33 kΩ
If no continuity, replace water temperature sensor.
3) Check power source and ground circuit for E.C.U.
O.K.
Water temperature sensor
Erase the self-diagnosis memory.
Perform driving test and then perform self-diagnosis again (Mode-III).
N.G.
1) Perform E.C.U. input/output signal inspection test.
2) If N.G., recheck the E.C.U. pin terminals damage or the connection of E.C.U. harness connector.
O.K.
INSPECTION END

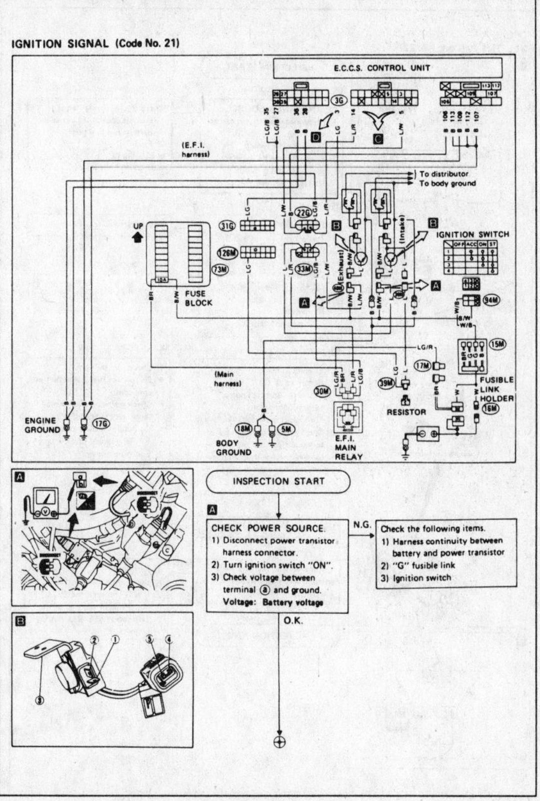
IGNITION SIGNAL (Code No. 21)
E.C.C.S. CONTROL UNIT
(E.F.I. harness)
To distributor
To body ground
IGNITION SWITCH
FUSE BLOCK
(Main harness)
ENGINE GROUND
BODY GROUND
E.F.I. MAIN RELAY
RESISTOR
FUSIBLE LINK HOLDER
INSPECTION START
CHECK POWER SOURCE.
1) Disconnect power transistor harness connector.
2) Turn ignition switch "ON".
3) Check voltage between terminal ⓐ and ground.
Voltage: Battery voltage
N.G.
Check the following items.
1) Harness continuity between battery and power transistor
2) "G" fusible link
3) Ignition switch
O.K.

ECCS DIAGNOSTIC CHART—1988 200SX WITH CA18ET ENGINE

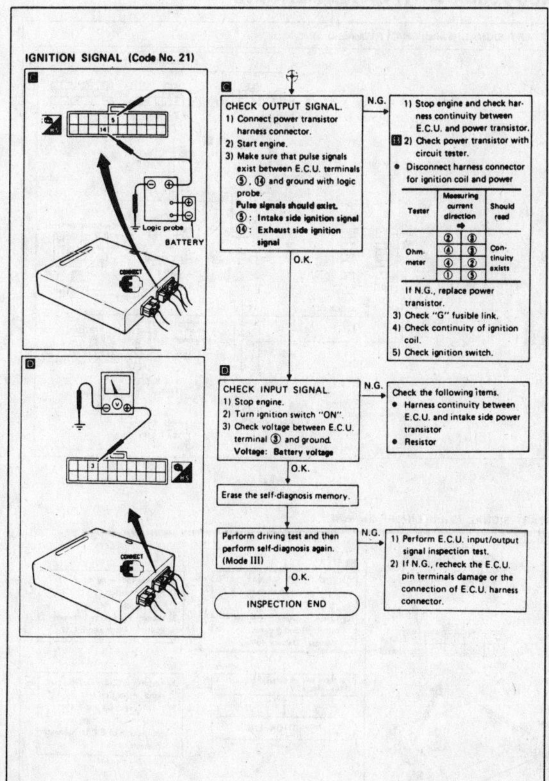
IGNITION SIGNAL (Code No. 21)
CHECK OUTPUT SIGNAL.
1) Connect power transistor harness connector.
2) Start engine.
3) Make sure that pulse signals exist between E.C.U. terminals ⑤, ⑭ and ground with logic probe.
Pulse signals should exist.
⑤: Intake side ignition signal
⑭: Exhaust side ignition signal
N.G.
1) Stop engine and check harness continuity between E.C.U. and power transistor.
2) Check power transistor with circuit tester.
• Disconnect harness connector for ignition coil and power transistor.
Tester
Measuring current direction
Should read
Ohm-meter
Continuity exists
If N.G., replace power transistor.
3) Check "G" fusible link.
4) Check continuity of ignition coil.
5) Check ignition switch.
Logic probe
BATTERY
O.K.
CHECK INPUT SIGNAL.
1) Stop engine.
2) Turn ignition switch "ON".
3) Check voltage between E.C.U. terminal ③ and ground.
Voltage: Battery voltage
N.G.
Check the following items.
• Harness continuity between E.C.U. and intake side power transistor
• Resistor
O.K.
Erase the self-diagnosis memory.
Perform driving test and then perform self-diagnosis again. (Mode III)
N.G.
1) Perform E.C.U. input/output signal inspection test.
2) If N.G., recheck the E.C.U. pin terminals damage or the connection of E.C.U. harness connector.
O.K.
INSPECTION END

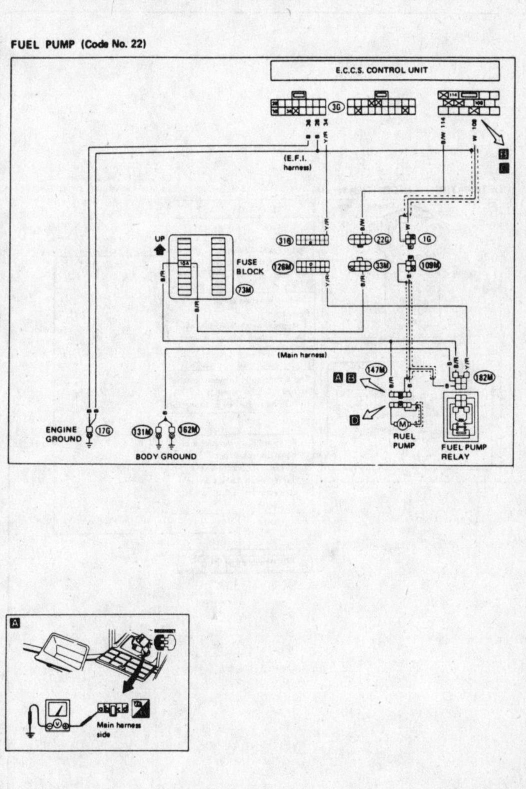
FUEL PUMP (Code No. 22)
E.C.C.S. CONTROL UNIT
(E.F.I. harness)
FUSE BLOCK
(Main harness)
ENGINE GROUND
BODY GROUND
FUEL PUMP
FUEL PUMP RELAY
Main harness side

ECCS DIAGNOSTIC CHART—1988 200SX WITH CA18ET ENGINE

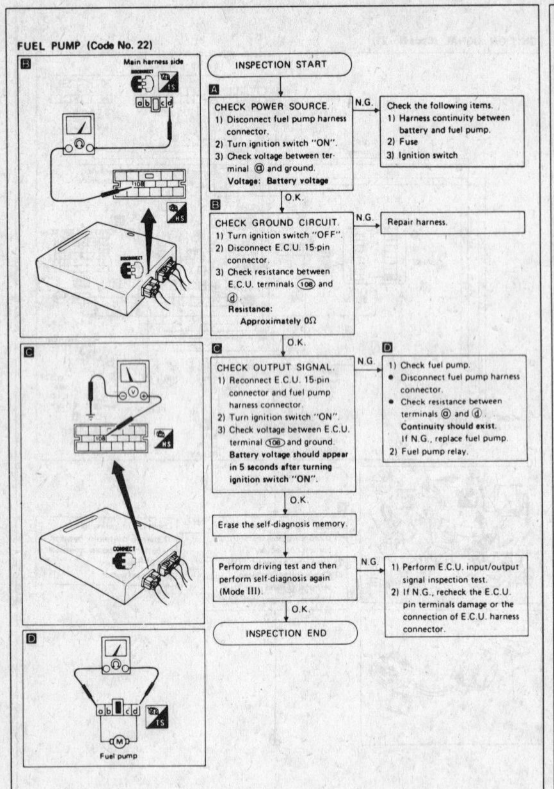
FUEL PUMP (Code No. 22)
Main harness side
Fuel pump
INSPECTION START
A
CHECK POWER SOURCE.
1) Disconnect fuel pump harness connector.
2) Turn ignition switch "ON".
3) Check voltage between terminal ⓐ and ground.
Voltage: Battery voltage
N.G.
Check the following items.
1) Harness continuity between battery and fuel pump.
2) Fuse
3) Ignition switch
O.K.
B
CHECK GROUND CIRCUIT.
1) Turn ignition switch "OFF".
2) Disconnect E.C.U. 15-pin connector.
3) Check resistance between E.C.U. terminals (108) and ⓓ.
Resistance:
Approximately 0Ω
N.G.
Repair harness.
O.K.
C
CHECK OUTPUT SIGNAL.
1) Reconnect E.C.U. 15-pin connector and fuel pump harness connector.
2) Turn ignition switch "ON".
3) Check voltage between E.C.U. terminal (108) and ground.
Battery voltage should appear in 5 seconds after turning ignition switch "ON".
N.G.
D
1) Check fuel pump.
• Disconnect fuel pump harness connector.
• Check resistance between terminals ⓐ and ⓓ.
Continuity should exist.
If N.G., replace fuel pump.
2) Fuel pump relay.
O.K.
Erase the self-diagnosis memory.
Perform driving test and then perform self-diagnosis again (Mode III).
N.G.
1) Perform E.C.U. input/output signal inspection test.
2) If N.G., recheck the E.C.U. pin terminals damage or the connection of E.C.U. harness connector.
O.K.
INSPECTION END

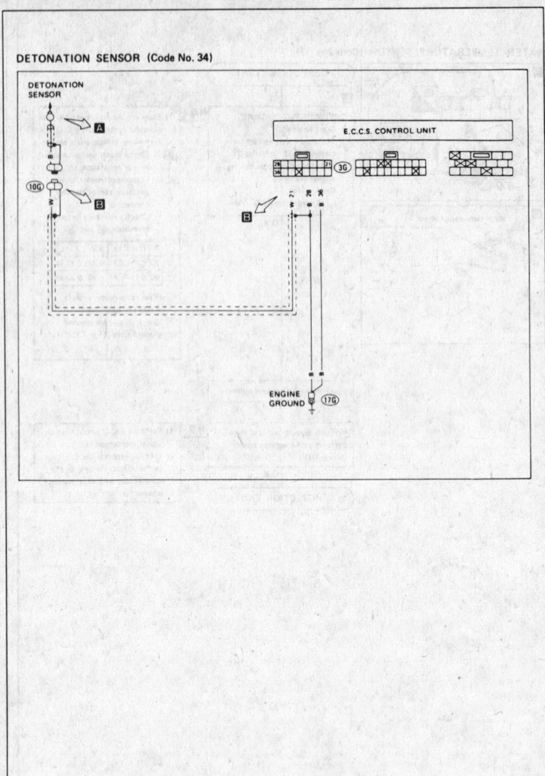
DETONATION SENSOR (Code No. 34)
DETONATION SENSOR
E.C.C.S. CONTROL UNIT
ENGINE GROUND

ECCS DIAGNOSTIC CHART—1988 200SX WITH CA18ET ENGINE

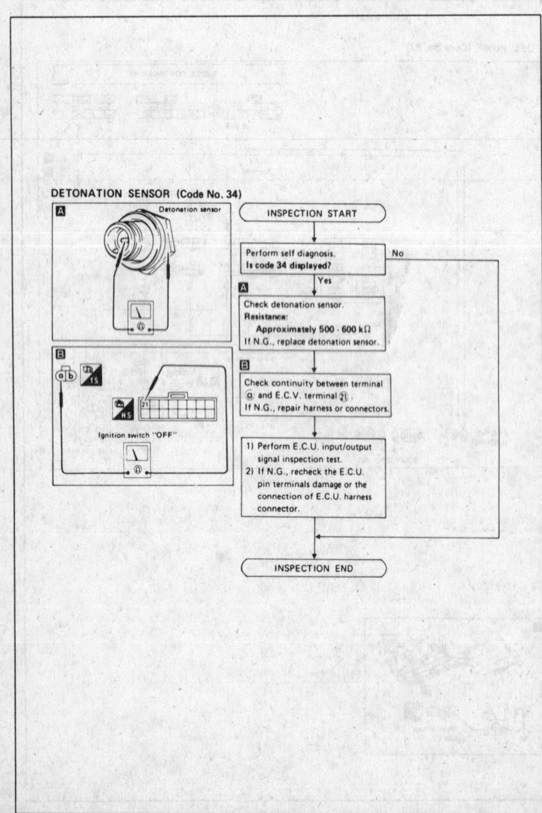
DETONATION SENSOR (Code No. 34)
Detonation sensor
Ignition switch "OFF"
INSPECTION START
Perform self diagnosis.
Is code 34 displayed?
No
Yes
A
Check detonation sensor.
Resistance:
Approximately 500 - 600 kΩ
If N.G., replace detonation sensor.
B
Check continuity between terminal ⓐ and E.C.V. terminal (21).
If N.G., repair harness or connectors.
1) Perform E.C.U. input/output signal inspection test.
2) If N.G., recheck the E.C.U. pin terminals damage or the connection of E.C.U. harness connector.
INSPECTION END

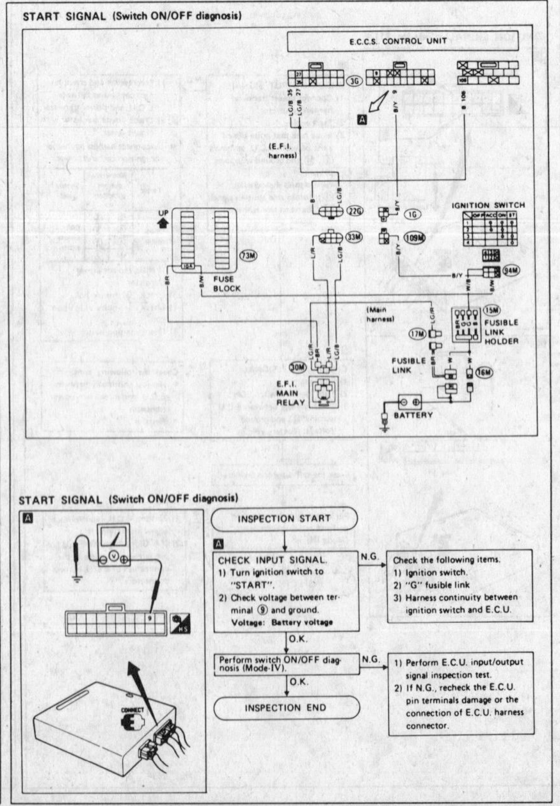
START SIGNAL (Switch ON/OFF diagnosis)
E.C.C.S. CONTROL UNIT
(E.F.I. harness)
IGNITION SWITCH
UP
FUSE BLOCK
(Main harness)
FUSIBLE LINK HOLDER
FUSIBLE LINK
E.F.I. MAIN RELAY
BATTERY
START SIGNAL (Switch ON/OFF diagnosis)
INSPECTION START
A
CHECK INPUT SIGNAL.
1) Turn ignition switch to "START".
2) Check voltage between terminal ⑨ and ground.
Voltage: Battery voltage
N.G.
Check the following items.
1) Ignition switch.
2) "G" fusible link
3) Harness continuity between ignition switch and E.C.U.
O.K.
Perform switch ON/OFF diagnosis (Mode-IV).
N.G.
1) Perform E.C.U. input/output signal inspection test.
2) If N.G., recheck the E.C.U. pin terminals damage or the connection of E.C.U. harness connector.
O.K.
INSPECTION END

ECCS DIAGNOSTIC CHART—1988 200SX WITH CA18ET ENGINE

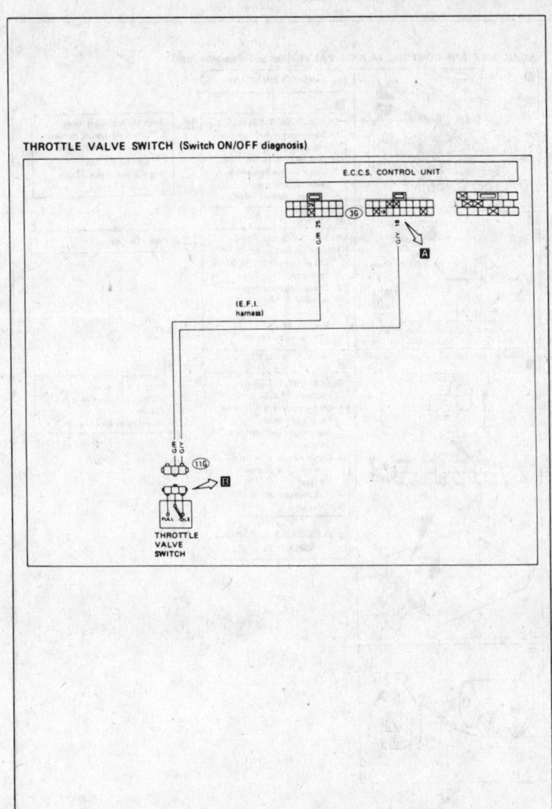
THROTTLE VALVE SWITCH (Switch ON/OFF diagnosis)
E.C.C.S. CONTROL UNIT
3G
(E.F.I. harness)
11G
THROTTLE VALVE SWITCH

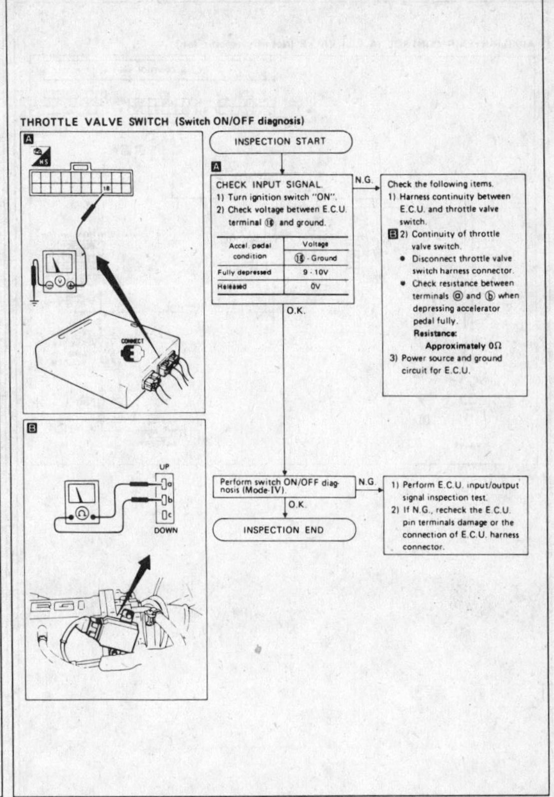
THROTTLE VALVE SWITCH (Switch ON/OFF diagnosis)
INSPECTION START
CHECK INPUT SIGNAL.
1) Turn ignition switch "ON".
2) Check voltage between E.C.U. terminal ⑱ and ground.
Accel. pedal condition | Voltage ⑱ - Ground
Fully depressed | 9 - 10V
Released | 0V
N.G.
Check the following items.
1) Harness continuity between E.C.U. and throttle valve switch.
2) Continuity of throttle valve switch.
• Disconnect throttle valve switch harness connector.
• Check resistance between terminals ⓐ and ⓑ when depressing accelerator pedal fully.
Resistance:
Approximately 0Ω
3) Power source and ground circuit for E.C.U.
O.K.
Perform switch ON/OFF diagnosis (Mode-IV).
N.G.
1) Perform E.C.U. input/output signal inspection test.
2) If N.G., recheck the E.C.U. pin terminals damage or the connection of E.C.U. harness connector.
O.K.
INSPECTION END
UP
DOWN

ECCS DIAGNOSTIC CHART—1988 200SX WITH CA18ET ENGINE

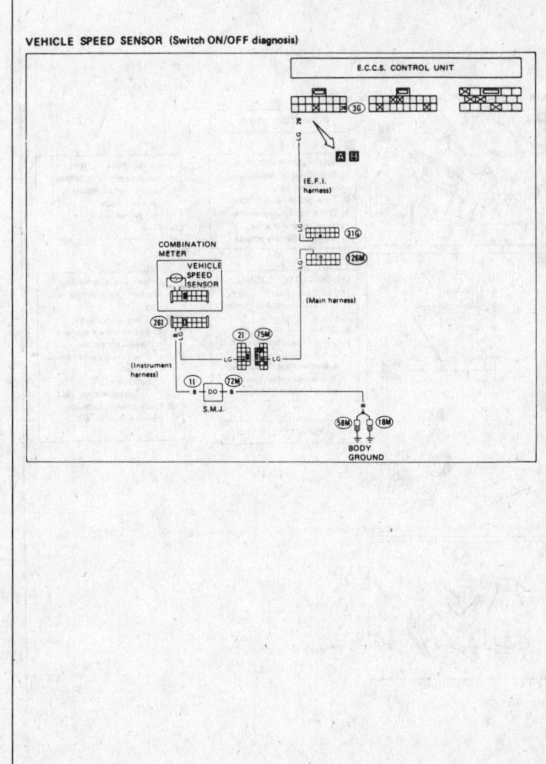
VEHICLE SPEED SENSOR (Switch ON/OFF diagnosis)
E.C.C.S. CONTROL UNIT
3G
(E.F.I. harness)
31G
126M
COMBINATION METER
VEHICLE SPEED SENSOR
(Main harness)
26I
21
75M
(Instrument harness)
11
72M
S.M.J.
58M
18M
BODY GROUND

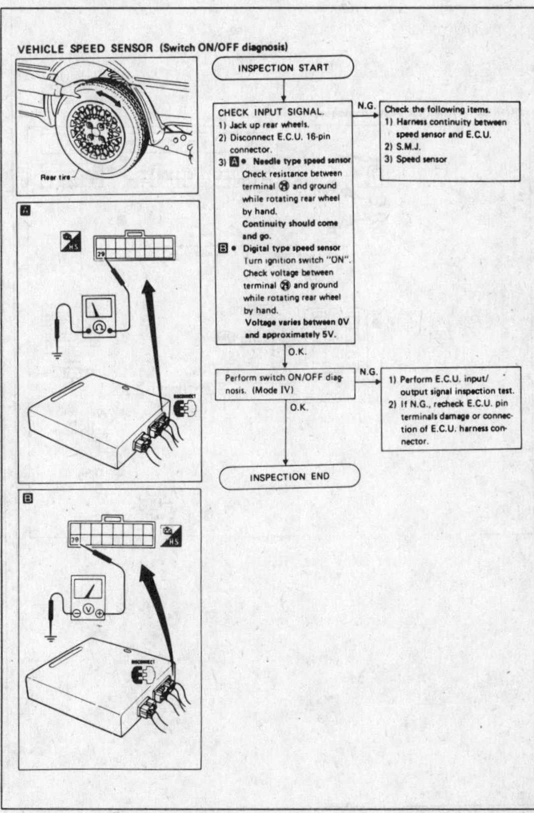
VEHICLE SPEED SENSOR (Switch ON/OFF diagnosis)
INSPECTION START
Rear tire
CHECK INPUT SIGNAL.
1) Jack up rear wheels.
2) Disconnect E.C.U. 16-pin connector.
3) A • Needle type speed sensor
Check resistance between terminal ㉙ and ground while rotating rear wheel by hand.
Continuity should come and go.
B • Digital type speed sensor
Turn ignition switch "ON".
Check voltage between terminal ㉙ and ground while rotating rear wheel by hand.
Voltage varies between 0V and approximately 5V.
N.G.
Check the following items.
1) Harness continuity between speed sensor and E.C.U.
2) S.M.J.
3) Speed sensor
O.K.
Perform switch ON/OFF diagnosis. (Mode IV)
N.G.
1) Perform E.C.U. input/output signal inspection test.
2) If N.G., recheck E.C.U. pin terminals damage or connection of E.C.U. harness connector.
O.K.
INSPECTION END

ECCS DIAGNOSTIC CHART—1988 200SX WITH CA18ET ENGINE

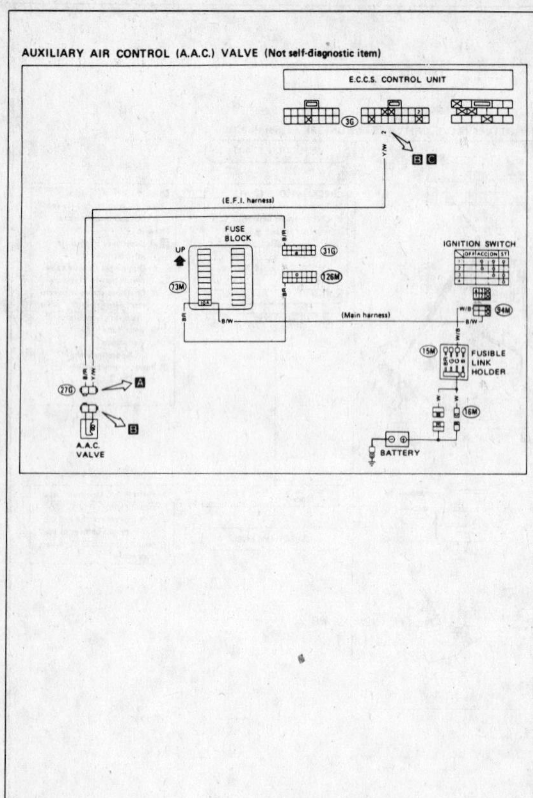
AUXILIARY AIR CONTROL (A.A.C.) VALVE (Not self-diagnostic item)
E.C.C.S. CONTROL UNIT
(E.F.I. harness)
FUSE BLOCK
UP
IGNITION SWITCH
(Main harness)
FUSIBLE LINK HOLDER
BATTERY
A.A.C. VALVE

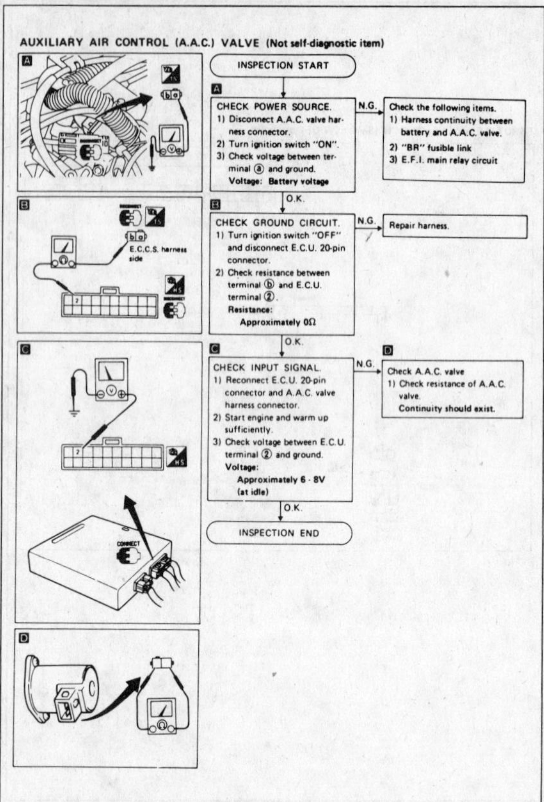
AUXILIARY AIR CONTROL (A.A.C.) VALVE (Not self-diagnostic item)
INSPECTION START
A
CHECK POWER SOURCE.
1) Disconnect A.A.C. valve harness connector.
2) Turn ignition switch "ON".
3) Check voltage between terminal ⓐ and ground.
Voltage: Battery voltage
N.G.
Check the following items.
1) Harness continuity between battery and A.A.C. valve.
2) "BR" fusible link
3) E.F.I. main relay circuit
O.K.
B
CHECK GROUND CIRCUIT.
1) Turn ignition switch "OFF" and disconnect E.C.U. 20-pin connector.
2) Check resistance between terminal ⓑ and E.C.U. terminal ②.
Resistance:
Approximately 0Ω
N.G.
Repair harness.
O.K.
C
CHECK INPUT SIGNAL.
1) Reconnect E.C.U. 20-pin connector and A.A.C. valve harness connector.
2) Start engine and warm up sufficiently.
3) Check voltage between E.C.U. terminal ② and ground.
Voltage:
Approximately 6 - 8V (at idle)
N.G.
D
Check A.A.C. valve
1) Check resistance of A.A.C. valve.
Continuity should exist.
O.K.
INSPECTION END
E.C.C.S. harness side

ECCS DIAGNOSTIC CHART—1988 200SX WITH CA18ET ENGINE

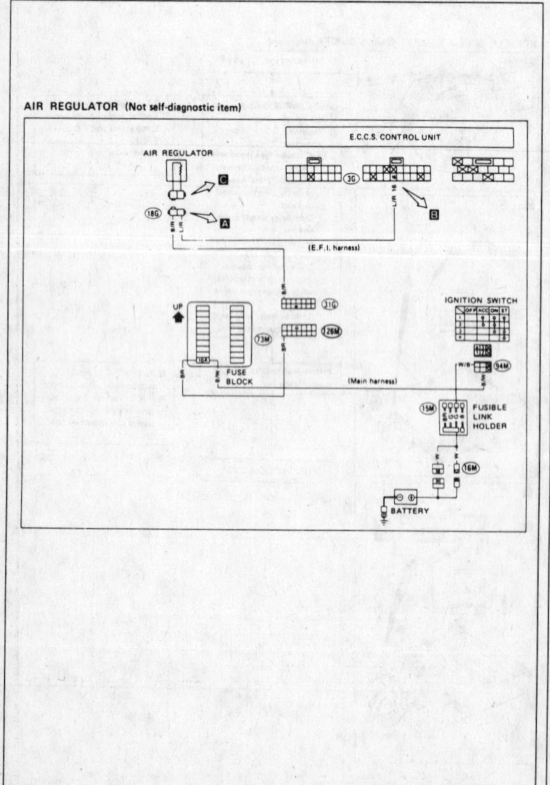
AIR REGULATOR (Not self-diagnostic item)
E.C.C.S. CONTROL UNIT
AIR REGULATOR
(E.F.I. harness)
IGNITION SWITCH
UP
FUSE BLOCK
(Main harness)
FUSIBLE LINK HOLDER
BATTERY

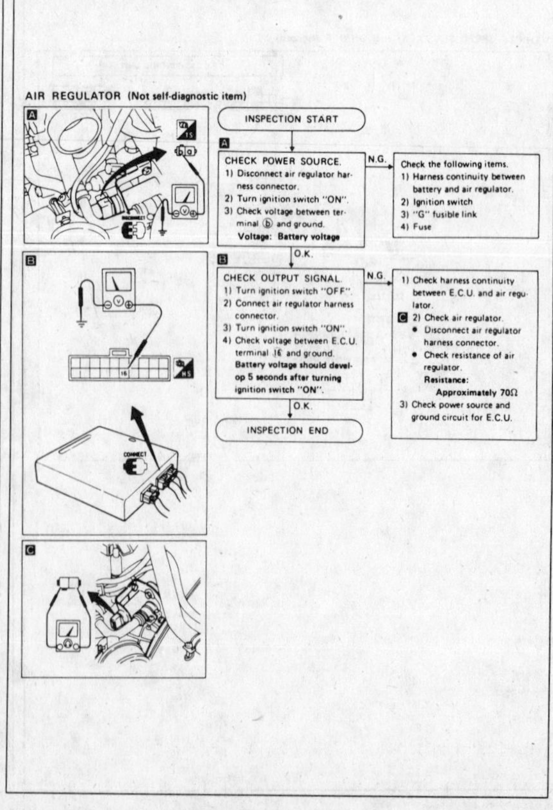
AIR REGULATOR (Not self-diagnostic item)
INSPECTION START
A
CHECK POWER SOURCE.
1) Disconnect air regulator harness connector.
2) Turn ignition switch "ON".
3) Check voltage between terminal ⓑ and ground.
Voltage: Battery voltage
N.G.
Check the following items.
1) Harness continuity between battery and air regulator.
2) Ignition switch
3) "G" fusible link
4) Fuse
O.K.
B
CHECK OUTPUT SIGNAL.
1) Turn ignition switch "OFF".
2) Connect air regulator harness connector.
3) Turn ignition switch "ON".
4) Check voltage between E.C.U. terminal ⑯ and ground.
Battery voltage should develop 5 seconds after turning ignition switch "ON".
N.G.
1) Check harness continuity between E.C.U. and air regulator.
C 2) Check air regulator.
• Disconnect air regulator harness connector.
• Check resistance of air regulator.
Resistance:
Approximately 70Ω
3) Check power source and ground circuit for E.C.U.
O.K.
INSPECTION END

ECCS DIAGNOSTIC CHART—1988 200SX WITH CA18ET ENGINE

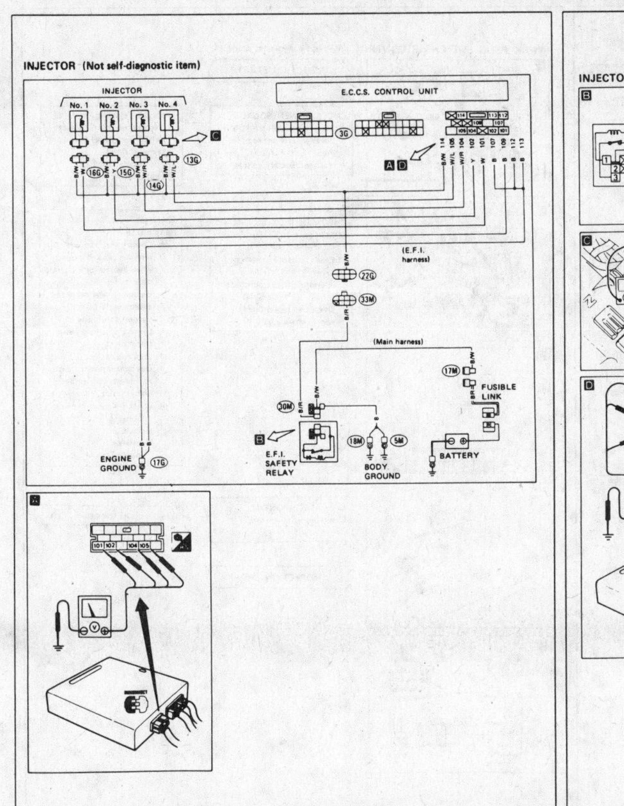
INJECTOR (Not self-diagnostic item)
INJECTOR
No. 1
No. 2
No. 3
No. 4
E.C.C.S. CONTROL UNIT
(E.F.I. harness)
(Main harness)
FUSIBLE LINK
ENGINE GROUND
E.F.I. SAFETY RELAY
BODY GROUND
BATTERY

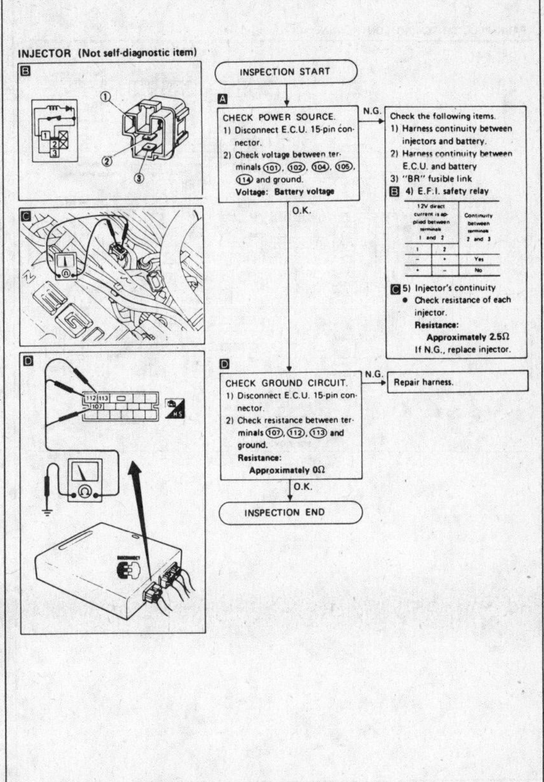
INJECTOR (Not self-diagnostic item)
INSPECTION START
A
CHECK POWER SOURCE.
1) Disconnect E.C.U. 15-pin connector.
2) Check voltage between terminals (101), (102), (104), (105), (114) and ground.
Voltage: Battery voltage
N.G.
Check the following items.
1) Harness continuity between injectors and battery.
2) Harness continuity between E.C.U. and battery
3) "BR" fusible link
B 4) E.F.I. safety relay
12V direct current is applied between terminals 1 and 2
Continuity between terminals 2 and 3
1
2
−
+
Yes
+
−
No
C 5) Injector's continuity
Check resistance of each injector.
Resistance:
Approximately 2.5Ω
If N.G., replace injector.
O.K.
D
CHECK GROUND CIRCUIT.
1) Disconnect E.C.U. 15-pin connector.
2) Check resistance between terminals (107), (112), (113) and ground.
Resistance:
Approximately 0Ω
N.G.
Repair harness.
O.K.
INSPECTION END

ECCS DIAGNOSTIC CHART—1988 200SX WITH CA18ET ENGINE

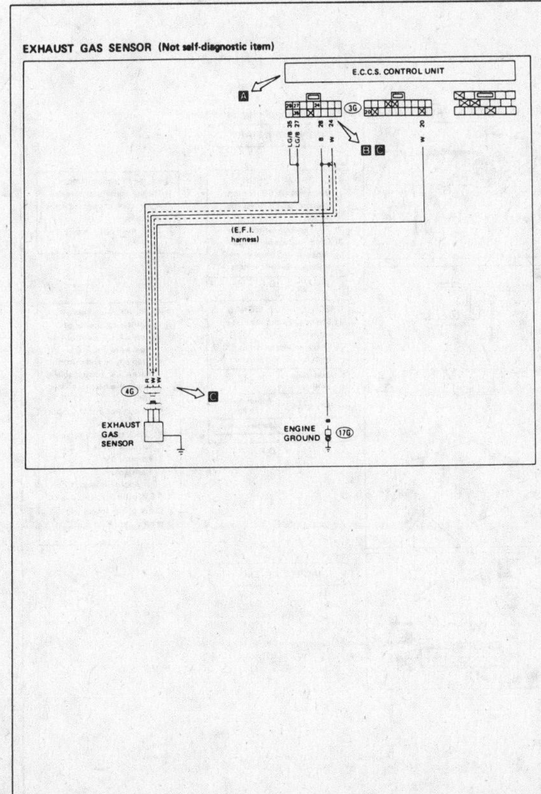
EXHAUST GAS SENSOR (Not self-diagnostic item)
E.C.C.S. CONTROL UNIT
(E.F.I. harness)
EXHAUST GAS SENSOR
ENGINE GROUND

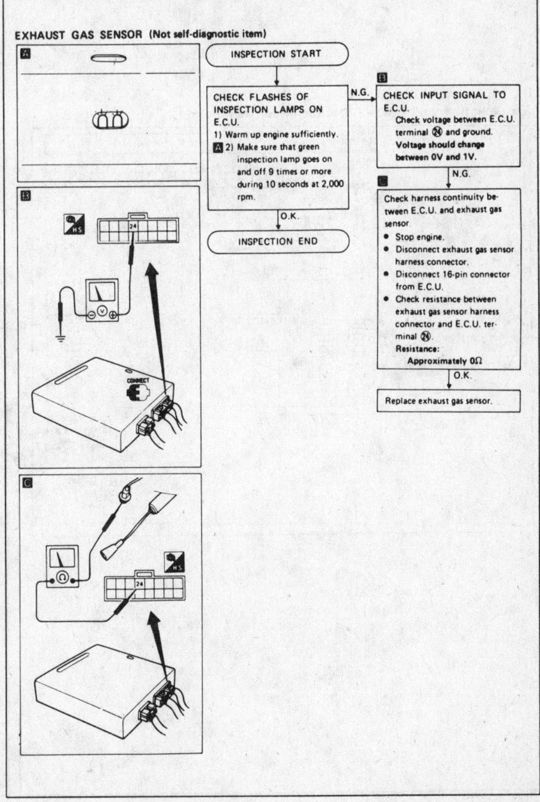
EXHAUST GAS SENSOR (Not self-diagnostic item)
INSPECTION START
CHECK FLASHES OF INSPECTION LAMPS ON E.C.U.
1) Warm up engine sufficiently.
A 2) Make sure that green inspection lamp goes on and off 9 times or more during 10 seconds at 2,000 rpm.
O.K.
INSPECTION END
N.G.
B
CHECK INPUT SIGNAL TO E.C.U.
Check voltage between E.C.U. terminal (24) and ground.
Voltage should change between 0V and 1V.
N.G.
C
Check harness continuity between E.C.U. and exhaust gas sensor.
Stop engine.
Disconnect exhaust gas sensor harness connector.
Disconnect 16-pin connector from E.C.U.
Check resistance between exhaust gas sensor harness connector and E.C.U. terminal (24).
Resistance:
Approximately 0Ω
O.K.
Replace exhaust gas sensor.

ECCS DIAGNOSTIC CHART—1988 200SX WITH CA18ET ENGINE

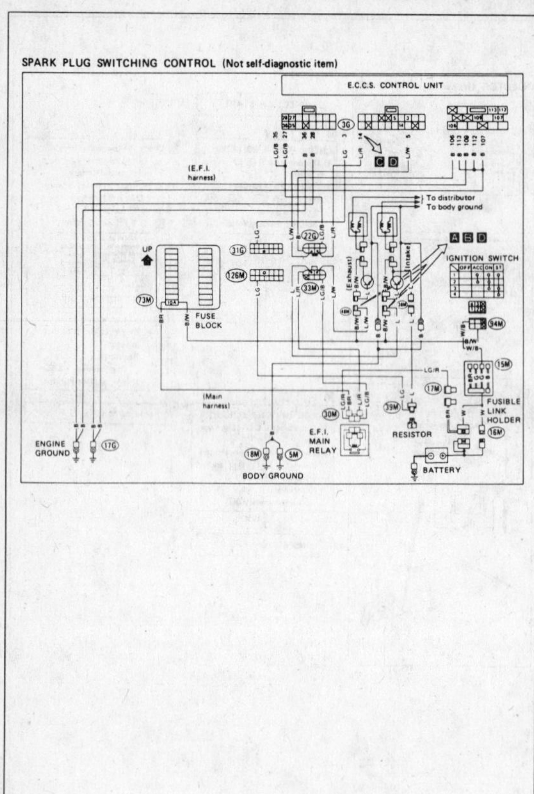
SPARK PLUG SWITCHING CONTROL (Not self-diagnostic item)
E.C.C.S. CONTROL UNIT
(E.F.I. harness)
To distributor
To body ground
IGNITION SWITCH
FUSE BLOCK
(Main harness)
ENGINE GROUND
E.F.I. MAIN RELAY
RESISTOR
FUSIBLE LINK HOLDER
BODY GROUND
BATTERY

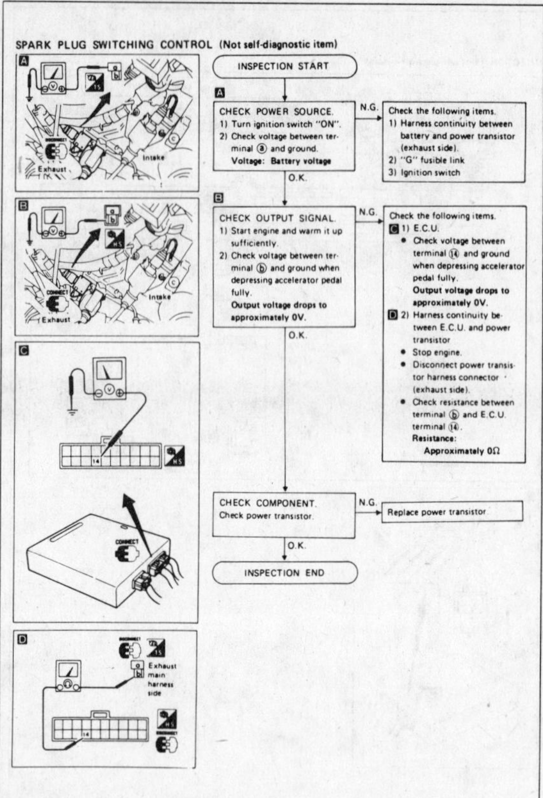
SPARK PLUG SWITCHING CONTROL (Not self-diagnostic item)
INSPECTION START
A
CHECK POWER SOURCE.
1) Turn ignition switch "ON".
2) Check voltage between terminal ⓐ and ground.
Voltage: Battery voltage
N.G.
Check the following items.
1) Harness continuity between battery and power transistor (exhaust side).
2) "G" fusible link
3) Ignition switch
O.K.
B
CHECK OUTPUT SIGNAL.
1) Start engine and warm it up sufficiently.
2) Check voltage between terminal ⓑ and ground when depressing accelerator pedal fully.
Output voltage drops to approximately 0V.
N.G.
Check the following items.
C 1) E.C.U.
• Check voltage between terminal ⑭ and ground when depressing accelerator pedal fully.
Output voltage drops to approximately 0V.
D 2) Harness continuity between E.C.U. and power transistor
• Stop engine.
• Disconnect power transistor harness connector (exhaust side).
• Check resistance between terminal ⓑ and E.C.U. terminal ⑭.
Resistance: Approximately 0Ω
O.K.
CHECK COMPONENT.
Check power transistor.
N.G.
Replace power transistor.
O.K.
INSPECTION END
Exhaust
Intake
Exhaust main harness side

ECCS DIAGNOSTIC CHART—1988 200SX WITH CA18ET ENGINE

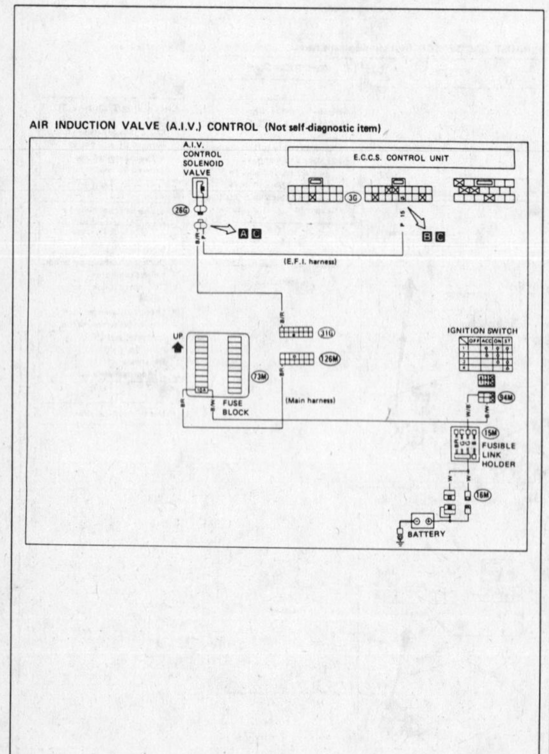
AIR INDUCTION VALVE (A.I.V.) CONTROL (Not self-diagnostic item)
A.I.V. CONTROL SOLENOID VALVE
E.C.C.S. CONTROL UNIT
(E.F.I. harness)
IGNITION SWITCH
FUSE BLOCK
(Main harness)
FUSIBLE LINK HOLDER
BATTERY

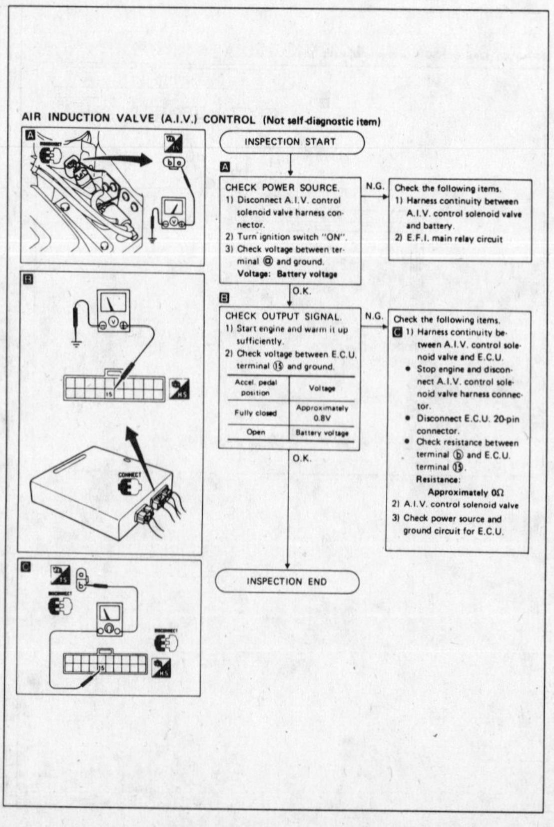
AIR INDUCTION VALVE (A.I.V.) CONTROL (Not self-diagnostic item)
INSPECTION START
A
CHECK POWER SOURCE.
1) Disconnect A.I.V. control solenoid valve harness connector.
2) Turn ignition switch "ON".
3) Check voltage between terminal ⓐ and ground.
Voltage: Battery voltage
N.G.
Check the following items.
1) Harness continuity between A.I.V. control solenoid valve and battery.
2) E.F.I. main relay circuit
O.K.
B
CHECK OUTPUT SIGNAL.
1) Start engine and warm it up sufficiently.
2) Check voltage between E.C.U. terminal ⑮ and ground.
Accel. pedal position | Voltage
Fully closed | Approximately 0.8V
Open | Battery voltage
N.G.
Check the following items.
C 1) Harness continuity between A.I.V. control solenoid valve and E.C.U.
• Stop engine and disconnect A.I.V. control solenoid valve harness connector.
• Disconnect E.C.U. 20-pin connector.
• Check resistance between terminal ⓑ and E.C.U. terminal ⑮.
Resistance: Approximately 0Ω
2) A.I.V. control solenoid valve
3) Check power source and ground circuit for E.C.U.
O.K.
INSPECTION END

ECCS DIAGNOSTIC CHART—1988 200SX WITH CA18ET ENGINE

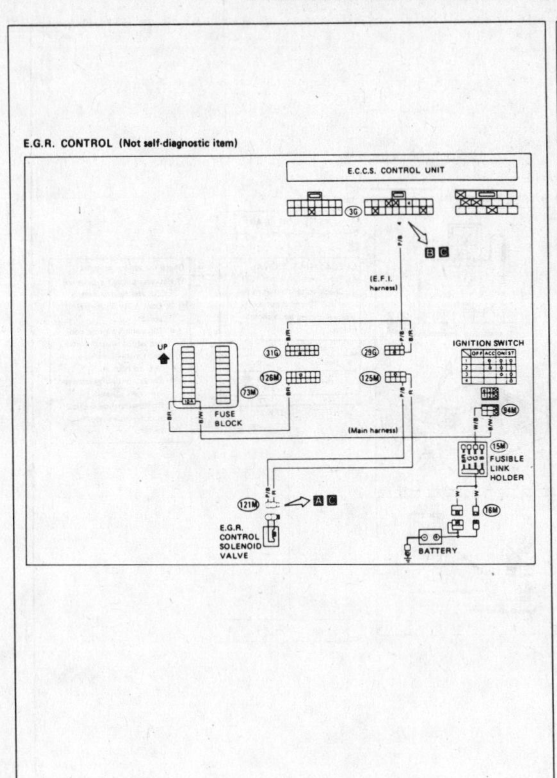
E.G.R. CONTROL (Not self-diagnostic item)
E.C.C.S. CONTROL UNIT
(E.F.I. harness)
UP
FUSE BLOCK
IGNITION SWITCH
(Main harness)
FUSIBLE LINK HOLDER
E.G.R. CONTROL SOLENOID VALVE
BATTERY

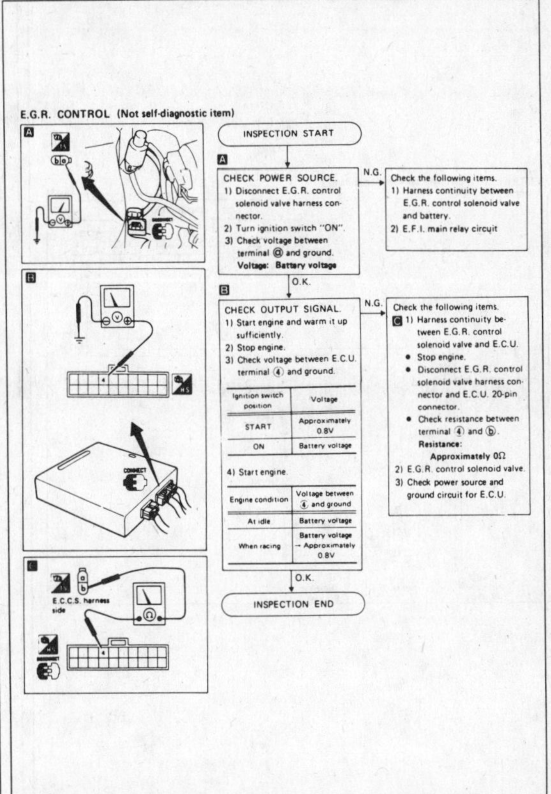
E.G.R. CONTROL (Not self-diagnostic item)
INSPECTION START
A
CHECK POWER SOURCE.
1) Disconnect E.G.R. control solenoid valve harness connector.
2) Turn ignition switch "ON".
3) Check voltage between terminal ⓐ and ground.
Voltage: Battery voltage
N.G.
Check the following items.
1) Harness continuity between E.G.R. control solenoid valve and battery.
2) E.F.I. main relay circuit
O.K.
B
CHECK OUTPUT SIGNAL.
1) Start engine and warm it up sufficiently.
2) Stop engine.
3) Check voltage between E.C.U. terminal ④ and ground.
Ignition switch position | Voltage
START | Approximately 0.8V
ON | Battery voltage
4) Start engine.
Engine condition | Voltage between ④ and ground
At idle | Battery voltage
When racing | Battery voltage → Approximately 0.8V
N.G.
Check the following items.
C 1) Harness continuity between E.G.R. control solenoid valve and E.C.U.
• Stop engine.
• Disconnect E.G.R. control solenoid valve harness connector and E.C.U. 20-pin connector.
• Check resistance between terminal ④ and ⓑ.
Resistance: Approximately 0Ω
2) E.G.R. control solenoid valve.
3) Check power source and ground circuit for E.C.U.
O.K.
INSPECTION END
E.C.C.S. harness side

ECCS DIAGNOSTIC CHART—1988 200SX WITH CA18ET ENGINE

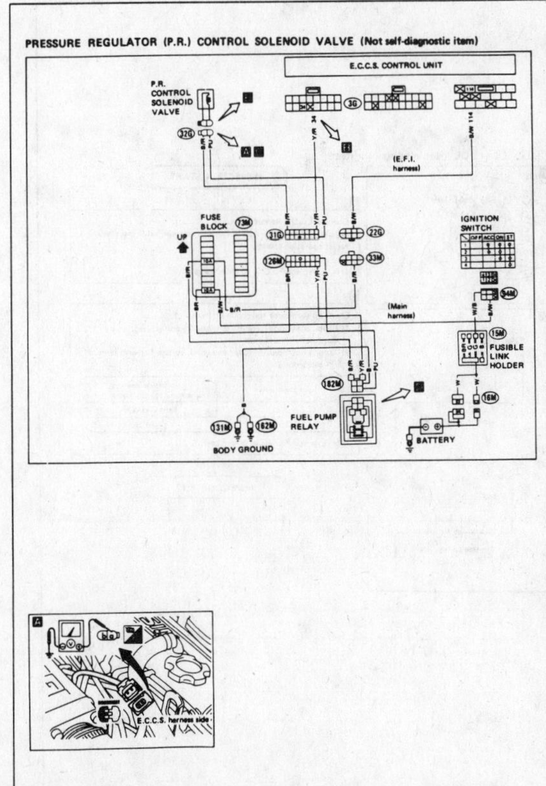
PRESSURE REGULATOR (P.R.) CONTROL SOLENOID VALVE (Not self-diagnostic item)
E.C.C.S. CONTROL UNIT
P.R. CONTROL SOLENOID VALVE
(E.F.I. harness)
FUSE BLOCK
UP
IGNITION SWITCH
(Main harness)
FUSIBLE LINK HOLDER
FUEL PUMP RELAY
BODY GROUND
BATTERY
E.C.C.S. harness side

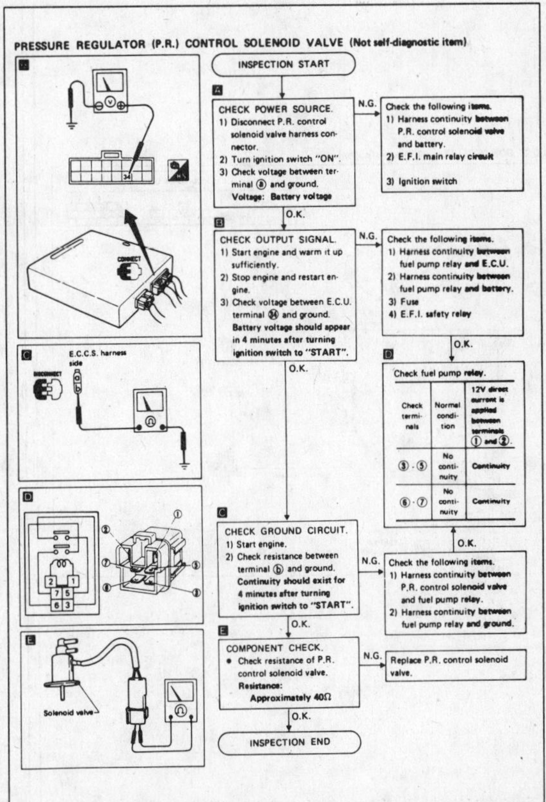
PRESSURE REGULATOR (P.R.) CONTROL SOLENOID VALVE (Not self-diagnostic item)
INSPECTION START
A
CHECK POWER SOURCE.
1) Disconnect P.R. control solenoid valve harness connector.
2) Turn ignition switch "ON".
3) Check voltage between terminal ⓐ and ground.
Voltage: Battery voltage
N.G.
Check the following items.
1) Harness continuity between P.R. control solenoid valve and battery.
2) E.F.I. main relay circuit
3) Ignition switch
O.K.
B
CHECK OUTPUT SIGNAL.
1) Start engine and warm it up sufficiently.
2) Stop engine and restart engine.
3) Check voltage between E.C.U. terminal ⑭ and ground.
Battery voltage should appear in 4 minutes after turning ignition switch to "START".
N.G.
Check the following items.
1) Harness continuity between fuel pump relay and E.C.U.
2) Harness continuity between fuel pump relay and battery.
3) Fuse
4) E.F.I. safety relay
O.K.
D
Check fuel pump relay.
Check terminals | Normal condition | 12V direct current is applied between terminals ① and ②.
③-⑤ | No continuity | Continuity
⑥-⑦ | No continuity | Continuity
O.K.
C
CHECK GROUND CIRCUIT.
1) Start engine.
2) Check resistance between terminal ⓑ and ground.
Continuity should exist for 4 minutes after turning ignition switch to "START".
N.G.
Check the following items.
1) Harness continuity between P.R. control solenoid valve and fuel pump relay.
2) Harness continuity between fuel pump relay and ground.
O.K.
E
COMPONENT CHECK.
• Check resistance of P.R. control solenoid valve.
Resistance: Approximately 40Ω
N.G.
Replace P.R. control solenoid valve.
O.K.
INSPECTION END
E.C.C.S. harness side
Solenoid valve

ECCS DIAGNOSTIC CHART – 1988 200SX WITH CA18ET ENGINE

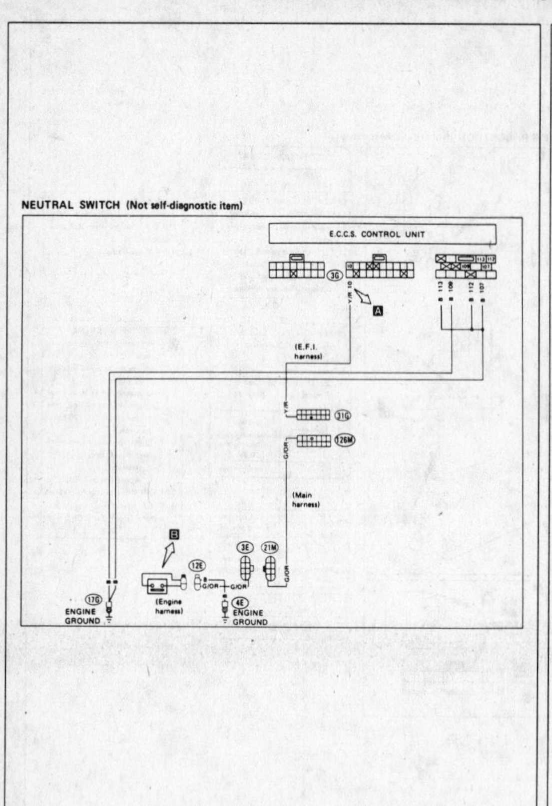
NEUTRAL SWITCH (Not self-diagnostic item)
E.C.C.S. CONTROL UNIT
3G
A
(E.F.I. harness)
31G
126M
(Main harness)
B
12E
3E
21M
G/OR
17G
ENGINE GROUND
(Engine harness)
4E
ENGINE GROUND

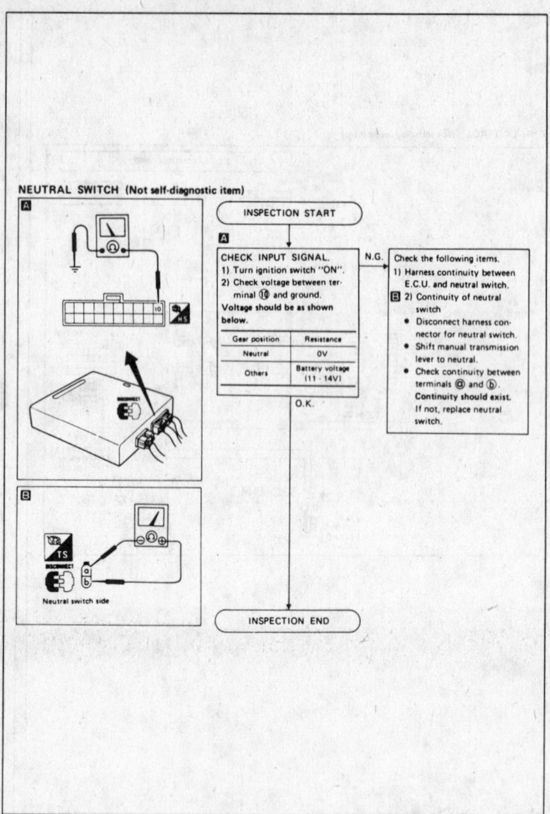
NEUTRAL SWITCH (Not self-diagnostic item)
A
B
TS
Neutral switch side
INSPECTION START
A
CHECK INPUT SIGNAL.
1) Turn ignition switch "ON".
2) Check voltage between terminal ⑩ and ground.
Voltage should be as shown below.
Gear position | Resistance
Neutral | 0V
Others | Battery voltage (11 - 14V)
N.G.
Check the following items.
1) Harness continuity between E.C.U. and neutral switch.
B 2) Continuity of neutral switch
• Disconnect harness connector for neutral switch.
• Shift manual transmission lever to neutral.
• Check continuity between terminals ⓐ and ⓑ.
Continuity should exist.
If not, replace neutral switch.
O.K.
INSPECTION END

ECCS DIAGNOSTIC CHART – 1988 200SX WITH CA18ET ENGINE

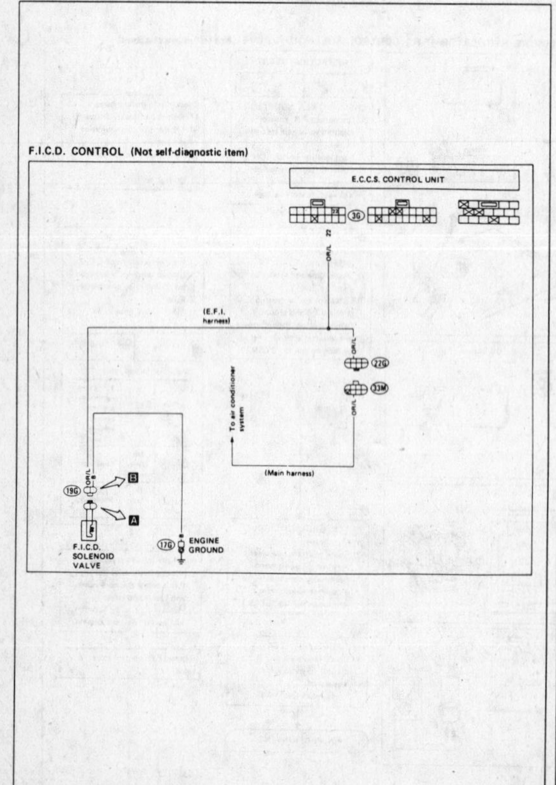
F.I.C.D. CONTROL (Not self-diagnostic item)
E.C.C.S. CONTROL UNIT
3G
(E.F.I. harness)
22G
33M
To air conditioner system
(Main harness)
19G
B
A
F.I.C.D. SOLENOID VALVE
17G
ENGINE GROUND

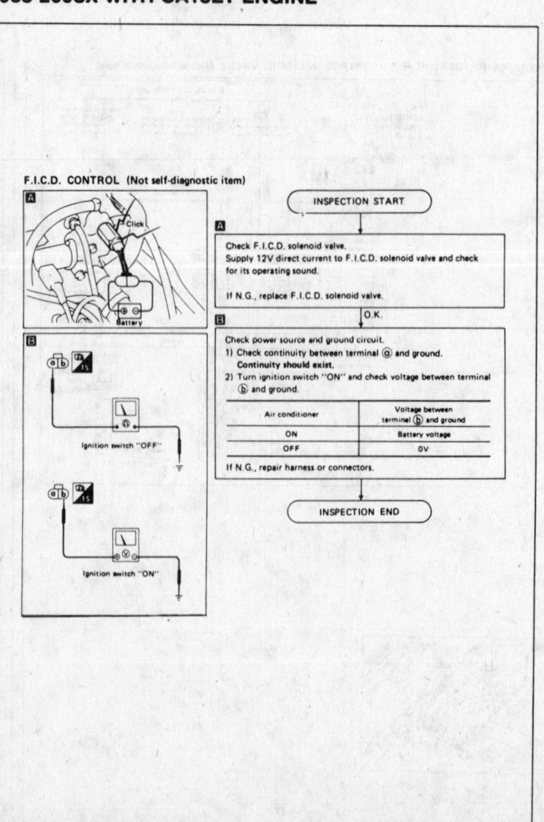
F.I.C.D. CONTROL (Not self-diagnostic item)
A
Click
Battery
B
Ignition switch "OFF"
Ignition switch "ON"
INSPECTION START
A
Check F.I.C.D. solenoid valve.
Supply 12V direct current to F.I.C.D. solenoid valve and check for its operating sound.
If N.G., replace F.I.C.D. solenoid valve.
O.K.
B
Check power source and ground circuit.
1) Check continuity between terminal ⓐ and ground.
Continuity should exist.
2) Turn ignition switch "ON" and check voltage between terminal ⓑ and ground.
Air conditioner | Voltage between terminal ⓑ and ground
ON | Battery voltage
OFF | 0V
If N.G., repair harness or connectors.
INSPECTION END

ECU SIGNAL INSPECTION—1988 200SX WITH CA18ET ENGINE

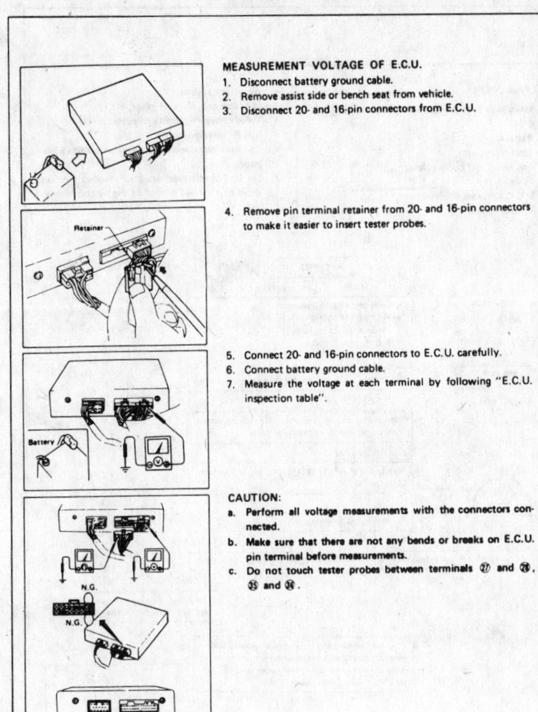
Retainer
Battery
N.G.
N.G.

MEASUREMENT VOLTAGE OF E.C.U.

1. Disconnect battery ground cable.
2. Remove assist side or bench seat from vehicle.
3. Disconnect 20- and 16-pin connectors from E.C.U.
4. Remove pin terminal retainer from 20- and 16-pin connectors to make it easier to insert tester probes.
5. Connect 20- and 16-pin connectors to E.C.U. carefully.
6. Connect battery ground cable.
7. Measure the voltage at each terminal by following "E.C.U. inspection table".

CAUTION:

a. **Perform all voltage measurements with the connectors connected.**
b. **Make sure that there are not any bends or breaks on E.C.U. pin terminal before measurements.**
c. Do not touch tester probes between terminals ㉗ and ㉘, ㉟ and ㊱.

E.C.U. inspection table

*Data are reference values.

TERMINAL NO.	ITEM	CONDITION	*DATA
2	A.A.C. valve	[Engine is running.] — At idle (after warming up)	6.0 - 8.0V
3	Ignition signal (from resistor)	Ignition switch "ON"	BATTERY VOLTAGE (11 - 14V)
4	E.G.R. control solenoid valve	[Engine is running.] — Engine is cold. [Water temperature is below 60°C (140°F).]	0.7 - 0.9V
		[Engine is running.] — After warming up [Water temperature is between 60°C (140°F) and 105°C (221°F).]	BATTERY VOLTAGE (11 - 14V)
5	Ignition signal (from intake side power transistor)	Engine is running.	0.4 - 2.2V Output voltage varies with engine speed.
8	Crank angle sensor (position signal)	Engine is running. **Do not turn engine at high speed under no-load.**	2.2 - 2.8V
9	Start signal	Ignition switch "START"	BATTERY VOLTAGE (11 - 14V)
10	Neutral signal	[Ignition switch "ON"] — Gear position: Neutral (M/T) : N or P range (A/T)	0V
		[Ignition switch "ON"] — Gear position: [Except neutral (M/T) Except N or P range (A/T)]	BATTERY VOLTAGE (11 - 14V)

ECU SIGNAL INPSECTION (CONT.)—1988 200SX WITH WITH CA18ET ENGINE

*Data are reference values.

TERMINAL NO.	ITEM	CONDITION	*DATA
14	Ignition signal (from exhaust side power transistor)	Engine is running. **Do not turn engine at high speed under no-load.**	0.4 - 2.2V (Output voltage varies with engine revolution.)
15	A.I.V. control solenoid valve	[Engine is running.] — At idle	0.7 - 0.9V
		[Engine is running.] — When depressing accelerator pedal [Water temperature is above 50°C (122°F).]	BATTERY VOLTAGE (11 - 14V)
16	Air regulator	[Ignition switch "ON"] — For 5 seconds after turning ignition switch "ON"	0.7 - 0.9V
		[Ignition switch "ON"] — 5 seconds after turning ignition switch "ON"	BATTERY VOLTAGE (11 - 14V)
17	Crank angle sensor (Reference signal)	Engine is running. **Do not turn engine at high speed under no-load.**	0.2 - 0.4V
18	Idle switch (⊖ side)	[Ignition switch "ON"] — Throttle valve: idle position	9 - 10V
		[Ignition switch "ON"] — Throttle valve: except idle position	0V
22	Air conditioner signal (Air conditioner equipped model)	[Ignition switch "ON"] — Air conditioner switch and heater fan switch "ON"	BATTERY VOLTAGE (11 - 14V)
		[Ignition switch "ON"] — Air conditioner switch "OFF"	0V
23	Water temperature sensor	Engine is running.	1.0 - 5.0V Output voltage varies with engine water temperature.

*Data are reference values.

TERMINAL NO.	ITEM	CONDITION	*DATA
24	Exhaust gas sensor	[Engine is running.] — After warming up sufficiently.	0 - Approximately 1.0V
25	Idle switch and full throttle switch (⊕ side)	Ignition switch "ON"	9 - 10V
27 35	Power source for E.C.U.	Ignition switch "ON"	BATTERY VOLTAGE (11 - 14V)
29	Vehicle speed sensor	[Ignition switch "ON"] — When rotating rear wheel slowly	Voltage varies between 0V and approximately 5V.
30	Air temperature sensor	Ignition switch "ON"	Approximately 3V [Air temperature is 20°C (68°F).] Output voltage varies with air temperature.
31	Air flow meter	Engine is running. **Do not turn engine at high speed under no-load.**	0 - Approximately 5V Output voltage varies with engine revolution.
34	Pressure regulator control solenoid valve (Fuel pump relay)	[Ignition switch "ON"] — For approximately 4 minutes after turning ignition switch to "START". [Water temperature is above 60°C (140°F).]	0.8 - 1.0V
		[Ignition switch "ON"] — Approximately 4 minutes after turning ignition switch to "START". [Water temperature is above 60°C (140°F).]	BATTERY VOLTAGE (11 - 14V)
		Ignition switch "ON" or "START". [Water temperature is below 60°C (140°F).]	

ECCS MIXTURE RATIO FEEDBACK SYSTEM INSPECTION—1988 200SX WITH CA18ET ENGINE

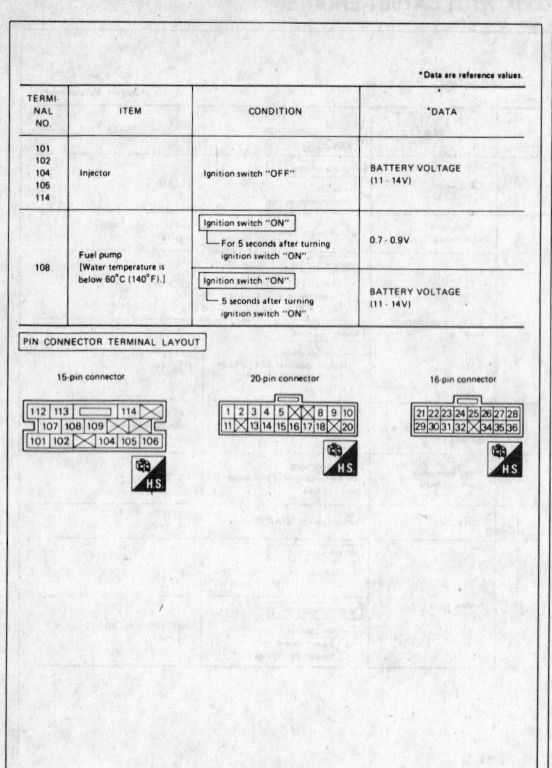

*Data are reference values.

TERMINAL NO.	ITEM	CONDITION	*DATA
101 102 104 105 114	Injector	Ignition switch "OFF"	BATTERY VOLTAGE (11 - 14V)
108	Fuel pump [Water temperature is below 60°C (140°F).]	Ignition switch "ON" — For 5 seconds after turning ignition switch "ON".	0.7 - 0.9V
		Ignition switch "ON" — 5 seconds after turning ignition switch "ON".	BATTERY VOLTAGE (11 - 14V)

PIN CONNECTOR TERMINAL LAYOUT
15-pin connector
112 113 114
107 108 109
101 102 104 105 106
20-pin connector
1 2 3 4 5 8 9 10
11 13 14 15 16 17 18 20
16-pin connector
21 22 23 24 25 26 27 28
29 30 31 32 34 35 36
HS

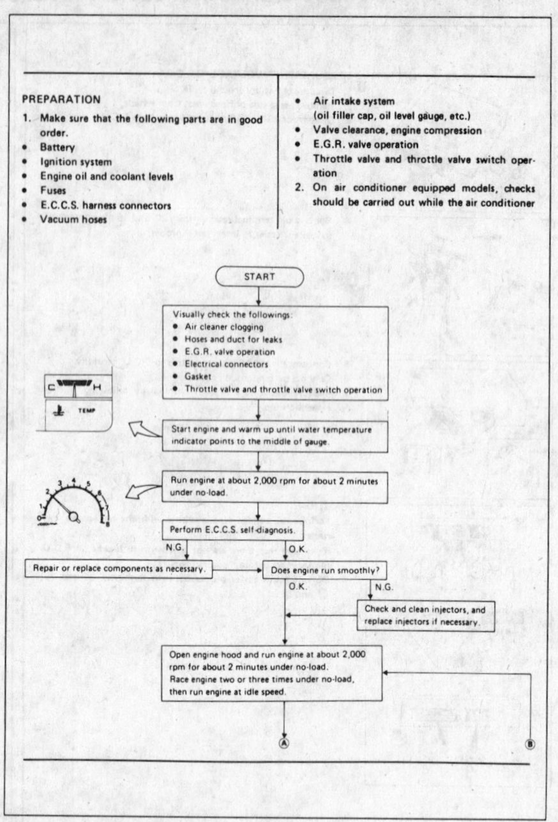

PREPARATION
1. Make sure that the following parts are in good order.
• Battery
• Ignition system
• Engine oil and coolant levels
• Fuses
• E.C.C.S. harness connectors
• Vacuum hoses
• Air intake system (oil filler cap, oil level gauge, etc.)
• Valve clearance, engine compression
• E.G.R. valve operation
• Throttle valve and throttle valve switch operation
2. On air conditioner equipped models, checks should be carried out while the air conditioner
START
Visually check the followings:
• Air cleaner clogging
• Hoses and duct for leaks
• E.G.R. valve operation
• Electrical connectors
• Gasket
• Throttle valve and throttle valve switch operation
C H
TEMP
Start engine and warm up until water temperature indicator points to the middle of gauge.
Run engine at about 2,000 rpm for about 2 minutes under no-load.
Perform E.C.C.S. self-diagnosis.
N.G.
O.K.
Repair or replace components as necessary.
Does engine run smoothly?
O.K.
N.G.
Check and clean injectors, and replace injectors if necessary.
Open engine hood and run engine at about 2,000 rpm for about 2 minutes under no-load.
Race engine two or three times under no-load, then run engine at idle speed.
A
B

ECCS MIXTURE RATIO FEEDBACK SYSTEM INPSECTION (CONT.) 1988 200SX WITH CA18ET ENGINE

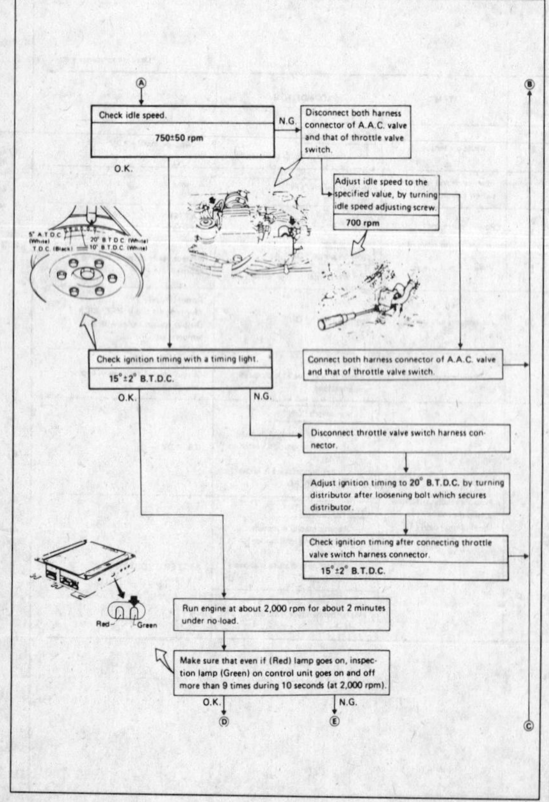

A
B
Check idle speed.
750±50 rpm
N.G.
Disconnect both harness connector of A.A.C. valve and that of throttle valve switch.
O.K.
Adjust idle speed to the specified value, by turning idle speed adjusting screw.
700 rpm
20° B.T.D.C. (White)
10° B.T.D.C. (White)
(Black)
Check ignition timing with a timing light.
15°±2° B.T.D.C.
Connect both harness connector of A.A.C. valve and that of throttle valve switch.
O.K.
N.G.
Disconnect throttle valve switch harness connector.
Adjust ignition timing to 20° B.T.D.C. by turning distributor after loosening bolt which secures distributor.
Check ignition timing after connecting throttle valve switch harness connector.
15°±2° B.T.D.C.
Red
Green
Run engine at about 2,000 rpm for about 2 minutes under no-load.
Make sure that even if (Red) lamp goes on, inspection lamp (Green) on control unit goes on and off more than 9 times during 10 seconds (at 2,000 rpm).
O.K.
N.G.
D
E
C

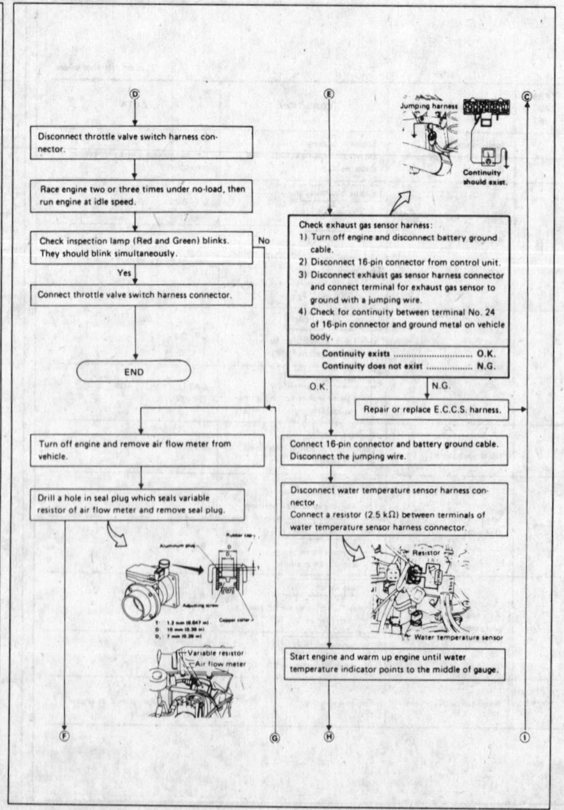

D
E
C
Disconnect throttle valve switch harness connector.
Race engine two or three times under no-load, then run engine at idle speed.
Check inspection lamp (Red and Green) blinks. They should blink simultaneously.
No
Yes
Connect throttle valve switch harness connector.
END
Jumping harness
Continuity should exist.
Check exhaust gas sensor harness:
1) Turn off engine and disconnect battery ground cable.
2) Disconnect 16-pin connector from control unit.
3) Disconnect exhaust gas sensor harness connector and connect terminal for exhaust gas sensor to ground with a jumping wire.
4) Check for continuity between terminal No. 24 of 16-pin connector and ground metal on vehicle body.
Continuity exists .... O.K.
Continuity does not exist .... N.G.
O.K.
N.G.
Repair or replace E.C.C.S. harness.
Turn off engine and remove air flow meter from vehicle.
Drill a hole in seal plug which seals variable resistor of air flow meter and remove seal plug.
Variable resistor
Air flow meter
Connect 16-pin connector and battery ground cable. Disconnect the jumping wire.
Disconnect water temperature sensor harness connector.
Connect a resistor (2.5 kΩ) between terminals of water temperature sensor harness connector.
Resistor
Water temperature sensor
Start engine and warm up engine until water temperature indicator points to the middle of gauge.
F
G
H
I

ECCS MIXTURE RATIO FEEDBACK SYSTEM INSPECTION (CONT.) – 1988 200SX WITH CA18ET ENGINE

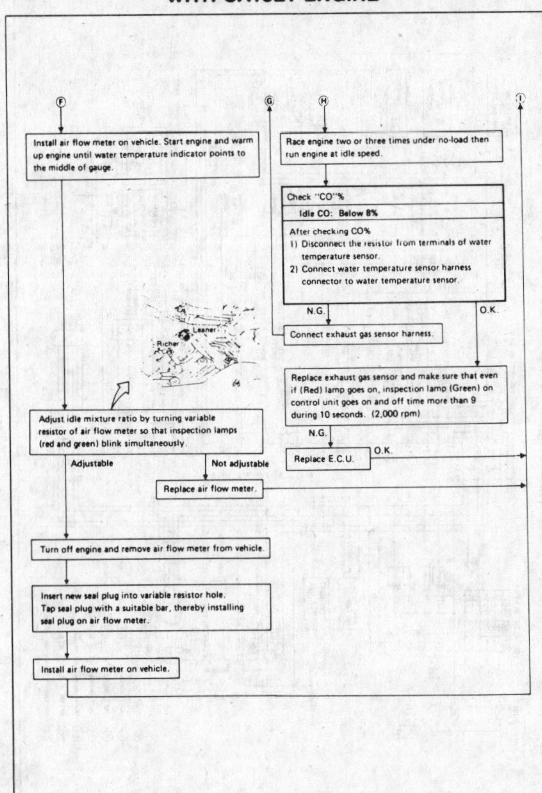
F
G
H
I
Install air flow meter on vehicle. Start engine and warm up engine until water temperature indicator points to the middle of gauge.
Race engine two or three times under no-load then run engine at idle speed.
Check "CO"%
Idle CO: Below 8%
After checking CO%
1) Disconnect the resistor from terminals of water temperature sensor.
2) Connect water temperature sensor harness connector to water temperature sensor.
N.G.
O.K.
Connect exhaust gas sensor harness.
Leaner
Richer
Replace exhaust gas sensor and make sure that even if (Red) lamp goes on, inspection lamp (Green) on control unit goes on and off time more than 9 during 10 seconds. (2,000 rpm)
N.G.
Replace E.C.U.
O.K.
Adjust idle mixture ratio by turning variable resistor of air flow meter so that inspection lamps (red and green) blink simultaneously.
Adjustable
Not adjustable
Replace air flow meter.
Turn off engine and remove air flow meter from vehicle.
Insert new seal plug into variable resistor hole. Tap seal plug with a suitable bar, thereby installing seal plug on air flow meter.
Install air flow meter on vehicle.

ECCS COMPONENTS LOCATION – 1989–90 240SX

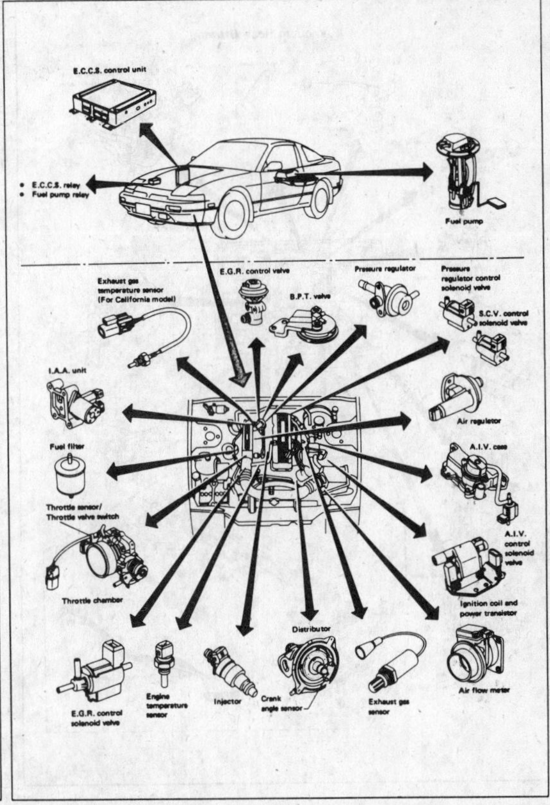
E.C.C.S. control unit
• E.C.C.S. relay
• Fuel pump relay
Fuel pump
Exhaust gas temperature sensor (For California model)
E.G.R. control valve
B.P.T. valve
Pressure regulator
Pressure regulator control solenoid valve
S.C.V. control solenoid valve
I.A.A. unit
Air regulator
Fuel filter
A.I.V. case
Throttle sensor/ Throttle valve switch
A.I.V. control solenoid valve
Throttle chamber
Ignition coil and power transistor
Distributor
Air flow meter
E.G.R. control solenoid valve
Engine temperature sensor
Injector
Crank angle sensor
Exhaust gas sensor

ECCS SYSTEM SCHEAMTIC – 1989–90 240SX

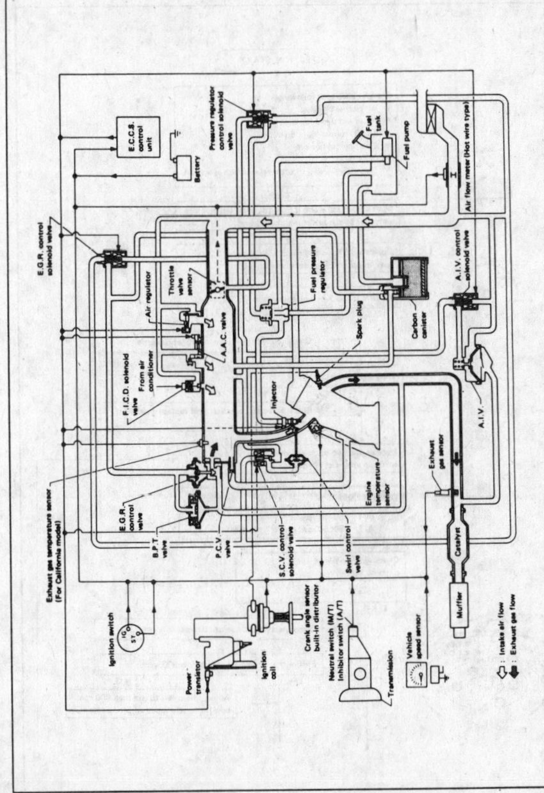
E.C.C.S. control unit
Battery
Pressure regulator control solenoid valve
Fuel tank
Fuel pump
Air flow meter (Hot wire type)
E.G.R. control solenoid valve
Throttle valve sensor
Air regulator
A.A.C. valve
Fuel pressure regulator
Carbon canister
A.I.V. control solenoid valve
Spark plug
F.I.C.D. solenoid valve
From air conditioner
Injector
A.I.V.
Exhaust gas sensor
Engine temperature sensor
E.G.R. control valve
B.P.T. valve
P.C.V. valve
S.C.V. control solenoid valve
Swirl control valve
Catalyst
Muffler
Exhaust gas temperature sensor (For California model)
Ignition switch
Power transistor
Ignition coil
Crank angle sensor built-in distributor
Neutral switch (M/T) Inhibitor switch (A/T)
Transmission
Vehicle speed sensor
Intake air flow
Exhaust gas flow

ECCS CONTROL SYSTEM CHART – 1989–90 240SX

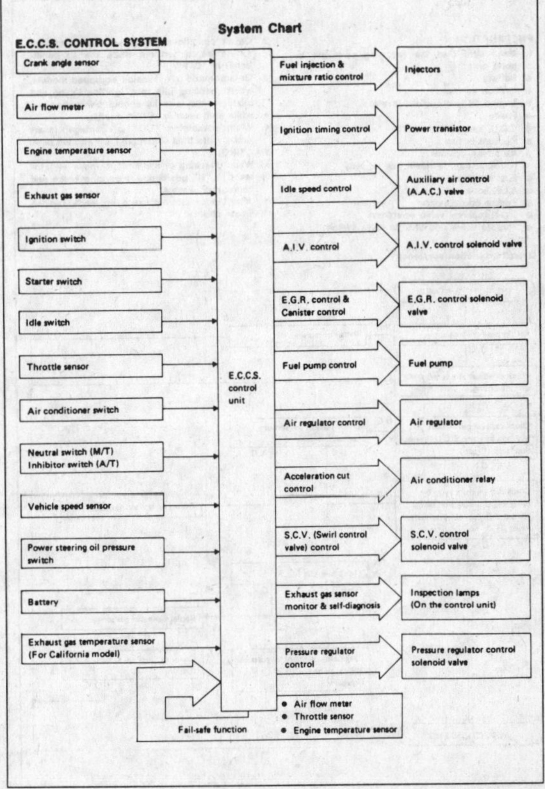
System Chart
E.C.C.S. CONTROL SYSTEM
Crank angle sensor
Air flow meter
Engine temperature sensor
Exhaust gas sensor
Ignition switch
Starter switch
Idle switch
Throttle sensor
Air conditioner switch
Neutral switch (M/T) Inhibitor switch (A/T)
Vehicle speed sensor
Power steering oil pressure switch
Battery
Exhaust gas temperature sensor (For California model)
E.C.C.S. control unit
Fuel injection & mixture ratio control
Ignition timing control
Idle speed control
A.I.V. control
E.G.R. control & Canister control
Fuel pump control
Air regulator control
Acceleration cut control
S.C.V. (Swirl control valve) control
Exhaust gas sensor monitor & self-diagnosis
Pressure regulator control
Injectors
Power transistor
Auxiliary air control (A.A.C.) valve
A.I.V. control solenoid valve
E.G.R. control solenoid valve
Fuel pump
Air regulator
Air conditioner relay
S.C.V. control solenoid valve
Inspection lamps (On the control unit)
Pressure regulator control solenoid valve
Fail-safe function
• Air flow meter
• Throttle sensor
• Engine temperature sensor

ECCS EMISSIONS CONTROL COMPONENTS LOCATION—1989–90 240SX

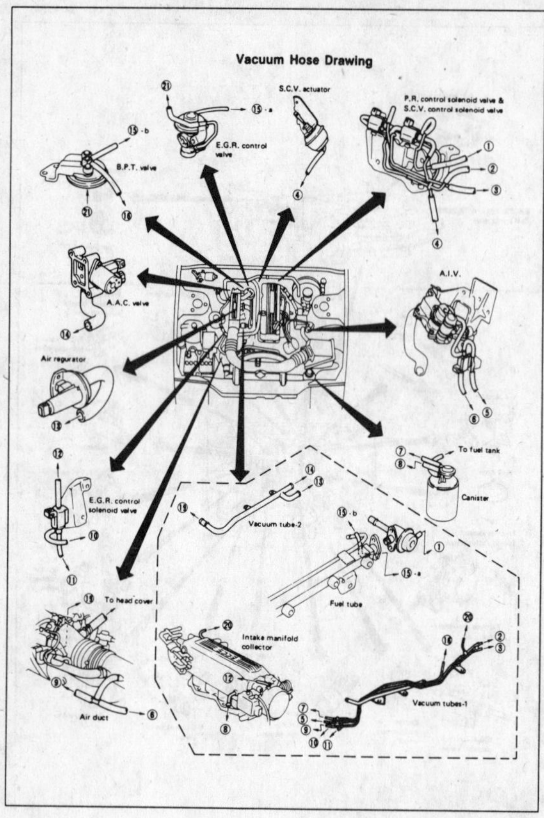
Vacuum Hose Drawing
S.C.V. actuator
P.R. control solenoid valve & S.C.V. control solenoid valve
E.G.R. control valve
B.P.T. valve
A.A.C. valve
Air regurator
A.I.V.
To fuel tank
Canister
E.G.R. control solenoid valve
Vacuum tubes-2
Fuel tube
To head cover
Intake manifold collector
Vacuum tubes-1
Air duct

ECCS CONTROL WIRING SCHEMATIC 1989–90 240SX

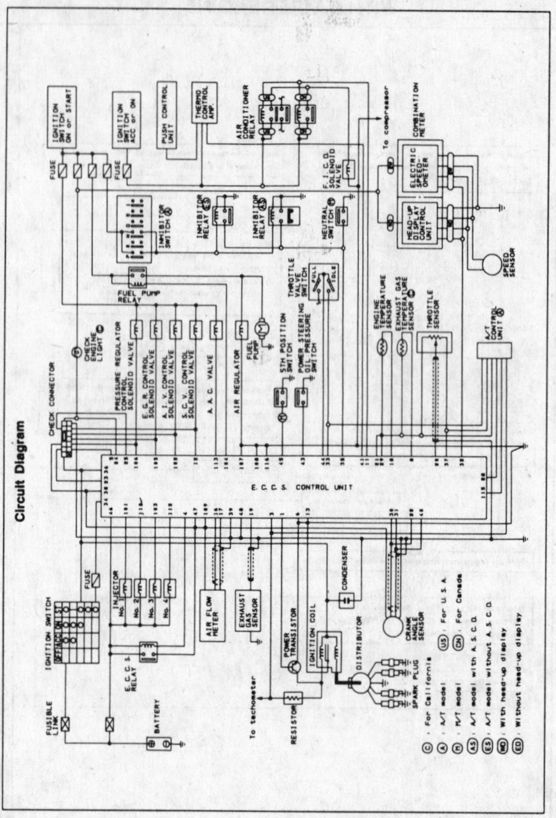
Circuit Diagram
E.C.C.S. CONTROL UNIT
IGNITION SWITCH
FUSIBLE LINK
BATTERY
FUSE
INJECTOR
AIR FLOW METER
EXHAUST GAS SENSOR
POWER TRANSISTOR
IGNITION COIL
CONDENSER
DISTRIBUTOR
SPARK PLUG
CRANK ANGLE SENSOR
RESISTOR
To tachometer
CHECK CONNECTOR
A.A.C. VALVE
AIR REGULATOR
FUEL PUMP RELAY
THROTTLE SENSOR
SPEED SENSOR
COMBINATION METER
To compressor
For California
A/T model
M/T model
A/T model with A.S.C.D.
A/T model without A.S.C.D.
With head-up display
Without head-up display
For U.S.A.
For Canada

ECCS IDLE SPEED, IGNITION TIMING AND IDLE MIXTURE RATIO INSPECTION 1989–90 240SX

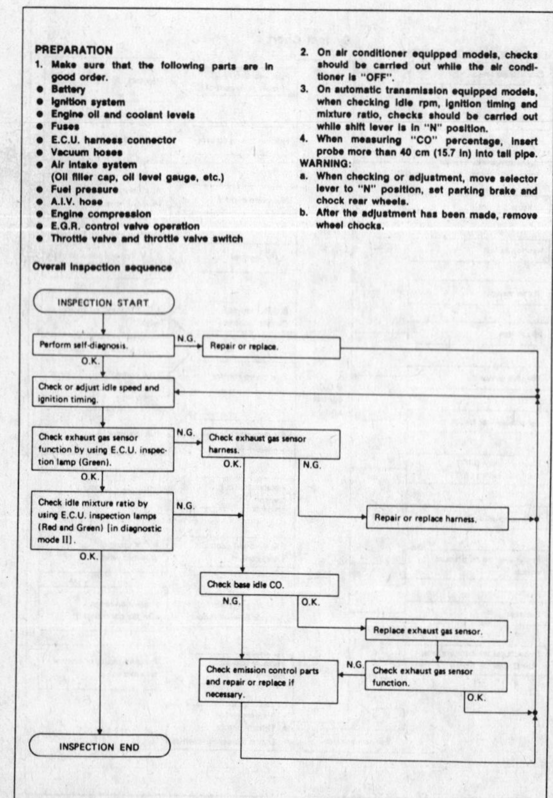

PREPARATION

1. Make sure that the following parts are in good order.
- Battery
- Ignition system
- Engine oil and coolant levels
- Fuses
- E.C.U. harness connector
- Vacuum hoses
- Air intake system (Oil filler cap, oil level gauge, etc.)
- Fuel pressure
- A.I.V. hose
- Engine compression
- E.G.R. control valve operation
- Throttle valve and throttle valve switch

2. On air conditioner equipped models, checks should be carried out while the air conditioner is "OFF".
3. On automatic transmission equipped models, when checking idle rpm, ignition timing and mixture ratio, checks should be carried out while shift lever is in "N" position.
4. When measuring "CO" percentage, insert probe more than 40 cm (15.7 in) into tail pipe.

WARNING:

a. When checking or adjustment, move selector lever to "N" position, set parking brake and chock rear wheels.
b. After the adjustment has been made, remove wheel chocks.

Overall inspection sequence

INSPECTION START
Perform self-diagnosis.
N.G.
Repair or replace.
O.K.
Check or adjust idle speed and ignition timing.
Check exhaust gas sensor function by using E.C.U. inspection lamp (Green).
N.G.
Check exhaust gas sensor harness.
O.K.
N.G.
O.K.
Check idle mixture ratio by using E.C.U. inspection lamps (Red and Green) [in diagnostic mode II].
N.G.
Repair or replace harness.
O.K.
Check base idle CO.
N.G.
O.K.
Replace exhaust gas sensor.
Check emission control parts and repair or replace if necessary.
N.G.
Check exhaust gas sensor function.
O.K.
INSPECTION END

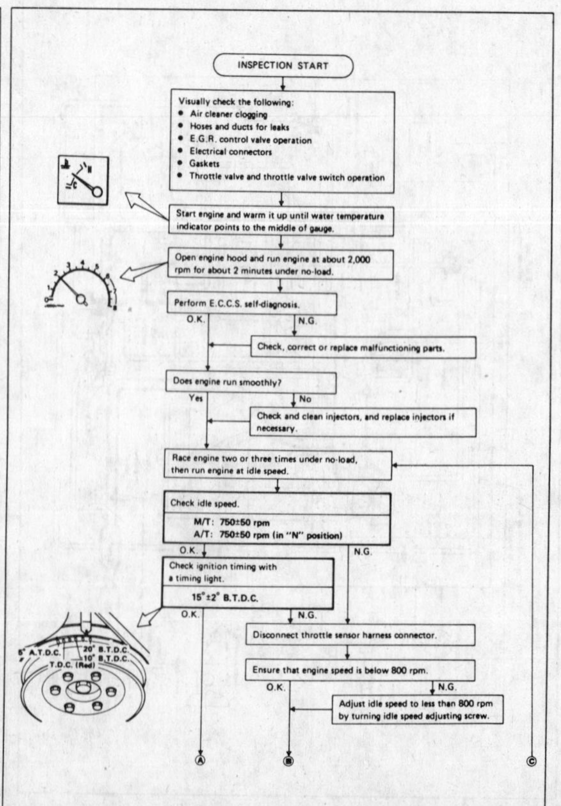
INSPECTION START
Visually check the following:
• Air cleaner clogging
• Hoses and ducts for leaks
• E.G.R. control valve operation
• Electrical connectors
• Gaskets
• Throttle valve and throttle valve switch operation
Start engine and warm it up until water temperature indicator points to the middle of gauge.
Open engine hood and run engine at about 2,000 rpm for about 2 minutes under no-load.
Perform E.C.C.S. self-diagnosis.
O.K.
N.G.
Check, correct or replace malfunctioning parts.
Does engine run smoothly?
Yes
No
Check and clean injectors, and replace injectors if necessary.
Race engine two or three times under no-load, then run engine at idle speed.
Check idle speed.
M/T: 750±50 rpm
A/T: 750±50 rpm (in "N" position)
O.K.
N.G.
Check ignition timing with a timing light.
15°±2° B.T.D.C.
O.K.
N.G.
Disconnect throttle sensor harness connector.
Ensure that engine speed is below 800 rpm.
O.K.
N.G.
Adjust idle speed to less than 800 rpm by turning idle speed adjusting screw.
5° A.T.D.C.
20° B.T.D.C.
10° B.T.D.C.
T.D.C. (Red)
A
B
C

ECCS IDLE SPEED, IGNITION TIMING AND IDLE MIXTURE RATIO INSPECTION (CONT.) 1989–90 240SX

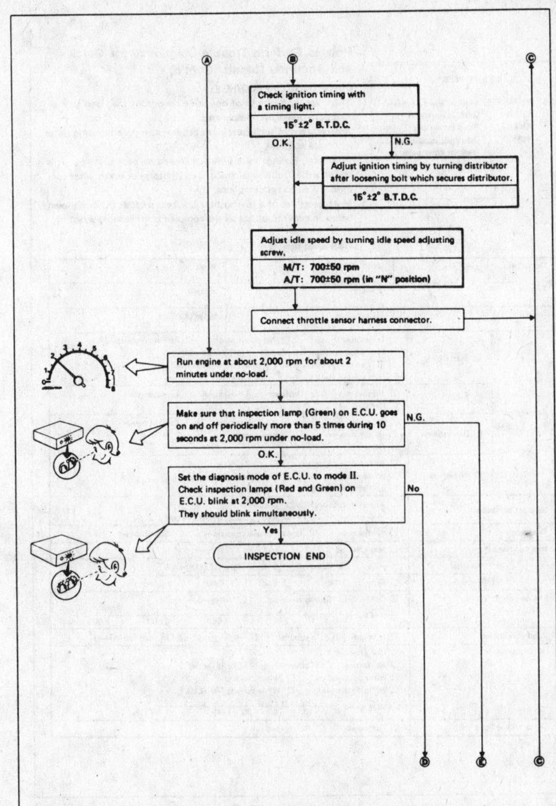
A
B
C
Check ignition timing with a timing light.
15°±2° B.T.D.C.
O.K.
N.G.
Adjust ignition timing by turning distributor after loosening bolt which secures distributor.
15°±2° B.T.D.C.
Adjust idle speed by turning idle speed adjusting screw.
M/T: 700±50 rpm
A/T: 700±50 rpm (in "N" position)
Connect throttle sensor harness connector.
Run engine at about 2,000 rpm for about 2 minutes under no-load.
Make sure that inspection lamp (Green) on E.C.U. goes on and off periodically more than 5 times during 10 seconds at 2,000 rpm under no-load.
N.G.
O.K.
Set the diagnosis mode of E.C.U. to mode II.
Check inspection lamps (Red and Green) on E.C.U. blink at 2,000 rpm.
They should blink simultaneously.
No
Yes
INSPECTION END
D
E
C

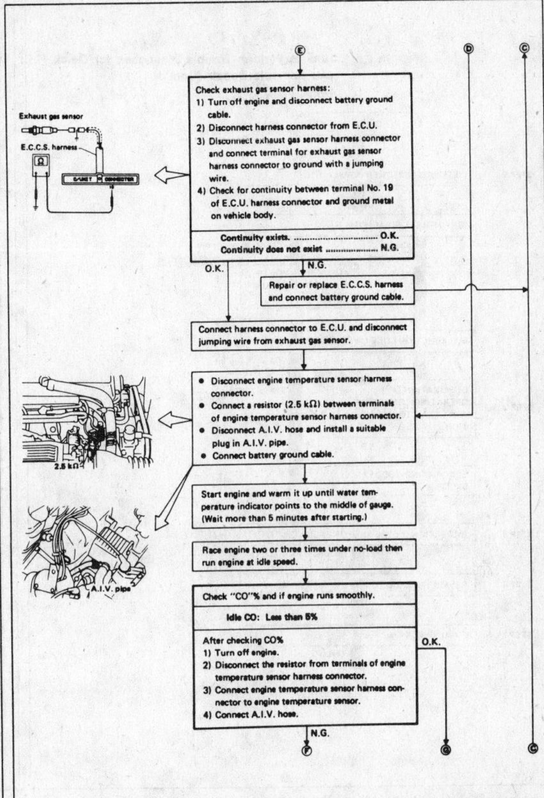
E
D
C
Exhaust gas sensor
E.C.C.S. harness
Check exhaust gas sensor harness:
1) Turn off engine and disconnect battery ground cable.
2) Disconnect harness connector from E.C.U.
3) Disconnect exhaust gas sensor harness connector and connect terminal for exhaust gas sensor harness connector to ground with a jumping wire.
4) Check for continuity between terminal No. 19 of E.C.U. harness connector and ground metal on vehicle body.
Continuity exists. .... O.K.
Continuity does not exist .... N.G.
O.K.
N.G.
Repair or replace E.C.C.S. harness and connect battery ground cable.
Connect harness connector to E.C.U. and disconnect jumping wire from exhaust gas sensor.
Disconnect engine temperature sensor harness connector.
Connect a resistor (2.5 kΩ) between terminals of engine temperature sensor harness connector.
Disconnect A.I.V. hose and install a suitable plug in A.I.V. pipe.
Connect battery ground cable.
2.5 kΩ
Start engine and warm it up until water temperature indicator points to the middle of gauge. (Wait more than 5 minutes after starting.)
Race engine two or three times under no-load then run engine at idle speed.
A.I.V. pipe
Check "CO"% and if engine runs smoothly.
Idle CO: Less than 5%
After checking CO%
1) Turn off engine.
2) Disconnect the resistor from terminals of engine temperature sensor harness connector.
3) Connect engine temperature sensor harness connector to engine temperature sensor.
4) Connect A.I.V. hose.
O.K.
N.G.
F
G
C

ECCS IDLE SPEED, IGNITION TIMING AND IDLE MIXTURE RATIO INSPECTION (CONT.)–1989–90 240SX

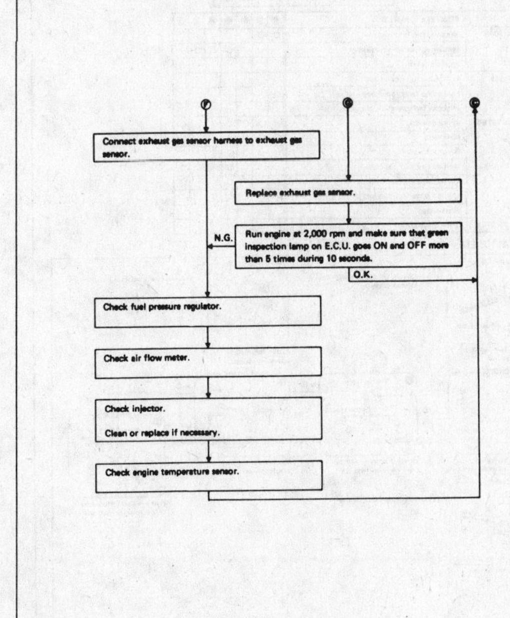
F
G
C
Connect exhaust gas sensor harness to exhaust gas sensor.
Replace exhaust gas sensor.
N.G.
Run engine at 2,000 rpm and make sure that green inspection lamp on E.C.U. goes ON and OFF more than 5 times during 10 seconds.
O.K.
Check fuel pressure regulator.
Check air flow meter.
Check injector.
Clean or replace if necessary.
Check engine temperature sensor.

ECCS DIAGNOSTIC CHART–1989–90 240SX

Sensors
Control unit
Actuators
INFO.
CAUSE

How to Perform Trouble Diagnoses for Quick and Accurate Repair

INTRODUCTION

The engine has an electronic control unit to control major systems such as fuel control, ignition control, idle speed control, etc. The control unit accepts input signals from sensors and instantly drives actuators. It is essential that both kinds of signals are proper and stable. At the same time, it is important that there are no conventional problems such as vacuum leaks, fouled spark plugs, or other problems with the engine.

It is much more difficult to diagnose a problem that occurs intermittently rather than continuously. Most intermittent problems are caused by poor electric connections or faulty wiring. In this case, careful checking of suspicious circuits may help prevent the replacement of good parts.

A visual check only may not find the cause of the problems. A road test with a circuit tester connected to a suspected circuit should be performed.

Before undertaking actual checks, take just a few minutes to talk with a customer who approaches with a driveability complaint. The customer is a very good supplier of information on such problems, especially intermittent ones. Through the talks with the customer, find out what symptoms are present and under what conditions they occur.

Start your diagnosis by looking for "conventional" problems first. This is one of the best ways to troubleshoot driveability problems on an electronically controlled engine vehicle.

ECCS DIAGNOSTIC CHART—1989–90 240SX

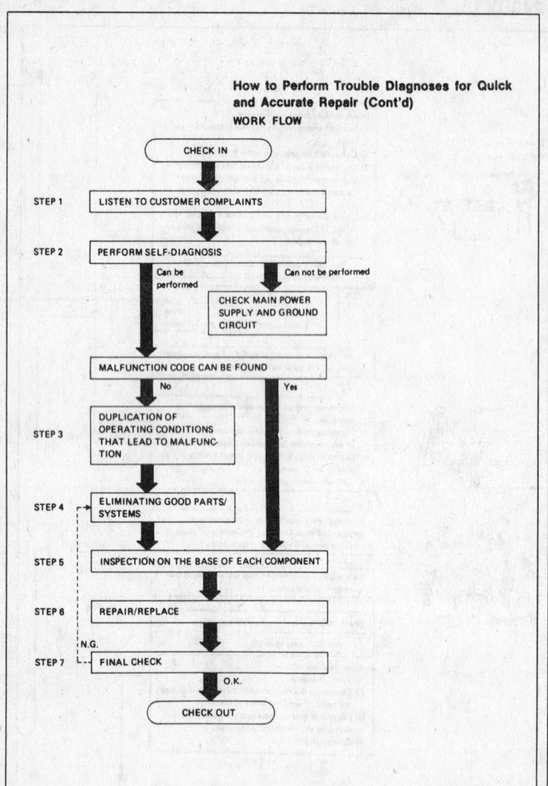
How to Perform Trouble Diagnoses for Quick and Accurate Repair (Cont'd)
WORK FLOW
CHECK IN
STEP 1
LISTEN TO CUSTOMER COMPLAINTS
STEP 2
PERFORM SELF-DIAGNOSIS
Can be performed
Can not be performed
CHECK MAIN POWER SUPPLY AND GROUND CIRCUIT
MALFUNCTION CODE CAN BE FOUND
No
Yes
STEP 3
DUPLICATION OF OPERATING CONDITIONS THAT LEAD TO MALFUNCTION
STEP 4
ELIMINATING GOOD PARTS/ SYSTEMS
STEP 5
INSPECTION ON THE BASE OF EACH COMPONENT
STEP 6
REPAIR/REPLACE
N.G.
STEP 7
FINAL CHECK
O.K.
CHECK OUT

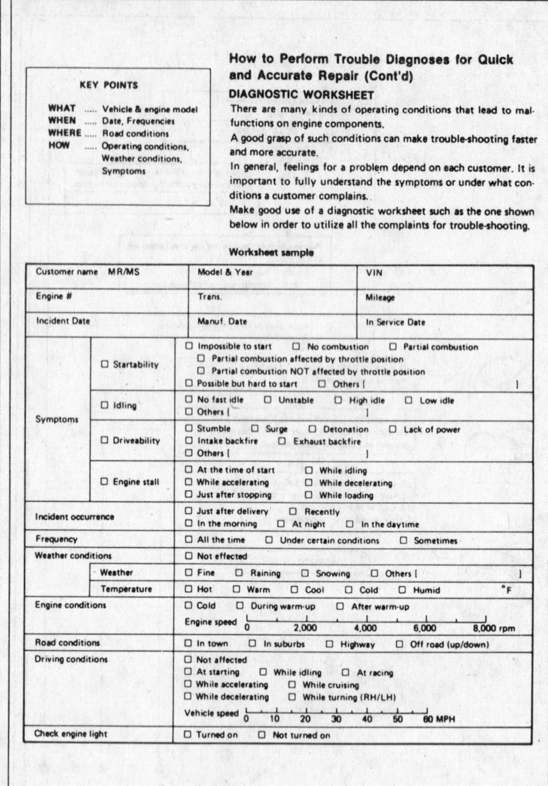

KEY POINTS

WHAT ..... Vehicle & engine model
WHEN ..... Date, Frequencies
WHERE ..... Road conditions
HOW ..... Operating conditions, Weather conditions, Symptoms

How to Perform Trouble Diagnoses for Quick and Accurate Repair (Cont'd)

DIAGNOSTIC WORKSHEET

There are many kinds of operating conditions that lead to malfunctions on engine components.
A good grasp of such conditions can make trouble-shooting faster and more accurate.
In general, feelings for a problem depend on each customer. It is important to fully understand the symptoms or under what conditions a customer complains.
Make good use of a diagnostic worksheet such as the one shown below in order to utilize all the complaints for trouble-shooting.

Worksheet sample

Customer name MR/MS		Model & Year	VIN
Engine #		Trans.	Mileage
Incident Date		Manuf. Date	In Service Date
Symptoms	☐ Startability	☐ Impossible to start ☐ No combustion ☐ Partial combustion ☐ Partial combustion affected by throttle position ☐ Partial combustion NOT affected by throttle position ☐ Possible but hard to start ☐ Others [ ]	
	☐ Idling	☐ No fast idle ☐ Unstable ☐ High idle ☐ Low idle ☐ Others [ ]	
	☐ Driveability	☐ Stumble ☐ Surge ☐ Detonation ☐ Lack of power ☐ Intake backfire ☐ Exhaust backfire ☐ Others [ ]	
	☐ Engine stall	☐ At the time of start ☐ While idling ☐ While accelerating ☐ While decelerating ☐ Just after stopping ☐ While loading	
Incident occurrence		☐ Just after delivery ☐ Recently ☐ In the morning ☐ At night ☐ In the daytime	
Frequency		☐ All the time ☐ Under certain conditions ☐ Sometimes	
Weather conditions		☐ Not effected	
	Weather	☐ Fine ☐ Raining ☐ Snowing ☐ Others [ ]	
	Temperature	☐ Hot ☐ Warm ☐ Cool ☐ Cold ☐ Humid °F	
Engine conditions		☐ Cold ☐ During warm-up ☐ After warm-up Engine speed 0 2,000 4,000 6,000 8,000 rpm	
Road conditions		☐ In town ☐ In suburbs ☐ Highway ☐ Off road (up/down)	
Driving conditions		☐ Not affected ☐ At starting ☐ While idling ☐ At racing ☐ While accelerating ☐ While cruising ☐ While decelerating ☐ While turning (RH/LH) Vehicle speed 0 10 20 30 40 50 60 MPH	
Check engine light		☐ Turned on ☐ Not turned on	

ECCS DIAGNOSTIC CHART—1989–90 240SX

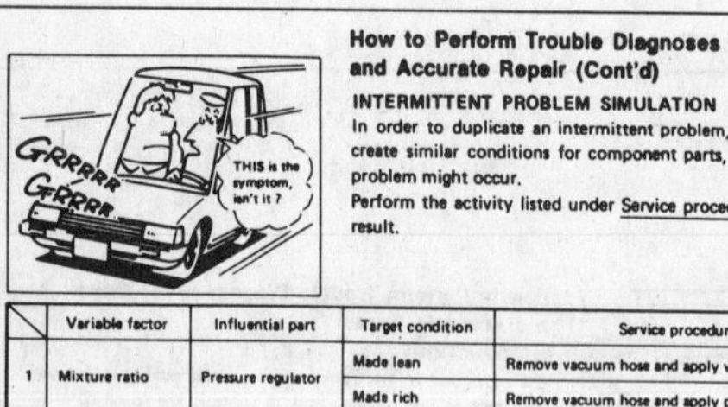
GRRRRR
GRRRR
THIS is the symptom, isn't it ?

How to Perform Trouble Diagnoses for Quick and Accurate Repair (Cont'd)

INTERMITTENT PROBLEM SIMULATION

In order to duplicate an intermittent problem, it is effective to create similar conditions for component parts, under which the problem might occur.
Perform the activity listed under Service procedure and note the result.

	Variable factor	Influential part	Target condition	Service procedure
1	Mixture ratio	Pressure regulator	Made lean	Remove vacuum hose and apply vacuum.
			Made rich	Remove vacuum hose and apply pressure.
2	Ignition timing	Distributor	Advanced	Rotate distributor clockwise.
			Retarded	Rotate distributor counterclockwise.
3	Mixture ratio feedback control	Exhaust gas sensor	Suspended	Disconnect exhaust gas sensor harness connector.
		Control unit	Operation check	Perform self-diagnosis (Mode I/II) at 2,000 rpm.
4	Idle speed	I.A.A. unit	Raised	Turn idle adjusting screw counterclockwise.
			Lowered	Turn idle adjusting screw clockwise.
5	Electric connection (Electric continuity)	Harness connectors and wires	Poor electric connection or faulty wiring	Tap or wiggle.
				Race engine rapidly. See if the torque reaction of the engine unit causes electric breaks.
6	Temperature	Control unit	Cooled	Cool with an icing spray or similar device.
			Warmed	Heat with a hair drier. [WARNING: Do not overheat the unit.]
7	Moisture	Electric parts	Damp	Wet. [WARNING: Do not directly pour water on components. Use a mist sprayer.]
8	Electric loads	Load switches	Loaded	Turn on head lights, air conditioner, rear defogger, etc.
9	Idle switch condition	Control unit	ON-OFF switching	Perform self-diagnosis (Mode IV).
10	Ignition spark	Timing light	Spark power check	Try to flash timing light for each cylinder.

Diagnostic Table

To assist with your trouble diagnoses, some typical diagnostic procedures for the following symptoms are described.

REMARKS

In the following pages, the numbers such as ●, ● in the above chart correspond to those in the service procedure described below.
Possible causes can be checked through the service procedure shown by the mark "○".

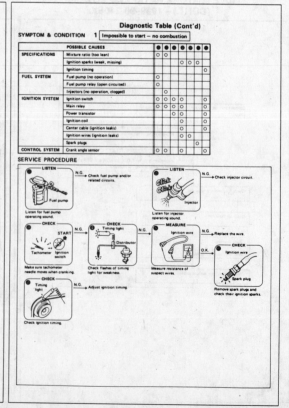

Diagnostic Table (Cont'd)

SYMPTOM & CONDITION 1 Impossible to start – no combustion

	POSSIBLE CAUSES	●	●	●	●	●	●	●
SPECIFICATIONS	Mixture ratio (too lean)	○	○					
	Ignition sparks (weak, missing)				○	○	○	
	Ignition timing							○
FUEL SYSTEM	Fuel pump (no operation)	○						
	Fuel pump relay (open circuited)	○						
	Injectors (no operation, clogged)		○					
IGNITION SYSTEM	Ignition switch	○	○	○	○			○
	Main relay	○	○	○	○			○
	Power transistor			○	○			○
	Ignition coil				○			○
	Center cable (ignition leaks)				○			○
	Ignition wires (ignition leaks)				○	○		
	Spark plugs						○	
CONTROL SYSTEM	Crank angle sensor	○	○		○			○

SERVICE PROCEDURE

● LISTEN — Fuel pump — N.G. → Check fuel pump and/or related circuits.
Listen for fuel pump operating sound.

● LISTEN — Click Click — Injector — N.G. → Check injector circuit.
Listen for injector operating sound.

● CHECK — START — Tachometer — Ignition switch — N.G. →
Make sure tachometer needle moves when cranking.

● CHECK — Timing light — Distributor — N.G. →
Check flashes of timing light for weakness.

● MEASURE — Ignition wire — N.G. → Replace the wire. O.K. →
Measure resistance of suspect wires.

● CHECK — Ignition wire — Spark plug
Remove spark plugs and check their ignition sparks.

● CHECK — Timing light — N.G. → Adjust ignition timing.
Check ignition timing.

ECCS DIAGNOSTIC CHART—1989–90 240SX

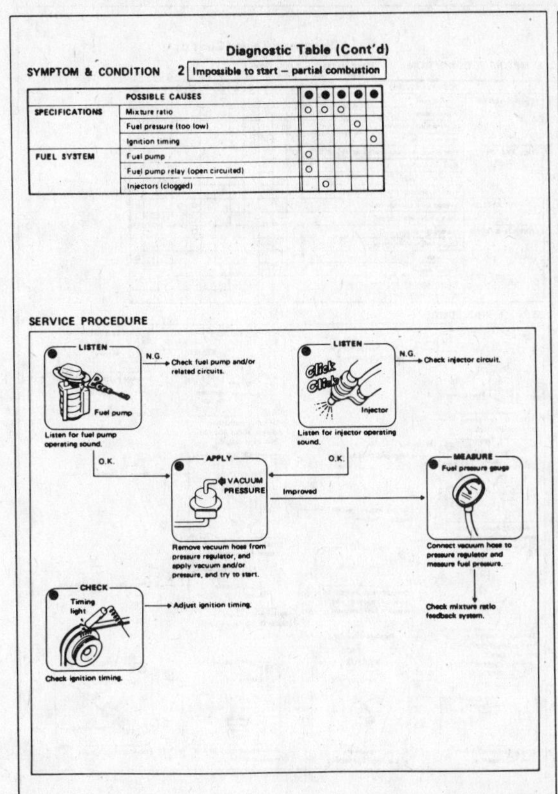

Diagnostic Table (Cont'd)

SYMPTOM & CONDITION 2 Impossible to start – partial combustion

POSSIBLE CAUSES		●	●	●	●	●
SPECIFICATIONS	Mixture ratio	○	○	○		
	Fuel pressure (too low)				○	
	Ignition timing					○
FUEL SYSTEM	Fuel pump	○				
	Fuel pump relay (open circuited)	○				
	Injectors (clogged)		○			

SERVICE PROCEDURE
LISTEN
Fuel pump
N.G.
Check fuel pump and/or related circuits.
Listen for fuel pump operating sound.
O.K.
LISTEN
Injector
N.G.
Check injector circuit.
Listen for injector operating sound.
O.K.
APPLY
VACUUM
PRESSURE
Improved
Remove vacuum hose from pressure regulator, and apply vacuum and/or pressure, and try to start.
MEASURE
Fuel pressure gauge
Connect vacuum hose to pressure regulator and measure fuel pressure.
Check mixture ratio feedback system.
CHECK
Timing light
Adjust ignition timing.
Check ignition timing.

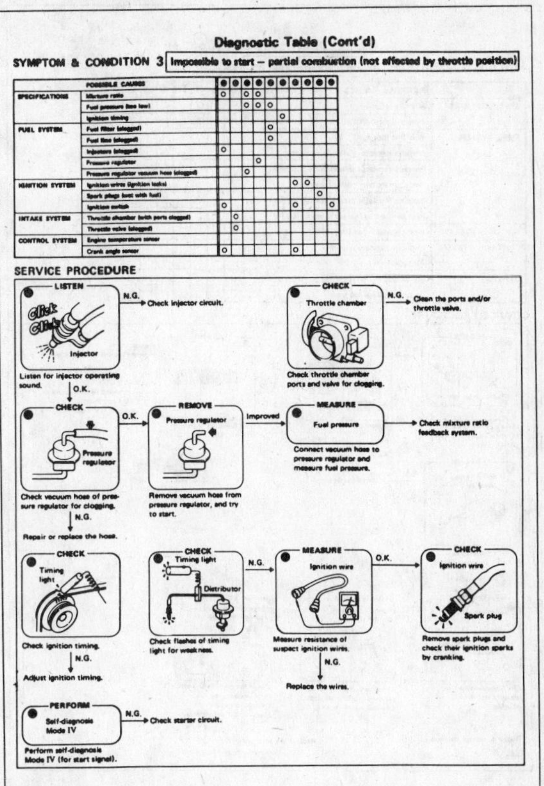

Diagnostic Table (Cont'd)

SYMPTOM & CONDITION 3 Impossible to start – partial combustion (not affected by throttle position)

POSSIBLE CAUSES		●	●	●	●	●	●	●	●	●	●
SPECIFICATIONS	Mixture ratio	○		○	○						
	Fuel pressure (too low)			○	○	○					
	Ignition timing						○				
FUEL SYSTEM	Fuel filter (clogged)					○					
	Fuel line (clogged)					○					
	Injectors (clogged)	○									
	Pressure regulator				○						
	Pressure regulator vacuum hose (clogged)			○							
IGNITION SYSTEM	Ignition wires (ignition leaks)							○	○		
	Spark plugs (wet with fuel)									○	
	Ignition switch	○						○			○
INTAKE SYSTEM	Throttle chamber (with ports clogged)		○								
	Throttle valve (clogged)		○								
CONTROL SYSTEM	Engine temperature sensor										
	Crank angle sensor	○						○			

SERVICE PROCEDURE
LISTEN
Injector
N.G.
Check injector circuit.
Listen for injector operating sound.
O.K.
CHECK
Throttle chamber
N.G.
Clean the ports and/or throttle valve.
Check throttle chamber ports and valve for clogging.
CHECK
Pressure regulator
O.K.
Check vacuum hose of pressure regulator for clogging.
N.G.
Repair or replace the hose.
REMOVE
Pressure regulator
Improved
Remove vacuum hose from pressure regulator, and try to start.
MEASURE
Fuel pressure
Check mixture ratio feedback system.
Connect vacuum hose to pressure regulator and measure fuel pressure.
CHECK
Timing light
Check ignition timing.
N.G.
Adjust ignition timing.
CHECK
Timing light
Distributor
N.G.
Check flashes of timing light for weakness.
MEASURE
Ignition wire
O.K.
Measure resistance of suspect ignition wires.
N.G.
Replace the wires.
CHECK
Ignition wire
Spark plug
Remove spark plugs and check their ignition sparks by cranking.
PERFORM
Self-diagnosis Mode IV
N.G.
Check starter circuit.
Perform self-diagnosis Mode IV (for start signal).

ECCS DIAGNOSTIC CHART—1989–90 240SX

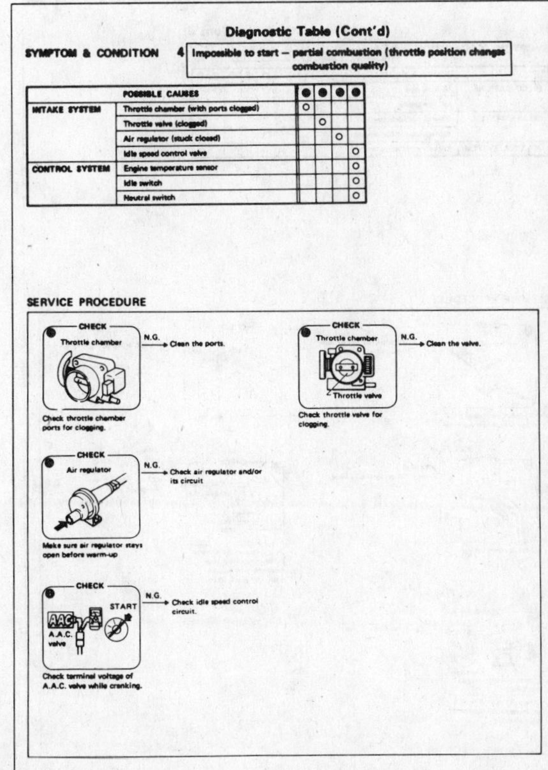

Diagnostic Table (Cont'd)

SYMPTOM & CONDITION 4 Impossible to start – partial combustion (throttle position changes combustion quality)

POSSIBLE CAUSES		●	●	●	●
INTAKE SYSTEM	Throttle chamber (with ports clogged)	○			
	Throttle valve (clogged)		○		
	Air regulator (stuck closed)			○	
	Idle speed control valve				○
CONTROL SYSTEM	Engine temperature sensor				○
	Idle switch				○
	Neutral switch				○

SERVICE PROCEDURE
CHECK
Throttle chamber
N.G.
Clean the ports.
Check throttle chamber ports for clogging.
CHECK
Throttle chamber
Throttle valve
N.G.
Clean the valve.
Check throttle valve for clogging.
CHECK
Air regulator
N.G.
Check air regulator and/or its circuit
Make sure air regulator stays open before warm-up
CHECK
START
AAC
A.A.C. valve
N.G.
Check idle speed control circuit.
Check terminal voltage of A.A.C. valve while cranking.

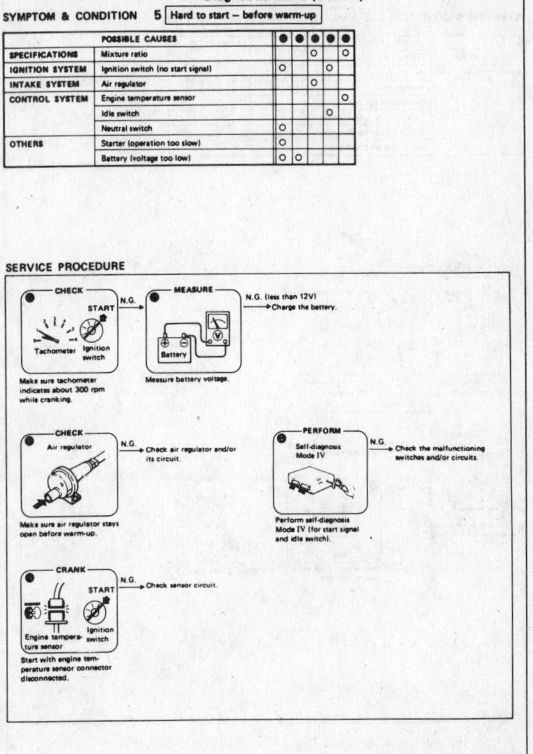

Diagnostic Table (Cont'd)

SYMPTOM & CONDITION 5 Hard to start – before warm-up

POSSIBLE CAUSES		●	●	●	●	●
SPECIFICATIONS	Mixture ratio			○		○
IGNITION SYSTEM	Ignition switch (no start signal)	○			○	
INTAKE SYSTEM	Air regulator			○		
CONTROL SYSTEM	Engine temperature sensor					○
	Idle switch				○	
	Neutral switch	○				
OTHERS	Starter (operation too slow)	○				
	Battery (voltage too low)	○	○			

SERVICE PROCEDURE
CHECK
START
Tachometer
Ignition switch
N.G.
Make sure tachometer indicates about 300 rpm while cranking.
MEASURE
Battery
N.G. (less than 12V)
Charge the battery.
Measure battery voltage.
CHECK
Air regulator
N.G.
Check air regulator and/or its circuit.
Make sure air regulator stays open before warm-up.
PERFORM
Self-diagnosis Mode IV
N.G.
Check the malfunctioning switches and/or circuits.
Perform self-diagnosis Mode IV (for start signal and idle switch).
CRANK
START
Engine temperature sensor
Ignition switch
N.G.
Check sensor circuit.
Start with engine temperature sensor connector disconnected.

ECCS DIAGNOSTIC CHART—1989–90 240SX

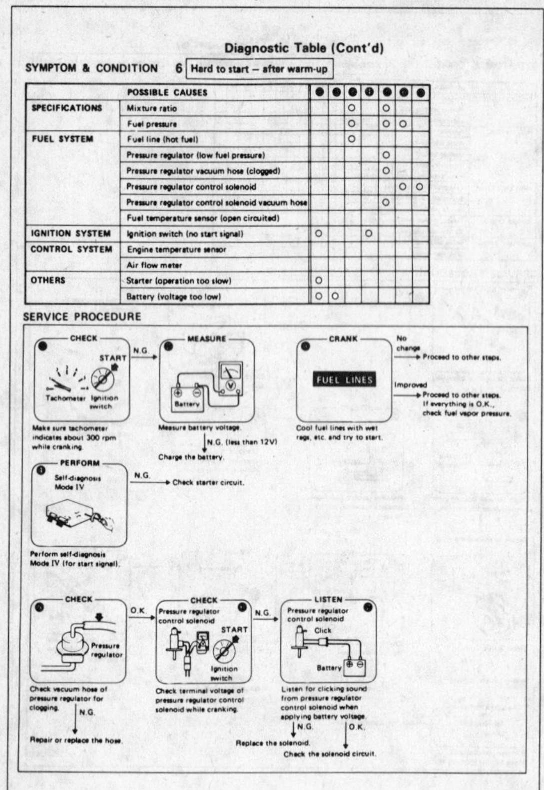

Diagnostic Table (Cont'd)

SYMPTOM & CONDITION 6 Hard to start – after warm-up

	POSSIBLE CAUSES	●	●	●	●	●	●	●
SPECIFICATIONS	Mixture ratio			O		O		
	Fuel pressure			O		O	O	
FUEL SYSTEM	Fuel line (hot fuel)			O				
	Pressure regulator (low fuel pressure)					O		
	Pressure regulator vacuum hose (clogged)					O		
	Pressure regulator control solenoid						O	O
	Pressure regulator control solenoid vacuum hose					O		
	Fuel temperature sensor (open circuited)							
IGNITION SYSTEM	Ignition switch (no start signal)	O			O			
CONTROL SYSTEM	Engine temperature sensor							
	Air flow meter							
OTHERS	Starter (operation too slow)	O						
	Battery (voltage too low)	O	O					

SERVICE PROCEDURE

CHECK
START
Tachometer
Ignition switch
N.G.
Make sure tachometer indicates about 300 rpm while cranking.
MEASURE
Battery
Measure battery voltage.
N.G. (less than 12V)
Charge the battery.
CRANK
FUEL LINES
No change
Proceed to other steps.
Improved
Proceed to other steps. If everything is O.K., check fuel vapor pressure.
Cool fuel lines with wet rags, etc. and try to start.
PERFORM
Self-diagnosis Mode IV
N.G.
Check starter circuit.
Perform self-diagnosis Mode IV (for start signal).
CHECK
Pressure regulator
Check vacuum hose of pressure regulator for clogging.
N.G.
Repair or replace the hose.
O.K.
CHECK
Pressure regulator control solenoid
START
Ignition switch
Check terminal voltage of pressure regulator control solenoid while cranking.
N.G.
LISTEN
Pressure regulator control solenoid
Click
Battery
Listen for clicking sound from pressure regulator control solenoid when applying battery voltage.
N.G.
O.K.
Replace the solenoid.
Check the solenoid circuit.

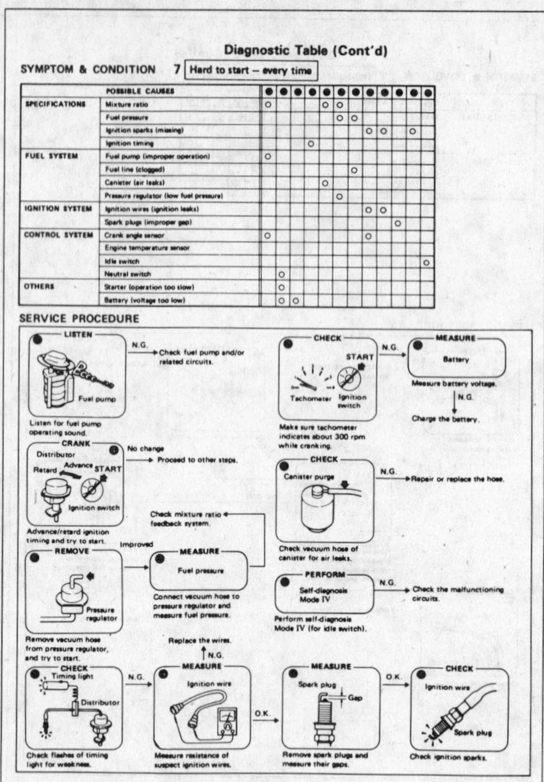

Diagnostic Table (Cont'd)

SYMPTOM & CONDITION 7 Hard to start – every time

	POSSIBLE CAUSES	●	●	●	●	●	●	●	●	●	●	●	●
SPECIFICATIONS	Mixture ratio	O				O	O						
	Fuel pressure						O	O					
	Ignition sparks (missing)								O	O		O	
	Ignition timing				O								
FUEL SYSTEM	Fuel pump (improper operation)	O											
	Fuel line (clogged)							O					
	Canister (air leaks)					O							
	Pressure regulator (low fuel pressure)						O						
IGNITION SYSTEM	Ignition wires (ignition leaks)								O	O			
	Spark plugs (improper gap)										O		
CONTROL SYSTEM	Crank angle sensor	O							O				
	Engine temperature sensor												
	Idle switch												O
	Neutral switch		O										
OTHERS	Starter (operation too slow)		O										
	Battery (voltage too low)		O	O									

SERVICE PROCEDURE

LISTEN
Fuel pump
N.G.
Check fuel pump and/or related circuits.
Listen for fuel pump operating sound.
CHECK
START
Tachometer
Ignition switch
N.G.
Make sure tachometer indicates about 300 rpm while cranking.
MEASURE
Battery
Measure battery voltage.
N.G.
Charge the battery.
CRANK
Distributor
Retard
Advance
START
Ignition switch
No change
Proceed to other steps.
Advance/retard ignition timing and try to start.
Improved
CHECK
Canister purge
N.G.
Repair or replace the hose.
Check vacuum hose of canister for air leaks.
Check mixture ratio feedback system.
REMOVE
Pressure regulator
Remove vacuum hose from pressure regulator, and try to start.
MEASURE
Fuel pressure
Connect vacuum hose to pressure regulator and measure fuel pressure.
PERFORM
Self-diagnosis Mode IV
N.G.
Check the malfunctioning circuits.
Perform self-diagnosis Mode IV (for idle switch).
Replace the wires.
N.G.
CHECK
Timing light
Distributor
Check flashes of timing light for weakness.
N.G.
MEASURE
Ignition wire
Measure resistance of suspect ignition wires.
O.K.
MEASURE
Spark plug
Gap
Remove spark plugs and measure their gaps.
O.K.
CHECK
Ignition wire
Spark plug
Check ignition sparks.

ECCS DIAGNOSTIC CHART—1989–90 240SX

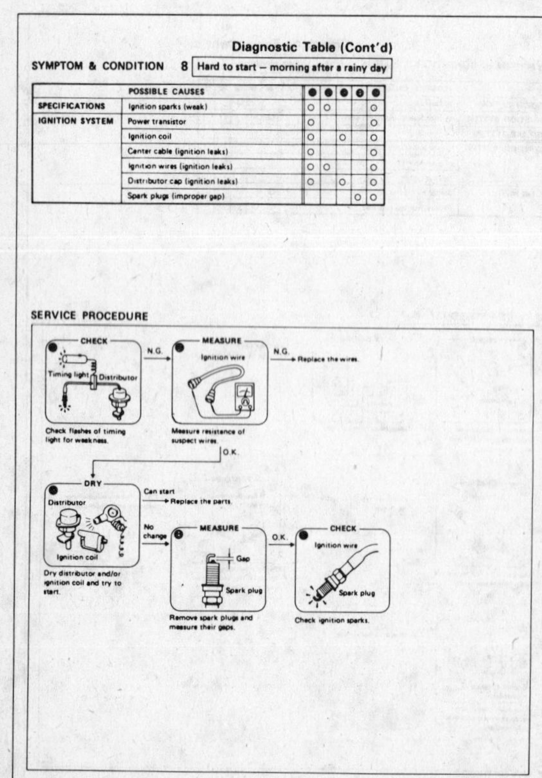

Diagnostic Table (Cont'd)

SYMPTOM & CONDITION 8 Hard to start – morning after a rainy day

	POSSIBLE CAUSES	●	●	●	●	●
SPECIFICATIONS	Ignition sparks (weak)	O	O			O
IGNITION SYSTEM	Power transistor	O				O
	Ignition coil	O		O		O
	Center cable (ignition leaks)	O				O
	Ignition wires (ignition leaks)	O	O			O
	Distributor cap (ignition leaks)	O		O		O
	Spark plugs (improper gap)				O	O

SERVICE PROCEDURE

CHECK
Timing light
Distributor
Check flashes of timing light for weakness.
N.G.
MEASURE
Ignition wire
N.G.
Replace the wires.
Measure resistance of suspect wires.
O.K.
DRY
Distributor
Ignition coil
Can start
Replace the parts.
No change
Dry distributor and/or ignition coil and try to start.
MEASURE
Gap
Spark plug
Remove spark plugs and measure their gaps.
O.K.
CHECK
Ignition wire
Spark plug
Check ignition sparks.

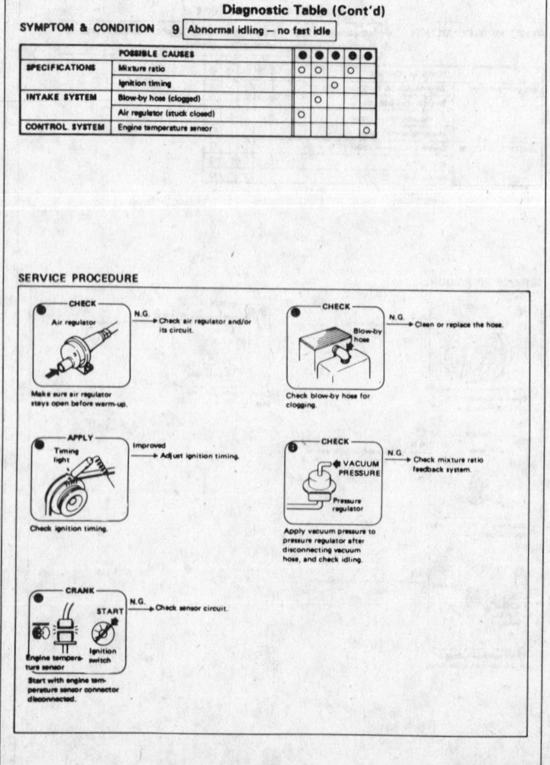

Diagnostic Table (Cont'd)

SYMPTOM & CONDITION 9 Abnormal idling – no fast idle

	POSSIBLE CAUSES	●	●	●	●	●
SPECIFICATIONS	Mixture ratio	O	O		O	
	Ignition timing			O		
INTAKE SYSTEM	Blow-by hose (clogged)		O			
	Air regulator (stuck closed)	O				
CONTROL SYSTEM	Engine temperature sensor					O

SERVICE PROCEDURE

CHECK
Air regulator
N.G.
Check air regulator and/or its circuit.
Make sure air regulator stays open before warm-up.
CHECK
Blow-by hose
N.G.
Clean or replace the hose.
Check blow-by hose for clogging.
APPLY
Timing light
Improved
Adjust ignition timing.
Check ignition timing.
CHECK
VACUUM PRESSURE
Pressure regulator
N.G.
Check mixture ratio feedback system.
Apply vacuum pressure to pressure regulator after disconnecting vacuum hose, and check idling.
CRANK
START
Ignition switch
Engine temperature sensor
N.G.
Check sensor circuit.
Start with engine temperature sensor connector disconnected.

ECCS DIAGNOSTIC CHART—1989-90 240SX

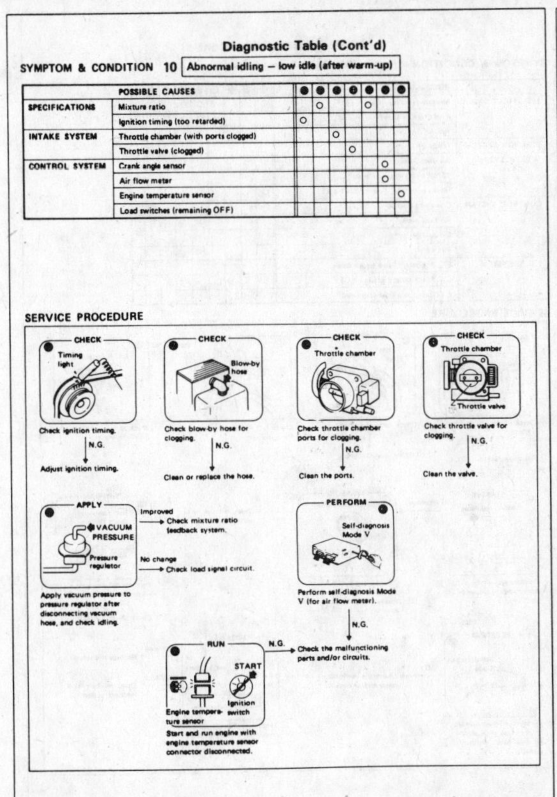

Diagnostic Table (Cont'd)

SYMPTOM & CONDITION 10 Abnormal idling – low idle (after warm-up)

POSSIBLE CAUSES		●	●	●	●	●	●	●
SPECIFICATIONS	Mixture ratio		○			○		
	Ignition timing (too retarded)	○						
INTAKE SYSTEM	Throttle chamber (with ports clogged)			○				
	Throttle valve (clogged)				○			
CONTROL SYSTEM	Crank angle sensor						○	
	Air flow meter						○	
	Engine temperature sensor							○
	Load switches (remaining OFF)							

SERVICE PROCEDURE

CHECK
Timing light
Check ignition timing.
N.G.
Adjust ignition timing.
CHECK
Blow-by hose
Check blow-by hose for clogging.
N.G.
Clean or replace the hose.
CHECK
Throttle chamber
Check throttle chamber ports for clogging.
N.G.
Clean the ports.
CHECK
Throttle chamber
Throttle valve
Check throttle valve for clogging.
N.G.
Clean the valve.
APPLY
VACUUM PRESSURE
Pressure regulator
Improved
Check mixture ratio feedback system.
No change
Check load signal circuit.
Apply vacuum pressure to pressure regulator after disconnecting vacuum hose, and check idling.
PERFORM
Self-diagnosis Mode V
Perform self-diagnosis Mode V (for air flow meter).
N.G.
Check the malfunctioning parts and/or circuits.
RUN
START
Engine temperature sensor
Ignition switch
Start and run engine with engine temperature sensor connector disconnected.
N.G.

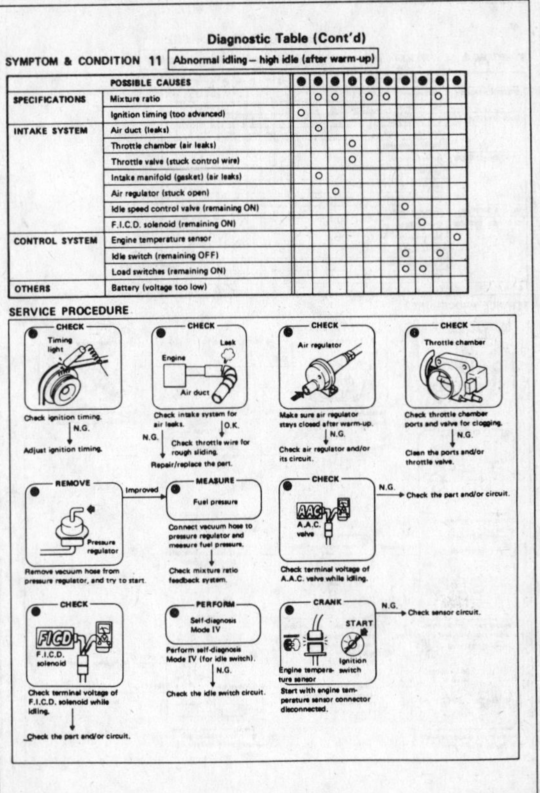

Diagnostic Table (Cont'd)

SYMPTOM & CONDITION 11 Abnormal idling – high idle (after warm-up)

POSSIBLE CAUSES		●	●	●	●	●	●	●	●	●	●
SPECIFICATIONS	Mixture ratio		○	○		○	○			○	
	Ignition timing (too advanced)	○									
INTAKE SYSTEM	Air duct (leaks)		○								
	Throttle chamber (air leaks)				○						
	Throttle valve (stuck control wire)				○						
	Intake manifold (gasket) (air leaks)		○								
	Air regulator (stuck open)			○							
	Idle speed control valve (remaining ON)							○			
	F.I.C.D. solenoid (remaining ON)								○		
CONTROL SYSTEM	Engine temperature sensor										○
	Idle switch (remaining OFF)							○		○	
	Load switches (remaining ON)							○	○		
OTHERS	Battery (voltage too low)										

SERVICE PROCEDURE

CHECK
Timing light
Check ignition timing.
N.G.
Adjust ignition timing.
CHECK
Engine
Leak
Air duct
Check intake system for air leaks.
N.G.
O.K.
Check throttle wire for rough sliding.
Repair/replace the part.
CHECK
Air regulator
Make sure air regulator stays closed after warm-up.
N.G.
Check air regulator and/or its circuit.
CHECK
Throttle chamber
Check throttle chamber ports and valve for clogging.
N.G.
Clean the ports and/or throttle valve.
REMOVE
Pressure regulator
Remove vacuum hose from pressure regulator, and try to start.
Improved
MEASURE
Fuel pressure
Connect vacuum hose to pressure regulator and measure fuel pressure.
Check mixture ratio feedback system.
CHECK
AAC
A.A.C. valve
Check terminal voltage of A.A.C. valve while idling.
N.G.
Check the part and/or circuit.
CHECK
FICD
F.I.C.D. solenoid
Check terminal voltage of F.I.C.D. solenoid while idling.
Check the part and/or circuit.
PERFORM
Self-diagnosis Mode IV
Perform self-diagnosis Mode IV (for idle switch).
N.G.
Check the idle switch circuit.
CRANK
START
Engine temperature sensor
Ignition switch
Start with engine temperature sensor connector disconnected.
N.G.
Check sensor circuit.

ECCS DIAGNOSTIC CHART—1989-90 240SX

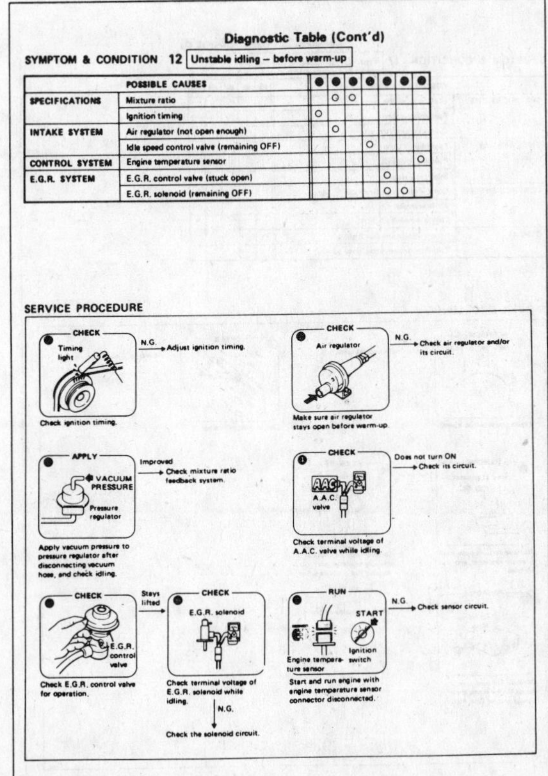

Diagnostic Table (Cont'd)

SYMPTOM & CONDITION 12 Unstable idling – before warm-up

POSSIBLE CAUSES		●	●	●	●	●	●	●
SPECIFICATIONS	Mixture ratio		○	○				
	Ignition timing	○						
INTAKE SYSTEM	Air regulator (not open enough)		○					
	Idle speed control valve (remaining OFF)				○			
CONTROL SYSTEM	Engine temperature sensor							○
E.G.R. SYSTEM	E.G.R. control valve (stuck open)					○		
	E.G.R. solenoid (remaining OFF)					○	○	

SERVICE PROCEDURE

CHECK
Timing light
N.G.
Adjust ignition timing.
Check ignition timing.
CHECK
Air regulator
N.G.
Check air regulator and/or its circuit.
Make sure air regulator stays open before warm-up.
APPLY
VACUUM PRESSURE
Pressure regulator
Improved
Check mixture ratio feedback system.
Apply vacuum pressure to pressure regulator after disconnecting vacuum hose, and check idling.
CHECK
AAC
A.A.C. valve
Does not turn ON
Check its circuit.
Check terminal voltage of A.A.C. valve while idling.
CHECK
E.G.R. control valve
Check E.G.R. control valve for operation.
Stays lifted
CHECK
E.G.R. solenoid
Check terminal voltage of E.G.R. solenoid while idling.
N.G.
Check the solenoid circuit.
RUN
START
Engine temperature sensor
Ignition switch
Start and run engine with engine temperature sensor connector disconnected.
N.G.
Check sensor circuit.

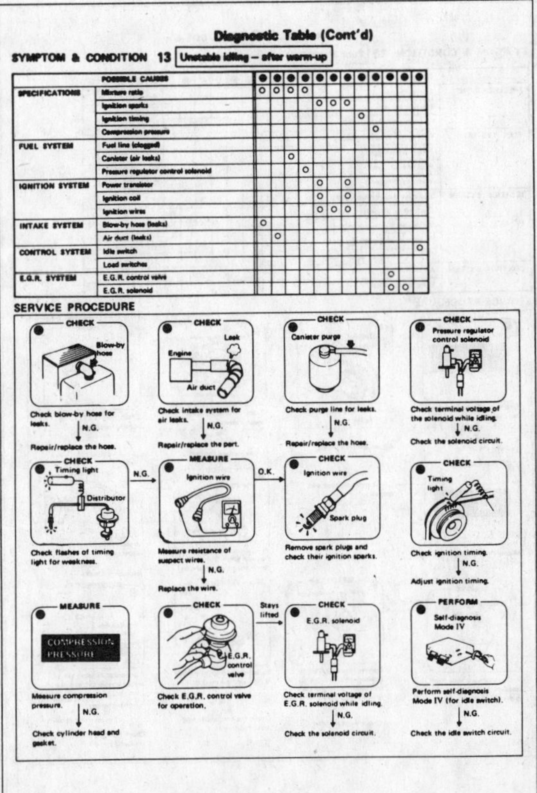

Diagnostic Table (Cont'd)

SYMPTOM & CONDITION 13 Unstable idling – after warm-up

POSSIBLE CAUSES		●	●	●	●	●	●	●	●	●	●	●	●
SPECIFICATIONS	Mixture ratio	○	○	○	○								
	Ignition sparks					○	○	○					
	Ignition timing								○				
	Compression pressure									○			
FUEL SYSTEM	Fuel line (clogged)												
	Canister (air leaks)			○									
	Pressure regulator control solenoid				○								
IGNITION SYSTEM	Power transistor					○		○					
	Ignition coil					○		○					
	Ignition wires					○	○	○					
INTAKE SYSTEM	Blow-by hose (leaks)	○											
	Air duct (leaks)		○										
CONTROL SYSTEM	Idle switch												○
	Load switches												
E.G.R. SYSTEM	E.G.R. control valve										○		
	E.G.R. solenoid										○	○	

SERVICE PROCEDURE

CHECK
Blow-by hose
Check blow-by hose for leaks.
N.G.
Repair/replace the hose.
CHECK
Engine
Leak
Air duct
Check intake system for air leaks.
N.G.
Repair/replace the part.
CHECK
Canister purge
Check purge line for leaks.
N.G.
Repair/replace the hose.
CHECK
Pressure regulator control solenoid
Check terminal voltage of the solenoid while idling.
N.G.
Check the solenoid circuit.
CHECK
Timing light
Distributor
Check flashes of timing light for weakness.
N.G.
MEASURE
Ignition wire
Measure resistance of suspect wires.
N.G.
Replace the wire.
O.K.
CHECK
Ignition wire
Spark plug
Remove spark plugs and check their ignition sparks.
CHECK
Timing light
Check ignition timing.
N.G.
Adjust ignition timing.
MEASURE
COMPRESSION PRESSURE
Measure compression pressure.
N.G.
Check cylinder head and gasket.
CHECK
E.G.R. control valve
Check E.G.R. control valve for operation.
Stays lifted
CHECK
E.G.R. solenoid
Check terminal voltage of E.G.R. solenoid while idling.
N.G.
Check the solenoid circuit.
PERFORM
Self-diagnosis Mode IV
Perform self-diagnosis Mode IV (for idle switch).
N.G.
Check the idle switch circuit.

ECCS DIAGNOSTIC CHART – 1989–90 240SX

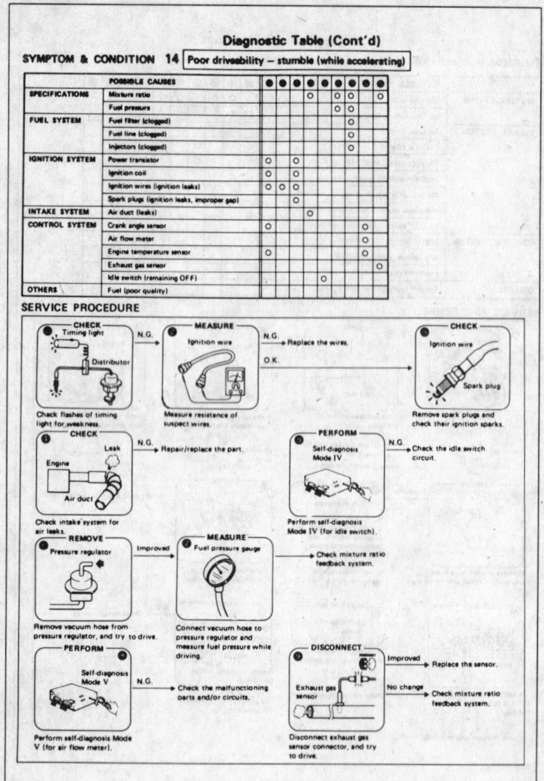

Diagnostic Table (Cont'd)

SYMPTOM & CONDITION 14 Poor driveability – stumble (while accelerating)

	POSSIBLE CAUSES	●	●	●	●	●	●	●	●	●
SPECIFICATIONS	Mixture ratio				○		○	○		○
	Fuel pressure						○	○		
FUEL SYSTEM	Fuel filter (clogged)							○		
	Fuel line (clogged)							○		
	Injectors (clogged)							○		
IGNITION SYSTEM	Power transistor	○		○						
	Ignition coil	○		○						
	Ignition wires (ignition leaks)	○	○	○						
	Spark plugs (ignition leaks, improper gap)			○						
INTAKE SYSTEM	Air duct (leaks)				○					
CONTROL SYSTEM	Crank angle sensor	○							○	
	Air flow meter								○	
	Engine temperature sensor	○							○	
	Exhaust gas sensor									○
	Idle switch (remaining OFF)					○				
OTHERS	Fuel (poor quality)									

SERVICE PROCEDURE

CHECK
Timing light
Distributor
Check flashes of timing light for weakness.
N.G.
MEASURE
Ignition wire
Measure resistance of suspect wires.
N.G.
Replace the wires.
O.K.
CHECK
Ignition wire
Spark plug
Remove spark plugs and check their ignition sparks.
CHECK
Leak
Engine
Air duct
Check intake system for air leaks.
N.G.
Repair/replace the part.
PERFORM
Self-diagnosis Mode IV
Perform self-diagnosis Mode IV (for idle switch).
N.G.
Check the idle switch circuit.
REMOVE
Pressure regulator
Remove vacuum hose from pressure regulator, and try to drive.
Improved
MEASURE
Fuel pressure gauge
Connect vacuum hose to pressure regulator and measure fuel pressure while driving.
Check mixture ratio feedback system.
PERFORM
Self-diagnosis Mode V
Perform self-diagnosis Mode V (for air flow meter).
N.G.
Check the malfunctioning parts and/or circuits.
DISCONNECT
Exhaust gas sensor
Disconnect exhaust gas sensor connector, and try to drive.
Improved
Replace the sensor.
No change
Check mixture ratio feedback system.

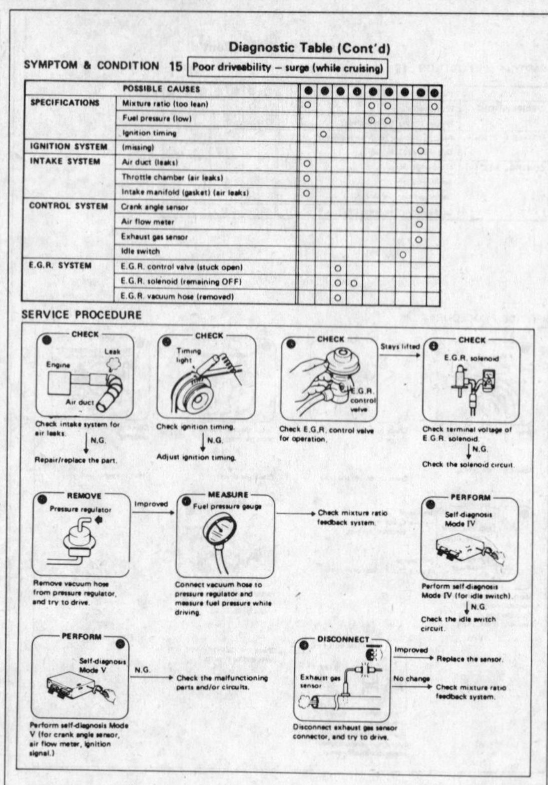

Diagnostic Table (Cont'd)

SYMPTOM & CONDITION 15 Poor driveability – surge (while cruising)

	POSSIBLE CAUSES	●	●	●	●	●	●	●	●	●
SPECIFICATIONS	Mixture ratio (too lean)	○				○	○			○
	Fuel pressure (low)					○	○			
	Ignition timing		○							
IGNITION SYSTEM	(missing)								○	
INTAKE SYSTEM	Air duct (leaks)	○								
	Throttle chamber (air leaks)	○								
	Intake manifold (gasket) (air leaks)	○								
CONTROL SYSTEM	Crank angle sensor								○	
	Air flow meter								○	
	Exhaust gas sensor								○	
	Idle switch							○		
E.G.R. SYSTEM	E.G.R. control valve (stuck open)			○						
	E.G.R. solenoid (remaining OFF)			○	○					
	E.G.R. vacuum hose (removed)			○						

SERVICE PROCEDURE

CHECK
Leak
Engine
Air duct
Check intake system for air leaks.
N.G.
Repair/replace the part.
CHECK
Timing light
Check ignition timing.
N.G.
Adjust ignition timing.
CHECK
E.G.R. control valve
Check E.G.R. control valve for operation.
Stays lifted
CHECK
E.G.R. solenoid
Check terminal voltage of E.G.R. solenoid.
N.G.
Check the solenoid circuit.
REMOVE
Pressure regulator
Remove vacuum hose from pressure regulator, and try to drive.
Improved
MEASURE
Fuel pressure gauge
Connect vacuum hose to pressure regulator and measure fuel pressure while driving.
Check mixture ratio feedback system.
PERFORM
Self-diagnosis Mode IV
Perform self-diagnosis Mode IV (for idle switch).
N.G.
Check the idle switch circuit.
PERFORM
Self-diagnosis Mode V
Perform self-diagnosis Mode V (for crank angle sensor, air flow meter, ignition signal.)
N.G.
Check the malfunctioning parts and/or circuits.
DISCONNECT
Exhaust gas sensor
Disconnect exhaust gas sensor connector, and try to drive.
Improved
Replace the sensor.
No change
Check mixture ratio feedback system.

ECCS DIAGNOSTIC CHART – 1989–90 240SX

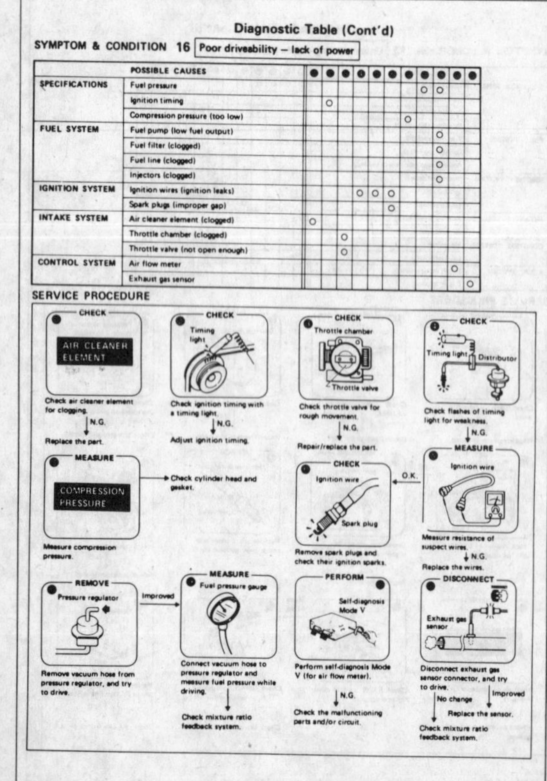

Diagnostic Table (Cont'd)

SYMPTOM & CONDITION 16 Poor driveability – lack of power

	POSSIBLE CAUSES	●	●	●	●	●	●	●	●	●	●	●
SPECIFICATIONS	Fuel pressure								○	○		
	Ignition timing		○									
	Compression pressure (too low)							○				
FUEL SYSTEM	Fuel pump (low fuel output)									○		
	Fuel filter (clogged)									○		
	Fuel line (clogged)									○		
	Injectors (clogged)									○		
IGNITION SYSTEM	Ignition wires (ignition leaks)				○	○	○					
	Spark plugs (improper gap)						○					
INTAKE SYSTEM	Air cleaner element (clogged)	○										
	Throttle chamber (clogged)			○								
	Throttle valve (not open enough)			○								
CONTROL SYSTEM	Air flow meter										○	
	Exhaust gas sensor											○

SERVICE PROCEDURE

CHECK
AIR CLEANER ELEMENT
Check air cleaner element for clogging.
N.G.
Replace the part.
CHECK
Timing light
Check ignition timing with a timing light.
N.G.
Adjust ignition timing.
CHECK
Throttle chamber
Throttle valve
Check throttle valve for rough movement.
N.G.
Repair/replace the part.
CHECK
Timing light
Distributor
Check flashes of timing light for weakness.
N.G.
MEASURE
COMPRESSION PRESSURE
Measure compression pressure.
Check cylinder head and gasket.
CHECK
Ignition wire
Spark plug
Remove spark plugs and check their ignition sparks.
O.K.
MEASURE
Ignition wire
Measure resistance of suspect wires.
N.G.
Replace the wires.
REMOVE
Pressure regulator
Remove vacuum hose from pressure regulator, and try to drive.
Improved
MEASURE
Fuel pressure gauge
Connect vacuum hose to pressure regulator and measure fuel pressure while driving.
Check mixture ratio feedback system.
PERFORM
Self-diagnosis Mode V
Perform self-diagnosis Mode V (for air flow meter).
N.G.
Check the malfunctioning parts and/or circuit.
DISCONNECT
Exhaust gas sensor
Disconnect exhaust gas sensor connector, and try to drive.
No change
Improved
Replace the sensor.
Check mixture ratio feedback system.

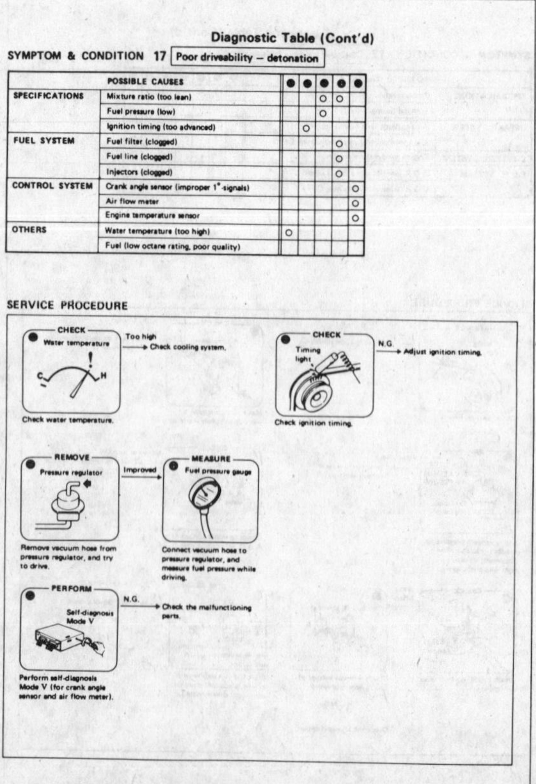

Diagnostic Table (Cont'd)

SYMPTOM & CONDITION 17 Poor driveability – detonation

	POSSIBLE CAUSES	●	●	●	●	●
SPECIFICATIONS	Mixture ratio (too lean)			○	○	
	Fuel pressure (low)			○		
	Ignition timing (too advanced)		○			
FUEL SYSTEM	Fuel filter (clogged)				○	
	Fuel line (clogged)				○	
	Injectors (clogged)				○	
CONTROL SYSTEM	Crank angle sensor (improper 1°-signals)					○
	Air flow meter					○
	Engine temperature sensor					○
OTHERS	Water temperature (too high)	○				
	Fuel (low octane rating, poor quality)					

SERVICE PROCEDURE

CHECK
Water temperature
Check water temperature.
Too high
Check cooling system.
CHECK
Timing light
Check ignition timing.
N.G.
Adjust ignition timing.
REMOVE
Pressure regulator
Remove vacuum hose from pressure regulator, and try to drive.
Improved
MEASURE
Fuel pressure gauge
Connect vacuum hose to pressure regulator, and measure fuel pressure while driving.
PERFORM
Self-diagnosis Mode V
Perform self-diagnosis Mode V (for crank angle sensor and air flow meter).
N.G.
Check the malfunctioning parts.

ECCS DIAGNOSTIC CHART—1989–90 240SX

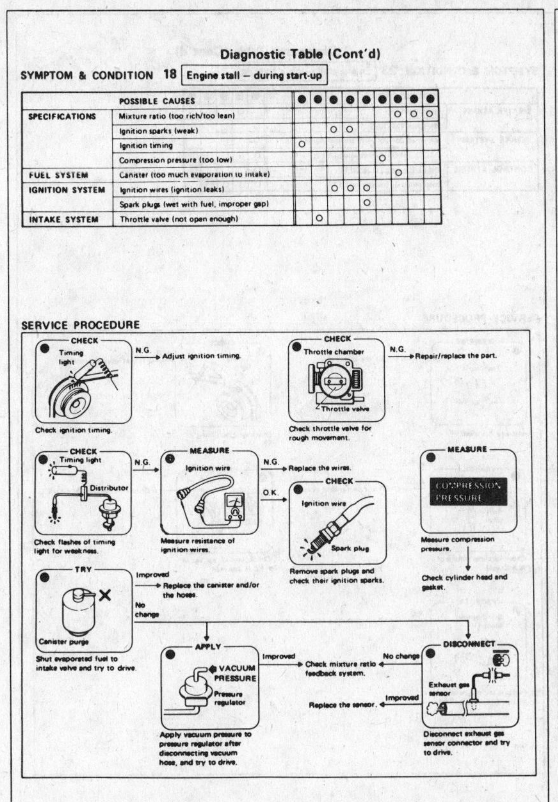

Diagnostic Table (Cont'd)

SYMPTOM & CONDITION 18 Engine stall – during start-up

	POSSIBLE CAUSES	●	●	●	●	●	●	●	●	●
SPECIFICATIONS	Mixture ratio (too rich/too lean)							○	○	○
	Ignition sparks (weak)			○	○					
	Ignition timing	○								
	Compression pressure (too low)						○			
FUEL SYSTEM	Canister (too much evaporation to intake)							○		
IGNITION SYSTEM	Ignition wires (ignition leaks)			○	○	○				
	Spark plugs (wet with fuel, improper gap)					○				
INTAKE SYSTEM	Throttle valve (not open enough)		○							

SERVICE PROCEDURE
CHECK
Timing light
N.G.
Adjust ignition timing.
Check ignition timing.
CHECK
Throttle chamber
Throttle valve
N.G.
Repair/replace the part.
Check throttle valve for rough movement.
CHECK
Timing light
Distributor
N.G.
Check flashes of timing light for weakness.
MEASURE
Ignition wire
N.G.
Replace the wires.
O.K.
Measure resistance of ignition wires.
CHECK
Ignition wire
Spark plug
Remove spark plugs and check their ignition sparks.
MEASURE
COMPRESSION PRESSURE
Measure compression pressure.
Check cylinder head and gasket.
TRY
Canister purge
Improved
Replace the canister and/or the hoses.
No change
Shut evaporated fuel to intake valve and try to drive.
APPLY
VACUUM PRESSURE
Pressure regulator
Improved
Check mixture ratio feedback system.
Apply vacuum pressure to pressure regulator after disconnecting vacuum hose, and try to drive.
DISCONNECT
Exhaust gas sensor
No change
Improved
Replace the sensor.
Disconnect exhaust gas sensor connector and try to drive.

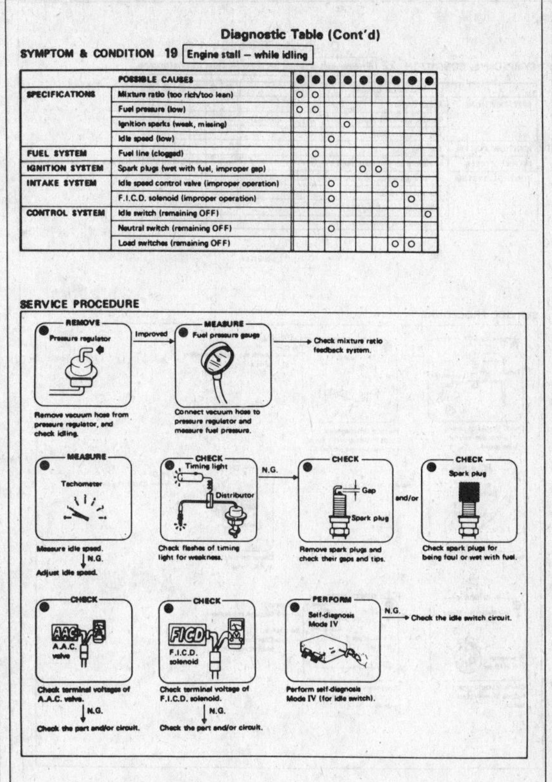

Diagnostic Table (Cont'd)

SYMPTOM & CONDITION 19 Engine stall – while idling

	POSSIBLE CAUSES	●	●	●	●	●	●	●	●	●
SPECIFICATIONS	Mixture ratio (too rich/too lean)	○	○							
	Fuel pressure (low)	○	○							
	Ignition sparks (weak, missing)				○					
	Idle speed (low)			○						
FUEL SYSTEM	Fuel line (clogged)		○							
IGNITION SYSTEM	Spark plugs (wet with fuel, improper gap)					○	○			
INTAKE SYSTEM	Idle speed control valve (improper operation)			○				○		
	F.I.C.D. solenoid (improper operation)			○					○	
CONTROL SYSTEM	Idle switch (remaining OFF)									○
	Neutral switch (remaining OFF)			○						
	Load switches (remaining OFF)							○	○	

SERVICE PROCEDURE
REMOVE
Pressure regulator
Improved
Remove vacuum hose from pressure regulator, and check idling.
MEASURE
Fuel pressure gauge
Check mixture ratio feedback system.
Connect vacuum hose to pressure regulator and measure fuel pressure.
MEASURE
Tachometer
Measure idle speed.
N.G.
Adjust idle speed.
CHECK
Timing light
Distributor
N.G.
Check flashes of timing light for weakness.
CHECK
Gap
Spark plug
and/or
Remove spark plugs and check their gaps and tips.
CHECK
Spark plug
Check spark plugs for being foul or wet with fuel.
CHECK
AAC
A.A.C. valve
Check terminal voltage of A.A.C. valve.
N.G.
Check the part and/or circuit.
CHECK
FICD
F.I.C.D. solenoid
Check terminal voltage of F.I.C.D. solenoid.
N.G.
Check the part and/or circuit.
PERFORM
Self-diagnosis Mode IV
N.G.
Check the idle switch circuit.
Perform self-diagnosis Mode IV (for idle switch).

ECCS DIAGNOSTIC CHART—1989–90 240SX

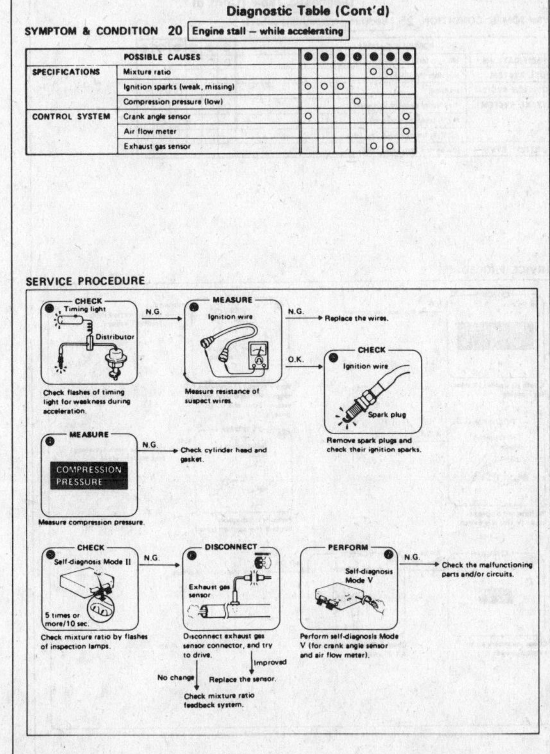

Diagnostic Table (Cont'd)

SYMPTOM & CONDITION 20 Engine stall – while accelerating

	POSSIBLE CAUSES	●	●	●	●	●	●	●
SPECIFICATIONS	Mixture ratio					○	○	
	Ignition sparks (weak, missing)	○	○	○				
	Compression pressure (low)				○			
CONTROL SYSTEM	Crank angle sensor	○						○
	Air flow meter							○
	Exhaust gas sensor					○	○	

SERVICE PROCEDURE
CHECK
Timing light
Distributor
N.G.
Check flashes of timing light for weakness during acceleration.
MEASURE
Ignition wire
N.G.
Replace the wires.
O.K.
Measure resistance of suspect wires.
CHECK
Ignition wire
Spark plug
Remove spark plugs and check their ignition sparks.
MEASURE
COMPRESSION PRESSURE
N.G.
Check cylinder head and gasket.
Measure compression pressure.
CHECK
Self-diagnosis Mode II
5 times or more/10 sec.
N.G.
Check mixture ratio by flashes of inspection lamps.
DISCONNECT
Exhaust gas sensor
Disconnect exhaust gas sensor connector, and try to drive.
Improved
Replace the sensor.
No change
Check mixture ratio feedback system.
PERFORM
Self-diagnosis Mode V
N.G.
Check the malfunctioning parts and/or circuits.
Perform self-diagnosis Mode V (for crank angle sensor and air flow meter).

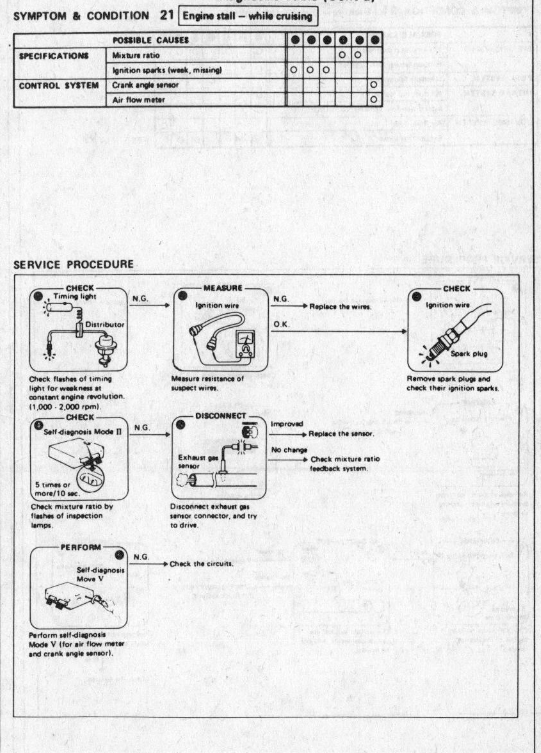

Diagnostic Table (Cont'd)

SYMPTOM & CONDITION 21 Engine stall – while cruising

	POSSIBLE CAUSES	●	●	●	●	●	●
SPECIFICATIONS	Mixture ratio				○	○	
	Ignition sparks (weak, missing)	○	○	○			
CONTROL SYSTEM	Crank angle sensor						○
	Air flow meter						○

SERVICE PROCEDURE
CHECK
Timing light
Distributor
N.G.
Check flashes of timing light for weakness at constant engine revolution. (1,000 - 2,000 rpm).
MEASURE
Ignition wire
N.G.
Replace the wires.
O.K.
Measure resistance of suspect wires.
CHECK
Ignition wire
Spark plug
Remove spark plugs and check their ignition sparks.
CHECK
Self-diagnosis Mode II
5 times or more/10 sec.
N.G.
Check mixture ratio by flashes of inspection lamps.
DISCONNECT
Exhaust gas sensor
Improved
Replace the sensor.
No change
Check mixture ratio feedback system.
Disconnect exhaust gas sensor connector, and try to drive.
PERFORM
Self-diagnosis Mode V
N.G.
Check the circuits.
Perform self-diagnosis Mode V (for air flow meter and crank angle sensor).

ECCS DIAGNOSTIC CHART—1989–90 240SX

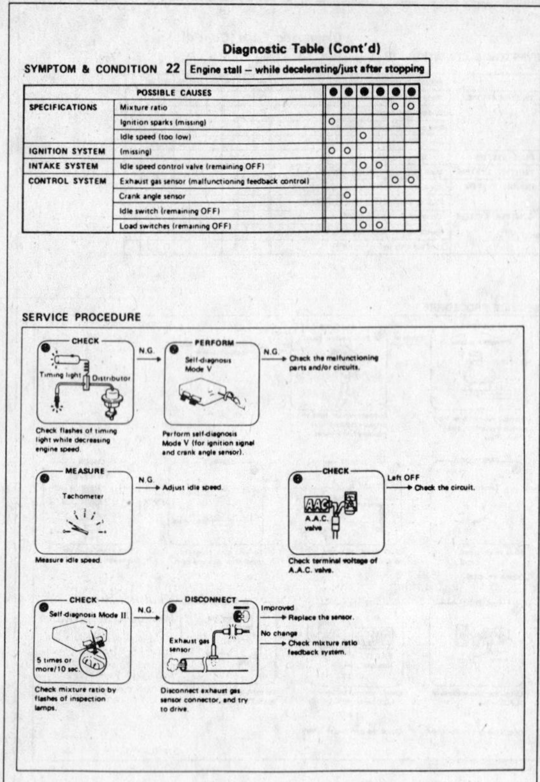

Diagnostic Table (Cont'd)

SYMPTOM & CONDITION 22 Engine stall – while decelerating/just after stopping

POSSIBLE CAUSES		●	●	●	●	●	●
SPECIFICATIONS	Mixture ratio					○	○
	Ignition sparks (missing)	○					
	Idle speed (too low)			○			
IGNITION SYSTEM	(missing)	○	○				
INTAKE SYSTEM	Idle speed control valve (remaining OFF)			○	○		
CONTROL SYSTEM	Exhaust gas sensor (malfunctioning feedback control)					○	○
	Crank angle sensor		○				
	Idle switch (remaining OFF)			○			
	Load switches (remaining OFF)			○	○		

SERVICE PROCEDURE

CHECK
Timing light
Distributor
N.G.
Check flashes of timing light while decreasing engine speed.
PERFORM
Self-diagnosis Mode V
N.G.
Check the malfunctioning parts and/or circuits.
Perform self-diagnosis Mode V (for ignition signal and crank angle sensor).
MEASURE
Tachometer
N.G.
Adjust idle speed.
Measure idle speed.
CHECK
AAC
A.A.C. valve
Left OFF
Check the circuit.
Check terminal voltage of A.A.C. valve.
CHECK
Self-diagnosis Mode II
5 times or more/10 sec.
N.G.
Check mixture ratio by flashes of inspection lamps.
DISCONNECT
Exhaust gas sensor
Improved
Replace the sensor.
No change
Check mixture ratio feedback system.
Disconnect exhaust gas sensor connector, and try to drive.

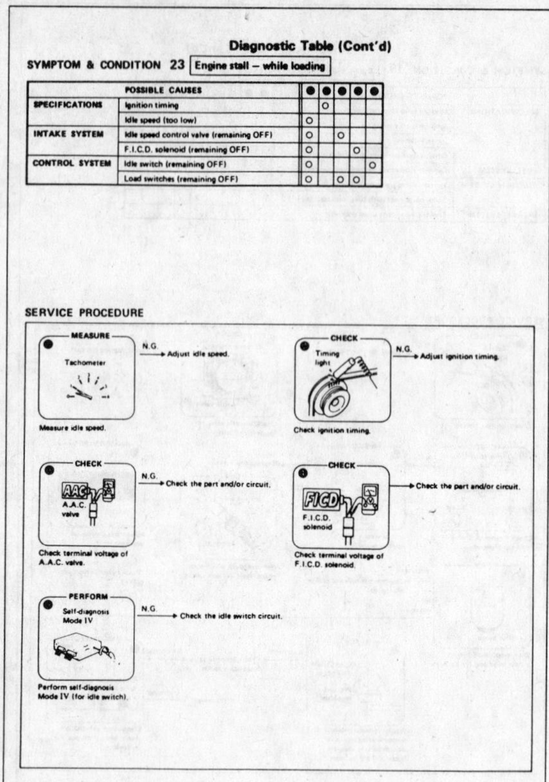

Diagnostic Table (Cont'd)

SYMPTOM & CONDITION 23 Engine stall – while loading

POSSIBLE CAUSES		●	●	●	●	●
SPECIFICATIONS	Ignition timing		○			
	Idle speed (too low)	○				
INTAKE SYSTEM	Idle speed control valve (remaining OFF)	○		○		
	F.I.C.D. solenoid (remaining OFF)	○			○	
CONTROL SYSTEM	Idle switch (remaining OFF)	○				○
	Load switches (remaining OFF)	○		○	○	

SERVICE PROCEDURE

MEASURE
Tachometer
N.G.
Adjust idle speed.
Measure idle speed.
CHECK
Timing light
N.G.
Adjust ignition timing.
Check ignition timing.
CHECK
AAC
A.A.C. valve
N.G.
Check the part and/or circuit.
Check terminal voltage of A.A.C. valve.
CHECK
FICD
F.I.C.D. solenoid
Check the part and/or circuit.
Check terminal voltage of F.I.C.D. solenoid.
PERFORM
Self-diagnosis Mode IV
N.G.
Check the idle switch circuit.
Perform self-diagnosis Mode IV (for idle switch).

ECCS DIAGNOSTIC CHART—1989–90 240SX

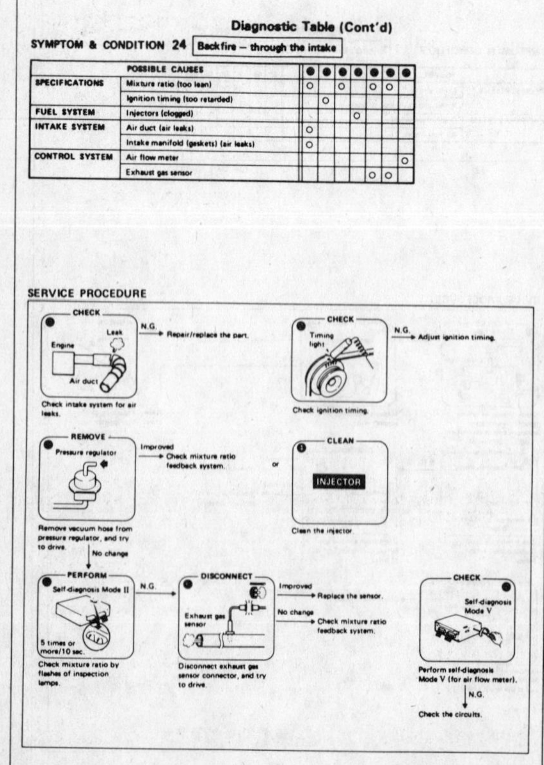

Diagnostic Table (Cont'd)

SYMPTOM & CONDITION 24 Backfire – through the intake

POSSIBLE CAUSES		●	●	●	●	●	●	●
SPECIFICATIONS	Mixture ratio (too lean)	○		○		○	○	
	Ignition timing (too retarded)		○					
FUEL SYSTEM	Injectors (clogged)				○			
INTAKE SYSTEM	Air duct (air leaks)	○						
	Intake manifold (gaskets) (air leaks)	○						
CONTROL SYSTEM	Air flow meter							○
	Exhaust gas sensor					○	○	

SERVICE PROCEDURE

CHECK
Engine
Leak
Air duct
N.G.
Repair/replace the part.
Check intake system for air leaks.
CHECK
Timing light
N.G.
Adjust ignition timing.
Check ignition timing.
REMOVE
Pressure regulator
Improved
Check mixture ratio feedback system.
or
Remove vacuum hose from pressure regulator, and try to drive.
No change
CLEAN
INJECTOR
Clean the injector.
PERFORM
Self-diagnosis Mode II
5 times or more/10 sec.
N.G.
Check mixture ratio by flashes of inspection lamps.
DISCONNECT
Exhaust gas sensor
Improved
Replace the sensor.
No change
Check mixture ratio feedback system.
Disconnect exhaust gas sensor connector, and try to drive.
CHECK
Self-diagnosis Mode V
Perform self-diagnosis Mode V (for air flow meter).
N.G.
Check the circuits.

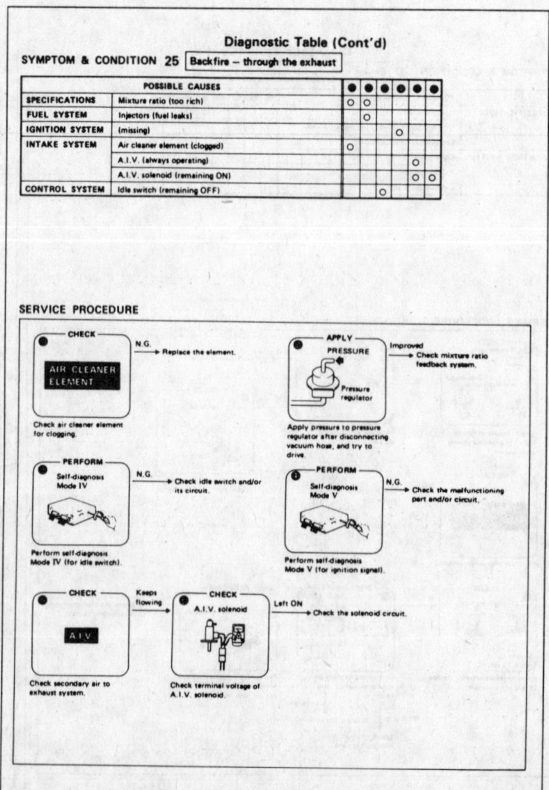

Diagnostic Table (Cont'd)

SYMPTOM & CONDITION 25 Backfire – through the exhaust

POSSIBLE CAUSES		●	●	●	●	●	●
SPECIFICATIONS	Mixture ratio (too rich)	○	○				
FUEL SYSTEM	Injectors (fuel leaks)		○				
IGNITION SYSTEM	(missing)				○		
INTAKE SYSTEM	Air cleaner element (clogged)	○					
	A.I.V. (always operating)					○	
	A.I.V. solenoid (remaining ON)					○	○
CONTROL SYSTEM	Idle switch (remaining OFF)			○			

SERVICE PROCEDURE

CHECK
AIR CLEANER ELEMENT
N.G.
Replace the element.
Check air cleaner element for clogging.
APPLY
PRESSURE
Pressure regulator
Improved
Check mixture ratio feedback system.
Apply pressure to pressure regulator after disconnecting vacuum hose, and try to drive.
PERFORM
Self-diagnosis Mode IV
N.G.
Check idle switch and/or its circuit.
Perform self-diagnosis Mode IV (for idle switch).
PERFORM
Self-diagnosis Mode V
N.G.
Check the malfunctioning part and/or circuit.
Perform self-diagnosis Mode V (for ignition signal).
CHECK
AIV
Keeps flowing
CHECK
A.I.V. solenoid
Left ON
Check the solenoid circuit.
Check secondary air to exhaust system.
Check terminal voltage of A.I.V. solenoid.

ECCS SELF-DIAGNOSTIC DESCRIPTION—1989–90 240SX

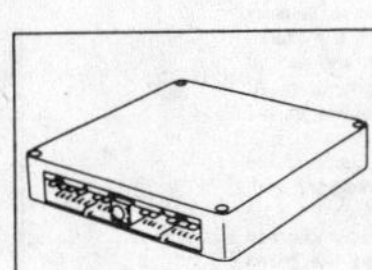

Self-diagnosis — Description

The self-diagnosis is useful to diagnose malfunctions in major sensors and actuators of the E.C.C.S. system. There are 5 modes in the self-diagnosis system.

1. **Mode I (Exhaust gas sensor monitor)**
   - During closed-loop operation:
     The green inspection lamp turns ON when a lean condition is detected and goes OFF under rich condition.
   - During open-loop operation condition:
     The green inspection lamp remains OFF or ON.
2. **Mode II (Mixture ratio feedback control monitor)**
   The green inspection lamp function is the same as Mode I.
   - During closed-loop operation:
     The red inspection lamp turns ON and OFF simultaneously with the green inspection lamp when the mixture ratio is controlled within the specified value.
   - During open-loop operation:
     The red inspection lamp remains ON or OFF.
3. **Mode III (Self-diagnostic system)**
   This mode is the same as the former self-diagnosis in self-diagnosis mode.
4. **Mode IV (Switches ON/OFF diagnostic system)**
   During this mode, the inspection lamps monitor the switch ON-OFF condition.
   - Idle switch
   - Starter switch
   - Vehicle speed sensor
5. **Mode V (Real-time diagnostic system)**
   The moment the malfunction is detected, the display will be presented immediately. That is, the condition at which the malfunction occurs can be found by observing the inspection lamps during driving test.

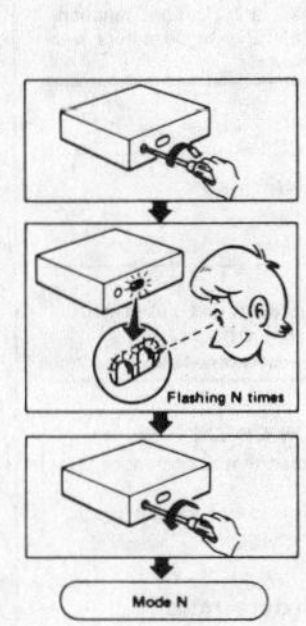
Flashing N times
Mode N

Self-diagnosis — Description (Cont'd)

HOW TO SWITCH THE DIAGNOSTIC MODES

1. Turn ignition switch "ON".
2. Turn diagnostic mode selector to E.C.U. (fully clockwise) and wait for inspection lamps to flash.
3. Count the number of flashes, and after the inspection lamps have flashed the number of the required mode, immediately turn diagnostic mode selector fully counterclockwise.

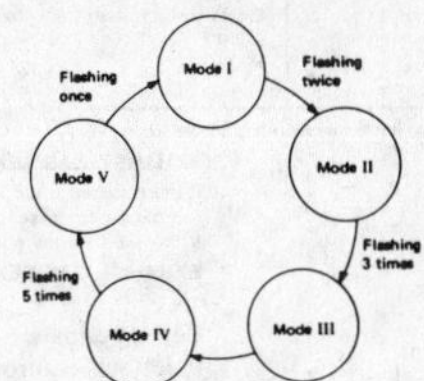
Mode I
Flashing twice
Mode II
Flashing 3 times
Mode III
Flashing 4 times
Mode IV
Flashing 5 times
Mode V
Flashing once

- **When the ignition switch is turned off during diagnosis in any mode and then turned on again (after power to the E.C.U. has dropped completely), the diagnosis will automatically return to Mode I.**
  The stored memory will be lost if:

1. Battery terminal is disconnected.
2. After selecting Mode III, Mode IV is selected.
   However, if the diagnostic mode selector is kept turned fully clockwise, it will continue to change in the order of Mode I → II → III → IV → V → I ... etc., and in this state the stored memory will not be erased.

This unit serves as an idle rpm feedback control. When the diagnostic mode selector is turned within the "diagnostic mode OFF" range, a target engine speed can be selected. Mark the original position of the selector before conducting self-diagnosis. Upon completion of self-diagnosis, return the selector to the previous position. Otherwise, engine speed may change before and after conducting self-diagnosis.

ECCS SELF-DIAGNOSTIC DESCRIPTION (CONT.)—1989–90 240SX

CHECK
CHECK ENGINE LIGHT

Self-diagnosis — Description (Cont'd)

CHECK ENGINE LIGHT (For California only)

This vehicle has a check engine light on the instrument panel. This light comes ON under the following conditions:

1) When ignition switch is turned "ON" (for bulb check).
2) When systems related to emission performance malfunction in Mode I (with engine running).
   - **This check engine light always illuminates and is synchronous with red L.E.D.**
   - **Malfunction systems related to emission performance can be detected by self-diagnosis, and they are clarified as self-diagnostic codes in Mode III.**
3) Check engine light will come "ON" only when malfunction is sensed.
   The check engine light will turn off when normal operation is resumed. Mode III memory must be cleared as the contents remain stored.

Code No.	Malfunction
12	Air flow meter circuit
13	Engine temperature sensor circuit
14	Vehicle speed sensor circuit
31	E.C.U. (E.C.C.S. control unit)
32	E.G.R. function
33	Exhaust gas sensor circuit
35	Exhaust gas temperature sensor circuit
43	Throttle sensor circuit
45	Injector leak

Use the following diagnostic flowchart to check and repair a malfunctioning system.

DIAGNOSIS START
↓
Turn ignition switch "ON" and make sure that check engine light comes "ON". — N.G. → Replace bulb.
↓ O.K.
Perform self-diagnosis and check which code is displayed in Mode III.
↓
Check electronic control system of affected code No. to locate faulty part.
↓
Repair or replace faulty part.
↓
Reinstall any part removed.
↓
Ⓐ

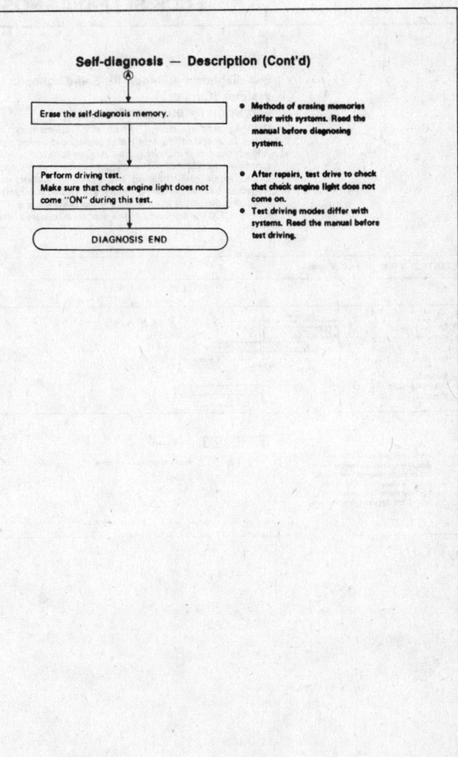
Self-diagnosis — Description (Cont'd)
Ⓐ
Erase the self-diagnosis memory.
• Methods of erasing memories differ with systems. Read the manual before diagnosing systems.
Perform driving test.
Make sure that check engine light does not come "ON" during this test.
• After repairs, test drive to check that check engine light does not come on.
• Test driving modes differ with systems. Read the manual before test driving.
DIAGNOSIS END

ECCS SELF-DIAGNOSTIC MODES I AND II 1989–90 240SX

Self-diagnosis — Mode I (Exhaust gas sensor monitor)

This mode checks the exhaust gas sensor for proper functioning. The operation of the E.C.U. L.E.D. in this mode differs with mixture ratio control conditions as follows:

Mode	L.E.D.	Engine stopped (Ignition switch "ON")	Engine running: Open loop condition	Engine running: Closed loop condition
Mode I (Monitor A)	Green	ON	*Remains ON or OFF	Blinks
	Red	ON	Except for California model • OFF	For California model • ON: when the CHECK ENGINE LIGHT ITEMS are stored in the E.C.U. • OFF: except for the above condition

*: Maintains conditions just before switching to open loop

EXHAUST GAS SENSOR FUNCTION CHECK

If the number of L.E.D. blinks is less than that specified, replace the exhaust gas sensor.
If the L.E.D. does not blink, check exhaust gas sensor circuit.

EXHAUST GAS SENSOR CIRCUIT CHECK

Self-diagnosis — Mode II (Mixture ratio feedback control monitor)

This mode checks, through the E.C.U. L.E.D., optimum control of the mixture ratio. The operation of the L.E.D., as shown below, differs with the control conditions of the mixture ratio (for example, richer or leaner mixture ratios, etc., which are controlled by the E.C.U.).

Mode	L.E.D.	Engine stopped (Ignition switch "ON")	Engine running: Open loop condition	Engine running: Closed loop condition		
Mode II (Monitor B)	Green	ON	*Remains ON or OFF	Blinks		
				Compensating mixture ratio		
	Red	OFF	*Remains ON or OFF (synchronous with green L.E.D.)	More than 5% rich	Between 5% lean and 5% rich	More than 5% lean
				OFF	Synchronized with green L.E.D.	Remains ON

*: Maintains conditions just before switching to open loop

If the red L.E.D. remains on or off during the closed-loop operation, the mixture ratio may not be controlled properly. Using the following procedures, check the related components or adjust the mixture ratio.

COMPONENT CHECK OR MIXTURE RATIO ADJUSTMENT

ECCS SELF-DIAGNOSTIC MODE III 1989–90 240SX

Self-diagnosis — Mode III (Self-diagnostic system)

The E.C.U. constantly monitors the function of these sensors and actuators, regardless of ignition key position. If a malfunction occurs, the information is stored in the E.C.U. and can be retrieved from the memory by turning on the diagnostic mode selector, located on the side of the E.C.U. When activated, the malfunction is indicated by flashing a red and a green L.E.D. (Light Emitting Diode), also located on the E.C.U. Since all the self-diagnostic results are stored in the E.C.U.'s memory even intermittent malfunctions can be diagnosed.

A malfunction is indicated by the number of both red and green flashing L.E.D.s. First, the red L.E.D. flashes and the green flashes follow. The red L.E.D. corresponds to units of ten and the green L.E.D. corresponds to units of one. For example, when the red L.E.D. flashes once and the green L.E.D. flashes twice, this signifies the number "12", showing that the air flow meter signal is malfunctioning. All problems are classified by code numbers in this way.

- **When the engine fails to start, crank it two or more seconds before beginning self-diagnosis.**
- **Before starting self-diagnosis, do not erase the stored memory before beginning self-diagnosis. If it is erased, the self-diagnosis function for intermittent malfunctions will be lost.**

DISPLAY CODE TABLE

Code No.	Detected items	California	Non-California
11	Crank angle sensor circuit	X	X
12	Air flow meter circuit	X	X
13	Engine temperature sensor circuit	X	X
14	Vehicle speed sensor circuit	X	X
21	Ignition signal missing in primary coil	X	X
31	E.C.U. (E.C.C.S. control unit)	X	X
32	E.G.R. function	X	–
33	Exhaust gas sensor circuit	X	X
35	Exhaust gas temperature sensor circuit	X	–
43	Throttle sensor circuit	X	X
45	Injector leak	X	–
55	No malfunction in the above circuit	X	X

X: Available –: Not available

ECCS SELF-DIAGNOSTIC MODE III (CONT.)—1989–90 240SX

Self-diagnosis — Mode III (Self-diagnostic system) (Cont'd)

RETENTION OF DIAGNOSTIC RESULTS

The diagnostic results will remain in E.C.U. memory until the starter is operated fifty times after a diagnostic item has been judged to be malfunctioning. The diagnostic result will then be cancelled automatically. If a diagnostic item which has been judged to be malfunctioning and stored in memory is again judged to be malfunctioning before the starter is operated fifty times, the second result will replace the previous one. It will be stored in E.C.U. memory until the starter is operated fifty times more.

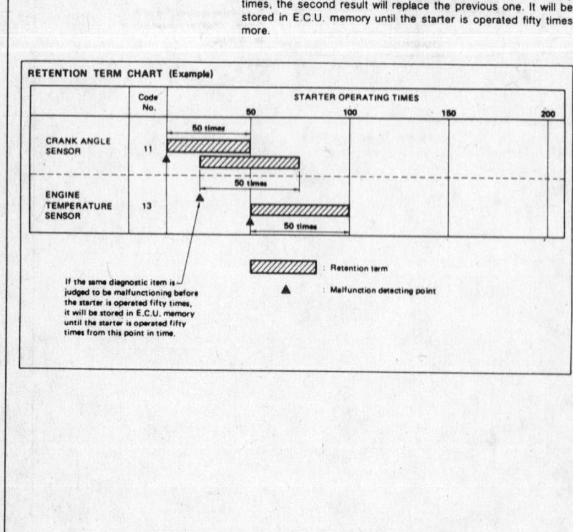
RETENTION TERM CHART (Example)
Code No.
STARTER OPERATING TIMES
50
100
150
200
50 times
CRANK ANGLE SENSOR
11
50 times
ENGINE TEMPERATURE SENSOR
13
50 times
If the same diagnostic item is judged to be malfunctioning before the starter is operated fifty times, it will be stored in E.C.U. memory until the starter is operated fifty times from this point in time.
: Retention term
: Malfunction detecting point

Self-diagnosis — Mode III (Self-diagnostic system) (Cont'd)

SELF-DIAGNOSTIC PROCEDURE

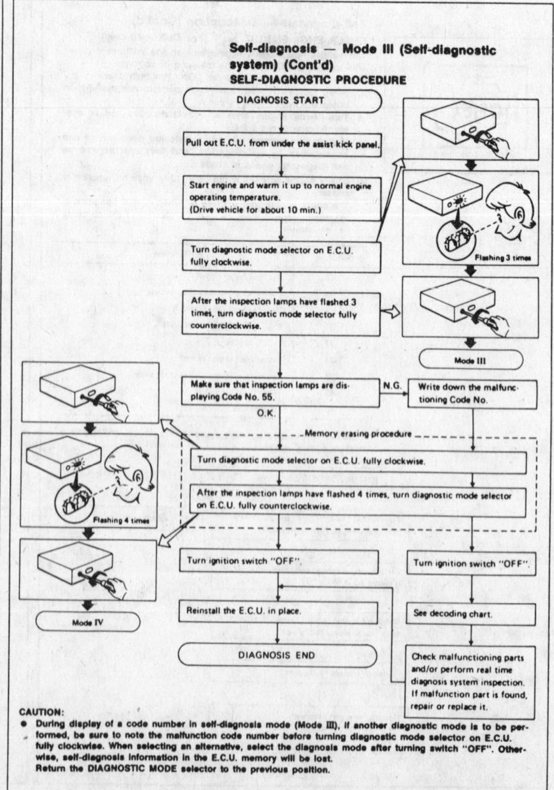
DIAGNOSIS START
Pull out E.C.U. from under the assist kick panel.
Start engine and warm it up to normal engine operating temperature. (Drive vehicle for about 10 min.)
Turn diagnostic mode selector on E.C.U. fully clockwise.
Flashing 3 times
After the inspection lamps have flashed 3 times, turn diagnostic mode selector fully counterclockwise.
Mode III
Make sure that inspection lamps are displaying Code No. 55.
N.G.
Write down the malfunctioning Code No.
O.K.
Memory erasing procedure
Turn diagnostic mode selector on E.C.U. fully clockwise.
After the inspection lamps have flashed 4 times, turn diagnostic mode selector on E.C.U. fully counterclockwise.
Flashing 4 times
Turn ignition switch "OFF".
Turn ignition switch "OFF".
Mode IV
Reinstall the E.C.U. in place.
See decoding chart.
DIAGNOSIS END
Check malfunctioning parts and/or perform real time diagnosis system inspection. If malfunction part is found, repair or replace it.

CAUTION:

- **During display of a code number in self-diagnosis mode (Mode III), if another diagnostic mode is to be performed, be sure to note the malfunction code number before turning diagnostic mode selector on E.C.U. fully clockwise. When selecting an alternative, select the diagnosis mode after turning switch "OFF". Otherwise, self-diagnosis information in the E.C.U. memory will be lost. Return the DIAGNOSTIC MODE selector to the previous position.**

ECCS SELF-DIAGNOSTIC MODE III DISPLAY CODES—1989-90 240SX

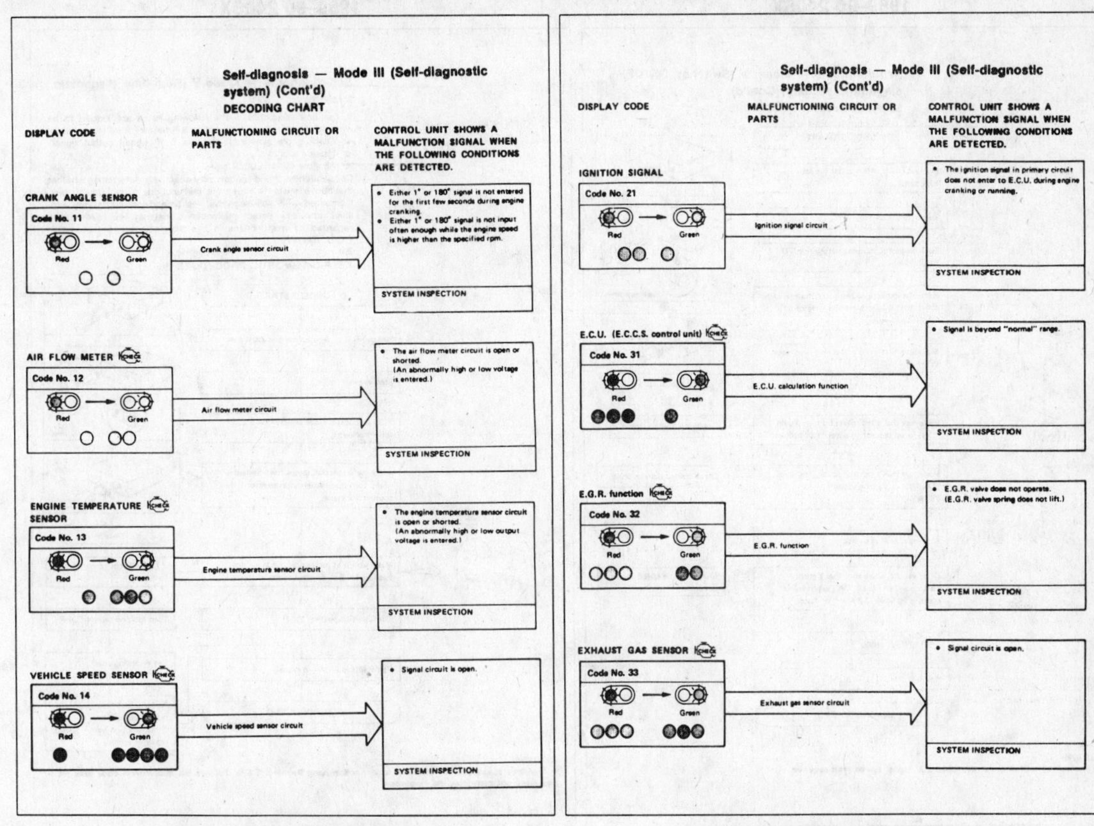

Self-diagnosis — Mode III (Self-diagnostic system) (Cont'd)
DECODING CHART
DISPLAY CODE
MALFUNCTIONING CIRCUIT OR PARTS
CONTROL UNIT SHOWS A MALFUNCTION SIGNAL WHEN THE FOLLOWING CONDITIONS ARE DETECTED.
CRANK ANGLE SENSOR
Code No. 11
Red
Green
Crank angle sensor circuit
• Either 1° or 180° signal is not entered for the first few seconds during engine cranking.
• Either 1° or 180° signal is not input often enough while the engine speed is higher than the specified rpm.
SYSTEM INSPECTION
AIR FLOW METER
Code No. 12
Air flow meter circuit
• The air flow meter circuit is open or shorted. (An abnormally high or low voltage is entered.)
ENGINE TEMPERATURE SENSOR
Code No. 13
Engine temperature sensor circuit
• The engine temperature sensor circuit is open or shorted. (An abnormally high or low output voltage is entered.)
VEHICLE SPEED SENSOR
Code No. 14
Vehicle speed sensor circuit
• Signal circuit is open.
Self-diagnosis — Mode III (Self-diagnostic system) (Cont'd)
IGNITION SIGNAL
Code No. 21
Ignition signal circuit
• The ignition signal in primary circuit does not enter to E.C.U. during engine cranking or running.
E.C.U. (E.C.C.S. control unit)
Code No. 31
E.C.U. calculation function
• Signal is beyond "normal" range.
E.G.R. function
Code No. 32
E.G.R. function
• E.G.R. valve does not operate. (E.G.R. valve spring does not lift.)
EXHAUST GAS SENSOR
Code No. 33
Exhaust gas sensor circuit
• Signal circuit is open.

ECCS SELF-DIAGNOSTIC MODE III DISPLAY CODES (CONT.) 1989-90 240SX

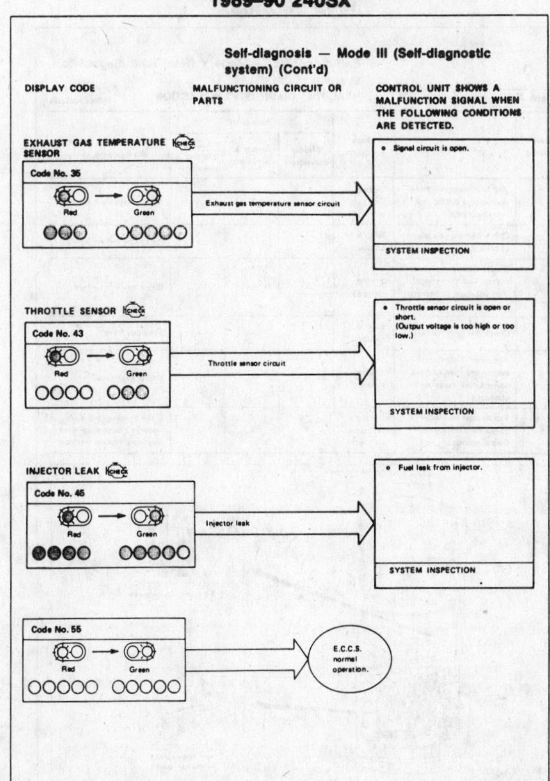

Self-diagnosis — Mode III (Self-diagnostic system) (Cont'd)
DISPLAY CODE
MALFUNCTIONING CIRCUIT OR PARTS
CONTROL UNIT SHOWS A MALFUNCTION SIGNAL WHEN THE FOLLOWING CONDITIONS ARE DETECTED.
EXHAUST GAS TEMPERATURE SENSOR
Code No. 35
Red
Green
Exhaust gas temperature sensor circuit
• Signal circuit is open.
SYSTEM INSPECTION
THROTTLE SENSOR
Code No. 43
Throttle sensor circuit
• Throttle sensor circuit is open or short. (Output voltage is too high or too low.)
INJECTOR LEAK
Code No. 45
Injector leak
• Fuel leak from injector.
Code No. 55
E.C.C.S. normal operation.

ECCS SELF-DIAGNOSTIC MODE IV 1989-90 240SX

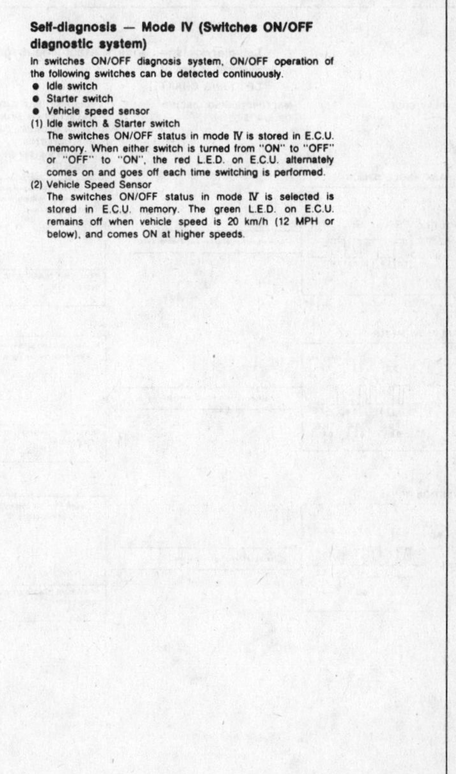

Self-diagnosis — Mode IV (Switches ON/OFF diagnostic system)
In switches ON/OFF diagnosis system, ON/OFF operation of the following switches can be detected continuously.
● Idle switch
● Starter switch
● Vehicle speed sensor
(1) Idle switch & Starter switch
The switches ON/OFF status in mode IV is stored in E.C.U. memory. When either switch is turned from "ON" to "OFF" or "OFF" to "ON", the red L.E.D. on E.C.U. alternately comes on and goes off each time switching is performed.
(2) Vehicle Speed Sensor
The switches ON/OFF status in mode IV is selected is stored in E.C.U. memory. The green L.E.D. on E.C.U. remains off when vehicle speed is 20 km/h (12 MPH or below), and comes ON at higher speeds.

ECCS SELF-DIAGNOSTIC MODE IV (CONT.) 1989–90 240SX

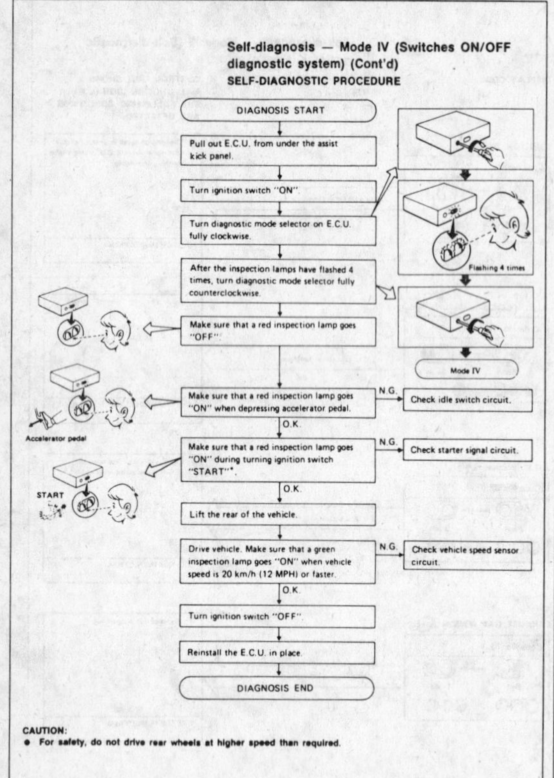
Self-diagnosis — Mode IV (Switches ON/OFF diagnostic system) (Cont'd)
SELF-DIAGNOSTIC PROCEDURE
DIAGNOSIS START
Pull out E.C.U. from under the assist kick panel.
Turn ignition switch "ON".
Turn diagnostic mode selector on E.C.U. fully clockwise.
After the inspection lamps have flashed 4 times, turn diagnostic mode selector fully counterclockwise.
Flashing 4 times
Mode IV
Make sure that a red inspection lamp goes "OFF".
Make sure that a red inspection lamp goes "ON" when depressing accelerator pedal.
N.G.
Check idle switch circuit.
O.K.
Accelerator pedal
Make sure that a red inspection lamp goes "ON" during turning ignition switch "START".
N.G.
Check starter signal circuit.
O.K.
START
Lift the rear of the vehicle.
Drive vehicle. Make sure that a green inspection lamp goes "ON" when vehicle speed is 20 km/h (12 MPH) or faster.
N.G.
Check vehicle speed sensor circuit.
O.K.
Turn ignition switch "OFF".
Reinstall the E.C.U. in place.
DIAGNOSIS END
CAUTION:
For safety, do not drive rear wheels at higher speed than required.

ECCS SELF-DIAGNOSTIC MODE V 1989–90 240SX

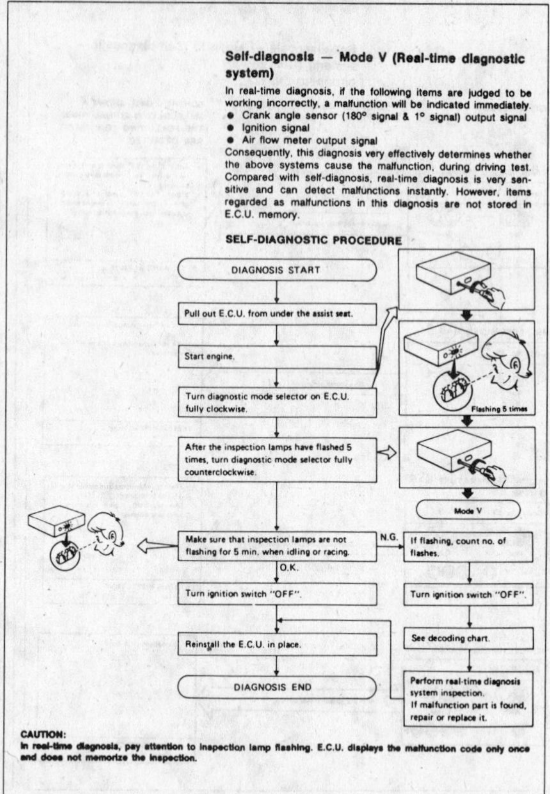
Self-diagnosis — Mode V (Real-time diagnostic system)
In real-time diagnosis, if the following items are judged to be working incorrectly, a malfunction will be indicated immediately.
Crank angle sensor (180° signal & 1° signal) output signal
Ignition signal
Air flow meter output signal
Consequently, this diagnosis very effectively determines whether the above systems cause the malfunction, during driving test. Compared with self-diagnosis, real-time diagnosis is very sensitive and can detect malfunctions instantly. However, items regarded as malfunctions in this diagnosis are not stored in E.C.U. memory.
SELF-DIAGNOSTIC PROCEDURE
DIAGNOSIS START
Pull out E.C.U. from under the assist seat.
Start engine.
Turn diagnostic mode selector on E.C.U. fully clockwise.
After the inspection lamps have flashed 5 times, turn diagnostic mode selector fully counterclockwise.
Flashing 5 times
Mode V
Make sure that inspection lamps are not flashing for 5 min. when idling or racing.
N.G.
If flashing, count no. of flashes.
Turn ignition switch "OFF".
See decoding chart.
Perform real-time diagnosis system inspection. If malfunction part is found, repair or replace it.
O.K.
Turn ignition switch "OFF".
Reinstall the E.C.U. in place.
DIAGNOSIS END
CAUTION:
In real-time diagnosis, pay attention to inspection lamp flashing. E.C.U. displays the malfunction code only once and does not memorize the inspection.

ECCS SELF-DIAGNOSTIC MODE V DECODING CHART 1989–90 240SX

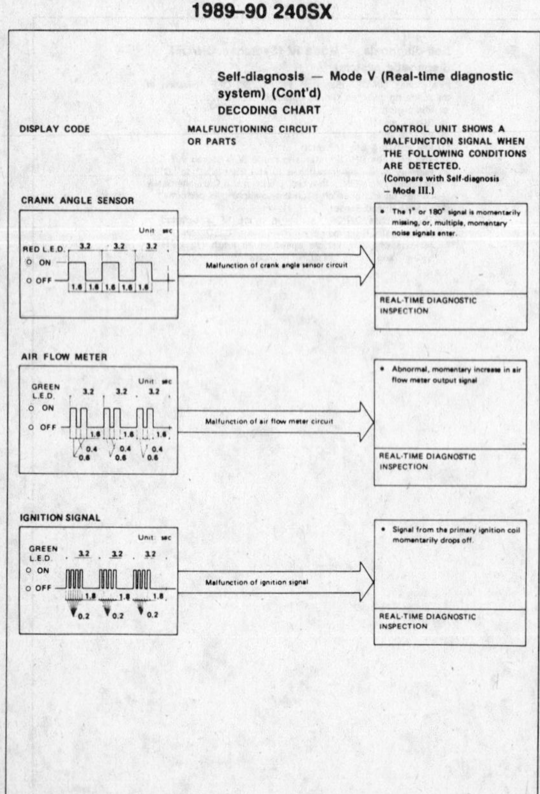
Self-diagnosis — Mode V (Real-time diagnostic system) (Cont'd)
DECODING CHART
DISPLAY CODE
MALFUNCTIONING CIRCUIT OR PARTS
CONTROL UNIT SHOWS A MALFUNCTION SIGNAL WHEN THE FOLLOWING CONDITIONS ARE DETECTED. (Compare with Self-diagnosis – Mode III.)
CRANK ANGLE SENSOR
Unit: sec
RED L.E.D.
ON
OFF
3.2 3.2 3.2
1.6 1.6 1.6 1.6 1.6
Malfunction of crank angle sensor circuit
The 1° or 180° signal is momentarily missing, or, multiple, momentary noise signals enter.
REAL-TIME DIAGNOSTIC INSPECTION
AIR FLOW METER
Unit: sec
GREEN L.E.D.
ON
OFF
3.2 3.2 3.2
1.6 1.6 1.6
0.4 0.4 0.4
0.6 0.6 0.6
Malfunction of air flow meter circuit
Abnormal, momentary increase in air flow meter output signal
REAL-TIME DIAGNOSTIC INSPECTION
IGNITION SIGNAL
Unit: sec
GREEN L.E.D.
ON
OFF
3.2 3.2 3.2
1.8 1.8 1.8
0.2 0.2 0.2
Malfunction of ignition signal
Signal from the primary ignition coil momentarily drops off.
REAL-TIME DIAGNOSTIC INSPECTION

ECCS SELF-DIAGNOSTIC MODE V INSPECTION 1989–90 240SX

Self-diagnosis — Mode V (Real-time diagnostic system) (Cont'd)

REAL-TIME DIAGNOSTIC INSPECTION

Crank Angle Sensor

X: Available
–: Not available

Check sequence	Check items	Check conditions	Check parts: Middle connectors	Check parts: Sensor & actuator	Check parts: E.C.U. harness connector	If malfunction, perform the following items.
1	Tap harness connector or component during real-time diagnosis.	During real-time diagnosis	X	X	X	Go to check item 2.
2	Check harness continuity at connector.	Engine stopped	X	–	–	Go to check item 3.
3	Disconnect harness connector, and then check dust adhesion to harness connector.	Engine stopped	X	–	X	Clean terminal surface.
4	Check pin terminal bend.	Engine stopped	–	–	X	Take out bend.
5	Reconnect harness connector and then recheck harness continuity at connector.	Engine stopped	X	–	–	Replace terminal.
6	Tap harness connector or component during real-time diagnosis.	During real-time diagnosis	X	X	X	If malfunction codes are displayed during real-time diagnosis, replace terminal.

E.C.U. harness connector

Crank angle sensor harness connector

ECCS SELF—DIAGNOSTIC MODE V INPSECTION—1989–90 240SX

Self-diagnosis — Mode V (Real-time diagnostic system) (Cont'd)

Air Flow Meter

X: Available
–: Not available

Check sequence	Check items	Check conditions	Check parts			If malfunction, perform the following items.
			Middle connectors	Sensor & actuator	E.C.U. harness connector	
1	Tap harness connector or component during real-time diagnosis.	During real-time diagnosis	X	X	X	Go to check item 2.
2	Check harness continuity at connector.	Engine stopped	X	–	–	Go to check item 3.
3	Disconnect harness connector, and then check dust adhesion to harness connector.	Engine stopped	X	–	X	Clean terminal surface.
4	Check pin terminal bend.	Engine stopped	–	–	X	Take out bend.
5	Reconnect harness connector and then recheck harness continuity at connector.	Engine stopped	X	–	–	Replace terminal.
6	Tap harness connector or component during real-time diagnosis.	During real-time diagnosis	X	X	X	If malfunction codes are displayed during real-time diagnosis, replace terminal.

E.C.U. harness connector
Air flow meter harness connector

Self-diagnosis — Mode V (Real-time diagnostic system) (Cont'd)

Ignition Signal

X: Available
–: Not available

Check sequence	Check items	Check conditions	Check parts			If malfunction, perform the following items.
			Middle connectors	Sensor & actuator	E.C.U. harness connector	
1	Tap harness connector or component during real-time diagnosis.	During real-time diagnosis	X	X	X	Go to check item 2.
2	Check harness continuity at connector.	Engine stopped	X	–	–	Go to check item 3.
3	Disconnect harness connector, and then check dust adhesion to harness connector.	Engine stopped	X	–	X	Clean terminal surface.
4	Check pin terminal bend.	Engine stopped	–	–	X	Take out bend.
5	Reconnect harness connector and then recheck harness continuity at connector.	Engine stopped	X	–	–	Replace terminal.
6	Tap harness connector or component during real-time diagnosis.	During real-time diagnosis	X	X	X	If malfunction codes are displayed during real-time diagnosis, replace terminal.

E.C.U. harness connector
Ignition coil and power transistor harness connector

ECCS SYSTEM CAUTION—1989–90 240SX

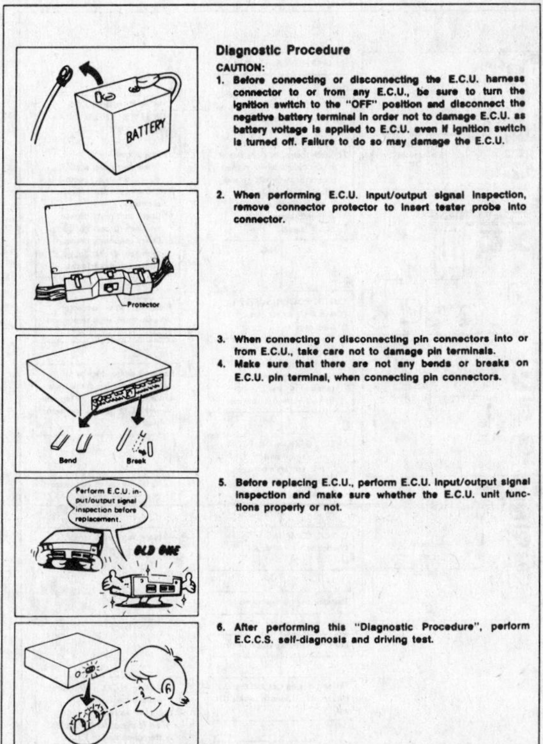
BATTERY
Protector
Bend
Break
Perform E.C.U. input/output signal inspection before replacement.
OLD ONE

Diagnostic Procedure

CAUTION:

1. **Before connecting or disconnecting the E.C.U. harness connector to or from any E.C.U., be sure to turn the ignition switch to the "OFF" position and disconnect the negative battery terminal in order not to damage E.C.U. as battery voltage is applied to E.C.U. even if ignition switch is turned off. Failure to do so may damage the E.C.U.**
2. **When performing E.C.U. input/output signal inspection, remove connector protector to insert tester probe into connector.**
3. **When connecting or disconnecting pin connectors into or from E.C.U., take care not to damage pin terminals.**
4. **Make sure that there are not any bends or breaks on E.C.U. pin terminal, when connecting pin connectors.**
5. **Before replacing E.C.U., perform E.C.U. input/output signal inspection and make sure whether the E.C.U. unit functions properly or not.**
6. **After performing this "Diagnostic Procedure", perform E.C.C.S. self-diagnosis and driving test.**

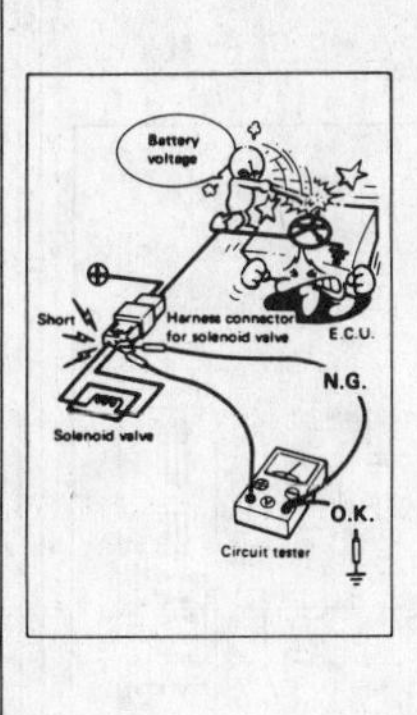
Battery voltage
Short
Harness connector for solenoid valve
E.C.U.
N.G.
Solenoid valve
O.K.
Circuit tester

Diagnostic Procedure (Cont'd)

7. **When measuring E.C.U. controlled components supply voltage with a circuit tester, separate one tester probe from the other.**
   If the two tester probes accidentally make contact with each other during measurement, the circuit will be shorted, resulting in damage to the control unit power transistor.

ECCS SELF-DIAGNOSTIC CHART—1989–90 240SX

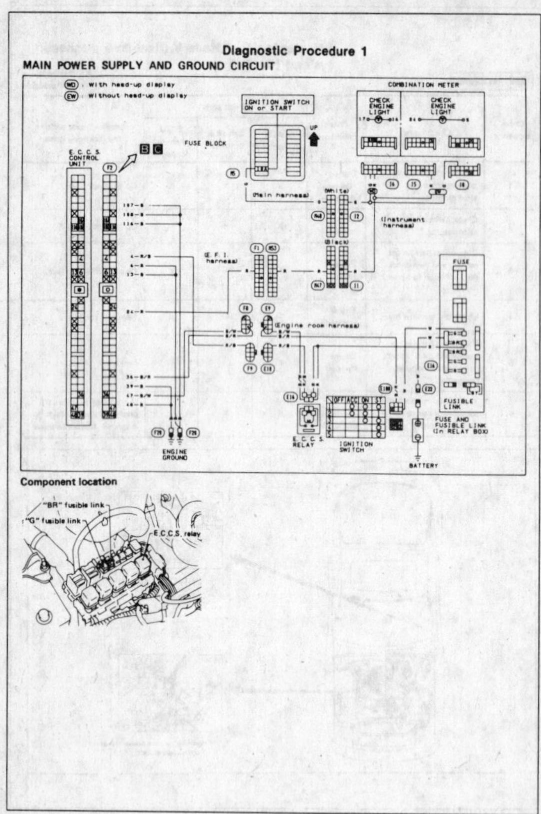
Diagnostic Procedure 1
MAIN POWER SUPPLY AND GROUND CIRCUIT
With head-up display
Without head-up display
E.C.C.S. CONTROL UNIT
FUSE BLOCK
IGNITION SWITCH ON or START
COMBINATION METER
CHECK ENGINE LIGHT
(Main harness)
(White)
(Black)
(Instrument harness)
(E.F.I. harness)
(Engine room harness)
FUSE
FUSIBLE LINK
FUSE AND FUSIBLE LINK (In RELAY BOX)
E.C.C.S. RELAY
IGNITION SWITCH
ENGINE GROUND
BATTERY
Component location
"BR" fusible link
"G" fusible link
E.C.C.S. relay

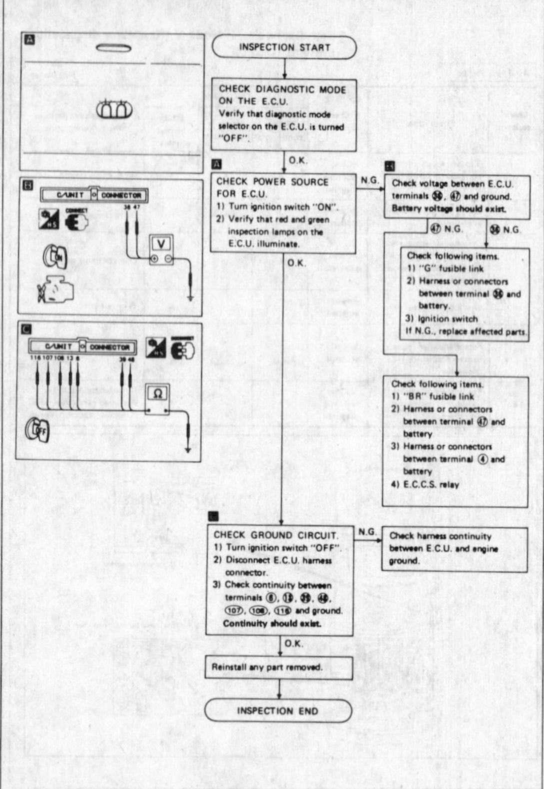
C/UNIT CONNECTOR
INSPECTION START
CHECK DIAGNOSTIC MODE ON THE E.C.U.
Verify that diagnostic mode selector on the E.C.U. is turned "OFF".
O.K.
A
CHECK POWER SOURCE FOR E.C.U.
1) Turn ignition switch "ON".
2) Verify that red and green inspection lamps on the E.C.U. illuminate.
N.G.
B
Check voltage between E.C.U. terminals ㊳, ㊼ and ground.
Battery voltage should exist.
㊼ N.G.
㊳ N.G.
Check following items.
1) "G" fusible link
2) Harness or connectors between terminal ㊳ and battery.
3) Ignition switch
If N.G., replace affected parts.
Check following items.
1) "BR" fusible link
2) Harness or connectors between terminal ㊼ and battery
3) Harness or connectors between terminal ④ and battery
4) E.C.C.S. relay
O.K.
C
CHECK GROUND CIRCUIT.
1) Turn ignition switch "OFF".
2) Disconnect E.C.U. harness connector.
3) Check continuity between terminals ⑧, ⑬, ㉟, ㊹, (107), (108), (116) and ground.
Continuity should exist.
N.G.
Check harness continuity between E.C.U. and engine ground.
O.K.
Reinstall any part removed.
INSPECTION END

ECCS SELF-DIAGNOSTIC CHART—1989–90 240SX

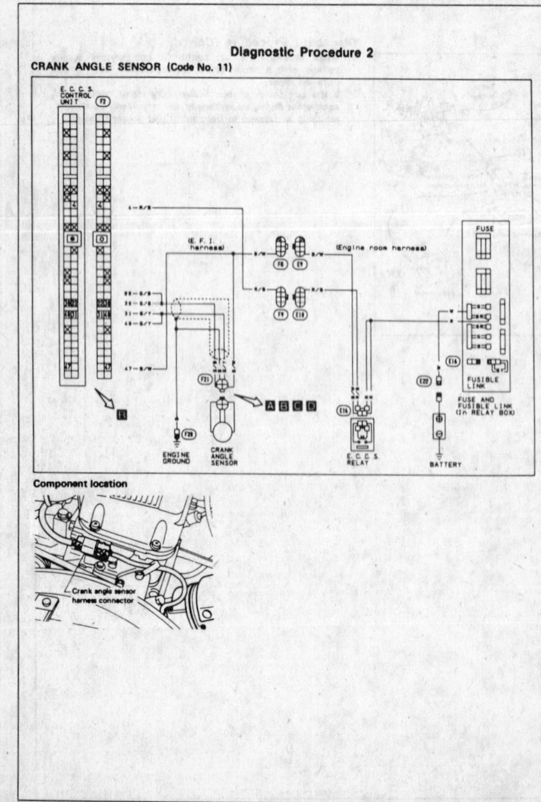
Diagnostic Procedure 2
CRANK ANGLE SENSOR (Code No. 11)
E.C.C.S. CONTROL UNIT
(E.F.I. harness)
(Engine room harness)
FUSE
FUSIBLE LINK
FUSE AND FUSIBLE LINK (In RELAY BOX)
ENGINE GROUND
CRANK ANGLE SENSOR
E.C.C.S. RELAY
BATTERY
Component location
Crank angle sensor harness connector

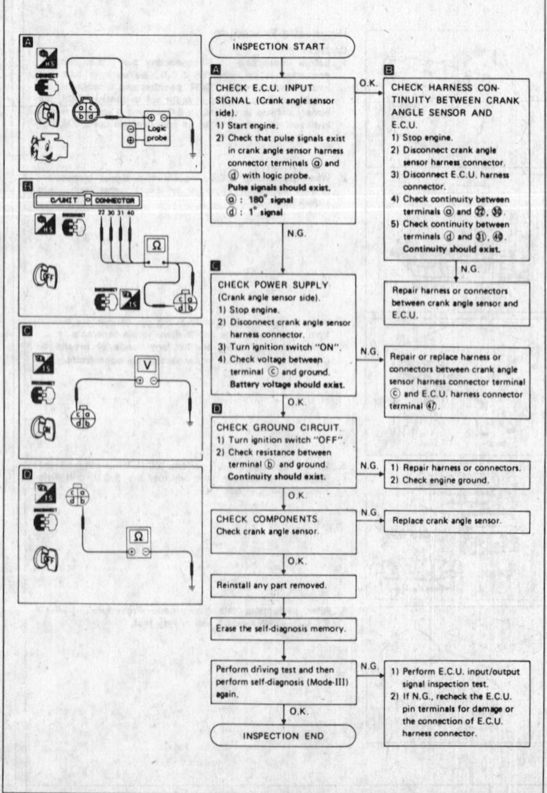
C/UNIT CONNECTOR
Logic probe
INSPECTION START
A
CHECK E.C.U. INPUT SIGNAL (Crank angle sensor side).
1) Start engine.
2) Check that pulse signals exist in crank angle sensor harness connector terminals ⓐ and ⓓ with logic probe.
Pulse signals should exist.
ⓐ : 180° signal
ⓓ : 1° signal
O.K.
B
CHECK HARNESS CONTINUITY BETWEEN CRANK ANGLE SENSOR AND E.C.U.
1) Stop engine.
2) Disconnect crank angle sensor harness connector.
3) Disconnect E.C.U. harness connector.
4) Check continuity between terminals ⓐ and ㉒, ㉚.
5) Check continuity between terminals ⓓ and ㉛, ㊵.
Continuity should exist.
N.G.
Repair harness or connectors between crank angle sensor and E.C.U.
N.G.
C
CHECK POWER SUPPLY (Crank angle sensor side).
1) Stop engine.
2) Disconnect crank angle sensor harness connector.
3) Turn ignition switch "ON".
4) Check voltage between terminal ⓒ and ground.
Battery voltage should exist.
N.G.
Repair or replace harness or connectors between crank angle sensor harness connector terminal ⓒ and E.C.U. harness connector terminal ㊼.
O.K.
D
CHECK GROUND CIRCUIT.
1) Turn ignition switch "OFF".
2) Check resistance between terminal ⓑ and ground.
Continuity should exist.
N.G.
1) Repair harness or connectors.
2) Check engine ground.
O.K.
CHECK COMPONENTS.
Check crank angle sensor.
N.G.
Replace crank angle sensor.
O.K.
Reinstall any part removed.
Erase the self-diagnosis memory.
Perform driving test and then perform self-diagnosis (Mode-III) again.
N.G.
1) Perform E.C.U. input/output signal inspection test.
2) If N.G., recheck the E.C.U. pin terminals for damage or the connection of E.C.U. harness connector.
O.K.
INSPECTION END

ECCS SELF-DIAGNOSTIC CHART—1989–90 240SX

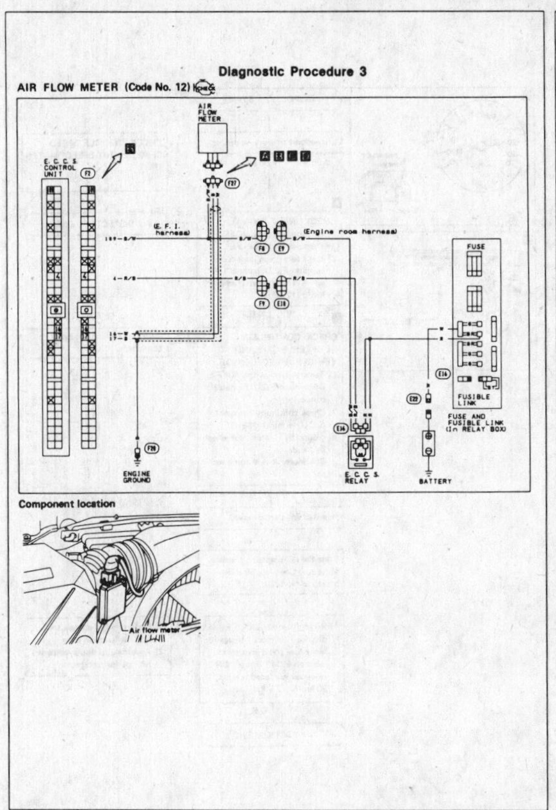
Diagnostic Procedure 3
AIR FLOW METER (Code No. 12)
E.C.C.S. CONTROL UNIT
AIR FLOW METER
(E.F.I. harness)
(Engine room harness)
FUSE
FUSIBLE LINK
FUSE AND FUSIBLE LINK (In RELAY BOX)
E.C.C.S. RELAY
BATTERY
ENGINE GROUND
Component location
Air flow meter

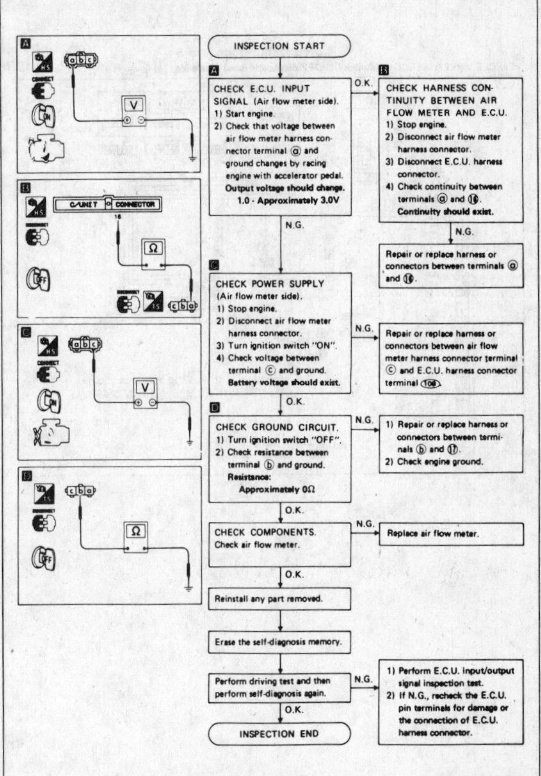
INSPECTION START
CHECK E.C.U. INPUT SIGNAL (Air flow meter side).
1) Start engine.
2) Check that voltage between air flow meter harness connector terminal ⓐ and ground changes by racing engine with accelerator pedal.
Output voltage should change.
1.0 - Approximately 3.0V
O.K.
CHECK HARNESS CONTINUITY BETWEEN AIR FLOW METER AND E.C.U.
1) Stop engine.
2) Disconnect air flow meter harness connector.
3) Disconnect E.C.U. harness connector.
4) Check continuity between terminals ⓐ and ⑯.
Continuity should exist.
N.G.
Repair or replace harness or connectors between terminals ⓐ and ⑯.
N.G.
CHECK POWER SUPPLY (Air flow meter side).
1) Stop engine.
2) Disconnect air flow meter harness connector.
3) Turn ignition switch "ON".
4) Check voltage between terminal ⓒ and ground.
Battery voltage should exist.
N.G.
Repair or replace harness or connectors between air flow meter harness connector terminal ⓒ and E.C.U. harness connector terminal ⑩⑧.
O.K.
CHECK GROUND CIRCUIT.
1) Turn ignition switch "OFF".
2) Check resistance between terminal ⓑ and ground.
Resistance:
Approximately 0Ω
N.G.
1) Repair or replace harness or connectors between terminals ⓑ and ⑪.
2) Check engine ground.
O.K.
CHECK COMPONENTS.
Check air flow meter.
N.G.
Replace air flow meter.
O.K.
Reinstall any part removed.
Erase the self-diagnosis memory.
Perform driving test and then perform self-diagnosis again.
N.G.
1) Perform E.C.U. input/output signal inspection test.
2) If N.G., recheck the E.C.U. pin terminals for damage or the connection of E.C.U. harness connector.
O.K.
INSPECTION END
C/UNIT CONNECTOR

ECCS SELF-DIAGNOSTIC CHART—1989–90 240SX

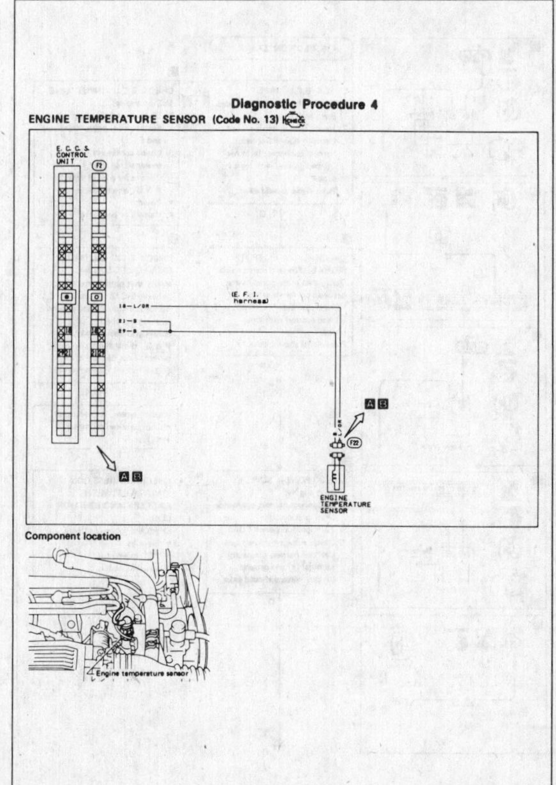
Diagnostic Procedure 4
ENGINE TEMPERATURE SENSOR (Code No. 13)
E.C.C.S. CONTROL UNIT
(E.F.I. harness)
ENGINE TEMPERATURE SENSOR
Component location
Engine temperature sensor

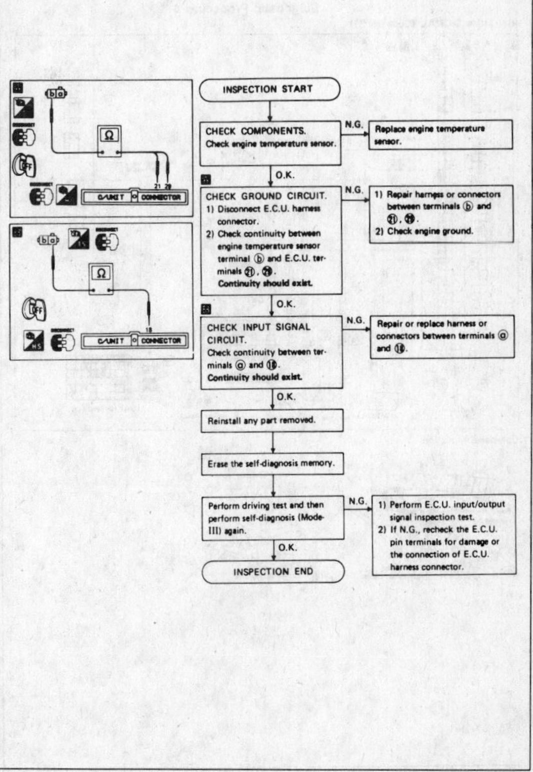
INSPECTION START
CHECK COMPONENTS.
Check engine temperature sensor.
N.G.
Replace engine temperature sensor.
O.K.
CHECK GROUND CIRCUIT.
1) Disconnect E.C.U. harness connector.
2) Check continuity between engine temperature sensor terminal ⓑ and E.C.U. terminals ㉑, ㉘.
Continuity should exist.
N.G.
1) Repair harness or connectors between terminals ⓑ and ㉑, ㉘.
2) Check engine ground.
O.K.
CHECK INPUT SIGNAL CIRCUIT.
Check continuity between terminals ⓐ and ⑱.
Continuity should exist.
N.G.
Repair or replace harness or connectors between terminals ⓐ and ⑱.
O.K.
Reinstall any part removed.
Erase the self-diagnosis memory.
Perform driving test and then perform self-diagnosis (Mode-III) again.
N.G.
1) Perform E.C.U. input/output signal inspection test.
2) If N.G., recheck the E.C.U. pin terminals for damage or the connection of E.C.U. harness connector.
O.K.
INSPECTION END
C/UNIT CONNECTOR

ECCS SELF-DIAGNOSTIC CHART – 1989-90 240SX

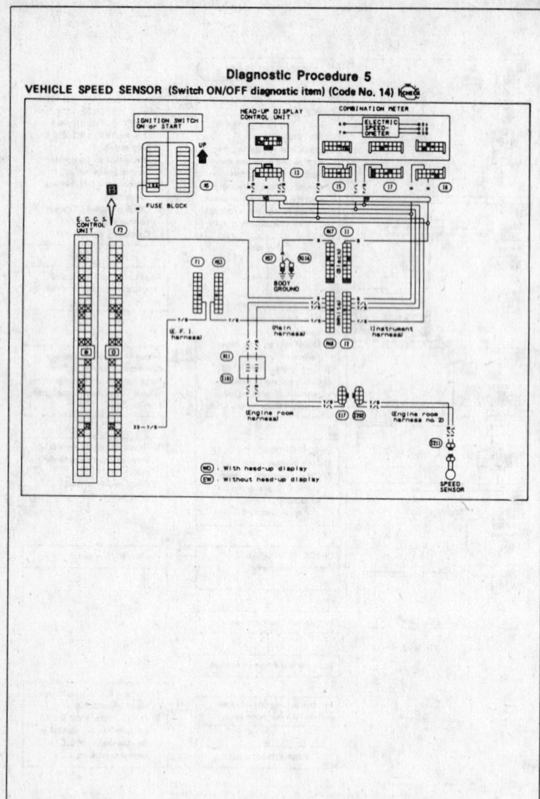
Diagnostic Procedure 5
VEHICLE SPEED SENSOR (Switch ON/OFF diagnostic item) (Code No. 14)
E.C.C.S. CONTROL UNIT
FUSE BLOCK
HEAD-UP DISPLAY CONTROL UNIT
COMBINATION METER
BODY GROUND
SPEED SENSOR
With head-up display
Without head-up display

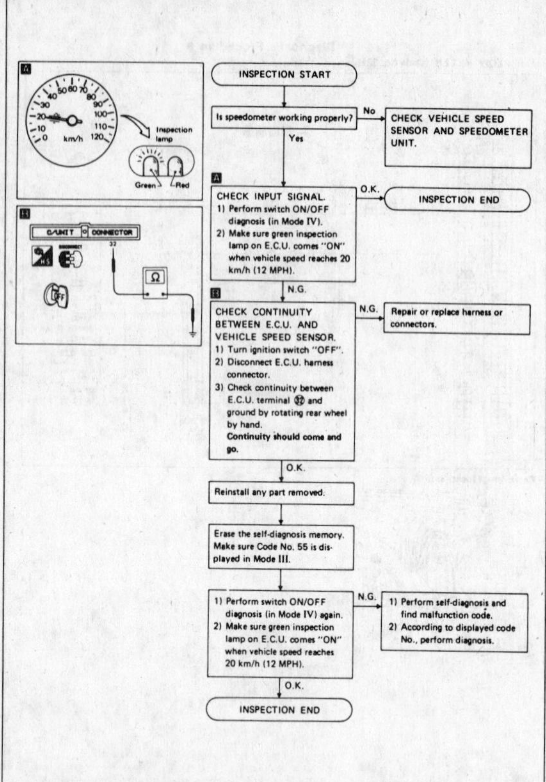
km/h
Inspection lamp
Green
Red
INSPECTION START
Is speedometer working properly?
No
CHECK VEHICLE SPEED SENSOR AND SPEEDOMETER UNIT.
Yes
CHECK INPUT SIGNAL.
1) Perform switch ON/OFF diagnosis (in Mode IV).
2) Make sure green inspection lamp on E.C.U. comes "ON" when vehicle speed reaches 20 km/h (12 MPH).
O.K.
INSPECTION END
N.G.
CHECK CONTINUITY BETWEEN E.C.U. AND VEHICLE SPEED SENSOR.
1) Turn ignition switch "OFF".
2) Disconnect E.C.U. harness connector.
3) Check continuity between E.C.U. terminal ㉜ and ground by rotating rear wheel by hand.
Continuity should come and go.
N.G.
Repair or replace harness or connectors.
O.K.
Reinstall any part removed.
Erase the self-diagnosis memory. Make sure Code No. 55 is displayed in Mode III.
1) Perform switch ON/OFF diagnosis (in Mode IV) again.
2) Make sure green inspection lamp on E.C.U. comes "ON" when vehicle speed reaches 20 km/h (12 MPH).
N.G.
1) Perform self-diagnosis and find malfunction code.
2) According to displayed code No., perform diagnosis.
O.K.
INSPECTION END

ECCS SELF-DIAGNOSTIC CHART – 1989-90 240SX

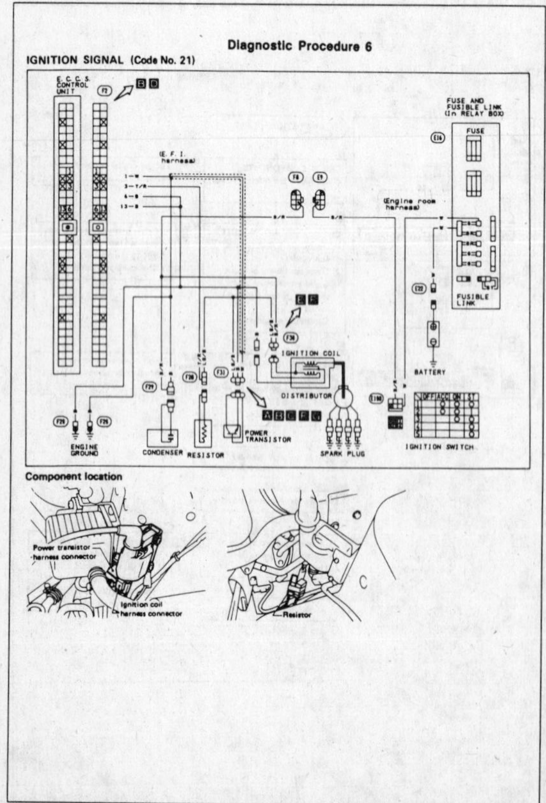
Diagnostic Procedure 6
IGNITION SIGNAL (Code No. 21)
E.C.C.S. CONTROL UNIT
E.F.I. harness
Engine room harness
FUSE AND FUSIBLE LINK (in RELAY BOX)
FUSE
FUSIBLE LINK
BATTERY
IGNITION COIL
DISTRIBUTOR
ENGINE GROUND
CONDENSER
RESISTOR
POWER TRANSISTOR
SPARK PLUG
IGNITION SWITCH
Component location
Power transistor harness connector
Ignition coil harness connector
Resistor

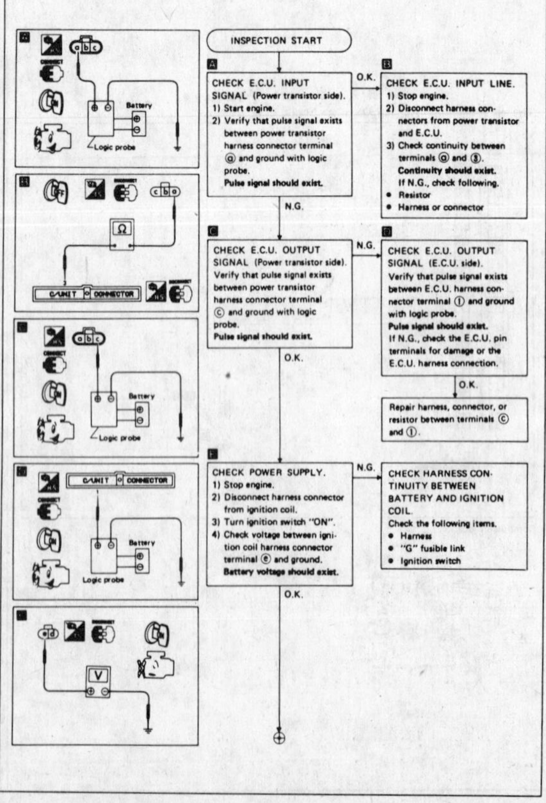
Battery
Logic probe
INSPECTION START
CHECK E.C.U. INPUT SIGNAL (Power transistor side).
1) Start engine.
2) Verify that pulse signal exists between power transistor harness connector terminal ⓐ and ground with logic probe.
Pulse signal should exist.
O.K.
CHECK E.C.U. INPUT LINE.
1) Stop engine.
2) Disconnect harness connectors from power transistor and E.C.U.
3) Check continuity between terminals ⓐ and ③.
Continuity should exist.
If N.G., check following.
• Resistor
• Harness or connector
N.G.
CHECK E.C.U. OUTPUT SIGNAL (Power transistor side).
Verify that pulse signal exists between power transistor harness connector terminal ⓒ and ground with logic probe.
Pulse signal should exist.
N.G.
CHECK E.C.U. OUTPUT SIGNAL (E.C.U. side).
Verify that pulse signal exists between E.C.U. harness connector terminal ① and ground with logic probe.
Pulse signal should exist.
If N.G., check the E.C.U. pin terminals for damage or the E.C.U. harness connection.
O.K.
Repair harness, connector, or resistor between terminals ⓒ and ①.
O.K.
CHECK POWER SUPPLY.
1) Stop engine.
2) Disconnect harness connector from ignition coil.
3) Turn ignition switch "ON".
4) Check voltage between ignition coil harness connector terminal ⓔ and ground.
Battery voltage should exist.
N.G.
CHECK HARNESS CONTINUITY BETWEEN BATTERY AND IGNITION COIL.
Check the following items.
• Harness
• "G" fusible link
• Ignition switch
O.K.

ECCS SELF-DIAGNOSTIC CHART—1989-90 240SX

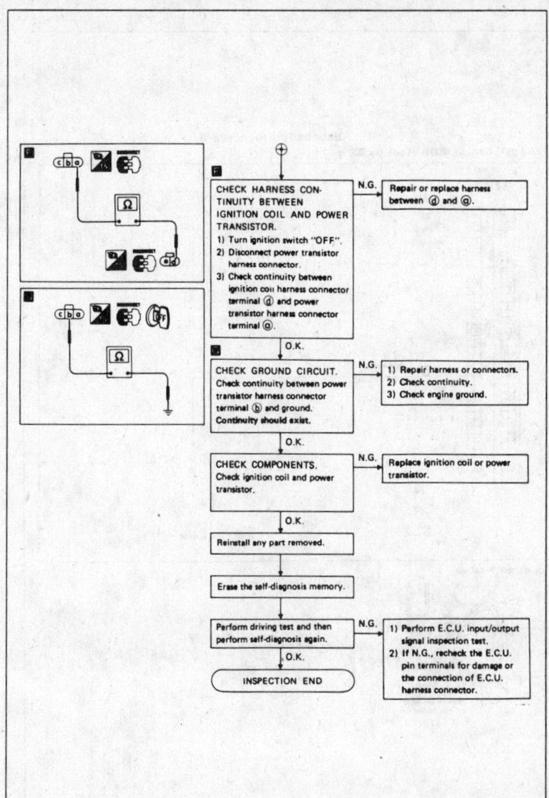
CHECK HARNESS CONTINUITY BETWEEN IGNITION COIL AND POWER TRANSISTOR.
1) Turn ignition switch "OFF".
2) Disconnect power transistor harness connector.
3) Check continuity between ignition coil harness connector terminal ⓓ and power transistor harness connector terminal ⓔ.
N.G.
Repair or replace harness between ⓓ and ⓔ.
O.K.
CHECK GROUND CIRCUIT.
Check continuity between power transistor harness connector terminal ⓑ and ground.
Continuity should exist.
N.G.
1) Repair harness or connectors.
2) Check continuity.
3) Check engine ground.
O.K.
CHECK COMPONENTS.
Check ignition coil and power transistor.
N.G.
Replace ignition coil or power transistor.
O.K.
Reinstall any part removed.
Erase the self-diagnosis memory.
Perform driving test and then perform self-diagnosis again.
N.G.
1) Perform E.C.U. input/output signal inspection test.
2) If N.G., recheck the E.C.U. pin terminals for damage or the connection of E.C.U. harness connector.
O.K.
INSPECTION END

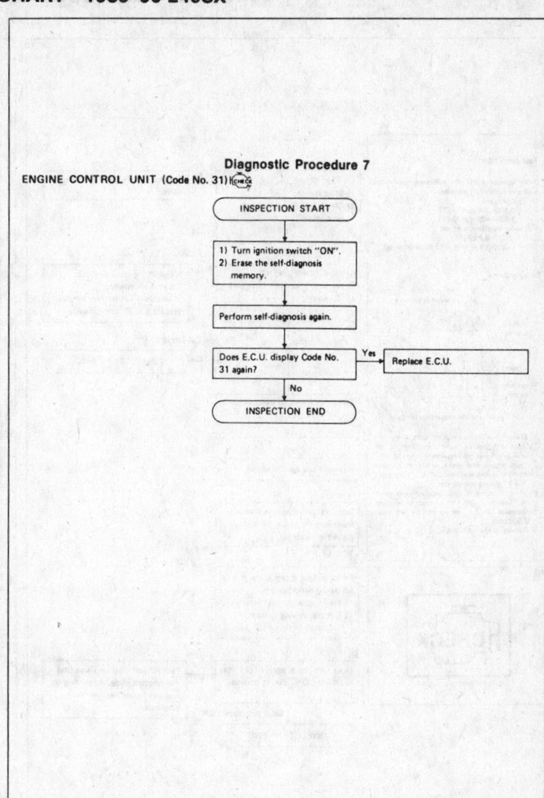
Diagnostic Procedure 7
ENGINE CONTROL UNIT (Code No. 31)
INSPECTION START
1) Turn ignition switch "ON".
2) Erase the self-diagnosis memory.
Perform self-diagnosis again.
Does E.C.U. display Code No. 31 again?
Yes
Replace E.C.U.
No
INSPECTION END

ECCS SELF-DIAGNOSTIC CHART—1989-90 240SX

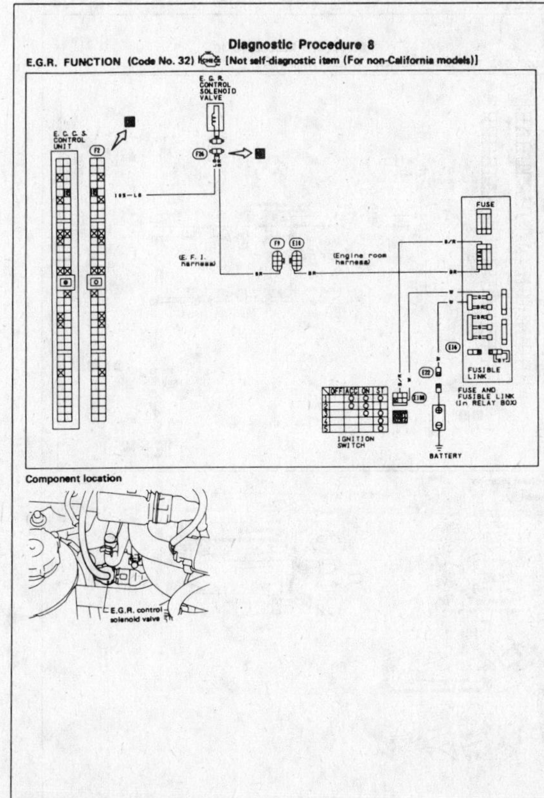
Diagnostic Procedure 8
E.G.R. FUNCTION (Code No. 32) [Not self-diagnostic item (For non-California models)]
E.G.R. CONTROL SOLENOID VALVE
E.C.C.S. CONTROL UNIT
(E.F.I. harness)
(Engine room harness)
FUSE
FUSIBLE LINK
FUSE AND FUSIBLE LINK (In RELAY BOX)
IGNITION SWITCH
BATTERY
Component location
E.G.R. control solenoid valve

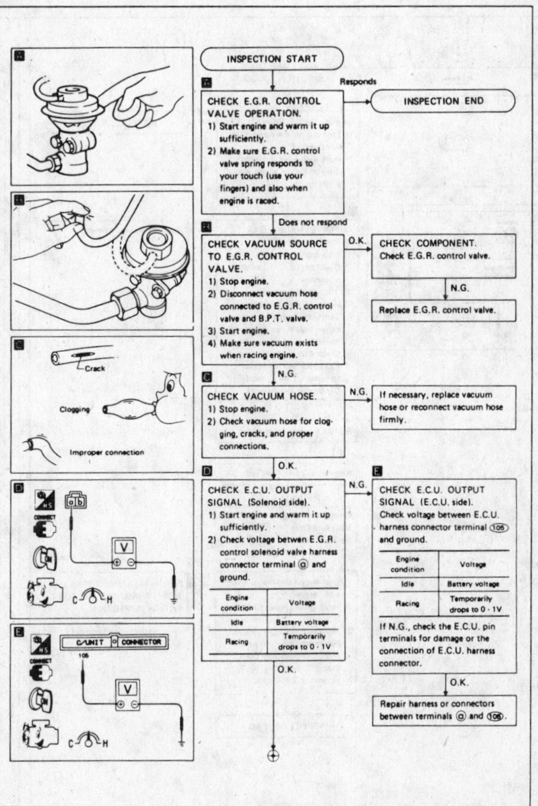
INSPECTION START
CHECK E.G.R. CONTROL VALVE OPERATION.
1) Start engine and warm it up sufficiently.
2) Make sure E.G.R. control valve spring responds to your touch (use your fingers) and also when engine is raced.
Responds
INSPECTION END
Does not respond
CHECK VACUUM SOURCE TO E.G.R. CONTROL VALVE.
1) Stop engine.
2) Disconnect vacuum hose connected to E.G.R. control valve and B.P.T. valve.
3) Start engine.
4) Make sure vacuum exists when racing engine.
O.K.
CHECK COMPONENT.
Check E.G.R. control valve.
N.G.
Replace E.G.R. control valve.
N.G.
CHECK VACUUM HOSE.
1) Stop engine.
2) Check vacuum hose for clogging, cracks, and proper connections.
N.G.
If necessary, replace vacuum hose or reconnect vacuum hose firmly.
Crack
Clogging
Improper connection
O.K.
CHECK E.C.U. OUTPUT SIGNAL (Solenoid side).
1) Start engine and warm it up sufficiently.
2) Check voltage betwen E.G.R. control solenoid valve harness connector terminal ⓐ and ground.
Engine condition | Voltage
Idle | Battery voltage
Racing | Temporarily drops to 0 - 1V
N.G.
CHECK E.C.U. OUTPUT SIGNAL (E.C.U. side).
Check voltage between E.C.U. harness connector terminal (105) and ground.
Engine condition | Voltage
Idle | Battery voltage
Racing | Temporarily drops to 0 - 1V
If N.G., check the E.C.U. pin terminals for damage or the connection of E.C.U. harness connector.
O.K.
Repair harness or connectors between terminals ⓐ and (105).
O.K.
C/UNIT CONNECTOR
105

ECCS SELF-DIAGNOSTIC CHART—1989–90 240SX

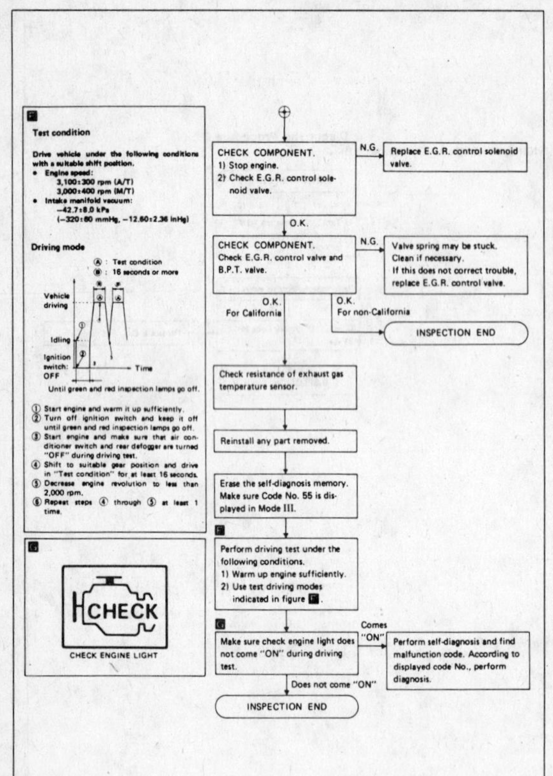
Test condition
Drive vehicle under the following conditions with a suitable shift position.
Engine speed: 3,100±300 rpm (A/T) 3,000±400 rpm (M/T)
Intake manifold vacuum: −42.7±8.0 kPa (−320±60 mmHg, −12.60±2.36 inHg)
Driving mode
Until green and red inspection lamps go off.
CHECK
CHECK ENGINE LIGHT
CHECK COMPONENT.
1) Stop engine.
2) Check E.G.R. control solenoid valve.
N.G.
Replace E.G.R. control solenoid valve.
O.K.
CHECK COMPONENT.
Check E.G.R. control valve and B.P.T. valve.
Valve spring may be stuck. Clean if necessary. If this does not correct trouble, replace E.G.R. control valve.
For California
For non-California
INSPECTION END
Check resistance of exhaust gas temperature sensor.
Reinstall any part removed.
Erase the self-diagnosis memory. Make sure Code No. 55 is displayed in Mode III.
Perform driving test under the following conditions.
1) Warm up engine sufficiently.
2) Use test driving modes indicated in figure.
Make sure check engine light does not come "ON" during driving test.
Comes "ON"
Perform self-diagnosis and find malfunction code. According to displayed code No., perform diagnosis.
Does not come "ON"
INSPECTION END

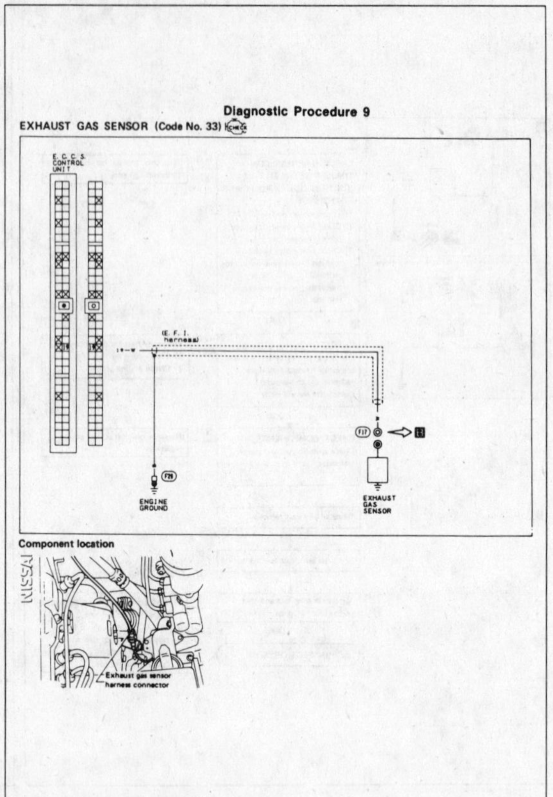
Diagnostic Procedure 9
EXHAUST GAS SENSOR (Code No. 33)
E.C.C.S. CONTROL UNIT
(E.F.I. harness)
ENGINE GROUND
EXHAUST GAS SENSOR
Component location
Exhaust gas sensor harness connector

ECCS SELF-DIAGNOSTIC CHART—1989–90 240SX

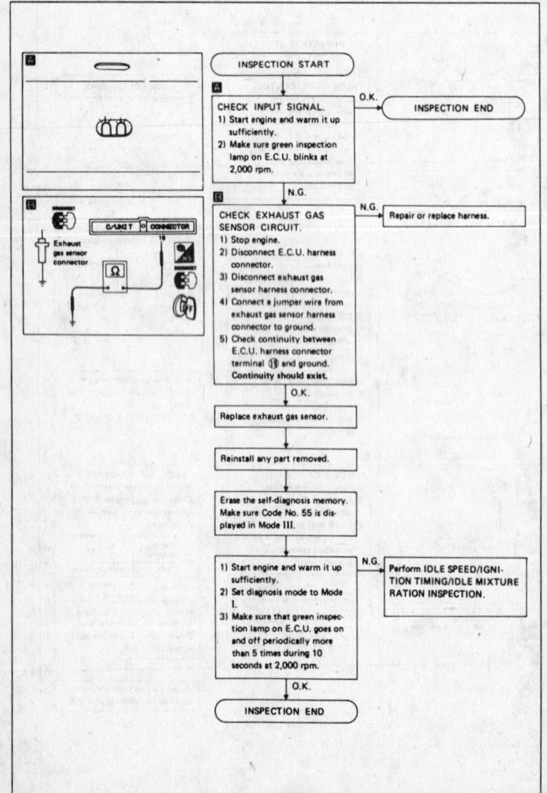
INSPECTION START
CHECK INPUT SIGNAL.
1) Start engine and warm it up sufficiently.
2) Make sure green inspection lamp on E.C.U. blinks at 2,000 rpm.
O.K.
INSPECTION END
N.G.
CHECK EXHAUST GAS SENSOR CIRCUIT.
1) Stop engine.
2) Disconnect E.C.U. harness connector.
3) Disconnect exhaust gas sensor harness connector.
4) Connect a jumper wire from exhaust gas sensor harness connector to ground.
5) Check continuity between E.C.U. harness connector terminal and ground. Continuity should exist.
Repair or replace harness.
Exhaust gas sensor connector
C/UNIT CONNECTOR
Replace exhaust gas sensor.
Reinstall any part removed.
Erase the self-diagnosis memory. Make sure Code No. 55 is displayed in Mode III.
1) Start engine and warm it up sufficiently.
2) Set diagnosis mode to Mode I.
3) Make sure that green inspection lamp on E.C.U. goes on and off periodically more than 5 times during 10 seconds at 2,000 rpm.
Perform IDLE SPEED/IGNITION TIMING/IDLE MIXTURE RATION INSPECTION.
INSPECTION END

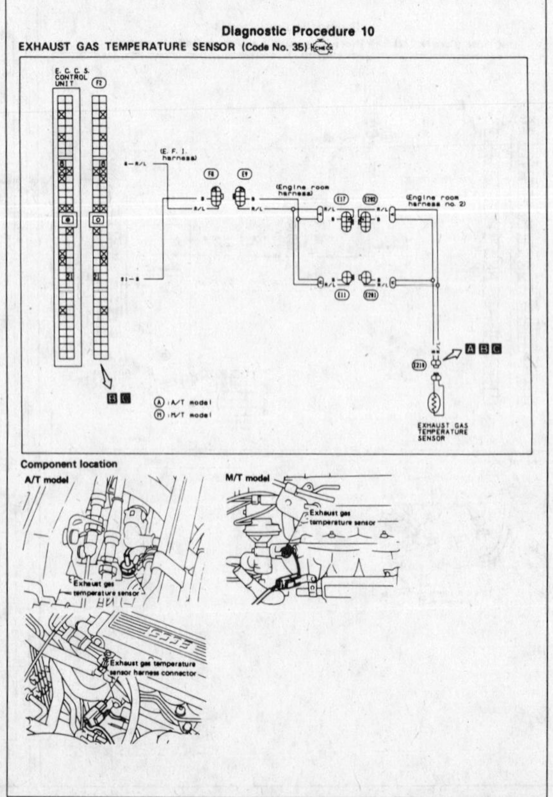
Diagnostic Procedure 10
EXHAUST GAS TEMPERATURE SENSOR (Code No. 35)
E.C.C.S. CONTROL UNIT
(E.F.I. harness)
(Engine room harness)
(Engine room harness no. 2)
A/T model
M/T model
EXHAUST GAS TEMPERATURE SENSOR
Component location
Exhaust gas temperature sensor
Exhaust gas temperature sensor harness connector

ECCS SELF-DIAGNOSTIC CHART – 1989–90 240SX

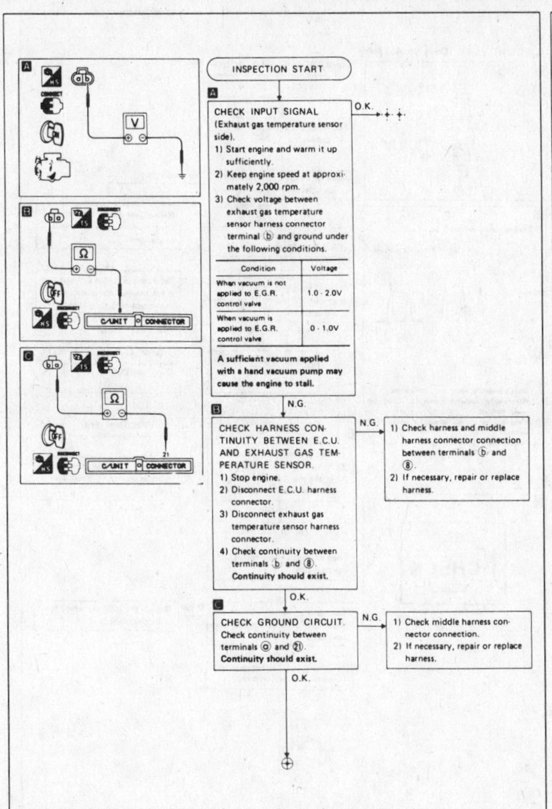

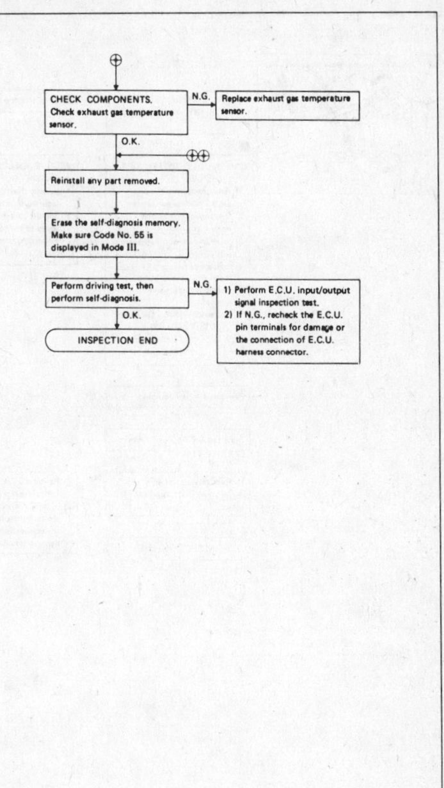

ECCS SELF-DIAGNOSTIC CHART – 1989–90 240SX

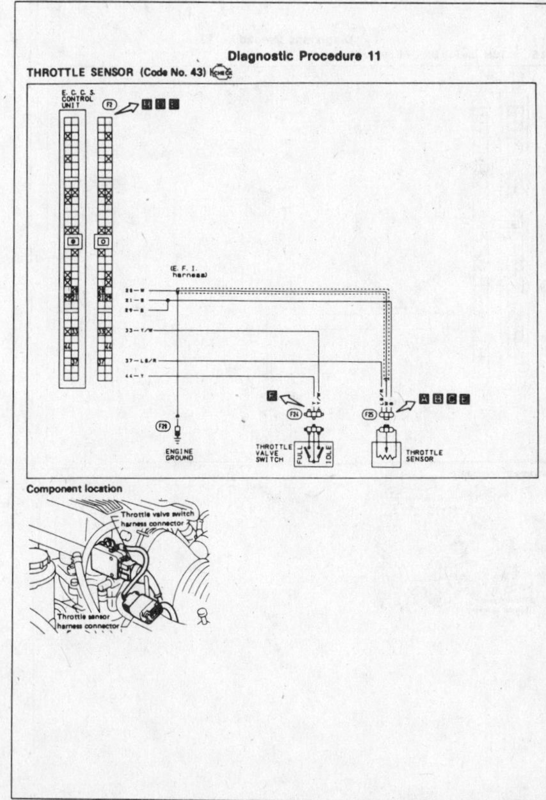

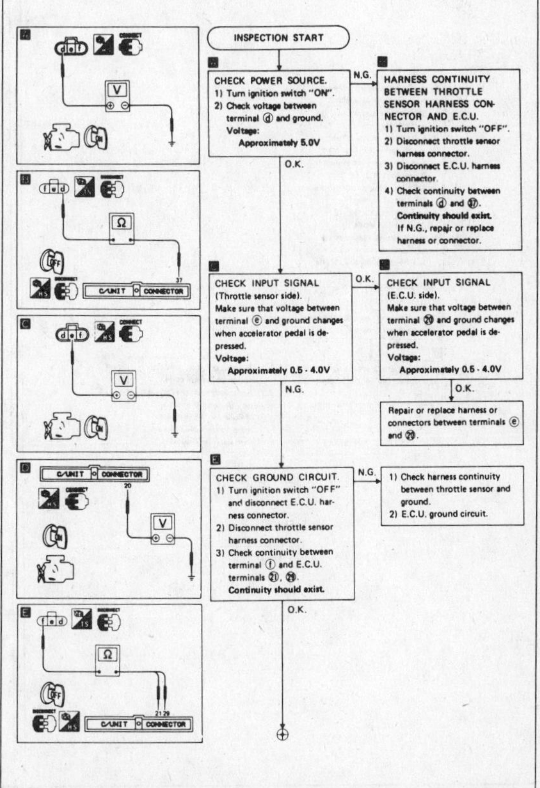

ECCS SELF-DIAGNOSTIC CHART—1989–90 240SX

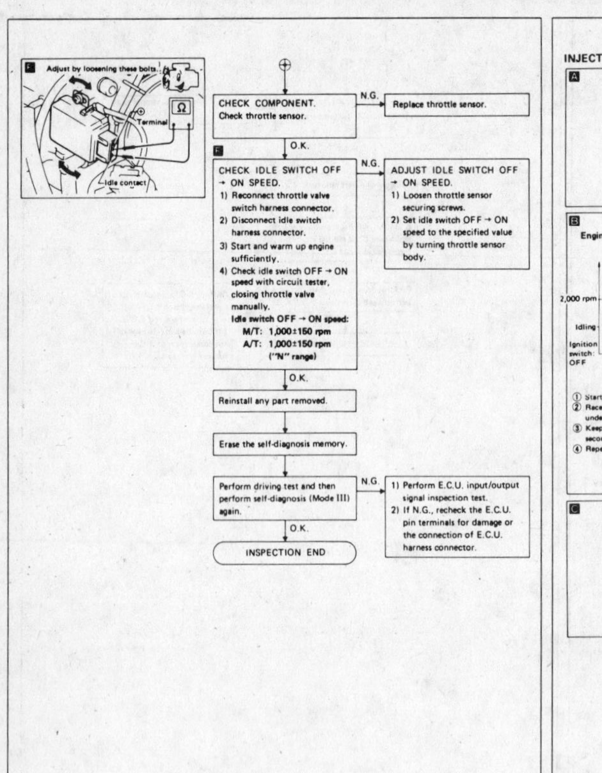

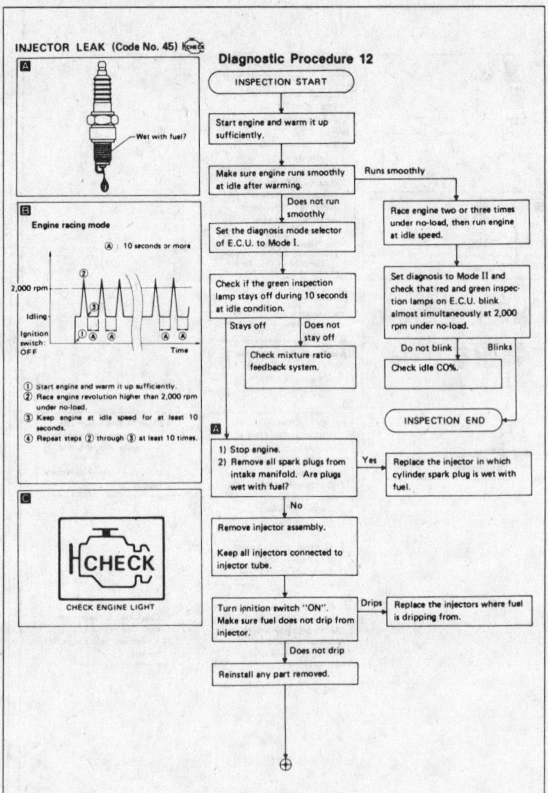

ECCS SELF-DIAGNOSTIC CHART—1989–90 240SX

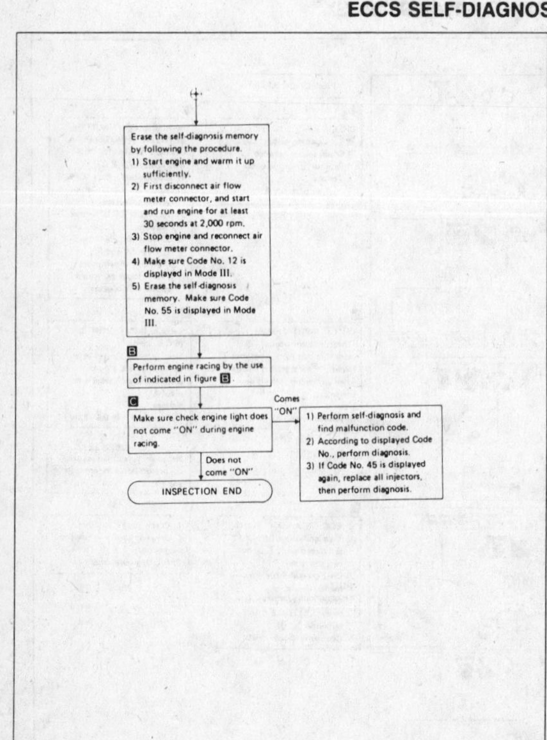

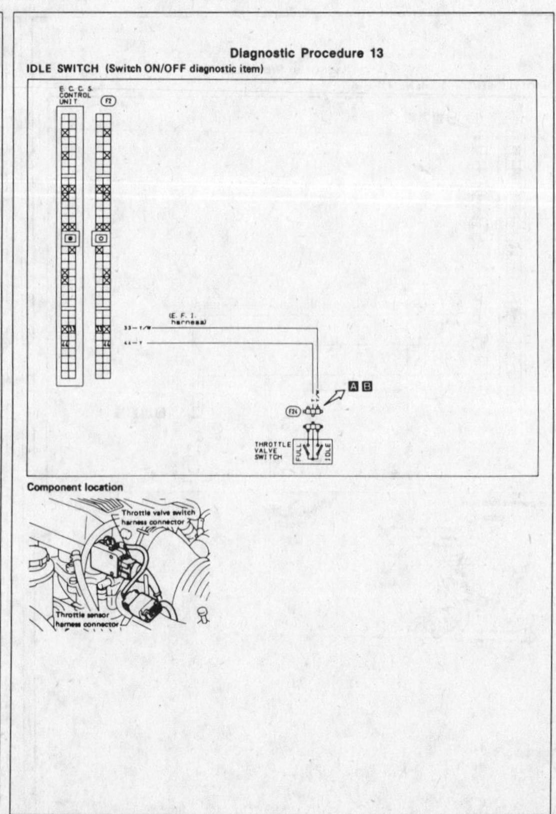

ECCS SELF-DIAGNOSTIC CHART—1989–90 240SX

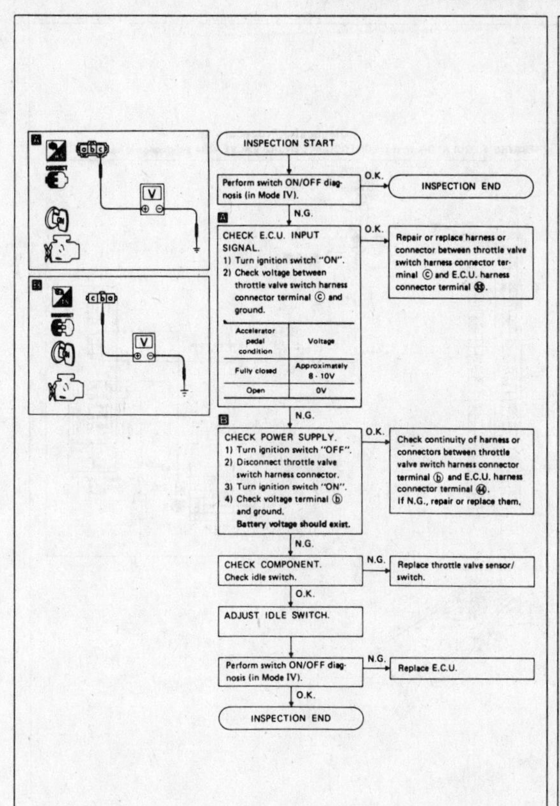
INSPECTION START
Perform switch ON/OFF diagnosis (in Mode IV).
O.K.
INSPECTION END
N.G.
CHECK E.C.U. INPUT SIGNAL.
1) Turn ignition switch "ON".
2) Check voltage between throttle valve switch harness connector terminal ⓒ and ground.
Accelerator pedal condition
Voltage
Fully closed
Approximately 8 - 10V
Open
0V
O.K.
Repair or replace harness or connector between throttle valve switch harness connector terminal ⓒ and E.C.U. harness connector terminal.
N.G.
CHECK POWER SUPPLY.
1) Turn ignition switch "OFF".
2) Disconnect throttle valve switch harness connector.
3) Turn ignition switch "ON".
4) Check voltage terminal ⓑ and ground.
Battery voltage should exist.
O.K.
Check continuity of harness or connectors between throttle valve switch harness connector terminal ⓑ and E.C.U. harness connector terminal.
If N.G., repair or replace them.
N.G.
CHECK COMPONENT.
Check idle switch.
N.G.
Replace throttle valve sensor/switch.
O.K.
ADJUST IDLE SWITCH.
Perform switch ON/OFF diagnosis (in Mode IV).
N.G.
Replace E.C.U.
O.K.
INSPECTION END

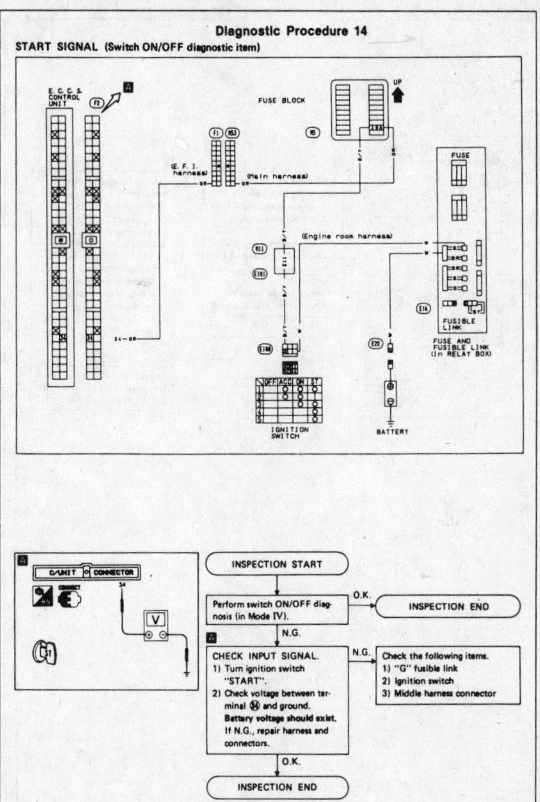
Diagnostic Procedure 14
START SIGNAL (Switch ON/OFF diagnostic item)
E.C.C.S. CONTROL UNIT
FUSE BLOCK
UP
FUSE
(E.F.I. harness)
(Main harness)
(Engine room harness)
FUSIBLE LINK
FUSE AND FUSIBLE LINK (In RELAY BOX)
IGNITION SWITCH
BATTERY
C/UNIT
CONNECTOR
INSPECTION START
Perform switch ON/OFF diagnosis (in Mode IV).
O.K.
INSPECTION END
N.G.
CHECK INPUT SIGNAL.
1) Turn ignition switch "START".
2) Check voltage between terminal and ground.
Battery voltage should exist.
If N.G., repair harness and connectors.
N.G.
Check the following items.
1) "G" fusible link
2) Ignition switch
3) Middle harness connector
O.K.
INSPECTION END

ECCS SELF-DIAGNOSTIC CHART—1989–90 240SX

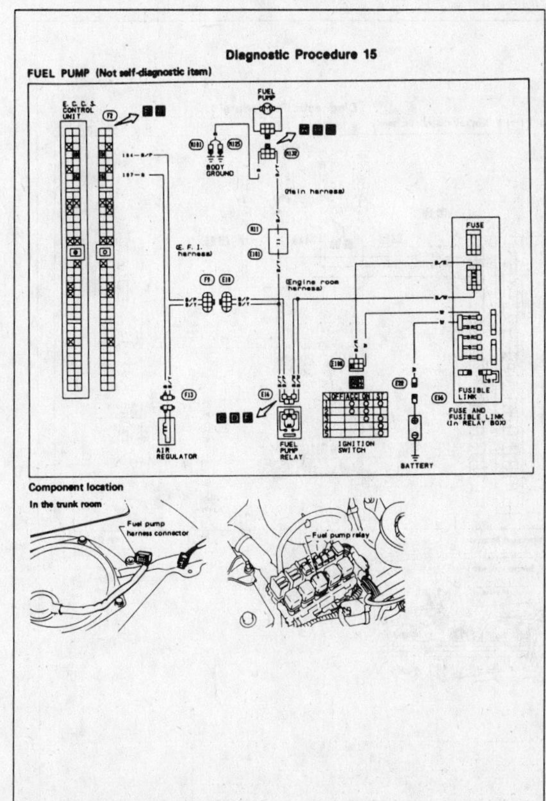
Diagnostic Procedure 15
FUEL PUMP (Not self-diagnostic item)
E.C.C.S. CONTROL UNIT
FUEL PUMP
BODY GROUND
(Main harness)
(E.F.I. harness)
(Engine room harness)
FUSE
FUSIBLE LINK
FUSE AND FUSIBLE LINK (In RELAY BOX)
AIR REGULATOR
FUEL PUMP RELAY
IGNITION SWITCH
BATTERY
Component location
In the trunk room
Fuel pump harness connector
Fuel pump relay

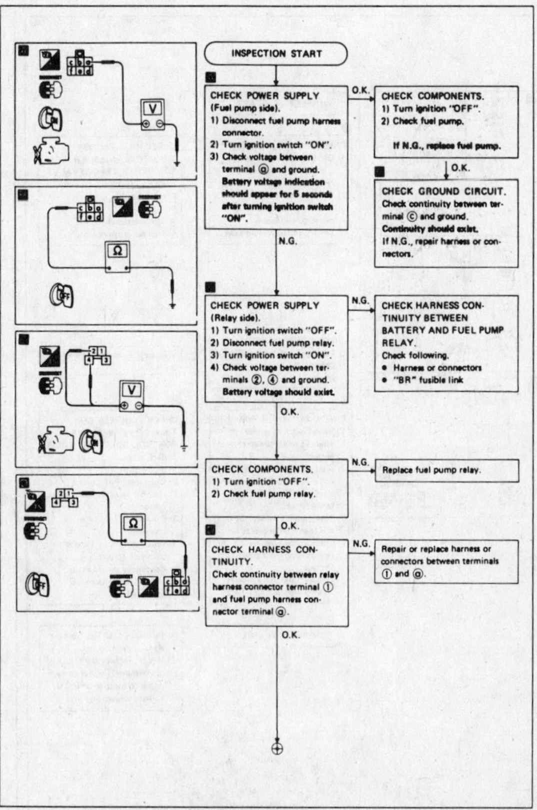
INSPECTION START
CHECK POWER SUPPLY (Fuel pump side).
1) Disconnect fuel pump harness connector.
2) Turn ignition switch "ON".
3) Check voltage between terminal ⓐ and ground.
Battery voltage indication should appear for 5 seconds after turning ignition switch "ON".
O.K.
CHECK COMPONENTS.
1) Turn ignition "OFF".
2) Check fuel pump.
If N.G., replace fuel pump.
O.K.
CHECK GROUND CIRCUIT.
Check continuity between terminal ⓒ and ground.
Continuity should exist.
If N.G., repair harness or connectors.
N.G.
CHECK POWER SUPPLY (Relay side).
1) Turn ignition switch "OFF".
2) Disconnect fuel pump relay.
3) Turn ignition switch "ON".
4) Check voltage between terminals ②, ④ and ground.
Battery voltage should exist.
N.G.
CHECK HARNESS CONTINUITY BETWEEN BATTERY AND FUEL PUMP RELAY.
Check following.
• Harness or connectors
• "BR" fusible link
O.K.
CHECK COMPONENTS.
1) Turn ignition "OFF".
2) Check fuel pump relay.
N.G.
Replace fuel pump relay.
O.K.
CHECK HARNESS CONTINUITY.
Check continuity between relay harness connector terminal ① and fuel pump harness connector terminal ⓐ.
N.G.
Repair or replace harness or connectors between terminals ① and ⓐ.
O.K.

ECCS SELF-DIAGNOSTIC CHART—1989–90 240SX

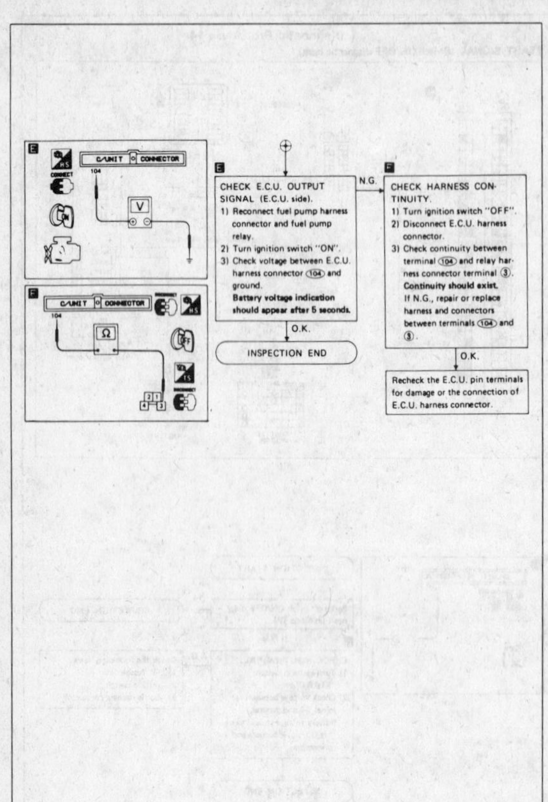
E
F
C/UNIT CONNECTOR
104
V
Ω
E
CHECK E.C.U. OUTPUT SIGNAL (E.C.U. side).
1) Reconnect fuel pump harness connector and fuel pump relay.
2) Turn ignition switch "ON".
3) Check voltage between E.C.U. harness connector (104) and ground.
Battery voltage indication should appear after 5 seconds.
N.G.
O.K.
INSPECTION END
F
CHECK HARNESS CONTINUITY.
1) Turn ignition switch "OFF".
2) Disconnect E.C.U. harness connector.
3) Check continuity between terminal (104) and relay harness connector terminal (3).
Continuity should exist.
If N.G., repair or replace harness and connectors between terminals (104) and (3).
O.K.
Recheck the E.C.U. pin terminals for damage or the connection of E.C.U. harness connector.

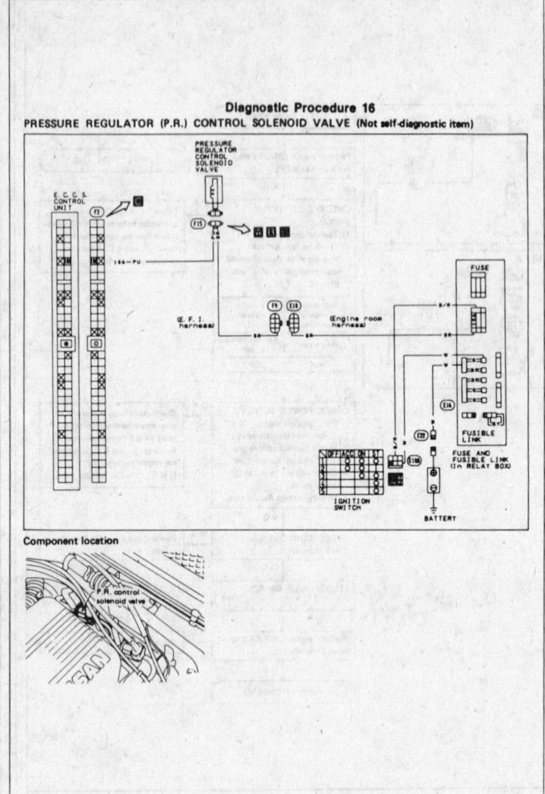
Diagnostic Procedure 16
PRESSURE REGULATOR (P.R.) CONTROL SOLENOID VALVE (Not self-diagnostic item)
PRESSURE REGULATOR CONTROL SOLENOID VALVE
E.C.C.S. CONTROL UNIT
FUSE
FUSIBLE LINK
FUSE AND FUSIBLE LINK (In RELAY BOX)
IGNITION SWITCH
BATTERY
Component location
P.R. control solenoid valve

ECCS SELF-DIAGNOSTIC CHART—1989–90 240SX

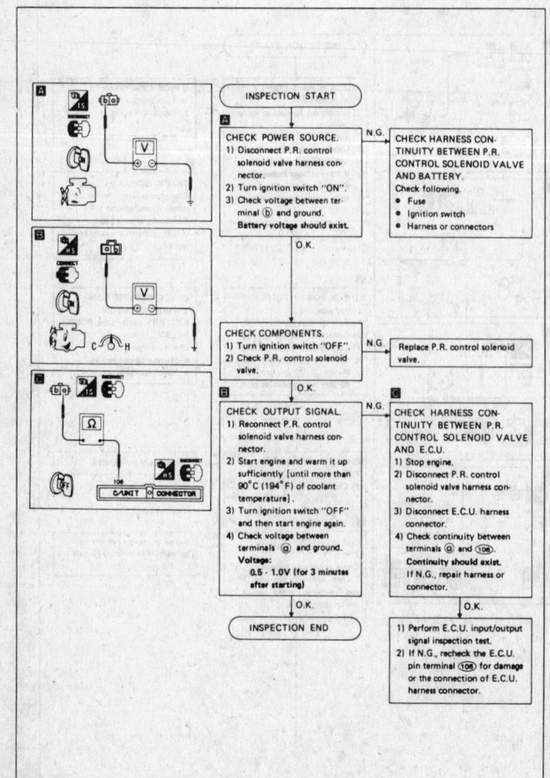
A
B
C
V
Ω
C/UNIT CONNECTOR
106
INSPECTION START
A
CHECK POWER SOURCE.
1) Disconnect P.R. control solenoid valve harness connector.
2) Turn ignition switch "ON".
3) Check voltage between terminal (b) and ground.
Battery voltage should exist.
N.G.
CHECK HARNESS CONTINUITY BETWEEN P.R. CONTROL SOLENOID VALVE AND BATTERY.
Check following.
• Fuse
• Ignition switch
• Harness or connectors
O.K.
CHECK COMPONENTS.
1) Turn ignition switch "OFF".
2) Check P.R. control solenoid valve.
N.G.
Replace P.R. control solenoid valve.
O.K.
B
CHECK OUTPUT SIGNAL.
1) Reconnect P.R. control solenoid valve harness connector.
2) Start engine and warm it up sufficiently [until more than 90°C (194°F) of coolant temperature].
3) Turn ignition switch "OFF" and then start engine again.
4) Check voltage between terminals (a) and ground.
Voltage:
0.5 - 1.0V (for 3 minutes after starting)
N.G.
C
CHECK HARNESS CONTINUITY BETWEEN P.R. CONTROL SOLENOID VALVE AND E.C.U.
1) Stop engine.
2) Disconnect P.R. control solenoid valve harness connector.
3) Disconnect E.C.U. harness connector.
4) Check continuity between terminals (a) and (106).
Continuity should exist.
If N.G., repair harness or connector.
O.K.
INSPECTION END
O.K.
1) Perform E.C.U. input/output signal inspection test.
2) If N.G., recheck the E.C.U. pin terminal (106) for damage or the connection of E.C.U. harness connector.

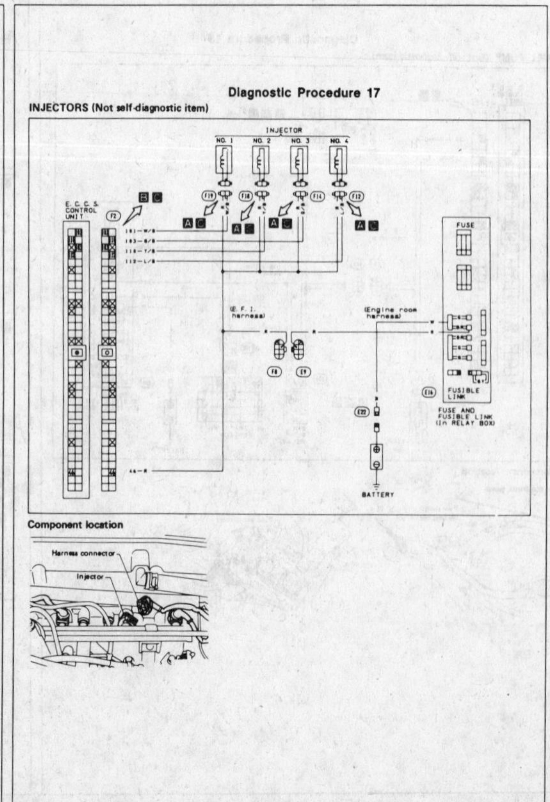
Diagnostic Procedure 17
INJECTORS (Not self-diagnostic item)
INJECTOR
NO. 1
NO. 2
NO. 3
NO. 4
E.C.C.S. CONTROL UNIT
FUSE
FUSIBLE LINK
FUSE AND FUSIBLE LINK (In RELAY BOX)
BATTERY
Component location
Harness connector
Injector

ECCS SELF-DIAGNOSTIC CHART—1989–90 240SX

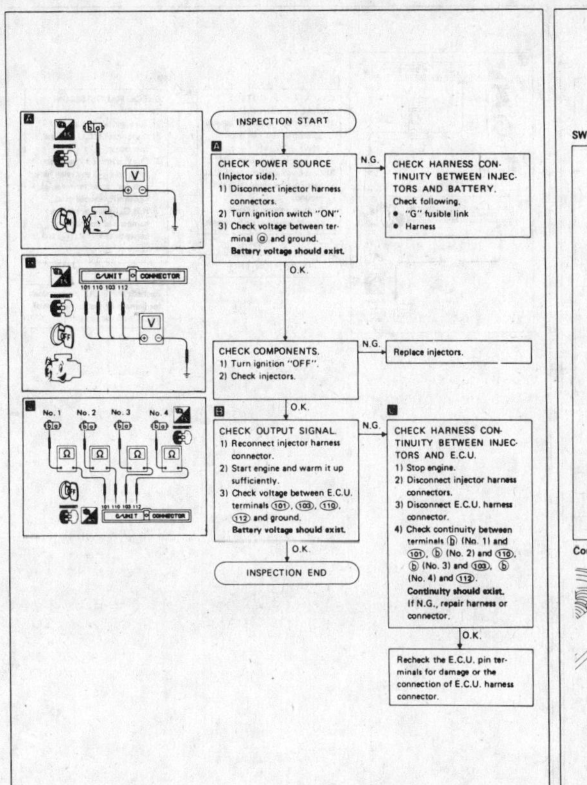
INSPECTION START
CHECK POWER SOURCE (Injector side).
1) Disconnect injector harness connectors.
2) Turn ignition switch "ON".
3) Check voltage between terminal ⓐ and ground.
Battery voltage should exist.
N.G.
CHECK HARNESS CONTINUITY BETWEEN INJECTORS AND BATTERY.
Check following.
• "G" fusible link
• Harness
O.K.
CHECK COMPONENTS.
1) Turn ignition "OFF".
2) Check injectors.
N.G.
Replace injectors.
O.K.
CHECK OUTPUT SIGNAL.
1) Reconnect injector harness connector.
2) Start engine and warm it up sufficiently.
3) Check voltage between E.C.U. terminals 101, 103, 110, 112 and ground.
Battery voltage should exist.
N.G.
CHECK HARNESS CONTINUITY BETWEEN INJECTORS AND E.C.U.
1) Stop engine.
2) Disconnect injector harness connectors.
3) Disconnect E.C.U. harness connector.
4) Check continuity between terminals ⓑ (No. 1) and 101, ⓑ (No. 2) and 110, ⓑ (No. 3) and 103, ⓑ (No. 4) and 112.
Continuity should exist.
If N.G., repair harness or connector.
O.K.
Recheck the E.C.U. pin terminals for damage or the connection of E.C.U. harness connector.
O.K.
INSPECTION END
No. 1
No. 2
No. 3
No. 4
C/UNIT
CONNECTOR

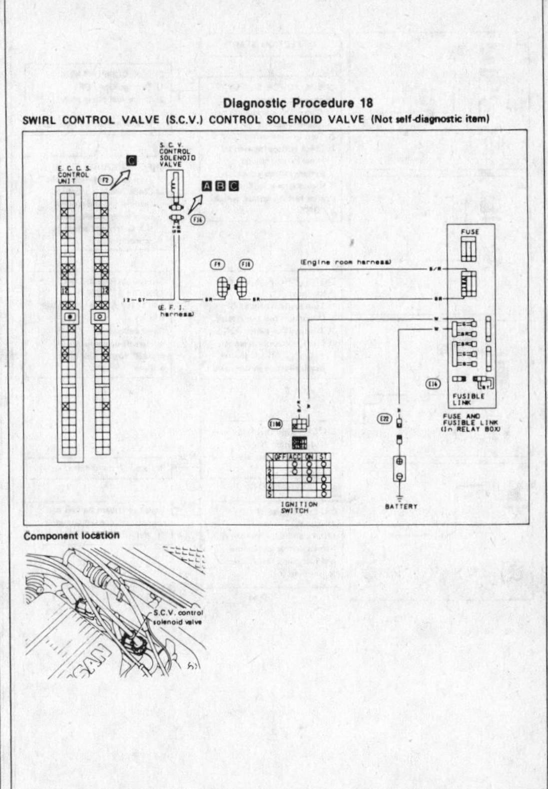
Diagnostic Procedure 18
SWIRL CONTROL VALVE (S.C.V.) CONTROL SOLENOID VALVE (Not self-diagnostic item)
E.C.C.S. CONTROL UNIT
S.C.V. CONTROL SOLENOID VALVE
(Engine room harness)
(E.F.I. harness)
FUSE
FUSIBLE LINK
FUSE AND FUSIBLE LINK (In RELAY BOX)
IGNITION SWITCH
BATTERY
Component location
S.C.V. control solenoid valve

ECCS SELF-DIAGNOSTIC CHART—1989–90 240SX

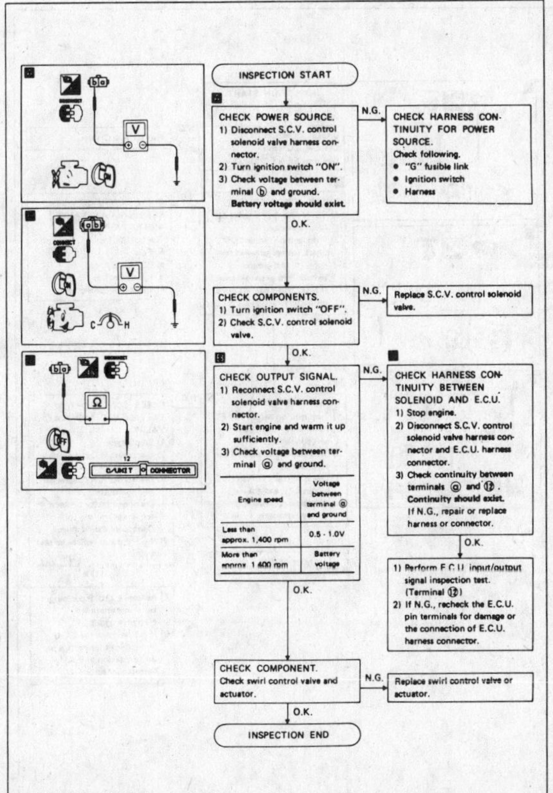
INSPECTION START
CHECK POWER SOURCE.
1) Disconnect S.C.V. control solenoid valve harness connector.
2) Turn ignition switch "ON".
3) Check voltage between terminal ⓑ and ground.
Battery voltage should exist.
N.G.
CHECK HARNESS CONTINUITY FOR POWER SOURCE.
Check following.
• "G" fusible link
• Ignition switch
• Harness
O.K.
CHECK COMPONENTS.
1) Turn ignition switch "OFF".
2) Check S.C.V. control solenoid valve.
N.G.
Replace S.C.V. control solenoid valve.
O.K.
CHECK OUTPUT SIGNAL.
1) Reconnect S.C.V. control solenoid valve harness connector.
2) Start engine and warm it up sufficiently.
3) Check voltage between terminal ⓐ and ground.
Engine speed
Voltage between terminal ⓐ and ground
Less than approx. 1,400 rpm
0.5 - 1.0V
More than approx. 1,600 rpm
Battery voltage
N.G.
CHECK HARNESS CONTINUITY BETWEEN SOLENOID AND E.C.U.
1) Stop engine.
2) Disconnect S.C.V. control solenoid valve harness connector and E.C.U. harness connector.
3) Check continuity between terminals ⓐ and 12.
Continuity should exist.
If N.G., repair or replace harness or connector.
O.K.
1) Perform E.C.U. input/output signal inspection test. (Terminal 12)
2) If N.G., recheck the E.C.U. pin terminals for damage or the connection of E.C.U. harness connector.
O.K.
CHECK COMPONENT.
Check swirl control valve and actuator.
N.G.
Replace swirl control valve or actuator.
O.K.
INSPECTION END
C/UNIT
CONNECTOR

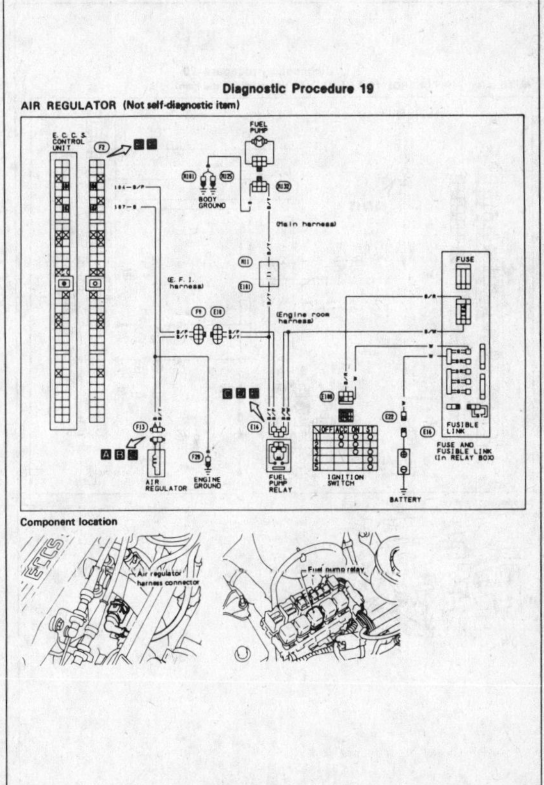
Diagnostic Procedure 19
AIR REGULATOR (Not self-diagnostic item)
E.C.C.S. CONTROL UNIT
FUEL PUMP
BODY GROUND
(Main harness)
(E.F.I. harness)
(Engine room harness)
FUSE
FUSIBLE LINK
FUSE AND FUSIBLE LINK (In RELAY BOX)
AIR REGULATOR
ENGINE GROUND
FUEL PUMP RELAY
IGNITION SWITCH
BATTERY
Component location
Air regulator harness connector
Fuel pump relay

ECCS SELF-DIAGNOSTIC CHART—1989-90 240SX

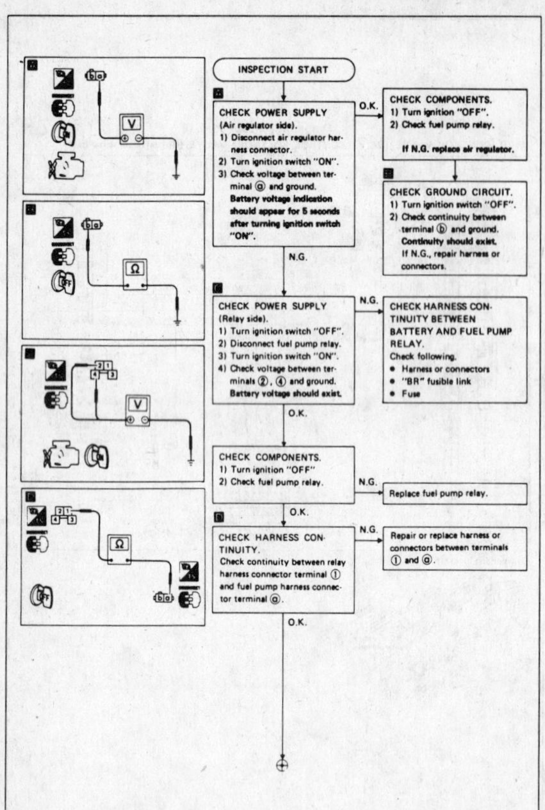
INSPECTION START
CHECK POWER SUPPLY (Air regulator side).
1) Disconnect air regulator harness connector.
2) Turn ignition switch "ON".
3) Check voltage between terminal ⓐ and ground.
Battery voltage indication should appear for 5 seconds after turning ignition switch "ON".
O.K.
CHECK COMPONENTS.
1) Turn ignition "OFF".
2) Check fuel pump relay.
If N.G. replace air regulator.
CHECK GROUND CIRCUIT.
1) Turn ignition switch "OFF".
2) Check continuity between terminal ⓑ and ground.
Continuity should exist.
If N.G., repair harness or connectors.
N.G.
CHECK POWER SUPPLY (Relay side).
1) Turn ignition switch "OFF".
2) Disconnect fuel pump relay.
3) Turn ignition switch "ON".
4) Check voltage between terminals ②, ④ and ground.
Battery voltage should exist.
N.G.
CHECK HARNESS CONTINUITY BETWEEN BATTERY AND FUEL PUMP RELAY.
Check following.
• Harness or connectors
• "BR" fusible link
• Fuse
O.K.
CHECK COMPONENTS.
1) Turn ignition "OFF"
2) Check fuel pump relay.
N.G.
Replace fuel pump relay.
O.K.
CHECK HARNESS CONTINUITY.
Check continuity between relay harness connector terminal ① and fuel pump harness connector terminal ⓐ.
N.G.
Repair or replace harness or connectors between terminals ① and ⓐ.
O.K.

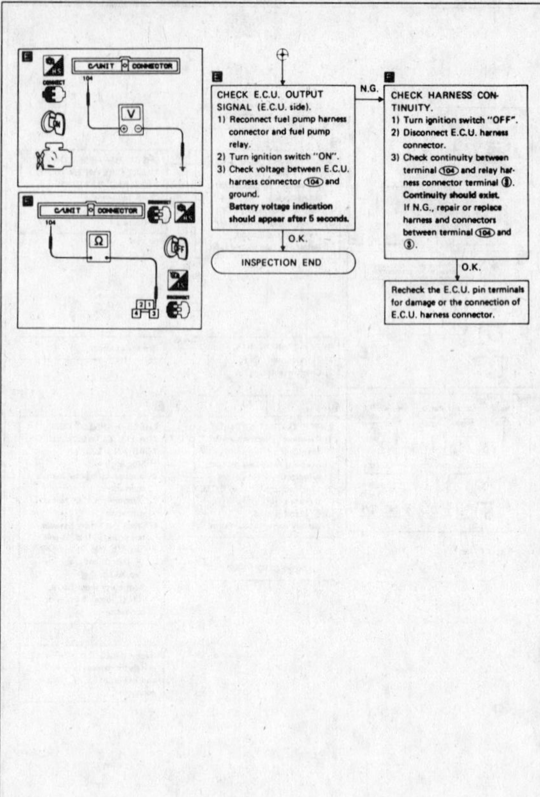
C/UNIT CONNECTOR
104
CHECK E.C.U. OUTPUT SIGNAL (E.C.U. side).
1) Reconnect fuel pump harness connector and fuel pump relay.
2) Turn ignition switch "ON".
3) Check voltage between E.C.U. harness connector (104) and ground.
Battery voltage indication should appear after 5 seconds.
N.G.
CHECK HARNESS CONTINUITY.
1) Turn ignition switch "OFF".
2) Disconnect E.C.U. harness connector.
3) Check continuity between terminal (104) and relay harness connector terminal ③.
Continuity should exist.
If N.G., repair or replace harness and connectors between terminal (104) and ③.
O.K.
Recheck the E.C.U. pin terminals for damage or the connection of E.C.U. harness connector.
O.K.
INSPECTION END

ECCS SELF-DIAGNOSTIC CHART—1989-90 240SX

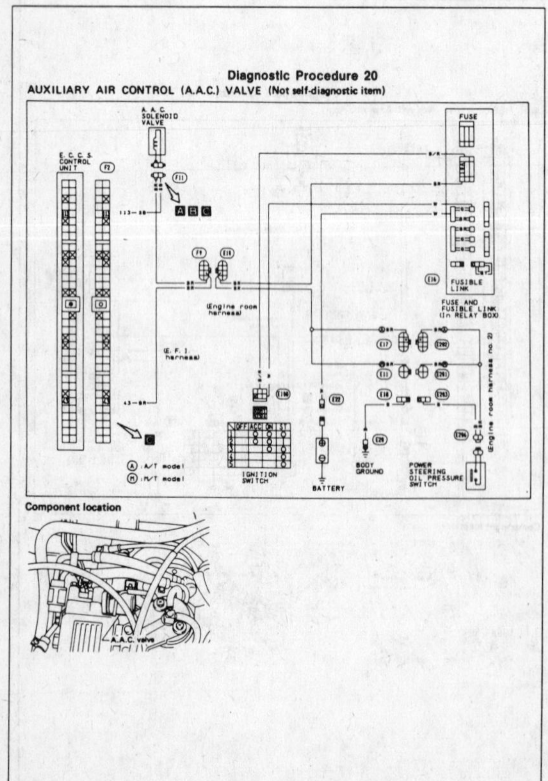
Diagnostic Procedure 20
AUXILIARY AIR CONTROL (A.A.C.) VALVE (Not self-diagnostic item)
A.A.C. SOLENOID VALVE
E.C.C.S. CONTROL UNIT
FUSE
FUSIBLE LINK
FUSE AND FUSIBLE LINK (In RELAY BOX)
(Engine room harness)
(E.F.I. harness)
IGNITION SWITCH
BATTERY
BODY GROUND
POWER STEERING OIL PRESSURE SWITCH
Ⓐ : A/T model
Ⓜ : M/T model
Component location
A.A.C. valve

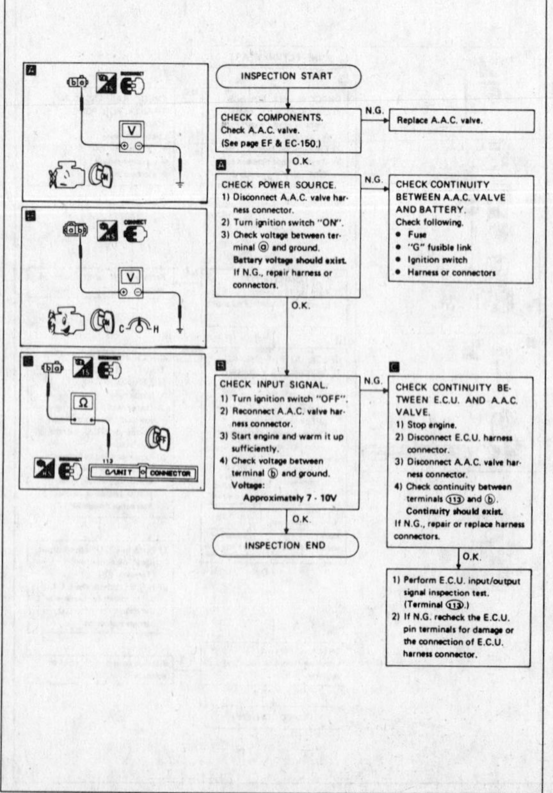
INSPECTION START
CHECK COMPONENTS.
Check A.A.C. valve.
(See page EF & EC-150.)
N.G.
Replace A.A.C. valve.
O.K.
CHECK POWER SOURCE.
1) Disconnect A.A.C. valve harness connector.
2) Turn ignition switch "ON".
3) Check voltage between terminal ⓐ and ground.
Battery voltage should exist.
If N.G., repair harness or connectors.
N.G.
CHECK CONTINUITY BETWEEN A.A.C. VALVE AND BATTERY.
Check following.
• Fuse
• "G" fusible link
• Ignition switch
• Harness or connectors
O.K.
CHECK INPUT SIGNAL.
1) Turn ignition switch "OFF".
2) Reconnect A.A.C. valve harness connector.
3) Start engine and warm it up sufficiently.
4) Check voltage between terminal ⓑ and ground.
Voltage:
Approximately 7 - 10V
N.G.
CHECK CONTINUITY BETWEEN E.C.U. AND A.A.C. VALVE.
1) Stop engine.
2) Disconnect E.C.U. harness connector.
3) Disconnect A.A.C. valve harness connector.
4) Check continuity between terminals (113) and ⓑ.
Continuity should exist.
If N.G., repair or replace harness connectors.
O.K.
1) Perform E.C.U. input/output signal inspection test. (Terminal (113).)
2) If N.G. recheck the E.C.U. pin terminals for damage or the connection of E.C.U. harness connector.
O.K.
INSPECTION END
C/UNIT CONNECTOR
113

ECCS SELF-DIAGNOSTIC CHART—1989–90 240SX

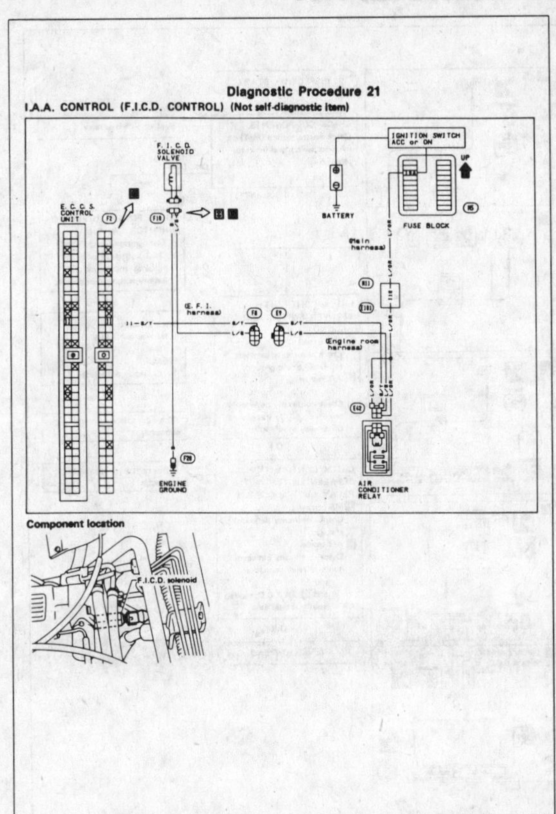

Diagnostic Procedure 21
I.A.A. CONTROL (F.I.C.D. CONTROL) (Not self-diagnostic item)
F.I.C.D. SOLENOID VALVE
E.C.C.S. CONTROL UNIT
IGNITION SWITCH ACC or ON
BATTERY
FUSE BLOCK
(Main harness)
(E.F.I. harness)
(Engine room harness)
ENGINE GROUND
AIR CONDITIONER RELAY
Component location
F.I.C.D. solenoid

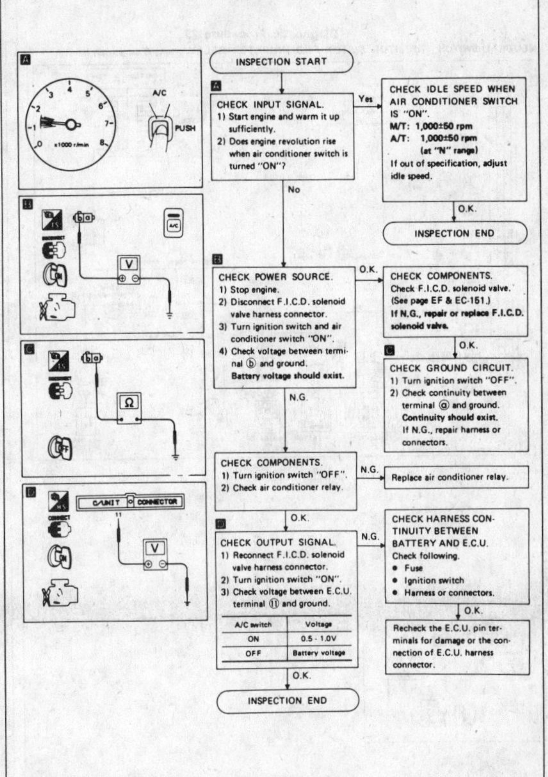

INSPECTION START
CHECK INPUT SIGNAL.
1) Start engine and warm it up sufficiently.
2) Does engine revolution rise when air conditioner switch is turned "ON"?
Yes
CHECK IDLE SPEED WHEN AIR CONDITIONER SWITCH IS "ON".
M/T: 1,000±50 rpm
A/T: 1,000±50 rpm (at "N" range)
If out of specification, adjust idle speed.
O.K.
INSPECTION END
No
CHECK POWER SOURCE.
1) Stop engine.
2) Disconnect F.I.C.D. solenoid valve harness connector.
3) Turn ignition switch and air conditioner switch "ON".
4) Check voltage between terminal ⓑ and ground.
Battery voltage should exist.
O.K.
CHECK COMPONENTS.
Check F.I.C.D. solenoid valve. (See page EF & EC-151.)
If N.G., repair or replace F.I.C.D. solenoid valve.
O.K.
CHECK GROUND CIRCUIT.
1) Turn ignition switch "OFF".
2) Check continuity between terminal ⓐ and ground.
Continuity should exist.
If N.G., repair harness or connectors.
N.G.
CHECK COMPONENTS.
1) Turn ignition switch "OFF".
2) Check air conditioner relay.
N.G.
Replace air conditioner relay.
O.K.
CHECK OUTPUT SIGNAL.
1) Reconnect F.I.C.D. solenoid valve harness connector.
2) Turn ignition switch "ON".
3) Check voltage between E.C.U. terminal ⑪ and ground.
A/C switch | Voltage
ON | 0.5 - 1.0V
OFF | Battery voltage
N.G.
CHECK HARNESS CONTINUITY BETWEEN BATTERY AND E.C.U.
Check following.
• Fuse
• Ignition switch
• Harness or connectors.
O.K.
Recheck the E.C.U. pin terminals for damage or the connection of E.C.U. harness connector.
O.K.
INSPECTION END
A/C
PUSH
x1000 r/min
C/UNIT CONNECTOR

ECCS SELF-DIAGNOSTIC CHART—1989–90 240SX

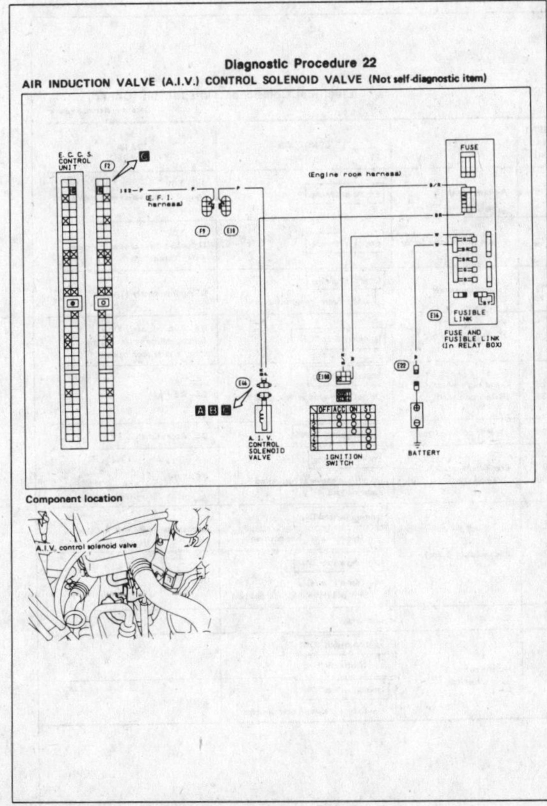

Diagnostic Procedure 22
AIR INDUCTION VALVE (A.I.V.) CONTROL SOLENOID VALVE (Not self-diagnostic item)
E.C.C.S. CONTROL UNIT
(E.F.I. harness)
(Engine room harness)
FUSE
FUSIBLE LINK
FUSE AND FUSIBLE LINK (in RELAY BOX)
A.I.V. CONTROL SOLENOID VALVE
IGNITION SWITCH
BATTERY
Component location
A.I.V. control solenoid valve

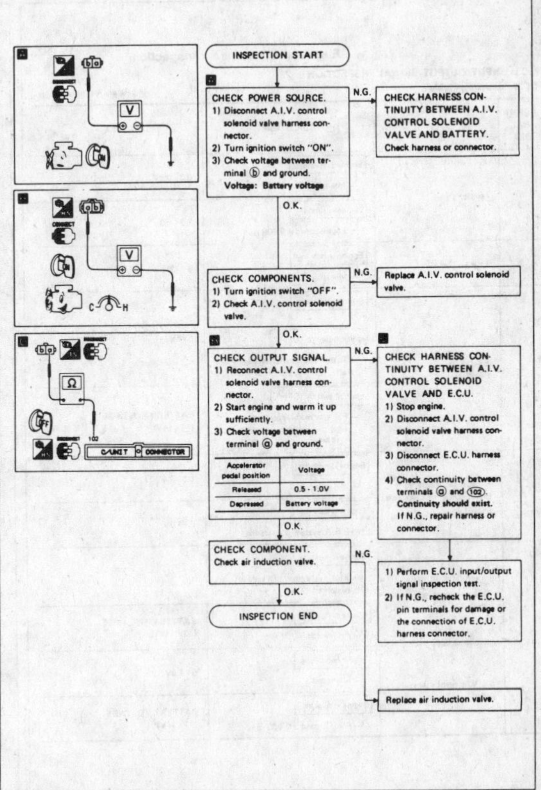

INSPECTION START
CHECK POWER SOURCE.
1) Disconnect A.I.V. control solenoid valve harness connector.
2) Turn ignition switch "ON".
3) Check voltage between terminal ⓑ and ground.
Voltage: Battery voltage
N.G.
CHECK HARNESS CONTINUITY BETWEEN A.I.V. CONTROL SOLENOID VALVE AND BATTERY.
Check harness or connector.
O.K.
CHECK COMPONENTS.
1) Turn ignition switch "OFF".
2) Check A.I.V. control solenoid valve.
N.G.
Replace A.I.V. control solenoid valve.
O.K.
CHECK OUTPUT SIGNAL.
1) Reconnect A.I.V. control solenoid valve harness connector.
2) Start engine and warm it up sufficiently.
3) Check voltage between terminal ⓐ and ground.
Accelerator pedal position | Voltage
Released | 0.5 - 1.0V
Depressed | Battery voltage
N.G.
CHECK HARNESS CONTINUITY BETWEEN A.I.V. CONTROL SOLENOID VALVE AND E.C.U.
1) Stop engine.
2) Disconnect A.I.V. control solenoid valve harness connector.
3) Disconnect E.C.U. harness connector.
4) Check continuity between terminals ⓐ and ⑩②.
Continuity should exist.
If N.G., repair harness or connector.
O.K.
CHECK COMPONENT.
Check air induction valve.
N.G.
1) Perform E.C.U. input/output signal inspection test.
2) If N.G., recheck the E.C.U. pin terminals for damage or the connection of E.C.U. harness connector.
O.K.
INSPECTION END
Replace air induction valve.
C/UNIT CONNECTOR

ECCS SELF-DIAGNOSTIC CHART—1989–90 240SX

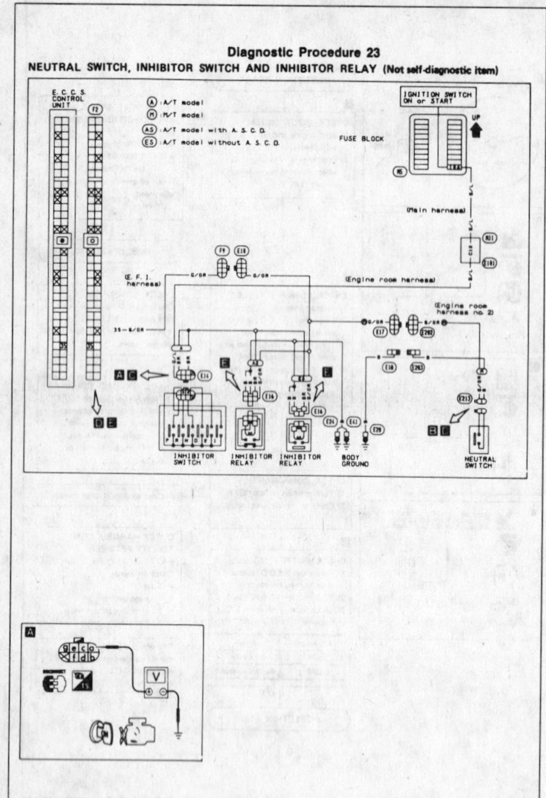
Diagnostic Procedure 23
NEUTRAL SWITCH, INHIBITOR SWITCH AND INHIBITOR RELAY (Not self-diagnostic item)
E.C.C.S. CONTROL UNIT
A/T model
M/T model
A/T model with A.S.C.D.
A/T model without A.S.C.D.
IGNITION SWITCH ON or START
FUSE BLOCK
(Main harness)
(Engine room harness)
(Engine room harness no. 2)
INHIBITOR SWITCH
INHIBITOR RELAY
INHIBITOR RELAY
BODY GROUND
NEUTRAL SWITCH

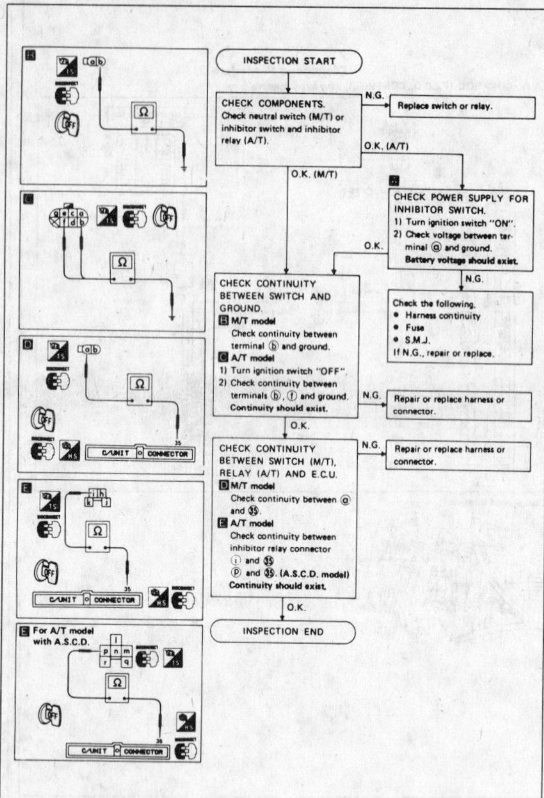
INSPECTION START
CHECK COMPONENTS.
Check neutral switch (M/T) or inhibitor switch and inhibitor relay (A/T).
N.G.
Replace switch or relay.
O.K. (A/T)
O.K. (M/T)
CHECK POWER SUPPLY FOR INHIBITOR SWITCH.
1) Turn ignition switch "ON".
2) Check voltage between terminal ⓐ and ground.
Battery voltage should exist.
O.K.
N.G.
Check the following.
• Harness continuity
• Fuse
• S.M.J.
If N.G., repair or replace.
CHECK CONTINUITY BETWEEN SWITCH AND GROUND.
M/T model
Check continuity between terminal ⓑ and ground.
A/T model
1) Turn ignition switch "OFF".
2) Check continuity between terminals ⓑ, ⓕ and ground.
Continuity should exist.
N.G.
Repair or replace harness or connector.
O.K.
CHECK CONTINUITY BETWEEN SWITCH (M/T), RELAY (A/T) AND E.C.U.
M/T model
Check continuity between ⓐ and 35.
A/T model
Check continuity between inhibitor relay connector ⓘ and 35
ⓟ and 35. (A.S.C.D. model)
Continuity should exist.
N.G.
Repair or replace harness or connector.
O.K.
INSPECTION END
For A/T model with A.S.C.D.
C/UNIT CONNECTOR

ECU SIGNAL INSPECTION—1989–90 240SX

Electrical Components Inspection

E.C.U. INPUT/OUTPUT SIGNAL INSPECTION
E.C.U. Inspection table

*Data are reference values.

TERMINAL NO.	ITEM	CONDITION	*DATA
1	Ignition signal	Engine is running. — Idle speed	0.3 - 0.6V
		Engine is running. — Engine speed is 2,000 rpm	1.2 - 1.5V
3	Ignition check	Engine is running. — Idle speed	9 - 12V
4	E.C.C.S. relay (Main relay)	Engine is running. ↓ Ignition switch "OFF" — Within approximately 1 second after turning ignition switch "OFF"	0 - 1V
		Ignition switch "OFF" — For approximately 1 second after turning ignition switch "OFF"	BATTERY VOLTAGE (11 - 14V)
8	Exhaust gas temperature sensor (Only for California model)	Engine is running. — Idle speed	1.0 - 2.0V
		Engine is running. — E.G.R. system is operating.	0 - 1.0V
11	Air conditioner relay	Engine is running. — Both A/C switch and blower switch are "ON"	0 - 1.0V
		Engine is running. — A/C switch is "OFF".	BATTERY VOLTAGE (11 - 14V)
12	S.C.V. control solenoid valve	Engine is running. — Idle speed	0 - 1.0V
		Engine is running. — Engine speed is 2,000 rpm.	BATTERY VOLTAGE (11 - 14V)

Electrical Components Inspection (Cont'd)

*Data are reference values.

TERMINAL NO.	ITEM	CONDITION	*DATA
16	Air flow meter	Engine is running.	1.0 - 3.0V Output voltage varies with engine revolution.
18	Engine temperature sensor	Engine is running.	1.0 - 5.0V Output voltage varies with engine water temperature.
19	Exhaust gas sensor	Engine is running. — After warming up sufficiently.	0 - Approximately 1.0V
20	Throttle sensor	Ignition switch "ON"	0.4 - Approximately 4V Output voltage varies with the throttle valve opening angle.
22 30	Crank angle sensor (Reference signal)	Engine is running. Do not run engine at high speed under no-load.	0.2 - 0.5V
28	Throttle opening signal	Ignition switch "ON"	0.3 - Approximately 3V
31 40	Crank angle sensor (Position signal)	Engine is running. Do not run engine at high speed under no-load.	2.0 - 3.0V
33	Idle switch (⊖ side)	Ignition switch "ON" — Throttle valve: idle position	Approximately 9 - 10V
		Ignition switch "ON" — Throttle valve: Any position except idle position	0V
34	Start signal	Cranking	8 - 12V
35	Neutral switch & Inhibitor switch	Ignition switch "ON" — Neutral/Parking	0V
		Ignition switch "ON" — Except the above gear position	6 - 7V

ECU SIGNAL INSPECTION (CONT.)—1989–90 240SX

Electrical Components Inspection (Cont'd)

*Data are reference values.

TERMINAL NO.	ITEM	CONDITION	*DATA
36	Ignition switch	Ignition switch "OFF"	0V
		Ignition switch "ON"	BATTERY VOLTAGE (11 - 14V)
37	Throttle sensor power supply	Ignition switch "ON"	Approximately 5V
38 47	Power supply for E.C.U.	Ignition switch "ON"	BATTERY VOLTAGE (11 - 14V)
41	Air conditioner switch	Engine is running. — Both air conditioner switch and blower switch are "ON".	0V
		Engine is running. — Air conditioner switch is "OFF".	BATTERY VOLTAGE (11 - 14V)
43	Power steering oil pressure switch	Engine is running. — Steering wheel is being turned.	0.1 - 0.3V
		Engine is running. — Steering wheel is not being turned.	8 - 9V
44	Idle switch (⊕ side)	Ignition switch "ON" — Throttle valve: idle position	Approximately 9 - 10V
		Ignition switch "ON" — Throttle valve: Except idle position	BATTERY VOLTAGE (11 - 14V)
45	5th position switch (M/T models)	Ignition switch "ON" — Gear is in 5th position.	0V
		Ignition switch "ON" — Gear is except in 5th position.	6 - 8V
46	Power supply (Back-up)	Ignition switch "OFF"	BATTERY VOLTAGE (11 - 14V)

Electrical Components Inspection (Cont'd)

*Data are reference values.

TERMINAL NO.	ITEM	CONDITION	*DATA
101	Injector No. 1	Engine is running.	BATTERY VOLTAGE (11 - 14V)
103	Injector No. 3		
110	Injector No. 2		
112	Injector No. 4		
102	A.I.V. control solenoid valve	Engine is running. — Idle speed	0 - 1.0V
		Engine is running. — Accelerator pedal is depressed. — After warming up	BATTERY VOLTAGE (11 - 14V)
104	Fuel pump relay	Ignition switch "ON" — For 5 seconds after turning ignition switch "ON" Engine is running.	0.7 - 0.9V
		Ignition switch "ON" — Within 5 seconds after turning ignition switch "ON"	BATTERY VOLTAGE (11 - 14V)
105	E.G.R. control solenoid valve	Engine is running. — Engine is cold. [Water temperature is below 60°C (140°F).]	0.7 - 0.9V
		Engine is running. (Racing) — After warming up [Water temperature is between 60°C (140°F) and 105°C (221°F).]	BATTERY VOLTAGE (11 - 14V)

ECU SIGNAL INSPECTION (CONT.)—1989–90 240SX

Electrical Components Inspection (Cont'd)

*Data are reference values.

TERMINAL NO.	ITEM	CONDITION	*DATA
106	Pressure regulator control solenoid valve	Stop and restart engine after warming it up. — Water temperature is above 90°C (194°F)	0 - 1.0V (for 3 minutes after ignition switch is turned off.)
			BATTERY VOLTAGE (After 3 minutes)
		Stop and restart engine after warming it up. — Water temperature is below 90°C (194°F)	BATTERY VOLTAGE (11 - 14V)
113	A.A.C. valve	Engine is running. — Idle speed	7 - 10V
		Engine is running. — Steering wheel is being turned. — Air conditioner is operating. — Rear defogger is "ON". — Headlamps are in high position.	4 - 7V

E.C.U. PIN CONNECTOR TERMINAL LAYOUT

[illegible]	1 2 3 4 5 6 7	15 16 17 18 19 20 21 22	31 32 33 34 35 36 37 38 39
[illegible]	8 9 10 11 12 13 14	23 24 25 26 27 28 29 30	40 41 42 43 44 45 46 47 48

H.S. CONNECT

ECCS COMPONENTS LOCATION—1988–89 300ZX

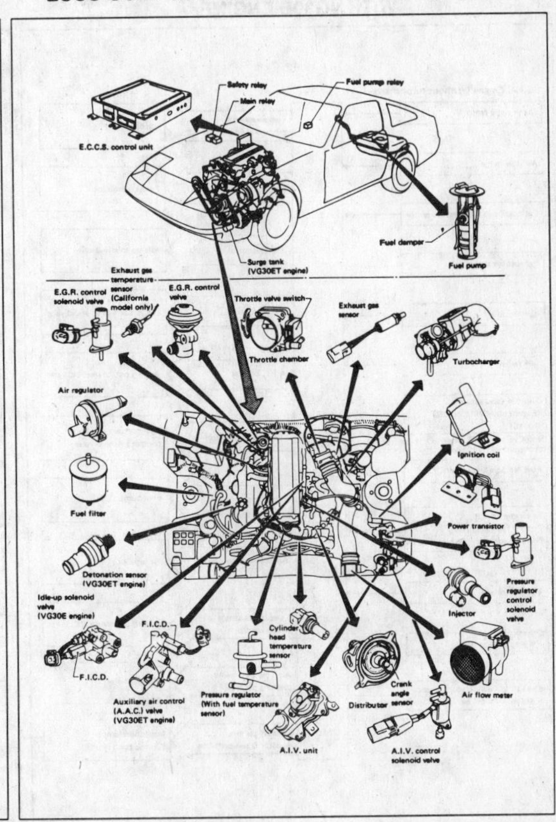
Safety relay
Fuel pump relay
Main relay
E.C.C.S. control unit
Fuel damper
Surge tank
(VG30ET engine)
Fuel pump
Exhaust gas temperature sensor (California model only)
E.G.R. control solenoid valve
E.G.R. control valve
Throttle valve switch
Exhaust gas sensor
Throttle chamber
Turbocharger
Air regulator
Ignition coil
Fuel filter
Power transistor
Detonation sensor
(VG30ET engine)
Pressure regulator control solenoid valve
Injector
Idle-up solenoid valve
(VG30E engine)
F.I.C.D.
Cylinder head temperature sensor
F.I.C.D.
Air flow meter
Auxiliary air control (A.A.C.) valve
(VG30ET engine)
Pressure regulator
(With fuel temperature sensor)
Distributor
Crank angle sensor
A.I.V. unit
A.I.V. control solenoid valve

ECCS SYSTEM SCHEMATIC—1988–89 300ZX WITH VG30E ENGINE

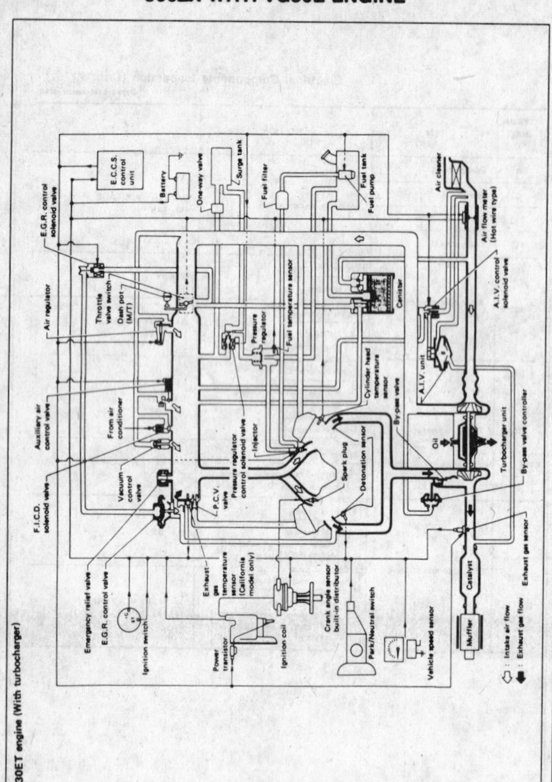
VG30ET engine (With turbocharger)
⇧ : Intake air flow
⬇ : Exhaust gas flow

ECCS SYSTEM SCHEMATIC—1988–89 300ZX WITH VG30ET ENGINE

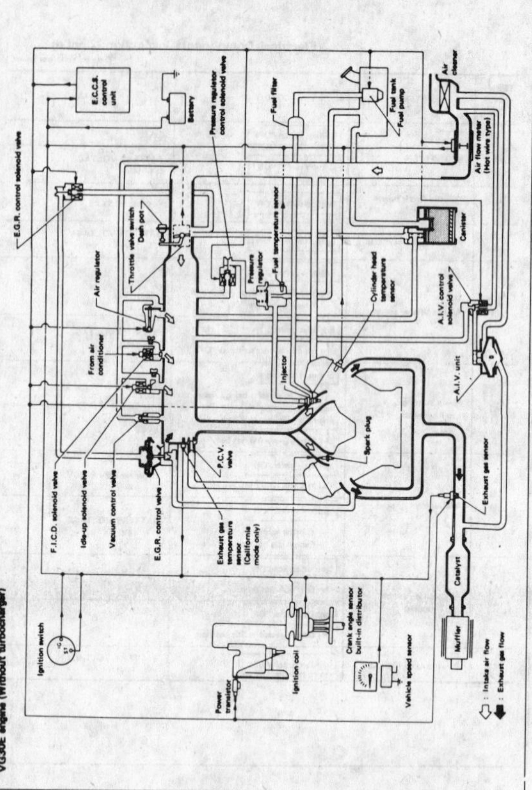
VG30E engine (Without turbocharger)
⇧ : Intake air flow
⬇ : Exhaust gas flow

ECCS CONTROL UNIT CHART—1988–89 300ZX WITH VG30E ENGINE

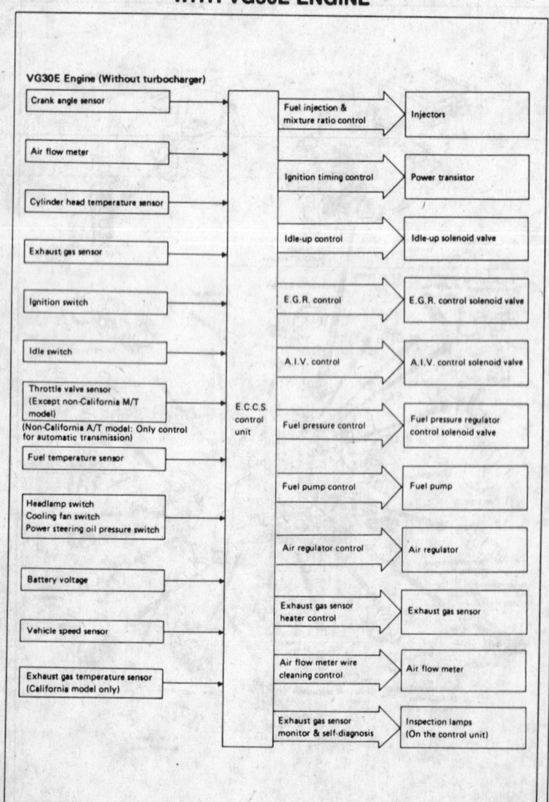
VG30E Engine (Without turbocharger)
Crank angle sensor
Air flow meter
Cylinder head temperature sensor
Exhaust gas sensor
Ignition switch
Idle switch
Throttle valve sensor (Except non-California M/T model)
(Non-California A/T model: Only control for automatic transmission)
Fuel temperature sensor
Headlamp switch
Cooling fan switch
Power steering oil pressure switch
Battery voltage
Vehicle speed sensor
Exhaust gas temperature sensor (California model only)
E.C.C.S. control unit
Fuel injection & mixture ratio control
Injectors
Ignition timing control
Power transistor
Idle-up control
Idle-up solenoid valve
E.G.R. control
E.G.R. control solenoid valve
A.I.V. control
A.I.V. control solenoid valve
Fuel pressure control
Fuel pressure regulator control solenoid valve
Fuel pump control
Fuel pump
Air regulator control
Air regulator
Exhaust gas sensor heater control
Exhaust gas sensor
Air flow meter wire cleaning control
Air flow meter
Exhaust gas sensor monitor & self-diagnosis
Inspection lamps (On the control unit)

ECCS CONTROL UNIT CHART—1988–89 300ZX WITH VG30ET ENGINE

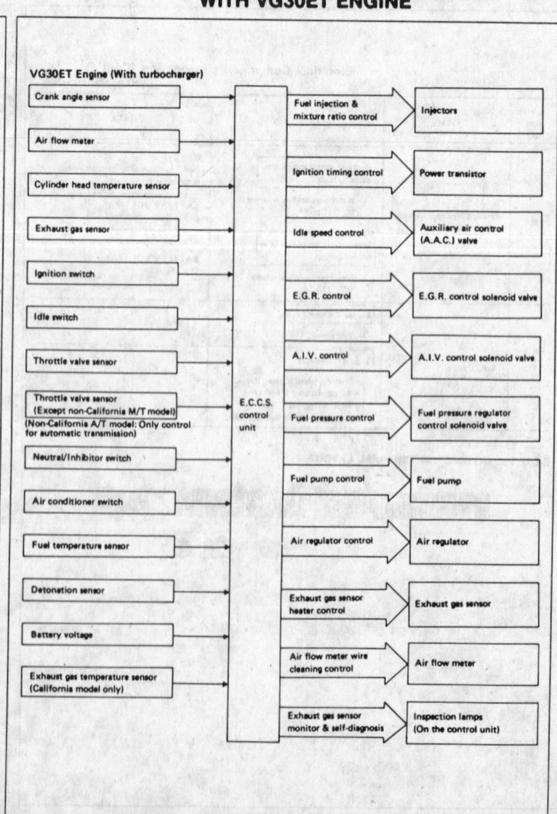
VG30ET Engine (With turbocharger)
Crank angle sensor
Air flow meter
Cylinder head temperature sensor
Exhaust gas sensor
Ignition switch
Idle switch
Throttle valve sensor
Throttle valve sensor (Except non-California M/T model)
(Non-California A/T model: Only control for automatic transmission)
Neutral/Inhibitor switch
Air conditioner switch
Fuel temperature sensor
Detonation sensor
Battery voltage
Exhaust gas temperature sensor (California model only)
E.C.C.S. control unit
Fuel injection & mixture ratio control
Injectors
Ignition timing control
Power transistor
Idle speed control
Auxiliary air control (A.A.C.) valve
E.G.R. control
E.G.R. control solenoid valve
A.I.V. control
A.I.V. control solenoid valve
Fuel pressure control
Fuel pressure regulator control solenoid valve
Fuel pump control
Fuel pump
Air regulator control
Air regulator
Exhaust gas sensor heater control
Exhaust gas sensor
Air flow meter wire cleaning control
Air flow meter
Exhaust gas sensor monitor & self-diagnosis
Inspection lamps (On the control unit)

ECCS FUEL FLOW SYSTEM SCHEMATIC 1988–89 300ZX

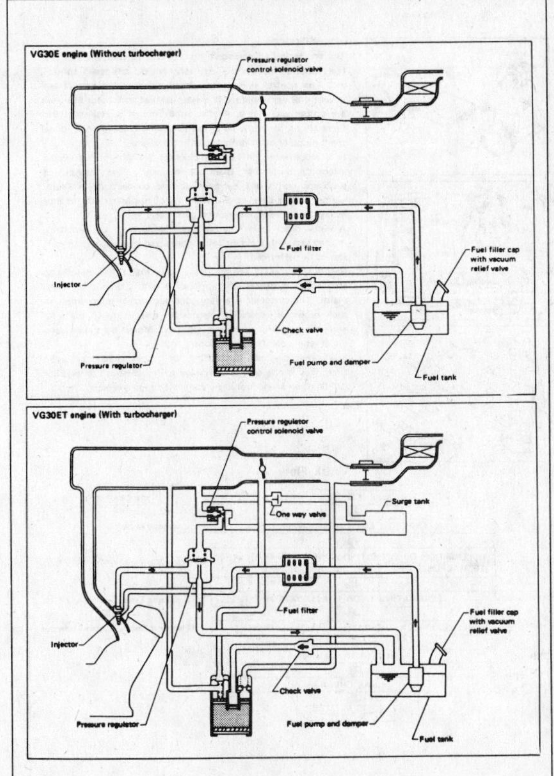
VG30E engine (Without turbocharger)
Pressure regulator control solenoid valve
Fuel filter
Fuel filler cap with vacuum relief valve
Injector
Check valve
Fuel pump and damper
Pressure regulator
Fuel tank
VG30ET engine (With turbocharger)
Pressure regulator control solenoid valve
Surge tank
One way valve
Fuel filter
Fuel filler cap with vacuum relief valve
Injector
Check valve
Fuel pump and damper
Pressure regulator
Fuel tank

ECCS AIR FLOW SYSTEM SCHEMATIC 1988–89 300ZX

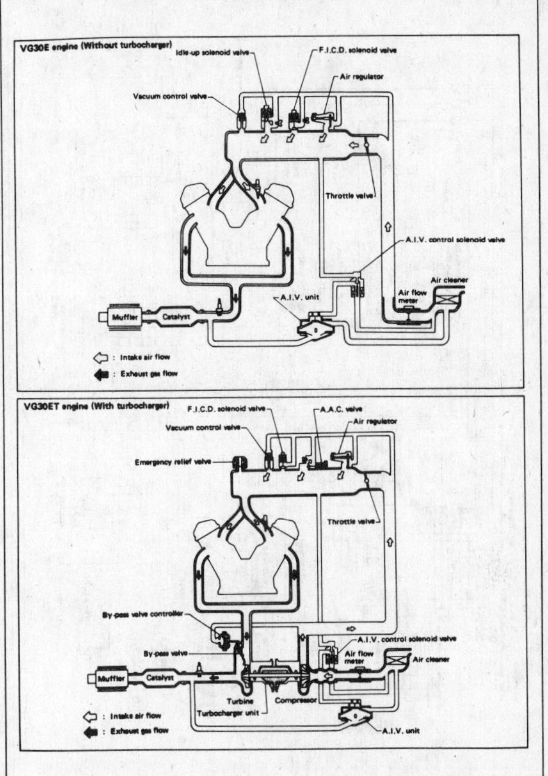
VG30E engine (Without turbocharger)
Idle-up solenoid valve
F.I.C.D. solenoid valve
Air regulator
Vacuum control valve
Throttle valve
A.I.V. control solenoid valve
Air cleaner
Air flow meter
A.I.V. unit
Muffler
Catalyst
: Intake air flow
: Exhaust gas flow
VG30ET engine (With turbocharger)
F.I.C.D. solenoid valve
A.A.C. valve
Air regulator
Vacuum control valve
Emergency relief valve
Throttle valve
By-pass valve controller
A.I.V. control solenoid valve
By-pass valve
Air flow meter
Air cleaner
Muffler
Catalyst
Turbine
Compressor
Turbocharger unit
A.I.V. unit
: Intake air flow
: Exhaust gas flow

ECCS SYSTEM WIRING SCHEMATIC 1988–89 300ZX

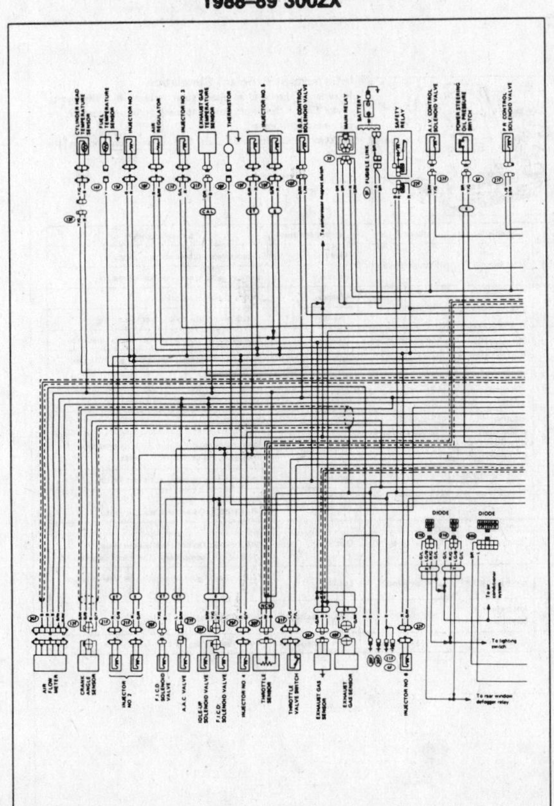

ECCS SYSTEM WIRING SCHEMATIC (CONT.) 1988–89 300ZX

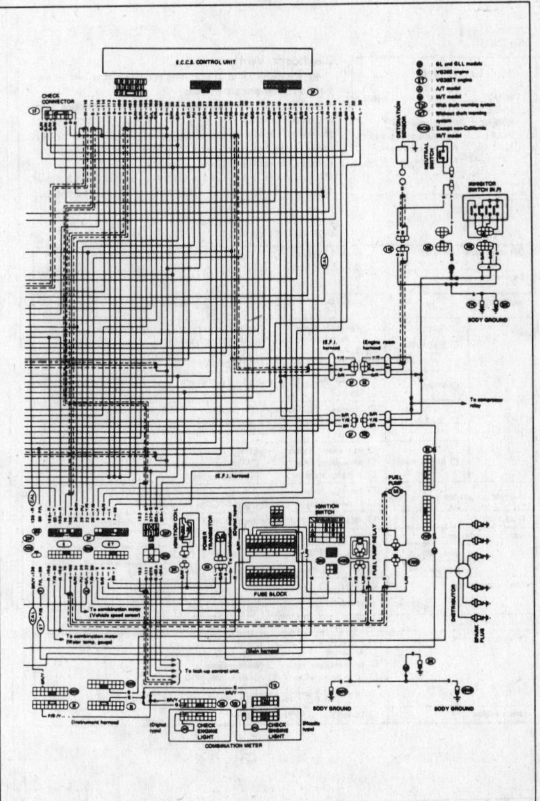
E.C.C.S. CONTROL UNIT
CHECK CONNECTOR
FUSE BLOCK
COMBINATION METER
BODY GROUND

ECCS CONTROL UNIT WIRING SCHEMATIC 1988–89 300ZX

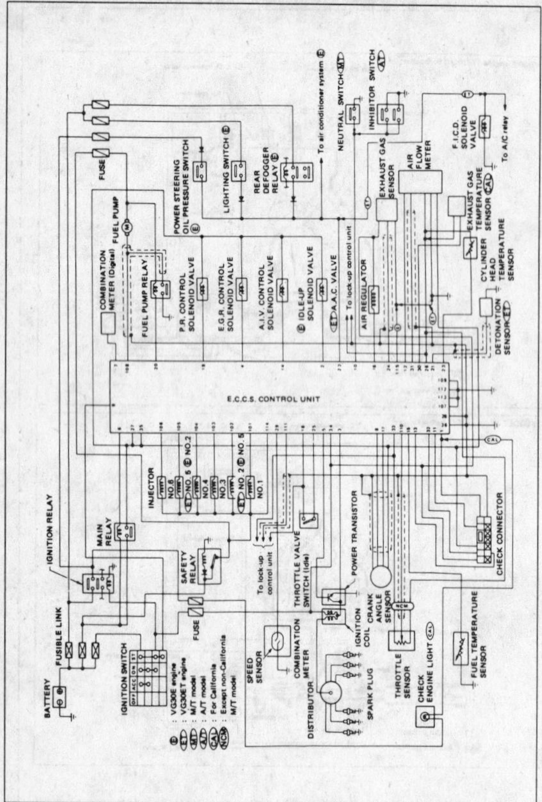

ECCS DIAGNOSTIC DESCRIPTION 1988–89 300ZX

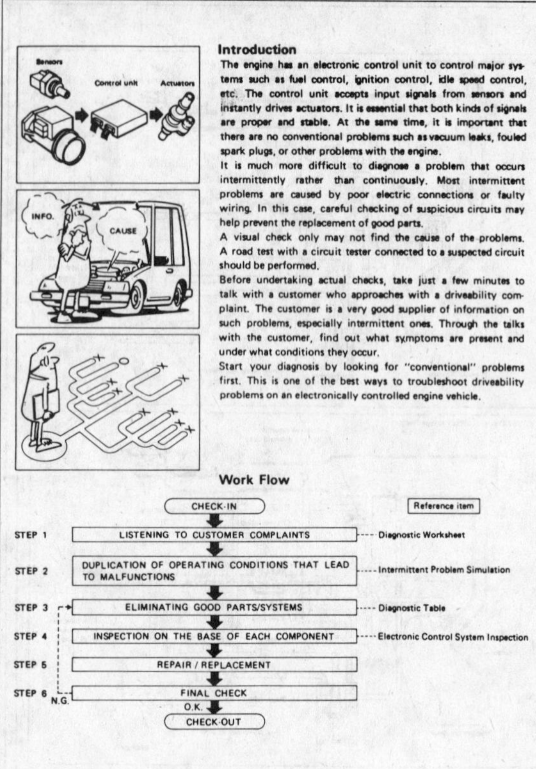
Sensors
Control unit
Actuators
INFO.
CAUSE

Introduction

The engine has an electronic control unit to control major systems such as fuel control, ignition control, idle speed control, etc. The control unit accepts input signals from sensors and instantly drives actuators. It is essential that both kinds of signals are proper and stable. At the same time, it is important that there are no conventional problems such as vacuum leaks, fouled spark plugs, or other problems with the engine.

It is much more difficult to diagnose a problem that occurs intermittently rather than continuously. Most intermittent problems are caused by poor electric connections or faulty wiring. In this case, careful checking of suspicious circuits may help prevent the replacement of good parts.

A visual check only may not find the cause of the problems. A road test with a circuit tester connected to a suspected circuit should be performed.

Before undertaking actual checks, take just a few minutes to talk with a customer who approaches with a driveability complaint. The customer is a very good supplier of information on such problems, especially intermittent ones. Through the talks with the customer, find out what symptoms are present and under what conditions they occur.

Start your diagnosis by looking for "conventional" problems first. This is one of the best ways to troubleshoot driveability problems on an electronically controlled engine vehicle.

Work Flow

CHECK-IN

		Reference item
STEP 1	LISTENING TO CUSTOMER COMPLAINTS	Diagnostic Worksheet
STEP 2	DUPLICATION OF OPERATING CONDITIONS THAT LEAD TO MALFUNCTIONS	Intermittent Problem Simulation
STEP 3	ELIMINATING GOOD PARTS/SYSTEMS	Diagnostic Table
STEP 4	INSPECTION ON THE BASE OF EACH COMPONENT	Electronic Control System Inspection
STEP 5	REPAIR / REPLACEMENT	
STEP 6	FINAL CHECK (N.G. → STEP 3; O.K. → CHECK-OUT)	

CHECK-OUT

ECCS DIAGNOSTIC DESCRIPTION (CONT.)—1988–89 300ZX

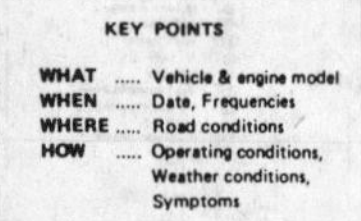

KEY POINTS

WHAT ..... Vehicle & engine model
WHEN ..... Date, Frequencies
WHERE ..... Road conditions
HOW ..... Operating conditions, Weather conditions, Symptoms

Diagnostic Worksheet

There are many kinds of operating conditions that lead to malfunctions on engine components.

A good grasp of such conditions can make troubleshooting faster and more accurate.

In general, feelings for a problem depend on each customer. It is important to fully understand the symptoms or under what conditions a customer complains.

Make good use of a diagnostic worksheet such as the one shown below in order to utilize all the complaints for troubleshooting.

WORKSHEET SAMPLE

Customer name MR/MS	Model & Year	VIN
Engine #	Trans.	Mileage
Incident Date	Manuf. Date	In Service Date
Symptoms	☐ Startability	☐ Impossible to start ☐ No combustion ☐ Partial combustion ☐ Partial combustion affected by throttle position ☐ Partial combustion NOT affected by throttle position ☐ Possible but hard to start ☐ Others [
	☐ Idling	☐ No fast idle ☐ Unstable ☐ High idle ☐ Low idle ☐ Others [ ]
	☐ Driveability	☐ Stumble ☐ Surge ☐ Detonation ☐ Lack of power ☐ Intake backfire ☐ Exhaust backfire ☐ Others [ ]
	☐ Engine stall	☐ At the time of start ☐ While idling ☐ While accelerating ☐ While decelerating ☐ Just after stopping ☐ While loading
Incident occurrence		☐ Just after delivery ☐ Recently ☐ In the morning ☐ At night ☐ In the daytime
Frequency		☐ All the time ☐ Under certain conditions ☐ Sometimes
Weather conditions		☐ Not effected
	Weather	☐ Fine ☐ Raining ☐ Snowing ☐ Others [ ]
	Temperature	☐ Hot ☐ Warm ☐ Cool ☐ Cold ☐ Humid °F
Engine conditions		☐ Cold ☐ During warm-up ☐ After warm-up Engine speed 0 2,000 4,000 6,000 8,000 rpm
Road conditions		☐ In town ☐ In suburbs ☐ Highway ☐ Off road (up/down)
Driving conditions		☐ Not affected ☐ At starting ☐ While idling ☐ At racing ☐ While accelerating ☐ While cruising ☐ While decelerating ☐ While turning (RH/LH) Vehicle speed 0 10 20 30 40 50 60 MPH
Check engine light		☐ Turned on ☐ Not turned on

GRRRRR
GRRRR
THIS is the symptom, isn't it ?

Intermittent Problem Simulation

In order to duplicate an intermittent problem, it is effective to create similar conditions for component parts, under which the problem might occur.

Perform the activity listed under Service procedure and note the result.

	Variable factor	Influential part	Target condition	Service procedure
1	Mixture ratio	Pressure regulator	Made lean	Remove vacuum hose and apply vacuum.
			Made rich	Remove vacuum hose and apply pressure.
2	Ignition timing	Distributor	Advanced	Rotate distributor clockwise.
			Retarded	Rotate distributor counterclockwise.
3	Mixture ratio feedback control	Exhaust gas sensor	Suspended	Disconnect exhaust gas sensor harness connector.
		Control unit	Operation check	Perform self-diagnosis (Mode I/II) at 2,000 rpm.
4	Idle speed	I.A.A. unit	Raised	Turn idle adjust screw counterclockwise.
			Lowered	Turn idle adjust screw clockwise.
5	Electric connection (Electric continuity)	Harness connectors and wires	Poor electric connection or faulty wiring	Tap or wiggle.
				Race engine rapidly. See if the torque reaction of the engine unit causes electric breaks.
6	Temperature	Control unit	Cooled	Cool with an icing spray or similar device.
			Warmed	Heat with a hair drier. [WARNING: Do not overheat the unit.]
7	Moisture	Electric parts	Damp	Wet [WARNING: Do not directly pour water on components. Use a mist sprayer.]
8	Electric loads	Load switches	Loaded	Turn on head lights, air conditioner, rear defogger, etc.
9	Idle switch condition	Control unit	ON-OFF switching	Perform self-diagnosis (Mode IV).
10	Ignition spark	Timing light	Spark power check	Try to flash timing light for each cylinder.

ECCS SPECIFICATIONS CHART—1988–89 300ZX

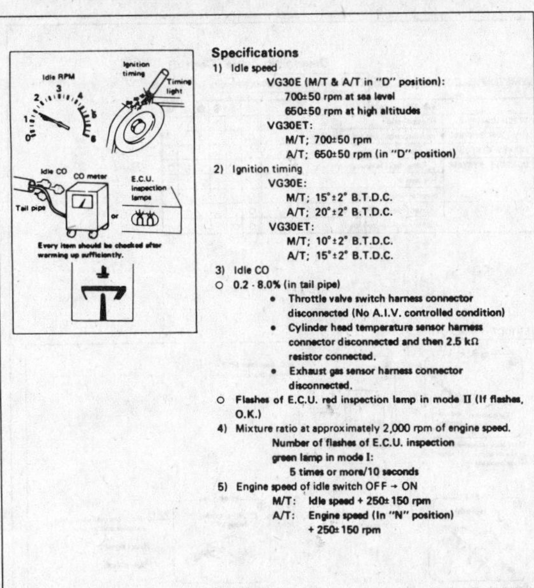
Idle RPM
Ignition timing
Timing light
Idle CO
CO meter
E.C.U. inspection lamps
Tail pipe
or
Every item should be checked after warming up sufficiently.

Specifications

1) Idle speed
   - VG30E (M/T & A/T in "D" position):
     - 700±50 rpm at sea level
     - 650±50 rpm at high altitudes
   - VG30ET:
     - M/T; 700±50 rpm
     - A/T; 650±50 rpm (in "D" position)
2) Ignition timing
   - VG30E:
     - M/T; 15°±2° B.T.D.C.
     - A/T; 20°±2° B.T.D.C.
   - VG30ET:
     - M/T; 10°±2° B.T.D.C.
     - A/T; 15°±2° B.T.D.C.
3) Idle CO
   - ○ 0.2 - 8.0% (in tail pipe)
     - Throttle valve switch harness connector disconnected (No A.I.V. controlled condition)
     - Cylinder head temperature sensor harness connector disconnected and then 2.5 kΩ resistor connected.
     - Exhaust gas sensor harness connector disconnected.
   - ○ Flashes of E.C.U. red inspection lamp in mode II (If flashes, O.K.)
4) Mixture ratio at approximately 2,000 rpm of engine speed.
   - Number of flashes of E.C.U. inspection green lamp in mode I:
     - 5 times or more/10 seconds
5) Engine speed of idle switch OFF → ON
   - M/T: Idle speed + 250±150 rpm
   - A/T: Engine speed (In "N" position) + 250±150 rpm

ECCS DIAGNOSTIC CHART—1988–89 300ZX

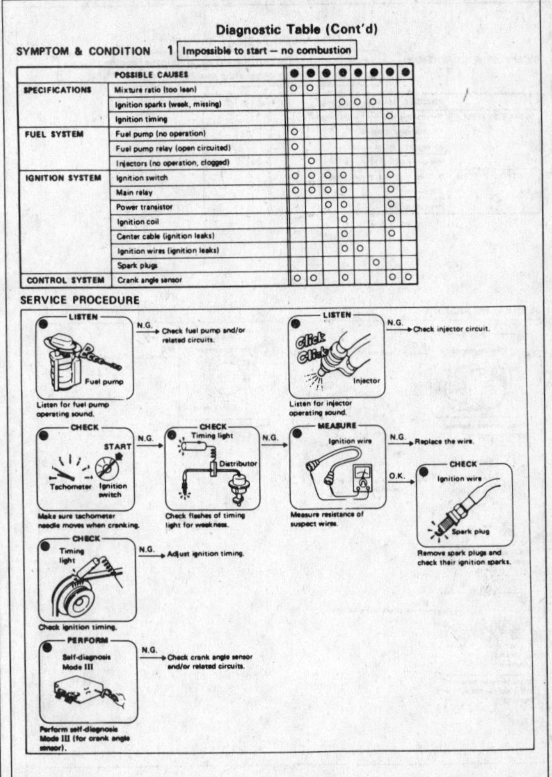

Diagnostic Table (Cont'd)

SYMPTOM & CONDITION 1 Impossible to start — no combustion

	POSSIBLE CAUSES	●	●	●	●	●	●	●	●
SPECIFICATIONS	Mixture ratio (too lean)	○	○						
	Ignition sparks (weak, missing)				○	○	○		
	Ignition timing							○	
FUEL SYSTEM	Fuel pump (no operation)	○							
	Fuel pump relay (open circuited)	○							
	Injectors (no operation, clogged)		○						
IGNITION SYSTEM	Ignition switch	○	○	○	○			○	
	Main relay	○	○	○	○			○	
	Power transistor			○	○			○	
	Ignition coil				○			○	
	Center cable (ignition leaks)				○			○	
	Ignition wires (ignition leaks)				○	○			
	Spark plugs						○		
CONTROL SYSTEM	Crank angle sensor	○	○		○			○	○

SERVICE PROCEDURE

LISTEN
Fuel pump
N.G. → Check fuel pump and/or related circuits.
Listen for fuel pump operating sound.
LISTEN
Click Click
Injector
N.G. → Check injector circuit.
Listen for injector operating sound.
CHECK
START
Tachometer
Ignition switch
Make sure tachometer needle moves when cranking.
N.G.
CHECK
Timing light
Distributor
Check flashes of timing light for weakness.
N.G.
MEASURE
Ignition wire
N.G. → Replace the wire.
Measure resistance of suspect wire.
O.K.
CHECK
Ignition wire
Spark plug
Remove spark plugs and check their ignition sparks.
CHECK
Timing light
N.G. → Adjust ignition timing.
Check ignition timing.
PERFORM
Self-diagnosis Mode III
N.G. → Check crank angle sensor and/or related circuits.
Perform self-diagnosis Mode III (for crank angle sensor).

ECCS DIAGNOSTIC CHART—1988–89 300ZX

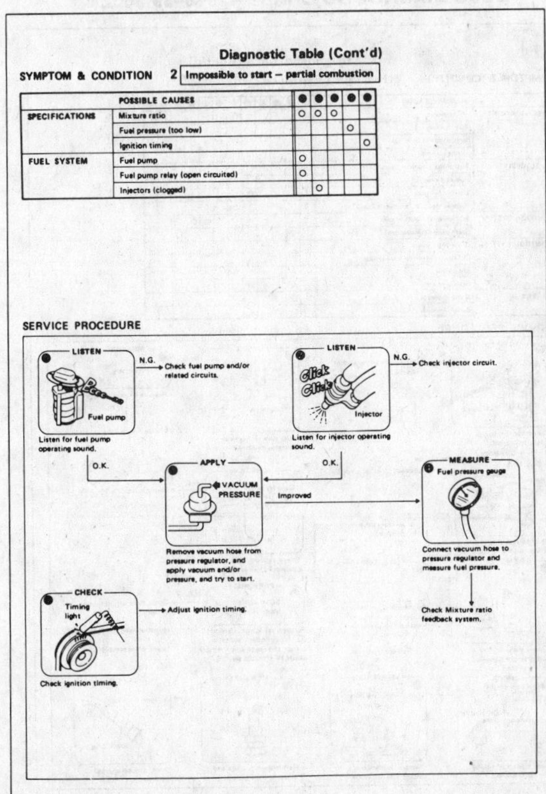

Diagnostic Table (Cont'd)

SYMPTOM & CONDITION 2 Impossible to start — partial combustion

	POSSIBLE CAUSES	●	●	●	●	●
SPECIFICATIONS	Mixture ratio	○	○	○		
	Fuel pressure (too low)				○	
	Ignition timing					○
FUEL SYSTEM	Fuel pump	○				
	Fuel pump relay (open circuited)	○				
	Injectors (clogged)		○			

SERVICE PROCEDURE

LISTEN
Fuel pump
N.G. → Check fuel pump and/or related circuits.
Listen for fuel pump operating sound.
LISTEN
Click Click
Injector
N.G. → Check injector circuit.
Listen for injector operating sound.
O.K.
APPLY
VACUUM PRESSURE
Remove vacuum hose from pressure regulator, and apply vacuum and/or pressure, and try to start.
O.K.
Improved
MEASURE
Fuel pressure gauge
Connect vacuum hose to pressure regulator and measure fuel pressure.
Check Mixture ratio feedback system.
CHECK
Timing light
→ Adjust ignition timing.
Check ignition timing.

ECCS DIAGNOSTIC CHART—1988–89 300ZX

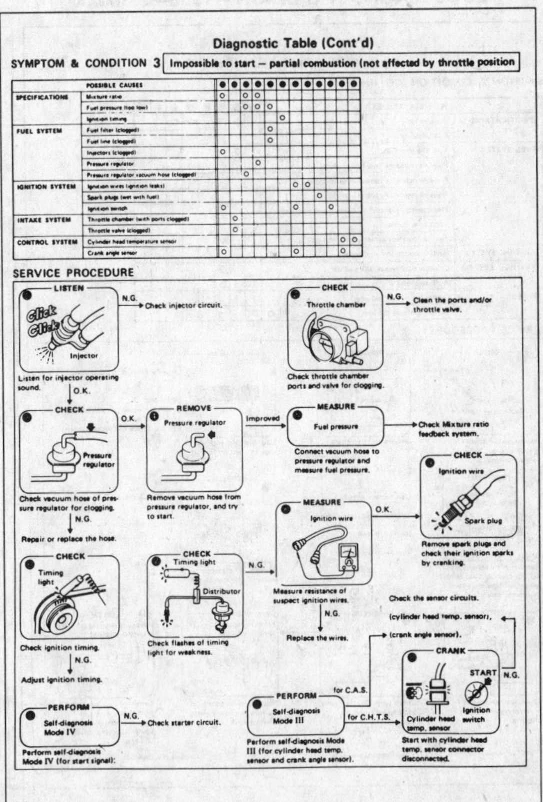

Diagnostic Table (Cont'd)

SYMPTOM & CONDITION 3 Impossible to start — partial combustion (not affected by throttle position)

	POSSIBLE CAUSES	●	●	●	●	●	●	●	●	●	●	●	●
SPECIFICATIONS	Mixture ratio	○		○	○								
	Fuel pressure (too low)			○	○	○							
	Ignition timing						○						
FUEL SYSTEM	Fuel filter (clogged)					○							
	Fuel line (clogged)					○							
	Injectors (clogged)	○											
	Pressure regulator				○								
	Pressure regulator vacuum hose (clogged)			○									
IGNITION SYSTEM	Ignition wires (ignition leaks)							○	○				
	Spark plugs (wet with fuel)									○			
	Ignition switch	○						○			○		
INTAKE SYSTEM	Throttle chamber (with ports clogged)		○										
	Throttle valve (clogged)		○										
CONTROL SYSTEM	Cylinder head temperature sensor											○	○
	Crank angle sensor	○						○				○	

SERVICE PROCEDURE

LISTEN
Click Click
Injector
N.G. → Check injector circuit.
Listen for injector operating sound.
O.K.
CHECK
Throttle chamber
N.G. → Clean the ports and/or throttle valve.
Check throttle chamber ports and valve for clogging.
CHECK
Pressure regulator
Check vacuum hose of pressure regulator for clogging.
N.G.
Repair or replace the hose.
O.K.
REMOVE
Pressure regulator
Remove vacuum hose from pressure regulator, and try to start.
Improved
MEASURE
Fuel pressure
→ Check Mixture ratio feedback system.
Connect vacuum hose to pressure regulator and measure fuel pressure.
CHECK
Ignition wire
Spark plug
Remove spark plugs and check their ignition sparks by cranking.
MEASURE
Ignition wire
O.K.
Measure resistance of suspect ignition wires.
N.G.
Replace the wires.
CHECK
Timing light
Check ignition timing.
N.G.
Adjust ignition timing.
CHECK
Timing light
Distributor
N.G.
Check flashes of timing light for weakness.
Check the sensor circuits.
(cylinder head temp. sensor),
(crank angle sensor).
CRANK
START
N.G.
Cylinder head temp. sensor
Ignition switch
Start with cylinder head temp. sensor connector disconnected.
PERFORM
Self-diagnosis Mode III
for C.A.S.
for C.H.T.S.
Perform self-diagnosis Mode III (for cylinder head temp. sensor and crank angle sensor).
PERFORM
Self-diagnosis Mode IV
N.G. → Check starter circuit.
Perform self-diagnosis Mode IV (for start signal).

ECCS DIAGNOSTIC CHART—1988–89 300ZX

Diagnostic Table (Cont'd)

SYMPTOM & CONDITION 4 Impossible to start – partial combustion (throttle position changes combustion quality)

	POSSIBLE CAUSES	●	●	●	●
INTAKE SYSTEM	Throttle chamber (with ports clogged)	○			
	Throttle valve (clogged)		○		
	Air regulator (stuck closed)			○	
	Idle speed control valve				○
CONTROL SYSTEM	Cylinder head temperature sensor				○
	Idle switch				○
	Neutral switch				○

SERVICE PROCEDURE

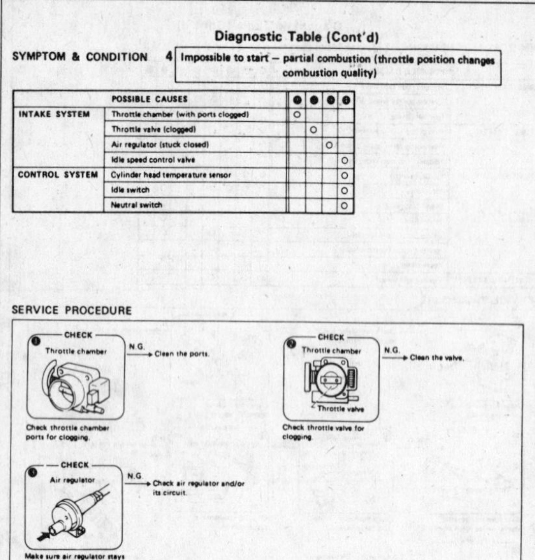
CHECK
Throttle chamber
N.G.
Clean the ports.
Check throttle chamber ports for clogging.
CHECK
Throttle chamber
Throttle valve
N.G.
Clean the valve.
Check throttle valve for clogging.
CHECK
Air regulator
N.G.
Check air regulator and/or its circuit.
Make sure air regulator stays open before warm-up
CHECK
START
Idle speed control valve
N.G.
Check idle speed control circuit.
Check terminal voltage of idle speed control valve while cranking.

ECCS DIAGNOSTIC CHART—1988–89 300ZX

Diagnostic Table (Cont'd)

SYMPTOM & CONDITION 5 Hard to start – before warm-up

	POSSIBLE CAUSES	●	●	●	●	●	●
SPECIFICATIONS	Mixture ratio			○			○
IGNITION SYSTEM	Ignition switch (no start signal)	○			○		
INTAKE SYSTEM	Air regulator			○			
CONTROL SYSTEM	Cylinder head temperature sensor					○	○
	Idle switch				○		
	Neutral switch	○					
OTHERS	Starter (operation too slow)	○					
	Battery (voltage too low)	○	○				

SERVICE PROCEDURE

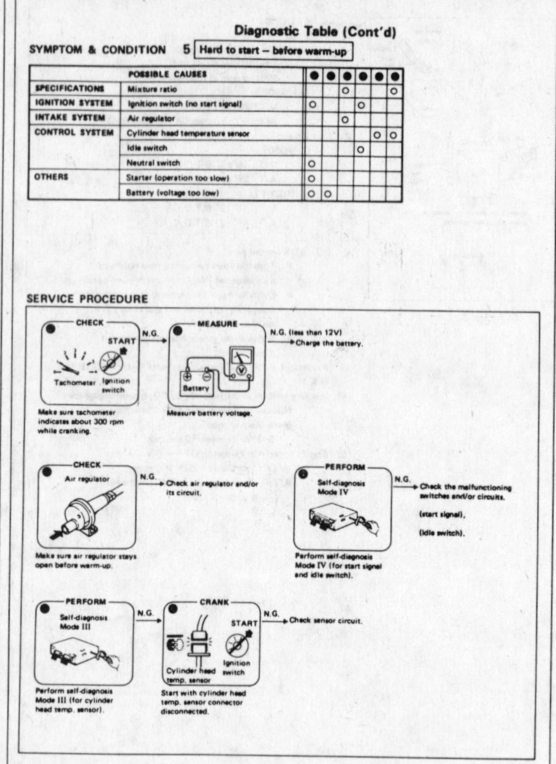
CHECK
START
Tachometer
Ignition switch
Make sure tachometer indicates about 300 rpm while cranking.
N.G.
MEASURE
Battery
Measure battery voltage.
N.G. (less than 12V)
Charge the battery.
CHECK
Air regulator
N.G.
Check air regulator and/or its circuit.
Make sure air regulator stays open before warm-up.
PERFORM
Self-diagnosis Mode IV
N.G.
Check the malfunctioning switches and/or circuits.
(start signal),
(idle switch).
Perform self-diagnosis Mode IV (for start signal and idle switch).
PERFORM
Self-diagnosis Mode III
Perform self-diagnosis Mode III (for cylinder head temp. sensor).
N.G.
CRANK
START
Cylinder head temp. sensor
Ignition switch
Start with cylinder head temp. sensor connector disconnected.
N.G.
Check sensor circuit.

ECCS DIAGNOSTIC CHART—1988–89 300ZX

Diagnostic Table (Cont'd)

SYMPTOM & CONDITION 6 Hard to start – after warm-up

	POSSIBLE CAUSES	●	●	●	●	●	●	●	●	●
SPECIFICATIONS	Mixture ratio			○			○	○		
	Fuel pressure			○			○	○	○	
FUEL SYSTEM	Fuel line (hot fuel)			○						
	Pressure regulator (low fuel pressure)						○			
	Pressure regulator vacuum hose (clogged)						○			
	Pressure regulator control solenoid								○	○
	Pressure regulator control solenoid vacuum hose						○			
	Surge tank (cracks)							○		
	Fuel temperature sensor (open circuited)				○					
IGNITION SYSTEM	Ignition switch (no start signal)	○				○				
CONTROL SYSTEM	Cylinder head temperature sensor				○					
	Air flow meter				○					
OTHERS	Starter (operation too slow)	○								
	Battery (voltage too low)	○	○							

SERVICE PROCEDURE

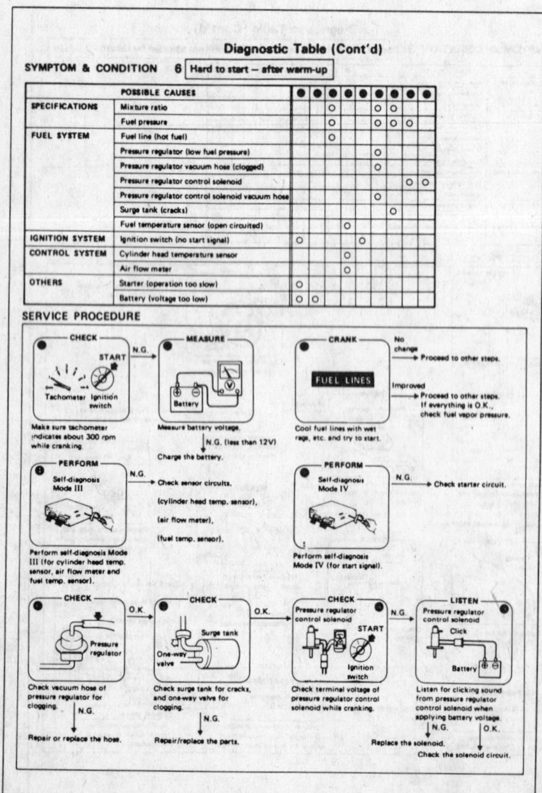
CHECK
START
Tachometer
Ignition switch
Make sure tachometer indicates about 300 rpm while cranking.
N.G.
MEASURE
Battery
Measure battery voltage.
N.G. (less than 12V)
Charge the battery.
CRANK
FUEL LINES
Cool fuel lines with wet rags, etc. and try to start.
No change
Proceed to other steps.
Improved
Proceed to other steps. If everything is O.K., check fuel vapor pressure.
PERFORM
Self-diagnosis Mode III
N.G.
Check sensor circuits.
(cylinder head temp. sensor),
(air flow meter),
(fuel temp. sensor).
Perform self-diagnosis Mode III (for cylinder head temp. sensor, air flow meter and fuel temp. sensor).
PERFORM
Self-diagnosis Mode IV
N.G.
Check starter circuit.
Perform self-diagnosis Mode IV (for start signal).
CHECK
Pressure regulator
Check vacuum hose of pressure regulator for clogging.
N.G.
Repair or replace the hose.
O.K.
CHECK
Surge tank
One-way valve
Check surge tank for cracks, and one-way valve for clogging.
N.G.
Repair/replace the parts.
O.K.
CHECK
Pressure regulator control solenoid
START
Ignition switch
Check terminal voltage of pressure regulator control solenoid while cranking.
N.G.
LISTEN
Pressure regulator control solenoid
Click
Battery
Listen for clicking sound from pressure regulator control solenoid when applying battery voltage.
N.G.
Replace the solenoid.
O.K.
Check the solenoid circuit.

ECCS DIAGNOSTIC CHART—1988–89 300ZX

Diagnostic Table (Cont'd)

SYMPTOM & CONDITION 7 Hard to start – every time

	POSSIBLE CAUSES	●	●	●	●	●	●	●	●	●	●	●	●
SPECIFICATIONS	Mixture ratio	○				○	○						
	Fuel pressure						○	○					
	Ignition sparks (missing)								○	○		○	
	Ignition timing				○								
FUEL SYSTEM	Fuel pump (improper operation)	○											
	Fuel line (clogged)							○					
	Canister (air leaks)					○							
	Pressure regulator (low fuel pressure)						○						
IGNITION SYSTEM	Ignition wires (ignition leaks)								○	○			
	Spark plugs (improper gap)										○		
CONTROL SYSTEM	Crank angle sensor	○							○				○
	Cylinder head temperature sensor												○
	Idle switch												○
	Neutral switch		○										
OTHERS	Starter (operation too slow)		○										
	Battery (voltage too low)		○	○									

SERVICE PROCEDURE

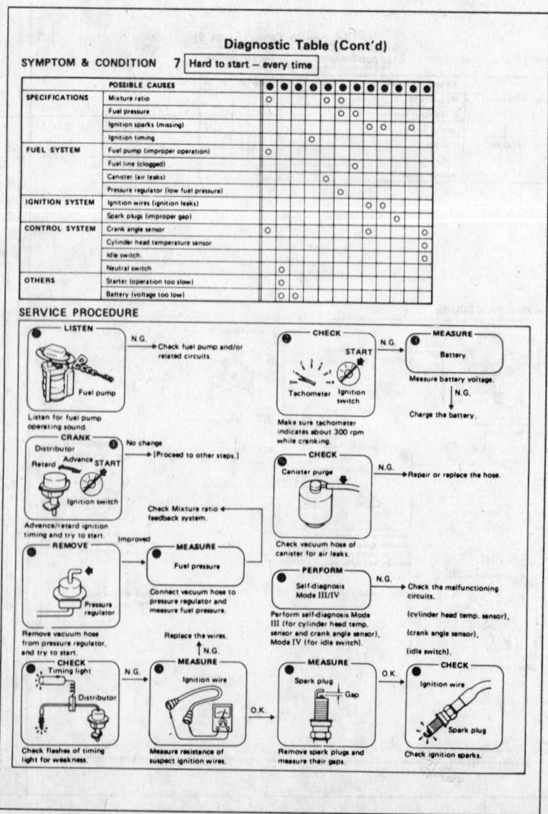
LISTEN
Fuel pump
Listen for fuel pump operating sound.
N.G.
Check fuel pump and/or related circuits.
CHECK
START
Tachometer
Ignition switch
Make sure tachometer indicates about 300 rpm while cranking.
N.G.
MEASURE
Battery
Measure battery voltage.
N.G.
Charge the battery.
CRANK
Distributor
Retard
Advance
START
Ignition switch
Advance/retard ignition timing and try to start.
No change
(Proceed to other steps.)
Improved
CHECK
Canister purge
N.G.
Repair or replace the hose.
Check vacuum hose of canister for air leaks.
Check Mixture ratio feedback system.
REMOVE
Pressure regulator
Remove vacuum hose from pressure regulator, and try to start.
MEASURE
Fuel pressure
Connect vacuum hose to pressure regulator and measure fuel pressure.
PERFORM
Self-diagnosis Mode III/IV
N.G.
Check the malfunctioning circuits.
(cylinder head temp. sensor),
(crank angle sensor),
(idle switch).
Perform self-diagnosis Mode III (for cylinder head temp. sensor and crank angle sensor), Mode IV (for idle switch).
CHECK
Timing light
Distributor
Check flashes of timing light for weakness.
N.G.
MEASURE
Ignition wire
Measure resistance of suspect ignition wires.
N.G.
Replace the wires.
O.K.
MEASURE
Spark plug
Gap
Remove spark plugs and measure their gaps.
O.K.
CHECK
Ignition wire
Spark plug
Check ignition sparks.

ECCS DIAGNOSTIC CHART—1988–89 300ZX

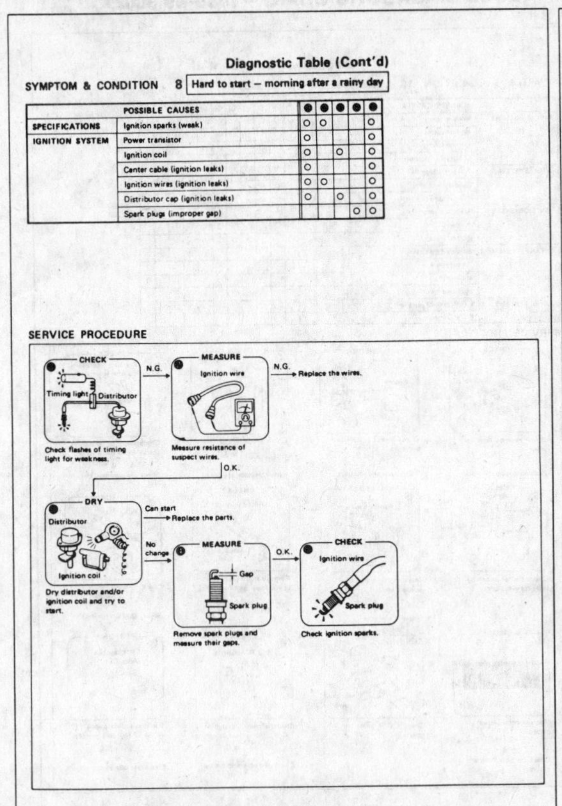

Diagnostic Table (Cont'd)

SYMPTOM & CONDITION 8 Hard to start – morning after a rainy day

	POSSIBLE CAUSES	●	●	●	●	●
SPECIFICATIONS	Ignition sparks (weak)	○	○			○
IGNITION SYSTEM	Power transistor	○				○
	Ignition coil	○		○		○
	Center cable (ignition leaks)	○				○
	Ignition wires (ignition leaks)	○	○			○
	Distributor cap (ignition leaks)	○		○		○
	Spark plugs (improper gap)				○	○

SERVICE PROCEDURE

CHECK
Timing light
Distributor
Check flashes of timing light for weakness.
N.G.
MEASURE
Ignition wire
Measure resistance of suspect wires.
N.G.
Replace the wires.
O.K.
DRY
Distributor
Ignition coil
Dry distributor and/or ignition coil and try to start.
Can start
Replace the parts.
No change
MEASURE
Gap
Spark plug
Remove spark plugs and measure their gaps.
O.K.
CHECK
Ignition wire
Spark plug
Check ignition sparks.

ECCS DIAGNOSTIC CHART—1988–89 300ZX

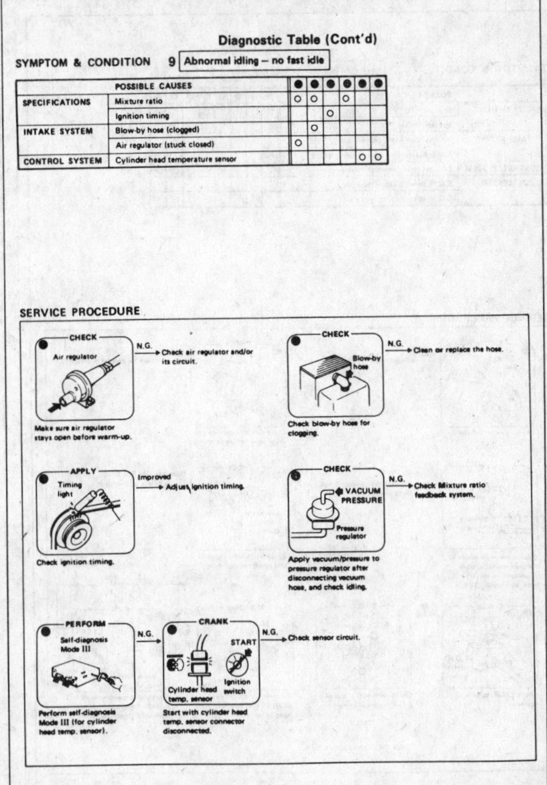

Diagnostic Table (Cont'd)

SYMPTOM & CONDITION 9 Abnormal idling – no fast idle

	POSSIBLE CAUSES	●	●	●	●	●	●
SPECIFICATIONS	Mixture ratio	○	○		○		
	Ignition timing			○			
INTAKE SYSTEM	Blow-by hose (clogged)		○				
	Air regulator (stuck closed)	○					
CONTROL SYSTEM	Cylinder head temperature sensor					○	○

SERVICE PROCEDURE

CHECK
Air regulator
Make sure air regulator stays open before warm-up.
N.G.
Check air regulator and/or its circuit.
CHECK
Blow-by hose
Check blow-by hose for clogging.
N.G.
Clean or replace the hose.
APPLY
Timing light
Check ignition timing.
Improved
Adjust ignition timing.
CHECK
VACUUM
PRESSURE
Pressure regulator
Apply vacuum/pressure to pressure regulator after disconnecting vacuum hose, and check idling.
N.G.
Check Mixture ratio feedback system.
PERFORM
Self-diagnosis Mode III
Perform self-diagnosis Mode III (for cylinder head temp. sensor).
N.G.
CRANK
START
Cylinder head temp. sensor
Ignition switch
Start with cylinder head temp. sensor connector disconnected.
N.G.
Check sensor circuit.

ECCS DIAGNOSTIC CHART—1988–89 300ZX

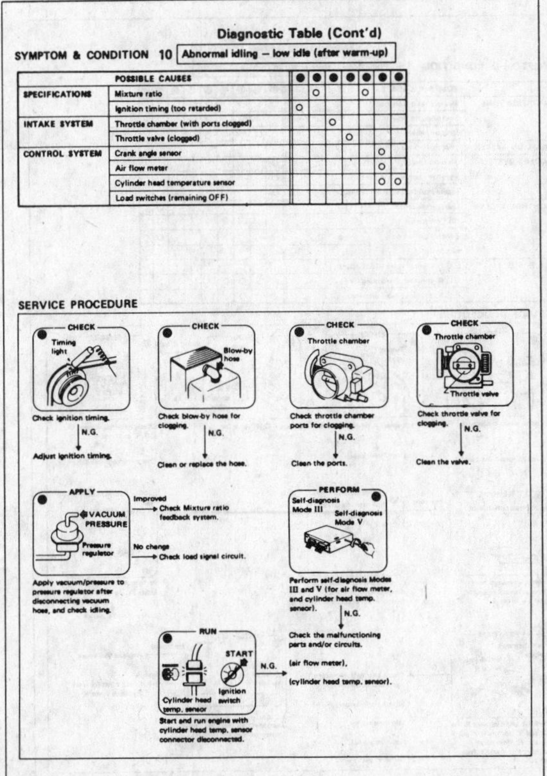

Diagnostic Table (Cont'd)

SYMPTOM & CONDITION 10 Abnormal idling – low idle (after warm-up)

	POSSIBLE CAUSES	●	●	●	●	●	●	●
SPECIFICATIONS	Mixture ratio		○			○		
	Ignition timing (too retarded)	○						
INTAKE SYSTEM	Throttle chamber (with ports clogged)			○				
	Throttle valve (clogged)				○			
CONTROL SYSTEM	Crank angle sensor						○	
	Air flow meter						○	
	Cylinder head temperature sensor						○	○
	Load switches (remaining OFF)							

SERVICE PROCEDURE

CHECK
Timing light
Check ignition timing.
N.G.
Adjust ignition timing.
CHECK
Blow-by hose
Check blow-by hose for clogging.
N.G.
Clean or replace the hose.
CHECK
Throttle chamber
Check throttle chamber ports for clogging.
N.G.
Clean the ports.
CHECK
Throttle chamber
Throttle valve
Check throttle valve for clogging.
N.G.
Clean the valve.
APPLY
VACUUM
PRESSURE
Pressure regulator
Apply vacuum/pressure to pressure regulator after disconnecting vacuum hose, and check idling.
Improved
Check Mixture ratio feedback system.
No change
Check load signal circuit.
PERFORM
Self-diagnosis Mode III
Self-diagnosis Mode V
Perform self-diagnosis Modes III and V (for air flow meter, and cylinder head temp. sensor).
N.G.
Check the malfunctioning parts and/or circuits.
(air flow meter),
(cylinder head temp. sensor).
RUN
START
Cylinder head temp. sensor
Ignition switch
Start and run engine with cylinder head temp. sensor connector disconnected.
N.G.

ECCS DIAGNOSTIC CHART—1988–89 300ZX

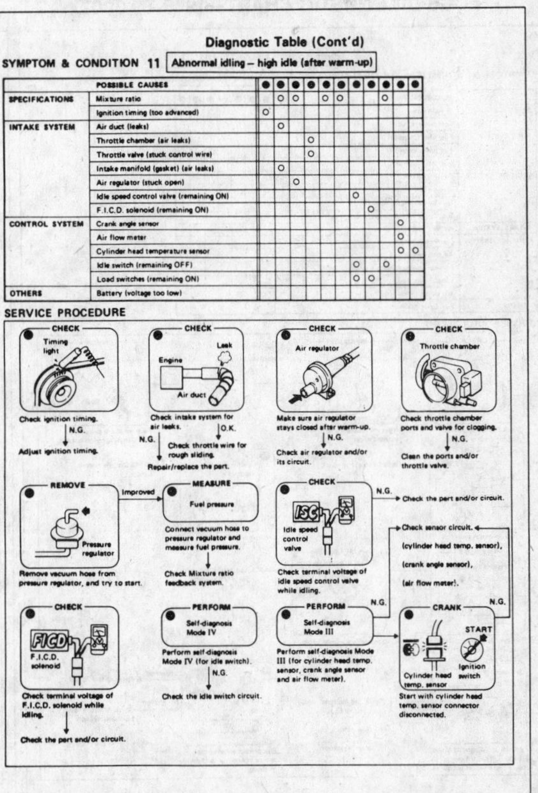

Diagnostic Table (Cont'd)

SYMPTOM & CONDITION 11 Abnormal idling – high idle (after warm-up)

	POSSIBLE CAUSES	●	●	●	●	●	●	●	●	●	●	●
SPECIFICATIONS	Mixture ratio		○	○		○	○			○		
	Ignition timing (too advanced)	○										
INTAKE SYSTEM	Air duct (leaks)		○									
	Throttle chamber (air leaks)				○							
	Throttle valve (stuck control wire)				○							
	Intake manifold (gasket) (air leaks)		○									
	Air regulator (stuck open)			○								
	Idle speed control valve (remaining ON)							○				
	F.I.C.D. solenoid (remaining ON)								○			
CONTROL SYSTEM	Crank angle sensor										○	
	Air flow meter										○	
	Cylinder head temperature sensor										○	○
	Idle switch (remaining OFF)							○		○		
	Load switches (remaining ON)							○	○			
OTHERS	Battery (voltage too low)											

SERVICE PROCEDURE

CHECK
Timing light
Check ignition timing.
N.G.
Adjust ignition timing.
CHECK
Engine
Leak
Air duct
Check intake system for air leaks.
N.G.
Repair/replace the part.
O.K.
Check throttle wire for rough sliding.
CHECK
Air regulator
Make sure air regulator stays closed after warm-up.
N.G.
Check air regulator and/or its circuit.
CHECK
Throttle chamber
Check throttle chamber ports and valve for clogging.
N.G.
Clean the ports and/or throttle valve.
REMOVE
Pressure regulator
Remove vacuum hose from pressure regulator, and try to start.
Improved
MEASURE
Fuel pressure
Connect vacuum hose to pressure regulator and measure fuel pressure.
Check Mixture ratio feedback system.
CHECK
ISC
Idle speed control valve
Check terminal voltage of idle speed control valve while idling.
N.G.
Check the part and/or circuit.
Check sensor circuit.
(cylinder head temp. sensor),
(crank angle sensor),
(air flow meter).
CHECK
FICD
F.I.C.D. solenoid
Check terminal voltage of F.I.C.D. solenoid while idling.
Check the part and/or circuit.
PERFORM
Self-diagnosis Mode IV
Perform self-diagnosis Mode IV (for idle switch).
N.G.
Check the idle switch circuit.
PERFORM
Self-diagnosis Mode III
Perform self-diagnosis Mode III (for cylinder head temp. sensor, crank angle sensor and air flow meter).
N.G.
CRANK
START
Cylinder head temp. sensor
Ignition switch
Start with cylinder head temp. sensor connector disconnected.
N.G.

ECCS DIAGNOSTIC CHART—1988–89 300ZX

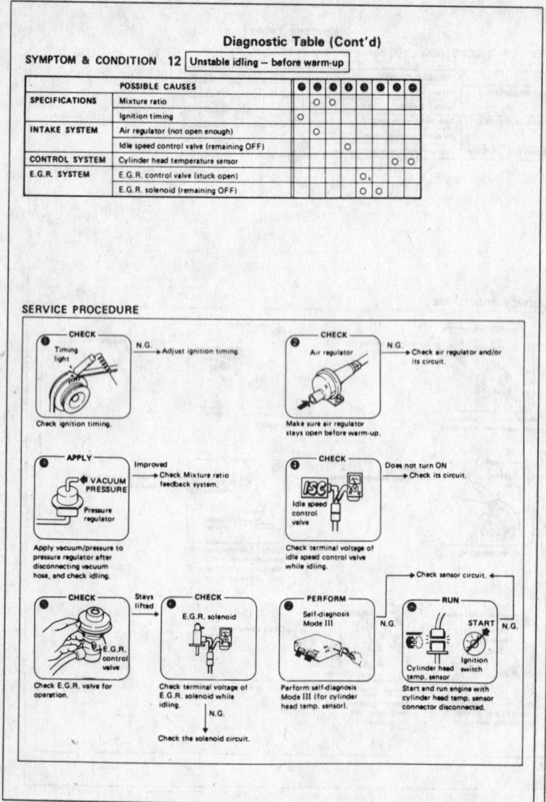

Diagnostic Table (Cont'd)

SYMPTOM & CONDITION 12 Unstable idling – before warm-up

POSSIBLE CAUSES		●	●	●	●	●	●	●	●
SPECIFICATIONS	Mixture ratio		O	O					
	Ignition timing	O							
INTAKE SYSTEM	Air regulator (not open enough)		O						
	Idle speed control valve (remaining OFF)				O				
CONTROL SYSTEM	Cylinder head temperature sensor							O	O
E.G.R. SYSTEM	E.G.R. control valve (stuck open)					O			
	E.G.R. solenoid (remaining OFF)					O	O		

SERVICE PROCEDURE
CHECK
Timing light
N.G. Adjust ignition timing.
Check ignition timing.
CHECK
Air regulator
N.G. Check air regulator and/or its circuit.
Make sure air regulator stays open before warm-up.
APPLY
VACUUM PRESSURE
Pressure regulator
Improved Check Mixture ratio feedback system.
Apply vacuum/pressure to pressure regulator after disconnecting vacuum hose, and check idling.
CHECK
ISC
Idle speed control valve
Does not turn ON Check its circuit.
Check terminal voltage of idle speed control valve while idling.
CHECK
E.G.R. control valve
Stays lifted
Check E.G.R. valve for operation.
CHECK
E.G.R. solenoid
Check terminal voltage of E.G.R. solenoid while idling.
N.G.
Check the solenoid circuit.
PERFORM
Self-diagnosis Mode III
N.G.
Perform self-diagnosis Mode III (for cylinder head temp. sensor).
Check sensor circuit.
RUN
START
Cylinder head temp. sensor
Ignition switch
N.G.
Start and run engine with cylinder head temp. sensor connector disconnected.

ECCS DIAGNOSTIC CHART—1988–89 300ZX

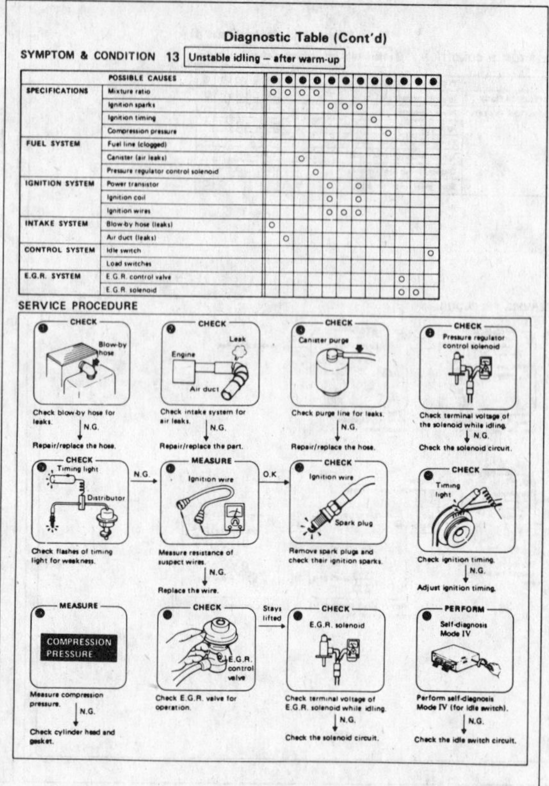

Diagnostic Table (Cont'd)

SYMPTOM & CONDITION 13 Unstable idling – after warm-up

POSSIBLE CAUSES		●	●	●	●	●	●	●	●	●	●	●	●
SPECIFICATIONS	Mixture ratio	O	O	O	O								
	Ignition sparks					O	O	O					
	Ignition timing								O				
	Compression pressure									O			
FUEL SYSTEM	Fuel line (clogged)												
	Canister (air leaks)			O									
	Pressure regulator control solenoid				O								
IGNITION SYSTEM	Power transistor					O		O					
	Ignition coil					O		O					
	Ignition wires					O	O	O					
INTAKE SYSTEM	Blow-by hose (leaks)	O											
	Air duct (leaks)		O										
CONTROL SYSTEM	Idle switch												O
	Load switches												
E.G.R. SYSTEM	E.G.R. control valve										O		
	E.G.R. solenoid										O	O	

SERVICE PROCEDURE
CHECK
Blow-by hose
Check blow-by hose for leaks.
N.G.
Repair/replace the hose.
CHECK
Leak
Engine
Air duct
Check intake system for air leaks.
N.G.
Repair/replace the part.
CHECK
Canister purge
Check purge line for leaks.
N.G.
Repair/replace the hose.
CHECK
Pressure regulator control solenoid
Check terminal voltage of the solenoid while idling.
N.G.
Check the solenoid circuit.
CHECK
Timing light
Distributor
Check flashes of timing light for weakness.
N.G.
MEASURE
Ignition wire
Measure resistance of suspect wires.
N.G.
Replace the wire.
O.K.
CHECK
Ignition wire
Spark plug
Remove spark plugs and check their ignition sparks.
CHECK
Timing light
Check ignition timing.
N.G.
Adjust ignition timing.
MEASURE
COMPRESSION PRESSURE
Measure compression pressure.
N.G.
Check cylinder head and gasket.
CHECK
E.G.R. control valve
Check E.G.R. valve for operation.
Stays lifted
CHECK
E.G.R. solenoid
Check terminal voltage of E.G.R. solenoid while idling.
N.G.
Check the solenoid circuit.
PERFORM
Self-diagnosis Mode IV
Perform self-diagnosis Mode IV (for idle switch).
N.G.
Check the idle switch circuit.

ECCS DIAGNOSTIC CHART—1988–89 300ZX

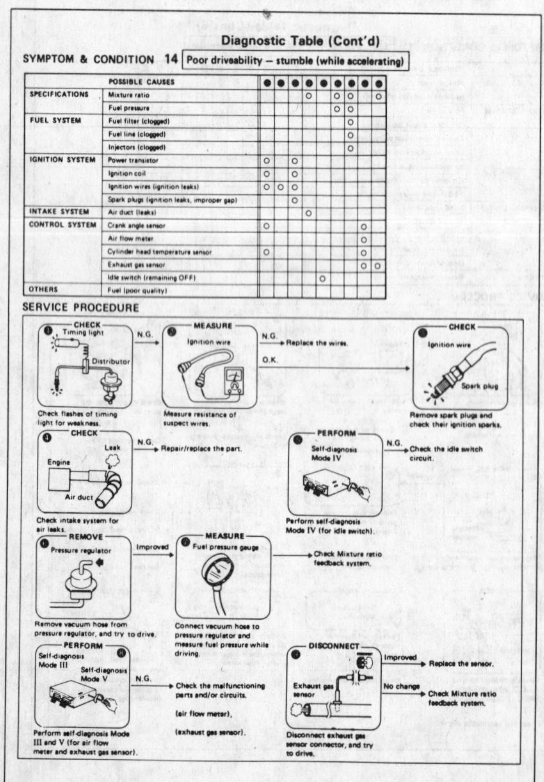

Diagnostic Table (Cont'd)

SYMPTOM & CONDITION 14 Poor driveability – stumble (while accelerating)

POSSIBLE CAUSES		●	●	●	●	●	●	●	●	●
SPECIFICATIONS	Mixture ratio				O		O	O		O
	Fuel pressure						O	O		
FUEL SYSTEM	Fuel filter (clogged)							O		
	Fuel line (clogged)							O		
	Injectors (clogged)							O		
IGNITION SYSTEM	Power transistor	O		O						
	Ignition coil	O		O						
	Ignition wires (ignition leaks)	O	O	O						
	Spark plugs (ignition leaks, improper gap)			O						
INTAKE SYSTEM	Air duct (leaks)				O					
CONTROL SYSTEM	Crank angle sensor	O							O	
	Air flow meter								O	
	Cylinder head temperature sensor	O							O	
	Exhaust gas sensor								O	O
	Idle switch (remaining OFF)					O				
OTHERS	Fuel (poor quality)									

SERVICE PROCEDURE
CHECK
Timing light
Distributor
N.G.
Check flashes of timing light for weakness.
MEASURE
Ignition wire
N.G. Replace the wires.
O.K.
Measure resistance of suspect wires.
CHECK
Ignition wire
Spark plug
Remove spark plugs and check their ignition sparks.
CHECK
Leak
Engine
Air duct
N.G. Repair/replace the part.
Check intake system for air leaks.
PERFORM
Self-diagnosis Mode IV
N.G. Check the idle switch circuit.
Perform self-diagnosis Mode IV (for idle switch).
REMOVE
Pressure regulator
Improved
Remove vacuum hose from pressure regulator, and try to drive.
MEASURE
Fuel pressure gauge
Check Mixture ratio feedback system.
Connect vacuum hose to pressure regulator and measure fuel pressure while driving.
PERFORM
Self-diagnosis Mode III
Self-diagnosis Mode V
N.G.
Check the malfunctioning parts and/or circuits.
(air flow meter),
(exhaust gas sensor).
Perform self-diagnosis Mode III and V (for air flow meter and exhaust gas sensor).
DISCONNECT
Exhaust gas sensor
Improved Replace the sensor.
No change Check Mixture ratio feedback system.
Disconnect exhaust gas sensor connector, and try to drive.

ECCS DIAGNOSTIC CHART—1988–89 300ZX

Diagnostic Table (Cont'd)

SYMPTOM & CONDITION 15 Poor driveability – surge (while cruising)

POSSIBLE CAUSES		●	●	●	●	●	●	●	●	●
SPECIFICATIONS	Mixture ratio (too lean)	O				O	O			O
	Fuel pressure (low)					O	O			
	Ignition timing		O							
IGNITION SYSTEM	(missing)								O	
INTAKE SYSTEM	Air duct (leaks)	O								
	Throttle chamber (air leaks)	O								
	Intake manifold (gasket) (air leaks)	O								
CONTROL SYSTEM	Crank angle sensor								O	
	Air flow meter								O	
	Exhaust gas sensor								O	O
	Idle switch							O		
E.G.R. SYSTEM	E.G.R. control valve (stuck open)			O						
	E.G.R. solenoid (remaining OFF)			O	O					
	E.G.R. vacuum hose (removed)			O						

SERVICE PROCEDURE

CHECK — Engine, Air duct, Leak

Check intake system for air leaks.
N.G. → Repair/replace the part.

CHECK — Timing light

Check ignition timing.
N.G. → Adjust ignition timing.

CHECK — E.G.R. control valve

Check E.G.R. valve for operation.
Stays lifted →

CHECK — E.G.R. solenoid

Check terminal voltage of E.G.R. solenoid.
N.G. → Check the solenoid circuit.

REMOVE — Pressure regulator

Remove vacuum hose from pressure regulator, and try to drive.
Improved →

MEASURE — Fuel pressure gauge

Connect vacuum hose to pressure regulator and measure fuel pressure while driving.
→ Check Mixture ratio feedback system.

PERFORM — Self-diagnosis Mode IV

Perform self-diagnosis Mode IV (for idle switch).
N.G. → Check the idle switch circuit.

PERFORM — Self-diagnosis Mode III, Self-diagnosis Mode V

Perform self-diagnosis Mode III and V (for crank angle sensor, air flow meter, ignition signal and exhaust gas sensor).
N.G. → Check the malfunctioning parts and/or circuits.
(crank angle sensor),
(air flow meter),
(ignition signal),
(exhaust gas sensor).

DISCONNECT — Exhaust gas sensor

Disconnect exhaust gas sensor connector, and try to drive.
Improved → Replace the sensor.
No change → Check mixture ratio feedback system.

ECCS DIAGNOSTIC CHART—1988–89 300ZX

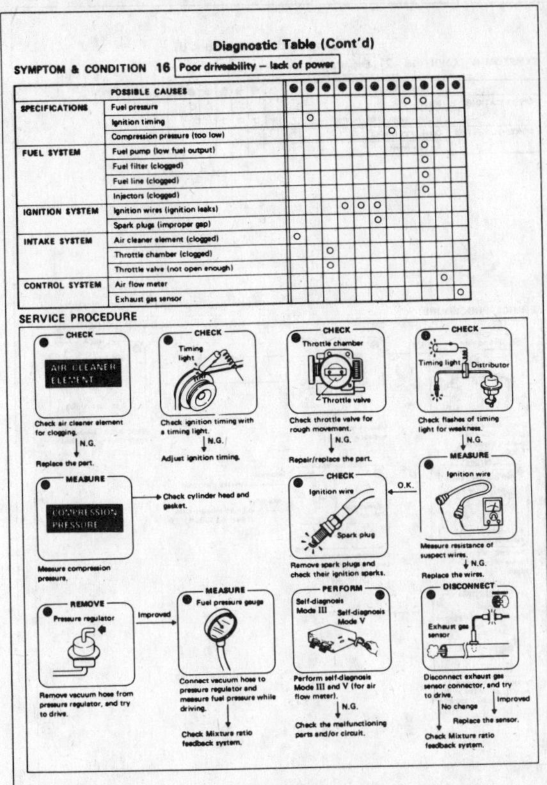

Diagnostic Table (Cont'd)

SYMPTOM & CONDITION 16 Poor driveability – lack of power

	POSSIBLE CAUSES	●	●	●	●	●	●	●	●	●	●	●
SPECIFICATIONS	Fuel pressure								○	○		
	Ignition timing		○									
	Compression pressure (too low)							○				
FUEL SYSTEM	Fuel pump (low fuel output)									○		
	Fuel filter (clogged)									○		
	Fuel line (clogged)									○		
	Injectors (clogged)									○		
IGNITION SYSTEM	Ignition wires (ignition leaks)				○	○	○					
	Spark plugs (improper gap)						○					
INTAKE SYSTEM	Air cleaner element (clogged)	○										
	Throttle chamber (clogged)			○								
	Throttle valve (not open enough)			○								
CONTROL SYSTEM	Air flow meter										○	
	Exhaust gas sensor											○

SERVICE PROCEDURE

CHECK
AIR CLEANER ELEMENT
Check air cleaner element for clogging.
N.G.
Replace the part.
CHECK
Timing light
Check ignition timing with a timing light.
N.G.
Adjust ignition timing.
CHECK
Throttle chamber
Throttle valve
Check throttle valve for rough movement.
N.G.
Repair/replace the part.
CHECK
Timing light
Distributor
Check flashes of timing light for weakness.
N.G.
MEASURE
Ignition wire
Measure resistance of suspect wires.
N.G.
Replace the wires.
O.K.
CHECK
Ignition wire
Spark plug
Remove spark plugs and check their ignition sparks.
MEASURE
COMPRESSION PRESSURE
Measure compression pressure.
Check cylinder head and gasket.
REMOVE
Pressure regulator
Remove vacuum hose from pressure regulator, and try to drive.
Improved
MEASURE
Fuel pressure gauge
Connect vacuum hose to pressure regulator and measure fuel pressure while driving.
Check Mixture ratio feedback system.
PERFORM
Self-diagnosis Mode III
Self-diagnosis Mode V
Perform self-diagnosis Mode III and V (for air flow meter).
N.G.
Check the malfunctioning parts and/or circuit.
DISCONNECT
Exhaust gas sensor
Disconnect exhaust gas sensor connector, and try to drive.
No change
Check Mixture ratio feedback system.
Improved
Replace the sensor.

ECCS DIAGNOSTIC CHART—1988–89 300ZX

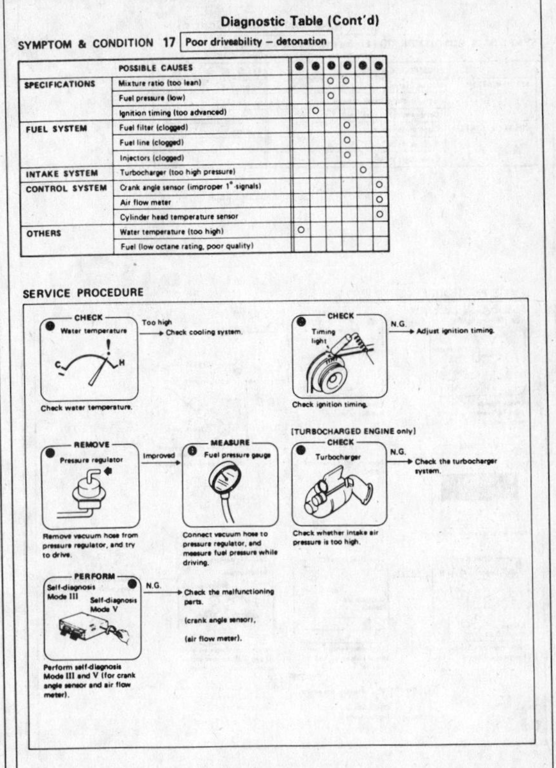

Diagnostic Table (Cont'd)

SYMPTOM & CONDITION 17 Poor driveability – detonation

	POSSIBLE CAUSES	●	●	●	●	●	●
SPECIFICATIONS	Mixture ratio (too lean)			○	○		
	Fuel pressure (low)			○			
	Ignition timing (too advanced)		○				
FUEL SYSTEM	Fuel filter (clogged)				○		
	Fuel line (clogged)				○		
	Injectors (clogged)				○		
INTAKE SYSTEM	Turbocharger (too high pressure)					○	
CONTROL SYSTEM	Crank angle sensor (improper 1°-signals)						○
	Air flow meter						○
	Cylinder head temperature sensor						○
OTHERS	Water temperature (too high)	○					
	Fuel (low octane rating, poor quality)						

SERVICE PROCEDURE

CHECK
Water temperature
C
H
Check water temperature.
Too high
Check cooling system.
CHECK
Timing light
Check ignition timing.
N.G.
Adjust ignition timing.
REMOVE
Pressure regulator
Remove vacuum hose from pressure regulator, and try to drive.
Improved
MEASURE
Fuel pressure gauge
Connect vacuum hose to pressure regulator, and measure fuel pressure while driving.
[TURBOCHARGED ENGINE only]
CHECK
Turbocharger
Check whether intake air pressure is too high.
N.G.
Check the turbocharger system.
PERFORM
Self-diagnosis Mode III
Self-diagnosis Mode V
Perform self-diagnosis Mode III and V (for crank angle sensor and air flow meter).
N.G.
Check the malfunctioning parts.
(crank angle sensor),
(air flow meter).

ECCS DIAGNOSTIC CHART—1988–89 300ZX

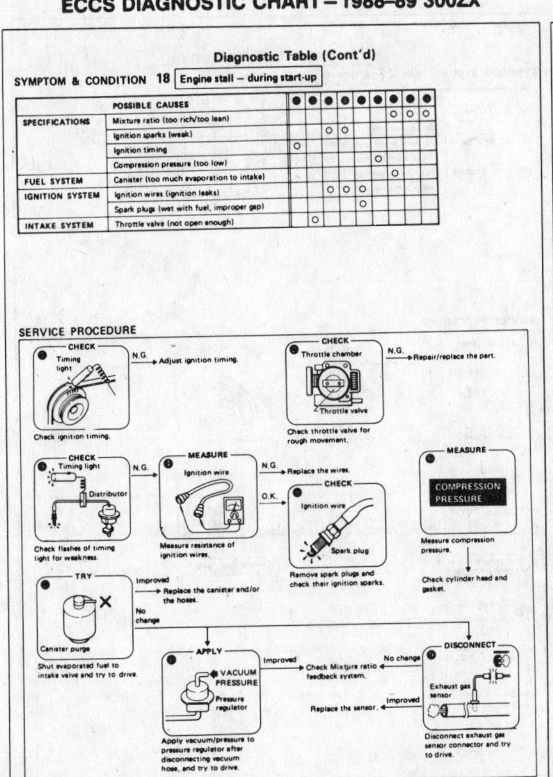

Diagnostic Table (Cont'd)

SYMPTOM & CONDITION 18 Engine stall – during start-up

	POSSIBLE CAUSES	●	●	●	●	●	●	●	●	●
SPECIFICATIONS	Mixture ratio (too rich/too lean)							○	○	○
	Ignition sparks (weak)			○	○					
	Ignition timing	○								
	Compression pressure (too low)						○			
FUEL SYSTEM	Canister (too much evaporation to intake)							○		
IGNITION SYSTEM	Ignition wires (ignition leaks)			○	○	○				
	Spark plugs (wet with fuel, improper gap)					○				
INTAKE SYSTEM	Throttle valve (not open enough)		○							

SERVICE PROCEDURE

CHECK
Timing light
Check ignition timing.
N.G.
Adjust ignition timing.
CHECK
Throttle chamber
Throttle valve
Check throttle valve for rough movement.
N.G.
Repair/replace the part.
CHECK
Timing light
Distributor
Check flashes of timing light for weakness.
N.G.
MEASURE
Ignition wire
Measure resistance of ignition wires.
N.G.
Replace the wires.
O.K.
CHECK
Ignition wire
Spark plug
Remove spark plugs and check their ignition sparks.
MEASURE
COMPRESSION PRESSURE
Measure compression pressure.
Check cylinder head and gasket.
TRY
Canister purge
Shut evaporated fuel to intake valve and try to drive.
Improved
Replace the canister and/or the hoses.
No change
APPLY
VACUUM PRESSURE
Pressure regulator
Apply vacuum/pressure to pressure regulator after disconnecting vacuum hose, and try to drive.
Improved
Check Mixture ratio feedback system.
No change
Replace the sensor.
Improved
DISCONNECT
Exhaust gas sensor
Disconnect exhaust gas sensor connector and try to drive.

ECCS DIAGNOSTIC CHART—1988–89 300ZX

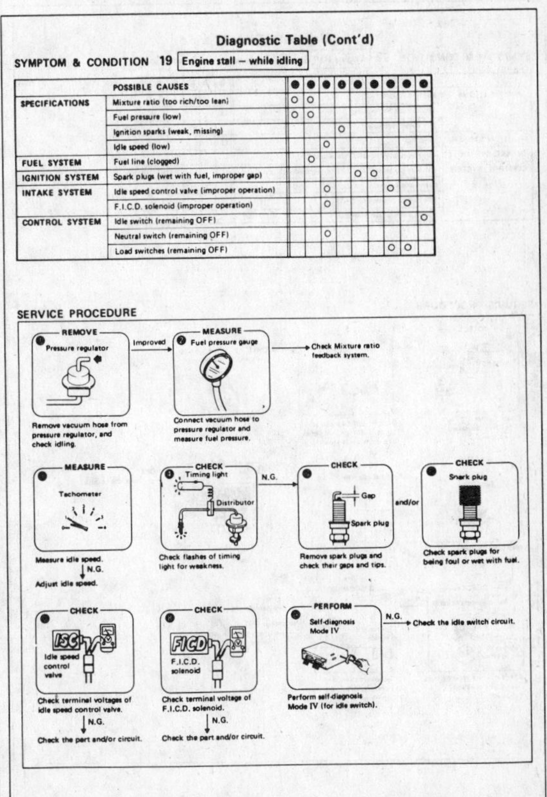

Diagnostic Table (Cont'd)

SYMPTOM & CONDITION 19 Engine stall – while idling

	POSSIBLE CAUSES	●	●	●	●	●	●	●	●	●
SPECIFICATIONS	Mixture ratio (too rich/too lean)	○	○							
	Fuel pressure (low)	○	○							
	Ignition sparks (weak, missing)				○					
	Idle speed (low)			○						
FUEL SYSTEM	Fuel line (clogged)		○							
IGNITION SYSTEM	Spark plugs (wet with fuel, improper gap)					○	○			
INTAKE SYSTEM	Idle speed control valve (improper operation)			○				○		
	F.I.C.D. solenoid (improper operation)			○					○	
CONTROL SYSTEM	Idle switch (remaining OFF)									○
	Neutral switch (remaining OFF)			○						
	Load switches (remaining OFF)							○	○	

SERVICE PROCEDURE

REMOVE
Pressure regulator
Remove vacuum hose from pressure regulator, and check idling.
Improved
MEASURE
Fuel pressure gauge
Connect vacuum hose to pressure regulator and measure fuel pressure.
Check Mixture ratio feedback system.
MEASURE
Tachometer
Measure idle speed.
N.G.
Adjust idle speed.
CHECK
Timing light
Distributor
Check flashes of timing light for weakness.
N.G.
CHECK
Gap
Spark plug
Remove spark plugs and check their gaps and tips.
and/or
CHECK
Snark plug
Check spark plugs for being foul or wet with fuel.
CHECK
ISC
Idle speed control valve
Check terminal voltages of idle speed control valve.
N.G.
Check the part and/or circuit.
CHECK
FICD
F.I.C.D. solenoid
Check terminal voltage of F.I.C.D. solenoid.
N.G.
Check the part and/or circuit.
PERFORM
Self-diagnosis Mode IV
Perform self-diagnosis Mode IV (for idle switch).
N.G.
Check the idle switch circuit.

ECCS DIAGNOSTIC CHART—1988–89 300ZX

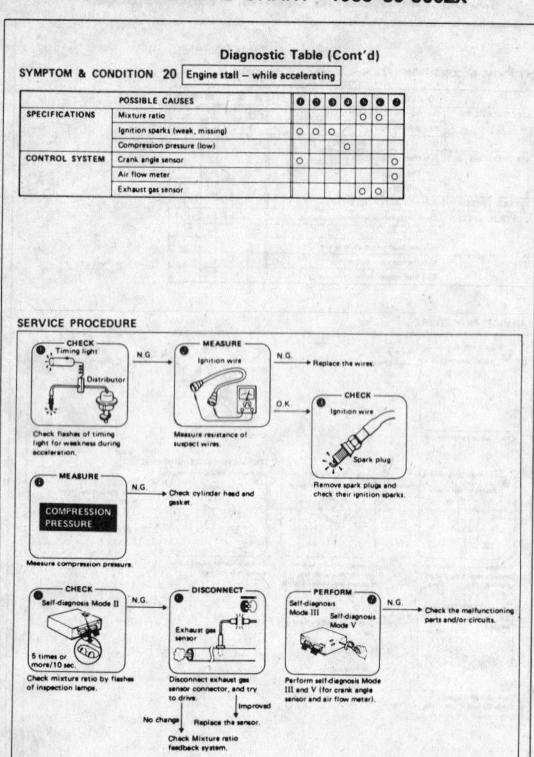

Diagnostic Table (Cont'd)

SYMPTOM & CONDITION 20 Engine stall – while accelerating

POSSIBLE CAUSES		1	2	3	4	5	6	7
SPECIFICATIONS	Mixture ratio					○	○	
	Ignition sparks (weak, missing)	○	○	○				
	Compression pressure (low)				○			
CONTROL SYSTEM	Crank angle sensor	○						○
	Air flow meter							○
	Exhaust gas sensor					○	○	

SERVICE PROCEDURE
CHECK
1 Timing light
Distributor
Check flashes of timing light for weakness during acceleration.
N.G.
MEASURE
2 Ignition wire
Measure resistance of suspect wires.
N.G.
Replace the wires.
O.K.
CHECK
3 Ignition wire
Spark plug
Remove spark plugs and check their ignition sparks.
MEASURE
4
COMPRESSION PRESSURE
Measure compression pressure.
N.G.
Check cylinder head and gasket.
CHECK
5 Self-diagnosis Mode II
5 times or more/10 sec.
Check mixture ratio by flashes of inspection lamps.
N.G.
DISCONNECT
6
Exhaust gas sensor
Disconnect exhaust gas sensor connector, and try to drive.
Improved
Replace the sensor.
No change
Check Mixture ratio feedback system.
PERFORM
7
Self-diagnosis Mode III
Self-diagnosis Mode V
Perform self-diagnosis Mode III and V (for crank angle sensor and air flow meter).
N.G.
Check the malfunctioning parts and/or circuits.

ECCS DIAGNOSTIC CHART—1988–89 300ZX

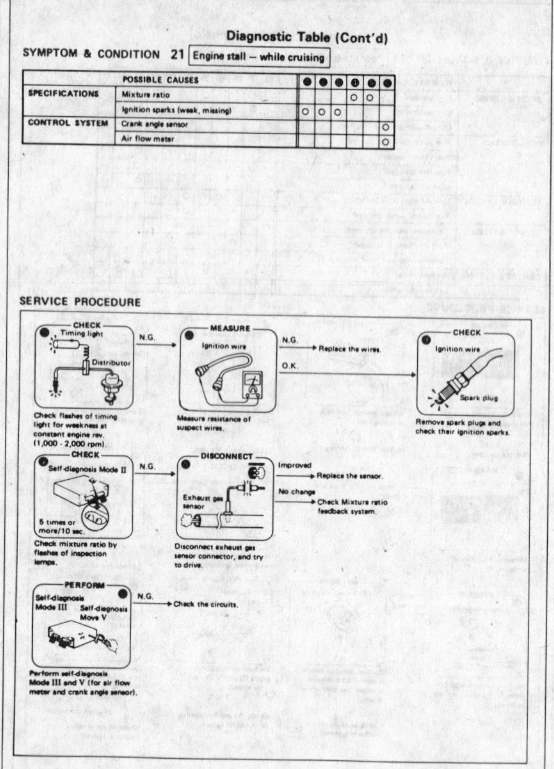

Diagnostic Table (Cont'd)

SYMPTOM & CONDITION 21 Engine stall – while cruising

POSSIBLE CAUSES		1	2	3	4	5	6
SPECIFICATIONS	Mixture ratio				○	○	
	Ignition sparks (weak, missing)	○	○	○			
CONTROL SYSTEM	Crank angle sensor						○
	Air flow meter						○

SERVICE PROCEDURE
CHECK
1 Timing light
Distributor
Check flashes of timing light for weakness at constant engine rev. (1,000 - 2,000 rpm).
N.G.
MEASURE
2 Ignition wire
Measure resistance of suspect wires.
N.G.
Replace the wires.
O.K.
CHECK
3 Ignition wire
Spark plug
Remove spark plugs and check their ignition sparks.
CHECK
4 Self-diagnosis Mode II
5 times or more/10 sec.
Check mixture ratio by flashes of inspection lamps.
N.G.
DISCONNECT
5
Exhaust gas sensor
Disconnect exhaust gas sensor connector, and try to drive.
Improved
Replace the sensor.
No change
Check Mixture ratio feedback system.
PERFORM
6
Self-diagnosis Mode III
Self-diagnosis Move V
Perform self-diagnosis Mode III and V (for air flow meter and crank angle sensor).
N.G.
Check the circuits.

ECCS DIAGNOSTIC CHART—1988–89 300ZX

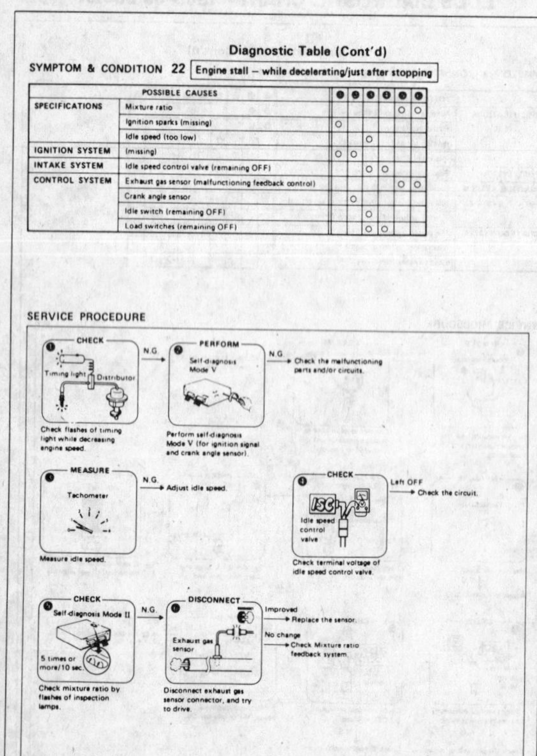

Diagnostic Table (Cont'd)

SYMPTOM & CONDITION 22 Engine stall – while decelerating/just after stopping

POSSIBLE CAUSES		1	2	3	4	5	6
SPECIFICATIONS	Mixture ratio					○	○
	Ignition sparks (missing)	○					
	Idle speed (too low)			○			
IGNITION SYSTEM	(missing)	○	○				
INTAKE SYSTEM	Idle speed control valve (remaining OFF)			○	○		
CONTROL SYSTEM	Exhaust gas sensor (malfunctioning feedback control)					○	○
	Crank angle sensor		○				
	Idle switch (remaining OFF)			○			
	Load switches (remaining OFF)			○	○		

SERVICE PROCEDURE
CHECK
1
Timing light
Distributor
Check flashes of timing light while decreasing engine speed.
N.G.
PERFORM
2 Self-diagnosis Mode V
Perform self-diagnosis Mode V (for ignition signal and crank angle sensor).
N.G.
Check the malfunctioning parts and/or circuits.
MEASURE
3 Tachometer
Measure idle speed.
N.G.
Adjust idle speed.
CHECK
4
ISC
Idle speed control valve
Check terminal voltage of idle speed control valve.
Left OFF
Check the circuit.
CHECK
5 Self-diagnosis Mode II
5 times or more/10 sec.
Check mixture ratio by flashes of inspection lamps.
N.G.
DISCONNECT
6
Exhaust gas sensor
Disconnect exhaust gas sensor connector, and try to drive.
Improved
Replace the sensor.
No change
Check Mixture ratio feedback system.

ECCS DIAGNOSTIC CHART—1988–89 300ZX

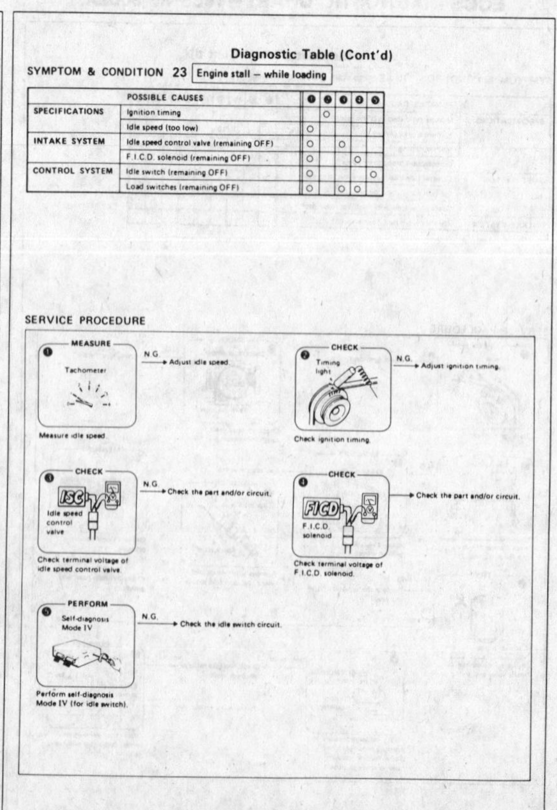

Diagnostic Table (Cont'd)

SYMPTOM & CONDITION 23 Engine stall – while loading

POSSIBLE CAUSES		1	2	3	4	5
SPECIFICATIONS	Ignition timing		○			
	Idle speed (too low)	○				
INTAKE SYSTEM	Idle speed control valve (remaining OFF)	○		○		
	F.I.C.D. solenoid (remaining OFF)	○			○	
CONTROL SYSTEM	Idle switch (remaining OFF)	○				○
	Load switches (remaining OFF)	○		○	○	

SERVICE PROCEDURE
MEASURE
1 Tachometer
Measure idle speed.
N.G.
Adjust idle speed.
CHECK
2 Timing light
Check ignition timing.
N.G.
Adjust ignition timing.
CHECK
3
ISC
Idle speed control valve
Check terminal voltage of idle speed control valve.
N.G.
Check the part and/or circuit.
CHECK
4
FICD
F.I.C.D. solenoid
Check terminal voltage of F.I.C.D. solenoid.
Check the part and/or circuit.
PERFORM
5 Self-diagnosis Mode IV
Perform self-diagnosis Mode IV (for idle switch).
N.G.
Check the idle switch circuit.

ECCS DIAGNOSTIC CHART—1988–89 300ZX

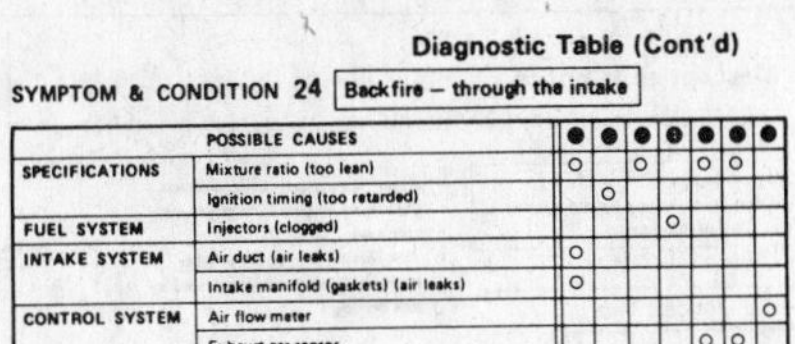

Diagnostic Table (Cont'd)

SYMPTOM & CONDITION 24 Backfire – through the intake

	POSSIBLE CAUSES	●	●	●	●	●	●	●
SPECIFICATIONS	Mixture ratio (too lean)	○		○		○	○	
	Ignition timing (too retarded)		○					
FUEL SYSTEM	Injectors (clogged)				○			
INTAKE SYSTEM	Air duct (air leaks)	○						
	Intake manifold (gaskets) (air leaks)	○						
CONTROL SYSTEM	Air flow meter							○
	Exhaust gas sensor					○	○	

SERVICE PROCEDURE

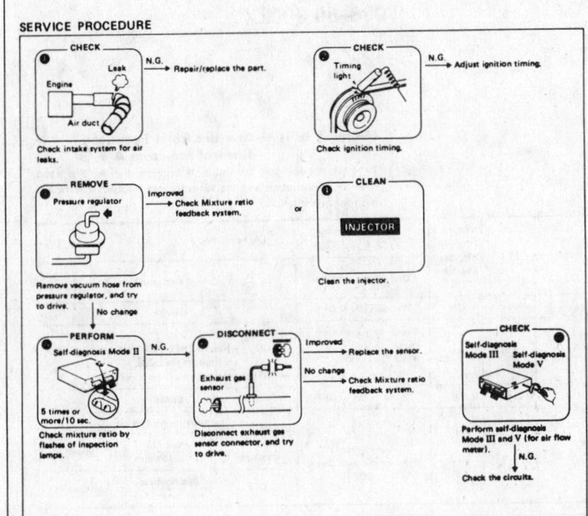
CHECK
Leak
Engine
Air duct
N.G. Repair/replace the part.
Check intake system for air leaks.
CHECK
Timing light
N.G. Adjust ignition timing.
Check ignition timing.
REMOVE
Pressure regulator
Improved Check Mixture ratio feedback system.
Remove vacuum hose from pressure regulator, and try to drive.
No change
or
CLEAN
INJECTOR
Clean the injector.
PERFORM
Self-diagnosis Mode II
5 times or more/10 sec.
Check mixture ratio by flashes of inspection lamps.
N.G.
DISCONNECT
Exhaust gas sensor
Improved Replace the sensor.
No change Check Mixture ratio feedback system.
Disconnect exhaust gas sensor connector, and try to drive.
CHECK
Self-diagnosis Mode III
Self-diagnosis Mode V
Perform self-diagnosis Mode III and V (for air flow meter).
N.G.
Check the circuits.

ECCS DIAGNOSTIC CHART—1988–89 300ZX

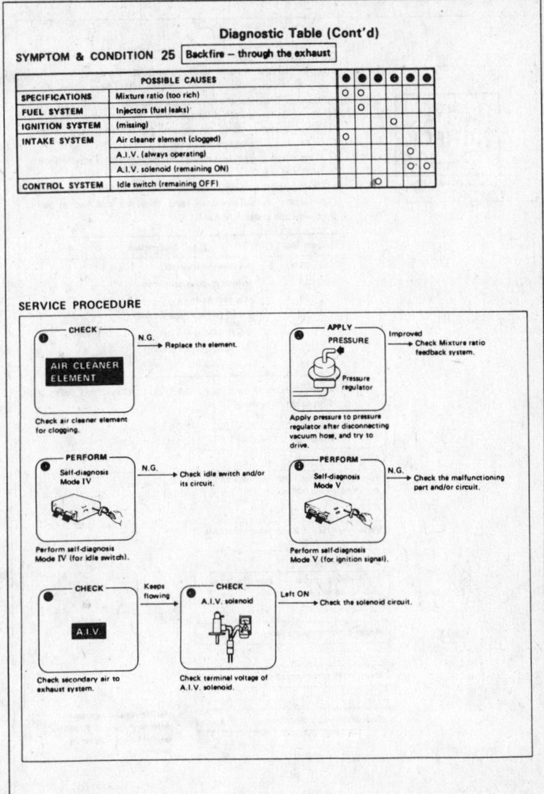

Diagnostic Table (Cont'd)

SYMPTOM & CONDITION 25 Backfire – through the exhaust

	POSSIBLE CAUSES	●	●	●	●	●	●
SPECIFICATIONS	Mixture ratio (too rich)	○	○				
FUEL SYSTEM	Injectors (fuel leaks)		○				
IGNITION SYSTEM	(missing)				○		
INTAKE SYSTEM	Air cleaner element (clogged)	○					
	A.I.V. (always operating)					○	
	A.I.V. solenoid (remaining ON)					○	○
CONTROL SYSTEM	Idle switch (remaining OFF)			○			

SERVICE PROCEDURE

CHECK
AIR CLEANER ELEMENT
N.G. Replace the element.
Check air cleaner element for clogging.
APPLY
PRESSURE
Pressure regulator
Improved Check Mixture ratio feedback system.
Apply pressure to pressure regulator after disconnecting vacuum hose, and try to drive.
PERFORM
Self-diagnosis Mode IV
N.G. Check idle switch and/or its circuit.
Perform self-diagnosis Mode IV (for idle switch).
PERFORM
Self-diagnosis Mode V
N.G. Check the malfunctioning part and/or circuit.
Perform self-diagnosis Mode V (for ignition signal).
CHECK
A.I.V.
Check secondary air to exhaust system.
Keeps flowing
CHECK
A.I.V. solenoid
Check terminal voltage of A.I.V. solenoid.
Left ON Check the solenoid circuit.

ECCS SELF-DIAGNOSTIC DESCRIPTION—1988–89 300ZX

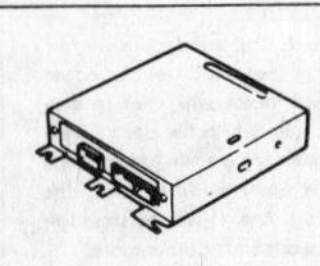

Description

The self-diagnosis is useful to diagnose malfunctions in major sensors and actuators of the E.C.C.S. system. There are 5 modes in the self-diagnosis system.

1. **Mode I – Mixture ratio feedback control monitor A**
   - During closed loop condition:
     The green inspection lamp turns ON when lean condition is detected and goes OFF by rich condition.
   - During open loop condition:
     The green inspection lamp remains ON or OFF.
2. **Mode II – Mixture ratio feedback control monitor B**
   The green inspection lamp function is the same as Mode I.
   - During closed loop condition:
     The red inspection lamp turns ON and OFF simultaneously with the green inspection lamp when the mixture ratio is controlled within the specified value.
   - During open loop condition:
     The red inspection lamp remains ON or OFF.
3. **Mode III – Self-diagnosis**
   This mode is the same as the former self-diagnosis in self-diagnosis mode.
4. **Mode IV – Switches ON/OFF diagnosis**
   During this mode, the inspection lamps monitor the switch ON-OFF condition.
   - Idle switch
   - Starter switch
   - Vehicle speed sensor
5. **Mode V – Real time diagnosis**
   The moment the malfunction is detected, the display will be presented immediately. That is, the condition at which the malfunction occurs can be found by observing the inspection lamps during driving test.

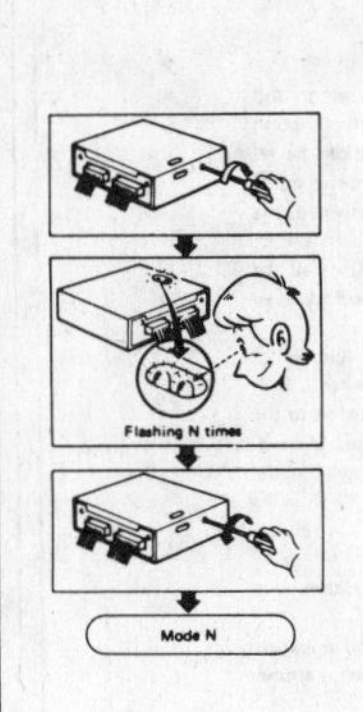
Flashing N times
Mode N

Description (Cont'd)

SWITCHING THE MODES

1. Turn ignition switch "ON".
2. Turn diagnostic mode selector on E.C.U. fully clockwise and wait the inspection lamps flash.
3. Count the number of the flashing time, and after the inspection lamps have flashed the number of the required mode, turn diagnostic mode selector fully counterclockwise immediately.

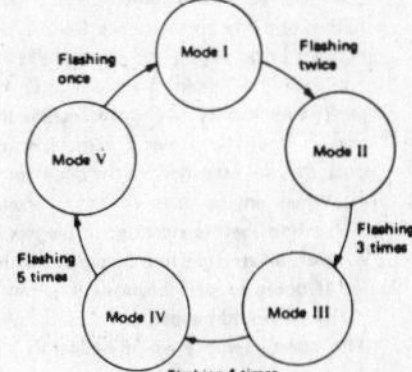
Mode I
Flashing twice
Mode II
Flashing 3 times
Mode III
Flashing 4 times
Mode IV
Flashing 5 times
Mode V
Flashing once

When the ignition switch is turned off during diagnosis, in each mode, and then turned back on again after the power to the E.C.U. has dropped off completely, the diagnosis will automatically return to Mode I.

The stored memory would be lost if:

1. Battery terminal is disconnected.
2. After selecting Mode III, Mode IV is selected.
   However, if the diagnostic mode selector is kept turned fully clockwise, it will continue to change in the order of Mode I → II → III → IV → V → I ... etc., and in this state the stored memory will not be erased.

ECCS SELF-DIAGNOSTIC DESCRIPTION (CONT.) 1988–89 300ZX

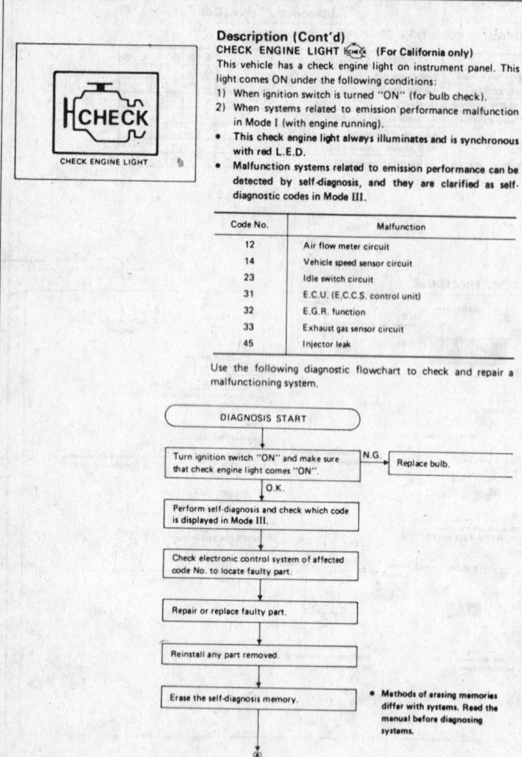

Description (Cont'd)

CHECK ENGINE LIGHT (For California only)

This vehicle has a check engine light on instrument panel. This light comes ON under the following conditions:

1) When ignition switch is turned "ON" (for bulb check).
2) When systems related to emission performance malfunction in Mode I (with engine running).

- **This check engine light always illuminates and is synchronous with red L.E.D.**
- **Malfunction systems related to emission performance can be detected by self-diagnosis, and they are clarified as self-diagnostic codes in Mode III.**

Code No.	Malfunction
12	Air flow meter circuit
14	Vehicle speed sensor circuit
23	Idle switch circuit
31	E.C.U. (E.C.C.S. control unit)
32	E.G.R. function
33	Exhaust gas sensor circuit
45	Injector leak

Use the following diagnostic flowchart to check and repair a malfunctioning system.

CHECK
CHECK ENGINE LIGHT
DIAGNOSIS START
Turn ignition switch "ON" and make sure that check engine light comes "ON".
N.G.
Replace bulb.
O.K.
Perform self-diagnosis and check which code is displayed in Mode III.
Check electronic control system of affected code No. to locate faulty part.
Repair or replace faulty part.
Reinstall any part removed.
Erase the self-diagnosis memory.
• Methods of erasing memories differ with systems. Read the manual before diagnosing systems.
Ⓐ

ECCS SELF-DIAGNOSTIC DESCRIPTION (CONT.) 1988–89 300ZX

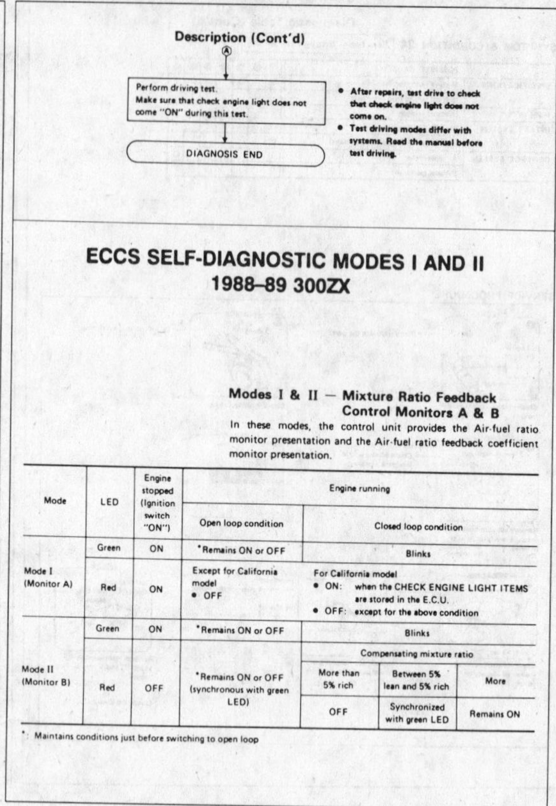

Description (Cont'd)

Ⓐ
Perform driving test.
Make sure that check engine light does not come "ON" during this test.
DIAGNOSIS END
• After repairs, test drive to check that check engine light does not come on.
• Test driving modes differ with systems. Read the manual before test driving.

ECCS SELF-DIAGNOSTIC MODES I AND II 1988–89 300ZX

Modes I & II — Mixture Ratio Feedback Control Monitors A & B

In these modes, the control unit provides the Air-fuel ratio monitor presentation and the Air-fuel ratio feedback coefficient monitor presentation.

Mode	LED	Engine stopped (Ignition switch "ON")	Engine running			
			Open loop condition	Closed loop condition		
Mode I (Monitor A)	Green	ON	*Remains ON or OFF	Blinks		
	Red	ON	Except for California model • OFF	For California model • ON: when the CHECK ENGINE LIGHT ITEMS are stored in the E.C.U. • OFF: except for the above condition		
Mode II (Monitor B)	Green	ON	*Remains ON or OFF	Blinks		
				Compensating mixture ratio		
				More than 5% rich	Between 5% lean and 5% rich	More
	Red	OFF	*Remains ON or OFF (synchronous with green LED)	OFF	Synchronized with green LED	Remains ON

*: Maintains conditions just before switching to open loop

ECCS SELF-DIAGNOSTIC MODE III—1988–89 300ZX

Mode III — Self-diagnostic System

The E.C.U. constantly monitors the function of these sensors and actuators, regardless of ignition key position. If a malfunction occurs, the information is stored in the E.C.U. and can be retrieved from the memory by turning on the diagnostic mode selector, located on the side of the E.C.U. When activated, the malfunction is indicated by flashing a red and a green L.E.D. (Light Emitting Diode), also located on the E.C.U. Since all the self-diagnostic results are stored in the E.C.U.'s memory even intermittent malfunctions can be diagnosed.

A malfunctioning part's group is indicated by the number of both the red and the green L.E.D.s flashing. First, the red L.E.D. flashes and the green flashes follow. The red L.E.D. refers to the number of tens while the green one refers to the number of units. For example, when the red L.E.D. flashes once and then the green one flashes twice, this means the number "12" showing the air flow meter signal is malfunctioning. In this way, all the problems are classified by the code numbers.

- **When engine fails to start, crank engine more than two seconds before starting self-diagnosis.**
- **Before starting self-diagnosis, do not erase stored memory. If doing so, self-diagnosis function for intermittent malfunctions would be lost.**

The stored memory would be lost if:

1. **Battery terminal is disconnected.**
2. **After selecting Mode III, Mode IV is selected.**

DISPLAY CODE TABLE

Code No.	Detected items	California	Non-California
11	Crank angle sensor ciruit	X	X
12	Air flow meter circuit	X	X
13	Cylinder head temperature sensor circuit	X	X
14	Vehicle speed sensor circuit	X	X
21	Ignition signal missing in primary coil	X	X
22	Fuel pump circuit	X	X
23	Idle switch circuit	X	X
31	E.C.U. (E.C.C.S. control unit)	X	X
32	E.G.R. function	X	–
33	Exhaust gas sensor circuit	X	X
34	Detonation sensor circuit [VG30ET]	X	X
35	Exhaust gas temperature circuit	X	–
42	Fuel temperature sensor circuit	X	X
43	Throttle sensor circuit	X	–
45	Injector leak	X	–
55	No malfunction in the above circuit	X	X

X: Available –: Not available

Mode III — Self-diagnostic System (Cont'd)

RETENTION OF DIAGNOSTIC RESULTS

The diagnostic result is retained in E.C.U. memory until the starter is operated fifty times after a diagnostic item is judged to be malfunctioning. The diagnostic result will then be cancelled automatically. If a diagnostic item which has been judged to be malfunctioning and stored in memory is again judged to be malfunctioning before the starter is operated fifty times, the second result will replace the previous one. It will be stored in E.C.U. memory until the starter is operated fifty times more.

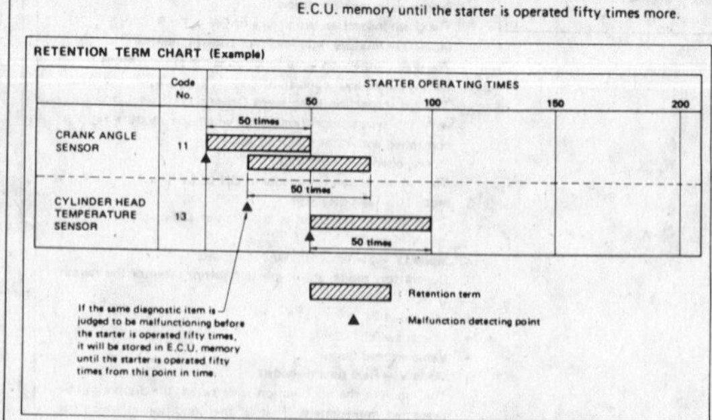
RETENTION TERM CHART (Example)
Code No.
STARTER OPERATING TIMES
50
100
150
200
50 times
CRANK ANGLE SENSOR
11
50 times
CYLINDER HEAD TEMPERATURE SENSOR
13
50 times
If the same diagnostic item is judged to be malfunctioning before the starter is operated fifty times, it will be stored in E.C.U. memory until the starter is operated fifty times from this point in time.
: Retention term
: Malfunction detecting point

ECCS SELF-DIAGNOSTIC MODE III (CONT.) 1988–89 300ZX

ECCS SELF-DIAGNOSTIC MODE III DISPLAY CODES 1988–89 300ZX

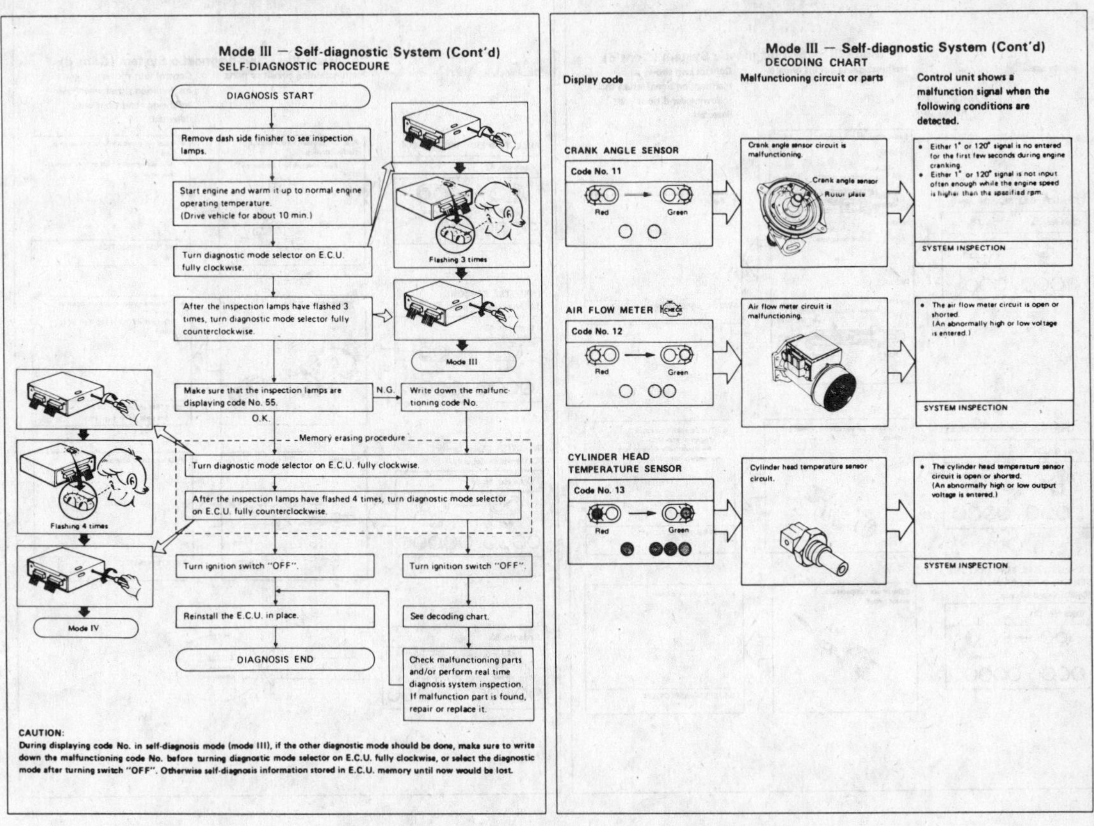
Mode III — Self-diagnostic System (Cont'd)
SELF-DIAGNOSTIC PROCEDURE
DIAGNOSIS START
Remove dash side finisher to see inspection lamps.
Start engine and warm it up to normal engine operating temperature. (Drive vehicle for about 10 min.)
Turn diagnostic mode selector on E.C.U. fully clockwise.
Flashing 3 times
After the inspection lamps have flashed 3 times, turn diagnostic mode selector fully counterclockwise.
Mode III
Make sure that the inspection lamps are displaying code No. 55.
N.G.
Write down the malfunctioning code No.
O.K.
Memory erasing procedure
Turn diagnostic mode selector on E.C.U. fully clockwise.
After the inspection lamps have flashed 4 times, turn diagnostic mode selector on E.C.U. fully counterclockwise.
Flashing 4 times
Turn ignition switch "OFF".
Turn ignition switch "OFF".
Mode IV
Reinstall the E.C.U. in place.
See decoding chart.
DIAGNOSIS END
Check malfunctioning parts and/or perform real time diagnosis system inspection. If malfunction part is found, repair or replace it.
CAUTION:
During displaying code No. in self-diagnosis mode (mode III), if the other diagnostic mode should be done, make sure to write down the malfunctioning code No. before turning diagnostic mode selector on E.C.U. fully clockwise, or select the diagnostic mode after turning switch "OFF". Otherwise self-diagnosis information stored in E.C.U. memory until now would be lost.
Mode III — Self-diagnostic System (Cont'd)
DECODING CHART
Display code
Malfunctioning circuit or parts
Control unit shows a malfunction signal when the following conditions are detected.
CRANK ANGLE SENSOR
Code No. 11
Red
Green
Crank angle sensor circuit is malfunctioning.
Crank angle sensor
Rotor plate
• Either 1° or 120° signal is no entered for the first few seconds during engine cranking.
• Either 1° or 120° signal is not input often enough while the engine speed is higher than the specified rpm.
SYSTEM INSPECTION
AIR FLOW METER
Code No. 12
Red
Green
Air flow meter circuit is malfunctioning.
• The air flow meter circuit is open or shorted. (An abnormally high or low voltage is entered.)
SYSTEM INSPECTION
CYLINDER HEAD TEMPERATURE SENSOR
Code No. 13
Red
Green
Cylinder head temperature sensor circuit.
• The cylinder head temperature sensor circuit is open or shorted. (An abnormally high or low output voltage is entered.)
SYSTEM INSPECTION

ECCS SELF-DIAGNOSTIC MODE III DISPLAY CODES (CONT.)—1988–89 300ZX

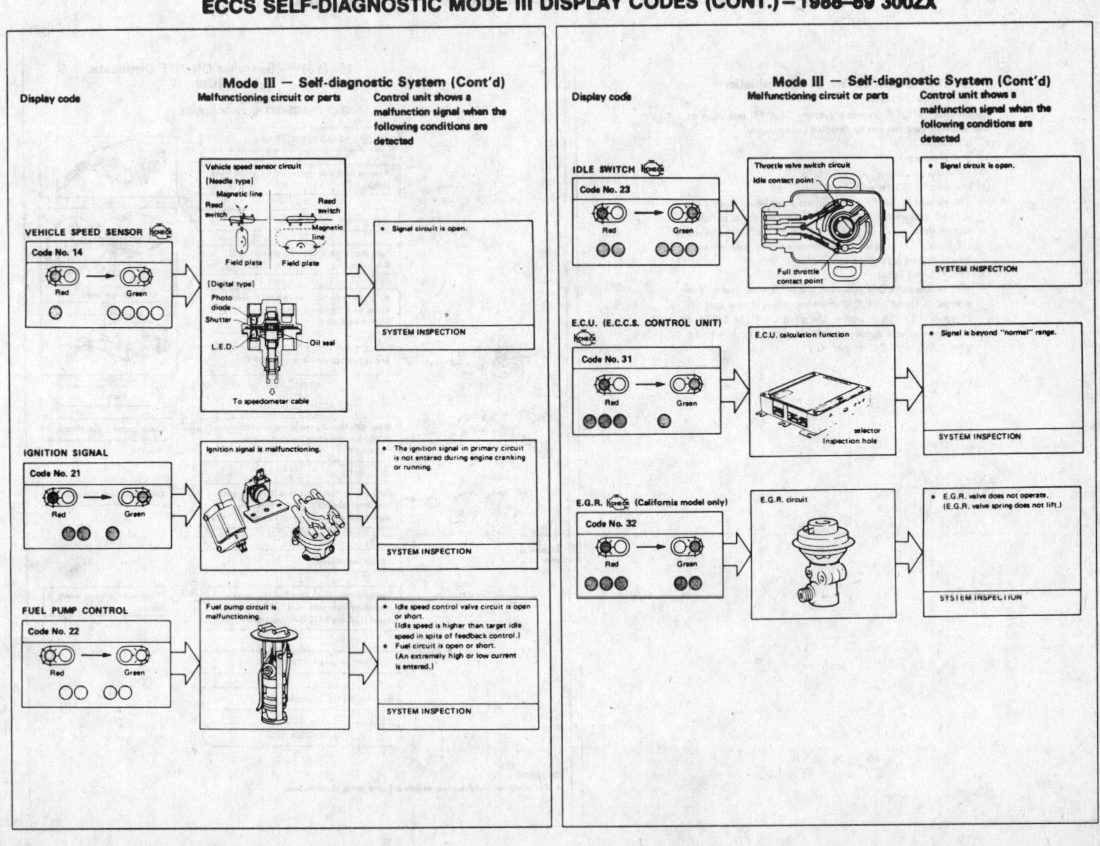
Mode III — Self-diagnostic System (Cont'd)
Display code
Malfunctioning circuit or parts
Control unit shows a malfunction signal when the following conditions are detected
VEHICLE SPEED SENSOR
Code No. 14
Red
Green
Vehicle speed sensor circuit
[Needle type]
Magnetic line
Reed switch
Reed switch
Magnetic line
Field plate
Field plate
[Digital type]
Photo diode
Shutter
L.E.D.
Oil seal
To speedometer cable
• Signal circuit is open.
SYSTEM INSPECTION
IGNITION SIGNAL
Code No. 21
Red
Green
Ignition signal is malfunctioning.
• The ignition signal in primary circuit is not entered during engine cranking or running.
SYSTEM INSPECTION
FUEL PUMP CONTROL
Code No. 22
Red
Green
Fuel pump circuit is malfunctioning.
• Idle speed control valve circuit is open or short. (Idle speed is higher than target idle speed in spite of feedback control.)
• Fuel circuit is open or short. (An extremely high or low current is entered.)
SYSTEM INSPECTION
Mode III — Self-diagnostic System (Cont'd)
Display code
Malfunctioning circuit or parts
Control unit shows a malfunction signal when the following conditions are detected
IDLE SWITCH
Code No. 23
Red
Green
Throttle valve switch circuit
Idle contact point
Full throttle contact point
• Signal circuit is open.
SYSTEM INSPECTION
E.C.U. (E.C.C.S. CONTROL UNIT)
Code No. 31
Red
Green
E.C.U. calculation function
selector
Inspection hole
• Signal is beyond "normal" range.
SYSTEM INSPECTION
E.G.R. (California model only)
Code No. 32
Red
Green
E.G.R. circuit
• E.G.R. valve does not operate. (E.G.R. valve spring does not lift.)
SYSTEM INSPECTION

ECCS SELF-DIAGNOSTIC MODE III DISPLAY CODES (CONT.)—1988–89 300ZX

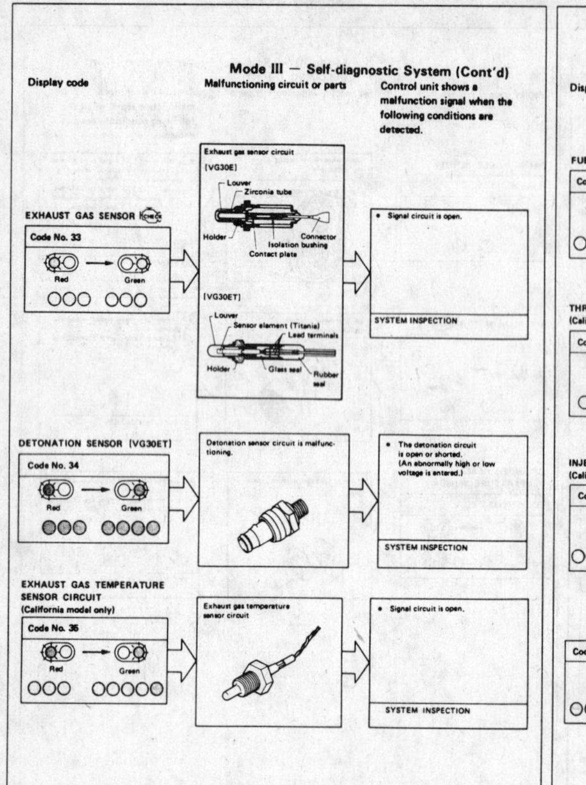
Mode III — Self-diagnostic System (Cont'd)
Display code
Malfunctioning circuit or parts
Control unit shows a malfunction signal when the following conditions are detected.
EXHAUST GAS SENSOR
Code No. 33
Red
Green
Exhaust gas sensor circuit
[VG30E]
Louver
Zirconia tube
Holder
Connector
Isolation bushing
Contact plate
[VG30ET]
Louver
Sensor element (Titania)
Lead terminals
Holder
Glass seal
Rubber seal
• Signal circuit is open.
SYSTEM INSPECTION
DETONATION SENSOR (VG30ET)
Code No. 34
Red
Green
Detonation sensor circuit is malfunctioning.
• The detonation circuit is open or shorted. (An abnormally high or low voltage is entered.)
SYSTEM INSPECTION
EXHAUST GAS TEMPERATURE SENSOR CIRCUIT
(California model only)
Code No. 35
Red
Green
Exhaust gas temperature sensor circuit
• Signal circuit is open.
SYSTEM INSPECTION

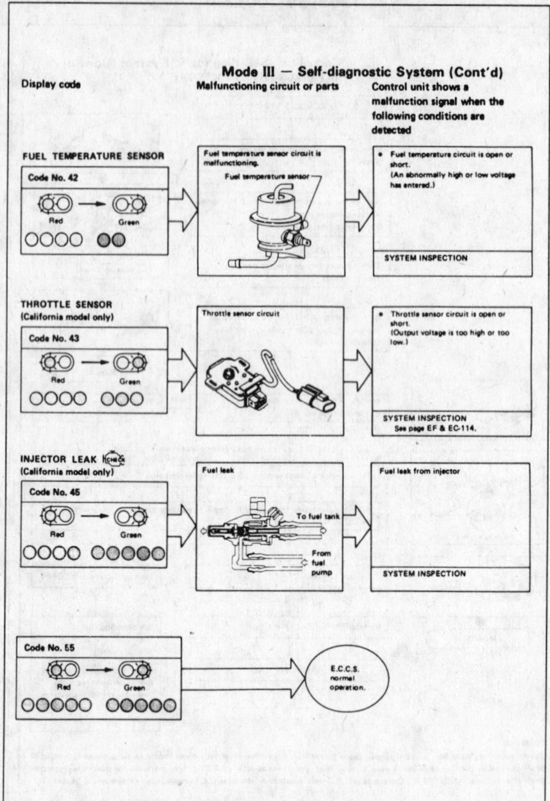
Mode III — Self-diagnostic System (Cont'd)
Display code
Malfunctioning circuit or parts
Control unit shows a malfunction signal when the following conditions are detected
FUEL TEMPERATURE SENSOR
Code No. 42
Red
Green
Fuel temperature sensor circuit is malfunctioning.
Fuel temperature sensor
• Fuel temperature circuit is open or short. (An abnormally high or low voltage has entered.)
SYSTEM INSPECTION
THROTTLE SENSOR
(California model only)
Code No. 43
Red
Green
Throttle sensor circuit
• Throttle sensor circuit is open or short. (Output voltage is too high or too low.)
SYSTEM INSPECTION
See page EF & EC-114.
INJECTOR LEAK
(California model only)
Code No. 45
Red
Green
Fuel leak
To fuel tank
From fuel pump
Fuel leak from injector
SYSTEM INSPECTION
Code No. 55
Red
Green
E.C.C.S. normal operation.

ECCS SELF-DIAGNOSTIC MODE IV—1988–89 300ZX

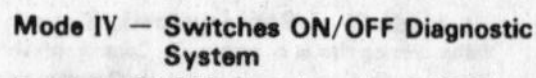

Mode IV — Switches ON/OFF Diagnostic System

In switches ON/OFF diagnosis system, ON/OFF operation of the following switches can be detected continuously.

- Idle switch
- Starter switch
- Vehicle speed sensor

(1) Idle switch & Starter switch
The switches ON/OFF status at the point when mode IV is selected is stored in E.C.U. memory. When either switch is turned from "ON" to "OFF" or "OFF" to "ON", the red L.E.D. on E.C.U. alternately comes on and goes off each time switching is detected.

(2) Vehicle Speed Sensor
The switches ON/OFF status at the point when mode IV is selected is stored in E.C.U. memory. When vehicle speed is 20 km/h (12 MPH) or slower, the green L.E.D. on E.C.U. is off. When vehicle speed exceeds 20 km/h (12 MPH), the green L.E.D. on E.C.U. comes "ON".

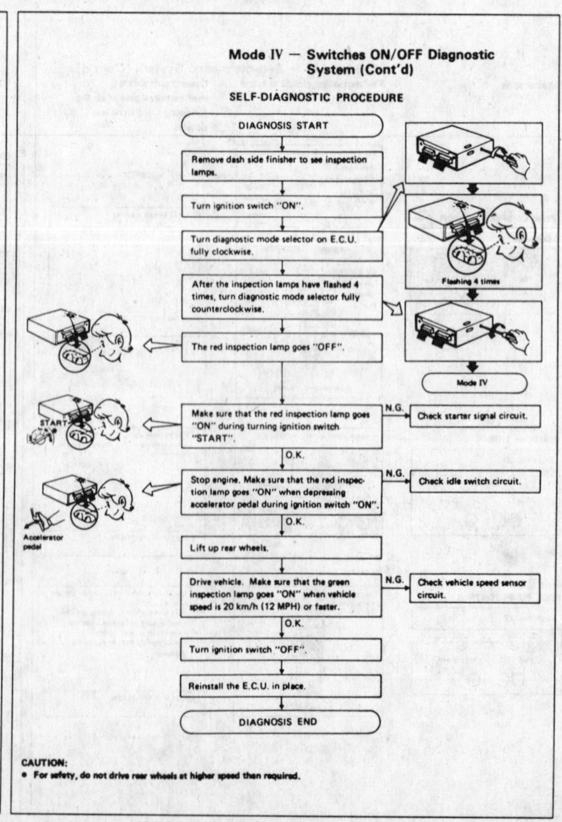
Mode IV — Switches ON/OFF Diagnostic System (Cont'd)
SELF-DIAGNOSTIC PROCEDURE
DIAGNOSIS START
Remove dash side finisher to see inspection lamps.
Turn ignition switch "ON".
Turn diagnostic mode selector on E.C.U. fully clockwise.
After the inspection lamps have flashed 4 times, turn diagnostic mode selector fully counterclockwise.
Flashing 4 times
The red inspection lamp goes "OFF".
Mode IV
Make sure that the red inspection lamp goes "ON" during turning ignition switch "START".
START
N.G.
Check starter signal circuit.
O.K.
Stop engine. Make sure that the red inspection lamp goes "ON" when depressing accelerator pedal during ignition switch "ON".
N.G.
Check idle switch circuit.
Accelerator pedal
O.K.
Lift up rear wheels.
Drive vehicle. Make sure that the green inspection lamp goes "ON" when vehicle speed is 20 km/h (12 MPH) or faster.
N.G.
Check vehicle speed sensor circuit.
O.K.
Turn ignition switch "OFF".
Reinstall the E.C.U. in place.
DIAGNOSIS END

CAUTION:

- **For safety, do not drive rear wheels at higher speed than required.**

ECCS SELF-DIAGNOSTIC MODE V 1988–89 300ZX

ECCS SELF-DIAGNOSTIC MODE V DECODING CHART 1988–89 300ZX

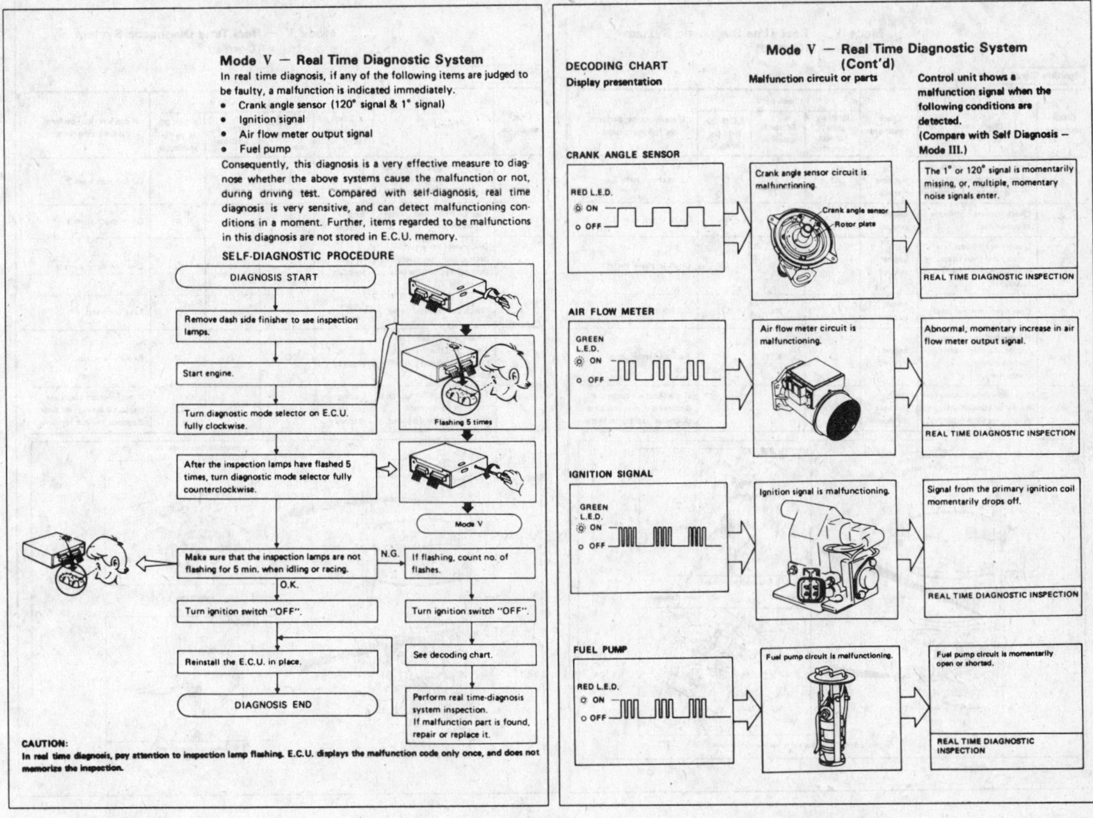
Mode V — Real Time Diagnostic System
In real time diagnosis, if any of the following items are judged to be faulty, a malfunction is indicated immediately.
• Crank angle sensor (120° signal & 1° signal)
• Ignition signal
• Air flow meter output signal
• Fuel pump
Consequently, this diagnosis is a very effective measure to diagnose whether the above systems cause the malfunction or not, during driving test. Compared with self-diagnosis, real time diagnosis is very sensitive, and can detect malfunctioning conditions in a moment. Further, items regarded to be malfunctions in this diagnosis are not stored in E.C.U. memory.
SELF-DIAGNOSTIC PROCEDURE
DIAGNOSIS START
Remove dash side finisher to see inspection lamps.
Start engine.
Turn diagnostic mode selector on E.C.U. fully clockwise.
Flashing 5 times
After the inspection lamps have flashed 5 times, turn diagnostic mode selector fully counterclockwise.
Mode V
Make sure that the inspection lamps are not flashing for 5 min. when idling or racing.
N.G.
If flashing, count no. of flashes.
O.K.
Turn ignition switch "OFF".
Turn ignition switch "OFF".
See decoding chart.
Reinstall the E.C.U. in place.
DIAGNOSIS END
Perform real time-diagnosis system inspection.
If malfunction part is found, repair or replace it.
CAUTION:
In real time diagnosis, pay attention to inspection lamp flashing. E.C.U. displays the malfunction code only once, and does not memorize the inspection.
Mode V — Real Time Diagnostic System (Cont'd)
DECODING CHART
Display presentation
Malfunction circuit or parts
Control unit shows a malfunction signal when the following conditions are detected. (Compare with Self Diagnosis — Mode III.)
CRANK ANGLE SENSOR
RED L.E.D.
ON
OFF
Crank angle sensor circuit is malfunctioning.
Crank angle sensor
Rotor plate
The 1° or 120° signal is momentarily missing, or, multiple, momentary noise signals enter.
REAL TIME DIAGNOSTIC INSPECTION
AIR FLOW METER
GREEN L.E.D.
ON
OFF
Air flow meter circuit is malfunctioning.
Abnormal, momentary increase in air flow meter output signal.
REAL TIME DIAGNOSTIC INSPECTION
IGNITION SIGNAL
GREEN L.E.D.
ON
OFF
Ignition signal is malfunctioning.
Signal from the primary ignition coil momentarily drops off.
REAL TIME DIAGNOSTIC INSPECTION
FUEL PUMP
RED L.E.D.
ON
OFF
Fuel pump circuit is malfunctioning.
Fuel pump circuit is momentarily open or shorted.
REAL TIME DIAGNOSTIC INSPECTION

ECCS SELF-DIAGNOSTIC MODE V INSPECTION—1988–89 300ZX

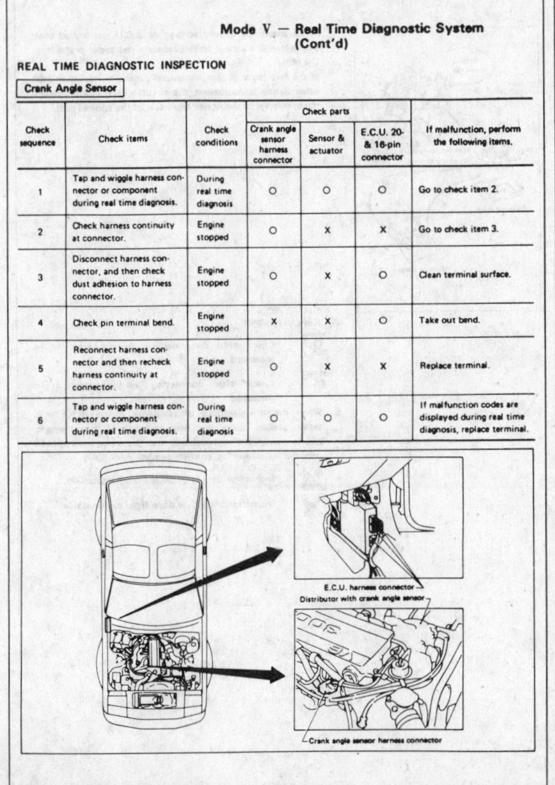

Mode V — Real Time Diagnostic System (Cont'd)

REAL TIME DIAGNOSTIC INSPECTION

Crank Angle Sensor

Check sequence	Check items	Check conditions	Check parts: Crank angle sensor harness connector	Check parts: Sensor & actuator	Check parts: E.C.U. 20- & 16-pin connector	If malfunction, perform the following items.
1	Tap and wiggle harness connector or component during real time diagnosis.	During real time diagnosis	O	O	O	Go to check item 2.
2	Check harness continuity at connector.	Engine stopped	O	X	X	Go to check item 3.
3	Disconnect harness connector, and then check dust adhesion to harness connector.	Engine stopped	O	X	O	Clean terminal surface.
4	Check pin terminal bend.	Engine stopped	X	X	O	Take out bend.
5	Reconnect harness connector and then recheck harness continuity at connector.	Engine stopped	O	X	X	Replace terminal.
6	Tap and wiggle harness connector or component during real time diagnosis.	During real time diagnosis	O	O	O	If malfunction codes are displayed during real time diagnosis, replace terminal.

E.C.U. harness connector
Distributor with crank angle sensor
Crank angle sensor harness connector

Mode V — Real Time Diagnostic System (Cont'd)

Air Flow Meter

Check sequence	Check items	Check conditions	Check parts: Air flow meter harness connector	Check parts: Sensor & actuator	Check parts: E.C.U. 20- & 16-pin connector	If malfunction, perform the following items.
1	Tap and wiggle harness connector or component during real time diagnosis.	During real time diagnosis	O	O	O	Go to check item 2.
2	Check harness continuity at connector.	Engine stopped	O	X	X	Go to check item 3.
3	Disconnect harness connector, and then check dust adhesion to harness connector.	Engine stopped	O	X	O	Clean terminal surface.
4	Check pin terminal bend.	Engine stopped	X	X	O	Take out bend.
5	Reconnect harness connector and then recheck harness continuity at connector.	Engine stopped	O	X	X	Replace terminal.
6	Tap and wiggle harness connector or component during real time diagnosis.	During real time diagnosis	O	O	O	If malfunction codes are displayed during real time diagnosis, replace terminal.

E.C.U. harness connector

Air flow meter harness connector

ECCS SELF-DIAGNOSTIC MODE V INSPECTION (CONT.)—1988–89 300ZX

Mode V — Real Time Diagnostic System (Cont'd)

Ignition Signal

Check sequence	Check items	Check conditions	Check parts: Ignition signal harness connector	Check parts: Sensor & actuator	Check parts: E.C.U. 20- & 16-pin connector	If malfunction, perform the following items.
1	Tap and wiggle harness connector or component during real time diagnosis.	During real time diagnosis	O	O	O	Go to check item 2.
2	Check harness continuity at connector.	Engine stopped	O	X	X	Go to check item 3.
3	Disconnect harness connector, and then check dust adhesion to harness connector.	Engine stopped	O	X	O	Clean terminal surface.
4	Check pin terminal bend.	Engine stopped	X	X	O	Take out bend.
5	Reconnect harness connector and then recheck harness continuity at connector.	Engine stopped	O	X	X	Replace terminal.
6	Tap and wiggle harness connector or component during real time diagnosis.	During real time diagnosis	O	O	O	If malfunction codes are displayed during real time diagnosis, replace terminal.

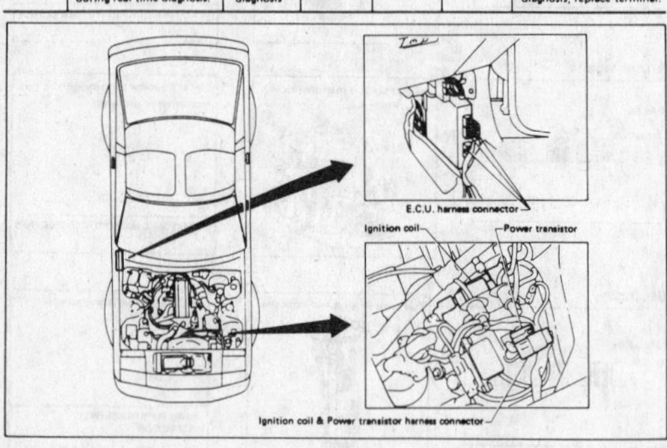
E.C.U. harness connector
Ignition coil
Power transistor
Ignition coil & Power transistor harness connector

Mode V — Real Time Diagnostic System (Cont'd)

Fuel pump

Check sequence	Check items	Check conditions	Check parts: Fuel pump harness connector	Check parts: Sensor & actuator	Check parts: E.C.U. 20- & 16-pin connector	If malfunction, perform the following items.
1	Tap and wiggle harness connector or component during real time diagnosis.	During real time diagnosis	O	O	O	Go to check item 2.
2	Check harness continuity at connector.	Engine stopped	O	X	X	Go to check item 3.
3	Disconnect harness connector, and then check dust adhesion to harness connector.	Engine stopped	O	X	O	Clean terminal surface.
4	Check pin terminal bend.	Engine stopped	X	X	O	Take out bend.
5	Reconnect harness connector and then recheck harness continuity at connector.	Engine stopped	O	X	X	Replace terminal.
6	Tap and wiggle harness connector or component during real time diagnosis.	During real time diagnosis	O	O	O	If malfunction codes are displayed during real time diagnosis, replace terminal.

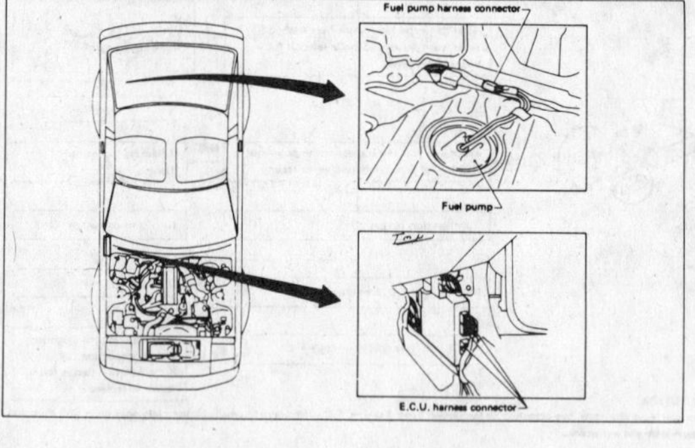
Fuel pump harness connector
Fuel pump
E.C.U. harness connector

ECCS SYSTEM CAUTION—1988–89 300ZX

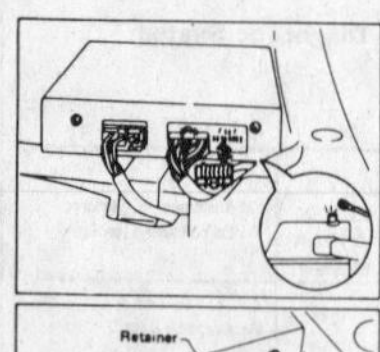

CAUTION:

1. Before connecting or disconnecting E.C.U. harness connector to or from any E.C.U., be sure to turn the ignition switch to the "OFF" position and disconnect the negative battery terminal in order not to damage E.C.U. as battery voltage is applied to E.C.U. even if ignition switch is turned off. Otherwise, there may be damage to the E.C.U.

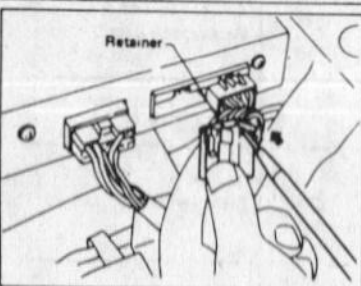
Retainer

2. When performing E.C.U. input/output signal inspection, remove pin terminal retainer from 20- and 16-pin connector to make it easier to insert tester probe into connector.

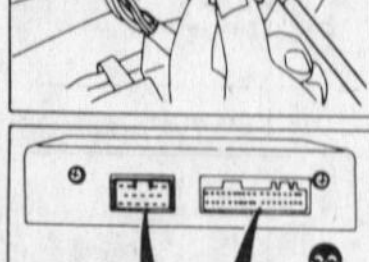
Bend
Break

3. When connecting pin connectors into E.C.U. or disconnecting them from E.C.U., take care not to damage pin terminal of E.C.U. (Bend or break).
4. Make sure that there are not any bends or breaks on E.C.U. pin terminal, when connecting pin connectors into E.C.U.

Perform E.C.U. input/output signal inspection before replacement.
OLD ONE

5. Before replacing E.C.U., perform E.C.U. input/output signal inspection and make sure whether E.C.U. functions properly or not.

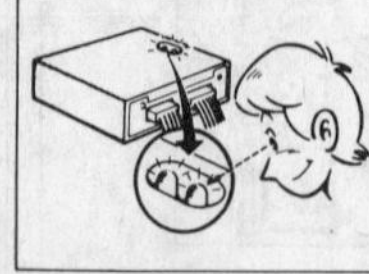

6. After performing this "ELECTRONIC CONTROL SYSTEM INSPECTION", perform E.C.C.S. self-diagnosis and driving test.

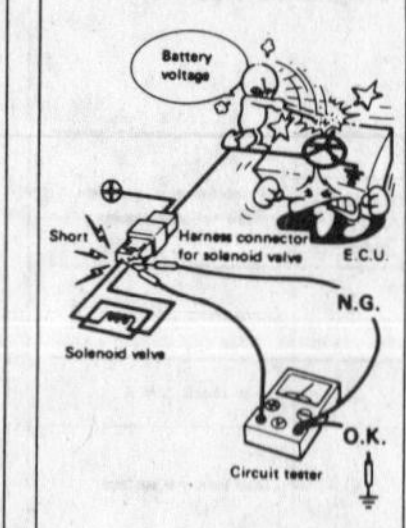
Battery voltage
Short
Harness connector for solenoid valve
E.C.U.
N.G.
Solenoid valve
O.K.
Circuit tester

7. When measuring supply voltage of E.C.U. controlled components with a circuit tester, separate one tester probe from the other.
   If the two tester probes accidentally make contact with each other during measurement, the circuit will be shorted, resulting in damage to the power transistor of the control unit.

8. Keys to symbols

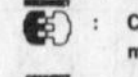

: Check after disconnecting the connector to be measured.

: Check after connecting the connector to be measured.

9. When measuring voltage or resistance at connector with tester probes, there are two methods of measurement; one is done from terminal side and the other from harness side. Before measuring, confirm symbol mark again.

: Inspection should be done from harness side.

: Inspection should be done from terminal side.

ECCS SELF-DIAGNOSTIC CHART–1988–89 300ZX

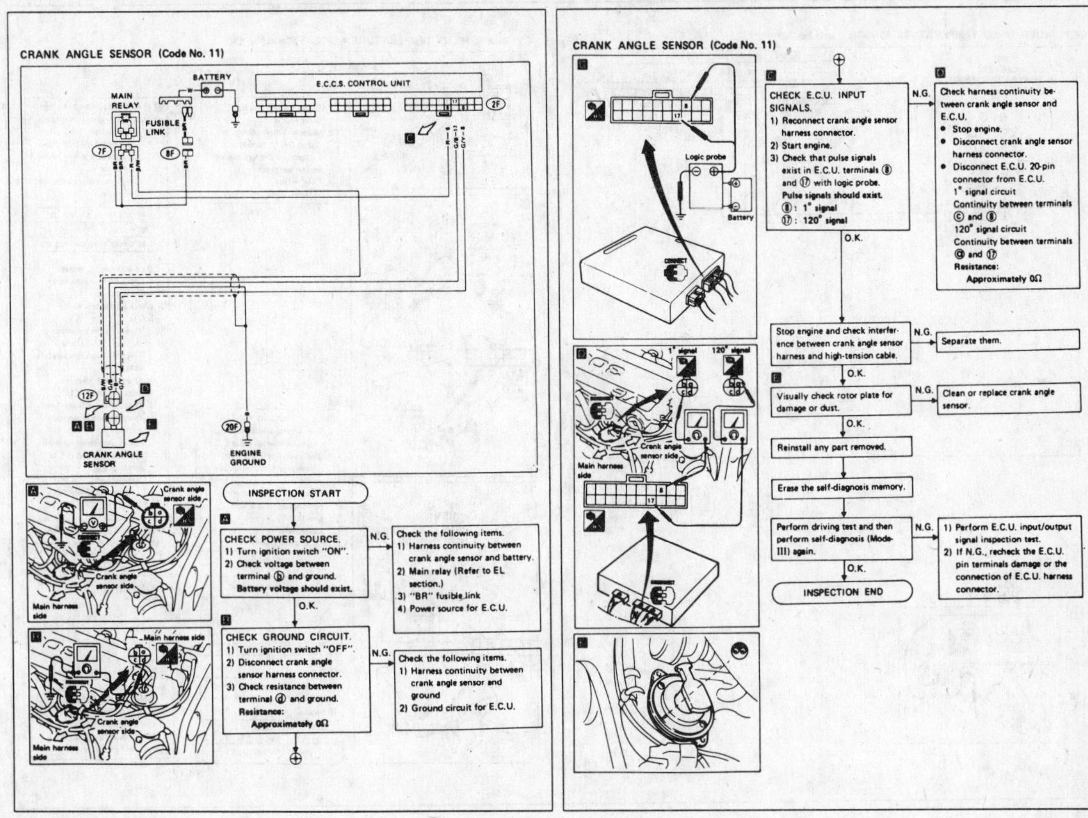
CRANK ANGLE SENSOR (Code No. 11)
BATTERY
MAIN RELAY
FUSIBLE LINK
E.C.C.S. CONTROL UNIT
CRANK ANGLE SENSOR
ENGINE GROUND
INSPECTION START
CHECK POWER SOURCE.
1) Turn ignition switch "ON".
2) Check voltage between terminal ⓑ and ground.
Battery voltage should exist.
N.G.
Check the following items.
1) Harness continuity between crank angle sensor and battery.
2) Main relay (Refer to EL section.)
3) "BR" fusible link
4) Power source for E.C.U.
O.K.
CHECK GROUND CIRCUIT.
1) Turn ignition switch "OFF".
2) Disconnect crank angle sensor harness connector.
3) Check resistance between terminal ⓓ and ground.
Resistance:
Approximately 0Ω
N.G.
Check the following items.
1) Harness continuity between crank angle sensor and ground
2) Ground circuit for E.C.U.
Crank angle sensor side
Main harness side
CRANK ANGLE SENSOR (Code No. 11)
Logic probe
Battery
CHECK E.C.U. INPUT SIGNALS.
1) Reconnect crank angle sensor harness connector.
2) Start engine.
3) Check that pulse signals exist in E.C.U. terminals ⑧ and ⑰ with logic probe.
Pulse signals should exist.
⑧: 1° signal
⑰: 120° signal
N.G.
Check harness continuity between crank angle sensor and E.C.U.
• Stop engine.
• Disconnect crank angle sensor harness connector.
• Disconnect E.C.U. 20-pin connector from E.C.U.
1° signal circuit
Continuity between terminals ⓒ and ⑧
120° signal circuit
Continuity between terminals ⓖ and ⑰
Resistance:
Approximately 0Ω
O.K.
Stop engine and check interference between crank angle sensor harness and high-tension cable.
N.G.
Separate them.
O.K.
Visually check rotor plate for damage or dust.
N.G.
Clean or replace crank angle sensor.
O.K.
Reinstall any part removed.
Erase the self-diagnosis memory.
Perform driving test and then perform self-diagnosis (Mode-III) again.
N.G.
1) Perform E.C.U. input/output signal inspection test.
2) If N.G., recheck the E.C.U. pin terminals damage or the connection of E.C.U. harness connector.
O.K.
INSPECTION END
1° signal
120° signal
Crank angle sensor side
Main harness side

ECCS SELF-DIAGNOSTIC CHART–1988–89 300ZX

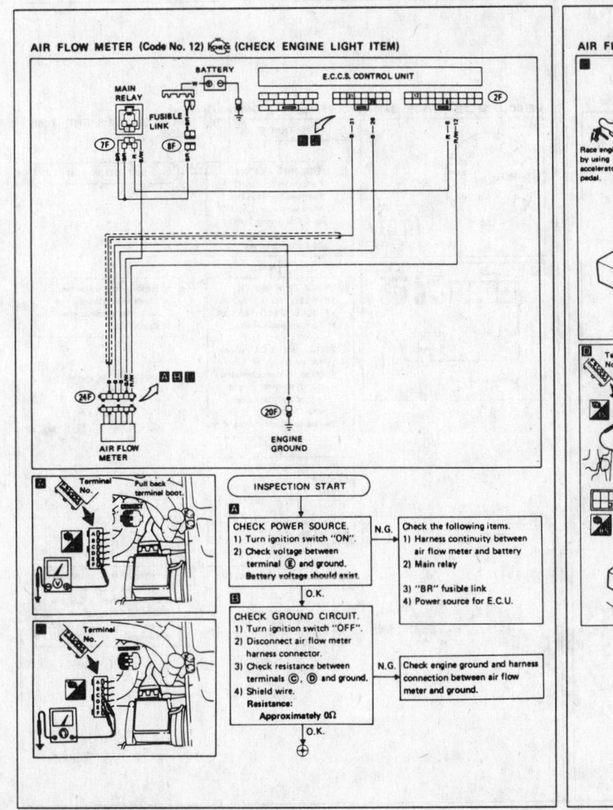
AIR FLOW METER (Code No. 12) (CHECK ENGINE LIGHT ITEM)
BATTERY
MAIN RELAY
FUSIBLE LINK
E.C.C.S. CONTROL UNIT
AIR FLOW METER
ENGINE GROUND
Terminal No.
Pull back terminal boot.
INSPECTION START
CHECK POWER SOURCE.
1) Turn ignition switch "ON".
2) Check voltage between terminal Ⓔ and ground.
Battery voltage should exist.
N.G.
Check the following items.
1) Harness continuity between air flow meter and battery
2) Main relay
3) "BR" fusible link
4) Power source for E.C.U.
O.K.
CHECK GROUND CIRCUIT.
1) Turn ignition switch "OFF".
2) Disconnect air flow meter harness connector.
3) Check resistance between terminals Ⓒ, Ⓓ and ground.
4) Shield wire.
Resistance:
Approximately 0Ω
N.G.
Check engine ground and harness connection between air flow meter and ground.
O.K.

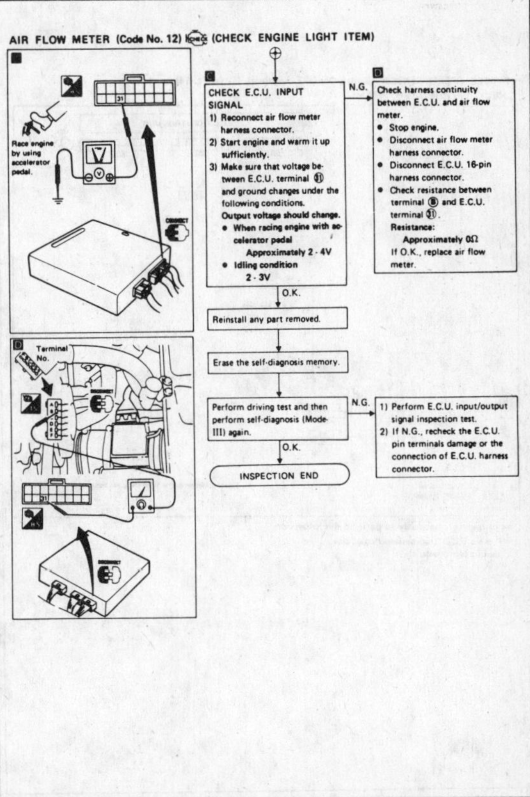
AIR FLOW METER (Code No. 12) (CHECK ENGINE LIGHT ITEM)
Race engine by using accelerator pedal.
CHECK E.C.U. INPUT SIGNAL
1) Reconnect air flow meter harness connector.
2) Start engine and warm it up sufficiently.
3) Make sure that voltage between E.C.U. terminal ㉛ and ground changes under the following conditions.
Output voltage should change.
• When racing engine with accelerator pedal
Approximately 2 - 4V
• Idling condition
2 - 3V
N.G.
Check harness continuity between E.C.U. and air flow meter.
• Stop engine.
• Disconnect air flow meter harness connector.
• Disconnect E.C.U. 16-pin harness connector.
• Check resistance between terminal Ⓑ and E.C.U. terminal ㉛.
Resistance:
Approximately 0Ω
If O.K., replace air flow meter.
O.K.
Reinstall any part removed.
Erase the self-diagnosis memory.
Perform driving test and then perform self-diagnosis (Mode-III) again.
N.G.
1) Perform E.C.U. input/output signal inspection test.
2) If N.G., recheck the E.C.U. pin terminals damage or the connection of E.C.U. harness connector.
O.K.
INSPECTION END
Terminal No.

ECCS SELF-DIAGNOSTIC CHART – 1988–89 300ZX

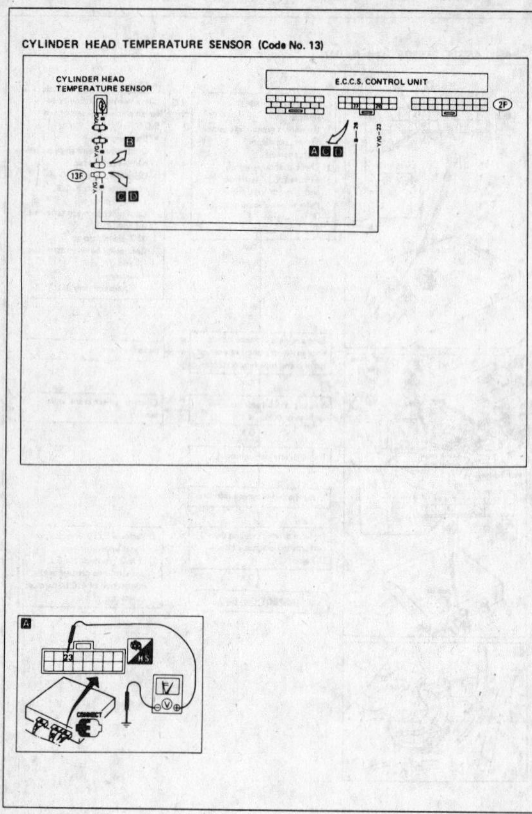
CYLINDER HEAD TEMPERATURE SENSOR (Code No. 13)
CYLINDER HEAD TEMPERATURE SENSOR
E.C.C.S. CONTROL UNIT
2F
13F

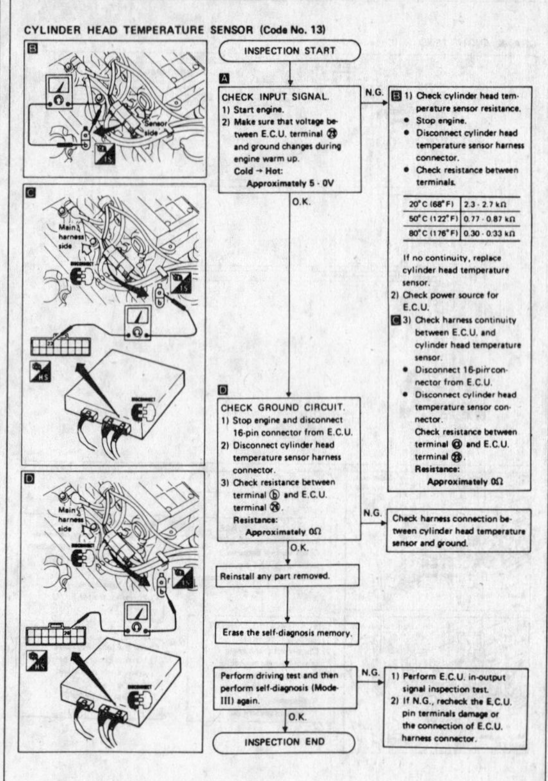
CYLINDER HEAD TEMPERATURE SENSOR (Code No. 13)
INSPECTION START
A CHECK INPUT SIGNAL.
1) Start engine.
2) Make sure that voltage between E.C.U. terminal and ground changes during engine warm up.
Cold → Hot:
Approximately 5 - 0V
N.G.
B 1) Check cylinder head temperature sensor resistance.
• Stop engine.
• Disconnect cylinder head temperature sensor harness connector.
• Check resistance between terminals.
20°C (68°F) 2.3 - 2.7 kΩ
50°C (122°F) 0.77 - 0.87 kΩ
80°C (176°F) 0.30 - 0.33 kΩ
If no continuity, replace cylinder head temperature sensor.
2) Check power source for E.C.U.
C 3) Check harness continuity between E.C.U. and cylinder head temperature sensor.
• Disconnect 16-pin connector from E.C.U.
• Disconnect cylinder head temperature sensor connector.
Check resistance between terminal ⓐ and E.C.U. terminal.
Resistance:
Approximately 0Ω
O.K.
D CHECK GROUND CIRCUIT.
1) Stop engine and disconnect 16-pin connector from E.C.U.
2) Disconnect cylinder head temperature sensor harness connector.
3) Check resistance between terminal ⓑ and E.C.U. terminal.
Resistance:
Approximately 0Ω
N.G.
Check harness connection between cylinder head temperature sensor and ground.
O.K.
Reinstall any part removed.
Erase the self-diagnosis memory.
Perform driving test and then perform self-diagnosis (Mode III) again.
N.G.
1) Perform E.C.U. in-output signal inspection test.
2) If N.G., recheck the E.C.U. pin terminals damage or the connection of E.C.U. harness connector.
O.K.
INSPECTION END
Sensor side
Main harness side

ECCS SELF-DIAGNOSTIC CHART – 1988–89 300ZX

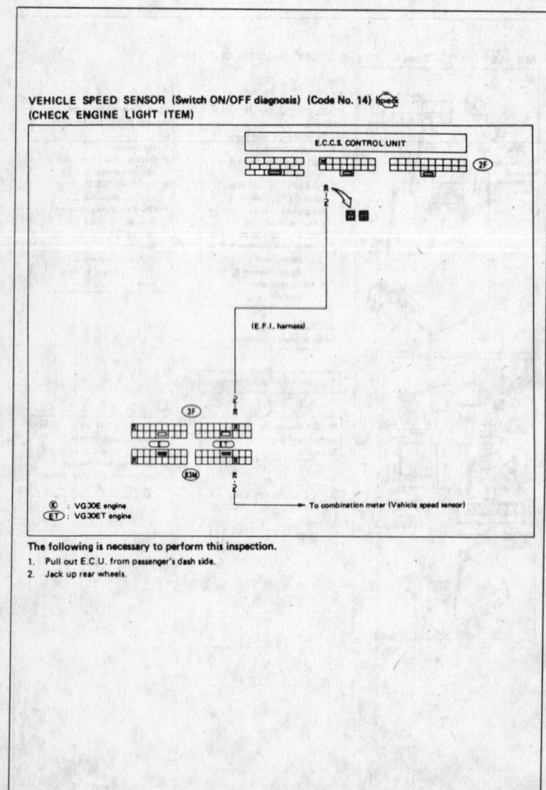
VEHICLE SPEED SENSOR (Switch ON/OFF diagnosis) (Code No. 14)
(CHECK ENGINE LIGHT ITEM)
E.C.C.S. CONTROL UNIT
2F
(E.F.I. harness)
3F
To combination meter (Vehicle speed sensor)
Ⓔ : VG30E engine
ⒺⓉ : VG30ET engine

The following is necessary to perform this inspection.
1. Pull out E.C.U. from passenger's dash side.
2. Jack up rear wheels.

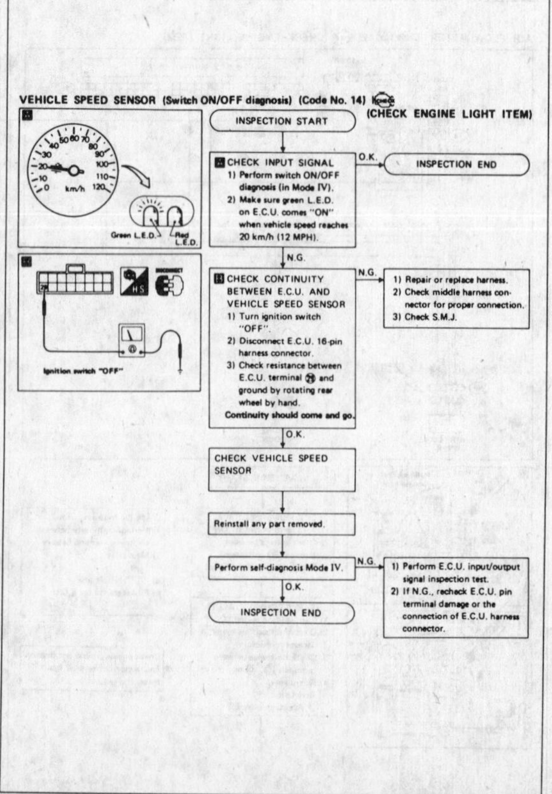
VEHICLE SPEED SENSOR (Switch ON/OFF diagnosis) (Code No. 14)
(CHECK ENGINE LIGHT ITEM)
INSPECTION START
A CHECK INPUT SIGNAL
1) Perform switch ON/OFF diagnosis (in Mode IV).
2) Make sure green L.E.D. on E.C.U. comes "ON" when vehicle speed reaches 20 km/h (12 MPH).
O.K.
INSPECTION END
N.G.
B CHECK CONTINUITY BETWEEN E.C.U. AND VEHICLE SPEED SENSOR
1) Turn ignition switch "OFF".
2) Disconnect E.C.U. 16-pin harness connector.
3) Check resistance between E.C.U. terminal and ground by rotating rear wheel by hand.
Continuity should come and go.
N.G.
1) Repair or replace harness.
2) Check middle harness connector for proper connection.
3) Check S.M.J.
O.K.
CHECK VEHICLE SPEED SENSOR
Reinstall any part removed.
Perform self-diagnosis Mode IV.
N.G.
1) Perform E.C.U. input/output signal inspection test.
2) If N.G., recheck E.C.U. pin terminal damage or the connection of E.C.U. harness connector.
O.K.
INSPECTION END
km/h
Green L.E.D.
Red L.E.D.
Ignition switch "OFF"

ECCS SELF-DIAGNOSTIC CHART—1988–89 300ZX

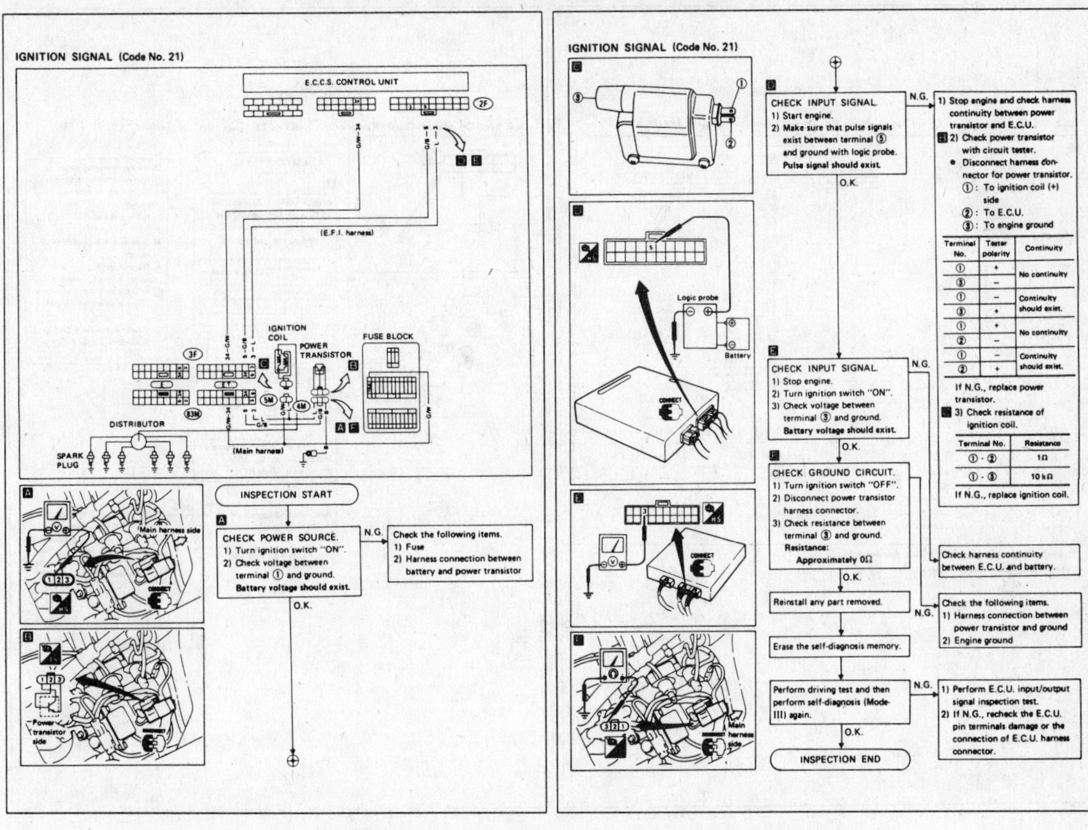
IGNITION SIGNAL (Code No. 21)
E.C.C.S. CONTROL UNIT
(E.F.I. harness)
IGNITION COIL
POWER TRANSISTOR
FUSE BLOCK
DISTRIBUTOR
SPARK PLUG
(Main harness)
Main harness side
Power transistor side
INSPECTION START
CHECK POWER SOURCE.
1) Turn ignition switch "ON".
2) Check voltage between terminal ① and ground.
Battery voltage should exist.
N.G.
Check the following items.
1) Fuse
2) Harness connection between battery and power transistor
O.K.
CHECK INPUT SIGNAL.
1) Start engine.
2) Make sure that pulse signals exist between terminal ⑤ and ground with logic probe.
Pulse signal should exist.
1) Stop engine and check harness continuity between power transistor and E.C.U.
2) Check power transistor with circuit tester.
• Disconnect harness connector for power transistor.
①: To ignition coil (+) side
②: To E.C.U.
③: To engine ground
Terminal No. | Tester polarity | Continuity
① + ③ − No continuity
① − ③ + Continuity should exist.
① + ② − No continuity
① − ② + Continuity should exist.
If N.G., replace power transistor.
3) Check resistance of ignition coil.
Terminal No. | Resistance
①-② | 1Ω
①-③ | 10 kΩ
If N.G., replace ignition coil.
Logic probe
Battery
CHECK INPUT SIGNAL.
1) Stop engine.
2) Turn ignition switch "ON".
3) Check voltage between terminal ③ and ground.
Battery voltage should exist.
CHECK GROUND CIRCUIT.
1) Turn ignition switch "OFF".
2) Disconnect power transistor harness connector.
3) Check resistance between terminal ③ and ground.
Resistance: Approximately 0Ω
Check harness continuity between E.C.U. and battery.
Check the following items.
1) Harness connection between power transistor and ground
2) Engine ground
Reinstall any part removed.
Erase the self-diagnosis memory.
Perform driving test and then perform self-diagnosis (Mode-III) again.
1) Perform E.C.U. input/output signal inspection test.
2) If N.G., recheck the E.C.U. pin terminals damage or the connection of E.C.U. harness connector.
INSPECTION END

ECCS SELF-DIAGNOSTIC CHART—1988–89 300ZX

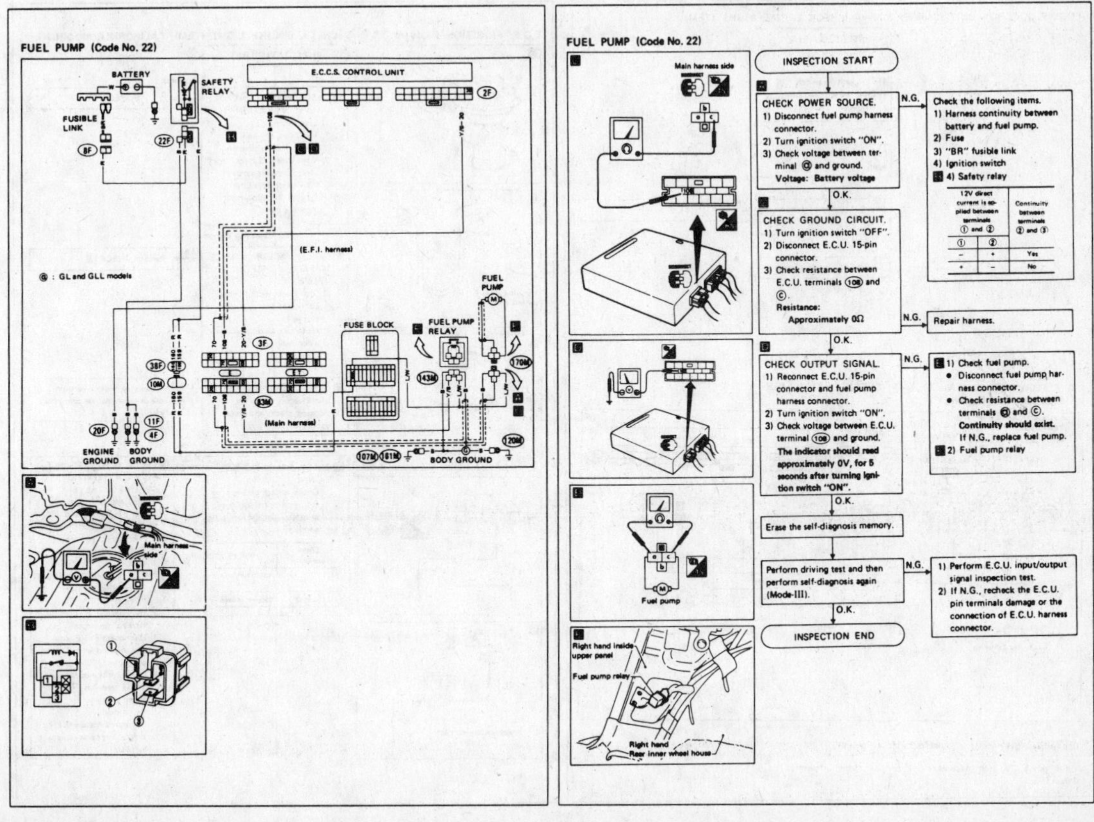
FUEL PUMP (Code No. 22)
BATTERY
SAFETY RELAY
FUSIBLE LINK
E.C.C.S. CONTROL UNIT
(E.F.I. harness)
Ⓢ: GL and GLL models
FUEL PUMP
FUSE BLOCK
FUEL PUMP RELAY
(Main harness)
ENGINE GROUND
BODY GROUND
Main harness side
INSPECTION START
CHECK POWER SOURCE.
1) Disconnect fuel pump harness connector.
2) Turn ignition switch "ON".
3) Check voltage between terminal ⓐ and ground.
Voltage: Battery voltage
N.G.
Check the following items.
1) Harness continuity between battery and fuel pump.
2) Fuse
3) "BR" fusible link
4) Ignition switch
4) Safety relay
12V direct current is applied between terminals ① and ② | Continuity between terminals ② and ③
Yes
No
O.K.
CHECK GROUND CIRCUIT.
1) Turn ignition switch "OFF".
2) Disconnect E.C.U. 15-pin connector.
3) Check resistance between E.C.U. terminals ⑩⑧ and ⓒ.
Resistance: Approximately 0Ω
Repair harness.
CHECK OUTPUT SIGNAL.
1) Reconnect E.C.U. 15-pin connector and fuel pump harness connector.
2) Turn ignition switch "ON".
3) Check voltage between E.C.U. terminal ⑩⑧ and ground.
The indicator should read approximately 0V, for 5 seconds after turning ignition switch "ON".
1) Check fuel pump.
• Disconnect fuel pump harness connector.
• Check resistance between terminals ⓐ and ⓒ.
Continuity should exist.
If N.G., replace fuel pump.
2) Fuel pump relay
Erase the self-diagnosis memory.
Perform driving test and then perform self-diagnosis again (Mode-III).
1) Perform E.C.U. input/output signal inspection test.
2) If N.G., recheck the E.C.U. pin terminals damage or the connection of E.C.U. harness connector.
INSPECTION END
Fuel pump
Right hand inside upper panel
Fuel pump relay
Right hand Rear inner wheel house

ECCS SELF-DIAGNOSTIC CHART—1988–89 300ZX

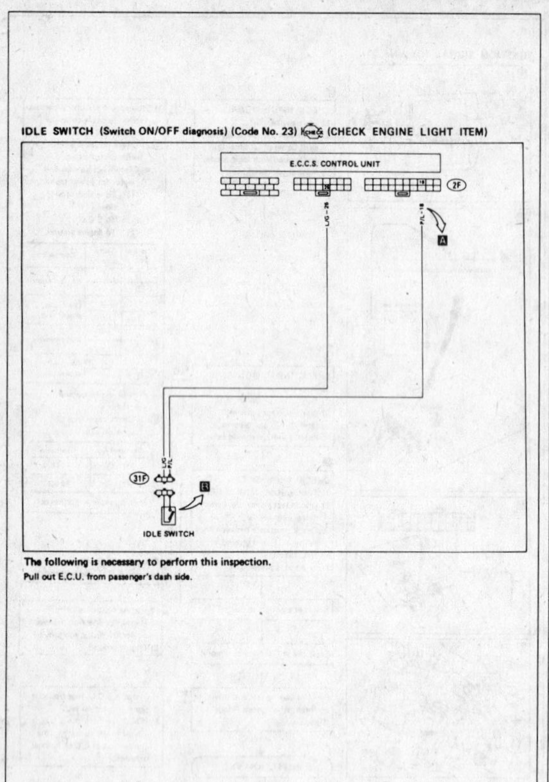
IDLE SWITCH (Switch ON/OFF diagnosis) (Code No. 23) (CHECK ENGINE LIGHT ITEM)
E.C.C.S. CONTROL UNIT
2F
31F
IDLE SWITCH
The following is necessary to perform this inspection.
Pull out E.C.U. from passenger's dash side.

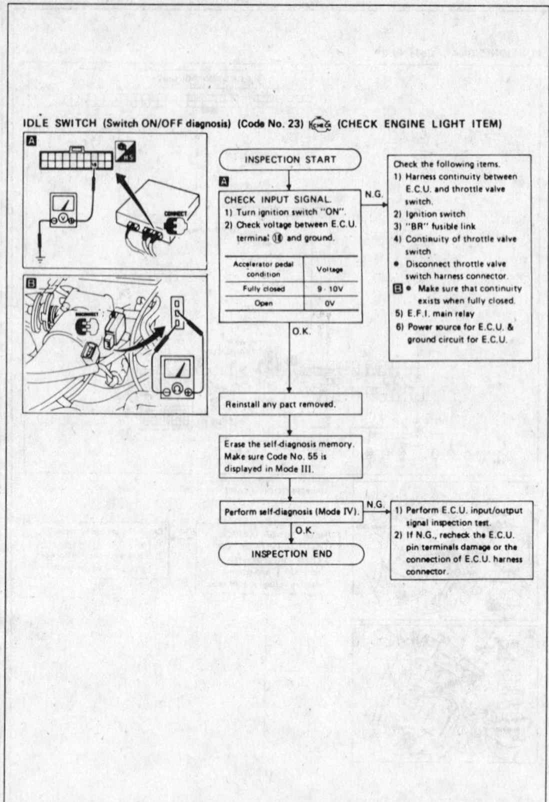
IDLE SWITCH (Switch ON/OFF diagnosis) (Code No. 23) (CHECK ENGINE LIGHT ITEM)
INSPECTION START
CHECK INPUT SIGNAL.
1) Turn ignition switch "ON".
2) Check voltage between E.C.U. terminal and ground.
Accelerator pedal condition
Voltage
Fully closed
9 - 10V
Open
0V
N.G.
Check the following items.
1) Harness continuity between E.C.U. and throttle valve switch.
2) Ignition switch
3) "BR" fusible link
4) Continuity of throttle valve switch
Disconnect throttle valve switch harness connector.
Make sure that continuity exists when fully closed.
5) E.F.I. main relay
6) Power source for E.C.U. & ground circuit for E.C.U.
O.K.
Reinstall any part removed.
Erase the self-diagnosis memory. Make sure Code No. 55 is displayed in Mode III.
Perform self-diagnosis (Mode IV).
N.G.
1) Perform E.C.U. input/output signal inspection test.
2) If N.G., recheck the E.C.U. pin terminals damage or the connection of E.C.U. harness connector.
O.K.
INSPECTION END

ECCS SELF-DIAGNOSTIC CHART—1988–89 300ZX

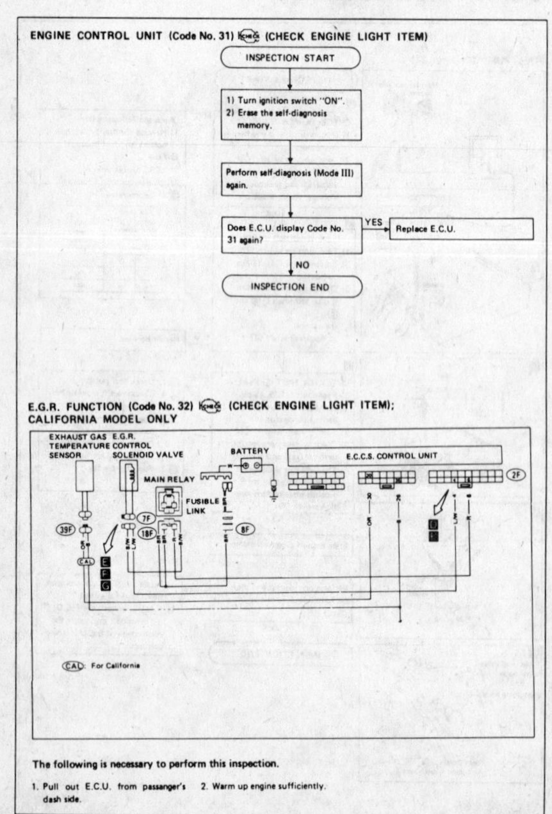
ENGINE CONTROL UNIT (Code No. 31) (CHECK ENGINE LIGHT ITEM)
INSPECTION START
1) Turn ignition switch "ON".
2) Erase the self-diagnosis memory.
Perform self-diagnosis (Mode III) again.
Does E.C.U. display Code No. 31 again?
YES
Replace E.C.U.
NO
INSPECTION END
E.G.R. FUNCTION (Code No. 32) (CHECK ENGINE LIGHT ITEM); CALIFORNIA MODEL ONLY
EXHAUST GAS TEMPERATURE SENSOR
E.G.R. CONTROL SOLENOID VALVE
BATTERY
MAIN RELAY
FUSIBLE LINK
E.C.C.S. CONTROL UNIT
2F
7F
18F
8F
39F
CAL For California
The following is necessary to perform this inspection.
1. Pull out E.C.U. from passanger's dash side.
2. Warm up engine sufficiently.

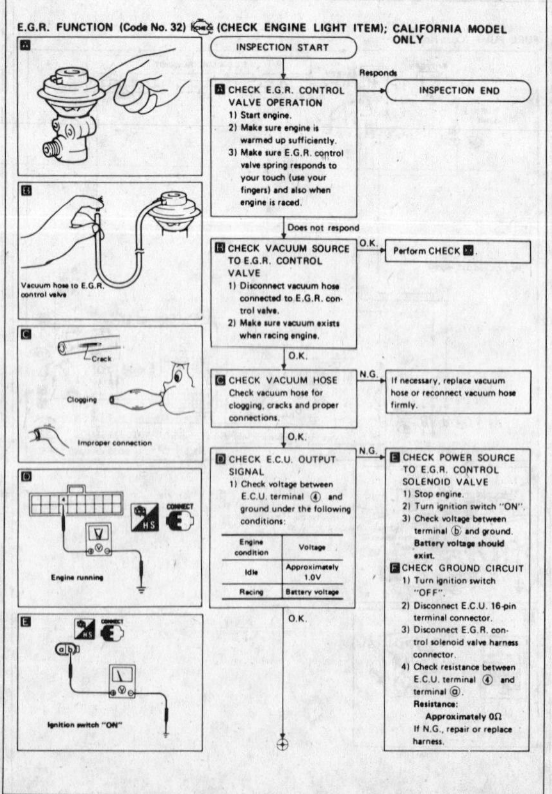
E.G.R. FUNCTION (Code No. 32) (CHECK ENGINE LIGHT ITEM); CALIFORNIA MODEL ONLY
INSPECTION START
CHECK E.G.R. CONTROL VALVE OPERATION
1) Start engine.
2) Make sure engine is warmed up sufficiently.
3) Make sure E.G.R. control valve spring responds to your touch (use your fingers) and also when engine is raced.
Responds
INSPECTION END
Does not respond
CHECK VACUUM SOURCE TO E.G.R. CONTROL VALVE
1) Disconnect vacuum hose connected to E.G.R. control valve.
2) Make sure vacuum exists when racing engine.
O.K.
Perform CHECK
Vacuum hose to E.G.R. control valve
O.K.
CHECK VACUUM HOSE
Check vacuum hose for clogging, cracks and proper connections.
N.G.
If necessary, replace vacuum hose or reconnect vacuum hose firmly.
Crack
Clogging
Improper connection
O.K.
CHECK E.C.U. OUTPUT SIGNAL
1) Check voltage between E.C.U. terminal ④ and ground under the following conditions:
Engine condition
Voltage
Idle
Approximately 1.0V
Racing
Battery voltage
N.G.
CHECK POWER SOURCE TO E.G.R. CONTROL SOLENOID VALVE
1) Stop engine.
2) Turn ignition switch "ON".
3) Check voltage between terminal ⓑ and ground. Battery voltage should exist.
CHECK GROUND CIRCUIT
1) Turn ignition switch "OFF".
2) Disconnect E.C.U. 16-pin terminal connector.
3) Disconnect E.G.R. control solenoid valve harness connector.
4) Check resistance between E.C.U. terminal ④ and terminal ⓐ.
Resistance: Approximately 0Ω
If N.G., repair or replace harness.
O.K.
Engine running
Ignition switch "ON"
CONNECT

ECCS SELF-DIAGNOSTIC CHART—1988–89 300ZX

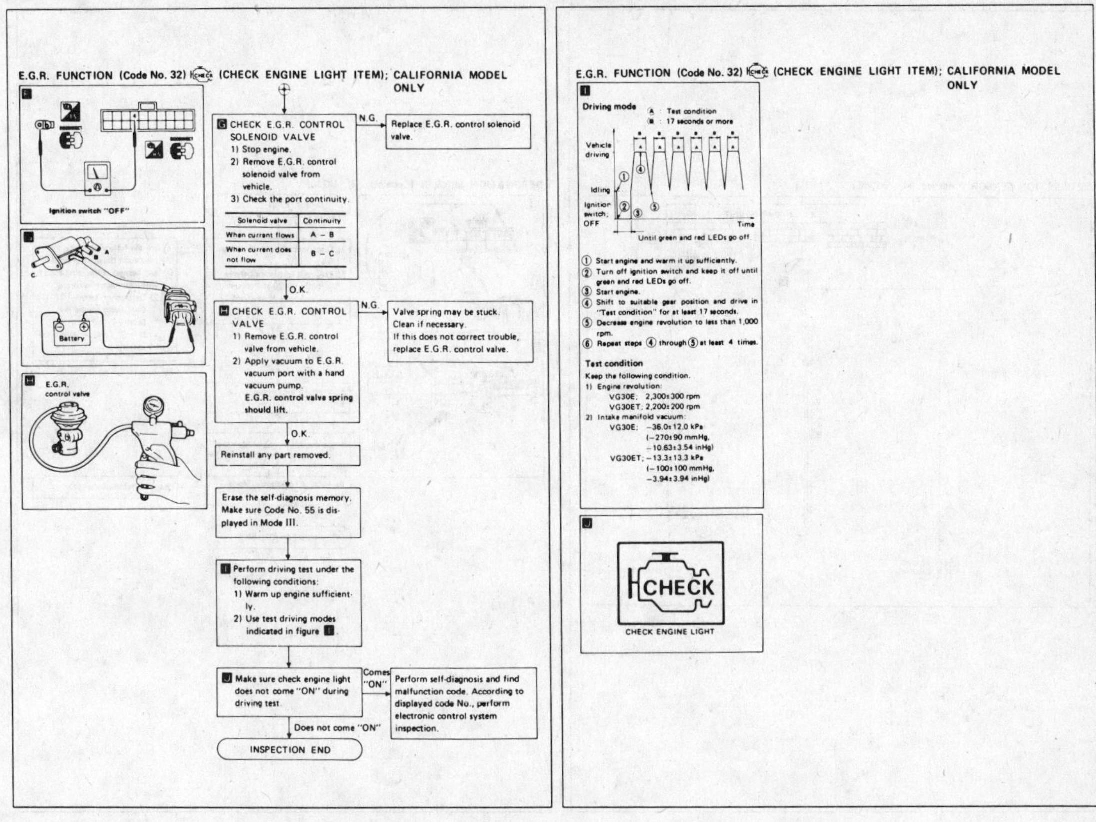

ECCS SELF-DIAGNOSTIC CHART—1988–89 300ZX

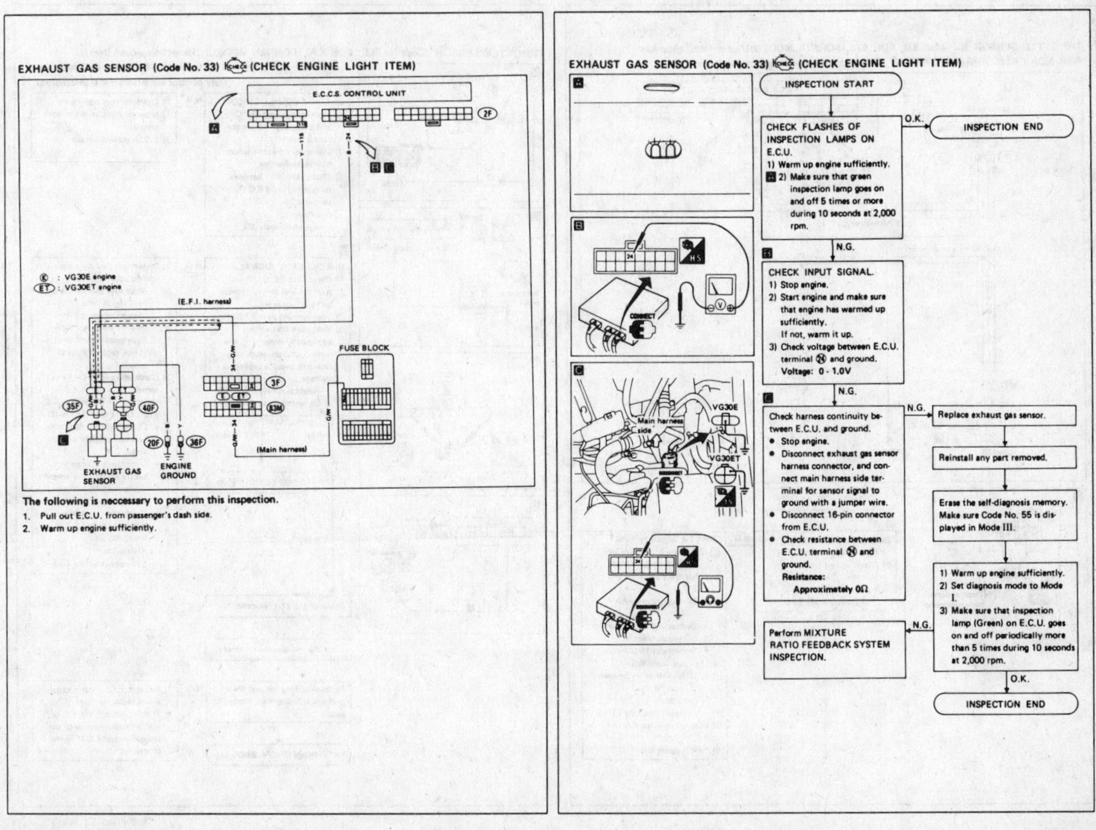

ECCS SELF-DIAGNOSTIC CHART—1988–89 300ZX

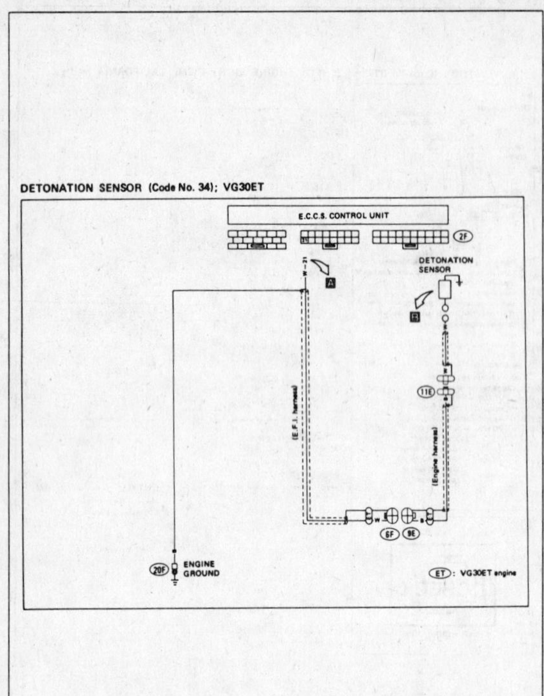
DETONATION SENSOR (Code No. 34); VG30ET
E.C.C.S. CONTROL UNIT
DETONATION SENSOR
(E.F.I. harness)
(Engine harness)
ENGINE GROUND
ET : VG30ET engine

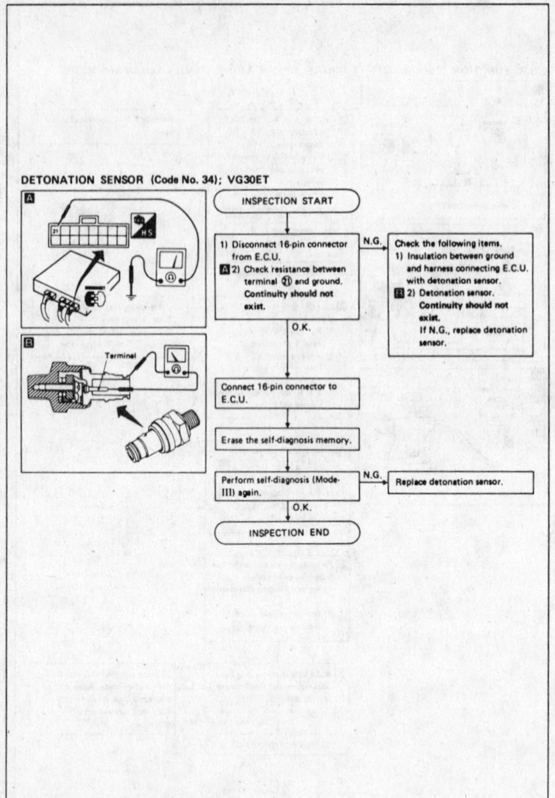
DETONATION SENSOR (Code No. 34); VG30ET
INSPECTION START
1) Disconnect 16-pin connector from E.C.U.
A 2) Check resistance between terminal 21 and ground. Continuity should not exist.
N.G.
Check the following items.
1) Insulation between ground and harness connecting E.C.U. with detonation sensor.
B 2) Detonation sensor. Continuity should not exist. If N.G., replace detonation sensor.
O.K.
Terminal
Connect 16-pin connector to E.C.U.
Erase the self-diagnosis memory.
Perform self-diagnosis (Mode III) again.
N.G.
Replace detonation sensor.
O.K.
INSPECTION END

ECCS SELF-DIAGNOSTIC CHART—1988–89 300ZX

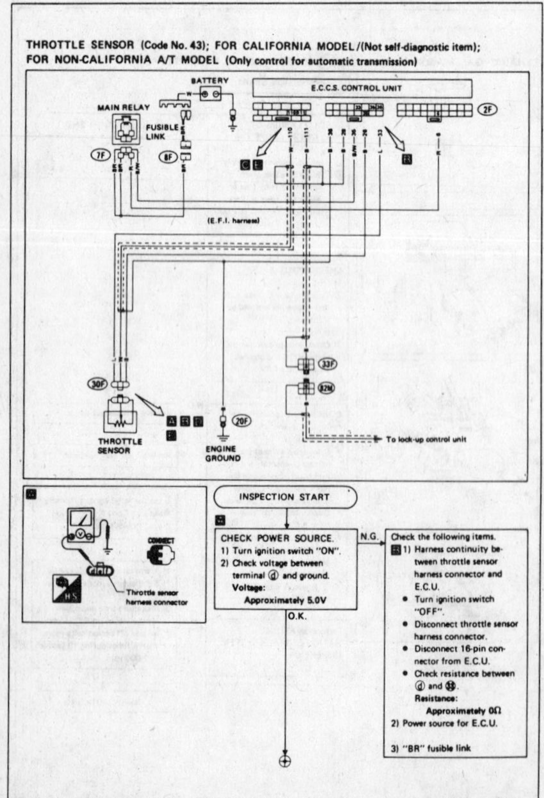
THROTTLE SENSOR (Code No. 43); FOR CALIFORNIA MODEL/(Not self-diagnostic item);
FOR NON-CALIFORNIA A/T MODEL (Only control for automatic transmission)
BATTERY
E.C.C.S. CONTROL UNIT
MAIN RELAY
FUSIBLE LINK
(E.F.I. harness)
THROTTLE SENSOR
ENGINE GROUND
To lock-up control unit
Throttle sensor harness connector
INSPECTION START
CHECK POWER SOURCE.
1) Turn ignition switch "ON".
2) Check voltage between terminal ⓓ and ground.
Voltage:
Approximately 5.0V
N.G.
O.K.
Check the following items.
1) Harness continuity between throttle sensor harness connector and E.C.U.
• Turn ignition switch "OFF".
• Disconnect throttle sensor harness connector.
• Disconnect 16-pin connector from E.C.U.
• Check resistance between ⓓ and 33.
Resistance:
Approximately 0Ω
2) Power source for E.C.U.
3) "BR" fusible link

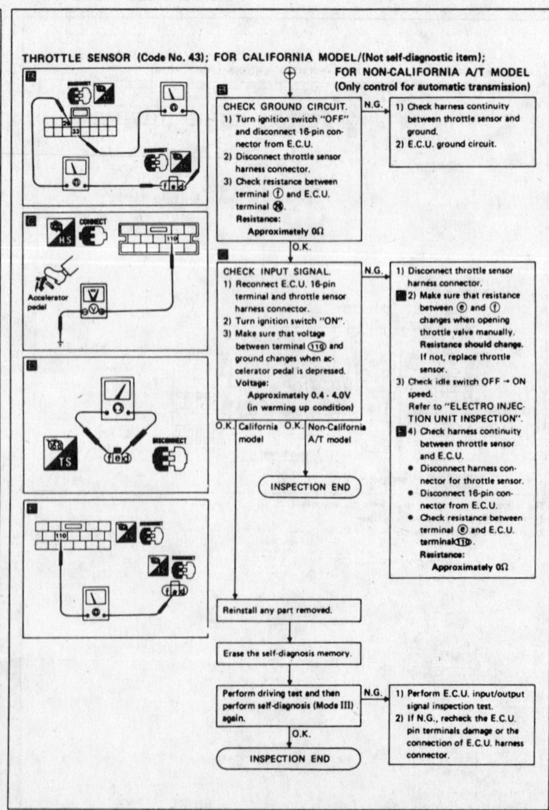
THROTTLE SENSOR (Code No. 43); FOR CALIFORNIA MODEL/(Not self-diagnostic item);
FOR NON-CALIFORNIA A/T MODEL
(Only control for automatic transmission)
CHECK GROUND CIRCUIT.
1) Turn ignition switch "OFF" and disconnect 16-pin connector from E.C.U.
2) Disconnect throttle sensor harness connector.
3) Check resistance between terminal ⓕ and E.C.U. terminal 36.
Resistance:
Approximately 0Ω
N.G.
1) Check harness continuity between throttle sensor and ground.
2) E.C.U. ground circuit.
O.K.
Accelerator pedal
CHECK INPUT SIGNAL.
1) Reconnect E.C.U. 16-pin terminal and throttle sensor harness connector.
2) Turn ignition switch "ON".
3) Make sure that voltage between terminal 110 and ground changes when accelerator pedal is depressed.
Voltage:
Approximately 0.4 - 4.0V
(in warming up condition)
N.G.
1) Disconnect throttle sensor harness connector.
2) Make sure that resistance between ⓔ and ⓕ changes when opening throttle valve manually. Resistance should change. If not, replace throttle sensor.
3) Check idle switch OFF → ON speed.
Refer to "ELECTRO INJECTION UNIT INSPECTION".
4) Check harness continuity between throttle sensor and E.C.U.
• Disconnect harness connector for throttle sensor.
• Disconnect 16-pin connector from E.C.U.
• Check resistance between terminal ⓔ and E.C.U. terminal 110.
Resistance:
Approximately 0Ω
O.K. California model
O.K. Non-California A/T model
INSPECTION END
Reinstall any part removed.
Erase the self-diagnosis memory.
Perform driving test and then perform self-diagnosis (Mode III) again.
N.G.
1) Perform E.C.U. input/output signal inspection test.
2) If N.G., recheck the E.C.U. pin terminals damage or the connection of E.C.U. harness connector.
O.K.
INSPECTION END

ECCS SELF-DIAGNOSTIC CHART—1988-89 300ZX

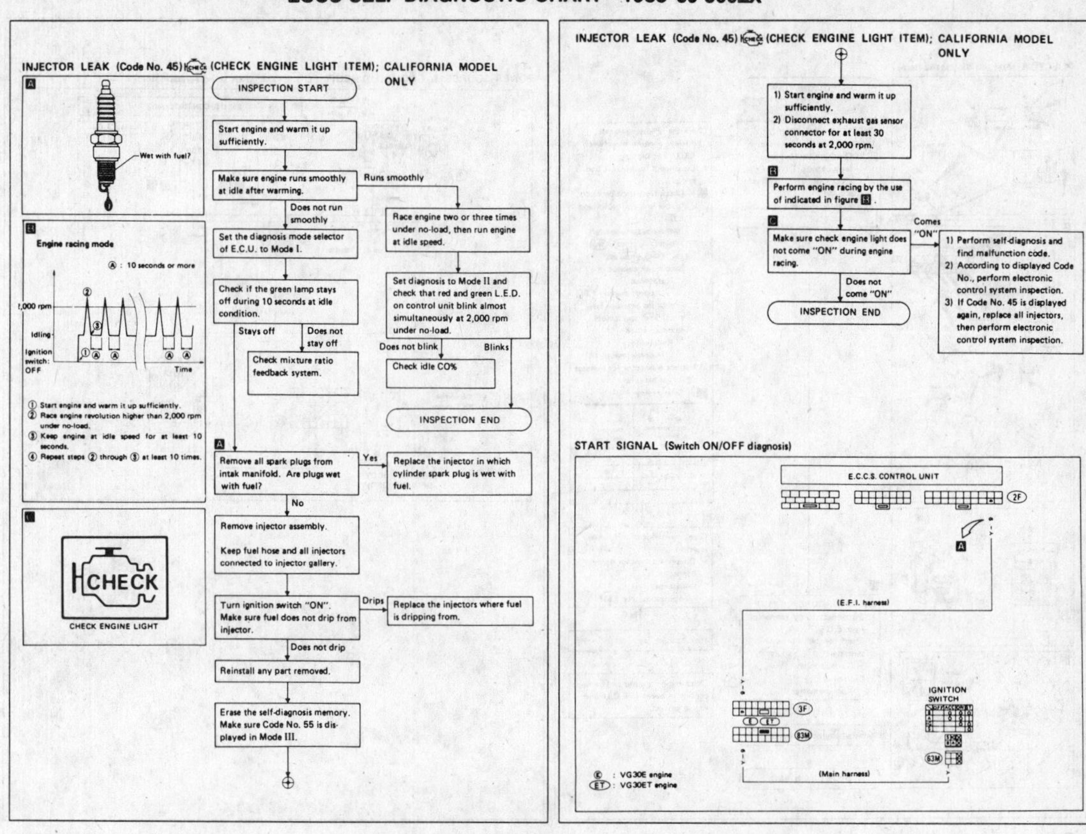
INJECTOR LEAK (Code No. 45) (CHECK ENGINE LIGHT ITEM); CALIFORNIA MODEL ONLY
A
Wet with fuel?
B
Engine racing mode
Ⓐ : 10 seconds or more
2,000 rpm
Idling
Ignition switch: OFF
Time
① Start engine and warm it up sufficiently.
② Race engine revolution higher than 2,000 rpm under no-load.
③ Keep engine at idle speed for at least 10 seconds.
④ Repeat steps ② through ③ at least 10 times.
C
CHECK
CHECK ENGINE LIGHT
INSPECTION START
Start engine and warm it up sufficiently.
Make sure engine runs smoothly at idle after warming.
Runs smoothly
Does not run smoothly
Race engine two or three times under no-load, then run engine at idle speed.
Set the diagnosis mode selector of E.C.U. to Mode I.
Set diagnosis to Mode II and check that red and green L.E.D. on control unit blink almost simultaneously at 2,000 rpm under no-load.
Check if the green lamp stays off during 10 seconds at idle condition.
Stays off
Does not stay off
Does not blink
Blinks
Check mixture ratio feedback system.
Check idle CO%
INSPECTION END
A
Remove all spark plugs from intak manifold. Are plugs wet with fuel?
Yes
Replace the injector in which cylinder spark plug is wet with fuel.
No
Remove injector assembly.
Keep fuel hose and all injectors connected to injector gallery.
Turn ignition switch "ON". Make sure fuel does not drip from injector.
Drips
Replace the injectors where fuel is dripping from.
Does not drip
Reinstall any part removed.
Erase the self-diagnosis memory. Make sure Code No. 55 is displayed in Mode III.
INJECTOR LEAK (Code No. 45) (CHECK ENGINE LIGHT ITEM); CALIFORNIA MODEL ONLY
1) Start engine and warm it up sufficiently.
2) Disconnect exhaust gas sensor connector for at least 30 seconds at 2,000 rpm.
B
Perform engine racing by the use of indicated in figure B.
C
Make sure check engine light does not come "ON" during engine racing.
Comes "ON"
1) Perform self-diagnosis and find malfunction code.
2) According to displayed Code No., perform electronic control system inspection.
3) If Code No. 45 is displayed again, replace all injectors, then perform electronic control system inspection.
Does not come "ON"
INSPECTION END
START SIGNAL (Switch ON/OFF diagnosis)
E.C.C.S. CONTROL UNIT
2F
A
(E.F.I. harness)
3F
63M
IGNITION SWITCH
(Main harness)
Ⓔ : VG30E engine
ET : VG30ET engine

ECCS SELF-DIAGNOSTIC CHART—1988-89 300ZX

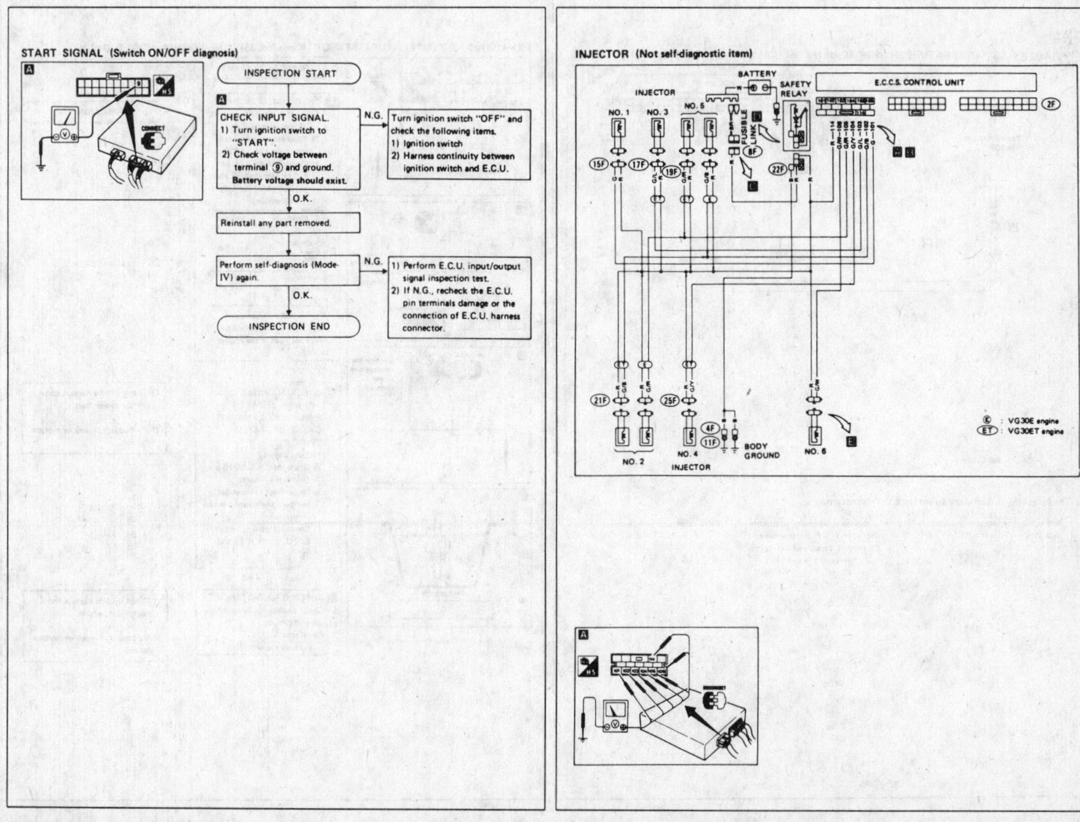
START SIGNAL (Switch ON/OFF diagnosis)
A
CONNECT
INSPECTION START
A
CHECK INPUT SIGNAL.
1) Turn ignition switch to "START".
2) Check voltage between terminal ⑨ and ground. Battery voltage should exist.
N.G.
Turn ignition switch "OFF" and check the following items.
1) Ignition switch
2) Harness continuity between ignition switch and E.C.U.
O.K.
Reinstall any part removed.
Perform self-diagnosis (Mode IV) again.
N.G.
1) Perform E.C.U. input/output signal inspection test.
2) If N.G., recheck the E.C.U. pin terminals damage or the connection of E.C.U. harness connector.
O.K.
INSPECTION END
INJECTOR (Not self-diagnostic item)
BATTERY
SAFETY RELAY
E.C.C.S. CONTROL UNIT
INJECTOR
NO. 1
NO. 3
NO. 5
FUSIBLE LINK
2F
15F
17F
19F
8F
22F
21F
29F
4F
11F
NO. 2
NO. 4
INJECTOR
BODY GROUND
NO. 6
Ⓔ : VG30E engine
ET : VG30ET engine
A

ECCS SELF-DIAGNOSTIC CHART—1988–89 300ZX

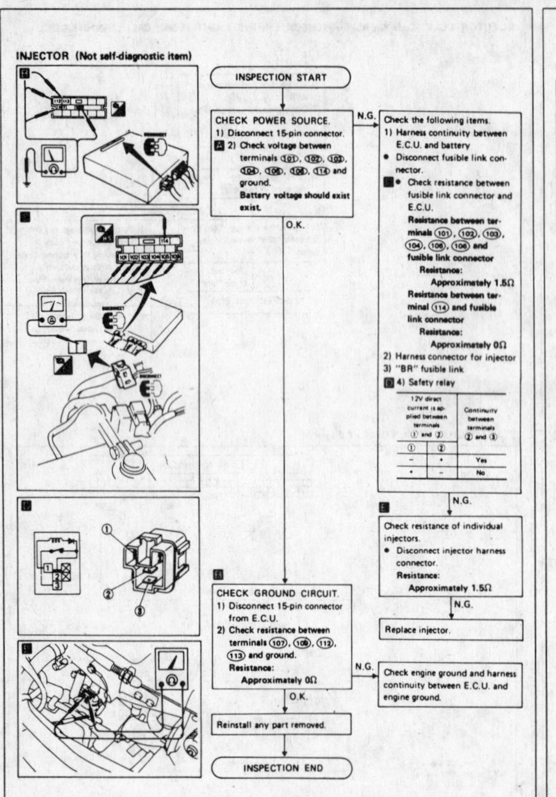
INJECTOR (Not self-diagnostic item)
INSPECTION START
CHECK POWER SOURCE.
1) Disconnect 15-pin connector.
2) Check voltage between terminals 101, 102, 103, 104, 105, 106, 114 and ground.
Battery voltage should exist.
N.G.
Check the following items.
1) Harness continuity between E.C.U. and battery
• Disconnect fusible link connector.
• Check resistance between fusible link connector and E.C.U.
Resistance between terminals 101, 102, 103, 104, 105, 106 and fusible link connector
Resistance: Approximately 1.5Ω
Resistance between terminal 114 and fusible link connector
Resistance: Approximately 0Ω
2) Harness connector for injector
3) "BR" fusible link
4) Safety relay
O.K.
N.G.
Check resistance of individual injectors.
• Disconnect injector harness connector.
Resistance: Approximately 1.5Ω
N.G.
Replace injector.
CHECK GROUND CIRCUIT.
1) Disconnect 15-pin connector from E.C.U.
2) Check resistance between terminals 107, 108, 112, 113 and ground.
Resistance: Approximately 0Ω
N.G.
Check engine ground and harness continuity between E.C.U. and engine ground.
O.K.
Reinstall any part removed.
INSPECTION END

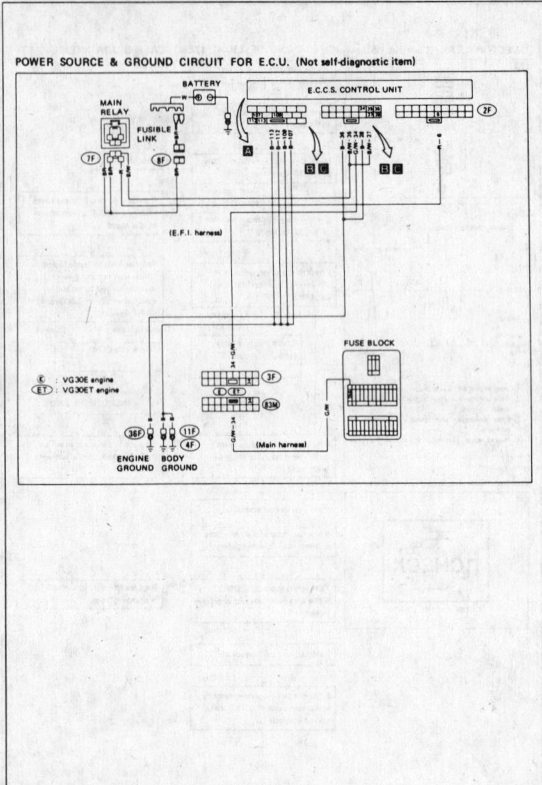
POWER SOURCE & GROUND CIRCUIT FOR E.C.U. (Not self-diagnostic item)
BATTERY
E.C.C.S. CONTROL UNIT
MAIN RELAY
FUSIBLE LINK
(E.F.I. harness)
FUSE BLOCK
E : VG30E engine
ET : VG30ET engine
(Main harness)
ENGINE GROUND
BODY GROUND

ECCS SELF-DIAGNOSTIC CHART—1988–89 300ZX

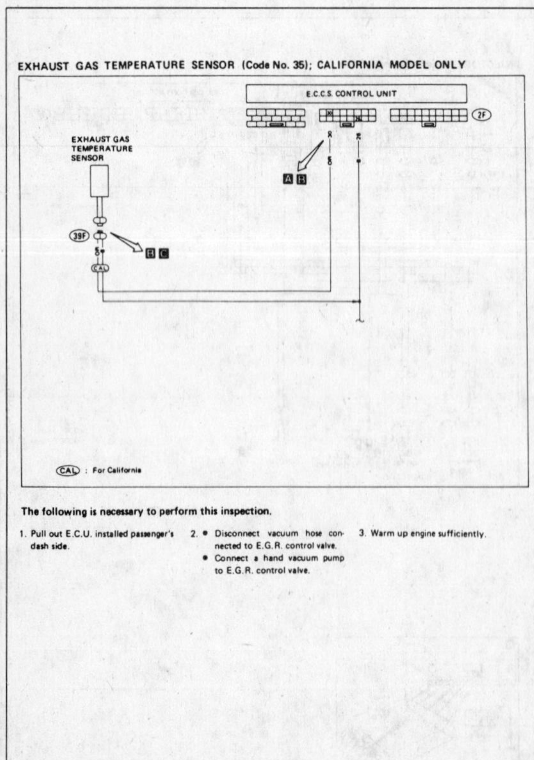
EXHAUST GAS TEMPERATURE SENSOR (Code No. 35); CALIFORNIA MODEL ONLY
E.C.C.S. CONTROL UNIT
EXHAUST GAS TEMPERATURE SENSOR
CAL : For California
The following is necessary to perform this inspection.
1. Pull out E.C.U. installed passenger's dash side.
2. • Disconnect vacuum hose connected to E.G.R. control valve.
• Connect a hand vacuum pump to E.G.R. control valve.
3. Warm up engine sufficiently.

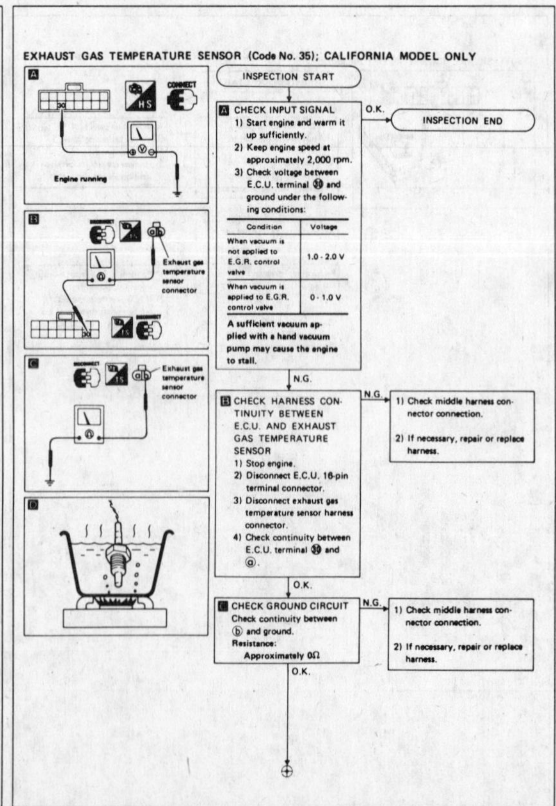
EXHAUST GAS TEMPERATURE SENSOR (Code No. 35); CALIFORNIA MODEL ONLY
INSPECTION START
Engine running
CHECK INPUT SIGNAL
1) Start engine and warm it up sufficiently.
2) Keep engine speed at approximately 2,000 rpm.
3) Check voltage between E.C.U. terminal 35 and ground under the following conditions:
Condition | Voltage
When vacuum is not applied to E.G.R. control valve | 1.0 - 2.0 V
When vacuum is applied to E.G.R. control valve | 0 - 1.0 V
A sufficient vacuum applied with a hand vacuum pump may cause the engine to stall.
O.K.
INSPECTION END
N.G.
Exhaust gas temperature sensor connector
CHECK HARNESS CONTINUITY BETWEEN E.C.U. AND EXHAUST GAS TEMPERATURE SENSOR
1) Stop engine.
2) Disconnect E.C.U. 16-pin terminal connector.
3) Disconnect exhaust gas temperature sensor harness connector.
4) Check continuity between E.C.U. terminal 35 and a.
N.G.
1) Check middle harness connector connection.
2) If necessary, repair or replace harness.
O.K.
CHECK GROUND CIRCUIT
Check continuity between b and ground.
Resistance: Approximately 0Ω
N.G.
1) Check middle harness connector connection.
2) If necessary, repair or replace harness.
O.K.

ECCS SELF-DIAGNOSTIC CHART—1988–89 300ZX

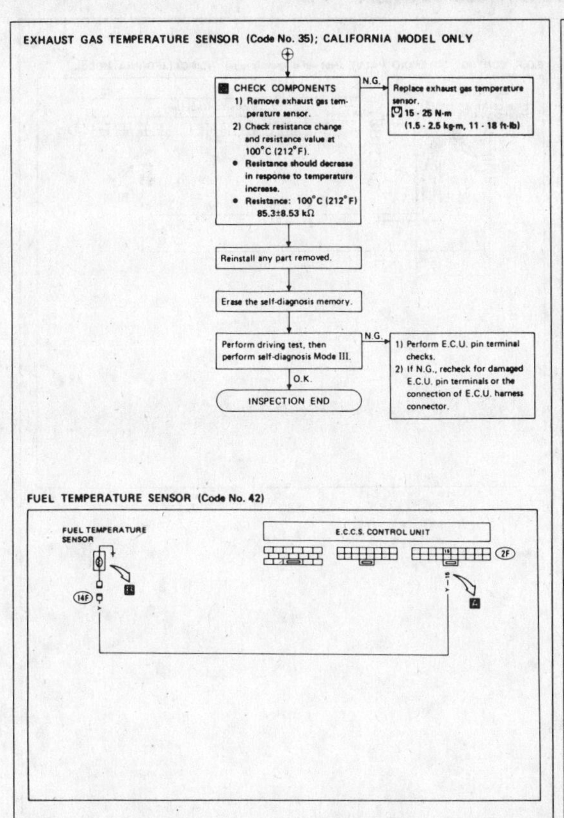
EXHAUST GAS TEMPERATURE SENSOR (Code No. 35); CALIFORNIA MODEL ONLY
CHECK COMPONENTS
1) Remove exhaust gas temperature sensor.
2) Check resistance change and resistance value at 100°C (212°F).
• Resistance should decrease in response to temperature increase.
• Resistance: 100°C (212°F) 85.3±8.53 kΩ
N.G.
Replace exhaust gas temperature sensor.
15 - 25 N·m (1.5 - 2.5 kg-m, 11 - 18 ft-lb)
Reinstall any part removed.
Erase the self-diagnosis memory.
Perform driving test, then perform self-diagnosis Mode III.
N.G.
1) Perform E.C.U. pin terminal checks.
2) If N.G., recheck for damaged E.C.U. pin terminals or the connection of E.C.U. harness connector.
O.K.
INSPECTION END
FUEL TEMPERATURE SENSOR (Code No. 42)
FUEL TEMPERATURE SENSOR
E.C.C.S. CONTROL UNIT
14F
2F

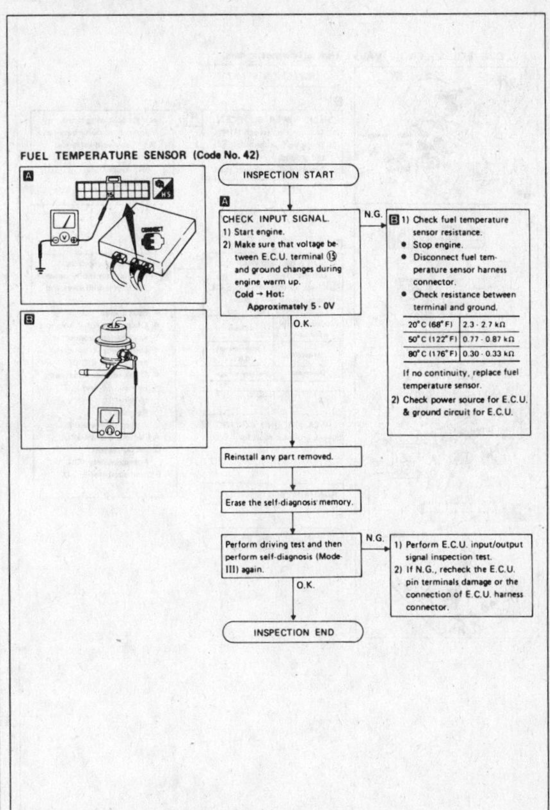
FUEL TEMPERATURE SENSOR (Code No. 42)
INSPECTION START
CHECK INPUT SIGNAL.
1) Start engine.
2) Make sure that voltage between E.C.U. terminal 15 and ground changes during engine warm up.
Cold → Hot:
Approximately 5 - 0V
N.G.
1) Check fuel temperature sensor resistance.
• Stop engine.
• Disconnect fuel temperature sensor harness connector.
• Check resistance between terminal and ground.
20°C (68°F) 2.3 - 2.7 kΩ
50°C (122°F) 0.77 - 0.87 kΩ
80°C (176°F) 0.30 - 0.33 kΩ
If no continuity, replace fuel temperature sensor.
2) Check power source for E.C.U. & ground circuit for E.C.U.
O.K.
Reinstall any part removed.
Erase the self-diagnosis memory.
Perform driving test and then perform self-diagnosis (Mode III) again.
N.G.
1) Perform E.C.U. input/output signal inspection test.
2) If N.G., recheck the E.C.U. pin terminals damage or the connection of E.C.U. harness connector.
O.K.
INSPECTION END

ECCS SELF-DIAGNOSTIC CHART—1988–89 300ZX

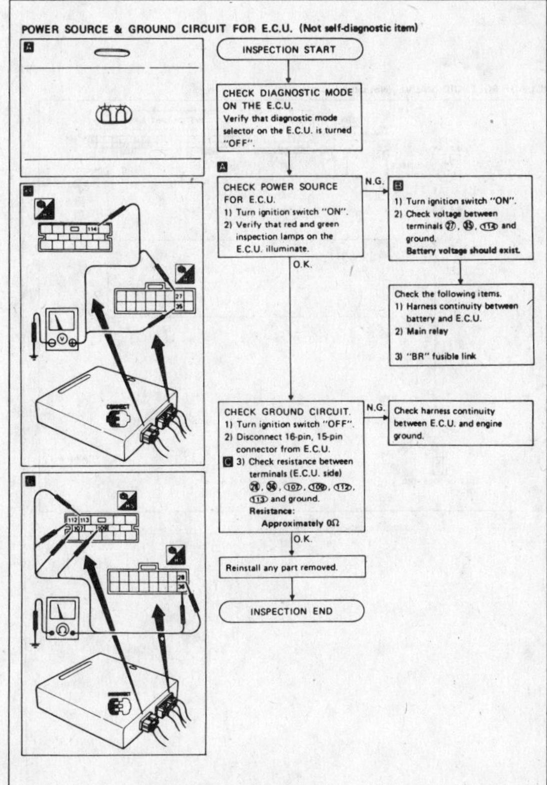
POWER SOURCE & GROUND CIRCUIT FOR E.C.U. (Not self-diagnostic item)
INSPECTION START
CHECK DIAGNOSTIC MODE ON THE E.C.U.
Verify that diagnostic mode selector on the E.C.U. is turned "OFF".
CHECK POWER SOURCE FOR E.C.U.
1) Turn ignition switch "ON".
2) Verify that red and green inspection lamps on the E.C.U. illuminate.
N.G.
1) Turn ignition switch "ON".
2) Check voltage between terminals 27, 35, 114 and ground.
Battery voltage should exist.
Check the following items.
1) Harness continuity between battery and E.C.U.
2) Main relay
3) "BR" fusible link
O.K.
CHECK GROUND CIRCUIT.
1) Turn ignition switch "OFF".
2) Disconnect 16-pin, 15-pin connector from E.C.U.
3) Check resistance between terminals (E.C.U. side) 26, 36, 107, 109, 112, 113 and ground.
Resistance:
Approximately 0Ω
N.G.
Check harness continuity between E.C.U. and engine ground.
O.K.
Reinstall any part removed.
INSPECTION END

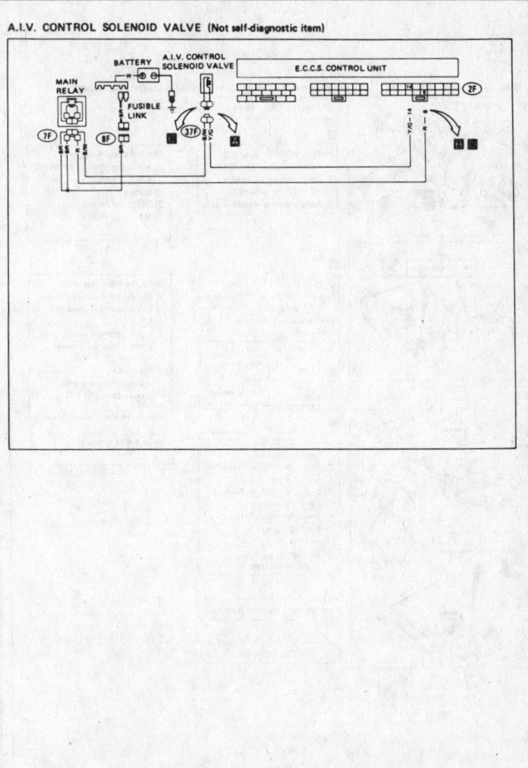
A.I.V. CONTROL SOLENOID VALVE (Not self-diagnostic item)
BATTERY
A.I.V. CONTROL SOLENOID VALVE
MAIN RELAY
FUSIBLE LINK
E.C.C.S. CONTROL UNIT
2F
7F
8F
37F

ECCS SELF-DIAGNOSTIC CHART—1988–89 300ZX

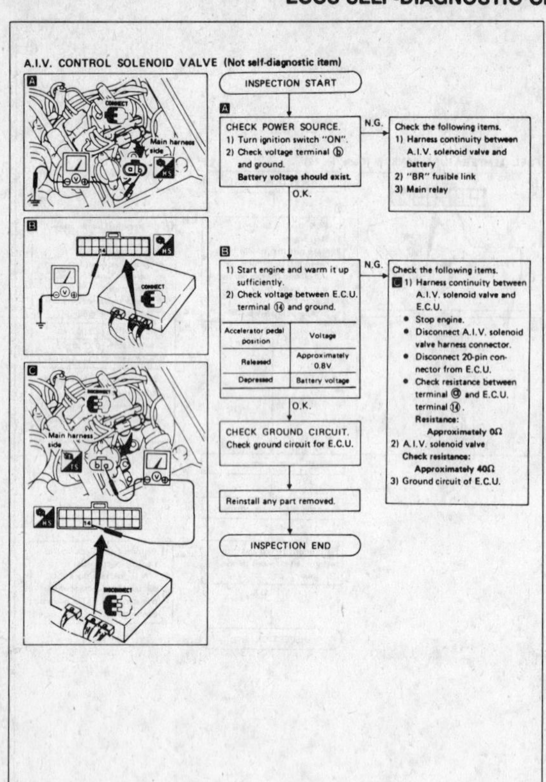
A.I.V. CONTROL SOLENOID VALVE (Not self-diagnostic item)
INSPECTION START
A
CHECK POWER SOURCE.
1) Turn ignition switch "ON".
2) Check voltage terminal ⓑ and ground.
Battery voltage should exist.
N.G.
Check the following items.
1) Harness continuity between A.I.V. solenoid valve and battery
2) "BR" fusible link
3) Main relay
O.K.
B
1) Start engine and warm it up sufficiently.
2) Check voltage between E.C.U. terminal ⑭ and ground.
Accelerator pedal position
Voltage
Released
Approximately 0.8V
Depressed
Battery voltage
N.G.
Check the following items.
1) Harness continuity between A.I.V. solenoid valve and E.C.U.
Stop engine.
Disconnect A.I.V. solenoid valve harness connector.
Disconnect 20-pin connector from E.C.U.
Check resistance between terminal ⓒ and E.C.U. terminal ⑭.
Resistance:
Approximately 0Ω
2) A.I.V. solenoid valve
Check resistance:
Approximately 40Ω
3) Ground circuit of E.C.U.
O.K.
CHECK GROUND CIRCUIT.
Check ground circuit for E.C.U.
Reinstall any part removed.
INSPECTION END
Main harness side
CONNECT
DISCONNECT

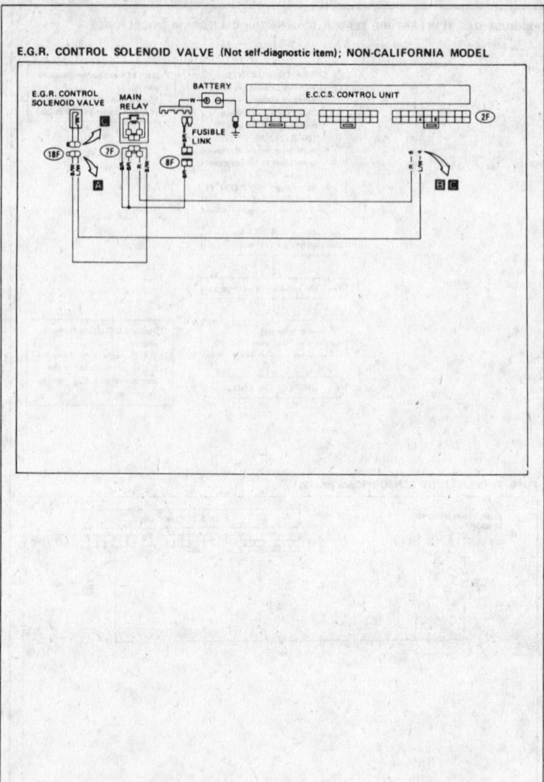
E.G.R. CONTROL SOLENOID VALVE (Not self-diagnostic item); NON-CALIFORNIA MODEL
E.G.R. CONTROL SOLENOID VALVE
MAIN RELAY
BATTERY
FUSIBLE LINK
E.C.C.S. CONTROL UNIT
18F
7F
8F
2F

ECCS SELF-DIAGNOSTIC CHART—1988–89 300ZX

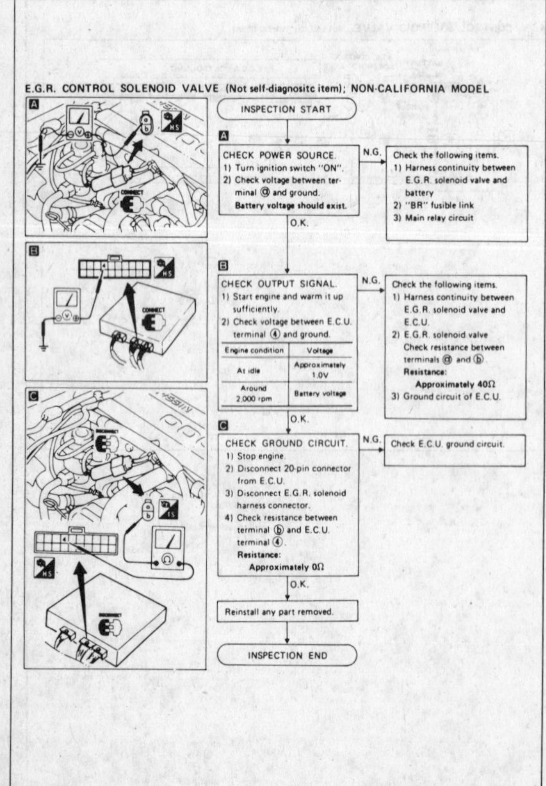
E.G.R. CONTROL SOLENOID VALVE (Not self-diagnostic item); NON-CALIFORNIA MODEL
INSPECTION START
A
CHECK POWER SOURCE.
1) Turn ignition switch "ON".
2) Check voltage between terminal ⓐ and ground.
Battery voltage should exist.
N.G.
Check the following items.
1) Harness continuity between E.G.R. solenoid valve and battery
2) "BR" fusible link
3) Main relay circuit
O.K.
B
CHECK OUTPUT SIGNAL.
1) Start engine and warm it up sufficiently.
2) Check voltage between E.C.U. terminal ④ and ground.
Engine condition
Voltage
At idle
Approximately 1.0V
Around 2,000 rpm
Battery voltage
N.G.
Check the following items.
1) Harness continuity between E.G.R. solenoid valve and E.C.U.
2) E.G.R. solenoid valve
Check resistance between terminals ⓐ and ⓑ.
Resistance:
Approximately 40Ω
3) Ground circuit of E.C.U.
O.K.
C
CHECK GROUND CIRCUIT.
1) Stop engine.
2) Disconnect 20-pin connector from E.C.U.
3) Disconnect E.G.R. solenoid harness connector.
4) Check resistance between terminal ⓑ and E.C.U. terminal ④.
Resistance:
Approximately 0Ω
N.G.
Check E.C.U. ground circuit.
O.K.
Reinstall any part removed.
INSPECTION END
CONNECT
DISCONNECT

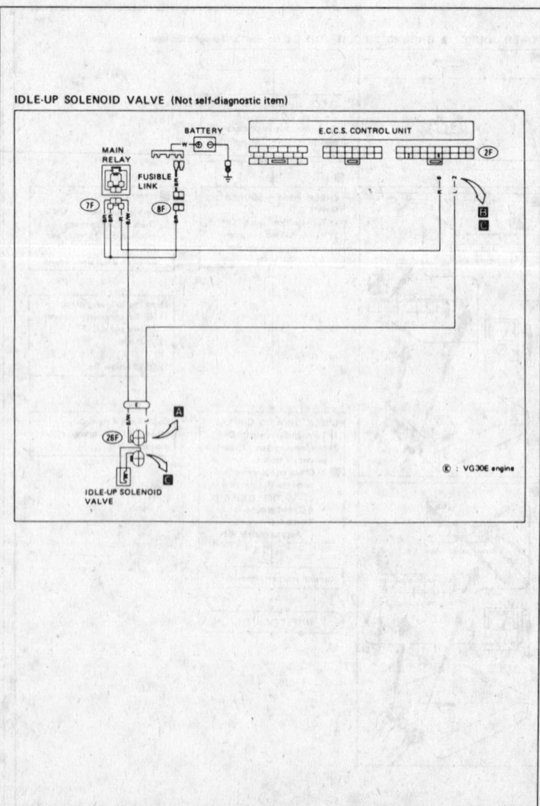
IDLE-UP SOLENOID VALVE (Not self-diagnostic item)
MAIN RELAY
BATTERY
FUSIBLE LINK
E.C.C.S. CONTROL UNIT
7F
8F
2F
28F
IDLE-UP SOLENOID VALVE
Ⓒ : VG30E engine

ECCS SELF-DIAGNOSTIC CHART—1988–89 300ZX

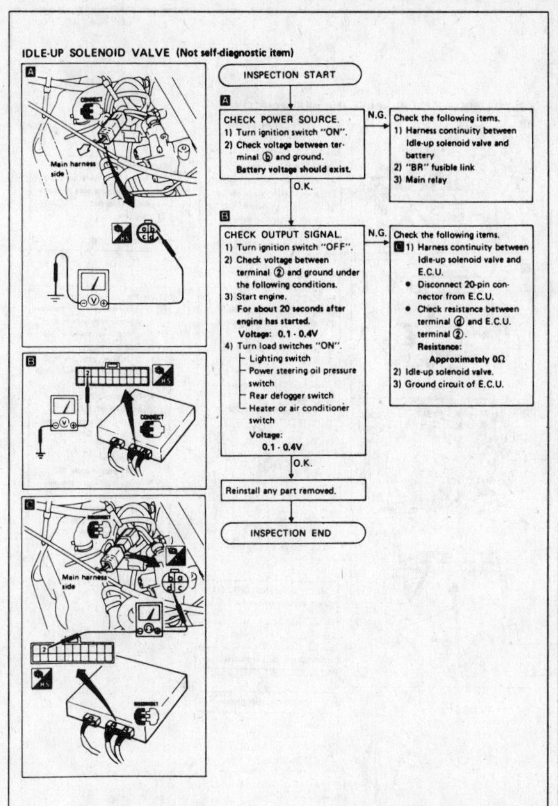
IDLE-UP SOLENOID VALVE (Not self-diagnostic item)
Main harness side
INSPECTION START
CHECK POWER SOURCE.
1) Turn ignition switch "ON".
2) Check voltage between terminal ⓑ and ground.
Battery voltage should exist.
N.G.
Check the following items.
1) Harness continuity between Idle-up solenoid valve and battery
2) "BR" fusible link
3) Main relay
O.K.
CHECK OUTPUT SIGNAL.
1) Turn ignition switch "OFF".
2) Check voltage between terminal ② and ground under the following conditions.
3) Start engine.
For about 20 seconds after engine has started.
Voltage: 0.1 - 0.4V
4) Turn load switches "ON".
Lighting switch
Power steering oil pressure switch
Rear defogger switch
Heater or air conditioner switch
Voltage:
0.1 - 0.4V
N.G.
Check the following items.
1) Harness continuity between Idle-up solenoid valve and E.C.U.
• Disconnect 20-pin connector from E.C.U.
• Check resistance between terminal ⓓ and E.C.U. terminal ②.
Resistance:
Approximately 0Ω
2) Idle-up solenoid valve.
3) Ground circuit of E.C.U.
O.K.
Reinstall any part removed.
INSPECTION END

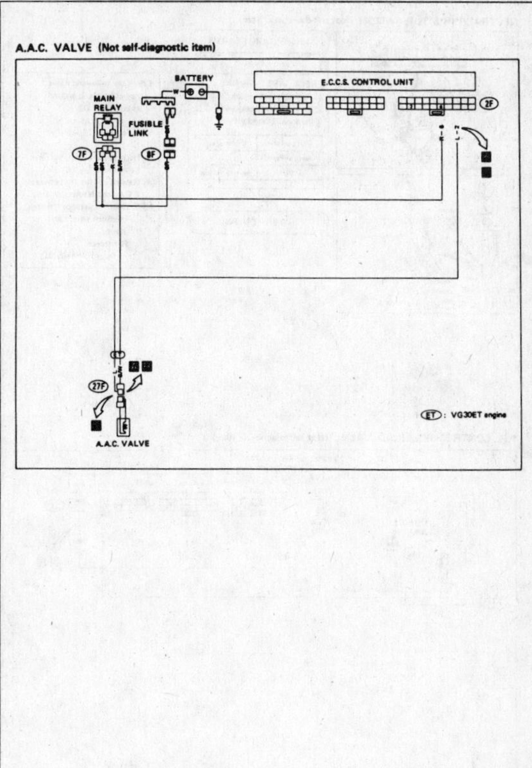
A.A.C. VALVE (Not self-diagnostic item)
BATTERY
E.C.C.S. CONTROL UNIT
MAIN RELAY
FUSIBLE LINK
A.A.C. VALVE
(ET): VG30ET engine

ECCS SELF-DIAGNOSTIC CHART—1988–89 300ZX

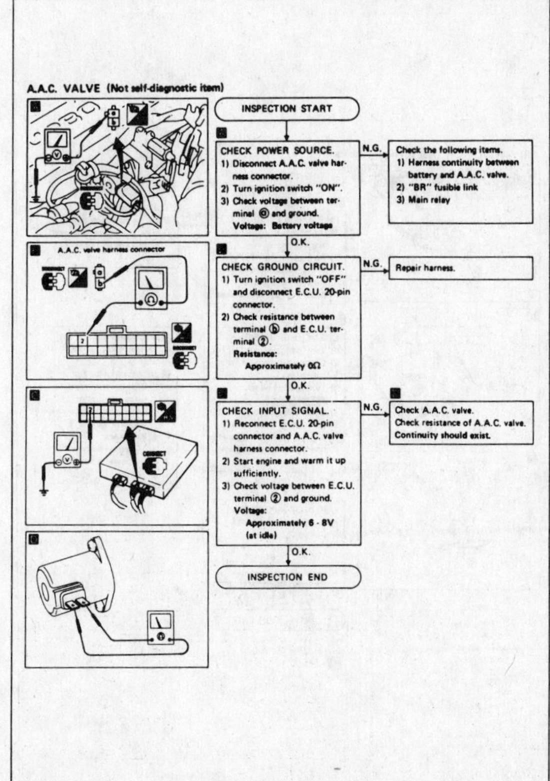
A.A.C. VALVE (Not self-diagnostic item)
A.A.C. valve harness connector
INSPECTION START
CHECK POWER SOURCE.
1) Disconnect A.A.C. valve harness connector.
2) Turn ignition switch "ON".
3) Check voltage between terminal ⓐ and ground.
Voltage: Battery voltage
N.G.
Check the following items.
1) Harness continuity between battery and A.A.C. valve.
2) "BR" fusible link
3) Main relay
O.K.
CHECK GROUND CIRCUIT.
1) Turn ignition switch "OFF" and disconnect E.C.U. 20-pin connector.
2) Check resistance between terminal ⓑ and E.C.U. terminal ②.
Resistance:
Approximately 0Ω
N.G.
Repair harness.
O.K.
CHECK INPUT SIGNAL.
1) Reconnect E.C.U. 20-pin connector and A.A.C. valve harness connector.
2) Start engine and warm it up sufficiently.
3) Check voltage between E.C.U. terminal ② and ground.
Voltage:
Approximately 6 - 8V
(at idle)
N.G.
Check A.A.C. valve.
Check resistance of A.A.C. valve.
Continuity should exist.
O.K.
INSPECTION END

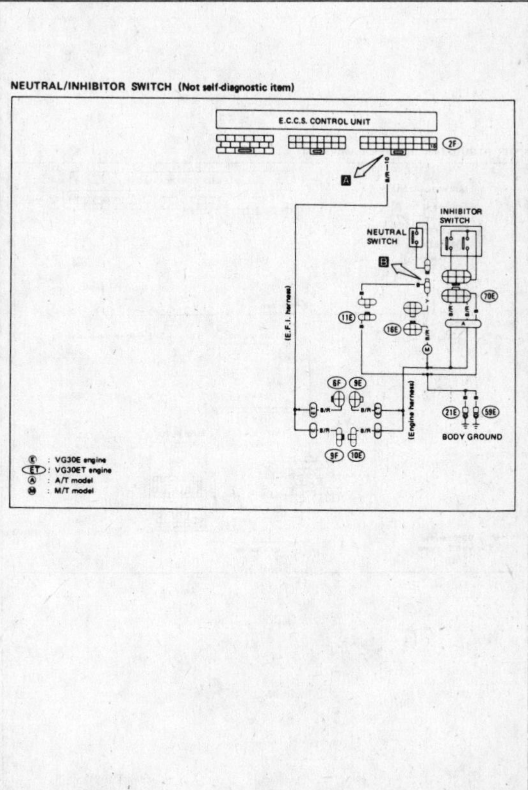
NEUTRAL/INHIBITOR SWITCH (Not self-diagnostic item)
E.C.C.S. CONTROL UNIT
INHIBITOR SWITCH
NEUTRAL SWITCH
(E.F.I. harness)
(Engine harness)
BODY GROUND
(E) : VG30E engine
(ET) : VG30ET engine
Ⓐ : A/T model
Ⓜ : M/T model

ECCS SELF-DIAGNOSTIC CHART—1988–89 300ZX

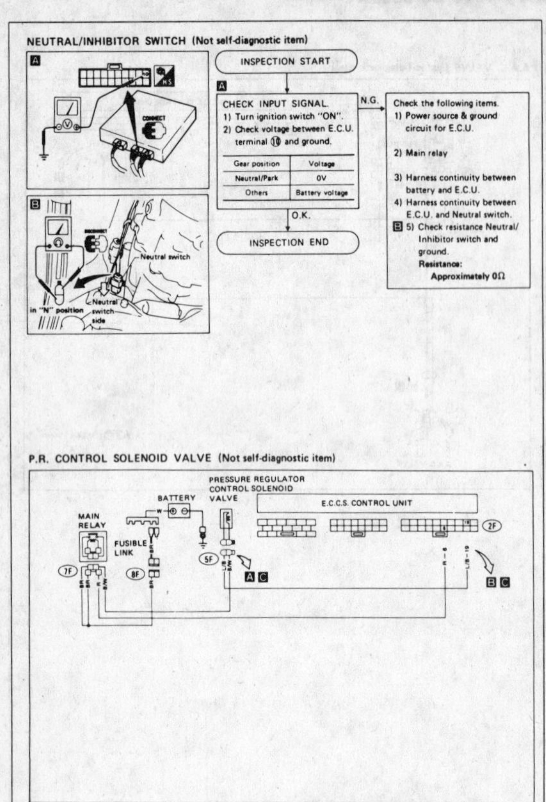
NEUTRAL/INHIBITOR SWITCH (Not self-diagnostic item)
INSPECTION START
CHECK INPUT SIGNAL.
1) Turn ignition switch "ON".
2) Check voltage between E.C.U. terminal ⑲ and ground.
Gear position | Voltage
Neutral/Park | 0V
Others | Battery voltage
N.G.
Check the following items.
1) Power source & ground circuit for E.C.U.
2) Main relay
3) Harness continuity between battery and E.C.U.
4) Harness continuity between E.C.U. and Neutral switch.
5) Check resistance Neutral/Inhibitor switch and ground.
Resistance:
Approximately 0Ω
O.K.
INSPECTION END
Neutral switch
in "N" position
Neutral switch side
P.R. CONTROL SOLENOID VALVE (Not self-diagnostic item)
PRESSURE REGULATOR CONTROL SOLENOID VALVE
BATTERY
E.C.C.S. CONTROL UNIT
MAIN RELAY
FUSIBLE LINK

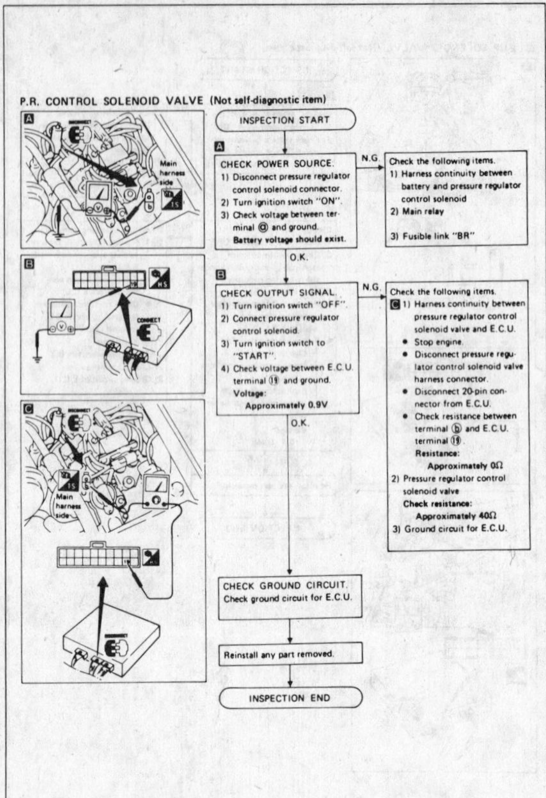
P.R. CONTROL SOLENOID VALVE (Not self-diagnostic item)
INSPECTION START
CHECK POWER SOURCE.
1) Disconnect pressure regulator control solenoid connector.
2) Turn ignition switch "ON".
3) Check voltage between terminal ⓐ and ground.
Battery voltage should exist.
N.G.
Check the following items.
1) Harness continuity between battery and pressure regulator control solenoid
2) Main relay
3) Fusible link "BR"
O.K.
CHECK OUTPUT SIGNAL.
1) Turn ignition switch "OFF".
2) Connect pressure regulator control solenoid.
3) Turn ignition switch to "START".
4) Check voltage between E.C.U. terminal ⑲ and ground.
Voltage:
Approximately 0.9V
N.G.
Check the following items.
1) Harness continuity between pressure regulator control solenoid valve and E.C.U.
Stop engine.
Disconnect pressure regulator control solenoid valve harness connector.
Disconnect 20-pin connector from E.C.U.
Check resistance between terminal ⓑ and E.C.U. terminal ⑲
Resistance:
Approximately 0Ω
2) Pressure regulator control solenoid valve
Check resistance:
Approximately 40Ω
3) Ground circuit for E.C.U.
O.K.
CHECK GROUND CIRCUIT.
Check ground circuit for E.C.U.
Reinstall any part removed.
INSPECTION END
Main harness side

ECCS SELF-DIAGNOSTIC CHART—1988–89 300ZX

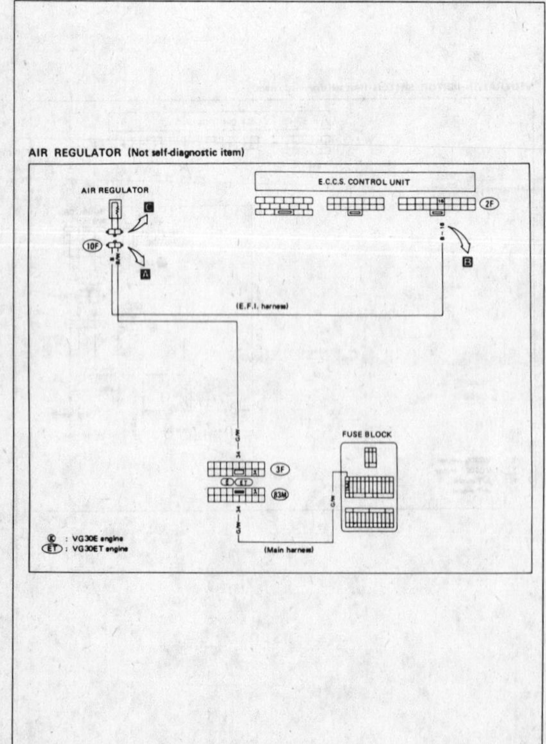
AIR REGULATOR (Not self-diagnostic item)
AIR REGULATOR
E.C.C.S. CONTROL UNIT
(E.F.I. harness)
FUSE BLOCK
(Main harness)
Ⓔ : VG30E engine
ⒺⓉ : VG30ET engine

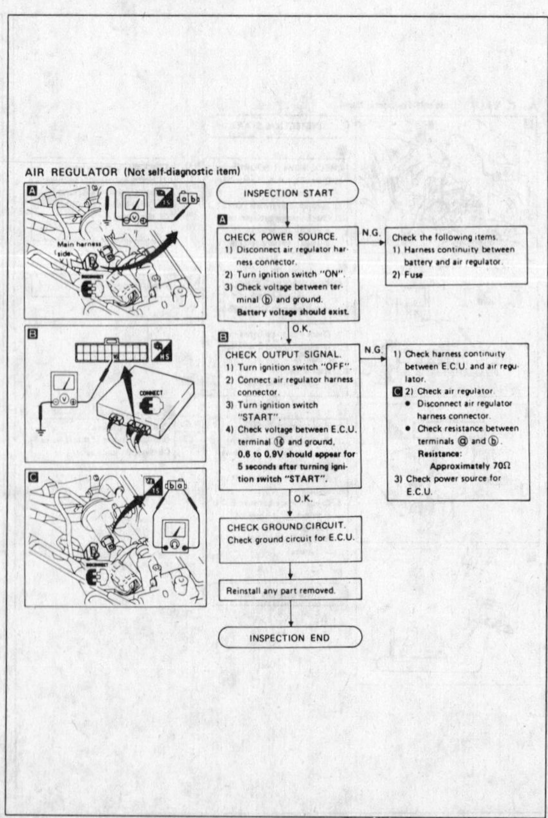
AIR REGULATOR (Not self-diagnostic item)
INSPECTION START
CHECK POWER SOURCE.
1) Disconnect air regulator harness connector.
2) Turn ignition switch "ON".
3) Check voltage between terminal ⓑ and ground.
Battery voltage should exist.
N.G.
Check the following items.
1) Harness continuity between battery and air regulator.
2) Fuse
O.K.
CHECK OUTPUT SIGNAL.
1) Turn ignition switch "OFF".
2) Connect air regulator harness connector.
3) Turn ignition switch "START".
4) Check voltage between E.C.U. terminal ⑯ and ground.
0.6 to 0.9V should appear for 5 seconds after turning ignition switch "START".
N.G.
1) Check harness continuity between E.C.U. and air regulator.
2) Check air regulator.
Disconnect air regulator harness connector.
Check resistance between terminals ⓐ and ⓑ.
Resistance:
Approximately 70Ω
3) Check power source for E.C.U.
O.K.
CHECK GROUND CIRCUIT.
Check ground circuit for E.C.U.
Reinstall any part removed.
INSPECTION END
Main harness side

ECU SIGNAL INSPECTION—1988–89 300ZX

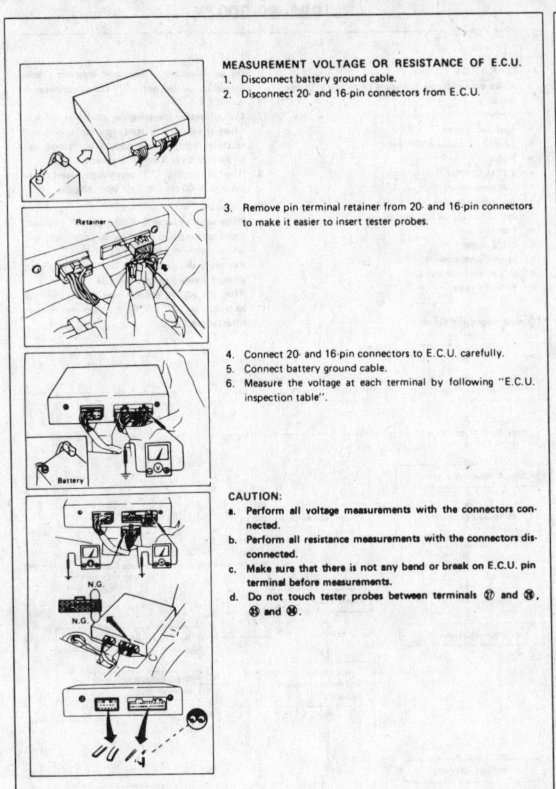
Retainer
Battery
N.G.
N.G.

MEASUREMENT VOLTAGE OR RESISTANCE OF E.C.U.

1. Disconnect battery ground cable.
2. Disconnect 20- and 16-pin connectors from E.C.U.
3. Remove pin terminal retainer from 20- and 16-pin connectors to make it easier to insert tester probes.
4. Connect 20- and 16-pin connectors to E.C.U. carefully.
5. Connect battery ground cable.
6. Measure the voltage at each terminal by following "E.C.U. inspection table".

CAUTION:

a. Perform all voltage measurements with the connectors connected.
b. Perform all resistance measurements with the connectors disconnected.
c. Make sure that there is not any bend or break on E.C.U. pin terminal before measurements.
d. Do not touch tester probes between terminals ㉗ and ㉘, ㉟ and ㊱.

E.C.U. inspection table

*Data are reference values.

TERMINAL NO.	ITEM	CONDITION	*DATA
2	Idle-up solenoid valve (VG30E)	Engine is running and gear position is in P or N (A/T). — For about 20 seconds after starting engine — Steering wheel is turned. — Blower and air conditioner switches are "ON". — Lighting switch is "ON".	0.1 - 0.4V
		Engine is running. — Except the conditions shown above	BATTERY VOLTAGE (11 - 14V)
	A.A.C. valve (VG30ET)	Engine is running. — Idle speed (after warm-up)	6.0 - 8.0V
3	Ignition check	Engine is running. — Idle speed	9 - 12V (Decreases as engine is revved up.)
4	E.G.R. control solenoid valve	Engine is running after being warmed up. — High engine revolution — Idle speed (Throttle valve switch "ON".)	Approximately 1.0V
		Engine is running. — Low engine revolution	BATTERY VOLTAGE (11 - 14V)
5	Ignition signal	Engine is running. — Idle speed	0.4 - 0.6V
		Engine is running. — Engine speed is 2,000 rpm.	1.2 - 1.5V
6	E.C.C.S. relay-1 (Main relay)	Engine is running. → Ignition switch "OFF" — For approximately 8 seconds after turning ignition switch "OFF"	0.7 - 0.9V
		Ignition switch "OFF" — Within approximately 8 seconds after turning ignition switch "OFF"	BATTERY VOLTAGE (11 - 14V)

ECU SIGNAL INSPECTION (CONT.)—1988–89 300ZX

*Data are reference values.

TERMINAL NO.	ITEM	CONDITION	*DATA
8	Crank angle sensor (Position signal)	Engine is running. Do not run engine at high speed under no-load.	2.5 - 2.7V
9	Start signal	Cranking	8 - 12V
10	Neutral switch (M/T) Inhibitor switch (A/T)	Ignition switch "ON" — Gear position is in Neutral or Parking.	0V
		Ignition switch "ON" — Any gear position except Neutral or Parking	BATTERY VOLTAGE (11 - 14V)
12	Air flow meter burn-off signal	Engine revolution is above 1,500 rpm and vehicle speed is more than 20 km/h (12MPH). → Ignition switch "OFF" — For 6 seconds after turning ignition switch "OFF"	0V
		Engine revolution is above 1,500 rpm and vehicle speed is more than 20 km/h (12 MPH). → Ignition switch "OFF" — For 1 second after the above 6 seconds have passed.	9.0 - 10.0V
14	A.I.V. control solenoid valve	Ignition switch "ON" — Release accelerator pedal. (Throttle valve switch "ON")	0.7 - 0.9V
		Ignition switch "ON" — Depress accelerator pedal. (Throttle valve switch "OFF")	BATTERY VOLTAGE (11 - 14V)
15	Fuel temperature sensor	Engine is running. — Idle speed	0.5V Output voltage varies with engine temperature.
16	Air regulator	Engine is running.	0.6 - 0.9V
17	Crank angle sensor (Reference signal)	Engine is running. Do not run engine at high speed under no-load.	0.2 - 0.4V

*Data are reference values.

TERMINAL NO.	ITEM	CONDITION	*DATA
18	Throttle valve switch (⊖ side)	Ignition switch "ON" — Release accelerator pedal. (Throttle valve switch "OFF")	9.0 - 10.0V
		Ignition switch "ON" — Depress accelerator pedal. (Throttle valve switch "ON")	0V
19	Pressure regulator control solenoid valve	Stop and restart engine after warming it up. — For 30 seconds	0.8 - 1.0V
		Stop and restart engine after warming it up. — After 3 minutes	BATTERY VOLTAGE (11 - 14V)
20	Fuel pump relay	Engine is running.	BATTERY VOLTAGE (11 - 14V)
22	Load signal	Engine is running and gear position is in P or N (A/T). — Steering wheel is turned. — Blower and air conditioner switchs are "ON". — Lighting switch is "ON".	BATTERY VOLTAGE (11 - 14V)
		Engine is running. — Except conditions shown above	0V
23	Cylinder head temperature sensor	Engine is running.	0 - 5.0V Output voltage varies with engine temperature.
24	Exhaust gas sensor	Engine is running. — After warming up sufficiently	0 - Approximately 1.0V
25	Idle switch (⊕side)	Ignition switch "ON"	9.0 - 10.0V
27 35	Power source for E.C.U.	Ignition switch "ON"	BATTERY VOLTAGE (11 - 14V)
29	Vehicle speed sensor	Ignition switch "ON" — While rotating rear wheel slowly	0 or 7.4V

ECU SIGNAL INSPECTION (CONT.) – 1988–89 300ZX

ECCS MIXTURE RATIO FEEDBACK SYSTEM INSPECTION 1988–89 300ZX

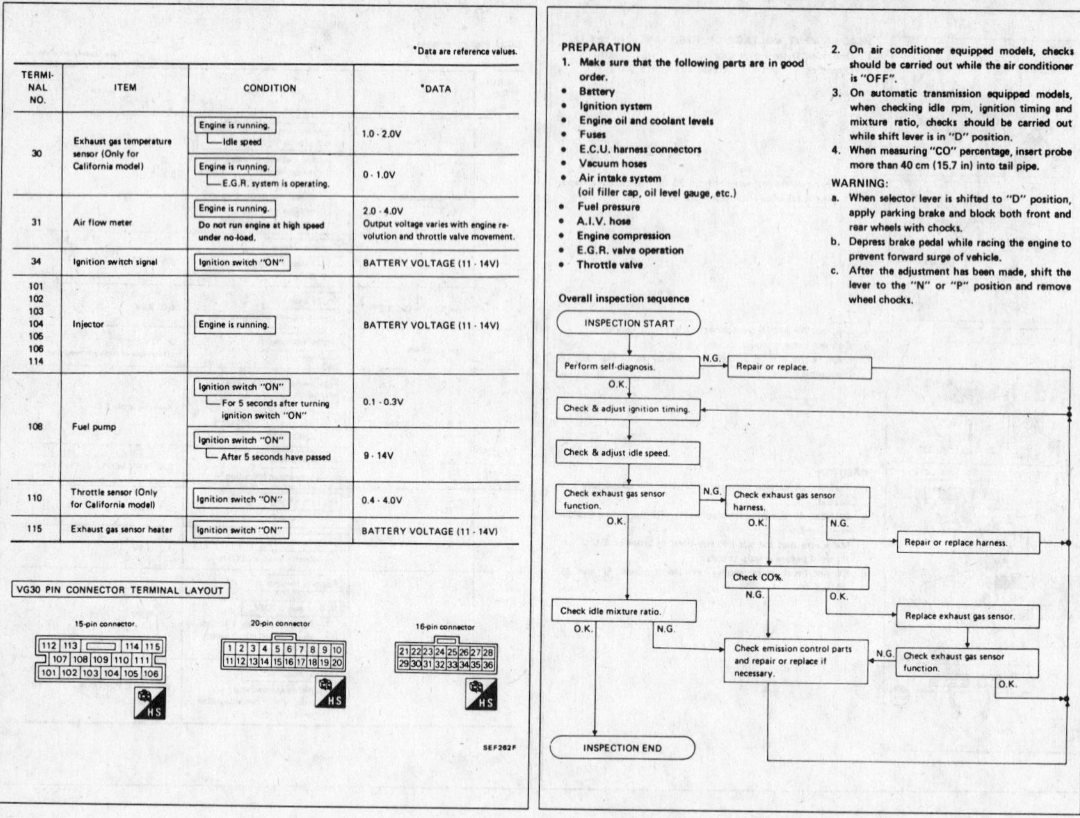

*Data are reference values.

TERMINAL NO.	ITEM	CONDITION	*DATA
30	Exhaust gas temperature sensor (Only for California model)	Engine is running. – Idle speed	1.0 - 2.0V
		Engine is running. – E.G.R. system is operating.	0 - 1.0V
31	Air flow meter	Engine is running. Do not run engine at high speed under no-load.	2.0 - 4.0V Output voltage varies with engine revolution and throttle valve movement.
34	Ignition switch signal	Ignition switch "ON"	BATTERY VOLTAGE (11 - 14V)
101 102 103 104 105 106 114	Injector	Engine is running.	BATTERY VOLTAGE (11 - 14V)
108	Fuel pump	Ignition switch "ON" – For 5 seconds after turning ignition switch "ON"	0.1 - 0.3V
		Ignition switch "ON" – After 5 seconds have passed	9 - 14V
110	Throttle sensor (Only for California model)	Ignition switch "ON"	0.4 - 4.0V
115	Exhaust gas sensor heater	Ignition switch "ON"	BATTERY VOLTAGE (11 - 14V)

VG30 PIN CONNECTOR TERMINAL LAYOUT
15-pin connector
112 113 114 115
107 108 109 110 111
101 102 103 104 105 106
20-pin connector
1 2 3 4 5 6 7 8 9 10
11 12 13 14 15 16 17 18 19 20
16-pin connector
21 22 23 24 25 26 27 28
29 30 31 32 33 34 35 36
HS
SEF282F

PREPARATION

1. Make sure that the following parts are in good order.
• Battery
• Ignition system
• Engine oil and coolant levels
• Fuses
• E.C.U. harness connectors
• Vacuum hoses
• Air intake system (oil filler cap, oil level gauge, etc.)
• Fuel pressure
• A.I.V. hose
• Engine compression
• E.G.R. valve operation
• Throttle valve
2. On air conditioner equipped models, checks should be carried out while the air conditioner is "OFF".
3. On automatic transmission equipped models, when checking idle rpm, ignition timing and mixture ratio, checks should be carried out while shift lever is in "D" position.
4. When measuring "CO" percentage, insert probe more than 40 cm (15.7 in) into tail pipe.

WARNING:

a. When selector lever is shifted to "D" position, apply parking brake and block both front and rear wheels with chocks.
b. Depress brake pedal while racing the engine to prevent forward surge of vehicle.
c. After the adjustment has been made, shift the lever to the "N" or "P" position and remove wheel chocks.

Overall inspection sequence

INSPECTION START
Perform self-diagnosis.
N.G.
Repair or replace.
O.K.
Check & adjust ignition timing.
Check & adjust idle speed.
Check exhaust gas sensor function.
N.G.
Check exhaust gas sensor harness.
O.K.
N.G.
Repair or replace harness.
Check CO%.
N.G.
O.K.
Replace exhaust gas sensor.
Check idle mixture ratio.
O.K.
N.G.
Check emission control parts and repair or replace if necessary.
N.G.
Check exhaust gas sensor function.
O.K.
INSPECTION END

ECCS MIXTURE RATIO FEEDBACK SYSTEM INSPECTION (CONT.) – 1988–89 300ZX

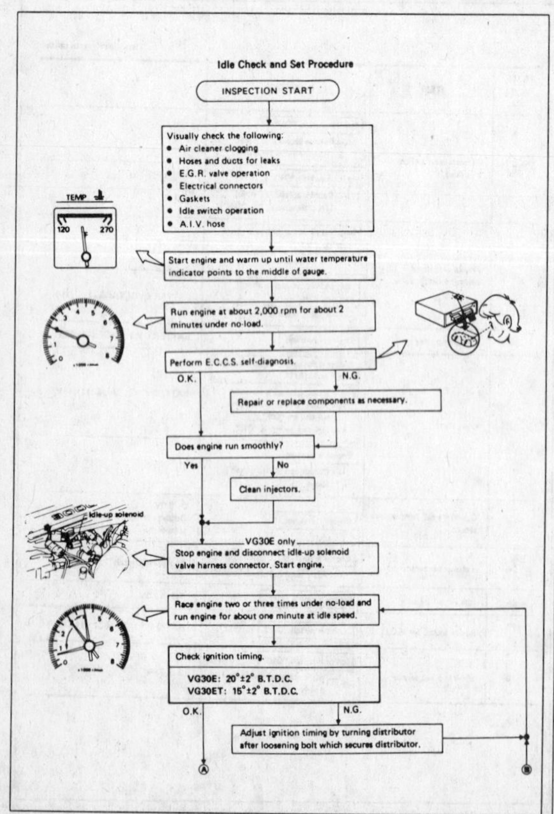

Idle Check and Set Procedure
INSPECTION START
Visually check the following:
• Air cleaner clogging
• Hoses and ducts for leaks
• E.G.R. valve operation
• Electrical connectors
• Gaskets
• Idle switch operation
• A.I.V. hose
TEMP
120
270
Start engine and warm up until water temperature indicator points to the middle of gauge.
Run engine at about 2,000 rpm for about 2 minutes under no-load.
Perform E.C.C.S. self-diagnosis.
O.K.
N.G.
Repair or replace components as necessary.
Does engine run smoothly?
Yes
No
Clean injectors.
Idle-up solenoid
VG30E only
Stop engine and disconnect idle-up solenoid valve harness connector. Start engine.
Race engine two or three times under no-load and run engine for about one minute at idle speed.
Check ignition timing.
VG30E: 20°±2° B.T.D.C.
VG30ET: 15°±2° B.T.D.C.
O.K.
N.G.
Adjust ignition timing by turning distributor after loosening bolt which secures distributor.
A
B

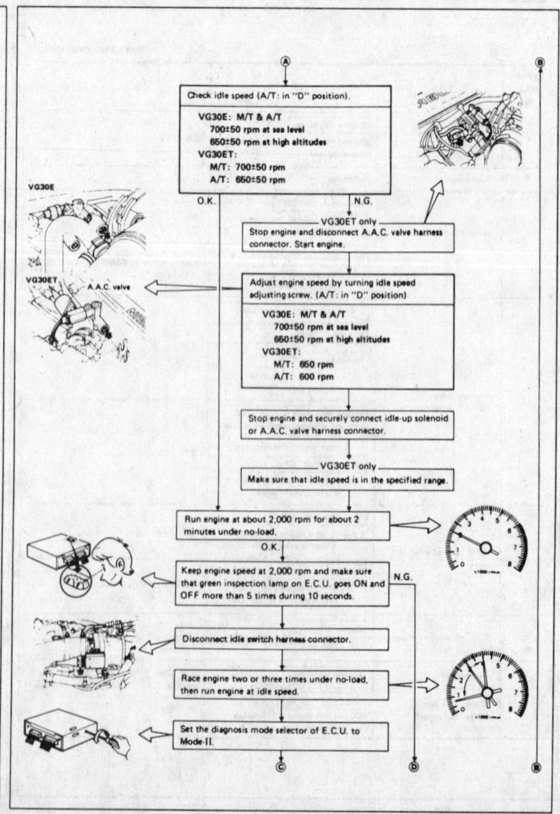

A
B
Check idle speed (A/T: in "D" position).
VG30E: M/T & A/T
700±50 rpm at sea level
650±50 rpm at high altitudes
VG30ET:
M/T: 700±50 rpm
A/T: 650±50 rpm
O.K.
N.G.
VG30E
VG30ET only
Stop engine and disconnect A.A.C. valve harness connector. Start engine.
VG30ET
A.A.C. valve
Adjust engine speed by turning idle speed adjusting screw. (A/T: in "D" position)
VG30E: M/T & A/T
700±50 rpm at sea level
650±50 rpm at high altitudes
VG30ET:
M/T: 650 rpm
A/T: 600 rpm
Stop engine and securely connect idle-up solenoid or A.A.C. valve harness connector.
VG30ET only
Make sure that idle speed is in the specified range.
Run engine at about 2,000 rpm for about 2 minutes under no-load.
O.K.
Keep engine speed at 2,000 rpm and make sure that green inspection lamp on E.C.U. goes ON and OFF more than 5 times during 10 seconds.
N.G.
Disconnect idle switch harness connector.
Race engine two or three times under no-load, then run engine at idle speed.
Set the diagnosis mode selector of E.C.U. to Mode-II.
C
D
B

ECCS MIXTURE RATIO FEEDBACK SYSTEM INSPECTION (CONT.)—1988–89 300ZX

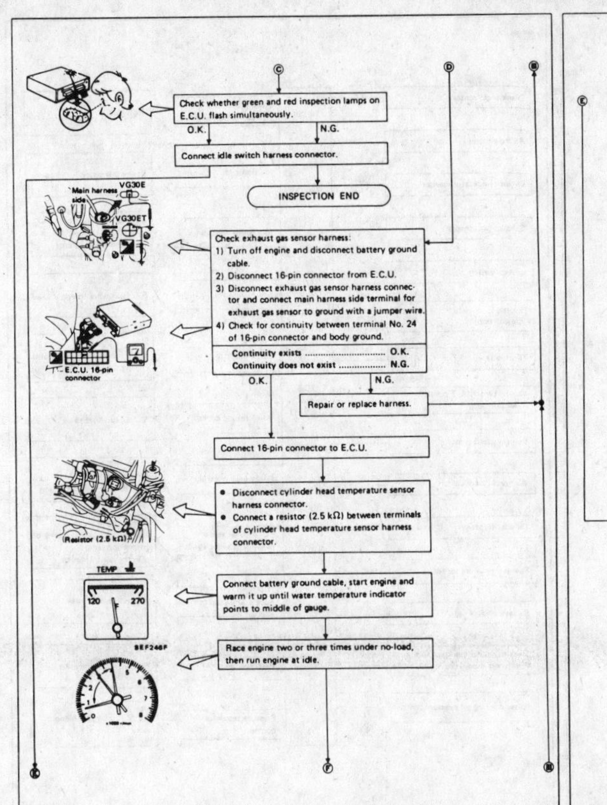
Check whether green and red inspection lamps on E.C.U. flash simultaneously.
O.K.
N.G.
Connect idle switch harness connector.
INSPECTION END
Main harness side
VG30E
VG30ET
Check exhaust gas sensor harness:
1) Turn off engine and disconnect battery ground cable.
2) Disconnect 16-pin connector from E.C.U.
3) Disconnect exhaust gas sensor harness connector and connect main harness side terminal for exhaust gas sensor to ground with a jumper wire.
4) Check for continuity between terminal No. 24 of 16-pin connector and body ground.
Continuity exists .... O.K.
Continuity does not exist .... N.G.
E.C.U. 16-pin connector
Repair or replace harness.
Connect 16-pin connector to E.C.U.
Disconnect cylinder head temperature sensor harness connector.
Connect a resistor (2.5 kΩ) between terminals of cylinder head temperature sensor harness connector.
Resistor (2.5 kΩ)
TEMP
120
270
Connect battery ground cable, start engine and warm it up until water temperature indicator points to middle of gauge.
SEF246P
Race engine two or three times under no-load, then run engine at idle.

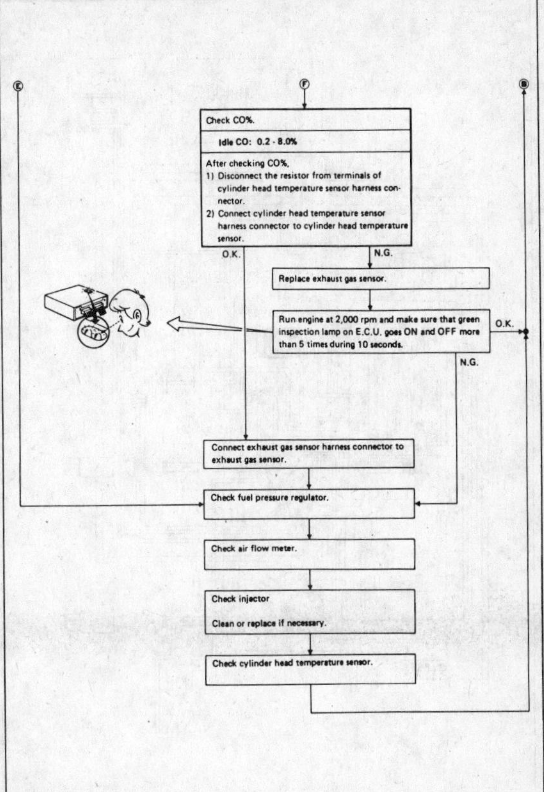
Check CO%.
Idle CO: 0.2 - 8.0%
After checking CO%,
1) Disconnect the resistor from terminals of cylinder head temperature sensor harness connector.
2) Connect cylinder head temperature sensor harness connector to cylinder head temperature sensor.
O.K.
N.G.
Replace exhaust gas sensor.
Run engine at 2,000 rpm and make sure that green inspection lamp on E.C.U. goes ON and OFF more than 5 times during 10 seconds.
O.K.
N.G.
Connect exhaust gas sensor harness connector to exhaust gas sensor.
Check fuel pressure regulator.
Check air flow meter.
Check injector
Clean or replace if necessary.
Check cylinder head temperature sensor.

ECCS COMPONENTS LOCATION—1990 300ZX

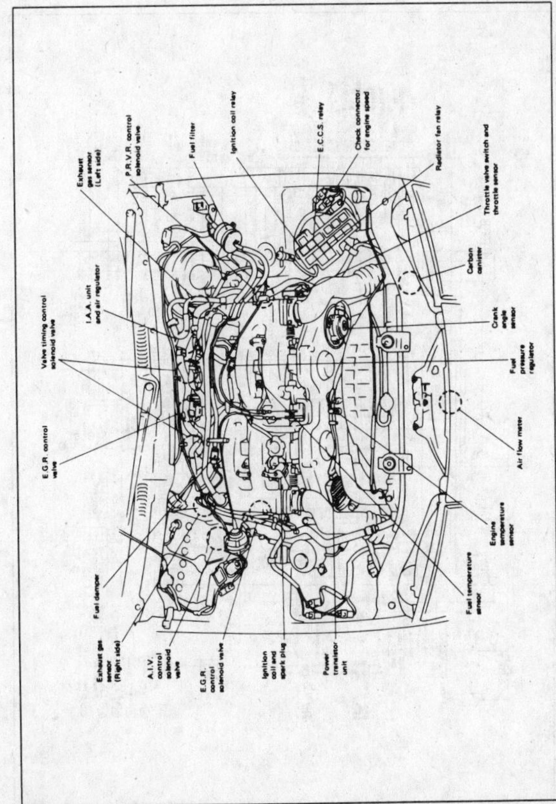
Exhaust gas sensor (Left side)
P.R.V.R. control solenoid valve
Fuel filter
Ignition coil relay
E.C.C.S. relay
Check connector for engine speed
Radiator fan relay
Throttle valve switch and throttle sensor
Carbon canister
Crank angle sensor
Fuel pressure regulator
Air flow meter
Engine temperature sensor
Fuel temperature sensor
Power transistor unit
Ignition coil and spark plug
E.G.R. control solenoid valve
A.I.V. control solenoid valve
Exhaust gas sensor (Right side)
Fuel damper
E.G.R. control valve
Valve timing control solenoid valve
I.A.A. unit and air regulator

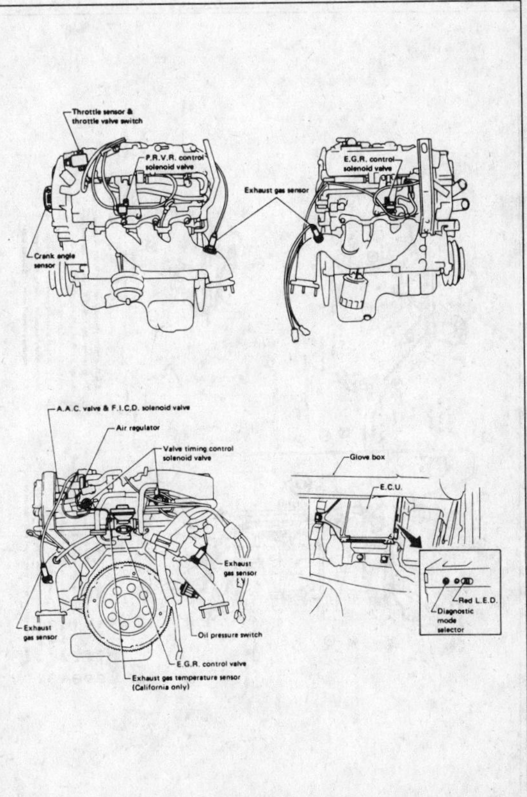
Throttle sensor & throttle valve switch
P.R.V.R. control solenoid valve
E.G.R. control solenoid valve
Exhaust gas sensor
Crank angle sensor
A.A.C. valve & F.I.C.D. solenoid valve
Air regulator
Valve timing control solenoid valve
Glove box
E.C.U.
Exhaust gas sensor
Exhaust gas sensor
Oil pressure switch
E.G.R. control valve
Exhaust gas temperature sensor (California only)
Red L.E.D.
Diagnostic mode selector

ECCS SYSTEM SCHEAMTIC—1990 300ZX

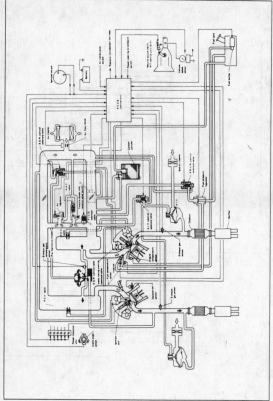

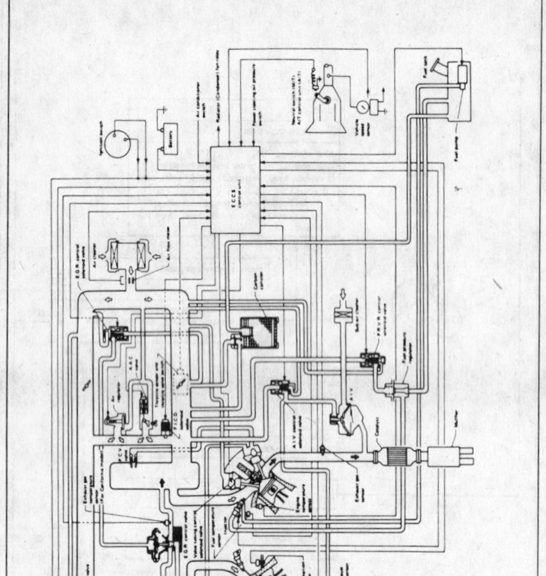

ECCS CONTROL SYSTEM CHART—1990 300ZX

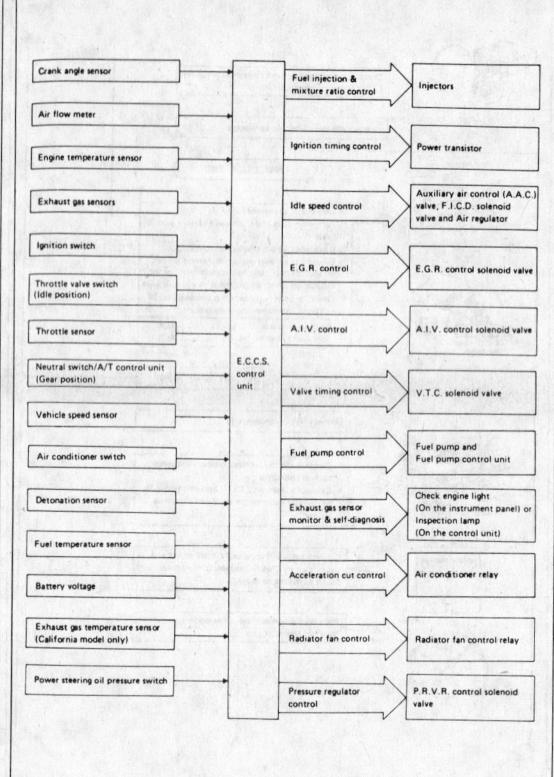
Crank angle sensor
Air flow meter
Engine temperature sensor
Exhaust gas sensors
Ignition switch
Throttle valve switch (Idle position)
Throttle sensor
Neutral switch/A/T control unit (Gear position)
Vehicle speed sensor
Air conditioner switch
Detonation sensor
Fuel temperature sensor
Battery voltage
Exhaust gas temperature sensor (California model only)
Power steering oil pressure switch
E.C.C.S. control unit
Fuel injection & mixture ratio control
Ignition timing control
Idle speed control
E.G.R. control
A.I.V. control
Valve timing control
Fuel pump control
Exhaust gas sensor monitor & self-diagnosis
Acceleration cut control
Radiator fan control
Pressure regulator control
Injectors
Power transistor
Auxiliary air control (A.A.C.) valve, F.I.C.D. solenoid valve and Air regulator
E.G.R. control solenoid valve
A.I.V. control solenoid valve
V.T.C. solenoid valve
Fuel pump and Fuel pump control unit
Check engine light (On the instrument panel) or Inspection lamp (On the control unit)
Air conditioner relay
Radiator fan control relay
P.R.V.R. control solenoid valve

ECCS VACUUM HOSE ROUTING—1990 300ZX

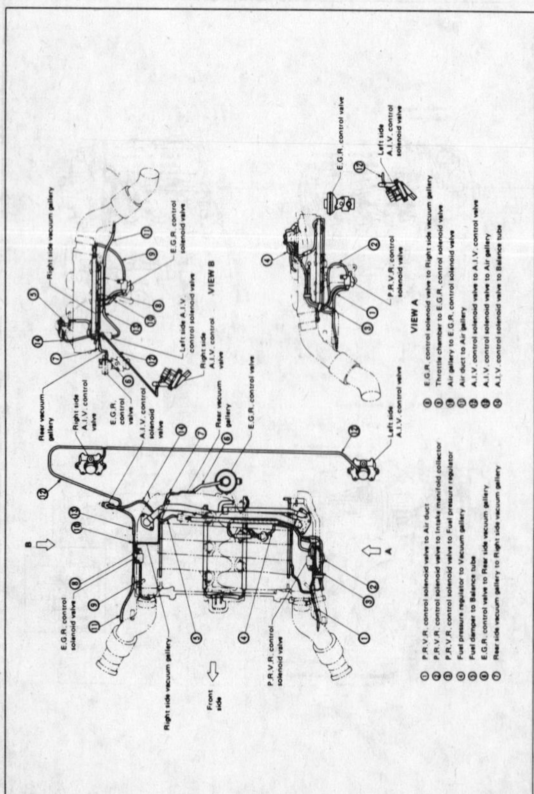

ECCS CONTROL UNIT WIRING SCHEMATIC 1990 300ZX

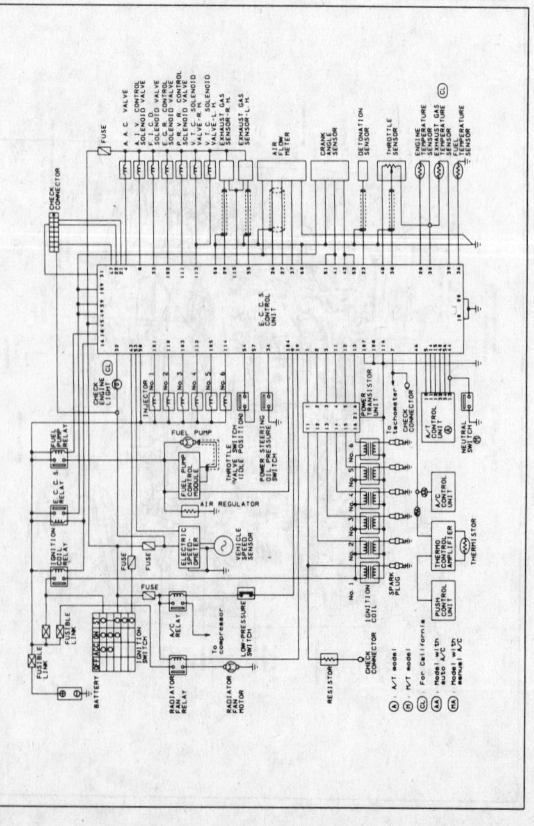

ECCS IDLE SPEED, IGNITION TIMING AND IDLE MIXTURE RATIO INSPECTION— 1990 300ZX

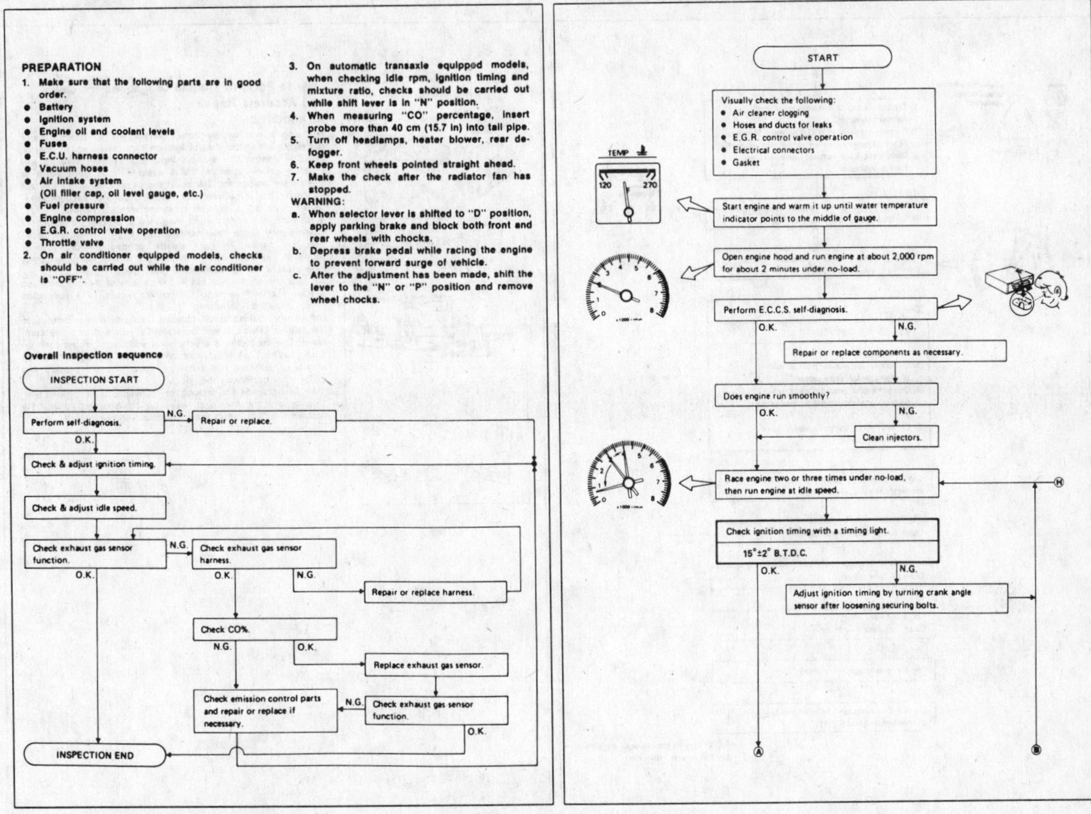
PREPARATION
1. Make sure that the following parts are in good order.
• Battery
• Ignition system
• Engine oil and coolant levels
• Fuses
• E.C.U. harness connector
• Vacuum hoses
• Air intake system
(Oil filler cap, oil level gauge, etc.)
• Fuel pressure
• Engine compression
• E.G.R. control valve operation
• Throttle valve
2. On air conditioner equipped models, checks should be carried out while the air conditioner is "OFF".
3. On automatic transaxle equipped models, when checking idle rpm, ignition timing and mixture ratio, checks should be carried out while shift lever is in "N" position.
4. When measuring "CO" percentage, insert probe more than 40 cm (15.7 in) into tail pipe.
5. Turn off headlamps, heater blower, rear defogger.
6. Keep front wheels pointed straight ahead.
7. Make the check after the radiator fan has stopped.
WARNING:
a. When selector lever is shifted to "D" position, apply parking brake and block both front and rear wheels with chocks.
b. Depress brake pedal while racing the engine to prevent forward surge of vehicle.
c. After the adjustment has been made, shift the lever to the "N" or "P" position and remove wheel chocks.
Overall inspection sequence
INSPECTION START
Perform self-diagnosis.
N.G.
Repair or replace.
O.K.
Check & adjust ignition timing.
Check & adjust idle speed.
Check exhaust gas sensor function.
N.G.
Check exhaust gas sensor harness.
O.K.
N.G.
Repair or replace harness.
O.K.
Check CO%.
N.G.
O.K.
Replace exhaust gas sensor.
Check emission control parts and repair or replace if necessary.
N.G.
Check exhaust gas sensor function.
O.K.
INSPECTION END
START
Visually check the following:
• Air cleaner clogging
• Hoses and ducts for leaks
• E.G.R. control valve operation
• Electrical connectors
• Gasket
TEMP
120
270
Start engine and warm it up until water temperature indicator points to the middle of gauge.
Open engine hood and run engine at about 2,000 rpm for about 2 minutes under no-load.
Perform E.C.C.S. self-diagnosis.
O.K.
N.G.
Repair or replace components as necessary.
Does engine run smoothly?
O.K.
N.G.
Clean injectors.
Race engine two or three times under no-load, then run engine at idle speed.
H
Check ignition timing with a timing light.
15°±2° B.T.D.C.
O.K.
N.G.
Adjust ignition timing by turning crank angle sensor after loosening securing bolts.
A
B

ECCS IDLE SPEED, IGNITION TIMING AND IDLE MIXTURE RATIO INSPECTION (CONT.)—1990 300ZX

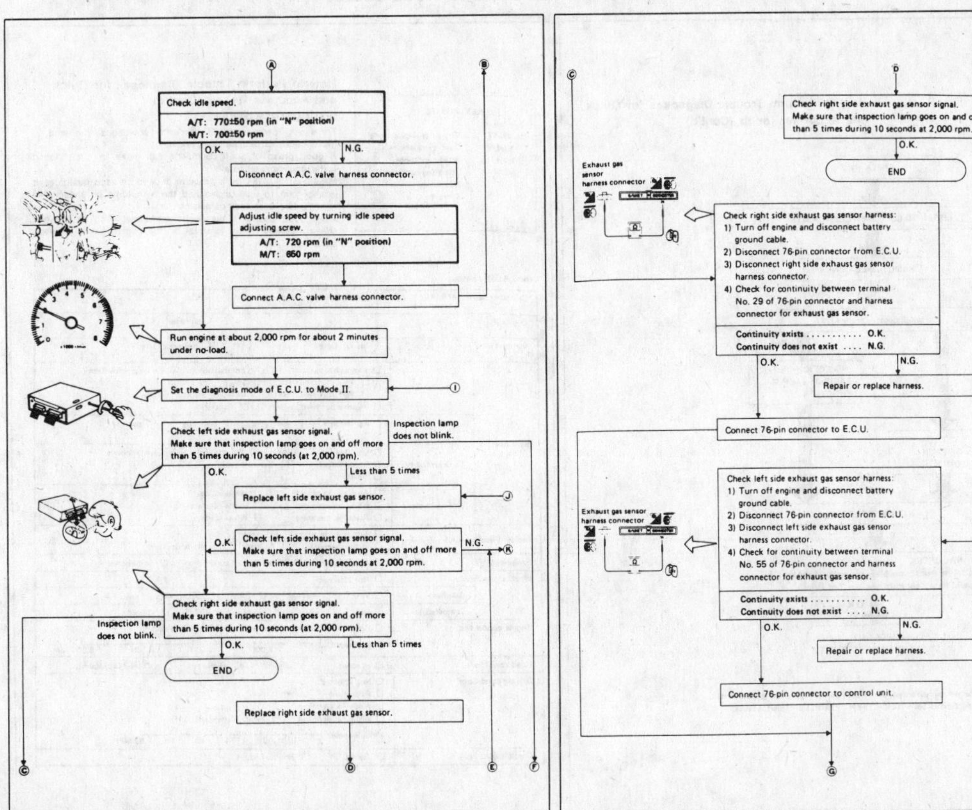
A
B
Check idle speed.
A/T: 770±50 rpm (in "N" position)
M/T: 700±50 rpm
O.K.
N.G.
Disconnect A.A.C. valve harness connector.
Adjust idle speed by turning idle speed adjusting screw.
A/T: 720 rpm (in "N" position)
M/T: 650 rpm
Connect A.A.C. valve harness connector.
Run engine at about 2,000 rpm for about 2 minutes under no-load.
Set the diagnosis mode of E.C.U. to Mode II.
I
Check left side exhaust gas sensor signal.
Make sure that inspection lamp goes on and off more than 5 times during 10 seconds (at 2,000 rpm).
Inspection lamp does not blink.
O.K.
Less than 5 times
Replace left side exhaust gas sensor.
J
O.K.
Check left side exhaust gas sensor signal.
Make sure that inspection lamp goes on and off more than 5 times during 10 seconds at 2,000 rpm.
N.G.
K
Check right side exhaust gas sensor signal.
Make sure that inspection lamp goes on and off more than 5 times during 10 seconds (at 2,000 rpm).
Inspection lamp does not blink.
O.K.
Less than 5 times
END
Replace right side exhaust gas sensor.
C
D
E
F
C
D
E
F
Check right side exhaust gas sensor signal.
Make sure that inspection lamp goes on and off more than 5 times during 10 seconds at 2,000 rpm.
N.G.
O.K.
END
Exhaust gas sensor harness connector
Check right side exhaust gas sensor harness:
1) Turn off engine and disconnect battery ground cable.
2) Disconnect 76-pin connector from E.C.U.
3) Disconnect right side exhaust gas sensor harness connector.
4) Check for continuity between terminal No. 29 of 76-pin connector and harness connector for exhaust gas sensor.
Continuity exists .... O.K.
Continuity does not exist .... N.G.
O.K.
N.G.
Repair or replace harness.
I
Connect 76-pin connector to E.C.U.
Exhaust gas sensor harness connector
Check left side exhaust gas sensor harness:
1) Turn off engine and disconnect battery ground cable.
2) Disconnect 76-pin connector from E.C.U.
3) Disconnect left side exhaust gas sensor harness connector.
4) Check for continuity between terminal No. 55 of 76-pin connector and harness connector for exhaust gas sensor.
Continuity exists .... O.K.
Continuity does not exist .... N.G.
O.K.
N.G.
Repair or replace harness.
I
Connect 76-pin connector to control unit.
G

ECCS IDLE SPEED, IGNITION TIMING AND IDLE MIXTURE RATIO INSPECTION (CONT.)—1990 300ZX

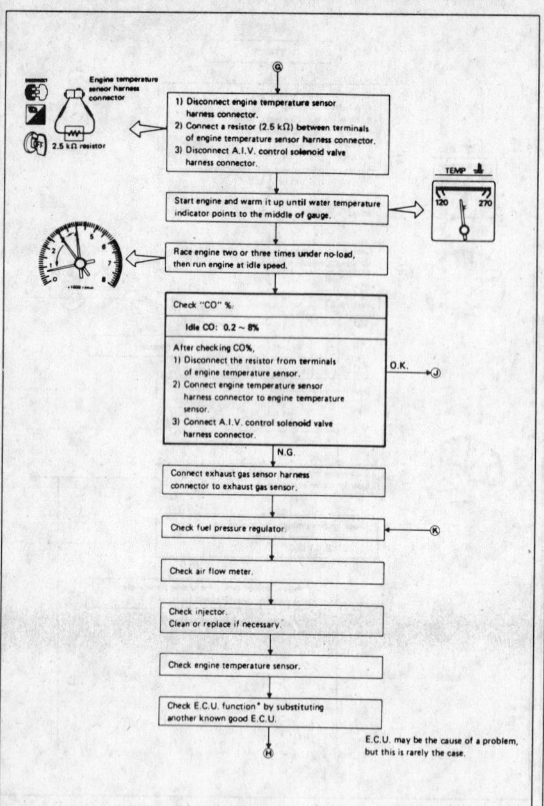
G
Engine temperature sensor harness connector
2.5 kΩ resistor
1) Disconnect engine temperature sensor harness connector.
2) Connect a resistor (2.5 kΩ) between terminals of engine temperature sensor harness connector.
3) Disconnect A.I.V. control solenoid valve harness connector.
TEMP
120
270
Start engine and warm it up until water temperature indicator points to the middle of gauge.
Race engine two or three times under no-load, then run engine at idle speed.
Check "CO" %.
Idle CO: 0.2 ~ 8%
After checking CO%,
1) Disconnect the resistor from terminals of engine temperature sensor.
2) Connect engine temperature sensor harness connector to engine temperature sensor.
3) Connect A.I.V. control solenoid valve harness connector.
O.K.
J
N.G.
Connect exhaust gas sensor harness connector to exhaust gas sensor.
Check fuel pressure regulator.
K
Check air flow meter.
Check injector.
Clean or replace if necessary.
Check engine temperature sensor.
Check E.C.U. function* by substituting another known good E.C.U.
E.C.U. may be the cause of a problem, but this is rarely the case.
H

ECCS DIAGNOSTIC CHART—1990 300ZX

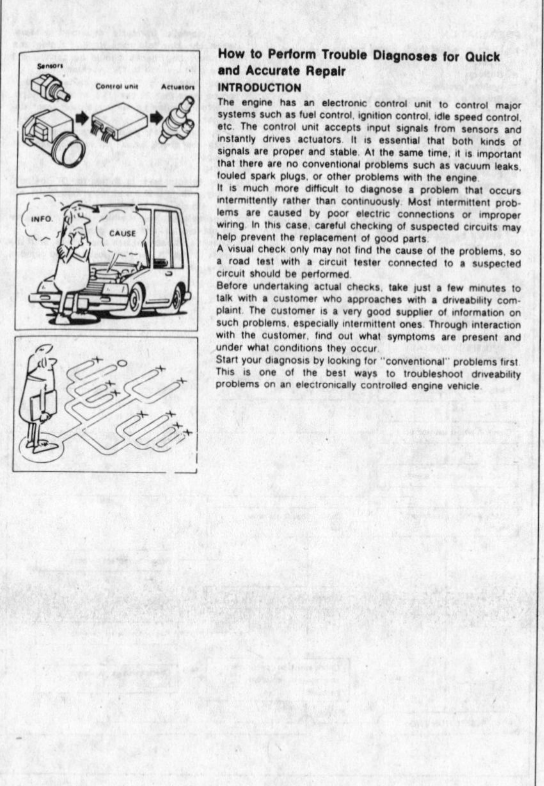
Sensors
Control unit
Actuators
INFO.
CAUSE

How to Perform Trouble Diagnoses for Quick and Accurate Repair

INTRODUCTION

The engine has an electronic control unit to control major systems such as fuel control, ignition control, idle speed control, etc. The control unit accepts input signals from sensors and instantly drives actuators. It is essential that both kinds of signals are proper and stable. At the same time, it is important that there are no conventional problems such as vacuum leaks, fouled spark plugs, or other problems with the engine.

It is much more difficult to diagnose a problem that occurs intermittently rather than continuously. Most intermittent problems are caused by poor electric connections or improper wiring. In this case, careful checking of suspected circuits may help prevent the replacement of good parts.

A visual check only may not find the cause of the problems, so a road test with a circuit tester connected to a suspected circuit should be performed.

Before undertaking actual checks, take just a few minutes to talk with a customer who approaches with a driveability complaint. The customer is a very good supplier of information on such problems, especially intermittent ones. Through interaction with the customer, find out what symptoms are present and under what conditions they occur.

Start your diagnosis by looking for "conventional" problems first. This is one of the best ways to troubleshoot driveability problems on an electronically controlled engine vehicle.

ECCS DIAGNOSTIC CHART—1990 300ZX

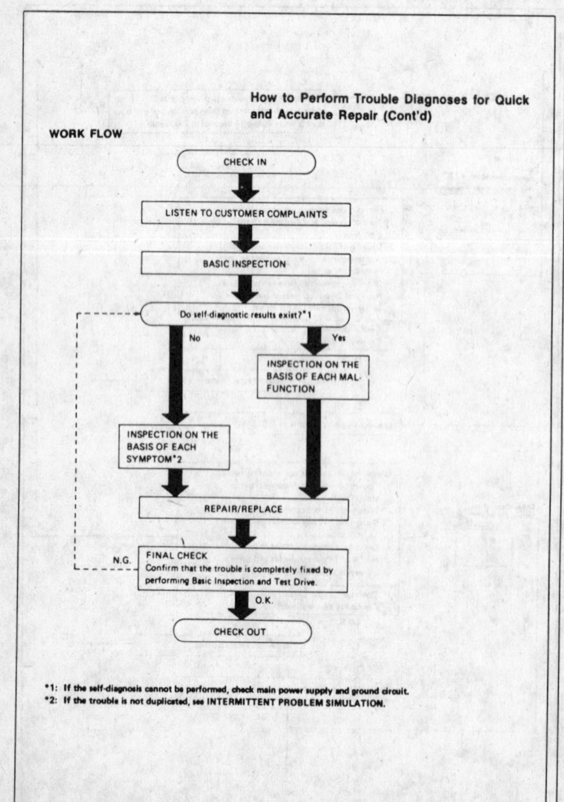
How to Perform Trouble Diagnoses for Quick and Accurate Repair (Cont'd)
WORK FLOW
CHECK IN
LISTEN TO CUSTOMER COMPLAINTS
BASIC INSPECTION
Do self-diagnostic results exist?*1
No
Yes
INSPECTION ON THE BASIS OF EACH MAL-FUNCTION
INSPECTION ON THE BASIS OF EACH SYMPTOM*2
REPAIR/REPLACE
N.G.
FINAL CHECK
Confirm that the trouble is completely fixed by performing Basic Inspection and Test Drive.
O.K.
CHECK OUT
*1: If the self-diagnosis cannot be performed, check main power supply and ground circuit.
*2: If the trouble is not duplicated, see INTERMITTENT PROBLEM SIMULATION.

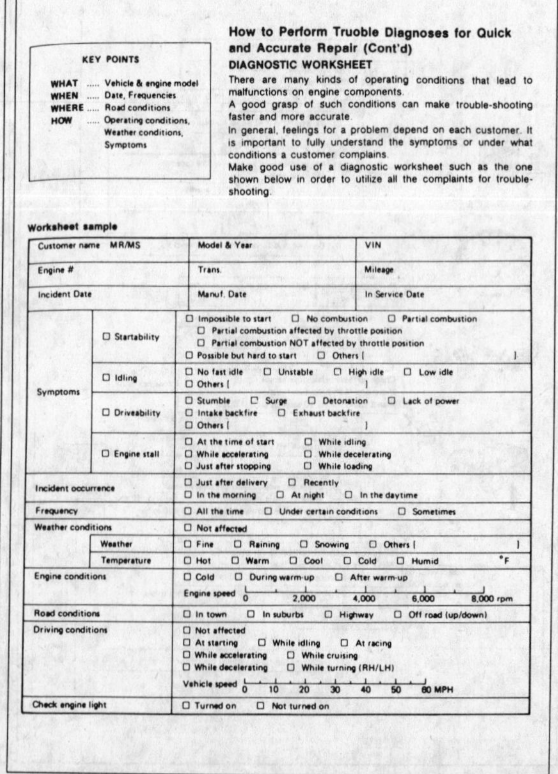

KEY POINTS

WHAT ..... Vehicle & engine model
WHEN ..... Date, Frequencies
WHERE ..... Road conditions
HOW ..... Operating conditions, Weather conditions, Symptoms

How to Perform Truoble Diagnoses for Quick and Accurate Repair (Cont'd)

DIAGNOSTIC WORKSHEET

There are many kinds of operating conditions that lead to malfunctions on engine components.

A good grasp of such conditions can make trouble-shooting faster and more accurate.

In general, feelings for a problem depend on each customer. It is important to fully understand the symptoms or under what conditions a customer complains.

Make good use of a diagnostic worksheet such as the one shown below in order to utilize all the complaints for trouble-shooting.

Worksheet sample

Customer name MR/MS	Model & Year	VIN
Engine #	Trans.	Mileage
Incident Date	Manuf. Date	In Service Date

Symptoms	☐ Startability	☐ Impossible to start ☐ No combustion ☐ Partial combustion ☐ Partial combustion affected by throttle position ☐ Partial combustion NOT affected by throttle position ☐ Possible but hard to start ☐ Others [ ]
	☐ Idling	☐ No fast idle ☐ Unstable ☐ High idle ☐ Low idle ☐ Others [ ]
	☐ Driveability	☐ Stumble ☐ Surge ☐ Detonation ☐ Lack of power ☐ Intake backfire ☐ Exhaust backfire ☐ Others [ ]
	☐ Engine stall	☐ At the time of start ☐ While idling ☐ While accelerating ☐ While decelerating ☐ Just after stopping ☐ While loading
Incident occurrence		☐ Just after delivery ☐ Recently ☐ In the morning ☐ At night ☐ In the daytime
Frequency		☐ All the time ☐ Under certain conditions ☐ Sometimes
Weather conditions		☐ Not affected
	Weather	☐ Fine ☐ Raining ☐ Snowing ☐ Others [ ]
	Temperature	☐ Hot ☐ Warm ☐ Cool ☐ Cold ☐ Humid °F
Engine conditions		☐ Cold ☐ During warm-up ☐ After warm-up Engine speed 0 2,000 4,000 6,000 8,000 rpm
Road conditions		☐ In town ☐ In suburbs ☐ Highway ☐ Off road (up/down)
Driving conditions		☐ Not affected ☐ At starting ☐ While idling ☐ At racing ☐ While accelerating ☐ While cruising ☐ While decelerating ☐ While turning (RH/LH) Vehicle speed 0 10 20 30 40 50 60 MPH
Check engine light		☐ Turned on ☐ Not turned on

ECCS DIAGNOSTIC CHART—1990 300ZX

GRRRRR
GRRRR
THIS is the symptom, isn't it ?

How to Perform Trouble Diagnoses for Quick and Accurate Repair (Cont'd)

INTERMITTENT PROBLEM SIMULATION

In order to duplicate an intermittent problem, it is effective to create similar conditions for component parts, under which the problem might occur.

Perform the activity listed under Service procedure and note the result.

	Variable factor	Influential part	Target condition	Service procedure
1	Mixture ratio	Pressure regulator	Made lean	Remove vacuum hose and apply vacuum.
			Made rich	Remove vacuum hose and apply pressure.
2	Ignition timing	Crank angle sensor	Advanced	Rotate distributor counter clockwise.
			Retarded	Rotate distributor clockwise.
3	Mixture ratio feedback control	Exhaust gas sensor	Suspended	Disconnect exhaust gas sensor harness connector.
		Control unit	Operation check	Perform self-diagnosis (Mode II) at 2,000 rpm.
4	Idle speed	A.A.C. valve	Raised	Turn idle adjusting screw counterclockwise.
			Lowered	Turn idle adjusting screw clockwise.
5	Electrical connection (Electric continuity)	Harness connectors and wires	Poor electrical connection or improper wiring	Tap or wiggle.
				Race engine rapidly. See if the torque reaction of the engine unit causes electric breaks.
6	Temperature	Control unit	Cooled	Cool with an icing spray or similar device.
			Warmed	Heat with a hair drier. [WARNING: Do not overheat the unit.]
7	Moisture	Electric parts	Damp	Wet. [WARNING: Do not directly pour water on components. Use a mist sprayer.]
8	Electric loads	Load switches	Loaded	Turn on head lights, air conditioner, rear defogger, etc.
9	Idle switch condition	Control unit	ON-OFF switching	Rotate throttle sensor body.
10	Ignition spark	Timing light	Spark power check	Try to flash timing light for each cylinder using ignition coil adapter (S.S.T.).

ECCS SELF-DIAGNOSTIC CHART—1990 300ZX

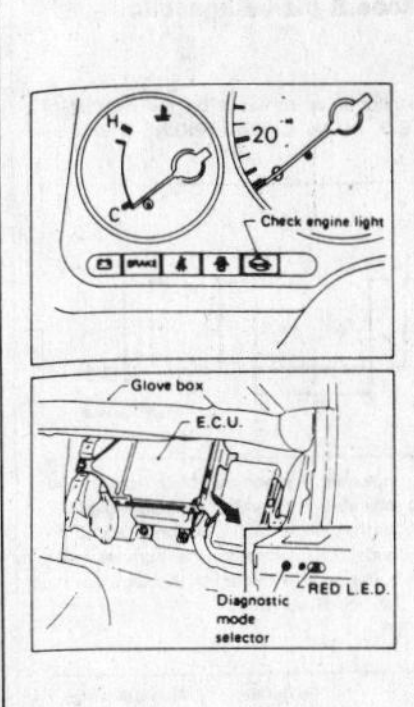
H
C
20
Check engine light
Glove box
E.C.U.
RED L.E.D.
Diagnostic mode selector

Self-diagnosis

CHECK ENGINE LIGHT

A check engine light has been adopted on the California, Federal and Canada models. This light blinks simultaneously with the RED L.E.D. on the E.C.U.

E.C.U. L.E.D.

In the E.C.U., the Green and Red L.E.D.'s have now been permanently changed to one RED L.E.D.

SELF-DIAGNOSTIC FUNCTION

Condition \ Mode		Mode I	Mode II
Ignition switch in "ON" position	Engine stopped	BULB CHECK	SELF-DIAGNOSTIC RESULTS
	Engine running	MALFUNCTION WARNING	EXHAUST GAS SENSOR MONITOR

ECCS SELF-DIAGNOSTIC CHART—1990 300ZX

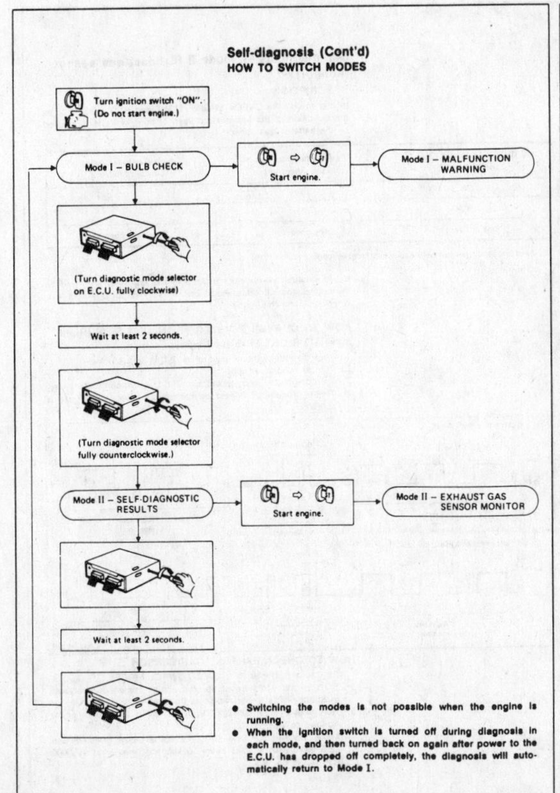
Self-diagnosis (Cont'd)
HOW TO SWITCH MODES
Turn ignition switch "ON". (Do not start engine.)
Mode I – BULB CHECK
Start engine.
Mode I – MALFUNCTION WARNING
(Turn diagnostic mode selector on E.C.U. fully clockwise)
Wait at least 2 seconds.
(Turn diagnostic mode selector fully counterclockwise.)
Mode II – SELF-DIAGNOSTIC RESULTS
Start engine.
Mode II – EXHAUST GAS SENSOR MONITOR
Wait at least 2 seconds.
● Switching the modes is not possible when the engine is running.
● When the ignition switch is turned off during diagnosis in each mode, and then turned back on again after power to the E.C.U. has dropped off completely, the diagnosis will automatically return to Mode I.

ECCS SELF-DIAGNOSTIC MODE I—1990 300ZX

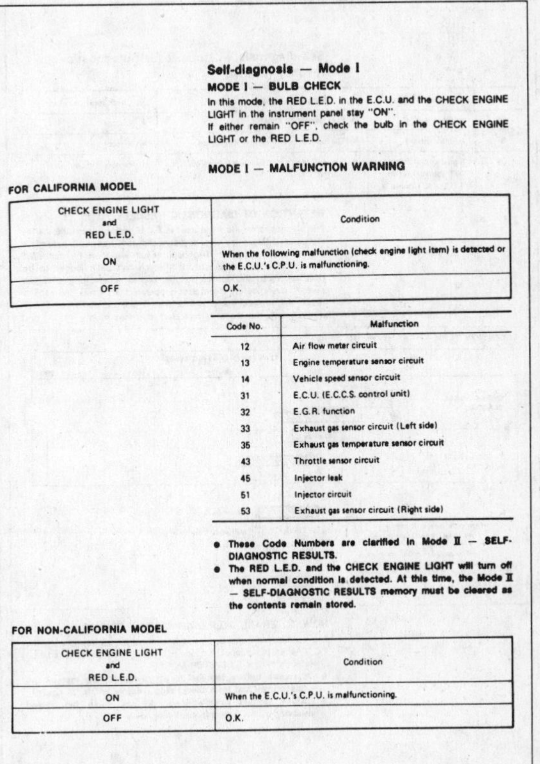

Self-diagnosis — Mode I

MODE I — BULB CHECK

In this mode, the RED L.E.D. in the E.C.U. and the CHECK ENGINE LIGHT in the instrument panel stay "ON".

If either remain "OFF", check the bulb in the CHECK ENGINE LIGHT or the RED L.E.D.

MODE I — MALFUNCTION WARNING

FOR CALIFORNIA MODEL

CHECK ENGINE LIGHT and RED L.E.D.	Condition
ON	When the following malfunction (check engine light item) is detected or the E.C.U.'s C.P.U. is malfunctioning.
OFF	O.K.

Code No.	Malfunction
12	Air flow meter circuit
13	Engine temperature sensor circuit
14	Vehicle speed sensor circuit
31	E.C.U. (E.C.C.S. control unit)
32	E.G.R. function
33	Exhaust gas sensor circuit (Left side)
35	Exhaust gas temperature sensor circuit
43	Throttle sensor circuit
45	Injector leak
51	Injector circuit
53	Exhaust gas sensor circuit (Right side)

- **These Code Numbers are clarified in Mode II — SELF-DIAGNOSTIC RESULTS.**
- **The RED L.E.D. and the CHECK ENGINE LIGHT will turn off when normal condition is detected. At this time, the Mode II — SELF-DIAGNOSTIC RESULTS memory must be cleared as the contents remain stored.**

FOR NON-CALIFORNIA MODEL

CHECK ENGINE LIGHT and RED L.E.D.	Condition
ON	When the E.C.U.'s C.P.U. is malfunctioning.
OFF	O.K.

ECCS SELF-DIAGNOSTIC MODE II – 1990 300ZX

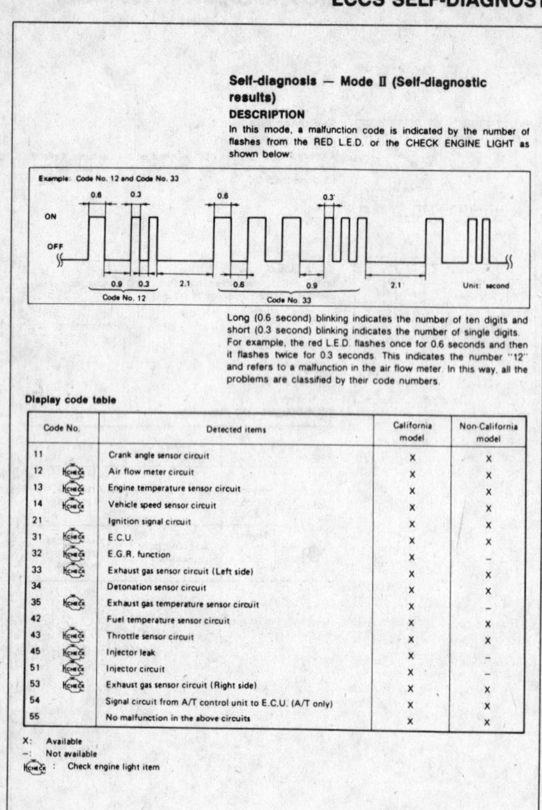

Self-diagnosis — Mode Ⅱ (Self-diagnostic results)

DESCRIPTION

In this mode, a malfunction code is indicated by the number of flashes from the RED L.E.D. or the CHECK ENGINE LIGHT as shown below:

Example: Code No. 12 and Code No. 33
0.6
0.3
0.6
0.3
ON
OFF
0.9
0.3
2.1
0.6
0.9
2.1
Unit: second
Code No. 12
Code No. 33

Long (0.6 second) blinking indicates the number of ten digits and short (0.3 second) blinking indicates the number of single digits. For example, the red L.E.D. flashes once for 0.6 seconds and then it flashes twice for 0.3 seconds. This indicates the number "12" and refers to a malfunction in the air flow meter. In this way, all the problems are classified by their code numbers.

Display code table

Code No.		Detected items	California model	Non-California model
11		Crank angle sensor circuit	X	X
12	CHECK	Air flow meter circuit	X	X
13	CHECK	Engine temperature sensor circuit	X	X
14	CHECK	Vehicle speed sensor circuit	X	X
21		Ignition signal circuit	X	X
31	CHECK	E.C.U.	X	X
32	CHECK	E.G.R. function	X	–
33	CHECK	Exhaust gas sensor circuit (Left side)	X	X
34		Detonation sensor circuit	X	X
35	CHECK	Exhaust gas temperature sensor circuit	X	–
42		Fuel temperature sensor circuit	X	X
43	CHECK	Throttle sensor circuit	X	X
45	CHECK	Injector leak	X	–
51	CHECK	Injector circuit	X	–
53	CHECK	Exhaust gas sensor circuit (Right side)	X	X
54		Signal circuit from A/T control unit to E.C.U. (A/T only)	X	X
55		No malfunction in the above circuits	X	X

X: Available
–: Not available
CHECK : Check engine light item

Self-diagnosis — Mode Ⅱ (Self-diagnostic results) (Cont'd)

Code No.	Detected items	Malfunction is detected when ...	Check item (remedy)
11	Crank angle sensor circuit	• Either 1° or 120° signal is not entered for the first few seconds during engine cranking. • Either 1° or 120° signal is not input often enough while the engine speed is higher than the specified rpm.	• Harness and connector (If harness and connector are normal, replace crank angle sensor.)
12	Air flow meter circuit	• The air flow meter circuit is open or shorted. (An abnormally high or low voltage is entered.)	• Harness and connector (If harness and connector are normal, replace air flow meter.)
13	Engine temperature sensor circuit	• The engine temperature sensor circuit is open or shorted. (An abnormally high or low output voltage is entered.)	• Harness and connector • Engine temperature sensor
14	Vehicle speed sensor circuit	• The vehicle speed sensor circuit is open or shorted.	• Harness and connector • Vehicle speed sensor (reed switch)
21	Ignition signal circuit	• The ignition signal in the primary circuit is not entered during engine cranking or running.	• Harness and connector • Power transistor unit
31	E.C.U.	• E.C.U. calculation function is malfunctioning.	(Replace E.C.C.S. control unit.)
32	E.G.R. function	• E.G.R. control valve does not operate. (E.G.R. control valve spring does not lift.)	• E.G.R. control valve • E.G.R. control solenoid valve
33	Exhaust gas sensor circuit (Left side)	• The exhaust gas sensor circuit is open or shorted. (An abnormally high or low output voltage is entered.)	• Harness and connector • Exhaust gas sensor • Fuel pressure • Injectors • Intake air leaks
53	Exhaust gas sensor circuit (Right side)		
34	Detonation sensor circuit	• The detonation circuit is open or shorted. (An abnormally high or low voltage is entered.)	• Harness and connector • Detonation sensor
35	Exhaust gas temperature sensor circuit	• The exhaust gas temperature sensor circuit is open or shorted. (An abnormally high or low voltage is entered.)	• Harness and connector • Exhaust gas temperature sensor
42	Fuel temperature sensor circuit	• The fuel temperature sensor circuit is open or shorted. (An abnormally high or low voltage is entered.)	• Harness and connector • Fuel temperature sensor
43	Throttle sensor circuit	• The throttle sensor circuit is open or shorted. (An abnormally high or low voltage is entered.)	• Harness and connector • Throttle sensor

ECCS SELF-DIAGNOSTIC MODE II (CONT.) – 1990 300ZX

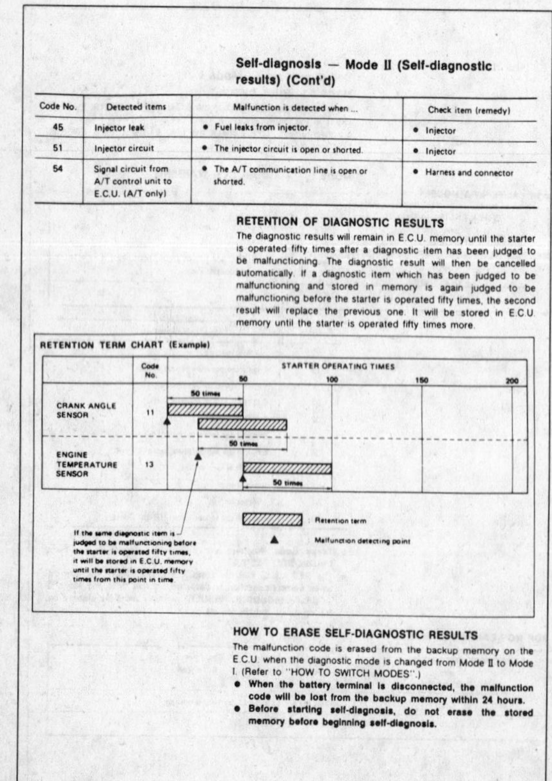

Self-diagnosis — Mode Ⅱ (Self-diagnostic results) (Cont'd)

Code No.	Detected items	Malfunction is detected when ...	Check item (remedy)
45	Injector leak	• Fuel leaks from injector.	• Injector
51	Injector circuit	• The injector circuit is open or shorted.	• Injector
54	Signal circuit from A/T control unit to E.C.U. (A/T only)	• The A/T communication line is open or shorted.	• Harness and connector

RETENTION OF DIAGNOSTIC RESULTS

The diagnostic results will remain in E.C.U. memory until the starter is operated fifty times after a diagnostic item has been judged to be malfunctioning. The diagnostic result will then be cancelled automatically. If a diagnostic item which has been judged to be malfunctioning and stored in memory is again judged to be malfunctioning before the starter is operated fifty times, the second result will replace the previous one. It will be stored in E.C.U. memory until the starter is operated fifty times more.

RETENTION TERM CHART (Example)
Code No.
STARTER OPERATING TIMES
50
100
150
200
CRANK ANGLE SENSOR
11
50 times
ENGINE TEMPERATURE SENSOR
13
50 times
50 times
: Retention term
Malfunction detecting point
If the same diagnostic item is judged to be malfunctioning before the starter is operated fifty times, it will be stored in E.C.U. memory until the starter is operated fifty times from this point in time.

HOW TO ERASE SELF-DIAGNOSTIC RESULTS

The malfunction code is erased from the backup memory on the E.C.U. when the diagnostic mode is changed from Mode Ⅱ to Mode I. (Refer to "HOW TO SWITCH MODES".)

- **When the battery terminal is disconnected, the malfunction code will be lost from the backup memory within 24 hours.**
- **Before starting self-diagnosis, do not erase the stored memory before beginning self-diagnosis.**

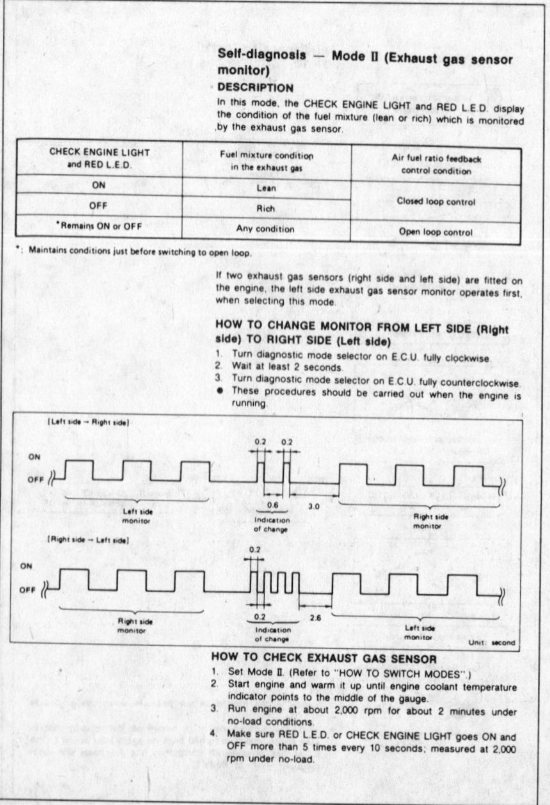

Self-diagnosis — Mode Ⅱ (Exhaust gas sensor monitor)

DESCRIPTION

In this mode, the CHECK ENGINE LIGHT and RED L.E.D. display the condition of the fuel mixture (lean or rich) which is monitored by the exhaust gas sensor.

CHECK ENGINE LIGHT and RED L.E.D.	Fuel mixture condition in the exhaust gas	Air fuel ratio feedback control condition
ON	Lean	Closed loop control
OFF	Rich	
*Remains ON or OFF	Any condition	Open loop control

*: Maintains conditions just before switching to open loop.

If two exhaust gas sensors (right side and left side) are fitted on the engine, the left side exhaust gas sensor monitor operates first, when selecting this mode.

HOW TO CHANGE MONITOR FROM LEFT SIDE (Right side) TO RIGHT SIDE (Left side)

1. Turn diagnostic mode selector on E.C.U. fully clockwise.
2. Wait at least 2 seconds.
3. Turn diagnostic mode selector on E.C.U. fully counterclockwise.
- These procedures should be carried out when the engine is running.

[Left side → Right side]
0.2
0.2
ON
OFF
0.6
3.0
Left side monitor
Indication of change
Right side monitor
[Right side → Left side]
0.2
ON
OFF
0.2
2.6
Right side monitor
Indication of change
Left side monitor
Unit: second

HOW TO CHECK EXHAUST GAS SENSOR

1. Set Mode Ⅱ. (Refer to "HOW TO SWITCH MODES".)
2. Start engine and warm it up until engine coolant temperature indicator points to the middle of the gauge.
3. Run engine at about 2,000 rpm for about 2 minutes under no-load conditions.
4. Make sure RED L.E.D. or CHECK ENGINE LIGHT goes ON and OFF more than 5 times every 10 seconds; measured at 2,000 rpm under no-load.

ECCS SELF-DIAGNOSTIC CHART – 1990 300ZX

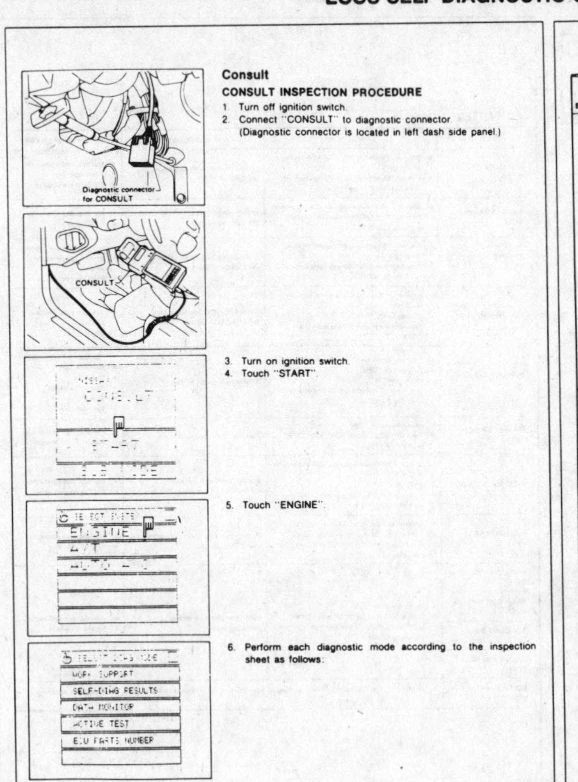
Diagnostic connector for CONSULT
CONSULT
ENGINE
SELF-DIAG RESULTS
DATA MONITOR
ACTIVE TEST
E.C.U PARTS NUMBER

Consult

CONSULT INSPECTION PROCEDURE

1. Turn off ignition switch.
2. Connect "CONSULT" to diagnostic connector. (Diagnostic connector is located in left dash side panel.)
3. Turn on ignition switch.
4. Touch "START".
5. Touch "ENGINE".
6. Perform each diagnostic mode according to the inspection sheet as follows.

	E.C.C.S. COMPONENT PARTS / MODE	WORK SUPPORT	SELF-DIAGNOSTIC RESULTS	DATA MONITOR	ACTIVE TEST
INPUT	Crank angle sensor		X	X	
	Air flow meter		X	X	
	Engine temperature sensor		X	X	X
	Exhaust gas sensors		X	X	
	Vehicle speed sensors		X	X	
	Throttle sensor	X	X	X	
	Fuel temperature sensor		X	X	
	Exhaust gas temperature sensor*		X	X	
	Detonation sensor		X		
	Ignition switch (start signal)			X	
	Air conditioner switch			X	
	Neutral switch			X	
	Power steering oil pressure switch			X	
	Battery			X	
	A/T signal		X	X	
OUTPUT	Injectors		X	X	X
	Power transistor (ignition signal)		X	X (Ignition timing)	X
	A.A.C. valve	X		X	X
	F.I.C.D. solenoid valve			X	X
	Valve timing control solenoid valve			X	X
	A.I.V. control solenoid valve			X	X
	P.R.V.R. control solenoid valve			X	X
	E.G.R. control solenoid valve			X	X
	Air conditioner relay			X	
	Fuel pump relay	X		X	X
	Radiator fan relay			X	X

*: The E.C.C.S. component part marked * is applicable to vehicles for California only.
X: Applicable

ECCS SELF-DIAGNOSTIC CHART – 1990 300ZX

FUNCTION

Diagnostic mode	Function
Work support	This mode enables a technician to adjust some devices faster and more accurately by following the indications on the CONSULT unit.
Self-diagnostic results	Self-diagnostic results can be read and erased quickly.
Data monitor	Input/Output data in the control unit can be read.
Active test	Mode in which CONSULT drives some actuators apart from the control units and also shifts some parameters in a specified range.
E.C.U. part numbers	E.C.U. part numbers can be read.

WORK SUPPORT MODE

WORK ITEM	CONDITION	USAGE
THROTTLE SENSOR ADJUSTMENT	CHECK THE THROTTLE SENSOR SIGNAL. ADJUST IT TO THE SPECIFIED VALUE BY ROTATING THE SENSOR BODY UNDER THE FOLLOWING CONDITIONS. • IGN SW "ON" • ENG NOT RUNNING • ACC PEDAL NOT PRESSED	When adjusting throttle sensor initial position,
IGNITION TIMING ADJUSTMENT*	• IGNITION TIMING FEEDBACK CONTROL WILL BE HELD BY TOUCHING "START". AFTER DOING SO, ADJUST IGNITION TIMING WITH A TIMING LIGHT BY TURNING THE CRANK ANGLE SENSOR.	When adjusting initial ignition timing.
AAC VALVE ADJUSTMENT	SET ENGINE RPM AT THE SPECIFIED VALUE UNDER THE FOLLOWING CONDITIONS. • ENGINE WARMED UP • NO-LOAD	When adjusting idle speed,
FUEL PRESSURE RELEASE	• FUEL PUMP WILL STOP BY TOUCHING "START" WHEN IDLING. CRANK A FEW TIMES AFTER ENGINE STALLS.	When releasing fuel pressure from fuel line,

*: The ignition timing feedback control is not adopted on model 300ZX, so it is not necessary to perform IGNITION TIMING ADJUSTMENT.

SELF-DIAGNOSTIC RESULTS MODE

DIAGNOSTIC ITEM	DIAGNOSTIC ITEM IS DETECTED WHEN....	CHECK ITEM (REMEDY)
CRANK ANGLE SENSOR	• Either 1° or 120° signal is not entered for the first few seconds during engine cranking. • Either 1° or 120° signal is not input often enough while the engine speed is higher than the specified rpm.	• Harness and connector (If harness and connector are normal, replace crank angle sensor.)
AIR FLOW METER	• The air flow meter circuit is open or shorted. (An abnormally high or low voltage is entered.)	• Harness and connector (If harness and connector are normal, replace air flow meter.)
ENGINE TEMPERATURE SENSOR	• The engine temperature sensor circuit is open or shorted. (An abnormally high or low output voltage is entered.)	• Harness and connector • Engine temperature sensor
VEHICLE SPEED SENSOR	• The vehicle speed sensor circuit is open or shorted.	• Harness and connector • Vehicle speed sensor (reed switch)
PRIMARY IGNITION SIGNAL	• The ignition signal in primary circuit is not entered during engine cranking or running.	• Harness and connector • Power transistor unit
CONTROL UNIT	• E.C.U. calculation function is malfunctioning.	(Replace E.C.C.S. control unit.)
EGR SYSTEM*	• E.G.R. control valve does not operate. (E.G.R. control valve spring does not lift.)	• E.G.R. control valve • E.G.R. control solenoid valve
EXHAUST GAS SENSOR EXHAUST GAS SENSOR (R)	• The exhaust gas sensor circuit is open or shorted. (An abnormally high or low output voltage is entered.)	• Harness and connector • Exhaust gas sensor • Fuel pressure • Injectors • Intake air leaks
DETONATION SENSOR	• The detonation circuit is open or shorted. (An abnormally high or low voltage is entered.)	• Harness and connector • Detonation sensor
EXHAUST GAS TEMPERATURE SENSOR	• The exhaust gas temperature sensor circuit is open or shorted. (An abnormally high or low voltage is entered.)	• Harness and connector • Exhaust gas temperature sensor
FUEL TEMPERATURE SENSOR	• The fuel temperature sensor circuit is open or shorted. (An abnormally high or low voltage is entered.)	• Harness and connector • Fuel temperature sensor
THROTTLE SENSOR	• The throttle sensor circuit is open or shorted. (An abnormally high or low voltage is entered.)	• Harness and connector • Throttle sensor
FUEL INJECTOR – FUEL LEAK*	• Fuel leaks from injector.	• Injector

ECCS SELF-DIAGNOSTIC CHART—1990 300ZX

DIAGNOSTIC ITEM	DIAGNOSTIC ITEM IS DETECTED WHEN....	CHECK ITEM (REMEDY)
FUEL INJECTOR – OPEN CIRCUIT*	• The injector circuit is open or shorted.	• Injector
A/T COMMUNICATION LINE	• The A/T communication line is open or shorted.	• Harness and connector

*: The diagnostic item marked * is applicable to vehicles for California only.

DATA MONITOR MODE

MONITOR ITEM	CONDITION		SPECIFICATION	CHECK ITEM WHEN OUTSIDE SPEC.
CAS, RPM (POS) CAS, RPM (REF)	• Tachometer: Connect • Run engine and compare tachometer indication with the CONSULT value.		Almost the same speed as the CONSULT value.	• Harness and connector • Crank angle sensor
AIR FLOW MTR	• Engine: After warming up, idle the engine • A/C switch "OFF" • Shift lever "N"		0.8 - 1.5V	• Harness and connector • Air flow meter
ENG TEMP SEN	• Engine: After warming up		More than 70°C (158°F)	• Harness and connector • Engine temperature sensor
EXH GAS SEN	• Engine: After warming up	Maintaining engine speed at 2,000 rpm	0 - 0.3V ↔ 0.6 - 1.0V	• Harness and connector • Exhaust gas sensor • Intake air leaks • Injectors
EXH GAS SEN-R				
M/R F/C MNT			LEAN ↔ RICH Changes more than 5 times during 10 seconds.	
M/R F/C MNT-R				
CAR SPEED SEN	• Turn drive wheels and compare speedometer indication with the CONSULT value		Almost the same speed as the CONSULT value	• Harness and connector • Vehicle speed sensor
BATTERY VOLT	• Ignition switch: ON (Engine stopped)		11 - 14V	• Battery • E.C.U. power supply circuit
THROTTLE SEN	• Ignition switch: ON (Engine stopped)	Throttle valve fully closed	0.4 - 0.5V	• Harness and connector • Throttle sensor • Throttle sensor adjustment
		Throttle valve fully opened	Approx. 4.0V	
FUEL TEMP SEN	• Engine: After warming up		20 - 60°C (68 - 140°F)	• Harness and connector • Fuel temp. sensor
EGR TEMP SEN*	• Engine: After warming up		Less than 4.5V	• Harness and connector • Exhaust gas temperature sensor
START SIGNAL	• Ignition switch: ON → START		OFF → ON	• Harness and connector • Starter switch
IDLE POSITION	• Ignition switch: ON (Engine stopped)	Throttle valve: Idle position	ON	• Harness and connector • Throttle sensor • Throttle sensor adjustment
		Throttle valve: Slightly open	OFF	
AIR COND SIG	• Engine: After warming up, idle the engine	A/C switch "OFF"	OFF	• Harness and connector • Air conditioner switch
		A/C switch "ON"	ON	
NEUTRAL SW	• Ignition switch: ON	Shift lever "P" or "N"	ON	• Harness and connector • Neutral switch
		Except above	OFF	
PW/ST SIGNAL	• Engine: After warming up, idle the engine	Steering wheel in neutral (forward direction)	OFF	• Harness and connector • Power steering oil pressure switch
		The steering wheel is turned	ON	

Remarks: The monitor item marked * is applicable to vehicles for California only.

ACTIVE TEST MODE

TEST ITEM	CONDITION	JUDGEMENT	CHECK ITEM (REMEDY)
FUEL INJECTION TEST	• Engine: Return to the original trouble condition • Change the amount of fuel injection with the CONSULT.	If trouble symptom disappears, see CHECK ITEM.	• Harness and connector • Fuel injectors • Exhaust gas sensors
AAC/V OPENING TEST	• Engine: After warming up, idle the engine. • Change the AAC valve opening percent with the CONSULT.	Engine speed changes according to the opening percent.	• Harness and connector • AAC valve
ENGINE TEMP TEST	• Engine: Return to the original trouble condition • Change the engine coolant temperature with the CONSULT.	If trouble symptom disappears, see CHECK ITEM.	• Harness and connector • Engine temperature sensor • Fuel injectors
IGN TIMING TEST	• Engine: Return to the original trouble condition • Timing light: Set • Retard the ignition timing with the CONSULT.	If trouble symptom disappears, see CHECK ITEM.	• Adjust initial ignition timing
POWER BALANCE TEST	• Engine: After warming up, idle the engine. • A/C switch "OFF" • Shift lever "N" • Cut off each injector signal one at a time with the CONSULT.	Engine runs rough or dies.	• Harness and connector • Compression • Injectors • Power transistor • Spark plugs • Ignition coils
RADIATOR FAN TEST	• Ignition switch: ON • Turn the radiator fan "ON" and "OFF" with the CONSULT.	Radiator fan moves and stops.	• Harness and connector • Radiator fan motor
FICD SOL/V TEST	• Engine: After warming up, idle the engine. • A/C switch "OFF" • Shift lever "N" • Turn the FICD solenoid valve "ON" with the CONSULT.	Engine speed will increase momentarily by approx. 200 rpm.	• Harness and connector • FICD solenoid valve
FUEL PUMP RLY TEST	• Ignition switch: ON (Engine stopped) • Turn the fuel pump relay "ON" and "OFF" with the CONSULT and listen to operating sound.	Fuel pump relay makes the operating sound.	• Harness and connector • Fuel pump relay
EGR CONT SOL/V TEST	• Ignition switch: ON • Turn solenoid valve "ON" and "OFF" with the CONSULT and listen to operating sound.	Each solenoid valve makes an operating sound.	• Harness and connector • Solenoid valve
PRVR CONT SOL/V TEST			
AIV CONT SOL/V TEST			
VALVE TIM SOL TEST			
SELF-LEARN CONT TEST	• In this test, the coefficient of self-learning control mixture ratio returns to the original coefficient by touching "CLEAR" on the screen.		

ECCS SELF-DIAGNOSTIC CHART—1990 300ZX

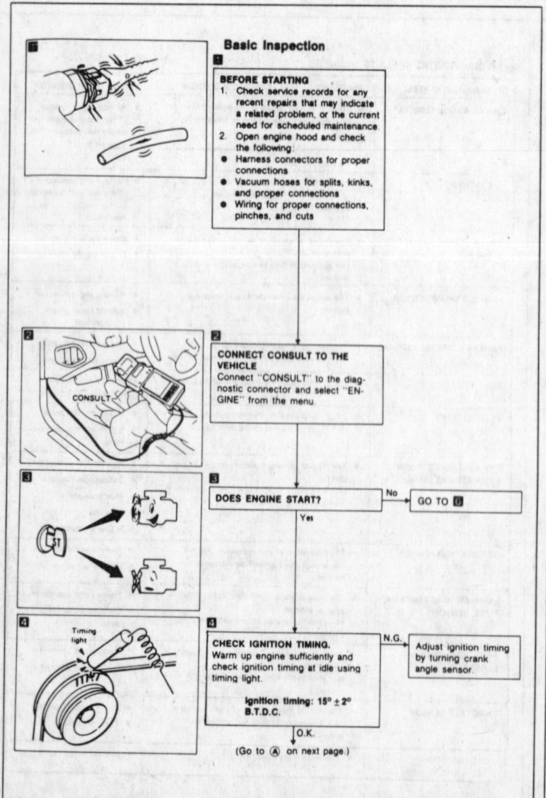
Basic Inspection
1
BEFORE STARTING
1. Check service records for any recent repairs that may indicate a related problem, or the current need for scheduled maintenance.
2. Open engine hood and check the following:
● Harness connectors for proper connections
● Vacuum hoses for splits, kinks, and proper connections
● Wiring for proper connections, pinches, and cuts
2
CONSULT
CONNECT CONSULT TO THE VEHICLE
Connect "CONSULT" to the diagnostic connector and select "ENGINE" from the menu.
3
DOES ENGINE START?
No
GO TO 6
Yes
4
Timing light
CHECK IGNITION TIMING.
Warm up engine sufficiently and check ignition timing at idle using timing light.
Ignition timing: 15° ± 2° B.T.D.C.
N.G.
Adjust ignition timing by turning crank angle sensor.
O.K.
(Go to Ⓐ on next page.)

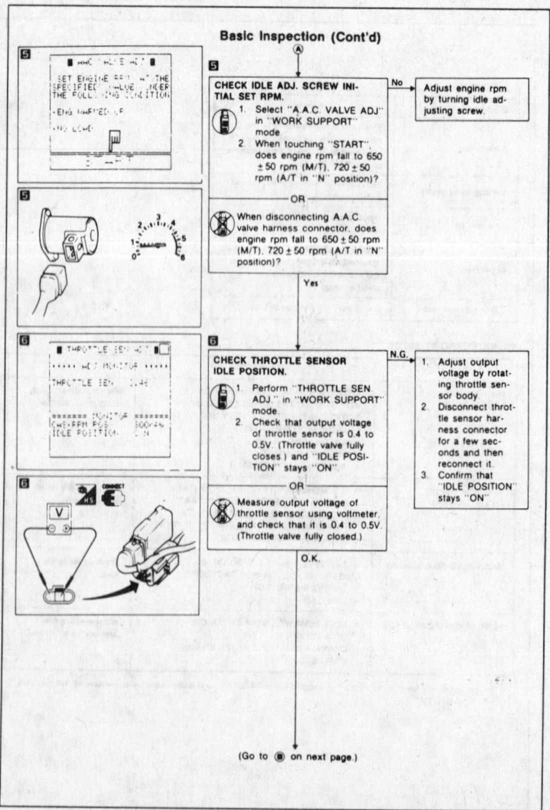
Basic Inspection (Cont'd)
Ⓐ
5
CHECK IDLE ADJ. SCREW INITIAL SET RPM.
1. Select "A.A.C. VALVE ADJ" in "WORK SUPPORT" mode
2. When touching "START", does engine rpm fall to 650 ± 50 rpm (M/T), 720 ± 50 rpm (A/T in "N" position)?
OR
When disconnecting A.A.C. valve harness connector, does engine rpm fall to 650 ± 50 rpm (M/T), 720 ± 50 rpm (A/T in "N" position)?
No
Adjust engine rpm by turning idle adjusting screw.
Yes
6
CHECK THROTTLE SENSOR IDLE POSITION.
1. Perform "THROTTLE SEN. ADJ." in "WORK SUPPORT" mode.
2. Check that output voltage of throttle sensor is 0.4 to 0.5V. (Throttle valve fully closes.) and "IDLE POSITION" stays "ON".
OR
Measure output voltage of throttle sensor using voltmeter, and check that it is 0.4 to 0.5V. (Throttle valve fully closed.)
N.G.
1. Adjust output voltage by rotating throttle sensor body.
2. Disconnect throttle sensor harness connector for a few seconds and then reconnect it.
3. Confirm that "IDLE POSITION" stays "ON".
O.K.
(Go to Ⓑ on next page.)

ECCS SELF-DIAGNOSTIC CHART—1990 300ZX

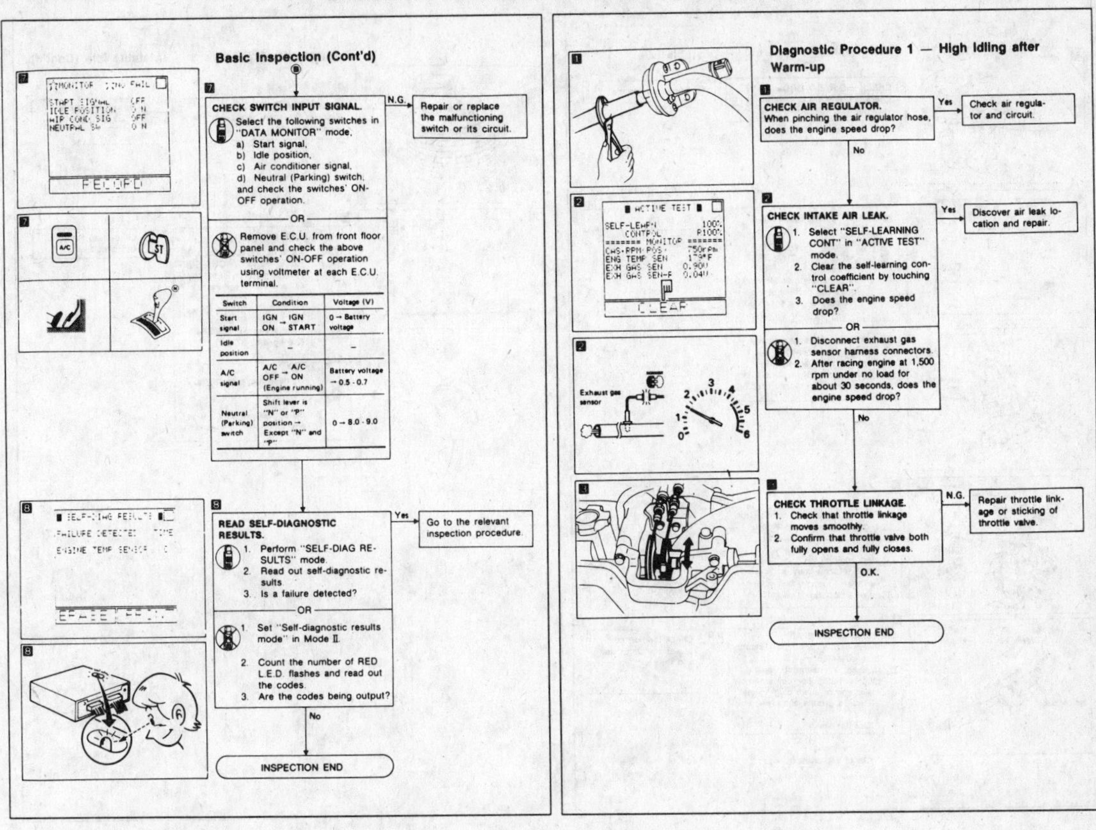
Basic Inspection (Cont'd)
CHECK SWITCH INPUT SIGNAL.
Select the following switches in "DATA MONITOR" mode.
a) Start signal,
b) Idle position,
c) Air conditioner signal,
d) Neutral (Parking) switch,
and check the switches' ON-OFF operation.
OR
Remove E.C.U. from front floor panel and check the above switches' ON-OFF operation using voltmeter at each E.C.U. terminal.
Switch | Condition | Voltage (V)
Start signal | IGN → START | 0 → Battery voltage
Idle position
A/C signal | A/C OFF → ON (Engine running) | Battery voltage → 0.5 - 0.7
Neutral (Parking) switch | Shift lever is "N" or "P" position → Except "N" and "P" | 0 → 8.0 - 9.0
N.G.
Repair or replace the malfunctioning switch or its circuit.
READ SELF-DIAGNOSTIC RESULTS.
1. Perform "SELF-DIAG RESULTS" mode.
2. Read out self-diagnostic results.
3. Is a failure detected?
OR
1. Set "Self-diagnostic results mode" in Mode II.
2. Count the number of RED L.E.D. flashes and read out the codes.
3. Are the codes being output?
Yes
Go to the relevant inspection procedure.
No
INSPECTION END
Diagnostic Procedure 1 — High Idling after Warm-up
CHECK AIR REGULATOR.
When pinching the air regulator hose, does the engine speed drop?
Yes
Check air regulator and circuit.
No
CHECK INTAKE AIR LEAK.
1. Select "SELF-LEARNING CONT" in "ACTIVE TEST" mode.
2. Clear the self-learning control coefficient by touching "CLEAR".
3. Does the engine speed drop?
OR
1. Disconnect exhaust gas sensor harness connectors.
2. After racing engine at 1,500 rpm under no load for about 30 seconds, does the engine speed drop?
Yes
Discover air leak location and repair.
No
Exhaust gas sensor
CHECK THROTTLE LINKAGE.
1. Check that throttle linkage moves smoothly.
2. Confirm that throttle valve both fully opens and fully closes.
N.G.
Repair throttle linkage or sticking of throttle valve.
O.K.
INSPECTION END

ECCS SELF-DIAGNOSTIC CHART—1990 300ZX

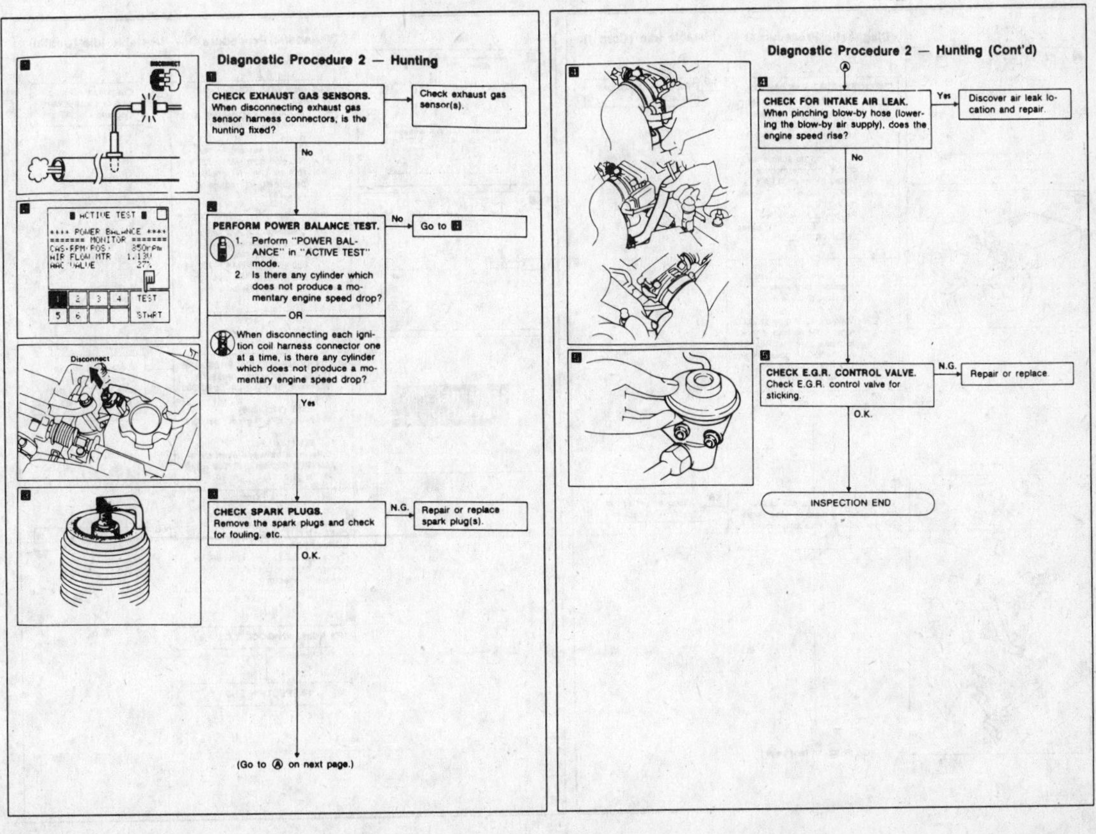
Diagnostic Procedure 2 — Hunting
CHECK EXHAUST GAS SENSORS.
When disconnecting exhaust gas sensor harness connectors, is the hunting fixed?
Check exhaust gas sensor(s).
No
PERFORM POWER BALANCE TEST.
1. Perform "POWER BALANCE" in "ACTIVE TEST" mode.
2. Is there any cylinder which does not produce a momentary engine speed drop?
OR
When disconnecting each ignition coil harness connector one at a time, is there any cylinder which does not produce a momentary engine speed drop?
No
Yes
Disconnect
CHECK SPARK PLUGS.
Remove the spark plugs and check for fouling, etc.
N.G.
Repair or replace spark plug(s).
O.K.
(Go to Ⓐ on next page.)
Diagnostic Procedure 2 — Hunting (Cont'd)
Ⓐ
CHECK FOR INTAKE AIR LEAK.
When pinching blow-by hose (lowering the blow-by air supply), does the engine speed rise?
Yes
Discover air leak location and repair.
No
CHECK E.G.R. CONTROL VALVE.
Check E.G.R. control valve for sticking.
N.G.
Repair or replace.
O.K.
INSPECTION END

ECCS SELF-DIAGNOSTIC CHART—1990 300ZX

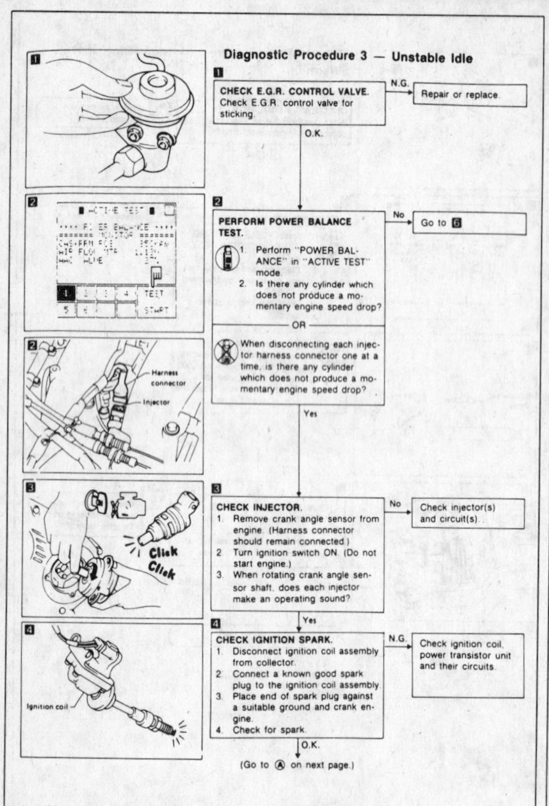
Diagnostic Procedure 3 — Unstable Idle
1
CHECK E.G.R. CONTROL VALVE.
Check E.G.R. control valve for sticking.
N.G.
Repair or replace.
O.K.
2
PERFORM POWER BALANCE TEST.
1. Perform "POWER BALANCE" in "ACTIVE TEST" mode.
2. Is there any cylinder which does not produce a momentary engine speed drop?
OR
When disconnecting each injector harness connector one at a time, is there any cylinder which does not produce a momentary engine speed drop?
No
Go to 6
Yes
Harness connector
Injector
3
CHECK INJECTOR.
1. Remove crank angle sensor from engine. (Harness connector should remain connected.)
2. Turn ignition switch ON. (Do not start engine.)
3. When rotating crank angle sensor shaft, does each injector make an operating sound?
No
Check injector(s) and circuit(s).
Click
Click
Yes
4
CHECK IGNITION SPARK.
1. Disconnect ignition coil assembly from collector.
2. Connect a known good spark plug to the ignition coil assembly.
3. Place end of spark plug against a suitable ground and crank engine.
4. Check for spark.
N.G.
Check ignition coil, power transistor unit and their circuits.
Ignition coil
O.K.
(Go to A on next page.)

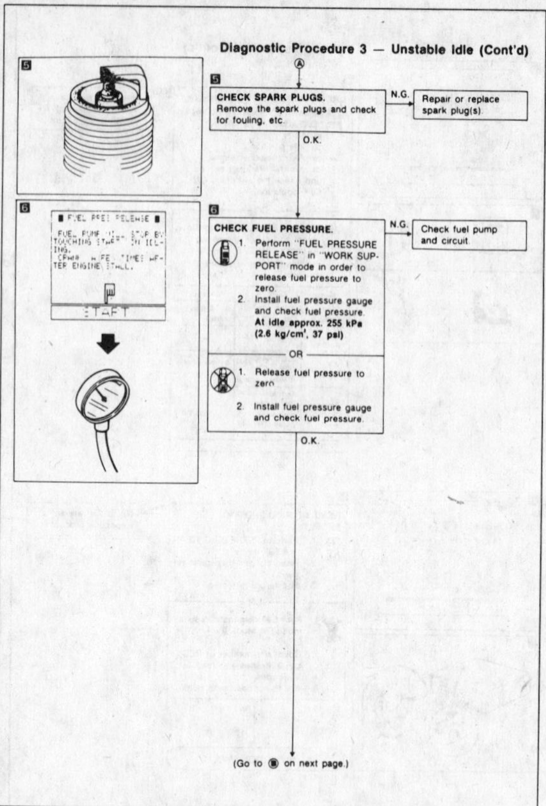
Diagnostic Procedure 3 — Unstable Idle (Cont'd)
A
5
CHECK SPARK PLUGS.
Remove the spark plugs and check for fouling, etc.
N.G.
Repair or replace spark plug(s).
O.K.
6
CHECK FUEL PRESSURE.
1. Perform "FUEL PRESSURE RELEASE" in "WORK SUPPORT" mode in order to release fuel pressure to zero.
2. Install fuel pressure gauge and check fuel pressure.
At idle approx. 255 kPa (2.6 kg/cm², 37 psi)
OR
1. Release fuel pressure to zero.
2. Install fuel pressure gauge and check fuel pressure.
N.G.
Check fuel pump and circuit.
START
O.K.
(Go to B on next page.)

ECCS SELF-DIAGNOSTIC CHART—1990 300ZX

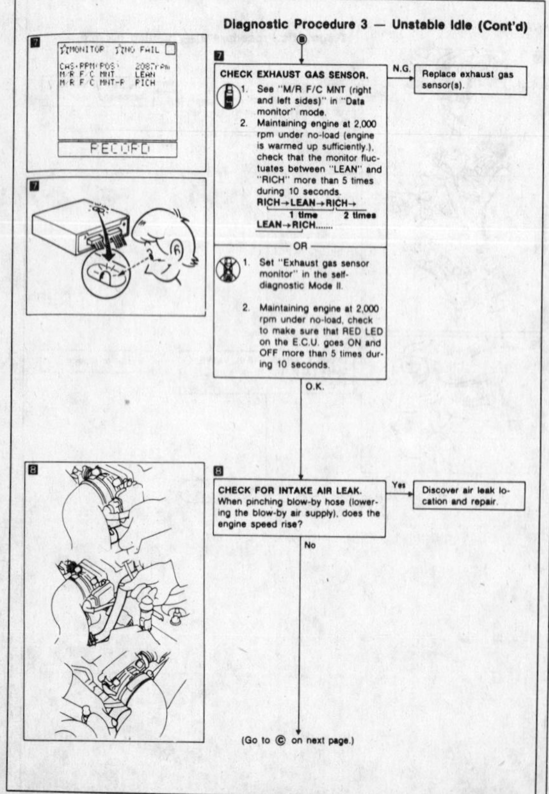
Diagnostic Procedure 3 — Unstable Idle (Cont'd)
B
7
CHECK EXHAUST GAS SENSOR.
1. See "M/R F/C MNT (right and left sides)" in "Data monitor" mode.
2. Maintaining engine at 2,000 rpm under no-load (engine is warmed up sufficiently.), check that the monitor fluctuates between "LEAN" and "RICH" more than 5 times during 10 seconds.
RICH→LEAN→RICH→
1 time 2 times
LEAN→RICH.......
OR
1. Set "Exhaust gas sensor monitor" in the self-diagnostic Mode II.
2. Maintaining engine at 2,000 rpm under no-load, check to make sure that RED LED on the E.C.U. goes ON and OFF more than 5 times during 10 seconds.
N.G.
Replace exhaust gas sensor(s).
RECORD
O.K.
8
CHECK FOR INTAKE AIR LEAK.
When pinching blow-by hose (lowering the blow-by air supply), does the engine speed rise?
Yes
Discover air leak location and repair.
No
(Go to C on next page.)

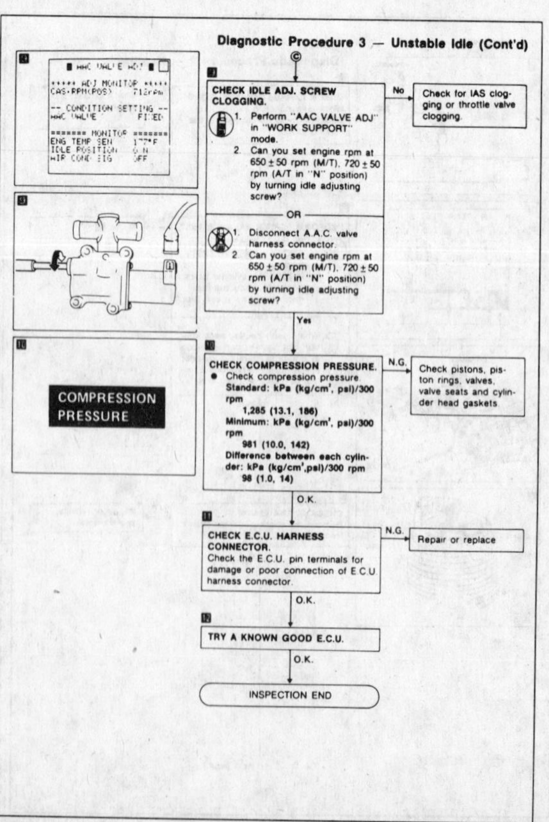
Diagnostic Procedure 3 — Unstable Idle (Cont'd)
C
9
CHECK IDLE ADJ. SCREW CLOGGING.
1. Perform "AAC VALVE ADJ" in "WORK SUPPORT" mode.
2. Can you set engine rpm at 650 ± 50 rpm (M/T), 720 ± 50 rpm (A/T in "N" position) by turning idle adjusting screw?
OR
1. Disconnect A.A.C. valve harness connector.
2. Can you set engine rpm at 650 ± 50 rpm (M/T), 720 ± 50 rpm (A/T in "N" position) by turning idle adjusting screw?
No
Check for IAS clogging or throttle valve clogging.
Yes
10
CHECK COMPRESSION PRESSURE.
• Check compression pressure.
Standard: kPa (kg/cm², psi)/300 rpm
1,285 (13.1, 186)
Minimum: kPa (kg/cm², psi)/300 rpm
981 (10.0, 142)
Difference between each cylinder: kPa (kg/cm², psi)/300 rpm
98 (1.0, 14)
N.G.
Check pistons, piston rings, valves, valve seats and cylinder head gaskets.
COMPRESSION PRESSURE
O.K.
11
CHECK E.C.U. HARNESS CONNECTOR.
Check the E.C.U. pin terminals for damage or poor connection of E.C.U harness connector.
N.G.
Repair or replace
O.K.
12
TRY A KNOWN GOOD E.C.U.
O.K.
INSPECTION END

ECCS SELF-DIAGNOSTIC CHART—1990 300ZX

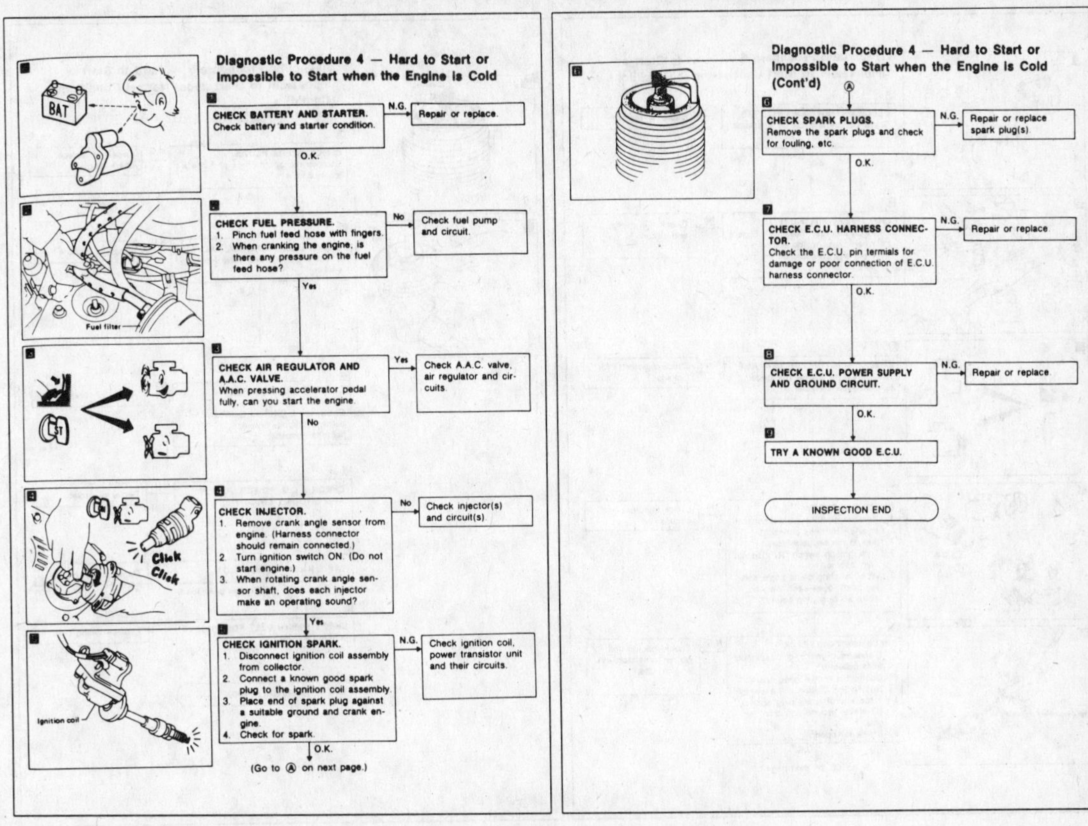
Diagnostic Procedure 4 — Hard to Start or Impossible to Start when the Engine is Cold
BAT
1
CHECK BATTERY AND STARTER.
Check battery and starter condition.
N.G.
Repair or replace.
O.K.
2
CHECK FUEL PRESSURE.
1. Pinch fuel feed hose with fingers.
2. When cranking the engine, is there any pressure on the fuel feed hose?
No
Check fuel pump and circuit.
Yes
Fuel filter
3
CHECK AIR REGULATOR AND A.A.C. VALVE.
When pressing accelerator pedal fully, can you start the engine.
Yes
Check A.A.C. valve, air regulator and circuits.
No
4
CHECK INJECTOR.
1. Remove crank angle sensor from engine. (Harness connector should remain connected.)
2. Turn ignition switch ON. (Do not start engine.)
3. When rotating crank angle sensor shaft, does each injector make an operating sound?
No
Check injector(s) and circuit(s).
Click
Click
Yes
5
CHECK IGNITION SPARK.
1. Disconnect ignition coil assembly from collector.
2. Connect a known good spark plug to the ignition coil assembly.
3. Place end of spark plug against a suitable ground and crank engine.
4. Check for spark.
N.G.
Check ignition coil, power transistor unit and their circuits.
Ignition coil
O.K.
(Go to Ⓐ on next page.)
Diagnostic Procedure 4 — Hard to Start or Impossible to Start when the Engine is Cold (Cont'd)
Ⓐ
6
CHECK SPARK PLUGS.
Remove the spark plugs and check for fouling, etc.
N.G.
Repair or replace spark plug(s)
O.K.
7
CHECK E.C.U. HARNESS CONNECTOR.
Check the E.C.U. pin termials for damage or poor connection of E.C.U. harness connector.
N.G.
Repair or replace.
O.K.
8
CHECK E.C.U. POWER SUPPLY AND GROUND CIRCUIT.
N.G.
Repair or replace.
O.K.
9
TRY A KNOWN GOOD E.C.U.
INSPECTION END

ECCS SELF-DIAGNOSTIC CHART—1990 300ZX

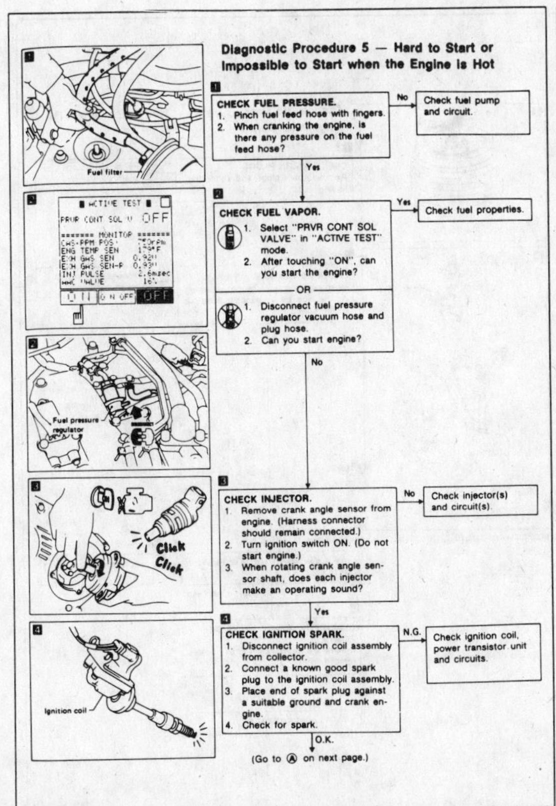
Diagnostic Procedure 5 — Hard to Start or Impossible to Start when the Engine is Hot
1
CHECK FUEL PRESSURE.
1. Pinch fuel feed hose with fingers.
2. When cranking the engine, is there any pressure on the fuel feed hose?
No
Check fuel pump and circuit.
Fuel filter
Yes
2
CHECK FUEL VAPOR.
1. Select "PRVR CONT SOL VALVE" in "ACTIVE TEST" mode.
2. After touching "ON", can you start the engine?
—OR—
1. Disconnect fuel pressure regulator vacuum hose and plug hose.
2. Can you start engine?
Yes
Check fuel properties.
ACTIVE TEST
PRVR CONT SOL V OFF
MONITOR
OFF
No
Fuel pressure regulator
3
CHECK INJECTOR.
1. Remove crank angle sensor from engine. (Harness connector should remain connected.)
2. Turn ignition switch ON. (Do not start engine.)
3. When rotating crank angle sensor shaft, does each injector make an operating sound?
No
Check injector(s) and circuit(s).
Click
Click
Yes
4
CHECK IGNITION SPARK.
1. Disconnect ignition coil assembly from collector.
2. Connect a known good spark plug to the ignition coil assembly.
3. Place end of spark plug against a suitable ground and crank engine.
4. Check for spark.
N.G.
Check ignition coil, power transistor unit and circuits.
Ignition coil
O.K.
(Go to Ⓐ on next page.)

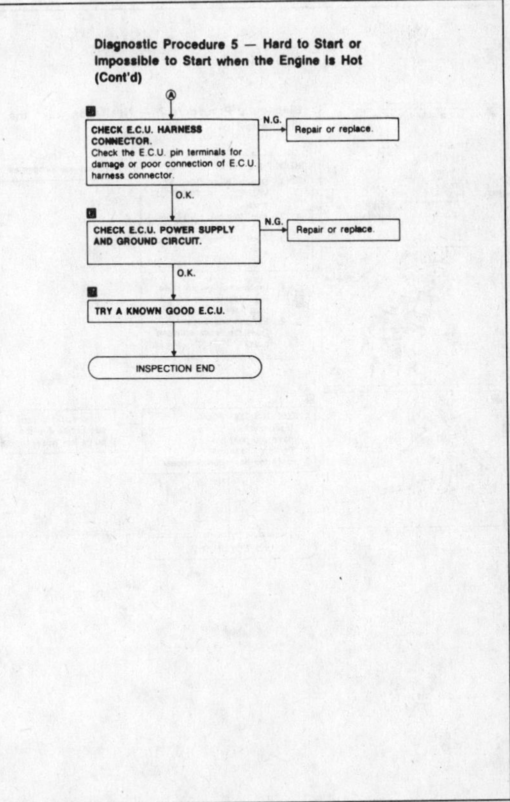
Diagnostic Procedure 5 — Hard to Start or Impossible to Start when the Engine is Hot (Cont'd)
Ⓐ
5
CHECK E.C.U. HARNESS CONNECTOR.
Check the E.C.U. pin terminals for damage or poor connection of E.C.U. harness connector.
N.G.
Repair or replace.
O.K.
6
CHECK E.C.U. POWER SUPPLY AND GROUND CIRCUIT.
N.G.
Repair or replace.
O.K.
7
TRY A KNOWN GOOD E.C.U.
INSPECTION END

ECCS SELF-DIAGNOSTIC CHART—1990 300ZX

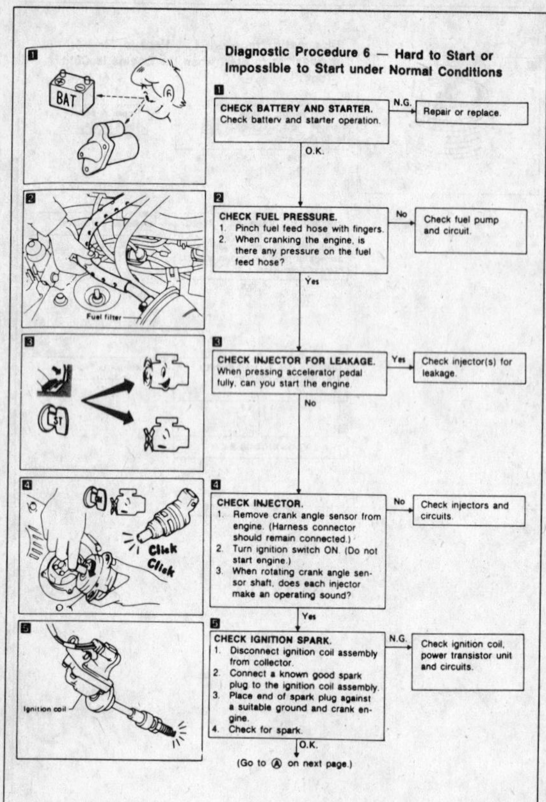
Diagnostic Procedure 6 — Hard to Start or Impossible to Start under Normal Conditions
1
BAT
CHECK BATTERY AND STARTER.
Check battery and starter operation.
N.G.
Repair or replace.
O.K.
2
CHECK FUEL PRESSURE.
1. Pinch fuel feed hose with fingers.
2. When cranking the engine, is there any pressure on the fuel feed hose?
No
Check fuel pump and circuit.
Yes
Fuel filter
3
CHECK INJECTOR FOR LEAKAGE.
When pressing accelerator pedal fully, can you start the engine.
Yes
Check injector(s) for leakage.
No
4
CHECK INJECTOR.
1. Remove crank angle sensor from engine. (Harness connector should remain connected.)
2. Turn ignition switch ON. (Do not start engine.)
3. When rotating crank angle sensor shaft, does each injector make an operating sound?
No
Check injectors and circuits.
Click
Click
Yes
5
CHECK IGNITION SPARK.
1. Disconnect ignition coil assembly from collector.
2. Connect a known good spark plug to the ignition coil assembly.
3. Place end of spark plug against a suitable ground and crank engine.
4. Check for spark.
N.G.
Check ignition coil, power transistor unit and circuits.
Ignition coil
O.K.
(Go to Ⓐ on next page.)

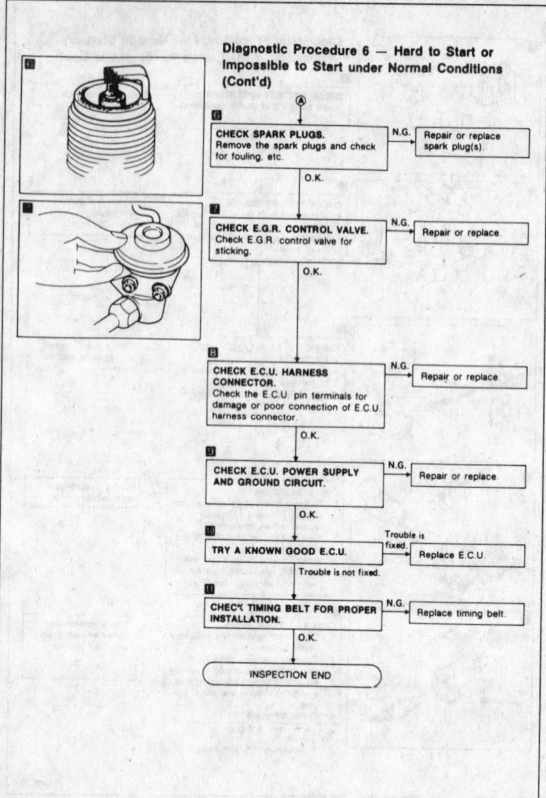
Diagnostic Procedure 6 — Hard to Start or Impossible to Start under Normal Conditions (Cont'd)
Ⓐ
6
CHECK SPARK PLUGS.
Remove the spark plugs and check for fouling, etc.
N.G.
Repair or replace spark plug(s).
O.K.
7
CHECK E.G.R. CONTROL VALVE.
Check E.G.R. control valve for sticking.
N.G.
Repair or replace.
O.K.
8
CHECK E.C.U. HARNESS CONNECTOR.
Check the E.C.U. pin terminals for damage or poor connection of E.C.U. harness connector.
N.G.
Repair or replace.
O.K.
9
CHECK E.C.U. POWER SUPPLY AND GROUND CIRCUIT.
N.G.
Repair or replace.
O.K.
10
TRY A KNOWN GOOD E.C.U.
Trouble is fixed.
Replace E.C.U.
Trouble is not fixed.
11
CHECK TIMING BELT FOR PROPER INSTALLATION.
N.G.
Replace timing belt.
O.K.
INSPECTION END

ECCS SELF-DIAGNOSTIC CHART—1990 300ZX

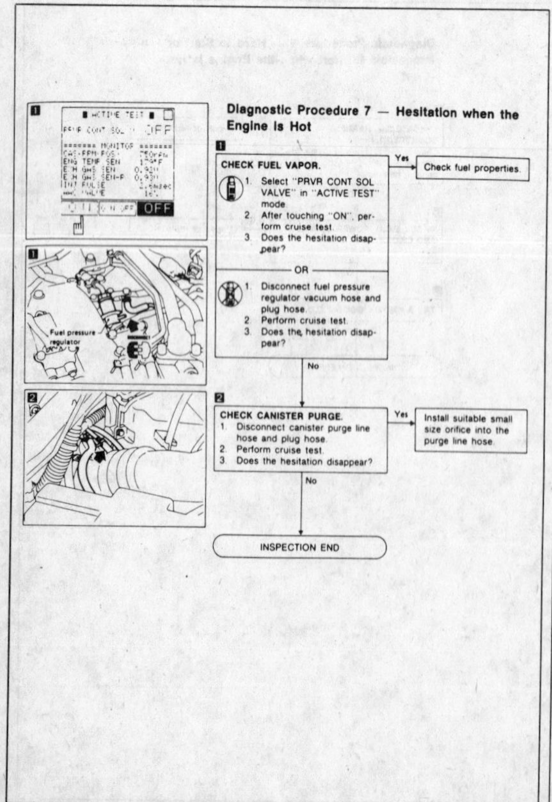
Diagnostic Procedure 7 — Hesitation when the Engine is Hot
1
ACTIVE TEST
OFF
OFF
CHECK FUEL VAPOR.
1. Select "PRVR CONT SOL VALVE" in "ACTIVE TEST" mode
2. After touching "ON", perform cruise test
3. Does the hesitation disappear?
Yes
Check fuel properties.
OR
1. Disconnect fuel pressure regulator vacuum hose and plug hose.
2. Perform cruise test.
3. Does the hesitation disappear?
Fuel pressure regulator
No
2
CHECK CANISTER PURGE.
1. Disconnect canister purge line hose and plug hose.
2. Perform cruise test.
3. Does the hesitation disappear?
Yes
Install suitable small size orifice into the purge line hose.
No
INSPECTION END

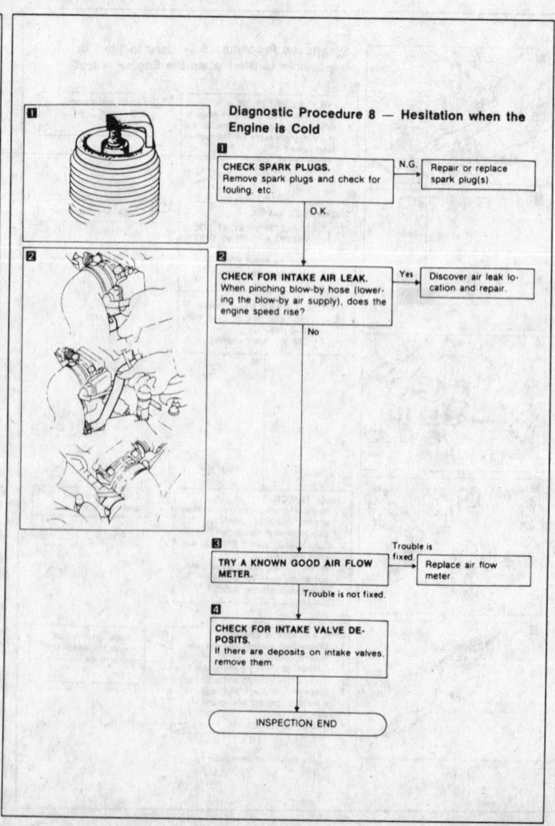
Diagnostic Procedure 8 — Hesitation when the Engine is Cold
1
CHECK SPARK PLUGS.
Remove spark plugs and check for fouling, etc.
N.G.
Repair or replace spark plug(s).
O.K.
2
CHECK FOR INTAKE AIR LEAK.
When pinching blow-by hose (lowering the blow-by air supply), does the engine speed rise?
Yes
Discover air leak location and repair.
No
3
TRY A KNOWN GOOD AIR FLOW METER.
Trouble is fixed.
Replace air flow meter.
Trouble is not fixed.
4
CHECK FOR INTAKE VALVE DEPOSITS.
If there are deposits on intake valves, remove them.
INSPECTION END

ECCS SELF-DIAGNOSTIC CHART—1990 300ZX

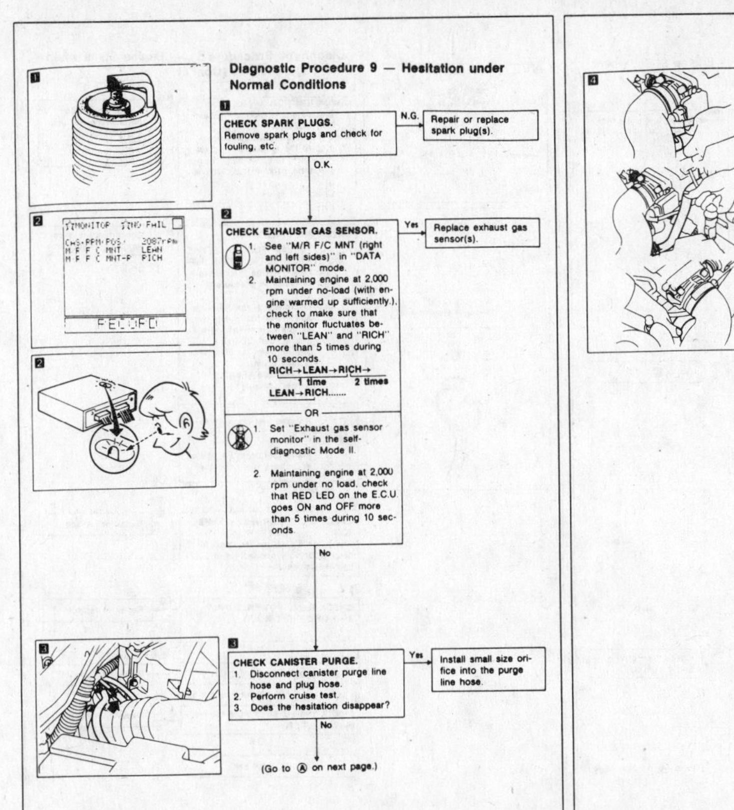
Diagnostic Procedure 9 — Hesitation under Normal Conditions
1
CHECK SPARK PLUGS.
Remove spark plugs and check for fouling, etc.
N.G.
Repair or replace spark plug(s).
O.K.
2
CHECK EXHAUST GAS SENSOR.
1. See "M/R F/C MNT (right and left sides)" in "DATA MONITOR" mode.
2. Maintaining engine at 2,000 rpm under no-load (with engine warmed up sufficiently.), check to make sure that the monitor fluctuates between "LEAN" and "RICH" more than 5 times during 10 seconds.
RICH→LEAN→RICH→
1 time
2 times
LEAN→RICH......
— OR —
1. Set "Exhaust gas sensor monitor" in the self-diagnostic Mode II.
2. Maintaining engine at 2,000 rpm under no load, check that RED LED on the E.C.U. goes ON and OFF more than 5 times during 10 seconds.
Yes
Replace exhaust gas sensor(s).
No
3
CHECK CANISTER PURGE.
1. Disconnect canister purge line hose and plug hose.
2. Perform cruise test.
3. Does the hesitation disappear?
Yes
Install small size orifice into the purge line hose.
No
(Go to Ⓐ on next page.)
MONITOR
RECORD

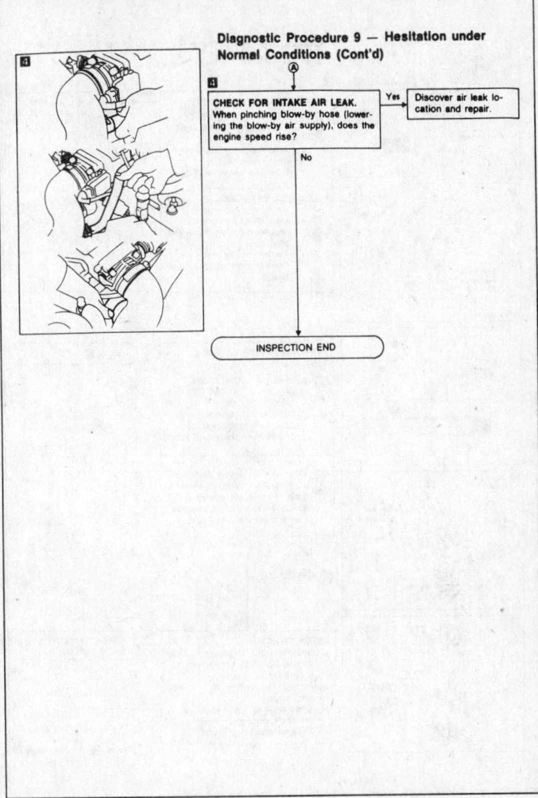
Diagnostic Procedure 9 — Hesitation under Normal Conditions (Cont'd)
Ⓐ
4
CHECK FOR INTAKE AIR LEAK.
When pinching blow-by hose (lowering the blow-by air supply), does the engine speed rise?
Yes
Discover air leak location and repair.
No
INSPECTION END

ECCS SELF-DIAGNOSTIC CHART—1990 300ZX

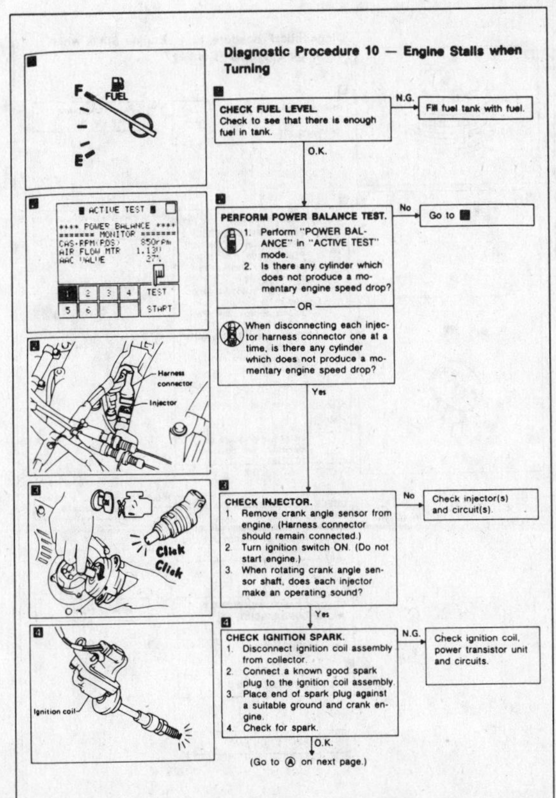
Diagnostic Procedure 10 — Engine Stalls when Turning
1
CHECK FUEL LEVEL.
Check to see that there is enough fuel in tank.
N.G.
Fill fuel tank with fuel.
O.K.
2
PERFORM POWER BALANCE TEST.
1. Perform "POWER BALANCE" in "ACTIVE TEST" mode.
2. Is there any cylinder which does not produce a momentary engine speed drop?
— OR —
When disconnecting each injector harness connector one at a time, is there any cylinder which does not produce a momentary engine speed drop?
No
Go to 5
Yes
3
CHECK INJECTOR.
1. Remove crank angle sensor from engine. (Harness connector should remain connected.)
2. Turn ignition switch ON. (Do not start engine.)
3. When rotating crank angle sensor shaft, does each injector make an operating sound?
No
Check injector(s) and circuit(s).
Yes
4
CHECK IGNITION SPARK.
1. Disconnect ignition coil assembly from collector.
2. Connect a known good spark plug to the ignition coil assembly.
3. Place end of spark plug against a suitable ground and crank engine.
4. Check for spark.
N.G.
Check ignition coil, power transistor unit and circuits.
O.K.
(Go to Ⓐ on next page.)
F
FUEL
E
ACTIVE TEST
POWER BALANCE
MONITOR
TEST
START
Harness connector
Injector
Click
Click
Ignition coil

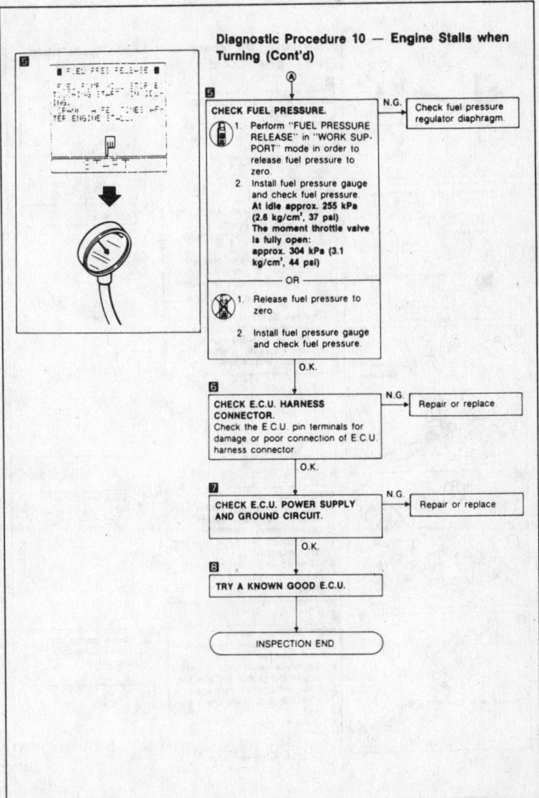
Diagnostic Procedure 10 — Engine Stalls when Turning (Cont'd)
Ⓐ
5
CHECK FUEL PRESSURE.
1. Perform "FUEL PRESSURE RELEASE" in "WORK SUPPORT" mode in order to release fuel pressure to zero.
2. Install fuel pressure gauge and check fuel pressure.
At idle approx. 255 kPa (2.6 kg/cm², 37 psi)
The moment throttle valve is fully open: approx. 304 kPa (3.1 kg/cm², 44 psi)
— OR —
1. Release fuel pressure to zero.
2. Install fuel pressure gauge and check fuel pressure.
N.G.
Check fuel pressure regulator diaphragm.
O.K.
6
CHECK E.C.U. HARNESS CONNECTOR.
Check the E.C.U. pin terminals for damage or poor connection of E.C.U. harness connector.
N.G.
Repair or replace.
O.K.
7
CHECK E.C.U. POWER SUPPLY AND GROUND CIRCUIT.
N.G.
Repair or replace
O.K.
8
TRY A KNOWN GOOD E.C.U.
INSPECTION END

ECCS SELF-DIAGNOSTIC CHART—1990 300ZX

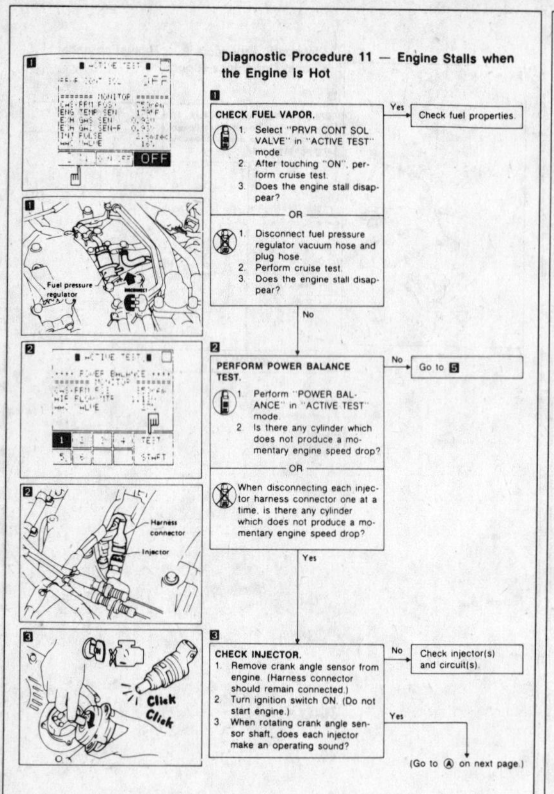
Diagnostic Procedure 11 — Engine Stalls when the Engine is Hot
1
CHECK FUEL VAPOR.
1. Select "PRVR CONT SOL VALVE" in "ACTIVE TEST" mode.
2. After touching "ON", perform cruise test.
3. Does the engine stall disappear?
OR
1. Disconnect fuel pressure regulator vacuum hose and plug hose.
2. Perform cruise test.
3. Does the engine stall disappear?
Yes
Check fuel properties.
No
Fuel pressure regulator
2
PERFORM POWER BALANCE TEST.
1. Perform "POWER BALANCE" in "ACTIVE TEST" mode.
2. Is there any cylinder which does not produce a momentary engine speed drop?
OR
When disconnecting each injector harness connector one at a time, is there any cylinder which does not produce a momentary engine speed drop?
No
Go to 5
Yes
Harness connector
Injector
3
CHECK INJECTOR.
1. Remove crank angle sensor from engine. (Harness connector should remain connected.)
2. Turn ignition switch ON. (Do not start engine.)
3. When rotating crank angle sensor shaft, does each injector make an operating sound?
No
Check injector(s) and circuit(s).
Yes
(Go to Ⓐ on next page.)
Click
Click

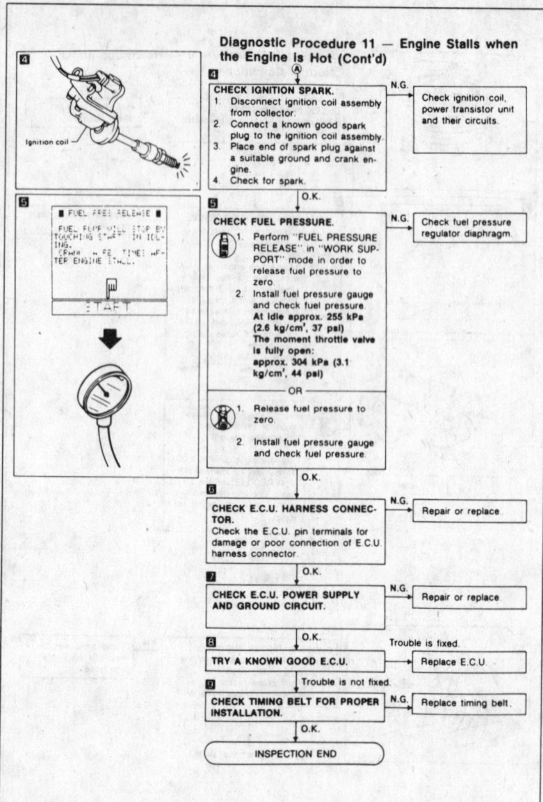
Diagnostic Procedure 11 — Engine Stalls when the Engine is Hot (Cont'd)
Ⓐ
4
CHECK IGNITION SPARK.
1. Disconnect ignition coil assembly from collector.
2. Connect a known good spark plug to the ignition coil assembly.
3. Place end of spark plug against a suitable ground and crank engine.
4. Check for spark.
N.G.
Check ignition coil, power transistor unit and their circuits.
O.K.
Ignition coil
5
CHECK FUEL PRESSURE.
1. Perform "FUEL PRESSURE RELEASE" in "WORK SUPPORT" mode in order to release fuel pressure to zero.
2. Install fuel pressure gauge and check fuel pressure.
At idle approx. 255 kPa (2.6 kg/cm², 37 psi)
The moment throttle valve is fully open:
approx. 304 kPa (3.1 kg/cm², 44 psi)
OR
1. Release fuel pressure to zero.
2. Install fuel pressure gauge and check fuel pressure.
N.G.
Check fuel pressure regulator diaphragm.
O.K.
START
6
CHECK E.C.U. HARNESS CONNECTOR.
Check the E.C.U. pin terminals for damage or poor connection of E.C.U. harness connector.
N.G.
Repair or replace.
O.K.
7
CHECK E.C.U. POWER SUPPLY AND GROUND CIRCUIT.
N.G.
Repair or replace.
O.K.
8
TRY A KNOWN GOOD E.C.U.
Trouble is fixed.
Replace E.C.U.
Trouble is not fixed.
9
CHECK TIMING BELT FOR PROPER INSTALLATION.
N.G.
Replace timing belt.
O.K.
INSPECTION END

ECCS SELF-DIAGNOSTIC CHART—1990 300ZX

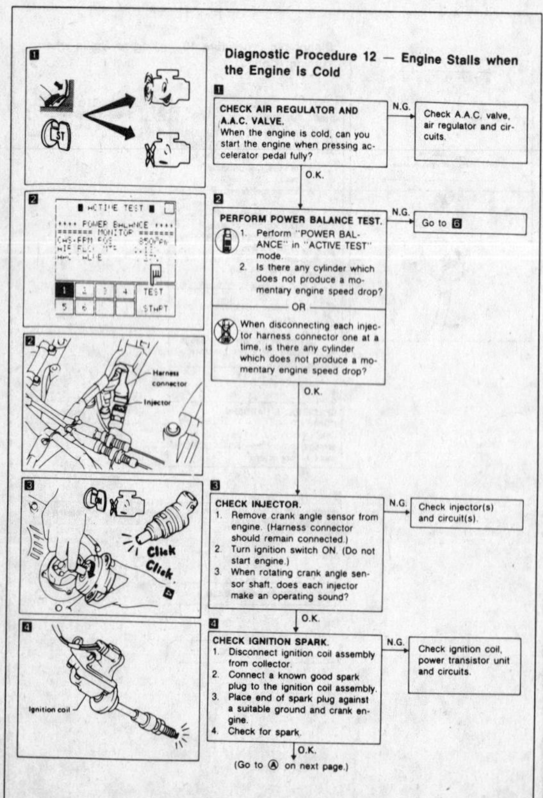
Diagnostic Procedure 12 — Engine Stalls when the Engine is Cold
1
CHECK AIR REGULATOR AND A.A.C. VALVE.
When the engine is cold, can you start the engine when pressing accelerator pedal fully?
N.G.
Check A.A.C. valve, air regulator and circuits.
O.K.
2
PERFORM POWER BALANCE TEST.
1. Perform "POWER BALANCE" in "ACTIVE TEST" mode.
2. Is there any cylinder which does not produce a momentary engine speed drop?
OR
When disconnecting each injector harness connector one at a time, is there any cylinder which does not produce a momentary engine speed drop?
N.G.
Go to 6
O.K.
Harness connector
Injector
3
CHECK INJECTOR.
1. Remove crank angle sensor from engine. (Harness connector should remain connected.)
2. Turn ignition switch ON. (Do not start engine.)
3. When rotating crank angle sensor shaft, does each injector make an operating sound?
N.G.
Check injector(s) and circuit(s).
O.K.
Click
Click
4
CHECK IGNITION SPARK.
1. Disconnect ignition coil assembly from collector.
2. Connect a known good spark plug to the ignition coil assembly.
3. Place end of spark plug against a suitable ground and crank engine.
4. Check for spark.
N.G.
Check ignition coil, power transistor unit and circuits.
O.K.
(Go to Ⓐ on next page.)
Ignition coil

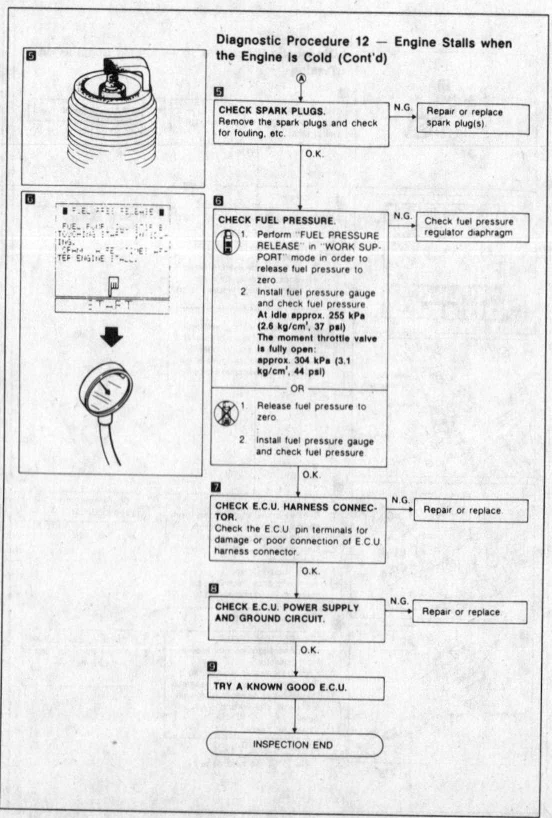
Diagnostic Procedure 12 — Engine Stalls when the Engine is Cold (Cont'd)
Ⓐ
5
CHECK SPARK PLUGS.
Remove the spark plugs and check for fouling, etc.
N.G.
Repair or replace spark plug(s).
O.K.
6
CHECK FUEL PRESSURE.
1. Perform "FUEL PRESSURE RELEASE" in "WORK SUPPORT" mode in order to release fuel pressure to zero.
2. Install fuel pressure gauge and check fuel pressure.
At idle approx. 255 kPa (2.6 kg/cm², 37 psi)
The moment throttle valve is fully open:
approx. 304 kPa (3.1 kg/cm², 44 psi)
OR
1. Release fuel pressure to zero.
2. Install fuel pressure gauge and check fuel pressure.
N.G.
Check fuel pressure regulator diaphragm
O.K.
START
7
CHECK E.C.U. HARNESS CONNECTOR.
Check the E.C.U. pin terminals for damage or poor connection of E.C.U. harness connector.
N.G.
Repair or replace.
O.K.
8
CHECK E.C.U. POWER SUPPLY AND GROUND CIRCUIT.
N.G.
Repair or replace.
O.K.
9
TRY A KNOWN GOOD E.C.U.
INSPECTION END

ECCS SELF-DIAGNOSTIC CHART—1990 300ZX

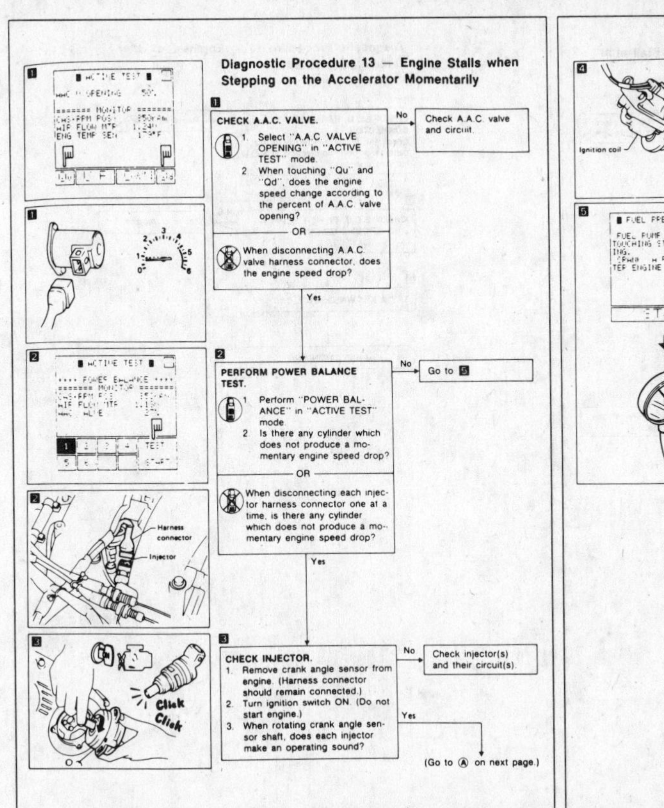
Diagnostic Procedure 13 — Engine Stalls when Stepping on the Accelerator Momentarily
CHECK A.A.C. VALVE.
1. Select "A.A.C. VALVE OPENING" in "ACTIVE TEST" mode.
2. When touching "Qu" and "Qd", does the engine speed change according to the percent of A.A.C. valve opening?
OR
When disconnecting A.A.C. valve harness connector, does the engine speed drop?
No
Check A.A.C. valve and circuit.
Yes
PERFORM POWER BALANCE TEST.
1. Perform "POWER BALANCE" in "ACTIVE TEST" mode.
2. Is there any cylinder which does not produce a momentary engine speed drop?
OR
When disconnecting each injector harness connector one at a time, is there any cylinder which does not produce a momentary engine speed drop?
No
Go to 5
Yes
Harness connector
Injector
CHECK INJECTOR.
1. Remove crank angle sensor from engine. (Harness connector should remain connected.)
2. Turn ignition switch ON. (Do not start engine.)
3. When rotating crank angle sensor shaft, does each injector make an operating sound?
No
Check injector(s) and their circuit(s).
Yes
(Go to Ⓐ on next page.)
Click
Click

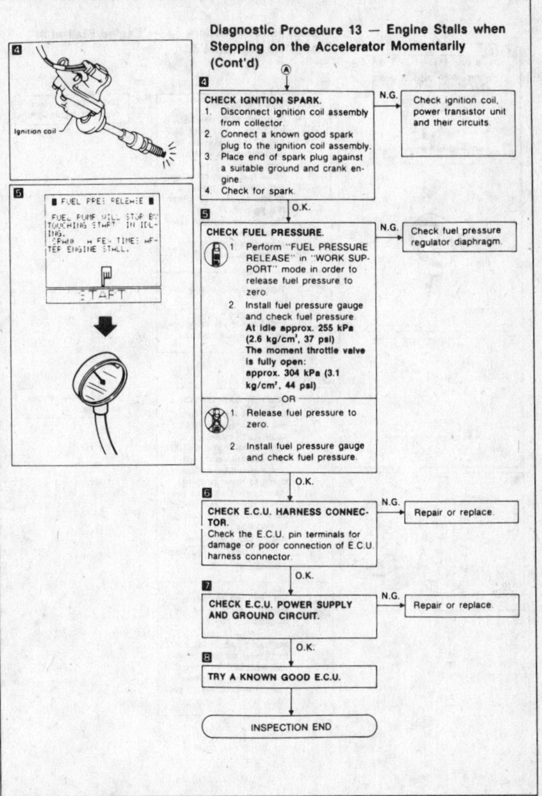
Diagnostic Procedure 13 — Engine Stalls when Stepping on the Accelerator Momentarily (Cont'd)
Ⓐ
CHECK IGNITION SPARK.
1. Disconnect ignition coil assembly from collector.
2. Connect a known good spark plug to the ignition coil assembly.
3. Place end of spark plug against a suitable ground and crank engine.
4. Check for spark.
N.G.
Check ignition coil, power transistor unit and their circuits.
Ignition coil
O.K.
CHECK FUEL PRESSURE.
1. Perform "FUEL PRESSURE RELEASE" in "WORK SUPPORT" mode in order to release fuel pressure to zero.
2. Install fuel pressure gauge and check fuel pressure.
At idle approx. 255 kPa (2.6 kg/cm², 37 psi)
The moment throttle valve is fully open: approx. 304 kPa (3.1 kg/cm², 44 psi)
OR
1. Release fuel pressure to zero.
2. Install fuel pressure gauge and check fuel pressure.
N.G.
Check fuel pressure regulator diaphragm.
START
O.K.
CHECK E.C.U. HARNESS CONNECTOR.
Check the E.C.U. pin terminals for damage or poor connection of E.C.U. harness connector.
N.G.
Repair or replace.
O.K.
CHECK E.C.U. POWER SUPPLY AND GROUND CIRCUIT.
N.G.
Repair or replace.
O.K.
TRY A KNOWN GOOD E.C.U.
INSPECTION END

ECCS SELF-DIAGNOSTIC CHART—1990 300ZX

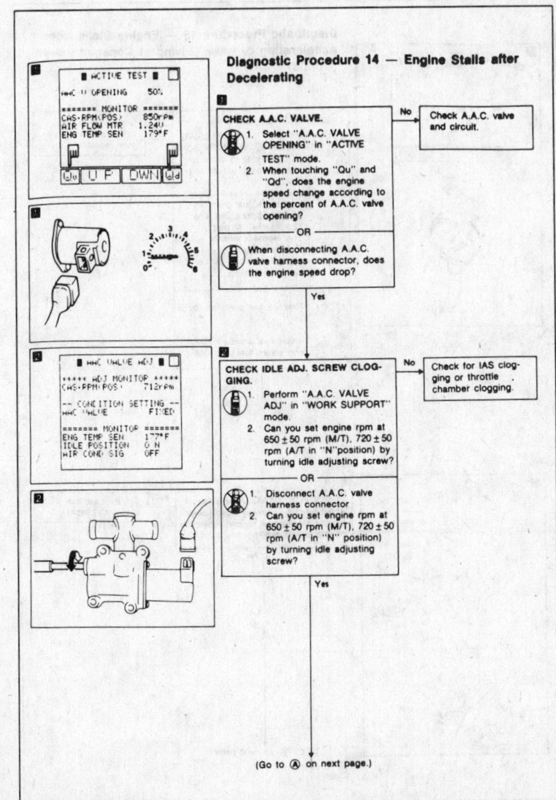
Diagnostic Procedure 14 — Engine Stalls after Decelerating
CHECK A.A.C. VALVE.
1. Select "A.A.C. VALVE OPENING" in "ACTIVE TEST" mode.
2. When touching "Qu" and "Qd", does the engine speed change according to the percent of A.A.C. valve opening?
OR
When disconnecting A.A.C. valve harness connector, does the engine speed drop?
No
Check A.A.C. valve and circuit.
Yes
CHECK IDLE ADJ. SCREW CLOGGING.
1. Perform "A.A.C. VALVE ADJ" in "WORK SUPPORT" mode.
2. Can you set engine rpm at 650±50 rpm (M/T), 720±50 rpm (A/T in "N" position) by turning idle adjusting screw?
OR
1. Disconnect A.A.C. valve harness connector.
2. Can you set engine rpm at 650±50 rpm (M/T), 720±50 rpm (A/T in "N" position) by turning idle adjusting screw?
No
Check for IAS clogging or throttle chamber clogging.
Yes
(Go to Ⓐ on next page.)

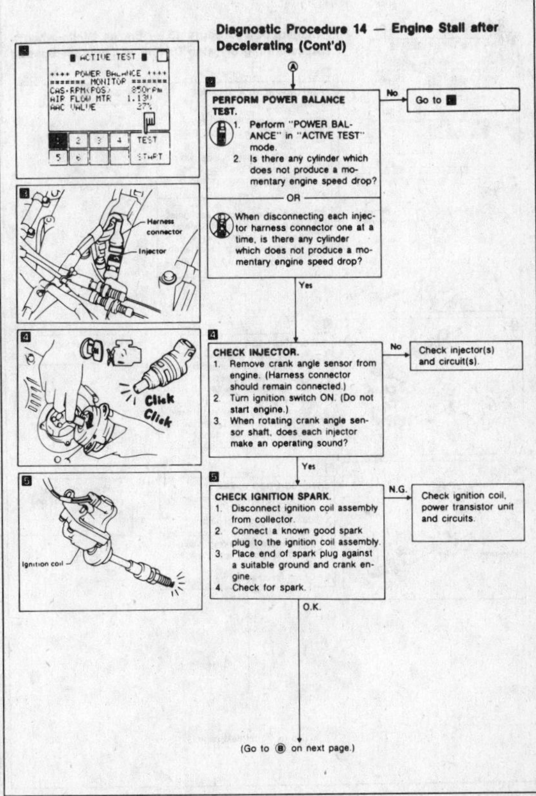
Diagnostic Procedure 14 — Engine Stall after Decelerating (Cont'd)
Ⓐ
PERFORM POWER BALANCE TEST.
1. Perform "POWER BALANCE" in "ACTIVE TEST" mode.
2. Is there any cylinder which does not produce a momentary engine speed drop?
OR
When disconnecting each injector harness connector one at a time, is there any cylinder which does not produce a momentary engine speed drop?
No
Go to 6
Yes
Harness connector
Injector
CHECK INJECTOR.
1. Remove crank angle sensor from engine. (Harness connector should remain connected.)
2. Turn ignition switch ON. (Do not start engine.)
3. When rotating crank angle sensor shaft, does each injector make an operating sound?
No
Check injector(s) and circuit(s).
Click
Click
Yes
CHECK IGNITION SPARK.
1. Disconnect ignition coil assembly from collector.
2. Connect a known good spark plug to the ignition coil assembly.
3. Place end of spark plug against a suitable ground and crank engine.
4. Check for spark.
N.G.
Check ignition coil, power transistor unit and circuits.
Ignition coil
O.K.
(Go to Ⓑ on next page.)

ECCS SELF-DIAGNOSTIC CHART—1990 300ZX

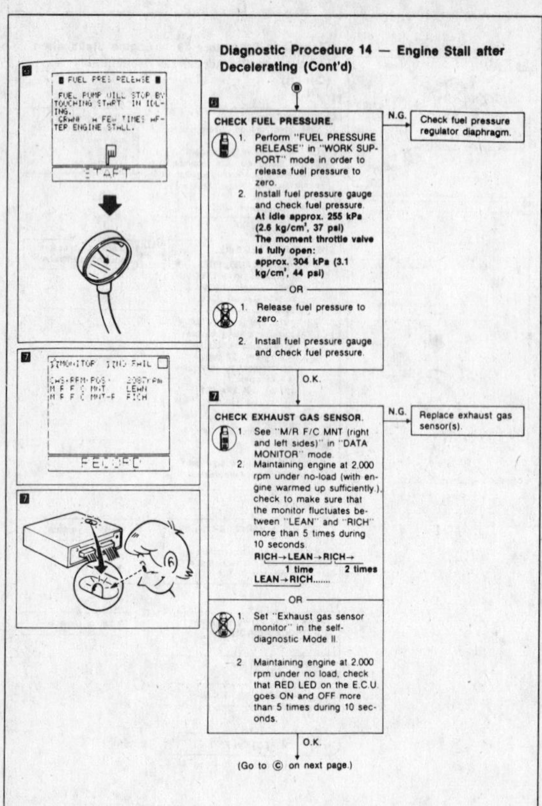
Diagnostic Procedure 14 — Engine Stall after Decelerating (Cont'd)
CHECK FUEL PRESSURE.
1. Perform "FUEL PRESSURE RELEASE" in "WORK SUPPORT" mode in order to release fuel pressure to zero.
2. Install fuel pressure gauge and check fuel pressure. At idle approx. 255 kPa (2.6 kg/cm², 37 psi) The moment throttle valve is fully open: approx. 304 kPa (3.1 kg/cm², 44 psi)
OR
1. Release fuel pressure to zero.
2. Install fuel pressure gauge and check fuel pressure.
N.G.
Check fuel pressure regulator diaphragm.
O.K.
CHECK EXHAUST GAS SENSOR.
1. See "M/R F/C MNT (right and left sides)" in "DATA MONITOR" mode.
2. Maintaining engine at 2,000 rpm under no-load (with engine warmed up sufficiently), check to make sure that the monitor fluctuates between "LEAN" and "RICH" more than 5 times during 10 seconds.
RICH→LEAN→RICH→
1 time
2 times
LEAN→RICH.......
OR
1. Set "Exhaust gas sensor monitor" in the self-diagnostic Mode II.
2. Maintaining engine at 2,000 rpm under no load, check that RED LED on the E.C.U. goes ON and OFF more than 5 times during 10 seconds.
N.G.
Replace exhaust gas sensor(s).
O.K.
(Go to © on next page.)

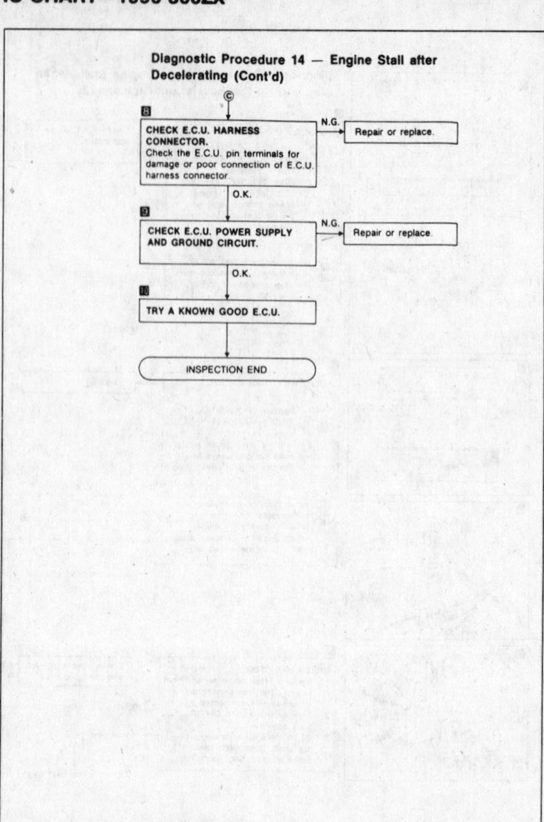
Diagnostic Procedure 14 — Engine Stall after Decelerating (Cont'd)
CHECK E.C.U. HARNESS CONNECTOR.
Check the E.C.U. pin terminals for damage or poor connection of E.C.U. harness connector.
N.G.
Repair or replace.
O.K.
CHECK E.C.U. POWER SUPPLY AND GROUND CIRCUIT.
N.G.
Repair or replace.
O.K.
TRY A KNOWN GOOD E.C.U.
INSPECTION END

ECCS SELF-DIAGNOSTIC CHART—1990 300ZX

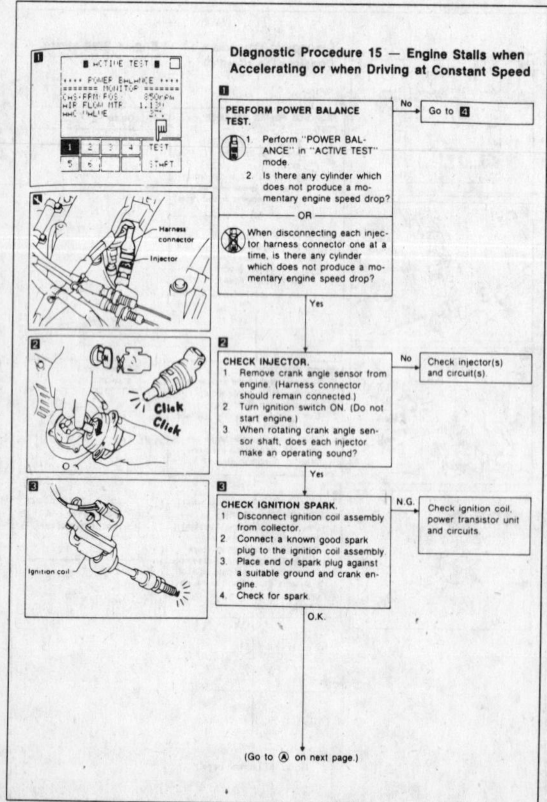
Diagnostic Procedure 15 — Engine Stalls when Accelerating or when Driving at Constant Speed
PERFORM POWER BALANCE TEST.
1. Perform "POWER BALANCE" in "ACTIVE TEST" mode.
2. Is there any cylinder which does not produce a momentary engine speed drop?
OR
When disconnecting each injector harness connector one at a time, is there any cylinder which does not produce a momentary engine speed drop?
No
Go to
Yes
Harness connector
Injector
CHECK INJECTOR.
1. Remove crank angle sensor from engine. (Harness connector should remain connected.)
2. Turn ignition switch ON. (Do not start engine.)
3. When rotating crank angle sensor shaft, does each injector make an operating sound?
No
Check injector(s) and circuit(s).
Click
Click
Yes
CHECK IGNITION SPARK.
1. Disconnect ignition coil assembly from collector.
2. Connect a known good spark plug to the ignition coil assembly.
3. Place end of spark plug against a suitable ground and crank engine.
4. Check for spark.
N.G.
Check ignition coil, power transistor unit and circuits.
Ignition coil
O.K.
(Go to Ⓐ on next page.)

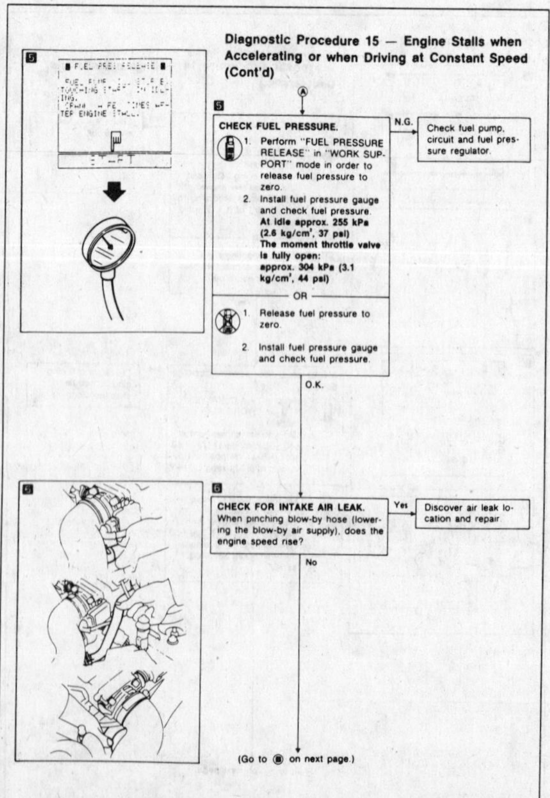
Diagnostic Procedure 15 — Engine Stalls when Accelerating or when Driving at Constant Speed (Cont'd)
CHECK FUEL PRESSURE.
1. Perform "FUEL PRESSURE RELEASE" in "WORK SUPPORT" mode in order to release fuel pressure to zero.
2. Install fuel pressure gauge and check fuel pressure. At idle approx. 255 kPa (2.6 kg/cm², 37 psi) The moment throttle valve is fully open: approx. 304 kPa (3.1 kg/cm², 44 psi)
OR
1. Release fuel pressure to zero.
2. Install fuel pressure gauge and check fuel pressure.
N.G.
Check fuel pump, circuit and fuel pressure regulator.
O.K.
CHECK FOR INTAKE AIR LEAK.
When pinching blow-by hose (lowering the blow-by air supply), does the engine speed rise?
Yes
Discover air leak location and repair.
No
(Go to Ⓑ on next page.)

ECCS SELF-DIAGNOSTIC CHART—1990 300ZX

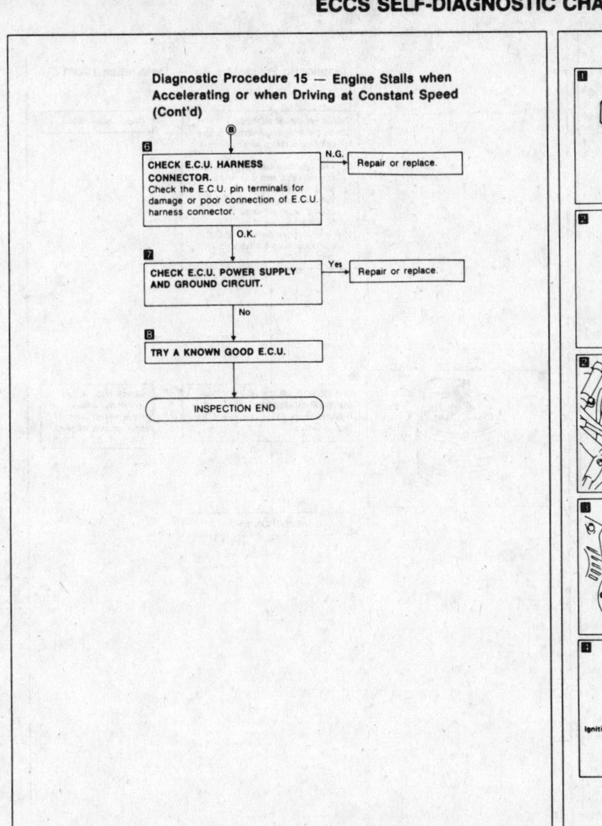
Diagnostic Procedure 15 — Engine Stalls when Accelerating or when Driving at Constant Speed (Cont'd)
B
6
CHECK E.C.U. HARNESS CONNECTOR.
Check the E.C.U. pin terminals for damage or poor connection of E.C.U. harness connector.
N.G.
Repair or replace.
O.K.
7
CHECK E.C.U. POWER SUPPLY AND GROUND CIRCUIT.
Yes
Repair or replace.
No
8
TRY A KNOWN GOOD E.C.U.
INSPECTION END

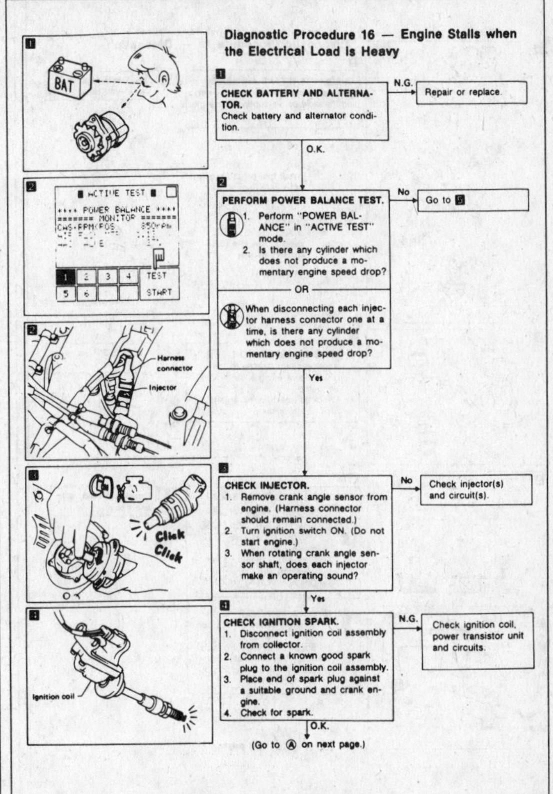
Diagnostic Procedure 16 — Engine Stalls when the Electrical Load is Heavy
BAT
1
CHECK BATTERY AND ALTERNATOR.
Check battery and alternator condition.
N.G.
Repair or replace.
O.K.
ACTIVE TEST
POWER BALANCE
MONITOR
TEST
START
2
PERFORM POWER BALANCE TEST.
1. Perform "POWER BALANCE" in "ACTIVE TEST" mode.
2. Is there any cylinder which does not produce a momentary engine speed drop?
OR
When disconnecting each injector harness connector one at a time, is there any cylinder which does not produce a momentary engine speed drop?
No
Go to 5
Harness connector
Injector
Yes
Click
Click
3
CHECK INJECTOR.
1. Remove crank angle sensor from engine. (Harness connector should remain connected.)
2. Turn ignition switch ON. (Do not start engine.)
3. When rotating crank angle sensor shaft, does each injector make an operating sound?
No
Check injector(s) and circuit(s).
Yes
Ignition coil
4
CHECK IGNITION SPARK.
1. Disconnect ignition coil assembly from collector.
2. Connect a known good spark plug to the ignition coil assembly.
3. Place end of spark plug against a suitable ground and crank engine.
4. Check for spark.
N.G.
Check ignition coil, power transistor unit and circuits.
O.K.
(Go to A on next page.)

ECCS SELF-DIAGNOSTIC CHART—1990 300ZX

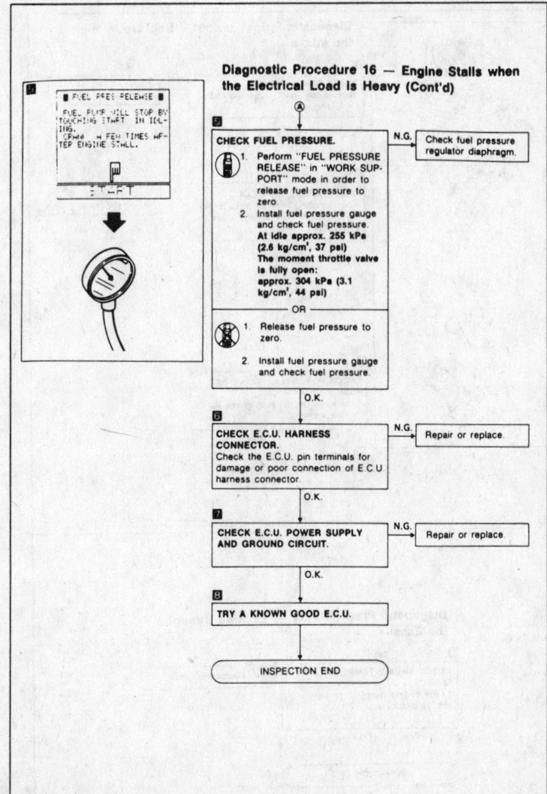
Diagnostic Procedure 16 — Engine Stalls when the Electrical Load is Heavy (Cont'd)
FUEL PRES RELEASE
FUEL PUMP WILL STOP BY TOUCHING START IN IDLING. CRANK A FEW TIMES AFTER ENGINE STALL.
START
A
5
CHECK FUEL PRESSURE.
1. Perform "FUEL PRESSURE RELEASE" in "WORK SUPPORT" mode in order to release fuel pressure to zero.
2. Install fuel pressure gauge and check fuel pressure.
At idle approx. 255 kPa (2.6 kg/cm², 37 psi)
The moment throttle valve is fully open:
approx. 304 kPa (3.1 kg/cm², 44 psi)
OR
1. Release fuel pressure to zero.
2. Install fuel pressure gauge and check fuel pressure.
N.G.
Check fuel pressure regulator diaphragm.
O.K.
6
CHECK E.C.U. HARNESS CONNECTOR.
Check the E.C.U. pin terminals for damage or poor connection of E.C.U. harness connector.
N.G.
Repair or replace.
O.K.
7
CHECK E.C.U. POWER SUPPLY AND GROUND CIRCUIT.
N.G.
Repair or replace.
O.K.
8
TRY A KNOWN GOOD E.C.U.
INSPECTION END

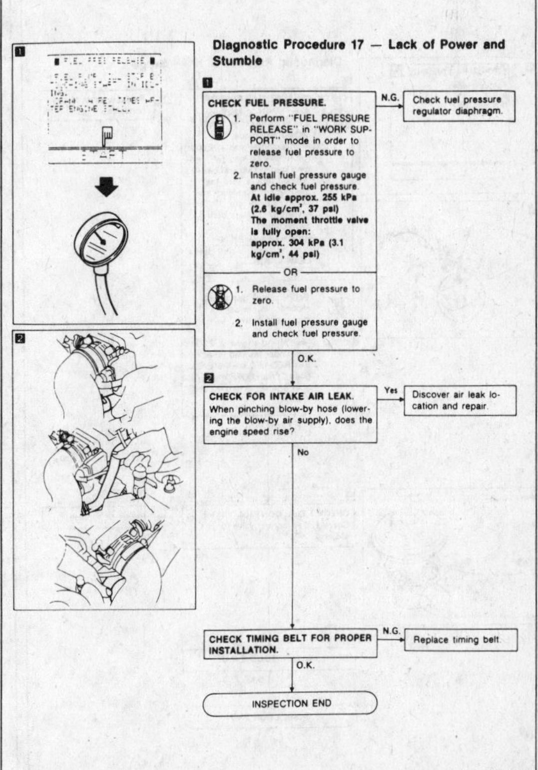
Diagnostic Procedure 17 — Lack of Power and Stumble
FUEL PRES RELEASE
START
1
CHECK FUEL PRESSURE.
1. Perform "FUEL PRESSURE RELEASE" in "WORK SUPPORT" mode in order to release fuel pressure to zero.
2. Install fuel pressure gauge and check fuel pressure.
At idle approx. 255 kPa (2.6 kg/cm², 37 psi)
The moment throttle valve is fully open:
approx. 304 kPa (3.1 kg/cm², 44 psi)
OR
1. Release fuel pressure to zero.
2. Install fuel pressure gauge and check fuel pressure.
N.G.
Check fuel pressure regulator diaphragm.
O.K.
2
CHECK FOR INTAKE AIR LEAK.
When pinching blow-by hose (lowering the blow-by air supply), does the engine speed rise?
Yes
Discover air leak location and repair.
No
CHECK TIMING BELT FOR PROPER INSTALLATION.
N.G.
Replace timing belt.
O.K.
INSPECTION END

ECCS SELF-DIAGNOSTIC CHART—1990 300ZX

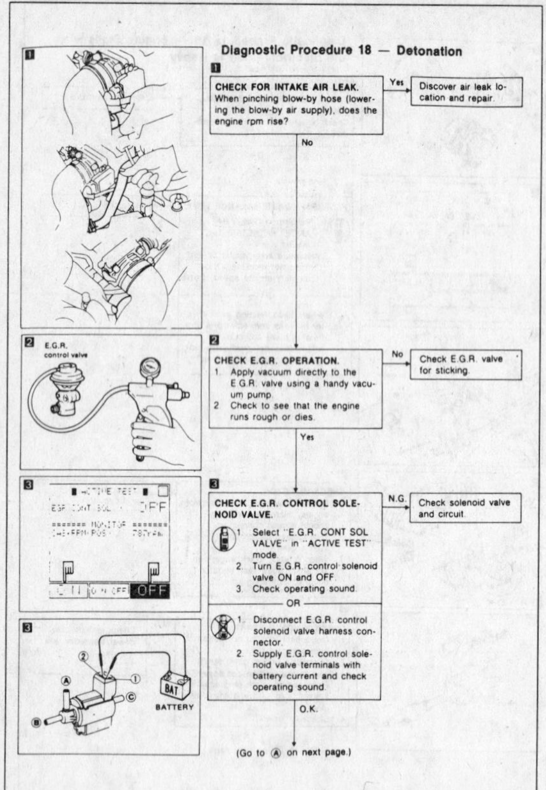
Diagnostic Procedure 18 — Detonation
1
CHECK FOR INTAKE AIR LEAK.
When pinching blow-by hose (lowering the blow-by air supply), does the engine rpm rise?
Yes
Discover air leak location and repair.
No
2
E.G.R. control valve
CHECK E.G.R. OPERATION.
1. Apply vacuum directly to the E.G.R. valve using a handy vacuum pump.
2. Check to see that the engine runs rough or dies.
No
Check E.G.R. valve for sticking.
Yes
3
ACTIVE TEST
EGR CONT SOL OFF
MONITOR
OFF
CHECK E.G.R. CONTROL SOLENOID VALVE.
1. Select "E.G.R. CONT SOL VALVE" in "ACTIVE TEST" mode.
2. Turn E.G.R. control solenoid valve ON and OFF.
3. Check operating sound.
OR
1. Disconnect E.G.R. control solenoid valve harness connector.
2. Supply E.G.R. control solenoid valve terminals with battery current and check operating sound.
N.G.
Check solenoid valve and circuit.
BAT
BATTERY
O.K.
(Go to A on next page.)

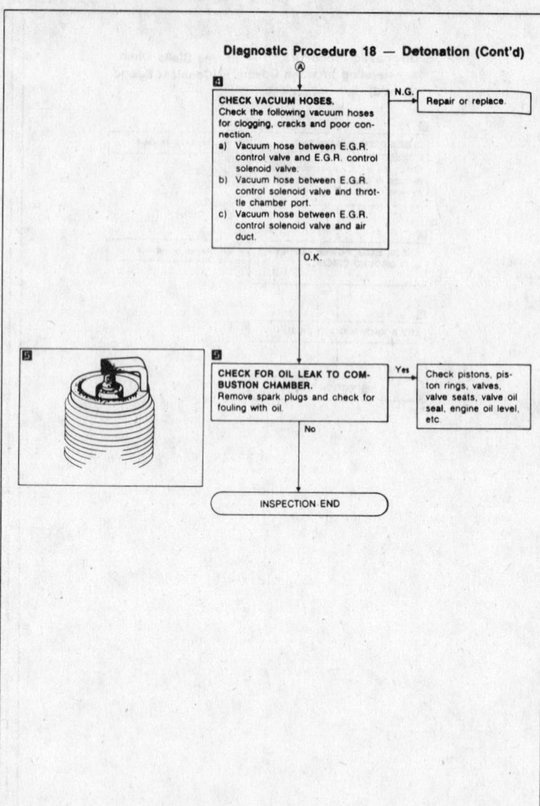
Diagnostic Procedure 18 — Detonation (Cont'd)
A
4
CHECK VACUUM HOSES.
Check the following vacuum hoses for clogging, cracks and poor connection.
a) Vacuum hose between E.G.R. control valve and E.G.R. control solenoid valve.
b) Vacuum hose between E.G.R. control solenoid valve and throttle chamber port.
c) Vacuum hose between E.G.R. control solenoid valve and air duct.
N.G.
Repair or replace.
O.K.
5
CHECK FOR OIL LEAK TO COMBUSTION CHAMBER.
Remove spark plugs and check for fouling with oil.
Yes
Check pistons, piston rings, valves, valve seats, valve oil seal, engine oil level, etc.
No
INSPECTION END

ECCS SELF-DIAGNOSTIC CHART—1990 300ZX

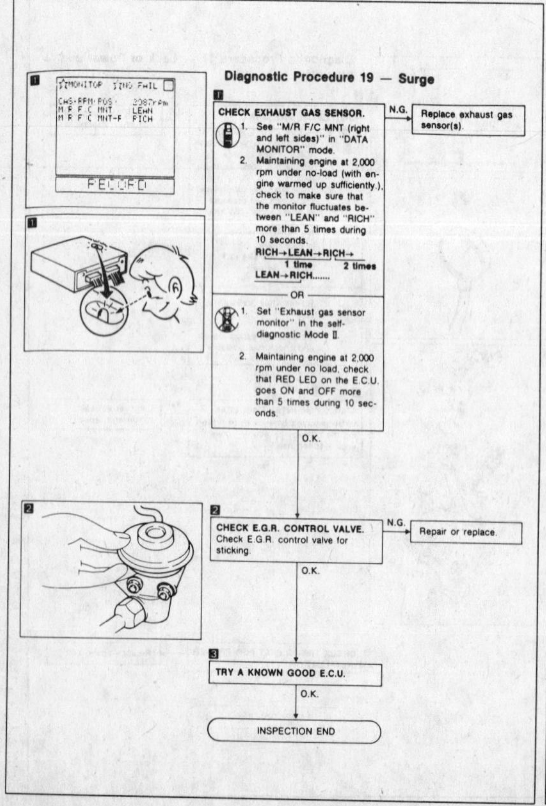
Diagnostic Procedure 19 — Surge
1
MONITOR
RECORD
CHECK EXHAUST GAS SENSOR.
1. See "M/R F/C MNT (right and left sides)" in "DATA MONITOR" mode.
2. Maintaining engine at 2,000 rpm under no-load (with engine warmed up sufficiently.), check to make sure that the monitor fluctuates between "LEAN" and "RICH" more than 5 times during 10 seconds.
RICH→LEAN→RICH→
1 time
2 times
LEAN→RICH.......
OR
1. Set "Exhaust gas sensor monitor" in the self-diagnostic Mode II.
2. Maintaining engine at 2,000 rpm under no load, check that RED LED on the E.C.U. goes ON and OFF more than 5 times during 10 seconds.
N.G.
Replace exhaust gas sensor(s).
O.K.
2
CHECK E.G.R. CONTROL VALVE.
Check E.G.R. control valve for sticking.
N.G.
Repair or replace.
O.K.
3
TRY A KNOWN GOOD E.C.U.
O.K.
INSPECTION END

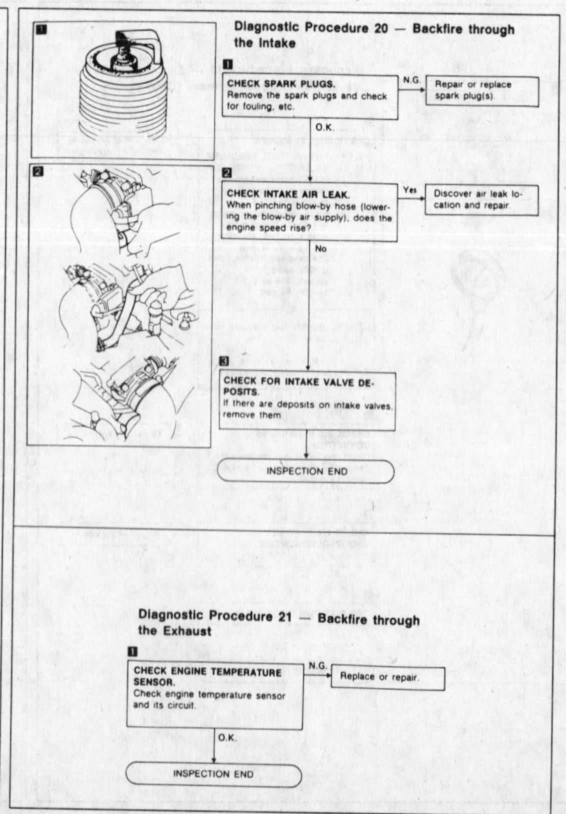
Diagnostic Procedure 20 — Backfire through the Intake
1
CHECK SPARK PLUGS.
Remove the spark plugs and check for fouling, etc.
N.G.
Repair or replace spark plug(s).
O.K.
2
CHECK INTAKE AIR LEAK.
When pinching blow-by hose (lowering the blow-by air supply), does the engine speed rise?
Yes
Discover air leak location and repair.
No
3
CHECK FOR INTAKE VALVE DEPOSITS.
If there are deposits on intake valves, remove them.
INSPECTION END
Diagnostic Procedure 21 — Backfire through the Exhaust
1
CHECK ENGINE TEMPERATURE SENSOR.
Check engine temperature sensor and its circuit.
N.G.
Replace or repair.
O.K.
INSPECTION END

ECCS SELF-DIAGNOSTIC CHART—1990 300ZX

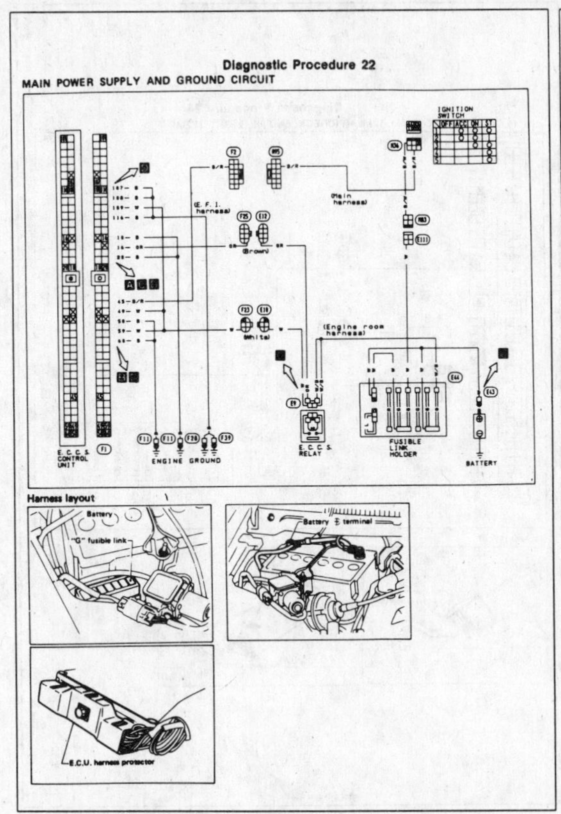
Diagnostic Procedure 22
MAIN POWER SUPPLY AND GROUND CIRCUIT
Harness layout
E.C.U. harness protector

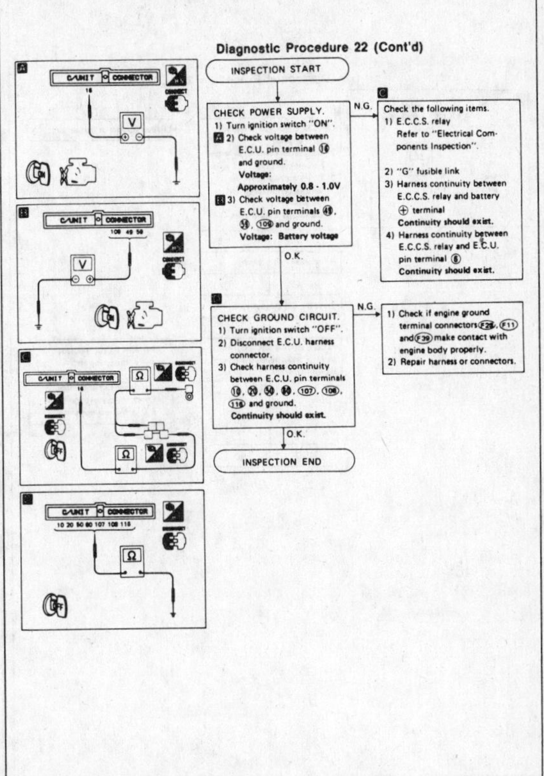
Diagnostic Procedure 22 (Cont'd)
INSPECTION START
CHECK POWER SUPPLY.
1) Turn ignition switch "ON".
2) Check voltage between E.C.U. pin terminal ⑯ and ground.
Voltage: Approximately 0.8 - 1.0V
3) Check voltage between E.C.U. pin terminals ㊾, 58, 108 and ground.
Voltage: Battery voltage
N.G.
Check the following items.
1) E.C.C.S. relay
Refer to "Electrical Components Inspection".
2) "G" fusible link
3) Harness continuity between E.C.C.S. relay and battery ⊕ terminal
Continuity should exist.
4) Harness continuity between E.C.C.S. relay and E.C.U. pin terminal ⑧
Continuity should exist.
O.K.
CHECK GROUND CIRCUIT.
1) Turn ignition switch "OFF".
2) Disconnect E.C.U. harness connector.
3) Check harness continuity between E.C.U. pin terminals ⑩, ⑳, 50, 60, 107, 108, 116 and ground.
Continuity should exist.
N.G.
1) Check if engine ground terminal connectors F28, F11 and F39 make contact with engine body properly.
2) Repair harness or connectors.
O.K.
INSPECTION END

ECCS SELF-DIAGNOSTIC CHART—1990 300ZX

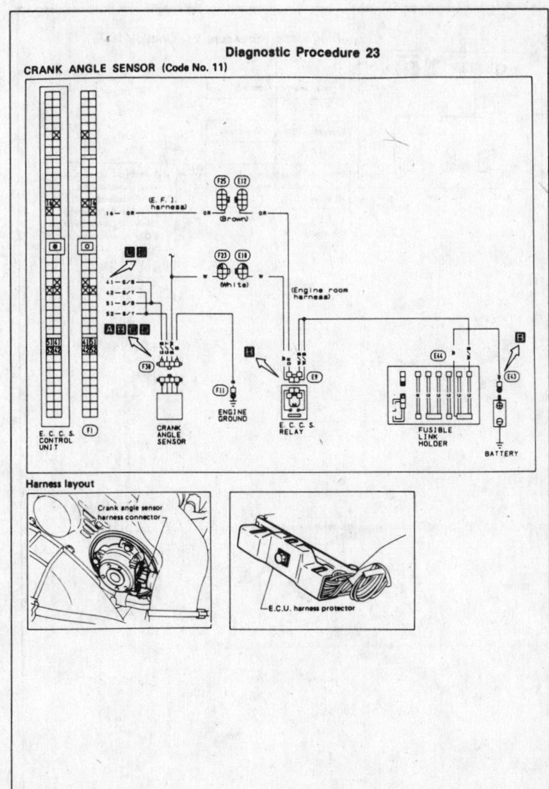
Diagnostic Procedure 23
CRANK ANGLE SENSOR (Code No. 11)
CRANK ANGLE SENSOR
ENGINE GROUND
E.C.C.S. RELAY
FUSIBLE LINK HOLDER
BATTERY
Harness layout
Crank angle sensor harness connector
E.C.U. harness protector

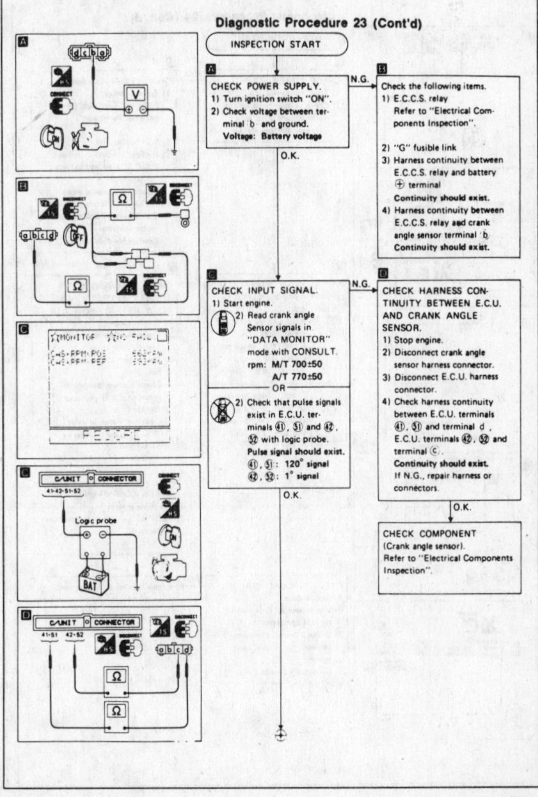
Diagnostic Procedure 23 (Cont'd)
INSPECTION START
CHECK POWER SUPPLY.
1) Turn ignition switch "ON".
2) Check voltage between terminal b and ground.
Voltage: Battery voltage
N.G.
Check the following items.
1) E.C.C.S. relay
Refer to "Electrical Components Inspection".
2) "G" fusible link
3) Harness continuity between E.C.C.S. relay and battery ⊕ terminal
Continuity should exist.
4) Harness continuity between E.C.C.S. relay and crank angle sensor terminal b
Continuity should exist.
O.K.
CHECK INPUT SIGNAL.
1) Start engine.
2) Read crank angle Sensor signals in "DATA MONITOR" mode with CONSULT.
rpm: M/T 700±50
A/T 770±50
—OR—
2) Check that pulse signals exist in E.C.U. terminals ㊶, 51 and ㊷, 52 with logic probe.
Pulse signal should exist.
㊶, 51: 120° signal
㊷, 52: 1° signal
N.G.
CHECK HARNESS CONTINUITY BETWEEN E.C.U. AND CRANK ANGLE SENSOR.
1) Stop engine.
2) Disconnect crank angle sensor harness connector.
3) Disconnect E.C.U. harness connector.
4) Check harness continuity between E.C.U. terminals ㊶, 51 and terminal d, E.C.U. terminals ㊷, 52 and terminal ⓒ.
Continuity should exist.
If N.G., repair harness or connectors.
O.K.
CHECK COMPONENT
(Crank angle sensor).
Refer to "Electrical Components Inspection".
Logic probe

ECCS SELF-DIAGNOSTIC CHART—1990 300ZX

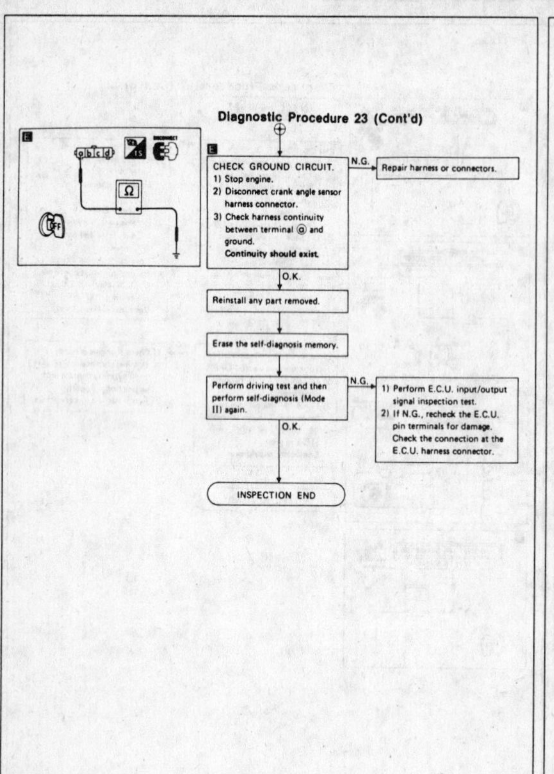
Diagnostic Procedure 23 (Cont'd)
E
CHECK GROUND CIRCUIT.
1) Stop engine.
2) Disconnect crank angle sensor harness connector.
3) Check harness continuity between terminal ⓐ and ground.
Continuity should exist.
N.G.
Repair harness or connectors.
O.K.
Reinstall any part removed.
Erase the self-diagnosis memory.
Perform driving test and then perform self-diagnosis (Mode II) again.
N.G.
1) Perform E.C.U. input/output signal inspection test.
2) If N.G., recheck the E.C.U. pin terminals for damage. Check the connection at the E.C.U. harness connector.
O.K.
INSPECTION END

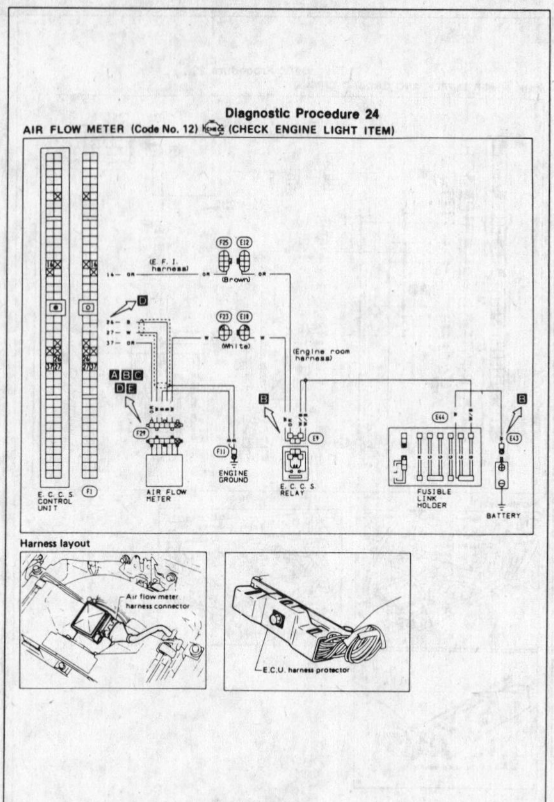
Diagnostic Procedure 24
AIR FLOW METER (Code No. 12) (CHECK ENGINE LIGHT ITEM)
(E.F.I. harness)
(Brown)
(White)
(Engine room harness)
E.C.C.S. CONTROL UNIT
AIR FLOW METER
ENGINE GROUND
E.C.C.S. RELAY
FUSIBLE LINK HOLDER
BATTERY
Harness layout
Air flow meter harness connector
E.C.U. harness protector

ECCS SELF-DIAGNOSTIC CHART—1990 300ZX

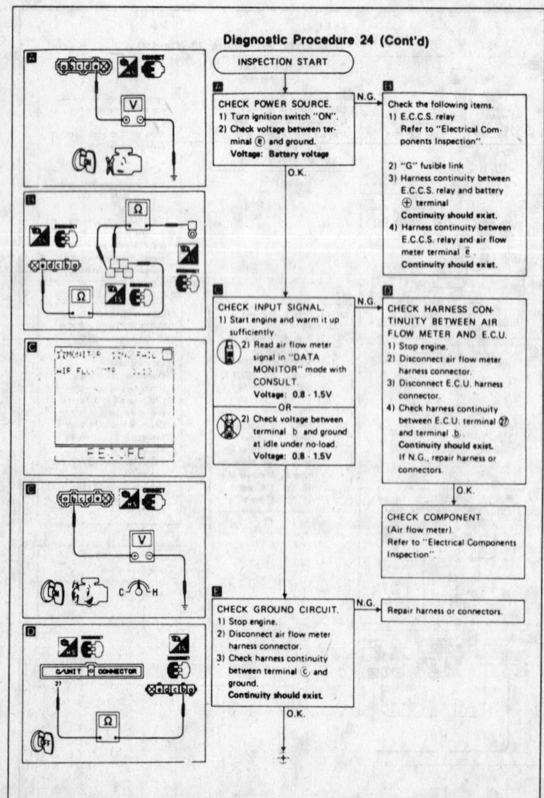
Diagnostic Procedure 24 (Cont'd)
INSPECTION START
A
CHECK POWER SOURCE.
1) Turn ignition switch "ON".
2) Check voltage between terminal ⓔ and ground.
Voltage: Battery voltage
N.G.
B
Check the following items.
1) E.C.C.S. relay
Refer to "Electrical Components Inspection".
2) "G" fusible link
3) Harness continuity between E.C.C.S. relay and battery ⊕ terminal
Continuity should exist.
4) Harness continuity between E.C.C.S. relay and air flow meter terminal ⓔ.
Continuity should exist.
O.K.
C
CHECK INPUT SIGNAL.
1) Start engine and warm it up sufficiently.
2) Read air flow meter signal in "DATA MONITOR" mode with CONSULT.
Voltage: 0.8 - 1.5V
OR
2) Check voltage between terminal ⓑ and ground at idle under no-load.
Voltage: 0.8 - 1.5V
N.G.
D
CHECK HARNESS CONTINUITY BETWEEN AIR FLOW METER AND E.C.U.
1) Stop engine.
2) Disconnect air flow meter harness connector.
3) Disconnect E.C.U. harness connector.
4) Check harness continuity between E.C.U. terminal ㉗ and terminal ⓑ.
Continuity should exist.
If N.G., repair harness or connectors.
O.K.
CHECK COMPONENT
(Air flow meter).
Refer to "Electrical Components Inspection".
E
CHECK GROUND CIRCUIT.
1) Stop engine.
2) Disconnect air flow meter harness connector.
3) Check harness continuity between terminal ⓒ and ground.
Continuity should exist.
N.G.
Repair harness or connectors.
O.K.
C/UNIT CONNECTOR
27

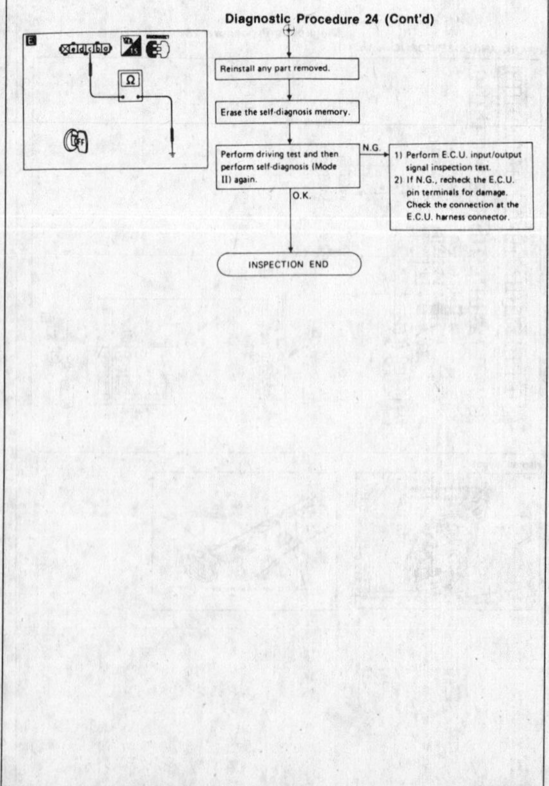
Diagnostic Procedure 24 (Cont'd)
E
Reinstall any part removed.
Erase the self-diagnosis memory.
Perform driving test and then perform self-diagnosis (Mode II) again.
N.G.
1) Perform E.C.U. input/output signal inspection test.
2) If N.G., recheck the E.C.U. pin terminals for damage. Check the connection at the E.C.U. harness connector.
O.K.
INSPECTION END

ECCS SELF-DIAGNOSTIC CHART—1990 300ZX

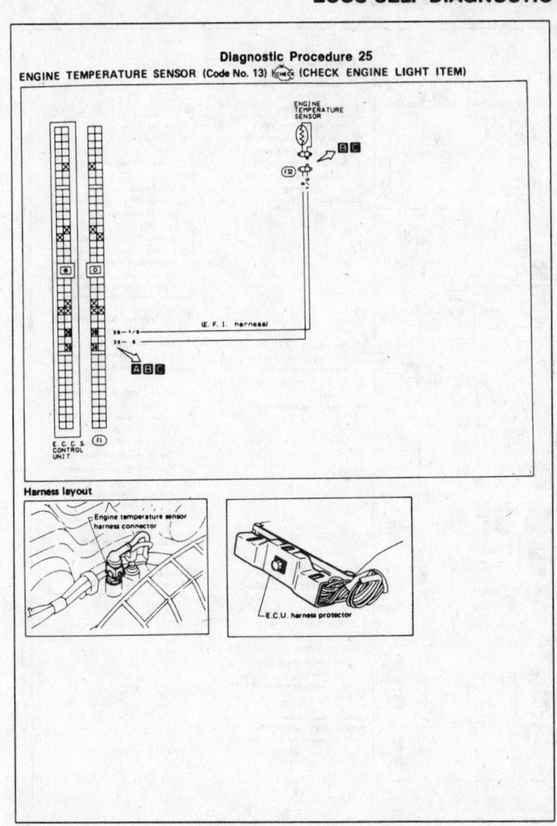
Diagnostic Procedure 25
ENGINE TEMPERATURE SENSOR (Code No. 13) (CHECK ENGINE LIGHT ITEM)
ENGINE TEMPERATURE SENSOR
E.C.C.S. CONTROL UNIT
(E.F.I. harness)
Harness layout
Engine temperature sensor harness connector
E.C.U. harness protector

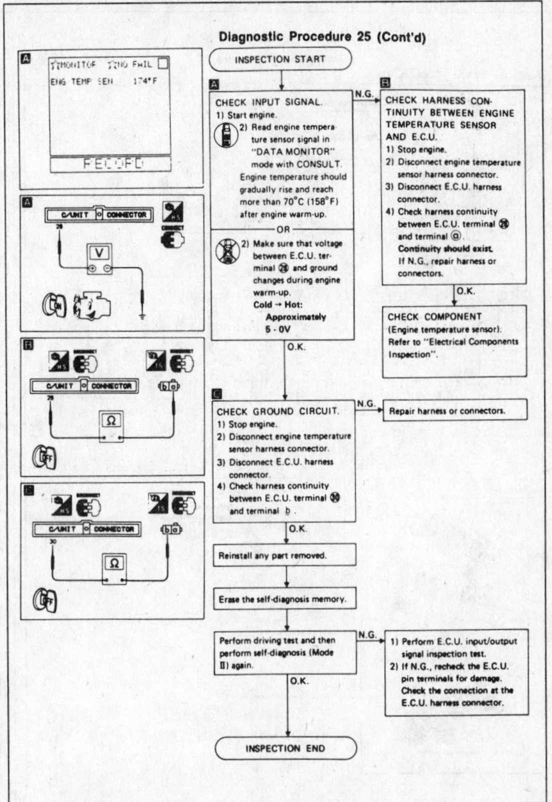
Diagnostic Procedure 25 (Cont'd)
INSPECTION START
CHECK INPUT SIGNAL.
1) Start engine.
2) Read engine temperature sensor signal in "DATA MONITOR" mode with CONSULT.
Engine temperature should gradually rise and reach more than 70°C (158°F) after engine warm-up.
OR
2) Make sure that voltage between E.C.U. terminal 28 and ground changes during engine warm-up.
Cold → Hot:
Approximately 5 - 0V
N.G.
CHECK HARNESS CONTINUITY BETWEEN ENGINE TEMPERATURE SENSOR AND E.C.U.
1) Stop engine.
2) Disconnect engine temperature sensor harness connector.
3) Disconnect E.C.U. harness connector.
4) Check harness continuity between E.C.U. terminal 28 and terminal a.
Continuity should exist.
If N.G., repair harness or connectors.
O.K.
CHECK COMPONENT
(Engine temperature sensor).
Refer to "Electrical Components Inspection".
O.K.
CHECK GROUND CIRCUIT.
1) Stop engine.
2) Disconnect engine temperature sensor harness connector.
3) Disconnect E.C.U. harness connector.
4) Check harness continuity between E.C.U. terminal 30 and terminal b.
N.G.
Repair harness or connectors.
O.K.
Reinstall any part removed.
Erase the self-diagnosis memory.
Perform driving test and then perform self-diagnosis (Mode II) again.
N.G.
1) Perform E.C.U. input/output signal inspection test.
2) If N.G., recheck the E.C.U. pin terminals for damage.
Check the connection at the E.C.U. harness connector.
O.K.
INSPECTION END
MONITOR NO FAIL
ENG TEMP SEN 174°F
RECORD
C/UNIT CONNECTOR

ECCS SELF-DIAGNOSTIC CHART—1990 300ZX

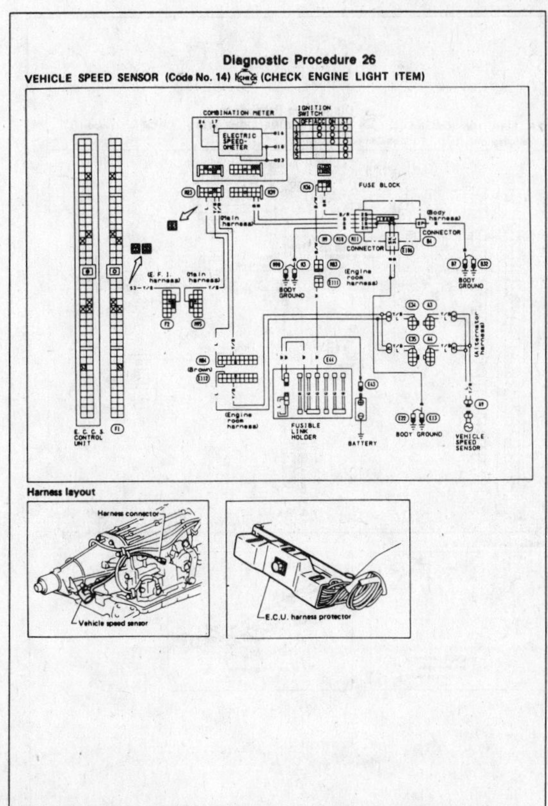
Diagnostic Procedure 26
VEHICLE SPEED SENSOR (Code No. 14) (CHECK ENGINE LIGHT ITEM)
COMBINATION METER
IGNITION SWITCH
ELECTRIC SPEEDOMETER
FUSE BLOCK
CONNECTOR
BODY GROUND
FUSIBLE LINK HOLDER
BATTERY
VEHICLE SPEED SENSOR
E.C.C.S. CONTROL UNIT
Harness layout
Harness connector
Vehicle speed sensor
E.C.U. harness protector

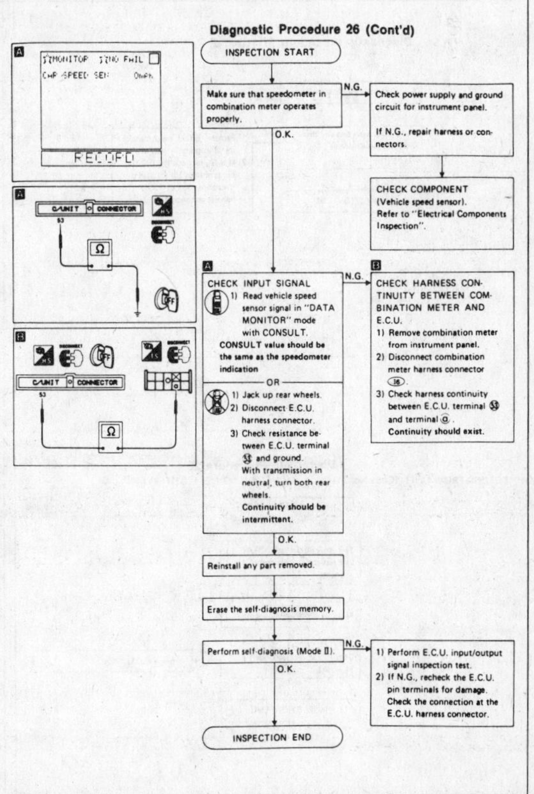
Diagnostic Procedure 26 (Cont'd)
INSPECTION START
Make sure that speedometer in combination meter operates properly.
N.G.
Check power supply and ground circuit for instrument panel.
If N.G., repair harness or connectors.
CHECK COMPONENT
(Vehicle speed sensor).
Refer to "Electrical Components Inspection".
O.K.
CHECK INPUT SIGNAL
1) Read vehicle speed sensor signal in "DATA MONITOR" mode with CONSULT.
CONSULT value should be the same as the speedometer indication
OR
1) Jack up rear wheels.
2) Disconnect E.C.U. harness connector.
3) Check resistance between E.C.U. terminal 53 and ground.
With transmission in neutral, turn both rear wheels.
Continuity should be intermittent.
N.G.
CHECK HARNESS CONTINUITY BETWEEN COMBINATION METER AND E.C.U.
1) Remove combination meter from instrument panel.
2) Disconnect combination meter harness connector.
3) Check harness continuity between E.C.U. terminal 53 and terminal a.
Continuity should exist.
O.K.
Reinstall any part removed.
Erase the self-diagnosis memory.
Perform self-diagnosis (Mode II).
N.G.
1) Perform E.C.U. input/output signal inspection test.
2) If N.G., recheck the E.C.U. pin terminals for damage.
Check the connection at the E.C.U. harness connector.
O.K.
INSPECTION END
MONITOR NO FAIL
RECORD
C/UNIT CONNECTOR

ECCS SELF-DIAGNOSTIC CHART—1990 300ZX

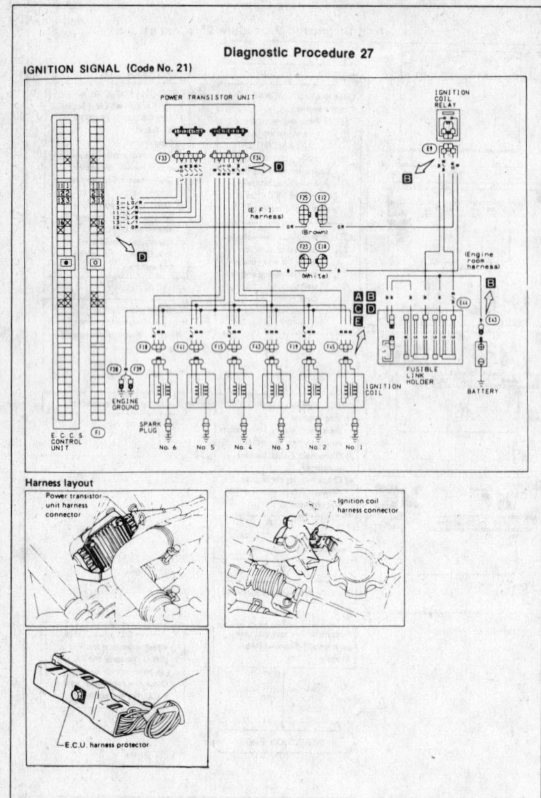
Diagnostic Procedure 27
IGNITION SIGNAL (Code No. 21)
POWER TRANSISTOR UNIT
IGNITION COIL RELAY
(E.F.I. harness)
(Brown)
(White)
(Engine room harness)
FUSIBLE LINK HOLDER
BATTERY
IGNITION COIL
ENGINE GROUND
SPARK PLUG
No. 6
No. 5
No. 4
No. 3
No. 2
No. 1
E.C.C.S. CONTROL UNIT
Harness layout
Power transistor unit harness connector
Ignition coil harness connector
E.C.U. harness protector

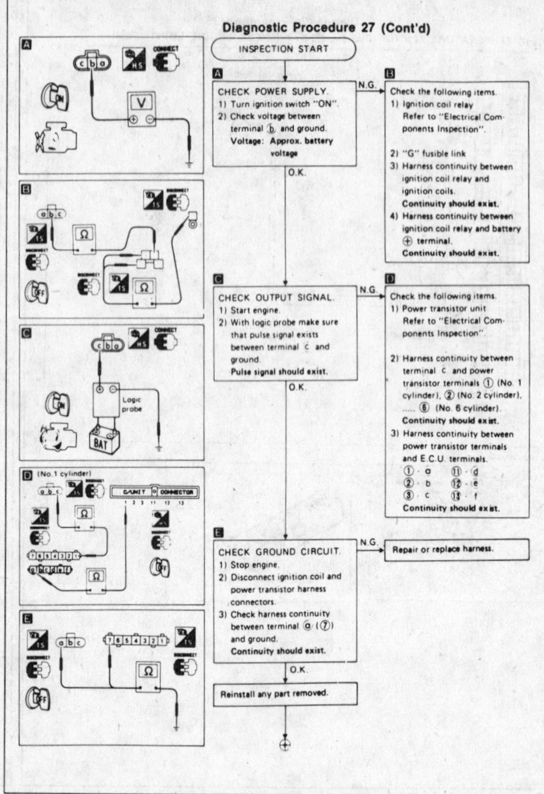
Diagnostic Procedure 27 (Cont'd)
INSPECTION START
A CHECK POWER SUPPLY.
1) Turn ignition switch "ON".
2) Check voltage between terminal ⓑ and ground.
Voltage: Approx. battery voltage
N.G.
B Check the following items.
1) Ignition coil relay Refer to "Electrical Components Inspection".
2) "G" fusible link
3) Harness continuity between ignition coil relay and ignition coils.
Continuity should exist.
4) Harness continuity between ignition coil relay and battery ⊕ terminal.
Continuity should exist.
O.K.
C CHECK OUTPUT SIGNAL.
1) Start engine.
2) With logic probe make sure that pulse signal exists between terminal c and ground.
Pulse signal should exist.
N.G.
D Check the following items.
1) Power transistor unit Refer to "Electrical Components Inspection".
2) Harness continuity between terminal c and power transistor terminals ① (No. 1 cylinder), ② (No. 2 cylinder), ... ⑥ (No. 6 cylinder).
Continuity should exist.
3) Harness continuity between power transistor terminals and E.C.U. terminals.
① · a ⑪ · d
② · b ⑫ · e
③ · c ⑬ · f
Continuity should exist.
O.K.
E CHECK GROUND CIRCUIT.
1) Stop engine.
2) Disconnect ignition coil and power transistor harness connectors.
3) Check harness continuity between terminal ⓐ (⑦) and ground.
Continuity should exist.
N.G.
Repair or replace harness.
O.K.
Reinstall any part removed.
CONNECT
DISCONNECT
Logic probe
BAT
(No. 1 cylinder)
CONNECTOR

ECCS SELF-DIAGNOSTIC CHART—1990 300ZX

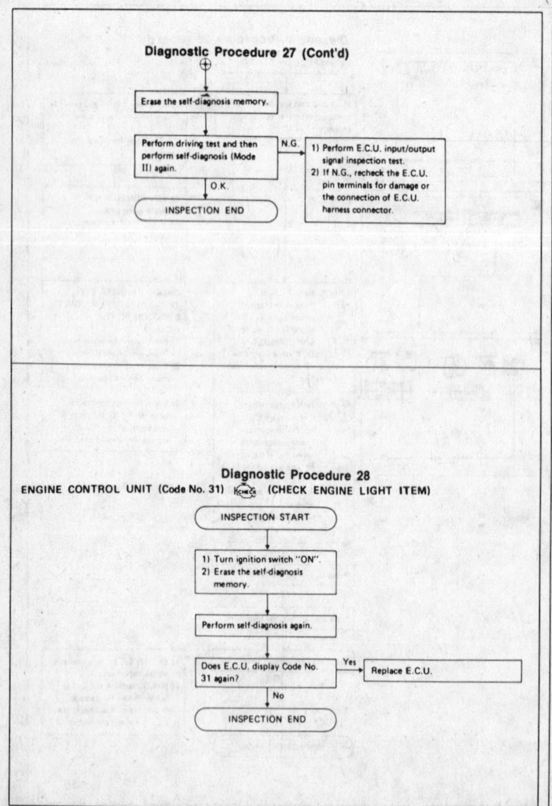
Diagnostic Procedure 27 (Cont'd)
Erase the self-diagnosis memory.
Perform driving test and then perform self-diagnosis (Mode III) again.
N.G.
1) Perform E.C.U. input/output signal inspection test.
2) If N.G., recheck the E.C.U. pin terminals for damage or the connection of E.C.U. harness connector.
O.K.
INSPECTION END
Diagnostic Procedure 28
ENGINE CONTROL UNIT (Code No. 31) (CHECK ENGINE LIGHT ITEM)
INSPECTION START
1) Turn ignition switch "ON".
2) Erase the self-diagnosis memory.
Perform self-diagnosis again.
Does E.C.U. display Code No. 31 again?
Yes
Replace E.C.U.
No
INSPECTION END

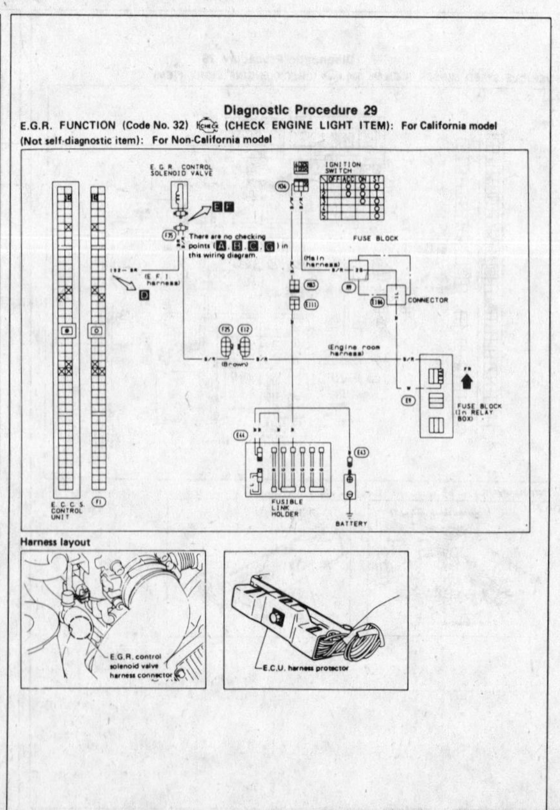
Diagnostic Procedure 29
E.G.R. FUNCTION (Code No. 32) (CHECK ENGINE LIGHT ITEM): For California model
(Not self-diagnostic item): For Non-California model
E.G.R. CONTROL SOLENOID VALVE
IGNITION SWITCH
There are no checking points (A, B, C, D) in this wiring diagram
FUSE BLOCK
(Main harness)
CONNECTOR
(E.F.I. harness)
(Brown)
(Engine room harness)
FUSE BLOCK (In RELAY BOX)
FUSIBLE LINK HOLDER
BATTERY
E.C.C.S. CONTROL UNIT
Harness layout
E.G.R. control solenoid valve harness connector
E.C.U. harness protector

ECCS SELF-DIAGNOSTIC CHART—1990 300ZX

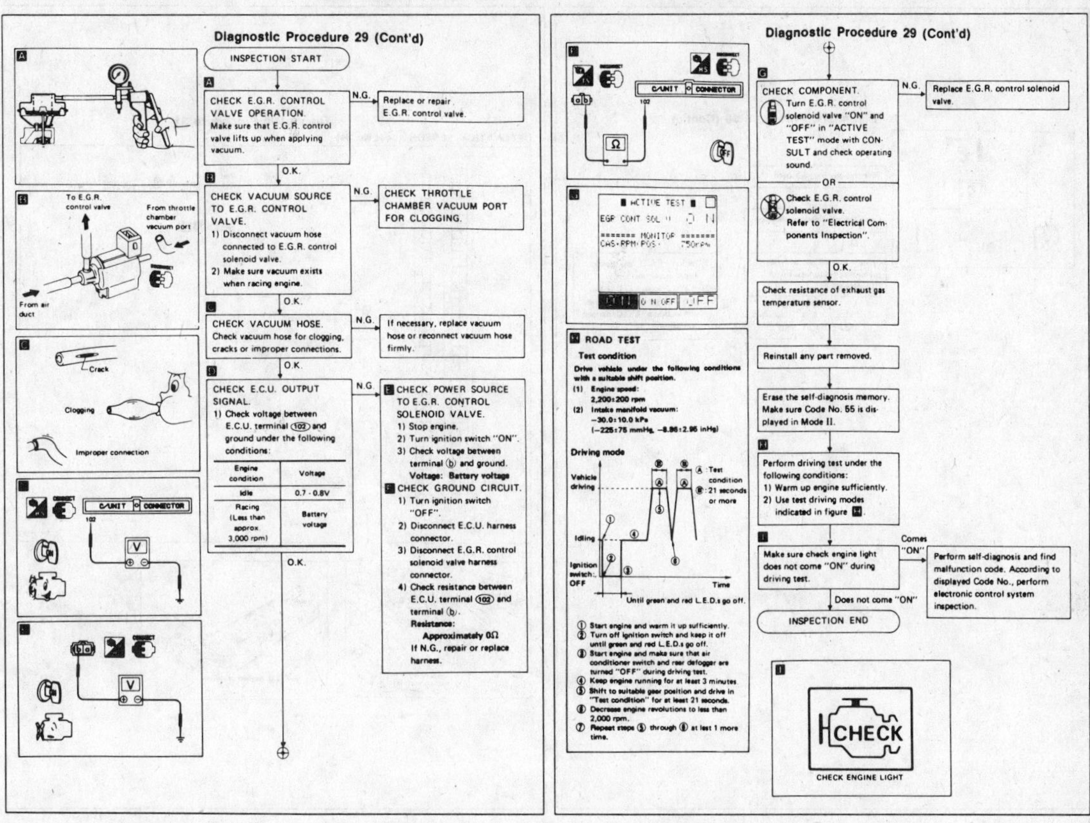
Diagnostic Procedure 29 (Cont'd)
INSPECTION START
CHECK E.G.R. CONTROL VALVE OPERATION.
Make sure that E.G.R. control valve lifts up when applying vacuum.
N.G.
Replace or repair E.G.R. control valve.
O.K.
CHECK VACUUM SOURCE TO E.G.R. CONTROL VALVE.
1) Disconnect vacuum hose connected to E.G.R. control solenoid valve.
2) Make sure vacuum exists when racing engine.
N.G.
CHECK THROTTLE CHAMBER VACUUM PORT FOR CLOGGING.
O.K.
CHECK VACUUM HOSE.
Check vacuum hose for clogging, cracks or improper connections.
N.G.
If necessary, replace vacuum hose or reconnect vacuum hose firmly.
O.K.
CHECK E.C.U. OUTPUT SIGNAL.
1) Check voltage between E.C.U. terminal (102) and ground under the following conditions:
Engine condition
Voltage
Idle
0.7 - 0.8V
Racing (Less than approx. 3,000 rpm)
Battery voltage
N.G.
CHECK POWER SOURCE TO E.G.R. CONTROL SOLENOID VALVE.
1) Stop engine.
2) Turn ignition switch "ON".
3) Check voltage between terminal (b) and ground.
Voltage: Battery voltage
CHECK GROUND CIRCUIT.
1) Turn ignition switch "OFF".
2) Disconnect E.C.U. harness connector.
3) Disconnect E.G.R. control solenoid valve harness connector.
4) Check resistance between E.C.U. terminal (102) and terminal (b).
Resistance: Approximately 0Ω
If N.G., repair or replace harness.
O.K.
To E.G.R. control valve
From throttle chamber vacuum port
From air duct
Crack
Clogging
Improper connection
C/UNIT
CONNECTOR
Diagnostic Procedure 29 (Cont'd)
CHECK COMPONENT.
Turn E.G.R. control solenoid valve "ON" and "OFF" in "ACTIVE TEST" mode with CONSULT and check operating sound.
OR
Check E.G.R. control solenoid valve.
Refer to "Electrical Components Inspection".
N.G.
Replace E.G.R. control solenoid valve.
O.K.
Check resistance of exhaust gas temperature sensor.
Reinstall any part removed.
Erase the self-diagnosis memory. Make sure Code No. 55 is displayed in Mode II.
Perform driving test under the following conditions:
1) Warm up engine sufficiently.
2) Use test driving modes indicated in figure.
Make sure check engine light does not come "ON" during driving test.
Comes "ON"
Perform self-diagnosis and find malfunction code. According to displayed Code No., perform electronic control system inspection.
Does not come "ON"
INSPECTION END
ACTIVE TEST
EGR CONT SOL
MONITOR
CAS-RPM(POS)
750rpm
ON OFF
ROAD TEST
Test condition
Drive vehicle under the following conditions with a suitable shift position.
(1) Engine speed: 2,200±200 rpm
(2) Intake manifold vacuum: −30.0±10.0 kPa (−225±75 mmHg, −8.86±2.95 inHg)
Driving mode
Vehicle driving
Idling
Ignition switch OFF
Time
Test condition
21 seconds or more
Until green and red L.E.D.s go off.
① Start engine and warm it up sufficiently.
② Turn off ignition switch and keep it off until green and red L.E.D.s go off.
③ Start engine and make sure that air conditioner switch and rear defogger are turned "OFF" during driving test.
④ Keep engine running for at least 3 minutes.
⑤ Shift to suitable gear position and drive in "Test condition" for at least 21 seconds.
⑥ Decrease engine revolutions to less than 2,000 rpm.
⑦ Repeat steps ⑤ through ⑥ at least 1 more time.
CHECK
CHECK ENGINE LIGHT

ECCS SELF-DIAGNOSTIC CHART—1990 300ZX

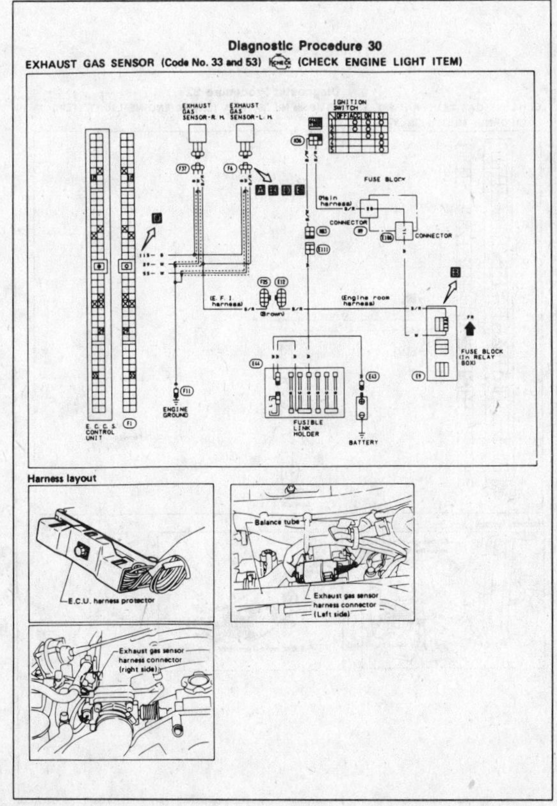
Diagnostic Procedure 30
EXHAUST GAS SENSOR (Code No. 33 and 53) (CHECK ENGINE LIGHT ITEM)
EXHAUST GAS SENSOR-R.H.
EXHAUST GAS SENSOR-L.H.
IGNITION SWITCH
FUSE BLOCK
CONNECTOR
(E.F.I. harness)
(Engine room harness)
FUSE BLOCK (J/L RELAY BOX)
ENGINE GROUND
E.C.C.S. CONTROL UNIT
FUSIBLE LINK HOLDER
BATTERY
Harness layout
E.C.U. harness protector
Balance tube
Exhaust gas sensor harness connector (Left side)
Exhaust gas sensor harness connector (right side)

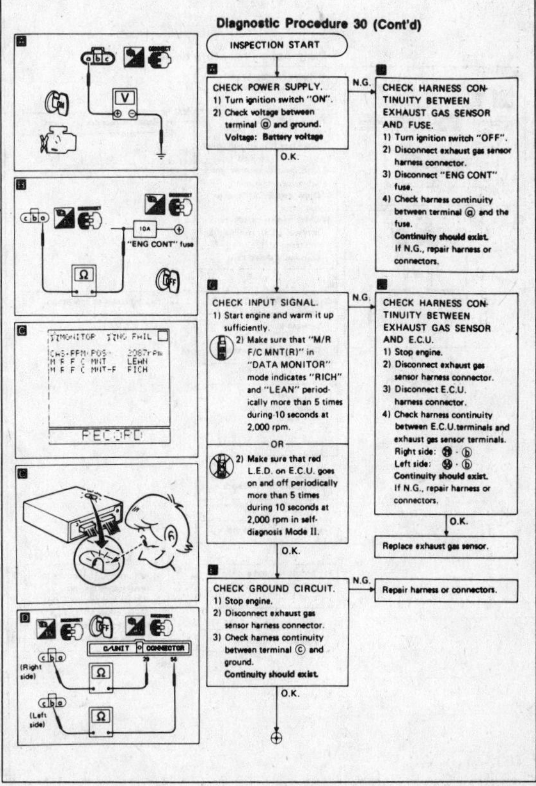
Diagnostic Procedure 30 (Cont'd)
INSPECTION START
CHECK POWER SUPPLY.
1) Turn ignition switch "ON".
2) Check voltage between terminal (a) and ground.
Voltage: Battery voltage
N.G.
CHECK HARNESS CONTINUITY BETWEEN EXHAUST GAS SENSOR AND FUSE.
1) Turn ignition switch "OFF".
2) Disconnect exhaust gas sensor harness connector.
3) Disconnect "ENG CONT" fuse.
4) Check harness continuity between terminal (a) and the fuse.
Continuity should exist.
If N.G., repair harness or connectors.
O.K.
CHECK INPUT SIGNAL.
1) Start engine and warm it up sufficiently.
2) Make sure that "M/R F/C MNT(R)" in "DATA MONITOR" mode indicates "RICH" and "LEAN" periodically more than 5 times during 10 seconds at 2,000 rpm.
OR
2) Make sure that red L.E.D. on E.C.U. goes on and off periodically more than 5 times during 10 seconds at 2,000 rpm in self-diagnosis Mode II.
N.G.
CHECK HARNESS CONTINUITY BETWEEN EXHAUST GAS SENSOR AND E.C.U.
1) Stop engine.
2) Disconnect exhaust gas sensor harness connector.
3) Disconnect E.C.U. harness connector.
4) Check harness continuity between E.C.U. terminals and exhaust gas sensor terminals.
Right side: (b)
Left side: (b)
Continuity should exist.
If N.G., repair harness or connectors.
O.K.
Replace exhaust gas sensor.
O.K.
CHECK GROUND CIRCUIT.
1) Stop engine.
2) Disconnect exhaust gas sensor harness connector.
3) Check harness continuity between terminal (c) and ground.
Continuity should exist.
N.G.
Repair harness or connectors.
O.K.
10A
"ENG CONT" fuse
MONITOR
CAS-RPM(POS)
2087rpm
M/R F/C MNT
LEAN
M/R F/C MNT-F
RICH
RECORD
C/UNIT
CONNECTOR
(Right side)
(Left side)

ECCS SELF-DIAGNOSTIC CHART—1990 300ZX

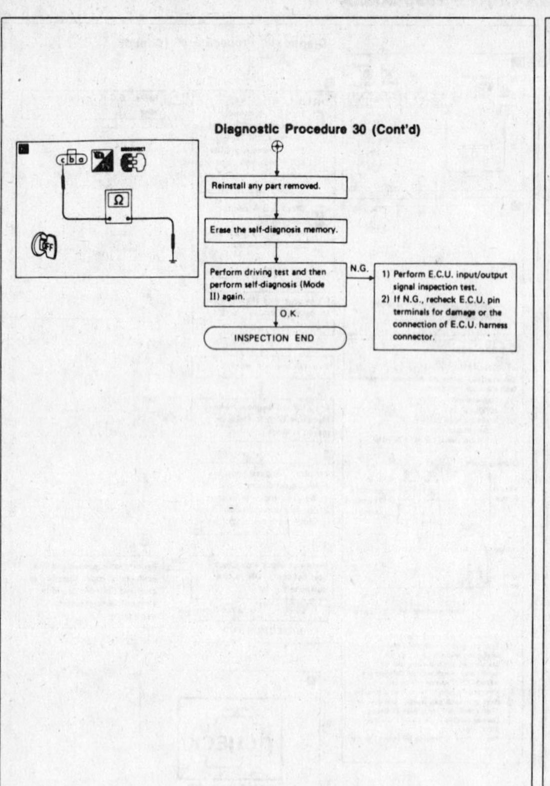
Diagnostic Procedure 30 (Cont'd)
Reinstall any part removed.
Erase the self-diagnosis memory.
Perform driving test and then perform self-diagnosis (Mode II) again.
N.G.
1) Perform E.C.U. input/output signal inspection test.
2) If N.G., recheck E.C.U. pin terminals for damage or the connection of E.C.U. harness connector.
O.K.
INSPECTION END

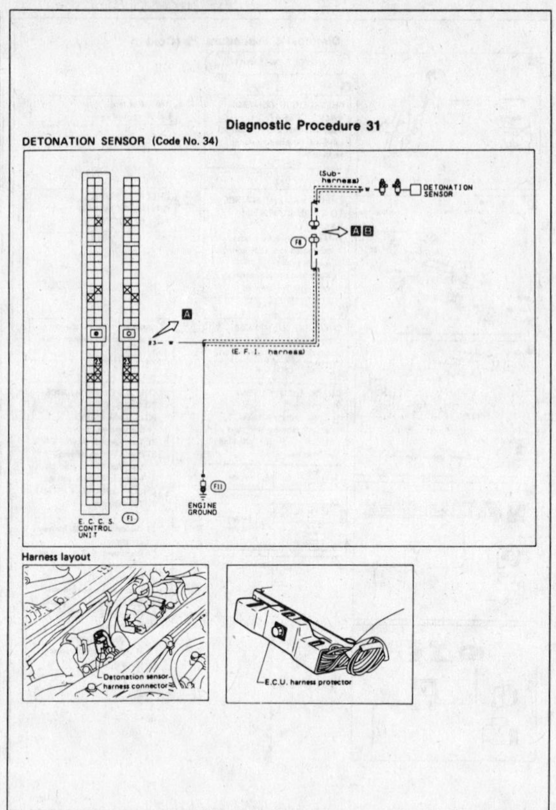
Diagnostic Procedure 31
DETONATION SENSOR (Code No. 34)
DETONATION SENSOR
E.C.C.S. CONTROL UNIT
ENGINE GROUND
Harness layout
Detonation sensor harness connector
E.C.U. harness protector

ECCS SELF-DIAGNOSTIC CHART—1990 300ZX

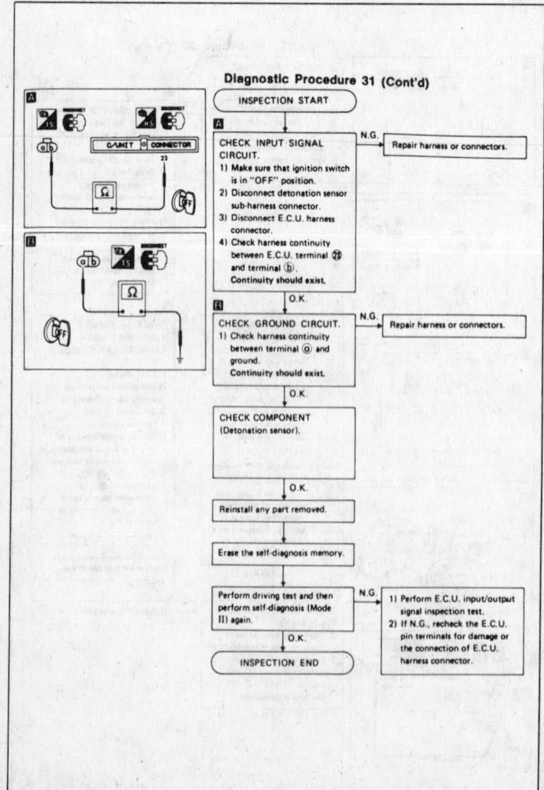
Diagnostic Procedure 31 (Cont'd)
INSPECTION START
A
CHECK INPUT SIGNAL CIRCUIT.
1) Make sure that ignition switch is in "OFF" position.
2) Disconnect detonation sensor sub-harness connector.
3) Disconnect E.C.U. harness connector.
4) Check harness continuity between E.C.U. terminal 23 and terminal b.
Continuity should exist.
N.G.
Repair harness or connectors.
O.K.
B
CHECK GROUND CIRCUIT.
1) Check harness continuity between terminal a and ground.
Continuity should exist.
N.G.
Repair harness or connectors.
O.K.
CHECK COMPONENT (Detonation sensor).
O.K.
Reinstall any part removed.
Erase the self-diagnosis memory.
Perform driving test and then perform self-diagnosis (Mode II) again.
N.G.
1) Perform E.C.U. input/output signal inspection test.
2) If N.G., recheck the E.C.U. pin terminals for damage or the connection of E.C.U. harness connector.
O.K.
INSPECTION END

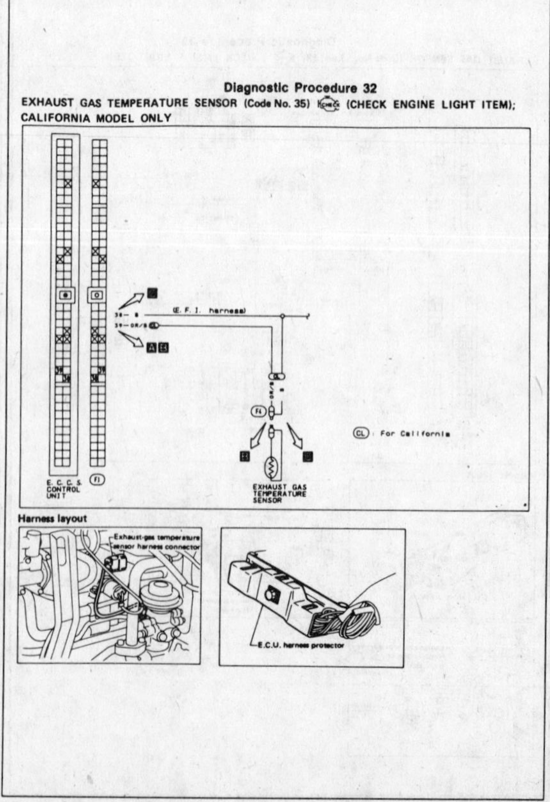
Diagnostic Procedure 32
EXHAUST GAS TEMPERATURE SENSOR (Code No. 35) (CHECK ENGINE LIGHT ITEM); CALIFORNIA MODEL ONLY
E.F.I. harness
CL: For California
EXHAUST GAS TEMPERATURE SENSOR
E.C.C.S. CONTROL UNIT
Harness layout
Exhaust-gas temperature sensor harness connector
E.C.U. harness protector

ECCS SELF-DIAGNOSTIC CHART—1990 300ZX

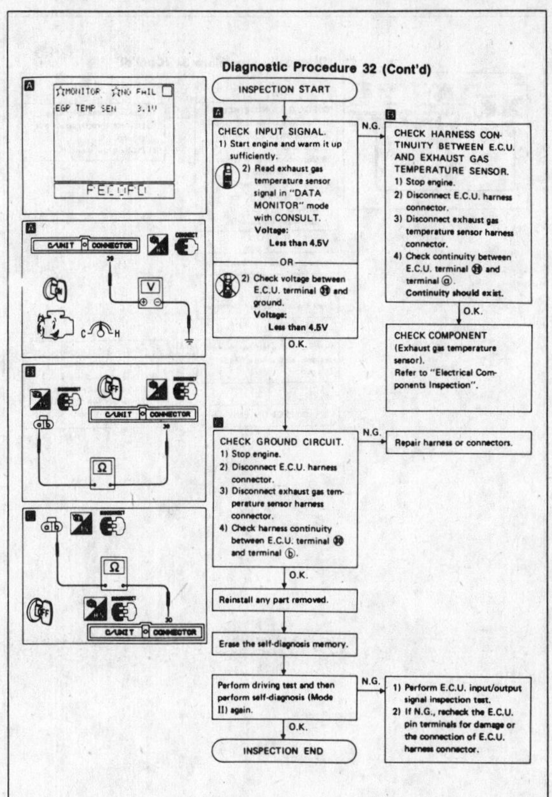
Diagnostic Procedure 32 (Cont'd)
INSPECTION START
CHECK INPUT SIGNAL.
1) Start engine and warm it up sufficiently.
2) Read exhaust gas temperature sensor signal in "DATA MONITOR" mode with CONSULT.
Voltage:
Less than 4.5V
OR
2) Check voltage between E.C.U. terminal 30 and ground.
Voltage:
Less than 4.5V
N.G.
CHECK HARNESS CONTINUITY BETWEEN E.C.U. AND EXHAUST GAS TEMPERATURE SENSOR.
1) Stop engine.
2) Disconnect E.C.U. harness connector.
3) Disconnect exhaust gas temperature sensor harness connector.
4) Check continuity between E.C.U. terminal 30 and terminal a.
Continuity should exist.
O.K.
CHECK COMPONENT
(Exhaust gas temperature sensor).
Refer to "Electrical Components Inspection".
O.K.
CHECK GROUND CIRCUIT.
1) Stop engine.
2) Disconnect E.C.U. harness connector.
3) Disconnect exhaust gas temperature sensor harness connector.
4) Check harness continuity between E.C.U. terminal 30 and terminal b.
N.G.
Repair harness or connectors.
O.K.
Reinstall any part removed.
Erase the self-diagnosis memory.
Perform driving test and then perform self-diagnosis (Mode II) again.
N.G.
1) Perform E.C.U. input/output signal inspection test.
2) If N.G., recheck the E.C.U. pin terminals for damage or the connection of E.C.U. harness connector.
O.K.
INSPECTION END
MONITOR NO FAIL
EGR TEMP SEN 3.1V
C/UNIT CONNECTOR
CONNECT
DISCONNECT
C H

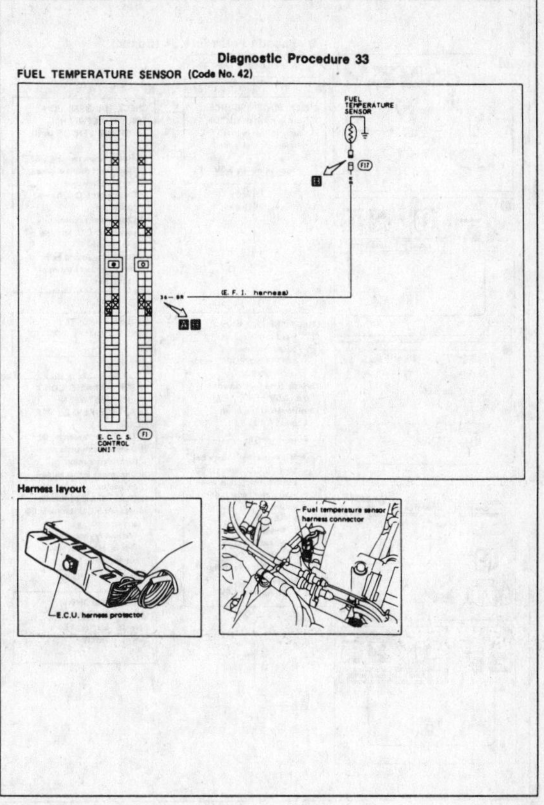
Diagnostic Procedure 33
FUEL TEMPERATURE SENSOR (Code No. 42)
FUEL TEMPERATURE SENSOR
F17
(E.F.I. harness)
E.C.C.S. CONTROL UNIT
F1
Harness layout
E.C.U. harness protector
Fuel temperature sensor harness connector

ECCS SELF-DIAGNOSTIC CHART—1990 300ZX

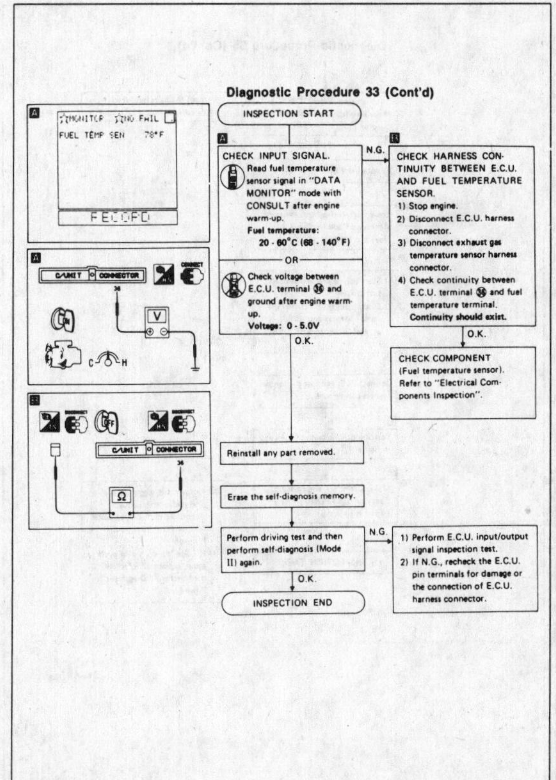
Diagnostic Procedure 33 (Cont'd)
INSPECTION START
CHECK INPUT SIGNAL.
Read fuel temperature sensor signal in "DATA MONITOR" mode with CONSULT after engine warm-up.
Fuel temperature:
20 - 60°C (68 - 140°F)
OR
Check voltage between E.C.U. terminal 36 and ground after engine warm-up.
Voltage: 0 - 5.0V
N.G.
CHECK HARNESS CONTINUITY BETWEEN E.C.U. AND FUEL TEMPERATURE SENSOR.
1) Stop engine.
2) Disconnect E.C.U. harness connector.
3) Disconnect exhaust gas temperature sensor harness connector.
4) Check continuity between E.C.U. terminal 36 and fuel temperature terminal.
Continuity should exist.
O.K.
CHECK COMPONENT
(Fuel temperature sensor).
Refer to "Electrical Components Inspection".
O.K.
Reinstall any part removed.
Erase the self-diagnosis memory.
Perform driving test and then perform self-diagnosis (Mode II) again.
N.G.
1) Perform E.C.U. input/output signal inspection test.
2) If N.G., recheck the E.C.U. pin terminals for damage or the connection of E.C.U. harness connector.
O.K.
INSPECTION END
MONITOR NO FAIL
FUEL TEMP SEN 78°F
C/UNIT CONNECTOR
C H

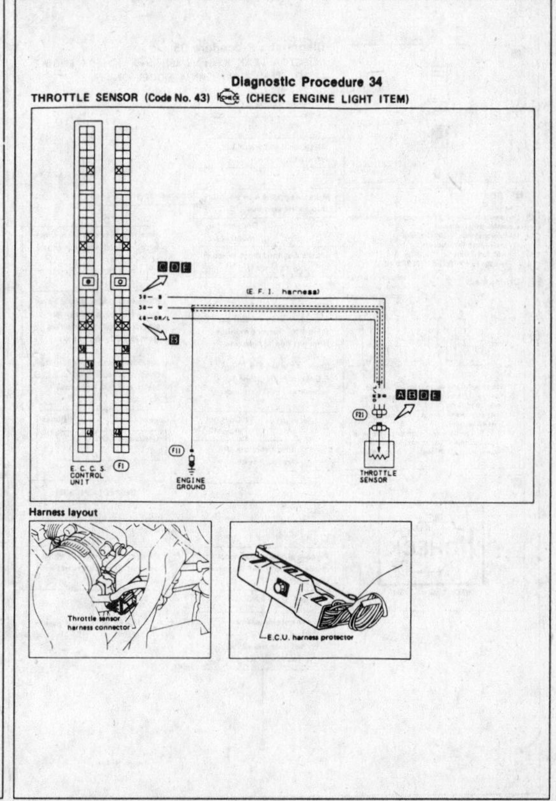
Diagnostic Procedure 34
THROTTLE SENSOR (Code No. 43) (CHECK ENGINE LIGHT ITEM)
(E.F.I. harness)
E.C.C.S. CONTROL UNIT
F1
F11
ENGINE GROUND
F21
THROTTLE SENSOR
Harness layout
Throttle sensor harness connector
E.C.U. harness protector

ECCS SELF-DIAGNOSTIC CHART—1990 300ZX

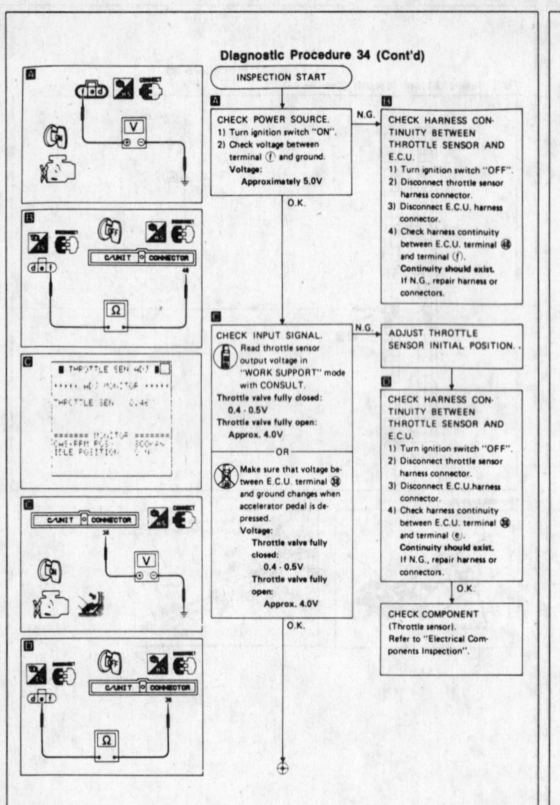
Diagnostic Procedure 34 (Cont'd)
INSPECTION START
CHECK POWER SOURCE.
1) Turn ignition switch "ON".
2) Check voltage between terminal (f) and ground.
Voltage:
Approximately 5.0V
N.G.
CHECK HARNESS CONTINUITY BETWEEN THROTTLE SENSOR AND E.C.U.
1) Turn ignition switch "OFF".
2) Disconnect throttle sensor harness connector.
3) Disconnect E.C.U. harness connector.
4) Check harness continuity between E.C.U. terminal 48 and terminal (f).
Continuity should exist.
If N.G., repair harness or connectors.
O.K.
CHECK INPUT SIGNAL.
Read throttle sensor output voltage in "WORK SUPPORT" mode with CONSULT.
Throttle valve fully closed:
0.4 - 0.5V
Throttle valve fully open:
Approx. 4.0V
OR
Make sure that voltage between E.C.U. terminal 38 and ground changes when accelerator pedal is depressed.
Voltage:
Throttle valve fully closed:
0.4 - 0.5V
Throttle valve fully open:
Approx. 4.0V
N.G.
ADJUST THROTTLE SENSOR INITIAL POSITION.
CHECK HARNESS CONTINUITY BETWEEN THROTTLE SENSOR AND E.C.U.
1) Turn ignition switch "OFF".
2) Disconnect throttle sensor harness connector.
3) Disconnect E.C.U. harness connector.
4) Check harness continuity between E.C.U. terminal 38 and terminal (e).
Continuity should exist.
If N.G., repair harness or connectors.
O.K.
CHECK COMPONENT (Throttle sensor).
Refer to "Electrical Components Inspection".
O.K.
C/UNIT CONNECTOR
THROTTLE SEN
MONITOR

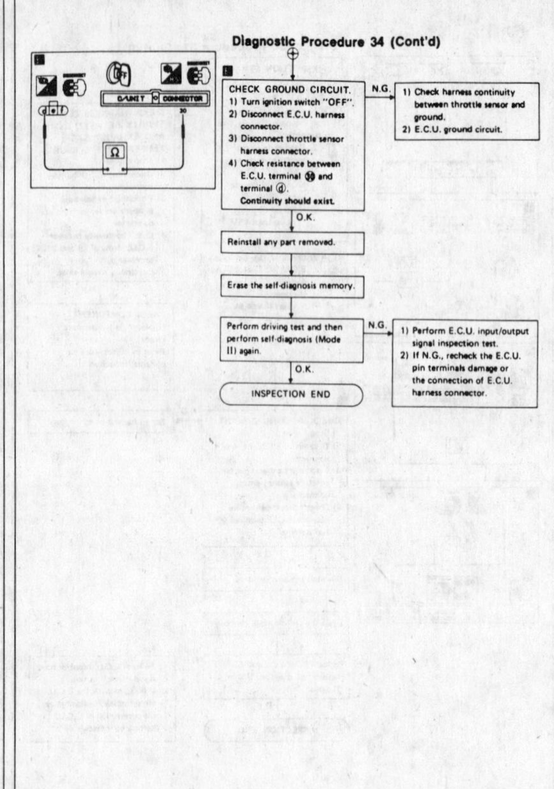
Diagnostic Procedure 34 (Cont'd)
C/UNIT CONNECTOR
CHECK GROUND CIRCUIT.
1) Turn ignition switch "OFF".
2) Disconnect E.C.U. harness connector.
3) Disconnect throttle sensor harness connector.
4) Check resistance between E.C.U. terminal 39 and terminal (d).
Continuity should exist.
N.G.
1) Check harness continuity between throttle sensor and ground.
2) E.C.U. ground circuit.
O.K.
Reinstall any part removed.
Erase the self-diagnosis memory.
Perform driving test and then perform self-diagnosis (Mode II) again.
N.G.
1) Perform E.C.U. input/output signal inspection test.
2) If N.G., recheck the E.C.U. pin terminals damage or the connection of E.C.U. harness connector.
O.K.
INSPECTION END

ECCS SELF-DIAGNOSTIC CHART—1990 300ZX

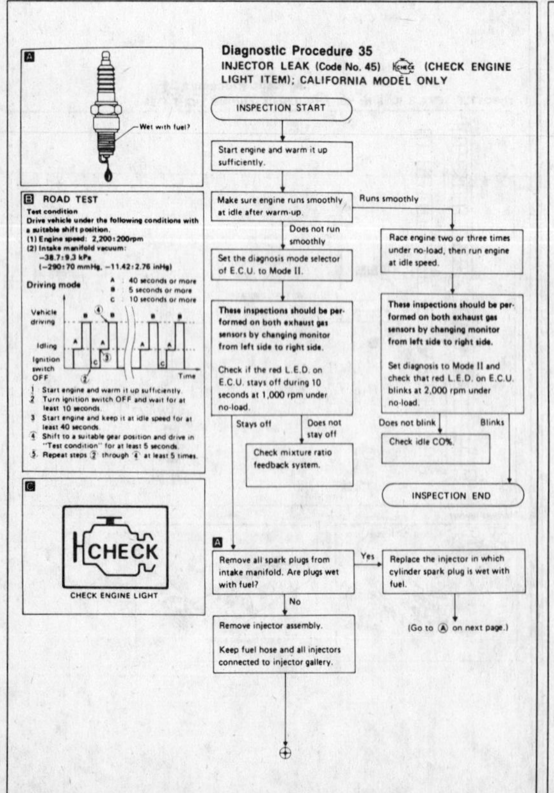
Diagnostic Procedure 35
INJECTOR LEAK (Code No. 45) (CHECK ENGINE LIGHT ITEM); CALIFORNIA MODEL ONLY
Wet with fuel?
INSPECTION START
Start engine and warm it up sufficiently.
Make sure engine runs smoothly at idle after warm-up.
Runs smoothly
Race engine two or three times under no-load, then run engine at idle speed.
Does not run smoothly
Set the diagnosis mode selector of E.C.U. to Mode II.
These inspections should be performed on both exhaust gas sensors by changing monitor from left side to right side.
Check if the red L.E.D. on E.C.U. stays off during 10 seconds at 1,000 rpm under no-load.
Set diagnosis to Mode II and check that red L.E.D. on E.C.U. blinks at 2,000 rpm under no-load.
Stays off
Does not stay off
Does not blink
Blinks
Check mixture ratio feedback system.
Check idle CO%.
INSPECTION END
ROAD TEST
Test condition
Drive vehicle under the following conditions with a suitable shift position.
(1) Engine speed: 2,200±200rpm
(2) Intake manifold vacuum: −38.7±9.3 kPa (−290±70 mmHg, −11.42±2.76 inHg)
Driving mode
A: 40 seconds or more
B: 5 seconds or more
C: 10 seconds or more
Vehicle driving
Idling
Ignition switch OFF
Time
1 Start engine and warm it up sufficiently.
2 Turn ignition switch OFF and wait for at least 10 seconds.
3 Start engine and keep it at idle speed for at least 40 seconds.
4 Shift to a suitable gear position and drive in "Test condition" for at least 5 seconds.
5 Repeat steps 2 through 4 at least 5 times.
CHECK
CHECK ENGINE LIGHT
Remove all spark plugs from intake manifold. Are plugs wet with fuel?
Yes
Replace the injector in which cylinder spark plug is wet with fuel.
No
Remove injector assembly.
Keep fuel hose and all injectors connected to injector gallery.
(Go to (A) on next page.)

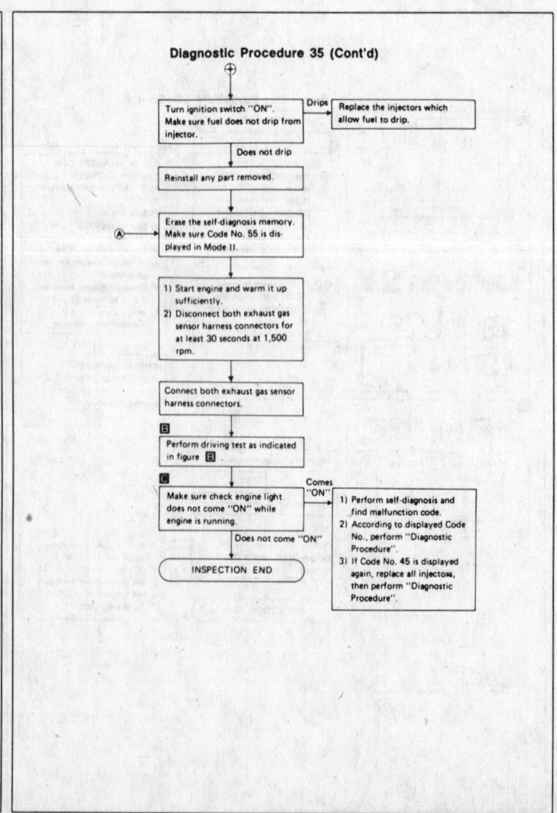
Diagnostic Procedure 35 (Cont'd)
Turn ignition switch "ON". Make sure fuel does not drip from injector.
Drips
Replace the injectors which allow fuel to drip.
Does not drip
Reinstall any part removed.
Erase the self-diagnosis memory. Make sure Code No. 55 is displayed in Mode II.
1) Start engine and warm it up sufficiently.
2) Disconnect both exhaust gas sensor harness connectors for at least 30 seconds at 1,500 rpm.
Connect both exhaust gas sensor harness connectors.
Perform driving test as indicated in figure B.
Make sure check engine light does not come "ON" while engine is running.
Comes "ON"
1) Perform self-diagnosis and find malfunction code.
2) According to displayed Code No., perform "Diagnostic Procedure".
3) If Code No. 45 is displayed again, replace all injectors, then perform "Diagnostic Procedure".
Does not come "ON"
INSPECTION END

ECCS SELF-DIAGNOSTIC CHART—1990 300ZX

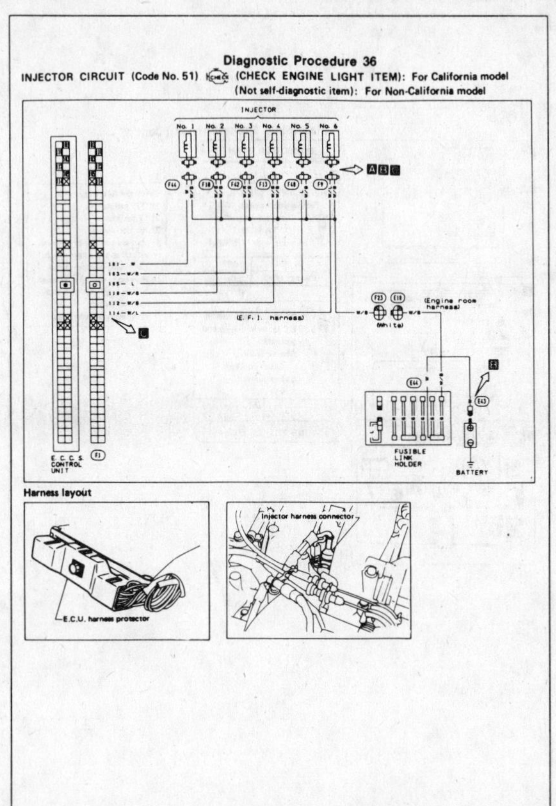
Diagnostic Procedure 36
INJECTOR CIRCUIT (Code No. 51) (CHECK ENGINE LIGHT ITEM): For California model
(Not self-diagnostic item): For Non-California model
INJECTOR
No. 1
No. 2
No. 3
No. 4
No. 5
No. 6
(E.F.I. harness)
(Engine room harness)
FUSIBLE LINK HOLDER
BATTERY
E.C.C.S. CONTROL UNIT
Harness layout
E.C.U. harness protector
Injector harness connector

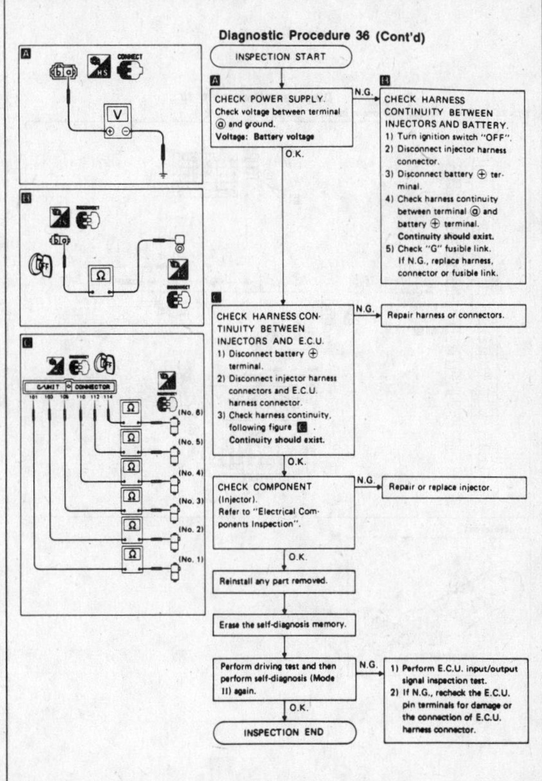
Diagnostic Procedure 36 (Cont'd)
INSPECTION START
CHECK POWER SUPPLY.
Check voltage between terminal ⓐ and ground.
Voltage: Battery voltage
N.G.
CHECK HARNESS CONTINUITY BETWEEN INJECTORS AND BATTERY.
1) Turn ignition switch "OFF".
2) Disconnect injector harness connector.
3) Disconnect battery ⊕ terminal.
4) Check harness continuity between terminal ⓐ and battery ⊕ terminal.
Continuity should exist.
5) Check "G" fusible link.
If N.G., replace harness, connector or fusible link.
O.K.
CHECK HARNESS CONTINUITY BETWEEN INJECTORS AND E.C.U.
1) Disconnect battery ⊕ terminal.
2) Disconnect injector harness connectors and E.C.U. harness connector.
3) Check harness continuity, following figure C.
Continuity should exist.
N.G.
Repair harness or connectors.
O.K.
CHECK COMPONENT (Injector).
Refer to "Electrical Components Inspection".
N.G.
Repair or replace injector.
O.K.
Reinstall any part removed.
Erase the self-diagnosis memory.
Perform driving test and then perform self-diagnosis (Mode II) again.
N.G.
1) Perform E.C.U. input/output signal inspection test.
2) If N.G., recheck the E.C.U. pin terminals for damage or the connection of E.C.U. harness connector.
O.K.
INSPECTION END
C/UNIT CONNECTOR
(No. 6)
(No. 5)
(No. 4)
(No. 3)
(No. 2)
(No. 1)

ECCS SELF-DIAGNOSTIC CHART—1990 300ZX

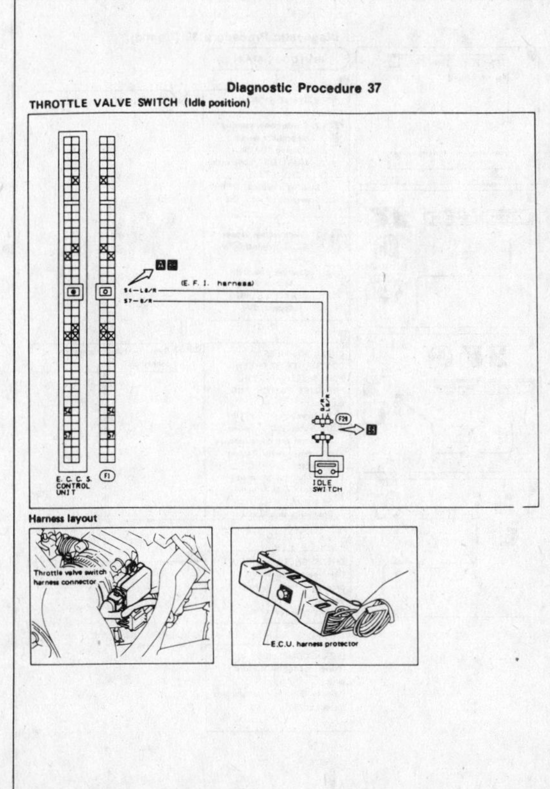
Diagnostic Procedure 37
THROTTLE VALVE SWITCH (Idle position)
(E.F.I. harness)
IDLE SWITCH
E.C.C.S. CONTROL UNIT
Harness layout
Throttle valve switch harness connector
E.C.U. harness protector

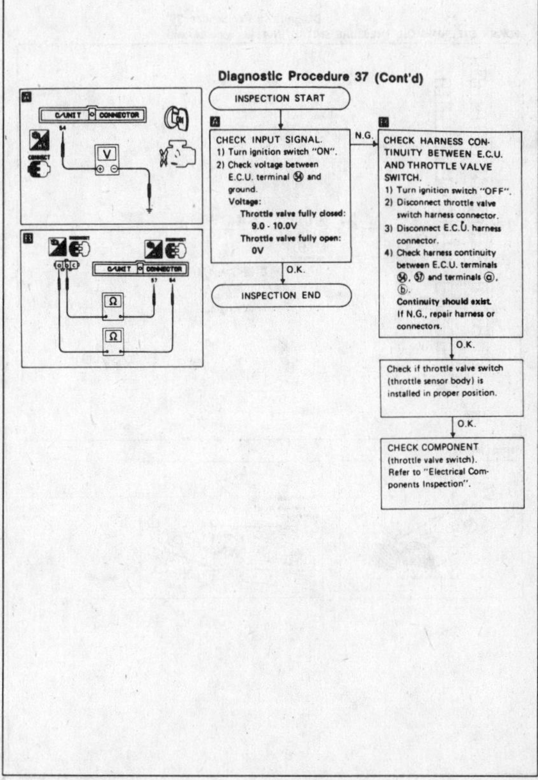
Diagnostic Procedure 37 (Cont'd)
INSPECTION START
CHECK INPUT SIGNAL.
1) Turn ignition switch "ON".
2) Check voltage between E.C.U. terminal 54 and ground.
Voltage:
Throttle valve fully closed: 9.0 - 10.0V
Throttle valve fully open: 0V
N.G.
CHECK HARNESS CONTINUITY BETWEEN E.C.U. AND THROTTLE VALVE SWITCH.
1) Turn ignition switch "OFF".
2) Disconnect throttle valve switch harness connector.
3) Disconnect E.C.U. harness connector.
4) Check harness continuity between E.C.U. terminals 54, 57 and terminals ⓐ, ⓑ.
Continuity should exist.
If N.G., repair harness or connectors.
O.K.
INSPECTION END
O.K.
Check if throttle valve switch (throttle sensor body) is installed in proper position.
O.K.
CHECK COMPONENT (throttle valve switch).
Refer to "Electrical Components Inspection".
C/UNIT CONNECTOR

ECCS SELF-DIAGNOSTIC CHART—1990 300ZX

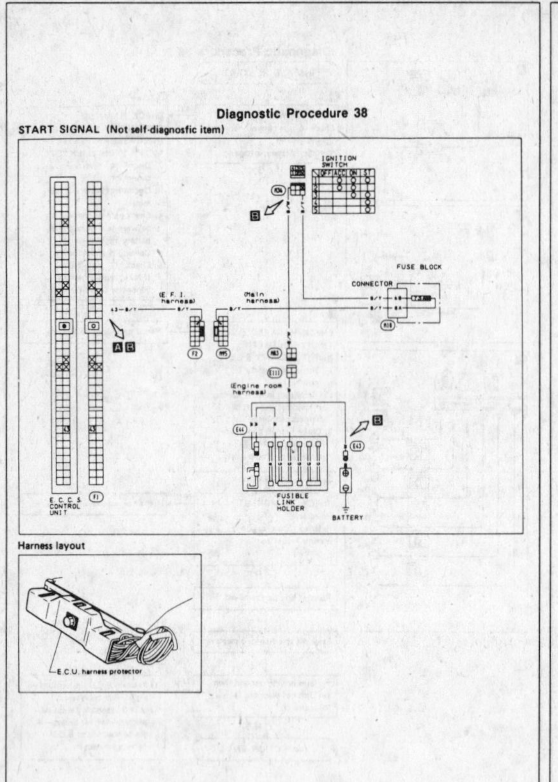
Diagnostic Procedure 38
START SIGNAL (Not self-diagnostic item)
IGNITION SWITCH
FUSE BLOCK
CONNECTOR
(E.F.I. harness)
(Main harness)
(Engine room harness)
FUSIBLE LINK HOLDER
BATTERY
E.C.C.S. CONTROL UNIT
Harness layout
E.C.U. harness protector

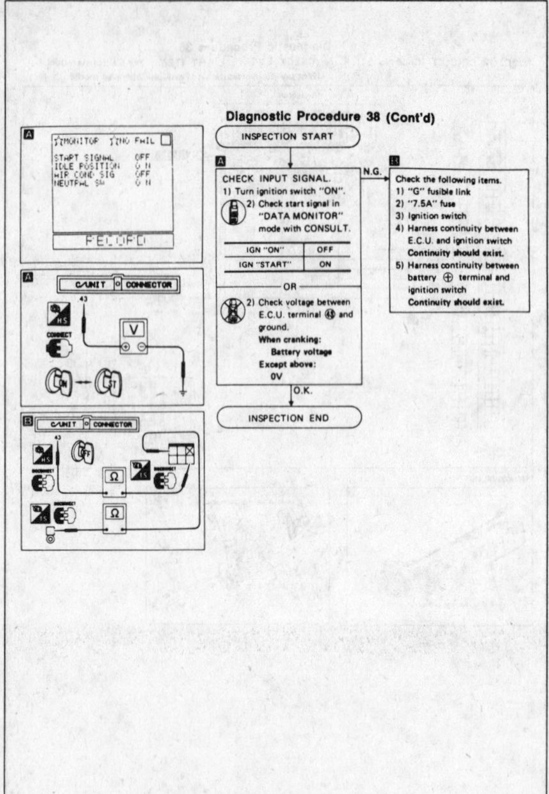
Diagnostic Procedure 38 (Cont'd)
MONITOR NO FAIL
START SIGNAL OFF
IDLE POSITION ON
AIR COND SIG OFF
NEUTRAL SW ON
RECORD
C/UNIT CONNECTOR
INSPECTION START
CHECK INPUT SIGNAL.
1) Turn ignition switch "ON".
2) Check start signal in "DATA MONITOR" mode with CONSULT.
IGN "ON" OFF
IGN "START" ON
OR
2) Check voltage between E.C.U. terminal 43 and ground.
When cranking: Battery voltage
Except above: 0V
N.G.
Check the following items.
1) "G" fusible link
2) "7.5A" fuse
3) Ignition switch
4) Harness continuity between E.C.U. and ignition switch
Continuity should exist.
5) Harness continuity between battery ⊕ terminal and ignition switch
Continuity should exist.
O.K.
INSPECTION END

ECCS SELF-DIAGNOSTIC CHART—1990 300ZX

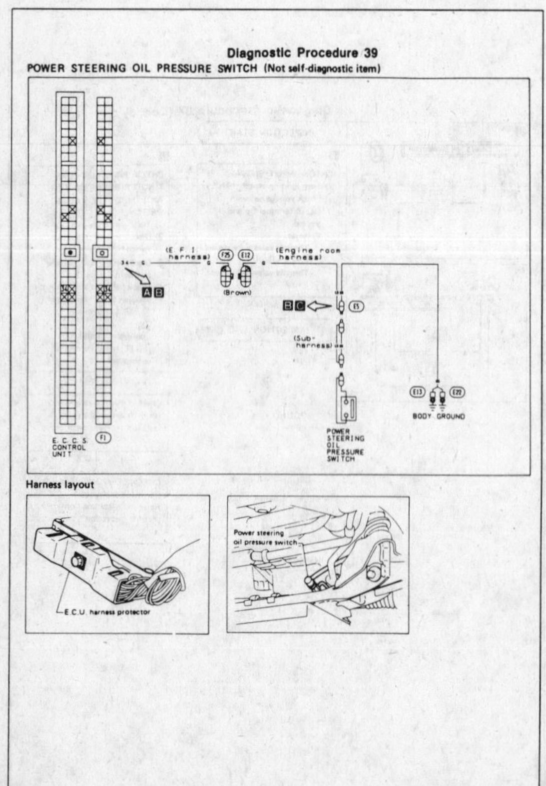
Diagnostic Procedure 39
POWER STEERING OIL PRESSURE SWITCH (Not self-diagnostic item)
(E.F.I. harness)
(Engine room harness)
(Brown)
(Sub-harness)
BODY GROUND
POWER STEERING OIL PRESSURE SWITCH
E.C.C.S. CONTROL UNIT
Harness layout
E.C.U. harness protector
Power steering oil pressure switch

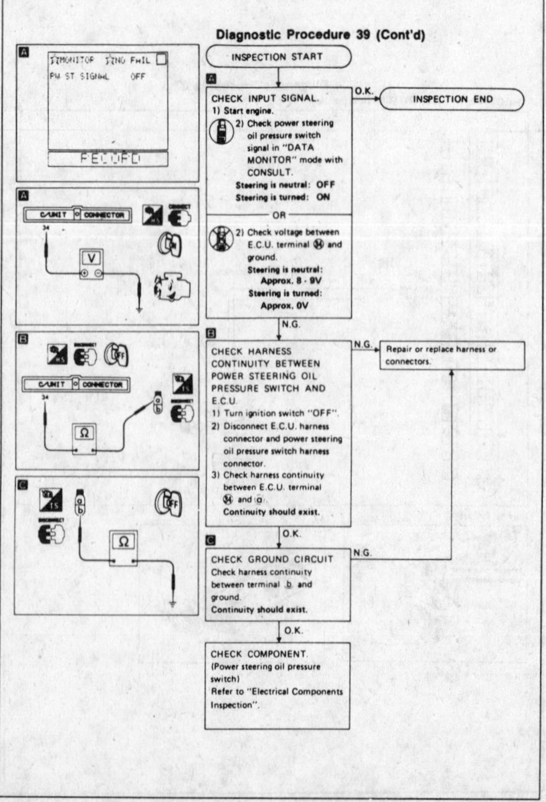
Diagnostic Procedure 39 (Cont'd)
MONITOR NO FAIL
PW ST SIGNAL OFF
RECORD
C/UNIT CONNECTOR
INSPECTION START
CHECK INPUT SIGNAL.
1) Start engine.
2) Check power steering oil pressure switch signal in "DATA MONITOR" mode with CONSULT.
Steering is neutral: OFF
Steering is turned: ON
OR
2) Check voltage between E.C.U. terminal 34 and ground.
Steering is neutral: Approx. 8 - 9V
Steering is turned: Approx. 0V
O.K.
INSPECTION END
N.G.
CHECK HARNESS CONTINUITY BETWEEN POWER STEERING OIL PRESSURE SWITCH AND E.C.U.
1) Turn ignition switch "OFF".
2) Disconnect E.C.U. harness connector and power steering oil pressure switch harness connector.
3) Check harness continuity between E.C.U. terminal 34 and a.
Continuity should exist.
N.G.
Repair or replace harness or connectors.
O.K.
CHECK GROUND CIRCUIT
Check harness continuity between terminal b and ground.
Continuity should exist.
N.G.
O.K.
CHECK COMPONENT.
(Power steering oil pressure switch)
Refer to "Electrical Components Inspection".

ECCS SELF-DIAGNOSTIC CHART—1990 300ZX

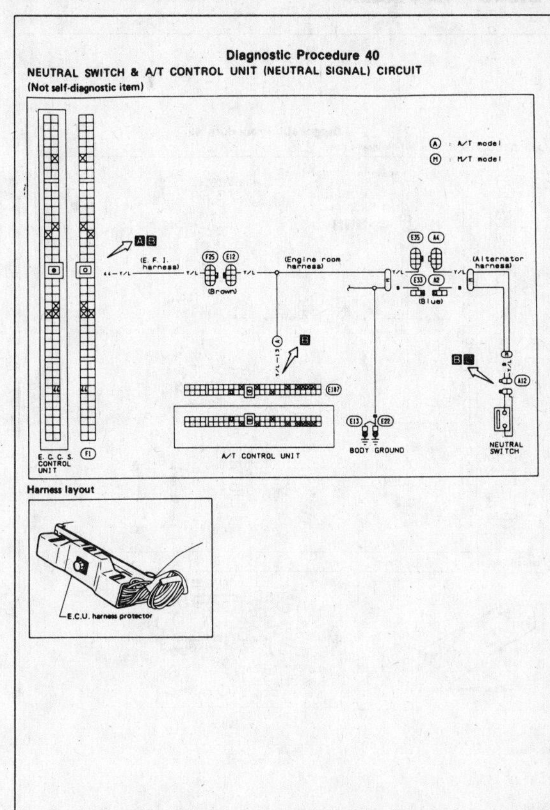
Diagnostic Procedure 40
NEUTRAL SWITCH & A/T CONTROL UNIT (NEUTRAL SIGNAL) CIRCUIT
(Not self-diagnostic item)
Ⓐ : A/T model
Ⓜ : M/T model
(E.F.I. harness)
(Brown)
(Engine room harness)
(Blue)
(Alternator harness)
E.C.C.S. CONTROL UNIT
A/T CONTROL UNIT
BODY GROUND
NEUTRAL SWITCH
Harness layout
E.C.U. harness protector

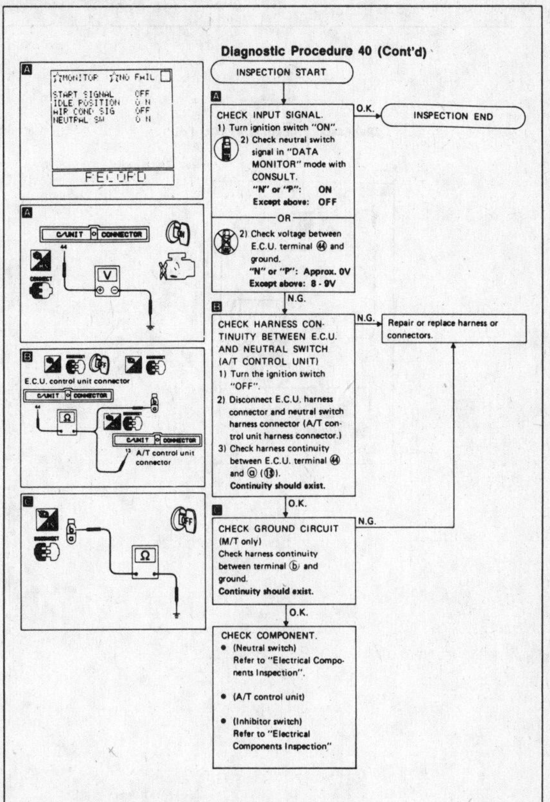
Diagnostic Procedure 40 (Cont'd)
INSPECTION START
CHECK INPUT SIGNAL.
1) Turn ignition switch "ON".
2) Check neutral switch signal in "DATA MONITOR" mode with CONSULT.
"N" or "P": ON
Except above: OFF
OR
2) Check voltage between E.C.U. terminal ㊹ and ground.
"N" or "P": Approx. 0V
Except above: 8 - 9V
O.K.
INSPECTION END
N.G.
CHECK HARNESS CONTINUITY BETWEEN E.C.U. AND NEUTRAL SWITCH (A/T CONTROL UNIT)
1) Turn the ignition switch "OFF".
2) Disconnect E.C.U. harness connector and neutral switch harness connector (A/T control unit harness connector.)
3) Check harness continuity between E.C.U. terminal ㊹ and ⓐ (⑬).
Continuity should exist.
N.G.
Repair or replace harness or connectors.
O.K.
CHECK GROUND CIRCUIT (M/T only)
Check harness continuity between terminal ⓑ and ground.
Continuity should exist.
N.G.
O.K.
CHECK COMPONENT.
• (Neutral switch) Refer to "Electrical Components Inspection".
• (A/T control unit)
• (Inhibitor switch) Refer to "Electrical Components Inspection"
MONITOR NO FAIL
START SIGNAL OFF
IDLE POSITION ON
AIR COND SIG OFF
NEUTRAL SW ON
RECORD
C/UNIT CONNECTOR
E.C.U. control unit connector
A/T control unit connector

ECCS SELF-DIAGNOSTIC CHART—1990 300ZX

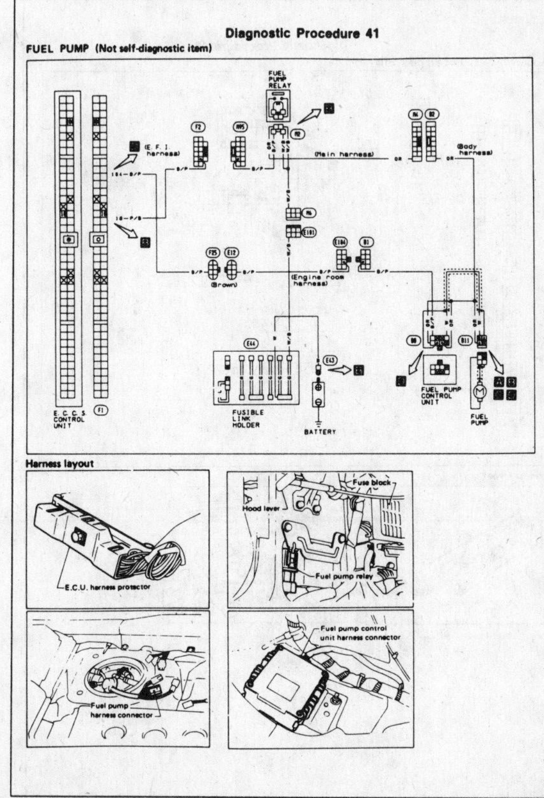
Diagnostic Procedure 41
FUEL PUMP (Not self-diagnostic item)
FUEL PUMP RELAY
(E.F.I. harness)
(Main harness)
(Body harness)
(Brown)
(Engine room harness)
FUSIBLE LINK HOLDER
BATTERY
FUEL PUMP CONTROL UNIT
FUEL PUMP
E.C.C.S. CONTROL UNIT
Harness layout
E.C.U. harness protector
Fuse block
Hood lever
Fuel pump relay
Fuel pump control unit harness connector
Fuel pump harness connector

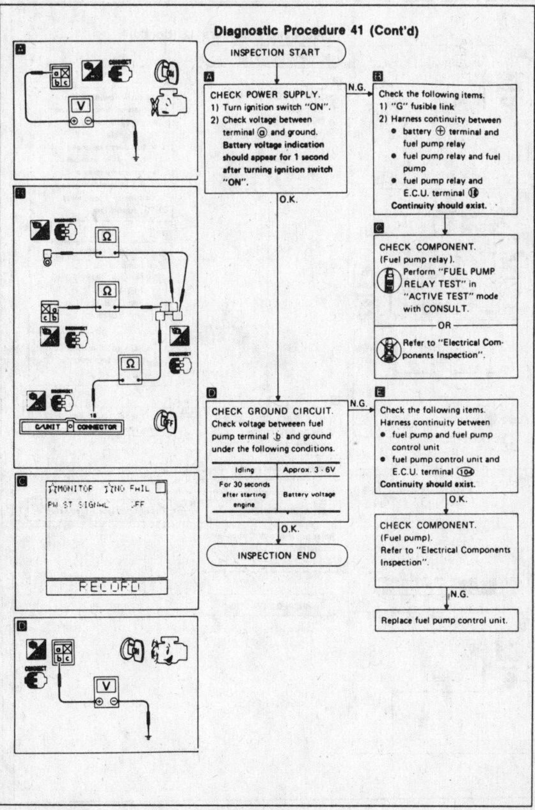
Diagnostic Procedure 41 (Cont'd)
INSPECTION START
CHECK POWER SUPPLY.
1) Turn ignition switch "ON".
2) Check voltage between terminal ⓐ and ground.
Battery voltage indication should appear for 1 second after turning ignition switch "ON".
N.G.
Check the following items.
1) "G" fusible link
2) Harness continuity between
• battery ⊕ terminal and fuel pump relay
• fuel pump relay and fuel pump
• fuel pump relay and E.C.U. terminal ⑱
Continuity should exist.
CHECK COMPONENT.
(Fuel pump relay).
Perform "FUEL PUMP RELAY TEST" in "ACTIVE TEST" mode with CONSULT.
OR
Refer to "Electrical Components Inspection".
O.K.
CHECK GROUND CIRCUIT.
Check voltage between fuel pump terminal ⓑ and ground under the following conditions.
Idling | Approx. 3 - 6V
For 30 seconds after starting engine | Battery voltage
N.G.
Check the following items.
Harness continuity between
• fuel pump and fuel pump control unit
• fuel pump control unit and E.C.U. terminal ⑩④
Continuity should exist.
O.K.
CHECK COMPONENT.
(Fuel pump).
Refer to "Electrical Components Inspection".
N.G.
Replace fuel pump control unit.
O.K.
INSPECTION END
MONITOR NO FAIL
FPCM ST SIGNAL OFF
RECORD
C/UNIT CONNECTOR

ECCS SELF-DIAGNOSTIC CHART—1990 300ZX

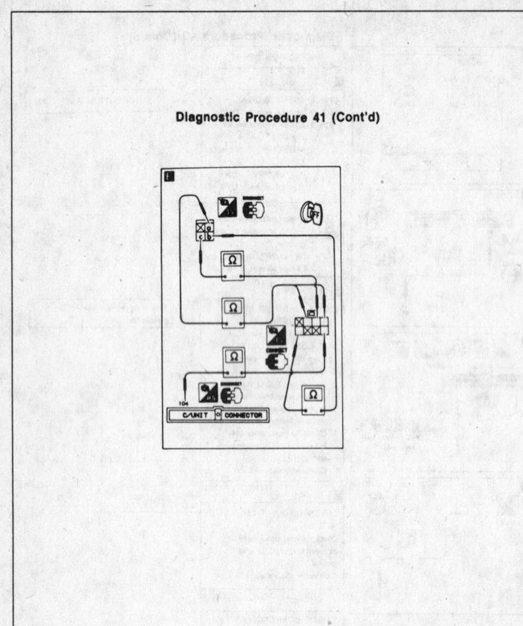
Diagnostic Procedure 41 (Cont'd)
C/UNIT
CONNECTOR

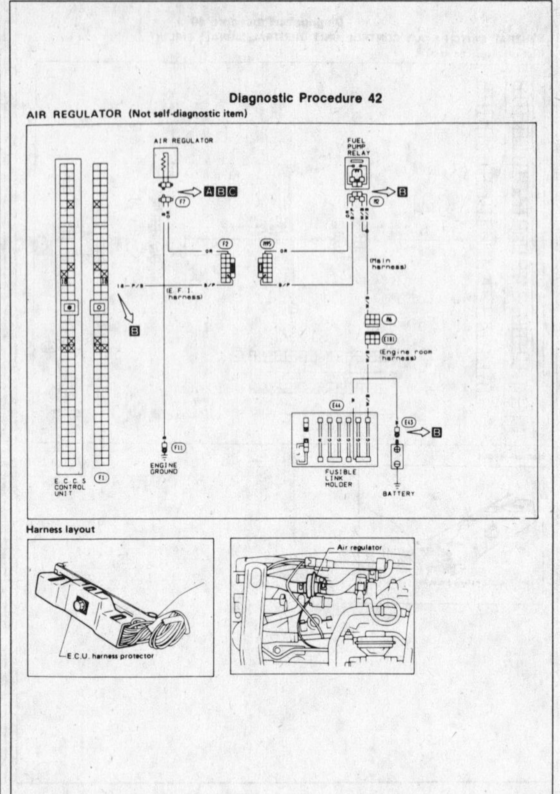
Diagnostic Procedure 42
AIR REGULATOR (Not self-diagnostic item)
AIR REGULATOR
FUEL PUMP RELAY
(E.F.I. harness)
(Main harness)
(Engine room harness)
ENGINE GROUND
E.C.C.S. CONTROL UNIT
FUSIBLE LINK HOLDER
BATTERY
Harness layout
E.C.U. harness protector
Air regulator

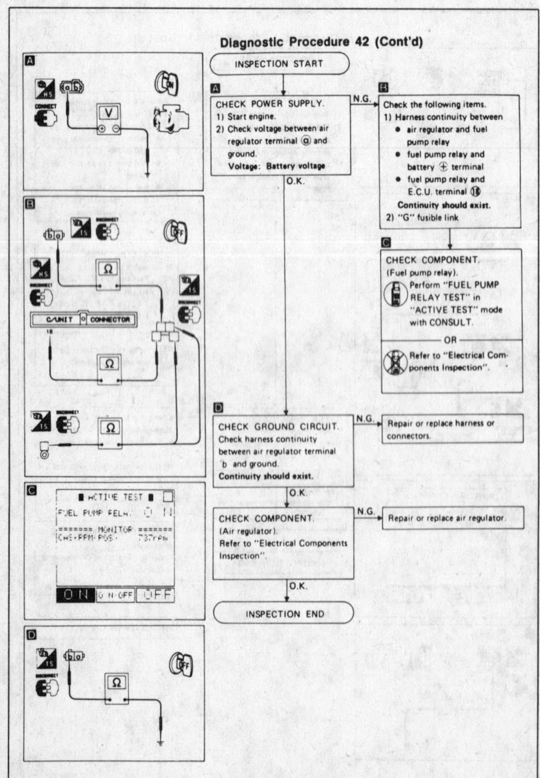
Diagnostic Procedure 42 (Cont'd)
INSPECTION START
A
CHECK POWER SUPPLY.
1) Start engine.
2) Check voltage between air regulator terminal ⓐ and ground.
Voltage: Battery voltage
N.G.
B
Check the following items.
1) Harness continuity between
• air regulator and fuel pump relay
• fuel pump relay and battery ⊕ terminal
• fuel pump relay and E.C.U. terminal ⑱
Continuity should exist.
2) "G" fusible link
C
CHECK COMPONENT.
(Fuel pump relay).
Perform "FUEL PUMP RELAY TEST" in "ACTIVE TEST" mode with CONSULT.
OR
Refer to "Electrical Components Inspection".
O.K.
D
CHECK GROUND CIRCUIT.
Check harness continuity between air regulator terminal ⓑ and ground.
Continuity should exist.
N.G.
Repair or replace harness or connectors.
O.K.
CHECK COMPONENT.
(Air regulator).
Refer to "Electrical Components Inspection".
N.G.
Repair or replace air regulator.
O.K.
INSPECTION END
C/UNIT
CONNECTOR
ACTIVE TEST
FUEL PUMP RELAY
MONITOR
ON
OFF

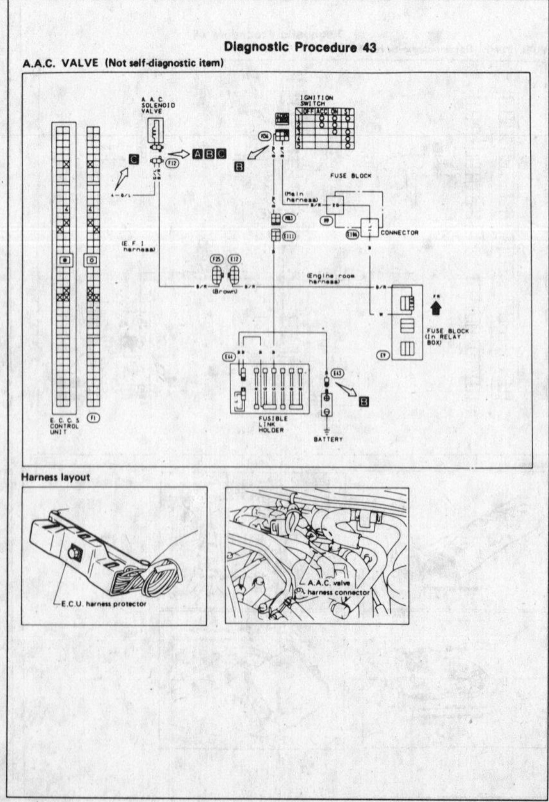
Diagnostic Procedure 43
A.A.C. VALVE (Not self-diagnostic item)
A.A.C. SOLENOID VALVE
IGNITION SWITCH
FUSE BLOCK
(Main harness)
(E.F.I. harness)
CONNECTOR
(Engine room harness)
(Brown)
FUSE BLOCK (In RELAY BOX)
E.C.C.S. CONTROL UNIT
FUSIBLE LINK HOLDER
BATTERY
Harness layout
E.C.U. harness protector
A.A.C. valve harness connector

ECCS SELF-DIAGNOSTIC CHART—1990 300ZX

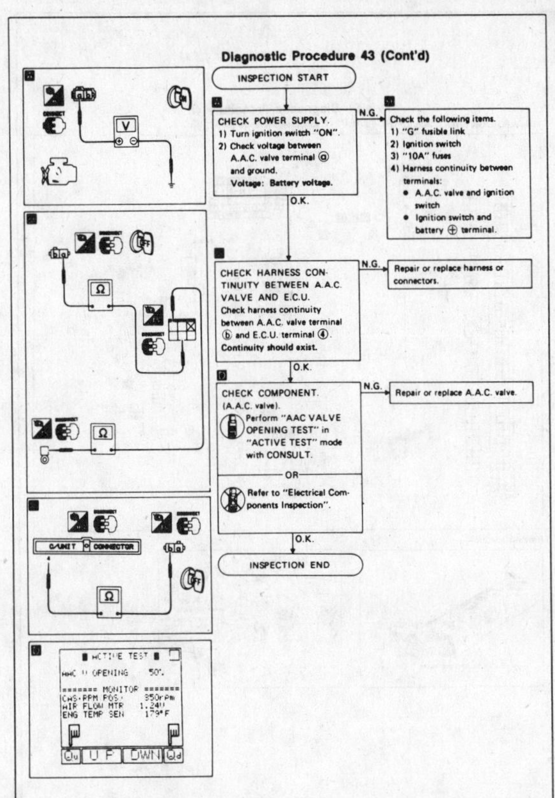
Diagnostic Procedure 43 (Cont'd)
INSPECTION START
CHECK POWER SUPPLY.
1) Turn ignition switch "ON".
2) Check voltage between A.A.C. valve terminal ⓐ and ground.
Voltage: Battery voltage.
N.G.
Check the following items.
1) "G" fusible link
2) Ignition switch
3) "10A" fuses
4) Harness continuity between terminals:
• A.A.C. valve and ignition switch
• Ignition switch and battery ⊕ terminal.
O.K.
CHECK HARNESS CONTINUITY BETWEEN A.A.C. VALVE AND E.C.U.
Check harness continuity between A.A.C. valve terminal ⓑ and E.C.U. terminal ④.
Continuity should exist.
Repair or replace harness or connectors.
CHECK COMPONENT. (A.A.C. valve).
Perform "AAC VALVE OPENING TEST" in "ACTIVE TEST" mode with CONSULT.
OR
Refer to "Electrical Components Inspection".
Repair or replace A.A.C. valve.
INSPECTION END

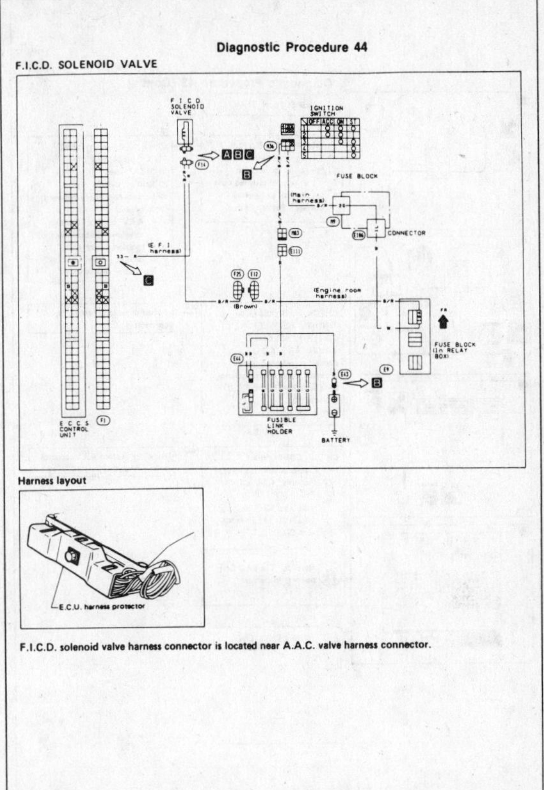
Diagnostic Procedure 44
F.I.C.D. SOLENOID VALVE
FUSE BLOCK
CONNECTOR
FUSE BLOCK (In RELAY BOX)
FUSIBLE LINK HOLDER
BATTERY
E.C.C.S. CONTROL UNIT
Harness layout
E.C.U. harness protector

F.I.C.D. solenoid valve harness connector is located near A.A.C. valve harness connector.

ECCS SELF-DIAGNOSTIC CHART—1990 300ZX

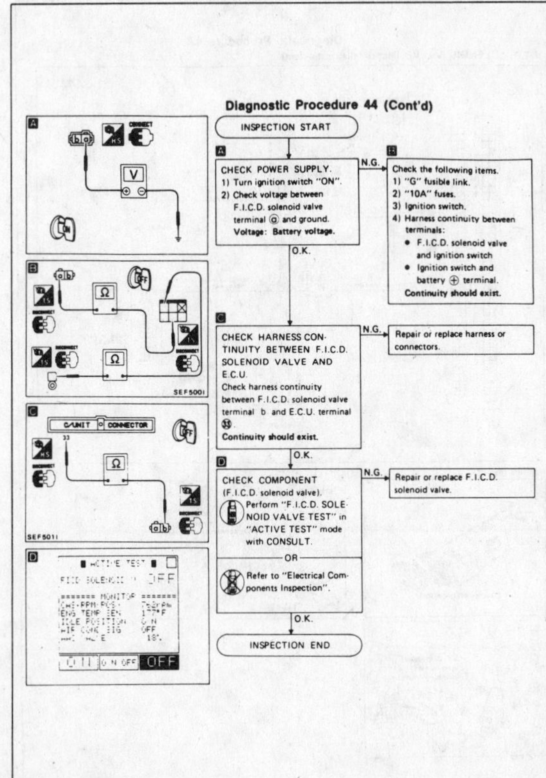
Diagnostic Procedure 44 (Cont'd)
INSPECTION START
CHECK POWER SUPPLY.
1) Turn ignition switch "ON".
2) Check voltage between F.I.C.D. solenoid valve terminal ⓐ and ground.
Voltage: Battery voltage.
N.G.
Check the following items.
1) "G" fusible link.
2) "10A" fuses.
3) Ignition switch.
4) Harness continuity between terminals:
• F.I.C.D. solenoid valve and ignition switch
• Ignition switch and battery ⊕ terminal.
Continuity should exist.
O.K.
CHECK HARNESS CONTINUITY BETWEEN F.I.C.D. SOLENOID VALVE AND E.C.U.
Check harness continuity between F.I.C.D. solenoid valve terminal b and E.C.U. terminal 33.
Continuity should exist.
Repair or replace harness or connectors.
CHECK COMPONENT (F.I.C.D. solenoid valve).
Perform "F.I.C.D. SOLENOID VALVE TEST" in "ACTIVE TEST" mode with CONSULT.
OR
Refer to "Electrical Components Inspection".
Repair or replace F.I.C.D. solenoid valve.
INSPECTION END
SEF500I
SEF501I

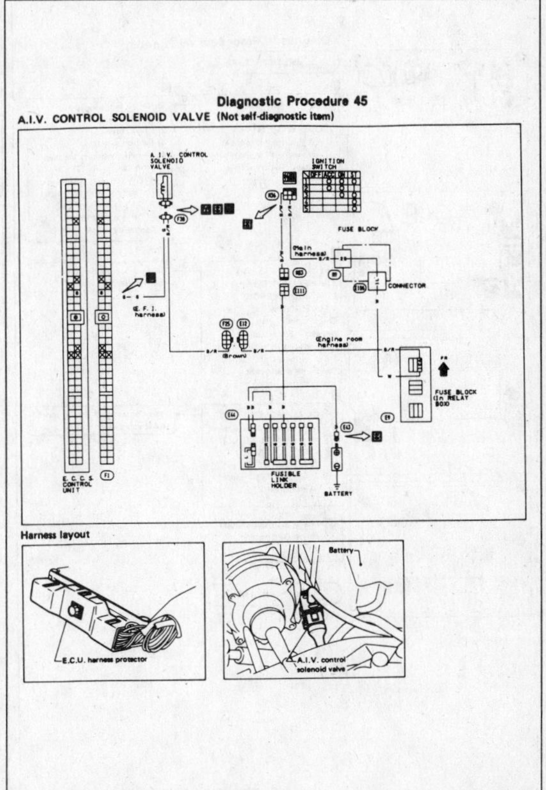
Diagnostic Procedure 45
A.I.V. CONTROL SOLENOID VALVE (Not self-diagnostic item)
FUSE BLOCK
CONNECTOR
FUSE BLOCK (In RELAY BOX)
FUSIBLE LINK HOLDER
BATTERY
E.C.C.S. CONTROL UNIT
Harness layout
E.C.U. harness protector
Battery
A.I.V. control solenoid valve

ECCS SELF-DIAGNOSTIC CHART—1990 300ZX

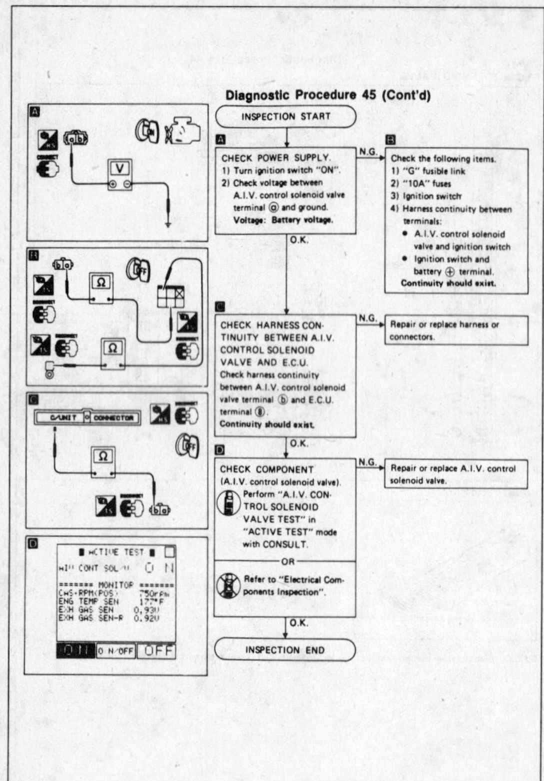
Diagnostic Procedure 45 (Cont'd)
INSPECTION START
A
CHECK POWER SUPPLY.
1) Turn ignition switch "ON".
2) Check voltage between A.I.V. control solenoid valve terminal ⓐ and ground.
Voltage: Battery voltage.
N.G.
B
Check the following items.
1) "G" fusible link
2) "10A" fuses
3) Ignition switch
4) Harness continuity between terminals:
• A.I.V. control solenoid valve and ignition switch
• Ignition switch and battery ⊕ terminal.
Continuity should exist.
O.K.
C
CHECK HARNESS CONTINUITY BETWEEN A.I.V. CONTROL SOLENOID VALVE AND E.C.U.
Check harness continuity between A.I.V. control solenoid valve terminal ⓑ and E.C.U. terminal ⑪.
Continuity should exist.
N.G.
Repair or replace harness or connectors.
O.K.
D
CHECK COMPONENT (A.I.V. control solenoid valve).
Perform "A.I.V. CONTROL SOLENOID VALVE TEST" in "ACTIVE TEST" mode with CONSULT.
OR
Refer to "Electrical Components Inspection".
N.G.
Repair or replace A.I.V. control solenoid valve.
O.K.
INSPECTION END
C/UNIT
CONNECTOR
ACTIVE TEST
MONITOR
CAS·RPM(POS) 750rpm
ENG TEMP SEN 177°F
EXH GAS SEN 0.93V
EXH GAS SEN-R 0.92V
ON
OFF

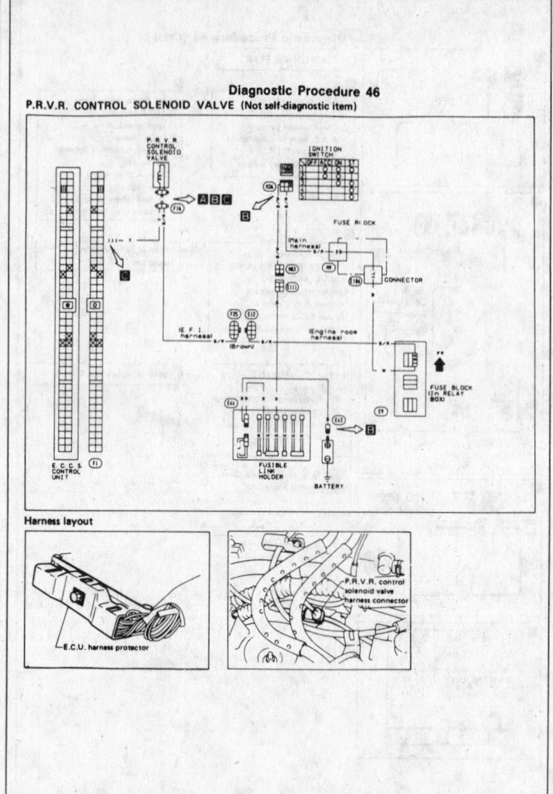
Diagnostic Procedure 46
P.R.V.R. CONTROL SOLENOID VALVE (Not self-diagnostic item)
P.R.V.R. CONTROL SOLENOID VALVE
IGNITION SWITCH
FUSE BLOCK
(Main harness)
CONNECTOR
(E.F.I. harness)
(Brown)
(Engine room harness)
FUSE BLOCK (In RELAY BOX)
E.C.C.S. CONTROL UNIT
FUSIBLE LINK HOLDER
BATTERY
Harness layout
E.C.U. harness protector
P.R.V.R. control solenoid valve harness connector

ECCS SELF-DIAGNOSTIC CHART—1990 300ZX

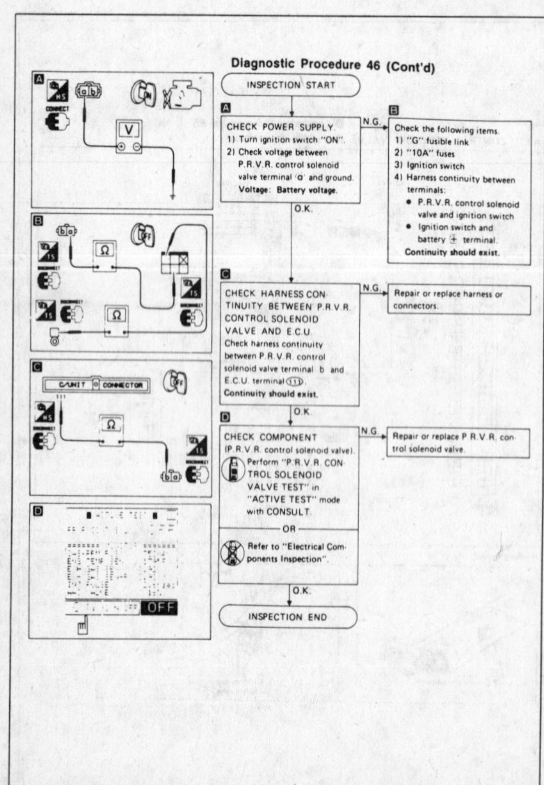
Diagnostic Procedure 46 (Cont'd)
INSPECTION START
A
CHECK POWER SUPPLY.
1) Turn ignition switch "ON".
2) Check voltage between P.R.V.R. control solenoid valve terminal ⓐ and ground.
Voltage: Battery voltage.
N.G.
B
Check the following items.
1) "G" fusible link
2) "10A" fuses
3) Ignition switch
4) Harness continuity between terminals:
• P.R.V.R. control solenoid valve and ignition switch
• Ignition switch and battery ⊕ terminal.
Continuity should exist.
O.K.
C
CHECK HARNESS CONTINUITY BETWEEN P.R.V.R. CONTROL SOLENOID VALVE AND E.C.U.
Check harness continuity between P.R.V.R. control solenoid valve terminal ⓑ and E.C.U. terminal ⑪.
Continuity should exist.
N.G.
Repair or replace harness or connectors.
O.K.
D
CHECK COMPONENT (P.R.V.R. control solenoid valve).
Perform "P.R.V.R. CONTROL SOLENOID VALVE TEST" in "ACTIVE TEST" mode with CONSULT.
OR
Refer to "Electrical Components Inspection".
N.G.
Repair or replace P.R.V.R. control solenoid valve.
O.K.
INSPECTION END
C/UNIT
CONNECTOR
OFF

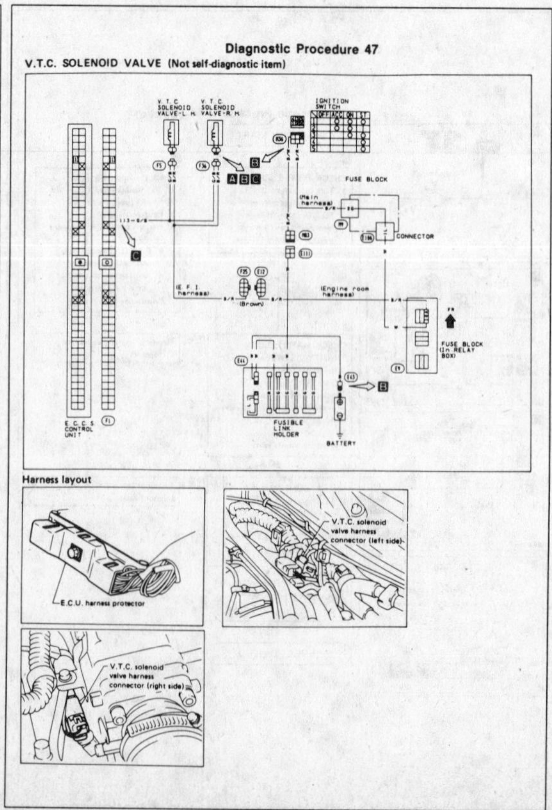
Diagnostic Procedure 47
V.T.C. SOLENOID VALVE (Not self-diagnostic item)
V.T.C. SOLENOID VALVE-L.H.
V.T.C. SOLENOID VALVE-R.H.
IGNITION SWITCH
FUSE BLOCK
(Main harness)
CONNECTOR
(E.F.I. harness)
(Brown)
(Engine room harness)
FUSE BLOCK (In RELAY BOX)
E.C.C.S. CONTROL UNIT
FUSIBLE LINK HOLDER
BATTERY
Harness layout
E.C.U. harness protector
V.T.C. solenoid valve harness connector (left side)
V.T.C. solenoid valve harness connector (right side)

ECCS SELF-DIAGNOSTIC CHART—1990 300ZX

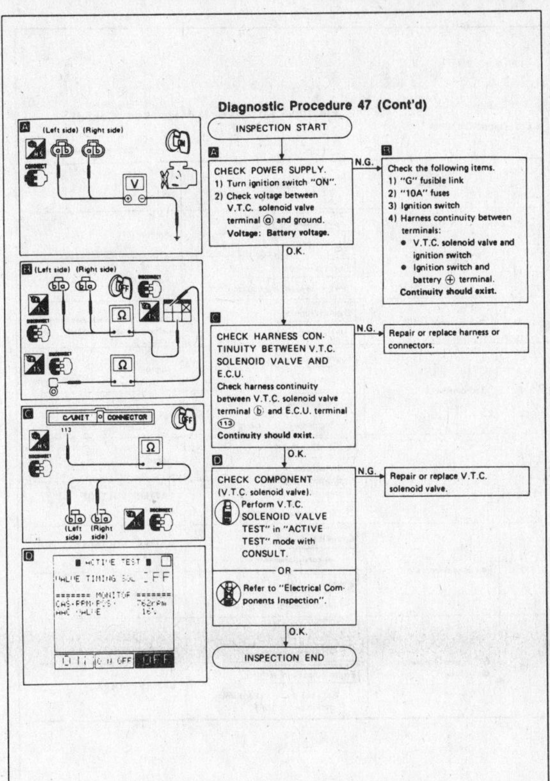
Diagnostic Procedure 47 (Cont'd)
INSPECTION START
CHECK POWER SUPPLY.
1) Turn ignition switch "ON".
2) Check voltage between V.T.C. solenoid valve terminal ⓐ and ground.
Voltage: Battery voltage.
N.G.
Check the following items.
1) "G" fusible link
2) "10A" fuses
3) Ignition switch
4) Harness continuity between terminals:
• V.T.C. solenoid valve and ignition switch
• Ignition switch and battery ⊕ terminal.
Continuity should exist.
O.K.
CHECK HARNESS CONTINUITY BETWEEN V.T.C. SOLENOID VALVE AND E.C.U.
Check harness continuity between V.T.C. solenoid valve terminal ⓑ and E.C.U. terminal ⑬
Continuity should exist.
N.G.
Repair or replace harness or connectors.
O.K.
CHECK COMPONENT (V.T.C. solenoid valve).
Perform V.T.C. SOLENOID VALVE TEST" in "ACTIVE TEST" mode with CONSULT.
OR
Refer to "Electrical Components Inspection".
N.G.
Repair or replace V.T.C. solenoid valve.
O.K.
INSPECTION END
(Left side) (Right side)
C/UNIT CONNECTOR
113
ACTIVE TEST
VALVE TIMING SOL OFF
MONITOR
OFF

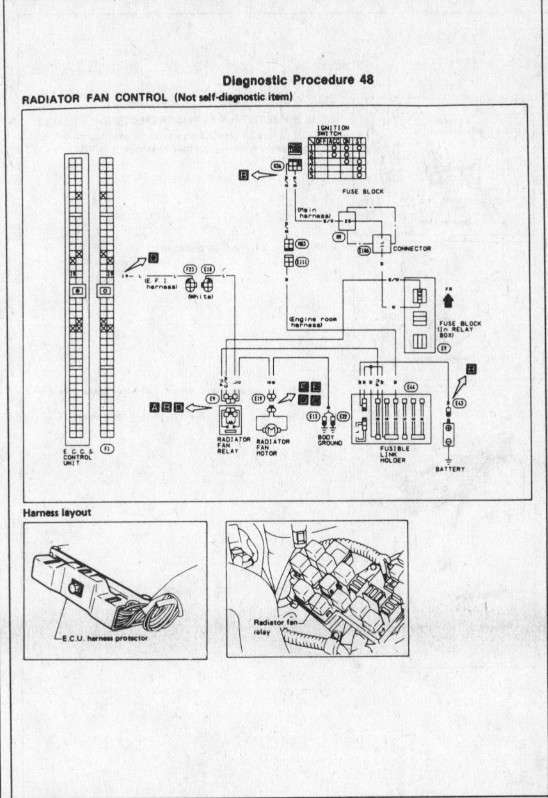
Diagnostic Procedure 48
RADIATOR FAN CONTROL (Not self-diagnostic item)
IGNITION SWITCH
FUSE BLOCK
CONNECTOR
FUSE BLOCK (In RELAY BOX)
E.C.C.S. CONTROL UNIT
RADIATOR FAN RELAY
RADIATOR FAN MOTOR
BODY GROUND
FUSIBLE LINK HOLDER
BATTERY
Harness layout
E.C.U. harness protector
Radiator fan relay

ECCS SELF-DIAGNOSTIC CHART—1990 300ZX

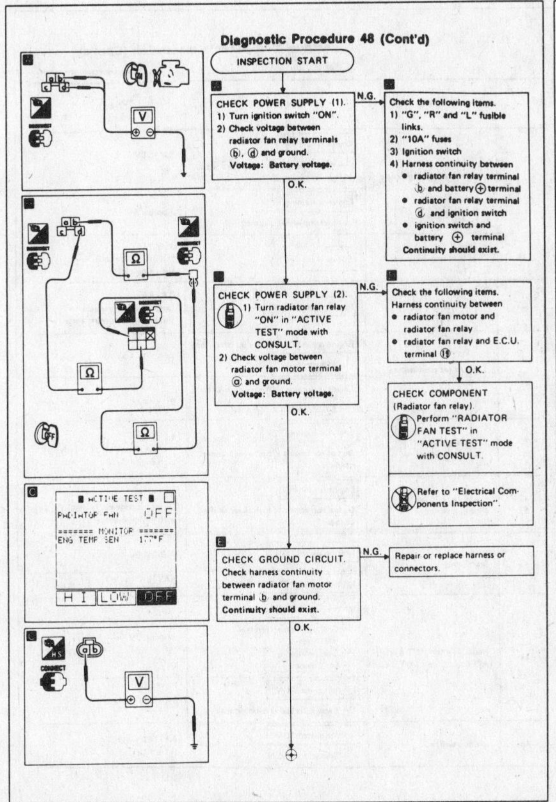
Diagnostic Procedure 48 (Cont'd)
INSPECTION START
CHECK POWER SUPPLY (1).
1) Turn ignition switch "ON".
2) Check voltage between radiator fan relay terminals ⓑ, ⓓ and ground.
Voltage: Battery voltage.
N.G.
Check the following items.
1) "G", "R" and "L" fusible links.
2) "10A" fuses
3) Ignition switch
4) Harness continuity between
• radiator fan relay terminal ⓑ and battery ⊕ terminal
• radiator fan relay terminal ⓓ and ignition switch
• ignition switch and battery ⊕ terminal
Continuity should exist.
O.K.
CHECK POWER SUPPLY (2).
1) Turn radiator fan relay "ON" in "ACTIVE TEST" mode with CONSULT.
2) Check voltage between radiator fan motor terminal ⓐ and ground.
Voltage: Battery voltage.
N.G.
Check the following items.
Harness continuity between
• radiator fan motor and radiator fan relay
• radiator fan relay and E.C.U. terminal ⑲
O.K.
CHECK COMPONENT (Radiator fan relay).
Perform "RADIATOR FAN TEST" in "ACTIVE TEST" mode with CONSULT.
Refer to "Electrical Components Inspection".
O.K.
CHECK GROUND CIRCUIT.
Check harness continuity between radiator fan motor terminal ⓑ and ground.
Continuity should exist.
N.G.
Repair or replace harness or connectors.
O.K.
ACTIVE TEST
RADIATOR FAN OFF
MONITOR
ENG TEMP SEN 177°F
HI LOW OFF

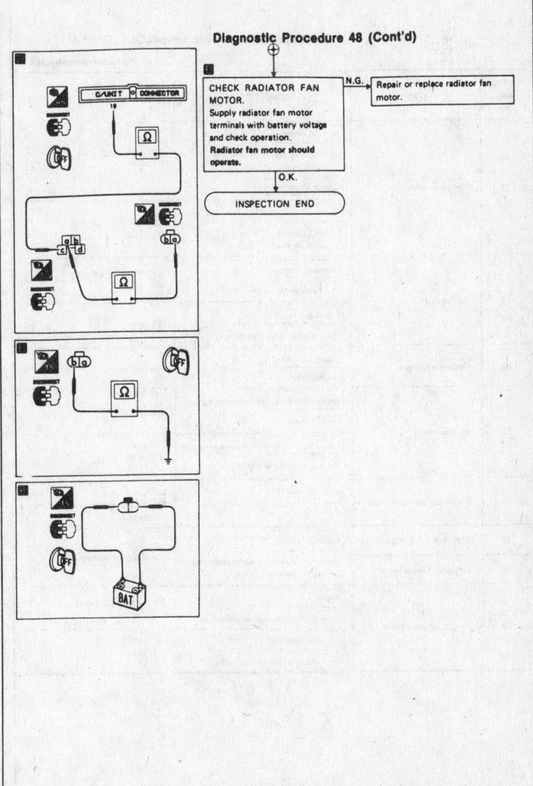
Diagnostic Procedure 48 (Cont'd)
CHECK RADIATOR FAN MOTOR.
Supply radiator fan motor terminals with battery voltage and check operation.
Radiator fan motor should operate.
N.G.
Repair or replace radiator fan motor.
O.K.
INSPECTION END
C/UNIT CONNECTOR
BAT

ECU SIGNAL INSPECTION—1990 300ZX

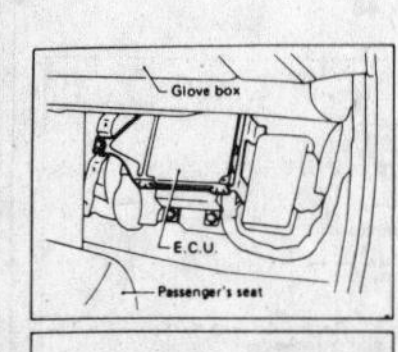
Glove box
E.C.U.
Passenger's seat

Electrical Components Inspection

E.C.U. INPUT/OUTPUT SIGNAL INSPECTION

1. E.C.U. is located behind front passenger side floor board. For this inspection, remove the front passenger side floor board.

E.C.U. harness protector

2. Remove E.C.U. harness protector.

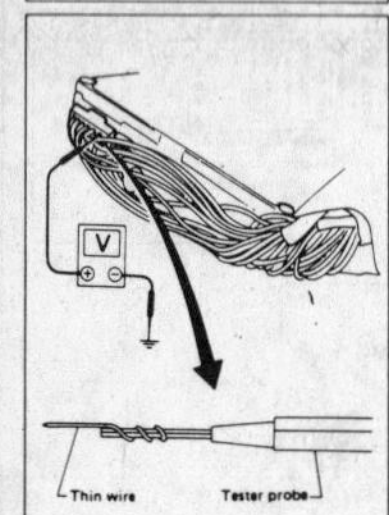
V
Thin wire
Tester probe

3. Perform all voltage measurements with the connectors connected.
Extend tester probe as shown to perform tests easily.

Electrical Components Inspection (Cont'd)

E.C.U. inspection table

*Data are reference values.

TERMINAL NO.	ITEM	CONDITION	*DATA
1 2 3 11 12 13	Ignition signal	Engine is running. └ Idle speed	Approx. 0.1V
		Engine is running. └ Engine speed is 2,000 rpm.	Approx. 0.14V
4	A.A.C. valve	Engine is running. └ Racing condition	Voltage briefly decreases from battery voltage (11 - 14V).
7	Tachometer	Engine is running. └ Idle speed	Approx. 0.7V
		Engine is running. └ Engine speed is 2,000 rpm.	Approx. 1.2V
8	A.I.V. control solenoid valve	Engine is running. └ Idle speed	Approx. 0V
		Engine is running. └ Engine speed is 2,000 rpm.	BATTERY VOLTAGE (11 - 14V)
9	Air conditioner relay	Engine is running. └ Air conditioner switch "OFF"	BATTERY VOLTAGE (11 - 14V)
		Engine is running. └ Air conditioner switch "ON"	Approx. 0V
16	E.C.U. power source (Self-shutoff)	Engine is running. └ Idle speed	0.8 - 1.0V
		Engine is not running. └ For a few seconds after turning ignition switch "OFF"	BATTERY VOLTAGE (11 - 14V)

ECU SIGNAL INSPECTION (CONT.)—1990 300ZX

Electrical Components Inspection (Cont'd)

*Data are reference values.

TERMINAL NO.	ITEM	CONDITION	*DATA
18	Fuel pump relay	Ignition switch "ON" └ For 5 seconds after turning ignition switch "ON"	0.7 - 0.9V
		Engine is running.	
		Ignition switch "ON" └ In 5 seconds after turning ignition switch "ON"	BATTERY VOLTAGE (11 - 14V)
19	Radiator fan	Engine is running. └ Radiator fan is not operating.	BATTERY VOLTAGE (11 - 14V)
		Engine is running. └ Radiator fan is operating.	0.1 - 0.3V
23	Detonation sensor	Engine is running. └ Idle speed	Approx. 2.5V
27	Air flow meter	Engine is running. (Warm-up condition) └ Idle speed	0.8 - 1.5V
		Engine is running. (Warm-up condition) └ Engine speed is 2,000 rpm.	1.0 - 1.6V
28	Engine temperature sensor	Engine is running.	0 - 5.0V Output voltage varies with engine temperature.
29 55	Right side exhaust gas sensor Left side exhaust gas sensor	Engine is running. └ After warming up sufficiently and engine speed is 2,000 rpm.	0 ⟷ approximately 1.0V
33	F.I.C.D. solenoid valve	Engine is running. └ A/C compressor is not operating.	BATTERY VOLTAGE (11 - 14V)
		Engine is running. └ A/C compressor is operating.	0.7 - 0.8V

Electrical Components Inspection (Cont'd)

*Data are reference values.

TERMINAL NO.	ITEM	CONDITION	*DATA
34	Power steering oil pressure switch	Engine is running. └ Steering wheel is in the "straight ahead" position.	8.0 - 9.0V
		Engine is running. └ Steering wheel is turned.	Approx. 0V
36	Fuel temperature sensor	Engine is running.	0 - 5.0V Output voltage varies with fuel temperature.
38	Throttle sensor	Ignition switch "ON"	0.4 - 4.0V Output voltage varies with throttle valve opening angle.
39	Exhaust gas temperature sensor	Engine is running. (Warm-up condition) └ Idle speed	Less than 4.5V
		Engine is running. (Warm-up condition) └ E.G.R. system is operating.	0 - 1.0V
41 51	Crank angle sensor (Reference signal)	Engine is running. Do not run engine at high speed under no-load.	1.2 - 1.4V Output voltage varies slightly with engine speed.
42 52	Crank angle sensor (Position signal)	Engine is running. Do not run engine at high speed under no-load.	2.5 - 2.7V Output voltage varies slightly with engine speed.
43	Start signal	Ignition switch "ON"	Approx. 0V
		Ignition switch "START"	BATTERY VOLTAGE (11 - 14V)
44	Neutral switch (M/T model) A/T control unit (A/T model)	Ignition switch "ON" └ Gear position is "Neutral" (M/T model). └ Gear position is "N" or "P" (A/T model).	Approx. 0V
		Ignition switch "ON" └ Except the above conditions	8.0 - 9.0V
45	Ignition switch	Ignition switch "ON" └ Engine stopped	BATTERY VOLTAGE (11 - 14V)

ECU SIGNAL INSPECTION (CONT.)—1990 300ZX

Electrical Components Inspection (Cont'd)

*Data are reference values.

TERMINAL NO.	ITEM	CONDITION	*DATA
46	Air conditioner switch	Engine is running. — Air conditioner switch "OFF"	BATTERY VOLTAGE (11 - 14V)
		Engine is running. — Air conditioner switch "ON"	0.5 - 0.7V
48	Power source for sensors	Ignition switch "ON" — Engine stopped	Approximately 5.0V
49	Battery source	Ignition switch "ON" — Engine stopped	BATTERY VOLTAGE (11 - 14V)
54	Idle switch	Ignition switch "ON" — Accelerator pedal is fully released (engine running).	9.0 - 10.0V
		Ignition switch "ON" — Accelerator pedal is depressed (engine running).	0V
57	Power source for idle switch	Ignition switch "ON" — Engine running	BATTERY VOLTAGE (11 - 14V)
59	Power supply	Ignition switch "ON" — Engine running	BATTERY VOLTAGE (11 - 14V)
101 103 105 110 112 114	Injectors	Ignition switch "OFF"	BATTERY VOLTAGE (11 - 14V)
102	E.G.R. control solenoid valve	Engine is running. (Warm-up condition) — Idle speed	0.7 - 0.8V
		Engine is running. (Warm-up condition) — Engine speed is 2,000 rpm.	BATTERY VOLTAGE (11 - 14V)

Electrical Components Inspection (Cont'd)

*Data are reference values.

TERMINAL NO.	ITEM	CONDITION	*DATA
104	Fuel pump voltage control	Ignition switch "ON" — Engine stopped	BATTERY VOLTAGE (11 - 14V)
		Engine is running. (Warm-up condition) — Idle speed	0V
		Engine is cranking.	Approx. 5.0V
111	P.R.V.R. control solenoid valve	Stop and restart engine after warming it up. — Fuel temperature is above 75°C (167°F)	0 - 1.0V (for 30 seconds after ignition switch is turned off.)
			BATTERY VOLTAGE (After 30 seconds)
		Stop and restart engine after warming it up. — Fuel temperature is below 75°C (167°F)	BATTERY VOLTAGE (11 - 14V)
113	Valve timing control solenoid valve	Engine is running. — Idle speed	BATTERY VOLTAGE (11 - 14V)
		Engine is running. — Engine speed is 3,000 rpm.	0.2 - 0.5V

E.C.U. HARNESS CONNECTOR TERMINAL LAYOUT

ECCS COMPONENTS LOCATION—1988 PULSAR

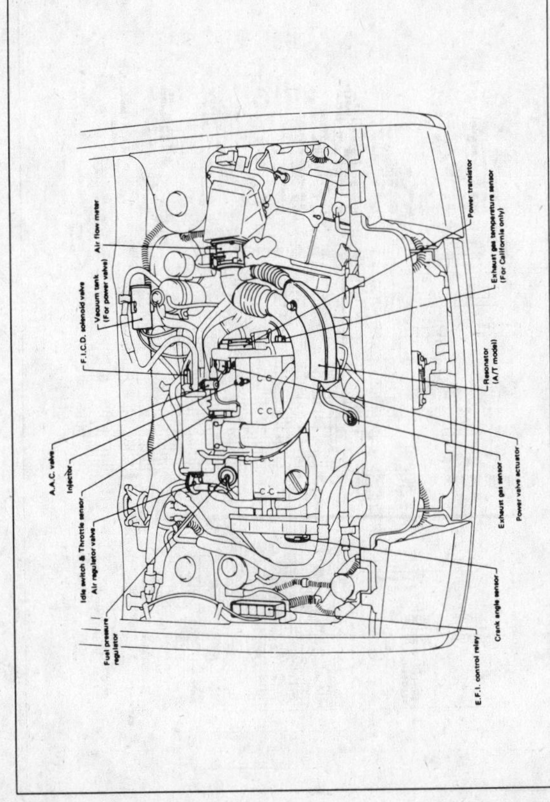
Air flow meter
F.I.C.D. solenoid valve
Vacuum tank (For power valve)
Power transistor
Exhaust gas temperature sensor (For California only)
Resonator (A/T model)
A.A.C. valve
Injector
Idle switch & Throttle sensor
Air regulator valve
Fuel pressure regulator
E.F.I. control relay
Crank angle sensor
Exhaust gas sensor
Power valve actuator

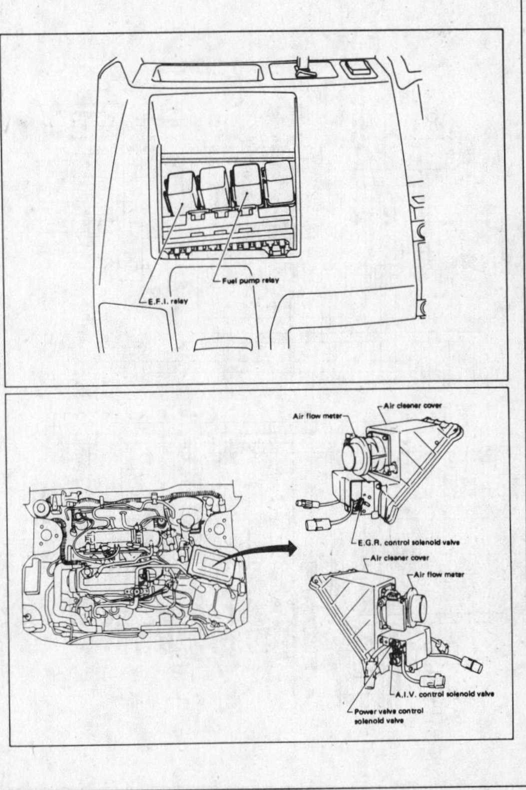
E.F.I. relay
Fuel pump relay
Air flow meter
Air cleaner cover
E.G.R. control solenoid valve
Air cleaner cover
Air flow meter
A.I.V. control solenoid valve
Power valve control solenoid valve

ECCS SYSTEM SCHEMATIC – 1988 PULSAR

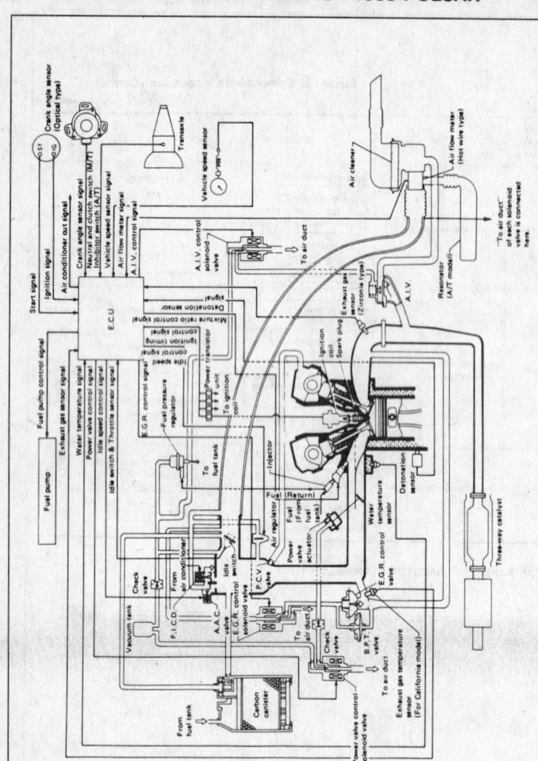

ECCS CONTROL SYSTEM CHART – 1988 PULSAR

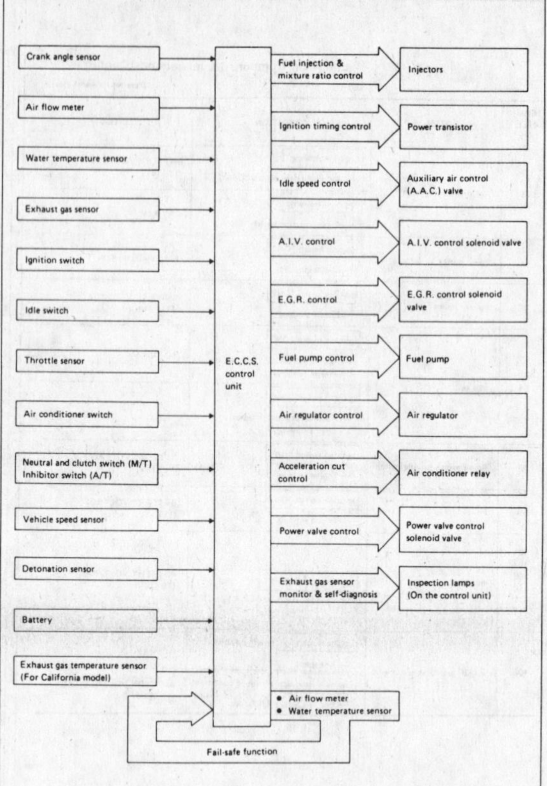
Crank angle sensor
Air flow meter
Water temperature sensor
Exhaust gas sensor
Ignition switch
Idle switch
Throttle sensor
Air conditioner switch
Neutral and clutch switch (M/T)
Inhibitor switch (A/T)
Vehicle speed sensor
Detonation sensor
Battery
Exhaust gas temperature sensor
(For California model)
E.C.C.S. control unit
Fuel injection & mixture ratio control
Ignition timing control
Idle speed control
A.I.V. control
E.G.R. control
Fuel pump control
Air regulator control
Acceleration cut control
Power valve control
Exhaust gas sensor monitor & self-diagnosis
Injectors
Power transistor
Auxiliary air control (A.A.C.) valve
A.I.V. control solenoid valve
E.G.R. control solenoid valve
Fuel pump
Air regulator
Air conditioner relay
Power valve control solenoid valve
Inspection lamps
(On the control unit)
Air flow meter
Water temperature sensor
Fail-safe function

ECCS CONTROL UNIT WIRING SCHEMATIC 1988 PULSAR

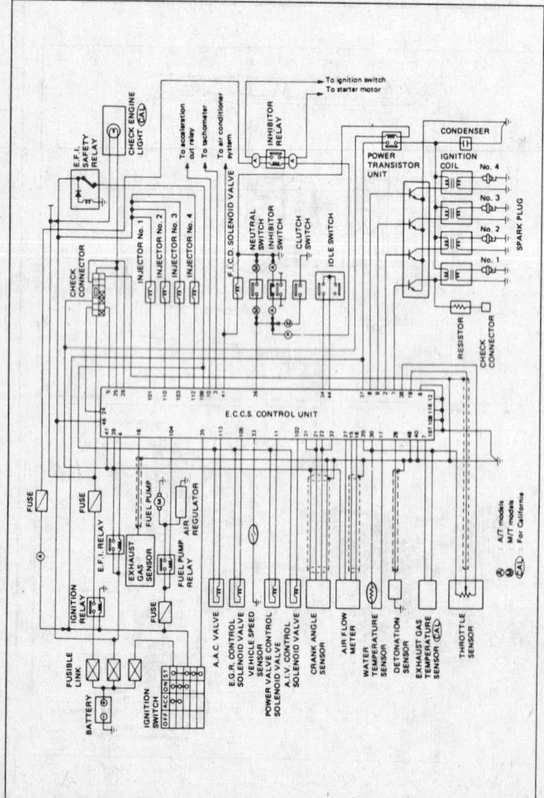

ECCS WIRING SCHEMATIC – 1988 PULSAR

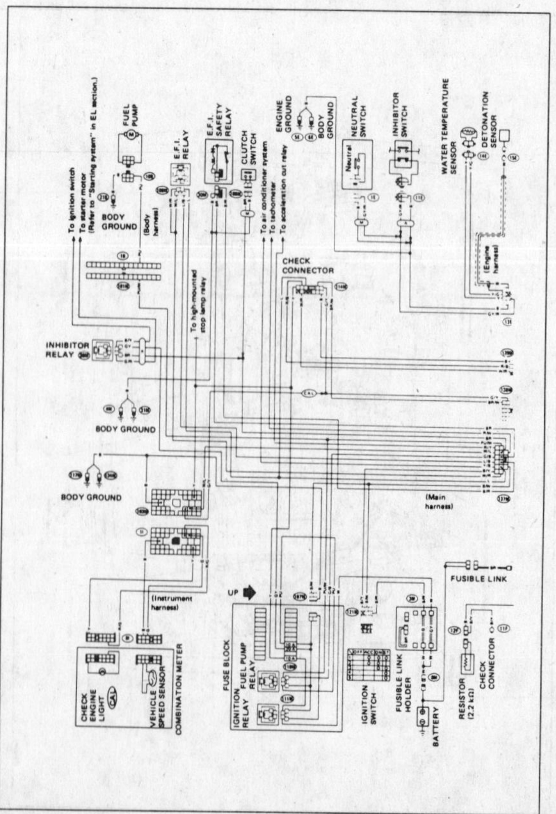

ECCS WIRING SCHEMATIC (CONT.)—1988 PULSAR

ENGINE GROUND
CONDENSER
BODY GROUND
F.I.C.D. SOLENOID VALVE
AIR REGULATOR
E.G.R. CONTROL SOLENOID VALVE
A.A.C. VALVE
POWER VALVE CONTROL SOLENOID VALVE
ENGINE GROUND
EXHAUST GAS SENSOR
THROTTLE SENSOR
AIR FLOW METER
(E.F.I. harness)
E.C.C.S. CONTROL UNIT
IDLE SWITCH
CRANK ANGLE SENSOR
(E.F.I. harness)
POWER TRANSISTOR RELAY
(E.F.I. sub harness)
ENGINE GROUND
No. 4 No. 3 No. 2 No. 1
IGNITION COIL
POWER TRANSISTOR UNIT
A.I.V. CONTROL SOLENOID VALVE
EXHAUST GAS TEMPERATURE SENSOR
No. 4 No. 3 No. 2 No. 1
INJECTOR
Ⓐ : A/T model
Ⓜ : M/T model
CAL : For California

ECCS SYSTEM DESCRIPTION—1988 PULSAR

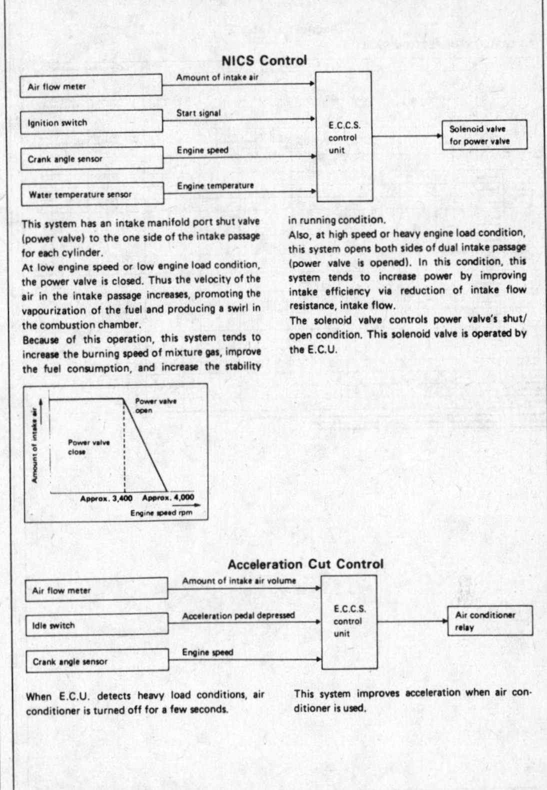

NICS Control

Air flow meter
Amount of intake air
Ignition switch
Start signal
Crank angle sensor
Engine speed
Water temperature sensor
Engine temperature
E.C.C.S. control unit
Solenoid valve for power valve

This system has an intake manifold port shut valve (power valve) to the one side of the intake passage for each cylinder.

At low engine speed or low engine load condition, the power valve is closed. Thus the velocity of the air in the intake passage increases, promoting the vapourization of the fuel and producing a swirl in the combustion chamber.

Because of this operation, this system tends to increase the burning speed of mixture gas, improve the fuel consumption, and increase the stability in running condition.

Also, at high speed or heavy engine load condition, this system opens both sides of dual intake passage (power valve is opened). In this condition, this system tends to increase power by improving intake efficiency via reduction of intake flow resistance, intake flow.

The solenoid valve controls power valve's shut/open condition. This solenoid valve is operated by the E.C.U.

Power valve open
Power valve close
Amount of intake air
Approx. 3,400
Approx. 4,000
Engine speed rpm

Acceleration Cut Control

Air flow meter
Amount of intake air volume
Idle switch
Acceleration pedal depressed
Crank angle sensor
Engine speed
E.C.C.S. control unit
Air conditioner relay

When E.C.U. detects heavy load conditions, air conditioner is turned off for a few seconds.

This system improves acceleration when air conditioner is used.

ECCS DIAGNOSTIC PROCEDURE—1988 PULSAR

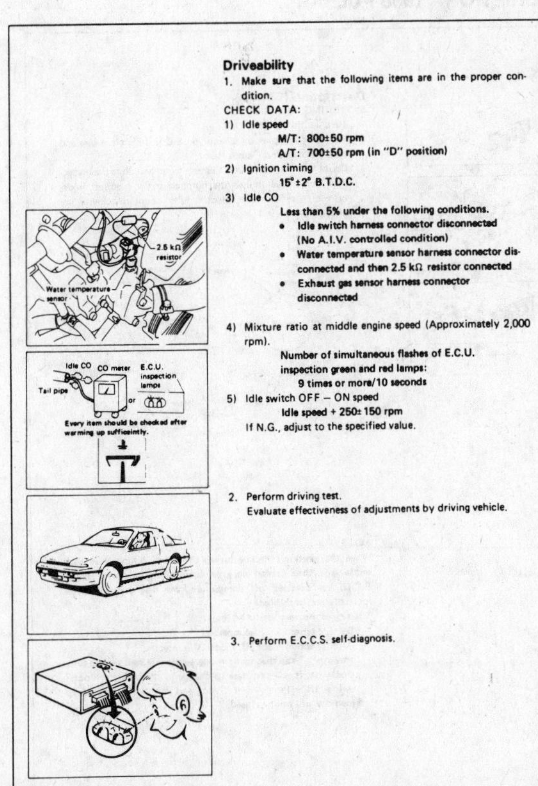

Driveability

1. Make sure that the following items are in the proper condition.

CHECK DATA:

1) Idle speed
   M/T: 800±50 rpm
   A/T: 700±50 rpm (in "D" position)
2) Ignition timing
   15°±2° B.T.D.C.
3) Idle CO
   Less than 5% under the following conditions.
   - Idle switch harness connector disconnected (No A.I.V. controlled condition)
   - Water temperature sensor harness connector disconnected and then 2.5 kΩ resistor connected
   - Exhaust gas sensor harness connector disconnected
4) Mixture ratio at middle engine speed (Approximately 2,000 rpm).
   Number of simultaneous flashes of E.C.U. inspection green and red lamps:
   9 times or more/10 seconds
5) Idle switch OFF – ON speed
   Idle speed + 250±150 rpm

If N.G., adjust to the specified value.

2.5 kΩ resistor
Water temperature sensor
Idle CO
CO meter
E.C.U. inspection lamps
Tail pipe
or
Every item should be checked after warming up sufficiently.

2. Perform driving test.
   Evaluate effectiveness of adjustments by driving vehicle.

3. Perform E.C.C.S. self-diagnosis.

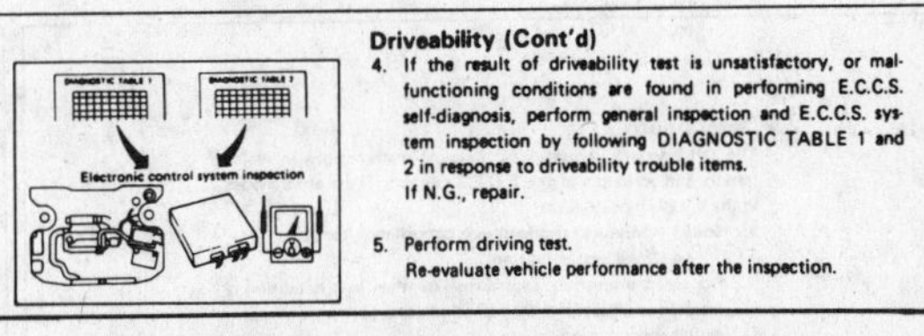

Driveability (Cont'd)

DIAGNOSTIC TABLE 1
DIAGNOSTIC TABLE 2
Electronic control system inspection

4. If the result of driveability test is unsatisfactory, or malfunctioning conditions are found in performing E.C.C.S. self-diagnosis, perform general inspection and E.C.C.S. system inspection by following DIAGNOSTIC TABLE 1 and 2 in response to driveability trouble items.
   If N.G., repair.

5. Perform driving test.
   Re-evaluate vehicle performance after the inspection.

ECCS DIAGNOSTIC TABLE 1—1988 PULSAR

Diagnostic Table 1

SYSTEM INSPECTION TABLE

System \ Sensor & actuator	Crank angle sensor	Air flow meter	Water temperature sensor	Ignition switch	Injector	Idle switch	Neutral switch or Inhibitor switch	Exhaust gas sensor	Throttle sensor	Air conditioner switch
Fuel injection & mixture ratio feedback control	O	O	O	O	O	O		O	O	
Ignition timing control	O	O	O	O		O				
A.I.V. control	O		O			O				
Fuel pump and air regulator control	O			O						
Idle speed control	O		O	O		O	O			O
E.G.R. control	O	O	O	O		O				
Power valve control	O	O	O	O						
Acceleration cut control	O	O				O				O

System \ Sensor & actuator	A.I.V. control solenoid valve	E.G.R. control solenoid valve	Power control valve solenoid valve	A.A.C. valve	Air regulator	Vehicle speed sensor	E.F.I. relay	Detonation sensor	Battery voltage
Fuel injection & mixture ratio feedback control							O		O
Ignition timing control							O	O	
A.I.V. control	O						O		
Fuel pump and air regulator control							O		
Idle speed control				O	O		O		O
E.G.R. control		O					O		
Power valve control			O				O		
Acceleration cut control							O		

This table indicates the inspection items for the E.C.C.S. control system. For each system, it is necessary to check sensors of actuators marked "O".

ECCS DIAGNOSTIC TABLE 2—1988 PULSAR

Diagnostic Table 2

DRIVEABILITY INSPECTION TABLE

TROUBLE ITEMS CONCERNED WITH DRIVEABILITY \ INSPECTION ITEM			GENERAL INSPECTION																		ECCS SYSTEM INSPECTION					
			FUEL FLOW SYSTEM				ELECTRIC SYSTEM					AIR FLOW SYSTEM							A A C VALVE		CRANK ANGLE SENSOR			VEHICLE SPEED SENSOR		
			Fuel level	Fuel pump	Fuel filter	Fuel line	Battery	Spark plug	Ignition coil	Alternator	Starter	Air cleaner	Air flow line	E G R B P T valve	A I V control valve	F I C D solenoid valve	Air regulator	P C V	Short	Open	180 signal noise	180° signal faults	1° signal faults	Short	Open	Poor connection
SURGE	ROAD/LOAD DRIVING	Heavy load			○	○		◎	○			○	○	○				○			○	○	○			
		Middle load			○	○		◎	○			○	○	◎				○			○	○	○			
		Light load			○	○		◎	○			○	○	◎				○			○	○	○			
	ACCELERATING DRIVING	Slow acceleration			○	○		◎	○			○	○	◎				○			○	○	○			
	DECELERATING DRIVING	Rapid deceleration						◎	○			○	○	○				○	○	○	○	○	○	○	○	○
		Slow deceleration						◎	○			○	○	○				○	○	○	○	○	○	○	○	○
HESITATION		Rapid acceleration			○	○		◎	○			○	○	○				○			○	○	○			
		Slow acceleration			○	○		◎	○			○	○	○				○			○	○	○			
STUMBLE		Rapid acceleration			○	○		◎	○			○	○	○				○			○	○	○			
		Slow acceleration			○	○		◎	○			○	○	○				○			○	○	○			
BACKFIRE					○	○			○				○					○			○	○	○			
AFTER FIRE								◎	○			○			○			○	○	○	○	○	○			
IDLE STABILITY								◎	○	○		○	○	◎		○	○	○	◎	◎	○	○	○	○	○	○
ENGINE STALL			○	○	○	○	○	◎	○	○		○	○	○		○	○	○	◎	◎	◎	◎	◎			
STARTABILITY			○	○			○	◎	○	○	○	○	○				○	○	○	○	◎	◎	◎			

This table indicates the inspection items for each type of symptom. It is necessary for each symptom to check sensors or actuators marked "◎" or "○". Items marked "◎" have a significant influence on driveability. Prior to items marked "○", check items marked "◎".
Improper mixture ratio, improper ignition condition, and an excess of E.G.R. volume can cause any symptom.

Diagnostic Table 2 (Cont'd)

ECCS SYSTEM INSPECTION																								
AIR FLOW METER *	WATER TEMPERATURE SENSOR			IDLE SWITCH		EXHAUST GAS SENSOR		INJECTOR	NEUTRAL SWITCH & INHIBITOR SWITCH	START SIGNAL		IGNITION SIGNAL		BATTERY VOLTAGE	FUEL PUMP CIRCUIT			E G R CONTROL SOLENOID VALVE		A I V CONTROL SOLENOID VALVE		POWER CONTROL VALVE SOLENOID VALVE (Include check valve)	DETONATION SENSOR	THROTTLE SENSOR
Poor connection	Short	Open	Poor connection	Short	Open	Short	Open	Clogging	Short	Short	Open	Short	Open	Low voltage	Short	Open	Poor connection	Short	Open	Short	Open	Short or open	Short or open	Short or open
◎	○	○	○			○	○	◎									○	◎	◎				○	
◎	○	○	○			○	○	◎									○	◎	◎				○	
◎	○	○	○			○	○	◎									○	◎	◎				○	
◎	○	○	○	○	○	○	○	◎									○	◎	◎				○	
◎	○	○	○	○	○	○	○	◎									○	○	○				○	
◎	○	○	○	○	○	○	○	◎									○	○	○				○	
◎	○	○	○	○	○	○	○	◎	○								○	○	○			○	○	◎
◎	○	○	○	○	○	○	○	◎	○								○	○	○			○	○	◎
◎	○	○	○	○	○			◎	○								○	○	○				○	◎
◎	○	○	○	○	○			◎	○								○	○	○				○	◎
◎	○		○	○	○	○	○	◎	○								○						○	○
◎		○	○	○	○	○	○	◎										◎	◎	○	○		○	○
○	○	○	○	○	○	○	○	◎	○								○	○	○				○	
◎	○	○	○	○	○	○	○	◎	○			○	○	○	○	○	○						○	
○	○	○	○					◎	○	○	○	○	○	○	○	○	○						○	

* If air flow meter circuit is short or open, the fail-safe system operates and engine revolution does not rise to approx. 2,000 rpm or more.

ECCS SELF-DIAGNOSTIC DESCRIPTION—1988 PULSAR

Description

The self-diagnosis is useful to diagnose malfunctions in major sensors and actuators of the E.C.C.S. system. There are 5 modes in the self-diagnosis system.

1. **Mode I – Mixture ratio feedback control monitor A**
   - During closed loop condition:
     The green inspection lamp turns ON when lean condition is detected and goes OFF by rich condition.
   - During open loop condition:
     The green inspection lamp remains ON or OFF.
2. **Mode II – Mixture ratio feedback control monitor B**
   The green inspection lamp function is the same as Mode I.
   - During closed loop condition:
     The red inspection lamp turns ON and OFF simultaneously with the green inspection lamp when the mixture ratio is controlled within the specified value.
   - During open loop condition:
     The red inspection lamp remains ON or OFF.
3. **Mode III – Self-diagnosis**
   This mode is the same as the former self-diagnosis in self-diagnosis mode.
4. **Mode IV – Switches ON/OFF diagnosis**
   During this mode, the inspection lamps monitor the switch ON-OFF condition.
   - Idle switch
   - Ignition switch "START"
   - Vehicle speed sensor
5. **Mode V – Real time diagnosis**
   The moment the malfunction is detected, the display will be presented immediately. That is, the condition at which the malfunction occurs can be found by observing the inspection lamps during driving test.

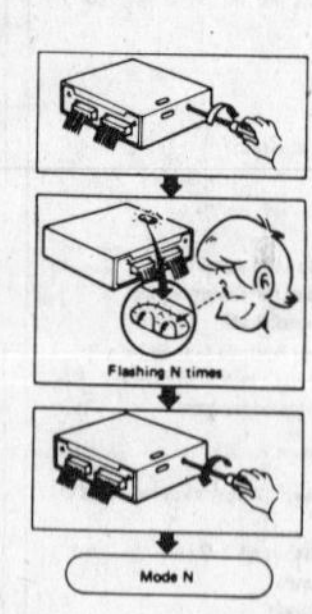
Flashing N times
Mode N

Description (Cont'd)

SWITCHING THE MODES

1. Turn ignition switch "ON".
2. Turn diagnostic mode selector on E.C.U. fully clockwise and wait the inspection lamps flash.
3. Count the number of the flashing time, and after the inspection lamps have flashed the number of the required mode, turn diagnostic mode selector fully counterclockwise immediately.

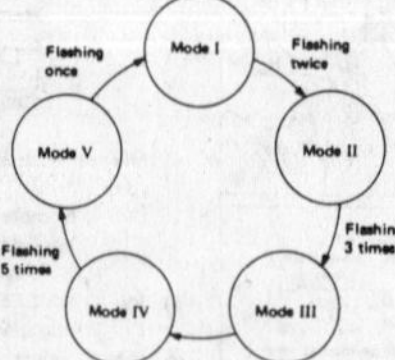
Mode I
Flashing twice
Mode II
Flashing 3 times
Mode III
Flashing 4 times
Mode IV
Flashing 5 times
Mode V
Flashing once

NOTE:
When the ignition switch is turned off during diagnosis, in each mode, and then turned back on again after the power to the E.C.U. has dropped off completely, the diagnosis will automatically return to Mode I.

The stored memory would be lost if:

1. Battery terminal is disconnected.
2. After selecting Mode III, Mode IV is selected.
   However, if the diagnostic mode selector is kept turned fully clockwise, it will continue to change in the order of Mode I → II → III → IV → V → I ... etc., and in this state the stored memory will not be erased.

ECCS SELF-DIAGNOSTIC DESCRIPTION (CONT.)–1988 PULSAR

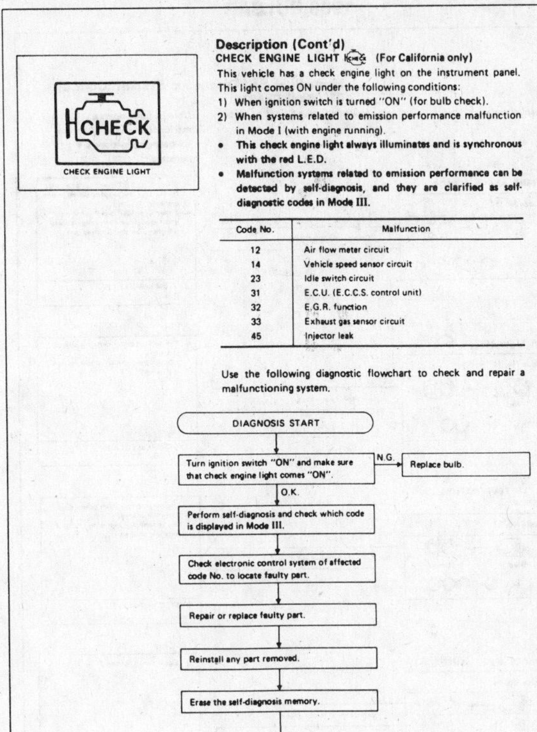
CHECK
CHECK ENGINE LIGHT

Description (Cont'd)

CHECK ENGINE LIGHT (For California only)

This vehicle has a check engine light on the instrument panel. This light comes ON under the following conditions:

1) When ignition switch is turned "ON" (for bulb check).
2) When systems related to emission performance malfunction in Mode I (with engine running).

- **This check engine light always illuminates and is synchronous with the red L.E.D.**
- **Malfunction systems related to emission performance can be detected by self-diagnosis, and they are clarified as self-diagnostic codes in Mode III.**

Code No.	Malfunction
12	Air flow meter circuit
14	Vehicle speed sensor circuit
23	Idle switch circuit
31	E.C.U. (E.C.C.S. control unit)
32	E.G.R. function
33	Exhaust gas sensor circuit
45	Injector leak

Use the following diagnostic flowchart to check and repair a malfunctioning system.

DIAGNOSIS START
Turn ignition switch "ON" and make sure that check engine light comes "ON".
N.G.
Replace bulb.
O.K.
Perform self-diagnosis and check which code is displayed in Mode III.
Check electronic control system of affected code No. to locate faulty part.
Repair or replace faulty part.
Reinstall any part removed.
Erase the self-diagnosis memory.
A

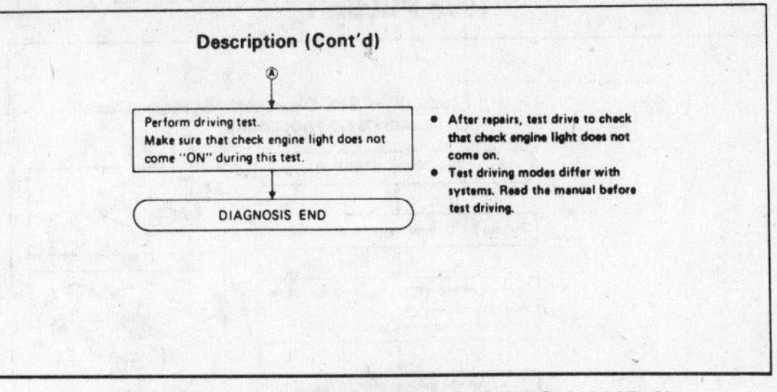

Description (Cont'd)

A
Perform driving test.
Make sure that check engine light does not come "ON" during this test.
DIAGNOSIS END
• After repairs, test drive to check that check engine light does not come on.
• Test driving modes differ with systems. Read the manual before test driving.

ECCS SELF-DIAGNOSTIC MODES I AND II 1988 PULSAR

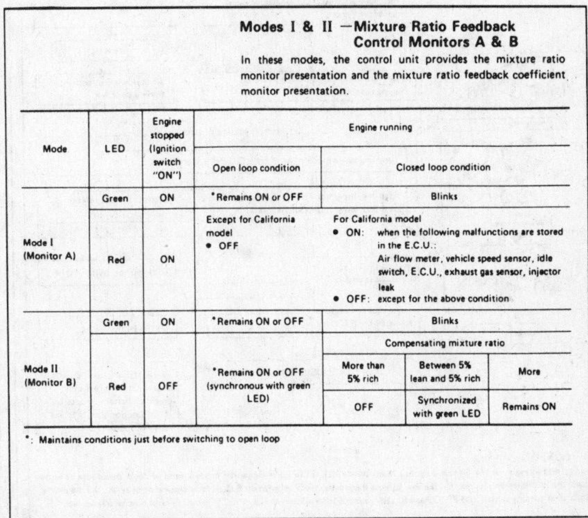

Modes I & II —Mixture Ratio Feedback Control Monitors A & B

In these modes, the control unit provides the mixture ratio monitor presentation and the mixture ratio feedback coefficient monitor presentation.

Mode	LED	Engine stopped (Ignition switch "ON")	Engine running			
			Open loop condition	Closed loop condition		
Mode I (Monitor A)	Green	ON	*Remains ON or OFF	Blinks		
	Red	ON	Except for California model • OFF	For California model • ON: when the following malfunctions are stored in the E.C.U.: Air flow meter, vehicle speed sensor, idle switch, E.C.U., exhaust gas sensor, injector leak • OFF: except for the above condition		
Mode II (Monitor B)	Green	ON	*Remains ON or OFF	Blinks		
	Red	OFF	*Remains ON or OFF (synchronous with green LED)	Compensating mixture ratio		
				More than 5% rich	Between 5% lean and 5% rich	More
				OFF	Synchronized with green LED	Remains ON

*: Maintains conditions just before switching to open loop

ECCS SELF-DIAGNOSTIC MODE III–1988 PULSAR

Mode III — Self-Diagnostic System

The E.C.U. constantly monitors the function of these sensors and actuators, regardless of ignition key position. If a malfunction occurs, the information is stored in the E.C.U. and can be retrieved from the memory by turning on the diagnostic mode selector, located on the side of the E.C.U. When activated, the malfunction is indicated by flashing a red and a green L.E.D. (Light Emitting Diode), also located on the E.C.U. Since all the self-diagnostic results are stored in the E.C.U.'s memory even intermittent malfunctions can be diagnosed.

A malfunctioning part's group is indicated by the number of both the red and the green L.E.D.s flashing. First, the red L.E.D. flashes and the green flashes follow. The red L.E.D. refers to the number of tens while the green one refers to the number of units. For example, when the red L.E.D. flashes once and then the green one flashes twice, this means the number "12" showing the air flow meter signal is malfunctioning. In this way, all the problems are classified by the code numbers.

- **When engine fails to start, crank engine more than two seconds before starting self-diagnosis.**
- **Before starting self-diagnosis, do not erase stored memory. If doing so, self-diagnosis function for intermittent malfunctions would be lost.**

The stored memory would be lost if:

1. **Battery terminal is disconnected.**
2. **After selecting Mode III, Mode IV is selected.**

DISPLAY CODE TABLE

X: Available
–: Not available

Code No.	Detected items	California	Non-California
11	Crank angle sensor circuit	X	X
12	Air flow meter circuit	X	X
13	Water temperature sensor circuit	X	X
14	Vehicle speed sensor circuit	X	X
21	Ignition signal circuit	X	X
23	Idle switch circuit	X	X
31	E.C.U. (E.C.C.S. control unit)	X	X
32	E.G.R. function	X	–
33	Exhaust gas sensor	X	X
34	Detonation sensor circuit	X	X
35	Exhaust gas temperature sensor circuit	X	–
43	Throttle sensor	X	X
45	Injector leak	X	–
55	No malfunctioning in the above circuit	X	X

Mode III — Self-Diagnostic System (Cont'd)

RETENTION OF DIAGNOSTIC RESULTS

The diagnostic result is retained in E.C.U. memory until the starter is operated fifty times after a diagnostic item is judged to be malfunctioning. The diagnostic result will then be cancelled automatically. If a diagnostic item which has been judged to be malfunctioning and stored in memory is again judged to be malfunctioning before the starter is operated fifty times, the second result will replace the previous one. It will be stored in E.C.U. memory until the starter is operated fifty times more.

RETENTION TERM CHART (Example)

Code No. — STARTER OPERATING TIMES — 50 — 100 — 150 — 200

CRANK ANGLE SENSOR — 11 — 50 times — 50 times

WATER TEMP. SENSOR — 13 — 50 times

: Retention term

▲ : Malfunction detecting point

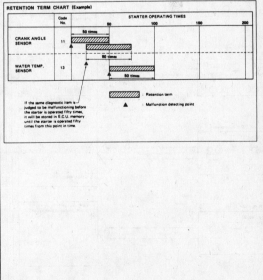

If the same diagnostic item is judged to be malfunctioning before the starter is operated fifty times, it will be stored in E.C.U. memory until the starter is operated fifty times from this point in time.

ECCS SELF-DIAGNOSTIC MODE III (CONT.) 1988 PULSAR

Mode III — Self-Diagnostic System (Cont'd)

SELF-DIAGNOSTIC PROCEDURE

DIAGNOSIS START

1. Pull out E.C.U. from under the assist seat.
2. Start engine and warm it up to normal engine operating temperature. (Drive vehicle for about 10 min.)
3. Turn diagnostic mode selector on E.C.U. fully clockwise. (Flashing 3 times)
4. After the inspection lamps have flashed 3 times, turn diagnostic mode selector fully counterclockwise. (Mode III)
5. Make sure that inspection lamps are displaying code No. 55.
   - N.G.: Write down the malfunctioning code No.
   - O.K.: continue.

Memory erasing procedure

6. Turn diagnostic mode selector on E.C.U. fully clockwise.
7. After the inspection lamps have flashed 4 times, turn diagnostic mode selector on E.C.U. fully counterclockwise. (Flashing 4 times; Mode IV)

O.K. path:

- Turn ignition switch "OFF".
- Reinstall the E.C.U. in place.
- DIAGNOSIS END

N.G. path:

- Turn ignition switch "OFF".
- See decoding chart.
- Check malfunctioning parts and/or perform real time diagnosis system inspection. If malfunction part is found, repair or replace it.

CAUTION:
During displaying code No. in self-diagnosis mode (mode III), if the other diagnostic mode should be done, make sure to write down the malfunctioning code No. before turning diagnostic mode selector on E.C.U. fully clockwise, or select the diagnostic mode after turning switch "OFF". Otherwise self-diagnosis information stored in E.C.U. memory until now would be lost.

ECCS SELF-DIAGNOSTIC MODE III DISPLAY CODES 1988 PULSAR

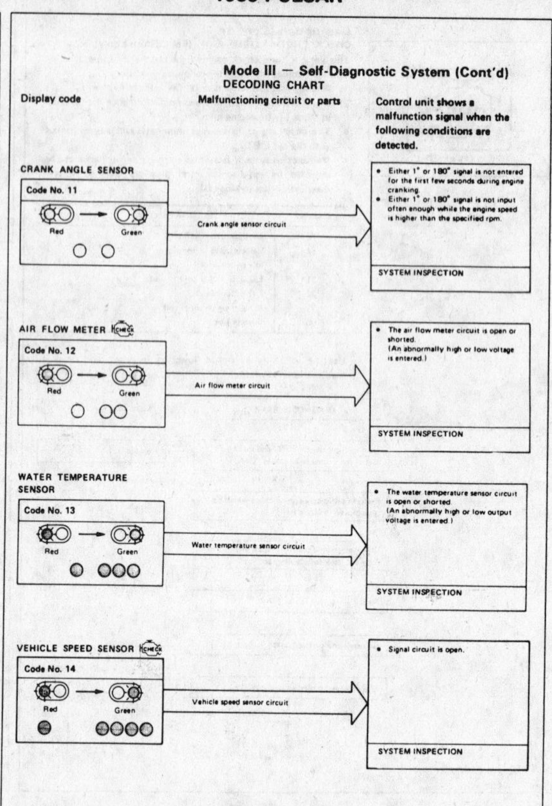
Mode III — Self-Diagnostic System (Cont'd)
DECODING CHART
Display code
Malfunctioning circuit or parts
Control unit shows a malfunction signal when the following conditions are detected.
CRANK ANGLE SENSOR
Code No. 11
Red
Green
Crank angle sensor circuit
• Either 1° or 180° signal is not entered for the first few seconds during engine cranking.
• Either 1° or 180° signal is not input often enough while the engine speed is higher than the specified rpm.
SYSTEM INSPECTION
AIR FLOW METER
Code No. 12
Red
Green
Air flow meter circuit
• The air flow meter circuit is open or shorted. (An abnormally high or low voltage is entered.)
SYSTEM INSPECTION
WATER TEMPERATURE SENSOR
Code No. 13
Red
Green
Water temperature sensor circuit
• The water temperature sensor circuit is open or shorted. (An abnormally high or low output voltage is entered.)
SYSTEM INSPECTION
VEHICLE SPEED SENSOR
Code No. 14
Red
Green
Vehicle speed sensor circuit
• Signal circuit is open.
SYSTEM INSPECTION

ECCS SELF-DIAGNOSTIC MODE III DISPLAY CODES (CONT.)—1988 PULSAR

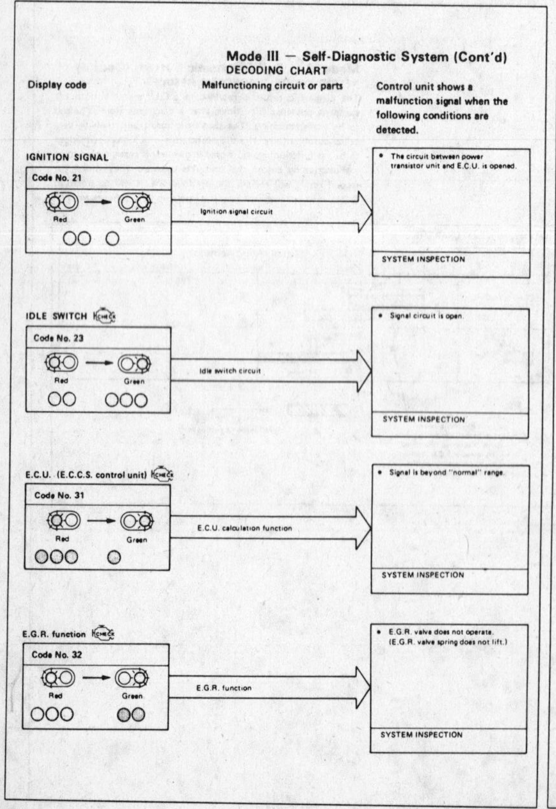
Mode III — Self-Diagnostic System (Cont'd)
DECODING CHART
Display code
Malfunctioning circuit or parts
Control unit shows a malfunction signal when the following conditions are detected.
IGNITION SIGNAL
Code No. 21
Red
Green
Ignition signal circuit
• The circuit between power transistor unit and E.C.U. is opened.
SYSTEM INSPECTION
IDLE SWITCH
Code No. 23
Red
Green
Idle switch circuit
• Signal circuit is open.
SYSTEM INSPECTION
E.C.U. (E.C.C.S. control unit)
Code No. 31
Red
Green
E.C.U. calculation function
• Signal is beyond "normal" range.
SYSTEM INSPECTION
E.G.R. function
Code No. 32
Red
Green
E.G.R. function
• E.G.R. valve does not operate. (E.G.R. valve spring does not lift.)
SYSTEM INSPECTION

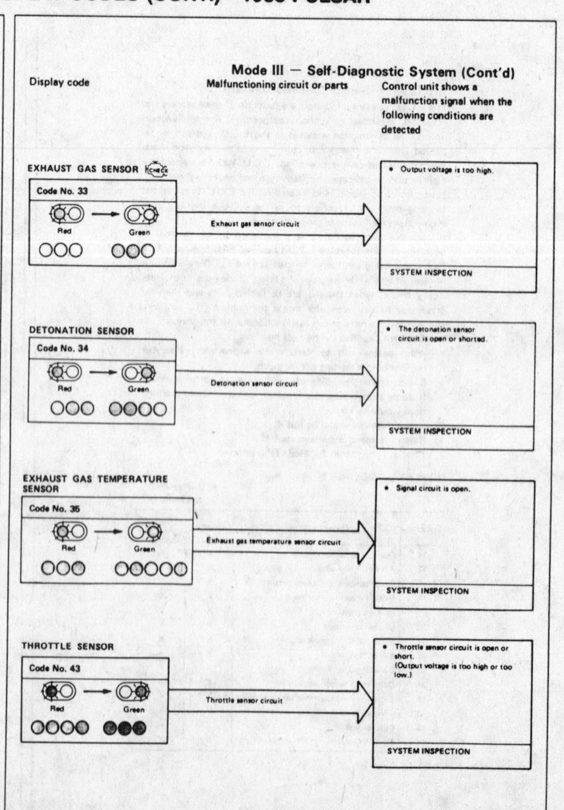
Mode III — Self-Diagnostic System (Cont'd)
Display code
Malfunctioning circuit or parts
Control unit shows a malfunction signal when the following conditions are detected
EXHAUST GAS SENSOR
Code No. 33
Red
Green
Exhaust gas sensor circuit
• Output voltage is too high.
SYSTEM INSPECTION
DETONATION SENSOR
Code No. 34
Red
Green
Detonation sensor circuit
• The detonation sensor circuit is open or shorted.
SYSTEM INSPECTION
EXHAUST GAS TEMPERATURE SENSOR
Code No. 35
Red
Green
Exhaust gas temperature sensor circuit
• Signal circuit is open.
SYSTEM INSPECTION
THROTTLE SENSOR
Code No. 43
Red
Green
Throttle sensor circuit
• Throttle sensor circuit is open or short. (Output voltage is too high or too low.)
SYSTEM INSPECTION

ECCS SELF-DIAGNOSTIC MODE III DISPLAY CODES (CONT.)—1988 PULSAR

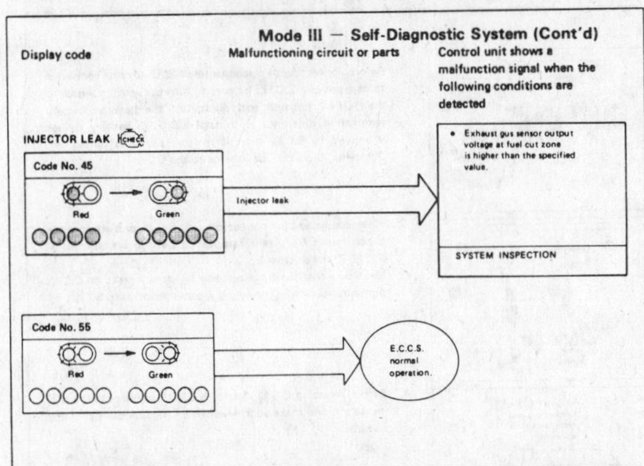
Mode III — Self-Diagnostic System (Cont'd)
Display code
Malfunctioning circuit or parts
Control unit shows a malfunction signal when the following conditions are detected
INJECTOR LEAK
Code No. 45
Red
Green
Injector leak
• Exhaust gas sensor output voltage at fuel cut zone is higher than the specified value.
SYSTEM INSPECTION
Code No. 55
Red
Green
E.C.C.S. normal operation.

ECCS SELF-DIAGNOSTIC MODE IV—1988 PULSAR

Mode IV — Switches ON/OFF Diagnostic System

In switches ON/OFF diagnosis system, ON/OFF operation of the following switches can be detected continuously.

- Idle switch
- Ignition switch "START"
- Vehicle speed sensor

(1) Idle switch & Ignition switch "START"
The switches ON/OFF status at the point when mode IV is selected is stored in E.C.U. memory. When either switch is turned from "ON" to "OFF" or "OFF" to "ON", the red L.E.D. on E.C.U. alternately comes on and goes off each time switching is detected.

(2) Vehicle Speed Sensor
The switches ON/OFF status at the point when mode IV is selected is stored in E.C.U. memory. When vehicle speed is 20 km/h (12 MPH) or slower, the green L.E.D. on E.C.U. is off. When vehicle speed exceeds 20 km/h (12 MPH), the green L.E.D. on E.C.U. comes "ON".

ECCS SELF-DIAGNOSTIC MODE IV (CONT.) 1988 PULSAR

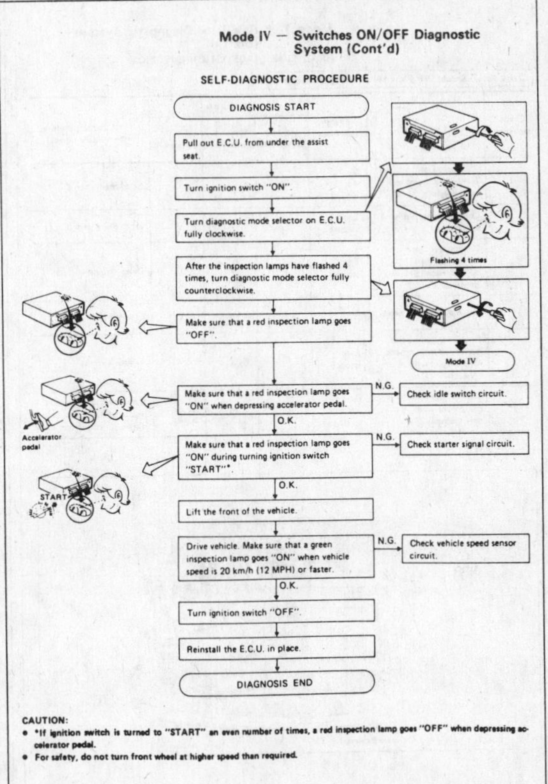
Mode IV — Switches ON/OFF Diagnostic System (Cont'd)
SELF-DIAGNOSTIC PROCEDURE
DIAGNOSIS START
Pull out E.C.U. from under the assist seat.
Turn ignition switch "ON".
Turn diagnostic mode selector on E.C.U. fully clockwise.
After the inspection lamps have flashed 4 times, turn diagnostic mode selector fully counterclockwise.
Flashing 4 times
Mode IV
Make sure that a red inspection lamp goes "OFF".
Make sure that a red inspection lamp goes "ON" when depressing accelerator pedal.
N.G.
Check idle switch circuit.
O.K.
Accelerator pedal
Make sure that a red inspection lamp goes "ON" during turning ignition switch "START"*.
N.G.
Check starter signal circuit.
O.K.
START
Lift the front of the vehicle.
Drive vehicle. Make sure that a green inspection lamp goes "ON" when vehicle speed is 20 km/h (12 MPH) or faster.
N.G.
Check vehicle speed sensor circuit.
O.K.
Turn ignition switch "OFF".
Reinstall the E.C.U. in place.
DIAGNOSIS END
CAUTION:
• *If ignition switch is turned to "START" an even number of times, a red inspection lamp goes "OFF" when depressing accelerator pedal.
• For safety, do not turn front wheel at higher speed than required.

ECCS SELF-DIAGNOSTIC MODE V—1988 PULSAR

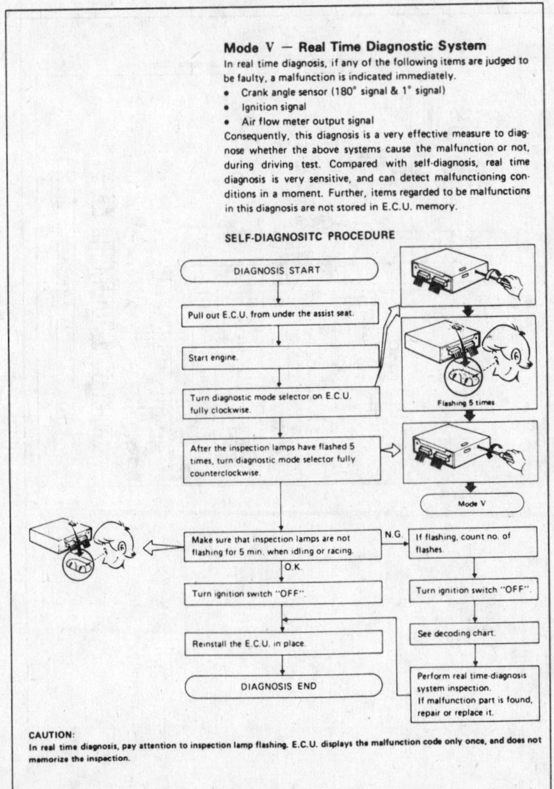
Mode V — Real Time Diagnostic System
In real time diagnosis, if any of the following items are judged to be faulty, a malfunction is indicated immediately.
• Crank angle sensor (180° signal & 1° signal)
• Ignition signal
• Air flow meter output signal
Consequently, this diagnosis is a very effective measure to diagnose whether the above systems cause the malfunction or not, during driving test. Compared with self-diagnosis, real time diagnosis is very sensitive, and can detect malfunctioning conditions in a moment. Further, items regarded to be malfunctions in this diagnosis are not stored in E.C.U. memory.
SELF-DIAGNOSITC PROCEDURE
DIAGNOSIS START
Pull out E.C.U. from under the assist seat.
Start engine.
Turn diagnostic mode selector on E.C.U. fully clockwise.
Flashing 5 times
After the inspection lamps have flashed 5 times, turn diagnostic mode selector fully counterclockwise.
Mode V
Make sure that inspection lamps are not flashing for 5 min. when idling or racing.
N.G.
If flashing, count no. of flashes.
O.K.
Turn ignition switch "OFF".
Turn ignition switch "OFF".
See decoding chart.
Reinstall the E.C.U. in place.
DIAGNOSIS END
Perform real time diagnosis system inspection.
If malfunction part is found, repair or replace it.
CAUTION:
In real time diagnosis, pay attention to inspection lamp flashing. E.C.U. displays the malfunction code only once, and does not memorize the inspection.

ECCS SELF-DIAGNOSTIC MODE V DECODING CHART 1988 PULSAR

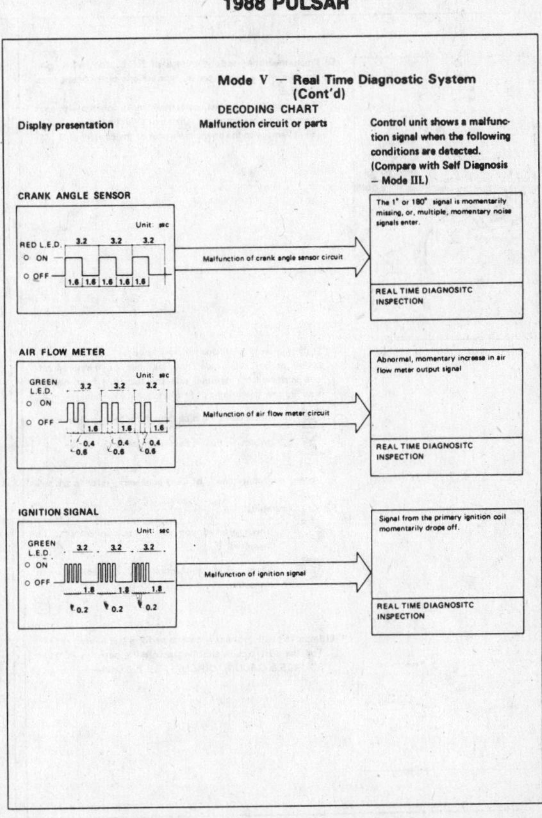
Mode V — Real Time Diagnostic System (Cont'd)
DECODING CHART
Display presentation
Malfunction circuit or parts
Control unit shows a malfunction signal when the following conditions are detected. (Compare with Self Diagnosis — Mode III.)
CRANK ANGLE SENSOR
Unit: sec
RED L.E.D.
ON
OFF
3.2
1.6
Malfunction of crank angle sensor circuit
The 1° or 180° signal is momentarily missing, or, multiple, momentary noise signals enter.
REAL TIME DIAGNOSITC INSPECTION
AIR FLOW METER
GREEN L.E.D.
ON
OFF
3.2
1.6
0.4
0.8
Malfunction of air flow meter circuit
Abnormal, momentary increase in air flow meter output signal
REAL TIME DIAGNOSITC INSPECTION
IGNITION SIGNAL
GREEN L.E.D.
ON
OFF
3.2
1.8
0.2
Malfunction of ignition signal
Signal from the primary ignition coil momentarily drops off.
REAL TIME DIAGNOSITC INSPECTION

ECCS SELF-DIAGNOSTIC MODE V INSPECTION 1988 PULSAR

Mode V — Real Time Diagnostic System (Cont'd)

REAL TIME DIAGNOSTIC INSPECTION

Crank Angle Sensor, Air Flow Meter and Ignition Signal

Check sequence	Check items	Check conditions	Check parts			If malfunction, perform the following items.
			Harness connectors	Sensor & actuator	E.C.U. connectors	
1	Tap harness connector or component during real time diagnosis.	During real time diagnosis	O	O	O	Go to check item 2.
2	Check harness continuity at connector.	Engine stopped	O	X	X	Go to check item 3.
3	Disconnect harness connector, and then check dust adhesion to harness connector.	Engine stopped	O	X	O	Clean terminal surface.
4	Check pin terminal bend.	Engine stopped	X	X	O	Take out bend.
5	Reconnect harness connector and then recheck harness continuity at connector.	Engine stopped	O	X	X	Replace terminal.
6	Tap harness connector or component during real time diagnosis.	During real time diagnosis	O	O	O	If malfunction codes are displayed during real time diagnosis, replace terminal.

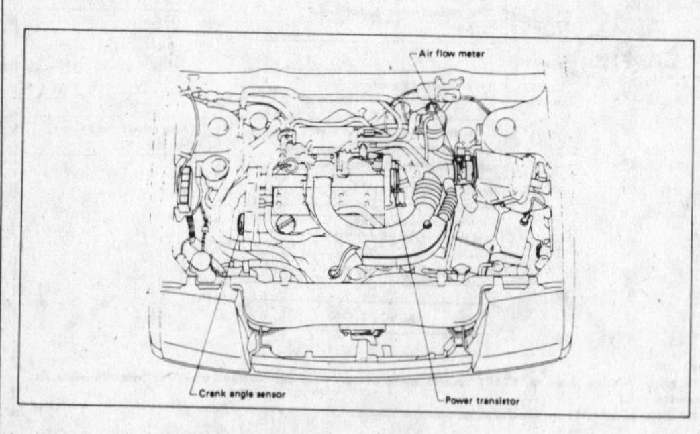
Air flow meter
Crank angle sensor
Power transistor

ECCS SELF-DIAGNOSTIC SYSTEM CAUTION 1988 PULSAR

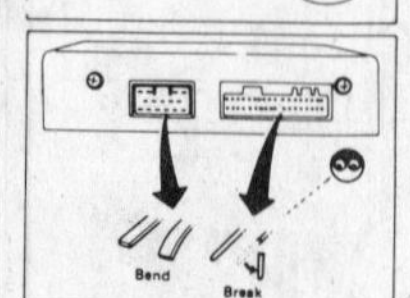

CAUTION:

1. Before connecting or disconnecting E.C.U. harness connector to or from any E.C.U., be sure to turn the ignition switch to the "OFF" position and disconnect the negative battery terminal in order not to damage E.C.U. as battery voltage is applied to E.C.U. even if ignition switch is turned off. Otherwise, there may be damage to the E.C.U.

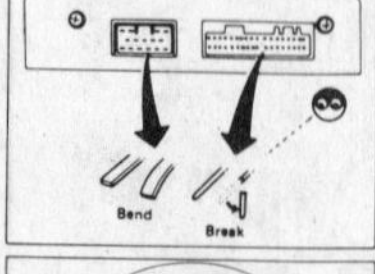
Bend
Break

2. When connecting pin connectors into E.C.U. or disconnecting them from E.C.U., take care not to damage pin terminal of E.C.U. (Bend or break).
3. Make sure that there are not any bends or breaks on E.C.U. pin terminal, when connecting pin connectors into E.C.U.

Perform E.C.U. input/output signal inspection before replacement
OLD ONE

4. Before replacing E.C.U., perform E.C.U. input/output signal inspection and make sure whether E.C.U. functions properly or not.

5. After performing this "ELECTRONIC CONTROL SYSTEM INSPECTION", perform E.C.C.S. self-diagnosis and driving test.

ECCS SELF-DIAGNOSTIC SYSTEM CAUTION (CONT.) 1988 PULSAR

6. When measuring supply voltage of E.C.U. controlled components with a circuit tester, separate one tester probe from the other.
If the two tester probes accidentally make contact with each other during measurement, the circuit will be shorted, resulting in damage to the power transistor of the control unit.
7. When measuring voltage or resistance at connector with tester probes, there are two methods of measurement; one is done from terminal side and the other from harness side. Before measuring, confirm symbol mark again.
HS : Inspection should be done from harness side.
TS : Inspection should be done from terminal side.
8. As for continuity check of joint connector, refer to EL section.
9. Key to symbols
DISCONNECT : Check after disconnecting the connector to be measured.
CONNECT : Check after connecting the connector to be measured.

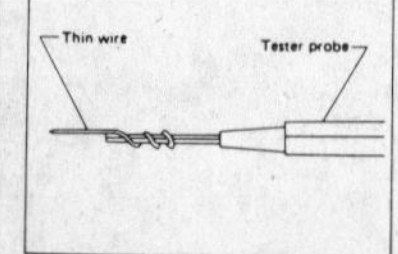
Thin wire
Tester probe

10. Improve tester probe as shown to perform test easily.
11. For the first trouble-shooting procedure, perform POWER SOURCE & GROUND CIRCUIT FOR E.C.U. check.

ECCS SELF-DIAGNOSTIC CHART—1988 PULSAR

POWER SOURCE & GROUND CIRCUIT FOR E.C.U. (Not self-diagnostic item)

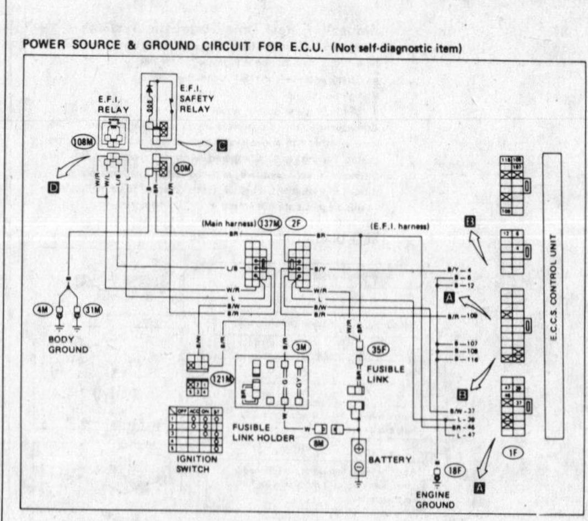
E.F.I. RELAY
E.F.I. SAFETY RELAY
(Main harness)
(E.F.I. harness)
BODY GROUND
FUSIBLE LINK
FUSIBLE LINK HOLDER
IGNITION SWITCH
BATTERY
ENGINE GROUND
E.C.C.S. CONTROL UNIT

ECCS SELF-DIAGNOSTIC CHART—1988 PULSAR

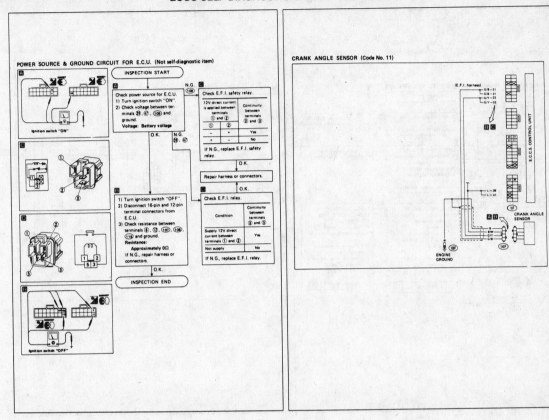

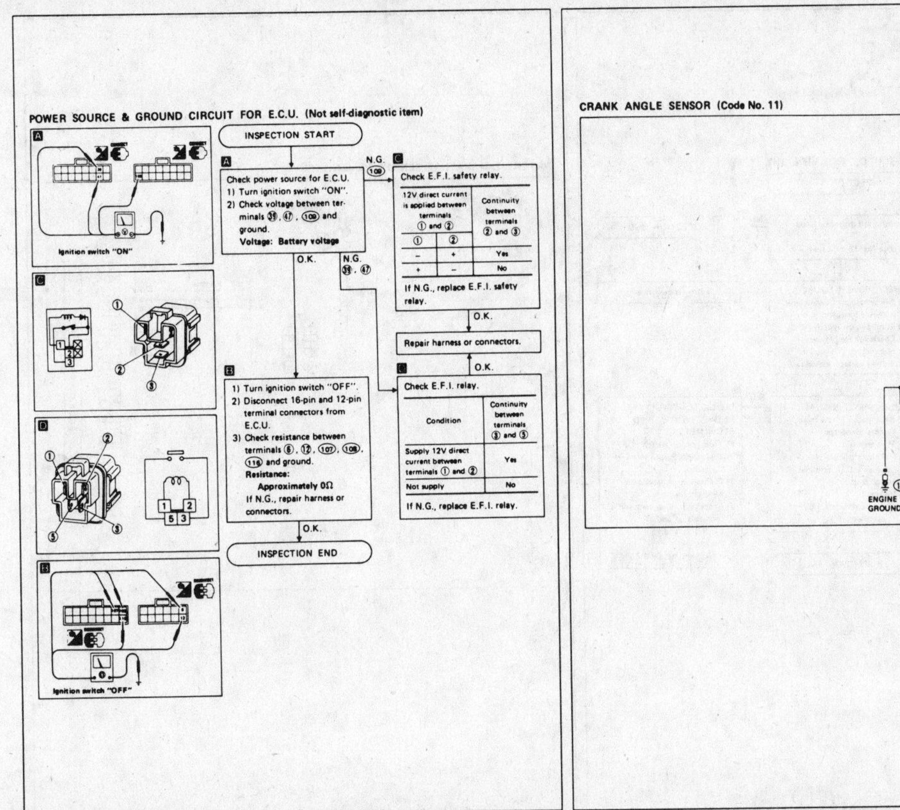
POWER SOURCE & GROUND CIRCUIT FOR E.C.U. (Not self-diagnostic item)
Ignition switch "ON"
Ignition switch "OFF"
INSPECTION START
Check power source for E.C.U.
1) Turn ignition switch "ON".
2) Check voltage between terminals ㉛, ㊼, ⑩⑧ and ground.
Voltage: Battery voltage
N.G.
Check E.F.I. safety relay.
12V direct current is applied between terminals ① and ②
Continuity between terminals ② and ③
Yes
No
If N.G., replace E.F.I. safety relay.
O.K.
Repair harness or connectors.
1) Turn ignition switch "OFF".
2) Disconnect 16-pin and 12-pin terminal connectors from E.C.U.
3) Check resistance between terminals ⑥, ⑫, ⑩⑦, ⑩⑧, ⑪⑥ and ground.
Resistance:
Approximately 0Ω
If N.G., repair harness or connectors.
Check E.F.I. relay.
Condition
Continuity between terminals ③ and ⑤
Supply 12V direct current between terminals ① and ②
Not supply
If N.G., replace E.F.I. relay.
INSPECTION END
CRANK ANGLE SENSOR (Code No. 11)
(E.F.I. harness)
E.C.C.S. CONTROL UNIT
CRANK ANGLE SENSOR
ENGINE GROUND

ECCS SELF-DIAGNOSTIC CHART—1988 PULSAR

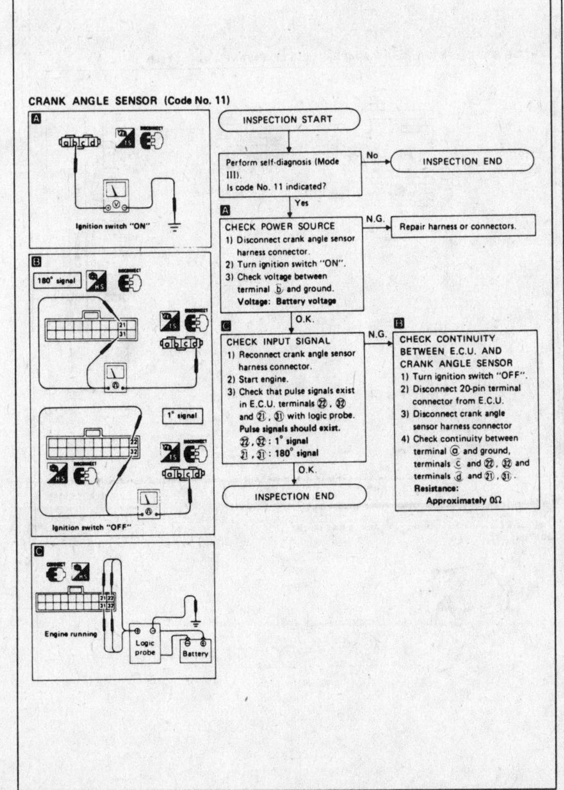
CRANK ANGLE SENSOR (Code No. 11)
Ignition switch "ON"
180° signal
1° signal
Ignition switch "OFF"
Engine running
Logic probe
Battery
INSPECTION START
Perform self-diagnosis (Mode III).
Is code No. 11 indicated?
No
INSPECTION END
Yes
CHECK POWER SOURCE
1) Disconnect crank angle sensor harness connector.
2) Turn ignition switch "ON".
3) Check voltage between terminal ⓑ and ground.
Voltage: Battery voltage
N.G.
Repair harness or connectors.
O.K.
CHECK INPUT SIGNAL
1) Reconnect crank angle sensor harness connector.
2) Start engine.
3) Check that pulse signals exist in E.C.U. terminals ㉒, ㉜ and ㉑, ㉛ with logic probe.
Pulse signals should exist.
㉒, ㉜ : 1° signal
㉑, ㉛ : 180° signal
CHECK CONTINUITY BETWEEN E.C.U. AND CRANK ANGLE SENSOR
1) Turn ignition switch "OFF".
2) Disconnect 20-pin terminal connector from E.C.U.
3) Disconnect crank angle sensor harness connector
4) Check continuity between terminal ⓐ and ground, terminals ⓒ and ㉒, ㉜ and terminals ⓓ and ㉑, ㉛.
Resistance:
Approximately 0Ω
INSPECTION END

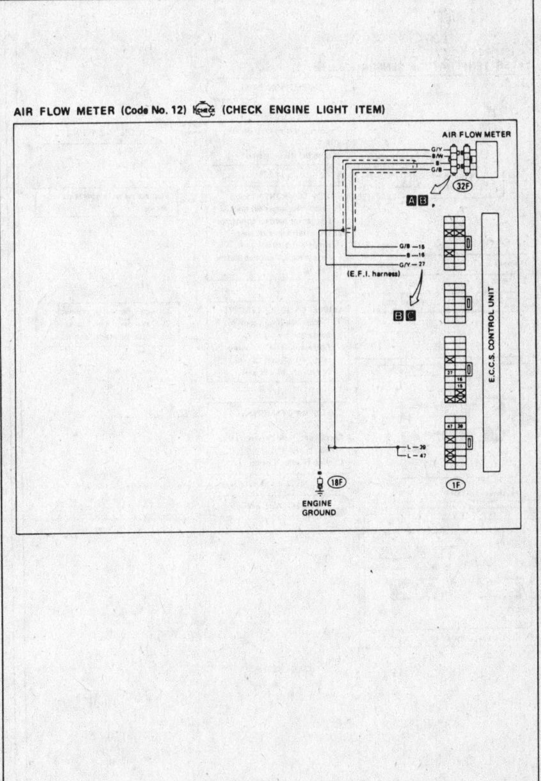
AIR FLOW METER (Code No. 12) (CHECK ENGINE LIGHT ITEM)
AIR FLOW METER
(E.F.I. harness)
E.C.C.S. CONTROL UNIT
ENGINE GROUND

ECCS SELF-DIAGNOSTIC CHART—1988 PULSAR

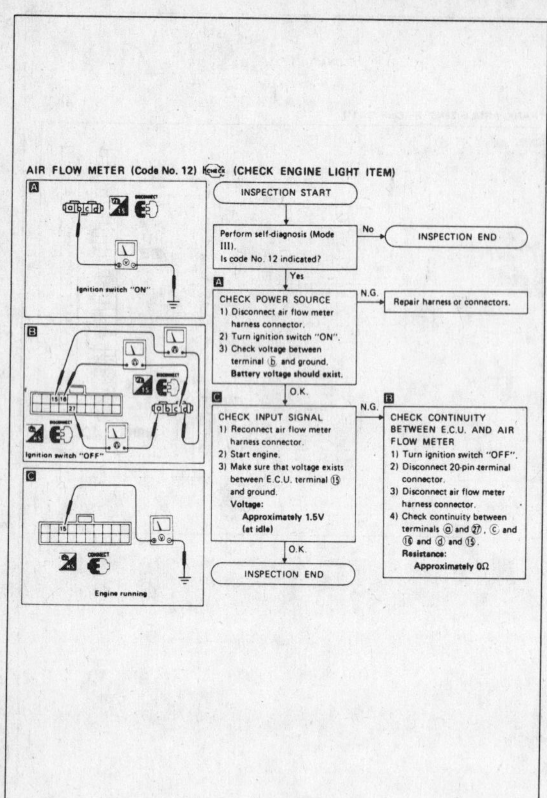
AIR FLOW METER (Code No. 12) (CHECK ENGINE LIGHT ITEM)
Ignition switch "ON"
Ignition switch "OFF"
Engine running
INSPECTION START
Perform self-diagnosis (Mode III).
Is code No. 12 indicated?
No
INSPECTION END
Yes
CHECK POWER SOURCE
1) Disconnect air flow meter harness connector.
2) Turn ignition switch "ON".
3) Check voltage between terminal ⓑ and ground.
Battery voltage should exist.
N.G.
Repair harness or connectors.
O.K.
CHECK INPUT SIGNAL
1) Reconnect air flow meter harness connector.
2) Start engine.
3) Make sure that voltage exists between E.C.U. terminal ⑮ and ground.
Voltage:
Approximately 1.5V
(at idle)
N.G.
CHECK CONTINUITY BETWEEN E.C.U. AND AIR FLOW METER
1) Turn ignition switch "OFF".
2) Disconnect 20-pin terminal connector.
3) Disconnect air flow meter harness connector.
4) Check continuity between terminals ⓐ and ㉗, ⓒ and ⑯ and ⓓ and ⑮.
Resistance:
Approximately 0Ω
O.K.
INSPECTION END

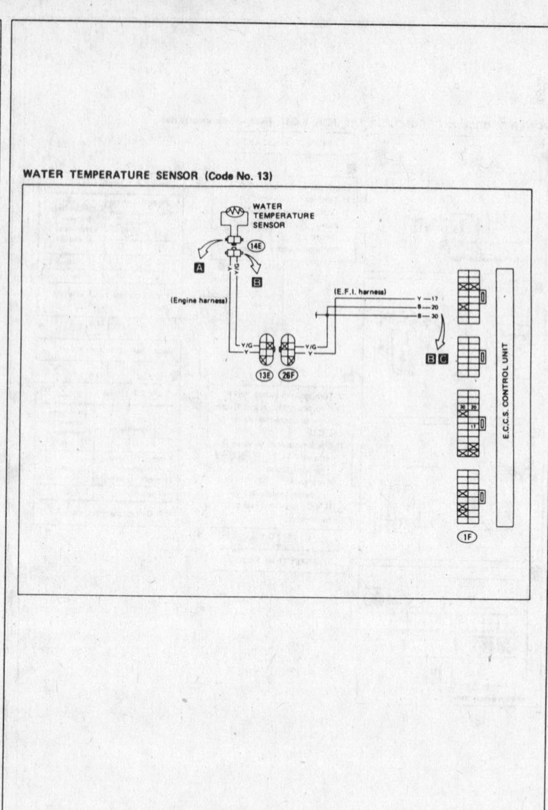
WATER TEMPERATURE SENSOR (Code No. 13)
WATER TEMPERATURE SENSOR
14E
(Engine harness)
(E.F.I. harness)
Y/G
13E
26F
1F
E.C.C.S. CONTROL UNIT

ECCS SELF-DIAGNOSTIC CHART—1988 PULSAR

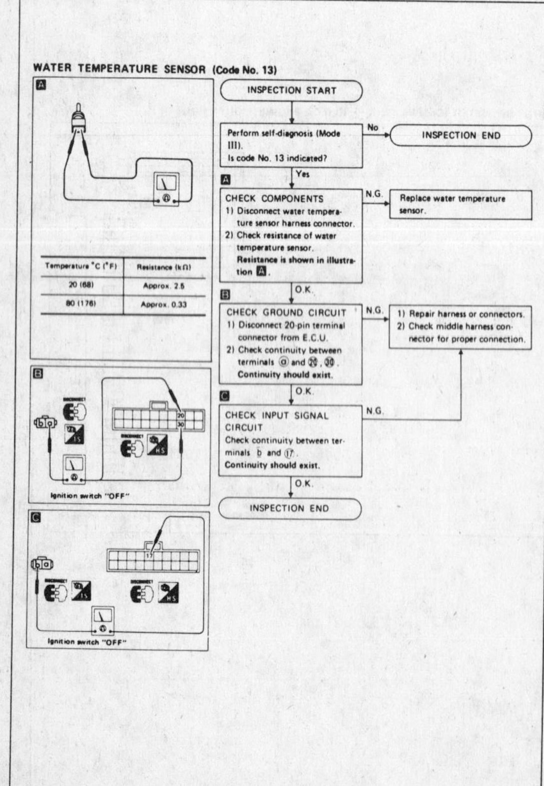
WATER TEMPERATURE SENSOR (Code No. 13)
INSPECTION START
Perform self-diagnosis (Mode III).
Is code No. 13 indicated?
No
INSPECTION END
Yes
CHECK COMPONENTS
1) Disconnect water temperature sensor harness connector.
2) Check resistance of water temperature sensor.
Resistance is shown in illustration A.
N.G.
Replace water temperature sensor.
O.K.
CHECK GROUND CIRCUIT
1) Disconnect 20-pin terminal connector from E.C.U.
2) Check continuity between terminals ⓐ and ⑳, ㉚.
Continuity should exist.
N.G.
1) Repair harness or connectors.
2) Check middle harness connector for proper connection.
O.K.
CHECK INPUT SIGNAL CIRCUIT
Check continuity between terminals ⓑ and ⑰.
Continuity should exist.
N.G.
O.K.
INSPECTION END
Ignition switch "OFF"
Ignition switch "OFF"

Temperature °C (°F)	Resistance (kΩ)
20 (68)	Approx. 2.5
80 (176)	Approx. 0.33

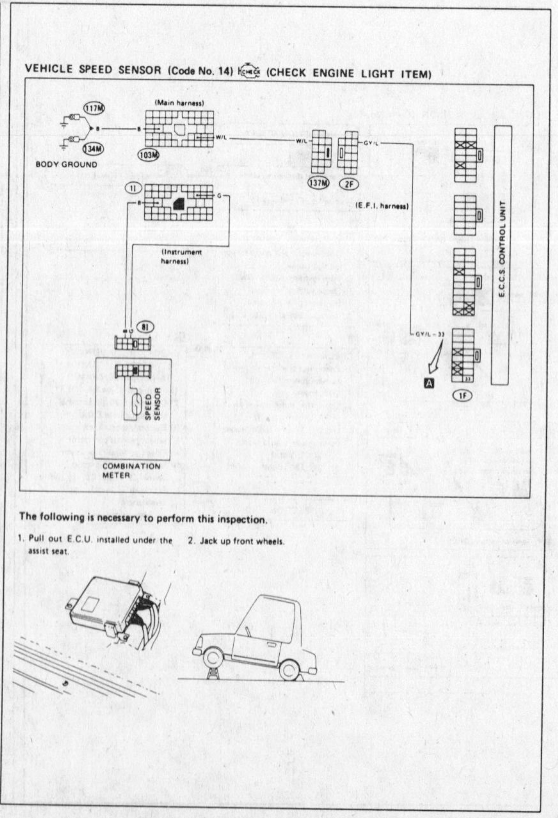
VEHICLE SPEED SENSOR (Code No. 14) (CHECK ENGINE LIGHT ITEM)
117M
134M
BODY GROUND
(Main harness)
103M
137M
2F
(E.F.I. harness)
(Instrument harness)
SPEED SENSOR
COMBINATION METER
E.C.C.S. CONTROL UNIT
1F
The following is necessary to perform this inspection.
1. Pull out E.C.U. installed under the assist seat.
2. Jack up front wheels.

ECCS SELF-DIAGNOSTIC CHART—1988 PULSAR

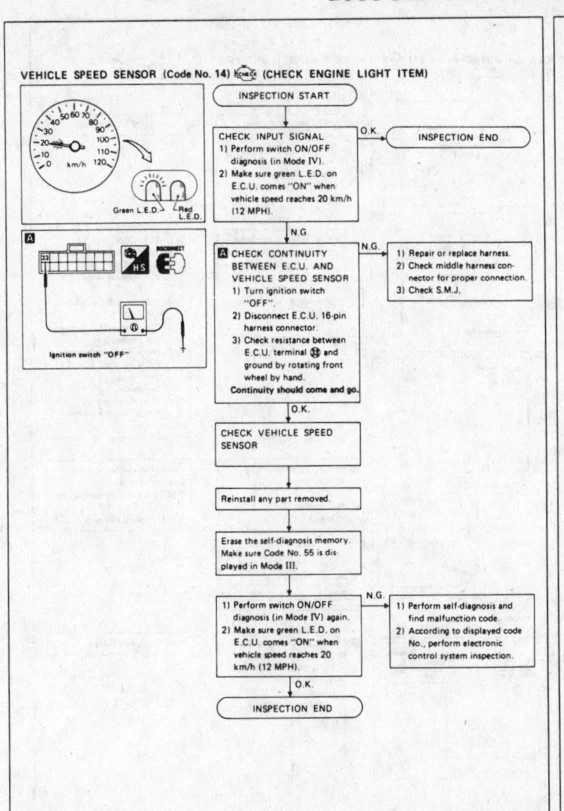
VEHICLE SPEED SENSOR (Code No. 14) (CHECK ENGINE LIGHT ITEM)
INSPECTION START
CHECK INPUT SIGNAL
1) Perform switch ON/OFF diagnosis (in Mode IV).
2) Make sure green L.E.D. on E.C.U. comes "ON" when vehicle speed reaches 20 km/h (12 MPH).
O.K.
INSPECTION END
N.G.
A CHECK CONTINUITY BETWEEN E.C.U. AND VEHICLE SPEED SENSOR
1) Turn ignition switch "OFF".
2) Disconnect E.C.U. 16-pin harness connector.
3) Check resistance between E.C.U. terminal 33 and ground by rotating front wheel by hand.
Continuity should come and go.
N.G.
1) Repair or replace harness.
2) Check middle harness connector for proper connection.
3) Check S.M.J.
O.K.
CHECK VEHICLE SPEED SENSOR
Reinstall any part removed.
Erase the self-diagnosis memory. Make sure Code No. 55 is displayed in Mode III.
1) Perform switch ON/OFF diagnosis (in Mode IV) again.
2) Make sure green L.E.D. on E.C.U. comes "ON" when vehicle speed reaches 20 km/h (12 MPH).
N.G.
1) Perform self-diagnosis and find malfunction code.
2) According to displayed code No., perform electronic control system inspection.
O.K.
INSPECTION END
km/h
Green L.E.D.
Red L.E.D.
Ignition switch "OFF"

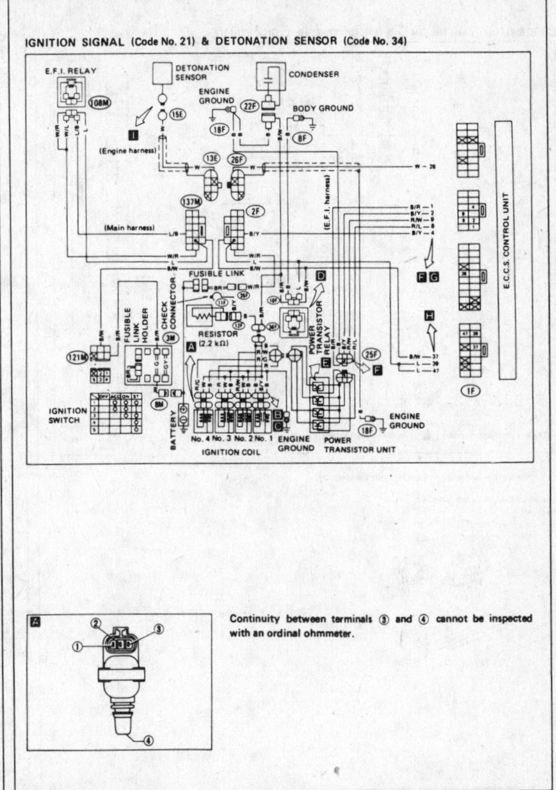
IGNITION SIGNAL (Code No. 21) & DETONATION SENSOR (Code No. 34)
E.F.I. RELAY
DETONATION SENSOR
CONDENSER
ENGINE GROUND
BODY GROUND
(Engine harness)
(Main harness)
(E.F.I. harness)
FUSIBLE LINK
FUSIBLE LINK HOLDER
CHECK CONNECTOR
RESISTOR (2.2 kΩ)
POWER TRANSISTOR RELAY
IGNITION SWITCH
BATTERY
No. 4 No. 3 No. 2 No. 1
IGNITION COIL
ENGINE GROUND
POWER TRANSISTOR UNIT
ENGINE GROUND
E.C.C.S. CONTROL UNIT
Continuity between terminals ③ and ④ cannot be inspected with an ordinal ohmmeter.

ECCS SELF-DIAGNOSTIC CHART—1988 PULSAR

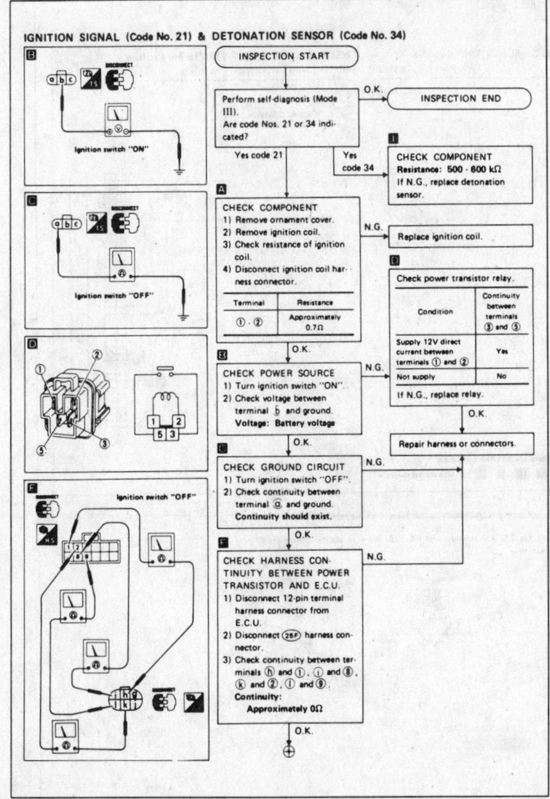
IGNITION SIGNAL (Code No. 21) & DETONATION SENSOR (Code No. 34)
INSPECTION START
Perform self-diagnosis (Mode III). Are code Nos. 21 or 34 indicated?
O.K.
INSPECTION END
Yes code 21
Yes code 34
I CHECK COMPONENT
Resistance: 500 - 600 kΩ
If N.G., replace detonation sensor.
A CHECK COMPONENT
1) Remove ornament cover.
2) Remove ignition coil.
3) Check resistance of ignition coil.
4) Disconnect ignition coil harness connector.
Terminal | Resistance
① - ② | Approximately 0.7Ω
N.G.
Replace ignition coil.
D Check power transistor relay.
Condition | Continuity between terminals ③ and ⑤
Supply 12V direct current between terminals ① and ② | Yes
Not supply | No
If N.G., replace relay.
O.K.
B CHECK POWER SOURCE
1) Turn ignition switch "ON".
2) Check voltage between terminal b and ground.
Voltage: Battery voltage
Repair harness or connectors.
C CHECK GROUND CIRCUIT
1) Turn ignition switch "OFF".
2) Check continuity between terminal c and ground.
Continuity should exist.
F CHECK HARNESS CONTINUITY BETWEEN POWER TRANSISTOR AND E.C.U.
1) Disconnect 12-pin terminal harness connector from E.C.U.
2) Disconnect 25F harness connector.
3) Check continuity between terminals ⓗ and ①, ⓘ and ⑧, ⓚ and ②, ⓛ and ⑨.
Continuity: Approximately 0Ω
Ignition switch "ON"
Ignition switch "OFF"

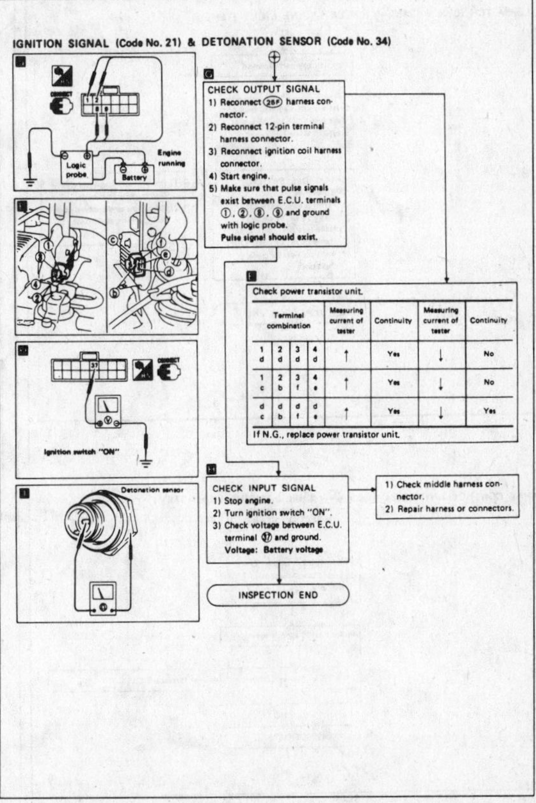
IGNITION SIGNAL (Code No. 21) & DETONATION SENSOR (Code No. 34)
G CHECK OUTPUT SIGNAL
1) Reconnect 25F harness connector.
2) Reconnect 12-pin terminal harness connector.
3) Reconnect ignition coil harness connector.
4) Start engine.
5) Make sure that pulse signals exist between E.C.U. terminals ①, ②, ⑧, ⑨ and ground with logic probe.
Pulse signal should exist.
Logic probe
Battery
Engine running
E Check power transistor unit.
Terminal combination | Measuring current of tester | Continuity | Measuring current of tester | Continuity
1 2 3 4 / d d d d | ↑ | Yes | ↓ | No
1 2 3 4 / c b f e | ↑ | Yes | ↓ | No
d d d d / c b f e | ↑ | Yes | ↓ | Yes
If N.G., replace power transistor unit.
Ignition switch "ON"
Detonation sensor
H CHECK INPUT SIGNAL
1) Stop engine.
2) Turn ignition switch "ON".
3) Check voltage between E.C.U. terminal 37 and ground.
Voltage: Battery voltage
1) Check middle harness connector.
2) Repair harness or connectors.
INSPECTION END

ECCS SELF-DIAGNOSTIC CHART—1988 PULSAR

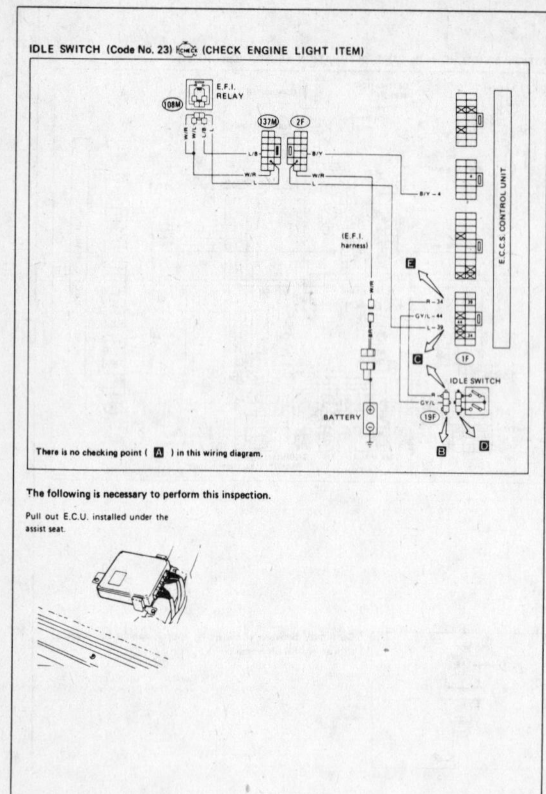
IDLE SWITCH (Code No. 23) (CHECK ENGINE LIGHT ITEM)
E.F.I. RELAY
108M
137M
2F
(E.F.I. harness)
E.C.C.S. CONTROL UNIT
IDLE SWITCH
BATTERY
1F
19F
There is no checking point (A) in this wiring diagram.
The following is necessary to perform this inspection.
Pull out E.C.U. installed under the assist seat.

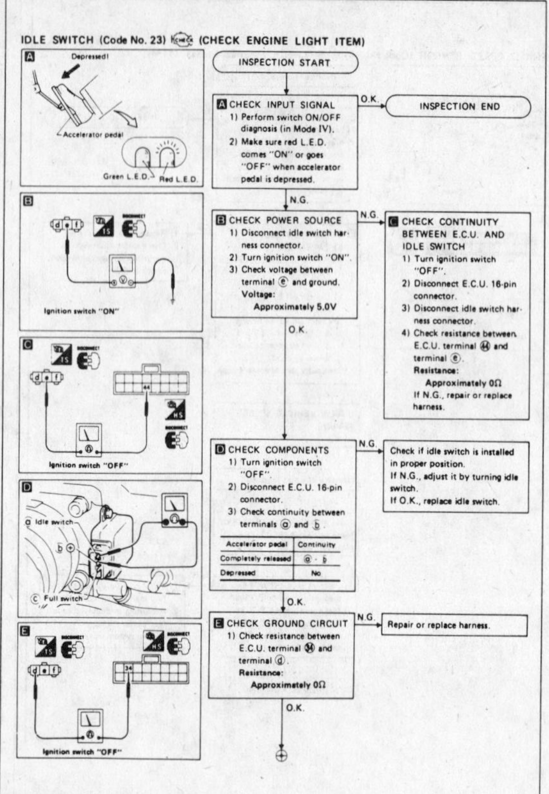
IDLE SWITCH (Code No. 23) (CHECK ENGINE LIGHT ITEM)
Depressed!
Accelerator pedal
Green L.E.D.
Red L.E.D.
Ignition switch "ON"
Ignition switch "OFF"
Idle switch
Full switch
Ignition switch "OFF"
INSPECTION START
A CHECK INPUT SIGNAL
1) Perform switch ON/OFF diagnosis (in Mode IV).
2) Make sure red L.E.D. comes "ON" or goes "OFF" when accelerator pedal is depressed.
O.K.
INSPECTION END
N.G.
B CHECK POWER SOURCE
1) Disconnect idle switch harness connector.
2) Turn ignition switch "ON".
3) Check voltage between terminal ⓔ and ground.
Voltage:
Approximately 5.0V
N.G.
C CHECK CONTINUITY BETWEEN E.C.U. AND IDLE SWITCH
1) Turn ignition switch "OFF".
2) Disconnect E.C.U. 16-pin connector.
3) Disconnect idle switch harness connector.
4) Check resistance between E.C.U. terminal ㊹ and terminal ⓔ.
Resistance:
Approximately 0Ω
If N.G., repair or replace harness.
O.K.
D CHECK COMPONENTS
1) Turn ignition switch "OFF".
2) Disconnect E.C.U. 16-pin connector.
3) Check continuity between terminals ⓐ and ⓑ.
Accelerator pedal | Continuity
Completely released | ⓐ - ⓑ
Depressed | No
N.G.
Check if idle switch is installed in proper position.
If N.G., adjust it by turning idle switch.
If O.K., replace idle switch.
O.K.
E CHECK GROUND CIRCUIT
1) Check resistance between E.C.U. terminal ㉞ and terminal ⓓ.
Resistance:
Approximately 0Ω
N.G.
Repair or replace harness.
O.K.

ECCS SELF-DIAGNOSTIC CHART—1988 PULSAR

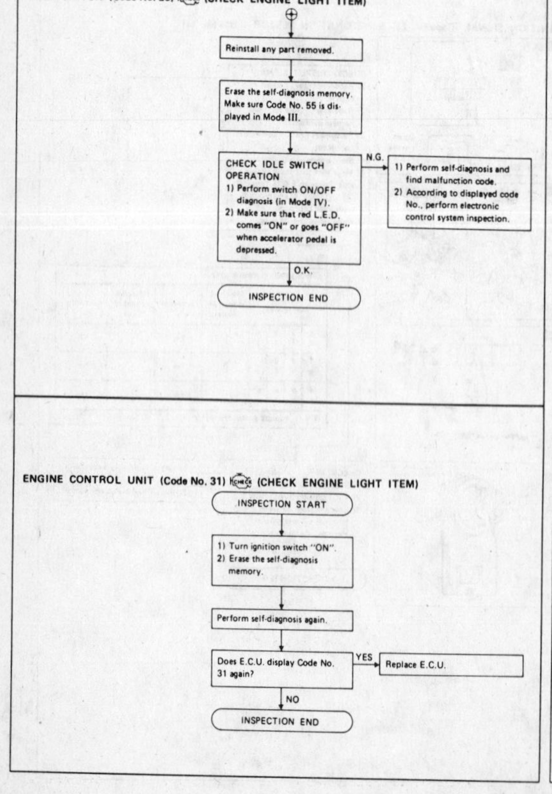
IDLE SWITCH (Code No. 23) (CHECK ENGINE LIGHT ITEM)
Reinstall any part removed.
Erase the self-diagnosis memory. Make sure Code No. 55 is displayed in Mode III.
CHECK IDLE SWITCH OPERATION
1) Perform switch ON/OFF diagnosis (in Mode IV).
2) Make sure that red L.E.D. comes "ON" or goes "OFF" when accelerator pedal is depressed.
N.G.
1) Perform self-diagnosis and find malfunction code.
2) According to displayed code No., perform electronic control system inspection.
O.K.
INSPECTION END
ENGINE CONTROL UNIT (Code No. 31) (CHECK ENGINE LIGHT ITEM)
INSPECTION START
1) Turn ignition switch "ON".
2) Erase the self-diagnosis memory.
Perform self-diagnosis again.
Does E.C.U. display Code No. 31 again?
YES
Replace E.C.U.
NO
INSPECTION END

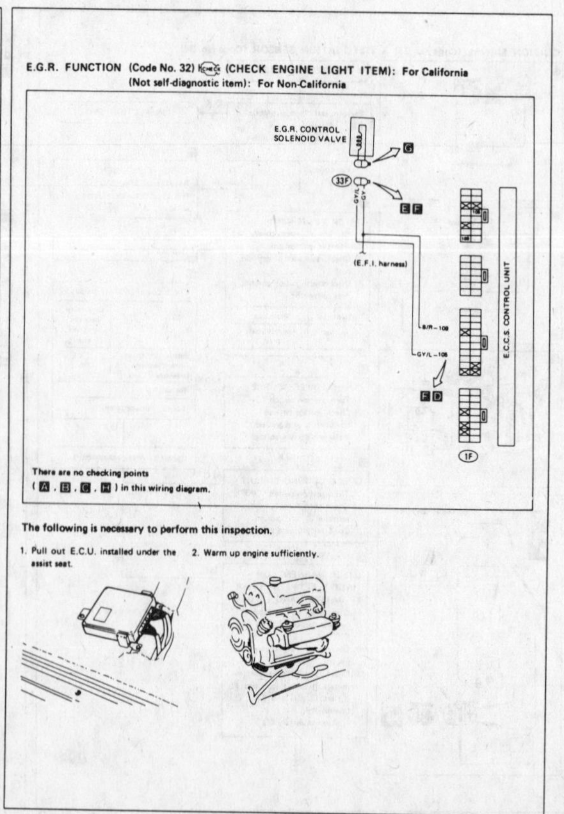
E.G.R. FUNCTION (Code No. 32) (CHECK ENGINE LIGHT ITEM): For California
(Not self-diagnostic item): For Non-California
E.G.R. CONTROL SOLENOID VALVE
33F
(E.F.I. harness)
E.C.C.S. CONTROL UNIT
1F
There are no checking points (A , B , C , H) in this wiring diagram.
The following is necessary to perform this inspection.
1. Pull out E.C.U. installed under the assist seat.
2. Warm up engine sufficiently.

ECCS SELF-DIAGNOSTIC CHART—1988 PULSAR

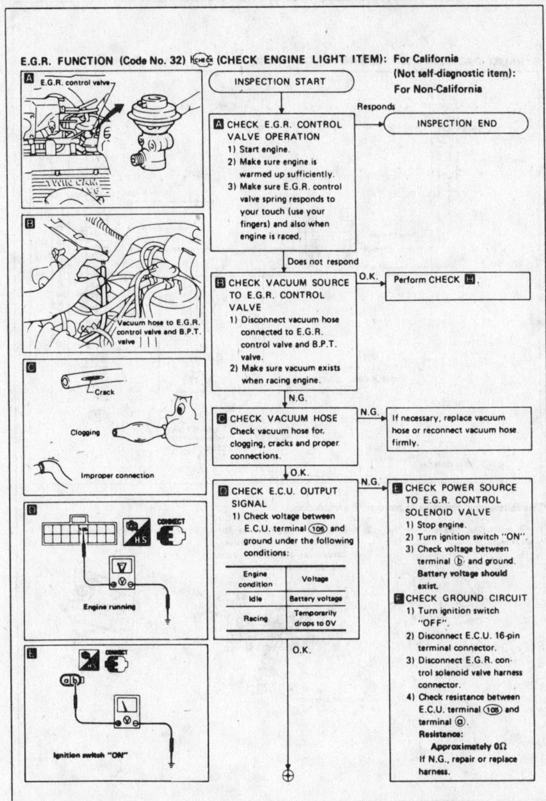
E.G.R. FUNCTION (Code No. 32) (CHECK ENGINE LIGHT ITEM): For California
(Not self-diagnostic item):
For Non-California
A
E.G.R. control valve
B
Vacuum hose to E.G.R. control valve and B.P.T. valve
C
Crack
Clogging
Improper connection
D
CONNECT
HS
Engine running
E
CONNECT
Ignition switch "ON"
INSPECTION START
A CHECK E.G.R. CONTROL VALVE OPERATION
1) Start engine.
2) Make sure engine is warmed up sufficiently.
3) Make sure E.G.R. control valve spring responds to your touch (use your fingers) and also when engine is raced.
Responds
INSPECTION END
Does not respond
B CHECK VACUUM SOURCE TO E.G.R. CONTROL VALVE
1) Disconnect vacuum hose connected to E.G.R. control valve and B.P.T. valve.
2) Make sure vacuum exists when racing engine.
O.K.
Perform CHECK H.
N.G.
C CHECK VACUUM HOSE
Check vacuum hose for clogging, cracks and proper connections.
N.G.
If necessary, replace vacuum hose or reconnect vacuum hose firmly.
O.K.
D CHECK E.C.U. OUTPUT SIGNAL
1) Check voltage between E.C.U. terminal (106) and ground under the following conditions:
Engine condition | Voltage
Idle | Battery voltage
Racing | Temporarily drops to 0V
N.G.
E CHECK POWER SOURCE TO E.G.R. CONTROL SOLENOID VALVE
1) Stop engine.
2) Turn ignition switch "ON".
3) Check voltage between terminal (b) and ground. Battery voltage should exist.
F CHECK GROUND CIRCUIT
1) Turn ignition switch "OFF".
2) Disconnect E.C.U. 16-pin terminal connector.
3) Disconnect E.G.R. control solenoid valve harness connector.
4) Check resistance between E.C.U. terminal (106) and terminal (a).
Resistance:
Approximately 0Ω
If N.G., repair or replace harness.
O.K.

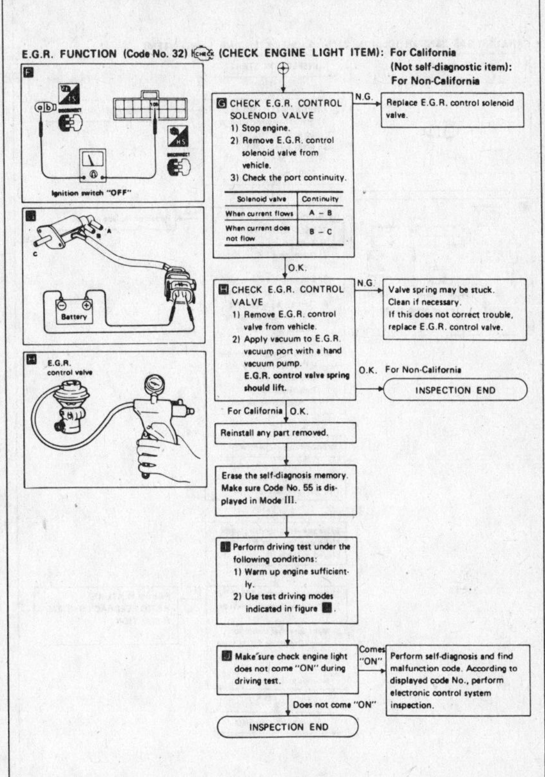
E.G.R. FUNCTION (Code No. 32) (CHECK ENGINE LIGHT ITEM): For California
(Not self-diagnostic item):
For Non-California
F
HS
Ignition switch "OFF"
G
A
B
C
Battery
H
E.G.R. control valve
G CHECK E.G.R. CONTROL SOLENOID VALVE
1) Stop engine.
2) Remove E.G.R. control solenoid valve from vehicle.
3) Check the port continuity.
Solenoid valve | Continuity
When current flows | A – B
When current does not flow | B – C
N.G.
Replace E.G.R. control solenoid valve.
O.K.
H CHECK E.G.R. CONTROL VALVE
1) Remove E.G.R. control valve from vehicle.
2) Apply vacuum to E.G.R. vacuum port with a hand vacuum pump.
E.G.R. control valve spring should lift.
N.G.
Valve spring may be stuck.
Clean if necessary.
If this does not correct trouble, replace E.G.R. control valve.
O.K. For Non-California
INSPECTION END
For California O.K.
Reinstall any part removed.
Erase the self-diagnosis memory. Make sure Code No. 55 is displayed in Mode III.
I Perform driving test under the following conditions:
1) Warm up engine sufficiently.
2) Use test driving modes indicated in figure I.
J Make sure check engine light does not come "ON" during driving test.
Comes "ON"
Perform self-diagnosis and find malfunction code. According to displayed code No., perform electronic control system inspection.
Does not come "ON"
INSPECTION END

ECCS SELF-DIAGNOSTIC CHART—1988 PULSAR

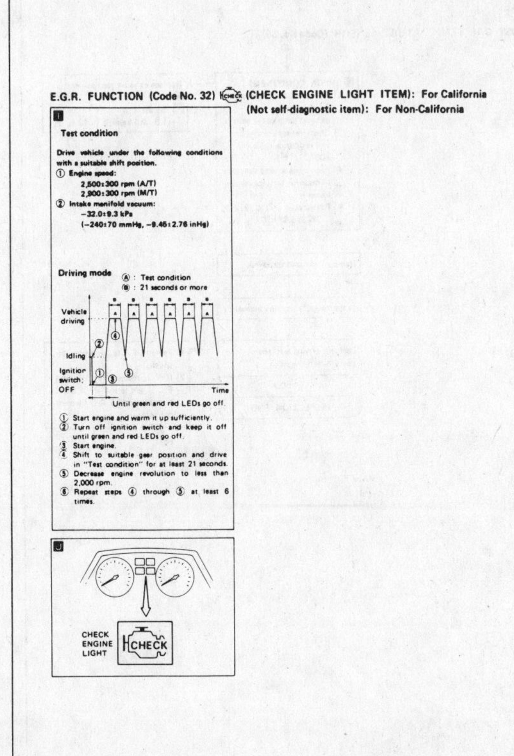
E.G.R. FUNCTION (Code No. 32) (CHECK ENGINE LIGHT ITEM): For California
(Not self-diagnostic item): For Non-California
I
Test condition
Drive vehicle under the following conditions with a suitable shift position.
① Engine speed:
2,500±300 rpm (A/T)
2,900±300 rpm (M/T)
② Intake manifold vacuum:
−32.0±9.3 kPa
(−240±70 mmHg, −9.45±2.76 inHg)
Driving mode
Ⓐ : Test condition
Ⓑ : 21 seconds or more
Vehicle driving
Idling
Ignition switch: OFF
Time
Until green and red LEDs go off.
① Start engine and warm it up sufficiently.
② Turn off ignition switch and keep it off until green and red LEDs go off.
③ Start engine.
④ Shift to suitable gear position and drive in "Test condition" for at least 21 seconds.
⑤ Decrease engine revolution to less than 2,000 rpm.
⑥ Repeat steps ④ through ⑤ at least 6 times.
J
CHECK ENGINE LIGHT
CHECK

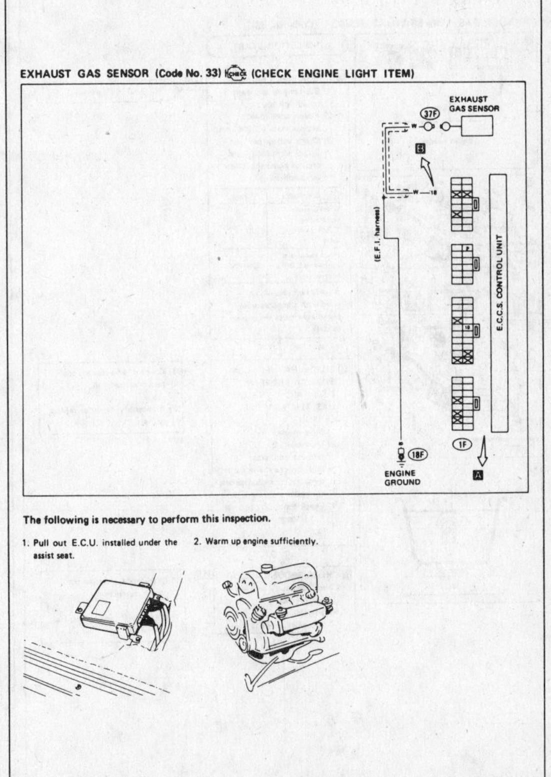
EXHAUST GAS SENSOR (Code No. 33) (CHECK ENGINE LIGHT ITEM)
EXHAUST GAS SENSOR
37F
W
B
W
18
(E.F.I. harness)
E.C.C.S. CONTROL UNIT
1F
18F
ENGINE GROUND
A
The following is necessary to perform this inspection.
1. Pull out E.C.U. installed under the assist seat.
2. Warm up engine sufficiently.

ECCS SELF-DIAGNOSTIC CHART—1988 PULSAR

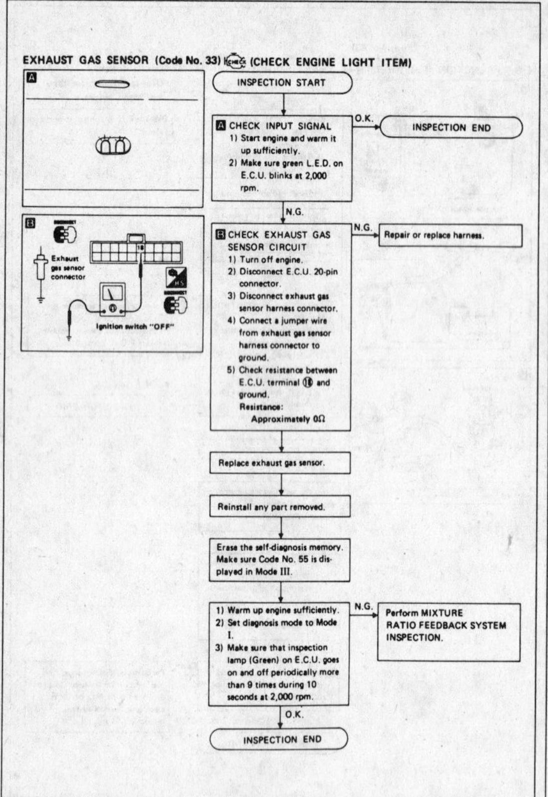
EXHAUST GAS SENSOR (Code No. 33) (CHECK ENGINE LIGHT ITEM)
INSPECTION START
A CHECK INPUT SIGNAL
1) Start engine and warm it up sufficiently.
2) Make sure green L.E.D. on E.C.U. blinks at 2,000 rpm.
O.K.
INSPECTION END
N.G.
B CHECK EXHAUST GAS SENSOR CIRCUIT
1) Turn off engine.
2) Disconnect E.C.U. 20-pin connector.
3) Disconnect exhaust gas sensor harness connector.
4) Connect a jumper wire from exhaust gas sensor harness connector to ground.
5) Check resistance between E.C.U. terminal ⑲ and ground.
Resistance: Approximately 0Ω
N.G.
Repair or replace harness.
Replace exhaust gas sensor.
Reinstall any part removed.
Erase the self-diagnosis memory. Make sure Code No. 55 is displayed in Mode III.
1) Warm up engine sufficiently.
2) Set diagnosis mode to Mode I.
3) Make sure that inspection lamp (Green) on E.C.U. goes on and off periodically more than 9 times during 10 seconds at 2,000 rpm.
N.G.
Perform MIXTURE RATIO FEEDBACK SYSTEM INSPECTION.
O.K.
INSPECTION END
Exhaust gas sensor connector
Ignition switch "OFF"

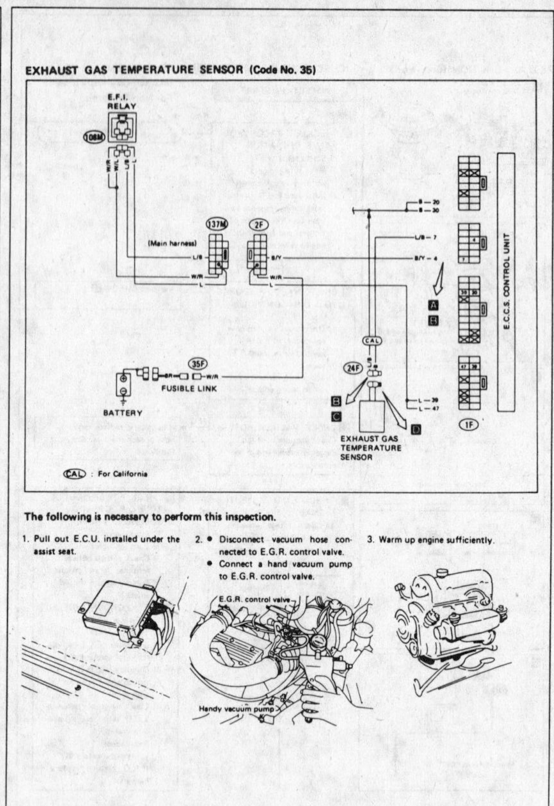
EXHAUST GAS TEMPERATURE SENSOR (Code No. 35)
E.F.I. RELAY
(Main harness)
BATTERY
FUSIBLE LINK
EXHAUST GAS TEMPERATURE SENSOR
E.C.C.S. CONTROL UNIT
CAL : For California
The following is necessary to perform this inspection.
1. Pull out E.C.U. installed under the assist seat.
2. • Disconnect vacuum hose connected to E.G.R. control valve.
• Connect a hand vacuum pump to E.G.R. control valve.
3. Warm up engine sufficiently.
E.G.R. control valve
Handy vacuum pump

ECCS SELF-DIAGNOSTIC CHART—1988 PULSAR

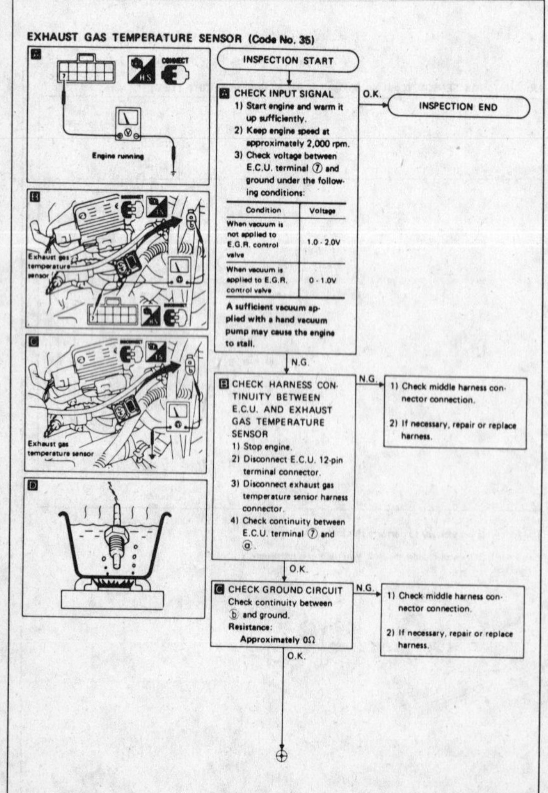
EXHAUST GAS TEMPERATURE SENSOR (Code No. 35)
INSPECTION START
A CHECK INPUT SIGNAL
1) Start engine and warm it up sufficiently.
2) Keep engine speed at approximately 2,000 rpm.
3) Check voltage between E.C.U. terminal ⑦ and ground under the following conditions:
Condition | Voltage
When vacuum is not applied to E.G.R. control valve | 1.0 - 2.0V
When vacuum is applied to E.G.R. control valve | 0 - 1.0V
A sufficient vacuum applied with a hand vacuum pump may cause the engine to stall.
O.K.
INSPECTION END
N.G.
B CHECK HARNESS CONTINUITY BETWEEN E.C.U. AND EXHAUST GAS TEMPERATURE SENSOR
1) Stop engine.
2) Disconnect E.C.U. 12-pin terminal connector.
3) Disconnect exhaust gas temperature sensor harness connector.
4) Check continuity between E.C.U. terminal ⑦ and ⓐ.
N.G.
1) Check middle harness connector connection.
2) If necessary, repair or replace harness.
O.K.
C CHECK GROUND CIRCUIT
Check continuity between ⓑ and ground.
Resistance: Approximately 0Ω
N.G.
1) Check middle harness connector connection.
2) If necessary, repair or replace harness.
O.K.
Engine running
Exhaust gas temperature sensor

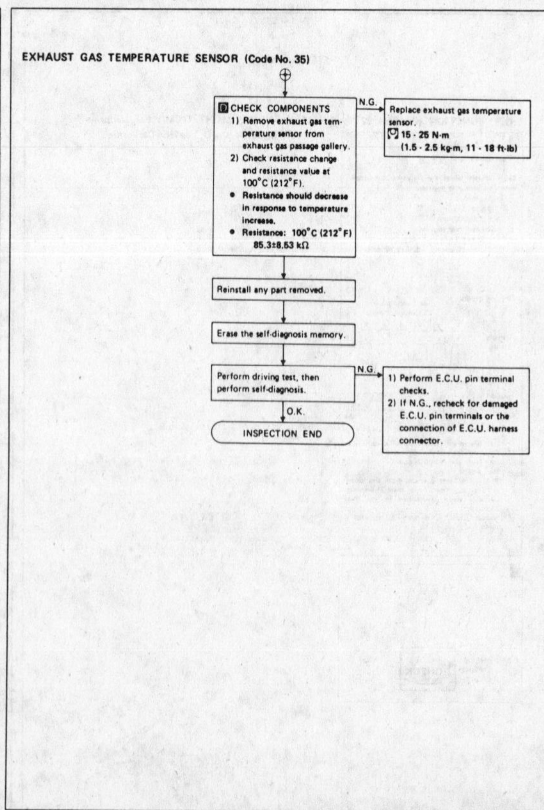
EXHAUST GAS TEMPERATURE SENSOR (Code No. 35)
D CHECK COMPONENTS
1) Remove exhaust gas temperature sensor from exhaust gas passage gallery.
2) Check resistance change and resistance value at 100°C (212°F).
• Resistance should decrease in response to temperature increase.
• Resistance: 100°C (212°F) 85.3±8.53 kΩ
N.G.
Replace exhaust gas temperature sensor.
15 - 25 N·m (1.5 - 2.5 kg-m, 11 - 18 ft-lb)
Reinstall any part removed.
Erase the self-diagnosis memory.
Perform driving test, then perform self-diagnosis.
N.G.
1) Perform E.C.U. pin terminal checks.
2) If N.G., recheck for damaged E.C.U. pin terminals or the connection of E.C.U. harness connector.
O.K.
INSPECTION END

ECCS SELF-DIAGNOSTIC CHART—1988 PULSAR

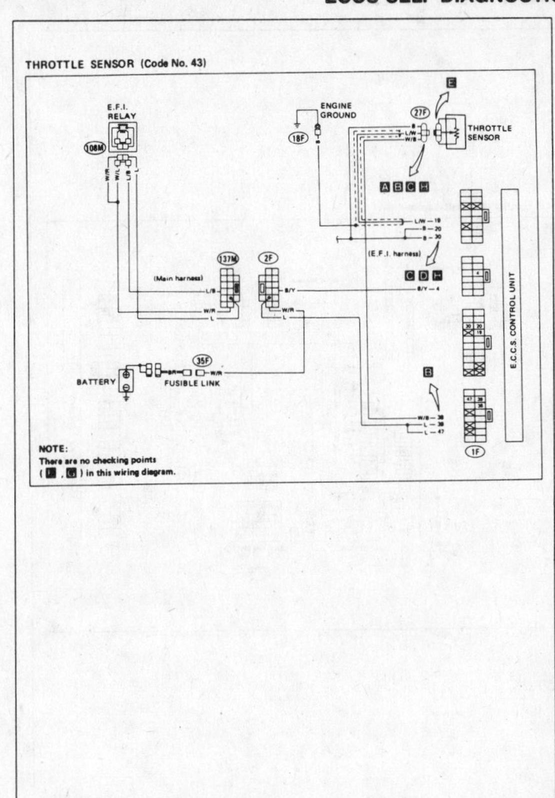
THROTTLE SENSOR (Code No. 43)
E.F.I. RELAY
ENGINE GROUND
THROTTLE SENSOR
(E.F.I. harness)
(Main harness)
BATTERY
FUSIBLE LINK
E.C.C.S. CONTROL UNIT
NOTE:
There are no checking points
(,) in this wiring diagram.

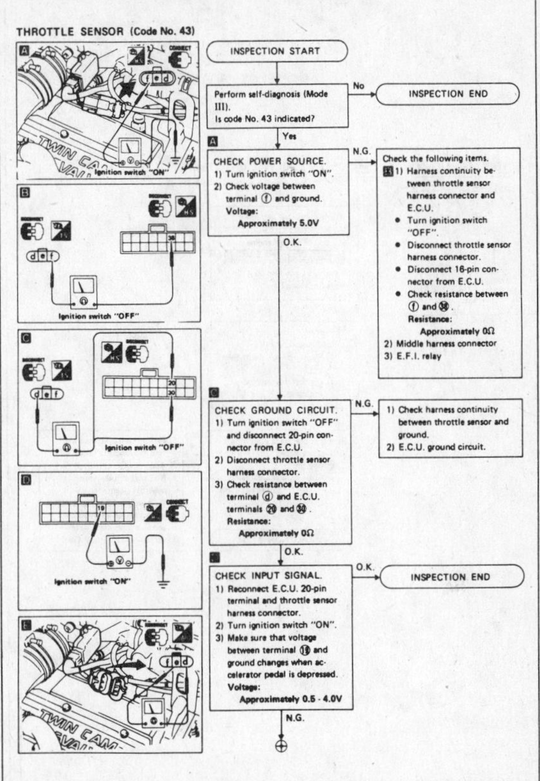
THROTTLE SENSOR (Code No. 43)
Ignition switch "ON"
Ignition switch "OFF"
Ignition switch "OFF"
Ignition switch "ON"
INSPECTION START
Perform self-diagnosis (Mode III).
Is code No. 43 indicated?
No
INSPECTION END
Yes
A
CHECK POWER SOURCE.
1) Turn ignition switch "ON".
2) Check voltage between terminal ⓕ and ground.
Voltage:
Approximately 5.0V
N.G.
Check the following items.
1) Harness continuity between throttle sensor harness connector and E.C.U.
• Turn ignition switch "OFF".
• Disconnect throttle sensor harness connector.
• Disconnect 16-pin connector from E.C.U.
• Check resistance between ⓕ and ㊳.
Resistance:
Approximately 0Ω
2) Middle harness connector
3) E.F.I. relay
O.K.
C
CHECK GROUND CIRCUIT.
1) Turn ignition switch "OFF" and disconnect 20-pin connector from E.C.U.
2) Disconnect throttle sensor harness connector.
3) Check resistance between terminal ⓓ and E.C.U. terminals ⑳ and ㉚.
Resistance:
Approximately 0Ω
N.G.
1) Check harness continuity between throttle sensor and ground.
2) E.C.U. ground circuit.
O.K.
CHECK INPUT SIGNAL.
1) Reconnect E.C.U. 20-pin terminal and throttle sensor harness connector.
2) Turn ignition switch "ON".
3) Make sure that voltage between terminal ⑲ and ground changes when accelerator pedal is depressed.
Voltage:
Approximately 0.5 - 4.0V
O.K.
INSPECTION END
N.G.

ECCS SELF-DIAGNOSTIC CHART—1988 PULSAR

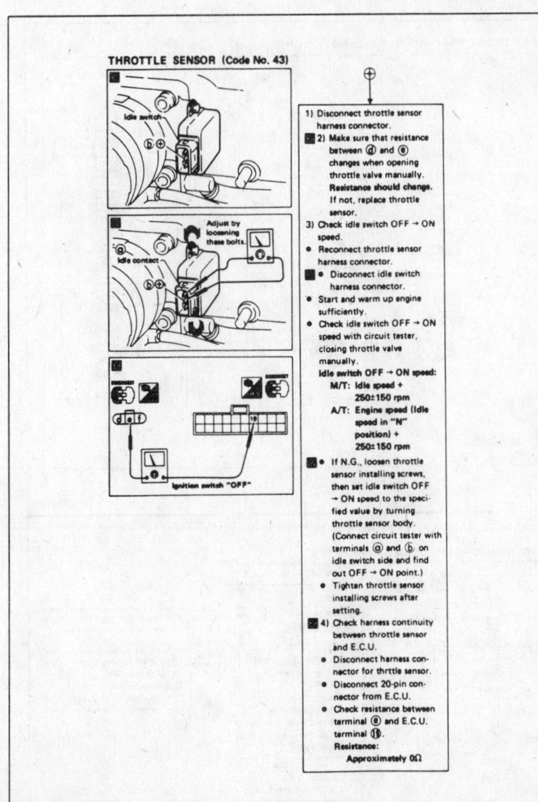
THROTTLE SENSOR (Code No. 43)
Idle switch
Adjust by loosening these bolts.
Idle contact
Ignition switch "OFF"
1) Disconnect throttle sensor harness connector.
2) Make sure that resistance between ⓓ and ⓔ changes when opening throttle valve manually.
Resistance should change.
If not, replace throttle sensor.
3) Check idle switch OFF → ON speed.
• Reconnect throttle sensor harness connector.
• Disconnect idle switch harness connector.
• Start and warm up engine sufficiently.
• Check idle switch OFF → ON speed with circuit tester, closing throttle valve manually.
Idle switch OFF → ON speed:
M/T: Idle speed + 250±150 rpm
A/T: Engine speed (Idle speed in "N" position) + 250±150 rpm
• If N.G., loosen throttle sensor installing screws, then set idle switch OFF → ON speed to the specified value by turning throttle sensor body. (Connect circuit tester with terminals ⓐ and ⓑ on idle switch side and find out OFF → ON point.)
• Tighten throttle sensor installing screws after setting.
4) Check harness continuity between throttle sensor and E.C.U.
• Disconnect harness connector for thrttle sensor.
• Disconnect 20-pin connector from E.C.U.
• Check resistance between terminal ⓔ and E.C.U. terminal ⑲.
Resistance:
Approximately 0Ω

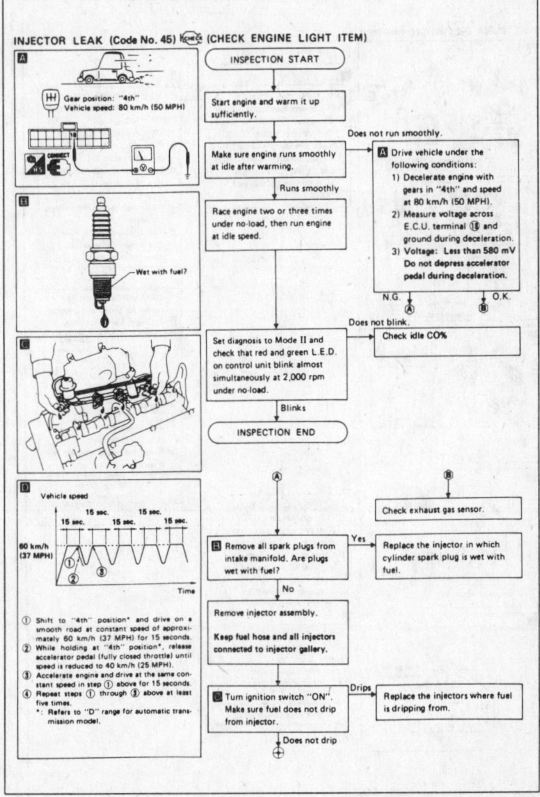
INJECTOR LEAK (Code No. 45) (CHECK ENGINE LIGHT ITEM)
Gear position: "4th"
Vehicle speed: 80 km/h (50 MPH)
Wet with fuel?
Vehicle speed
15 sec.
60 km/h (37 MPH)
Time
① Shift to "4th" position* and drive on a smooth road at constant speed of approximately 60 km/h (37 MPH) for 15 seconds.
② While holding at "4th" position*, release accelerator pedal (fully closed throttle) until speed is reduced to 40 km/h (25 MPH).
③ Accelerate engine and drive at the same constant speed in step ① above for 15 seconds.
④ Repeat steps ① through ③ above at least five times.
*: Refers to "D" range for automatic transmission model.
INSPECTION START
Start engine and warm it up sufficiently.
Make sure engine runs smoothly at idle after warming.
Does not run smoothly.
A Drive vehicle under the following conditions:
1) Decelerate engine with gears in "4th" and speed at 80 km/h (50 MPH).
2) Measure voltage across E.C.U. terminal ⑲ and ground during deceleration.
3) Voltage: Less than 580 mV
Do not depress accelerator pedal during deceleration.
N.G.
O.K.
Runs smoothly
Race engine two or three times under no-load, then run engine at idle speed.
Set diagnosis to Mode II and check that red and green L.E.D. on control unit blink almost simultaneously at 2,000 rpm under no-load.
Does not blink.
Check idle CO%
Blinks
INSPECTION END
Check exhaust gas sensor.
B Remove all spark plugs from intake manifold. Are plugs wet with fuel?
Yes
Replace the injector in which cylinder spark plug is wet with fuel.
No
Remove injector assembly.
Keep fuel hose and all injectors connected to injector gallery.
C Turn ignition switch "ON". Make sure fuel does not drip from injector.
Drips
Replace the injectors where fuel is dripping from.
Does not drip

ECCS SELF-DIAGNOSTIC CHART—1988 PULSAR

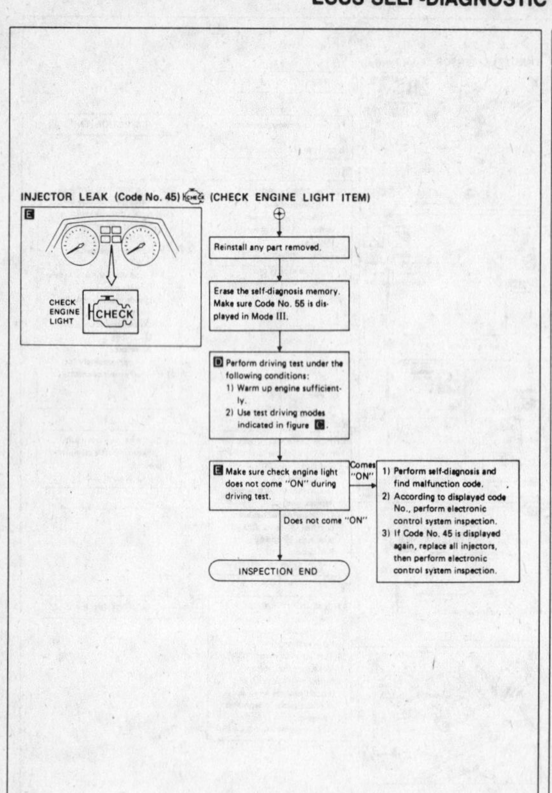
INJECTOR LEAK (Code No. 45) (CHECK ENGINE LIGHT ITEM)
E
CHECK ENGINE LIGHT
CHECK
Reinstall any part removed.
Erase the self-diagnosis memory. Make sure Code No. 55 is displayed in Mode III.
D Perform driving test under the following conditions:
1) Warm up engine sufficiently.
2) Use test driving modes indicated in figure D.
E Make sure check engine light does not come "ON" during driving test.
Comes "ON"
1) Perform self-diagnosis and find malfunction code.
2) According to displayed code No., perform electronic control system inspection.
3) If Code No. 45 is displayed again, replace all injectors, then perform electronic control system inspection.
Does not come "ON"
INSPECTION END

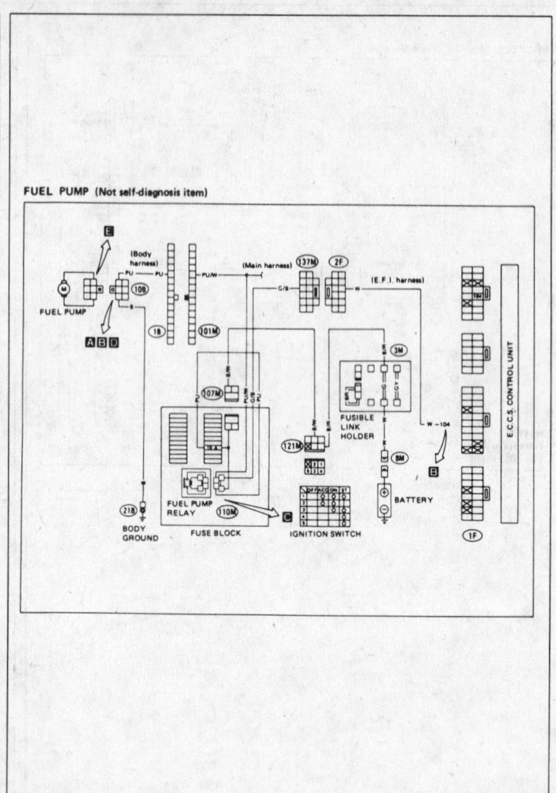
FUEL PUMP (Not self-diagnosis item)
FUEL PUMP
(Body harness)
(Main harness)
(E.F.I. harness)
A B D
E
FUSIBLE LINK HOLDER
BATTERY
FUEL PUMP RELAY
BODY GROUND
FUSE BLOCK
IGNITION SWITCH
E.C.C.S. CONTROL UNIT
B
C

ECCS SELF-DIAGNOSTIC CHART—1988 PULSAR

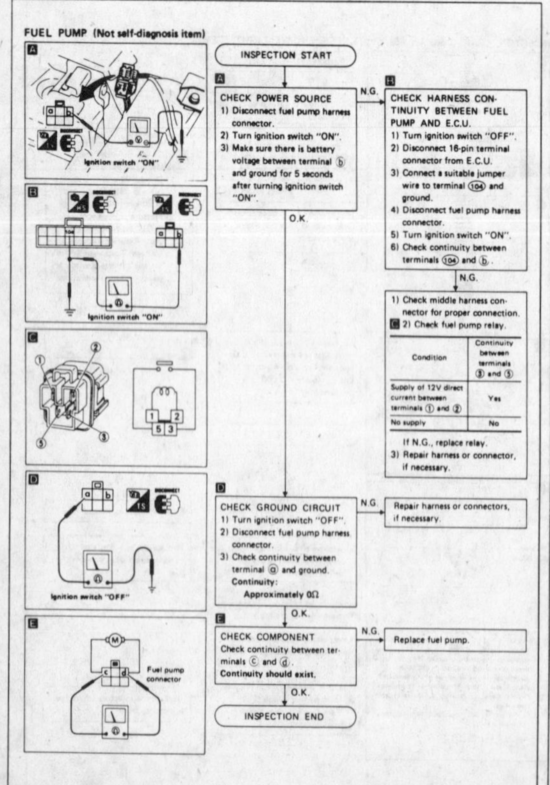
FUEL PUMP (Not self-diagnosis item)
A
Ignition switch "ON"
B
Ignition switch "ON"
C
D
Ignition switch "OFF"
E
Fuel pump connector
INSPECTION START
A
CHECK POWER SOURCE
1) Disconnect fuel pump harness connector.
2) Turn ignition switch "ON".
3) Make sure there is battery voltage between terminal ⓑ and ground for 5 seconds after turning ignition switch "ON".
N.G.
B
CHECK HARNESS CONTINUITY BETWEEN FUEL PUMP AND E.C.U.
1) Turn ignition switch "OFF".
2) Disconnect 16-pin terminal connector from E.C.U.
3) Connect a suitable jumper wire to terminal (104) and ground.
4) Disconnect fuel pump harness connector.
5) Turn ignition switch "ON".
6) Check continuity between terminals (104) and ⓑ.
N.G.
1) Check middle harness connector for proper connection.
C 2) Check fuel pump relay.
Condition | Continuity between terminals ③ and ⑤
Supply of 12V direct current between terminals ① and ② | Yes
No supply | No
If N.G., replace relay.
3) Repair harness or connector, if necessary.
O.K.
D
CHECK GROUND CIRCUIT
1) Turn ignition switch "OFF".
2) Disconnect fuel pump harness connector.
3) Check continuity between terminal ⓐ and ground.
Continuity: Approximately 0Ω
N.G.
Repair harness or connectors, if necessary.
O.K.
E
CHECK COMPONENT
Check continuity between terminals ⓒ and ⓓ.
Continuity should exist.
N.G.
Replace fuel pump.
O.K.
INSPECTION END

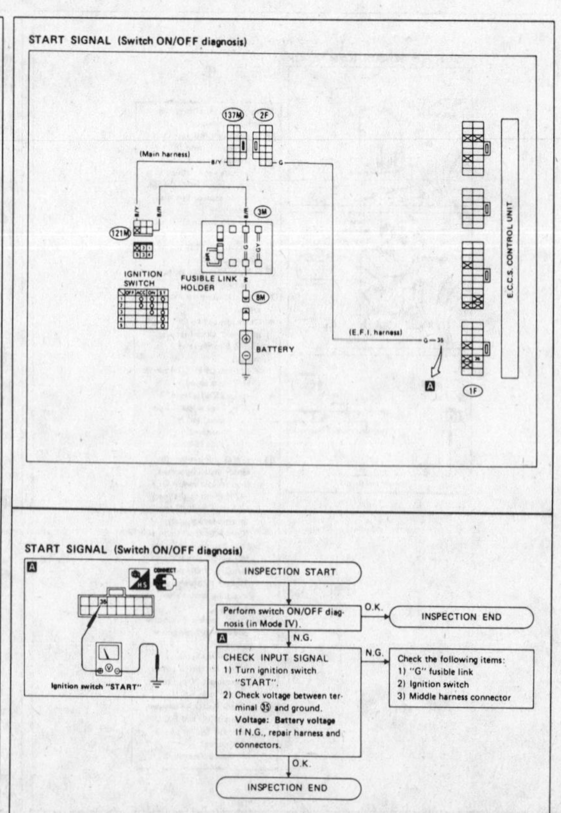
START SIGNAL (Switch ON/OFF diagnosis)
(Main harness)
IGNITION SWITCH
FUSIBLE LINK HOLDER
BATTERY
(E.F.I. harness)
E.C.C.S. CONTROL UNIT
A
START SIGNAL (Switch ON/OFF diagnosis)
A
Ignition switch "START"
INSPECTION START
Perform switch ON/OFF diagnosis (in Mode IV).
O.K.
INSPECTION END
N.G.
A
CHECK INPUT SIGNAL
1) Turn ignition switch "START".
2) Check voltage between terminal (35) and ground.
Voltage: Battery voltage
If N.G., repair harness and connectors.
N.G.
Check the following items:
1) "G" fusible link
2) Ignition switch
3) Middle harness connector
O.K.
INSPECTION END

ECCS SELF-DIAGNOSTIC CHART—1988 PULSAR

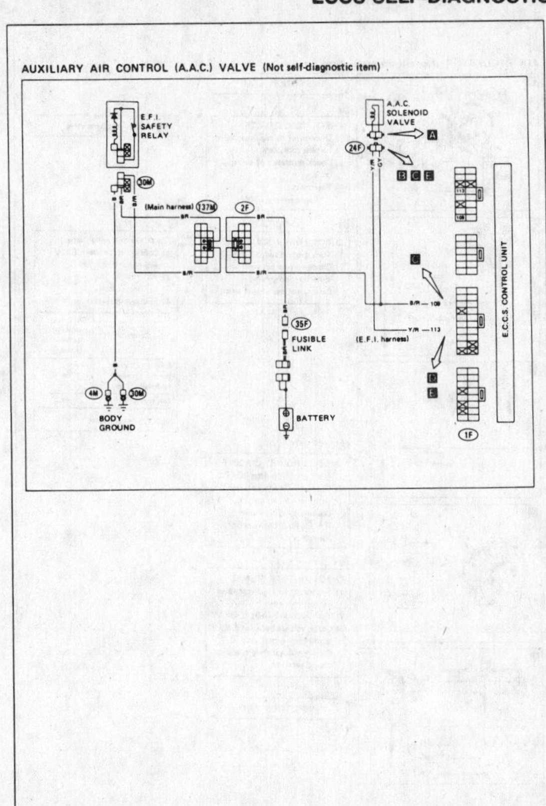
AUXILIARY AIR CONTROL (A.A.C.) VALVE (Not self-diagnostic item)
E.F.I. SAFETY RELAY
A.A.C. SOLENOID VALVE
(Main harness)
FUSIBLE LINK
(E.F.I. harness)
BATTERY
BODY GROUND
E.C.C.S. CONTROL UNIT

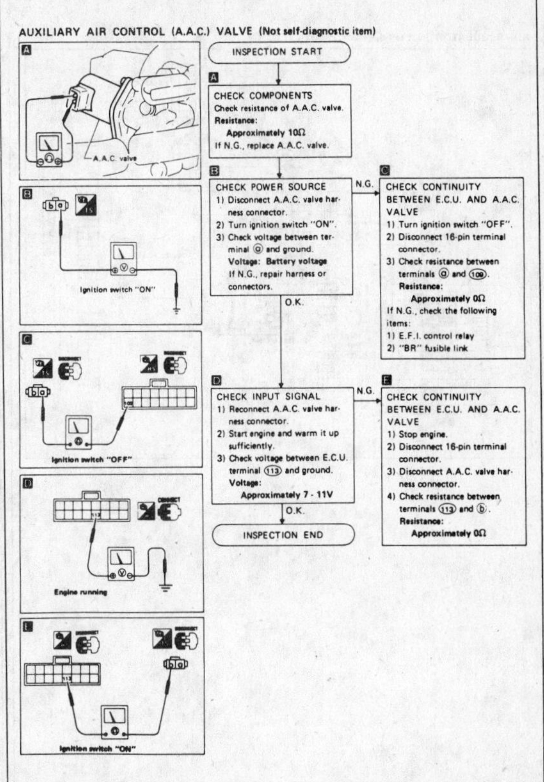
AUXILIARY AIR CONTROL (A.A.C.) VALVE (Not self-diagnostic item)
INSPECTION START
CHECK COMPONENTS
Check resistance of A.A.C. valve.
Resistance:
Approximately 10Ω
If N.G., replace A.A.C. valve.
A.A.C. valve
CHECK POWER SOURCE
1) Disconnect A.A.C. valve harness connector.
2) Turn ignition switch "ON".
3) Check voltage between terminal ⓐ and ground.
Voltage: Battery voltage
If N.G., repair harness or connectors.
O.K.
N.G.
CHECK CONTINUITY BETWEEN E.C.U. AND A.A.C. VALVE
1) Turn ignition switch "OFF".
2) Disconnect 16-pin terminal connector.
3) Check resistance between terminals ⓐ and (108).
Resistance:
Approximately 0Ω
If N.G., check the following items:
1) E.F.I. control relay
2) "BR" fusible link
Ignition switch "ON"
CHECK INPUT SIGNAL
1) Reconnect A.A.C. valve harness connector.
2) Start engine and warm it up sufficiently.
3) Check voltage between E.C.U. terminal (113) and ground.
Voltage:
Approximately 7 - 11V
O.K.
INSPECTION END
N.G.
CHECK CONTINUITY BETWEEN E.C.U. AND A.A.C. VALVE
1) Stop engine.
2) Disconnect 16-pin terminal connector.
3) Disconnect A.A.C. valve harness connector.
4) Check resistance between terminals (113) and ⓑ.
Resistance:
Approximately 0Ω
Ignition switch "OFF"
Engine running
Ignition switch "ON"

ECCS SELF-DIAGNOSTIC CHART—1988 PULSAR

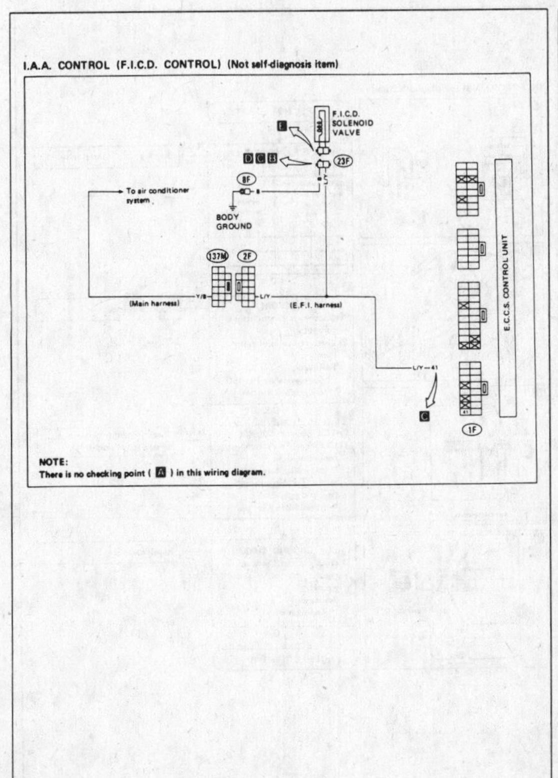
I.A.A. CONTROL (F.I.C.D. CONTROL) (Not self-diagnosis item)
F.I.C.D. SOLENOID VALVE
To air conditioner system
BODY GROUND
(Main harness)
(E.F.I. harness)
E.C.C.S. CONTROL UNIT
NOTE:
There is no checking point (A) in this wiring diagram.

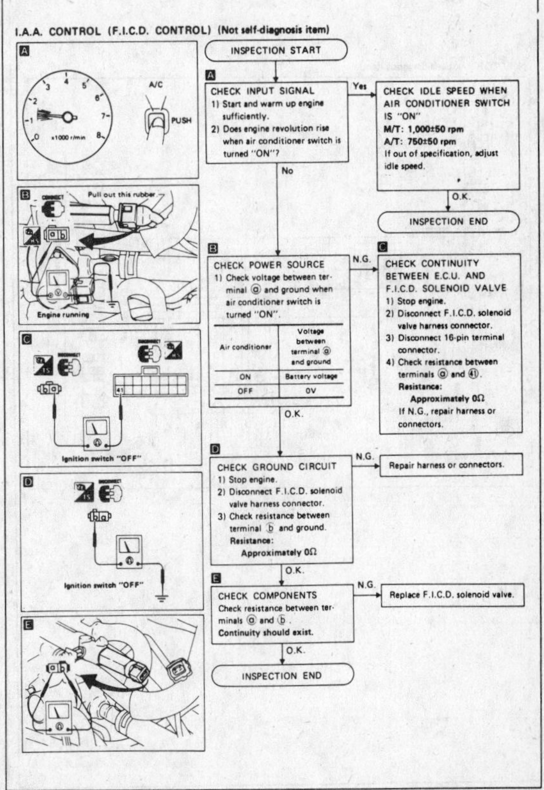
I.A.A. CONTROL (F.I.C.D. CONTROL) (Not self-diagnosis item)
INSPECTION START
CHECK INPUT SIGNAL
1) Start and warm up engine sufficiently.
2) Does engine revolution rise when air conditioner switch is turned "ON"?
Yes
No
CHECK IDLE SPEED WHEN AIR CONDITIONER SWITCH IS "ON"
M/T: 1,000±50 rpm
A/T: 750±50 rpm
If out of specification, adjust idle speed.
O.K.
INSPECTION END
A/C
PUSH
Pull out this rubber
Engine running
CHECK POWER SOURCE
1) Check voltage between terminal ⓐ and ground when air conditioner switch is turned "ON".
Air conditioner | Voltage between terminal ⓐ and ground
ON | Battery voltage
OFF | 0V
O.K.
N.G.
CHECK CONTINUITY BETWEEN E.C.U. AND F.I.C.D. SOLENOID VALVE
1) Stop engine.
2) Disconnect F.I.C.D. solenoid valve harness connector.
3) Disconnect 16-pin terminal connector.
4) Check resistance between terminals ⓐ and (41).
Resistance:
Approximately 0Ω
If N.G., repair harness or connectors.
Ignition switch "OFF"
CHECK GROUND CIRCUIT
1) Stop engine.
2) Disconnect F.I.C.D. solenoid valve harness connector.
3) Check resistance between terminal ⓑ and ground.
Resistance:
Approximately 0Ω
N.G.
Repair harness or connectors.
O.K.
CHECK COMPONENTS
Check resistance between terminals ⓐ and ⓑ.
Continuity should exist.
N.G.
Replace F.I.C.D. solenoid valve.
O.K.
INSPECTION END

ECCS SELF-DIAGNOSTIC CHART—1988 PULSAR

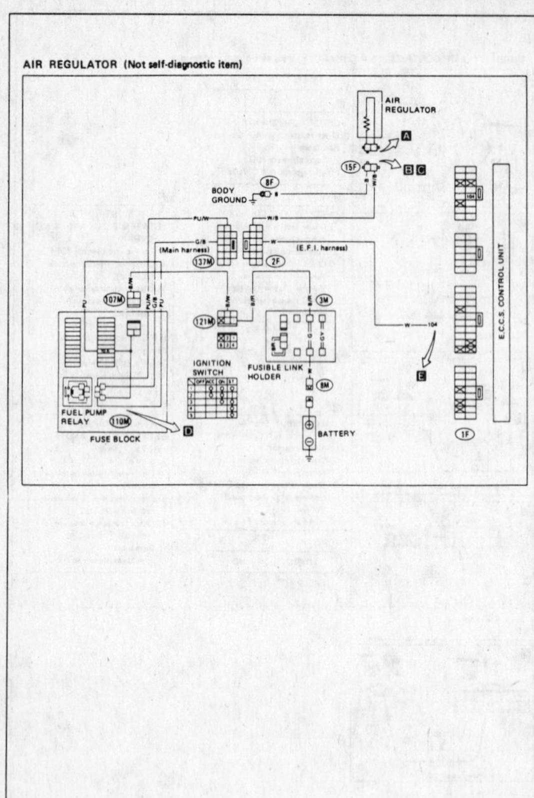
AIR REGULATOR (Not self-diagnostic item)
AIR REGULATOR
BODY GROUND
(Main harness)
(E.F.I. harness)
IGNITION SWITCH
FUSIBLE LINK HOLDER
FUEL PUMP RELAY
FUSE BLOCK
BATTERY
E.C.C.S. CONTROL UNIT

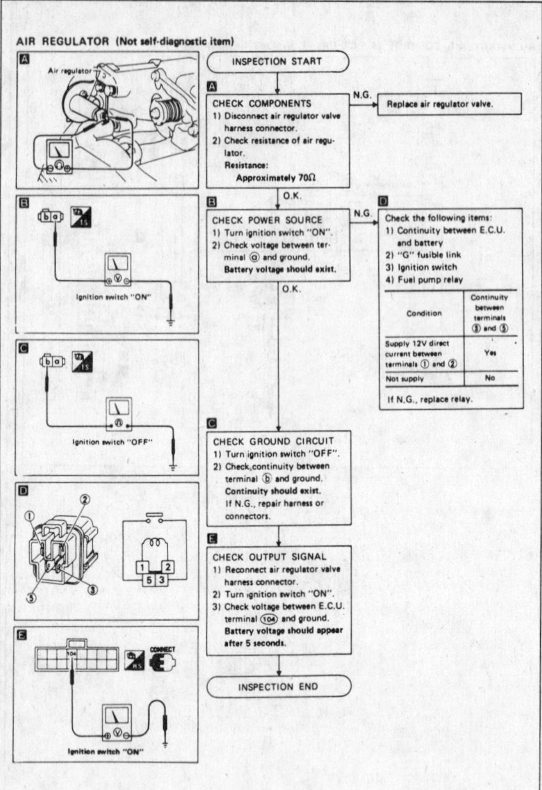
AIR REGULATOR (Not self-diagnostic item)
Air regulator
Ignition switch "ON"
Ignition switch "OFF"
CONNECT
Ignition switch "ON"
INSPECTION START
A
CHECK COMPONENTS
1) Disconnect air regulator valve harness connector.
2) Check resistance of air regulator.
Resistance:
Approximately 70Ω
N.G.
Replace air regulator valve.
O.K.
B
CHECK POWER SOURCE
1) Turn ignition switch "ON".
2) Check voltage between terminal ⓐ and ground.
Battery voltage should exist.
N.G.
D
Check the following items:
1) Continuity between E.C.U. and battery
2) "G" fusible link
3) Ignition switch
4) Fuel pump relay
Condition | Continuity between terminals ③ and ⑤
Supply 12V direct current between terminals ① and ② | Yes
Not supply | No
If N.G., replace relay.
O.K.
C
CHECK GROUND CIRCUIT
1) Turn ignition switch "OFF".
2) Check continuity between terminal ⓑ and ground.
Continuity should exist.
If N.G., repair harness or connectors.
E
CHECK OUTPUT SIGNAL
1) Reconnect air regulator valve harness connector.
2) Turn ignition switch "ON".
3) Check voltage between E.C.U. terminal (104) and ground.
Battery voltage should appear after 5 seconds.
INSPECTION END

ECCS SELF-DIAGNOSTIC CHART—1988 PULSAR

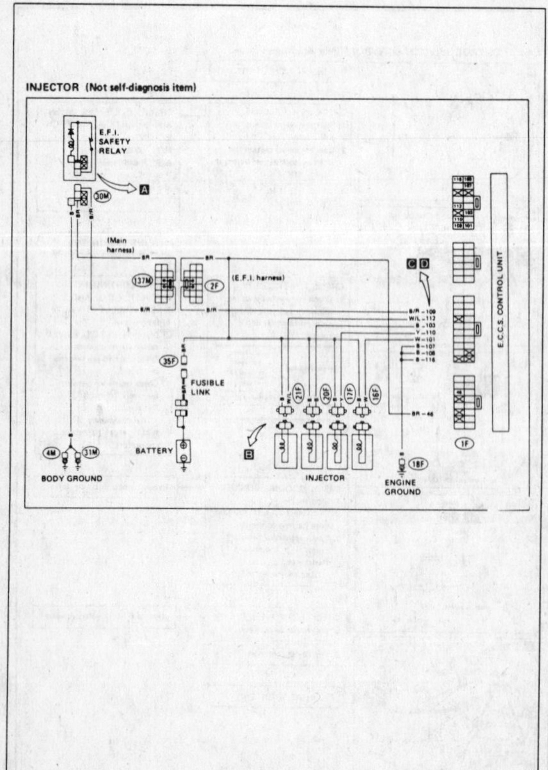
INJECTOR (Not self-diagnosis item)
E.F.I. SAFETY RELAY
(Main harness)
(E.F.I. harness)
FUSIBLE LINK
BATTERY
BODY GROUND
INJECTOR
ENGINE GROUND
E.C.C.S. CONTROL UNIT

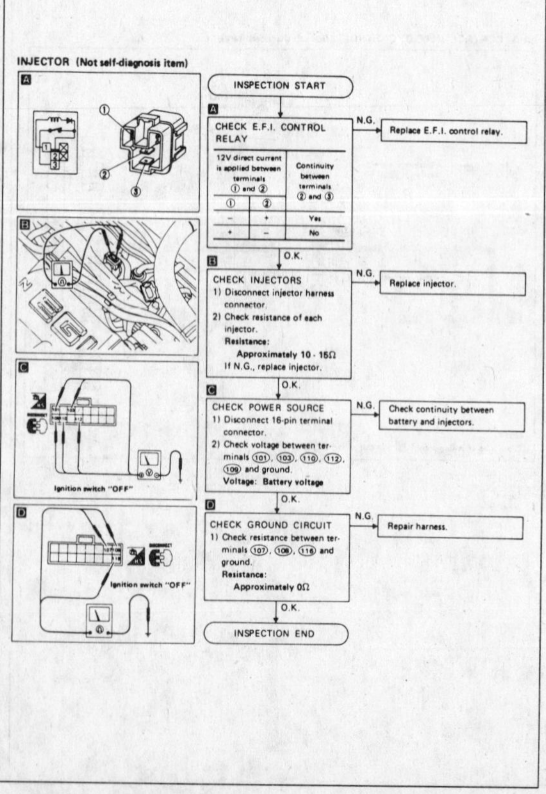
INJECTOR (Not self-diagnosis item)
Ignition switch "OFF"
Ignition switch "OFF"
INSPECTION START
A
CHECK E.F.I. CONTROL RELAY
12V direct current is applied between terminals ① and ② | Continuity between terminals ② and ③
① | ② |
– | + | Yes
+ | – | No
N.G.
Replace E.F.I. control relay.
O.K.
B
CHECK INJECTORS
1) Disconnect injector harness connector.
2) Check resistance of each injector.
Resistance:
Approximately 10 - 15Ω
If N.G., replace injector.
N.G.
Replace injector.
O.K.
C
CHECK POWER SOURCE
1) Disconnect 16-pin terminal connector.
2) Check voltage between terminals (101), (103), (110), (112), (109) and ground.
Voltage: Battery voltage
N.G.
Check continuity between battery and injectors.
O.K.
D
CHECK GROUND CIRCUIT
1) Check resistance between terminals (107), (108), (116) and ground.
Resistance:
Approximately 0Ω
N.G.
Repair harness.
O.K.
INSPECTION END

ECCS SELF-DIAGNOSTIC CHART—1988 PULSAR

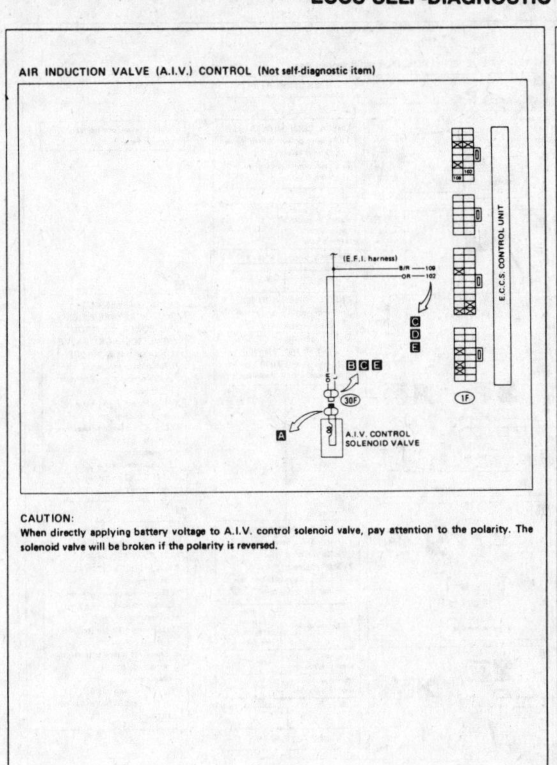
AIR INDUCTION VALVE (A.I.V.) CONTROL (Not self-diagnostic item)
(E.F.I. harness)
E.C.C.S. CONTROL UNIT
A.I.V. CONTROL SOLENOID VALVE
30F
1F
CAUTION:
When directly applying battery voltage to A.I.V. control solenoid valve, pay attention to the polarity. The solenoid valve will be broken if the polarity is reversed.

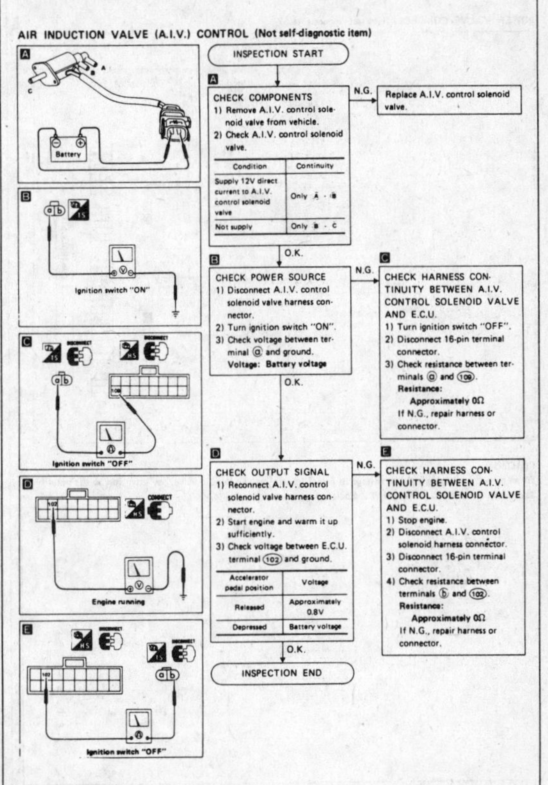
AIR INDUCTION VALVE (A.I.V.) CONTROL (Not self-diagnostic item)
Battery
Ignition switch "ON"
Ignition switch "OFF"
Engine running
Ignition switch "OFF"
INSPECTION START
A CHECK COMPONENTS
1) Remove A.I.V. control solenoid valve from vehicle.
2) Check A.I.V. control solenoid valve.
Condition | Continuity
Supply 12V direct current to A.I.V. control solenoid valve | Only A - B
Not supply | Only B - C
N.G. Replace A.I.V. control solenoid valve.
O.K.
B CHECK POWER SOURCE
1) Disconnect A.I.V. control solenoid valve harness connector.
2) Turn ignition switch "ON".
3) Check voltage between terminal ⓐ and ground.
Voltage: Battery voltage
N.G.
C CHECK HARNESS CONTINUITY BETWEEN A.I.V. CONTROL SOLENOID VALVE AND E.C.U.
1) Turn ignition switch "OFF".
2) Disconnect 16-pin terminal connector.
3) Check resistance between terminals ⓐ and (109).
Resistance: Approximately 0Ω
If N.G., repair harness or connector.
O.K.
D CHECK OUTPUT SIGNAL
1) Reconnect A.I.V. control solenoid valve harness connector.
2) Start engine and warm it up sufficiently.
3) Check voltage between E.C.U. terminal (102) and ground.
Accelerator pedal position | Voltage
Released | Approximately 0.8V
Depressed | Battery voltage
N.G.
E CHECK HARNESS CONTINUITY BETWEEN A.I.V. CONTROL SOLENOID VALVE AND E.C.U.
1) Stop engine.
2) Disconnect A.I.V. control solenoid harness connector.
3) Disconnect 16-pin terminal connector.
4) Check resistance between terminals ⓑ and (102).
Resistance: Approximately 0Ω
If N.G., repair harness or connector.
O.K.
INSPECTION END

ECCS SELF-DIAGNOSTIC CHART—1988 PULSAR

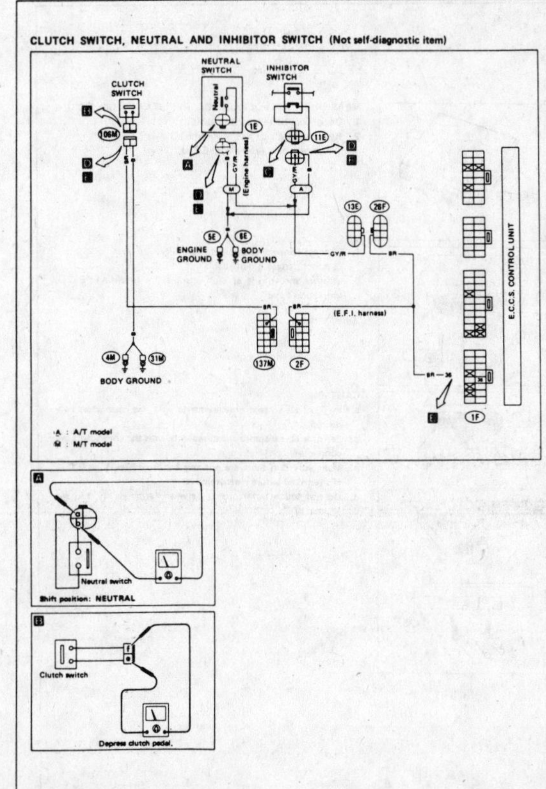
CLUTCH SWITCH, NEUTRAL AND INHIBITOR SWITCH (Not self-diagnostic item)
CLUTCH SWITCH
NEUTRAL SWITCH
INHIBITOR SWITCH
ENGINE GROUND
BODY GROUND
(E.F.I. harness)
E.C.C.S. CONTROL UNIT
BODY GROUND
A : A/T model
M : M/T model
Neutral switch
Shift position: NEUTRAL
Clutch switch
Depress clutch pedal.

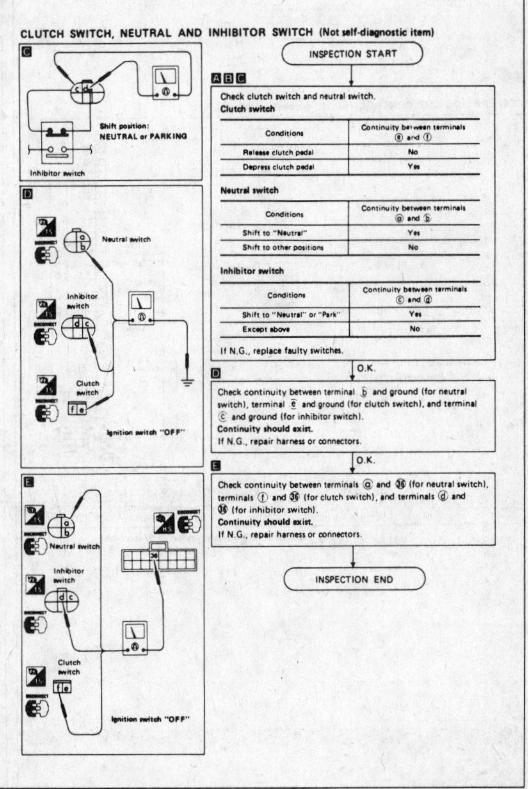
CLUTCH SWITCH, NEUTRAL AND INHIBITOR SWITCH (Not self-diagnostic item)
Shift position: NEUTRAL or PARKING
Inhibitor switch
Neutral switch
Clutch switch
Ignition switch "OFF"
INSPECTION START
A B C
Check clutch switch and neutral switch.
Clutch switch
Conditions | Continuity between terminals ⓔ and ⓕ
Release clutch pedal | No
Depress clutch pedal | Yes
Neutral switch
Conditions | Continuity between terminals ⓐ and ⓑ
Shift to "Neutral" | Yes
Shift to other positions | No
Inhibitor switch
Conditions | Continuity between terminals ⓒ and ⓓ
Shift to "Neutral" or "Park" | Yes
Except above | No
If N.G., replace faulty switches.
O.K.
D
Check continuity between terminal ⓑ and ground (for neutral switch), terminal ⓔ and ground (for clutch switch), and terminal ⓒ and ground (for inhibitor switch).
Continuity should exist.
If N.G., repair harness or connectors.
O.K.
E
Check continuity between terminals ⓐ and (36) (for neutral switch), terminals ⓕ and (36) (for clutch switch), and terminals ⓓ and (36) (for inhibitor switch).
Continuity should exist.
If N.G., repair harness or connectors.
INSPECTION END

ECCS SELF-DIAGNOSTIC CHART—1988 PULSAR

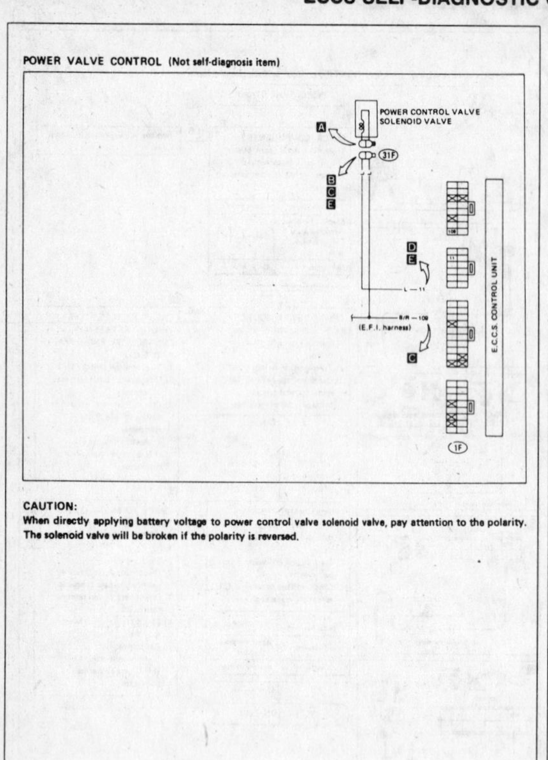
POWER VALVE CONTROL (Not self-diagnosis item)
POWER CONTROL VALVE SOLENOID VALVE
31F
L—11
B/R—109
(E.F.I. harness)
E.C.C.S. CONTROL UNIT
1F
CAUTION:
When directly applying battery voltage to power control valve solenoid valve, pay attention to the polarity. The solenoid valve will be broken if the polarity is reversed.

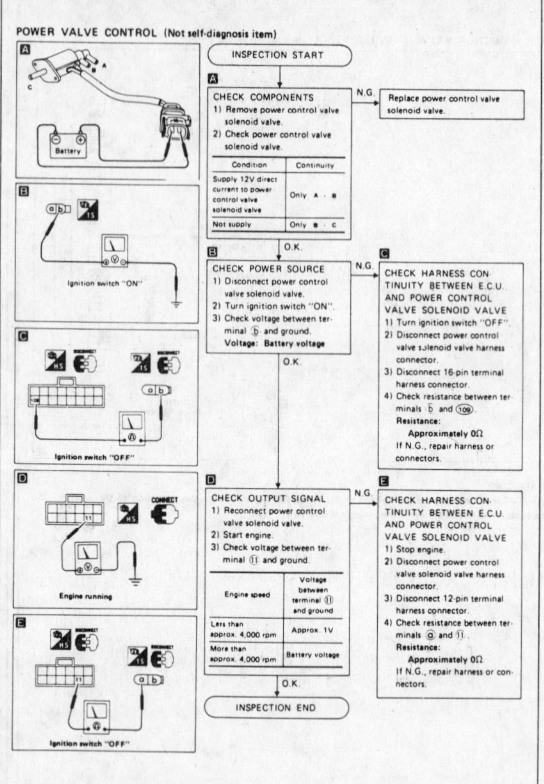
POWER VALVE CONTROL (Not self-diagnosis item)
INSPECTION START
A CHECK COMPONENTS
1) Remove power control valve solenoid valve.
2) Check power control valve solenoid valve.
Condition | Continuity
Supply 12V direct current to power control valve solenoid valve | Only A - B
Not supply | Only B - C
N.G. Replace power control valve solenoid valve.
O.K.
B CHECK POWER SOURCE
1) Disconnect power control valve solenoid valve.
2) Turn ignition switch "ON".
3) Check voltage between terminal ⓑ and ground.
Voltage: Battery voltage
N.G.
C CHECK HARNESS CONTINUITY BETWEEN E.C.U. AND POWER CONTROL VALVE SOLENOID VALVE
1) Turn ignition switch "OFF".
2) Disconnect power control valve solenoid valve harness connector.
3) Disconnect 16-pin terminal harness connector.
4) Check resistance between terminals ⓑ and (109).
Resistance:
Approximately 0Ω
If N.G., repair harness or connectors.
O.K.
D CHECK OUTPUT SIGNAL
1) Reconnect power control valve solenoid valve.
2) Start engine.
3) Check voltage between terminal ⑪ and ground.
Engine speed | Voltage between terminal ⑪ and ground
Less than approx. 4,000 rpm | Approx. 1V
More than approx. 4,000 rpm | Battery voltage
N.G.
E CHECK HARNESS CONTINUITY BETWEEN E.C.U. AND POWER CONTROL VALVE SOLENOID VALVE
1) Stop engine.
2) Disconnect power control valve solenoid valve harness connector.
3) Disconnect 12-pin terminal harness connector.
4) Check resistance between terminals ⓐ and ⑪.
Resistance:
Approximately 0Ω
If N.G., repair harness or connectors.
O.K.
INSPECTION END
Battery
Ignition switch "ON"
Ignition switch "OFF"
Engine running
Ignition switch "OFF"

ECCS SELF-DIAGNOSTIC CHART—1988 PULSAR

ACCELERATION CUT CONTROL (Not self-diagnostic item)

(Main harness) (E.F.I. harness)
R/Y—10
B—6
B—12
137M 2F
B—107
B—108
B—116
To thermo control amplifier
To air conditioner switch
R—34
GY/L—44
143M
ACCELERATION CUT RELAY
E.C.C.S. CONTROL UNIT
1F
19F IDLE SWITCH
18F
ENGINE GROUND

For inspection of this system, refer to HA section.

ECU SIGNAL INSPECTION—1988 PULSAR

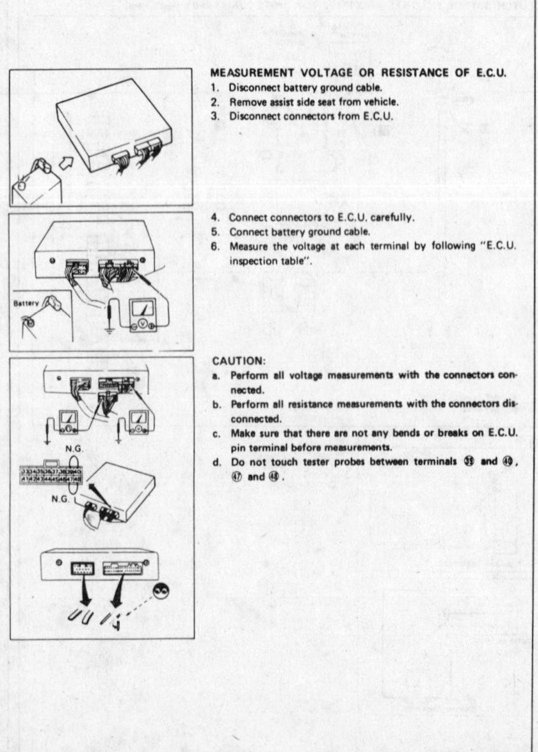
MEASUREMENT VOLTAGE OR RESISTANCE OF E.C.U.
1. Disconnect battery ground cable.
2. Remove assist side seat from vehicle.
3. Disconnect connectors from E.C.U.
4. Connect connectors to E.C.U. carefully.
5. Connect battery ground cable.
6. Measure the voltage at each terminal by following "E.C.U. inspection table".
Battery
CAUTION:
a. Perform all voltage measurements with the connectors connected.
b. Perform all resistance measurements with the connectors disconnected.
c. Make sure that there are not any bends or breaks on E.C.U. pin terminal before measurements.
d. Do not touch tester probes between terminals ㊴ and ㊵, ㊼ and ㊽.
N.G.
N.G.

ECU SIGNAL INSPECTION (CONT.)—1988 PULSAR

E.C.U. inspection table

*Data are reference values.

TERMINAL NO.	ITEM	CONDITION	*DATA
1	Ignition signal for No. 1 cylinder	Engine is running.	0 - 1.0V
2	Ignition signal for No. 2 cylinder		
8	Ignition signal for No. 3 cylinder		
9	Ignition signal for No. 4 cylinder		
7	Exhaust gas temperature sensor	Engine is running. — Idle speed	1.0 - 2.0V
		Engine is running. — E.G.R. system is operating.	0 - 1.0V
10	Air conditioner cut signal	Engine is running. — Idle speed	0 - 1.0V
		Engine is running. — Sudden racing	BATTERY VOLTAGE (11 - 14V)
11	Power valve control solenoid valve	Engine is running. — Engine speed is less than approximately 4,000 rpm.	Approximately 1V
		Engine is running. — Engine speed is more than approximately 4,000 rpm.	BATTERY VOLTAGE (11 - 14V)
15	Air flow meter	Engine is running. Do not run engine at high speed under no-load.	1.0 - 3.0V Output voltage varies with engine revolution.
17	Water temperature sensor	Engine is running.	1.0 - 5.0V Output voltage varies with engine water temperature.

*Data are reference values.

TERMINAL NO.	ITEM	CONDITION	*DATA
18	Exhaust gas sensor	Engine is running. — After warming up sufficiently	0 - Approximately 1.0V
19	Throttle sensor	Ignition switch "ON"	0.4 - 4.0V Output voltage varies with the throttle valve opening angle.
21 31	Crank angle sensor (Reference signal)	Engine is running. Do not run engine at high speed under no-load.	0.6 - 0.7V
22 32	Crank angle sensor (Position signal)	Engine is running. Do not run engine at high speed under no-load.	2.0 - 2.5V
34	Idle switch (⊖ side)	Ignition switch "ON" — Throttle valve: idle position	Approximately 9 - 10V
		Ignition switch "ON" — Throttle valve: Any position except idle position	0V
35	Start signal	Cranking	8 - 12V
36	Neutral switch & Inhibitor switch	Ignition switch "ON" — Neutral/Parking	0V
		Ignition switch "ON" — Except the above gear position	BATTERY VOLTAGE (11 - 14V)
37	Ignition switch	Ignition switch "OFF"	0V
		Ignition switch "ON"	BATTERY VOLTAGE (11 - 14V)
41	Air conditioner	Engine is running. — Both air conditioner switch and blower switch are "ON".	BATTERY VOLTAGE (11 - 14V)

ECU SIGNAL INSPECTION (CONT.)—1988 PULSAR

*Data are reference values.

TERMINAL NO.	ITEM	CONDITION	*DATA
44	Idle switch (⊕ side)	Ignition switch "ON" — Throttle valve: idle position	Approximately 9 - 11V
		Ignition switch "ON" — Throttle valve: Except idle position	BATTERY VOLTAGE (11 - 14V)
46	Power source (Back-up)	Ignition switch "OFF"	BATTERY VOLTAGE (11 - 14V)
39 47	Power source for E.C.U.	Ignition switch "ON"	BATTERY VOLTAGE (11 - 14V)
101	Injector No. 1	Engine is running.	BATTERY VOLTAGE (11 - 14V)
103	Injector No. 3		
110	Injector No. 2		
112	Injector No. 4		
102	A.I.V. control solenoid	Engine is running. — Idle speed	0 - 1.0V
		Engine is running. — Accelerator pedal is depressed. — After warming up	BATTERY VOLTAGE (11 - 14V)
104	Fuel pump relay	Ignition switch "ON" — For 5 seconds after turning ignition switch "ON"; Engine is running.	0.7 - 0.9V
		Ignition switch "ON" — In 5 seconds after turning ignition switch "ON"	BATTERY VOLTAGE (11 - 14V)

*Data are reference values

TERMINAL NO.	ITEM	CONDITION	*DATA
105	E.G.R. control solenoid valve	Engine is running. — Engine is cold. [Water temperature is below 60°C (140°F).]	0.7 - 0.9V
		Engine is running. — After warming up [Water temperature is between 65°C (149°F) and 105°C (221°F).]	BATTERY VOLTAGE (11 - 14V)
113	A.A.C. valve	Engine is running. — Idle speed	8 - 10V
		Engine is running. — Steering wheel is turned. — Air conditioner is operating. — Rear defogger is "ON". — Head lights are in high position.	7 - 8V

E.C.U. PIN CONNECTOR TERMINAL LAYOUT

ECCS MIXTURE RATIO FEEDBACK SYSTEM INSPECTION – 1988 PULSAR

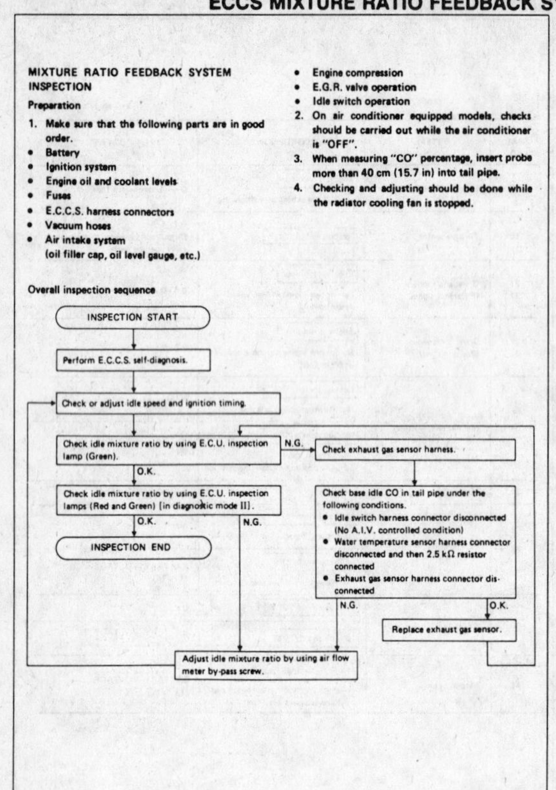

MIXTURE RATIO FEEDBACK SYSTEM INSPECTION

Preparation

1. Make sure that the following parts are in good order.
 - Battery
 - Ignition system
 - Engine oil and coolant levels
 - Fuses
 - E.C.C.S. harness connectors
 - Vacuum hoses
 - Air intake system (oil filler cap, oil level gauge, etc.)
 - Engine compression
 - E.G.R. valve operation
 - Idle switch operation
2. On air conditioner equipped models, checks should be carried out while the air conditioner is "OFF".
3. When measuring "CO" percentage, insert probe more than 40 cm (15.7 in) into tail pipe.
4. Checking and adjusting should be done while the radiator cooling fan is stopped.

Overall inspection sequence

INSPECTION START
Perform E.C.C.S. self-diagnosis.
Check or adjust idle speed and ignition timing.
Check idle mixture ratio by using E.C.U. inspection lamp (Green).
N.G.
Check exhaust gas sensor harness.
O.K.
Check idle mixture ratio by using E.C.U. inspection lamps (Red and Green) [in diagnostic mode II].
O.K.
N.G.
INSPECTION END
Check base idle CO in tail pipe under the following conditions.
• Idle switch harness connector disconnected (No A.I.V. controlled condition)
• Water temperature sensor harness connector disconnected and then 2.5 kΩ resistor connected
• Exhaust gas sensor harness connector dis-connected
N.G.
O.K.
Replace exhaust gas sensor.
Adjust idle mixture ratio by using air flow meter by-pass screw.

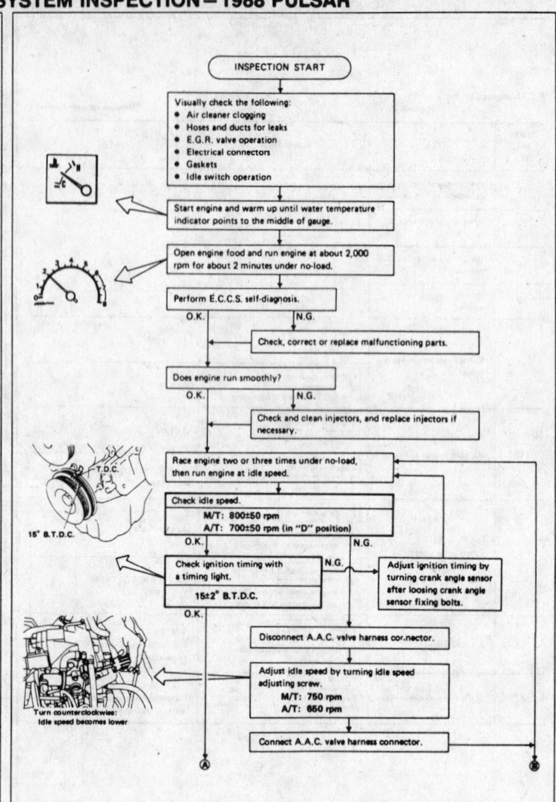
INSPECTION START
Visually check the following:
• Air cleaner clogging
• Hoses and ducts for leaks
• E.G.R. valve operation
• Electrical connectors
• Gaskets
• Idle switch operation
Start engine and warm up until water temperature indicator points to the middle of gauge.
Open engine food and run engine at about 2,000 rpm for about 2 minutes under no-load.
Perform E.C.C.S. self-diagnosis.
O.K.
N.G.
Check, correct or replace malfunctioning parts.
Does engine run smoothly?
O.K.
N.G.
Check and clean injectors, and replace injectors if necessary.
T.D.C.
15° B.T.D.C.
Race engine two or three times under no-load, then run engine at idle speed.
Check idle speed.
M/T: 800±50 rpm
A/T: 700±50 rpm (in "D" position)
O.K.
N.G.
Check ignition timing with a timing light.
15±2° B.T.D.C.
N.G.
Adjust ignition timing by turning crank angle sensor after loosing crank angle sensor fixing bolts.
O.K.
Disconnect A.A.C. valve harness cor.nector.
Adjust idle speed by turning idle speed adjusting screw.
M/T: 750 rpm
A/T: 650 rpm
Turn counterclockwise: Idle speed becomes lower
Connect A.A.C. valve harness connector.
A
B

ECCS MIXTURE RATIO FEEDBACK SYSTEM INSPECTION (CONT.) – 1988 PULSAR

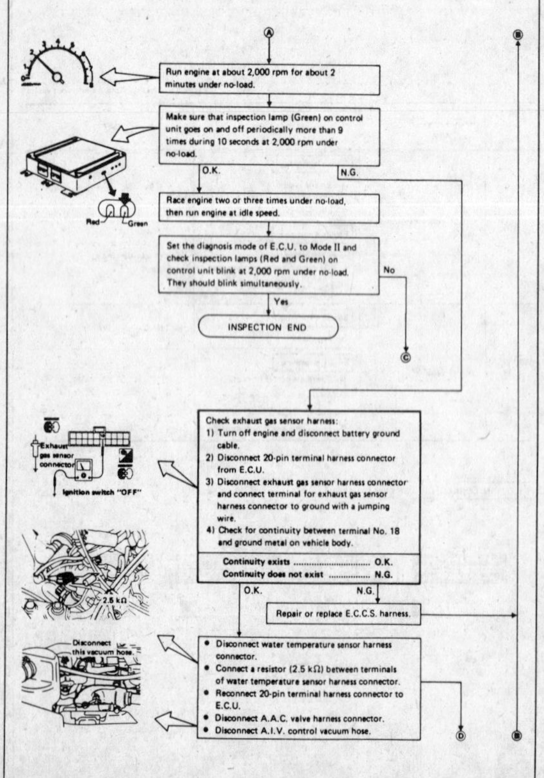
A
B
Run engine at about 2,000 rpm for about 2 minutes under no-load.
Make sure that inspection lamp (Green) on control unit goes on and off periodically more than 9 times during 10 seconds at 2,000 rpm under no-load.
Red
Green
O.K.
N.G.
Race engine two or three times under no-load, then run engine at idle speed.
Set the diagnosis mode of E.C.U. to Mode II and check inspection lamps (Red and Green) on control unit blink at 2,000 rpm under no-load. They should blink simultaneously.
No
Yes
INSPECTION END
C
Exhaust gas sensor connector
Ignition switch "OFF"
Check exhaust gas sensor harness:
1) Turn off engine and disconnect battery ground cable.
2) Disconnect 20-pin terminal harness connector from E.C.U.
3) Disconnect exhaust gas sensor harness connector and connect terminal for exhaust gas sensor harness connector to ground with a jumping wire.
4) Check for continuity between terminal No. 18 and ground metal on vehicle body.
Continuity exists .... O.K.
Continuity does not exist .... N.G.
O.K.
N.G.
2.5 kΩ
Repair or replace E.C.C.S. harness.
Disconnect this vacuum hose.
• Disconnect water temperature sensor harness connector.
• Connect a resistor (2.5 kΩ) between terminals of water temperature sensor harness connector.
• Reconnect 20-pin terminal harness connector to E.C.U.
• Disconnect A.A.C. valve harness connector.
• Disconnect A.I.V. control vacuum hose.
D
B

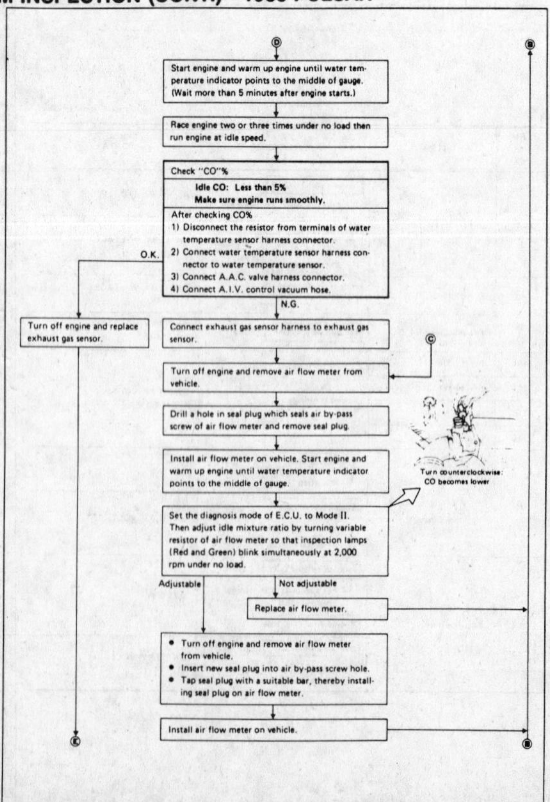
D
B
Start engine and warm up engine until water tem-perature indicator points to the middle of gauge. (Wait more than 5 minutes after engine starts.)
Race engine two or three times under no load then run engine at idle speed.
Check "CO"%
Idle CO: Less than 5%
Make sure engine runs smoothly.
After checking CO%
1) Disconnect the resistor from terminals of water temperature sensor harness connector.
2) Connect water temperature sensor harness con-nector to water temperature sensor.
3) Connect A.A.C. valve harness connector.
4) Connect A.I.V. control vacuum hose.
O.K.
N.G.
Turn off engine and replace exhaust gas sensor.
Connect exhaust gas sensor harness to exhaust gas sensor.
C
Turn off engine and remove air flow meter from vehicle.
Drill a hole in seal plug which seals air by-pass screw of air flow meter and remove seal plug.
Install air flow meter on vehicle. Start engine and warm up engine until water temperature indicator points to the middle of gauge.
Turn counterclockwise: CO becomes lower
Set the diagnosis mode of E.C.U. to Mode II. Then adjust idle mixture ratio by turning variable resistor of air flow meter so that inspection lamps (Red and Green) blink simultaneously at 2,000 rpm under no load.
Adjustable
Not adjustable
Replace air flow meter.
• Turn off engine and remove air flow meter from vehicle.
• Insert new seal plug into air by-pass screw hole.
• Tap seal plug with a suitable bar, thereby install-ing seal plug on air flow meter.
Install air flow meter on vehicle.
E
B

ECCS MIXTURE RATIO FEEDBACK SYSTEM INSPECTION (CONT.)—1988 PULSAR

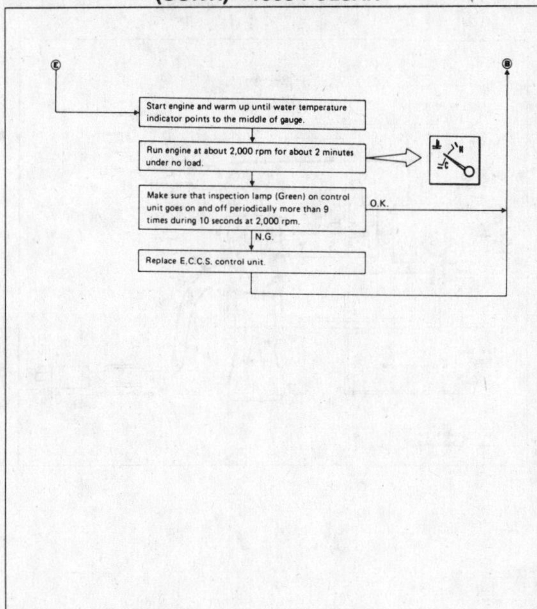
E
B
Start engine and warm up until water temperature indicator points to the middle of gauge.
Run engine at about 2,000 rpm for about 2 minutes under no load.
Make sure that inspection lamp (Green) on control unit goes on and off periodically more than 9 times during 10 seconds at 2,000 rpm.
O.K.
N.G.
Replace E.C.C.S. control unit.

ECCS COMPONENTS LOCATION—1989 PULSAR

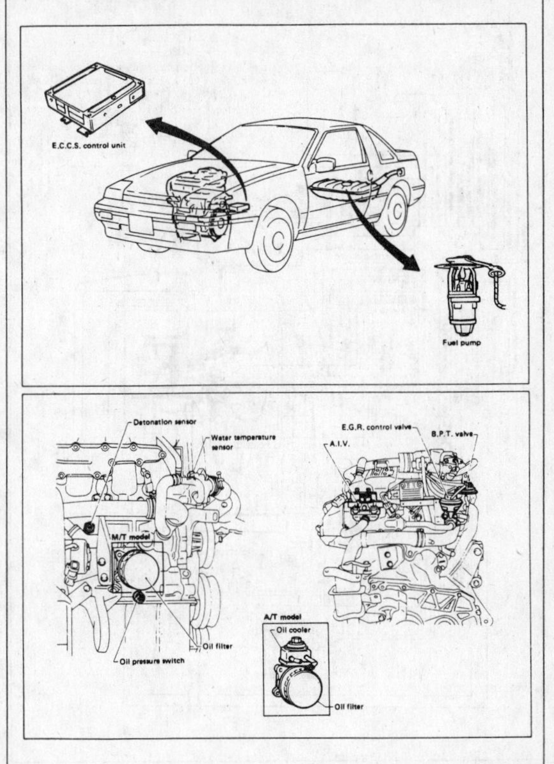
E.C.C.S. control unit
Fuel pump
Detonation sensor
Water temperature sensor
E.G.R. control valve
B.P.T. valve
A.I.V.
M/T model
Oil filter
Oil pressure switch
A/T model
Oil cooler
Oil filter

ECCS COMPONENTS LOCATION (CONT.)—1989 PULSAR

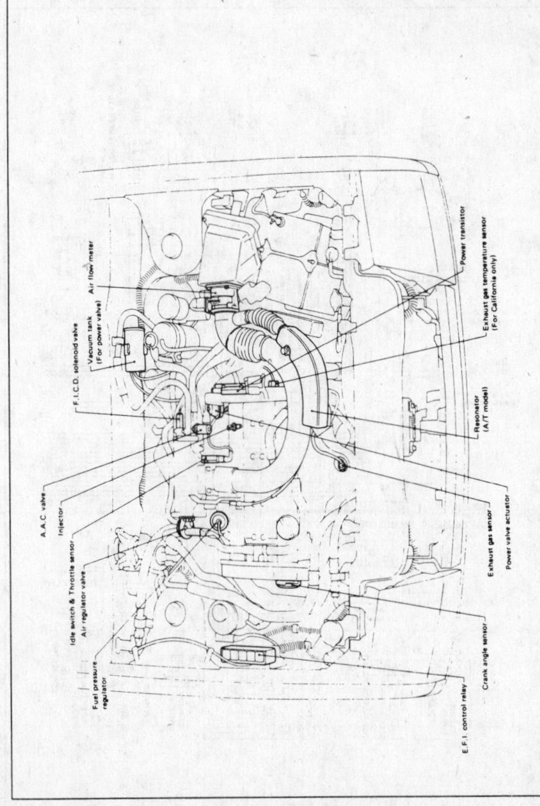
A.A.C. valve
Injector
Idle switch & Throttle sensor
Air regulator valve
Fuel pressure regulator
F.I.C.D. solenoid valve
Vacuum tank (For power valve)
Air flow meter
Power transistor
Exhaust gas temperature sensor (For California only)
Resonator (A/T model)
Exhaust gas sensor
Power valve actuator
E.F.I. control relay
Crank angle sensor

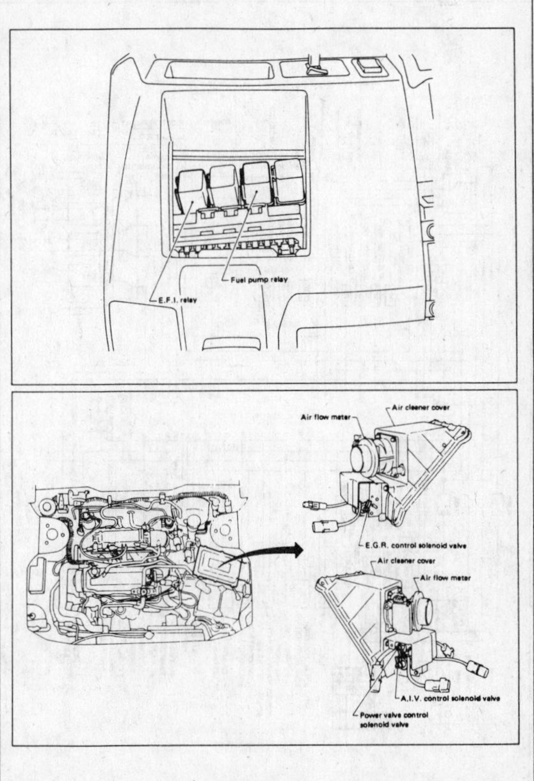
E.F.I. relay
Fuel pump relay
Air flow meter
Air cleaner cover
E.G.R. control solenoid valve
Air cleaner cover
Air flow meter
A.I.V. control solenoid valve
Power valve control solenoid valve

ECCS SYSTEM SCHEMATIC—1989 PULSAR

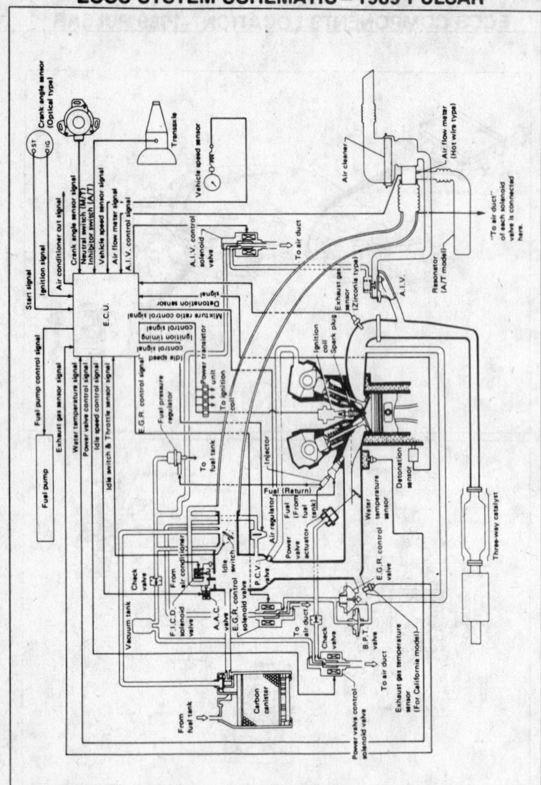

ECCS AIR FLOW SYSTEM SCHEMATIC—1989 PULSAR

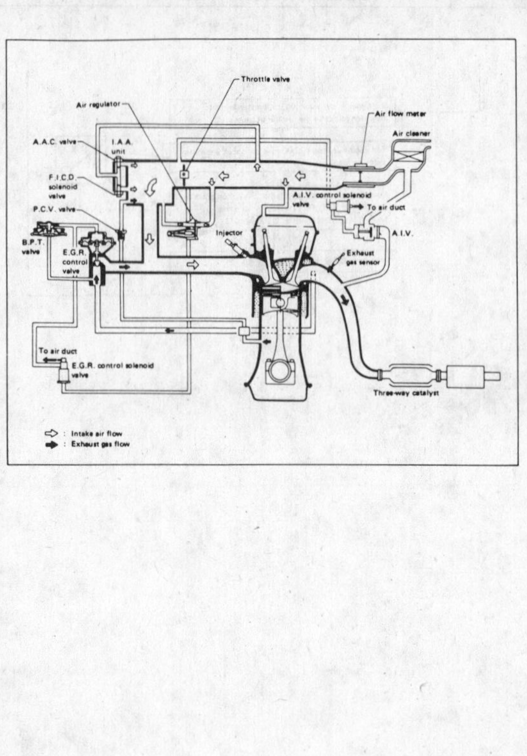
Throttle valve
Air regulator
Air flow meter
Air cleaner
A.A.C. valve
I.A.A. unit
F.I.C.D. solenoid valve
A.I.V. control solenoid valve
To air duct
P.C.V. valve
Injector
A.I.V.
B.P.T. valve
E.G.R. control valve
Exhaust gas sensor
To air duct
E.G.R. control solenoid valve
Three-way catalyst
: Intake air flow
: Exhaust gas flow

ECCS CONTROL UNIT WIRING SCHEMATIC—1989 PULSAR

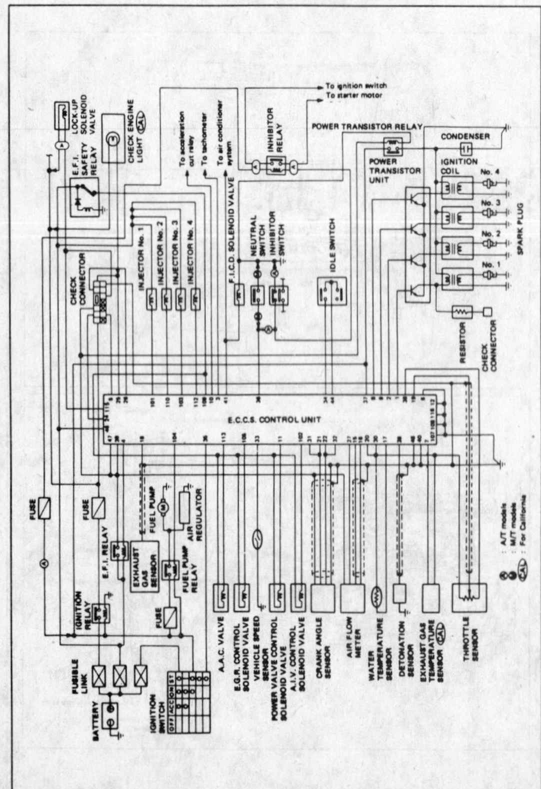

ECCS WIRING SCHEAMTIC—1989 PULSAR

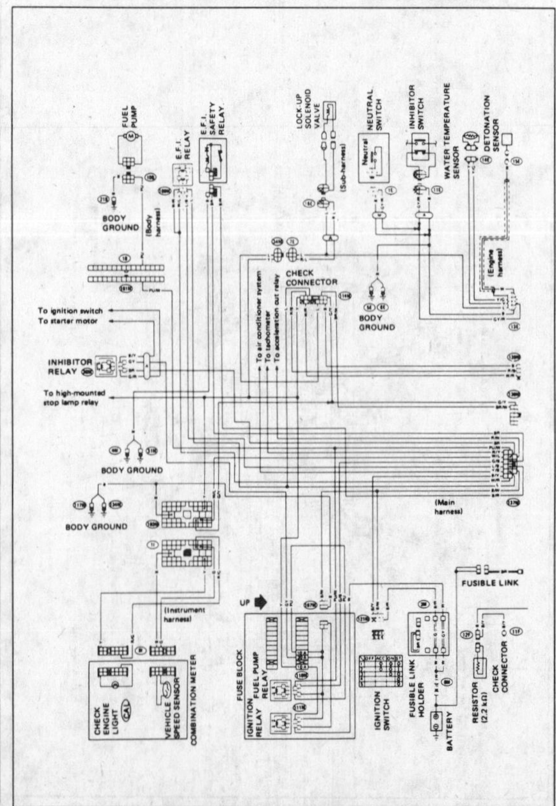

ECCS WIRING SCHEAMTIC (CONT.)—1989 PULSAR

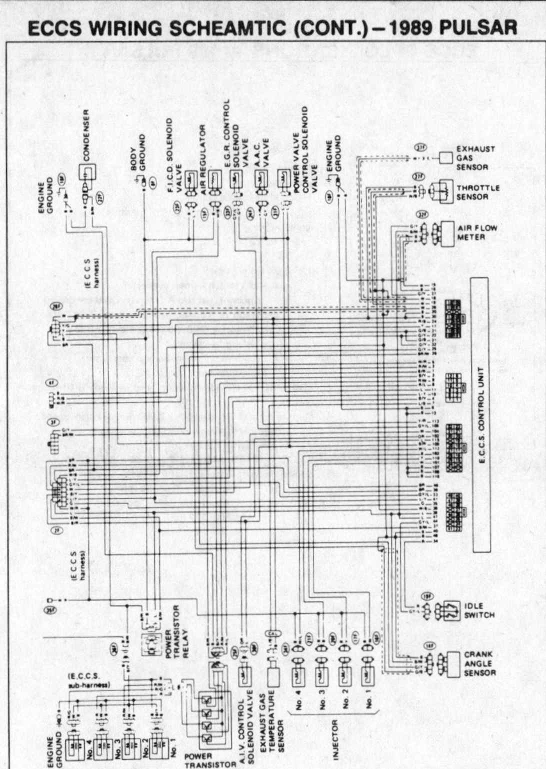
ENGINE GROUND
CONDENSER
BODY GROUND
F.I.C.D. SOLENOID VALVE
AIR REGULATOR
E.G.R. CONTROL SOLENOID VALVE
A.A.C. VALVE
POWER VALVE CONTROL SOLENOID VALVE
ENGINE GROUND
EXHAUST GAS SENSOR
THROTTLE SENSOR
AIR FLOW METER
(E.C.C.S. harness)
E.C.C.S. CONTROL UNIT
IDLE SWITCH
CRANK ANGLE SENSOR
POWER TRANSISTOR RELAY
(E.C.C.S. sub-harness)
ENGINE GROUND
IGNITION COIL
POWER TRANSISTOR UNIT
A.I.V. CONTROL SOLENOID VALVE
EXHAUST GAS TEMPERATURE SENSOR
INJECTOR
No. 4
No. 3
No. 2
No. 1
A/T model
M/T model
CAL : For California

ECCS DESCRIPTION—1989 PULSAR

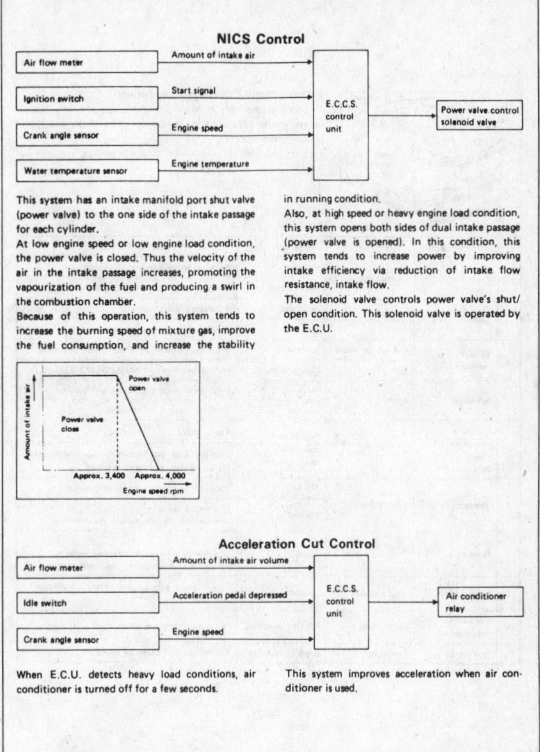

NICS Control

Air flow meter
Amount of intake air
Ignition switch
Start signal
Crank angle sensor
Engine speed
Water temperature sensor
Engine temperature
E.C.C.S. control unit
Power valve control solenoid valve

This system has an intake manifold port shut valve (power valve) to the one side of the intake passage for each cylinder.

At low engine speed or low engine load condition, the power valve is closed. Thus the velocity of the air in the intake passage increases, promoting the vapourization of the fuel and producing a swirl in the combustion chamber.

Because of this operation, this system tends to increase the burning speed of mixture gas, improve the fuel consumption, and increase the stability in running condition.

Also, at high speed or heavy engine load condition, this system opens both sides of dual intake passage (power valve is opened). In this condition, this system tends to increase power by improving intake efficiency via reduction of intake flow resistance, intake flow.

The solenoid valve controls power valve's shut/open condition. This solenoid valve is operated by the E.C.U.

Amount of intake air
Power valve open
Power valve close
Approx. 3,400
Approx. 4,000
Engine speed rpm

Acceleration Cut Control

Air flow meter
Amount of intake air volume
Idle switch
Acceleration pedal depressed
Crank angle sensor
Engine speed
E.C.C.S. control unit
Air conditioner relay

When E.C.U. detects heavy load conditions, air conditioner is turned off for a few seconds.

This system improves acceleration when air conditioner is used.

ECCS DIAGNOSTIC PROCEDURE (CONT.)—1989 PULSAR

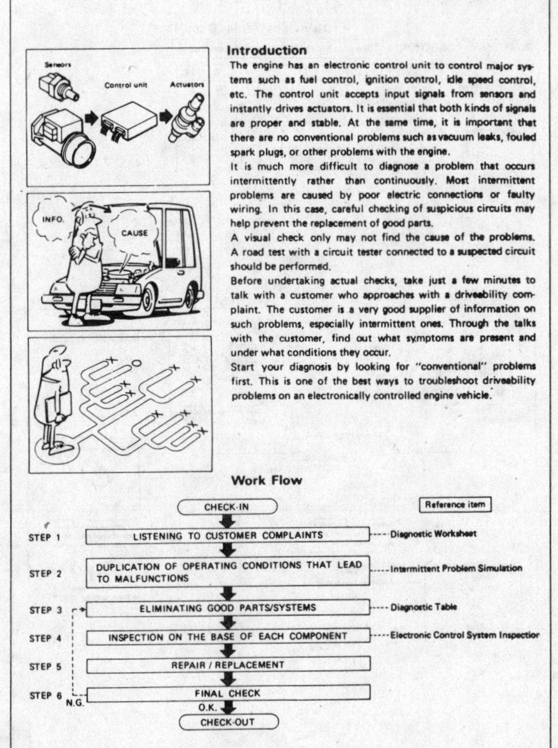
Sensors
Control unit
Actuators
INFO.
CAUSE

Introduction

The engine has an electronic control unit to control major systems such as fuel control, ignition control, idle speed control, etc. The control unit accepts input signals from sensors and instantly drives actuators. It is essential that both kinds of signals are proper and stable. At the same time, it is important that there are no conventional problems such as vacuum leaks, fouled spark plugs, or other problems with the engine.

It is much more difficult to diagnose a problem that occurs intermittently rather than continuously. Most intermittent problems are caused by poor electric connections or faulty wiring. In this case, careful checking of suspicious circuits may help prevent the replacement of good parts.

A visual check only may not find the cause of the problems. A road test with a circuit tester connected to a suspected circuit should be performed.

Before undertaking actual checks, take just a few minutes to talk with a customer who approaches with a driveability complaint. The customer is a very good supplier of information on such problems, especially intermittent ones. Through the talks with the customer, find out what symptoms are present and under what conditions they occur.

Start your diagnosis by looking for "conventional" problems first. This is one of the best ways to troubleshoot driveability problems on an electronically controlled engine vehicle.

Work Flow

CHECK-IN
Reference item
STEP 1 LISTENING TO CUSTOMER COMPLAINTS ---- Diagnostic Worksheet
STEP 2 DUPLICATION OF OPERATING CONDITIONS THAT LEAD TO MALFUNCTIONS ---- Intermittent Problem Simulation
STEP 3 ELIMINATING GOOD PARTS/SYSTEMS ---- Diagnostic Table
STEP 4 INSPECTION ON THE BASE OF EACH COMPONENT ---- Electronic Control System Inspection
STEP 5 REPAIR / REPLACEMENT
STEP 6 FINAL CHECK
N.G.
O.K.
CHECK-OUT

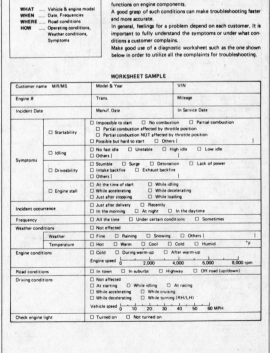

KEY POINTS

WHAT ..... Vehicle & engine model
WHEN ..... Date, Frequencies
WHERE ..... Road conditions
HOW ..... Operating conditions, Weather conditions, Symptoms

Diagnostic Worksheet

There are many kinds of operating conditions that lead to malfunctions on engine components.

A good grasp of such conditions can make troubleshooting faster and more accurate.

In general, feelings for a problem depend on each customer. It is important to fully understand the symptoms or under what conditions a customer complains.

Make good use of a diagnostic worksheet such as the one shown below in order to utilize all the complaints for troubleshooting.

WORKSHEET SAMPLE

Customer name MR/MS		Model & Year	VIN
Engine #		Trans.	Mileage
Incident Date		Manuf. Date	In Service Date
Symptoms	☐ Startability	☐ Impossible to start ☐ No combustion ☐ Partial combustion ☐ Partial combustion affected by throttle position ☐ Partial combustion NOT affected by throttle position ☐ Possible but hard to start ☐ Others [ ]	
	☐ Idling	☐ No fast idle ☐ Unstable ☐ High idle ☐ Low idle ☐ Others [ ]	
	☐ Driveability	☐ Stumble ☐ Surge ☐ Detonation ☐ Lack of power ☐ Intake backfire ☐ Exhaust backfire ☐ Others [ ]	
	☐ Engine stall	☐ At the time of start ☐ While idling ☐ While accelerating ☐ While decelerating ☐ Just after stopping ☐ While loading	
Incident occurrence		☐ Just after delivery ☐ Recently ☐ In the morning ☐ At night ☐ In the daytime	
Frequency		☐ All the time ☐ Under certain conditions ☐ Sometimes	
Weather conditions		☐ Not effected	
	Weather	☐ Fine ☐ Raining ☐ Snowing ☐ Others [ ]	
	Temperature	☐ Hot ☐ Warm ☐ Cool ☐ Cold ☐ Humid °F	
Engine conditions		☐ Cold ☐ During warm-up ☐ After warm-up Engine speed 0 2,000 4,000 6,000 8,000 rpm	
Road conditions		☐ In town ☐ In suburbs ☐ Highway ☐ Off road (up/down)	
Driving conditions		☐ Not affected ☐ At starting ☐ While idling ☐ At racing ☐ While accelerating ☐ While cruising ☐ While decelerating ☐ While turning (RH/LH) Vehicle speed 0 10 20 30 40 50 60 MPH	
Check engine light		☐ Turned on ☐ Not turned on	

ECCS DIAGNOSTIC PROCEDURE (CONT.)—1989 PULSAR

Intermittent Problem Simulation

In order to duplicate an intermittent problem, it is effective to create similar conditions for component parts, under which the problem might occur.

Perform the activity listed under Service procedure and note the result.

	Variable factor	Influential part	Target condition	Service procedure
1	Mixture ratio	Pressure regulator	Made lean	Remove vacuum hose and apply vacuum.
			Made rich	Remove vacuum hose and apply pressure.
2	Ignition timing	Distributor	Advanced	Rotate distributor clockwise.
			Retarded	Rotate distributor counterclockwise.
3	Mixture ratio feedback control	Exhaust gas sensor	Suspended	Disconnect exhaust gas sensor harness connector.
		Control unit	Operation check	Perform self-diagnosis (Mode I/II) at 2,000 rpm.
4	Idle speed	I.A.A. unit	Raised	Turn idle adjust screw counterclockwise.
			Lowered	Turn idle adjust screw clockwise.
5	Electric connection (Electric continuity)	Harness connectors and wires	Poor electric connection or faulty wiring	Tap or wiggle.
				Race engine rapidly. See if the torque reaction of the engine unit causes electric breaks.
6	Temperature	Control unit	Cooled	Cool with an icing spray or similar device.
			Warmed	Heat with a hair drier. [WARNING: Do not overheat the unit.]
7	Moisture	Electric parts	Damp	Wet [WARNING: Do not directly pour water on components. Use a mist sprayer.]
8	Electric loads	Load switches	Loaded	Turn on head lights, air conditioner, rear defogger, etc.
9	Idle switch condition	Control unit	ON-OFF switching	Perform self-diagnosis (Mode IV).
10	Ignition spark	Timing light	Spark power check	Try to flash timing light for each cylinder.

ECCS SPECIFICATIONS—1989 PULSAR

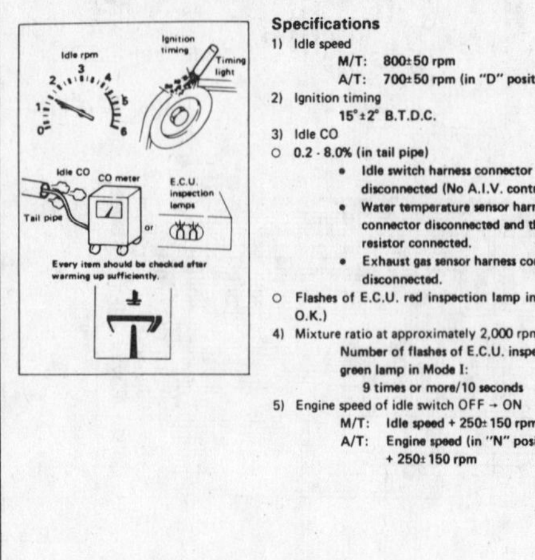
Idle rpm
Ignition timing
Timing light
Idle CO
CO meter
E.C.U. inspection lamps
Tail pipe
or
Every item should be checked after warming up sufficiently.

Specifications

1) Idle speed
   M/T: 800±50 rpm
   A/T: 700±50 rpm (in "D" position)
2) Ignition timing
   15°±2° B.T.D.C.
3) Idle CO
   ○ 0.2 - 8.0% (in tail pipe)
   - Idle switch harness connector disconnected (No A.I.V. controlled condition)
   - Water temperature sensor harness connector disconnected and then 2.5 kΩ resistor connected.
   - Exhaust gas sensor harness connector disconnected.

   ○ Flashes of E.C.U. red inspection lamp in Mode II (If flashes, O.K.)
4) Mixture ratio at approximately 2,000 rpm of engine speed.
   Number of flashes of E.C.U. inspection green lamp in Mode I:
   9 times or more/10 seconds
5) Engine speed of idle switch OFF → ON
   M/T: Idle speed + 250±150 rpm
   A/T: Engine speed (in "N" position) + 250±150 rpm

ECCS DIAGNOSTIC CHART—1989 PULSAR

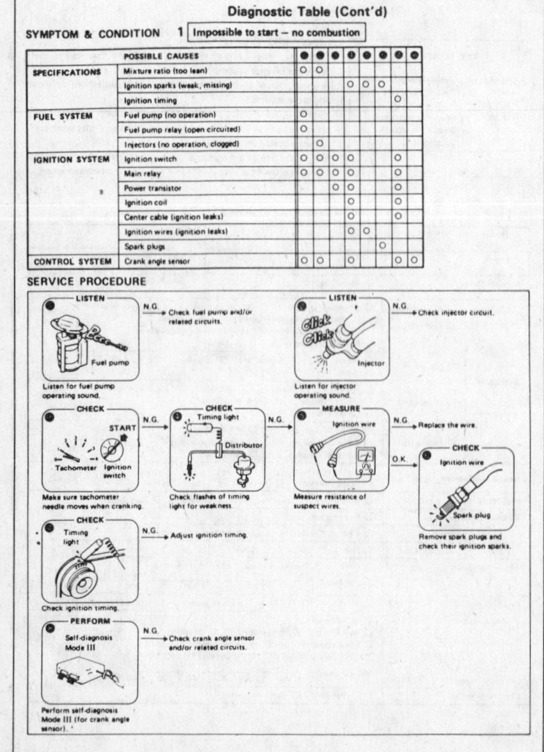

Diagnostic Table (Cont'd)

SYMPTOM & CONDITION 1 Impossible to start — no combustion

	POSSIBLE CAUSES	●	●	●	●	●	●	●	●
SPECIFICATIONS	Mixture ratio (too lean)	○	○						
	Ignition sparks (weak, missing)				○	○	○		
	Ignition timing							○	
FUEL SYSTEM	Fuel pump (no operation)	○							
	Fuel pump relay (open circuited)	○							
	Injectors (no operation, clogged)		○						
IGNITION SYSTEM	Ignition switch	○	○	○	○			○	
	Main relay	○	○	○	○			○	
	Power transistor			○	○			○	
	Ignition coil				○			○	
	Center cable (ignition leaks)				○			○	
	Ignition wires (ignition leaks)				○	○			
	Spark plugs						○		
CONTROL SYSTEM	Crank angle sensor	○	○		○			○	○

SERVICE PROCEDURE

LISTEN
Fuel pump
N.G. → Check fuel pump and/or related circuits.
Listen for fuel pump operating sound.
LISTEN
Click Click
Injector
N.G. → Check injector circuit.
Listen for injector operating sound.
CHECK
START
Tachometer
Ignition switch
N.G.
Make sure tachometer needle moves when cranking.
CHECK
Timing light
Distributor
N.G.
Check flashes of timing light for weakness.
MEASURE
Ignition wire
N.G. → Replace the wire.
O.K.
Measure resistance of suspect wires.
CHECK
Ignition wire
Spark plug
Remove spark plugs and check their ignition sparks.
CHECK
Timing light
N.G. → Adjust ignition timing.
Check ignition timing.
PERFORM
Self-diagnosis Mode III
N.G. → Check crank angle sensor and/or related circuits.
Perform self-diagnosis Mode III (for crank angle sensor).

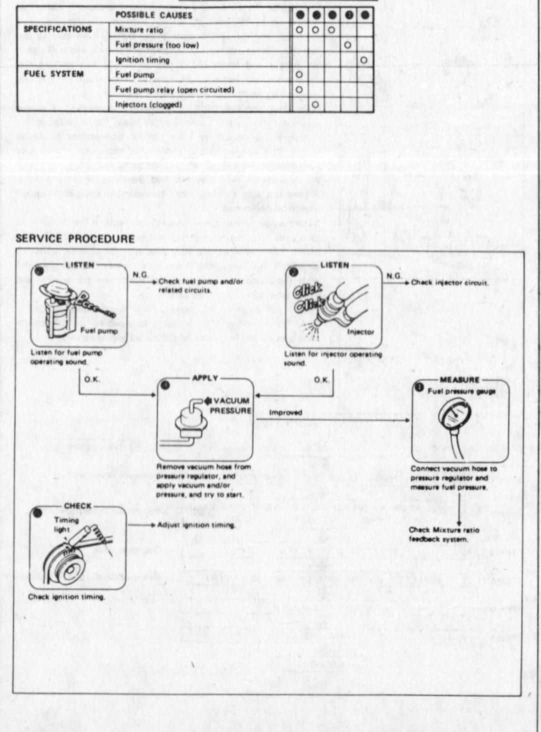

Diagnostic Table (Cont'd)

SYMPTOM & CONDITION 2 Impossible to start — partial combustion

	POSSIBLE CAUSES	●	●	●	●	●
SPECIFICATIONS	Mixture ratio	○	○	○		
	Fuel pressure (too low)				○	
	Ignition timing					○
FUEL SYSTEM	Fuel pump	○				
	Fuel pump relay (open circuited)	○				
	Injectors (clogged)		○			

SERVICE PROCEDURE

LISTEN
Fuel pump
N.G. → Check fuel pump and/or related circuits.
Listen for fuel pump operating sound.
LISTEN
Click Click
Injector
N.G. → Check injector circuit.
Listen for injector operating sound.
O.K.
O.K.
APPLY
VACUUM PRESSURE
Improved
Remove vacuum hose from pressure regulator, and apply vacuum and/or pressure, and try to start.
MEASURE
Fuel pressure gauge
Connect vacuum hose to pressure regulator and measure fuel pressure.
Check Mixture ratio feedback system.
CHECK
Timing light
Adjust ignition timing.
Check ignition timing.

ECCS DIAGNOSTIC CHART—1989 PULSAR

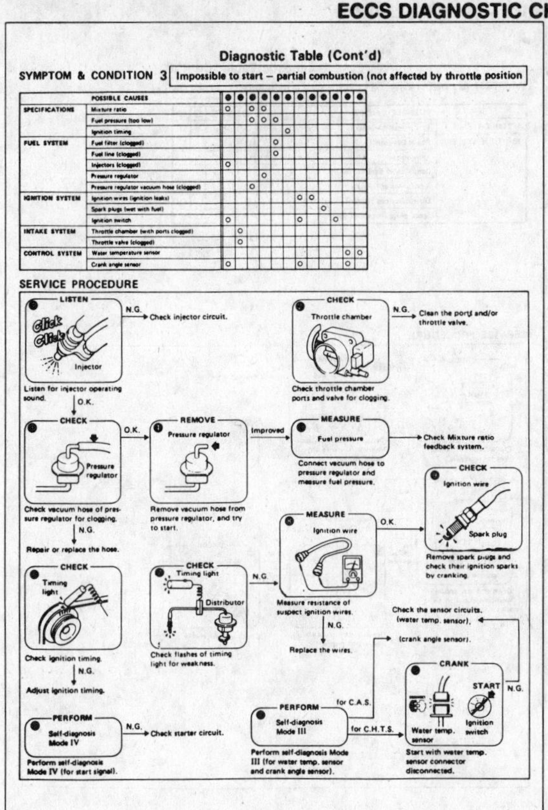

Diagnostic Table (Cont'd)

SYMPTOM & CONDITION 3 Impossible to start – partial combustion (not affected by throttle position

	POSSIBLE CAUSES	●	●	●	●	●	●	●	●	●	●	●	●
SPECIFICATIONS	Mixture ratio	O		O	O								
	Fuel pressure (too low)			O	O	O							
	Ignition timing						O						
FUEL SYSTEM	Fuel filter (clogged)					O							
	Fuel line (clogged)					O							
	Injectors (clogged)	O											
	Pressure regulator				O								
	Pressure regulator vacuum hose (clogged)			O									
IGNITION SYSTEM	Ignition wires (ignition leaks)							O	O				
	Spark plugs (wet with fuel)									O			
	Ignition switch	O						O			O		
INTAKE SYSTEM	Throttle chamber (with ports clogged)		O										
	Throttle valve (clogged)		O										
CONTROL SYSTEM	Water temperature sensor											O	O
	Crank angle sensor	O						O				O	

SERVICE PROCEDURE

LISTEN
Injector
N.G. → Check injector circuit.
Listen for injector operating sound.
O.K.
CHECK
Pressure regulator
Check vacuum hose of pressure regulator for clogging.
N.G.
Repair or replace the hose.
O.K.
REMOVE
Pressure regulator
Remove vacuum hose from pressure regulator, and try to start.
Improved
MEASURE
Fuel pressure
Connect vacuum hose to pressure regulator and measure fuel pressure.
→ Check Mixture ratio feedback system.
CHECK
Throttle chamber
Check throttle chamber ports and valve for clogging.
N.G. → Clean the porty and/or throttle valve.
CHECK
Timing light
Check ignition timing.
N.G.
Adjust ignition timing.
CHECK
Timing light
Distributor
Check flashes of timing light for weakness.
N.G.
MEASURE
Ignition wire
Measure resistance of suspect ignition wires.
N.G.
Replace the wires.
O.K.
CHECK
Ignition wire
Spark plug
Remove spark plugs and check their ignition sparks by cranking.
Check the sensor circuits.
(water temp. sensor).
(crank angle sensor).
PERFORM
Self-diagnosis Mode IV
N.G. → Check starter circuit.
Perform self-diagnosis Mode IV (for start signal).
PERFORM
Self-diagnosis Mode III
for C.A.S.
for C.H.T.S.
Perform self-diagnosis Mode III (for water temp. sensor and crank angle sensor).
CRANK
START
Water temp. sensor
Ignition switch
N.G.
Start with water temp. sensor connector disconnected.

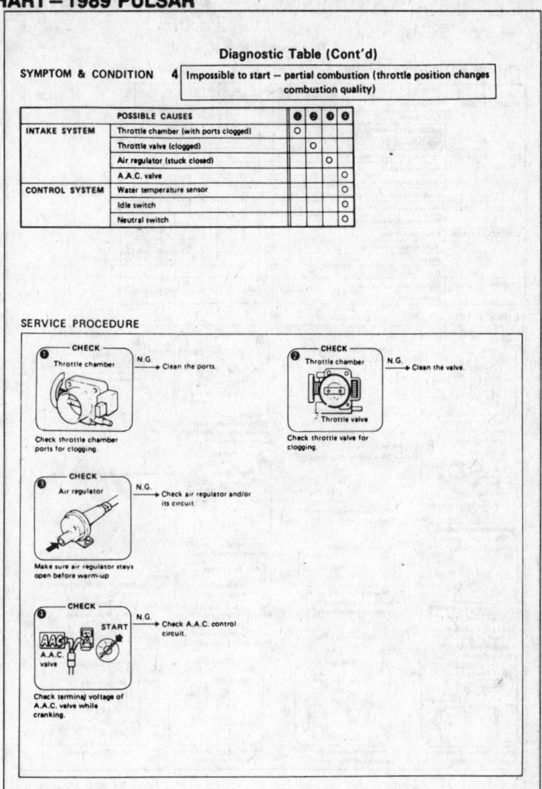

Diagnostic Table (Cont'd)

SYMPTOM & CONDITION 4 Impossible to start – partial combustion (throttle position changes combustion quality)

	POSSIBLE CAUSES	❶	❷	❸	❹
INTAKE SYSTEM	Throttle chamber (with ports clogged)	O			
	Throttle valve (clogged)		O		
	Air regulator (stuck closed)			O	
	A.A.C. valve				O
CONTROL SYSTEM	Water temperature sensor				O
	Idle switch				O
	Neutral switch				O

SERVICE PROCEDURE

CHECK
Throttle chamber
N.G. → Clean the ports.
Check throttle chamber ports for clogging.
CHECK
Throttle chamber
Throttle valve
N.G. → Clean the valve.
Check throttle valve for clogging.
CHECK
Air regulator
N.G. → Check air regulator and/or its circuit.
Make sure air regulator stays open before warm-up
CHECK
START
AAC
A.A.C. valve
N.G. → Check A.A.C. control circuit.
Check terminal voltage of A.A.C. valve while cranking.

ECCS DIAGNOSTIC CHART—1989 PULSAR

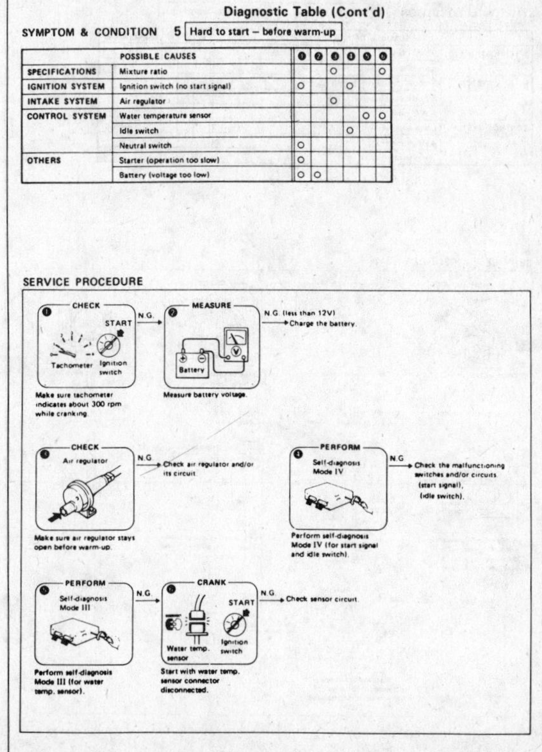

Diagnostic Table (Cont'd)

SYMPTOM & CONDITION 5 Hard to start – before warm-up

	POSSIBLE CAUSES	❶	❷	❸	❹	❺	❻
SPECIFICATIONS	Mixture ratio			O			O
IGNITION SYSTEM	Ignition switch (no start signal)	O			O		
INTAKE SYSTEM	Air regulator			O			
CONTROL SYSTEM	Water temperature sensor					O	O
	Idle switch				O		
	Neutral switch	O					
OTHERS	Starter (operation too slow)	O					
	Battery (voltage too low)	O	O				

SERVICE PROCEDURE

CHECK
START
Tachometer
Ignition switch
Make sure tachometer indicates about 300 rpm while cranking.
N.G.
MEASURE
Battery
Measure battery voltage.
N.G. (less than 12V) → Charge the battery.
CHECK
Air regulator
N.G. → Check air regulator and/or its circuit.
Make sure air regulator stays open before warm-up.
PERFORM
Self-diagnosis Mode IV
N.G → Check the malfunctioning switches and/or circuits (start signal), (idle switch).
Perform self-diagnosis Mode IV (for start signal and idle switch).
PERFORM
Self-diagnosis Mode III
N.G.
Perform self-diagnosis Mode III (for water temp. sensor).
CRANK
START
Water temp. sensor
Ignition switch
N.G. → Check sensor circuit.
Start with water temp. sensor connector disconnected.

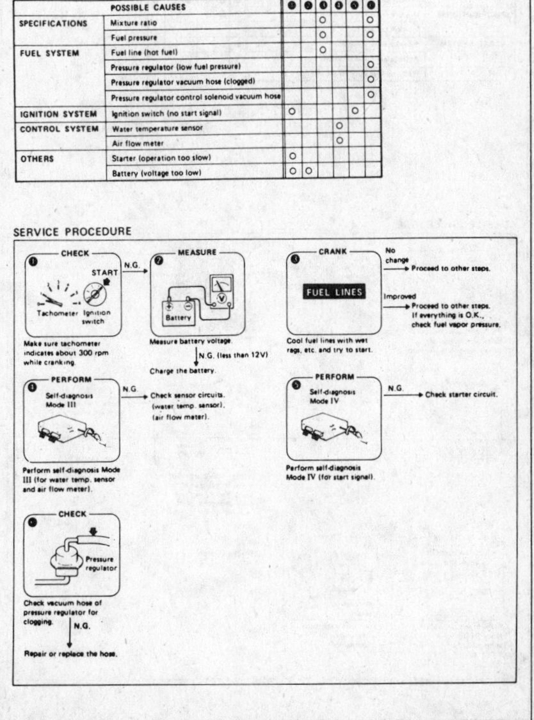

Diagnostic Table (Cont'd)

SYMPTOM & CONDITION 6 Hard to start – after warm-up

	POSSIBLE CAUSES	❶	❷	❸	❹	❺	❻
SPECIFICATIONS	Mixture ratio			O			O
	Fuel pressure			O			O
FUEL SYSTEM	Fuel line (hot fuel)			O			
	Pressure regulator (low fuel pressure)						O
	Pressure regulator vacuum hose (clogged)						O
	Pressure regulator control solenoid vacuum hose						O
IGNITION SYSTEM	Ignition switch (no start signal)	O				O	
CONTROL SYSTEM	Water temperature sensor				O		
	Air flow meter				O		
OTHERS	Starter (operation too slow)	O					
	Battery (voltage too low)	O	O				

SERVICE PROCEDURE

CHECK
START
Tachometer
Ignition switch
Make sure tachometer indicates about 300 rpm while cranking.
N.G.
MEASURE
Battery
Measure battery voltage.
N.G. (less than 12V)
Charge the battery.
CRANK
FUEL LINES
Cool fuel lines with wet rags, etc. and try to start.
No change → Proceed to other steps.
Improved → Proceed to other steps. If everything is O.K., check fuel vapor pressure.
PERFORM
Self-diagnosis Mode III
N.G. → Check sensor circuits. (water temp. sensor). (air flow meter).
Perform self-diagnosis Mode III (for water temp. sensor and air flow meter).
PERFORM
Self-diagnosis Mode IV
N.G. → Check starter circuit.
Perform self-diagnosis Mode IV (for start signal).
CHECK
Pressure regulator
Check vacuum hose of pressure regulator for clogging.
N.G.
Repair or replace the hose.

ECCS DIAGNOSTIC CHART—1989 PULSAR

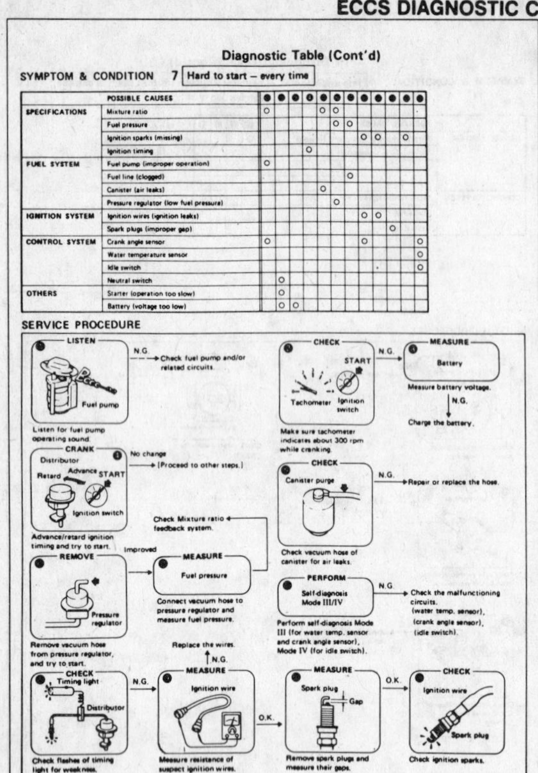

Diagnostic Table (Cont'd)

SYMPTOM & CONDITION 7 Hard to start – every time

	POSSIBLE CAUSES	1	2	3	4	5	6	7	8	9	10	11	12
SPECIFICATIONS	Mixture ratio	○				○	○						
	Fuel pressure						○	○					
	Ignition sparks (missing)								○	○		○	
	Ignition timing				○								
FUEL SYSTEM	Fuel pump (improper operation)	○											
	Fuel line (clogged)							○					
	Canister (air leaks)					○							
	Pressure regulator (low fuel pressure)						○						
IGNITION SYSTEM	Ignition wires (ignition leaks)								○	○			
	Spark plugs (improper gap)										○		
CONTROL SYSTEM	Crank angle sensor	○							○				○
	Water temperature sensor												○
	Idle switch												○
	Neutral switch		○										
OTHERS	Starter (operation too slow)		○										
	Battery (voltage too low)		○	○									

SERVICE PROCEDURE

LISTEN
Fuel pump
N.G.
Check fuel pump and/or related circuits.
Listen for fuel pump operating sound.
CHECK
START
Tachometer
Ignition switch
Make sure tachometer indicates about 300 rpm while cranking.
N.G.
MEASURE
Battery
Measure battery voltage.
N.G.
Charge the battery.
CRANK
Distributor
Advance
Retard
START
Ignition switch
Advance/retard ignition timing and try to start.
No change
(Proceed to other steps.)
CHECK
Canister purge
Check vacuum hose of canister for air leaks.
N.G.
Repair or replace the hose.
REMOVE
Pressure regulator
Remove vacuum hose from pressure regulator, and try to start.
Improved
MEASURE
Fuel pressure
Connect vacuum hose to pressure regulator and measure fuel pressure.
Check Mixture ratio feedback system.
PERFORM
Self-diagnosis Mode III/IV
Perform self-diagnosis Mode III (for water temp. sensor and crank angle sensor), Mode IV (for idle switch).
N.G.
Check the malfunctioning circuits. (water temp. sensor), (crank angle sensor), (idle switch).
CHECK
Timing light
Distributor
Check flashes of timing light for weakness.
N.G.
MEASURE
Ignition wire
Measure resistance of suspect ignition wires.
N.G.
Replace the wires.
O.K.
MEASURE
Spark plug
Gap
Remove spark plugs and measure their gaps.
O.K.
CHECK
Ignition wire
Spark plug
Check ignition sparks.

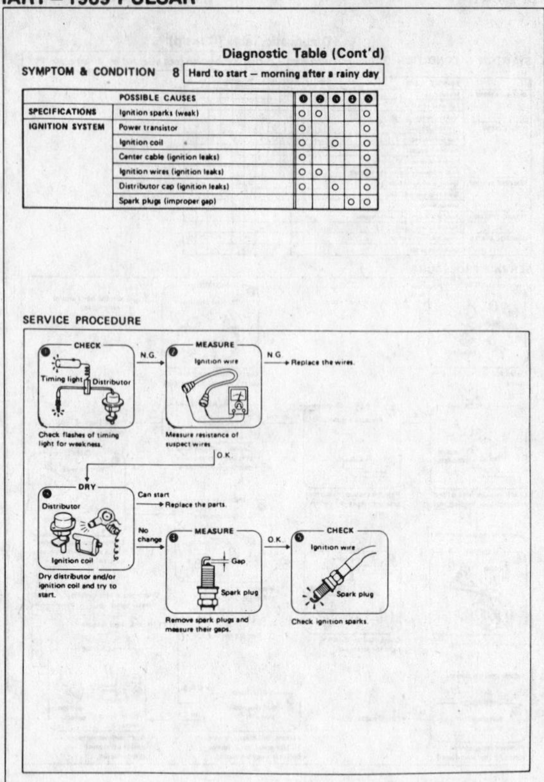

Diagnostic Table (Cont'd)

SYMPTOM & CONDITION 8 Hard to start – morning after a rainy day

	POSSIBLE CAUSES	1	2	3	4	5
SPECIFICATIONS	Ignition sparks (weak)	○	○			○
IGNITION SYSTEM	Power transistor	○				○
	Ignition coil	○		○		○
	Center cable (ignition leaks)	○				○
	Ignition wires (ignition leaks)	○	○			○
	Distributor cap (ignition leaks)	○		○		○
	Spark plugs (improper gap)				○	○

SERVICE PROCEDURE

CHECK
Timing light
Distributor
Check flashes of timing light for weakness.
N.G.
MEASURE
Ignition wire
Measure resistance of suspect wires.
N.G.
Replace the wires.
O.K.
DRY
Distributor
Ignition coil
Dry distributor and/or ignition coil and try to start.
Can start
Replace the parts.
No change
MEASURE
Gap
Spark plug
Remove spark plugs and measure their gaps.
O.K.
CHECK
Ignition wire
Spark plug
Check ignition sparks.

ECCS DIAGNOSTIC CHART—1989 PULSAR

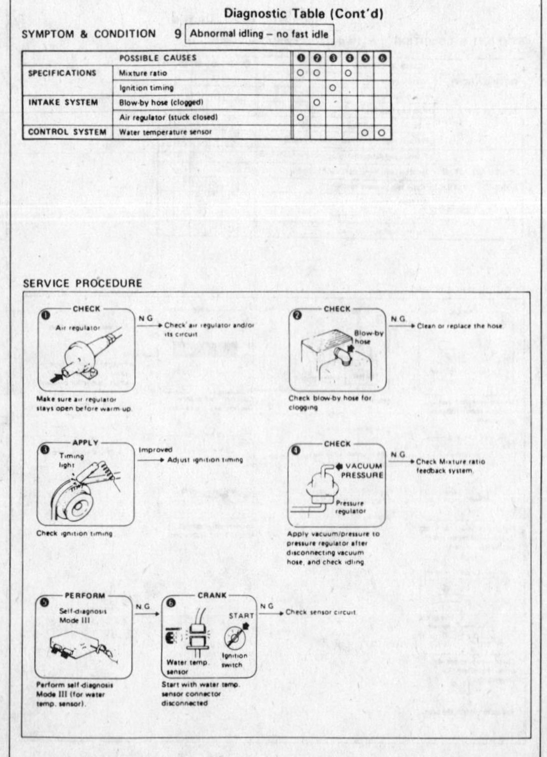

Diagnostic Table (Cont'd)

SYMPTOM & CONDITION 9 Abnormal idling – no fast idle

	POSSIBLE CAUSES	1	2	3	4	5	6
SPECIFICATIONS	Mixture ratio	○	○		○		
	Ignition timing			○			
INTAKE SYSTEM	Blow-by hose (clogged)		○				
	Air regulator (stuck closed)	○					
CONTROL SYSTEM	Water temperature sensor					○	○

SERVICE PROCEDURE

CHECK
Air regulator
N.G.
Check air regulator and/or its circuit
Make sure air regulator stays open before warm-up.
CHECK
Blow-by hose
N.G.
Clean or replace the hose.
Check blow-by hose for clogging
APPLY
Timing light
Improved
Adjust ignition timing
Check ignition timing
CHECK
VACUUM PRESSURE
Pressure regulator
N.G.
Check Mixture ratio feedback system.
Apply vacuum/pressure to pressure regulator after disconnecting vacuum hose, and check idling
PERFORM
Self-diagnosis Mode III
Perform self-diagnosis Mode III (for water temp. sensor).
N.G.
CRANK
START
Water temp. sensor
Ignition switch
Start with water temp. sensor connector disconnected
N.G.
Check sensor circuit.

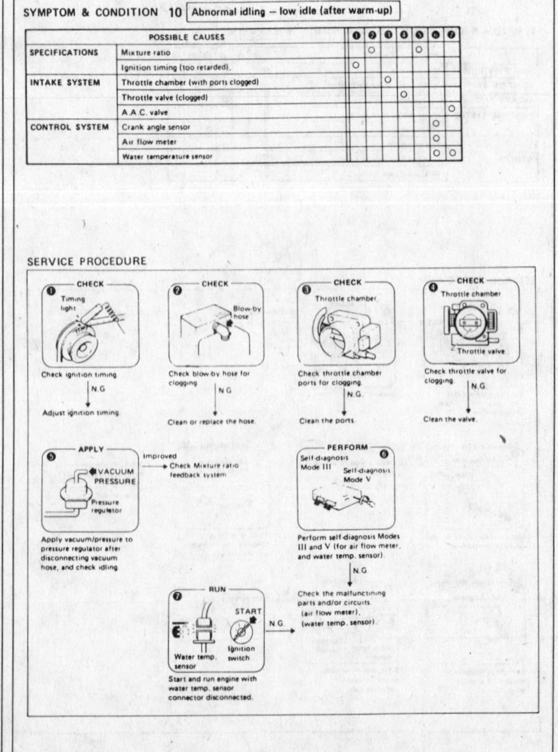

SYMPTOM & CONDITION 10 Abnormal idling – low idle (after warm-up)

	POSSIBLE CAUSES	1	2	3	4	5	6	7
SPECIFICATIONS	Mixture ratio		○			○		
	Ignition timing (too retarded).	○						
INTAKE SYSTEM	Throttle chamber (with ports clogged)			○				
	Throttle valve (clogged)				○			
	A.A.C. valve							○
CONTROL SYSTEM	Crank angle sensor						○	
	Air flow meter						○	
	Water temperature sensor						○	○

SERVICE PROCEDURE

CHECK
Timing light
Check ignition timing.
N.G.
Adjust ignition timing.
CHECK
Blow-by hose
Check blow-by hose for clogging
N.G.
Clean or replace the hose.
CHECK
Throttle chamber
Check throttle chamber ports for clogging.
N.G.
Clean the ports.
CHECK
Throttle chamber
Throttle valve
Check throttle valve for clogging.
N.G.
Clean the valve.
APPLY
VACUUM PRESSURE
Pressure regulator
Improved
Check Mixture ratio feedback system
Apply vacuum/pressure to pressure regulator after disconnecting vacuum hose, and check idling.
PERFORM
Self-diagnosis Mode III
Self-diagnosis Mode V
Perform self-diagnosis Modes III and V (for air flow meter, and water temp. sensor).
N.G.
Check the malfunctining parts and/or circuits. (air flow meter), (water temp. sensor).
RUN
START
Water temp. sensor
Ignition switch
N.G.
Start and run engine with water temp. sensor connector disconnected.

ECCS DIAGNOSTIC CHART—1989 PULSAR

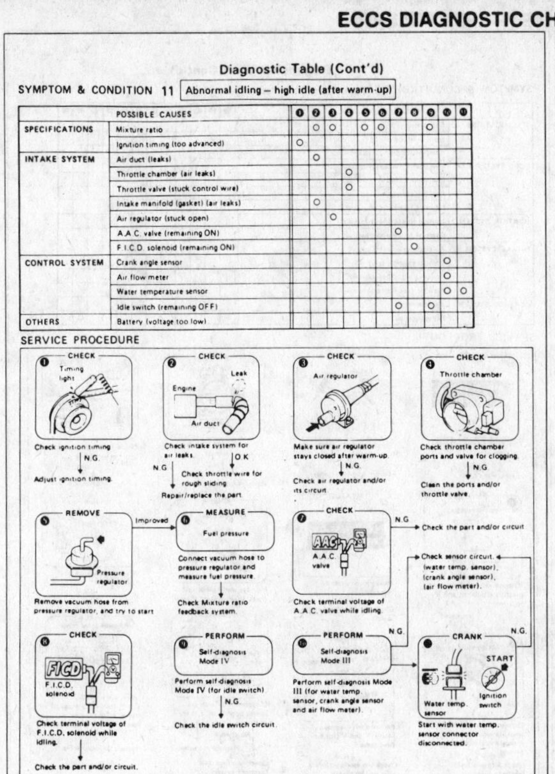

Diagnostic Table (Cont'd)

SYMPTOM & CONDITION 11 Abnormal idling – high idle (after warm-up)

	POSSIBLE CAUSES	1	2	3	4	5	6	7	8	9	10	11
SPECIFICATIONS	Mixture ratio		○	○		○	○			○		
	Ignition timing (too advanced)	○										
INTAKE SYSTEM	Air duct (leaks)		○									
	Throttle chamber (air leaks)				○							
	Throttle valve (stuck control wire)				○							
	Intake manifold (gasket) (air leaks)		○									
	Air regulator (stuck open)			○								
	A.A.C. valve (remaining ON)							○				
	F.I.C.D. solenoid (remaining ON)								○			
CONTROL SYSTEM	Crank angle sensor										○	
	Air flow meter										○	
	Water temperature sensor										○	○
	Idle switch (remaining OFF)							○		○		
OTHERS	Battery (voltage too low)											

SERVICE PROCEDURE

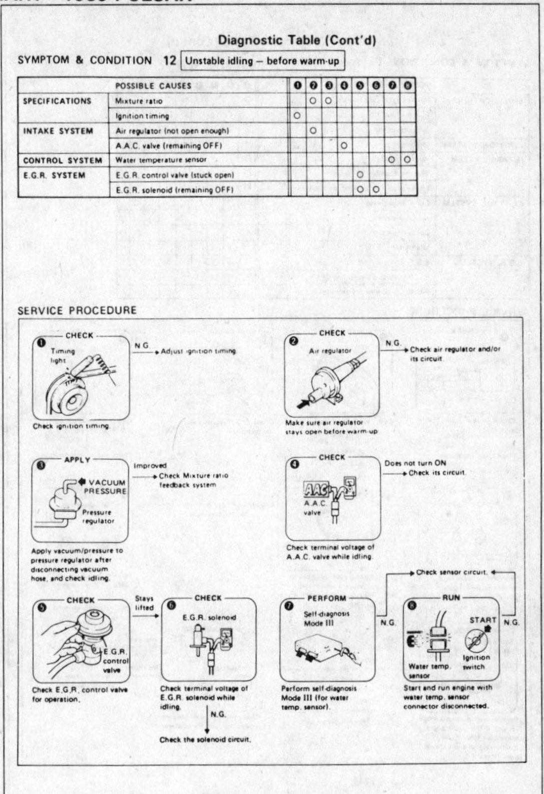

Diagnostic Table (Cont'd)

SYMPTOM & CONDITION 12 Unstable idling – before warm-up

	POSSIBLE CAUSES	1	2	3	4	5	6	7	8
SPECIFICATIONS	Mixture ratio		○	○					
	Ignition timing	○							
INTAKE SYSTEM	Air regulator (not open enough)		○						
	A.A.C. valve (remaining OFF)				○				
CONTROL SYSTEM	Water temperature sensor							○	○
E.G.R. SYSTEM	E.G.R. control valve (stuck open)					○			
	E.G.R. solenoid (remaining OFF)					○	○		

SERVICE PROCEDURE

ECCS DIAGNOSTIC CHART—1989 PULSAR

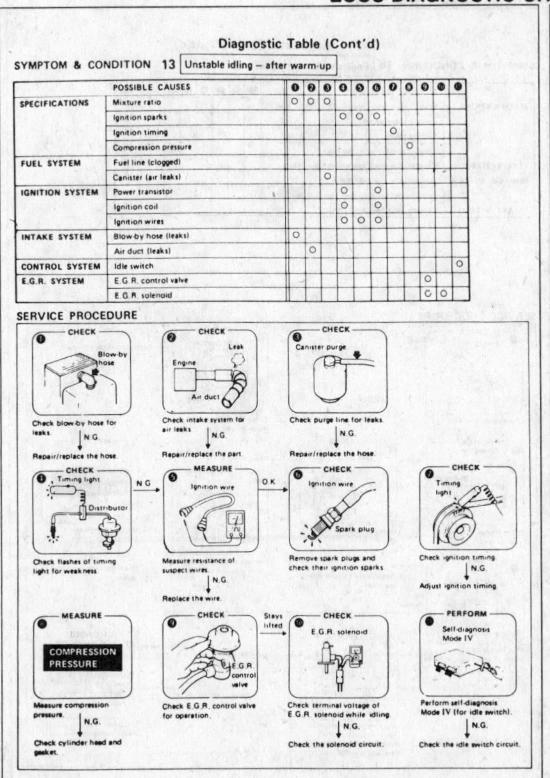

Diagnostic Table (Cont'd)

SYMPTOM & CONDITION 13 Unstable idling – after warm-up

	POSSIBLE CAUSES	1	2	3	4	5	6	7	8	9	10	11
SPECIFICATIONS	Mixture ratio	○	○	○								
	Ignition sparks				○	○	○					
	Ignition timing							○				
	Compression pressure								○			
FUEL SYSTEM	Fuel line (clogged)											
	Canister (air leaks)			○								
IGNITION SYSTEM	Power transistor				○		○					
	Ignition coil				○		○					
	Ignition wires				○	○	○					
INTAKE SYSTEM	Blow-by hose (leaks)	○										
	Air duct (leaks)		○									
CONTROL SYSTEM	Idle switch											○
E.G.R. SYSTEM	E.G.R. control valve									○		
	E.G.R. solenoid									○	○	

SERVICE PROCEDURE

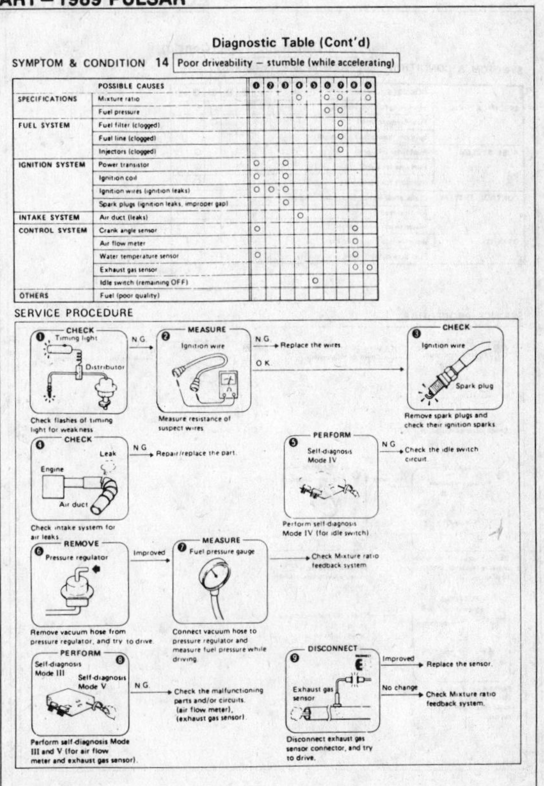

Diagnostic Table (Cont'd)

SYMPTOM & CONDITION 14 Poor driveability – stumble (while accelerating)

	POSSIBLE CAUSES	1	2	3	4	5	6	7	8	9
SPECIFICATIONS	Mixture ratio				○		○	○		○
	Fuel pressure						○	○		
FUEL SYSTEM	Fuel filter (clogged)							○		
	Fuel line (clogged)							○		
	Injectors (clogged)							○		
IGNITION SYSTEM	Power transistor	○		○						
	Ignition coil	○		○						
	Ignition wires (ignition leaks)	○	○	○						
	Spark plugs (ignition leaks, improper gap)			○						
INTAKE SYSTEM	Air duct (leaks)				○					
CONTROL SYSTEM	Crank angle sensor	○							○	
	Air flow meter								○	
	Water temperature sensor	○							○	
	Exhaust gas sensor								○	○
	Idle switch (remaining OFF)					○				
OTHERS	Fuel (poor quality)									

SERVICE PROCEDURE

ECCS DIAGNOSTIC CHART—1989 PULSAR

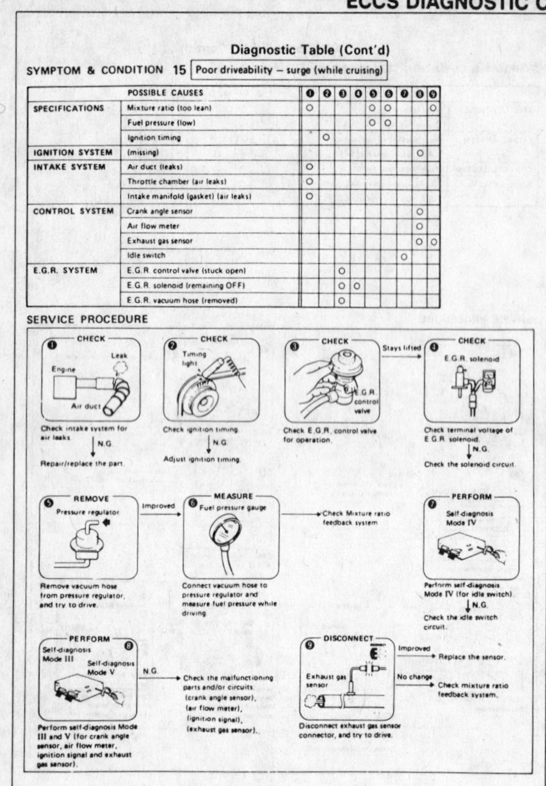

Diagnostic Table (Cont'd)

SYMPTOM & CONDITION 15 Poor driveability – surge (while cruising)

	POSSIBLE CAUSES	1	2	3	4	5	6	7	8	9
SPECIFICATIONS	Mixture ratio (too lean)	O				O	O			O
	Fuel pressure (low)					O	O			
	Ignition timing		O							
IGNITION SYSTEM	(missing)								O	
INTAKE SYSTEM	Air duct (leaks)	O								
	Throttle chamber (air leaks)	O								
	Intake manifold (gasket) (air leaks)	O								
CONTROL SYSTEM	Crank angle sensor								O	
	Air flow meter								O	
	Exhaust gas sensor								O	O
	Idle switch							O		
E.G.R. SYSTEM	E.G.R. control valve (stuck open)			O						
	E.G.R. solenoid (remaining OFF)			O	O					
	E.G.R. vacuum hose (removed)			O						

SERVICE PROCEDURE

1 CHECK
Engine
Leak
Air duct
Check intake system for air leaks.
N.G.
Repair/replace the part.
2 CHECK
Timing light
Check ignition timing.
N.G.
Adjust ignition timing.
3 CHECK
E.G.R. control valve
Check E.G.R. control valve for operation.
Stays lifted
4 CHECK
E.G.R. solenoid
Check terminal voltage of E.G.R. solenoid.
N.G.
Check the solenoid circuit.
5 REMOVE
Pressure regulator
Remove vacuum hose from pressure regulator, and try to drive.
Improved
6 MEASURE
Fuel pressure gauge
Connect vacuum hose to pressure regulator and measure fuel pressure while driving
Check Mixture ratio feedback system
7 PERFORM
Self-diagnosis Mode IV
Perform self-diagnosis Mode IV (for idle switch).
N.G.
Check the idle switch circuit.
8 PERFORM
Self-diagnosis Mode III
Self-diagnosis Mode V
Perform self-diagnosis Mode III and V (for crank angle sensor, air flow meter, ignition signal and exhaust gas sensor).
N.G.
Check the malfunctioning parts and/or circuits (crank angle sensor), (air flow meter), (ignition signal), (exhaust gas sensor).
9 DISCONNECT
Exhaust gas sensor
Disconnect exhaust gas sensor connector, and try to drive.
Improved
Replace the sensor.
No change
Check mixture ratio feedback system.

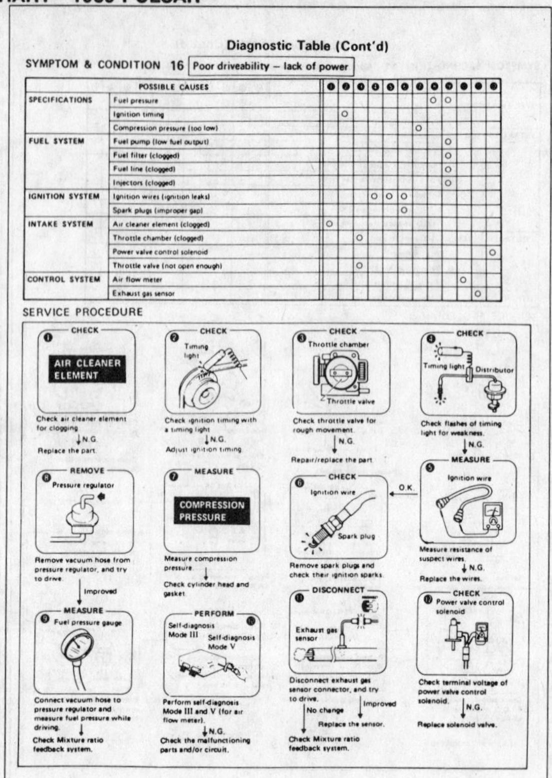

Diagnostic Table (Cont'd)

SYMPTOM & CONDITION 16 Poor driveability – lack of power

	POSSIBLE CAUSES	1	2	3	4	5	6	7	8	9	10	11	12
SPECIFICATIONS	Fuel pressure								O	O			
	Ignition timing		O										
	Compression pressure (too low)							O					
FUEL SYSTEM	Fuel pump (low fuel output)									O			
	Fuel filter (clogged)									O			
	Fuel line (clogged)									O			
	Injectors (clogged)									O			
IGNITION SYSTEM	Ignition wires (ignition leaks)				O	O	O						
	Spark plugs (improper gap)						O						
INTAKE SYSTEM	Air cleaner element (clogged)	O											
	Throttle chamber (clogged)			O									
	Power valve control solenoid												O
	Throttle valve (not open enough)			O									
CONTROL SYSTEM	Air flow meter										O		
	Exhaust gas sensor											O	

SERVICE PROCEDURE

1 CHECK
AIR CLEANER ELEMENT
Check air cleaner element for clogging.
N.G.
Replace the part.
2 CHECK
Timing light
Check ignition timing with a timing light.
N.G.
Adjust ignition timing.
3 CHECK
Throttle chamber
Throttle valve
Check throttle valve for rough movement.
N.G.
Repair/replace the part.
4 CHECK
Timing light
Distributor
Check flashes of timing light for weakness.
N.G.
5 MEASURE
Ignition wire
Measure resistance of suspect wires.
N.G.
Replace the wires.
O.K.
6 CHECK
Ignition wire
Spark plug
Remove spark plugs and check their ignition sparks.
7 MEASURE
COMPRESSION PRESSURE
Measure compression pressure.
Check cylinder head and gasket.
8 REMOVE
Pressure regulator
Remove vacuum hose from pressure regulator, and try to drive.
Improved
9 MEASURE
Fuel pressure gauge
Connect vacuum hose to pressure regulator and measure fuel pressure while driving.
Check Mixture ratio feedback system.
10 PERFORM
Self-diagnosis Mode III
Self-diagnosis Mode V
Perform self-diagnosis Mode III and V (for air flow meter).
N.G.
Check the malfunctioning parts and/or circuit.
11 DISCONNECT
Exhaust gas sensor
Disconnect exhaust gas sensor connector, and try to drive.
No change
Check Mixture ratio feedback system.
Improved
Replace the sensor.
12 CHECK
Power valve control solenoid
Check terminal voltage of power valve control solenoid.
N.G.
Replace solenoid valve.

ECCS DIAGNOSTIC CHART—1989 PULSAR

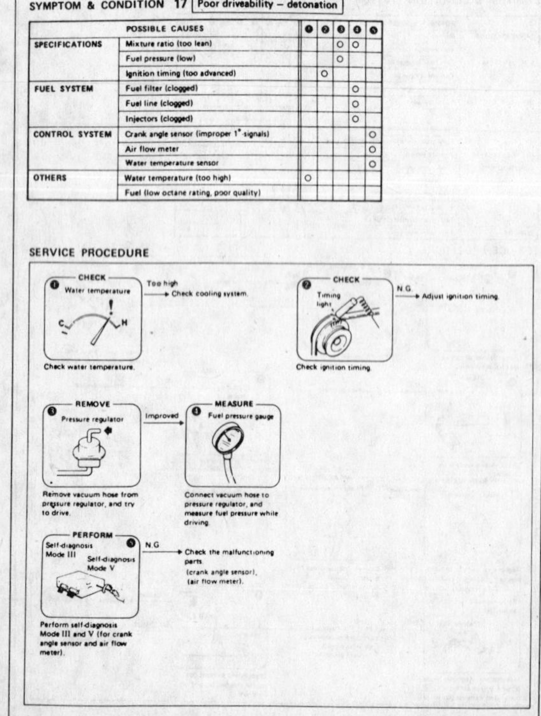

Diagnostic Table (Cont'd)

SYMPTOM & CONDITION 17 Poor driveability – detonation

	POSSIBLE CAUSES	1	2	3	4	5
SPECIFICATIONS	Mixture ratio (too lean)			O	O	
	Fuel pressure (low)			O		
	Ignition timing (too advanced)		O			
FUEL SYSTEM	Fuel filter (clogged)				O	
	Fuel line (clogged)				O	
	Injectors (clogged)				O	
CONTROL SYSTEM	Crank angle sensor (improper 1° signals)					O
	Air flow meter					O
	Water temperature sensor					O
OTHERS	Water temperature (too high)	O				
	Fuel (low octane rating, poor quality)					

SERVICE PROCEDURE

1 CHECK
Water temperature
C
H
Check water temperature.
Too high
Check cooling system.
2 CHECK
Timing light
Check ignition timing.
N.G.
Adjust ignition timing.
3 REMOVE
Pressure regulator
Remove vacuum hose from pressure regulator, and try to drive.
Improved
4 MEASURE
Fuel pressure gauge
Connect vacuum hose to pressure regulator, and measure fuel pressure while driving.
5 PERFORM
Self-diagnosis Mode III
Self-diagnosis Mode V
Perform self-diagnosis Mode III and V (for crank angle sensor and air flow meter).
N.G.
Check the malfunctioning parts. (crank angle sensor), (air flow meter).

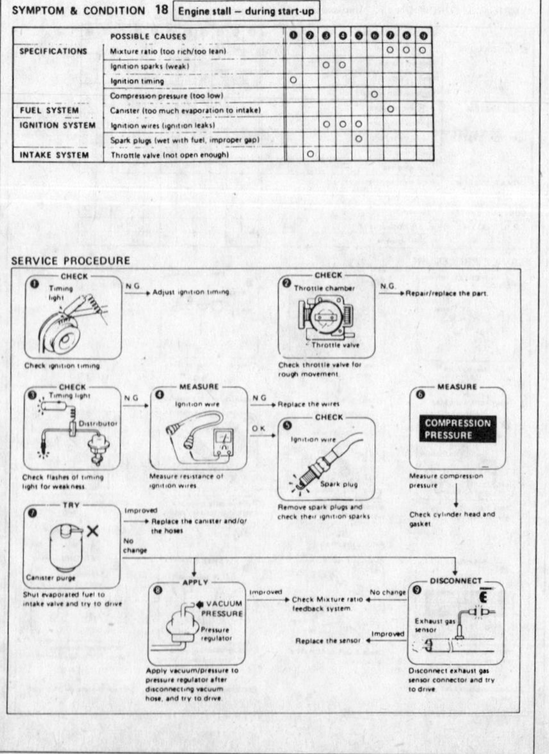

Diagnostic Table (Cont'd)

SYMPTOM & CONDITION 18 Engine stall – during start-up

	POSSIBLE CAUSES	1	2	3	4	5	6	7	8	9
SPECIFICATIONS	Mixture ratio (too rich/too lean)							O	O	O
	Ignition sparks (weak)			O	O					
	Ignition timing	O								
	Compression pressure (too low)						O			
FUEL SYSTEM	Canister (too much evaporation to intake)							O		
IGNITION SYSTEM	Ignition wires (ignition leaks)			O	O	O				
	Spark plugs (wet with fuel, improper gap)					O				
INTAKE SYSTEM	Throttle valve (not open enough)		O							

SERVICE PROCEDURE

1 CHECK
Timing light
Check ignition timing.
N.G.
Adjust ignition timing.
2 CHECK
Throttle chamber
Throttle valve
Check throttle valve for rough movement.
N.G.
Repair/replace the part.
3 CHECK
Timing light
Distributor
Check flashes of timing light for weakness.
N.G.
4 MEASURE
Ignition wire
Measure resistance of ignition wires.
N.G.
Replace the wires.
O.K.
5 CHECK
Ignition wire
Spark plug
Remove spark plugs and check their ignition sparks.
6 MEASURE
COMPRESSION PRESSURE
Measure compression pressure.
Check cylinder head and gasket.
7 TRY
Canister purge
Shut evaporated fuel to intake valve and try to drive.
Improved
Replace the canister and/or the hoses.
No change
8 APPLY
VACUUM PRESSURE
Pressure regulator
Apply vacuum/pressure to pressure regulator after disconnecting vacuum hose, and try to drive.
Improved
Check Mixture ratio feedback system.
No change
9 DISCONNECT
Exhaust gas sensor
Disconnect exhaust gas sensor connector and try to drive.
Improved
Replace the sensor.

ECCS DIAGNOSTIC CHART—1989 PULSAR

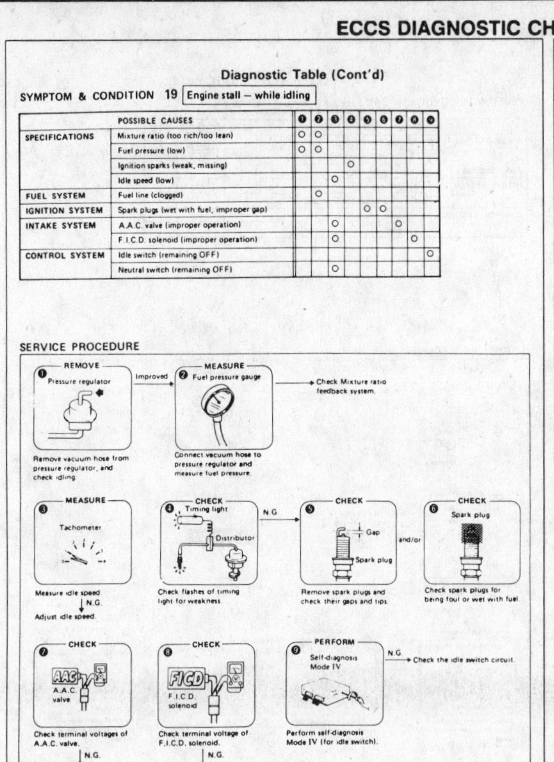

Diagnostic Table (Cont'd)

SYMPTOM & CONDITION 19 Engine stall – while idling

POSSIBLE CAUSES		1	2	3	4	5	6	7	8	9
SPECIFICATIONS	Mixture ratio (too rich/too lean)	O	O							
	Fuel pressure (low)	O	O							
	Ignition sparks (weak, missing)				O					
	Idle speed (low)			O						
FUEL SYSTEM	Fuel line (clogged)		O							
IGNITION SYSTEM	Spark plugs (wet with fuel, improper gap)					O	O			
INTAKE SYSTEM	A.A.C. valve (improper operation)			O				O		
	F.I.C.D. solenoid (improper operation)			O					O	
CONTROL SYSTEM	Idle switch (remaining OFF)									O
	Neutral switch (remaining OFF)			O						

SERVICE PROCEDURE
1 REMOVE
Pressure regulator
Remove vacuum hose from pressure regulator, and check idling.
Improved
2 MEASURE
Fuel pressure gauge
Connect vacuum hose to pressure regulator and measure fuel pressure.
Check Mixture ratio feedback system.
3 MEASURE
Tachometer
Measure idle speed.
N.G.
Adjust idle speed.
4 CHECK
Timing light
Distributor
Check flashes of timing light for weakness.
N.G.
5 CHECK
Gap
Spark plug
Remove spark plugs and check their gaps and tips.
and/or
6 CHECK
Spark plug
Check spark plugs for being foul or wet with fuel.
7 CHECK
AAC
A.A.C. valve
Check terminal voltages of A.A.C. valve.
N.G.
Check the part and/or circuit.
8 CHECK
FICD
F.I.C.D. solenoid
Check terminal voltage of F.I.C.D. solenoid.
N.G.
Check the part and/or circuit.
9 PERFORM
Self-diagnosis Mode IV
Perform self-diagnosis Mode IV (for idle switch).
N.G.
Check the idle switch circuit.

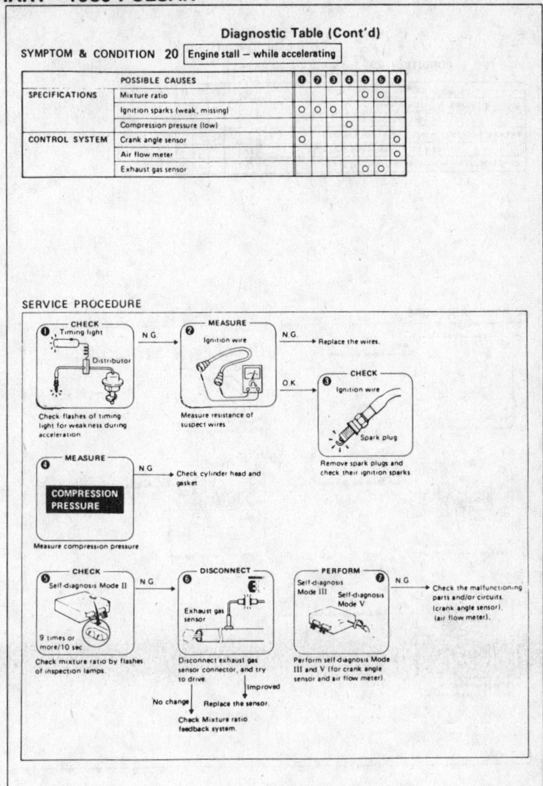

Diagnostic Table (Cont'd)

SYMPTOM & CONDITION 20 Engine stall – while accelerating

POSSIBLE CAUSES		1	2	3	4	5	6	7
SPECIFICATIONS	Mixture ratio					O	O	
	Ignition sparks (weak, missing)	O	O	O				
	Compression pressure (low)				O			
CONTROL SYSTEM	Crank angle sensor	O						O
	Air flow meter							O
	Exhaust gas sensor					O	O	

SERVICE PROCEDURE
1 CHECK
Timing light
Distributor
Check flashes of timing light for weakness during acceleration
N.G.
2 MEASURE
Ignition wire
Measure resistance of suspect wires
N.G.
Replace the wires.
O.K.
3 CHECK
Ignition wire
Spark plug
Remove spark plugs and check their ignition sparks
4 MEASURE
COMPRESSION PRESSURE
Measure compression pressure
N.G.
Check cylinder head and gasket
5 CHECK
Self-diagnosis Mode II
9 times or more/10 sec
Check mixture ratio by flashes of inspection lamps.
N.G.
6 DISCONNECT
Exhaust gas sensor
Disconnect exhaust gas sensor connector, and try to drive.
Improved
Replace the sensor.
No change
Check Mixture ratio feedback system.
7 PERFORM
Self-diagnosis Mode III
Self-diagnosis Mode V
Perform self-diagnosis Mode III and V (for crank angle sensor and air flow meter).
N.G.
Check the malfunctioning parts and/or circuits. (crank angle sensor), (air flow meter).

ECCS DIAGNOSTIC CHART—1989 PULSAR

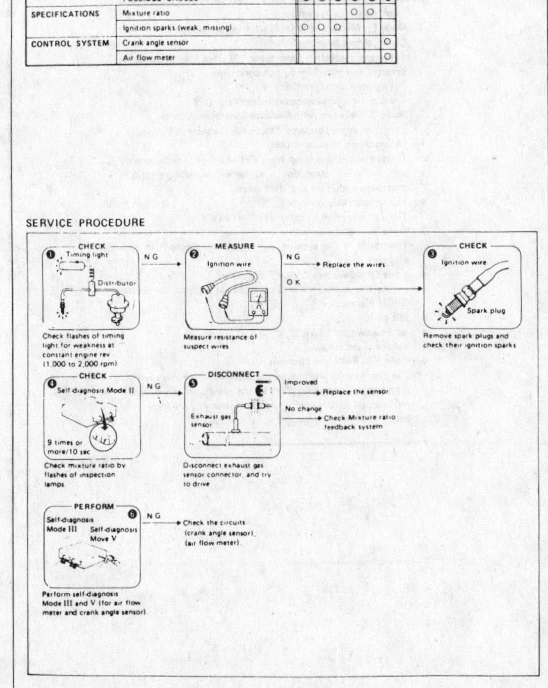

Diagnostic Table (Cont'd)

SYMPTOM & CONDITION 21 Engine stall – while cruising

POSSIBLE CAUSES		1	2	3	4	5	6
SPECIFICATIONS	Mixture ratio				O	O	
	Ignition sparks (weak, missing)	O	O	O			
CONTROL SYSTEM	Crank angle sensor						O
	Air flow meter						O

SERVICE PROCEDURE
1 CHECK
Timing light
Distributor
Check flashes of timing light for weakness at constant engine rev (1,000 to 2,000 rpm)
N.G.
2 MEASURE
Ignition wire
Measure resistance of suspect wires
N.G.
Replace the wires
O.K.
3 CHECK
Ignition wire
Spark plug
Remove spark plugs and check their ignition sparks
4 CHECK
Self-diagnosis Mode II
9 times or more/10 sec
Check mixture ratio by flashes of inspection lamps.
N.G.
5 DISCONNECT
Exhaust gas sensor
Disconnect exhaust gas sensor connector, and try to drive
Improved
Replace the sensor
No change
Check Mixture ratio feedback system
6 PERFORM
Self-diagnosis Mode III
Self-diagnosis Move V
Perform self-diagnosis Mode III and V (for air flow meter and crank angle sensor).
N.G.
Check the circuits (crank angle sensor), (air flow meter).

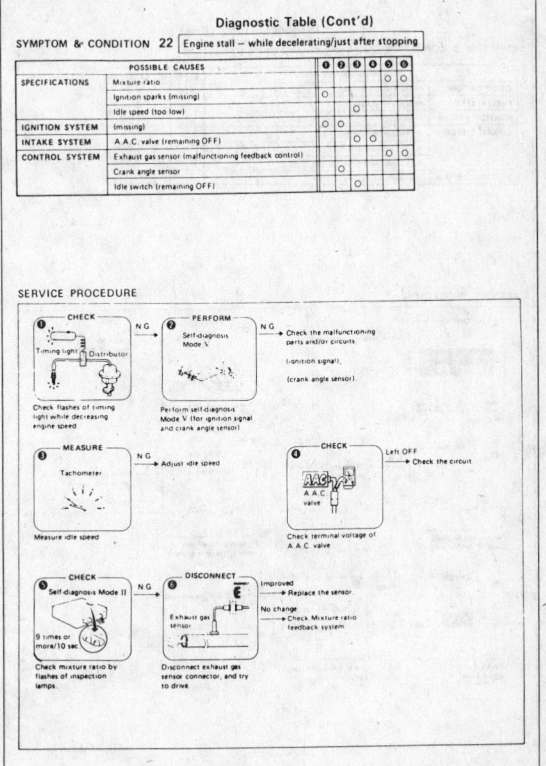

Diagnostic Table (Cont'd)

SYMPTOM & CONDITION 22 Engine stall – while decelerating/just after stopping

POSSIBLE CAUSES		1	2	3	4	5	6
SPECIFICATIONS	Mixture ratio					O	O
	Ignition sparks (missing)	O					
	Idle speed (too low)			O			
IGNITION SYSTEM	(missing)	O	O				
INTAKE SYSTEM	A.A.C. valve (remaining OFF)			O	O		
CONTROL SYSTEM	Exhaust gas sensor (malfunctioning feedback control)					O	O
	Crank angle sensor		O				
	Idle switch (remaining OFF)			O			

SERVICE PROCEDURE
1 CHECK
Timing light
Distributor
Check flashes of timing light while decreasing engine speed
N.G.
2 PERFORM
Self-diagnosis Mode V
Perform self-diagnosis Mode V (for ignition signal and crank angle sensor)
N.G.
Check the malfunctioning parts and/or circuits.
(ignition signal),
(crank angle sensor).
3 MEASURE
Tachometer
Measure idle speed
N.G.
Adjust idle speed
4 CHECK
AAC
A.A.C. valve
Check terminal voltage of A.A.C. valve
Left OFF
Check the circuit
5 CHECK
Self-diagnosis Mode II
9 times or more/10 sec.
Check mixture ratio by flashes of inspection lamps.
N.G.
6 DISCONNECT
Exhaust gas sensor
Disconnect exhaust gas sensor connector, and try to drive.
Improved
Replace the sensor.
No change
Check Mixture ratio feedback system

ECCS DIAGNOSTIC CHART – 1989 PULSAR

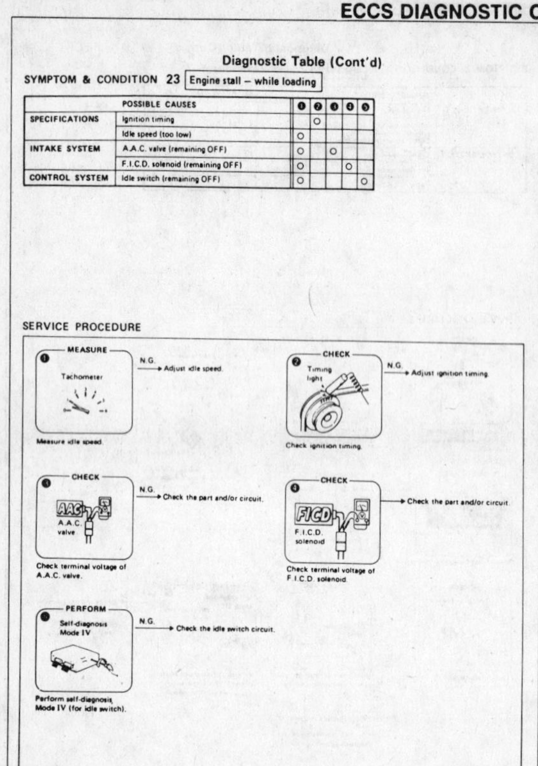

Diagnostic Table (Cont'd)

SYMPTOM & CONDITION 23 Engine stall – while loading

POSSIBLE CAUSES		1	2	3	4	5
SPECIFICATIONS	Ignition timing		○			
	Idle speed (too low)	○				
INTAKE SYSTEM	A.A.C. valve (remaining OFF)	○		○		
	F.I.C.D. solenoid (remaining OFF)	○			○	
CONTROL SYSTEM	Idle switch (remaining OFF)	○				○

SERVICE PROCEDURE
1 MEASURE
Tachometer
N.G. → Adjust idle speed.
Measure idle speed.
2 CHECK
Timing light
N.G. → Adjust ignition timing.
Check ignition timing.
3 CHECK
AAC
A.A.C. valve
N.G. → Check the part and/or circuit.
Check terminal voltage of A.A.C. valve.
4 CHECK
FICD
F.I.C.D. solenoid
→ Check the part and/or circuit.
Check terminal voltage of F.I.C.D. solenoid.
5 PERFORM
Self-diagnosis Mode IV
N.G. → Check the idle switch circuit.
Perform self-diagnosis Mode IV (for idle switch).

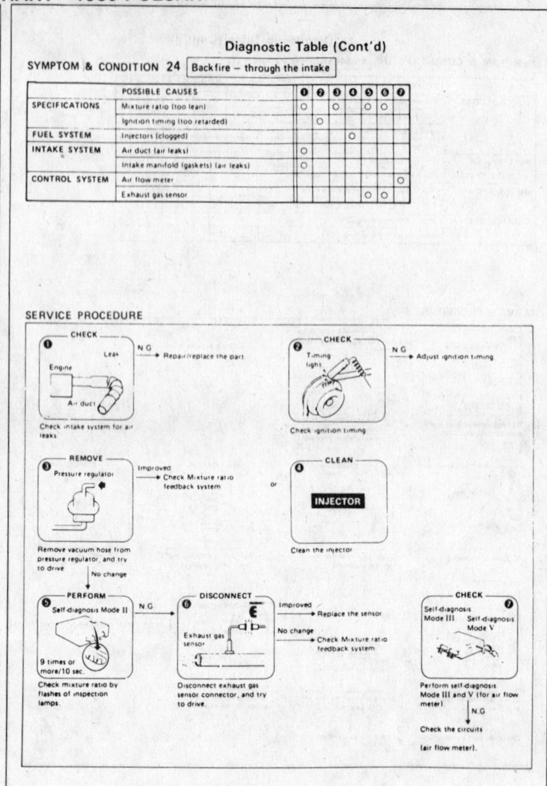

Diagnostic Table (Cont'd)

SYMPTOM & CONDITION 24 Backfire – through the intake

POSSIBLE CAUSES		1	2	3	4	5	6	7
SPECIFICATIONS	Mixture ratio (too lean)	○		○		○	○	
	Ignition timing (too retarded)		○					
FUEL SYSTEM	Injectors (clogged)				○			
INTAKE SYSTEM	Air duct (air leaks)	○						
	Intake manifold (gaskets) (air leaks)	○						
CONTROL SYSTEM	Air flow meter							○
	Exhaust gas sensor					○	○	

SERVICE PROCEDURE
1 CHECK
Leak
Engine
Air duct
N.G. → Repair/replace the part.
Check intake system for air leaks.
2 CHECK
Timing light
N.G. → Adjust ignition timing.
Check ignition timing.
3 REMOVE
Pressure regulator
Improved → Check Mixture ratio feedback system.
or
Remove vacuum hose from pressure regulator, and try to drive.
No change
4 CLEAN
INJECTOR
Clean the injector.
5 PERFORM
Self-diagnosis Mode II
9 times or more/10 sec.
N.G.
Check mixture ratio by flashes of inspection lamps.
6 DISCONNECT
Exhaust gas sensor
Improved → Replace the sensor.
No change → Check Mixture ratio feedback system.
Disconnect exhaust gas sensor connector, and try to drive.
7 CHECK
Self-diagnosis Mode III
Self-diagnosis Mode V
Perform self-diagnosis Mode III and V (for air flow meter).
N.G.
Check the circuits (air flow meter).

ECCS DIAGNOSTIC CHART – 1989 PULSAR

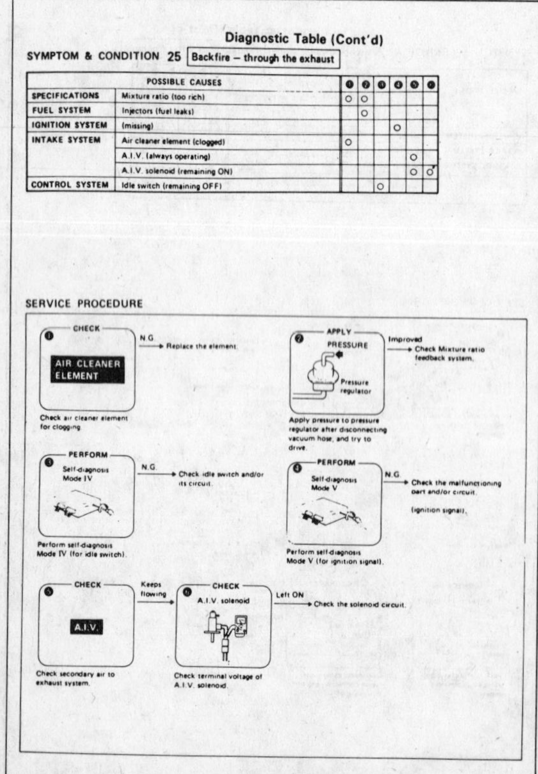

Diagnostic Table (Cont'd)

SYMPTOM & CONDITION 25 Backfire – through the exhaust

POSSIBLE CAUSES		1	2	3	4	5	6
SPECIFICATIONS	Mixture ratio (too rich)	○	○				
FUEL SYSTEM	Injectors (fuel leaks)		○				
IGNITION SYSTEM	(missing)				○		
INTAKE SYSTEM	Air cleaner element (clogged)	○					
	A.I.V. (always operating)					○	
	A.I.V. solenoid (remaining ON)					○	○
CONTROL SYSTEM	Idle switch (remaining OFF)			○			

SERVICE PROCEDURE
1 CHECK
AIR CLEANER ELEMENT
N.G. → Replace the element.
Check air cleaner element for clogging.
2 APPLY PRESSURE
Pressure regulator
Improved → Check Mixture ratio feedback system.
Apply pressure to pressure regulator after disconnecting vacuum hose, and try to drive.
3 PERFORM
Self-diagnosis Mode IV
N.G. → Check idle switch and/or its circuit.
Perform self-diagnosis Mode IV (for idle switch).
4 PERFORM
Self-diagnosis Mode V
N.G. → Check the malfunctioning part and/or circuit. (ignition signal).
Perform self-diagnosis Mode V (for ignition signal).
5 CHECK
A.I.V.
Keeps flowing
Check secondary air to exhaust system.
6 CHECK
A.I.V. solenoid
Left ON → Check the solenoid circuit.
Check terminal voltage of A.I.V. solenoid.

ECCS SELF-DIAGNOSTIC DESCRIPTION – 1989 PULSAR

Description

The self-diagnosis is useful to diagnose malfunctions in major sensors and actuators of the E.C.C.S. system. There are 5 modes in the self-diagnosis system.

1. **Mode I – Mixture ratio feedback control monitor A**
   - During closed loop condition:
     The green inspection lamp turns ON when lean condition is detected and goes OFF by rich condition.
   - During open loop condition:
     The green inspection lamp remains ON or OFF.
2. **Mode II – Mixture ratio feedback control monitor B**
   The green inspection lamp function is the same as Mode I.
   - During closed loop condition:
     The red inspection lamp turns ON and OFF simultaneously with the green inspection lamp when the mixture ratio is controlled within the specified value.
   - During open loop condition:
     The red inspection lamp remains ON or OFF.
3. **Mode III – Self-diagnosis**
   This mode is the same as the former self-diagnosis in self-diagnosis mode.
4. **Mode IV – Switches ON/OFF diagnosis**
   During this mode, the inspection lamps monitor the switch ON-OFF condition.
   - Idle switch
   - Ignition switch "START"
   - Vehicle speed sensor
5. **Mode V – Real time diagnosis**
   The moment the malfunction is detected, the display will be presented immediately. That is, the condition at which the malfunction occurs can be found by observing the inspection lamps during driving test.

ECCS SELF-DIAGNOSTIC DESCRIPTION (CONT.)—1989 PULSAR

Flashing N times

Mode N

Description (Cont'd)

SWITCHING THE MODES

1. Turn ignition switch "ON".
2. Turn diagnostic mode selector on E.C.U. fully clockwise and wait the inspection lamps flash.
3. Count the number of the flashing time, and after the inspection lamps have flashed the number of the required mode, turn diagnostic mode selector fully counterclockwise immediately.

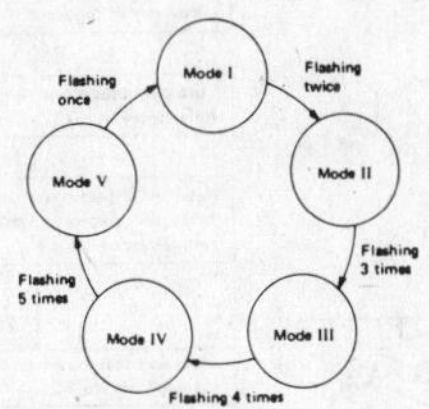
Mode I
Flashing twice
Mode II
Flashing 3 times
Mode III
Flashing 4 times
Mode IV
Flashing 5 times
Mode V
Flashing once

NOTE:
When the ignition switch is turned off during diagnosis, in each mode, and then turned back on again after the power to the E.C.U. has dropped off completely, the diagnosis will automatically return to Mode I.

The stored memory would be lost if:

1. Battery terminal is disconnected.
2. After selecting Mode III, Mode IV is selected.
   However, if the diagnostic mode selector is kept turned fully clockwise, it will continue to change in the order of Mode I → II → III → IV → V → I ... etc., and in this state the stored memory will not be erased.

CHECK
CHECK ENGINE LIGHT

Description (Cont'd)

CHECK ENGINE LIGHT (For California only)

This vehicle has a check engine light on the instrument panel. This light comes ON under the following conditions:

1) When ignition switch is turned "ON" (for bulb check).
2) When systems related to emission performance malfunction in Mode I (with engine running).
   - **This check engine light always illuminates and is synchronous with the red L.E.D.**
   - **Malfunction systems related to emission performance can be detected by self-diagnosis, and they are clarified as self-diagnostic codes in Mode III.**
3) Check engine light will be "ON" only when malfunction is sensed.
   The check engine light will turn off when normal operation is resumed. Mode III memory must be cleared as the contents remain stored.

Code No.	Malfunction
12	Air flow meter circuit
13	Water temperature sensor circuit
14	Vehicle speed sensor circuit
23	Idle switch circuit
31	E.C.U. (E.C.C.S. control unit)
32	E.G.R. function
33	Exhaust gas sensor circuit
35	Exhaust gas temperature sensor circuit
43	Throttle sensor circuit
45	Injector leak

Use the following diagnostic flowchart to check and repair a malfunctioning system.

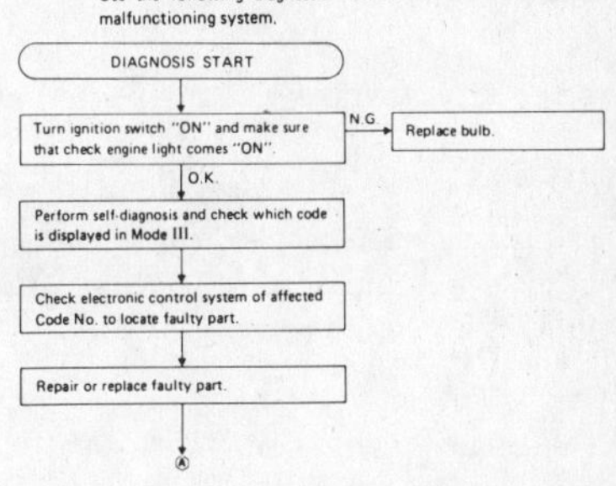
DIAGNOSIS START
Turn ignition switch "ON" and make sure that check engine light comes "ON".
N.G.
Replace bulb.
O.K.
Perform self-diagnosis and check which code is displayed in Mode III.
Check electronic control system of affected Code No. to locate faulty part.
Repair or replace faulty part.
Ⓐ

ECCS SELF-DIAGNOSTIC DESCRIPTION (CONT.)—1989 PULSAR

Description (Cont'd)

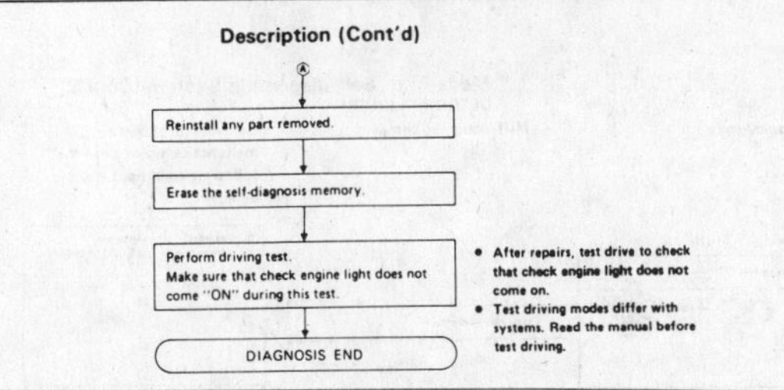
Ⓐ
Reinstall any part removed.
Erase the self-diagnosis memory.
Perform driving test.
Make sure that check engine light does not come "ON" during this test.
DIAGNOSIS END

- **After repairs, test drive to check that check engine light does not come on.**
- **Test driving modes differ with systems. Read the manual before test driving.**

ECCS SELF-DIAGNOSTIC MODES I AND II—1989 PULSAR

Modes I & II—Mixture Ratio Feedback Control Monitors A & B

In these modes, the control unit provides the mixture ratio monitor presentation and the mixture ratio feedback coefficient monitor presentation.

Mode	LED	Engine stopped (Ignition switch "ON")	Engine running: Open loop condition	Engine running: Closed loop condition		
Mode I (Monitor A)	Green	ON	*Remains ON or OFF	Blinks		
	Red	ON	Except for California model • OFF	For California model • ON: when the following malfunctions are stored in the E.C.U.: Air flow meter, water temperature sensor, vehicle speed sensor, idle switch, E.C.U., E.G.R. function, exhaust gas sensor, exhaust gas temperature sensor, throttle sensor, injector leak • OFF: except for the above condition		
Mode II (Monitor B)	Green	ON	*Remains ON or OFF	Blinks		
				Compensating mixture ratio		
				More than 5% rich	Between 5% lean and 5% rich	More
	Red	OFF	*Remains ON or OFF (synchronous with green LED)	OFF	Synchronized with green LED	Remains ON

*: Maintains conditions just before switching to open loop

ECCS SELF-DIAGNOSTIC MODE III—1989 PULSAR

Mode III — Self-diagnostic System

The E.C.U. constantly monitors the function of these sensors and actuators, regardless of ignition key position. If a malfunction occurs, the information is stored in the E.C.U. and can be retrieved from the memory by turning on the diagnostic mode selector, located on the side of the E.C.U. When activated, the malfunction is indicated by flashing a red and a green L.E.D. (Light Emitting Diode), also located on the E.C.U. Since all the self-diagnostic results are stored in the E.C.U.'s memory even intermittent malfunctions can be diagnosed.

A malfunctioning part's group is indicated by the number of both the red and the green L.E.D.s flashing. First, the red L.E.D. flashes and the green flashes follow. The red L.E.D. refers to the number of tens while the green one refers to the number of units. For example, when the red L.E.D. flashes once and then the green one flashes twice, this means the number "12" showing the air flow meter signal is malfunctioning. In this way, all the problems are classified by the code numbers.

- **When engine fails to start, crank engine more than two seconds before starting self-diagnosis.**
- **Before starting self-diagnosis, do not erase stored memory. If doing so, self-diagnosis function for intermittent malfunctions would be lost.**

The stored memory would be lost if:

1. **Battery terminal is disconnected.**
2. **After selecting Mode III, Mode IV is selected.**

DISPLAY CODE TABLE

X: Available
—: Not available

Code No.	Detected items	California	Non-California
11	Crank angle sensor circuit	X	X
12	Air flow meter circuit	X	X
13	Water temperature sensor circuit	X	X
14	Vehicle speed sensor circuit	X	X
21	Ignition signal circuit	X	X
23	Idle switch circuit	X	X
31	E.C.U. (E.C.C.S. control unit)	X	X
32	E.G.R. function	X	—
33	Exhaust gas sensor	X	X
34	Detonation sensor circuit	X	X
35	Exhaust gas temperature sensor circuit	X	—
43	Throttle sensor	X	X
45	Injector leak	X	—
55	No malfunctioning in the above circuit	X	X

ECCS SELF-DIAGNOSTIC MODE III (CONT.)—1989 PULSAR

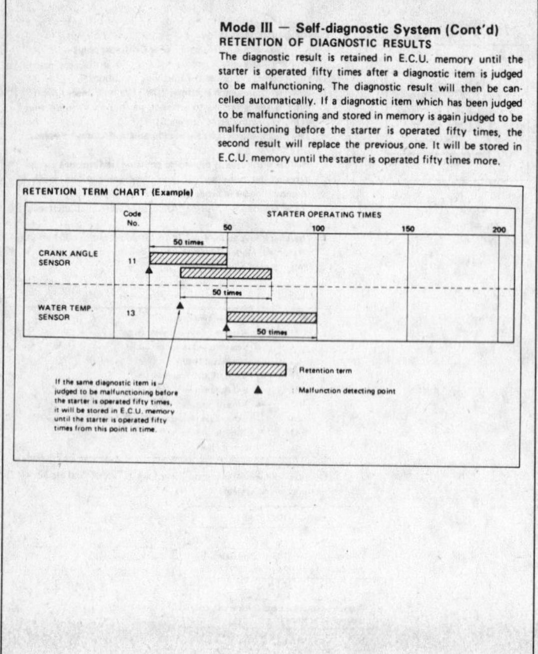

Mode III — Self-diagnostic System (Cont'd)
RETENTION OF DIAGNOSTIC RESULTS

The diagnostic result is retained in E.C.U. memory until the starter is operated fifty times after a diagnostic item is judged to be malfunctioning. The diagnostic result will then be cancelled automatically. If a diagnostic item which has been judged to be malfunctioning and stored in memory is again judged to be malfunctioning before the starter is operated fifty times, the second result will replace the previous one. It will be stored in E.C.U. memory until the starter is operated fifty times more.

RETENTION TERM CHART (Example)
Code No.
STARTER OPERATING TIMES
50
100
150
200
CRANK ANGLE SENSOR
11
50 times
WATER TEMP. SENSOR
13
50 times
50 times
Retention term
Malfunction detecting point
If the same diagnostic item is judged to be malfunctioning before the starter is operated fifty times, it will be stored in E.C.U. memory until the starter is operated fifty times from this point in time.

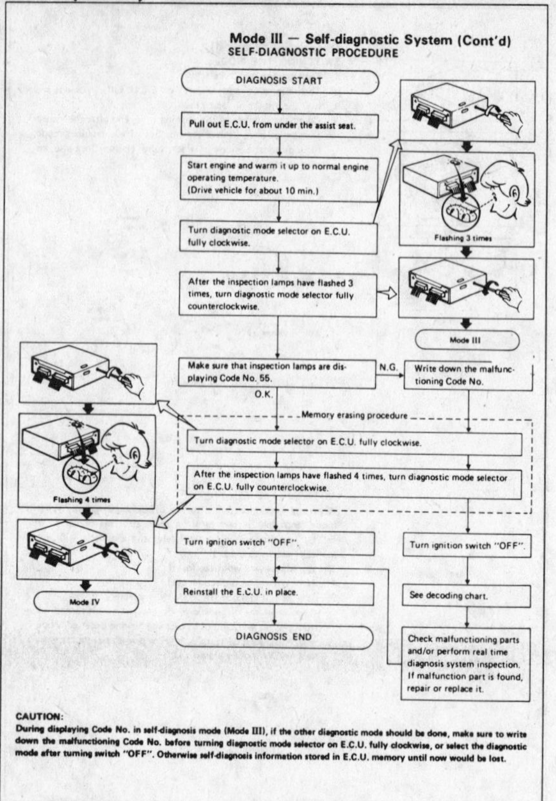

Mode III — Self-diagnostic System (Cont'd)
SELF-DIAGNOSTIC PROCEDURE

DIAGNOSIS START
Pull out E.C.U. from under the assist seat.
Start engine and warm it up to normal engine operating temperature.
(Drive vehicle for about 10 min.)
Turn diagnostic mode selector on E.C.U. fully clockwise.
Flashing 3 times
After the inspection lamps have flashed 3 times, turn diagnostic mode selector fully counterclockwise.
Mode III
Make sure that inspection lamps are displaying Code No. 55.
N.G.
Write down the malfunctioning Code No.
O.K.
Memory erasing procedure
Turn diagnostic mode selector on E.C.U. fully clockwise.
Flashing 4 times
After the inspection lamps have flashed 4 times, turn diagnostic mode selector on E.C.U. fully counterclockwise.
Turn ignition switch "OFF".
Turn ignition switch "OFF".
Mode IV
Reinstall the E.C.U. in place.
See decoding chart.
DIAGNOSIS END
Check malfunctioning parts and/or perform real time diagnosis system inspection. If malfunction part is found, repair or replace it.

CAUTION:
During displaying Code No. in self-diagnosis mode (Mode III), if the other diagnostic mode should be done, make sure to write down the malfunctioning Code No. before turning diagnostic mode selector on E.C.U. fully clockwise, or select the diagnostic mode after turning switch "OFF". Otherwise self-diagnosis information stored in E.C.U. memory until now would be lost.

ECCS SELF-DIAGNOSTIC MODE III DISPLAY CODES—1989 PULSAR

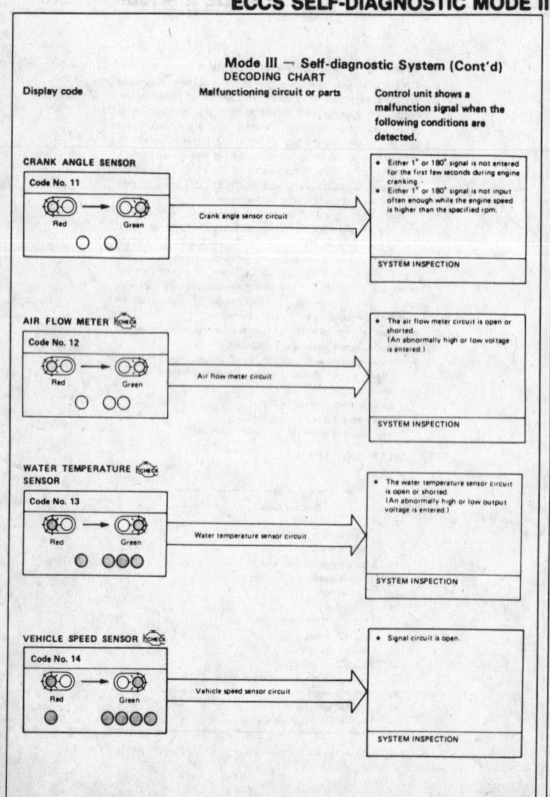

Mode III — Self-diagnostic System (Cont'd)
DECODING CHART

Display code	Malfunctioning circuit or parts	Control unit shows a malfunction signal when the following conditions are detected.
CRANK ANGLE SENSOR — Code No. 11 (Red → Green)	Crank angle sensor circuit	• Either 1° or 180° signal is not entered for the first few seconds during engine cranking. • Either 1° or 180° signal is not input often enough while the engine speed is higher than the specified rpm. SYSTEM INSPECTION
AIR FLOW METER CHECK — Code No. 12 (Red → Green)	Air flow meter circuit	• The air flow meter circuit is open or shorted. (An abnormally high or low voltage is entered.) SYSTEM INSPECTION
WATER TEMPERATURE SENSOR CHECK — Code No. 13 (Red → Green)	Water temperature sensor circuit	• The water temperature sensor circuit is open or shorted. (An abnormally high or low output voltage is entered.) SYSTEM INSPECTION
VEHICLE SPEED SENSOR CHECK — Code No. 14 (Red → Green)	Vehicle speed sensor circuit	• Signal circuit is open. SYSTEM INSPECTION

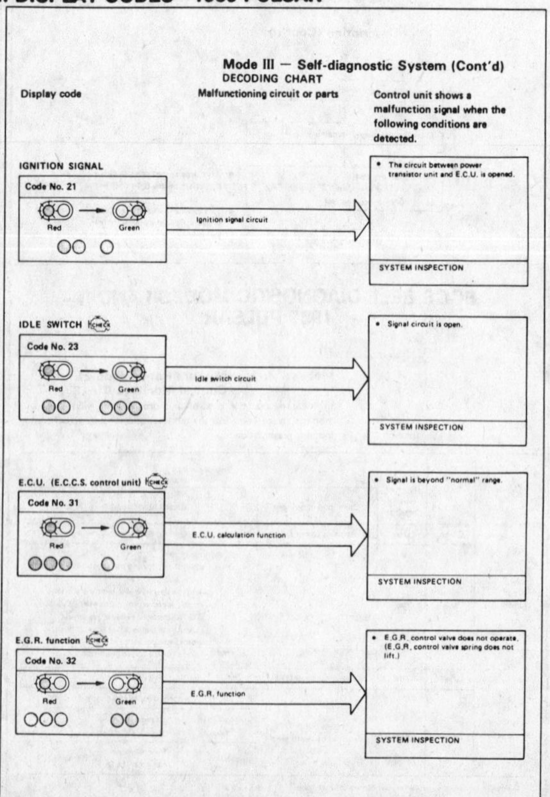

Mode III — Self-diagnostic System (Cont'd)
DECODING CHART

Display code	Malfunctioning circuit or parts	Control unit shows a malfunction signal when the following conditions are detected.
IGNITION SIGNAL — Code No. 21 (Red → Green)	Ignition signal circuit	• The circuit between power transistor unit and E.C.U. is opened. SYSTEM INSPECTION
IDLE SWITCH CHECK — Code No. 23 (Red → Green)	Idle switch circuit	• Signal circuit is open. SYSTEM INSPECTION
E.C.U. (E.C.C.S. control unit) CHECK — Code No. 31 (Red → Green)	E.C.U. calculation function	• Signal is beyond "normal" range. SYSTEM INSPECTION
E.G.R. function CHECK — Code No. 32 (Red → Green)	E.G.R. function	• E.G.R. control valve does not operate. (E.G.R. control valve spring does not lift.) SYSTEM INSPECTION

ECCS SELF-DIAGNOSTIC MODE III DISPLAY CODES (CONT.)—1989 PULSAR

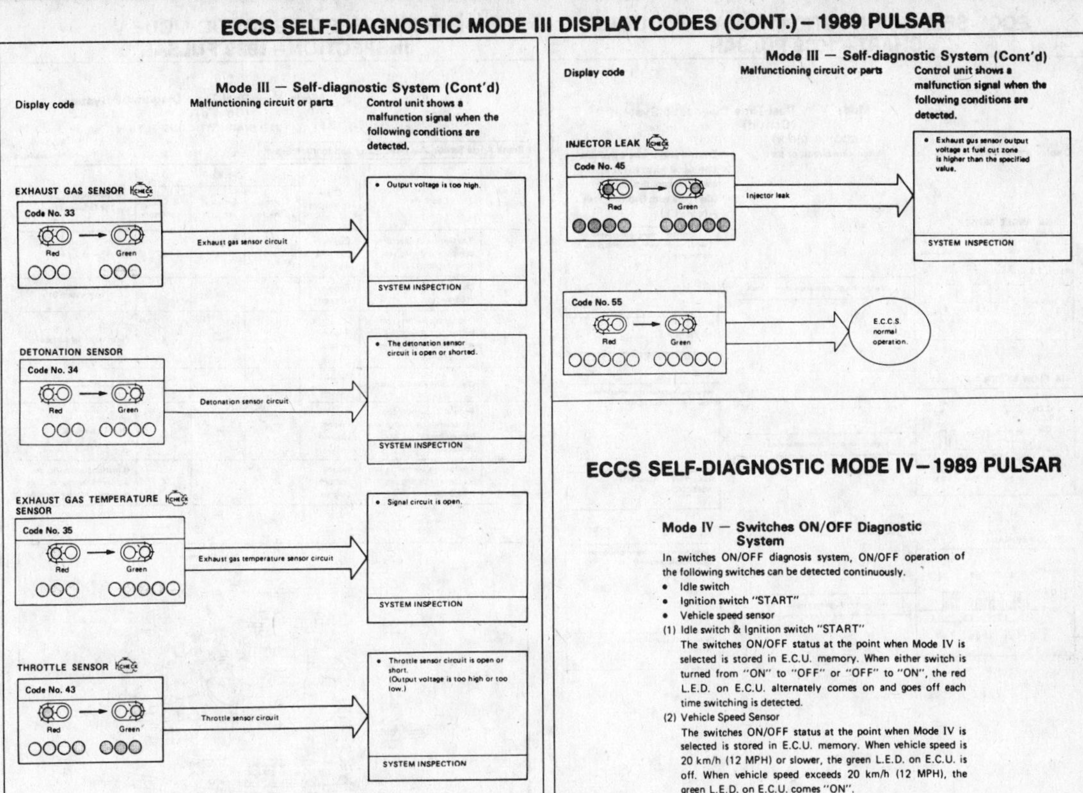
Mode III — Self-diagnostic System (Cont'd)
Display code
Malfunctioning circuit or parts
Control unit shows a malfunction signal when the following conditions are detected.
EXHAUST GAS SENSOR
Code No. 33
Red
Green
Exhaust gas sensor circuit
• Output voltage is too high.
SYSTEM INSPECTION
DETONATION SENSOR
Code No. 34
Detonation sensor circuit
• The detonation sensor circuit is open or shorted.
SYSTEM INSPECTION
EXHAUST GAS TEMPERATURE SENSOR
Code No. 35
Exhaust gas temperature sensor circuit
• Signal circuit is open.
SYSTEM INSPECTION
THROTTLE SENSOR
Code No. 43
Throttle sensor circuit
• Throttle sensor circuit is open or short. (Output voltage is too high or too low.)
SYSTEM INSPECTION
Mode III — Self-diagnostic System (Cont'd)
INJECTOR LEAK
Code No. 45
Injector leak
• Exhaust gas sensor output voltage at fuel cut zone is higher than the specified value.
SYSTEM INSPECTION
Code No. 55
E.C.C.S. normal operation.

ECCS SELF-DIAGNOSTIC MODE IV—1989 PULSAR

Mode IV — Switches ON/OFF Diagnostic System

In switches ON/OFF diagnosis system, ON/OFF operation of the following switches can be detected continuously.

- Idle switch
- Ignition switch "START"
- Vehicle speed sensor

(1) Idle switch & Ignition switch "START"
The switches ON/OFF status at the point when Mode IV is selected is stored in E.C.U. memory. When either switch is turned from "ON" to "OFF" or "OFF" to "ON", the red L.E.D. on E.C.U. alternately comes on and goes off each time switching is detected.

(2) Vehicle Speed Sensor
The switches ON/OFF status at the point when Mode IV is selected is stored in E.C.U. memory. When vehicle speed is 20 km/h (12 MPH) or slower, the green L.E.D. on E.C.U. is off. When vehicle speed exceeds 20 km/h (12 MPH), the green L.E.D. on E.C.U. comes "ON".

ECCS SELF-DIAGNOSTIC MODE IV (CONT.)—1989 PULSAR

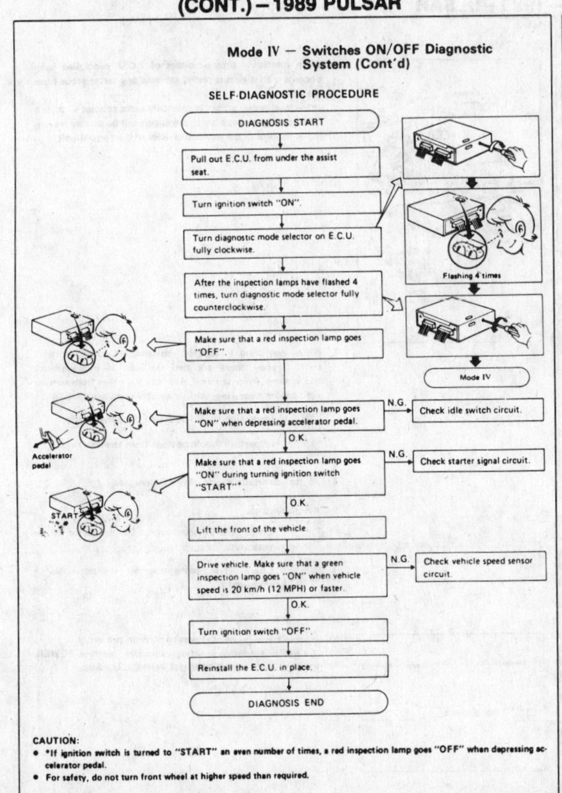
Mode IV — Switches ON/OFF Diagnostic System (Cont'd)
SELF-DIAGNOSTIC PROCEDURE
DIAGNOSIS START
Pull out E.C.U. from under the assist seat.
Turn ignition switch "ON".
Turn diagnostic mode selector on E.C.U. fully clockwise.
After the inspection lamps have flashed 4 times, turn diagnostic mode selector fully counterclockwise.
Flashing 4 times
Mode IV
Make sure that a red inspection lamp goes "OFF".
Make sure that a red inspection lamp goes "ON" when depressing accelerator pedal.
N.G.
Check idle switch circuit.
O.K.
Accelerator pedal
Make sure that a red inspection lamp goes "ON" during turning ignition switch "START"*.
N.G.
Check starter signal circuit.
O.K.
START
Lift the front of the vehicle.
Drive vehicle. Make sure that a green inspection lamp goes "ON" when vehicle speed is 20 km/h (12 MPH) or faster.
N.G.
Check vehicle speed sensor circuit.
O.K.
Turn ignition switch "OFF".
Reinstall the E.C.U. in place.
DIAGNOSIS END

CAUTION:
- *If ignition switch is turned to "START" an even number of times, a red inspection lamp goes "OFF" when depressing accelerator pedal.
- For safety, do not turn front wheel at higher speed than required.

ECCS SELF-DIAGNOSTIC MODE V—1989 PULSAR

Mode V — Real Time Diagnostic System

In real time diagnosis, if any of the following items are judged to be faulty, a malfunction is indicated immediately.

- Crank angle sensor (180° signal & 1° signal)
- Ignition signal
- Air flow meter output signal

Consequently, this diagnosis is a very effective measure to diagnose whether the above systems cause the malfunction or not, during driving test. Compared with self-diagnosis, real time diagnosis is very sensitive, and can detect malfunctioning conditions in a moment. Further, items regarded to be malfunctions in this diagnosis are not stored in E.C.U. memory.

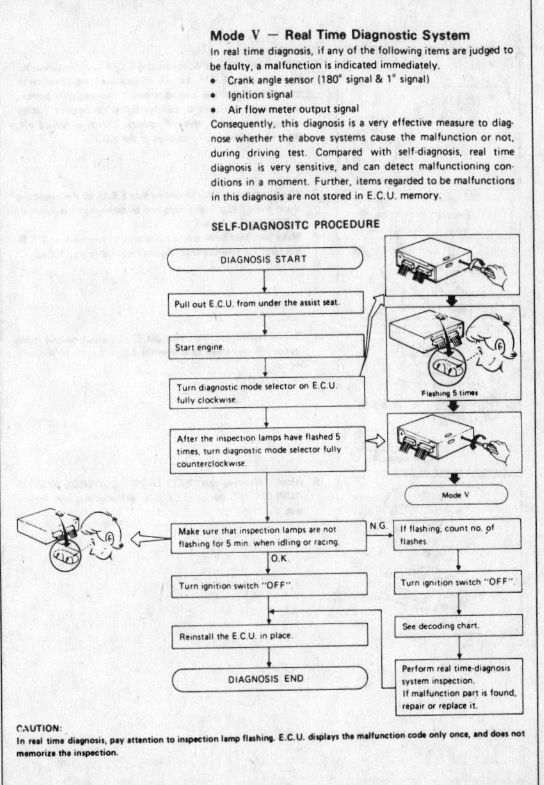
SELF-DIAGNOSITC PROCEDURE
DIAGNOSIS START
Pull out E.C.U. from under the assist seat.
Start engine.
Turn diagnostic mode selector on E.C.U. fully clockwise.
Flashing 5 times
After the inspection lamps have flashed 5 times, turn diagnostic mode selector fully counterclockwise.
Mode V
Make sure that inspection lamps are not flashing for 5 min. when idling or racing.
N.G.
If flashing, count no. of flashes.
O.K.
Turn ignition switch "OFF".
Turn ignition switch "OFF".
See decoding chart.
Reinstall the E.C.U. in place.
Perform real time-diagnosis system inspection. If malfunction part is found, repair or replace it.
DIAGNOSIS END

CAUTION:
In real time diagnosis, pay attention to inspection lamp flashing. E.C.U. displays the malfunction code only once, and does not memorize the inspection.

ECCS SELF-DIAGNOSTIC MODE V DECODING CHART—1989 PULSAR

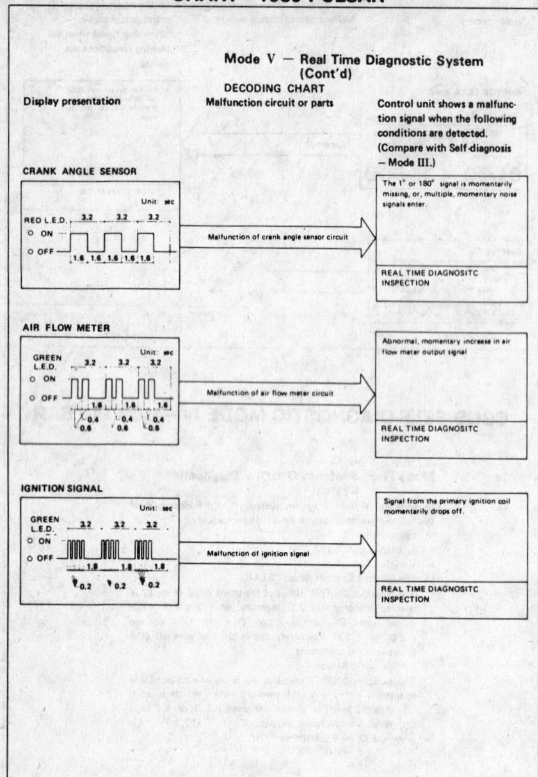
Mode V — Real Time Diagnostic System (Cont'd)
DECODING CHART
Display presentation
Malfunction circuit or parts
Control unit shows a malfunction signal when the following conditions are detected. (Compare with Self-diagnosis – Mode III.)
CRANK ANGLE SENSOR
Unit: sec
RED L.E.D.
ON
OFF
3.2 3.2 3.2
1.6 1.6 1.6 1.6 1.6
Malfunction of crank angle sensor circuit
The 1° or 180° signal is momentarily missing, or, multiple, momentary noise signals enter.
REAL TIME DIAGNOSITC INSPECTION
AIR FLOW METER
Unit: sec
GREEN L.E.D.
ON
OFF
3.2 3.2 3.2
1.6 1.6 1.6
0.4 0.4 0.4
0.6 0.6 0.6
Malfunction of air flow meter circuit
Abnormal, momentary increase in air flow meter output signal
REAL TIME DIAGNOSITC INSPECTION
IGNITION SIGNAL
Unit: sec
GREEN L.E.D.
ON
OFF
3.2 3.2 3.2
1.8 1.8 1.8
0.2 0.2 0.2
Malfunction of ignition signal
Signal from the primary ignition coil momentarily drops off.
REAL TIME DIAGNOSITC INSPECTION

ECCS SELF-DIAGNOSTIC MODE V INSPECTION—1989 PULSAR

Mode V — Real Time Diagnostic System (Cont'd)

REAL TIME DIAGNOSTIC INSPECTION

Crank Angle Sensor, Air Flow Meter and Ignition Signal

X: Available
–: Not available

Check sequence	Check items	Check conditions	Check parts: Harness connectors	Check parts: Sensor & actuator	Check parts: E.C.U. connectors	If malfunction, perform the following items.
1	Tap harness connector or component during real time diagnosis.	During real time diagnosis	X	X	X	Go to check item 2.
2	Check harness continuity at connector.	Engine stopped	X	–	–	Go to check item 3.
3	Disconnect harness connector, and then check dust adhesion to harness connector.	Engine stopped	X	–	X	Clean terminal surface.
4	Check pin terminal bend.	Engine stopped	–	–	X	Take out bend.
5	Reconnect harness connector and then recheck harness continuity at connector.	Engine stopped	X	–	–	Replace terminal.
6	Tap harness connector or component during real time diagnosis.	During real time diagnosis	X	X	X	If malfunction codes are displayed during real time diagnosis, replace terminal.

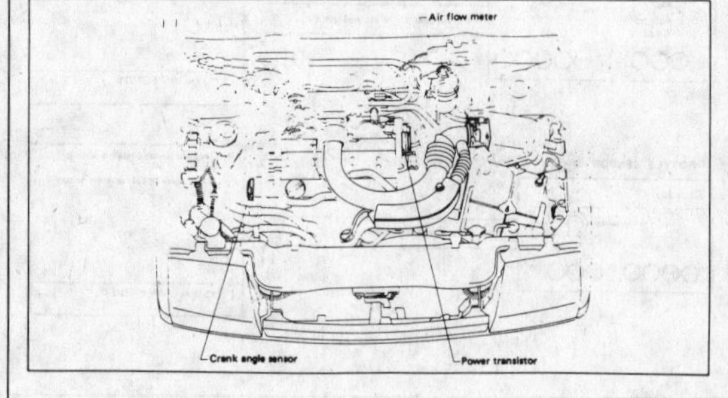
Air flow meter
Crank angle sensor
Power transistor

ECCS CAUTION—1989 PULSAR

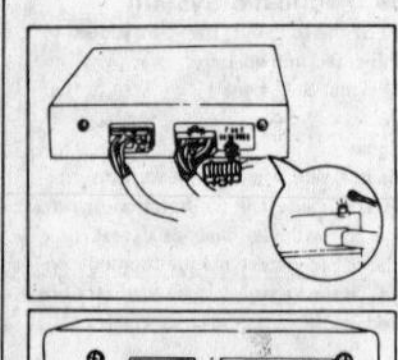

CAUTION:

1. **Before connecting or disconnecting E.C.U. harness connector to or from any E.C.U., be sure to turn the ignition switch to the "OFF" position and disconnect the negative battery terminal in order not to damage E.C.U. as battery voltage is applied to E.C.U. even if ignition switch is turned off. Otherwise, there may be damage to the E.C.U.**

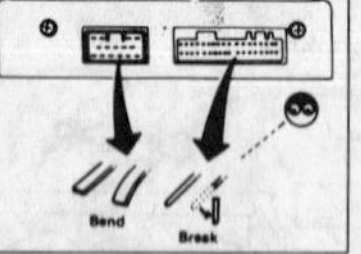
Bend
Break

2. **When connecting pin connectors into E.C.U. or disconnecting them from E.C.U., take care not to damage pin terminal of E.C.U. (Bend or break).**
3. **Make sure that there are not any bends or breaks on E.C.U. pin terminal, when connecting pin connectors into E.C.U.**

Perform E.C.U. input/output signal inspection before replacement.
OLD ONE

4. **Before replacing E.C.U., perform E.C.U. input/output signal inspection and make sure whether E.C.U. functions properly or not.**

5. **After performing this "ELECTRONIC CONTROL SYSTEM INSPECTION", perform E.C.C.S. self-diagnosis and driving test.**

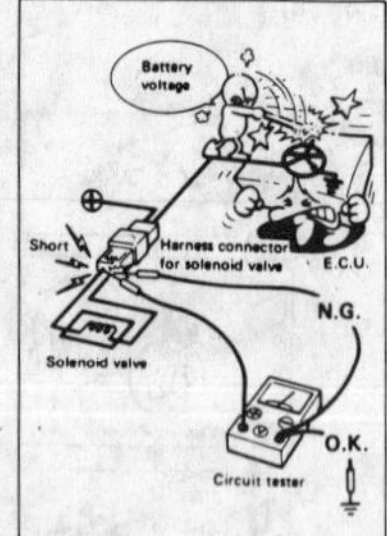
Battery voltage
Short
Harness connector for solenoid valve
E.C.U.
N.G.
Solenoid valve
O.K.
Circuit tester

6. **When measuring supply voltage of E.C.U. controlled components with a circuit tester, separate one tester probe from the other.**
   If the two tester probes accidentally make contact with each other during measurement, the circuit will be shorted, resulting in damage to the power transistor of the control unit.
7. **When measuring voltage or resistance at connector with tester probes, there are two methods of measurement; one is done from terminal side and the other from harness side. Before measuring, confirm symbol mark again.**

   HS : **Inspection should be done from harness side.**

   TS : **Inspection should be done from terminal side.**
8. **As for continuity check of joint connector.**
9. **Key to symbols**

   DISCONNECT : **Check after disconnecting the connector to be measured.**

   CONNECT : **Check after connecting the connector to be measured.**

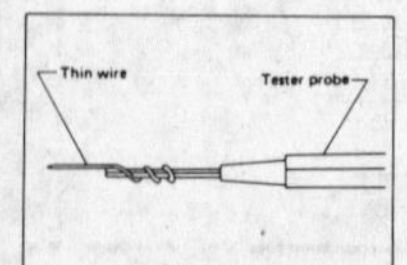
Thin wire
Tester probe

10. **Improve tester probe as shown to perform test easily.**
11. **For the first trouble-shooting procedure, perform POWER SOURCE & GROUND CIRCUIT FOR E.C.U. check.**

ECCS SELF-DIAGNOSTIC CHART – 1989 PULSAR

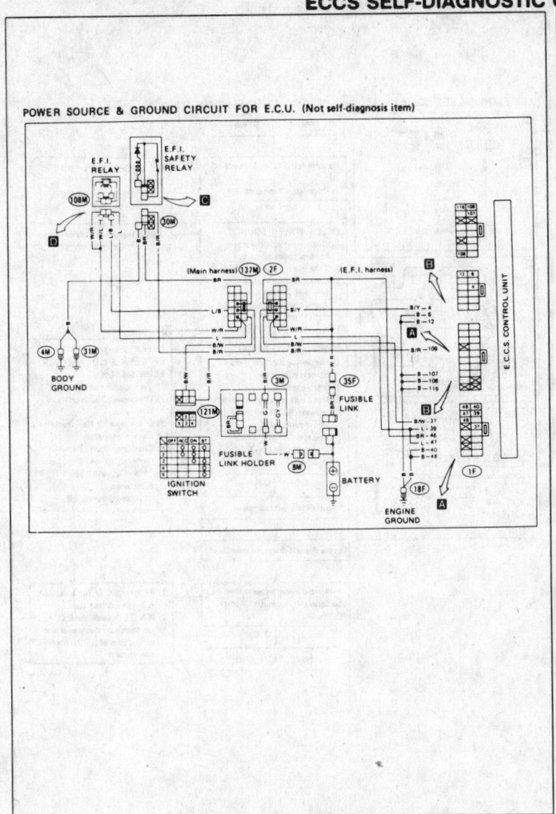
POWER SOURCE & GROUND CIRCUIT FOR E.C.U. (Not self-diagnosis item)
E.F.I. RELAY
E.F.I. SAFETY RELAY
(Main harness)
(E.F.I. harness)
BODY GROUND
FUSIBLE LINK
FUSIBLE LINK HOLDER
IGNITION SWITCH
BATTERY
ENGINE GROUND
E.C.C.S. CONTROL UNIT

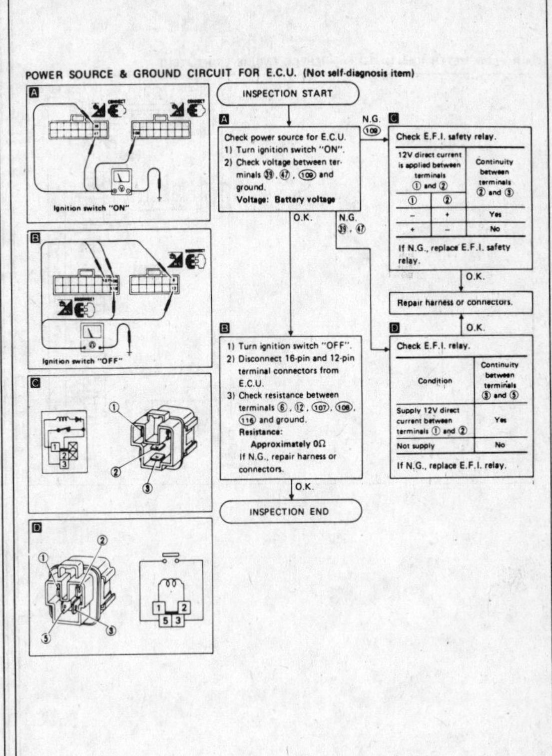
POWER SOURCE & GROUND CIRCUIT FOR E.C.U. (Not self-diagnosis item)
Ignition switch "ON"
Ignition switch "OFF"
INSPECTION START
A
Check power source for E.C.U.
1) Turn ignition switch "ON".
2) Check voltage between terminals 39, 47, 109 and ground.
Voltage: Battery voltage
N.G. 109
C
Check E.F.I. safety relay.
12V direct current is applied between terminals 1 and 2
Continuity between terminals 2 and 3
1 | 2
– | + | Yes
+ | – | No
If N.G., replace E.F.I. safety relay.
O.K.
Repair harness or connectors.
O.K.
N.G. 39, 47
D
Check E.F.I. relay.
Condition
Continuity between terminals 3 and 5
Supply 12V direct current between terminals 1 and 2 | Yes
Not supply | No
If N.G., replace E.F.I. relay.
O.K.
B
1) Turn ignition switch "OFF".
2) Disconnect 16-pin and 12-pin terminal connectors from E.C.U.
3) Check resistance between terminals 6, 12, 107, 108, 116 and ground.
Resistance:
Approximately 0Ω
If N.G., repair harness or connectors.
O.K.
INSPECTION END

ECCS SELF-DIAGNOSTIC CHART – 1989 PULSAR

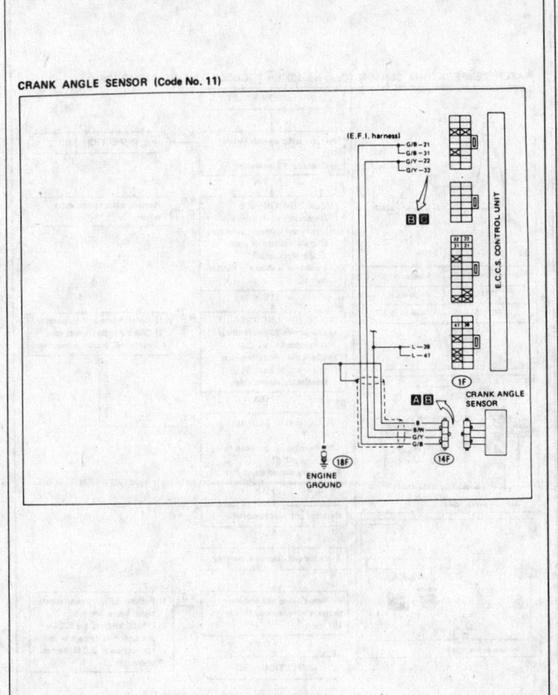
CRANK ANGLE SENSOR (Code No. 11)
(E.F.I. harness)
E.C.C.S. CONTROL UNIT
CRANK ANGLE SENSOR
ENGINE GROUND

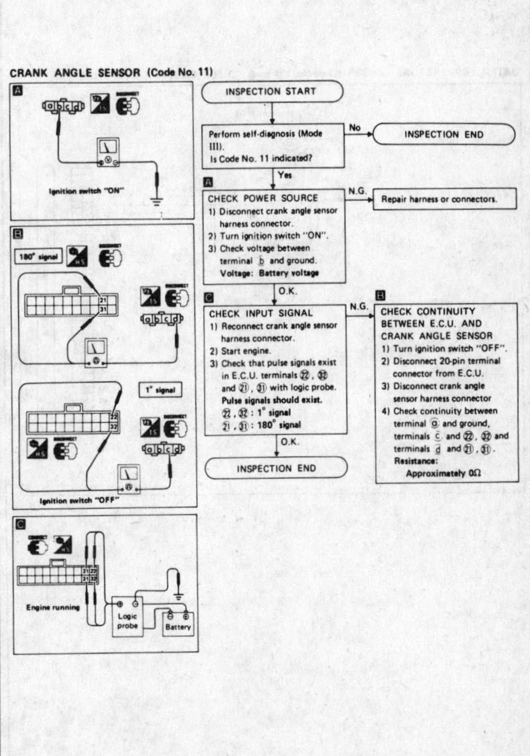
CRANK ANGLE SENSOR (Code No. 11)
Ignition switch "ON"
180° signal
1° signal
Ignition switch "OFF"
Engine running
Logic probe
Battery
INSPECTION START
Perform self-diagnosis (Mode III).
Is Code No. 11 indicated?
No
INSPECTION END
Yes
A
CHECK POWER SOURCE
1) Disconnect crank angle sensor harness connector.
2) Turn ignition switch "ON".
3) Check voltage between terminal b and ground.
Voltage: Battery voltage
N.G.
Repair harness or connectors.
O.K.
C
CHECK INPUT SIGNAL
1) Reconnect crank angle sensor harness connector.
2) Start engine.
3) Check that pulse signals exist in E.C.U. terminals 22, 32 and 21, 31 with logic probe.
Pulse signals should exist.
22, 32 : 1° signal
21, 31 : 180° signal
O.K.
INSPECTION END
N.G.
B
CHECK CONTINUITY BETWEEN E.C.U. AND CRANK ANGLE SENSOR
1) Turn ignition switch "OFF".
2) Disconnect 20-pin terminal connector from E.C.U.
3) Disconnect crank angle sensor harness connector
4) Check continuity between terminal a and ground, terminals c and 22, 32 and terminals d and 21, 31.
Resistance:
Approximately 0Ω

ECCS SELF-DIAGNOSTIC CHART—1989 PULSAR

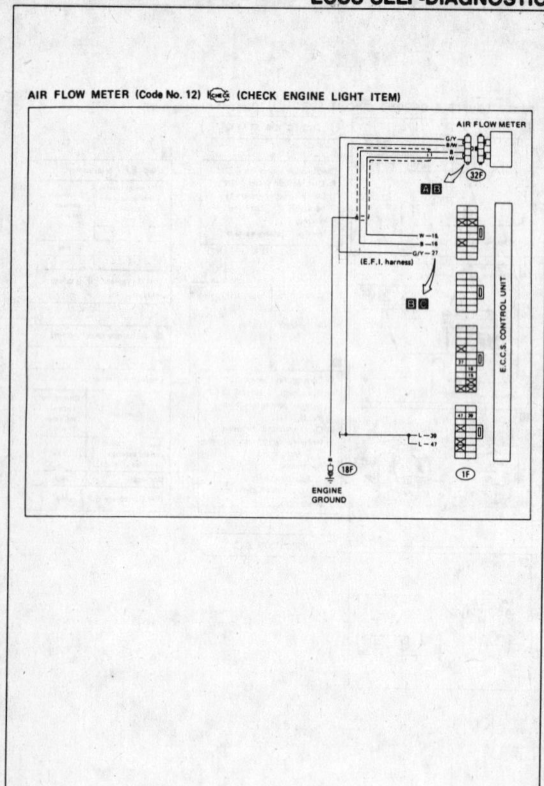
AIR FLOW METER (Code No. 12) (CHECK ENGINE LIGHT ITEM)
AIR FLOW METER
(E.F.I. harness)
E.C.C.S. CONTROL UNIT
ENGINE GROUND

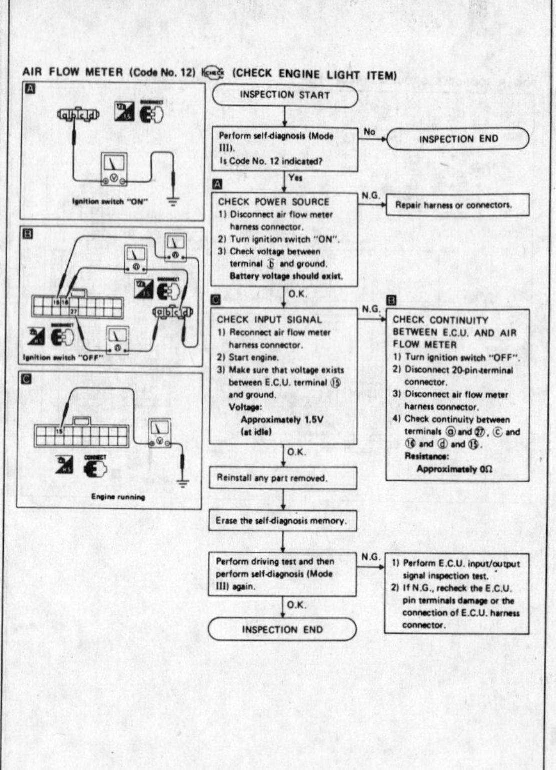
AIR FLOW METER (Code No. 12) (CHECK ENGINE LIGHT ITEM)
Ignition switch "ON"
Ignition switch "OFF"
Engine running
INSPECTION START
Perform self-diagnosis (Mode III).
Is Code No. 12 indicated?
No
INSPECTION END
Yes
CHECK POWER SOURCE
1) Disconnect air flow meter harness connector.
2) Turn ignition switch "ON".
3) Check voltage between terminal ⓑ and ground.
Battery voltage should exist.
N.G.
Repair harness or connectors.
O.K.
CHECK INPUT SIGNAL
1) Reconnect air flow meter harness connector.
2) Start engine.
3) Make sure that voltage exists between E.C.U. terminal ⑮ and ground.
Voltage:
Approximately 1.5V (at idle)
N.G.
CHECK CONTINUITY BETWEEN E.C.U. AND AIR FLOW METER
1) Turn ignition switch "OFF".
2) Disconnect 20-pin-terminal connector.
3) Disconnect air flow meter harness connector.
4) Check continuity between terminals ⓐ and ㉗, ⓒ and ⑯ and ⓓ and ⑮.
Resistance:
Approximately 0Ω
O.K.
Reinstall any part removed.
Erase the self-diagnosis memory.
Perform driving test and then perform self-diagnosis (Mode III) again.
N.G.
1) Perform E.C.U. input/output signal inspection test.
2) If N.G., recheck the E.C.U. pin terminals damage or the connection of E.C.U. harness connector.
O.K.
INSPECTION END

ECCS SELF-DIAGNOSTIC CHART—1989 PULSAR

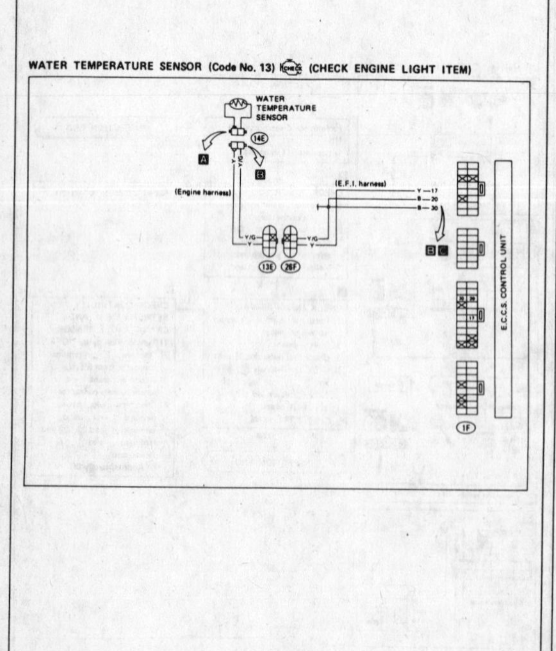
WATER TEMPERATURE SENSOR (Code No. 13) (CHECK ENGINE LIGHT ITEM)
WATER TEMPERATURE SENSOR
(Engine harness)
(E.F.I. harness)
E.C.C.S. CONTROL UNIT

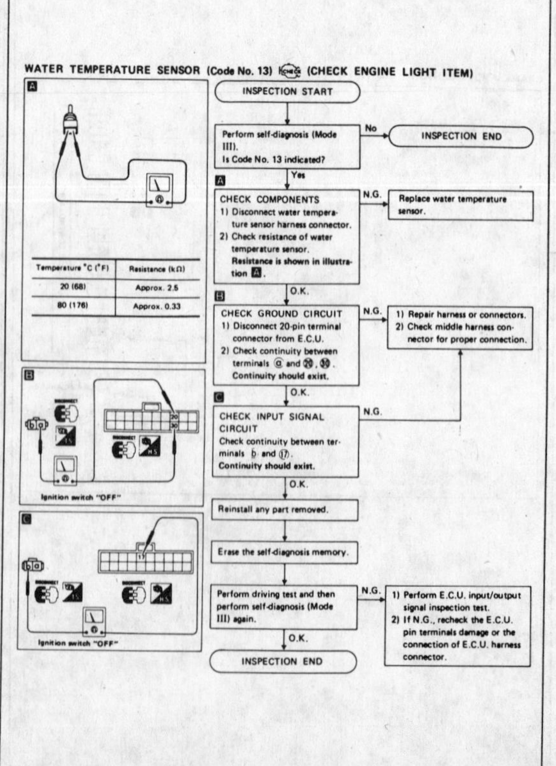
WATER TEMPERATURE SENSOR (Code No. 13) (CHECK ENGINE LIGHT ITEM)
Ignition switch "OFF"
INSPECTION START
Perform self-diagnosis (Mode III).
Is Code No. 13 indicated?
No
INSPECTION END
Yes
CHECK COMPONENTS
1) Disconnect water temperature sensor harness connector.
2) Check resistance of water temperature sensor.
Resistance is shown in illustration A.
N.G.
Replace water temperature sensor.
O.K.
CHECK GROUND CIRCUIT
1) Disconnect 20-pin terminal connector from E.C.U.
2) Check continuity between terminals ⓐ and ⑳, ㉚.
Continuity should exist.
N.G.
1) Repair harness or connectors.
2) Check middle harness connector for proper connection.
O.K.
CHECK INPUT SIGNAL CIRCUIT
Check continuity between terminals ⓑ and ⑰.
Continuity should exist.
N.G.
O.K.
Reinstall any part removed.
Erase the self-diagnosis memory.
Perform driving test and then perform self-diagnosis (Mode III) again.
N.G.
1) Perform E.C.U. input/output signal inspection test.
2) If N.G., recheck the E.C.U. pin terminals damage or the connection of E.C.U. harness connector.
O.K.
INSPECTION END

Temperature °C (°F)	Resistance (kΩ)
20 (68)	Approx. 2.5
80 (176)	Approx. 0.33

ECCS SELF-DIAGNOSTIC CHART—1989 PULSAR

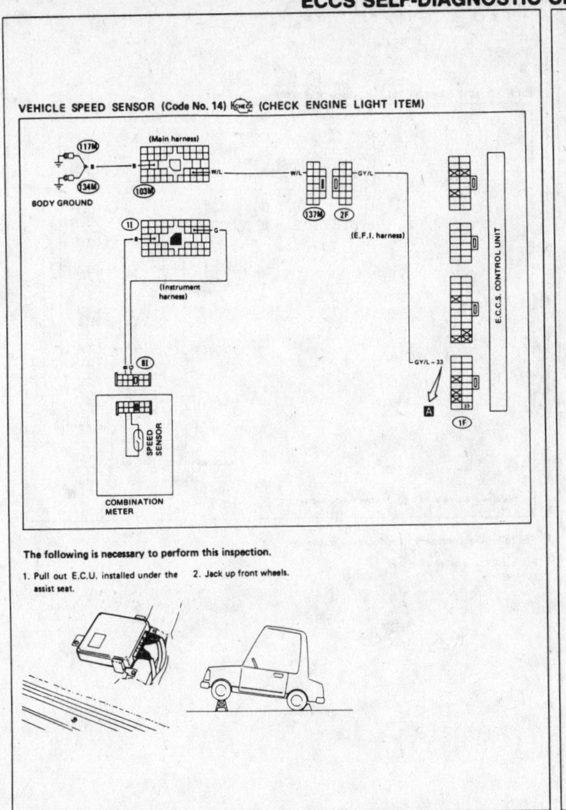
VEHICLE SPEED SENSOR (Code No. 14) (CHECK ENGINE LIGHT ITEM)
(Main harness)
BODY GROUND
(E.F.I. harness)
(Instrument harness)
SPEED SENSOR
COMBINATION METER
E.C.C.S. CONTROL UNIT
The following is necessary to perform this inspection.
1. Pull out E.C.U. installed under the assist seat.
2. Jack up front wheels.

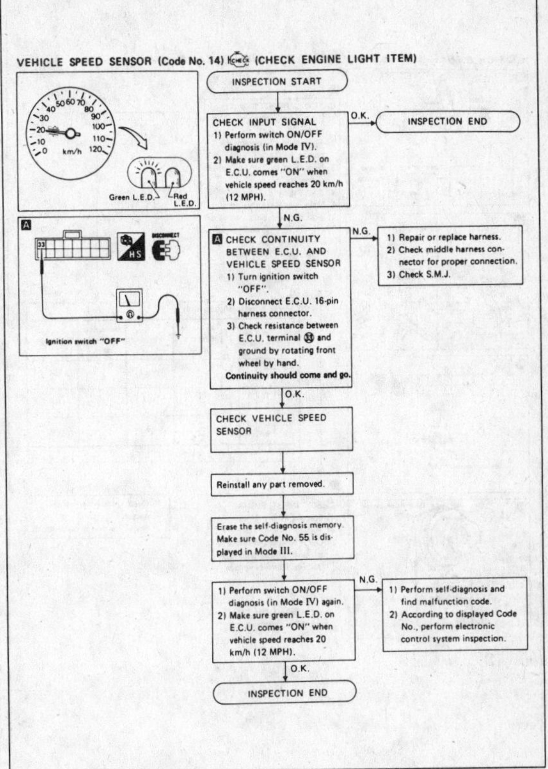
VEHICLE SPEED SENSOR (Code No. 14) (CHECK ENGINE LIGHT ITEM)
INSPECTION START
CHECK INPUT SIGNAL
1) Perform switch ON/OFF diagnosis (in Mode IV).
2) Make sure green L.E.D. on E.C.U. comes "ON" when vehicle speed reaches 20 km/h (12 MPH).
O.K.
INSPECTION END
N.G.
CHECK CONTINUITY BETWEEN E.C.U. AND VEHICLE SPEED SENSOR
1) Turn ignition switch "OFF".
2) Disconnect E.C.U. 16-pin harness connector.
3) Check resistance between E.C.U. terminal 33 and ground by rotating front wheel by hand.
Continuity should come and go.
N.G.
1) Repair or replace harness.
2) Check middle harness connector for proper connection.
3) Check S.M.J.
O.K.
CHECK VEHICLE SPEED SENSOR
Reinstall any part removed.
Erase the self-diagnosis memory. Make sure Code No. 55 is displayed in Mode III.
1) Perform switch ON/OFF diagnosis (in Mode IV) again.
2) Make sure green L.E.D. on E.C.U. comes "ON" when vehicle speed reaches 20 km/h (12 MPH).
N.G.
1) Perform self-diagnosis and find malfunction code.
2) According to displayed Code No., perform electronic control system inspection.
O.K.
INSPECTION END
Green L.E.D.
Red L.E.D.
Ignition switch "OFF"

ECCS SELF-DIAGNOSTIC CHART—1989 PULSAR

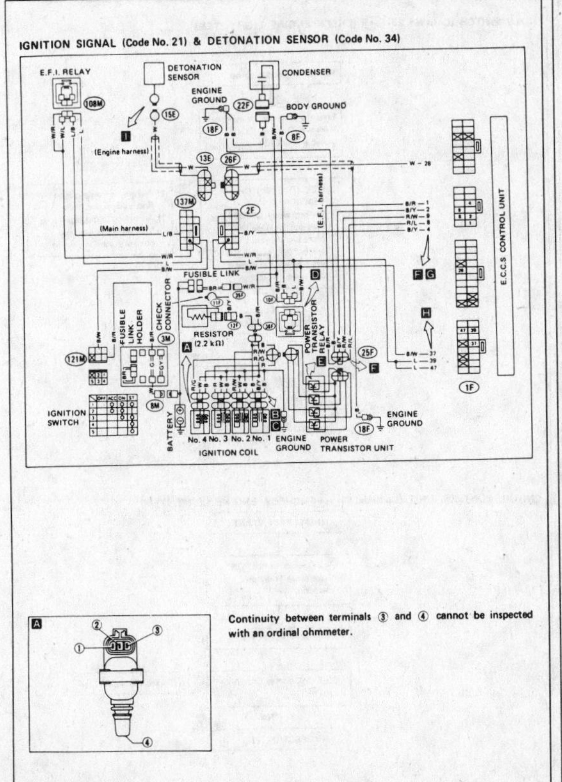
IGNITION SIGNAL (Code No. 21) & DETONATION SENSOR (Code No. 34)
E.F.I. RELAY
DETONATION SENSOR
CONDENSER
ENGINE GROUND
BODY GROUND
(Engine harness)
(Main harness)
(E.F.I. harness)
FUSIBLE LINK
FUSIBLE LINK HOLDER
CHECK CONNECTOR
RESISTOR (2.2 kΩ)
POWER TRANSISTOR RELAY
IGNITION SWITCH
BATTERY
No. 4 No. 3 No. 2 No. 1
IGNITION COIL
ENGINE GROUND
POWER TRANSISTOR UNIT
E.C.C.S. CONTROL UNIT
Continuity between terminals ③ and ④ cannot be inspected with an ordinal ohmmeter.

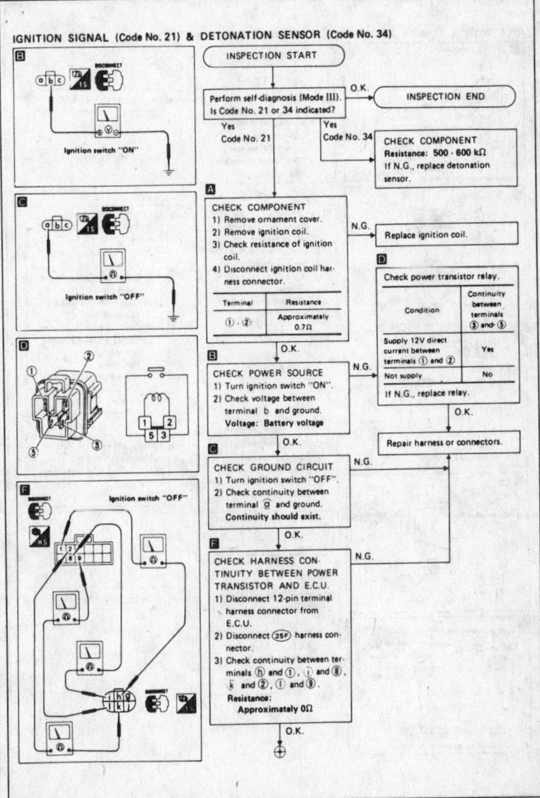
IGNITION SIGNAL (Code No. 21) & DETONATION SENSOR (Code No. 34)
INSPECTION START
Perform self-diagnosis (Mode III). Is Code No. 21 or 34 indicated?
O.K.
INSPECTION END
Yes Code No. 21
Yes Code No. 34
CHECK COMPONENT
Resistance: 500 - 600 kΩ
If N.G., replace detonation sensor.
CHECK COMPONENT
1) Remove ornament cover.
2) Remove ignition coil.
3) Check resistance of ignition coil.
4) Disconnect ignition coil harness connector.
Terminal | Resistance
① - ② | Approximately 0.7Ω
N.G.
Replace ignition coil.
Check power transistor relay.
Condition | Continuity between terminals ③ and ⑤
Supply 12V direct current between terminals ① and ② | Yes
Not supply | No
If N.G., replace relay.
O.K.
CHECK POWER SOURCE
1) Turn ignition switch "ON".
2) Check voltage between terminal ⓑ and ground.
Voltage: Battery voltage
N.G.
Repair harness or connectors.
O.K.
CHECK GROUND CIRCUIT
1) Turn ignition switch "OFF".
2) Check continuity between terminal ⑨ and ground.
Continuity should exist.
N.G.
O.K.
CHECK HARNESS CONTINUITY BETWEEN POWER TRANSISTOR AND E.C.U.
1) Disconnect 12-pin terminal harness connector from E.C.U.
2) Disconnect 25F harness connector.
3) Check continuity between terminals ⓗ and ①, ⓘ and ⑧, ⓚ and ②, ⓛ and ⑨.
Resistance: Approximately 0Ω
N.G.
O.K.
Ignition switch "ON"
Ignition switch "OFF"

ECCS SELF-DIAGNOSTIC CHART—1989 PULSAR

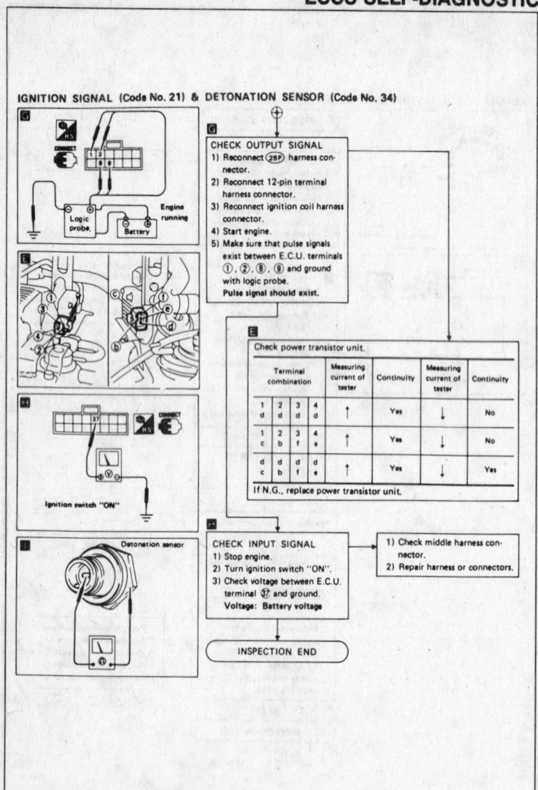
IGNITION SIGNAL (Code No. 21) & DETONATION SENSOR (Code No. 34)
G
Logic probe
Battery
Engine running
CHECK OUTPUT SIGNAL
1) Reconnect (26F) harness connector.
2) Reconnect 12-pin terminal harness connector.
3) Reconnect ignition coil harness connector.
4) Start engine.
5) Make sure that pulse signals exist between E.C.U. terminals ①, ②, ⑧, ⑨ and ground with logic probe.
Pulse signal should exist.
E
Check power transistor unit.
Terminal combination
Measuring current of tester
Continuity
Measuring current of tester
Continuity
1 2 3 4 / d d d d — ↑ — Yes — ↓ — No
1 2 3 4 / c b f e — ↑ — Yes — ↓ — No
d d d d / c b f e — ↑ — Yes — ↓ — Yes
If N.G., replace power transistor unit.
H
Ignition switch "ON"
J
Detonation sensor
CHECK INPUT SIGNAL
1) Stop engine.
2) Turn ignition switch "ON".
3) Check voltage between E.C.U. terminal ㊲ and ground.
Voltage: Battery voltage
1) Check middle harness connector.
2) Repair harness or connectors.
INSPECTION END

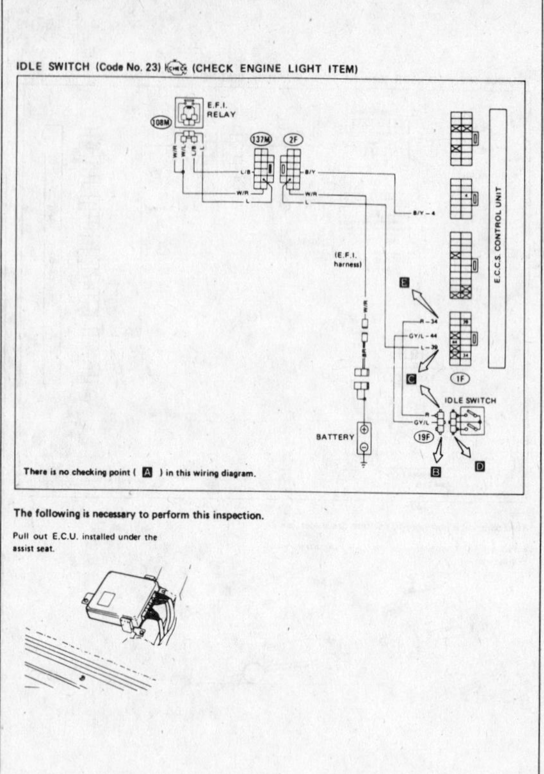
IDLE SWITCH (Code No. 23) (CHECK ENGINE LIGHT ITEM)
E.F.I. RELAY
108M
137M
2F
E.C.C.S. CONTROL UNIT
(E.F.I. harness)
BATTERY
IDLE SWITCH
1F
19F
There is no checking point (A) in this wiring diagram.
The following is necessary to perform this inspection.
Pull out E.C.U. installed under the assist seat.

ECCS SELF-DIAGNOSTIC CHART—1989 PULSAR

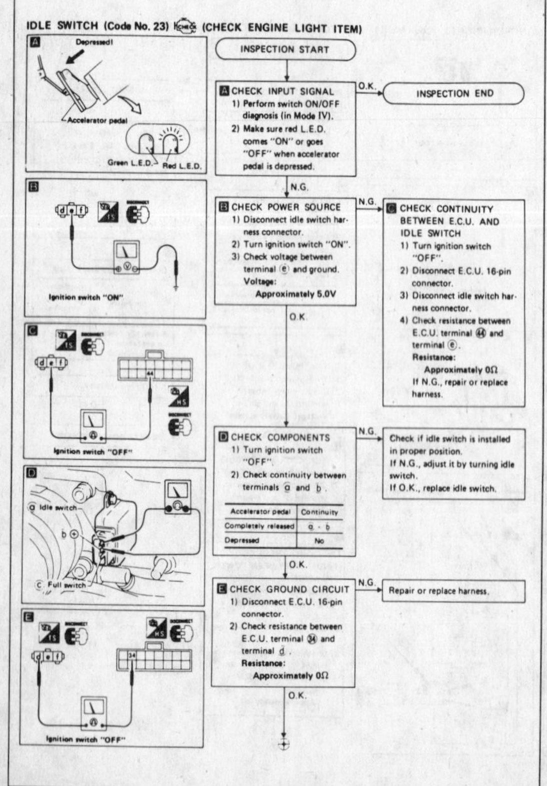
IDLE SWITCH (Code No. 23) (CHECK ENGINE LIGHT ITEM)
INSPECTION START
A
Depressed!
Accelerator pedal
Green L.E.D.
Red L.E.D.
A CHECK INPUT SIGNAL
1) Perform switch ON/OFF diagnosis (in Mode IV).
2) Make sure red L.E.D. comes "ON" or goes "OFF" when accelerator pedal is depressed.
O.K.
INSPECTION END
N.G.
B CHECK POWER SOURCE
1) Disconnect idle switch harness connector.
2) Turn ignition switch "ON".
3) Check voltage between terminal ⓔ and ground.
Voltage: Approximately 5.0V
N.G.
C CHECK CONTINUITY BETWEEN E.C.U. AND IDLE SWITCH
1) Turn ignition switch "OFF".
2) Disconnect E.C.U. 16-pin connector.
3) Disconnect idle switch harness connector.
4) Check resistance between E.C.U. terminal ㊹ and terminal ⓔ.
Resistance: Approximately 0Ω
If N.G., repair or replace harness.
Ignition switch "ON"
Ignition switch "OFF"
O.K.
D CHECK COMPONENTS
1) Turn ignition switch "OFF".
2) Check continuity between terminals ⓐ and ⓑ.
Accelerator pedal — Continuity
Completely released — a - b
Depressed — No
N.G.
Check if idle switch is installed in proper position.
If N.G., adjust it by turning idle switch.
If O.K., replace idle switch.
Idle switch
Full switch
O.K.
E CHECK GROUND CIRCUIT
1) Disconnect E.C.U. 16-pin connector.
2) Check resistance between E.C.U. terminal ㉞ and terminal ⓓ.
Resistance: Approximately 0Ω
N.G.
Repair or replace harness.
O.K.
Ignition switch "OFF"

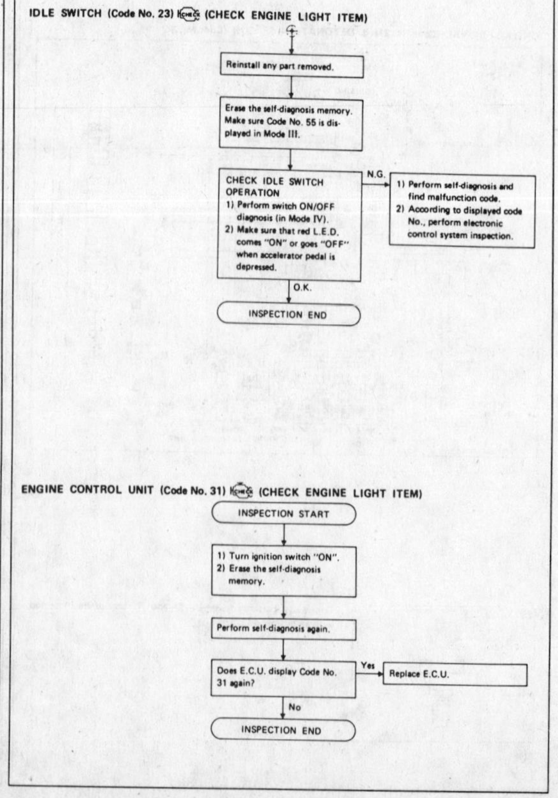
IDLE SWITCH (Code No. 23) (CHECK ENGINE LIGHT ITEM)
Reinstall any part removed.
Erase the self-diagnosis memory. Make sure Code No. 55 is displayed in Mode III.
CHECK IDLE SWITCH OPERATION
1) Perform switch ON/OFF diagnosis (in Mode IV).
2) Make sure that red L.E.D. comes "ON" or goes "OFF" when accelerator pedal is depressed.
N.G.
1) Perform self-diagnosis and find malfunction code.
2) According to displayed code No., perform electronic control system inspection.
O.K.
INSPECTION END
ENGINE CONTROL UNIT (Code No. 31) (CHECK ENGINE LIGHT ITEM)
INSPECTION START
1) Turn ignition switch "ON".
2) Erase the self-diagnosis memory.
Perform self-diagnosis again.
Does E.C.U. display Code No. 31 again?
Yes
Replace E.C.U.
No
INSPECTION END

ECCS SELF-DIAGNOSTIC CHART—1989 PULSAR

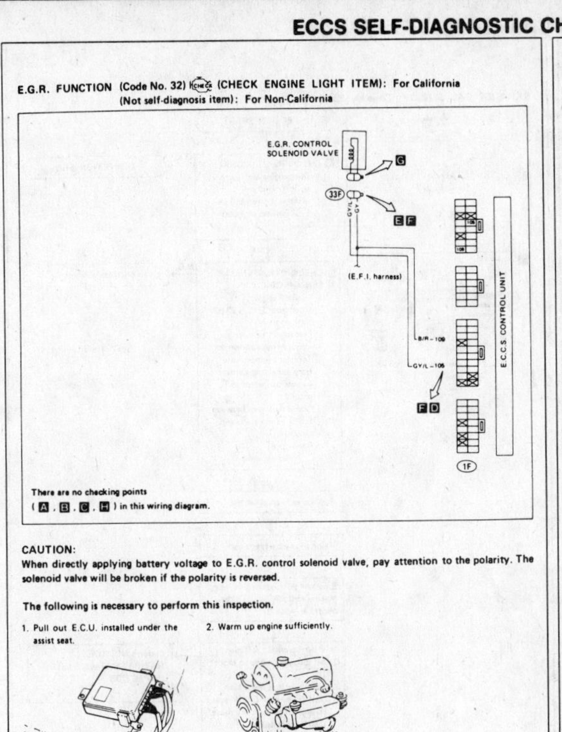
E.G.R. FUNCTION (Code No. 32) (CHECK ENGINE LIGHT ITEM): For California
(Not self-diagnosis item): For Non-California
E.G.R. CONTROL SOLENOID VALVE
G
E F
(E.F.I. harness)
B/R—106
GY/L—105
F D
E.C.C.S. CONTROL UNIT
1F
There are no checking points (A, B, C, H) in this wiring diagram.
CAUTION:
When directly applying battery voltage to E.G.R. control solenoid valve, pay attention to the polarity. The solenoid valve will be broken if the polarity is reversed.
The following is necessary to perform this inspection.
1. Pull out E.C.U. installed under the assist seat.
2. Warm up engine sufficiently.

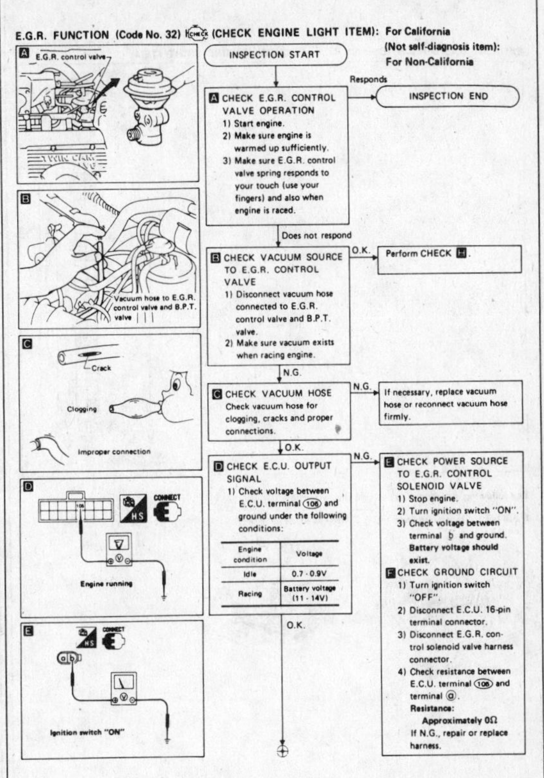
E.G.R. FUNCTION (Code No. 32) (CHECK ENGINE LIGHT ITEM): For California
(Not self-diagnosis item): For Non-California
A E.G.R. control valve
INSPECTION START
Responds
INSPECTION END
A CHECK E.G.R. CONTROL VALVE OPERATION
1) Start engine.
2) Make sure engine is warmed up sufficiently.
3) Make sure E.G.R. control valve spring responds to your touch (use your fingers) and also when engine is raced.
Does not respond
B
Vacuum hose to E.G.R. control valve and B.P.T. valve
B CHECK VACUUM SOURCE TO E.G.R. CONTROL VALVE
1) Disconnect vacuum hose connected to E.G.R. control valve and B.P.T. valve.
2) Make sure vacuum exists when racing engine.
O.K.
Perform CHECK H.
N.G.
C
Crack
Clogging
Improper connection
C CHECK VACUUM HOSE
Check vacuum hose for clogging, cracks and proper connections.
N.G.
If necessary, replace vacuum hose or reconnect vacuum hose firmly.
O.K.
D
CONNECT
Engine running
D CHECK E.C.U. OUTPUT SIGNAL
1) Check voltage between E.C.U. terminal 106 and ground under the following conditions:
Engine condition | Voltage
Idle | 0.7 - 0.9V
Racing | Battery voltage (11 - 14V)
O.K.
N.G.
E CHECK POWER SOURCE TO E.G.R. CONTROL SOLENOID VALVE
1) Stop engine.
2) Turn ignition switch "ON".
3) Check voltage between terminal b and ground. Battery voltage should exist.
F CHECK GROUND CIRCUIT
1) Turn ignition switch "OFF".
2) Disconnect E.C.U. 16-pin terminal connector.
3) Disconnect E.G.R. control solenoid valve harness connector.
4) Check resistance between E.C.U. terminal 106 and terminal a.
Resistance: Approximately 0Ω
If N.G., repair or replace harness.
E
CONNECT
Ignition switch "ON"

ECCS SELF-DIAGNOSTIC CHART—1989 PULSAR

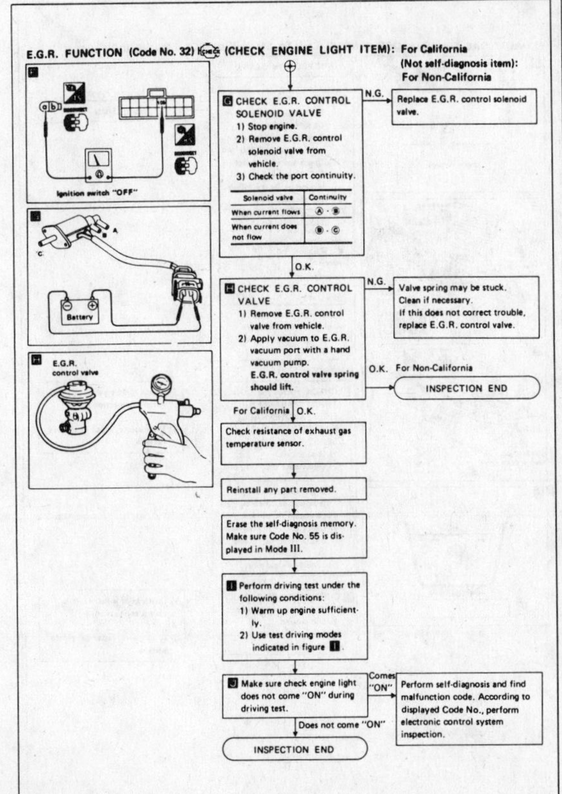
E.G.R. FUNCTION (Code No. 32) (CHECK ENGINE LIGHT ITEM): For California
(Not self-diagnosis item): For Non-California
F
Ignition switch "OFF"
G CHECK E.G.R. CONTROL SOLENOID VALVE
1) Stop engine.
2) Remove E.G.R. control solenoid valve from vehicle.
3) Check the port continuity.
Solenoid valve | Continuity
When current flows | A - B
When current does not flow | B - C
N.G.
Replace E.G.R. control solenoid valve.
O.K.
G
Battery
H E.G.R. control valve
H CHECK E.G.R. CONTROL VALVE
1) Remove E.G.R. control valve from vehicle.
2) Apply vacuum to E.G.R. vacuum port with a hand vacuum pump.
E.G.R. control valve spring should lift.
N.G.
Valve spring may be stuck. Clean if necessary. If this does not correct trouble, replace E.G.R. control valve.
O.K. For Non-California
INSPECTION END
For California O.K.
Check resistance of exhaust gas temperature sensor.
Reinstall any part removed.
Erase the self-diagnosis memory. Make sure Code No. 55 is displayed in Mode III.
I Perform driving test under the following conditions:
1) Warm up engine sufficiently.
2) Use test driving modes indicated in figure I.
J Make sure check engine light does not come "ON" during driving test.
Comes "ON"
Perform self-diagnosis and find malfunction code. According to displayed Code No., perform electronic control system inspection.
Does not come "ON"
INSPECTION END

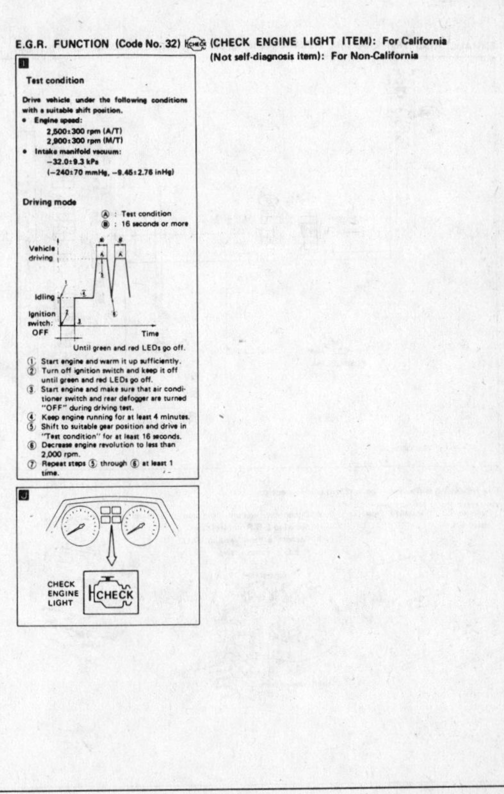
E.G.R. FUNCTION (Code No. 32) (CHECK ENGINE LIGHT ITEM): For California
(Not self-diagnosis item): For Non-California
I
Test condition
Drive vehicle under the following conditions with a suitable shift position.
• Engine speed:
2,500±300 rpm (A/T)
2,900±300 rpm (M/T)
• Intake manifold vacuum:
−32.0±9.3 kPa
(−240±70 mmHg, −9.45±2.76 inHg)
Driving mode
A : Test condition
B : 16 seconds or more
Vehicle driving
Idling
Ignition switch: OFF
Time
Until green and red LEDs go off.
1 Start engine and warm it up sufficiently.
2 Turn off ignition switch and keep it off until green and red LEDs go off.
3 Start engine and make sure that air conditioner switch and rear defogger are turned "OFF" during driving test.
4 Keep engine running for at least 4 minutes.
5 Shift to suitable gear position and drive in "Test condition" for at least 16 seconds.
6 Decrease engine revolution to less than 2,000 rpm.
7 Repeat steps 5 through 6 at least 1 time.
J
CHECK ENGINE LIGHT
CHECK

ECCS SELF-DIAGNOSTIC CHART—1989 PULSAR

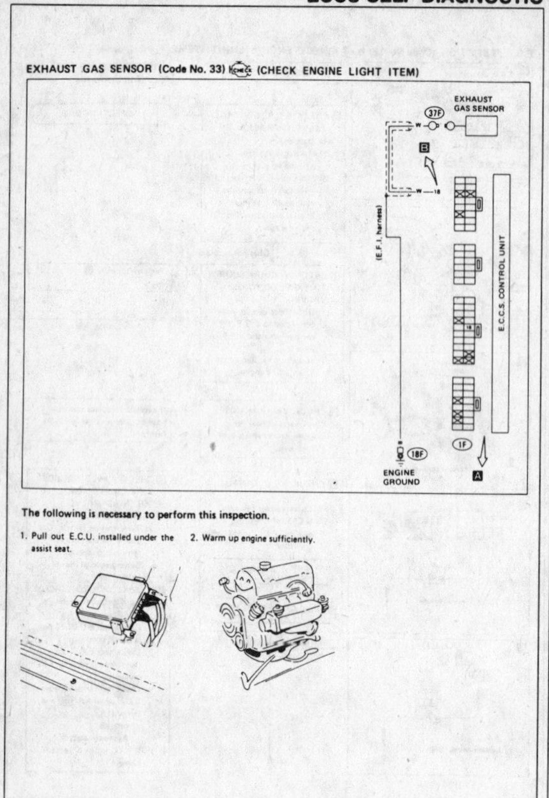
EXHAUST GAS SENSOR (Code No. 33) (CHECK ENGINE LIGHT ITEM)
EXHAUST GAS SENSOR
37F
(E.F.I. harness)
E.C.C.S. CONTROL UNIT
18F
1F
ENGINE GROUND
The following is necessary to perform this inspection.
1. Pull out E.C.U. installed under the assist seat.
2. Warm up engine sufficiently.

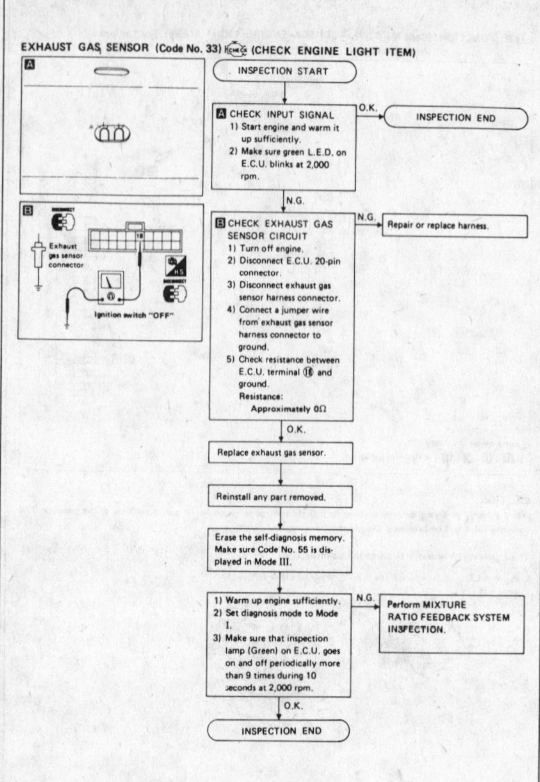
EXHAUST GAS SENSOR (Code No. 33) (CHECK ENGINE LIGHT ITEM)
INSPECTION START
A CHECK INPUT SIGNAL
1) Start engine and warm it up sufficiently.
2) Make sure green L.E.D. on E.C.U. blinks at 2,000 rpm.
O.K.
INSPECTION END
N.G.
B CHECK EXHAUST GAS SENSOR CIRCUIT
1) Turn off engine.
2) Disconnect E.C.U. 20-pin connector.
3) Disconnect exhaust gas sensor harness connector.
4) Connect a jumper wire from exhaust gas sensor harness connector to ground.
5) Check resistance between E.C.U. terminal ⑱ and ground.
Resistance: Approximately 0Ω
N.G.
Repair or replace harness.
O.K.
Replace exhaust gas sensor.
Reinstall any part removed.
Erase the self-diagnosis memory. Make sure Code No. 55 is displayed in Mode III.
1) Warm up engine sufficiently.
2) Set diagnosis mode to Mode I.
3) Make sure that inspection lamp (Green) on E.C.U. goes on and off periodically more than 9 times during 10 seconds at 2,000 rpm.
N.G.
Perform MIXTURE RATIO FEEDBACK SYSTEM INSPECTION.
O.K.
INSPECTION END
Exhaust gas sensor connector
Ignition switch "OFF"

ECCS SELF-DIAGNOSTIC CHART—1989 PULSAR

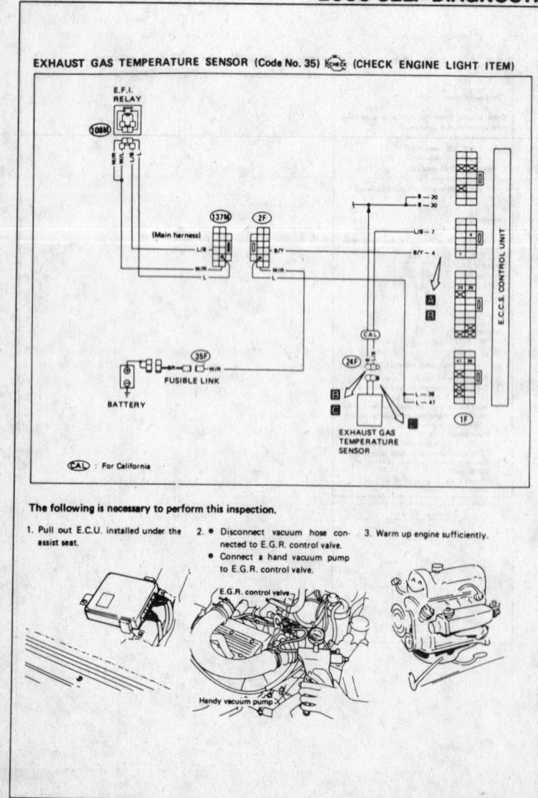
EXHAUST GAS TEMPERATURE SENSOR (Code No. 35) (CHECK ENGINE LIGHT ITEM)
E.F.I. RELAY
108M
137M
2F
(Main harness)
E.C.C.S. CONTROL UNIT
35F
FUSIBLE LINK
BATTERY
24F
1F
EXHAUST GAS TEMPERATURE SENSOR
CAL : For California
The following is necessary to perform this inspection.
1. Pull out E.C.U. installed under the assist seat.
2. • Disconnect vacuum hose connected to E.G.R. control valve.
• Connect a hand vacuum pump to E.G.R. control valve.
3. Warm up engine sufficiently.
E.G.R. control valve
Handy vacuum pump

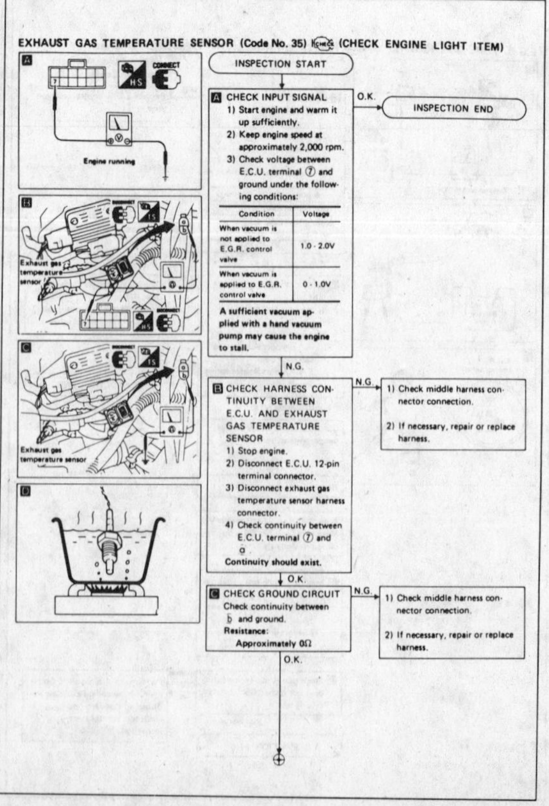
EXHAUST GAS TEMPERATURE SENSOR (Code No. 35) (CHECK ENGINE LIGHT ITEM)
INSPECTION START
A CHECK INPUT SIGNAL
1) Start engine and warm it up sufficiently.
2) Keep engine speed at approximately 2,000 rpm.
3) Check voltage between E.C.U. terminal ⑦ and ground under the following conditions:
Condition | Voltage
When vacuum is not applied to E.G.R. control valve | 1.0 - 2.0V
When vacuum is applied to E.G.R. control valve | 0 - 1.0V
A sufficient vacuum applied with a hand vacuum pump may cause the engine to stall.
O.K.
INSPECTION END
N.G.
B CHECK HARNESS CONTINUITY BETWEEN E.C.U. AND EXHAUST GAS TEMPERATURE SENSOR
1) Stop engine.
2) Disconnect E.C.U. 12-pin terminal connector.
3) Disconnect exhaust gas temperature sensor harness connector.
4) Check continuity between E.C.U. terminal ⑦ and a.
Continuity should exist.
N.G.
1) Check middle harness connector connection.
2) If necessary, repair or replace harness.
O.K.
C CHECK GROUND CIRCUIT
Check continuity between b and ground.
Resistance: Approximately 0Ω
N.G.
1) Check middle harness connector connection.
2) If necessary, repair or replace harness.
O.K.
Engine running
Exhaust gas temperature sensor

ECCS SELF-DIAGNOSTIC CHART—1989 PULSAR

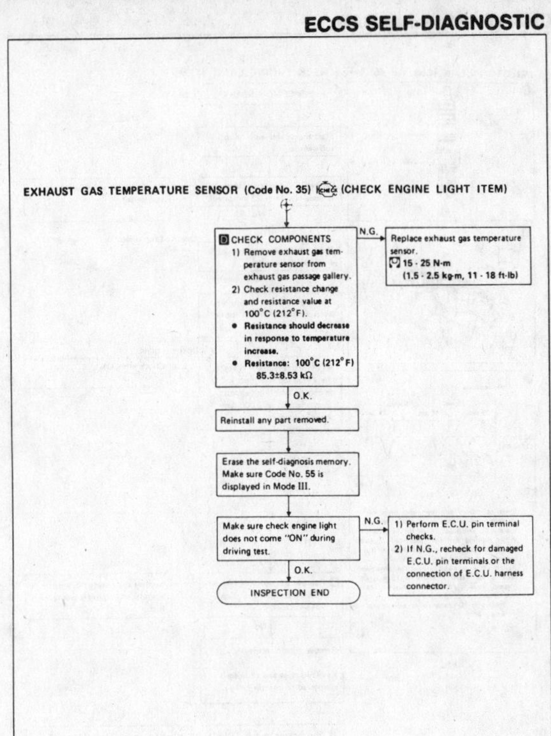
EXHAUST GAS TEMPERATURE SENSOR (Code No. 35) (CHECK ENGINE LIGHT ITEM)
D CHECK COMPONENTS
1) Remove exhaust gas temperature sensor from exhaust gas passage gallery.
2) Check resistance change and resistance value at 100°C (212°F).
• Resistance should decrease in response to temperature increase.
• Resistance: 100°C (212°F) 85.3±8.53 kΩ
N.G.
Replace exhaust gas temperature sensor.
15 - 25 N-m (1.5 - 2.5 kg-m, 11 - 18 ft-lb)
O.K.
Reinstall any part removed.
Erase the self-diagnosis memory. Make sure Code No. 55 is displayed in Mode III.
Make sure check engine light does not come "ON" during driving test.
N.G.
1) Perform E.C.U. pin terminal checks.
2) If N.G., recheck for damaged E.C.U. pin terminals or the connection of E.C.U. harness connector.
O.K.
INSPECTION END

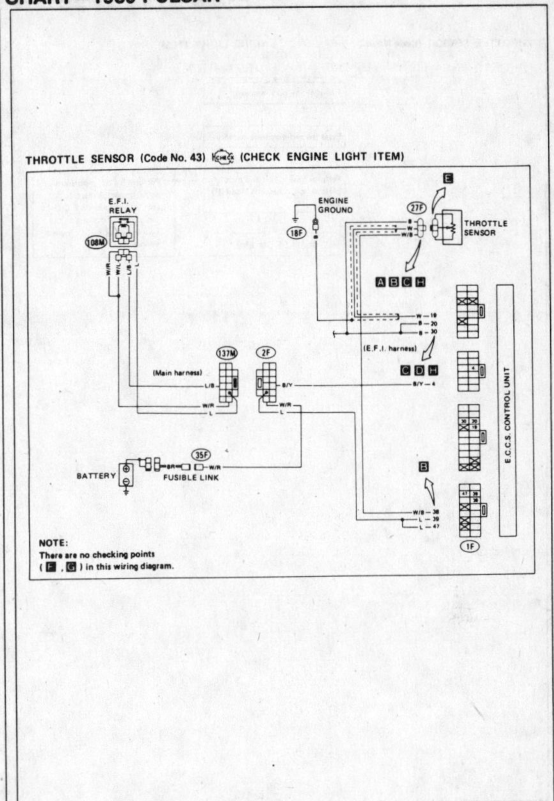
THROTTLE SENSOR (Code No. 43) (CHECK ENGINE LIGHT ITEM)
E.F.I. RELAY
ENGINE GROUND
THROTTLE SENSOR
(E.F.I. harness)
(Main harness)
BATTERY
FUSIBLE LINK
E.C.C.S. CONTROL UNIT
NOTE:
There are no checking points (F, G) in this wiring diagram.

ECCS SELF-DIAGNOSTIC CHART—1989 PULSAR

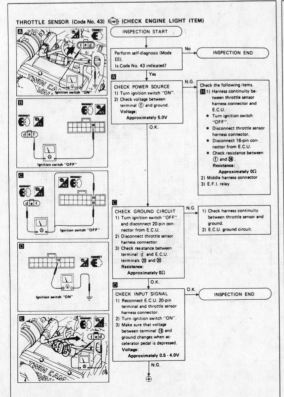

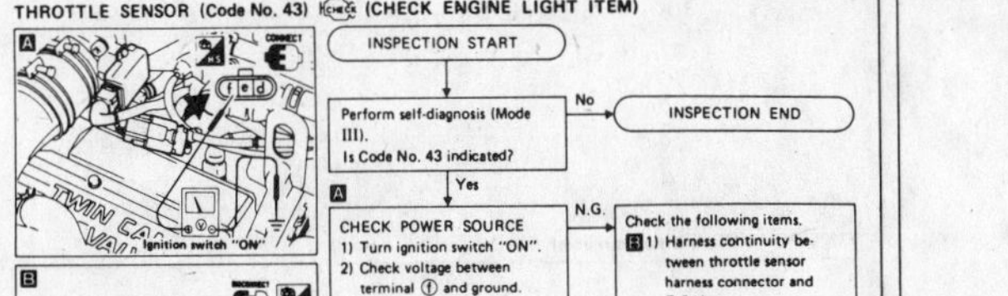
THROTTLE SENSOR (Code No. 43) (CHECK ENGINE LIGHT ITEM)
INSPECTION START
Perform self-diagnosis (Mode III). Is Code No. 43 indicated?
No
INSPECTION END
Yes
Ignition switch "ON"

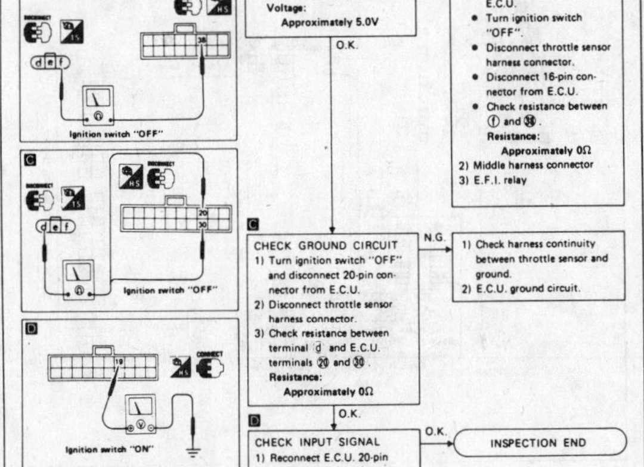
A CHECK POWER SOURCE
1) Turn ignition switch "ON".
2) Check voltage between terminal ① and ground.
Voltage: Approximately 5.0V
N.G.
Check the following items.
B 1) Harness continuity between throttle sensor harness connector and E.C.U.
• Turn ignition switch "OFF".
• Disconnect throttle sensor harness connector.
• Disconnect 16-pin connector from E.C.U.
• Check resistance between ① and ㊳.
Resistance: Approximately 0Ω
2) Middle harness connector
3) E.F.I. relay
O.K.
C CHECK GROUND CIRCUIT
1) Turn ignition switch "OFF" and disconnect 20-pin connector from E.C.U.
2) Disconnect throttle sensor harness connector.
3) Check resistance between terminal ⓓ and E.C.U. terminals ⑳ and ㉚.
Resistance: Approximately 0Ω
N.G.
1) Check harness continuity between throttle sensor and ground.
2) E.C.U. ground circuit.
O.K.
D CHECK INPUT SIGNAL
1) Reconnect E.C.U. 20-pin terminal and throttle sensor harness connector.
2) Turn ignition switch "ON".
3) Make sure that voltage between terminal ⑲ and ground changes when accelerator pedal is depressed.
Voltage: Approximately 0.5 - 4.0V
O.K.
INSPECTION END
N.G.
Ignition switch "OFF"
Ignition switch "ON"

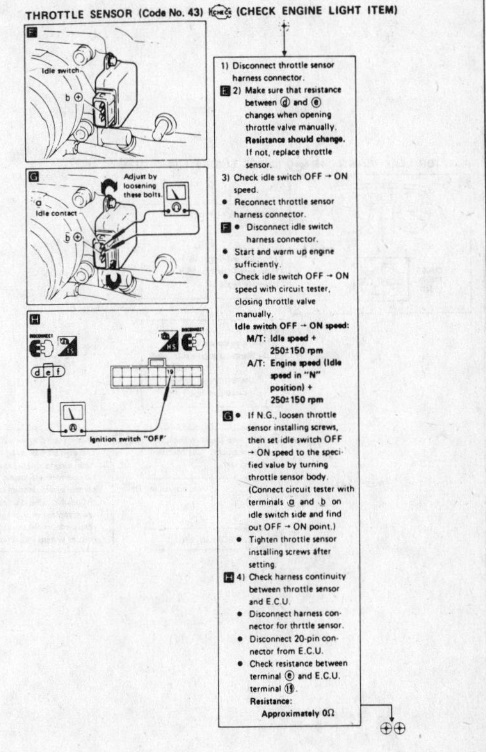
THROTTLE SENSOR (Code No. 43) (CHECK ENGINE LIGHT ITEM)
Idle switch
Adjust by loosening these bolts.
Idle contact
Ignition switch "OFF"
1) Disconnect throttle sensor harness connector.
E 2) Make sure that resistance between ⓓ and ⓔ changes when opening throttle valve manually.
Resistance should change.
If not, replace throttle sensor.
3) Check idle switch OFF → ON speed.
• Reconnect throttle sensor harness connector.
F • Disconnect idle switch harness connector.
• Start and warm up engine sufficiently.
• Check idle switch OFF → ON speed with circuit tester, closing throttle valve manually.
Idle switch OFF → ON speed:
M/T: Idle speed + 250±150 rpm
A/T: Engine speed (Idle speed in "N" position) + 250±150 rpm
G • If N.G., loosen throttle sensor installing screws, then set idle switch OFF → ON speed to the specified value by turning throttle sensor body. (Connect circuit tester with terminals ⓐ and ⓑ on idle switch side and find out OFF → ON point.)
• Tighten throttle sensor installing screws after setting.
H 4) Check harness continuity between throttle sensor and E.C.U.
• Disconnect harness connector for thrttle sensor.
• Disconnect 20-pin connector from E.C.U.
• Check resistance between terminal ⓔ and E.C.U. terminal ⑲.
Resistance: Approximately 0Ω

ECCS SELF-DIAGNOSTIC CHART—1989 PULSAR

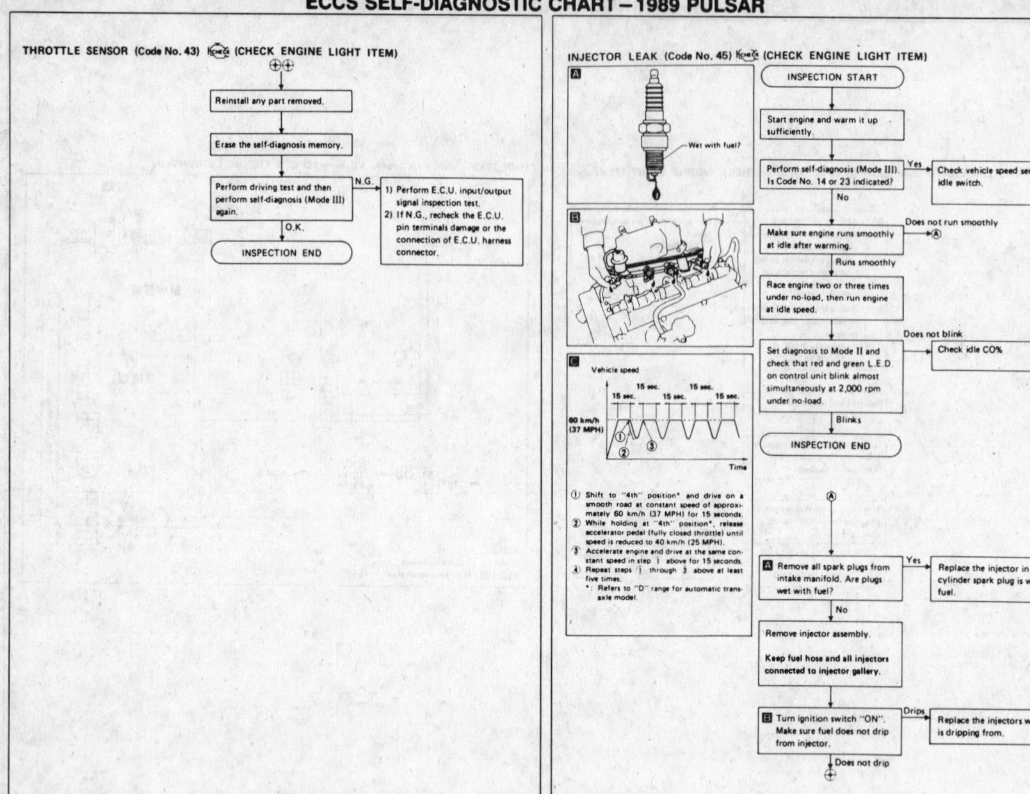
THROTTLE SENSOR (Code No. 43) (CHECK ENGINE LIGHT ITEM)
Reinstall any part removed.
Erase the self-diagnosis memory.
Perform driving test and then perform self-diagnosis (Mode III) again.
N.G.
1) Perform E.C.U. input/output signal inspection test.
2) If N.G., recheck the E.C.U. pin terminals damage or the connection of E.C.U. harness connector.
O.K.
INSPECTION END
INJECTOR LEAK (Code No. 45) (CHECK ENGINE LIGHT ITEM)
Wet with fuel?
Vehicle speed
15 sec.
60 km/h (37 MPH)
Time
① Shift to "4th" position* and drive on a smooth road at constant speed of approximately 60 km/h (37 MPH) for 15 seconds.
② While holding at "4th" position*, release accelerator pedal (fully closed throttle) until speed is reduced to 40 km/h (25 MPH).
③ Accelerate engine and drive at the same constant speed in step ① above for 15 seconds.
④ Repeat steps ① through ③ above at least five times.
*: Refers to "D" range for automatic transaxle model.
INSPECTION START
Start engine and warm it up sufficiently.
Perform self-diagnosis (Mode III). Is Code No. 14 or 23 indicated?
Yes
Check vehicle speed sensor or idle switch.
No
Make sure engine runs smoothly at idle after warming.
Does not run smoothly
Runs smoothly
Race engine two or three times under no-load, then run engine at idle speed.
Does not blink
Set diagnosis to Mode II and check that red and green L.E.D. on control unit blink almost simultaneously at 2,000 rpm under no-load.
Check idle CO%
Blinks
INSPECTION END
Remove all spark plugs from intake manifold. Are plugs wet with fuel?
Yes
Replace the injector in which cylinder spark plug is wet with fuel.
No
Remove injector assembly.
Keep fuel hose and all injectors connected to injector gallery.
Turn ignition switch "ON". Make sure fuel does not drip from injector.
Drips
Replace the injectors where fuel is dripping from.
Does not drip

ECCS SELF-DIAGNOSTIC CHART—1989 PULSAR

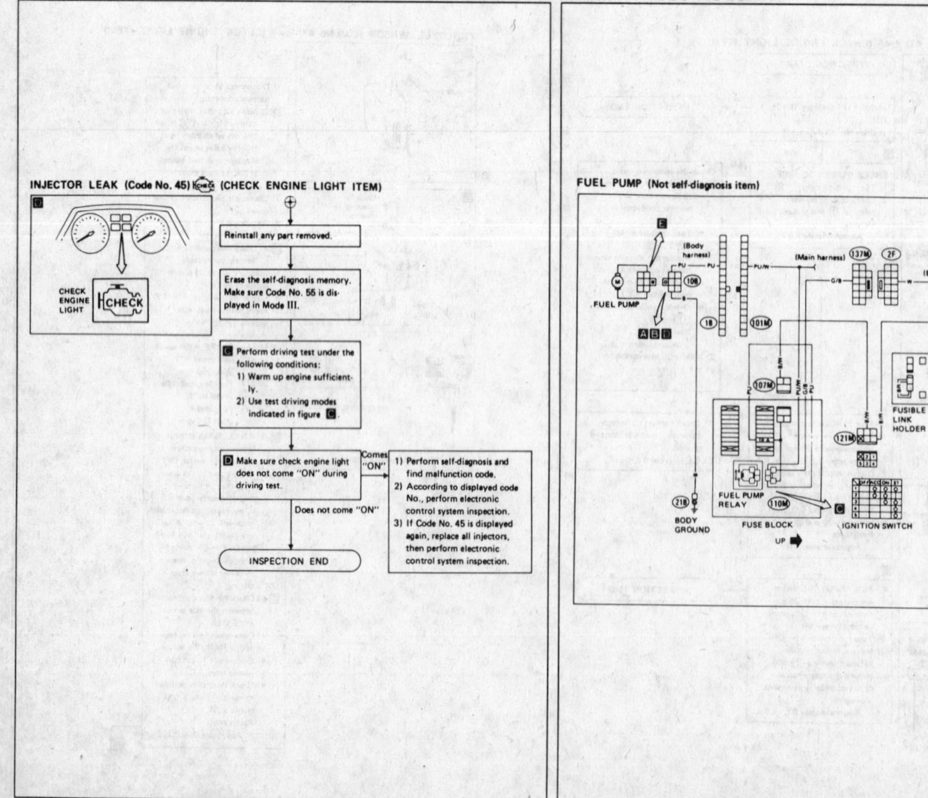
INJECTOR LEAK (Code No. 45) (CHECK ENGINE LIGHT ITEM)
CHECK ENGINE LIGHT
CHECK
Reinstall any part removed.
Erase the self-diagnosis memory. Make sure Code No. 55 is displayed in Mode III.
Perform driving test under the following conditions:
1) Warm up engine sufficiently.
2) Use test driving modes indicated in figure.
Make sure check engine light does not come "ON" during driving test.
Comes "ON"
1) Perform self-diagnosis and find malfunction code.
2) According to displayed code No., perform electronic control system inspection.
3) If Code No. 45 is displayed again, replace all injectors, then perform electronic control system inspection.
Does not come "ON"
INSPECTION END
FUEL PUMP (Not self-diagnosis item)
FUEL PUMP
(Body harness)
(Main harness)
(E.F.I. harness)
FUSIBLE LINK HOLDER
BATTERY
FUEL PUMP RELAY
BODY GROUND
FUSE BLOCK
UP
IGNITION SWITCH
E.C.C.S. CONTROL UNIT

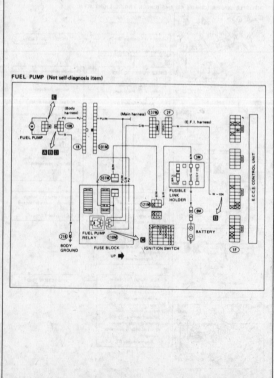

ECCS SELF-DIAGNOSTIC CHART—1989 PULSAR

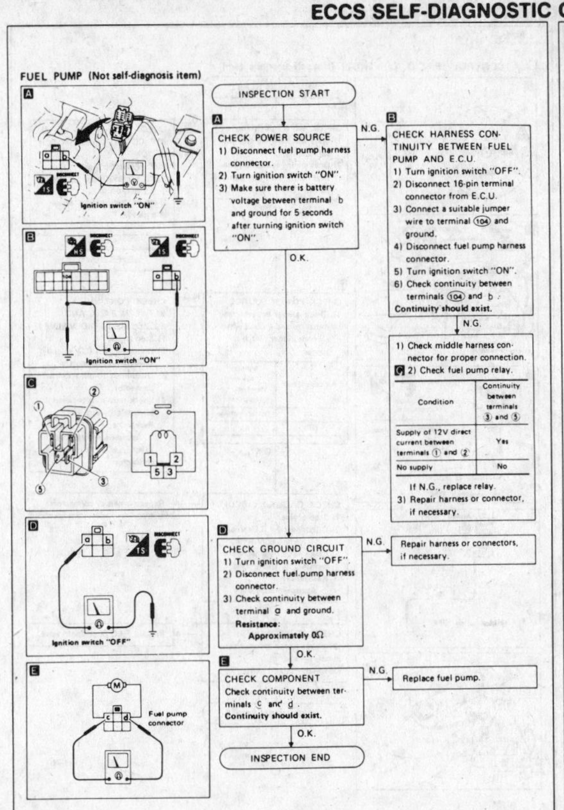
FUEL PUMP (Not self-diagnosis item)
Ignition switch "ON"
Ignition switch "ON"
Ignition switch "OFF"
Fuel pump connector
INSPECTION START
CHECK POWER SOURCE
1) Disconnect fuel pump harness connector.
2) Turn ignition switch "ON".
3) Make sure there is battery voltage between terminal b and ground for 5 seconds after turning ignition switch "ON".
N.G.
CHECK HARNESS CONTINUITY BETWEEN FUEL PUMP AND E.C.U.
1) Turn ignition switch "OFF".
2) Disconnect 16-pin terminal connector from E.C.U.
3) Connect a suitable jumper wire to terminal 104 and ground.
4) Disconnect fuel pump harness connector.
5) Turn ignition switch "ON".
6) Check continuity between terminals 104 and b.
Continuity should exist.
N.G.
1) Check middle harness connector for proper connection.
2) Check fuel pump relay.
Condition
Continuity between terminals 3 and 5
Supply of 12V direct current between terminals 1 and 2
Yes
No supply
No
If N.G., replace relay.
3) Repair harness or connector, if necessary.
O.K.
CHECK GROUND CIRCUIT
1) Turn ignition switch "OFF".
2) Disconnect fuel pump harness connector.
3) Check continuity between terminal a and ground.
Resistance: Approximately 0Ω
N.G.
Repair harness or connectors, if necessary.
O.K.
CHECK COMPONENT
Check continuity between terminals c and d.
Continuity should exist.
N.G.
Replace fuel pump.
O.K.
INSPECTION END

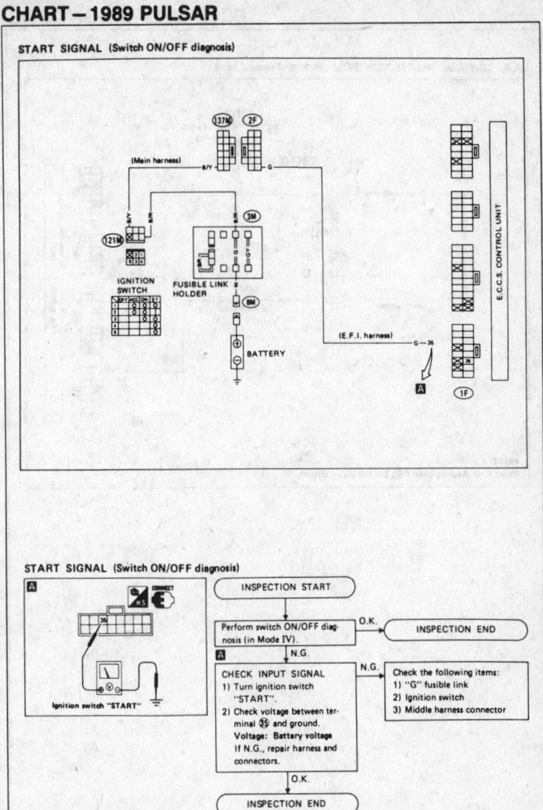
START SIGNAL (Switch ON/OFF diagnosis)
(Main harness)
IGNITION SWITCH
FUSIBLE LINK HOLDER
BATTERY
(E.F.I. harness)
E.C.C.S. CONTROL UNIT
START SIGNAL (Switch ON/OFF diagnosis)
Ignition switch "START"
INSPECTION START
Perform switch ON/OFF diagnosis (in Mode IV).
O.K.
INSPECTION END
N.G.
CHECK INPUT SIGNAL
1) Turn ignition switch "START".
2) Check voltage between terminal 35 and ground.
Voltage: Battery voltage
If N.G., repair harness and connectors.
N.G.
Check the following items:
1) "G" fusible link
2) Ignition switch
3) Middle harness connector
O.K.
INSPECTION END

ECCS SELF-DIAGNOSTIC CHART—1989 PULSAR

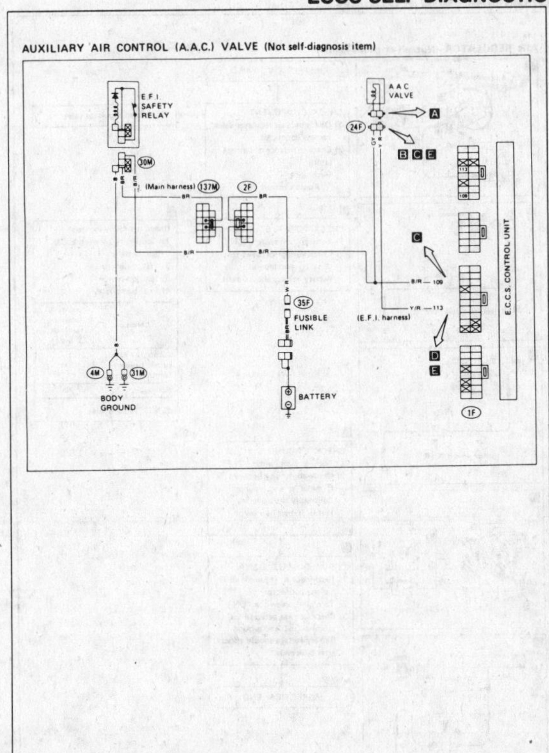
AUXILIARY AIR CONTROL (A.A.C.) VALVE (Not self-diagnosis item)
E.F.I. SAFETY RELAY
(Main harness)
A.A.C. VALVE
FUSIBLE LINK
BODY GROUND
BATTERY
(E.F.I. harness)
E.C.C.S. CONTROL UNIT

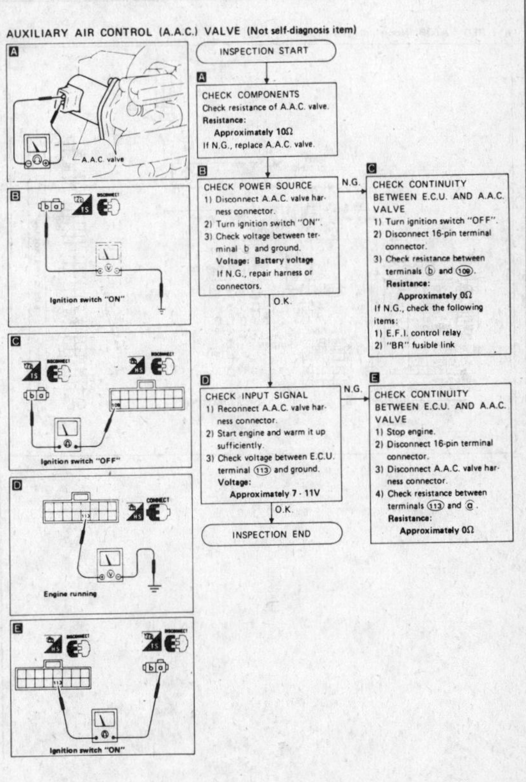
AUXILIARY AIR CONTROL (A.A.C.) VALVE (Not self-diagnosis item)
A.A.C. valve
Ignition switch "ON"
Ignition switch "OFF"
Engine running
Ignition switch "ON"
INSPECTION START
CHECK COMPONENTS
Check resistance of A.A.C. valve.
Resistance: Approximately 10Ω
If N.G., replace A.A.C. valve.
CHECK POWER SOURCE
1) Disconnect A.A.C. valve harness connector.
2) Turn ignition switch "ON".
3) Check voltage between terminal b and ground.
Voltage: Battery voltage
If N.G., repair harness or connectors.
N.G.
CHECK CONTINUITY BETWEEN E.C.U. AND A.A.C. VALVE
1) Turn ignition switch "OFF".
2) Disconnect 16-pin terminal connector.
3) Check resistance between terminals b and 109.
Resistance: Approximately 0Ω
If N.G., check the following items:
1) E.F.I. control relay
2) "BR" fusible link
O.K.
CHECK INPUT SIGNAL
1) Reconnect A.A.C. valve harness connector.
2) Start engine and warm it up sufficiently.
3) Check voltage between E.C.U. terminal 113 and ground.
Voltage: Approximately 7 - 11V
N.G.
CHECK CONTINUITY BETWEEN E.C.U. AND A.A.C. VALVE
1) Stop engine.
2) Disconnect 16-pin terminal connector.
3) Disconnect A.A.C. valve harness connector.
4) Check resistance between terminals 113 and a.
Resistance: Approximately 0Ω
O.K.
INSPECTION END

ECCS SELF-DIAGNOSTIC CHART—1989 PULSAR

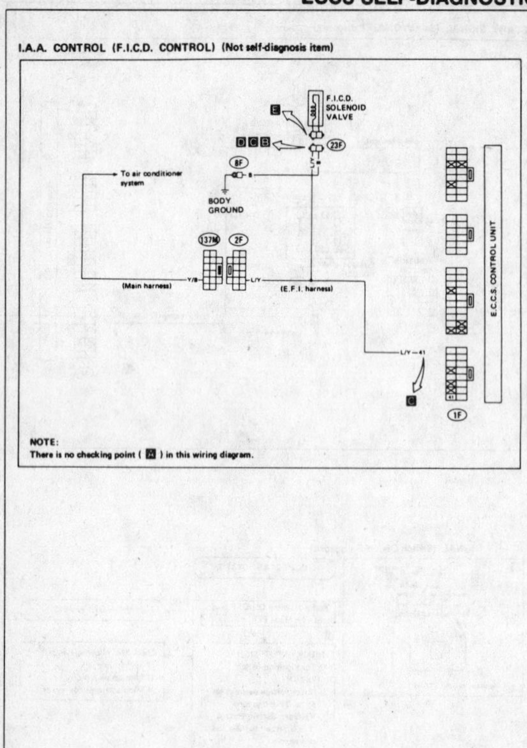
I.A.A. CONTROL (F.I.C.D. CONTROL) (Not self-diagnosis item)
F.I.C.D. SOLENOID VALVE
To air conditioner system
BODY GROUND
(Main harness)
(E.F.I. harness)
E.C.C.S. CONTROL UNIT
NOTE:
There is no checking point (A) in this wiring diagram.

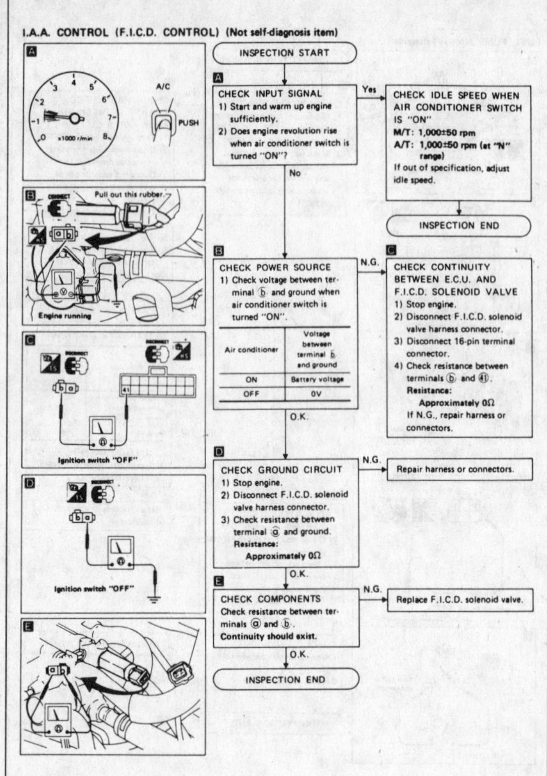
I.A.A. CONTROL (F.I.C.D. CONTROL) (Not self-diagnosis item)
INSPECTION START
A CHECK INPUT SIGNAL
1) Start and warm up engine sufficiently.
2) Does engine revolution rise when air conditioner switch is turned "ON"?
Yes
No
CHECK IDLE SPEED WHEN AIR CONDITIONER SWITCH IS "ON"
M/T: 1,000±50 rpm
A/T: 1,000±50 rpm (at "N" range)
If out of specification, adjust idle speed.
INSPECTION END
B CHECK POWER SOURCE
1) Check voltage between terminal ⓑ and ground when air conditioner switch is turned "ON".
Air conditioner | Voltage between terminal ⓑ and ground
ON | Battery voltage
OFF | 0V
N.G.
C CHECK CONTINUITY BETWEEN E.C.U. AND F.I.C.D. SOLENOID VALVE
1) Stop engine.
2) Disconnect F.I.C.D. solenoid valve harness connector.
3) Disconnect 16-pin terminal connector.
4) Check resistance between terminals ⓑ and ㊶.
Resistance: Approximately 0Ω
If N.G., repair harness or connectors.
O.K.
D CHECK GROUND CIRCUIT
1) Stop engine.
2) Disconnect F.I.C.D. solenoid valve harness connector.
3) Check resistance between terminal ⓐ and ground.
Resistance: Approximately 0Ω
N.G.
Repair harness or connectors.
O.K.
E CHECK COMPONENTS
Check resistance between terminals ⓐ and ⓑ.
Continuity should exist.
N.G.
Replace F.I.C.D. solenoid valve.
O.K.
INSPECTION END
A/C
PUSH
Pull out this rubber.
Engine running
Ignition switch "OFF"
Ignition switch "OFF"

ECCS SELF-DIAGNOSTIC CHART—1989 PULSAR

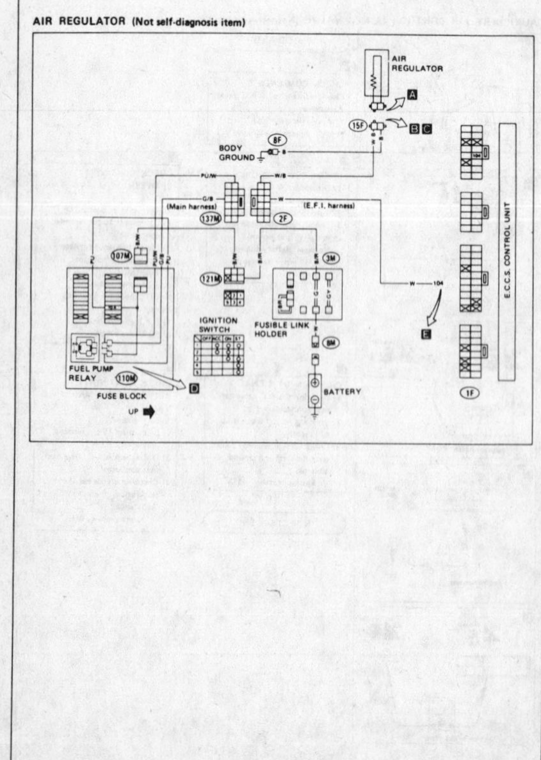
AIR REGULATOR (Not self-diagnosis item)
AIR REGULATOR
BODY GROUND
(Main harness)
(E.F.I. harness)
IGNITION SWITCH
FUSIBLE LINK HOLDER
FUEL PUMP RELAY
FUSE BLOCK
UP
BATTERY
E.C.C.S. CONTROL UNIT

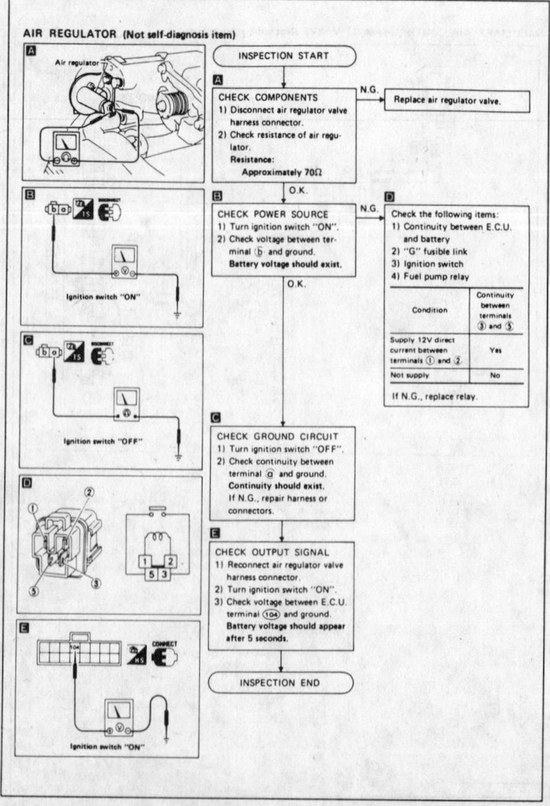
AIR REGULATOR (Not self-diagnosis item)
INSPECTION START
A CHECK COMPONENTS
1) Disconnect air regulator valve harness connector.
2) Check resistance of air regulator.
Resistance: Approximately 70Ω
N.G.
Replace air regulator valve.
O.K.
B CHECK POWER SOURCE
1) Turn ignition switch "ON".
2) Check voltage between terminal ⓑ and ground.
Battery voltage should exist.
N.G.
D Check the following items:
1) Continuity between E.C.U. and battery
2) "G" fusible link
3) Ignition switch
4) Fuel pump relay
Condition | Continuity between terminals ③ and ⑤
Supply 12V direct current between terminals ① and ② | Yes
Not supply | No
If N.G., replace relay.
O.K.
C CHECK GROUND CIRCUIT
1) Turn ignition switch "OFF".
2) Check continuity between terminal ⓐ and ground.
Continuity should exist.
If N.G., repair harness or connectors.
E CHECK OUTPUT SIGNAL
1) Reconnect air regulator valve harness connector.
2) Turn ignition switch "ON".
3) Check voltage between E.C.U. terminal 104 and ground.
Battery voltage should appear after 5 seconds.
INSPECTION END
Air regulator
Ignition switch "ON"
Ignition switch "OFF"
Ignition switch "ON"

ECCS SELF-DIAGNOSTIC CHART—1989 PULSAR

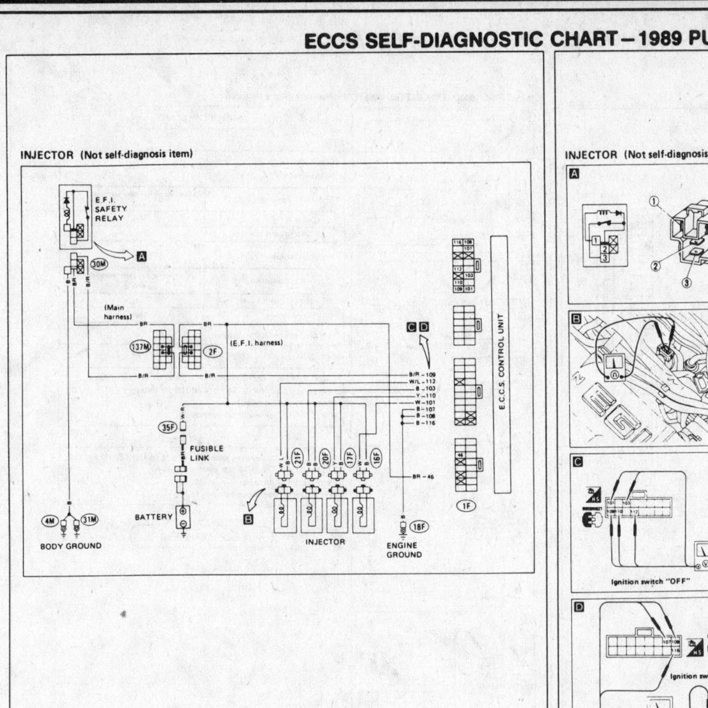
INJECTOR (Not self-diagnosis item)
E.F.I. SAFETY RELAY
(Main harness)
(E.F.I. harness)
FUSIBLE LINK
BATTERY
BODY GROUND
INJECTOR
ENGINE GROUND
E.C.C.S. CONTROL UNIT

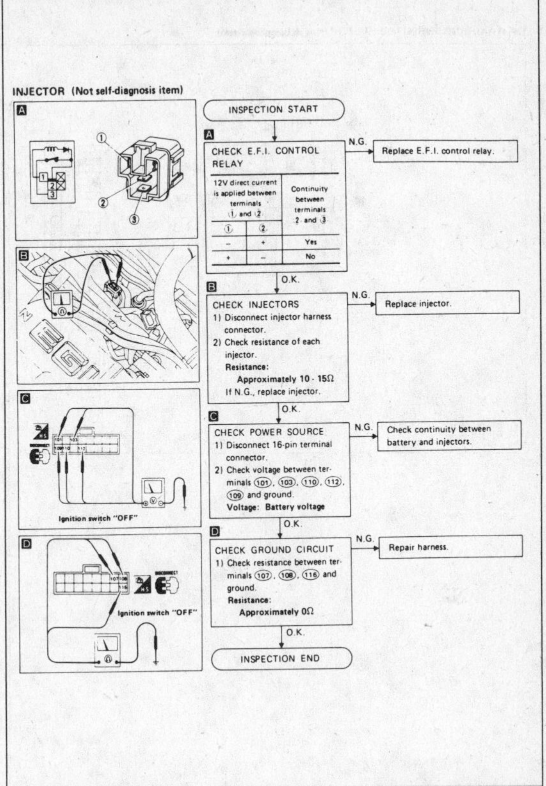
INJECTOR (Not self-diagnosis item)
INSPECTION START
CHECK E.F.I. CONTROL RELAY
12V direct current is applied between terminals ① and ②
Continuity between terminals ② and ③
Replace E.F.I. control relay.
CHECK INJECTORS
1) Disconnect injector harness connector.
2) Check resistance of each injector.
Resistance: Approximately 10 - 15Ω
If N.G., replace injector.
Replace injector.
CHECK POWER SOURCE
1) Disconnect 16-pin terminal connector.
2) Check voltage between terminals 101, 103, 110, 112, 109 and ground.
Voltage: Battery voltage
Check continuity between battery and injectors.
CHECK GROUND CIRCUIT
1) Check resistance between terminals 107, 108, 116 and ground.
Resistance: Approximately 0Ω
Repair harness.
INSPECTION END
Ignition switch "OFF"

ECCS SELF-DIAGNOSTIC CHART—1989 PULSAR

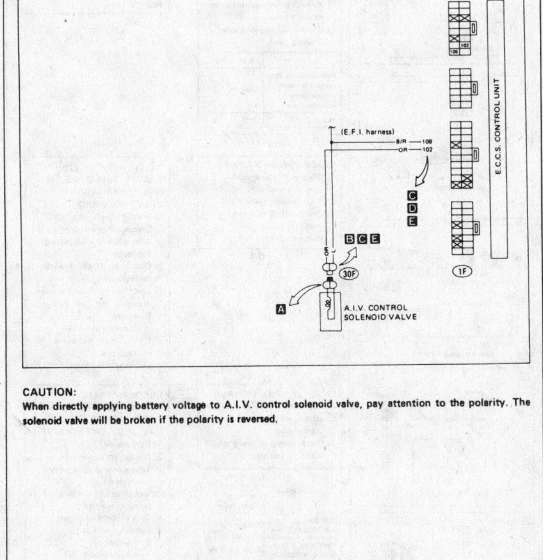
AIR INDUCTION VALVE (A.I.V.) CONTROL (Not self-diagnosis item)
(E.F.I. harness)
A.I.V. CONTROL SOLENOID VALVE
E.C.C.S. CONTROL UNIT

CAUTION:
When directly applying battery voltage to A.I.V. control solenoid valve, pay attention to the polarity. The solenoid valve will be broken if the polarity is reversed.

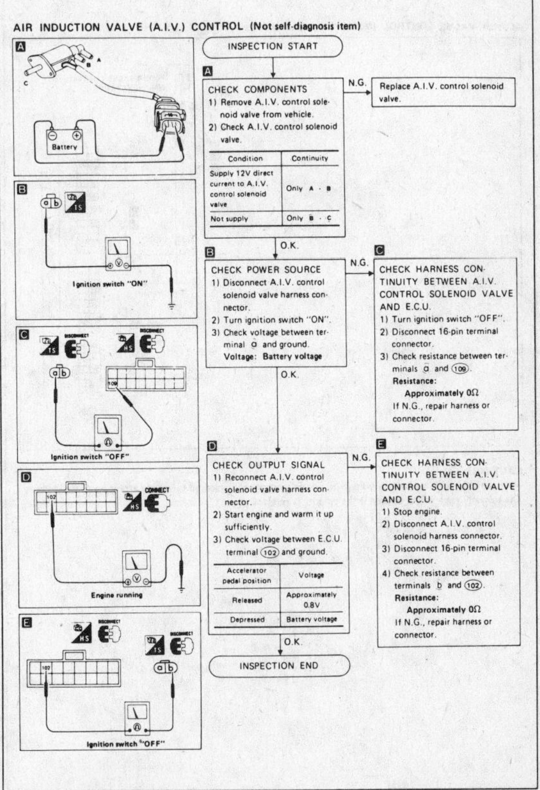
AIR INDUCTION VALVE (A.I.V.) CONTROL (Not self-diagnosis item)
INSPECTION START
CHECK COMPONENTS
1) Remove A.I.V. control solenoid valve from vehicle.
2) Check A.I.V. control solenoid valve.
Condition / Continuity
Supply 12V direct current to A.I.V. control solenoid valve / Only A - B
Not supply / Only B - C
Replace A.I.V. control solenoid valve.
CHECK POWER SOURCE
1) Disconnect A.I.V. control solenoid valve harness connector.
2) Turn ignition switch "ON".
3) Check voltage between terminal a and ground.
Voltage: Battery voltage
CHECK HARNESS CONTINUITY BETWEEN A.I.V. CONTROL SOLENOID VALVE AND E.C.U.
1) Turn ignition switch "OFF".
2) Disconnect 16-pin terminal connector.
3) Check resistance between terminals a and 109.
Resistance: Approximately 0Ω
If N.G., repair harness or connector.
CHECK OUTPUT SIGNAL
1) Reconnect A.I.V. control solenoid valve harness connector.
2) Start engine and warm it up sufficiently.
3) Check voltage between E.C.U. terminal 102 and ground.
Accelerator pedal position / Voltage
Released / Approximately 0.8V
Depressed / Battery voltage
CHECK HARNESS CONTINUITY BETWEEN A.I.V. CONTROL SOLENOID VALVE AND E.C.U.
1) Stop engine.
2) Disconnect A.I.V. control solenoid valve harness connector.
3) Disconnect 16-pin terminal connector.
4) Check resistance between terminals b and 102.
Resistance: Approximately 0Ω
If N.G., repair harness or connector.
INSPECTION END
Battery
Ignition switch "ON"
Ignition switch "OFF"
Engine running

ECCS SELF-DIAGNOSTIC CHART – 1989 PULSAR

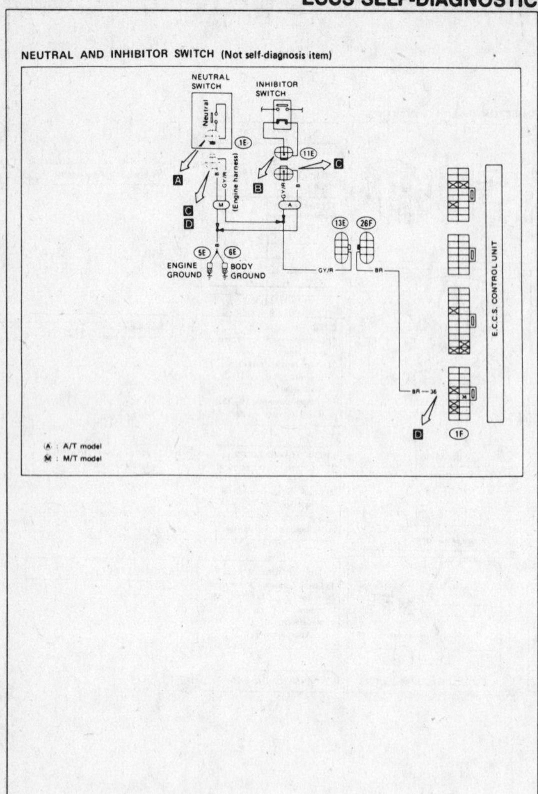
NEUTRAL AND INHIBITOR SWITCH (Not self-diagnosis item)
NEUTRAL SWITCH
INHIBITOR SWITCH
ENGINE GROUND
BODY GROUND
E.C.C.S. CONTROL UNIT
Ⓐ : A/T model
Ⓜ : M/T model

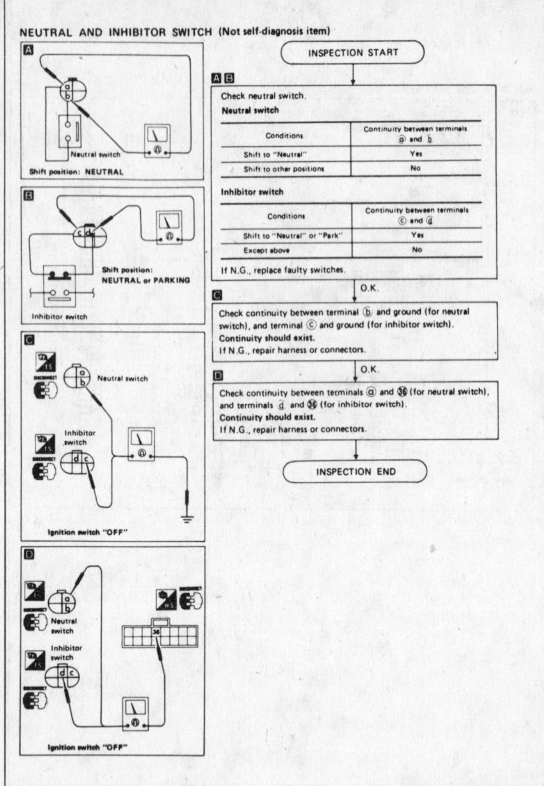
NEUTRAL AND INHIBITOR SWITCH (Not self-diagnosis item)
Neutral switch
Shift position: NEUTRAL
Shift position: NEUTRAL or PARKING
Inhibitor switch
Ignition switch "OFF"
INSPECTION START
Check neutral switch.
Neutral switch
Conditions | Continuity between terminals ⓐ and ⓑ
Shift to "Neutral" | Yes
Shift to other positions | No
Inhibitor switch
Conditions | Continuity between terminals ⓒ and ⓓ
Shift to "Neutral" or "Park" | Yes
Except above | No
If N.G., replace faulty switches.
O.K.
Check continuity between terminal ⓑ and ground (for neutral switch), and terminal ⓒ and ground (for inhibitor switch).
Continuity should exist.
If N.G., repair harness or connectors.
O.K.
Check continuity between terminals ⓐ and ㊱ (for neutral switch), and terminals ⓓ and ㊱ (for inhibitor switch).
Continuity should exist.
If N.G., repair harness or connectors.
INSPECTION END

ECCS SELF-DIAGNOSTIC CHART – 1989 PULSAR

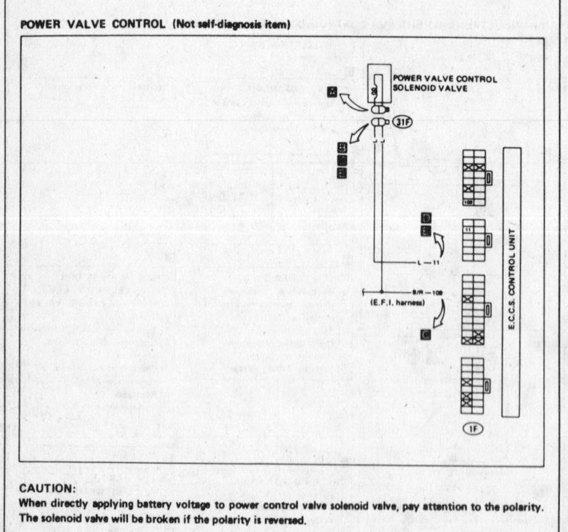
POWER VALVE CONTROL (Not self-diagnosis item)
POWER VALVE CONTROL SOLENOID VALVE
(E.F.I. harness)
E.C.C.S. CONTROL UNIT

CAUTION:
When directly applying battery voltage to power control valve solenoid valve, pay attention to the polarity. The solenoid valve will be broken if the polarity is reversed.

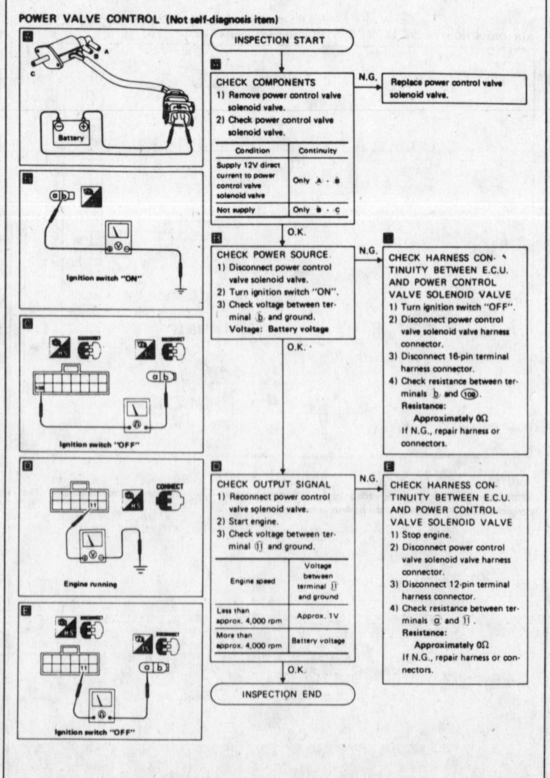
POWER VALVE CONTROL (Not self-diagnosis item)
Battery
Ignition switch "ON"
Ignition switch "OFF"
Engine running
INSPECTION START
CHECK COMPONENTS
1) Remove power control valve solenoid valve.
2) Check power control valve solenoid valve.
Condition | Continuity
Supply 12V direct current to power control valve solenoid valve | Only A · B
Not supply | Only B · C
N.G.
Replace power control valve solenoid valve.
O.K.
CHECK POWER SOURCE
1) Disconnect power control valve solenoid valve.
2) Turn ignition switch "ON".
3) Check voltage between terminal ⓑ and ground.
Voltage: Battery voltage
N.G.
CHECK HARNESS CONTINUITY BETWEEN E.C.U. AND POWER CONTROL VALVE SOLENOID VALVE
1) Turn ignition switch "OFF".
2) Disconnect power control valve solenoid valve harness connector.
3) Disconnect 16-pin terminal harness connector.
4) Check resistance between terminals ⓑ and ⑩⑨.
Resistance:
Approximately 0Ω
If N.G., repair harness or connectors.
O.K.
CHECK OUTPUT SIGNAL
1) Reconnect power control valve solenoid valve.
2) Start engine.
3) Check voltage between terminal ⑪ and ground.
Engine speed | Voltage between terminal ⑪ and ground
Less than approx. 4,000 rpm | Approx. 1V
More than approx. 4,000 rpm | Battery voltage
N.G.
CHECK HARNESS CONTINUITY BETWEEN E.C.U. AND POWER CONTROL VALVE SOLENOID VALVE
1) Stop engine.
2) Disconnect power control valve solenoid valve harness connector.
3) Disconnect 12-pin terminal harness connector.
4) Check resistance between terminals ⓐ and ⑪.
Resistance:
Approximately 0Ω
If N.G., repair harness or connectors.
O.K.
INSPECTION END

ECCS SELF-DIAGNOSTIC CHART—1989 PULSAR

ACCELERATION CUT CONTROL (Not self-diagnosis item)

(Main harness) (E.F.I. harness)

To thermo control amplifier
To air conditioner switch

137M 2F 143M 1F 19F 18F

ACCELERATION CUT RELAY

IDLE SWITCH

ENGINE GROUND

E.C.C.S. CONTROL UNIT

ECU SIGNAL INSPECTION—1989 PULSAR

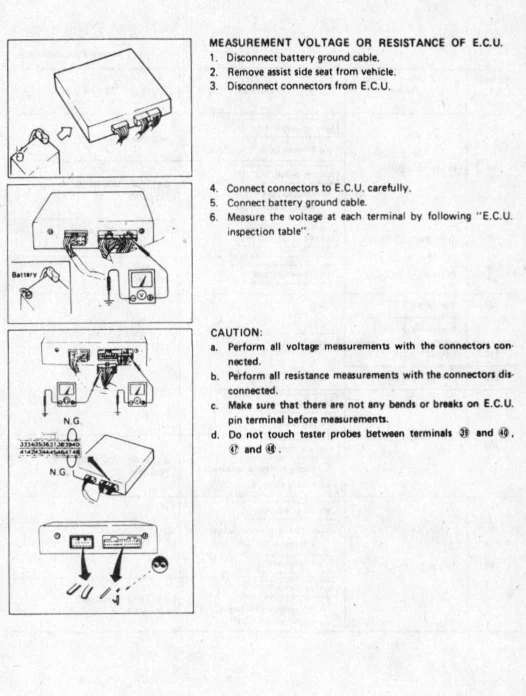
Battery
N.G.
N.G.

MEASUREMENT VOLTAGE OR RESISTANCE OF E.C.U.

1. Disconnect battery ground cable.
2. Remove assist side seat from vehicle.
3. Disconnect connectors from E.C.U.
4. Connect connectors to E.C.U. carefully.
5. Connect battery ground cable.
6. Measure the voltage at each terminal by following "E.C.U. inspection table".

CAUTION:

a. Perform all voltage measurements with the connectors connected.
b. Perform all resistance measurements with the connectors disconnected.
c. Make sure that there are not any bends or breaks on E.C.U. pin terminal before measurements.
d. Do not touch tester probes between terminals ㊴ and ㊵, ㊼ and ㊽.

ECU SIGNAL INSPECTION (CONT.)—1989 PULSAR

E.C.U. inspection table

*Data are reference values.

TERMINAL NO.	ITEM	CONDITION	*DATA
1	Ignition signal for No. 1 cylinder	Engine is running.	0 - 1.0V
2	Ignition signal for No. 2 cylinder		
8	Ignition signal for No. 3 cylinder		
9	Ignition signal for No. 4 cylinder		
7	Exhaust gas temperature sensor	Engine is running. — Idle speed	1.0 - 2.0V
		Engine is running. — E.G.R. system is operating.	0 - 1.0V
10	Air conditioner cut signal	Engine is running. — Idle speed	0 - 1.0V
		Engine is running. — Sudden racing	BATTERY VOLTAGE (11 - 14V)
11	Power valve control solenoid valve	Engine is running. — Engine speed is less than approximately 4,000 rpm.	Approximately 1V
		Engine is running. — Engine speed is more than approximately 4,000 rpm.	BATTERY VOLTAGE (11 - 14V)
15	Air flow meter	Engine is running. Do not run engine at high speed under no-load.	1.0 - 3.0V Output voltage varies with engine revolution.
17	Water temperature sensor	Engine is running.	1.0 - 5.0V Output voltage varies with engine water temperature.

*Data are reference values.

TERMINAL NO.	ITEM	CONDITION	*DATA
18	Exhaust gas sensor	Engine is running. — After warming up sufficiently	0 - Approximately 1.0V
19	Throttle sensor	Ignition switch "ON"	0.4 - 4.0V Output voltage varies with the throttle valve opening angle.
21 31	Crank angle sensor (Reference signal)	Engine is running. Do not run engine at high speed under no-load.	0.6 - 0.7V
22 32	Crank angle sensor (Position signal)	Engine is running. Do not run engine at high speed under no-load.	2.0 - 2.5V
34	Idle switch (⊖ side)	Ignition switch "ON" — Throttle valve: Idle position	Approximately 9 - 10V
		Ignition switch "ON" — Throttle valve: Any position except idle position	0V
35	Start signal	Cranking	8 - 12V
36	Neutral switch & Inhibitor switch	Ignition switch "ON" — Neutral/Parking	0V
		Ignition switch "ON" — Except the above gear position	BATTERY VOLTAGE (11 - 14V)
37	Ignition switch	Ignition switch "OFF"	0V
		Ignition switch "ON"	BATTERY VOLTAGE (11 - 14V)
41	Air conditioner	Engine is running. — Both air conditioner switch and blower switch are "ON".	BATTERY VOLTAGE (11 - 14V)

ECU SIGNAL INSPECTION (CONT.)—1989 PULSAR

*Data are reference values.

TERMINAL NO.	ITEM	CONDITION	*DATA
44	Idle switch (⊕ side)	Ignition switch "ON" — Throttle valve: Idle position	Approximately 9 - 11V
		Ignition switch "ON" — Throttle valve: Except idle position	BATTERY VOLTAGE (11 - 14V)
46	Power source (Back-up)	Ignition switch "OFF"	BATTERY VOLTAGE (11 - 14V)
39 47	Power source for E.C.U.	Ignition switch "ON"	BATTERY VOLTAGE (11 - 14V)
101	Injector No. 1	Engine is running.	BATTERY VOLTAGE (11 - 14V)
103	Injector No. 3		
110	Injector No. 2		
112	Injector No. 4		
102	A.I.V. control solenoid valve	Engine is running. — Idle speed	0 - 1.0V
		Engine is running. — Accelerator pedal is depressed. — After warming up	BATTERY VOLTAGE (11 - 14V)
104	Fuel pump relay	Ignition switch "ON" — For 5 seconds after turning ignition switch "ON"	0.7 - 0.9V
		Engine is running.	
		Ignition switch "ON" — In 5 seconds after turning ignition switch "ON"	BATTERY VOLTAGE (11 - 14V)

*Data are reference values

TERMINAL NO.	ITEM	CONDITION	*DATA
105	E.G.R. control solenoid valve	Engine is running. — Engine is cold. [Water temperature is below 60°C (140°F).]	0.7 - 0.9V
		Engine is running. — After warming up [Water temperature is between 65°C (149°F) and 105°C (221°F).]	BATTERY VOLTAGE (11 - 14V)
113	A.A.C. valve	Engine is running. — Idle speed	8 - 10V
		Engine is running. — Steering wheel is turned. — Air conditioner is operating. — Rear defogger is "ON". — Head lights are in high position.	7 - 8V

E.C.U. PIN CONNECTOR TERMINAL LAYOUT

101	102	103	104	105	⊠	107	108
109	110	⊠	112	113	⊠	115	116

1	2	3	4	5	6
7	8	9	10	11	12

⊠	⊠	15	16	17	18	19	20	21	22
⊠	24	25	26	27	28	⊠	30	31	32

33	34	35	36	37	38	39	40
41	⊠	⊠	44	⊠	46	47	48

ECCS MIXTURE RATIO FEEDBACK SYSTEM INSPECTION—1989 PULSAR

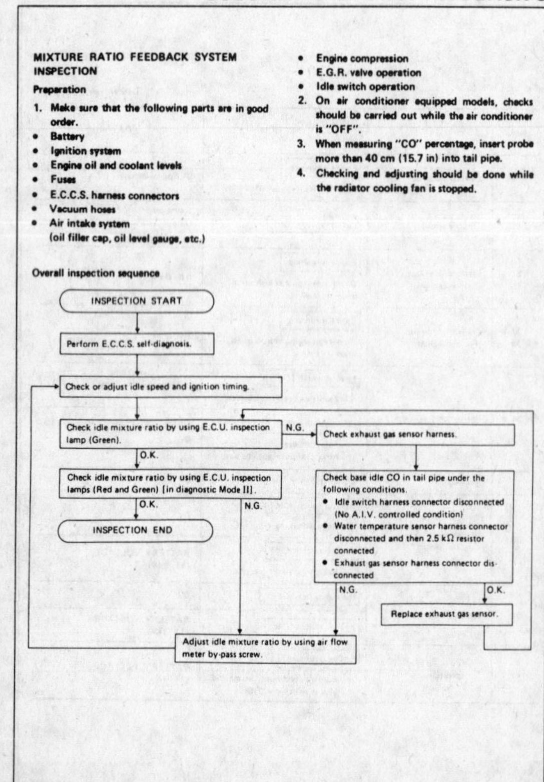

MIXTURE RATIO FEEDBACK SYSTEM INSPECTION

Preparation

1. **Make sure that the following parts are in good order.**
   - **Battery**
   - **Ignition system**
   - **Engine oil and coolant levels**
   - **Fuses**
   - **E.C.C.S. harness connectors**
   - **Vacuum hoses**
   - **Air intake system (oil filler cap, oil level gauge, etc.)**
   - **Engine compression**
   - **E.G.R. valve operation**
   - **Idle switch operation**
2. **On air conditioner equipped models, checks should be carried out while the air conditioner is "OFF".**
3. **When measuring "CO" percentage, insert probe more than 40 cm (15.7 in) into tail pipe.**
4. **Checking and adjusting should be done while the radiator cooling fan is stopped.**

Overall inspection sequence

INSPECTION START
Perform E.C.C.S. self-diagnosis.
Check or adjust idle speed and ignition timing.
Check idle mixture ratio by using E.C.U. inspection lamp (Green).
N.G.
Check exhaust gas sensor harness.
O.K.
Check idle mixture ratio by using E.C.U. inspection lamps (Red and Green) [in diagnostic Mode II].
O.K.
N.G.
INSPECTION END
Check base idle CO in tail pipe under the following conditions.
• Idle switch harness connector disconnected (No A.I.V. controlled condition)
• Water temperature sensor harness connector disconnected and then 2.5 kΩ resistor connected
• Exhaust gas sensor harness connector dis-connected
N.G.
O.K.
Replace exhaust gas sensor.
Adjust idle mixture ratio by using air flow meter by-pass screw.

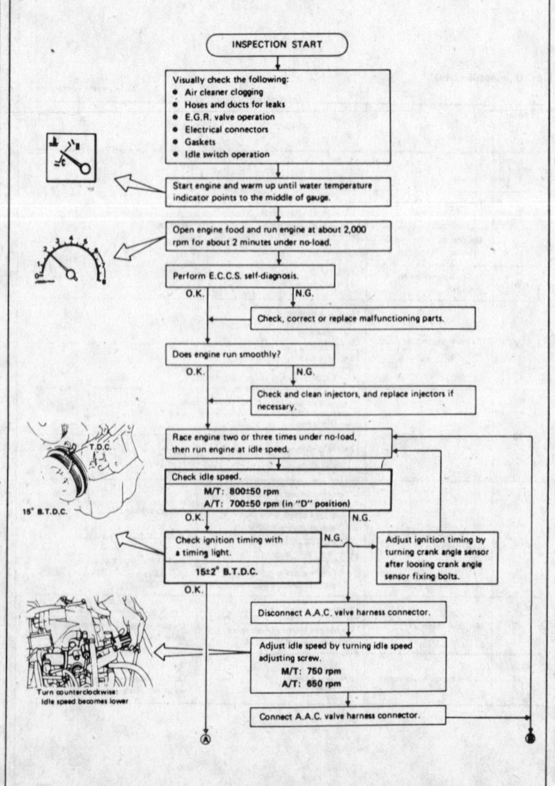

INSPECTION START
Visually check the following:
• Air cleaner clogging
• Hoses and ducts for leaks
• E.G.R. valve operation
• Electrical connectors
• Gaskets
• Idle switch operation
Start engine and warm up until water temperature indicator points to the middle of gauge.
Open engine food and run engine at about 2,000 rpm for about 2 minutes under no-load.
Perform E.C.C.S. self-diagnosis.
O.K.
N.G.
Check, correct or replace malfunctioning parts.
Does engine run smoothly?
O.K.
N.G.
Check and clean injectors, and replace injectors if necessary.
T.D.C.
15° B.T.D.C.
Race engine two or three times under no-load, then run engine at idle speed.
Check idle speed.
M/T: 800±50 rpm
A/T: 700±50 rpm (in "D" position)
O.K.
N.G.
Check ignition timing with a timing light.
15±2° B.T.D.C.
N.G.
Adjust ignition timing by turning crank angle sensor after loosing crank angle sensor fixing bolts.
O.K.
Disconnect A.A.C. valve harness connector.
Adjust idle speed by turning idle speed adjusting screw.
M/T: 750 rpm
A/T: 650 rpm
Turn counterclockwise: Idle speed becomes lower
Connect A.A.C. valve harness connector.
A

ECCS MIXTURE RATIO FEEDBACK SYSTEM INSPECTION (CONT.)—1989 PULSAR

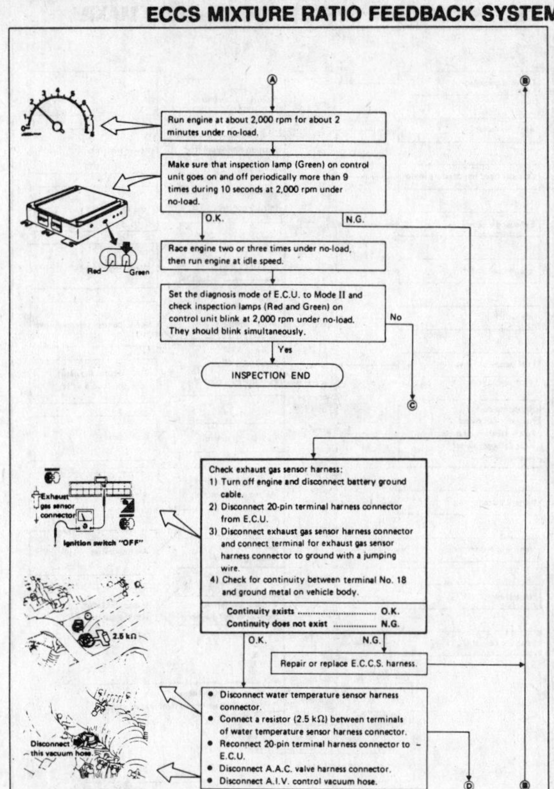
A
B
Run engine at about 2,000 rpm for about 2 minutes under no-load.
Make sure that inspection lamp (Green) on control unit goes on and off periodically more than 9 times during 10 seconds at 2,000 rpm under no-load.
Red
Green
O.K.
N.G.
Race engine two or three times under no-load, then run engine at idle speed.
Set the diagnosis mode of E.C.U. to Mode II and check inspection lamps (Red and Green) on control unit blink at 2,000 rpm under no-load. They should blink simultaneously.
No
Yes
INSPECTION END
C
Check exhaust gas sensor harness:
1) Turn off engine and disconnect battery ground cable.
2) Disconnect 20-pin terminal harness connector from E.C.U.
3) Disconnect exhaust gas sensor harness connector and connect terminal for exhaust gas sensor harness connector to ground with a jumping wire.
4) Check for continuity between terminal No. 18 and ground metal on vehicle body.
Continuity exists .... O.K.
Continuity does not exist .... N.G.
Exhaust gas sensor connector
Ignition switch "OFF"
2.5 kΩ
O.K.
N.G.
Repair or replace E.C.C.S. harness.
• Disconnect water temperature sensor harness connector.
• Connect a resistor (2.5 kΩ) between terminals of water temperature sensor harness connector.
• Reconnect 20-pin terminal harness connector to E.C.U.
• Disconnect A.A.C. valve harness connector.
• Disconnect A.I.V. control vacuum hose.
Disconnect this vacuum hose.
D
B

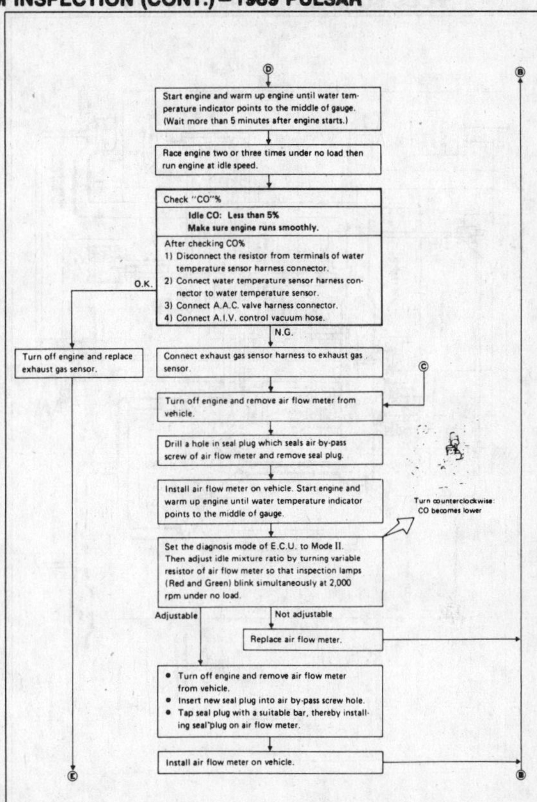
D
B
Start engine and warm up engine until water temperature indicator points to the middle of gauge. (Wait more than 5 minutes after engine starts.)
Race engine two or three times under no load then run engine at idle speed.
Check "CO"%
Idle CO: Less than 5%
Make sure engine runs smoothly.
After checking CO%
1) Disconnect the resistor from terminals of water temperature sensor harness connector.
2) Connect water temperature sensor harness connector to water temperature sensor.
3) Connect A.A.C. valve harness connector.
4) Connect A.I.V. control vacuum hose.
O.K.
N.G.
Turn off engine and replace exhaust gas sensor.
Connect exhaust gas sensor harness to exhaust gas sensor.
C
Turn off engine and remove air flow meter from vehicle.
Drill a hole in seal plug which seals air by-pass screw of air flow meter and remove seal plug.
Install air flow meter on vehicle. Start engine and warm up engine until water temperature indicator points to the middle of gauge.
Turn counterclockwise: CO becomes lower
Set the diagnosis mode of E.C.U. to Mode II. Then adjust idle mixture ratio by turning variable resistor of air flow meter so that inspection lamps (Red and Green) blink simultaneously at 2,000 rpm under no load.
Adjustable
Not adjustable
Replace air flow meter.
• Turn off engine and remove air flow meter from vehicle.
• Insert new seal plug into air by-pass screw hole.
• Tap seal plug with a suitable bar, thereby installing seal plug on air flow meter.
Install air flow meter on vehicle.
E
B

ECCS MIXTURE RATIO FEEDBACK SYSTEM INSPECTION (CONT.)—1989 PULSAR

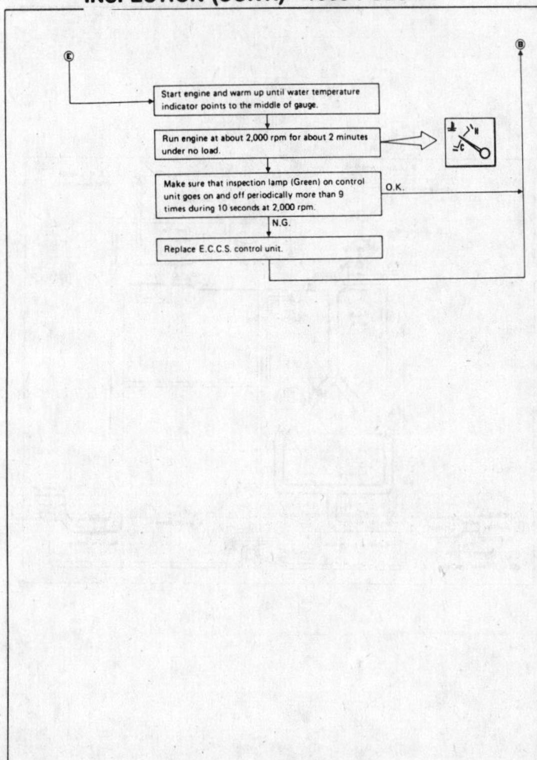
E
B
Start engine and warm up until water temperature indicator points to the middle of gauge.
Run engine at about 2,000 rpm for about 2 minutes under no load.
H
C
Make sure that inspection lamp (Green) on control unit goes on and off periodically more than 9 times during 10 seconds at 2,000 rpm.
O.K.
N.G.
Replace E.C.C.S. control unit.

ECCS COMPONENTS LOCATION—1988 MAXIMA

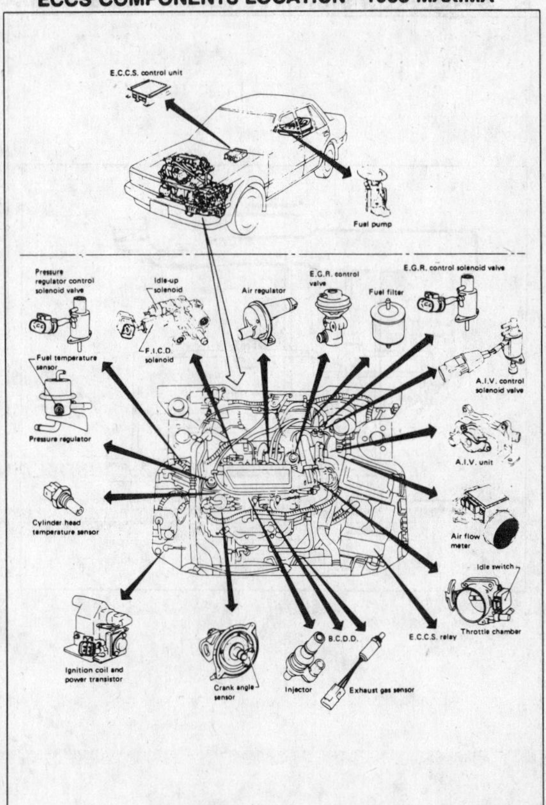
E.C.C.S. control unit
Fuel pump
Pressure regulator control solenoid valve
Idle-up solenoid
F.I.C.D. solenoid
Air regulator
E.G.R. control valve
Fuel filter
E.G.R. control solenoid valve
A.I.V. control solenoid valve
Fuel temperature sensor
Pressure regulator
A.I.V. unit
Cylinder head temperature sensor
Air flow meter
Idle switch
Throttle chamber
E.C.C.S. relay
B.C.D.D.
Ignition coil and power transistor
Crank angle sensor
Injector
Exhaust gas sensor

ECCS SCHEMATIC—1988 MAXIMA

ECCS CONTROL SYSTEM CHART—1988 MAXIMA

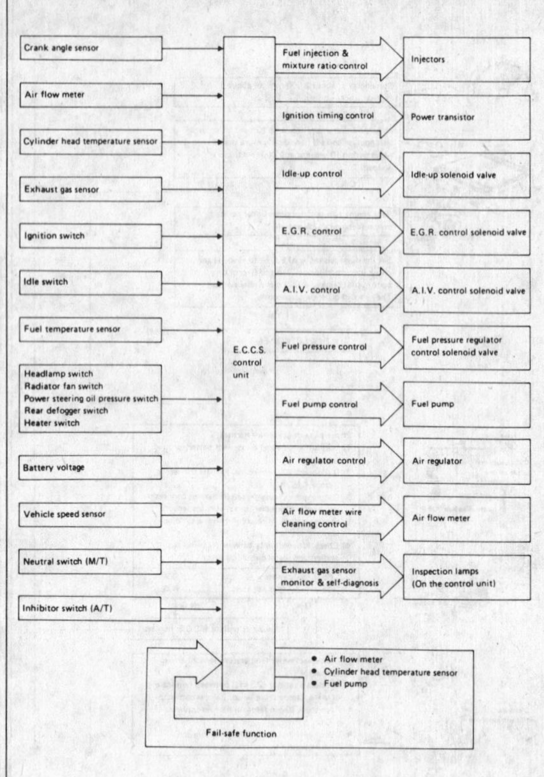
Crank angle sensor
Air flow meter
Cylinder head temperature sensor
Exhaust gas sensor
Ignition switch
Idle switch
Fuel temperature sensor
Headlamp switch
Radiator fan switch
Power steering oil pressure switch
Rear defogger switch
Heater switch
Battery voltage
Vehicle speed sensor
Neutral switch (M/T)
Inhibitor switch (A/T)
E.C.C.S. control unit
Fuel injection & mixture ratio control
Ignition timing control
Idle-up control
E.G.R. control
A.I.V. control
Fuel pressure control
Fuel pump control
Air regulator control
Air flow meter wire cleaning control
Exhaust gas sensor monitor & self-diagnosis
Injectors
Power transistor
Idle-up solenoid valve
E.G.R. control solenoid valve
A.I.V. control solenoid valve
Fuel pressure regulator control solenoid valve
Fuel pump
Air regulator
Air flow meter
Inspection lamps (On the control unit)
Air flow meter
Cylinder head temperature sensor
Fuel pump
Fail-safe function

ECCS FUEL FLOW SYSTEM SCHEAMTIC—1988 MAXIMA

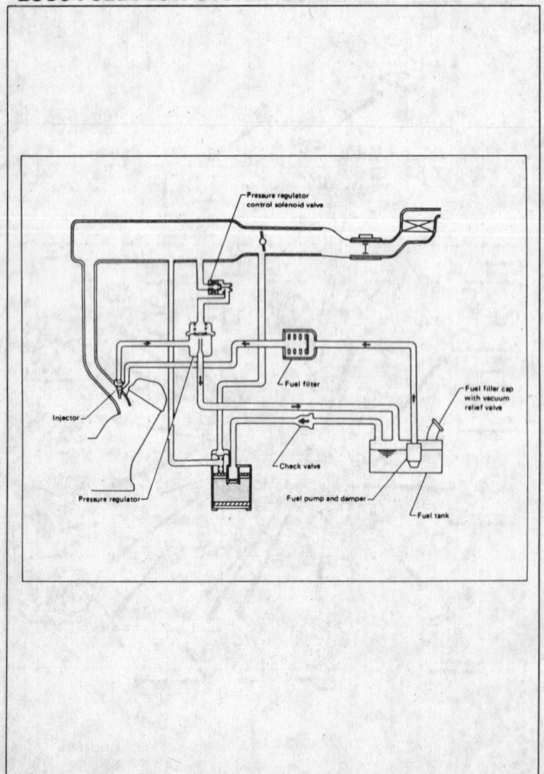
Pressure regulator control solenoid valve
Fuel filter
Fuel filler cap with vacuum relief valve
Injector
Check valve
Pressure regulator
Fuel pump and damper
Fuel tank

ECCS AIR FLOW SYSTEM SCHEMATIC
1988 MAXIMA

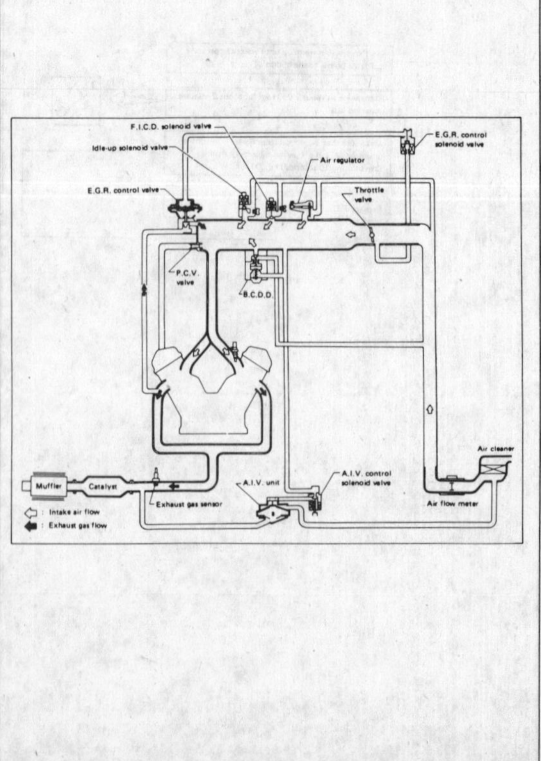
F.I.C.D. solenoid valve
Idle-up solenoid valve
E.G.R. control solenoid valve
Air regulator
E.G.R. control valve
Throttle valve
P.C.V. valve
B.C.D.D.
Air cleaner
Muffler
Catalyst
Exhaust gas sensor
A.I.V. unit
A.I.V. control solenoid valve
Air flow meter
Intake air flow
Exhaust gas flow

ECCS CONTROL UNIT WIRING SCHEMATIC 1988 MAXIMA

ECCS WIRING SCHEMATIC—1988 MAXIMA

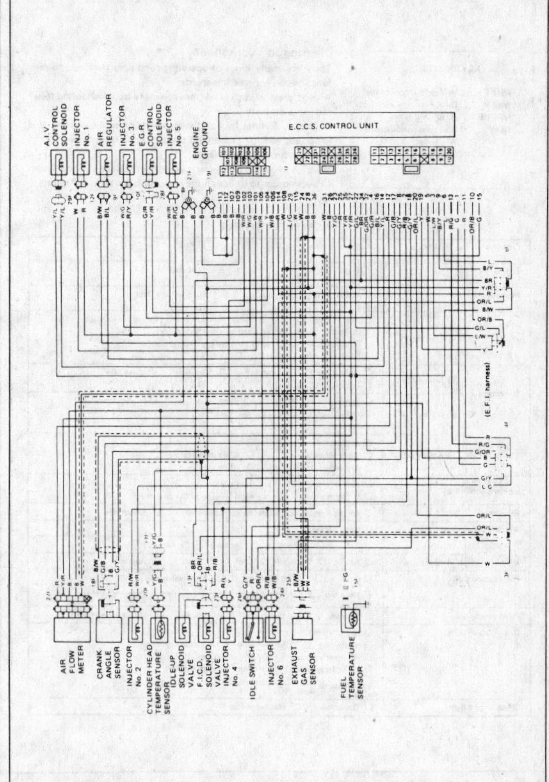
E.C.C.S. CONTROL UNIT
A.I.V. CONTROL SOLENOID
INJECTOR No. 1
AIR REGULATOR
INJECTOR No. 3
E.G.R. CONTROL SOLENOID
INJECTOR No. 5
ENGINE GROUND
AIR FLOW METER
CRANK ANGLE SENSOR
CYLINDER HEAD TEMPERATURE SENSOR
INJECTOR No. 2
IDLE-UP SOLENOID VALVE
F.I.C.D. SOLENOID VALVE
INJECTOR No. 4
IDLE SWITCH
INJECTOR No. 6
EXHAUST GAS SENSOR
FUEL TEMPERATURE SENSOR

ECCS WIRING SCHEMATIC (CONT.)—1988 MAXIMA

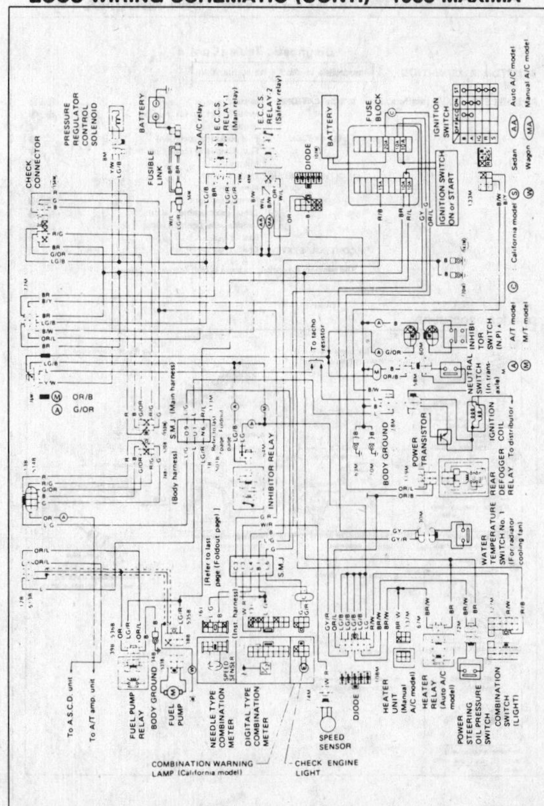
CHECK CONNECTOR
PRESSURE REGULATOR CONTROL SOLENOID
BATTERY
FUSIBLE LINK
E.C.C.S. RELAY-1 (Main relay)
E.C.C.S. RELAY-2 (Safety relay)
BATTERY
FUSE BLOCK
IGNITION SWITCH
DIODE
OR/B
G/OR
INHIBITOR RELAY
BODY GROUND
POWER TRANSISTOR
IGNITION COIL
FUEL PUMP RELAY
FUEL PUMP
NEEDLE TYPE COMBINATION METER
DIGITAL TYPE COMBINATION METER
SPEED SENSOR
DIODE
HEATER UNIT
POWER STEERING OIL PRESSURE SWITCH
COMBINATION SWITCH (LIGHT)
COMBINATION WARNING LAMP (California model)
CHECK ENGINE LIGHT

ECCS DIAGNOSTIC PROCEDURE—1988 MAXIMA

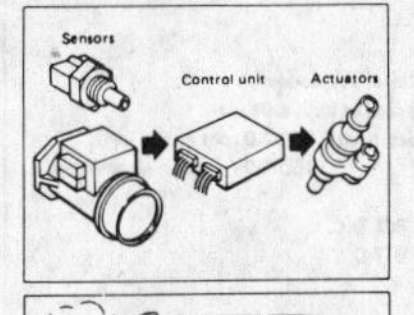
Sensors
Control unit
Actuators

INFO.
CAUSE

Introduction

The engine has an electronic control unit to control major systems such as fuel control, ignition control, idle speed control, etc. The control unit accepts input signals from sensors and instantly drives actuators. It is essential that both kinds of signals are proper and stable. At the same time, it is indispensable that there are no conventional problems such as vacuum leaks, fouled spark plugs, or other problems with the engine.

It is much more difficult to diagnose a problem that occurs intermittently rather than continuously. Most intermittent problems are caused by poor electric connections or faulty wiring. In this case, careful checking of suspicious circuits may help prevent the replacement of good parts.

A visual check only may not find the cause of the problems. A road test with a circuit tester connected to a suspected circuit should be performed.

Before undertaking actual checks, take just a few minutes to talk with a customer who approaches with a driveability complaint. The customer is a very good supplier of information on such problems, especially intermittent ones. Through the talks with the customer, find out what symptoms are present and under what conditions they occur.

Start your diagnosis by looking for "conventional" problems first. This is one of the best ways to troubleshoot driveability problems on an electronically controlled engine vehicle.

Work Flow

		Reference item
	CHECK-IN	
STEP 1	LISTENING TO CUSTOMER COMPLAINTS	Diagnostic Worksheet
STEP 2	DUPLICATION OF OPERATING CONDITIONS THAT LEAD TO MALFUNCTIONS	Intermittent Problem Simulation
STEP 3	ELIMINATING GOOD PARTS/SYSTEMS	Diagnostic Table
STEP 4	INSPECTION ON THE BASE OF EACH COMPONENT	Electronic Control System Inspection
STEP 5	REPAIR / REPLACEMENT	
STEP 6	FINAL CHECK (N.G. → STEP 3; O.K. ↓)	
	CHECK-OUT	

ECCS DIAGNOSTIC PROCEDURE (CONT.)—1988 MAXIMA

KEY POINTS

WHAT ..... Vehicle & engine model
WHEN ..... Date, Frequencies
WHERE ..... Road conditions
HOW ..... Operating conditions, Weather conditions, Symptoms

Diagnostic Worksheet

There are many kinds of operating conditions that lead to malfunctions on engine components.
A good grasp of such conditions can make troubleshooting faster and more accurate.
In general, feelings for a problem depend on each customer. It is important to fully understand the symptoms or under what conditions a customer complains..
Make good use of a diagnostic worksheet such as the one shown below in order to utilize all the complaints for troubleshooting.

WORKSHEET SAMPLE

Customer name MR/MS		Model & Year	VIN
Engine #		Trans.	Mileage
Incident Date		Manuf. Date	In Service Date
Symptoms	☐ Startability	☐ Impossible to start ☐ No combustion ☐ Partial combustion ☐ Partial combustion affected by throttle position ☐ Partial combustion NOT affected by throttle position ☐ Possible but hard to start ☐ Others [ ]	
	☐ Idling	☐ No fast idle ☐ Unstable ☐ High idle ☐ Low idle ☐ Others [ ]	
	☐ Driveability	☐ Stumble ☐ Surge ☐ Detonation ☐ Lack of power ☐ Intake backfire ☐ Exhaust backfire ☐ Others [ ]	
	☐ Engine stall	☐ At the time of start ☐ While idling ☐ While accelerating ☐ While decelerating ☐ Just after stopping ☐ While loading	
Incident occurrence		☐ Just after delivery ☐ Recently ☐ In the morning ☐ At night ☐ In the daytime	
Frequency		☐ All the time ☐ Under certain conditions ☐ Sometimes	
Weather conditions		☐ Not effected	
	Weather	☐ Fine ☐ Raining ☐ Snowing ☐ Others [ ]	
	Temperature	☐ Hot ☐ Warm ☐ Cool ☐ Cold ☐ Humid °F Engine speed 0 2,000 4,000 6,000 8,000 rpm	
Road conditions		☐ In town ☐ In suburbs ☐ Highway ☐ Off road (up/down)	
Driving conditions		☐ Not affected ☐ At starting ☐ While idling ☐ At racing ☐ While accelerating ☐ While cruising ☐ While decelerating ☐ While turning (RH/LH) Vehicle speed 0 10 20 30 40 50 60 MPH	
Check engine lights		☐ Turned on ☐ Not turned on	

Intermittent Problem Simulation

In order to duplicate an intermittent problem, it is effective to create similar conditions for component parts, under which the problem might occur.

	Variable factor	Influential part	Target condition	Service procedure
1	Mixture ratio	Pressure regulator	Made lean	Remove vacuum hose and apply vacuum.
			Made rich	Remove vacuum hose and apply pressure.
2	Ignition timing	Distributor	Advanced	Rotate distributor clockwise.
			Retarded	Rotate distributor counterclockwise.
3	Mixture ratio feedback control	Exhaust gas sensor	Suspended	Disconnect exhaust gas sensor harness connector.
		Control unit	Operation check	Perform self-diagnosis (Mode I/II) at 2,000 rpm.
4	Idle speed	I.A.A. unit	Raised	Turn idle adjust screw counterclockwise.
			Lowered	Turn idle adjust screw clockwise.
5	Electric connection (Electric continuity)	Harness connectors and wires	Poor electric connection or faulty wiring	Tap or wiggle.
				Race engine rapidly. See if the torque reaction of the engine unit causes electric breaks.
6	Characteristics against thermal condition	Control unit	Cooled	Cool with an icing spray or similar device.
			Warmed	Heat with a hand drier. [WARNING: Do not overheat the unit.]
7	Characteristics against moisture condition	Electric parts	Damp	Wet [WARNING: Do not directly pour water on components. Use a mist sprayer.]
8	Electric loads	Load switches	Loaded	Turn on head lights, air conditioner, rear defogger, etc.
9	Idle switch condition	Control unit	ON-OFF switching	Perform self-diagnosis (Mode IV).
10	Ignition spark	Timing light	Spark power check	Try to flash timing light for each cylinder.

ECCS SPECIFICATIONS—1988 MAXIMA

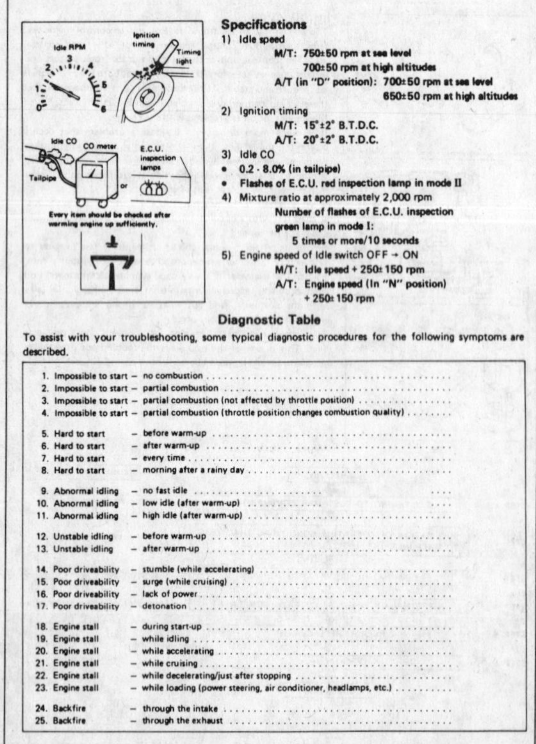
Idle RPM
Ignition timing
Timing light
Idle CO
CO meter
E.C.U. inspection lamps
Tailpipe
or
Every item should be checked after warming engine up sufficiently.

Specifications

1) Idle speed
M/T: 750±50 rpm at sea level
700±50 rpm at high altitudes
A/T (in "D" position): 700±50 rpm at sea level
650±50 rpm at high altitudes
2) Ignition timing
M/T: 15°±2° B.T.D.C.
A/T: 20°±2° B.T.D.C.
3) Idle CO
- 0.2 - 8.0% (in tailpipe)
- Flashes of E.C.U. red inspection lamp in mode II

4) Mixture ratio at approximately 2,000 rpm
Number of flashes of E.C.U. inspection green lamp in mode I:
5 times or more/10 seconds
5) Engine speed of Idle switch OFF → ON
M/T: Idle speed + 250± 150 rpm
A/T: Engine speed (In "N" position) + 250± 150 rpm

Diagnostic Table

To assist with your troubleshooting, some typical diagnostic procedures for the following symptoms are described.

1. Impossible to start – no combustion
2. Impossible to start – partial combustion
3. Impossible to start – partial combustion (not affected by throttle position)
4. Impossible to start – partial combustion (throttle position changes combustion quality)
5. Hard to start – before warm-up
6. Hard to start – after warm-up
7. Hard to start – every time
8. Hard to start – morning after a rainy day
9. Abnormal idling – no fast idle
10. Abnormal idling – low idle (after warm-up)
11. Abnormal idling – high idle (after warm-up)
12. Unstable idling – before warm-up
13. Unstable idling – after warm-up
14. Poor driveability – stumble (while accelerating)
15. Poor driveability – surge (while cruising)
16. Poor driveability – lack of power
17. Poor driveability – detonation
18. Engine stall – during start-up
19. Engine stall – while idling
20. Engine stall – while accelerating
21. Engine stall – while cruising
22. Engine stall – while decelerating/just after stopping
23. Engine stall – while loading (power steering, air conditioner, headlamps, etc.)
24. Backfire – through the intake
25. Backfire – through the exhaust

ECCS DIAGNOSTIC CHART—1988 MAXIMA

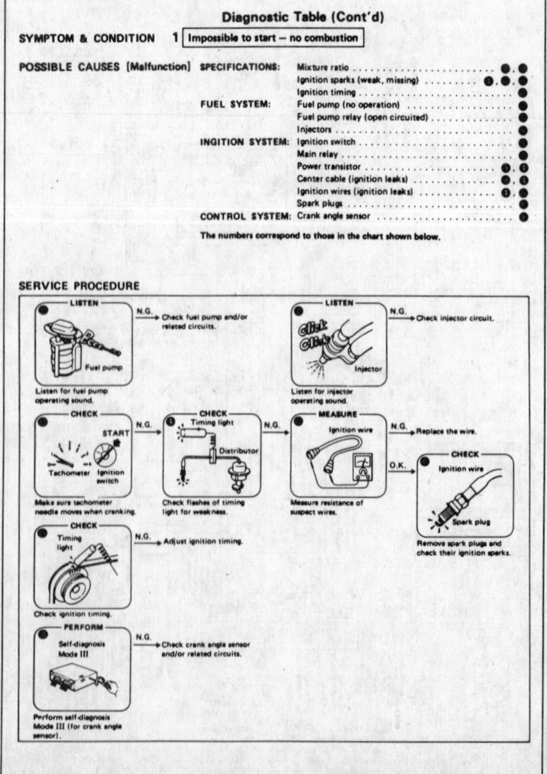

Diagnostic Table (Cont'd)

SYMPTOM & CONDITION 1 **Impossible to start – no combustion**

POSSIBLE CAUSES [Malfunction]

SPECIFICATIONS: Mixture ratio; Ignition sparks (weak, missing); Ignition timing
FUEL SYSTEM: Fuel pump (no operation); Fuel pump relay (open circuited); Injectors
INGITION SYSTEM: Ignition switch; Main relay; Power transistor; Center cable (ignition leaks); Ignition wires (ignition leaks); Spark plugs
CONTROL SYSTEM: Crank angle sensor

The numbers correspond to those in the chart shown below.

SERVICE PROCEDURE

LISTEN
Fuel pump
N.G. → Check fuel pump and/or related circuits.
Listen for fuel pump operating sound.
LISTEN
Click Click
Injector
N.G. → Check injector circuit.
Listen for injector operating sound.
CHECK
START
Tachometer
Ignition switch
N.G.
Make sure tachometer needle moves when cranking.
CHECK
Timing light
Distributor
N.G.
Check flashes of timing light for weakness.
MEASURE
Ignition wire
N.G. → Replace the wire.
O.K.
Measure resistance of suspect wires.
CHECK
Ignition wire
Spark plug
Remove spark plugs and check their ignition sparks.
CHECK
Timing light
N.G. → Adjust ignition timing.
Check ignition timing.
PERFORM
Self-diagnosis Mode III
N.G. → Check crank angle sensor and/or related circuits.
Perform self-diagnosis Mode III (for crank angle sensor).

ECCS DIAGNOSTIC CHART—1988 MAXIMA

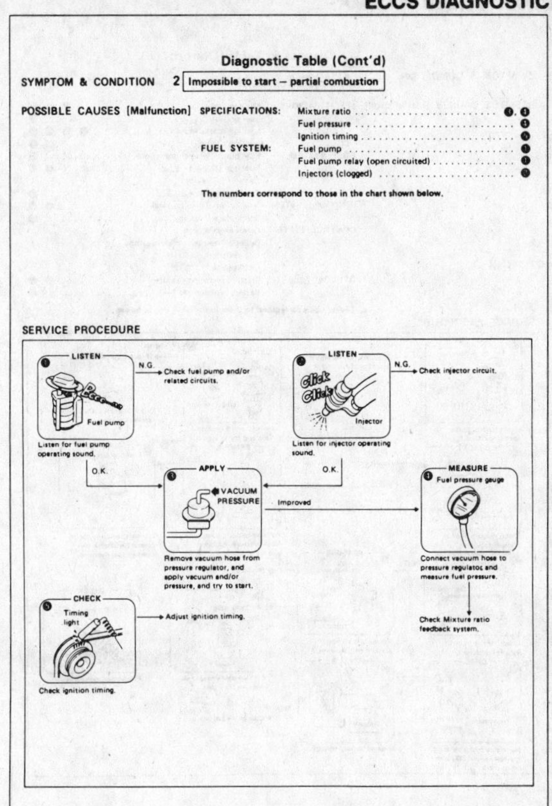
Diagnostic Table (Cont'd)
SYMPTOM & CONDITION 2 Impossible to start – partial combustion
POSSIBLE CAUSES [Malfunction]
SPECIFICATIONS: Mixture ratio
Fuel pressure
Ignition timing
FUEL SYSTEM: Fuel pump
Fuel pump relay (open circuited)
Injectors (clogged)
The numbers correspond to those in the chart shown below.
SERVICE PROCEDURE
LISTEN
Fuel pump
N.G.
Check fuel pump and/or related circuits.
Listen for fuel pump operating sound.
O.K.
LISTEN
Injector
N.G.
Check injector circuit.
Listen for injector operating sound.
O.K.
APPLY
VACUUM PRESSURE
Remove vacuum hose from pressure regulator, and apply vacuum and/or pressure, and try to start.
Improved
MEASURE
Fuel pressure gauge
Connect vacuum hose to pressure regulator, and measure fuel pressure.
Check Mixture ratio feedback system.
CHECK
Timing light
Adjust ignition timing.
Check ignition timing.

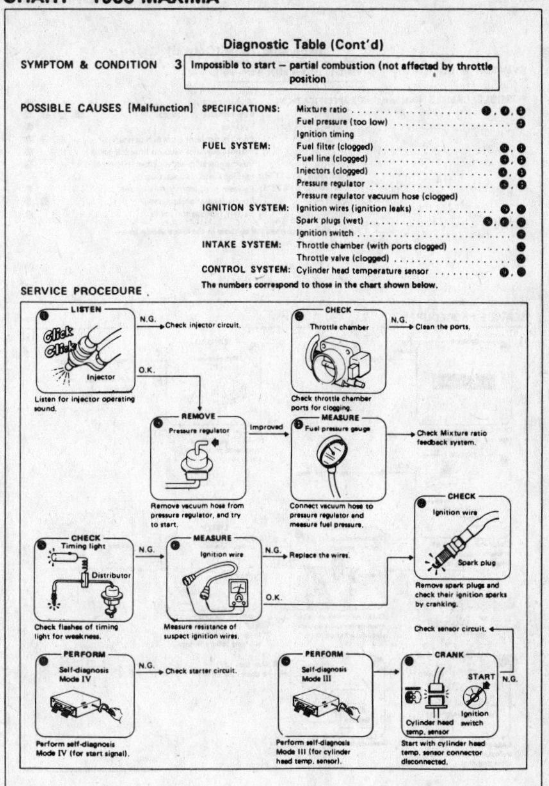
Diagnostic Table (Cont'd)
SYMPTOM & CONDITION 3 Impossible to start – partial combustion (not affected by throttle position
POSSIBLE CAUSES [Malfunction]
SPECIFICATIONS: Mixture ratio
Fuel pressure (too low)
Ignition timing
FUEL SYSTEM: Fuel filter (clogged)
Fuel line (clogged)
Injectors (clogged)
Pressure regulator
Pressure regulator vacuum hose (clogged)
IGNITION SYSTEM: Ignition wires (ignition leaks)
Spark plugs (wet)
Ignition switch
INTAKE SYSTEM: Throttle chamber (with ports clogged)
Throttle valve (clogged)
CONTROL SYSTEM: Cylinder head temperature sensor
The numbers correspond to those in the chart shown below.
SERVICE PROCEDURE
LISTEN
Injector
N.G.
Check injector circuit.
O.K.
Listen for injector operating sound.
CHECK
Throttle chamber
N.G.
Clean the ports.
Check throttle chamber ports for clogging.
REMOVE
Pressure regulator
Remove vacuum hose from pressure regulator, and try to start.
Improved
MEASURE
Fuel pressure gauge
Connect vacuum hose to pressure regulator and measure fuel pressure.
Check Mixture ratio feedback system.
CHECK
Timing light
Distributor
Check flashes of timing light for weakness.
N.G.
MEASURE
Ignition wire
Measure resistance of suspect ignition wires.
N.G.
Replace the wires.
O.K.
CHECK
Ignition wire
Spark plug
Remove spark plugs and check their ignition sparks by cranking.
Check sensor circuit.
PERFORM
Self-diagnosis Mode IV
N.G.
Check starter circuit.
Perform self-diagnosis Mode IV (for start signal).
PERFORM
Self-diagnosis Mode III
Perform self-diagnosis Mode III (for cylinder head temp. sensor).
CRANK
START
N.G.
Cylinder head temp. sensor
Ignition switch
Start with cylinder head temp. sensor connector disconnected.

ECCS DIAGNOSTIC CHART—1988 MAXIMA

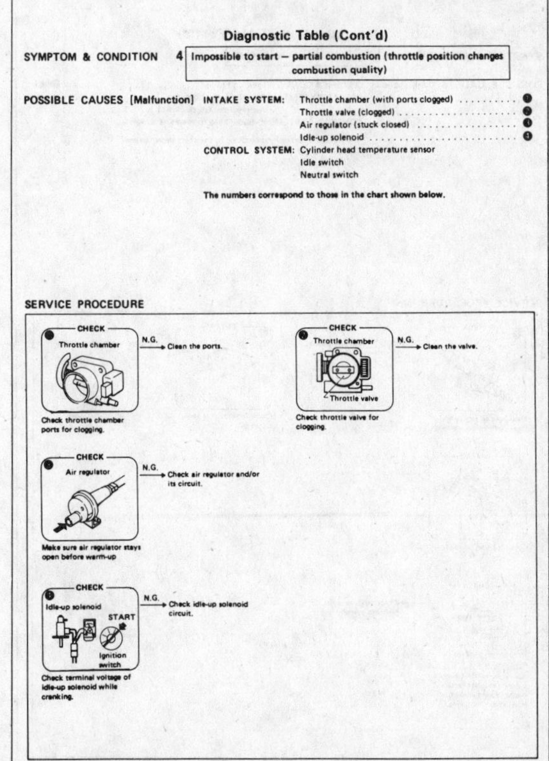
Diagnostic Table (Cont'd)
SYMPTOM & CONDITION 4 Impossible to start – partial combustion (throttle position changes combustion quality)
POSSIBLE CAUSES [Malfunction]
INTAKE SYSTEM: Throttle chamber (with ports clogged)
Throttle valve (clogged)
Air regulator (stuck closed)
Idle-up solenoid
CONTROL SYSTEM: Cylinder head temperature sensor
Idle switch
Neutral switch
The numbers correspond to those in the chart shown below.
SERVICE PROCEDURE
CHECK
Throttle chamber
N.G.
Clean the ports.
Check throttle chamber ports for clogging.
CHECK
Throttle chamber
Throttle valve
N.G.
Clean the valve.
Check throttle valve for clogging.
CHECK
Air regulator
N.G.
Check air regulator and/or its circuit.
Make sure air regulator stays open before warm-up
CHECK
Idle-up solenoid
START
Ignition switch
N.G.
Check idle-up solenoid circuit.
Check terminal voltage of idle-up solenoid while cranking.

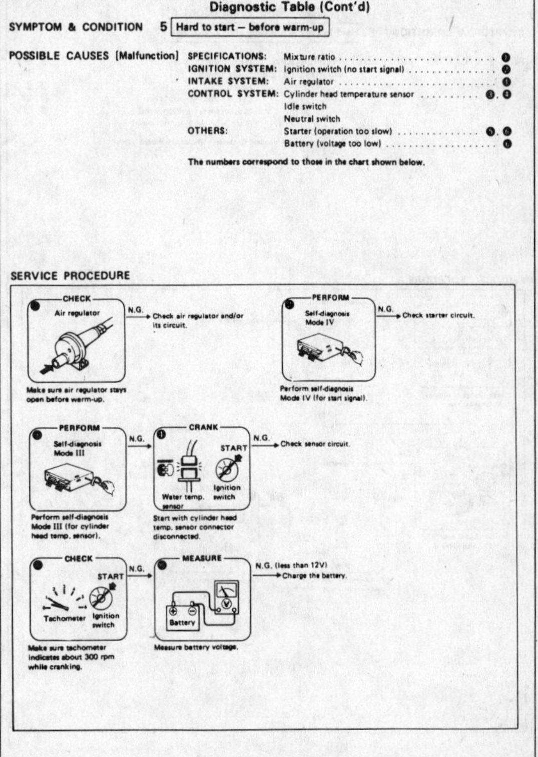
Diagnostic Table (Cont'd)
SYMPTOM & CONDITION 5 Hard to start – before warm-up
POSSIBLE CAUSES [Malfunction]
SPECIFICATIONS: Mixture ratio
IGNITION SYSTEM: Ignition switch (no start signal)
INTAKE SYSTEM: Air regulator
CONTROL SYSTEM: Cylinder head temperature sensor
Idle switch
Neutral switch
OTHERS: Starter (operation too slow)
Battery (voltage too low)
The numbers correspond to those in the chart shown below.
SERVICE PROCEDURE
CHECK
Air regulator
N.G.
Check air regulator and/or its circuit.
Make sure air regulator stays open before warm-up.
PERFORM
Self-diagnosis Mode IV
N.G.
Check starter circuit.
Perform self-diagnosis Mode IV (for start signal).
PERFORM
Self-diagnosis Mode III
N.G.
Perform self-diagnosis Mode III (for cylinder head temp. sensor).
CRANK
START
Water temp. sensor
Ignition switch
N.G.
Check sensor circuit.
Start with cylinder head temp. sensor connector disconnected.
CHECK
START
Tachometer
Ignition switch
N.G.
Make sure tachometer indicates about 300 rpm while cranking.
MEASURE
Battery
N.G. (less than 12V)
Charge the battery.
Measure battery voltage.

ECCS DIAGNOSTIC CHART—1988 MAXIMA

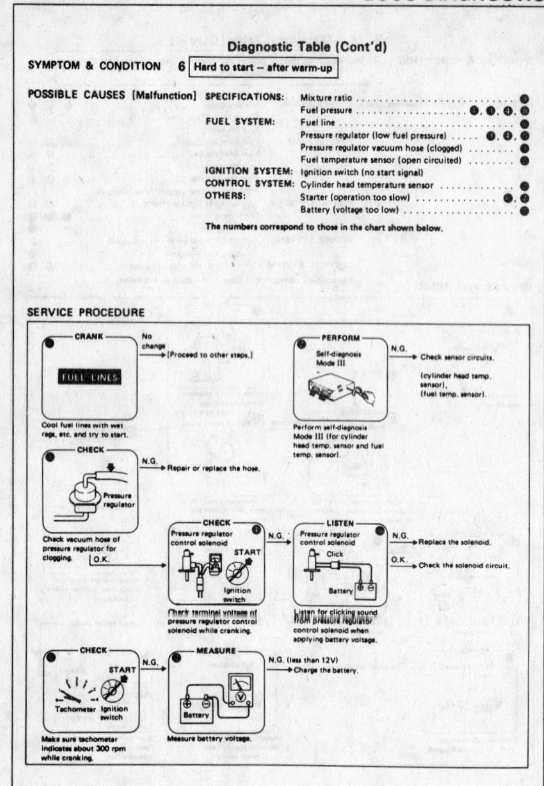
Diagnostic Table (Cont'd)
SYMPTOM & CONDITION 6 Hard to start – after warm-up
POSSIBLE CAUSES [Malfunction]
SPECIFICATIONS: Mixture ratio
Fuel pressure
FUEL SYSTEM: Fuel line
Pressure regulator (low fuel pressure)
Pressure regulator vacuum hose (clogged)
Fuel temperature sensor (open circuited)
IGNITION SYSTEM: Ignition switch (no start signal)
CONTROL SYSTEM: Cylinder head temperature sensor
OTHERS: Starter (operation too slow)
Battery (voltage too low)
The numbers correspond to those in the chart shown below.
SERVICE PROCEDURE
CRANK
No change
[Proceed to other steps.]
FUEL LINES
Cool fuel lines with wet rags, etc. and try to start.
PERFORM
Self-diagnosis Mode III
N.G.
Check sensor circuits.
(cylinder head temp. sensor), (fuel temp. sensor).
Perform self-diagnosis Mode III (for cylinder head temp. sensor and fuel temp. sensor).
CHECK
N.G.
Repair or replace the hose.
Pressure regulator
Check vacuum hose of pressure regulator for clogging.
O.K.
CHECK
Pressure regulator control solenoid
START
Ignition switch
Check terminal voltage of pressure regulator control solenoid while cranking.
N.G.
LISTEN
Pressure regulator control solenoid
Click
Battery
N.G.
Replace the solenoid.
O.K.
Check the solenoid circuit.
Listen for clicking sound from pressure regulator control solenoid when applying battery voltage.
CHECK
START
Tachometer
Ignition switch
Make sure tachometer indicates about 300 rpm while cranking.
N.G.
MEASURE
Battery
Measure battery voltage.
N.G. (less than 12V)
Charge the battery.

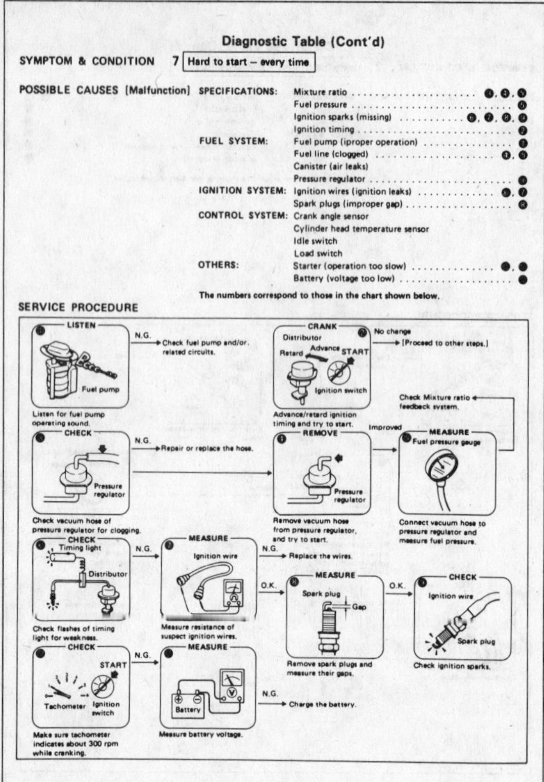
Diagnostic Table (Cont'd)
SYMPTOM & CONDITION 7 Hard to start – every time
POSSIBLE CAUSES [Malfunction]
SPECIFICATIONS: Mixture ratio
Fuel pressure
Ignition sparks (missing)
Ignition timing
FUEL SYSTEM: Fuel pump (improper operation)
Fuel line (clogged)
Canister (air leaks)
Pressure regulator
IGNITION SYSTEM: Ignition wires (ignition leaks)
Spark plugs (improper gap)
CONTROL SYSTEM: Crank angle sensor
Cylinder head temperature sensor
Idle switch
Load switch
OTHERS: Starter (operation too slow)
Battery (voltage too low)
The numbers correspond to those in the chart shown below.
SERVICE PROCEDURE
LISTEN
Fuel pump
N.G.
Check fuel pump and/or related circuits.
Listen for fuel pump operating sound.
CRANK
Distributor
Retard
Advance
START
Ignition switch
No change
[Proceed to other steps.]
Advance/retard ignition timing and try to start.
Check Mixture ratio feedback system.
Improved
CHECK
N.G.
Repair or replace the hose.
Pressure regulator
Check vacuum hose of pressure regulator for clogging.
REMOVE
Pressure regulator
Remove vacuum hose from pressure regulator, and try to start.
MEASURE
Fuel pressure gauge
Connect vacuum hose to pressure regulator and measure fuel pressure.
CHECK
Timing light
Distributor
Check flashes of timing light for weakness.
N.G.
MEASURE
Ignition wire
Measure resistance of suspect ignition wires.
N.G.
Replace the wires.
O.K.
MEASURE
Spark plug
Gap
Remove spark plugs and measure their gaps.
O.K.
CHECK
Ignition wire
Spark plug
Check ignition sparks.
CHECK
START
Tachometer
Ignition switch
Make sure tachometer indicates about 300 rpm while cranking.
N.G.
MEASURE
Battery
Measure battery voltage.
N.G.
Charge the battery.

ECCS DIAGNOSTIC CHART—1988 MAXIMA

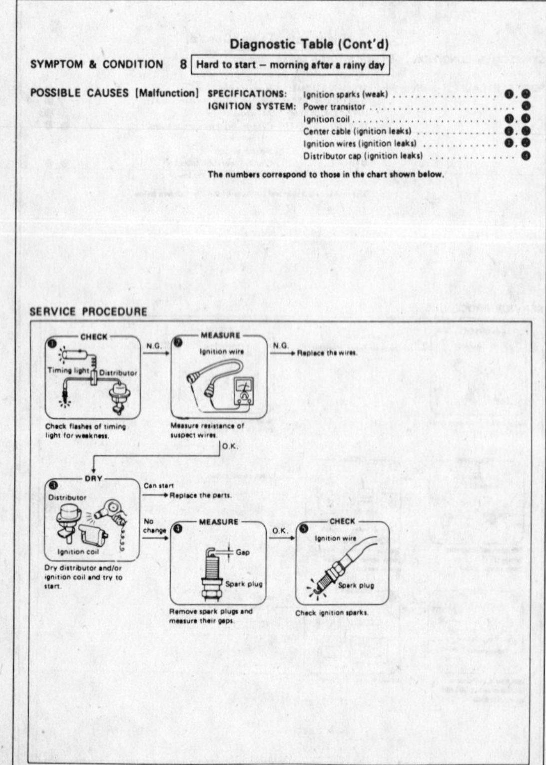
Diagnostic Table (Cont'd)
SYMPTOM & CONDITION 8 Hard to start – morning after a rainy day
POSSIBLE CAUSES [Malfunction]
SPECIFICATIONS: Ignition sparks (weak)
IGNITION SYSTEM: Power transistor
Ignition coil
Center cable (ignition leaks)
Ignition wires (ignition leaks)
Distributor cap (ignition leaks)
The numbers correspond to those in the chart shown below.
SERVICE PROCEDURE
CHECK
Timing light
Distributor
Check flashes of timing light for weakness.
N.G.
MEASURE
Ignition wire
Measure resistance of suspect wires.
N.G.
Replace the wires.
O.K.
DRY
Distributor
Ignition coil
Dry distributor and/or ignition coil and try to start.
Can start
Replace the parts.
No change
MEASURE
Gap
Spark plug
Remove spark plugs and measure their gaps.
O.K.
CHECK
Ignition wire
Spark plug
Check ignition sparks.

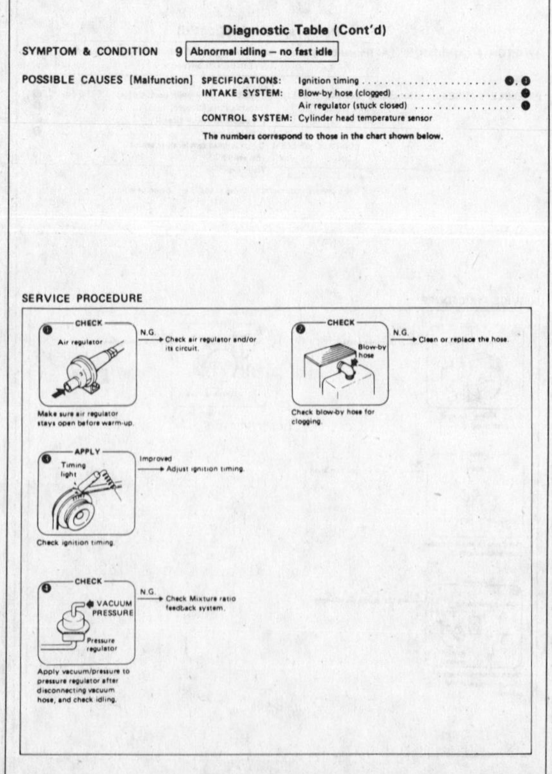
Diagnostic Table (Cont'd)
SYMPTOM & CONDITION 9 Abnormal idling – no fast idle
POSSIBLE CAUSES [Malfunction]
SPECIFICATIONS: Ignition timing
INTAKE SYSTEM: Blow-by hose (clogged)
Air regulator (stuck closed)
CONTROL SYSTEM: Cylinder head temperature sensor
The numbers correspond to those in the chart shown below.
SERVICE PROCEDURE
CHECK
Air regulator
N.G.
Check air regulator and/or its circuit.
Make sure air regulator stays open before warm-up.
CHECK
Blow-by hose
N.G.
Clean or replace the hose.
Check blow-by hose for clogging.
APPLY
Timing light
Improved
Adjust ignition timing.
Check ignition timing.
CHECK
VACUUM PRESSURE
Pressure regulator
N.G.
Check Mixture ratio feedback system.
Apply vacuum/pressure to pressure regulator after disconnecting vacuum hose, and check idling.

ECCS DIAGNOSTIC CHART—1988 MAXIMA

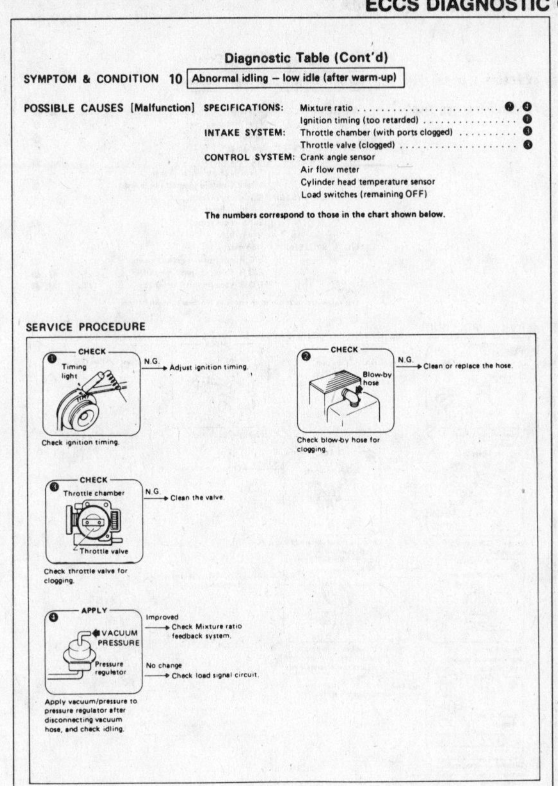

Diagnostic Table (Cont'd)

SYMPTOM & CONDITION 10 Abnormal idling – low idle (after warm-up)

POSSIBLE CAUSES [Malfunction]

SPECIFICATIONS:	Mixture ratio	❷, ❹
	Ignition timing (too retarded)	❶
INTAKE SYSTEM:	Throttle chamber (with ports clogged)	❸
	Throttle valve (clogged)	❸
CONTROL SYSTEM:	Crank angle sensor	
	Air flow meter	
	Cylinder head temperature sensor	
	Load switches (remaining OFF)	

The numbers correspond to those in the chart shown below.

SERVICE PROCEDURE

❶ CHECK
Timing light
N.G. → Adjust ignition timing.
Check ignition timing.
❷ CHECK
Blow-by hose
N.G. → Clean or replace the hose.
Check blow-by hose for clogging.
❸ CHECK
Throttle chamber
Throttle valve
N.G. → Clean the valve.
Check throttle valve for clogging.
❹ APPLY
VACUUM PRESSURE
Pressure regulator
Improved → Check Mixture ratio feedback system.
No change → Check load signal circuit.
Apply vacuum/pressure to pressure regulator after disconnecting vacuum hose, and check idling.

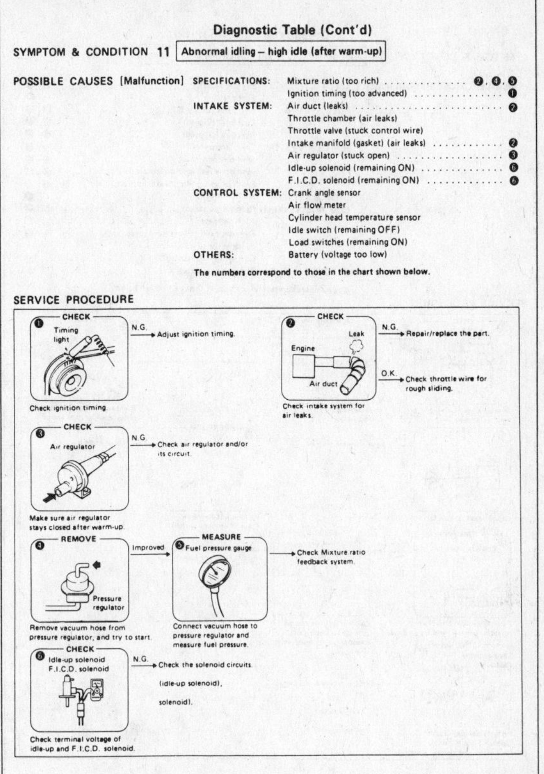

Diagnostic Table (Cont'd)

SYMPTOM & CONDITION 11 Abnormal idling – high idle (after warm-up)

POSSIBLE CAUSES [Malfunction]

SPECIFICATIONS:	Mixture ratio (too rich)	❷, ❹, ❺
	Ignition timing (too advanced)	❶
INTAKE SYSTEM:	Air duct (leaks)	❷
	Throttle chamber (air leaks)	
	Throttle valve (stuck control wire)	
	Intake manifold (gasket) (air leaks)	❷
	Air regulator (stuck open)	❸
	Idle-up solenoid (remaining ON)	❻
	F.I.C.D. solenoid (remaining ON)	❻
CONTROL SYSTEM:	Crank angle sensor	
	Air flow meter	
	Cylinder head temperature sensor	
	Idle switch (remaining OFF)	
	Load switches (remaining ON)	
OTHERS:	Battery (voltage too low)	

The numbers correspond to those in the chart shown below.

SERVICE PROCEDURE

❶ CHECK
Timing light
N.G. → Adjust ignition timing.
Check ignition timing.
❷ CHECK
Engine
Leak
Air duct
N.G. → Repair/replace the part.
O.K. → Check throttle wire for rough sliding.
Check intake system for air leaks.
❸ CHECK
Air regulator
N.G. → Check air regulator and/or its circuit.
Make sure air regulator stays closed after warm-up.
❹ REMOVE
Pressure regulator
Improved
Remove vacuum hose from pressure regulator, and try to start.
❺ MEASURE
Fuel pressure gauge
→ Check Mixture ratio feedback system.
Connect vacuum hose to pressure regulator and measure fuel pressure.
❻ CHECK
Idle-up solenoid
F.I.C.D. solenoid
N.G. → Check the solenoid circuits (idle-up solenoid), solenoid).
Check terminal voltage of idle-up and F.I.C.D. solenoid.

ECCS DIAGNOSTIC CHART—1988 MAXIMA

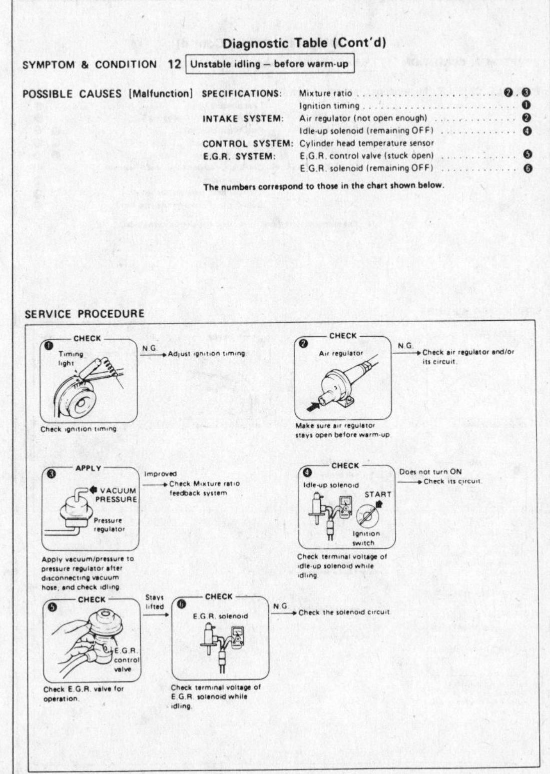

Diagnostic Table (Cont'd)

SYMPTOM & CONDITION 12 Unstable idling – before warm-up

POSSIBLE CAUSES [Malfunction]

SPECIFICATIONS:	Mixture ratio	❷, ❸
	Ignition timing	❶
INTAKE SYSTEM:	Air regulator (not open enough)	❷
	Idle-up solenoid (remaining OFF)	❹
CONTROL SYSTEM:	Cylinder head temperature sensor	
E.G.R. SYSTEM:	E.G.R. control valve (stuck open)	❺
	E.G.R. solenoid (remaining OFF)	❻

The numbers correspond to those in the chart shown below.

SERVICE PROCEDURE

❶ CHECK
Timing light
N.G. → Adjust ignition timing.
Check ignition timing.
❷ CHECK
Air regulator
N.G. → Check air regulator and/or its circuit.
Make sure air regulator stays open before warm-up.
❸ APPLY
VACUUM PRESSURE
Pressure regulator
Improved → Check Mixture ratio feedback system.
Apply vacuum/pressure to pressure regulator after disconnecting vacuum hose, and check idling.
❹ CHECK
Idle-up solenoid
START
Ignition switch
Does not turn ON → Check its circuit.
Check terminal voltage of idle-up solenoid while idling.
❺ CHECK
E.G.R. control valve
Stays lifted
Check E.G.R. valve for operation.
❻ CHECK
E.G.R. solenoid
N.G. → Check the solenoid circuit.
Check terminal voltage of E.G.R. solenoid while idling.

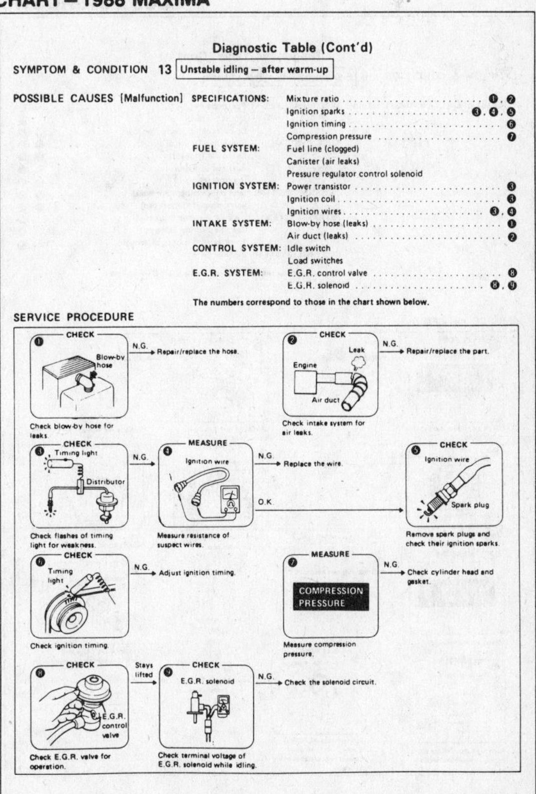

Diagnostic Table (Cont'd)

SYMPTOM & CONDITION 13 Unstable idling – after warm-up

POSSIBLE CAUSES [Malfunction]

SPECIFICATIONS:	Mixture ratio	❶, ❷
	Ignition sparks	❸, ❹, ❺
	Ignition timing	❻
	Compression pressure	❼
FUEL SYSTEM:	Fuel line (clogged)	
	Canister (air leaks)	
	Pressure regulator control solenoid	
IGNITION SYSTEM:	Power transistor	❸
	Ignition coil	❸
	Ignition wires	❸, ❹
INTAKE SYSTEM:	Blow-by hose (leaks)	❶
	Air duct (leaks)	❷
CONTROL SYSTEM:	Idle switch	
	Load switches	
E.G.R. SYSTEM:	E.G.R. control valve	❽
	E.G.R. solenoid	❽, ❾

The numbers correspond to those in the chart shown below.

SERVICE PROCEDURE

❶ CHECK
Blow-by hose
N.G. → Repair/replace the hose.
Check blow-by hose for leaks.
❷ CHECK
Engine
Leak
Air duct
N.G. → Repair/replace the part.
Check intake system for air leaks.
❸ CHECK
Timing light
Distributor
N.G.
Check flashes of timing light for weakness.
❹ MEASURE
Ignition wire
N.G. → Replace the wire.
O.K.
Measure resistance of suspect wires.
❺ CHECK
Ignition wire
Spark plug
Remove spark plugs and check their ignition sparks.
❻ CHECK
Timing light
N.G. → Adjust ignition timing.
Check ignition timing.
❼ MEASURE
COMPRESSION PRESSURE
N.G. → Check cylinder head and gasket.
Measure compression pressure.
❽ CHECK
E.G.R. control valve
Stays lifted
Check E.G.R. valve for operation.
❾ CHECK
E.G.R. solenoid
N.G. → Check the solenoid circuit.
Check terminal voltage of E.G.R. solenoid while idling.

ECCS DIAGNOSTIC CHART—1988 MAXIMA

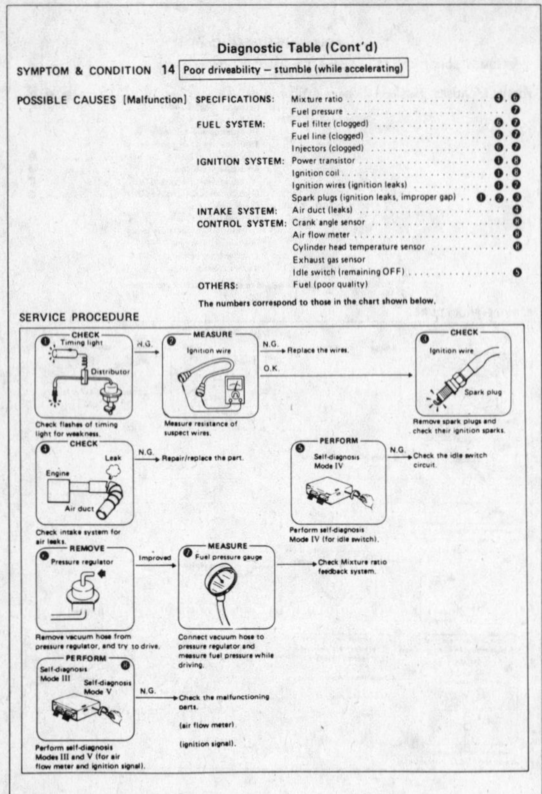
Diagnostic Table (Cont'd)
SYMPTOM & CONDITION 14 Poor driveability – stumble (while accelerating)
POSSIBLE CAUSES [Malfunction]
SPECIFICATIONS: Mixture ratio ❹, ❻
Fuel pressure ❼
FUEL SYSTEM: Fuel filter (clogged) ❻, ❼
Fuel line (clogged) ❻, ❼
Injectors (clogged) ❻, ❼
IGNITION SYSTEM: Power transistor ❶, ❽
Ignition coil ❶, ❽
Ignition wires (ignition leaks) ❶, ❷
Spark plugs (ignition leaks, improper gap) ❶, ❷, ❸
INTAKE SYSTEM: Air duct (leaks) ❹
CONTROL SYSTEM: Crank angle sensor ❽
Air flow meter ❽
Cylinder head temperature sensor ❽
Exhaust gas sensor
Idle switch (remaining OFF) ❺
OTHERS: Fuel (poor quality)
The numbers correspond to those in the chart shown below.
SERVICE PROCEDURE
❶ CHECK
Timing light
Distributor
Check flashes of timing light for weakness.
N.G.
❷ MEASURE
Ignition wire
Measure resistance of suspect wires.
N.G.
Replace the wires.
O.K.
❸ CHECK
Ignition wire
Spark plug
Remove spark plugs and check their ignition sparks.
❹ CHECK
Leak
Engine
Air duct
Check intake system for air leaks.
N.G.
Repair/replace the part.
❺ PERFORM
Self-diagnosis Mode IV
N.G.
Check the idle switch circuit.
Perform self-diagnosis Mode IV (for idle switch).
❻ REMOVE
Pressure regulator
Remove vacuum hose from pressure regulator, and try to drive.
Improved
❼ MEASURE
Fuel pressure gauge
Connect vacuum hose to pressure regulator and measure fuel pressure while driving.
Check Mixture ratio feedback system.
❽ PERFORM
Self-diagnosis Mode III
Self-diagnosis Mode V
N.G.
Check the malfunctioning parts.
(air flow meter).
(ignition signal).
Perform self-diagnosis Modes III and V (for air flow meter and ignition signal).

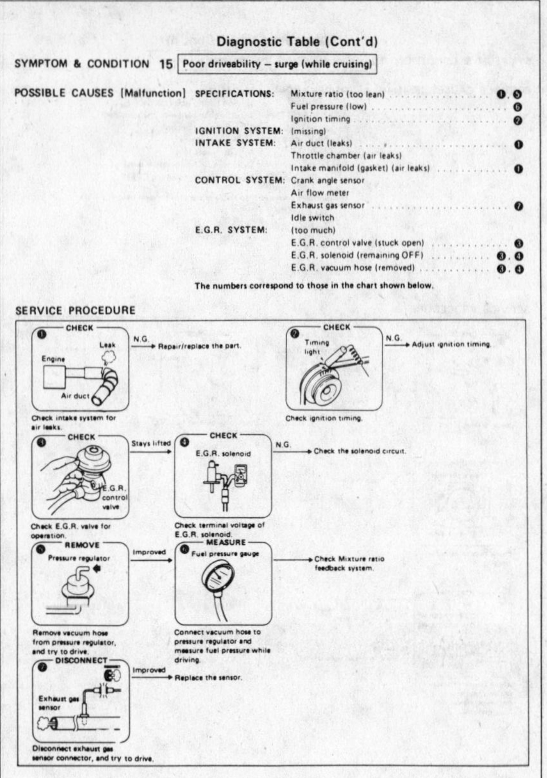
Diagnostic Table (Cont'd)
SYMPTOM & CONDITION 15 Poor driveability – surge (while cruising)
POSSIBLE CAUSES [Malfunction]
SPECIFICATIONS: Mixture ratio (too lean) ❶, ❺
Fuel pressure (low) ❻
Ignition timing ❷
IGNITION SYSTEM: (missing)
INTAKE SYSTEM: Air duct (leaks) ❶
Throttle chamber (air leaks)
Intake manifold (gasket) (air leaks) ❶
CONTROL SYSTEM: Crank angle sensor
Air flow meter
Exhaust gas sensor ❼
Idle switch
E.G.R. SYSTEM: (too much)
E.G.R. control valve (stuck open) ❸
E.G.R. solenoid (remaining OFF) ❸, ❹
E.G.R. vacuum hose (removed) ❸, ❹
The numbers correspond to those in the chart shown below.
SERVICE PROCEDURE
❶ CHECK
Leak
Engine
Air duct
Check intake system for air leaks.
N.G.
Repair/replace the part.
❷ CHECK
Timing light
Check ignition timing.
N.G.
Adjust ignition timing.
❸ CHECK
E.G.R. control valve
Check E.G.R. valve for operation.
Stays lifted
❹ CHECK
E.G.R. solenoid
Check terminal voltage of E.G.R. solenoid.
N.G.
Check the solenoid circuit.
❺ REMOVE
Pressure regulator
Remove vacuum hose from pressure regulator, and try to drive.
Improved
❻ MEASURE
Fuel pressure gauge
Connect vacuum hose to pressure regulator and measure fuel pressure while driving.
Check Mixture ratio feedback system.
❼ DISCONNECT
Exhaust gas sensor
Disconnect exhaust gas sensor connector, and try to drive.
Improved
Replace the sensor.

ECCS DIAGNOSTIC CHART—1988 MAXIMA

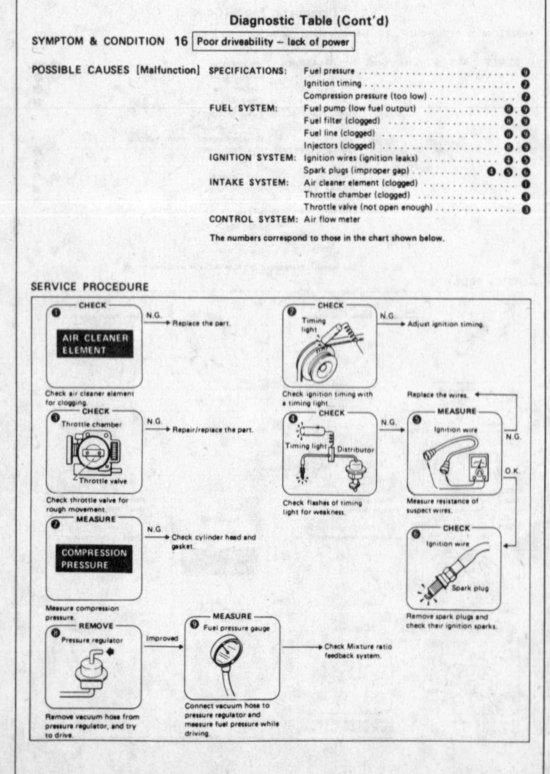
Diagnostic Table (Cont'd)
SYMPTOM & CONDITION 16 Poor driveability – lack of power
POSSIBLE CAUSES [Malfunction]
SPECIFICATIONS: Fuel pressure ❾
Ignition timing ❷
Compression pressure (too low) ❼
FUEL SYSTEM: Fuel pump (low fuel output) ❽, ❾
Fuel filter (clogged) ❽, ❾
Fuel line (clogged) ❽, ❾
Injectors (clogged) ❽, ❾
IGNITION SYSTEM: Ignition wires (ignition leaks) ❹, ❺
Spark plugs (improper gap) ❹, ❺, ❻
INTAKE SYSTEM: Air cleaner element (clogged) ❶
Throttle chamber (clogged) ❸
Throttle valve (not open enough) ❸
CONTROL SYSTEM: Air flow meter
The numbers correspond to those in the chart shown below.
SERVICE PROCEDURE
❶ CHECK
AIR CLEANER ELEMENT
Check air cleaner element for clogging.
N.G.
Replace the part.
❷ CHECK
Timing light
Check ignition timing with a timing light.
N.G.
Adjust ignition timing.
❸ CHECK
Throttle chamber
Throttle valve
Check throttle valve for rough movement.
N.G.
Repair/replace the part.
❹ CHECK
Timing light
Distributor
Check flashes of timing light for weakness.
N.G.
❺ MEASURE
Ignition wire
Measure resistance of suspect wires.
N.G.
Replace the wires.
O.K.
❻ CHECK
Ignition wire
Spark plug
Remove spark plugs and check their ignition sparks.
❼ MEASURE
COMPRESSION PRESSURE
Measure compression pressure.
N.G.
Check cylinder head and gasket.
❽ REMOVE
Pressure regulator
Remove vacuum hose from pressure regulator, and try to drive.
Improved
❾ MEASURE
Fuel pressure gauge
Connect vacuum hose to pressure regulator and measure fuel pressure while driving.
Check Mixture ratio feedback system.

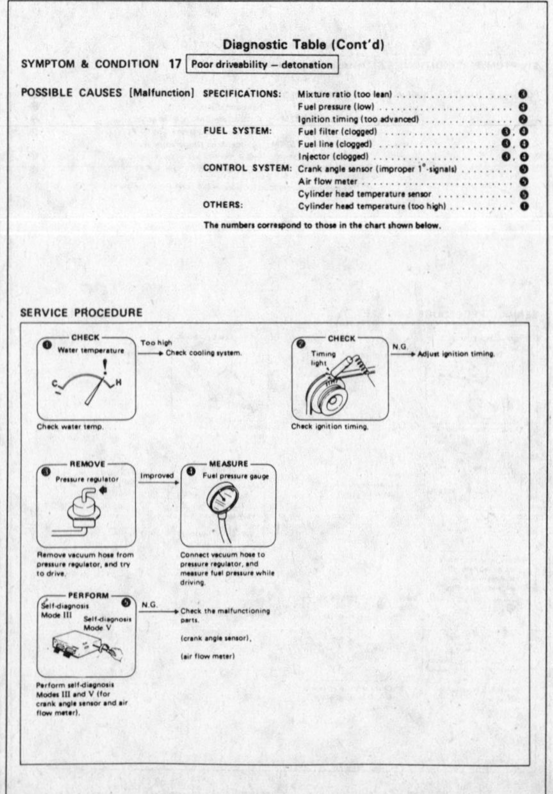
Diagnostic Table (Cont'd)
SYMPTOM & CONDITION 17 Poor driveability – detonation
POSSIBLE CAUSES [Malfunction]
SPECIFICATIONS: Mixture ratio (too lean) ❸
Fuel pressure (low) ❹
Ignition timing (too advanced) ❷
FUEL SYSTEM: Fuel filter (clogged) ❸, ❹
Fuel line (clogged) ❸, ❹
Injector (clogged) ❸, ❹
CONTROL SYSTEM: Crank angle sensor (improper 1°-signals) ❺
Air flow meter ❺
Cylinder head temperature sensor ❺
OTHERS: Cylinder head temperature (too high) ❶
The numbers correspond to those in the chart shown below.
SERVICE PROCEDURE
❶ CHECK
Water temperature
C
H
Check water temp.
Too high
Check cooling system.
❷ CHECK
Timing light
Check ignition timing.
N.G.
Adjust ignition timing.
❸ REMOVE
Pressure regulator
Remove vacuum hose from pressure regulator, and try to drive.
Improved
❹ MEASURE
Fuel pressure gauge
Connect vacuum hose to pressure regulator, and measure fuel pressure while driving.
❺ PERFORM
Self-diagnosis Mode III
Self-diagnosis Mode V
Perform self-diagnosis Modes III and V (for crank angle sensor and air flow meter).
N.G.
Check the malfunctioning parts.
(crank angle sensor),
(air flow meter)

ECCS DIAGNOSTIC CHART—1988 MAXIMA

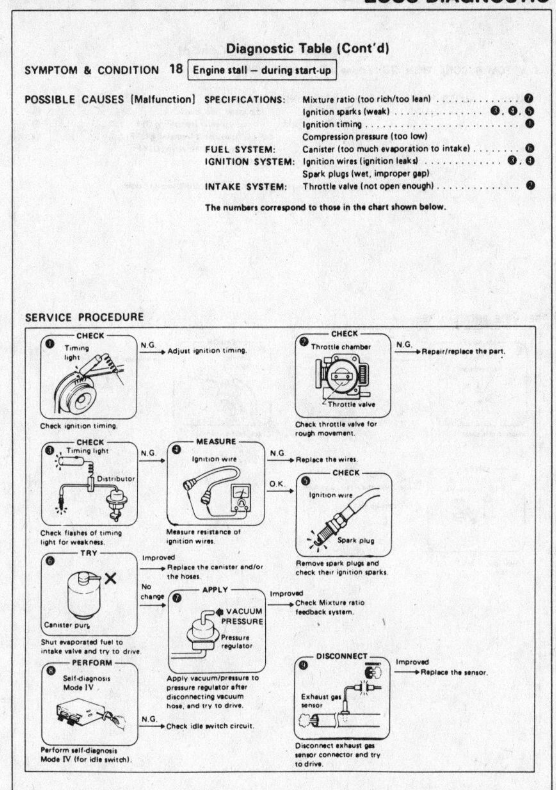
Diagnostic Table (Cont'd)
SYMPTOM & CONDITION 18 Engine stall – during start-up
POSSIBLE CAUSES [Malfunction] SPECIFICATIONS: Mixture ratio (too rich/too lean) 7
Ignition sparks (weak) 3, 4, 5
Ignition timing 1
Compression pressure (too low)
FUEL SYSTEM: Canister (too much evaporation to intake) 6
IGNITION SYSTEM: Ignition wires (ignition leaks) 3, 4
Spark plugs (wet, improper gap)
INTAKE SYSTEM: Throttle valve (not open enough) 2
The numbers correspond to those in the chart shown below.
SERVICE PROCEDURE
CHECK Timing light
N.G. Adjust ignition timing.
Check ignition timing.
CHECK Throttle chamber
N.G. Repair/replace the part.
Throttle valve
Check throttle valve for rough movement.
CHECK Timing light Distributor
Check flashes of timing light for weakness.
MEASURE Ignition wire
Measure resistance of ignition wires.
N.G. Replace the wires.
O.K.
CHECK Ignition wire Spark plug
Remove spark plugs and check their ignition sparks.
TRY
Canister purg
Shut evaporated fuel to intake valve and try to drive.
Improved Replace the canister and/or the hoses.
No change
APPLY VACUUM PRESSURE Pressure regulator
Apply vacuum/pressure to pressure regulator after disconnecting vacuum hose, and try to drive.
Improved Check Mixture ratio feedback system.
PERFORM Self-diagnosis Mode IV
N.G. Check idle switch circuit.
Perform self-diagnosis Mode IV (for idle switch).
DISCONNECT Exhaust gas sensor
Improved Replace the sensor.
Disconnect exhaust gas sensor connector and try to drive.

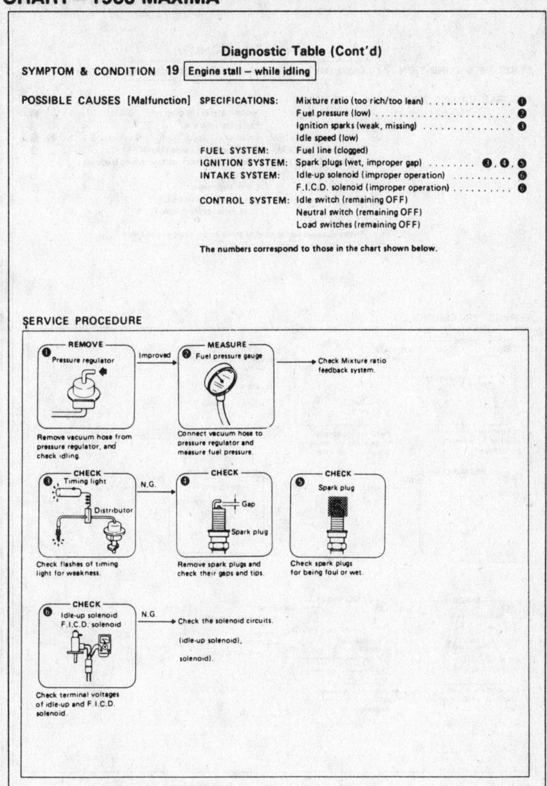
Diagnostic Table (Cont'd)
SYMPTOM & CONDITION 19 Engine stall – while idling
POSSIBLE CAUSES [Malfunction] SPECIFICATIONS: Mixture ratio (too rich/too lean) 1
Fuel pressure (low) 2
Ignition sparks (weak, missing) 3
Idle speed (low)
FUEL SYSTEM: Fuel line (clogged)
IGNITION SYSTEM: Spark plugs (wet, improper gap) 3, 4, 5
INTAKE SYSTEM: Idle-up solenoid (improper operation) 6
F.I.C.D. solenoid (improper operation) 6
CONTROL SYSTEM: Idle switch (remaining OFF)
Neutral switch (remaining OFF)
Load switches (remaining OFF)
The numbers correspond to those in the chart shown below.
SERVICE PROCEDURE
REMOVE Pressure regulator
Remove vacuum hose from pressure regulator, and check idling.
Improved
MEASURE Fuel pressure gauge
Connect vacuum hose to pressure regulator and measure fuel pressure.
Check Mixture ratio feedback system.
CHECK Timing light Distributor
Check flashes of timing light for weakness.
N.G.
CHECK Gap Spark plug
Remove spark plugs and check their gaps and tips.
CHECK Spark plug
Check spark plugs for being foul or wet.
CHECK Idle-up solenoid F.I.C.D. solenoid
N.G. Check the solenoid circuits. (idle-up solenoid), solenoid).
Check terminal voltages of idle-up and F.I.C.D. solenoid.

ECCS DIAGNOSTIC CHART—1988 MAXIMA

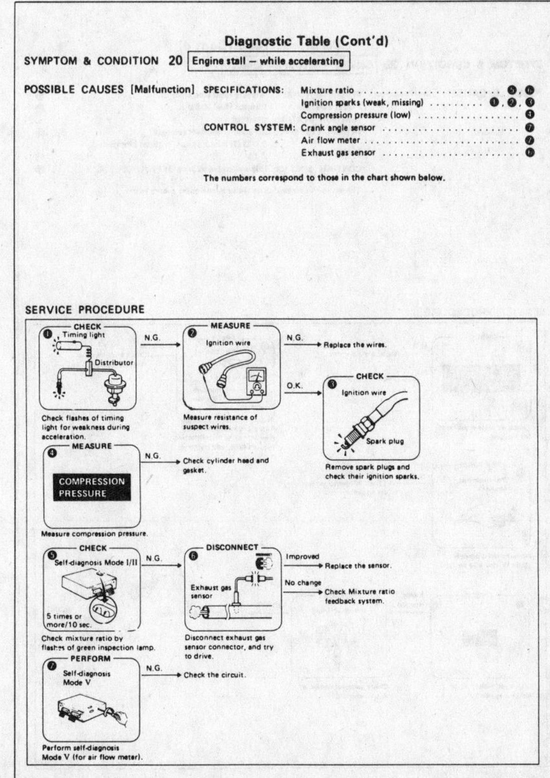
Diagnostic Table (Cont'd)
SYMPTOM & CONDITION 20 Engine stall – while accelerating
POSSIBLE CAUSES [Malfunction] SPECIFICATIONS: Mixture ratio 5, 6
Ignition sparks (weak, missing) 1, 2, 3
Compression pressure (low) 4
CONTROL SYSTEM: Crank angle sensor 7
Air flow meter 7
Exhaust gas sensor 6
The numbers correspond to those in the chart shown below.
SERVICE PROCEDURE
CHECK Timing light Distributor
Check flashes of timing light for weakness during acceleration.
N.G.
MEASURE Ignition wire
Measure resistance of suspect wires.
N.G. Replace the wires.
O.K.
CHECK Ignition wire Spark plug
Remove spark plugs and check their ignition sparks.
MEASURE COMPRESSION PRESSURE
Measure compression pressure.
N.G. Check cylinder head and gasket.
CHECK Self-diagnosis Mode I/II
5 times or more/10 sec.
Check mixture ratio by flashes of green inspection lamp.
DISCONNECT Exhaust gas sensor
Improved Replace the sensor.
No change Check Mixture ratio feedback system.
Disconnect exhaust gas sensor connector, and try to drive.
PERFORM Self-diagnosis Mode V
N.G. Check the circuit.
Perform self-diagnosis Mode V (for air flow meter).

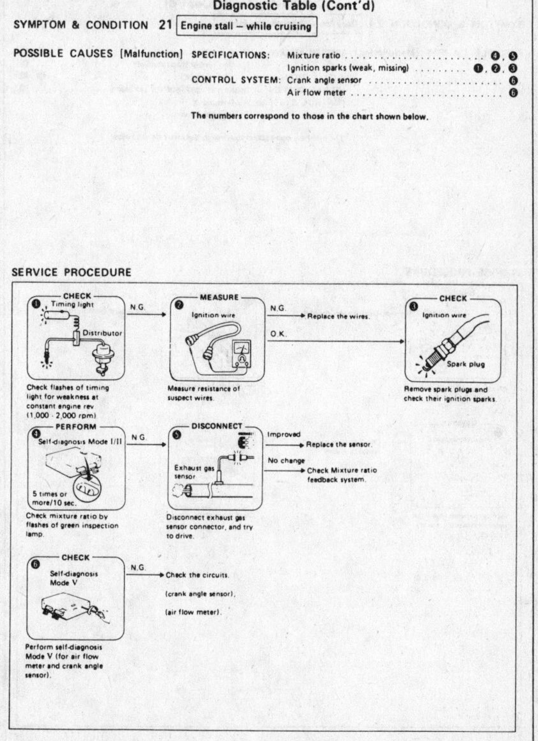
Diagnostic Table (Cont'd)
SYMPTOM & CONDITION 21 Engine stall – while cruising
POSSIBLE CAUSES [Malfunction] SPECIFICATIONS: Mixture ratio 4, 5
Ignition sparks (weak, missing) 1, 2, 3
CONTROL SYSTEM: Crank angle sensor 6
Air flow meter 6
The numbers correspond to those in the chart shown below.
SERVICE PROCEDURE
CHECK Timing light Distributor
Check flashes of timing light for weakness at constant engine rev. (1,000 - 2,000 rpm)
N.G.
MEASURE Ignition wire
Measure resistance of suspect wires.
N.G. Replace the wires.
O.K.
CHECK Ignition wire Spark plug
Remove spark plugs and check their ignition sparks.
PERFORM Self-diagnosis Mode I/II
5 times or more/10 sec.
Check mixture ratio by flashes of green inspection lamp.
N.G.
DISCONNECT Exhaust gas sensor
Improved Replace the sensor.
No change Check Mixture ratio feedback system.
Disconnect exhaust gas sensor connector, and try to drive.
CHECK Self-diagnosis Mode V
N.G. Check the circuits. (crank angle sensor), (air flow meter).
Perform self-diagnosis Mode V (for air flow meter and crank angle sensor).

ECCS DIAGNOSTIC CHART—1988 MAXIMA

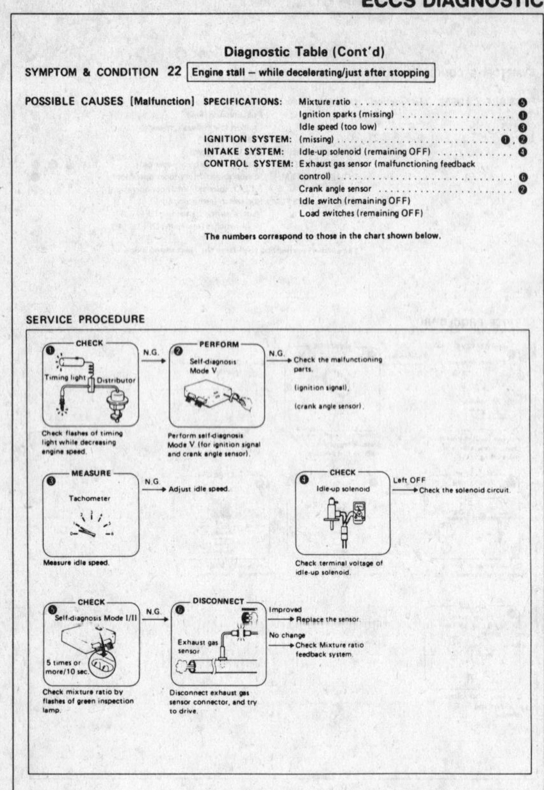
Diagnostic Table (Cont'd)
SYMPTOM & CONDITION 22 Engine stall – while decelerating/just after stopping
POSSIBLE CAUSES [Malfunction]
SPECIFICATIONS: Mixture ratio 5
Ignition sparks (missing) 1
Idle speed (too low) 3
IGNITION SYSTEM: (missing) 1, 2
INTAKE SYSTEM: Idle-up solenoid (remaining OFF) 4
CONTROL SYSTEM: Exhaust gas sensor (malfunctioning feedback control) 6
Crank angle sensor 2
Idle switch (remaining OFF)
Load switches (remaining OFF)
The numbers correspond to those in the chart shown below.
SERVICE PROCEDURE
CHECK
Timing light
Distributor
N.G.
PERFORM
Self-diagnosis Mode V
N.G.
Check the malfunctioning parts. (ignition signal), (crank angle sensor).
Check flashes of timing light while decreasing engine speed.
Perform self-diagnosis Mode V (for ignition signal and crank angle sensor).
MEASURE
Tachometer
N.G.
Adjust idle speed.
Measure idle speed.
CHECK
Idle-up solenoid
Left OFF
Check the solenoid circuit.
Check terminal voltage of idle-up solenoid.
CHECK
Self-diagnosis Mode I/II
5 times or more/10 sec.
N.G.
DISCONNECT
Exhaust gas sensor
Improved
Replace the sensor.
No change
Check Mixture ratio feedback system.
Check mixture ratio by flashes of green inspection lamp.
Disconnect exhaust gas sensor connector, and try to drive.

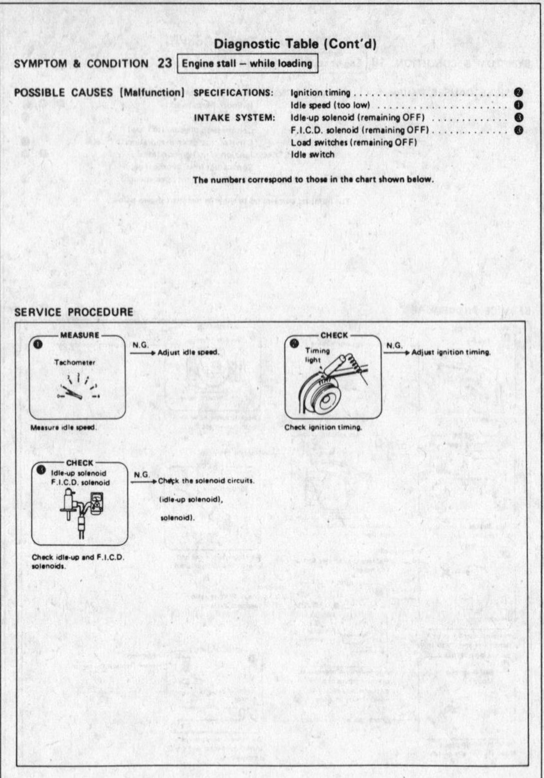
Diagnostic Table (Cont'd)
SYMPTOM & CONDITION 23 Engine stall – while loading
POSSIBLE CAUSES [Malfunction]
SPECIFICATIONS: Ignition timing 2
Idle speed (too low) 1
INTAKE SYSTEM: Idle-up solenoid (remaining OFF) 3
F.I.C.D. solenoid (remaining OFF) 3
Load switches (remaining OFF)
Idle switch
The numbers correspond to those in the chart shown below.
SERVICE PROCEDURE
MEASURE
Tachometer
N.G.
Adjust idle speed.
Measure idle speed.
CHECK
Timing light
N.G.
Adjust ignition timing.
Check ignition timing.
CHECK
Idle-up solenoid
F.I.C.D. solenoid
N.G.
Check the solenoid circuits. (idle-up solenoid), solenoid).
Check idle-up and F.I.C.D. solenoids.

ECCS DIAGNOSTIC CHART—1988 MAXIMA

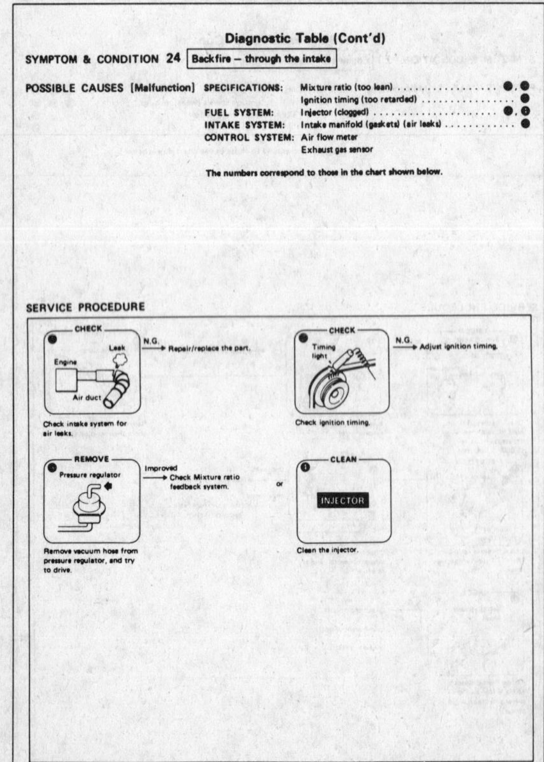
Diagnostic Table (Cont'd)
SYMPTOM & CONDITION 24 Backfire – through the intake
POSSIBLE CAUSES [Malfunction]
SPECIFICATIONS: Mixture ratio (too lean) ....
Ignition timing (too retarded) ....
FUEL SYSTEM: Injector (clogged) ....
INTAKE SYSTEM: Intake manifold (gaskets) (air leaks) ....
CONTROL SYSTEM: Air flow meter
Exhaust gas sensor
The numbers correspond to those in the chart shown below.
SERVICE PROCEDURE
CHECK
Engine
Leak
Air duct
N.G.
Repair/replace the part.
Check intake system for air leaks.
CHECK
Timing light
N.G.
Adjust ignition timing.
Check ignition timing.
REMOVE
Pressure regulator
Improved
Check Mixture ratio feedback system.
Remove vacuum hose from pressure regulator, and try to drive.
or
CLEAN
INJECTOR
Clean the injector.

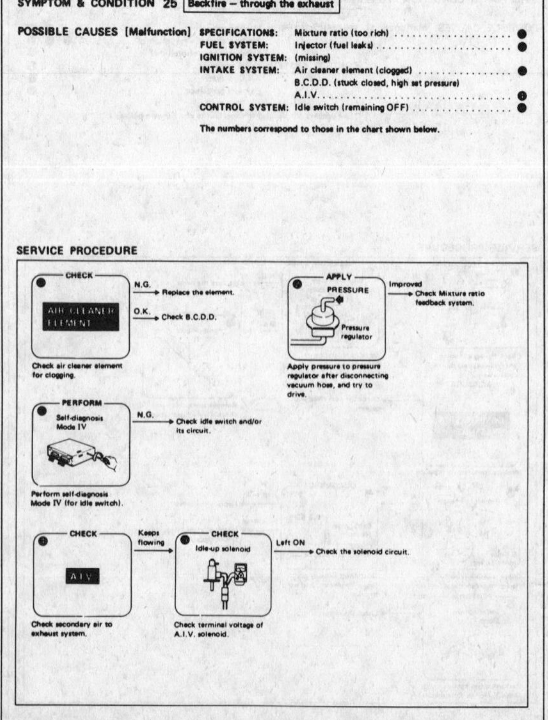
Diagnostic Table (Cont'd)
SYMPTOM & CONDITION 25 Backfire – through the exhaust
POSSIBLE CAUSES [Malfunction]
SPECIFICATIONS: Mixture ratio (too rich) ....
FUEL SYSTEM: Injector (fuel leaks) ....
IGNITION SYSTEM: (missing)
INTAKE SYSTEM: Air cleaner element (clogged) ....
B.C.D.D. (stuck closed, high set pressure)
A.I.V. ....
CONTROL SYSTEM: Idle switch (remaining OFF) ....
The numbers correspond to those in the chart shown below.
SERVICE PROCEDURE
CHECK
AIR CLEANER ELEMENT
N.G.
Replace the element.
O.K.
Check B.C.D.D.
Check air cleaner element for clogging.
APPLY
PRESSURE
Pressure regulator
Improved
Check Mixture ratio feedback system.
Apply pressure to pressure regulator after disconnecting vacuum hose, and try to drive.
PERFORM
Self-diagnosis Mode IV
N.G.
Check idle switch and/or its circuit.
Perform self-diagnosis Mode IV (for idle switch).
CHECK
A.I.V.
Keeps flowing
CHECK
Idle-up solenoid
Left ON
Check the solenoid circuit.
Check secondary air to exhaust system.
Check terminal voltage of A.I.V. solenoid.

ECCS SELF-DIAGNOSTIC DESCRIPTION—1988 MAXIMA

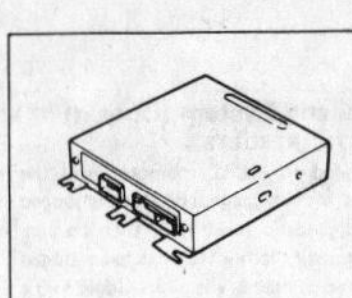

Description

The self-diagnosis is useful to diagnose malfunctions in major sensors and actuators of the E.C.C.S. system. There are 5 modes in the self-diagnosis system.

1. **Mode I – Mixture ratio feedback control monitor A**
   - During closed loop condition:
     The green inspection lamp turns ON when lean condition is detected and goes OFF by rich condition.
   - During open loop condition:
     The green inspection lamp remains ON or OFF.
2. **Mode II – Mixture ratio feedback control monitor B**
   The green inspection lamp function is the same as Mode I.
   - During closed loop condition:
     The red inspection lamp turns ON and OFF simultaneously with the green inspection lamp when the mixture ratio is controlled within the specified value.
   - During open loop condition:
     The red inspection lamp remains ON or OFF.
3. **Mode III – Self-diagnosis**
   This mode is the same as the former self-diagnosis in self-diagnosis mode.
4. **Mode IV – Switches ON/OFF diagnosis**
   During this mode, the inspection lamps monitor the switch ON-OFF condition.
   - Idle switch
   - Starter switch
   - Vehicle speed sensor
5. **Mode V – Real time diagnosis**
   The moment the malfunction is detected, the display will be presented immediately. That is, the condition at which the malfunction occurs can be found by observing the inspection lamps during driving test.

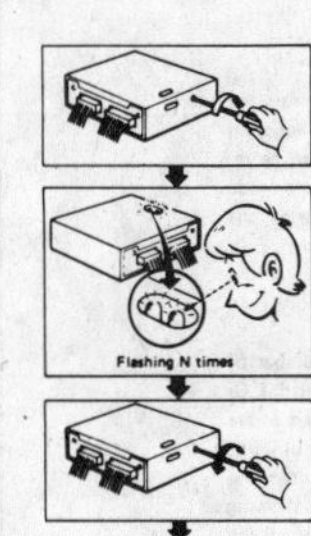
Flashing N times
Mode N

Description (Cont'd)

SWITCHING THE MODES

1. Turn ignition switch "ON".
2. Turn diagnostic mode selector on E.C.U. fully clockwise and wait the inspection lamps flash.
3. Count the number of the flashing time, and after the inspection lamps have flashed the number of the required mode, turn diagnostic mode selector fully counterclockwise immediately.

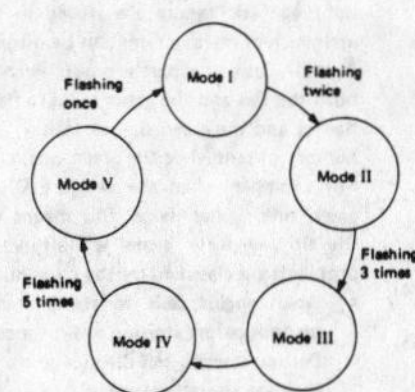
Mode I
Flashing twice
Mode II
Flashing 3 times
Mode III
Flashing 4 times
Mode IV
Flashing 5 times
Mode V
Flashing once

When the ignition switch is turned off during diagnosis, in each mode, and then turned back on again after the power to the E.C.U. has dropped off completely, the diagnosis will automatically return to Mode I.

The stored memory would be lost if:

1. Battery terminal is disconnected.
2. After selecting Mode III, Mode IV is selected.
   However, if the diagnostic mode selector is kept turned fully clockwise, it will continue to change in the order of Mode I → II → III → IV → V → I ... etc., and in this state the stored memory will not be erased.

ECCS SELF-DIAGNOSTIC DESCRIPTION (CONT.) 1988 MAXIMA

CHECK
CHECK ENGINE LIGHT

Description (Cont'd)

CHECK ENGINE LIGHT (For California only)

This vehicle has a check engine light on instrument panel. This light comes ON under the following conditions:

1) When ignition switch is turned "ON" (for bulb check).
2) When systems related to emission performance malfunction in Mode I (with engine running).

- **This check engine light always illuminates and is synchronous with red L.E.D.**
- **Malfunction systems related to emission performance can be detected by self-diagnosis, and they are clarified as self-diagnostic codes in Mode III.**

Code No.	Malfunction
12	Air flow meter circuit
14	Vehicle speed sensor circuit
23	Idle switch circuit
31	E.C.U. (E.C.C.S. control unit)
33	Exhaust gas sensor circuit

Use the following diagnostic flowchart to check and repair a malfunctioning system.

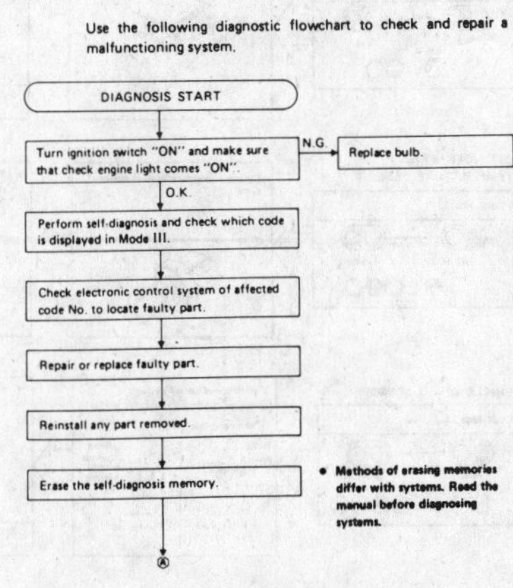
DIAGNOSIS START
Turn ignition switch "ON" and make sure that check engine light comes "ON".
N.G.
Replace bulb.
O.K.
Perform self-diagnosis and check which code is displayed in Mode III.
Check electronic control system of affected code No. to locate faulty part.
Repair or replace faulty part.
Reinstall any part removed.
Erase the self-diagnosis memory.
• Methods of erasing memories differ with systems. Read the manual before diagnosing systems.
A

ECCS SELF-DIAGNOSTIC DESCRIPTION (CONT.) 1988 MAXIMA

Description (Cont'd)

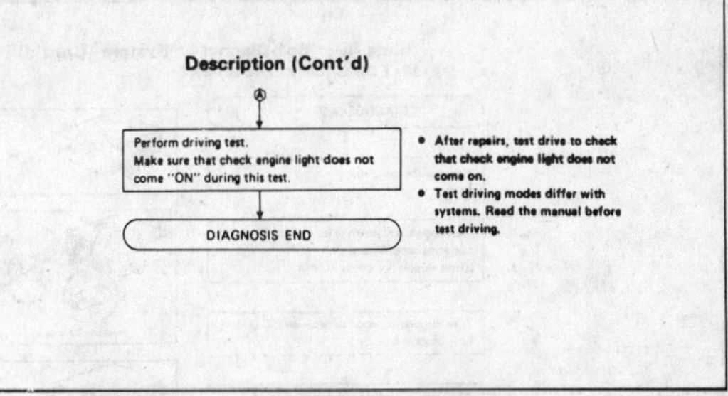
A
Perform driving test.
Make sure that check engine light does not come "ON" during this test.
DIAGNOSIS END
• After repairs, test drive to check that check engine light does not come on.
• Test driving modes differ with systems. Read the manual before test driving.

ECCS SELF-DIAGNOSTIC MODES I AND II 1988 MAXIMA

Modes I & II —Mixture Ratio Feedback Control Monitors A & B

In these modes, the control unit provides the Air-fuel ratio monitor presentation and the Air-fuel ratio feedback coefficient monitor presentation.

Mode	LED	Engine stopped (Ignition switch "ON")	Engine running			
			Open loop condition	Closed loop condition		
Mode I (Monitor A)	Green	ON	*Remains ON or OFF	Blinks		
	Red	ON	Except for California model • OFF	For California model • ON: when CHECK ENGINE LIGHT ITEMS are stored in the E.C.U. • OFF: except for the above condition		
Mode II (Monitor B)	Green	ON	*Remains ON or OFF	Blinks		
	Red	OFF	*Remains ON or OFF (synchronous with green LED)	Compensating mixture ratio		
				More than 5% rich	Between 5% lean and 5% rich	More
				OFF	Synchronized with green LED	Remains ON

*: Maintains conditions just before switching to open loop

ECCS SELF-DIAGNOSTIC MODE III—1988 MAXIMA

Mode III — Self-Diagnostic System

The E.C.U. constantly monitors the function of these sensors and actuators, regardless of ignition key position. If a malfunction occurs, the information is stored in the E.C.U. and can be retrieved from the memory by turning on the diagnostic mode selector, located on the side of the E.C.U. When activated, the malfunction is indicated by flashing a red and a green L.E.D. (Light Emitting Diode), also located on the E.C.U. Since all the self-diagnostic results are stored in the E.C.U.'s memory even intermittent malfunctions can be diagnosed.

A malfunctioning part's group is indicated by the number of both the red and the green L.E.D.s flashing. First, the red L.E.D. flashes and the green flashes follow. The red L.E.D. refers to the number of tens while the green one refers to the number of units. For example, when the red L.E.D. flashes once and then the green one flashes twice, this means the number "12" showing the air flow meter signal is malfunctioning. In this way, all the problems are classified by the code numbers.

- **When engine fails to start, crank engine more than two seconds before starting self-diagnosis.**
- **Before starting self-diagnosis, do not erase stored memory. If doing so, self-diagnosis function for intermittent malfunctions would be lost.**

The stored memory would be lost if:

1. **Battery terminal is disconnected.**
2. **After selecting Mode III, Mode IV is selected.**

DISPLAY CODE TABLE

Code No.	Detected items
11	Crank angle sensor circuit
12	Air flow meter circuit
13	Cylinder head temperature sensor circuit
14	Vehicle speed sensor circuit
21	Ignition signal missing in primary coil
22	Fuel pump circuit
23	Idle switch circuit
31	E.C.U. (E.C.C.S. control unit)
33	Exhaust gas sensor circuit
42	Fuel temperature sensor circuit
55	No malfunction in the above circuit

Mode III — Self-Diagnostic System (Cont'd)

RETENTION OF DIAGNOSTIC RESULTS

The diagnostic result is retained in E.C.U. memory until the starter is operated fifty times after a diagnostic item is judged to be malfunctioning. The diagnostic result will then be cancelled automatically. If a diagnostic item which has been judged to be malfunctioning and stored in memory is again judged to be malfunctioning before the starter is operated fifty times, the second result will replace the previous one. It will be stored in E.C.U. memory until the starter is operated fifty times more.

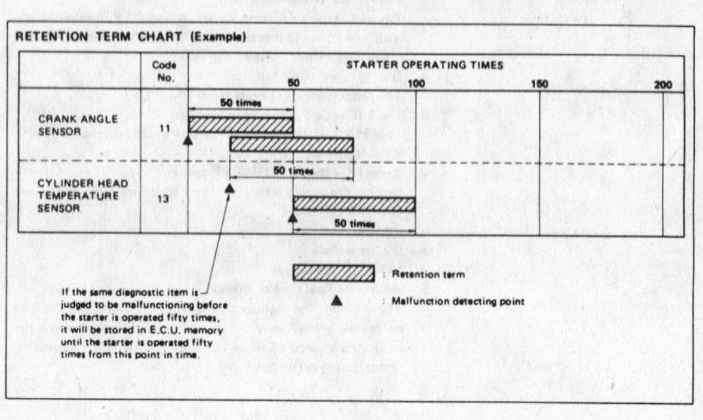
RETENTION TERM CHART (Example)
Code No.
STARTER OPERATING TIMES
50
100
150
200
CRANK ANGLE SENSOR
11
50 times
CYLINDER HEAD TEMPERATURE SENSOR
13
50 times
50 times
If the same diagnostic item is judged to be malfunctioning before the starter is operated fifty times, it will be stored in E.C.U. memory until the starter is operated fifty times from this point in time.
: Retention term
: Malfunction detecting point

ECCS SELF-DIAGNOSTIC MODE III (CONT.) 1988 MAXIMA

Mode III — Self-Diagnostic System (Cont'd)

SELF-DIAGNOSTIC PROCEDURE

DIAGNOSIS START

Remove the assist seat.

Start engine and warm it up to normal engine operating temperature. (Drive vehicle for about 10 min.)

Turn diagnostic mode selector on E.C.U. fully clockwise.

Flashing 3 times

After the inspection lamps have flashed 3 times, turn diagnostic mode selector fully counterclockwise.

Mode III

Make sure that inspection lamps are displaying code No. 55.

N.G. → Write down the malfunctioning code No.

O.K.

Memory erasing procedure

Turn diagnostic mode selector on E.C.U. fully clockwise.

Flashing 4 times

After the inspection lamps have flashed 4 times, turn diagnostic mode selector on E.C.U. fully counterclockwise.

Mode IV

Turn ignition switch "OFF".

Reinstall the E.C.U. in place.

DIAGNOSIS END

(N.G. branch:) Turn ignition switch "OFF".

See decoding chart.

Check malfunctioning parts and/or perform real time diagnosis system inspection. If malfunction part is found, repair or replace it.

CAUTION:

During displaying code No. in self-diagnosis mode (mode III), if the other diagnostic mode should be done, make sure to write down the malfunctioning code No. before turning diagnostic mode selector on E.C.U. fully clockwise, or select the diagnostic mode after turning switch "OFF". Otherwise self-diagnosis information stored in E.C.U. memory until now would be lost.

ECCS SELF-DIAGNOSTIC MODE III DISPLAY CODES 1988 MAXIMA

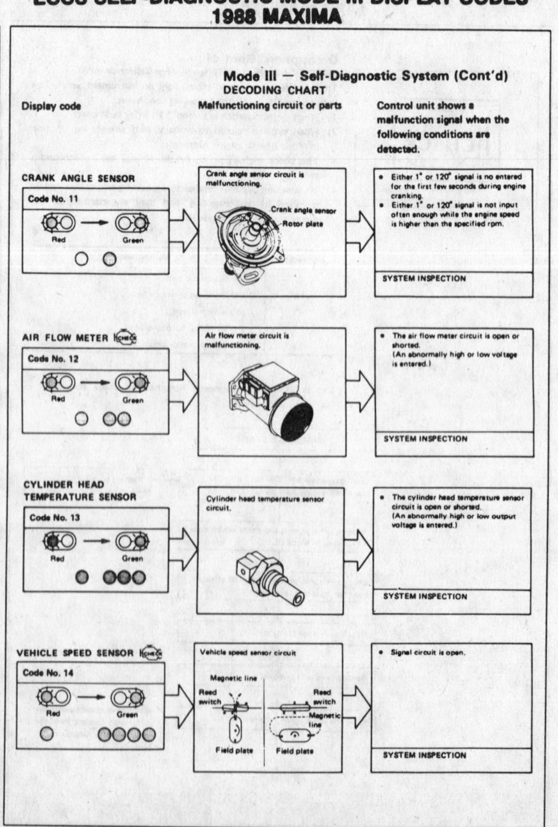
Mode III — Self-Diagnostic System (Cont'd)
DECODING CHART
Display code
Malfunctioning circuit or parts
Control unit shows a malfunction signal when the following conditions are detected.
CRANK ANGLE SENSOR
Code No. 11
Red
Green
Crank angle sensor circuit is malfunctioning.
Crank angle sensor
Rotor plate
• Either 1° or 120° signal is no entered for the first few seconds during engine cranking.
• Either 1° or 120° signal is not input often enough while the engine speed is higher than the specified rpm.
SYSTEM INSPECTION
AIR FLOW METER
Code No. 12
Red
Green
Air flow meter circuit is malfunctioning.
• The air flow meter circuit is open or shorted. (An abnormally high or low voltage is entered.)
SYSTEM INSPECTION
CYLINDER HEAD TEMPERATURE SENSOR
Code No. 13
Red
Green
Cylinder head temperature sensor circuit.
• The cylinder head temperature sensor circuit is open or shorted. (An abnormally high or low output voltage is entered.)
SYSTEM INSPECTION
VEHICLE SPEED SENSOR
Code No. 14
Red
Green
Vehicle speed sensor circuit
Magnetic line
Reed switch
Reed switch
Magnetic line
Field plate
Field plate
• Signal circuit is open.
SYSTEM INSPECTION

ECCS SELF-DIAGNOSTIC MODE III DISPLAY CODES—1988 MAXIMA

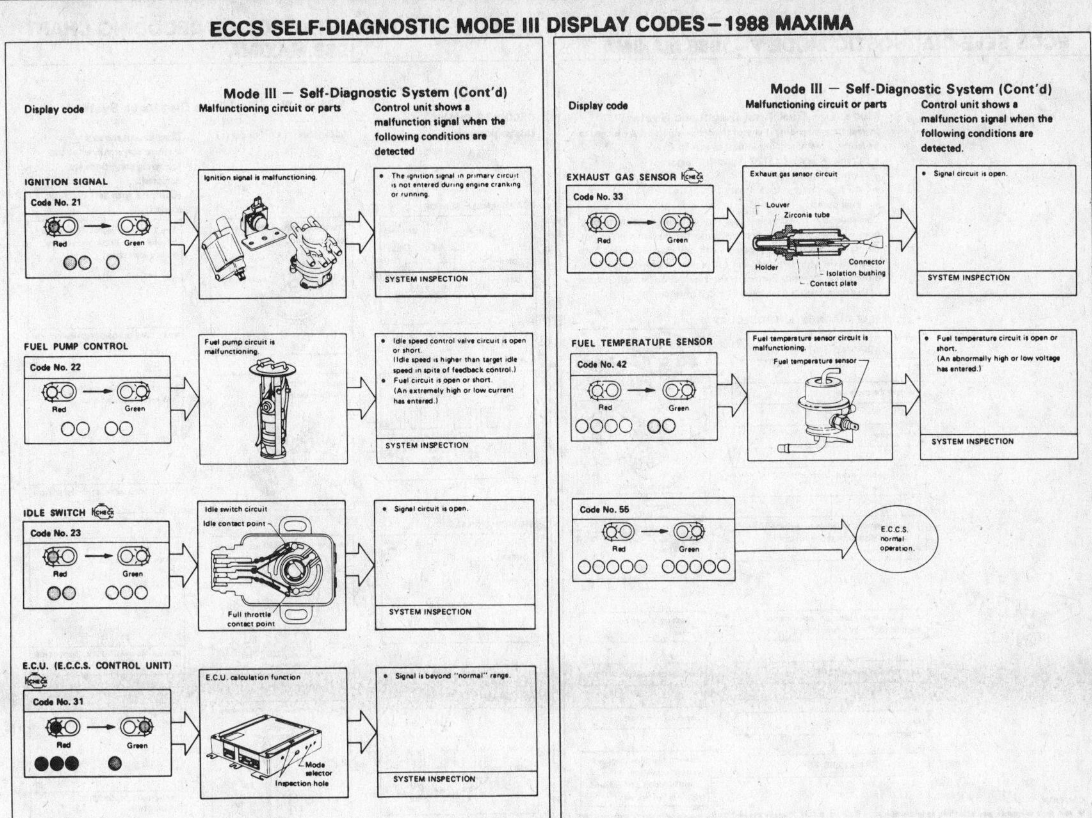
Mode III — Self-Diagnostic System (Cont'd)
Display code
Malfunctioning circuit or parts
Control unit shows a malfunction signal when the following conditions are detected
IGNITION SIGNAL
Code No. 21
Red
Green
Ignition signal is malfunctioning.
The ignition signal in primary circuit is not entered during engine cranking or running.
SYSTEM INSPECTION
FUEL PUMP CONTROL
Code No. 22
Fuel pump circuit is malfunctioning.
Idle speed control valve circuit is open or short. (Idle speed is higher than target idle speed in spite of feedback control.)
Fuel circuit is open or short. (An extremely high or low current has entered.)
IDLE SWITCH
Code No. 23
Idle switch circuit
Idle contact point
Full throttle contact point
Signal circuit is open.
E.C.U. (E.C.C.S. CONTROL UNIT)
Code No. 31
E.C.U. calculation function
Mode selector
Inspection hole
Signal is beyond "normal" range.
EXHAUST GAS SENSOR
Code No. 33
Exhaust gas sensor circuit
Louver
Zirconia tube
Connector
Holder
Isolation bushing
Contact plate
Signal circuit is open.
FUEL TEMPERATURE SENSOR
Code No. 42
Fuel temperature sensor circuit is malfunctioning.
Fuel temperature sensor
Fuel temperature circuit is open or short. (An abnormally high or low voltage has entered.)
Code No. 55
E.C.C.S. normal operation.

ECCS SELF-DIAGNOSTIC MODE IV—1988 MAXIMA

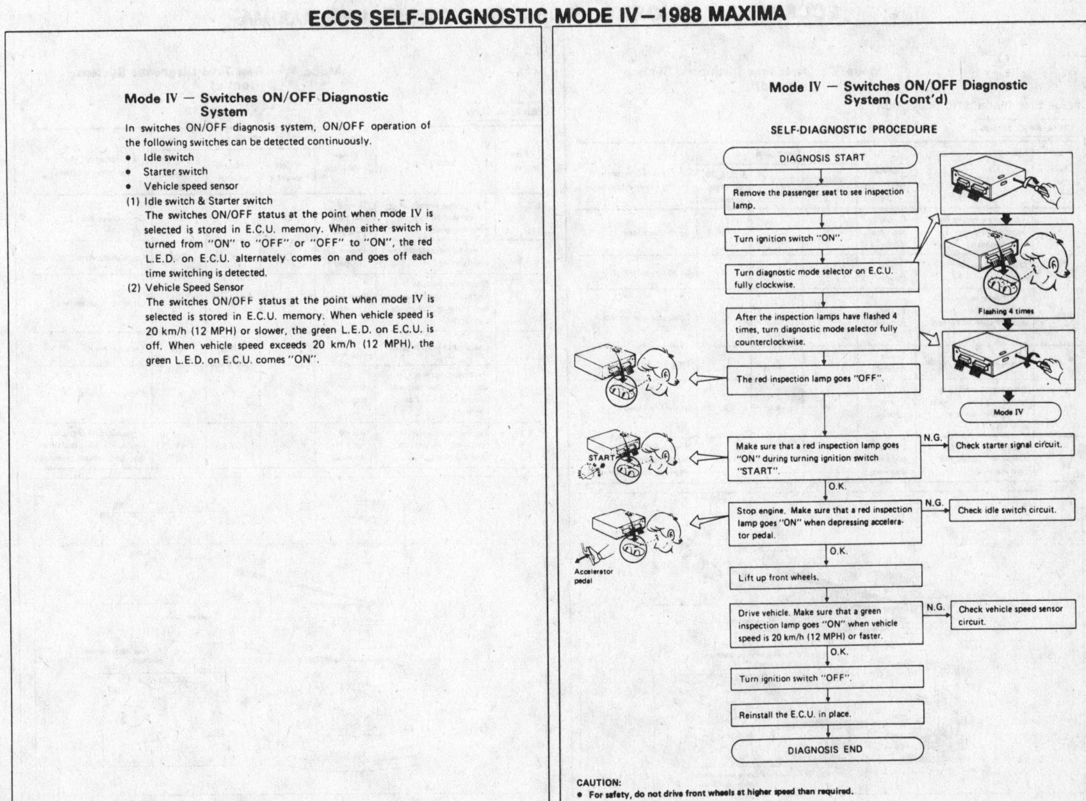

Mode IV — Switches ON/OFF Diagnostic System

In switches ON/OFF diagnosis system, ON/OFF operation of the following switches can be detected continuously.

- Idle switch
- Starter switch
- Vehicle speed sensor

(1) Idle switch & Starter switch
The switches ON/OFF status at the point when mode IV is selected is stored in E.C.U. memory. When either switch is turned from "ON" to "OFF" or "OFF" to "ON", the red L.E.D. on E.C.U. alternately comes on and goes off each time switching is detected.

(2) Vehicle Speed Sensor
The switches ON/OFF status at the point when mode IV is selected is stored in E.C.U. memory. When vehicle speed is 20 km/h (12 MPH) or slower, the green L.E.D. on E.C.U. is off. When vehicle speed exceeds 20 km/h (12 MPH), the green L.E.D. on E.C.U. comes "ON".

Mode IV — Switches ON/OFF Diagnostic System (Cont'd)

SELF-DIAGNOSTIC PROCEDURE

DIAGNOSIS START
Remove the passenger seat to see inspection lamp.
Turn ignition switch "ON".
Turn diagnostic mode selector on E.C.U. fully clockwise.
Flashing 4 times
After the inspection lamps have flashed 4 times, turn diagnostic mode selector fully counterclockwise.
Mode IV
The red inspection lamp goes "OFF".
Make sure that a red inspection lamp goes "ON" during turning ignition switch "START".
START
N.G.
Check starter signal circuit.
O.K.
Stop engine. Make sure that a red inspection lamp goes "ON" when depressing accelerator pedal.
N.G.
Check idle switch circuit.
Accelerator pedal
O.K.
Lift up front wheels.
Drive vehicle. Make sure that a green inspection lamp goes "ON" when vehicle speed is 20 km/h (12 MPH) or faster.
N.G.
Check vehicle speed sensor circuit.
O.K.
Turn ignition switch "OFF".
Reinstall the E.C.U. in place.
DIAGNOSIS END

CAUTION:
- For safety, do not drive front wheels at higher speed than required.

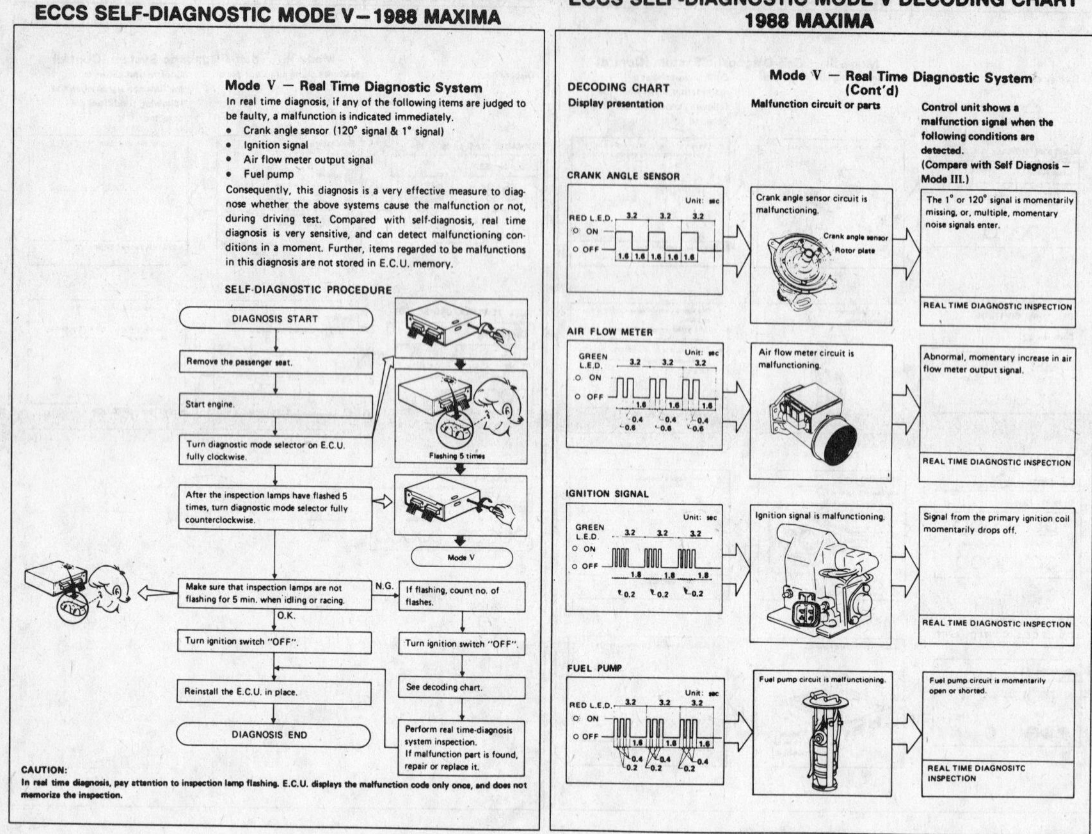

ECCS SELF-DIAGNOSTIC MODE V—1988 MAXIMA

Mode V — Real Time Diagnostic System

In real time diagnosis, if any of the following items are judged to be faulty, a malfunction is indicated immediately.

- Crank angle sensor (120° signal & 1° signal)
- Ignition signal
- Air flow meter output signal
- Fuel pump

Consequently, this diagnosis is a very effective measure to diagnose whether the above systems cause the malfunction or not, during driving test. Compared with self-diagnosis, real time diagnosis is very sensitive, and can detect malfunctioning conditions in a moment. Further, items regarded to be malfunctions in this diagnosis are not stored in E.C.U. memory.

SELF-DIAGNOSTIC PROCEDURE

DIAGNOSIS START

Remove the passenger seat.

Start engine.

Turn diagnostic mode selector on E.C.U. fully clockwise.

Flashing 5 times

After the inspection lamps have flashed 5 times, turn diagnostic mode selector fully counterclockwise.

Mode V

Make sure that inspection lamps are not flashing for 5 min. when idling or racing.

N.G. → If flashing, count no. of flashes. → Turn ignition switch "OFF". → See decoding chart. → Perform real time-diagnosis system inspection. If malfunction part is found, repair or replace it.

O.K. → Turn ignition switch "OFF". → Reinstall the E.C.U. in place.

DIAGNOSIS END

CAUTION:
In real time diagnosis, pay attention to inspection lamp flashing. E.C.U. displays the malfunction code only once, and does not memorize the inspection.

ECCS SELF-DIAGNOSTIC MODE V DECODING CHART 1988 MAXIMA

Mode V — Real Time Diagnostic System (Cont'd)

DECODING CHART

Display presentation	Malfunction circuit or parts	Control unit shows a malfunction signal when the following conditions are detected. (Compare with Self Diagnosis – Mode III.)
CRANK ANGLE SENSOR (RED L.E.D.; Unit: sec; 3.2, 3.2, 3.2; 1.6, 1.6, 1.6, 1.6, 1.6)	Crank angle sensor circuit is malfunctioning. (Crank angle sensor; Rotor plate)	The 1° or 120° signal is momentarily missing, or, multiple, momentary noise signals enter. REAL TIME DIAGNOSTIC INSPECTION
AIR FLOW METER (GREEN L.E.D.; Unit: sec; 3.2, 3.2, 3.2; 1.6, 1.6, 1.6; 0.4, 0.4, 0.4; 0.6, 0.6, 0.6)	Air flow meter circuit is malfunctioning.	Abnormal, momentary increase in air flow meter output signal. REAL TIME DIAGNOSTIC INSPECTION
IGNITION SIGNAL (GREEN L.E.D.; Unit: sec; 3.2, 3.2, 3.2; 1.8, 1.8, 1.8; 0.2, 0.2, 0.2)	Ignition signal is malfunctioning.	Signal from the primary ignition coil momentarily drops off. REAL TIME DIAGNOSTIC INSPECTION
FUEL PUMP (RED L.E.D.; Unit: sec; 3.2, 3.2, 3.2; 1.6, 1.6, 1.6; 0.4, 0.4, 0.4; 0.2, 0.2, 0.2)	Fuel pump circuit is malfunctioning.	Fuel pump circuit is momentarily open or shorted. REAL TIME DIAGNOSITC INSPECTION

ECCS SELF-DIAGNOSTIC MODE V INSPECTION—1988 MAXIMA

Mode V — Real Time Diagnostic System (Cont'd)

REAL TIME DIAGNOSTIC INSPECTION

Crank Angle Sensor

Check sequence	Check items	Check conditions	Check parts: Middle connector	Sensor & actuator	E.C.U. 20- & 16-pin connector	If malfunction, perform the following items.
1	Tap harness connector or component during real time diagnosis.	During real time diagnosis	O	O	O	Go to check item 2.
2	Check harness continuity at connector.	Engine stopped	O	X	X	Go to check item 3.
3	Disconnect harness connector, and then check dust adhesion to harness connector.	Engine stopped	O	X	O	Clean terminal surface.
4	Check pin terminal bend.	Engine stopped	X	X	O	Take out bend.
5	Reconnect harness connector and then recheck harness continuity at connector.	Engine stopped	O	X	X	Replace terminal.
6	Tap harness connector or component during real time diagnosis.	During real time diagnosis	O	O	O	If malfunction codes are displayed during real time diagnosis, replace terminal.

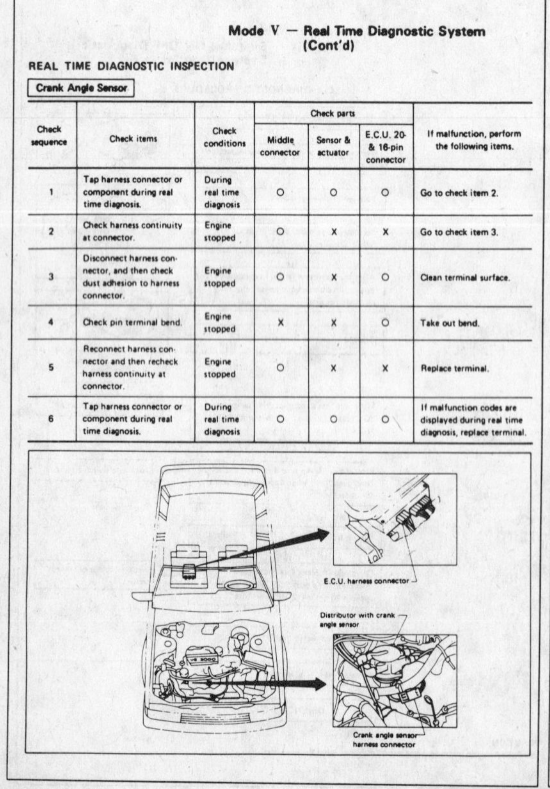
E.C.U. harness connector
Distributor with crank angle sensor
Crank angle sensor harness connector

Mode V — Real Time Diagnostic System (Cont'd)

Air Flow Meter

Check sequence	Check items	Check conditions	Check parts: Middle connector	Sensor & actuator	E.C.U. 20- & 16-pin connector	If malfunction, perform the following items.
1	Tap harness connector or component during real time diagnosis.	During real time diagnosis	O	O	O	Go to check item 2.
2	Check harness continuity at connector.	Engine stopped	O	X	X	Go to check item 3.
3	Disconnect harness connector, and then check dust adhesion to harness connector.	Engine stopped	O	X	O	Clean terminal surface.
4	Check pin terminal bend.	Engine stopped	X	X	O	Take out bend.
5	Reconnect harness connector and then recheck harness continuity at connector.	Engine stopped	O	X	X	Replace terminal.
6	Tap harness connector or component during real time diagnosis.	During real time diagnosis	O	O	O	If malfunction codes are displayed during real time diagnosis, replace terminal.

E.C.U. harness connector

Air flow meter

Air flow meter harness connector

ECCS SELF-DIAGNOSTIC MODE V INSPECTION (CONT.)—1988 MAXIMA

Mode V — Real Time Diagnostic System (Cont'd)

Ignition Signal

Check sequence	Check items	Check conditions	Check parts: Middle connector	Sensor & actuator	E.C.U. 20- & 16-pin connector	If malfunction, perform the following items.
1	Tap harness connector or component during real time diagnosis.	During real time diagnosis	O	O	O	Go to check item 2.
2	Check harness continuity at connector.	Engine stopped	O	X	X	Go to check item 3.
3	Disconnect harness connector, and then check dust adhesion to harness connector.	Engine stopped	O	X	O	Clean terminal surface.
4	Check pin terminal bend.	Engine stopped	X	X	O	Take out bend.
5	Reconnect harness connector and then recheck harness continuity at connector.	Engine stopped	O	X	X	Replace terminal.
6	Tap harness connector or component during real time diagnosis.	During real time diagnosis	O	O	O	If malfunction codes are displayed during real time diagnosis, replace terminal.

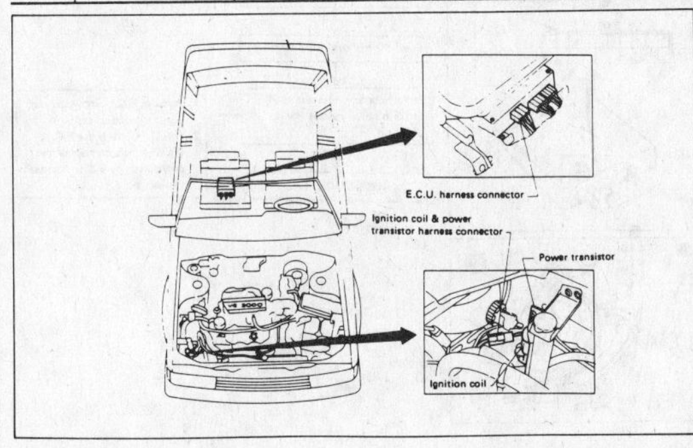
E.C.U. harness connector
Ignition coil & power transistor harness connector
Power transistor
Ignition coil

Mode V — Real Time Diagnostic System (Cont'd)

Fuel pump

Check sequence	Check items	Check conditions	Check parts: Middle connector	Sensor & actuator	E.C.U. 20- & 16-pin connector	If malfunction, perform the following items.
1	Tap harness connector or component during real time diagnosis.	During real time diagnosis	O	O	O	Go to check item 2.
2	Check harness continuity at connector.	Engine stopped	O	X	X	Go to check item 3.
3	Disconnect harness connector, and then check dust adhesion to harness connector.	Engine stopped	O	X	O	Clean terminal surface.
4	Check pin terminal bend.	Engine stopped	X	X	O	Take out bend.
5	Reconnect harness connector and then recheck harness continuity at connector.	Engine stopped	O	X	X	Replace terminal.
6	Tap harness connector or component during real time diagnosis.	During real time diagnosis	O	O	O	If malfunction codes are displayed during real time diagnosis, replace terminal.

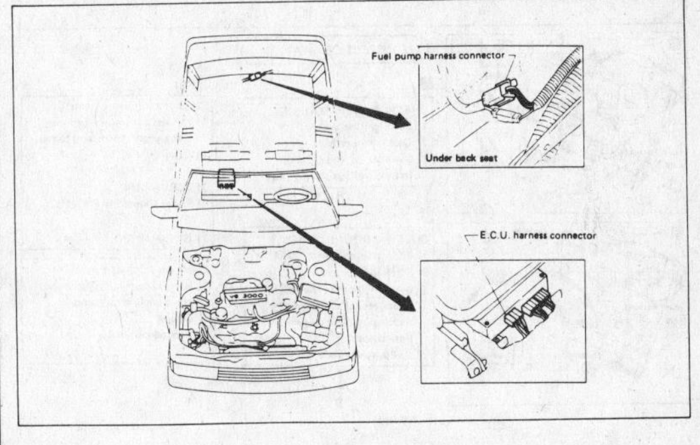
Fuel pump harness connector
Under back seat
E.C.U. harness connector

ECCS CAUTION—1988 MAXIMA

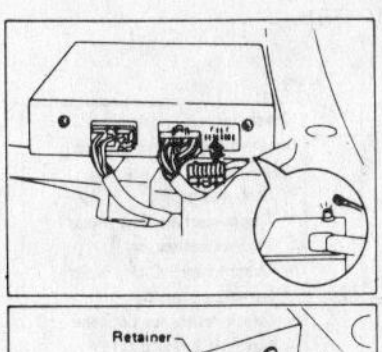

CAUTION:

1. Before connecting or disconnecting E.C.U. harness connector to or from any E.C.U., be sure to turn the ignition switch to the "OFF" position and disconnect the negative battery terminal in order not to damage E.C.U. as battery voltage is applied to E.C.U. even if ignition switch is turned off. Otherwise, there may be damage to the E.C.U.

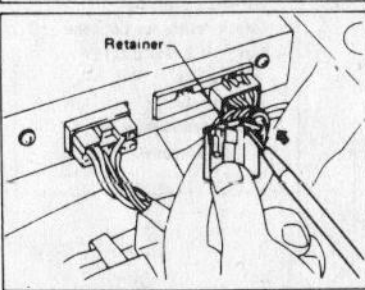
Retainer

2. When performing E.C.U. input/output signal inspection, remove pin terminal retainer from 20-and 16-pin connector to make it easier to insert tester probe into connector.

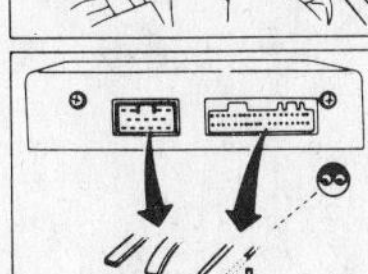
Bend
Break

3. When connecting pin connectors into E.C.U. or disconnecting them from E.C.U., take care not to damage pin terminal of E.C.U. (Bend or break).
4. Make sure that there are not any bends or breaks on E.C.U. pin terminal, when connecting pin connectors into E.C.U.

Perform E.C.U. input/output signal inspection before replacement.
OLD ONE

5. Before replacing E.C.U., perform E.C.U. input/output signal inspection and make sure whether E.C.U. functions properly or not.

6. After performing this "ELECTRONIC CONTROL SYSTEM INSPECTION", perform E.C.C.S. self-diagnosis and driving test.

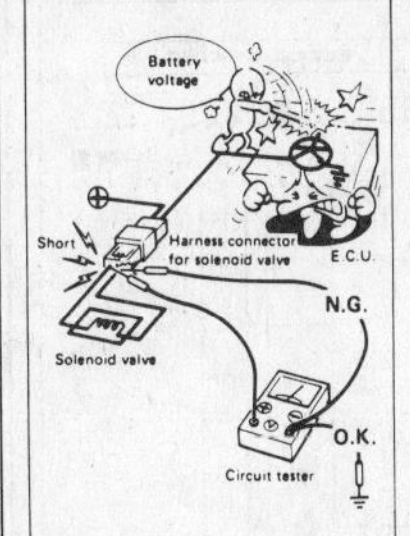
Battery voltage
Short
Harness connector for solenoid valve
E.C.U.
N.G.
Solenoid valve
O.K.
Circuit tester

7. When measuring supply voltage of E.C.U. controlled components with a circuit tester, separate one tester probe from the other.
   If the two tester probes accidentally make contact with each other during measurement, the circuit will be shorted, resulting in damage to the power transistor of the control unit.
8. Keys to symbols

    : Check after disconnecting the connector to be measured.

   Check after connecting the connector to be measured.
9. When measuring voltage or resistance at connector with tester probes, there are two methods of measurement; one is done from terminal side and the other from harness side. Before measuring, confirm symbol mark again.

   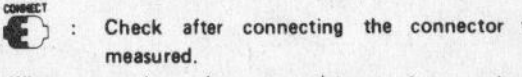 : Inspection should be done from harness side.

    : Inspection should be done from terminal side.
10. As for continuity check of joint connector.

ECCS SELF-DIAGNOSTIC CHART—1988 MAXIMA

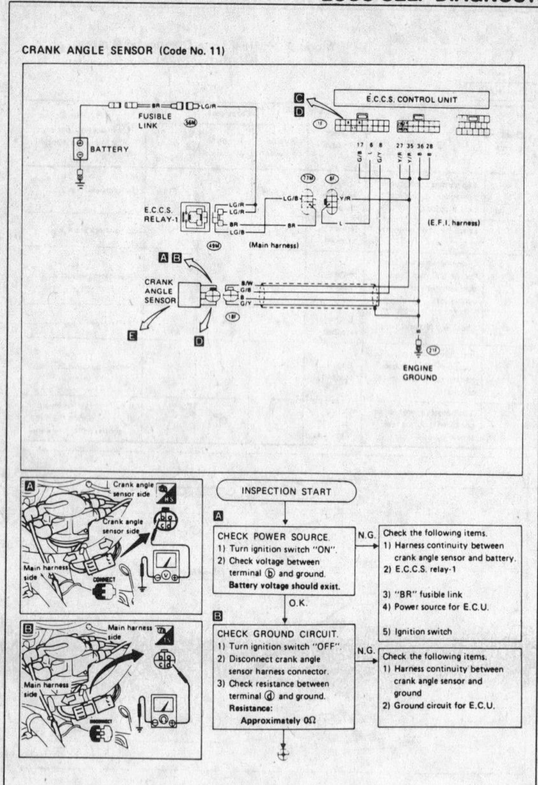
CRANK ANGLE SENSOR (Code No. 11)
E.C.C.S. CONTROL UNIT
FUSIBLE LINK
BATTERY
E.C.C.S. RELAY-1
(Main harness)
(E.F.I. harness)
CRANK ANGLE SENSOR
ENGINE GROUND
INSPECTION START
CHECK POWER SOURCE.
1) Turn ignition switch "ON".
2) Check voltage between terminal ⓑ and ground.
Battery voltage should exist.
N.G.
Check the following items.
1) Harness continuity between crank angle sensor and battery.
2) E.C.C.S. relay-1
3) "BR" fusible link
4) Power source for E.C.U.
5) Ignition switch
O.K.
CHECK GROUND CIRCUIT.
1) Turn ignition switch "OFF".
2) Disconnect crank angle sensor harness connector.
3) Check resistance between terminal ⓓ and ground.
Resistance:
Approximately 0Ω
N.G.
Check the following items.
1) Harness continuity between crank angle sensor and ground
2) Ground circuit for E.C.U.
Crank angle sensor side
Crank angle sensor side
Main harness side
Main harness side

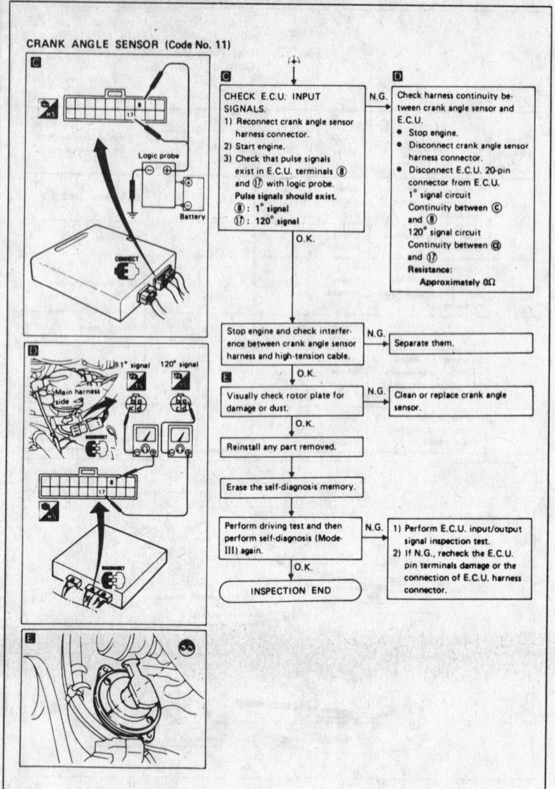
CRANK ANGLE SENSOR (Code No. 11)
Logic probe
Battery
CHECK E.C.U. INPUT SIGNALS.
1) Reconnect crank angle sensor harness connector.
2) Start engine.
3) Check that pulse signals exist in E.C.U. terminals ⑧ and ⑰ with logic probe.
Pulse signals should exist.
⑧: 1° signal
⑰: 120° signal
N.G.
Check harness continuity between crank angle sensor and E.C.U.
• Stop engine.
• Disconnect crank angle sensor harness connector.
• Disconnect E.C.U. 20-pin connector from E.C.U.
1° signal circuit
Continuity between ⓒ and ⑧
120° signal circuit
Continuity between ⓐ and ⑰
Resistance:
Approximately 0Ω
O.K.
Stop engine and check interference between crank angle sensor harness and high-tension cable.
N.G.
Separate them.
O.K.
Visually check rotor plate for damage or dust.
N.G.
Clean or replace crank angle sensor.
O.K.
Reinstall any part removed.
Erase the self-diagnosis memory.
Perform driving test and then perform self-diagnosis (Mode-III) again.
N.G.
1) Perform E.C.U. input/output signal inspection test.
2) If N.G., recheck the E.C.U. pin terminals damage or the connection of E.C.U. harness connector.
O.K.
INSPECTION END
1° signal
120° signal
Main harness side

ECCS SELF-DIAGNOSTIC CHART—1988 MAXIMA

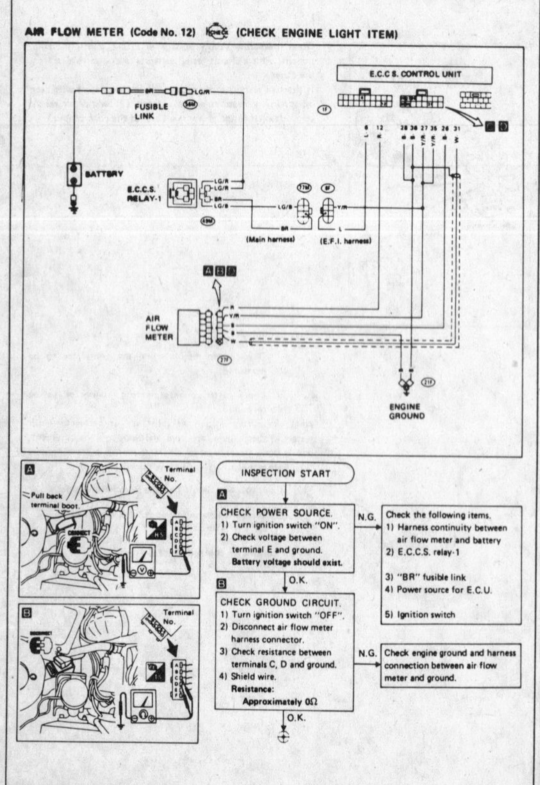
AIR FLOW METER (Code No. 12) (CHECK ENGINE LIGHT ITEM)
E.C.C.S. CONTROL UNIT
FUSIBLE LINK
BATTERY
E.C.C.S. RELAY-1
(Main harness)
(E.F.I. harness)
AIR FLOW METER
ENGINE GROUND
INSPECTION START
CHECK POWER SOURCE.
1) Turn ignition switch "ON".
2) Check voltage between terminal E and ground.
Battery voltage should exist.
N.G.
Check the following items.
1) Harness continuity between air flow meter and battery
2) E.C.C.S. relay-1
3) "BR" fusible link
4) Power source for E.C.U.
5) Ignition switch
O.K.
CHECK GROUND CIRCUIT.
1) Turn ignition switch "OFF".
2) Disconnect air flow meter harness connector.
3) Check resistance between terminals C, D and ground.
4) Shield wire.
Resistance:
Approximately 0Ω
N.G.
Check engine ground and harness connection between air flow meter and ground.
O.K.
Pull back terminal boot
Terminal No.
Terminal No.

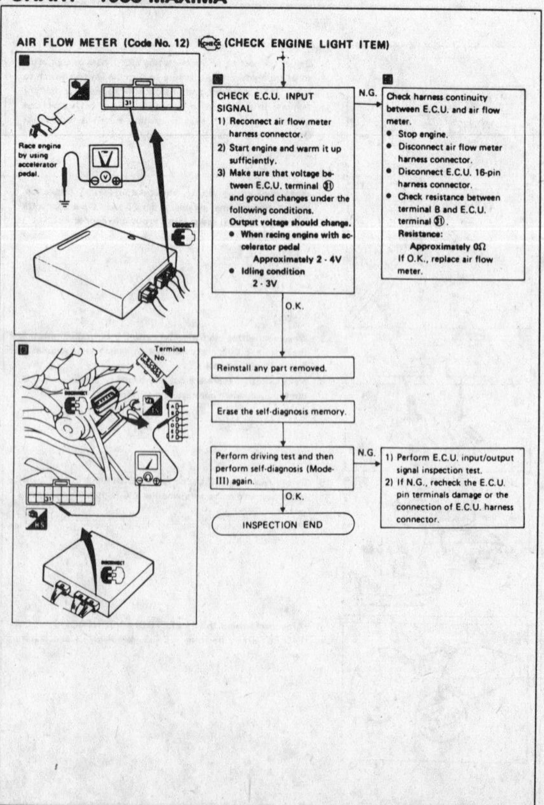
AIR FLOW METER (Code No. 12) (CHECK ENGINE LIGHT ITEM)
Race engine by using accelerator pedal.
CHECK E.C.U. INPUT SIGNAL
1) Reconnect air flow meter harness connector.
2) Start engine and warm it up sufficiently.
3) Make sure that voltage between E.C.U. terminal ㉛ and ground changes under the following conditions.
Output voltage should change.
• When racing engine with accelerator pedal
Approximately 2 - 4V
• Idling condition
2 - 3V
N.G.
Check harness continuity between E.C.U. and air flow meter.
• Stop engine.
• Disconnect air flow meter harness connector.
• Disconnect E.C.U. 16-pin harness connector.
• Check resistance between terminal B and E.C.U. terminal ㉛.
Resistance:
Approximately 0Ω
If O.K., replace air flow meter.
O.K.
Reinstall any part removed.
Erase the self-diagnosis memory.
Perform driving test and then perform self-diagnosis (Mode-III) again.
N.G.
1) Perform E.C.U. input/output signal inspection test.
2) If N.G., recheck the E.C.U. pin terminals damage or the connection of E.C.U. harness connector.
O.K.
INSPECTION END
Terminal No.

ECCS SELF-DIAGNOSTIC CHART—1988 MAXIMA

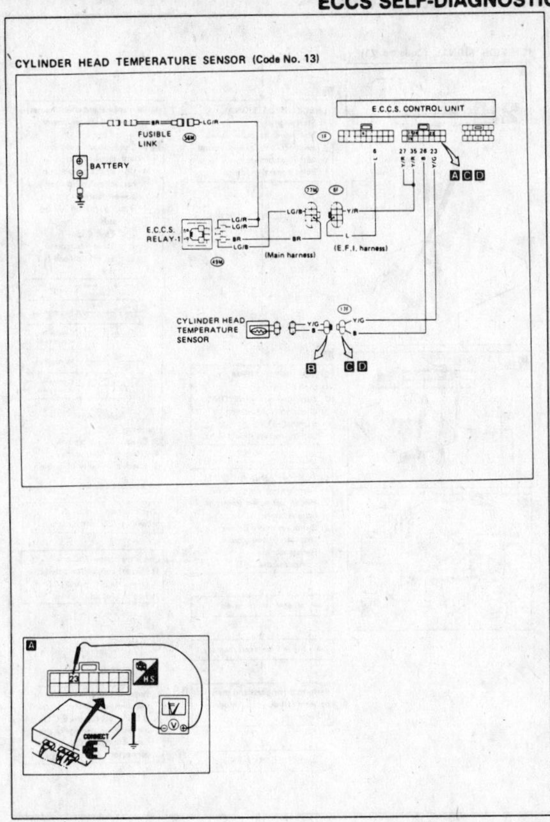
CYLINDER HEAD TEMPERATURE SENSOR (Code No. 13)
E.C.C.S. CONTROL UNIT
FUSIBLE LINK
BATTERY
E.C.C.S. RELAY-1
(Main harness)
(E.F.I. harness)
CYLINDER HEAD TEMPERATURE SENSOR

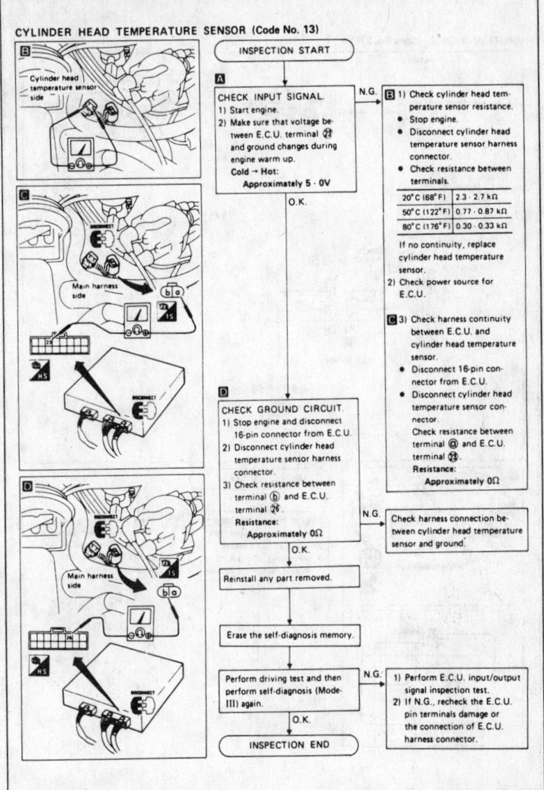
CYLINDER HEAD TEMPERATURE SENSOR (Code No. 13)
Cylinder head temperature sensor side
Main harness side
INSPECTION START
A
CHECK INPUT SIGNAL.
1) Start engine.
2) Make sure that voltage between E.C.U. terminal 23 and ground changes during engine warm up.
Cold → Hot:
Approximately 5 - 0V
N.G.
B 1) Check cylinder head temperature sensor resistance.
• Stop engine.
• Disconnect cylinder head temperature sensor harness connector.
• Check resistance between terminals.
20°C (68°F) 2.3 - 2.7 kΩ
50°C (122°F) 0.77 - 0.87 kΩ
80°C (176°F) 0.30 - 0.33 kΩ
If no continuity, replace cylinder head temperature sensor.
2) Check power source for E.C.U.
C 3) Check harness continuity between E.C.U. and cylinder head temperature sensor.
• Disconnect 16-pin connector from E.C.U.
• Disconnect cylinder head temperature sensor connector.
Check resistance between terminal ⓐ and E.C.U. terminal 23.
Resistance:
Approximately 0Ω
O.K.
D
CHECK GROUND CIRCUIT.
1) Stop engine and disconnect 16-pin connector from E.C.U.
2) Disconnect cylinder head temperature sensor harness connector.
3) Check resistance between terminal ⓑ and E.C.U. terminal 26.
Resistance:
Approximately 0Ω
N.G.
Check harness connection between cylinder head temperature sensor and ground.
O.K.
Reinstall any part removed.
Erase the self-diagnosis memory.
Perform driving test and then perform self-diagnosis (Mode-III) again.
N.G.
1) Perform E.C.U. input/output signal inspection test.
2) If N.G., recheck the E.C.U. pin terminals damage or the connection of E.C.U. harness connector.
O.K.
INSPECTION END

ECCS SELF-DIAGNOSTIC CHART—1988 MAXIMA

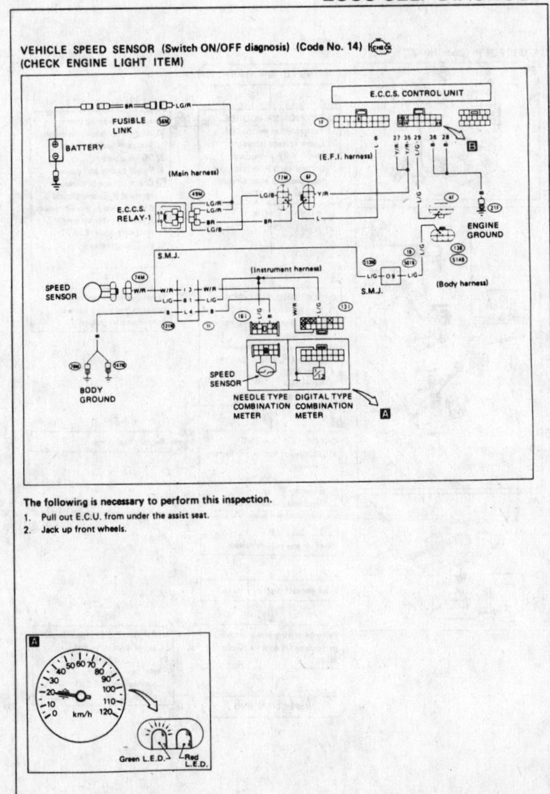
VEHICLE SPEED SENSOR (Switch ON/OFF diagnosis) (Code No. 14)
(CHECK ENGINE LIGHT ITEM)
E.C.C.S. CONTROL UNIT
FUSIBLE LINK
BATTERY
(Main harness)
(E.F.I. harness)
E.C.C.S. RELAY-1
ENGINE GROUND
S.M.J.
(Instrument harness)
(Body harness)
SPEED SENSOR
BODY GROUND
SPEED SENSOR
NEEDLE TYPE COMBINATION METER
DIGITAL TYPE COMBINATION METER
The following is necessary to perform this inspection.
1. Pull out E.C.U. from under the assist seat.
2. Jack up front wheels.
km/h
Green L.E.D.
Red L.E.D.

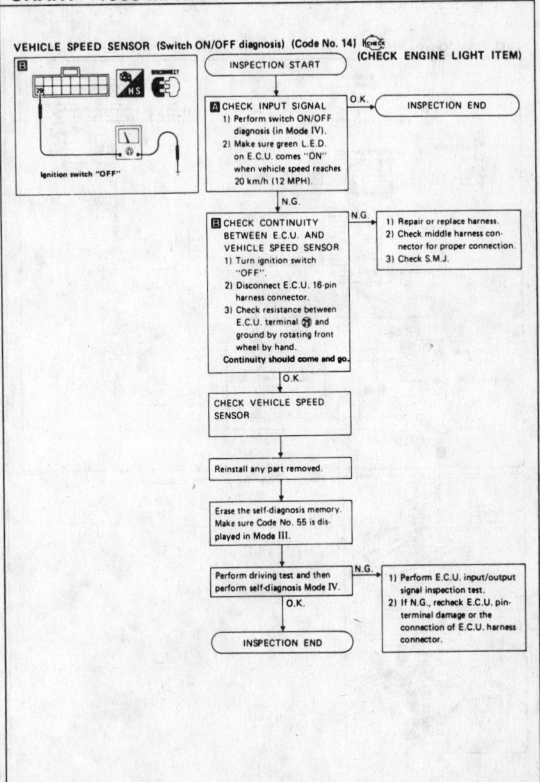
VEHICLE SPEED SENSOR (Switch ON/OFF diagnosis) (Code No. 14)
(CHECK ENGINE LIGHT ITEM)
Ignition switch "OFF"
INSPECTION START
A CHECK INPUT SIGNAL
1) Perform switch ON/OFF diagnosis (in Mode IV).
2) Make sure green L.E.D. on E.C.U. comes "ON" when vehicle speed reaches 20 km/h (12 MPH).
O.K.
INSPECTION END
N.G.
B CHECK CONTINUITY BETWEEN E.C.U. AND VEHICLE SPEED SENSOR
1) Turn ignition switch "OFF".
2) Disconnect E.C.U. 16-pin harness connector.
3) Check resistance between E.C.U. terminal 29 and ground by rotating front wheel by hand.
Continuity should come and go.
N.G.
1) Repair or replace harness.
2) Check middle harness connector for proper connection.
3) Check S.M.J.
O.K.
CHECK VEHICLE SPEED SENSOR
Reinstall any part removed.
Erase the self-diagnosis memory. Make sure Code No. 55 is displayed in Mode III.
Perform driving test and then perform self-diagnosis Mode IV.
N.G.
1) Perform E.C.U. input/output signal inspection test.
2) If N.G., recheck E.C.U. pin-terminal damage or the connection of E.C.U. harness connector.
O.K.
INSPECTION END

ECCS SELF-DIAGNOSTIC CHART—1988 MAXIMA

IGNITION SIGNAL (Code No. 21)

FUSIBLE LINK HOLDER
IGNITION SWITCH
E.C.C.S. CONTROL UNIT
(E.F.I. harness)
BATTERY
ENGINE GROUND
IGNITION COIL
POWER TRANSISTOR
S.M.J.
(Body harness)
(Main harness)
To distributor
BODY GROUND

A
Main harness side

INSPECTION START

A
CHECK POWER SOURCE.
1) Turn ignition switch "ON".
2) Check voltage between terminal ① and ground.
Battery voltage should exist.

N.G. → Check the following items.
1) Harness connection between battery and power transistor
2) "G" fusible link
3) Ignition switch

O.K.

B
To distributor

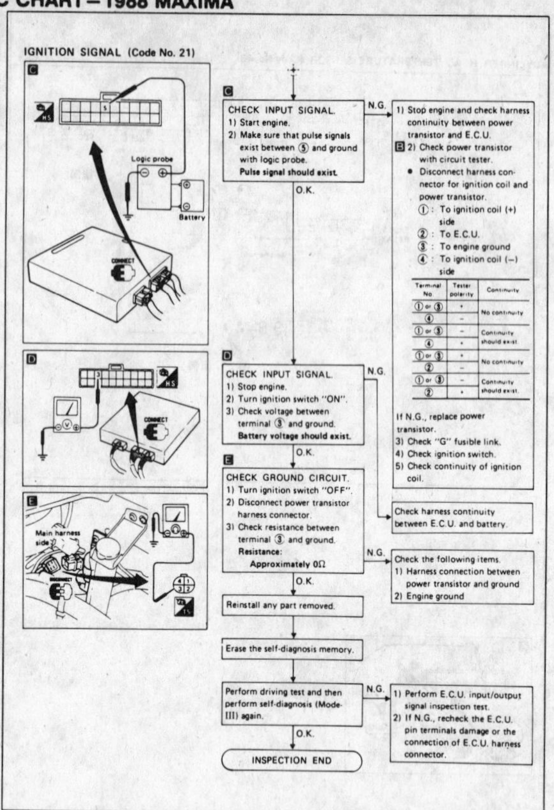
IGNITION SIGNAL (Code No. 21)
C
Logic probe
Battery
D
E
Main harness side
C
CHECK INPUT SIGNAL.
1) Start engine.
2) Make sure that pulse signals exist between ⑤ and ground with logic probe.
Pulse signal should exist.
N.G.
1) Stop engine and check harness continuity between power transistor and E.C.U.
B 2) Check power transistor with circuit tester.
• Disconnect harness connector for ignition coil and power transistor.
①: To ignition coil (+) side
②: To E.C.U.
③: To engine ground
④: To ignition coil (−) side
Terminal No. | Tester polarity | Continuity
① or ③ | + | No continuity
④ | − |
① or ③ | − | Continuity should exist
④ | + |
① or ③ | + | No continuity
② | − |
① or ③ | − | Continuity should exist.
② | + |
If N.G., replace power transistor.
3) Check "G" fusible link.
4) Check ignition switch.
5) Check continuity of ignition coil.
O.K.
D
CHECK INPUT SIGNAL.
1) Stop engine.
2) Turn ignition switch "ON".
3) Check voltage between terminal ③ and ground.
Battery voltage should exist.
N.G.
Check harness continuity between E.C.U. and battery.
O.K.
E
CHECK GROUND CIRCUIT.
1) Turn ignition switch "OFF".
2) Disconnect power transistor harness connector.
3) Check resistance between terminal ③ and ground.
Resistance:
Approximately 0Ω
N.G.
Check the following items.
1) Harness connection between power transistor and ground
2) Engine ground
O.K.
Reinstall any part removed.
Erase the self-diagnosis memory.
Perform driving test and then perform self-diagnosis (Mode-III) again.
N.G.
1) Perform E.C.U. input/output signal inspection test.
2) If N.G., recheck the E.C.U. pin terminals damage or the connection of E.C.U. harness connector.
O.K.
INSPECTION END

ECCS SELF-DIAGNOSTIC CHART—1988 MAXIMA

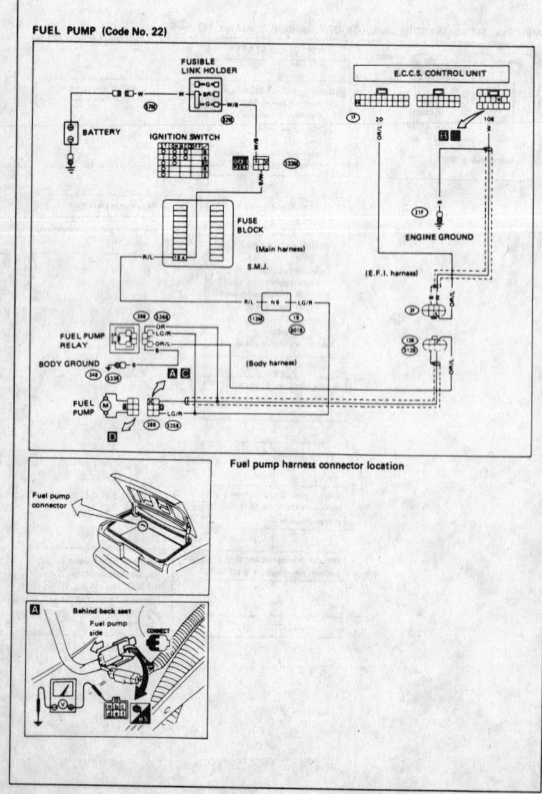
FUEL PUMP (Code No. 22)
FUSIBLE LINK HOLDER
E.C.C.S. CONTROL UNIT
BATTERY
IGNITION SWITCH
FUSE BLOCK
(Main harness)
S.M.J.
(E.F.I. harness)
ENGINE GROUND
FUEL PUMP RELAY
BODY GROUND
(Body harness)
FUEL PUMP
Fuel pump harness connector location
Fuel pump connector
Behind back seat
Fuel pump side

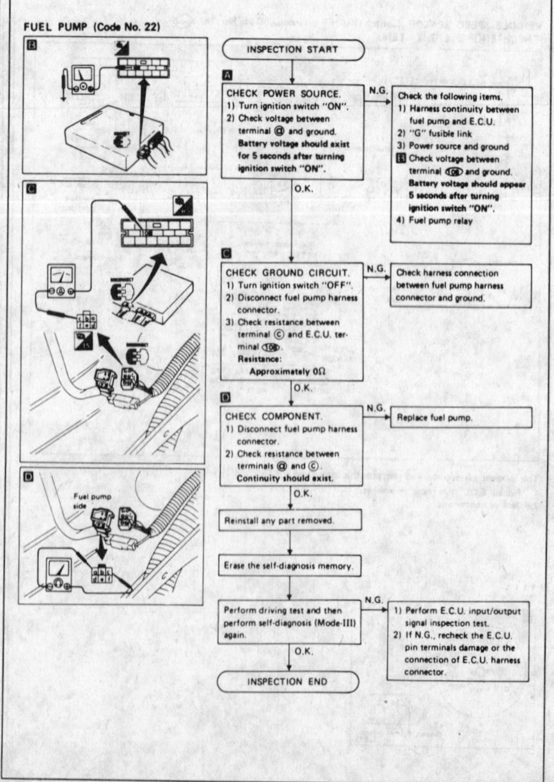
FUEL PUMP (Code No. 22)
B
C
D
Fuel pump side
INSPECTION START
A
CHECK POWER SOURCE.
1) Turn ignition switch "ON".
2) Check voltage between terminal ⓐ and ground.
Battery voltage should exist for 5 seconds after turning ignition switch "ON".
N.G.
Check the following items.
1) Harness continuity between fuel pump and E.C.U.
2) "G" fusible link
3) Power source and ground
B Check voltage between terminal (108) and ground.
Battery voltage should appear 5 seconds after turning ignition switch "ON".
4) Fuel pump relay
O.K.
C
CHECK GROUND CIRCUIT.
1) Turn ignition switch "OFF".
2) Disconnect fuel pump harness connector.
3) Check resistance between terminal ⓒ and E.C.U. terminal (108).
Resistance:
Approximately 0Ω
N.G.
Check harness connection between fuel pump harness connector and ground.
O.K.
D
CHECK COMPONENT.
1) Disconnect fuel pump harness connector.
2) Check resistance between terminals ⓐ and ⓒ.
Continuity should exist.
N.G.
Replace fuel pump.
O.K.
Reinstall any part removed.
Erase the self-diagnosis memory.
Perform driving test and then perform self-diagnosis (Mode-III) again.
N.G.
1) Perform E.C.U. input/output signal inspection test.
2) If N.G., recheck the E.C.U. pin terminals damage or the connection of E.C.U. harness connector.
O.K.
INSPECTION END

ECCS SELF-DIAGNOSTIC CHART—1988 MAXIMA

IDLE SWITCH (Switch ON/OFF diagnosis) (Code No. 23) (CHECK ENGINE LIGHT ITEM)

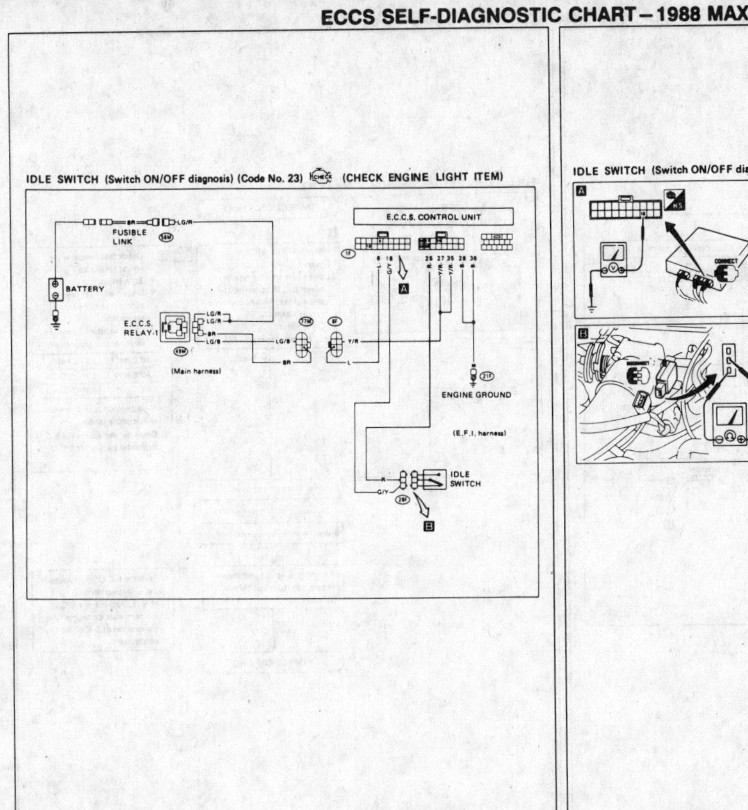
FUSIBLE LINK
BATTERY
E.C.C.S. RELAY-1
(Main harness)
E.C.C.S. CONTROL UNIT
ENGINE GROUND
(E.F.I. harness)
IDLE SWITCH
A
B

IDLE SWITCH (Switch ON/OFF diagnosis) (Code No. 23) (CHECK ENGINE LIGHT ITEM)

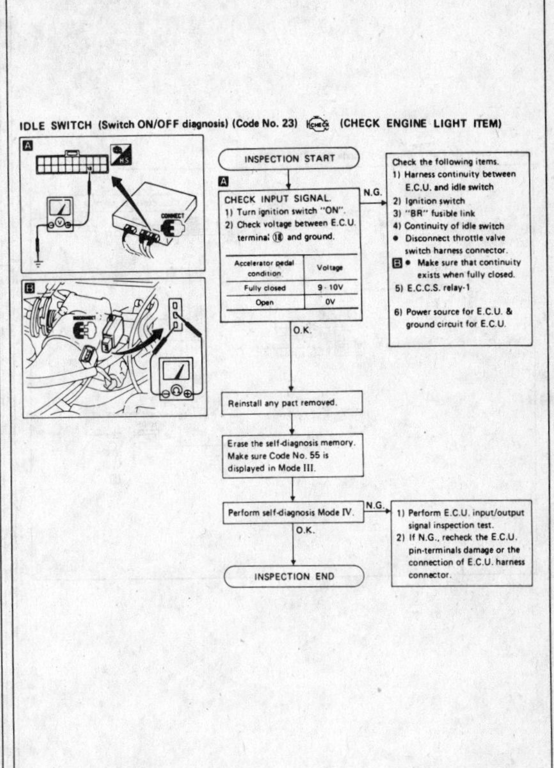
INSPECTION START
A
CHECK INPUT SIGNAL.
1) Turn ignition switch "ON".
2) Check voltage between E.C.U. terminal 18 and ground.
Accelerator pedal condition | Voltage
Fully closed | 9 - 10V
Open | 0V
N.G.
Check the following items.
1) Harness continuity between E.C.U. and idle switch
2) Ignition switch
3) "BR" fusible link
4) Continuity of idle switch
• Disconnect throttle valve switch harness connector.
B • Make sure that continuity exists when fully closed.
5) E.C.C.S. relay-1
6) Power source for E.C.U. & ground circuit for E.C.U.
O.K.
Reinstall any part removed.
Erase the self-diagnosis memory. Make sure Code No. 55 is displayed in Mode III.
Perform self-diagnosis Mode IV.
N.G.
1) Perform E.C.U. input/output signal inspection test.
2) If N.G., recheck the E.C.U. pin-terminals damage or the connection of E.C.U. harness connector.
O.K.
INSPECTION END

ECCS SELF-DIAGNOSTIC CHART—1988 MAXIMA

ENGINE CONTROL UNIT (Code No. 31) (CHECK ENGINE LIGHT ITEM)

INSPECTION START

1) Turn ignition switch "ON".
2) Erase the self-diagnosis memory.

Perform self-diagnosis again.

Does E.C.U. display Code No. 31 again? — YES → Replace E.C.U.

NO

INSPECTION END

EXHAUST GAS SENSOR (Code No. 33) (CHECK ENGINE LIGHT ITEM)

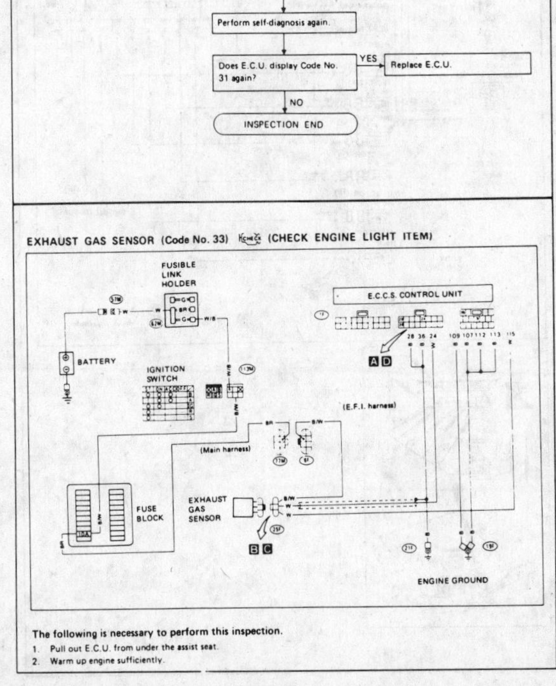
FUSIBLE LINK HOLDER
BATTERY
IGNITION SWITCH
E.C.C.S. CONTROL UNIT
A D
(E.F.I. harness)
(Main harness)
FUSE BLOCK
EXHAUST GAS SENSOR
B C
ENGINE GROUND

The following is necessary to perform this inspection.

1. Pull out E.C.U. from under the assist seat.
2. Warm up engine sufficiently.

EXHAUST GAS SENSOR (Code No. 33) (CHECK ENGINE LIGHT ITEM)

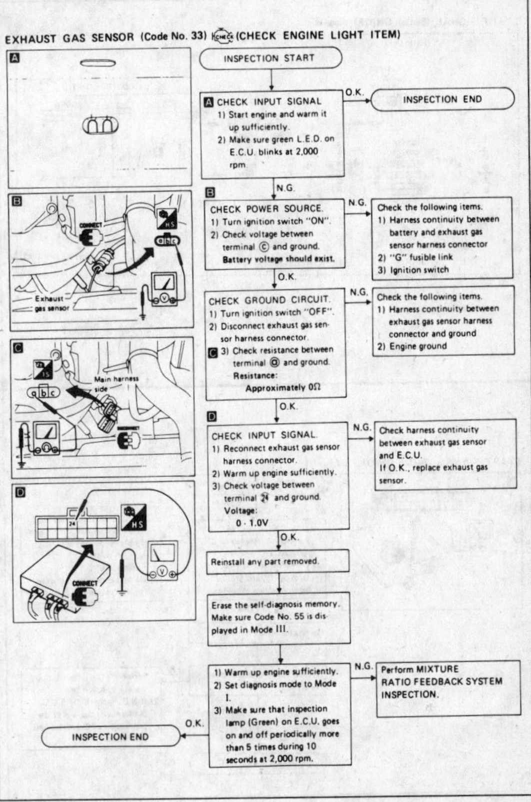
INSPECTION START
A CHECK INPUT SIGNAL
1) Start engine and warm it up sufficiently.
2) Make sure green L.E.D. on E.C.U. blinks at 2,000 rpm.
O.K.
INSPECTION END
N.G.
B
CHECK POWER SOURCE.
1) Turn ignition switch "ON".
2) Check voltage between terminal © and ground.
Battery voltage should exist.
N.G.
Check the following items.
1) Harness continuity between battery and exhaust gas sensor harness connector
2) "G" fusible link
3) Ignition switch
O.K.
CHECK GROUND CIRCUIT.
1) Turn ignition switch "OFF".
2) Disconnect exhaust gas sensor harness connector.
C 3) Check resistance between terminal Ⓑ and ground.
Resistance:
Approximately 0Ω
N.G.
Check the following items.
1) Harness continuity between exhaust gas sensor harness connector and ground
2) Engine ground
O.K.
D
CHECK INPUT SIGNAL.
1) Reconnect exhaust gas sensor harness connector.
2) Warm up engine sufficiently.
3) Check voltage between terminal 24 and ground.
Voltage:
0 - 1.0V
N.G.
Check harness continuity between exhaust gas sensor and E.C.U.
If O.K., replace exhaust gas sensor.
O.K.
Reinstall any part removed.
Erase the self-diagnosis memory. Make sure Code No. 55 is displayed in Mode III.
1) Warm up engine sufficiently.
2) Set diagnosis mode to Mode I.
3) Make sure that inspection lamp (Green) on E.C.U. goes on and off periodically more than 5 times during 10 seconds at 2,000 rpm.
N.G.
Perform MIXTURE RATIO FEEDBACK SYSTEM INSPECTION.
O.K.
INSPECTION END
Exhaust gas sensor
Main harness side
CONNECT

ECCS SELF-DIAGNOSTIC CHART—1988 MAXIMA

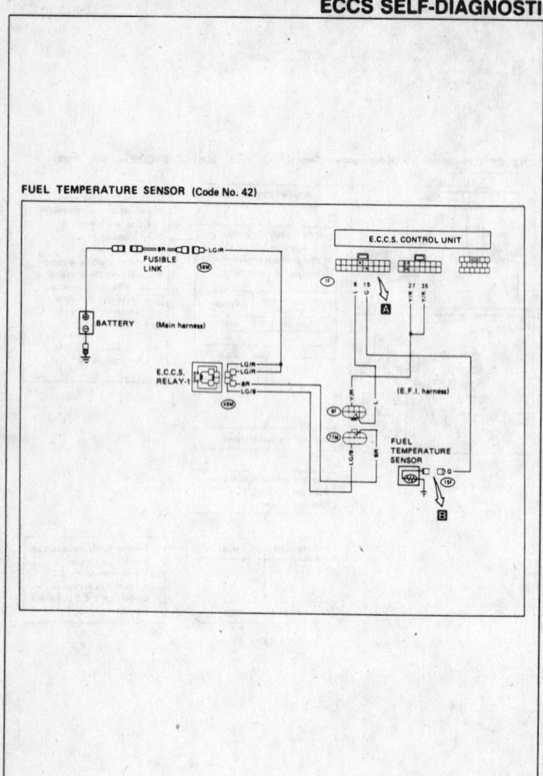
FUEL TEMPERATURE SENSOR (Code No. 42)
E.C.C.S. CONTROL UNIT
FUSIBLE LINK
BATTERY
(Main harness)
E.C.C.S. RELAY-1
(E.F.I. harness)
FUEL TEMPERATURE SENSOR

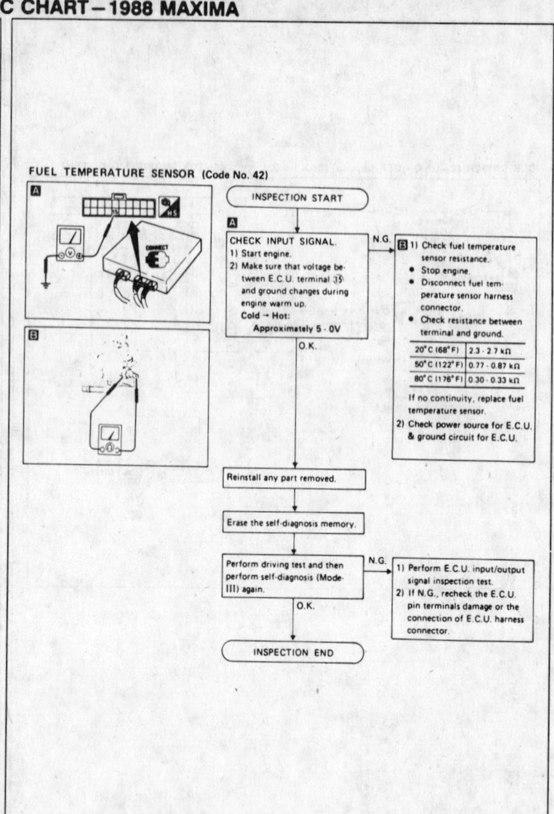
FUEL TEMPERATURE SENSOR (Code No. 42)
INSPECTION START
CHECK INPUT SIGNAL.
1) Start engine.
2) Make sure that voltage between E.C.U. terminal 35 and ground changes during engine warm up.
Cold → Hot:
Approximately 5 - 0V
N.G.
1) Check fuel temperature sensor resistance.
• Stop engine.
• Disconnect fuel temperature sensor harness connector.
• Check resistance between terminal and ground.
20°C (68°F) 2.3 - 2.7 kΩ
50°C (122°F) 0.77 - 0.87 kΩ
80°C (176°F) 0.30 - 0.33 kΩ
If no continuity, replace fuel temperature sensor.
2) Check power source for E.C.U. & ground circuit for E.C.U.
O.K.
Reinstall any part removed.
Erase the self-diagnosis memory.
Perform driving test and then perform self-diagnosis (Mode-III) again.
N.G.
1) Perform E.C.U. input/output signal inspection test.
2) If N.G., recheck the E.C.U. pin terminals damage or the connection of E.C.U. harness connector.
O.K.
INSPECTION END

ECCS SELF-DIAGNOSTIC CHART—1988 MAXIMA

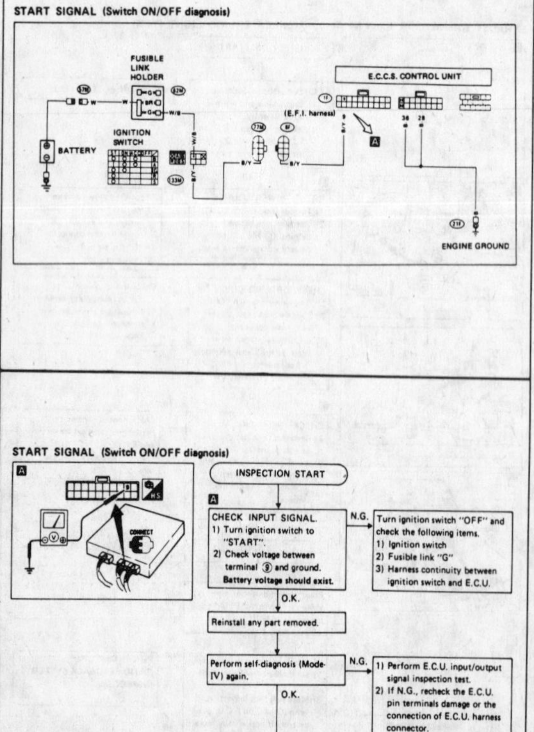
START SIGNAL (Switch ON/OFF diagnosis)
FUSIBLE LINK HOLDER
E.C.C.S. CONTROL UNIT
(E.F.I. harness)
IGNITION SWITCH
BATTERY
ENGINE GROUND
START SIGNAL (Switch ON/OFF diagnosis)
INSPECTION START
CHECK INPUT SIGNAL.
1) Turn ignition switch to "START".
2) Check voltage between terminal 9 and ground.
Battery voltage should exist.
N.G.
Turn ignition switch "OFF" and check the following items.
1) Ignition switch
2) Fusible link "G"
3) Harness continuity between ignition switch and E.C.U.
O.K.
Reinstall any part removed.
Perform self-diagnosis (Mode-IV) again.
N.G.
1) Perform E.C.U. input/output signal inspection test.
2) If N.G., recheck the E.C.U. pin terminals damage or the connection of E.C.U. harness connector.
O.K.
INSPECTION END

ECCS DIAGNOSTIC CHART—1988 MAXIMA

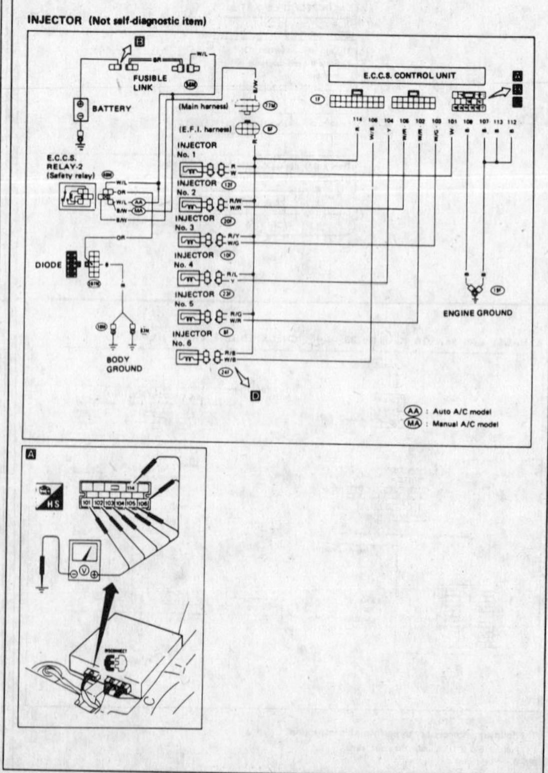
INJECTOR (Not self-diagnostic item)
FUSIBLE LINK
BATTERY
(Main harness)
(E.F.I. harness)
E.C.C.S. CONTROL UNIT
E.C.C.S. RELAY-2 (Safety relay)
INJECTOR No. 1
INJECTOR No. 2
INJECTOR No. 3
INJECTOR No. 4
INJECTOR No. 5
INJECTOR No. 6
DIODE
BODY GROUND
ENGINE GROUND
AA : Auto A/C model
MA : Manual A/C model

ECCS DIAGNOSTIC CHART—1988 MAXIMA

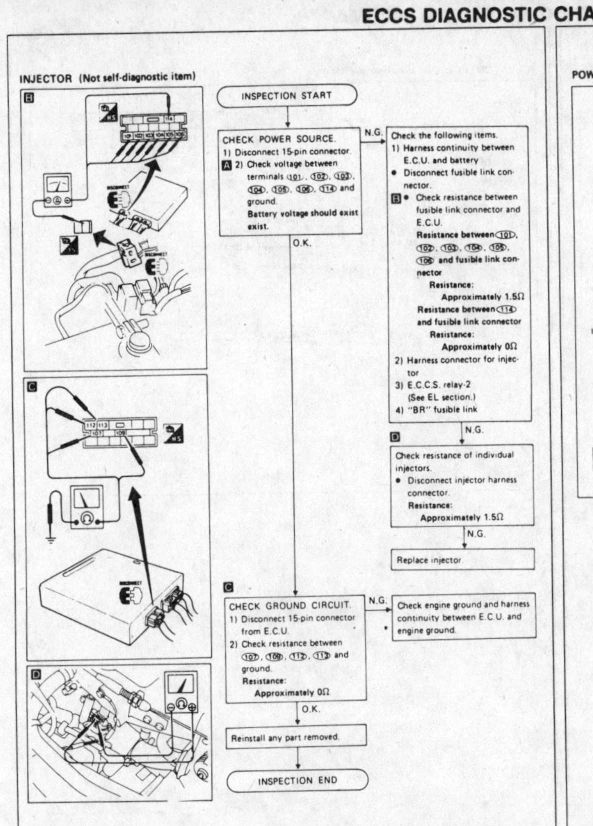
INJECTOR (Not self-diagnostic item)
INSPECTION START
CHECK POWER SOURCE.
1) Disconnect 15-pin connector.
A 2) Check voltage between terminals 101, 102, 103, 104, 105, 106, 114 and ground.
Battery voltage should exist.
N.G.
Check the following items.
1) Harness continuity between E.C.U. and battery
• Disconnect fusible link connector.
B • Check resistance between fusible link connector and E.C.U.
Resistance between 101, 102, 103, 104, 105, 106 and fusible link connector
Resistance: Approximately 1.5Ω
Resistance between 114 and fusible link connector
Resistance: Approximately 0Ω
2) Harness connector for injector
3) E.C.C.S. relay-2 (See EL section.)
4) "BR" fusible link
N.G.
D
Check resistance of individual injectors.
• Disconnect injector harness connector.
Resistance: Approximately 1.5Ω
N.G.
Replace injector
O.K.
C
CHECK GROUND CIRCUIT.
1) Disconnect 15-pin connector from E.C.U.
2) Check resistance between 107, 109, 112, 113 and ground.
Resistance: Approximately 0Ω
N.G.
Check engine ground and harness continuity between E.C.U. and engine ground.
O.K.
Reinstall any part removed.
INSPECTION END

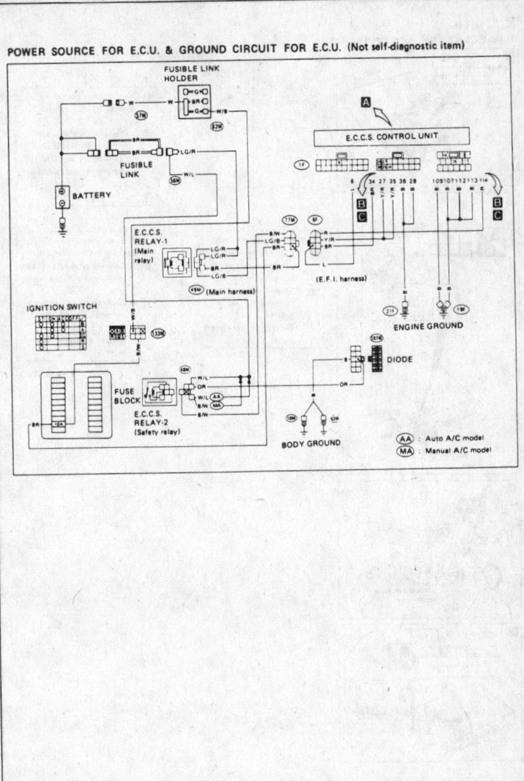
POWER SOURCE FOR E.C.U. & GROUND CIRCUIT FOR E.C.U. (Not self-diagnostic item)
FUSIBLE LINK HOLDER
E.C.C.S. CONTROL UNIT
FUSIBLE LINK
BATTERY
E.C.C.S. RELAY-1 (Main relay)
(Main harness)
(E.F.I. harness)
IGNITION SWITCH
ENGINE GROUND
DIODE
FUSE BLOCK
E.C.C.S. RELAY-2 (Safety relay)
BODY GROUND
AA : Auto A/C model
MA : Manual A/C model

ECCS DIAGNOSTIC CHART—1988 MAXIMA

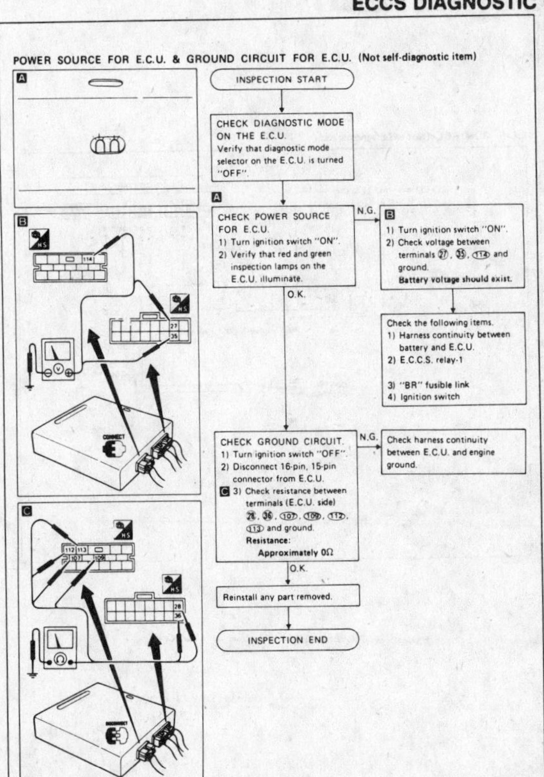
POWER SOURCE FOR E.C.U. & GROUND CIRCUIT FOR E.C.U. (Not self-diagnostic item)
INSPECTION START
CHECK DIAGNOSTIC MODE ON THE E.C.U.
Verify that diagnostic mode selector on the E.C.U. is turned "OFF".
A
CHECK POWER SOURCE FOR E.C.U.
1) Turn ignition switch "ON".
2) Verify that red and green inspection lamps on the E.C.U. illuminate.
N.G.
B
1) Turn ignition switch "ON".
2) Check voltage between terminals 27, 35, 114 and ground.
Battery voltage should exist.
Check the following items.
1) Harness continuity between battery and E.C.U.
2) E.C.C.S. relay-1
3) "BR" fusible link
4) Ignition switch
O.K.
CHECK GROUND CIRCUIT.
1) Turn ignition switch "OFF".
2) Disconnect 16-pin, 15-pin connector from E.C.U.
C 3) Check resistance between terminals (E.C.U. side) 28, 36, 107, 109, 112, 113 and ground.
Resistance: Approximately 0Ω
N.G.
Check harness continuity between E.C.U. and engine ground.
O.K.
Reinstall any part removed.
INSPECTION END

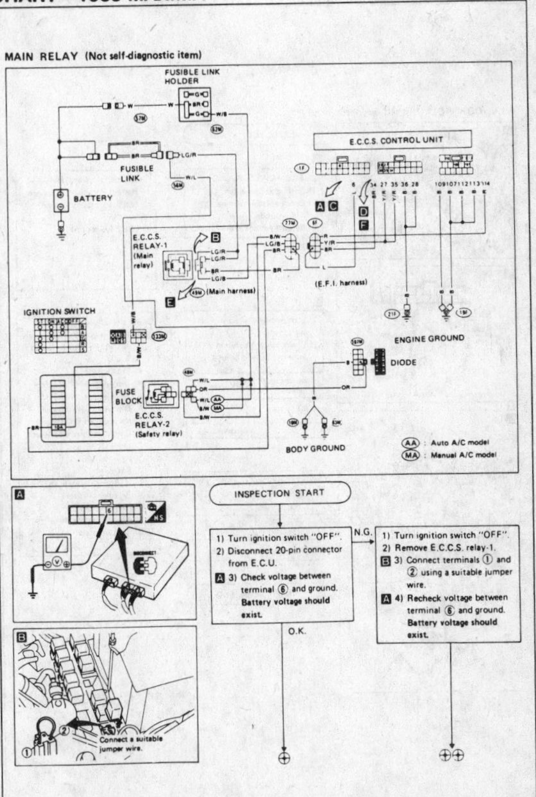
MAIN RELAY (Not self-diagnostic item)
FUSIBLE LINK HOLDER
E.C.C.S. CONTROL UNIT
FUSIBLE LINK
BATTERY
E.C.C.S. RELAY-1 (Main relay)
(Main harness)
(E.F.I. harness)
IGNITION SWITCH
ENGINE GROUND
DIODE
FUSE BLOCK
E.C.C.S. RELAY-2 (Safety relay)
BODY GROUND
AA : Auto A/C model
MA : Manual A/C model
INSPECTION START
1) Turn ignition switch "OFF".
2) Disconnect 20-pin connector from E.C.U.
A 3) Check voltage between terminal 6 and ground.
Battery voltage should exist.
N.G.
1) Turn ignition switch "OFF".
2) Remove E.C.C.S. relay-1.
B 3) Connect terminals 1 and 2 using a suitable jumper wire.
A 4) Recheck voltage between terminal 6 and ground.
Battery voltage should exist.
O.K.
Connect a suitable jumper wire.

ECCS DIAGNOSTIC CHART—1988 MAXIMA

MAIN RELAY (Not self-diagnostic item)

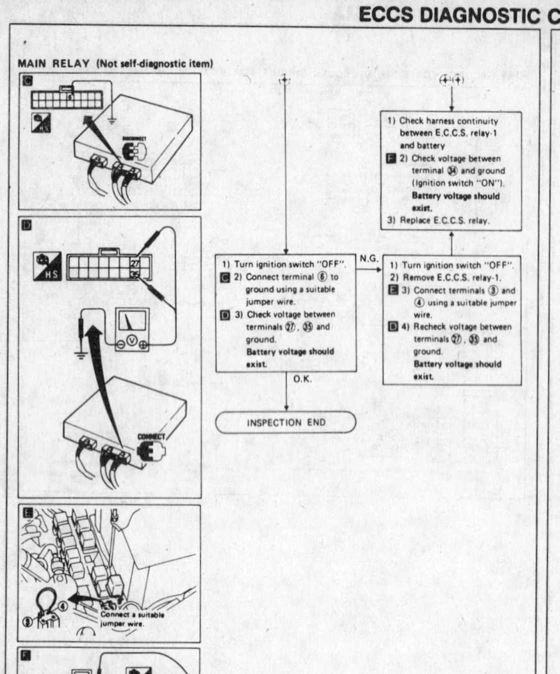
1) Check harness continuity between E.C.C.S. relay-1 and battery
F 2) Check voltage between terminal 34 and ground (Ignition switch "ON"). Battery voltage should exist.
3) Replace E.C.C.S. relay.
1) Turn ignition switch "OFF".
C 2) Connect terminal 6 to ground using a suitable jumper wire.
D 3) Check voltage between terminals 27, 35 and ground. Battery voltage should exist.
N.G.
1) Turn ignition switch "OFF".
2) Remove E.C.C.S. relay-1.
E 3) Connect terminals 3 and 4 using a suitable jumper wire.
D 4) Recheck voltage between terminals 27, 35 and ground. Battery voltage should exist.
O.K.
INSPECTION END
CONNECT
Connect a suitable jumper wire.

A.I.V. CONTROL (Not self-diagnostic item)

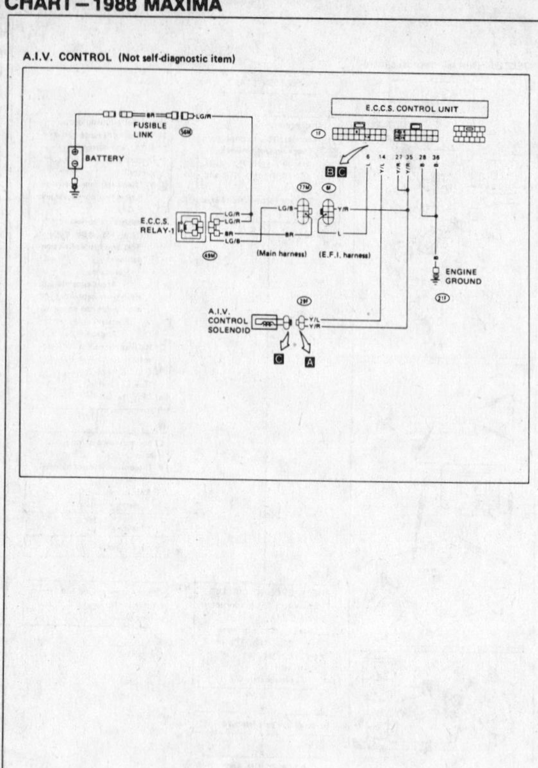
E.C.C.S. CONTROL UNIT
FUSIBLE LINK
BATTERY
E.C.C.S. RELAY-1
(Main harness)
(E.F.I. harness)
ENGINE GROUND
A.I.V. CONTROL SOLENOID

ECCS DIAGNOSTIC CHART—1988 MAXIMA

A.I.V. CONTROL (Not self-diagnostic item)

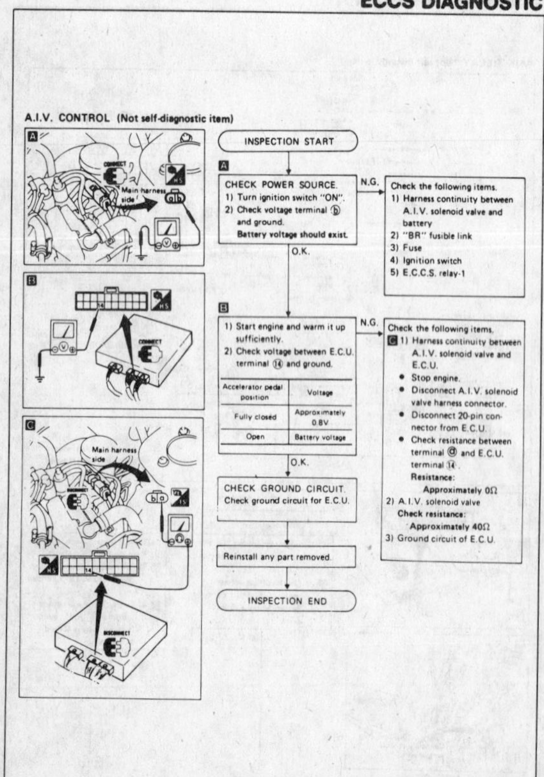
INSPECTION START
A
CHECK POWER SOURCE.
1) Turn ignition switch "ON".
2) Check voltage terminal ⓑ and ground.
Battery voltage should exist.
N.G.
Check the following items.
1) Harness continuity between A.I.V. solenoid valve and battery
2) "BR" fusible link
3) Fuse
4) Ignition switch
5) E.C.C.S. relay-1
O.K.
B
1) Start engine and warm it up sufficiently.
2) Check voltage between E.C.U. terminal 14 and ground.
Accelerator pedal position | Voltage
Fully closed | Approximately 0.8V
Open | Battery voltage
N.G.
Check the following items.
C 1) Harness continuity between A.I.V. solenoid valve and E.C.U.
• Stop engine.
• Disconnect A.I.V. solenoid valve harness connector.
• Disconnect 20-pin connector from E.C.U.
• Check resistance between terminal ⓐ and E.C.U. terminal 14.
Resistance: Approximately 0Ω
2) A.I.V. solenoid valve
Check resistance: Approximately 40Ω
3) Ground circuit of E.C.U.
O.K.
CHECK GROUND CIRCUIT.
Check ground circuit for E.C.U.
Reinstall any part removed.
INSPECTION END
Main harness side
CONNECT

E.G.R. CONTROL (Not self-diagnostic item)

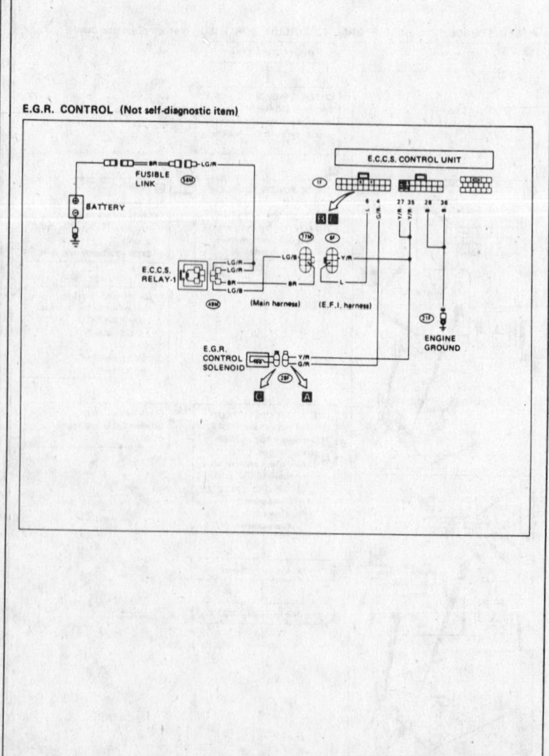
E.C.C.S. CONTROL UNIT
FUSIBLE LINK
BATTERY
E.C.C.S. RELAY-1
(Main harness)
(E.F.I. harness)
ENGINE GROUND
E.G.R. CONTROL SOLENOID

ECCS DIAGNOSTIC CHART—1988 MAXIMA

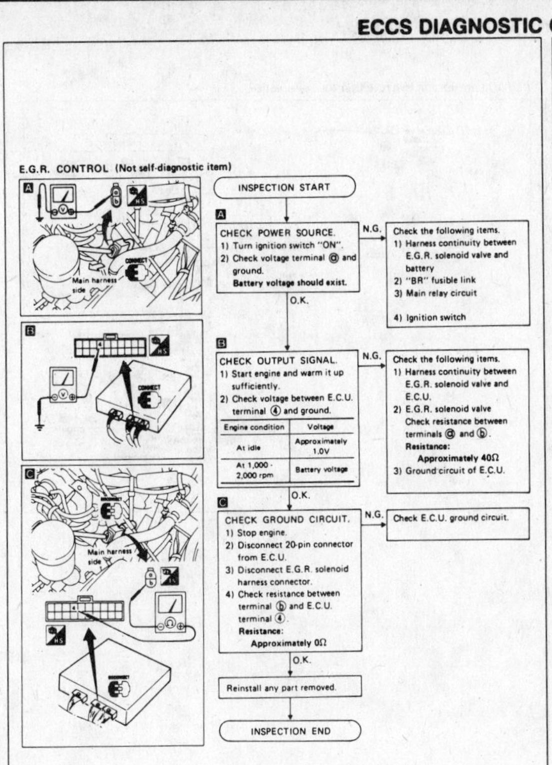
E.G.R. CONTROL (Not self-diagnostic item)
INSPECTION START
A
CHECK POWER SOURCE.
1) Turn ignition switch "ON".
2) Check voltage terminal ⓐ and ground.
Battery voltage should exist.
N.G.
Check the following items.
1) Harness continuity between E.G.R. solenoid valve and battery
2) "BR" fusible link
3) Main relay circuit
4) Ignition switch
O.K.
B
CHECK OUTPUT SIGNAL.
1) Start engine and warm it up sufficiently.
2) Check voltage between E.C.U. terminal ④ and ground.
Engine condition | Voltage
At idle | Approximately 1.0V
At 1,000 - 2,000 rpm | Battery voltage
N.G.
Check the following items.
1) Harness continuity between E.G.R. solenoid valve and E.C.U.
2) E.G.R. solenoid valve
Check resistance between terminals ⓐ and ⓑ.
Resistance:
Approximately 40Ω
3) Ground circuit of E.C.U.
O.K.
C
CHECK GROUND CIRCUIT.
1) Stop engine.
2) Disconnect 20-pin connector from E.C.U.
3) Disconnect E.G.R. solenoid harness connector.
4) Check resistance between terminal ⓑ and E.C.U. terminal ④.
Resistance:
Approximately 0Ω
N.G.
Check E.C.U. ground circuit.
O.K.
Reinstall any part removed.
INSPECTION END
Main harness side
CONNECT
DISCONNECT

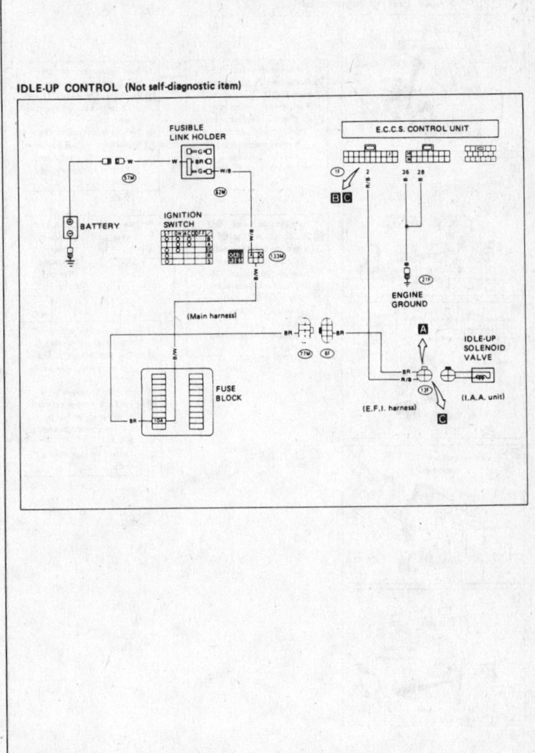
IDLE-UP CONTROL (Not self-diagnostic item)
FUSIBLE LINK HOLDER
E.C.C.S. CONTROL UNIT
BATTERY
IGNITION SWITCH
FUSE BLOCK
(Main harness)
ENGINE GROUND
IDLE-UP SOLENOID VALVE
(I.A.A. unit)
(E.F.I. harness)

ECCS DIAGNOSTIC CHART—1988 MAXIMA

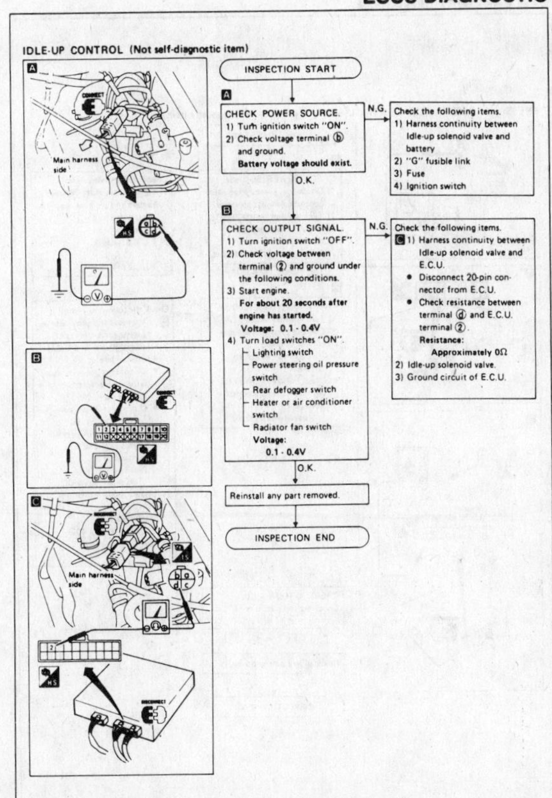
IDLE-UP CONTROL (Not self-diagnostic item)
INSPECTION START
A
CHECK POWER SOURCE.
1) Turn ignition switch "ON".
2) Check voltage terminal ⓑ and ground.
Battery voltage should exist.
N.G.
Check the following items.
1) Harness continuity between Idle-up solenoid valve and battery
2) "G" fusible link
3) Fuse
4) Ignition switch
O.K.
B
CHECK OUTPUT SIGNAL.
1) Turn ignition switch "OFF".
2) Check voltage between terminal ② and ground under the following conditions.
3) Start engine.
For about 20 seconds after engine has started.
Voltage: 0.1 - 0.4V
4) Turn load switches "ON".
Lighting switch
Power steering oil pressure switch
Rear defogger switch
Heater or air conditioner switch
Radiator fan switch
Voltage:
0.1 - 0.4V
N.G.
Check the following items.
C 1) Harness continuity between Idle-up solenoid valve and E.C.U.
• Disconnect 20-pin connector from E.C.U.
• Check resistance between terminal ⓓ and E.C.U. terminal ②.
Resistance:
Approximately 0Ω
2) Idle-up solenoid valve.
3) Ground circuit of E.C.U.
O.K.
Reinstall any part removed.
INSPECTION END
Main harness side
CONNECT
DISCONNECT

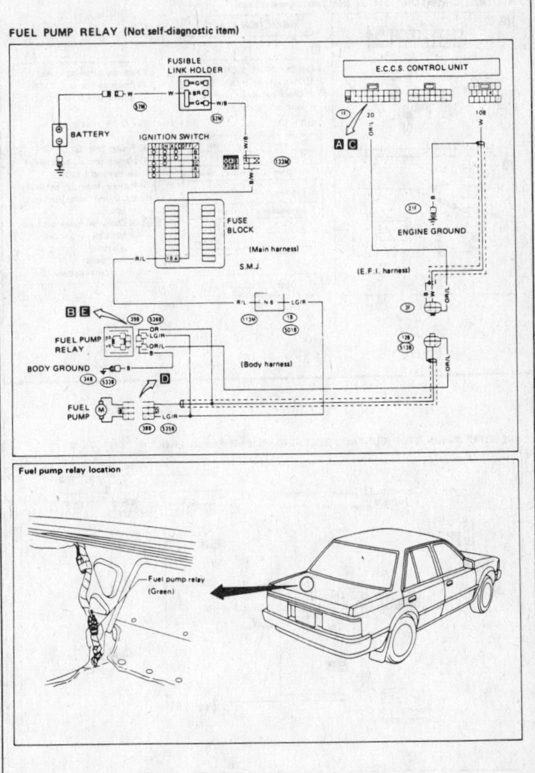
FUEL PUMP RELAY (Not self-diagnostic item)
FUSIBLE LINK HOLDER
E.C.C.S. CONTROL UNIT
BATTERY
IGNITION SWITCH
FUSE BLOCK
(Main harness)
S.M.J.
ENGINE GROUND
(E.F.I. harness)
FUEL PUMP RELAY
BODY GROUND
(Body harness)
FUEL PUMP
Fuel pump relay location
Fuel pump relay (Green)

ECCS DIAGNOSTIC CHART—1988 MAXIMA

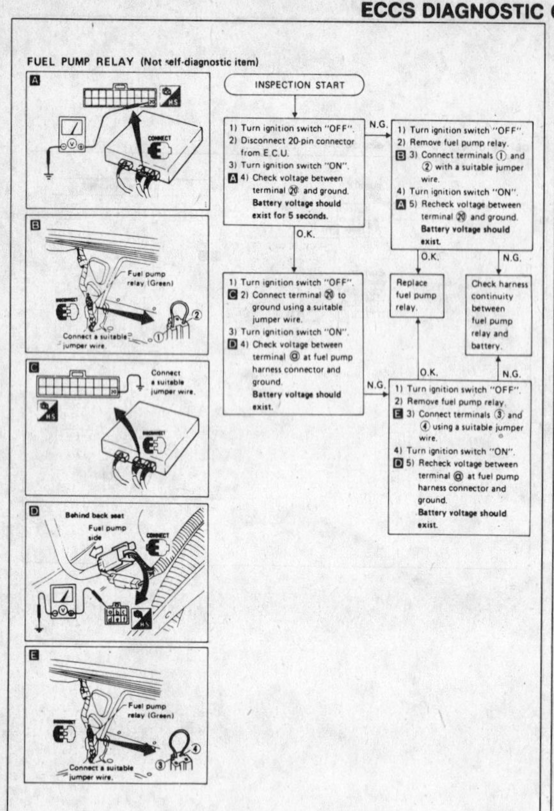
FUEL PUMP RELAY (Not self-diagnostic item)
INSPECTION START
1) Turn ignition switch "OFF". 2) Disconnect 20-pin connector from E.C.U. 3) Turn ignition switch "ON". A 4) Check voltage between terminal ⑳ and ground. Battery voltage should exist for 5 seconds.
N.G.
1) Turn ignition switch "OFF". 2) Remove fuel pump relay. B 3) Connect terminals ① and ② with a suitable jumper wire. 4) Turn ignition switch "ON". A 5) Recheck voltage between terminal ⑳ and ground. Battery voltage should exist.
O.K.
N.G.
Replace fuel pump relay.
Check harness continuity between fuel pump relay and battery.
1) Turn ignition switch "OFF". C 2) Connect terminal ⑳ to ground using a suitable jumper wire. 3) Turn ignition switch "ON". D 4) Check voltage between terminal ⓐ at fuel pump harness connector and ground. Battery voltage should exist.
1) Turn ignition switch "OFF". 2) Remove fuel pump relay. E 3) Connect terminals ③ and ④ using a suitable jumper wire. 4) Turn ignition switch "ON". D 5) Recheck voltage between terminal ⓐ at fuel pump harness connector and ground. Battery voltage should exist.
Fuel pump relay (Green)
Connect a suitable jumper wire.
Behind back seat
Fuel pump side

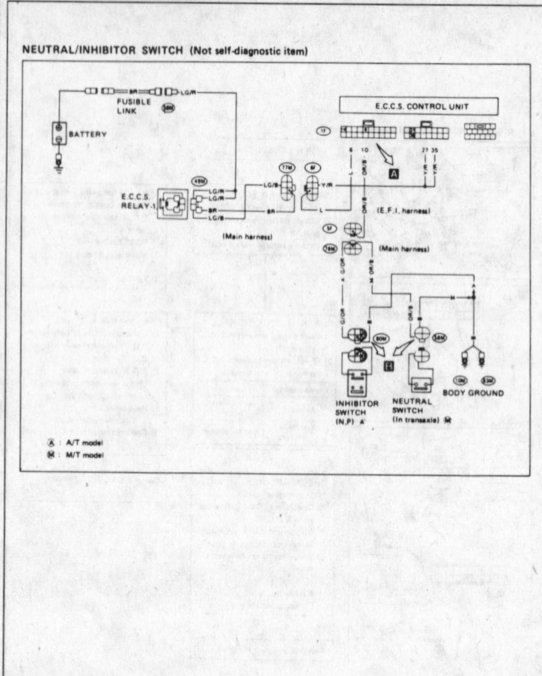
NEUTRAL/INHIBITOR SWITCH (Not self-diagnostic item)
FUSIBLE LINK
BATTERY
E.C.C.S. CONTROL UNIT
E.C.C.S. RELAY-1
(E.F.I. harness)
(Main harness)
INHIBITOR SWITCH (N,P) Ⓐ
NEUTRAL SWITCH (In transaxle) Ⓜ
BODY GROUND
Ⓐ : A/T model
Ⓜ : M/T model

ECCS DIAGNOSTIC CHART—1988 MAXIMA

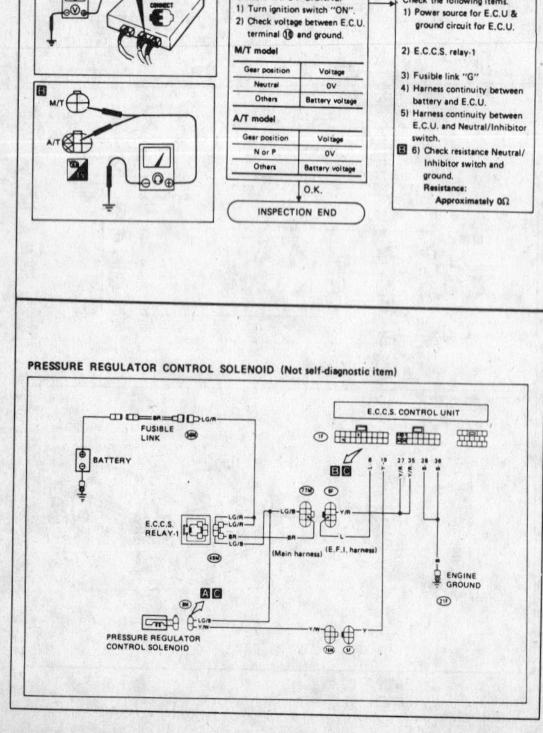
NEUTRAL/INHIBITOR SWITCH (Not self-diagnostic item)
INSPECTION START
CHECK INPUT SIGNAL. 1) Turn ignition switch "ON". 2) Check voltage between E.C.U. terminal ⑱ and ground.
M/T model
Gear position | Voltage
Neutral | 0V
Others | Battery voltage
A/T model
Gear position | Voltage
N or P | 0V
Others | Battery voltage
N.G.
Check the following items. 1) Power source for E.C.U & ground circuit for E.C.U. 2) E.C.C.S. relay-1 3) Fusible link "G" 4) Harness continuity between battery and E.C.U. 5) Harness continuity between E.C.U. and Neutral/Inhibitor switch. B 6) Check resistance Neutral/Inhibitor switch and ground. Resistance: Approximately 0Ω
O.K.
INSPECTION END
M/T
A/T
PRESSURE REGULATOR CONTROL SOLENOID (Not self-diagnostic item)
FUSIBLE LINK
BATTERY
E.C.C.S. CONTROL UNIT
E.C.C.S. RELAY-1
(Main harness)
(E.F.I. harness)
ENGINE GROUND
PRESSURE REGULATOR CONTROL SOLENOID

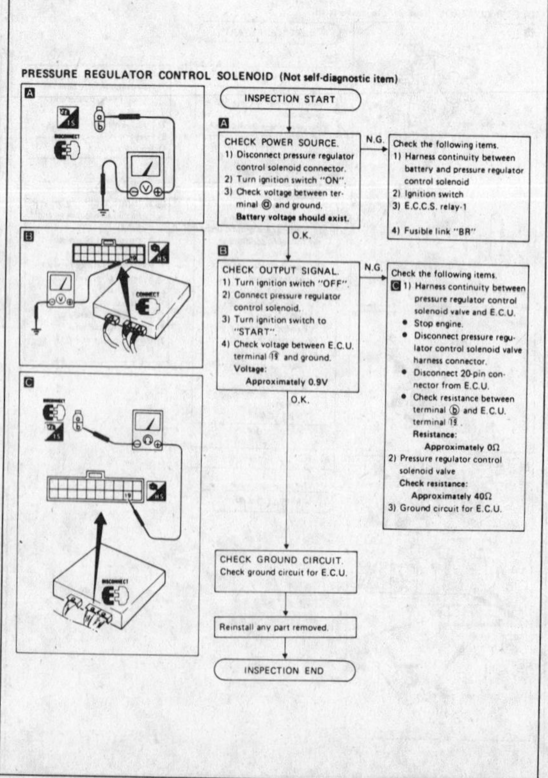
PRESSURE REGULATOR CONTROL SOLENOID (Not self-diagnostic item)
INSPECTION START
CHECK POWER SOURCE. 1) Disconnect pressure regulator control solenoid connector. 2) Turn ignition switch "ON". 3) Check voltage between terminal ⓐ and ground. Battery voltage should exist.
N.G.
Check the following items. 1) Harness continuity between battery and pressure regulator control solenoid 2) Ignition switch 3) E.C.C.S. relay-1 4) Fusible link "BR"
O.K.
CHECK OUTPUT SIGNAL. 1) Turn ignition switch "OFF". 2) Connect pressure regulator control solenoid. 3) Turn ignition switch to "START". 4) Check voltage between E.C.U. terminal ⑲ and ground. Voltage: Approximately 0.9V
N.G.
Check the following items. C 1) Harness continuity between pressure regulator control solenoid valve and E.C.U. • Stop engine. • Disconnect pressure regulator control solenoid valve harness connector. • Disconnect 20-pin connector from E.C.U. • Check resistance between terminal ⓑ and E.C.U. terminal ⑲. Resistance: Approximately 0Ω 2) Pressure regulator control solenoid valve Check resistance: Approximately 40Ω 3) Ground circuit for E.C.U.
O.K.
CHECK GROUND CIRCUIT. Check ground circuit for E.C.U.
Reinstall any part removed.
INSPECTION END

ECCS DIAGNOSTIC CHART—1988 MAXIMA

AIR REGULATOR (Not self-diagnostic item)

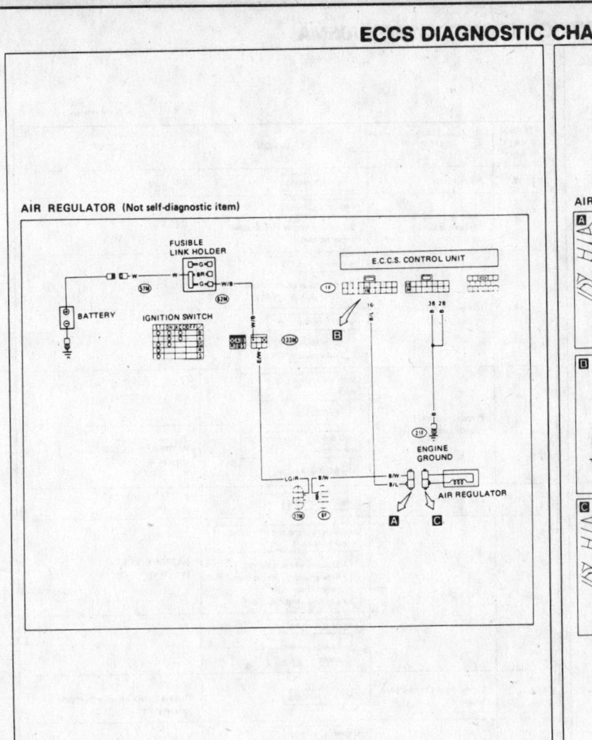
FUSIBLE
LINK HOLDER
BATTERY
IGNITION SWITCH
E.C.C.S. CONTROL UNIT
ENGINE
GROUND
AIR REGULATOR

AIR REGULATOR (Not self-diagnostic item)

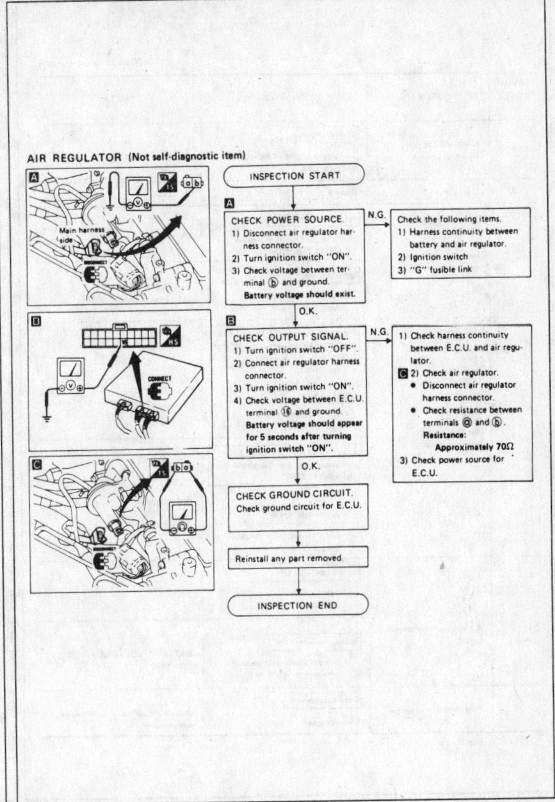
Main harness side
INSPECTION START
A
CHECK POWER SOURCE.
1) Disconnect air regulator harness connector.
2) Turn ignition switch "ON".
3) Check voltage between terminal ⓑ and ground.
Battery voltage should exist.
N.G.
Check the following items.
1) Harness continuity between battery and air regulator.
2) Ignition switch
3) "G" fusible link
O.K.
B
CHECK OUTPUT SIGNAL.
1) Turn ignition switch "OFF".
2) Connect air regulator harness connector.
3) Turn ignition switch "ON".
4) Check voltage between E.C.U. terminal ⑯ and ground.
Battery voltage should appear for 5 seconds after turning ignition switch "ON".
N.G.
1) Check harness continuity between E.C.U. and air regulator.
C 2) Check air regulator.
• Disconnect air regulator harness connector.
• Check resistance between terminals ⓐ and ⓑ.
Resistance:
Approximately 70Ω
3) Check power source for E.C.U.
O.K.
CHECK GROUND CIRCUIT.
Check ground circuit for E.C.U.
Reinstall any part removed.
INSPECTION END

ECU SIGNAL INSPECTION—1988 MAXIMA

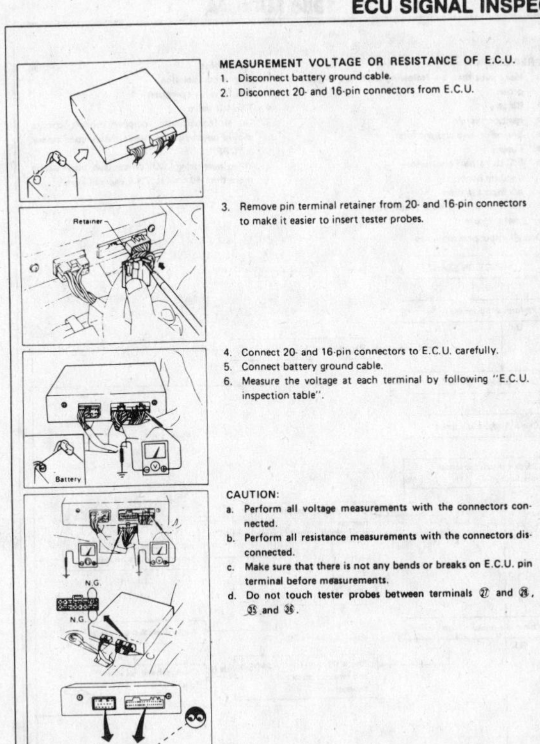

MEASUREMENT VOLTAGE OR RESISTANCE OF E.C.U.

1. Disconnect battery ground cable.
2. Disconnect 20- and 16-pin connectors from E.C.U.
3. Remove pin terminal retainer from 20- and 16-pin connectors to make it easier to insert tester probes.
4. Connect 20- and 16-pin connectors to E.C.U. carefully.
5. Connect battery ground cable.
6. Measure the voltage at each terminal by following "E.C.U. inspection table".

CAUTION:

a. Perform all voltage measurements with the connectors connected.
b. Perform all resistance measurements with the connectors disconnected.
c. Make sure that there is not any bends or breaks on E.C.U. pin terminal before measurements.
d. Do not touch tester probes between terminals ㉗ and ㉘, ㉟ and ㊱.

E.C.U. inspection table

*Data are reference values.

TERMINAL NO.	ITEM	CONDITION	*DATA
2	Idle-up solenoid	Engine is running and gear position is in P or N (A/T). — For about 20 seconds after starting engine — Steering wheel is turned. — Blower and air conditioner switches are "ON". — Lighting switch is "ON".	0.1 - 0.4V
		Engine is running. — Except conditions shown above	BATTERY VOLTAGE (11 - 14V)
3	Ignition check	Engine is running. — Idle speed	9 - 12V (Decreases as engine is revved up.)
4	E.G.R. control solenoid	Engine is running after being warmed up. — High engine revolution — Idle speed (Idle switch is "ON".)	Approx. 1.0V
		Engine is running. — Low engine revolution	BATTERY VOLTAGE (11 - 14V)
5	Ignition signal	Engine is running. — Idle speed	0.4 - 0.6V
		Engine is running. — Engine speed is 2,000 rpm.	1.2 - 1.5V
6	E.C.C.S. relay-1 (Main relay)	Engine is running. ⇩ Ignition switch "OFF" — For approximately 6 seconds after turning ignition switch "OFF"	0.7 - 0.9V
		Ignition switch "OFF" — Within approximately 6 seconds after turning ignition switch "OFF"	BATTERY VOLTAGE (11 - 14V)

ECU SIGNAL INPSECTION (CONT)—1988 MAXIMA

*Data are reference values.

TERMINAL NO.	ITEM	CONDITION	*DATA
8	Crank angle sensor (Position signal)	Engine is running. Do not run engine at high speed under no-load.	2.5 - 2.7V
9	Start signal	Cranking	8 - 12V
10	Neutral switch or Inhibitor switch	Ignition switch "ON" — Gear position is in Neutral (M/T), N or P (A/T).	0V
		Ignition switch "ON" — Any gear position except Neutral (M/T), or N or P (A/T)	BATTERY VOLTAGE (11 - 14V)
12	Air flow meter burn-off signal	Engine revolutions are above 1,500 rpm and vehicle speed is greater than 20 km/h (12 MPH). ⇩ Ignition switch "OFF" — For 6 seconds after turning ignition switch "OFF"	0V
		Engine revolution are above 1,500 rpm and vehicle speed is greater than 20 km/h (12 MPH). ⇩ Ignition switch "OFF" — For 1 second after the above 6 seconds have passed.	9.0 - 10.0V
14	A.I.V. control solenoid	Ignition switch "OFF" — Release accelerator pedal. (Idle switch "ON")	0.7 - 0.9V
		Ignition switch "ON" — Depress accelerator pedal. (Idle switch "OFF")	BATTERY VOLTAGE (11 - 14V)
15	Fuel temperature sensor	Engine is running. — Idle speed	0 - 5V Output voltage varies with engine temperature.

*Data are reference values.

TERMINAL NO.	ITEM	CONDITION	*DATA
16	Air regulator	Engine is running.	0.6 - 0.9V
17	Crank angle sensor (Reference signal)	Engine is running. Do not run engine at high speed under no-load.	0.2 - 0.4V
18	Idle switch (⊖ side)	Ignition switch "ON" — Release accelerator pedal. (Idle switch "OFF")	8.0 - 10.0V
		Ignition switch "ON" — Depress accelerator pedal. (Idle switch "ON")	0V
19	Pressure regulator control solenoid	Stop and restart engine after warming it up. — For 30 seconds	0.8 - 1.0V
		Stop and restart engine after warming it up. — After 3 minutes	BATTERY VOLTAGE (11 - 14V)
20	Fuel pump relay	Engine is running.	BATTERY VOLTAGE (11 - 14V)
22	Load signal	Engine is running and gear position is in P or N (A/T). — Steering wheel is turned. — Blower and air conditioner switches are "ON". — Lighting switch is "ON".	BATTERY VOLTAGE (11 - 14V)
		Engine is running. — Except conditions shown above	0V
23	Cylinder head temperature sensor	Engine is running.	0 - 5.0V Output voltage varies with engine temperature.
24	Exhaust gas sensor	Engine is running. — After warming up sufficiently	0 - Approximately 1.0V
25	Idle switch (⊕ side)	Ignition switch "ON"	9.0 - 10.0V

ECU SIGNAL INPSECTION (CONT)—1988 MAXIMA

*Data are reference values.

TERMINAL NO.	ITEM	CONDITION	*DATA
27 35	Power source for E.C.U.	Ignition switch "ON"	BATTERY VOLTAGE (11 - 14V)
29	Vehicle speed sensor	Ignition switch "ON" — While rotating front wheel slowly	0 or 7.4V
31	Air flow meter	Engine is running. Do not run engine at high speed under no-load.	2.0 - 4.0V Output voltage varies with engine revolution and throttle valve movement.
34	Ignition switch signal	Ignition switch "ON"	BATTERY VOLTAGE (11 - 14V)
101 102 103 104 105 106 114	Injector	Engine is running.	BATTERY VOLTAGE (11 - 14V)
108	Fuel pump	Ignition switch "ON" — For 5 seconds after turning ignition switch "ON"	0.1 - 0.3V
		Ignition switch "ON" — Within 5 seconds after turning ignition switch "ON"	BATTERY VOLTAGE (11 - 14V)

VG30 PIN CONNECTOR TERMINAL LAYOUT

20-pin connector

1	2	3	4	5	6	7	8	9	10
11	12	13	14	15	16	17	18	19	20

HS

16-pin connector

21	22	23	24	25	26	27	28
29	30	31	32	33	34	35	36

HS

15-pin connector

112	113		114	115	
107	108	109	110	111	
101	102	103	104	105	106

HS

ECCS MIXTURE RATIO FEEDBACK SYSTEM INSPECTION 1988 MAXIMA

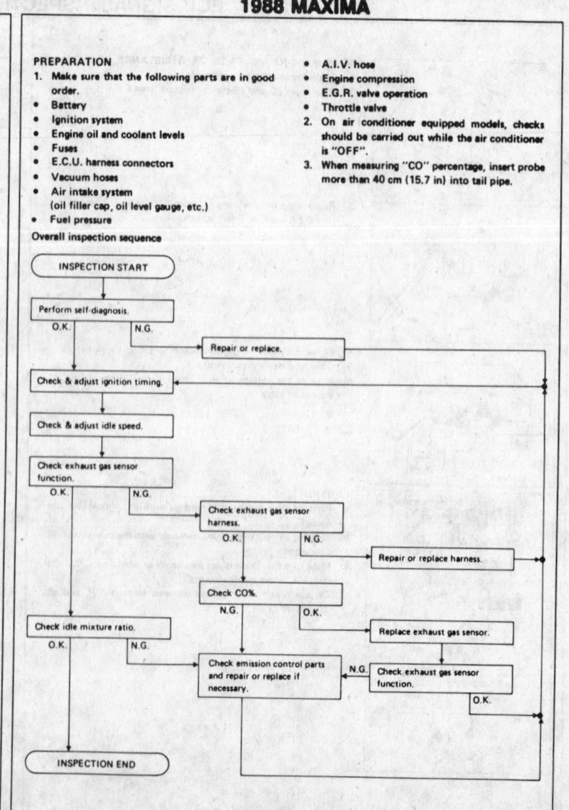

PREPARATION

1. Make sure that the following parts are in good order.
   - Battery
   - Ignition system
   - Engine oil and coolant levels
   - Fuses
   - E.C.U. harness connectors
   - Vacuum hoses
   - Air intake system (oil filler cap, oil level gauge, etc.)
   - Fuel pressure
   - A.I.V. hose
   - Engine compression
   - E.G.R. valve operation
   - Throttle valve
2. On air conditioner equipped models, checks should be carried out while the air conditioner is "OFF".
3. When measuring "CO" percentage, insert probe more than 40 cm (15.7 in) into tail pipe.

Overall inspection sequence

INSPECTION START
Perform self-diagnosis.
O.K.
N.G.
Repair or replace.
Check & adjust ignition timing.
Check & adjust idle speed.
Check exhaust gas sensor function.
O.K.
N.G.
Check exhaust gas sensor harness.
O.K.
N.G.
Repair or replace harness.
Check CO%.
N.G.
O.K.
Replace exhaust gas sensor.
Check idle mixture ratio.
O.K.
N.G.
Check emission control parts and repair or replace if necessary.
N.G.
Check exhaust gas sensor function.
O.K.
INSPECTION END

ECCS MIXTURE RATIO FEEDBACK SYSTEM INSPECTION (CONT.)—1988 MAXIMA

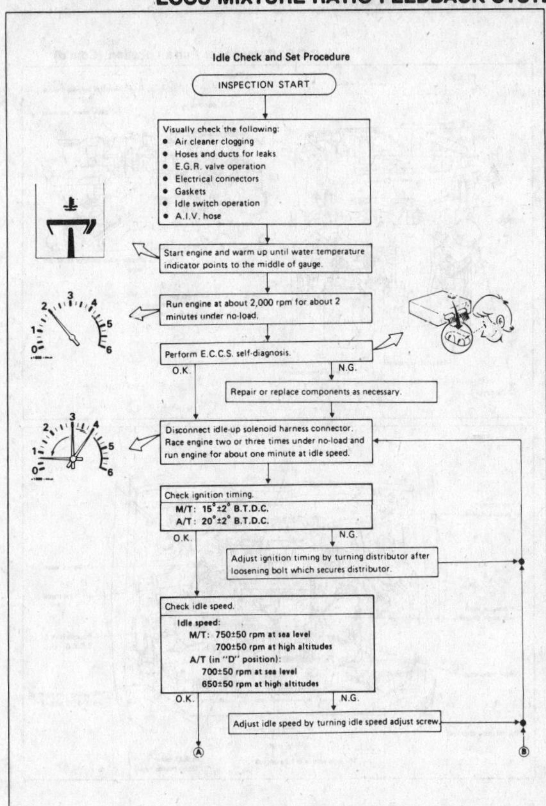
Idle Check and Set Procedure
INSPECTION START
Visually check the following:
• Air cleaner clogging
• Hoses and ducts for leaks
• E.G.R. valve operation
• Electrical connectors
• Gaskets
• Idle switch operation
• A.I.V. hose
Start engine and warm up until water temperature indicator points to the middle of gauge.
Run engine at about 2,000 rpm for about 2 minutes under no-load.
Perform E.C.C.S. self-diagnosis.
O.K.
N.G.
Repair or replace components as necessary.
Disconnect idle-up solenoid harness connector. Race engine two or three times under no-load and run engine for about one minute at idle speed.
Check ignition timing.
M/T: 15°±2° B.T.D.C.
A/T: 20°±2° B.T.D.C.
O.K.
N.G.
Adjust ignition timing by turning distributor after loosening bolt which secures distributor.
Check idle speed.
Idle speed:
M/T: 750±50 rpm at sea level
700±50 rpm at high altitudes
A/T (in "D" position):
700±50 rpm at sea level
650±50 rpm at high altitudes
O.K.
N.G.
Adjust idle speed by turning idle speed adjust screw.
A
B

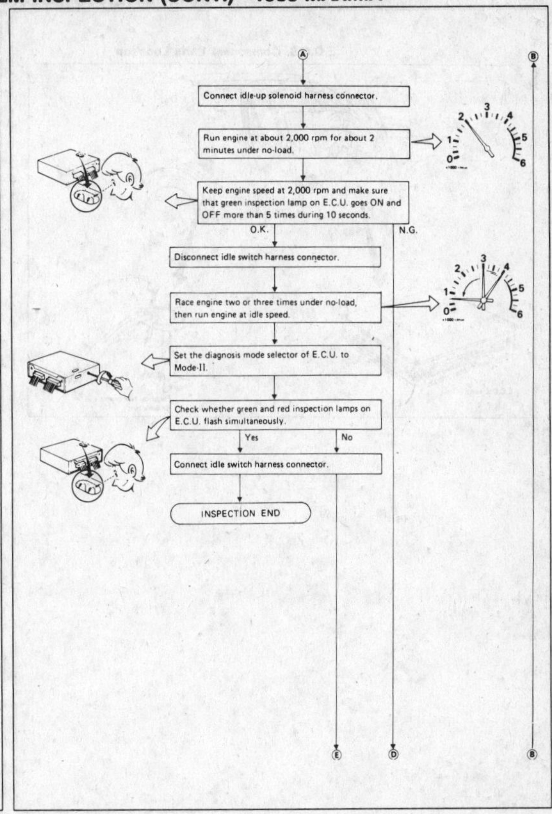
A
B
Connect idle-up solenoid harness connector.
Run engine at about 2,000 rpm for about 2 minutes under no-load.
Keep engine speed at 2,000 rpm and make sure that green inspection lamp on E.C.U. goes ON and OFF more than 5 times during 10 seconds.
O.K.
N.G.
Disconnect idle switch harness connector.
Race engine two or three times under no-load, then run engine at idle speed.
Set the diagnosis mode selector of E.C.U. to Mode-II.
Check whether green and red inspection lamps on E.C.U. flash simultaneously.
Yes
No
Connect idle switch harness connector.
INSPECTION END
E
D
B

ECCS MIXTURE RATIO FEEDBACK SYSTEM INSPECTION (CONT.)—1988 MAXIMA

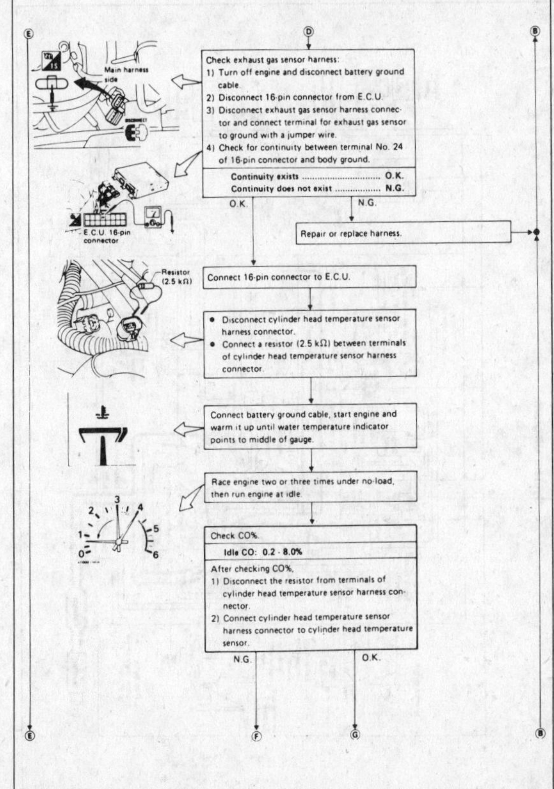
E
D
B
Main harness side
Check exhaust gas sensor harness:
1) Turn off engine and disconnect battery ground cable.
2) Disconnect 16-pin connector from E.C.U.
3) Disconnect exhaust gas sensor harness connector and connect terminal for exhaust gas sensor to ground with a jumper wire.
4) Check for continuity between terminal No. 24 of 16-pin connector and body ground.
Continuity exists .... O.K.
Continuity does not exist .... N.G.
O.K.
N.G.
E.C.U. 16-pin connector
Repair or replace harness.
Resistor (2.5 kΩ)
Connect 16-pin connector to E.C.U.
• Disconnect cylinder head temperature sensor harness connector.
• Connect a resistor (2.5 kΩ) between terminals of cylinder head temperature sensor harness connector.
Connect battery ground cable, start engine and warm it up until water temperature indicator points to middle of gauge.
Race engine two or three times under no-load, then run engine at idle.
Check CO%.
Idle CO: 0.2 - 8.0%
After checking CO%,
1) Disconnect the resistor from terminals of cylinder head temperature sensor harness connector.
2) Connect cylinder head temperature sensor harness connector to cylinder head temperature sensor.
N.G.
O.K.
E
F
G
B

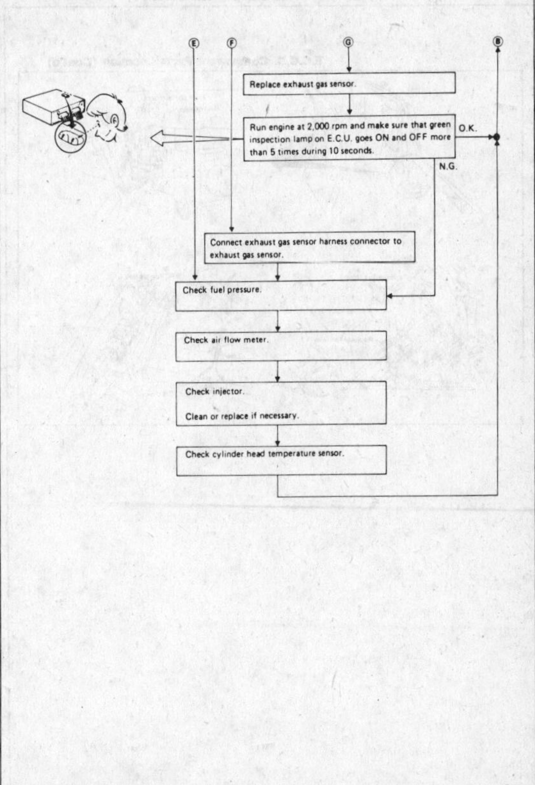
E
F
G
B
Replace exhaust gas sensor.
Run engine at 2,000 rpm and make sure that green inspection lamp on E.C.U. goes ON and OFF more than 5 times during 10 seconds.
O.K.
N.G.
Connect exhaust gas sensor harness connector to exhaust gas sensor.
Check fuel pressure.
Check air flow meter.
Check injector.
Clean or replace if necessary.
Check cylinder head temperature sensor.

ECCS COMPONENTS LOCATION—1989–90 MAXIMA

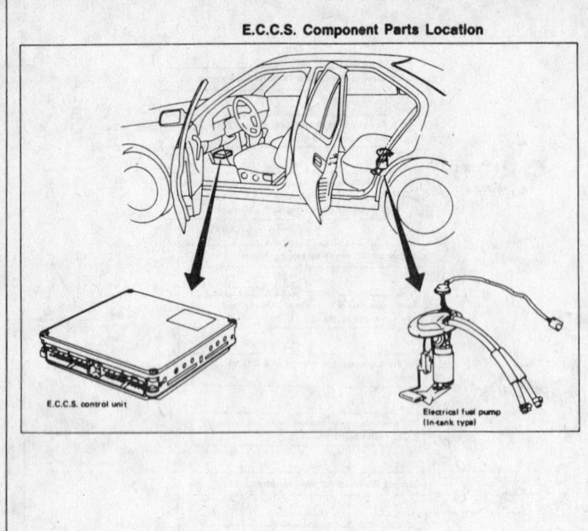
E.C.C.S. Component Parts Location
E.C.C.S. control unit
Electrical fuel pump
(In-tank type)

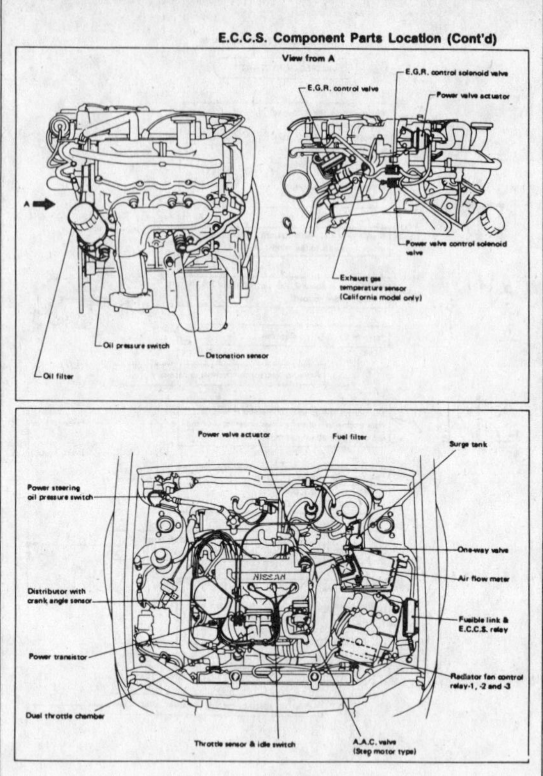
E.C.C.S. Component Parts Location (Cont'd)
View from A
A
E.G.R. control solenoid valve
E.G.R. control valve
Power valve actuator
Power valve control solenoid valve
Exhaust gas temperature sensor (California model only)
Oil pressure switch
Detonation sensor
Oil filter
Power valve actuator
Fuel filter
Surge tank
Power steering oil pressure switch
One-way valve
Air flow meter
Distributor with crank angle sensor
Fusible link & E.C.C.S. relay
Power transistor
Radiator fan control relay-1, -2 and -3
Dual throttle chamber
Throttle sensor & idle switch
A.A.C. valve (Step motor type)

ECCS COMPONENTS LOCATION (CONT.)—1989–90 MAXIMA

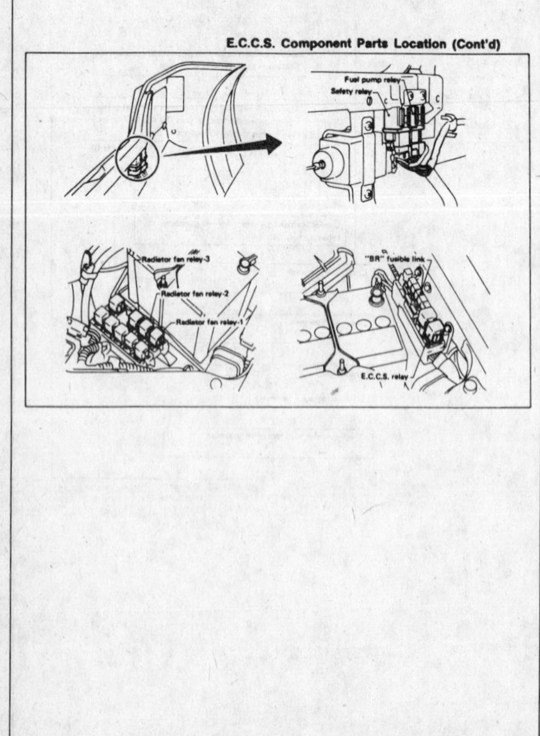
E.C.C.S. Component Parts Location (Cont'd)
Fuel pump relay
Safety relay
Radiator fan relay-3
Radiator fan relay-2
Radiator fan relay-1
"BR" fusible link
E.C.C.S. relay

ECCS SCHEMATIC—1989–90 MAXIMA

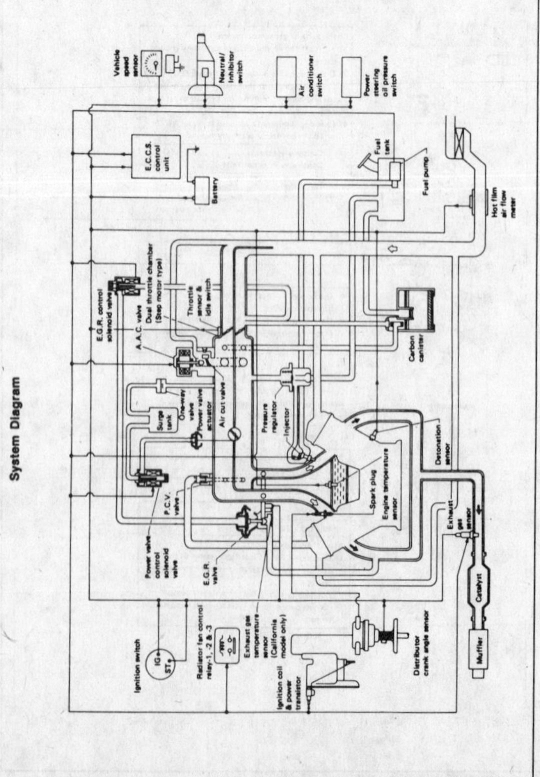
System Diagram
Vehicle speed sensor
Neutral/Inhibitor switch
Air conditioner switch
Power steering oil pressure switch
E.C.C.S. control unit
Battery
Fuel tank
Fuel pump
Hot film air flow meter
E.G.R. control solenoid valve
A.A.C. valve Dual throttle chamber (Step motor type)
Throttle sensor & idle switch
Carbon canister
Surge tank
One-way valve
Power valve actuator
Air cut valve
Pressure regulator
Injector
Detonation sensor
P.C.V. valve
Spark plug
Engine temperature sensor
Exhaust gas sensor
Power valve control solenoid valve
E.G.R. valve
Catalyst
Ignition switch
Radiator fan control relay-1, -2 & -3
Exhaust gas temperature sensor (California model only)
Ignition coil & power transistor
Distributor crank angle sensor
Muffler

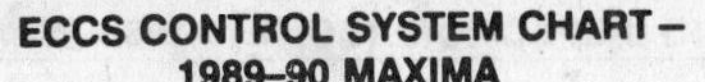

ECCS CONTROL SYSTEM CHART— 1989–90 MAXIMA

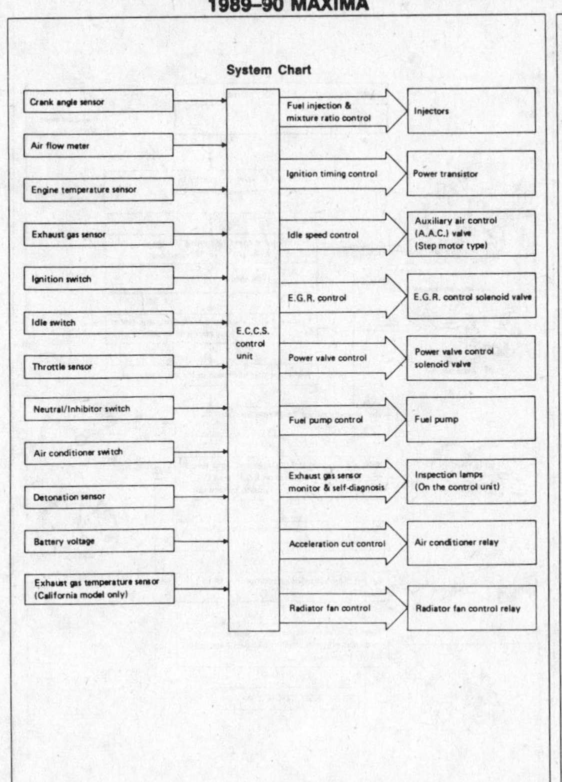
System Chart
Crank angle sensor
Air flow meter
Engine temperature sensor
Exhaust gas sensor
Ignition switch
Idle switch
Throttle sensor
Neutral/Inhibitor switch
Air conditioner switch
Detonation sensor
Battery voltage
Exhaust gas temperature sensor (California model only)
E.C.C.S. control unit
Fuel injection & mixture ratio control
Ignition timing control
Idle speed control
E.G.R. control
Power valve control
Fuel pump control
Exhaust gas sensor monitor & self-diagnosis
Acceleration cut control
Radiator fan control
Injectors
Power transistor
Auxiliary air control (A.A.C.) valve (Step motor type)
E.G.R. control solenoid valve
Power valve control solenoid valve
Fuel pump
Inspection lamps (On the control unit)
Air conditioner relay
Radiator fan control relay

ECCS VACUUM HOSE ROUTING— 1989–90 MAXIMA

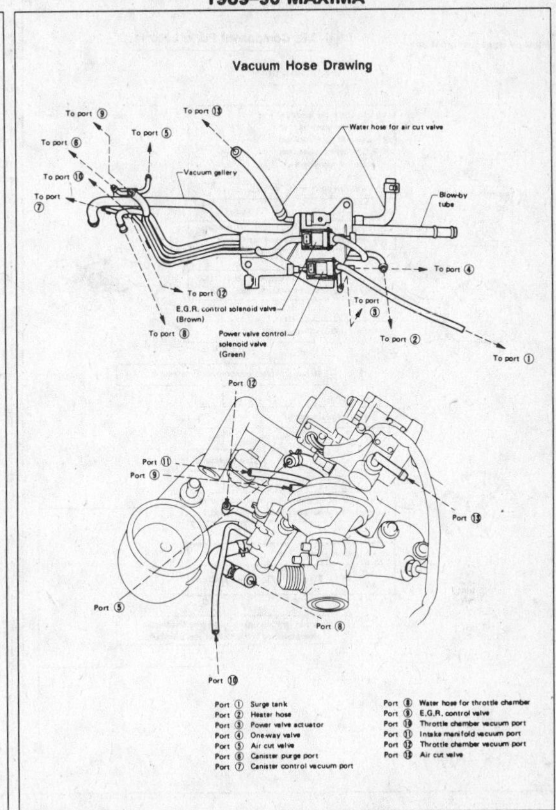
Vacuum Hose Drawing
Water hose for air cut valve
Vacuum gallery
Blow-by tube
E.G.R. control solenoid valve (Brown)
Power valve control solenoid valve (Green)
Port ① Surge tank
Port ② Heater hose
Port ③ Power valve actuator
Port ④ One-way valve
Port ⑤ Air cut valve
Port ⑥ Canister purge port
Port ⑦ Canister control vacuum port
Port ⑧ Water hose for throttle chamber
Port ⑨ E.G.R. control valve
Port ⑩ Throttle chamber vacuum port
Port ⑪ Intake manifold vacuum port
Port ⑫ Throttle chamber vacuum port
Port ⑬ Air cut valve

ECCS CONTROL UNIT WIRING SCHEMATIC—1989–90 MAXIMA

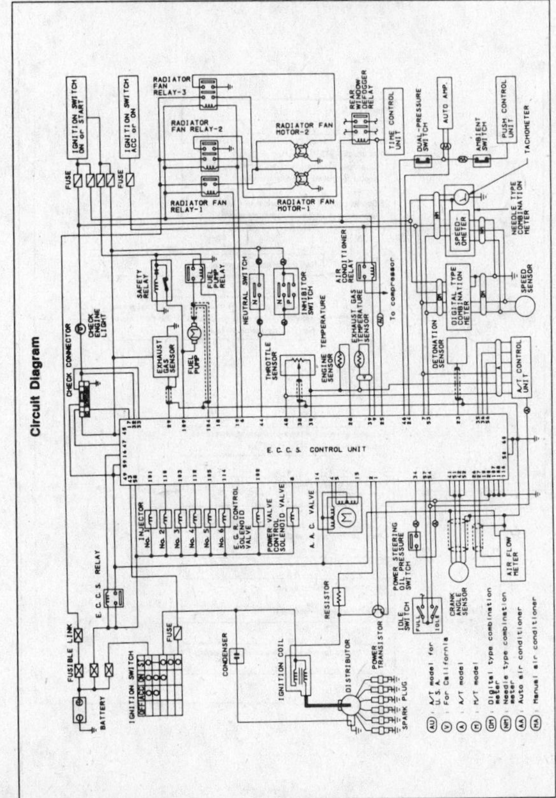
Circuit Diagram
E.C.C.S. CONTROL UNIT
BATTERY
FUSIBLE LINK
IGNITION SWITCH
E.C.C.S. RELAY
FUSE
RADIATOR FAN RELAY-3
RADIATOR FAN RELAY-2
RADIATOR FAN RELAY-1
RADIATOR FAN MOTOR-2
RADIATOR FAN MOTOR-1
CHECK CONNECTOR
CHECK ENGINE LIGHT
SAFETY RELAY
FUEL PUMP RELAY
EXHAUST GAS SENSOR
FUEL PUMP
NEUTRAL SWITCH
INHIBITOR SWITCH
THROTTLE SENSOR
ENGINE TEMPERATURE SENSOR
EXHAUST GAS TEMPERATURE SENSOR
AIR CONDITIONER RELAY
To compressor
DETONATION SENSOR
TACHOMETER
SPEED SENSOR
A/T CONTROL UNIT
INJECTOR
POWER VALVE CONTROL SOLENOID VALVE
A.A.C. VALVE
RESISTOR
CONDENSER
IGNITION COIL
DISTRIBUTOR
POWER TRANSISTOR
SPARK PLUG
POWER STEERING OIL PRESSURE SWITCH
IDLE SWITCH
CRANK ANGLE SENSOR
AIR FLOW METER

ECCS IDLE SPEED, IGNITION TIMING AND IDLE MIXTURE RATIO INSPECTION— 1989–90 MAXIMA

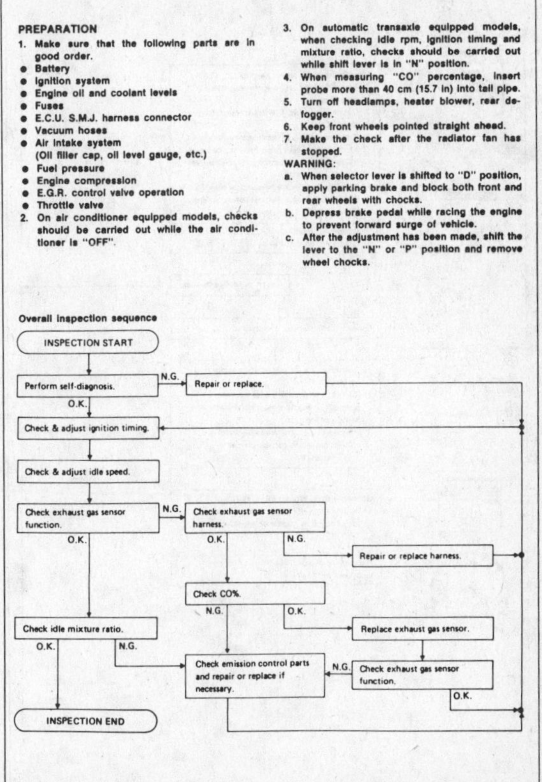

PREPARATION

1. Make sure that the following parts are in good order.
   - Battery
   - Ignition system
   - Engine oil and coolant levels
   - Fuses
   - E.C.U. S.M.J. harness connector
   - Vacuum hoses
   - Air intake system (Oil filler cap, oil level gauge, etc.)
   - Fuel pressure
   - Engine compression
   - E.G.R. control valve operation
   - Throttle valve
2. On air conditioner equipped models, checks should be carried out while the air conditioner is "OFF".
3. On automatic transaxle equipped models, when checking idle rpm, ignition timing and mixture ratio, checks should be carried out while shift lever is in "N" position.
4. When measuring "CO" percentage, insert probe more than 40 cm (15.7 in) into tail pipe.
5. Turn off headlamps, heater blower, rear defogger.
6. Keep front wheels pointed straight ahead.
7. Make the check after the radiator fan has stopped.

WARNING:

a. When selector lever is shifted to "D" position, apply parking brake and block both front and rear wheels with chocks.
b. Depress brake pedal while racing the engine to prevent forward surge of vehicle.
c. After the adjustment has been made, shift the lever to the "N" or "P" position and remove wheel chocks.

Overall inspection sequence
INSPECTION START
Perform self-diagnosis.
N.G.
Repair or replace.
O.K.
Check & adjust ignition timing.
Check & adjust idle speed.
Check exhaust gas sensor function.
Check exhaust gas sensor harness.
Repair or replace harness.
Check CO%.
Replace exhaust gas sensor.
Check idle mixture ratio.
Check emission control parts and repair or replace if necessary.
Check exhaust gas sensor function.
INSPECTION END

ECCS IDLE SPEED, IGNITION TIMING AND IDLE MIXTURE RATIO INSPECTION (CONT.)—1989–90 MAXIMA

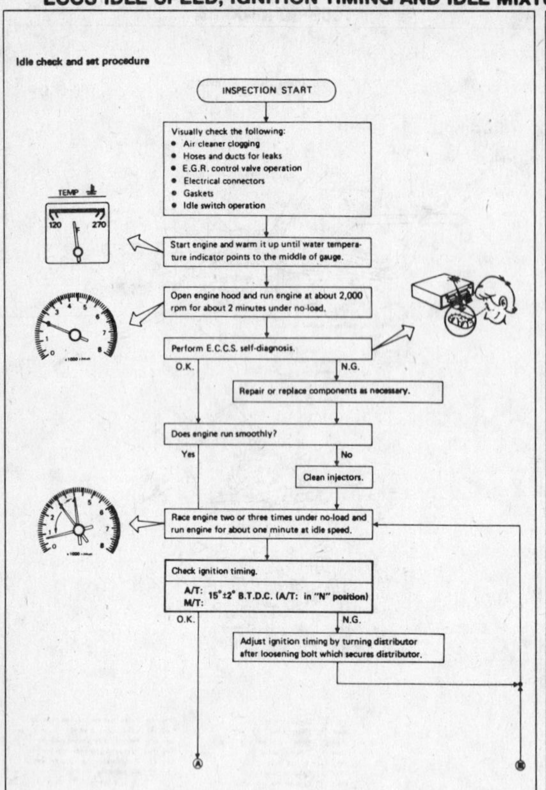
Idle check and set procedure
INSPECTION START
Visually check the following:
• Air cleaner clogging
• Hoses and ducts for leaks
• E.G.R. control valve operation
• Electrical connectors
• Gaskets
• Idle switch operation
TEMP
120
270
Start engine and warm it up until water temperature indicator points to the middle of gauge.
Open engine hood and run engine at about 2,000 rpm for about 2 minutes under no-load.
Perform E.C.C.S. self-diagnosis.
O.K.
N.G.
Repair or replace components as necessary.
Does engine run smoothly?
Yes
No
Clean injectors.
Race engine two or three times under no-load and run engine for about one minute at idle speed.
Check ignition timing.
A/T:
M/T:
15°±2° B.T.D.C. (A/T: in "N" position)
O.K.
N.G.
Adjust ignition timing by turning distributor after loosening bolt which secures distributor.
A
B

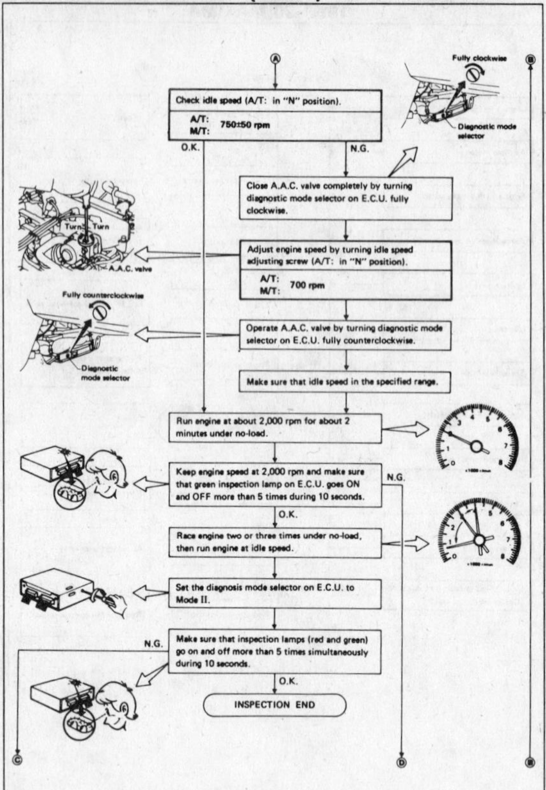
A
B
Fully clockwise
Diagnostic mode selector
Check idle speed (A/T: in "N" position).
A/T:
M/T:
750±50 rpm
O.K.
N.G.
Close A.A.C. valve completely by turning diagnostic mode selector on E.C.U. fully clockwise.
Turn
Turn
A.A.C. valve
Adjust engine speed by turning idle speed adjusting screw (A/T: in "N" position).
A/T:
M/T:
700 rpm
Fully counterclockwise
Operate A.A.C. valve by turning diagnostic mode selector on E.C.U. fully counterclockwise.
Diagnostic mode selector
Make sure that idle speed in the specified range.
Run engine at about 2,000 rpm for about 2 minutes under no-load.
Keep engine speed at 2,000 rpm and make sure that green inspection lamp on E.C.U. goes ON and OFF more than 5 times during 10 seconds.
N.G.
O.K.
Race engine two or three times under no-load, then run engine at idle speed.
Set the diagnosis mode selector on E.C.U. to Mode II.
N.G.
Make sure that inspection lamps (red and green) go on and off more than 5 times simultaneously during 10 seconds.
O.K.
INSPECTION END
C
D
B

ECCS IDLE SPEED, IGNITION TIMING AND IDLE MIXTURE RATIO INSPECTION (CONT.)—1989–90 MAXIMA

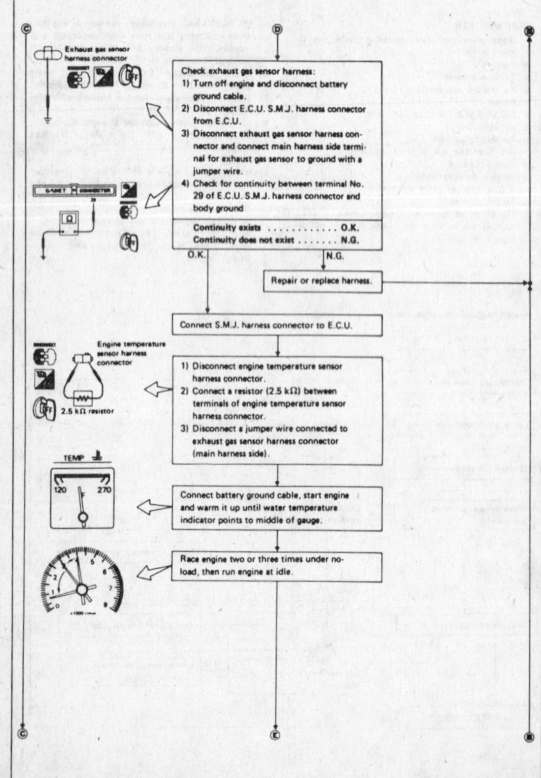
C
D
B
Exhaust gas sensor harness connector
Check exhaust gas sensor harness:
1) Turn off engine and disconnect battery ground cable.
2) Disconnect E.C.U. S.M.J. harness connector from E.C.U.
3) Disconnect exhaust gas sensor harness connector and connect main harness side terminal for exhaust gas sensor to ground with a jumper wire.
4) Check for continuity between terminal No. 29 of E.C.U. S.M.J. harness connector and body ground
Continuity exists .... O.K.
Continuity does not exist .... N.G.
O.K.
N.G.
Repair or replace harness.
Connect S.M.J. harness connector to E.C.U.
Engine temperature sensor harness connector
2.5 kΩ resistor
1) Disconnect engine temperature sensor harness connector.
2) Connect a resistor (2.5 kΩ) between terminals of engine temperature sensor harness connector.
3) Disconnect a jumper wire connected to exhaust gas sensor harness connector (main harness side).
TEMP
120
270
Connect battery ground cable, start engine and warm it up until water temperature indicator points to middle of gauge.
Race engine two or three times under no-load, then run engine at idle.
C
E
B

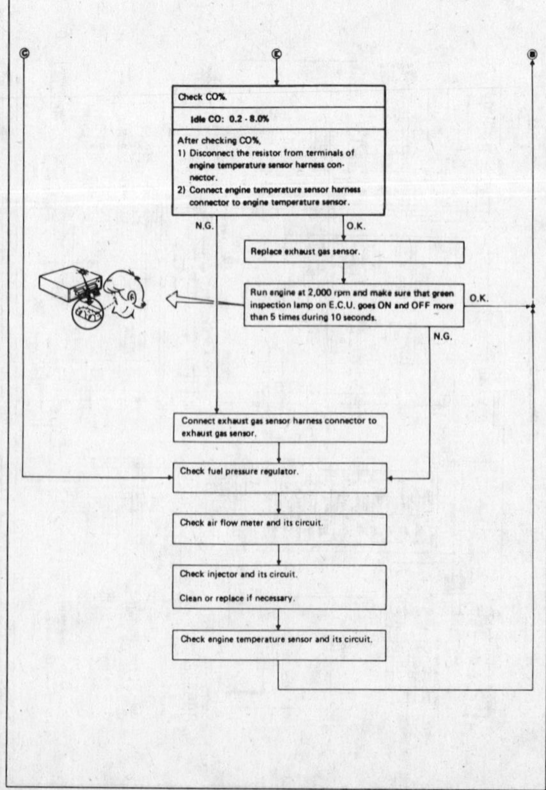
C
E
B
Check CO%.
Idle CO: 0.2 - 8.0%
After checking CO%,
1) Disconnect the resistor from terminals of engine temperature sensor harness connector.
2) Connect engine temperature sensor harness connector to engine temperature sensor.
N.G.
O.K.
Replace exhaust gas sensor.
Run engine at 2,000 rpm and make sure that green inspection lamp on E.C.U. goes ON and OFF more than 5 times during 10 seconds.
O.K.
N.G.
Connect exhaust gas sensor harness connector to exhaust gas sensor.
Check fuel pressure regulator.
Check air flow meter and its circuit.
Check injector and its circuit.
Clean or replace if necessary.
Check engine temperature sensor and its circuit.

ECCS DIAGNOSTIC PROCEDURE (CONT.)—1989-90 MAXIMA

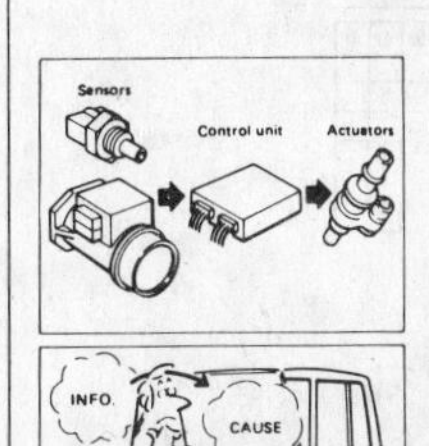
Sensors
Control unit
Actuators

How to Perform Trouble Diagnoses for Quick and Accurate Repair

INTRODUCTION

The engine has an electronic control unit to control major systems such as fuel control, ignition control, idle speed control, etc. The control unit accepts input signals from sensors and instantly drives actuators. It is essential that both kinds of signals are proper and stable. At the same time, it is important that there are no conventional problems such as vacuum leaks, fouled spark plugs, or other problems with the engine.

It is much more difficult to diagnose a problem that occurs intermittently rather than continuously. Most intermittent problems are caused by poor electric connections or improper wiring. In this case, careful checking of suspected circuits may help prevent the replacement of good parts.

A visual check only may not find the cause of the problems, so a road test with a circuit tester connected to a suspected circuit should be performed.

Before undertaking actual checks, take just a few minutes to talk with a customer who approaches with a driveability complaint. The customer is a very good supplier of information on such problems, especially intermittent ones. Through interaction with the customer, find out what symptoms are present and under what conditions they occur.

Start your diagnosis by looking for "conventional" problems first. This is one of the best ways to troubleshoot driveability problems on an electronically controlled engine vehicle.

How to Perform Trouble Diagnoses for Quick and Accurate Repair (Cont'd)

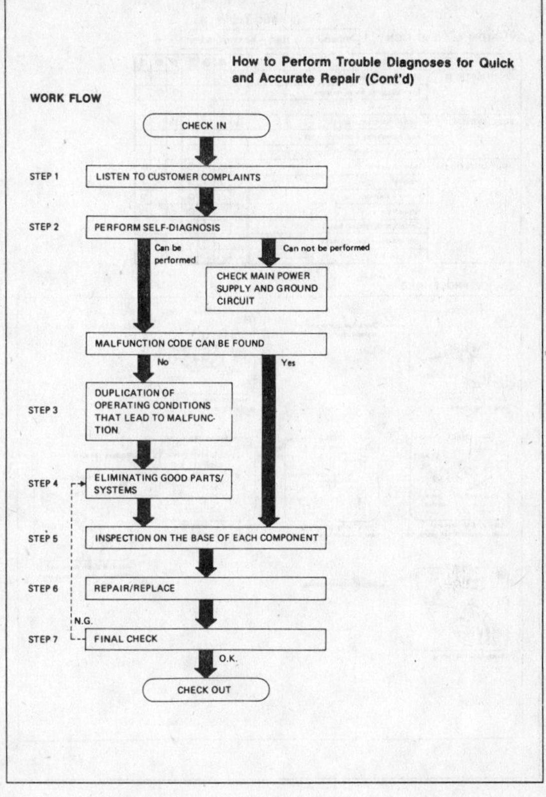
WORK FLOW
CHECK IN
STEP 1 LISTEN TO CUSTOMER COMPLAINTS
STEP 2 PERFORM SELF-DIAGNOSIS
Can be performed
Can not be performed
CHECK MAIN POWER SUPPLY AND GROUND CIRCUIT
MALFUNCTION CODE CAN BE FOUND
No
Yes
STEP 3 DUPLICATION OF OPERATING CONDITIONS THAT LEAD TO MALFUNCTION
STEP 4 ELIMINATING GOOD PARTS/SYSTEMS
STEP 5 INSPECTION ON THE BASE OF EACH COMPONENT
STEP 6 REPAIR/REPLACE
N.G.
STEP 7 FINAL CHECK
O.K.
CHECK OUT

ECCS DIAGNOSTIC PROCEDURE (CONT.)—1989-90 MAXIMA

KEY POINTS

WHAT ..... Vehicle & engine model
WHEN ..... Date, Frequencies
WHERE ..... Road conditions
HOW ..... Operating conditions, Weather conditions, Symptoms

How to Perform Trouble Diagnoses for Quick and Accurate Repair (Cont'd)

DIAGNOSTIC WORKSHEET

There are many kinds of operating conditions that lead to malfunctions on engine components.

A good grasp of such conditions can make trouble-shooting faster and more accurate.

In general, feelings for a problem depend on each customer. It is important to fully understand the symptoms or under what conditions a customer complains.

Make good use of a diagnostic worksheet such as the one shown below in order to utilize all the complaints for trouble-shooting.

Worksheet sample

Customer name MR/MS		Model & Year	VIN
Engine #		Trans.	Mileage
Incident Date		Manuf. Date	In Service Date
Symptoms	☐ Startability	☐ Impossible to start ☐ No combustion ☐ Partial combustion ☐ Partial combustion affected by throttle position ☐ Partial combustion NOT affected by throttle position ☐ Possible but hard to start ☐ Others [ ]	
	☐ Idling	☐ No fast idle ☐ Unstable ☐ High idle ☐ Low idle ☐ Others [ ]	
	☐ Driveability	☐ Stumble ☐ Surge ☐ Detonation ☐ Lack of power ☐ Intake backfire ☐ Exhaust backfire ☐ Others [ ]	
	☐ Engine stall	☐ At the time of start ☐ While idling ☐ While accelerating ☐ While decelerating ☐ Just after stopping ☐ While loading	
Incident occurrence		☐ Just after delivery ☐ Recently ☐ In the morning ☐ At night ☐ In the daytime	
Frequency		☐ All the time ☐ Under certain conditions ☐ Sometimes	
Weather conditions		☐ Not effected	
	Weather	☐ Fine ☐ Raining ☐ Snowing ☐ Others [ ]	
	Temperature	☐ Hot ☐ Warm ☐ Cool ☐ Cold ☐ Humid °F	
Engine conditions		☐ Cold ☐ During warm-up ☐ After warm-up Engine speed 0 2,000 4,000 6,000 8,000 rpm	
Road conditions		☐ In town ☐ In suburbs ☐ Highway ☐ Off road (up/down)	
Driving conditions		☐ Not affected ☐ At starting ☐ While idling ☐ At racing ☐ While accelerating ☐ While cruising ☐ While decelerating ☐ While turning (RH/LH) Vehicle speed 0 10 20 30 40 50 60 MPH	
Check engine light		☐ Turned on ☐ Not turned on	

How to Perform Trouble Diagnoses for Quick and Accurate Repair (Cont'd)

INTERMITTENT PROBLEM SIMULATION

In order to duplicate an intermittent problem, it is effective to create similar conditions for component parts, under which the problem might occur.

Perform the activity listed under Service procedure and note the result.

GRRRRR
GRRRR
THIS is the symptom, isn't it ?

	Variable factor	Influential part	Target condition	Service procedure
1	Mixture ratio	Pressure regulator	Made lean	Remove vacuum hose and apply vacuum.
			Made rich	Remove vacuum hose and apply pressure.
2	Ignition timing	Distributor	Advanced	Rotate distributor clockwise.
			Retarded	Rotate distributor counterclockwise.
3	Mixture ratio feedback control	Exhaust gas sensor	Suspended	Disconnect exhaust gas sensor harness connector.
		Control unit	Operation check	Perform self-diagnosis (Mode I/II) at 2,000 rpm.
4	Idle speed	A.A.C. valve	Raised	Turn idle adjusting screw counterclockwise.
			Lowered	Turn idle adjusting screw clockwise.
5	Electric connection (Electric continuity)	Harness connectors and wires	Poor electric connection or faulty wiring	Tap or wiggle.
				Race engine rapidly. See if the torque reaction of the engine unit causes electric breaks.
6	Temperature	Control unit	Cooled	Cool with an icing spray or similar device.
			Warmed	Heat with a hair drier. [WARNING: Do not overheat the unit.]
7	Moisture	Electric parts	Damp	Wet. [WARNING: Do not directly pour water on components. Use a mist sprayer.]
8	Electric loads	Load switches	Loaded	Turn on head lights, air conditioner, rear defogger, etc.
9	Idle switch condition	Control unit	ON-OFF switching	Perform self-diagnosis (Mode IV).
10	Ignition spark	Timing light	Spark power check	Try to flash timing light for each cylinder.

ECCS DIAGNOSTIC CHART—1989–90 MAXIMA

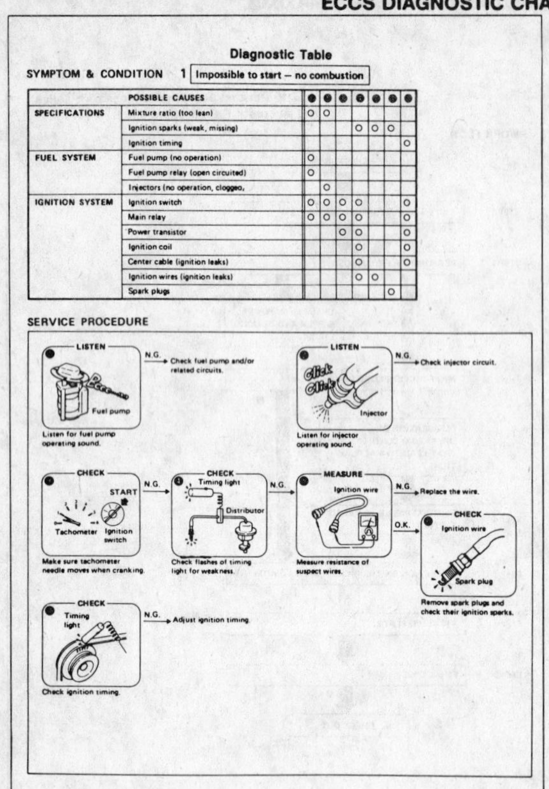

Diagnostic Table

SYMPTOM & CONDITION 1 Impossible to start – no combustion

POSSIBLE CAUSES		●	●	●	●	●	●	●
SPECIFICATIONS	Mixture ratio (too lean)	○	○					
	Ignition sparks (weak, missing)				○	○	○	
	Ignition timing							○
FUEL SYSTEM	Fuel pump (no operation)	○						
	Fuel pump relay (open circuited)	○						
	Injectors (no operation, cloggeo,		○					
IGNITION SYSTEM	Ignition switch	○	○	○	○			○
	Main relay	○	○	○	○			○
	Power transistor			○	○			○
	Ignition coil				○			○
	Center cable (ignition leaks)				○			○
	Ignition wires (ignition leaks)				○	○		
	Spark plugs						○	

SERVICE PROCEDURE

LISTEN
Fuel pump
N.G. → Check fuel pump and/or related circuits.
Listen for fuel pump operating sound.
LISTEN
Click Click
Injector
N.G. → Check injector circuit.
Listen for injector operating sound.
CHECK
START
Tachometer
Ignition switch
Make sure tachometer needle moves when cranking.
N.G.
CHECK
Timing light
Distributor
Check flashes of timing light for weakness.
N.G.
MEASURE
Ignition wire
N.G. → Replace the wire.
Measure resistance of suspect wires.
O.K.
CHECK
Ignition wire
Spark plug
Remove spark plugs and check their ignition sparks.
CHECK
Timing light
N.G. → Adjust ignition timing.
Check ignition timing.

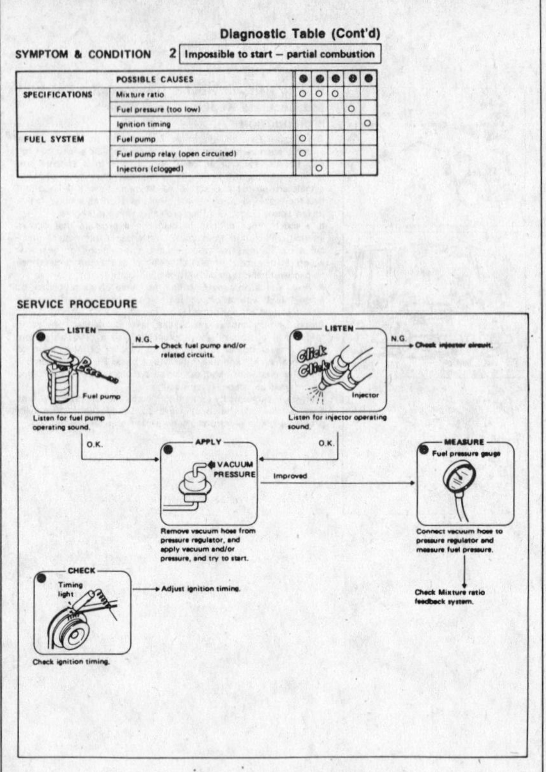

Diagnostic Table (Cont'd)

SYMPTOM & CONDITION 2 Impossible to start – partial combustion

POSSIBLE CAUSES		●	●	●	●	●
SPECIFICATIONS	Mixture ratio	○	○	○		
	Fuel pressure (too low)				○	
	Ignition timing					○
FUEL SYSTEM	Fuel pump	○				
	Fuel pump relay (open circuited)	○				
	Injectors (clogged)		○			

SERVICE PROCEDURE

LISTEN
Fuel pump
N.G. → Check fuel pump and/or related circuits.
Listen for fuel pump operating sound.
LISTEN
Click Click
Injector
N.G. → Check injector circuit.
Listen for injector operating sound.
O.K.
O.K.
APPLY
VACUUM
PRESSURE
Remove vacuum hose from pressure regulator, and apply vacuum and/or pressure, and try to start.
Improved
MEASURE
Fuel pressure gauge
Connect vacuum hose to pressure regulator and measure fuel pressure.
Check Mixture ratio feedback system.
CHECK
Timing light
→ Adjust ignition timing.
Check ignition timing.

ECCS DIAGNOSTIC CHART—1989–90 MAXIMA

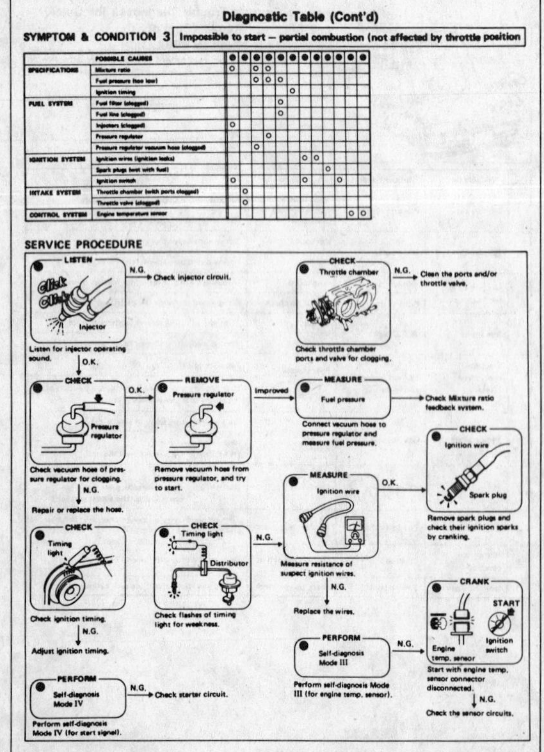

Diagnostic Table (Cont'd)

SYMPTOM & CONDITION 3 Impossible to start – partial combustion (not affected by throttle position

POSSIBLE CAUSES		●	●	●	●	●	●	●	●	●	●	●	●
SPECIFICATIONS	Mixture ratio	○		○	○								
	Fuel pressure (too low)			○	○	○							
	Ignition timing						○						
FUEL SYSTEM	Fuel filter (clogged)					○							
	Fuel line (clogged)					○							
	Injectors (clogged)	○											
	Pressure regulator				○								
	Pressure regulator vacuum hose (clogged)			○									
IGNITION SYSTEM	Ignition wires (ignition leaks)							○	○				
	Spark plugs (wet with fuel)									○			
	Ignition switch	○						○			○		
INTAKE SYSTEM	Throttle chamber (with ports clogged)		○										
	Throttle valve (clogged)		○										
CONTROL SYSTEM	Engine temperature sensor											○	○

SERVICE PROCEDURE

LISTEN
Click Click
Injector
N.G. → Check injector circuit.
Listen for injector operating sound.
O.K.
CHECK
Throttle chamber
N.G. → Clean the ports and/or throttle valve.
Check throttle chamber ports and valve for clogging.
CHECK
Pressure regulator
Check vacuum hose of pressure regulator for clogging.
N.G.
Repair or replace the hose.
O.K.
REMOVE
Pressure regulator
Remove vacuum hose from pressure regulator, and try to start.
Improved
MEASURE
Fuel pressure
→ Check Mixture ratio feedback system.
Connect vacuum hose to pressure regulator and measure fuel pressure.
CHECK
Ignition wire
Spark plug
Remove spark plugs and check their ignition sparks by cranking.
O.K.
MEASURE
Ignition wire
Measure resistance of suspect ignition wires.
N.G.
Replace the wires.
CHECK
Timing light
Check ignition timing.
N.G.
Adjust ignition timing.
CHECK
Timing light
Distributor
Check flashes of timing light for weakness.
N.G.
CRANK
START
Engine temp. sensor
Ignition switch
Start with engine temp. sensor connector disconnected.
N.G.
Check the sensor circuits.
PERFORM
Self-diagnosis Mode III
N.G.
Perform self-diagnosis Mode III (for engine temp. sensor).
PERFORM
Self-diagnosis Mode IV
N.G. → Check starter circuit.
Perform self-diagnosis Mode IV (for start signal).

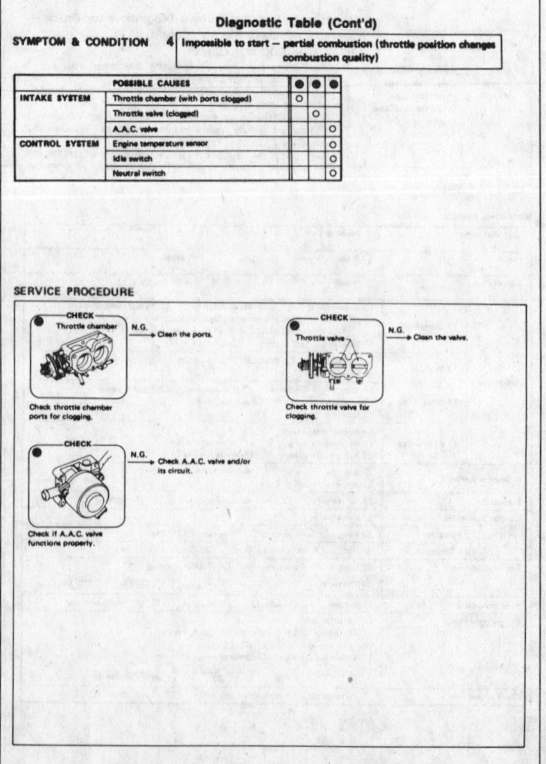

Diagnostic Table (Cont'd)

SYMPTOM & CONDITION 4 Impossible to start – partial combustion (throttle position changes combustion quality)

POSSIBLE CAUSES		●	●	●
INTAKE SYSTEM	Throttle chamber (with ports clogged)	○		
	Throttle valve (clogged)		○	
	A.A.C. valve			○
CONTROL SYSTEM	Engine temperature sensor			○
	Idle switch			○
	Neutral switch			○

SERVICE PROCEDURE

CHECK
Throttle chamber
N.G. → Clean the ports.
Check throttle chamber ports for clogging.
CHECK
Throttle valve
N.G. → Clean the valve.
Check throttle valve for clogging.
CHECK
N.G. → Check A.A.C. valve and/or its circuit.
Check if A.A.C. valve functions properly.

ECCS DIAGNOSTIC CHART—1989–90 MAXIMA

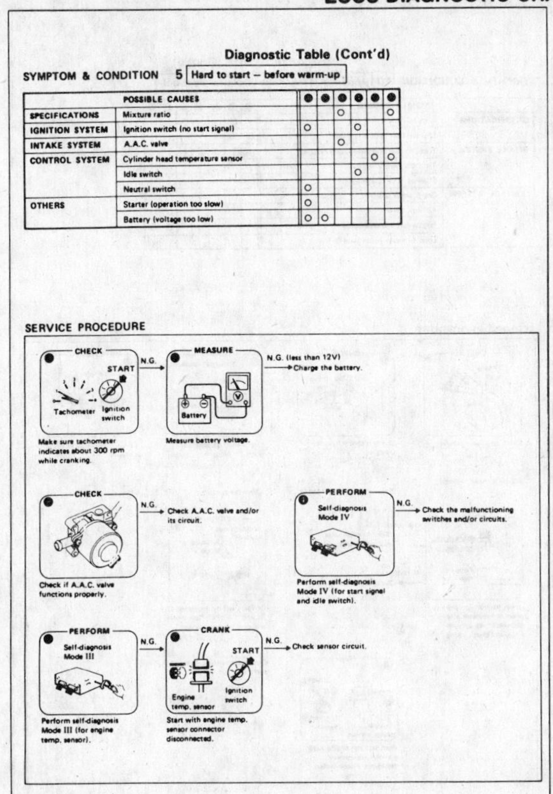

Diagnostic Table (Cont'd)

SYMPTOM & CONDITION 5 Hard to start – before warm-up

POSSIBLE CAUSES		●	●	●	●	●	●
SPECIFICATIONS	Mixture ratio			O			O
IGNITION SYSTEM	Ignition switch (no start signal)	O			O		
INTAKE SYSTEM	A.A.C. valve			O			
CONTROL SYSTEM	Cylinder head temperature sensor					O	O
	Idle switch				O		
	Neutral switch	O					
OTHERS	Starter (operation too slow)	O					
	Battery (voltage too low)	O	O				

SERVICE PROCEDURE
CHECK
START
Tachometer
Ignition switch
Make sure tachometer indicates about 300 rpm while cranking.
N.G.
MEASURE
Battery
Measure battery voltage.
N.G. (less than 12V)
Charge the battery.
CHECK
Check if A.A.C. valve functions properly.
N.G.
Check A.A.C. valve and/or its circuit.
PERFORM
Self-diagnosis Mode IV
Perform self-diagnosis Mode IV (for start signal and idle switch).
N.G.
Check the malfunctioning switches and/or circuits.
PERFORM
Self-diagnosis Mode III
Perform self-diagnosis Mode III (for engine temp. sensor).
N.G.
CRANK
START
Engine temp. sensor
Ignition switch
Start with engine temp. sensor connector disconnected.
N.G.
Check sensor circuit.

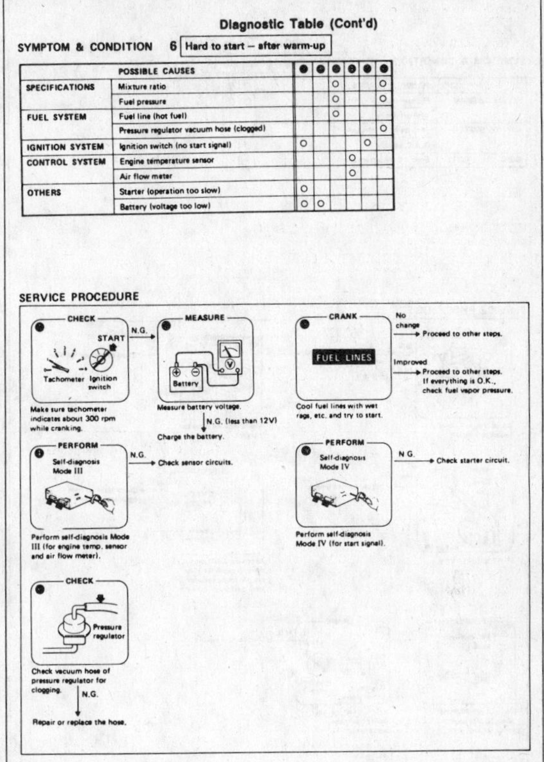

Diagnostic Table (Cont'd)

SYMPTOM & CONDITION 6 Hard to start – after warm-up

POSSIBLE CAUSES		●	●	●	●	●	●
SPECIFICATIONS	Mixture ratio			O			O
	Fuel pressure			O			O
FUEL SYSTEM	Fuel line (hot fuel)			O			
	Pressure regulator vacuum hose (clogged)						O
IGNITION SYSTEM	Ignition switch (no start signal)	O				O	
CONTROL SYSTEM	Engine temperature sensor				O		
	Air flow meter				O		
OTHERS	Starter (operation too slow)	O					
	Battery (voltage too low)	O	O				

SERVICE PROCEDURE
CHECK
START
Tachometer
Ignition switch
Make sure tachometer indicates about 300 rpm while cranking.
N.G.
MEASURE
Battery
Measure battery voltage.
N.G. (less than 12V)
Charge the battery.
CRANK
FUEL LINES
Cool fuel lines with wet rags, etc. and try to start.
No change
Proceed to other steps.
Improved
Proceed to other steps. If everything is O.K., check fuel vapor pressure.
PERFORM
Self-diagnosis Mode III
Perform self-diagnosis Mode III (for engine temp. sensor and air flow meter).
N.G.
Check sensor circuits.
PERFORM
Self-diagnosis Mode IV
Perform self-diagnosis Mode IV (for start signal).
N.G.
Check starter circuit.
CHECK
Pressure regulator
Check vacuum hose of pressure regulator for clogging.
N.G.
Repair or replace the hose.

ECCS DIAGNOSTIC CHART—1989–90 MAXIMA

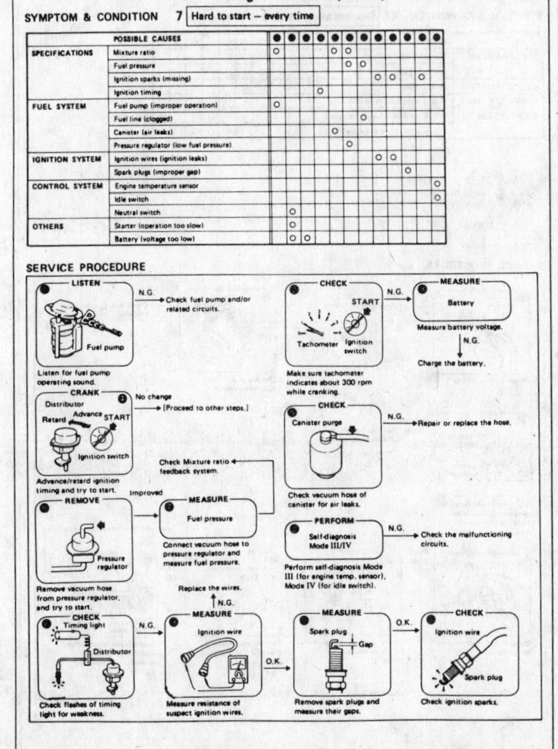

Diagnostic Table (Cont'd)

SYMPTOM & CONDITION 7 Hard to start – every time

POSSIBLE CAUSES		●	●	●	●	●	●	●	●	●	●	●	●
SPECIFICATIONS	Mixture ratio	O				O	O						
	Fuel pressure						O	O					
	Ignition sparks (missing)								O	O		O	
	Ignition timing				O								
FUEL SYSTEM	Fuel pump (improper operation)	O											
	Fuel line (clogged)							O					
	Canister (air leaks)					O							
	Pressure regulator (low fuel pressure)						O						
IGNITION SYSTEM	Ignition wires (ignition leaks)								O	O			
	Spark plugs (improper gap)										O		
CONTROL SYSTEM	Engine temperature sensor												O
	Idle switch												O
	Neutral switch		O										
OTHERS	Starter (operation too slow)		O										
	Battery (voltage too low)		O	O									

SERVICE PROCEDURE
LISTEN
Fuel pump
Listen for fuel pump operating sound.
N.G.
Check fuel pump and/or related circuits.
CHECK
START
Tachometer
Ignition switch
Make sure tachometer indicates about 300 rpm while cranking.
N.G.
MEASURE
Battery
Measure battery voltage.
N.G.
Charge the battery.
CRANK
Distributor
Advance
Retard
START
Ignition switch
Advance/retard ignition timing and try to start.
No change
[Proceed to other steps.]
Improved
Check Mixture ratio feedback system.
CHECK
Canister purge
Check vacuum hose of canister for air leaks.
N.G.
Repair or replace the hose.
REMOVE
Pressure regulator
Remove vacuum hose from pressure regulator, and try to start.
MEASURE
Fuel pressure
Connect vacuum hose to pressure regulator and measure fuel pressure.
PERFORM
Self-diagnosis Mode III/IV
Perform self-diagnosis Mode III (for engine temp. sensor), Mode IV (for idle switch).
N.G.
Check the malfunctioning circuits.
CHECK
Timing light
Distributor
Check flashes of timing light for weakness.
N.G.
MEASURE
Ignition wire
Measure resistance of suspect ignition wires.
N.G.
Replace the wires.
O.K.
MEASURE
Spark plug
Gap
Remove spark plugs and measure their gaps.
O.K.
CHECK
Ignition wire
Spark plug
Check ignition sparks.

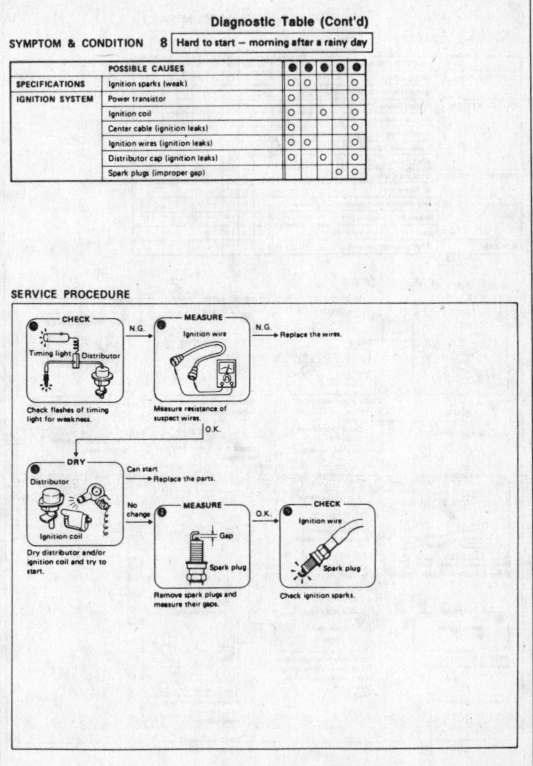

Diagnostic Table (Cont'd)

SYMPTOM & CONDITION 8 Hard to start – morning after a rainy day

POSSIBLE CAUSES		●	●	●	●	●
SPECIFICATIONS	Ignition sparks (weak)	O	O			O
IGNITION SYSTEM	Power transistor	O				O
	Ignition coil	O		O		O
	Center cable (ignition leaks)	O				O
	Ignition wires (ignition leaks)	O	O			O
	Distributor cap (ignition leaks)	O		O		O
	Spark plugs (improper gap)				O	O

SERVICE PROCEDURE
CHECK
Timing light
Distributor
Check flashes of timing light for weakness.
N.G.
MEASURE
Ignition wire
Measure resistance of suspect wires.
N.G.
Replace the wires.
O.K.
DRY
Distributor
Ignition coil
Dry distributor and/or ignition coil and try to start.
Can start
Replace the parts.
No change
MEASURE
Gap
Spark plug
Remove spark plugs and measure their gaps.
O.K.
CHECK
Ignition wire
Spark plug
Check ignition sparks.

ECCS DIAGNOSTIC CHART—1989–90 MAXIMA

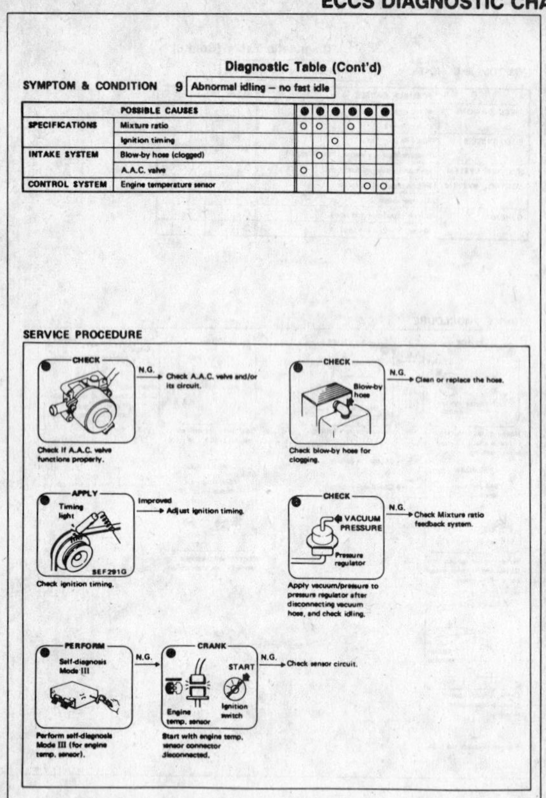

Diagnostic Table (Cont'd)

SYMPTOM & CONDITION 9 Abnormal idling – no fast idle

	POSSIBLE CAUSES	●	●	●	●	●	●
SPECIFICATIONS	Mixture ratio	○	○		○		
	Ignition timing			○			
INTAKE SYSTEM	Blow-by hose (clogged)		○				
	A.A.C. valve	○					
CONTROL SYSTEM	Engine temperature sensor					○	○

SERVICE PROCEDURE

CHECK
N.G.
Check A.A.C. valve and/or its circuit.
Check if A.A.C. valve functions properly.
CHECK
Blow-by hose
N.G.
Clean or replace the hose.
Check blow-by hose for clogging.
APPLY
Timing light
SEF291G
Improved
Adjust ignition timing.
Check ignition timing.
CHECK
VACUUM PRESSURE
Pressure regulator
N.G.
Check Mixture ratio feedback system.
Apply vacuum/pressure to pressure regulator after disconnecting vacuum hose, and check idling.
PERFORM
Self-diagnosis Mode III
N.G.
Perform self-diagnosis Mode III (for engine temp. sensor).
CRANK
START
Engine temp. sensor
Ignition switch
N.G.
Check sensor circuit.
Start with engine temp. sensor connector disconnected.

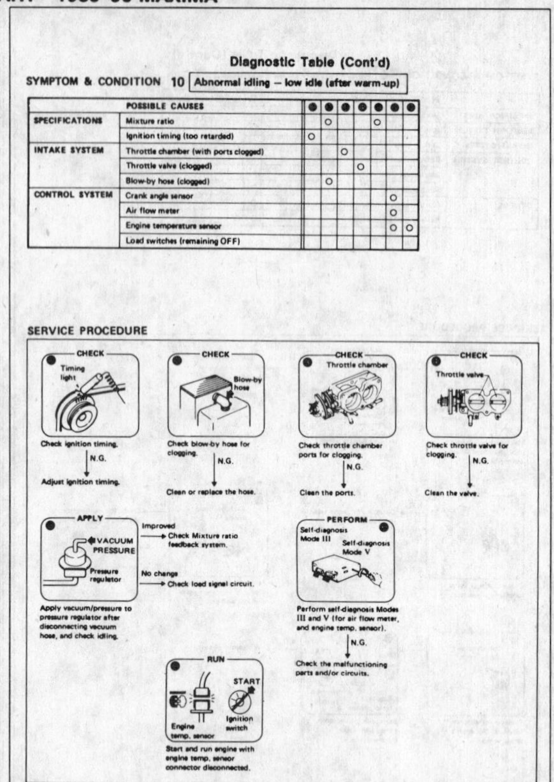

Diagnostic Table (Cont'd)

SYMPTOM & CONDITION 10 Abnormal idling – low idle (after warm-up)

	POSSIBLE CAUSES	●	●	●	●	●	●	●
SPECIFICATIONS	Mixture ratio		○			○		
	Ignition timing (too retarded)	○						
INTAKE SYSTEM	Throttle chamber (with ports clogged)			○				
	Throttle valve (clogged)				○			
	Blow-by hose (clogged)		○					
CONTROL SYSTEM	Crank angle sensor						○	
	Air flow meter						○	
	Engine temperature sensor						○	○
	Load switches (remaining OFF)							

SERVICE PROCEDURE

CHECK
Timing light
Check ignition timing.
N.G.
Adjust ignition timing.
CHECK
Blow-by hose
Check blow-by hose for clogging.
N.G.
Clean or replace the hose.
CHECK
Throttle chamber
Check throttle chamber ports for clogging.
N.G.
Clean the ports.
CHECK
Throttle valve
Check throttle valve for clogging.
N.G.
Clean the valve.
APPLY
VACUUM PRESSURE
Pressure regulator
Improved
Check Mixture ratio feedback system.
No change
Check load signal circuit.
Apply vacuum/pressure to pressure regulator after disconnecting vacuum hose, and check idling.
PERFORM
Self-diagnosis Mode III
Self-diagnosis Mode V
Perform self-diagnosis Modes III and V (for air flow meter, and engine temp. sensor).
N.G.
Check the malfunctioning parts and/or circuits.
RUN
START
Engine temp. sensor
Ignition switch
Start and run engine with engine temp. sensor connector disconnected.

ECCS DIAGNOSTIC CHART—1989–90 MAXIMA

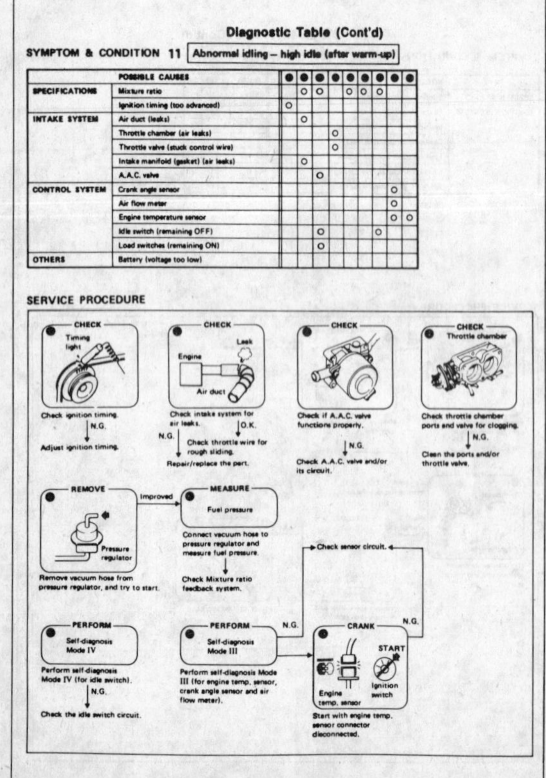

Diagnostic Table (Cont'd)

SYMPTOM & CONDITION 11 Abnormal idling – high idle (after warm-up)

	POSSIBLE CAUSES	●	●	●	●	●	●	●	●	●
SPECIFICATIONS	Mixture ratio		○	○		○	○	○		
	Ignition timing (too advanced)	○								
INTAKE SYSTEM	Air duct (leaks)		○							
	Throttle chamber (air leaks)				○					
	Throttle valve (stuck control wire)				○					
	Intake manifold (gasket) (air leaks)		○							
	A.A.C. valve			○						
CONTROL SYSTEM	Crank angle sensor								○	
	Air flow meter								○	
	Engine temperature sensor								○	○
	Idle switch (remaining OFF)			○				○		
	Load switches (remaining ON)			○						
OTHERS	Battery (voltage too low)									

SERVICE PROCEDURE

CHECK
Timing light
Check ignition timing.
N.G.
Adjust ignition timing.
CHECK
Engine
Leak
Air duct
Check intake system for air leaks.
N.G.
Repair/replace the part.
O.K.
Check throttle wire for rough sliding.
CHECK
Check if A.A.C. valve functions properly.
N.G.
Check A.A.C. valve and/or its circuit.
CHECK
Throttle chamber
Check throttle chamber ports and valve for clogging.
N.G.
Clean the ports and/or throttle valve.
REMOVE
Pressure regulator
Remove vacuum hose from pressure regulator, and try to start.
Improved
MEASURE
Fuel pressure
Connect vacuum hose to pressure regulator and measure fuel pressure.
Check Mixture ratio feedback system.
PERFORM
Self-diagnosis Mode IV
Perform self-diagnosis Mode IV (for idle switch).
N.G.
Check the idle switch circuit.
PERFORM
Self-diagnosis Mode III
Perform self-diagnosis Mode III (for engine temp. sensor, crank angle sensor and air flow meter).
N.G.
Check sensor circuit.
CRANK
START
Engine temp. sensor
Ignition switch
N.G.
Start with engine temp. sensor connector disconnected.

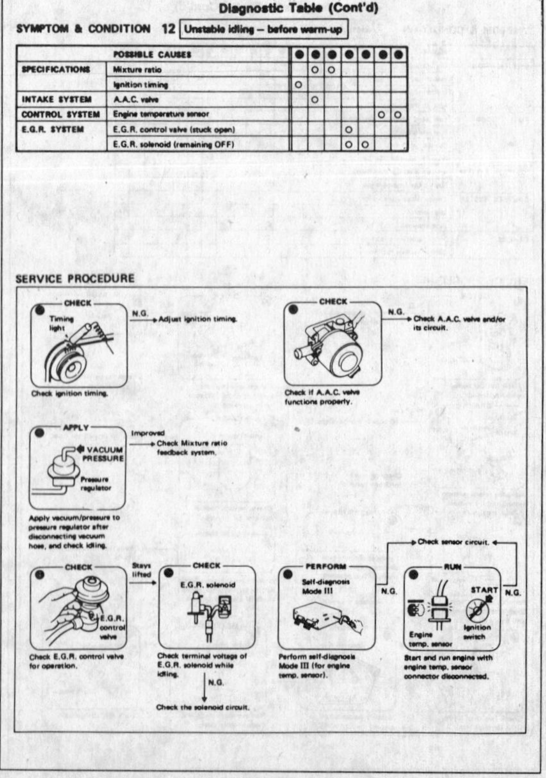

Diagnostic Table (Cont'd)

SYMPTOM & CONDITION 12 Unstable idling – before warm-up

	POSSIBLE CAUSES	●	●	●	●	●	●	●
SPECIFICATIONS	Mixture ratio		○	○				
	Ignition timing	○						
INTAKE SYSTEM	A.A.C. valve		○					
CONTROL SYSTEM	Engine temperature sensor						○	○
E.G.R. SYSTEM	E.G.R. control valve (stuck open)				○			
	E.G.R. solenoid (remaining OFF)				○	○		

SERVICE PROCEDURE

CHECK
Timing light
N.G.
Adjust ignition timing.
Check ignition timing.
CHECK
N.G.
Check A.A.C. valve and/or its circuit.
Check if A.A.C. valve functions properly.
APPLY
VACUUM PRESSURE
Pressure regulator
Improved
Check Mixture ratio feedback system.
Apply vacuum/pressure to pressure regulator after disconnecting vacuum hose, and check idling.
CHECK
E.G.R. control valve
Check E.G.R. control valve for operation.
Stays lifted
CHECK
E.G.R. solenoid
Check terminal voltage of E.G.R. solenoid while idling.
N.G.
Check the solenoid circuit.
PERFORM
Self-diagnosis Mode III
Perform self-diagnosis Mode III (for engine temp. sensor).
N.G.
Check sensor circuit.
RUN
START
Engine temp. sensor
Ignition switch
N.G.
Start and run engine with engine temp. sensor connector disconnected.

ECCS DIAGNOSTIC CHART—1989–90 MAXIMA

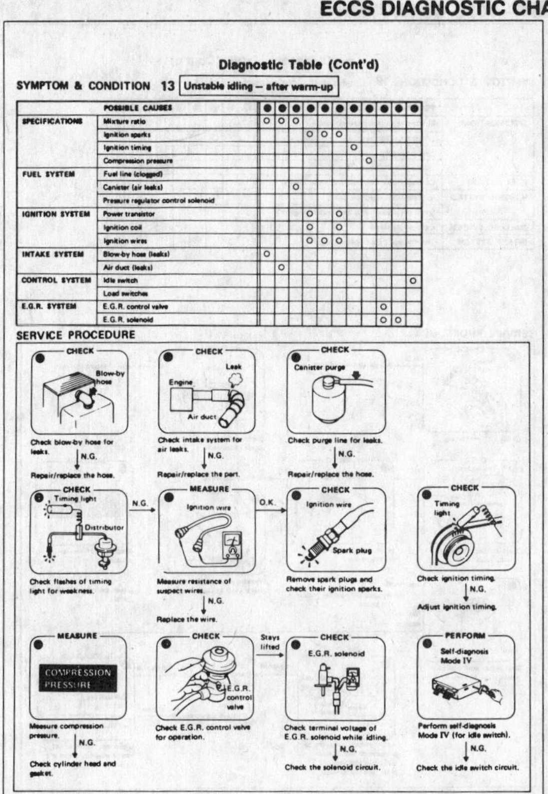

Diagnostic Table (Cont'd)

SYMPTOM & CONDITION 13 Unstable idling – after warm-up

	POSSIBLE CAUSES	●	●	●	●	●	●	●	●	●	●	●
SPECIFICATIONS	Mixture ratio	○	○	○								
	Ignition sparks				○	○	○					
	Ignition timing							○				
	Compression pressure								○			
FUEL SYSTEM	Fuel line (clogged)											
	Canister (air leaks)			○								
	Pressure regulator control solenoid											
IGNITION SYSTEM	Power transistor				○		○					
	Ignition coil				○		○					
	Ignition wires				○	○	○					
INTAKE SYSTEM	Blow-by hose (leaks)	○										
	Air duct (leaks)		○									
CONTROL SYSTEM	Idle switch											○
	Load switches											
E.G.R. SYSTEM	E.G.R. control valve									○		
	E.G.R. solenoid									○	○	

SERVICE PROCEDURE

CHECK
Blow-by hose
Check blow-by hose for leaks.
N.G.
Repair/replace the hose.
CHECK
Engine
Leak
Air duct
Check intake system for air leaks.
N.G.
Repair/replace the part.
CHECK
Canister purge
Check purge line for leaks.
N.G.
Repair/replace the hose.
CHECK
Timing light
Distributor
Check flashes of timing light for weakness.
N.G.
MEASURE
Ignition wire
Measure resistance of suspect wires.
N.G.
Replace the wire.
O.K.
CHECK
Ignition wire
Spark plug
Remove spark plugs and check their ignition sparks.
CHECK
Timing light
Check ignition timing.
N.G.
Adjust ignition timing.
MEASURE
COMPRESSION PRESSURE
Measure compression pressure.
N.G.
Check cylinder head and gasket.
CHECK
E.G.R. control valve
Check E.G.R. control valve for operation.
Stays lifted
CHECK
E.G.R. solenoid
Check terminal voltage of E.G.R. solenoid while idling.
N.G.
Check the solenoid circuit.
PERFORM
Self-diagnosis Mode IV
Perform self-diagnosis Mode IV (for idle switch).
N.G.
Check the idle switch circuit.

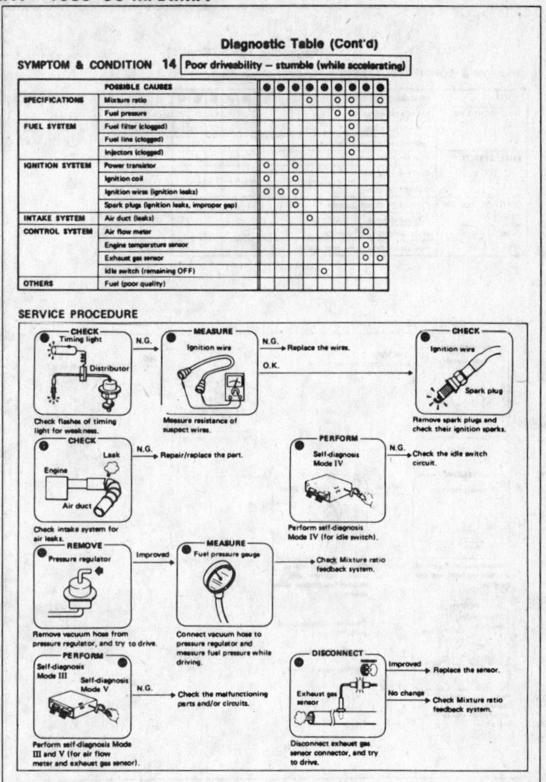

Diagnostic Table (Cont'd)

SYMPTOM & CONDITION 14 Poor driveability – stumble (while accelerating)

	POSSIBLE CAUSES	●	●	●	●	●	●	●	●	●
SPECIFICATIONS	Mixture ratio				○		○	○		○
	Fuel pressure						○	○		
FUEL SYSTEM	Fuel filter (clogged)							○		
	Fuel line (clogged)							○		
	Injectors (clogged)							○		
IGNITION SYSTEM	Power transistor	○		○						
	Ignition coil	○		○						
	Ignition wires (ignition leaks)	○	○	○						
	Spark plugs (ignition leaks, improper gap)			○						
INTAKE SYSTEM	Air duct (leaks)				○					
CONTROL SYSTEM	Air flow meter								○	
	Engine temperature sensor								○	
	Exhaust gas sensor								○	○
	Idle switch (remaining OFF)					○				
OTHERS	Fuel (poor quality)									

SERVICE PROCEDURE

CHECK
Timing light
Distributor
Check flashes of timing light for weakness.
N.G.
MEASURE
Ignition wire
Measure resistance of suspect wires.
N.G.
Replace the wires.
O.K.
CHECK
Ignition wire
Spark plug
Remove spark plugs and check their ignition sparks.
CHECK
Engine
Leak
Air duct
Check intake system for air leaks.
N.G.
Repair/replace the part.
PERFORM
Self-diagnosis Mode IV
Perform self-diagnosis Mode IV (for idle switch).
N.G.
Check the idle switch circuit.
REMOVE
Pressure regulator
Remove vacuum hose from pressure regulator, and try to drive.
Improved
MEASURE
Fuel pressure gauge
Connect vacuum hose to pressure regulator and measure fuel pressure while driving.
Check Mixture ratio feedback system.
PERFORM
Self-diagnosis Mode III
Self-diagnosis Mode V
Perform self-diagnosis Mode III and V (for air flow meter and exhaust gas sensor).
N.G.
Check the malfunctioning parts and/or circuits.
DISCONNECT
Exhaust gas sensor
Disconnect exhaust gas sensor connector, and try to drive.
Improved
Replace the sensor.
No change
Check Mixture ratio feedback system.

ECCS DIAGNOSTIC CHART—1989–90 MAXIMA

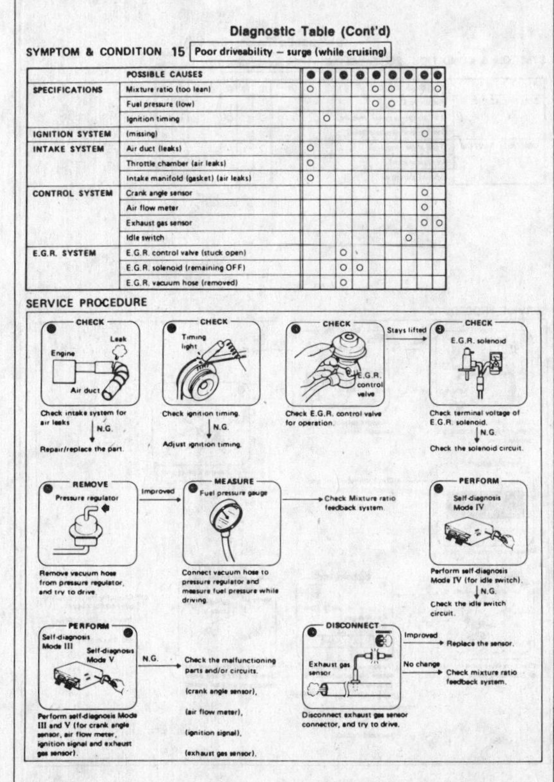

Diagnostic Table (Cont'd)

SYMPTOM & CONDITION 15 Poor driveability – surge (while cruising)

	POSSIBLE CAUSES	●	●	●	●	●	●	●	●	●
SPECIFICATIONS	Mixture ratio (too lean)	○				○	○			○
	Fuel pressure (low)					○	○			
	Ignition timing		○							
IGNITION SYSTEM	(missing)								○	
INTAKE SYSTEM	Air duct (leaks)	○								
	Throttle chamber (air leaks)	○								
	Intake manifold (gasket) (air leaks)	○								
CONTROL SYSTEM	Crank angle sensor								○	
	Air flow meter								○	
	Exhaust gas sensor								○	○
	Idle switch							○		
E.G.R. SYSTEM	E.G.R. control valve (stuck open)			○						
	E.G.R. solenoid (remaining OFF)			○	○					
	E.G.R. vacuum hose (removed)			○						

SERVICE PROCEDURE

CHECK
Engine
Leak
Air duct
Check intake system for air leaks
N.G.
Repair/replace the part.
CHECK
Timing light
Check ignition timing.
N.G.
Adjust ignition timing.
CHECK
E.G.R. control valve
Check E.G.R. control valve for operation.
Stays lifted
CHECK
E.G.R. solenoid
Check terminal voltage of E.G.R. solenoid.
N.G.
Check the solenoid circuit.
REMOVE
Pressure regulator
Remove vacuum hose from pressure regulator, and try to drive.
Improved
MEASURE
Fuel pressure gauge
Connect vacuum hose to pressure regulator and measure fuel pressure while driving.
Check Mixture ratio feedback system.
PERFORM
Self-diagnosis Mode IV
Perform self-diagnosis Mode IV (for idle switch).
N.G.
Check the idle switch circuit.
PERFORM
Self-diagnosis Mode III
Self-diagnosis Mode V
Perform self-diagnosis Mode III and V (for crank angle sensor, air flow meter, ignition signal and exhaust gas sensor).
N.G.
Check the malfunctioning parts and/or circuits.
(crank angle sensor),
(air flow meter),
(ignition signal),
(exhaust gas sensor).
DISCONNECT
Exhaust gas sensor
Disconnect exhaust gas sensor connector, and try to drive.
Improved
Replace the sensor.
No change
Check mixture ratio feedback system.

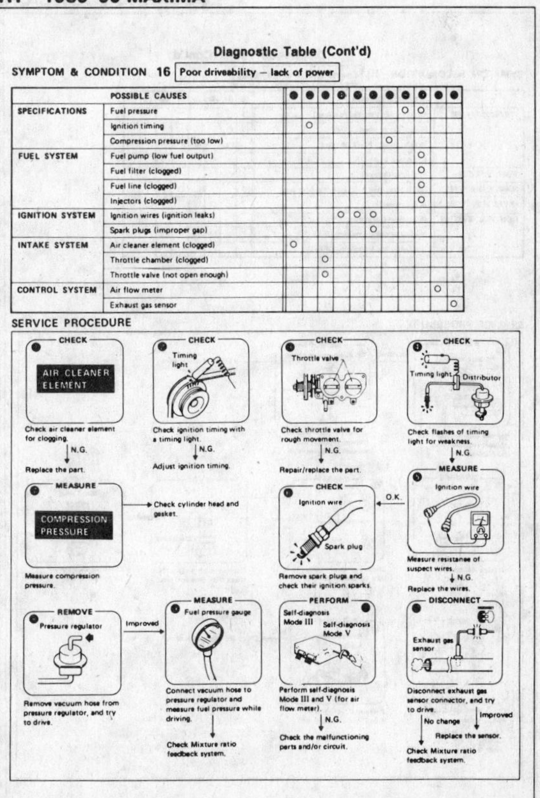

Diagnostic Table (Cont'd)

SYMPTOM & CONDITION 16 Poor driveability – lack of power

	POSSIBLE CAUSES	●	●	●	●	●	●	●	●	●	●	●
SPECIFICATIONS	Fuel pressure								○	○		
	Ignition timing		○									
	Compression pressure (too low)							○				
FUEL SYSTEM	Fuel pump (low fuel output)									○		
	Fuel filter (clogged)									○		
	Fuel line (clogged)									○		
	Injectors (clogged)									○		
IGNITION SYSTEM	Ignition wires (ignition leaks)				○	○	○					
	Spark plugs (improper gap)						○					
INTAKE SYSTEM	Air cleaner element (clogged)	○										
	Throttle chamber (clogged)			○								
	Throttle valve (not open enough)			○								
CONTROL SYSTEM	Air flow meter										○	
	Exhaust gas sensor											○

SERVICE PROCEDURE

CHECK
AIR CLEANER ELEMENT
Check air cleaner element for clogging.
N.G.
Replace the part.
CHECK
Timing light
Check ignition timing with a timing light.
N.G.
Adjust ignition timing.
CHECK
Throttle valve
Check throttle valve for rough movement.
N.G.
Repair/replace the part.
CHECK
Timing light
Distributor
Check flashes of timing light for weakness.
N.G.
MEASURE
COMPRESSION PRESSURE
Measure compression pressure.
Check cylinder head and gasket.
CHECK
Ignition wire
Spark plug
Remove spark plugs and check their ignition sparks.
O.K.
MEASURE
Ignition wire
Measure resistance of suspect wires.
N.G.
Replace the wires.
REMOVE
Pressure regulator
Remove vacuum hose from pressure regulator, and try to drive.
Improved
MEASURE
Fuel pressure gauge
Connect vacuum hose to pressure regulator and measure fuel pressure while driving.
Check Mixture ratio feedback system.
PERFORM
Self-diagnosis Mode III
Self-diagnosis Mode V
Perform self-diagnosis Mode III and V (for air flow meter).
N.G.
Check the malfunctioning parts and/or circuit.
DISCONNECT
Exhaust gas sensor
Disconnect exhaust gas sensor connector, and try to drive.
No change
Improved
Replace the sensor.
Check Mixture ratio feedback system.

ECCS DIAGNOSTIC CHART—1989-90 MAXIMA

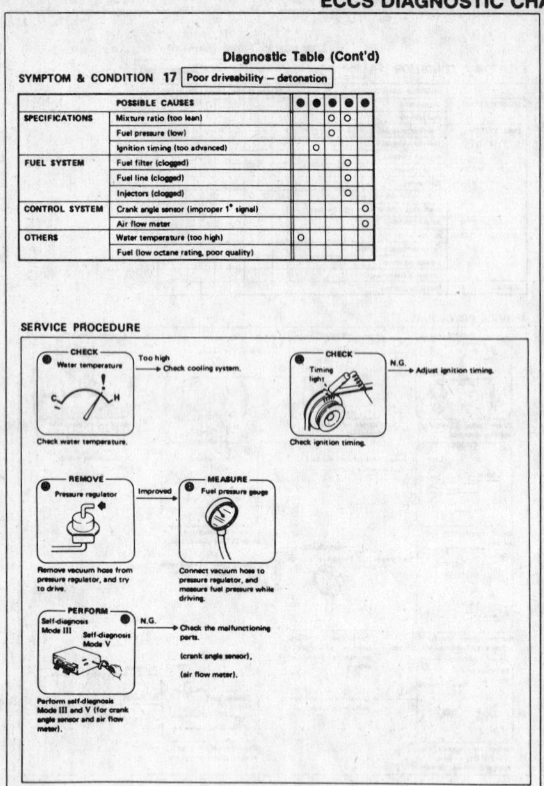

Diagnostic Table (Cont'd)

SYMPTOM & CONDITION 17 Poor driveability – detonation

POSSIBLE CAUSES		●	●	●	●	●
SPECIFICATIONS	Mixture ratio (too lean)			○	○	
	Fuel pressure (low)			○		
	Ignition timing (too advanced)		○			
FUEL SYSTEM	Fuel filter (clogged)				○	
	Fuel line (clogged)				○	
	Injectors (clogged)				○	
CONTROL SYSTEM	Crank angle sensor (improper 1° signal)					○
	Air flow meter					○
OTHERS	Water temperature (too high)	○				
	Fuel (low octane rating, poor quality)					

SERVICE PROCEDURE

CHECK
Water temperature
C
H
Too high
Check cooling system.
Check water temperature.
CHECK
Timing light
N.G.
Adjust ignition timing.
Check ignition timing.
REMOVE
Pressure regulator
Improved
MEASURE
Fuel pressure gauge
Remove vacuum hose from pressure regulator, and try to drive.
Connect vacuum hose to pressure regulator, and measure fuel pressure while driving.
PERFORM
Self-diagnosis Mode III
Self-diagnosis Mode V
N.G.
Check the malfunctioning parts.
(crank angle sensor),
(air flow meter).
Perform self-diagnosis Mode III and V (for crank angle sensor and air flow meter).

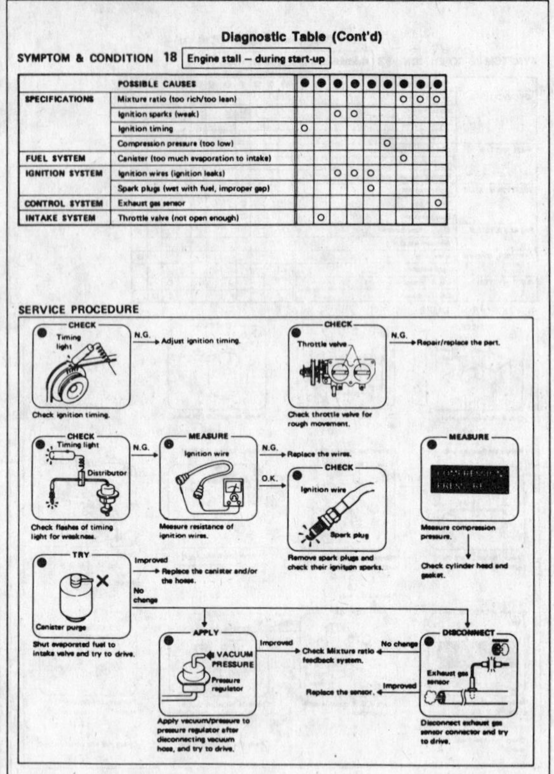

Diagnostic Table (Cont'd)

SYMPTOM & CONDITION 18 Engine stall – during start-up

POSSIBLE CAUSES		●	●	●	●	●	●	●	●	●
SPECIFICATIONS	Mixture ratio (too rich/too lean)							○	○	○
	Ignition sparks (weak)			○	○					
	Ignition timing	○								
	Compression pressure (too low)						○			
FUEL SYSTEM	Canister (too much evaporation to intake)							○		
IGNITION SYSTEM	Ignition wires (ignition leaks)			○	○	○				
	Spark plugs (wet with fuel, improper gap)					○				
CONTROL SYSTEM	Exhaust gas sensor									○
INTAKE SYSTEM	Throttle valve (not open enough)		○							

SERVICE PROCEDURE

CHECK
Timing light
N.G.
Adjust ignition timing.
Check ignition timing.
CHECK
Throttle valve
N.G.
Repair/replace the part.
Check throttle valve for rough movement.
CHECK
Timing light
Distributor
N.G.
Check flashes of timing light for weakness.
MEASURE
Ignition wire
N.G.
Replace the wires.
O.K.
Measure resistance of ignition wires.
CHECK
Ignition wire
Spark plug
Remove spark plugs and check their ignition sparks.
MEASURE
Measure compression pressure.
Check cylinder head and gasket.
TRY
Canister purge
Improved
Replace the canister and/or the hoses.
No change
Shut evaporated fuel to intake valve and try to drive.
APPLY
VACUUM
PRESSURE
Pressure regulator
Improved
Check Mixture ratio feedback system.
Apply vacuum/pressure to pressure regulator after disconnecting vacuum hose, and try to drive.
DISCONNECT
Exhaust gas sensor
No change
Improved
Replace the sensor.
Disconnect exhaust gas sensor connector and try to drive.

ECCS DIAGNOSTIC CHART—1989-90 MAXIMA

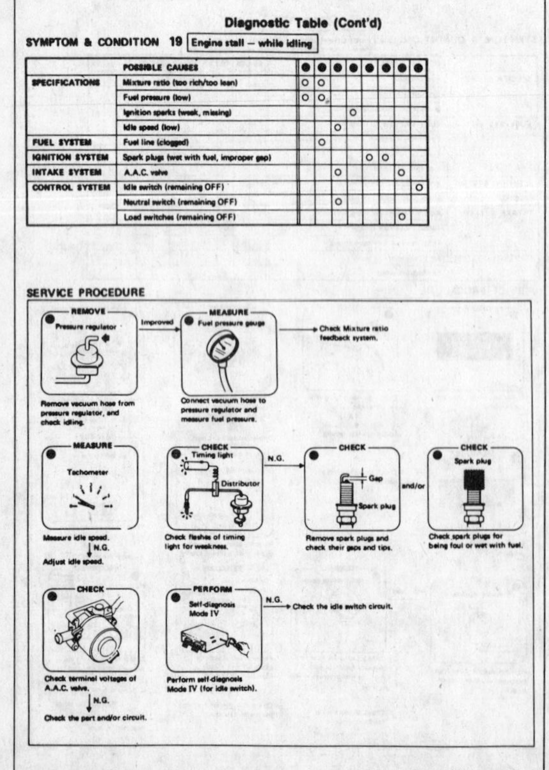

Diagnostic Table (Cont'd)

SYMPTOM & CONDITION 19 Engine stall – while idling

POSSIBLE CAUSES		●	●	●	●	●	●	●	●
SPECIFICATIONS	Mixture ratio (too rich/too lean)	○	○						
	Fuel pressure (low)	○	○						
	Ignition sparks (weak, missing)				○				
	Idle speed (low)			○					
FUEL SYSTEM	Fuel line (clogged)		○						
IGNITION SYSTEM	Spark plugs (wet with fuel, improper gap)					○	○		
INTAKE SYSTEM	A.A.C. valve			○				○	
CONTROL SYSTEM	Idle switch (remaining OFF)								○
	Neutral switch (remaining OFF)			○					
	Load switches (remaining OFF)							○	

SERVICE PROCEDURE

REMOVE
Pressure regulator
Improved
MEASURE
Fuel pressure gauge
Check Mixture ratio feedback system.
Remove vacuum hose from pressure regulator, and check idling.
Connect vacuum hose to pressure regulator and measure fuel pressure.
MEASURE
Tachometer
Measure idle speed.
N.G.
Adjust idle speed.
CHECK
Timing light
Distributor
N.G.
Check flashes of timing light for weakness.
CHECK
Gap
Spark plug
and/or
Remove spark plugs and check their gaps and tips.
CHECK
Spark plug
Check spark plugs for being foul or wet with fuel.
CHECK
Check terminal voltages of A.A.C. valve.
N.G.
Check the part and/or circuit.
PERFORM
Self-diagnosis Mode IV
N.G.
Check the idle switch circuit.
Perform self-diagnosis Mode IV (for idle switch).

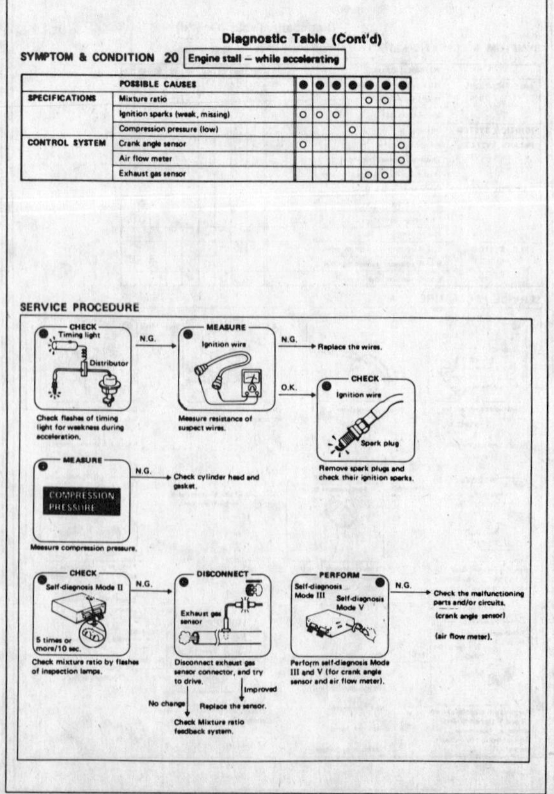

Diagnostic Table (Cont'd)

SYMPTOM & CONDITION 20 Engine stall – while accelerating

POSSIBLE CAUSES		●	●	●	●	●	●	●
SPECIFICATIONS	Mixture ratio					○	○	
	Ignition sparks (weak, missing)	○	○	○				
	Compression pressure (low)				○			
CONTROL SYSTEM	Crank angle sensor	○						○
	Air flow meter							○
	Exhaust gas sensor					○	○	

SERVICE PROCEDURE

CHECK
Timing light
Distributor
N.G.
Check flashes of timing light for weakness during acceleration.
MEASURE
Ignition wire
N.G.
Replace the wires.
O.K.
Measure resistance of suspect wires.
CHECK
Ignition wire
Spark plug
Remove spark plugs and check their ignition sparks.
MEASURE
COMPRESSION PRESSURE
N.G.
Check cylinder head and gasket.
Measure compression pressure.
CHECK
Self-diagnosis Mode II
5 times or more/10 sec.
N.G.
Check mixture ratio by flashes of inspection lamps.
DISCONNECT
Exhaust gas sensor
Disconnect exhaust gas sensor connector, and try to drive.
Improved
Replace the sensor.
No change
Check Mixture ratio feedback system.
PERFORM
Self-diagnosis Mode III
Self-diagnosis Mode V
N.G.
Check the malfunctioning parts and/or circuits.
(crank angle sensor)
(air flow meter).
Perform self-diagnosis Mode III and V (for crank angle sensor and air flow meter).

ECCS DIAGNOSTIC CHART—1989–90 MAXIMA

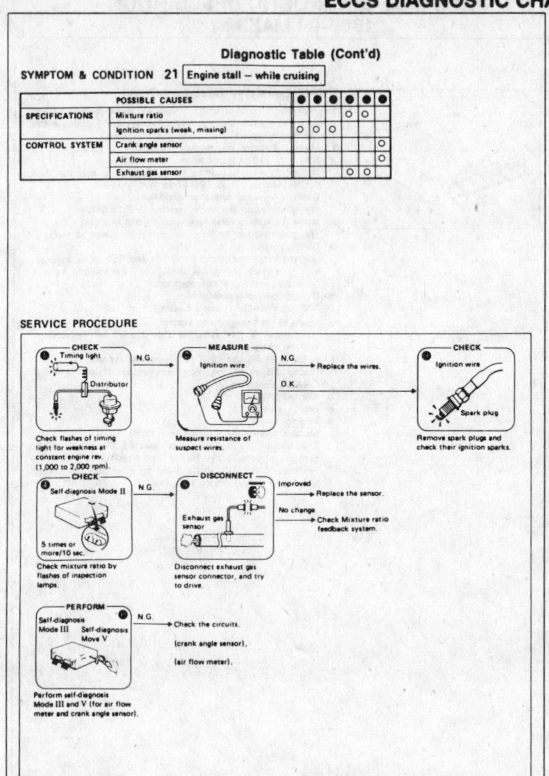

Diagnostic Table (Cont'd)

SYMPTOM & CONDITION 21 Engine stall – while cruising

POSSIBLE CAUSES		●	●	●	●	●	●
SPECIFICATIONS	Mixture ratio				○	○	
	Ignition sparks (weak, missing)	○	○	○			
CONTROL SYSTEM	Crank angle sensor						○
	Air flow meter						○
	Exhaust gas sensor				○	○	

SERVICE PROCEDURE

CHECK
Timing light
Distributor
N.G.
Check flashes of timing light for weakness at constant engine rev. (1,000 to 2,000 rpm).
MEASURE
Ignition wire
N.G.
Replace the wires.
O.K.
Measure resistance of suspect wires.
CHECK
Ignition wire
Spark plug
Remove spark plugs and check their ignition sparks.
CHECK
Self-diagnosis Mode II
5 times or more/10 sec.
N.G.
Check mixture ratio by flashes of inspection lamps.
DISCONNECT
Exhaust gas sensor
Improved
Replace the sensor.
No change
Check Mixture ratio feedback system.
Disconnect exhaust gas sensor connector, and try to drive.
PERFORM
Self-diagnosis Mode III
Self-diagnosis Move V
N.G.
Check the circuits.
(crank angle sensor),
(air flow meter).
Perform self-diagnosis Mode III and V (for air flow meter and crank angle sensor).

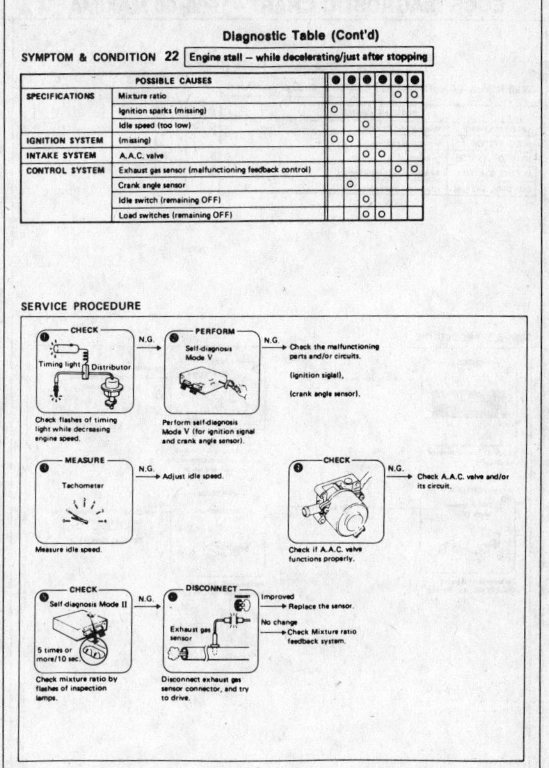

Diagnostic Table (Cont'd)

SYMPTOM & CONDITION 22 Engine stall – while decelerating/just after stopping

POSSIBLE CAUSES		●	●	●	●	●	●
SPECIFICATIONS	Mixture ratio					○	○
	Ignition sparks (missing)	○					
	Idle speed (too low)			○			
IGNITION SYSTEM	(missing)	○	○				
INTAKE SYSTEM	A.A.C. valve			○	○		
CONTROL SYSTEM	Exhaust gas sensor (malfunctioning feedback control)					○	○
	Crank angle sensor		○				
	Idle switch (remaining OFF)			○			
	Load switches (remaining OFF)			○	○		

SERVICE PROCEDURE

CHECK
Timing light
Distributor
N.G.
Check flashes of timing light while decreasing engine speed.
PERFORM
Self-diagnosis Mode V
N.G.
Check the malfunctioning parts and/or circuits.
(ignition siglal),
(crank angle sensor).
Perform self-diagnosis Mode V (for ignition signal and crank angle sensor).
MEASURE
Tachometer
N.G.
Adjust idle speed.
Measure idle speed.
CHECK
N.G.
Check A.A.C. valve and/or its circuit.
Check if A.A.C. valve functions properly.
CHECK
Self-diagnosis Mode II
5 times or more/10 sec.
N.G.
Check mixture ratio by flashes of inspection lamps.
DISCONNECT
Exhaust gas sensor
Improved
Replace the sensor.
No change
Check Mixture ratio feedback system.
Disconnect exhaust gas sensor connector, and try to drive.

ECCS DIAGNOSTIC CHART—1989–90 MAXIMA

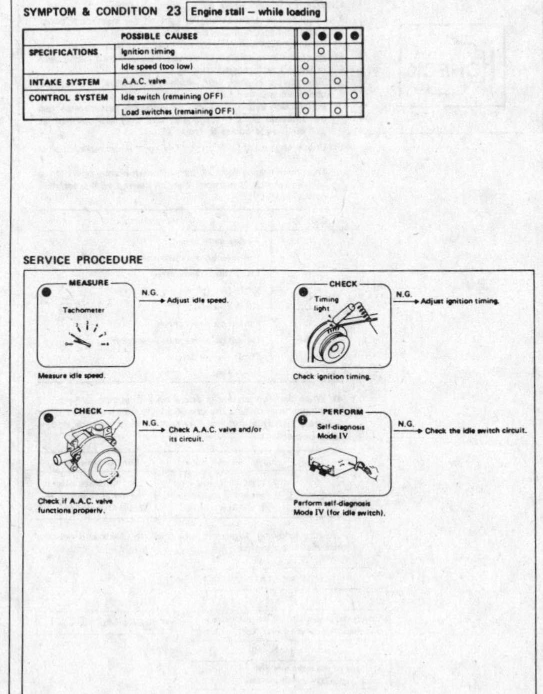

Diagnostic Table (Cont'd)

SYMPTOM & CONDITION 23 Engine stall – while loading

POSSIBLE CAUSES		●	●	●	●
SPECIFICATIONS	Ignition timing		○		
	Idle speed (too low)	○			
INTAKE SYSTEM	A.A.C. valve	○		○	
CONTROL SYSTEM	Idle switch (remaining OFF)	○			○
	Load switches (remaining OFF)	○		○	

SERVICE PROCEDURE

MEASURE
Tachometer
N.G.
Adjust idle speed.
Measure idle speed.
CHECK
Timing light
N.G.
Adjust ignition timing.
Check ignition timing.
CHECK
N.G.
Check A.A.C. valve and/or its circuit.
Check if A.A.C. valve functions properly.
PERFORM
Self-diagnosis Mode IV
N.G.
Check the idle switch circuit.
Perform self-diagnosis Mode IV (for idle switch).

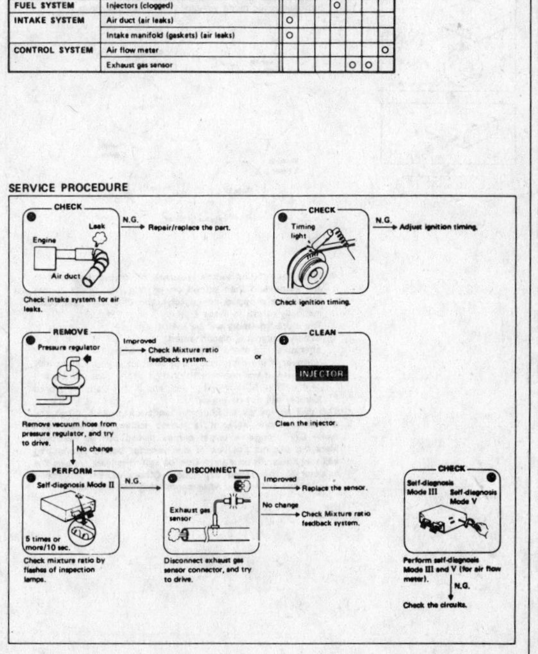

Diagnostic Table (Cont'd)

SYMPTOM & CONDITION 24 Backfire – through the intake

POSSIBLE CAUSES		●	●	●	●	●	●	●
SPECIFICATIONS	Mixture ratio (too lean)	○		○		○	○	
	Ignition timing (too retarded)		○					
FUEL SYSTEM	Injectors (clogged)				○			
INTAKE SYSTEM	Air duct (air leaks)	○						
	Intake manifold (gaskets) (air leaks)	○						
CONTROL SYSTEM	Air flow meter							○
	Exhaust gas sensor					○	○	

SERVICE PROCEDURE

CHECK
Engine
Leak
Air duct
N.G.
Repair/replace the part.
Check intake system for air leaks.
CHECK
Timing light
N.G.
Adjust ignition timing.
Check ignition timing.
REMOVE
Pressure regulator
Improved
Check Mixture ratio feedback system.
or
Remove vacuum hose from pressure regulator, and try to drive.
No change
CLEAN
INJECTOR
Clean the injector.
PERFORM
Self-diagnosis Mode II
5 times or more/10 sec.
N.G.
Check mixture ratio by flashes of inspection lamps.
DISCONNECT
Exhaust gas sensor
Improved
Replace the sensor.
No change
Check Mixture ratio feedback system.
Disconnect exhaust gas sensor connector, and try to drive.
CHECK
Self-diagnosis Mode III
Self-diagnosis Mode V
Perform self-diagnosis Mode III and V (for air flow meter).
N.G.
Check the circuits.

ECCS DIAGNOSTIC CHART—1989–90 MAXIMA

Diagnostic Table (Cont'd)

SYMPTOM & CONDITION 25 Backfire – through the exhaust

POSSIBLE CAUSES		●	●	●	●
SPECIFICATIONS	Mixture ratio (too rich)	○	○		
FUEL SYSTEM	Injectors (fuel leaks)		○		
IGNITION SYSTEM	(missing)				○
INTAKE SYSTEM	Air cleaner element (clogged)	○			
CONTROL SYSTEM	Idle switch (remaining OFF)			○	

SERVICE PROCEDURE

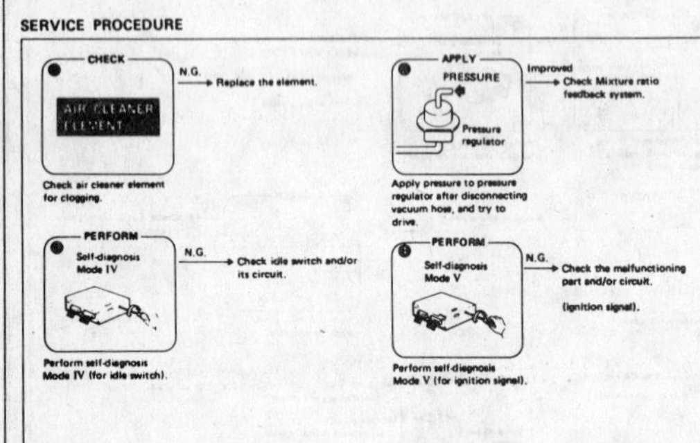
CHECK
AIR CLEANER ELEMENT
N.G.
Replace the element.
Check air cleaner element for clogging.
APPLY PRESSURE
Pressure regulator
Improved
Check Mixture ratio feedback system.
Apply pressure to pressure regulator after disconnecting vacuum hose, and try to drive.
PERFORM
Self-diagnosis Mode IV
N.G.
Check idle switch and/or its circuit.
Perform self-diagnosis Mode IV (for idle switch).
PERFORM
Self-diagnosis Mode V
N.G.
Check the malfunctioning part and/or circuit.
(ignition signal).
Perform self-diagnosis Mode V (for ignition signal).

ECCS SELF-DIAGNOSTIC DESCRIPTION—1989–90 MAXIMA

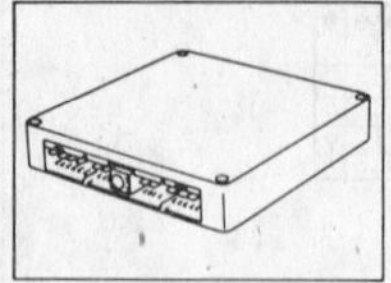

Self-diagnosis — Description

The self-diagnosis is useful to diagnose malfunctions in major sensors and actuators of the E.C.C.S. There are 5 modes in the self-diagnosis system.

1. **Mode I (Exhaust gas sensor monitor)**
   - During closed-loop operation:
     The green inspection lamp turns ON when a lean condition is detected and goes OFF under rich condition.
   - During open-loop operation condition:
     The green inspection lamp remains OFF or ON.
2. **Mode II (Mixture ratio feedback control monitor)**
   The green inspection lamp function is the same as Mode I.
   - During closed-loop operation:
     The red inspection lamp turns ON and OFF simultaneously with the green inspection lamp when the mixture ratio is controlled within the specified value.
   - During open-loop operation:
     The red inspection lamp remains ON or OFF.
3. **Mode III (Self-diagnostic system)**
   This mode is the same as the former self-diagnosis in self-diagnosis mode.
4. **Mode IV (Switches ON/OFF diagnostic system)**
   During this mode, the inspection lamps monitor the switch ON-OFF condition.
   - Idle switch
   - Starter switch
   - Vehicle speed sensor
5. **Mode V (Real-time diagnostic system)**
   The moment the malfunction is detected, the display will be presented immediately. That is, the condition at which the malfunction occurs can be found by observing the inspection lamps during driving test.

ECCS SELF-DIAGNOSTIC DESCRIPTION (CONT.)—1989–90 MAXIMA

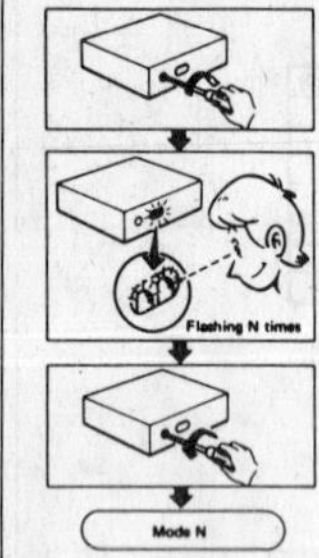
Flashing N times
Mode N

Self-diagnosis — Description (Cont'd)

HOW TO SWITCH THE DIAGNOSTIC MODES

1. Turn ignition switch "ON".
2. Turn diagnostic mode selector to E.C.U. (fully clockwise) and wait for inspection lamps to flash.
3. Count the number of flashes, and after the inspection lamps have flashed the number of the required mode, immediately turn diagnostic mode selector fully counterclockwise.

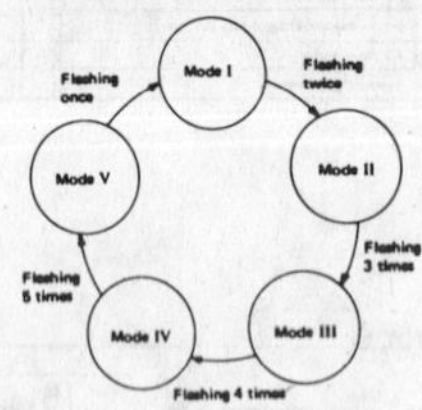
Mode I
Flashing twice
Mode II
Flashing 3 times
Mode III
Flashing 4 times
Mode IV
Flashing 5 times
Mode V
Flashing once

- **When the ignition switch is turned off during diagnosis in any mode and then turned on again (after power to the E.C.U. has dropped completely), the diagnosis will automatically return to Mode I.**
  The stored memory will be lost if:
1. Battery terminal is disconnected.
2. After selecting Mode III, Mode IV is selected.
   However, if the diagnostic mode selector is kept turned fully clockwise, it will continue to change in the order of Mode I → II → III → IV → V → I ... etc., and in this state the stored memory will not be erased.

This unit serves as an idle rpm feedback control. When the diagnostic mode selector is turned within the "diagnostic mode OFF" range, a target engine speed can be selected. Mark the original position of the selector before conducting self-diagnosis. Upon completion of self-diagnosis, return the selector to the previous position. Otherwise, engine speed may change before and after conducting self-diagnosis.

CHECK
CHECK ENGINE LIGHT

Self-diagnosis — Description (Cont'd)

CHECK ENGINE LIGHT

This vehicle has a check engine light on the instrument panel. This light comes ON under the following conditions:

1) When ignition switch is turned "ON" (for bulb check).
2) When systems related to emission performance malfunction in Mode I (with engine running).
   - **This check engine light always illuminates and is synchronous with red L.E.D.**
   - **Malfunction systems related to emission performance can be detected by self-diagnosis, and they are clarified as self-diagnostic codes in Mode III.**
3) Check engine light will come "ON" only when malfunction is sensed.
   The check engine light will turn off when normal operation is resumed. Mode III memory must be cleared as the contents remain stored.

Code No.	Malfunction
12	Air flow meter circuit
13	Engine temperature sensor circuit
14	Vehicle speed sensor circuit
31	E.C.U. (E.C.C.S. control unit)
32	E.G.R. function
33	Exhaust gas sensor circuit
35	Exhaust gas temperature sensor circuit
43	Throttle sensor circuit
45	Injector leak

4) When back-up system is operating with engine running.
Conditions under which the check engine light illuminates differ between California and non-California models, as indicated in table below:

	California model	Non-California model
Condition	Light illuminates when any one of conditions 1), 2) and 3) is satisfied.	Light illuminates when any one of conditions 1), 2) and 4) is satisfied.

Use the following diagnostic flow chart to check and repair a malfunctioning system.

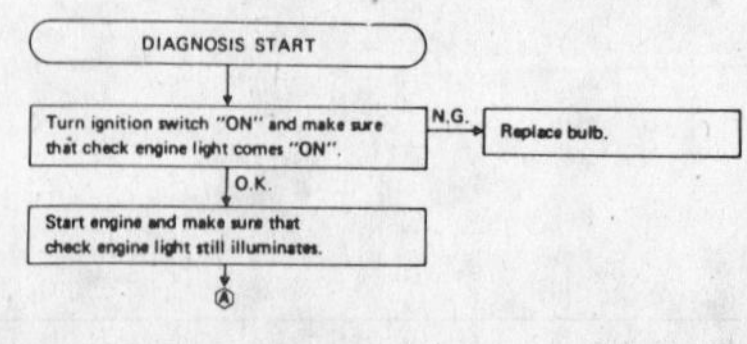
DIAGNOSIS START
Turn ignition switch "ON" and make sure that check engine light comes "ON".
N.G.
Replace bulb.
O.K.
Start engine and make sure that check engine light still illuminates.
Ⓐ

ECCS SELF-DIAGNOSTIC DESCRIPTION (CONT.)—1989-90 MAXIMA

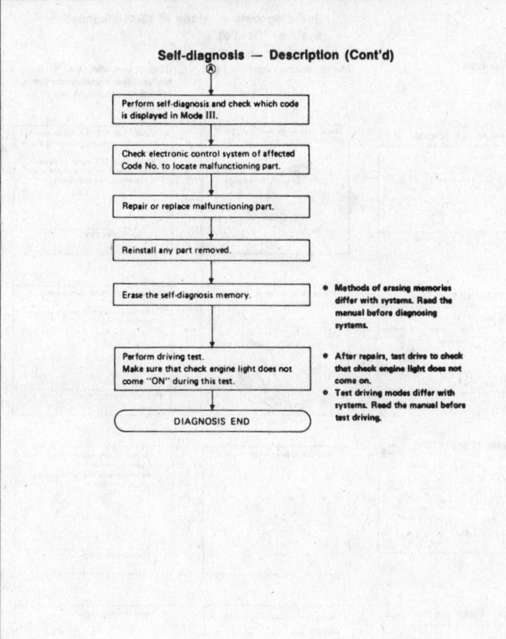
Self-diagnosis — Description (Cont'd)
Ⓐ
Perform self-diagnosis and check which code is displayed in Mode III.
Check electronic control system of affected Code No. to locate malfunctioning part.
Repair or replace malfunctioning part.
Reinstall any part removed.
Erase the self-diagnosis memory.
Perform driving test.
Make sure that check engine light does not come "ON" during this test.
DIAGNOSIS END
• Methods of erasing memories differ with systems. Read the manual before diagnosing systems.
• After repairs, test drive to check that check engine light does not come on.
• Test driving modes differ with systems. Read the manual before test driving.

ECCS SELF-DIAGNOSTIC MODE I AND II—1989-90 MAXIMA

Self-diagnosis — Mode I (Exhaust gas sensor monitor)

This mode checks the exhaust gas sensor for proper functioning. The operation of the E.C.U. L.E.D. in this mode differs with mixture ratio control conditions as follows:

Mode	L.E.D.	Engine stopped (Ignition switch "ON")	Engine running	
			Open loop condition	Closed loop condition
Mode I (Monitor A)	Green	ON	*Remains ON or OFF	Blinks
	Red	ON	Except for California model • When back-up system is operating	For California model • ON: a. when the CHECK ENGINE LIGHT ITEMS are stored in the E.C.U. b. when back-up system is operating • OFF: except for the above conditions

*: Maintains conditions just before switching to open loop

EXHAUST GAS SENSOR FUNCTION CHECK

If the number of L.E.D. blinks is less than that specified, replace the exhaust gas sensor.

If the L.E.D. does not blink, check exhaust gas sensor circuit.

EXHAUST GAS SENSOR CIRCUIT CHECK

Self-diagnosis — Mode II (Mixture ratio feedback control monitor)

This mode checks, through the E.C.U. L.E.D., optimum control of the mixture ratio. The operation of the L.E.D., as shown below, differs with the control conditions of the mixture ratio (for example, richer or leaner mixture ratios, etc., which are controlled by the E.C.U.).

Mode	L.E.D.	Engine stopped (Ignition switch "ON")	Engine running			
			Open loop condition	Closed loop condition		
Mode II (Monitor B)	Green	ON	*Remains ON or OFF	Blinks		
	Red	OFF	*Remains ON or OFF (synchronous with green L.E.D.)	Compensating mixture ratio		
				More than 5% rich	Between 5% lean and 5% rich	More
				OFF	Synchronized with green L.E.D.	Remains ON

*: Maintains conditions just before switching to open loop

If the red L.E.D. remains on or off during the closed-loop operation, the mixture ratio may not be controlled properly. Using the following procedures, check the related components or adjust the mixture ratio.

COMPONENT CHECK OR MIXTURE RATIO ADJUSTMENT

ECCS SELF-DIAGNOSTIC MODE III—1989-90 MAXIMA

Self-diagnosis — Mode III (Self-diagnostic system)

The E.C.U. constantly monitors the function of these sensors and actuators, regardless of ignition key position. If a malfunction occurs, the information is stored in the E.C.U. and can be retrieved from the memory by turning on the diagnostic mode selector, located on the side of the E.C.U. When activated, the malfunction is indicated by flashing a red and a green L.E.D. (Light Emitting Diode), also located on the E.C.U. Since all the self-diagnostic results are stored in the E.C.U.'s memory even intermittent malfunctions can be diagnosed.

A malfunction is indicated by the number of both red and green flashing L.E.D.s. First, the red L.E.D. flashes and the green flashes follow. The red L.E.D. corresponds to units of ten and the green L.E.D. corresponds to units of one. For example, when the red L.E.D. flashes once and the green L.E.D. flashes twice, this signifies the number "12", showing that the air flow meter signal is malfunctioning. All problems are classified by code numbers in this way.

- **When the engine fails to start, crank it two or more seconds before beginning self-diagnosis.**
- **Before starting self-diagnosis, do not erase the stored memory before beginning self-diagnosis. If it is erased, the self-diagnosis function for intermittent malfunctions will be lost.**

The stored memory would be lost if:
1. **Battery terminal is disconnected.**
2. **After selecting Mode III, Mode IV is selected.**

DISPLAY CODE TABLE

Code No.	Detected items	California	Non-California
11	Crank angle sensor ciruit	X	X
12	Air flow meter circuit	X	X
13	Engine temperature sensor circuit	X	X
14	Vehicle speed sensor circuit	X	X
21	Ignition signal missing in primary coil	X	X
22	Fuel pump circuit	X	X
31	E.C.U. (E.C.C.S. control unit)	X	X
32	E.G.R. function	X	–
33	Exhaust gas sensor circuit	X	X
34	Detonation sensor circuit	X	X
35	Exhaust gas temperature sensor circuit	X	–
43	Throttle sensor circuit	X	X
45	Injector leak	X	–
55	No malfunction in the above circuit	X	X

X: Available –: Not available

Self-diagnosis — Mode III (Self-diagnostic system) (Cont'd)

RETENTION OF DIAGNOSTIC RESULTS

The diagnostic results will remain in E.C.U. memory until the starter is operated fifty times after a diagnostic item has been judged to be malfunctioning. The diagnostic result will then be cancelled automatically. If a diagnostic item which has been judged to be malfunctioning and stored in memory is again judged to be malfunctioning before the starter is operated fifty times, the second result will replace the previous one. It will be stored in E.C.U. memory until the starter is operated fifty times more.

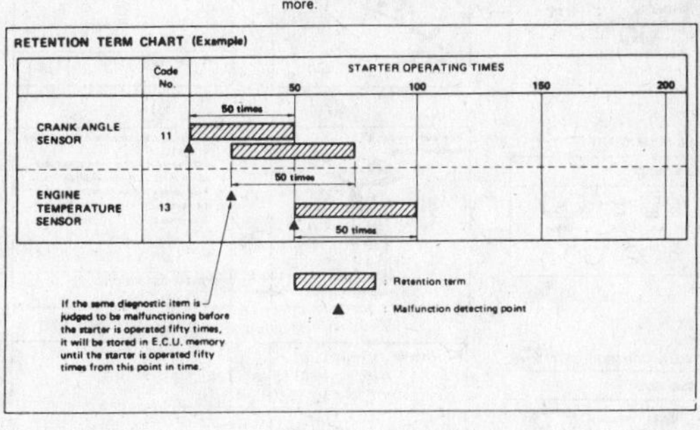
RETENTION TERM CHART (Example)
STARTER OPERATING TIMES
Code No.
50
100
150
200
CRANK ANGLE SENSOR
11
50 times
ENGINE TEMPERATURE SENSOR
13
50 times
50 times
Retention term
Malfunction detecting point
If the same diagnostic item is judged to be malfunctioning before the starter is operated fifty times, it will be stored in E.C.U. memory until the starter is operated fifty times from this point in time.

ECCS SELF-DIAGNOSTIC MODE III (CONT.)—1989–90 MAXIMA

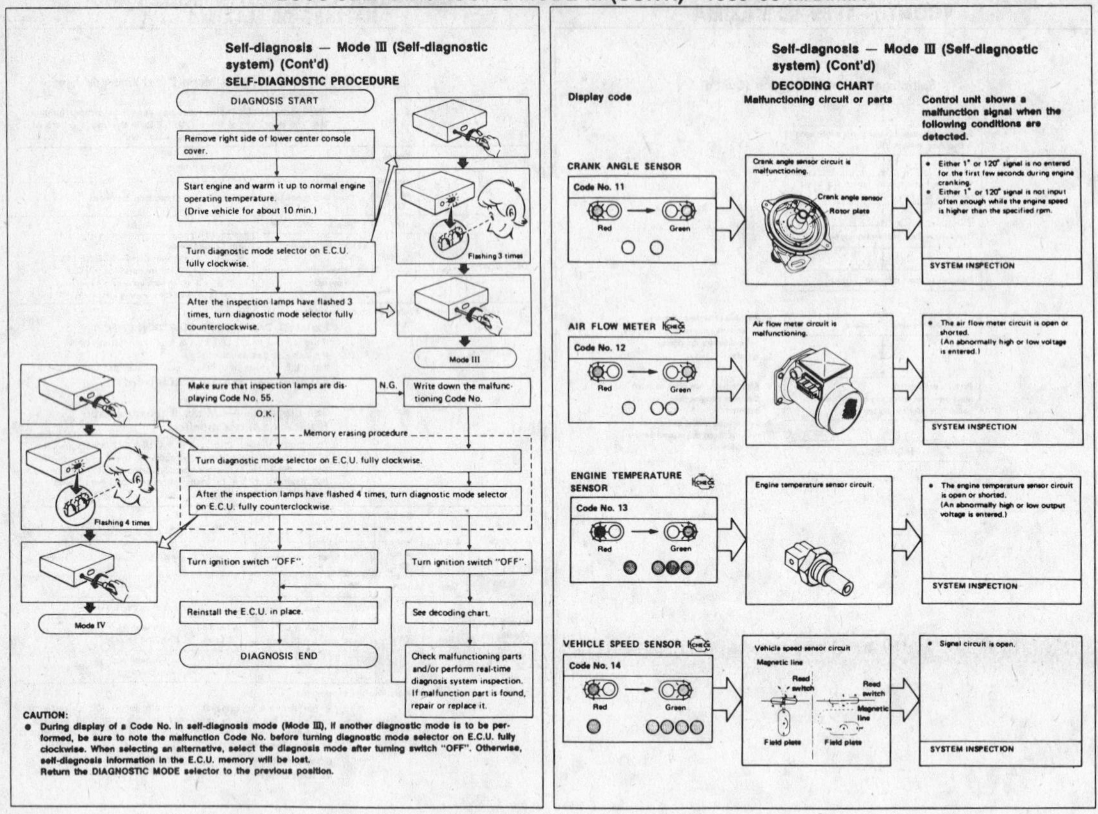
Self-diagnosis — Mode III (Self-diagnostic system) (Cont'd)
SELF-DIAGNOSTIC PROCEDURE
DIAGNOSIS START
Remove right side of lower center console cover.
Start engine and warm it up to normal engine operating temperature. (Drive vehicle for about 10 min.)
Turn diagnostic mode selector on E.C.U. fully clockwise.
Flashing 3 times
After the inspection lamps have flashed 3 times, turn diagnostic mode selector fully counterclockwise.
Mode III
Make sure that inspection lamps are displaying Code No. 55.
N.G.
Write down the malfunctioning Code No.
O.K.
Memory erasing procedure
Turn diagnostic mode selector on E.C.U. fully clockwise.
Flashing 4 times
After the inspection lamps have flashed 4 times, turn diagnostic mode selector on E.C.U. fully counterclockwise.
Turn ignition switch "OFF".
Turn ignition switch "OFF".
Mode IV
Reinstall the E.C.U. in place.
See decoding chart.
DIAGNOSIS END
Check malfunctioning parts and/or perform real-time diagnosis system inspection. If malfunction part is found, repair or replace it.
CAUTION:
During display of a Code No. in self-diagnosis mode (Mode III), if another diagnostic mode is to be performed, be sure to note the malfunction Code No. before turning diagnostic mode selector on E.C.U. fully clockwise. When selecting an alternative, select the diagnosis mode after turning switch "OFF". Otherwise, self-diagnosis information in the E.C.U. memory will be lost. Return the DIAGNOSTIC MODE selector to the previous position.
Self-diagnosis — Mode III (Self-diagnostic system) (Cont'd)
DECODING CHART
Display code
Malfunctioning circuit or parts
Control unit shows a malfunction signal when the following conditions are detected.
CRANK ANGLE SENSOR
Code No. 11
Red
Green
Crank angle sensor circuit is malfunctioning.
Crank angle sensor
Rotor plate
Either 1° or 120° signal is not entered for the first few seconds during engine cranking.
Either 1° or 120° signal is not input often enough while the engine speed is higher than the specified rpm.
SYSTEM INSPECTION
AIR FLOW METER
Code No. 12
Air flow meter circuit is malfunctioning.
The air flow meter circuit is open or shorted. (An abnormally high or low voltage is entered.)
SYSTEM INSPECTION
ENGINE TEMPERATURE SENSOR
Code No. 13
Engine temperature sensor circuit.
The engine temperature sensor circuit is open or shorted. (An abnormally high or low output voltage is entered.)
SYSTEM INSPECTION
VEHICLE SPEED SENSOR
Code No. 14
Vehicle speed sensor circuit
Magnetic line
Reed switch
Reed switch
Magnetic line
Field plate
Field plate
Signal circuit is open.
SYSTEM INSPECTION

ECCS SELF-DIAGNOSTIC MODE III DISPLAY CODES (CONT.)—1989–90 MAXIMA

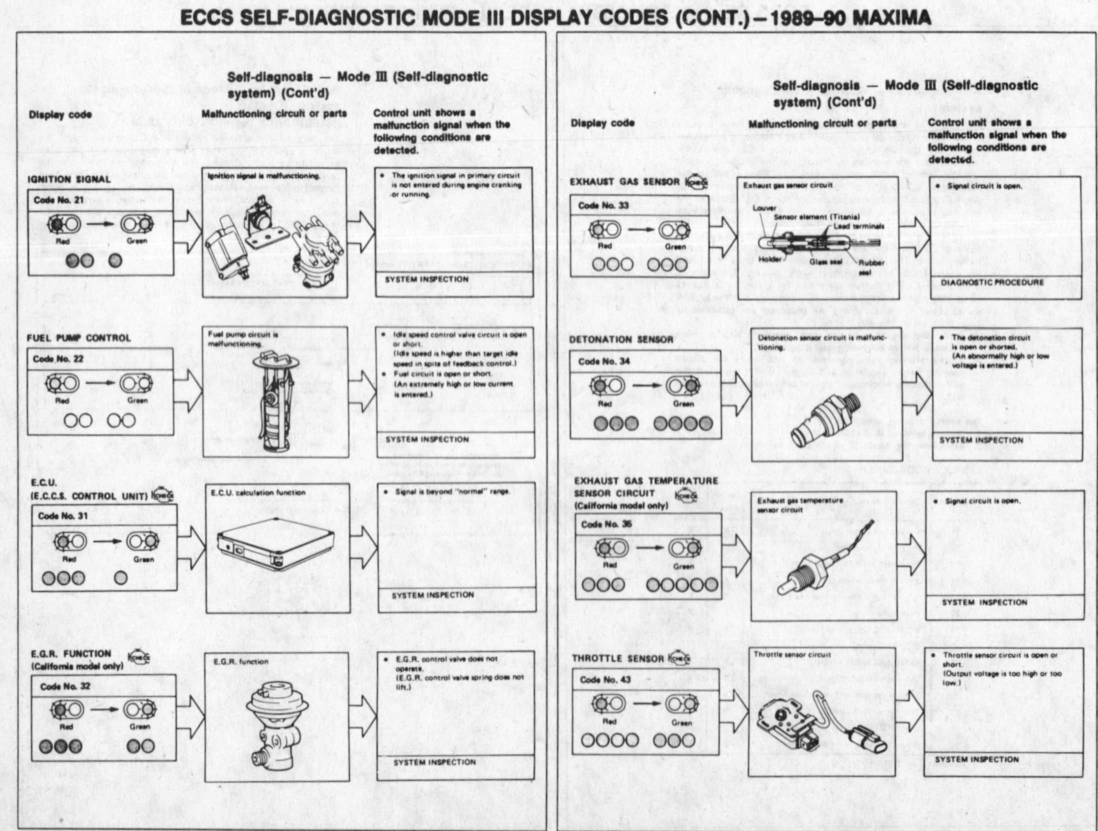
Self-diagnosis — Mode III (Self-diagnostic system) (Cont'd)
Display code
Malfunctioning circuit or parts
Control unit shows a malfunction signal when the following conditions are detected.
IGNITION SIGNAL
Code No. 21
Red
Green
Ignition signal is malfunctioning.
The ignition signal in primary circuit is not entered during engine cranking or running.
SYSTEM INSPECTION
FUEL PUMP CONTROL
Code No. 22
Fuel pump circuit is malfunctioning.
Idle speed control valve circuit is open or short. (Idle speed is higher than target idle speed in spite of feedback control.)
Fuel circuit is open or short. (An extremely high or low current is entered.)
SYSTEM INSPECTION
E.C.U. (E.C.C.S. CONTROL UNIT)
Code No. 31
E.C.U. calculation function
Signal is beyond "normal" range.
SYSTEM INSPECTION
E.G.R. FUNCTION (California model only)
Code No. 32
E.G.R. function
E.G.R. control valve does not operate. (E.G.R. control valve spring does not lift.)
SYSTEM INSPECTION
EXHAUST GAS SENSOR
Code No. 33
Exhaust gas sensor circuit
Louver
Sensor element (Titania)
Lead terminals
Holder
Glass seal
Rubber seal
Signal circuit is open.
DIAGNOSTIC PROCEDURE
DETONATION SENSOR
Code No. 34
Detonation sensor circuit is malfunctioning.
The detonation circuit is open or shorted. (An abnormally high or low voltage is entered.)
SYSTEM INSPECTION
EXHAUST GAS TEMPERATURE SENSOR CIRCUIT (California model only)
Code No. 35
Exhaust gas temperature sensor circuit
Signal circuit is open.
SYSTEM INSPECTION
THROTTLE SENSOR
Code No. 43
Throttle sensor circuit
Throttle sensor circuit is open or short. (Output voltage is too high or too low.)
SYSTEM INSPECTION

ECCS SELF-DIAGNOSTIC MODE III DISPLAY CODES (CONT.)—1989–90 MAXIMA

Self-diagnosis — Mode III (Self-diagnostic system) (Cont'd)

Display code	Malfunctioning circuit or parts	Control unit shows a malfunction signal when the following conditions are detected.
INJECTOR LEAK (California model only) Code No. 45 Red → Green	Fuel leak (O-ring)	Fuel leak from injector SYSTEM INSPECTION
Code No. 55 Red → Green	E.C.C.S. normal operation.	

ECCS SELF-DIAGNOSTIC MODE IV—1989–90 MAXIMA

Self-diagnosis — Mode IV (Switches ON/OFF diagnostic system)

In switches ON/OFF diagnostic system, ON/OFF operation of the following switches can be detected continuously.

- Idle switch
- Starter switch
- Vehicle speed sensor

(1) Idle switch & Starter switch
The switches ON/OFF status in Mode IV is stored in E.C.U. memory. When either switch is turned from "ON" to "OFF" or "OFF" to "ON", the red L.E.D. on E.C.U. alternately comes on and goes off each time switching is performed.

(2) Vehicle Speed Sensor
The switches ON/OFF status in Mode IV is selected is stored in E.C.U. memory. The green L.E.D. on E.C.U. remains off when vehicle speed is 20 km/h (12 MPH or below), and comes ON at higher speeds.

ECCS SELF-DIAGNOSTIC MODE IV (CONT.)—1989–90 MAXIMA

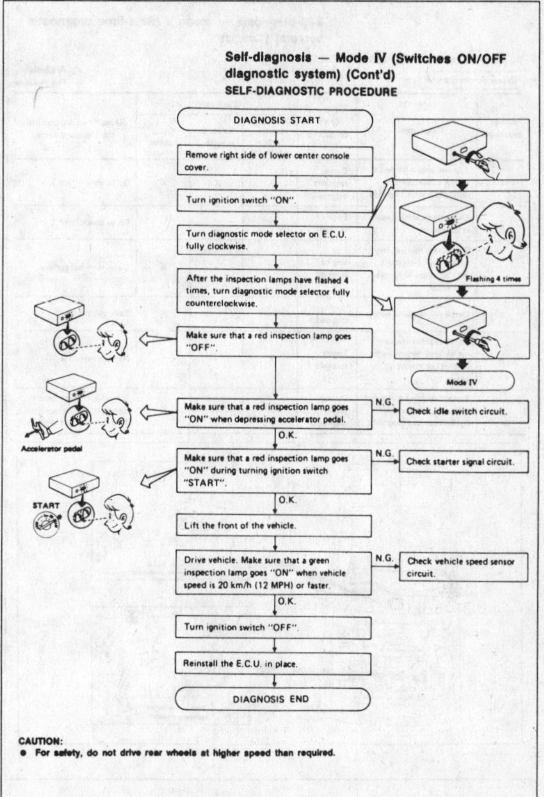
Self-diagnosis — Mode IV (Switches ON/OFF diagnostic system) (Cont'd)
SELF-DIAGNOSTIC PROCEDURE
DIAGNOSIS START
Remove right side of lower center console cover.
Turn ignition switch "ON".
Turn diagnostic mode selector on E.C.U. fully clockwise.
After the inspection lamps have flashed 4 times, turn diagnostic mode selector fully counterclockwise.
Flashing 4 times
Make sure that a red inspection lamp goes "OFF".
Mode IV
Make sure that a red inspection lamp goes "ON" when depressing accelerator pedal.
N.G.
Check idle switch circuit.
O.K.
Accelerator pedal
Make sure that a red inspection lamp goes "ON" during turning ignition switch "START".
Check starter signal circuit.
START
Lift the front of the vehicle.
Drive vehicle. Make sure that a green inspection lamp goes "ON" when vehicle speed is 20 km/h (12 MPH) or faster.
Check vehicle speed sensor circuit.
Turn ignition switch "OFF".
Reinstall the E.C.U. in place.
DIAGNOSIS END

CAUTION:
- **For safety, do not drive rear wheels at higher speed than required.**

ECCS SELF-DIAGNOSTIC MODE V—1989–90 MAXIMA

ECCS SELF-DIAGNOSTIC MODE V DECODING CHART—1989–90 MAXIMA

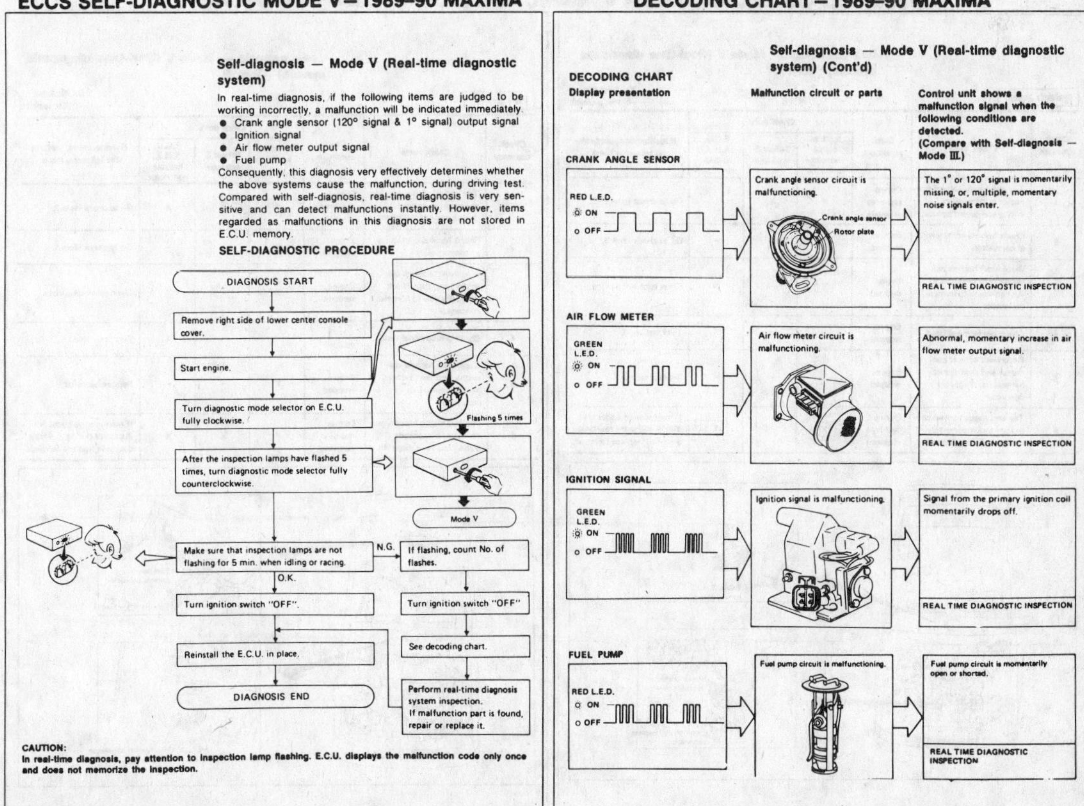
Self-diagnosis — Mode V (Real-time diagnostic system)
In real-time diagnosis, if the following items are judged to be working incorrectly, a malfunction will be indicated immediately.
● Crank angle sensor (120° signal & 1° signal) output signal
● Ignition signal
● Air flow meter output signal
● Fuel pump
Consequently, this diagnosis very effectively determines whether the above systems cause the malfunction, during driving test. Compared with self-diagnosis, real-time diagnosis is very sensitive and can detect malfunctions instantly. However, items regarded as malfunctions in this diagnosis are not stored in E.C.U. memory.
SELF-DIAGNOSTIC PROCEDURE
DIAGNOSIS START
Remove right side of lower center console cover.
Start engine.
Turn diagnostic mode selector on E.C.U. fully clockwise.
Flashing 5 times
After the inspection lamps have flashed 5 times, turn diagnostic mode selector fully counterclockwise.
Mode V
Make sure that inspection lamps are not flashing for 5 min. when idling or racing.
N.G.
If flashing, count No. of flashes.
O.K.
Turn ignition switch "OFF".
Turn ignition switch "OFF".
See decoding chart.
Reinstall the E.C.U. in place.
DIAGNOSIS END
Perform real-time diagnosis system inspection. If malfunction part is found, repair or replace it.
CAUTION:
In real-time diagnosis, pay attention to inspection lamp flashing. E.C.U. displays the malfunction code only once and does not memorize the inspection.
Self-diagnosis — Mode V (Real-time diagnostic system) (Cont'd)
DECODING CHART
Display presentation
Malfunction circuit or parts
Control unit shows a malfunction signal when the following conditions are detected. (Compare with Self-diagnosis — Mode III.)
CRANK ANGLE SENSOR
RED L.E.D. ON OFF
Crank angle sensor circuit is malfunctioning.
Crank angle sensor
Rotor plate
The 1° or 120° signal is momentarily missing, or, multiple, momentary noise signals enter.
REAL TIME DIAGNOSTIC INSPECTION
AIR FLOW METER
GREEN L.E.D. ON OFF
Air flow meter circuit is malfunctioning.
Abnormal, momentary increase in air flow meter output signal.
REAL TIME DIAGNOSTIC INSPECTION
IGNITION SIGNAL
GREEN L.E.D. ON OFF
Ignition signal is malfunctioning.
Signal from the primary ignition coil momentarily drops off.
REAL TIME DIAGNOSTIC INSPECTION
FUEL PUMP
RED L.E.D. ON OFF
Fuel pump circuit is malfunctioning.
Fuel pump circuit is momentarily open or shorted.
REAL TIME DIAGNOSTIC INSPECTION

ECCS SELF-DIAGNOSTIC MODE V INSPECTION (CONT.) – 1989–90 MAXIMA

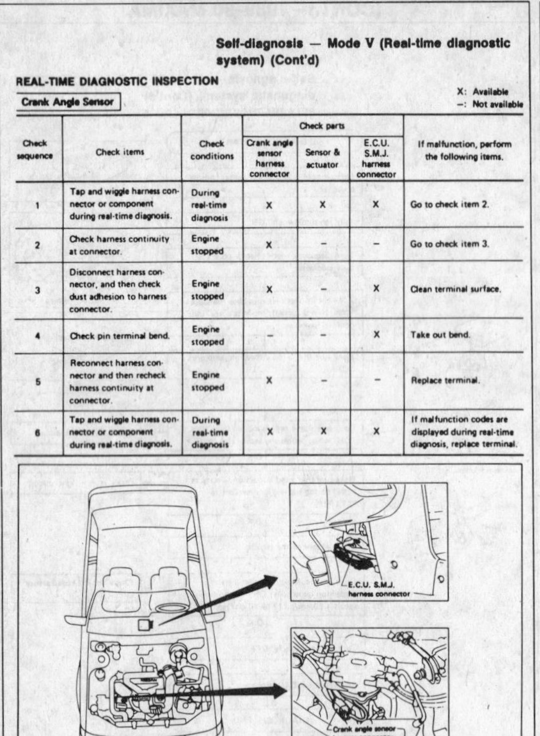

Self-diagnosis — Mode V (Real-time diagnostic system) (Cont'd)

REAL-TIME DIAGNOSTIC INSPECTION

Crank Angle Sensor

X: Available
–: Not available

Check sequence	Check items	Check conditions	Check parts: Crank angle sensor harness connector	Check parts: Sensor & actuator	Check parts: E.C.U. S.M.J. harness connector	If malfunction, perform the following items.
1	Tap and wiggle harness connector or component during real-time diagnosis.	During real-time diagnosis	X	X	X	Go to check item 2.
2	Check harness continuity at connector.	Engine stopped	X	–	–	Go to check item 3.
3	Disconnect harness connector, and then check dust adhesion to harness connector.	Engine stopped	X	–	X	Clean terminal surface.
4	Check pin terminal bend.	Engine stopped	–	–	X	Take out bend.
5	Reconnect harness connector and then recheck harness continuity at connector.	Engine stopped	X	–	–	Replace terminal.
6	Tap and wiggle harness connector or component during real-time diagnosis.	During real-time diagnosis	X	X	X	If malfunction codes are displayed during real-time diagnosis, replace terminal.

E.C.U. S.M.J. harness connector
Crank angle sensor harness connector

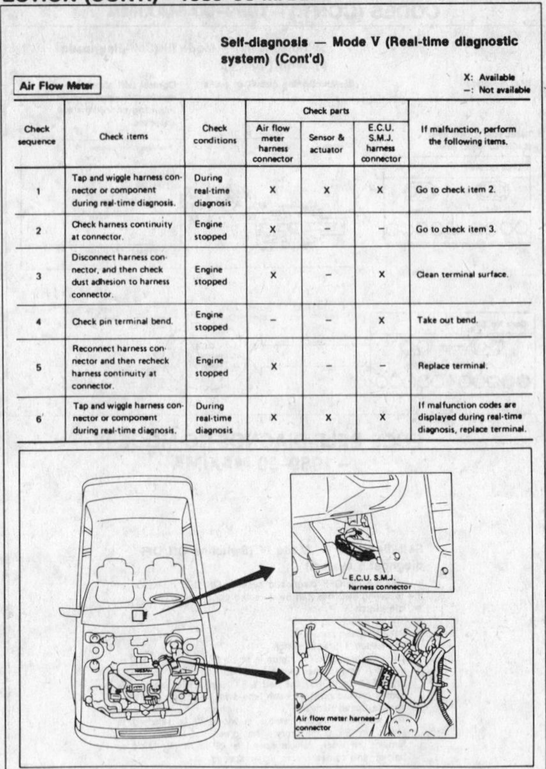

Self-diagnosis — Mode V (Real-time diagnostic system) (Cont'd)

Air Flow Meter

X: Available
–: Not available

Check sequence	Check items	Check conditions	Check parts: Air flow meter harness connector	Check parts: Sensor & actuator	Check parts: E.C.U. S.M.J. harness connector	If malfunction, perform the following items.
1	Tap and wiggle harness connector or component during real-time diagnosis.	During real-time diagnosis	X	X	X	Go to check item 2.
2	Check harness continuity at connector.	Engine stopped	X	–	–	Go to check item 3.
3	Disconnect harness connector, and then check dust adhesion to harness connector.	Engine stopped	X	–	X	Clean terminal surface.
4	Check pin terminal bend.	Engine stopped	–	–	X	Take out bend.
5	Reconnect harness connector and then recheck harness continuity at connector.	Engine stopped	X	–	–	Replace terminal.
6	Tap and wiggle harness connector or component during real-time diagnosis.	During real-time diagnosis	X	X	X	If malfunction codes are displayed during real-time diagnosis, replace terminal.

E.C.U. S.M.J. harness connector
Air flow meter harness connector

ECCS SELF-DIAGNOSTIC MODE V INSPECTION (CONT.) – 1989–90 MAXIMA

Self-diagnosis — Mode V (Real-time diagnostic system) (Cont'd)

Ignition Signal

X: Available
–: Not available

Check sequence	Check items	Check conditions	Check parts: Ignition signal harness connector	Check parts: Sensor & actuator	Check parts: E.C.U. S.M.J. harness connector	If malfunction, perform the following items.
1	Tap and wiggle harness connector or component during real-time diagnosis.	During real-time diagnosis	X	X	X	Go to check item 2.
2	Check harness continuity at connector.	Engine stopped	X	–	–	Go to check item 3.
3	Disconnect harness connector, and then check dust adhesion to harness connector.	Engine stopped	X	–	X	Clean terminal surface.
4	Check pin terminal bend.	Engine stopped	–	–	X	Take out bend.
5	Reconnect harness connector and then recheck harness continuity at connector.	Engine stopped	X	–	–	Replace terminal.
6	Tap and wiggle harness connector or component during real-time diagnosis.	During real-time diagnosis	X	X	X	If malfunction codes are displayed during real-time diagnosis, replace terminal.

E.C.U. S.M.J. harness connector
Power transistor harness connector
Ignition coil harness connector

Self-diagnosis — Mode V (Real-time diagnostic system) (Cont'd)

Fuel pump

X: Available
–: Not available

Check sequence	Check items	Check conditions	Check parts: Fuel pump harness connector	Check parts: Sensor & actuator	Check parts: E.C.U. S.M.J. harness connector	If malfunction, perform the following items.
1	Tap and wiggle harness connector or component during real-time diagnosis.	During real-time diagnosis	X	X	X	Go to check item 2.
2	Check harness continuity at connector.	Engine stopped	X	–	–	Go to check item 3.
3	Disconnect harness connector, and then check dust adhesion to harness connector.	Engine stopped	X	–	X	Clean terminal surface.
4	Check pin terminal bend.	Engine stopped	–	–	X	Take out bend.
5	Reconnect harness connector and then recheck harness continuity at connector.	Engine stopped	X	–	–	Replace terminal.
6	Tap and wiggle harness connector or component during real-time diagnosis.	During real-time diagnosis	X	X	X	If malfunction codes are displayed during real-time diagnosis, replace terminal.

Fuel pump harness connector
E.C.U. S.M.J. harness connector

ECCS CAUTION—1989-90 MAXIMA

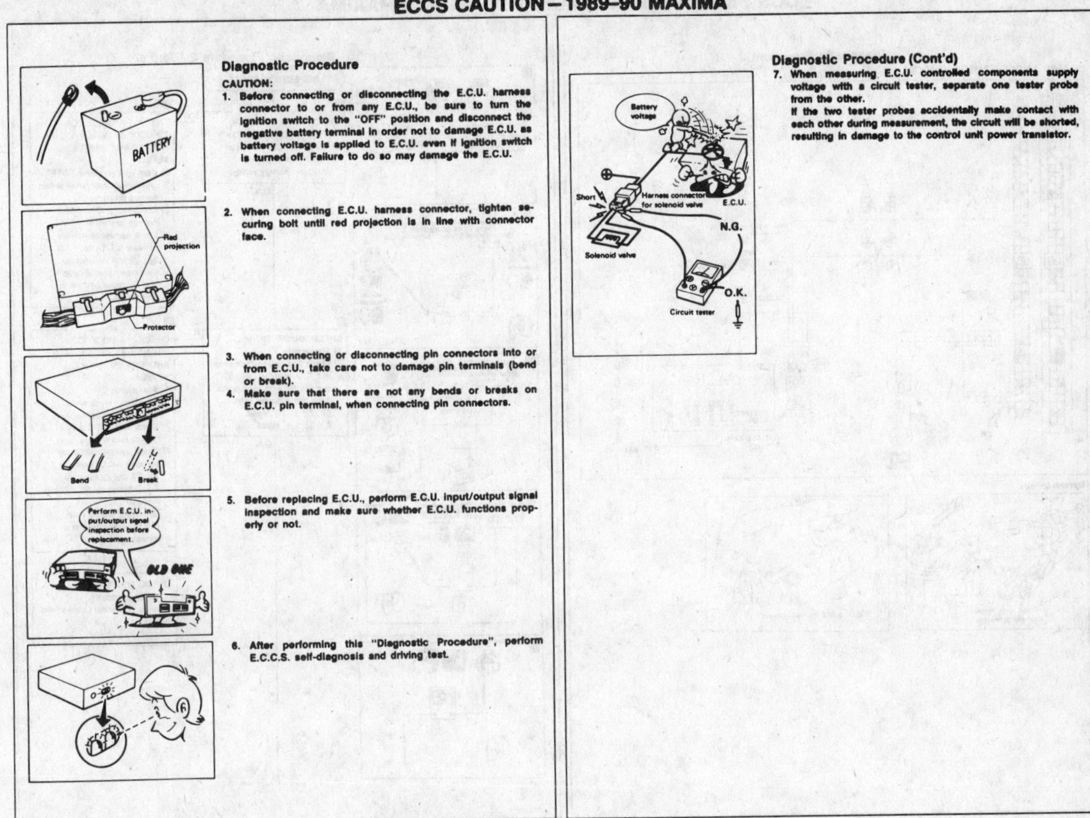
BATTERY
Red projection
Protector
Bend
Break
Perform E.C.U. input/output signal inspection before replacement.
OLD ONE
Battery voltage
Short
Harness connector for solenoid valve
E.C.U.
Solenoid valve
N.G.
O.K.
Circuit tester

Diagnostic Procedure

CAUTION:

1. Before connecting or disconnecting the E.C.U. harness connector to or from any E.C.U., be sure to turn the ignition switch to the "OFF" position and disconnect the negative battery terminal in order not to damage E.C.U. as battery voltage is applied to E.C.U. even if ignition switch is turned off. Failure to do so may damage the E.C.U.
2. When connecting E.C.U. harness connector, tighten securing bolt until red projection is in line with connector face.
3. When connecting or disconnecting pin connectors into or from E.C.U., take care not to damage pin terminals (bend or break).
4. Make sure that there are not any bends or breaks on E.C.U. pin terminal, when connecting pin connectors.
5. Before replacing E.C.U., perform E.C.U. input/output signal inspection and make sure whether E.C.U. functions properly or not.
6. After performing this "Diagnostic Procedure", perform E.C.C.S. self-diagnosis and driving test.

Diagnostic Procedure (Cont'd)

7. When measuring E.C.U. controlled components supply voltage with a circuit tester, separate one tester probe from the other.
If the two tester probes accidentally make contact with each other during measurement, the circuit will be shorted, resulting in damage to the control unit power transistor.

ECCS SELF-DIAGNOSTIC CHART—1989-90 MAXIMA

Diagnostic Procedure 1

MAIN POWER SUPPLY AND GROUND CIRCUIT

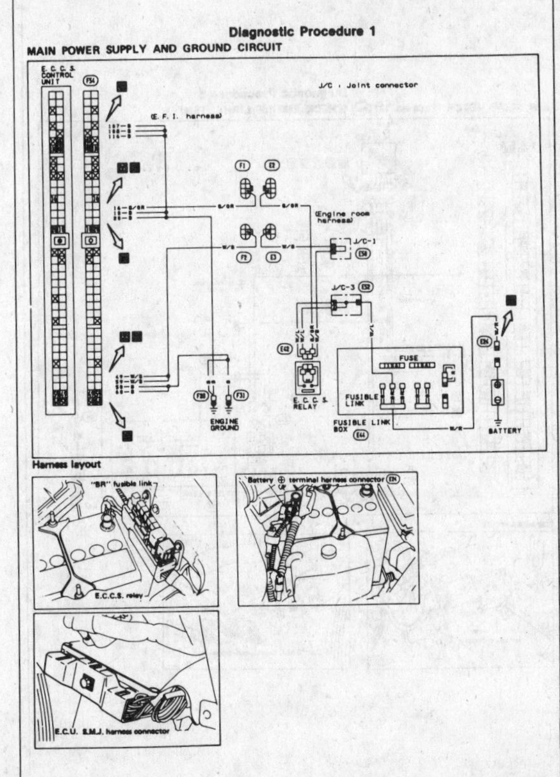
E.C.C.S. CONTROL UNIT
J/C : Joint connector
(E.F.I. harness)
(Engine room harness)
J/C-1
J/C-3
FUSE
FUSIBLE LINK
E.C.C.S. RELAY
FUSIBLE LINK BOX
ENGINE GROUND
BATTERY
Harness layout
"BR" fusible link
E.C.C.S. relay
Battery ⊕ terminal harness connector
E.C.U. S.M.J. harness connector

Diagnostic Procedure 1 (Cont'd)

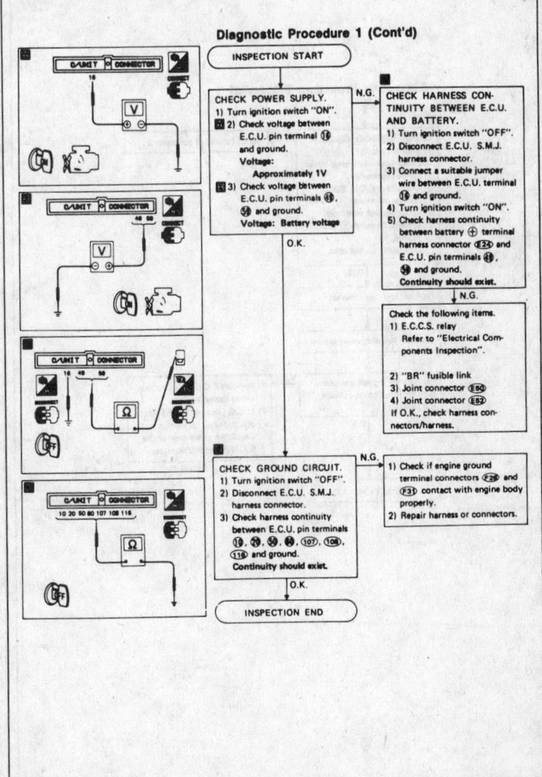
C/UNIT CONNECTOR
16
49 58
16 49 58
10 20 50 60 107 108 116

INSPECTION START

CHECK POWER SUPPLY.
1) Turn ignition switch "ON".
2) Check voltage between E.C.U. pin terminal ⑯ and ground.
Voltage: Approximately 1V
3) Check voltage between E.C.U. pin terminals ㊾, 58 and ground.
Voltage: Battery voltage

N.G. → CHECK HARNESS CONTINUITY BETWEEN E.C.U. AND BATTERY.
1) Turn ignition switch "OFF".
2) Disconnect E.C.U. S.M.J. harness connector.
3) Connect a suitable jumper wire between E.C.U. terminal ⑯ and ground.
4) Turn ignition switch "ON".
5) Check harness continuity between battery ⊕ terminal harness connector (E24) and E.C.U. pin terminals ㊾, 58 and ground.
Continuity should exist.

N.G. → Check the following items.
1) E.C.C.S. relay
Refer to "Electrical Components Inspection".
2) "BR" fusible link
3) Joint connector (E60)
4) Joint connector (E62)
If O.K., check harness connectors/harness.

O.K. → CHECK GROUND CIRCUIT.
1) Turn ignition switch "OFF".
2) Disconnect E.C.U. S.M.J. harness connector.
3) Check harness continuity between E.C.U. pin terminals ⑩, ⑳, ㊿, 60, 107, 108, 116 and ground.
Continuity should exist.

N.G. → 1) Check if engine ground terminal connectors (F28) and (F31) contact with engine body properly.
2) Repair harness or connectors.

O.K. → INSPECTION END

ECCS SELF-DIAGNOSTIC CHART—1989–90 MAXIMA

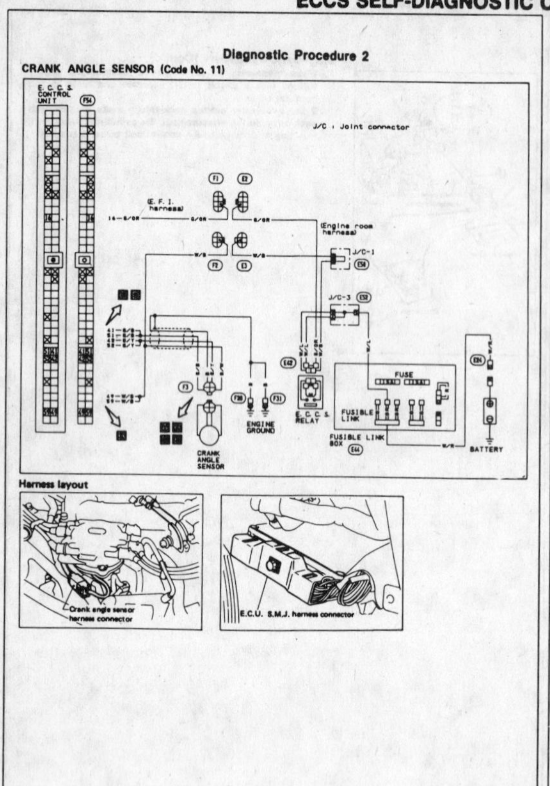
Diagnostic Procedure 2
CRANK ANGLE SENSOR (Code No. 11)
E.C.C.S. CONTROL UNIT
J/C : Joint connector
(E.F.I. harness)
(Engine room harness)
J/C-1
J/C-3
FUSE
E.C.C.S. RELAY
FUSIBLE LINK
FUSIBLE LINK BOX
BATTERY
ENGINE GROUND
CRANK ANGLE SENSOR
Harness layout
Crank angle sensor harness connector
E.C.U. S.M.J. harness connector

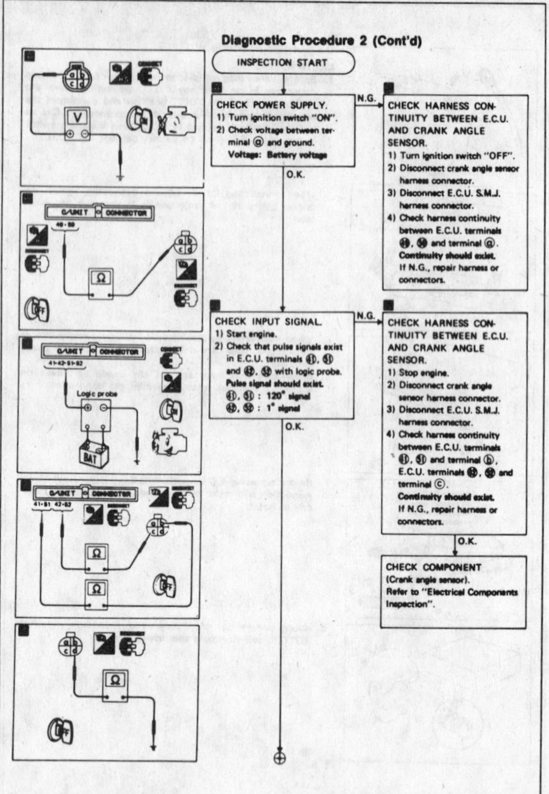
Diagnostic Procedure 2 (Cont'd)
INSPECTION START
CHECK POWER SUPPLY.
1) Turn ignition switch "ON".
2) Check voltage between terminal ⓐ and ground.
Voltage: Battery voltage
N.G.
CHECK HARNESS CONTINUITY BETWEEN E.C.U. AND CRANK ANGLE SENSOR.
1) Turn ignition switch "OFF".
2) Disconnect crank angle sensor harness connector.
3) Disconnect E.C.U. S.M.J. harness connector.
4) Check harness continuity between E.C.U. terminals ㊽, ㊾ and terminal ⓐ.
Continuity should exist.
If N.G., repair harness or connectors.
O.K.
CHECK INPUT SIGNAL.
1) Start engine.
2) Check that pulse signals exist in E.C.U. terminals ㊶, ㊿ and ㊷, ㊿ with logic probe.
Pulse signal should exist.
㊶, ㊿ : 120° signal
㊷, ㊿ : 1° signal
N.G.
CHECK HARNESS CONTINUITY BETWEEN E.C.U. AND CRANK ANGLE SENSOR.
1) Stop engine.
2) Disconnect crank angle sensor harness connector.
3) Disconnect E.C.U. S.M.J. harness connector.
4) Check harness continuity between E.C.U. terminals ㊶, ㊿ and terminal ⓑ, E.C.U. terminals ㊷, ㊿ and terminal ⓒ.
Continuity should exist.
If N.G., repair harness or connectors.
O.K.
CHECK COMPONENT
(Crank angle sensor).
Refer to "Electrical Components Inspection".
Logic probe
BAT

ECCS SELF-DIAGNOSTIC CHART—1989–90 MAXIMA

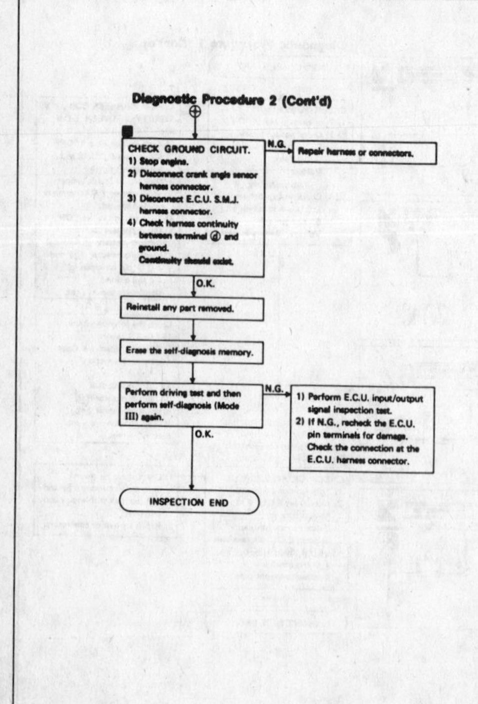
Diagnostic Procedure 2 (Cont'd)
CHECK GROUND CIRCUIT.
1) Stop engine.
2) Disconnect crank angle sensor harness connector.
3) Disconnect E.C.U. S.M.J. harness connector.
4) Check harness continuity between terminal ⓓ and ground.
Continuity should exist.
N.G.
Repair harness or connectors.
O.K.
Reinstall any part removed.
Erase the self-diagnosis memory.
Perform driving test and then perform self-diagnosis (Mode III) again.
N.G.
1) Perform E.C.U. input/output signal inspection test.
2) If N.G., recheck the E.C.U. pin terminals for damage. Check the connection at the E.C.U. harness connector.
O.K.
INSPECTION END

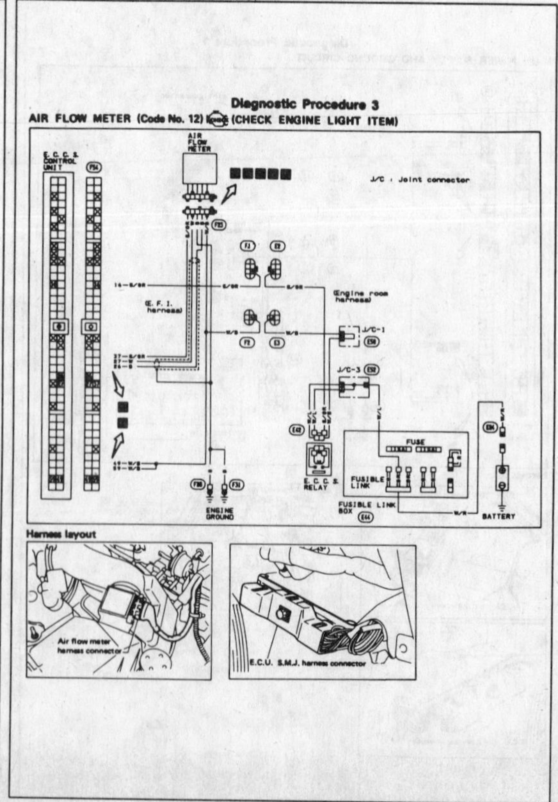
Diagnostic Procedure 3
AIR FLOW METER (Code No. 12) (CHECK ENGINE LIGHT ITEM)
E.C.C.S. CONTROL UNIT
AIR FLOW METER
J/C : Joint connector
(E.F.I. harness)
(Engine room harness)
J/C-1
J/C-3
FUSE
E.C.C.S. RELAY
FUSIBLE LINK
FUSIBLE LINK BOX
BATTERY
ENGINE GROUND
Harness layout
Air flow meter harness connector
E.C.U. S.M.J. harness connector

ECCS SELF-DIAGNOSTIC CHART—1989–90 MAXIMA

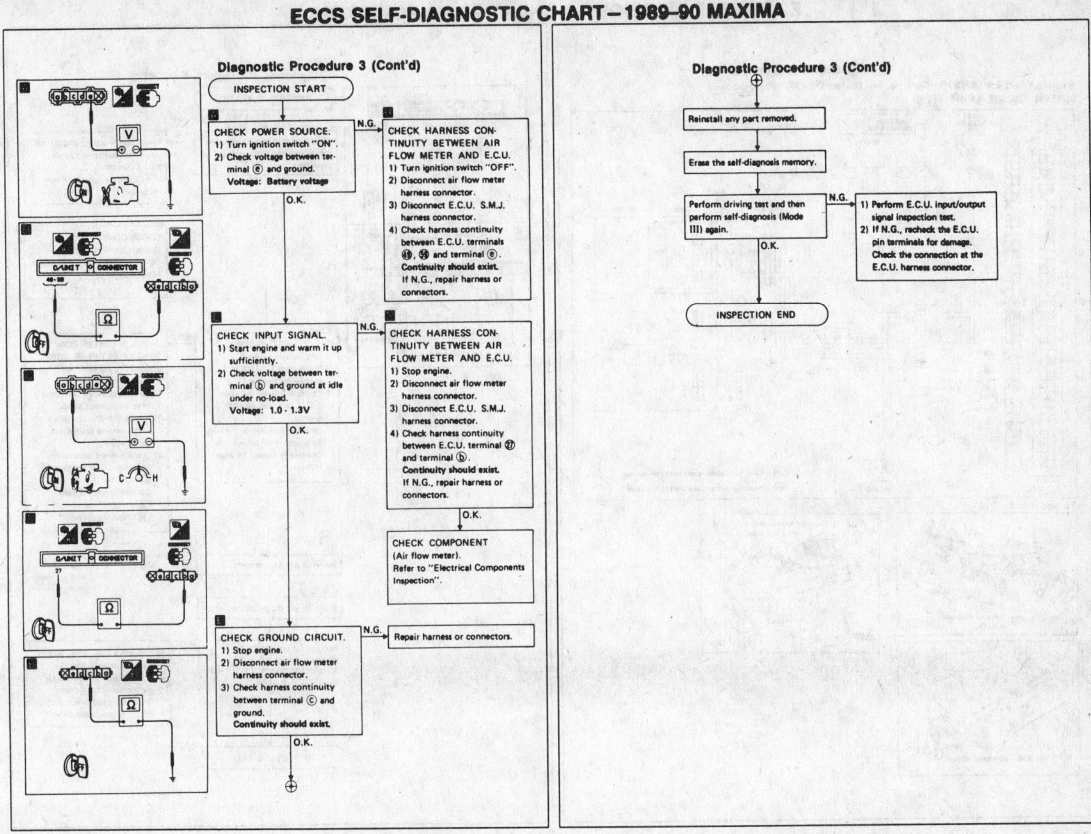
Diagnostic Procedure 3 (Cont'd)
INSPECTION START
CHECK POWER SOURCE.
1) Turn ignition switch "ON".
2) Check voltage between terminal ⓔ and ground.
Voltage: Battery voltage
N.G.
CHECK HARNESS CONTINUITY BETWEEN AIR FLOW METER AND E.C.U.
1) Turn ignition switch "OFF".
2) Disconnect air flow meter harness connector.
3) Disconnect E.C.U. S.M.J. harness connector.
4) Check harness continuity between E.C.U. terminals ㊽, ㊾ and terminal ⓔ.
Continuity should exist.
If N.G., repair harness or connectors.
O.K.
CHECK INPUT SIGNAL.
1) Start engine and warm it up sufficiently.
2) Check voltage between terminal ⓑ and ground at idle under no-load.
Voltage: 1.0 - 1.3V
N.G.
CHECK HARNESS CONTINUITY BETWEEN AIR FLOW METER AND E.C.U.
1) Stop engine.
2) Disconnect air flow meter harness connector.
3) Disconnect E.C.U. S.M.J. harness connector.
4) Check harness continuity between E.C.U. terminal ㉗ and terminal ⓑ.
Continuity should exist.
If N.G., repair harness or connectors.
O.K.
CHECK COMPONENT
(Air flow meter).
Refer to "Electrical Components Inspection".
O.K.
CHECK GROUND CIRCUIT.
1) Stop engine.
2) Disconnect air flow meter harness connector.
3) Check harness continuity between terminal ⓒ and ground.
Continuity should exist.
N.G.
Repair harness or connectors.
O.K.
C/UNIT CONNECTOR
CONNECTOR
Diagnostic Procedure 3 (Cont'd)
Reinstall any part removed.
Erase the self-diagnosis memory.
Perform driving test and then perform self-diagnosis (Mode III) again.
N.G.
1) Perform E.C.U. input/output signal inspection test.
2) If N.G., recheck the E.C.U. pin terminals for damage. Check the connection at the E.C.U. harness connector.
O.K.
INSPECTION END

ECCS SELF-DIAGNOSTIC CHART—1989–90 MAXIMA

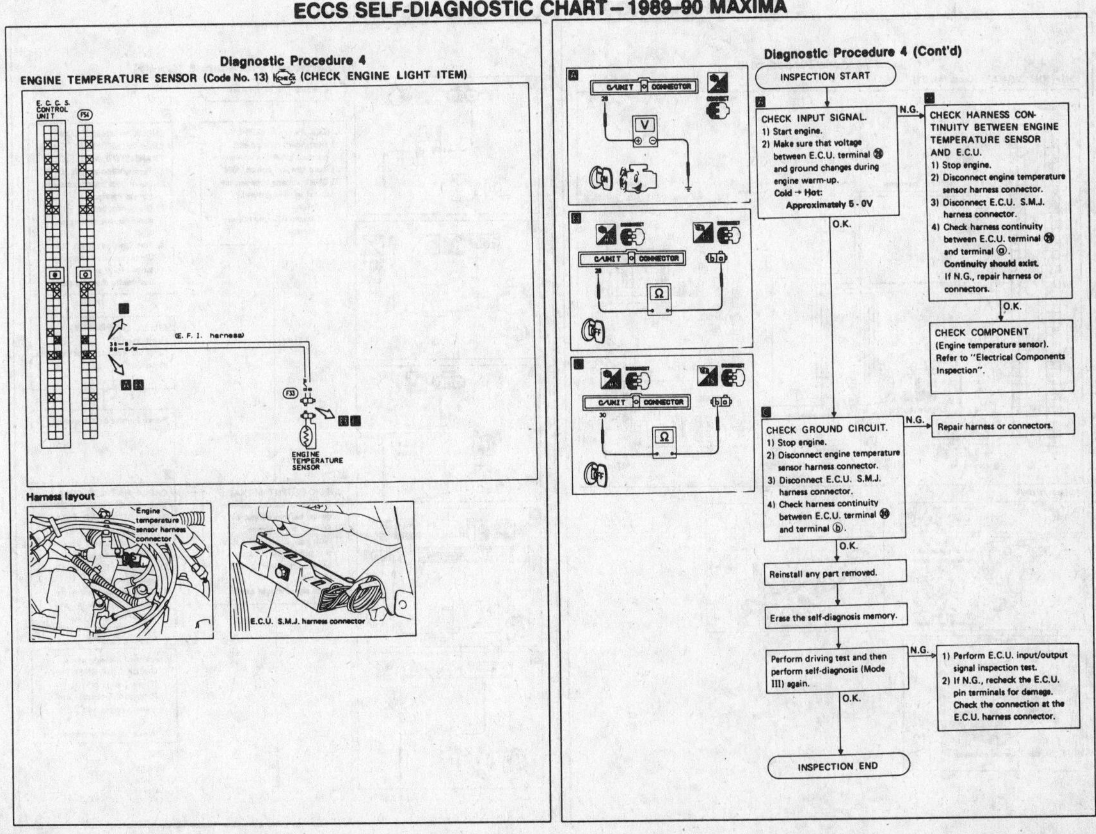
Diagnostic Procedure 4
ENGINE TEMPERATURE SENSOR (Code No. 13) (CHECK ENGINE LIGHT ITEM)
E.C.C.S. CONTROL UNIT
(F54)
(E.F.I. harness)
(F33)
ENGINE TEMPERATURE SENSOR
Harness layout
Engine temperature sensor harness connector
E.C.U. S.M.J. harness connector
Diagnostic Procedure 4 (Cont'd)
INSPECTION START
CHECK INPUT SIGNAL.
1) Start engine.
2) Make sure that voltage between E.C.U. terminal ㉘ and ground changes during engine warm-up.
Cold → Hot:
Approximately 5 - 0V
N.G.
CHECK HARNESS CONTINUITY BETWEEN ENGINE TEMPERATURE SENSOR AND E.C.U.
1) Stop engine.
2) Disconnect engine temperature sensor harness connector.
3) Disconnect E.C.U. S.M.J. harness connector.
4) Check harness continuity between E.C.U. terminal ㉘ and terminal ⓐ.
Continuity should exist.
If N.G., repair harness or connectors.
O.K.
CHECK COMPONENT
(Engine temperature sensor).
Refer to "Electrical Components Inspection".
O.K.
CHECK GROUND CIRCUIT.
1) Stop engine.
2) Disconnect engine temperature sensor harness connector.
3) Disconnect E.C.U. S.M.J. harness connector.
4) Check harness continuity between E.C.U. terminal ㉚ and terminal ⓑ.
N.G.
Repair harness or connectors.
O.K.
Reinstall any part removed.
Erase the self-diagnosis memory.
Perform driving test and then perform self-diagnosis (Mode III) again.
N.G.
1) Perform E.C.U. input/output signal inspection test.
2) If N.G., recheck the E.C.U. pin terminals for damage. Check the connection at the E.C.U. harness connector.
O.K.
INSPECTION END
C/UNIT CONNECTOR

ECCS SELF-DIAGNOSTIC CHART—1989–90 MAXIMA

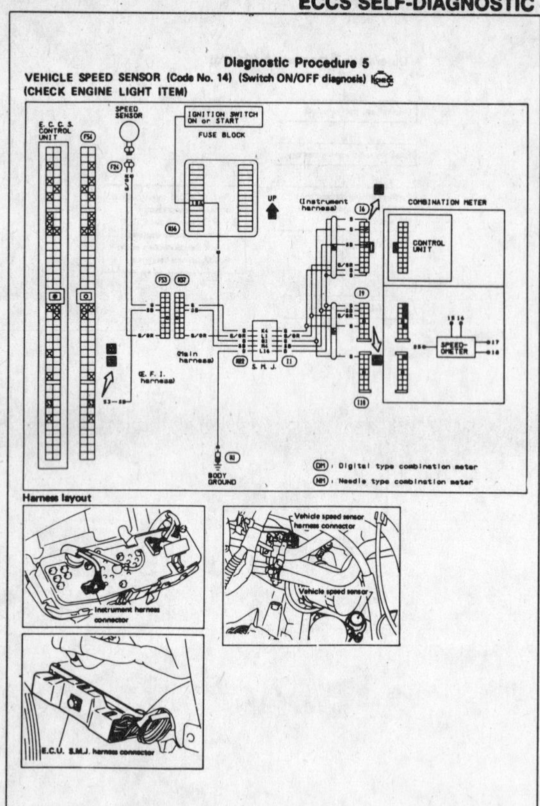
Diagnostic Procedure 5
VEHICLE SPEED SENSOR (Code No. 14) (Switch ON/OFF diagnosis)
(CHECK ENGINE LIGHT ITEM)
E.C.C.S. CONTROL UNIT
SPEED SENSOR
IGNITION SWITCH ON or START
FUSE BLOCK
COMBINATION METER
CONTROL UNIT
SPEED-OMETER
S.M.J.
BODY GROUND
Digital type combination meter
Needle type combination meter
Harness layout
Instrument harness connector
Vehicle speed sensor harness connector
Vehicle speed sensor
E.C.U. S.M.J. harness connector

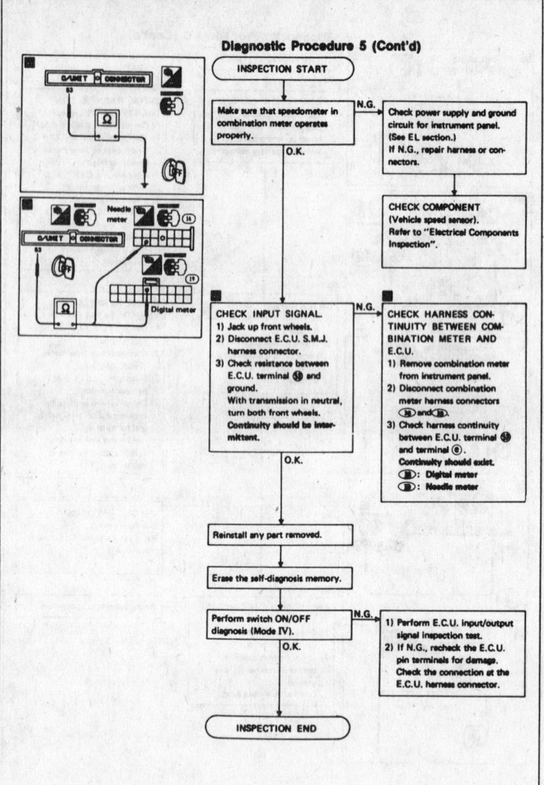
Diagnostic Procedure 5 (Cont'd)
INSPECTION START
Make sure that speedometer in combination meter operates properly.
N.G.
Check power supply and ground circuit for instrument panel. (See EL section.) If N.G., repair harness or connectors.
CHECK COMPONENT (Vehicle speed sensor). Refer to "Electrical Components Inspection".
O.K.
CHECK INPUT SIGNAL.
1) Jack up front wheels.
2) Disconnect E.C.U. S.M.J. harness connector.
3) Check resistance between E.C.U. terminal 53 and ground. With transmission in neutral, turn both front wheels. Continuity should be intermittent.
N.G.
CHECK HARNESS CONTINUITY BETWEEN COMBINATION METER AND E.C.U.
1) Remove combination meter from instrument panel.
2) Disconnect combination meter harness connectors.
3) Check harness continuity between E.C.U. terminal 53 and terminal. Continuity should exist.
Digital meter
Needle meter
O.K.
Reinstall any part removed.
Erase the self-diagnosis memory.
Perform switch ON/OFF diagnosis (Mode IV).
N.G.
1) Perform E.C.U. input/output signal inspection test.
2) If N.G., recheck the E.C.U. pin terminals for damage. Check the connection at the E.C.U. harness connector.
O.K.
INSPECTION END

ECCS SELF-DIAGNOSTIC CHART—1989–90 MAXIMA

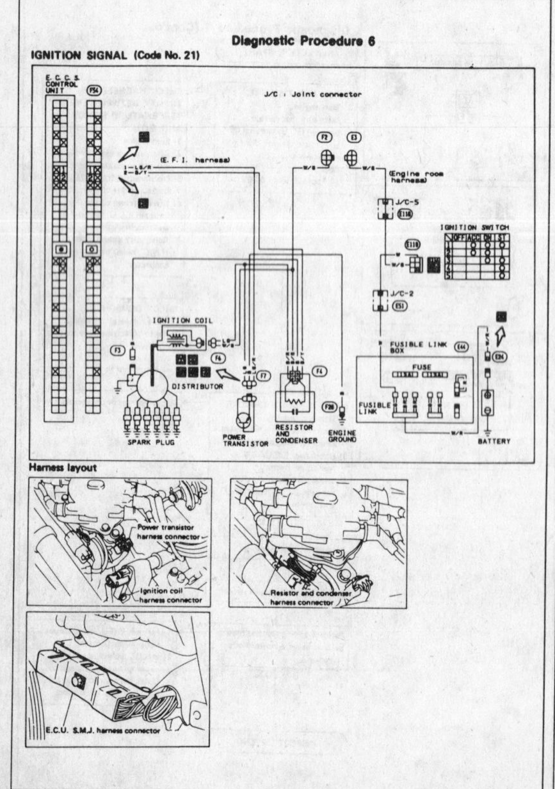
Diagnostic Procedure 6
IGNITION SIGNAL (Code No. 21)
E.C.C.S. CONTROL UNIT
J/C : Joint connector
(E.F.I. harness)
(Engine room harness)
IGNITION SWITCH
IGNITION COIL
DISTRIBUTOR
SPARK PLUG
POWER TRANSISTOR
RESISTOR AND CONDENSER
ENGINE GROUND
FUSIBLE LINK BOX
FUSE
FUSIBLE LINK
BATTERY
Harness layout
Power transistor harness connector
Ignition coil harness connector
Resistor and condenser harness connector
E.C.U. S.M.J. harness connector

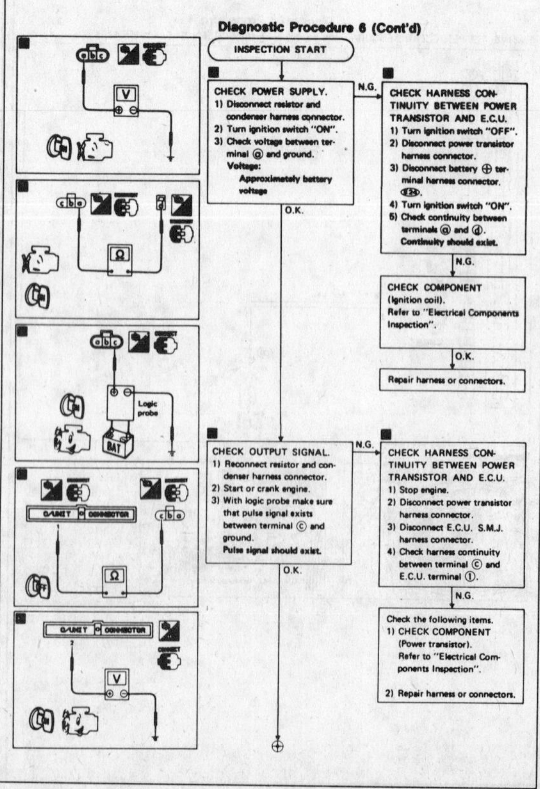
Diagnostic Procedure 6 (Cont'd)
INSPECTION START
CHECK POWER SUPPLY.
1) Disconnect resistor and condenser harness connector.
2) Turn ignition switch "ON".
3) Check voltage between terminal ⓐ and ground.
Voltage: Approximately battery voltage
N.G.
CHECK HARNESS CONTINUITY BETWEEN POWER TRANSISTOR AND E.C.U.
1) Turn ignition switch "OFF".
2) Disconnect power transistor harness connector.
3) Disconnect battery ⊕ terminal harness connector.
4) Turn ignition switch "ON".
5) Check continuity between terminals ⓒ and ⓓ. Continuity should exist.
N.G.
CHECK COMPONENT (Ignition coil). Refer to "Electrical Components Inspection".
O.K.
Repair harness or connectors.
O.K.
CHECK OUTPUT SIGNAL.
1) Reconnect resistor and condenser harness connector.
2) Start or crank engine.
3) With logic probe make sure that pulse signal exists between terminal ⓒ and ground.
Pulse signal should exist.
N.G.
CHECK HARNESS CONTINUITY BETWEEN POWER TRANSISTOR AND E.C.U.
1) Stop engine.
2) Disconnect power transistor harness connector.
3) Disconnect E.C.U. S.M.J. harness connector.
4) Check harness continuity between terminal ⓒ and E.C.U. terminal ①.
N.G.
Check the following items.
1) CHECK COMPONENT (Power transistor). Refer to "Electrical Components Inspection".
2) Repair harness or connectors.
O.K.
Logic probe

ECCS SELF-DIAGNOSTIC CHART—1989-90 MAXIMA

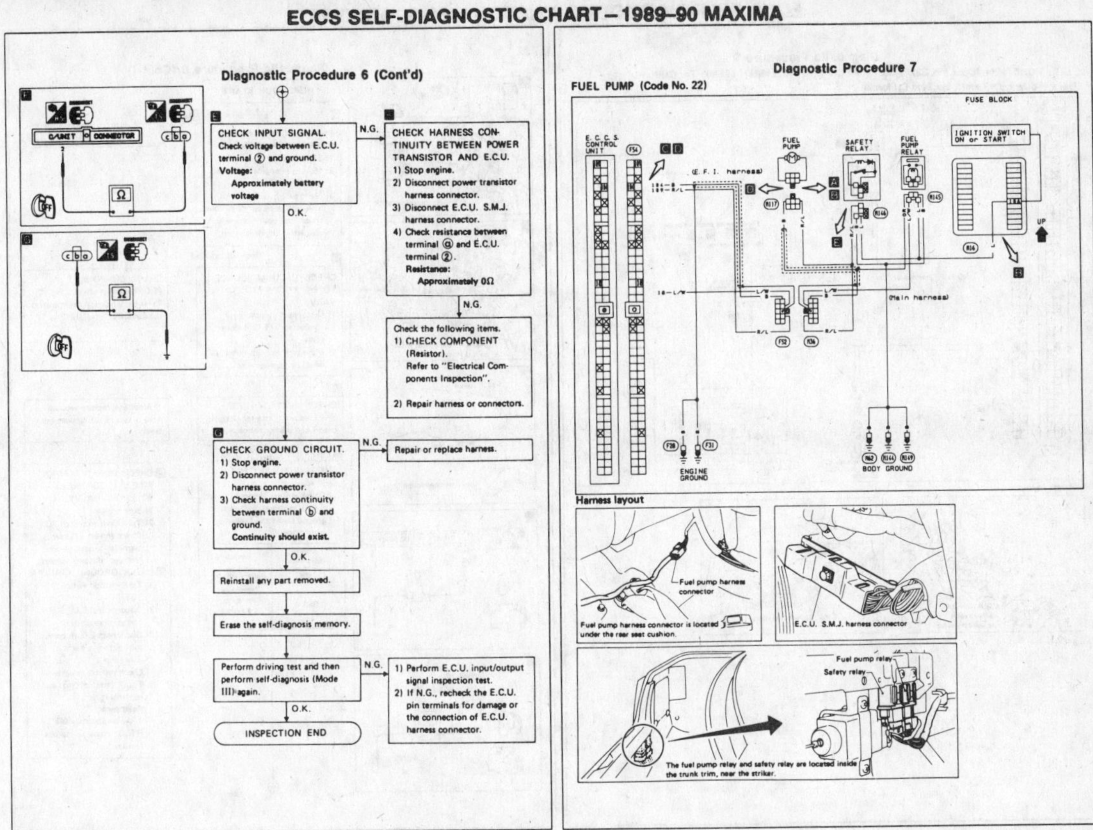
Diagnostic Procedure 6 (Cont'd)
CHECK INPUT SIGNAL.
Check voltage between E.C.U. terminal ② and ground.
Voltage:
Approximately battery voltage
N.G.
O.K.
CHECK HARNESS CONTINUITY BETWEEN POWER TRANSISTOR AND E.C.U.
1) Stop engine.
2) Disconnect power transistor harness connector.
3) Disconnect E.C.U. S.M.J. harness connector.
4) Check resistance between terminal ⓐ and E.C.U. terminal ②.
Resistance:
Approximately 0Ω
Check the following items.
1) CHECK COMPONENT (Resistor).
Refer to "Electrical Components Inspection".
2) Repair harness or connectors.
CHECK GROUND CIRCUIT.
1) Stop engine.
2) Disconnect power transistor harness connector.
3) Check harness continuity between terminal ⓑ and ground.
Continuity should exist.
Repair or replace harness.
Reinstall any part removed.
Erase the self-diagnosis memory.
Perform driving test and then perform self-diagnosis (Mode III) again.
1) Perform E.C.U. input/output signal inspection test.
2) If N.G., recheck the E.C.U. pin terminals for damage or the connection of E.C.U. harness connector.
INSPECTION END
C/UNIT CONNECTOR
Diagnostic Procedure 7
FUEL PUMP (Code No. 22)
E.C.C.S. CONTROL UNIT
FUEL PUMP
SAFETY RELAY
FUEL PUMP RELAY
FUSE BLOCK
IGNITION SWITCH ON or START
(E.F.I. harness)
(Main harness)
UP
ENGINE GROUND
BODY GROUND
Harness layout
Fuel pump harness connector
Fuel pump harness connector is located under the rear seat cushion.
E.C.U. S.M.J. harness connector
Fuel pump relay
Safety relay
The fuel pump relay and safety relay are located inside the trunk trim, near the striker.

ECCS SELF-DIAGNOSTIC CHART—1989-90 MAXIMA

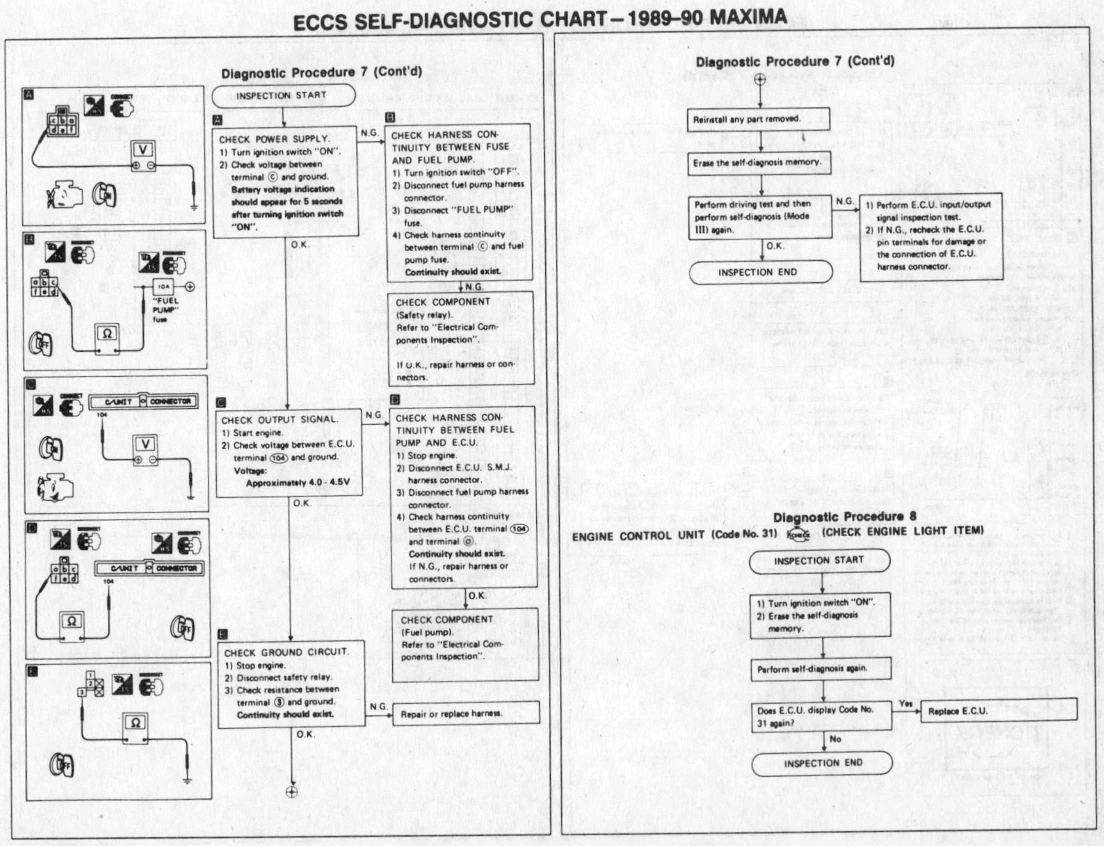
Diagnostic Procedure 7 (Cont'd)
INSPECTION START
CHECK POWER SUPPLY.
1) Turn ignition switch "ON".
2) Check voltage between terminal ⓒ and ground.
Battery voltage indication should appear for 5 seconds after turning ignition switch "ON".
N.G.
O.K.
CHECK HARNESS CONTINUITY BETWEEN FUSE AND FUEL PUMP.
1) Turn ignition switch "OFF".
2) Disconnect fuel pump harness connector.
3) Disconnect "FUEL PUMP" fuse.
4) Check harness continuity between terminal ⓒ and fuel pump fuse.
Continuity should exist.
CHECK COMPONENT (Safety relay).
Refer to "Electrical Components Inspection".
If O.K., repair harness or connectors.
CHECK OUTPUT SIGNAL.
1) Start engine.
2) Check voltage between E.C.U. terminal (104) and ground.
Voltage:
Approximately 4.0 - 4.5V
CHECK HARNESS CONTINUITY BETWEEN FUEL PUMP AND E.C.U.
1) Stop engine.
2) Disconnect E.C.U. S.M.J. harness connector.
3) Disconnect fuel pump harness connector.
4) Check harness continuity between E.C.U. terminal (104) and terminal ⓞ.
Continuity should exist.
If N.G., repair harness or connectors.
CHECK COMPONENT (Fuel pump).
Refer to "Electrical Components Inspection".
CHECK GROUND CIRCUIT.
1) Stop engine.
2) Disconnect safety relay.
3) Check resistance between terminal ③ and ground.
Continuity should exist.
Repair or replace harness.
"FUEL PUMP" fuse
10A
C/UNIT CONNECTOR
104
Diagnostic Procedure 7 (Cont'd)
Reinstall any part removed.
Erase the self-diagnosis memory.
Perform driving test and then perform self-diagnosis (Mode III) again.
1) Perform E.C.U. input/output signal inspection test.
2) If N.G., recheck the E.C.U. pin terminals for damage or the connection of E.C.U. harness connector.
INSPECTION END
Diagnostic Procedure 8
ENGINE CONTROL UNIT (Code No. 31) (CHECK ENGINE LIGHT ITEM)
INSPECTION START
1) Turn ignition switch "ON".
2) Erase the self-diagnosis memory.
Perform self-diagnosis again.
Does E.C.U. display Code No. 31 again?
Yes
Replace E.C.U.
No
INSPECTION END

ECCS SELF-DIAGNOSTIC CHART—1989–90 MAXIMA

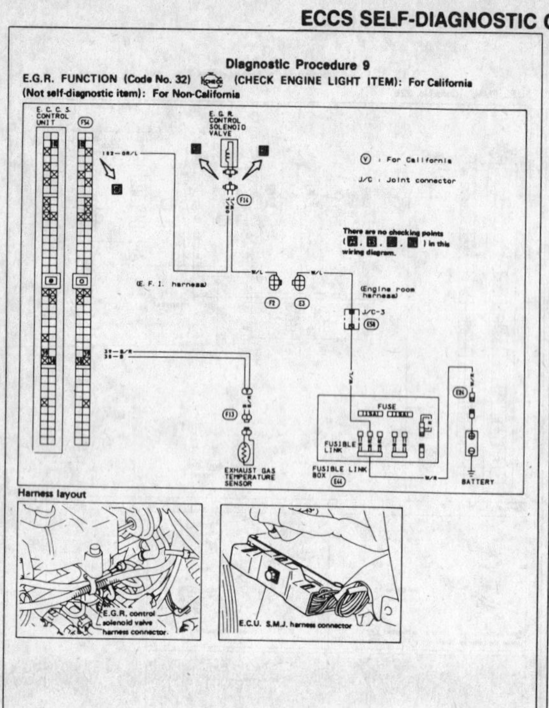
Diagnostic Procedure 9
E.G.R. FUNCTION (Code No. 32) (CHECK ENGINE LIGHT ITEM): For California
(Not self-diagnostic item): For Non-California
E.C.C.S. CONTROL UNIT
E.G.R. CONTROL SOLENOID VALVE
Ⓥ : For California
J/C : Joint connector
There are no checking points in this wiring diagram.
(E.F.I. harness)
(Engine room harness)
J/C-3
FUSE
FUSIBLE LINK
FUSIBLE LINK BOX
BATTERY
EXHAUST GAS TEMPERATURE SENSOR
Harness layout
E.G.R. control solenoid valve harness connector.
E.C.U. S.M.J. harness connector

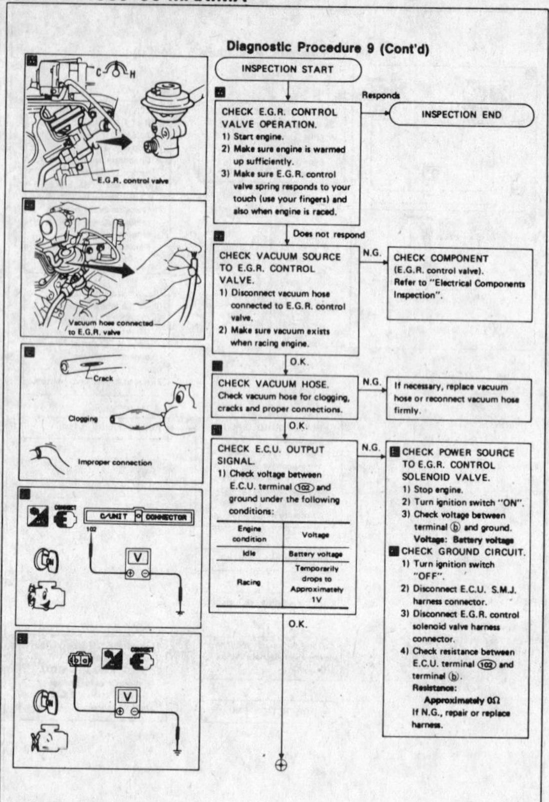
Diagnostic Procedure 9 (Cont'd)
INSPECTION START
CHECK E.G.R. CONTROL VALVE OPERATION.
1) Start engine.
2) Make sure engine is warmed up sufficiently.
3) Make sure E.G.R. control valve spring responds to your touch (use your fingers) and also when engine is raced.
Responds
INSPECTION END
Does not respond
CHECK VACUUM SOURCE TO E.G.R. CONTROL VALVE.
1) Disconnect vacuum hose connected to E.G.R. control valve.
2) Make sure vacuum exists when racing engine.
N.G.
CHECK COMPONENT (E.G.R. control valve). Refer to "Electrical Components Inspection".
O.K.
CHECK VACUUM HOSE.
Check vacuum hose for clogging, cracks and proper connections.
N.G.
If necessary, replace vacuum hose or reconnect vacuum hose firmly.
O.K.
CHECK E.C.U. OUTPUT SIGNAL.
1) Check voltage between E.C.U. terminal 102 and ground under the following conditions:
Engine condition | Voltage
Idle | Battery voltage
Racing | Temporarily drops to Approximately 1V
N.G.
CHECK POWER SOURCE TO E.G.R. CONTROL SOLENOID VALVE.
1) Stop engine.
2) Turn ignition switch "ON".
3) Check voltage between terminal ⓑ and ground.
Voltage: Battery voltage
CHECK GROUND CIRCUIT.
1) Turn ignition switch "OFF".
2) Disconnect E.C.U. S.M.J. harness connector.
3) Disconnect E.G.R. control solenoid valve harness connector.
4) Check resistance between E.C.U. terminal 102 and terminal ⓑ.
Resistance: Approximately 0Ω
If N.G., repair or replace harness.
O.K.
E.G.R. control valve
Vacuum hose connected to E.G.R. valve
Crack
Clogging
Improper connection
CONNECT
C/UNIT CONNECTOR

ECCS SELF-DIAGNOSTIC CHART—1989–90 MAXIMA

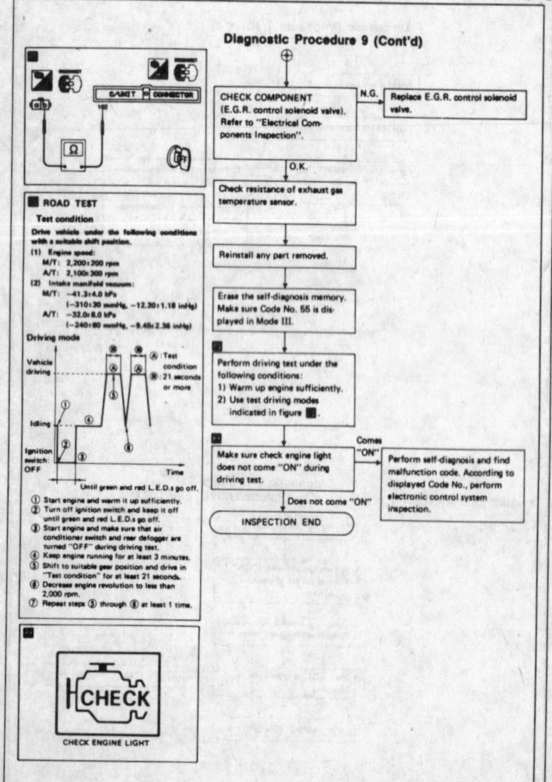
Diagnostic Procedure 9 (Cont'd)
C/UNIT CONNECTOR
CHECK COMPONENT (E.G.R. control solenoid valve). Refer to "Electrical Components Inspection".
N.G.
Replace E.G.R. control solenoid valve.
O.K.
Check resistance of exhaust gas temperature sensor.
Reinstall any part removed.
Erase the self-diagnosis memory. Make sure Code No. 55 is displayed in Mode III.
Perform driving test under the following conditions:
1) Warm up engine sufficiently.
2) Use test driving modes indicated in figure.
Make sure check engine light does not come "ON" during driving test.
Comes "ON"
Perform self-diagnosis and find malfunction code. According to displayed Code No., perform electronic control system inspection.
Does not come "ON"
INSPECTION END
ROAD TEST
Test condition
Drive vehicle under the following conditions with a suitable shift position.
(1) Engine speed:
M/T: 2,200±200 rpm
A/T: 2,100±300 rpm
(2) Intake manifold vacuum:
M/T: −41.3±4.0 kPa (−310±30 mmHg, −12.20±1.18 inHg)
A/T: −32.0±8.0 kPa (−240±60 mmHg, −9.45±2.36 inHg)
Driving mode
Vehicle driving
Idling
Ignition switch OFF
Time
Ⓐ: Test condition
Ⓑ: 21 seconds or more
Until green and red L.E.D.s go off.
① Start engine and warm it up sufficiently.
② Turn off ignition switch and keep it off until green and red L.E.D.s go off.
③ Start engine and make sure that air conditioner switch and rear defogger are turned "OFF" during driving test.
④ Keep engine running for at least 3 minutes.
⑤ Shift to suitable gear position and drive in "Test condition" for at least 21 seconds.
⑥ Decrease engine revolution to less than 2,000 rpm.
⑦ Repeat steps ⑤ through ⑥ at least 1 time.
CHECK
CHECK ENGINE LIGHT

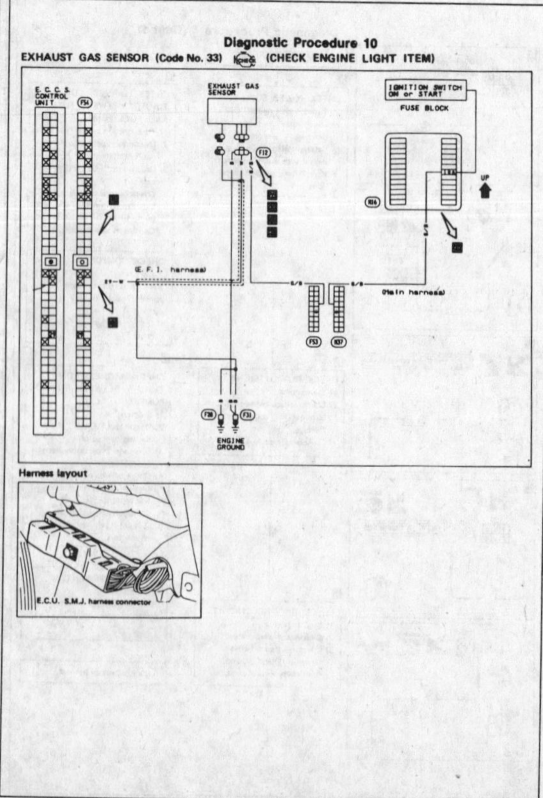
Diagnostic Procedure 10
EXHAUST GAS SENSOR (Code No. 33) (CHECK ENGINE LIGHT ITEM)
E.C.C.S. CONTROL UNIT
EXHAUST GAS SENSOR
IGNITION SWITCH ON or START
FUSE BLOCK
(E.F.I. harness)
(Main harness)
ENGINE GROUND
Harness layout
E.C.U. S.M.J. harness connector

ECCS SELF-DIAGNOSTIC CHART—1989–90 MAXIMA

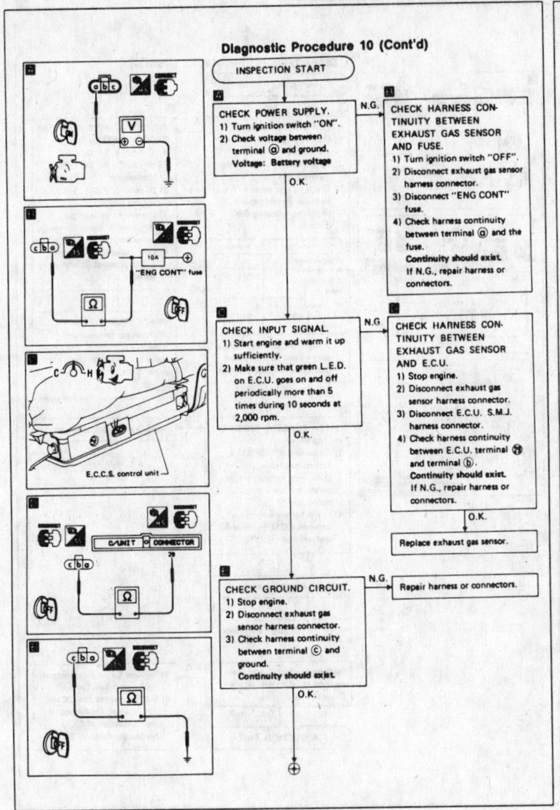
Diagnostic Procedure 10 (Cont'd)
INSPECTION START
CHECK POWER SUPPLY.
1) Turn ignition switch "ON".
2) Check voltage between terminal ⓐ and ground.
Voltage: Battery voltage
N.G.
CHECK HARNESS CONTINUITY BETWEEN EXHAUST GAS SENSOR AND FUSE.
1) Turn ignition switch "OFF".
2) Disconnect exhaust gas sensor harness connector.
3) Disconnect "ENG CONT" fuse.
4) Check harness continuity between terminal ⓐ and the fuse.
Continuity should exist.
If N.G., repair harness or connectors.
O.K.
CHECK INPUT SIGNAL.
1) Start engine and warm it up sufficiently.
2) Make sure that green L.E.D. on E.C.U. goes on and off periodically more than 5 times during 10 seconds at 2,000 rpm.
N.G.
CHECK HARNESS CONTINUITY BETWEEN EXHAUST GAS SENSOR AND E.C.U.
1) Stop engine.
2) Disconnect exhaust gas sensor harness connector.
3) Disconnect E.C.U. S.M.J. harness connector.
4) Check harness continuity between E.C.U. terminal 29 and terminal ⓑ.
Continuity should exist.
If N.G., repair harness or connectors.
O.K.
Replace exhaust gas sensor.
O.K.
CHECK GROUND CIRCUIT.
1) Stop engine.
2) Disconnect exhaust gas sensor harness connector.
3) Check harness continuity between terminal ⓒ and ground.
Continuity should exist.
N.G.
Repair harness or connectors.
O.K.
V
10A
"ENG CONT" fuse
Ω
C/UNIT CONNECTOR
29
E.C.C.S. control unit

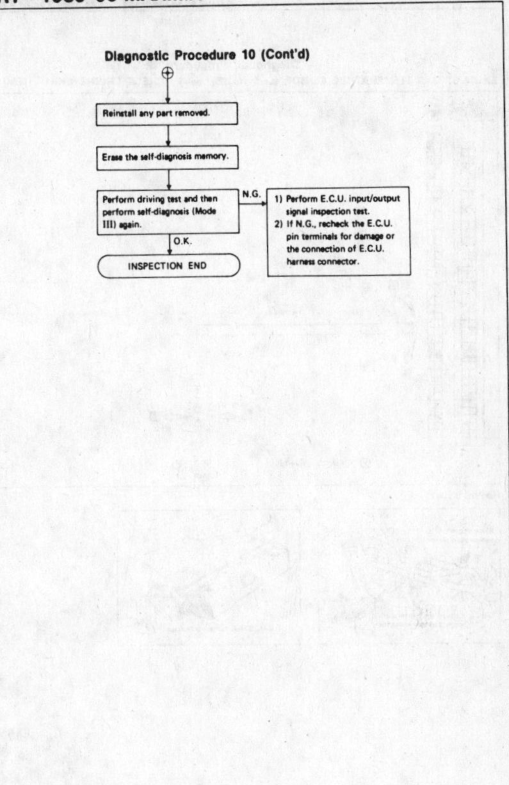
Diagnostic Procedure 10 (Cont'd)
Reinstall any part removed.
Erase the self-diagnosis memory.
Perform driving test and then perform self-diagnosis (Mode III) again.
N.G.
1) Perform E.C.U. input/output signal inspection test.
2) If N.G., recheck the E.C.U. pin terminals for damage or the connection of E.C.U. harness connector.
O.K.
INSPECTION END

ECCS SELF-DIAGNOSTIC CHART—1989–90 MAXIMA

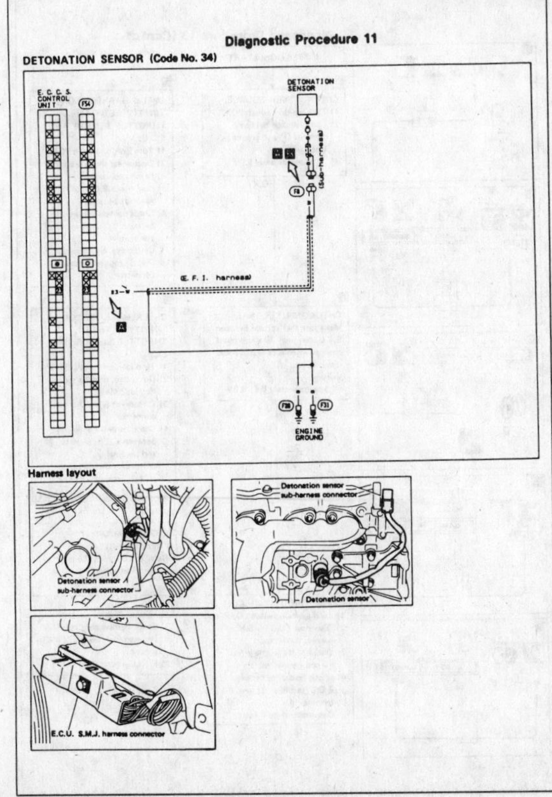
Diagnostic Procedure 11
DETONATION SENSOR (Code No. 34)
E.C.C.S. CONTROL UNIT
DETONATION SENSOR
(Sub-harness)
(E.F.I. harness)
ENGINE GROUND
Harness layout
Detonation sensor sub-harness connector
Detonation sensor
E.C.U. S.M.J. harness connector

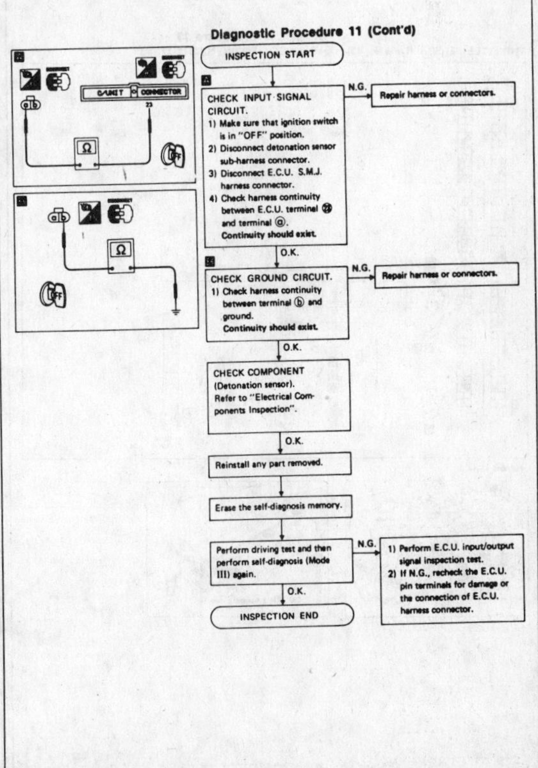
Diagnostic Procedure 11 (Cont'd)
INSPECTION START
CHECK INPUT SIGNAL CIRCUIT.
1) Make sure that ignition switch is in "OFF" position.
2) Disconnect detonation sensor sub-harness connector.
3) Disconnect E.C.U. S.M.J. harness connector.
4) Check harness continuity between E.C.U. terminal 23 and terminal ⓐ.
Continuity should exist.
N.G.
Repair harness or connectors.
O.K.
CHECK GROUND CIRCUIT.
1) Check harness continuity between terminal ⓑ and ground.
Continuity should exist.
N.G.
Repair harness or connectors.
O.K.
CHECK COMPONENT (Detonation sensor).
Refer to "Electrical Components Inspection".
O.K.
Reinstall any part removed.
Erase the self-diagnosis memory.
Perform driving test and then perform self-diagnosis (Mode III) again.
N.G.
1) Perform E.C.U. input/output signal inspection test.
2) If N.G., recheck the E.C.U. pin terminals for damage or the connection of E.C.U. harness connector.
O.K.
INSPECTION END
C/UNIT CONNECTOR
23
Ω

ECCS SELF-DIAGNOSTIC CHART—1989–90 MAXIMA

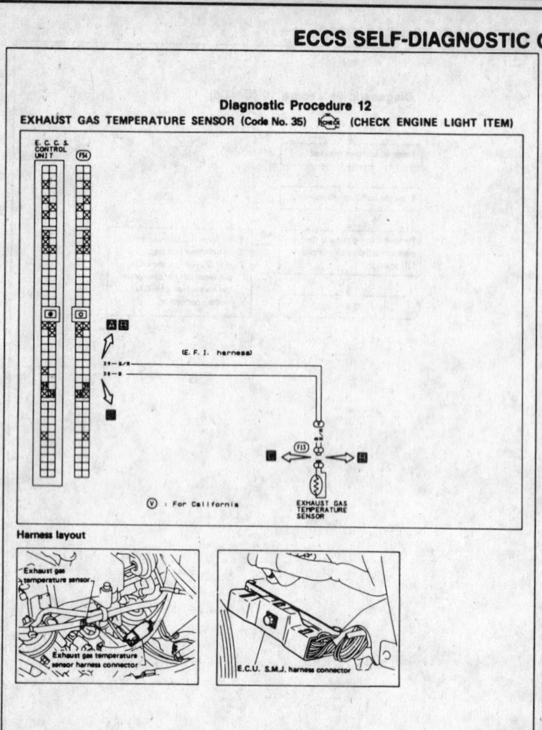
Diagnostic Procedure 12
EXHAUST GAS TEMPERATURE SENSOR (Code No. 35) (CHECK ENGINE LIGHT ITEM)
E.C.C.S. CONTROL UNIT
F54
(E.F.I. harness)
F13
EXHAUST GAS TEMPERATURE SENSOR
For California
Harness layout
Exhaust gas temperature sensor
Exhaust gas temperature sensor harness connector
E.C.U. S.M.J. harness connector

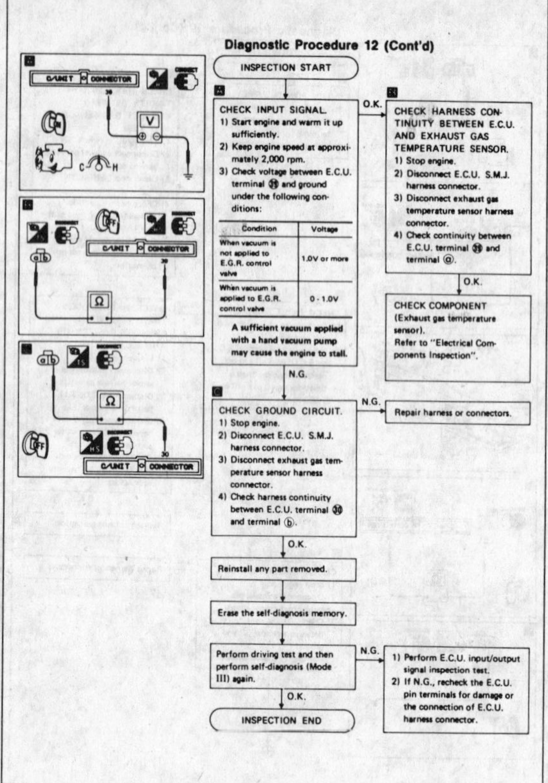
Diagnostic Procedure 12 (Cont'd)
INSPECTION START
CHECK INPUT SIGNAL.
1) Start engine and warm it up sufficiently.
2) Keep engine speed at approximately 2,000 rpm.
3) Check voltage between E.C.U. terminal 35 and ground under the following conditions:
Condition | Voltage
When vacuum is not applied to E.G.R. control valve | 1.0V or more
When vacuum is applied to E.G.R. control valve | 0 - 1.0V
A sufficient vacuum applied with a hand vacuum pump may cause the engine to stall.
O.K.
CHECK HARNESS CONTINUITY BETWEEN E.C.U. AND EXHAUST GAS TEMPERATURE SENSOR.
1) Stop engine.
2) Disconnect E.C.U. S.M.J. harness connector.
3) Disconnect exhaust gas temperature sensor harness connector.
4) Check continuity between E.C.U. terminal 35 and terminal ⓐ.
O.K.
CHECK COMPONENT (Exhaust gas temperature sensor). Refer to "Electrical Components Inspection".
N.G.
CHECK GROUND CIRCUIT.
1) Stop engine.
2) Disconnect E.C.U. S.M.J. harness connector.
3) Disconnect exhaust gas temperature sensor harness connector.
4) Check harness continuity between E.C.U. terminal 30 and terminal ⓑ.
N.G.
Repair harness or connectors.
O.K.
Reinstall any part removed.
Erase the self-diagnosis memory.
Perform driving test and then perform self-diagnosis (Mode III) again.
N.G.
1) Perform E.C.U. input/output signal inspection test.
2) If N.G., recheck the E.C.U. pin terminals for damage or the connection of E.C.U. harness connector.
O.K.
INSPECTION END
C/UNIT CONNECTOR

ECCS SELF-DIAGNOSTIC CHART—1989–90 MAXIMA

Diagnostic Procedure 13

THROTTLE SENSOR (Code No. 43) (CHECK ENGINE LIGHT ITEM)

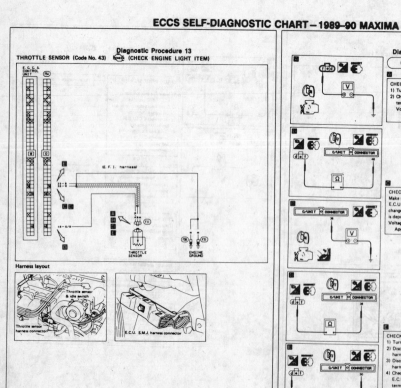

E.C.C.S. CONTROL UNIT

F54

(E.F.I. harness)

F11

THROTTLE SENSOR

F30 F31

ENGINE GROUND

Harness layout

Throttle sensor & idle switch

Throttle sensor harness connector

E.C.U. S.M.J. harness connector

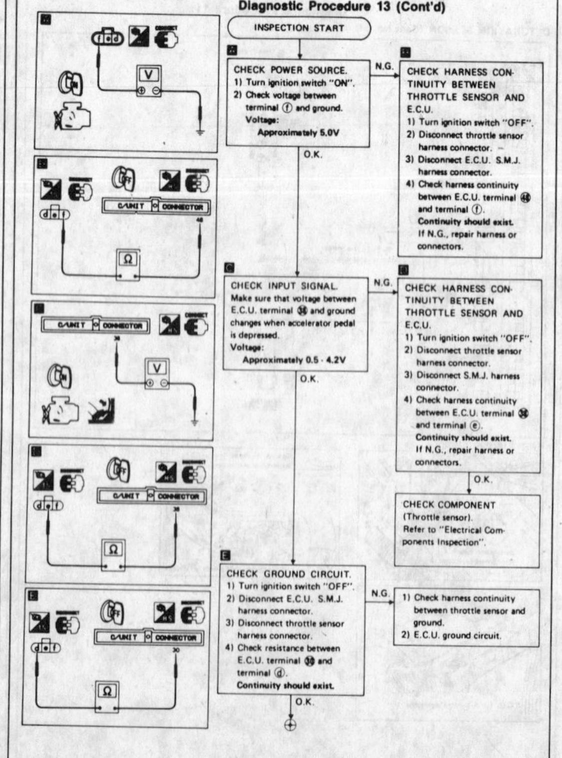
Diagnostic Procedure 13 (Cont'd)
INSPECTION START
CHECK POWER SOURCE.
1) Turn ignition switch "ON".
2) Check voltage between terminal ⓕ and ground.
Voltage: Approximately 5.0V
N.G.
CHECK HARNESS CONTINUITY BETWEEN THROTTLE SENSOR AND E.C.U.
1) Turn ignition switch "OFF".
2) Disconnect throttle sensor harness connector.
3) Disconnect E.C.U. S.M.J. harness connector.
4) Check harness continuity between E.C.U. terminal 48 and terminal ⓕ.
Continuity should exist.
If N.G., repair harness or connectors.
O.K.
CHECK INPUT SIGNAL.
Make sure that voltage between E.C.U. terminal 38 and ground changes when accelerator pedal is depressed.
Voltage: Approximately 0.5 - 4.2V
N.G.
CHECK HARNESS CONTINUITY BETWEEN THROTTLE SENSOR AND E.C.U.
1) Turn ignition switch "OFF".
2) Disconnect throttle sensor harness connector.
3) Disconnect S.M.J. harness connector.
4) Check harness continuity between E.C.U. terminal 38 and terminal ⓔ.
Continuity should exist.
If N.G., repair harness or connectors.
O.K.
CHECK COMPONENT (Throttle sensor). Refer to "Electrical Components Inspection".
O.K.
CHECK GROUND CIRCUIT.
1) Turn ignition switch "OFF".
2) Disconnect E.C.U. S.M.J. harness connector.
3) Disconnect throttle sensor harness connector.
4) Check resistance between E.C.U. terminal 30 and terminal ⓓ.
Continuity should exist.
N.G.
1) Check harness continuity between throttle sensor and ground.
2) E.C.U. ground circuit.
O.K.
C/UNIT CONNECTOR

ECCS SELF-DIAGNOSTIC CHART—1989–90 MAXIMA

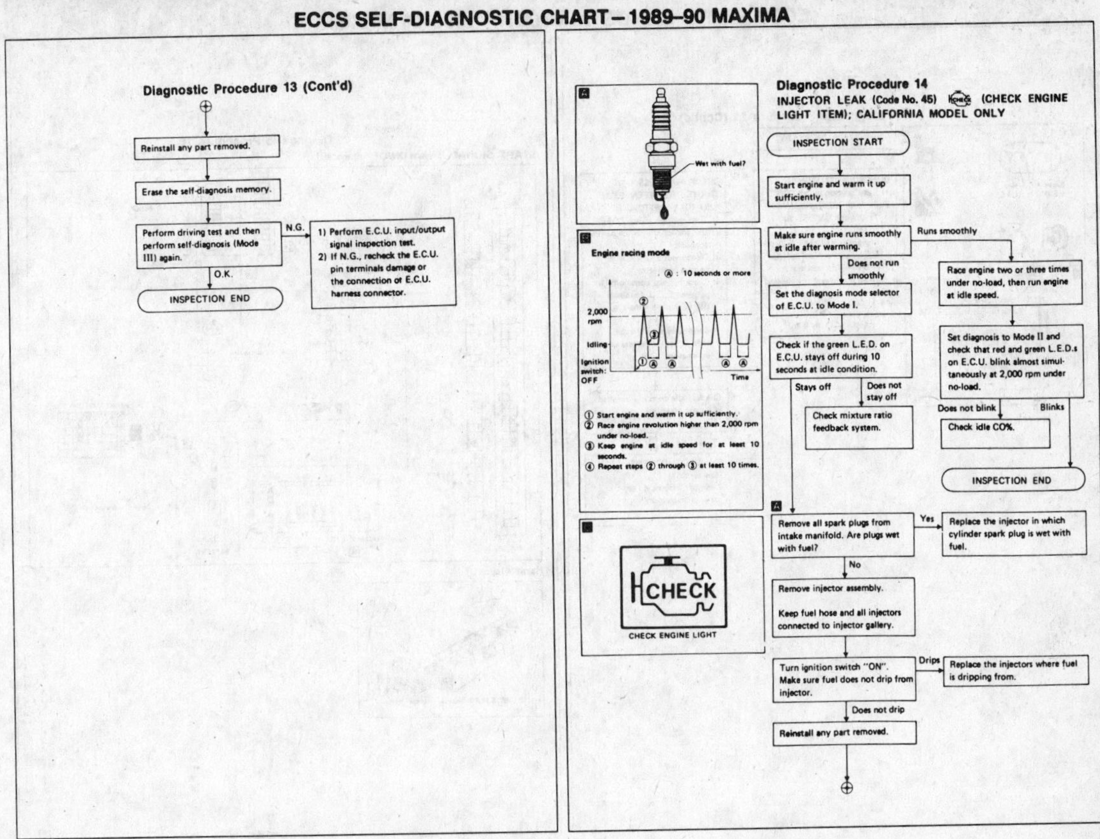
Diagnostic Procedure 13 (Cont'd)
Reinstall any part removed.
Erase the self-diagnosis memory.
Perform driving test and then perform self-diagnosis (Mode III) again.
N.G.
1) Perform E.C.U. input/output signal inspection test.
2) If N.G., recheck the E.C.U. pin terminals damage or the connection of E.C.U. harness connector.
O.K.
INSPECTION END
Wet with fuel?
Engine racing mode
Ⓐ : 10 seconds or more
2,000 rpm
Idling
Ignition switch: OFF
Time
① Start engine and warm it up sufficiently.
② Race engine revolution higher than 2,000 rpm under no-load.
③ Keep engine at idle speed for at least 10 seconds.
④ Repeat steps ② through ③ at least 10 times.
CHECK
CHECK ENGINE LIGHT
Diagnostic Procedure 14
INJECTOR LEAK (Code No. 45) (CHECK ENGINE LIGHT ITEM); CALIFORNIA MODEL ONLY
INSPECTION START
Start engine and warm it up sufficiently.
Make sure engine runs smoothly at idle after warming.
Runs smoothly
Does not run smoothly
Set the diagnosis mode selector of E.C.U. to Mode I.
Check if the green L.E.D. on E.C.U. stays off during 10 seconds at idle condition.
Stays off
Does not stay off
Check mixture ratio feedback system.
Race engine two or three times under no-load, then run engine at idle speed.
Set diagnosis to Mode II and check that red and green L.E.D.s on E.C.U. blink almost simultaneously at 2,000 rpm under no-load.
Does not blink
Blinks
Check idle CO%.
INSPECTION END
Remove all spark plugs from intake manifold. Are plugs wet with fuel?
Yes
Replace the injector in which cylinder spark plug is wet with fuel.
No
Remove injector assembly.
Keep fuel hose and all injectors connected to injector gallery.
Turn ignition switch "ON". Make sure fuel does not drip from injector.
Drips
Replace the injectors where fuel is dripping from.
Does not drip
Reinstall any part removed.

ECCS SELF-DIAGNOSTIC CHART—1989–90 MAXIMA

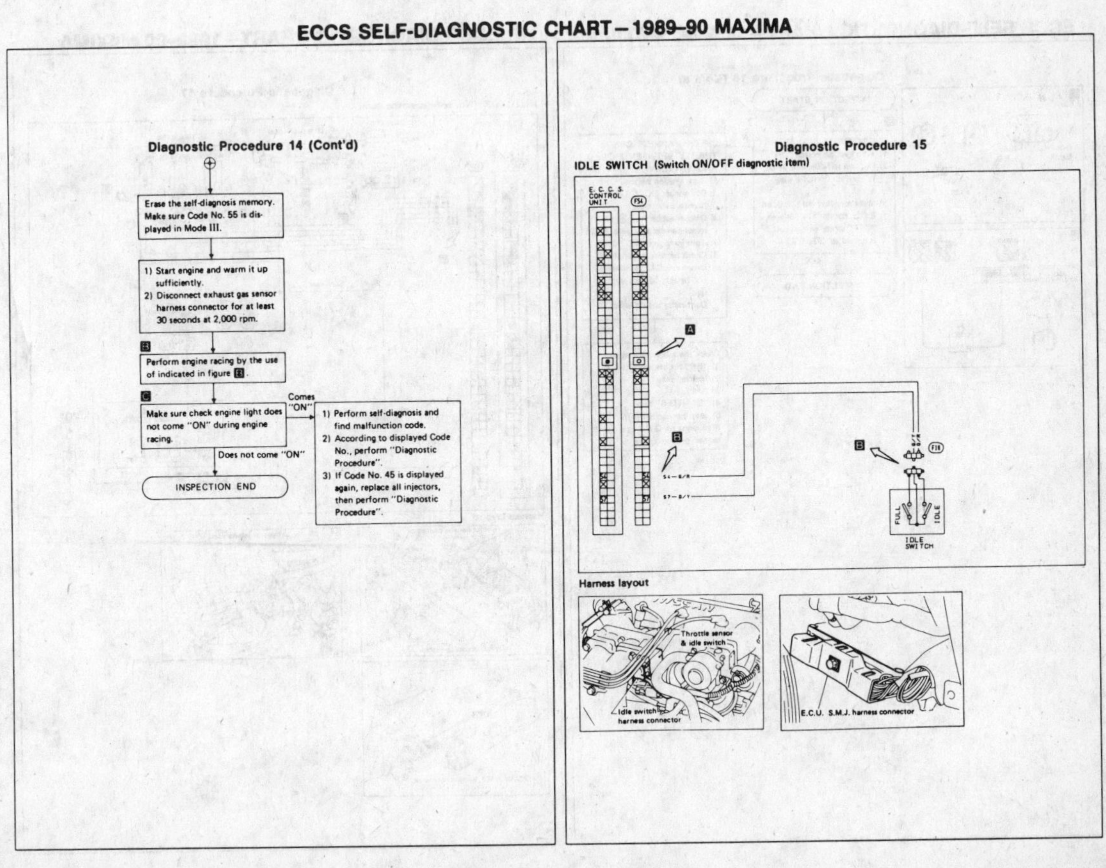
Diagnostic Procedure 14 (Cont'd)
Erase the self-diagnosis memory. Make sure Code No. 55 is displayed in Mode III.
1) Start engine and warm it up sufficiently.
2) Disconnect exhaust gas sensor harness connector for at least 30 seconds at 2,000 rpm.
Perform engine racing by the use of indicated in figure B.
Make sure check engine light does not come "ON" during engine racing.
Comes "ON"
1) Perform self-diagnosis and find malfunction code.
2) According to displayed Code No., perform "Diagnostic Procedure".
3) If Code No. 45 is displayed again, replace all injectors, then perform "Diagnostic Procedure".
Does not come "ON"
INSPECTION END
Diagnostic Procedure 15
IDLE SWITCH (Switch ON/OFF diagnostic item)
E.C.C.S. CONTROL UNIT
F54
F19
FULL
IDLE
IDLE SWITCH
Harness layout
Throttle sensor & idle switch
Idle switch harness connector
E.C.U. S.M.J. harness connector

ECCS SELF-DIAGNOSTIC CHART – 1989–90 MAXIMA

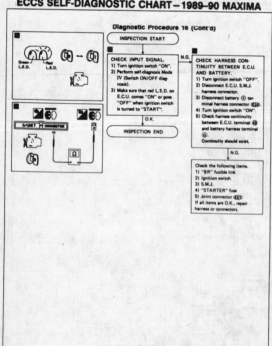

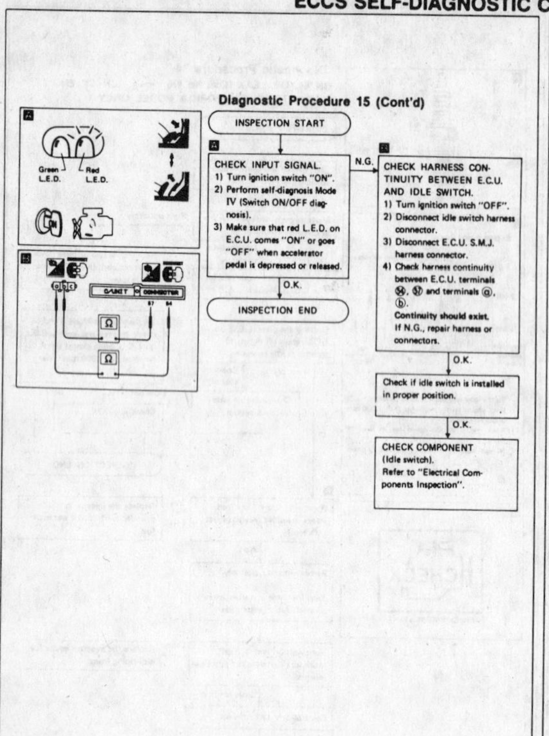
Diagnostic Procedure 15 (Cont'd)
Green L.E.D.
Red L.E.D.
INSPECTION START
CHECK INPUT SIGNAL.
1) Turn ignition switch "ON".
2) Perform self-diagnosis Mode IV (Switch ON/OFF diagnosis).
3) Make sure that red L.E.D. on E.C.U. comes "ON" or goes "OFF" when accelerator pedal is depressed or released.
O.K.
INSPECTION END
N.G.
CHECK HARNESS CONTINUITY BETWEEN E.C.U. AND IDLE SWITCH.
1) Turn ignition switch "OFF".
2) Disconnect idle switch harness connector.
3) Disconnect E.C.U. S.M.J. harness connector.
4) Check harness continuity between E.C.U. terminals 54, 57 and terminals ⓐ, ⓑ.
Continuity should exist.
If N.G., repair harness or connectors.
O.K.
Check if idle switch is installed in proper position.
O.K.
CHECK COMPONENT (Idle switch).
Refer to "Electrical Components Inspection".

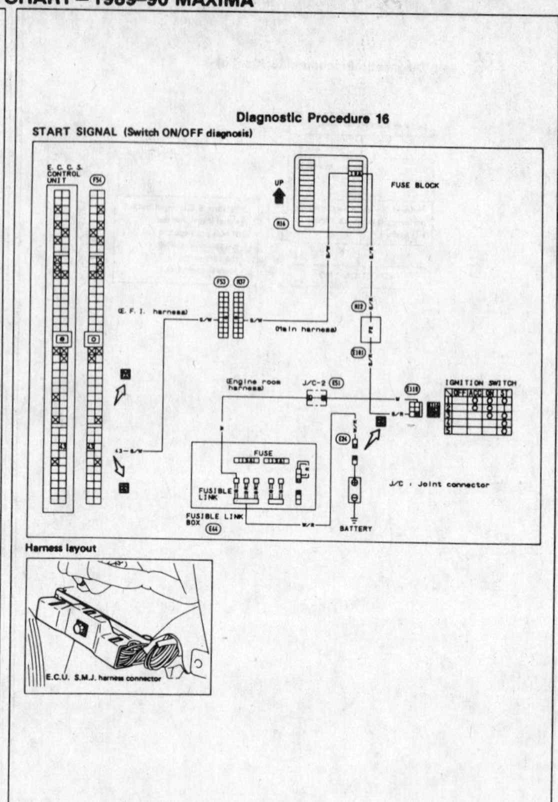
Diagnostic Procedure 16
START SIGNAL (Switch ON/OFF diagnosis)
E.C.C.S. CONTROL UNIT
UP
FUSE BLOCK
(E.F.I. harness)
(Main harness)
(Engine room harness)
J/C-2
IGNITION SWITCH
FUSE
FUSIBLE LINK
FUSIBLE LINK BOX
BATTERY
J/C : Joint connector
Harness layout
E.C.U. S.M.J. harness connector

ECCS SELF-DIAGNOSTIC CHART – 1989–90 MAXIMA

Diagnostic Procedure 16 (Cont'd)

Green L.E.D. / Red L.E.D.

INSPECTION START

CHECK INPUT SIGNAL.

1) Turn ignition switch "ON".
2) Perform self-diagnosis Mode IV (Switch ON/OFF diagnosis).
3) Make sure that red L.E.D. on E.C.U. comes "ON" or goes "OFF" when ignition switch is turned to "START".

O.K. → INSPECTION END

N.G. →

CHECK HARNESS CONTINUITY BETWEEN E.C.U. AND BATTERY.

1) Turn ignition switch "OFF".
2) Disconnect E.C.U. S.M.J. harness connector.
3) Disconnect battery ⊕ terminal harness connector (E24).
4) Turn ignition switch "ON".
5) Check harness continuity between E.C.U. terminal 43 and battery harness terminal ⓐ.

Continuity should exist.

N.G. →

Check the following items.

1) "BR" fusible link
2) Ignition switch
3) S.M.J.
4) "STARTER" fuse
5) Joint connector (E51)

If all items are O.K., repair harness or connectors.

ECCS DIAGNOSTIC CHART – 1989–90 MAXIMA

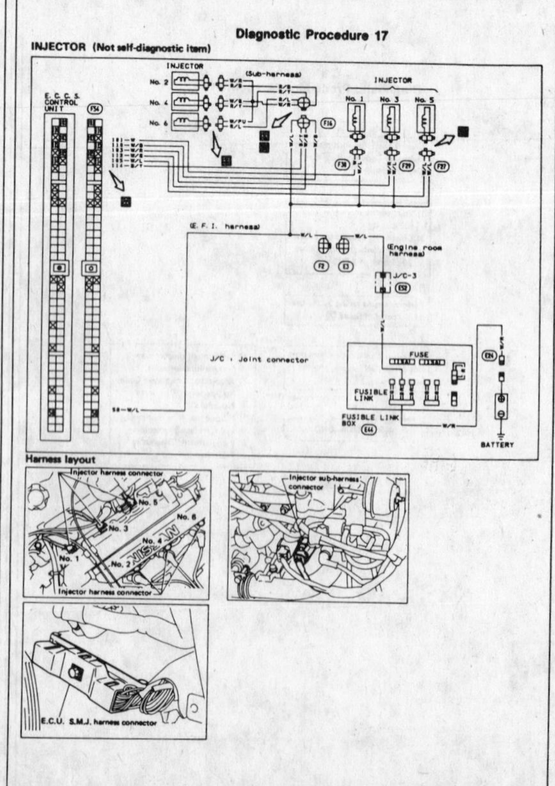
Diagnostic Procedure 17
INJECTOR (Not self-diagnostic item)
E.C.C.S. CONTROL UNIT
INJECTOR
No. 2
No. 4
No. 6
(Sub-harness)
No. 1
No. 3
No. 5
(E.F.I. harness)
(Engine room harness)
J/C-3
J/C : Joint connector
FUSE
FUSIBLE LINK
FUSIBLE LINK BOX
BATTERY
Harness layout
Injector harness connector
Injector sub-harness connector
E.C.U. S.M.J. harness connector

ECCS DIAGNOSTIC CHART—1989–90 MAXIMA

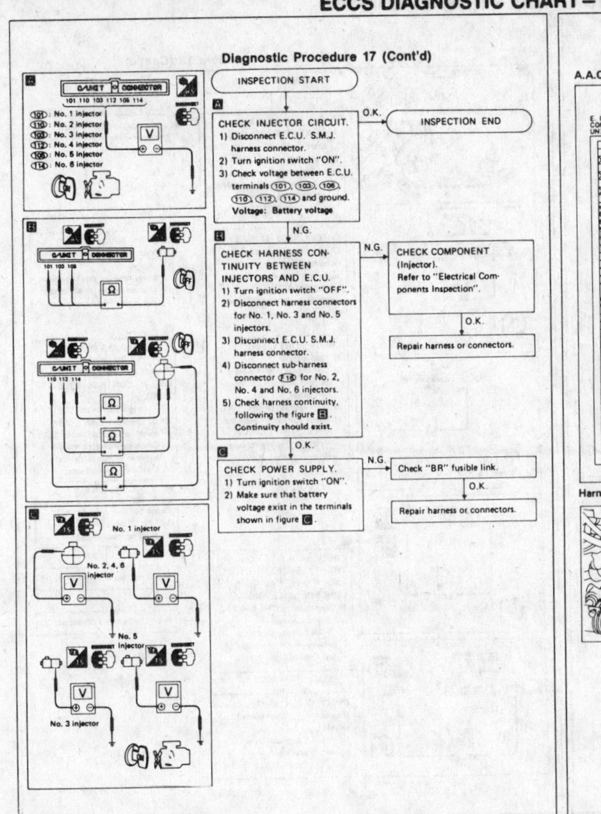
Diagnostic Procedure 17 (Cont'd)
INSPECTION START
CHECK INJECTOR CIRCUIT.
1) Disconnect E.C.U. S.M.J. harness connector.
2) Turn ignition switch "ON".
3) Check voltage between E.C.U. terminals 101, 103, 106, 110, 112, 114 and ground.
Voltage: Battery voltage
O.K.
INSPECTION END
N.G.
CHECK HARNESS CONTINUITY BETWEEN INJECTORS AND E.C.U.
1) Turn ignition switch "OFF".
2) Disconnect harness connectors for No. 1, No. 3 and No. 5 injectors.
3) Disconnect E.C.U. S.M.J. harness connector.
4) Disconnect sub-harness connector F14 for No. 2, No. 4 and No. 6 injectors.
5) Check harness continuity, following the figure B.
Continuity should exist.
N.G.
CHECK COMPONENT (Injector).
Refer to "Electrical Components Inspection".
O.K.
Repair harness or connectors.
O.K.
CHECK POWER SUPPLY.
1) Turn ignition switch "ON".
2) Make sure that battery voltage exist in the terminals shown in figure C.
N.G.
Check "BR" fusible link.
O.K.
Repair harness or connectors.
C/UNIT CONNECTOR
101 110 103 112 106 114
101 No. 1 injector
110 No. 2 injector
103 No. 3 injector
112 No. 4 injector
106 No. 5 injector
114 No. 6 injector
No. 1 injector
No. 2, 4, 6 injector
No. 5 injector
No. 3 injector

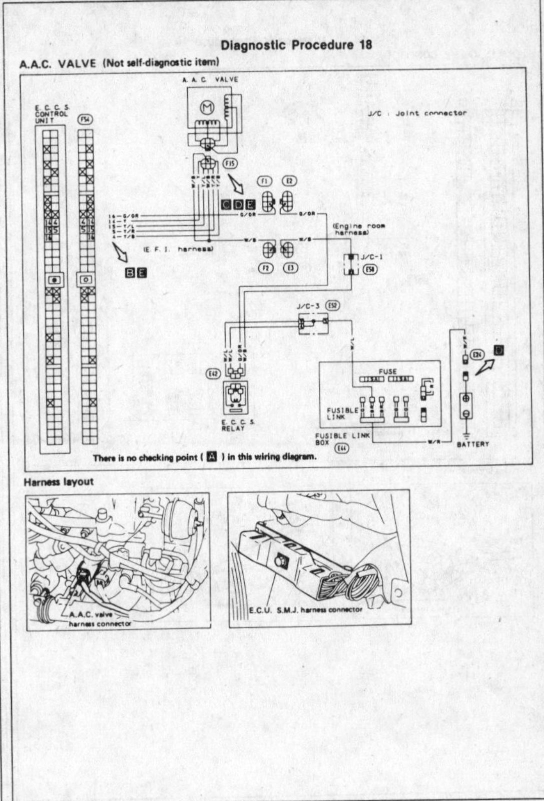
Diagnostic Procedure 18
A.A.C. VALVE (Not self-diagnostic item)
A.A.C. VALVE
J/C : Joint connector
E.C.C.S. CONTROL UNIT
(E.F.I. harness)
(Engine room harness)
J/C-1
J/C-3
E.C.C.S. RELAY
FUSE
FUSIBLE LINK
FUSIBLE LINK BOX
BATTERY
There is no checking point (A) in this wiring diagram.
Harness layout
A.A.C. valve harness connector
E.C.U. S.M.J. harness connector

ECCS DIAGNOSTIC CHART—1989–90 MAXIMA

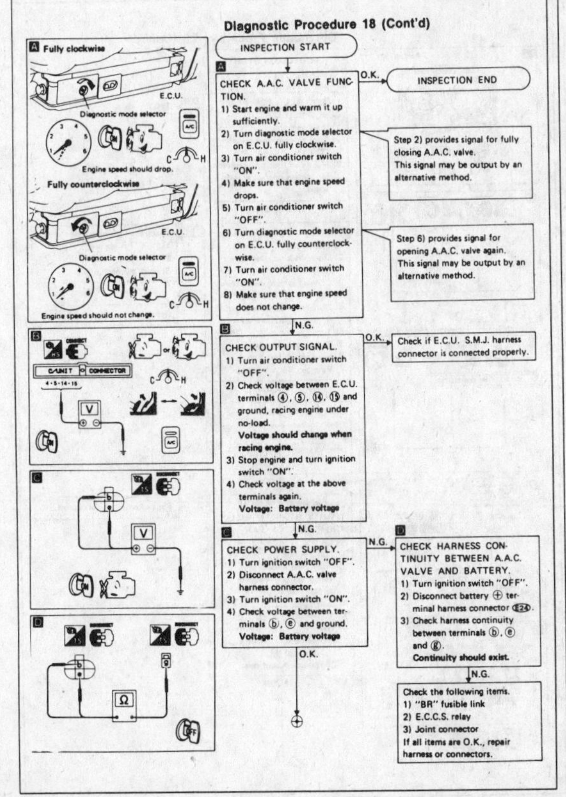
Diagnostic Procedure 18 (Cont'd)
INSPECTION START
CHECK A.A.C. VALVE FUNCTION.
1) Start engine and warm it up sufficiently.
2) Turn diagnostic mode selector on E.C.U. fully clockwise.
3) Turn air conditioner switch "ON".
4) Make sure that engine speed drops.
5) Turn air conditioner switch "OFF".
6) Turn diagnostic mode selector on E.C.U. fully counterclockwise.
7) Turn air conditioner switch "ON".
8) Make sure that engine speed does not change.
O.K.
INSPECTION END
Step 2) provides signal for fully closing A.A.C. valve. This signal may be output by an alternative method.
Step 6) provides signal for opening A.A.C. valve again. This signal may be output by an alternative method.
N.G.
CHECK OUTPUT SIGNAL.
1) Turn air conditioner switch "OFF".
2) Check voltage between E.C.U. terminals 4, 5, 14, 15 and ground, racing engine under no-load.
Voltage should change when racing engine.
3) Stop engine and turn ignition switch "ON".
4) Check voltage at the above terminals again.
Voltage: Battery voltage
O.K.
Check if E.C.U. S.M.J. harness connector is connected properly.
N.G.
CHECK POWER SUPPLY.
1) Turn ignition switch "OFF".
2) Disconnect A.A.C. valve harness connector.
3) Turn ignition switch "ON".
4) Check voltage between terminals b, e and ground.
Voltage: Battery voltage
O.K.
N.G.
CHECK HARNESS CONTINUITY BETWEEN A.A.C. VALVE AND BATTERY.
1) Turn ignition switch "OFF".
2) Disconnect battery ⊕ terminal harness connector E24.
3) Check harness continuity between terminals b, e and g.
Continuity should exist.
N.G.
Check the following items.
1) "BR" fusible link
2) E.C.C.S. relay
3) Joint connector
If all items are O.K., repair harness or connectors.
Fully clockwise
E.C.U.
Diagnostic mode selector
Engine speed should drop.
Fully counterclockwise
Engine speed should not change.
C/UNIT CONNECTOR
4·5·14·15

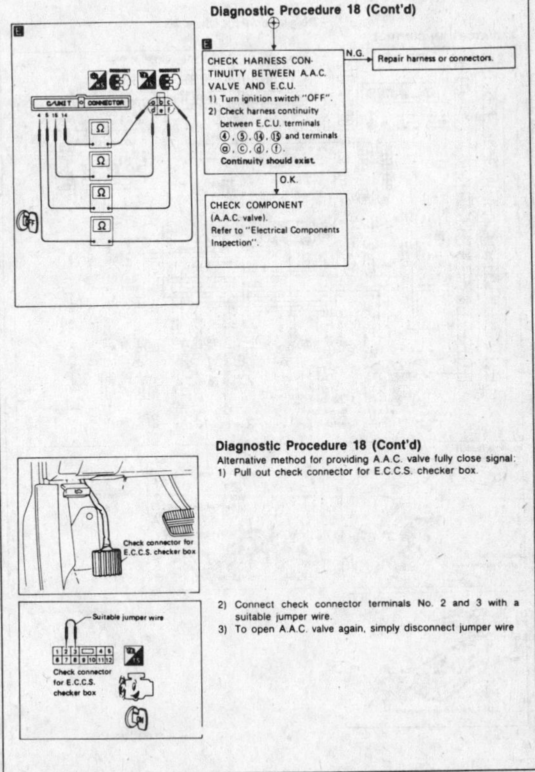
Diagnostic Procedure 18 (Cont'd)
CHECK HARNESS CONTINUITY BETWEEN A.A.C. VALVE AND E.C.U.
1) Turn ignition switch "OFF".
2) Check harness continuity between E.C.U. terminals 4, 5, 14, 15 and terminals a, c, d, f.
Continuity should exist.
N.G.
Repair harness or connectors.
O.K.
CHECK COMPONENT (A.A.C. valve).
Refer to "Electrical Components Inspection".
C/UNIT CONNECTOR
4 5 15 14
Diagnostic Procedure 18 (Cont'd)
Alternative method for providing A.A.C. valve fully close signal:
1) Pull out check connector for E.C.C.S. checker box.
Check connector for E.C.C.S. checker box
Suitable jumper wire
Check connector for E.C.C.S. checker box
2) Connect check connector terminals No. 2 and 3 with a suitable jumper wire.
3) To open A.A.C. valve again, simply disconnect jumper wire

ECCS SELF-DIAGNOSTIC CHART – 1989–90 MAXIMA

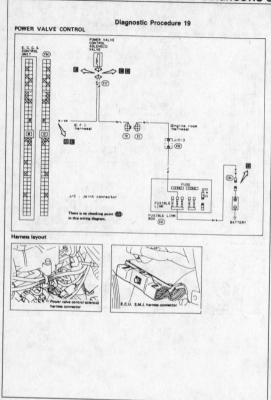

Diagnostic Procedure 19

POWER VALVE CONTROL

E.C.C.S. CONTROL UNIT

POWER VALVE CONTROL SOLENOID VALVE

(E.F.I. harness)

(Engine room harness)

J/C-3

FUSE

FUSIBLE LINK

FUSIBLE LINK BOX

BATTERY

J/C : Joint connector

There is no checking point in this wiring diagram.

Harness layout

Power valve control solenoid harness connector

E.C.U. S.M.J. harness connector

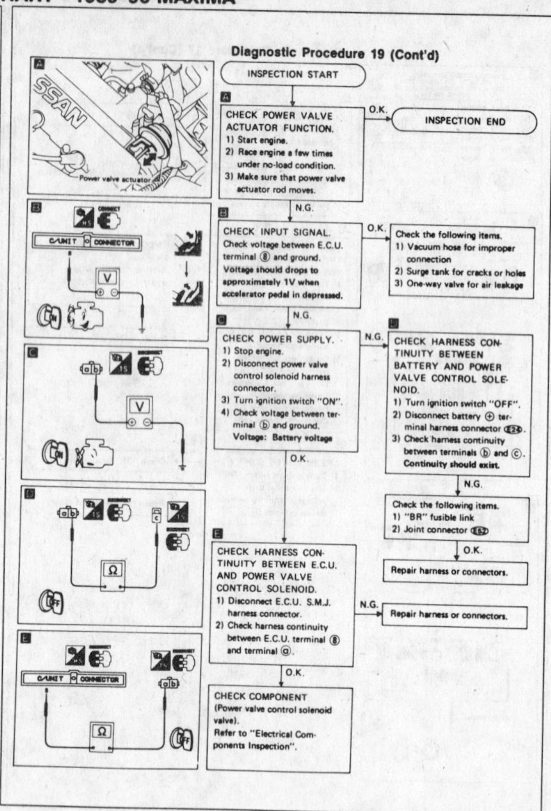
Diagnostic Procedure 19 (Cont'd)
Power valve actuator
INSPECTION START
CHECK POWER VALVE ACTUATOR FUNCTION.
1) Start engine.
2) Race engine a few times under no-load condition.
3) Make sure that power valve actuator rod moves.
O.K.
INSPECTION END
N.G.
CHECK INPUT SIGNAL.
Check voltage between E.C.U. terminal ⑧ and ground.
Voltage should drops to approximately 1V when accelerator pedal in depressed.
O.K.
Check the following items.
1) Vacuum hose for improper connection
2) Surge tank for cracks or holes
3) One-way valve for air leakage
N.G.
CHECK POWER SUPPLY.
1) Stop engine.
2) Disconnect power valve control solenoid harness connector.
3) Turn ignition switch "ON".
4) Check voltage between terminal ⓑ and ground.
Voltage: Battery voltage
N.G.
CHECK HARNESS CONTINUITY BETWEEN BATTERY AND POWER VALVE CONTROL SOLENOID.
1) Turn ignition switch "OFF".
2) Disconnect battery ⊕ terminal harness connector.
3) Check harness continuity between terminals ⓑ and ⓒ.
Continuity should exist.
N.G.
Check the following items.
1) "BR" fusible link
2) Joint connector
O.K.
Repair harness or connectors.
O.K.
CHECK HARNESS CONTINUITY BETWEEN E.C.U. AND POWER VALVE CONTROL SOLENOID.
1) Disconnect E.C.U. S.M.J. harness connector.
2) Check harness continuity between E.C.U. terminal ⑧ and terminal ⓐ.
N.G.
Repair harness or connectors.
O.K.
CHECK COMPONENT
(Power valve control solenoid valve).
Refer to "Electrical Components Inspection".
C/UNIT CONNECTOR

ECCS SELF-DIAGNOSTIC CHART – 1989–90 MAXIMA

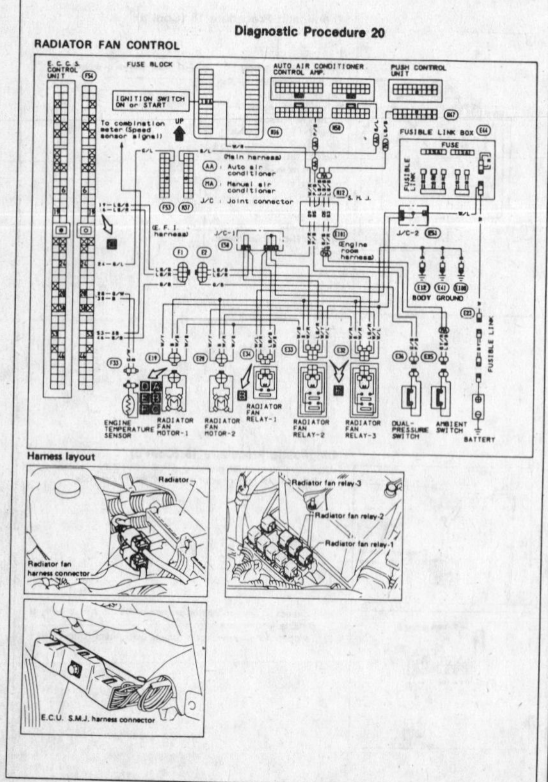
Diagnostic Procedure 20
RADIATOR FAN CONTROL
E.C.C.S. CONTROL UNIT
FUSE BLOCK
IGNITION SWITCH ON or START
To combination meter (Speed sensor signal)
AUTO AIR CONDITIONER CONTROL AMP.
PUSH CONTROL UNIT
FUSIBLE LINK BOX
(Main harness)
(AA) : Auto air conditioner
(MA) : Manual air conditioner
J/C : Joint connector
(E.F.I. harness)
(Engine room harness)
J/C-1
J/C-2
BODY GROUND
ENGINE TEMPERATURE SENSOR
RADIATOR FAN MOTOR-1
RADIATOR FAN MOTOR-2
RADIATOR FAN RELAY-1
RADIATOR FAN RELAY-2
RADIATOR FAN RELAY-3
DUAL-PRESSURE SWITCH
AMBIENT SWITCH
BATTERY
Harness layout
Radiator
Radiator fan harness connector
Radiator fan relay-3
Radiator fan relay-2
Radiator fan relay-1
E.C.U. S.M.J. harness connector

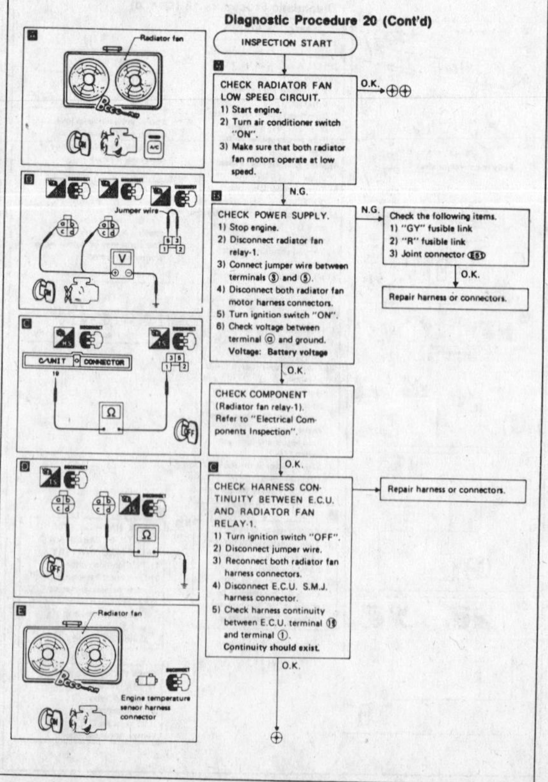
Diagnostic Procedure 20 (Cont'd)
Radiator fan
INSPECTION START
CHECK RADIATOR FAN LOW SPEED CIRCUIT.
1) Start engine.
2) Turn air conditioner switch "ON".
3) Make sure that both radiator fan motors operate at low speed.
O.K.
N.G.
CHECK POWER SUPPLY.
1) Stop engine.
2) Disconnect radiator fan relay-1.
3) Connect jumper wire between terminals ③ and ⑤.
4) Disconnect both radiator fan motor harness connectors.
5) Turn ignition switch "ON".
6) Check voltage between terminal ⓐ and ground.
Voltage: Battery voltage
Jumper wire
N.G.
Check the following items.
1) "GY" fusible link
2) "R" fusible link
3) Joint connector
O.K.
Repair harness or connectors.
O.K.
CHECK COMPONENT
(Radiator fan relay-1).
Refer to "Electrical Components Inspection".
O.K.
CHECK HARNESS CONTINUITY BETWEEN E.C.U. AND RADIATOR FAN RELAY-1.
1) Turn ignition switch "OFF".
2) Disconnect jumper wire.
3) Reconnect both radiator fan harness connectors.
4) Disconnect E.C.U. S.M.J. harness connector.
5) Check harness continuity between E.C.U. terminal ⑪ and terminal ①.
Continuity should exist.
Repair harness or connectors.
O.K.
C/UNIT CONNECTOR
Engine temperature sensor harness connector

ECCS SELF-DIAGNOSTIC CHART—1989-90 MAXIMA

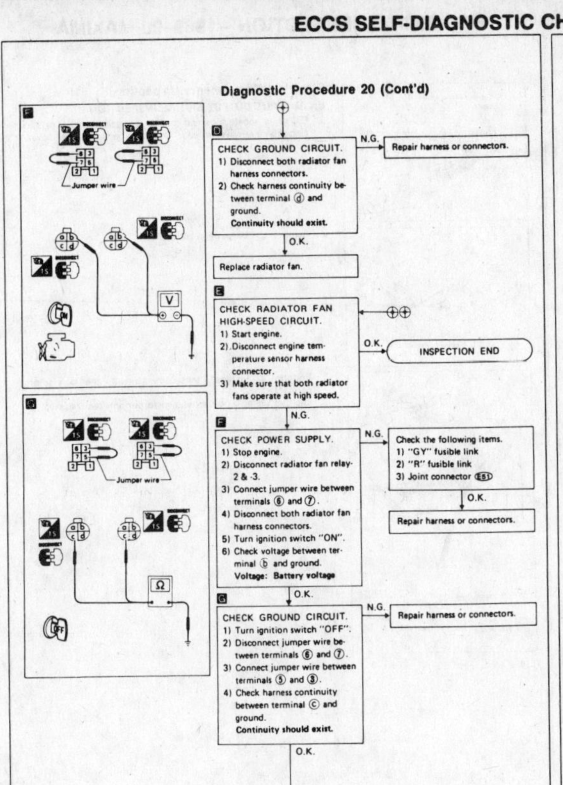
Diagnostic Procedure 20 (Cont'd)
Jumper wire
CHECK GROUND CIRCUIT.
1) Disconnect both radiator fan harness connectors.
2) Check harness continuity between terminal ⓓ and ground.
Continuity should exist.
N.G.
Repair harness or connectors.
O.K.
Replace radiator fan.
CHECK RADIATOR FAN HIGH-SPEED CIRCUIT.
1) Start engine.
2) Disconnect engine temperature sensor harness connector.
3) Make sure that both radiator fans operate at high speed.
O.K.
INSPECTION END
N.G.
CHECK POWER SUPPLY.
1) Stop engine.
2) Disconnect radiator fan relay-2 & -3.
3) Connect jumper wire between terminals ⑥ and ⑦.
4) Disconnect both radiator fan harness connectors.
5) Turn ignition switch "ON".
6) Check voltage between terminal ⓑ and ground.
Voltage: Battery voltage
N.G.
Check the following items.
1) "GY" fusible link
2) "R" fusible link
3) Joint connector
O.K.
Repair harness or connectors.
O.K.
CHECK GROUND CIRCUIT.
1) Turn ignition switch "OFF".
2) Disconnect jumper wire between terminals ⑥ and ⑦.
3) Connect jumper wire between terminals ⑤ and ⑧.
4) Check harness continuity between terminal ⓒ and ground.
Continuity should exist.
N.G.
Repair harness or connectors.
O.K.

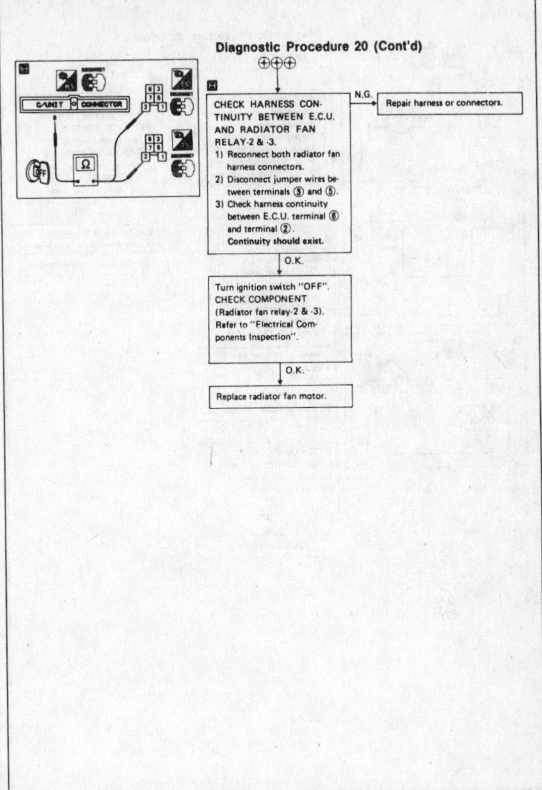
Diagnostic Procedure 20 (Cont'd)
C.M.I.T CONNECTOR
CHECK HARNESS CONTINUITY BETWEEN E.C.U. AND RADIATOR FAN RELAY-2 & -3.
1) Reconnect both radiator fan harness connectors.
2) Disconnect jumper wires between terminals ⑤ and ⑧.
3) Check harness continuity between E.C.U. terminal ⑥ and terminal ②.
Continuity should exist.
N.G.
Repair harness or connectors.
O.K.
Turn ignition switch "OFF".
CHECK COMPONENT
(Radiator fan relay-2 & -3).
Refer to "Electrical Components Inspection".
O.K.
Replace radiator fan motor.

ECCS SELF-DIAGNOSTIC CHART—1989-90 MAXIMA

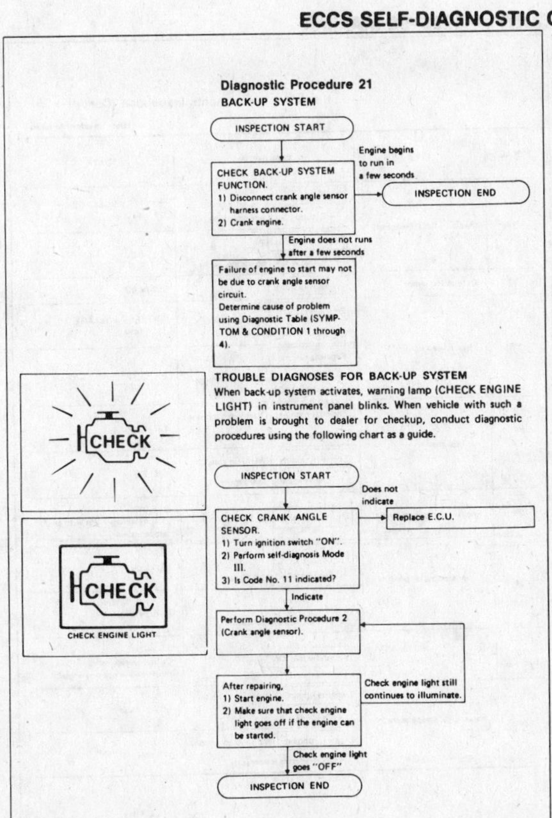
Diagnostic Procedure 21
BACK-UP SYSTEM
INSPECTION START
CHECK BACK-UP SYSTEM FUNCTION.
1) Disconnect crank angle sensor harness connector.
2) Crank engine.
Engine begins to run in a few seconds
INSPECTION END
Engine does not runs after a few seconds
Failure of engine to start may not be due to crank angle sensor circuit.
Determine cause of problem using Diagnostic Table (SYMPTOM & CONDITION 1 through 4).
TROUBLE DIAGNOSES FOR BACK-UP SYSTEM
When back-up system activates, warning lamp (CHECK ENGINE LIGHT) in instrument panel blinks. When vehicle with such a problem is brought to dealer for checkup, conduct diagnostic procedures using the following chart as a guide.
CHECK
CHECK
CHECK ENGINE LIGHT
INSPECTION START
CHECK CRANK ANGLE SENSOR.
1) Turn ignition switch "ON".
2) Perform self-diagnosis Mode III.
3) Is Code No. 11 indicated?
Does not indicate
Replace E.C.U.
Indicate
Perform Diagnostic Procedure 2 (Crank angle sensor).
After repairing,
1) Start engine.
2) Make sure that check engine light goes off if the engine can be started.
Check engine light still continues to illuminate.
Check engine light goes "OFF"
INSPECTION END

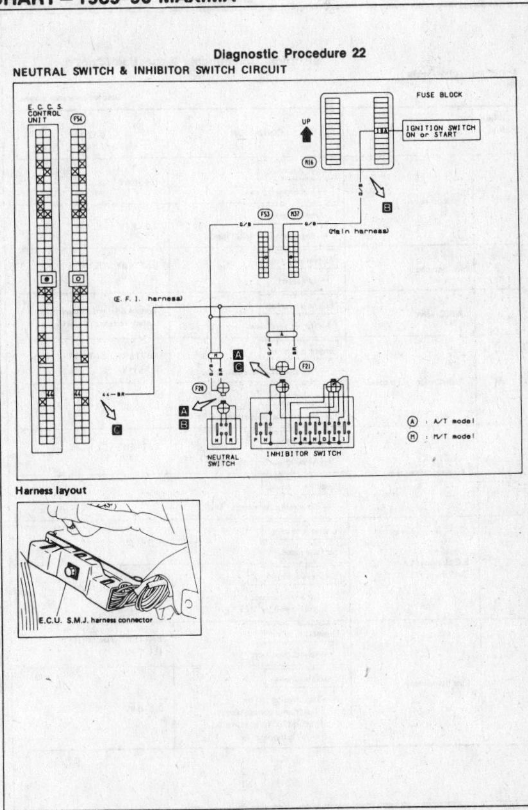
Diagnostic Procedure 22
NEUTRAL SWITCH & INHIBITOR SWITCH CIRCUIT
E.C.C.S. CONTROL UNIT
FUSE BLOCK
UP
IGNITION SWITCH ON or START
(Main harness)
(E.F.I. harness)
NEUTRAL SWITCH
INHIBITOR SWITCH
Ⓐ : A/T model
Ⓜ : M/T model
Harness layout
E.C.U. S.M.J. harness connector

ECCS SELF-DIAGNOSTIC CHART—1989-90 MAXIMA

ECU SIGNAL INPSECTION—1989-90 MAXIMA

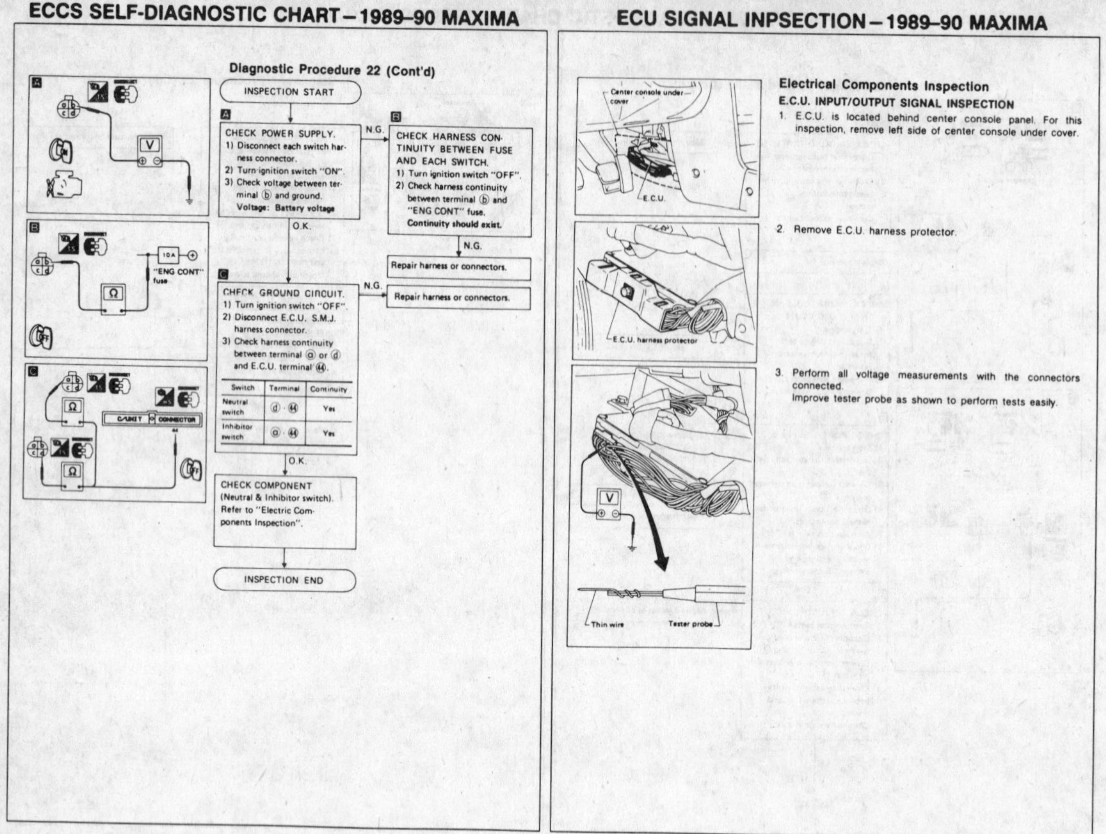
Diagnostic Procedure 22 (Cont'd)
INSPECTION START
A CHECK POWER SUPPLY.
1) Disconnect each switch harness connector.
2) Turn ignition switch "ON".
3) Check voltage between terminal ⓑ and ground.
Voltage: Battery voltage
N.G.
B CHECK HARNESS CONTINUITY BETWEEN FUSE AND EACH SWITCH.
1) Turn ignition switch "OFF".
2) Check harness continuity between terminal ⓑ and "ENG CONT" fuse.
Continuity should exist.
N.G.
Repair harness or connectors.
O.K.
C CHECK GROUND CIRCUIT.
1) Turn ignition switch "OFF".
2) Disconnect E.C.U. S.M.J. harness connector.
3) Check harness continuity between terminal ⓐ or ⓓ and E.C.U. terminal ㊹.
Switch | Terminal | Continuity
Neutral switch | ⓓ - ㊹ | Yes
Inhibitor switch | ⓐ - ㊹ | Yes
N.G.
Repair harness or connectors.
O.K.
CHECK COMPONENT (Neutral & Inhibitor switch).
Refer to "Electric Components Inspection".
INSPECTION END
10A
"ENG CONT" fuse
V
Ω
Electrical Components Inspection
E.C.U. INPUT/OUTPUT SIGNAL INSPECTION
1. E.C.U. is located behind center console panel. For this inspection, remove left side of center console under cover.
Center console under cover
E.C.U.
2. Remove E.C.U. harness protector.
E.C.U. harness protector
3. Perform all voltage measurements with the connectors connected.
Improve tester probe as shown to perform tests easily.
Thin wire
Tester probe

ECU SIGNAL INSPECTION (CONT.)—1989-90 MAXIMA

Electrical Components Inspection (Cont'd)

E.C.U. inspection table

*Data are reference values.

TERMINAL NO.	ITEM	CONDITION	*DATA
1	Ignition signal	Engine is running. — Idle speed	0.4 - 0.6V
		Engine is running. — Engine speed is 4,000 rpm.	1.9 - 2.1V
2	Ignition check	Engine is running. — Idle speed	BATTERY VOLTAGE (11 - 14V)
4, 5, 14, 15	A.A.C. valve	Engine is running. — Racing condition	Voltage briefly decreases from battery voltage (11 - 14V).
6	Radiator fan (High speed)	Engine is running. — Radiator fan is not operating.	BATTERY VOLTAGE (11 - 14V)
		Engine is running. — Radiator fan is operating.	0.7 - 0.8V
9	Air conditioner relay	Engine is running. — Air conditioner switch "OFF"	BATTERY VOLTAGE (11 - 14V)
		Engine is running. — Air conditioner switch "ON"	0.7 - 0.8V
16	E.C.U. power source (Self-shutoff)	Engine is running. — Idle speed	0.8 - 1.0V
		Engine is not running. — For a few seconds after turning ignition switch "OFF"	BATTERY VOLTAGE (11 - 14V)
18	Fuel pump relay	Engine is running. — Normal condition	BATTERY VOLTAGE (11 - 14V)
		Engine is running. — Abnormal condition [Fuel pump voltage control circuit (E.C.U. terminal No. ⑩④) is inoperative.]	0.7 - 0.8V

Electrical Components Inspection (Cont'd)

*Data are reference values.

TERMINAL NO.	ITEM	CONDITION	*DATA
19	Radiator fan (Low speed)	Engine is running. — Radiator fan is not operating.	BATTERY VOLTAGE (11 - 14V)
		Engine is running. — Radiator fan is operating.	0.7 - 0.8V
23	Detonation sensor	Engine is running.	3.0 - 4.0V
25	Electric load signal	Engine is running. — Electric load signal is "OFF".	BATTERY VOLTAGE (11 - 14V)
		Engine is running. — Electric load signal (Rear defogger switch) is "ON".	0.1 - 0.3V
27	Air flow meter	Engine is running. (Warm-up condition) — Idle speed	1.0 - 1.3V
		Engine is running. (Warm-up condition) — Engine speed is 3,000 rpm.	1.8 - 2.0V
28	Engine temperature sensor	Engine is running.	0 - 5.0V Output voltage varies with engine temperature.
34	Power steering oil pressure switch	Engine is running. — Steering wheel stays straight.	8.0 - 9.0V
		Engine is running. — Steering wheel is turned.	0 - 0.2V
38	Throttle sensor	Ignition switch "ON"	0.5 - 4.2V Output voltage varies with throttle valve opening angle.
39	Exhaust gas temperature sensor	Engine is running. (Warm-up condition) — Idle speed	1.0V or more
		Engine is running. (Warm-up condition) — E.G.R. system is operating.	0 - 1.0V

ECU SIGNAL INSPECTION (CONT.) – 1989–90 MAXIMA

Electrical Components Inspection (Cont'd)

*Data are reference values.

TERMINAL NO.	ITEM	CONDITION	*DATA
41 51	Crank angle sensor (Reference signal)	Engine is running. Do not run engine at high speed under no-load.	0.2 - 0.4V Output voltage slightly varies with engine speed.
42 52	Crank angle sensor (Position signal)	Engine is running. Do not run engine at high speed under no-load.	2.5 - 2.7V Output voltage slightly varies with engine speed.
43	Start signal	Ignition switch "ON"	BATTERY VOLTAGE (11 - 14V)
		Ignition switch "START"	0.7 - 0.8V
44	Neutral switch (M/T model) Inhibitor switch (A/T model)	Ignition switch "ON" – Gear position is "Neutral" (M/T model). – Gear position is "N" or "P" (A/T model).	BATTERY VOLTAGE (11 - 14V)
		Ignition switch "ON" – Except the above conditions	0V
45	Ignition switch	Ignition switch "ON" – Engine stopped	BATTERY VOLTAGE (11 - 14V)
46	Air conditioner switch	Engine is running. – Air conditioner switch "OFF"	8.0 - 9.0V
		Engine is running. – Air conditioner switch "ON"	0.5 - 0.7V
48	Power source for sensors	Ignition switch "ON" – Engine stopped	Approximately 5.0V
49	Battery source	Ignition switch "ON" – Engine stopped	BATTERY VOLTAGE (11 - 14V)
53	Vehicle speed sensor	Ignition switch "ON" – Engine stopped – While rotating front wheel by hand	0 or 7.0 - 9.0V
54	Idle switch	Ignition switch "ON" – Accelerator pedal is fully released (engine stopped).	9.0 - 10.0V
		Ignition switch "ON" – Accelerator pedal is depressed (engine stopped).	0V

Electrical Components Inspection (Cont'd)

*Data are reference values.

TERMINAL NO.	ITEM	CONDITION	*DATA
57	Power source for idle switch	Ignition switch "ON" – Engine stopped	8.0 - 9.0V
59	Power supply	Ignition switch "ON" – Engine speed	BATTERY VOLTAGE (11 - 14V)
101 103 105 109 110 112 114	Injectors	Ignition switch "OFF"	BATTERY VOLTAGE (11 - 14V)
102	E.G.R. control solenoid valve	Engine is running. (Warm-up condition) – Idle speed – Engine speed is approximately 3,400 rpm or more.	0.7 - 0.8V
		Engine is running. (Warm-up condition) – Engine speed is between idle and approximately 3,400 rpm.	BATTERY VOLTAGE (11 - 14V)
104	Fuel pump voltage control	Ignition switch "ON" – Engine stopped	BATTERY VOLTAGE (11 - 14V)
		Engine is running. – For 30 seconds after engine begins to run.	4.0 - 4.5V
		Engine is running. – Racing (up to 4,000 rpm)	2.0 - 4.5V
113	Power valve control solenoid valve	Engine is running. – Idle speed	BATTERY VOLTAGE (11 - 14V)
		Engine is running. – Racing (up to 4,000 rpm)	0.7 - 0.8V

E.C.U. S.M.J. HARNESS CONNECTOR TERMINAL LAYOUT

EECS COMPONENTS LOCATION – 1988–89 STANZA

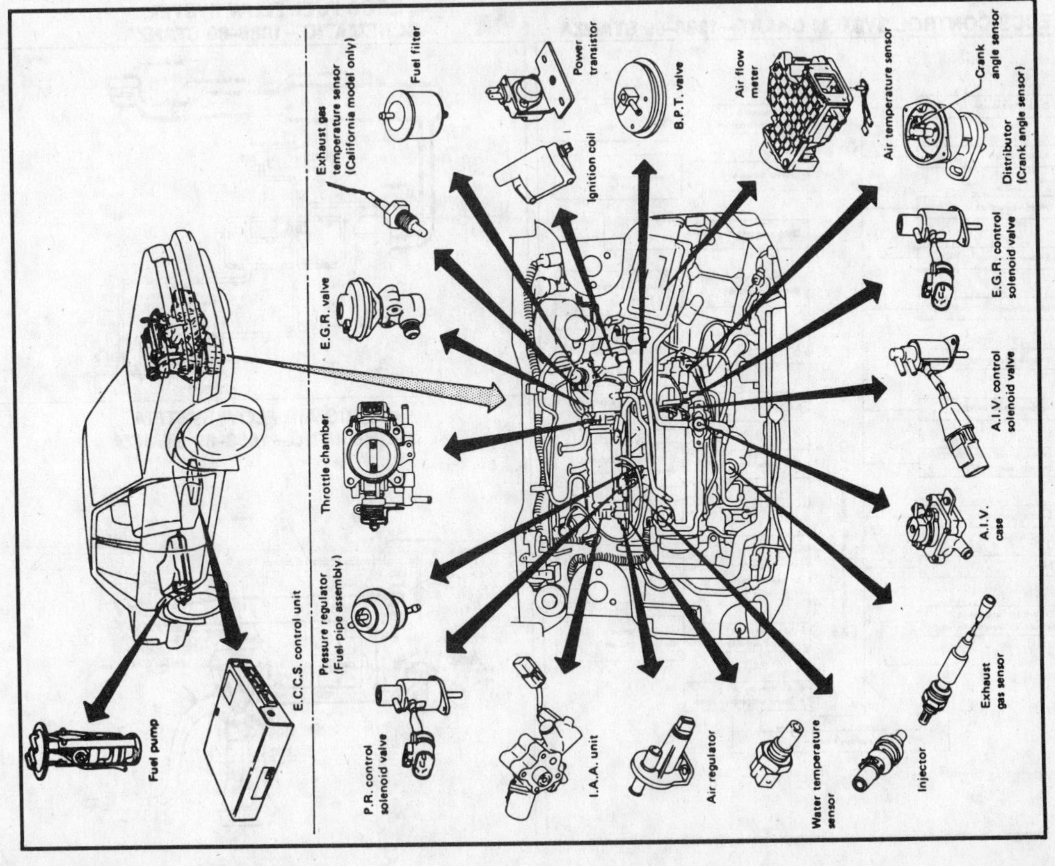
Exhaust gas temperature sensor (California model only)
Fuel filter
Power transistor
Ignition coil
B.P.T. valve
Air flow meter
Air temperature sensor
Crank angle sensor
Distributor (Crank angle sensor)
E.G.R. control solenoid valve
E.G.R. valve
A.I.V. control solenoid valve
Throttle chamber
A.I.V. case
Pressure regulator (Fuel pipe assembly)
E.C.C.S. control unit
Exhaust gas sensor
Fuel pump
P.R. control solenoid valve
I.A.A. unit
Air regulator
Water temperature sensor
Injector

ECCS SCHEMATIC—1988–89 STANZA

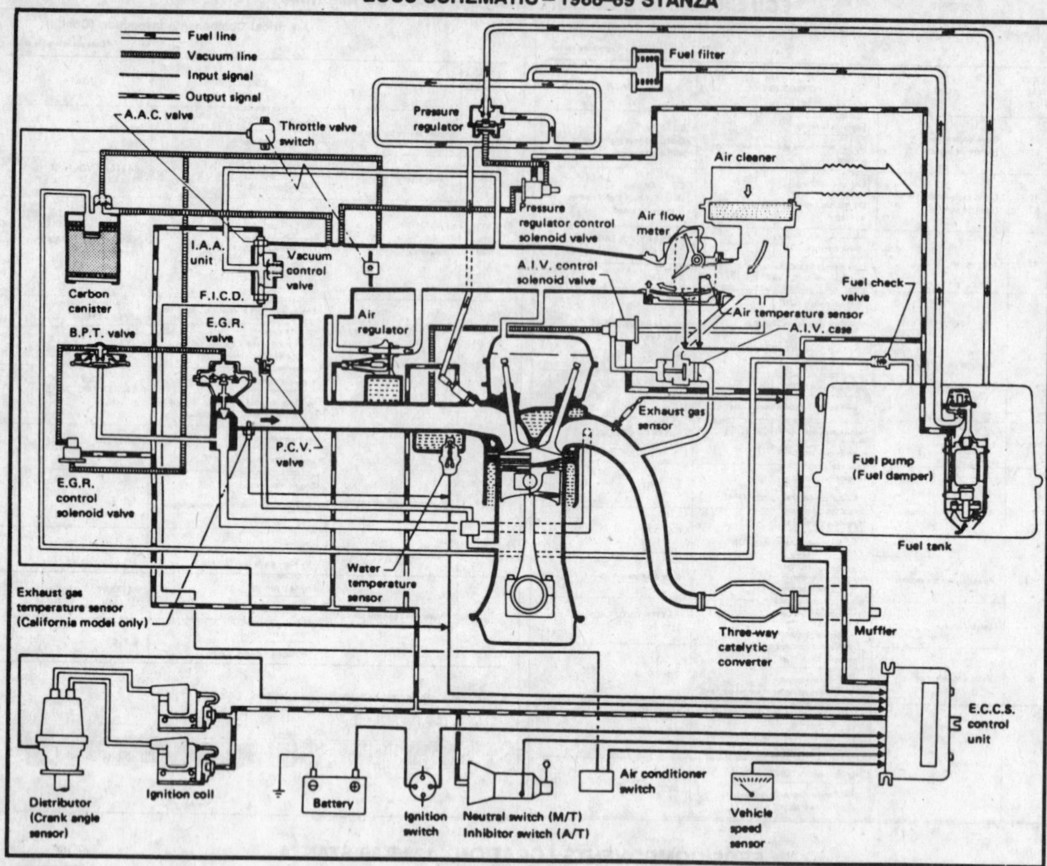
Fuel line
Vacuum line
Input signal
Output signal
A.A.C. valve
Throttle valve switch
Pressure regulator
Fuel filter
Air cleaner
Pressure regulator control solenoid valve
Air flow meter
I.A.A. unit
Vacuum control valve
Carbon canister
F.I.C.D.
A.I.V. control solenoid valve
Fuel check valve
E.G.R. valve
Air regulator
Air temperature sensor
A.I.V. case
B.P.T. valve
Exhaust gas sensor
P.C.V. valve
Fuel pump (Fuel damper)
E.G.R. control solenoid valve
Fuel tank
Water temperature sensor
Exhaust gas temperature sensor (California model only)
Three-way catalytic converter
Muffler
E.C.C.S. control unit
Air conditioner switch
Distributor (Crank angle sensor)
Ignition coil
Battery
Ignition switch
Neutral switch (M/T) Inhibitor switch (A/T)
Vehicle speed sensor

ECCS CONTROL SYSTEM CHART—1988–89 STANZA

Input		Control	Output
Crank angle sensor	E.C.C.S. control unit	Fuel injection & mixture ratio control	Injectors
Air flow meter		Ignition timing control	Power transistor
Water temperature sensor		Spark plug switching control	
Exhaust gas sensor		Idle speed control	Auxiliary air control (A.A.C.) valve
Ignition switch		A.I.V. control	A.I.V. control solenoid valve
Throttle valve switch		E.G.R. control	E.G.R. control solenoid valve
Air conditioner switch		Fuel pressure control	Fuel pressure regulator control solenoid valve
Clutch-Neutral switch (M/T) Inhibitor switch (A/T)		Fuel pump control	Fuel pump
Vehicle speed sensor	Barometric pressure sensor (Installed in E.C.U.) (Except for California)	Air regulator control	Air regulator
Battery		Exhaust gas sensor monitor & self-diagnosis	Inspection lamps (On the control unit)
Exhaust gas temperature sensor (California model only)			

Fail-safe function
• Air flow meter
• Water temperature sensor

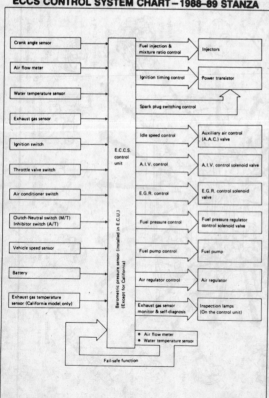

ECCS FUEL FLOW SYSTEM SCHEMATIC—1988–89 STANZA

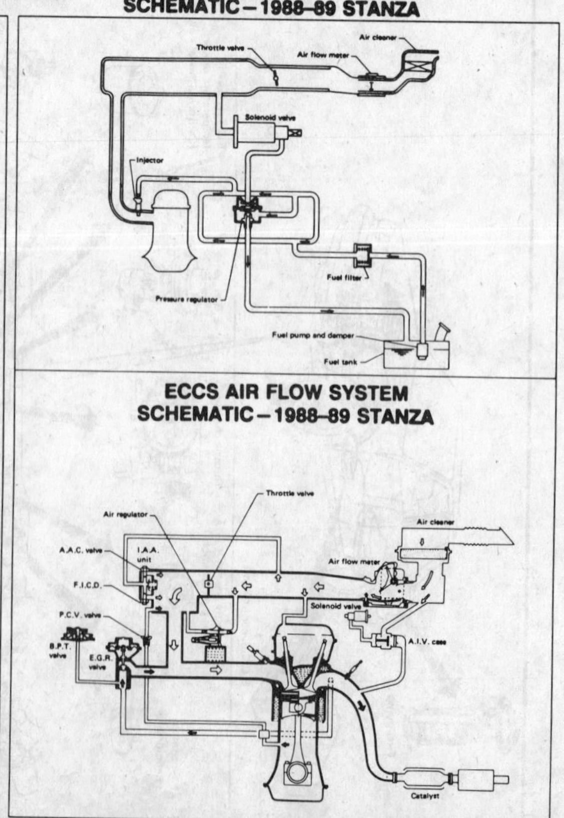
Throttle valve
Air cleaner
Air flow meter
Solenoid valve
Injector
Fuel filter
Pressure regulator
Fuel pump and damper
Fuel tank
ECCS AIR FLOW SYSTEM SCHEMATIC—1988–89 STANZA
Throttle valve
Air regulator
Air cleaner
A.A.C. valve
I.A.A. unit
Air flow meter
F.I.C.D.
Solenoid valve
P.C.V. valve
B.P.T. valve
E.G.R. valve
A.I.V. case
Catalyst

ECCS WIRING SCHEMATIC–1988–89 STANZA

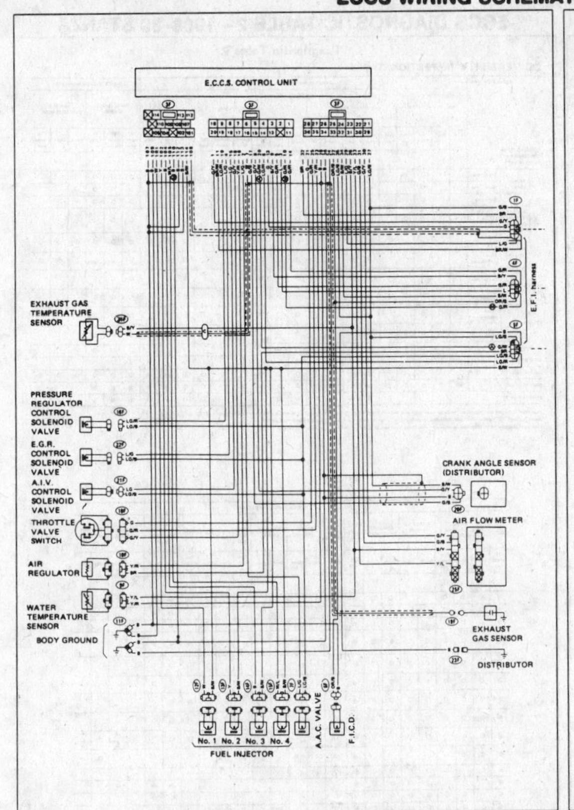

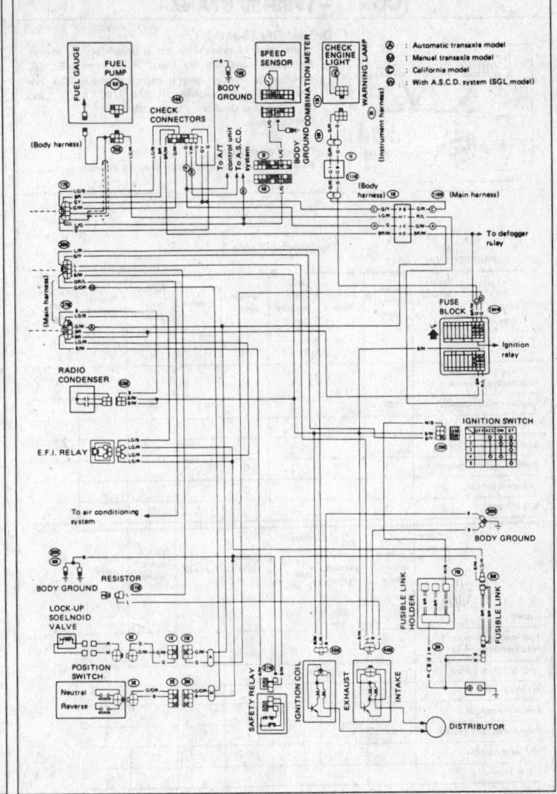

ECCS CONTROL UNIT WIRING SCHEMATIC–1988–89 STANZA

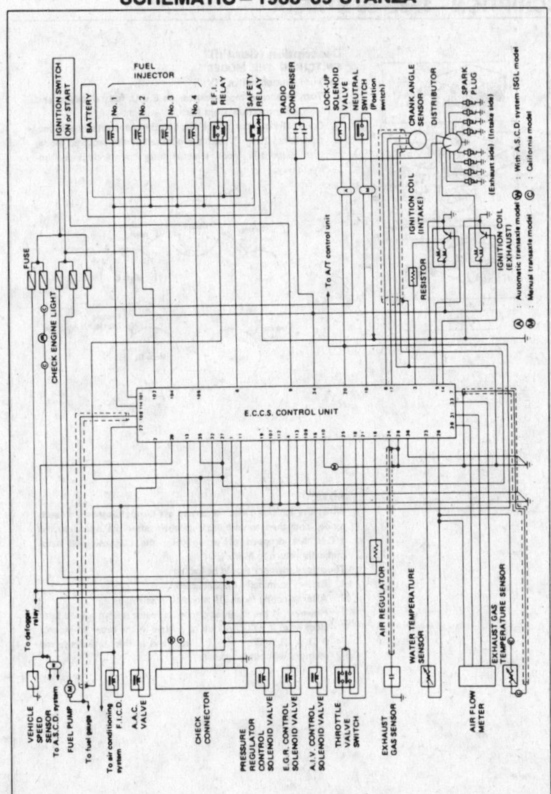

ECCS DIAGNOSTIC PROCEDURE–1988–89 STANZA

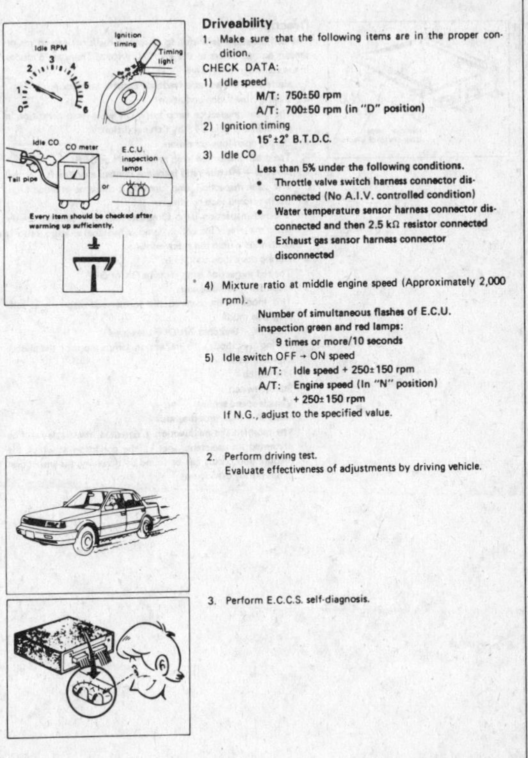

Driveability

1. Make sure that the following items are in the proper condition.

CHECK DATA:

1) Idle speed
   M/T: 750±50 rpm
   A/T: 700±50 rpm (in "D" position)
2) Ignition timing
   15°±2° B.T.D.C.
3) Idle CO
   Less than 5% under the following conditions.
   - Throttle valve switch harness connector disconnected (No A.I.V. controlled condition)
   - Water temperature sensor harness connector disconnected and then 2.5 kΩ resistor connected
   - Exhaust gas sensor harness connector disconnected
4) Mixture ratio at middle engine speed (Approximately 2,000 rpm).
   Number of simultaneous flashes of E.C.U. inspection green and red lamps:
   9 times or more/10 seconds
5) Idle switch OFF → ON speed
   M/T: Idle speed + 250±150 rpm
   A/T: Engine speed (In "N" position) + 250±150 rpm

If N.G., adjust to the specified value.

2. Perform driving test.
Evaluate effectiveness of adjustments by driving vehicle.

3. Perform E.C.C.S. self-diagnosis.

ECCS DIAGNOSTIC PROCEDURE (CONT.)—1988–89 STANZA

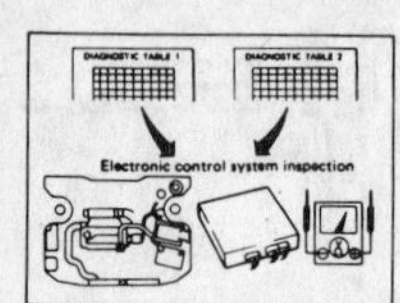
DIAGNOSTIC TABLE 1
DIAGNOSTIC TABLE 2
Electronic control system inspection

Driveability (Cont'd)

4. If the result of driveability test is unsatisfactory, or malfunctioning conditions are found in performing E.C.C.S. self-diagnosis, perform general inspection and E.C.C.S. system inspection by following DIAGNOSTIC TABLE 1 and 2 in response to driveability trouble items.
   If N.G., repair.
5. Perform driving test.
   Re-evaluate vehicle performance after the inspection.

Diagnostic Table 1

SYSTEM INSPECTION TABLE

System \ Sensor & actuator	Crank angle sensor	Air flow meter	Water temperature sensor	Ignition switch	Injector	Throttle valve switch	Neutral switch	Exhaust gas sensor
Fuel injection & mixture ratio feedback control	○	○	○	○	○	○	○	○
Ignition timing control	○	○	○	○		○		
A.I.V. control	○		○			○		
Fuel pump control	○			○				
Pressure regulator control	○		○	○	○			
Idle speed control	○		○	○		○	○	
E.G.R. control	○		○	○		○		

System \ Sensor & actuator	A.I.V. control solenoid valve	E.G.R. control solenoid valve	Pressure regulator control solenoid valve	A.A.C. valve	Air regulator	Vehicle speed sensor	E.F.I. relay
Fuel injection & mixture ratio feedback control							○
Ignition timing control							○
A.I.V. control	○						○
Fuel pump control							○
Pressure regulator control			○				○
Idle speed control				○	○	○	○
E.G.R. control		○					○

This table indicates the inspection items for the E.C.C.S. control system. For each system, it is necessary to check sensors or actuators marked "○".

ECCS DIAGNOSTIC TABLE 2—1988–89 STANZA

Diagnostic Table 2

DRIVEABILITY INSPECTION TABLE

GENERAL INSPECTION

Trouble items concerned with driveability \ Inspection item			Fuel flow system: Fuel level	Fuel pump	Fuel filter	Fuel line	Electric system: Battery	Spark plug	Ignition wire	Alternator	Starter	Air flow system: Air cleaner	Air flow line	E.G.R. B.P.T. valve	A.I.V. control valve	F.I.C.D. solenoid valve	Air regulator	F.C.V.	A.A.C. valve: Short	Open
SURGE	ROAD/LOAD DRIVING	Heavy load			○	○		●	○			○	○	○				○		
		Middle load			○	○		●	○			○	○	●				○		
		Light load			○	○		●	○			○	○	●				○		
	ACCELERATING DRIVING	Slow acceleration			○	○		●	○			○	○	●				○		
	DECELERATING DRIVING	Rapid deceleration						●	○			○	○	○				○	○	○
		Slow deceleration						●	○			○	○	○				○	○	○
HESITATION		Rapid acceleration			○	○		●	○			○	○	○				○		
		Slow acceleration			○	○		●	○			○	○	○				○		
STUMBLE		Rapid acceleration			○	○		●	○			○	○	○				○		
		Slow acceleration			○	○		●	○			○	○	○				○		
BACKFIRE					○	○			○				○					○		
AFTER FIRE								●	○			○			○			○	○	○
IDLE STABILITY								●	○	○		○	○	●		○	○	○	●	●
ENGINE STALL			○	○	○	○	○	●	○	○		○	○	○		○	○	○	●	●
STARTABILITY			○	○			○	●	○	○	○	○	○				○	○	○	○

E.C.C.S. SYSTEM INSPECTION

Trouble items concerned with driveability			Crank angle sensor: 180° signal noise	180° signal faults	1° signal faults	Vehicle speed sensor: Short	Open	Poor connection	Air flow meter: Poor connection	Water temperature sensor: Short	Open	Poor connection	Throttle valve switch: Short	Open	Exhaust gas sensor: Short	Open	Injector: Clogging
SURGE	ROAD/LOAD DRIVING	Heavy load	○	○	○				●	○	○	○			○	○	●
		Middle load	○	○	○				●	○	○	○			○	○	●
		Light load	○	○	○				●	○	○	○			○	○	●
	ACCELERATING DRIVING	Slow acceleration	○	○	○				●	○	○	○	○	○	○	○	●
	DECELERATING DRIVING	Rapid deceleration	○	○	○	○	○	○	●	○	○	○	○	○	○	○	●
		Slow deceleration	○	○	○	○	○	○	●	○	○	○	○	○	○	○	●
HESITATION		Rapid acceleration	○	○	○				●	○	○	○	○	○	○	○	●
		Slow acceleration	○	○	○				●	○	○	○	○	○	○	○	●
STUMBLE		Rapid acceleration	○	○	○				●	○	○	○	○	○			●
		Slow acceleration	○	○	○				●	○	○	○	○	○			●
BACKFIRE			○	○	○				●	○		○	○	○	○	○	●
AFTER FIRE			○	○	○				●		○	○	○	○	○	○	●
IDLE STABILITY			○	○	○	○	○	○	○	○	○	○	○	○	○	○	●
ENGINE STALL			●	●	●				●	○	○	○	○	○	○	○	●
STARTABILITY			●	●	●				○	○	○	○					●

E.C.C.S. SYSTEM INSPECTION (continued)

Trouble items concerned with driveability			Neutral switch: Short	Start signal: Short	Open	Ignition signal: Short	Open	Battery voltage: Low voltage	Fuel pump circuit: Short	Open	Poor connection	E.G.R. control solenoid valve: Short	Open	A.I.V. control solenoid valve: Short	Open	P.R. control solenoid valve: Short	Open
SURGE	ROAD/LOAD DRIVING	Heavy load									○	●	●				
		Middle load									○	●	●				
		Light load									○	●	●				
	ACCELERATING DRIVING	Slow acceleration									○	●	●				
	DECELERATING DRIVING	Rapid deceleration									○	○	○				
		Slow deceleration									○	○	○				
HESITATION		Rapid acceleration	○								○	○	○				
		Slow acceleration	○								○	○	○				
STUMBLE		Rapid acceleration	○								○	○	○			○	○
		Slow acceleration	○								○	○	○			○	○
BACKFIRE			○								○						
AFTER FIRE												●	●	○	○		
IDLE STABILITY			○								○	○	○				
ENGINE STALL			○			○	○	○	○	○	○					○	○
STARTABILITY			○	○	○	○	○	○	○	○	○					○	○

This table indicates the inspection items for each type of symptom. It is necessary for each symptom to check sensors or actuators marked "●" or "○".
Items marked "●" have a significant influence on driveability. Prior to items marked "○", check items marked "●".
Improper mixture ratio, improper ignition condition, and an excess of E.G.R. volume can cause any symptom.

ECCS SELF-DIAGNOSTIC DESCRIPTION—1988–89 STANZA

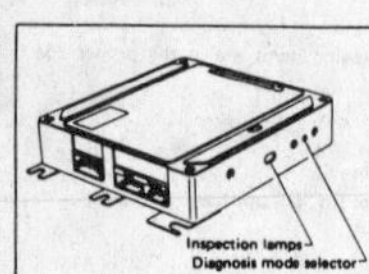
Inspection lamps
Diagnosis mode selector

Description

The self-diagnosis is useful to diagnose malfunctions in major sensors and actuators of the E.C.C.S. system. There are 5 modes in the self-diagnosis system.

1. **Mode I – Mixture ratio feedback control monitor A**
   - During closed loop condition:
     The green inspection lamp turns ON when lean condition is detected and goes OFF by rich condition.
   - During open loop condition:
     The green inspection lamp remains ON or OFF.
2. **Mode II – Mixture ratio feedback control monitor B**
   The green inspection lamp function is the same as Mode I.
   - During closed loop condition:
     The red inspection lamp turns ON and OFF simultaneously with the green inspection lamp when the mixture ratio is controlled within the specified value.
   - During open loop condition:
     The red inspection lamp remains ON or OFF.
3. **Mode III – Self-diagnosis**
   This mode is the same as the former self-diagnosis in self-diagnosis mode.
4. **Mode IV – Switches ON/OFF diagnosis**
   During this mode, the inspection lamps monitor the switch ON-OFF condition.
   - Idle switch
   - Starter switch
   - Vehicle speed sensor
5. **Mode V – Real time diagnosis**
   The moment the malfunction is detected, the display will be presented immediately. That is, the condition at which the malfunction occurs can be found by observing the inspection lamps during driving test.

Flashing N times
Mode N

Description (Cont'd)

SWITCHING THE MODES

1. Turn ignition switch "ON".
2. Turn diagnostic mode selector on E.C.U. fully clockwise and wait the inspection lamps flash.
3. Count the number of the flashing time, and after the inspection lamps have flashed the number of the required mode, turn diagnostic mode selector fully counterclockwise immediately.

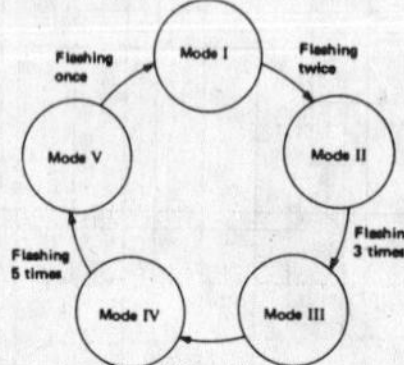
Mode I
Flashing once
Flashing twice
Mode II
Flashing 3 times
Mode III
Flashing 4 times
Mode IV
Flashing 5 times
Mode V

NOTE:
When the ignition switch is turned off during diagnosis, in each mode, and then turned back on again after the power to the E.C.U. has dropped off completely, the diagnosis will automatically return to Mode I.
The stored memory would be lost if:
1. Battery terminal is disconnected.
2. After selecting Mode III, Mode IV is selected.
   However, if the diagnostic mode selector is kept turned fully clockwise, it will continue to change in the order of Mode I → II → III → IV → V → I ... etc., and in this state the stored memory will not be erased.

ECCS SELF-DIAGNOSTIC DESCRIPTION (CONT.) – 1988–89 STANZA

CHECK

CHECK ENGINE LIGHT

Description (Cont'd)

CHECK ENGINE LIGHT (For California only)

This vehicle has a check engine light on instrument panel. This light comes ON under the following conditions:

1) When ignition switch is turned "ON" (for bulb check).
2) When systems related to emission performance malfunction in Mode I (with engine running).

- **This check engine light always illuminates and is synchronous with red L.E.D.**
- **Malfunction systems related to emission performance can be detected by self-diagnosis, and they are clarified as self-diagnostic codes in Mode III.**

Code No.	Malfunction
12	Air flow meter circuit
14	Vehicle speed sensor circuit
23	Idle switch circuit
24	Full switch circuit
31	Engine control unit
32	E.G.R. function
33	Exhaust gas sensor circuit
45	Injector leak

Use the following diagnostic flowchart to check and repair a malfunctioning system.

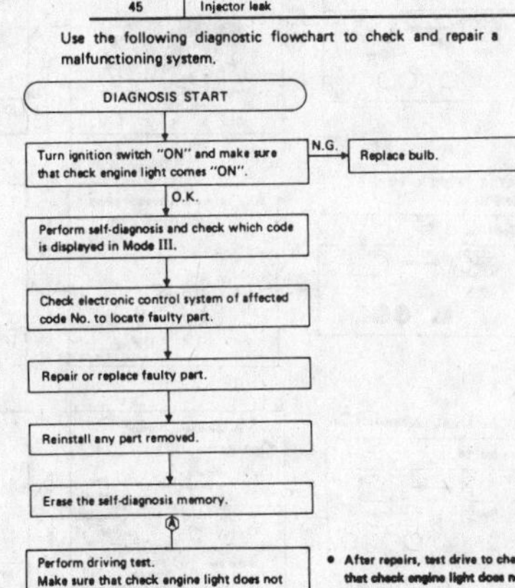
DIAGNOSIS START
Turn ignition switch "ON" and make sure that check engine light comes "ON".
N.G.
Replace bulb.
O.K.
Perform self-diagnosis and check which code is displayed in Mode III.
Check electronic control system of affected code No. to locate faulty part.
Repair or replace faulty part.
Reinstall any part removed.
Erase the self-diagnosis memory.
A
Perform driving test.
Make sure that check engine light does not come "ON" during this test.
DIAGNOSIS END

- **After repairs, test drive to check that check engine light does not come on.**
- **Test driving modes differ with systems. Read the manual before test driving.**

ECCS SELF-DIAGNOSTIC MODES I AND II – 1988–89 STANZA

Modes I & II — Mixture Ratio Feedback Control Monitors A & B

In these modes, the control unit provides the Air-fuel ratio monitor presentation and the Air-fuel ratio feedback coefficient monitor presentation.

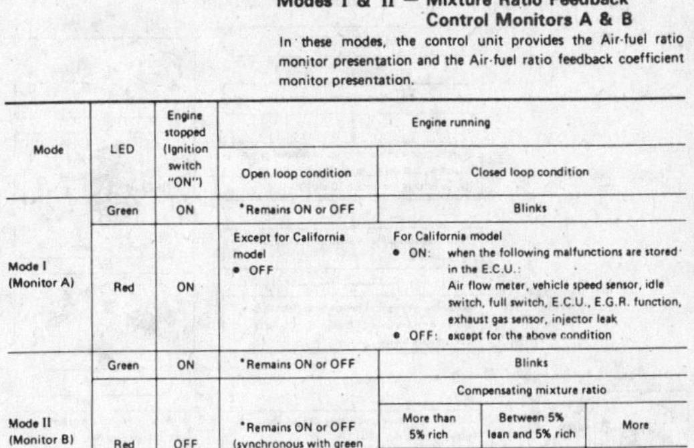

Mode	LED	Engine stopped (Ignition switch "ON")	Engine running			
			Open loop condition	Closed loop condition		
Mode I (Monitor A)	Green	ON	*Remains ON or OFF	Blinks		
	Red	ON	Except for California model • OFF	For California model • ON: when the following malfunctions are stored in the E.C.U.: Air flow meter, vehicle speed sensor, idle switch, full switch, E.C.U., E.G.R. function, exhaust gas sensor, injector leak • OFF: except for the above condition		
Mode II (Monitor B)	Green	ON	*Remains ON or OFF	Blinks		
	Red	OFF	*Remains ON or OFF (synchronous with green LED)	Compensating mixture ratio		
				More than 5% rich	Between 5% lean and 5% rich	More
				OFF	Synchronized with green LED	Remains ON

*: Maintains conditions just before switching to open loop

ECCS SELF-DIAGNOSTIC MODE III – 1988–89 STANZA

Mode III — Self-Diagnostic System

The E.C.U. constantly monitors the function of these sensors and actuators, regardless of ignition key position. If a malfunction occurs, the information is stored in the E.C.U. and can be retrieved from the memory by turning on the diagnostic mode selector, located on the side of the E.C.U. When activated, the malfunction is indicated by flashing a red and a green L.E.D. (Light Emitting Diode), also located on the E.C.U. Since all the self-diagnostic results are stored in the E.C.U.'s memory even intermittent malfunctions can be diagnosed.

A malfunctioning part's group is indicated by the number of both the red and the green L.E.D.s flashing. First, the red L.E.D. flashes and the green flashes follow. The red L.E.D. refers to the number of tens while the green one refers to the number of units. For example, when the red L.E.D. flashes once and then the green one flashes twice, this means the number "12" showing the air flow meter signal is malfunctioning. In this way, all the problems are classified by the code numbers.

- **When engine fails to start, crank engine more than two seconds before starting self-diagnosis.**
- **Before starting self-diagnosis, do not erase stored memory. If doing so, self-diagnosis function for intermittent malfunctions would be lost.**

The stored memory would be lost if:

1. **Battery terminal is disconnected.**
2. **After selecting Mode III, Mode IV is selected.**

DISPLAY CODE TABLE

Code No.	Detected items
11	Crank angle sensor circuit
12	Air flow meter circuit
13	Water temperature sensor circuit
14	Vehicle speed sensor circuit
21	Ignition signal missing in primary coil
22	Fuel pump circuit
23	Idle switch circuit
24	Full switch circuit
31	E.C.U.
32*	E.G.R. function
33	Exhaust gas sensor circuit
35*	Exhaust gas temperature sensor circuit
41	Air temperature sensor circuit
45*	Injector leak
55	No malfunctioning in the above circuit

*: For California only

Mode III — Self-Diagnostic System (Cont'd)

RETENTION OF DIAGNOSTIC RESULTS

The diagnostic result is retained in E.C.U. memory until the starter is operated fifty times after a diagnostic item is judged to be malfunctioning. The diagnostic result will then be cancelled automatically. If a diagnostic item which has been judged to be malfunctioning and stored in memory is again judged to be malfunctioning before the starter is operated fifty times, the second result will replace the previous one. It will be stored in E.C.U. memory until the starter is operated fifty times more.

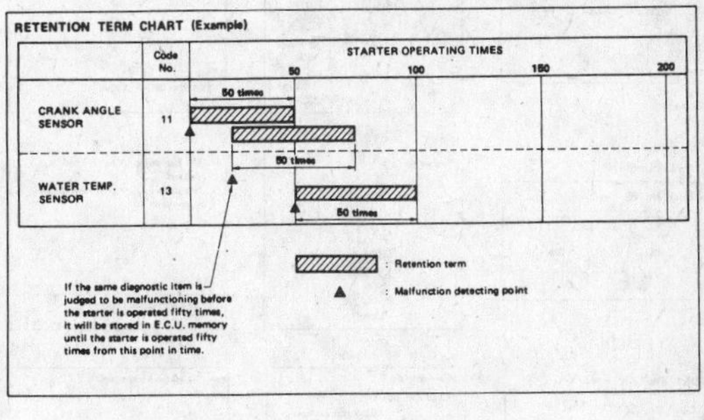
RETENTION TERM CHART (Example)
Code No.
STARTER OPERATING TIMES
50
100
150
200
CRANK ANGLE SENSOR
11
50 times
WATER TEMP. SENSOR
13
50 times
50 times
: Retention term
: Malfunction detecting point
If the same diagnostic item is judged to be malfunctioning before the starter is operated fifty times, it will be stored in E.C.U. memory until the starter is operated fifty times from this point in time.

ECCS SELF-DIAGNOSTIC MODE III (CONT.)—1988–89 STANZA

ECCS SELF-DIAGNOSTIC MODE III DISPLAY CODES—1988–89 STANZA

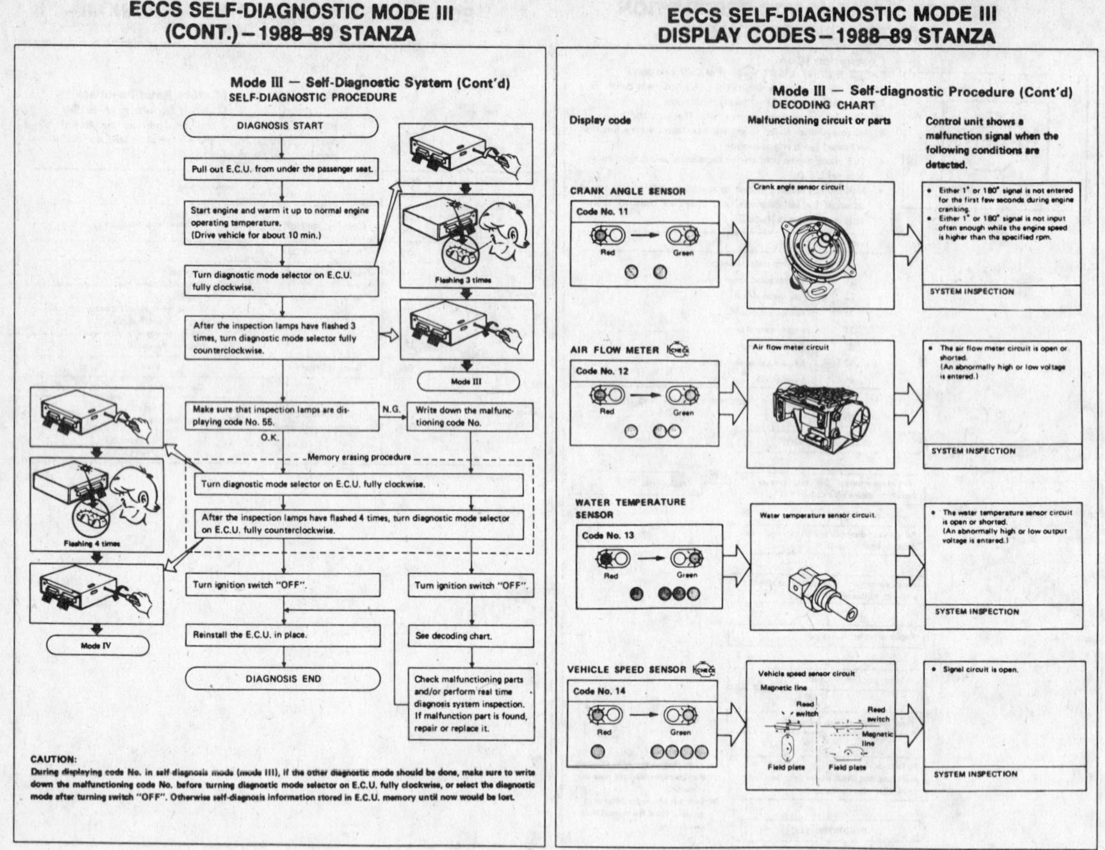
Mode III — Self-Diagnostic System (Cont'd)
SELF-DIAGNOSTIC PROCEDURE
DIAGNOSIS START
Pull out E.C.U. from under the passenger seat.
Start engine and warm it up to normal engine operating temperature. (Drive vehicle for about 10 min.)
Turn diagnostic mode selector on E.C.U. fully clockwise.
Flashing 3 times
After the inspection lamps have flashed 3 times, turn diagnostic mode selector fully counterclockwise.
Mode III
Make sure that inspection lamps are displaying code No. 55.
N.G.
Write down the malfunctioning code No.
O.K.
Memory erasing procedure
Turn diagnostic mode selector on E.C.U. fully clockwise.
After the inspection lamps have flashed 4 times, turn diagnostic mode selector on E.C.U. fully counterclockwise.
Flashing 4 times
Turn ignition switch "OFF".
Turn ignition switch "OFF".
Reinstall the E.C.U. in place.
See decoding chart.
Mode IV
DIAGNOSIS END
Check malfunctioning parts and/or perform real time diagnosis system inspection. If malfunction part is found, repair or replace it.
CAUTION:
During displaying code No. in self diagnosis mode (mode III), if the other diagnostic mode should be done, make sure to write down the malfunctioning code No. before turning diagnostic mode selector on E.C.U. fully clockwise, or select the diagnostic mode after turning switch "OFF". Otherwise self-diagnosis information stored in E.C.U. memory until now would be lost.
Mode III — Self-diagnostic Procedure (Cont'd)
DECODING CHART
Display code
Malfunctioning circuit or parts
Control unit shows a malfunction signal when the following conditions are detected.
CRANK ANGLE SENSOR
Code No. 11
Red
Green
Crank angle sensor circuit
Either 1° or 180° signal is not entered for the first few seconds during engine cranking.
Either 1° or 180° signal is not input often enough while the engine speed is higher than the specified rpm.
SYSTEM INSPECTION
AIR FLOW METER
Code No. 12
Air flow meter circuit
The air flow meter circuit is open or shorted. (An abnormally high or low voltage is entered.)
SYSTEM INSPECTION
WATER TEMPERATURE SENSOR
Code No. 13
Water temperature sensor circuit.
The water temperature sensor circuit is open or shorted. (An abnormally high or low output voltage is entered.)
SYSTEM INSPECTION
VEHICLE SPEED SENSOR
Code No. 14
Vehicle speed sensor circuit
Magnetic line
Reed switch
Reed switch
Magnetic line
Field plate
Field plate
Signal circuit is open.
SYSTEM INSPECTION

ECCS SELF-DIAGNOSTIC MODE III DISPLAY CODES (CONT.)—1988–89 STANZA

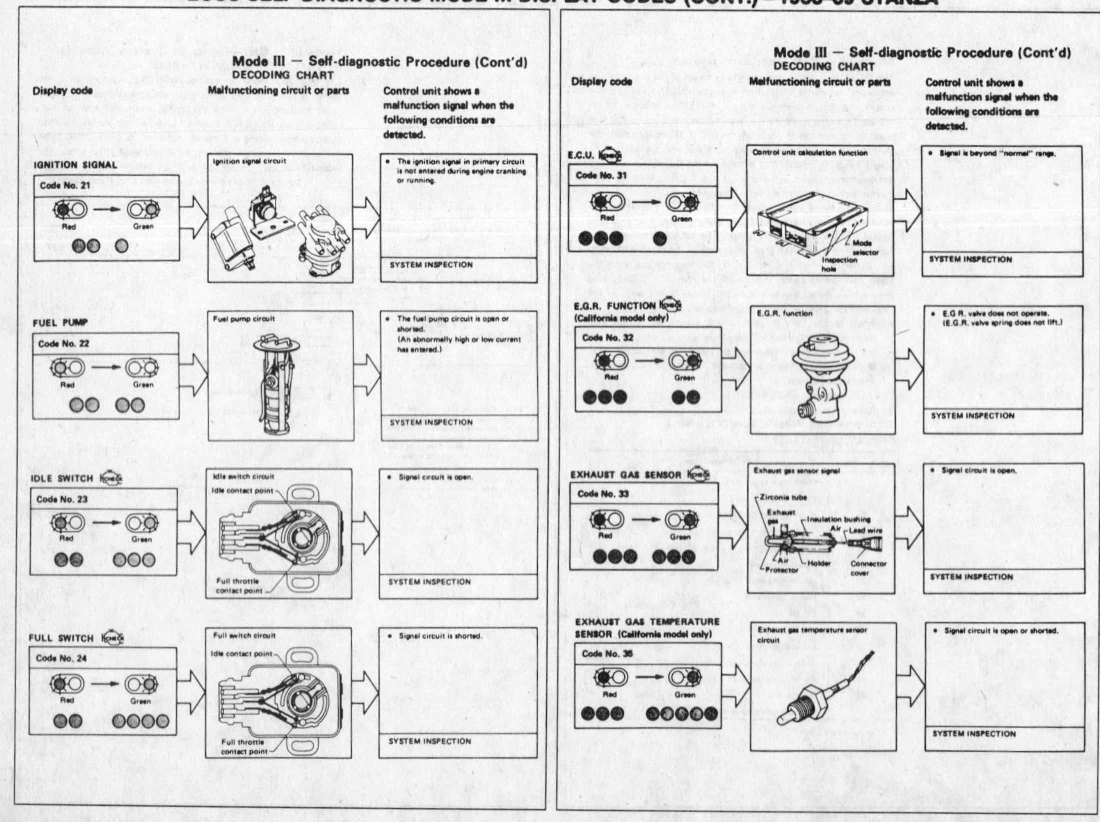
Mode III — Self-diagnostic Procedure (Cont'd)
DECODING CHART
Display code
Malfunctioning circuit or parts
Control unit shows a malfunction signal when the following conditions are detected.
IGNITION SIGNAL
Code No. 21
Red
Green
Ignition signal circuit
The ignition signal in primary circuit is not entered during engine cranking or running.
SYSTEM INSPECTION
FUEL PUMP
Code No. 22
Fuel pump circuit
The fuel pump circuit is open or shorted. (An abnormally high or low current has entered.)
SYSTEM INSPECTION
IDLE SWITCH
Code No. 23
Idle switch circuit
Idle contact point
Full throttle contact point
Signal circuit is open.
SYSTEM INSPECTION
FULL SWITCH
Code No. 24
Full switch circuit
Idle contact point
Full throttle contact point
Signal circuit is shorted.
SYSTEM INSPECTION
Mode III — Self-diagnostic Procedure (Cont'd)
DECODING CHART
Display code
Malfunctioning circuit or parts
Control unit shows a malfunction signal when the following conditions are detected.
E.C.U.
Code No. 31
Control unit calculation function
Mode selector
Inspection hole
Signal is beyond "normal" range.
SYSTEM INSPECTION
E.G.R. FUNCTION (California model only)
Code No. 32
E.G.R. function
E.G.R. valve does not operate. (E.G.R. valve spring does not lift.)
SYSTEM INSPECTION
EXHAUST GAS SENSOR
Code No. 33
Exhaust gas sensor signal
Zirconia tube
Exhaust gas
Insulation bushing
Air
Lead wire
Air
Protector
Holder
Connector cover
Signal circuit is open.
SYSTEM INSPECTION
EXHAUST GAS TEMPERATURE SENSOR (California model only)
Code No. 35
Exhaust gas temperature sensor circuit
Signal circuit is open or shorted.
SYSTEM INSPECTION

ECCS SELF-DIAGNOSTIC MODE IV—1988–89 STANZA

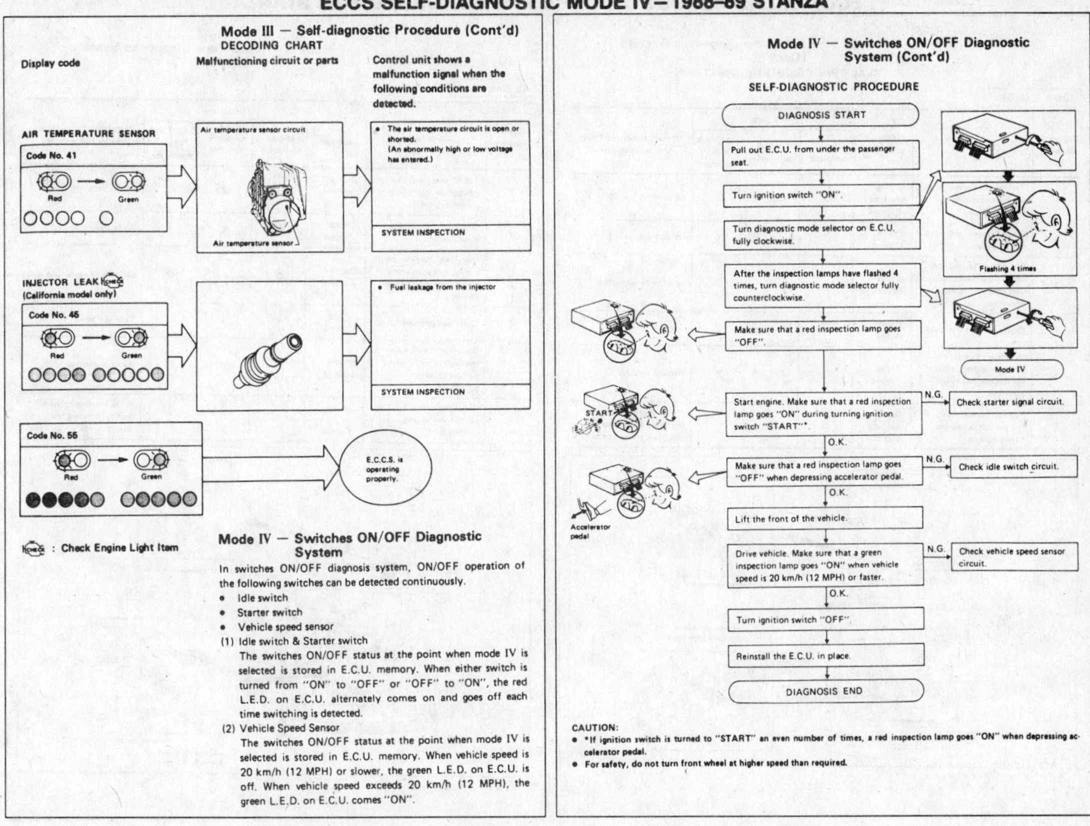
Mode III — Self-diagnostic Procedure (Cont'd)
DECODING CHART
Display code
Malfunctioning circuit or parts
Control unit shows a malfunction signal when the following conditions are detected.
AIR TEMPERATURE SENSOR
Code No. 41
Red
Green
Air temperature sensor circuit
Air temperature sensor
The air temperature circuit is open or shorted. (An abnormally high or low voltage has entered.)
SYSTEM INSPECTION
INJECTOR LEAK
(California model only)
Code No. 45
Red
Green
Fuel leakage from the injector
SYSTEM INSPECTION
Code No. 55
Red
Green
E.C.C.S. is operating properly.
: Check Engine Light Item
Mode IV — Switches ON/OFF Diagnostic System
In switches ON/OFF diagnosis system, ON/OFF operation of the following switches can be detected continuously.
Idle switch
Starter switch
Vehicle speed sensor
(1) Idle switch & Starter switch
The switches ON/OFF status at the point when mode IV is selected is stored in E.C.U. memory. When either switch is turned from "ON" to "OFF" or "OFF" to "ON", the red L.E.D. on E.C.U. alternately comes on and goes off each time switching is detected.
(2) Vehicle Speed Sensor
The switches ON/OFF status at the point when mode IV is selected is stored in E.C.U. memory. When vehicle speed is 20 km/h (12 MPH) or slower, the green L.E.D. on E.C.U. is off. When vehicle speed exceeds 20 km/h (12 MPH), the green L.E.D. on E.C.U. comes "ON".
Mode IV — Switches ON/OFF Diagnostic System (Cont'd)
SELF-DIAGNOSTIC PROCEDURE
DIAGNOSIS START
Pull out E.C.U. from under the passenger seat.
Turn ignition switch "ON".
Turn diagnostic mode selector on E.C.U. fully clockwise.
Flashing 4 times
After the inspection lamps have flashed 4 times, turn diagnostic mode selector fully counterclockwise.
Make sure that a red inspection lamp goes "OFF".
Mode IV
Start engine. Make sure that a red inspection lamp goes "ON" during turning ignition switch "START"*.
START
N.G.
Check starter signal circuit.
O.K.
Make sure that a red inspection lamp goes "OFF" when depressing accelerator pedal.
N.G.
Check idle switch circuit.
Accelerator pedal
O.K.
Lift the front of the vehicle.
Drive vehicle. Make sure that a green inspection lamp goes "ON" when vehicle speed is 20 km/h (12 MPH) or faster.
N.G.
Check vehicle speed sensor circuit.
O.K.
Turn ignition switch "OFF".
Reinstall the E.C.U. in place.
DIAGNOSIS END
CAUTION:
*If ignition switch is turned to "START" an even number of times, a red inspection lamp goes "ON" when depressing accelerator pedal.
For safety, do not turn front wheel at higher speed than required.

ECCS SELF-DIAGNOSTIC MODE V—1988–89 STANZA

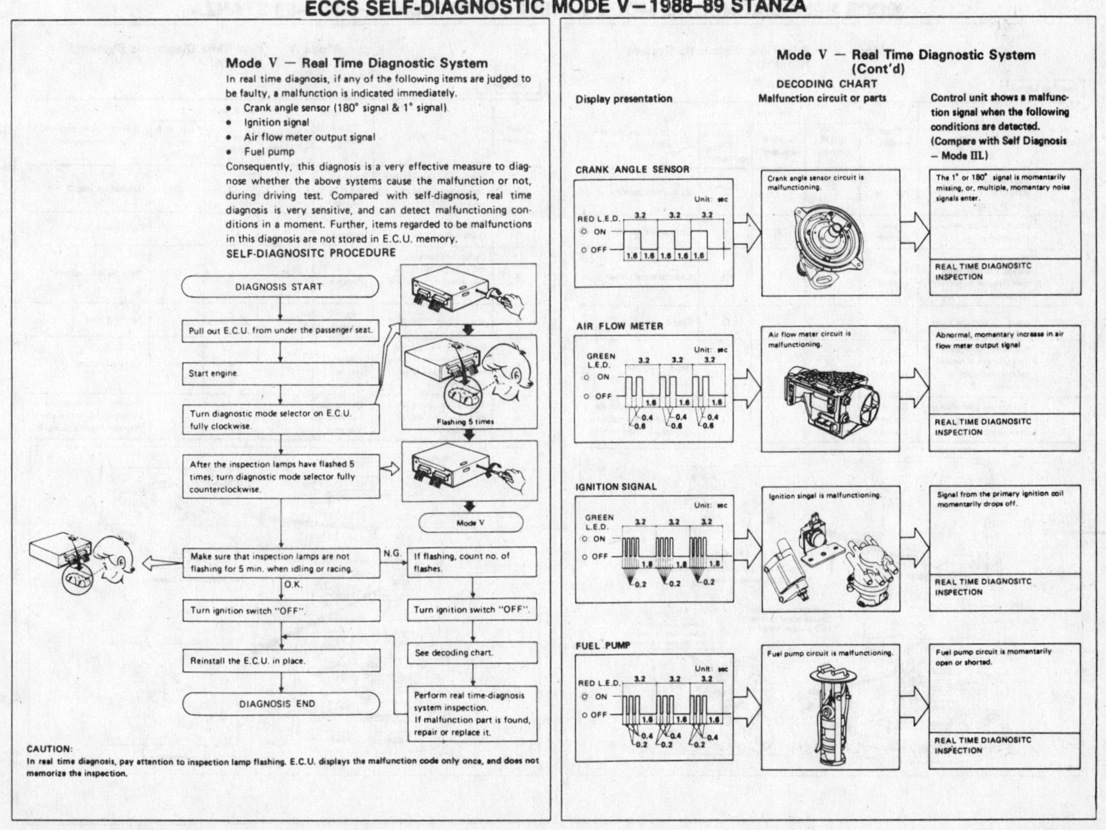
Mode V — Real Time Diagnostic System
In real time diagnosis, if any of the following items are judged to be faulty, a malfunction is indicated immediately.
Crank angle sensor (180° signal & 1° signal)
Ignition signal
Air flow meter output signal
Fuel pump
Consequently, this diagnosis is a very effective measure to diagnose whether the above systems cause the malfunction or not, during driving test. Compared with self-diagnosis, real time diagnosis is very sensitive, and can detect malfunctioning conditions in a moment. Further, items regarded to be malfunctions in this diagnosis are not stored in E.C.U. memory.
SELF-DIAGNOSITC PROCEDURE
DIAGNOSIS START
Pull out E.C.U. from under the passenger seat.
Start engine.
Turn diagnostic mode selector on E.C.U. fully clockwise.
Flashing 5 times
After the inspection lamps have flashed 5 times, turn diagnostic mode selector fully counterclockwise.
Mode V
Make sure that inspection lamps are not flashing for 5 min. when idling or racing.
N.G.
If flashing, count no. of flashes.
O.K.
Turn ignition switch "OFF".
Turn ignition switch "OFF".
See decoding chart.
Reinstall the E.C.U. in place.
DIAGNOSIS END
Perform real time-diagnosis system inspection. If malfunction part is found, repair or replace it.
CAUTION:
In real time diagnosis, pay attention to inspection lamp flashing. E.C.U. displays the malfunction code only once, and does not memorize the inspection.
Mode V — Real Time Diagnostic System (Cont'd)
DECODING CHART
Display presentation
Malfunction circuit or parts
Control unit shows a malfunction signal when the following conditions are detected. (Compare with Self Diagnosis – Mode III.)
CRANK ANGLE SENSOR
Unit: sec
RED L.E.D.
ON
OFF
3.2 3.2 3.2
1.6 1.6 1.6 1.6 1.6 1.6
Crank angle sensor circuit is malfunctioning.
The 1° or 180° signal is momentarily missing, or, multiple, momentary noise signals enter.
REAL TIME DIAGNOSITC INSPECTION
AIR FLOW METER
Unit: sec
GREEN L.E.D.
ON
OFF
3.2 3.2 3.2
1.6 1.6 1.6
0.4 0.4 0.4
0.8 0.8 0.8
Air flow meter circuit is malfunctioning.
Abnormal, momentary increase in air flow meter output signal
REAL TIME DIAGNOSITC INSPECTION
IGNITION SIGNAL
Unit: sec
GREEN L.E.D.
ON
OFF
3.2 3.2 3.2
1.8 1.8 1.8
0.2 0.2 0.2
Ignition singal is malfunctioning.
Signal from the primary ignition coil momentarily drops off.
REAL TIME DIAGNOSITC INSPECTION
FUEL PUMP
Unit: sec
RED L.E.D.
ON
OFF
3.2 3.2 3.2
1.6 1.6 1.6
0.4 0.4 0.4
0.2 0.2 0.2
Fuel pump circuit is malfunctioning.
Fuel pump circuit is momentarily open or shorted.
REAL TIME DIAGNOSITC INSPECTION

ECCS SELF-DIAGNOSTIC MODE V INSPECTION – 1988–89 STANZA

Mode V — Real Time Diagnostic System (Cont'd)

REAL TIME DIAGNOSTIC INSPECTION

Crank Angle Sensor

Check sequence	Check items	Check conditions	Check parts			If malfunction, perform the following items.
			Middle connector	Sensor & actuator	E.C.U. 20 & 16 pin connector	
1	Tap harness connector or component during real time diagnosis.	During real time diagnosis	O	O	O	Go to check item 2.
2	Check harness continuity at connector.	Engine stopped	O	X	X	Go to check item 3.
3	Disconnect harness connector, and then check dust adhesion to harness connector.	Engine stopped	O	X	O	Clean terminal surface.
4	Check pin terminal bend.	Engine stopped	X	X	O	Take out bend.
5	Reconnect harness connector and then recheck harness continuity at connector.	Engine stopped	O	X	X	Replace terminal.
6	Tap harness connector or component during real time diagnosis.	During real time diagnosis	O	O	O	If malfunction codes are displayed during real time diagnosis, replace terminal.

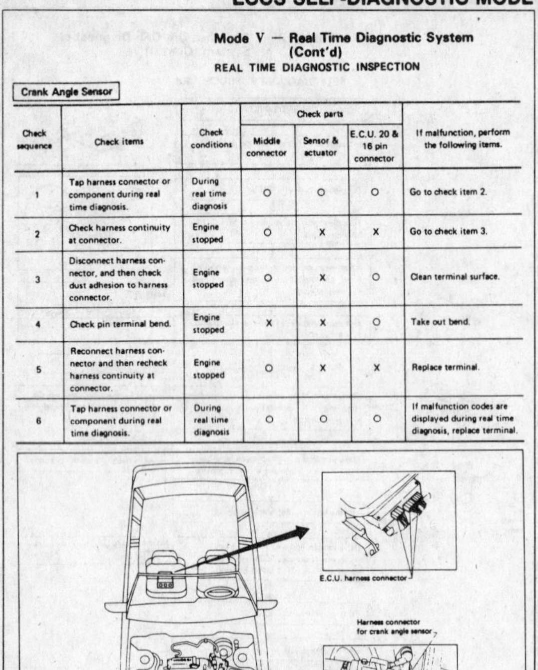
E.C.U. harness connector
Harness connector for crank angle sensor

Mode V — Real Time Diagnostic System (Cont'd)

Air Flow Meter

Check sequence	Check items	Check conditions	Check parts			If malfunction, perform the following items.
			Middle connector	Sensor & actuator	E.C.U. 20 & 16 pin connector	
1	Tap harness connector or component during real time diagnosis.	During real time diagnosis	O	O	O	Go to check item 2.
2	Check harness continuity at connector.	Engine stopped	O	X	X	Go to check item 3.
3	Disconnect harness connector, and then check dust adhesion to harness connector.	Engine stopped	O	X	O	Clean terminal surface.
4	Check pin terminal bend.	Engine stopped	X	X	O	Take out bend.
5	Reconnect harness connector and then recheck harness continuity at connector.	Engine stopped	O	X	X	Replace terminal.
6	Tap harness connector or component during real time diagnosis.	During real time diagnosis	O	O	O	If malfunction codes are displayed during real time diagnosis, replace terminal.

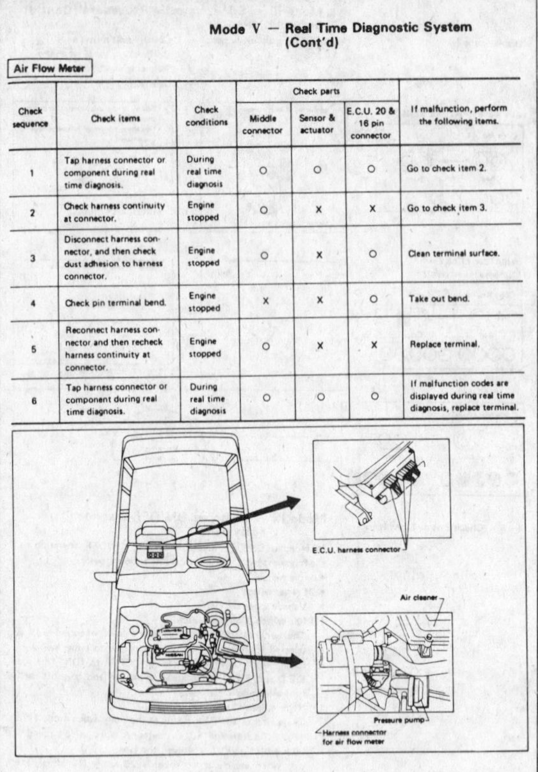
E.C.U. harness connector
Air cleaner
Pressure pump
Harness connector for air flow meter

ECCS SELF-DIAGNOSTIC MODE V INSPECTION (CONT.) – 1988–89 STANZA

Mode V — Real Time Diagnostic System (Cont'd)

Ignition Signal

Check sequence	Check items	Check conditions	Check parts			If malfunction, perform the following items.
			Middle connector	Sensor & actuator	E.C.U. 20 & 16 pin connector	
1	Tap harness connector or component during real time diagnosis.	During real time diagnosis	O	O	O	Go to check item 2.
2	Check harness continuity at connector.	Engine stopped	O	X	X	Go to check item 3.
3	Disconnect harness connector, and then check dust adhesion to harness connector.	Engine stopped	O	X	O	Clean terminal surface.
4	Check pin terminal bend.	Engine stopped	X	X	O	Take out bend.
5	Reconnect harness connector and then recheck harness continuity at connector.	Engine stopped	O	X	X	Replace terminal.
6	Tap harness connector or component during real time diagnosis.	During real time diagnosis	O	O	O	If malfunction codes are displayed during real time diagnosis, replace terminal.

E.C.U. harness connector
Harness connector for intake side of ignition coil & power transistor
Harness connector for exhaust side of ignition coil & power transistor

Mode V — Real Time Diagnostic System (Cont'd)

Fuel Pump

Check sequence	Check items	Check conditions	Check parts			If malfunction, perform the following items.
			Middle connector	Sensor & actuator	E.C.U. 20 & 16 pin connector	
1	Tap harness connector or component during real time diagnosis.	During real time diagnosis	O	O	O	Go to check item 2.
2	Check harness continuity at connector.	Engine stopped	O	X	X	Go to check item 3.
3	Disconnect harness connector, and then check dust adhesion to harness connector.	Engine stopped	O	X	O	Clean terminal surface.
4	Check pin terminal bend.	Engine stopped	X	X	O	Take out bend.
5	Reconnect harness connector and then recheck harness continuity at connector.	Engine stopped	O	X	X	Replace terminal.
6	Tap harness connector or component during real time diagnosis.	During real time diagnosis	O	O	O	If malfunction codes are displayed during real time diagnosis, replace terminal.

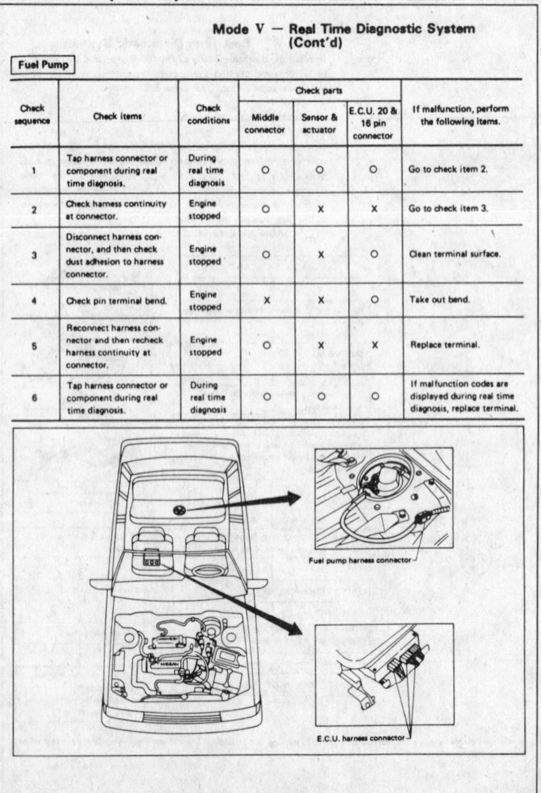
Fuel pump harness connector
E.C.U. harness connector

ECCS CAUTION—1988-89 STANZA

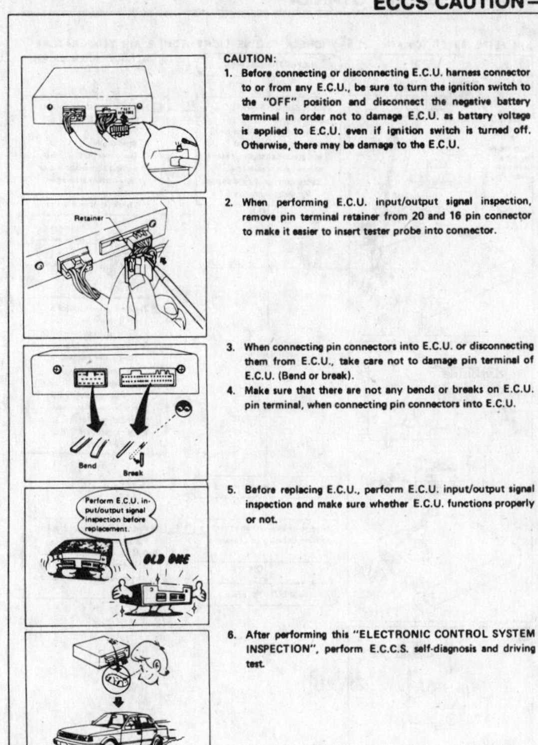
Retainer
Bend
Break
Perform E.C.U. in-put/output signal inspection before replacement.
OLD ONE

CAUTION:

1. Before connecting or disconnecting E.C.U. harness connector to or from any E.C.U., be sure to turn the ignition switch to the "OFF" position and disconnect the negative battery terminal in order not to damage E.C.U. as battery voltage is applied to E.C.U. even if ignition switch is turned off. Otherwise, there may be damage to the E.C.U.
2. When performing E.C.U. input/output signal inspection, remove pin terminal retainer from 20 and 16 pin connector to make it easier to insert tester probe into connector.
3. When connecting pin connectors into E.C.U. or disconnecting them from E.C.U., take care not to damage pin terminal of E.C.U. (Bend or break).
4. Make sure that there are not any bends or breaks on E.C.U. pin terminal, when connecting pin connectors into E.C.U.
5. Before replacing E.C.U., perform E.C.U. input/output signal inspection and make sure whether E.C.U. functions properly or not.
6. After performing this "ELECTRONIC CONTROL SYSTEM INSPECTION", perform E.C.C.S. self-diagnosis and driving test.

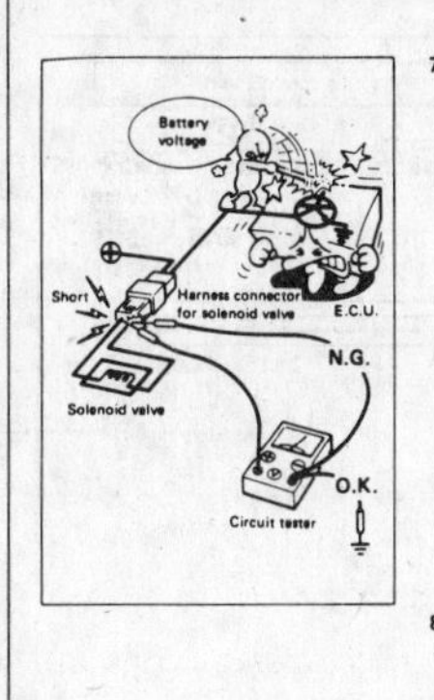
Battery voltage
Short
Harness connector for solenoid valve
E.C.U.
N.G.
Solenoid valve
O.K.
Circuit tester

7. When measuring supply voltage of E.C.U. controlled components with a circuit tester, separate one tester probe from the other.
   If the two tester probes accidentally make contact with each other during measurement, the circuit will be shorted, resulting in damage to the power transistor of the control unit.
8. Keys to symbols
   - DISCONNECT : Check after disconnecting the connector to be measured.
   - CONNECT : Check after connecting the connector to be measured.
9. When measuring voltage or resistance at connector with tester probes, there are two methods of measurement; one is done from terminal side and the other from harness side. Before measuring, confirm symbol mark again.
   - H.S. : Inspection should be done from harness side.
   - T.S. : Inspection should be done from terminal side.

ECCS SELF-DIAGNOSTIC CHART—1988-89 STANZA

CRANK ANGLE SENSOR (Code No. 11)

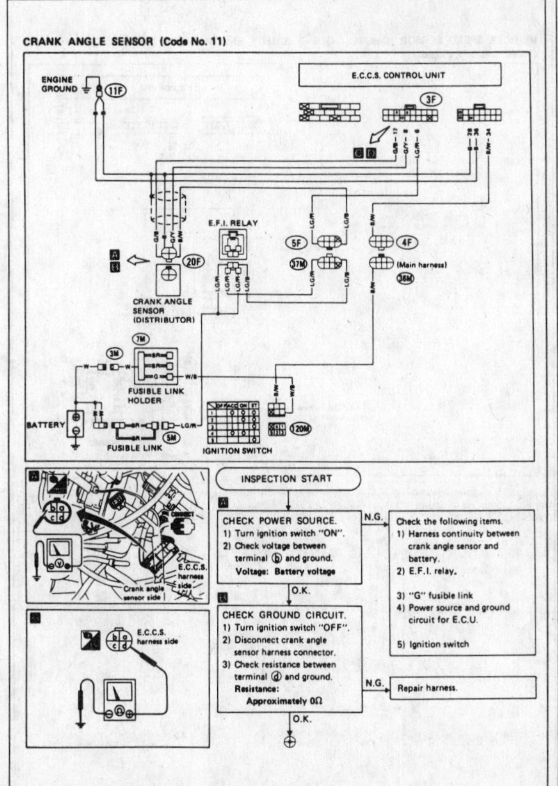
ENGINE GROUND
E.C.C.S. CONTROL UNIT
E.F.I. RELAY
(Main harness)
CRANK ANGLE SENSOR (DISTRIBUTOR)
FUSIBLE LINK HOLDER
BATTERY
FUSIBLE LINK
IGNITION SWITCH
E.C.C.S. harness side
Crank angle sensor side
INSPECTION START
A
CHECK POWER SOURCE.
1) Turn ignition switch "ON".
2) Check voltage between terminal ⓑ and ground.
Voltage: Battery voltage
N.G.
Check the following items.
1) Harness continuity between crank angle sensor and battery.
2) E.F.I. relay.
3) "G" fusible link
4) Power source and ground circuit for E.C.U.
5) Ignition switch
O.K.
B
CHECK GROUND CIRCUIT.
1) Turn ignition switch "OFF".
2) Disconnect crank angle sensor harness connector.
3) Check resistance between terminal ⓓ and ground.
Resistance:
Approximately 0Ω
N.G.
Repair harness.
O.K.

CRANK ANGLE SENSOR (Code No. 11)

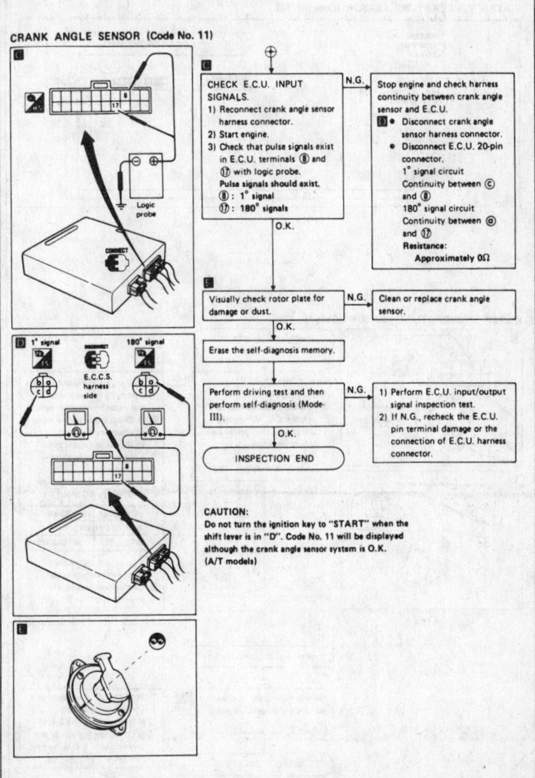
Logic probe
1° signal
180° signal
E.C.C.S. harness side
C
CHECK E.C.U. INPUT SIGNALS.
1) Reconnect crank angle sensor harness connector.
2) Start engine.
3) Check that pulse signals exist in E.C.U. terminals ⑧ and ⑰ with logic probe.
Pulse signals should exist.
⑧ : 1° signal
⑰ : 180° signals
N.G.
Stop engine and check harness continuity between crank angle sensor and E.C.U.
D • Disconnect crank angle sensor harness connector.
• Disconnect E.C.U. 20-pin connector.
1° signal circuit
Continuity between ⓒ and ⑧
180° signal circuit
Continuity between ⓐ and ⑰
Resistance:
Approximately 0Ω
O.K.
E
Visually check rotor plate for damage or dust.
N.G.
Clean or replace crank angle sensor.
O.K.
Erase the self-diagnosis memory.
Perform driving test and then perform self-diagnosis (Mode-III).
N.G.
1) Perform E.C.U. input/output signal inspection test.
2) If N.G., recheck the E.C.U. pin terminal damage or the connection of E.C.U. harness connector.
O.K.
INSPECTION END
CAUTION:
Do not turn the ignition key to "START" when the shift lever is in "D". Code No. 11 will be displayed although the crank angle sensor system is O.K. (A/T models)

ECCS SELF-DIAGNOSTIC CHART—1988–89 STANZA

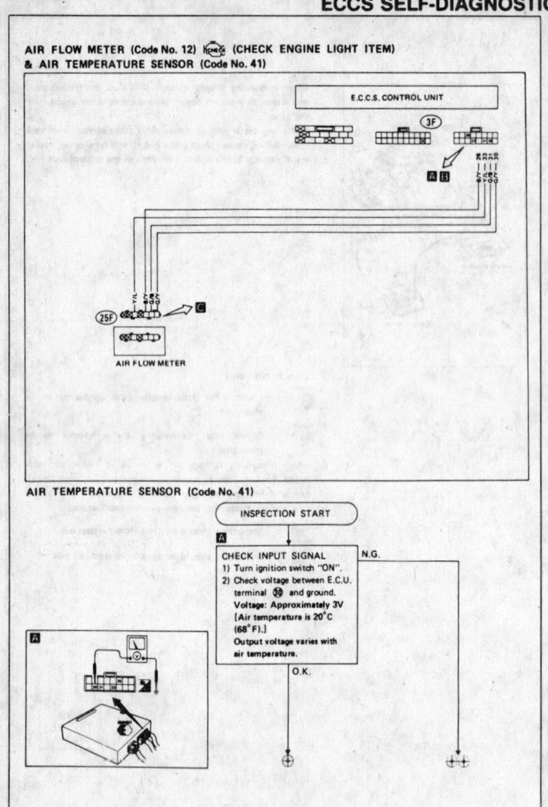
AIR FLOW METER (Code No. 12) (CHECK ENGINE LIGHT ITEM) & AIR TEMPERATURE SENSOR (Code No. 41)
E.C.C.S. CONTROL UNIT
3F
25F
AIR FLOW METER
AIR TEMPERATURE SENSOR (Code No. 41)
INSPECTION START
CHECK INPUT SIGNAL.
1) Turn ignition switch "ON".
2) Check voltage between E.C.U. terminal 30 and ground.
Voltage: Approximately 3V
[Air temperature is 20°C (68°F).]
Output voltage varies with air temperature.
N.G.
O.K.

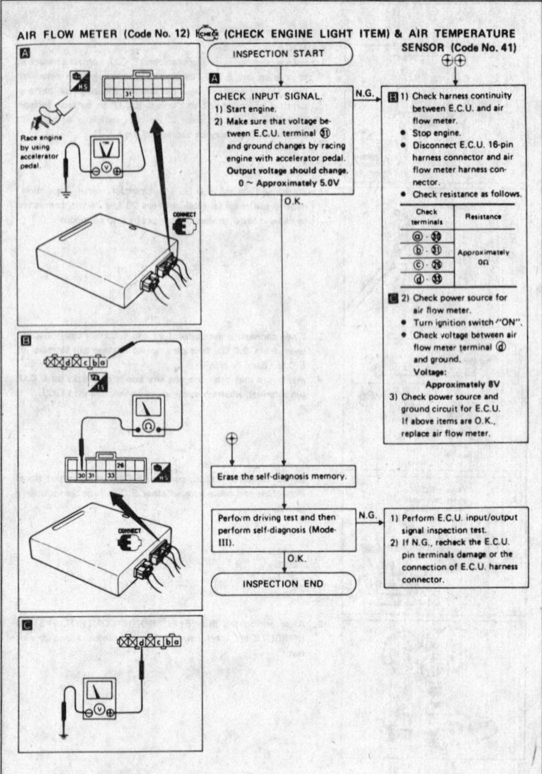
AIR FLOW METER (Code No. 12) (CHECK ENGINE LIGHT ITEM) & AIR TEMPERATURE SENSOR (Code No. 41)
INSPECTION START
Race engine by using accelerator pedal.
CONNECT
CHECK INPUT SIGNAL.
1) Start engine.
2) Make sure that voltage between E.C.U. terminal 31 and ground changes by racing engine with accelerator pedal.
Output voltage should change.
0 ~ Approximately 5.0V
N.G.
O.K.
1) Check harness continuity between E.C.U. and air flow meter.
Stop engine.
Disconnect E.C.U. 16-pin harness connector and air flow meter harness connector.
Check resistance as follows.
Check terminals
Resistance
Approximately 0Ω
2) Check power source for air flow meter.
Turn ignition switch "ON".
Check voltage between air flow meter terminal ⓓ and ground.
Voltage: Approximately 8V
3) Check power source and ground circuit for E.C.U.
If above items are O.K., replace air flow meter.
Erase the self-diagnosis memory.
Perform driving test and then perform self-diagnosis (Mode-III).
N.G.
1) Perform E.C.U. input/output signal inspection test.
2) If N.G., recheck the E.C.U. pin terminals damage or the connection of E.C.U. harness connector.
O.K.
INSPECTION END

ECCS SELF-DIAGNOSTIC CHART—1988–89 STANZA

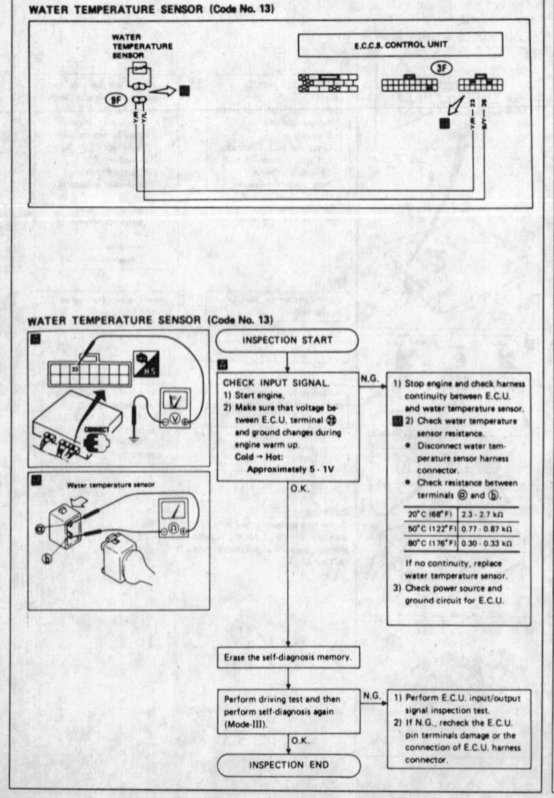
WATER TEMPERATURE SENSOR (Code No. 13)
WATER TEMPERATURE SENSOR
E.C.C.S. CONTROL UNIT
3F
9F
WATER TEMPERATURE SENSOR (Code No. 13)
INSPECTION START
CONNECT
Water temperature sensor
CHECK INPUT SIGNAL.
1) Start engine.
2) Make sure that voltage between E.C.U. terminal 23 and ground changes during engine warm up.
Cold → Hot:
Approximately 5 - 1V
N.G.
O.K.
1) Stop engine and check harness continuity between E.C.U. and water temperature sensor.
2) Check water temperature sensor resistance.
Disconnect water temperature sensor harness connector.
Check resistance between terminals ⓐ and ⓑ.
20°C (68°F) 2.3 - 2.7 kΩ
50°C (122°F) 0.77 - 0.87 kΩ
80°C (176°F) 0.30 - 0.33 kΩ
If no continuity, replace water temperature sensor.
3) Check power source and ground circuit for E.C.U.
Erase the self-diagnosis memory.
Perform driving test and then perform self-diagnosis again (Mode-III).
N.G.
1) Perform E.C.U. input/output signal inspection test.
2) If N.G., recheck the E.C.U. pin terminals damage or the connection of E.C.U. harness connector.
O.K.
INSPECTION END

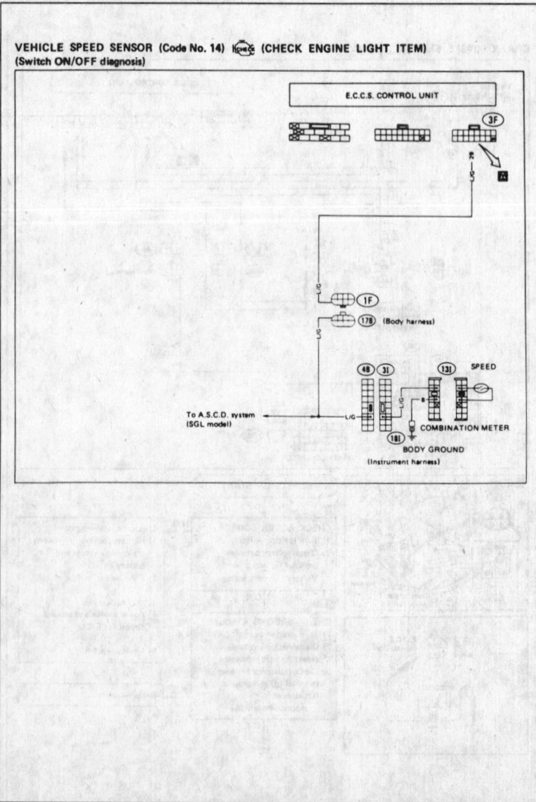
VEHICLE SPEED SENSOR (Code No. 14) (CHECK ENGINE LIGHT ITEM)
(Switch ON/OFF diagnosis)
E.C.C.S. CONTROL UNIT
3F
1F
17B
(Body harness)
48
31
13I
SPEED
To A.S.C.D. system (SGL model)
COMBINATION METER
18I
BODY GROUND
(Instrument harness)

ECCS SELF-DIAGNOSTIC CHART–1988–89 STANZA

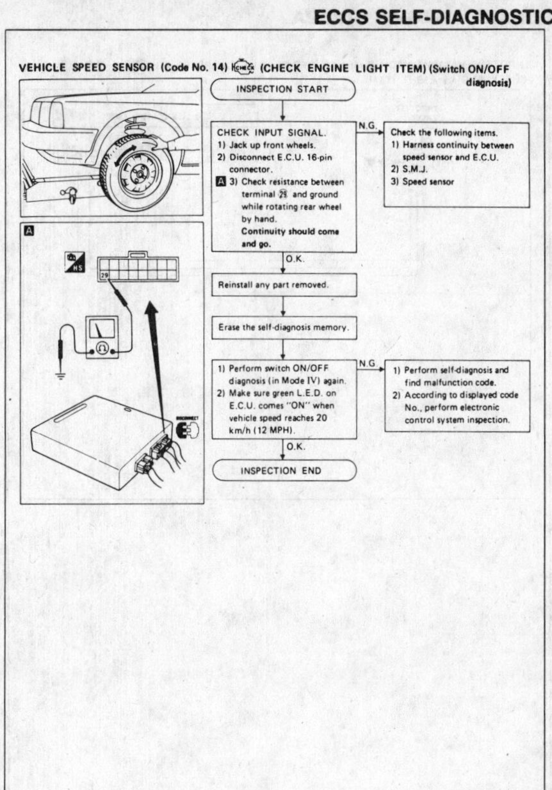
VEHICLE SPEED SENSOR (Code No. 14) (CHECK ENGINE LIGHT ITEM) (Switch ON/OFF diagnosis)
INSPECTION START
CHECK INPUT SIGNAL.
1) Jack up front wheels.
2) Disconnect E.C.U. 16-pin connector.
A 3) Check resistance between terminal 29 and ground while rotating rear wheel by hand.
Continuity should come and go.
N.G.
Check the following items.
1) Harness continuity between speed sensor and E.C.U.
2) S.M.J.
3) Speed sensor
O.K.
Reinstall any part removed.
Erase the self-diagnosis memory.
1) Perform switch ON/OFF diagnosis (in Mode IV) again.
2) Make sure green L.E.D. on E.C.U. comes "ON" when vehicle speed reaches 20 km/h (12 MPH).
N.G.
1) Perform self-diagnosis and find malfunction code.
2) According to displayed code No., perform electronic control system inspection.
O.K.
INSPECTION END
A

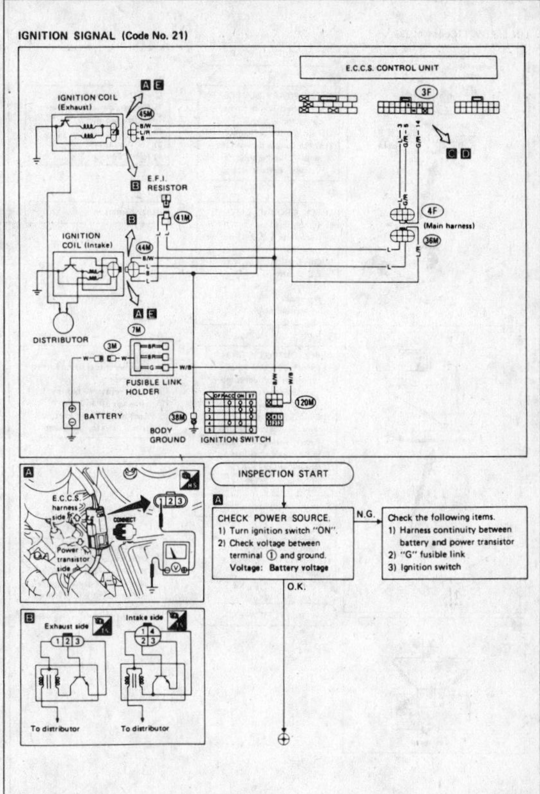
IGNITION SIGNAL (Code No. 21)
E.C.C.S. CONTROL UNIT
IGNITION COIL (Exhaust)
E.F.I. RESISTOR
IGNITION COIL (Intake)
(Main harness)
DISTRIBUTOR
FUSIBLE LINK HOLDER
BATTERY
BODY GROUND
IGNITION SWITCH
INSPECTION START
A
CHECK POWER SOURCE.
1) Turn ignition switch "ON".
2) Check voltage between terminal ① and ground.
Voltage: Battery voltage
N.G.
Check the following items.
1) Harness continuity between battery and power transistor
2) "G" fusible link
3) Ignition switch
O.K.
E.C.C.S. harness side
Power transistor side
B
Exhaust side
Intake side
To distributor
To distributor

ECCS SELF-DIAGNOSTIC CHART–1988–89 STANZA

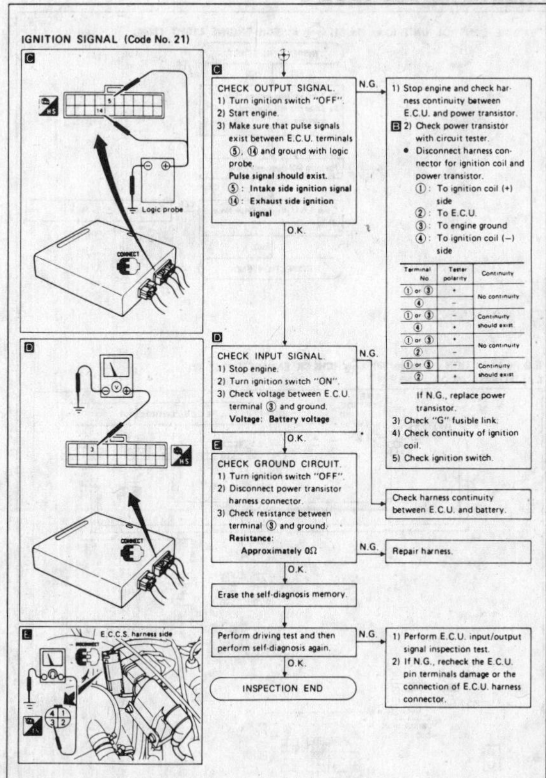
IGNITION SIGNAL (Code No. 21)
C
Logic probe
C
CHECK OUTPUT SIGNAL.
1) Turn ignition switch "OFF".
2) Start engine.
3) Make sure that pulse signals exist between E.C.U. terminals ⑤, ⑭ and ground with logic probe.
Pulse signal should exist.
⑤ : Intake side ignition signal
⑭ : Exhaust side ignition signal
N.G.
1) Stop engine and check harness continuity between E.C.U. and power transistor.
B 2) Check power transistor with circuit tester.
• Disconnect harness connector for ignition coil and power transistor.
①: To ignition coil (+) side
②: To E.C.U.
③: To engine ground
④: To ignition coil (−) side
If N.G., replace power transistor.
3) Check "G" fusible link.
4) Check continuity of ignition coil.
5) Check ignition switch.
O.K.
D
CHECK INPUT SIGNAL.
1) Stop engine.
2) Turn ignition switch "ON".
3) Check voltage between E.C.U. terminal ③ and ground.
Voltage: Battery voltage
N.G.
O.K.
E
CHECK GROUND CIRCUIT.
1) Turn ignition switch "OFF".
2) Disconnect power transistor harness connector.
3) Check resistance between terminal ③ and ground.
Resistance: Approximately 0Ω
Check harness continuity between E.C.U. and battery.
N.G.
Repair harness.
O.K.
Erase the self-diagnosis memory.
Perform driving test and then perform self-diagnosis again.
N.G.
1) Perform E.C.U. input/output signal inspection test.
2) If N.G., recheck the E.C.U. pin terminals damage or the connection of E.C.U. harness connector.
O.K.
INSPECTION END
E.C.C.S. harness side

Terminal No.	Tester polarity	Continuity
① or ③	+	No continuity
④	−	
① or ③	−	Continuity should exist
④	+	
① or ③	+	No continuity
②	−	
① or ③	−	Continuity should exist
②	+	

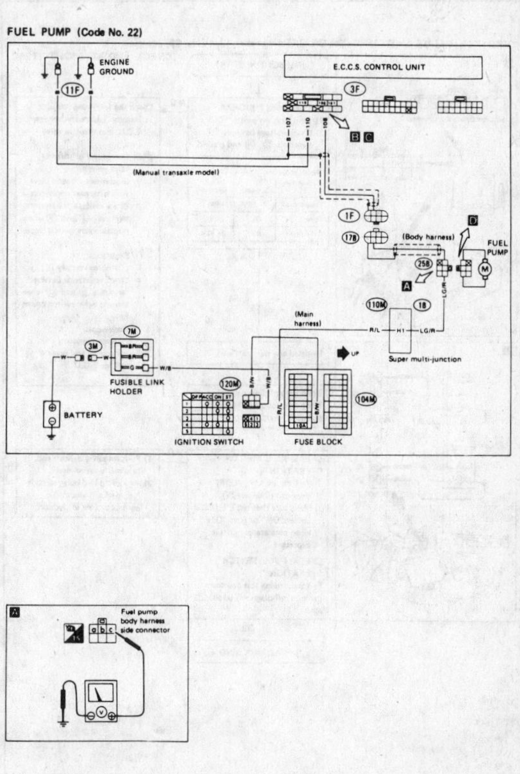
FUEL PUMP (Code No. 22)
ENGINE GROUND
E.C.C.S. CONTROL UNIT
(Manual transaxle model)
(Body harness)
FUEL PUMP
(Main harness)
Super multi-junction
FUSIBLE LINK HOLDER
BATTERY
IGNITION SWITCH
FUSE BLOCK
A
Fuel pump body harness side connector

ECCS SELF-DIAGNOSTIC CHART—1988–89 STANZA

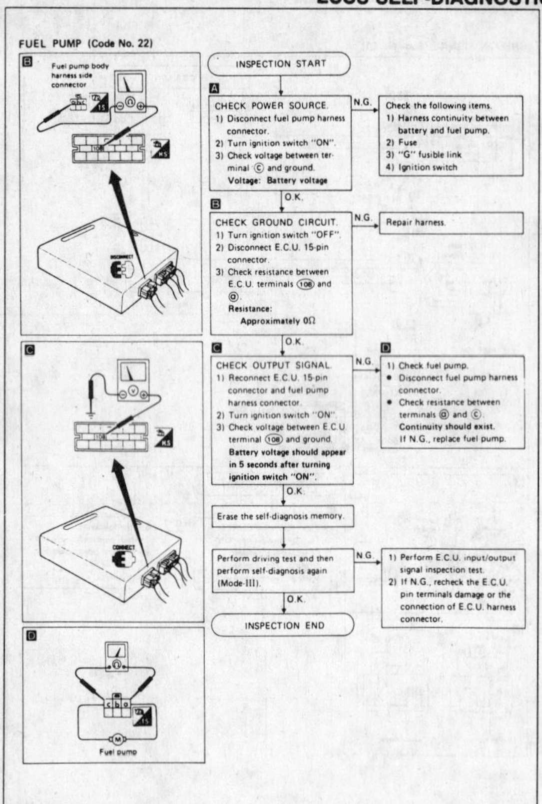
FUEL PUMP (Code No. 22)
INSPECTION START
CHECK POWER SOURCE.
1) Disconnect fuel pump harness connector.
2) Turn ignition switch "ON".
3) Check voltage between terminal © and ground.
Voltage: Battery voltage
N.G.
Check the following items.
1) Harness continuity between battery and fuel pump.
2) Fuse
3) "G" fusible link
4) Ignition switch
O.K.
CHECK GROUND CIRCUIT.
1) Turn ignition switch "OFF".
2) Disconnect E.C.U. 15-pin connector.
3) Check resistance between E.C.U. terminals 108 and ©.
Resistance:
Approximately 0Ω
Repair harness.
CHECK OUTPUT SIGNAL.
1) Reconnect E.C.U. 15-pin connector and fuel pump harness connector.
2) Turn ignition switch "ON".
3) Check voltage between E.C.U. terminal 108 and ground.
Battery voltage should appear in 5 seconds after turning ignition switch "ON".
1) Check fuel pump.
• Disconnect fuel pump harness connector.
• Check resistance between terminals ⓐ and ©.
Continuity should exist.
If N.G., replace fuel pump.
Erase the self-diagnosis memory.
Perform driving test and then perform self-diagnosis again (Mode-III).
1) Perform E.C.U. input/output signal inspection test.
2) If N.G., recheck the E.C.U. pin terminals damage or the connection of E.C.U. harness connector.
INSPECTION END
Fuel pump body harness side connector
DISCONNECT
CONNECT
Fuel pump

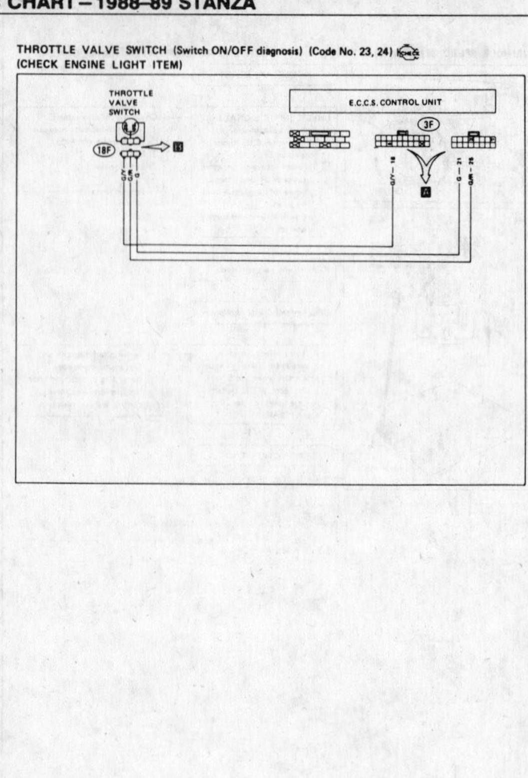
THROTTLE VALVE SWITCH (Switch ON/OFF diagnosis) (Code No. 23, 24)
(CHECK ENGINE LIGHT ITEM)
THROTTLE VALVE SWITCH
E.C.C.S. CONTROL UNIT
18F
3F

ECCS SELF-DIAGNOSTIC CHART—1988–89 STANZA

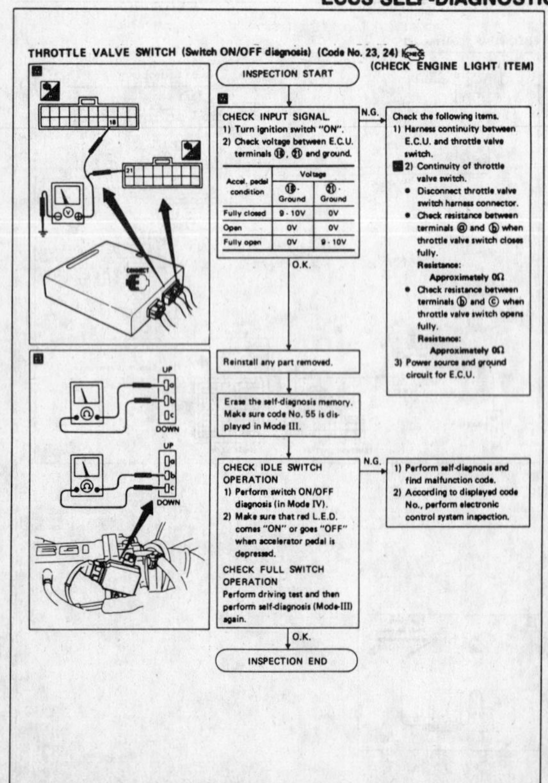
THROTTLE VALVE SWITCH (Switch ON/OFF diagnosis) (Code No. 23, 24)
(CHECK ENGINE LIGHT ITEM)
INSPECTION START
CHECK INPUT SIGNAL.
1) Turn ignition switch "ON".
2) Check voltage between E.C.U. terminals 18, 21 and ground.
Accel. pedal condition | Voltage 18 - Ground | 21 - Ground
Fully closed | 9 - 10V | 0V
Open | 0V | 0V
Fully open | 0V | 9 - 10V
N.G.
Check the following items.
1) Harness continuity between E.C.U. and throttle valve switch.
2) Continuity of throttle valve switch.
• Disconnect throttle valve switch harness connector.
• Check resistance between terminals ⓐ and ⓑ when throttle valve switch closes fully.
Resistance:
Approximately 0Ω
• Check resistance between terminals ⓑ and © when throttle valve switch opens fully.
Resistance:
Approximately 0Ω
3) Power source and ground circuit for E.C.U.
O.K.
Reinstall any part removed.
Erase the self-diagnosis memory.
Make sure code No. 55 is displayed in Mode III.
CHECK IDLE SWITCH OPERATION
1) Perform switch ON/OFF diagnosis (in Mode IV).
2) Make sure that red L.E.D. comes "ON" or goes "OFF" when accelerator pedal is depressed.
CHECK FULL SWITCH OPERATION
Perform driving test and then perform self-diagnosis (Mode-III) again.
1) Perform self-diagnosis and find malfunction code.
2) According to displayed code No., perform electronic control system inspection.
INSPECTION END
UP
DOWN

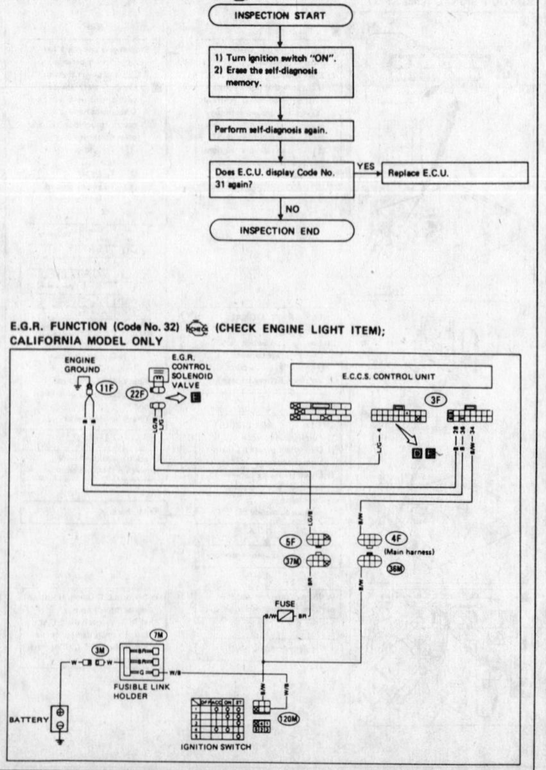
ENGINE CONTROL UNIT (Code No. 31) (CHECK ENGINE LIGHT ITEM)
INSPECTION START
1) Turn ignition switch "ON".
2) Erase the self-diagnosis memory.
Perform self-diagnosis again.
Does E.C.U. display Code No. 31 again?
YES
Replace E.C.U.
NO
INSPECTION END
E.G.R. FUNCTION (Code No. 32) (CHECK ENGINE LIGHT ITEM);
CALIFORNIA MODEL ONLY
ENGINE GROUND
E.G.R. CONTROL SOLENOID VALVE
E.C.C.S. CONTROL UNIT
(Main harness)
FUSE
FUSIBLE LINK HOLDER
BATTERY
IGNITION SWITCH

ECCS SELF-DIAGNOSTIC CHART—1988–89 STANZA

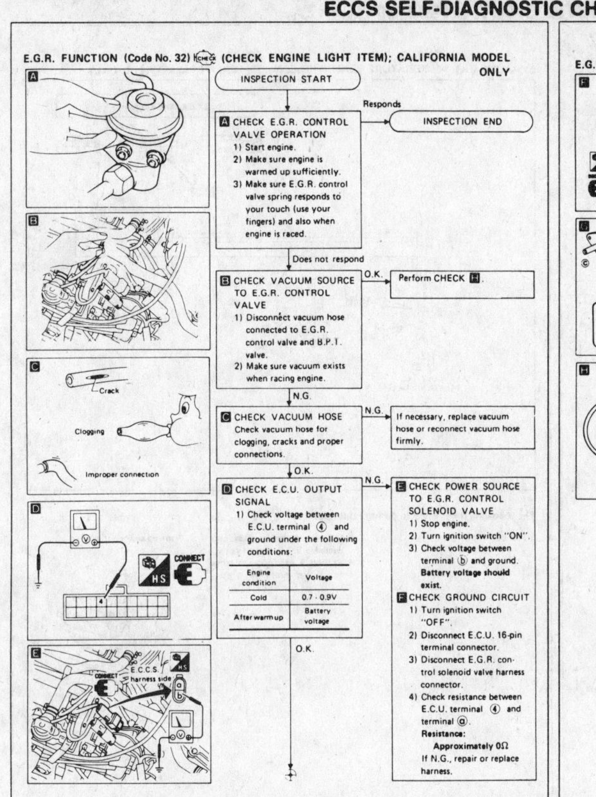
E.G.R. FUNCTION (Code No. 32) (CHECK ENGINE LIGHT ITEM); CALIFORNIA MODEL ONLY
INSPECTION START
Responds
INSPECTION END
A CHECK E.G.R. CONTROL VALVE OPERATION
1) Start engine.
2) Make sure engine is warmed up sufficiently.
3) Make sure E.G.R. control valve spring responds to your touch (use your fingers) and also when engine is raced.
Does not respond
B CHECK VACUUM SOURCE TO E.G.R. CONTROL VALVE
1) Disconnect vacuum hose connected to E.G.R. control valve and B.P.T. valve.
2) Make sure vacuum exists when racing engine.
O.K.
Perform CHECK H.
N.G.
C CHECK VACUUM HOSE
Check vacuum hose for clogging, cracks and proper connections.
N.G.
If necessary, replace vacuum hose or reconnect vacuum hose firmly.
O.K.
D CHECK E.C.U. OUTPUT SIGNAL
1) Check voltage between E.C.U. terminal ④ and ground under the following conditions:
Engine condition | Voltage
Cold | 0.7 - 0.9V
After warm up | Battery voltage
N.G.
E CHECK POWER SOURCE TO E.G.R. CONTROL SOLENOID VALVE
1) Stop engine.
2) Turn ignition switch "ON".
3) Check voltage between terminal ⓑ and ground. Battery voltage should exist.
F CHECK GROUND CIRCUIT
1) Turn ignition switch "OFF".
2) Disconnect E.C.U. 16-pin terminal connector.
3) Disconnect E.G.R. control solenoid valve harness connector.
4) Check resistance between E.C.U. terminal ④ and terminal ⓐ.
Resistance: Approximately 0Ω
If N.G., repair or replace harness.
O.K.
Crack
Clogging
Improper connection
CONNECT
HS
E.C.C.S. harness side

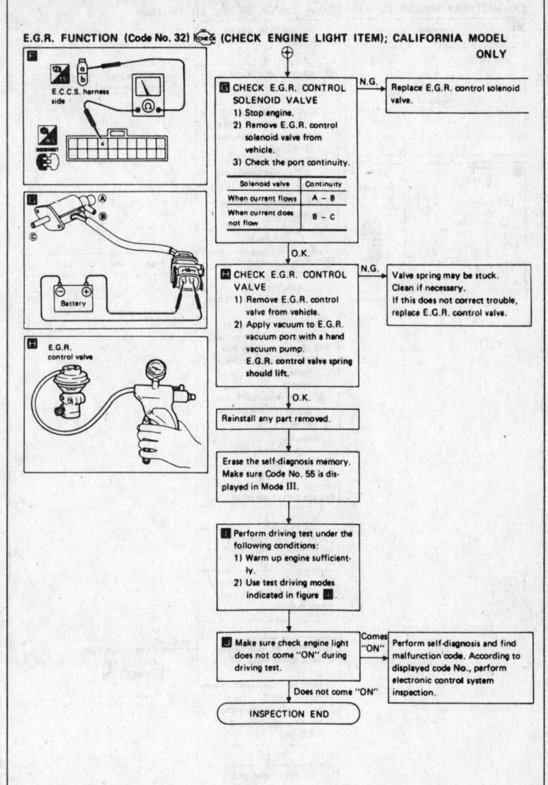
E.G.R. FUNCTION (Code No. 32) (CHECK ENGINE LIGHT ITEM); CALIFORNIA MODEL ONLY
E.C.C.S. harness side
G CHECK E.G.R. CONTROL SOLENOID VALVE
1) Stop engine.
2) Remove E.G.R. control solenoid valve from vehicle.
3) Check the port continuity.
Solenoid valve | Continuity
When current flows | A – B
When current does not flow | B – C
N.G.
Replace E.G.R. control solenoid valve.
O.K.
Battery
H CHECK E.G.R. CONTROL VALVE
1) Remove E.G.R. control valve from vehicle.
2) Apply vacuum to E.G.R. vacuum port with a hand vacuum pump.
E.G.R. control valve spring should lift.
N.G.
Valve spring may be stuck. Clean if necessary.
If this does not correct trouble, replace E.G.R. control valve.
E.G.R. control valve
O.K.
Reinstall any part removed.
Erase the self-diagnosis memory. Make sure Code No. 55 is displayed in Mode III.
I Perform driving test under the following conditions:
1) Warm up engine sufficiently.
2) Use test driving modes indicated in figure I.
J Make sure check engine light does not come "ON" during driving test.
Comes "ON"
Perform self-diagnosis and find malfunction code. According to displayed code No., perform electronic control system inspection.
Does not come "ON"
INSPECTION END

ECCS SELF-DIAGNOSTIC CHART—1988–89 STANZA

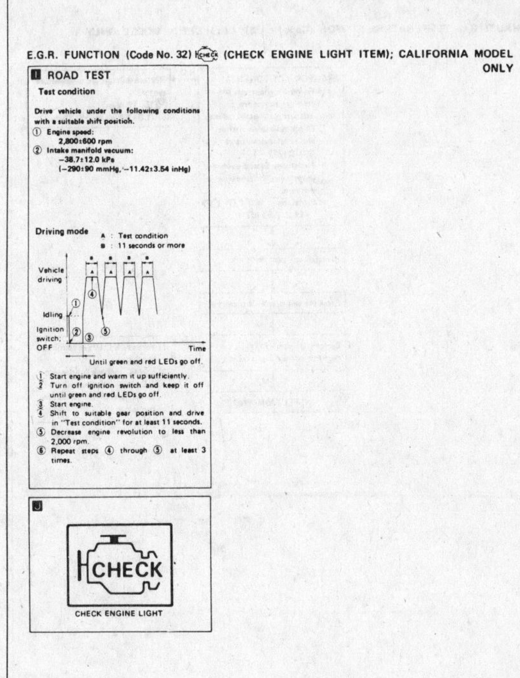
E.G.R. FUNCTION (Code No. 32) (CHECK ENGINE LIGHT ITEM); CALIFORNIA MODEL ONLY
I ROAD TEST
Test condition
Drive vehicle under the following conditions with a suitable shift position.
① Engine speed: 2,800±600 rpm
② Intake manifold vacuum: −38.7±12.0 kPa (−290±90 mmHg, −11.42±3.54 inHg)
Driving mode
A : Test condition
B : 11 seconds or more
Vehicle driving
Idling
Ignition switch: OFF
Time
Until green and red LEDs go off.
① Start engine and warm it up sufficiently.
② Turn off ignition switch and keep it off until green and red LEDs go off.
③ Start engine.
④ Shift to suitable gear position and drive in "Test condition" for at least 11 seconds.
⑤ Decrease engine revolution to less than 2,000 rpm.
⑥ Repeat steps ④ through ⑤ at least 3 times.
J
CHECK
CHECK ENGINE LIGHT

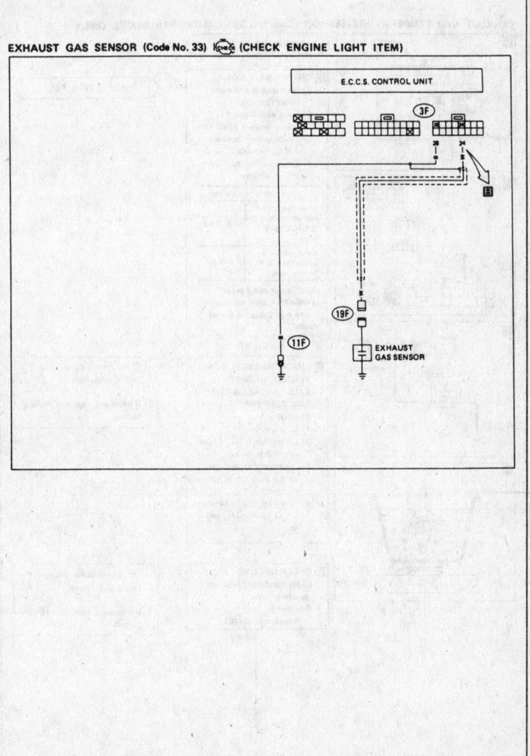
EXHAUST GAS SENSOR (Code No. 33) (CHECK ENGINE LIGHT ITEM)
E.C.C.S. CONTROL UNIT
3F
19F
11F
EXHAUST GAS SENSOR

ECCS SELF-DIAGNOSTIC CHART—1988–89 STANZA

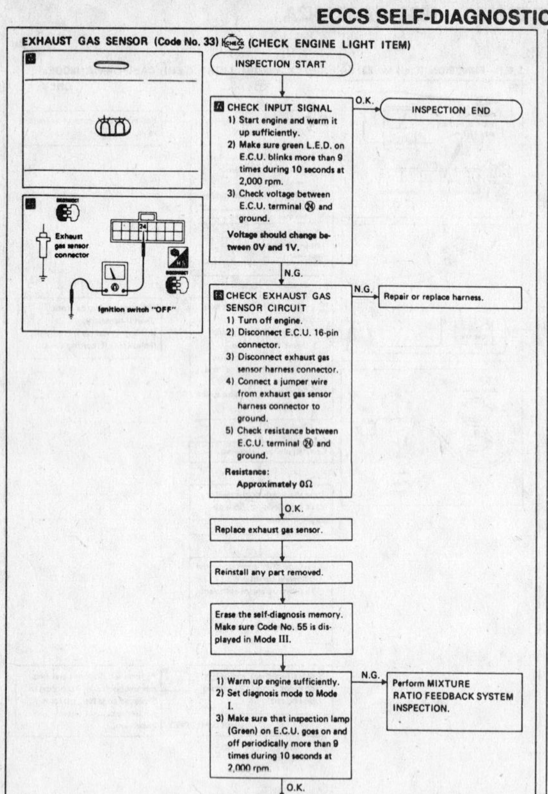
EXHAUST GAS SENSOR (Code No. 33) (CHECK ENGINE LIGHT ITEM)
INSPECTION START
A CHECK INPUT SIGNAL
1) Start engine and warm it up sufficiently.
2) Make sure green L.E.D. on E.C.U. blinks more than 9 times during 10 seconds at 2,000 rpm.
3) Check voltage between E.C.U. terminal 24 and ground.
Voltage should change between 0V and 1V.
O.K.
INSPECTION END
N.G.
B CHECK EXHAUST GAS SENSOR CIRCUIT
1) Turn off engine.
2) Disconnect E.C.U. 16-pin connector.
3) Disconnect exhaust gas sensor harness connector.
4) Connect a jumper wire from exhaust gas sensor harness connector to ground.
5) Check resistance between E.C.U. terminal 24 and ground.
Resistance:
Approximately 0Ω
N.G.
Repair or replace harness.
O.K.
Replace exhaust gas sensor.
Reinstall any part removed.
Erase the self-diagnosis memory. Make sure Code No. 55 is displayed in Mode III.
1) Warm up engine sufficiently.
2) Set diagnosis mode to Mode I.
3) Make sure that inspection lamp (Green) on E.C.U. goes on and off periodically more than 9 times during 10 seconds at 2,000 rpm.
N.G.
Perform MIXTURE RATIO FEEDBACK SYSTEM INSPECTION.
O.K.
INSPECTION END
Exhaust gas sensor connector
Ignition switch "OFF"

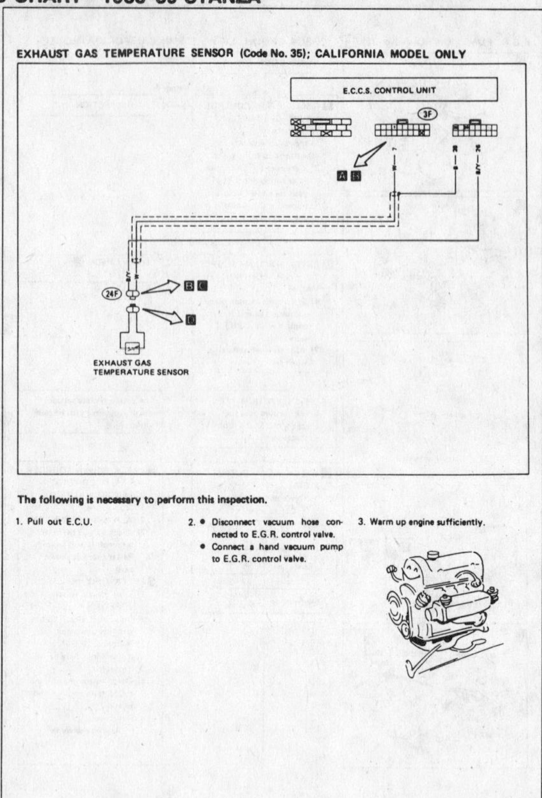
EXHAUST GAS TEMPERATURE SENSOR (Code No. 35); CALIFORNIA MODEL ONLY
E.C.C.S. CONTROL UNIT
3F
24F
EXHAUST GAS TEMPERATURE SENSOR
The following is necessary to perform this inspection.
1. Pull out E.C.U.
2. • Disconnect vacuum hose connected to E.G.R. control valve.
• Connect a hand vacuum pump to E.G.R. control valve.
3. Warm up engine sufficiently.

ECCS SELF-DIAGNOSTIC CHART—1988–89 STANZA

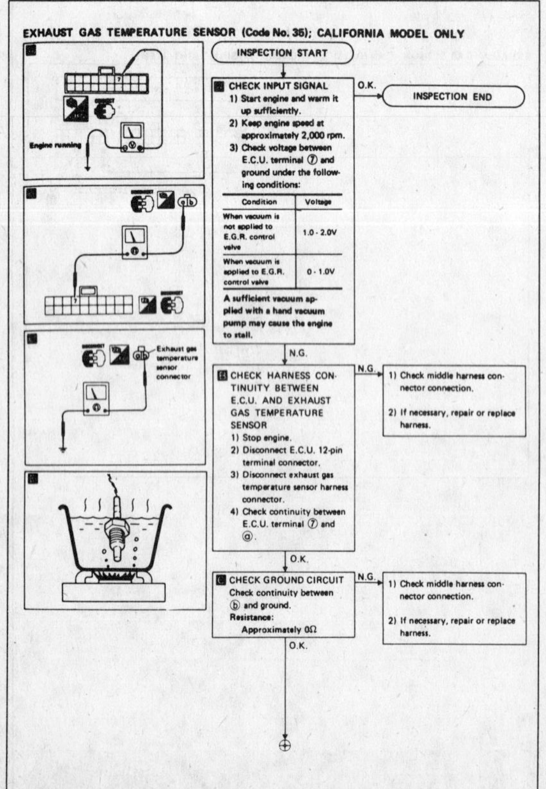
EXHAUST GAS TEMPERATURE SENSOR (Code No. 35); CALIFORNIA MODEL ONLY
INSPECTION START
A CHECK INPUT SIGNAL
1) Start engine and warm it up sufficiently.
2) Keep engine speed at approximately 2,000 rpm.
3) Check voltage between E.C.U. terminal 7 and ground under the following conditions:
Condition | Voltage
When vacuum is not applied to E.G.R. control valve | 1.0 - 2.0V
When vacuum is applied to E.G.R. control valve | 0 - 1.0V
A sufficient vacuum applied with a hand vacuum pump may cause the engine to stall.
O.K.
INSPECTION END
N.G.
B CHECK HARNESS CONTINUITY BETWEEN E.C.U. AND EXHAUST GAS TEMPERATURE SENSOR
1) Stop engine.
2) Disconnect E.C.U. 12-pin terminal connector.
3) Disconnect exhaust gas temperature sensor harness connector.
4) Check continuity between E.C.U. terminal 7 and a.
N.G.
1) Check middle harness connector connection.
2) If necessary, repair or replace harness.
O.K.
C CHECK GROUND CIRCUIT
Check continuity between b and ground.
Resistance:
Approximately 0Ω
N.G.
1) Check middle harness connector connection.
2) If necessary, repair or replace harness.
O.K.
Engine running
Exhaust gas temperature sensor connector

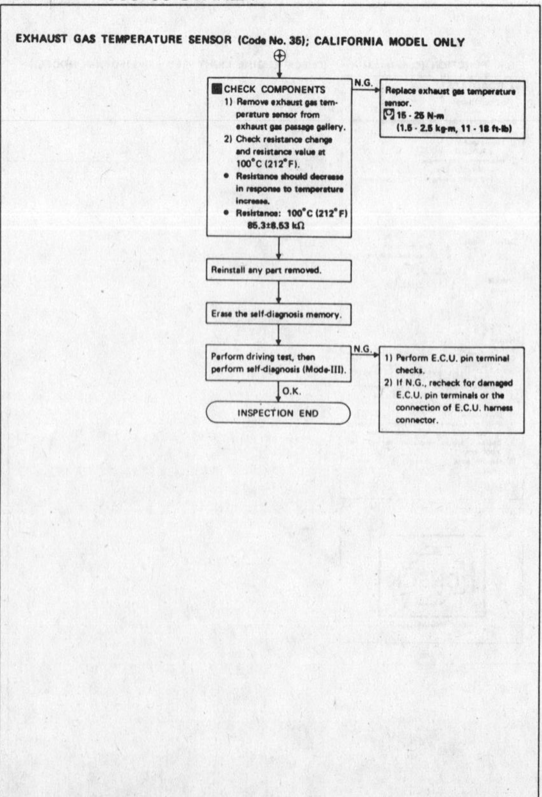
EXHAUST GAS TEMPERATURE SENSOR (Code No. 35); CALIFORNIA MODEL ONLY
D CHECK COMPONENTS
1) Remove exhaust gas temperature sensor from exhaust gas passage gallery.
2) Check resistance change and resistance value at 100°C (212°F).
• Resistance should decrease in response to temperature increase.
• Resistance: 100°C (212°F) 85.3±8.53 kΩ
N.G.
Replace exhaust gas temperature sensor.
15 - 25 N-m (1.5 - 2.5 kg-m, 11 - 18 ft-lb)
Reinstall any part removed.
Erase the self-diagnosis memory.
Perform driving test, then perform self-diagnosis (Mode-III).
N.G.
1) Perform E.C.U. pin terminal checks.
2) If N.G., recheck for damaged E.C.U. pin terminals or the connection of E.C.U. harness connector.
O.K.
INSPECTION END

ECCS SELF-DIAGNOSTIC CHART—1988–89 STANZA

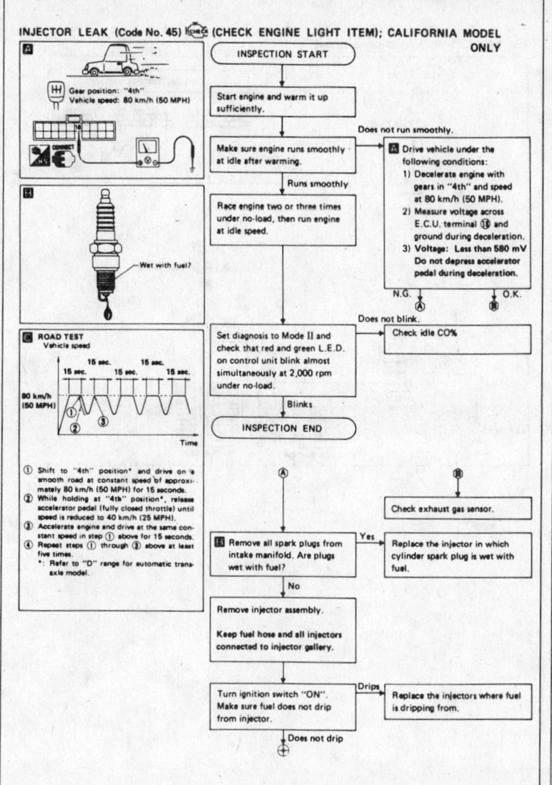
INJECTOR LEAK (Code No. 45) (CHECK ENGINE LIGHT ITEM); CALIFORNIA MODEL ONLY
A
Gear position: "4th"
Vehicle speed: 80 km/h (50 MPH)
CONNECT
INSPECTION START
Start engine and warm it up sufficiently.
Does not run smoothly.
Make sure engine runs smoothly at idle after warming.
A Drive vehicle under the following conditions:
1) Decelerate engine with gears in "4th" and speed at 80 km/h (50 MPH).
2) Measure voltage across E.C.U. terminal and ground during deceleration.
3) Voltage: Less than 580 mV
Do not depress accelerator pedal during deceleration.
Runs smoothly
B
Wet with fuel?
Race engine two or three times under no-load, then run engine at idle speed.
N.G.
O.K.
Does not blink.
C ROAD TEST
Vehicle speed
15 sec.
80 km/h (50 MPH)
Time
Set diagnosis to Mode II and check that red and green L.E.D. on control unit blink almost simultaneously at 2,000 rpm under no-load.
Check idle CO%
Blinks
INSPECTION END
① Shift to "4th" position* and drive on a smooth road at constant speed of approximately 80 km/h (50 MPH) for 15 seconds.
② While holding at "4th" position*, release accelerator pedal (fully closed throttle) until speed is reduced to 40 km/h (25 MPH).
③ Accelerate engine and drive at the same constant speed in step ① above for 15 seconds.
④ Repeat steps ① through ③ above at least five times.
*: Refer to "D" range for automatic transaxle model.
Check exhaust gas sensor.
B Remove all spark plugs from intake manifold. Are plugs wet with fuel?
Yes
Replace the injector in which cylinder spark plug is wet with fuel.
No
Remove injector assembly.
Keep fuel hose and all injectors connected to injector gallery.
Turn ignition switch "ON". Make sure fuel does not drip from injector.
Drips
Replace the injectors where fuel is dripping from.
Does not drip

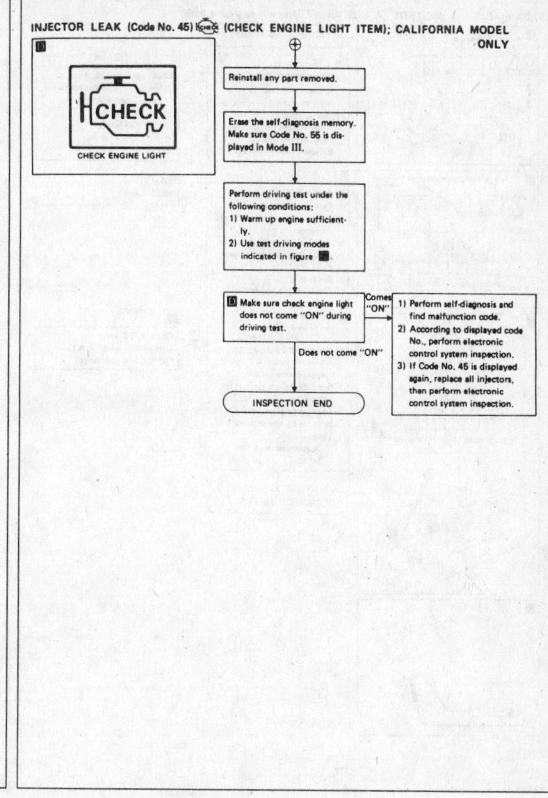
INJECTOR LEAK (Code No. 45) (CHECK ENGINE LIGHT ITEM); CALIFORNIA MODEL ONLY
D
CHECK
CHECK ENGINE LIGHT
Reinstall any part removed.
Erase the self-diagnosis memory. Make sure Code No. 55 is displayed in Mode III.
Perform driving test under the following conditions:
1) Warm up engine sufficiently.
2) Use test driving modes indicated in figure C.
D Make sure check engine light does not come "ON" during driving test.
Comes "ON"
1) Perform self-diagnosis and find malfunction code.
2) According to displayed code No., perform electronic control system inspection.
3) If Code No. 45 is displayed again, replace all injectors, then perform electronic control system inspection.
Does not come "ON"
INSPECTION END

ECCS SELF-DIAGNOSTIC CHART—1988–89 STANZA

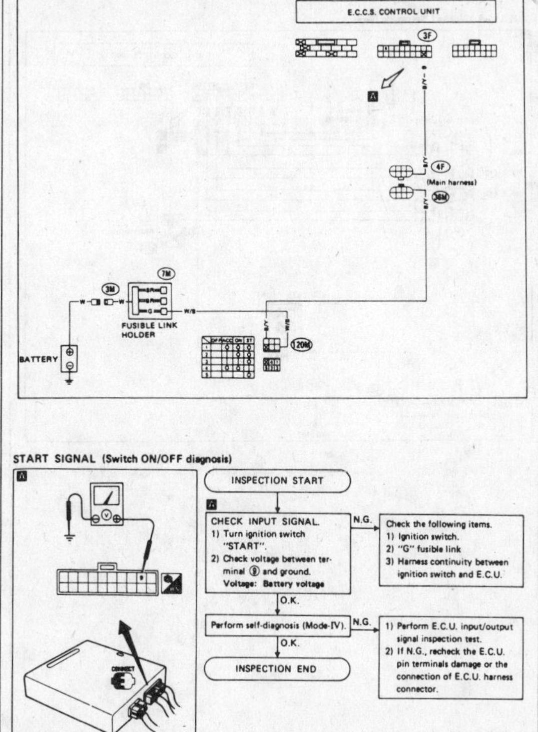
START SIGNAL (Switch ON/OFF diagnosis)
E.C.C.S. CONTROL UNIT
3F
4F
(Main harness)
36M
7M
3M
FUSIBLE LINK HOLDER
BATTERY
120M
START SIGNAL (Switch ON/OFF diagnosis)
A
CONNECT
INSPECTION START
A
CHECK INPUT SIGNAL.
1) Turn ignition switch "START".
2) Check voltage between terminal ⑨ and ground.
Voltage: Battery voltage
N.G.
Check the following items.
1) Ignition switch.
2) "G" fusible link
3) Harness continuity between ignition switch and E.C.U.
O.K.
Perform self-diagnosis (Mode-IV).
N.G.
1) Perform E.C.U. input/output signal inspection test.
2) If N.G., recheck the E.C.U. pin terminals damage or the connection of E.C.U. harness connector.
O.K.
INSPECTION END

ECCS DIAGNOSTIC CHART—1988–89 STANZA

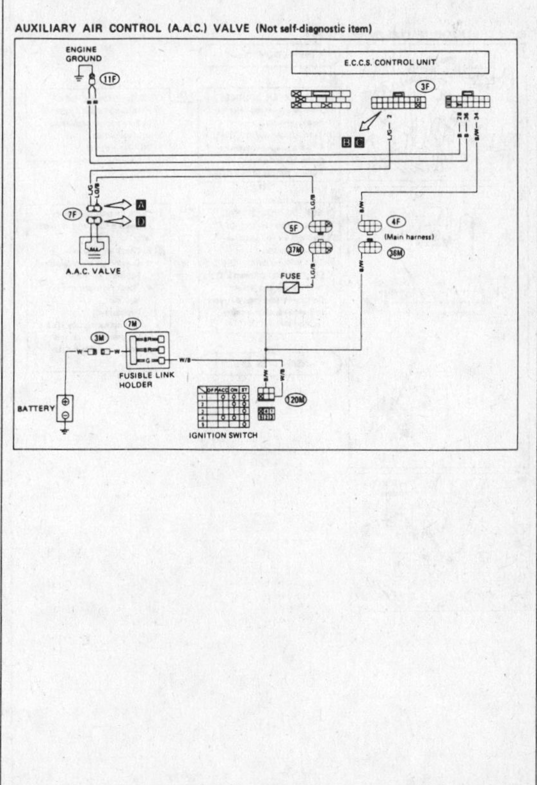
AUXILIARY AIR CONTROL (A.A.C.) VALVE (Not self-diagnostic item)
ENGINE GROUND
11F
E.C.C.S. CONTROL UNIT
3F
7F
A.A.C. VALVE
5F
37M
4F
(Main harness)
36M
FUSE
7M
3M
FUSIBLE LINK HOLDER
BATTERY
120M
IGNITION SWITCH

ECCS DIAGNOSTIC CHART – 1988–89 STANZA

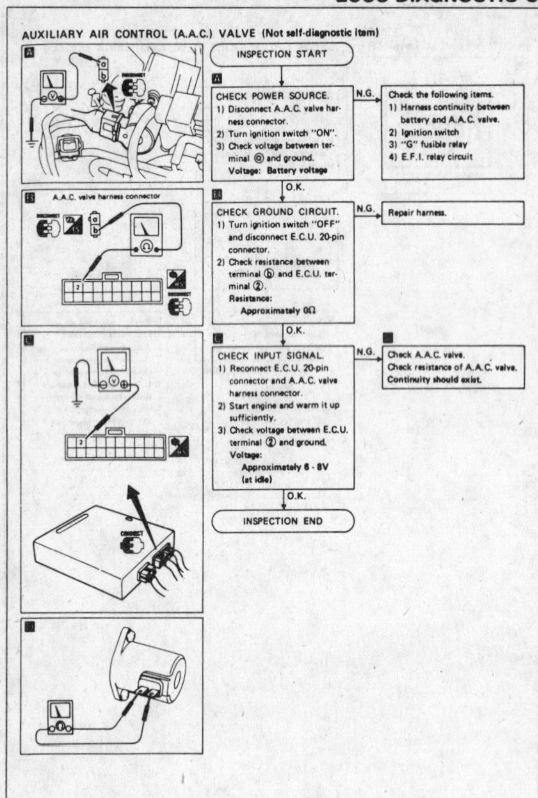
AUXILIARY AIR CONTROL (A.A.C.) VALVE (Not self-diagnostic item)
INSPECTION START
CHECK POWER SOURCE.
1) Disconnect A.A.C. valve harness connector.
2) Turn ignition switch "ON".
3) Check voltage between terminal ⓒ and ground.
Voltage: Battery voltage
N.G.
Check the following items.
1) Harness continuity between battery and A.A.C. valve.
2) Ignition switch
3) "G" fusible relay
4) E.F.I. relay circuit
O.K.
A.A.C. valve harness connector
CHECK GROUND CIRCUIT.
1) Turn ignition switch "OFF" and disconnect E.C.U. 20-pin connector.
2) Check resistance between terminal ⓑ and E.C.U. terminal ②.
Resistance:
Approximately 0Ω
Repair harness.
CHECK INPUT SIGNAL.
1) Reconnect E.C.U. 20-pin connector and A.A.C. valve harness connector.
2) Start engine and warm it up sufficiently.
3) Check voltage between E.C.U. terminal ② and ground.
Voltage:
Approximately 6 - 8V
(at idle)
Check A.A.C. valve.
Check resistance of A.A.C. valve.
Continuity should exist.
INSPECTION END
CONNECT

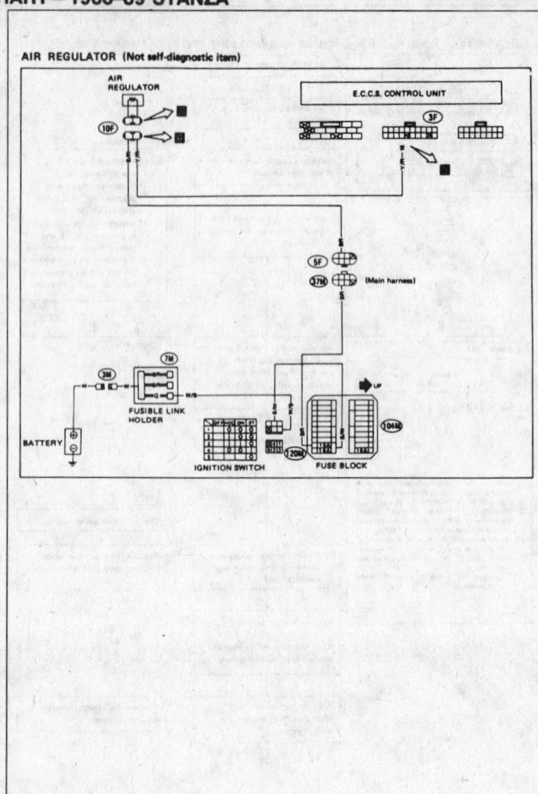
AIR REGULATOR (Not self-diagnostic item)
AIR REGULATOR
E.C.C.S. CONTROL UNIT
10F
3F
5F
37M
(Main harness)
3M
7M
FUSIBLE LINK HOLDER
BATTERY
IGNITION SWITCH
FUSE BLOCK
120M
104M
UP

ECCS DIAGNOSTIC CHART – 1988–89 STANZA

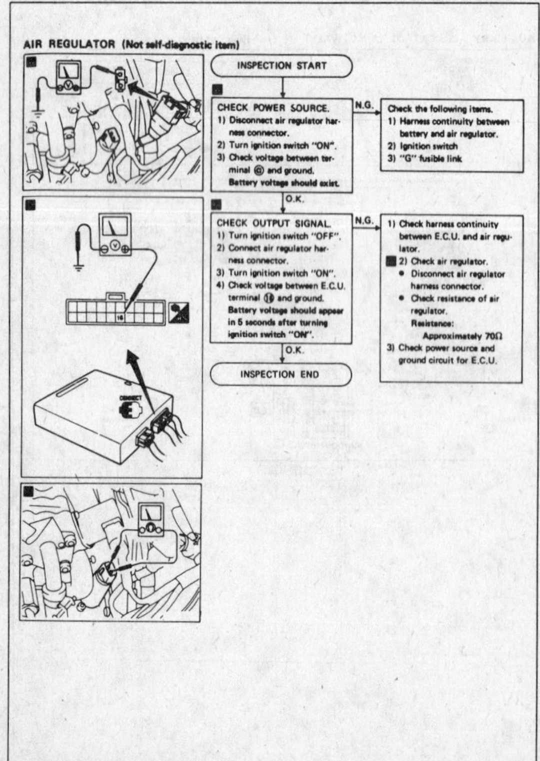
AIR REGULATOR (Not self-diagnostic item)
INSPECTION START
CHECK POWER SOURCE.
1) Disconnect air regulator harness connector.
2) Turn ignition switch "ON".
3) Check voltage between terminal ⓐ and ground.
Battery voltage should exist.
N.G.
Check the following items.
1) Harness continuity between battery and air regulator.
2) Ignition switch
3) "G" fusible link
O.K.
CHECK OUTPUT SIGNAL.
1) Turn ignition switch "OFF".
2) Connect air regulator harness connector.
3) Turn ignition switch "ON".
4) Check voltage between E.C.U. terminal ⑯ and ground.
Battery voltage should appear in 5 seconds after turning ignition switch "ON".
1) Check harness continuity between E.C.U. and air regulator.
2) Check air regulator.
• Disconnect air regulator harness connector.
• Check resistance of air regulator.
Resistance:
Approximately 70Ω
3) Check power source and ground circuit for E.C.U.
INSPECTION END
CONNECT

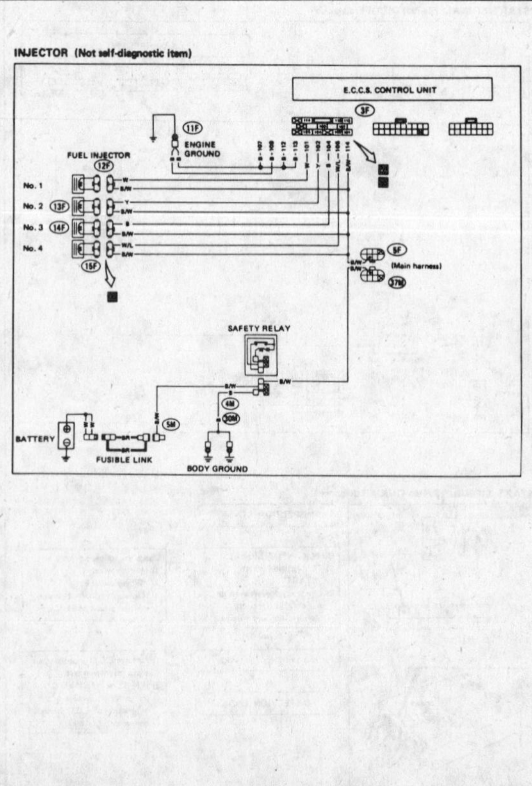
INJECTOR (Not self-diagnostic item)
E.C.C.S. CONTROL UNIT
3F
11F
ENGINE GROUND
FUEL INJECTOR
12F
No. 1
No. 2
13F
No. 3
14F
No. 4
15F
5F
(Main harness)
37M
SAFETY RELAY
4M
30M
5M
BATTERY
FUSIBLE LINK
BODY GROUND

ECCS DIAGNOSTIC CHART—1988–89 STANZA

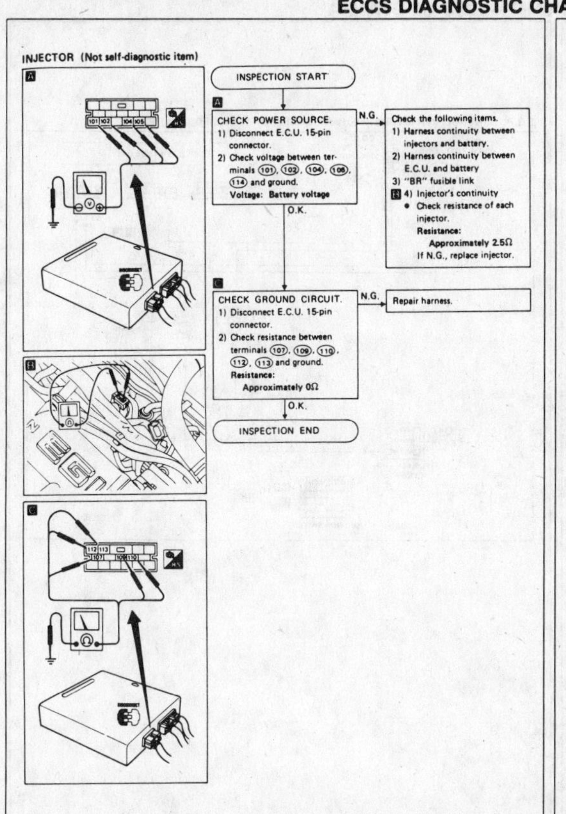
INJECTOR (Not self-diagnostic item)
INSPECTION START
A
CHECK POWER SOURCE.
1) Disconnect E.C.U. 15-pin connector.
2) Check voltage between terminals (101), (102), (104), (106), (114) and ground.
Voltage: Battery voltage
N.G.
Check the following items.
1) Harness continuity between injectors and battery.
2) Harness continuity between E.C.U. and battery
3) "BR" fusible link
B 4) Injector's continuity
• Check resistance of each injector.
Resistance:
Approximately 2.5Ω
If N.G., replace injector.
O.K.
C
CHECK GROUND CIRCUIT.
1) Disconnect E.C.U. 15-pin connector.
2) Check resistance between terminals (107), (109), (110), (112), (113) and ground.
Resistance:
Approximately 0Ω
N.G.
Repair harness.
O.K.
INSPECTION END

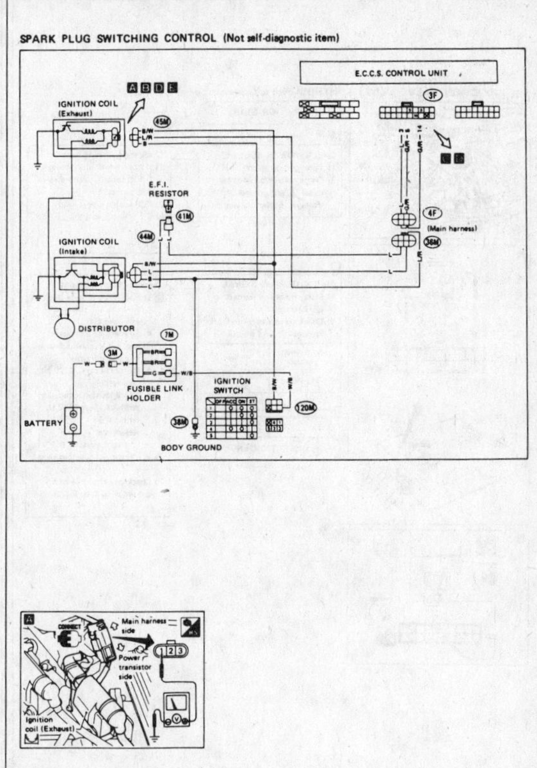
SPARK PLUG SWITCHING CONTROL (Not self-diagnostic item)
E.C.C.S. CONTROL UNIT
IGNITION COIL (Exhaust)
E.F.I. RESISTOR
IGNITION COIL (Intake)
DISTRIBUTOR
(Main harness)
FUSIBLE LINK HOLDER
IGNITION SWITCH
BATTERY
BODY GROUND
Main harness side
Power transistor side
Ignition coil (Exhaust)

ECCS DIAGNOSTIC CHART—1988–89 STANZA

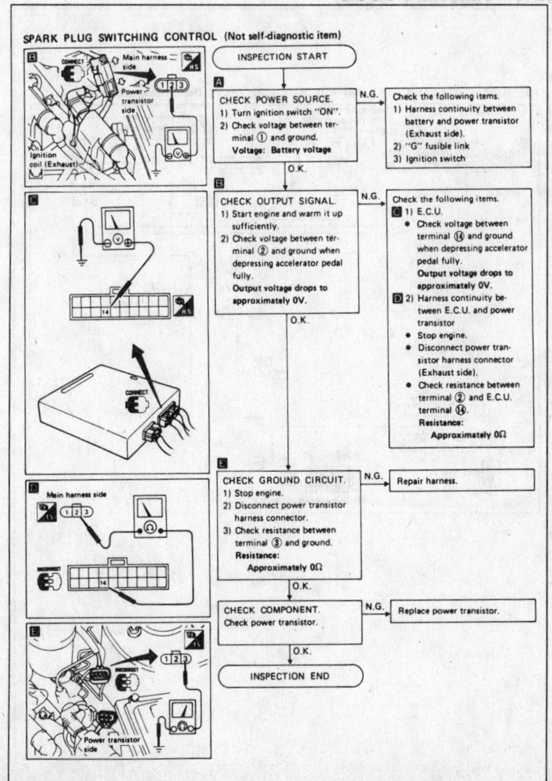
SPARK PLUG SWITCHING CONTROL (Not self-diagnostic item)
INSPECTION START
A
CHECK POWER SOURCE.
1) Turn ignition switch "ON".
2) Check voltage between terminal ① and ground.
Voltage: Battery voltage
N.G.
Check the following items.
1) Harness continuity between battery and power transistor (Exhaust side).
2) "G" fusible link
3) Ignition switch
O.K.
B
CHECK OUTPUT SIGNAL.
1) Start engine and warm it up sufficiently.
2) Check voltage between terminal ② and ground when depressing accelerator pedal fully.
Output voltage drops to approximately 0V.
N.G.
Check the following items.
C 1) E.C.U.
• Check voltage between terminal ⑭ and ground when depressing accelerator pedal fully.
Output voltage drops to approximately 0V.
D 2) Harness continuity between E.C.U. and power transistor
• Stop engine.
• Disconnect power transistor harness connector (Exhaust side).
• Check resistance between terminal ② and E.C.U. terminal ⑭.
Resistance:
Approximately 0Ω
O.K.
E
CHECK GROUND CIRCUIT.
1) Stop engine.
2) Disconnect power transistor harness connector.
3) Check resistance between terminal ③ and ground.
Resistance:
Approximately 0Ω
N.G.
Repair harness.
O.K.
CHECK COMPONENT.
Check power transistor.
N.G.
Replace power transistor.
O.K.
INSPECTION END
Main harness side
Power transistor side
Ignition coil (Exhaust)

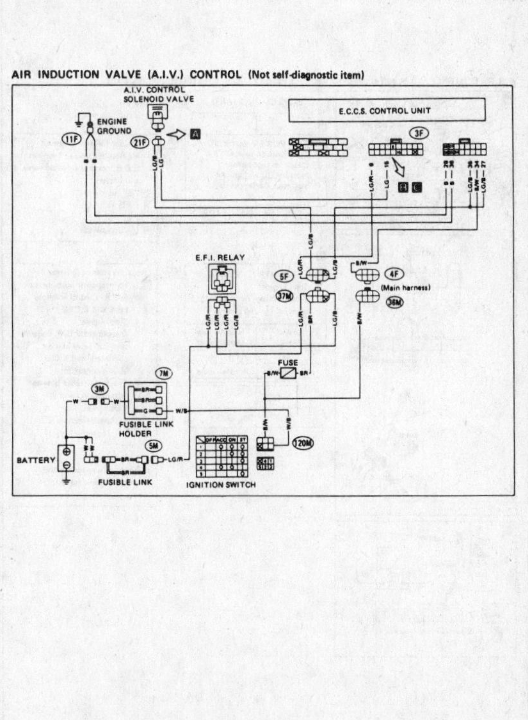
AIR INDUCTION VALVE (A.I.V.) CONTROL (Not self-diagnostic item)
A.I.V. CONTROL SOLENOID VALVE
ENGINE GROUND
E.C.C.S. CONTROL UNIT
E.F.I. RELAY
(Main harness)
FUSE
FUSIBLE LINK HOLDER
BATTERY
FUSIBLE LINK
IGNITION SWITCH

ECCS DIAGNOSTIC CHART—1988–89 STANZA

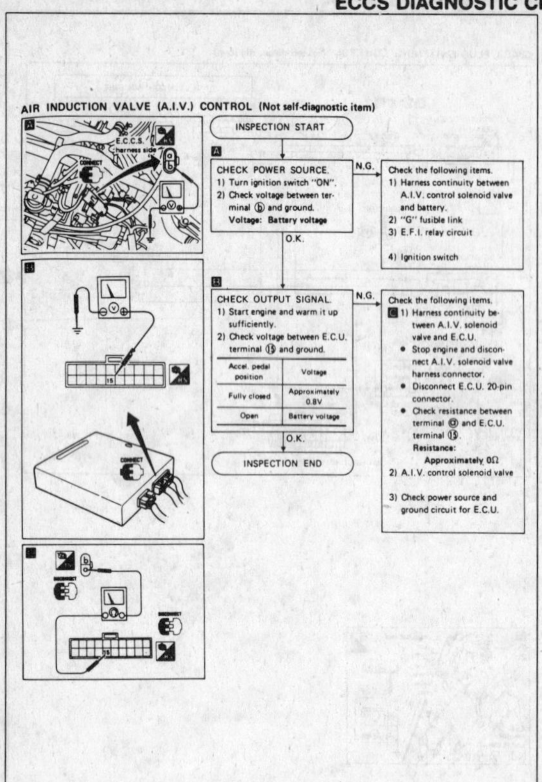
AIR INDUCTION VALVE (A.I.V.) CONTROL (Not self-diagnostic item)
INSPECTION START
CHECK POWER SOURCE.
1) Turn ignition switch "ON".
2) Check voltage between terminal ⓑ and ground.
Voltage: Battery voltage
N.G.
Check the following items.
1) Harness continuity between A.I.V. control solenoid valve and battery.
2) "G" fusible link
3) E.F.I. relay circuit
4) Ignition switch
O.K.
CHECK OUTPUT SIGNAL.
1) Start engine and warm it up sufficiently.
2) Check voltage between E.C.U. terminal ⑮ and ground.
Accel. pedal position | Voltage
Fully closed | Approximately 0.8V
Open | Battery voltage
N.G.
Check the following items.
1) Harness continuity between A.I.V. solenoid valve and E.C.U.
• Stop engine and disconnect A.I.V. solenoid valve harness connector.
• Disconnect E.C.U. 20-pin connector.
• Check resistance between terminal ⓑ and E.C.U. terminal ⑮.
Resistance: Approximately 0Ω
2) A.I.V. control solenoid valve
3) Check power source and ground circuit for E.C.U.
O.K.
INSPECTION END

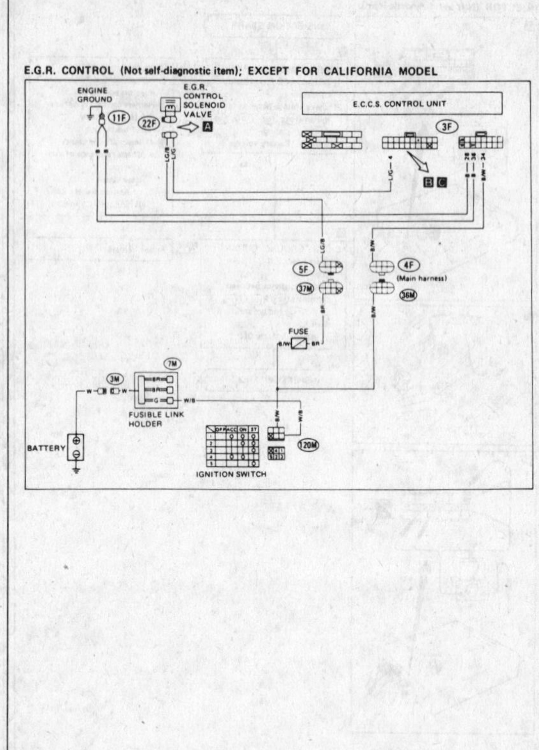
E.G.R. CONTROL (Not self-diagnostic item); EXCEPT FOR CALIFORNIA MODEL
ENGINE GROUND
E.G.R. CONTROL SOLENOID VALVE
E.C.C.S. CONTROL UNIT
(Main harness)
FUSE
FUSIBLE LINK HOLDER
BATTERY
IGNITION SWITCH

ECCS DIAGNOSTIC CHART—1988–89 STANZA

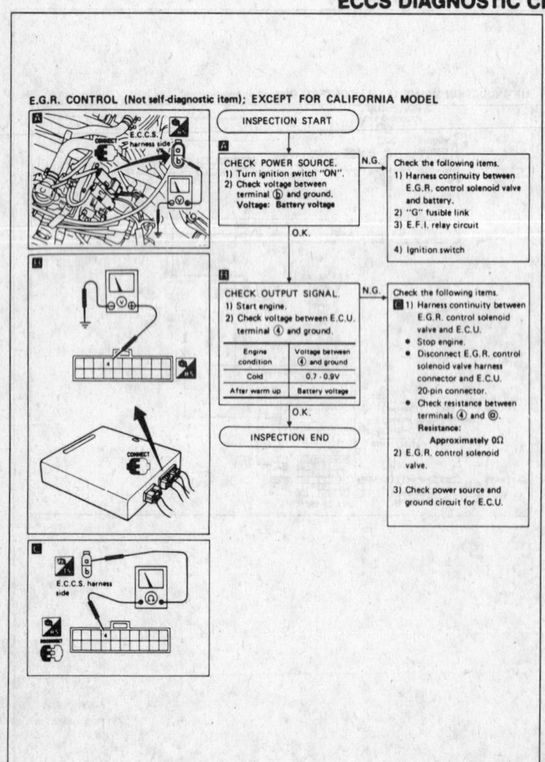
E.G.R. CONTROL (Not self-diagnostic item); EXCEPT FOR CALIFORNIA MODEL
INSPECTION START
CHECK POWER SOURCE.
1) Turn ignition switch "ON".
2) Check voltage between terminal ⓑ and ground.
Voltage: Battery voltage
N.G.
Check the following items.
1) Harness continuity between E.G.R. control solenoid valve and battery.
2) "G" fusible link
3) E.F.I. relay circuit
4) Ignition switch
O.K.
CHECK OUTPUT SIGNAL.
1) Start engine.
2) Check voltage between E.C.U. terminal ④ and ground.
Engine condition | Voltage between ④ and ground
Cold | 0.7 - 0.9V
After warm up | Battery voltage
N.G.
Check the following items.
1) Harness continuity between E.G.R. control solenoid valve and E.C.U.
• Stop engine.
• Disconnect E.G.R. control solenoid valve harness connector and E.C.U. 20-pin connector.
• Check resistance between terminals ④ and ⓑ.
Resistance: Approximately 0Ω
2) E.G.R. control solenoid valve.
3) Check power source and ground circuit for E.C.U.
O.K.
INSPECTION END
E.C.C.S. harness side

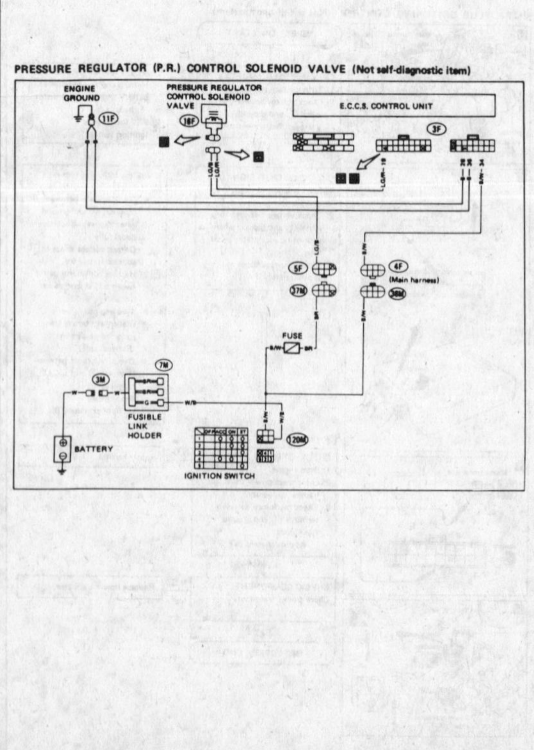
PRESSURE REGULATOR (P.R.) CONTROL SOLENOID VALVE (Not self-diagnostic item)
ENGINE GROUND
PRESSURE REGULATOR CONTROL SOLENOID VALVE
E.C.C.S. CONTROL UNIT
(Main harness)
FUSE
FUSIBLE LINK HOLDER
BATTERY
IGNITION SWITCH

ECCS DIAGNOSTIC CHART—1988-89 STANZA

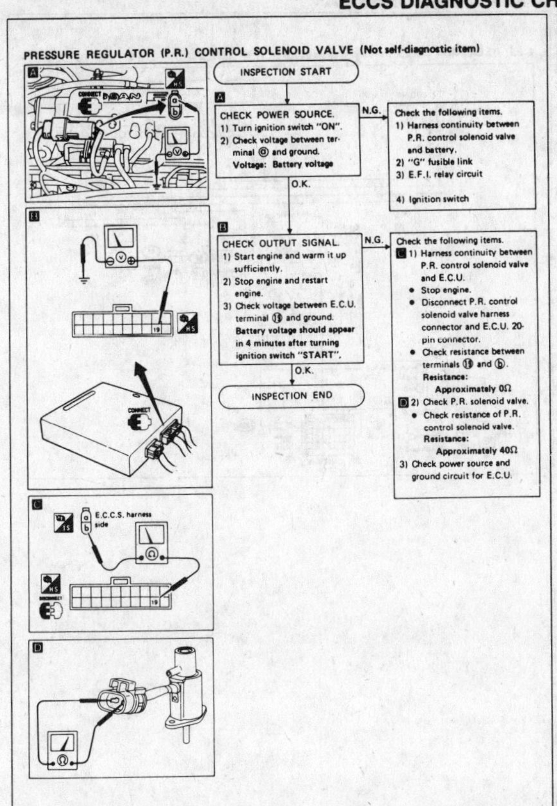
PRESSURE REGULATOR (P.R.) CONTROL SOLENOID VALVE (Not self-diagnostic item)
INSPECTION START
A
CHECK POWER SOURCE.
1) Turn ignition switch "ON".
2) Check voltage between terminal ⓐ and ground.
Voltage: Battery voltage
N.G.
Check the following items.
1) Harness continuity between P.R. control solenoid valve and battery.
2) "G" fusible link
3) E.F.I. relay circuit
4) Ignition switch
O.K.
B
CHECK OUTPUT SIGNAL.
1) Start engine and warm it up sufficiently.
2) Stop engine and restart engine.
3) Check voltage between E.C.U. terminal ⑲ and ground.
Battery voltage should appear in 4 minutes after turning ignition switch "START".
N.G.
Check the following items.
C 1) Harness continuity between P.R. control solenoid valve and E.C.U.
• Stop engine.
• Disconnect P.R. control solenoid valve harness connector and E.C.U. 20-pin connector.
• Check resistance between terminals ⑲ and ⓑ.
Resistance: Approximately 0Ω
D 2) Check P.R. solenoid valve.
• Check resistance of P.R. control solenoid valve.
Resistance: Approximately 40Ω
3) Check power source and ground circuit for E.C.U.
O.K.
INSPECTION END
E.C.C.S. harness side

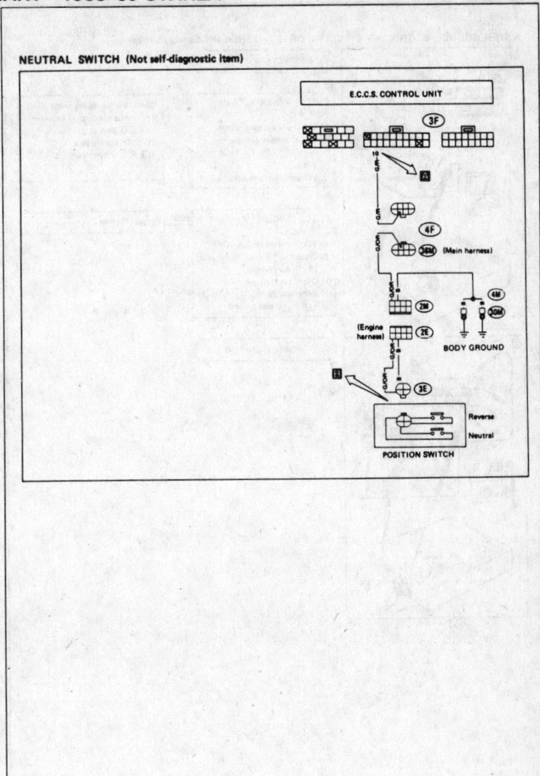
NEUTRAL SWITCH (Not self-diagnostic item)
E.C.C.S. CONTROL UNIT
3F
4F
38M (Main harness)
2M
4M
30M
(Engine harness)
2E
BODY GROUND
3E
Reverse
Neutral
POSITION SWITCH

ECCS DIAGNOSTIC CHART—1988-89 STANZA

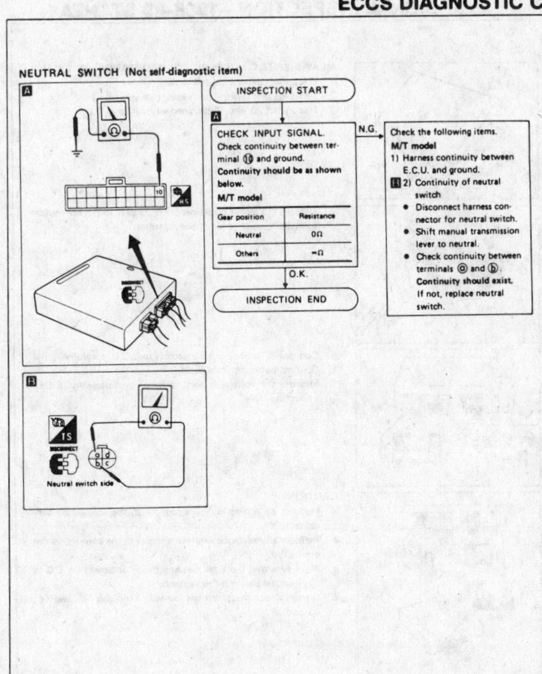
NEUTRAL SWITCH (Not self-diagnostic item)
INSPECTION START
A
CHECK INPUT SIGNAL.
Check continuity between terminal ⑩ and ground.
Continuity should be as shown below.
M/T model
Gear position | Resistance
Neutral | 0Ω
Others | ∞Ω
N.G.
Check the following items.
M/T model
1) Harness continuity between E.C.U. and ground.
B 2) Continuity of neutral switch
• Disconnect harness connector for neutral switch.
• Shift manual transmission lever to neutral.
• Check continuity between terminals ⓒ and ⓓ.
Continuity should exist.
If not, replace neutral switch.
O.K.
INSPECTION END
Neutral switch side

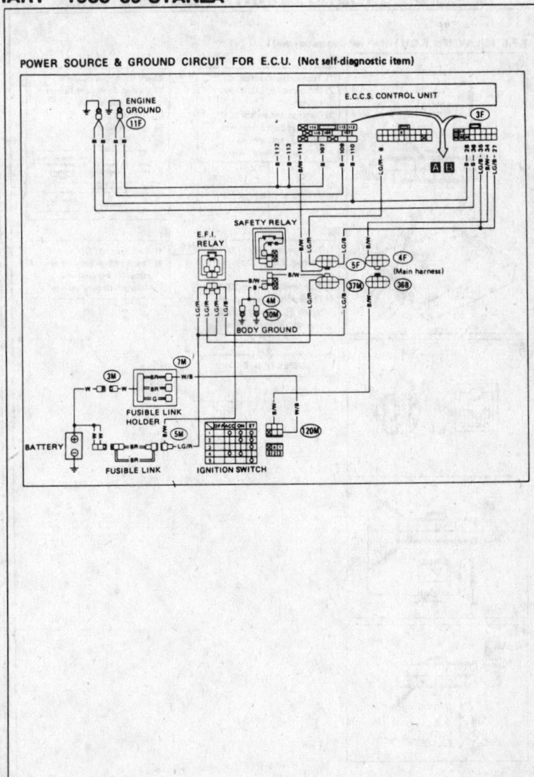
POWER SOURCE & GROUND CIRCUIT FOR E.C.U. (Not self-diagnostic item)
E.C.C.S. CONTROL UNIT
ENGINE GROUND
11F
3F
SAFETY RELAY
E.F.I. RELAY
5F
4F
37M
36B
(Main harness)
4M
30M
BODY GROUND
7M
3M
FUSIBLE LINK HOLDER
5M
120M
BATTERY
FUSIBLE LINK
IGNITION SWITCH

ECCS DIAGNOSTIC CHART—1988–89 STANZA

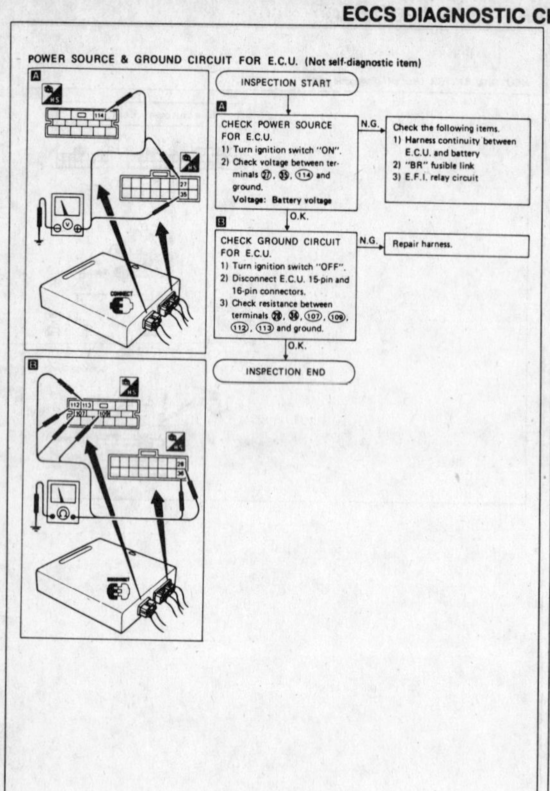
POWER SOURCE & GROUND CIRCUIT FOR E.C.U. (Not self-diagnostic item)
INSPECTION START
A
CHECK POWER SOURCE FOR E.C.U.
1) Turn ignition switch "ON".
2) Check voltage between terminals 27, 35, 114 and ground.
Voltage: Battery voltage
N.G.
Check the following items.
1) Harness continuity between E.C.U. and battery
2) "BR" fusible link
3) E.F.I. relay circuit
O.K.
B
CHECK GROUND CIRCUIT FOR E.C.U.
1) Turn ignition switch "OFF".
2) Disconnect E.C.U. 15-pin and 16-pin connectors.
3) Check resistance between terminals 26, 38, 107, 109, 112, 113 and ground.
N.G.
Repair harness.
O.K.
INSPECTION END
CONNECT

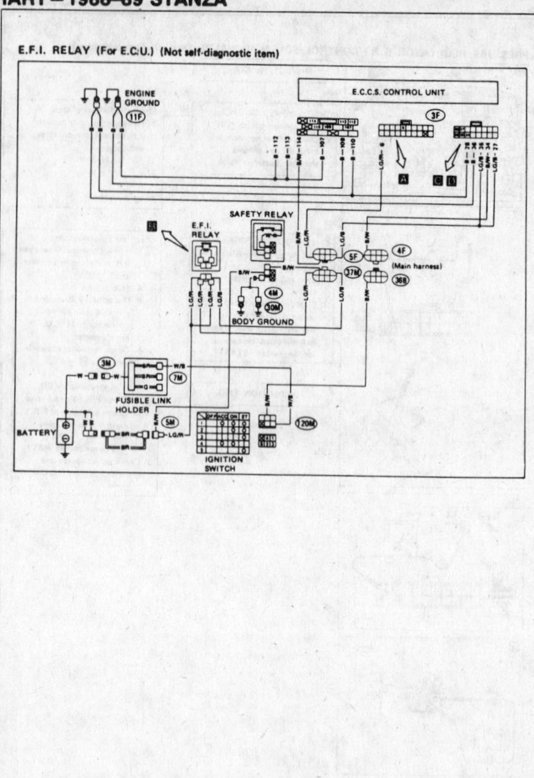
E.F.I. RELAY (For E.C.U.) (Not self-diagnostic item)
ENGINE GROUND
11F
E.C.C.S. CONTROL UNIT
3F
SAFETY RELAY
E.F.I. RELAY
5F
4F
37M
(Main harness)
36B
3M
4M
20M
BODY GROUND
7M
FUSIBLE LINK HOLDER
BATTERY
5M
120M
IGNITION SWITCH

ECCS DIAGNOSTIC CHART—1988–89 STANZA

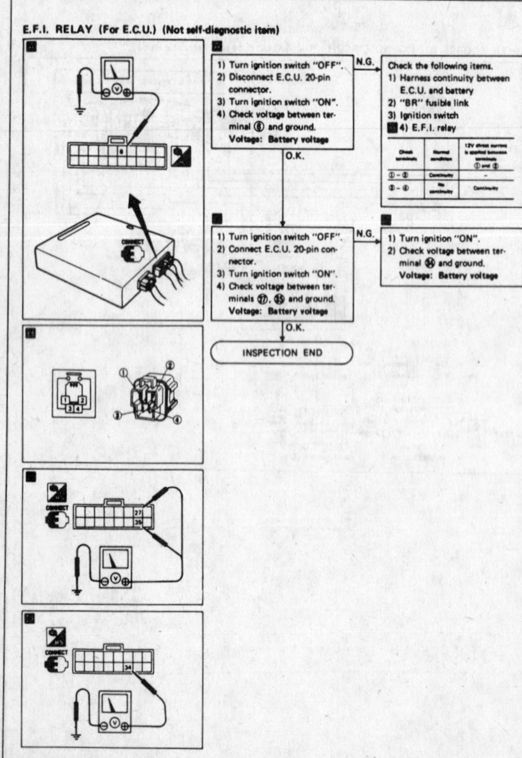
E.F.I. RELAY (For E.C.U.) (Not self-diagnostic item)
1) Turn ignition switch "OFF".
2) Disconnect E.C.U. 20-pin connector.
3) Turn ignition switch "ON".
4) Check voltage between terminal 6 and ground.
Voltage: Battery voltage
N.G.
Check the following items.
1) Harness continuity between E.C.U. and battery
2) "BR" fusible link
3) Ignition switch
4) E.F.I. relay
O.K.
1) Turn ignition switch "OFF".
2) Connect E.C.U. 20-pin connector.
3) Turn ignition switch "ON".
4) Check voltage between terminals 27, 35 and ground.
Voltage: Battery voltage
N.G.
1) Turn ignition "ON".
2) Check voltage between terminal 34 and ground.
Voltage: Battery voltage
O.K.
INSPECTION END
CONNECT

ECU SIGNAL INSPECTION—1988–89 STANZA

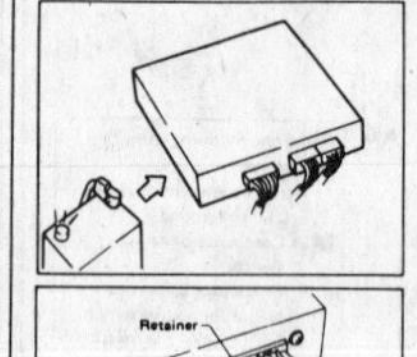

MEASUREMENT VOLTAGE OR RESISTANCE OF E.C.U.

1. Disconnect battery ground cable.
2. Remove assist side or bench seat from vehicle.
3. Disconnect 20- and 16-pin connectors from E.C.U.

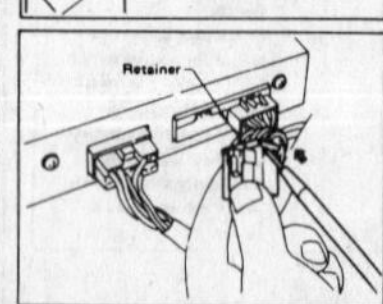
Retainer

4. Remove pin terminal retainer from 20- and 16-pin connectors to make it easier to insert tester probes.

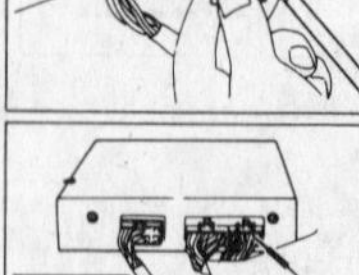
Battery

5. Connect 20- and 16-pin connectors to E.C.U. carefully.
6. Connect battery ground cable.
7. Measure the voltage at each terminal by following "E.C.U. inspection table".

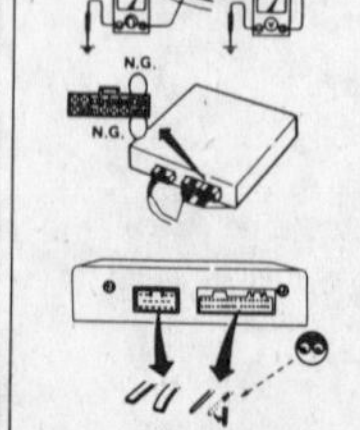
N.G.
N.G.

CAUTION:

a. Perform all voltage measurements with the connectors connected.
b. Perform all resistance measurements with the connectors disconnected.
c. Make sure that there are not any bends or breaks on E.C.U. pin terminal before measurements.
d. Do not touch tester probes between terminals 27 and 28, 35 and 36.

ECU SIGNAL INSPECTION (CONT.)—1988–89 STANZA

E.C.U. inspection table

*Data are reference values.

TERMINAL NO.	ITEM	CONDITION	*DATA
2	A.A.C. valve	Engine is running. At idle (after warming up)	6.0 - 8.0V
3	Ignition signal (from resistor)	Ignition switch "ON"	BATTERY VOLTAGE (11 - 14V)
4	E.G.R. control solenoid valve	Engine is running. Engine is cold. [Water temperature is below 60°C (140°F).]	0.7 - 0.9V
		Engine is running. After warming up [Water temperature is between 60°C (140°F) and 105°C (221°F).]	BATTERY VOLTAGE (11 - 14V)
5	Ignition signal (from intake side power transistor)	Engine is running.	0.4 - 2.2V Output voltage varies with engine speed.
6	E.F.I. relay	Ignition switch "ON"	0.8 - 1.0V
7	Exhaust gas temperature sensor	Engine is running. E.G.R. system is operated.	0 - 1.0V
		Engine is running. E.G.R. system is not operated.	1.0 - 2.0V
8	Crank angle sensor (position signal)	Engine is running Do not turn engine at high speed under no-load.	2.2 - 2.8V
9	Start signal	Ignition switch "START"	BATTERY VOLTAGE (11 - 14V)
10	Neutral signal	Ignition switch "ON" Gear position: Neutral (M/T) : N or P range (A/T)	0V
		Ignition switch "ON" Gear position: [Except neutral (M/T) Except N or P range (A/T)]	BATTERY VOLTAGE (11 - 14V)

*Data are reference values.

TERMINAL NO.	ITEM	CONDITION	*DATA
14	Ignition signal (from exhaust side power transistor)	Engine is running. Do not turn engine at high speed under no-load.	0.4 - 2.2V (Output voltage varies with engine revolution.)
15	A.I.V. control solenoid valve	Engine is running. At idle	0.7 - 0.9V
		Engine is running. When depressing accelerator pedal [Water temperature is above 50°C (122°F).]	BATTERY VOLTAGE (11 - 14V)
16	Air regulator	Ignition switch "ON" For 5 seconds after turning ignition switch "ON"	0.7 - 0.9V
		Ignition switch "ON" After 5 seconds have passed	BATTERY VOLTAGE (11 - 14V)
17	Crank angle sensor (Reference signal)	Engine is running. Do not turn engine at high speed under no-load.	0.2 - 0.4V
18	Idle switch (⊖ side)	Ignition switch "ON" Throttle valve: idle position	9 - 10V
		Ignition switch "ON" Throttle valve: except idle position	0V

ECU SIGNAL INSPECTION (CONT.)—1988–89 STANZA

*Data are reference values.

TERMINAL NO.	ITEM	CONDITION	*DATA
19	Pressure regulator control solenoid valve	Ignition switch "ON" For approximately 4 minutes after turning ignition switch to "START". [Water temperature is above 60°C (140°F).]	0.7 - 0.9V
		Ignition switch "ON" Approximately 4 minutes after turning ignition switch to "START". [Water temperature is above 60°C (140°F).]	BATTERY VOLTAGE (11 - 14V)
		Ignition switch "ON" or "START". [Water temperature is below 60°C (140°F).]	
21	Full throttle switch (⊖ side)	Ignition switch "ON" Throttle valve: fully open	9 - 10V
		Ignition switch "ON" Throttle valve: Any position except full throttle	0V
22	Air conditioner signal (Air conditioner equipped model)	Ignition switch "ON" Air conditioner switch and heater fan switch "ON"	BATTERY VOLTAGE (11 - 14V)
		Ignition switch "ON" Air conditioner switch "OFF"	0V
23	Water temperature sensor	Engine is running.	1.0 - 5.0V Output voltage varies with engine water temperature.
24	Exhaust gas sensor	Engine is running. After warming up sufficiently.	0 - Approximately 1.0V
25	Idle switch and full throttle switch (⊕ side)	Ignition switch "ON"	9 - 10V
27 35	Power source for E.C.U.	Ignition switch "ON"	BATTERY VOLTAGE (11 - 14V)

*Data are reference values.

TERMINAL NO.	ITEM	CONDITION	*DATA
29	Vehicle speed sensor	Ignition switch "ON" When rotating front wheel slowly	Voltage varies between 0V and approximately 5V.
30	Air temperature sensor	Ignition switch "ON"	Approximately 3V [Air temperature is 20°C (68°F).] Output voltage varies with air temperature.
31	Air flow meter	Engine is running. Do not turn engine at high speed under no-load.	2 - 4V Output voltage varies with engine revolution.
33	Power source for air flow meter	Ignition switch "ON"	8V
34	Ignition switch signal	Ignition switch "ON"	BATTERY VOLTAGE (11 - 14V)
101 102 104 105 114	Injector	Ignition switch "OFF"	BATTERY VOLTAGE (11 - 14V)
108	Fuel pump	Ignition switch "ON" For 5 seconds after turning ignition switch "ON".	0.7 - 0.9V
		Ignition switch "ON" After 5 seconds have passed	BATTERY VOLTAGE (9 - 14V)

PIN CONNECTOR TERMINAL LAYOUT

15-pin connector

112	113		114	
	107	108	109	110
101	102		104	105

HS

20-pin connector

1	2	3	4	5	6		8	9	10
11	12	13	14	15	16	17	18	19	20

HS

16-pin connector

21	22	23	24	25	26	27	28
29	30	31	32	33	34	35	36

HS

ECCS MIXTURE RATIO FEEDBACK SYSTEM INSPECTION—1988–89 STANZA

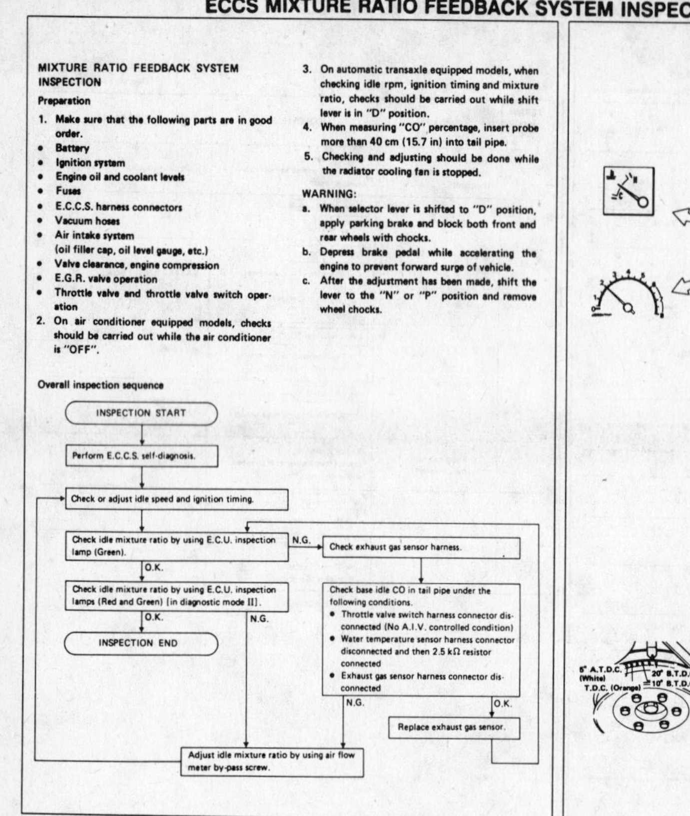

MIXTURE RATIO FEEDBACK SYSTEM INSPECTION

Preparation

1. Make sure that the following parts are in good order.
   - Battery
   - Ignition system
   - Engine oil and coolant levels
   - Fuses
   - E.C.C.S. harness connectors
   - Vacuum hoses
   - Air intake system (oil filler cap, oil level gauge, etc.)
   - Valve clearance, engine compression
   - E.G.R. valve operation
   - Throttle valve and throttle valve switch operation
2. On air conditioner equipped models, checks should be carried out while the air conditioner is "OFF".
3. On automatic transaxle equipped models, when checking idle rpm, ignition timing and mixture ratio, checks should be carried out while shift lever is in "D" position.
4. When measuring "CO" percentage, insert probe more than 40 cm (15.7 in) into tail pipe.
5. Checking and adjusting should be done while the radiator cooling fan is stopped.

WARNING:

a. When selector lever is shifted to "D" position, apply parking brake and block both front and rear wheels with chocks.
b. Depress brake pedal while accelerating the engine to prevent forward surge of vehicle.
c. After the adjustment has been made, shift the lever to the "N" or "P" position and remove wheel chocks.

Overall inspection sequence

INSPECTION START
Perform E.C.C.S. self-diagnosis.
Check or adjust idle speed and ignition timing.
Check idle mixture ratio by using E.C.U. inspection lamp (Green).
N.G.
Check exhaust gas sensor harness.
O.K.
Check idle mixture ratio by using E.C.U. inspection lamps (Red and Green) [in diagnostic mode II].
O.K.
N.G.
INSPECTION END
Check base idle CO in tail pipe under the following conditions.
• Throttle valve switch harness connector disconnected (No A.I.V. controlled condition)
• Water temperature sensor harness connector disconnected and then 2.5 kΩ resistor connected
• Exhaust gas sensor harness connector disconnected
N.G.
O.K.
Replace exhaust gas sensor.
Adjust idle mixture ratio by using air flow meter by-pass screw.

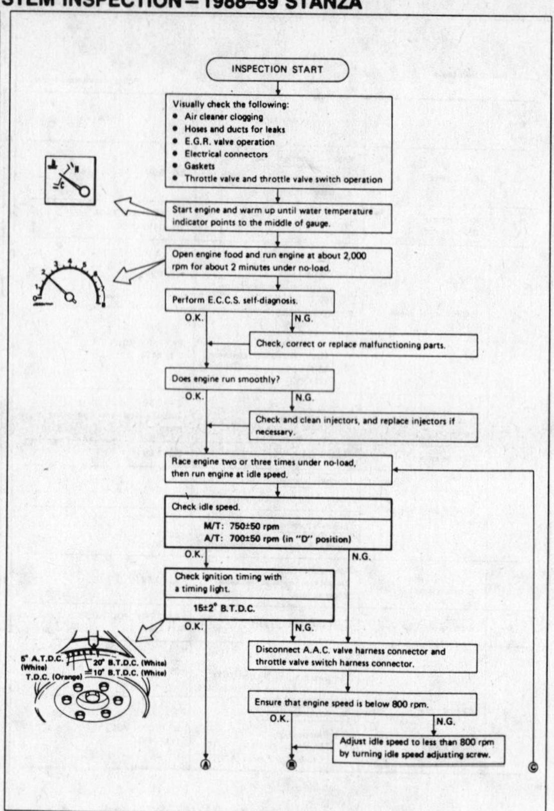

INSPECTION START
Visually check the following:
• Air cleaner clogging
• Hoses and ducts for leaks
• E.G.R. valve operation
• Electrical connectors
• Gaskets
• Throttle valve and throttle valve switch operation
Start engine and warm up until water temperature indicator points to the middle of gauge.
Open engine food and run engine at about 2,000 rpm for about 2 minutes under no-load.
Perform E.C.C.S. self-diagnosis.
O.K.
N.G.
Check, correct or replace malfunctioning parts.
Does engine run smoothly?
O.K.
N.G.
Check and clean injectors, and replace injectors if necessary.
Race engine two or three times under no-load, then run engine at idle speed.
Check idle speed.
M/T: 750±50 rpm
A/T: 700±50 rpm (in "D" position)
O.K.
N.G.
Check ignition timing with a timing light.
15±2° B.T.D.C.
O.K.
N.G.
5° A.T.D.C. (White)
20° B.T.D.C. (White)
10° B.T.D.C. (White)
T.D.C. (Orange)
Disconnect A.A.C. valve harness connector and throttle valve switch harness connector.
Ensure that engine speed is below 800 rpm.
O.K.
N.G.
Adjust idle speed to less than 800 rpm by turning idle speed adjusting screw.
A
B
C

ECCS MIXTURE RATIO FEEDBACK SYSTEM INSPECTION (CONT.)—1988–89 STANZA

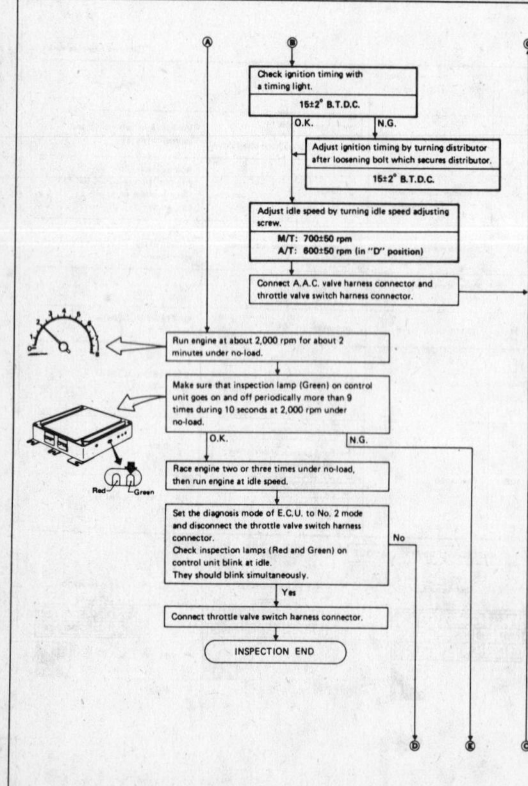

A
B
C
Check ignition timing with a timing light.
15±2° B.T.D.C.
O.K.
N.G.
Adjust ignition timing by turning distributor after loosening bolt which secures distributor.
15±2° B.T.D.C.
Adjust idle speed by turning idle speed adjusting screw.
M/T: 700±50 rpm
A/T: 600±50 rpm (in "D" position)
Connect A.A.C. valve harness connector and throttle valve switch harness connector.
Run engine at about 2,000 rpm for about 2 minutes under no-load.
Make sure that inspection lamp (Green) on control unit goes on and off periodically more than 9 times during 10 seconds at 2,000 rpm under no-load.
O.K.
N.G.
Red
Green
Race engine two or three times under no-load, then run engine at idle speed.
Set the diagnosis mode of E.C.U. to No. 2 mode and disconnect the throttle valve switch harness connector.
Check inspection lamps (Red and Green) on control unit blink at idle.
They should blink simultaneously.
No
Yes
Connect throttle valve switch harness connector.
INSPECTION END
D
E
C

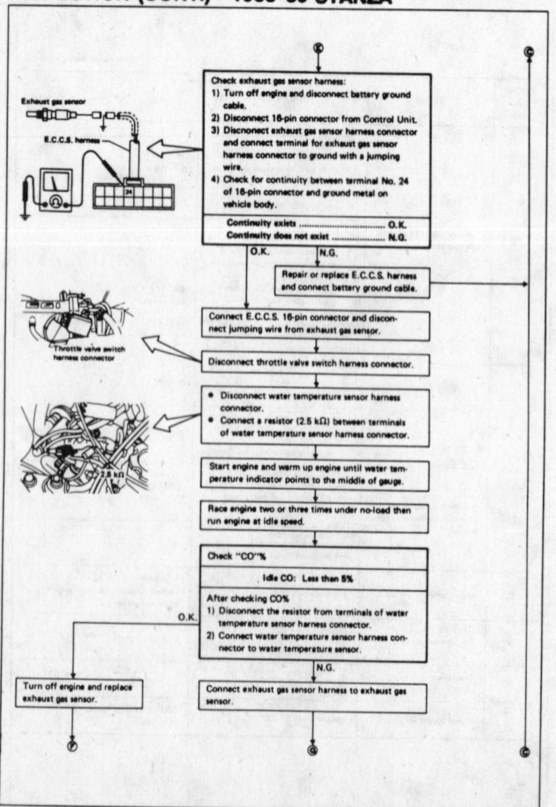

E
C
Check exhaust gas sensor harness:
1) Turn off engine and disconnect battery ground cable.
2) Disconnect 16-pin connector from Control Unit.
3) Disconnect exhaust gas sensor harness connector and connect terminal for exhaust gas sensor harness connector to ground with a jumping wire.
4) Check for continuity between terminal No. 24 of 16-pin connector and ground metal on vehicle body.
Continuity exists .... O.K.
Continuity does not exist .... N.G.
Exhaust gas sensor
E.C.C.S. harness
24
O.K.
N.G.
Repair or replace E.C.C.S. harness and connect battery ground cable.
Connect E.C.C.S. 16-pin connector and disconnect jumping wire from exhaust gas sensor.
Throttle valve switch harness connector
Disconnect throttle valve switch harness connector.
• Disconnect water temperature sensor harness connector.
• Connect a resistor (2.5 kΩ) between terminals of water temperature sensor harness connector.
2.5 kΩ
Start engine and warm up engine until water temperature indicator points to the middle of gauge.
Race engine two or three times under no-load then run engine at idle speed.
Check "CO"%
Idle CO: Less than 5%
After checking CO%
1) Disconnect the resistor from terminals of water temperature sensor harness connector.
2) Connect water temperature sensor harness connector to water temperature sensor.
O.K.
N.G.
Turn off engine and replace exhaust gas sensor.
Connect exhaust gas sensor harness to exhaust gas sensor.
F
G
C

ECCS MIXTURE RATIO FEEDBACK SYSTEM INSPECTION (CONT.)—1988–89 STANZA

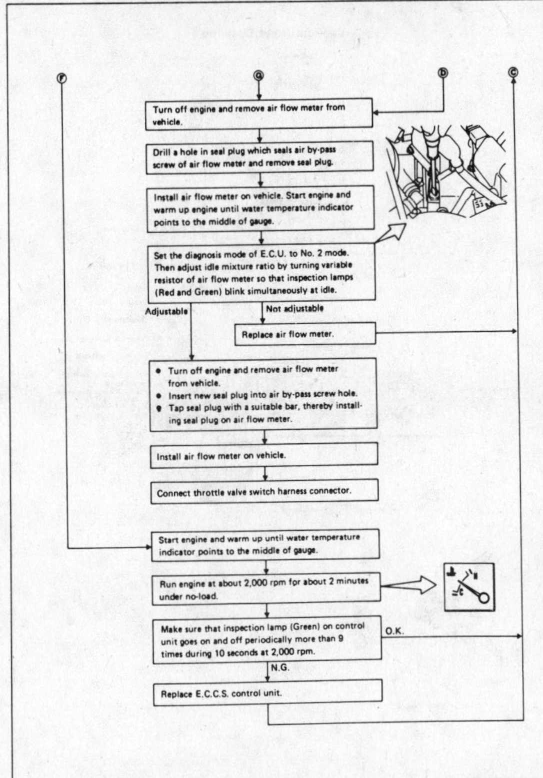
Turn off engine and remove air flow meter from vehicle.
Drill a hole in seal plug which seals air by-pass screw of air flow meter and remove seal plug.
Install air flow meter on vehicle. Start engine and warm up engine until water temperature indicator points to the middle of gauge.
Set the diagnosis mode of E.C.U. to No. 2 mode. Then adjust idle mixture ratio by turning variable resistor of air flow meter so that inspection lamps (Red and Green) blink simultaneously at idle.
Adjustable
Not adjustable
Replace air flow meter.
Turn off engine and remove air flow meter from vehicle.
Insert new seal plug into air by-pass screw hole.
Tap seal plug with a suitable bar, thereby installing seal plug on air flow meter.
Install air flow meter on vehicle.
Connect throttle valve switch harness connector.
Start engine and warm up until water temperature indicator points to the middle of gauge.
Run engine at about 2,000 rpm for about 2 minutes under no-load.
Make sure that inspection lamp (Green) on control unit goes on and off periodically more than 9 times during 10 seconds at 2,000 rpm.
O.K.
N.G.
Replace E.C.C.S. control unit.

ECCS COMPONENTS LOCATION—1990 STANZA

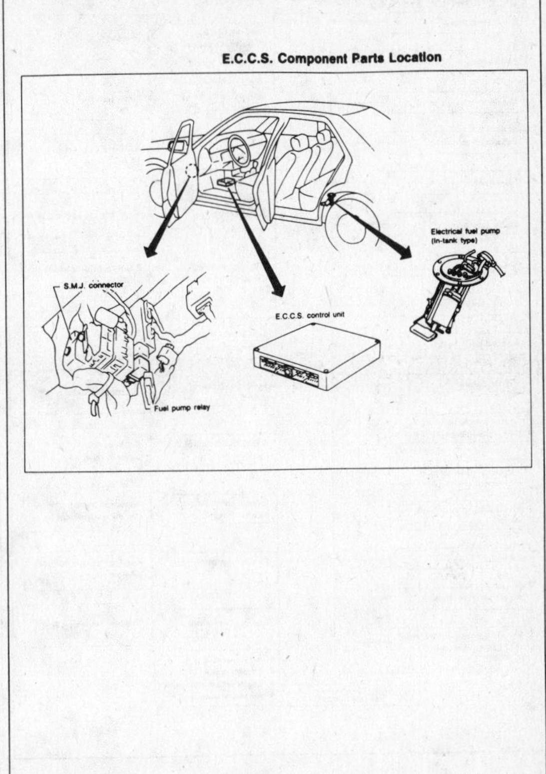
E.C.C.S. Component Parts Location
Electrical fuel pump (In-tank type)
S.M.J. connector
E.C.C.S. control unit
Fuel pump relay

ECCS COMPONENTS LOCATION (CONT.)—1990 STANZA

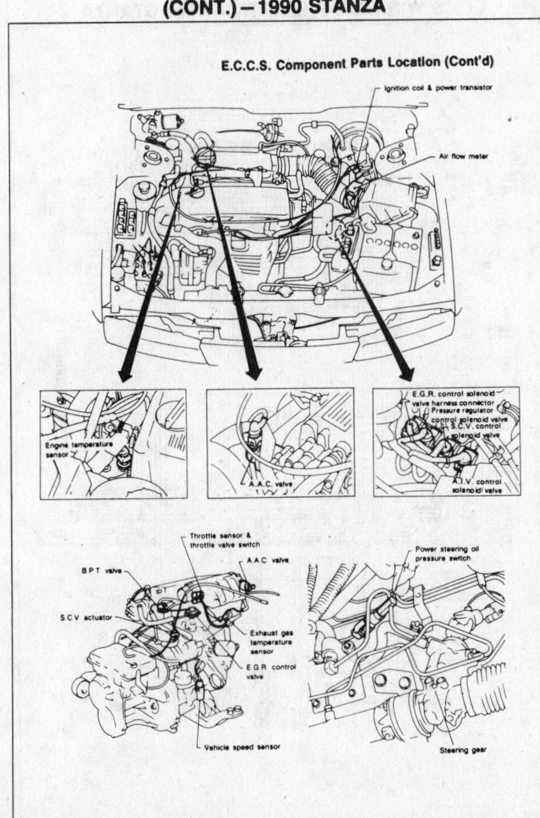
E.C.C.S. Component Parts Location (Cont'd)
Ignition coil & power transistor
Air flow meter
Engine temperature sensor
A.A.C. valve
E.G.R. control solenoid valve harness connector
Pressure regulator control solenoid valve
S.C.V. control solenoid valve
A.I.V. control solenoid valve
Throttle sensor & throttle valve switch
A.A.C. valve
B.P.T. valve
S.C.V. actuator
Exhaust gas temperature sensor
E.G.R. control valve
Vehicle speed sensor
Power steering oil pressure switch
Steering gear

ECCS SCHEMATIC—1990 STANZA

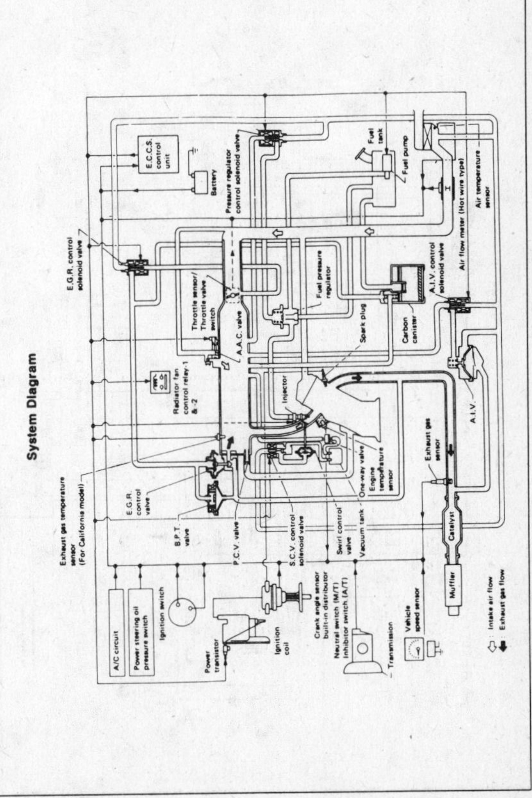
System Diagram
E.C.C.S. control unit
Battery
Fuel tank
Fuel pump
Air flow meter (Hot wire type)
Air temperature sensor
Pressure regulator control solenoid valve
E.G.R. control solenoid valve
Throttle sensor/Throttle valve switch
A.A.C. valve
Fuel pressure regulator
Spark plug
Carbon canister
A.I.V. control solenoid valve
Radiator fan control relay 1 & 2
Injector
A.I.V.
Exhaust gas sensor
Exhaust gas temperature sensor (For California model)
E.G.R. control valve
B.P.T. valve
P.C.V. valve
S.C.V. control solenoid valve
Swirl control valve
Vacuum tank
One-way valve
Engine temperature sensor
Catalyst
Muffler
A/C circuit
Power steering oil pressure switch
Ignition switch
Power transistor
Ignition coil
Crank angle sensor built-into distributor
Neutral switch (M/T) Inhibitor switch (A/T)
Transmission
Vehicle speed sensor
Intake air flow
Exhaust gas flow

ECCS CONTROL SYSTEM CHART—1990 STANZA

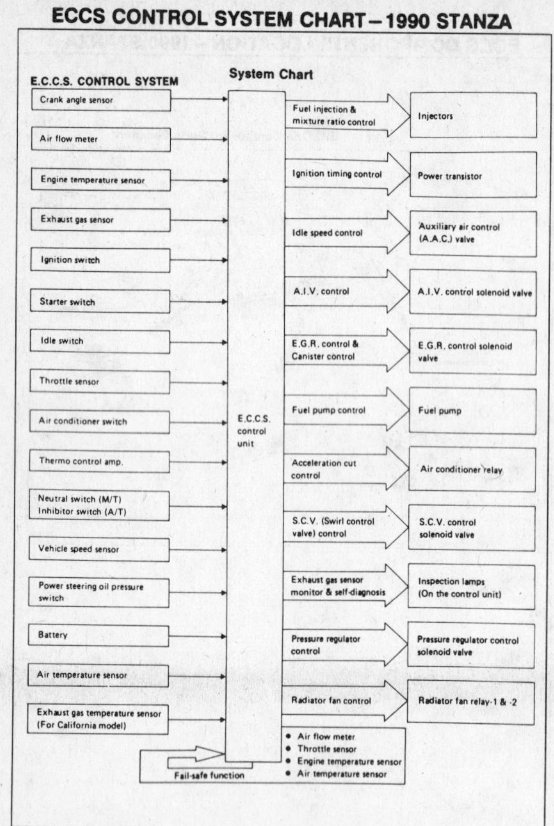
System Chart
E.C.C.S. CONTROL SYSTEM
Crank angle sensor
Air flow meter
Engine temperature sensor
Exhaust gas sensor
Ignition switch
Starter switch
Idle switch
Throttle sensor
Air conditioner switch
Thermo control amp.
Neutral switch (M/T)
Inhibitor switch (A/T)
Vehicle speed sensor
Power steering oil pressure switch
Battery
Air temperature sensor
Exhaust gas temperature sensor (For California model)
E.C.C.S. control unit
Fuel injection & mixture ratio control
Ignition timing control
Idle speed control
A.I.V. control
E.G.R. control & Canister control
Fuel pump control
Acceleration cut control
S.C.V. (Swirl control valve) control
Exhaust gas sensor monitor & self-diagnosis
Pressure regulator control
Radiator fan control
Injectors
Power transistor
Auxiliary air control (A.A.C.) valve
A.I.V. control solenoid valve
E.G.R. control solenoid valve
Fuel pump
Air conditioner relay
S.C.V. control solenoid valve
Inspection lamps (On the control unit)
Pressure regulator control solenoid valve
Radiator fan relay-1 & -2
Air flow meter
Throttle sensor
Engine temperature sensor
Air temperature sensor
Fail-safe function

ECCS VACUUM HOSE ROUTING—1990 STANZA

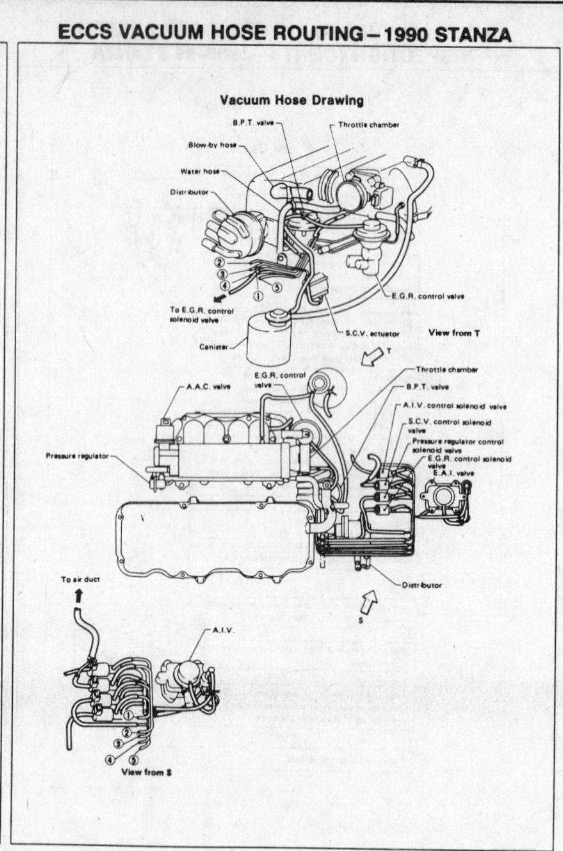
Vacuum Hose Drawing
B.P.T. valve
Throttle chamber
Blow-by hose
Water hose
Distributor
E.G.R. control valve
To E.G.R. control solenoid valve
S.C.V. actuator
View from T
Canister
A.A.C. valve
E.G.R. control valve
Throttle chamber
B.P.T. valve
A.I.V. control solenoid valve
S.C.V. control solenoid valve
Pressure regulator control solenoid valve
E.G.R. control solenoid valve
E.A.I. valve
Pressure regulator
Distributor
To air duct
A.I.V.
View from S

ECCS CONTROL UNIT WIRING—1990 STANZA

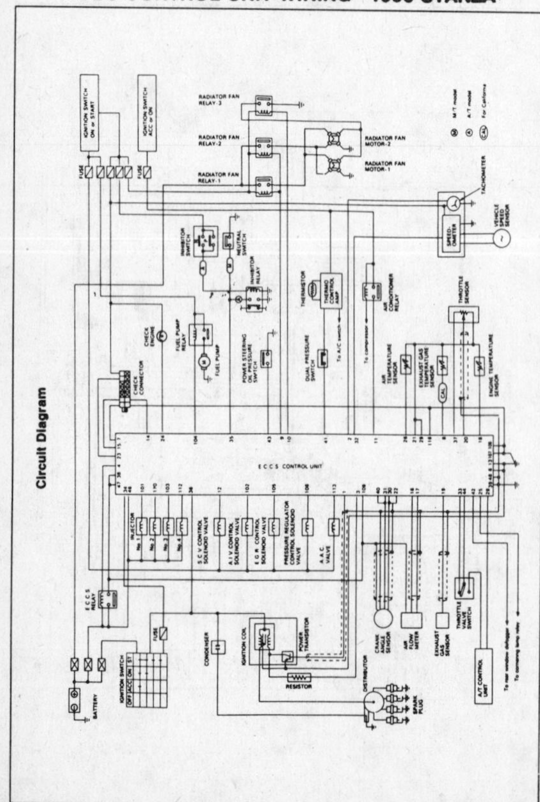
Circuit Diagram
E.C.C.S. CONTROL UNIT

ECCS WIRING SCHEMATIC—1990 STANZA

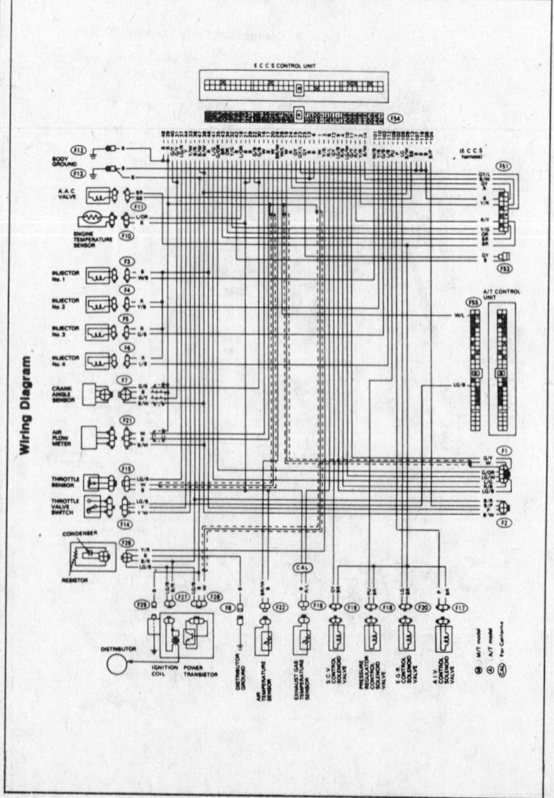
Wiring Diagram
E.C.C.S. CONTROL UNIT

ECCS WIRING SCHEMATIC (CONT.)—1990 STANZA

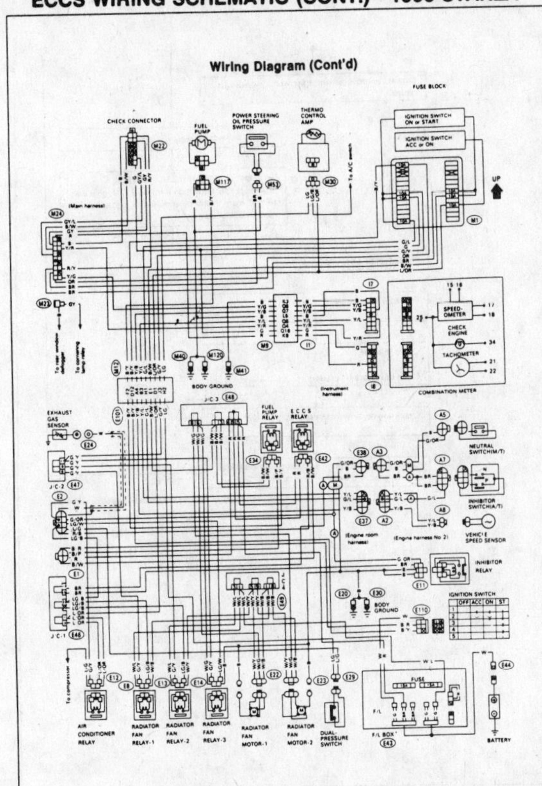
Wiring Diagram (Cont'd)
FUSE BLOCK
CHECK CONNECTOR
FUEL PUMP
POWER STEERING OIL PRESSURE SWITCH
THERMO CONTROL AMP
IGNITION SWITCH ON or START
IGNITION SWITCH ACC or ON
UP
SPEEDOMETER
CHECK ENGINE
TACHOMETER
COMBINATION METER
BODY GROUND
EXHAUST GAS SENSOR
FUEL PUMP RELAY
E.C.C.S RELAY
NEUTRAL SWITCH(M/T)
INHIBITOR SWITCH(A/T)
VEHICLE SPEED SENSOR
INHIBITOR RELAY
IGNITION SWITCH
BODY GROUND
AIR CONDITIONER RELAY
RADIATOR FAN RELAY-1
RADIATOR FAN RELAY-2
RADIATOR FAN RELAY-3
RADIATOR FAN MOTOR-1
RADIATOR FAN MOTOR-2
DUAL-PRESSURE SWITCH
F/L BOX
BATTERY

ECCS IDLE SPEED, IGINTION TIMING AND IDLE MIXTURE RATIO INSPECTION—1990 STANZA

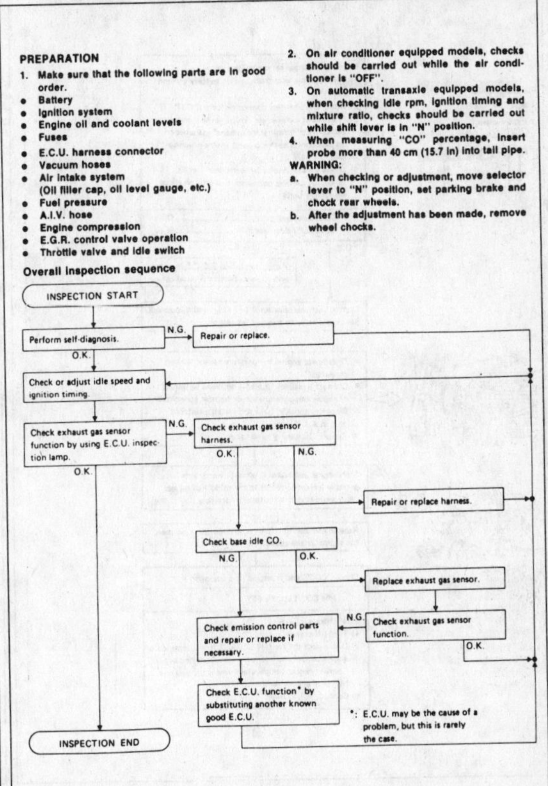
PREPARATION
1. Make sure that the following parts are in good order.
• Battery
• Ignition system
• Engine oil and coolant levels
• Fuses
• E.C.U. harness connector
• Vacuum hoses
• Air intake system (Oil filler cap, oil level gauge, etc.)
• Fuel pressure
• A.I.V. hose
• Engine compression
• E.G.R. control valve operation
• Throttle valve and idle switch
2. On air conditioner equipped models, checks should be carried out while the air conditioner is "OFF".
3. On automatic transaxle equipped models, when checking idle rpm, ignition timing and mixture ratio, checks should be carried out while shift lever is in "N" position.
4. When measuring "CO" percentage, insert probe more than 40 cm (15.7 in) into tail pipe.
WARNING:
a. When checking or adjustment, move selector lever to "N" position, set parking brake and chock rear wheels.
b. After the adjustment has been made, remove wheel chocks.
Overall inspection sequence
INSPECTION START
Perform self-diagnosis.
N.G.
Repair or replace.
O.K.
Check or adjust idle speed and ignition timing.
Check exhaust gas sensor function by using E.C.U. inspection lamp.
N.G.
Check exhaust gas sensor harness.
O.K.
N.G.
O.K.
Repair or replace harness.
Check base idle CO.
N.G.
O.K.
Replace exhaust gas sensor.
Check emission control parts and repair or replace if necessary.
N.G.
Check exhaust gas sensor function.
O.K.
Check E.C.U. function* by substituting another known good E.C.U.
*: E.C.U. may be the cause of a problem, but this is rarely the case.
INSPECTION END

ECCS IDLE SPEED, IGINTION TIMING AND IDLE MIXTURE RATIO INSPECTION (CONT.)—1990 STANZA

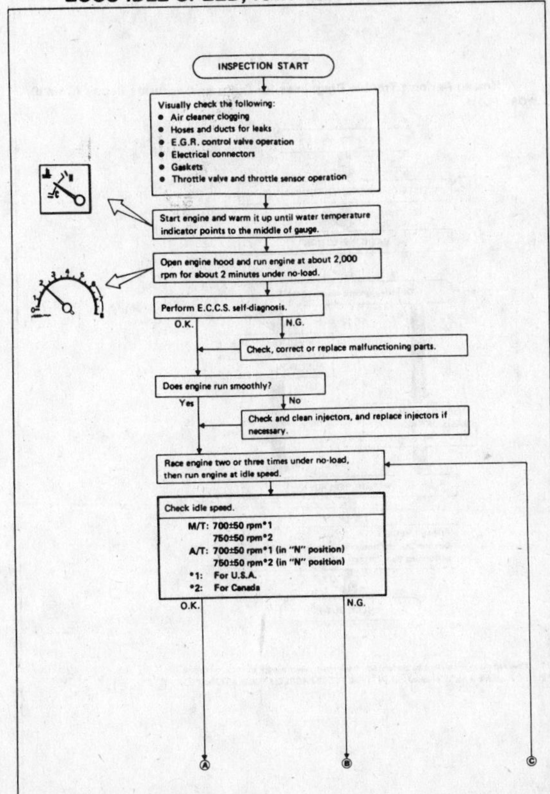
INSPECTION START
Visually check the following:
• Air cleaner clogging
• Hoses and ducts for leaks
• E.G.R. control valve operation
• Electrical connectors
• Gaskets
• Throttle valve and throttle sensor operation
Start engine and warm it up until water temperature indicator points to the middle of gauge.
Open engine hood and run engine at about 2,000 rpm for about 2 minutes under no-load.
Perform E.C.C.S. self-diagnosis.
O.K.
N.G.
Check, correct or replace malfunctioning parts.
Does engine run smoothly?
Yes
No
Check and clean injectors, and replace injectors if necessary.
Race engine two or three times under no-load, then run engine at idle speed.
Check idle speed.
M/T: 700±50 rpm*1
750±50 rpm*2
A/T: 700±50 rpm*1 (in "N" position)
750±50 rpm*2 (in "N" position)
*1: For U.S.A.
*2: For Canada
O.K.
N.G.
A
B
C

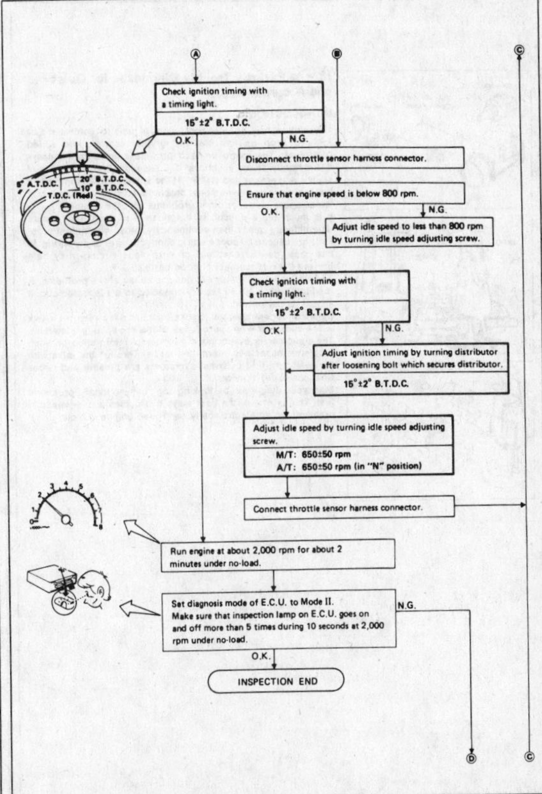
A
B
C
Check ignition timing with a timing light.
15°±2° B.T.D.C.
O.K.
N.G.
5° A.T.D.C.
20° B.T.D.C.
10° B.T.D.C.
T.D.C. (Red)
Disconnect throttle sensor harness connector.
Ensure that engine speed is below 800 rpm.
O.K.
N.G.
Adjust idle speed to less than 800 rpm by turning idle speed adjusting screw.
Check ignition timing with a timing light.
15°±2° B.T.D.C.
O.K.
N.G.
Adjust ignition timing by turning distributor after loosening bolt which secures distributor.
15°±2° B.T.D.C.
Adjust idle speed by turning idle speed adjusting screw.
M/T: 650±50 rpm
A/T: 650±50 rpm (in "N" position)
Connect throttle sensor harness connector.
Run engine at about 2,000 rpm for about 2 minutes under no-load.
Set diagnosis mode of E.C.U. to Mode II.
Make sure that inspection lamp on E.C.U. goes on and off more than 5 times during 10 seconds at 2,000 rpm under no-load.
N.G.
O.K.
INSPECTION END
D
C

ECCS IDLE SPEED, IGINTION TIMING AND IDLE MIXTURE RATIO INSPECTION (CONT.) – 1990 STANZA

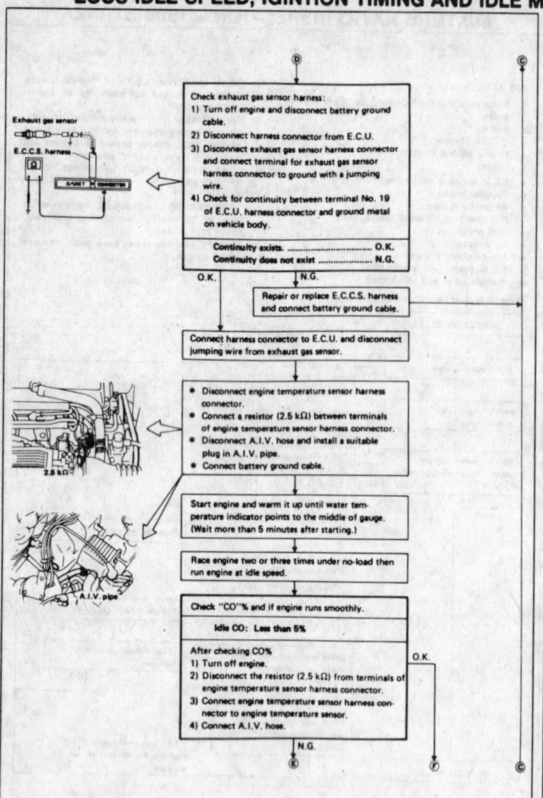
D
C
Exhaust gas sensor
E.C.C.S. harness
Check exhaust gas sensor harness:
1) Turn off engine and disconnect battery ground cable.
2) Disconnect harness connector from E.C.U.
3) Disconnect exhaust gas sensor harness connector and connect terminal for exhaust gas sensor harness connector to ground with a jumping wire.
4) Check for continuity between terminal No. 19 of E.C.U. harness connector and ground metal on vehicle body.
Continuity exists. .... O.K.
Continuity does not exist .... N.G.
O.K.
N.G.
Repair or replace E.C.C.S. harness and connect battery ground cable.
Connect harness connector to E.C.U. and disconnect jumping wire from exhaust gas sensor.
2.5 kΩ
• Disconnect engine temperature sensor harness connector.
• Connect a resistor (2.5 kΩ) between terminals of engine temperature sensor harness connector.
• Disconnect A.I.V. hose and install a suitable plug in A.I.V. pipe.
• Connect battery ground cable.
Start engine and warm it up until water temperature indicator points to the middle of gauge. (Wait more than 5 minutes after starting.)
Race engine two or three times under no-load then run engine at idle speed.
A.I.V. pipe
Check "CO"% and if engine runs smoothly.
Idle CO: Less than 5%
After checking CO%
1) Turn off engine.
2) Disconnect the resistor (2.5 kΩ) from terminals of engine temperature sensor harness connector.
3) Connect engine temperature sensor harness connector to engine temperature sensor.
4) Connect A.I.V. hose.
O.K.
N.G.
E
F
C

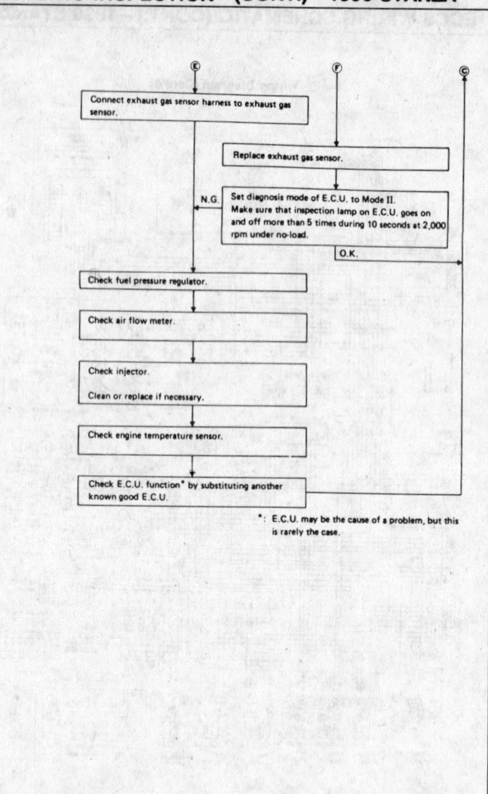
E
F
C
Connect exhaust gas sensor harness to exhaust gas sensor.
Replace exhaust gas sensor.
N.G.
Set diagnosis mode of E.C.U. to Mode II.
Make sure that inspection lamp on E.C.U. goes on and off more than 5 times during 10 seconds at 2,000 rpm under no-load.
O.K.
Check fuel pressure regulator.
Check air flow meter.
Check injector.
Clean or replace if necessary.
Check engine temperature sensor.
Check E.C.U. function* by substituting another known good E.C.U.
*: E.C.U. may be the cause of a problem, but this is rarely the case.

ECCS DIAGNOSTIC DESCRIPTION – 1990 STANZA

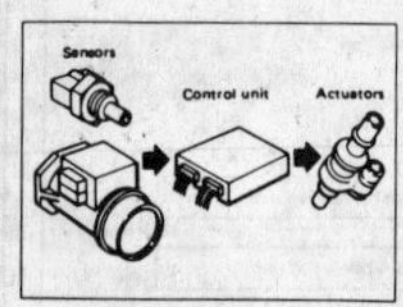
Sensors
Control unit
Actuators

INFO.
CAUSE

How to Perform Trouble Diagnoses for Quick and Accurate Repair

INTRODUCTION

The engine has an electronic control unit to control major systems such as fuel control, ignition control, idle speed control, etc. The control unit accepts input signals from sensors and instantly drives actuators. It is essential that both kinds of signals are proper and stable. At the same time, it is important that there are no conventional problems such as vacuum leaks, fouled spark plugs, or other problems with the engine.

It is much more difficult to diagnose a problem that occurs intermittently rather than continuously. Most intermittent problems are caused by poor electric connections or faulty wiring. In this case, careful checking of suspicious circuits may help prevent the replacement of good parts.

A visual check only may not find the cause of the problems. A road test with a circuit tester connected to a suspected circuit should be performed.

Before undertaking actual checks, take just a few minutes to talk with a customer who approaches with a driveability complaint. The customer is a very good supplier of information on such problems, especially intermittent ones. Through the talks with the customer, find out what symptoms are present and under what conditions they occur.

Start your diagnosis by looking for "conventional" problems first. This is one of the best ways to troubleshoot driveability problems on an electronically controlled engine vehicle.

How to Perform Trouble Diagnoses for Quick and Accurate Repair (Cont'd)

WORK FLOW

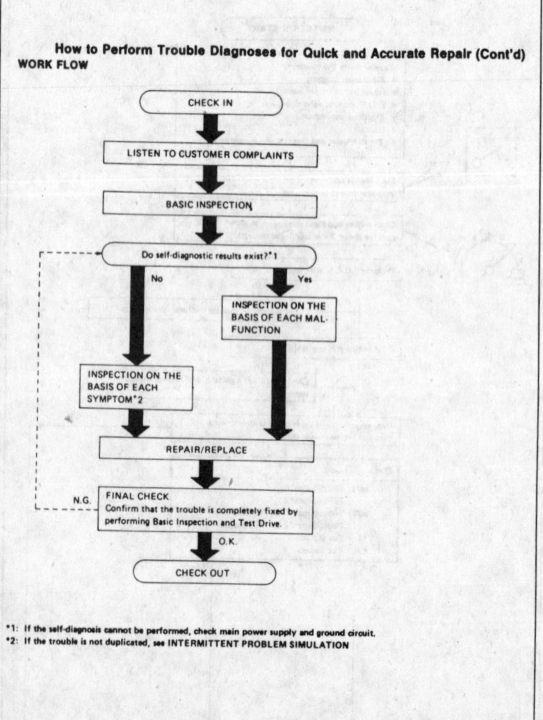
CHECK IN
LISTEN TO CUSTOMER COMPLAINTS
BASIC INSPECTION
Do self-diagnostic results exist?*1
No
Yes
INSPECTION ON THE BASIS OF EACH MALFUNCTION
INSPECTION ON THE BASIS OF EACH SYMPTOM*2
REPAIR/REPLACE
N.G.
FINAL CHECK
Confirm that the trouble is completely fixed by performing Basic Inspection and Test Drive.
O.K.
CHECK OUT
*1: If the self-diagnosis cannot be performed, check main power supply and ground circuit.
*2: If the trouble is not duplicated, see INTERMITTENT PROBLEM SIMULATION

ECCS DIAGNOSTIC DESCRIPTION (CONT.)—1990 STANZA

How to Perform Trouble Diagnoses for Quick and Accurate Repair (Cont'd)

KEY POINTS

- WHAT ..... Vehicle & engine model
- WHEN ..... Date, Frequencies
- WHERE ..... Road conditions
- HOW ..... Operating conditions, Weather conditions, Symptoms

DIAGNOSTIC WORKSHEET

There are many kinds of operating conditions that lead to malfunctions on engine components.
A good grasp of such conditions can make trouble-shooting faster and more accurate.
In general, feelings for a problem depend on each customer. It is important to fully understand the symptoms or under what conditions a customer complains.
Make good use of a diagnostic worksheet such as the one shown below in order to utilize all the complaints for trouble-shooting.

Worksheet sample

Customer name MR/MS		Model & Year	VIN
Engine #		Trans.	Mileage
Incident Date		Manuf. Date	In Service Date

Symptoms	☐ Startability	☐ Impossible to start ☐ No combustion ☐ Partial combustion ☐ Partial combustion affected by throttle position ☐ Partial combustion NOT affected by throttle position ☐ Possible but hard to start ☐ Others [
	☐ Idling	☐ No fast idle ☐ Unstable ☐ High idle ☐ Low idle ☐ Others [ ]
	☐ Driveability	☐ Stumble ☐ Surge ☐ Detonation ☐ Lack of power ☐ Intake backfire ☐ Exhaust backfire ☐ Others [ ]
	☐ Engine stall	☐ At the time of start ☐ While idling ☐ While accelerating ☐ While decelerating ☐ Just after stopping ☐ While loading
Incident occurrence		☐ Just after delivery ☐ Recently ☐ In the morning ☐ At night ☐ In the daytime
Frequency		☐ All the time ☐ Under certain conditions ☐ Sometimes
Weather conditions		☐ Note effected
	Weather	☐ Fine ☐ Raining ☐ Snowing ☐ Others [ ]
	Temperature	☐ Hot ☐ Warm ☐ Cool ☐ Cold ☐ Humid °F
Engine conditions		☐ Cold During warm-up ☐ After warm-up Engine speed 0 2,000 4,000 6,000 8,000 rpm
Road conditions		☐ In town ☐ In suburbs ☐ Highway ☐ Off road (up/down)
Driving conditions		☐ Not affected ☐ At starting ☐ While idling ☐ At racing ☐ While accelerating ☐ While cruising ☐ While decelerating ☐ While turning (RH/LH) Vehicle speed 0 10 20 30 40 50 60 MPH
Check engine light		☐ Turned on ☐ Not turned on

How to Perform Trouble Diagnoses for Quick and Accurate Repair (Cont'd)

GRRRRR
GRRRR
THIS is the symptom, isn't it?

INTERMITTENT PROBLEM SIMULATION

In order to duplicate an intermittent problem, it is effective to create similar conditions for component parts, under which the problem might occur.
Perform the activity listed under Service procedure and note the result.

	Variable factor	Influential part	Target condition	Service procedure
1	Mixture ratio	Pressure regulator	Made lean	Remove vacuum hose and apply vacuum.
			Made rich	Remove vacuum hose and apply pressure.
2	Ignition timing	Distributor	Advanced	Rotate distributor clockwise.
			Retarded	Rotate distributor counterclockwise.
3	Mixture ratio feed-back control	Exhaust gas sensor	Suspended	Disconnect exhaust gas sensor harness connector.
		Control unit	Operation check	Perform self-diagnosis (Mode I/II) at 2,000 rpm.
4	Idle speed	I.A.A. unit	Raised	Turn idle adjusting screw counterclockwise.
			Lowered	Turn idle adjusting screw clockwise.
5	Electric connection (Electric continuity)	Harness connectors and wires	Poor electric connection or faulty wiring	Tap or wiggle.
				Race engine rapidly. See if the torque reaction of the engine unit causes electric breaks.
6	Temperature	Control unit	Cooled	Cool with an icing spray or similar device.
			Warmed	Heat with a hair drier. **[WARNING: Do not overheat the unit.]**
7	Moisture	Electric parts	Damp	Wet. **[WARNING: Do not directly pour water on components. Use a mist sprayer.]**
8	Electric loads	Load switches	Loaded	Turn on headlights, air conditioner, rear defogger, etc.
9	Idle switch condition	Control unit	ON-OFF switching	Perform self-diagnosis (Mode IV).
10	Ignition spark	Timing light	Spark power check	Try to flash timing light for each cylinder.

ECCS SELF-DIAGNOSTIC DESCRIPTION—1990 STANZA

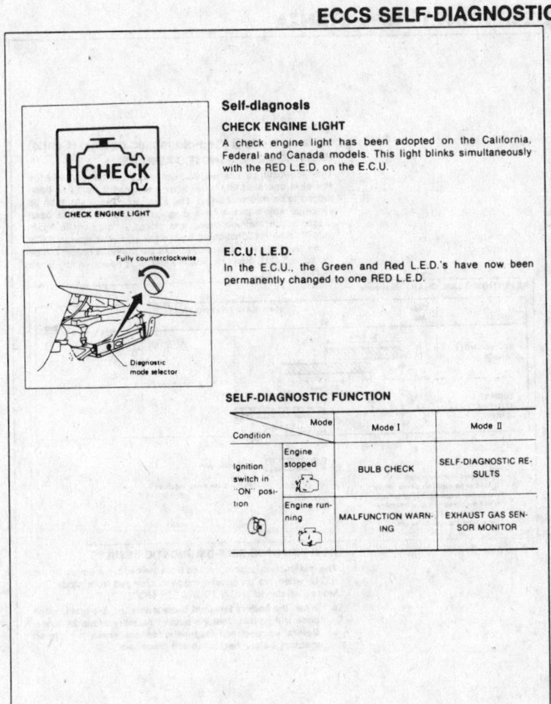
CHECK
CHECK ENGINE LIGHT
Fully counterclockwise
Diagnostic mode selector

Self-diagnosis

CHECK ENGINE LIGHT

A check engine light has been adopted on the California, Federal and Canada models. This light blinks simultaneously with the RED L.E.D. on the E.C.U.

E.C.U. L.E.D.

In the E.C.U., the Green and Red L.E.D.'s have now been permanently changed to one RED L.E.D.

SELF-DIAGNOSTIC FUNCTION

Condition \ Mode		Mode I	Mode II
Ignition switch in "ON" position	Engine stopped	BULB CHECK	SELF-DIAGNOSTIC RESULTS
	Engine running	MALFUNCTION WARNING	EXHAUST GAS SENSOR MONITOR

Self-diagnosis (Cont'd)

HOW TO SWITCH MODES

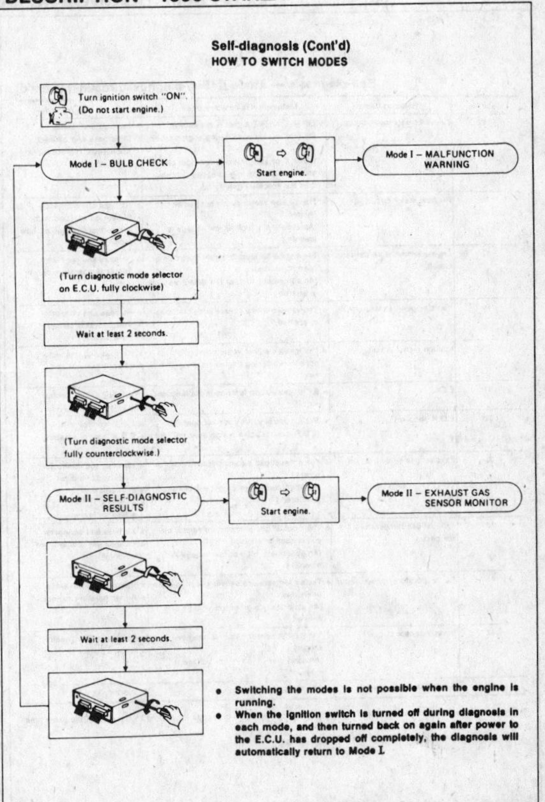
Turn ignition switch "ON".
(Do not start engine.)
Mode I – BULB CHECK
Start engine.
Mode I – MALFUNCTION WARNING
(Turn diagnostic mode selector on E.C.U. fully clockwise)
Wait at least 2 seconds.
(Turn diagnostic mode selector fully counterclockwise.)
Mode II – SELF-DIAGNOSTIC RESULTS
Start engine.
Mode II – EXHAUST GAS SENSOR MONITOR
Wait at least 2 seconds.

- **Switching the modes is not possible when the engine is running.**
- **When the ignition switch is turned off during diagnosis in each mode, and then turned back on again after power to the E.C.U. has dropped off completely, the diagnosis will automatically return to Mode I.**

ECCS SELF-DIAGNOSTIC MODE I—1990 STANZA

Self-diagnosis — Mode I

MODE I — BULB CHECK

In this mode, the RED L.E.D. in the E.C.U. and the CHECK ENGINE LIGHT in the instrument panel stay "ON".
If either remain "OFF", check the bulb in the CHECK ENGINE LIGHT or the RED L.E.D.

MODE I — MALFUNCTION WARNING

FOR CALIFORNIA MODEL

CHECK ENGINE LIGHT AND RED L.E.D.	CONDITION
ON	WHEN THE FOLLOWING MALFUNCTION (CHECK ENGINE LIGHT ITEM) IS DETECTED OR THE E.C.U.'S C.P.U. IS MALFUNCTIONING.
OFF	O.K.

CODE NO.	MALFUNCTION
12	AIR FLOW METER CIRCUIT
13	ENGINE TEMPERATURE SENSOR CIRCUIT
14	VEHICLE SPEED SENSOR CIRCUIT
31	E.C.U.(E.C.C.S. CONTROL UNIT)
32	E.G.R. FUNCTION
33	EXHAUST GAS SENSOR CIRCUIT
35	EXHAUST GAS TEMPERATURE SENSOR CIRCUIT
43	THROTTLE SENSOR CIRCUIT
45	INJECTOR LEAK

- **These Code Numbers are clarified in Mode II — SELF-DIAGNOSTIC RESULTS.**
- **The RED L.E.D. and the CHECK ENGINE LIGHT will turn off when normal condition is detected. At this time, the Mode II — SELF-DIAGNOSTIC RESULTS memory must be cleared as the contents remain stored.**

FOR NON-CALIFORNIA MODEL

RED L.E.D.	CONDITION
ON	WHEN THE E.C.U.'S C.P.U. IS MALFUNCTIONING.
OFF	O.K.

ECCS SELF-DIAGNOSTIC II—1990 STANZA

Self-diagnosis — Mode II (Self-diagnostic results)

DESCRIPTION

In this mode, a malfunction code is indicated by the number of flashes from the RED L.E.D. or the CHECK ENGINE LIGHT as shown below:

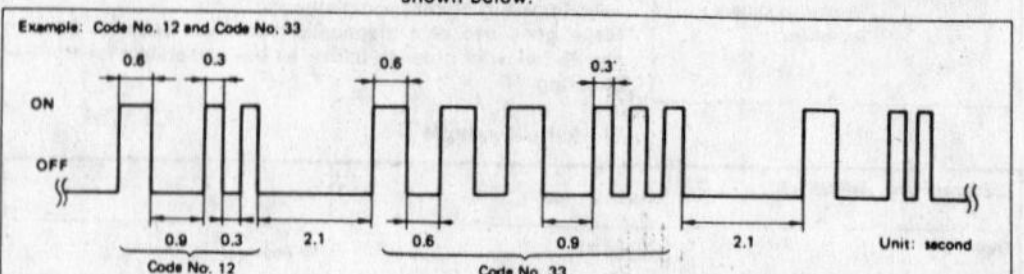
Example: Code No. 12 and Code No. 33
0.6
0.3
0.6
0.3
ON
OFF
0.9
0.3
2.1
0.6
0.9
2.1
Unit: second
Code No. 12
Code No. 33

Long (0.6 second) blinking indicates the number of ten digits and short (0.3 second) blinking indicates the number of single digits.
For example, the red L.E.D. flashes once for 0.6 seconds and then it flashes twice for 0.3 seconds. This indicates the number "12" and refers to a malfunction in the air flow meter. In this way, all the problems are classified by their code numbers.

Display code table

Code No.		Detected items	California model	Non-California model
11		Crank angle sensor circuit	X	X
12	CHECK	Air flow meter circuit	X	X
13	CHECK	Engine temperature sensor circuit	X	X
14	CHECK	Vehicle speed sensor circuit	X	X
21		Ignition signal circuit	X	X
31	CHECK	E.C.U.	X	X
32	CHECK	E.G.R. function	X	—
33	CHECK	Exhaust gas sensor circuit	X	X
35	CHECK	Exhaust gas temperat..re sensor circuit	X	—
41		Air temperature sensor circuit	X	X
43	CHECK	Throttle sensor circuit	X	X
45	CHECK	Injector leak	X	—
55		No malfunction in the above circuits	X	X

X: Available
—: Not available
CHECK: Check engine light item

ECCS SELF-DIAGNOSTIC MODE II (CONT.)—1990 STANZA

Self-diagnosis — Mode II (Self-diagnostic results) (Cont'd)

Code No.	Detected items	Malfunction is detected when ...	Check item (remedy)
*11	Crank angle sensor circuit	• Either 1° or 180° signal is not entered for the first few seconds during engine cranking. • Either 1° or 180° signal is not input often enough while the engine speed is higher than the specified rpm.	• Harness and connector (If harness and connector are normal, replace crank angle sensor.)
12	Air flow meter circuit	• The air flow meter circuit is open or shorted. (An abnormally high or low voltage is entered.)	• Harness and connector (If harness and connector are normal, replace air flow meter.)
13	Engine temperature sensor circuit	• The engine temperature sensor circuit is open or shorted. (An abnormally high or low output voltage is entered.)	• Harness and connector • Engine temperature sensor
14	Vehicle speed sensor circuit	• The vehicle speed sensor circuit is open or shorted.	• Harness and connector • Vehicle speed sensor (reed switch)
*21	Ignition signal circuit	• The ignition signal in the primary circuit is not entered during engine cranking or running.	• Harness and connector • power transistor unit
31	E.C.U.	• E.C.U. calculation function is malfunctioning.	(Replace E.C.C.S. control unit.)
32	E.G.R. function	• E.G.R. control valve does not operate. (E.G.R. control valve spring does not lift.)	• E.G.R. control valve • E.G.R. control solenoid valve
33	Exhaust gas sensor circuit	• The exhaust gas sensor circuit is open or shorted. (An abnormally high or low output voltage is entered.)	• Harness and connector • Exhaust gas sensor • Fuel pressure • Injectors • Intake air leaks
35	Exhaust gas temperature sensor circuit	• The exhaust gas temperature sensor circuit is open or shorted. (An abnormally high or low voltage is entered.)	• Harness and connector • Exhaust gas temperature sensor
41	Air temperature sensor circuit	• The air temperature sensor circuit is open or shorted. (An abnormally high or low voltage is entered.)	• Harness and connector • Air temperature sensor
43	Throttle sensor circuit	• The throttle sensor circuit is open or shorted. (An abnormally high or low voltage is entered.)	• Harness and connector • Throttle sensor
45	injector leak	• Fuel leaks from injector.	• Injector

*: Check items causing a malfunction of crank angle sensor circuit first, if both code No. 11 and 21 come out at the same time.

Self-diagnosis — Mode II (Self-diagnostic results) (Cont'd)

RETENTION OF DIAGNOSTIC RESULTS

The diagnostic results will remain in E.C.U. memory until the starter is operated fifty times after a diagnostic item has been judged to be malfunctioning. The diagnostic result will then be canceled automatically. If a diagnostic item which has been judged to be malfunctioning and stored in memory is again judged to be malfunctioning before the starter is operated fifty times, the second result will replace the previous one. It will be stored in E.C.U. memory until the starter is operated fifty times more.

RETENTION TERM CHART (Example)

	Code No.	STARTER OPERATING TIMES: 50	100	150	200
CRANK ANGLE SENSOR	11	50 times			
ENGINE TEMPERATURE SENSOR	13	50 times / 50 times			

If the same diagnostic item is judged to be malfunctioning before the starter is operated fifty times, it will be stored in E.C.U. memory until the starter is operated fifty times from this point in time.

▨ : Retention term
▲ Malfunction detecting point

HOW TO ERASE SELF-DIAGNOSTIC RESULTS

The malfunction code is erased from the backup memory on the E.C.U. when the diagnostic mode is changed from Mode II to Mode I. (Refer to "HOW TO SWITCH MODES".)

- **When the battery terminal is disconnected, the malfunction code will be lost from the backup memory within 24 hours.**
- **Before starting self-diagnosis, do not erase the stored memory before beginning self-diagnosis.**

ECCS SELF-DIAGNOSTIC MODE II (CONT.)—1990 STANZA

Self-diagnosis — Mode II (Exhaust gas sensor monitor)

DESCRIPTION

In this mode, the CHECK ENGINE LIGHT and RED L.E.D. display the condition of the fuel mixture (lean or rich) which is monitored by the exhaust gas sensor.

CHECK ENGINE LIGHT and RED L.E.D.	Fuel mixture condition in the exhaust gas	Air fuel ratio feedback control condition
ON	Lean	Closed loop control
OFF	Rich	
*Remains ON or OFF	Any condition	Open loop control

*: Maintains conditions just before switching to open loop.

HOW TO CHECK EXHAUST GAS SENSOR

1. Set Mode II. (Refer to "HOW TO SWITCH MODES".)
2. Start engine and warm it up until engine coolant temperature indicator points to the middle of the gauge.
3. Run engine at about 2,000 rpm for about 2 minutes under no-load conditions.
4. Make sure RED L.E.D. or CHECK ENGINE LIGHT goes ON and OFF more than 5 times every 10 seconds; measured at 2,000 rpm under no-load.

ECCS SELF-DIAGNOSTIC CHART—1990 STANZA

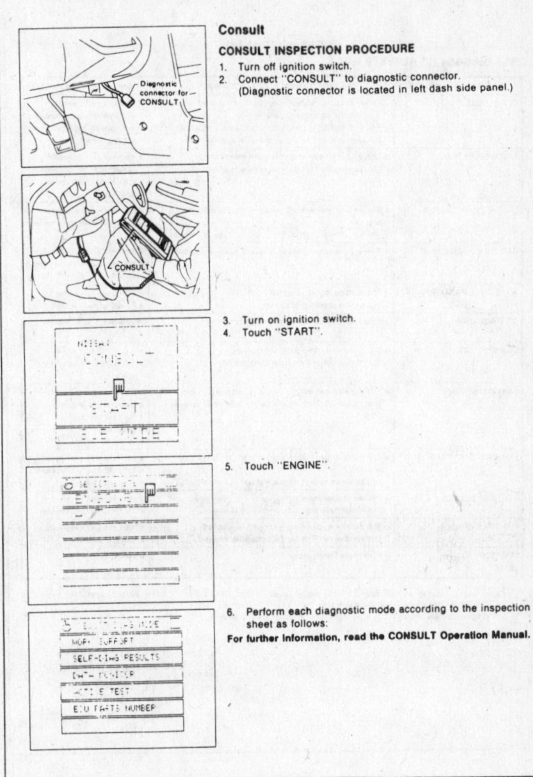
Diagnostic connector for CONSULT
CONSULT
START

Consult

CONSULT INSPECTION PROCEDURE

1. Turn off ignition switch.
2. Connect "CONSULT" to diagnostic connector. (Diagnostic connector is located in left dash side panel.)
3. Turn on ignition switch.
4. Touch "START".
5. Touch "ENGINE".
6. Perform each diagnostic mode according to the inspection sheet as follows:

For further information, read the CONSULT Operation Manual.

ECCS SELF-DIAGNOSTIC CHART—1990 STANZA

Consult (Cont'd)

E.C.C.S. COMPONENT PARTS APPLICATION

	E.C.C.S. COMPONENT PARTS \ MODE	WORK SUPPORT	SELF-DIAGNOSTIC RESULTS	DATA MONITOR	ACTIVE TEST
INPUT	Crank angle sensor		X	X	
	Air flow meter		X	X	
	Engine temperature sensor		X	X	X
	Exhaust gas sensors		X	X	
	Vehicle speed sensors		X	X	
	Throttle sensor	X	X	X	
	Exhaust gas temperature sensor*		X	X	
	Ignition switch (start signal)			X	
	Air conditioner switch			X	
	Neutral switch			X	
	Power steering oil pressure switch			X	
	Battery			X	
	Air temperature sensor			X	
OUTPUT	Injectors		X	X	X
	Power transistor (ignition signal)		X	X (Ignition timing)	X
	A.A.C. valve	X		X	X
	Swirl control solenoid valve			X	X
	A.I.V. control solenoid valve			X	X
	P.R.V.R. control solenoid valve			X	X
	E.G.R. control solenoid valve			X	X
	Air conditioner relay			X	
	Fuel pump relay	X		X	X
	Radiator fan relay			X	X

*: The E.C.C.S. component part marked * is applicable to vehicles for California only.
X: Applicable

CAUTION:

a. **When servicing fuel system after performing "Fuel Pressure Release" in "Work Support" mode, turn off ignition key with CONSULT set in "Work Support" mode.**
b. **Be sure to set gears to "Neutral" before conducting "Active Test" on idle adjustment or A.A.C. valve.**

Consult (Cont'd)

FUNCTION

Diagnostic mode	Function
Work support	This mode enables a technician to adjust some devices faster and more accurately by following the indications on the CONSULT unit.
Self-diagnostic results	Self-diagnostic results can be read and erased quickly.
Data monitor	Input/Output data in the control unit can be read.
Active test	Mode in which CONSULT drives some actuators apart from the control units and also shifts some parameters in a specified range.
E.C.U. part numbers	E.C.U. part numbers can be read.

WORK SUPPORT MODE

WORK ITEM	CONDITION	USAGE
THROTTLE SENSOR ADJUSTMENT	CHECK THE THROTTLE SENSOR SIGNAL. ADJUST IT TO THE SPECIFIED VALUE BY ROTATING THE SENSOR BODY UNDER THE FOLLOWING CONDITIONS. • IGN SW "ON" • ENG NOT RUNNING • ACC PEDAL NOT PRESSED	When adjusting throttle sensor initial position.
IGNITION TIMING ADJUSTMENT*	• IGNITION TIMING FEEDBACK CONTROL WILL BE HELD BY TOUCHING "START". AFTER DOING SO, ADJUST IGNITION TIMING WITH A TIMING LIGHT BY TURNING THE CRANK ANGLE SENSOR.	When adjusting initial ignition timing.
AAC VALVE ADJUSTMENT	SET ENGINE RPM AT THE SPECIFIED VALUE UNDER THE FOLLOWING CONDITIONS. • ENGINE WARMED UP • NO-LOAD	When adjusting idle speed.
FUEL PRESSURE RELEASE	• FUEL PUMP WILL STOP BY TOUCHING "START" WHEN IDLING. CRANK A FEW TIMES AFTER ENGINE STALLS.	When releasing fuel pressure from fuel line.

ECCS SELF-DIAGNOSTIC CHART – 1990 STANZA

Consult (Cont'd)

SELF-DIAGNOSTIC RESULTS MODE

DIAGNOSTIC ITEM	DIAGNOSTIC ITEM IS DETECTED WHEN ...	CHECK ITEM (REMEDY)
CRANK ANGLE SENSOR*	• Either 1° or 180° signal is not entered for the first few seconds during engine cranking. • Either 1° or 180° signal is not input often enough while the engine speed is higher than the specified rpm.	• Harness and connector (If harness and connector are normal, replace crank angle sensor.)
AIR FLOW METER	• The air flow meter circuit is open or shorted. (An abnormally high or low voltage is entered.)	• Harness and connector (If harness and connector are normal, replace air flow meter.)
ENGINE TEMP SENSOR	• The engine temperature sensor circuit is open or shorted. (An abnormally high or low output voltage is entered.)	• Harness and connector • Engine temperature sensor
CAR SPEED SENSOR	• The vehicle speed sensor circuit is open or shorted.	• Harness and connector • Vehicle speed sensor (reed switch)
IGN SIGNAL-PRIMARY*	• The ignition signal in primary circuit is not entered during engine cranking or running.	• Harness and connector • Power transistor unit
CONTROL UNIT	• E.C.U. calculation function is malfunctioning.	(Replace E.C.C.S. control unit.)
EGR SYSTEM*, **	• E.G.R. control valve does not operate. (E.G.R. control valve spring does not lift.)	• E.G.R. control valve • E.G.R. control solenoid valve
EXH GAS SENSOR	• The exhaust gas sensor circuit is open or shorted. (An abnormally high or low output voltage is entered.)	• Harness and connector • Exhaust gas sensor • Fuel pressure • Injectors • Intake air leaks
EXH GAS TEMP SENSOR**	• The exhaust gas temperature sensor circuit is open or shorted. (An abnormally high or low voltage is entered.)	• Harness and connector • Exhaust gas temperature sensor
AIR TEMP SENSOR	• The air temperature sensor circuit is open or shorted. (An abnormally high or low voltage is entered.)	• Harness and connector • Air temperature sensor
THROTTLE SENSOR	• The throttle sensor circuit is open or shorted. (An abnormally high or low voltage is entered.)	• Harness and connector • Throttle sensor
INJECTOR LEAK**	• Fuel leaks from injector.	• Injector

*: Check items causing a malfuction of crank angle senser circuit first, if both "CRANK ANGLE SENSOR" and "IGN SIGNAL-PRIMARY" come out at the same time.

**: The diagnostic item marked ** is applicable to vehicles for California only.

Consult (Cont'd)

DATA MONITOR MODE

MONITOR ITEM	CONDITION		SPECIFICATION	CHECK ITEM WHEN OUTSIDE SPEC.
CAS RPM (REF)	• Tachometer: Connect • Run engine and compare tachometer indication with the CONSULT value.		Almost the same speed as the CONSULT value.	• Harness and connector • Crank angle sensor
AIR FLOW MTR	• Engine: After warming up, idle the engine • A/C switch "OFF" • Shift lever "N" • No load	Idle	1.3 - 1.8V	• Harness and connector • Air flow meter
		2,000 rpm	1.8 - 2.2V	
ENG TEMP SEN	• Engine: After warming up		More than 70°C (158°F)	• Harness and connector • Engine temperature sensor
EXH GAS SEN	• Engine: After warming up	Maintaining engine speed at 2,000 rpm	0 - 0.3V ←→ 0.6 - 1.0V	• Harness and connector • Exhaust gas sensor • Intake air leaks • Injectors
M/R F/C MNT			LEAN ←→ RICH Changes more than 5 times during 10 seconds.	
CAR SPEED SEN	• Turn drive wheels and compare speedometer indication with the CONSULT value		Almost the same speed as the CONSULT value	• Harness and connector • Vehicle speed sensor
BATTERY VOLT	• Ignition switch: ON (Engine stopped)		11 - 14V	• Battery • E.C.U. power supply circuit
THROTTLE SEN	• Ignition switch: ON (Engine stopped)	Throttle valve fully closed	0.4 - 0.5V	• Harness and connector • Throttle sensor • Throttle sensor adjustment
		Throttle valve fully opened	Approx. 4.0V	
AIR TEMP SEN	• Engine: After warming up		20 - 60°C (68 - 140°F)	• Harness and connector • Fuel temp. sensor
EGR TEMP SEN*	• Engine: After warming up		Less than 4.5V	• Harness and connector • Exhaust gas temperature sensor
START SIGNAL	• Ignition switch: ON → START		OFF → ON	• Harness and connector • Starter switch
IDLE POSITION	• Ignition switch: ON (Engine stopped)	Throttle valve: Idle position	ON	• Harness and connector • Throttle sensor • Throttle sensor adjustment
		Throttle valve: Slightly open	OFF	
AIR COND SIG	• Engine: After warming up, idle the engine	A/C switch "OFF"	OFF	• Harness and connector • Air conditioner switch
		A/C switch "ON"	ON	
NEUTRAL SW	• Ignition switch: ON	Shift lever "P" or "N"	ON	• Harness and connector • Neutral switch
		Except above	OFF	
PW/ST SIGNAL	• Engine: After warming up, idle the engine	Steering wheel in neutral (forward direction)	OFF	• Harness and connector • Power steering oil pressure switch
		The steering wheel is turned	ON	
INJ PULSE	Idle		2.9 - 3.6 msec.	• Harness and connector • Injector • Air flow meter
	2,000 rpm		2.6 - 3.3 msec.	
IGN TIMING	Idle		15 deg.	• Harness and connector • Crank angle sensor
	2,000 rpm		More than 25 deg.	
AAC VALVE	Idle		15° - 40°	• Harness and connector • AAC valve
	2,000 rpm		—	

Remarks: The monitor item marked * is applicable to vehicles for California only.
Specifications are reference values.

ECCS SELF-DIAGNOSTIC CHART – 1990 STANZA

Consult (Cont'd)

ACTIVE TEST MODE

TEST ITEM	CONDITION	JUDGEMENT	CHECK ITEM (REMEDY)
FUEL INJECTION TEST	• Engine: Return to the original trouble condition • Change the amount of fuel injection with the CONSULT.	If trouble symptom disappears, see CHECK ITEM.	• Harness and connector • Fuel injectors • Exhaust gas sensors
AAC/V OPENING TEST	• Engine: After warming up, idle the engine. • Change the AAC valve opening percent with the CONSULT.	Engine speed changes according to the opening percent.	• Harness and connector • AAC valve
ENGINE TEMP TEST	• Engine: Return to the original trouble condition • Change the engine coolant temperature with the CONSULT.	If trouble symptom disappears, see CHECK ITEM.	• Harness and connector • Engine temperature sensor • Fuel injectors
IGN TIMING TEST	• Engine: Return to the original trouble condition • Timing light: Set • Retard the ignition timing with the CONSULT.	If trouble symptom disappears, see CHECK ITEM.	• Adjust initial ignition timing
POWER BALANCE TEST	• Engine: After warming up, idle the engine. • A/C switch "OFF" • Shift lever "N" • Cut off each injector signal one at a time with the CONSULT.	Engine runs rough or dies.	• Harness and connector • Compression • Injectors • Power transistor • Spark plugs • Ignition coils
RADIATOR FAN TEST	• Ignition switch: ON • Turn the radiator fan "ON" and "OFF" with the CONSULT.	Radiator fan moves and stops.	• Harness and connector • Radiator fan motor
FUEL PUMP RLY TEST	• Ignition switch: ON (Engine stopped) • Turn the fuel pump relay "ON" and "OFF" with the CONSULT and listen to operating sound.	Fuel pump relay makes the operating sound.	• Harness and connector • Fuel pump relay
EGR CONT SOLV/V TEST PRVR CONT SOL/V TEST AIV CONT SOL/V TEST SWIRL CONT SOL/V TEST	• Ignition switch: ON • Turn solenoid valve "ON" and "OFF" with the CONSULT and listen to operating sound.	Each solenoid valve makes an operating sound.	• Harness and connector • Solenoid valve
SELF-LEARN CONT TEST	• In this test, the coefficient of self-learning control mixture ratio returns to the original coefficient by touching "CLEAR" on the screen.		

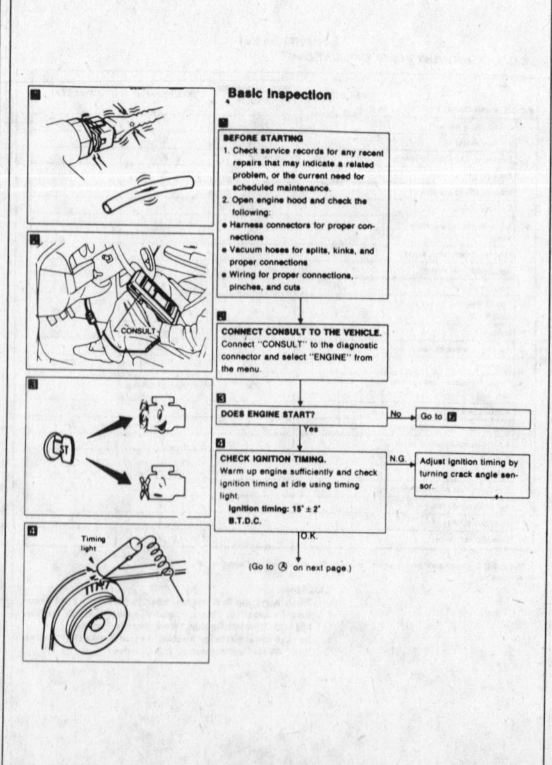
CONSULT
Timing light
Basic Inspection
BEFORE STARTING
1. Check service records for any recent repairs that may indicate a related problem, or the current need for scheduled maintenance.
2. Open engine hood and check the following:
• Harness connectors for proper connections
• Vacuum hoses for splits, kinks, and proper connections
• Wiring for proper connections, pinches, and cuts
CONNECT CONSULT TO THE VEHICLE.
Connect "CONSULT" to the diagnostic connector and select "ENGINE" from the menu.
DOES ENGINE START?
No
Go to
Yes
CHECK IGNITION TIMING.
Warm up engine sufficiently and check ignition timing at idle using timing light.
Ignition timing: 15° ± 2° B.T.D.C.
N.G.
Adjust ignition timing by turning crack angle sensor.
O.K.
(Go to Ⓐ on next page.)

ECCS SELF-DIAGNOSTIC CHART—1990 STANZA

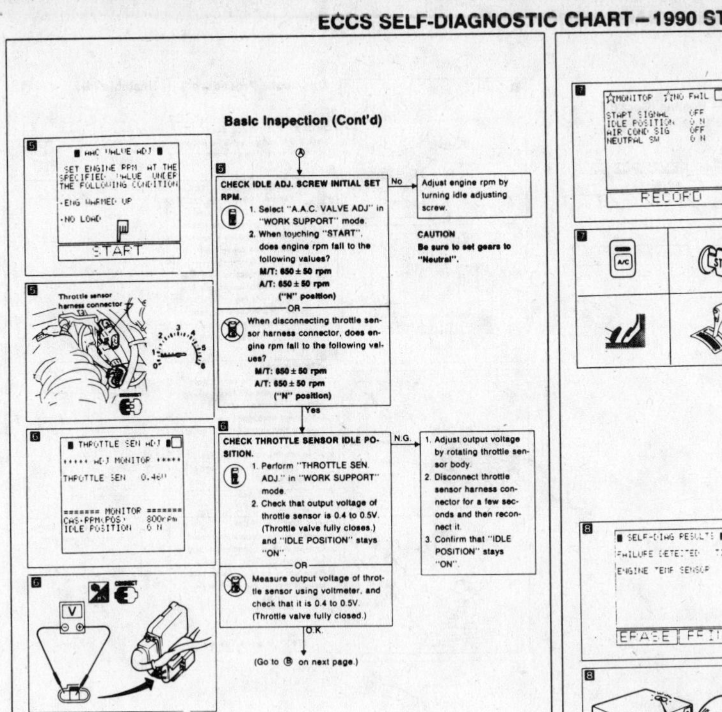
Basic Inspection (Cont'd)
A
5
CHECK IDLE ADJ. SCREW INITIAL SET RPM.
1. Select "A.A.C. VALVE ADJ" in "WORK SUPPORT" mode.
2. When touching "START", does engine rpm fall to the following values?
M/T: 650 ± 50 rpm
A/T: 650 ± 50 rpm
("N" position)
OR
When disconnecting throttle sensor harness connector, does engine rpm fall to the following values?
M/T: 650 ± 50 rpm
A/T: 650 ± 50 rpm
("N" position)
No
Adjust engine rpm by turning idle adjusting screw.
CAUTION
Be sure to set gears to "Neutral".
Yes
6
CHECK THROTTLE SENSOR IDLE POSITION.
1. Perform "THROTTLE SEN. ADJ." in "WORK SUPPORT" mode.
2. Check that output voltage of throttle sensor is 0.4 to 0.5V. (Throttle valve fully closes.) and "IDLE POSITION" stays "ON".
OR
Measure output voltage of throttle sensor using voltmeter, and check that it is 0.4 to 0.5V. (Throttle valve fully closed.)
N.G.
1. Adjust output voltage by rotating throttle sensor body.
2. Disconnect throttle sensor harness connector for a few seconds and then reconnect it.
3. Confirm that "IDLE POSITION" stays "ON".
O.K.
(Go to B on next page.)
START
Throttle sensor harness connector
V

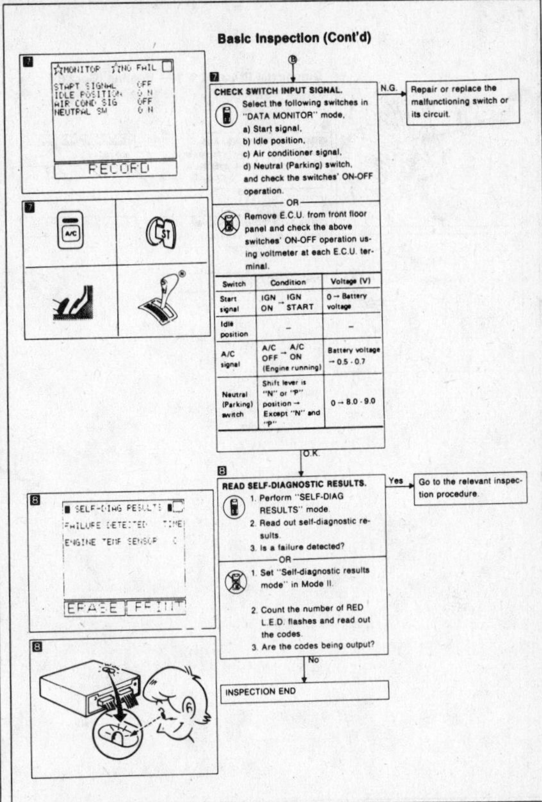
Basic Inspection (Cont'd)
B
7
CHECK SWITCH INPUT SIGNAL.
Select the following switches in "DATA MONITOR" mode,
a) Start signal,
b) Idle position,
c) Air conditioner signal,
d) Neutral (Parking) switch,
and check the switches' ON-OFF operation.
OR
Remove E.C.U. from front floor panel and check the above switches' ON-OFF operation using voltmeter at each E.C.U. terminal.
N.G.
Repair or replace the malfunctioning switch or its circuit.
O.K.
8
READ SELF-DIAGNOSTIC RESULTS.
1. Perform "SELF-DIAG RESULTS" mode.
2. Read out self-diagnostic results.
3. Is a failure detected?
OR
1. Set "Self-diagnostic results mode" in Mode II.
2. Count the number of RED L.E.D. flashes and read out the codes.
3. Are the codes being output?
Yes
Go to the relevant inspection procedure.
No
INSPECTION END
RECORD
ERASE
A/C

Switch	Condition	Voltage (V)
Start signal	IGN ON → IGN START	0 → Battery voltage
Idle position	–	–
A/C signal	A/C OFF → A/C ON (Engine running)	Battery voltage → 0.5 - 0.7
Neutral (Parking) switch	Shift lever is "N" or "P" position → Except "N" and "P"	0 → 8.0 - 9.0

ECCS SELF-DIAGNOSTIC CHART—1990 STANZA

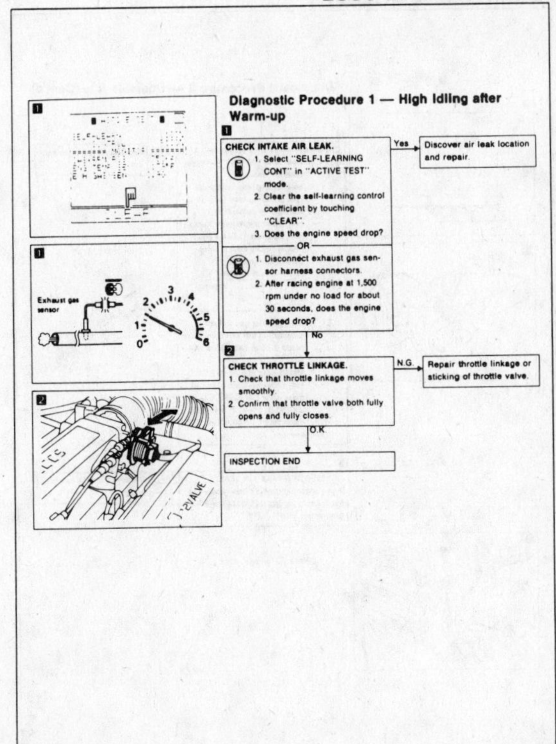
Diagnostic Procedure 1 — High Idling after Warm-up
1
CHECK INTAKE AIR LEAK.
1. Select "SELF-LEARNING CONT" in "ACTIVE TEST" mode.
2. Clear the self-learning control coefficient by touching "CLEAR".
3. Does the engine speed drop?
OR
1. Disconnect exhaust gas sensor harness connectors.
2. After racing engine at 1,500 rpm under no load for about 30 seconds, does the engine speed drop?
Yes
Discover air leak location and repair.
No
2
CHECK THROTTLE LINKAGE.
1. Check that throttle linkage moves smoothly.
2. Confirm that throttle valve both fully opens and fully closes.
N.G.
Repair throttle linkage or sticking of throttle valve.
O.K.
INSPECTION END
Exhaust gas sensor

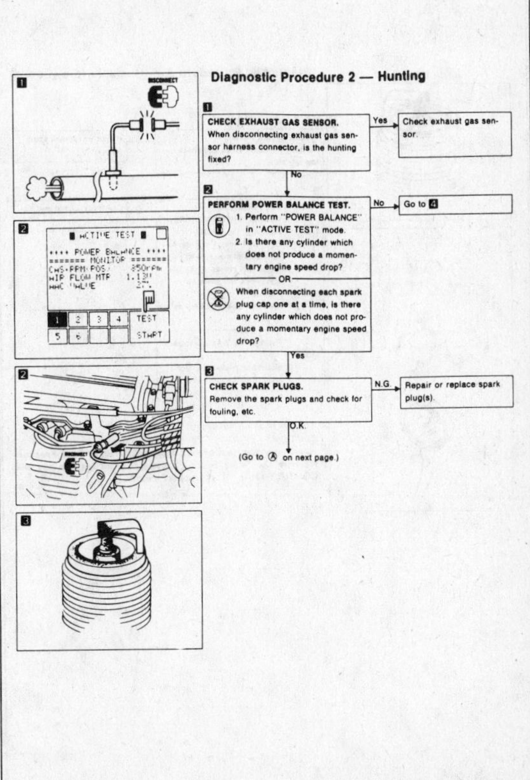
Diagnostic Procedure 2 — Hunting
1
CHECK EXHAUST GAS SENSOR.
When disconnecting exhaust gas sensor harness connector, is the hunting fixed?
Yes
Check exhaust gas sensor.
No
2
PERFORM POWER BALANCE TEST.
1. Perform "POWER BALANCE" in "ACTIVE TEST" mode.
2. Is there any cylinder which does not produce a momentary engine speed drop?
OR
When disconnecting each spark plug cap one at a time, is there any cylinder which does not produce a momentary engine speed drop?
No
Go to 4
Yes
3
CHECK SPARK PLUGS.
Remove the spark plugs and check for fouling, etc.
N.G.
Repair or replace spark plug(s).
O.K.
(Go to A on next page.)
DISCONNECT
TEST
START

ECCS SELF-DIAGNOSTIC CHART—1990 STANZA

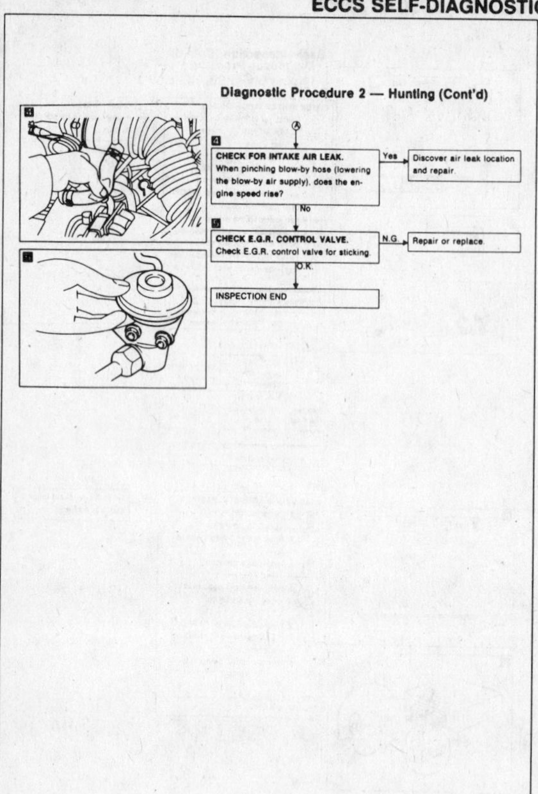
Diagnostic Procedure 2 — Hunting (Cont'd)
A
CHECK FOR INTAKE AIR LEAK.
When pinching blow-by hose (lowering the blow-by air supply), does the engine speed rise?
Yes
Discover air leak location and repair.
No
CHECK E.G.R. CONTROL VALVE.
Check E.G.R. control valve for sticking.
N.G.
Repair or replace.
O.K.
INSPECTION END

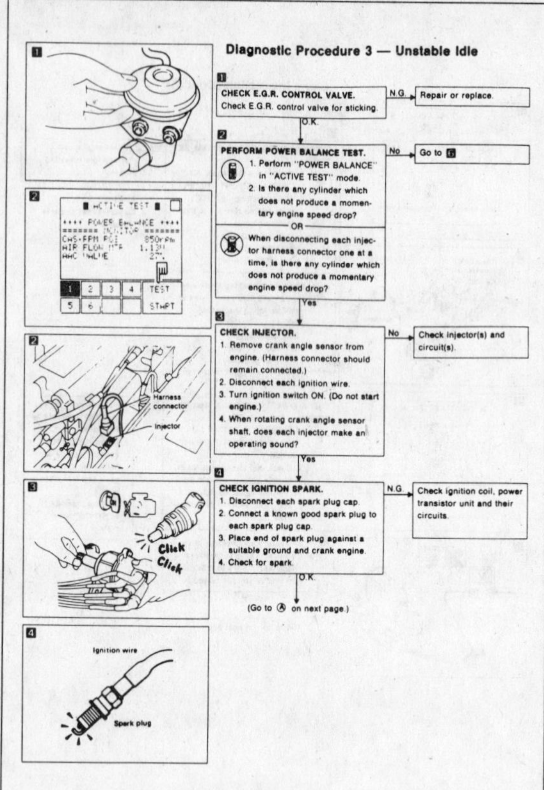
Diagnostic Procedure 3 — Unstable Idle
CHECK E.G.R. CONTROL VALVE.
Check E.G.R. control valve for sticking.
N.G.
Repair or replace.
O.K.
PERFORM POWER BALANCE TEST.
1. Perform "POWER BALANCE" in "ACTIVE TEST" mode.
2. Is there any cylinder which does not produce a momentary engine speed drop?
OR
When disconnecting each injector harness connector one at a time, is there any cylinder which does not produce a momentary engine speed drop?
No
Go to
Yes
CHECK INJECTOR.
1. Remove crank angle sensor from engine. (Harness connector should remain connected.)
2. Disconnect each ignition wire.
3. Turn ignition switch ON. (Do not start engine.)
4. When rotating crank angle sensor shaft, does each injector make an operating sound?
No
Check injector(s) and circuit(s).
Yes
CHECK IGNITION SPARK.
1. Disconnect each spark plug cap.
2. Connect a known good spark plug to each spark plug cap.
3. Place end of spark plug against a suitable ground and crank engine.
4. Check for spark.
N.G.
Check ignition coil, power transistor unit and their circuits.
O.K.
(Go to A on next page.)
ACTIVE TEST
POWER BALANCE
TEST
START
Harness connector
Injector
Click
Click
Ignition wire
Spark plug

ECCS SELF-DIAGNOSTIC CHART—1990 STANZA

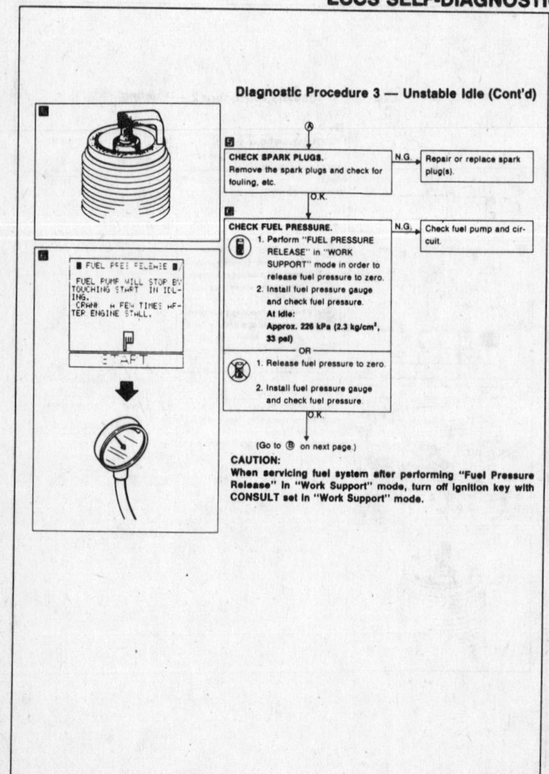
Diagnostic Procedure 3 — Unstable Idle (Cont'd)
A
CHECK SPARK PLUGS.
Remove the spark plugs and check for fouling, etc.
N.G.
Repair or replace spark plug(s).
O.K.
CHECK FUEL PRESSURE.
1. Perform "FUEL PRESSURE RELEASE" in "WORK SUPPORT" mode in order to release fuel pressure to zero.
2. Install fuel pressure gauge and check fuel pressure.
At idle:
Approx. 226 kPa (2.3 kg/cm², 33 psi)
OR
1. Release fuel pressure to zero.
2. Install fuel pressure gauge and check fuel pressure.
N.G.
Check fuel pump and circuit.
O.K.
(Go to B on next page.)
CAUTION:
When servicing fuel system after performing "Fuel Pressure Release" in "Work Support" mode, turn off ignition key with CONSULT set in "Work Support" mode.
FUEL PRES RELEASE
START

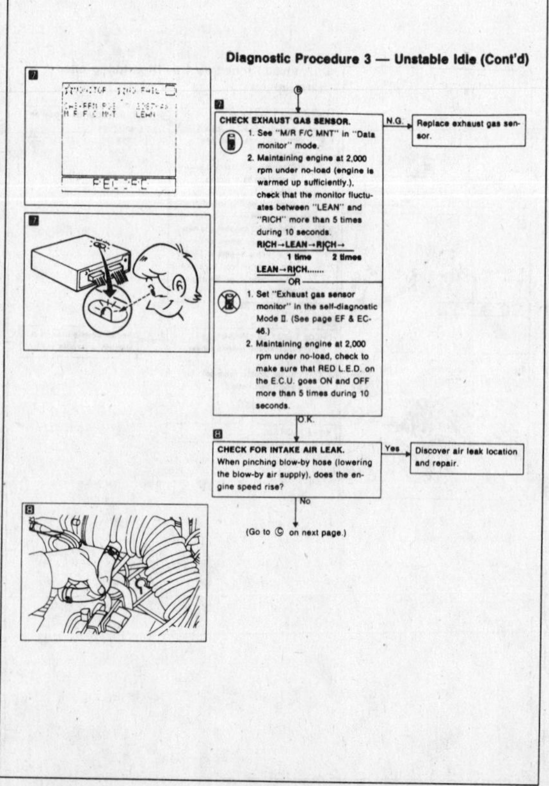
Diagnostic Procedure 3 — Unstable Idle (Cont'd)
B
CHECK EXHAUST GAS SENSOR.
1. See "M/R F/C MNT" in "Data monitor" mode.
2. Maintaining engine at 2,000 rpm under no-load (engine is warmed up sufficiently.), check that the monitor fluctuates between "LEAN" and "RICH" more than 5 times during 10 seconds.
RICH→LEAN→RICH→
1 time 2 times
LEAN→RICH.......
OR
1. Set "Exhaust gas sensor monitor" in the self-diagnostic Mode II. (See page EF & EC-46.)
2. Maintaining engine at 2,000 rpm under no-load, check to make sure that RED L.E.D. on the E.C.U. goes ON and OFF more than 5 times during 10 seconds.
N.G.
Replace exhaust gas sensor.
O.K.
CHECK FOR INTAKE AIR LEAK.
When pinching blow-by hose (lowering the blow-by air supply), does the engine speed rise?
Yes
Discover air leak location and repair.
No
(Go to C on next page.)

ECCS SELF-DIAGNOSTIC CHART—1990 STANZA

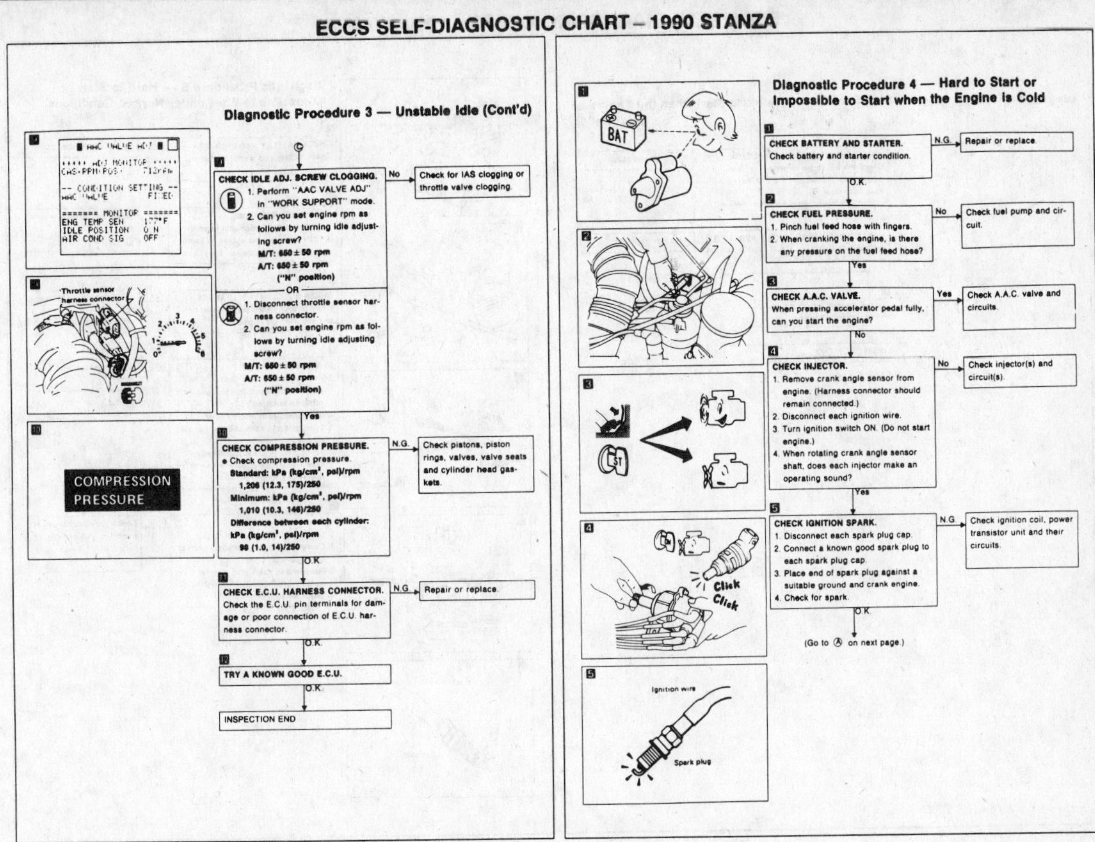
Diagnostic Procedure 3 — Unstable Idle (Cont'd)
++++++ ADJ MONITOR ++++++
CAS·RPM·POS· 712rpm
-- CONDITION SETTING --
====== MONITOR ======
ENG TEMP SEN 177°F
IDLE POSITION ON
AIR COND SIG OFF
Throttle sensor harness connector
COMPRESSION PRESSURE
9
CHECK IDLE ADJ. SCREW CLOGGING.
1. Perform "AAC VALVE ADJ" in "WORK SUPPORT" mode.
2. Can you set engine rpm as follows by turning idle adjusting screw?
M/T: 650 ± 50 rpm
A/T: 650 ± 50 rpm ("N" position)
OR
1. Disconnect throttle sensor harness connector.
2. Can you set engine rpm as follows by turning idle adjusting screw?
M/T: 650 ± 50 rpm
A/T: 650 ± 50 rpm ("N" position)
No
Check for IAS clogging or throttle valve clogging.
Yes
10
CHECK COMPRESSION PRESSURE.
• Check compression pressure.
Standard: kPa (kg/cm², psi)/rpm 1,206 (12.3, 175)/250
Minimum: kPa (kg/cm², psi)/rpm 1,010 (10.3, 146)/250
Difference between each cylinder: kPa (kg/cm², psi)/rpm 98 (1.0, 14)/250
N.G.
Check pistons, piston rings, valves, valve seats and cylinder head gaskets.
O.K.
11
CHECK E.C.U. HARNESS CONNECTOR.
Check the E.C.U. pin terminals for damage or poor connection of E.C.U. harness connector.
N.G.
Repair or replace.
O.K.
12
TRY A KNOWN GOOD E.C.U.
O.K.
INSPECTION END
Diagnostic Procedure 4 — Hard to Start or Impossible to Start when the Engine is Cold
BAT
1
CHECK BATTERY AND STARTER.
Check battery and starter condition.
N.G.
Repair or replace.
O.K.
2
CHECK FUEL PRESSURE.
1. Pinch fuel feed hose with fingers.
2. When cranking the engine, is there any pressure on the fuel feed hose?
No
Check fuel pump and circuit.
Yes
3
CHECK A.A.C. VALVE.
When pressing accelerator pedal fully, can you start the engine?
Yes
Check A.A.C. valve and circuits.
No
4
CHECK INJECTOR.
1. Remove crank angle sensor from engine. (Harness connector should remain connected.)
2. Disconnect each ignition wire.
3. Turn ignition switch ON. (Do not start engine.)
4. When rotating crank angle sensor shaft, does each injector make an operating sound?
No
Check injector(s) and circuit(s).
Yes
5
CHECK IGNITION SPARK.
1. Disconnect each spark plug cap.
2. Connect a known good spark plug to each spark plug cap.
3. Place end of spark plug against a suitable ground and crank engine.
4. Check for spark.
N.G.
Check ignition coil, power transistor unit and their circuits.
O.K.
(Go to Ⓐ on next page.)
Click
Click
Ignition wire
Spark plug

ECCS SELF-DIAGNOSTIC CHART—1990 STANZA

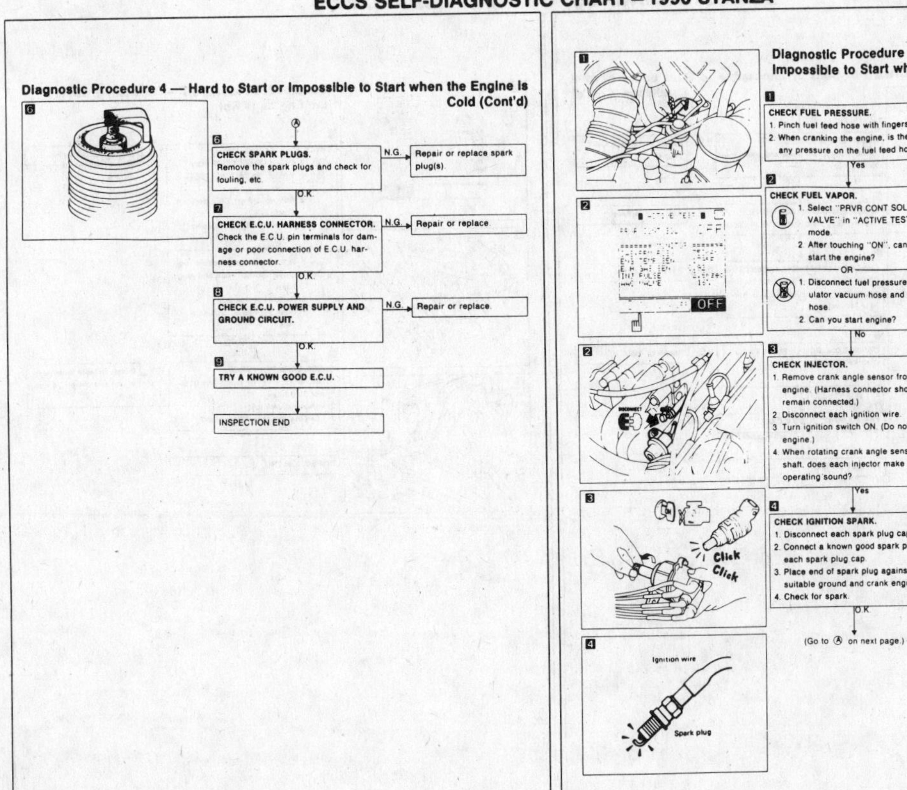
Diagnostic Procedure 4 — Hard to Start or Impossible to Start when the Engine is Cold (Cont'd)
Ⓐ
6
CHECK SPARK PLUGS.
Remove the spark plugs and check for fouling, etc.
N.G.
Repair or replace spark plug(s).
O.K.
7
CHECK E.C.U. HARNESS CONNECTOR.
Check the E.C.U. pin terminals for damage or poor connection of E.C.U. harness connector.
N.G.
Repair or replace.
O.K.
8
CHECK E.C.U. POWER SUPPLY AND GROUND CIRCUIT.
N.G.
Repair or replace.
O.K.
9
TRY A KNOWN GOOD E.C.U.
INSPECTION END

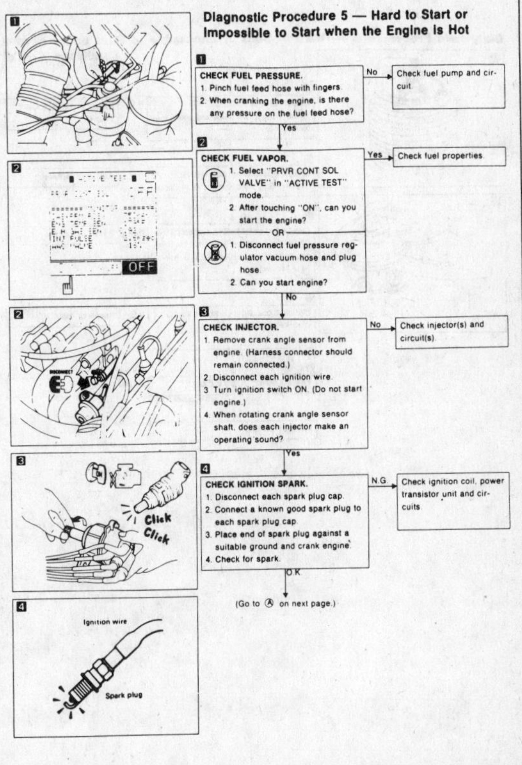
Diagnostic Procedure 5 — Hard to Start or Impossible to Start when the Engine is Hot
1
CHECK FUEL PRESSURE.
1. Pinch fuel feed hose with fingers.
2. When cranking the engine, is there any pressure on the fuel feed hose?
No
Check fuel pump and circuit.
Yes
2
CHECK FUEL VAPOR.
1. Select "PRVR CONT SOL VALVE" in "ACTIVE TEST" mode.
2. After touching "ON", can you start the engine?
OR
1. Disconnect fuel pressure regulator vacuum hose and plug hose.
2. Can you start engine?
Yes
Check fuel properties.
No
3
CHECK INJECTOR.
1. Remove crank angle sensor from engine. (Harness connector should remain connected.)
2. Disconnect each ignition wire.
3. Turn ignition switch ON. (Do not start engine.)
4. When rotating crank angle sensor shaft, does each injector make an operating sound?
No
Check injector(s) and circuit(s).
Yes
4
CHECK IGNITION SPARK.
1. Disconnect each spark plug cap.
2. Connect a known good spark plug to each spark plug cap.
3. Place end of spark plug against a suitable ground and crank engine.
4. Check for spark.
N.G.
Check ignition coil, power transistor unit and circuits.
O.K.
(Go to Ⓐ on next page.)
OFF
Click
Click
Ignition wire
Spark plug

ECCS SELF-DIAGNOSTIC CHART—1990 STANZA

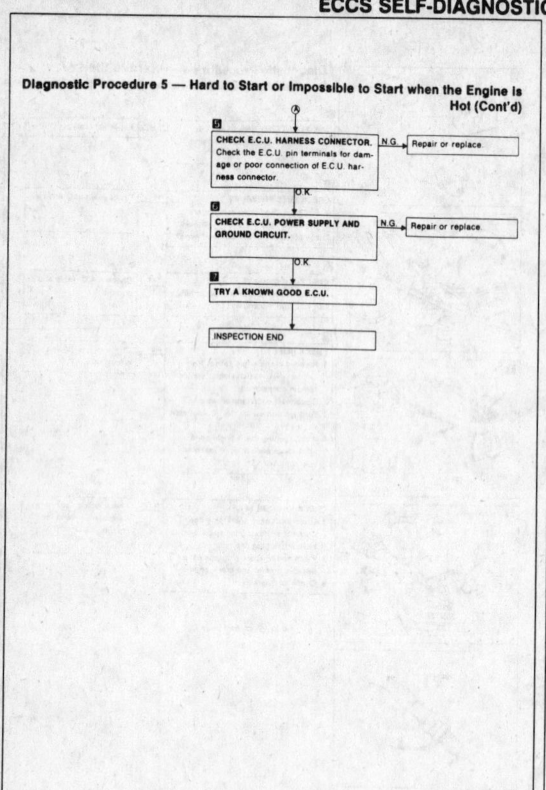
Diagnostic Procedure 5 — Hard to Start or Impossible to Start when the Engine is Hot (Cont'd)
A
5
CHECK E.C.U. HARNESS CONNECTOR.
Check the E.C.U. pin terminals for damage or poor connection of E.C.U. harness connector.
N.G.
Repair or replace.
O.K.
6
CHECK E.C.U. POWER SUPPLY AND GROUND CIRCUIT.
N.G.
Repair or replace.
O.K.
7
TRY A KNOWN GOOD E.C.U.
INSPECTION END

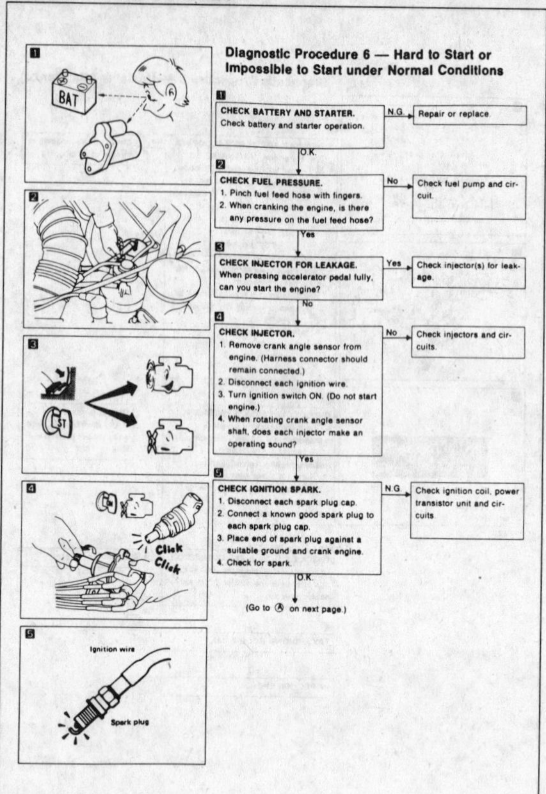
BAT
Click
Click
Ignition wire
Spark plug
Diagnostic Procedure 6 — Hard to Start or Impossible to Start under Normal Conditions
1
CHECK BATTERY AND STARTER.
Check battery and starter operation.
N.G.
Repair or replace.
O.K.
2
CHECK FUEL PRESSURE.
1. Pinch fuel feed hose with fingers.
2. When cranking the engine, is there any pressure on the fuel feed hose?
No
Check fuel pump and circuit.
Yes
3
CHECK INJECTOR FOR LEAKAGE.
When pressing accelerator pedal fully, can you start the engine?
Yes
Check injector(s) for leakage.
No
4
CHECK INJECTOR.
1. Remove crank angle sensor from engine. (Harness connector should remain connected.)
2. Disconnect each ignition wire.
3. Turn ignition switch ON. (Do not start engine.)
4. When rotating crank angle sensor shaft, does each injector make an operating sound?
No
Check injectors and circuits.
Yes
5
CHECK IGNITION SPARK.
1. Disconnect each spark plug cap.
2. Connect a known good spark plug to each spark plug cap.
3. Place end of spark plug against a suitable ground and crank engine.
4. Check for spark.
N.G.
Check ignition coil, power transistor unit and circuits.
O.K.
(Go to A on next page.)

ECCS SELF-DIAGNOSTIC CHART—1990 STANZA

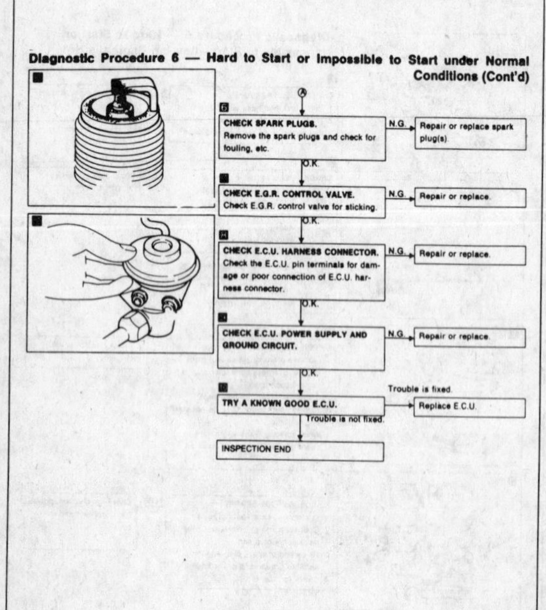
Diagnostic Procedure 6 — Hard to Start or Impossible to Start under Normal Conditions (Cont'd)
A
6
CHECK SPARK PLUGS.
Remove the spark plugs and check for fouling, etc.
N.G.
Repair or replace spark plug(s).
O.K.
7
CHECK E.G.R. CONTROL VALVE.
Check E.G.R. control valve for sticking.
N.G.
Repair or replace.
O.K.
8
CHECK E.C.U. HARNESS CONNECTOR.
Check the E.C.U. pin terminals for damage or poor connection of E.C.U. harness connector.
N.G.
Repair or replace.
O.K.
9
CHECK E.C.U. POWER SUPPLY AND GROUND CIRCUIT.
N.G.
Repair or replace.
O.K.
10
TRY A KNOWN GOOD E.C.U.
Trouble is fixed.
Replace E.C.U.
Trouble is not fixed.
INSPECTION END

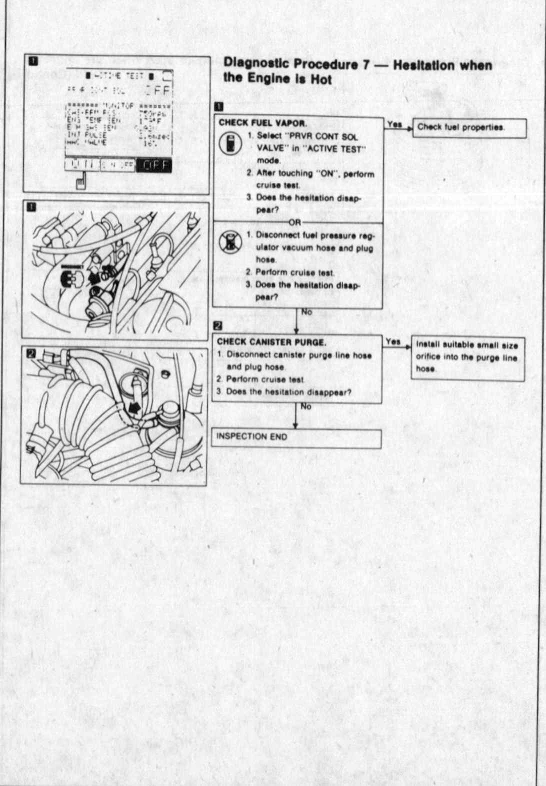
OFF
Diagnostic Procedure 7 — Hesitation when the Engine is Hot
1
CHECK FUEL VAPOR.
1. Select "PRVR CONT SOL VALVE" in "ACTIVE TEST" mode.
2. After touching "ON", perform cruise test.
3. Does the hesitation disappear?
OR
1. Disconnect fuel pressure regulator vacuum hose and plug hose.
2. Perform cruise test.
3. Does the hesitation disappear?
Yes
Check fuel properties.
No
2
CHECK CANISTER PURGE.
1. Disconnect canister purge line hose and plug hose.
2. Perform cruise test.
3. Does the hesitation disappear?
Yes
Install suitable small size orifice into the purge line hose.
No
INSPECTION END

ECCS SELF-DIAGNOSTIC CHART—1990 STANZA

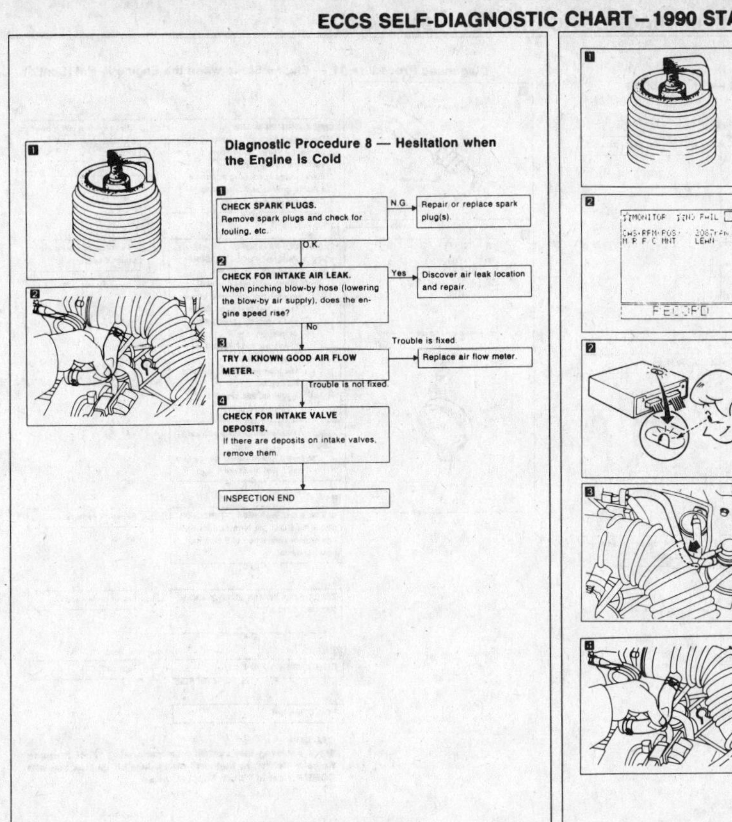
Diagnostic Procedure 8 — Hesitation when the Engine is Cold
1
CHECK SPARK PLUGS.
Remove spark plugs and check for fouling, etc.
N.G.
Repair or replace spark plug(s).
O.K.
2
CHECK FOR INTAKE AIR LEAK.
When pinching blow-by hose (lowering the blow-by air supply), does the engine speed rise?
Yes
Discover air leak location and repair.
No
3
TRY A KNOWN GOOD AIR FLOW METER.
Trouble is fixed
Replace air flow meter.
Trouble is not fixed.
4
CHECK FOR INTAKE VALVE DEPOSITS.
If there are deposits on intake valves, remove them
INSPECTION END

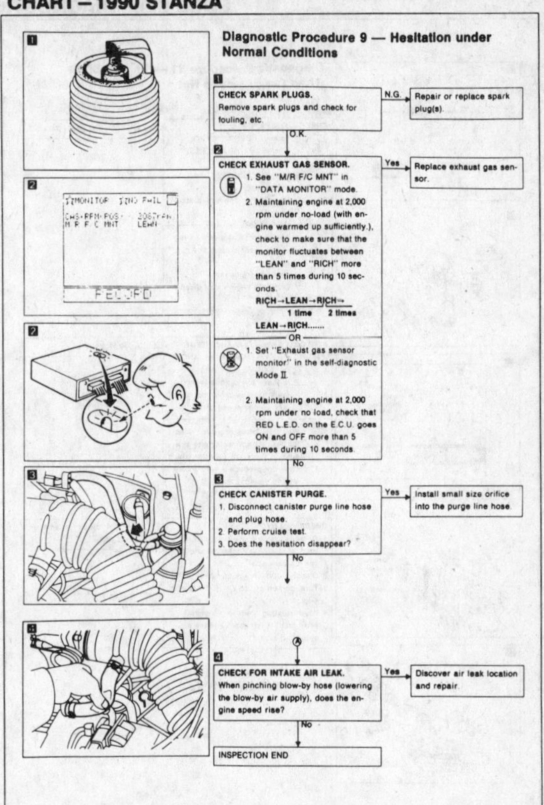
Diagnostic Procedure 9 — Hesitation under Normal Conditions
1
CHECK SPARK PLUGS.
Remove spark plugs and check for fouling, etc.
N.G.
Repair or replace spark plug(s).
O.K.
2
CHECK EXHAUST GAS SENSOR.
1. See "M/R F/C MNT" in "DATA MONITOR" mode.
2. Maintaining engine at 2,000 rpm under no-load (with engine warmed up sufficiently.), check to make sure that the monitor fluctuates between "LEAN" and "RICH" more than 5 times during 10 seconds.
RICH→LEAN→RICH→
1 time 2 times
LEAN→RICH.......
OR
1. Set "Exhaust gas sensor monitor" in the self-diagnostic Mode II
2. Maintaining engine at 2,000 rpm under no load, check that RED L.E.D. on the E.C.U. goes ON and OFF more than 5 times during 10 seconds.
Yes
Replace exhaust gas sensor.
No
3
CHECK CANISTER PURGE.
1. Disconnect canister purge line hose and plug hose.
2. Perform cruise test.
3. Does the hesitation disappear?
Yes
Install small size orifice into the purge line hose.
No
A
4
CHECK FOR INTAKE AIR LEAK.
When pinching blow-by hose (lowering the blow-by air supply), does the engine speed rise?
Yes
Discover air leak location and repair.
No
INSPECTION END
RECORD

ECCS SELF-DIAGNOSTIC CHART—1990 STANZA

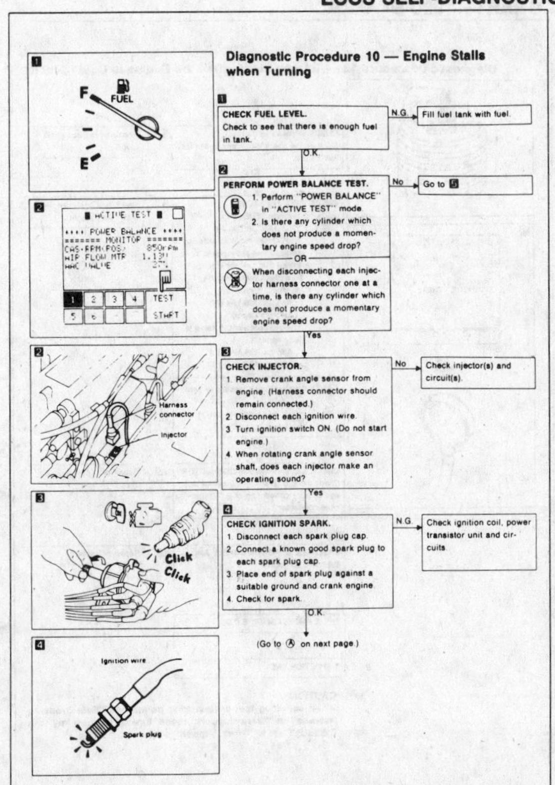
Diagnostic Procedure 10 — Engine Stalls when Turning
1
CHECK FUEL LEVEL.
Check to see that there is enough fuel in tank.
N.G.
Fill fuel tank with fuel.
O.K.
2
PERFORM POWER BALANCE TEST.
1. Perform "POWER BALANCE" in "ACTIVE TEST" mode.
2. Is there any cylinder which does not produce a momentary engine speed drop?
No
Go to 5
OR
When disconnecting each injector harness connector one at a time, is there any cylinder which does not produce a momentary engine speed drop?
Yes
3
CHECK INJECTOR.
1. Remove crank angle sensor from engine. (Harness connector should remain connected.)
2. Disconnect each ignition wire.
3. Turn ignition switch ON. (Do not start engine.)
4. When rotating crank angle sensor shaft, does each injector make an operating sound?
No
Check injector(s) and circuit(s).
Yes
4
CHECK IGNITION SPARK.
1. Disconnect each spark plug cap.
2. Connect a known good spark plug to each spark plug cap.
3. Place end of spark plug against a suitable ground and crank engine.
4. Check for spark.
N.G.
Check ignition coil, power transistor unit and circuits.
O.K.
(Go to A on next page.)
FUEL
F
E
ACTIVE TEST
POWER BALANCE
MONITOR
TEST
START
Harness connector
Injector
Click
Click
Ignition wire
Spark plug

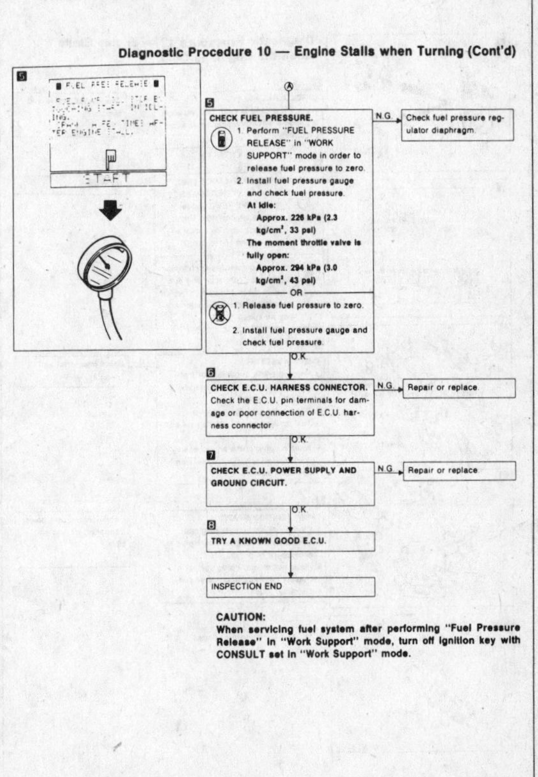
Diagnostic Procedure 10 — Engine Stalls when Turning (Cont'd)
A
5
CHECK FUEL PRESSURE.
1. Perform "FUEL PRESSURE RELEASE" in "WORK SUPPORT" mode in order to release fuel pressure to zero.
2. Install fuel pressure gauge and check fuel pressure.
At idle:
Approx. 226 kPa (2.3 kg/cm², 33 psi)
The moment throttle valve is fully open:
Approx. 294 kPa (3.0 kg/cm², 43 psi)
OR
1. Release fuel pressure to zero.
2. Install fuel pressure gauge and check fuel pressure.
N.G.
Check fuel pressure regulator diaphragm.
O.K.
6
CHECK E.C.U. HARNESS CONNECTOR.
Check the E.C.U. pin terminals for damage or poor connection of E.C.U. harness connector.
N.G.
Repair or replace.
O.K.
7
CHECK E.C.U. POWER SUPPLY AND GROUND CIRCUIT.
N.G.
Repair or replace.
O.K.
8
TRY A KNOWN GOOD E.C.U.
INSPECTION END
CAUTION:
When servicing fuel system after performing "Fuel Pressure Release" in "Work Support" mode, turn off ignition key with CONSULT set in "Work Support" mode.
START

ECCS SELF-DIAGNOSTIC CHART—1990 STANZA

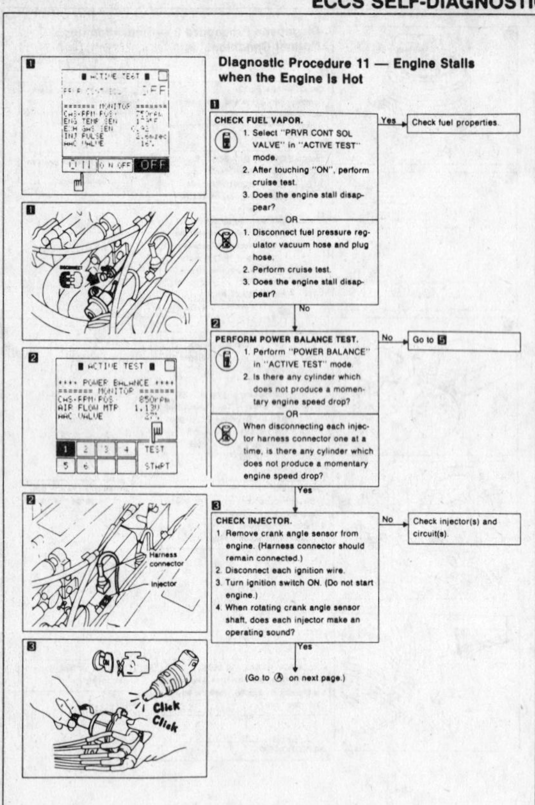
Diagnostic Procedure 11 — Engine Stalls when the Engine is Hot
CHECK FUEL VAPOR.
1. Select "PRVR CONT SOL VALVE" in "ACTIVE TEST" mode.
2. After touching "ON", perform cruise test.
3. Does the engine stall disappear?
OR
1. Disconnect fuel pressure regulator vacuum hose and plug hose.
2. Perform cruise test.
3. Does the engine stall disappear?
Yes
Check fuel properties.
No
PERFORM POWER BALANCE TEST.
1. Perform "POWER BALANCE" in "ACTIVE TEST" mode.
2. Is there any cylinder which does not produce a momentary engine speed drop?
OR
When disconnecting each injector harness connector one at a time, is there any cylinder which does not produce a momentary engine speed drop?
No
Go to 5
Yes
CHECK INJECTOR.
1. Remove crank angle sensor from engine. (Harness connector should remain connected.)
2. Disconnect each ignition wire.
3. Turn ignition switch ON. (Do not start engine.)
4. When rotating crank angle sensor shaft, does each injector make an operating sound?
No
Check injector(s) and circuit(s).
Yes
(Go to Ⓐ on next page.)
Harness connector
Injector
Click
Click

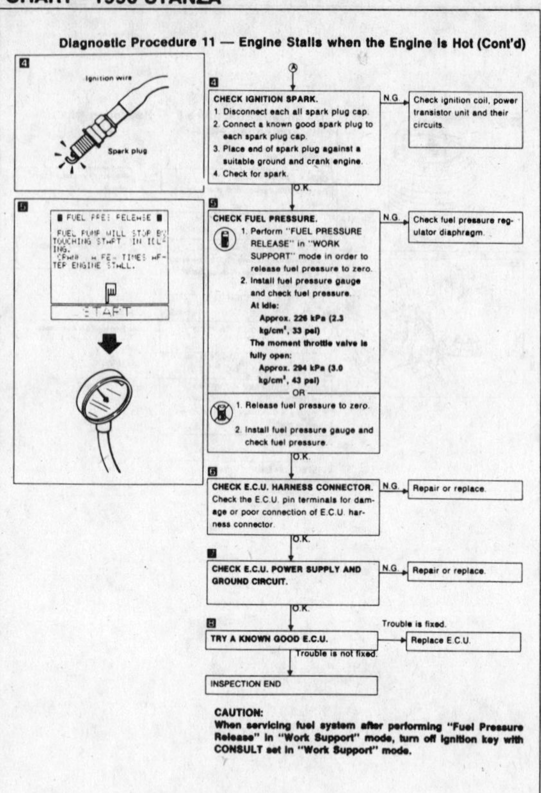
Diagnostic Procedure 11 — Engine Stalls when the Engine is Hot (Cont'd)
Ignition wire
Spark plug
CHECK IGNITION SPARK.
1. Disconnect each all spark plug cap.
2. Connect a known good spark plug to each spark plug cap.
3. Place end of spark plug against a suitable ground and crank engine.
4. Check for spark.
N.G.
Check ignition coil, power transistor unit and their circuits.
O.K.
CHECK FUEL PRESSURE.
1. Perform "FUEL PRESSURE RELEASE" in "WORK SUPPORT" mode in order to release fuel pressure to zero.
2. Install fuel pressure gauge and check fuel pressure.
At idle:
Approx. 226 kPa (2.3 kg/cm², 33 psi)
The moment throttle valve is fully open:
Approx. 294 kPa (3.0 kg/cm², 43 psi)
OR
1. Release fuel pressure to zero.
2. Install fuel pressure gauge and check fuel pressure.
N.G.
Check fuel pressure regulator diaphragm.
O.K.
CHECK E.C.U. HARNESS CONNECTOR.
Check the E.C.U. pin terminals for damage or poor connection of E.C.U. harness connector.
N.G.
Repair or replace.
O.K.
CHECK E.C.U. POWER SUPPLY AND GROUND CIRCUIT.
N.G.
Repair or replace.
O.K.
TRY A KNOWN GOOD E.C.U.
Trouble is fixed.
Replace E.C.U.
Trouble is not fixed.
INSPECTION END
CAUTION:
When servicing fuel system after performing "Fuel Pressure Release" in "Work Support" mode, turn off ignition key with CONSULT set in "Work Support" mode.

ECCS SELF-DIAGNOSTIC CHART—1990 STANZA

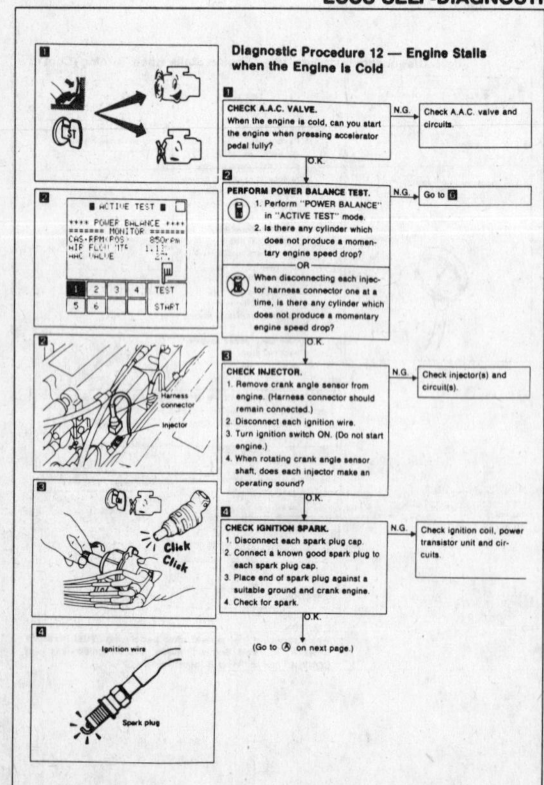
Diagnostic Procedure 12 — Engine Stalls when the Engine is Cold
CHECK A.A.C. VALVE.
When the engine is cold, can you start the engine when pressing accelerator pedal fully?
N.G.
Check A.A.C. valve and circuits.
O.K.
PERFORM POWER BALANCE TEST.
1. Perform "POWER BALANCE" in "ACTIVE TEST" mode.
2. Is there any cylinder which does not produce a momentary engine speed drop?
OR
When disconnecting each injector harness connector one at a time, is there any cylinder which does not produce a momentary engine speed drop?
N.G.
Go to 6
O.K.
CHECK INJECTOR.
1. Remove crank angle sensor from engine. (Harness connector should remain connected.)
2. Disconnect each ignition wire.
3. Turn ignition switch ON. (Do not start engine.)
4. When rotating crank angle sensor shaft, does each injector make an operating sound?
N.G.
Check injector(s) and circuit(s).
O.K.
CHECK IGNITION SPARK.
1. Disconnect each spark plug cap.
2. Connect a known good spark plug to each spark plug cap.
3. Place end of spark plug against a suitable ground and crank engine.
4. Check for spark.
N.G.
Check ignition coil, power transistor unit and circuits.
O.K.
(Go to Ⓐ on next page.)
Harness connector
Injector
Click
Click
Ignition wire
Spark plug

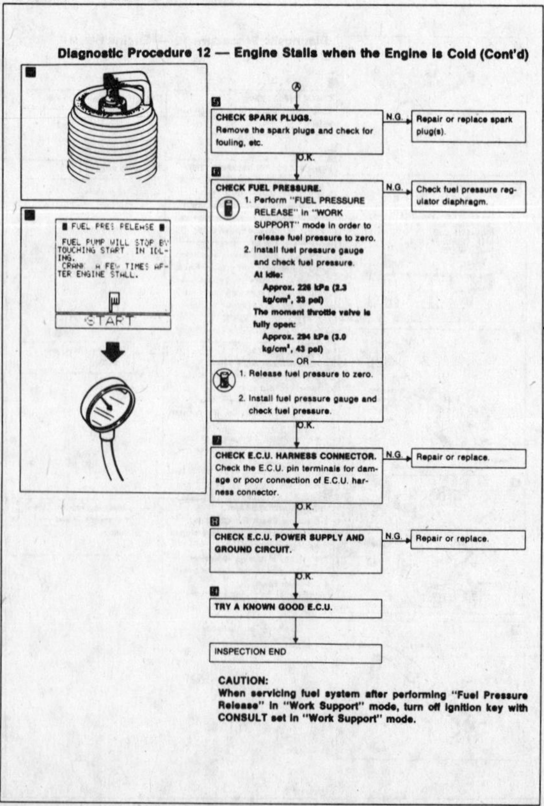
Diagnostic Procedure 12 — Engine Stalls when the Engine is Cold (Cont'd)
CHECK SPARK PLUGS.
Remove the spark plugs and check for fouling, etc.
N.G.
Repair or replace spark plug(s).
O.K.
CHECK FUEL PRESSURE.
1. Perform "FUEL PRESSURE RELEASE" in "WORK SUPPORT" mode in order to release fuel pressure to zero.
2. Install fuel pressure gauge and check fuel pressure.
At idle:
Approx. 226 kPa (2.3 kg/cm², 33 psi)
The moment throttle valve is fully open:
Approx. 294 kPa (3.0 kg/cm², 43 psi)
OR
1. Release fuel pressure to zero.
2. Install fuel pressure gauge and check fuel pressure.
N.G.
Check fuel pressure regulator diaphragm.
O.K.
CHECK E.C.U. HARNESS CONNECTOR.
Check the E.C.U. pin terminals for damage or poor connection of E.C.U. harness connector.
N.G.
Repair or replace.
O.K.
CHECK E.C.U. POWER SUPPLY AND GROUND CIRCUIT.
N.G.
Repair or replace.
O.K.
TRY A KNOWN GOOD E.C.U.
INSPECTION END
CAUTION:
When servicing fuel system after performing "Fuel Pressure Release" in "Work Support" mode, turn off ignition key with CONSULT set in "Work Support" mode.

ECCS SELF-DIAGNOSTIC CHART—1990 STANZA

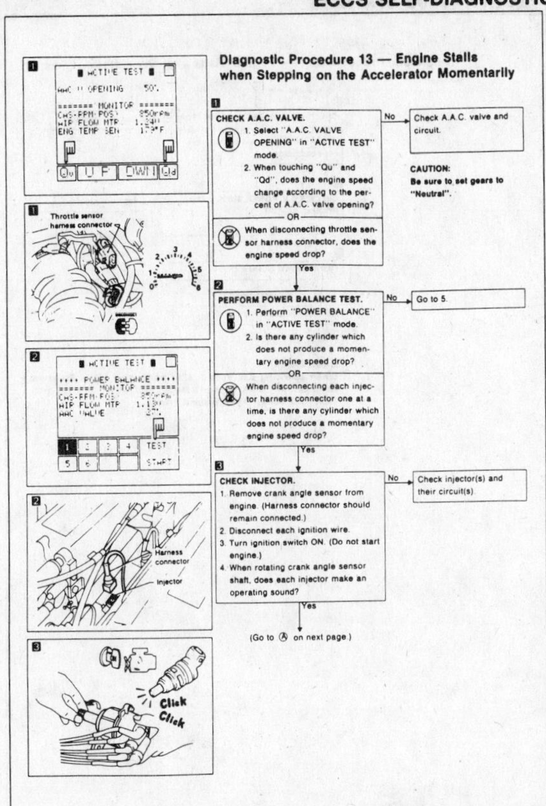
Diagnostic Procedure 13 — Engine Stalls when Stepping on the Accelerator Momentarily
ACTIVE TEST
AAC V OPENING 50%
MONITOR
CAS-RPM(POS) 850rpm
AIR FLOW MTR 1.24V
ENG TEMP SEN 179°F
Qu UP DWN Qd
Throttle sensor harness connector
1 CHECK A.A.C. VALVE.
1. Select "A.A.C. VALVE OPENING" in "ACTIVE TEST" mode.
2. When touching "Qu" and "Qd", does the engine speed change according to the percent of A.A.C. valve opening?
OR
When disconnecting throttle sensor harness connector, does the engine speed drop?
No
Check A.A.C. valve and circuit.
CAUTION:
Be sure to set gears to "Neutral".
Yes
2 PERFORM POWER BALANCE TEST.
1. Perform "POWER BALANCE" in "ACTIVE TEST" mode.
2. Is there any cylinder which does not produce a momentary engine speed drop?
OR
When disconnecting each injector harness connector one at a time, is there any cylinder which does not produce a momentary engine speed drop?
No
Go to 5.
POWER BALANCE
AAC VALVE 34%
TEST
START
Yes
3 CHECK INJECTOR.
1. Remove crank angle sensor from engine. (Harness connector should remain connected.)
2. Disconnect each ignition wire.
3. Turn ignition switch ON. (Do not start engine.)
4. When rotating crank angle sensor shaft, does each injector make an operating sound?
No
Check injector(s) and their circuit(s).
Harness connector
Injector
Click
Click
Yes
(Go to Ⓐ on next page.)

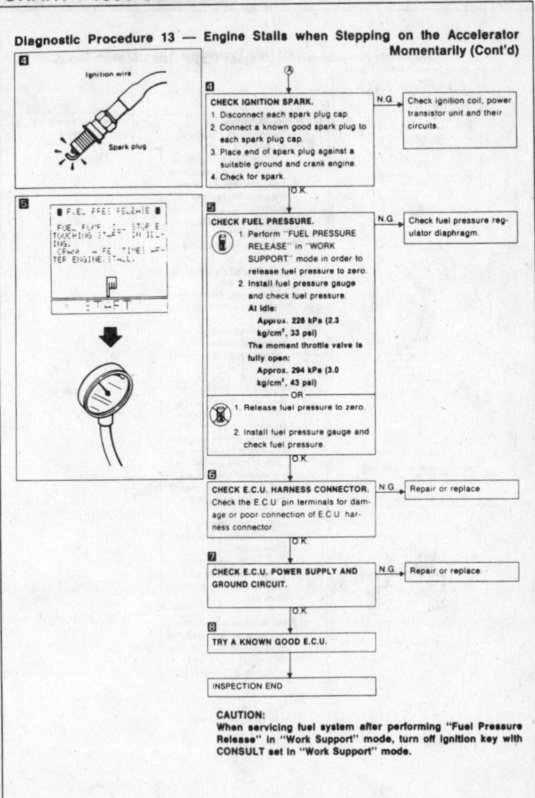
Diagnostic Procedure 13 — Engine Stalls when Stepping on the Accelerator Momentarily (Cont'd)
Ignition wire
Spark plug
Ⓐ
4 CHECK IGNITION SPARK.
1. Disconnect each spark plug cap.
2. Connect a known good spark plug to each spark plug cap.
3. Place end of spark plug against a suitable ground and crank engine.
4. Check for spark.
N.G.
Check ignition coil, power transistor unit and their circuits.
O.K.
FUEL PRES RELEASE
START
5 CHECK FUEL PRESSURE.
1. Perform "FUEL PRESSURE RELEASE" in "WORK SUPPORT" mode in order to release fuel pressure to zero.
2. Install fuel pressure gauge and check fuel pressure.
At idle:
Approx. 226 kPa (2.3 kg/cm², 33 psi)
The moment throttle valve is fully open:
Approx. 294 kPa (3.0 kg/cm², 43 psi)
OR
1. Release fuel pressure to zero.
2. Install fuel pressure gauge and check fuel pressure.
N.G.
Check fuel pressure regulator diaphragm.
O.K.
6 CHECK E.C.U. HARNESS CONNECTOR.
Check the E.C.U. pin terminals for damage or poor connection of E.C.U. harness connector.
N.G.
Repair or replace.
O.K.
7 CHECK E.C.U. POWER SUPPLY AND GROUND CIRCUIT.
N.G.
Repair or replace.
O.K.
8 TRY A KNOWN GOOD E.C.U.
INSPECTION END
CAUTION:
When servicing fuel system after performing "Fuel Pressure Release" in "Work Support" mode, turn off ignition key with CONSULT set in "Work Support" mode.

ECCS SELF-DIAGNOSTIC CHART—1990 STANZA

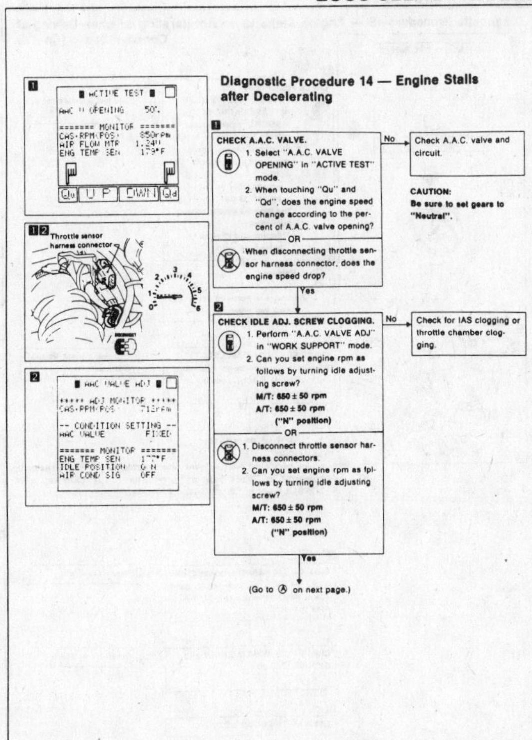
Diagnostic Procedure 14 — Engine Stalls after Decelerating
ACTIVE TEST
AAC V OPENING 50%
MONITOR
CAS-RPM(POS) 850rpm
AIR FLOW MTR 1.24V
ENG TEMP SEN 179°F
Qu UP DWN Qd
Throttle sensor harness connector
1 CHECK A.A.C. VALVE.
1. Select "A.A.C. VALVE OPENING" in "ACTIVE TEST" mode.
2. When touching "Qu" and "Qd", does the engine speed change according to the percent of A.A.C. valve opening?
OR
When disconnecting throttle sensor harness connector, does the engine speed drop?
No
Check A.A.C. valve and circuit.
CAUTION:
Be sure to set gears to "Neutral".
Yes
2 CHECK IDLE ADJ. SCREW CLOGGING.
1. Perform "A.A.C. VALVE ADJ" in "WORK SUPPORT" mode.
2. Can you set engine rpm as follows by turning idle adjusting screw?
M/T: 650 ± 50 rpm
A/T: 650 ± 50 rpm
("N" position)
OR
1. Disconnect throttle sensor harness connectors.
2. Can you set engine rpm as follows by turning idle adjusting screw?
M/T: 650 ± 50 rpm
A/T: 650 ± 50 rpm
("N" position)
No
Check for IAS clogging or throttle chamber clogging.
AAC VALVE ADJ
CONDITION SETTING
AAC VALVE FIXED
MONITOR
IDLE POSITION ON
AIR COND SIG OFF
Yes
(Go to Ⓐ on next page.)

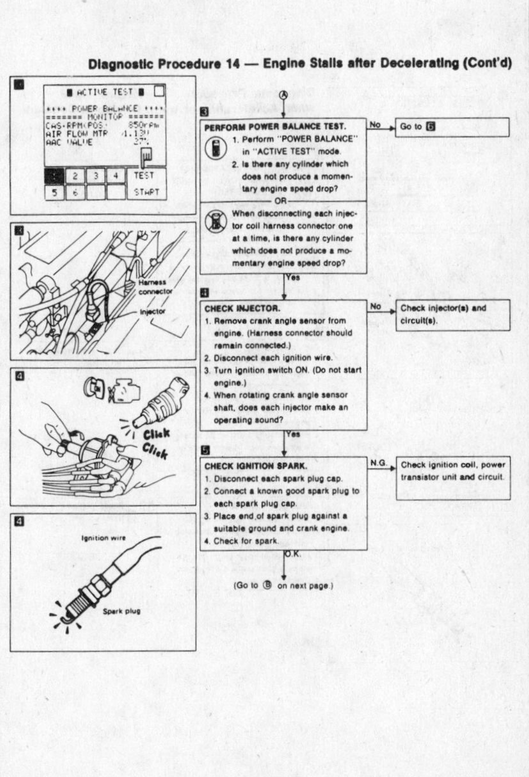
Diagnostic Procedure 14 — Engine Stalls after Decelerating (Cont'd)
Ⓐ
ACTIVE TEST
POWER BALANCE
TEST
START
3 PERFORM POWER BALANCE TEST.
1. Perform "POWER BALANCE" in "ACTIVE TEST" mode.
2. Is there any cylinder which does not produce a momentary engine speed drop?
OR
When disconnecting each injector coil harness connector one at a time, is there any cylinder which does not produce a momentary engine speed drop?
No
Go to 5
Yes
Harness connector
Injector
4 CHECK INJECTOR.
1. Remove crank angle sensor from engine. (Harness connector should remain connected.)
2. Disconnect each ignition wire.
3. Turn ignition switch ON. (Do not start engine.)
4. When rotating crank angle sensor shaft, does each injector make an operating sound?
No
Check injector(s) and circuit(s).
Click
Click
Yes
5 CHECK IGNITION SPARK.
1. Disconnect each spark plug cap.
2. Connect a known good spark plug to each spark plug cap.
3. Place end of spark plug against a suitable ground and crank engine.
4. Check for spark.
N.G.
Check ignition coil, power transistor unit and circuit.
Ignition wire
Spark plug
O.K.
(Go to Ⓑ on next page.)

ECCS SELF-DIAGNOSTIC CHART—1990 STANZA

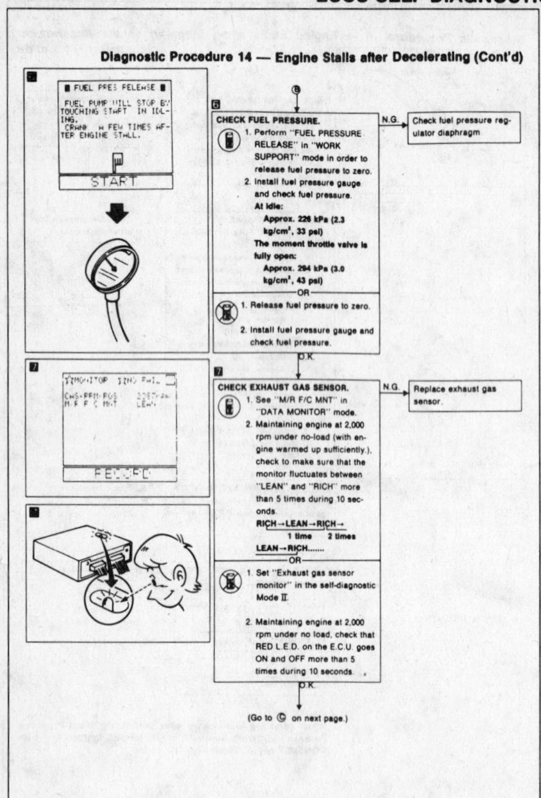
Diagnostic Procedure 14 — Engine Stalls after Decelerating (Cont'd)
FUEL PRES RELEASE
FUEL PUMP WILL STOP BY TOUCHING START IN IDLING. CRANK A FEW TIMES AFTER ENGINE STALL.
START
RECORD
B
6
CHECK FUEL PRESSURE.
1. Perform "FUEL PRESSURE RELEASE" in "WORK SUPPORT" mode in order to release fuel pressure to zero.
2. Install fuel pressure gauge and check fuel pressure.
At idle:
Approx. 226 kPa (2.3 kg/cm², 33 psi)
The moment throttle valve is fully open:
Approx. 294 kPa (3.0 kg/cm², 43 psi)
OR
1. Release fuel pressure to zero.
2. Install fuel pressure gauge and check fuel pressure.
N.G.
Check fuel pressure regulator diaphragm
O.K.
7
CHECK EXHAUST GAS SENSOR.
1. See "M/R F/C MNT" in "DATA MONITOR" mode.
2. Maintaining engine at 2,000 rpm under no-load (with engine warmed up sufficiently.), check to make sure that the monitor fluctuates between "LEAN" and "RICH" more than 5 times during 10 seconds.
RICH→LEAN→RICH→
1 time 2 times
LEAN→RICH.......
OR
1. Set "Exhaust gas sensor monitor" in the self-diagnostic Mode II.
2. Maintaining engine at 2,000 rpm under no load, check that RED L.E.D. on the E.C.U. goes ON and OFF more than 5 times during 10 seconds.
N.G.
Replace exhaust gas sensor.
O.K.
(Go to C on next page.)

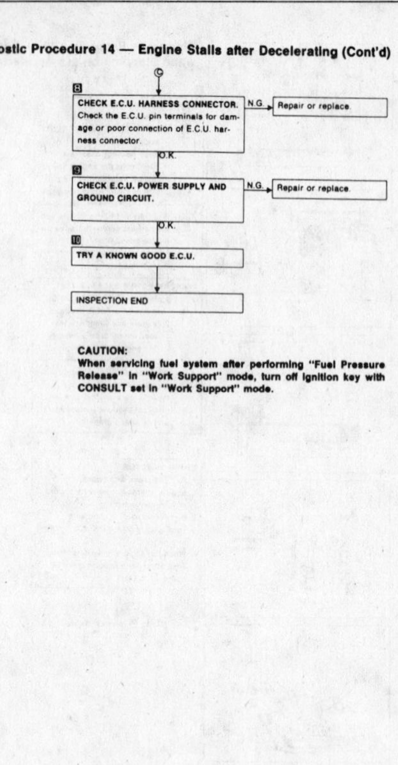
Diagnostic Procedure 14 — Engine Stalls after Decelerating (Cont'd)
C
8
CHECK E.C.U. HARNESS CONNECTOR.
Check the E.C.U. pin terminals for damage or poor connection of E.C.U. harness connector.
N.G.
Repair or replace.
O.K.
9
CHECK E.C.U. POWER SUPPLY AND GROUND CIRCUIT.
N.G.
Repair or replace.
O.K.
10
TRY A KNOWN GOOD E.C.U.
INSPECTION END
CAUTION:
When servicing fuel system after performing "Fuel Pressure Release" in "Work Support" mode, turn off ignition key with CONSULT set in "Work Support" mode.

ECCS SELF-DIAGNOSTIC CHART—1990 STANZA

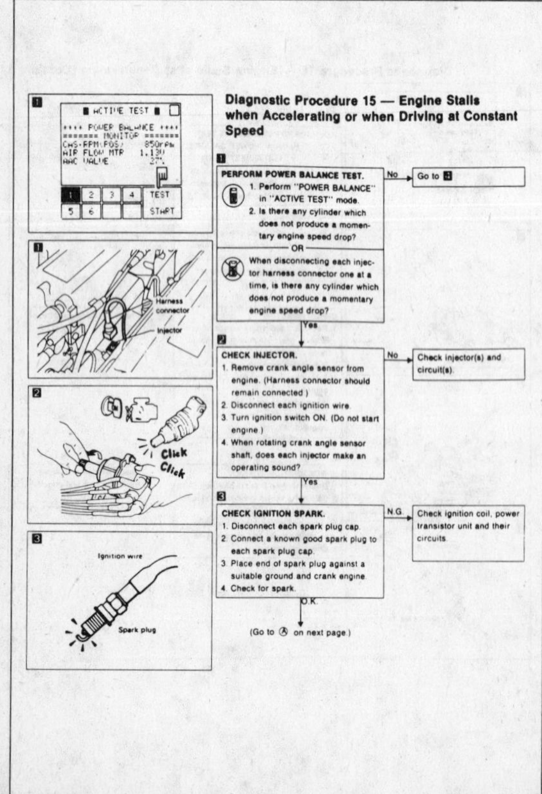
ACTIVE TEST
POWER BALANCE
MONITOR
CKS-RPM-POS 850rpm
AIR FLOW MTR 1.13V
AAC VALVE 27%
TEST
START
Harness connector
Injector
Click
Click
Ignition wire
Spark plug
Diagnostic Procedure 15 — Engine Stalls when Accelerating or when Driving at Constant Speed
1
PERFORM POWER BALANCE TEST.
1. Perform "POWER BALANCE" in "ACTIVE TEST" mode.
2. Is there any cylinder which does not produce a momentary engine speed drop?
OR
When disconnecting each injector harness connector one at a time, is there any cylinder which does not produce a momentary engine speed drop?
No
Go to 4
Yes
2
CHECK INJECTOR.
1. Remove crank angle sensor from engine. (Harness connector should remain connected.)
2. Disconnect each ignition wire.
3. Turn ignition switch ON. (Do not start engine.)
4. When rotating crank angle sensor shaft, does each injector make an operating sound?
No
Check injector(s) and circuit(s).
Yes
3
CHECK IGNITION SPARK.
1. Disconnect each spark plug cap.
2. Connect a known good spark plug to each spark plug cap.
3. Place end of spark plug against a suitable ground and crank engine.
4. Check for spark.
N.G.
Check ignition coil, power transistor unit and their circuits.
O.K.
(Go to A on next page.)

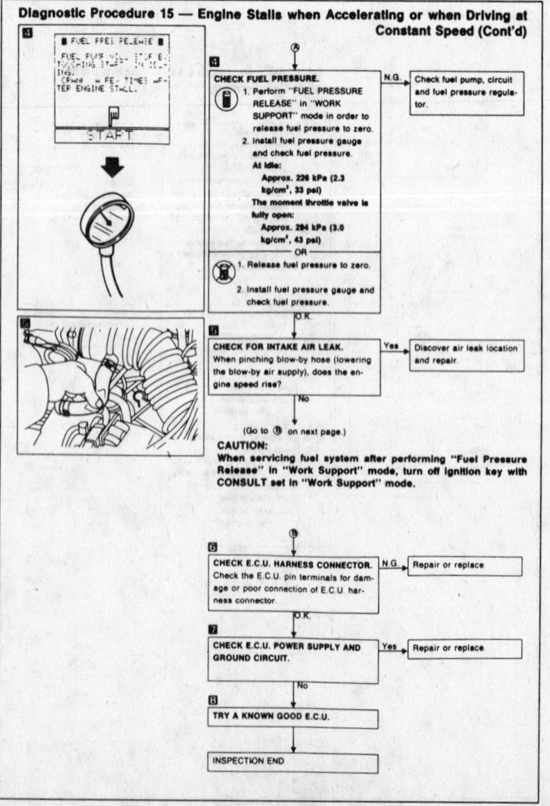
Diagnostic Procedure 15 — Engine Stalls when Accelerating or when Driving at Constant Speed (Cont'd)
FUEL PRES RELEASE
FUEL PUMP WILL STOP BY TOUCHING START IN IDLING. CRANK A FEW TIMES AFTER ENGINE STALL.
START
A
4
CHECK FUEL PRESSURE.
1. Perform "FUEL PRESSURE RELEASE" in "WORK SUPPORT" mode in order to release fuel pressure to zero.
2. Install fuel pressure gauge and check fuel pressure.
At idle:
Approx. 226 kPa (2.3 kg/cm², 33 psi)
The moment throttle valve is fully open:
Approx. 294 kPa (3.0 kg/cm², 43 psi)
OR
1. Release fuel pressure to zero.
2. Install fuel pressure gauge and check fuel pressure.
N.G.
Check fuel pump, circuit and fuel pressure regulator.
O.K.
5
CHECK FOR INTAKE AIR LEAK.
When pinching blow-by hose (lowering the blow-by air supply), does the engine speed rise?
Yes
Discover air leak location and repair.
No
(Go to B on next page.)
CAUTION:
When servicing fuel system after performing "Fuel Pressure Release" in "Work Support" mode, turn off ignition key with CONSULT set in "Work Support" mode.
B
6
CHECK E.C.U. HARNESS CONNECTOR.
Check the E.C.U. pin terminals for damage or poor connection of E.C.U. harness connector.
N.G.
Repair or replace.
O.K.
7
CHECK E.C.U. POWER SUPPLY AND GROUND CIRCUIT.
Yes
Repair or replace.
No
8
TRY A KNOWN GOOD E.C.U.
INSPECTION END

ECCS SELF-DIAGNOSTIC CHART—1990 STANZA

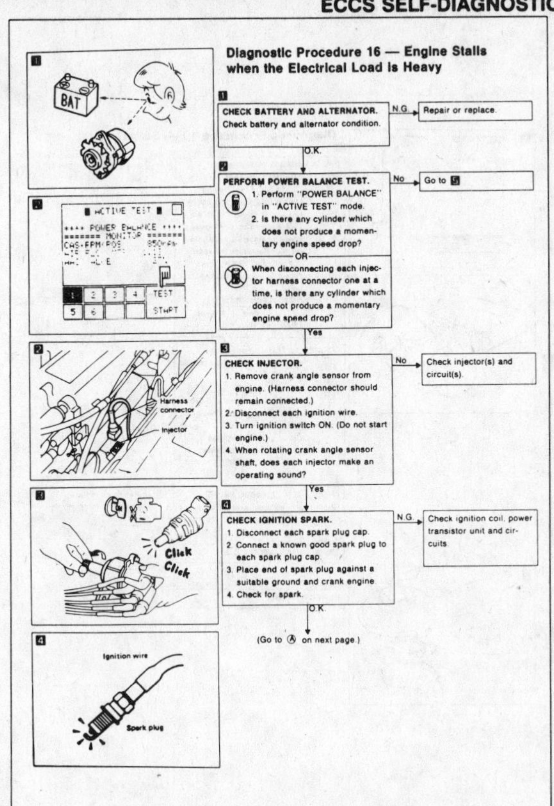
Diagnostic Procedure 16 — Engine Stalls when the Electrical Load is Heavy
1
BAT
CHECK BATTERY AND ALTERNATOR.
Check battery and alternator condition.
N.G.
Repair or replace.
O.K.
2
PERFORM POWER BALANCE TEST.
1. Perform "POWER BALANCE" in "ACTIVE TEST" mode.
2. Is there any cylinder which does not produce a momentary engine speed drop?
OR
When disconnecting each injector harness connector one at a time, is there any cylinder which does not produce a momentary engine speed drop?
No
Go to 5
ACTIVE TEST
POWER BALANCE
MONITOR
TEST
START
Yes
3
CHECK INJECTOR.
1. Remove crank angle sensor from engine. (Harness connector should remain connected.)
2. Disconnect each ignition wire.
3. Turn ignition switch ON. (Do not start engine.)
4. When rotating crank angle sensor shaft, does each injector make an operating sound?
No
Check injector(s) and circuit(s).
Harness connector
Injector
Click
Click
Yes
4
CHECK IGNITION SPARK.
1. Disconnect each spark plug cap.
2. Connect a known good spark plug to each spark plug cap.
3. Place end of spark plug against a suitable ground and crank engine.
4. Check for spark.
N.G.
Check ignition coil, power transistor unit and circuits.
O.K.
(Go to Ⓐ on next page.)
Ignition wire
Spark plug

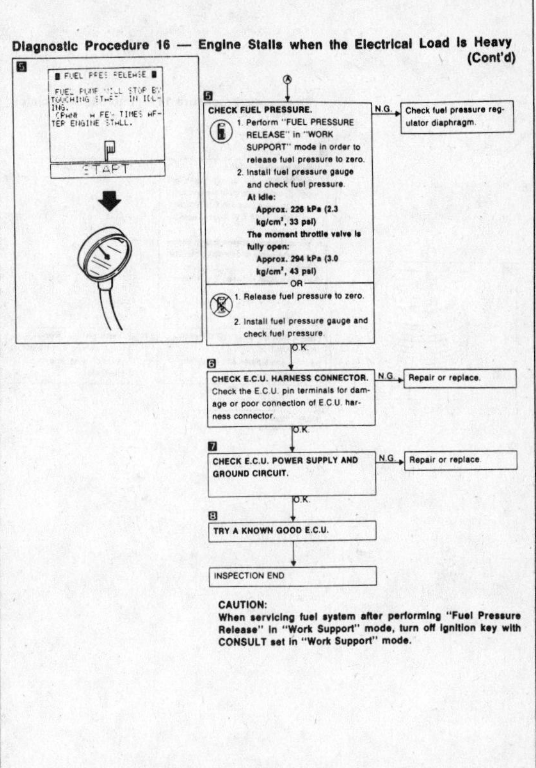
Diagnostic Procedure 16 — Engine Stalls when the Electrical Load is Heavy (Cont'd)
5
FUEL PRES RELEASE
FUEL PUMP WILL STOP BY TOUCHING START IN IDLING. CRANK A FEW TIMES AFTER ENGINE STALL.
START
Ⓐ
5
CHECK FUEL PRESSURE.
1. Perform "FUEL PRESSURE RELEASE" in "WORK SUPPORT" mode in order to release fuel pressure to zero.
2. Install fuel pressure gauge and check fuel pressure.
At idle:
Approx. 226 kPa (2.3 kg/cm², 33 psi)
The moment throttle valve is fully open:
Approx. 294 kPa (3.0 kg/cm², 43 psi)
OR
1. Release fuel pressure to zero.
2. Install fuel pressure gauge and check fuel pressure.
N.G.
Check fuel pressure regulator diaphragm.
O.K.
6
CHECK E.C.U. HARNESS CONNECTOR.
Check the E.C.U. pin terminals for damage or poor connection of E.C.U. harness connector.
N.G.
Repair or replace.
O.K.
7
CHECK E.C.U. POWER SUPPLY AND GROUND CIRCUIT.
N.G.
Repair or replace.
O.K.
8
TRY A KNOWN GOOD E.C.U.
INSPECTION END
CAUTION:
When servicing fuel system after performing "Fuel Pressure Release" in "Work Support" mode, turn off ignition key with CONSULT set in "Work Support" mode.

ECCS SELF-DIAGNOSTIC CHART—1990 STANZA

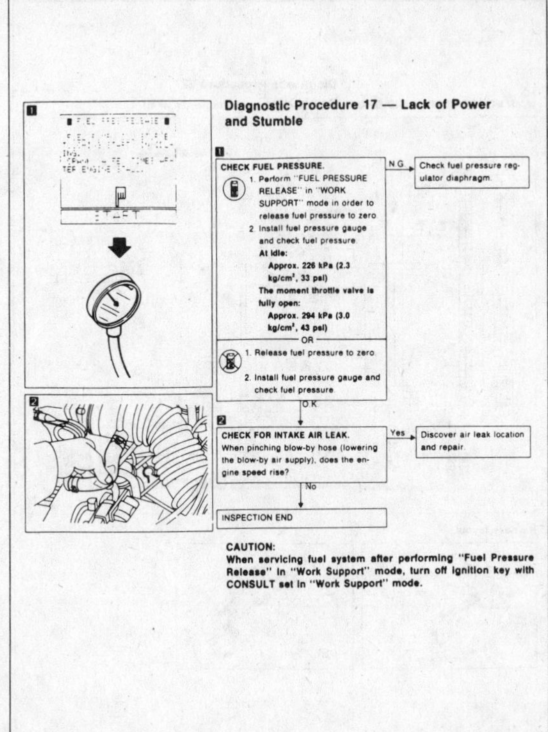
Diagnostic Procedure 17 — Lack of Power and Stumble
1
FUEL PRES RELEASE
START
1
CHECK FUEL PRESSURE.
1. Perform "FUEL PRESSURE RELEASE" in "WORK SUPPORT" mode in order to release fuel pressure to zero.
2. Install fuel pressure gauge and check fuel pressure.
At idle:
Approx. 226 kPa (2.3 kg/cm², 33 psi)
The moment throttle valve is fully open:
Approx. 294 kPa (3.0 kg/cm², 43 psi)
OR
1. Release fuel pressure to zero.
2. Install fuel pressure gauge and check fuel pressure.
N.G.
Check fuel pressure regulator diaphragm.
O.K.
2
CHECK FOR INTAKE AIR LEAK.
When pinching blow-by hose (lowering the blow-by air supply), does the engine speed rise?
Yes
Discover air leak location and repair.
No
INSPECTION END
CAUTION:
When servicing fuel system after performing "Fuel Pressure Release" in "Work Support" mode, turn off ignition key with CONSULT set in "Work Support" mode.

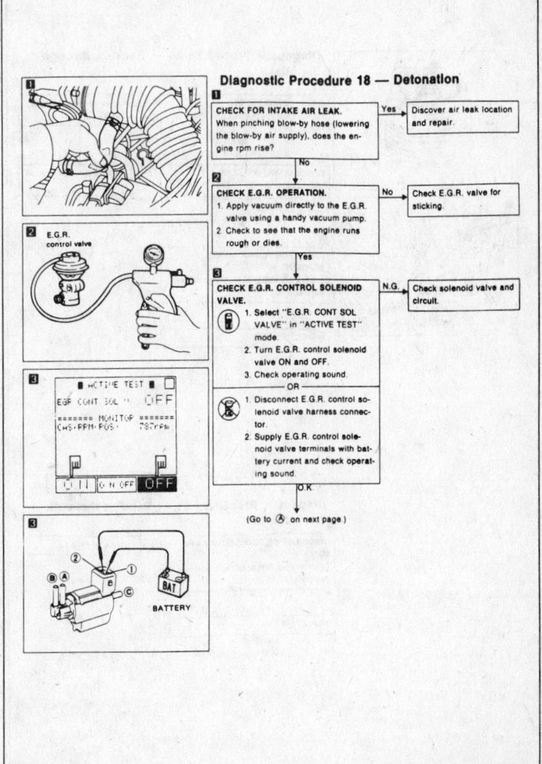
Diagnostic Procedure 18 — Detonation
1
CHECK FOR INTAKE AIR LEAK.
When pinching blow-by hose (lowering the blow-by air supply), does the engine rpm rise?
Yes
Discover air leak location and repair.
No
2
CHECK E.G.R. OPERATION.
1. Apply vacuum directly to the E.G.R. valve using a handy vacuum pump.
2. Check to see that the engine runs rough or dies.
No
Check E.G.R. valve for sticking.
E.G.R. control valve
Yes
3
CHECK E.G.R. CONTROL SOLENOID VALVE.
1. Select "E.G.R. CONT SOL VALVE" in "ACTIVE TEST" mode.
2. Turn E.G.R. control solenoid valve ON and OFF.
3. Check operating sound.
OR
1. Disconnect E.G.R. control solenoid valve harness connector.
2. Supply E.G.R. control solenoid valve terminals with battery current and check operating sound.
N.G.
Check solenoid valve and circuit.
ACTIVE TEST
EGR CONT SOL V OFF
MONITOR
ON
OFF
O.K.
(Go to Ⓐ on next page.)
BAT
BATTERY

ECCS SELF-DIAGNOSTIC CHART—1990 STANZA

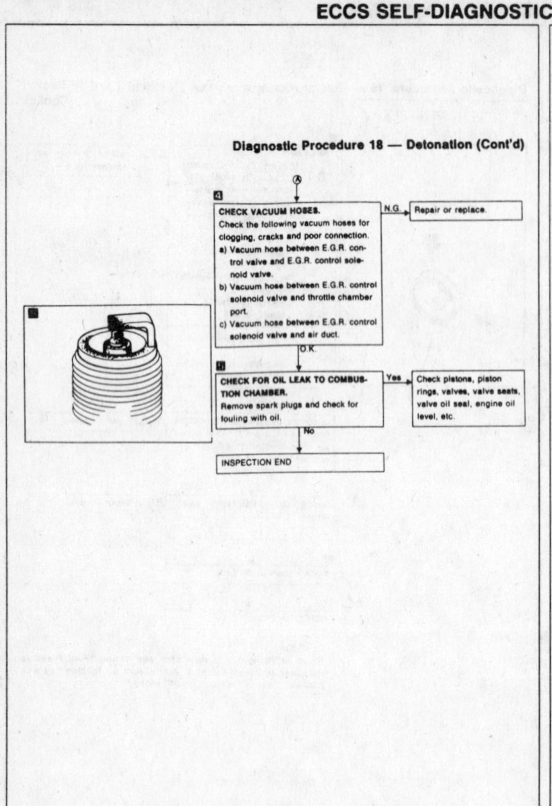
Diagnostic Procedure 18 — Detonation (Cont'd)
A
4
CHECK VACUUM HOSES.
Check the following vacuum hoses for clogging, cracks and poor connection.
a) Vacuum hose between E.G.R. control valve and E.G.R. control solenoid valve.
b) Vacuum hose between E.G.R. control solenoid valve and throttle chamber port.
c) Vacuum hose between E.G.R. control solenoid valve and air duct.
N.G.
Repair or replace.
O.K.
5
CHECK FOR OIL LEAK TO COMBUSTION CHAMBER.
Remove spark plugs and check for fouling with oil.
Yes
Check pistons, piston rings, valves, valve seats, valve oil seal, engine oil level, etc.
No
INSPECTION END

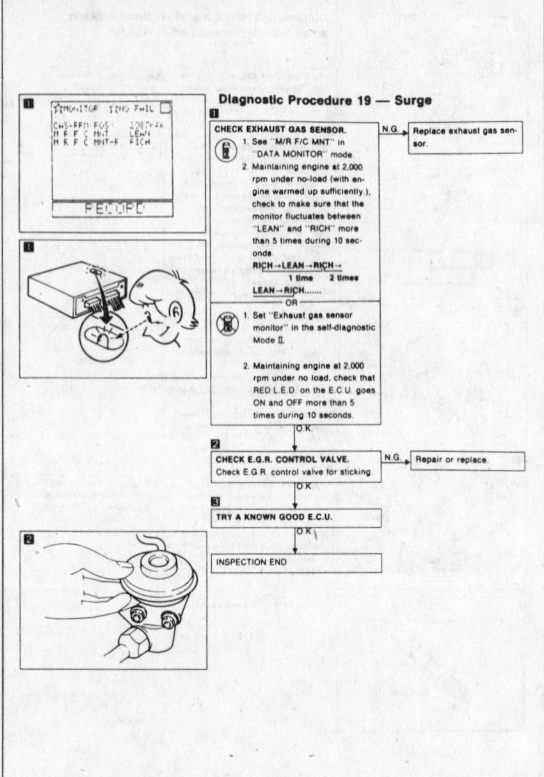
Diagnostic Procedure 19 — Surge
1
CHECK EXHAUST GAS SENSOR.
1. See "M/R F/C MNT" in "DATA MONITOR" mode.
2. Maintaining engine at 2,000 rpm under no-load (with engine warmed up sufficiently.), check to make sure that the monitor fluctuates between "LEAN" and "RICH" more than 5 times during 10 seconds.
RICH→LEAN→RICH→
1 time 2 times
LEAN→RICH.......
OR
1. Set "Exhaust gas sensor monitor" in the self-diagnostic Mode II.
2. Maintaining engine at 2,000 rpm under no load, check that RED L.E.D. on the E.C.U. goes ON and OFF more than 5 times during 10 seconds.
N.G.
Replace exhaust gas sensor.
O.K.
2
CHECK E.G.R. CONTROL VALVE.
Check E.G.R. control valve for sticking.
N.G.
Repair or replace.
O.K.
3
TRY A KNOWN GOOD E.C.U.
O.K.
INSPECTION END
RECORD

ECCS SELF-DIAGNOSTIC CHART—1990 STANZA

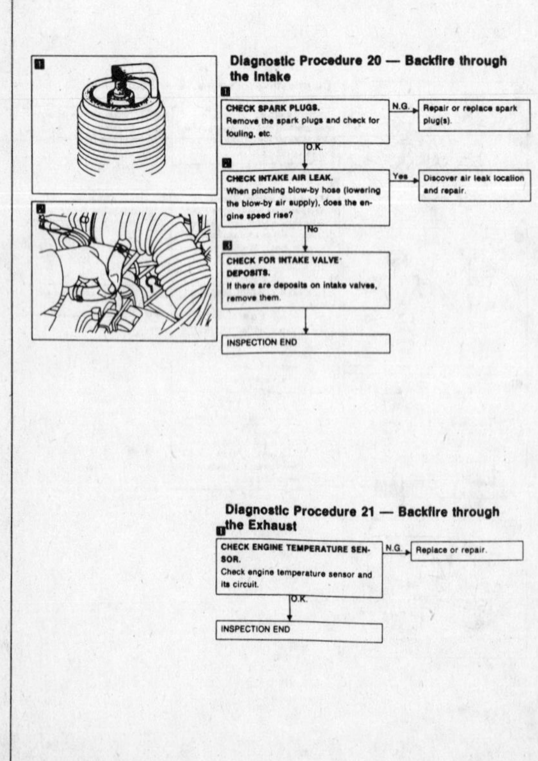
Diagnostic Procedure 20 — Backfire through the Intake
1
CHECK SPARK PLUGS.
Remove the spark plugs and check for fouling, etc.
N.G.
Repair or replace spark plug(s).
O.K.
2
CHECK INTAKE AIR LEAK.
When pinching blow-by hose (lowering the blow-by air supply), does the engine speed rise?
Yes
Discover air leak location and repair.
No
3
CHECK FOR INTAKE VALVE DEPOSITS.
If there are deposits on intake valves, remove them.
INSPECTION END
Diagnostic Procedure 21 — Backfire through the Exhaust
1
CHECK ENGINE TEMPERATURE SENSOR.
Check engine temperature sensor and its circuit.
N.G.
Replace or repair.
O.K.
INSPECTION END

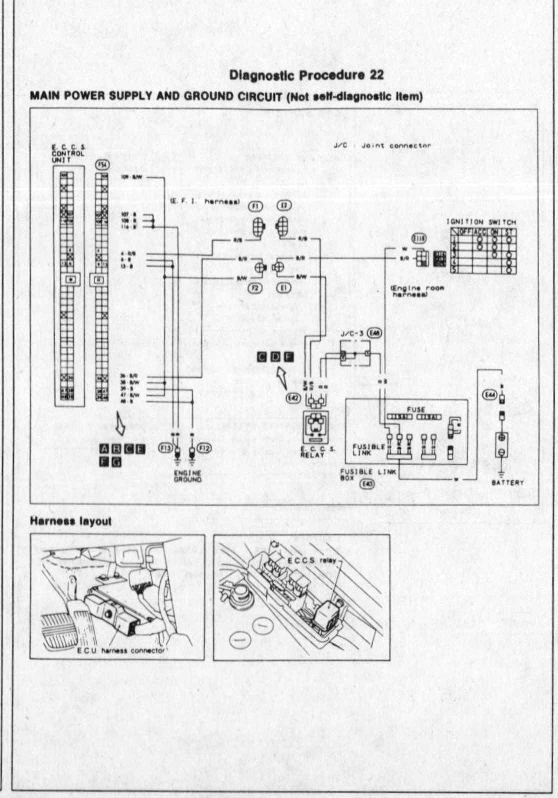
Diagnostic Procedure 22
MAIN POWER SUPPLY AND GROUND CIRCUIT (Not self-diagnostic item)
J/C : Joint connector
E.C.C.S. CONTROL UNIT
(E.F.I. harness)
IGNITION SWITCH
(Engine room harness)
J/C-3
E.C.C.S. RELAY
FUSE
FUSIBLE LINK
FUSIBLE LINK BOX
BATTERY
ENGINE GROUND
Harness layout
E.C.C.S. relay
E.C.U. harness connector

ECCS DIAGNOSTIC CHART—1990 STANZA

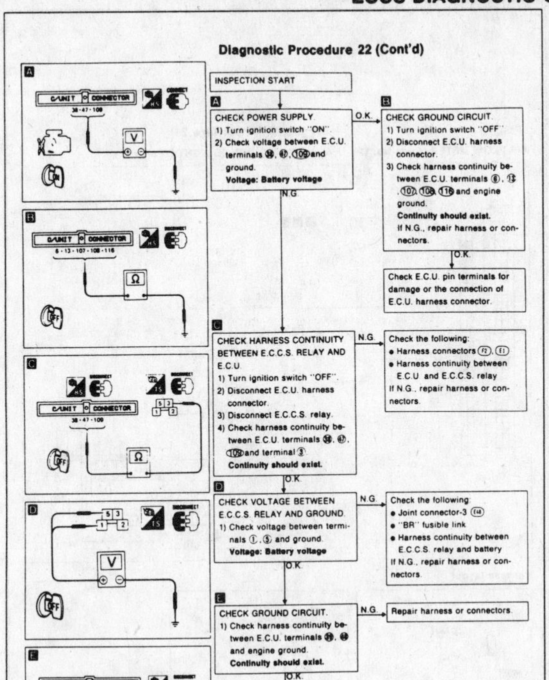

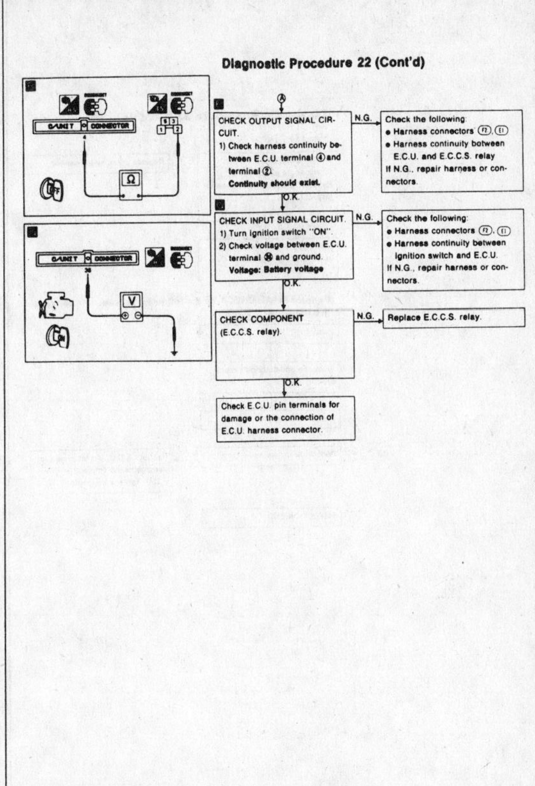

ECCS SELF-DIAGNOSTIC CHART—1990 STANZA

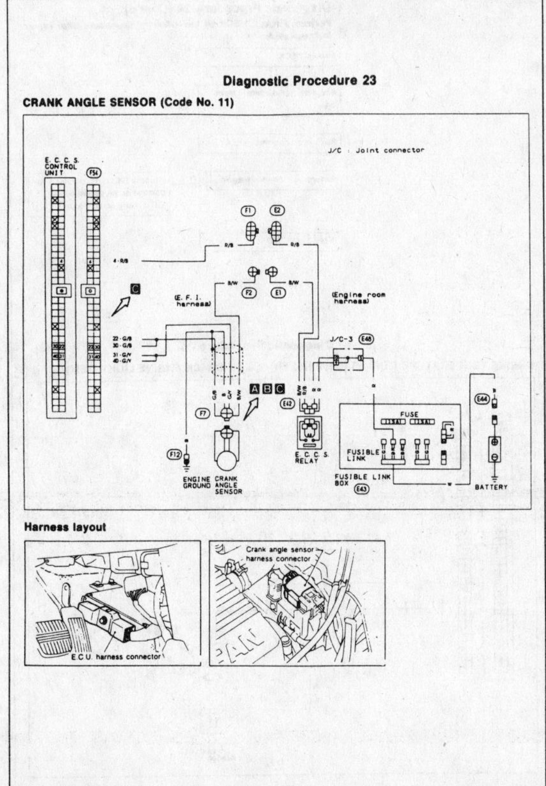

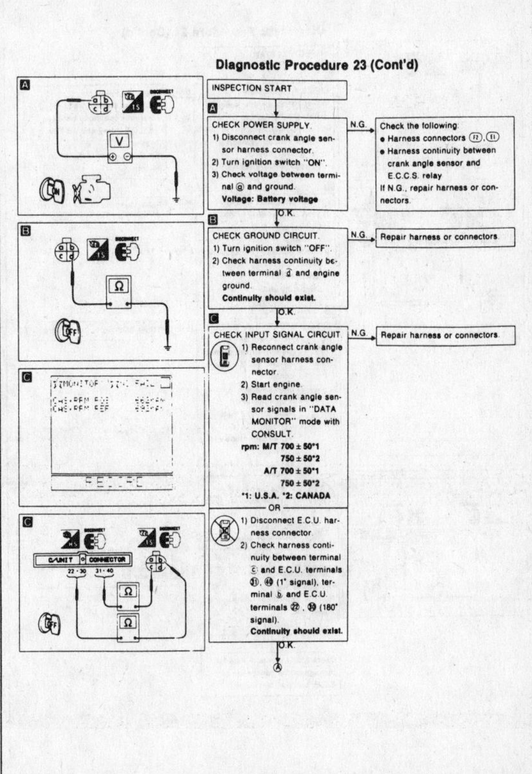

ECCS SELF-DIAGNOSTIC CHART—1990 STANZA

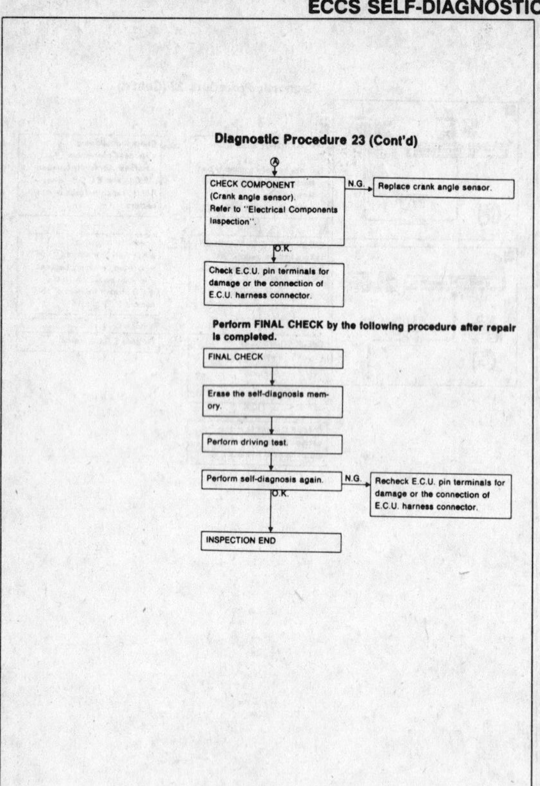
Diagnostic Procedure 23 (Cont'd)
CHECK COMPONENT (Crank angle sensor). Refer to "Electrical Components Inspection".
N.G.
Replace crank angle sensor.
O.K.
Check E.C.U. pin terminals for damage or the connection of E.C.U. harness connector.
Perform FINAL CHECK by the following procedure after repair is completed.
FINAL CHECK
Erase the self-diagnosis memory.
Perform driving test.
Perform self-diagnosis again.
N.G.
Recheck E.C.U. pin terminals for damage or the connection of E.C.U. harness connector.
O.K.
INSPECTION END

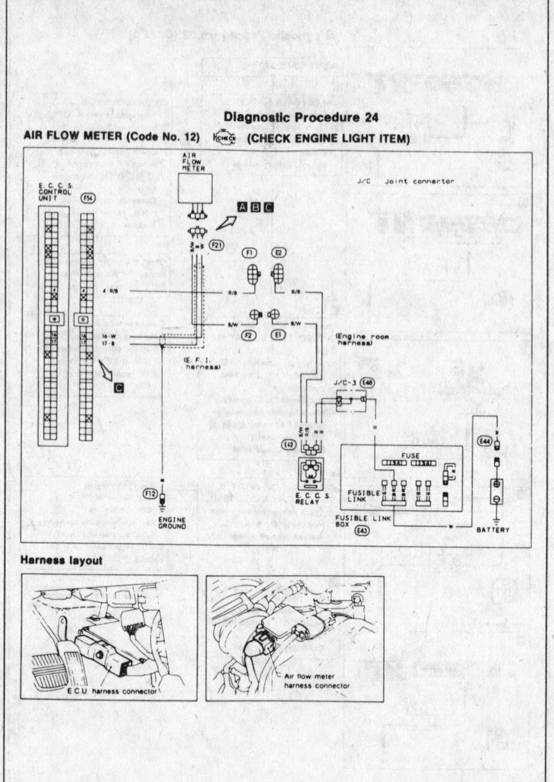
Diagnostic Procedure 24
AIR FLOW METER (Code No. 12)
(CHECK ENGINE LIGHT ITEM)
E.C.C.S. CONTROL UNIT
AIR FLOW METER
J/C Joint connector
(Engine room harness)
(E.F.I. harness)
J/C-3
ENGINE GROUND
E.C.C.S. RELAY
FUSE
FUSIBLE LINK
FUSIBLE LINK BOX
BATTERY
Harness layout
E.C.U. harness connector
Air flow meter harness connector

ECCS SELF-DIAGNOSTIC CHART—1990 STANZA

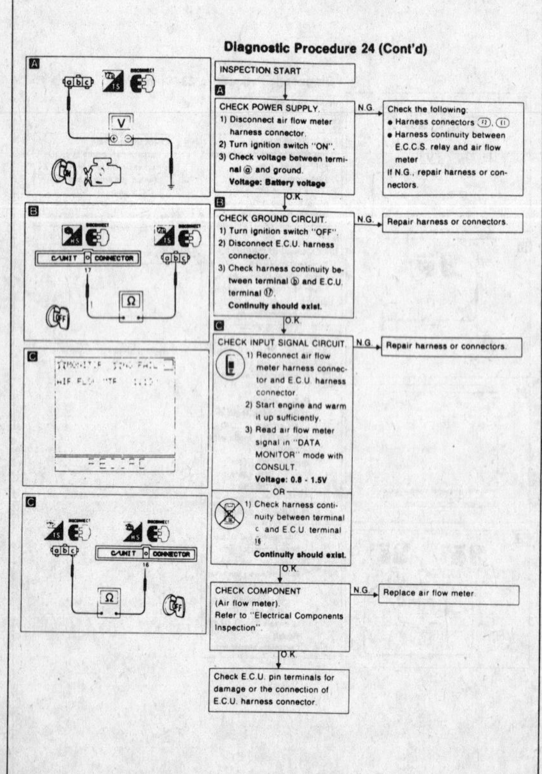
Diagnostic Procedure 24 (Cont'd)
INSPECTION START
CHECK POWER SUPPLY.
1) Disconnect air flow meter harness connector.
2) Turn ignition switch "ON".
3) Check voltage between terminal ⓐ and ground.
Voltage: Battery voltage
N.G.
Check the following:
• Harness connectors F2, E1
• Harness continuity between E.C.C.S. relay and air flow meter
If N.G., repair harness or connectors.
O.K.
CHECK GROUND CIRCUIT
1) Turn ignition switch "OFF".
2) Disconnect E.C.U. harness connector.
3) Check harness continuity between terminal ⓑ and E.C.U. terminal ⑰.
Continuity should exist.
N.G.
Repair harness or connectors.
O.K.
CHECK INPUT SIGNAL CIRCUIT.
1) Reconnect air flow meter harness connector and E.C.U. harness connector.
2) Start engine and warm it up sufficiently.
3) Read air flow meter signal in "DATA MONITOR" mode with CONSULT.
Voltage: 0.8 - 1.5V
—OR—
1) Check harness continuity between terminal c and E.C.U. terminal 16.
Continuity should exist.
N.G.
Repair harness or connectors.
O.K.
CHECK COMPONENT (Air flow meter). Refer to "Electrical Components Inspection".
N.G.
Replace air flow meter.
O.K.
Check E.C.U. pin terminals for damage or the connection of E.C.U. harness connector.
C/UNIT
CONNECTOR

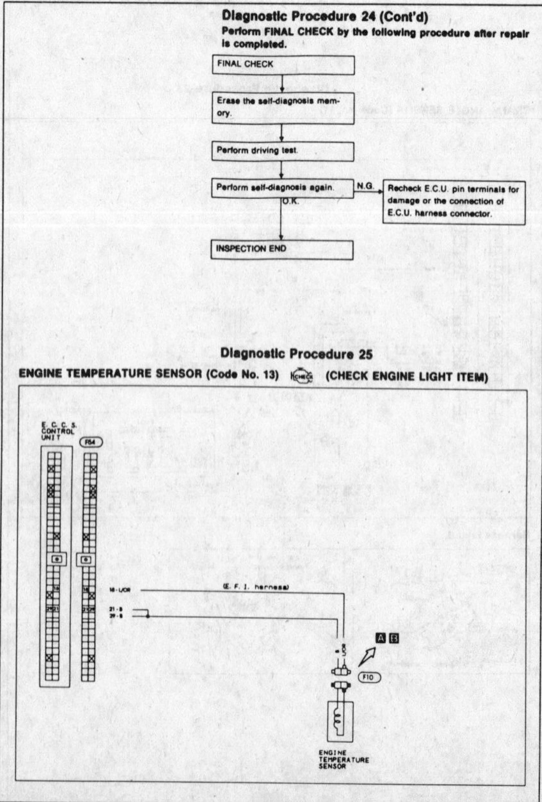
Diagnostic Procedure 24 (Cont'd)
Perform FINAL CHECK by the following procedure after repair is completed.
FINAL CHECK
Erase the self-diagnosis memory.
Perform driving test.
Perform self-diagnosis again.
N.G.
Recheck E.C.U. pin terminals for damage or the connection of E.C.U. harness connector.
O.K.
INSPECTION END
Diagnostic Procedure 25
ENGINE TEMPERATURE SENSOR (Code No. 13)
(CHECK ENGINE LIGHT ITEM)
E.C.C.S. CONTROL UNIT
(E.F.I. harness)
ENGINE TEMPERATURE SENSOR

ECCS SELF-DIAGNOSTIC CHART—1990 STANZA

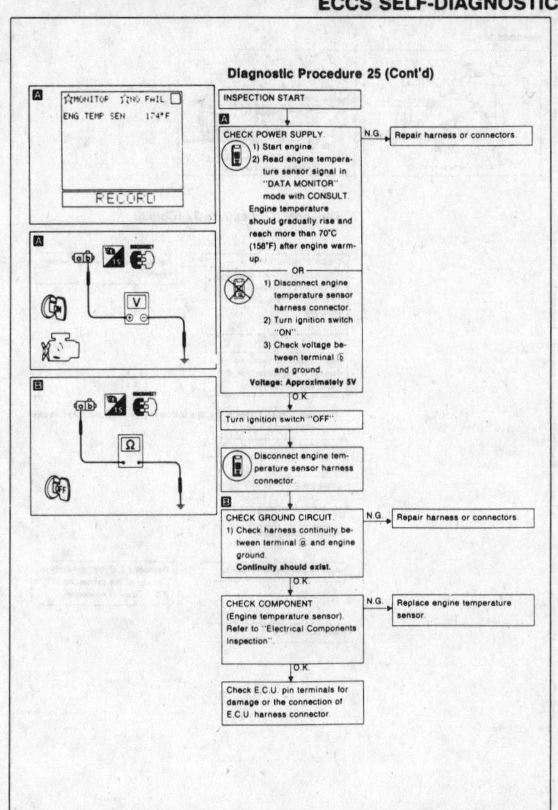
Diagnostic Procedure 25 (Cont'd)
MONITOR NO FAIL
ENG TEMP SEN 174°F
RECORD
INSPECTION START
CHECK POWER SUPPLY.
1) Start engine
2) Read engine temperature sensor signal in "DATA MONITOR" mode with CONSULT.
Engine temperature should gradually rise and reach more than 70°C (158°F) after engine warm-up.
OR
1) Disconnect engine temperature sensor harness connector.
2) Turn ignition switch "ON".
3) Check voltage between terminal ⓑ and ground.
Voltage: Approximately 5V
N.G.
Repair harness or connectors.
O.K.
Turn ignition switch "OFF".
Disconnect engine temperature sensor harness connector.
CHECK GROUND CIRCUIT.
1) Check harness continuity between terminal ⓐ and engine ground.
Continuity should exist.
N.G.
Repair harness or connectors.
O.K.
CHECK COMPONENT
(Engine temperature sensor).
Refer to "Electrical Components Inspection".
N.G.
Replace engine temperature sensor.
O.K.
Check E.C.U. pin terminals for damage or the connection of E.C.U. harness connector.

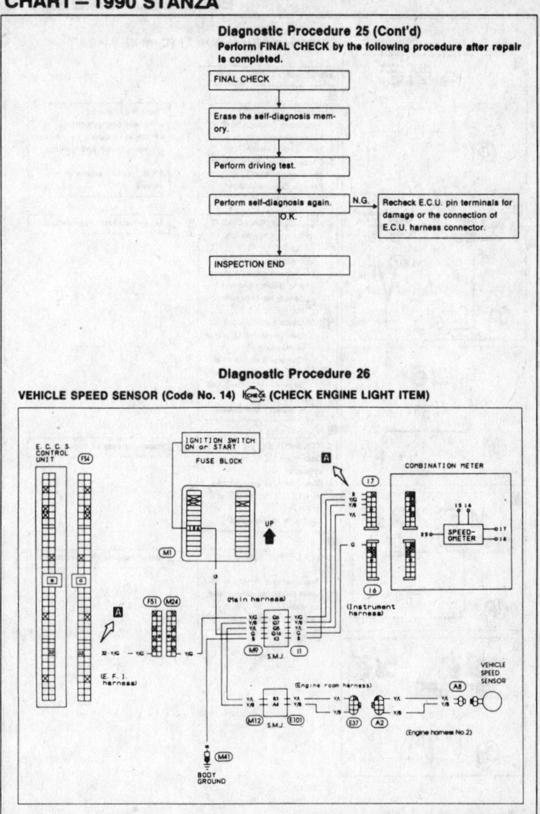
Diagnostic Procedure 25 (Cont'd)
Perform FINAL CHECK by the following procedure after repair is completed.
FINAL CHECK
Erase the self-diagnosis memory.
Perform driving test.
Perform self-diagnosis again.
N.G.
Recheck E.C.U. pin terminals for damage or the connection of E.C.U. harness connector.
O.K.
INSPECTION END
Diagnostic Procedure 26
VEHICLE SPEED SENSOR (Code No. 14) (CHECK ENGINE LIGHT ITEM)
E.C.C.S. CONTROL UNIT
IGNITION SWITCH ON or START
FUSE BLOCK
COMBINATION METER
SPEEDOMETER
(Main harness)
(Instrument harness)
S.M.J.
(E.F.I. harness)
(Engine room harness)
VEHICLE SPEED SENSOR
(Engine harness No.2)
BODY GROUND

ECCS SELF-DIAGNOSTIC CHART—1990 STANZA

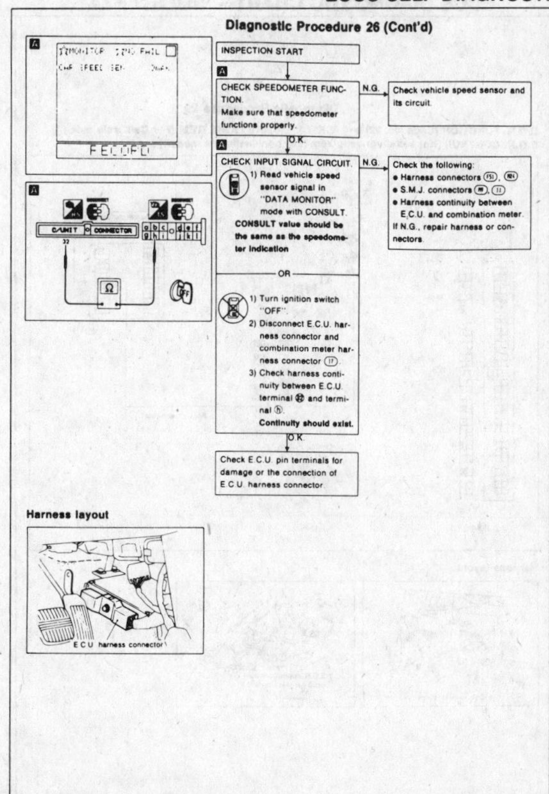
Diagnostic Procedure 26 (Cont'd)
MONITOR NO FAIL
RECORD
INSPECTION START
CHECK SPEEDOMETER FUNCTION.
Make sure that speedometer functions properly.
N.G.
Check vehicle speed sensor and its circuit.
O.K.
CHECK INPUT SIGNAL CIRCUIT.
1) Read vehicle speed sensor signal in "DATA MONITOR" mode with CONSULT.
CONSULT value should be the same as the speedometer indication
OR
1) Turn ignition switch "OFF".
2) Disconnect E.C.U. harness connector and combination meter harness connector Ⓘ⑰.
3) Check harness continuity between E.C.U. terminal ㊳ and terminal ⑤.
Continuity should exist.
N.G.
Check the following:
• Harness connectors
• S.M.J. connectors
• Harness continuity between E.C.U. and combination meter.
If N.G., repair harness or connectors
O.K.
Check E.C.U. pin terminals for damage or the connection of E.C.U. harness connector.
C/UNIT CONNECTOR
Harness layout
E.C.U. harness connector

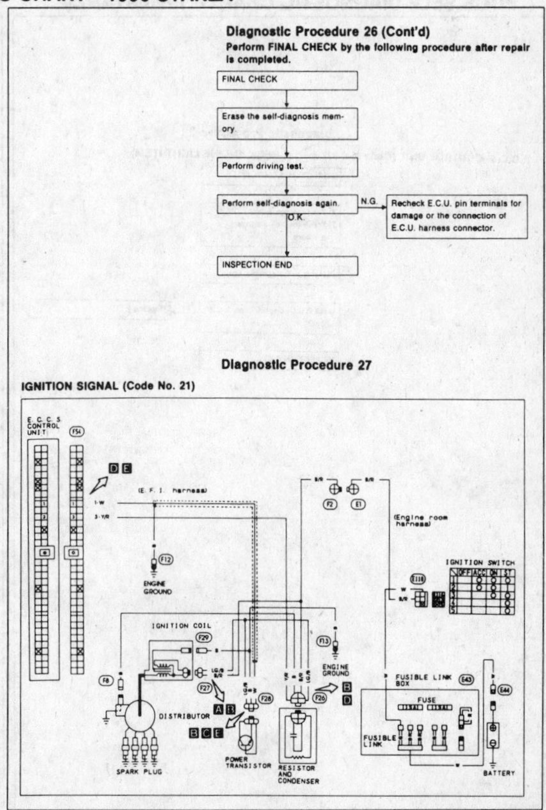
Diagnostic Procedure 26 (Cont'd)
Perform FINAL CHECK by the following procedure after repair is completed.
FINAL CHECK
Erase the self-diagnosis memory.
Perform driving test.
Perform self-diagnosis again.
N.G.
Recheck E.C.U. pin terminals for damage or the connection of E.C.U. harness connector.
O.K.
INSPECTION END
Diagnostic Procedure 27
IGNITION SIGNAL (Code No. 21)
E.C.C.S. CONTROL UNIT
(E.F.I. harness)
(Engine room harness)
ENGINE GROUND
IGNITION SWITCH
IGNITION COIL
DISTRIBUTOR
SPARK PLUG
POWER TRANSISTOR
RESISTOR AND CONDENSER
FUSIBLE LINK BOX
FUSE
FUSIBLE LINK
BATTERY

ECCS SELF-DIAGNOSTIC CHART—1990 STANZA

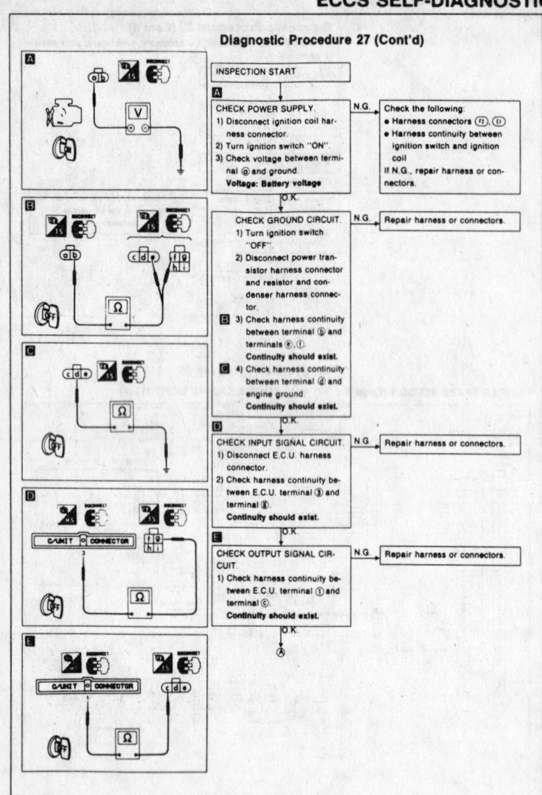
Diagnostic Procedure 27 (Cont'd)
INSPECTION START
A
CHECK POWER SUPPLY.
1) Disconnect ignition coil harness connector.
2) Turn ignition switch "ON".
3) Check voltage between terminal ⓐ and ground.
Voltage: Battery voltage
N.G.
Check the following:
• Harness connectors (F2), (I1)
• Harness continuity between ignition switch and ignition coil
If N.G., repair harness or connectors.
O.K.
CHECK GROUND CIRCUIT.
1) Turn ignition switch "OFF".
2) Disconnect power transistor harness connector and resistor and condenser harness connector.
B 3) Check harness continuity between terminal ⓑ and terminals ⓔ, ⓕ.
Continuity should exist.
C 4) Check harness continuity between terminal ⓓ and engine ground
Continuity should exist.
N.G.
Repair harness or connectors.
O.K.
D
CHECK INPUT SIGNAL CIRCUIT.
1) Disconnect E.C.U. harness connector.
2) Check harness continuity between E.C.U. terminal ③ and terminal ⓔ.
Continuity should exist.
N.G.
Repair harness or connectors.
O.K.
E
CHECK OUTPUT SIGNAL CIRCUIT.
1) Check harness continuity between E.C.U. terminal ① and terminal ⓒ.
Continuity should exist.
N.G.
Repair harness or connectors.
O.K.
A
C/UNIT CONNECTOR

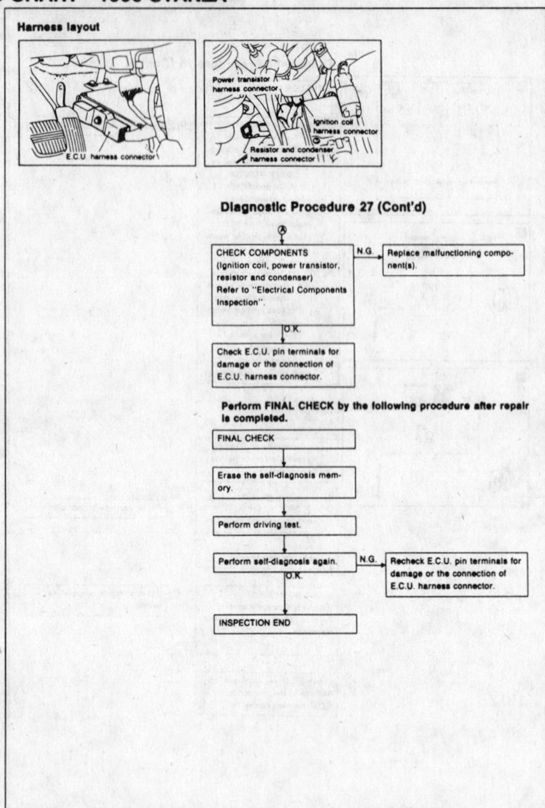
Harness layout
E.C.U. harness connector
Power transistor harness connector
Ignition coil harness connector
Resistor and condenser harness connector
Diagnostic Procedure 27 (Cont'd)
A
CHECK COMPONENTS
(Ignition coil, power transistor, resistor and condenser)
Refer to "Electrical Components Inspection".
N.G.
Replace malfunctioning component(s).
O.K.
Check E.C.U. pin terminals for damage or the connection of E.C.U. harness connector.
Perform FINAL CHECK by the following procedure after repair is completed.
FINAL CHECK
Erase the self-diagnosis memory.
Perform driving test.
Perform self-diagnosis again.
N.G.
Recheck E.C.U. pin terminals for damage or the connection of E.C.U. harness connector.
O.K.
INSPECTION END

ECCS SELF-DIAGNOSTIC CHART—1990 STANZA

Diagnostic Procedure 28

E.C.C.S. CONTROL UNIT (Code No. 31) (CHECK ENGINE LIGHT ITEM)

INSPECTION START
↓
1) Turn ignition switch "ON".
2) Erase the self-diagnosis memory.
↓
Perform self-diagnosis again.
↓
Does E.C.U. display Code No. 31 again? — Yes → Replace E.C.U.
↓ No
INSPECTION END

ECCS DIAGNOSTIC CHART—1990 STANZA

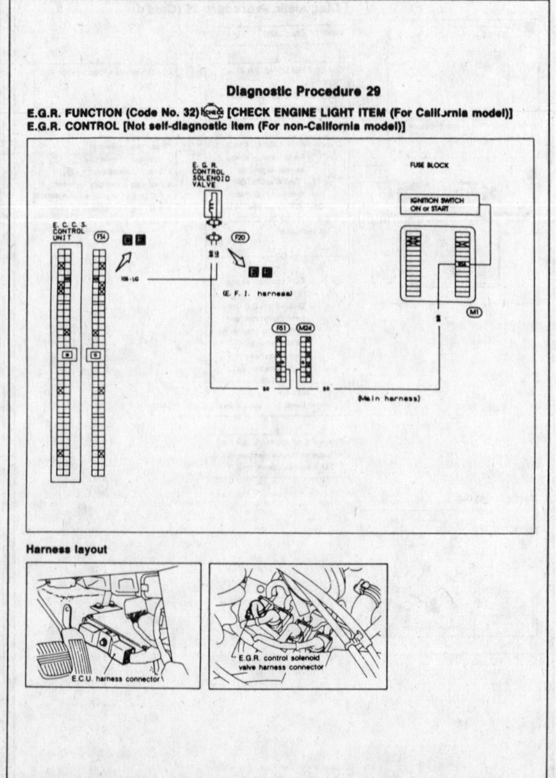
Diagnostic Procedure 29
E.G.R. FUNCTION (Code No. 32) [CHECK ENGINE LIGHT ITEM (For California model)]
E.G.R. CONTROL [Not self-diagnostic item (For non-California model)]
E.G.R. CONTROL SOLENOID VALVE
E.C.C.S. CONTROL UNIT
FUSE BLOCK
IGNITION SWITCH ON or START
(E.F.I. harness)
(Main harness)
Harness layout
E.C.U. harness connector
E.G.R. control solenoid valve harness connector

ECCS DIAGNOSTIC CHART—1990 STANZA

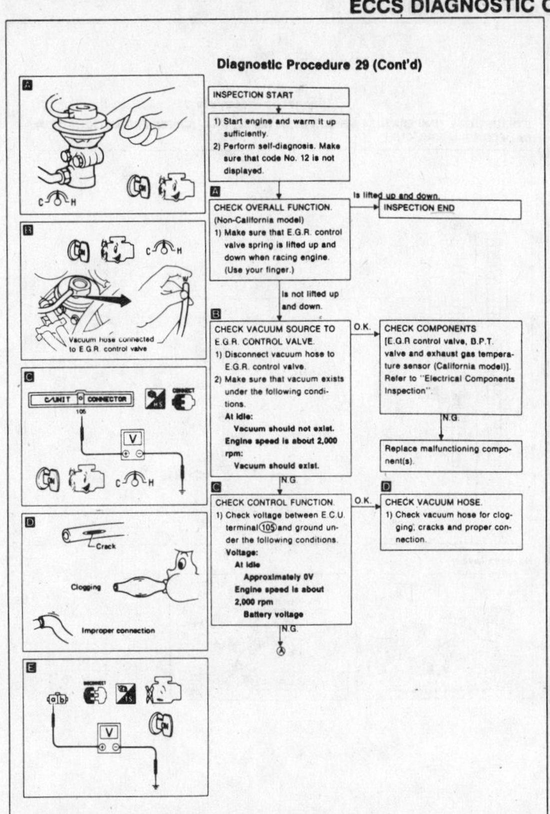
Diagnostic Procedure 29 (Cont'd)
INSPECTION START
1) Start engine and warm it up sufficiently.
2) Perform self-diagnosis. Make sure that code No. 12 is not displayed.
A
CHECK OVERALL FUNCTION.
(Non-California model)
1) Make sure that E.G.R. control valve spring is lifted up and down when racing engine. (Use your finger.)
Is lifted up and down.
INSPECTION END
Is not lifted up and down.
B
CHECK VACUUM SOURCE TO E.G.R. CONTROL VALVE.
1) Disconnect vacuum hose to E.G.R. control valve.
2) Make sure that vacuum exists under the following conditions.
At idle:
Vacuum should not exist.
Engine speed is about 2,000 rpm:
Vacuum should exist.
O.K.
CHECK COMPONENTS
[E.G.R control valve, B.P.T. valve and exhaust gas temperature sensor (California model)].
Refer to "Electrical Components Inspection".
N.G.
Replace malfunctioning component(s).
N.G.
C
CHECK CONTROL FUNCTION.
1) Check voltage between E.C.U. terminal (105) and ground under the following conditions.
Voltage:
At idle
Approximately 0V
Engine speed is about 2,000 rpm
Battery voltage
O.K.
D
CHECK VACUUM HOSE.
1) Check vacuum hose for clogging, cracks and proper connection.
N.G.
Ⓐ
Vacuum hose connected to E.G.R. control valve
C/UNIT CONNECTOR
105
V
Crack
Clogging
Improper connection

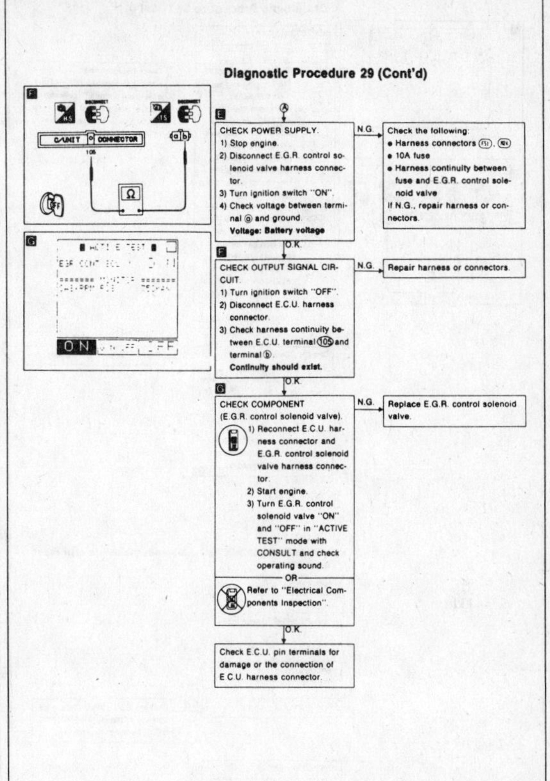
Diagnostic Procedure 29 (Cont'd)
C/UNIT CONNECTOR
105
Ⓐ
E
CHECK POWER SUPPLY.
1) Stop engine.
2) Disconnect E.G.R. control solenoid valve harness connector.
3) Turn ignition switch "ON".
4) Check voltage between terminal ⓐ and ground.
Voltage: Battery voltage
N.G.
Check the following:
• Harness connectors (F51), (R4)
• 10A fuse
• Harness continuity between fuse and E.G.R. control solenoid valve
If N.G., repair harness or connectors.
O.K.
F
CHECK OUTPUT SIGNAL CIRCUIT.
1) Turn ignition switch "OFF".
2) Disconnect E.C.U. harness connector.
3) Check harness continuity between E.C.U. terminal (105) and terminal ⓑ.
Continuity should exist.
N.G.
Repair harness or connectors.
O.K.
G
CHECK COMPONENT
(E.G.R. control solenoid valve).
1) Reconnect E.C.U. harness connector and E.G.R. control solenoid valve harness connector.
2) Start engine.
3) Turn E.G.R. control solenoid valve "ON" and "OFF" in "ACTIVE TEST" mode with CONSULT and check operating sound.
OR
Refer to "Electrical Components Inspection".
N.G.
Replace E.G.R. control solenoid valve.
O.K.
Check E.C.U. pin terminals for damage or the connection of E.C.U. harness connector.
ON

ECCS DIAGNOSTIC CHART—1990 STANZA

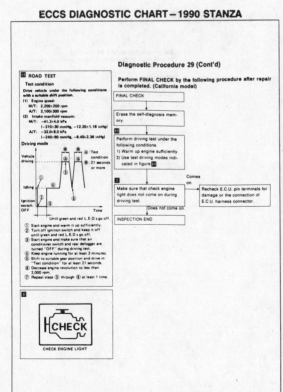

Diagnostic Procedure 29 (Cont'd)

H ROAD TEST

Test condition

Drive vehicle under the following conditions with a suitable shift position.

(1) Engine speed:
- **M/T: 2,200±200 rpm**
- **A/T: 2,100±300 rpm**

(2) Intake manifold vacuum:
- **M/T: −41.3±4.0 kPa (−310±30 mmHg, −12.20±1.18 inHg)**
- **A/T: −32.0±8.0 kPa (−240±60 mmHg, −9.45±2.36 inHg)**

Driving mode

Vehicle driving — Idling — Ignition switch: OFF — Time

Ⓐ: Test condition
Ⓑ: 21 seconds or more

Until green and red L.E.D.s go off.

① Start engine and warm it up sufficiently.
② Turn off ignition switch and keep it off until green and red L.E.D.s go off.
③ Start engine and make sure that air conditioner switch and rear defogger are turned "OFF" during driving test.
④ Keep engine running for at least 3 minutes.
⑤ Shift to suitable gear position and drive in "Test condition" for at least 21 seconds.
⑥ Decrease engine revolution to less than 2,000 rpm.
⑦ Repeat steps ⑤ through ⑥ at least 1 time.

I

CHECK

CHECK ENGINE LIGHT

Perform FINAL CHECK by the following procedure after repair is completed. (California model)

FINAL CHECK
↓
Erase the self-diagnosis memory.
↓
H Perform driving test under the following conditions.
1) Warm up engine sufficiently.
2) Use test driving modes indicated in figure **H**.
↓
I Make sure that check engine light does not come on during driving test.
- Comes on. → Recheck E.C.U. pin terminals for damage or the connection of E.C.U. harness connector.
- Does not come on. → INSPECTION END

ECCS SELF-DIAGNOSTIC CHART—1990 STANZA

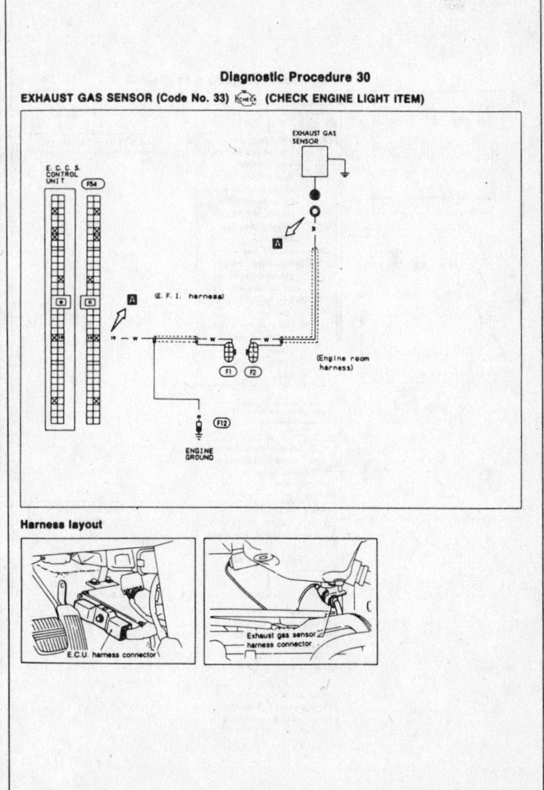
Diagnostic Procedure 30
EXHAUST GAS SENSOR (Code No. 33) (CHECK ENGINE LIGHT ITEM)
EXHAUST GAS SENSOR
E.C.C.S. CONTROL UNIT
F54
(E.F.I. harness)
F1
F2
(Engine room harness)
F12
ENGINE GROUND
Harness layout
E.C.U. harness connector
Exhaust gas sensor harness connector

ECCS SELF-DIAGNOSTIC CHART—1990 STANZA

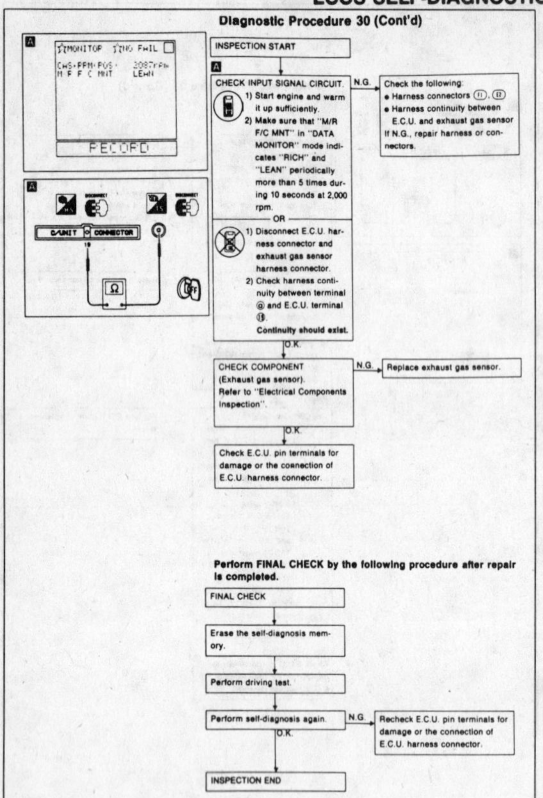
Diagnostic Procedure 30 (Cont'd)
MONITOR NO FAIL
CHS•RPM•POS 3087rpm
M/R F/C MNT LEAN
RECORD
C/UNIT CONNECTOR
INSPECTION START
CHECK INPUT SIGNAL CIRCUIT.
1) Start engine and warm it up sufficiently.
2) Make sure that "M/R F/C MNT" in "DATA MONITOR" mode indicates "RICH" and "LEAN" periodically more than 5 times during 10 seconds at 2,000 rpm.
OR
1) Disconnect E.C.U. harness connector and exhaust gas sensor harness connector.
2) Check harness continuity between terminal ⓐ and E.C.U. terminal ⑲.
Continuity should exist.
N.G.
Check the following:
• Harness connectors (F1), (F2)
• Harness continuity between E.C.U. and exhaust gas sensor
If N.G., repair harness or connectors.
O.K.
CHECK COMPONENT
(Exhaust gas sensor).
Refer to "Electrical Components Inspection".
N.G.
Replace exhaust gas sensor.
O.K.
Check E.C.U. pin terminals for damage or the connection of E.C.U. harness connector.
Perform FINAL CHECK by the following procedure after repair is completed.
FINAL CHECK
Erase the self-diagnosis memory.
Perform driving test.
Perform self-diagnosis again.
N.G.
Recheck E.C.U. pin terminals for damage or the connection of E.C.U. harness connector.
O.K.
INSPECTION END

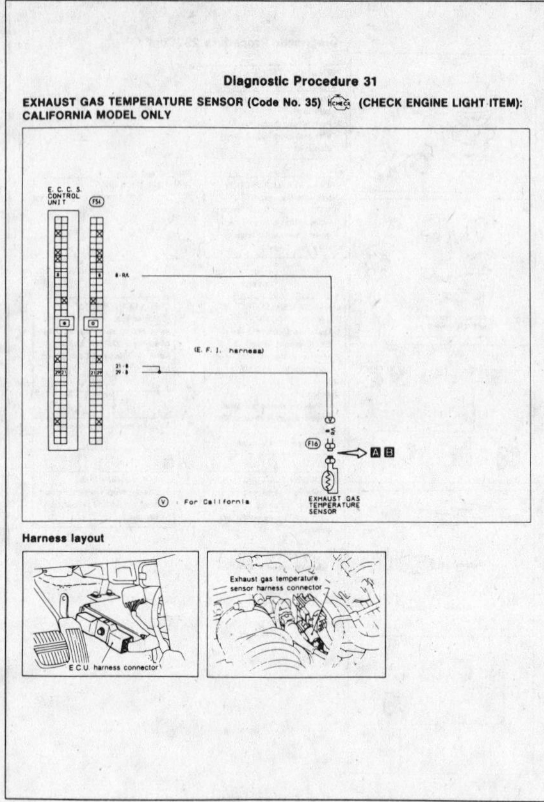
Diagnostic Procedure 31
EXHAUST GAS TEMPERATURE SENSOR (Code No. 35) (CHECK ENGINE LIGHT ITEM):
CALIFORNIA MODEL ONLY
E.C.C.S. CONTROL UNIT
(F54)
(E.F.I. harness)
(F16)
EXHAUST GAS TEMPERATURE SENSOR
Ⓥ: For California
Harness layout
Exhaust gas temperature sensor harness connector
E.C.U. harness connector

ECCS SELF-DIAGNOSTIC CHART—1990 STANZA

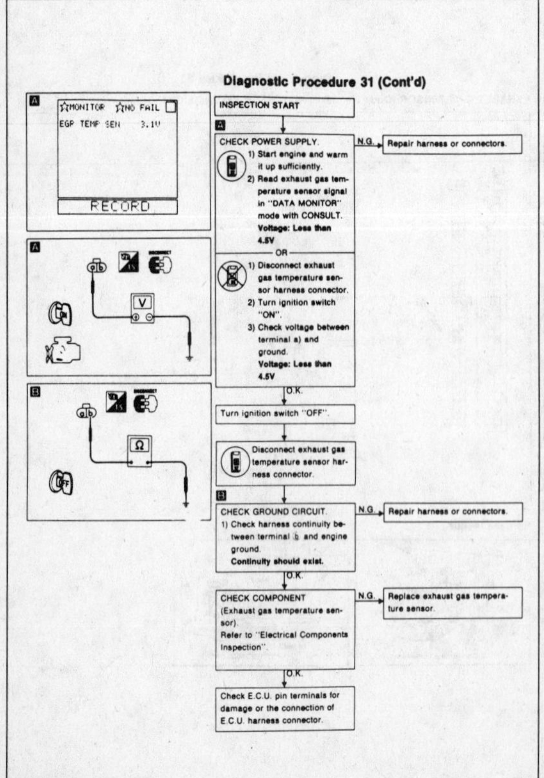
Diagnostic Procedure 31 (Cont'd)
MONITOR NO FAIL
EGP TEMP SEN 3.1V
RECORD
INSPECTION START
CHECK POWER SUPPLY.
1) Start engine and warm it up sufficiently.
2) Read exhaust gas temperature sensor signal in "DATA MONITOR" mode with CONSULT.
Voltage: Less than 4.5V
OR
1) Disconnect exhaust gas temperature sensor harness connector.
2) Turn ignition switch "ON".
3) Check voltage between terminal ⓐ and ground.
Voltage: Less than 4.5V
N.G.
Repair harness or connectors.
O.K.
Turn ignition switch "OFF".
Disconnect exhaust gas temperature sensor harness connector.
CHECK GROUND CIRCUIT.
1) Check harness continuity between terminal ⓑ and engine ground.
Continuity should exist.
N.G.
Repair harness or connectors.
O.K.
CHECK COMPONENT
(Exhaust gas temperature sensor).
Refer to "Electrical Components Inspection".
N.G.
Replace exhaust gas temperature sensor.
O.K.
Check E.C.U. pin terminals for damage or the connection of E.C.U. harness connector.

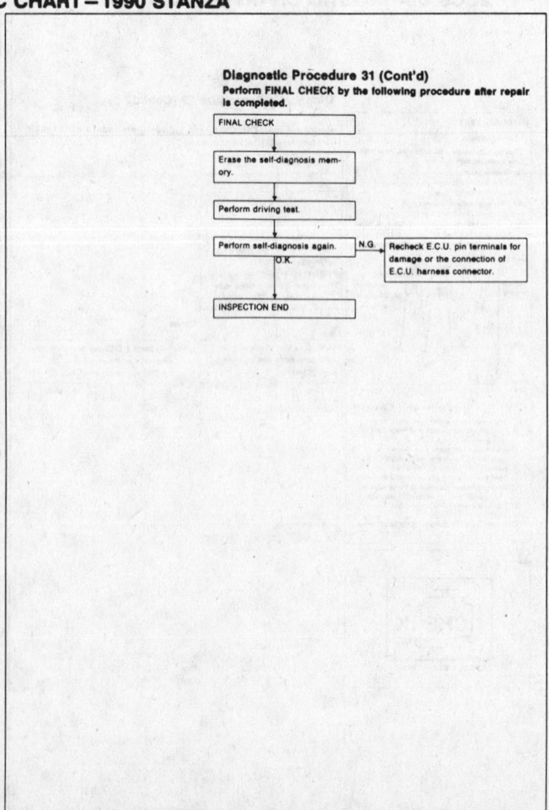
Diagnostic Procedure 31 (Cont'd)
Perform FINAL CHECK by the following procedure after repair is completed.
FINAL CHECK
Erase the self-diagnosis memory.
Perform driving test.
Perform self-diagnosis again.
N.G.
Recheck E.C.U. pin terminals for damage or the connection of E.C.U. harness connector.
O.K.
INSPECTION END

ECCS SELF-DIAGNOSTIC CHART—1990 STANZA

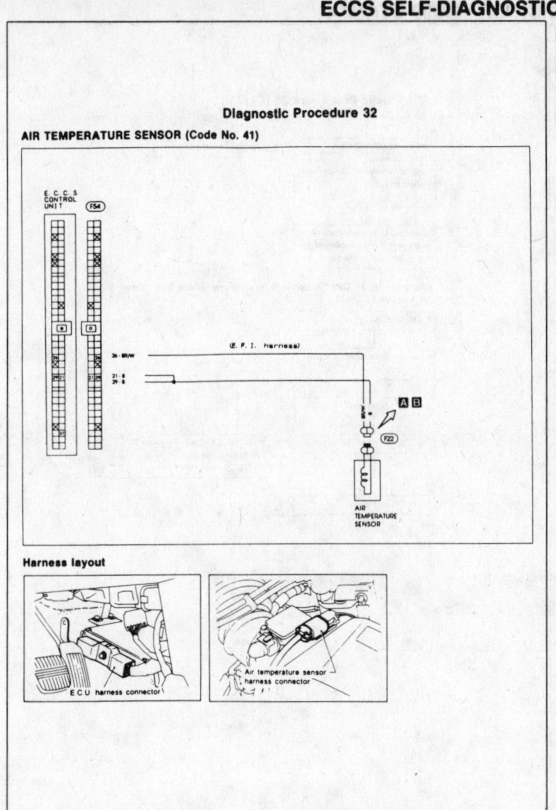
Diagnostic Procedure 32
AIR TEMPERATURE SENSOR (Code No. 41)
E.C.C.S. CONTROL UNIT
F54
(E. F. I. harness)
A B
F22
AIR TEMPERATURE SENSOR
Harness layout
E.C.U. harness connector
Air temperature sensor harness connector

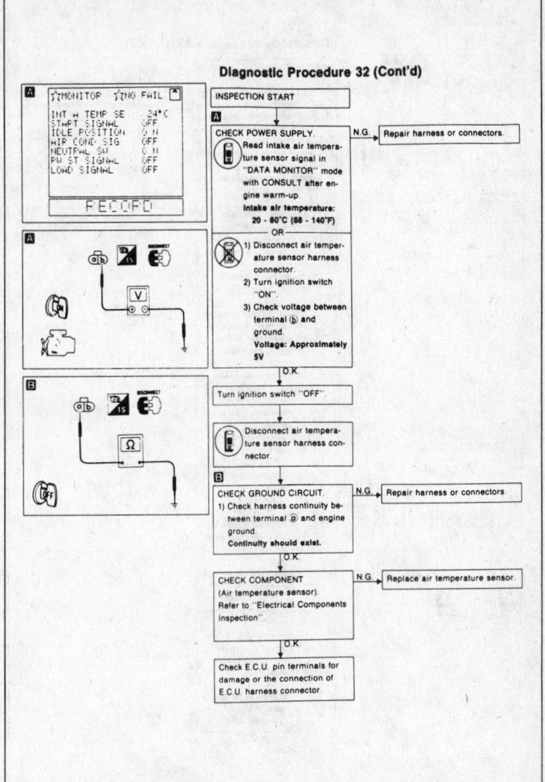
Diagnostic Procedure 32 (Cont'd)
MONITOR NO FAIL
INT A TEMP SE 24°C
START SIGNAL OFF
IDLE POSITION ON
AIR COND SIG OFF
NEUTRAL SW ON
PW ST SIGNAL OFF
LOAD SIGNAL OFF
RECORD
INSPECTION START
CHECK POWER SUPPLY.
Read intake air temperature sensor signal in "DATA MONITOR" mode with CONSULT after engine warm-up.
Intake air temperature: 20 - 80°C (68 - 140°F)
OR
1) Disconnect air temperature sensor harness connector.
2) Turn ignition switch "ON".
3) Check voltage between terminal (b) and ground.
Voltage: Approximately 5V
N.G.
Repair harness or connectors.
O.K.
Turn ignition switch "OFF".
Disconnect air temperature sensor harness connector.
CHECK GROUND CIRCUIT.
1) Check harness continuity between terminal (a) and engine ground.
Continuity should exist.
N.G.
Repair harness or connectors.
O.K.
CHECK COMPONENT
(Air temperature sensor).
Refer to "Electrical Components Inspection".
N.G.
Replace air temperature sensor.
O.K.
Check E.C.U. pin terminals for damage or the connection of E.C.U. harness connector.

ECCS SELF-DIAGNOSTIC CHART—1990 STANZA

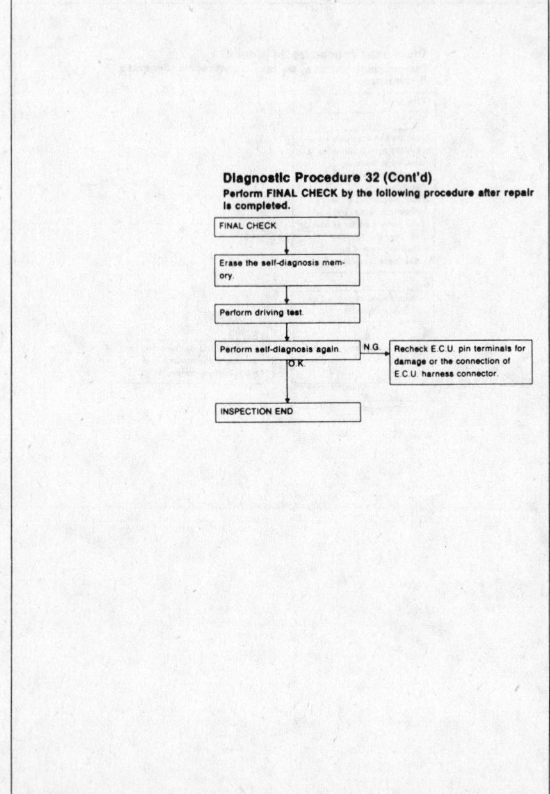
Diagnostic Procedure 32 (Cont'd)
Perform FINAL CHECK by the following procedure after repair is completed.
FINAL CHECK
Erase the self-diagnosis memory.
Perform driving test.
Perform self-diagnosis again.
N.G.
Recheck E.C.U. pin terminals for damage or the connection of E.C.U. harness connector.
O.K.
INSPECTION END

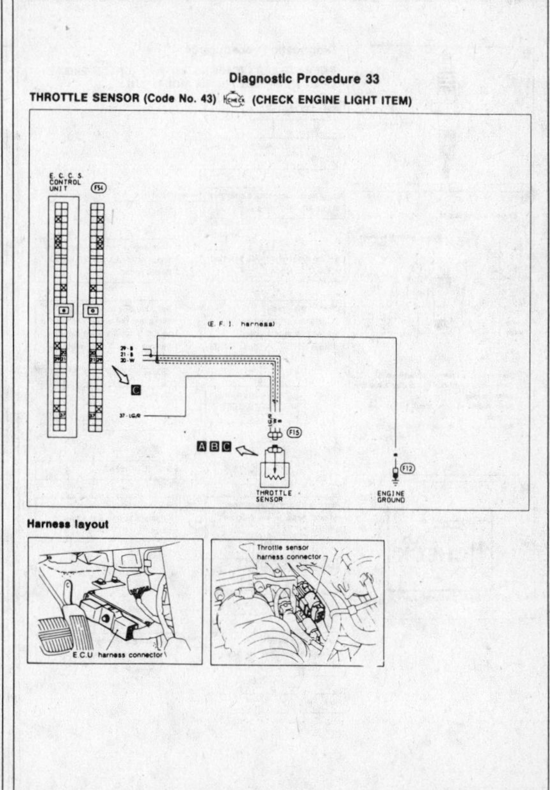
Diagnostic Procedure 33
THROTTLE SENSOR (Code No. 43) (CHECK ENGINE LIGHT ITEM)
E.C.C.S. CONTROL UNIT
F54
(E. F. I. harness)
C
A B C
F15
F12
THROTTLE SENSOR
ENGINE GROUND
Harness layout
E.C.U. harness connector
Throttle sensor harness connector

ECCS SELF-DIAGNOSTIC CHART—1990 STANZA

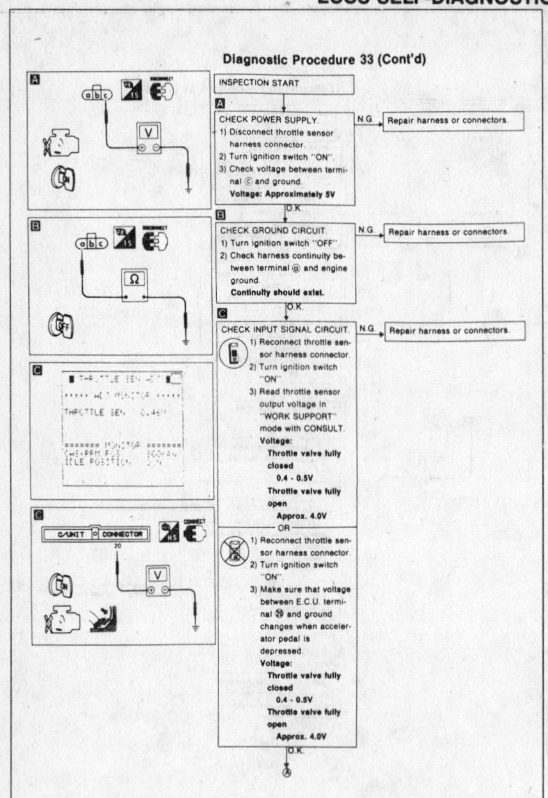
Diagnostic Procedure 33 (Cont'd)
INSPECTION START
A
CHECK POWER SUPPLY.
1) Disconnect throttle sensor harness connector.
2) Turn ignition switch "ON".
3) Check voltage between terminal © and ground.
Voltage: Approximately 5V
N.G.
Repair harness or connectors.
O.K.
B
CHECK GROUND CIRCUIT.
1) Turn ignition switch "OFF".
2) Check harness continuity between terminal ⓐ and engine ground.
Continuity should exist.
N.G.
Repair harness or connectors.
O.K.
C
CHECK INPUT SIGNAL CIRCUIT.
1) Reconnect throttle sensor harness connector.
2) Turn ignition switch "ON".
3) Read throttle sensor output voltage in "WORK SUPPORT" mode with CONSULT.
Voltage:
Throttle valve fully closed
0.4 - 0.5V
Throttle valve fully open
Approx. 4.0V
N.G.
Repair harness or connectors.
OR
1) Reconnect throttle sensor harness connector.
2) Turn ignition switch "ON".
3) Make sure that voltage between E.C.U. terminal ⑳ and ground changes when accelerator pedal is depressed.
Voltage:
Throttle valve fully closed
0.4 - 0.5V
Throttle valve fully open
Approx. 4.0V
O.K.
Ⓐ
a b c
V
Ω
C/UNIT CONNECTOR
20
THROTTLE SEN ADJ
THROTTLE SEN
MONITOR
IDLE POSITION

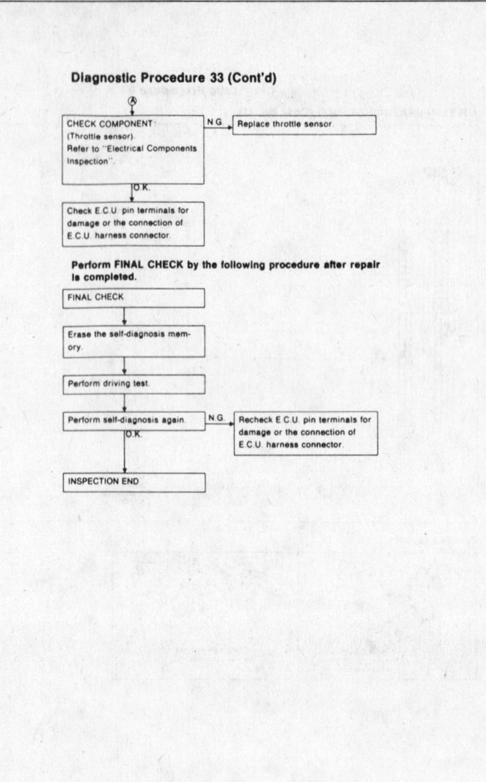
Diagnostic Procedure 33 (Cont'd)
Ⓐ
CHECK COMPONENT
(Throttle sensor).
Refer to "Electrical Components Inspection".
N.G.
Replace throttle sensor.
O.K.
Check E.C.U. pin terminals for damage or the connection of E.C.U. harness connector.
Perform FINAL CHECK by the following procedure after repair is completed.
FINAL CHECK
Erase the self-diagnosis memory.
Perform driving test.
Perform self-diagnosis again.
N.G.
Recheck E.C.U. pin terminals for damage or the connection of E.C.U. harness connector.
O.K.
INSPECTION END

ECCS SELF-DIAGNOSTIC CHART—1990 STANZA

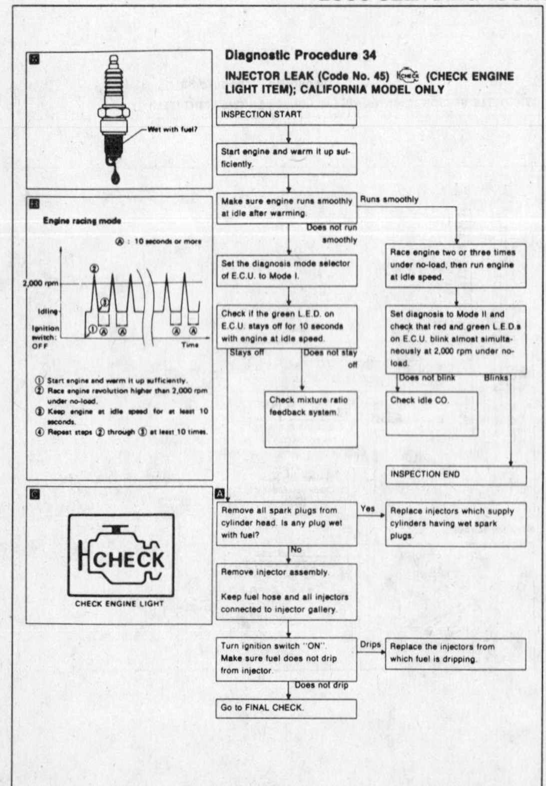
Diagnostic Procedure 34
INJECTOR LEAK (Code No. 45) (CHECK ENGINE LIGHT ITEM); CALIFORNIA MODEL ONLY
INSPECTION START
Start engine and warm it up sufficiently.
Make sure engine runs smoothly at idle after warming.
Runs smoothly
Does not run smoothly
Set the diagnosis mode selector of E.C.U. to Mode I.
Check if the green L.E.D. on E.C.U. stays off for 10 seconds with engine at idle speed.
Stays off
Does not stay off
Check mixture ratio feedback system.
Race engine two or three times under no-load, then run engine at idle speed.
Set diagnosis to Mode II and check that red and green L.E.D.s on E.C.U. blink almost simultaneously at 2,000 rpm under no-load.
Does not blink
Blinks
Check idle CO.
INSPECTION END
A
Remove all spark plugs from cylinder head. Is any plug wet with fuel?
Yes
Replace injectors which supply cylinders having wet spark plugs.
No
Remove injector assembly.
Keep fuel hose and all injectors connected to injector gallery.
Turn ignition switch "ON". Make sure fuel does not drip from injector.
Drips
Replace the injectors from which fuel is dripping.
Does not drip
Go to FINAL CHECK.
Wet with fuel?
Engine racing mode
Ⓐ : 10 seconds or more
2,000 rpm
Idling
Ignition switch: OFF
Time
① Start engine and warm it up sufficiently.
② Race engine revolution higher than 2,000 rpm under no-load.
③ Keep engine at idle speed for at least 10 seconds.
④ Repeat steps ② through ③ at least 10 times.
CHECK
CHECK ENGINE LIGHT

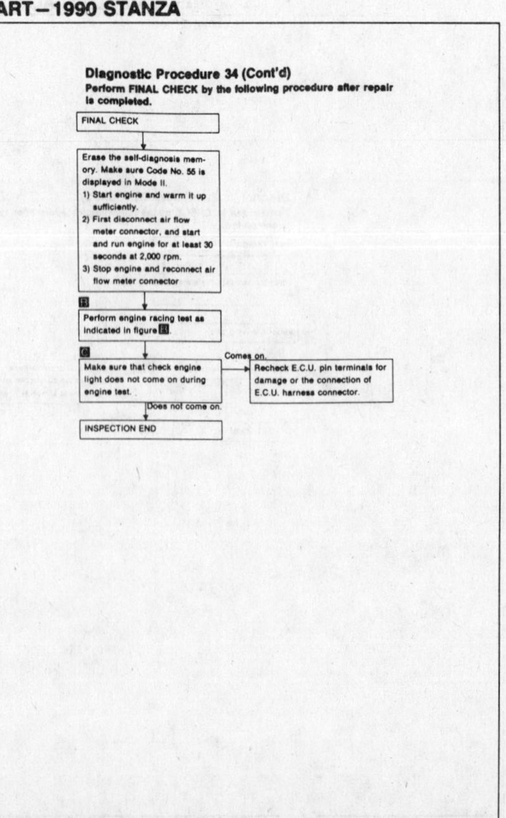
Diagnostic Procedure 34 (Cont'd)
Perform FINAL CHECK by the following procedure after repair is completed.
FINAL CHECK
Erase the self-diagnosis memory. Make sure Code No. 55 is displayed in Mode II.
1) Start engine and warm it up sufficiently.
2) First disconnect air flow meter connector, and start and run engine for at least 30 seconds at 2,000 rpm.
3) Stop engine and reconnect air flow meter connector.
B
Perform engine racing test as indicated in figure B.
C
Make sure that check engine light does not come on during engine test.
Comes on.
Recheck E.C.U. pin terminals for damage or the connection of E.C.U. harness connector.
Does not come on.
INSPECTION END

ECCS DIAGNOSTIC CHART—1990 STANZA

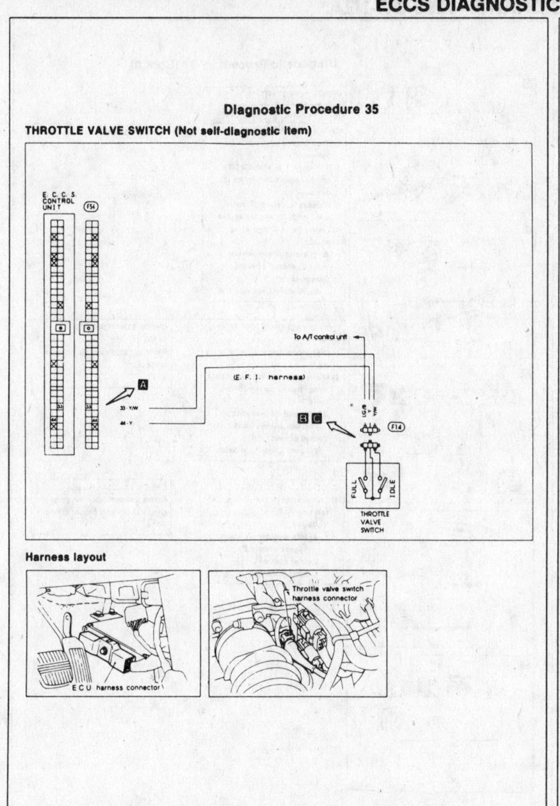
Diagnostic Procedure 35
THROTTLE VALVE SWITCH (Not self-diagnostic item)
E.C.C.S. CONTROL UNIT
To A/T control unit
(E.F.I. harness)
THROTTLE VALVE SWITCH
FULL
IDLE
Harness layout
Throttle valve switch harness connector
E.C.U. harness connector

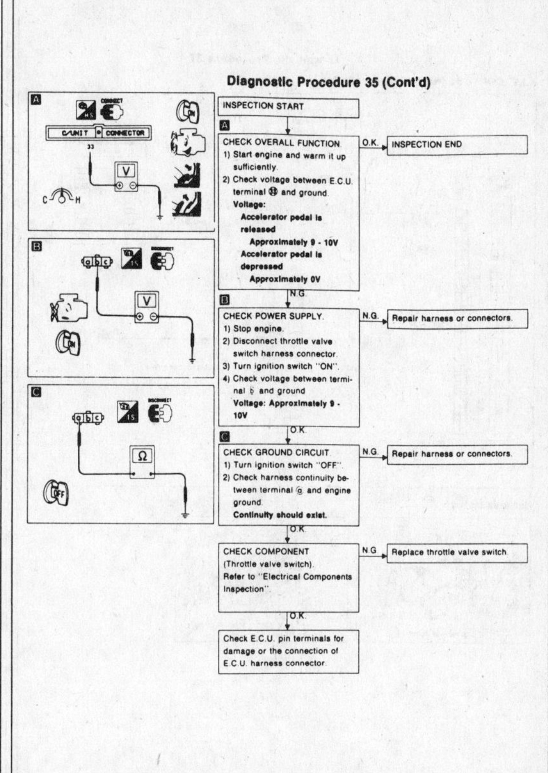
Diagnostic Procedure 35 (Cont'd)
INSPECTION START
A
CHECK OVERALL FUNCTION.
1) Start engine and warm it up sufficiently.
2) Check voltage between E.C.U. terminal ㉝ and ground.
Voltage:
Accelerator pedal is released
Approximately 9 - 10V
Accelerator pedal is depressed
Approximately 0V
O.K.
INSPECTION END
N.G.
B
CHECK POWER SUPPLY.
1) Stop engine.
2) Disconnect throttle valve switch harness connector.
3) Turn ignition switch "ON".
4) Check voltage between terminal ⓑ and ground
Voltage: Approximately 9 - 10V
N.G.
Repair harness or connectors.
O.K.
C
CHECK GROUND CIRCUIT.
1) Turn ignition switch "OFF".
2) Check harness continuity between terminal ⓒ and engine ground.
Continuity should exist.
N.G.
Repair harness or connectors.
O.K.
CHECK COMPONENT
(Throttle valve switch).
Refer to "Electrical Components Inspection".
N.G
Replace throttle valve switch.
O.K.
Check E.C.U. pin terminals for damage or the connection of E.C.U. harness connector.

ECCS DIAGNOSTIC CHART—1990 STANZA

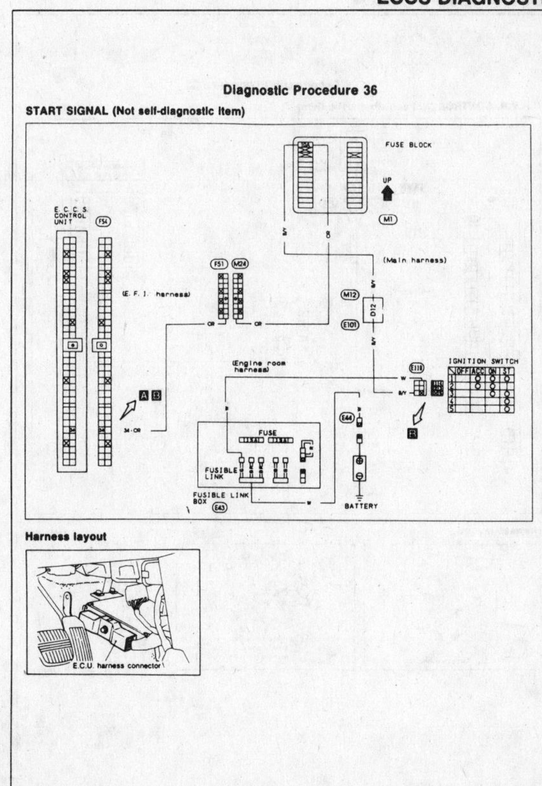
Diagnostic Procedure 36
START SIGNAL (Not self-diagnostic item)
FUSE BLOCK
E.C.C.S. CONTROL UNIT
(Main harness)
(E.F.I. harness)
(Engine room harness)
IGNITION SWITCH
FUSE
FUSIBLE LINK
FUSIBLE LINK BOX
BATTERY
Harness layout
E.C.U. harness connector

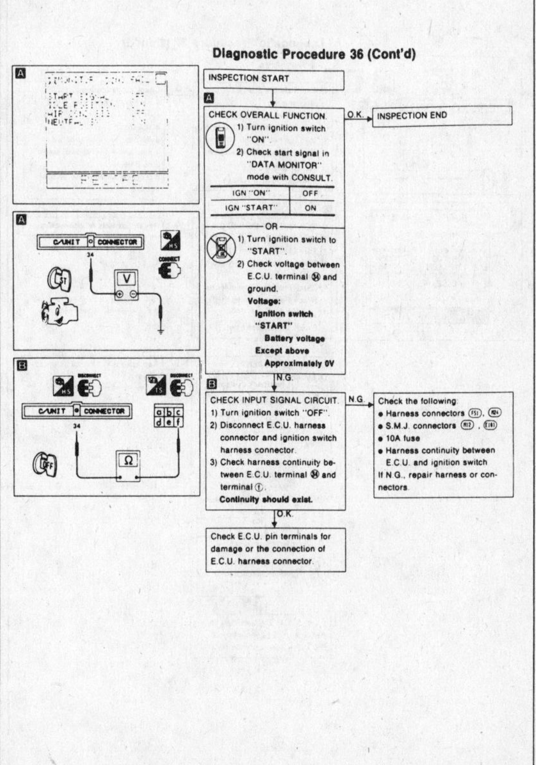
Diagnostic Procedure 36 (Cont'd)
INSPECTION START
A
CHECK OVERALL FUNCTION.
1) Turn ignition switch "ON".
2) Check start signal in "DATA MONITOR" mode with CONSULT.
IGN "ON" | OFF
IGN "START" | ON
OR
1) Turn ignition switch to "START".
2) Check voltage between E.C.U. terminal ㉞ and ground.
Voltage:
Ignition switch "START"
Battery voltage
Except above
Approximately 0V
O.K.
INSPECTION END
N.G.
B
CHECK INPUT SIGNAL CIRCUIT.
1) Turn ignition switch "OFF".
2) Disconnect E.C.U. harness connector and ignition switch harness connector.
3) Check harness continuity between E.C.U. terminal ㉞ and terminal ⓕ.
Continuity should exist.
N.G.
Check the following:
• Harness connectors (F51), (M24)
• S.M.J. connectors (M12), (E101)
• 10A fuse
• Harness continuity between E.C.U. and ignition switch
If N.G., repair harness or connectors.
O.K.
Check E.C.U. pin terminals for damage or the connection of E.C.U. harness connector.

ECCS DIAGNOSTIC CHART—1990 STANZA

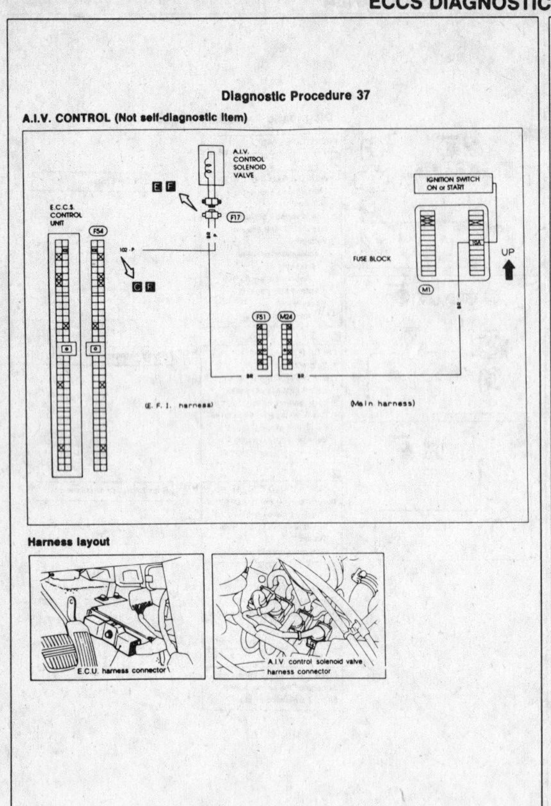
Diagnostic Procedure 37
A.I.V. CONTROL (Not self-diagnostic item)
A.I.V. CONTROL SOLENOID VALVE
E.C.C.S. CONTROL UNIT
IGNITION SWITCH ON or START
FUSE BLOCK
UP
(E. F. I. harness)
(Main harness)
Harness layout
E.C.U. harness connector
A.I.V. control solenoid valve harness connector

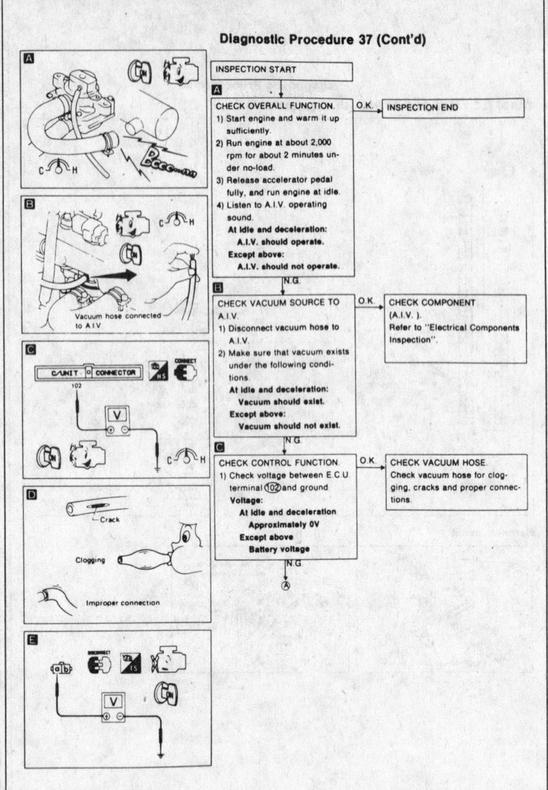
Diagnostic Procedure 37 (Cont'd)
INSPECTION START
CHECK OVERALL FUNCTION.
1) Start engine and warm it up sufficiently.
2) Run engine at about 2,000 rpm for about 2 minutes under no-load.
3) Release accelerator pedal fully, and run engine at idle.
4) Listen to A.I.V. operating sound.
At idle and deceleration:
A.I.V. should operate.
Except above:
A.I.V. should not operate.
O.K.
INSPECTION END
N.G.
CHECK VACUUM SOURCE TO A.I.V.
1) Disconnect vacuum hose to A.I.V.
2) Make sure that vacuum exists under the following conditions.
At idle and deceleration:
Vacuum should exist.
Except above:
Vacuum should not exist.
O.K.
CHECK COMPONENT (A.I.V.).
Refer to "Electrical Components Inspection".
N.G.
CHECK CONTROL FUNCTION.
1) Check voltage between E.C.U. terminal 102 and ground.
Voltage:
At idle and deceleration
Approximately 0V
Except above
Battery voltage
O.K.
CHECK VACUUM HOSE.
Check vacuum hose for clogging, cracks and proper connections.
N.G.
Vacuum hose connected to A.I.V.
C/UNIT CONNECTOR
Crack
Clogging
Improper connection

ECCS DIAGNOSTIC CHART—1990 STANZA

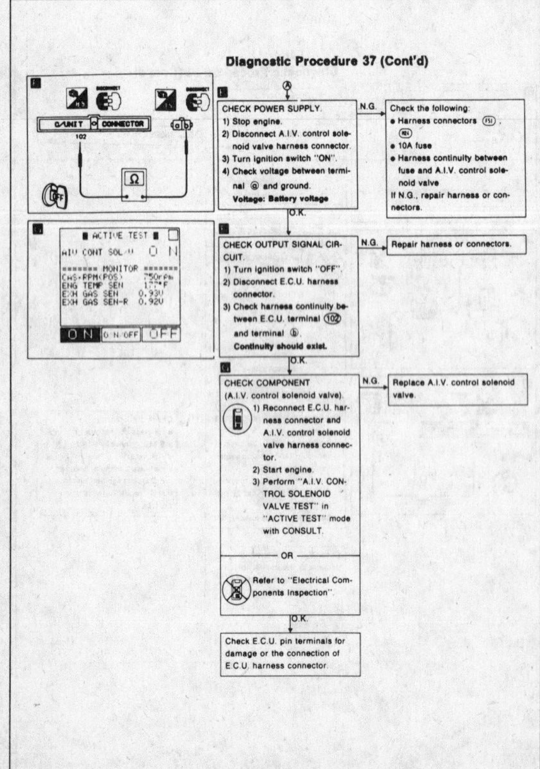
Diagnostic Procedure 37 (Cont'd)
CHECK POWER SUPPLY.
1) Stop engine.
2) Disconnect A.I.V. control solenoid valve harness connector.
3) Turn ignition switch "ON".
4) Check voltage between terminal ⓐ and ground.
Voltage: Battery voltage
N.G.
Check the following:
• Harness connectors F51, R61
• 10A fuse
• Harness continuity between fuse and A.I.V. control solenoid valve
If N.G., repair harness or connectors.
O.K.
CHECK OUTPUT SIGNAL CIRCUIT.
1) Turn ignition switch "OFF".
2) Disconnect E.C.U. harness connector.
3) Check harness continuity between E.C.U. terminal 102 and terminal ⓑ.
Continuity should exist.
N.G.
Repair harness or connectors.
O.K.
CHECK COMPONENT
(A.I.V. control solenoid valve).
1) Reconnect E.C.U. harness connector and A.I.V. control solenoid valve harness connector.
2) Start engine.
3) Perform "A.I.V. CONTROL SOLENOID VALVE TEST" in "ACTIVE TEST" mode with CONSULT.
OR
Refer to "Electrical Components Inspection".
N.G.
Replace A.I.V. control solenoid valve.
O.K.
Check E.C.U. pin terminals for damage or the connection of E.C.U. harness connector.
C/UNIT CONNECTOR
ACTIVE TEST
AIV CONT SOL-V
MONITOR
CAS·RPM(POS)
ENG TEMP SEN
EXH GAS SEN
EXH GAS SEN-R
ON
ON/OFF
OFF

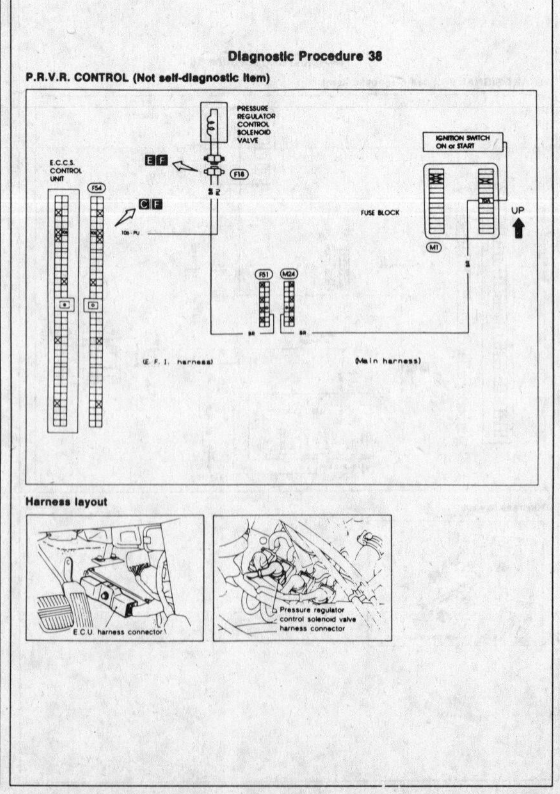
Diagnostic Procedure 38
P.R.V.R. CONTROL (Not self-diagnostic item)
PRESSURE REGULATOR CONTROL SOLENOID VALVE
E.C.C.S. CONTROL UNIT
IGNITION SWITCH ON or START
FUSE BLOCK
UP
(E. F. I. harness)
(Main harness)
Harness layout
E.C.U. harness connector
Pressure regulator control solenoid valve harness connector

ECCS DIAGNOSTIC CHART—1990 STANZA

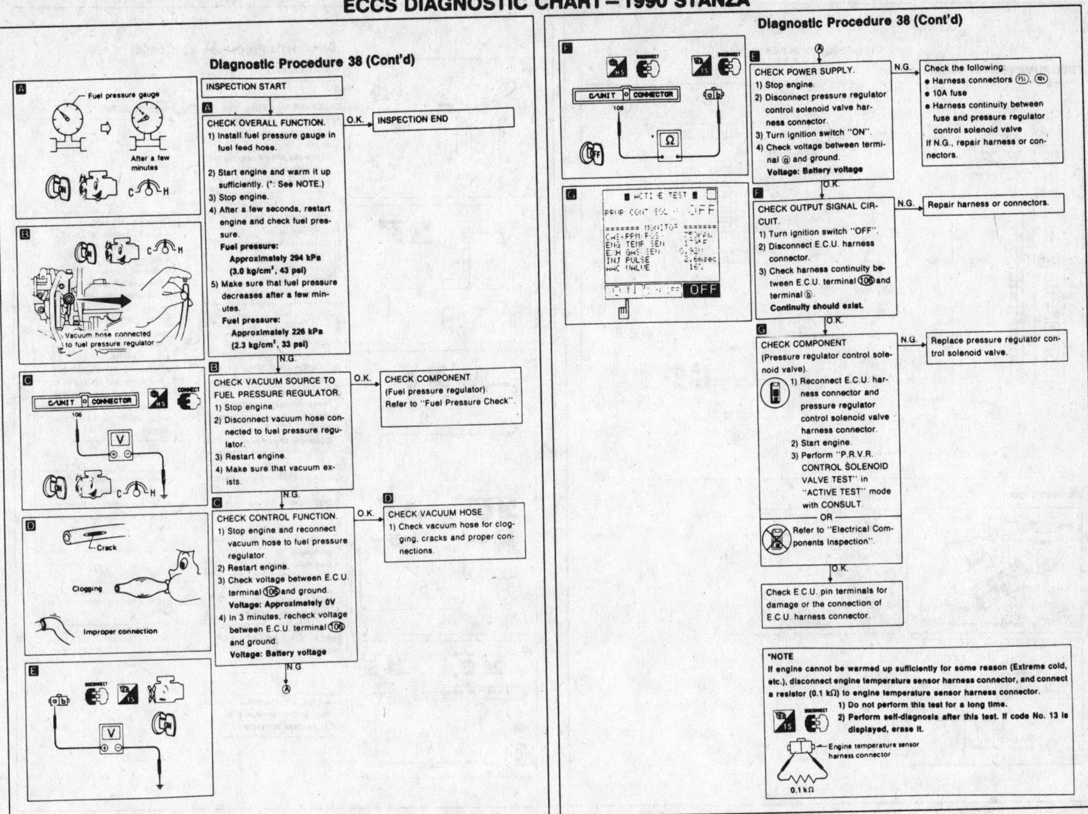
Diagnostic Procedure 38 (Cont'd)
Fuel pressure gauge
After a few minutes
Vacuum hose connected to fuel pressure regulator
C/UNIT CONNECTOR
106
Crack
Clogging
Improper connection
INSPECTION START
CHECK OVERALL FUNCTION.
1) Install fuel pressure gauge in fuel feed hose.
2) Start engine and warm it up sufficiently. (*: See NOTE.)
3) Stop engine.
4) After a few seconds, restart engine and check fuel pressure.
Fuel pressure:
Approximately 294 kPa
(3.0 kg/cm², 43 psi)
5) Make sure that fuel pressure decreases after a few minutes.
Fuel pressure:
Approximately 226 kPa
(2.3 kg/cm², 33 psi)
O.K.
INSPECTION END
N.G.
CHECK VACUUM SOURCE TO FUEL PRESSURE REGULATOR.
1) Stop engine.
2) Disconnect vacuum hose connected to fuel pressure regulator.
3) Restart engine.
4) Make sure that vacuum exists.
CHECK COMPONENT
(Fuel pressure regulator)
Refer to "Fuel Pressure Check".
CHECK CONTROL FUNCTION.
1) Stop engine and reconnect vacuum hose to fuel pressure regulator.
2) Restart engine.
3) Check voltage between E.C.U. terminal 106 and ground.
Voltage: Approximately 0V
4) In 3 minutes, recheck voltage between E.C.U. terminal 106 and ground.
Voltage: Battery voltage
CHECK VACUUM HOSE.
1) Check vacuum hose for clogging, cracks and proper connections.
Diagnostic Procedure 38 (Cont'd)
ACTIVE TEST
OFF
CHECK POWER SUPPLY.
1) Stop engine.
2) Disconnect pressure regulator control solenoid valve harness connector.
3) Turn ignition switch "ON".
4) Check voltage between terminal ⓐ and ground.
Voltage: Battery voltage
Check the following.
• Harness connectors
• 10A fuse
• Harness continuity between fuse and pressure regulator control solenoid valve
If N.G., repair harness or connectors.
CHECK OUTPUT SIGNAL CIRCUIT.
1) Turn ignition switch "OFF".
2) Disconnect E.C.U. harness connector.
3) Check harness continuity between E.C.U. terminal 106 and terminal ⓑ.
Continuity should exist.
Repair harness or connectors.
CHECK COMPONENT
(Pressure regulator control solenoid valve).
1) Reconnect E.C.U. harness connector and pressure regulator control solenoid valve harness connector.
2) Start engine.
3) Perform "P.R.V.R. CONTROL SOLENOID VALVE TEST" in "ACTIVE TEST" mode with CONSULT.
OR
Refer to "Electrical Components Inspection".
Replace pressure regulator control solenoid valve.
Check E.C.U. pin terminals for damage or the connection of E.C.U. harness connector.
*NOTE
If engine cannot be warmed up sufficiently for some reason (Extreme cold, etc.), disconnect engine temperature sensor harness connector, and connect a resistor (0.1 kΩ) to engine temperature sensor harness connector.
1) Do not perform this test for a long time.
2) Perform self-diagnosis after this test. If code No. 13 is displayed, erase it.
Engine temperature sensor harness connector
0.1 kΩ

ECCS DIAGNOSTIC CHART—1990 STANZA

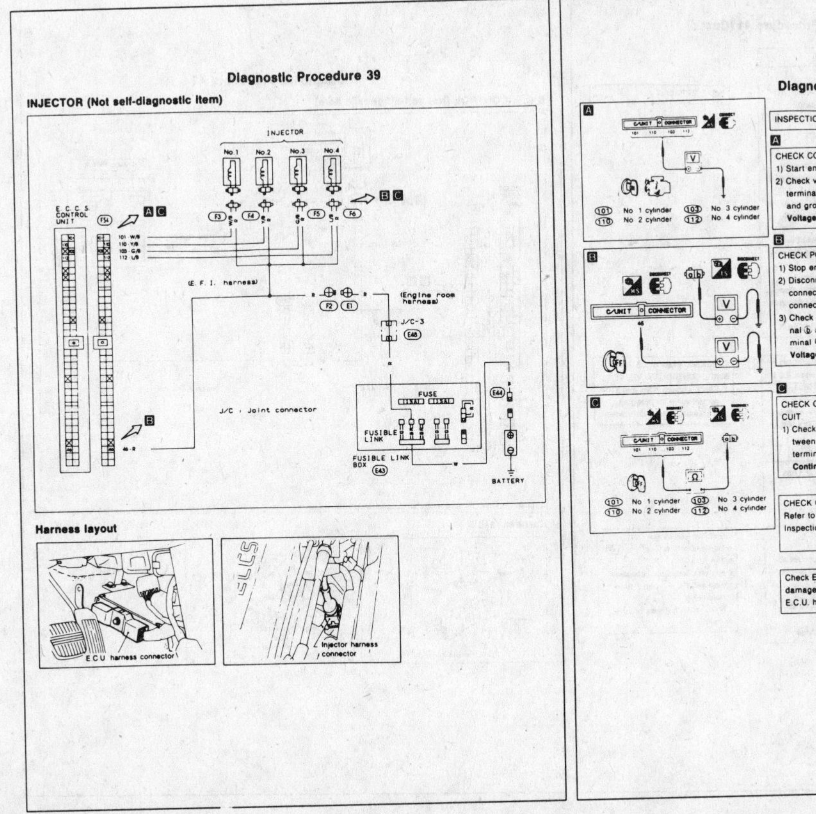
Diagnostic Procedure 39
INJECTOR (Not self-diagnostic item)
INJECTOR
No.1
No.2
No.3
No.4
E.C.C.S. CONTROL UNIT
(E.F.I. harness)
(Engine room harness)
J/C-3
FUSE
FUSIBLE LINK
FUSIBLE LINK BOX
J/C : Joint connector
BATTERY
Harness layout
ECU harness connector
Injector harness connector

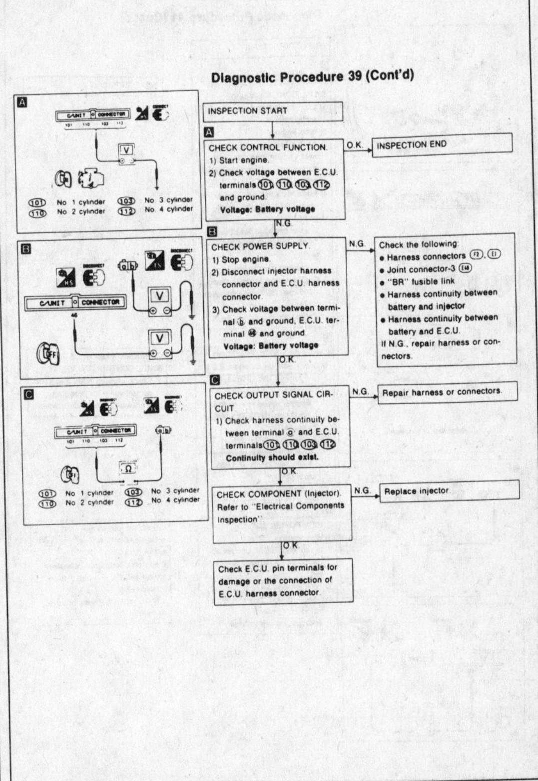
Diagnostic Procedure 39 (Cont'd)
C/UNIT CONNECTOR
No 1 cylinder
No 2 cylinder
No 3 cylinder
No 4 cylinder
INSPECTION START
CHECK CONTROL FUNCTION.
1) Start engine.
2) Check voltage between E.C.U. terminals 101, 110, 103, 112 and ground.
Voltage: Battery voltage
O.K.
INSPECTION END
N.G.
CHECK POWER SUPPLY.
1) Stop engine.
2) Disconnect injector harness connector and E.C.U. harness connector.
3) Check voltage between terminal ⓑ and ground, E.C.U. terminal 46 and ground.
Voltage: Battery voltage
Check the following.
• Harness connectors
• Joint connector-3
• "BR" fusible link
• Harness continuity between battery and injector
• Harness continuity between battery and E.C.U.
If N.G., repair harness or connectors.
CHECK OUTPUT SIGNAL CIRCUIT
1) Check harness continuity between terminal ⓐ and E.C.U. terminals 101, 110, 103, 112.
Continuity should exist.
Repair harness or connectors.
CHECK COMPONENT (Injector).
Refer to "Electrical Components Inspection".
Replace injector
Check E.C.U. pin terminals for damage or the connection of E.C.U. harness connector.

ECCS DIAGNOSTIC CHART—1990 STANZA

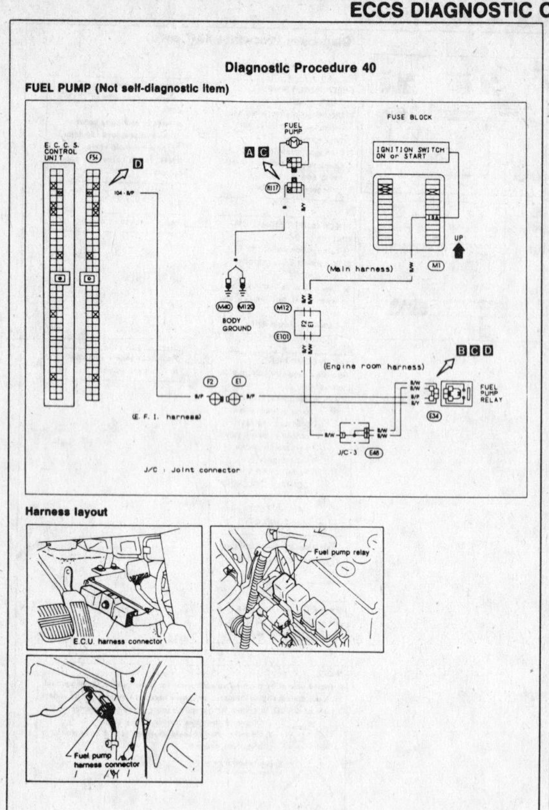
Diagnostic Procedure 40
FUEL PUMP (Not self-diagnostic item)
E.C.C.S. CONTROL UNIT
FUEL PUMP
FUSE BLOCK
IGNITION SWITCH ON or START
UP
(Main harness)
BODY GROUND
(Engine room harness)
FUEL PUMP RELAY
(E.F.I. harness)
J/C : Joint connector
J/C-3
Harness layout
E.C.U. harness connector
Fuel pump relay
Fuel pump harness connector

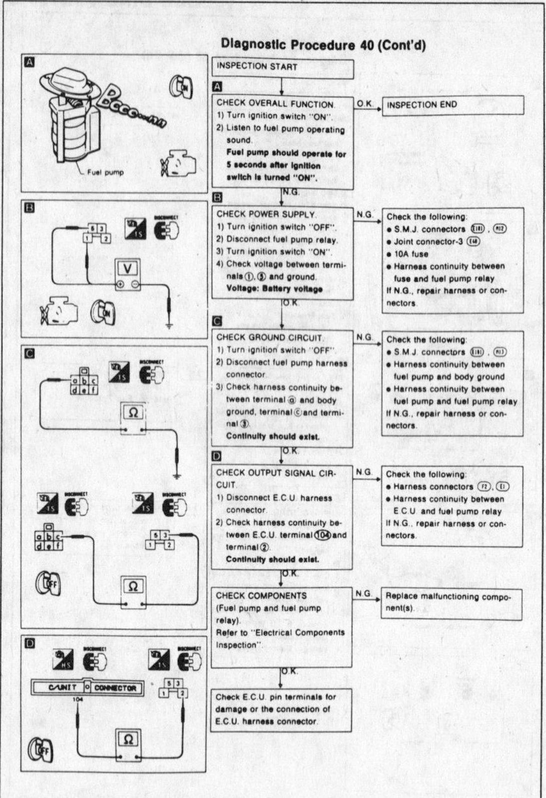
Diagnostic Procedure 40 (Cont'd)
Fuel pump
INSPECTION START
CHECK OVERALL FUNCTION.
1) Turn ignition switch "ON".
2) Listen to fuel pump operating sound.
Fuel pump should operate for 5 seconds after ignition switch is turned "ON".
O.K.
INSPECTION END
N.G.
CHECK POWER SUPPLY.
1) Turn ignition switch "OFF".
2) Disconnect fuel pump relay.
3) Turn ignition switch "ON".
4) Check voltage between terminals ①, ⑤ and ground.
Voltage: Battery voltage
Check the following:
• S.M.J. connectors
• Joint connector-3
• 10A fuse
• Harness continuity between fuse and fuel pump relay
If N.G., repair harness or connectors.
CHECK GROUND CIRCUIT.
1) Turn ignition switch "OFF".
2) Disconnect fuel pump harness connector.
3) Check harness continuity between terminal ⓐ and body ground, terminal ⓒ and terminal ③.
Continuity should exist.
Check the following:
• S.M.J. connectors
• Harness continuity between fuel pump and body ground
• Harness continuity between fuel pump and fuel pump relay
If N.G., repair harness or connectors.
CHECK OUTPUT SIGNAL CIRCUIT.
1) Disconnect E.C.U. harness connector.
2) Check harness continuity between E.C.U. terminal 104 and terminal ②.
Continuity should exist.
Check the following:
• Harness connectors
• Harness continuity between E.C.U. and fuel pump relay
If N.G., repair harness or connectors.
CHECK COMPONENTS
(Fuel pump and fuel pump relay).
Refer to "Electrical Components Inspection".
Replace malfunctioning component(s).
Check E.C.U. pin terminals for damage or the connection of E.C.U. harness connector.
C/UNIT CONNECTOR

ECCS DIAGNOSTIC CHART—1990 STANZA

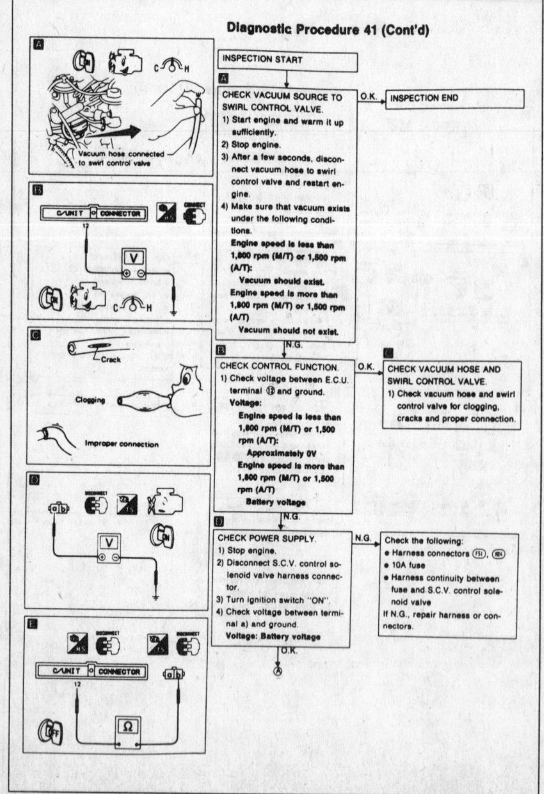
Diagnostic Procedure 41 (Cont'd)
Vacuum hose connected to swirl control valve
INSPECTION START
CHECK VACUUM SOURCE TO SWIRL CONTROL VALVE.
1) Start engine and warm it up sufficiently.
2) Stop engine.
3) After a few seconds, disconnect vacuum hose to swirl control valve and restart engine.
4) Make sure that vacuum exists under the following conditions.
Engine speed is less than 1,800 rpm (M/T) or 1,500 rpm (A/T):
Vacuum should exist.
Engine speed is more than 1,800 rpm (M/T) or 1,500 rpm (A/T)
Vacuum should not exist.
O.K.
INSPECTION END
N.G.
CHECK CONTROL FUNCTION.
1) Check voltage between E.C.U. terminal ⑫ and ground.
Voltage:
Engine speed is less than 1,800 rpm (M/T) or 1,500 rpm (A/T):
Approximately 0V
Engine speed is more than 1,800 rpm (M/T) or 1,500 rpm (A/T)
Battery voltage
CHECK VACUUM HOSE AND SWIRL CONTROL VALVE.
1) Check vacuum hose and swirl control valve for clogging, cracks and proper connection.
Crack
Clogging
Improper connection
CHECK POWER SUPPLY.
1) Stop engine.
2) Disconnect S.C.V. control solenoid valve harness connector.
3) Turn ignition switch "ON".
4) Check voltage between terminal a) and ground.
Voltage: Battery voltage
Check the following:
• Harness connectors
• 10A fuse
• Harness continuity between fuse and S.C.V. control solenoid valve
If N.G., repair harness or connectors.
C/UNIT CONNECTOR

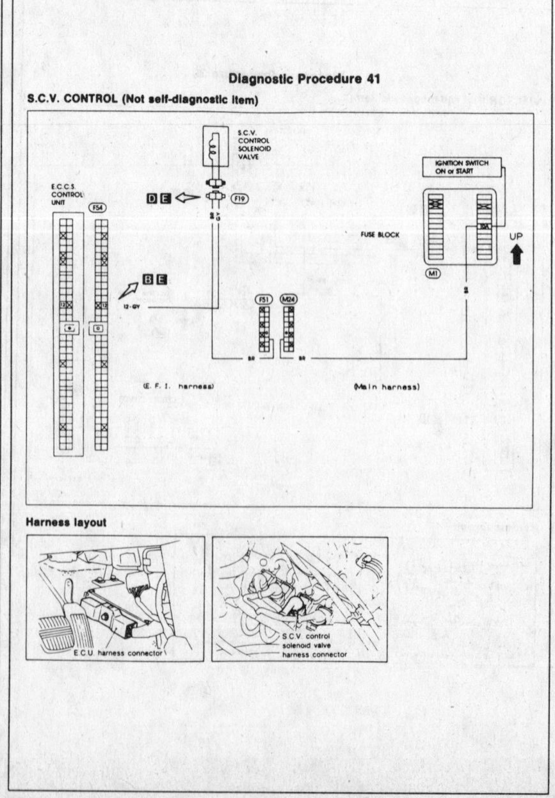
Diagnostic Procedure 41
S.C.V. CONTROL (Not self-diagnostic item)
S.C.V. CONTROL SOLENOID VALVE
E.C.C.S. CONTROL UNIT
IGNITION SWITCH ON or START
FUSE BLOCK
UP
(E.F.I. harness)
(Main harness)
Harness layout
E.C.U. harness connector
S.C.V. control solenoid valve harness connector

ECCS DIAGNOSTIC CHART—1990 STANZA

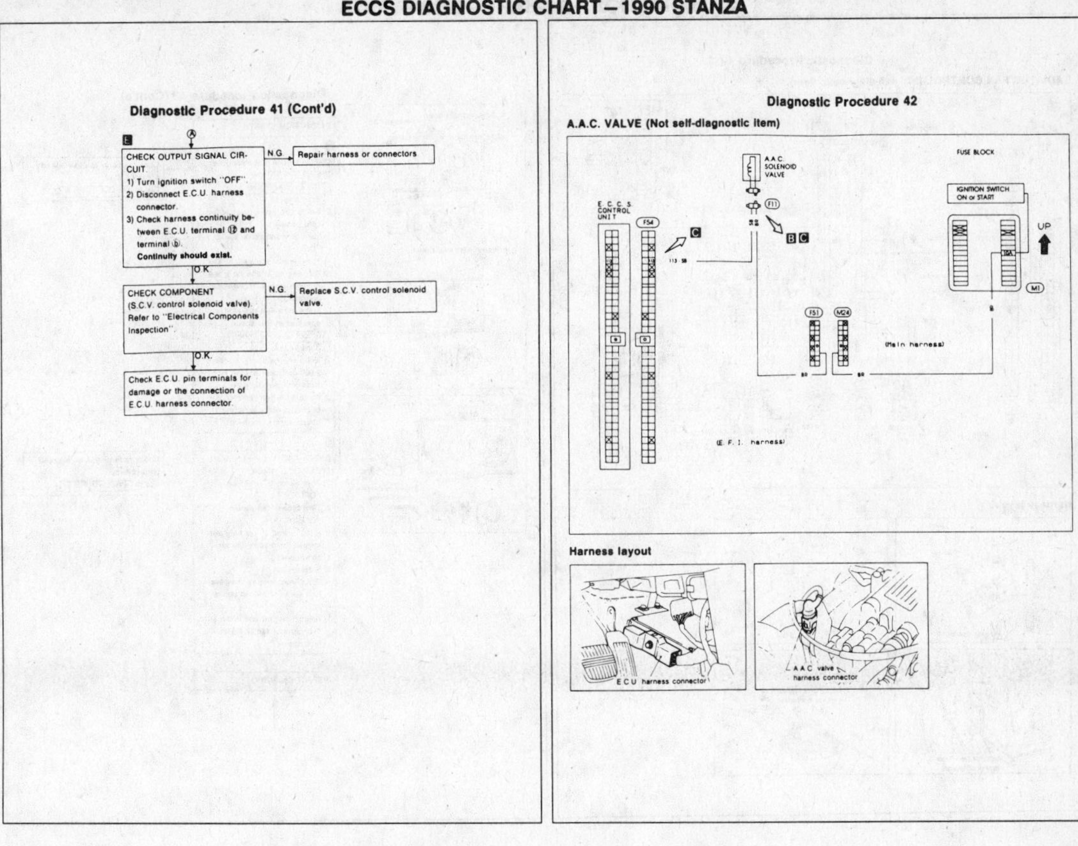
Diagnostic Procedure 41 (Cont'd)
E
CHECK OUTPUT SIGNAL CIR-CUIT
1) Turn ignition switch "OFF".
2) Disconnect E.C.U. harness connector.
3) Check harness continuity be-tween E.C.U. terminal and terminal .
Continuity should exist.
N.G.
Repair harness or connectors
O.K.
CHECK COMPONENT
(S.C.V. control solenoid valve).
Refer to "Electrical Components Inspection".
N.G.
Replace S.C.V. control solenoid valve.
O.K.
Check E.C.U. pin terminals for damage or the connection of E.C.U. harness connector.
Diagnostic Procedure 42
A.A.C. VALVE (Not self-diagnostic item)
A.A.C. SOLENOID VALVE
FUSE BLOCK
IGNITION SWITCH ON or START
E.C.C.S. CONTROL UNIT
UP
(Main harness)
(E.F.I. harness)
Harness layout
E.C.U. harness connector
A.A.C. valve harness connector

ECCS DIAGNOSTIC CHART—1990 STANZA

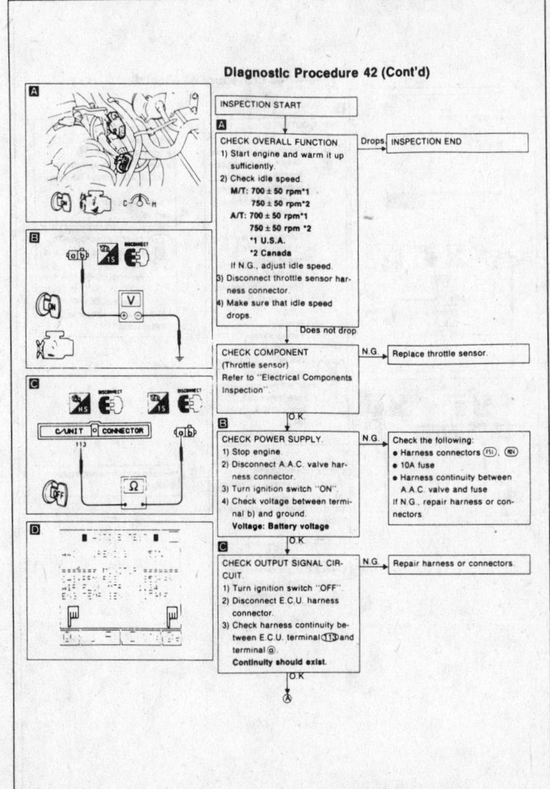
Diagnostic Procedure 42 (Cont'd)
A
B
C
C/UNIT CONNECTOR
D
INSPECTION START
A
CHECK OVERALL FUNCTION.
1) Start engine and warm it up sufficiently.
2) Check idle speed.
M/T: 700 ± 50 rpm*1
750 ± 50 rpm*2
A/T: 700 ± 50 rpm*1
750 ± 50 rpm *2
*1 U.S.A.
*2 Canada
If N.G., adjust idle speed.
3) Disconnect throttle sensor harness connector.
4) Make sure that idle speed drops.
Drops
INSPECTION END
Does not drop
CHECK COMPONENT
(Throttle sensor)
Refer to "Electrical Components Inspection".
N.G.
Replace throttle sensor.
O.K.
B
CHECK POWER SUPPLY.
1) Stop engine.
2) Disconnect A.A.C. valve harness connector.
3) Turn ignition switch "ON".
4) Check voltage between terminal b) and ground.
Voltage: Battery voltage
N.G.
Check the following:
• Harness connectors
• 10A fuse
• Harness continuity between A.A.C. valve and fuse
If N.G., repair harness or connectors.
O.K.
C
CHECK OUTPUT SIGNAL CIRCUIT.
1) Turn ignition switch "OFF".
2) Disconnect E.C.U. harness connector.
3) Check harness continuity between E.C.U. terminal 113 and terminal a.
Continuity should exist.
N.G.
Repair harness or connectors.
O.K.

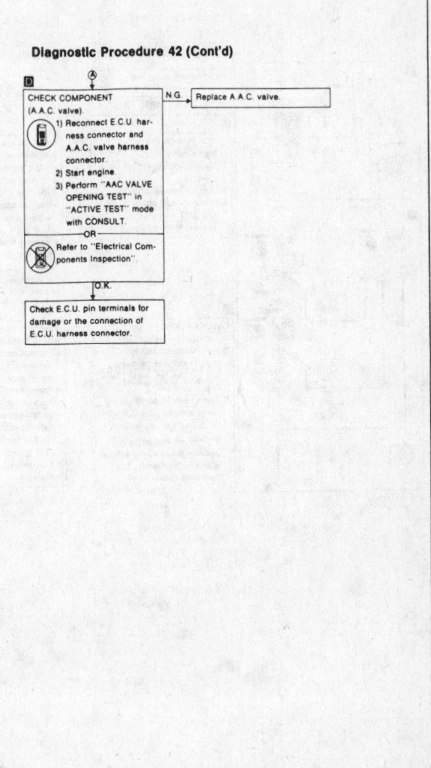
Diagnostic Procedure 42 (Cont'd)
D
CHECK COMPONENT
(A.A.C. valve).
1) Reconnect E.C.U. harness connector and A.A.C. valve harness connector.
2) Start engine.
3) Perform "AAC VALVE OPENING TEST" in "ACTIVE TEST" mode with CONSULT.
OR
Refer to "Electrical Components Inspection".
N.G.
Replace A.A.C. valve.
O.K.
Check E.C.U. pin terminals for damage or the connection of E.C.U. harness connector.

ECCS DIAGNOSTIC CHART—1990 STANZA

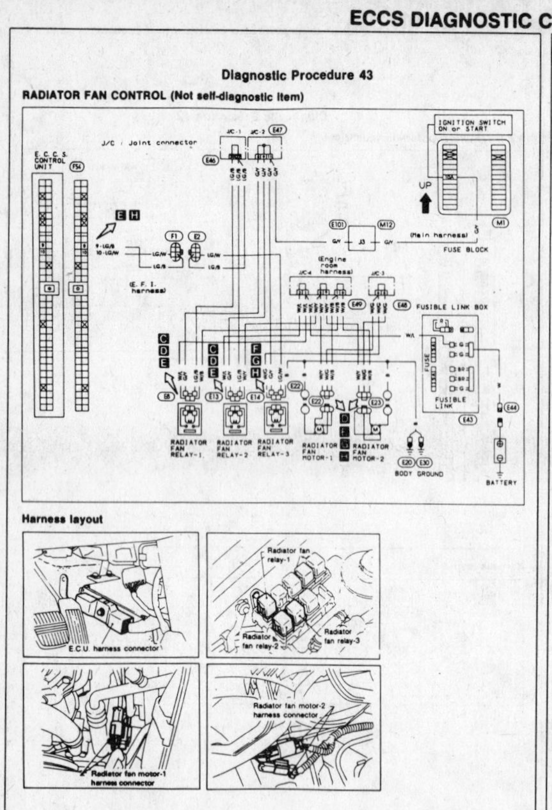
Diagnostic Procedure 43
RADIATOR FAN CONTROL (Not self-diagnostic item)
J/C : Joint connector
E.C.C.S. CONTROL UNIT
IGNITION SWITCH ON or START
UP
Main harness
FUSE BLOCK
Engine room harness
FUSIBLE LINK BOX
E.F.I. harness
RADIATOR FAN RELAY-1
RADIATOR FAN RELAY-2
RADIATOR FAN RELAY-3
RADIATOR FAN MOTOR-1
RADIATOR FAN MOTOR-2
FUSIBLE LINK
BODY GROUND
BATTERY
Harness layout
E.C.U. harness connector
Radiator fan relay-1
Radiator fan relay-2
Radiator fan relay-3
Radiator fan motor-1 harness connector
Radiator fan motor-2 harness connector

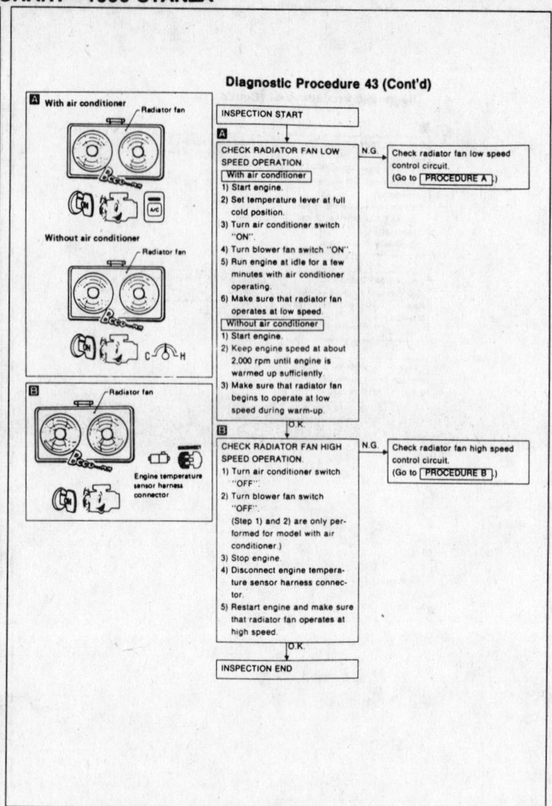
Diagnostic Procedure 43 (Cont'd)
A With air conditioner
Radiator fan
Without air conditioner
B
Engine temperature sensor harness connector
INSPECTION START
A
CHECK RADIATOR FAN LOW SPEED OPERATION.
With air conditioner
1) Start engine.
2) Set temperature lever at full cold position.
3) Turn air conditioner switch "ON".
4) Turn blower fan switch "ON".
5) Run engine at idle for a few minutes with air conditioner operating.
6) Make sure that radiator fan operates at low speed.
Without air conditioner
1) Start engine.
2) Keep engine speed at about 2,000 rpm until engine is warmed up sufficiently.
3) Make sure that radiator fan begins to operate at low speed during warm-up.
N.G.
Check radiator fan low speed control circuit.
(Go to PROCEDURE A.)
O.K.
B
CHECK RADIATOR FAN HIGH SPEED OPERATION.
1) Turn air conditioner switch "OFF".
2) Turn blower fan switch "OFF".
(Step 1) and 2) are only performed for model with air conditioner.)
3) Stop engine.
4) Disconnect engine temperature sensor harness connector.
5) Restart engine and make sure that radiator fan operates at high speed.
N.G.
Check radiator fan high speed control circuit.
(Go to PROCEDURE B.)
O.K.
INSPECTION END

ECCS DIAGNOSTIC CHART—1990 STANZA

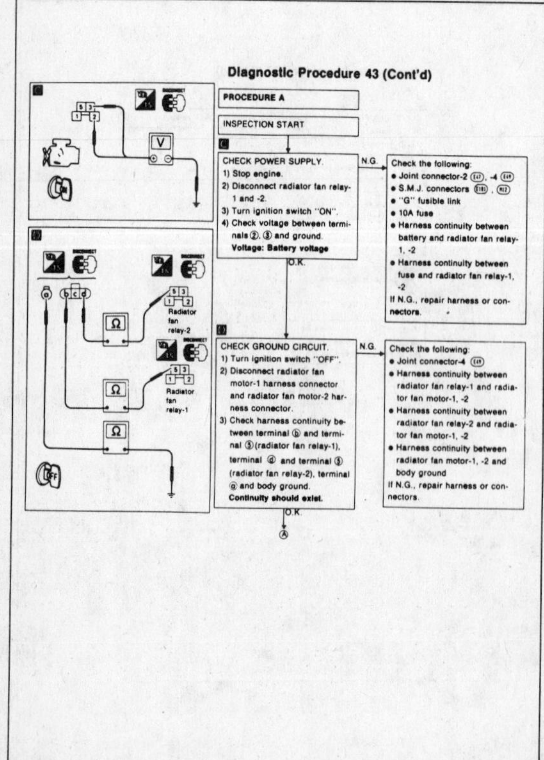
Diagnostic Procedure 43 (Cont'd)
PROCEDURE A
INSPECTION START
C
CHECK POWER SUPPLY.
1) Stop engine.
2) Disconnect radiator fan relay-1 and -2.
3) Turn ignition switch "ON".
4) Check voltage between terminals ②, ③ and ground.
Voltage: Battery voltage
N.G.
Check the following:
• Joint connector-2 (E47), -4 (E49)
• S.M.J. connectors (E101), (M12)
• "G" fusible link
• 10A fuse
• Harness continuity between battery and radiator fan relay-1, -2
• Harness continuity between fuse and radiator fan relay-1, -2
If N.G., repair harness or connectors.
O.K.
D
CHECK GROUND CIRCUIT.
1) Turn ignition switch "OFF".
2) Disconnect radiator fan motor-1 harness connector and radiator fan motor-2 harness connector.
3) Check harness continuity between terminal ⓑ and terminal ⑤(radiator fan relay-1), terminal ⓓ and terminal ⑤ (radiator fan relay-2), terminal ⓐ and body ground.
Continuity should exist.
N.G.
Check the following:
• Joint connector-4 (E49)
• Harness continuity between radiator fan relay-1 and radiator fan motor-1, -2
• Harness continuity between radiator fan relay-2 and radiator fan motor-1, -2
• Harness continuity between radiator fan motor-1, -2 and body ground
If N.G., repair harness or connectors.
O.K.
Ⓐ
Radiator fan relay-2
Radiator fan relay-1

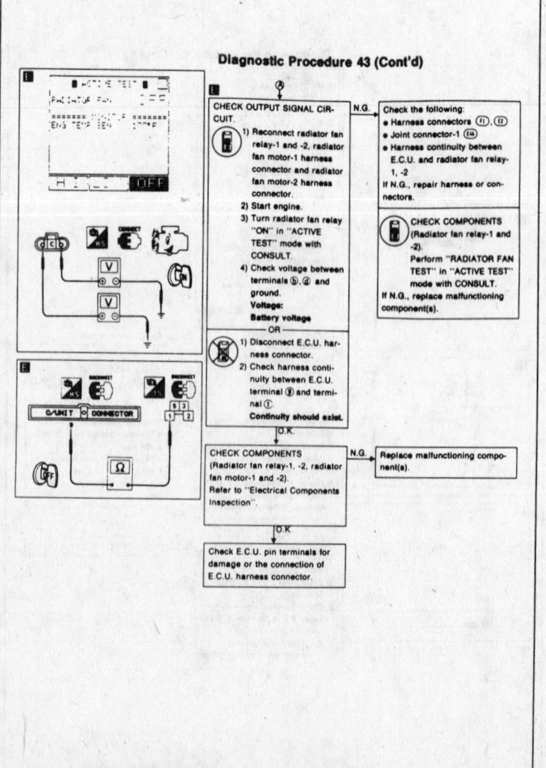
Diagnostic Procedure 43 (Cont'd)
Ⓐ
E
CHECK OUTPUT SIGNAL CIRCUIT.
1) Reconnect radiator fan relay-1 and -2, radiator fan motor-1 harness connector and radiator fan motor-2 harness connector.
2) Start engine.
3) Turn radiator fan relay "ON" in "ACTIVE TEST" mode with CONSULT.
4) Check voltage between terminals ⓑ, ⓓ and ground.
Voltage:
Battery voltage
OR
1) Disconnect E.C.U. harness connector.
2) Check harness continuity between E.C.U. terminal ⑨ and terminal ①.
Continuity should exist.
N.G.
Check the following:
• Harness connectors (F1), (E2)
• Joint connector-1 (E14)
• Harness continuity between E.C.U. and radiator fan relay-1, -2
If N.G., repair harness or connectors.
CHECK COMPONENTS
(Radiator fan relay-1 and -2).
Perform "RADIATOR FAN TEST" in "ACTIVE TEST" mode with CONSULT.
If N.G., replace malfunctioning component(s).
O.K.
CHECK COMPONENTS
(Radiator fan relay-1, -2, radiator fan motor-1 and -2).
Refer to "Electrical Components Inspection".
N.G.
Replace malfunctioning component(s).
O.K.
Check E.C.U. pin terminals for damage or the connection of E.C.U. harness connector.
C/UNIT CONNECTOR
OFF

ECCS DIAGNOSTIC CHART—1990 STANZA

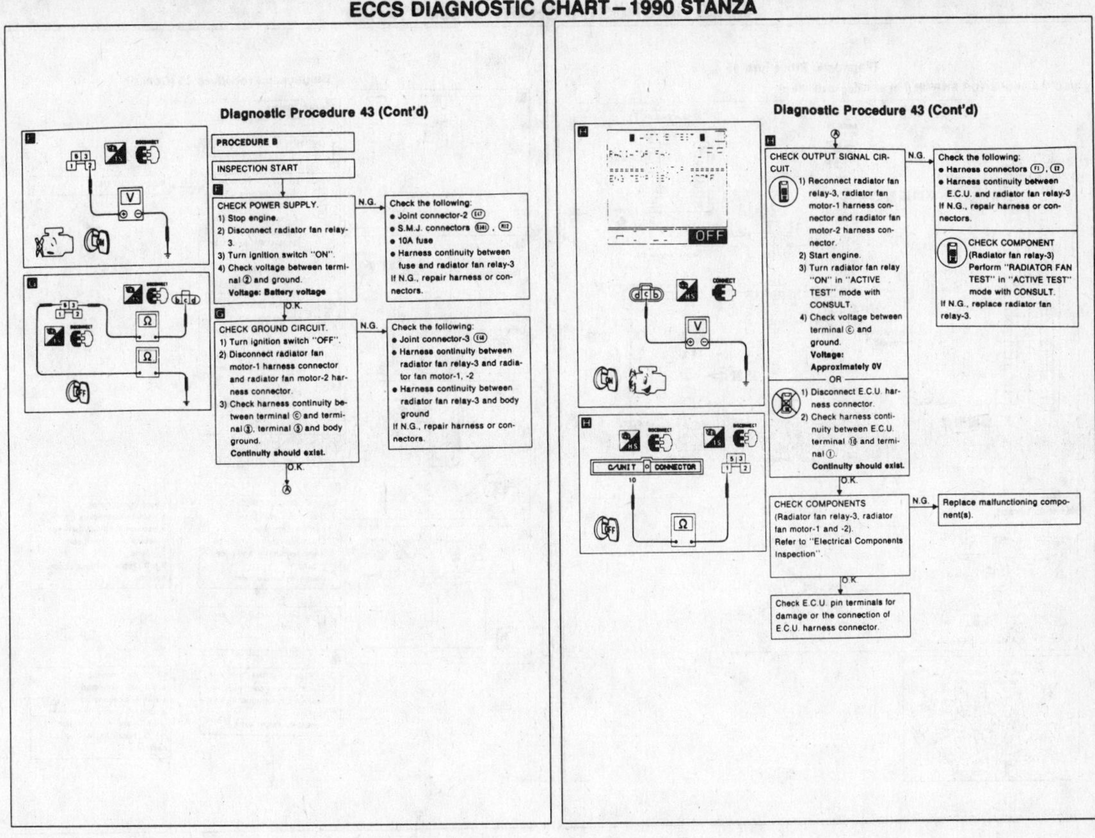
Diagnostic Procedure 43 (Cont'd)
PROCEDURE B
INSPECTION START
CHECK POWER SUPPLY.
1) Stop engine.
2) Disconnect radiator fan relay-3.
3) Turn ignition switch "ON".
4) Check voltage between terminal ② and ground.
Voltage: Battery voltage
N.G.
Check the following:
• Joint connector-2
• S.M.J. connectors
• 10A fuse
• Harness continuity between fuse and radiator fan relay-3
If N.G., repair harness or connectors.
O.K.
CHECK GROUND CIRCUIT.
1) Turn ignition switch "OFF".
2) Disconnect radiator fan motor-1 harness connector and radiator fan motor-2 harness connector.
3) Check harness continuity between terminal ⓒ and terminal ③, terminal ⑤ and body ground.
Continuity should exist.
N.G.
Check the following:
• Joint connector-3
• Harness continuity between radiator fan relay-3 and radiator fan motor-1, -2
• Harness continuity between radiator fan relay-3 and body ground
If N.G., repair harness or connectors.
O.K.
Diagnostic Procedure 43 (Cont'd)
OFF
CHECK OUTPUT SIGNAL CIRCUIT.
1) Reconnect radiator fan relay-3, radiator fan motor-1 harness connector and radiator fan motor-2 harness connector.
2) Start engine.
3) Turn radiator fan relay "ON" in "ACTIVE TEST" mode with CONSULT.
4) Check voltage between terminal ⓒ and ground.
Voltage: Approximately 0V
OR
1) Disconnect E.C.U. harness connector.
2) Check harness continuity between E.C.U. terminal ⑩ and terminal ①.
Continuity should exist.
N.G.
Check the following:
• Harness connectors
• Harness continuity between E.C.U. and radiator fan relay-3
If N.G., repair harness or connectors.
CHECK COMPONENT
(Radiator fan relay-3)
Perform "RADIATOR FAN TEST" in "ACTIVE TEST" mode with CONSULT.
If N.G., replace radiator fan relay-3.
C/UNIT CONNECTOR
O.K.
CHECK COMPONENTS
(Radiator fan relay-3, radiator fan motor-1 and -2).
Refer to "Electrical Components Inspection".
N.G.
Replace malfunctioning component(s).
O.K.
Check E.C.U. pin terminals for damage or the connection of E.C.U. harness connector.

ECCS DIAGNOSTIC CHART—1990 STANZA

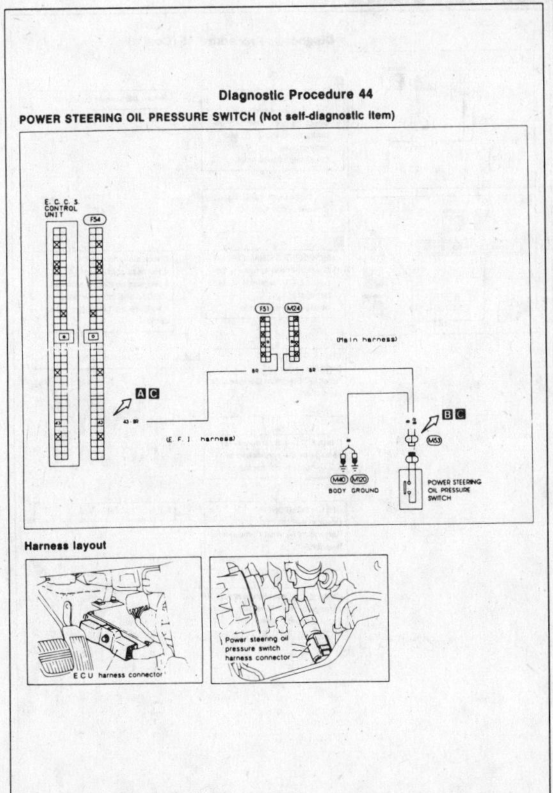
Diagnostic Procedure 44
POWER STEERING OIL PRESSURE SWITCH (Not self-diagnostic item)
E.C.C.S. CONTROL UNIT
F54
F51
M24
(Main harness)
(E.F.I. harness)
BODY GROUND
POWER STEERING OIL PRESSURE SWITCH
Harness layout
E.C.U. harness connector
Power steering oil pressure switch harness connector

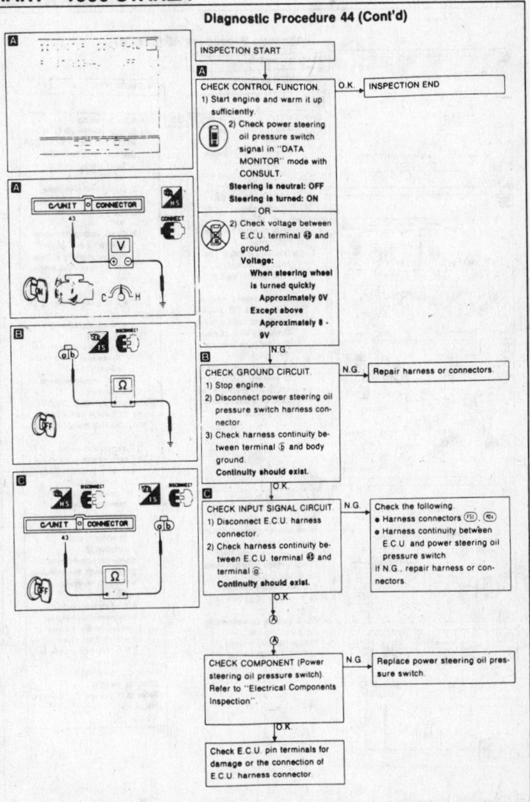
Diagnostic Procedure 44 (Cont'd)
INSPECTION START
CHECK CONTROL FUNCTION.
1) Start engine and warm it up sufficiently.
2) Check power steering oil pressure switch signal in "DATA MONITOR" mode with CONSULT.
Steering is neutral: OFF
Steering is turned: ON
OR
2) Check voltage between E.C.U. terminal ㊸ and ground.
Voltage:
When steering wheel is turned quickly
Approximately 0V
Except above
Approximately 8 - 9V
O.K.
INSPECTION END
C/UNIT CONNECTOR
N.G.
CHECK GROUND CIRCUIT.
1) Stop engine.
2) Disconnect power steering oil pressure switch harness connector.
3) Check harness continuity between terminal ⓑ and body ground.
Continuity should exist.
N.G.
Repair harness or connectors.
O.K.
CHECK INPUT SIGNAL CIRCUIT.
1) Disconnect E.C.U. harness connector.
2) Check harness continuity between E.C.U. terminal ㊸ and terminal ⓐ.
Continuity should exist.
N.G.
Check the following:
• Harness connectors
• Harness continuity between E.C.U. and power steering oil pressure switch
If N.G., repair harness or connectors.
O.K.
CHECK COMPONENT (Power steering oil pressure switch).
Refer to "Electrical Components Inspection".
N.G.
Replace power steering oil pressure switch.
O.K.
Check E.C.U. pin terminals for damage or the connection of E.C.U. harness connector.

ECCS DIAGNOSTIC CHART—1990 STANZA

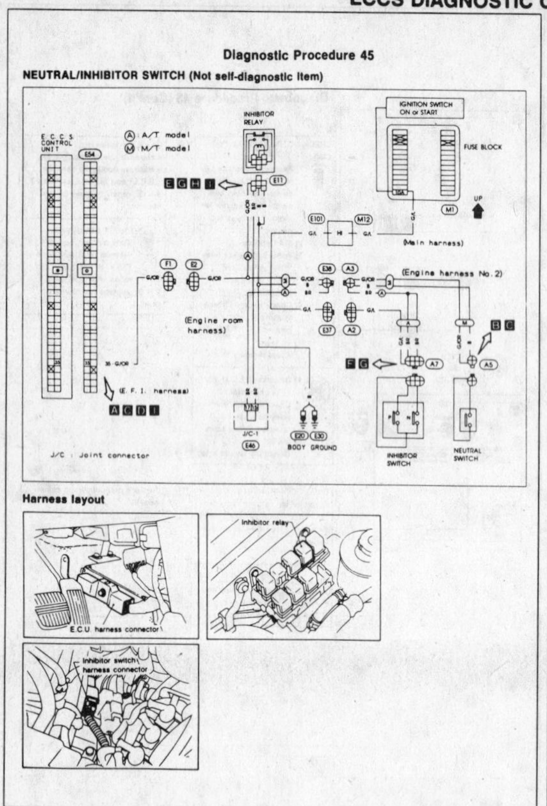
Diagnostic Procedure 45
NEUTRAL/INHIBITOR SWITCH (Not self-diagnostic item)
E.C.C.S. CONTROL UNIT
(A) A/T model
(M) M/T model
INHIBITOR RELAY
IGNITION SWITCH ON or START
FUSE BLOCK
(Main harness)
(Engine harness No. 2)
(Engine room harness)
(E.F.I. harness)
J/C-1
BODY GROUND
INHIBITOR SWITCH
NEUTRAL SWITCH
J/C : Joint connector
Harness layout
E.C.U. harness connector
Inhibitor relay
Inhibitor switch harness connector

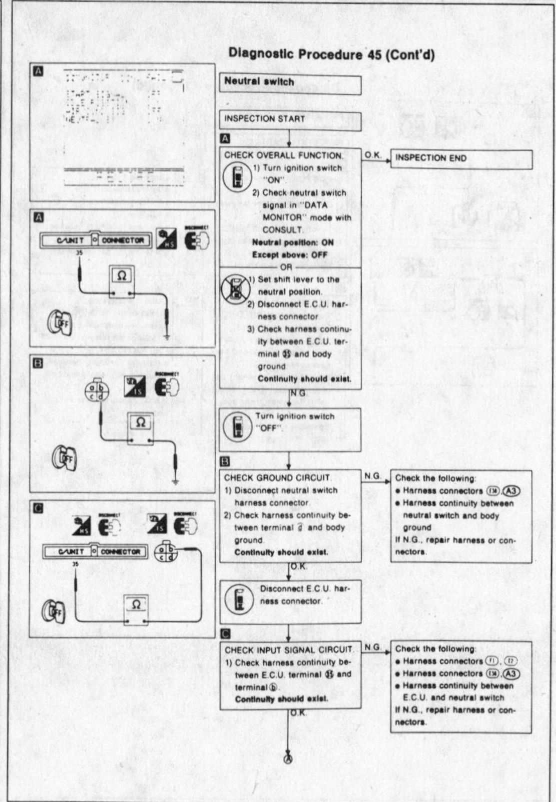
Diagnostic Procedure 45 (Cont'd)
Neutral switch
INSPECTION START
A
CHECK OVERALL FUNCTION.
1) Turn ignition switch "ON".
2) Check neutral switch signal in "DATA MONITOR" mode with CONSULT.
Neutral position: ON
Except above: OFF
OR
1) Set shift lever to the neutral position.
2) Disconnect E.C.U. harness connector.
3) Check harness continuity between E.C.U. terminal (35) and body ground.
Continuity should exist.
O.K.
INSPECTION END
N.G.
Turn ignition switch "OFF".
B
CHECK GROUND CIRCUIT.
1) Disconnect neutral switch harness connector.
2) Check harness continuity between terminal (d) and body ground.
Continuity should exist.
N.G.
Check the following:
• Harness connectors (I38), (A3)
• Harness continuity between neutral switch and body ground
If N.G., repair harness or connectors.
O.K.
Disconnect E.C.U. harness connector.
C
CHECK INPUT SIGNAL CIRCUIT.
1) Check harness continuity between E.C.U. terminal (35) and terminal (b).
Continuity should exist.
N.G.
Check the following:
• Harness connectors (F1), (I2)
• Harness connectors (I38), (A3)
• Harness continuity between E.C.U. and neutral switch
If N.G., repair harness or connectors.
O.K.
(A)
C/UNIT CONNECTOR
35

ECCS DIAGNOSTIC CHART—1990 STANZA

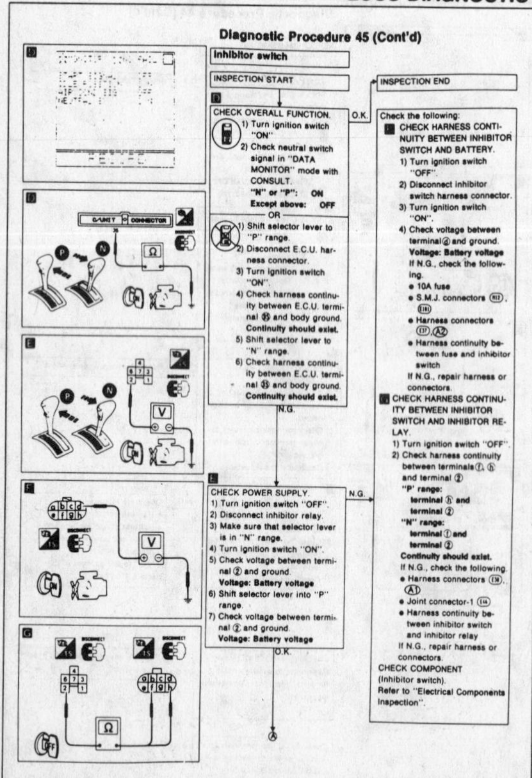
Diagnostic Procedure 45 (Cont'd)
Inhibitor switch
INSPECTION START
D
CHECK OVERALL FUNCTION.
1) Turn ignition switch "ON"
2) Check neutral switch signal in "DATA MONITOR" mode with CONSULT.
"N" or "P": ON
Except above: OFF
OR
1) Shift selector lever to "P" range.
2) Disconnect E.C.U. harness connector.
3) Turn ignition switch "ON".
4) Check harness continuity between E.C.U. terminal (35) and body ground.
Continuity should exist.
5) Shift selector lever to "N" range.
6) Check harness continuity between E.C.U. terminal (35) and body ground.
Continuity should exist.
O.K.
INSPECTION END
N.G.
E
CHECK POWER SUPPLY.
1) Turn ignition switch "OFF".
2) Disconnect inhibitor relay.
3) Make sure that selector lever is in "N" range.
4) Turn ignition switch "ON".
5) Check voltage between terminal (2) and ground.
Voltage: Battery voltage
6) Shift selector lever into "P" range.
7) Check voltage between terminal (2) and ground.
Voltage: Battery voltage
N.G.
Check the following:
F CHECK HARNESS CONTINUITY BETWEEN INHIBITOR SWITCH AND BATTERY.
1) Turn ignition switch "OFF".
2) Disconnect inhibitor switch harness connector.
3) Turn ignition switch "ON".
4) Check voltage between terminal (a) and ground.
Voltage: Battery voltage
If N.G., check the following.
• 10A fuse
• S.M.J. connectors (M12), (E101)
• Harness connectors (E37), (A2)
• Harness continuity between fuse and inhibitor switch
If N.G., repair harness or connectors.
G CHECK HARNESS CONTINUITY BETWEEN INHIBITOR SWITCH AND INHIBITOR RELAY.
1) Turn ignition switch "OFF".
2) Check harness continuity between terminals (f), (h) and terminal (2)
"P" range:
terminal (h) and terminal (2)
"N" range:
terminal (f) and terminal (2)
Continuity should exist.
If N.G., check the following.
• Harness connectors (E38), (A1)
• Joint connector-1 (E46)
• Harness continuity between inhibitor switch and inhibitor relay
If N.G., repair harness or connectors.
CHECK COMPONENT (Inhibitor switch).
Refer to "Electrical Components Inspection".
O.K.
(A)
C/UNIT CONNECTOR

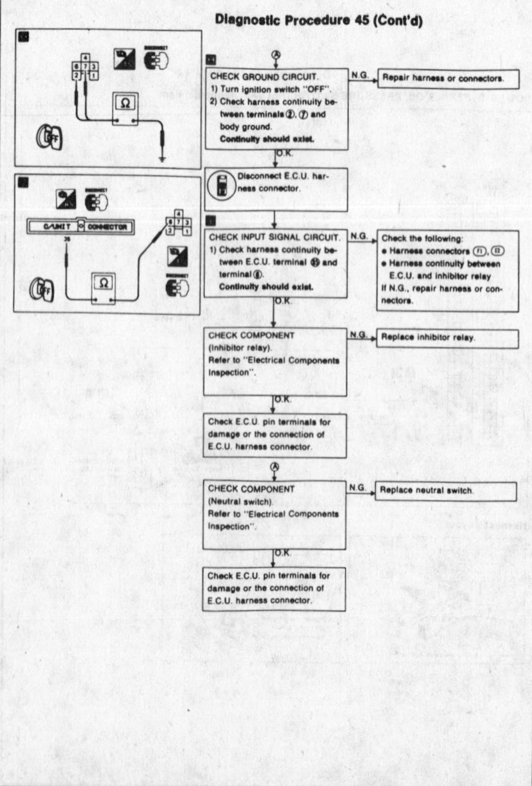
Diagnostic Procedure 45 (Cont'd)
(A)
H
CHECK GROUND CIRCUIT.
1) Turn ignition switch "OFF".
2) Check harness continuity between terminals (2), (7) and body ground.
Continuity should exist.
N.G.
Repair harness or connectors.
O.K.
Disconnect E.C.U. harness connector.
I
CHECK INPUT SIGNAL CIRCUIT.
1) Check harness continuity between E.C.U. terminal (35) and terminal (6).
Continuity should exist.
N.G.
Check the following:
• Harness connectors (F1), (I2)
• Harness continuity between E.C.U. and inhibitor relay
If N.G., repair harness or connectors.
O.K.
CHECK COMPONENT (Inhibitor relay).
Refer to "Electrical Components Inspection".
N.G.
Replace inhibitor relay.
O.K.
Check E.C.U. pin terminals for damage or the connection of E.C.U. harness connector.
(A)
CHECK COMPONENT (Neutral switch).
Refer to "Electrical Components Inspection".
N.G.
Replace neutral switch.
O.K.
Check E.C.U. pin terminals for damage or the connection of E.C.U. harness connector.
C/UNIT CONNECTOR

ECU SIGNAL INSPECTION—1990 STANZA

Electrical Components Inspection

E.C.U. INPUT/OUTPUT SIGNAL INSPECTION

E.C.U. Inspection table

*Date are reference values.

TER-MINAL NO.	ITEM	CONDITION	*DATA
1	Ignition signal	Engine is running. — Idle speed	0.3 - 0.6V
		Engine is running. — Engine speed is 2,000 rpm	Approximately 1.0V
3	Ignition check	Engine is running. — Idle speed	9 - 12V
4	E.C.C.S. relay (Main relay)	Engine is running. → Ignition switch "OFF" — Within approximately 1 second after turning ignition switch "OFF"	0 - 1V
		Ignition switch "OFF" — For approximately 1 second after turning ignition switch "OFF"	BATTERY VOLTAGE (11 - 14V)
8	Exhaust gas temperature sensor (Only for California model)	Engine is running. — Idle speed	1.0 - 2.0V
		Engine is running. — E.G.R. system is operating.	0 - 1.0V
11	Air conditioner relay	Engine is running. — Both A/C switch and blower switch are "ON".	0 - 1.0V
		Engine is running — A/C switch is "OFF".	BATTERY VOLTAGE (11 - 14V)
12	S.C.V. control solenoid valve	Engine is running. — Idle speed	0 - 1.0V
		Engine is running. — Engine speed is 2,000 rpm.	BATTERY VOLTAGE (11 - 14V)
16	Air flow meter	Engine is running.	1.0 - 3.0V Output voltage varies with engine revolution.

Electrical Components Inspection (Cont'd)

*Date are reference values.

TER-MINAL NO.	ITEM	CONDITION	*DATA
18	Engine temperature sensor	Engine is running.	1.0 - 5.0V Output voltage varies with engine water temperature.
19	Exhaust gas sensor	Engine is running. — After warming up sufficiently.	0 - Approximately 1.0V
20	Throttle sensor	Ignition switch "ON"	0.4 - Approximately 4V Output voltage varies with the throttle valve opening angle.
22 30	Crank angle sensor (Reference signal)	Engine is running. **Do not run engine at high speed under no-load.**	0.2 - 0.5V
26	Air temperature sensor	Ignition switch "ON" — Temperature of intake air is 20°C (68°F)	Approximately 1.0 - 1.5V
		Ignition switch "ON" — Temperature of intake air is 80°C (176°F)	Approximately 0.3V
31 40	Crank angle sensor (Position signal)	Engine is running. **Do not run engine at high speed under no-load.**	2.0 - 3.0V
33	Idle switch (⊖side)	Ignition switch "ON" — Throttle valve: idle posion	Approximately 9 - 10V
		Ignition switch "ON" — Throttle valve: Any position except idle position	0V
34	Start signal	Cranking	8 - 12V
35	Neutral switch & Inhibitor switch	Ignition switch "ON" — Neutral/Parking	0V
		Ignition switch "ON" — Except the above gear position	Approximately 6V
36	Ignition switch	Ignition switch "OFF"	0V
		Ignition switch "ON"	BATTERY VOLTAGE (11 - 14V)
37	Throttle sensor power supply	Ignition switch "ON"	Approximately 5V
38 47	Power supply for E.C.U.	Ignition switch "ON"	BATTERY VOLTAGE (11 - 14V)

ECU SIGNAL INSPECTION (CONT.)—1990 STANZA

Electrical Components Inspection (Cont'd)

*Date are reference values.

TER-MINAL NO.	ITEM	CONDITION	*DATA
41	Air conditioner switch	Engine is running. — Both air conditioner switch and blower switch are "ON".	0V
		Engine is running. — Air conditioner switch is "OFF".	BATTERY VOLTAGE (11 - 14V)
43	Power steering oil pressure switch	Engine is running. — Steering wheel is being turned.	0V
		Engine is running. — Steering wheel is not being turned.	8 - 9V
44	Idle switch (⊕ side)	Ignition switch "ON" — Throttle valve: Idle position	Approximately 9 - 10V
		Ignition switch "ON" — Throttle valve: Except idle position	BATTERY VOLTAGE (11 - 14V)
45	Thermo control amp.	Engine is running Air conditioner is operating. — Evaporator outlet air temperature is between 3.0 - 8.0°C (37 - 46°F)	Approximately 8 - 9V
		Engine is running. Air conditioner is operating. — Evaporator outlet air temperature is over 8°C (46°F)	Approximately. 0V
46	Power supply (Back-up)	Ignition switch "OFF"	BATTERY VOLTAGE (11 - 14V)
101	Injector No. 1	Engine is running	BATTERY VOLTAGE (11 - 14V)
103	Injector No. 3		
110	Injector No. 2		
112	Injector No. 4		
102	A.I.V. control solenoid valve	Engine is running. — Idle speed	0 - 1.0V
		Engine is running. — Accelerator pedal is depressed. After warning up	BATTERY VOLTAGE (11 - 14V)

Electrical Components Inspection (Cont'd)

*Date are reference values.

TER-MINAL NO.	ITEM	CONDITION	*DATA
104	Fuel pump relay	Ignition switch "ON" — For 5 seconds after turning ignition switch "ON" Engine is running.	0.7 - 0.9V
		Ignition switch "ON" — Within 5 seconds after turning ignition switch "ON"	BATTERY VOLTAGE (11 - 14V)
105	E.G.R. control solenoid valve	Engine is running. — Engine is cold. [Water temperature is below 60°C (140°F).]	0.7 - 0.9V
		Engine is running. — After warning up [Water temperature is between 60°C (140°F) and 105°C (221°F)]	BATTERY VOLTAGE (11 - 14V)
106	Pressure regulator control solenoid valve	Stop and restart engine after warming it up. — Water temperature is above 90°C (194°F)	0 - 1.0V (for 3 minutes after ignition switch is turned off.) / BATTERY VOLTAGE (After 3 minutes)
		Stop and restart engine after warming it up. — Water temperature is below 90°C (194°F)	BATTERY VOLTAGE (11 - 14V)
113	A.A.C. valve	Engine is running. — Idle speed	7 - 10V
		Engine is running. — Steering wheel is being turned. — Air conditioner is operating — Rear defogger is "ON". — Headlamp are in high position.	4 - 7V

E.C.U. HARNESS CONNECTOR TERMINAL LAYOUT

101 102 103 104 105 106 107 108	1 2 3 4 5 6 7	O	15 16 17 18 19 20 21 22	31 32 33 34 35 36 37 38 39
109 110 111 112 113 114 115 116	8 9 10 11 12 13 14		23 24 25 26 27 28 29 30	40 41 42 43 44 45 46 47 48

H.S. CONNECT

ECU SIGNAL INSPECTION (CONT.) – 1990 STANZA

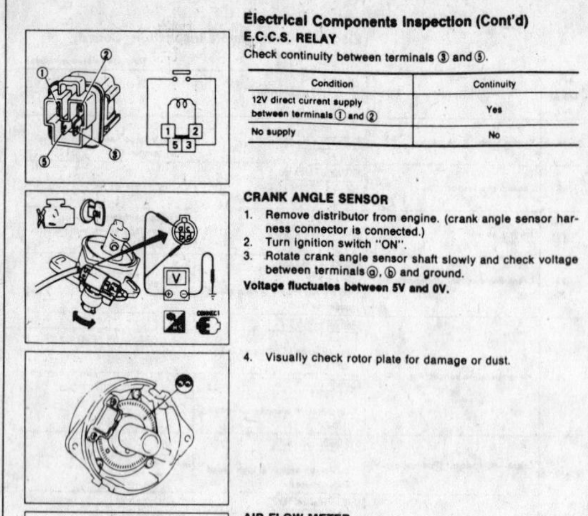

Electrical Components Inspection (Cont'd)

E.C.C.S. RELAY

Check continuity between terminals ③ and ⑤.

Condition	Continuity
12V direct current supply between terminals ① and ②	Yes
No supply	No

CRANK ANGLE SENSOR

1. Remove distributor from engine. (crank angle sensor harness connector is connected.)
2. Turn ignition switch "ON".
3. Rotate crank angle sensor shaft slowly and check voltage between terminals ⓐ, ⓑ and ground.

Voltage fluctuates between 5V and 0V.

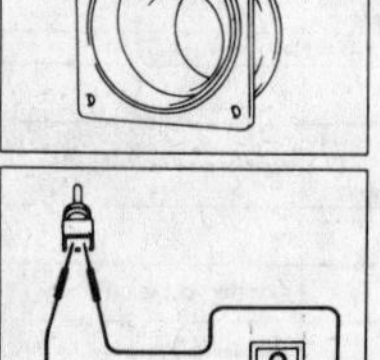

4. Visually check rotor plate for damage or dust.

AIR FLOW METER

- Visually check hot wire air passage for dust.

ENGINE TEMPERATURE SENSOR

Check engine temperature sensor resistance.

Temperature °C (°F)	Resistance kΩ
20 (68)	2.1 - 2.9
80 (176)	0.30 - 0.33

ECCS COMPONENTS LOCATION – 1990 AXXESS

E.C.C.S. Component Parts Location

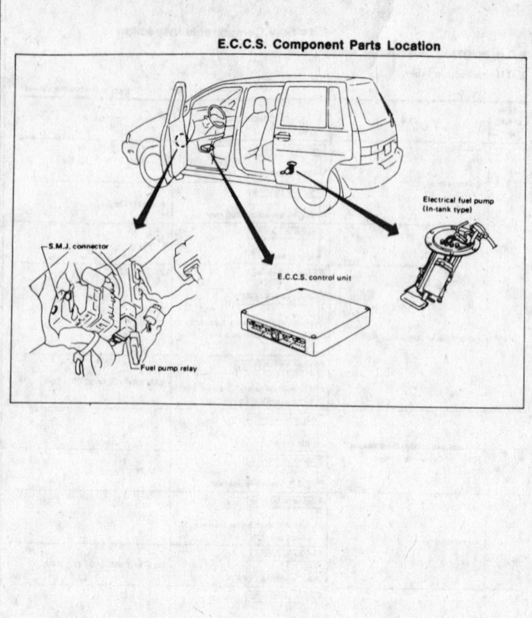
Electrical fuel pump
(In-tank type)
S.M.J. connector
E.C.C.S. control unit
Fuel pump relay

ECCS COMPONENTS LOCATION – 1990 AXXESS

E.C.C.S. Component Parts Location (Cont'd)

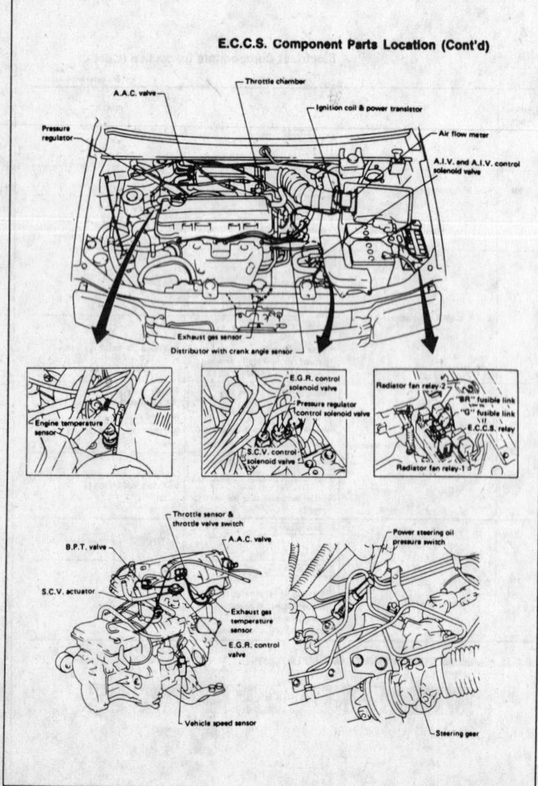
A.A.C. valve
Throttle chamber
Ignition coil & power transistor
Pressure regulator
Air flow meter
A.I.V. and A.I.V. control solenoid valve
Exhaust gas sensor
Distributor with crank angle sensor
Engine temperature sensor
E.G.R. control solenoid valve
Pressure regulator control solenoid valve
S.C.V. control solenoid valve
Radiator fan relay-2
"BR" fusible link
"G" fusible link
E.C.C.S. relay
Radiator fan relay-1
Throttle sensor & throttle valve switch
A.A.C. valve
B.P.T. valve
S.C.V. actuator
Exhaust gas temperature sensor
E.G.R. control valve
Vehicle speed sensor
Power steering oil pressure switch
Steering gear

ECCS SCHEMATIC – 1990 AXXESS

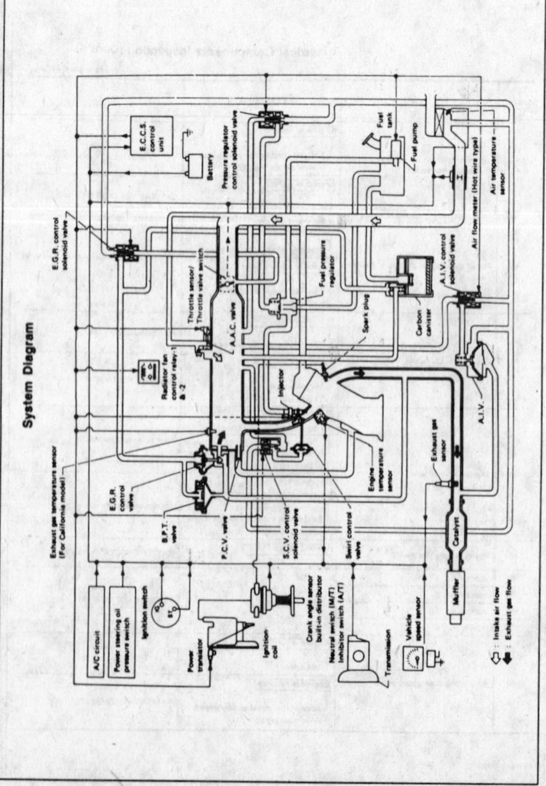
System Diagram
E.C.C.S. control unit
Battery
Fuel tank
Fuel pump
Air temperature sensor
Air flow meter (Hot wire type)
A.I.V. control solenoid valve
Carbon canister
Spark plug
Fuel pressure regulator
A.A.C. valve
Injector
E.G.R. control valve
P.C.V. valve
S.C.V. control solenoid valve
Swirl control valve
Engine temperature sensor
Exhaust gas sensor
A.I.V.
Catalyst
Muffler
Ignition coil
Ignition switch
Power transistor
Transmission
Vehicle speed sensor
A/C circuit
Intake air flow
Exhaust gas flow

ECCS CONTROL SYSTEM CHART—1990 AXXESS

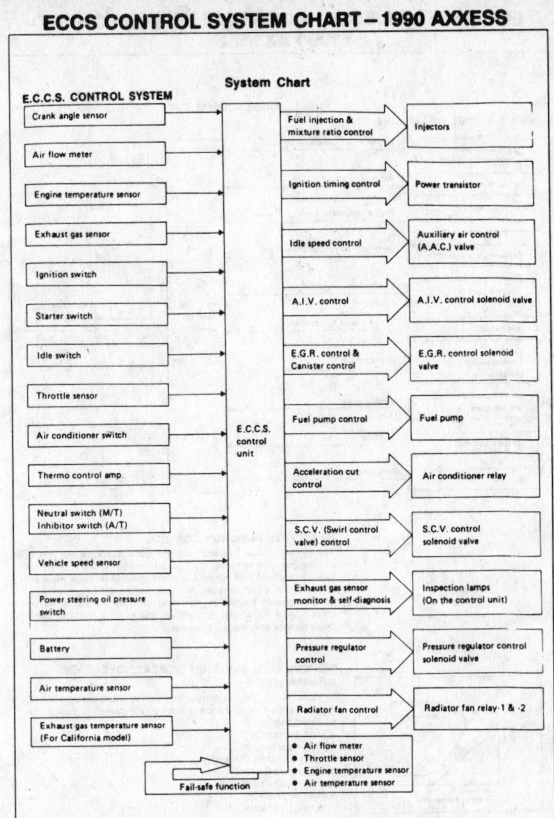
System Chart
E.C.C.S. CONTROL SYSTEM
Crank angle sensor
Air flow meter
Engine temperature sensor
Exhaust gas sensor
Ignition switch
Starter switch
Idle switch
Throttle sensor
Air conditioner switch
Thermo control amp.
Neutral switch (M/T)
Inhibitor switch (A/T)
Vehicle speed sensor
Power steering oil pressure switch
Battery
Air temperature sensor
Exhaust gas temperature sensor (For California model)
E.C.C.S. control unit
Fuel injection & mixture ratio control
Injectors
Ignition timing control
Power transistor
Idle speed control
Auxiliary air control (A.A.C.) valve
A.I.V. control
A.I.V. control solenoid valve
E.G.R. control & Canister control
E.G.R. control solenoid valve
Fuel pump control
Fuel pump
Acceleration cut control
Air conditioner relay
S.C.V. (Swirl control valve) control
S.C.V. control solenoid valve
Exhaust gas sensor monitor & self-diagnosis
Inspection lamps (On the control unit)
Pressure regulator control
Pressure regulator control solenoid valve
Radiator fan control
Radiator fan relay-1 & -2
Air flow meter
Throttle sensor
Engine temperature sensor
Air temperature sensor
Fail-safe function

ECCS VACUUM HOSE ROUTING—1990 AXXESS

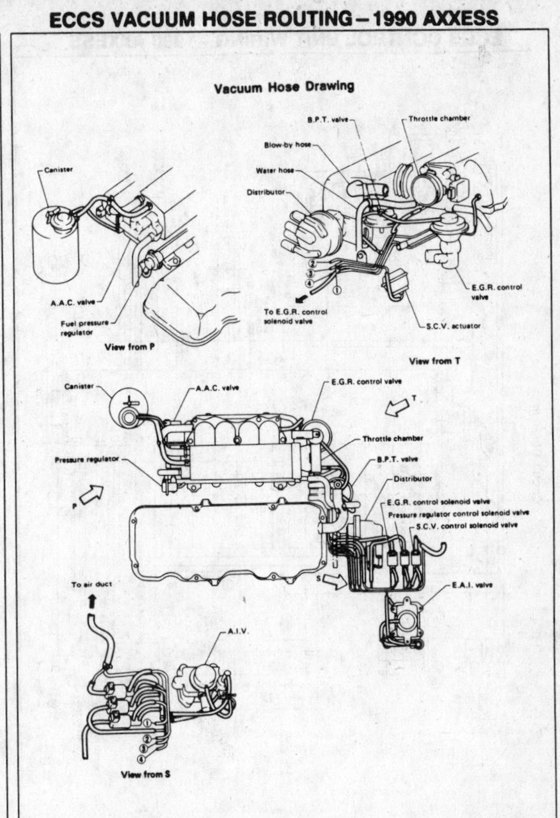
Vacuum Hose Drawing
B.P.T. valve
Throttle chamber
Blow-by hose
Water hose
Distributor
Canister
A.A.C. valve
Fuel pressure regulator
To E.G.R. control solenoid valve
E.G.R. control valve
S.C.V. actuator
View from P
View from T
E.G.R. control valve
Throttle chamber
Pressure regulator
B.P.T. valve
Distributor
E.G.R. control solenoid valve
Pressure regulator control solenoid valve
S.C.V. control solenoid valve
E.A.I. valve
To air duct
A.I.V.
View from S

ECCS WIRING SCHEMATIC—1900 AXXESS

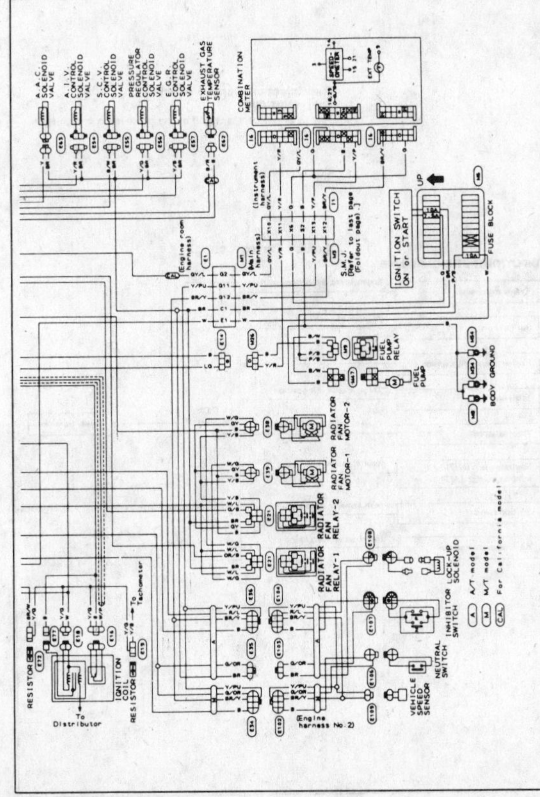
COMBINATION METER
FUSE BLOCK
IGNITION SWITCH ON or START
FUEL PUMP RELAY
FUEL PUMP
BODY GROUND
RADIATOR FAN MOTOR-1
RADIATOR FAN MOTOR-2
RADIATOR FAN RELAY-2
RADIATOR FAN RELAY-1
LOCK-UP SOLENOID
INHIBITOR SWITCH
NEUTRAL SWITCH
VEHICLE SPEED SENSOR
RESISTOR
IGNITION COIL
To Distributor
To Tachometer

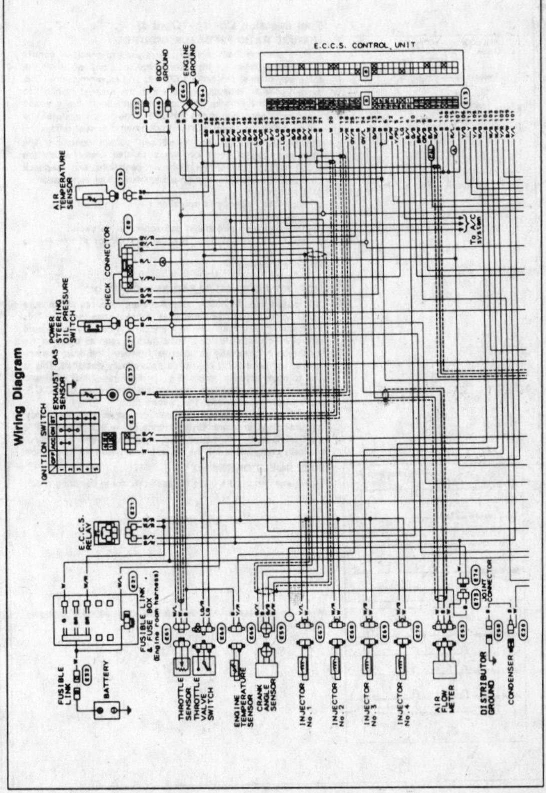
Wiring Diagram
E.C.C.S. CONTROL UNIT
AIR TEMPERATURE SENSOR
CHECK CONNECTOR
POWER STEERING OIL PRESSURE SWITCH
EXHAUST GAS SENSOR
IGNITION SWITCH
E.C.C.S. RELAY
FUSIBLE LINK
BATTERY
THROTTLE SENSOR
ENGINE TEMPERATURE SENSOR
CRANK ANGLE SENSOR
INJECTOR No.1
INJECTOR No.2
INJECTOR No.3
INJECTOR No.4
AIR FLOW METER
DISTRIBUTOR GROUND
CONDENSER

ECCS CONTROL UNIT WIRING—1990 AXXESS

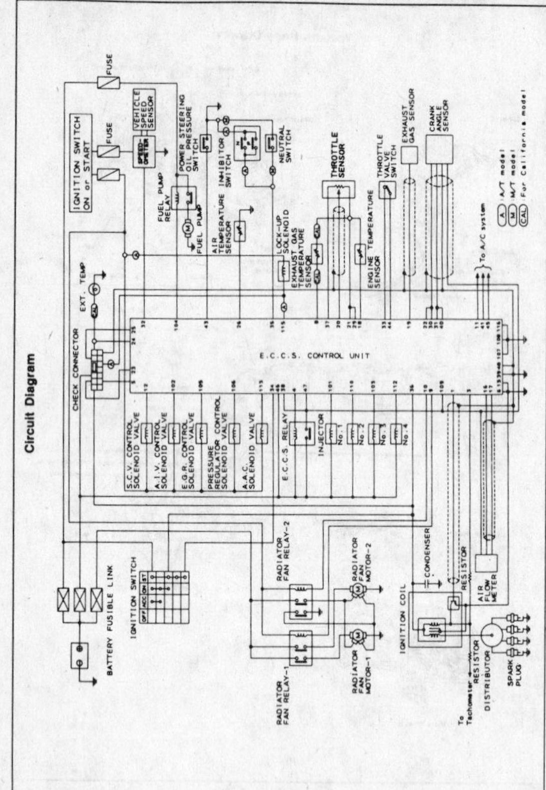
Circuit Diagram
E.C.C.S. CONTROL UNIT
BATTERY FUSIBLE LINK
IGNITION SWITCH
RADIATOR FAN RELAY-1
RADIATOR FAN RELAY-2
RADIATOR FAN MOTOR-1
RADIATOR FAN MOTOR-2
IGNITION COIL
CONDENSER
RESISTOR
AIR FLOW METER
DISTRIBUTOR
SPARK PLUG
THROTTLE SENSOR
CRANK ANGLE SENSOR
EXHAUST GAS SENSOR
CHECK CONNECTOR
FUEL PUMP
FUEL PUMP RELAY
FUSE
A/T model
M/T model
For California model

ECCS FUEL INJECTION CONTROL SYSTEM CHART—1990 AXXESS

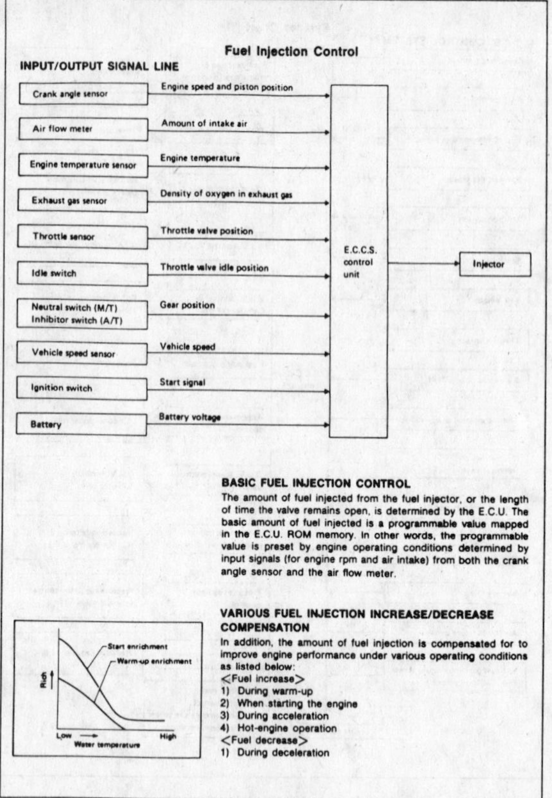
Fuel Injection Control
INPUT/OUTPUT SIGNAL LINE
Crank angle sensor
Engine speed and piston position
Air flow meter
Amount of intake air
Engine temperature sensor
Engine temperature
Exhaust gas sensor
Density of oxygen in exhaust gas
Throttle sensor
Throttle valve position
Idle switch
Throttle valve idle position
Neutral switch (M/T)
Inhibitor switch (A/T)
Gear position
Vehicle speed sensor
Vehicle speed
Ignition switch
Start signal
Battery
Battery voltage
E.C.C.S. control unit
Injector
Start enrichment
Warm-up enrichment
Rich
Low
High
Water temperature

BASIC FUEL INJECTION CONTROL

The amount of fuel injected from the fuel injector, or the length of time the valve remains open, is determined by the E.C.U. The basic amount of fuel injected is a programmable value mapped in the E.C.U. ROM memory. In other words, the programmable value is preset by engine operating conditions determined by input signals (for engine rpm and air intake) from both the crank angle sensor and the air flow meter.

VARIOUS FUEL INJECTION INCREASE/DECREASE COMPENSATION

In addition, the amount of fuel injection is compensated for to improve engine performance under various operating conditions as listed below:

<Fuel increase>
1) During warm-up
2) When starting the engine
3) During acceleration
4) Hot-engine operation

<Fuel decrease>
1) During deceleration

ECCS FUEL INJECTION CONTROL SYSTEM CHART (CONT.)—1990 AXXESS

CLOSED LOOP CONTROL — E.C.C.S. control unit; Feedback signal; Injection pulse; Exhaust gas sensor; Injector; Combustion; Fuel injection; Engine

Fuel Injection Control (Cont'd)

MIXTURE RATIO FEEDBACK CONTROL

Mixture ratio feedback system is designed to precisely control the mixture ratio to the stoichiometric point so that the three-way catalyst can reduce CO, HC and NOx emissions. This system uses an exhaust gas sensor in the exhaust manifold to check the air-fuel ratio. The control unit adjusts the injection pulse width according to the sensor voltage so the mixture ratio will be within the range of the stoichiometric air-fuel ratio.

This stage refers to the closed-loop control condition. The open-loop control condition refers to that under which the E.C.U. detects any of the following conditions and feedback control stops in order to maintain stabilized fuel combustion.

1) Deceleration
2) High-load, high-speed operation
3) Engine idling
4) Malfunctioning of exhaust gas sensor or its circuit
5) Insufficient activation of exhaust gas sensor at low engine temperature
6) Engine starting

MIXTURE RATIO SELF-LEARNING CONTROL

The mixture ratio feedback control system monitors the mixture ratio signal transmitted from the exhaust gas sensor. This feedback signal is then sent to the E.C.U. to control the amount of fuel injection to provide a basic mixture ratio as close to the theoretical mixture ratio as possible. However, the basic mixture ratio is not necessarily controlled as originally designed. This is due to manufacturing errors (e.g., air flow meter hot wire) and changes during operation (injector clogging, etc.) of E.C.C.S. parts which directly affect the mixture ratio.

Accordingly, a difference between the basic and theoretical mixture ratios is quantitatively monitored in this system. It is then computed in terms of "fuel injection duration" to automatically compensate for the difference between the two ratios.

FUEL INJECTION TIMING

Fuel is injected once a cycle for each cylinder in the firing order.

Injection pulse — No. 1 cylinder; No. 2 cylinder; No. 3 cylinder; No. 4 cylinder; 1 engine cycle

When engine starts, fuel is injected into all four cylinders simultaneously twice a cycle.

No. 1 cylinder; No. 2 cylinder; No. 3 cylinder; No. 4 cylinder; 1 engine cycle

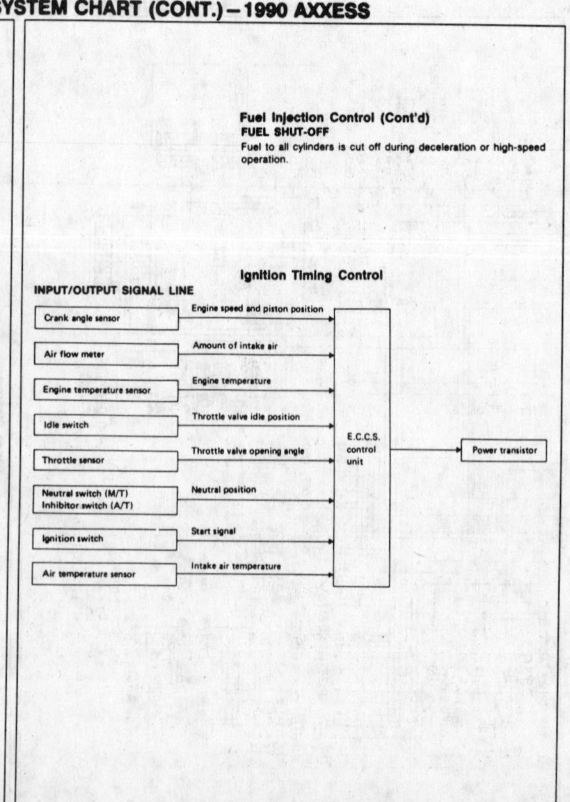
Ignition Timing Control
INPUT/OUTPUT SIGNAL LINE
Crank angle sensor
Engine speed and piston position
Air flow meter
Amount of intake air
Engine temperature sensor
Engine temperature
Idle switch
Throttle valve idle position
Throttle sensor
Throttle valve opening angle
Neutral switch (M/T)
Inhibitor switch (A/T)
Neutral position
Ignition switch
Start signal
Air temperature sensor
Intake air temperature
E.C.C.S. control unit
Power transistor

Fuel Injection Control (Cont'd)

FUEL SHUT-OFF

Fuel to all cylinders is cut off during deceleration or high-speed operation.

ECCS IGNITION TIMING CONTROL DESCRIPTION—1990 AXXESS

Ignition Timing Control (Cont'd)

SYSTEM DESCRIPTION

The ignition timing is controlled by the E.C.U. in order to maintain the best air-fuel ratio in response to every running condition of the engine.

The ignition timing data is stored in the ROM located in the E.C.U., in the form of the map shown below.

The E.C.U. detects information such as the injection pulse width and crank angle sensor signal which varies every moment. Then responding to this information, ignition signals are transmitted to the power transistor.

e.g N: 1,800 rpm, Tp: 1.50 msec
A °B.T.D.C.

In addition to this,

1 At starting
2 During warm-up
3 At idle
4 At low battery voltage
5 During swirl control valve operates
6 During hot engine operation
7 At acceleration
8 When intake air temperature is extremely high

the ignition timing is revised by the E.C.U. according to the other data stored in the ROM.

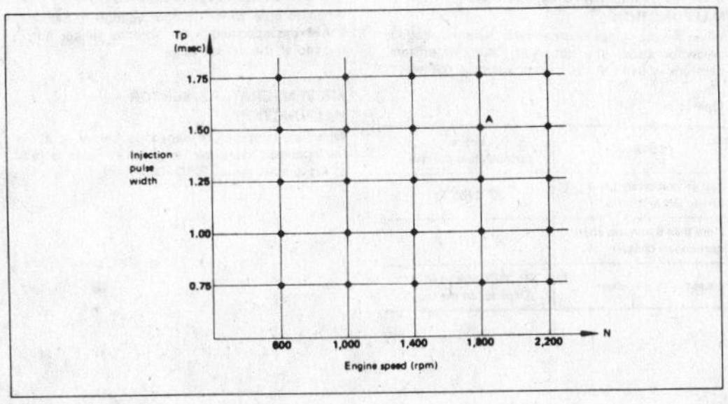

ECCS IDLE SPEED CONTROL DESCRIPTION—1990 AXXESS

Idle Speed Control

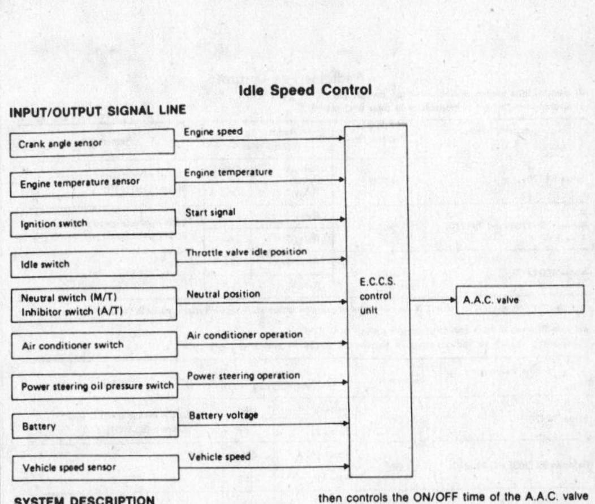
INPUT/OUTPUT SIGNAL LINE
Crank angle sensor
Engine speed
Engine temperature sensor
Engine temperature
Ignition switch
Start signal
Idle switch
Throttle valve idle position
Neutral switch (M/T)
Inhibitor switch (A/T)
Neutral position
Air conditioner switch
Air conditioner operation
Power steering oil pressure switch
Power steering operation
Battery
Battery voltage
Vehicle speed sensor
Vehicle speed
E.C.C.S. control unit
A.A.C. valve

SYSTEM DESCRIPTION

This system automatically controls engine idle speed to a specified level. Idle speed is controlled through fine adjustment of the amount of air which by-passes the throttle valve via A.A.C. valve. The A.A.C. valve repeats ON/OFF operation according to the signal sent from the E.C.U. The crank angle sensor detects the actual engine speed and sends a signal to the E.C.U. The E.C.U. then controls the ON/OFF time of the A.A.C. valve so that engine speed coincides with the target value memorized in ROM. The target engine speed is the lowest speed at which the engine can operate steadily. The optimum value stored in the ROM is determined by taking into consideration various engine conditions, such as noise and vibration transmitted to the compartment, fuel consumption, and engine load.

ECCS FUEL PUMP AND AIR INDUCTION VALVE CONTROL DESCRIPTION—1990 AXXESS

Fuel Pump Control

INPUT/OUTPUT SIGNAL LINE

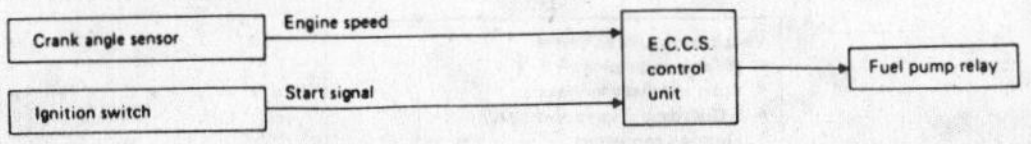

Crank angle sensor → Engine speed → E.C.C.S. control unit
Ignition switch → Start signal → E.C.C.S. control unit
E.C.C.S. control unit → Fuel pump relay

SYSTEM DESCRIPTION

The E.C.U. activates the fuel pump for several seconds after the ignition switch is turned on to improve engine startability. If the E.C.U. receives a 1° signal from the crank angle sensor, it knows that the engine is rotating, and causes the pump to perform. If the 1° signal is not received when the ignition switch is on, the engine stalls. The E.C.U. stops pump operation and prevents battery discharging, thereby improving safety. The E.C.U. does not directly drive the fuel pump. It controls the ON/OFF fuel pump relay, which in turn controls the fuel pump.

Condition	Fuel pump operation
Ignition switch is tunred to ON.	Operates for 5 seconds
Engine running and cranking	Operates
When engine is stopped	Stops in 1 second
Except as shown above	Stops

Air Induction Valve (A.I.V.) Control

INPUT/OUTPUT SIGNAL LINE

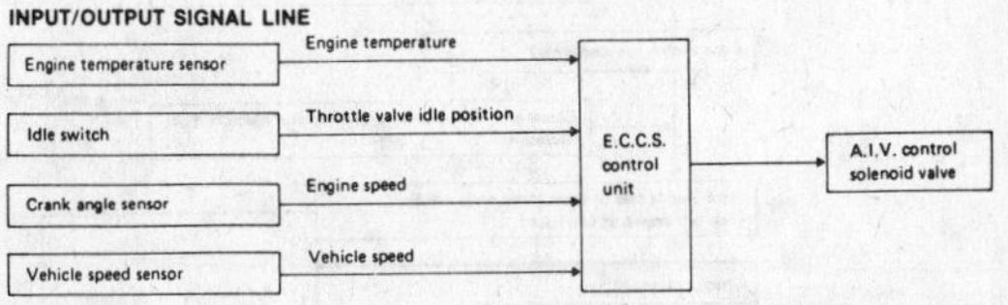

Engine temperature sensor → Engine temperature → E.C.C.S. control unit
Idle switch → Throttle valve idle position → E.C.C.S. control unit
Crank angle sensor → Engine speed → E.C.C.S. control unit
Vehicle speed sensor → Vehicle speed → E.C.C.S. control unit
E.C.C.S. control unit → A.I.V. control solenoid valve

SYSTEM DESCRIPTION

The air induction system is designed to send secondary air to the exhaust manifold, utilizing the vacuum caused by exhaust pulsation in the exhaust manifold.

The exhaust pressure in the exhaust manifold usually pulsates in response to the opening and closing of the exhaust valve and decreases below atmospheric pressure periodically.

If a secondary air intake pipe is opened to the atmosphere under vacuum conditions, secondary air can be drawn into the exhaust manifold in proportion to the vacuum.

The air induction valve is controlled by the E.C.C.S. control unit, corresponding to the engine temperature. When the engine is cold, the A.I.V. control system operates to reduce HC and CO.

In extremely cold conditions, A.I.V. control system does not operate to reduce after-burning. This system also operates during deceleration for the purpose of blowing off water around the air induction valve.

Engine condition	Water temperature °C (°F)	A.I.V. control solenoid valve	A.I.V. control system
Idle or deceleration	Between 28 (82) and 115 (239)	ON	Operates

ECCS FUEL PRESSURE REGULATOR AND SWIRL CONTROL VALVE DESCRIPTION—1990 AXXESS

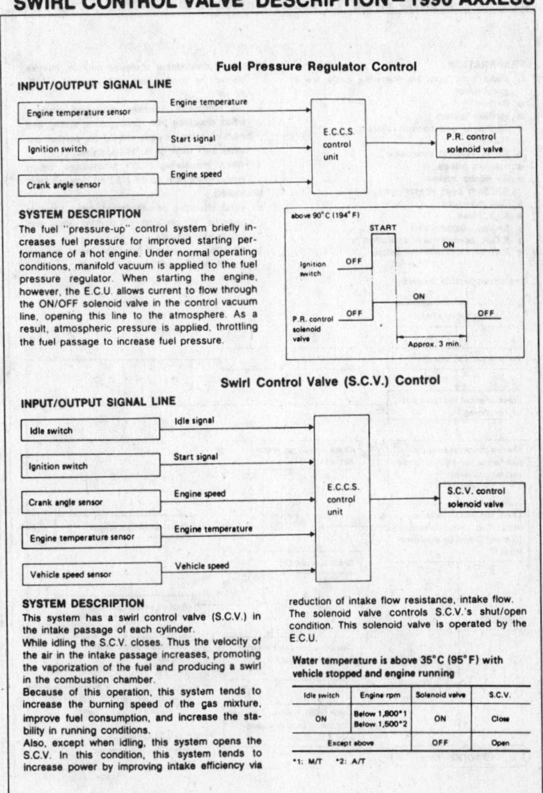
INPUT/OUTPUT SIGNAL LINE
Engine temperature sensor
Engine temperature
Ignition switch
Start signal
Crank angle sensor
Engine speed
E.C.C.S. control unit
P.R. control solenoid valve
above 90°C (194°F)
START
ON
OFF
Ignition switch
P.R. control solenoid valve
Approx. 3 min.
INPUT/OUTPUT SIGNAL LINE
Idle switch
Idle signal
Ignition switch
Start signal
Crank angle sensor
Engine speed
Engine temperature sensor
Engine temperature
Vehicle speed sensor
Vehicle speed
E.C.C.S. control unit
S.C.V. control solenoid valve

Fuel Pressure Regulator Control

SYSTEM DESCRIPTION

The fuel "pressure-up" control system briefly increases fuel pressure for improved starting performance of a hot engine. Under normal operating conditions, manifold vacuum is applied to the fuel pressure regulator. When starting the engine, however, the E.C.U. allows current to flow through the ON/OFF solenoid valve in the control vacuum line, opening this line to the atmosphere. As a result, atmospheric pressure is applied, throttling the fuel passage to increase fuel pressure.

Swirl Control Valve (S.C.V.) Control

SYSTEM DESCRIPTION

This system has a swirl control valve (S.C.V.) in the intake passage of each cylinder.

While idling the S.C.V. closes. Thus the velocity of the air in the intake passage increases, promoting the vaporization of the fuel and producing a swirl in the combustion chamber.

Because of this operation, this system tends to increase the burning speed of the gas mixture, improve fuel consumption, and increase the stability in running conditions.

Also, except when idling, this system opens the S.C.V. In this condition, this system tends to increase power by improving intake efficiency via reduction of intake flow resistance, intake flow.

The solenoid valve controls S.C.V.'s shut/open condition. This solenoid valve is operated by the E.C.U.

Water temperature is above 35°C (95°F) with vehicle stopped and engine running

Idle switch	Engine rpm	Solenoid valve	S.C.V.
ON	Below 1,800*1 Below 1,500*2	ON	Close
Except above		OFF	Open

*1: M/T *2: A/T

ECCS RADIATOR FAN AND FAIL SAFE SYSTEM INSPECTION—1990 AXXESS

Radiator Fan Control

Air conditioner switch and blower fan switch are "ON"
[Evaporator outlet air temperature is over 8°C (46°F).]

Engine temperature °C (°F)	Radiator fan	Radiator fan control relay		Remarks
Below 94 (201)	*HIGH	Relay-1	ON	*OFF if vehicle speed is above 80 km/h (50 MPH).
		Relay-2	ON	
Between 95 (203) and 99 (210)	*HIGH	Relay-1	ON	*LOW if vehicle speed is above 80 km/h (50 MPH).
		Relay-2	ON	
Above 100 (212)	HIGH	Relay-1	ON	—
		Relay-2	ON	

The radiator fan operates at HIGHT if the self-diagnosing engine temperature sensor system results in "N.G.".

Air conditioner switch and blower fan switch are "ON"
[Evaporator outlet air temperature is between 3°C (37°F) and 8°C (46°F).]

Engine temperature °C (°F)	Radiator fan	Radiator fan control relay		Remarks
Below 94 (201)	*LOW	Relay-1	ON	*OFF if vehicle speed is above 80 km/h (50 MPH).
		Relay-2	OFF	
Between 95 (203) and 99 (210)	LOW	Relay-1	ON	—
		Relay-2	OFF	
Above 100 (212)	HIGH	Relay-1	ON	—
		Relay-2	ON	

The radiator fan operates at HIGH if the self-diagnosing engine temperature sensor system results in "N.G.".

Fail-safe System

AIR FLOW METER MALFUNCTION

If the air flow meter output voltage is above or below the specified value, the E.C.U. senses an air flow meter malfunction. In case of a malfunction, the throttle sensor substitutes for the air flow meter.

Though air flow meter is malfunctioning, it is possible to drive the vehicle and start the engine. But engine speed will not rise more than 2,400 rpm in order to inform the driver of fail-safe system operation while driving.

Operation

System	Fixed condition
E.G.R. control system	OFF
Idle speed control system	A duty ratio is fixed at the preprogrammed value.
Fuel injection control system	Fuel is shut off above 2,400 rpm. (Engine speed does not exceed 2,400 rpm.)

ENGINE TEMPERATURE SENSOR MALFUNCTION

When engine temperature sensor output voltage is below or above the specified value, water temperature is fixed at the preset value as follows:

Operation

Condition	Engine temperature decided
Just as ignition switch is turned ON or Start	20°C (68°F)
More than 6 minutes after ignition ON or Start	80°C (176°F)
Except as shown above	20 - 80°C (68 - 176°F) (Depends on the time)

THROTTLE SENSOR MALFUNCTION

When throttle sensor output voltage is below or above the specified value, throttle sensor output is fixed at the preset value.

AIR TEMPERATURE SENSOR MALFUNCTION

When air temperature sensor is below or above the specified value, air temperature value is fixed at the preset value [20°C (68°F)].

ECCS IDLE SPEED, IGNITION TIMING AND IDLE MIXTURE RATIO INSPECTION— 1990 AXXESS

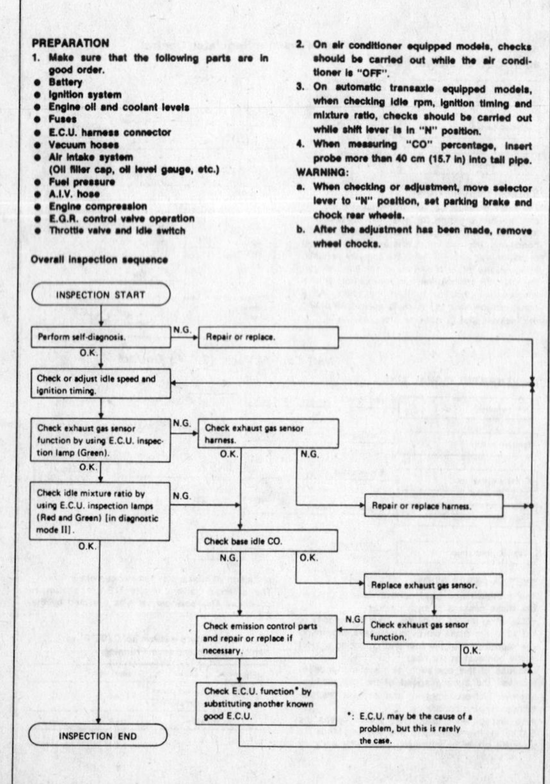

PREPARATION

1. Make sure that the following parts are in good order.
   - Battery
   - Ignition system
   - Engine oil and coolant levels
   - Fuses
   - E.C.U. harness connector
   - Vacuum hoses
   - Air intake system (Oil filler cap, oil level gauge, etc.)
   - Fuel pressure
   - A.I.V. hose
   - Engine compression
   - E.G.R. control valve operation
   - Throttle valve and idle switch
2. On air conditioner equipped models, checks should be carried out while the air conditioner is "OFF".
3. On automatic transaxle equipped models, when checking idle rpm, ignition timing and mixture ratio, checks should be carried out while shift lever is in "N" position.
4. When measuring "CO" percentage, insert probe more than 40 cm (15.7 in) into tail pipe.

WARNING:

a. When checking or adjustment, move selector lever to "N" position, set parking brake and chock rear wheels.
b. After the adjustment has been made, remove wheel chocks.

Overall inspection sequence

INSPECTION START
Perform self-diagnosis.
N.G.
Repair or replace.
O.K.
Check or adjust idle speed and ignition timing.
Check exhaust gas sensor function by using E.C.U. inspection lamp (Green).
N.G.
Check exhaust gas sensor harness.
O.K.
N.G.
O.K.
Check idle mixture ratio by using E.C.U. inspection lamps (Red and Green) [in diagnostic mode II].
N.G.
Repair or replace harness.
O.K.
Check base idle CO.
N.G.
O.K.
Replace exhaust gas sensor.
Check emission control parts and repair or replace if necessary.
N.G.
Check exhaust gas sensor function.
O.K.
Check E.C.U. function* by substituting another known good E.C.U.
*: E.C.U. may be the cause of a problem, but this is rarely the case.
INSPECTION END

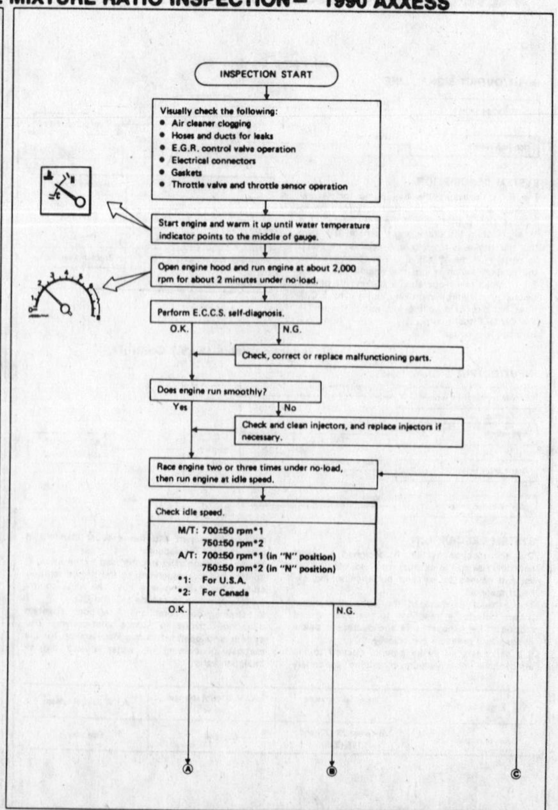

INSPECTION START
Visually check the following:
Air cleaner clogging
Hoses and ducts for leaks
E.G.R. control valve operation
Electrical connectors
Gaskets
Throttle valve and throttle sensor operation
Start engine and warm it up until water temperature indicator points to the middle of gauge.
Open engine hood and run engine at about 2,000 rpm for about 2 minutes under no-load.
Perform E.C.C.S. self-diagnosis.
O.K.
N.G.
Check, correct or replace malfunctioning parts.
Does engine run smoothly?
Yes
No
Check and clean injectors, and replace injectors if necessary.
Race engine two or three times under no-load, then run engine at idle speed.
Check idle speed.
M/T: 700±50 rpm*1
750±50 rpm*2
A/T: 700±50 rpm*1 (in "N" position)
750±50 rpm*2 (in "N" position)
*1: For U.S.A.
*2: For Canada
O.K.
N.G.
A
B
C

ECCS IDLE SPEED, IGNITION TIMING AND IDLE MIXTURE RATIO INSPECTION (CONT.)—1990 AXXESS

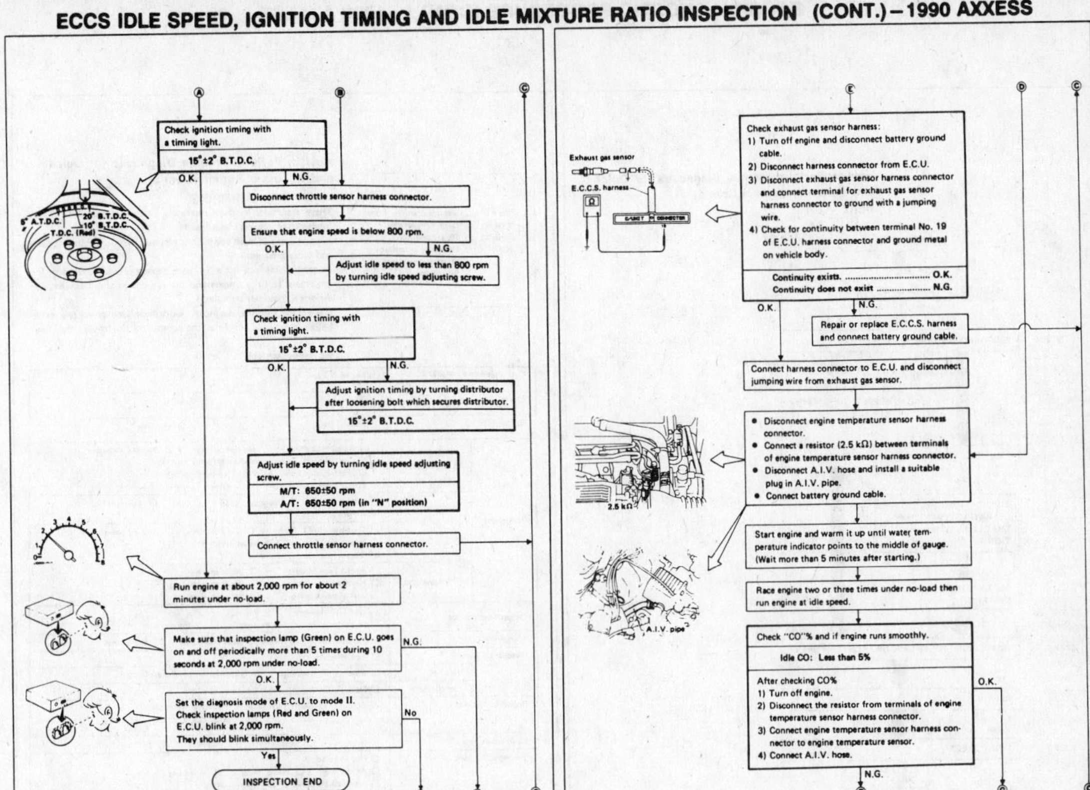
Check ignition timing with a timing light.
15°±2° B.T.D.C.
O.K.
N.G.
5° A.T.D.C.
20° B.T.D.C.
10° B.T.D.C.
T.D.C. (Red)
Disconnect throttle sensor harness connector.
Ensure that engine speed is below 800 rpm.
Adjust idle speed to less than 800 rpm by turning idle speed adjusting screw.
Check ignition timing with a timing light.
15°±2° B.T.D.C.
Adjust ignition timing by turning distributor after loosening bolt which secures distributor.
15°±2° B.T.D.C.
Adjust idle speed by turning idle speed adjusting screw.
M/T: 650±50 rpm
A/T: 650±50 rpm (in "N" position)
Connect throttle sensor harness connector.
Run engine at about 2,000 rpm for about 2 minutes under no-load.
Make sure that inspection lamp (Green) on E.C.U. goes on and off periodically more than 5 times during 10 seconds at 2,000 rpm under no-load.
Set the diagnosis mode of E.C.U. to mode II. Check inspection lamps (Red and Green) on E.C.U. blink at 2,000 rpm. They should blink simultaneously.
No
Yes
INSPECTION END
Exhaust gas sensor
E.C.C.S. harness
Check exhaust gas sensor harness:
1) Turn off engine and disconnect battery ground cable.
2) Disconnect harness connector from E.C.U.
3) Disconnect exhaust gas sensor harness connector and connect terminal for exhaust gas sensor harness connector to ground with a jumping wire.
4) Check for continuity between terminal No. 19 of E.C.U. harness connector and ground metal on vehicle body.
Continuity exists. .... O.K.
Continuity does not exist .... N.G.
Repair or replace E.C.C.S. harness and connect battery ground cable.
Connect harness connector to E.C.U. and disconnect jumping wire from exhaust gas sensor.
Disconnect engine temperature sensor harness connector.
Connect a resistor (2.5 kΩ) between terminals of engine temperature sensor harness connector.
Disconnect A.I.V. hose and install a suitable plug in A.I.V. pipe.
Connect battery ground cable.
2.5 kΩ
Start engine and warm it up until water temperature indicator points to the middle of gauge. (Wait more than 5 minutes after starting.)
Race engine two or three times under no-load then run engine at idle speed.
A.I.V. pipe
Check "CO"% and if engine runs smoothly.
Idle CO: Less than 5%
After checking CO%
1) Turn off engine.
2) Disconnect the resistor from terminals of engine temperature sensor harness connector.
3) Connect engine temperature sensor harness connector to engine temperature sensor.
4) Connect A.I.V. hose.

ECCS IDLE SPEED, IGNITION TIMING AND IDLE MIXTURE RATIO INSPECTION —1990 AXXESS

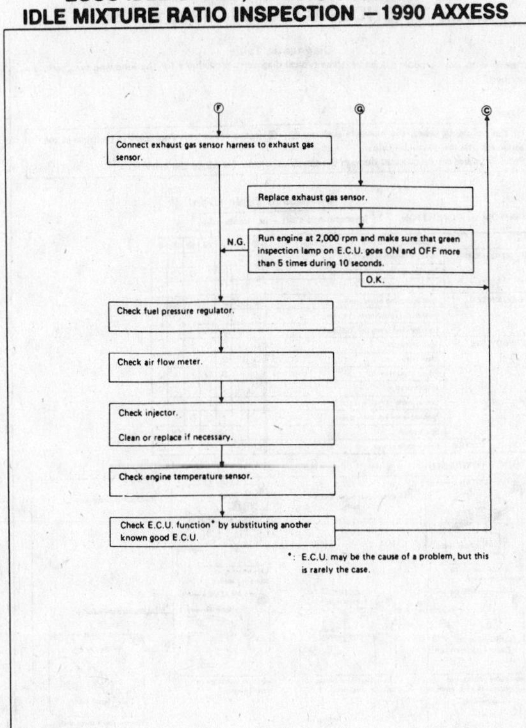
Connect exhaust gas sensor harness to exhaust gas sensor.
Replace exhaust gas sensor.
N.G.
Run engine at 2,000 rpm and make sure that green inspection lamp on E.C.U. goes ON and OFF more than 5 times during 10 seconds.
O.K.
Check fuel pressure regulator.
Check air flow meter.
Check injector.
Clean or replace if necessary.
Check engine temperature sensor.
Check E.C.U. function* by substituting another known good E.C.U.
*: E.C.U. may be the cause of a problem, but this is rarely the case.

ECCS DIAGNOSTIC PROCEDURE—1990 AXXESS

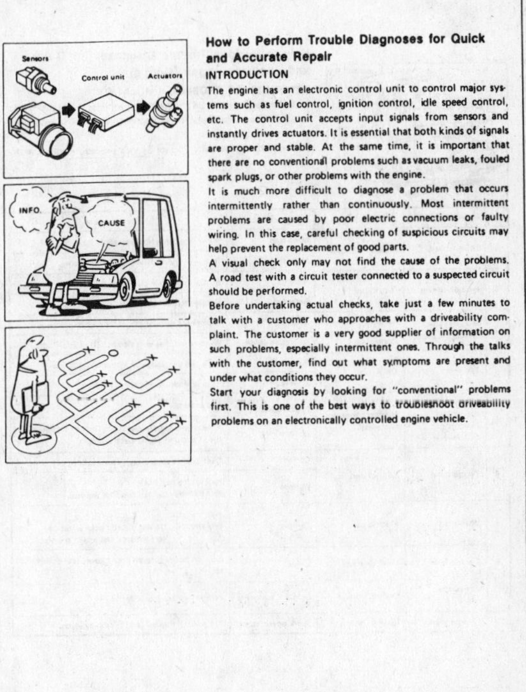
Sensors
Control unit
Actuators
INFO.
CAUSE

How to Perform Trouble Diagnoses for Quick and Accurate Repair

INTRODUCTION

The engine has an electronic control unit to control major systems such as fuel control, ignition control, idle speed control, etc. The control unit accepts input signals from sensors and instantly drives actuators. It is essential that both kinds of signals are proper and stable. At the same time, it is important that there are no conventional problems such as vacuum leaks, fouled spark plugs, or other problems with the engine.

It is much more difficult to diagnose a problem that occurs intermittently rather than continuously. Most intermittent problems are caused by poor electric connections or faulty wiring. In this case, careful checking of suspicious circuits may help prevent the replacement of good parts.

A visual check only may not find the cause of the problems. A road test with a circuit tester connected to a suspected circuit should be performed.

Before undertaking actual checks, take just a few minutes to talk with a customer who approaches with a driveability complaint. The customer is a very good supplier of information on such problems, especially intermittent ones. Through the talks with the customer, find out what symptoms are present and under what conditions they occur.

Start your diagnosis by looking for "conventional" problems first. This is one of the best ways to troubleshoot driveability problems on an electronically controlled engine vehicle.

ECCS DIAGNOSTIC PROCEDURE (CONT.)—1990 AXXESS

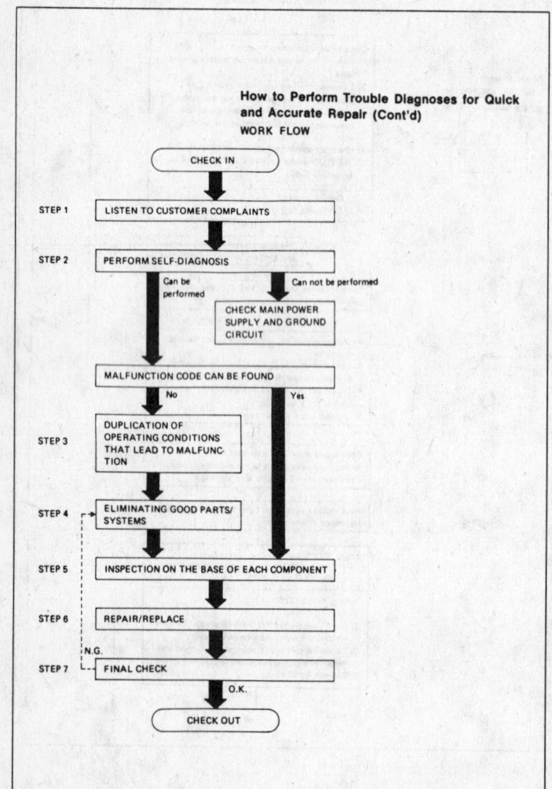
How to Perform Trouble Diagnoses for Quick and Accurate Repair (Cont'd)
WORK FLOW
CHECK IN
STEP 1
LISTEN TO CUSTOMER COMPLAINTS
STEP 2
PERFORM SELF-DIAGNOSIS
Can be performed
Can not be performed
CHECK MAIN POWER SUPPLY AND GROUND CIRCUIT
MALFUNCTION CODE CAN BE FOUND
No
Yes
STEP 3
DUPLICATION OF OPERATING CONDITIONS THAT LEAD TO MALFUNCTION
STEP 4
ELIMINATING GOOD PARTS/SYSTEMS
STEP 5
INSPECTION ON THE BASE OF EACH COMPONENT
STEP 6
REPAIR/REPLACE
N.G.
STEP 7
FINAL CHECK
O.K.
CHECK OUT

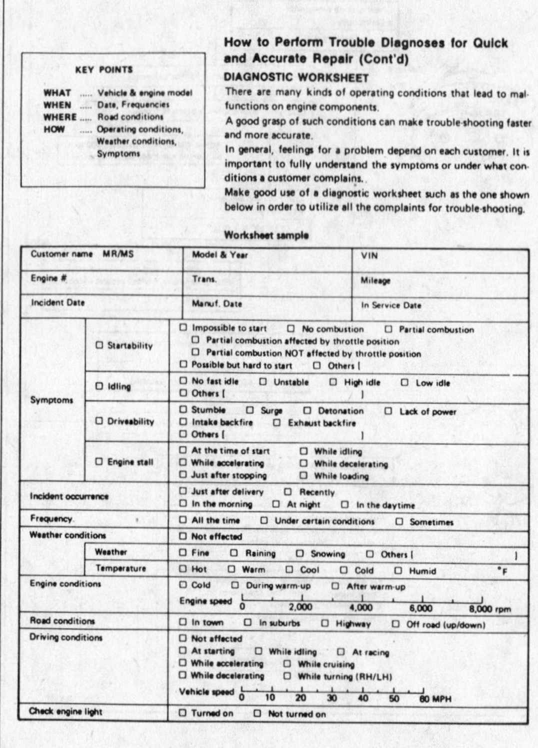

KEY POINTS	
WHAT	Vehicle & engine model
WHEN	Date, Frequencies
WHERE	Road conditions
HOW	Operating conditions, Weather conditions, Symptoms

How to Perform Trouble Diagnoses for Quick and Accurate Repair (Cont'd)

DIAGNOSTIC WORKSHEET

There are many kinds of operating conditions that lead to malfunctions on engine components.
A good grasp of such conditions can make trouble-shooting faster and more accurate.
In general, feelings for a problem depend on each customer. It is important to fully understand the symptoms or under what conditions a customer complains.
Make good use of a diagnostic worksheet such as the one shown below in order to utilize all the complaints for trouble-shooting.

Worksheet sample

Customer name MR/MS		Model & Year	VIN
Engine #		Trans.	Mileage
Incident Date		Manuf. Date	In Service Date
Symptoms	☐ Startability	☐ Impossible to start ☐ No combustion ☐ Partial combustion ☐ Partial combustion affected by throttle position ☐ Partial combustion NOT affected by throttle position ☐ Possible but hard to start ☐ Others [	
	☐ Idling	☐ No fast idle ☐ Unstable ☐ High idle ☐ Low idle ☐ Others [ ]	
	☐ Driveability	☐ Stumble ☐ Surge ☐ Detonation ☐ Lack of power ☐ Intake backfire ☐ Exhaust backfire ☐ Others [ ]	
	☐ Engine stall	☐ At the time of start ☐ While idling ☐ While accelerating ☐ While decelerating ☐ Just after stopping ☐ While loading	
Incident occurrence		☐ Just after delivery ☐ Recently ☐ In the morning ☐ At night ☐ In the daytime	
Frequency		☐ All the time ☐ Under certain conditions ☐ Sometimes	
Weather conditions		☐ Not effected	
	Weather	☐ Fine ☐ Raining ☐ Snowing ☐ Others [ ]	
	Temperature	☐ Hot ☐ Warm ☐ Cool ☐ Cold ☐ Humid °F	
Engine conditions		☐ Cold ☐ During warm-up ☐ After warm-up Engine speed 0 2,000 4,000 6,000 8,000 rpm	
Road conditions		☐ In town ☐ In suburbs ☐ Highway ☐ Off road (up/down)	
Driving conditions		☐ Not affected ☐ At starting ☐ While idling ☐ At racing ☐ While accelerating ☐ While cruising ☐ While decelerating ☐ While turning (RH/LH) Vehicle speed 0 10 20 30 40 50 60 MPH	
Check engine light		☐ Turned on ☐ Not turned on	

ECCS DIAGNOSTIC PROCEDURE —1990 AXXESS

How to Perform Trouble Diagnoses for Quick and Accurate Repair (Cont'd)

INTERMITTENT PROBLEM SIMULATION

In order to duplicate an intermittent problem, it is effective to create similar conditions for component parts, under which the problem might occur.
Perform the activity listed under Service procedure and note the result.

	Variable factor	Influential part	Target condition	Service procedure
1	Mixture ratio	Pressure regulator	Made lean	Remove vacuum hose and apply vacuum.
			Made rich	Remove vacuum hose and apply pressure.
2	Ignition timing	Distributor	Advanced	Rotate distributor clockwise.
			Retarded	Rotate distributor counterclockwise.
3	Mixture ratio feedback control	Exhaust gas sensor	Suspended	Disconnect exhaust gas sensor harness connector.
		Control unit	Operation check	Perform self-diagnosis (Mode I/II) at 2,000 rpm.
4	Idle speed	I.A.A. unit	Raised	Turn idle adjusting screw counterclockwise.
			Lowered	Turn idle adjusting screw clockwise.
5	Electric connection (Electric continuity)	Harness connectors and wires	Poor electric connection or faulty wiring	Tap or wiggle.
				Race engine rapidly. See if the torque reaction of the engine unit causes electric breaks.
6	Temperature	Control unit	Cooled	Cool with an icing spray or similar device.
			Warmed	Heat with a hair drier. [WARNING: Do not overheat the unit.]
7	Moisture	Electric parts	Damp	Wet. [WARNING: Do not directly pour water on components. Use a mist sprayer.]
8	Electric loads	Load switches	Loaded	Turn on head lights, air conditioner, rear defogger, etc.
9	Idle switch condition	Control unit	ON-OFF switching	Perform self-diagnosis (Mode IV).
10	Ignition spark	Timing light	Spark power check	Try to flash timing light for each cylinder.

ECCS DIAGNOSTIC CHART—1990 AXXESS

Diagnostic Table

To assist with your trouble diagnoses, some typical diagnostic procedures for the following symptoms are described.

REMARKS

In the following pages, the numbers such as ●, ● in the above chart correspond to those in the service procedure described below.
Possible causes can be checked through the service procedure shown by the mark "○".

Diagnostic Table (Cont'd)

SYMPTOM & CONDITION 1 Impossible to start — no combustion

	POSSIBLE CAUSES	●	●	●	●	●	●	●
SPECIFICATIONS	Mixture ratio (too lean)	○	○					
	Ignition sparks (weak, missing)				○	○	○	
	Ignition timing							○
FUEL SYSTEM	Fuel pump (no operation)	○						
	Fuel pump relay (open circuited)	○						
	Injectors (no operation, clogged)		○					
IGNITION SYSTEM	Ignition switch	○	○	○	○			○
	Main relay	○	○	○	○			○
	Power transistor			○	○			○
	Ignition coil				○			○
	Center cable (ignition leaks)				○			○
	Ignition wires (ignition leaks)				○	○		
	Spark plugs						○	
CONTROL SYSTEM	Crank angle sensor	○	○		○			○

SERVICE PROCEDURE

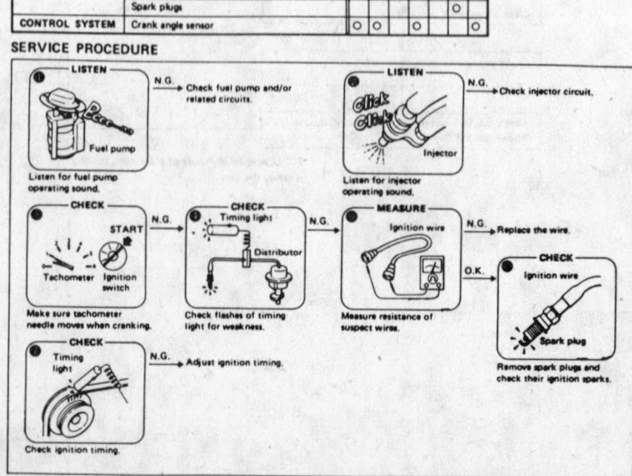
LISTEN
Fuel pump
N.G.
Check fuel pump and/or related circuits.
Listen for fuel pump operating sound.
LISTEN
Click
Click
Injector
N.G.
Check injector circuit.
Listen for injector operating sound.
CHECK
START
Tachometer
Ignition switch
N.G.
Make sure tachometer needle moves when cranking.
CHECK
Timing light
Distributor
N.G.
Check flashes of timing light for weakness.
MEASURE
Ignition wire
N.G.
Replace the wire.
O.K.
Measure resistance of suspect wires.
CHECK
Ignition wire
Spark plug
Remove spark plugs and check their ignition sparks.
CHECK
Timing light
N.G.
Adjust ignition timing.
Check ignition timing.

ECCS DIAGNOSTIC CHART—1990 AXXESS

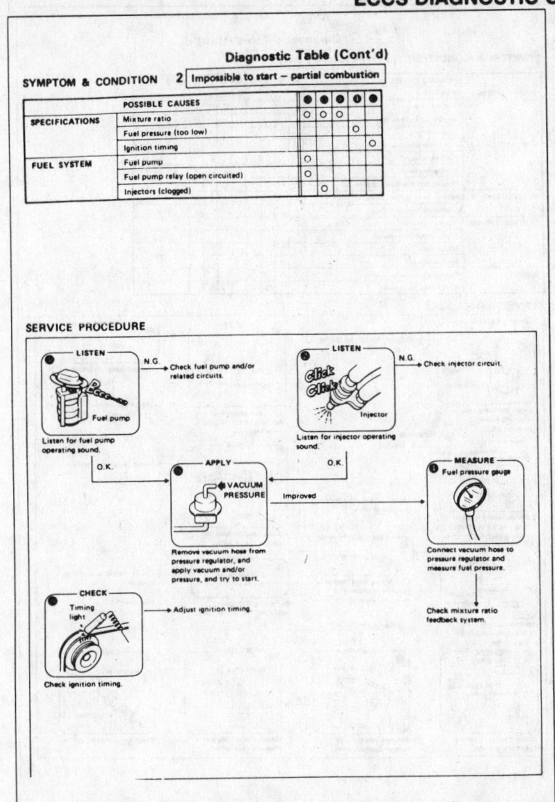

Diagnostic Table (Cont'd)

SYMPTOM & CONDITION 2 Impossible to start – partial combustion

	POSSIBLE CAUSES	1	2	3	4	5
SPECIFICATIONS	Mixture ratio	○	○	○		
	Fuel pressure (too low)				○	
	Ignition timing					○
FUEL SYSTEM	Fuel pump	○				
	Fuel pump relay (open circuited)	○				
	Injectors (clogged)		○			

SERVICE PROCEDURE
LISTEN
Fuel pump
N.G.
Check fuel pump and/or related circuits.
Listen for fuel pump operating sound.
O.K.
LISTEN
Injector
N.G.
Check injector circuit.
Listen for injector operating sound.
O.K.
APPLY
VACUUM PRESSURE
Remove vacuum hose from pressure regulator, and apply vacuum and/or pressure, and try to start.
Improved
MEASURE
Fuel pressure gauge
Connect vacuum hose to pressure regulator and measure fuel pressure.
Check mixture ratio feedback system.
CHECK
Timing light
Adjust ignition timing.
Check ignition timing.

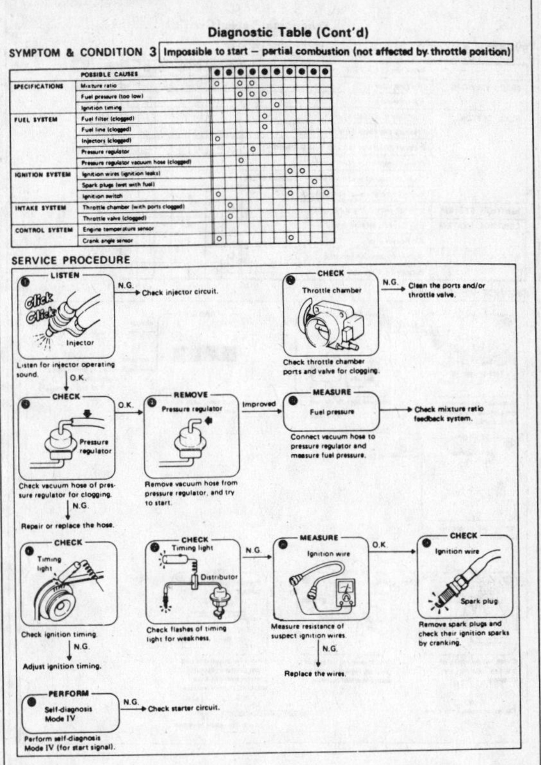

Diagnostic Table (Cont'd)

SYMPTOM & CONDITION 3 Impossible to start – partial combustion (not affected by throttle position)

	POSSIBLE CAUSES	1	2	3	4	5	6	7	8	9	10
SPECIFICATIONS	Mixture ratio	○		○	○						
	Fuel pressure (too low)			○	○	○					
	Ignition timing						○				
FUEL SYSTEM	Fuel filter (clogged)					○					
	Fuel line (clogged)					○					
	Injectors (clogged)	○									
	Pressure regulator				○						
	Pressure regulator vacuum hose (clogged)			○							
IGNITION SYSTEM	Ignition wires (ignition leaks)							○	○		
	Spark plugs (wet with fuel)									○	
	Ignition switch	○						○			○
INTAKE SYSTEM	Throttle chamber (with ports clogged)		○								
	Throttle valve (clogged)		○								
CONTROL SYSTEM	Engine temperature sensor										
	Crank angle sensor	○						○			

SERVICE PROCEDURE
LISTEN
Injector
N.G.
Check injector circuit.
Listen for injector operating sound.
O.K.
CHECK
Throttle chamber
N.G.
Clean the ports and/or throttle valve.
Check throttle chamber ports and valve for clogging.
CHECK
Pressure regulator
Check vacuum hose of pressure regulator for clogging.
N.G.
Repair or replace the hose.
O.K.
REMOVE
Pressure regulator
Remove vacuum hose from pressure regulator, and try to start.
Improved
MEASURE
Fuel pressure
Connect vacuum hose to pressure regulator and measure fuel pressure.
Check mixture ratio feedback system.
CHECK
Timing light
Check ignition timing.
N.G.
Adjust ignition timing.
CHECK
Timing light
Distributor
Check flashes of timing light for weakness.
N.G.
MEASURE
Ignition wire
Measure resistance of suspect ignition wires.
N.G.
Replace the wires.
O.K.
CHECK
Ignition wire
Spark plug
Remove spark plugs and check their ignition sparks by cranking.
PERFORM
Self-diagnosis Mode IV
N.G.
Check starter circuit.
Perform self-diagnosis Mode IV (for start signal).

ECCS DIAGNOSTIC CHART—1990 AXXESS

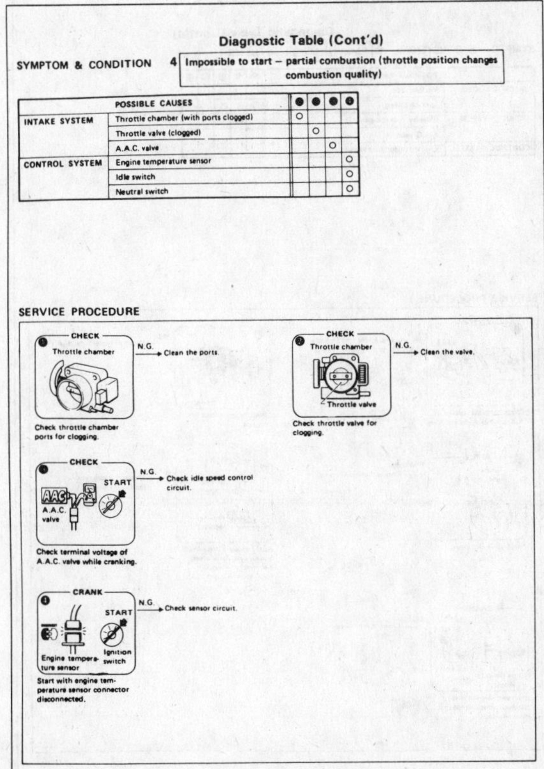

Diagnostic Table (Cont'd)

SYMPTOM & CONDITION 4 Impossible to start – partial combustion (throttle position changes combustion quality)

	POSSIBLE CAUSES	1	2	3	4
INTAKE SYSTEM	Throttle chamber (with ports clogged)	○			
	Throttle valve (clogged)		○		
	A.A.C. valve			○	
CONTROL SYSTEM	Engine temperature sensor				○
	Idle switch				○
	Neutral switch				○

SERVICE PROCEDURE
CHECK
Throttle chamber
N.G.
Clean the ports.
Check throttle chamber ports for clogging.
CHECK
Throttle chamber
Throttle valve
N.G.
Clean the valve.
Check throttle valve for clogging.
CHECK
START
A.A.C. valve
N.G.
Check idle speed control circuit.
Check terminal voltage of A.A.C. valve while cranking.
CRANK
START
Engine temperature sensor
Ignition switch
N.G.
Check sensor circuit.
Start with engine temperature sensor connector disconnected.

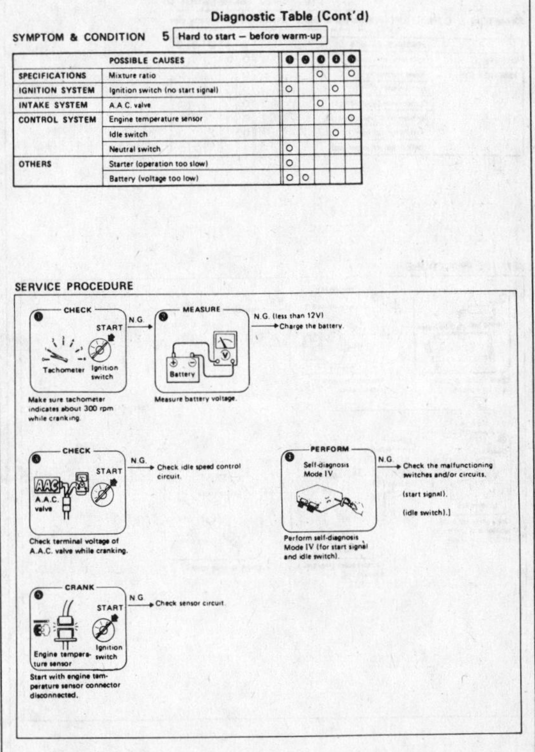

Diagnostic Table (Cont'd)

SYMPTOM & CONDITION 5 Hard to start – before warm-up

	POSSIBLE CAUSES	1	2	3	4	5
SPECIFICATIONS	Mixture ratio			○		○
IGNITION SYSTEM	Ignition switch (no start signal)	○			○	
INTAKE SYSTEM	A.A.C. valve			○		
CONTROL SYSTEM	Engine temperature sensor					○
	Idle switch				○	
	Neutral switch	○				
OTHERS	Starter (operation too slow)	○				
	Battery (voltage too low)	○	○			

SERVICE PROCEDURE
CHECK
START
Tachometer
Ignition switch
Make sure tachometer indicates about 300 rpm while cranking.
N.G.
MEASURE
Battery
Measure battery voltage.
N.G. (less than 12V)
Charge the battery.
CHECK
START
A.A.C. valve
N.G.
Check idle speed control circuit.
Check terminal voltage of A.A.C. valve while cranking.
PERFORM
Self-diagnosis Mode IV
N.G.
Check the malfunctioning switches and/or circuits.
(start signal).
(idle switch).]
Perform self-diagnosis Mode IV (for start signal and idle switch).
CRANK
START
Engine temperature sensor
Ignition switch
N.G.
Check sensor circuit.
Start with engine temperature sensor connector disconnected.

ECCS DIAGNOSTIC CHART—1990 AXXESS

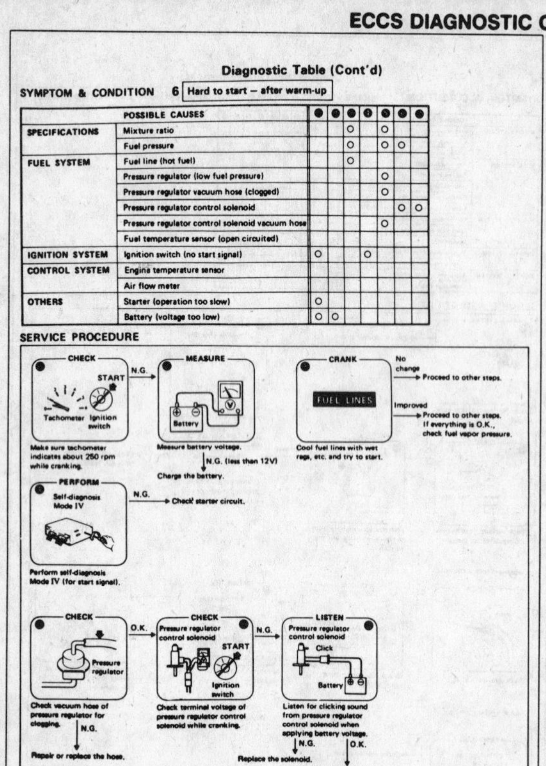

Diagnostic Table (Cont'd)

SYMPTOM & CONDITION 6 Hard to start – after warm-up

	POSSIBLE CAUSES	●	●	●	●	●	●	●
SPECIFICATIONS	Mixture ratio			O		O		
	Fuel pressure			O		O	O	
FUEL SYSTEM	Fuel line (hot fuel)			O				
	Pressure regulator (low fuel pressure)					O		
	Pressure regulator vacuum hose (clogged)					O		
	Pressure regulator control solenoid						O	O
	Pressure regulator control solenoid vacuum hose					O		
	Fuel temperature sensor (open circuited)							
IGNITION SYSTEM	Ignition switch (no start signal)	O			O			
CONTROL SYSTEM	Engine temperature sensor							
	Air flow meter							
OTHERS	Starter (operation too slow)	O						
	Battery (voltage too low)	O	O					

SERVICE PROCEDURE

CHECK
START
Tachometer
Ignition switch
Make sure tachometer indicates about 250 rpm while cranking.
N.G.
MEASURE
Battery
Measure battery voltage.
N.G. (less than 12V)
Charge the battery.
CRANK
FUEL LINES
Cool fuel lines with wet rags, etc. and try to start.
No change → Proceed to other steps.
Improved → Proceed to other steps. If everything is O.K., check fuel vapor pressure.
PERFORM
Self-diagnosis Mode IV
Perform self-diagnosis Mode IV (for start signal).
N.G. → Check starter circuit.
CHECK
Pressure regulator
Check vacuum hose of pressure regulator for clogging.
N.G.
Repair or replace the hose.
O.K.
CHECK
Pressure regulator control solenoid
START
Ignition switch
Check terminal voltage of pressure regulator control solenoid while cranking.
N.G.
LISTEN
Pressure regulator control solenoid
Click
Battery
Listen for clicking sound from pressure regulator control solenoid when applying battery voltage.
N.G.
Replace the solenoid.
O.K.
Check the solenoid circuit.

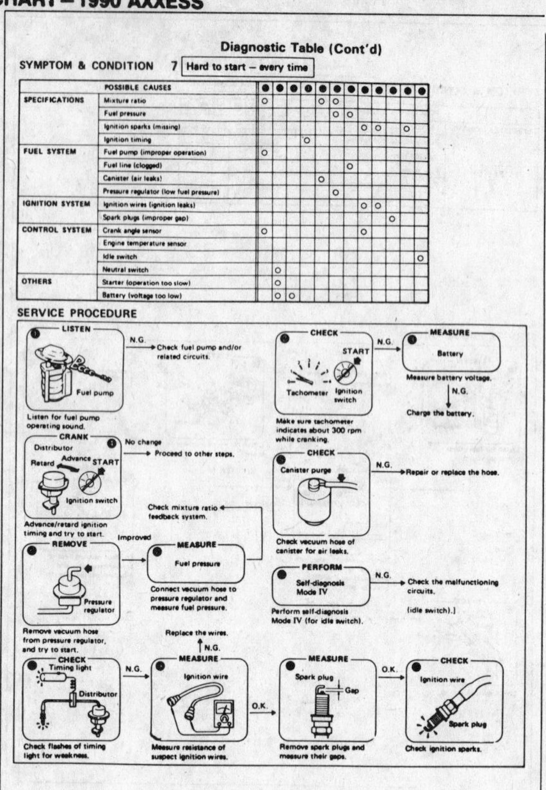

Diagnostic Table (Cont'd)

SYMPTOM & CONDITION 7 Hard to start – every time

	POSSIBLE CAUSES	●	●	●	●	●	●	●	●	●	●	●	●
SPECIFICATIONS	Mixture ratio	O				O	O						
	Fuel pressure						O	O					
	Ignition sparks (missing)								O	O		O	
	Ignition timing				O								
FUEL SYSTEM	Fuel pump (improper operation)	O											
	Fuel line (clogged)							O					
	Canister (air leaks)					O							
	Pressure regulator (low fuel pressure)						O						
IGNITION SYSTEM	Ignition wires (ignition leaks)								O	O			
	Spark plugs (improper gap)										O		
CONTROL SYSTEM	Crank angle sensor	O							O				
	Engine temperature sensor												
	Idle switch												O
	Neutral switch		O										
OTHERS	Starter (operation too slow)		O										
	Battery (voltage too low)		O	O									

SERVICE PROCEDURE

LISTEN
Fuel pump
Listen for fuel pump operating sound.
N.G. → Check fuel pump and/or related circuits.
CHECK
START
Tachometer
Ignition switch
Make sure tachometer indicates about 300 rpm while cranking.
N.G.
MEASURE
Battery
Measure battery voltage.
N.G.
Charge the battery.
CRANK
Distributor
Retard
Advance
START
Ignition switch
Advance/retard ignition timing and try to start.
No change → Proceed to other steps.
CHECK
Canister purge
Check vacuum hose of canister for air leaks.
N.G. → Repair or replace the hose.
Check mixture ratio feedback system.
REMOVE
Pressure regulator
Remove vacuum hose from pressure regulator, and try to start.
Improved
MEASURE
Fuel pressure
Connect vacuum hose to pressure regulator and measure fuel pressure.
PERFORM
Self-diagnosis Mode IV
Perform self-diagnosis Mode IV (for idle switch).
N.G. → Check the malfunctioning circuits.
(idle switch).]
Replace the wires.
N.G.
CHECK
Timing light
Distributor
Check flashes of timing light for weakness.
N.G.
MEASURE
Ignition wire
Measure resistance of suspect ignition wires.
O.K.
MEASURE
Spark plug
Gap
Remove spark plugs and measure their gaps.
O.K.
CHECK
Ignition wire
Spark plug
Check ignition sparks.

ECCS DIAGNOSTIC CHART—1990 AXXESS

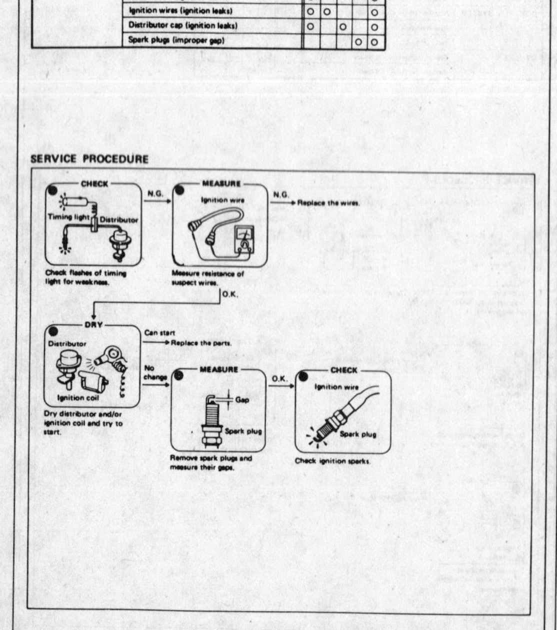

Diagnostic Table (Cont'd)

SYMPTOM & CONDITION 8 Hard to start – morning after a rainy day

	POSSIBLE CAUSES	●	●	●	●	●
SPECIFICATIONS	Ignition sparks (weak)	O	O			O
IGNITION SYSTEM	Power transistor	O				O
	Ignition coil	O		O		O
	Center cable (ignition leaks)	O				O
	Ignition wires (ignition leaks)	O	O			O
	Distributor cap (ignition leaks)	O		O		O
	Spark plugs (improper gap)				O	O

SERVICE PROCEDURE

CHECK
Timing light
Distributor
Check flashes of timing light for weakness.
N.G.
MEASURE
Ignition wire
Measure resistance of suspect wires.
N.G. → Replace the wires.
O.K.
DRY
Distributor
Ignition coil
Dry distributor and/or ignition coil and try to start.
Can start → Replace the parts.
No change
MEASURE
Gap
Spark plug
Remove spark plugs and measure their gaps.
O.K.
CHECK
Ignition wire
Spark plug
Check ignition sparks.

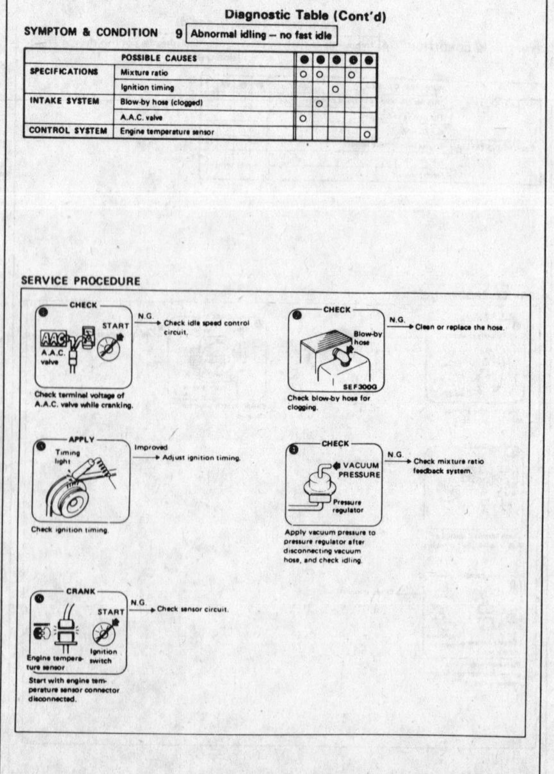

Diagnostic Table (Cont'd)

SYMPTOM & CONDITION 9 Abnormal idling – no fast idle

	POSSIBLE CAUSES	●	●	●	●	●
SPECIFICATIONS	Mixture ratio	O	O		O	
	Ignition timing			O		
INTAKE SYSTEM	Blow-by hose (clogged)		O			
	A.A.C. valve	O				
CONTROL SYSTEM	Engine temperature sensor					O

SERVICE PROCEDURE

CHECK
AAC
A.A.C. valve
START
Check terminal voltage of A.A.C. valve while cranking.
N.G. → Check idle speed control circuit.
CHECK
Blow-by hose
SEF300G
Check blow-by hose for clogging.
N.G. → Clean or replace the hose.
APPLY
Timing light
Check ignition timing.
Improved → Adjust ignition timing.
CHECK
VACUUM PRESSURE
Pressure regulator
Apply vacuum pressure to pressure regulator after disconnecting vacuum hose, and check idling.
N.G. → Check mixture ratio feedback system.
CRANK
START
Engine temperature sensor
Ignition switch
Start with engine temperature sensor connector disconnected.
N.G. → Check sensor circuit.

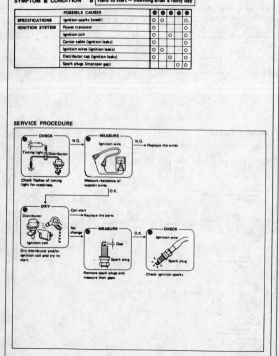

ECCS DIAGNOSTIC CHART—1990 AXXESS

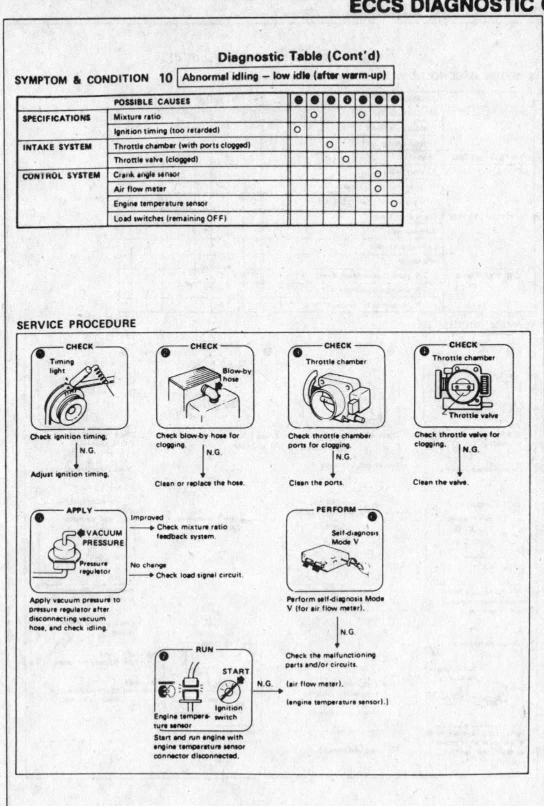

Diagnostic Table (Cont'd)

SYMPTOM & CONDITION 10 Abnormal idling – low idle (after warm-up)

	POSSIBLE CAUSES	1	2	3	4	5	6	7
SPECIFICATIONS	Mixture ratio		O			O		
	Ignition timing (too retarded)	O						
INTAKE SYSTEM	Throttle chamber (with ports clogged)			O				
	Throttle valve (clogged)				O			
CONTROL SYSTEM	Crank angle sensor						O	
	Air flow meter						O	
	Engine temperature sensor							O
	Load switches (remaining OFF)							

SERVICE PROCEDURE
CHECK
Timing light
Check ignition timing.
N.G.
Adjust ignition timing.
CHECK
Blow-by hose
Check blow-by hose for clogging.
N.G.
Clean or replace the hose.
CHECK
Throttle chamber
Check throttle chamber ports for clogging.
N.G.
Clean the ports.
CHECK
Throttle chamber
Throttle valve
Check throttle valve for clogging.
N.G.
Clean the valve.
APPLY
VACUUM PRESSURE
Pressure regulator
Apply vacuum pressure to pressure regulator after disconnecting vacuum hose, and check idling.
Improved
Check mixture ratio feedback system.
No change
Check load signal circuit.
PERFORM
Self-diagnosis Mode V
Perform self-diagnosis Mode V (for air flow meter).
N.G.
Check the malfunctioning parts and/or circuits.
(air flow meter).
(engine temperature sensor).]
RUN
START
Engine temperature sensor
Ignition switch
Start and run engine with engine temperature sensor connector disconnected.
N.G.

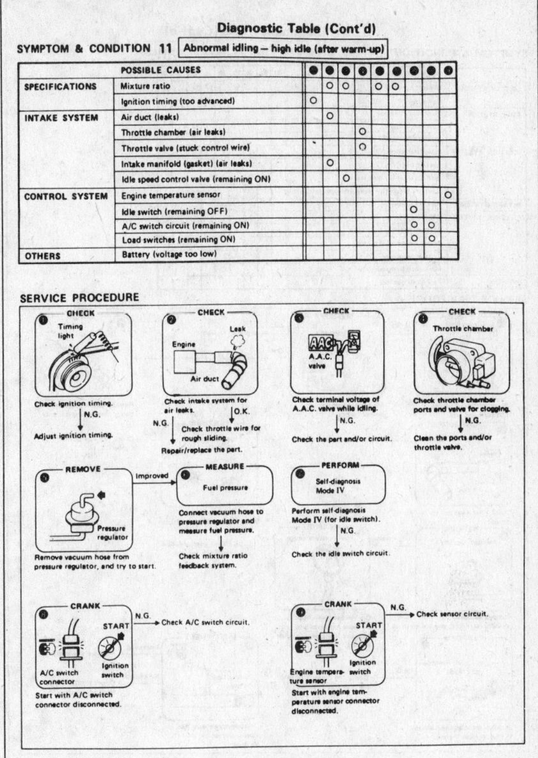

Diagnostic Table (Cont'd)

SYMPTOM & CONDITION 11 Abnormal idling – high idle (after warm-up)

	POSSIBLE CAUSES	1	2	3	4	5	6	7	8	9
SPECIFICATIONS	Mixture ratio		O	O		O	O			
	Ignition timing (too advanced)	O								
INTAKE SYSTEM	Air duct (leaks)		O							
	Throttle chamber (air leaks)				O					
	Throttle valve (stuck control wire)				O					
	Intake manifold (gasket) (air leaks)		O							
	Idle speed control valve (remaining ON)			O						
CONTROL SYSTEM	Engine temperature sensor									O
	Idle switch (remaining OFF)							O		
	A/C switch circuit (remaining ON)							O	O	
	Load switches (remaining ON)							O	O	
OTHERS	Battery (voltage too low)									

SERVICE PROCEDURE
CHECK
Timing light
Check ignition timing.
N.G.
Adjust ignition timing.
CHECK
Leak
Engine
Air duct
Check intake system for air leaks.
N.G.
Repair/replace the part.
O.K.
Check throttle wire for rough sliding.
CHECK
A.A.C. valve
Check terminal voltage of A.A.C. valve while idling.
N.G.
Check the part and/or circuit.
CHECK
Throttle chamber
Check throttle chamber ports and valve for clogging.
N.G.
Clean the ports and/or throttle valve.
REMOVE
Pressure regulator
Remove vacuum hose from pressure regulator, and try to start.
Improved
MEASURE
Fuel pressure
Connect vacuum hose to pressure regulator and measure fuel pressure.
Check mixture ratio feedback system.
PERFORM
Self-diagnosis Mode IV
Perform self-diagnosis Mode IV (for idle switch).
N.G.
Check the idle switch circuit.
CRANK
START
A/C switch connector
Ignition switch
Start with A/C switch connector disconnected.
N.G.
Check A/C switch circuit.
CRANK
START
Engine temperature sensor
Ignition switch
Start with engine temperature sensor connector disconnected.
N.G.
Check sensor circuit.

ECCS DIAGNOSTIC CHART—1990 AXXESS

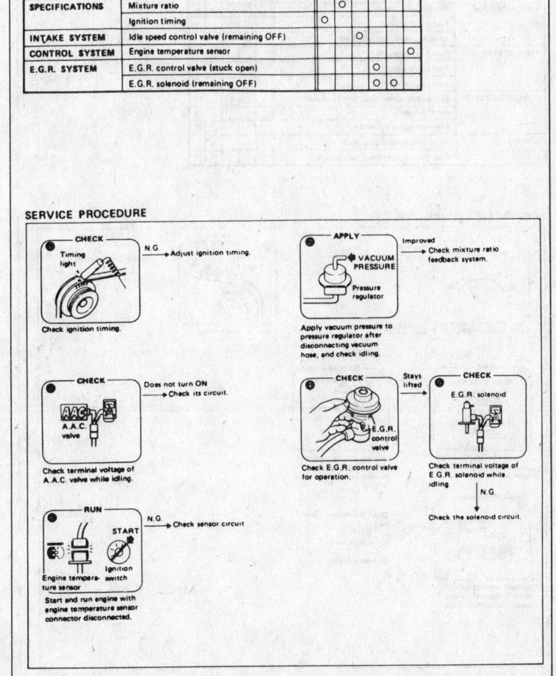

Diagnostic Table (Cont'd)

SYMPTOM & CONDITION 12 Unstable idling – before warm-up

	POSSIBLE CAUSES	1	2	3	4	5	6
SPECIFICATIONS	Mixture ratio		O				
	Ignition timing	O					
INTAKE SYSTEM	Idle speed control valve (remaining OFF)			O			
CONTROL SYSTEM	Engine temperature sensor						O
E.G.R. SYSTEM	E.G.R. control valve (stuck open)				O		
	E.G.R. solenoid (remaining OFF)				O	O	

SERVICE PROCEDURE
CHECK
Timing light
Check ignition timing.
N.G.
Adjust ignition timing.
APPLY
VACUUM PRESSURE
Pressure regulator
Apply vacuum pressure to pressure regulator after disconnecting vacuum hose, and check idling.
Improved
Check mixture ratio feedback system.
CHECK
A.A.C. valve
Check terminal voltage of A.A.C. valve while idling.
Does not turn ON
Check its circuit.
CHECK
E.G.R. control valve
Check E.G.R. control valve for operation.
Stays lifted
CHECK
E.G.R. solenoid
Check terminal voltage of E.G.R. solenoid while idling.
N.G.
Check the solenoid circuit.
RUN
START
Engine temperature sensor
Ignition switch
Start and run engine with engine temperature sensor connector disconnected.
N.G.
Check sensor circuit

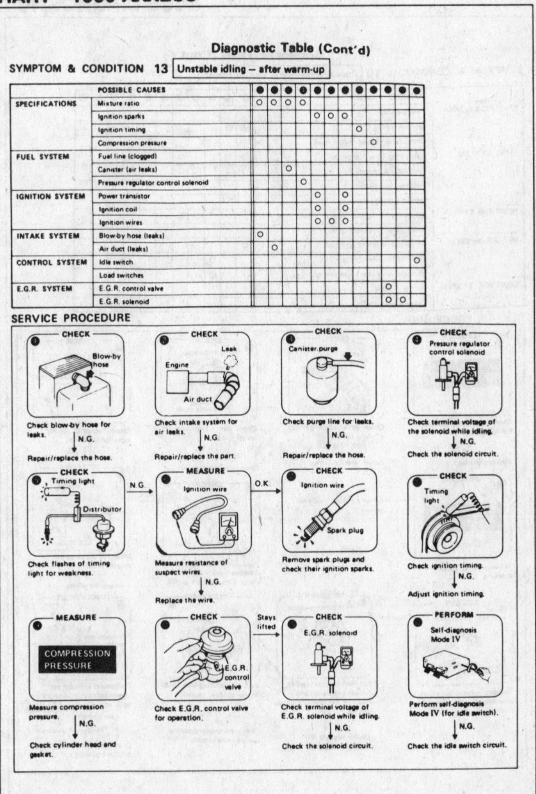

Diagnostic Table (Cont'd)

SYMPTOM & CONDITION 13 Unstable idling – after warm-up

	POSSIBLE CAUSES	1	2	3	4	5	6	7	8	9	10	11	12
SPECIFICATIONS	Mixture ratio	O	O	O	O								
	Ignition sparks					O	O	O					
	Ignition timing								O				
	Compression pressure									O			
FUEL SYSTEM	Fuel line (clogged)												
	Canister (air leaks)			O									
	Pressure regulator control solenoid				O								
IGNITION SYSTEM	Power transistor					O		O					
	Ignition coil					O		O					
	Ignition wires					O	O	O					
INTAKE SYSTEM	Blow-by hose (leaks)	O											
	Air duct (leaks)		O										
CONTROL SYSTEM	Idle switch												O
	Load switches												
E.G.R. SYSTEM	E.G.R. control valve										O		
	E.G.R. solenoid										O	O	

SERVICE PROCEDURE
CHECK
Blow-by hose
Check blow-by hose for leaks.
N.G.
Repair/replace the hose.
CHECK
Leak
Engine
Air duct
Check intake system for air leaks.
N.G.
Repair/replace the part.
CHECK
Canister purge
Check purge line for leaks.
N.G.
Repair/replace the hose.
CHECK
Pressure regulator control solenoid
Check terminal voltage of the solenoid while idling.
N.G.
Check the solenoid circuit.
CHECK
Timing light
Distributor
Check flashes of timing light for weakness.
N.G.
MEASURE
Ignition wire
Measure resistance of suspect wires.
N.G.
Replace the wire.
O.K.
CHECK
Ignition wire
Spark plug
Remove spark plugs and check their ignition sparks.
CHECK
Timing light
Check ignition timing.
N.G.
Adjust ignition timing.
MEASURE
COMPRESSION PRESSURE
Measure compression pressure.
N.G.
Check cylinder head and gasket.
CHECK
E.G.R. control valve
Check E.G.R. control valve for operation.
Stays lifted
CHECK
E.G.R. solenoid
Check terminal voltage of E.G.R. solenoid while idling.
N.G.
Check the solenoid circuit.
PERFORM
Self-diagnosis Mode IV
Perform self-diagnosis Mode IV (for idle switch).
N.G.
Check the idle switch circuit.

ECCS DIAGNOSTIC CHART—1990 AXXESS

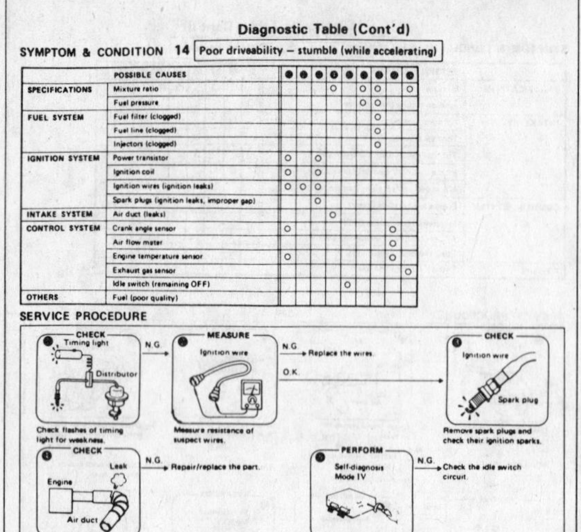

Diagnostic Table (Cont'd)

SYMPTOM & CONDITION 14 Poor driveability – stumble (while accelerating)

	POSSIBLE CAUSES	●	●	●	●	●	●	●	●	●
SPECIFICATIONS	Mixture ratio				○		○	○		○
	Fuel pressure						○	○		
FUEL SYSTEM	Fuel filter (clogged)							○		
	Fuel line (clogged)							○		
	Injectors (clogged)							○		
IGNITION SYSTEM	Power transistor	○		○						
	Ignition coil	○		○						
	Ignition wires (ignition leaks)	○	○	○						
	Spark plugs (ignition leaks, improper gap)			○						
INTAKE SYSTEM	Air duct (leaks)				○					
CONTROL SYSTEM	Crank angle sensor	○							○	
	Air flow meter								○	
	Engine temperature sensor	○							○	
	Exhaust gas sensor									○
	Idle switch (remaining OFF)					○				
OTHERS	Fuel (poor quality)									

SERVICE PROCEDURE

CHECK
Timing light
Distributor
Check flashes of timing light for weakness.
N.G.
MEASURE
Ignition wire
Measure resistance of suspect wires.
N.G
Replace the wires.
O.K.
CHECK
Ignition wire
Spark plug
Remove spark plugs and check their ignition sparks.
CHECK
Leak
Engine
Air duct
Check intake system for air leaks.
N.G.
Repair/replace the part.
PERFORM
Self-diagnosis Mode IV
Perform self-diagnosis Mode IV (for idle switch).
N.G.
Check the idle switch circuit.
REMOVE
Pressure regulator
Remove vacuum hose from pressure regulator, and try to drive.
Improved
MEASURE
Fuel pressure gauge
Connect vacuum hose to pressure regulator and measure fuel pressure while driving.
Check mixture ratio feedback system.
PERFORM
Self-diagnosis Mode V
Perform self-diagnosis Mode V (for air flow meter).
N.G.
Check the malfunctioning parts and/or circuits.
DISCONNECT
Exhaust gas sensor
Disconnect exhaust gas sensor connector, and try to drive.
Improved
Replace the sensor.
No change
Check mixture ratio feedback system.

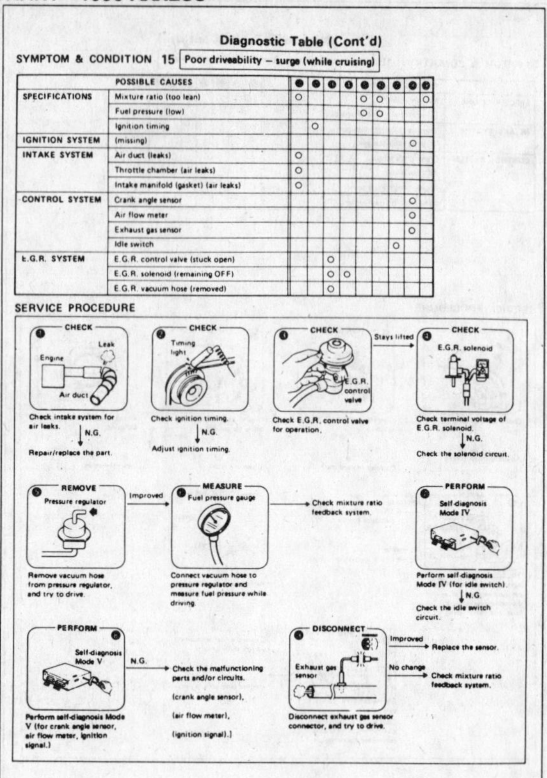

Diagnostic Table (Cont'd)

SYMPTOM & CONDITION 15 Poor driveability – surge (while cruising)

	POSSIBLE CAUSES	●	●	●	●	●	●	●	●	●
SPECIFICATIONS	Mixture ratio (too lean)	○				○	○			○
	Fuel pressure (low)					○	○			
	Ignition timing		○							
IGNITION SYSTEM	(missing)								○	
INTAKE SYSTEM	Air duct (leaks)	○								
	Throttle chamber (air leaks)	○								
	Intake manifold (gasket) (air leaks)	○								
CONTROL SYSTEM	Crank angle sensor								○	
	Air flow meter								○	
	Exhaust gas sensor								○	
	Idle switch							○		
E.G.R. SYSTEM	E.G.R. control valve (stuck open)			○						
	E.G.R. solenoid (remaining OFF)			○	○					
	E.G.R. vacuum hose (removed)			○						

SERVICE PROCEDURE

CHECK
Leak
Engine
Air duct
Check intake system for air leaks.
N.G.
Repair/replace the part.
CHECK
Timing light
Check ignition timing.
N.G.
Adjust ignition timing.
CHECK
E.G.R. control valve
Check E.G.R. control valve for operation.
Stays lifted
CHECK
E.G.R. solenoid
Check terminal voltage of E.G.R. solenoid.
N.G.
Check the solenoid circuit.
REMOVE
Pressure regulator
Remove vacuum hose from pressure regulator, and try to drive.
Improved
MEASURE
Fuel pressure gauge
Connect vacuum hose to pressure regulator and measure fuel pressure while driving.
Check mixture ratio feedback system.
PERFORM
Self-diagnosis Mode IV
Perform self-diagnosis Mode IV (for idle switch).
N.G.
Check the idle switch circuit.
PERFORM
Self-diagnosis Mode V
Perform self-diagnosis Mode V (for crank angle sensor, air flow meter, ignition signal.)
N.G.
Check the malfunctioning parts and/or circuits.
(crank angle sensor),
(air flow meter),
(ignition signal).]
DISCONNECT
Exhaust gas sensor
Disconnect exhaust gas sensor connector, and try to drive.
Improved
Replace the sensor.
No change
Check mixture ratio feedback system.

ECCS DIAGNOSTIC CHART—1990 AXXESS

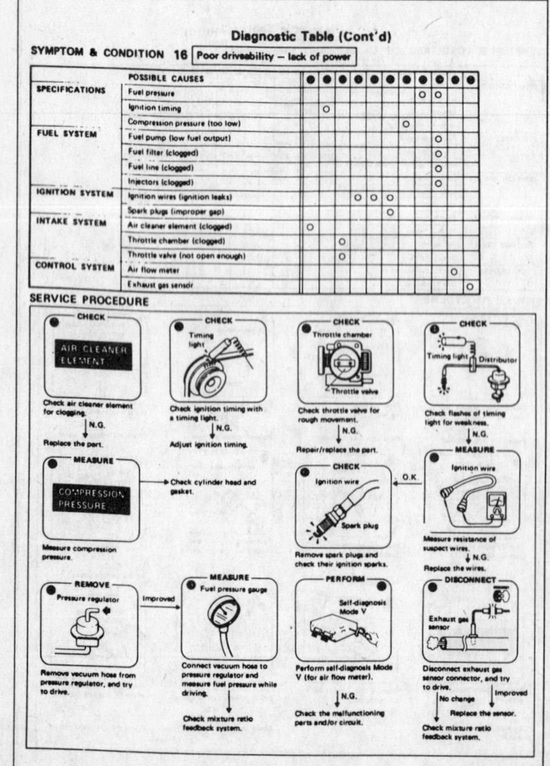

Diagnostic Table (Cont'd)

SYMPTOM & CONDITION 16 Poor driveability – lack of power

	POSSIBLE CAUSES	●	●	●	●	●	●	●	●	●	●	●
SPECIFICATIONS	Fuel pressure								○	○		
	Ignition timing		○									
	Compression pressure (too low)							○				
FUEL SYSTEM	Fuel pump (low fuel output)									○		
	Fuel filter (clogged)									○		
	Fuel line (clogged)									○		
	Injectors (clogged)									○		
IGNITION SYSTEM	Ignition wires (ignition leaks)				○	○	○					
	Spark plugs (improper gap)						○					
INTAKE SYSTEM	Air cleaner element (clogged)	○										
	Throttle chamber (clogged)			○								
	Throttle valve (not open enough)			○								
CONTROL SYSTEM	Air flow meter										○	
	Exhaust gas sensor											○

SERVICE PROCEDURE

CHECK
AIR CLEANER ELEMENT
Check air cleaner element for clogging.
N.G.
Replace the part.
CHECK
Timing light
Check ignition timing with a timing light.
N.G.
Adjust ignition timing.
CHECK
Throttle chamber
Throttle valve
Check throttle valve for rough movement.
N.G.
Repair/replace the part.
CHECK
Timing light
Distributor
Check flashes of timing light for weakness.
N.G.
MEASURE
COMPRESSION PRESSURE
Measure compression pressure.
Check cylinder head and gasket.
CHECK
Ignition wire
Spark plug
Remove spark plugs and check their ignition sparks.
O.K.
MEASURE
Ignition wire
Measure resistance of suspect wires.
N.G.
Replace the wires.
REMOVE
Pressure regulator
Remove vacuum hose from pressure regulator, and try to drive.
Improved
MEASURE
Fuel pressure gauge
Connect vacuum hose to pressure regulator and measure fuel pressure while driving.
Check mixture ratio feedback system.
PERFORM
Self-diagnosis Mode V
Perform self-diagnosis Mode V (for air flow meter).
N.G.
Check the malfunctioning parts and/or circuit.
DISCONNECT
Exhaust gas sensor
Disconnect exhaust gas sensor connector, and try to drive.
No change
Improved
Replace the sensor.
Check mixture ratio feedback system.

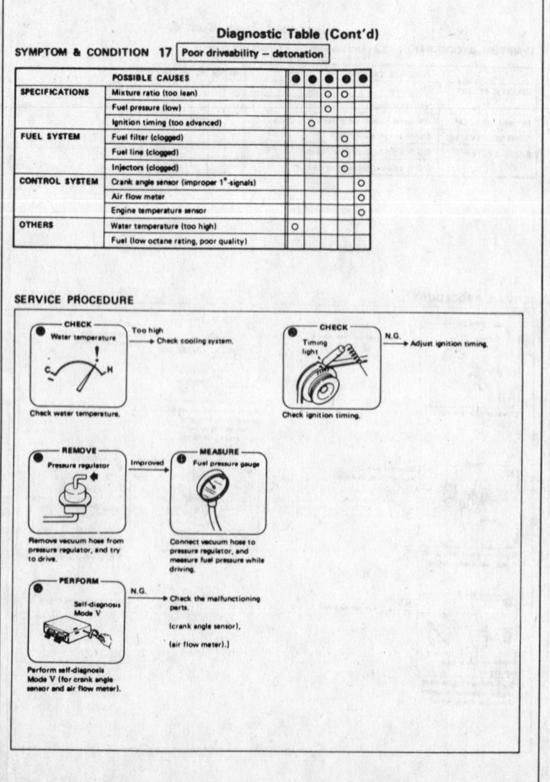

Diagnostic Table (Cont'd)

SYMPTOM & CONDITION 17 Poor driveability – detonation

	POSSIBLE CAUSES	●	●	●	●	●
SPECIFICATIONS	Mixture ratio (too lean)			○	○	
	Fuel pressure (low)			○		
	Ignition timing (too advanced)		○			
FUEL SYSTEM	Fuel filter (clogged)				○	
	Fuel line (clogged)				○	
	Injectors (clogged)				○	
CONTROL SYSTEM	Crank angle sensor (improper 1°-signals)					○
	Air flow meter					○
	Engine temperature sensor					○
OTHERS	Water temperature (too high)	○				
	Fuel (low octane rating, poor quality)					

SERVICE PROCEDURE

CHECK
Water temperature
C
H
Check water temperature.
Too high
Check cooling system.
CHECK
Timing light
Check ignition timing.
N.G.
Adjust ignition timing.
REMOVE
Pressure regulator
Remove vacuum hose from pressure regulator, and try to drive.
Improved
MEASURE
Fuel pressure gauge
Connect vacuum hose to pressure regulator, and measure fuel pressure while driving.
PERFORM
Self-diagnosis Mode V
Perform self-diagnosis Mode V (for crank angle sensor and air flow meter).
N.G.
Check the malfunctioning parts.
(crank angle sensor),
(air flow meter).]

ECCS DIAGNOSTIC CHART—1990 AXXESS

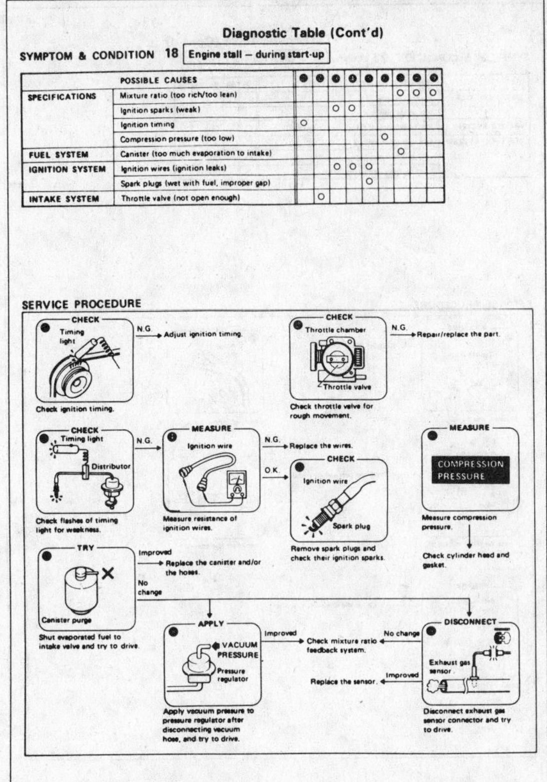

Diagnostic Table (Cont'd)

SYMPTOM & CONDITION 18 Engine stall – during start-up

	POSSIBLE CAUSES	1	2	3	4	5	6	7	8	9
SPECIFICATIONS	Mixture ratio (too rich/too lean)							O	O	O
	Ignition sparks (weak)			O	O					
	Ignition timing	O								
	Compression pressure (too low)						O			
FUEL SYSTEM	Canister (too much evaporation to intake)							O		
IGNITION SYSTEM	Ignition wires (ignition leaks)			O	O	O				
	Spark plugs (wet with fuel, improper gap)					O				
INTAKE SYSTEM	Throttle valve (not open enough)		O							

SERVICE PROCEDURE
CHECK
Timing light
N.G. → Adjust ignition timing.
Check ignition timing.
CHECK
Throttle chamber
Throttle valve
N.G. → Repair/replace the part.
Check throttle valve for rough movement.
CHECK
Timing light
Distributor
N.G.
Check flashes of timing light for weakness.
MEASURE
Ignition wire
N.G. → Replace the wires.
O.K.
Measure resistance of ignition wires.
CHECK
Ignition wire
Spark plug
Remove spark plugs and check their ignition sparks.
MEASURE
COMPRESSION PRESSURE
Measure compression pressure.
Check cylinder head and gasket.
TRY
Improved → Replace the canister and/or the hoses.
No change
Canister purge
Shut evaporated fuel to intake valve and try to drive.
APPLY
VACUUM PRESSURE
Pressure regulator
Improved → Check mixture ratio feedback system.
Apply vacuum pressure to pressure regulator after disconnecting vacuum hose, and try to drive.
DISCONNECT
Exhaust gas sensor
No change
Improved
Replace the sensor.
Disconnect exhaust gas sensor connector and try to drive.

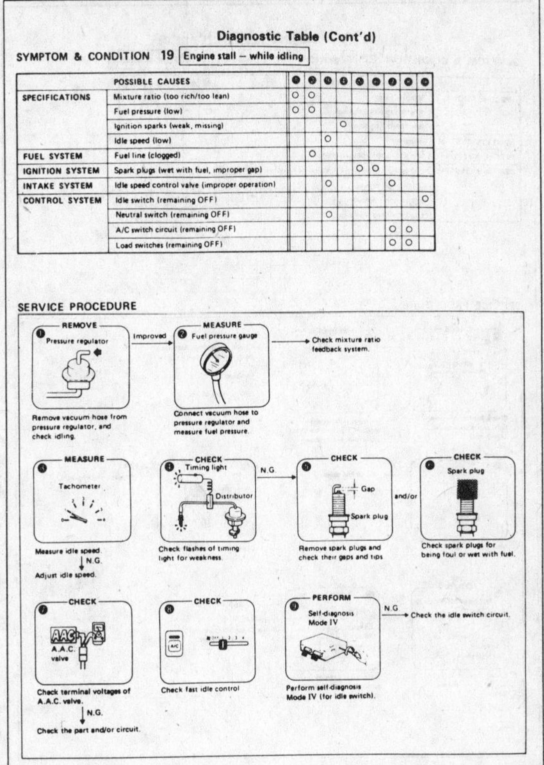

Diagnostic Table (Cont'd)

SYMPTOM & CONDITION 19 Engine stall – while idling

	POSSIBLE CAUSES	1	2	3	4	5	6	7	8	9
SPECIFICATIONS	Mixture ratio (too rich/too lean)	O	O							
	Fuel pressure (low)	O	O							
	Ignition sparks (weak, missing)				O					
	Idle speed (low)			O						
FUEL SYSTEM	Fuel line (clogged)		O							
IGNITION SYSTEM	Spark plugs (wet with fuel, improper gap)					O	O			
INTAKE SYSTEM	Idle speed control valve (improper operation)			O				O		
CONTROL SYSTEM	Idle switch (remaining OFF)									O
	Neutral switch (remaining OFF)			O						
	A/C switch circuit (remaining OFF)							O	O	
	Load switches (remaining OFF)							O	O	

SERVICE PROCEDURE
REMOVE
Pressure regulator
Improved
Remove vacuum hose from pressure regulator, and check idling.
MEASURE
Fuel pressure gauge
→ Check mixture ratio feedback system.
Connect vacuum hose to pressure regulator and measure fuel pressure.
MEASURE
Tachometer
Measure idle speed.
N.G.
Adjust idle speed.
CHECK
Timing light
Distributor
N.G.
Check flashes of timing light for weakness.
CHECK
Gap
Spark plug
Remove spark plugs and check their gaps and tips.
and/or
CHECK
Spark plug
Check spark plugs for being foul or wet with fuel.
CHECK
A.A.C. valve
Check terminal voltages of A.A.C. valve.
N.G.
Check the part and/or circuit.
CHECK
Check fast idle control
PERFORM
Self-diagnosis Mode IV
N.G. → Check the idle switch circuit.
Perform self-diagnosis Mode IV (for idle switch).

ECCS DIAGNOSTIC CHART—1990 AXXESS

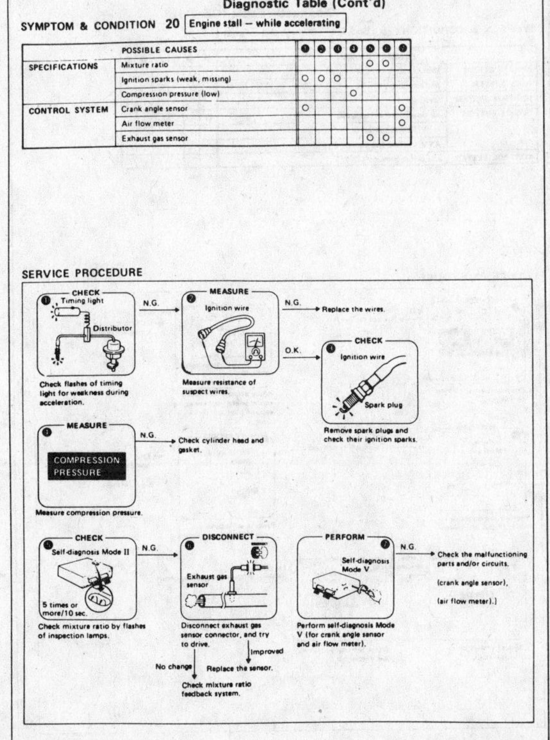

Diagnostic Table (Cont'd)

SYMPTOM & CONDITION 20 Engine stall – while accelerating

	POSSIBLE CAUSES	1	2	3	4	5	6	7
SPECIFICATIONS	Mixture ratio					O	O	
	Ignition sparks (weak, missing)	O	O	O				
	Compression pressure (low)				O			
CONTROL SYSTEM	Crank angle sensor	O						O
	Air flow meter							O
	Exhaust gas sensor					O	O	

SERVICE PROCEDURE
CHECK
Timing light
Distributor
N.G.
Check flashes of timing light for weakness during acceleration.
MEASURE
Ignition wire
N.G. → Replace the wires.
O.K.
Measure resistance of suspect wires.
CHECK
Ignition wire
Spark plug
Remove spark plugs and check their ignition sparks.
MEASURE
COMPRESSION PRESSURE
N.G. → Check cylinder head and gasket.
Measure compression pressure.
CHECK
Self-diagnosis Mode II
5 times or more/10 sec.
N.G.
Check mixture ratio by flashes of inspection lamps.
DISCONNECT
Exhaust gas sensor
Disconnect exhaust gas sensor connector, and try to drive.
Improved
No change
Replace the sensor.
Check mixture ratio feedback system.
PERFORM
Self-diagnosis Mode V
N.G. → Check the malfunctioning parts and/or circuits.
(crank angle sensor),
(air flow meter).)
Perform self-diagnosis Mode V (for crank angle sensor and air flow meter).

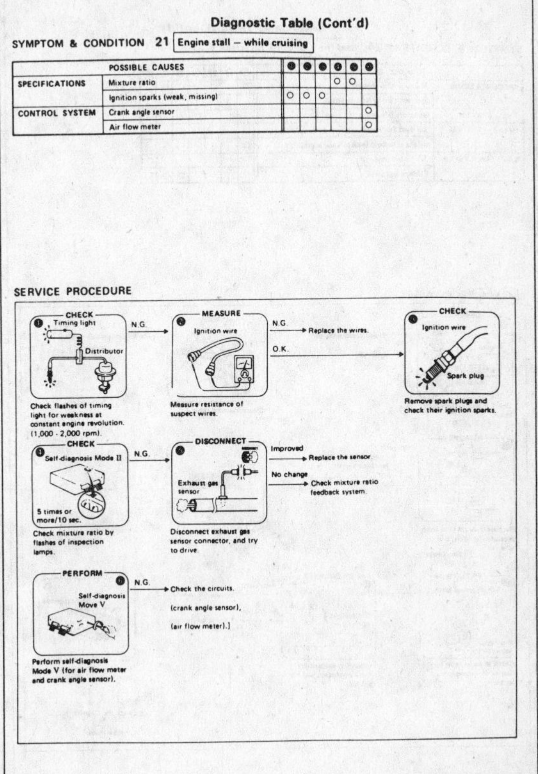

Diagnostic Table (Cont'd)

SYMPTOM & CONDITION 21 Engine stall – while cruising

	POSSIBLE CAUSES	1	2	3	4	5	6
SPECIFICATIONS	Mixture ratio				O	O	
	Ignition sparks (weak, missing)	O	O	O			
CONTROL SYSTEM	Crank angle sensor						O
	Air flow meter						O

SERVICE PROCEDURE
CHECK
Timing light
Distributor
N.G.
Check flashes of timing light for weakness at constant engine revolution. (1,000 - 2,000 rpm).
MEASURE
Ignition wire
N.G. → Replace the wires.
O.K.
Measure resistance of suspect wires.
CHECK
Ignition wire
Spark plug
Remove spark plugs and check their ignition sparks.
CHECK
Self-diagnosis Mode II
5 times or more/10 sec.
N.G.
Check mixture ratio by flashes of inspection lamps.
DISCONNECT
Exhaust gas sensor
Improved → Replace the sensor.
No change → Check mixture ratio feedback system.
Disconnect exhaust gas sensor connector, and try to drive.
PERFORM
Self-diagnosis Move V
N.G. → Check the circuits.
(crank angle sensor),
(air flow meter).)
Perform self-diagnosis Mode V (for air flow meter and crank angle sensor).

ECCS DIAGNOSTIC CHART – 1990 AXXESS

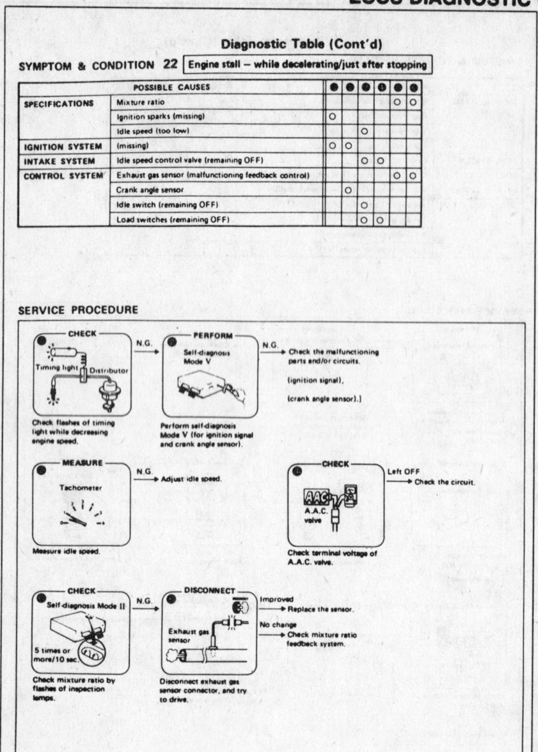

Diagnostic Table (Cont'd)

SYMPTOM & CONDITION 22 Engine stall – while decelerating/just after stopping

POSSIBLE CAUSES		●	●	●	●	●	●
SPECIFICATIONS	Mixture ratio					O	O
	Ignition sparks (missing)	O					
	Idle speed (too low)			O			
IGNITION SYSTEM	(missing)	O	O				
INTAKE SYSTEM	Idle speed control valve (remaining OFF)			O	O		
CONTROL SYSTEM	Exhaust gas sensor (malfunctioning feedback control)					O	O
	Crank angle sensor		O				
	Idle switch (remaining OFF)			O			
	Load switches (remaining OFF)			O	O		

SERVICE PROCEDURE
CHECK
Timing light
Distributor
Check flashes of timing light while decreasing engine speed.
N.G.
PERFORM
Self-diagnosis Mode V
Perform self-diagnosis Mode V (for ignition signal and crank angle sensor).
N.G.
Check the malfunctioning parts and/or circuits. (ignition signal), (crank angle sensor).)
MEASURE
Tachometer
Measure idle speed.
N.G.
Adjust idle speed.
CHECK
A.A.C. valve
Check terminal voltage of A.A.C. valve.
Left OFF
Check the circuit.
CHECK
Self-diagnosis Mode II
5 times or more/10 sec.
Check mixture ratio by flashes of inspection lamps.
N.G.
DISCONNECT
Exhaust gas sensor
Disconnect exhaust gas sensor connector, and try to drive.
Improved
Replace the sensor.
No change
Check mixture ratio feedback system.

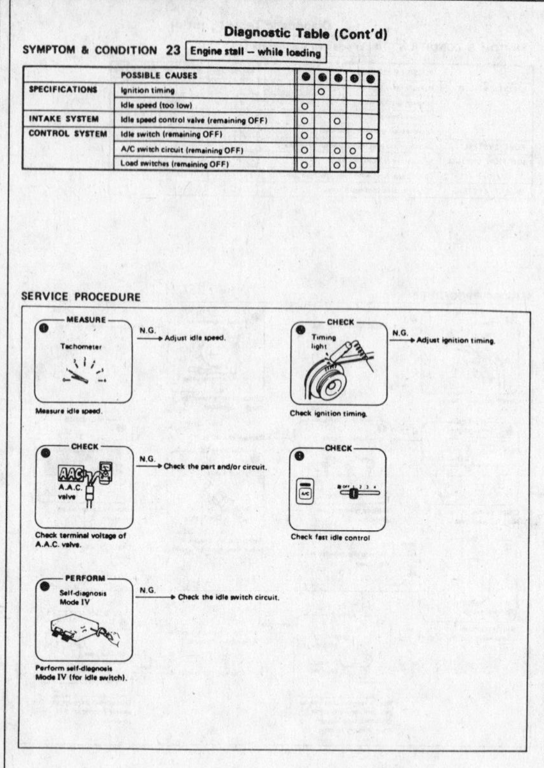

Diagnostic Table (Cont'd)

SYMPTOM & CONDITION 23 Engine stall – while loading

POSSIBLE CAUSES		●	●	●	●	●
SPECIFICATIONS	Ignition timing		O			
	Idle speed (too low)	O				
INTAKE SYSTEM	Idle speed control valve (remaining OFF)	O		O		
CONTROL SYSTEM	Idle switch (remaining OFF)	O				O
	A/C switch circuit (remaining OFF)	O		O	O	
	Load switches (remaining OFF)	O		O	O	

SERVICE PROCEDURE
MEASURE
Tachometer
Measure idle speed.
N.G.
Adjust idle speed.
CHECK
Timing light
Check ignition timing.
N.G.
Adjust ignition timing.
CHECK
A.A.C. valve
Check terminal voltage of A.A.C. valve.
N.G.
Check the part and/or circuit.
CHECK
Check fast idle control
PERFORM
Self-diagnosis Mode IV
Perform self-diagnosis Mode IV (for idle switch).
N.G.
Check the idle switch circuit.

ECCS DIAGNOSTIC CHART – 1990 AXXESS

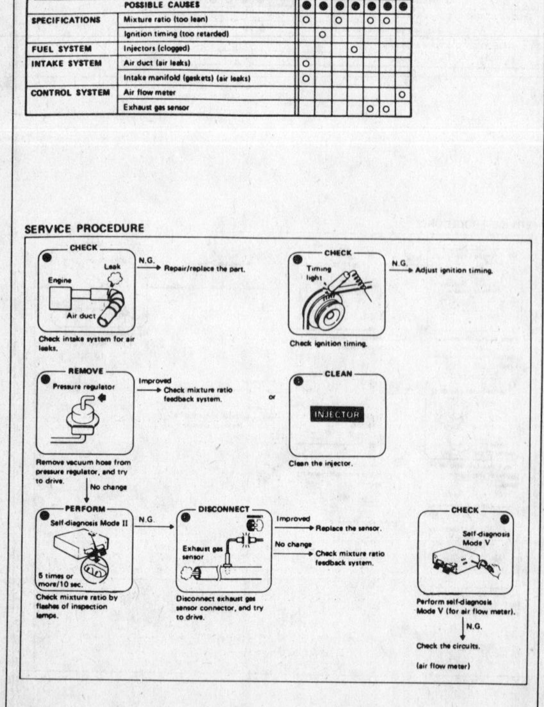

Diagnostic Table (Cont'd)

SYMPTOM & CONDITION 24 Backfire – through the intake

POSSIBLE CAUSES		●	●	●	●	●	●	●
SPECIFICATIONS	Mixture ratio (too lean)	O		O		O	O	
	Ignition timing (too retarded)		O					
FUEL SYSTEM	Injectors (clogged)				O			
INTAKE SYSTEM	Air duct (air leaks)	O						
	Intake manifold (gaskets) (air leaks)	O						
CONTROL SYSTEM	Air flow meter							O
	Exhaust gas sensor					O	O	

SERVICE PROCEDURE
CHECK
Engine
Leak
Air duct
Check intake system for air leaks.
N.G.
Repair/replace the part.
CHECK
Timing light
Check ignition timing.
N.G.
Adjust ignition timing.
REMOVE
Pressure regulator
Remove vacuum hose from pressure regulator, and try to drive.
Improved
Check mixture ratio feedback system.
or
CLEAN
INJECTOR
Clean the injector.
No change
PERFORM
Self-diagnosis Mode II
5 times or more/10 sec.
Check mixture ratio by flashes of inspection lamps.
N.G.
DISCONNECT
Exhaust gas sensor
Disconnect exhaust gas sensor connector, and try to drive.
Improved
Replace the sensor.
No change
Check mixture ratio feedback system.
CHECK
Self-diagnosis Mode V
Perform self-diagnosis Mode V (for air flow meter).
N.G.
Check the circuits. (air flow meter)

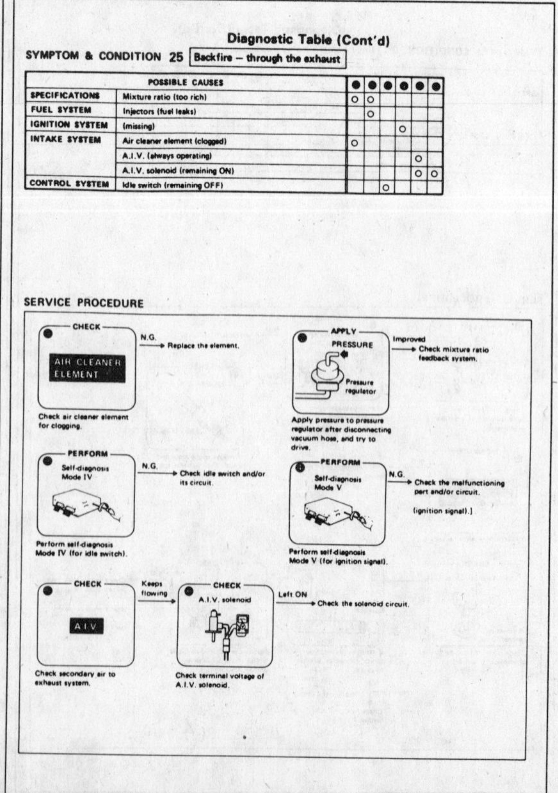

Diagnostic Table (Cont'd)

SYMPTOM & CONDITION 25 Backfire – through the exhaust

POSSIBLE CAUSES		●	●	●	●	●	●
SPECIFICATIONS	Mixture ratio (too rich)	O	O				
FUEL SYSTEM	Injectors (fuel leaks)		O				
IGNITION SYSTEM	(missing)				O		
INTAKE SYSTEM	Air cleaner element (clogged)	O					
	A.I.V. (always operating)					O	
	A.I.V. solenoid (remaining ON)					O	O
CONTROL SYSTEM	Idle switch (remaining OFF)			O			

SERVICE PROCEDURE
CHECK
AIR CLEANER ELEMENT
Check air cleaner element for clogging.
N.G.
Replace the element.
APPLY PRESSURE
Pressure regulator
Apply pressure to pressure regulator after disconnecting vacuum hose, and try to drive.
Improved
Check mixture ratio feedback system.
PERFORM
Self-diagnosis Mode IV
Perform self-diagnosis Mode IV (for idle switch).
N.G.
Check idle switch and/or its circuit.
PERFORM
Self-diagnosis Mode V
Perform self-diagnosis Mode V (for ignition signal).
N.G.
Check the malfunctioning part and/or circuit. (ignition signal).)
CHECK
A.I.V.
Check secondary air to exhaust system.
Keeps flowing
CHECK
A.I.V. solenoid
Check terminal voltage of A.I.V. solenoid.
Left ON
Check the solenoid circuit.

ECCS SELF-DIAGNOSTIC DESCRIPTION—1990 AXXESS

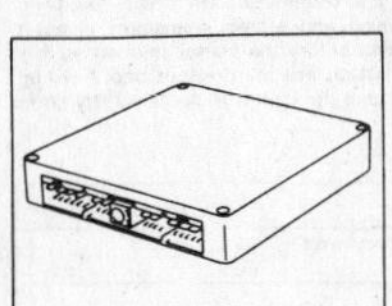

Self-diagnosis — Description

The self-diagnosis is useful to diagnose malfunctions in major sensors and actuators of the E.C.C.S. system. There are 5 modes in the self-diagnosis system.

1. **Mode I (Exhaust gas sensor monitor)**
   - During closed-loop operation:
     The green inspection lamp turns ON when a lean condition is detected and goes OFF under rich condition.
   - During open-loop operation condition:
     The green inspection lamp remains OFF or ON.
2. **Mode II (Mixture ratio feedback control monitor)**
   The green inspection lamp function is the same as Mode I.
   - During closed-loop operation:
     The red inspection lamp turns ON and OFF simultaneously with the green inspection lamp when the mixture ratio is controlled within the specified value.
   - During open-loop operation:
     The red inspection lamp remains ON or OFF.
3. **Mode III (Self-diagnostic system)**
   This mode is the same as the former self-diagnosis in self-diagnosis mode.
4. **Mode IV (Switches ON/OFF diagnostic system)**
   During this mode, the inspection lamps monitor the switch ON-OFF condition.
   - Idle switch
   - Starter switch
   - Vehicle speed sensor
5. **Mode V (Real-time diagnostic system)**
   The moment the malfunction is detected, the display will be presented immediately. That is, the condition at which the malfunction occurs can be found by observing the inspection lamps during driving test.

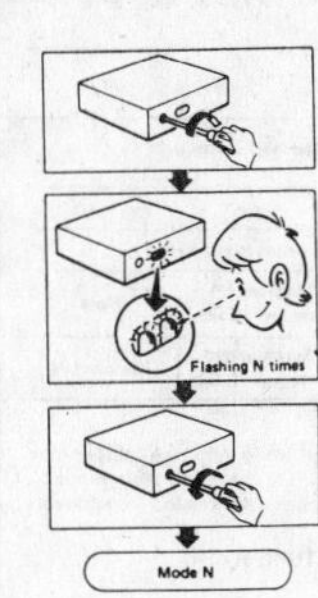
Flashing N times
Mode N

Self-diagnosis — Description (Cont'd)

HOW TO SWITCH THE DIAGNOSTIC MODES

1. Turn ignition switch "ON".
2. Turn diagnostic mode selector to E.C.U. (fully clockwise) and wait for inspection lamps to flash.
3. Count the number of flashes, and after the inspection lamps have flashed the number of the required mode, immediately turn diagnostic mode selector fully counterclockwise.

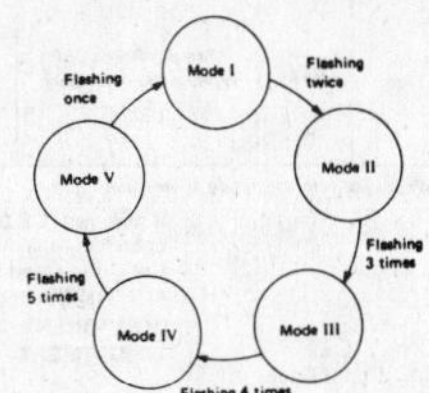
Mode I
Flashing twice
Mode II
Flashing 3 times
Mode III
Flashing 4 times
Mode IV
Flashing 5 times
Mode V
Flashing once

- **When the ignition switch is turned off during diagnosis in any mode and then turned on again (after power to the E.C.U. has dropped completely), the diagnosis will automatically return to Mode I.**

The stored memory will be lost if:

1. Battery terminal is disconnected.
2. After selecting Mode III, Mode IV is selected.
   However, if the diagnostic mode selector is kept turned fully clockwise, it will continue to change in the order of Mode I → II → III → IV → V → I ... etc., and in this state the stored memory will not be erased.

This unit serves as an idle rpm feedback control. When the diagnostic mode selector is turned within the "diagnostic mode OFF" range, a target engine speed can be selected. Mark the original position of the selector before conducting self-diagnosis. Upon completion of self-diagnosis, return the selector to the previous position. Otherwise, engine speed may change before and after conducting self-diagnosis.

ECCS SELF-DIAGNOSTIC DESCRIPTION (CONT.)—1990 AXXESS

CHECK
CHECK ENGINE LIGHT

Self-diagnosis — Description (Cc:nt'd)

CHECK ENGINE LIGHT (For ' 'ɔrnia only)

This vehicle has a check engine light o., the instrument panel. This light comes ON under the following conditions:

1) When ignition switch is turned "ON" (for bulb check).
2) When systems related to emission performance malfunction in Mode I (with engine running).
   - **This check engine light always illuminates and is synchronous with red L.E.D.**
   - **Malfunction systems related to emission performance can be detected by self-diagnosis, and they are clarified as self-diagnostic codes in Mode III.**
3) Check engine light will come "ON" only when malfunction is sensed.
   The check engine light will turn off when normal operation is resumed. Mode III memory must be cleared as the contents remain stored.

Code No.	Malfunction
12	Air flow meter circuit
13	Engine temperature sensor circuit
14	Vehicle speed sensor circuit
31	E.C.U. (E.C.C.S. control unit)
32	E.G.R. function
33	Exhaust gas sensor circuit
35	Exhaust gas temperature sensor circuit
43	Throttle sensor circuit
45	Injector leak

Use the following diagnostic flowchart to check and repair a malfunctioning system.

DIAGNOSIS START
↓
Turn ignition switch "ON" and make sure that check engine light comes "ON". — N.G. → Replace bulb.
↓ O.K.
Perform self-diagnosis and check which code is displayed in Mode III.
↓
Check electronic control system of affected code No. to locate faulty part.
↓
Repair or replace faulty part.
↓
Reinstall any part removed.
↓
Ⓐ

Self-diagnosis — Description (Cont'd)

Ⓐ
↓
Erase the self-diagnosis memory.
- Methods of erasing memories differ with systems. Read the manual before diagnosing systems.

↓
Perform driving test.
Make sure that check engine light does not come "ON" during this test.
- After repairs, test drive to check that check engine light does not come on.
- Test driving modes differ with systems. Read the manual before test driving.

↓
DIAGNOSIS END

ECCS SELF-DIAGNOSTIC MODES I AND II—1990 AXXESS

Self-diagnosis — Mode I (Exhaust gas sensor monitor)

This mode checks the exhaust gas sensor for proper functioning. The operation of the E.C.U. L.E.D. in this mode differs with mixture ratio control conditions as follows:

Mode	L.E.D.	Engine stopped (Ignition switch "ON")	Engine running	
			Open loop condition	Closed loop condition
Mode I (Monitor A)	Green	ON	*Remains ON or OFF	Blinks
	Red	ON	Except for California model • OFF	For California model • ON: when the CHECK ENGINE LIGHT ITEMS are stored in the E.C.U. • OFF: except for the above condition

*: Maintains conditions just before switching to open loop

EXHAUST GAS SENSOR FUNCTION CHECK

If the number of L.E.D. blinks is less than that specified, replace the exhaust gas sensor.
If the L.E.D. does not blink, check exhaust gas sensor circuit.

EXHAUST GAS SENSOR CIRCUIT CHECK

ECCS SELF-DIAGNOSTIC MODE III—1990 AXXESS

Self-diagnosis — Mode II (Mixture ratio feedback control monitor)

This mode checks, through the E.C.U. L.E.D., optimum control of the mixture ratio. The operation of the L.E.D., as shown below, differs with the control conditions of the mixture ratio (for example, richer or leaner mixture ratios, etc., which are controlled by the E.C.U.).

Mode	L.E.D.	Engine stopped (Ignition switch "ON")	Engine running			
			Open loop condition	Closed loop condition		
Mode II (Monitor B)	Green	ON	*Remains ON or OFF	Blinks		
	Red	OFF	*Remains ON or OFF (synchronous with green L.E.D.)	Compensating mixture ratio		
				More than 5% rich	Between 5% lean and 5% rich	More
				OFF	Synchronized with green L.E.D.	Remains ON

*: Maintains conditions just before switching to open loop

If the red L.E.D. remains on or off during the closed-loop operation, the mixture ratio may not be controlled properly. Using the following procedures, check the related components or adjust the mixture ratio.

COMPONENT CHECK OR MIXTURE RATIO ADJUSTMENT

Self-diagnosis — Mode III (Self-diagnostic system)

The E.C.U. constantly monitors the function of these sensors and actuators, regardless of ignition key position. If a malfunction occurs, the information is stored in the E.C.U. and can be retrieved from the memory by turning on the diagnostic mode selector, located on the side of the E.C.U. When activated, the malfunction is indicated by flashing a red and a green L.E.D. (Light Emitting Diode), also located on the E.C.U. Since all the self-diagnostic results are stored in the E.C.U.'s memory even intermittent malfunctions can be diagnosed.

A malfunction is indicated by the number of both red and green flashing L.E.D.s. First, the red L.E.D. flashes and the green flashes follow. The red L.E.D. corresponds to units of ten and the green L.E.D. corresponds to units of one. For example, when the red L.E.D. flashes once and the green L.E.D. flashes twice, this signifies the number "12", showing that the air flow meter signal is malfunctioning. All problems are classified by code numbers in this way.

- **When the engine fails to start, crank it two or more seconds before beginning self-diagnosis.**
- **Before starting self-diagnosis, do not erase the stored memory before beginning self-diagnosis. If it is erased, the self-diagnosis function for intermittent malfunctions will be lost.**

Self-diagnosis — Mode III (Self-diagnostic system) (Cont'd)

RETENTION OF DIAGNOSTIC RESULTS

The diagnostic results will remain in E.C.U. memory until the starter is operated fifty times after a diagnostic item has been judged to be malfunctioning. The diagnostic result will then be cancelled automatically. If a diagnostic item which has been judged to be malfunctioning and stored in memory is again judged to be malfunctioning before the starter is operated fifty times, the second result will replace the previous one. It will be stored in E.C.U. memory until the starter is operated fifty times more.

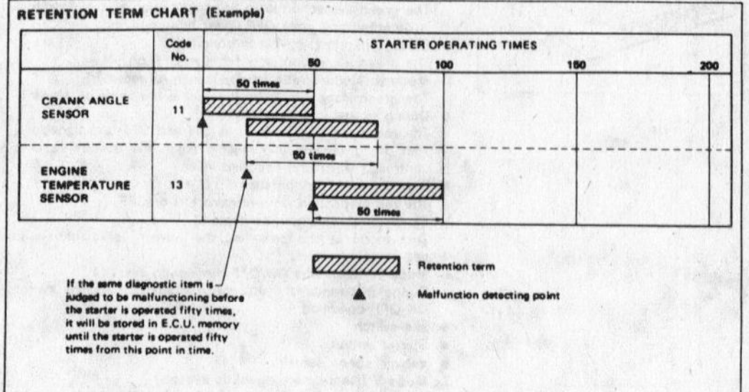
RETENTION TERM CHART (Example)
Code No.
STARTER OPERATING TIMES
50
100
150
200
50 times
CRANK ANGLE SENSOR
11
50 times
ENGINE TEMPERATURE SENSOR
13
50 times
: Retention term
: Malfunction detecting point
If the same diagnostic item is judged to be malfunctioning before the starter is operated fifty times, it will be stored in E.C.U. memory until the starter is operated fifty times from this point in time.

DISPLAY CODE TABLE

Code No.	Detected items	California	Non-California
11	Crank angle sensor circuit	X	X
12	Air flow meter circuit	X	X
13	Engine temperature sensor circuit	X	X
14	Vehicle speed sensor circuit	X	X
21	Ignition signal missing in primary coil	X	X
31	E.C.U. (E.C.C.S. control unit)	X	X
32	E.G.R. function	X	–
33	Exhaust gas sensor circuit	X	X
35	Exhaust gas temperature sensor circuit	X	–
41	Air temperature sensor circuit	X	X
43	Throttle sensor circuit	X	X
45	Injector leak	X	–
55	No malfunction in the above circuits	X	X

X: Available –: Not available

ECCS SELF-DIAGNOSTIC MODE III (CONT.)—1990 AXXESS

ECCS SELF-DIAGNOSTIC MODE III DISPLAY CODES—1990 AXXESS

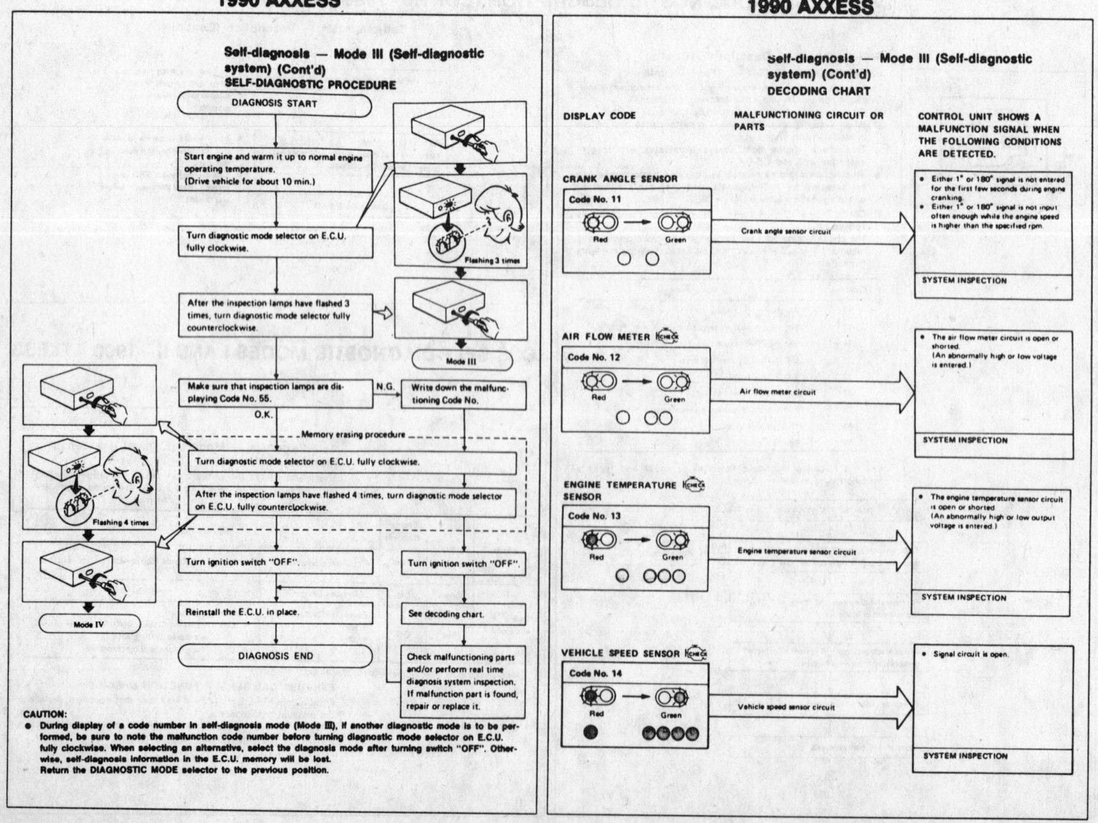
Self-diagnosis — Mode III (Self-diagnostic system) (Cont'd)
SELF-DIAGNOSTIC PROCEDURE
DIAGNOSIS START
Start engine and warm it up to normal engine operating temperature. (Drive vehicle for about 10 min.)
Turn diagnostic mode selector on E.C.U. fully clockwise.
Flashing 3 times
After the inspection lamps have flashed 3 times, turn diagnostic mode selector fully counterclockwise.
Mode III
Make sure that inspection lamps are displaying Code No. 55.
N.G.
Write down the malfunctioning Code No.
O.K.
Memory erasing procedure
Turn diagnostic mode selector on E.C.U. fully clockwise.
After the inspection lamps have flashed 4 times, turn diagnostic mode selector on E.C.U. fully counterclockwise.
Flashing 4 times
Turn ignition switch "OFF".
Turn ignition switch "OFF".
Reinstall the E.C.U. in place.
See decoding chart.
Mode IV
DIAGNOSIS END
Check malfunctioning parts and/or perform real time diagnosis system inspection. If malfunction part is found, repair or replace it.
CAUTION:
During display of a code number in self-diagnosis mode (Mode III), if another diagnostic mode is to be performed, be sure to note the malfunction code number before turning diagnostic mode selector on E.C.U. fully clockwise. When selecting an alternative, select the diagnosis mode after turning switch "OFF". Otherwise, self-diagnosis information in the E.C.U. memory will be lost.
Return the DIAGNOSTIC MODE selector to the previous position.
Self-diagnosis — Mode III (Self-diagnostic system) (Cont'd)
DECODING CHART
DISPLAY CODE
MALFUNCTIONING CIRCUIT OR PARTS
CONTROL UNIT SHOWS A MALFUNCTION SIGNAL WHEN THE FOLLOWING CONDITIONS ARE DETECTED.
CRANK ANGLE SENSOR
Code No. 11
Red
Green
Crank angle sensor circuit
Either 1° or 180° signal is not entered for the first few seconds during engine cranking.
Either 1° or 180° signal is not input often enough while the engine speed is higher than the specified rpm.
SYSTEM INSPECTION
AIR FLOW METER
Code No. 12
Red
Green
Air flow meter circuit
The air flow meter circuit is open or shorted. (An abnormally high or low voltage is entered.)
SYSTEM INSPECTION
ENGINE TEMPERATURE SENSOR
Code No. 13
Red
Green
Engine temperature sensor circuit
The engine temperature sensor circuit is open or shorted (An abnormally high or low output voltage is entered.)
SYSTEM INSPECTION
VEHICLE SPEED SENSOR
Code No. 14
Red
Green
Vehicle speed sensor circuit
Signal circuit is open.
SYSTEM INSPECTION

ECCS SELF-DIAGNOSTIC MODE III DISPLAY CODES (CONT.)—1990 AXXESS

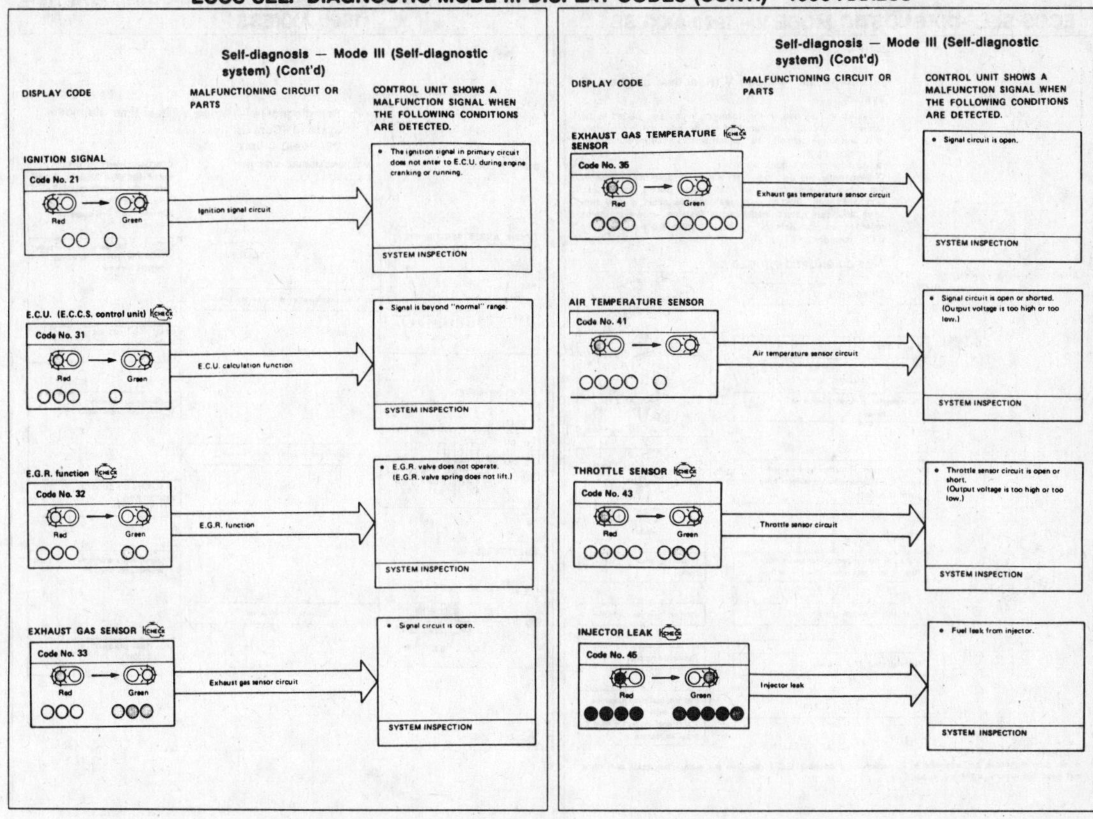
Self-diagnosis — Mode III (Self-diagnostic system) (Cont'd)
DISPLAY CODE
MALFUNCTIONING CIRCUIT OR PARTS
CONTROL UNIT SHOWS A MALFUNCTION SIGNAL WHEN THE FOLLOWING CONDITIONS ARE DETECTED.
IGNITION SIGNAL
Code No. 21
Red
Green
Ignition signal circuit
• The ignition signal in primary circuit does not enter to E.C.U. during engine cranking or running.
SYSTEM INSPECTION
E.C.U. (E.C.C.S. control unit)
Code No. 31
E.C.U. calculation function
• Signal is beyond "normal" range.
SYSTEM INSPECTION
E.G.R. function
Code No. 32
E.G.R. function
• E.G.R. valve does not operate. (E.G.R. valve spring does not lift.)
SYSTEM INSPECTION
EXHAUST GAS SENSOR
Code No. 33
Exhaust gas sensor circuit
• Signal circuit is open.
SYSTEM INSPECTION
Self-diagnosis — Mode III (Self-diagnostic system) (Cont'd)
DISPLAY CODE
MALFUNCTIONING CIRCUIT OR PARTS
CONTROL UNIT SHOWS A MALFUNCTION SIGNAL WHEN THE FOLLOWING CONDITIONS ARE DETECTED.
EXHAUST GAS TEMPERATURE SENSOR
Code No. 35
Exhaust gas temperature sensor circuit
• Signal circuit is open.
SYSTEM INSPECTION
AIR TEMPERATURE SENSOR
Code No. 41
Air temperature sensor circuit
• Signal circuit is open or shorted. (Output voltage is too high or too low.)
SYSTEM INSPECTION
THROTTLE SENSOR
Code No. 43
Throttle sensor circuit
• Throttle sensor circuit is open or short. (Output voltage is too high or too low.)
SYSTEM INSPECTION
INJECTOR LEAK
Code No. 45
Injector leak
• Fuel leak from injector.
SYSTEM INSPECTION

ECCS SELF-DIAGNOSTIC MODE III DISPLAY CODES—1990 AXXESS

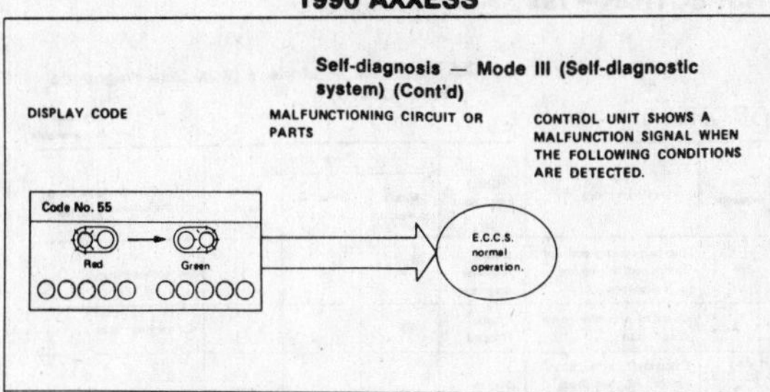
Self-diagnosis — Mode III (Self-diagnostic system) (Cont'd)
DISPLAY CODE
MALFUNCTIONING CIRCUIT OR PARTS
CONTROL UNIT SHOWS A MALFUNCTION SIGNAL WHEN THE FOLLOWING CONDITIONS ARE DETECTED.
Code No. 55
Red
Green
E.C.C.S. normal operation.

ECCS SELF-DIAGNOSTIC MODE IV—1990 AXXESS

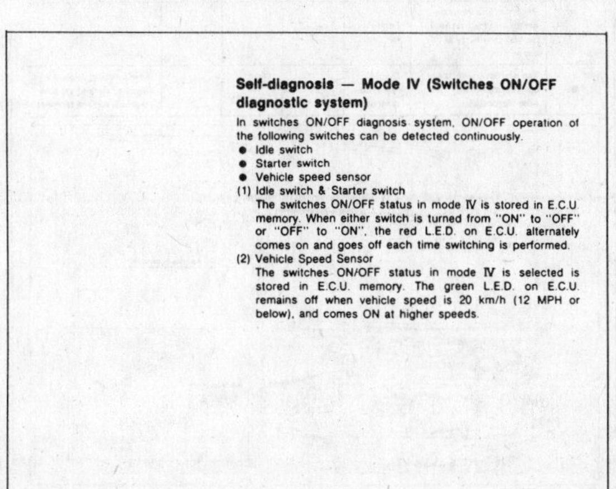

Self-diagnosis — Mode IV (Switches ON/OFF diagnostic system)

In switches ON/OFF diagnosis system, ON/OFF operation of the following switches can be detected continuously.

- Idle switch
- Starter switch
- Vehicle speed sensor

(1) Idle switch & Starter switch
The switches ON/OFF status in mode IV is stored in E.C.U. memory. When either switch is turned from "ON" to "OFF" or "OFF" to "ON", the red L.E.D. on E.C.U. alternately comes on and goes off each time switching is performed.

(2) Vehicle Speed Sensor
The switches ON/OFF status in mode IV is selected is stored in E.C.U. memory. The green L.E.D. on E.C.U. remains off when vehicle speed is 20 km/h (12 MPH or below), and comes ON at higher speeds.

ECCS SELF-DIAGNOSTIC MODE IV (CONT.)—1990 AXXESS

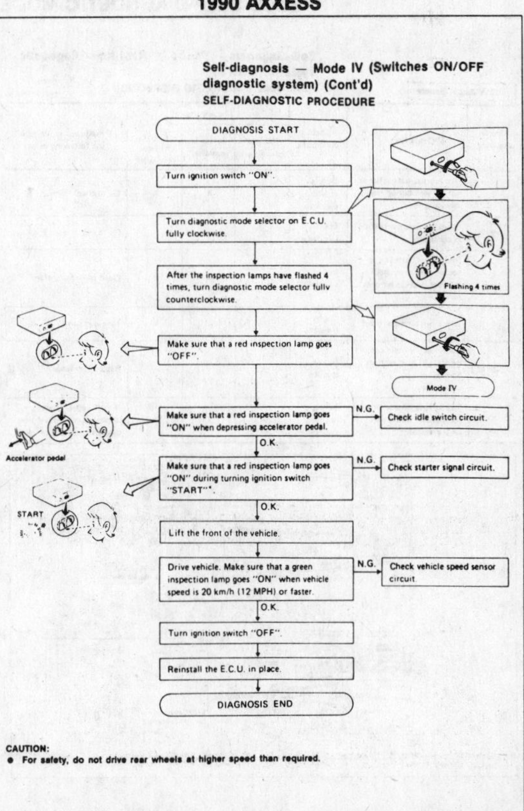
Self-diagnosis — Mode IV (Switches ON/OFF diagnostic system) (Cont'd)
SELF-DIAGNOSTIC PROCEDURE
DIAGNOSIS START
Turn ignition switch "ON".
Turn diagnostic mode selector on E.C.U. fully clockwise.
After the inspection lamps have flashed 4 times, turn diagnostic mode selector fully counterclockwise.
Flashing 4 times
Make sure that a red inspection lamp goes "OFF".
Mode IV
Make sure that a red inspection lamp goes "ON" when depressing accelerator pedal.
N.G.
Check idle switch circuit.
O.K.
Accelerator pedal
Make sure that a red inspection lamp goes "ON" during turning ignition switch "START".
N.G.
Check starter signal circuit.
O.K.
START
Lift the front of the vehicle.
Drive vehicle. Make sure that a green inspection lamp goes "ON" when vehicle speed is 20 km/h (12 MPH) or faster.
N.G.
Check vehicle speed sensor circuit.
O.K.
Turn ignition switch "OFF".
Reinstall the E.C.U. in place.
DIAGNOSIS END
CAUTION:
• For safety, do not drive rear wheels at higher speed than required.

ECCS SELF-DIAGNOSTIC MODE V—1990 AXXESS

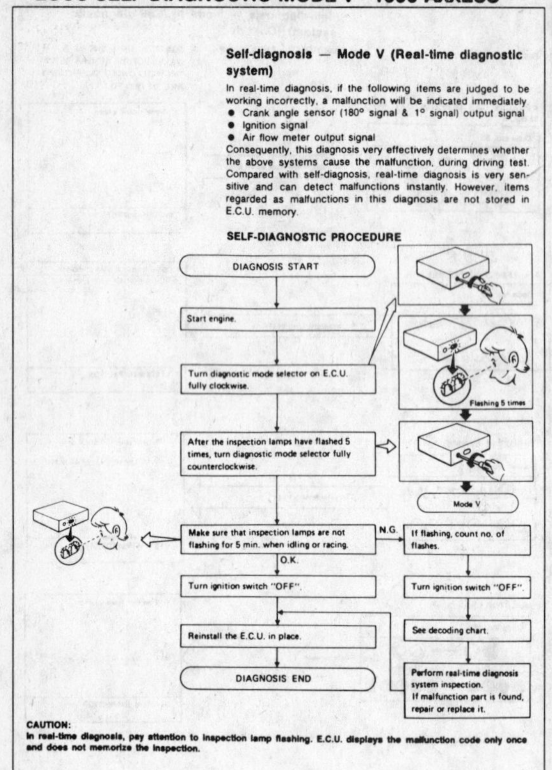
Self-diagnosis — Mode V (Real-time diagnostic system)
In real-time diagnosis, if the following items are judged to be working incorrectly, a malfunction will be indicated immediately
● Crank angle sensor (180° signal & 1° signal) output signal
● Ignition signal
● Air flow meter output signal
Consequently, this diagnosis very effectively determines whether the above systems cause the malfunction, during driving test. Compared with self-diagnosis, real-time diagnosis is very sensitive and can detect malfunctions instantly. However, items regarded as malfunctions in this diagnosis are not stored in E.C.U. memory.
SELF-DIAGNOSTIC PROCEDURE
DIAGNOSIS START
Start engine.
Turn diagnostic mode selector on E.C.U. fully clockwise.
Flashing 5 times
After the inspection lamps have flashed 5 times, turn diagnostic mode selector fully counterclockwise.
Mode V
Make sure that inspection lamps are not flashing for 5 min. when idling or racing.
N.G.
O.K.
If flashing, count no. of flashes.
Turn ignition switch "OFF".
Turn ignition switch "OFF".
See decoding chart.
Reinstall the E.C.U. in place.
Perform real-time diagnosis system inspection.
If malfunction part is found, repair or replace it.
DIAGNOSIS END
CAUTION:
In real-time diagnosis, pay attention to inspection lamp flashing. E.C.U. displays the malfunction code only once and does not memorize the inspection.

ECCS SELF-DIAGNOSTIC MODE V DECODING CHART—1990 AXXESS

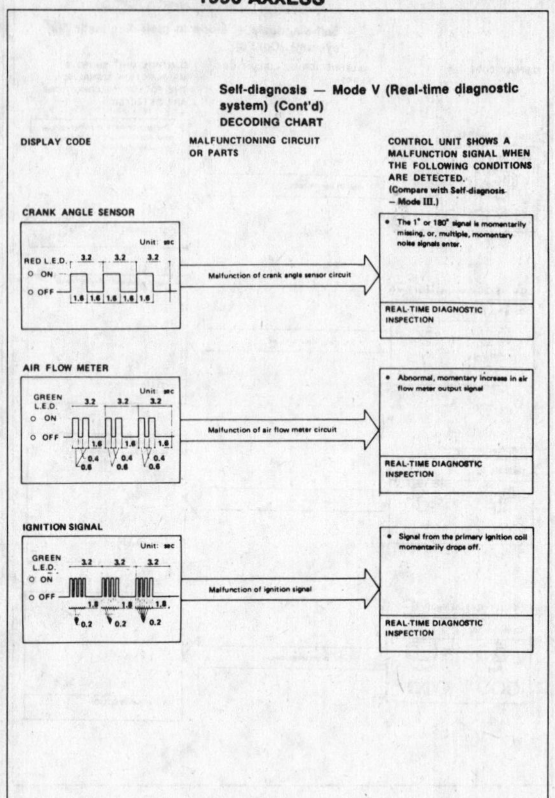
Self-diagnosis — Mode V (Real-time diagnostic system) (Cont'd)
DECODING CHART
DISPLAY CODE
MALFUNCTIONING CIRCUIT OR PARTS
CONTROL UNIT SHOWS A MALFUNCTION SIGNAL WHEN THE FOLLOWING CONDITIONS ARE DETECTED. (Compare with Self-diagnosis — Mode III.)
CRANK ANGLE SENSOR
Unit: sec
RED L.E.D.
ON
OFF
3.2 3.2 3.2
1.6 1.6 1.6 1.6 1.6
Malfunction of crank angle sensor circuit
• The 1° or 180° signal is momentarily missing, or, multiple, momentary noise signals enter.
REAL-TIME DIAGNOSTIC INSPECTION
AIR FLOW METER
Unit: sec
GREEN L.E.D.
ON
OFF
3.2 3.2 3.2
1.6 1.6 1.6
0.4 0.4 0.4
0.6 0.6 0.6
Malfunction of air flow meter circuit
• Abnormal, momentary increase in air flow meter output signal
REAL-TIME DIAGNOSTIC INSPECTION
IGNITION SIGNAL
Unit: sec
GREEN L.E.D.
ON
OFF
3.2 3.2 3.2
1.8 1.8 1.8
0.2 0.2 0.2
Malfunction of ignition signal
• Signal from the primary ignition coil momentarily drops off.
REAL-TIME DIAGNOSTIC INSPECTION

ECCS SELF-DIAGNOSTIC MODE V INSPECTION—1990 AXXESS

Self-diagnosis — Mode V (Real-time diagnostic system) (Cont'd)

REAL-TIME DIAGNOSTIC INSPECTION

X: Available
—: Not available

Crank Angle Sensor

Check sequence	Check items	Check conditions	Check parts: Middle connectors	Sensor & actuator	E.C.U. harness connector	If malfunction, perform the following items.
1	Tap harness connector or component during real-time diagnosis.	During real-time diagnosis	X	X	X	Go to check item 2.
2	Check harness continuity at connector.	Engine stopped	X	—	—	Go to check item 3.
3	Disconnect harness connector, and then check dust adhesion to harness connector.	Engine stopped	X	—	X	Clean terminal surface.
4	Check pin terminal bend.	Engine stopped	—	—	X	Take out bend.
5	Reconnect harness connector and then recheck harness continuity at connector.	Engine stopped	X	—	—	Replace terminal.
6	Tap harness connector or component during real-time diagnosis.	During real-time diagnosis	X	X	X	If malfunction codes are displayed during real-time diagnosis, replace terminal.

E.C.U. S.M.J. harness connector

Crank angle sensor harness connector

Self-diagnosis — Mode V (Real-time diagnostic system) (Cont'd)

X: Available
—: Not available

Air Flow Meter

Check sequence	Check items	Check conditions	Check parts: Middle connectors	Sensor & actuator	E.C.U. harness connector	If malfunction, perform the following items.
1	Tap harness connector or component during real-time diagnosis.	During real-time diagnosis	X	X	X	Go to check item 2.
2	Check harness continuity at connector.	Engine stopped	X	—	—	Go to check item 3.
3	Disconnect harness connector, and then check dust adhesion to harness connector.	Engine stopped	X	—	X	Clean terminal surface.
4	Check pin terminal bend.	Engine stopped	—	—	X	Take out bend.
5	Reconnect harness connector and then recheck harness continuity at connector.	Engine stopped	X	—	—	Replace terminal.
6	Tap harness connector or component during real-time diagnosis.	During real-time diagnosis	X	X	X	If malfunction codes are displayed during real-time diagnosis, replace terminal.

E.C.U. S.M.J. harness connector

Air flow meter

ECCS SELF-DIAGNOSTIC MODE V INSPECTION—1990 AXXESS

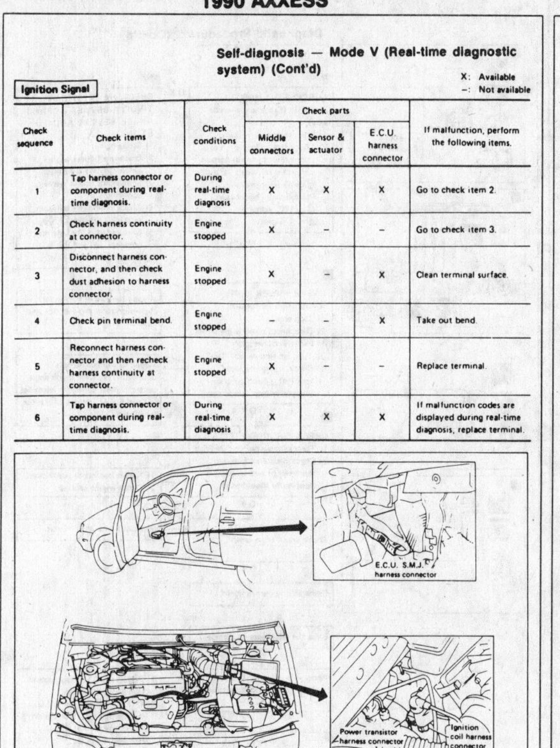

Self-diagnosis — Mode V (Real-time diagnostic system) (Cont'd)

X: Available
–: Not available

Ignition Signal

Check sequence	Check items	Check conditions	Check parts: Middle connectors	Check parts: Sensor & actuator	Check parts: E.C.U. harness connector	If malfunction, perform the following items.
1	Tap harness connector or component during real-time diagnosis.	During real-time diagnosis	X	X	X	Go to check item 2.
2	Check harness continuity at connector.	Engine stopped	X	–	–	Go to check item 3.
3	Disconnect harness connector, and then check dust adhesion to harness connector.	Engine stopped	X	–	X	Clean terminal surface.
4	Check pin terminal bend.	Engine stopped	–	–	X	Take out bend.
5	Reconnect harness connector and then recheck harness continuity at connector.	Engine stopped	X	–	–	Replace terminal.
6	Tap harness connector or component during real-time diagnosis.	During real-time diagnosis	X	X	X	If malfunction codes are displayed during real-time diagnosis, replace terminal.

E.C.U. S.M.J. harness connector
Power transistor harness connector
Ignition coil harness connector

ECCS CAUTION—1990 AXXESS

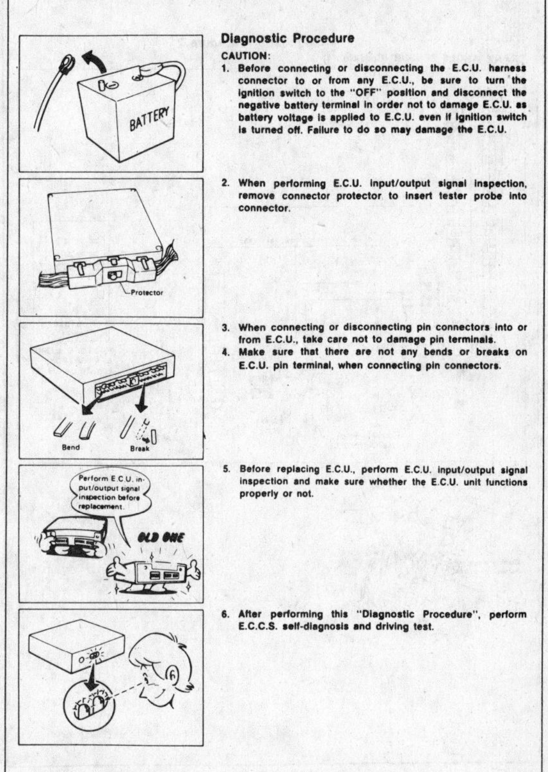

BATTERY
Protector
Bend
Break
Perform E.C.U. input/output signal inspection before replacement.
OLD ONE

Diagnostic Procedure

CAUTION:

1. Before connecting or disconnecting the E.C.U. harness connector to or from any E.C.U., be sure to turn the ignition switch to the "OFF" position and disconnect the negative battery terminal in order not to damage E.C.U. as battery voltage is applied to E.C.U. even if ignition switch is turned off. Failure to do so may damage the E.C.U.
2. When performing E.C.U. input/output signal inspection, remove connector protector to insert tester probe into connector.
3. When connecting or disconnecting pin connectors into or from E.C.U., take care not to damage pin terminals.
4. Make sure that there are not any bends or breaks on E.C.U. pin terminal, when connecting pin connectors.
5. Before replacing E.C.U., perform E.C.U. input/output signal inspection and make sure whether the E.C.U. unit functions properly or not.
6. After performing this "Diagnostic Procedure", perform E.C.C.S. self-diagnosis and driving test.

ECCS CAUTION (CONT.)—1990 AXXESS

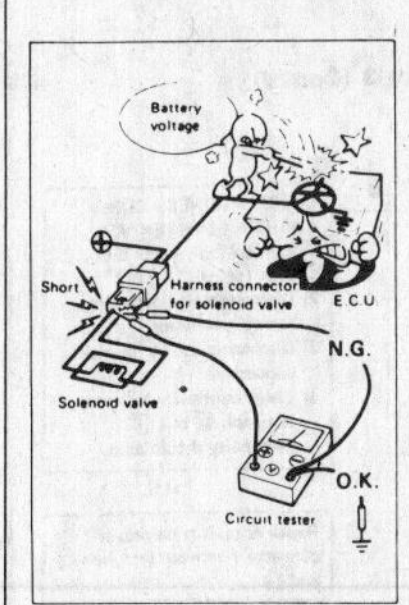

Battery voltage
Short
Harness connector for solenoid valve
E.C.U.
N.G.
Solenoid valve
O.K.
Circuit tester

Diagnostic Procedure (Cont'd)

7. When measuring E.C.U. controlled components supply voltage with a circuit tester, separate one tester probe from the other.
If the two tester probes accidentally make contact with each other during measurement, the circuit will be shorted, resulting in damage to the control unit power transistor.

ECCS SELF-DIAGNOSTIC CHART—1990 AXXESS

Diagnostic Procedure 1

MAIN POWER SUPPLY AND GROUND CIRCUIT

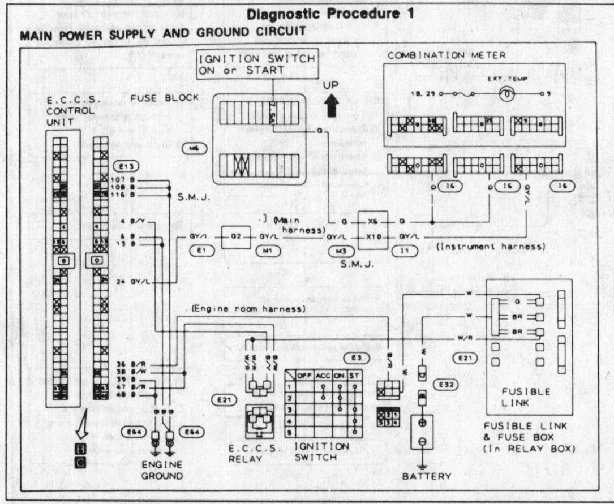

IGNITION SWITCH ON or START
COMBINATION METER
E.C.C.S. CONTROL UNIT
FUSE BLOCK
UP
S.M.J.
(Main harness)
(Instrument harness)
S.M.J.
(Engine room harness)
ENGINE GROUND
E.C.C.S. RELAY
IGNITION SWITCH
OFF ACC ON ST
BATTERY
FUSIBLE LINK
FUSIBLE LINK & FUSE BOX (In RELAY BOX)

ECCS SELF-DIAGNOSTIC CHART—1990 AXXESS

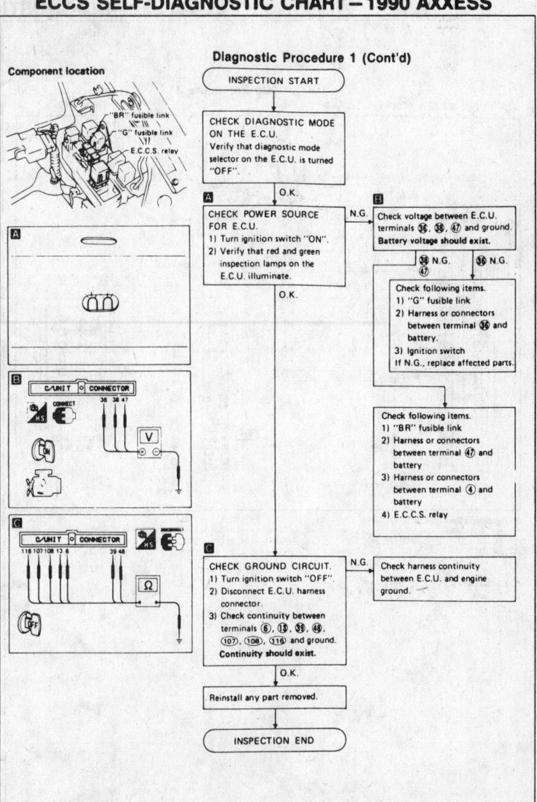

Component location
"BR" fusible link
"G" fusible link
E.C.C.S. relay
C/UNIT CONNECTOR
Diagnostic Procedure 1 (Cont'd)
INSPECTION START
CHECK DIAGNOSTIC MODE ON THE E.C.U.
Verify that diagnostic mode selector on the E.C.U. is turned "OFF".
O.K.
A
CHECK POWER SOURCE FOR E.C.U.
1) Turn ignition switch "ON".
2) Verify that red and green inspection lamps on the E.C.U. illuminate.
N.G.
B
Check voltage between E.C.U. terminals 36, 38, 47 and ground.
Battery voltage should exist.
38 47 N.G.
36 N.G.
Check following items.
1) "G" fusible link
2) Harness or connectors between terminal 36 and battery.
3) Ignition switch
If N.G., replace affected parts.
Check following items.
1) "BR" fusible link
2) Harness or connectors between terminal 47 and battery
3) Harness or connectors between terminal 4 and battery
4) E.C.C.S. relay
O.K.
C
CHECK GROUND CIRCUIT.
1) Turn ignition switch "OFF".
2) Disconnect E.C.U. harness connector.
3) Check continuity between terminals 6, 13, 39, 48, 107, 108, 116 and ground.
Continuity should exist.
N.G.
Check harness continuity between E.C.U. and engine ground.
O.K.
Reinstall any part removed.
INSPECTION END

ECCS SELF-DIAGNOSTIC CHART—1990 AXXESS

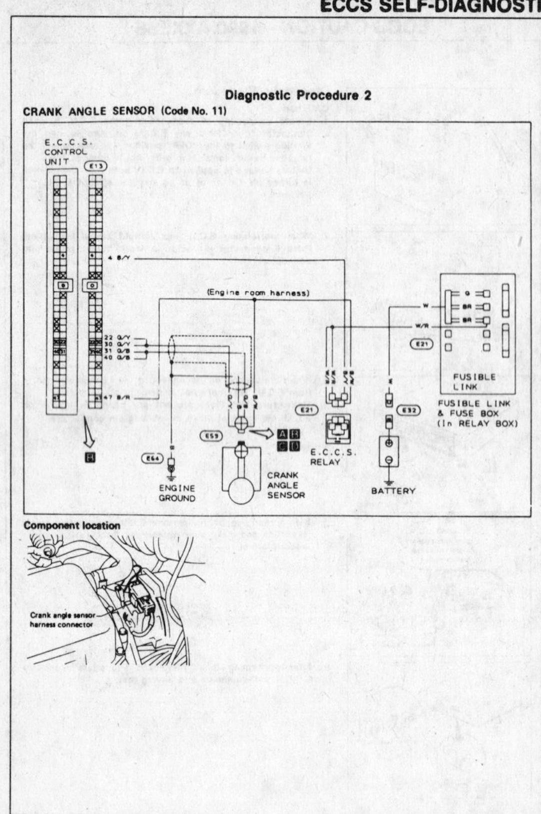
Diagnostic Procedure 2
CRANK ANGLE SENSOR (Code No. 11)
E.C.C.S. CONTROL UNIT
E13
4 B/Y
(Engine room harness)
22 G/Y
30 G/Y
31 G/B
40 G/B
47 B/R
W
W/R
E21
E32
E53
E64
FUSIBLE LINK
FUSIBLE LINK & FUSE BOX (In RELAY BOX)
E.C.C.S. RELAY
BATTERY
ENGINE GROUND
CRANK ANGLE SENSOR
Component location
Crank angle sensor harness connector

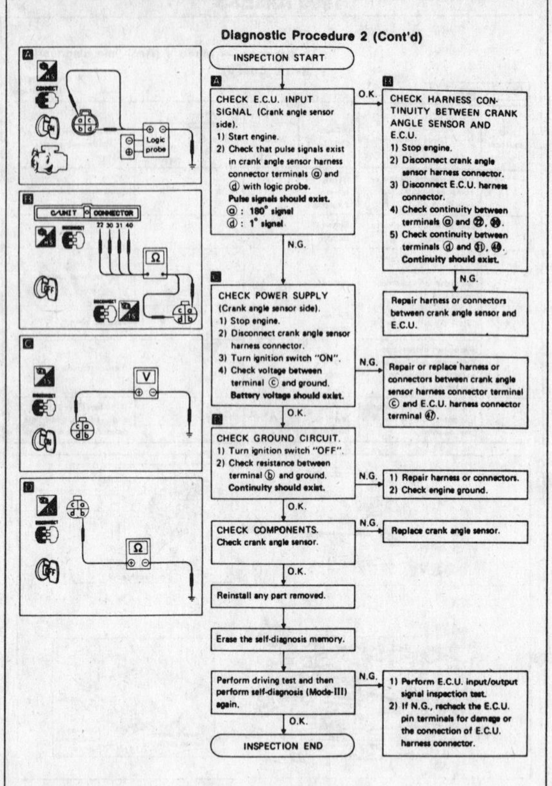
Diagnostic Procedure 2 (Cont'd)
INSPECTION START
CHECK E.C.U. INPUT SIGNAL (Crank angle sensor side).
1) Start engine.
2) Check that pulse signals exist in crank angle sensor harness connector terminals ⓐ and ⓓ with logic probe.
Pulse signals should exist.
ⓐ : 180° signal
ⓓ : 1° signal
O.K.
CHECK HARNESS CONTINUITY BETWEEN CRANK ANGLE SENSOR AND E.C.U.
1) Stop engine.
2) Disconnect crank angle sensor harness connector.
3) Disconnect E.C.U. harness connector.
4) Check continuity between terminals ⓐ and ㉒, ㉚.
5) Check continuity between terminals ⓓ and ㉛, ㊵.
Continuity should exist.
N.G.
Repair harness or connectors between crank angle sensor and E.C.U.
N.G.
CHECK POWER SUPPLY (Crank angle sensor side).
1) Stop engine.
2) Disconnect crank angle sensor harness connector.
3) Turn ignition switch "ON".
4) Check voltage between terminal ⓒ and ground.
Battery voltage should exist.
N.G.
Repair or replace harness or connectors between crank angle sensor harness connector terminal ⓒ and E.C.U. harness connector terminal ㊼.
O.K.
CHECK GROUND CIRCUIT.
1) Turn ignition switch "OFF".
2) Check resistance between terminal ⓑ and ground.
Continuity should exist.
N.G.
1) Repair harness or connectors.
2) Check engine ground.
O.K.
CHECK COMPONENTS.
Check crank angle sensor.
N.G.
Replace crank angle sensor.
O.K.
Reinstall any part removed.
Erase the self-diagnosis memory.
Perform driving test and then perform self-diagnosis (Mode-III) again.
N.G.
1) Perform E.C.U. input/output signal inspection test.
2) If N.G., recheck the E.C.U. pin terminals for damage or the connection of E.C.U. harness connector.
O.K.
INSPECTION END
Logic probe
C/UNIT CONNECTOR
22 30 31 40

ECCS SELF-DIAGNOSTIC CHART—1990 AXXESS

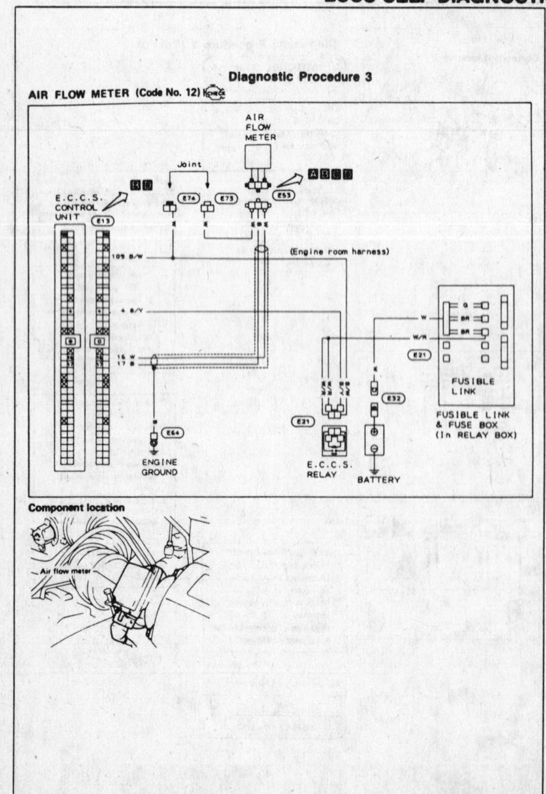
Diagnostic Procedure 3
AIR FLOW METER (Code No. 12)
AIR FLOW METER
Joint
E.C.C.S. CONTROL UNIT
E13
E74
E73
E53
108 B/W
(Engine room harness)
4 B/Y
16
17
W
W/R
E21
E32
E64
FUSIBLE LINK
FUSIBLE LINK & FUSE BOX (In RELAY BOX)
E.C.C.S. RELAY
BATTERY
ENGINE GROUND
Component location
Air flow meter

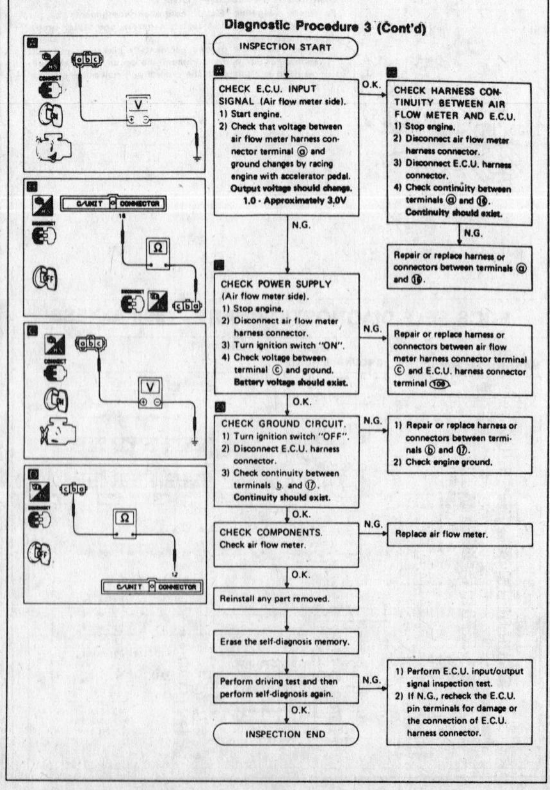
Diagnostic Procedure 3 (Cont'd)
INSPECTION START
CHECK E.C.U. INPUT SIGNAL (Air flow meter side).
1) Start engine.
2) Check that voltage between air flow meter harness connector terminal ⓐ and ground changes by racing engine with accelerator pedal.
Output voltage should change.
1.0 - Approximately 3.0V
O.K.
CHECK HARNESS CONTINUITY BETWEEN AIR FLOW METER AND E.C.U.
1) Stop engine.
2) Disconnect air flow meter harness connector.
3) Disconnect E.C.U. harness connector.
4) Check continuity between terminals ⓐ and ⑯.
Continuity should exist.
N.G.
Repair or replace harness or connectors between terminals ⓐ and ⑯.
N.G.
CHECK POWER SUPPLY (Air flow meter side).
1) Stop engine.
2) Disconnect air flow meter harness connector.
3) Turn ignition switch "ON".
4) Check voltage between terminal ⓒ and ground.
Battery voltage should exist.
N.G.
Repair or replace harness or connectors between air flow meter harness connector terminal ⓒ and E.C.U. harness connector terminal (108).
O.K.
CHECK GROUND CIRCUIT.
1) Turn ignition switch "OFF".
2) Disconnect E.C.U. harness connector.
3) Check continuity between terminals b and ⑰.
Continuity should exist.
N.G.
1) Repair or replace harness or connectors between terminals ⓑ and ⑰.
2) Check engine ground.
O.K.
CHECK COMPONENTS.
Check air flow meter.
N.G.
Replace air flow meter.
O.K.
Reinstall any part removed.
Erase the self-diagnosis memory.
Perform driving test and then perform self-diagnosis again.
N.G.
1) Perform E.C.U. input/output signal inspection test.
2) If N.G., recheck the E.C.U. pin terminals for damage or the connection of E.C.U. harness connector.
O.K.
INSPECTION END
C/UNIT CONNECTOR

ECCS SELF-DIAGNOSTIC CHART—1990 AXXESS

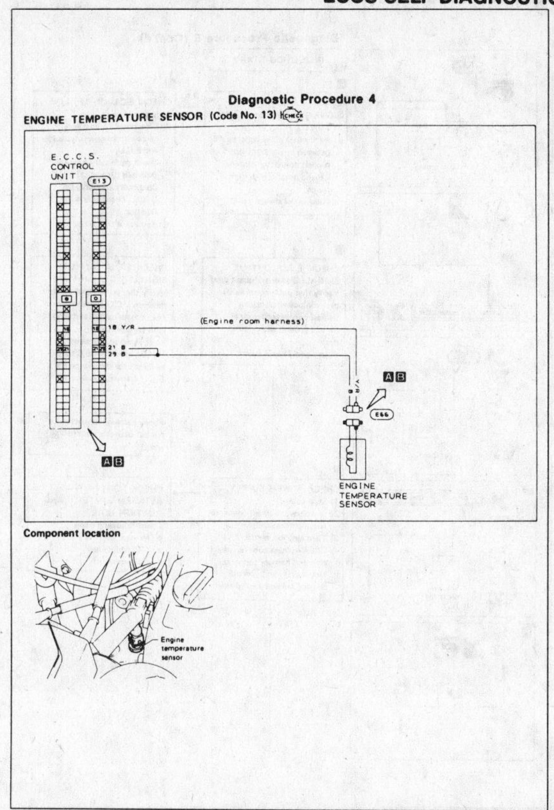
Diagnostic Procedure 4
ENGINE TEMPERATURE SENSOR (Code No. 13)
E.C.C.S. CONTROL UNIT
E13
18 Y/R
21 B
29 B
(Engine room harness)
E66
ENGINE TEMPERATURE SENSOR
Component location
Engine temperature sensor

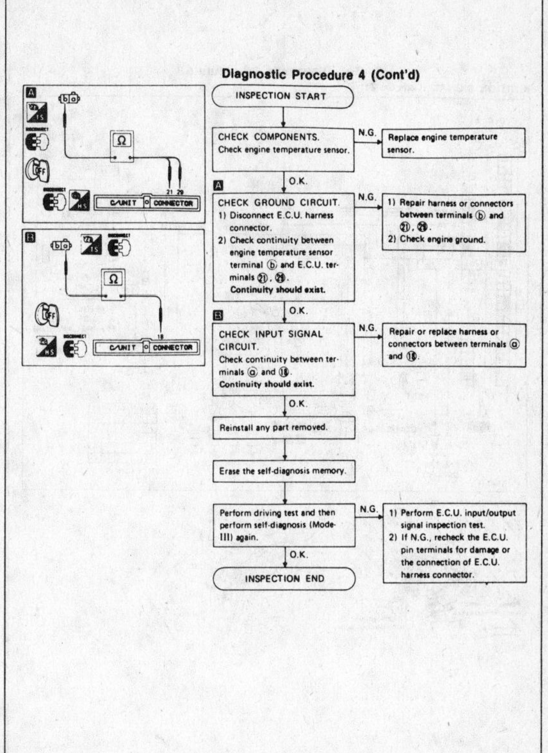
Diagnostic Procedure 4 (Cont'd)
INSPECTION START
CHECK COMPONENTS.
Check engine temperature sensor.
N.G.
Replace engine temperature sensor.
O.K.
CHECK GROUND CIRCUIT.
1) Disconnect E.C.U. harness connector.
2) Check continuity between engine temperature sensor terminal ⓑ and E.C.U. terminals ㉑, ㉙.
Continuity should exist.
N.G.
1) Repair harness or connectors between terminals ⓑ and ㉑, ㉙.
2) Check engine ground.
O.K.
CHECK INPUT SIGNAL CIRCUIT.
Check continuity between terminals ⓐ and ⑱.
Continuity should exist.
N.G.
Repair or replace harness or connectors between terminals ⓐ and ⑱.
O.K.
Reinstall any part removed.
Erase the self-diagnosis memory.
Perform driving test and then perform self-diagnosis (Mode III) again.
N.G.
1) Perform E.C.U. input/output signal inspection test.
2) If N.G., recheck the E.C.U. pin terminals for damage or the connection of E.C.U. harness connector.
O.K.
INSPECTION END
C/UNIT CONNECTOR

ECCS SELF-DIAGNOSTIC CHART—1990 AXXESS

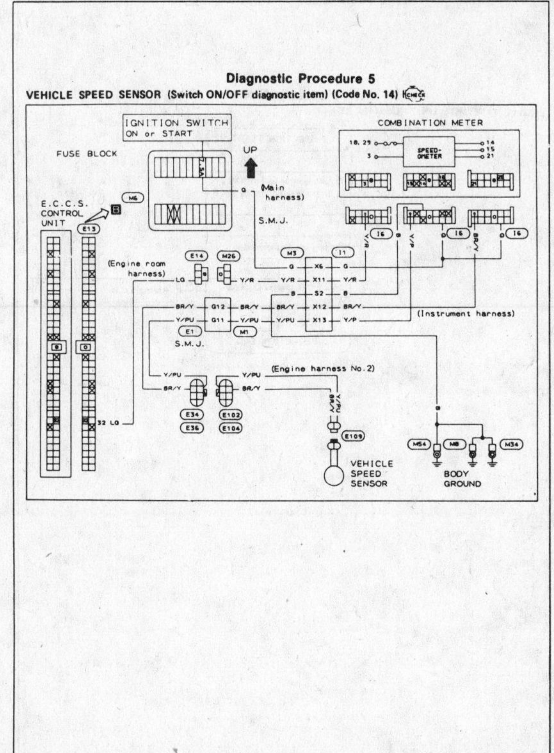
Diagnostic Procedure 5
VEHICLE SPEED SENSOR (Switch ON/OFF diagnostic item) (Code No. 14)
IGNITION SWITCH ON or START
FUSE BLOCK
COMBINATION METER
SPEEDOMETER
UP
(Main harness)
S.M.J.
E.C.C.S. CONTROL UNIT
E13
(Engine room harness)
(Instrument harness)
(Engine harness No. 2)
32 LG
VEHICLE SPEED SENSOR
BODY GROUND

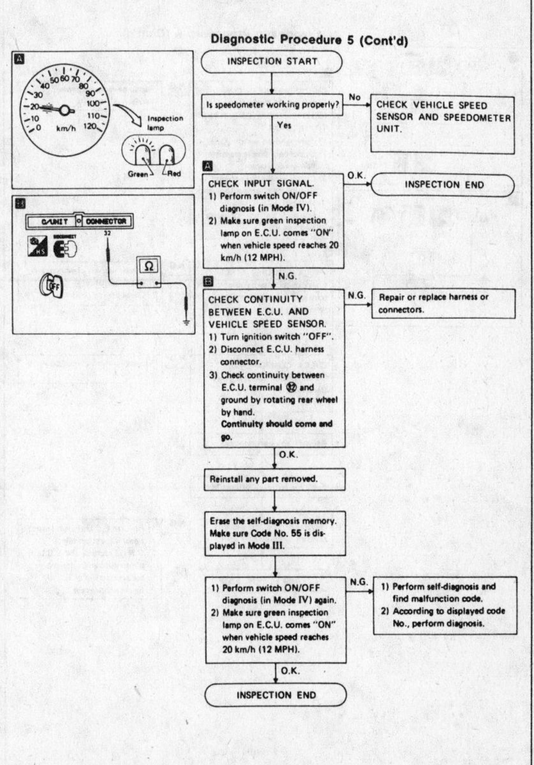
Diagnostic Procedure 5 (Cont'd)
INSPECTION START
Is speedometer working properly?
No
CHECK VEHICLE SPEED SENSOR AND SPEEDOMETER UNIT.
Yes
CHECK INPUT SIGNAL.
1) Perform switch ON/OFF diagnosis (in Mode IV).
2) Make sure green inspection lamp on E.C.U. comes "ON" when vehicle speed reaches 20 km/h (12 MPH).
O.K.
INSPECTION END
N.G.
CHECK CONTINUITY BETWEEN E.C.U. AND VEHICLE SPEED SENSOR.
1) Turn ignition switch "OFF".
2) Disconnect E.C.U. harness connector.
3) Check continuity between E.C.U. terminal ㉜ and ground by rotating rear wheel by hand.
Continuity should come and go.
N.G.
Repair or replace harness or connectors.
O.K.
Reinstall any part removed.
Erase the self-diagnosis memory. Make sure Code No. 55 is displayed in Mode III.
1) Perform switch ON/OFF diagnosis (in Mode IV) again.
2) Make sure green inspection lamp on E.C.U. comes "ON" when vehicle speed reaches 20 km/h (12 MPH).
N.G.
1) Perform self-diagnosis and find malfunction code.
2) According to displayed code No., perform diagnosis.
O.K.
INSPECTION END
Inspection lamp
Green
Red
C/UNIT CONNECTOR

ECCS SELF-DIAGNOSTIC CHART—1990 AXXESS

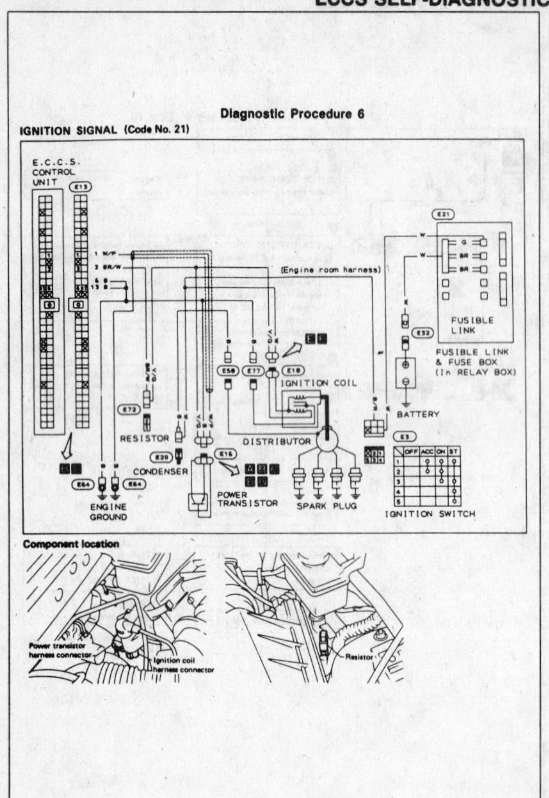
Diagnostic Procedure 6
IGNITION SIGNAL (Code No. 21)
E.C.C.S. CONTROL UNIT
(Engine room harness)
FUSIBLE LINK
FUSIBLE LINK & FUSE BOX (In RELAY BOX)
IGNITION COIL
BATTERY
RESISTOR
DISTRIBUTOR
CONDENSER
POWER TRANSISTOR
SPARK PLUG
IGNITION SWITCH
ENGINE GROUND
Component location
Power transistor harness connector
Ignition coil harness connector
Resistor

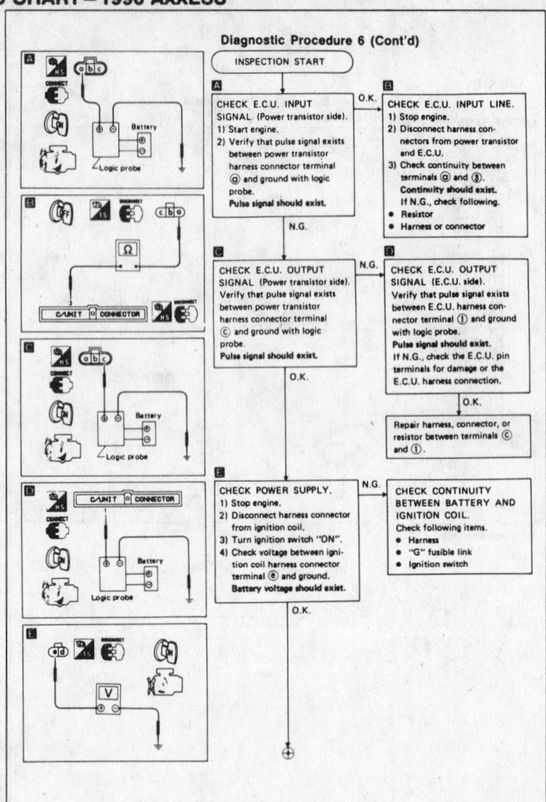
Diagnostic Procedure 6 (Cont'd)
INSPECTION START
A CHECK E.C.U. INPUT SIGNAL (Power transistor side).
1) Start engine.
2) Verify that pulse signal exists between power transistor harness connector terminal ⓐ and ground with logic probe.
Pulse signal should exist.
O.K.
B CHECK E.C.U. INPUT LINE.
1) Stop engine.
2) Disconnect harness connectors from power transistor and E.C.U.
3) Check continuity between terminals ⓐ and ③.
Continuity should exist.
If N.G., check following.
• Resistor
• Harness or connector
N.G.
C CHECK E.C.U. OUTPUT SIGNAL (Power transistor side).
Verify that pulse signal exists between power transistor harness connector terminal ⓒ and ground with logic probe.
Pulse signal should exist.
N.G.
D CHECK E.C.U. OUTPUT SIGNAL (E.C.U. side).
Verify that pulse signal exists between E.C.U. harness connector terminal ① and ground with logic probe.
Pulse signal should exist.
If N.G., check the E.C.U. pin terminals for damage or the E.C.U. harness connection.
O.K.
Repair harness, connector, or resistor between terminals ⓒ and ①.
O.K.
E CHECK POWER SUPPLY.
1) Stop engine.
2) Disconnect harness connector from ignition coil.
3) Turn ignition switch "ON".
4) Check voltage between ignition coil harness connector terminal ⓔ and ground.
Battery voltage should exist.
N.G.
CHECK CONTINUITY BETWEEN BATTERY AND IGNITION COIL.
Check following items.
• Harness
• "G" fusible link
• Ignition switch
O.K.
Battery
Logic probe
C/UNIT CONNECTOR

ECCS SELF-DIAGNOSTIC CHART—1990 AXXESS

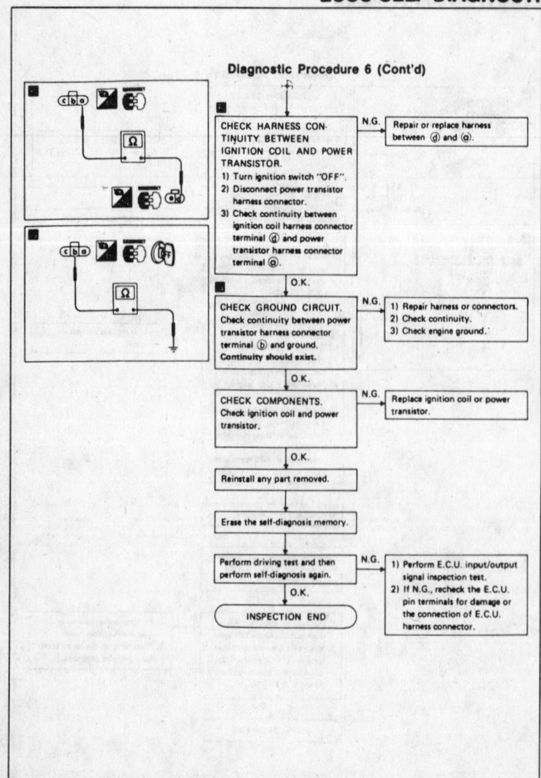
Diagnostic Procedure 6 (Cont'd)
F CHECK HARNESS CONTINUITY BETWEEN IGNITION COIL AND POWER TRANSISTOR.
1) Turn ignition switch "OFF".
2) Disconnect power transistor harness connector.
3) Check continuity between ignition coil harness connector terminal ⓓ and power transistor harness connector terminal ⓐ.
N.G.
Repair or replace harness between ⓓ and ⓐ.
O.K.
G CHECK GROUND CIRCUIT.
Check continuity between power transistor harness connector terminal ⓑ and ground.
Continuity should exist.
N.G.
1) Repair harness or connectors.
2) Check continuity.
3) Check engine ground.
O.K.
CHECK COMPONENTS.
Check ignition coil and power transistor.
N.G.
Replace ignition coil or power transistor.
O.K.
Reinstall any part removed.
Erase the self-diagnosis memory.
Perform driving test and then perform self-diagnosis again.
N.G.
1) Perform E.C.U. input/output signal inspection test.
2) If N.G., recheck the E.C.U. pin terminals for damage or the connection of E.C.U. harness connector.
O.K.
INSPECTION END

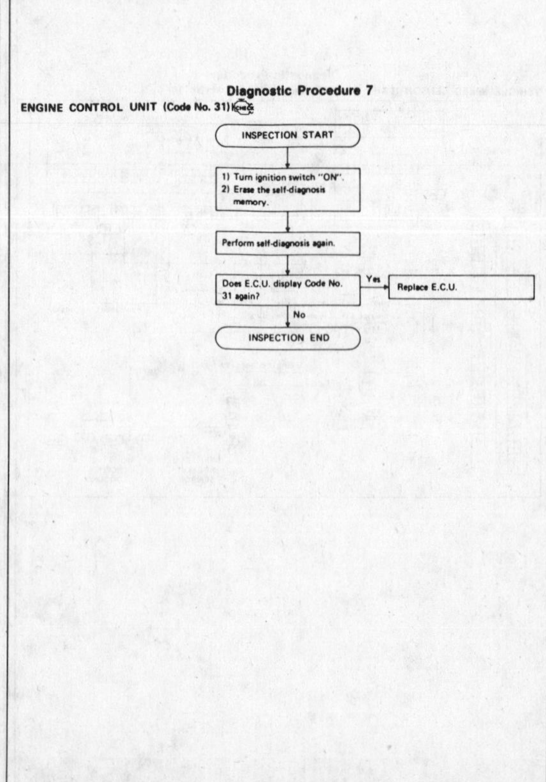
Diagnostic Procedure 7
ENGINE CONTROL UNIT (Code No. 31)
INSPECTION START
1) Turn ignition switch "ON".
2) Erase the self-diagnosis memory.
Perform self-diagnosis again.
Does E.C.U. display Code No. 31 again?
Yes
Replace E.C.U.
No
INSPECTION END

ECCS DIAGNOSTIC CHART—1990 AXXESS

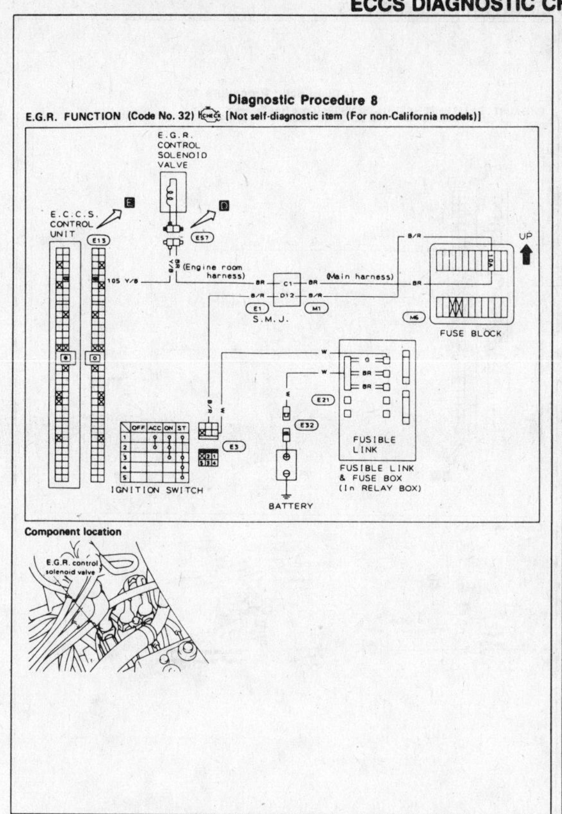
Diagnostic Procedure 8
E.G.R. FUNCTION (Code No. 32) [Not self-diagnostic item (For non-California models)]
E.G.R. CONTROL SOLENOID VALVE
E.C.C.S. CONTROL UNIT
(Engine room harness)
(Main harness)
S.M.J.
UP
FUSE BLOCK
FUSIBLE LINK
FUSIBLE LINK & FUSE BOX (In RELAY BOX)
IGNITION SWITCH
BATTERY
Component location
E.G.R. control solenoid valve

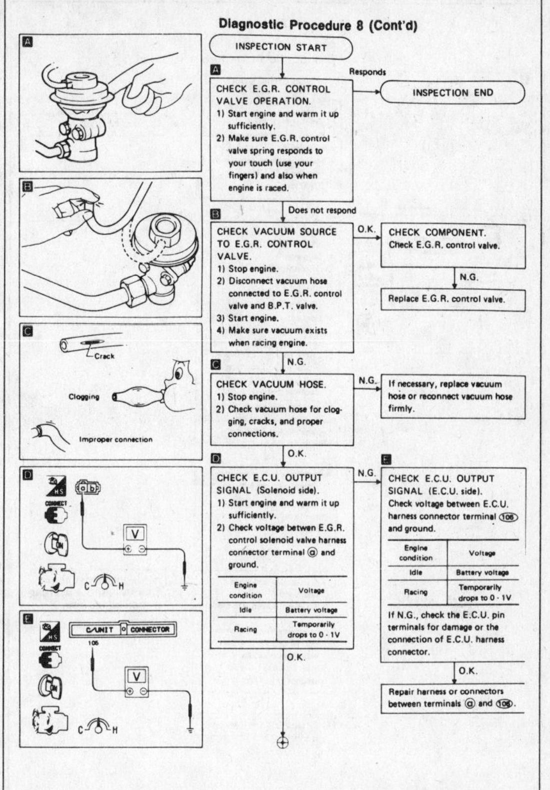
Diagnostic Procedure 8 (Cont'd)
INSPECTION START
CHECK E.G.R. CONTROL VALVE OPERATION.
1) Start engine and warm it up sufficiently.
2) Make sure E.G.R. control valve spring responds to your touch (use your fingers) and also when engine is raced.
Responds
INSPECTION END
Does not respond
CHECK VACUUM SOURCE TO E.G.R. CONTROL VALVE.
1) Stop engine.
2) Disconnect vacuum hose connected to E.G.R. control valve and B.P.T. valve.
3) Start engine.
4) Make sure vacuum exists when racing engine.
O.K.
CHECK COMPONENT. Check E.G.R. control valve.
N.G.
Replace E.G.R. control valve.
N.G.
CHECK VACUUM HOSE.
1) Stop engine.
2) Check vacuum hose for clogging, cracks, and proper connections.
If necessary, replace vacuum hose or reconnect vacuum hose firmly.
Crack
Clogging
Improper connection
CHECK E.C.U. OUTPUT SIGNAL (Solenoid side).
1) Start engine and warm it up sufficiently.
2) Check voltage betwen E.G.R. control solenoid valve harness connector terminal ⓐ and ground.
Engine condition | Voltage
Idle | Battery voltage
Racing | Temporarily drops to 0 - 1V
CHECK E.C.U. OUTPUT SIGNAL (E.C.U. side).
Check voltage between E.C.U. harness connector terminal 106 and ground.
Engine condition | Voltage
Idle | Battery voltage
Racing | Temporarily drops to 0 - 1V
If N.G., check the E.C.U. pin terminals for damage or the connection of E.C.U. harness connector.
O.K.
Repair harness or connectors between terminals ⓐ and 106.
C/UNIT CONNECTOR

ECCS DIAGNOSTIC CHART—1990 AXXESS

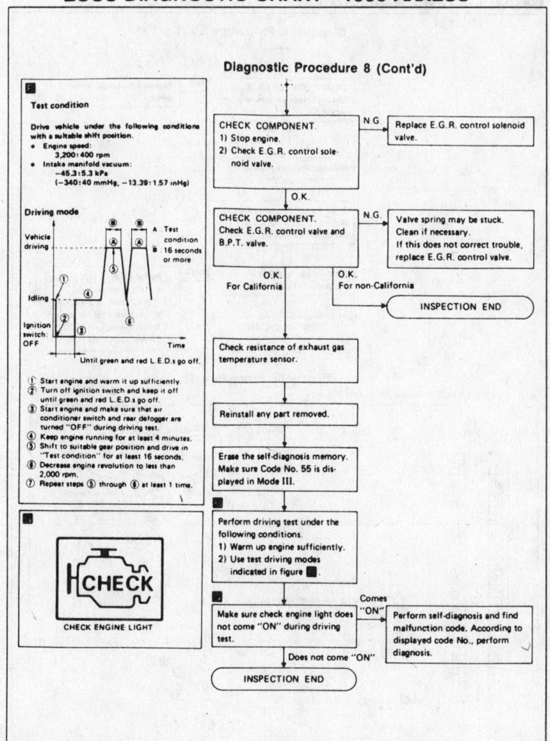
Diagnostic Procedure 8 (Cont'd)
Test condition
Drive vehicle under the following conditions with a suitable shift position.
• Engine speed: 3,200±400 rpm
• Intake manifold vacuum: −45.3±5.3 kPa (−340±40 mmHg, −13.39±1.57 inHg)
Driving mode
A Test condition
B 16 seconds or more
Vehicle driving
Idling
Ignition switch OFF
Time
Until green and red L.E.D.s go off.
① Start engine and warm it up sufficiently.
② Turn off ignition switch and keep it off until green and red L.E.D.s go off.
③ Start engine and make sure that air conditioner switch and rear defogger are turned "OFF" during driving test.
④ Keep engine running for at least 4 minutes.
⑤ Shift to suitable gear position and drive in "Test condition" for at least 16 seconds.
⑥ Decrease engine revolution to less than 2,000 rpm.
⑦ Repeat steps ⑤ through ⑥ at least 1 time.
CHECK
CHECK ENGINE LIGHT
CHECK COMPONENT.
1) Stop engine.
2) Check E.G.R. control solenoid valve.
N.G.
Replace E.G.R. control solenoid valve.
O.K.
CHECK COMPONENT. Check E.G.R. control valve and B.P.T. valve.
N.G.
Valve spring may be stuck. Clean if necessary. If this does not correct trouble, replace E.G.R. control valve.
O.K. For California
O.K. For non-California
INSPECTION END
Check resistance of exhaust gas temperature sensor.
Reinstall any part removed.
Erase the self-diagnosis memory. Make sure Code No. 55 is displayed in Mode III.
Perform driving test under the following conditions.
1) Warm up engine sufficiently.
2) Use test driving modes indicated in figure.
Make sure check engine light does not come "ON" during driving test.
Comes "ON"
Perform self-diagnosis and find malfunction code. According to displayed code No., perform diagnosis.
Does not come "ON"
INSPECTION END

ECCS SELF-DIAGNOSTIC CHART—1990 AXXESS

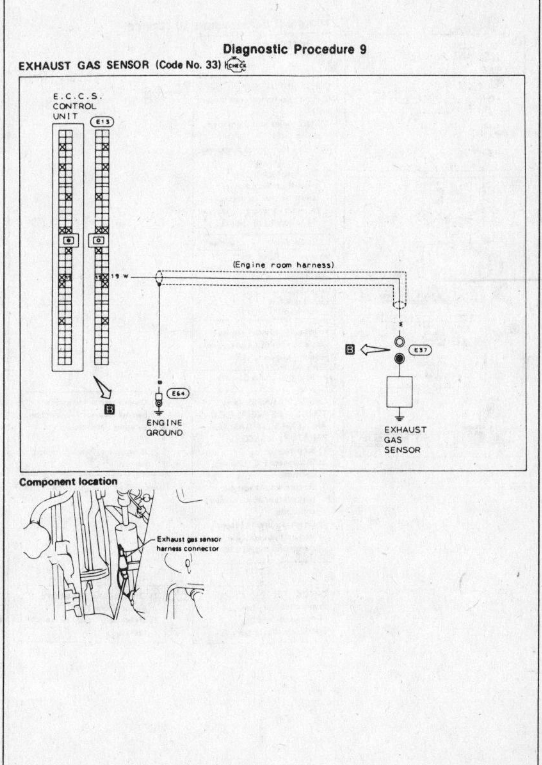
Diagnostic Procedure 9
EXHAUST GAS SENSOR (Code No. 33)
E.C.C.S. CONTROL UNIT
(Engine room harness)
ENGINE GROUND
EXHAUST GAS SENSOR
Component location
Exhaust gas sensor harness connector

ECCS SELF-DIAGNOSTIC CHART—1990 AXXESS

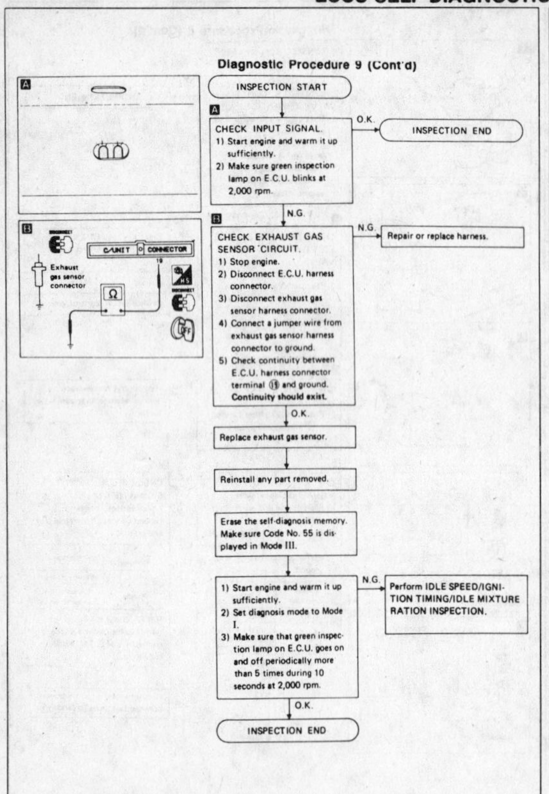
Diagnostic Procedure 9 (Cont'd)
INSPECTION START
CHECK INPUT SIGNAL.
1) Start engine and warm it up sufficiently.
2) Make sure green inspection lamp on E.C.U. blinks at 2,000 rpm.
O.K.
INSPECTION END
N.G.
CHECK EXHAUST GAS SENSOR CIRCUIT.
1) Stop engine.
2) Disconnect E.C.U. harness connector.
3) Disconnect exhaust gas sensor harness connector.
4) Connect a jumper wire from exhaust gas sensor harness connector to ground.
5) Check continuity between E.C.U. harness connector terminal ⑲ and ground.
Continuity should exist.
N.G.
Repair or replace harness.
O.K.
Replace exhaust gas sensor.
Reinstall any part removed.
Erase the self-diagnosis memory. Make sure Code No. 55 is displayed in Mode III.
1) Start engine and warm it up sufficiently.
2) Set diagnosis mode to Mode I.
3) Make sure that green inspection lamp on E.C.U. goes on and off periodically more than 5 times during 10 seconds at 2,000 rpm.
N.G.
Perform IDLE SPEED/IGNITION TIMING/IDLE MIXTURE RATION INSPECTION.
O.K.
INSPECTION END
Exhaust gas sensor connector
C/UNIT
CONNECTOR

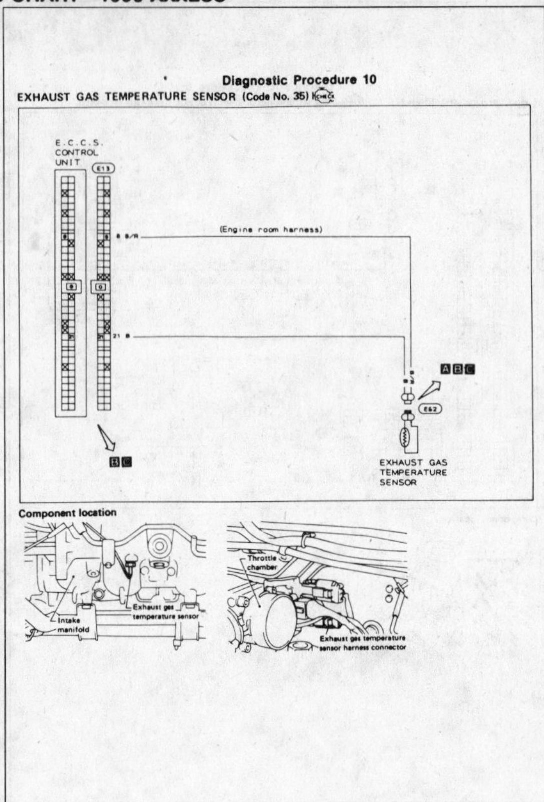
Diagnostic Procedure 10
EXHAUST GAS TEMPERATURE SENSOR (Code No. 35)
E.C.C.S. CONTROL UNIT
E13
(Engine room harness)
8 B/R
21 B
E62
EXHAUST GAS TEMPERATURE SENSOR
Component location
Intake manifold
Exhaust gas temperature sensor
Throttle chamber
Exhaust gas temperature sensor harness connector

ECCS SELF-DIAGNOSTIC CHART—1990 AXXESS

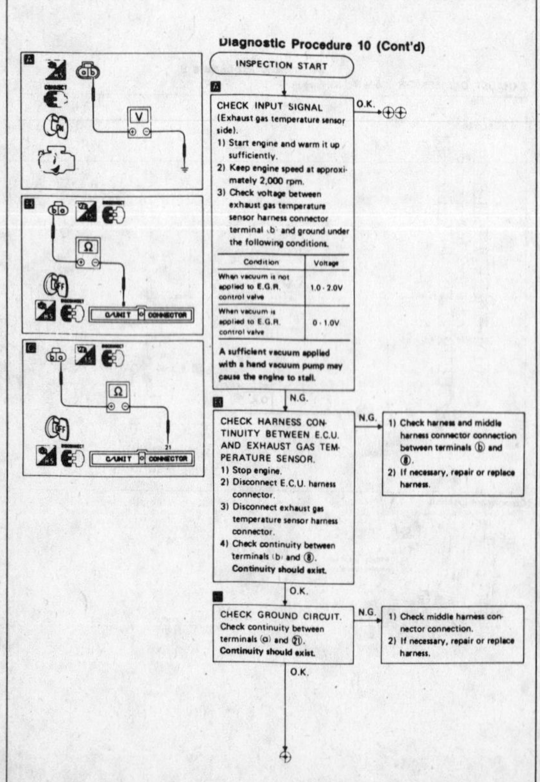
Diagnostic Procedure 10 (Cont'd)
INSPECTION START
CHECK INPUT SIGNAL (Exhaust gas temperature sensor side).
1) Start engine and warm it up sufficiently.
2) Keep engine speed at approximately 2,000 rpm.
3) Check voltage between exhaust gas temperature sensor harness connector terminal ⓑ and ground under the following conditions.
Condition | Voltage
When vacuum is not applied to E.G.R. control valve | 1.0 - 2.0V
When vacuum is applied to E.G.R. control valve | 0 - 1.0V
A sufficient vacuum applied with a hand vacuum pump may cause the engine to stall.
O.K.
N.G.
CHECK HARNESS CONTINUITY BETWEEN E.C.U. AND EXHAUST GAS TEMPERATURE SENSOR.
1) Stop engine.
2) Disconnect E.C.U. harness connector.
3) Disconnect exhaust gas temperature sensor harness connector.
4) Check continuity between terminals ⓑ and ⑧.
Continuity should exist.
N.G.
1) Check harness and middle harness connector connection between terminals ⓑ and ⑧.
2) If necessary, repair or replace harness.
O.K.
CHECK GROUND CIRCUIT.
Check continuity between terminals ⓐ and ㉑.
Continuity should exist.
N.G.
1) Check middle harness connector connection.
2) If necessary, repair or replace harness.
O.K.
C/UNIT
CONNECTOR

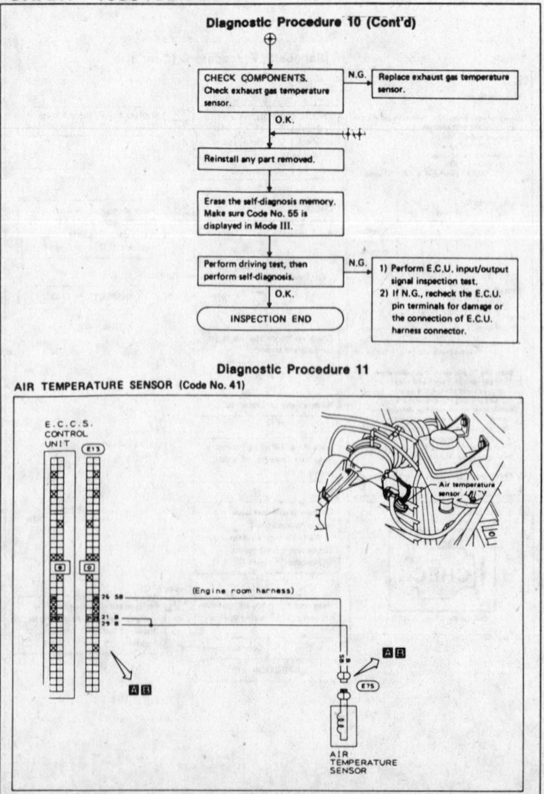
Diagnostic Procedure 10 (Cont'd)
CHECK COMPONENTS.
Check exhaust gas temperature sensor.
N.G.
Replace exhaust gas temperature sensor.
O.K.
Reinstall any part removed.
Erase the self-diagnosis memory. Make sure Code No. 55 is displayed in Mode III.
Perform driving test, then perform self-diagnosis.
N.G.
1) Perform E.C.U. input/output signal inspection test.
2) If N.G., recheck the E.C.U. pin terminals for damage or the connection of E.C.U. harness connector.
O.K.
INSPECTION END
Diagnostic Procedure 11
AIR TEMPERATURE SENSOR (Code No. 41)
E.C.C.S. CONTROL UNIT
E13
(Engine room harness)
Air temperature sensor
E75
AIR TEMPERATURE SENSOR

ECCS SELF-DIAGNOSTIC CHART—1990 AXXESS

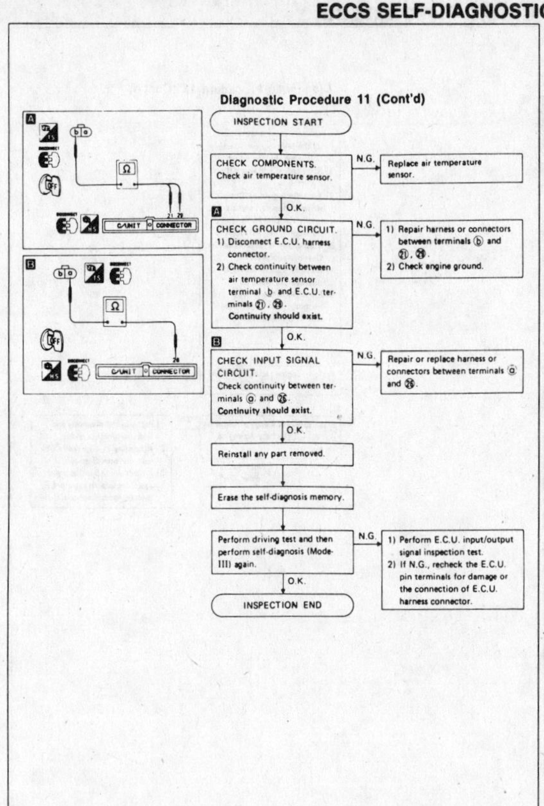
Diagnostic Procedure 11 (Cont'd)
INSPECTION START
CHECK COMPONENTS. Check air temperature sensor.
N.G. Replace air temperature sensor.
O.K.
CHECK GROUND CIRCUIT. 1) Disconnect E.C.U. harness connector. 2) Check continuity between air temperature sensor terminal ⓑ and E.C.U. terminals ㉑, ㉙. Continuity should exist.
N.G. 1) Repair harness or connectors between terminals ⓑ and ㉑, ㉙. 2) Check engine ground.
O.K.
CHECK INPUT SIGNAL CIRCUIT. Check continuity between terminals ⓐ and ㉖. Continuity should exist.
N.G. Repair or replace harness or connectors between terminals ⓐ and ㉖.
O.K.
Reinstall any part removed.
Erase the self-diagnosis memory.
Perform driving test and then perform self-diagnosis (Mode III) again.
N.G. 1) Perform E.C.U. input/output signal inspection test. 2) If N.G., recheck the E.C.U. pin terminals for damage or the connection of E.C.U. harness connector.
O.K.
INSPECTION END

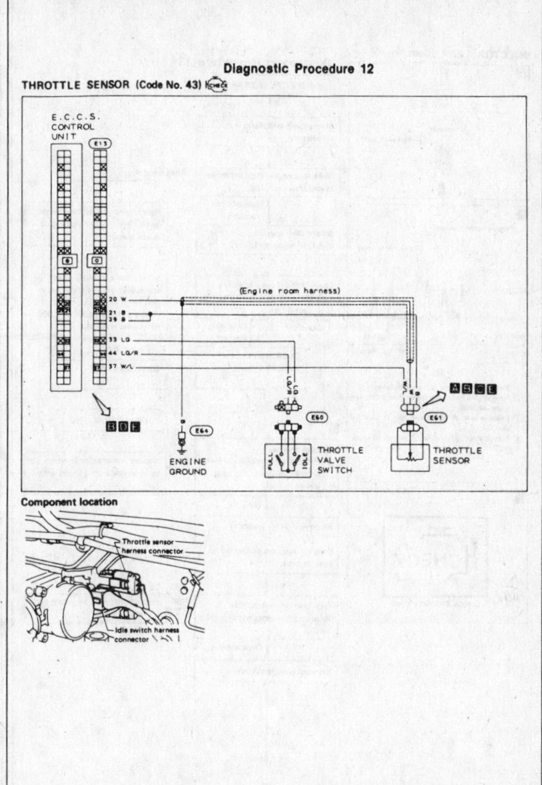
Diagnostic Procedure 12
THROTTLE SENSOR (Code No. 43)
E.C.C.S. CONTROL UNIT
(Engine room harness)
ENGINE GROUND
THROTTLE VALVE SWITCH
THROTTLE SENSOR
Component location
Throttle sensor harness connector
Idle switch harness connector

ECCS SELF-DIAGNOSTIC CHART—1990 AXXESS

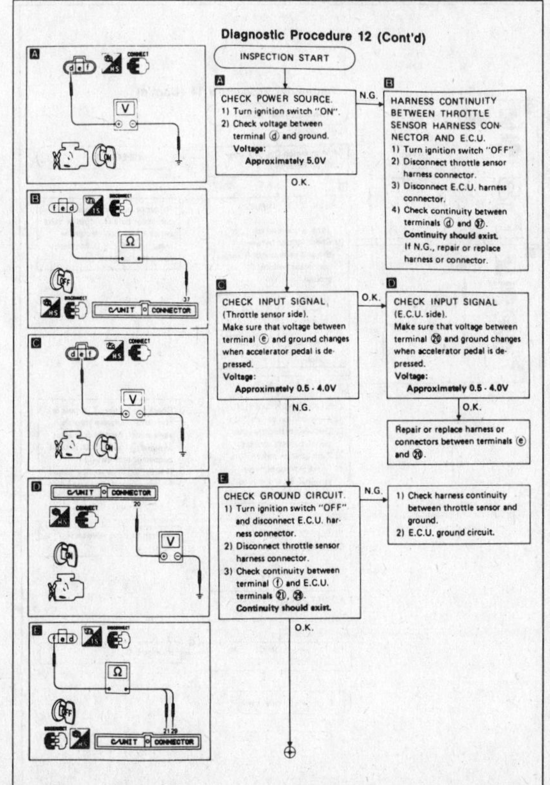
Diagnostic Procedure 12 (Cont'd)
INSPECTION START
CHECK POWER SOURCE. 1) Turn ignition switch "ON". 2) Check voltage between terminal ⓓ and ground. Voltage: Approximately 5.0V
N.G. HARNESS CONTINUITY BETWEEN THROTTLE SENSOR HARNESS CONNECTOR AND E.C.U. 1) Turn ignition switch "OFF". 2) Disconnect throttle sensor harness connector. 3) Disconnect E.C.U. harness connector. 4) Check continuity between terminals ⓓ and ㊲. Continuity should exist. If N.G., repair or replace harness or connector.
O.K.
CHECK INPUT SIGNAL (Throttle sensor side). Make sure that voltage between terminal ⓔ and ground changes when accelerator pedal is depressed. Voltage: Approximately 0.5 - 4.0V
O.K. CHECK INPUT SIGNAL (E.C.U. side). Make sure that voltage between terminal ⑳ and ground changes when accelerator pedal is depressed. Voltage: Approximately 0.5 - 4.0V
O.K. Repair or replace harness or connectors between terminals ⓔ and ⑳.
N.G.
CHECK GROUND CIRCUIT. 1) Turn ignition switch "OFF" and disconnect E.C.U. harness connector. 2) Disconnect throttle sensor harness connector. 3) Check continuity between terminal ⓕ and E.C.U. terminals ㉑, ㉙. Continuity should exist.
N.G. 1) Check harness continuity between throttle sensor and ground. 2) E.C.U. ground circuit.
O.K.

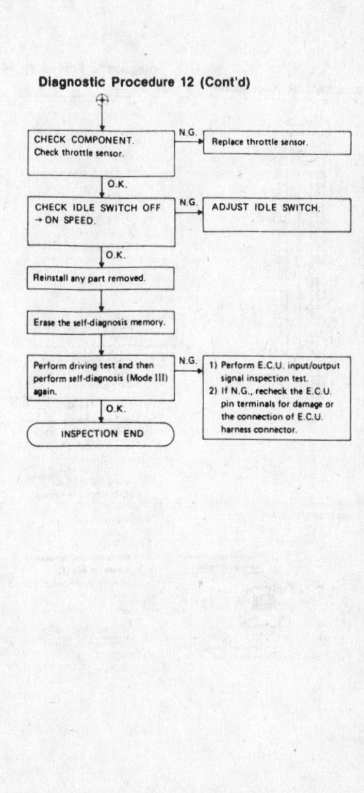
Diagnostic Procedure 12 (Cont'd)
CHECK COMPONENT. Check throttle sensor.
N.G. Replace throttle sensor.
O.K.
CHECK IDLE SWITCH OFF → ON SPEED.
N.G. ADJUST IDLE SWITCH.
O.K.
Reinstall any part removed.
Erase the self-diagnosis memory.
Perform driving test and then perform self-diagnosis (Mode III) again.
N.G. 1) Perform E.C.U. input/output signal inspection test. 2) If N.G., recheck the E.C.U. pin terminals for damage or the connection of E.C.U. harness connector.
O.K.
INSPECTION END

ECCS SELF-DIAGNOSTIC CHART—1990 AXXESS

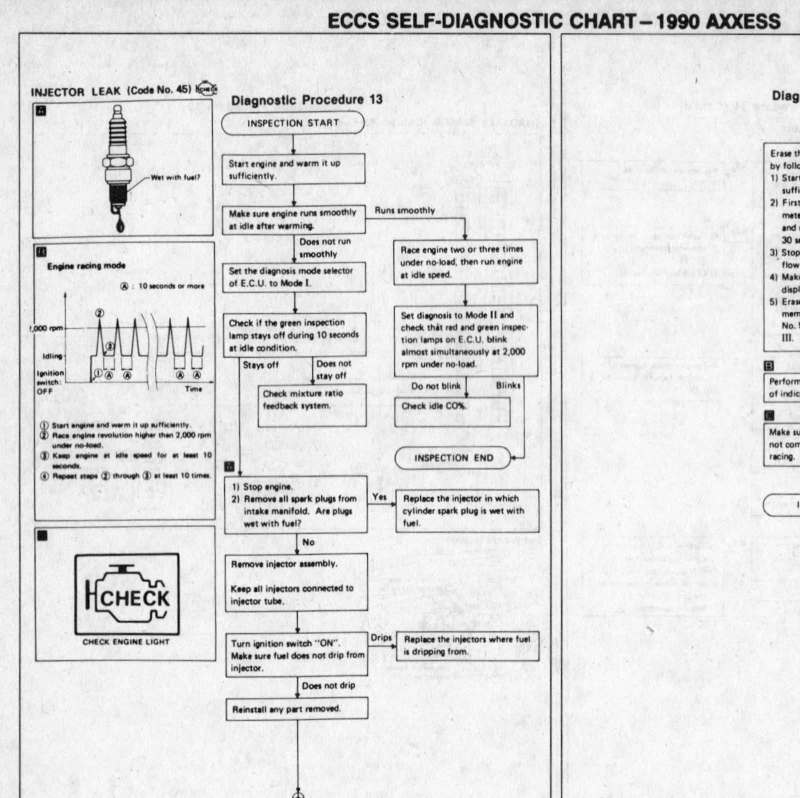
INJECTOR LEAK (Code No. 45)
Wet with fuel?
Engine racing mode
Ⓐ : 10 seconds or more
2,000 rpm
Idling
Ignition switch: OFF
Time
① Start engine and warm it up sufficiently.
② Race engine revolution higher than 2,000 rpm under no-load.
③ Keep engine at idle speed for at least 10 seconds.
④ Repeat steps ② through ③ at least 10 times.
CHECK
CHECK ENGINE LIGHT
Diagnostic Procedure 13
INSPECTION START
Start engine and warm it up sufficiently.
Make sure engine runs smoothly at idle after warming.
Runs smoothly
Does not run smoothly
Set the diagnosis mode selector of E.C.U. to Mode I.
Race engine two or three times under no-load, then run engine at idle speed.
Check if the green inspection lamp stays off during 10 seconds at idle condition.
Stays off
Does not stay off
Check mixture ratio feedback system.
Set diagnosis to Mode II and check that red and green inspection lamps on E.C.U. blink almost simultaneously at 2,000 rpm under no-load.
Do not blink
Blinks
Check idle CO%.
INSPECTION END
1) Stop engine.
2) Remove all spark plugs from intake manifold. Are plugs wet with fuel?
Yes
Replace the injector in which cylinder spark plug is wet with fuel.
No
Remove injector assembly.
Keep all injectors connected to injector tube.
Turn ignition switch "ON". Make sure fuel does not drip from injector.
Drips
Replace the injectors where fuel is dripping from.
Does not drip
Reinstall any part removed.

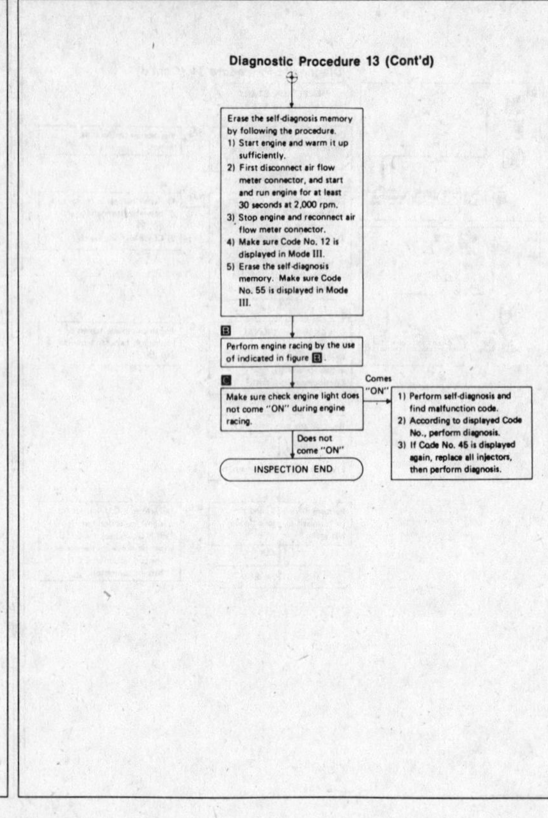
Diagnostic Procedure 13 (Cont'd)
Erase the self-diagnosis memory by following the procedure.
1) Start engine and warm it up sufficiently.
2) First disconnect air flow meter connector, and start and run engine for at least 30 seconds at 2,000 rpm.
3) Stop engine and reconnect air flow meter connector.
4) Make sure Code No. 12 is displayed in Mode III.
5) Erase the self-diagnosis memory. Make sure Code No. 55 is displayed in Mode III.
Perform engine racing by the use of indicated in figure
Make sure check engine light does not come "ON" during engine racing.
Comes "ON"
1) Perform self-diagnosis and find malfunction code.
2) According to displayed Code No., perform diagnosis.
3) If Code No. 45 is displayed again, replace all injectors, then perform diagnosis.
Does not come "ON"
INSPECTION END

ECCS SELF-DIAGNOSTIC CHART—1990 AXXESS

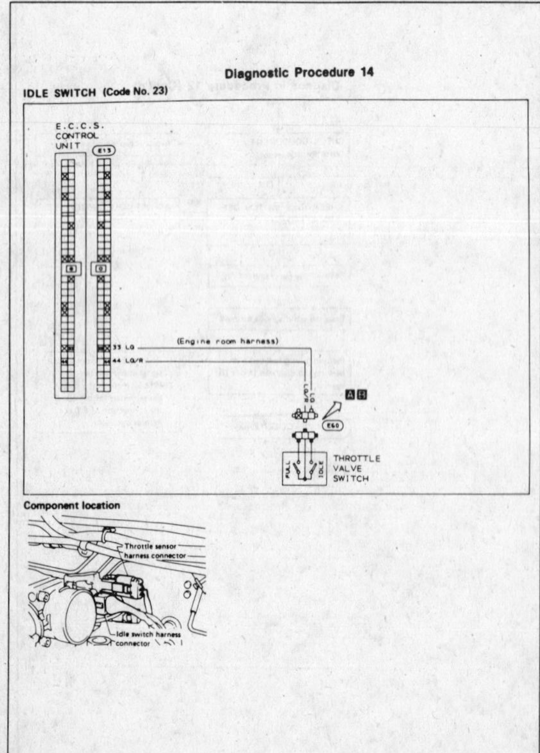
Diagnostic Procedure 14
IDLE SWITCH (Code No. 23)
E.C.C.S. CONTROL UNIT
(Engine room harness)
33 LG
44 LG/R
THROTTLE VALVE SWITCH
Component location
Throttle sensor harness connector
Idle switch harness connector

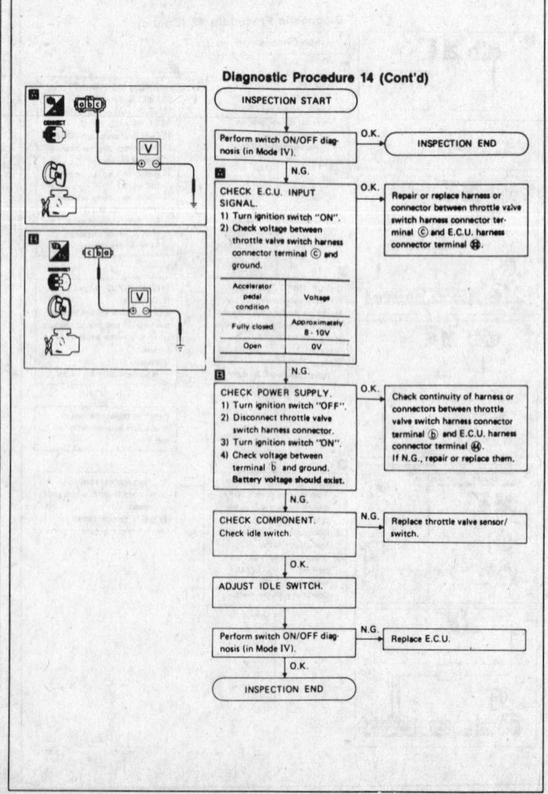
Diagnostic Procedure 14 (Cont'd)
INSPECTION START
Perform switch ON/OFF diagnosis (in Mode IV).
O.K.
INSPECTION END
N.G.
CHECK E.C.U. INPUT SIGNAL.
1) Turn ignition switch "ON".
2) Check voltage between throttle valve switch harness connector terminal Ⓒ and ground.
Accelerator pedal condition
Voltage
Fully closed
Approximately 8 - 10V
Open
0V
O.K.
Repair or replace harness or connector between throttle valve switch harness connector terminal Ⓒ and E.C.U. harness connector terminal 33.
N.G.
CHECK POWER SUPPLY.
1) Turn ignition switch "OFF".
2) Disconnect throttle valve switch harness connector.
3) Turn ignition switch "ON".
4) Check voltage between terminal ⓑ and ground. Battery voltage should exist.
O.K.
Check continuity of harness or connectors between throttle valve switch harness connector terminal ⓑ and E.C.U. harness connector terminal 44.
If N.G., repair or replace them.
N.G.
CHECK COMPONENT. Check idle switch.
N.G.
Replace throttle valve sensor/switch.
O.K.
ADJUST IDLE SWITCH.
Perform switch ON/OFF diagnosis (in Mode IV).
N.G.
Replace E.C.U.
O.K.
INSPECTION END

ECCS SELF-DIAGNOSTIC CHART—1990 AXXESS

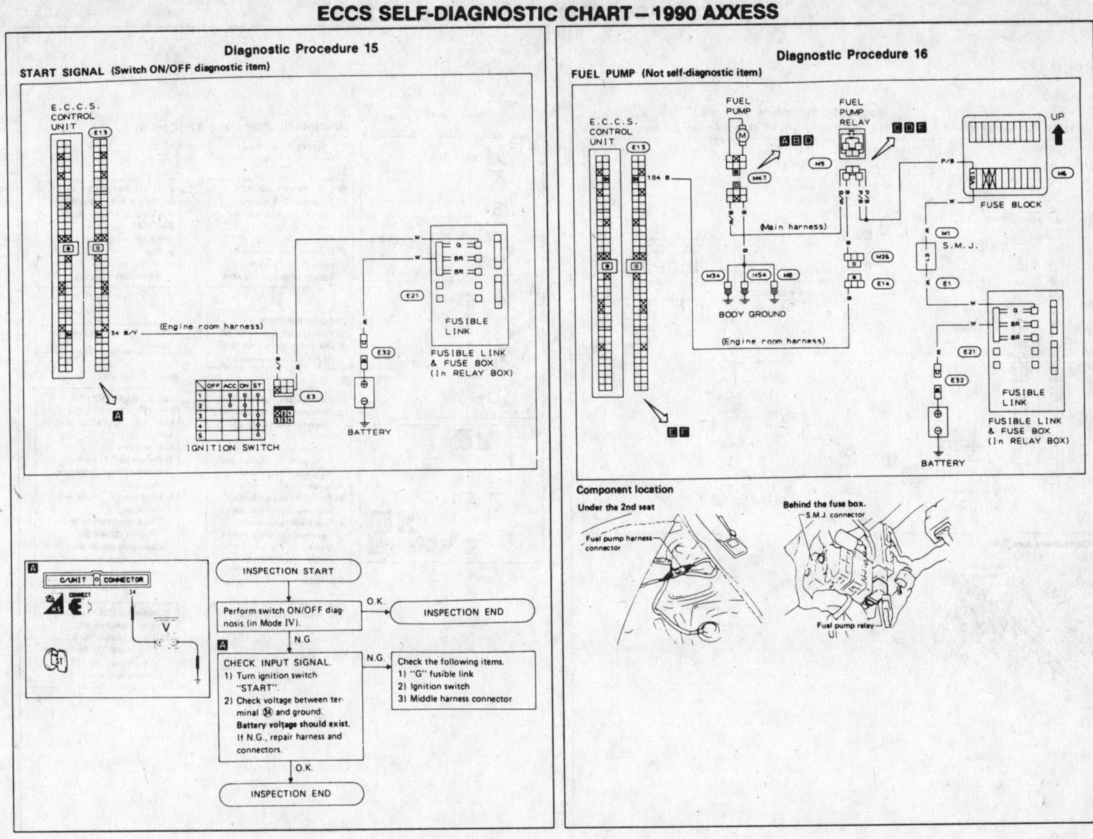
Diagnostic Procedure 15
START SIGNAL (Switch ON/OFF diagnostic item)
E.C.C.S. CONTROL UNIT
(Engine room harness)
FUSIBLE LINK
FUSIBLE LINK & FUSE BOX (In RELAY BOX)
BATTERY
IGNITION SWITCH
OFF ACC ON ST
C/UNIT CONNECTOR
INSPECTION START
Perform switch ON/OFF diagnosis (in Mode IV).
O.K.
INSPECTION END
N.G.
CHECK INPUT SIGNAL.
1) Turn ignition switch "START".
2) Check voltage between terminal ㉞ and ground.
Battery voltage should exist.
If N.G., repair harness and connectors.
N.G.
Check the following items.
1) "G" fusible link
2) Ignition switch
3) Middle harness connector
O.K.
INSPECTION END
Diagnostic Procedure 16
FUEL PUMP (Not self-diagnostic item)
E.C.C.S. CONTROL UNIT
FUEL PUMP
FUEL PUMP RELAY
UP
FUSE BLOCK
(Main harness)
S.M.J.
BODY GROUND
(Engine room harness)
FUSIBLE LINK
FUSIBLE LINK & FUSE BOX (In RELAY BOX)
BATTERY
Component location
Under the 2nd seat
Fuel pump harness connector
Behind the fuse box.
S.M.J. connector
Fuel pump relay

ECCS DIAGNOSTIC CHART—1990 AXXESS

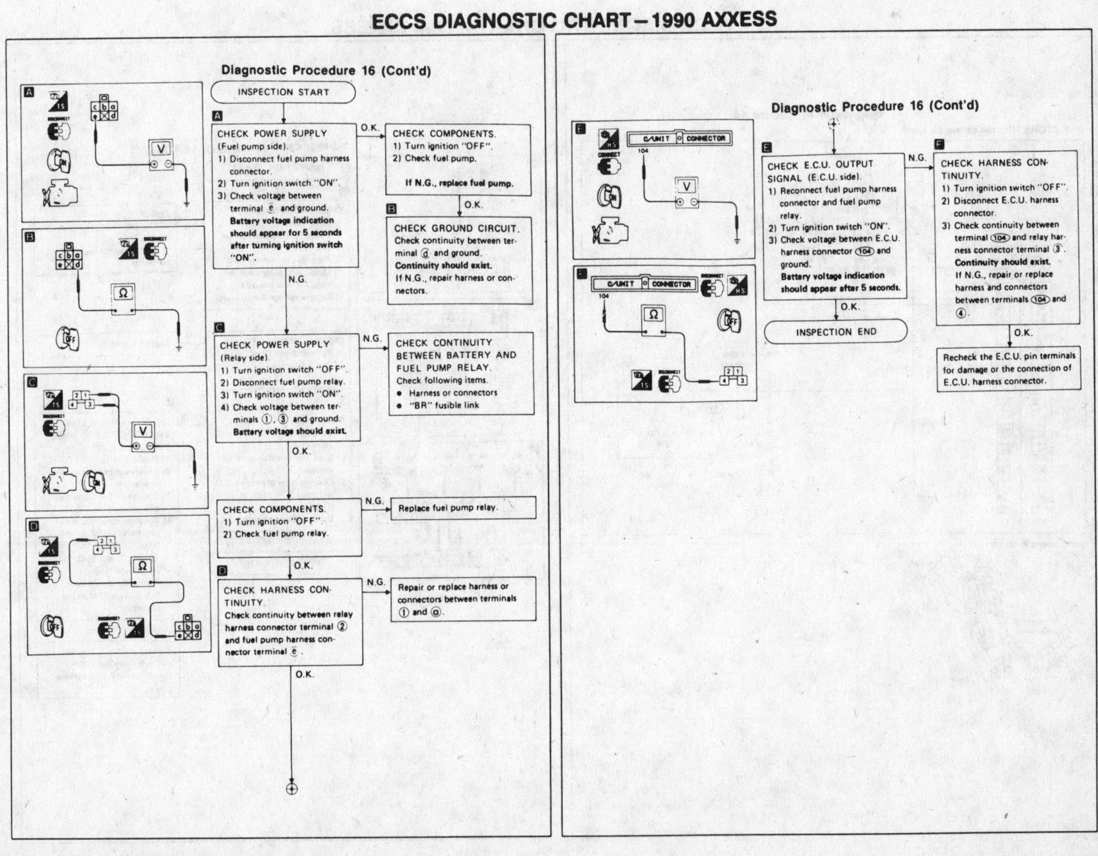
Diagnostic Procedure 16 (Cont'd)
INSPECTION START
CHECK POWER SUPPLY (Fuel pump side).
1) Disconnect fuel pump harness connector.
2) Turn ignition switch "ON".
3) Check voltage between terminal ⓔ and ground.
Battery voltage indication should appear for 5 seconds after turning ignition switch "ON".
O.K.
CHECK COMPONENTS.
1) Turn ignition "OFF".
2) Check fuel pump.
If N.G., replace fuel pump.
O.K.
CHECK GROUND CIRCUIT.
Check continuity between terminal ⓓ and ground.
Continuity should exist.
If N.G., repair harness or connectors.
N.G.
CHECK POWER SUPPLY (Relay side).
1) Turn ignition switch "OFF".
2) Disconnect fuel pump relay.
3) Turn ignition switch "ON".
4) Check voltage between terminals ①, ③ and ground.
Battery voltage should exist.
N.G.
CHECK CONTINUITY BETWEEN BATTERY AND FUEL PUMP RELAY.
Check following items.
• Harness or connectors
• "BR" fusible link
O.K.
CHECK COMPONENTS.
1) Turn ignition "OFF".
2) Check fuel pump relay.
N.G.
Replace fuel pump relay.
O.K.
CHECK HARNESS CONTINUITY.
Check continuity between relay harness connector terminal ② and fuel pump harness connector terminal ⓔ.
N.G.
Repair or replace harness or connectors between terminals ① and ⓐ.
O.K.
Diagnostic Procedure 16 (Cont'd)
C/UNIT CONNECTOR
CHECK E.C.U. OUTPUT SIGNAL (E.C.U. side).
1) Reconnect fuel pump harness connector and fuel pump relay.
2) Turn ignition switch "ON".
3) Check voltage between E.C.U. harness connector ⑩④ and ground.
Battery voltage indication should appear after 5 seconds.
O.K.
INSPECTION END
N.G.
CHECK HARNESS CONTINUITY.
1) Turn ignition switch "OFF".
2) Disconnect E.C.U. harness connector.
3) Check continuity between terminal ⑩④ and relay harness connector terminal ③.
Continuity should exist.
If N.G., repair or replace harness and connectors between terminals ⑩④ and ④.
O.K.
Recheck the E.C.U. pin terminals for damage or the connection of E.C.U. harness connector.

ECCS DIAGNOSTIC CHART—1990 AXXESS

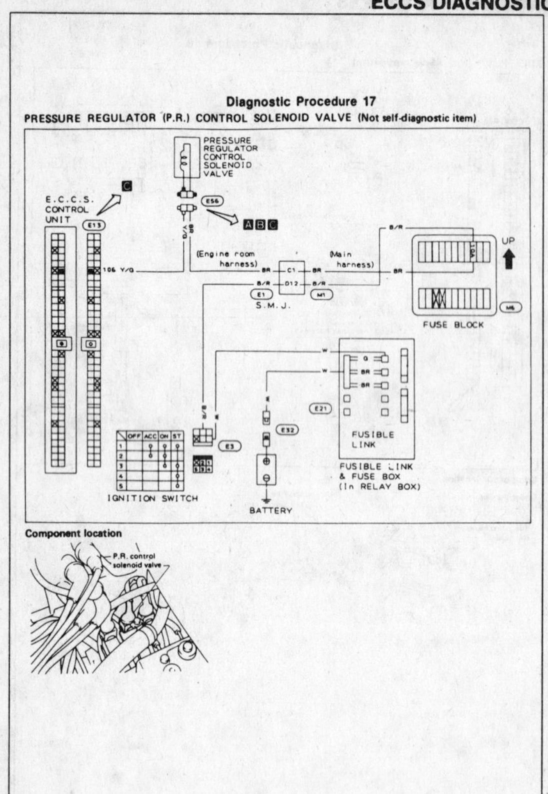
Diagnostic Procedure 17
PRESSURE REGULATOR (P.R.) CONTROL SOLENOID VALVE (Not self-diagnostic item)
PRESSURE REGULATOR CONTROL SOLENOID VALVE
E.C.C.S. CONTROL UNIT
S.M.J.
FUSE BLOCK
UP
IGNITION SWITCH
FUSIBLE LINK
FUSIBLE LINK & FUSE BOX (In RELAY BOX)
BATTERY
Component location
P.R. control solenoid valve

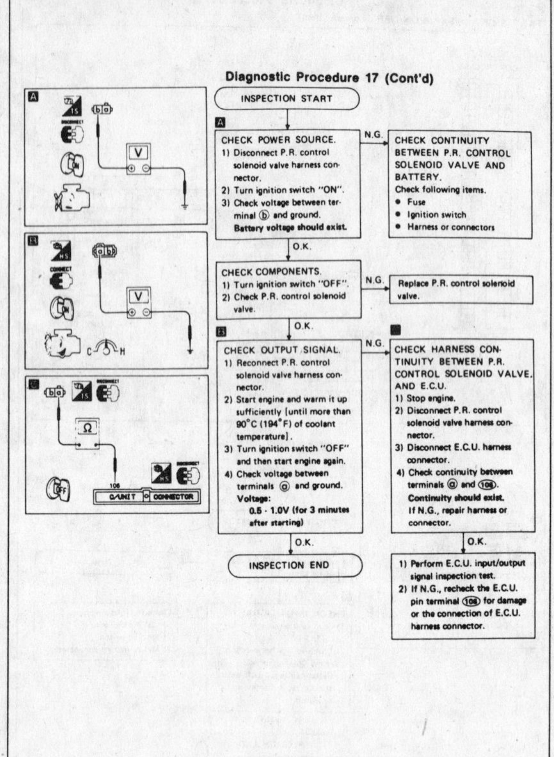
Diagnostic Procedure 17 (Cont'd)
INSPECTION START
CHECK POWER SOURCE.
1) Disconnect P.R. control solenoid valve harness connector.
2) Turn ignition switch "ON".
3) Check voltage between terminal ⓑ and ground.
Battery voltage should exist.
N.G.
CHECK CONTINUITY BETWEEN P.R. CONTROL SOLENOID VALVE AND BATTERY.
Check following items.
• Fuse
• Ignition switch
• Harness or connectors
O.K.
CHECK COMPONENTS.
1) Turn ignition switch "OFF".
2) Check P.R. control solenoid valve.
N.G.
Replace P.R. control solenoid valve.
O.K.
CHECK OUTPUT SIGNAL.
1) Reconnect P.R. control solenoid valve harness connector.
2) Start engine and warm it up sufficiently [until more than 90°C (194°F) of coolant temperature].
3) Turn ignition switch "OFF" and then start engine again.
4) Check voltage between terminals ⓐ and ground.
Voltage:
0.5 - 1.0V (for 3 minutes after starting)
N.G.
CHECK HARNESS CONTINUITY BETWEEN P.R. CONTROL SOLENOID VALVE AND E.C.U.
1) Stop engine.
2) Disconnect P.R. control solenoid valve harness connector.
3) Disconnect E.C.U. harness connector.
4) Check continuity between terminals ⓐ and 106.
Continuity should exist.
If N.G., repair harness or connector.
O.K.
INSPECTION END
O.K.
1) Perform E.C.U. input/output signal inspection test.
2) If N.G., recheck the E.C.U. pin terminal 106 for damage or the connection of E.C.U. harness connector.

ECCS DIAGNOSTIC CHART—1990 AXXESS

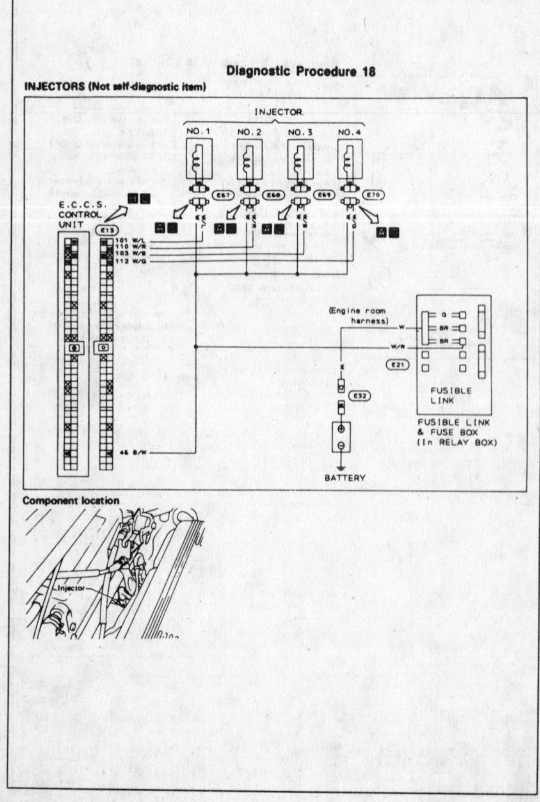
Diagnostic Procedure 18
INJECTORS (Not self-diagnostic item)
INJECTOR
NO. 1
NO. 2
NO. 3
NO. 4
E.C.C.S. CONTROL UNIT
(Engine room harness)
FUSIBLE LINK
FUSIBLE LINK & FUSE BOX (In RELAY BOX)
BATTERY
Component location
Injector

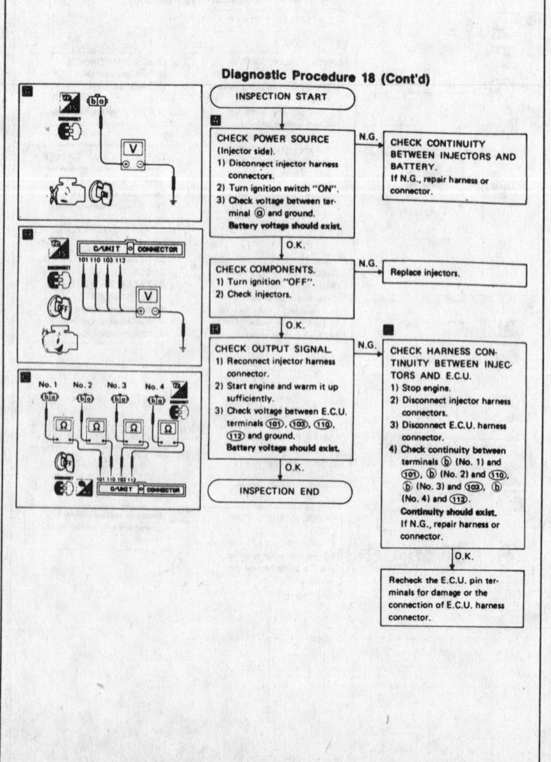
Diagnostic Procedure 18 (Cont'd)
INSPECTION START
CHECK POWER SOURCE (Injector side).
1) Disconnect injector harness connectors.
2) Turn ignition switch "ON".
3) Check voltage between terminal ⓐ and ground.
Battery voltage should exist.
N.G.
CHECK CONTINUITY BETWEEN INJECTORS AND BATTERY.
If N.G., repair harness or connector.
O.K.
CHECK COMPONENTS.
1) Turn ignition "OFF".
2) Check injectors.
N.G.
Replace injectors.
O.K.
CHECK OUTPUT SIGNAL.
1) Reconnect injector harness connector.
2) Start engine and warm it up sufficiently.
3) Check voltage between E.C.U. terminals 101, 103, 110, 112 and ground.
Battery voltage should exist.
N.G.
CHECK HARNESS CONTINUITY BETWEEN INJECTORS AND E.C.U.
1) Stop engine.
2) Disconnect injector harness connectors.
3) Disconnect E.C.U. harness connector.
4) Check continuity between terminals ⓑ (No. 1) and 101, ⓑ (No. 2) and 110, ⓑ (No. 3) and 103, ⓑ (No. 4) and 112.
Continuity should exist.
If N.G., repair harness or connector.
O.K.
INSPECTION END
O.K.
Recheck the E.C.U. pin terminals for damage or the connection of E.C.U. harness connector.

ECCS DIAGNOSTIC CHART—1990 AXXESS

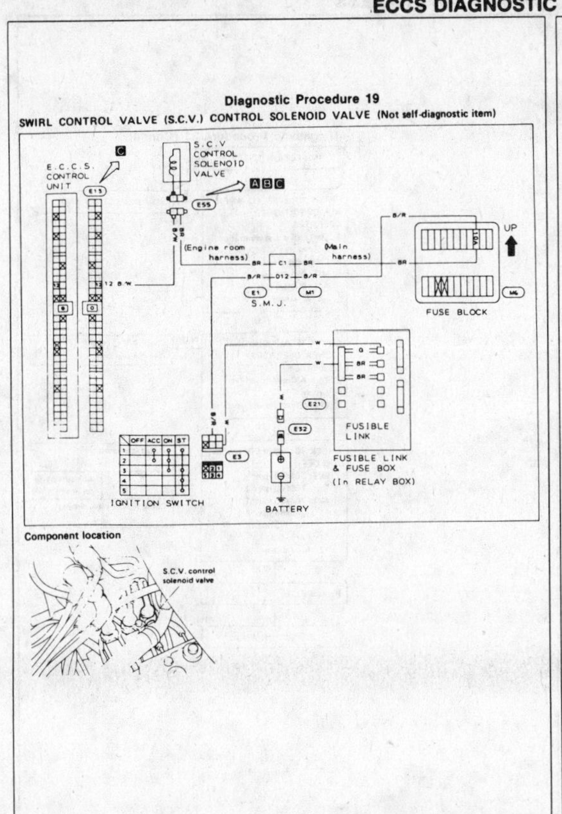
Diagnostic Procedure 19
SWIRL CONTROL VALVE (S.C.V.) CONTROL SOLENOID VALVE (Not self-diagnostic item)
E.C.C.S. CONTROL UNIT
S.C.V CONTROL SOLENOID VALVE
(Engine room harness)
(Main harness)
S.M.J.
FUSE BLOCK
UP
FUSIBLE LINK
FUSIBLE LINK & FUSE BOX (In RELAY BOX)
BATTERY
IGNITION SWITCH
Component location
S.C.V. control solenoid valve

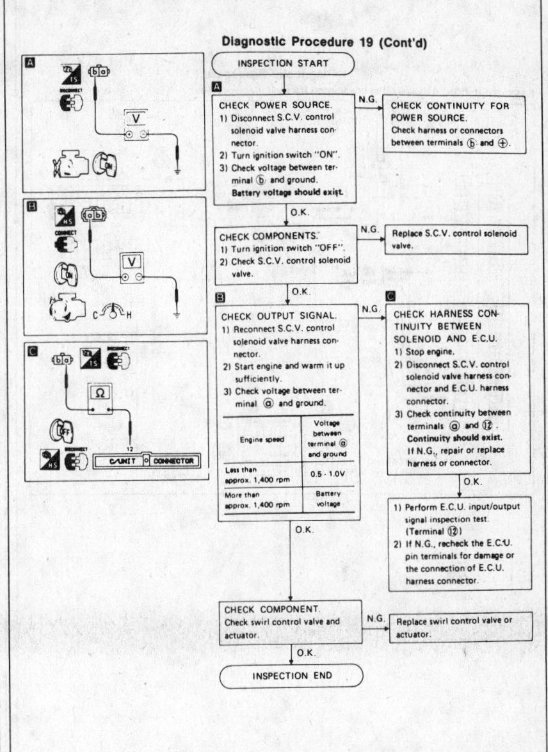
Diagnostic Procedure 19 (Cont'd)
INSPECTION START
CHECK POWER SOURCE.
1) Disconnect S.C.V. control solenoid valve harness connector.
2) Turn ignition switch "ON".
3) Check voltage between terminal ⓑ and ground.
Battery voltage should exist.
N.G.
CHECK CONTINUITY FOR POWER SOURCE.
Check harness or connectors between terminals ⓑ and ⊕.
O.K.
CHECK COMPONENTS.
1) Turn ignition switch "OFF".
2) Check S.C.V. control solenoid valve.
N.G.
Replace S.C.V. control solenoid valve.
O.K.
CHECK OUTPUT SIGNAL.
1) Reconnect S.C.V. control solenoid valve harness connector.
2) Start engine and warm it up sufficiently.
3) Check voltage between terminal ⓐ and ground.
Engine speed
Voltage between terminal ⓐ and ground
Less than approx. 1,400 rpm
0.5 - 1.0V
More than approx. 1,400 rpm
Battery voltage
N.G.
CHECK HARNESS CONTINUITY BETWEEN SOLENOID AND E.C.U.
1) Stop engine.
2) Disconnect S.C.V. control solenoid valve harness connector and E.C.U. harness connector.
3) Check continuity between terminals ⓐ and ⑫.
Continuity should exist.
If N.G., repair or replace harness or connector.
O.K.
1) Perform E.C.U. input/output signal inspection test. (Terminal ⑫)
2) If N.G., recheck the E.C.U. pin terminals for damage or the connection of E.C.U. harness connector.
O.K.
CHECK COMPONENT.
Check swirl control valve and actuator.
N.G.
Replace swirl control valve or actuator.
O.K.
INSPECTION END
C/UNIT CONNECTOR

ECCS DIAGNOSTIC CHART—1990 AXXESS

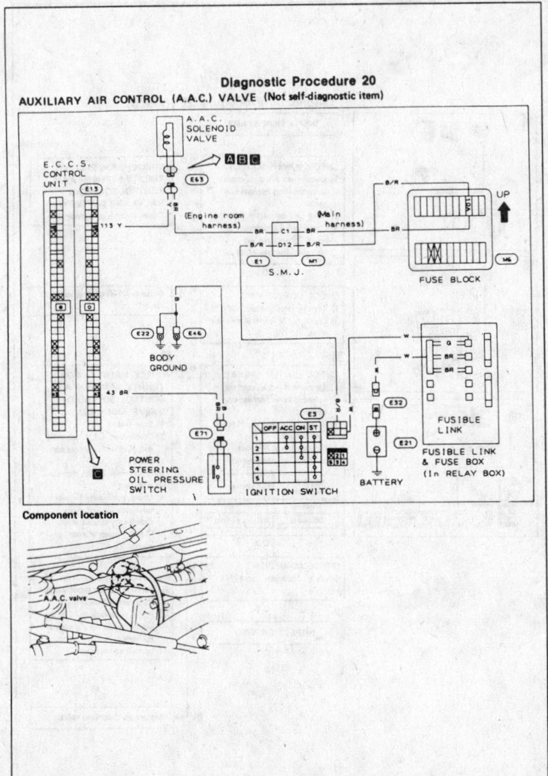
Diagnostic Procedure 20
AUXILIARY AIR CONTROL (A.A.C.) VALVE (Not self-diagnostic item)
E.C.C.S. CONTROL UNIT
A.A.C. SOLENOID VALVE
(Engine room harness)
(Main harness)
S.M.J.
FUSE BLOCK
UP
BODY GROUND
POWER STEERING OIL PRESSURE SWITCH
IGNITION SWITCH
BATTERY
FUSIBLE LINK
FUSIBLE LINK & FUSE BOX (In RELAY BOX)
Component location
A.A.C. valve

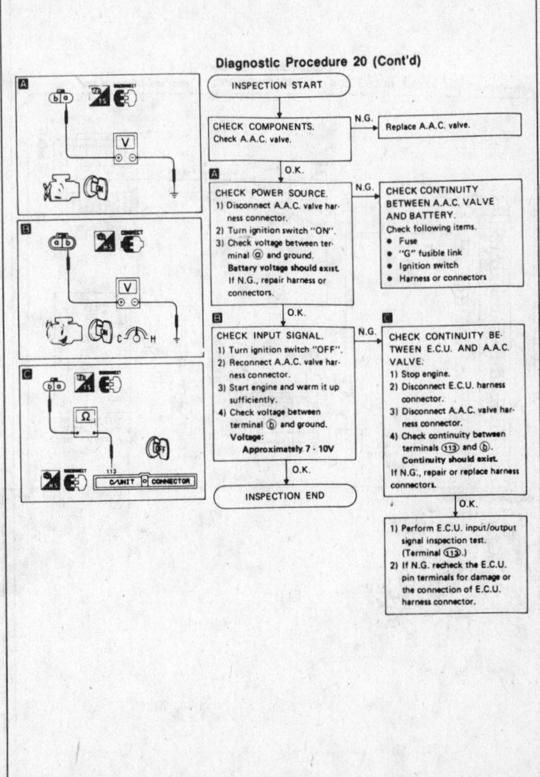
Diagnostic Procedure 20 (Cont'd)
INSPECTION START
CHECK COMPONENTS.
Check A.A.C. valve.
N.G.
Replace A.A.C. valve.
O.K.
CHECK POWER SOURCE.
1) Disconnect A.A.C. valve harness connector.
2) Turn ignition switch "ON".
3) Check voltage between terminal ⓐ and ground.
Battery voltage should exist.
If N.G., repair harness or connectors.
N.G.
CHECK CONTINUITY BETWEEN A.A.C. VALVE AND BATTERY.
Check following items.
Fuse
"G" fusible link
Ignition switch
Harness or connectors
O.K.
CHECK INPUT SIGNAL.
1) Turn ignition switch "OFF".
2) Reconnect A.A.C. valve harness connector.
3) Start engine and warm it up sufficiently.
4) Check voltage between terminal ⓑ and ground.
Voltage:
Approximately 7 - 10V
N.G.
CHECK CONTINUITY BETWEEN E.C.U. AND A.A.C. VALVE.
1) Stop engine.
2) Disconnect E.C.U. harness connector.
3) Disconnect A.A.C. valve harness connector.
4) Check continuity between terminals ⑪③ and ⓑ.
Continuity should exist.
If N.G., repair or replace harness connectors.
O.K.
1) Perform E.C.U. input/output signal inspection test. (Terminal ⑪③.)
2) If N.G. recheck the E.C.U. pin terminals for damage or the connection of E.C.U. harness connector.
O.K.
INSPECTION END
C/UNIT CONNECTOR

ECCS DIAGNOSTIC CHART—1990 AXXESS

Diagnostic Procedure 21

FAST IDLE CONTROL (With air conditioner) (Not self-diagnositc item)

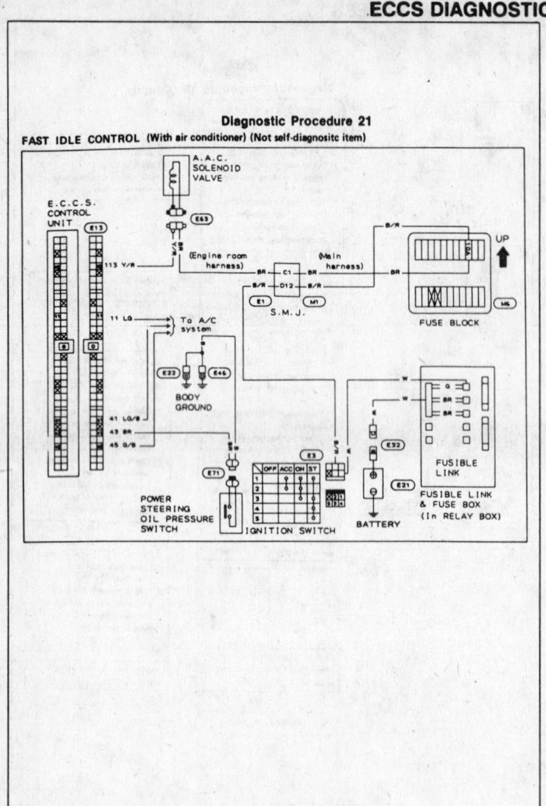
A.A.C. SOLENOID VALVE
E.C.C.S. CONTROL UNIT
E13
E53
113 Y/R
(Engine room harness)
(Main harness)
C1
D12
S.M.J.
E1
M1
UP
FUSE BLOCK
M6
11 LG
To A/C system
E22
E46
BODY GROUND
41 LG/R
43 BR
45 G/R
E71
E3
E32
E21
FUSIBLE LINK
FUSIBLE LINK & FUSE BOX (In RELAY BOX)
POWER STEERING OIL PRESSURE SWITCH
IGNITION SWITCH
BATTERY

Diagnostic Procedure 21 (Cont'd)

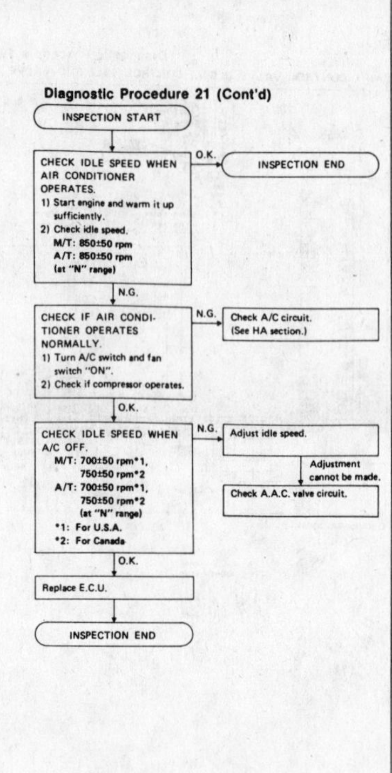
INSPECTION START
CHECK IDLE SPEED WHEN AIR CONDITIONER OPERATES.
1) Start engine and warm it up sufficiently.
2) Check idle speed.
M/T: 850±50 rpm
A/T: 850±50 rpm
(at "N" range)
O.K.
INSPECTION END
N.G.
CHECK IF AIR CONDITIONER OPERATES NORMALLY.
1) Turn A/C switch and fan switch "ON".
2) Check if compressor operates.
N.G.
Check A/C circuit. (See HA section.)
O.K.
CHECK IDLE SPEED WHEN A/C OFF.
M/T: 700±50 rpm*1, 750±50 rpm*2
A/T: 700±50 rpm*1, 750±50 rpm*2
(at "N" range)
*1: For U.S.A.
*2: For Canada
N.G.
Adjust idle speed.
Adjustment cannot be made.
Check A.A.C. valve circuit.
O.K.
Replace E.C.U.
INSPECTION END

ECCS DIAGNOSTIC CHART—1990 AXXESS

Diagnostic Procedure 22

AIR INDUCTION VALVE (A.I.V.) CONTROL SOLENOID VALVE (Not self-diagnostic item)

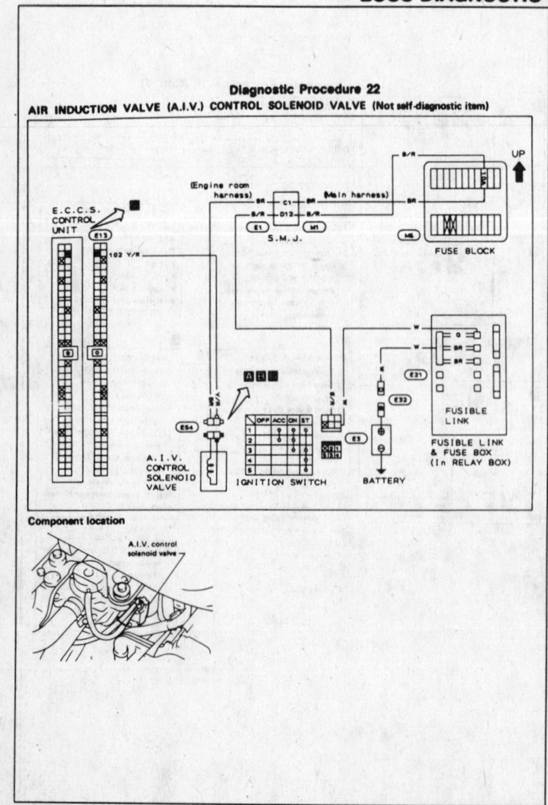
UP
(Engine room harness)
(Main harness)
C1
D12
S.M.J.
E1
M1
M6
FUSE BLOCK
E.C.C.S. CONTROL UNIT
E13
102 Y/R
E21
E32
E3
E54
FUSIBLE LINK
FUSIBLE LINK & FUSE BOX (In RELAY BOX)
A.I.V. CONTROL SOLENOID VALVE
IGNITION SWITCH
BATTERY
Component location
A.I.V. control solenoid valve

Diagnostic Procedure 22 (Cont'd)

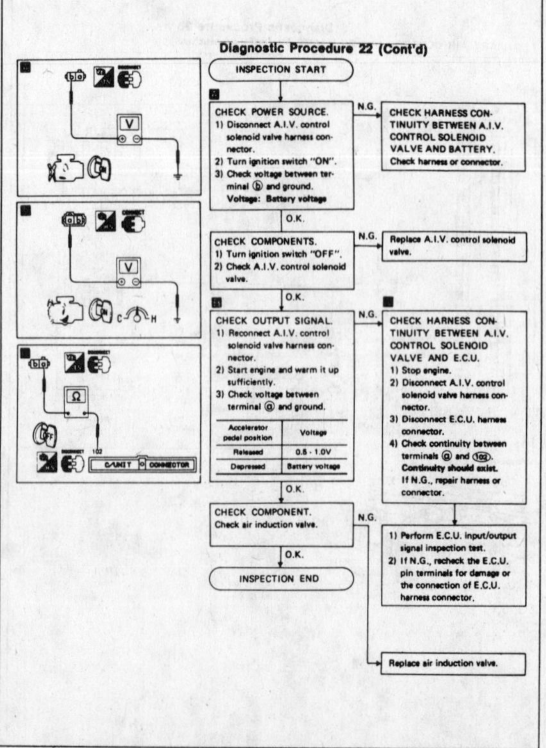
INSPECTION START
CHECK POWER SOURCE.
1) Disconnect A.I.V. control solenoid valve harness connector.
2) Turn ignition switch "ON".
3) Check voltage between terminal ⓑ and ground.
Voltage: Battery voltage
N.G.
CHECK HARNESS CONTINUITY BETWEEN A.I.V. CONTROL SOLENOID VALVE AND BATTERY.
Check harness or connector.
O.K.
CHECK COMPONENTS.
1) Turn ignition switch "OFF".
2) Check A.I.V. control solenoid valve.
N.G.
Replace A.I.V. control solenoid valve.
O.K.
CHECK OUTPUT SIGNAL.
1) Reconnect A.I.V. control solenoid valve harness connector.
2) Start engine and warm it up sufficiently.
3) Check voltage between terminal ⓐ and ground.
Accelerator pedal position | Voltage
Released | 0.5 - 1.0V
Depressed | Battery voltage
N.G.
CHECK HARNESS CONTINUITY BETWEEN A.I.V. CONTROL SOLENOID VALVE AND E.C.U.
1) Stop engine.
2) Disconnect A.I.V. control solenoid valve harness connector.
3) Disconnect E.C.U. harness connector.
4) Check continuity between terminals ⓐ and (102).
Continuity should exist.
If N.G., repair harness or connector.
1) Perform E.C.U. input/output signal inspection test.
2) If N.G., recheck the E.C.U. pin terminals for damage or the connection of E.C.U. harness connector.
O.K.
CHECK COMPONENT.
Check air induction valve.
N.G.
Replace air induction valve.
O.K.
INSPECTION END
102
C/UNIT CONNECTOR

ECCS DIAGNOSTIC CHART—1990 AXXESS

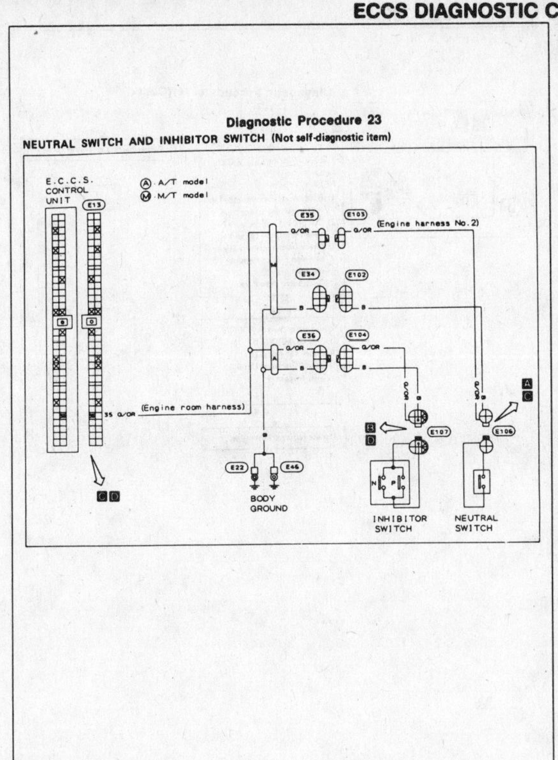
Diagnostic Procedure 23
NEUTRAL SWITCH AND INHIBITOR SWITCH (Not self-diagnostic item)
E.C.C.S. CONTROL UNIT
A A/T model
M M/T model
(Engine harness No. 2)
(Engine room harness)
BODY GROUND
INHIBITOR SWITCH
NEUTRAL SWITCH

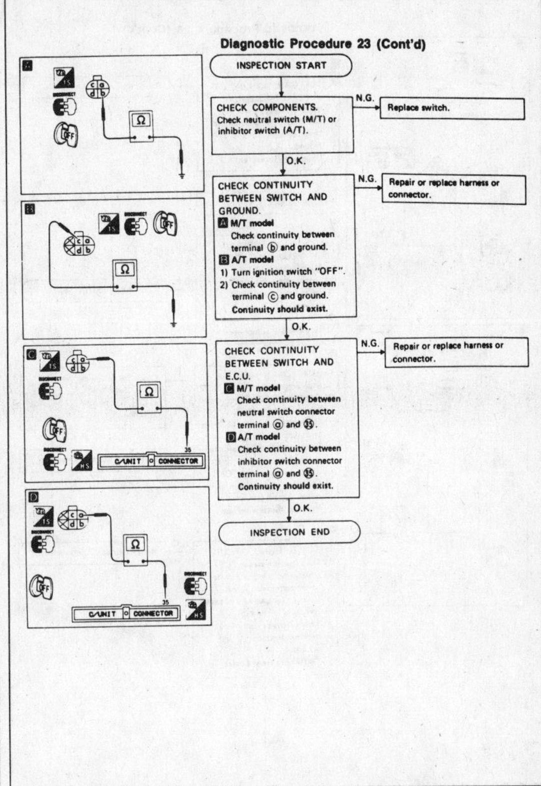
Diagnostic Procedure 23 (Cont'd)
INSPECTION START
CHECK COMPONENTS. Check neutral switch (M/T) or inhibitor switch (A/T).
N.G.
Replace switch.
O.K.
CHECK CONTINUITY BETWEEN SWITCH AND GROUND.
A M/T model
Check continuity between terminal ⓑ and ground.
B A/T model
1) Turn ignition switch "OFF".
2) Check continuity between terminal ⓒ and ground.
Continuity should exist.
N.G.
Repair or replace harness or connector.
O.K.
CHECK CONTINUITY BETWEEN SWITCH AND E.C.U.
C M/T model
Check continuity between neutral switch connector terminal ⓐ and ㉟.
D A/T model
Check continuity between inhibitor switch connector terminal ⓐ and ㉟.
Continuity should exist.
N.G.
Repair or replace harness or connector.
O.K.
INSPECTION END
C/UNIT CONNECTOR

ECCS DIAGNOSTIC CHART—1990 AXXESS

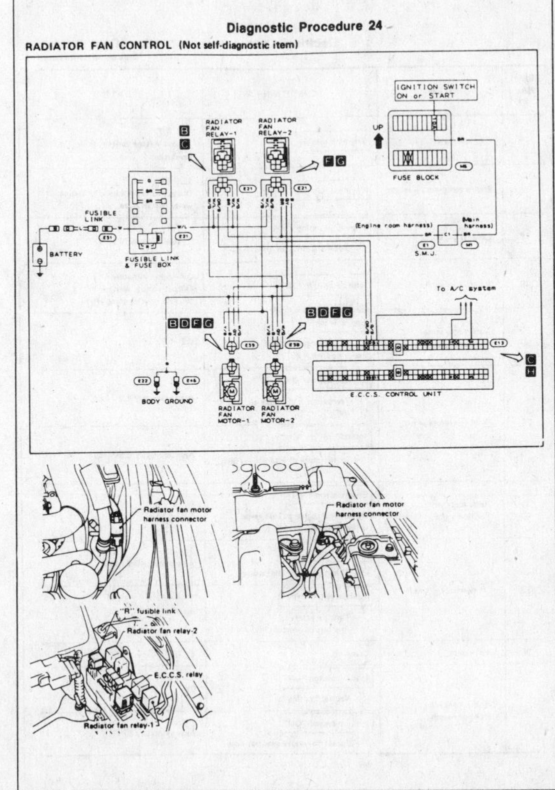
Diagnostic Procedure 24
RADIATOR FAN CONTROL (Not self-diagnostic item)
IGNITION SWITCH ON or START
RADIATOR FAN RELAY-1
RADIATOR FAN RELAY-2
FUSE BLOCK
FUSIBLE LINK
BATTERY
FUSIBLE LINK & FUSE BOX
(Engine room harness)
(Main harness)
S.M.J.
To A/C system
BODY GROUND
RADIATOR FAN MOTOR-1
RADIATOR FAN MOTOR-2
E.C.C.S. CONTROL UNIT
Radiator fan motor harness connector
Radiator fan motor harness connector
"R" fusible link
Radiator fan relay-2
E.C.C.S. relay
Radiator fan relay-1

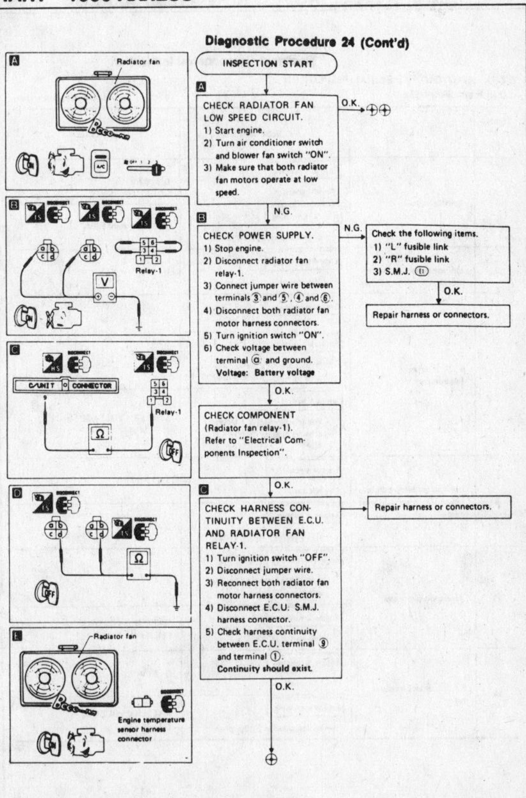
Diagnostic Procedure 24 (Cont'd)
INSPECTION START
A
CHECK RADIATOR FAN LOW SPEED CIRCUIT.
1) Start engine.
2) Turn air conditioner switch and blower fan switch "ON".
3) Make sure that both radiator fan motors operate at low speed.
O.K.
N.G.
B
CHECK POWER SUPPLY.
1) Stop engine.
2) Disconnect radiator fan relay-1.
3) Connect jumper wire between terminals ③ and ⑤, ④ and ⑥.
4) Disconnect both radiator fan motor harness connectors.
5) Turn ignition switch "ON".
6) Check voltage between terminal ① and ground.
Voltage: Battery voltage
N.G.
Check the following items.
1) "L" fusible link
2) "R" fusible link
3) S.M.J. Ⓔ
O.K.
Repair harness or connectors.
O.K.
CHECK COMPONENT (Radiator fan relay-1).
Refer to "Electrical Components Inspection".
O.K.
C
CHECK HARNESS CONTINUITY BETWEEN E.C.U. AND RADIATOR FAN RELAY-1.
1) Turn ignition switch "OFF".
2) Disconnect jumper wire.
3) Reconnect both radiator fan motor harness connectors.
4) Disconnect E.C.U. S.M.J. harness connector.
5) Check harness continuity between E.C.U. terminal ⑨ and terminal ①.
Continuity should exist.
Repair harness or connectors.
O.K.
Radiator fan
Relay-1
C/UNIT CONNECTOR
Engine temperature sensor harness connector

ECCS DIAGNOSTIC CHART—1990 AXXESS

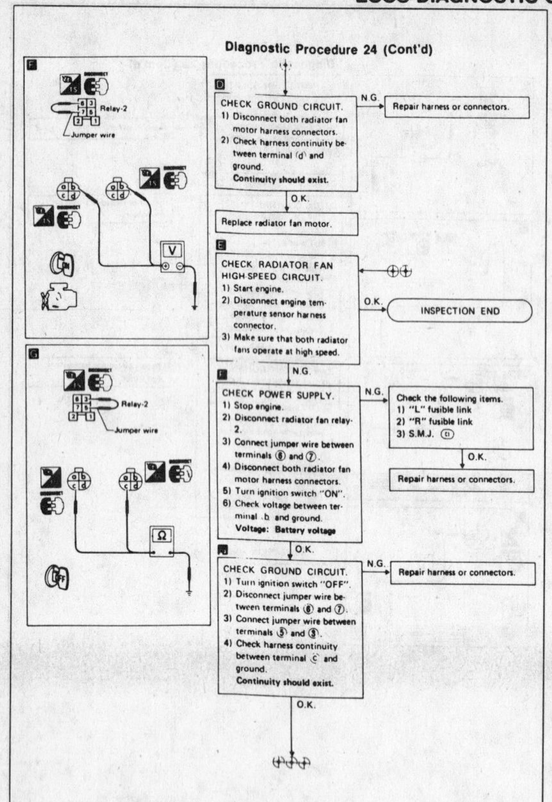
Diagnostic Procedure 24 (Cont'd)
Relay-2
Jumper wire
CHECK GROUND CIRCUIT.
1) Disconnect both radiator fan motor harness connectors.
2) Check harness continuity between terminal (d) and ground.
Continuity should exist.
N.G.
Repair harness or connectors.
O.K.
Replace radiator fan motor.
CHECK RADIATOR FAN HIGH-SPEED CIRCUIT.
1) Start engine.
2) Disconnect engine temperature sensor harness connector.
3) Make sure that both radiator fans operate at high speed.
O.K.
INSPECTION END
N.G.
CHECK POWER SUPPLY.
1) Stop engine.
2) Disconnect radiator fan relay-2.
3) Connect jumper wire between terminals ⑥ and ⑦.
4) Disconnect both radiator fan motor harness connectors.
5) Turn ignition switch "ON".
6) Check voltage between terminal (b) and ground.
Voltage: Battery voltage
Check the following items.
1) "L" fusible link
2) "R" fusible link
3) S.M.J.
O.K.
Repair harness or connectors.
O.K.
CHECK GROUND CIRCUIT.
1) Turn ignition switch "OFF".
2) Disconnect jumper wire between terminals ⑥ and ⑦.
3) Connect jumper wire between terminals ⑤ and ③.
4) Check harness continuity between terminal (c) and ground.
Continuity should exist.
N.G.
Repair harness or connectors.
O.K.

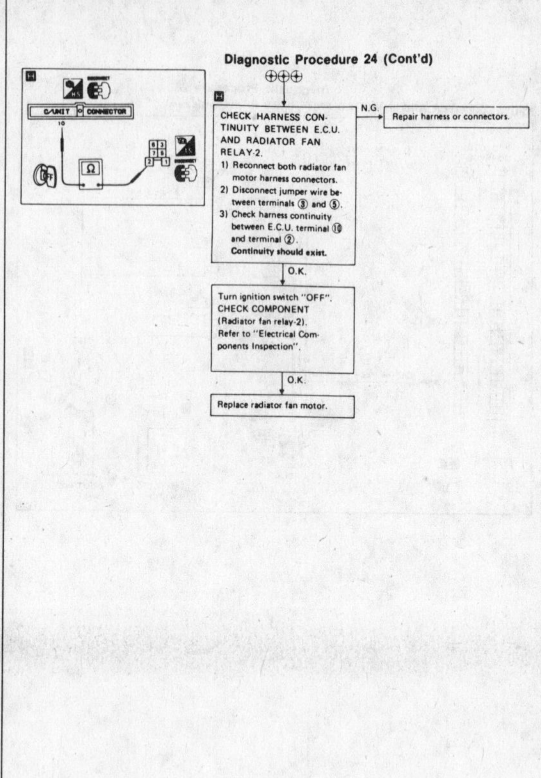
Diagnostic Procedure 24 (Cont'd)
C/UNIT CONNECTOR
10
CHECK HARNESS CONTINUITY BETWEEN E.C.U. AND RADIATOR FAN RELAY-2.
1) Reconnect both radiator fan motor harness connectors.
2) Disconnect jumper wire between terminals ③ and ⑤.
3) Check harness continuity between E.C.U. terminal ⑩ and terminal ②.
Continuity should exist.
N.G.
Repair harness or connectors.
O.K.
Turn ignition switch "OFF".
CHECK COMPONENT (Radiator fan relay-2).
Refer to "Electrical Components Inspection".
O.K.
Replace radiator fan motor.

ECU SIGNAL INSPECTION—1990 AXXESS

Electrical Components Inspection

E.C.U. INPUT/OUTPUT SIGNAL INSPECTION
E.C.U. Inspection table

*Data are reference values.

TERMINAL NO.	ITEM	CONDITION	*DATA
1	Ignition signal	Engine is running. – Idle speed	0.3 - 0.6V
		Engine is running. – Engine speed is 2,000 rpm	Approximately 1.0V
3	Ignition check	Engine is running. – Idle speed	9 - 12V
4	E.C.C.S. relay (Main relay)	Engine is running. → Ignition switch "OFF" – Within approximately 1 second after turning ignition switch "OFF"	0 - 1V
		Ignition switch "OFF" – For approximately 1 second after turning ignition switch "OFF"	BATTERY VOLTAGE (11 - 14V)
8	Exhaust gas temperature sensor (Only for California model)	Engine is running. – Idle speed	1.0 - 2.0V
		Engine is running. – E.G.R. system is operating.	0 - 1.0V
11	Air conditioner relay	Engine is running. – Both A/C switch and blower switch are "ON"	0 - 1.0V
		Engine is running. – A/C switch is "OFF".	BATTERY VOLTAGE (11 - 14V)
12	S.C.V. control solenoid valve	Engine is running. – Idle speed	0 - 1.0V
		Engine is running. – Engine speed is 2,000 rpm.	BATTERY VOLTAGE (11 - 14V)

ECU SIGNAL INSPECTION (CONT.)—1990 AXXESS

Electrical Components Inspection (Cont'd)

*Data are reference values.

TERMINAL NO.	ITEM	CONDITION	*DATA
16	Air flow meter	Engine is running.	1.0 - 3.0V Output voltage varies with engine revolution.
18	Engine temperature sensor	Engine is running.	1.0 - 5.0V Output voltage varies with engine water temperature.
19	Exhaust gas sensor	Engine is running. – After warming up sufficiently.	0 - Approximately 1.0V
20	Throttle sensor	Ignition switch "ON"	0.4 - Approximately 4V Output voltage varies with the throttle valve opening angle.
22 30	Crank angle sensor (Reference signal)	Engine is running. Do not run engine at high speed under no-load.	0.2 - 0.5V
26	Air temperature sensor	Ignition switch "ON" – Temperature of intake air is 20°C (68°F)	Approximately 1.0 - 1.5V
		Ignition switch "ON" – Temperature of intake air is 80°C (176°F)	Approximately 0.3V
31 40	Crank angle sensor (Position signal)	Engine is running. Do not run engine at high speed under no-load.	2.0 - 3.0V
33	Idle switch (⊖ side)	Ignition switch "ON" – Throttle valve: idle position	Approximately 9 - 10V
		Ignition switch "ON" – Throttle valve: Any position except idle position	0V
34	Start signal	Cranking	8 - 12V
36	Neutral switch & Inhibitor switch	Ignition switch "ON" – Neutral/Parking	0V
		Ignition switch "ON" – Except the above gear position	Approximately 12V

ECU SIGNAL INSPECTION (CONT.)—1990 AXXESS

Electrical Components Inspection (Cont'd)

*Data are reference values.

TERMINAL NO.	ITEM	CONDITION	*DATA
36	Ignition switch	Ignition switch "OFF"	0V
		Ignition switch "ON"	BATTERY VOLTAGE (11 - 14V)
37	Throttle sensor power supply	Ignition switch "ON"	Approximately 5V
38 47	Power supply for E.C.U.	Ignition switch "ON"	BATTERY VOLTAGE (11 - 14V)
41	Air conditioner switch	Engine is running. └ Both air conditioner switch and blower switch are "ON".	0V
		Engine is running. └ Air conditioner switch is "OFF".	BATTERY VOLTAGE (11 - 14V)
43	Power steering oil pressure switch	Engine is running. └ Steering wheel is being turned.	0V
		Engine is running. └ Steering wheel is not being turned.	8 - 9V
44	Idle switch (⊕ side)	Ignition switch "ON" └ Throttle valve: idle position	Approximately 9 - 10V
		Ignition switch "ON" └ Throttle valve: Except idle position	BATTERY VOLTAGE (11 - 14V)
45	Thermo control amp.	Engine is running. Air conditioner is operating. └ Evaporator outlet air temperature is between 3.0 - 8.0°C (37 - 46°F)	Approximately 8 - 9V
		Engine is running. Air conditioner is operating. └ Evaporator outlet air temperature is over 8°C (46°F)	Approximately 0V
46	Power supply (Back-up)	Ignition switch "OFF"	BATTERY VOLTAGE (11 - 14V)

Electrical Components Inspection (Cont'd)

*Data are reference values.

TERMINAL NO.	ITEM	CONDITION	*DATA
101	Injector No. 1	Engine is running.	BATTERY VOLTAGE (11 - 14V)
103	Injector No. 3		
110	Injector No. 2		
112	Injector No. 4		
102	A.I.V. control solenoid valve	Engine is running. └ Idle speed	0 - 1.0V
		Engine is running. ├ Accelerator pedal is depressed. └ After warming up	BATTERY VOLTAGE (11 - 14V)
104	Fuel pump relay	Ignition switch "ON" └ For 5 seconds after turning ignition switch "ON" Engine is running.	0.7 - 0.9V
		Ignition switch "ON" └ Within 5 seconds after turning ignition switch "ON"	BATTERY VOLTAGE (11 - 14V)
105	E.G.R. control solenoid valve	Engine is running. └ Engine is cold. [Water temperature is below 60°C (140°F).]	0.7 - 0.9V
		Engine is running. └ After warming up [Water temperature is between 60°C (140°F) and 105°C (221°F).]	BATTERY VOLTAGE (11 - 14V)

ECU SIGNAL INSPECTION (CONT.)—1990 AXXESS

Electrical Components Inspection (Cont'd)

*Data are reference values.

TERMINAL NO.	ITEM	CONDITION	*DATA
106	Pressure regulator control solenoid valve	Stop and restart engine after warming it up. └ Water temperature is above 90°C (194°F)	0 - 1.0V (for 3 minutes after ignition switch is turned off.)
			BATTERY VOLTAGE (After 3 minutes)
		Stop and restart engine after warming it up. └ Water temperature is below 90°C (194°F)	BATTERY VOLTAGE (11 - 14V)
113	A.A.C. valve	Engine is running. └ Idle speed	7 - 10V
		Engine is running. ├ Steering wheel is being turned. ├ Air conditioner is operating. ├ Rear defogger is "ON". └ Headlamps are in high position.	4 - 7V

E.C.U. PIN CONNECTOR TERMINAL LAYOUT

1 2 3 4 5 6 7
8 9 10 11 12 13 14
15 16 17 18 19 20 21 22
23 24 25 26 27 28 29 30
31 32 33 34 35 36 37 38 39
40 41 42 43 44 45 46 47 48

H.S
CONNECT

ECCS SPECIFICATIONS—1990 AXXESS

General Specifications

IGNITION TIMING B.T.D.C.	15° ± 2°
IDLE SPEED rpm	M/T 700±50*1, 750±50*2 A/T 700±50*1, 750±50*2 (in "N" position)

*1: For U.S.A.
*2: For Canada

Inspection and Adjustment

ENGINE TEMPERATURE SENSOR Thermistor resistance kΩ	20°C (68°F)	80°C (176°F)
	2.1 - 2.9	0.30 - 0.33
AIR TEMPERATURE SENSOR Thermistor resistance kΩ	20°C (68°F)	80°C (176°F)
	2.1 - 2.9	0.27 - 0.38
IDLE SWITCH Engine speed when idle switch is changed from "OFF" to "ON" rpm	M/T 1,000±150 A/T 1,000±150 (in "N" position)	
FUEL PRESSURE at idling (Measuring point: between fuel filter and fuel pipe) Vacuum hose is connected kPa (kg/cm², psi)	Approximately 226 (2.3, 33)	
Vacuum hose is disconnected kPa (kg/cm², psi)	Approximately 294 (3.0, 43)	
FUEL INJECTOR Coil resistance Ω	Approximately 10 - 15	
EXHAUST GAS TEMPERATURE SENSOR Thermistor resistance kΩ	100°C (212°F)	
	85.3±8.53	

ECCS COMPONENTS LOCATION— 1990 TRUCK/PATHFINDER WITH VG30E ENGINE

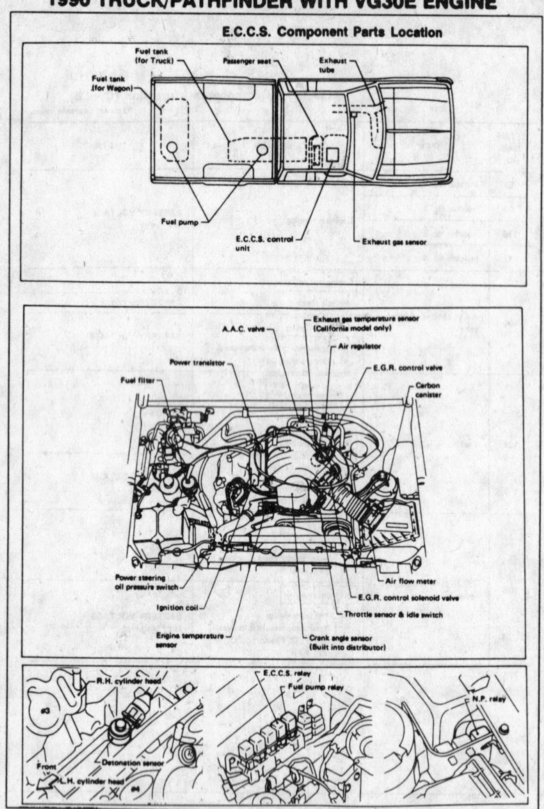
E.C.C.S. Component Parts Location
Fuel tank (for Truck)
Fuel tank (for Wagon)
Passenger seat
Exhaust tube
Fuel pump
E.C.C.S. control unit
Exhaust gas sensor
A.A.C. valve
Exhaust gas temperature sensor (California model only)
Air regulator
Power transistor
E.G.R. control valve
Fuel filter
Carbon canister
Power steering oil pressure switch
Air flow meter
Ignition coil
E.G.R. control solenoid valve
Throttle sensor & idle switch
Engine temperature sensor
Crank angle sensor (Built into distributor)
R.H. cylinder head
Detonation sensor
Front
L.H. cylinder head
E.C.C.S. relay
Fuel pump relay
N.P. relay

ECCS CONTROL SYSTEM CHART— 1990 TRUCK/PATHFINDER WITH VG30E ENGINE

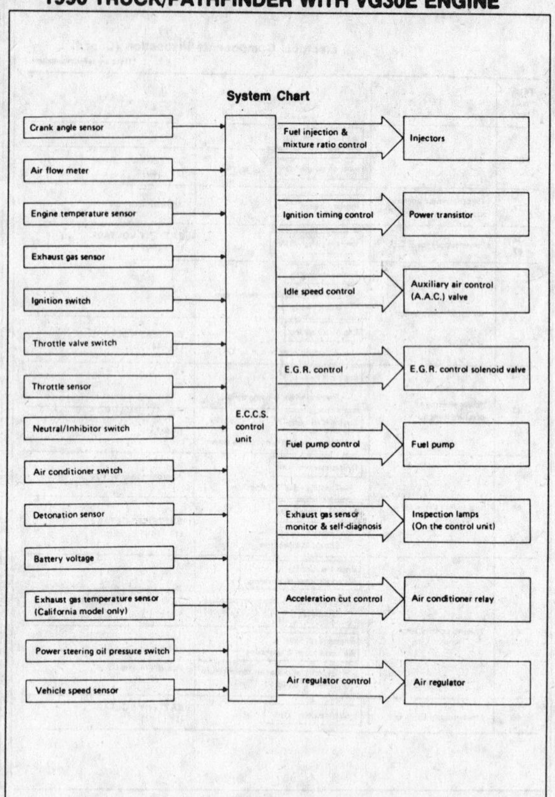
System Chart
Crank angle sensor
Air flow meter
Engine temperature sensor
Exhaust gas sensor
Ignition switch
Throttle valve switch
Throttle sensor
Neutral/Inhibitor switch
Air conditioner switch
Detonation sensor
Battery voltage
Exhaust gas temperature sensor (California model only)
Power steering oil pressure switch
Vehicle speed sensor
E.C.C.S. control unit
Fuel injection & mixture ratio control
Ignition timing control
Idle speed control
E.G.R. control
Fuel pump control
Exhaust gas sensor monitor & self-diagnosis
Acceleration cut control
Air regulator control
Injectors
Power transistor
Auxiliary air control (A.A.C.) valve
E.G.R. control solenoid valve
Fuel pump
Inspection lamps (On the control unit)
Air conditioner relay
Air regulator

ECCS SYSTEM SCHEMATIC— 1990 TRUCK/PATHFINDER WITH VG30E ENGINE

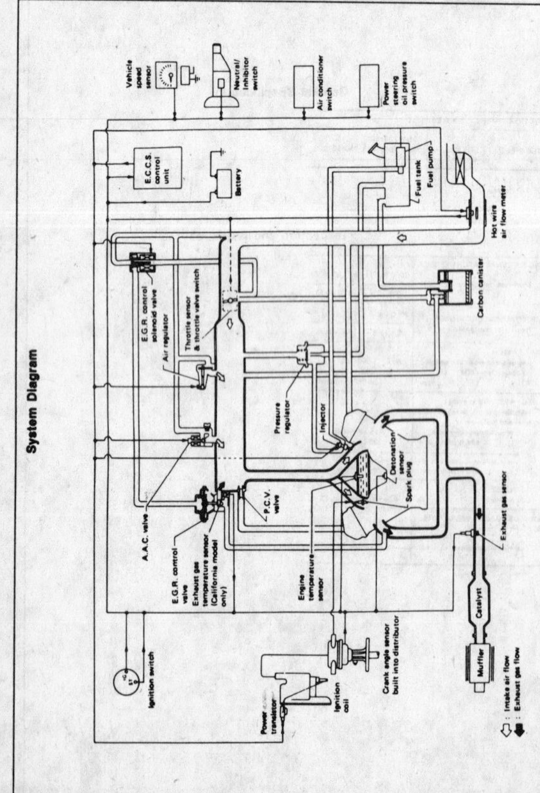
System Diagram
Vehicle speed sensor
Neutral/Inhibitor switch
Air conditioner switch
Power steering oil pressure switch
E.C.C.S. control unit
Battery
Fuel tank
Fuel pump
Hot wire air flow meter
Carbon canister
E.G.R. control solenoid valve
Air regulator
Throttle sensor & throttle valve switch
Pressure regulator
Injector
Detonation sensor
Spark plug
P.C.V. valve
Exhaust gas sensor
A.A.C. valve
Engine temperature sensor
Catalyst
Muffler
Ignition switch
Power transistor
Ignition coil
Crank angle sensor built into distributor
Intake air flow
Exhaust gas flow

ECCS VACUUM HOSE ROUTING— 1990 TRUCK/PATHFINDER WITH VG30E ENGINE

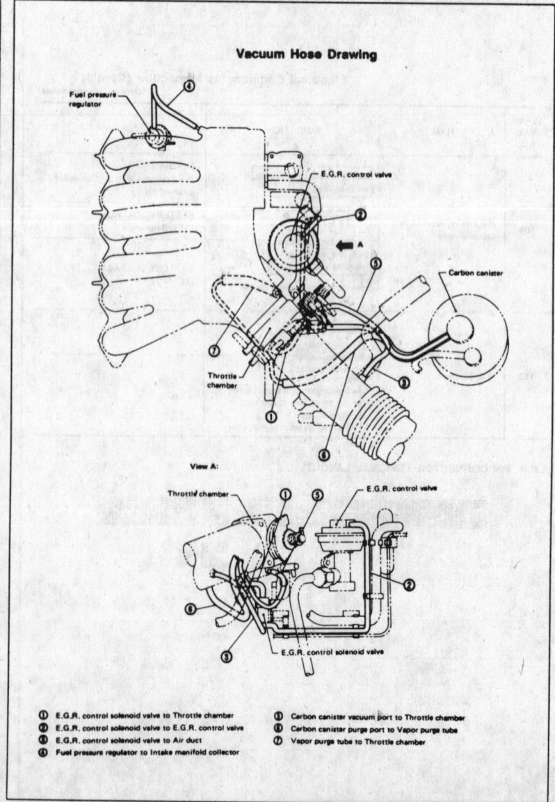
Vacuum Hose Drawing
Fuel pressure regulator
E.G.R. control valve
Carbon canister
Throttle chamber
View A:
Throttle chamber
E.G.R. control valve
E.G.R. control solenoid valve
① E.G.R. control solenoid valve to Throttle chamber
② E.G.R. control solenoid valve to E.G.R. control valve
③ E.G.R. control solenoid valve to Air duct
④ Fuel pressure regulator to Intake manifold collector
⑤ Carbon canister vacuum port to Throttle chamber
⑥ Carbon canister purge port to Vapor purge tube
⑦ Vapor purge tube to Throttle chamber

ECCS WIRING SCHEMATIC—1990 TRUCK/PATHFINDER WITH VG30E ENGINE

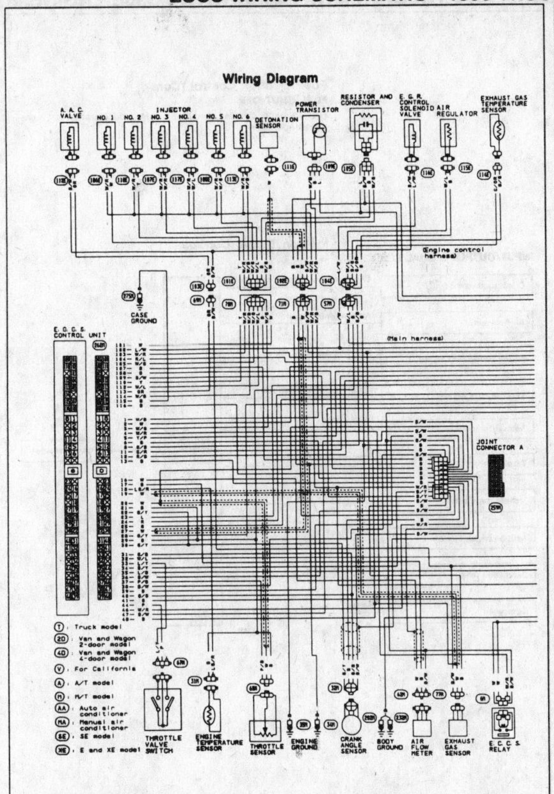
Wiring Diagram
E.C.C.S. CONTROL UNIT
JOINT CONNECTOR A
(Main harness)
(Engine control harness)
CASE GROUND
Truck model
Van and Wagon 2-door model
Van and Wagon 4-door model
For California
A/T model
M/T model
Auto air conditioner
Manual air conditioner
SE model
E and XE model
THROTTLE VALVE SWITCH
ENGINE TEMPERATURE SENSOR
THROTTLE SENSOR
ENGINE GROUND
CRANK ANGLE SENSOR
BODY GROUND
AIR FLOW METER
EXHAUST GAS SENSOR
E.C.C.S. RELAY

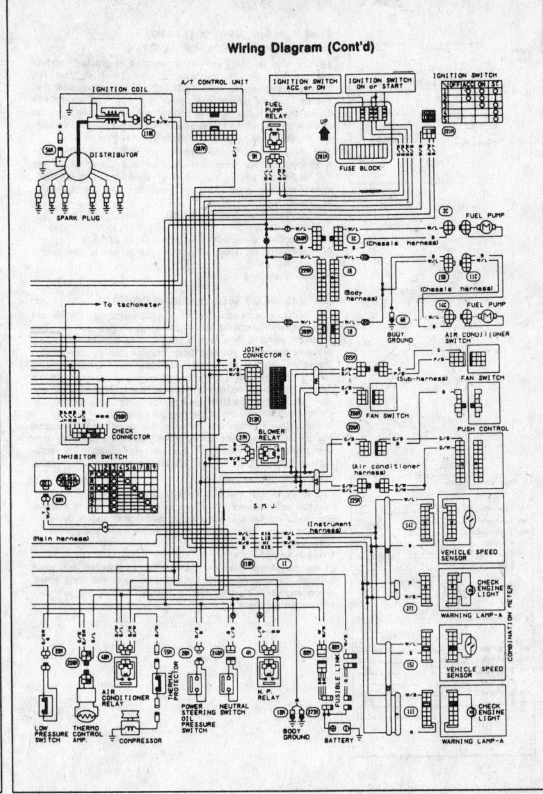
Wiring Diagram (Cont'd)
IGNITION COIL
A/T CONTROL UNIT
DISTRIBUTOR
SPARK PLUG
FUEL PUMP RELAY
FUSE BLOCK
FUEL PUMP
(Chassis harness)
(Body harness)
To tachometer
JOINT CONNECTOR C
CHECK CONNECTOR
INHIBITOR SWITCH
BLOWER RELAY
FAN SWITCH
PUSH CONTROL
(Air conditioner harness)
(Instrument harness)
(Main harness)
VEHICLE SPEED SENSOR
CHECK ENGINE LIGHT
WARNING LAMP-A
COMBINATION METER
LOW PRESSURE SWITCH
THERMO CONTROL AMP.
AIR CONDITIONER RELAY
COMPRESSOR
POWER STEERING OIL PRESSURE SWITCH
NEUTRAL SWITCH
N.P. RELAY
BODY GROUND
BATTERY
FUSIBLE LINK

ECCS CONTROL UNIT WIRING—1990 TRUCK/PATHFINDER WITH VG30E ENGINE

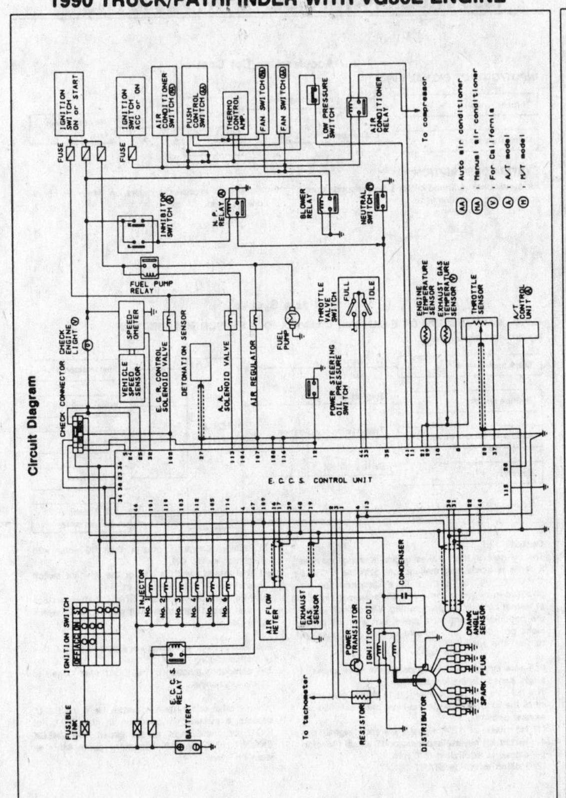
Circuit Diagram
E.C.C.S. CONTROL UNIT
IGNITION SWITCH
FUSE
FUEL PUMP RELAY
CHECK CONNECTOR
VEHICLE SPEED SENSOR
DETONATION SENSOR
AIR REGULATOR
FUEL PUMP
THROTTLE VALVE SWITCH
INJECTOR
AIR FLOW METER
EXHAUST GAS SENSOR
POWER TRANSISTOR
IGNITION COIL
CONDENSER
CRANK ANGLE SENSOR
SPARK PLUG
DISTRIBUTOR
RESISTOR
To tachometer
BATTERY
FUSIBLE LINK
To compressor

ECCS FUEL INJECTION CONTROL DESCRIPTION—1990 TRUCK/PATHFINDER WITH VG30E ENGINE.

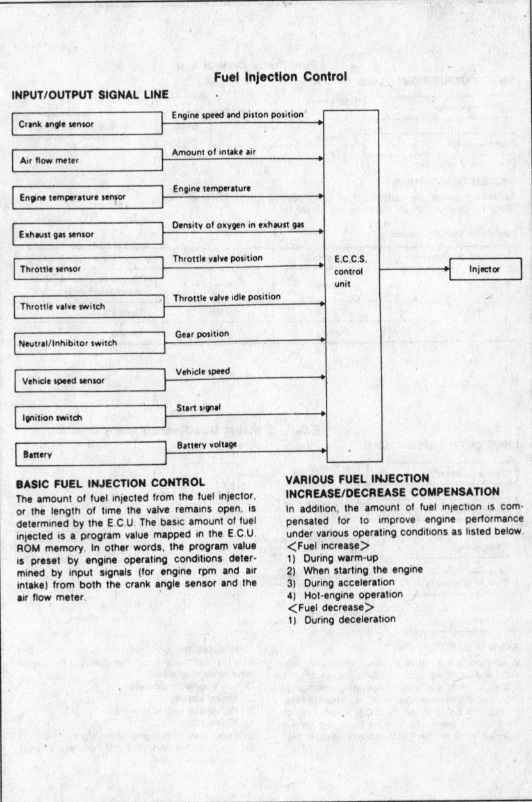
Fuel Injection Control
INPUT/OUTPUT SIGNAL LINE
Crank angle sensor
Engine speed and piston position
Air flow meter
Amount of intake air
Engine temperature sensor
Engine temperature
Exhaust gas sensor
Density of oxygen in exhaust gas
Throttle sensor
Throttle valve position
Throttle valve switch
Throttle valve idle position
Neutral/Inhibitor switch
Gear position
Vehicle speed sensor
Vehicle speed
Ignition switch
Start signal
Battery
Battery voltage
E.C.C.S. control unit
Injector

BASIC FUEL INJECTION CONTROL

The amount of fuel injected from the fuel injector, or the length of time the valve remains open, is determined by the E.C.U. The basic amount of fuel injected is a program value mapped in the E.C.U. ROM memory. In other words, the program value is preset by engine operating conditions determined by input signals (for engine rpm and air intake) from both the crank angle sensor and the air flow meter.

VARIOUS FUEL INJECTION INCREASE/DECREASE COMPENSATION

In addition, the amount of fuel injection is compensated for to improve engine performance under various operating conditions as listed below.

<Fuel increase>

1) During warm-up
2) When starting the engine
3) During acceleration
4) Hot-engine operation

<Fuel decrease>

1) During deceleration

ECCS FUEL INJECTION CONTROL DESCRIPTION (CONT.) – 1990 TRUCK/ PATHFINDER WITH VG30E ENGINE.

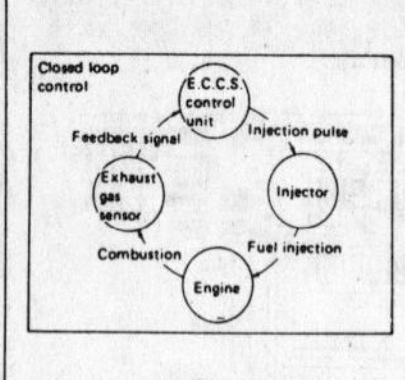
Closed loop control
E.C.C.S. control unit
Injection pulse
Feedback signal
Exhaust gas sensor
Injector
Combustion
Fuel injection
Engine

Fuel Injection Control (Cont'd)

MIXTURE RATIO FEEDBACK CONTROL

Mixture ratio feedback system is designed to precisely control the mixture ratio to the stoichiometric point so that the three-way catalyst can reduce CO, HC and NOx emissions. This system uses an exhaust gas sensor in the exhaust manifold to check the air-fuel ratio. The control unit adjusts the injection pulse width according to the sensor voltage so the mixture ratio will be within the range of the stoichiometric air-fuel ratio.

This stage refers to the closed-loop control condition. The open-loop control condition refers to that under which the E.C.U. detects any of the following conditions and feedback control stops in order to maintain stabilized fuel combustion.

1) Deceleration
2) High-load, high-speed operation
3) Engine idling
4) Malfunction of exhaust gas sensor or its circuit
5) Insufficient activation of exhaust gas sensor at low engine temperature
6) Engine starting

MIXTURE RATIO SELF-LEARNING CONTROL

The mixture ratio feedback control system monitors the mixture ratio signal transmitted from the exhaust gas sensor. This feedback signal is then sent to the E.C.U. to control the amount of fuel injection to provide a basic mixture ratio as close to the theoretical mixture ratio as possible. However, the basic mixture ratio is not necessarily controlled as originally designed. This is due to manufacturing errors (e.g., air flow meter hot wire) and changes during operation (injector clogging, etc.) of E.C.C.S. parts which directly affect the mixture ratio.

Accordingly, a difference between the basic and theoretical mixture ratios is quantitatively monitored in this system. It is then computed in terms of "fuel injection duration" to automatically compensate for the difference between the two ratios.

FUEL INJECTION TIMING

Two types of fuel injection systems are used — simultaneous injection and sequential injection. In the former, fuel is injected into all six cylinders simultaneously twice each engine cycle. In other words, pulse signals of the same width are simultaneously transmitted from the E.C.U. to the six injectors two times for each engine cycle.

In the sequential injection system, fuel is injected into each cylinder during each engine cycle according to the firing order. When engine is starting, fuel is injected into all six cylinders simultaneously twice a cycle.

• Simultaneous injection
No. 1 cylinder
No. 2 cylinder
No. 3 cylinder
No. 4 cylinder
No. 5 cylinder
No. 6 cylinder
1 engine cycle

• Sequential injection
No. 1 cylinder
No. 2 cylinder
No. 3 cylinder
No. 4 cylinder
No. 5 cylinder
No. 6 cylinder
1 engine cycle

Fuel Injection Control (Cont'd)

FUEL SHUT-OFF

Fuel to each cylinder is cut off during deceleration or high-speed operation.

Ignition Timing Control

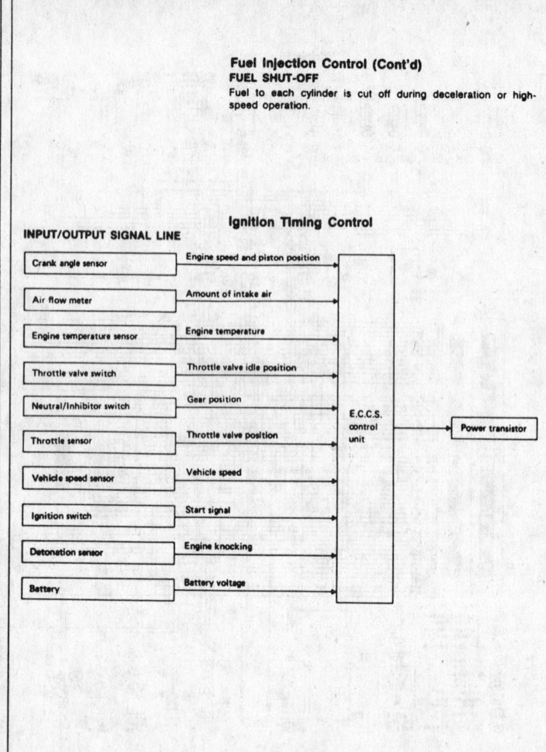
INPUT/OUTPUT SIGNAL LINE
Crank angle sensor
Engine speed and piston position
Air flow meter
Amount of intake air
Engine temperature sensor
Engine temperature
Throttle valve switch
Throttle valve idle position
Neutral/Inhibitor switch
Gear position
Throttle sensor
Throttle valve position
Vehicle speed sensor
Vehicle speed
Ignition switch
Start signal
Detonation sensor
Engine knocking
Battery
Battery voltage
E.C.C.S. control unit
Power transistor

ECCS FUEL PUMP CONTROL DESCRIPTION – 1990 TRUCK/PATHFINDER WITH VG30E ENGINE

Fuel Pump Control

INPUT/OUTPUT SIGNAL LINE

Crank angle sensor — Engine speed → E.C.C.S. control unit → Fuel pump
Ignition switch — Start signal → E.C.C.S. control unit

SYSTEM DESCRIPTION

To reduce power consumption, fuel pump relay ON-OFF operation controls the fuel pump as follows:

Fuel pump ON-OFF control

Ignition switch position	Engine condition	Fuel pump relay	Fuel pump operation
ON	Stopped	ON → OFF	Operates for a few seconds after ignition switch turns to "ON"
	Starting	ON	Operates
	Running	ON	Operates

E.G.R. (Exhaust Gas Recirculation) Control

INPUT/OUTPUT SIGNAL LINE

Crank angle sensor — Engine speed
Engine temperature sensor — Engine temperature
Throttle sensor — Throttle valve idle position
Throttle valve switch — Throttle valve idle position
Ignition switch — Start signal
→ E.C.C.S. control unit → E.G.R. control solenoid valve

SYSTEM DESCRIPTION

In addition, a system is provided which precisely cuts and controls port vacuum applied to the E.G.R. valve to suit engine operating conditions. This cut-and-control operation is accomplished through the E.C.U. When the E.C.U. detects any of the following conditions, current flows through the solenoid valve in the E.G.R. control vacuum line. This causes the port vacuum to be discharged into the atmosphere so that the E.G.R. control valve remains closed.

1) Low engine temperature
2) Engine starting
3) High-speed engine operation
4) Engine idling
5) Excessively high engine temperature
6) C.P.U. malfunction of E.C.U. and crank angle sensor malfunction

ECCS ACCELERATION CUT CONTROL AND FAIL SAFE SYSTEMS DESCRIPTION – 1990 TRUCK/PATHFINDER WITH VG30E ENGINE

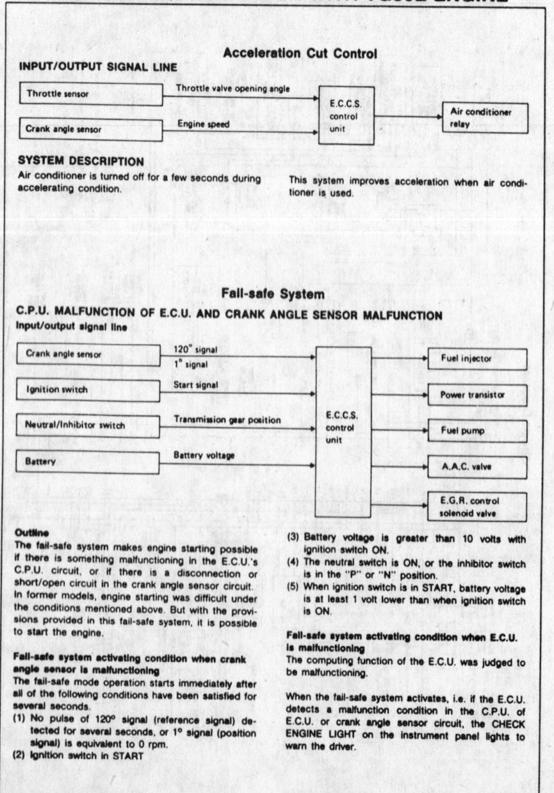
INPUT/OUTPUT SIGNAL LINE
Throttle sensor
Throttle valve opening angle
Crank angle sensor
Engine speed
E.C.C.S. control unit
Air conditioner relay
Input/output signal line
Crank angle sensor
120° signal
1° signal
Ignition switch
Start signal
Neutral/Inhibitor switch
Transmission gear position
Battery
Battery voltage
E.C.C.S. control unit
Fuel injector
Power transistor
Fuel pump
A.A.C. valve
E.G.R. control solenoid valve

Acceleration Cut Control

SYSTEM DESCRIPTION

Air conditioner is turned off for a few seconds during accelerating condition.

This system improves acceleration when air conditioner is used.

Fail-safe System

C.P.U. MALFUNCTION OF E.C.U. AND CRANK ANGLE SENSOR MALFUNCTION

Outline

The fail-safe system makes engine starting possible if there is something malfunctioning in the E.C.U.'s C.P.U. circuit, or if there is a disconnection or short/open circuit in the crank angle sensor circuit. In former models, engine starting was difficult under the conditions mentioned above. But with the provisions provided in this fail-safe system, it is possible to start the engine.

Fail-safe system activating condition when crank angle sensor is malfunctioning

The fail-safe mode operation starts immediately after all of the following conditions have been satisfied for several seconds.

(1) No pulse of 120° signal (reference signal) detected for several seconds, or 1° signal (position signal) is equivalent to 0 rpm.
(2) Ignition switch in START
(3) Battery voltage is greater than 10 volts with ignition switch ON.
(4) The neutral switch is ON, or the inhibitor switch is in the "P" or "N" position.
(5) When ignition switch is in START, battery voltage is at least 1 volt lower than when ignition switch is ON.

Fail-safe system activating condition when E.C.U. is malfunctioning

The computing function of the E.C.U. was judged to be malfunctioning.

When the fail-safe system activates, i.e. if the E.C.U. detects a malfunction condition in the C.P.U. of E.C.U. or crank angle sensor circuit, the CHECK ENGINE LIGHT on the instrument panel lights to warn the driver.

ECCS IGNITION TIMING CONTROL DESCRIPTION— 1990 TRUCK/PATHFINDER WITH VG30E ENGINE

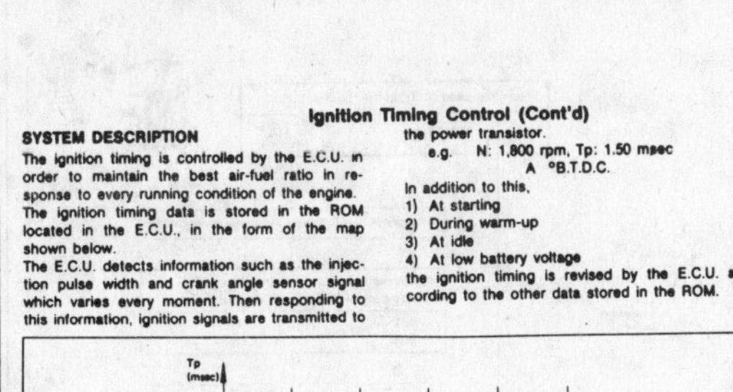

Ignition Timing Control (Cont'd)

SYSTEM DESCRIPTION

The ignition timing is controlled by the E.C.U. in order to maintain the best air-fuel ratio in response to every running condition of the engine.

The ignition timing data is stored in the ROM located in the E.C.U., in the form of the map shown below.

The E.C.U. detects information such as the injection pulse width and crank angle sensor signal which varies every moment. Then responding to this information, ignition signals are transmitted to the power transistor.

e.g. N: 1,800 rpm, Tp: 1.50 msec
A °B.T.D.C.

In addition to this,

1) At starting
2) During warm-up
3) At idle
4) At low battery voltage

the ignition timing is revised by the E.C.U. according to the other data stored in the ROM.

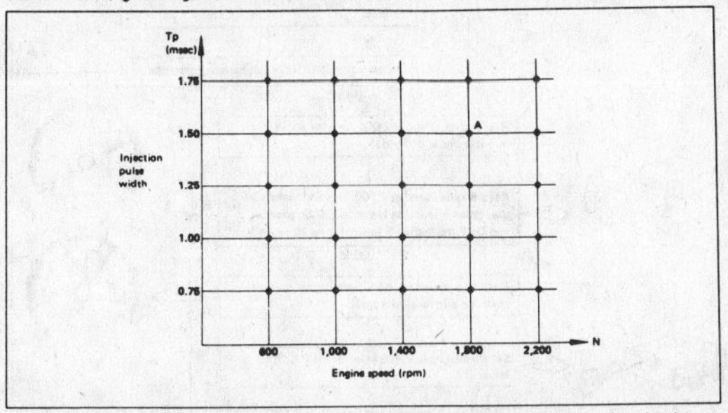
Tp (msec)
Injection pulse width
1.75
1.50
1.25
1.00
0.75
A
600
1,000
1,400
1,800
2,200
N
Engine speed (rpm)

The retard system by detonation sensor is designed only for emergencies. The basic ignition timing is pre-programmed within the anti-knocking zone, even if recommended fuel is used under dry conditions. Consequently, the retard system does not operate under normal driving conditions.

However, if engine knocking occurs, the detonation sensor monitors the condition and the signal is transmitted to the E.C.C.S. control unit. After receiving it, the control unit retards the ignition timing to avoid the knocking condition.

ECCS IDLE SPEED CONTROL DESCRIPTION— 1990 TRUCK/PATHFINDER WITH VG30E ENGINE

Idle Speed Control

INPUT/OUTPUT SIGNAL LINE

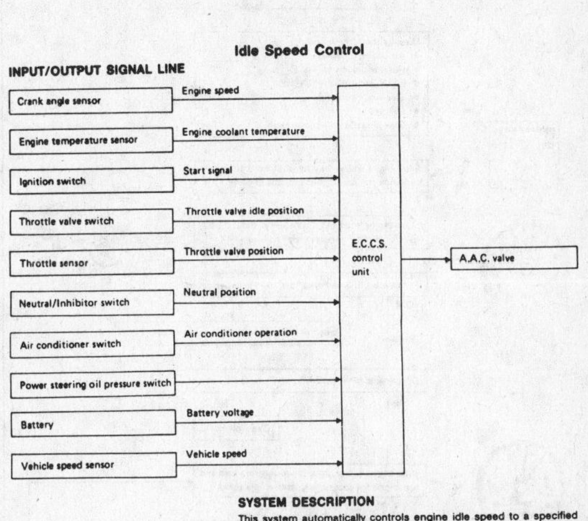
Crank angle sensor
Engine speed
Engine temperature sensor
Engine coolant temperature
Ignition switch
Start signal
Throttle valve switch
Throttle valve idle position
Throttle sensor
Throttle valve position
Neutral/Inhibitor switch
Neutral position
Air conditioner switch
Air conditioner operation
Power steering oil pressure switch
Battery
Battery voltage
Vehicle speed sensor
Vehicle speed
E.C.C.S. control unit
A.A.C. valve

SYSTEM DESCRIPTION

This system automatically controls engine idle speed to a specified level. Idle speed is controlled through fine adjustment of the amount of air which by-passes the throttle valve via A.A.C. valve. The A.A.C. valve changes the opening of the air by-pass passage to control the amount of auxiliary air. The opening of the valve is varied to allow for optimum control of the engine idling speed. The crank angle sensor detects the actual engine speed and sends a signal to the E.C.U. The E.C.U. then controls the ON/OFF time of the A.A.C. valve so that engine speed coincides with the target value memorized in ROM. The target engine speed is the lowest speed at which the engine can operate steadily. The optimum value stored in the ROM is determined by taking into consideration various engine conditions, such as warming up and during deceleration, fuel consumption, and engine load (air conditioner, electrical load).

ECCS FAIL SAFE SYSTEM CONTROL (CONT.)— 1990 TRUCK/PATHFINDER WITH VG30E ENGINE

Fail-safe System (Cont'd)

Engine control, with fail-safe system, operates when E.C.U. or crank angle sensor is malfunctioning

When the fail-safe system is operating, fuel injection, ignition timing, fuel pump operation, engine idle speed, and E.G.R. operation, are controlled under certain limitations.

Cancellation of fail-safe system when E.C.U. or crank angle sensor is malfunctioning

Activation of the fail-safe system is canceled each time the ignition switch is turned OFF. The system is reactivated if all of the above-mentioned activating conditions are satisfied after turning the ignition switch from OFF to ON.

AIR FLOW METER MALFUNCTION

If the air flow meter output voltage is above or below the specified value, the E.C.U. senses an air flow meter malfunction. In case of a malfunction, the throttle sensor substites for the air flow meter.

Though air flow meter is malfunctioning, it is possible to drive the vehicle and start the engine. But engine speed will not rise more than 3,000 rpm in order to inform the driver of fail-safe system operation while driving.

Operation

Engine condition	Starter switch	Fail-safe system	Fail-safe functioning
Stopped	ANY	Does not operate	–
Cranking	ON	Operates	Engine will be started by a pre-determined injection pulse on E.C.U.
Running	OFF		Engine speed will not rise above 3,000 rpm

ENGINE TEMPERATURE SENSOR MALFUNCTION

When engine temperature sensor output voltage is below or above the specified value, water temperature is fixed at the preset value as follows:

Operation

Condition	Engine temperature decided
Just as ignition switch is turned ON or Start	20°C (68°F)
More than 6 minutes after ignition ON or Start	80°C (176°F)
Except as shown above	20 - 80°C (68 - 176°F) (Depends on the time)

DETONATION SENSOR MALFUNCTION

When the output signal of the detonation sensor is abnormal, the E.C.U. judges it to be malfunctioning. When detonation sensor is malfunctioning, ignition timing will retard according to operating conditions.

THROTTLE SENSOR MALFUNCTION

When throttle sensor output voltage is below or above the specified value, throttle sensor output is fixed at the preset value.

ECCS IDLE SPEED, IGNITION TIMING AND IDLE MIXTURE RATIO INSPECTION— 1990 TRUCK/PATHFINDER WITH VG30E ENGINE

PREPARATION

1. **Make sure that the following parts are in good order.**
   - **Battery**
   - **Ignition system**
   - **Engine oil and coolant levels**
   - **Fuses**
   - **E.C.U. S.M.J. harness connector**
   - **Vacuum hoses**
   - **Air intake system (Oil filler cap, oil level gauge, etc.)**
   - **Fuel pressure**
   - **Engine compression**
   - **E.G.R. control valve operation**
   - **Throttle valve**
2. **On air conditioner equipped models, checks should be carried out while the air conditioner is "OFF".**
3. **On automatic transmission equipped models, when checking idle rpm, ignition timing and mixture ratio, checks should be carried out while shift lever is in "N" position.**
4. **When measuring "CO" percentage, insert probe more than 40 cm (15.7 in) into tail pipe.**
5. **Turn off headlamps, heater blower, rear defogger.**
6. **Keep front wheels pointed straight ahead.**
7. **Make the check after the radiator fan has stopped.**

WARNING:

a. **When selector lever is shifted to "D" position, apply parking brake and block both front and rear wheels with chocks.**
b. **Depress brake pedal while racing the engine to prevent forward surge of vehicle.**
c. **After the adjustment has been made, shift the lever to the "N" or "P" position and remove wheel chocks.**

Overall inspection sequence

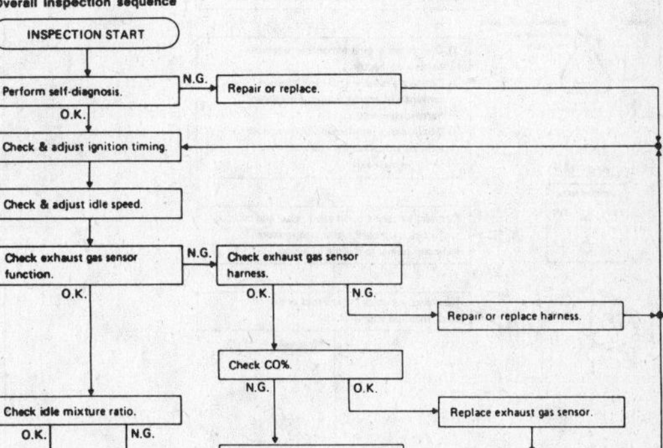
INSPECTION START
Perform self-diagnosis.
N.G.
Repair or replace.
O.K.
Check & adjust ignition timing.
Check & adjust idle speed.
Check exhaust gas sensor function.
N.G.
Check exhaust gas sensor harness.
O.K.
O.K.
N.G.
Repair or replace harness.
Check CO%.
N.G.
O.K.
Replace exhaust gas sensor.
Check idle mixture ratio.
O.K.
N.G.
Check emission control parts and repair or replace if necessary.
N.G.
Check exhaust gas sensor function.
O.K.
INSPECTION END

ECCS IDLE SPEED, IGNITION TIMING AND IDLE MIXTURE RATIO INSPECTION (CONT.)—1990 TRUCK/PATHFINDER WITH VG30E ENGINE

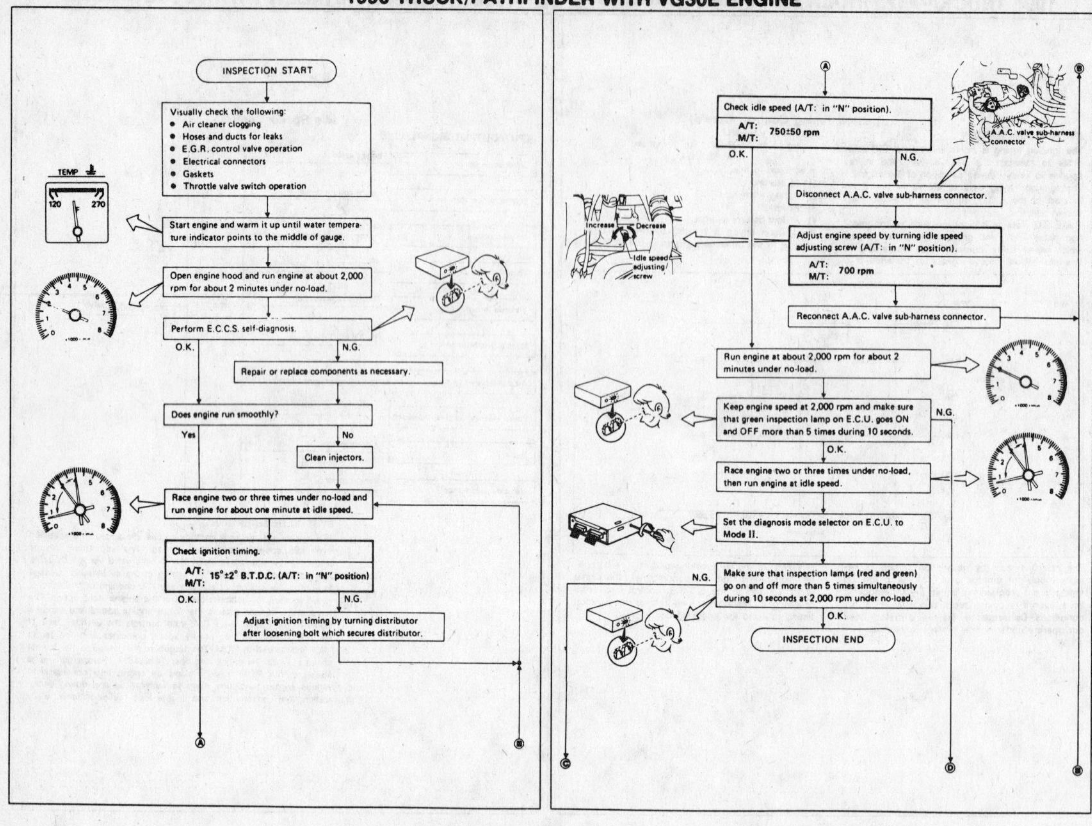
INSPECTION START
Visually check the following:
• Air cleaner clogging
• Hoses and ducts for leaks
• E.G.R. control valve operation
• Electrical connectors
• Gaskets
• Throttle valve switch operation
TEMP
120
270
Start engine and warm it up until water temperature indicator points to the middle of gauge.
Open engine hood and run engine at about 2,000 rpm for about 2 minutes under no-load.
Perform E.C.C.S. self-diagnosis.
O.K.
N.G.
Repair or replace components as necessary.
Does engine run smoothly?
Yes
No
Clean injectors.
Race engine two or three times under no-load and run engine for about one minute at idle speed.
Check ignition timing.
A/T:
M/T: 15°±2° B.T.D.C. (A/T: in "N" position)
O.K.
N.G.
Adjust ignition timing by turning distributor after loosening bolt which secures distributor.
A
B
Check idle speed (A/T: in "N" position).
A/T:
M/T: 750±50 rpm
O.K.
N.G.
A.A.C. valve sub-harness connector
Disconnect A.A.C. valve sub-harness connector.
Increase
Decrease
Idle speed adjusting screw
Adjust engine speed by turning idle speed adjusting screw (A/T: in "N" position).
A/T:
M/T: 700 rpm
Reconnect A.A.C. valve sub-harness connector.
Run engine at about 2,000 rpm for about 2 minutes under no-load.
Keep engine speed at 2,000 rpm and make sure that green inspection lamp on E.C.U. goes ON and OFF more than 5 times during 10 seconds.
N.G.
O.K.
Race engine two or three times under no-load, then run engine at idle speed.
Set the diagnosis mode selector on E.C.U. to Mode II.
N.G.
Make sure that inspection lamps (red and green) go on and off more than 5 times simultaneously during 10 seconds at 2,000 rpm under no-load.
O.K.
INSPECTION END
C
D
E

ECCS IDLE SPEED, IGNITION TIMING AND IDLE MIXTURE RATIO INSPECTION (CONT.)—1990 TRUCK/PATHFINDER WITH VG30E ENGINE

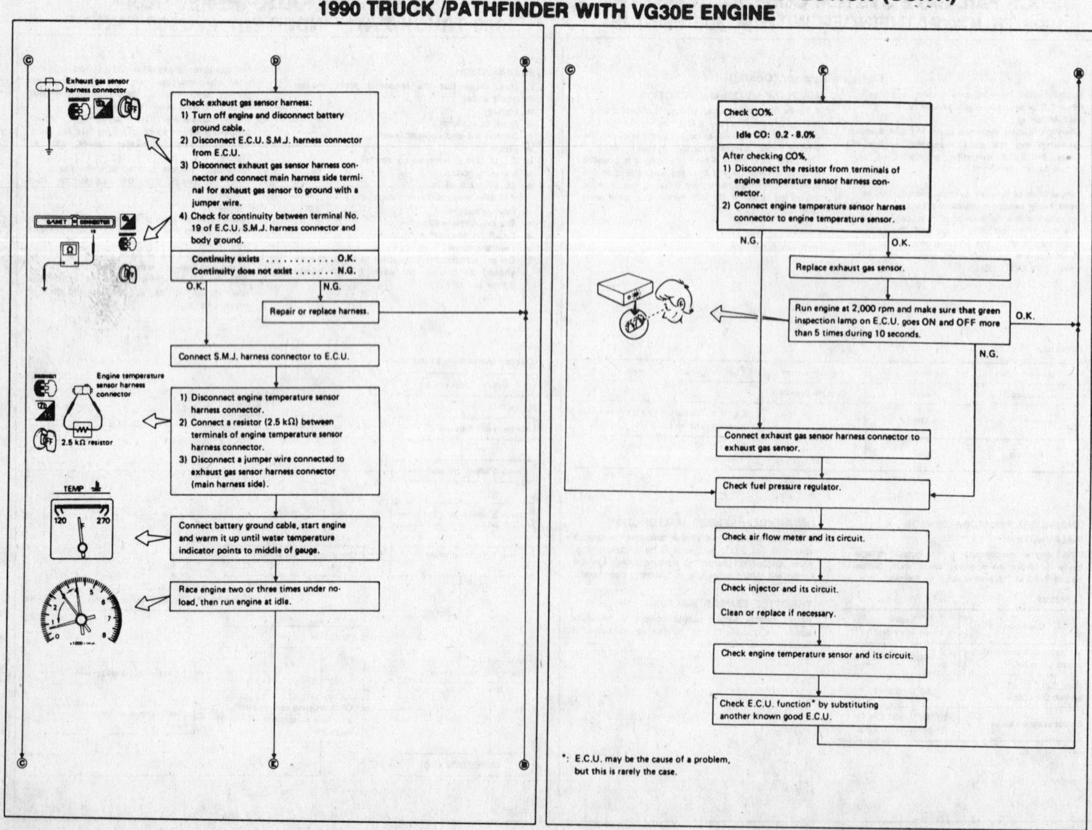
C
D
E
Exhaust gas sensor harness connector
Check exhaust gas sensor harness:
1) Turn off engine and disconnect battery ground cable.
2) Disconnect E.C.U. S.M.J. harness connector from E.C.U.
3) Disconnect exhaust gas sensor harness connector and connect main harness side terminal for exhaust gas sensor to ground with a jumper wire.
4) Check for continuity between terminal No. 19 of E.C.U. S.M.J. harness connector and body ground.
Continuity exists .... O.K.
Continuity does not exist .... N.G.
O.K.
N.G.
Repair or replace harness.
Connect S.M.J. harness connector to E.C.U.
Engine temperature sensor harness connector
2.5 kΩ resistor
1) Disconnect engine temperature sensor harness connector.
2) Connect a resistor (2.5 kΩ) between terminals of engine temperature sensor harness connector.
3) Disconnect a jumper wire connected to exhaust gas sensor harness connector (main harness side).
TEMP
120
270
Connect battery ground cable, start engine and warm it up until water temperature indicator points to middle of gauge.
Race engine two or three times under no-load, then run engine at idle.
Check CO%.
Idle CO: 0.2 - 8.0%
After checking CO%,
1) Disconnect the resistor from terminals of engine temperature sensor harness connector.
2) Connect engine temperature sensor harness connector to engine temperature sensor.
N.G.
O.K.
Replace exhaust gas sensor.
Run engine at 2,000 rpm and make sure that green inspection lamp on E.C.U. goes ON and OFF more than 5 times during 10 seconds.
O.K.
N.G.
Connect exhaust gas sensor harness connector to exhaust gas sensor.
Check fuel pressure regulator.
Check air flow meter and its circuit.
Check injector and its circuit.
Clean or replace if necessary.
Check engine temperature sensor and its circuit.
Check E.C.U. function* by substituting another known good E.C.U.
*: E.C.U. may be the cause of a problem, but this is rarely the case.

ECCS DIAGNOSTIC DESCRIPTION – 1990 TRUCK/PATHFINDER WITH VG30E ENGINE

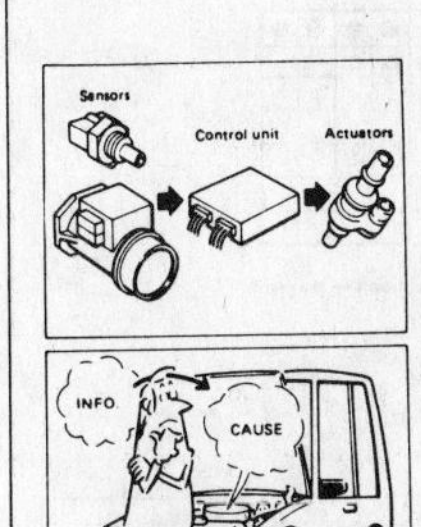
Sensors
Control unit
Actuators

How to Perform Trouble Diagnoses for Quick and Accurate Repair

INTRODUCTION

The engine has an electronic control unit to control major systems such as fuel control, ignition control, idle speed control, etc. The control unit accepts input signals from sensors and instantly drives actuators. It is essential that both kinds of signals are proper and stable. At the same time, it is important that there are no conventional problems such as vacuum leaks, fouled spark plugs, or other problems with the engine.

It is much more difficult to diagnose a problem that occurs intermittently rather than continuously. Most intermittent problems are caused by poor electric connections or improper wiring. In this case, careful checking of suspected circuits may help prevent the replacement of good parts.

A visual check only may not find the cause of the problems, so a road test with a circuit tester connected to a suspected circuit should be performed.

Before undertaking actual checks, take just a few minutes to talk with a customer who approaches with a driveability complaint. The customer is a very good supplier of information on such problems, especially intermittent ones. Through interaction with the customer, find out what symptoms are present and under what conditions they occur.

Start your diagnosis by looking for "conventional" problems first. This is one of the best ways to troubleshoot driveability problems on an electronically controlled engine vehicle.

How to Perform Trouble Diagnoses for Quick and Accurate Repair (Cont'd)

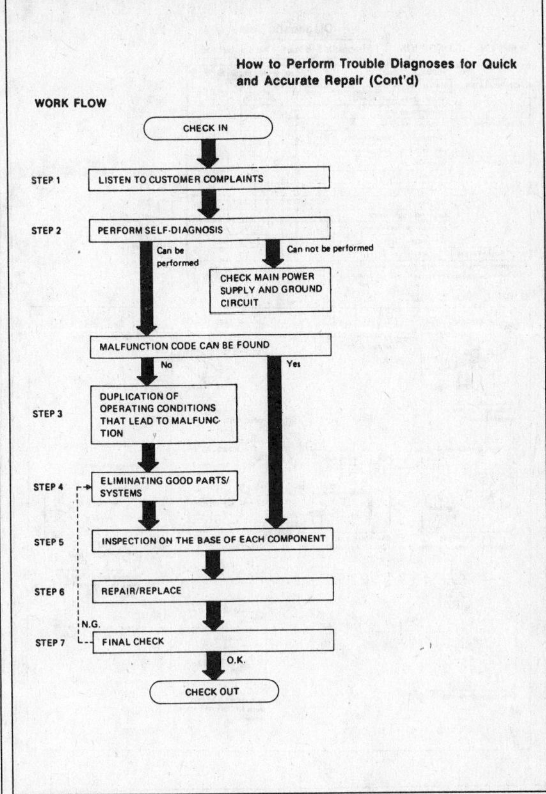
WORK FLOW
CHECK IN
STEP 1
LISTEN TO CUSTOMER COMPLAINTS
STEP 2
PERFORM SELF-DIAGNOSIS
Can be performed
Can not be performed
CHECK MAIN POWER SUPPLY AND GROUND CIRCUIT
MALFUNCTION CODE CAN BE FOUND
No
Yes
STEP 3
DUPLICATION OF OPERATING CONDITIONS THAT LEAD TO MALFUNCTION
STEP 4
ELIMINATING GOOD PARTS/SYSTEMS
STEP 5
INSPECTION ON THE BASE OF EACH COMPONENT
STEP 6
REPAIR/REPLACE
N.G.
STEP 7
FINAL CHECK
O.K.
CHECK OUT

ECCS DIAGNOSTIC DESCRIPTION (CONT.) – 1990 TRUCK/PATHFINDER WITH VG30E ENGINE

KEY POINTS

WHAT ..... Vehicle & engine model
WHEN ..... Date, Frequencies
WHERE ..... Road conditions
HOW ..... Operating conditions, Weather conditions, Symptoms

How to Perform Trouble Diagnoses for Quick and Accurate Repair (Cont'd)

DIAGNOSTIC WORKSHEET

There are many kinds of operating conditions that lead to malfunctions on engine components.

A good grasp of such conditions can make trouble-shooting faster and more accurate.

In general, feelings for a problem depend on each customer. It is important to fully understand the symptoms or under what conditions a customer complains.

Make good use of a diagnostic worksheet such as the one shown below in order to utilize all the complaints for trouble-shooting.

Worksheet sample

Customer name MR/MS		Model & Year	VIN
Engine #		Trans.	Mileage
Incident Date		Manuf. Date	In Service Date
Symptoms	☐ Startability	☐ Impossible to start ☐ No combustion ☐ Partial combustion ☐ Partial combustion affected by throttle position ☐ Partial combustion NOT affected by throttle position ☐ Possible but hard to start ☐ Others [ ]	
	☐ Idling	☐ No fast idle ☐ Unstable ☐ High idle ☐ Low idle ☐ Others [ ]	
	☐ Driveability	☐ Stumble ☐ Surge ☐ Detonation ☐ Lack of power ☐ Intake backfire ☐ Exhaust backfire ☐ Others [ ]	
	☐ Engine stall	☐ At the time of start ☐ While idling ☐ While accelerating ☐ While decelerating ☐ Just after stopping ☐ While loading	
Incident occurrence		☐ Just after delivery ☐ Recently ☐ In the morning ☐ At night ☐ In the daytime	
Frequency		☐ All the time ☐ Under certain conditions ☐ Sometimes	
Weather conditions		☐ Not affected	
	Weather	☐ Fine ☐ Raining ☐ Snowing ☐ Others [ ]	
	Temperature	☐ Hot ☐ Warm ☐ Cool ☐ Cold ☐ Humid °F	
Engine conditions		☐ Cold ☐ During warm-up ☐ After warm-up Engine speed 0 2,000 4,000 6,000 8,000 rpm	
Road conditions		☐ In town ☐ In suburbs ☐ Highway ☐ Off road (up/down)	
Driving conditions		☐ Not affected ☐ At starting ☐ While idling ☐ At racing ☐ While accelerating ☐ While cruising ☐ While decelerating ☐ While turning (RH/LH) Vehicle speed 0 10 20 30 40 50 60 MPH	
Check engine light		☐ Turned on ☐ Not turned on	

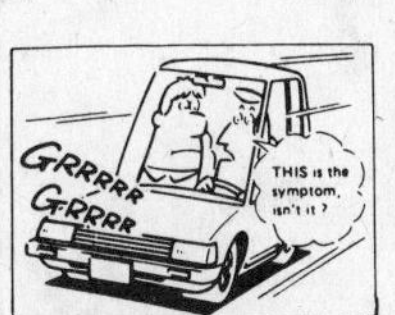
GRRRRR
GRRRR
THIS is the symptom, isn't it ?

How to Perform Trouble Diagnoses for Quick and Accurate Repair (Cont'd)

INTERMITTENT PROBLEM SIMULATION

In order to duplicate an intermittent problem, it is effective to create similar conditions for component parts, under which the problem might occur.

Perform the activity listed under Service procedure and note the result.

	Variable factor	Influential part	Target condition	Service procedure
1	Mixture ratio	Pressure regulator	Made lean	Remove vacuum hose and apply vacuum.
			Made rich	Remove vacuum hose and apply pressure.
2	Ignition timing	Crank angle sensor	Advanced	Rotate distributor counter clockwise.
			Retarded	Rotate distributor clockwise.
3	Mixture ratio feedback control	Exhaust gas sensor	Suspended	Disconnect exhaust gas sensor harness connector.
		Control unit	Operation check	Perform self-diagnosis (Mode I/II) at 2,000 rpm.
4	Idle speed	A.A.C. valve	Raised	Turn idle adjusting screw counterclockwise.
			Lowered	Turn idle adjusting screw clockwise.
5	Electrical connection (Electric continuity)	Harness connectors and wires	Poor electrical connection or improper wiring	Tap or wiggle.
				Race engine rapidly. See if the torque reaction of the engine unit causes electric breaks.
6	Temperature	Control unit	Cooled	Cool with an icing spray or similar device.
			Warmed	Heat with a hair drier. [WARNING: Do not overheat the unit.]
7	Moisture	Electric parts	Damp	Wet. [WARNING: Do not directly pour water on components. Use a mist sprayer.]
8	Electric loads	Load switches	Loaded	Turn on head lights, air conditioner, rear defogger, etc.
9	Idle switch condition	Control unit	ON-OFF switching	Rotate throttle sensor body.
10	Ignition spark	Timing light	Spark power check	Try to flash timing light for each cylinder using ignition coil adapter (S.S.T.).

ECCS DIAGNOSTIC CHART—1990 TRUCK/PATHFINDER WITH VG30E ENGINE

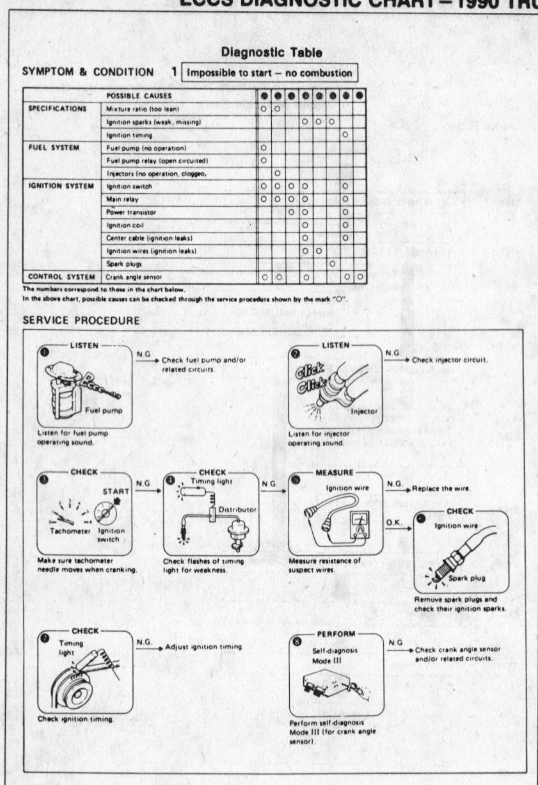

Diagnostic Table

SYMPTOM & CONDITION 1 Impossible to start – no combustion

	POSSIBLE CAUSES	1	2	3	4	5	6	7	8
SPECIFICATIONS	Mixture ratio (too lean)	O	O						
	Ignition sparks (weak, missing)				O	O	O		
	Ignition timing							O	
FUEL SYSTEM	Fuel pump (no operation)	O							
	Fuel pump relay (open circuited)	O							
	Injectors (no operation, clogged)		O						
IGNITION SYSTEM	Ignition switch	O	O	O	O			O	
	Main relay	O	O	O	O			O	
	Power transistor			O	O			O	
	Ignition coil				O			O	
	Center cable (ignition leaks)				O			O	
	Ignition wires (ignition leaks)				O	O			
	Spark plugs						O		
CONTROL SYSTEM	Crank angle sensor	O	O		O			O	O

The numbers correspond to those in the chart below.
In the above chart, possible causes can be checked through the service procedure shown by the mark "O".

SERVICE PROCEDURE

1 LISTEN
Fuel pump
Listen for fuel pump operating sound.
N.G. → Check fuel pump and/or related circuits.
2 LISTEN
Click Click
Injector
Listen for injector operating sound.
N.G. → Check injector circuit.
3 CHECK
START
Tachometer
Ignition switch
Make sure tachometer needle moves when cranking.
N.G.
4 CHECK
Timing light
Distributor
Check flashes of timing light for weakness.
N.G.
5 MEASURE
Ignition wire
Measure resistance of suspect wires.
N.G. → Replace the wire.
O.K.
6 CHECK
Ignition wire
Spark plug
Remove spark plugs and check their ignition sparks.
7 CHECK
Timing light
Check ignition timing.
N.G. → Adjust ignition timing.
8 PERFORM
Self-diagnosis Mode III
Perform self-diagnosis Mode III (for crank angle sensor).
N.G. → Check crank angle sensor and/or related circuits.

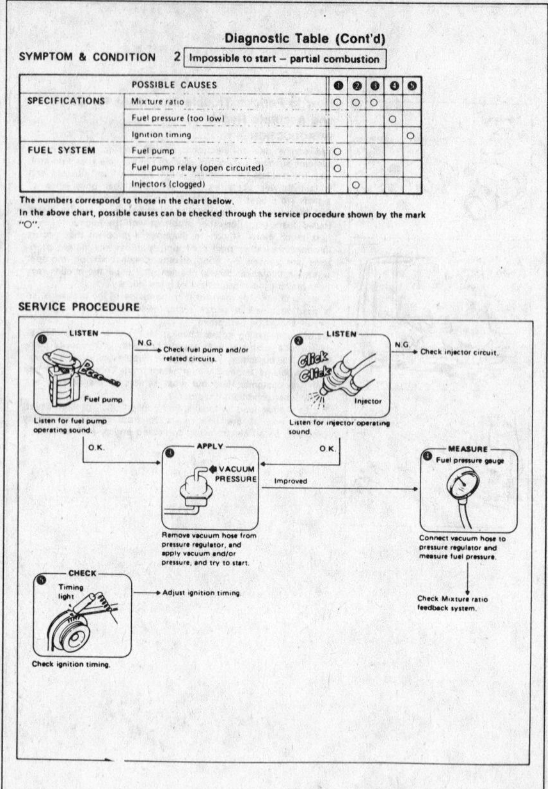

Diagnostic Table (Cont'd)

SYMPTOM & CONDITION 2 Impossible to start – partial combustion

	POSSIBLE CAUSES	1	2	3	4	5
SPECIFICATIONS	Mixture ratio	O	O	O		
	Fuel pressure (too low)				O	
	Ignition timing					O
FUEL SYSTEM	Fuel pump	O				
	Fuel pump relay (open circuited)	O				
	Injectors (clogged)		O			

The numbers correspond to those in the chart below.
In the above chart, possible causes can be checked through the service procedure shown by the mark "O".

SERVICE PROCEDURE

1 LISTEN
Fuel pump
Listen for fuel pump operating sound.
N.G. → Check fuel pump and/or related circuits.
O.K.
2 LISTEN
Click Click
Injector
Listen for injector operating sound.
N.G. → Check injector circuit.
O.K.
3 APPLY
VACUUM PRESSURE
Remove vacuum hose from pressure regulator, and apply vacuum and/or pressure, and try to start.
Improved
4 MEASURE
Fuel pressure gauge
Connect vacuum hose to pressure regulator and measure fuel pressure.
Check Mixture ratio feedback system.
5 CHECK
Timing light
Check ignition timing.
→ Adjust ignition timing.

ECCS DIAGNOSTIC CHART—1990 TRUCK/PATHFINDER WITH VG30E ENGINE

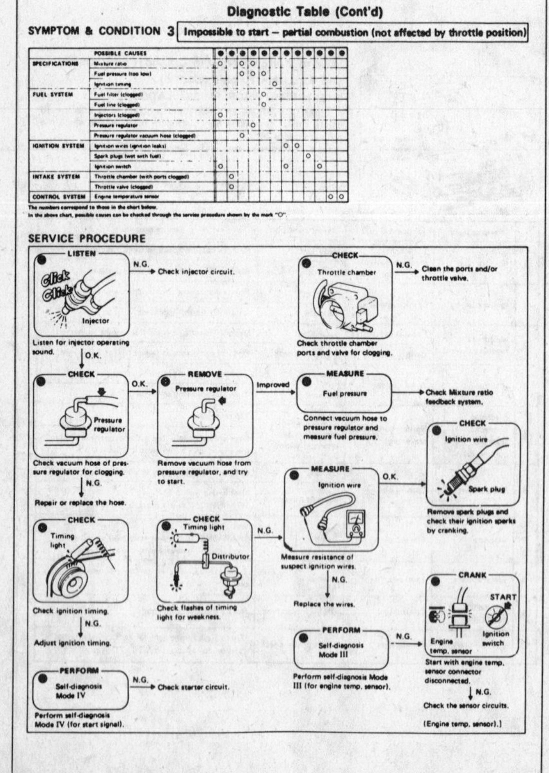

Diagnostic Table (Cont'd)

SYMPTOM & CONDITION 3 Impossible to start – partial combustion (not affected by throttle position)

	POSSIBLE CAUSES	1	2	3	4	5	6	7	8	9	10	11	12
SPECIFICATIONS	Mixture ratio	O		O	O								
	Fuel pressure (too low)			O	O	O							
	Ignition timing						O						
FUEL SYSTEM	Fuel filter (clogged)					O							
	Fuel line (clogged)					O							
	Injectors (clogged)	O											
	Pressure regulator				O								
	Pressure regulator vacuum hose (clogged)			O									
IGNITION SYSTEM	Ignition wires (ignition leaks)							O	O				
	Spark plugs (wet with fuel)									O			
	Ignition switch	O						O			O		
INTAKE SYSTEM	Throttle chamber (with ports clogged)		O										
	Throttle valve (clogged)		O										
CONTROL SYSTEM	Engine temperature sensor											O	O

The numbers correspond to those in the chart below.
In the above chart, possible causes can be checked through the service procedure shown by the mark "O".

SERVICE PROCEDURE

LISTEN
Click Click
Injector
Listen for injector operating sound.
N.G. → Check injector circuit.
O.K.
CHECK
Throttle chamber
Check throttle chamber ports and valve for clogging.
N.G. → Clean the ports and/or throttle valve.
CHECK
Pressure regulator
Check vacuum hose of pressure regulator for clogging.
N.G.
Repair or replace the hose.
O.K.
REMOVE
Pressure regulator
Remove vacuum hose from pressure regulator, and try to start.
Improved
MEASURE
Fuel pressure
Connect vacuum hose to pressure regulator and measure fuel pressure.
→ Check Mixture ratio feedback system.
CHECK
Ignition wire
Spark plug
Remove spark plugs and check their ignition sparks by cranking.
MEASURE
Ignition wire
Measure resistance of suspect ignition wires.
O.K.
N.G.
Replace the wires.
CHECK
Timing light
Check ignition timing.
N.G.
Adjust ignition timing.
CHECK
Timing light
Distributor
Check flashes of timing light for weakness.
N.G.
CRANK
START
Engine temp. sensor
Ignition switch
Start with engine temp. sensor connector disconnected.
N.G.
Check the sensor circuits.
(Engine temp. sensor).]
PERFORM
Self-diagnosis Mode III
Perform self-diagnosis Mode III (for engine temp. sensor).
N.G.
PERFORM
Self-diagnosis Mode IV
Perform self-diagnosis Mode IV (for start signal).
N.G. → Check starter circuit.

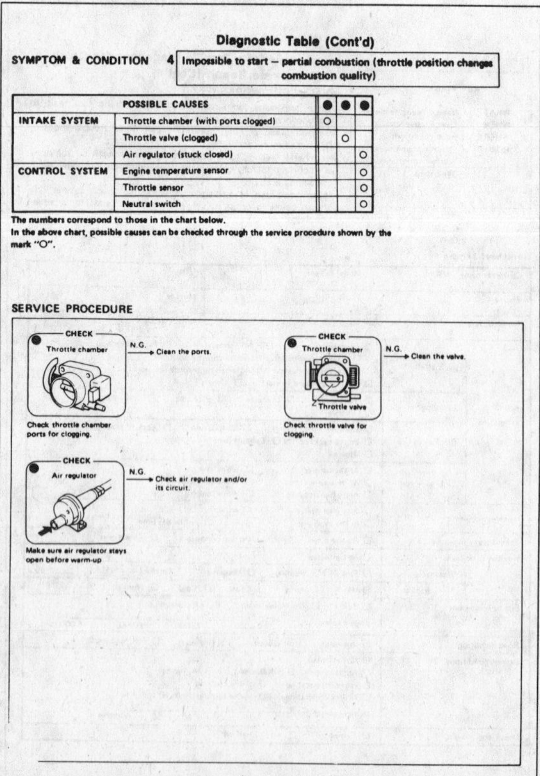

Diagnostic Table (Cont'd)

SYMPTOM & CONDITION 4 Impossible to start – partial combustion (throttle position changes combustion quality)

	POSSIBLE CAUSES	1	2	3
INTAKE SYSTEM	Throttle chamber (with ports clogged)	O		
	Throttle valve (clogged)		O	
	Air regulator (stuck closed)			O
CONTROL SYSTEM	Engine temperature sensor			O
	Throttle sensor			O
	Neutral switch			O

The numbers correspond to those in the chart below.
In the above chart, possible causes can be checked through the service procedure shown by the mark "O".

SERVICE PROCEDURE

CHECK
Throttle chamber
Check throttle chamber ports for clogging.
N.G. → Clean the ports.
CHECK
Throttle chamber
Throttle valve
Check throttle valve for clogging.
N.G. → Clean the valve.
CHECK
Air regulator
Make sure air regulator stays open before warm-up
N.G. → Check air regulator and/or its circuit.

ECCS DIAGNOSTIC CHART—1990 TRUCK/PATHFINDER WITH VG30E ENGINE

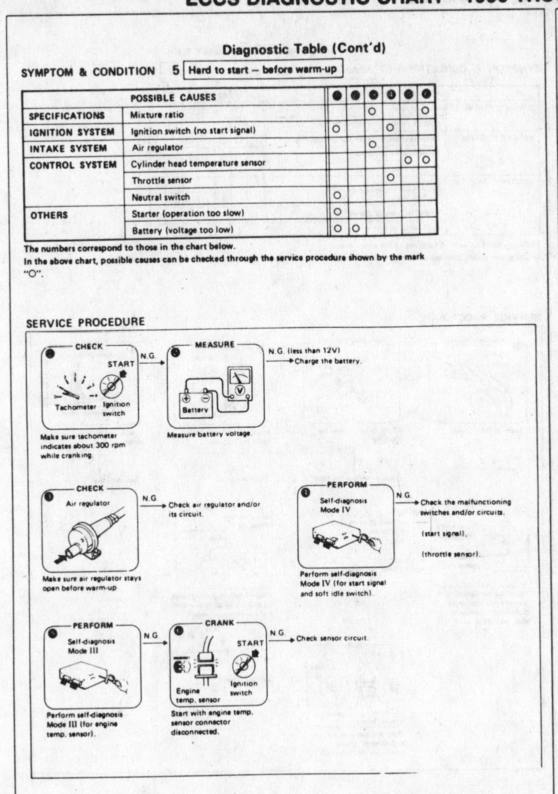

Diagnostic Table (Cont'd)

SYMPTOM & CONDITION 5 Hard to start – before warm-up

POSSIBLE CAUSES		❶	❷	❸	❹	❺	❻
SPECIFICATIONS	Mixture ratio			O			O
IGNITION SYSTEM	Ignition switch (no start signal)	O			O		
INTAKE SYSTEM	Air regulator			O			
CONTROL SYSTEM	Cylinder head temperature sensor					O	O
	Throttle sensor				O		
	Neutral switch	O					
OTHERS	Starter (operation too slow)	O					
	Battery (voltage too low)	O	O				

The numbers correspond to those in the chart below.
In the above chart, possible causes can be checked through the service procedure shown by the mark "O".

SERVICE PROCEDURE

CHECK
START
Tachometer
Ignition switch
Make sure tachometer indicates about 300 rpm while cranking.
N.G.
MEASURE
Battery
Measure battery voltage.
N.G. (less than 12V)
Charge the battery.
CHECK
Air regulator
Make sure air regulator stays open before warm-up
N.G.
Check air regulator and/or its circuit.
PERFORM
Self-diagnosis Mode IV
Perform self-diagnosis Mode IV (for start signal and soft idle switch).
N.G.
Check the malfunctioning switches and/or circuits.
(start signal).
(throttle sensor).
PERFORM
Self-diagnosis Mode III
Perform self-diagnosis Mode III (for engine temp. sensor).
N.G.
CRANK
START
Engine temp. sensor
Ignition switch
Start with engine temp. sensor connector disconnected.
N.G.
Check sensor circuit.

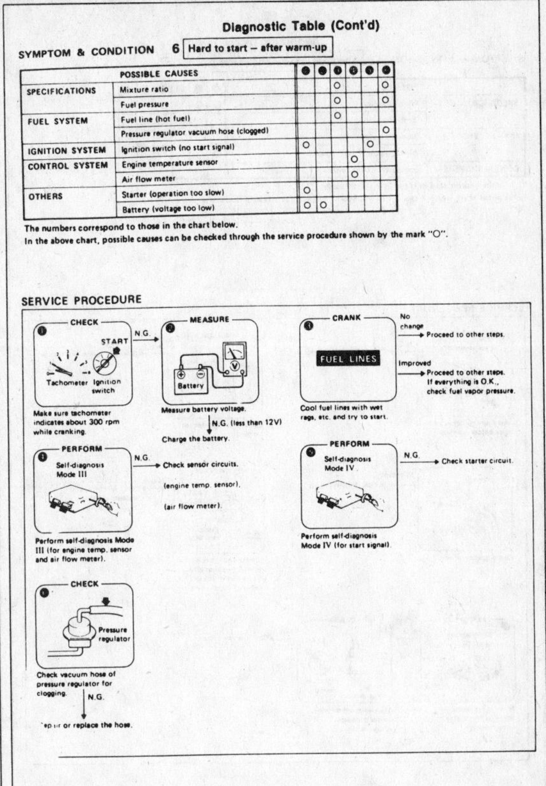

Diagnostic Table (Cont'd)

SYMPTOM & CONDITION 6 Hard to start – after warm-up

POSSIBLE CAUSES		❶	❷	❸	❹	❺	❻
SPECIFICATIONS	Mixture ratio			O			O
	Fuel pressure			O			O
FUEL SYSTEM	Fuel line (hot fuel)			O			
	Pressure regulator vacuum hose (clogged)						O
IGNITION SYSTEM	Ignition switch (no start signal)	O				O	
CONTROL SYSTEM	Engine temperature sensor				O		
	Air flow meter				O		
OTHERS	Starter (operation too slow)	O					
	Battery (voltage too low)	O	O				

The numbers correspond to those in the chart below.
In the above chart, possible causes can be checked through the service procedure shown by the mark "O".

SERVICE PROCEDURE

CHECK
START
Tachometer
Ignition switch
Make sure tachometer indicates about 300 rpm while cranking.
N.G.
MEASURE
Battery
Measure battery voltage.
N.G. (less than 12V)
Charge the battery.
CRANK
FUEL LINES
Cool fuel lines with wet rags, etc. and try to start.
No change
Proceed to other steps.
Improved
Proceed to other steps. If everything is O.K., check fuel vapor pressure.
PERFORM
Self-diagnosis Mode III
Perform self-diagnosis Mode III (for engine temp. sensor and air flow meter).
N.G.
Check sensor circuits.
(engine temp. sensor).
(air flow meter).
PERFORM
Self-diagnosis Mode IV
Perform self-diagnosis Mode IV (for start signal).
N.G.
Check starter circuit.
CHECK
Pressure regulator
Check vacuum hose of pressure regulator for clogging.
N.G.
'ep ir or replace the hose.

ECCS DIAGNOSTIC CHART—1990 TRUCK/PATHFINDER WITH VG30E ENGINE

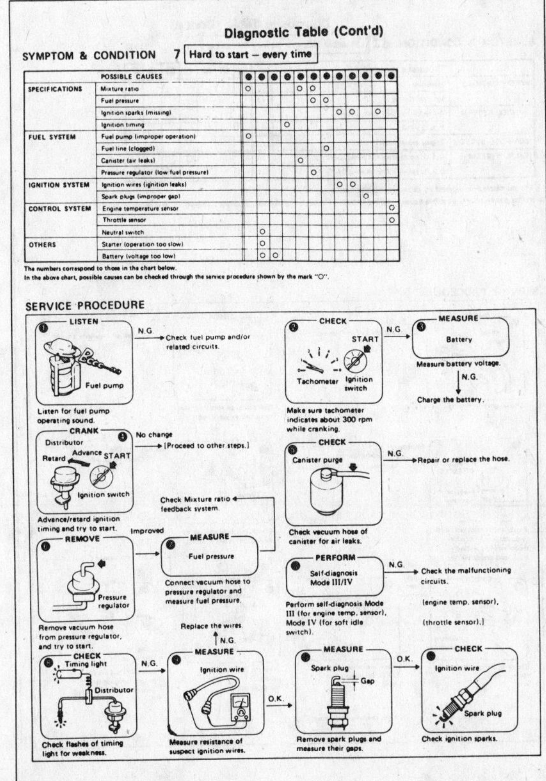

Diagnostic Table (Cont'd)

SYMPTOM & CONDITION 7 Hard to start – every time

POSSIBLE CAUSES		❶	❷	❸	❹	❺	❻	❼	❽	❾	❿	⓫	⓬
SPECIFICATIONS	Mixture ratio	O				O	O						
	Fuel pressure						O	O					
	Ignition sparks (missing)								O	O		O	
	Ignition timing				O								
FUEL SYSTEM	Fuel pump (improper operation)	O											
	Fuel line (clogged)							O					
	Canister (air leaks)					O							
	Pressure regulator (low fuel pressure)						O						
IGNITION SYSTEM	Ignition wires (ignition leaks)								O	O			
	Spark plugs (improper gap)										O		
CONTROL SYSTEM	Engine temperature sensor												O
	Throttle sensor												O
	Neutral switch		O										
OTHERS	Starter (operation too slow)		O										
	Battery (voltage too low)		O	O									

The numbers correspond to those in the chart below.
In the above chart, possible causes can be checked through the service procedure shown by the mark "O".

SERVICE PROCEDURE

LISTEN
Fuel pump
Listen for fuel pump operating sound.
N.G.
Check fuel pump and/or related circuits.
CHECK
START
Tachometer
Ignition switch
Make sure tachometer indicates about 300 rpm while cranking.
N.G.
MEASURE
Battery
Measure battery voltage.
N.G.
Charge the battery.
CRANK
Distributor
Retard
Advance
START
Ignition switch
Advance/retard ignition timing and try to start.
No change
[Proceed to other steps.]
CHECK
Canister purge
Check vacuum hose of canister for air leaks.
N.G.
Repair or replace the hose.
Check Mixture ratio feedback system.
Improved
REMOVE
Pressure regulator
Remove vacuum hose from pressure regulator, and try to start.
MEASURE
Fuel pressure
Connect vacuum hose to pressure regulator and measure fuel pressure.
PERFORM
Self-diagnosis Mode III/IV
Perform self-diagnosis Mode III (for engine temp. sensor), Mode IV (for soft idle switch).
N.G.
Check the malfunctioning circuits.
(engine temp. sensor),
(throttle sensor).]
Replace the wires.
N.G.
CHECK
Timing light
Distributor
Check flashes of timing light for weakness.
N.G.
MEASURE
Ignition wire
Measure resistance of suspect ignition wires.
O.K.
MEASURE
Spark plug
Gap
Remove spark plugs and measure their gaps.
O.K.
CHECK
Ignition wire
Spark plug
Check ignition sparks.

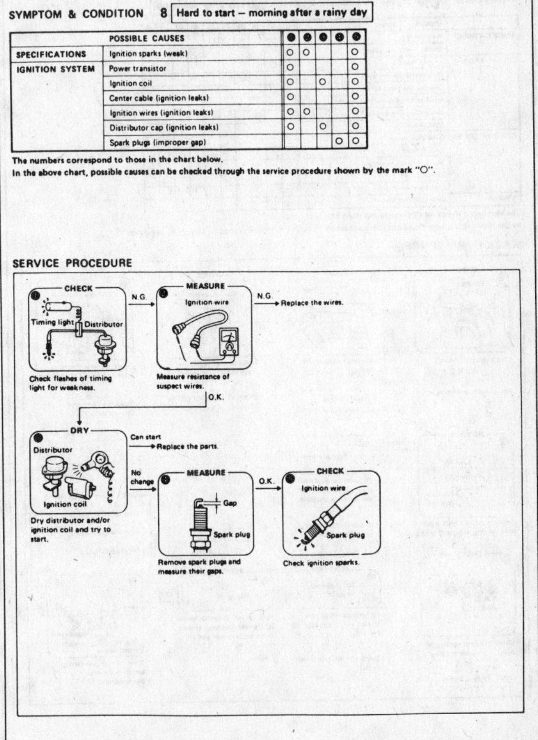

Diagnostic Table (Cont'd)

SYMPTOM & CONDITION 8 Hard to start – morning after a rainy day

POSSIBLE CAUSES		❶	❷	❸	❹	❺
SPECIFICATIONS	Ignition sparks (weak)	O	O			O
IGNITION SYSTEM	Power transistor	O				O
	Ignition coil	O		O		O
	Center cable (ignition leaks)	O				O
	Ignition wires (ignition leaks)	O	O			O
	Distributor cap (ignition leaks)	O		O		O
	Spark plugs (improper gap)				O	O

The numbers correspond to those in the chart below.
In the above chart, possible causes can be checked through the service procedure shown by the mark "O".

SERVICE PROCEDURE

CHECK
Timing light
Distributor
Check flashes of timing light for weakness.
N.G.
MEASURE
Ignition wire
Measure resistance of suspect wires.
N.G.
Replace the wires.
O.K.
DRY
Distributor
Ignition coil
Dry distributor and/or ignition coil and try to start.
Can start
Replace the parts.
No change
MEASURE
Gap
Spark plug
Remove spark plugs and measure their gaps.
O.K.
CHECK
Ignition wire
Spark plug
Check ignition sparks.

ECCS DIAGNOSTIC CHART—1990 TRUCK/PATHFINDER WITH VG30E ENGINE

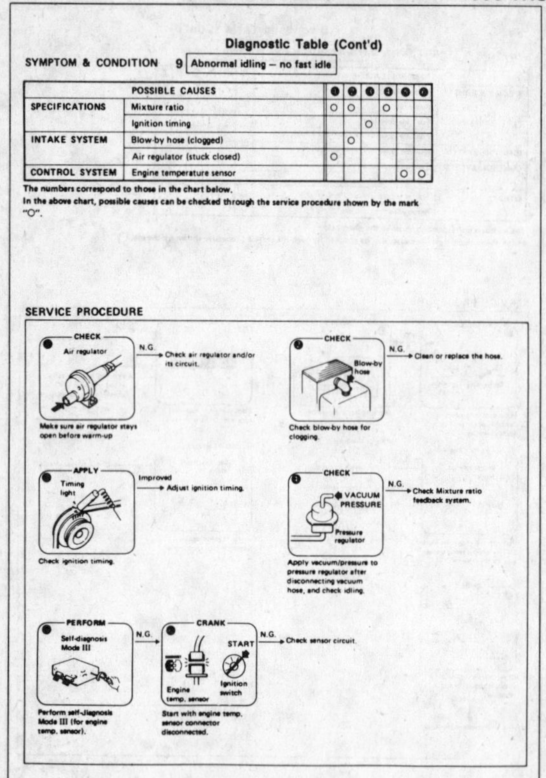

Diagnostic Table (Cont'd)

SYMPTOM & CONDITION 9 Abnormal idling – no fast idle

POSSIBLE CAUSES		1	2	3	4	5	6
SPECIFICATIONS	Mixture ratio	O	O		O		
	Ignition timing			O			
INTAKE SYSTEM	Blow-by hose (clogged)		O				
	Air regulator (stuck closed)	O					
CONTROL SYSTEM	Engine temperature sensor					O	O

The numbers correspond to those in the chart below.
In the above chart, possible causes can be checked through the service procedure shown by the mark "O".

SERVICE PROCEDURE

CHECK
Air regulator
N.G.
Check air regulator and/or its circuit.
Make sure air regulator stays open before warm-up
CHECK
Blow-by hose
N.G.
Clean or replace the hose.
Check blow-by hose for clogging.
APPLY
Timing light
Improved
Adjust ignition timing.
Check ignition timing.
CHECK
VACUUM PRESSURE
Pressure regulator
N.G.
Check Mixture ratio feedback system.
Apply vacuum/pressure to pressure regulator after disconnecting vacuum hose, and check idling.
PERFORM
Self-diagnosis Mode III
N.G.
Perform self-diagnosis Mode III (for engine temp. sensor).
CRANK
START
Engine temp. sensor
Ignition switch
N.G.
Check sensor circuit.
Start with engine temp. sensor connector disconnected.

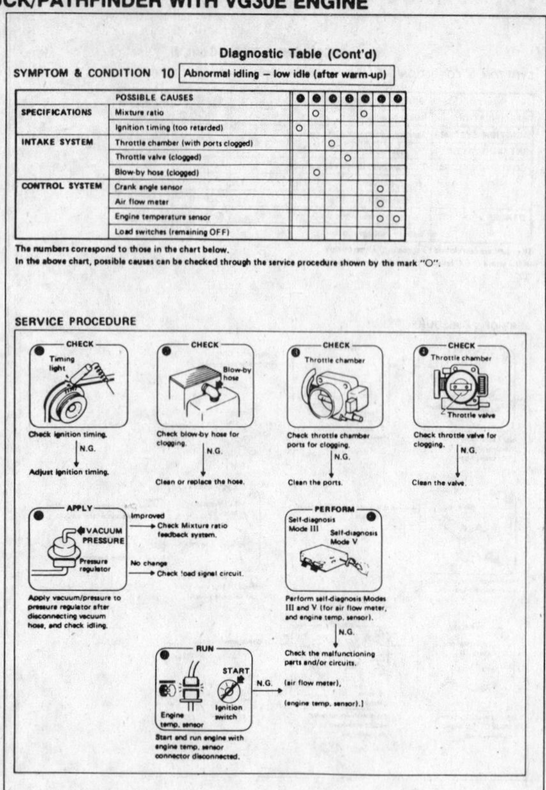

Diagnostic Table (Cont'd)

SYMPTOM & CONDITION 10 Abnormal idling – low idle (after warm-up)

POSSIBLE CAUSES		1	2	3	4	5	6	7
SPECIFICATIONS	Mixture ratio		O			O		
	Ignition timing (too retarded)	O						
INTAKE SYSTEM	Throttle chamber (with ports clogged)			O				
	Throttle valve (clogged)				O			
	Blow-by hose (clogged)		O					
CONTROL SYSTEM	Crank angle sensor						O	
	Air flow meter						O	
	Engine temperature sensor						O	O
	Load switches (remaining OFF)							

The numbers correspond to those in the chart below.
In the above chart, possible causes can be checked through the service procedure shown by the mark "O".

SERVICE PROCEDURE

CHECK
Timing light
Check ignition timing.
N.G.
Adjust ignition timing.
CHECK
Blow-by hose
Check blow-by hose for clogging.
N.G.
Clean or replace the hose.
CHECK
Throttle chamber
Check throttle chamber ports for clogging.
N.G.
Clean the ports.
CHECK
Throttle chamber
Throttle valve
Check throttle valve for clogging.
N.G.
Clean the valve.
APPLY
VACUUM PRESSURE
Pressure regulator
Improved
Check Mixture ratio feedback system.
No change
Check load signal circuit.
Apply vacuum/pressure to pressure regulator after disconnecting vacuum hose, and check idling.
PERFORM
Self-diagnosis Mode III
Self-diagnosis Mode V
Perform self-diagnosis Modes III and V (for air flow meter, and engine temp. sensor).
N.G.
Check the malfunctioning parts and/or circuits.
RUN
START
Engine temp. sensor
Ignition switch
N.G.
(air flow meter),
(engine temp. sensor).]
Start and run engine with engine temp. sensor connector disconnected.

ECCS DIAGNOSTIC CHART—1990 TRUCK/PATHFINDER WITH VG30E ENGINE

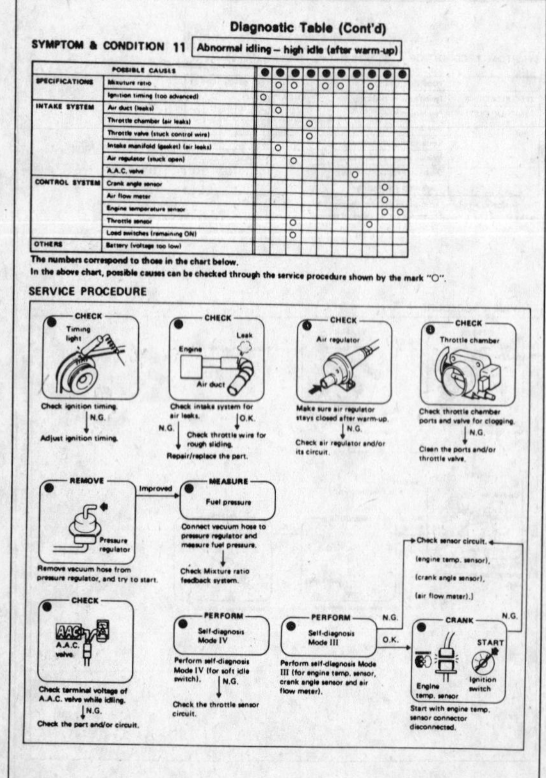

Diagnostic Table (Cont'd)

SYMPTOM & CONDITION 11 Abnormal idling – high idle (after warm-up)

POSSIBLE CAUSES		1	2	3	4	5	6	7	8	9	10
SPECIFICATIONS	Mixuture ratio		O	O		O	O		O		
	Ignition timing (too advanced)	O									
INTAKE SYSTEM	Air duct (leaks)		O								
	Throttle chamber (air leaks)				O						
	Throttle valve (stuck control wire)				O						
	Intake manifold (gasket) (air leaks)		O								
	Air regulator (stuck open)			O							
	A.A.C. valve							O			
CONTROL SYSTEM	Crank angle sensor									O	
	Air flow meter									O	
	Engine temperature sensor									O	O
	Throttle sensor			O					O		
	Load switches (remaining ON)			O							
OTHERS	Battery (voltage too low)										

The numbers correspond to those in the chart below.
In the above chart, possible causes can be checked through the service procedure shown by the mark "O".

SERVICE PROCEDURE

CHECK
Timing light
Check ignition timing.
N.G.
Adjust ignition timing.
CHECK
Engine
Leak
Air duct
Check intake system for air leaks.
O.K.
N.G.
Check throttle wire for rough sliding.
Repair/replace the part.
CHECK
Air regulator
Make sure air regulator stays closed after warm-up.
N.G.
Check air regulator and/or its circuit.
CHECK
Throttle chamber
Check throttle chamber ports and valve for clogging.
N.G.
Clean the ports and/or throttle valve.
REMOVE
Pressure regulator
Remove vacuum hose from pressure regulator, and try to start.
Improved
MEASURE
Fuel pressure
Connect vacuum hose to pressure regulator and measure fuel pressure.
Check Mixture ratio feedback system.
CHECK
A.A.C. valve
Check terminal voltage of A.A.C. valve while idling.
N.G.
Check the part and/or circuit.
PERFORM
Self-diagnosis Mode IV
Perform self-diagnosis Mode IV (for soft idle switch).
N.G.
Check the throttle sensor circuit.
PERFORM
Self-diagnosis Mode III
Perform self-diagnosis Mode III (for engine temp. sensor, crank angle sensor and air flow meter).
N.G.
O.K.
CRANK
START
Engine temp. sensor
Ignition switch
N.G.
Start with engine temp. sensor connector disconnected.
Check sensor circuit.
(engine temp. sensor),
(crank angle sensor),
(air flow meter).]

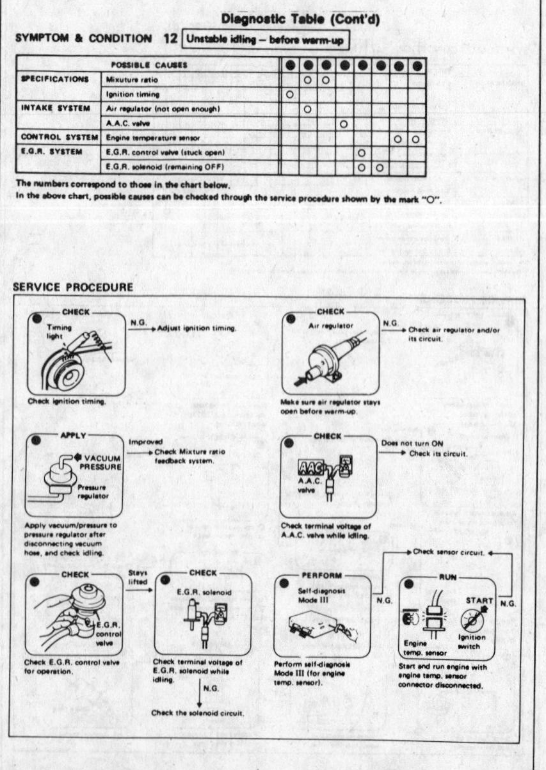

Diagnostic Table (Cont'd)

SYMPTOM & CONDITION 12 Unstable idling – before warm-up

POSSIBLE CAUSES		1	2	3	4	5	6	7	8
SPECIFICATIONS	Mixuture ratio		O	O					
	Ignition timing	O							
INTAKE SYSTEM	Air regulator (not open enough)		O						
	A.A.C. valve				O				
CONTROL SYSTEM	Engine temperature sensor							O	O
E.G.R. SYSTEM	E.G.R. control valve (stuck open)					O			
	E.G.R. solenoid (remaining OFF)					O	O		

The numbers correspond to those in the chart below.
In the above chart, possible causes can be checked through the service procedure shown by the mark "O".

SERVICE PROCEDURE

CHECK
Timing light
N.G.
Adjust ignition timing.
Check ignition timing.
CHECK
Air regulator
N.G.
Check air regulator and/or its circuit.
Make sure air regulator stays open before warm-up.
APPLY
VACUUM PRESSURE
Pressure regulator
Improved
Check Mixture ratio feedback system.
Apply vacuum/pressure to pressure regulator after disconnecting vacuum hose, and check idling.
CHECK
A.A.C. valve
Does not turn ON
Check its circuit.
Check terminal voltage of A.A.C. valve while idling.
CHECK
E.G.R. control valve
Stays lifted
Check E.G.R. control valve for operation.
CHECK
E.G.R. solenoid
Check terminal voltage of E.G.R. solenoid while idling.
N.G.
Check the solenoid circuit.
PERFORM
Self-diagnosis Mode III
Perform self-diagnosis Mode III (for engine temp. sensor).
N.G.
Check sensor circuit.
RUN
START
Engine temp. sensor
Ignition switch
N.G.
Start and run engine with engine temp. sensor connector disconnected.

ECCS DIAGNOSTIC CHART – 1990 TRUCK/PATHFINDER WITH VG30E ENGINE

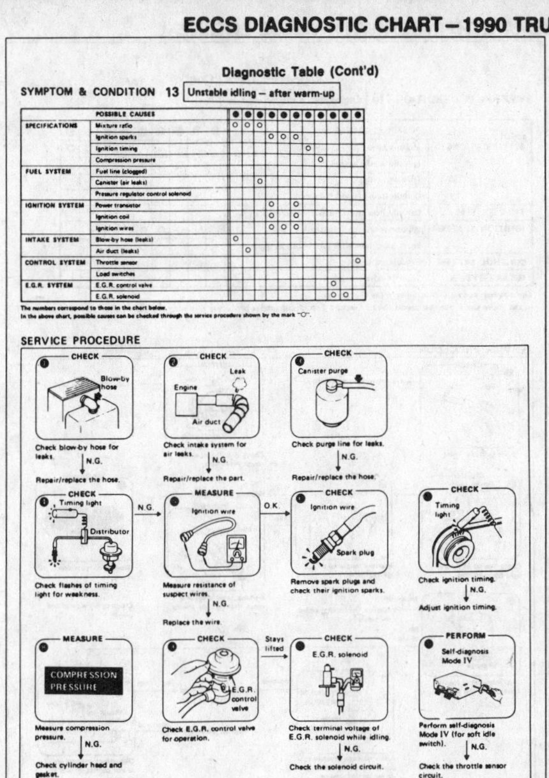

Diagnostic Table (Cont'd)

SYMPTOM & CONDITION 13 Unstable idling – after warm-up

	POSSIBLE CAUSES	●	●	●	●	●	●	●	●	●	●	●
SPECIFICATIONS	Mixture ratio	O	O	O								
	Ignition sparks				O	O	O					
	Ignition timing							O				
	Compression pressure								O			
FUEL SYSTEM	Fuel line (clogged)											
	Canister (air leaks)			O								
	Pressure regulator control solenoid											
IGNITION SYSTEM	Power transistor				O		O					
	Ignition coil				O		O					
	Ignition wires				O	O	O					
INTAKE SYSTEM	Blow-by hose (leaks)	O										
	Air duct (leaks)		O									
CONTROL SYSTEM	Throttle sensor											O
	Load switches											
E.G.R. SYSTEM	E.G.R. control valve									O		
	E.G.R. solenoid									O	O	

The numbers correspond to those in the chart below.
In the above chart, possible causes can be checked through the service procedure shown by the mark "O".

SERVICE PROCEDURE

CHECK — Blow-by hose. Check blow-by hose for leaks. N.G. → Repair/replace the hose.

CHECK — Engine, Leak, Air duct. Check intake system for air leaks. N.G. → Repair/replace the part.

CHECK — Canister purge. Check purge line for leaks. N.G. → Repair/replace the hose.

CHECK — Timing light, Distributor. Check flashes of timing light for weakness. N.G. →

MEASURE — Ignition wire. Measure resistance of suspect wires. N.G. → Replace the wire. O.K. →

CHECK — Ignition wire, Spark plug. Remove spark plugs and check their ignition sparks.

CHECK — Timing light. Check ignition timing. N.G. → Adjust ignition timing.

MEASURE — COMPRESSION PRESSURE. Measure compression pressure. N.G. → Check cylinder head and gasket.

CHECK — E.G.R. control valve. Check E.G.R. control valve for operation. Stays lifted →

CHECK — E.G.R. solenoid. Check terminal voltage of E.G.R. solenoid while idling. N.G. → Check the solenoid circuit.

PERFORM — Self-diagnosis Mode IV. Perform self-diagnosis Mode IV (for soft idle switch). N.G. → Check the throttle sensor circuit.

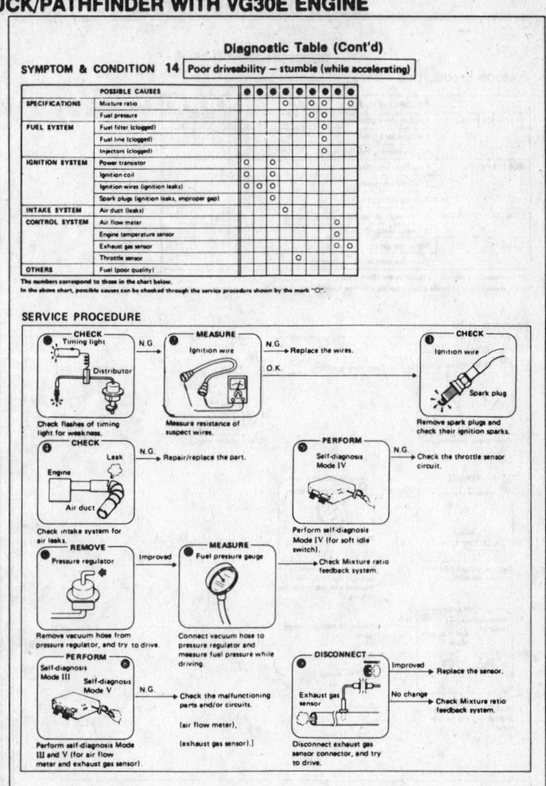

Diagnostic Table (Cont'd)

SYMPTOM & CONDITION 14 Poor driveability – stumble (while accelerating)

	POSSIBLE CAUSES	●	●	●	●	●	●	●	●	●
SPECIFICATIONS	Mixture ratio				O		O	O		O
	Fuel pressure						O	O		
FUEL SYSTEM	Fuel filter (clogged)							O		
	Fuel line (clogged)							O		
	Injectors (clogged)							O		
IGNITION SYSTEM	Power transistor	O		O						
	Ignition coil	O		O						
	Ignition wires (ignition leaks)	O	O	O						
	Spark plugs (ignition leaks, improper gap)			O						
INTAKE SYSTEM	Air duct (leaks)				O					
CONTROL SYSTEM	Air flow meter								O	
	Engine temperature sensor								O	
	Exhaust gas sensor								O	O
	Throttle sensor					O				
OTHERS	Fuel (poor quality)									

The numbers correspond to those in the chart below.
In the above chart, possible causes can be checked through the service procedure shown by the mark "O".

SERVICE PROCEDURE

CHECK — Timing light, Distributor. Check flashes of timing light for weakness. N.G. →

MEASURE — Ignition wire. Measure resistance of suspect wires. N.G. → Replace the wires. O.K. →

CHECK — Ignition wire, Spark plug. Remove spark plugs and check their ignition sparks.

CHECK — Engine, Leak, Air duct. Check intake system for air leaks. N.G. → Repair/replace the part.

PERFORM — Self-diagnosis Mode IV. Perform self-diagnosis Mode IV (for soft idle switch). N.G. → Check the throttle sensor circuit.

REMOVE — Pressure regulator. Remove vacuum hose from pressure regulator, and try to drive. Improved →

MEASURE — Fuel pressure gauge. Connect vacuum hose to pressure regulator and measure fuel pressure while driving. → Check Mixture ratio feedback system.

PERFORM — Self-diagnosis Mode III, Self-diagnosis Mode V. Perform self-diagnosis Mode III and V (for air flow meter and exhaust gas sensor). N.G. → Check the malfunctioning parts and/or circuits. (air flow meter), (exhaust gas sensor).]

DISCONNECT — Exhaust gas sensor. Disconnect exhaust gas sensor connector, and try to drive. Improved → Replace the sensor. No change → Check Mixture ratio feedback system.

ECCS DIAGNOSTIC CHART – 1990 TRUCK/PATHFINDER WITH VG30E ENGINE

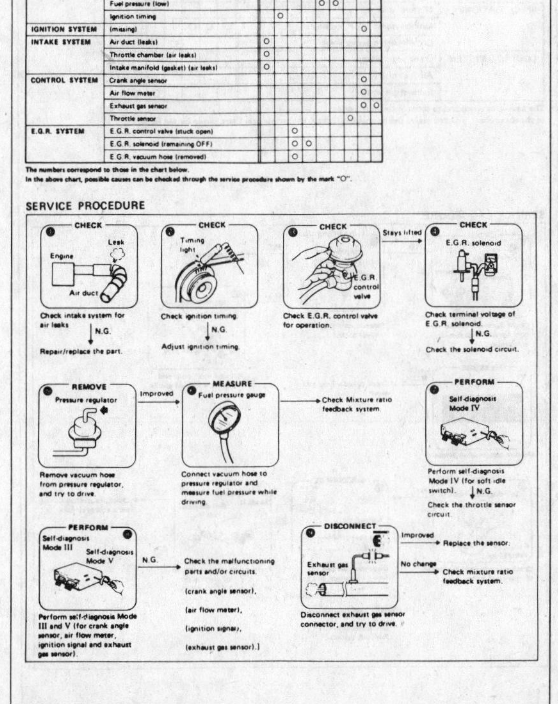

Diagnostic Table (Cont'd)

SYMPTOM & CONDITION 15 Poor driveability – surge (while cruising)

	POSSIBLE CAUSES	●	●	●	●	●	●	●	●	●
SPECIFICATIONS	Mixture ratio (too lean)	O				O	O			O
	Fuel pressure (low)					O	O			
	Ignition timing		O							
IGNITION SYSTEM	(missing)								O	
INTAKE SYSTEM	Air duct (leaks)	O								
	Throttle chamber (air leaks)	O								
	Intake manifold (gasket) (air leaks)	O								
CONTROL SYSTEM	Crank angle sensor								O	
	Air flow meter								O	
	Exhaust gas sensor								O	O
	Throttle sensor							O		
E.G.R. SYSTEM	E.G.R. control valve (stuck open)			O						
	E.G.R. solenoid (remaining OFF)			O	O					
	E.G.R. vacuum hose (removed)			O						

The numbers correspond to those in the chart below.
In the above chart, possible causes can be checked through the service procedure shown by the mark "O".

SERVICE PROCEDURE

CHECK — Engine, Leak, Air duct. Check intake system for air leaks. N.G. → Repair/replace the part.

CHECK — Timing light. Check ignition timing. N.G. → Adjust ignition timing.

CHECK — E.G.R. control valve. Check E.G.R. control valve for operation. Stays lifted →

CHECK — E.G.R. solenoid. Check terminal voltage of E.G.R. solenoid. N.G. → Check the solenoid circuit.

REMOVE — Pressure regulator. Remove vacuum hose from pressure regulator, and try to drive. Improved →

MEASURE — Fuel pressure gauge. Connect vacuum hose to pressure regulator and measure fuel pressure while driving. → Check Mixture ratio feedback system.

PERFORM — Self-diagnosis Mode IV. Perform self-diagnosis Mode IV (for soft idle switch). N.G. → Check the throttle sensor circuit.

PERFORM — Self-diagnosis Mode III, Self-diagnosis Mode V. Perform self-diagnosis Mode III and V (for crank angle sensor, air flow meter, ignition signal and exhaust gas sensor). N.G. → Check the malfunctioning parts and/or circuits. (crank angle sensor), (air flow meter), (ignition signal), (exhaust gas sensor).]

DISCONNECT — Exhaust gas sensor. Disconnect exhaust gas sensor connector, and try to drive. Improved → Replace the sensor. No change → Check mixture ratio feedback system.

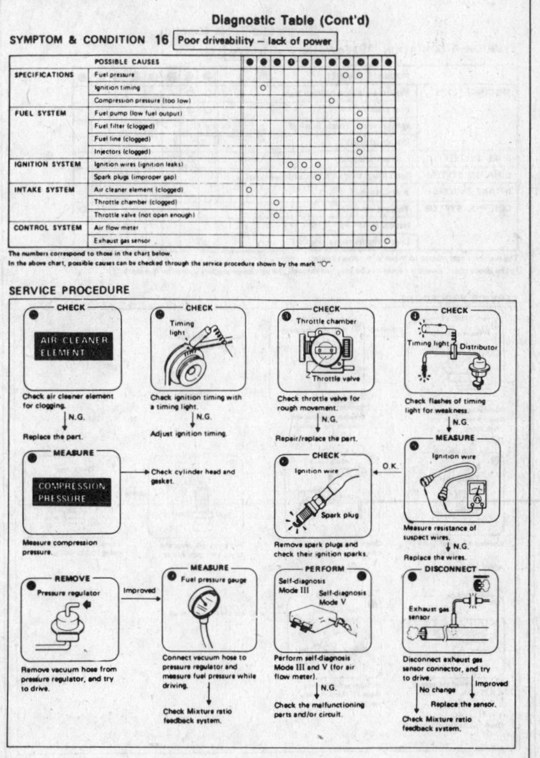

Diagnostic Table (Cont'd)

SYMPTOM & CONDITION 16 Poor driveability – lack of power

	POSSIBLE CAUSES	●	●	●	●	●	●	●	●	●	●	●
SPECIFICATIONS	Fuel pressure								O	O		
	Ignition timing		O									
	Compression pressure (too low)							O				
FUEL SYSTEM	Fuel pump (low fuel output)									O		
	Fuel filter (clogged)									O		
	Fuel line (clogged)									O		
	Injectors (clogged)									O		
IGNITION SYSTEM	Ignition wires (ignition leaks)				O	O	O					
	Spark plugs (improper gap)						O					
INTAKE SYSTEM	Air cleaner element (clogged)	O										
	Throttle chamber (clogged)			O								
	Throttle valve (not open enough)			O								
CONTROL SYSTEM	Air flow meter										O	
	Exhaust gas sensor											O

The numbers correspond to those in the chart below.
In the above chart, possible causes can be checked through the service procedure shown by the mark "O".

SERVICE PROCEDURE

CHECK — AIR CLEANER ELEMENT. Check air cleaner element for clogging. N.G. → Replace the part.

CHECK — Timing light. Check ignition timing with a timing light. N.G. → Adjust ignition timing.

CHECK — Throttle chamber, Throttle valve. Check throttle valve for rough movement. N.G. → Repair/replace the part.

CHECK — Timing light, Distributor. Check flashes of timing light for weakness. N.G. →

MEASURE — COMPRESSION PRESSURE. Measure compression pressure. → Check cylinder head and gasket.

MEASURE — Ignition wire. Measure resistance of suspect wires. N.G. → Replace the wires. O.K. →

CHECK — Ignition wire, Spark plug. Remove spark plugs and check their ignition sparks.

REMOVE — Pressure regulator. Remove vacuum hose from pressure regulator, and try to drive. Improved →

MEASURE — Fuel pressure gauge. Connect vacuum hose to pressure regulator and measure fuel pressure while driving. → Check Mixture ratio feedback system.

PERFORM — Self-diagnosis Mode III, Self-diagnosis Mode V. Perform self-diagnosis Mode III and V (for air flow meter). N.G. → Check the malfunctioning parts and/or circuit.

DISCONNECT — Exhaust gas sensor. Disconnect exhaust gas sensor connector, and try to drive. Improved → Replace the sensor. No change → Check Mixture ratio feedback system.

ECCS DIAGNOSTIC CHART—1990 TRUCK/PATHFINDER WITH VG30E ENGINE

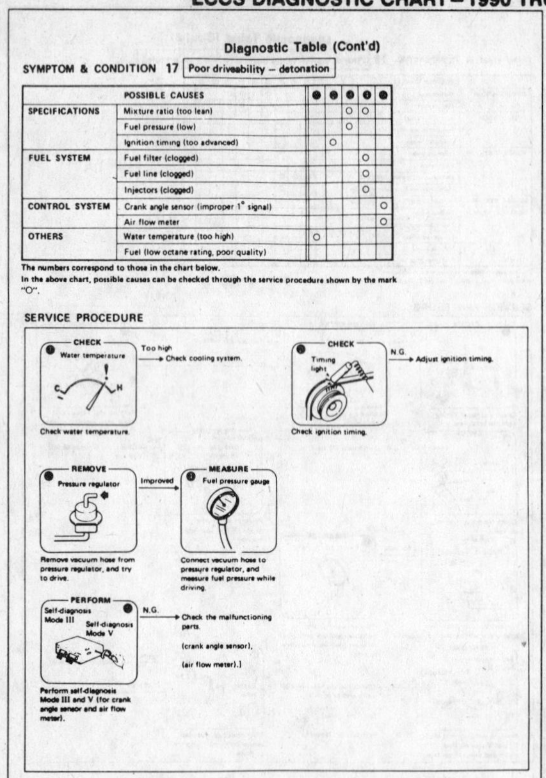

Diagnostic Table (Cont'd)

SYMPTOM & CONDITION 17 Poor driveability – detonation

	POSSIBLE CAUSES	●	●	●	●	●
SPECIFICATIONS	Mixture ratio (too lean)			○	○	
	Fuel pressure (low)			○		
	Ignition timing (too advanced)		○			
FUEL SYSTEM	Fuel filter (clogged)				○	
	Fuel line (clogged)				○	
	Injectors (clogged)				○	
CONTROL SYSTEM	Crank angle sensor (improper 1° signal)					○
	Air flow meter					○
OTHERS	Water temperature (too high)	○				
	Fuel (low octane rating, poor quality)					

The numbers correspond to those in the chart below.
In the above chart, possible causes can be checked through the service procedure shown by the mark "O".

SERVICE PROCEDURE

CHECK
Water temperature
C
H
Too high
Check cooling system.
Check water temperature.
CHECK
Timing light
N.G.
Adjust ignition timing.
Check ignition timing.
REMOVE
Pressure regulator
Improved
MEASURE
Fuel pressure gauge
Remove vacuum hose from pressure regulator, and try to drive.
Connect vacuum hose to pressure regulator, and measure fuel pressure while driving.
PERFORM
Self-diagnosis Mode III
Self-diagnosis Mode V
N.G.
Check the malfunctioning parts.
(crank angle sensor),
(air flow meter).)
Perform self-diagnosis Mode III and V (for crank angle sensor and air flow meter).

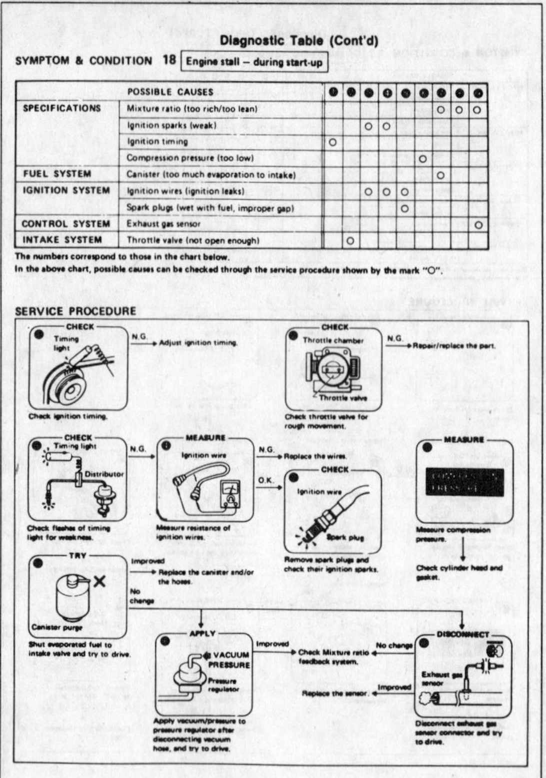

Diagnostic Table (Cont'd)

SYMPTOM & CONDITION 18 Engine stall – during start-up

	POSSIBLE CAUSES	●	●	●	●	●	●	●	●	●
SPECIFICATIONS	Mixture ratio (too rich/too lean)							○	○	○
	Ignition sparks (weak)			○	○					
	Ignition timing	○								
	Compression pressure (too low)						○			
FUEL SYSTEM	Canister (too much evaporation to intake)							○		
IGNITION SYSTEM	Ignition wires (ignition leaks)			○	○	○				
	Spark plugs (wet with fuel, improper gap)					○				
CONTROL SYSTEM	Exhaust gas sensor									○
INTAKE SYSTEM	Throttle valve (not open enough)		○							

The numbers correspond to those in the chart below.
In the above chart, possible causes can be checked through the service procedure shown by the mark "O".

SERVICE PROCEDURE

CHECK
Timing light
N.G.
Adjust ignition timing.
Check ignition timing.
CHECK
Throttle chamber
Throttle valve
N.G.
Repair/replace the part.
Check throttle valve for rough movement.
CHECK
Timing light
Distributor
N.G.
Check flashes of timing light for weakness.
MEASURE
Ignition wire
N.G.
Replace the wires.
O.K.
Measure resistance of ignition wires.
CHECK
Ignition wire
Spark plug
Remove spark plugs and check their ignition sparks.
MEASURE
Measure compression pressure.
Check cylinder head and gasket.
TRY
Canister purge
Improved
Replace the canister and/or the hoses.
No change
Shut evaporated fuel to intake valve and try to drive.
APPLY
VACUUM PRESSURE
Pressure regulator
Apply vacuum/pressure to pressure regulator after disconnecting vacuum hose, and try to drive.
Improved
Check Mixture ratio feedback system.
No change
DISCONNECT
Exhaust gas sensor
Improved
Replace the sensor.
Disconnect exhaust gas sensor connector and try to drive.

ECCS DIAGNOSTIC CHART—1990 TRUCK/PATHFINDER WITH VG30E ENGINE

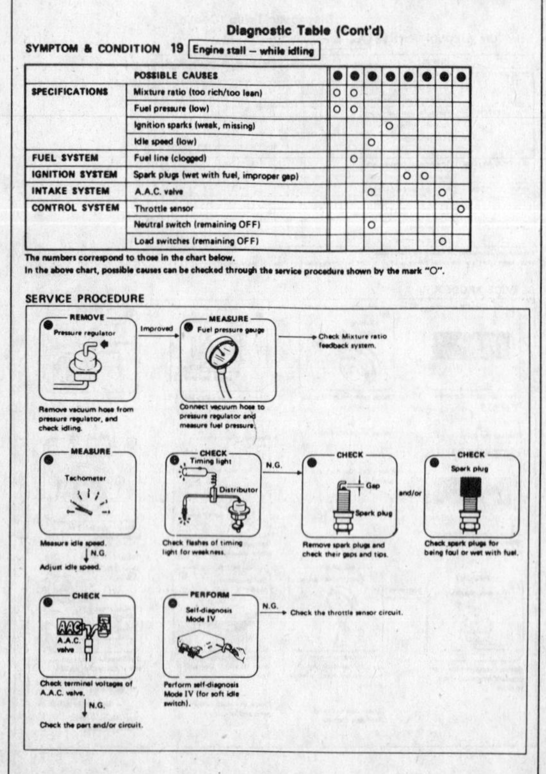

Diagnostic Table (Cont'd)

SYMPTOM & CONDITION 19 Engine stall – while idling

	POSSIBLE CAUSES	●	●	●	●	●	●	●	●
SPECIFICATIONS	Mixture ratio (too rich/too lean)	○	○						
	Fuel pressure (low)	○	○						
	Ignition sparks (weak, missing)				○				
	Idle speed (low)			○					
FUEL SYSTEM	Fuel line (clogged)		○						
IGNITION SYSTEM	Spark plugs (wet with fuel, improper gap)					○	○		
INTAKE SYSTEM	A.A.C. valve			○				○	
CONTROL SYSTEM	Throttle sensor								○
	Neutral switch (remaining OFF)			○					
	Load switches (remaining OFF)							○	

The numbers correspond to those in the chart below.
In the above chart, possible causes can be checked through the service procedure shown by the mark "O".

SERVICE PROCEDURE

REMOVE
Pressure regulator
Improved
MEASURE
Fuel pressure gauge
Check Mixture ratio feedback system.
Remove vacuum hose from pressure regulator, and check idling.
Connect vacuum hose to pressure regulator and measure fuel pressure.
MEASURE
Tachometer
Measure idle speed.
N.G.
Adjust idle speed.
CHECK
Timing light
Distributor
N.G.
Check flashes of timing light for weakness.
CHECK
Gap
Spark plug
Remove spark plugs and check their gaps and tips.
and/or
CHECK
Spark plug
Check spark plugs for being foul or wet with fuel.
CHECK
A.A.C. valve
Check terminal voltages of A.A.C. valve.
N.G.
Check the part and/or circuit.
PERFORM
Self-diagnosis Mode IV
N.G.
Check the throttle sensor circuit.
Perform self-diagnosis Mode IV (for soft idle switch).

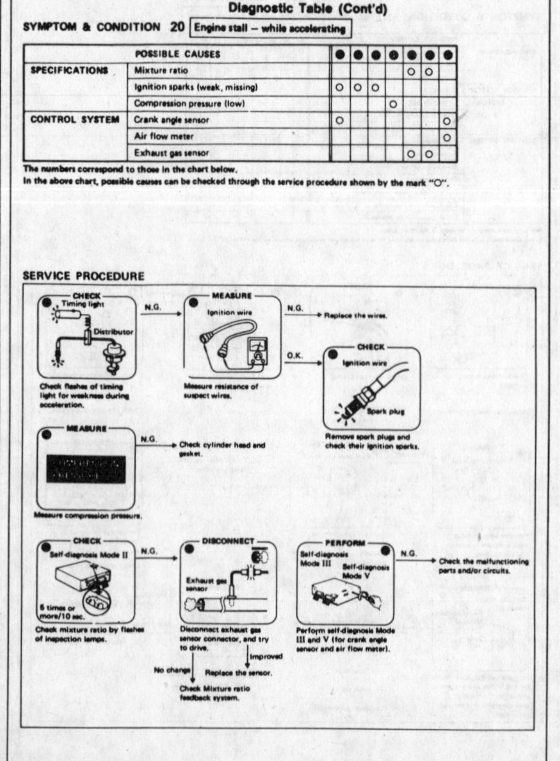

Diagnostic Table (Cont'd)

SYMPTOM & CONDITION 20 Engine stall – while accelerating

	POSSIBLE CAUSES	●	●	●	●	●	●	●
SPECIFICATIONS	Mixture ratio					○	○	
	Ignition sparks (weak, missing)	○	○	○				
	Compression pressure (low)				○			
CONTROL SYSTEM	Crank angle sensor	○						○
	Air flow meter							○
	Exhaust gas sensor					○	○	

The numbers correspond to those in the chart below.
In the above chart, possible causes can be checked through the service procedure shown by the mark "O".

SERVICE PROCEDURE

CHECK
Timing light
Distributor
N.G.
Check flashes of timing light for weakness during acceleration.
MEASURE
Ignition wire
N.G.
Replace the wires.
O.K.
Measure resistance of suspect wires.
CHECK
Ignition wire
Spark plug
Remove spark plugs and check their ignition sparks.
MEASURE
N.G.
Check cylinder head and gasket.
Measure compression pressure.
CHECK
Self-diagnosis Mode II
5 times or more/10 sec.
N.G.
Check mixture ratio by flashes of inspection lamps.
DISCONNECT
Exhaust gas sensor
Disconnect exhaust gas sensor connector, and try to drive.
Improved
Replace the sensor.
No change
Check Mixture ratio feedback system.
PERFORM
Self-diagnosis Mode III
Self-diagnosis Mode V
N.G.
Check the malfunctioning parts and/or circuits.
Perform self-diagnosis Mode III and V (for crank angle sensor and air flow meter).

ECCS DIAGNOSTIC CHART—1990 TRUCK/PATHFINDER WITH VG30E ENGINE

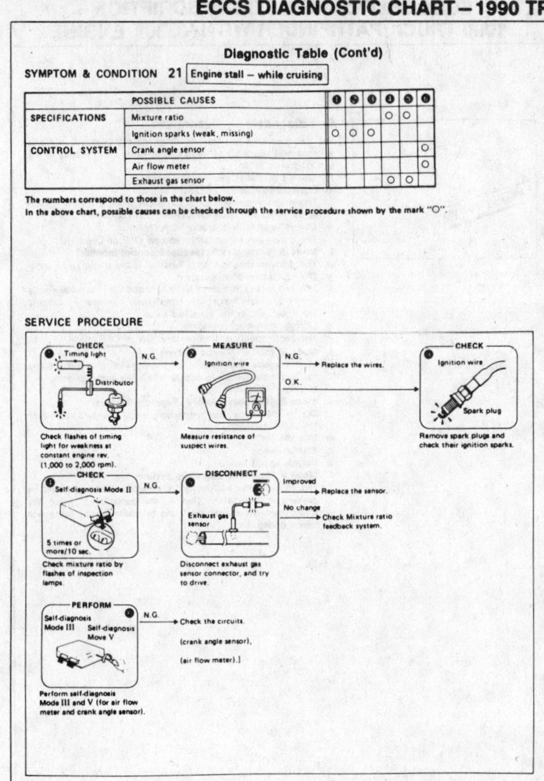

Diagnostic Table (Cont'd)

SYMPTOM & CONDITION 21 **Engine stall – while cruising**

POSSIBLE CAUSES		1	2	3	4	5	6
SPECIFICATIONS	Mixture ratio				O	O	
	Ignition sparks (weak, missing)	O	O	O			
CONTROL SYSTEM	Crank angle sensor						O
	Air flow meter						O
	Exhaust gas sensor				O	O	

The numbers correspond to those in the chart below.
In the above chart, possible causes can be checked through the service procedure shown by the mark "O".

SERVICE PROCEDURE

CHECK
Timing light
Distributor
N.G.
MEASURE
Ignition wire
N.G.
Replace the wires.
O.K.
CHECK
Ignition wire
Spark plug
Check flashes of timing light for weakness at constant engine rev. (1,000 to 2,000 rpm).
Measure resistance of suspect wires.
Remove spark plugs and check their ignition sparks.
CHECK
Self-diagnosis Mode II
N.G.
DISCONNECT
Improved
Replace the sensor.
No change
Check Mixture ratio feedback system.
Exhaust gas sensor
5 times or more/10 sec.
Check mixture ratio by flashes of inspection lamps.
Disconnect exhaust gas sensor connector, and try to drive.
PERFORM
Self-diagnosis Mode III
Self-diagnosis Move V
N.G.
Check the circuits.
(crank angle sensor),
(air flow meter).)
Perform self-diagnosis Mode III and V (for air flow meter and crank angle sensor).

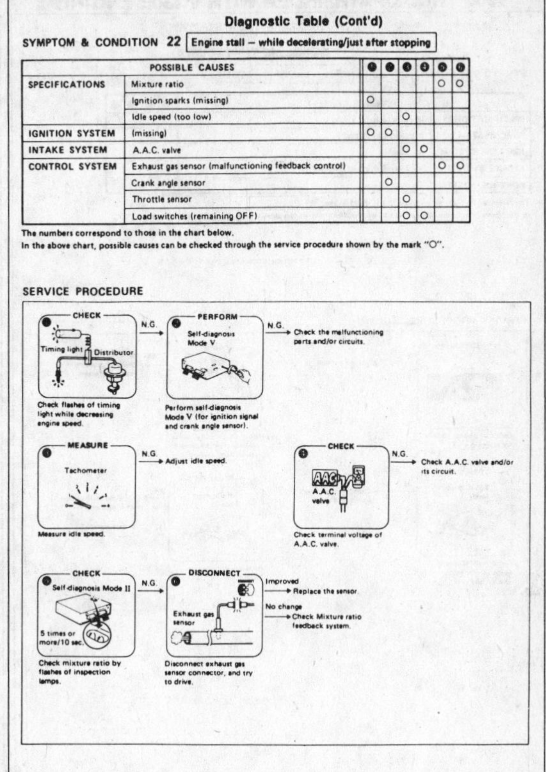

Diagnostic Table (Cont'd)

SYMPTOM & CONDITION 22 **Engine stall – while decelerating/just after stopping**

POSSIBLE CAUSES		1	2	3	4	5	6
SPECIFICATIONS	Mixture ratio					O	O
	Ignition sparks (missing)	O					
	Idle speed (too low)			O			
IGNITION SYSTEM	(missing)	O	O				
INTAKE SYSTEM	A.A.C. valve			O	O		
CONTROL SYSTEM	Exhaust gas sensor (malfunctioning feedback control)					O	O
	Crank angle sensor		O				
	Throttle sensor			O			
	Load switches (remaining OFF)			O	O		

The numbers correspond to those in the chart below.
In the above chart, possible causes can be checked through the service procedure shown by the mark "O".

SERVICE PROCEDURE

CHECK
Timing light
Distributor
N.G.
PERFORM
Self-diagnosis Mode V
N.G.
Check the malfunctioning parts and/or circuits.
Check flashes of timing light while decreasing engine speed.
Perform self-diagnosis Mode V (for ignition signal and crank angle sensor).
MEASURE
Tachometer
N.G.
Adjust idle speed.
Measure idle speed.
CHECK
AAC
A.A.C. valve
N.G.
Check A.A.C. valve and/or its circuit.
Check terminal voltage of A.A.C. valve.
CHECK
Self-diagnosis Mode II
N.G.
DISCONNECT
Improved
Replace the sensor.
No change
Check Mixture ratio feedback system.
Exhaust gas sensor
5 times or more/10 sec.
Check mixture ratio by flashes of inspection lamps.
Disconnect exhaust gas sensor connector, and try to drive.

ECCS DIAGNOSTIC CHART—1990 TRUCK/PATHFINDER WITH VG30E ENGINE

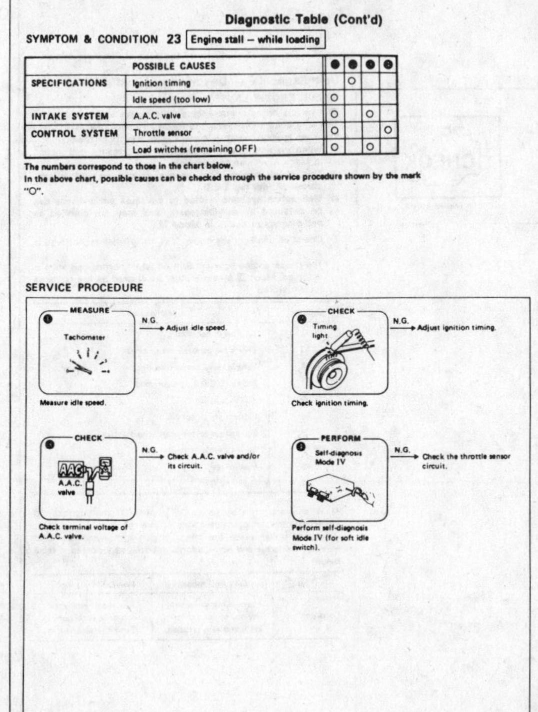

Diagnostic Table (Cont'd)

SYMPTOM & CONDITION 23 **Engine stall – while loading**

POSSIBLE CAUSES		1	2	3	4
SPECIFICATIONS	Ignition timing		O		
	Idle speed (too low)	O			
INTAKE SYSTEM	A.A.C. valve	O		O	
CONTROL SYSTEM	Throttle sensor	O			O
	Load switches (remaining OFF)	O		O	

The numbers correspond to those in the chart below.
In the above chart, possible causes can be checked through the service procedure shown by the mark "O".

SERVICE PROCEDURE

MEASURE
Tachometer
N.G.
Adjust idle speed.
Measure idle speed.
CHECK
Timing light
N.G.
Adjust ignition timing.
Check ignition timing.
CHECK
AAC
A.A.C. valve
N.G.
Check A.A.C. valve and/or its circuit.
Check terminal voltage of A.A.C. valve.
PERFORM
Self-diagnosis Mode IV
N.G.
Check the throttle sensor circuit.
Perform self-diagnosis Mode IV (for soft idle switch).

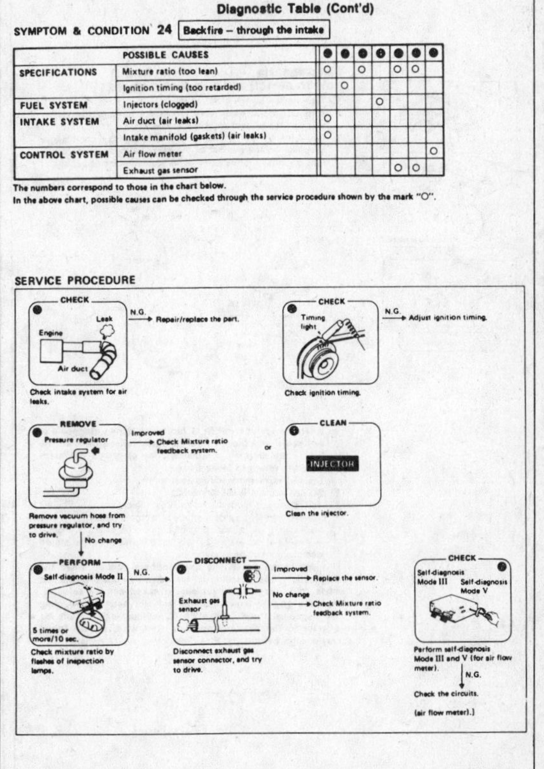

Diagnostic Table (Cont'd)

SYMPTOM & CONDITION 24 **Backfire – through the intake**

POSSIBLE CAUSES		1	2	3	4	5	6	7
SPECIFICATIONS	Mixture ratio (too lean)	O		O		O	O	
	Ignition timing (too retarded)		O					
FUEL SYSTEM	Injectors (clogged)				O			
INTAKE SYSTEM	Air duct (air leaks)	O						
	Intake manifold (gaskets) (air leaks)	O						
CONTROL SYSTEM	Air flow meter							O
	Exhaust gas sensor					O	O	

The numbers correspond to those in the chart below.
In the above chart, possible causes can be checked through the service procedure shown by the mark "O".

SERVICE PROCEDURE

CHECK
Engine
Leak
Air duct
N.G.
Repair/replace the part.
Check intake system for air leaks.
CHECK
Timing light
N.G.
Adjust ignition timing.
Check ignition timing.
REMOVE
Pressure regulator
Improved
Check Mixture ratio feedback system.
or
CLEAN
INJECTOR
Clean the injector.
Remove vacuum hose from pressure regulator, and try to drive.
No change
PERFORM
Self-diagnosis Mode II
N.G.
DISCONNECT
Exhaust gas sensor
Improved
Replace the sensor.
No change
Check Mixture ratio feedback system.
CHECK
Self-diagnosis Mode III
Self-diagnosis Mode V
5 times or more/10 sec.
Check mixture ratio by flashes of inspection lamps.
Disconnect exhaust gas sensor connector, and try to drive.
Perform self-diagnosis Mode III and V (for air flow meter).
N.G.
Check the circuits.
(air flow meter).)

ECCS DIAGNOSTIC CHART — 1990 TRUCK/PATHFINDER WITH VG30E ENGINE

Diagnostic Table (Cont'd)

SYMPTOM & CONDITION 25 Backfire – through the exhaust

POSSIBLE CAUSES					
SPECIFICATIONS	Mixture ratio (too rich)	O	O		
FUEL SYSTEM	Injectors (fuel leaks)		O		
IGNITION SYSTEM	(missing)				O
INTAKE SYSTEM	Air cleaner element (clogged)	O			
CONTROL SYSTEM	Throttle sensor			O	

The numbers correspond to those in the chart below.
In the above chart, possible causes can be checked through the service procedure shown by the mark "O".

SERVICE PROCEDURE

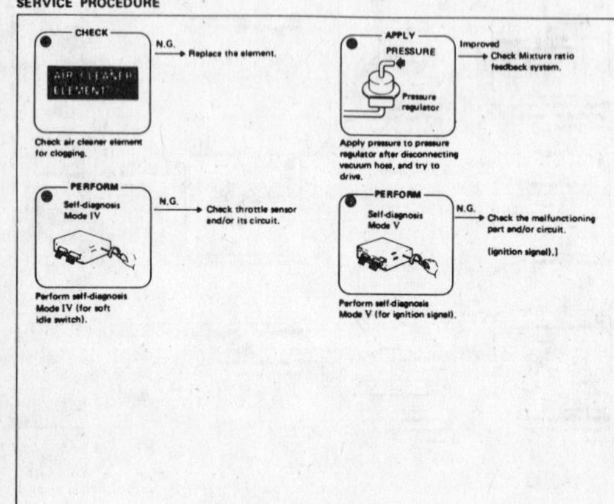
CHECK
AIR CLEANER ELEMENT
N.G.
Replace the element.
Check air cleaner element for clogging.
APPLY PRESSURE
Pressure regulator
Improved
Check Mixture ratio feedback system.
Apply pressure to pressure regulator after disconnecting vacuum hose, and try to drive.
PERFORM
Self-diagnosis Mode IV
N.G.
Check throttle sensor and/or its circuit.
Perform self-diagnosis Mode IV (for soft idle switch).
PERFORM
Self-diagnosis Mode V
N.G.
Check the malfunctioning part and/or circuit.
(ignition signal).]
Perform self-diagnosis Mode V (for ignition signal).

ECCS SELF-DIAGNOSTIC DESCRIPTION — 1990 TRUCK/PATHFINDER WITH VG30E ENGINE

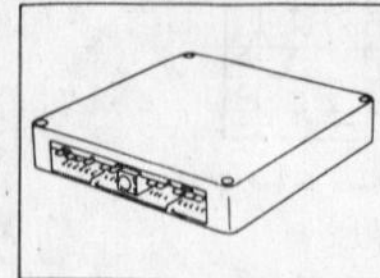

Self-diagnosis — Description

The self-diagnosis is useful to diagnose malfunctions in major sensors and actuators of the E.C.C.S. There are 5 modes in the self-diagnosis system.

1. **Mode I (Exhaust gas sensor monitor)**
   - During closed-loop operation:
     The green inspection lamp turns ON when a lean condition is detected and goes OFF under rich condition.
   - During open-loop operation condition:
     The green inspection lamp remains OFF or ON.
2. **Mode II (Mixture ratio feedback control monitor)**
   The green inspection lamp function is the same as Mode I.
   - During closed-loop operation:
     The red inspection lamp turns ON and OFF simultaneously with the green inspection lamp when the mixture ratio is controlled within the specified value.
   - During open-loop operation:
     The red inspection lamp remains ON or OFF.
3. **Mode III (Self-diagnostic system)**
   In this mode the number of both green and red L.E.D.'s flashing indicates the group to which the malfunctioning part belongs.
4. **Mode IV (Switches ON/OFF diagnostic system)**
   During this mode, the inspection lamps monitor the switch ON-OFF condition.
   - Soft idle switch
   - Starter switch
   - Vehicle speed sensor
5. **Mode V (Real-time diagnostic system)**
   The moment the malfunction is detected, the display will be presented immediately. That is, the condition at which the malfunction occurs can be found by observing the inspection lamps during driving test.

ECCS SELF-DIAGNOSTIC DESCRIPTION (CONT.) — 1990 TRUCK/PATHFINDER WITH VG30E ENGINE

Flashing N times
Mode N

Self-diagnosis — Description (Cont'd)

HOW TO SWITCH THE DIAGNOSTIC MODES

1. Turn ignition switch "ON".
2. Turn diagnostic mode selector to E.C.U. (fully clockwise) and wait for inspection lamps to flash.
3. Count the number of flashes, and after the inspection lamps have flashed the number of the required mode, immediately turn diagnostic mode selector fully counterclockwise.

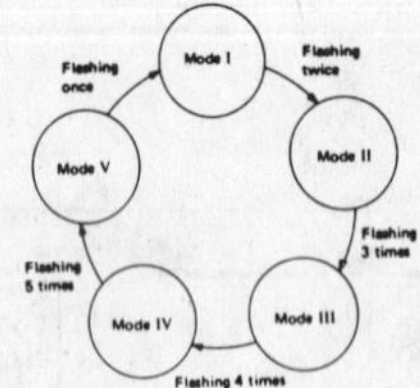
Mode I
Flashing once
Flashing twice
Mode II
Flashing 3 times
Mode III
Flashing 4 times
Mode IV
Flashing 5 times
Mode V

- **When the ignition switch is turned off during diagnosis in any mode and then turned on again (after power to the E.C.U. has dropped completely), the diagnosis will automatically return to Mode I.**
  The stored memory will be lost if:

1. Battery terminal is disconnected.
2. After selecting Mode III, Mode IV is selected.
   However, if the diagnostic mode selector is kept turned fully clockwise, it will continue to change in the order of Mode I → II → III → IV → V → I ... etc., and in this state the stored memory will not be erased.

This unit serves as an idle rpm feedback control. When the diagnostic mode selector is turned within the "diagnostic mode OFF" range, a target engine speed can be selected. Mark the original position of the selector before conducting self-diagnosis. Upon completion of self-diagnosis, return the selector to the previous position. Otherwise, engine speed may change before and after conducting self-diagnosis.

CHECK
CHECK ENGINE LIGHT

Self-diagnosis — Description (Cont'd)

CHECK ENGINE LIGHT

This vehicle has a check engine light on the instrument panel. This light comes ON under the following conditions:

1) When ignition switch is turned "ON" (for bulb check).
2) When systems related to emission performance malfunction in Mode I (with engine running).
   - **This check engine light always illuminates and is synchronous with red L.E.D.**
   - **Malfunction systems related to emission performance can be detected by self-diagnosis, and they are clarified as self-diagnostic codes in Mode III.**
3) Check engine light will come "ON" only when malfunction is sensed.
   The check engine light will turn off when normal operation is resumed. Mode III memory must be cleared as the contents remain stored.

Code No.	Malfunction
12	Air flow meter circuit
13	Engine temperature sensor circuit
14	Vehicle speed sensor circuit
31	E.C.U. (E.C.C.S. control unit)
32	E.G.R. function
33	Exhaust gas sensor circuit
35	Exhaust gas temperature sensor circuit
43	Throttle sensor circuit
45	Injector leak
51	Injector circuit

4) When crank angle sensor or C.P.U. of E.C.U. malfunctions and fail-safe system operates during engine rotation.

Conditions under which the check engine light illuminates differ between California and non-California models, as indicated in table below:

	California model	Non-California model
Condition	Light illuminates when any one of conditions 1), 2), 3) and 4) is satisfied.	Light illuminates when any one of conditions 1), 2) and 4) is satisfied.

ECCS SELF-DIAGNOSTIC DESCRIPTION (CONT.)—1990 TRUCK/PATHFINDER WITH VG30E ENGINE

Self-diagnosis — Description (Cont'd)

Use the following diagnostic flow chart to check and repair a malfunctioning system.

DIAGNOSIS START

↓

Turn ignition switch "ON" and make sure that check engine light comes "ON". — N.G. → Replace bulb.

↓ O.K.

Start engine and make sure that check engine light still illuminates.

↓

Perform self-diagnosis and check which code is displayed in Mode III.

↓

Check electronic control system of affected Code No. to locate malfunctioning part.

↓

Repair or replace malfunctioning part.

↓

Reinstall any part removed.

↓

Erase the self-diagnosis memory.

- **Methods of erasing memories differ with systems. Read the manual before diagnosing systems.**

↓

Perform driving test.
Make sure that check engine light does not come "ON" during this test.

- **After repairs, test drive to check that check engine light does not come on.**
- **Test driving modes differ with systems. Read the manual before test driving.**

↓

DIAGNOSIS END

ECCS SELF-DIAGNOSTIC MODES I AND II—1990 TRUCK/PATHFINDER WITH VG30E ENGINE

Self-diagnosis — Mode I (Exhaust gas sensor monitor)

This mode checks the exhaust gas sensor for proper functioning. The operation of the E.C.U. L.E.D. in this mode differs with mixture ratio control conditions as follows:

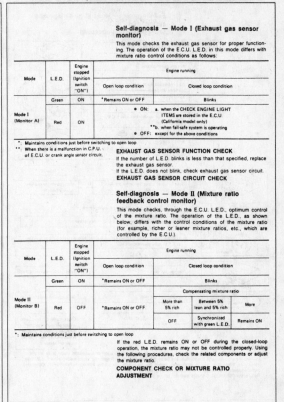

Mode	L.E.D.	Engine stopped (Ignition switch "ON")	Engine running	
			Open loop condition	Closed loop condition
Mode I (Monitor A)	Green	ON	*Remains ON or OFF	Blinks
	Red	ON	● ON: a. when the CHECK ENGINE LIGHT ITEMS are stored in the E.C.U. (California model only) **b. when fail-safe system is operating ● OFF: except for the above conditions	

*: Maintains conditions just before switching to open loop
**: When there is a malfunction in C.P.U. of E.C.U. or crank angle sensor circuit.

EXHAUST GAS SENSOR FUNCTION CHECK

If the number of L.E.D. blinks is less than that specified, replace the exhaust gas sensor.
If the L.E.D. does not blink, check exhaust gas sensor circuit.

EXHAUST GAS SENSOR CIRCUIT CHECK

Self-diagnosis — Mode II (Mixture ratio feedback control monitor)

This mode checks, through the E.C.U. L.E.D., optimum control of the mixture ratio. The operation of the L.E.D., as shown below, differs with the control conditions of the mixture ratio (for example, richer or leaner mixture ratios, etc., which are controlled by the E.C.U.).

Mode	L.E.D.	Engine stopped (Ignition switch "ON")	Engine running			
			Open loop condition	Closed loop condition		
Mode II (Monitor B)	Green	ON	*Remains ON or OFF	Blinks		
	Red	OFF	*Remains ON or OFF	Compensating mixture ratio		
				More than 5% rich	Between 5% lean and 5% rich	More
				OFF	Synchronized with green L.E.D.	Remains ON

*: Maintains conditions just before switching to open loop

If the red L.E.D. remains ON or OFF during the closed-loop operation, the mixture ratio may not be controlled properly. Using the following procedures, check the related components or adjust the mixture ratio.

COMPONENT CHECK OR MIXTURE RATIO ADJUSTMENT

ECCS SELF-DIAGNOSTIC MODE III—1990 TRUCK/PATHFINDER WITH VG30E ENGINE

Self-diagnosis — Mode III (Self-diagnostic system)

The E.C.U. constantly monitors the function of these sensors and actuators, regardless of ignition key position. If a malfunction occurs, the information is stored in the E.C.U. and can be retrieved from the memory by turning on the diagnostic mode selector, located on the side of the E.C.U. When activated, the malfunction is indicated by flashing a red and a green L.E.D. (Light Emitting Diode), also located on the E.C.U. Since all the self-diagnostic results are stored in the E.C.U.'s memory even intermittent malfunctions can be diagnosed.

A malfunction is indicated by the number of both red and green flashing L.E.D.s. First, the red L.E.D. flashes and the green flashes follow. The red L.E.D. corresponds to units of ten and the green L.E.D. corresponds to units of one. For example, when the red L.E.D. flashes once and the green L.E.D. flashes twice, this signifies the number "12", showing that the air flow meter signal is malfunctioning. All problems are classified by code numbers in this way.

- **When the engine fails to start, crank it two or more seconds before beginning self-diagnosis.**
- **Before starting self-diagnosis, do not erase the stored memory before beginning self-diagnosis. If it is erased, the self-diagnosis function for intermittent malfunctions will be lost.**

The stored memory would be lost if:

1. **Battery terminal is disconnected.**
2. **After selecting Mode III, Mode IV is selected.**

DISPLAY CODE TABLE

Code No.	Detected items	California	Non-California
11	Crank angle sensor ciruit	X	X
12	Air flow meter circuit	X	X
13	Engine temperature sensor circuit	X	X
14	Vehicle speed sensor circuit	X	X
21	Ignition signal missing in primary coil	X	X
31	E.C.U. (E.C.C.S. control unit)	X	X
32	E.G.R. function	X	–
33	Exhaust gas sensor circuit	X	X
34	Detonation sensor circuit	X	X
35	Exhaust gas temperature sensor circuit	X	–
43	Throttle sensor circuit	X	X
45	Injector leak	X	–
51	Injector circuit	X	–
55	No malfunction in the above circuit	X	X

X: Available –: Not available

Self-diagnosis — Mode III (Self-diagnostic system) (Cont'd)

RETENTION OF DIAGNOSTIC RESULTS

The diagnostic results will remain in E.C.U. memory until the starter is operated fifty times after a diagnostic item has been judged to be malfunctioning. The diagnostic result will then be cancelled automatically. If a diagnostic item which has been judged to be malfunctioning and stored in memory is again judged to be malfunctioning before the starter is operated fifty times, the second result will replace the previous one. It will be stored in E.C.U. memory until the starter is operated fifty times more.

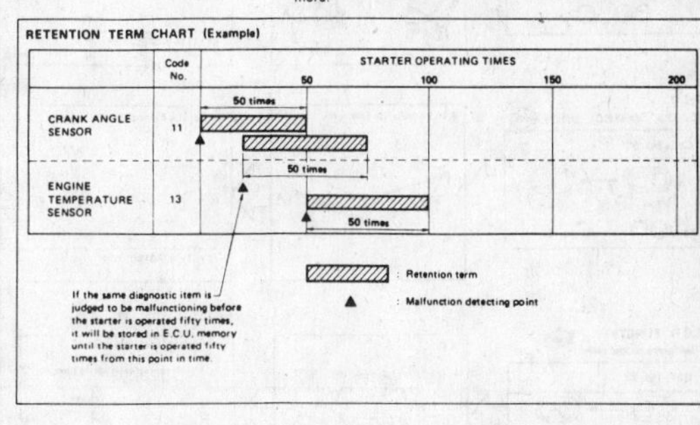
RETENTION TERM CHART (Example)
Code No.
STARTER OPERATING TIMES
50
100
150
200
CRANK ANGLE SENSOR
11
50 times
ENGINE TEMPERATURE SENSOR
13
50 times
50 times
If the same diagnostic item is judged to be malfunctioning before the starter is operated fifty times, it will be stored in E.C.U. memory until the starter is operated fifty times from this point in time.
: Retention term
: Malfunction detecting point

ECCS SELF-DIAGNOSTIC MODE III (CONT.)—1990 TRUCK/PATHFINDER WITH VG30E ENGINE

ECCS SELF-DIAGNOSTIC MODE III DISPLAY CODES—1990 TRUCK/PATHFINDER WITH VG30E ENGINE

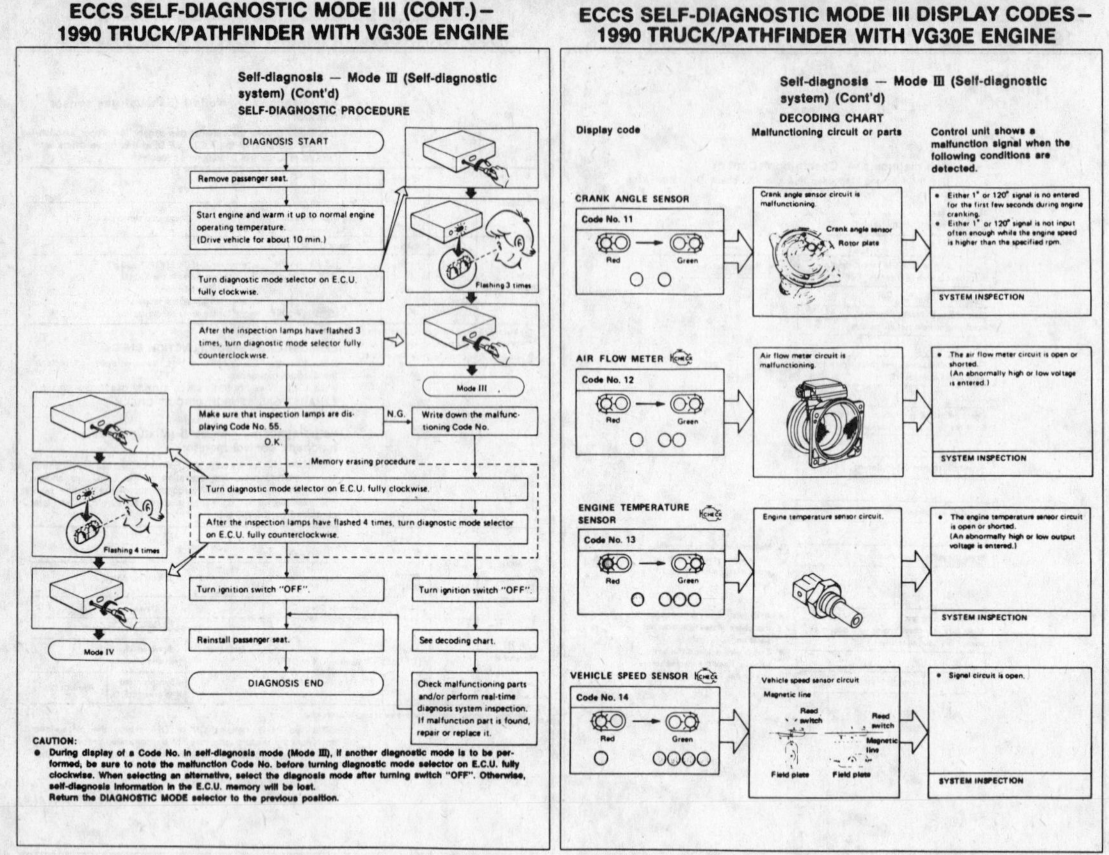
Self-diagnosis — Mode III (Self-diagnostic system) (Cont'd)
SELF-DIAGNOSTIC PROCEDURE
DIAGNOSIS START
Remove passenger seat.
Start engine and warm it up to normal engine operating temperature. (Drive vehicle for about 10 min.)
Turn diagnostic mode selector on E.C.U. fully clockwise.
Flashing 3 times
After the inspection lamps have flashed 3 times, turn diagnostic mode selector fully counterclockwise.
Mode III
Make sure that inspection lamps are displaying Code No. 55.
N.G.
Write down the malfunctioning Code No.
O.K.
Memory erasing procedure
Turn diagnostic mode selector on E.C.U. fully clockwise.
After the inspection lamps have flashed 4 times, turn diagnostic mode selector on E.C.U. fully counterclockwise.
Flashing 4 times
Turn ignition switch "OFF".
Turn ignition switch "OFF".
Reinstall passenger seat.
See decoding chart.
Mode IV
DIAGNOSIS END
Check malfunctioning parts and/or perform real-time diagnosis system inspection. If malfunction part is found, repair or replace it.
CAUTION:
During display of a Code No. in self-diagnosis mode (Mode III), if another diagnostic mode is to be performed, be sure to note the malfunction Code No. before turning diagnostic mode selector on E.C.U. fully clockwise. When selecting an alternative, select the diagnosis mode after turning switch "OFF". Otherwise, self-diagnosis information in the E.C.U. memory will be lost.
Return the DIAGNOSTIC MODE selector to the previous position.
Self-diagnosis — Mode III (Self-diagnostic system) (Cont'd)
DECODING CHART
Display code
Malfunctioning circuit or parts
Control unit shows a malfunction signal when the following conditions are detected.
CRANK ANGLE SENSOR
Code No. 11
Red
Green
Crank angle sensor circuit is malfunctioning.
Crank angle sensor
Rotor plate
Either 1° or 120° signal is no entered for the first few seconds during engine cranking.
Either 1° or 120° signal is not input often enough while the engine speed is higher than the specified rpm.
SYSTEM INSPECTION
AIR FLOW METER
Code No. 12
Red
Green
Air flow meter circuit is malfunctioning.
The air flow meter circuit is open or shorted. (An abnormally high or low voltage is entered.)
SYSTEM INSPECTION
ENGINE TEMPERATURE SENSOR
Code No. 13
Red
Green
Engine temperature sensor circuit.
The engine temperature sensor circuit is open or shorted. (An abnormally high or low output voltage is entered.)
SYSTEM INSPECTION
VEHICLE SPEED SENSOR
Code No. 14
Red
Green
Vehicle speed sensor circuit
Magnetic line
Reed switch
Reed switch
Magnetic line
Field plate
Field plate
Signal circuit is open.
SYSTEM INSPECTION

ECCS SELF-DIAGNOSTIC MODE III DISPLAY CODES (CONT.)—1990 TRUCK/PATHFINDER WITH VG30E ENGINE

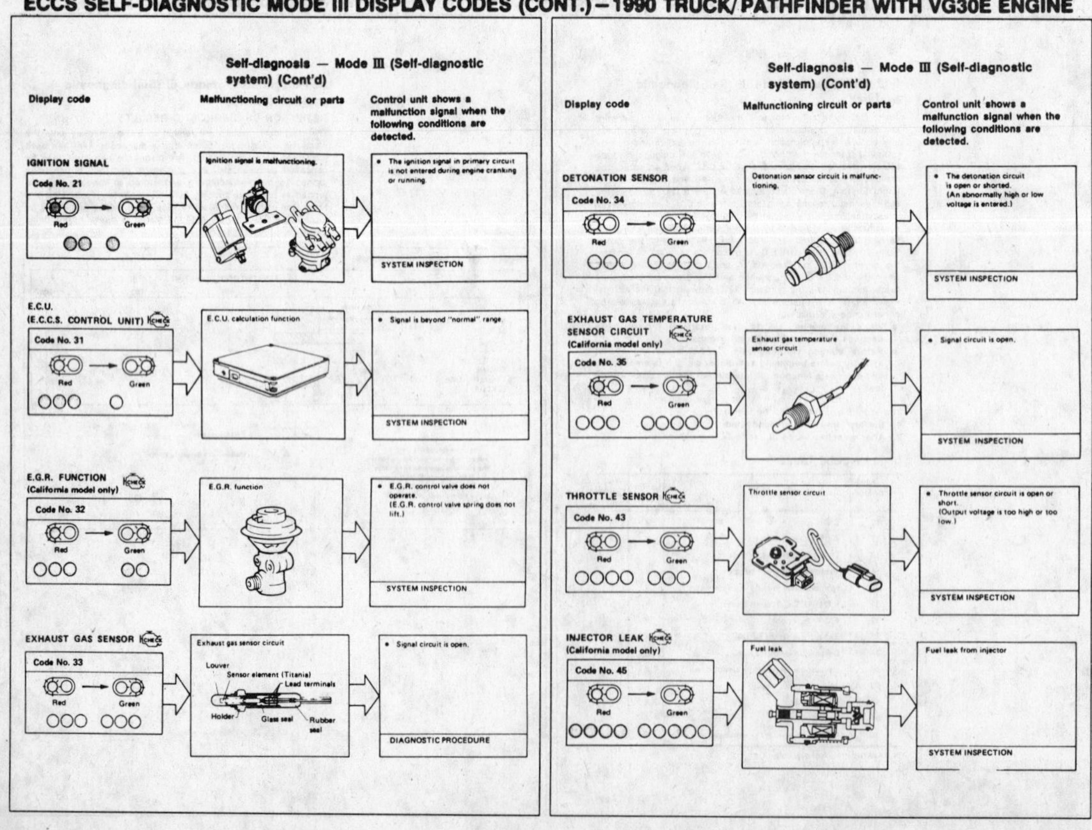
Self-diagnosis — Mode III (Self-diagnostic system) (Cont'd)
Display code
Malfunctioning circuit or parts
Control unit shows a malfunction signal when the following conditions are detected.
IGNITION SIGNAL
Code No. 21
Red
Green
Ignition signal is malfunctioning.
The ignition signal in primary circuit is not entered during engine cranking or running.
SYSTEM INSPECTION
E.C.U. (E.C.C.S. CONTROL UNIT)
Code No. 31
Red
Green
E.C.U. calculation function
Signal is beyond "normal" range.
SYSTEM INSPECTION
E.G.R. FUNCTION (California model only)
Code No. 32
Red
Green
E.G.R. function
E.G.R. control valve does not operate. (E.G.R. control valve spring does not lift.)
SYSTEM INSPECTION
EXHAUST GAS SENSOR
Code No. 33
Red
Green
Exhaust gas sensor circuit
Louver
Sensor element (Titania)
Lead terminals
Holder
Glass seal
Rubber seal
Signal circuit is open.
DIAGNOSTIC PROCEDURE
Self-diagnosis — Mode III (Self-diagnostic system) (Cont'd)
Display code
Malfunctioning circuit or parts
Control unit shows a malfunction signal when the following conditions are detected.
DETONATION SENSOR
Code No. 34
Red
Green
Detonation sensor circuit is malfunctioning.
The detonation circuit is open or shorted. (An abnormally high or low voltage is entered.)
SYSTEM INSPECTION
EXHAUST GAS TEMPERATURE SENSOR CIRCUIT (California model only)
Code No. 35
Red
Green
Exhaust gas temperature sensor circuit
Signal circuit is open.
SYSTEM INSPECTION
THROTTLE SENSOR
Code No. 43
Red
Green
Throttle sensor circuit
Throttle sensor circuit is open or short. (Output voltage is too high or too low.)
SYSTEM INSPECTION
INJECTOR LEAK (California model only)
Code No. 45
Red
Green
Fuel leak
Fuel leak from injector
SYSTEM INSPECTION

ECCS SELF-DIAGNOSTIC MODE III DISPLAY CODES— 1990 TRUCK/ PATHFINDER WITH VG30E ENGINE

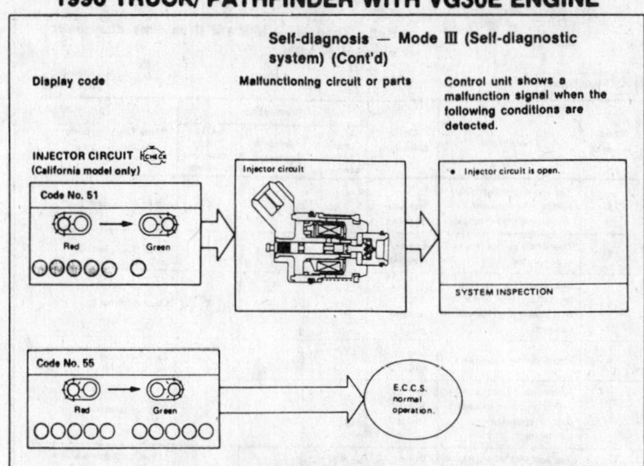
Self-diagnosis — Mode III (Self-diagnostic system) (Cont'd)
Display code
Malfunctioning circuit or parts
Control unit shows a malfunction signal when the following conditions are detected.
INJECTOR CIRCUIT
(California model only)
Code No. 51
Red
Green
Injector circuit
• Injector circuit is open.
SYSTEM INSPECTION
Code No. 55
Red
Green
E.C.C.S. normal operation.

ECCS SELF-DIAGNOSTIC MODE IV— 1990 TRUCK/PATHFINDER WITH VG30E ENGINE

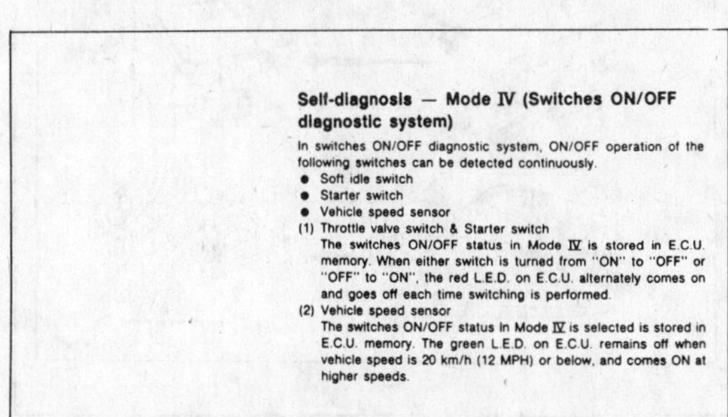

Self-diagnosis — Mode IV (Switches ON/OFF diagnostic system)

In switches ON/OFF diagnostic system, ON/OFF operation of the following switches can be detected continuously.

- Soft idle switch
- Starter switch
- Vehicle speed sensor

(1) Throttle valve switch & Starter switch
The switches ON/OFF status in Mode IV is stored in E.C.U. memory. When either switch is turned from "ON" to "OFF" or "OFF" to "ON", the red L.E.D. on E.C.U. alternately comes on and goes off each time switching is performed.

(2) Vehicle speed sensor
The switches ON/OFF status in Mode IV is selected is stored in E.C.U. memory. The green L.E.D. on E.C.U. remains off when vehicle speed is 20 km/h (12 MPH) or below, and comes ON at higher speeds.

ECCS SELF-DIAGNOSTIC MODE IV (CONT.)— 1990 TRUCK/PATHFINDER WITH VG30E ENGINE

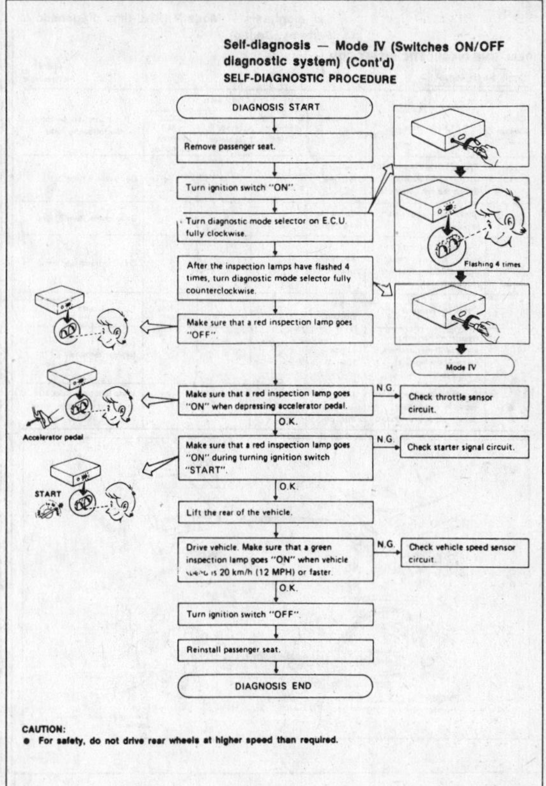
Self-diagnosis — Mode IV (Switches ON/OFF diagnostic system) (Cont'd)
SELF-DIAGNOSTIC PROCEDURE
DIAGNOSIS START
Remove passenger seat.
Turn ignition switch "ON".
Turn diagnostic mode selector on E.C.U. fully clockwise.
After the inspection lamps have flashed 4 times, turn diagnostic mode selector fully counterclockwise.
Flashing 4 times
Mode IV
Make sure that a red inspection lamp goes "OFF".
Make sure that a red inspection lamp goes "ON" when depressing accelerator pedal.
N.G.
Check throttle sensor circuit.
O.K.
Accelerator pedal
Make sure that a red inspection lamp goes "ON" during turning ignition switch "START".
N.G.
Check starter signal circuit.
O.K.
START
Lift the rear of the vehicle.
Drive vehicle. Make sure that a green inspection lamp goes "ON" when vehicle speed is 20 km/h (12 MPH) or faster.
N.G.
Check vehicle speed sensor circuit.
O.K.
Turn ignition switch "OFF".
Reinstall passenger seat.
DIAGNOSIS END
CAUTION:
• For safety, do not drive rear wheels at higher speed than required.

ECCS SELF-DIAGNOSTIC MODE V— 1990 TRUCK/PATHFINDER WITH VG30E ENGINE

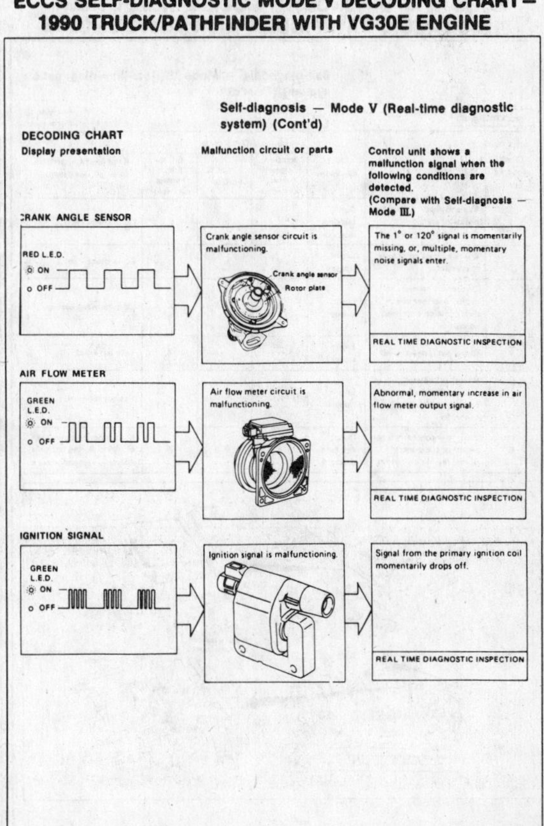
Self-diagnosis — Mode V (Real-time diagnostic system)
In real-time diagnosis, if the following items are judged to be working incorrectly, a malfunction will be indicated immediately.
• Crank angle sensor (120° signal & 1° signal) output signal
• Ignition signal
• Air flow meter output signal
Consequently, this diagnosis very effectively determines whether the above systems cause the malfunction, during driving test. Compared with self-diagnosis, real-time diagnosis is very sensitive and can detect malfunctions instantly. However, items regarded as malfunctions in this diagnosis are not stored in E.C.U. memory.
SELF-DIAGNOSTIC PROCEDURE
DIAGNOSIS START
Remove passenger seat.
Start engine.
Turn diagnostic mode selector on E.C.U. fully clockwise.
Flashing 5 times
After the inspection lamps have flashed 5 times, turn diagnostic mode selector fully counterclockwise.
Mode V
Make sure that inspection lamps are not flashing for 5 min. when idling or racing.
N.G.
If flashing, count No. of flashes.
O.K.
Turn ignition switch "OFF".
Turn ignition switch "OFF".
See decoding chart.
Reinstall passenger seat.
Perform real-time diagnosis system inspection. If malfunction part is found, repair or replace it.
DIAGNOSIS END
CAUTION:
In real-time diagnosis, pay attention to inspection lamp flashing. E.C.U. displays the malfunction code only once and does not memorize the inspection.

ECCS SELF-DIAGNOSTIC MODE V DECODING CHART— 1990 TRUCK/PATHFINDER WITH VG30E ENGINE

Self-diagnosis — Mode V (Real-time diagnostic system) (Cont'd)
DECODING CHART
Display presentation
Malfunction circuit or parts
Control unit shows a malfunction signal when the following conditions are detected. (Compare with Self-diagnosis — Mode III.)
CRANK ANGLE SENSOR
RED L.E.D.
ON
OFF
Crank angle sensor circuit is malfunctioning.
Crank angle sensor
Rotor plate
The 1° or 120° signal is momentarily missing, or, multiple, momentary noise signals enter.
REAL TIME DIAGNOSTIC INSPECTION
AIR FLOW METER
GREEN L.E.D.
ON
OFF
Air flow meter circuit is malfunctioning.
Abnormal, momentary increase in air flow meter output signal.
REAL TIME DIAGNOSTIC INSPECTION
IGNITION SIGNAL
GREEN L.E.D.
ON
OFF
Ignition signal is malfunctioning.
Signal from the primary ignition coil momentarily drops off.
REAL TIME DIAGNOSTIC INSPECTION

ECCS SELF-DIAGNOSTIC MODE V INSPECTION—1990 TRUCK/PATHFINDER WITH VG30E ENGINE

Self-diagnosis — Mode V (Real-time diagnostic system) (Cont'd)

REAL-TIME DIAGNOSTIC INSPECTION

Crank Angle Sensor

X: Available
–: Not available

Check sequence	Check items	Check conditions	Check parts: Crank angle sensor harness connector	Check parts: Sensor & actuator	Check parts: E.C.U. S.M.J. harness connector	If malfunction, perform the following items.
1	Tap and wiggle harness connector or component during real-time diagnosis.	During real-time diagnosis	X	X	X	Go to check item 2.
2	Check harness continuity at connector.	Engine stopped	X	–	–	Go to check item 3.
3	Disconnect harness connector, and then check dust adhesion to harness connector.	Engine stopped	X	–	X	Clean terminal surface.
4	Check pin terminal bend.	Engine stopped	–	–	X	Take out bend.
5	Reconnect harness connector and then recheck harness continuity at connector.	Engine stopped	X	–	–	Replace terminal.
6	Tap and wiggle harness connector or component during real-time diagnosis.	During real-time diagnosis	X	X	X	If malfunction codes are displayed during real-time diagnosis, replace terminal.

E.C.U.
Crank angle sensor harness connector

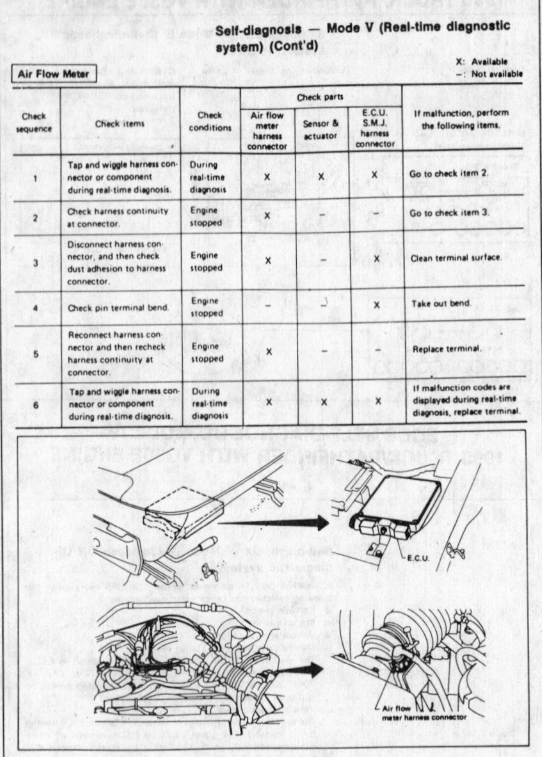
E.C.U.
Air flow meter harness connector

Self-diagnosis — Mode V (Real-time diagnostic system) (Cont'd)

Air Flow Meter

X: Available
–: Not available

Check sequence	Check items	Check conditions	Check parts: Air flow meter harness connector	Check parts: Sensor & actuator	Check parts: E.C.U. S.M.J. harness connector	If malfunction, perform the following items.
1	Tap and wiggle harness connector or component during real-time diagnosis.	During real-time diagnosis	X	X	X	Go to check item 2.
2	Check harness continuity at connector.	Engine stopped	X	–	–	Go to check item 3.
3	Disconnect harness connector, and then check dust adhesion to harness connector.	Engine stopped	X	–	X	Clean terminal surface.
4	Check pin terminal bend.	Engine stopped	–	–	X	Take out bend.
5	Reconnect harness connector and then recheck harness continuity at connector.	Engine stopped	X	–	–	Replace terminal.
6	Tap and wiggle harness connector or component during real-time diagnosis.	During real-time diagnosis	X	X	X	If malfunction codes are displayed during real-time diagnosis, replace terminal.

ECCS SELF-DIAGNOSTIC MODE V INSPECTION—1990 TRUCK/ PATHFINDER WITH VG30E ENGINE

Self-diagnosis — Mode V (Real-time diagnostic system) (Cont'd)

Ignition Signal

X: Available
–: Not available

Check sequence	Check items	Check conditions	Check parts: Ignition signal harness connector	Check parts: Sensor & actuator	Check parts: E.C.U. S.M.J. harness connector	If malfunction, perform the following items.
1	Tap and wiggle harness connector or component during real-time diagnosis.	During real-time diagnosis	X	X	X	Go to check item 2.
2	Check harness continuity at connector.	Engine stopped	X	–	–	Go to check item 3.
3	Disconnect harness connector, and then check dust adhesion to harness connector.	Engine stopped	X	–	X	Clean terminal surface.
4	Check pin terminal bend.	Engine stopped	–	–	X	Take out bend.
5	Reconnect harness connector and then recheck harness continuity at connector.	Engine stopped	X	–	–	Replace terminal.
6	Tap and wiggle harness connector or component during real-time diagnosis.	During real-time diagnosis	X	X	X	If malfunction codes are displayed during real-time diagnosis, replace terminal.

E.C.U.
Power transistor harness connector
Ignition coil harness connector

ECCS CAUTIONS—1990 TRUCK/PATHFINDER WITH VG30E ENGINE

Diagnostic Procedure

CAUTION:

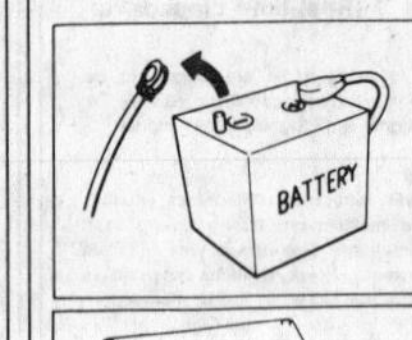
BATTERY

1. **Before connecting or disconnecting the E.C.U. harness connector to or from any E.C.U., be sure to turn the ignition switch to the "OFF" position and disconnect the negative battery terminal in order not to damage E.C.U. as battery voltage is applied to E.C.U. even if ignition switch is turned off. Failure to do so may damage the E.C.U.**

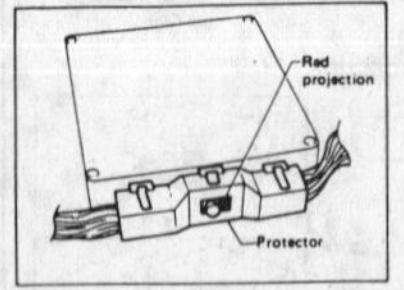
Red projection
Protector

2. **When connecting E.C.U. harness connector, tighten securing bolt until red projection is in line with connector face.**

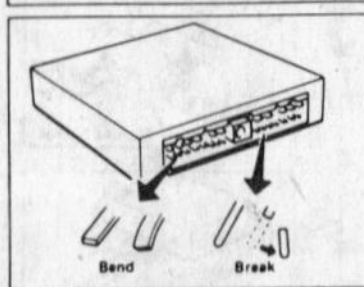
Bend
Break

3. **When connecting or disconnecting pin connectors into or from E.C.U., take care not to damage pin terminals (bend or break).**
4. **Make sure that there are not any bends or breaks on E.C.U. pin terminal, when connecting pin connectors.**

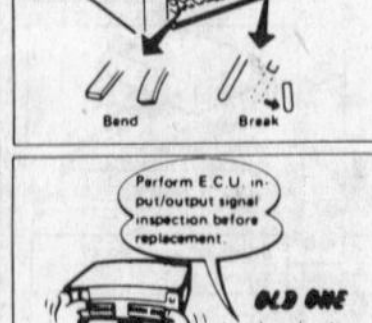
Perform E.C.U. input/output signal inspection before replacement.
OLD ONE

5. **Before replacing E.C.U., perform E.C.U. input/output signal inspection and make sure whether E.C.U. functions properly or not.**

6. **After performing this "Diagnostic Procedure", perform E.C.C.S. self-diagnosis and driving test.**

ECCS CAUTIONS (CONT.)—1990 TRUCK/PATHFINDER WITH VG30E ENGINE

Diagnostic Procedure (Cont'd)

7. When measuring E.C.U. controlled components supply voltage with a circuit tester, separate one tester probe from the other.
If the two tester probes accidentally make contact with each other during measurement, the circuit will be shorted, resulting in damage to the control unit power transistor.

Battery voltage
Short
Harness connector for solenoid valve
E.C.U.
N.G.
Solenoid valve
O.K.
Circuit tester

ECCS DIAGNOSTIC CHART—1990 TRUCK/PATHFINDER WITH VG30E ENGINE

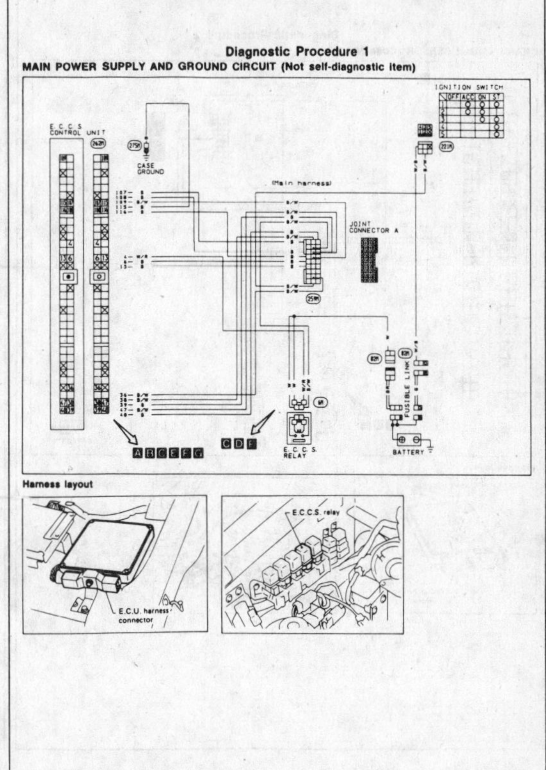
Diagnostic Procedure 1
MAIN POWER SUPPLY AND GROUND CIRCUIT (Not self-diagnostic item)
IGNITION SWITCH
E.C.C.S. CONTROL UNIT
CASE GROUND
(Main harness)
JOINT CONNECTOR A
FUSIBLE LINK
E.C.C.S. RELAY
BATTERY
A B C E F G
C D F
Harness layout
E.C.U. harness connector
E.C.C.S. relay

ECCS DIAGNOSTIC CHART—1990 TRUCK/PATHFINDER WITH VG30E ENGINE

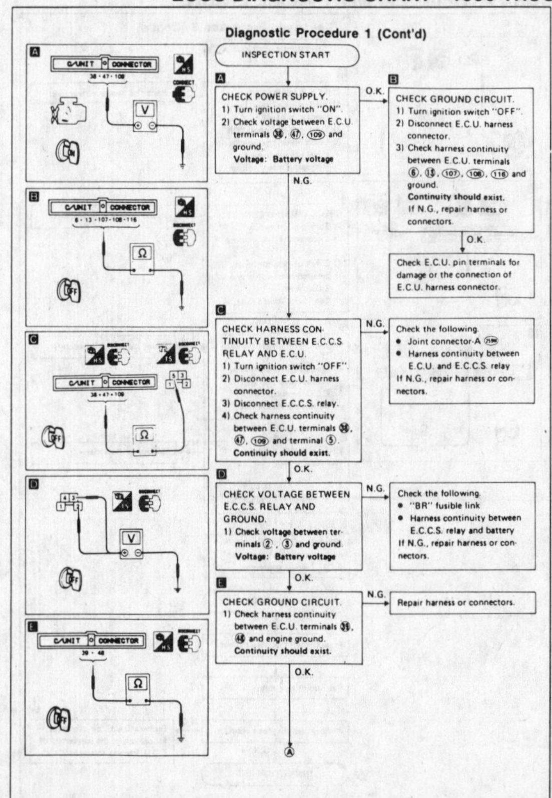
Diagnostic Procedure 1 (Cont'd)
INSPECTION START
A
CHECK POWER SUPPLY.
1) Turn ignition switch "ON".
2) Check voltage between E.C.U. terminals 38, 47, 109 and ground.
Voltage: Battery voltage
O.K.
N.G.
B
CHECK GROUND CIRCUIT.
1) Turn ignition switch "OFF".
2) Disconnect E.C.U. harness connector.
3) Check harness continuity between E.C.U. terminals 6, 13, 107, 108, 116 and ground.
Continuity should exist.
If N.G., repair harness or connectors.
O.K.
Check E.C.U. pin terminals for damage or the connection of E.C.U. harness connector.
C
CHECK HARNESS CONTINUITY BETWEEN E.C.C.S. RELAY AND E.C.U.
1) Turn ignition switch "OFF".
2) Disconnect E.C.U. harness connector.
3) Disconnect E.C.C.S. relay.
4) Check harness continuity between E.C.U. terminals 38, 47, 109 and terminal 5.
Continuity should exist.
N.G.
Check the following.
• Joint connector-A
• Harness continuity between E.C.U. and E.C.C.S. relay
If N.G., repair harness or connectors.
O.K.
D
CHECK VOLTAGE BETWEEN E.C.C.S. RELAY AND GROUND.
1) Check voltage between terminals 2, 3 and ground.
Voltage: Battery voltage
N.G.
Check the following.
• "BR" fusible link
• Harness continuity between E.C.C.S. relay and battery
If N.G., repair harness or connectors.
O.K.
E
CHECK GROUND CIRCUIT.
1) Check harness continuity between E.C.U. terminals 39, 48 and engine ground.
Continuity should exist.
N.G.
Repair harness or connectors.
O.K.
A
C/UNIT CONNECTOR
38 · 47 · 109
6 · 13 · 107 · 108 · 116
38 · 47 · 109
39 · 48

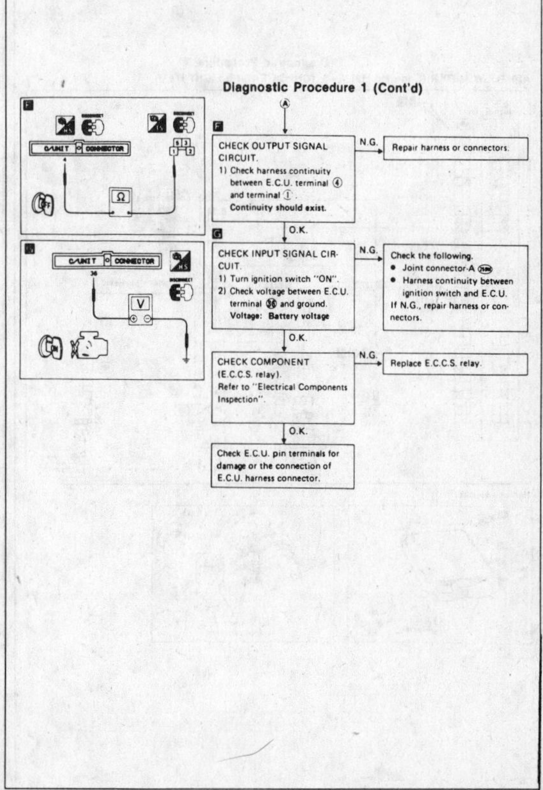
Diagnostic Procedure 1 (Cont'd)
A
F
CHECK OUTPUT SIGNAL CIRCUIT.
1) Check harness continuity between E.C.U. terminal 4 and terminal 1.
Continuity should exist.
N.G.
Repair harness or connectors.
O.K.
G
CHECK INPUT SIGNAL CIRCUIT.
1) Turn ignition switch "ON".
2) Check voltage between E.C.U. terminal 36 and ground.
Voltage: Battery voltage
N.G.
Check the following.
• Joint connector-A
• Harness continuity between ignition switch and E.C.U.
If N.G., repair harness or connectors.
O.K.
CHECK COMPONENT (E.C.C.S. relay).
Refer to "Electrical Components Inspection".
N.G.
Replace E.C.C.S. relay.
O.K.
Check E.C.U. pin terminals for damage or the connection of E.C.U. harness connector.
C/UNIT CONNECTOR
36

ECCS SELF-DIAGNOSTIC CHART—1990 TRUCK/PATHFINDER WITH VG30E ENGINE

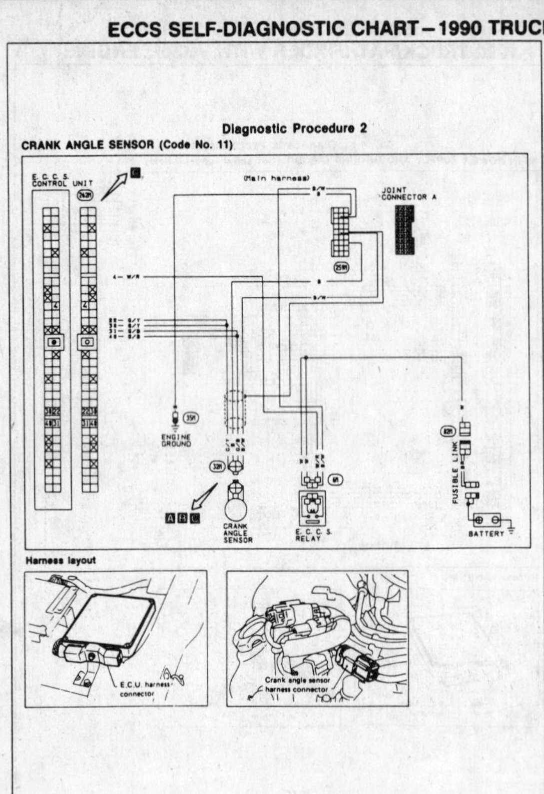
Diagnostic Procedure 2
CRANK ANGLE SENSOR (Code No. 11)
E.C.C.S. CONTROL UNIT
(Main harness)
JOINT CONNECTOR A
ENGINE GROUND
CRANK ANGLE SENSOR
E.C.C.S. RELAY
FUSIBLE LINK
BATTERY
Harness layout
E.C.U. harness connector
Crank angle sensor harness connector

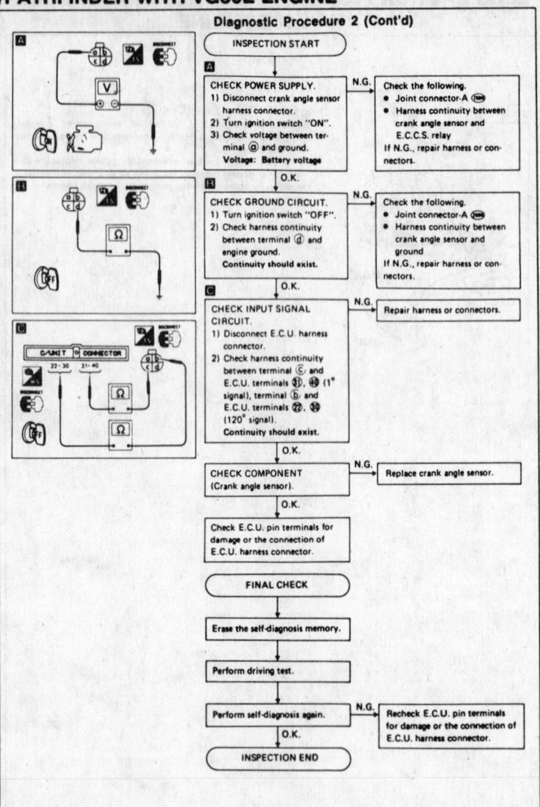
Diagnostic Procedure 2 (Cont'd)
INSPECTION START
A
CHECK POWER SUPPLY.
1) Disconnect crank angle sensor harness connector.
2) Turn ignition switch "ON".
3) Check voltage between terminal ⓐ and ground.
Voltage: Battery voltage
N.G.
Check the following.
• Joint connector-A
• Harness continuity between crank angle sensor and E.C.C.S. relay
If N.G., repair harness or connectors.
O.K.
B
CHECK GROUND CIRCUIT.
1) Turn ignition switch "OFF".
2) Check harness continuity between terminal ⓓ and engine ground.
Continuity should exist.
N.G.
Check the following.
• Joint connector-A
• Harness continuity between crank angle sensor and ground
If N.G., repair harness or connectors.
O.K.
C
CHECK INPUT SIGNAL CIRCUIT.
1) Disconnect E.C.U. harness connector.
2) Check harness continuity between terminal ⓒ and E.C.U. terminals 31, 40 (1° signal), terminal ⓑ and E.C.U. terminals 22, 30 (120° signal).
Continuity should exist.
N.G.
Repair harness or connectors.
C/UNIT CONNECTOR
22-30
31-40
O.K.
CHECK COMPONENT (Crank angle sensor).
N.G.
Replace crank angle sensor.
O.K.
Check E.C.U. pin terminals for damage or the connection of E.C.U. harness connector.
FINAL CHECK
Erase the self-diagnosis memory.
Perform driving test.
Perform self-diagnosis again.
N.G.
Recheck E.C.U. pin terminals for damage or the connection of E.C.U. harness connector.
O.K.
INSPECTION END

ECCS SELF-DIAGNOSTIC CHART—1990 TRUCK/PATHFINDER WITH VG30E ENGINE

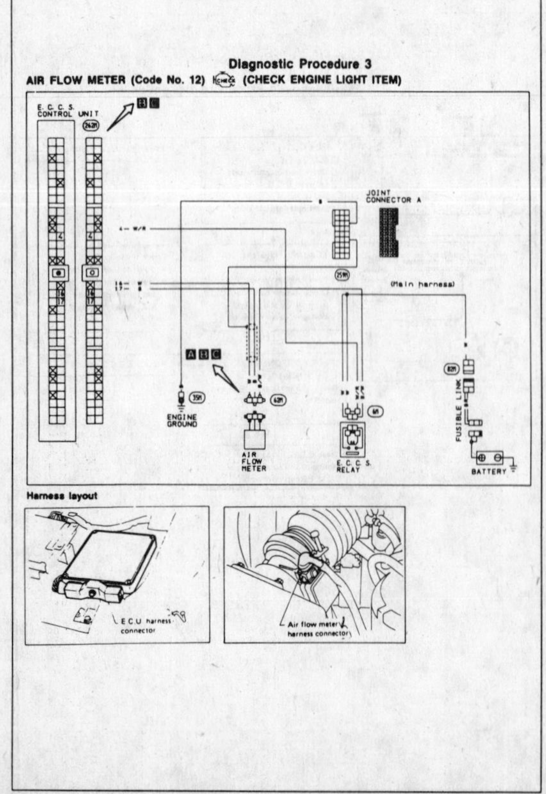
Diagnostic Procedure 3
AIR FLOW METER (Code No. 12) (CHECK ENGINE LIGHT ITEM)
E.C.C.S. CONTROL UNIT
JOINT CONNECTOR A
(Main harness)
ENGINE GROUND
AIR FLOW METER
E.C.C.S. RELAY
FUSIBLE LINK
BATTERY
Harness layout
E.C.U. harness connector
Air flow meter harness connector

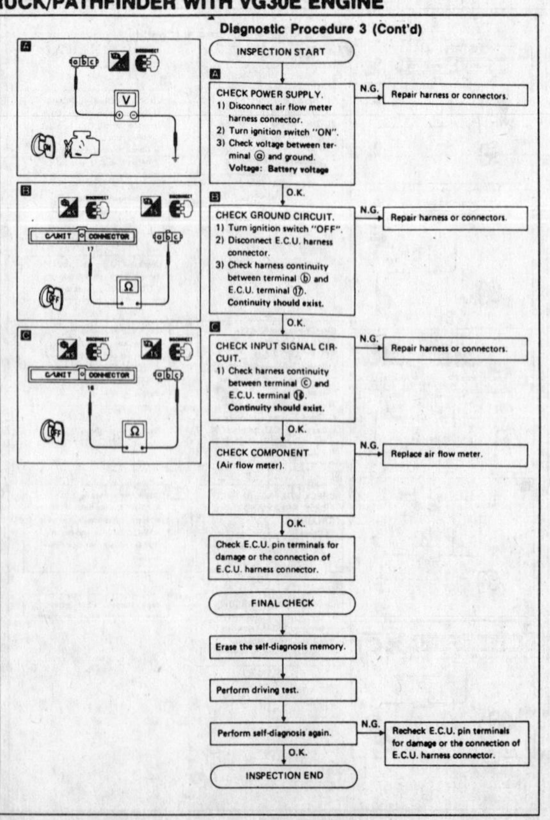
Diagnostic Procedure 3 (Cont'd)
INSPECTION START
A
CHECK POWER SUPPLY.
1) Disconnect air flow meter harness connector.
2) Turn ignition switch "ON".
3) Check voltage between terminal ⓐ and ground.
Voltage: Battery voltage
N.G.
Repair harness or connectors.
O.K.
B
CHECK GROUND CIRCUIT.
1) Turn ignition switch "OFF".
2) Disconnect E.C.U. harness connector.
3) Check harness continuity between terminal ⓑ and E.C.U. terminal 17.
Continuity should exist.
N.G.
Repair harness or connectors.
C/UNIT CONNECTOR
O.K.
C
CHECK INPUT SIGNAL CIRCUIT.
1) Check harness continuity between terminal ⓒ and E.C.U. terminal 16.
Continuity should exist.
N.G.
Repair harness or connectors.
O.K.
CHECK COMPONENT (Air flow meter).
N.G.
Replace air flow meter.
O.K.
Check E.C.U. pin terminals for damage or the connection of E.C.U. harness connector.
FINAL CHECK
Erase the self-diagnosis memory.
Perform driving test.
Perform self-diagnosis again.
N.G.
Recheck E.C.U. pin terminals for damage or the connection of E.C.U. harness connector.
O.K.
INSPECTION END

ECCS SELF-DIAGNOSTIC CHART—1990 TRUCK/PATHFINDER WITH VG30E ENGINE

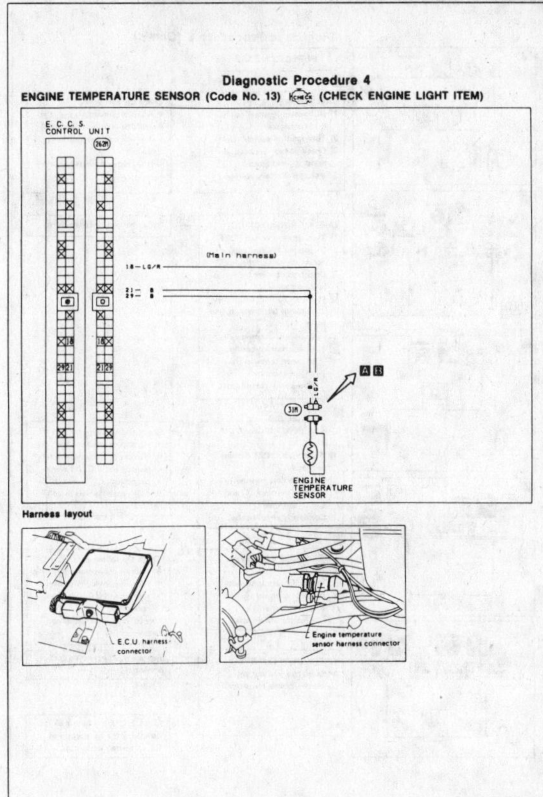
Diagnostic Procedure 4
ENGINE TEMPERATURE SENSOR (Code No. 13) (CHECK ENGINE LIGHT ITEM)
E.C.C.S. CONTROL UNIT
(Main harness)
ENGINE TEMPERATURE SENSOR
Harness layout
E.C.U. harness connector
Engine temperature sensor harness connector

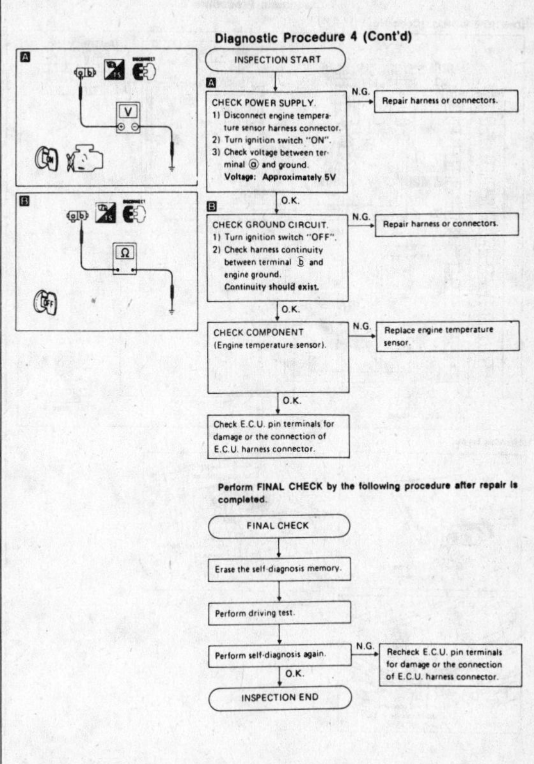
Diagnostic Procedure 4 (Cont'd)
INSPECTION START
CHECK POWER SUPPLY.
1) Disconnect engine temperature sensor harness connector.
2) Turn ignition switch "ON".
3) Check voltage between terminal ⓐ and ground.
Voltage: Approximately 5V
N.G.
Repair harness or connectors.
O.K.
CHECK GROUND CIRCUIT.
1) Turn ignition switch "OFF".
2) Check harness continuity between terminal ⓑ and engine ground.
Continuity should exist.
N.G.
Repair harness or connectors.
O.K.
CHECK COMPONENT
(Engine temperature sensor).
N.G.
Replace engine temperature sensor.
O.K.
Check E.C.U. pin terminals for damage or the connection of E.C.U. harness connector.
Perform FINAL CHECK by the following procedure after repair is completed.
FINAL CHECK
Erase the self-diagnosis memory.
Perform driving test.
Perform self-diagnosis again.
N.G.
Recheck E.C.U. pin terminals for damage or the connection of E.C.U. harness connector.
O.K.
INSPECTION END

ECCS SELF-DIAGNOSTIC CHART—1990 TRUCK/PATHFINDER WITH VG30E ENGINE

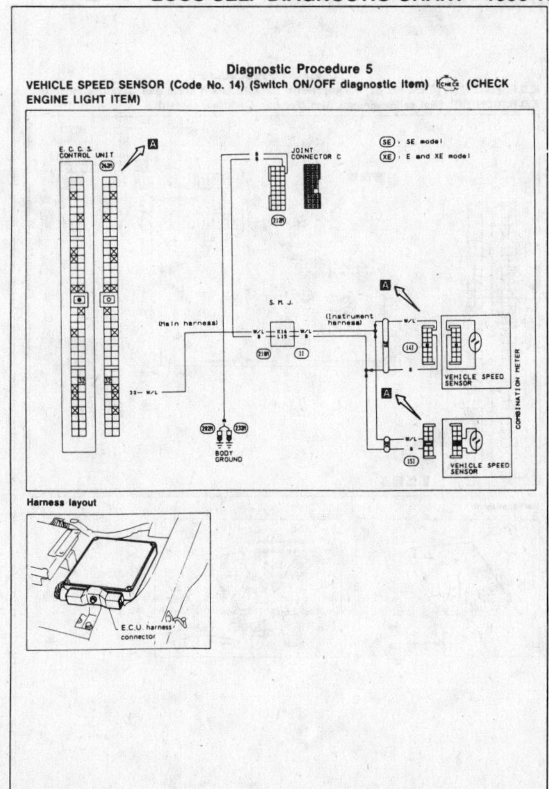
Diagnostic Procedure 5
VEHICLE SPEED SENSOR (Code No. 14) (Switch ON/OFF diagnostic item) (CHECK ENGINE LIGHT ITEM)
E.C.C.S. CONTROL UNIT
JOINT CONNECTOR C
SE : SE model
XE : E and XE model
S.M.J.
(Main harness)
(Instrument harness)
VEHICLE SPEED SENSOR
COMBINATION METER
BODY GROUND
Harness layout
E.C.U. harness connector

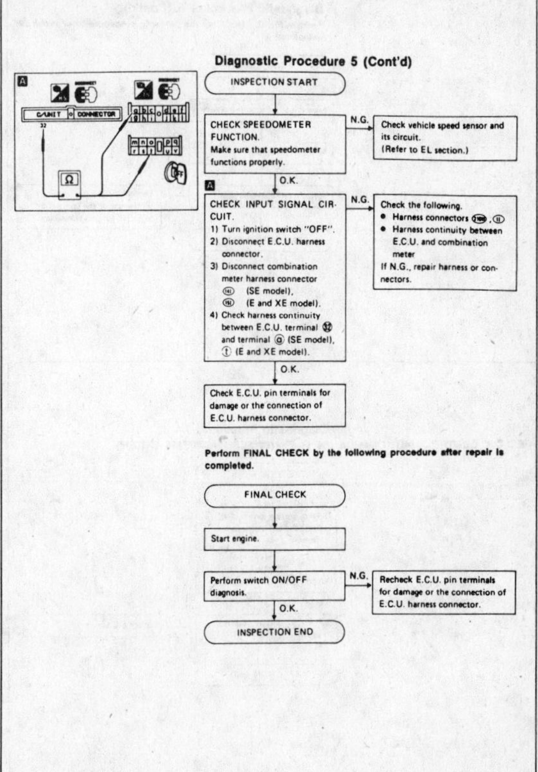
Diagnostic Procedure 5 (Cont'd)
INSPECTION START
CHECK SPEEDOMETER FUNCTION.
Make sure that speedometer functions properly.
N.G.
Check vehicle speed sensor and its circuit.
(Refer to EL section.)
O.K.
CHECK INPUT SIGNAL CIRCUIT.
1) Turn ignition switch "OFF".
2) Disconnect E.C.U. harness connector.
3) Disconnect combination meter harness connector
(SE model),
(E and XE model).
4) Check harness continuity between E.C.U. terminal and terminal ⓐ (SE model), (E and XE model).
N.G.
Check the following.
• Harness connectors
• Harness continuity between E.C.U. and combination meter
If N.G., repair harness or connectors.
O.K.
Check E.C.U. pin terminals for damage or the connection of E.C.U. harness connector.
Perform FINAL CHECK by the following procedure after repair is completed.
FINAL CHECK
Start engine.
Perform switch ON/OFF diagnosis.
N.G.
Recheck E.C.U. pin terminals for damage or the connection of E.C.U. harness connector.
O.K.
INSPECTION END

ECCS SELF-DIAGNOSTIC CHART—1990 TRUCK/PATHFINDER WITH VG30E ENGINE

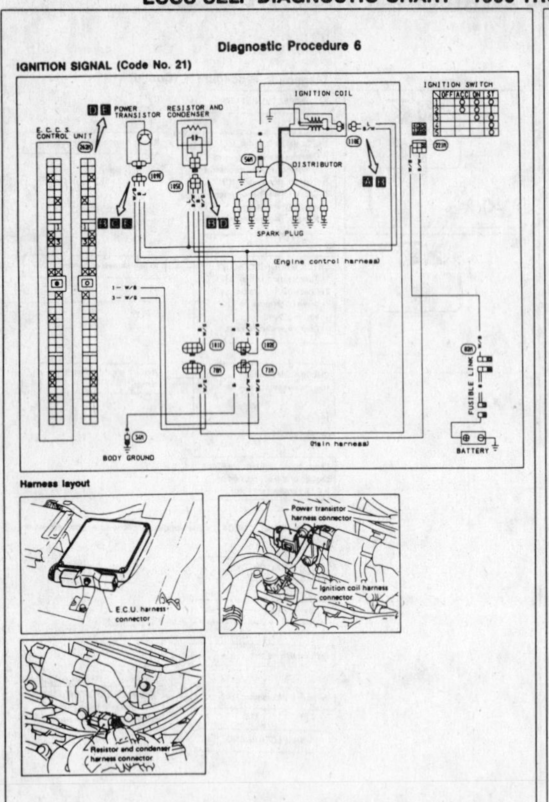
Diagnostic Procedure 6
IGNITION SIGNAL (Code No. 21)
E.C.C.S CONTROL UNIT
POWER TRANSISTOR
RESISTOR AND CONDENSER
IGNITION COIL
IGNITION SWITCH
DISTRIBUTOR
SPARK PLUG
(Engine control harness)
(Main harness)
FUSIBLE LINK
BATTERY
BODY GROUND
Harness layout
Power transistor harness connector
Ignition coil harness connector
E.C.U. harness connector
Resistor and condenser harness connector

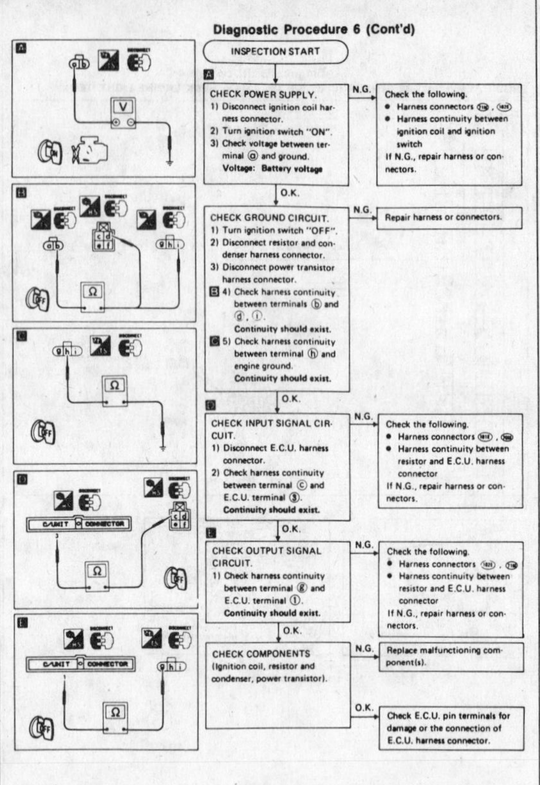
Diagnostic Procedure 6 (Cont'd)
INSPECTION START
A
CHECK POWER SUPPLY.
1) Disconnect ignition coil harness connector.
2) Turn ignition switch "ON".
3) Check voltage between terminal ⓐ and ground.
Voltage: Battery voltage
N.G.
Check the following.
• Harness connectors
• Harness continuity between ignition coil and ignition switch
If N.G., repair harness or connectors.
O.K.
CHECK GROUND CIRCUIT.
1) Turn ignition switch "OFF".
2) Disconnect resistor and condenser harness connector.
3) Disconnect power transistor harness connector.
B 4) Check harness continuity between terminals ⓑ and ⓓ, ⓘ.
Continuity should exist.
C 5) Check harness continuity between terminal ⓗ and engine ground.
Continuity should exist.
N.G.
Repair harness or connectors.
O.K.
D
CHECK INPUT SIGNAL CIRCUIT.
1) Disconnect E.C.U. harness connector.
2) Check harness continuity between terminal ⓒ and E.C.U. terminal ③.
Continuity should exist.
N.G.
Check the following.
• Harness connectors
• Harness continuity between resistor and E.C.U. harness connector
If N.G., repair harness or connectors.
O.K.
E
CHECK OUTPUT SIGNAL CIRCUIT.
1) Check harness continuity between terminal ⓖ and E.C.U. terminal ①.
Continuity should exist.
N.G.
Check the following.
• Harness connectors
• Harness continuity between resistor and E.C.U. harness connector
If N.G., repair harness or connectors.
O.K.
CHECK COMPONENTS
(Ignition coil, resistor and condenser, power transistor).
N.G.
Replace malfunctioning component(s).
O.K.
Check E.C.U. pin terminals for damage or the connection of E.C.U. harness connector.
C/UNIT
CONNECTOR

ECCS SELF-DIAGNOSTIC CHART—1990 TRUCK/PATHFINDER WITH VG30E ENGINE

Diagnostic Procedure 6 (Cont'd)

Perform FINAL CHECK by the following procedure after repair is completed.

FINAL CHECK
↓
Erase the self-diagnosis memory.
↓
Perform driving test.
↓
Perform self-diagnosis again. — N.G. → Recheck E.C.U. pin terminals for damage or the connection of E.C.U. harness connector.
↓ O.K.
INSPECTION END

Diagnostic Procedure 7

E.C.C.S. CONTROL UNIT (Code No. 31) (CHECK ENGINE LIGHT ITEM)

INSPECTION START
↓
1) Turn ignition switch "ON".
2) Erase the self-diagnosis memory.
↓
Perform self-diagnosis again.
↓
Does E.C.U. display Code No. 31 again? — Yes → Replace E.C.U.
↓ No
INSPECTION END

ECCS DIAGNOSTIC CHART—1990 TRUCK/PATHFINDER WITH VG30E ENGINE

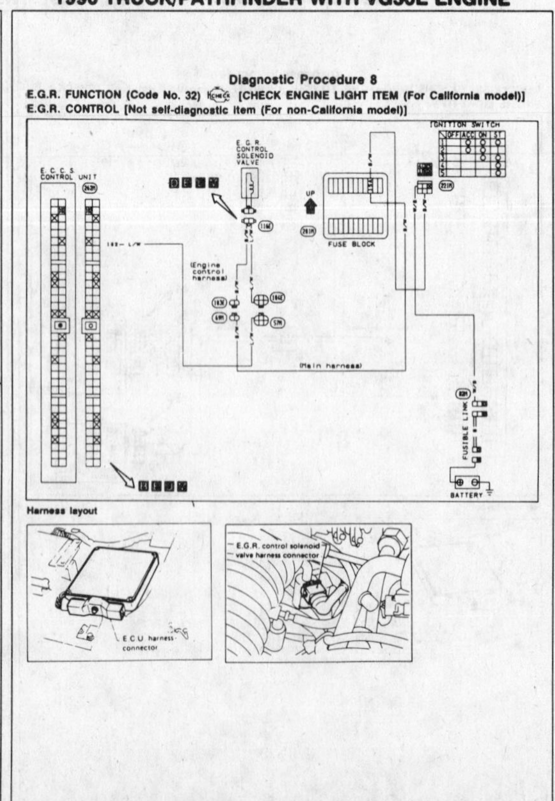
Diagnostic Procedure 8
E.G.R. FUNCTION (Code No. 32) [CHECK ENGINE LIGHT ITEM (For California model)]
E.G.R. CONTROL [Not self-diagnostic item (For non-California model)]
IGNITION SWITCH
E.C.C.S CONTROL UNIT
E.G.R. CONTROL SOLENOID VALVE
UP
FUSE BLOCK
(Engine control harness)
(Main harness)
FUSIBLE LINK
BATTERY
Harness layout
E.C.U. harness connector
E.G.R. control solenoid valve harness connector

ECCS DIAGNOSTIC CHART—1990 TRUCK/PATHFINDER WITH VG30E ENGINE

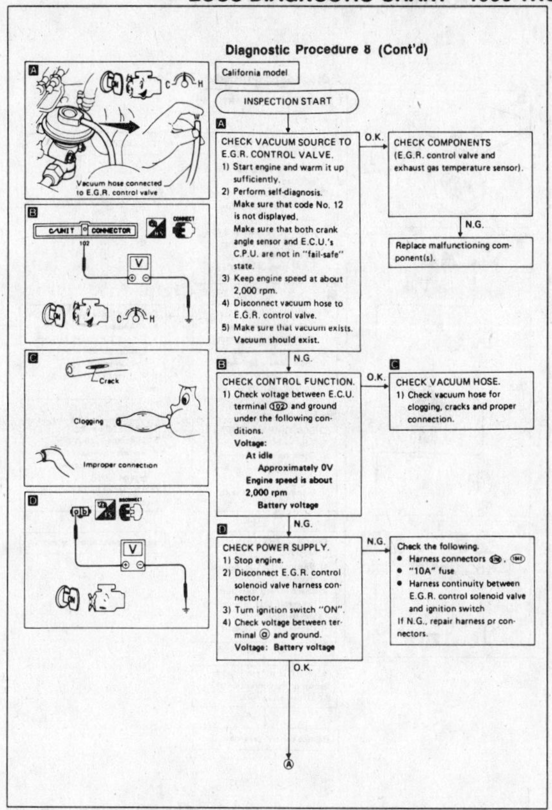
Diagnostic Procedure 8 (Cont'd)
California model
INSPECTION START
Vacuum hose connected to E.G.R. control valve
Crack
Clogging
Improper connection
CHECK VACUUM SOURCE TO E.G.R. CONTROL VALVE.
1) Start engine and warm it up sufficiently.
2) Perform self-diagnosis. Make sure that code No. 12 is not displayed. Make sure that both crank angle sensor and E.C.U.'s C.P.U. are not in "fail-safe" state.
3) Keep engine speed at about 2,000 rpm.
4) Disconnect vacuum hose to E.G.R. control valve.
5) Make sure that vacuum exists. Vacuum should exist.
O.K.
CHECK COMPONENTS (E.G.R. control valve and exhaust gas temperature sensor).
N.G.
Replace malfunctioning component(s).
N.G.
CHECK CONTROL FUNCTION.
1) Check voltage between E.C.U. terminal 102 and ground under the following conditions.
Voltage:
At idle
Approximately 0V
Engine speed is about 2,000 rpm
Battery voltage
O.K.
CHECK VACUUM HOSE.
1) Check vacuum hose for clogging, cracks and proper connection.
N.G.
CHECK POWER SUPPLY.
1) Stop engine.
2) Disconnect E.G.R. control solenoid valve harness connector.
3) Turn ignition switch "ON".
4) Check voltage between terminal ⓐ and ground.
Voltage: Battery voltage
N.G.
Check the following.
• Harness connectors
• "10A" fuse
• Harness continuity between E.G.R. control solenoid valve and ignition switch
If N.G., repair harness or connectors.
O.K.
Ⓐ

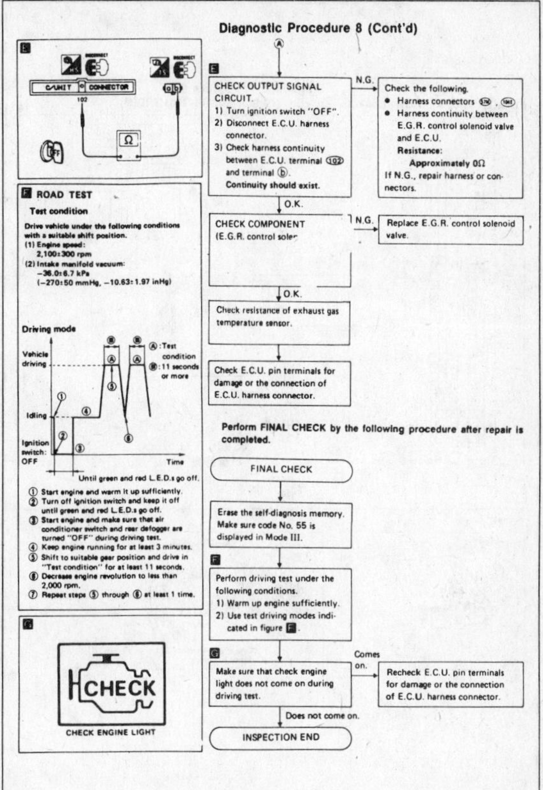
Diagnostic Procedure 8 (Cont'd)
Ⓐ
CHECK OUTPUT SIGNAL CIRCUIT.
1) Turn ignition switch "OFF".
2) Disconnect E.C.U. harness connector.
3) Check harness continuity between E.C.U. terminal 102 and terminal ⓑ.
Continuity should exist.
N.G.
Check the following.
• Harness connectors
• Harness continuity between E.G.R. control solenoid valve and E.C.U.
Resistance:
Approximately 0Ω
If N.G., repair harness or connectors.
O.K.
CHECK COMPONENT (E.G.R. control solenoid valve).
N.G.
Replace E.G.R. control solenoid valve.
O.K.
Check resistance of exhaust gas temperature sensor.
Check E.C.U. pin terminals for damage or the connection of E.C.U. harness connector.
ROAD TEST
Test condition
Drive vehicle under the following conditions with a suitable shift position.
(1) Engine speed: 2,100±300 rpm
(2) Intake manifold vacuum: −36.0±6.7 kPa (−270±50 mmHg, −10.63±1.97 inHg)
Driving mode
Vehicle driving
Idling
Ignition switch: OFF
Time
Ⓐ: Test condition
Ⓑ: 11 seconds or more
Until green and red L.E.D.s go off.
① Start engine and warm it up sufficiently.
② Turn off ignition switch and keep it off until green and red L.E.D.s go off.
③ Start engine and make sure that air conditioner switch and rear defogger are turned "OFF" during driving test.
④ Keep engine running for at least 3 minutes.
⑤ Shift to suitable gear position and drive in "Test condition" for at least 11 seconds.
⑥ Decrease engine revolution to less than 2,000 rpm.
⑦ Repeat steps ⑤ through ⑥ at least 1 time.
CHECK
CHECK ENGINE LIGHT
Perform FINAL CHECK by the following procedure after repair is completed.
FINAL CHECK
Erase the self-diagnosis memory. Make sure code No. 55 is displayed in Mode III.
Perform driving test under the following conditions.
1) Warm up engine sufficiently.
2) Use test driving modes indicated in figure.
Make sure that check engine light does not come on during driving test.
Comes on.
Recheck E.C.U. pin terminals for damage or the connection of E.C.U. harness connector.
Does not come on.
INSPECTION END

ECCS DIAGNOSTIC CHART—1990 TRUCK/PATHFINDER WITH VG30E ENGINE

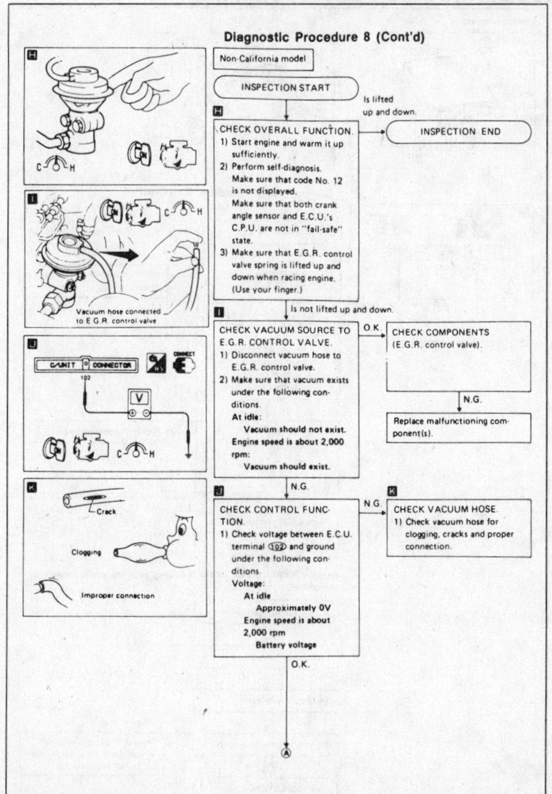
Diagnostic Procedure 8 (Cont'd)
Non-California model
INSPECTION START
CHECK OVERALL FUNCTION.
1) Start engine and warm it up sufficiently.
2) Perform self-diagnosis. Make sure that code No. 12 is not displayed. Make sure that both crank angle sensor and E.C.U.'s C.P.U. are not in "fail-safe" state.
3) Make sure that E.G.R. control valve spring is lifted up and down when racing engine. (Use your finger.)
Is lifted up and down.
INSPECTION END
Is not lifted up and down.
Vacuum hose connected to E.G.R. control valve
Crack
Clogging
Improper connection
CHECK VACUUM SOURCE TO E.G.R. CONTROL VALVE.
1) Disconnect vacuum hose to E.G.R. control valve.
2) Make sure that vacuum exists under the following conditions.
At idle:
Vacuum should not exist.
Engine speed is about 2,000 rpm:
Vacuum should exist.
O.K.
CHECK COMPONENTS (E.G.R. control valve).
N.G.
Replace malfunctioning component(s).
N.G.
CHECK CONTROL FUNCTION.
1) Check voltage between E.C.U. terminal 102 and ground under the following conditions.
Voltage:
At idle
Approximately 0V
Engine speed is about 2,000 rpm
Battery voltage
N.G.
CHECK VACUUM HOSE.
1) Check vacuum hose for clogging, cracks and proper connection.
O.K.
Ⓐ

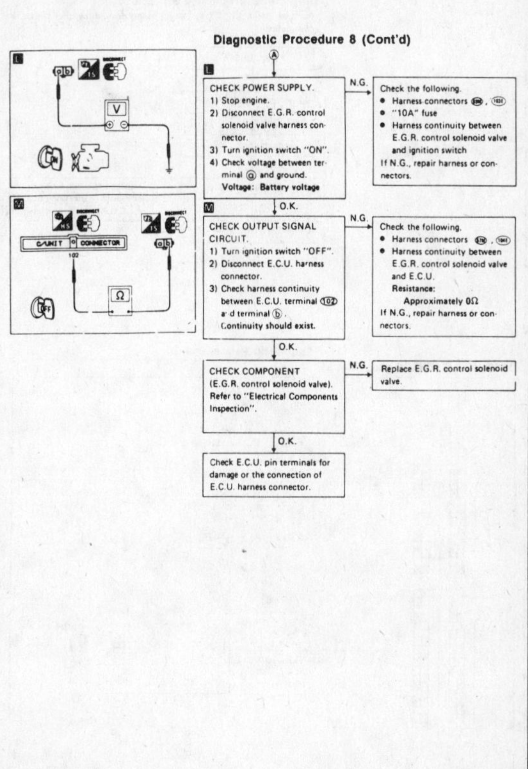
Diagnostic Procedure 8 (Cont'd)
Ⓐ
CHECK POWER SUPPLY.
1) Stop engine.
2) Disconnect E.G.R. control solenoid valve harness connector.
3) Turn ignition switch "ON".
4) Check voltage between terminal ⓐ and ground.
Voltage: Battery voltage
N.G.
Check the following.
• Harness connectors
• "10A" fuse
• Harness continuity between E.G.R. control solenoid valve and ignition switch
If N.G., repair harness or connectors.
O.K.
CHECK OUTPUT SIGNAL CIRCUIT.
1) Turn ignition switch "OFF".
2) Disconnect E.C.U. harness connector.
3) Check harness continuity between E.C.U. terminal 102 and terminal ⓑ.
Continuity should exist.
N.G.
Check the following.
• Harness connectors
• Harness continuity between E.G.R. control solenoid valve and E.C.U.
Resistance:
Approximately 0Ω
If N.G., repair harness or connectors.
O.K.
CHECK COMPONENT (E.G.R. control solenoid valve). Refer to "Electrical Components Inspection".
N.G.
Replace E.G.R. control solenoid valve.
O.K.
Check E.C.U. pin terminals for damage or the connection of E.C.U. harness connector.

ECCS SELF-DIAGNOSTIC CHART—1990 TRUCK/PATHFINDER WITH VG30E ENGINE

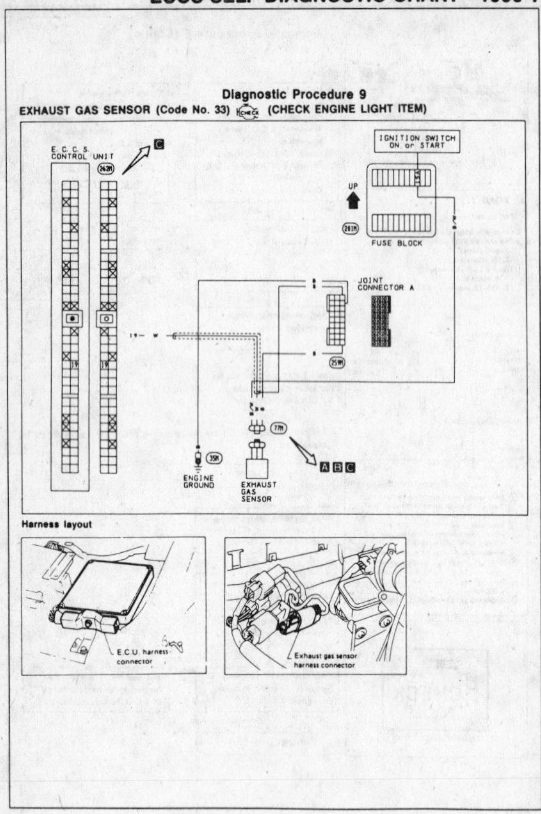
Diagnostic Procedure 9
EXHAUST GAS SENSOR (Code No. 33) (CHECK ENGINE LIGHT ITEM)
E.C.C.S. CONTROL UNIT
IGNITION SWITCH ON or START
UP
FUSE BLOCK
JOINT CONNECTOR A
ENGINE GROUND
EXHAUST GAS SENSOR
Harness layout
E.C.U. harness connector
Exhaust gas sensor harness connector

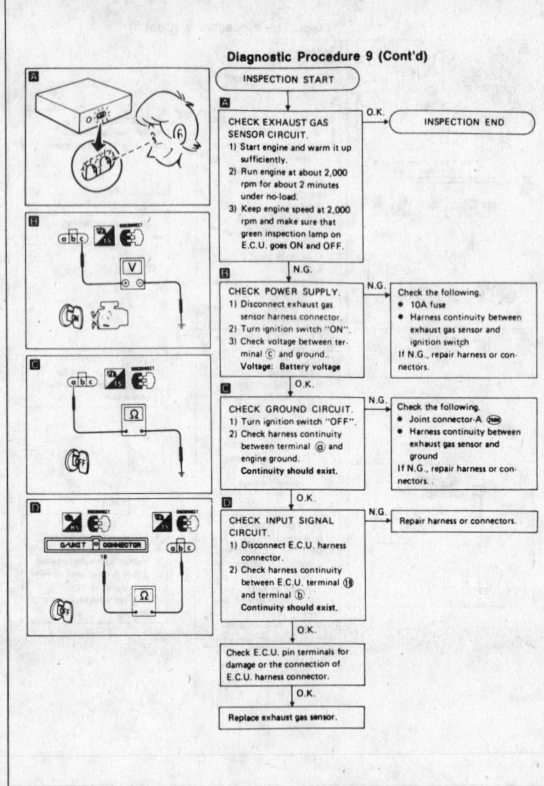
Diagnostic Procedure 9 (Cont'd)
INSPECTION START
CHECK EXHAUST GAS SENSOR CIRCUIT.
1) Start engine and warm it up sufficiently.
2) Run engine at about 2,000 rpm for about 2 minutes under no-load.
3) Keep engine speed at 2,000 rpm and make sure that green inspection lamp on E.C.U. goes ON and OFF.
O.K.
INSPECTION END
N.G.
CHECK POWER SUPPLY.
1) Disconnect exhaust gas sensor harness connector.
2) Turn ignition switch "ON".
3) Check voltage between terminal ⓒ and ground.
Voltage: Battery voltage
N.G.
Check the following.
• 10A fuse
• Harness continuity between exhaust gas sensor and ignition switch
If N.G., repair harness or connectors.
O.K.
CHECK GROUND CIRCUIT.
1) Turn ignition switch "OFF".
2) Check harness continuity between terminal ⓐ and engine ground.
Continuity should exist.
N.G.
Check the following.
• Joint connector-A
• Harness continuity between exhaust gas sensor and ground
If N.G., repair harness or connectors.
O.K.
CHECK INPUT SIGNAL CIRCUIT.
1) Disconnect E.C.U. harness connector.
2) Check harness continuity between E.C.U. terminal ⑲ and terminal ⓑ.
Continuity should exist.
N.G.
Repair harness or connectors.
O.K.
Check E.C.U. pin terminals for damage or the connection of E.C.U. harness connector.
O.K.
Replace exhaust gas sensor.

ECCS SELF-DIAGNOSTIC CHART—1990 TRUCK/PATHFINDER WITH VG30E ENGINE

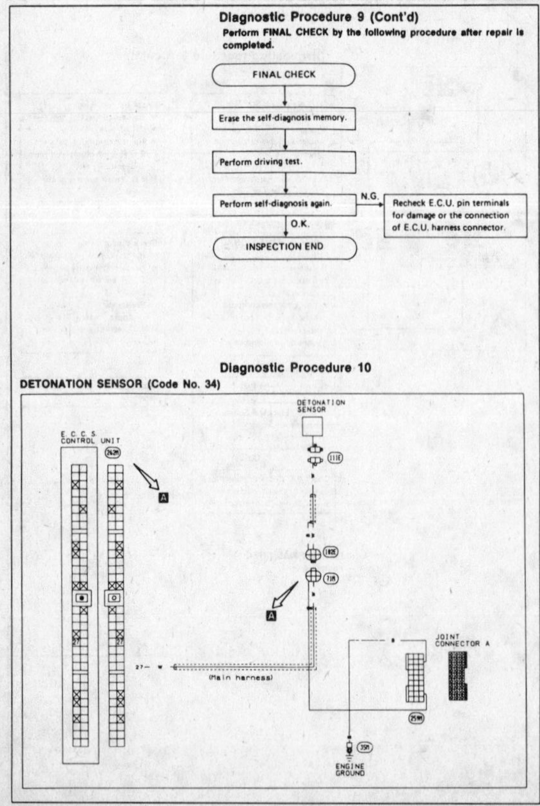
Diagnostic Procedure 9 (Cont'd)
Perform FINAL CHECK by the following procedure after repair is completed.
FINAL CHECK
Erase the self-diagnosis memory.
Perform driving test.
Perform self-diagnosis again.
N.G.
Recheck E.C.U. pin terminals for damage or the connection of E.C.U. harness connector.
O.K.
INSPECTION END
Diagnostic Procedure 10
DETONATION SENSOR (Code No. 34)
DETONATION SENSOR
E.C.C.S. CONTROL UNIT
JOINT CONNECTOR A
(Main harness)
ENGINE GROUND

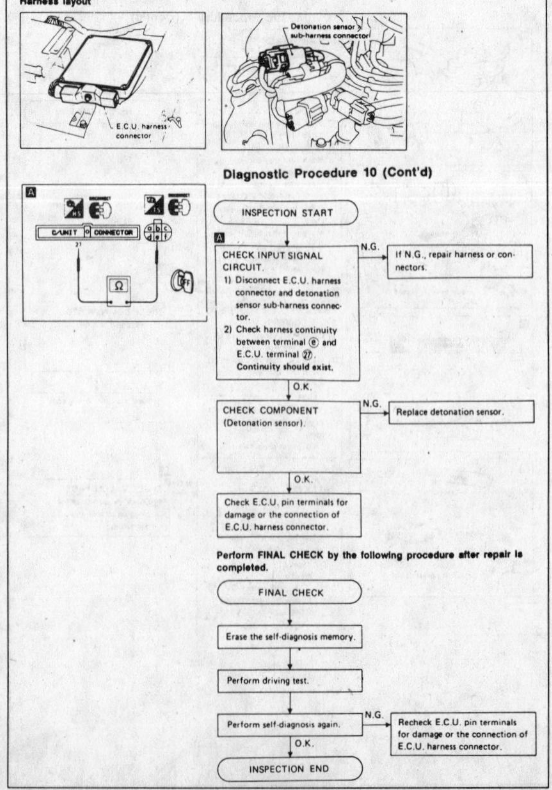
Harness layout
E.C.U. harness connector
Detonation sensor sub-harness connector
Diagnostic Procedure 10 (Cont'd)
INSPECTION START
CHECK INPUT SIGNAL CIRCUIT.
1) Disconnect E.C.U. harness connector and detonation sensor sub-harness connector.
2) Check harness continuity between terminal ⓔ and E.C.U. terminal ㉗.
Continuity should exist.
N.G.
If N.G., repair harness or connectors.
O.K.
CHECK COMPONENT (Detonation sensor).
N.G.
Replace detonation sensor.
O.K.
Check E.C.U. pin terminals for damage or the connection of E.C.U. harness connector.
Perform FINAL CHECK by the following procedure after repair is completed.
FINAL CHECK
Erase the self-diagnosis memory.
Perform driving test.
Perform self-diagnosis again.
N.G.
Recheck E.C.U. pin terminals for damage or the connection of E.C.U. harness connector.
O.K.
INSPECTION END

ECCS SELF-DIAGNOSTIC CHART—1990 TRUCK/PATHFINDER WITH VG30E ENGINE

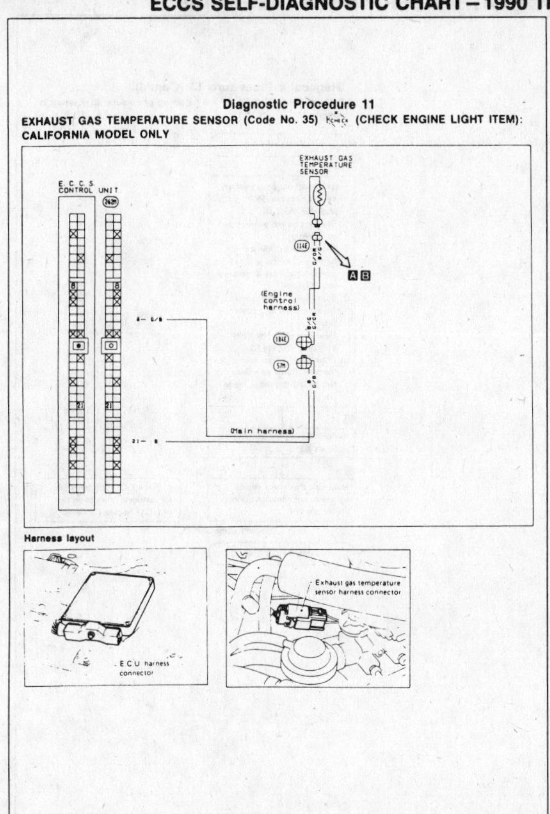
Diagnostic Procedure 11
EXHAUST GAS TEMPERATURE SENSOR (Code No. 35) (CHECK ENGINE LIGHT ITEM):
CALIFORNIA MODEL ONLY
E.C.C.S. CONTROL UNIT
EXHAUST GAS TEMPERATURE SENSOR
(Engine control harness)
(Main harness)
Harness layout
E.C.U. harness connector
Exhaust gas temperature sensor harness connector

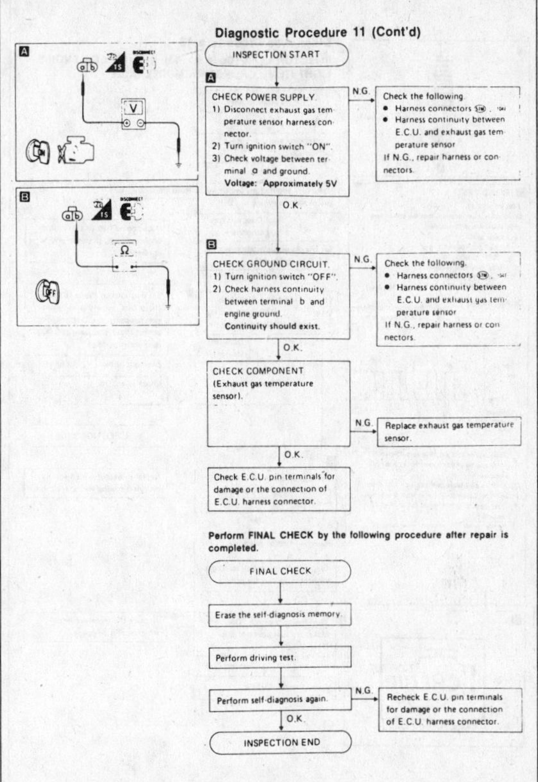
Diagnostic Procedure 11 (Cont'd)
INSPECTION START
CHECK POWER SUPPLY.
1) Disconnect exhaust gas temperature sensor harness connector.
2) Turn ignition switch "ON".
3) Check voltage between terminal ⓐ and ground.
Voltage: Approximately 5V
N.G.
Check the following.
• Harness connectors
• Harness continuity between E.C.U. and exhaust gas temperature sensor
If N.G., repair harness or connectors.
O.K.
CHECK GROUND CIRCUIT.
1) Turn ignition switch "OFF".
2) Check harness continuity between terminal ⓑ and engine ground.
Continuity should exist.
N.G.
Check the following.
• Harness connectors
• Harness continuity between E.C.U. and exhaust gas temperature sensor
If N.G., repair harness or connectors.
O.K.
CHECK COMPONENT
(Exhaust gas temperature sensor).
N.G.
Replace exhaust gas temperature sensor.
O.K.
Check E.C.U. pin terminals for damage or the connection of E.C.U. harness connector.
Perform FINAL CHECK by the following procedure after repair is completed.
FINAL CHECK
Erase the self-diagnosis memory.
Perform driving test.
Perform self-diagnosis again.
N.G.
Recheck E.C.U. pin terminals for damage or the connection of E.C.U. harness connector.
O.K.
INSPECTION END

ECCS SELF-DIAGNOSTIC CHART—1990 TRUCK/PATHFINDER WITH VG30E ENGINE

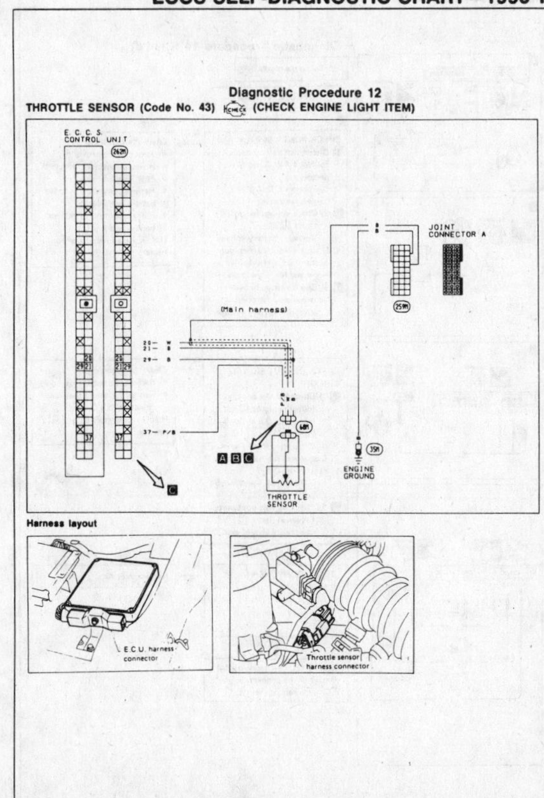
Diagnostic Procedure 12
THROTTLE SENSOR (Code No. 43) (CHECK ENGINE LIGHT ITEM)
E.C.C.S. CONTROL UNIT
JOINT CONNECTOR A
(Main harness)
ENGINE GROUND
THROTTLE SENSOR
Harness layout
E.C.U. harness connector
Throttle sensor harness connector

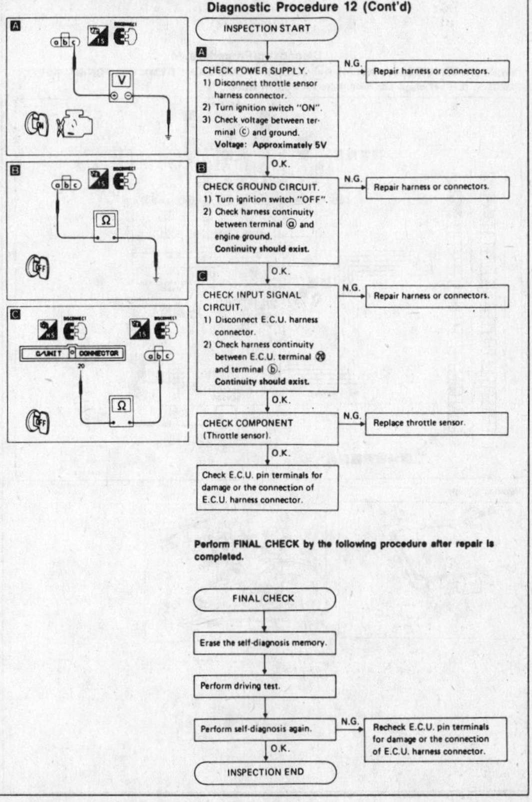
Diagnostic Procedure 12 (Cont'd)
INSPECTION START
CHECK POWER SUPPLY.
1) Disconnect throttle sensor harness connector.
2) Turn ignition switch "ON".
3) Check voltage between terminal ⓒ and ground.
Voltage: Approximately 5V
N.G.
Repair harness or connectors.
O.K.
CHECK GROUND CIRCUIT.
1) Turn ignition switch "OFF".
2) Check harness continuity between terminal ⓐ and engine ground.
Continuity should exist.
N.G.
Repair harness or connectors.
O.K.
CHECK INPUT SIGNAL CIRCUIT.
1) Disconnect E.C.U. harness connector.
2) Check harness continuity between E.C.U. terminal ⑳ and terminal ⓑ.
Continuity should exist.
N.G.
Repair harness or connectors.
C/UNIT CONNECTOR
O.K.
CHECK COMPONENT
(Throttle sensor).
N.G.
Replace throttle sensor.
O.K.
Check E.C.U. pin terminals for damage or the connection of E.C.U. harness connector.
Perform FINAL CHECK by the following procedure after repair is completed.
FINAL CHECK
Erase the self-diagnosis memory.
Perform driving test.
Perform self-diagnosis again.
N.G.
Recheck E.C.U. pin terminals for damage or the connection of E.C.U. harness connector.
O.K.
INSPECTION END

ECCS SELF-DIAGNOSTIC CHART—1990 TRUCK/PATHFINDER WITH VG30E ENGINE

A

Wet with fuel?

B ROAD TEST

Test condition

Drive vehicle under the following conditions with a suitable shift position.

(1) Engine speed:
 M/T: 2,600±600 rpm
 A/T: 2,500±700 rpm

(2) Intake manifold vacuum:
 −46.7±9.3 kPa
 (−350±70 mmHg, −13.78±2.76 inHg)

Driving mode

Ⓐ : 60 seconds or more
Ⓑ : 5 seconds or more
Ⓒ : 10 seconds or more

Vehicle driving
Idling
Ignition switch OFF
Time

① Start engine and warm it up sufficiently.
② Keep engine at idle speed for at least 60 seconds.
③ Shift to a suitable gear position and drive in "Test condition" for at least 5 seconds.
④ Keep engine at idle speed for at least 10 seconds.
⑤ Repeat steps ② through ④ at least 10 times.

C

CHECK

CHECK ENGINE LIGHT

Diagnostic Procedure 13

INJECTOR LEAK (Code No. 45) (CHECK ENGINE LIGHT ITEM): CALIFORNIA MODEL ONLY

INSPECTION START

↓

Start engine and warm it up sufficiently.

↓

Make sure engine runs smoothly at idle after warming.
- Runs smoothly. → Race engine two or three times under no-load, then run engine at idle speed.
  ↓
  Set diagnosis to Mode II and check that red and green inspection lamps on E.C.U. blink almost simultaneously at 2,000 rpm under no-load.
  - Does not blink. → Check idle CO%.
  - Blinks. → INSPECTION END
- Does not run smoothly. ↓

Set the diagnosis mode selector of E.C.U. to Mode I.

↓

Check if the green inspection lamp on E.C.U. stays off during 10 seconds at idle condition.
- Does not stay off. → Check mixture ratio feedback system.
- Stay off. ↓

A Remove all spark plugs from cylinder head. Is any plug wet with fuel?
- Yes → Replace injector which supply cylinders having wet spark plugs.
- No ↓

Remove injector assembly.

Keep fuel hoses and all injectors connected to injector gallery.

↓

Turn ignition switch "ON". Makesure fuel does not drip from injector.
- Drips. → Replace the injectors from which fuel is dripping.
- Does not drip. ↓

Go to FINAL CHECK.

Diagnostic Procedure 13 (Cont'd)

Perform FINAL CHECK by the following procedure after repair is completed.

FINAL CHECK

↓

Erase the self-diagnosis memory. Make sure code No. 55 is displayed in Mode III.
1) Start engine and warm it up sufficiently.
2) Disconnect air flow meter harness connector and run engine for at least 30 seconds at 2,000 rpm.
3) Stop engine and reconnect air flow meter harness connector.
4) Make sure Code No. 12 is displayed in Mode III.
5) Erase the self-diagnosis memory. Make sure Code No. 55 is displayed in Mode III.

↓

B Perform driving test as indicated in figure **B**.

↓

C Make sure that check engine light does not come on during engine test.
- Comes on. → Recheck E.C.U. pin terminals for damage or the connection of E.C.U. harness connector.
- Does not come on. ↓

INSPECTION END

ECCS DIAGNOSTIC CHART—1990 TRUCK/PATHFINDER WITH VG30E ENGINE

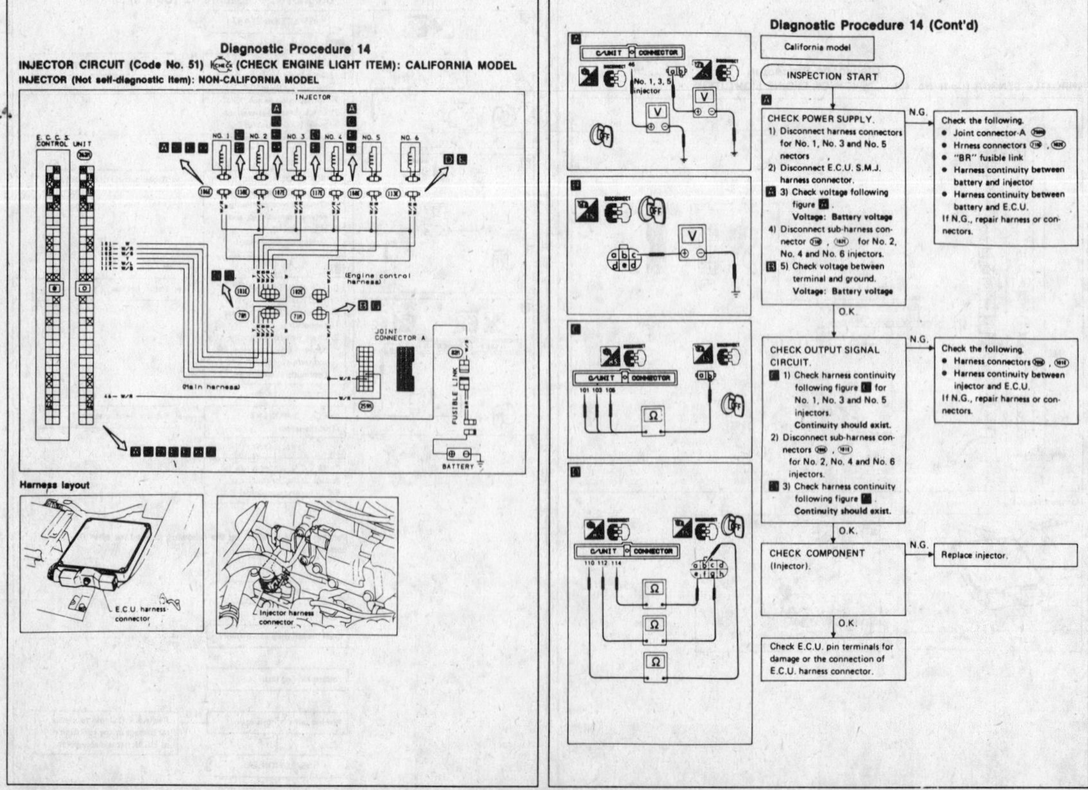
Diagnostic Procedure 14
INJECTOR CIRCUIT (Code No. 51) (CHECK ENGINE LIGHT ITEM): CALIFORNIA MODEL
INJECTOR (Not self-diagnostic Item): NON-CALIFORNIA MODEL
INJECTOR
E.C.C.S. CONTROL UNIT
NO. 1
NO. 2
NO. 3
NO. 4
NO. 5
NO. 6
(Engine control harness)
JOINT CONNECTOR A
(Main harness)
FUSIBLE LINK
BATTERY
Harness layout
E.C.U. harness connector
Injector harness connector
C/UNIT CONNECTOR
No. 1, 3, 5 injector
Diagnostic Procedure 14 (Cont'd)
California model
INSPECTION START
CHECK POWER SUPPLY.
1) Disconnect harness connectors for No. 1, No. 3 and No. 5 nectors
2) Disconnect E.C.U. S.M.J. harness connector.
3) Check voltage following figure.
Voltage: Battery voltage
4) Disconnect sub-harness connector for No. 2, No. 4 and No. 6 injectors.
5) Check voltage between terminal and ground.
Voltage: Battery voltage
N.G.
Check the following.
Joint connector-A
Hrness connectors
"BR" fusible link
Harness continuity between battery and injector
Harness continuity between battery and E.C.U.
If N.G., repair harness or connectors.
O.K.
CHECK OUTPUT SIGNAL CIRCUIT.
1) Check harness continuity following figure for No. 1, No. 3 and No. 5 injectors.
Continuity should exist.
2) Disconnect sub-harness connectors for No. 2, No. 4 and No. 6 injectors.
3) Check harness continuity following figure.
Continuity should exist.
N.G.
Check the following.
Harness connectors
Harness continuity between injector and E.C.U.
If N.G., repair harness or connectors.
O.K.
CHECK COMPONENT (Injector).
N.G.
Replace injector.
O.K.
Check E.C.U. pin terminals for damage or the connection of E.C.U. harness connector.

ECCS DIAGNOSTIC CHART—1990 TRUCK/PATHFINDER WITH VG30E ENGINE

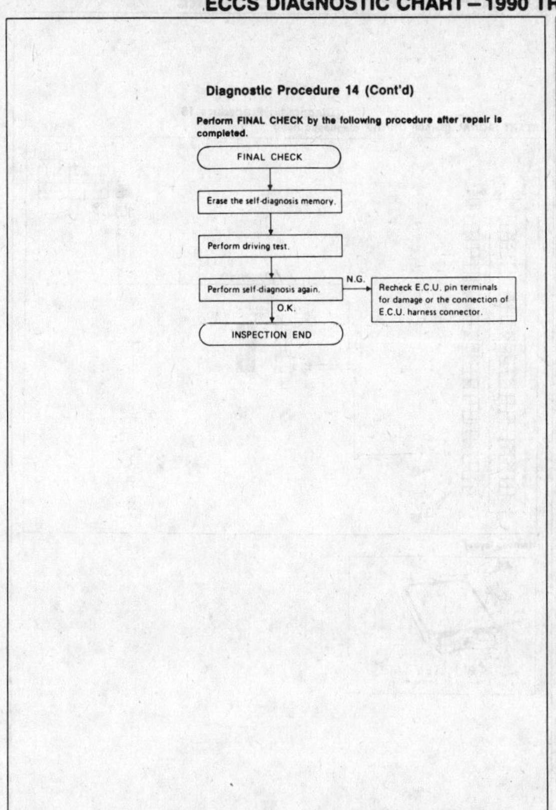
Diagnostic Procedure 14 (Cont'd)
Perform FINAL CHECK by the following procedure after repair is completed.
FINAL CHECK
Erase the self-diagnosis memory.
Perform driving test.
Perform self-diagnosis again.
N.G.
Recheck E.C.U. pin terminals for damage or the connection of E.C.U. harness connector.
O.K.
INSPECTION END

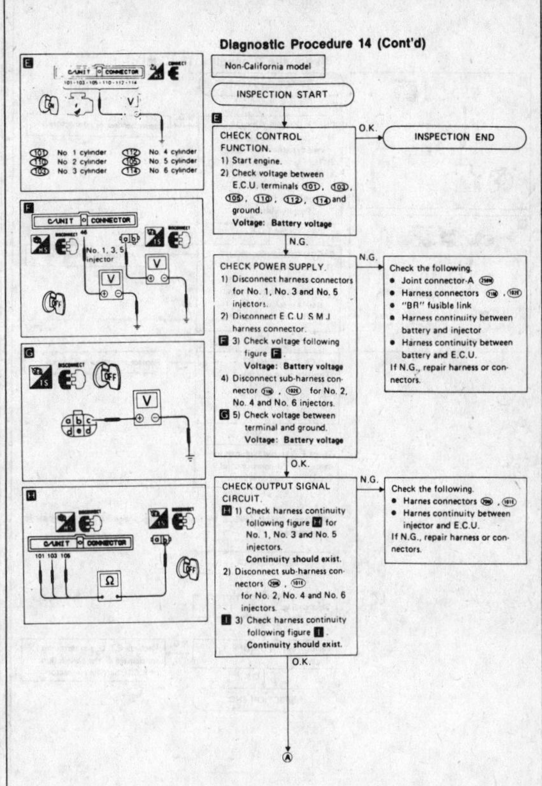
Diagnostic Procedure 14 (Cont'd)
Non-California model
INSPECTION START
CHECK CONTROL FUNCTION.
1) Start engine.
2) Check voltage between E.C.U. terminals 101, 103, 105, 110, 112, 114 and ground.
Voltage: Battery voltage
O.K.
INSPECTION END
N.G.
CHECK POWER SUPPLY.
1) Disconnect harness connectors for No. 1, No. 3 and No. 5 injectors.
2) Disconnect E.C.U. S.M.J. harness connector.
3) Check voltage following figure F.
Voltage: Battery voltage
4) Disconnect sub-harness connector, for No. 2, No. 4 and No. 6 injectors.
5) Check voltage between terminal and ground.
Voltage: Battery voltage
Check the following.
Joint connector-A
Harness connectors
"BR" fusible link
Harness continuity between battery and injector
Harness continuity between battery and E.C.U.
If N.G., repair harness or connectors.
CHECK OUTPUT SIGNAL CIRCUIT.
1) Check harness continuity following figure H for No. 1, No. 3 and No. 5 injectors.
Continuity should exist.
2) Disconnect sub-harness connectors, for No. 2, No. 4 and No. 6 injectors.
3) Check harness continuity following figure I.
Continuity should exist.
Check the following.
Harnes connectors
Harnes continuity between injector and E.C.U.
If N.G., repair harness or connectors.
No 1 cylinder
No 2 cylinder
No 3 cylinder
No 4 cylinder
No 5 cylinder
No 6 cylinder
No. 1, 3, 5 injector
C/UNIT CONNECTOR
DISCONNECT
OFF

ECCS DIAGNOSTIC CHART—1990 TRUCK/PATHFINDER WITH VG30E ENGINE

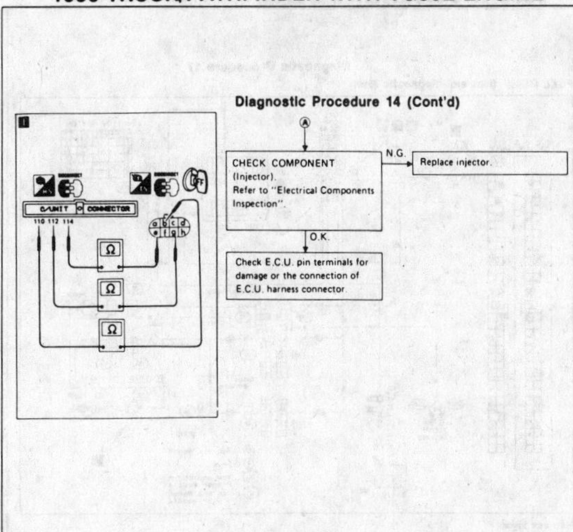
Diagnostic Procedure 14 (Cont'd)
CHECK COMPONENT (Injector). Refer to "Electrical Components Inspection".
N.G.
Replace injector.
O.K.
Check E.C.U. pin terminals for damage or the connection of E.C.U. harness connector.
C/UNIT CONNECTOR
110 112 114

ECCS SELF-DIAGNOSTIC CHART—1990 TRUCK/PATHFINDER WITH VG30E ENGINE

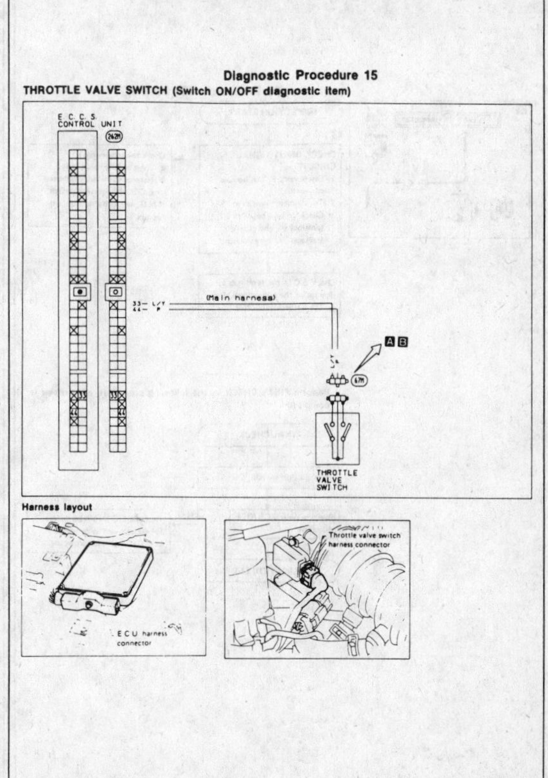
Diagnostic Procedure 15
THROTTLE VALVE SWITCH (Switch ON/OFF diagnostic item)
E.C.C.S. CONTROL UNIT
(Main harness)
THROTTLE VALVE SWITCH
Harness layout
E.C.U. harness connector
Throttle valve switch harness connector

ECCS SELF-DIAGNOSTIC CHART—1990 TRUCK/PATHFINDER WITH VG30E ENGINE

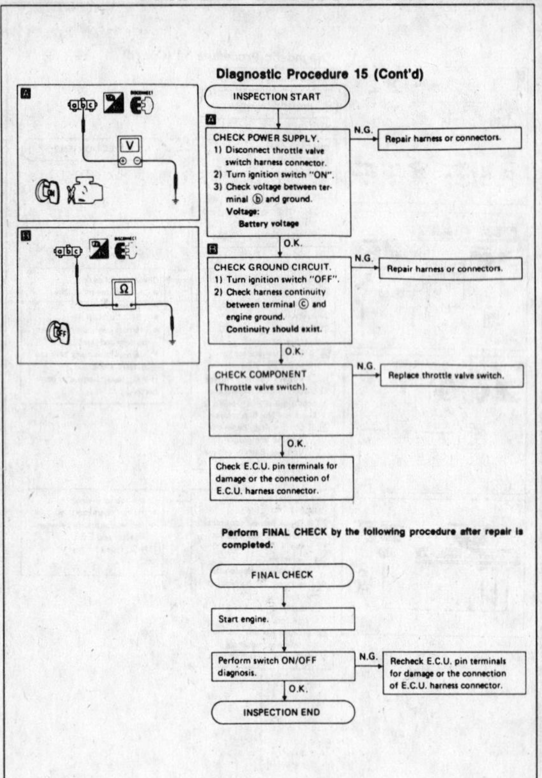
Diagnostic Procedure 15 (Cont'd)
INSPECTION START
CHECK POWER SUPPLY.
1) Disconnect throttle valve switch harness connector.
2) Turn ignition switch "ON".
3) Check voltage between terminal ⓑ and ground.
Voltage: Battery voltage
N.G.
Repair harness or connectors.
O.K.
CHECK GROUND CIRCUIT.
1) Turn ignition switch "OFF".
2) Check harness continuity between terminal ⓒ and engine ground.
Continuity should exist.
N.G.
Repair harness or connectors.
O.K.
CHECK COMPONENT
(Throttle valve switch).
N.G.
Replace throttle valve switch.
O.K.
Check E.C.U. pin terminals for damage or the connection of E.C.U. harness connector.
Perform FINAL CHECK by the following procedure after repair is completed.
FINAL CHECK
Start engine.
Perform switch ON/OFF diagnosis.
N.G.
Recheck E.C.U. pin terminals for damage or the connection of E.C.U. harness connector.
O.K.
INSPECTION END

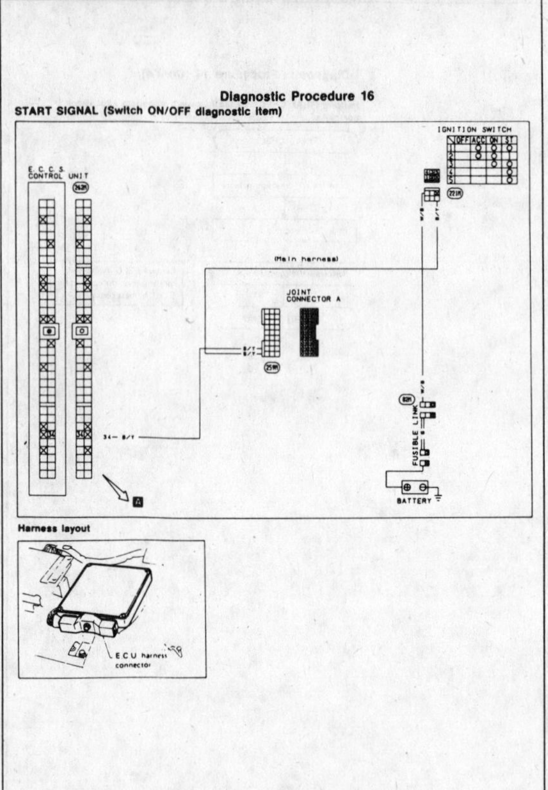
Diagnostic Procedure 16
START SIGNAL (Switch ON/OFF diagnostic item)
E.C.C.S. CONTROL UNIT
IGNITION SWITCH
Main harness
JOINT CONNECTOR A
FUSIBLE LINK
BATTERY
Harness layout
E.C.U. harness connector

ECCS SELF-DIAGNOSTIC CHART—1990 TRUCK/PATHFINDER WITH VG30E ENGINE

Diagnostic Procedure 16 (Cont'd)

INSPECTION START

CHECK INPUT SIGNAL CIRCUIT.
1) Disconnect E.C.U. harness connector.
2) Turn ignition switch to "ST".
3) Check voltage between E.C.U. terminal ⓗ and ground.
Voltage: Battery voltage

N.G. → Check the following.
- Joint connector-A
- Harness continuity between E.C.U. and ignition switch

If N.G., repair harness or connectors.

O.K.

Check E.C.U. pin terminals for damage or the connection of E.C.U. harness connector.

Perform FINAL CHECK by the following procedure after repair is completed.

FINAL CHECK

Turn ignition switch "ON".

Perform switch ON/OFF diagnosis.

N.G. → Recheck E.C.U. pin terminals for damage or the connection of E.C.U. harness connector.

O.K.

INSPECTION END

ECCS DIAGNOSTIC CHART—1990 TRUCK/PATHFINDER WITH VG30E ENGINE

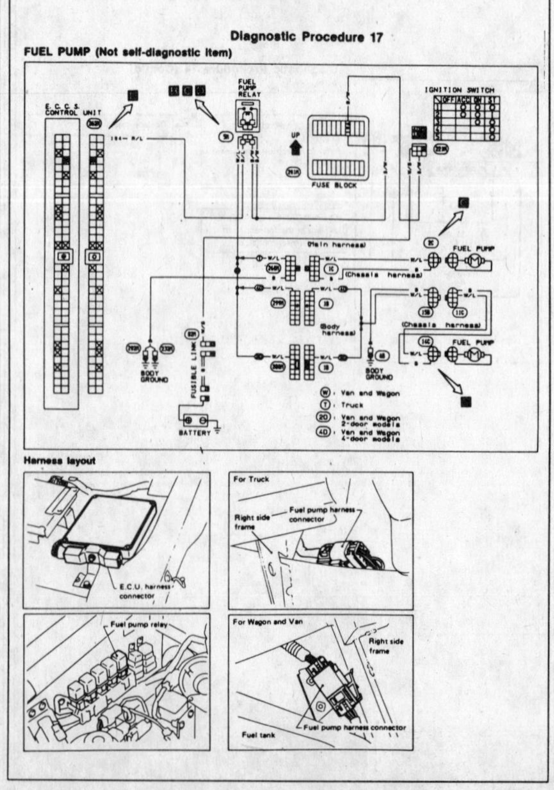
Diagnostic Procedure 17
FUEL PUMP (Not self-diagnostic item)
E.C.C.S. CONTROL UNIT
FUEL PUMP RELAY
IGNITION SWITCH
FUSE BLOCK
Main harness
Chassis harness
Body harness
FUEL PUMP
BODY GROUND
FUSIBLE LINK
BATTERY
Ⓦ: Van and Wagon
Ⓣ: Truck
2D: Van and Wagon 2-door models
4D: Van and Wagon 4-door models
Harness layout
E.C.U. harness connector
For Truck
Right side frame
Fuel pump harness connector
Fuel pump relay
For Wagon and Van
Right side frame
Fuel tank
Fuel pump harness connector

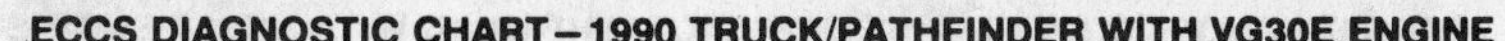

ECCS DIAGNOSTIC CHART—1990 TRUCK/PATHFINDER WITH VG30E ENGINE

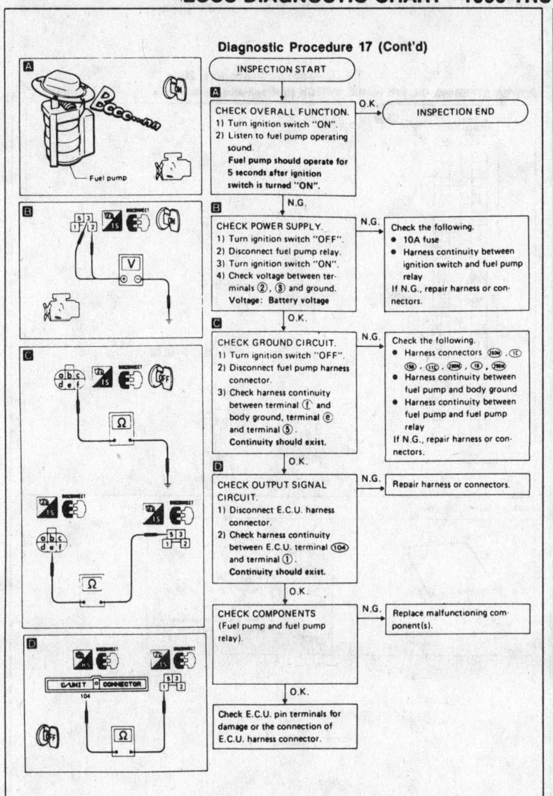
Diagnostic Procedure 17 (Cont'd)
INSPECTION START
A
CHECK OVERALL FUNCTION.
1) Turn ignition switch "ON".
2) Listen to fuel pump operating sound.
Fuel pump should operate for 5 seconds after ignition switch is turned "ON".
O.K.
INSPECTION END
N.G.
B
CHECK POWER SUPPLY.
1) Turn ignition switch "OFF".
2) Disconnect fuel pump relay.
3) Turn ignition switch "ON".
4) Check voltage between terminals ②, ⑤ and ground.
Voltage: Battery voltage
N.G.
Check the following.
• 10A fuse
• Harness continuity between ignition switch and fuel pump relay
If N.G., repair harness or connectors.
O.K.
C
CHECK GROUND CIRCUIT.
1) Turn ignition switch "OFF".
2) Disconnect fuel pump harness connector.
3) Check harness continuity between terminal ⓕ and body ground, terminal ⓔ and terminal ⑤.
Continuity should exist.
N.G.
Check the following.
• Harness connectors
• Harness continuity between fuel pump and body ground
• Harness continuity between fuel pump and fuel pump relay
If N.G., repair harness or connectors.
O.K.
D
CHECK OUTPUT SIGNAL CIRCUIT.
1) Disconnect E.C.U. harness connector.
2) Check harness continuity between E.C.U. terminal 104 and terminal ①.
Continuity should exist.
N.G.
Repair harness or connectors.
O.K.
CHECK COMPONENTS
(Fuel pump and fuel pump relay).
N.G.
Replace malfunctioning component(s).
O.K.
Check E.C.U. pin terminals for damage or the connection of E.C.U. harness connector.
Fuel pump
DISCONNECT
C/UNIT CONNECTOR
104

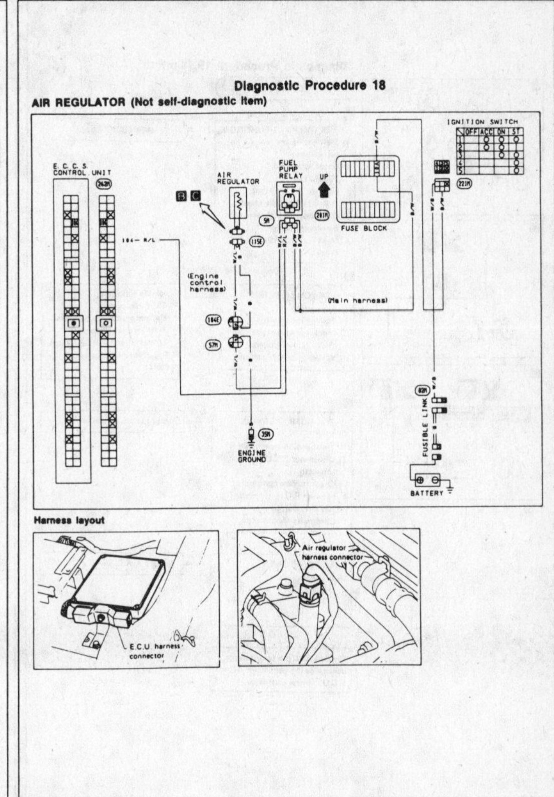
Diagnostic Procedure 18
AIR REGULATOR (Not self-diagnostic item)
E.C.C.S. CONTROL UNIT
AIR REGULATOR
FUEL PUMP RELAY
UP
FUSE BLOCK
IGNITION SWITCH
(Engine control harness)
(Main harness)
ENGINE GROUND
FUSIBLE LINK
BATTERY
Harness layout
E.C.U. harness connector
Air regulator harness connector

ECCS DIAGNOSTIC CHART—1990 TRUCK/PATHFINDER WITH VG30E ENGINE

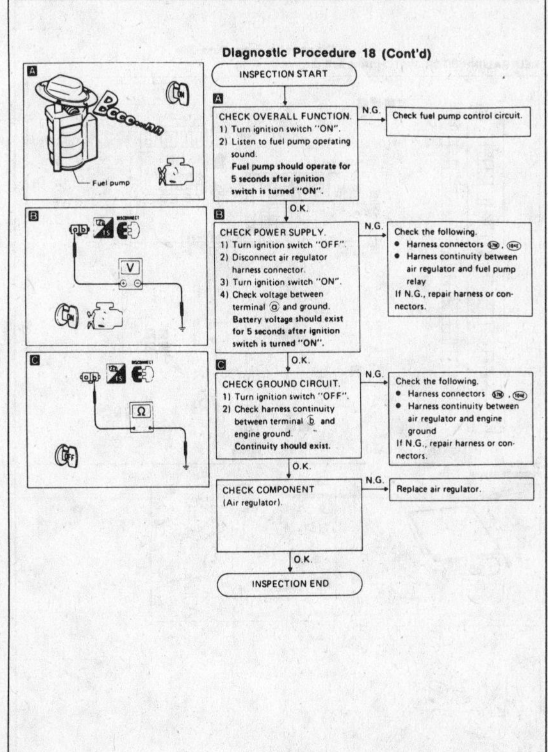
Diagnostic Procedure 18 (Cont'd)
INSPECTION START
A
CHECK OVERALL FUNCTION.
1) Turn ignition switch "ON".
2) Listen to fuel pump operating sound.
Fuel pump should operate for 5 seconds after ignition switch is turned "ON".
N.G.
Check fuel pump control circuit.
O.K.
B
CHECK POWER SUPPLY.
1) Turn ignition switch "OFF".
2) Disconnect air regulator harness connector.
3) Turn ignition switch "ON".
4) Check voltage between terminal ⓐ and ground.
Battery voltage should exist for 5 seconds after ignition switch is turned "ON".
N.G.
Check the following.
• Harness connectors
• Harness continuity between air regulator and fuel pump relay
If N.G., repair harness or connectors.
O.K.
C
CHECK GROUND CIRCUIT.
1) Turn ignition switch "OFF".
2) Check harness continuity between terminal ⓑ and engine ground.
Continuity should exist.
N.G.
Check the following.
• Harness connectors
• Harness continuity between air regulator and engine ground
If N.G., repair harness or connectors.
O.K.
CHECK COMPONENT
(Air regulator).
N.G.
Replace air regulator.
O.K.
INSPECTION END
Fuel pump
DISCONNECT

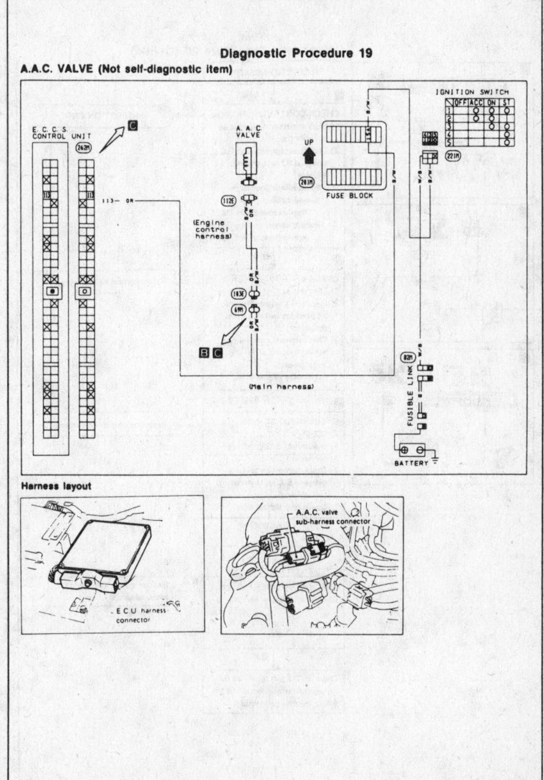
Diagnostic Procedure 19
A.A.C. VALVE (Not self-diagnostic item)
E.C.C.S. CONTROL UNIT
A.A.C. VALVE
UP
FUSE BLOCK
IGNITION SWITCH
(Engine control harness)
(Main harness)
FUSIBLE LINK
BATTERY
Harness layout
E.C.U harness connector
A.A.C. valve sub-harness connector

ECCS DIAGNOSTIC CHART—1990 TRUCK/PATHFINDER WITH VG30E ENGINE

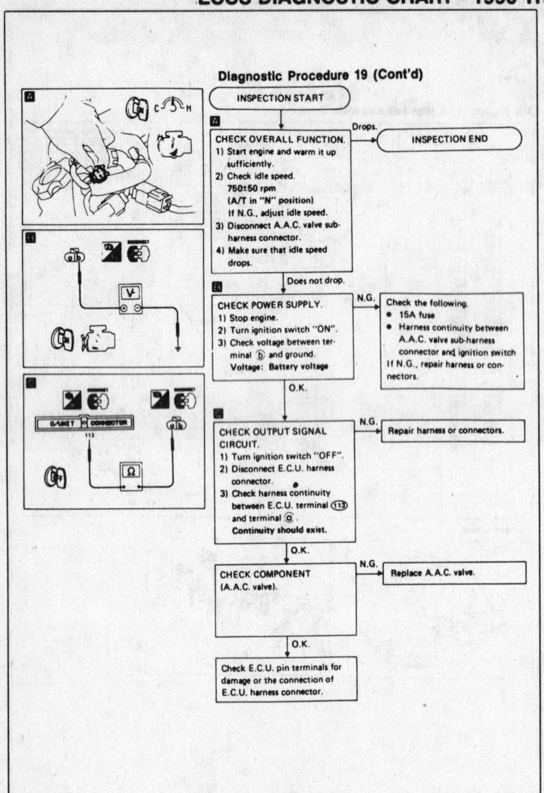
Diagnostic Procedure 19 (Cont'd)
INSPECTION START
CHECK OVERALL FUNCTION.
1) Start engine and warm it up sufficiently.
2) Check idle speed.
750±50 rpm
(A/T in "N" position)
If N.G., adjust idle speed.
3) Disconnect A.A.C. valve sub-harness connector.
4) Make sure that idle speed drops.
Drops.
INSPECTION END
Does not drop.
CHECK POWER SUPPLY.
1) Stop engine.
2) Turn ignition switch "ON".
3) Check voltage between terminal ⓑ and ground.
Voltage: Battery voltage
N.G.
Check the following.
• 15A fuse
• Harness continuity between A.A.C. valve sub-harness connector and ignition switch
If N.G., repair harness or connectors.
O.K.
CHECK OUTPUT SIGNAL CIRCUIT.
1) Turn ignition switch "OFF".
2) Disconnect E.C.U. harness connector.
3) Check harness continuity between E.C.U. terminal ⑪③ and terminal ⓒ.
Continuity should exist.
N.G.
Repair harness or connectors.
O.K.
CHECK COMPONENT (A.A.C. valve).
N.G.
Replace A.A.C. valve.
O.K.
Check E.C.U. pin terminals for damage or the connection of E.C.U. harness connector.
113

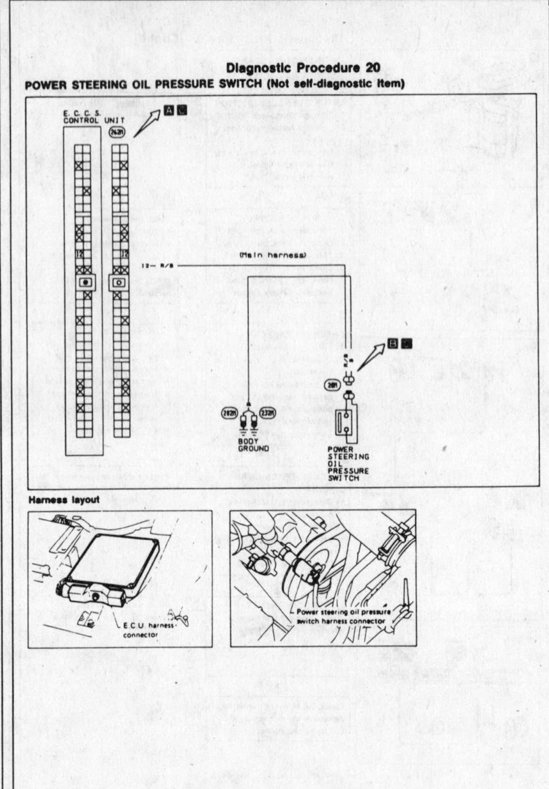
Diagnostic Procedure 20
POWER STEERING OIL PRESSURE SWITCH (Not self-diagnostic item)
E.C.C.S. CONTROL UNIT
(Main harness)
12 — R/B
BODY GROUND
POWER STEERING OIL PRESSURE SWITCH
Harness layout
E.C.U. harness connector
Power steering oil pressure switch harness connector

ECCS DIAGNOSTIC CHART—1990 TRUCK/PATHFINDER WITH VG30E ENGINE

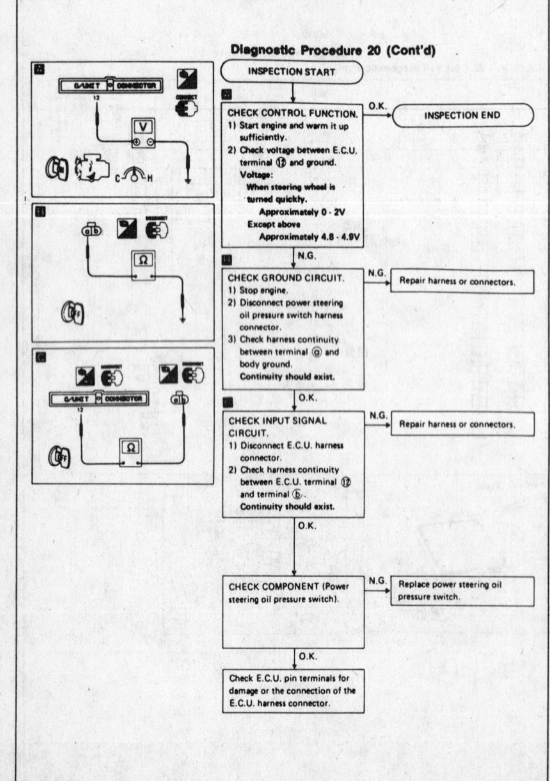
Diagnostic Procedure 20 (Cont'd)
INSPECTION START
CHECK CONTROL FUNCTION.
1) Start engine and warm it up sufficiently.
2) Check voltage between E.C.U. terminal ⑫ and ground.
Voltage:
When steering wheel is turned quickly.
Approximately 0 - 2V
Except above
Approximately 4.8 - 4.9V
O.K.
INSPECTION END
N.G.
CHECK GROUND CIRCUIT.
1) Stop engine.
2) Disconnect power steering oil pressure switch harness connector.
3) Check harness continuity between terminal ⓐ and body ground.
Continuity should exist.
N.G.
Repair harness or connectors.
O.K.
CHECK INPUT SIGNAL CIRCUIT.
1) Disconnect E.C.U. harness connector.
2) Check harness continuity between E.C.U. terminal ⑫ and terminal ⓑ.
Continuity should exist.
N.G.
Repair harness or connectors.
O.K.
CHECK COMPONENT (Power steering oil pressure switch).
N.G.
Replace power steering oil pressure switch.
O.K.
Check E.C.U. pin terminals for damage or the connection of the E.C.U. harness connector.

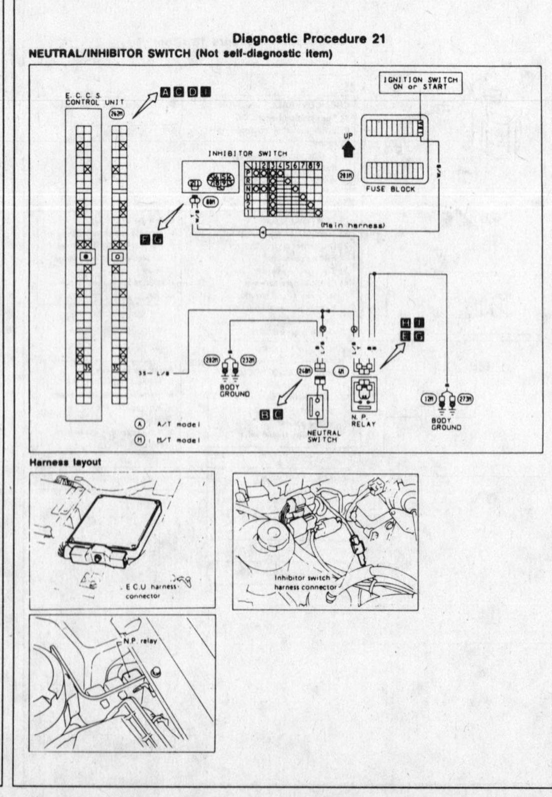
Diagnostic Procedure 21
NEUTRAL/INHIBITOR SWITCH (Not self-diagnostic item)
E.C.C.S. CONTROL UNIT
IGNITION SWITCH ON or START
INHIBITOR SWITCH
UP
FUSE BLOCK
(Main harness)
35 — L/B
BODY GROUND
NEUTRAL SWITCH
N.P. RELAY
Ⓐ A/T model
Ⓜ M/T model
Harness layout
E.C.U. harness connector
Inhibitor switch harness connector
N.P. relay

ECCS DIAGNOSTIC CHART—1990 TRUCK/PATHFINDER WITH VG30E ENGINE

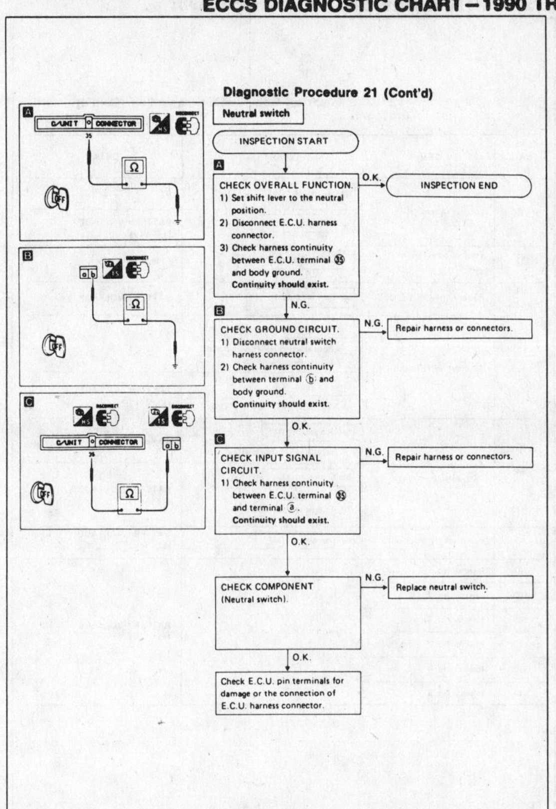

Diagnostic Procedure 21 (Cont'd)
Neutral switch
INSPECTION START
A
CHECK OVERALL FUNCTION.
1) Set shift lever to the neutral position.
2) Disconnect E.C.U. harness connector.
3) Check harness continuity between E.C.U. terminal 35 and body ground.
Continuity should exist.
O.K.
INSPECTION END
N.G.
B
CHECK GROUND CIRCUIT.
1) Disconnect neutral switch harness connector.
2) Check harness continuity between terminal b and body ground.
Continuity should exist.
N.G.
Repair harness or connectors.
O.K.
C
CHECK INPUT SIGNAL CIRCUIT.
1) Check harness continuity between E.C.U. terminal 35 and terminal a.
Continuity should exist.
N.G.
Repair harness or connectors.
O.K.
CHECK COMPONENT (Neutral switch).
N.G.
Replace neutral switch.
O.K.
Check E.C.U. pin terminals for damage or the connection of E.C.U. harness connector.
C/UNIT
CONNECTOR
35

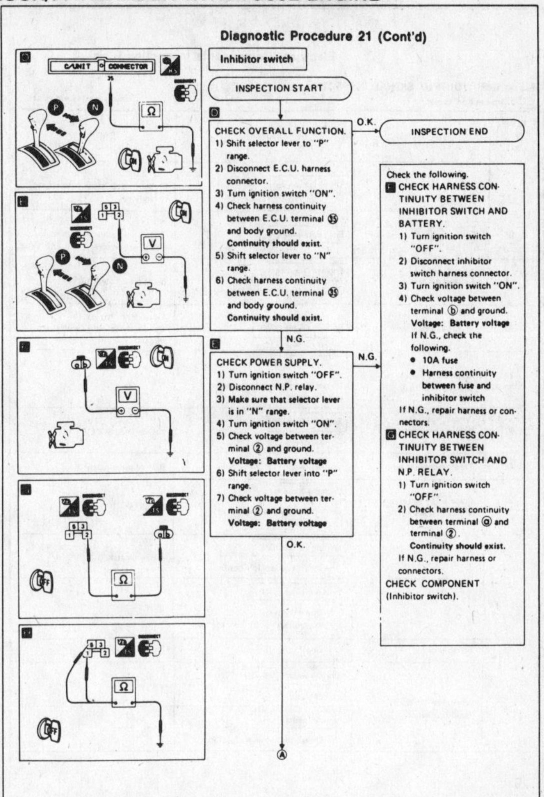

Diagnostic Procedure 21 (Cont'd)
Inhibitor switch
INSPECTION START
D
CHECK OVERALL FUNCTION.
1) Shift selector lever to "P" range.
2) Disconnect E.C.U. harness connector.
3) Turn ignition switch "ON".
4) Check harness continuity between E.C.U. terminal 35 and body ground.
Continuity should exist.
5) Shift selector lever to "N" range.
6) Check harness continuity between E.C.U. terminal 35 and body ground.
Continuity should exist.
O.K.
INSPECTION END
N.G.
E
CHECK POWER SUPPLY.
1) Turn ignition switch "OFF".
2) Disconnect N.P. relay.
3) Make sure that selector lever is in "N" range.
4) Turn ignition switch "ON".
5) Check voltage between terminal 2 and ground.
Voltage: Battery voltage
6) Shift selector lever into "P" range.
7) Check voltage between terminal 2 and ground.
Voltage: Battery voltage
N.G.
O.K.
A
Check the following.
F CHECK HARNESS CONTINUITY BETWEEN INHIBITOR SWITCH AND BATTERY.
1) Turn ignition switch "OFF".
2) Disconnect inhibitor switch harness connector.
3) Turn ignition switch "ON".
4) Check voltage between terminal b and ground.
Voltage: Battery voltage
If N.G., check the following.
• 10A fuse
• Harness continuity between fuse and inhibitor switch
If N.G., repair harness or connectors.
G CHECK HARNESS CONTINUITY BETWEEN INHIBITOR SWITCH AND N.P. RELAY.
1) Turn ignition switch "OFF".
2) Check harness continuity between terminal a and terminal 2.
Continuity should exist.
If N.G., repair harness or connectors.
CHECK COMPONENT (Inhibitor switch).
C/UNIT
CONNECTOR
35

ECCS DIAGNOSTIC CHART—1990 TRUCK/PATHFINDER WITH VG30E ENGINE

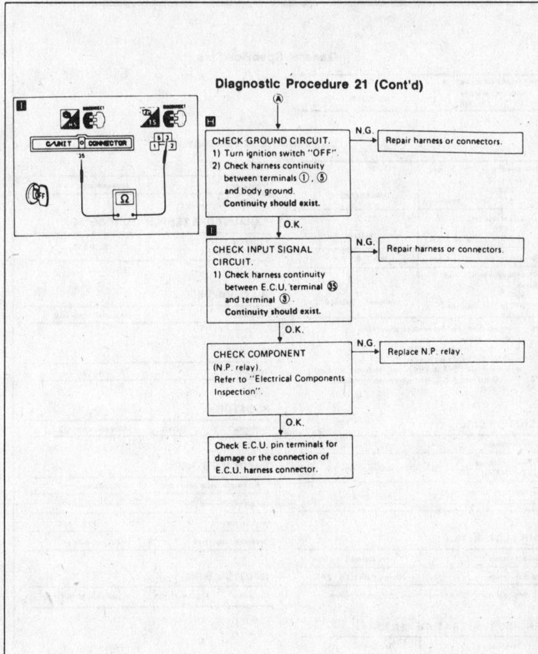

Diagnostic Procedure 21 (Cont'd)
A
H
CHECK GROUND CIRCUIT.
1) Turn ignition switch "OFF".
2) Check harness continuity between terminals 1, 5 and body ground.
Continuity should exist.
N.G.
Repair harness or connectors.
O.K.
I
CHECK INPUT SIGNAL CIRCUIT.
1) Check harness continuity between E.C.U. terminal 35 and terminal 3.
Continuity should exist.
N.G.
Repair harness or connectors.
O.K.
CHECK COMPONENT (N.P. relay).
Refer to "Electrical Components Inspection".
N.G.
Replace N.P. relay.
O.K.
Check E.C.U. pin terminals for damage or the connection of E.C.U. harness connector.
C/UNIT
CONNECTOR
35

ECU SIGNAL INSPECTION—1990 TRUCK/PATHFINDER WITH VG30E ENGINE

Electrical Components Inspection

E.C.U. INPUT/OUTPUT SIGNAL INSPECTION

E.C.U. Inspection table

*Data are reference values.

TERMINAL NO.	ITEM	CONDITION	*DATA
1	Ignition signal	Engine is running. — Idle speed	0.5 - 0.6V
		Engine is running. — Engine speed is 2,000 rpm.	1.2 - 1.3V
2	Tachometer	Engine is running. — Idle speed	Approximately 1.0V
		Engine is running. — Engine speed is 2,000 rpm.	2.7 - 2.9V
3	Ignition check	Engine is running. — Idle speed	9 - 12V
4	E.C.U. power source (Self-shutoff)	Engine is running. — Idle speed	0 - 1V
		Engine is not running. — For a few seconds after turning ignition switch "OFF"	BATTERY VOLTAGE (11 - 14V)
8	Exhaust gas temperature sensor (Only for California model)	Engine is running. — Idle speed	1.0 - 2.0V
		Engine is running. (Racing) — After warming up	0 - 1.0V
11	Air conditioner relay	Engine is ruuning. — Both A/C switch and blower switch are "ON".	0 - 1.0V
		Engine is running. — A/C switch is "OFF".	BATTERY VOLTAGE (11 - 14V)
12	Power steering oil pressure switch	Engine is running. — Steering wheel is being turned.	0 - 2.0V
		Engine is running. — Steering wheel is not being turned.	4.8 - 4.9V

ECU SIGNAL INSPECTION (CONT.)—1990 TRUCK/PATHFINDER WITH VG30E ENGINE

Electrical Components Inspection (Cont'd)

E.C.U. INPUT/OUTPUT SIGNAL INSPECTION

E.C.U. Inspection table

*Data are reference values.

TERMINAL NO.	ITEM	CONDITION	*DATA
16	Air flow meter	Engine is running.	1.0 - 3.0V Output voltage varies with engine revolution.
18	Engine temperature sensor	Engine is running.	1.0 - 3.0V Output voltage varies with engine water temperature.
19	Exhaust gas sensor	Engine is running. — After warming up sufficiently.	0 - Approximately 1.0V
20	Throttle sensor	Ignition switch "ON"	0.4 - Approximately 4V Output voltage varies with the throttle valve opening angle.
22 30	Crank angle sensor (Reference signal)	Engine is running. **Do not run engine at high speed under no-load.**	0.2 - 0.5V
27	Detonation sensor	Engine is running. — Idle speed	Approximately 2.5V
28	Throttle opening signal	Ignition switch "ON"	0.3 - Approximately 3V
31 40	Crank angle sensor (Position signal)	Engine is running. **Do not run engine at high speed under no-load.**	2.0 - 3.0V
33	Throttle valve switch (⊖ side)	Ignition switch "ON" — Throttle valve: Idle position	Approximately 8 - 10V
		Ignition switch "ON" — Throttle valve: Any position except idle position	0V
34	Start signal	Cranking	8 - 12V
35	Neutral switch & Inhibitor switch	Ignition switch "ON" — Neutral/Parking	0V
		Ignition switch "ON" — Except the above gear position	6 - 7V

Electrical Components Inspection (Cont'd)

*Data are reference values.

TERMINAL NO.	ITEM	CONDITION	*DATA
36	Ignition switch	Ignition switch "OFF"	0V
		Ignition switch "ON"	BATTERY VOLTAGE (11 - 14V)
37	Throttle sensor power supply	Ignition switch "ON"	Approximately 5V
38 47	Power supply for E.C.U.	Ignition switch "ON"	BATTERY VOLTAGE
41	Air conditioner switch	Engine is running. — Both air conditioner switch and blower switch are "ON".	0V
		Engine is running. — Air conditioner switch is "OFF".	BATTERY VOLTAGE (11 - 14V)
44	Throttle valve switch (⊕ side)	Ignition switch "ON" — Throttle valve: Idle position	Approximately 9 - 10V
		Ignition switch "ON" — Throttle valve: Except idle position	BATTERY VOLTAGE (11 - 14V)
46	Power supply (Back-up)	Ignition switch "OFF"	BATTERY VOLTAGE (11 - 14V)
101	Injector No. 1	Engine is running.	BATTERY VOLTAGE (11 - 14V)
103	Injector No. 3		
105	Injector No. 5		
110	Injector No. 2		
112	Injector No. 4		
114	Injector No. 6		

ECU SIGNAL INSPECTION (CONT.)—1990 TRUCK/PATHFINDER WITH VG30E ENGINE

Electrical Components Inspection (Cont'd)

*Data are reference values.

TERMINAL NO.	ITEM	CONDITION	*DATA
102	E.G.R.control solenoid valve	Engine is running. (Warm-up condition) — Idle speed	0.7 - 0.9V
		Engine is running. (Warm-up condition) — Engine speed is 2,000 rpm.	BATTERY VOLTAGE (11 - 14V)
		Engine is running. (Warm-up condition) — Engine speed is above 3,100 rpm. (A/T model) — Engine speed is above 2,600 rpm. (M/T model)	0.8 - 0.9V
104	Fuel pump relay	Ignition switch "ON" — For 5 seconds after turning ignition switch "ON"; Engine is running.	0.7 - 0.9V
		Ignition switch "ON" — Within 5 seconds after turning ignition switch "ON"	BATTERY VOLTAGE (11 - 14V)
113	A.A.C. valve	Engine is running. — Idle speed	7 - 10V
		Engine is running. — Steering wheel is being turned. — Air conditioner is operating. — Rear defogger is "ON". — Headlamps are in high position.	4 - 7V

E.C.U. HARNESS CONNECTOR TERMINAL LAYOUT

101	102	103	104	105	106	107	108	1	2	3	4	5	6	7	15	16	17	18	19	20	21	22	31	32	33	34	35	36	37	38	39
109	110	111	112	113	114	115	116	8	9	10	11	12	13	14	23	24	25	26	27	28	29	30	40	41	42	43	44	45	46	47	48

HS CONNECT

ECCS SPECIFICATIONS—1990 TRUCK/PATHFINDER WITH VG30E ENGINE

General Specifications

PRESSURE REGULATOR Regulated pressure kPa (kg/cm², psi)	299.1 (3.05, 43.4)

Inspection and Adjustment

Idle speed*1 rpm No-load*2 M/T A/T (in "N" position)	750±50 (700)*3
Air conditioner: ON M/T A/T (in "N" position)	800±50
Ignition timing degree	15°±2° B.T.D.C.
Throttle valve switch touch speed rpm M/T A/T (in "N" position)	Idle speed + 250±150*3

*1: Feedback controlled and needs no adjustments
*2: Under the following conditions:
- Air conditioner switch: OFF
- Steering wheel: Kept straight
- Electric load: OFF (Lights, heater, fan & rear defogger)

*3: ( ): Disconnect A.A.C. valve sub-harness connector.

IGNITION COIL

Primary voltage	V	12
Primary resistance [at 20°C (68°F)]	Ω	Approximately 1.0
Secondary resistance [at 20°C (68°F)]	kΩ	Approximately 10

AIR FLOW METER

Supply voltage	V	Battery voltage (11 - 14)
Output voltage	V	Approximately 1.5 - 2.0*

*: Engine is warmed up sufficiently and idling under no-load.

ENGINE TEMPERATURE SENSOR

Temperature °C (°F)	Resistance kΩ
20 (68)	2.1 - 2.9
50 (122)	0.68 - 1.00
80 (176)	0.30 - 0.33

FUEL PUMP

Resistance	Ω	Approximately 1.5

EXHAUST GAS TEMPERATURE SENSOR

Resistance [at 100°C (212°F)]	kΩ	85.3±8.53

A.A.C. VALVE

Resistance	Ω	Approximately 10.0

INJECTOR

Resistance	Ω	10 - 14

RESISTOR

Resistance	kΩ	Approximately 2.2

THROTTLE SENSOR

Accelerator pedal conditions	Resistance kΩ
Completely released	Approximately 1
Partially released	1 - 9
Completely depressed	Approximately 9

IGNITION WIRE

Resistance	kΩ/m (kΩ/ft)	Less than 30 (9.1)

ECCS COMPONENTS LOCATION—1990 TRUCK/PATHFINDER WITH KA24E ENGINE

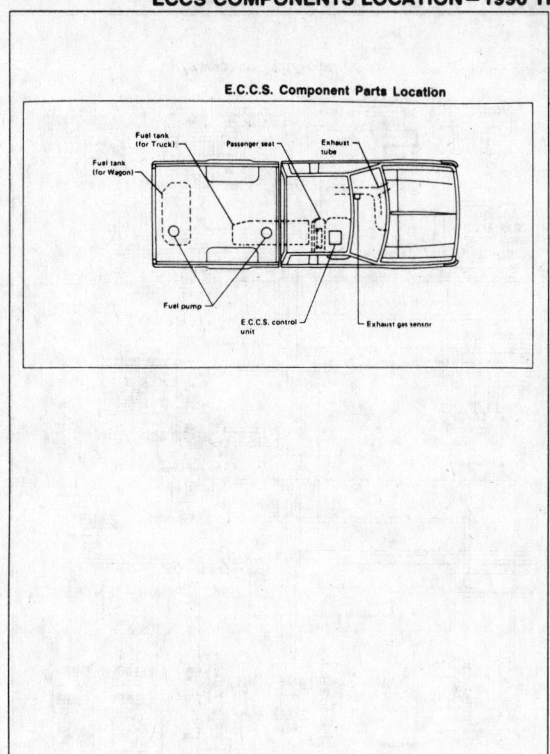
E.C.C.S. Component Parts Location
Fuel tank (for Truck)
Passenger seat
Exhaust tube
Fuel tank (for Wagon)
Fuel pump
E.C.C.S. control unit
Exhaust gas sensor

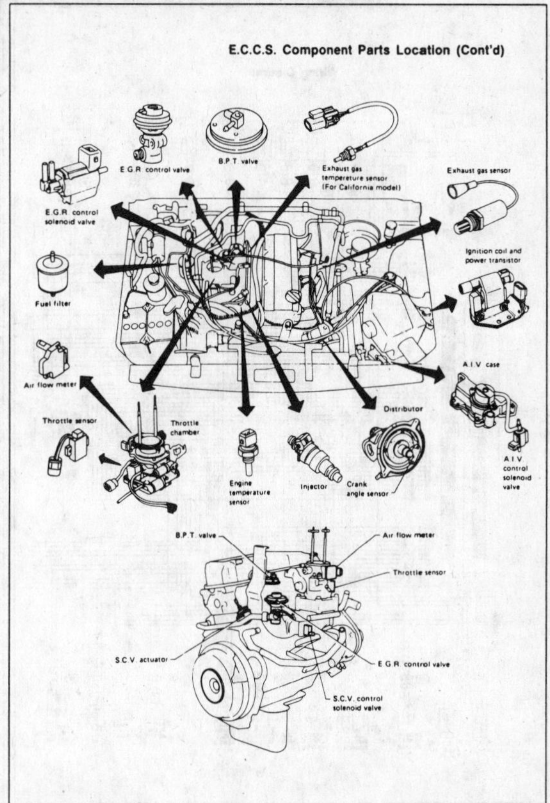
E.C.C.S. Component Parts Location (Cont'd)
E.G.R. control valve
B.P.T. valve
Exhaust gas temperature sensor (For California model)
Exhaust gas sensor
E.G.R. control solenoid valve
Ignition coil and power transistor
Fuel filter
A.I.V. case
Air flow meter
Throttle sensor
Throttle chamber
Distributor
Engine temperature sensor
Injector
Crank angle sensor
A.I.V. control solenoid valve
B.P.T. valve
Air flow meter
Throttle sensor
S.C.V. actuator
E.G.R. control valve
S.C.V. control solenoid valve

ECCS SCHEMATIC—1990 TRUCK/PATHFINDER WITH KA24E ENGINE

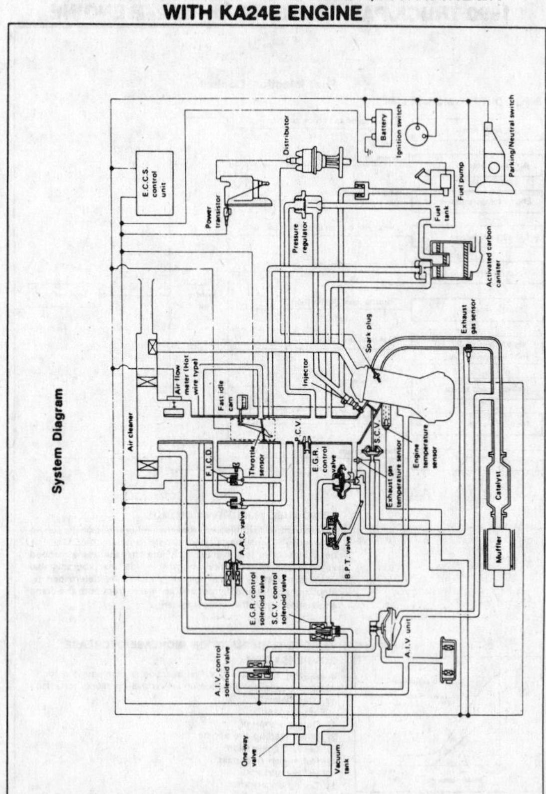
System Diagram
E.C.C.S. control unit
Power transistor
Distributor
Battery
Ignition switch
Parking/Neutral switch
Fuel pump
Fuel tank
Pressure regulator
Activated carbon canister
Exhaust gas sensor
Spark plug
Injector
Air flow meter (Hot wire type)
Fast idle cam
Air cleaner
F.I.C.D.
Throttle sensor
P.C.V.
E.G.R. control valve
S.C.V.
Exhaust gas temperature sensor
Engine temperature sensor
Catalyst
Muffler
A.A.C. valve
B.P.T. valve
E.G.R. control solenoid valve
S.C.V. control solenoid valve
A.I.V. control solenoid valve
A.I.V. unit
One-way valve
Vacuum tank

ECCS CONTROL SYSTEM CHART—1990 TRUCK/PATHFINDER WITH KA24E ENGINE

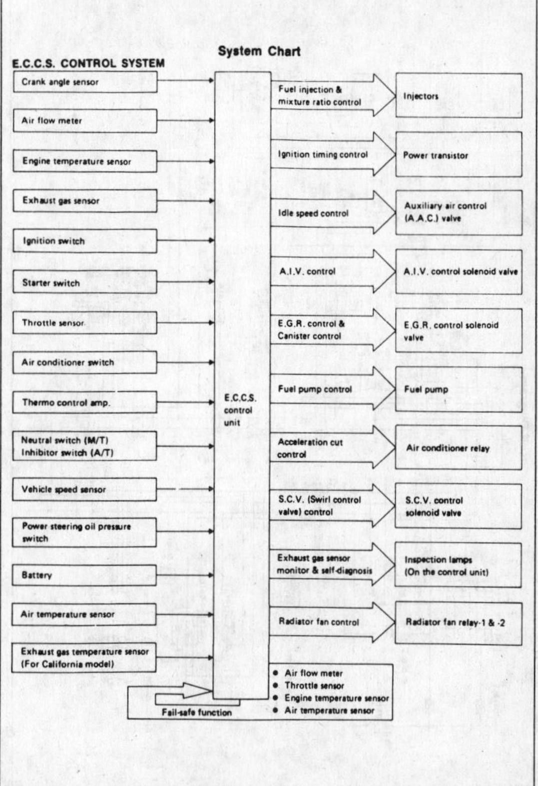
System Chart
E.C.C.S. CONTROL SYSTEM
Crank angle sensor
Air flow meter
Engine temperature sensor
Exhaust gas sensor
Ignition switch
Starter switch
Throttle sensor
Air conditioner switch
Thermo control amp.
Neutral switch (M/T) Inhibitor switch (A/T)
Vehicle speed sensor
Power steering oil pressure switch
Battery
Air temperature sensor
Exhaust gas temperature sensor (For California model)
E.C.C.S. control unit
Fuel injection & mixture ratio control
Ignition timing control
Idle speed control
A.I.V. control
E.G.R. control & Canister control
Fuel pump control
Acceleration cut control
S.C.V. (Swirl control valve) control
Exhaust gas sensor monitor & self-diagnosis
Radiator fan control
Injectors
Power transistor
Auxiliary air control (A.A.C.) valve
A.I.V. control solenoid valve
E.G.R. control solenoid valve
Fuel pump
Air conditioner relay
S.C.V. control solenoid valve
Inspection lamps (On the control unit)
Radiator fan relay-1 & -2
Air flow meter
Throttle sensor
Engine temperature sensor
Air temperature sensor
Fail-safe function

ECCS WIRING SCHEMATIC—1990 TRUCK/PATHFINDER WITH KA24E ENGINE

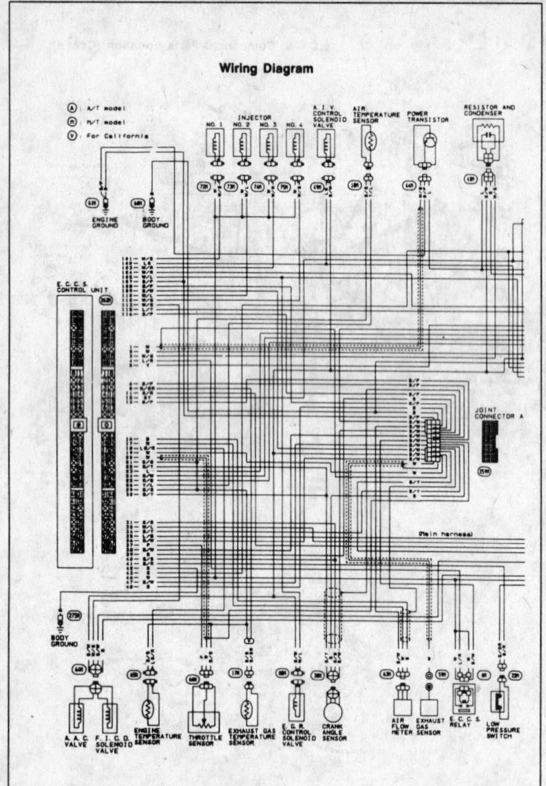
Wiring Diagram
Ⓐ : A/T model
Ⓜ : M/T model
Ⓥ : For California
INJECTOR
NO. 1
NO. 2
NO. 3
NO. 4
A.I.V. CONTROL SOLENOID VALVE
AIR TEMPERATURE SENSOR
POWER TRANSISTOR
RESISTOR AND CONDENSER
ENGINE GROUND
BODY GROUND
E.C.C.S. CONTROL UNIT
JOINT CONNECTOR A
(Main harness)
A.A.C. VALVE
F.I.C.D. SOLENOID VALVE
ENGINE TEMPERATURE SENSOR
THROTTLE SENSOR
EXHAUST GAS TEMPERATURE SENSOR
E.G.R. CONTROL SOLENOID VALVE
CRANK ANGLE SENSOR
AIR FLOW METER
EXHAUST GAS SENSOR
E.C.C.S. RELAY
LOW PRESSURE SWITCH

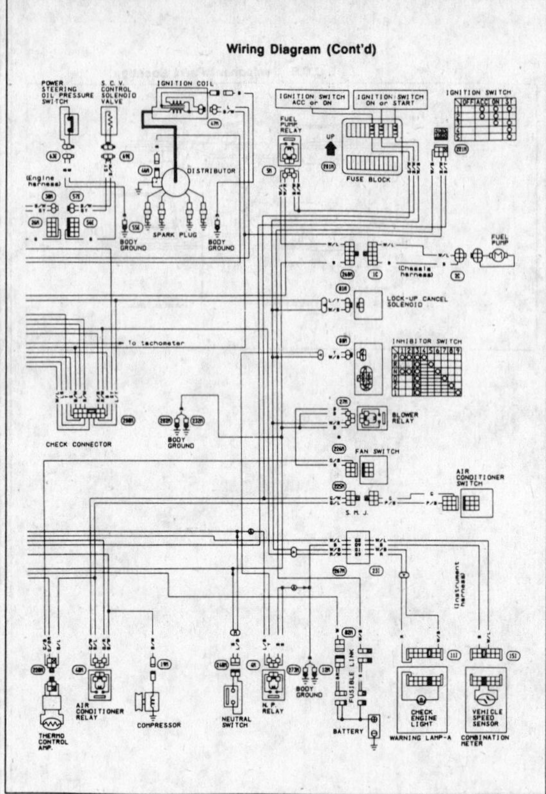
Wiring Diagram (Cont'd)
POWER STEERING OIL PRESSURE SWITCH
S.C.V. CONTROL SOLENOID VALVE
IGNITION COIL
IGNITION SWITCH ACC or ON
IGNITION SWITCH ON or START
IGNITION SWITCH
FUEL PUMP RELAY
DISTRIBUTOR
SPARK PLUG
FUSE BLOCK
BODY GROUND
FUEL PUMP
(Chassis harness)
LOCK-UP CANCEL SOLENOID
To tachometer
INHIBITOR SWITCH
BLOWER RELAY
FAN SWITCH
AIR CONDITIONER SWITCH
CHECK CONNECTOR
(Instrument harness)
THERMO CONTROL AMP.
AIR CONDITIONER RELAY
COMPRESSOR
NEUTRAL SWITCH
N.P. RELAY
FUSIBLE LINK
BATTERY
CHECK ENGINE LIGHT
WARNING LAMP-A
VEHICLE SPEED SENSOR
COMBINATION METER

ECCS CONTROL UNIT WIRING SCHEMATIC—1990 TRUCK/PATHFINDER WITH KA24E ENGINE

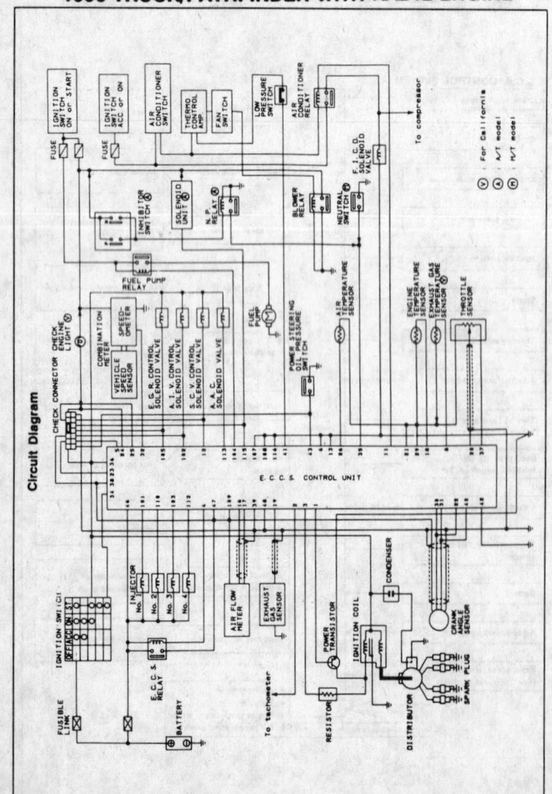
Circuit Diagram
IGNITION SWITCH ON or START
IGNITION SWITCH ACC or ON
AIR CONDITIONER SWITCH
THERMO CONTROL AMP.
FAN SWITCH
LOW PRESSURE SWITCH
AIR CONDITIONER RELAY
To compressor
Ⓥ : For California
Ⓐ : A/T model
Ⓜ : M/T model
FUSE
INHIBITOR SWITCH Ⓐ
SOLENOID UNIT Ⓐ
N.P. RELAY Ⓜ
BLOWER RELAY
NEUTRAL SWITCH Ⓜ
F.I.C.D. SOLENOID VALVE
FUEL PUMP RELAY
FUEL PUMP
POWER STEERING OIL PRESSURE SWITCH
AIR TEMPERATURE SENSOR
ENGINE TEMPERATURE SENSOR
EXHAUST GAS TEMPERATURE SENSOR
THROTTLE SENSOR
CHECK CONNECTOR
CHECK ENGINE LIGHT
COMBINATION METER
SPEEDOMETER
VEHICLE SPEED SENSOR
E.G.R. CONTROL SOLENOID VALVE
A.I.V. CONTROL SOLENOID VALVE
S.C.V. CONTROL SOLENOID VALVE
A.A.C. SOLENOID VALVE
E.C.C.S. CONTROL UNIT
INJECTOR
AIR FLOW METER
EXHAUST GAS SENSOR
POWER TRANSISTOR
IGNITION COIL
CONDENSER
CRANK ANGLE SENSOR
IGNITION SWITCH
E.C.C.S. RELAY
FUSIBLE LINK
BATTERY
To tachometer
RESISTOR
DISTRIBUTOR
SPARK PLUG

ECCS FUEL INJECTION CONTROL DESCRIPTION—1990 TRUCK/PATHFINDER WITH KA24E ENGINE

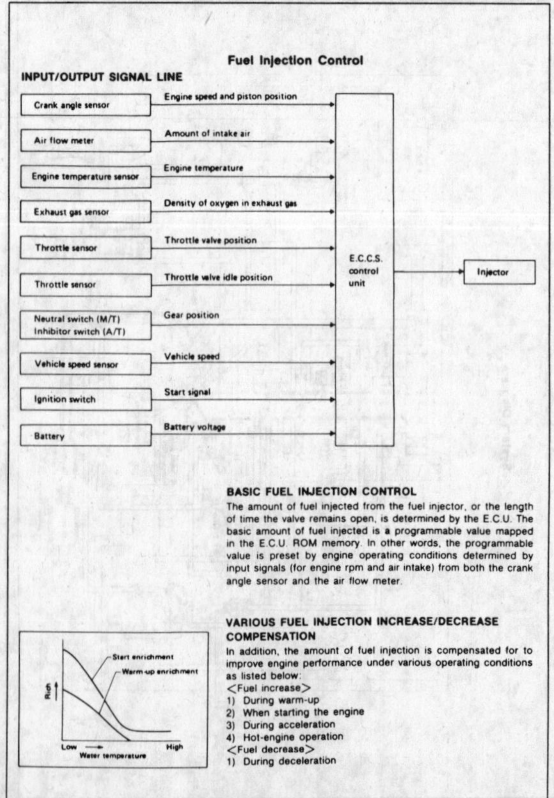
Fuel Injection Control
INPUT/OUTPUT SIGNAL LINE
Crank angle sensor
Engine speed and piston position
Air flow meter
Amount of intake air
Engine temperature sensor
Engine temperature
Exhaust gas sensor
Density of oxygen in exhaust gas
Throttle sensor
Throttle valve position
Throttle sensor
Throttle valve idle position
Neutral switch (M/T)
Inhibitor switch (A/T)
Gear position
Vehicle speed sensor
Vehicle speed
Ignition switch
Start signal
Battery
Battery voltage
E.C.C.S. control unit
Injector
Start enrichment
Warm-up enrichment
Rich
Low
High
Water temperature

BASIC FUEL INJECTION CONTROL

The amount of fuel injected from the fuel injector, or the length of time the valve remains open, is determined by the E.C.U. The basic amount of fuel injected is a programmable value mapped in the E.C.U. ROM memory. In other words, the programmable value is preset by engine operating conditions determined by input signals (for engine rpm and air intake) from both the crank angle sensor and the air flow meter.

VARIOUS FUEL INJECTION INCREASE/DECREASE COMPENSATION

In addition, the amount of fuel injection is compensated for to improve engine performance under various operating conditions as listed below:

<Fuel increase>
1) During warm-up
2) When starting the engine
3) During acceleration
4) Hot-engine operation

<Fuel decrease>
1) During deceleration

ECCS FUEL INJECTION CONTROL DESCRIPTION (CONT.)—1990 TRUCK/ PATHFINDER WITH KA24E ENGINE

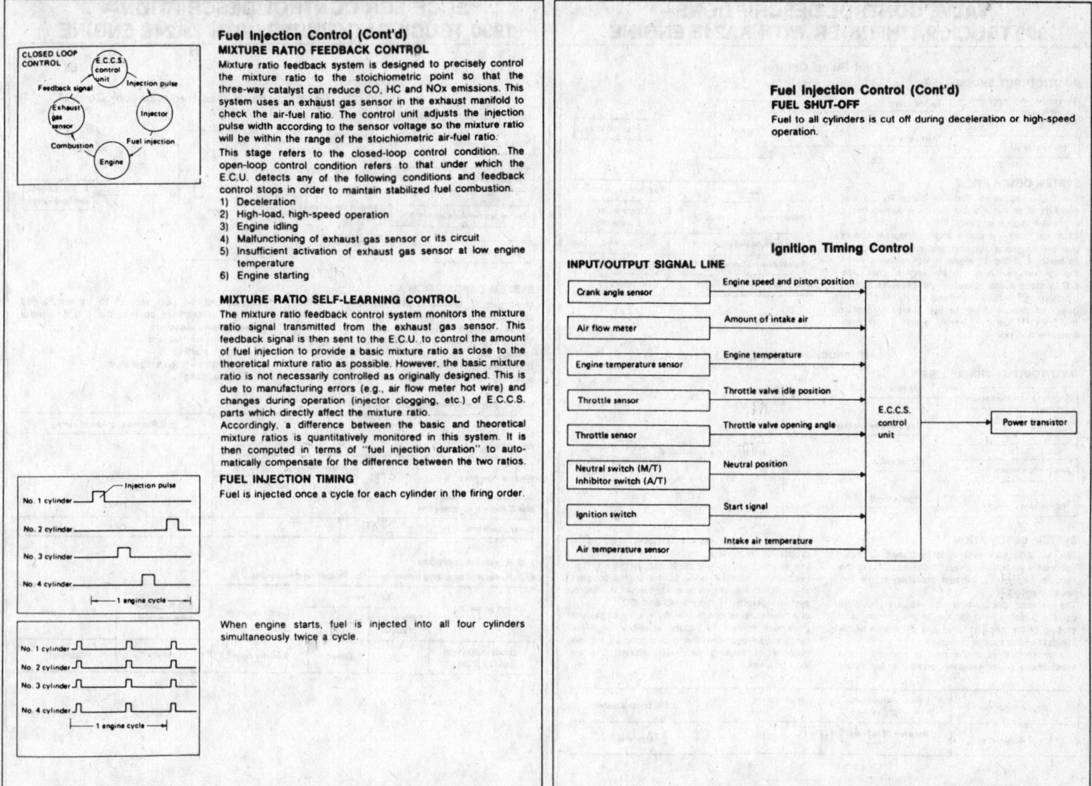
CLOSED LOOP CONTROL
E.C.C.S. control unit
Injection pulse
Feedback signal
Exhaust gas sensor
Injector
Combustion
Fuel injection
Engine
Injection pulse
No. 1 cylinder
No. 2 cylinder
No. 3 cylinder
No. 4 cylinder
1 engine cycle
INPUT/OUTPUT SIGNAL LINE
Crank angle sensor
Engine speed and piston position
Air flow meter
Amount of intake air
Engine temperature sensor
Engine temperature
Throttle sensor
Throttle valve idle position
Throttle sensor
Throttle valve opening angle
Neutral switch (M/T)
Inhibitor switch (A/T)
Neutral position
Ignition switch
Start signal
Air temperature sensor
Intake air temperature
E.C.C.S. control unit
Power transistor

Fuel Injection Control (Cont'd)

MIXTURE RATIO FEEDBACK CONTROL

Mixture ratio feedback system is designed to precisely control the mixture ratio to the stoichiometric point so that the three-way catalyst can reduce CO, HC and NOx emissions. This system uses an exhaust gas sensor in the exhaust manifold to check the air-fuel ratio. The control unit adjusts the injection pulse width according to the sensor voltage so the mixture ratio will be within the range of the stoichiometric air-fuel ratio.

This stage refers to the closed-loop control condition. The open-loop control condition refers to that under which the E.C.U. detects any of the following conditions and feedback control stops in order to maintain stabilized fuel combustion.

1) Deceleration
2) High-load, high-speed operation
3) Engine idling
4) Malfunctioning of exhaust gas sensor or its circuit
5) Insufficient activation of exhaust gas sensor at low engine temperature
6) Engine starting

MIXTURE RATIO SELF-LEARNING CONTROL

The mixture ratio feedback control system monitors the mixture ratio signal transmitted from the exhaust gas sensor. This feedback signal is then sent to the E.C.U. to control the amount of fuel injection to provide a basic mixture ratio as close to the theoretical mixture ratio as possible. However, the basic mixture ratio is not necessarily controlled as originally designed. This is due to manufacturing errors (e.g., air flow meter hot wire) and changes during operation (injector clogging, etc.) of E.C.C.S. parts which directly affect the mixture ratio.

Accordingly, a difference between the basic and theoretical mixture ratios is quantitatively monitored in this system. It is then computed in terms of "fuel injection duration" to automatically compensate for the difference between the two ratios.

FUEL INJECTION TIMING

Fuel is injected once a cycle for each cylinder in the firing order.

When engine starts, fuel is injected into all four cylinders simultaneously twice a cycle.

Fuel Injection Control (Cont'd)

FUEL SHUT-OFF

Fuel to all cylinders is cut off during deceleration or high-speed operation.

Ignition Timing Control

ECCS IGNITION TIMING CONTROL DESCRIPTION—1990 TRUCK/PATHFINDER WITH KA24E ENGINE

Ignition Timing Control (Cont'd)

SYSTEM DESCRIPTION

The ignition timing is controlled by the E.C.U. in order to maintain the best air-fuel ratio in response to every running condition of the engine. The ignition timing data is stored in the ROM located in the E.C.U., in the form of the map shown below.

The E.C.U. detects information such as the injection pulse width and crank angle sensor signal which varies every moment. Then responding to this information, ignition signals are transmitted to the power transistor.

e.g. N: 1,800 rpm, Tp: 1.50 msec
A °B.T.D.C.

In addition to this,

1 At starting
2 During warm-up
3 At idle
4 At low battery voltage
5 During swirl control valve operates
6 During hot engine operation
7 At acceleration
8 When intake air temperature is extremely high the ignition timing is revised by the E.C.U. according to the other data stored in the ROM.

Tp (msec)
1.75
1.50
1.25
1.00
0.75
Injection pulse width
A
800 1,000 1,400 1,800 2,200 N
Engine speed (rpm)

ECCS IDLE SPEED CONTROL DESCRIPTION—1990 TRUCK/PATHFINDER WITH KA24E ENGINE

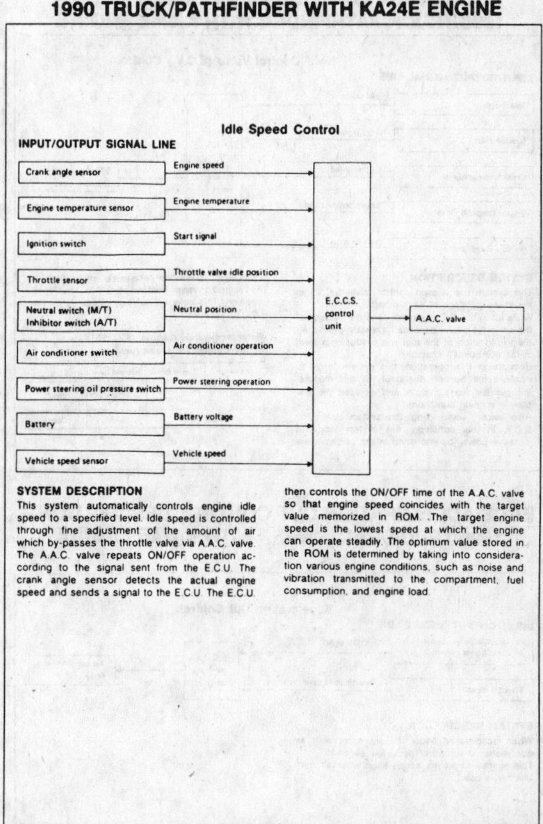
INPUT/OUTPUT SIGNAL LINE
Crank angle sensor
Engine speed
Engine temperature sensor
Engine temperature
Ignition switch
Start signal
Throttle sensor
Throttle valve idle position
Neutral switch (M/T)
Inhibitor switch (A/T)
Neutral position
Air conditioner switch
Air conditioner operation
Power steering oil pressure switch
Power steering operation
Battery
Battery voltage
Vehicle speed sensor
Vehicle speed
E.C.C.S. control unit
A.A.C. valve

Idle Speed Control

SYSTEM DESCRIPTION

This system automatically controls engine idle speed to a specified level. Idle speed is controlled through fine adjustment of the amount of air which by-passes the throttle valve via A.A.C. valve. The A.A.C. valve repeats ON/OFF operation according to the signal sent from the E.C.U. The crank angle sensor detects the actual engine speed and sends a signal to the E.C.U. The E.C.U. then controls the ON/OFF time of the A.A.C. valve so that engine speed coincides with the target value memorized in ROM. The target engine speed is the lowest speed at which the engine can operate steadily. The optimum value stored in the ROM is determined by taking into consideration various engine conditions, such as noise and vibration transmitted to the compartment, fuel consumption, and engine load.

ECCS FUEL PUMP AND AIR INDUCTION VALVE CONTROL DESCRIPTIONS—1990 TRUCK/PATHFINDER WITH KA24E ENGINE

Fuel Pump Control

INPUT/OUTPUT SIGNAL LINE

Crank angle sensor — Engine speed → E.C.C.S. control unit → Fuel pump relay
Ignition switch — Start signal → E.C.C.S. control unit

SYSTEM DESCRIPTION

The E.C.U. activates the fuel pump for several seconds after the ignition switch is turned on to improve engine startability. If the E.C.U. receives a 1° signal from the crank angle sensor, it knows that the engine is rotating, and causes the pump to perform. If the 1° signal is not received when the ignition switch is on, the engine stalls. The E.C.U. stops pump operation and prevents battery discharging, thereby improving safety. The E.C.U. does not directly drive the fuel pump. It controls the ON/OFF fuel pump relay, which in turn controls the fuel pump.

Condition	Fuel pump operation
Ignition switch is tunred to ON.	Operates for 5 seconds
Engine running and cranking	Operates
When engine is stopped	Stops in 1 second
Except as shown above	Stops

Air Induction Valve (A.I.V.) Control

INPUT/OUTPUT SIGNAL LINE

Engine temperature sensor — Engine temperature → E.C.C.S. control unit
Throttle sensor — Throttle valve idle position → E.C.C.S. control unit
Crank angle sensor — Engine speed → E.C.C.S. control unit
Vehicle speed sensor — Vehicle speed → E.C.C.S. control unit
E.C.C.S. control unit → A.I.V. control solenoid valve

SYSTEM DESCRIPTION

The air induction system is designed to send secondary air to the exhaust manifold, utilizing the vacuum caused by exhaust pulsation in the exhaust manifold.

The exhaust pressure in the exhaust manifold usually pulsates in response to the opening and closing of the exhaust valve and decreases below atmospheric pressure periodically.

If a secondary air intake pipe is opened to the atmosphere under vacuum conditions, secondary air can be drawn into the exhaust manifold in proportion to the vacuum.

The air induction valve is controlled by the E.C.C.S. control unit, corresponding to the engine temperature. When the engine is cold, the A.I.V. control system operates to reduce HC and CO.

In extremely cold conditions, A.I.V. control system does not operate to reduce after-burning. This system also operates during deceleration for the purpose of blowing off water around the air induction valve.

Engine condition	Water temperature °C (°F)	A.I.V. control solenoid valve	A.I.V. control system
Idle or deceleration	Between 28 (82) and 115 (239)	ON	Operates

ECCS EGR CONTROL DESCRIPTION—1990 TRUCK/PATHFINDER WITH KA24E ENGINE

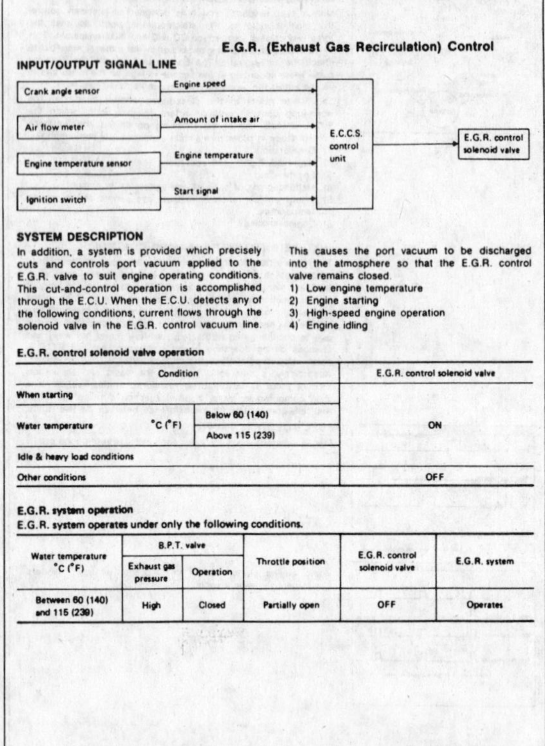

E.G.R. (Exhaust Gas Recirculation) Control

INPUT/OUTPUT SIGNAL LINE

Crank angle sensor
Engine speed
Air flow meter
Amount of intake air
Engine temperature sensor
Engine temperature
Ignition switch
Start signal
E.C.C.S. control unit
E.G.R. control solenoid valve

SYSTEM DESCRIPTION

In addition, a system is provided which precisely cuts and controls port vacuum applied to the E.G.R. valve to suit engine operating conditions. This cut-and-control operation is accomplished through the E.C.U. When the E.C.U. detects any of the following conditions, current flows through the solenoid valve in the E.G.R. control vacuum line. This causes the port vacuum to be discharged into the atmosphere so that the E.G.R. control valve remains closed.

1) Low engine temperature
2) Engine starting
3) High-speed engine operation
4) Engine idling

E.G.R. control solenoid valve operation

Condition			E.G.R. control solenoid valve
When starting			ON
Water temperature	°C (°F)	Below 60 (140)	ON
		Above 115 (239)	ON
Idle & heavy load conditions			ON
Other conditions			OFF

E.G.R. system operation

E.G.R. system operates under only the following conditions.

Water temperature °C (°F)	B.P.T. valve: Exhaust gas pressure	B.P.T. valve: Operation	Throttle position	E.G.R. control solenoid valve	E.G.R. system
Between 60 (140) and 115 (239)	High	Closed	Partially open	OFF	Operates

ECCS SWIRL CONTROL VALVE DESCRIPTION—1990 TRUCK/PATHFINDER WITH KA24E ENGINE

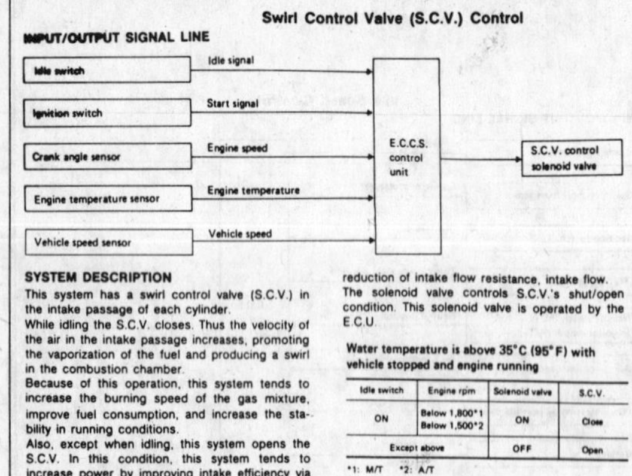

Swirl Control Valve (S.C.V.) Control

INPUT/OUTPUT SIGNAL LINE

Idle switch
Idle signal
Ignition switch
Start signal
Crank angle sensor
Engine speed
Engine temperature sensor
Engine temperature
Vehicle speed sensor
Vehicle speed
E.C.C.S. control unit
S.C.V. control solenoid valve

SYSTEM DESCRIPTION

This system has a swirl control valve (S.C.V.) in the intake passage of each cylinder.

While idling the S.C.V. closes. Thus the velocity of the air in the intake passage increases, promoting the vaporization of the fuel and producing a swirl in the combustion chamber.

Because of this operation, this system tends to increase the burning speed of the gas mixture, improve fuel consumption, and increase the stability in running conditions.

Also, except when idling, this system opens the S.C.V. In this condition, this system tends to increase power by improving intake efficiency via reduction of intake flow resistance, intake flow.

The solenoid valve controls S.C.V.'s shut/open condition. This solenoid valve is operated by the E.C.U.

Water temperature is above 35°C (95°F) with vehicle stopped and engine running

Idle switch	Engine rpm	Solenoid valve	S.C.V.
ON	Below 1,800*1 Below 1,500*2	ON	Close
Except above		OFF	Open

*1: M/T *2: A/T

ECCS ACCELERATION CUT CONTROL DESCRIPTION—1990 TRUCK/PATHFINDER WITH KA24E ENGINE

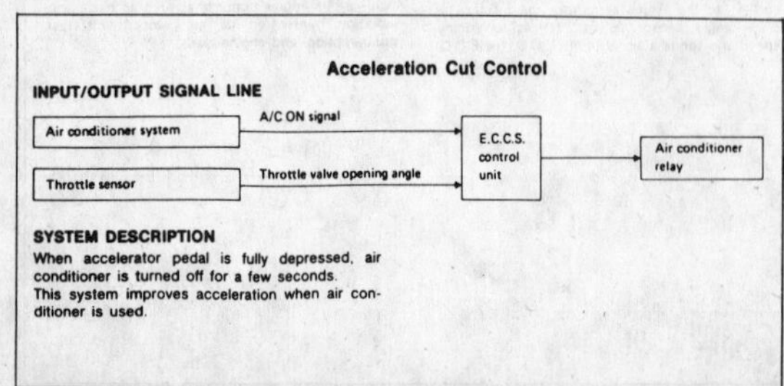

Acceleration Cut Control

INPUT/OUTPUT SIGNAL LINE

Air conditioner system
A/C ON signal
Throttle sensor
Throttle valve opening angle
E.C.C.S. control unit
Air conditioner relay

SYSTEM DESCRIPTION

When accelerator pedal is fully depressed, air conditioner is turned off for a few seconds.

This system improves acceleration when air conditioner is used.

ECCS IDLE SPEED, IGNITION TIMING AND IDLE MIXTURE RATIO INSPECTION—1990 TRUCK/PATHFINDER WITH KA24E ENGINE

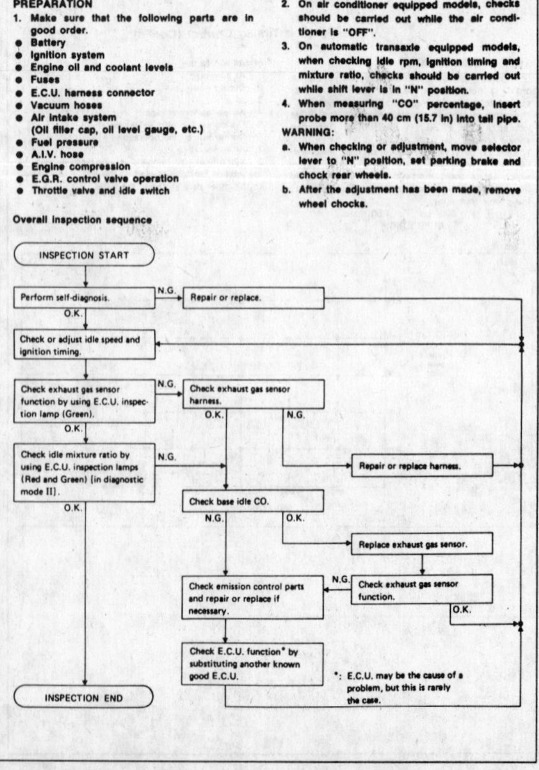

PREPARATION

1. Make sure that the following parts are in good order.
   - Battery
   - Ignition system
   - Engine oil and coolant levels
   - Fuses
   - E.C.U. harness connector
   - Vacuum hoses
   - Air intake system (Oil filler cap, oil level gauge, etc.)
   - Fuel pressure
   - A.I.V. hose
   - Engine compression
   - E.G.R. control valve operation
   - Throttle valve and idle switch
2. On air conditioner equipped models, checks should be carried out while the air conditioner is "OFF".
3. On automatic transaxle equipped models, when checking idle rpm, ignition timing and mixture ratio, checks should be carried out while shift lever is in "N" position.
4. When measuring "CO" percentage, insert probe more than 40 cm (15.7 in) into tail pipe.

WARNING:

a. When checking or adjustment, move selector lever to "N" position, set parking brake and chock rear wheels.
b. After the adjustment has been made, remove wheel chocks.

Overall inspection sequence

INSPECTION START
Perform self-diagnosis.
N.G.
Repair or replace.
O.K.
Check or adjust idle speed and ignition timing.
Check exhaust gas sensor function by using E.C.U. inspection lamp (Green).
N.G.
Check exhaust gas sensor harness.
O.K.
N.G.
Check idle mixture ratio by using E.C.U. inspection lamps (Red and Green) [in diagnostic mode II].
Repair or replace harness.
Check base idle CO.
Replace exhaust gas sensor.
Check emission control parts and repair or replace if necessary.
Check exhaust gas sensor function.
Check E.C.U. function* by substituting another known good E.C.U.
*: E.C.U. may be the cause of a problem, but this is rarely the case.
INSPECTION END

ECCS IDLE SPEED, IGNITION TIMING AND IDLE MIXTURE RATIO INSPECTION (CONT.)— 1990 TRUCK/PATHFINDER WITH KA24E ENGINE

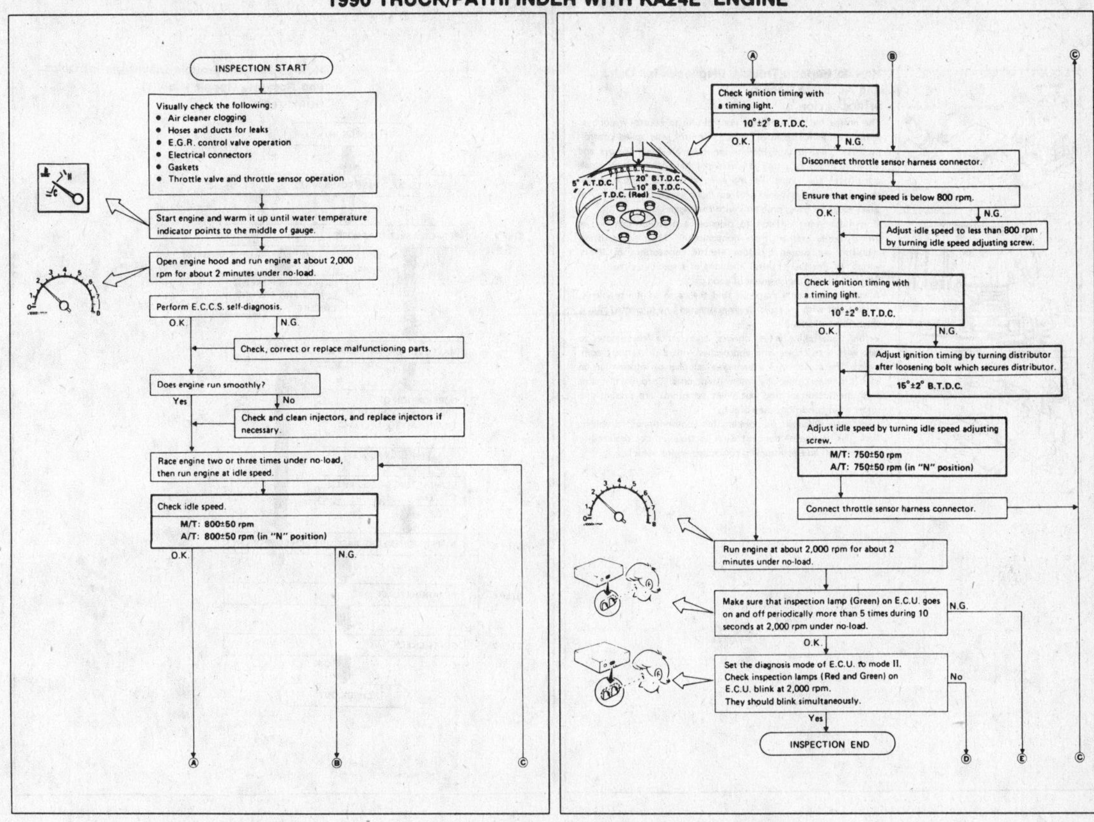
INSPECTION START
Visually check the following:
Air cleaner clogging
Hoses and ducts for leaks
E.G.R. control valve operation
Electrical connectors
Gaskets
Throttle valve and throttle sensor operation
Start engine and warm it up until water temperature indicator points to the middle of gauge.
Open engine hood and run engine at about 2,000 rpm for about 2 minutes under no-load.
Perform E.C.C.S. self-diagnosis.
O.K.
N.G.
Check, correct or replace malfunctioning parts.
Does engine run smoothly?
Yes
No
Check and clean injectors, and replace injectors if necessary.
Race engine two or three times under no-load, then run engine at idle speed.
Check idle speed.
M/T: 800±50 rpm
A/T: 800±50 rpm (in "N" position)
Check ignition timing with a timing light.
10°±2° B.T.D.C.
5° A.T.D.C.
20° B.T.D.C.
10° B.T.D.C.
T.D.C. (Red)
Disconnect throttle sensor harness connector.
Ensure that engine speed is below 800 rpm.
Adjust idle speed to less than 800 rpm by turning idle speed adjusting screw.
Adjust ignition timing by turning distributor after loosening bolt which secures distributor.
15°±2° B.T.D.C.
Adjust idle speed by turning idle speed adjusting screw.
M/T: 750±50 rpm
A/T: 750±50 rpm (in "N" position)
Connect throttle sensor harness connector.
Run engine at about 2,000 rpm for about 2 minutes under no-load.
Make sure that inspection lamp (Green) on E.C.U. goes on and off periodically more than 5 times during 10 seconds at 2,000 rpm under no-load.
Set the diagnosis mode of E.C.U. to mode II. Check inspection lamps (Red and Green) on E.C.U. blink at 2,000 rpm. They should blink simultaneously.
INSPECTION END

ECCS IDLE SPEED, IGNITION TIMING AND IDLE MIXTURE RATIO INSPECTION (CONT.)— 1990 TRUCK/PATHFINDER WITH KA24E ENGINE

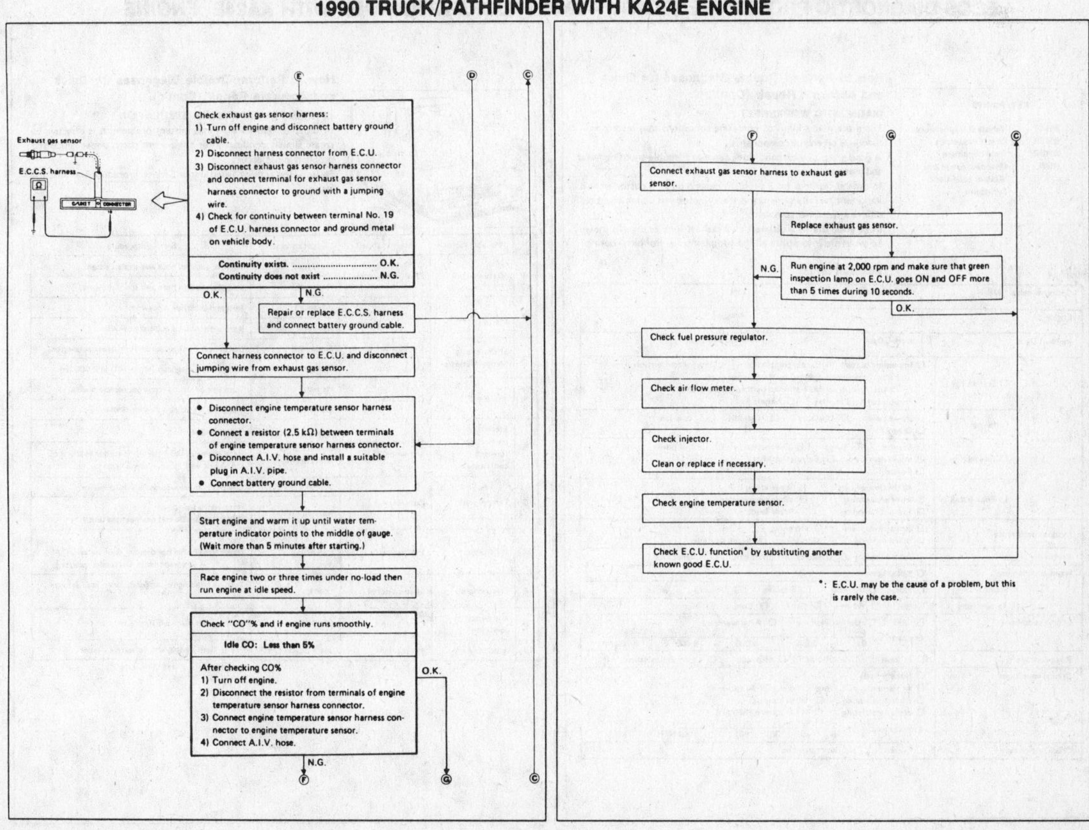
Check exhaust gas sensor harness:
1) Turn off engine and disconnect battery ground cable.
2) Disconnect harness connector from E.C.U.
3) Disconnect exhaust gas sensor harness connector and connect terminal for exhaust gas sensor harness connector to ground with a jumping wire.
4) Check for continuity between terminal No. 19 of E.C.U. harness connector and ground metal on vehicle body.
Continuity exists. .... O.K.
Continuity does not exist .... N.G.
Exhaust gas sensor
E.C.C.S. harness
Repair or replace E.C.C.S. harness and connect battery ground cable.
Connect harness connector to E.C.U. and disconnect jumping wire from exhaust gas sensor.
Disconnect engine temperature sensor harness connector.
Connect a resistor (2.5 kΩ) between terminals of engine temperature sensor harness connector.
Disconnect A.I.V. hose and install a suitable plug in A.I.V. pipe.
Connect battery ground cable.
Start engine and warm it up until water temperature indicator points to the middle of gauge. (Wait more than 5 minutes after starting.)
Race engine two or three times under no-load then run engine at idle speed.
Check "CO"% and if engine runs smoothly.
Idle CO: Less than 5%
After checking CO%
1) Turn off engine.
2) Disconnect the resistor from terminals of engine temperature sensor harness connector.
3) Connect engine temperature sensor harness connector to engine temperature sensor.
4) Connect A.I.V. hose.
Connect exhaust gas sensor harness to exhaust gas sensor.
Replace exhaust gas sensor.
Run engine at 2,000 rpm and make sure that green inspection lamp on E.C.U. goes ON and OFF more than 5 times during 10 seconds.
Check fuel pressure regulator.
Check air flow meter.
Check injector.
Clean or replace if necessary.
Check engine temperature sensor.
Check E.C.U. function* by substituting another known good E.C.U.
*: E.C.U. may be the cause of a problem, but this is rarely the case.

ECCS DIAGNOSTIC PROCEDURE—1990 TRUCK/PATHFINDER WITH KA24E ENGINE

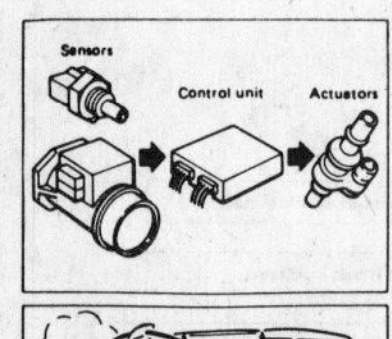
Sensors
Control unit
Actuators

INFO.
CAUSE

How to Perform Trouble Diagnoses for Quick and Accurate Repair

INTRODUCTION

The engine has an electronic control unit to control major systems such as fuel control, ignition control, idle speed control, etc. The control unit accepts input signals from sensors and instantly drives actuators. It is essential that both kinds of signals are proper and stable. At the same time, it is important that there are no conventional problems such as vacuum leaks, fouled spark plugs, or other problems with the engine.

It is much more difficult to diagnose a problem that occurs intermittently rather than continuously. Most intermittent problems are caused by poor electric connections or faulty wiring. In this case, careful checking of suspicious circuits may help prevent the replacement of good parts.

A visual check only may not find the cause of the problems. A road test with a circuit tester connected to a suspected circuit should be performed.

Before undertaking actual checks, take just a few minutes to talk with a customer who approaches with a driveability complaint. The customer is a very good supplier of information on such problems, especially intermittent ones. Through the talks with the customer, find out what symptoms are present and under what conditions they occur.

Start your diagnosis by looking for "conventional" problems first. This is one of the best ways to troubleshoot driveability problems on an electronically controlled engine vehicle.

How to Perform Trouble Diagnoses for Quick and Accurate Repair (Cont'd)

WORK FLOW

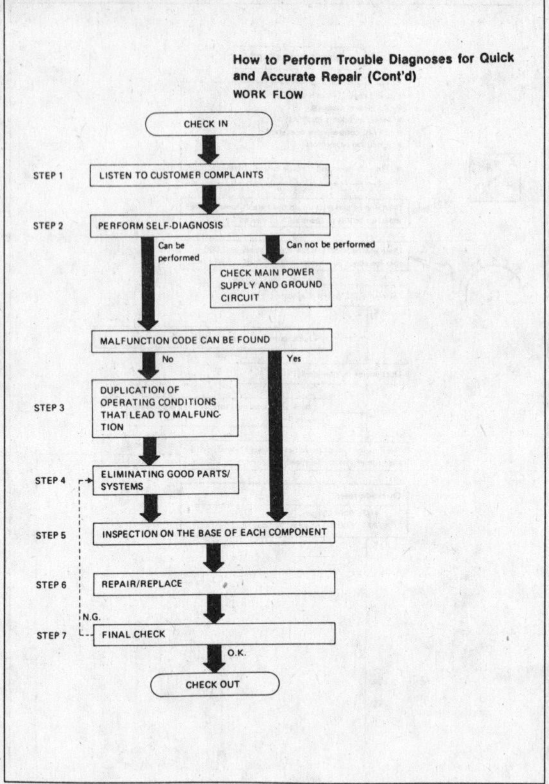
CHECK IN
STEP 1
LISTEN TO CUSTOMER COMPLAINTS
STEP 2
PERFORM SELF-DIAGNOSIS
Can be performed
Can not be performed
CHECK MAIN POWER SUPPLY AND GROUND CIRCUIT
MALFUNCTION CODE CAN BE FOUND
No
Yes
STEP 3
DUPLICATION OF OPERATING CONDITIONS THAT LEAD TO MALFUNCTION
STEP 4
ELIMINATING GOOD PARTS/ SYSTEMS
STEP 5
INSPECTION ON THE BASE OF EACH COMPONENT
STEP 6
REPAIR/REPLACE
N.G.
STEP 7
FINAL CHECK
O.K.
CHECK OUT

ECCS DIAGNOSTIC PROCEDURE (CONT.)—1990 TRUCK/PATHFINDER WITH KA24E ENGINE

KEY POINTS

WHAT ..... Vehicle & engine model
WHEN ..... Date, Frequencies
WHERE ..... Road conditions
HOW ..... Operating conditions, Weather conditions, Symptoms

How to Perform Trouble Diagnoses for Quick and Accurate Repair (Cont'd)

DIAGNOSTIC WORKSHEET

There are many kinds of operating conditions that lead to malfunctions on engine components.

A good grasp of such conditions can make trouble-shooting faster and more accurate.

In general, feelings for a problem depend on each customer. It is important to fully understand the symptoms or under what conditions a customer complains.

Make good use of a diagnostic worksheet such as the one shown below in order to utilize all the complaints for trouble-shooting.

Worksheet sample

Customer name MR/MS	Model & Year	VIN
Engine #	Trans.	Mileage
Incident Date	Manuf. Date	In Service Date

Symptoms	☐ Startability	☐ Impossible to start ☐ No combustion ☐ Partial combustion ☐ Partial combustion affected by throttle position ☐ Partial combustion NOT affected by throttle position ☐ Possible but hard to start ☐ Others [ ]
	☐ Idling	☐ No fast idle ☐ Unstable ☐ High idle ☐ Low idle ☐ Others [ ]
	☐ Driveability	☐ Stumble ☐ Surge ☐ Detonation ☐ Lack of power ☐ Intake backfire ☐ Exhaust backfire ☐ Others [ ]
	☐ Engine stall	☐ At the time of start ☐ While idling ☐ While accelerating ☐ While decelerating ☐ Just after stopping ☐ While loading
Incident occurrence		☐ Just after delivery ☐ Recently ☐ In the morning ☐ At night ☐ In the daytime
Frequency		☐ All the time ☐ Under certain conditions ☐ Sometimes
Weather conditions		☐ Not effected
	Weather	☐ Fine ☐ Raining ☐ Snowing ☐ Others [ ]
	Temperature	☐ Hot ☐ Warm ☐ Cool ☐ Cold ☐ Humid °F
Engine conditions		☐ Cold ☐ During warm-up ☐ After warm-up Engine speed 0 2,000 4,000 6,000 8,000 rpm
Road conditions		☐ In town ☐ In suburbs ☐ Highway ☐ Off road (up/down)
Driving conditions		☐ Not affected ☐ At starting ☐ While idling ☐ At racing ☐ While accelerating ☐ While cruising ☐ While decelerating ☐ While turning (RH/LH) Vehicle speed 0 10 20 30 40 50 60 MPH
Check engine light		☐ Turned on ☐ Not turned on

GRRRRR
GRRRR
THIS is the symptom, isn't it ?

How to Perform Trouble Diagnoses for Quick and Accurate Repair (Cont'd)

INTERMITTENT PROBLEM SIMULATION

In order to duplicate an intermittent problem, it is effective to create similar conditions for component parts, under which the problem might occur.

Perform the activity listed under Service procedure and note the result.

	Variable factor	Influential part	Target condition	Service procedure
1	Mixture ratio	Pressure regulator	Made lean	Remove vacuum hose and apply vacuum.
			Made rich	Remove vacuum hose and apply pressure.
2	Ignition timing	Distributor	Advanced	Rotate distributor clockwise.
			Retarded	Rotate distributor counterclockwise.
3	Mixture ratio feedback control	Exhaust gas sensor	Suspended	Disconnect exhaust gas sensor harness connector.
		Control unit	Operation check	Perform self-diagnosis (Mode I/II) at 2,000 rpm.
4	Idle speed	I.A.A. unit	Raised	Turn idle adjusting screw counterclockwise.
			Lowered	Turn idle adjusting screw clockwise.
5	Electric connection (Electric continuity)	Harness connectors and wires	Poor electric connection or faulty wiring	Tap or wiggle.
				Race engine rapidly. See if the torque reaction of the engine unit causes electric breaks.
6	Temperature	Control unit	Cooled	Cool with an icing spray or similar device.
			Warmed	Heat with a hair drier. [WARNING: Do not overheat the unit.]
7	Moisture	Electric parts	Damp	Wet. [WARNING: Do not directly pour water on components. Use a mist sprayer.]
8	Electric loads	Load switches	Loaded	Turn on head lights, air conditioner, rear defogger, etc.
9	Idle switch condition	Control unit	ON-OFF switching	Perform self-diagnosis (Mode IV).
10	Ignition spark	Timing light	Spark power check	Try to flash timing light for each cylinder.

ECCS DIAGNOSTIC CHART—1990 TRUCK/PATHFINDER WITH KA24E ENGINE

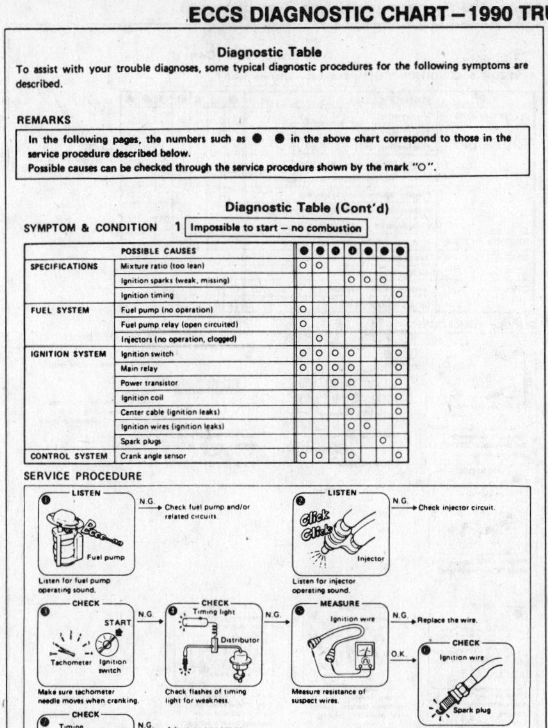

Diagnostic Table

To assist with your trouble diagnoses, some typical diagnostic procedures for the following symptoms are described.

REMARKS

In the following pages, the numbers such as ● ● in the above chart correspond to those in the service procedure described below.
Possible causes can be checked through the service procedure shown by the mark "O".

Diagnostic Table (Cont'd)

SYMPTOM & CONDITION 1 Impossible to start – no combustion

	POSSIBLE CAUSES	1	2	3	4	5	6	7
SPECIFICATIONS	Mixture ratio (too lean)	O	O					
	Ignition sparks (weak, missing)				O	O	O	
	Ignition timing							O
FUEL SYSTEM	Fuel pump (no operation)	O						
	Fuel pump relay (open circuited)	O						
	Injectors (no operation, clogged)		O					
IGNITION SYSTEM	Ignition switch	O	O	O	O			O
	Main relay	O	O	O	O			O
	Power transistor			O	O			O
	Ignition coil				O			O
	Center cable (ignition leaks)				O			O
	Ignition wires (ignition leaks)				O	O		
	Spark plugs						O	
CONTROL SYSTEM	Crank angle sensor	O	O		O			O

SERVICE PROCEDURE

LISTEN
Fuel pump
N.G.
Check fuel pump and/or related circuits.
Listen for fuel pump operating sound.
LISTEN
Click Click
Injector
N.G.
Check injector circuit.
Listen for injector operating sound.
CHECK
START
Tachometer
Ignition switch
N.G.
Make sure tachometer needle moves when cranking.
CHECK
Timing light
Distributor
N.G.
Check flashes of timing light for weakness.
MEASURE
Ignition wire
N.G.
Replace the wire.
O.K.
Measure resistance of suspect wires.
CHECK
Ignition wire
Spark plug
Remove spark plugs and check their ignition sparks.
CHECK
Timing light
N.G.
Adjust ignition timing.
Check ignition timing.

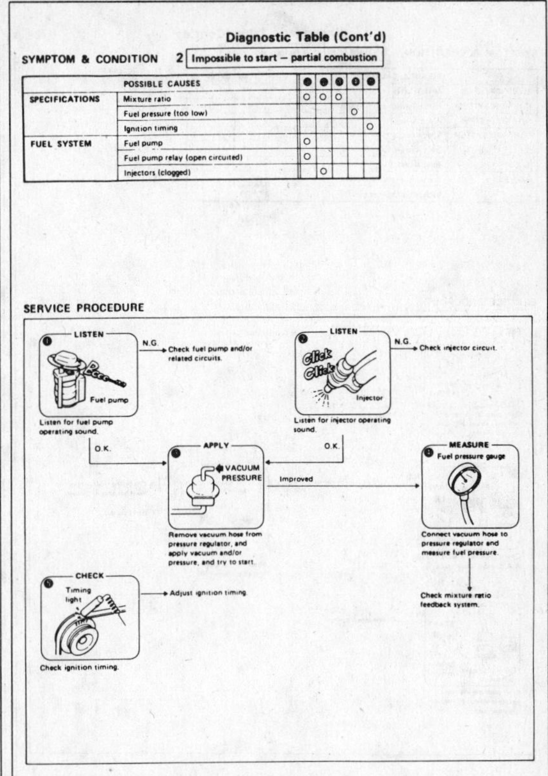

Diagnostic Table (Cont'd)

SYMPTOM & CONDITION 2 Impossible to start – partial combustion

	POSSIBLE CAUSES	1	2	3	4	5
SPECIFICATIONS	Mixture ratio	O	O	O		
	Fuel pressure (too low)				O	
	Ignition timing					O
FUEL SYSTEM	Fuel pump	O				
	Fuel pump relay (open circuited)	O				
	Injectors (clogged)		O			

SERVICE PROCEDURE

LISTEN
Fuel pump
N.G.
Check fuel pump and/or related circuits.
Listen for fuel pump operating sound.
O.K.
LISTEN
Click Click
Injector
N.G.
Check injector circuit.
Listen for injector operating sound.
O.K.
APPLY
VACUUM PRESSURE
Improved
Remove vacuum hose from pressure regulator, and apply vacuum and/or pressure, and try to start.
MEASURE
Fuel pressure gauge
Connect vacuum hose to pressure regulator and measure fuel pressure.
Check mixture ratio feedback system.
CHECK
Timing light
Adjust ignition timing.
Check ignition timing.

ECCS DIAGNOSTIC CHART—1990 TRUCK/PATHFINDER WITH KA24E ENGINE

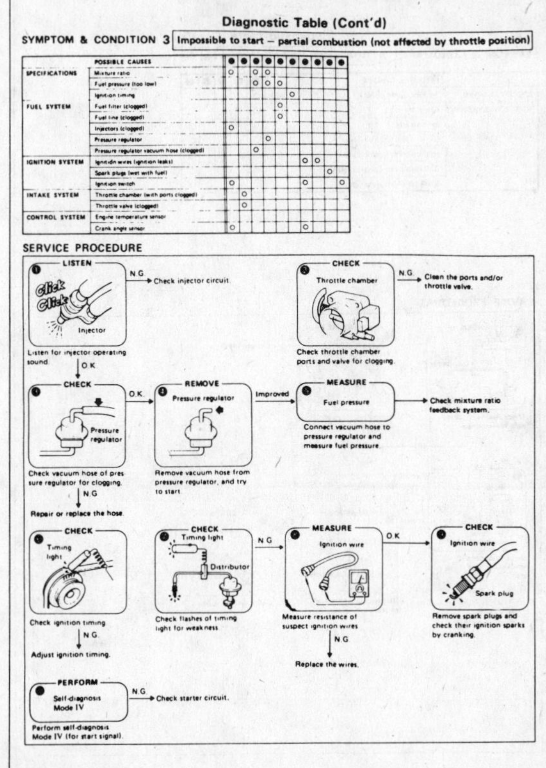

Diagnostic Table (Cont'd)

SYMPTOM & CONDITION 3 Impossible to start – partial combustion (not affected by throttle position)

	POSSIBLE CAUSES	1	2	3	4	5	6	7	8	9	10
SPECIFICATIONS	Mixture ratio	O		O	O						
	Fuel pressure (too low)			O	O	O					
	Ignition timing						O				
FUEL SYSTEM	Fuel filter (clogged)					O					
	Fuel line (clogged)					O					
	Injectors (clogged)	O									
	Pressure regulator				O						
	Pressure regulator vacuum hose (clogged)			O							
IGNITION SYSTEM	Ignition wires (ignition leaks)							O	O		
	Spark plugs (wet with fuel)									O	
	Ignition switch	O						O			O
INTAKE SYSTEM	Throttle chamber (with ports clogged)		O								
	Throttle valve (clogged)		O								
CONTROL SYSTEM	Engine temperature sensor										
	Crank angle sensor	O						O			

SERVICE PROCEDURE

LISTEN
Click Click
Injector
N.G.
Check injector circuit.
Listen for injector operating sound.
O.K.
CHECK
Throttle chamber
N.G.
Clean the ports and/or throttle valve.
Check throttle chamber ports and valve for clogging.
CHECK
Pressure regulator
O.K.
Check vacuum hose of pressure regulator for clogging.
N.G.
Repair or replace the hose.
REMOVE
Pressure regulator
Improved
Remove vacuum hose from pressure regulator, and try to start.
MEASURE
Fuel pressure
Check mixture ratio feedback system.
Connect vacuum hose to pressure regulator and measure fuel pressure.
CHECK
Timing light
Check ignition timing.
N.G.
Adjust ignition timing.
CHECK
Timing light
Distributor
N.G.
Check flashes of timing light for weakness.
MEASURE
Ignition wire
O.K.
Measure resistance of suspect ignition wires.
N.G.
Replace the wires.
CHECK
Ignition wire
Spark plug
Remove spark plugs and check their ignition sparks by cranking.
PERFORM
Self-diagnosis Mode IV
N.G.
Check starter circuit.
Perform self-diagnosis Mode IV (for start signal).

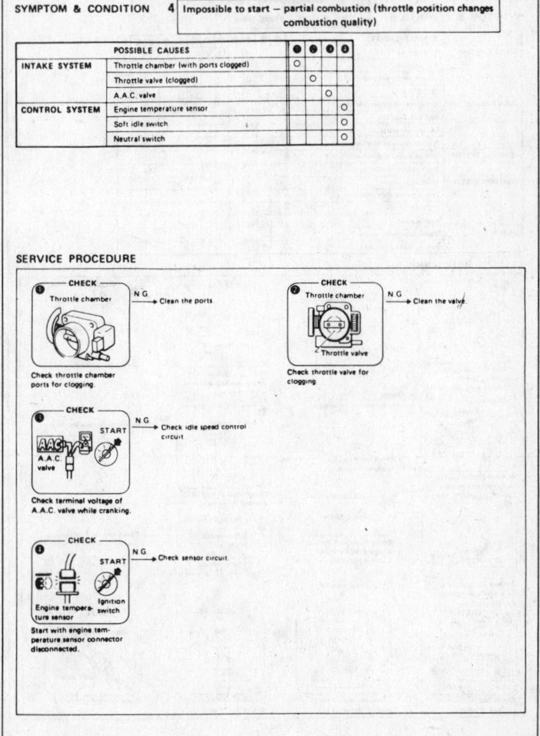

Diagnostic Table (Cont'd)

SYMPTOM & CONDITION 4 Impossible to start – partial combustion (throttle position changes combustion quality)

	POSSIBLE CAUSES	1	2	3	4
INTAKE SYSTEM	Throttle chamber (with ports clogged)	O			
	Throttle valve (clogged)		O		
	A.A.C. valve			O	
CONTROL SYSTEM	Engine temperature sensor				O
	Soft idle switch				O
	Neutral switch				O

SERVICE PROCEDURE

CHECK
Throttle chamber
N.G.
Clean the ports.
Check throttle chamber ports for clogging.
CHECK
Throttle chamber
Throttle valve
N.G.
Clean the valve.
Check throttle valve for clogging.
CHECK
AAC
A.A.C. valve
START
N.G.
Check idle speed control circuit.
Check terminal voltage of A.A.C. valve while cranking.
CHECK
START
Engine temperature sensor
Ignition switch
N.G.
Check sensor circuit.
Start with engine temperature sensor connector disconnected.

ECCS DIAGNOSTIC CHART—1990 TRUCK/PATHFINDER WITH KA24E ENGINE

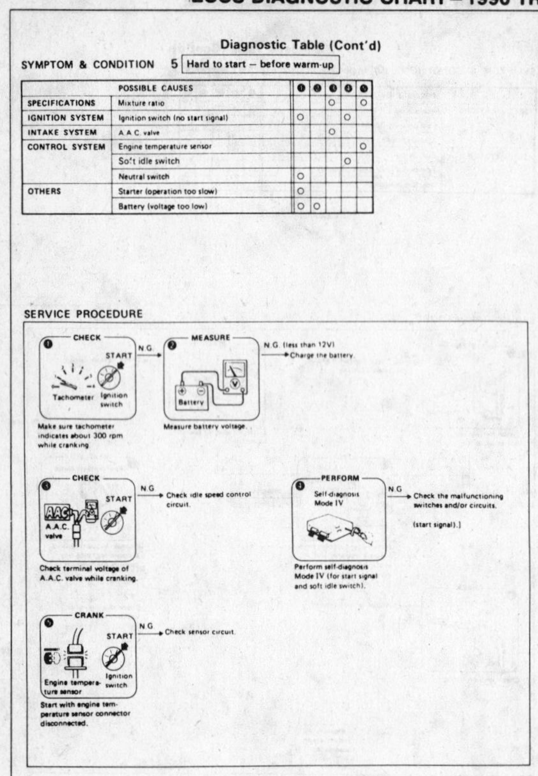

Diagnostic Table (Cont'd)

SYMPTOM & CONDITION 5 Hard to start – before warm-up

POSSIBLE CAUSES		1	2	3	4	5
SPECIFICATIONS	Mixture ratio			○		○
IGNITION SYSTEM	Ignition switch (no start signal)	○			○	
INTAKE SYSTEM	A.A.C. valve			○		
CONTROL SYSTEM	Engine temperature sensor					○
	Soft idle switch				○	
	Neutral switch	○				
OTHERS	Starter (operation too slow)	○				
	Battery (voltage too low)	○	○			

SERVICE PROCEDURE

CHECK
START
Tachometer
Ignition switch
Make sure tachometer indicates about 300 rpm while cranking.
N.G.
MEASURE
Battery
Measure battery voltage.
N.G. (less than 12V)
Charge the battery.
CHECK
START
A.A.C. valve
Check terminal voltage of A.A.C. valve while cranking.
N.G
Check idle speed control circuit.
PERFORM
Self-diagnosis Mode IV
Perform self-diagnosis Mode IV (for start signal and soft idle switch).
N.G
Check the malfunctioning switches and/or circuits.
(start signal).]
CRANK
START
Engine temperature sensor
Ignition switch
Start with engine temperature sensor connector disconnected.
N.G
Check sensor circuit.

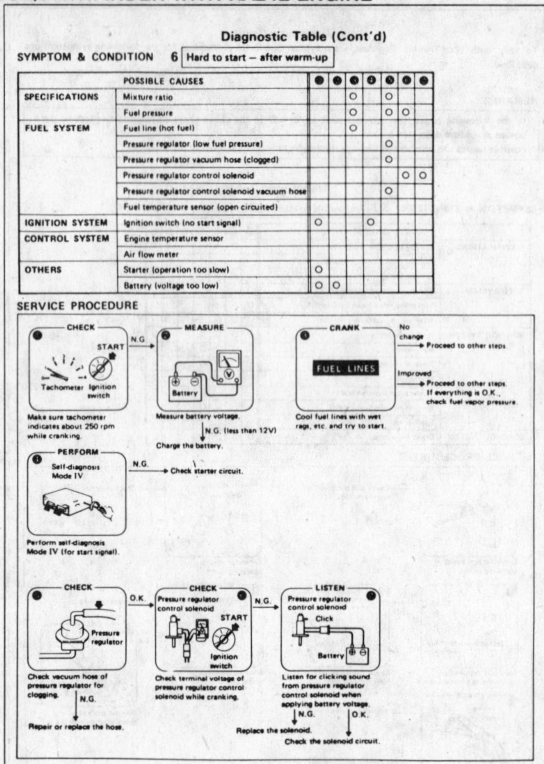

Diagnostic Table (Cont'd)

SYMPTOM & CONDITION 6 Hard to start – after warm-up

POSSIBLE CAUSES		1	2	3	4	5	6	7
SPECIFICATIONS	Mixture ratio			○		○		
	Fuel pressure			○		○	○	
FUEL SYSTEM	Fuel line (hot fuel)			○				
	Pressure regulator (low fuel pressure)					○		
	Pressure regulator vacuum hose (clogged)					○		
	Pressure regulator control solenoid						○	○
	Pressure regulator control solenoid vacuum hose					○		
	Fuel temperature sensor (open circuited)							
IGNITION SYSTEM	Ignition switch (no start signal)	○			○			
CONTROL SYSTEM	Engine temperature sensor							
	Air flow meter							
OTHERS	Starter (operation too slow)	○						
	Battery (voltage too low)	○	○					

SERVICE PROCEDURE

CHECK
START
Tachometer
Ignition switch
Make sure tachometer indicates about 250 rpm while cranking.
N.G
MEASURE
Battery
Measure battery voltage.
N.G. (less than 12V)
Charge the battery.
CRANK
FUEL LINES
Cool fuel lines with wet rags, etc. and try to start.
No change
Proceed to other steps.
Improved
Proceed to other steps. If everything is O.K., check fuel vapor pressure.
PERFORM
Self-diagnosis Mode IV
Perform self-diagnosis Mode IV (for start signal).
N.G.
Check starter circuit.
CHECK
Pressure regulator
Check vacuum hose of pressure regulator for clogging.
N.G.
Repair or replace the hose.
O.K.
CHECK
Pressure regulator control solenoid
START
Ignition switch
Check terminal voltage of pressure regulator control solenoid while cranking.
N.G.
LISTEN
Pressure regulator control solenoid
Click
Battery
Listen for clicking sound from pressure regulator control solenoid when applying battery voltage.
N.G.
O.K.
Replace the solenoid.
Check the solenoid circuit.

ECCS DIAGNOSTIC CHART—1990 TRUCK/PATHFINDER WITH KA24E ENGINE

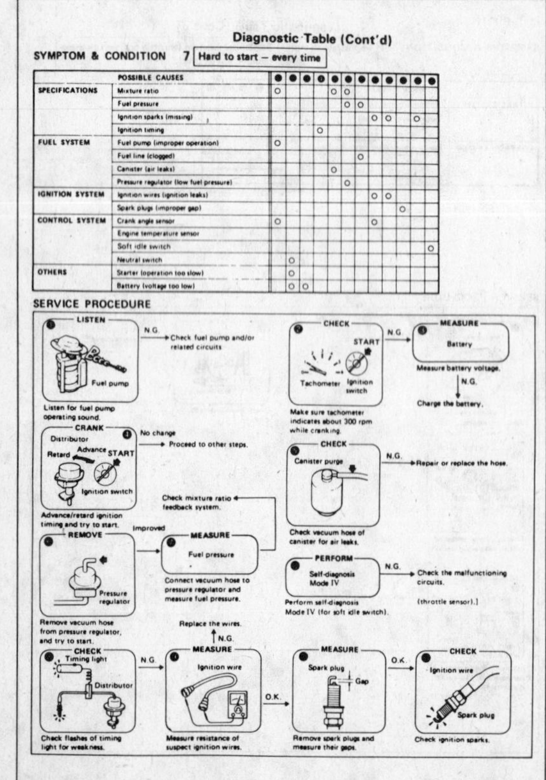

Diagnostic Table (Cont'd)

SYMPTOM & CONDITION 7 Hard to start – every time

POSSIBLE CAUSES		1	2	3	4	5	6	7	8	9	10	11	12
SPECIFICATIONS	Mixture ratio	○				○	○						
	Fuel pressure						○	○					
	Ignition sparks (missing)								○	○		○	
	Ignition timing				○								
FUEL SYSTEM	Fuel pump (improper operation)	○											
	Fuel line (clogged)							○					
	Canister (air leaks)					○							
	Pressure regulator (low fuel pressure)						○						
IGNITION SYSTEM	Ignition wires (ignition leaks)								○	○			
	Spark plugs (improper gap)										○		
CONTROL SYSTEM	Crank angle sensor	○							○				
	Engine temperature sensor												
	Soft idle switch												○
	Neutral switch		○										
OTHERS	Starter (operation too slow)		○										
	Battery (voltage too low)		○	○									

SERVICE PROCEDURE

LISTEN
Fuel pump
Listen for fuel pump operating sound.
N.G.
Check fuel pump and/or related circuits
CHECK
START
Tachometer
Ignition switch
Make sure tachometer indicates about 300 rpm while cranking.
N.G
MEASURE
Battery
Measure battery voltage.
N.G.
Charge the battery.
CRANK
Distributor
Retard
Advance
START
Ignition switch
Advance/retard ignition timing and try to start.
No change
Proceed to other steps.
CHECK
Canister purge
Check vacuum hose of canister for air leaks.
N.G.
Repair or replace the hose.
Check mixture ratio feedback system.
Improved
REMOVE
Pressure regulator
Remove vacuum hose from pressure regulator, and try to start.
MEASURE
Fuel pressure
Connect vacuum hose to pressure regulator and measure fuel pressure.
PERFORM
Self-diagnosis Mode IV
Perform self-diagnosis Mode IV (for soft idle switch).
N.G.
Check the malfunctioning circuits.
(throttle sensor).]
Replace the wires.
N.G.
CHECK
Timing light
Distributor
Check flashes of timing light for weakness.
N.G.
MEASURE
Ignition wire
Measure resistance of suspect ignition wires.
O.K.
MEASURE
Spark plug
Gap
Remove spark plugs and measure their gaps.
O.K.
CHECK
Ignition wire
Spark plug
Check ignition sparks.

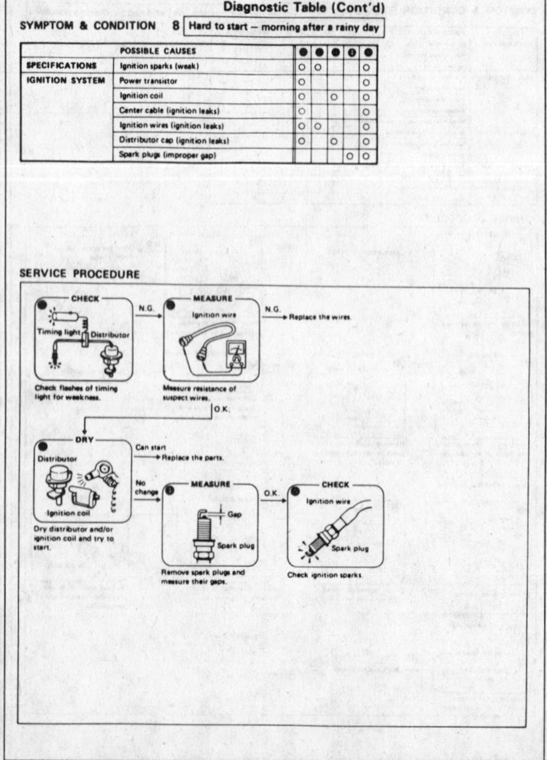

Diagnostic Table (Cont'd)

SYMPTOM & CONDITION 8 Hard to start – morning after a rainy day

POSSIBLE CAUSES		1	2	3	4	5
SPECIFICATIONS	Ignition sparks (weak)	○	○			○
IGNITION SYSTEM	Power transistor	○				○
	Ignition coil	○		○		○
	Center cable (ignition leaks)	○				○
	Ignition wires (ignition leaks)	○	○			○
	Distributor cap (ignition leaks)	○		○		○
	Spark plugs (improper gap)				○	○

SERVICE PROCEDURE

CHECK
Timing light
Distributor
Check flashes of timing light for weakness.
N.G.
MEASURE
Ignition wire
Measure resistance of suspect wires.
N.G.
Replace the wires.
O.K.
DRY
Distributor
Ignition coil
Dry distributor and/or ignition coil and try to start.
Can start
Replace the parts.
No change
MEASURE
Gap
Spark plug
Remove spark plugs and measure their gaps.
O.K.
CHECK
Ignition wire
Spark plug
Check ignition sparks.

ECCS DIAGNOSTIC CHART—1990 TRUCK/PATHFINDER WITH KA24E ENGINE

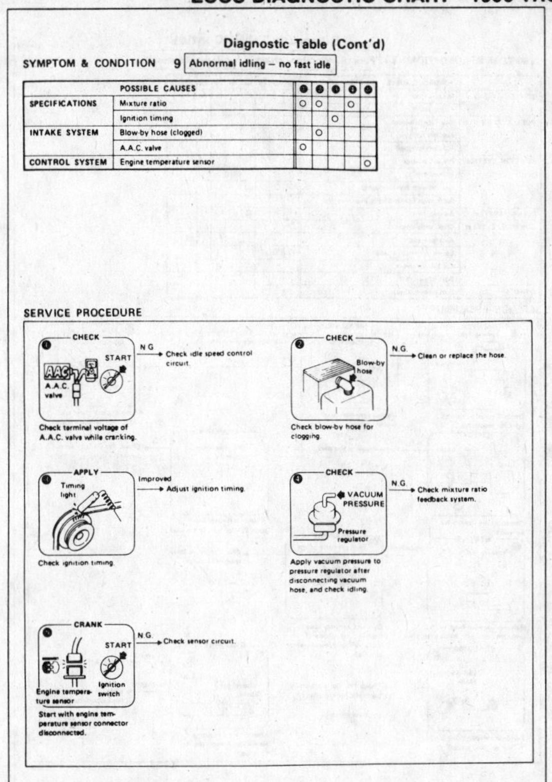

Diagnostic Table (Cont'd)

SYMPTOM & CONDITION 9 Abnormal idling – no fast idle

	POSSIBLE CAUSES	1	2	3	4	5
SPECIFICATIONS	Mixture ratio	○	○		○	
	Ignition timing			○		
INTAKE SYSTEM	Blow-by hose (clogged)		○			
	A.A.C. valve	○				
CONTROL SYSTEM	Engine temperature sensor					○

SERVICE PROCEDURE

1 CHECK
AAC
A.A.C. valve
START
N.G. → Check idle speed control circuit.
Check terminal voltage of A.A.C. valve while cranking.
2 CHECK
Blow-by hose
N.G. → Clean or replace the hose.
Check blow-by hose for clogging.
3 APPLY
Timing light
Improved → Adjust ignition timing.
Check ignition timing.
4 CHECK
VACUUM PRESSURE
Pressure regulator
N.G. → Check mixture ratio feedback system.
Apply vacuum pressure to pressure regulator after disconnecting vacuum hose, and check idling.
5 CRANK
START
Engine temperature sensor
Ignition switch
N.G. → Check sensor circuit.
Start with engine temperature sensor connector disconnected.

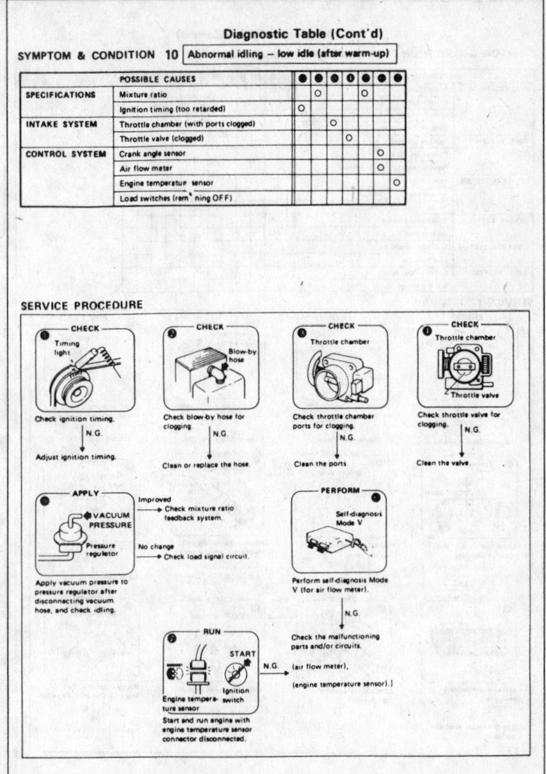

Diagnostic Table (Cont'd)

SYMPTOM & CONDITION 10 Abnormal idling – low idle (after warm-up)

	POSSIBLE CAUSES	1	2	3	4	5	6	7
SPECIFICATIONS	Mixture ratio		○			○		
	Ignition timing (too retarded)	○						
INTAKE SYSTEM	Throttle chamber (with ports clogged)			○				
	Throttle valve (clogged)				○			
CONTROL SYSTEM	Crank angle sensor						○	
	Air flow meter						○	
	Engine temperatur sensor							○
	Load switches (rem ning OFF)							

SERVICE PROCEDURE

1 CHECK
Timing light
Check ignition timing.
N.G. → Adjust ignition timing.
2 CHECK
Blow-by hose
Check blow-by hose for clogging.
N.G. → Clean or replace the hose.
3 CHECK
Throttle chamber
Check throttle chamber ports for clogging.
N.G. → Clean the ports
4 CHECK
Throttle chamber
Throttle valve
Check throttle valve for clogging.
N.G. → Clean the valve.
5 APPLY
VACUUM PRESSURE
Pressure regulator
Improved → Check mixture ratio feedback system.
No change → Check load signal circuit.
Apply vacuum pressure to pressure regulator after disconnecting vacuum hose, and check idling.
6 PERFORM
Self-diagnosis Mode V
Perform self-diagnosis Mode V (for air flow meter).
N.G. → Check the malfunctioning parts and/or circuits.
7 RUN
START
Engine temperature sensor
Ignition switch
N.G. → (air flow meter),
(engine temperature sensor).]
Start and run engine with engine temperature sensor connector disconnected.

ECCS DIAGNOSTIC CHART—1990 TRUCK/PATHFINDER WITH KA24E ENGINE

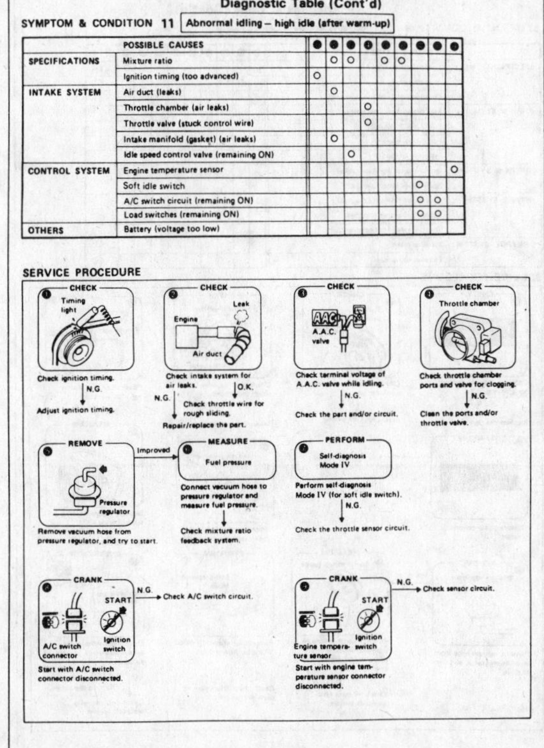

Diagnostic Table (Cont'd)

SYMPTOM & CONDITION 11 Abnormal idling – high idle (after warm-up)

	POSSIBLE CAUSES	1	2	3	4	5	6	7	8	9
SPECIFICATIONS	Mixture ratio		○	○		○	○			
	Ignition timing (too advanced)	○								
INTAKE SYSTEM	Air duct (leaks)		○							
	Throttle chamber (air leaks)				○					
	Throttle valve (stuck control wire)				○					
	Intake manifold (gasket) (air leaks)		○							
	Idle speed control valve (remaining ON)			○						
CONTROL SYSTEM	Engine temperature sensor									○
	Soft idle switch							○		
	A/C switch circuit (remaining ON)							○	○	
	Load switches (remaining ON)							○	○	
OTHERS	Battery (voltage too low)									

SERVICE PROCEDURE

1 CHECK
Timing light
Check ignition timing.
N.G. → Adjust ignition timing.
2 CHECK
Engine
Leak
Air duct
Check intake system for air leaks.
N.G. → Repair/replace the part.
O.K. → Check throttle wire for rough sliding.
3 CHECK
AAC
A.A.C. valve
Check terminal voltage of A.A.C. valve while idling.
N.G. → Check the part and/or circuit.
4 CHECK
Throttle chamber
Check throttle chamber ports and valve for clogging.
N.G. → Clean the ports and/or throttle valve.
5 REMOVE
Pressure regulator
Remove vacuum hose from pressure regulator, and try to start.
Improved → 6 MEASURE
Fuel pressure
Connect vacuum hose to pressure regulator and measure fuel pressure.
Check mixture ratio feedback system.
7 PERFORM
Self-diagnosis Mode IV
Perform self-diagnosis Mode IV (for soft idle switch).
N.G. → Check the throttle sensor circuit.
8 CRANK
START
A/C switch connector
Ignition switch
N.G. → Check A/C switch circuit.
Start with A/C switch connector disconnected.
9 CRANK
START
Engine temperature sensor
Ignition switch
N.G. → Check sensor circuit.
Start with engine temperature sensor connector disconnected.

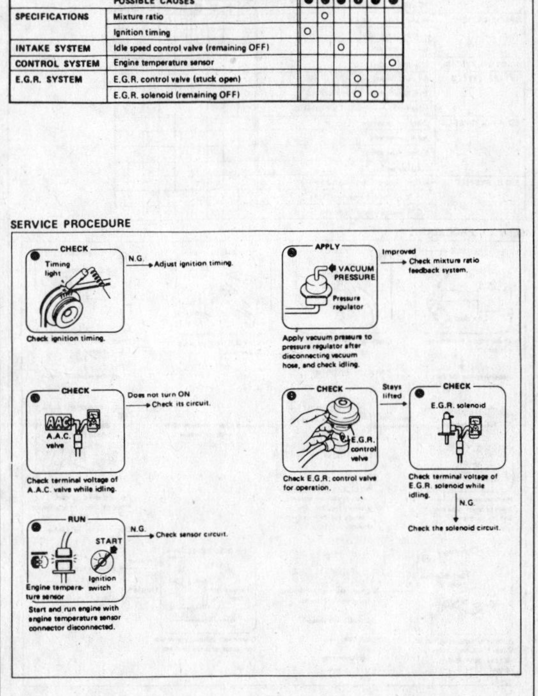

Diagnostic Table (Cont'd)

SYMPTOM & CONDITION 12 Unstable idling – before warm-up

	POSSIBLE CAUSES	1	2	3	4	5	6
SPECIFICATIONS	Mixture ratio		○				
	Ignition timing	○					
INTAKE SYSTEM	Idle speed control valve (remaining OFF)			○			
CONTROL SYSTEM	Engine temperature sensor						○
E.G.R. SYSTEM	E.G.R. control valve (stuck open)				○		
	E.G.R. solenoid (remaining OFF)				○	○	

SERVICE PROCEDURE

1 CHECK
Timing light
N.G. → Adjust ignition timing.
Check ignition timing.
2 APPLY
VACUUM PRESSURE
Pressure regulator
Improved → Check mixture ratio feedback system.
Apply vacuum pressure to pressure regulator after disconnecting vacuum hose, and check idling.
3 CHECK
AAC
A.A.C. valve
Does not turn ON → Check its circuit.
Check terminal voltage of A.A.C. valve while idling.
4 CHECK
E.G.R. control valve
Check E.G.R. control valve for operation.
Stays lifted → 5 CHECK
E.G.R. solenoid
Check terminal voltage of E.G.R. solenoid while idling.
N.G. → Check the solenoid circuit.
6 RUN
START
Engine temperature sensor
Ignition switch
N.G. → Check sensor circuit.
Start and run engine with engine temperature sensor connector disconnected.

ECCS DIAGNOSTIC CHART—1990 TRUCK/PATHFINDER WITH KA24E ENGINE

Diagnostic Table (Cont'd)

SYMPTOM & CONDITION 13 Unstable idling – after warm-up

	POSSIBLE CAUSES	●	●	●	●	●	●	●	●	●	●	●	●
SPECIFICATIONS	Mixture ratio	O	O	O	O								
	Ignition sparks					O	O	O					
	Ignition timing								O				
	Compression pressure									O			
FUEL SYSTEM	Fuel line (clogged)												
	Canister (air leaks)			O									
	Pressure regulator control solenoid				O								
IGNITION SYSTEM	Power transistor					O		O					
	Ignition coil					O		O					
	Ignition wires					O	O	O					
INTAKE SYSTEM	Blow-by hose (leaks)	O											
	Air duct (leaks)		O										
CONTROL SYSTEM	Soft idle switch												O
	Load switches												
E.G.R. SYSTEM	E.G.R. control valve										O		
	E.G.R. solenoid										O	O	

SERVICE PROCEDURE

CHECK — Blow-by hose
Check blow-by hose for leaks. N.G. → Repair/replace the hose.

CHECK — Engine, Leak, Air duct
Check intake system for air leaks. N.G. → Repair/replace the part.

CHECK — Canister purge
Check purge line for leaks. N.G. → Repair/replace the hose.

CHECK — Pressure regulator control solenoid
Check terminal voltage of the solenoid while idling. N.G. → Check the solenoid circuit.

CHECK — Timing light, Distributor
Check flashes of timing light for weakness. N.G. →

MEASURE — Ignition wire
Measure resistance of suspect wires. N.G. → Replace the wire. O.K. →

CHECK — Ignition wire, Spark plug
Remove spark plugs and check their ignition sparks.

CHECK — Timing light
Check ignition timing. N.G. → Adjust ignition timing.

MEASURE — COMPRESSION PRESSURE
Measure compression pressure. N.G. → Check cylinder head and gasket.

CHECK — E.G.R. control valve
Check E.G.R. control valve for operation. Stays lifted →

CHECK — E.G.R. solenoid
Check terminal voltage of E.G.R. solenoid while idling. N.G. → Check the solenoid circuit.

PERFORM — Self-diagnosis Mode IV
Perform self-diagnosis Mode IV (for soft idle switch). N.G. → Check the throttle sensor circuit.

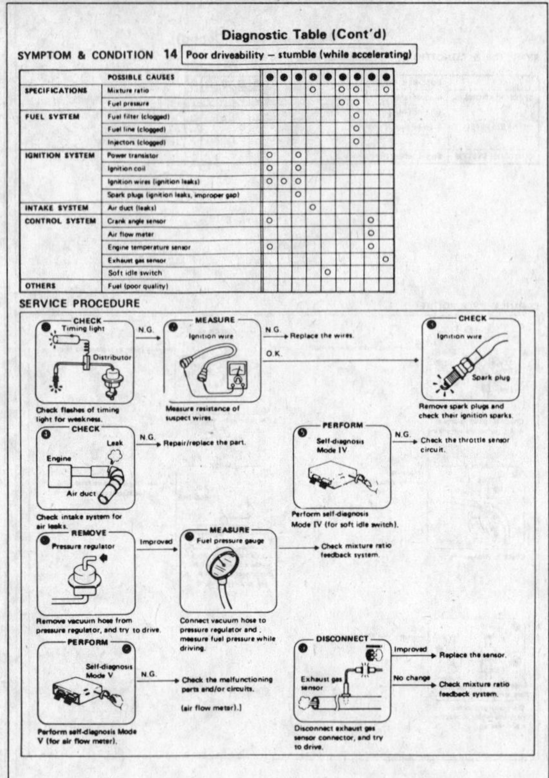

Diagnostic Table (Cont'd)

SYMPTOM & CONDITION 14 Poor driveability – stumble (while accelerating)

	POSSIBLE CAUSES	●	●	●	●	●	●	●	●	●
SPECIFICATIONS	Mixture ratio				O		O	O		O
	Fuel pressure						O	O		
FUEL SYSTEM	Fuel filter (clogged)							O		
	Fuel line (clogged)							O		
	Injectors (clogged)							O		
IGNITION SYSTEM	Power transistor	O		O						
	Ignition coil	O		O						
	Ignition wires (ignition leaks)	O	O	O						
	Spark plugs (ignition leaks, improper gap)			O						
INTAKE SYSTEM	Air duct (leaks)				O					
CONTROL SYSTEM	Crank angle sensor	O							O	
	Air flow meter								O	
	Engine temperature sensor	O							O	
	Exhaust gas sensor									O
	Soft idle switch					O				
OTHERS	Fuel (poor quality)									

SERVICE PROCEDURE

CHECK
Timing light
Distributor
Check flashes of timing light for weakness.
N.G.
MEASURE
Ignition wire
Measure resistance of suspect wires.
N.G. Replace the wires.
O.K.
CHECK
Ignition wire
Spark plug
Remove spark plugs and check their ignition sparks.
CHECK
Engine
Leak
Air duct
Check intake system for air leaks.
N.G. Repair/replace the part.
PERFORM
Self-diagnosis Mode IV
Perform self-diagnosis Mode IV (for soft idle switch).
N.G. Check the throttle sensor circuit.
REMOVE
Pressure regulator
Remove vacuum hose from pressure regulator, and try to drive.
Improved
MEASURE
Fuel pressure gauge
Connect vacuum hose to pressure regulator and measure fuel pressure while driving.
Check mixture ratio feedback system.
PERFORM
Self-diagnosis Mode V
Perform self-diagnosis Mode V (for air flow meter).
N.G. Check the malfunctioning parts and/or circuits.
(air flow meter).]
DISCONNECT
Exhaust gas sensor
Disconnect exhaust gas sensor connector, and try to drive.
Improved Replace the sensor.
No change Check mixture ratio feedback system.

ECCS DIAGNOSTIC CHART—1990 TRUCK/PATHFINDER WITH KA24E ENGINE

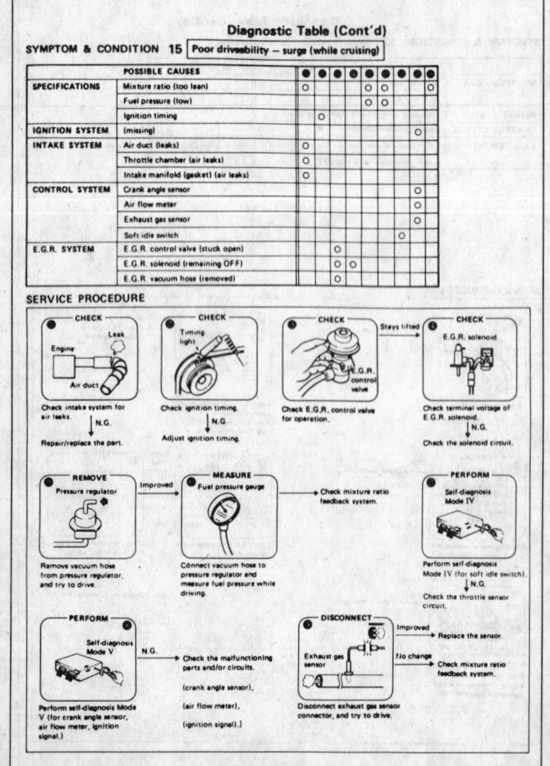

Diagnostic Table (Cont'd)

SYMPTOM & CONDITION 15 Poor driveability – surge (while cruising)

	POSSIBLE CAUSES	●	●	●	●	●	●	●	●	●
SPECIFICATIONS	Mixture ratio (too lean)	O				O	O			O
	Fuel pressure (low)					O	O			
	Ignition timing		O							
IGNITION SYSTEM	(missing)								O	
INTAKE SYSTEM	Air duct (leaks)	O								
	Throttle chamber (air leaks)	O								
	Intake manifold (gasket) (air leaks)	O								
CONTROL SYSTEM	Crank angle sensor								O	
	Air flow meter								O	
	Exhaust gas sensor								O	
	Soft idle switch							O		
E.G.R. SYSTEM	E.G.R. control valve (stuck open)			O						
	E.G.R. solenoid (remaining OFF)			O	O					
	E.G.R. vacuum hose (removed)			O						

SERVICE PROCEDURE

CHECK
Engine
Leak
Air duct
Check intake system for air leaks.
N.G.
Repair/replace the part.
CHECK
Timing light
Check ignition timing.
N.G.
Adjust ignition timing.
CHECK
E.G.R. control valve
Check E.G.R. control valve for operation.
Stays lifted
CHECK
E.G.R. solenoid
Check terminal voltage of E.G.R. solenoid.
N.G.
Check the solenoid circuit.
REMOVE
Pressure regulator
Remove vacuum hose from pressure regulator, and try to drive.
Improved
MEASURE
Fuel pressure gauge
Connect vacuum hose to pressure regulator and measure fuel pressure while driving.
Check mixture ratio feedback system.
PERFORM
Self-diagnosis Mode IV
Perform self-diagnosis Mode IV (for soft idle switch).
N.G.
Check the throttle sensor circuit.
PERFORM
Self-diagnosis Mode V
Perform self-diagnosis Mode V (for crank angle sensor, air flow meter, ignition signal.)
N.G.
Check the malfunctioning parts and/or circuits.
(crank angle sensor),
(air flow meter),
(ignition signal).]
DISCONNECT
Exhaust gas sensor
Disconnect exhaust gas sensor connector, and try to drive.
Improved
Replace the sensor.
No change
Check mixture ratio feedback system.

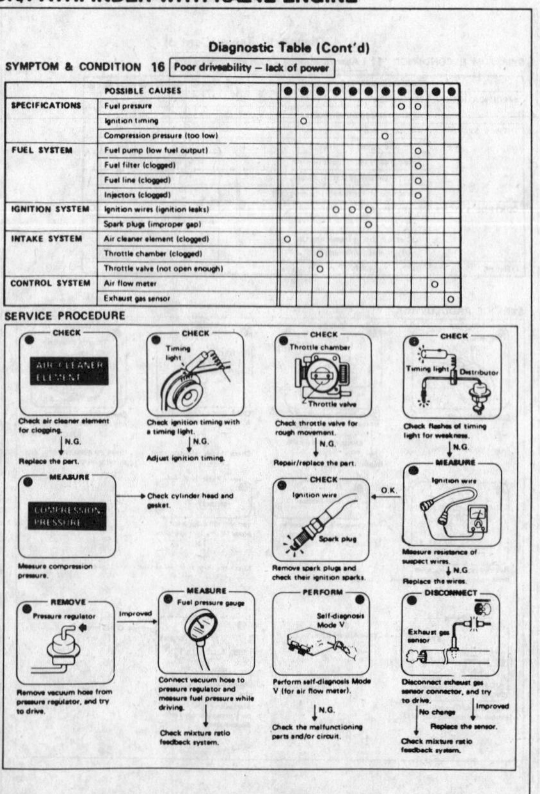

Diagnostic Table (Cont'd)

SYMPTOM & CONDITION 16 Poor driveability – lack of power

	POSSIBLE CAUSES	●	●	●	●	●	●	●	●	●	●	●
SPECIFICATIONS	Fuel pressure								O	O		
	Ignition timing		O									
	Compression pressure (too low)							O				
FUEL SYSTEM	Fuel pump (low fuel output)									O		
	Fuel filter (clogged)									O		
	Fuel line (clogged)									O		
	Injectors (clogged)									O		
IGNITION SYSTEM	Ignition wires (ignition leaks)				O	O	O					
	Spark plugs (improper gap)						O					
INTAKE SYSTEM	Air cleaner element (clogged)	O										
	Throttle chamber (clogged)			O								
	Throttle valve (not open enough)			O								
CONTROL SYSTEM	Air flow meter										O	
	Exhaust gas sensor											O

SERVICE PROCEDURE

CHECK
AIR CLEANER ELEMENT
Check air cleaner element for clogging.
N.G.
Replace the part.
CHECK
Timing light
Check ignition timing with a timing light.
N.G.
Adjust ignition timing.
CHECK
Throttle chamber
Throttle valve
Check throttle valve for rough movement.
N.G.
Repair/replace the part.
CHECK
Timing light
Distributor
Check flashes of timing light for weakness.
N.G.
MEASURE
COMPRESSION PRESSURE
Measure compression pressure.
Check cylinder head and gasket.
CHECK
Ignition wire
Spark plug
Remove spark plugs and check their ignition sparks.
O.K.
MEASURE
Ignition wire
Measure resistance of suspect wires.
N.G.
Replace the wires.
REMOVE
Pressure regulator
Remove vacuum hose from pressure regulator, and try to drive.
Improved
MEASURE
Fuel pressure gauge
Connect vacuum hose to pressure regulator and measure fuel pressure while driving.
Check mixture ratio feedback system.
PERFORM
Self-diagnosis Mode V
Perform self-diagnosis Mode V (for air flow meter).
N.G.
Check the malfunctioning parts and/or circuit.
DISCONNECT
Exhaust gas sensor
Disconnect exhaust gas sensor connector, and try to drive.
No change
Improved
Replace the sensor.
Check mixture ratio feedback system.

ECCS DIAGNOSTIC CHART—1990 TRUCK/PATHFINDER WITH KA24E ENGINE

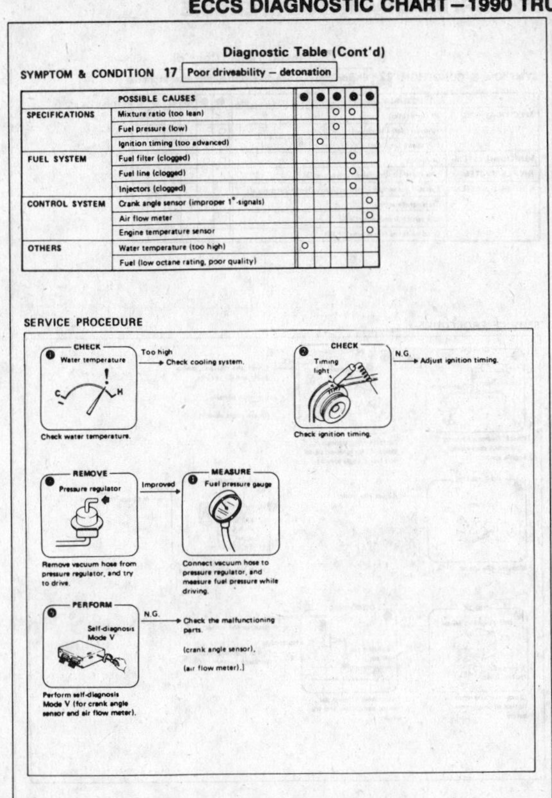

Diagnostic Table (Cont'd)

SYMPTOM & CONDITION 17 Poor driveability – detonation

	POSSIBLE CAUSES	1	2	3	4	5
SPECIFICATIONS	Mixture ratio (too lean)			O	O	
	Fuel pressure (low)			O		
	Ignition timing (too advanced)		O			
FUEL SYSTEM	Fuel filter (clogged)				O	
	Fuel line (clogged)				O	
	Injectors (clogged)				O	
CONTROL SYSTEM	Crank angle sensor (improper 1°-signals)					O
	Air flow meter					O
	Engine temperature sensor					O
OTHERS	Water temperature (too high)	O				
	Fuel (low octane rating, poor quality)					

SERVICE PROCEDURE

CHECK
Water temperature
Too high → Check cooling system.
Check water temperature.
CHECK
Timing light
N.G. → Adjust ignition timing.
Check ignition timing.
REMOVE
Pressure regulator
Improved
MEASURE
Fuel pressure gauge
Remove vacuum hose from pressure regulator, and try to drive.
Connect vacuum hose to pressure regulator, and measure fuel pressure while driving.
PERFORM
Self-diagnosis Mode V
N.G. → Check the malfunctioning parts.
(crank angle sensor),
(air flow meter).)
Perform self-diagnosis Mode V (for crank angle sensor and air flow meter).

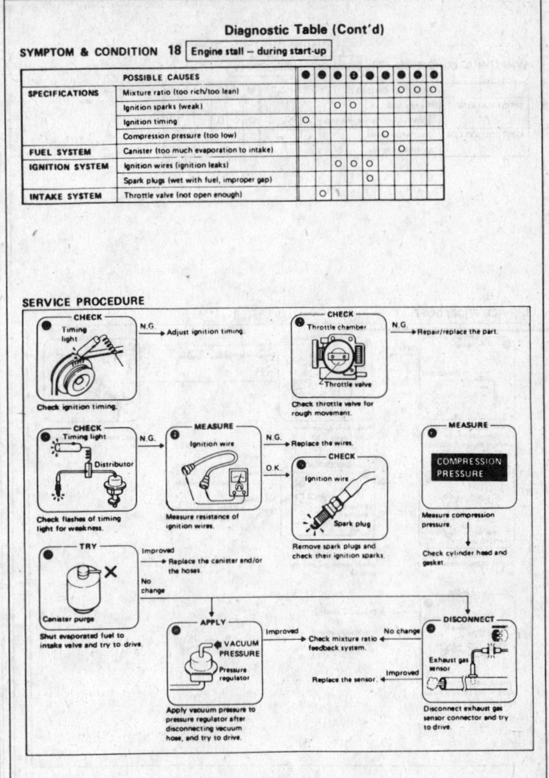

Diagnostic Table (Cont'd)

SYMPTOM & CONDITION 18 Engine stall – during start-up

	POSSIBLE CAUSES	1	2	3	4	5	6	7	8	9
SPECIFICATIONS	Mixture ratio (too rich/too lean)							O	O	O
	Ignition sparks (weak)			O	O					
	Ignition timing	O								
	Compression pressure (too low)						O			
FUEL SYSTEM	Canister (too much evaporation to intake)							O		
IGNITION SYSTEM	Ignition wires (ignition leaks)			O	O	O				
	Spark plugs (wet with fuel, improper gap)					O				
INTAKE SYSTEM	Throttle valve (not open enough)		O							

SERVICE PROCEDURE

CHECK
Timing light
N.G. → Adjust ignition timing.
Check ignition timing.
CHECK
Throttle chamber
N.G. → Repair/replace the part.
Throttle valve
Check throttle valve for rough movement.
CHECK
Timing light
Distributor
N.G.
MEASURE
Ignition wire
N.G. → Replace the wires.
O.K.
CHECK
Ignition wire
Spark plug
MEASURE
COMPRESSION PRESSURE
Check flashes of timing light for weakness.
Measure resistance of ignition wires.
Remove spark plugs and check their ignition sparks.
Measure compression pressure.
Check cylinder head and gasket.
TRY
Canister purge
Improved → Replace the canister and/or the hoses.
No change
Shut evaporated fuel to intake valve and try to drive.
APPLY
VACUUM PRESSURE
Pressure regulator
Improved → Check mixture ratio feedback system.
No change
DISCONNECT
Exhaust gas sensor
Replace the sensor. Improved
Apply vacuum pressure to pressure regulator after disconnecting vacuum hose, and try to drive.
Disconnect exhaust gas sensor connector and try to drive.

ECCS DIAGNOSTIC CHART—1990 TRUCK/PATHFINDER WITH KA24E ENGINE

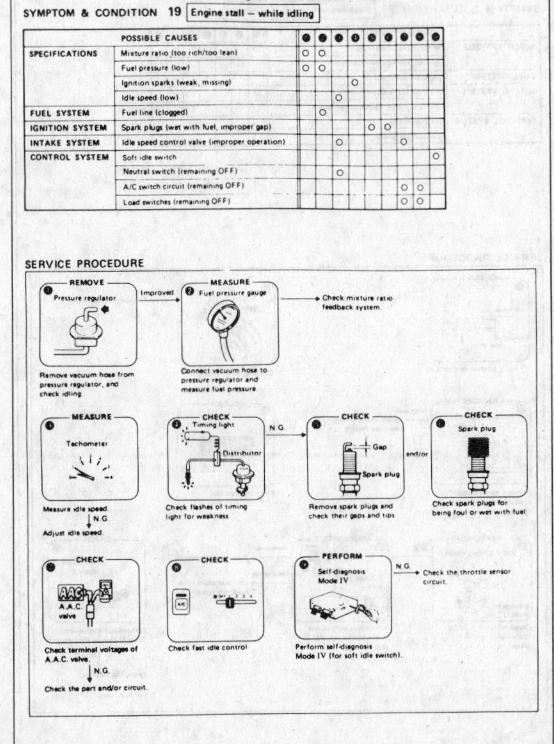

Diagnostic Table (Cont'd)

SYMPTOM & CONDITION 19 Engine stall – while idling

	POSSIBLE CAUSES	1	2	3	4	5	6	7	8	9
SPECIFICATIONS	Mixture ratio (too rich/too lean)	O	O							
	Fuel pressure (low)	O	O							
	Ignition sparks (weak, missing)				O					
	Idle speed (low)			O						
FUEL SYSTEM	Fuel line (clogged)		O							
IGNITION SYSTEM	Spark plugs (wet with fuel, improper gap)					O	O			
INTAKE SYSTEM	Idle speed control valve (improper operation)			O				O		
CONTROL SYSTEM	Soft idle switch									O
	Neutral switch (remaining OFF)			O						
	A/C switch circuit (remaining OFF)							O	O	
	Load switches (remaining OFF)							O	O	

SERVICE PROCEDURE

REMOVE
Pressure regulator
Improved
MEASURE
Fuel pressure gauge
→ Check mixture ratio feedback system.
Remove vacuum hose from pressure regulator, and check idling.
Connect vacuum hose to pressure regulator and measure fuel pressure.
MEASURE
Tachometer
Measure idle speed.
N.G.
Adjust idle speed.
CHECK
Timing light
Distributor
N.G.
CHECK
Gap
Spark plug
and/or
CHECK
Spark plug
Check flashes of timing light for weakness.
Remove spark plugs and check their gaps and tips.
Check spark plugs for being foul or wet with fuel.
CHECK
A.A.C. valve
Check terminal voltages of A.A.C. valve.
N.G.
Check the part and/or circuit.
CHECK
Check fast idle control
PERFORM
Self-diagnosis Mode IV
N.G. → Check the throttle sensor circuit.
Perform self-diagnosis Mode IV (for soft idle switch).

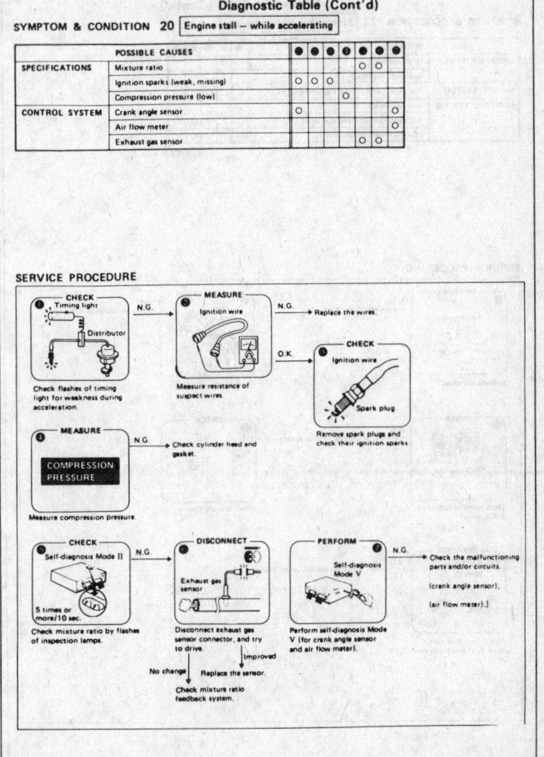

Diagnostic Table (Cont'd)

SYMPTOM & CONDITION 20 Engine stall – while accelerating

	POSSIBLE CAUSES	1	2	3	4	5	6	7
SPECIFICATIONS	Mixture ratio					O	O	
	Ignition sparks (weak, missing)	O	O	O				
	Compression pressure (low)				O			
CONTROL SYSTEM	Crank angle sensor	O						O
	Air flow meter							O
	Exhaust gas sensor					O	O	

SERVICE PROCEDURE

CHECK
Timing light
Distributor
N.G.
MEASURE
Ignition wire
N.G. → Replace the wires.
O.K.
CHECK
Ignition wire
Spark plug
Check flashes of timing light for weakness during acceleration.
Measure resistance of suspect wires.
Remove spark plugs and check their ignition sparks.
MEASURE
COMPRESSION PRESSURE
N.G. → Check cylinder head and gasket.
Measure compression pressure.
CHECK
Self-diagnosis Mode II
5 times or more/10 sec.
N.G.
DISCONNECT
Exhaust gas sensor
PERFORM
Self-diagnosis Mode V
N.G. → Check the malfunctioning parts and/or circuits.
(crank angle sensor),
(air flow meter).)
Check mixture ratio by flashes of inspection lamps.
Disconnect exhaust gas sensor connector, and try to drive.
Perform self-diagnosis Mode V (for crank angle sensor and air flow meter).
Improved
No change
Replace the sensor.
Check mixture ratio feedback system.

ECCS DIAGNOSTIC CHART—1990 TRUCK/PATHFINDER WITH KA24E ENGINE

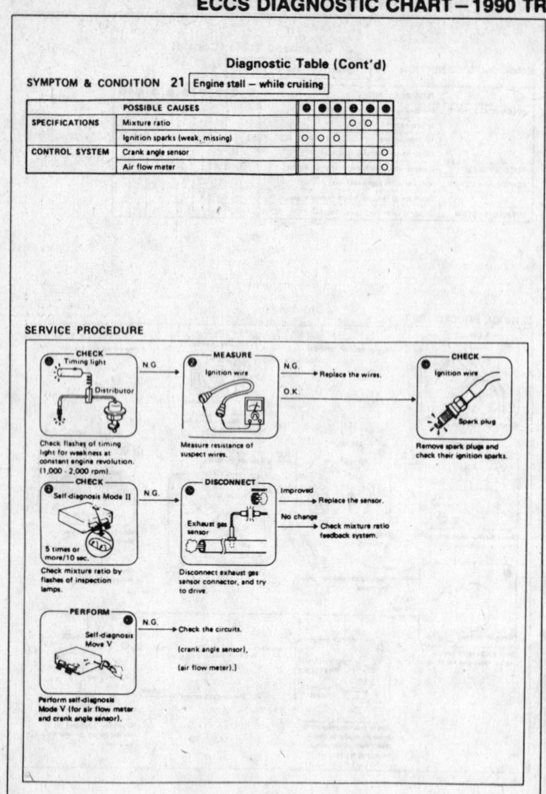

Diagnostic Table (Cont'd)

SYMPTOM & CONDITION 21 Engine stall – while cruising

POSSIBLE CAUSES		❶	❷	❸	❹	❺	❻
SPECIFICATIONS	Mixture ratio				○	○	
	Ignition sparks (weak, missing)	○	○	○			
CONTROL SYSTEM	Crank angle sensor						○
	Air flow meter						○

SERVICE PROCEDURE
CHECK
Timing light
Distributor
Check flashes of timing light for weakness at constant engine revolution. (1,000 - 2,000 rpm).
N.G.
MEASURE
Ignition wire
Measure resistance of suspect wires.
N.G.
Replace the wires.
O.K.
CHECK
Ignition wire
Spark plug
Remove spark plugs and check their ignition sparks.
CHECK
Self-diagnosis Mode II
5 times or more/10 sec.
Check mixture ratio by flashes of inspection lamps.
N.G.
DISCONNECT
Exhaust gas sensor
Disconnect exhaust gas sensor connector, and try to drive.
Improved
Replace the sensor.
No change
Check mixture ratio feedback system.
PERFORM
Self-diagnosis Move V
Perform self-diagnosis Mode V (for air flow meter and crank angle sensor).
N.G.
Check the circuits.
(crank angle sensor),
(air flow meter).]

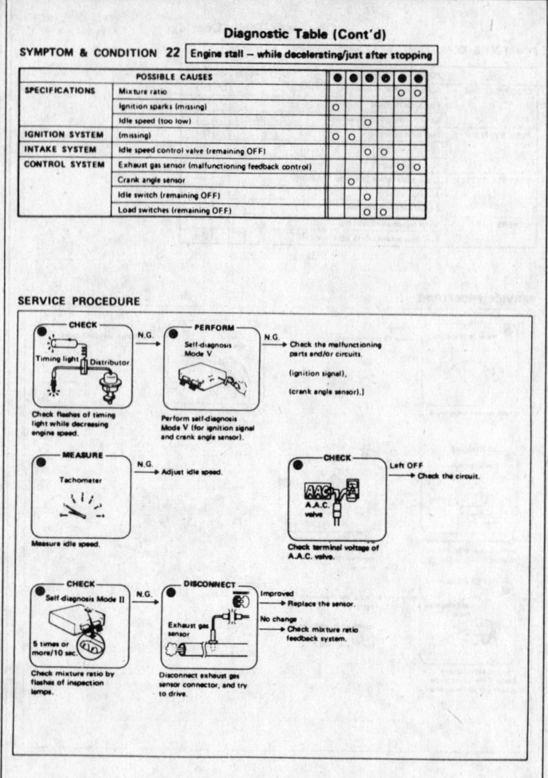

Diagnostic Table (Cont'd)

SYMPTOM & CONDITION 22 Engine stall – while decelerating/just after stopping

POSSIBLE CAUSES		❶	❷	❸	❹	❺	❻
SPECIFICATIONS	Mixture ratio					○	○
	Ignition sparks (missing)	○					
	Idle speed (too low)			○			
IGNITION SYSTEM	(missing)	○	○				
INTAKE SYSTEM	Idle speed control valve (remaining OFF)			○	○		
CONTROL SYSTEM	Exhaust gas sensor (malfunctioning feedback control)					○	○
	Crank angle sensor		○				
	Idle switch (remaining OFF)			○			
	Load switches (remaining OFF)			○	○		

SERVICE PROCEDURE
CHECK
Timing light
Distributor
Check flashes of timing light while decreasing engine speed.
N.G.
PERFORM
Self-diagnosis Mode V
Perform self-diagnosis Mode V (for ignition signal and crank angle sensor).
N.G.
Check the malfunctioning parts and/or circuits.
(ignition signal),
(crank angle sensor).]
MEASURE
Tachometer
Measure idle speed.
N.G.
Adjust idle speed.
CHECK
A.A.C. valve
Check terminal voltage of A.A.C. valve.
Left OFF
Check the circuit.
CHECK
Self-diagnosis Mode II
5 times or more/10 sec.
Check mixture ratio by flashes of inspection lamps.
N.G.
DISCONNECT
Exhaust gas sensor
Disconnect exhaust gas sensor connector, and try to drive.
Improved
Replace the sensor.
No change
Check mixture ratio feedback system.

ECCS DIAGNOSTIC CHART—1990 TRUCK/PATHFINDER WITH KA24E ENGINE

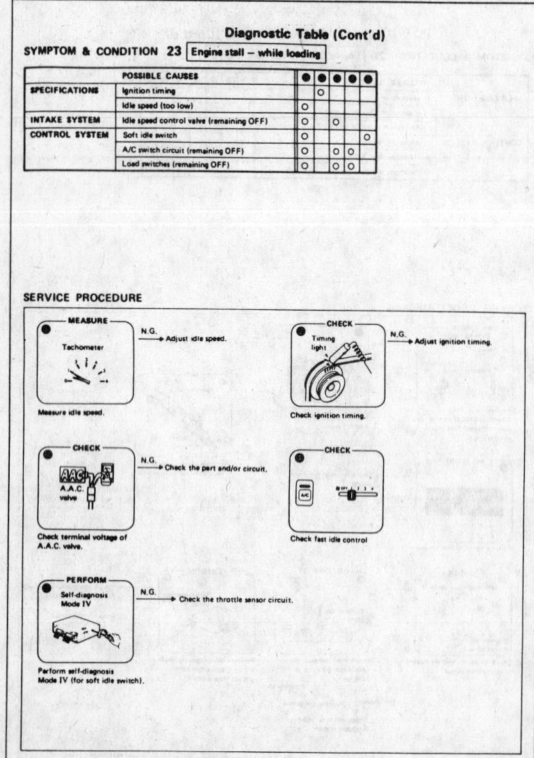

Diagnostic Table (Cont'd)

SYMPTOM & CONDITION 23 Engine stall – while loading

POSSIBLE CAUSES		❶	❷	❸	❹	❺
SPECIFICATIONS	Ignition timing		○			
	Idle speed (too low)	○				
INTAKE SYSTEM	Idle speed control valve (remaining OFF)	○		○		
CONTROL SYSTEM	Soft idle switch	○				○
	A/C switch circuit (remaining OFF)	○		○	○	
	Load switches (remaining OFF)	○		○	○	

SERVICE PROCEDURE
MEASURE
Tachometer
Measure idle speed.
N.G.
Adjust idle speed.
CHECK
Timing light
Check ignition timing.
N.G.
Adjust ignition timing.
CHECK
A.A.C. valve
Check terminal voltage of A.A.C. valve.
N.G.
Check the part and/or circuit.
CHECK
A/C
Check fast idle control
PERFORM
Self-diagnosis Mode IV
Perform self-diagnosis Mode IV (for soft idle switch).
N.G.
Check the throttle sensor circuit.

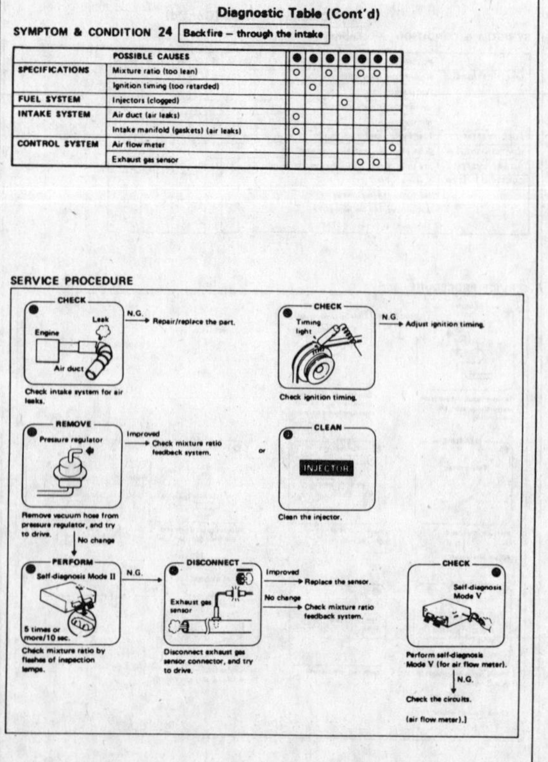

Diagnostic Table (Cont'd)

SYMPTOM & CONDITION 24 Backfire – through the intake

POSSIBLE CAUSES		❶	❷	❸	❹	❺	❻	❼
SPECIFICATIONS	Mixture ratio (too lean)	○		○		○	○	
	Ignition timing (too retarded)		○					
FUEL SYSTEM	Injectors (clogged)				○			
INTAKE SYSTEM	Air duct (air leaks)	○						
	Intake manifold (gaskets) (air leaks)	○						
CONTROL SYSTEM	Air flow meter							○
	Exhaust gas sensor					○	○	

SERVICE PROCEDURE
CHECK
Engine
Leak
Air duct
Check intake system for air leaks.
N.G.
Repair/replace the part.
CHECK
Timing light
Check ignition timing.
N.G.
Adjust ignition timing.
REMOVE
Pressure regulator
Remove vacuum hose from pressure regulator, and try to drive.
Improved
Check mixture ratio feedback system.
or
CLEAN
INJECTOR
Clean the injector.
No change
PERFORM
Self-diagnosis Mode II
5 times or more/10 sec.
Check mixture ratio by flashes of inspection lamps.
N.G.
DISCONNECT
Exhaust gas sensor
Disconnect exhaust gas sensor connector, and try to drive.
Improved
Replace the sensor.
No change
Check mixture ratio feedback system.
CHECK
Self-diagnosis Mode V
Perform self-diagnosis Mode V (for air flow meter).
N.G.
Check the circuits.
(air flow meter).]

ECCS DIAGNOSTIC CHART—1990 TRUCK/PATHFINDER WITH KA24E ENGINE

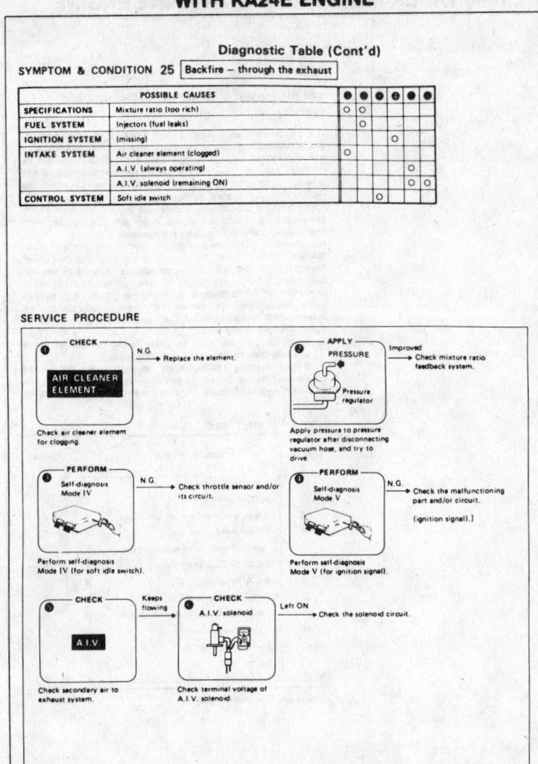

Diagnostic Table (Cont'd)

SYMPTOM & CONDITION 25 Backfire – through the exhaust

POSSIBLE CAUSES		❶	❷	❸	❹	❺	❻
SPECIFICATIONS	Mixture ratio (too rich)	○	○				
FUEL SYSTEM	Injectors (fuel leaks)		○				
IGNITION SYSTEM	(missing)				○		
INTAKE SYSTEM	Air cleaner element (clogged)	○					
	A.I.V. (always operating)					○	
	A.I.V. solenoid (remaining ON)					○	○
CONTROL SYSTEM	Soft idle switch			○			

SERVICE PROCEDURE
❶ CHECK
AIR CLEANER ELEMENT
N.G. → Replace the element.
Check air cleaner element for clogging.
❷ APPLY PRESSURE
Pressure regulator
Improved → Check mixture ratio feedback system.
Apply pressure to pressure regulator after disconnecting vacuum hose, and try to drive.
❸ PERFORM
Self-diagnosis Mode IV
N.G. → Check throttle sensor and/or its circuit.
Perform self-diagnosis Mode IV (for soft idle switch).
❹ PERFORM
Self-diagnosis Mode V
N.G. → Check the malfunctioning part and/or circuit.
(ignition signal).]
Perform self-diagnosis Mode V (for ignition signal).
❺ CHECK
A.I.V.
Keeps flowing
Check secondary air to exhaust system.
❻ CHECK
A.I.V. solenoid
Left ON → Check the solenoid circuit.
Check terminal voltage of A.I.V. solenoid.

ECCS SELF-DIAGNOSTIC DESCRIPTION—1990 TRUCK/PATHFINDER WITH KA24E ENGINE

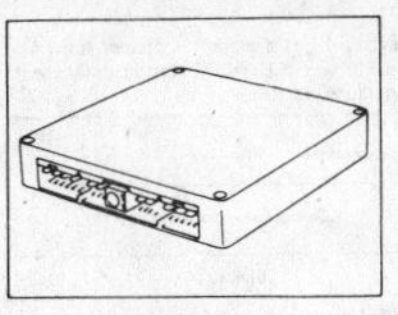

Self-diagnosis — Description

The self-diagnosis is useful to diagnose malfunctions in major sensors and actuators of the E.C.C.S. system. There are 5 modes in the self-diagnosis system.

1. **Mode I (Exhaust gas sensor monitor)**
   - During closed-loop operation:
     The green inspection lamp turns ON when a lean condition is detected and goes OFF under rich condition.
   - During open-loop operation condition:
     The green inspection lamp remains OFF or ON.
2. **Mode II (Mixture ratio feedback control monitor)**
   The green inspection lamp function is the same as Mode I.
   - During closed-loop operation:
     The red inspection lamp turns ON and OFF simultaneously with the green inspection lamp when the mixture ratio is controlled within the specified value.
   - During open-loop operation:
     The red inspection lamp remains ON or OFF.
3. **Mode III (Self-diagnostic system)**
   In this mode the number of both green and red L.E.D.'s flashing indicates the group to which the malfunctioning part belongs.
4. **Mode IV (Switches ON/OFF diagnostic system)**
   During this mode, the inspection lamps monitor the switch ON-OFF condition.
   - Soft idle switch
   - Starter switch
   - Vehicle speed sensor
5. **Mode V (Real-time diagnostic system)**
   The moment the malfunction is detected, the display will be presented immediately. That is, the condition at which the malfunction occurs can be found by observing the inspection lamps during driving test.

ECCS SELF-DIAGNOSTIC DESCRIPTION—1990 TRUCK/PATHFINDER WITH KA24E ENGINE

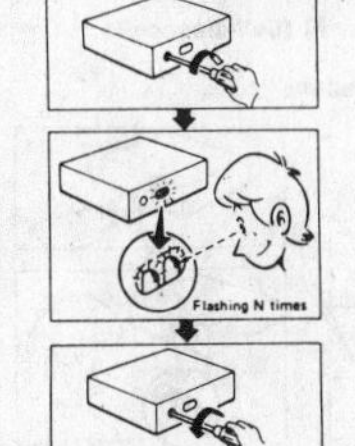

Flashing N times
Mode N

Self-diagnosis — Description (Cont'd)

HOW TO SWITCH THE DIAGNOSTIC MODES

1. Turn ignition switch "ON".
2. Turn diagnostic mode selector to E.C.U. (fully clockwise) and wait for inspection lamps to flash.
3. Count the number of flashes, and after the inspection lamps have flashed the number of the required mode, immediately turn diagnostic mode selector fully counterclockwise.

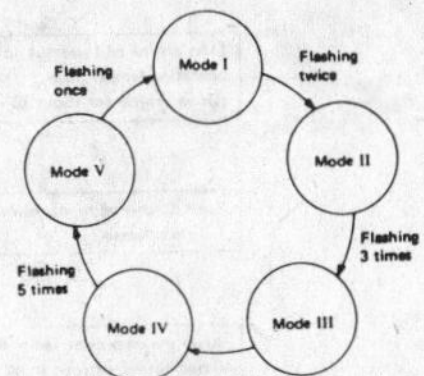

Mode I
Flashing twice
Mode II
Flashing 3 times
Mode III
Flashing 4 times
Mode IV
Flashing 5 times
Mode V
Flashing once

- **When the ignition switch is turned off during diagnosis in any mode and then turned on again (after power to the E.C.U. has dropped completely), the diagnosis will automatically return to Mode I.**
  The stored memory will be lost if:
1. Battery terminal is disconnected.
2. After selecting Mode III, Mode IV is selected.
   However, if the diagnostic mode selector is kept turned fully clockwise, it will continue to change in the order of Mode I → II → III → IV → V → I ... etc., and in this state the stored memory will not be erased.

This unit serves as an idle rpm feedback control. When the diagnostic mode selector is turned within the "diagnostic mode OFF" range, a target engine speed can be selected. Mark the original position of the selector before conducting self-diagnosis. Upon completion of self-diagnosis, return the selector to the previous position. Otherwise, engine speed may change before and after conducting self-diagnosis.

CHECK

CHECK ENGINE LIGHT

CHECK ENGINE LIGHT (For California only)

This vehicle has a check engine light on the instrument panel. This light comes ON under the following conditions:

1) When ignition switch is turned "ON" (for bulb check).
2) When systems related to emission performance malfunction in Mode I (with engine running).

- **This check engine light always illuminates and is synchronous with red L.E.D.**
- **Malfunction systems related to emission performance can be detected by self-diagnosis, and they are clarified as self-diagnostic codes in Mode III.**

Self-diagnosis — Description (Cont'd)

3) Check engine light will come "ON" only when malfunction is sensed.
   The check engine light will turn off when normal operation is resumed. Mode III memory must be cleared as the contents remain stored.

Code No.	Malfunction
12	Air flow meter circuit
13	Engine temperature sensor circuit
14	Vehicle speed sensor circuit
31	E.C.U. (E.C.C.S. control unit)
32	E.G.R. function
33	Exhaust gas sensor circuit
35	Exhaust gas temperature sensor circuit
43	Throttle sensor circuit
45	Injector leak

Use the following diagnostic flowchart to check and repair a malfunctioning system.

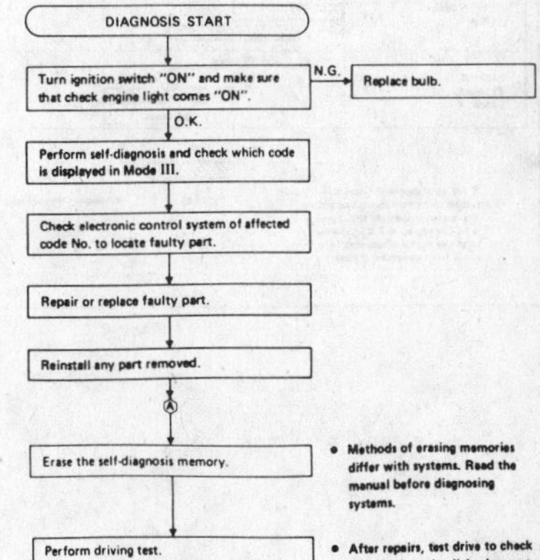

DIAGNOSIS START
Turn ignition switch "ON" and make sure that check engine light comes "ON".
N.G.
Replace bulb.
O.K.
Perform self-diagnosis and check which code is displayed in Mode III.
Check electronic control system of affected code No. to locate faulty part.
Repair or replace faulty part.
Reinstall any part removed.
A
Erase the self-diagnosis memory.
Perform driving test.
Make sure that check engine light does not come "ON" during this test.
DIAGNOSIS END
• Methods of erasing memories differ with systems. Read the manual before diagnosing systems.
• After repairs, test drive to check that check engine light does not come on.
• Test driving modes differ with systems. Read the manual before test driving.

ECCS SELF-DIAGNOSTIC MODES I AND II— 1990 TRUCK/PATHFINDER WITH KA24E ENGINE

Self-diagnosis — Mode I (Exhaust gas sensor monitor)

This mode checks the exhaust gas sensor for proper functioning. The operation of the E.C.U. L.E.D. in this mode differs with mixture ratio control conditions as follows:

Mode	L.E.D.	Engine stopped (Ignition switch "ON")	Engine running	
			Open loop condition	Closed loop condition
Mode I (Monitor A)	Green	ON	*Remains ON or OFF	Blinks
	Red	ON	Except for California model • OFF	For California model • ON: when the CHECK ENGINE LIGHT ITEMS are stored in the E.C.U. • OFF: except for the above condition

*: Maintains conditions just before switching to open loop

EXHAUST GAS SENSOR FUNCTION CHECK

If the number of L.E.D. blinks is less than that specified, replace the exhaust gas sensor.
If the L.E.D. does not blink, check exhaust gas sensor circuit.

EXHAUST GAS SENSOR CIRCUIT CHECK

Self-diagnosis — Mode II (Mixture ratio feedback control monitor)

This mode checks, through the E.C.U. L.E.D., optimum control of the mixture ratio. The operation of the L.E.D., as shown below, differs with the control conditions of the mixture ratio (for example, richer or leaner mixture ratios, etc., which are controlled by the E.C.U.).

Mode	L.E.D.	Engine stopped (Ignition switch "ON")	Engine running			
			Open loop condition	Closed loop condition		
Mode II (Monitor B)	Green	ON	*Remains ON or OFF	Blinks		
				Compensating mixture ratio		
				More than 5% rich	Between 5% lean and 5% rich	More
	Red	OFF	*Remains ON or OFF (synchronous with green L.E.D.)	OFF	Synchronized with green L.E.D.	Remains ON

*: Maintains conditions just before switching to open loop

If the red L.E.D. remains on or off during the closed-loop operation, the mixture ratio may not be controlled properly. Using the following procedures, check the related components or adjust the mixture ratio.

COMPONENT CHECK OR MIXTURE RATIO ADJUSTMENT

ECCS SELF-DIAGNOSTIC MODE III— 1990 TRUCK/PATHFINDER WITH KA24E ENGINE

Self-diagnosis — Mode III (Self-diagnostic system)

The E.C.U. constantly monitors the function of these sensors and actuators, regardless of ignition key position. If a malfunction occurs, the information is stored in the E.C.U. and can be retrieved from the memory by turning on the diagnostic mode selector, located on the side of the E.C.U. When activated, the malfunction is indicated by flashing a red and a green L.E.D. (Light Emitting Diode), also located on the E.C.U. Since all the self-diagnostic results are stored in the E.C.U.'s memory even intermittent malfunctions can be diagnosed.

A malfunction is indicated by the number of both red and green flashing L.E.D.s. First, the red L.E.D. flashes and the green flashes follow. The red L.E.D. corresponds to units of ten and the green L.E.D. corresponds to units of one. For example, when the red L.E.D. flashes once and the green L.E.D. flashes twice, this signifies the number "12", showing that the air flow meter signal is malfunctioning. All problems are classified by code numbers in this way.

- **When the engine fails to start, crank it two or more seconds before beginning self-diagnosis.**
- **Before starting self-diagnosis, do not erase the stored memory before beginning self-diagnosis. If it is erased, the self-diagnosis function for intermittent malfunctions will be lost.**

DISPLAY CODE TABLE

Code No.	Detected items	California	Non-California
11	Crank angle sensor circuit	X	X
12	Air flow meter circuit	X	X
13	Engine temperature sensor circuit	X	X
14	Vehicle speed sensor circuit	X	X
21	Ignition signal missing in primary coil	X	X
31	E.C.U. (E.C.C.S. control unit)	X	X
32	E.G.R. function	X	–
33	Exhaust gas sensor circuit	X	X
35	Exhaust gas temperature sensor circuit	X	–
41	Air temperature sensor circuit	X	X
43	Throttle sensor circuit	X	X
45	Injector leak	X	–
55	No malfunction in the above circuits	X	X

X: Available —: Not available

ECCS SELF-DIAGNOSTIC MODE III (CONT.)—1990 TRUCK/PATHFINDER WITH KA24E ENGINE

Self-diagnosis — Mode III (Self-diagnostic system) (Cont'd)

RETENTION OF DIAGNOSTIC RESULTS

The diagnostic results will remain in E.C.U. memory until the starter is operated fifty times after a diagnostic item has been judged to be malfunctioning. The diagnostic result will then be cancelled automatically. If a diagnostic item which has been judged to be malfunctioning and stored in memory is again judged to be malfunctioning before the starter is operated fifty times, the second result will replace the previous one. It will be stored in E.C.U. memory until the starter is operated fifty times more.

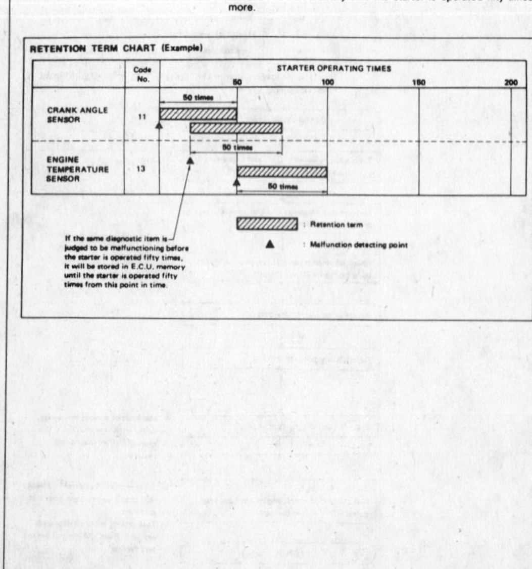
RETENTION TERM CHART (Example)
Code No.
STARTER OPERATING TIMES
50
100
150
200
CRANK ANGLE SENSOR
11
ENGINE TEMPERATURE SENSOR
13
50 times
50 times
50 times
: Retention term
: Malfunction detecting point
If the same diagnostic item is judged to be malfunctioning before the starter is operated fifty times, it will be stored in E.C.U. memory until the starter is operated fifty times from this point in time.

Self-diagnosis — Mode III (Self-diagnostic system) (Cont'd)

SELF-DIAGNOSTIC PROCEDURE

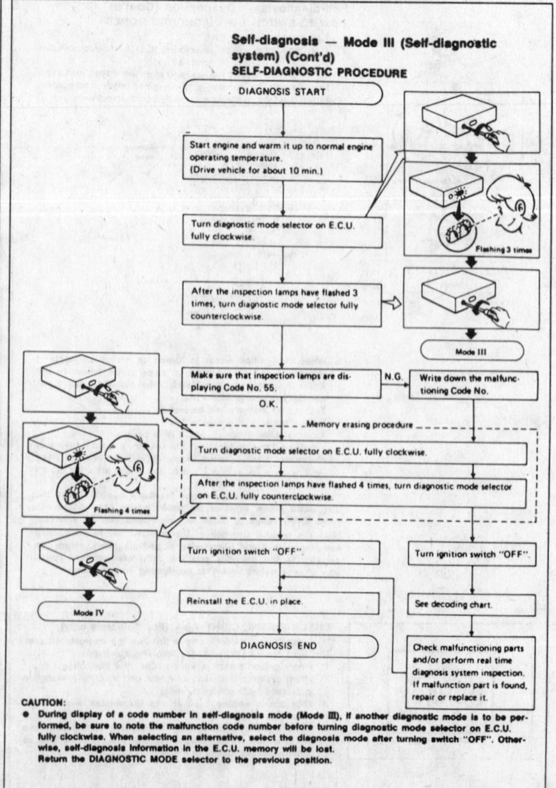
DIAGNOSIS START
Start engine and warm it up to normal engine operating temperature. (Drive vehicle for about 10 min.)
Turn diagnostic mode selector on E.C.U. fully clockwise.
Flashing 3 times
After the inspection lamps have flashed 3 times, turn diagnostic mode selector fully counterclockwise.
Mode III
Make sure that inspection lamps are displaying Code No. 55.
N.G.
Write down the malfunctioning Code No.
O.K.
Memory erasing procedure
Turn diagnostic mode selector on E.C.U. fully clockwise.
After the inspection lamps have flashed 4 times, turn diagnostic mode selector on E.C.U. fully counterclockwise.
Flashing 4 times
Turn ignition switch "OFF".
Turn ignition switch "OFF".
Reinstall the E.C.U. in place.
See decoding chart.
Mode IV
DIAGNOSIS END
Check malfunctioning parts and/or perform real time diagnosis system inspection. If malfunction part is found, repair or replace it.

CAUTION:

- **During display of a code number in self-diagnosis mode (Mode III), if another diagnostic mode is to be performed, be sure to note the malfunction code number before turning diagnostic mode selector on E.C.U. fully clockwise. When selecting an alternative, select the diagnosis mode after turning switch "OFF". Otherwise, self-diagnosis information in the E.C.U. memory will be lost.
Return the DIAGNOSTIC MODE selector to the previous position.**

ECCS SELF-DIAGNOSTIC MODE III DISPLAY CODES—1990 TRUCK/PATHFINDER WITH KA24E ENGINE

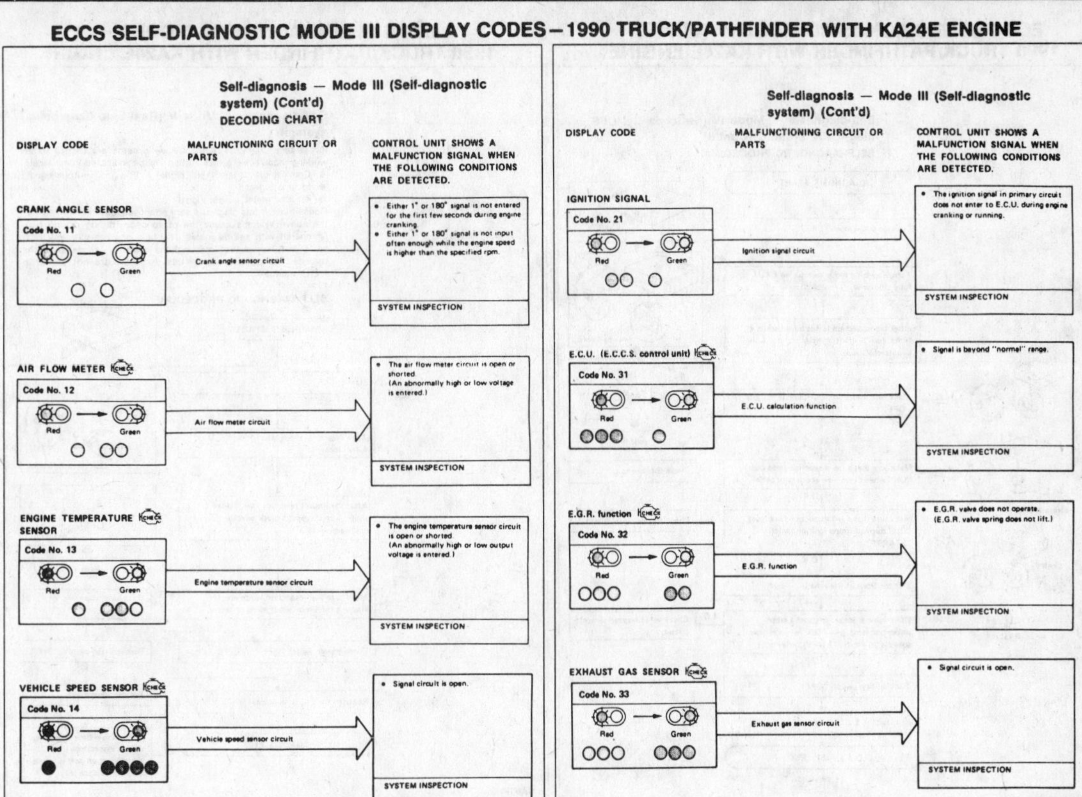
Self-diagnosis — Mode III (Self-diagnostic system) (Cont'd)
DECODING CHART
DISPLAY CODE
MALFUNCTIONING CIRCUIT OR PARTS
CONTROL UNIT SHOWS A MALFUNCTION SIGNAL WHEN THE FOLLOWING CONDITIONS ARE DETECTED.
CRANK ANGLE SENSOR
Code No. 11
Red
Green
Crank angle sensor circuit
• Either 1° or 180° signal is not entered for the first few seconds during engine cranking.
• Either 1° or 180° signal is not input often enough while the engine speed is higher than the specified rpm.
SYSTEM INSPECTION
AIR FLOW METER
Code No. 12
Air flow meter circuit
• The air flow meter circuit is open or shorted. (An abnormally high or low voltage is entered.)
SYSTEM INSPECTION
ENGINE TEMPERATURE SENSOR
Code No. 13
Engine temperature sensor circuit
• The engine temperature sensor circuit is open or shorted. (An abnormally high or low output voltage is entered.)
SYSTEM INSPECTION
VEHICLE SPEED SENSOR
Code No. 14
Vehicle speed sensor circuit
• Signal circuit is open.
SYSTEM INSPECTION
Self-diagnosis — Mode III (Self-diagnostic system) (Cont'd)
DISPLAY CODE
MALFUNCTIONING CIRCUIT OR PARTS
CONTROL UNIT SHOWS A MALFUNCTION SIGNAL WHEN THE FOLLOWING CONDITIONS ARE DETECTED.
IGNITION SIGNAL
Code No. 21
Ignition signal circuit
• The ignition signal in primary circuit does not enter to E.C.U. during engine cranking or running.
SYSTEM INSPECTION
E.C.U. (E.C.C.S. control unit)
Code No. 31
E.C.U. calculation function
• Signal is beyond "normal" range.
SYSTEM INSPECTION
E.G.R. function
Code No. 32
E.G.R. function
• E.G.R. valve does not operate. (E.G.R. valve spring does not lift.)
SYSTEM INSPECTION
EXHAUST GAS SENSOR
Code No. 33
Exhaust gas sensor circuit
• Signal circuit is open.
SYSTEM INSPECTION

ECCS SELF-DIAGNOSTIC MODE III DISPLAY CODES (CONT.)—1990 TRUCK/ PATHFINDER WITH KA24E ENGINE

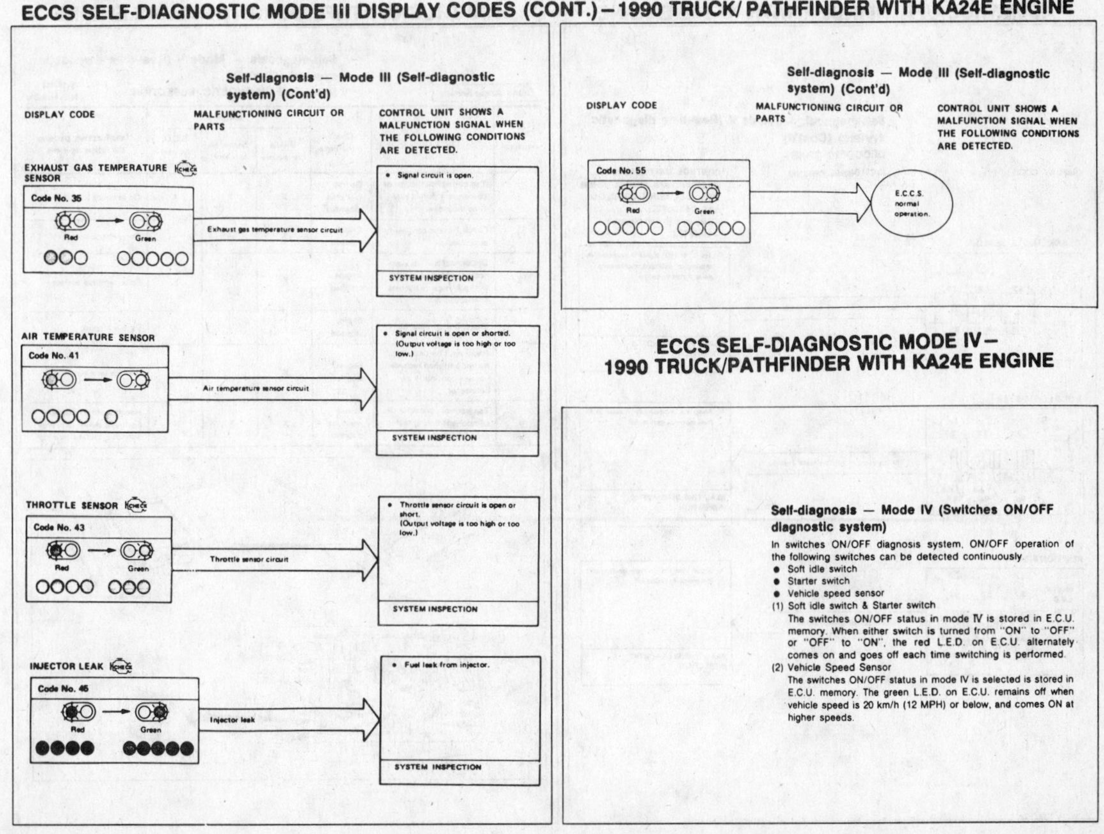
Self-diagnosis — Mode III (Self-diagnostic system) (Cont'd)
DISPLAY CODE
MALFUNCTIONING CIRCUIT OR PARTS
CONTROL UNIT SHOWS A MALFUNCTION SIGNAL WHEN THE FOLLOWING CONDITIONS ARE DETECTED.
EXHAUST GAS TEMPERATURE SENSOR
Code No. 35
Red
Green
Exhaust gas temperature sensor circuit
• Signal circuit is open.
SYSTEM INSPECTION
AIR TEMPERATURE SENSOR
Code No. 41
Air temperature sensor circuit
• Signal circuit is open or shorted. (Output voltage is too high or too low.)
SYSTEM INSPECTION
THROTTLE SENSOR
Code No. 43
Throttle sensor circuit
• Throttle sensor circuit is open or short. (Output voltage is too high or too low.)
SYSTEM INSPECTION
INJECTOR LEAK
Code No. 45
Injector leak
• Fuel leak from injector.
SYSTEM INSPECTION
Self-diagnosis — Mode III (Self-diagnostic system) (Cont'd)
DISPLAY CODE
MALFUNCTIONING CIRCUIT OR PARTS
CONTROL UNIT SHOWS A MALFUNCTION SIGNAL WHEN THE FOLLOWING CONDITIONS ARE DETECTED.
Code No. 55
E.C.C.S. normal operation.

ECCS SELF-DIAGNOSTIC MODE IV—1990 TRUCK/PATHFINDER WITH KA24E ENGINE

Self-diagnosis — Mode IV (Switches ON/OFF diagnostic system)

In switches ON/OFF diagnosis system, ON/OFF operation of the following switches can be detected continuously.

- Soft idle switch
- Starter switch
- Vehicle speed sensor

(1) Soft idle switch & Starter switch
The switches ON/OFF status in mode IV is stored in E.C.U. memory. When either switch is turned from "ON" to "OFF" or "OFF" to "ON", the red L.E.D. on E.C.U. alternately comes on and goes off each time switching is performed.

(2) Vehicle Speed Sensor
The switches ON/OFF status in mode IV is selected is stored in E.C.U. memory. The green L.E.D. on E.C.U. remains off when vehicle speed is 20 km/h (12 MPH) or below, and comes ON at higher speeds.

ECCS SELF-DIAGNOSTIC MODE IV (CONT.)—1990 TRUCK/PATHFINDER WITH KA24E ENGINE

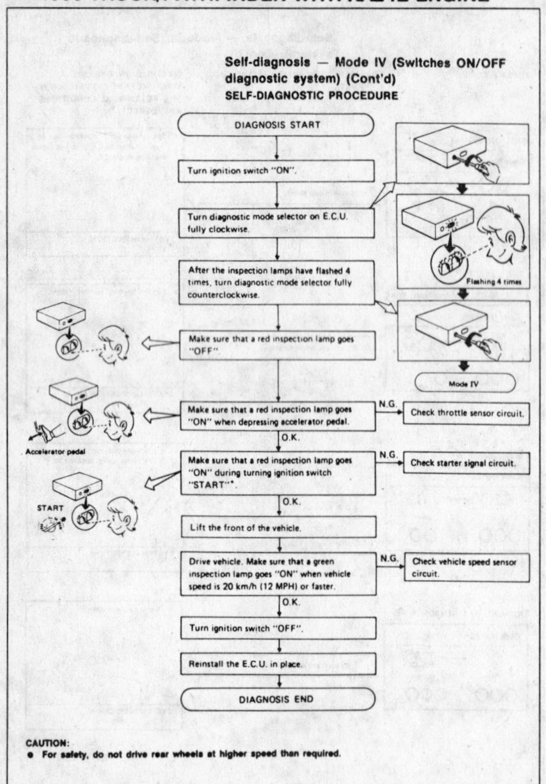
Self-diagnosis — Mode IV (Switches ON/OFF diagnostic system) (Cont'd)
SELF-DIAGNOSTIC PROCEDURE
DIAGNOSIS START
Turn ignition switch "ON".
Turn diagnostic mode selector on E.C.U. fully clockwise.
After the inspection lamps have flashed 4 times, turn diagnostic mode selector fully counterclockwise.
Flashing 4 times
Make sure that a red inspection lamp goes "OFF".
Mode IV
Make sure that a red inspection lamp goes "ON" when depressing accelerator pedal.
N.G.
Check throttle sensor circuit.
O.K.
Accelerator pedal
Make sure that a red inspection lamp goes "ON" during turning ignition switch "START".
N.G.
Check starter signal circuit.
START
O.K.
Lift the front of the vehicle.
Drive vehicle. Make sure that a green inspection lamp goes "ON" when vehicle speed is 20 km/h (12 MPH) or faster.
N.G.
Check vehicle speed sensor circuit.
O.K.
Turn ignition switch "OFF".
Reinstall the E.C.U. in place.
DIAGNOSIS END
CAUTION:
• For safety, do not drive rear wheels at higher speed than required.

ECCS SELF-DIAGNOSTIC MODE V—1990 TRUCK/PATHFINDER WITH KA24E ENGINE

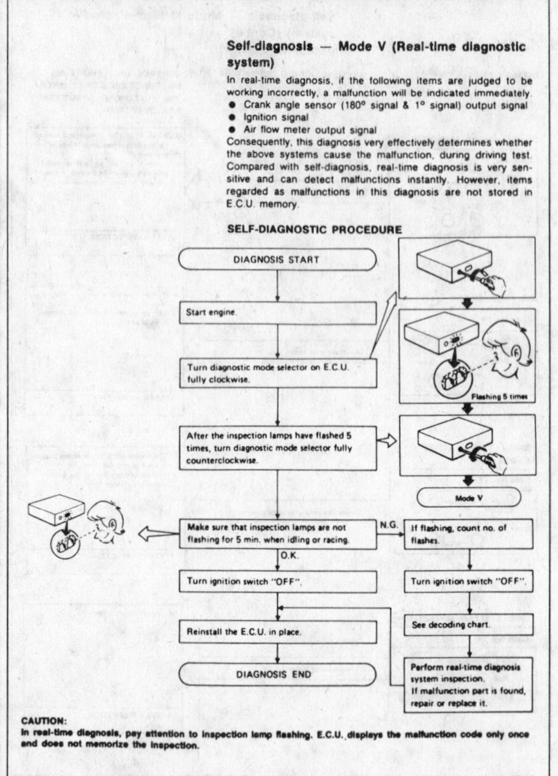
Self-diagnosis — Mode V (Real-time diagnostic system)
In real-time diagnosis, if the following items are judged to be working incorrectly, a malfunction will be indicated immediately.
• Crank angle sensor (180° signal & 1° signal) output signal
• Ignition signal
• Air flow meter output signal
Consequently, this diagnosis very effectively determines whether the above systems cause the malfunction, during driving test. Compared with self-diagnosis, real-time diagnosis is very sensitive and can detect malfunctions instantly. However, items regarded as malfunctions in this diagnosis are not stored in E.C.U. memory.
SELF-DIAGNOSTIC PROCEDURE
DIAGNOSIS START
Start engine.
Turn diagnostic mode selector on E.C.U. fully clockwise.
Flashing 5 times
After the inspection lamps have flashed 5 times, turn diagnostic mode selector fully counterclockwise.
Mode V
Make sure that inspection lamps are not flashing for 5 min. when idling or racing.
N.G.
If flashing, count no. of flashes.
O.K.
Turn ignition switch "OFF".
Turn ignition switch "OFF".
See decoding chart.
Reinstall the E.C.U. in place.
Perform real-time diagnosis system inspection.
If malfunction part is found, repair or replace it.
DIAGNOSIS END
CAUTION:
In real-time diagnosis, pay attention to inspection lamp flashing. E.C.U. displays the malfunction code only once and does not memorize the inspection.

ECCS SELF-DIAGNOSTIC MODE V DECODING CHART—1990 TRUCK/PATHFINDER WITH KA24E ENGINE

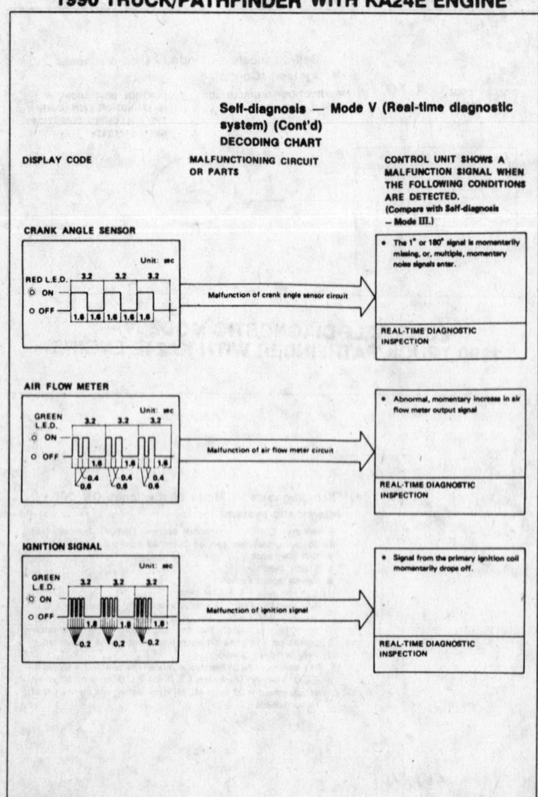
Self-diagnosis — Mode V (Real-time diagnostic system) (Cont'd)
DECODING CHART
DISPLAY CODE
MALFUNCTIONING CIRCUIT OR PARTS
CONTROL UNIT SHOWS A MALFUNCTION SIGNAL WHEN THE FOLLOWING CONDITIONS ARE DETECTED. (Compare with Self-diagnosis – Mode III.)
CRANK ANGLE SENSOR
Unit: sec
RED L.E.D.
ON
OFF
3.2
1.6
Malfunction of crank angle sensor circuit
• The 1° or 180° signal is momentarily missing, or, multiple, momentary noise signals enter.
REAL-TIME DIAGNOSTIC INSPECTION
AIR FLOW METER
Unit: sec
GREEN L.E.D.
ON
OFF
3.2
1.6
0.4
0.8
Malfunction of air flow meter circuit
• Abnormal, momentary increase in air flow meter output signal
REAL-TIME DIAGNOSTIC INSPECTION
IGNITION SIGNAL
Unit: sec
GREEN L.E.D.
ON
OFF
3.2
1.8
0.2
Malfunction of ignition signal
• Signal from the primary ignition coil momentarily drops off.
REAL-TIME DIAGNOSTIC INSPECTION

ECCS SELF-DIAGNOSTIC MODE V INSPECTION—1990 TRUCK/PATHFINDER WITH KA24E ENGINE

Self-diagnosis — Mode V (Real-time diagnostic system) (Cont'd)
REAL-TIME DIAGNOSTIC INSPECTION

Crank Angle Sensor

X: Available
–: Not available

Check sequence	Check items	Check conditions	Check parts: Middle connectors	Sensor & actuator	E.C.U. harness connector	If malfunction, perform the following items.
1	Tap harness connector or component during real-time diagnosis.	During real-time diagnosis	X	X	X	Go to check item 2.
2	Check harness continuity at connector.	Engine stopped	X	–	–	Go to check item 3.
3	Disconnect harness connector, and then check dust adhesion to harness connector.	Engine stopped	X	–	X	Clean terminal surface.
4	Check pin terminal bend.	Engine stopped	–	–	X	Take out bend.
5	Reconnect harness connector and then recheck harness continuity at connector.	Engine stopped	X	–	–	Replace terminal.
6	Tap harness connector or component during real-time diagnosis.	During real-time diagnosis	X	X	X	If malfunction codes are displayed during real-time diagnosis, replace terminal.

E.C.U. harness connector
Distributor
Crank angle sensor

ECCS SELF-DIAGNOSTIC MODE V INSPECTION—1990 TRUCK/PATHFINDER WITH KA24E ENGINE

Self-diagnosis — Mode V (Real-time diagnostic system) (Cont'd)

X: Available
–: Not available

Air Flow Meter

Check sequence	Check items	Check conditions	Check parts			If malfunction, perform the following items.
			Middle connectors	Sensor & actuator	E.C.U. harness connector	
1	Tap harness connector or component during real-time diagnosis.	During real-time diagnosis	X	X	X	Go to check item 2.
2	Check harness continuity at connector.	Engine stopped	X	–	–	Go to check item 3.
3	Disconnect harness connector, and then check dust adhesion to harness connector.	Engine stopped	X	–	X	Clean terminal surface.
4	Check pin terminal bend.	Engine stopped	–	–	X	Take out bend.
5	Reconnect harness connector and then recheck harness continuity at connector.	Engine stopped	X	–	–	Replace terminal.
6	Tap harness connector or component during real-time diagnosis.	During real-time diagnosis	X	X	X	If malfunction codes are displayed during real-time diagnosis, replace terminal.

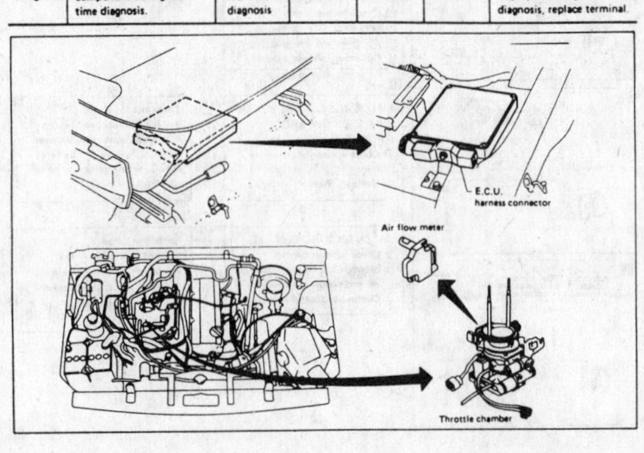
E.C.U.
harness connector
Air flow meter
Throttle chamber

Self-diagnosis — Mode V (Real-time diagnostic system) (Cont'd)

X: Available
–: Not available

Ignition Signal

Check sequence	Check items	Check conditions	Check parts			If malfunction, perform the following items.
			Middle connectors	Sensor & actuator	E.C.U. harness connector	
1	Tap harness connector or component during real-time diagnosis.	During real-time diagnosis	X	X	X	Go to check item 2.
2	Check harness continuity at connector.	Engine stopped	X	–	–	Go to check item 3.
3	Disconnect harness connector, and then check dust adhesion to harness connector.	Engine stopped	X	–	X	Clean terminal surface.
4	Check pin terminal bend.	Engine stopped	–	–	X	Take out bend.
5	Reconnect harness connector and then recheck harness continuity at connector.	Engine stopped	X	–	–	Replace terminal.
6	Tap harness connector or component during real-time diagnosis.	During real-time diagnosis	X	X	X	If malfunction codes are displayed during real-time diagnosis, replace terminal.

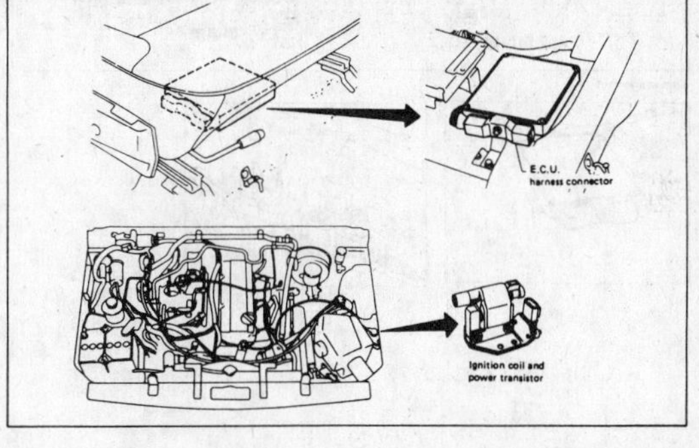
E.C.U.
harness connector
Ignition coil and
power transistor

ECCS CAUTION—1990 TRUCK/PATHFINDER WITH KA24E ENGINE

Diagnostic Procedure

CAUTION:

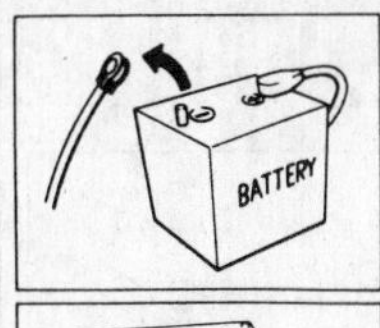
BATTERY

1. **Before connecting or disconnecting the E.C.U. harness connector to or from any E.C.U., be sure to turn the ignition switch to the "OFF" position and disconnect the negative battery terminal in order not to damage E.C.U. as battery voltage is applied to E.C.U. even if ignition switch is turned off. Failure to do so may damage the E.C.U.**

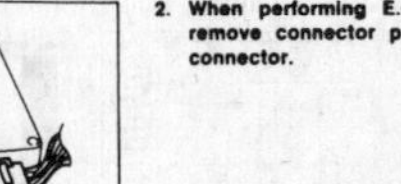

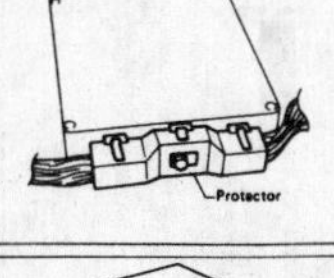
Protector

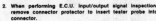

2. **When performing E.C.U. input/output signal inspection, remove connector protector to insert tester probe into connector.**

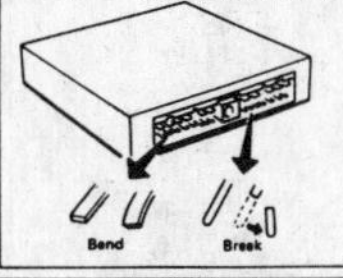
Bend
Break

3. **When connecting or disconnecting pin connectors into or from E.C.U., take care not to damage pin terminals.**
4. **Make sure that there are not any bends or breaks on E.C.U. pin terminal, when connecting pin connectors.**

Perform E.C.U. input/output signal inspection before replacement.
OLD ONE

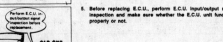

5. **Before replacing E.C.U., perform E.C.U. input/output signal inspection and make sure whether the E.C.U. unit functions properly or not.**

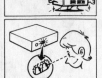

6. **After performing this "Diagnostic Procedure", perform E.C.C.S. self-diagnosis and driving test.**

Diagnostic Procedure (Cont'd)

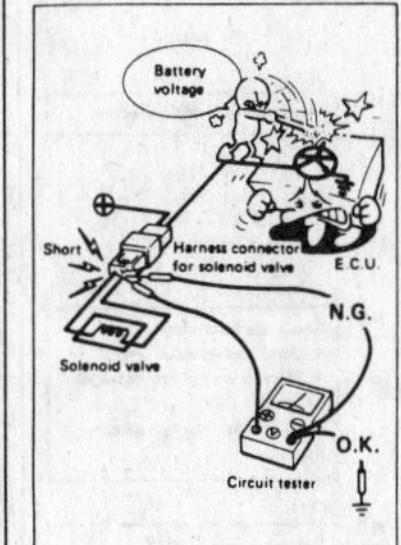
Battery voltage
Short
Harness connector for solenoid valve
E.C.U.
N.G.
Solenoid valve
O.K.
Circuit tester

7. **When measuring E.C.U. controlled components supply voltage with a circuit tester, separate one tester probe from the other.**
   If the two tester probes accidentally make contact with each other during measurement, the circuit will be shorted, resulting in damage to the control unit power transistor.

ECCS DIAGNOSTIC CHART—1990 TRUCK/PATHFINDER WITH KA24E ENGINE

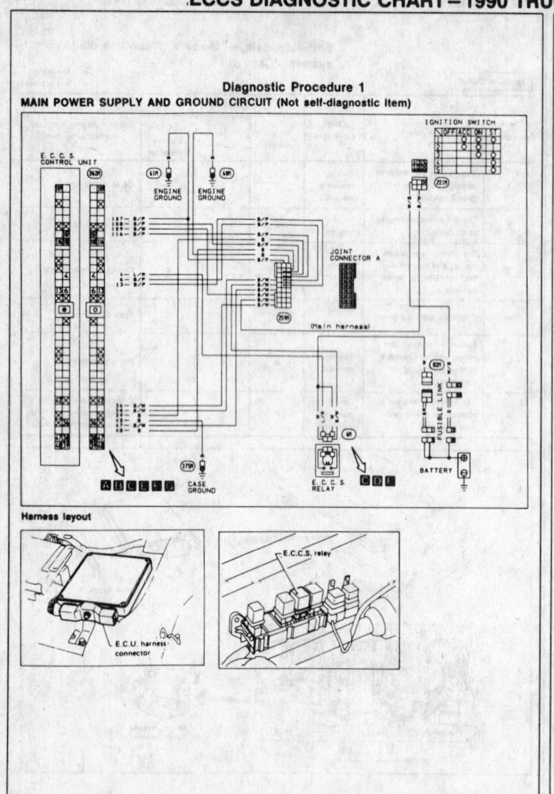
Diagnostic Procedure 1
MAIN POWER SUPPLY AND GROUND CIRCUIT (Not self-diagnostic item)
E.C.C.S. CONTROL UNIT
ENGINE GROUND
ENGINE GROUND
IGNITION SWITCH
JOINT CONNECTOR A
(Main harness)
FUSIBLE LINK
BATTERY
CASE GROUND
E.C.C.S. RELAY
Harness layout
E.C.U. harness connector
E.C.C.S. relay

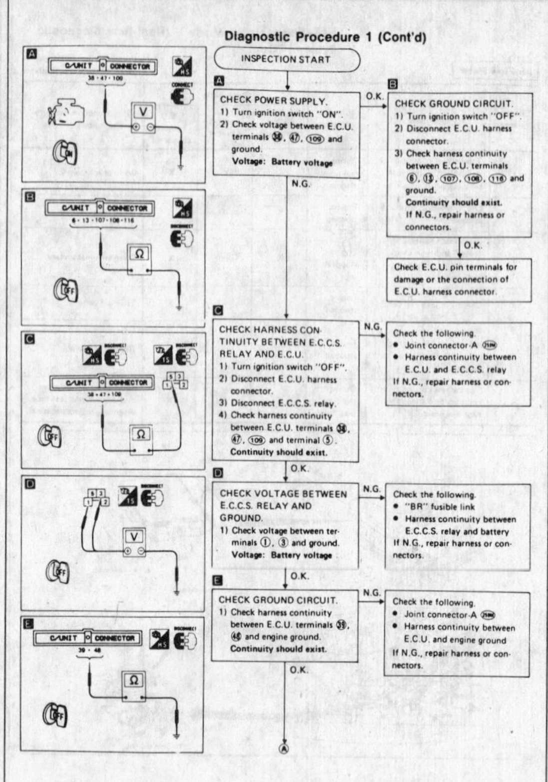
Diagnostic Procedure 1 (Cont'd)
C/UNIT CONNECTOR
CONNECT
DISCONNECT
INSPECTION START
A
CHECK POWER SUPPLY.
1) Turn ignition switch "ON".
2) Check voltage between E.C.U. terminals 38, 47, 109 and ground.
Voltage: Battery voltage
O.K.
B
CHECK GROUND CIRCUIT.
1) Turn ignition switch "OFF".
2) Disconnect E.C.U. harness connector.
3) Check harness continuity between E.C.U. terminals 6, 13, 107, 108, 116 and ground.
Continuity should exist.
If N.G., repair harness or connectors.
O.K.
Check E.C.U. pin terminals for damage or the connection of E.C.U. harness connector.
N.G.
C
CHECK HARNESS CONTINUITY BETWEEN E.C.C.S. RELAY AND E.C.U.
1) Turn ignition switch "OFF".
2) Disconnect E.C.U. harness connector.
3) Disconnect E.C.C.S. relay.
4) Check harness continuity between E.C.U. terminals 38, 47, 109 and terminal 5.
Continuity should exist.
N.G.
Check the following.
• Joint connector-A
• Harness continuity between E.C.U. and E.C.C.S. relay
If N.G., repair harness or connectors.
O.K.
D
CHECK VOLTAGE BETWEEN E.C.C.S. RELAY AND GROUND.
1) Check voltage between terminals 1, 3 and ground.
Voltage: Battery voltage
N.G.
Check the following.
• "BR" fusible link
• Harness continuity between E.C.C.S. relay and battery
If N.G., repair harness or connectors.
O.K.
E
CHECK GROUND CIRCUIT.
1) Check harness continuity between E.C.U. terminals 39, 48 and engine ground.
Continuity should exist.
N.G.
Check the following.
• Joint connector-A
• Harness continuity between E.C.U. and engine ground
If N.G., repair harness or connectors.
O.K.
A

ECCS DIAGNOSTIC CHART—1990 TRUCK/PATHFINDER WITH KA24E ENGINE

Diagnostic Procedure 1 (Cont'd)

Ⓐ

F

CHECK OUTPUT SIGNAL CIRCUIT.
1) Check harness continuity between E.C.U. terminal ④ and terminal ②.
Continuity should exist.

N.G. → Repair harness or connectors.

O.K.

G

CHECK INPUT SIGNAL CIRCUIT.
1) Turn ignition switch "ON".
2) Check voltage between E.C.U. terminal ㊱ and ground.
Voltage: Battery voltage

N.G. → Check the following.
- Joint connector-A
- Harness continuity between ignition switch and E.C.U.

If N.G., repair harness or connectors.

O.K.

CHECK COMPONENT (E.C.C.S. relay).

N.G. → Replace E.C.C.S. relay.

O.K.

Check E.C.U. pin terminals for damage or the connection of E.C.U. harness connector.

C/UNIT CONNECTOR

ECCS SELF-DIAGNOSTIC CHART—1990 TRUCK/PATHFINDER WITH KA24E ENGINE

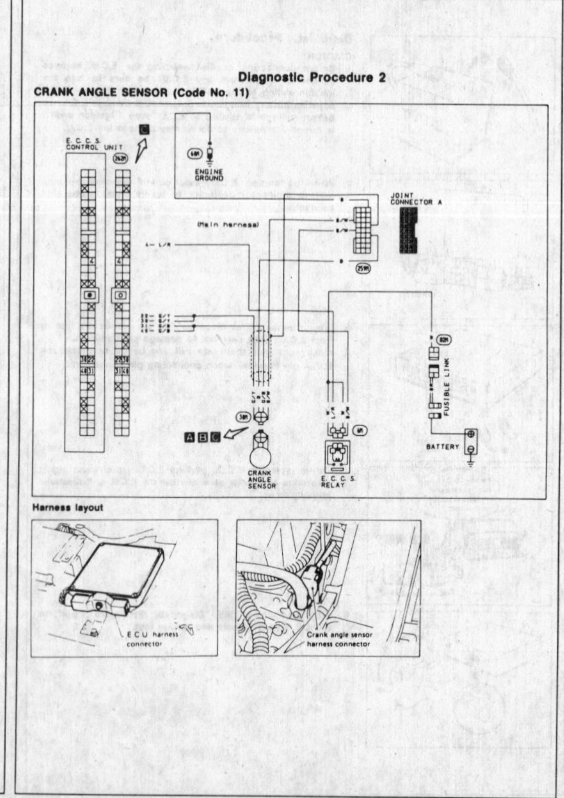
Diagnostic Procedure 2
CRANK ANGLE SENSOR (Code No. 11)
E.C.C.S. CONTROL UNIT
ENGINE GROUND
JOINT CONNECTOR A
(Main harness)
FUSIBLE LINK
BATTERY
CRANK ANGLE SENSOR
E.C.C.S. RELAY
Harness layout
E.C.U. harness connector
Crank angle sensor harness connector

ECCS SELF-DIAGNOSTIC CHART—1990 TRUCK/PATHFINDER WITH KA24E ENGINE

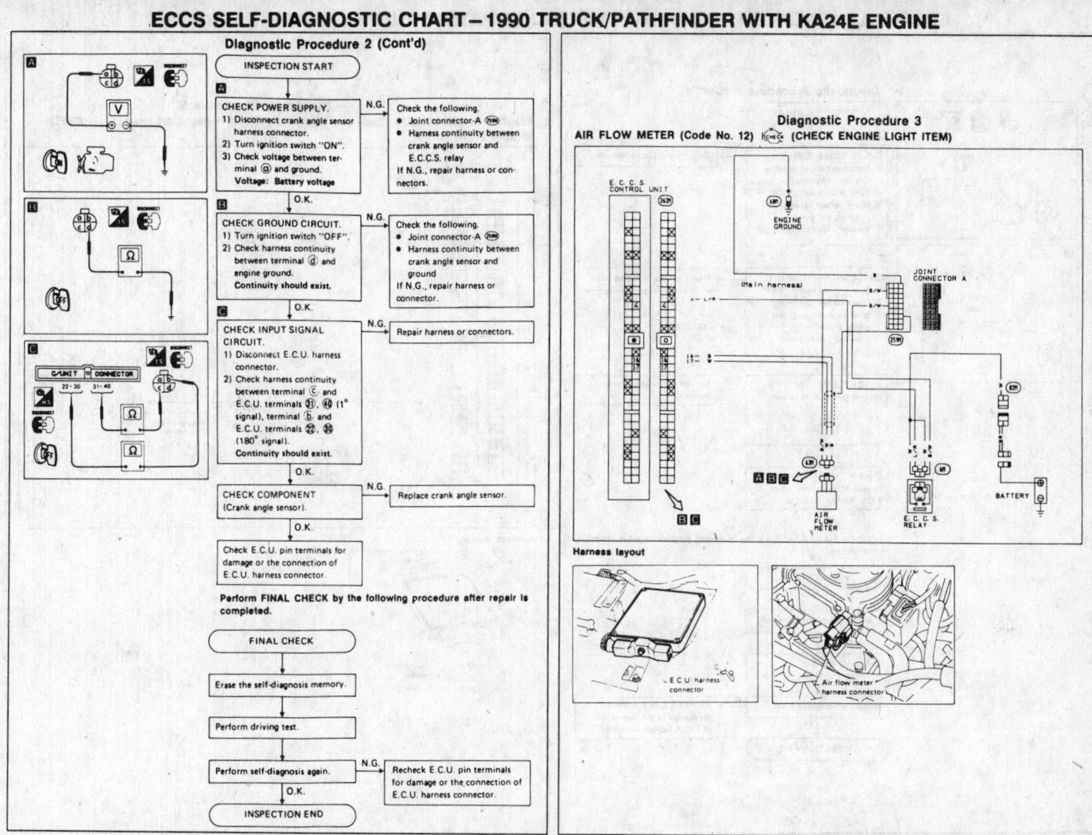
Diagnostic Procedure 2 (Cont'd)
INSPECTION START
A CHECK POWER SUPPLY.
1) Disconnect crank angle sensor harness connector.
2) Turn ignition switch "ON".
3) Check voltage between terminal ⓐ and ground.
Voltage: Battery voltage
N.G. Check the following.
• Joint connector-A
• Harness continuity between crank angle sensor and E.C.C.S. relay
If N.G., repair harness or connectors.
O.K.
B CHECK GROUND CIRCUIT.
1) Turn ignition switch "OFF".
2) Check harness continuity between terminal ⓓ and engine ground.
Continuity should exist.
N.G. Check the following.
• Joint connector-A
• Harness continuity between crank angle sensor and ground
If N.G., repair harness or connector.
O.K.
C CHECK INPUT SIGNAL CIRCUIT.
1) Disconnect E.C.U. harness connector.
2) Check harness continuity between terminal ⓒ and E.C.U. terminals ㉛, ㊵ (1° signal), terminal ⓑ and E.C.U. terminals ㉒, ㉚ (180° signal).
Continuity should exist.
N.G. Repair harness or connectors.
O.K.
CHECK COMPONENT (Crank angle sensor).
N.G. Replace crank angle sensor.
O.K.
Check E.C.U. pin terminals for damage or the connection of E.C.U. harness connector.
Perform FINAL CHECK by the following procedure after repair is completed.
FINAL CHECK
Erase the self-diagnosis memory.
Perform driving test.
Perform self-diagnosis again.
N.G. Recheck E.C.U. pin terminals for damage or the connection of E.C.U. harness connector.
O.K.
INSPECTION END
Diagnostic Procedure 3
AIR FLOW METER (Code No. 12) (CHECK ENGINE LIGHT ITEM)
E.C.C.S. CONTROL UNIT
ENGINE GROUND
JOINT CONNECTOR A
(Main harness)
AIR FLOW METER
E.C.C.S. RELAY
BATTERY
Harness layout
E.C.U. harness connector
Air flow meter harness connector

ECCS SELF-DIAGNOSTIC CHART—1990 TRUCK/PATHFINDER WITH KA24E ENGINE

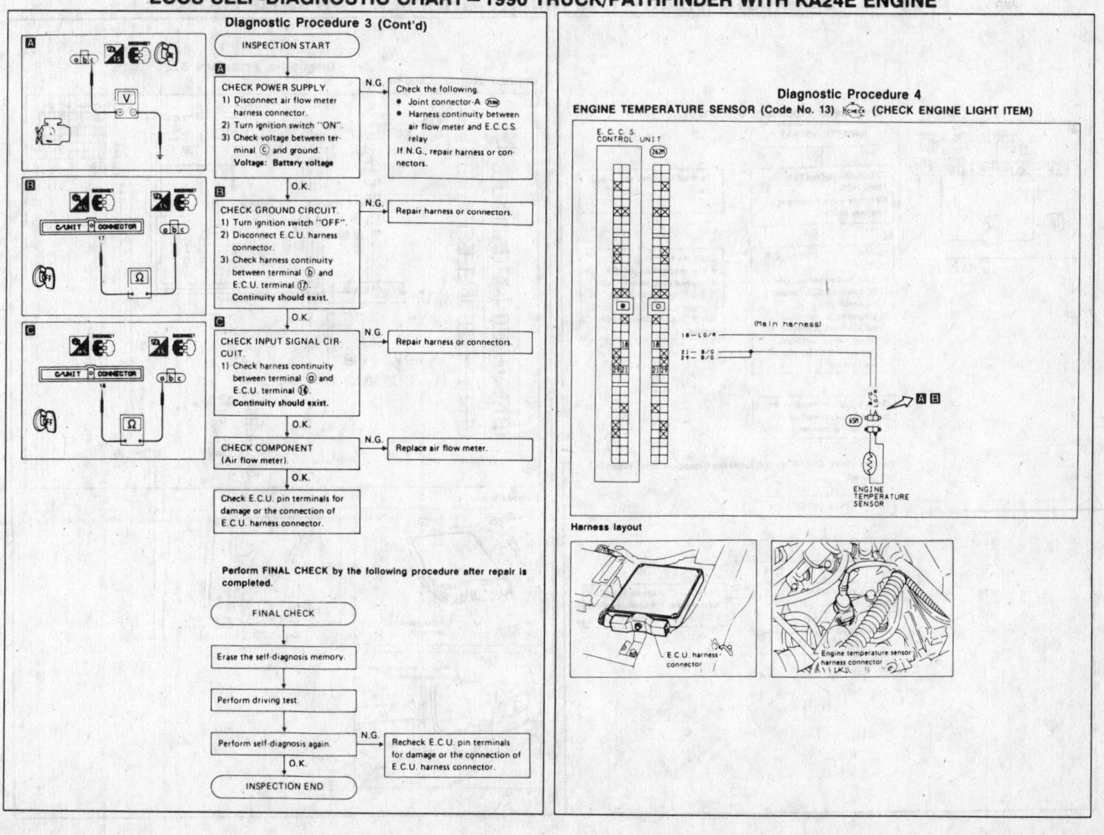
Diagnostic Procedure 3 (Cont'd)
INSPECTION START
A CHECK POWER SUPPLY.
1) Disconnect air flow meter harness connector.
2) Turn ignition switch "ON".
3) Check voltage between terminal ⓒ and ground.
Voltage: Battery voltage
N.G. Check the following.
• Joint connector-A
• Harness continuity between air flow meter and E.C.C.S. relay
If N.G., repair harness or connectors.
O.K.
B CHECK GROUND CIRCUIT.
1) Turn ignition switch "OFF".
2) Disconnect E.C.U. harness connector.
3) Check harness continuity between terminal ⓑ and E.C.U. terminal ⑰.
Continuity should exist.
N.G. Repair harness or connectors.
O.K.
C CHECK INPUT SIGNAL CIRCUIT.
1) Check harness continuity between terminal ⓐ and E.C.U. terminal ⑯.
Continuity should exist.
N.G. Repair harness or connectors.
O.K.
CHECK COMPONENT (Air flow meter).
N.G. Replace air flow meter.
O.K.
Check E.C.U. pin terminals for damage or the connection of E.C.U. harness connector.
Perform FINAL CHECK by the following procedure after repair is completed.
FINAL CHECK
Erase the self-diagnosis memory.
Perform driving test.
Perform self-diagnosis again.
N.G. Recheck E.C.U. pin terminals for damage or the connection of E.C.U. harness connector.
O.K.
INSPECTION END
Diagnostic Procedure 4
ENGINE TEMPERATURE SENSOR (Code No. 13) (CHECK ENGINE LIGHT ITEM)
E.C.C.S. CONTROL UNIT
(Main harness)
ENGINE TEMPERATURE SENSOR
Harness layout
E.C.U. harness connector
Engine temperature sensor harness connector

ECCS SELF-DIAGNOSTIC CHART—1990 TRUCK/PATHFINDER WITH KA24E ENGINE

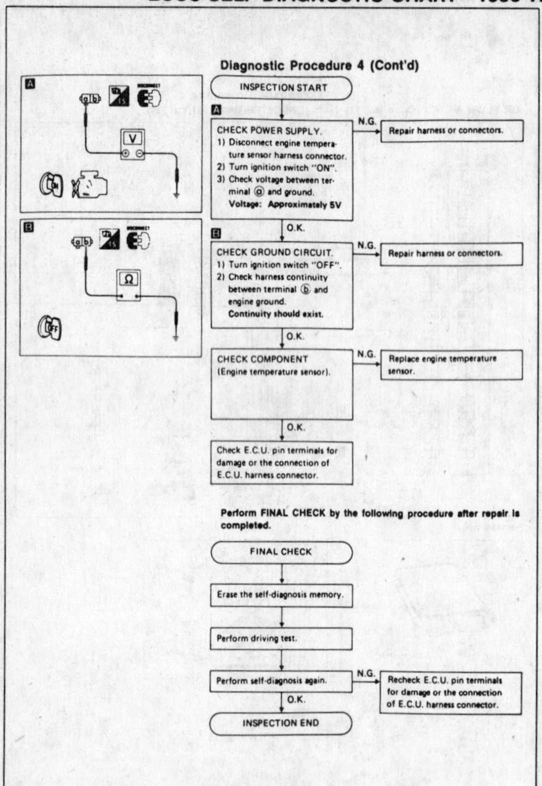
Diagnostic Procedure 4 (Cont'd)
INSPECTION START
A
CHECK POWER SUPPLY.
1) Disconnect engine temperature sensor harness connector.
2) Turn ignition switch "ON".
3) Check voltage between terminal ⓐ and ground.
Voltage: Approximately 5V
N.G.
Repair harness or connectors.
O.K.
B
CHECK GROUND CIRCUIT.
1) Turn ignition switch "OFF".
2) Check harness continuity between terminal ⓑ and engine ground.
Continuity should exist.
N.G.
Repair harness or connectors.
O.K.
CHECK COMPONENT
(Engine temperature sensor).
N.G.
Replace engine temperature sensor.
O.K.
Check E.C.U. pin terminals for damage or the connection of E.C.U. harness connector.
Perform FINAL CHECK by the following procedure after repair is completed.
FINAL CHECK
Erase the self-diagnosis memory.
Perform driving test.
Perform self-diagnosis again.
N.G.
Recheck E.C.U. pin terminals for damage or the connection of E.C.U. harness connector.
O.K.
INSPECTION END

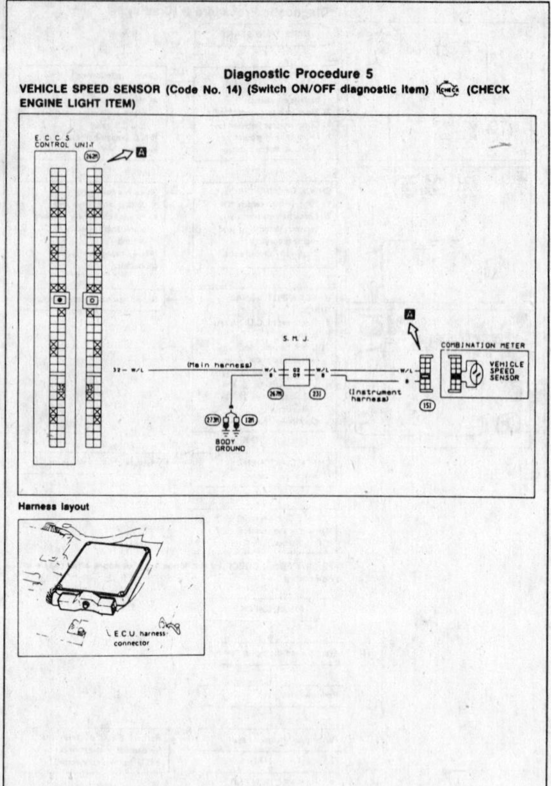
Diagnostic Procedure 5
VEHICLE SPEED SENSOR (Code No. 14) (Switch ON/OFF diagnostic item) (CHECK ENGINE LIGHT ITEM)
E.C.C.S. CONTROL UNIT
S.M.J.
COMBINATION METER
VEHICLE SPEED SENSOR
(Main harness)
(Instrument harness)
BODY GROUND
Harness layout
E.C.U. harness connector

ECCS SELF-DIAGNOSTIC CHART—1990 TRUCK/PATHFINDER WITH KA24E ENGINE

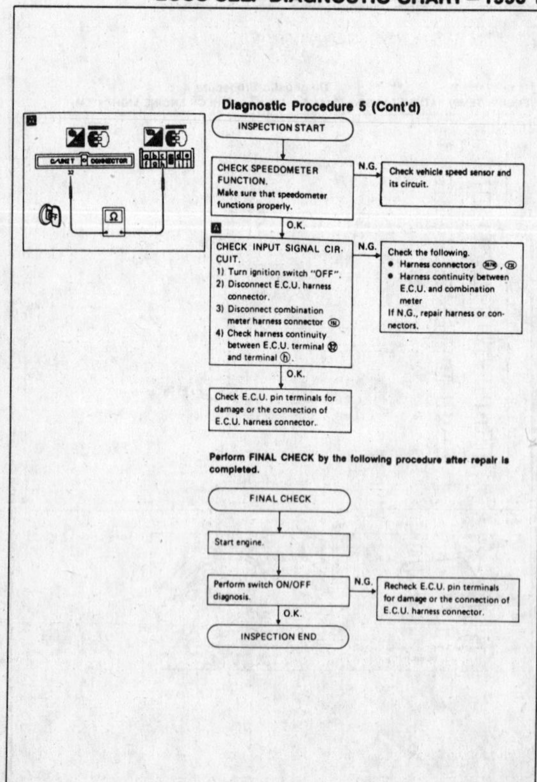
Diagnostic Procedure 5 (Cont'd)
INSPECTION START
CHECK SPEEDOMETER FUNCTION.
Make sure that speedometer functions properly.
N.G.
Check vehicle speed sensor and its circuit.
O.K.
A
CHECK INPUT SIGNAL CIRCUIT.
1) Turn ignition switch "OFF".
2) Disconnect E.C.U. harness connector.
3) Disconnect combination meter harness connector.
4) Check harness continuity between E.C.U. terminal and terminal ⓗ.
N.G.
Check the following.
• Harness connectors
• Harness continuity between E.C.U. and combination meter
If N.G., repair harness or connectors.
O.K.
Check E.C.U. pin terminals for damage or the connection of E.C.U. harness connector.
Perform FINAL CHECK by the following procedure after repair is completed.
FINAL CHECK
Start engine.
Perform switch ON/OFF diagnosis.
N.G.
Recheck E.C.U. pin terminals for damage or the connection of E.C.U. harness connector.
O.K.
INSPECTION END

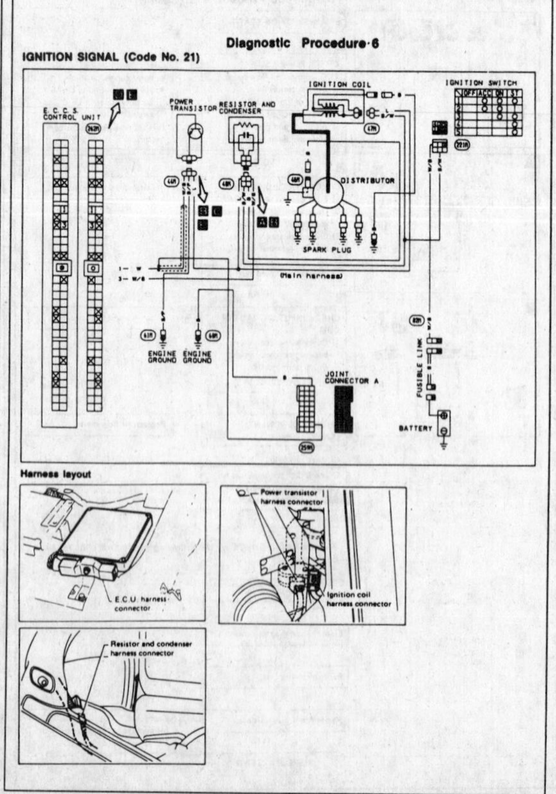
Diagnostic Procedure 6
IGNITION SIGNAL (Code No. 21)
E.C.C.S. CONTROL UNIT
POWER TRANSISTOR
RESISTOR AND CONDENSER
IGNITION COIL
IGNITION SWITCH
DISTRIBUTOR
SPARK PLUG
(Main harness)
ENGINE GROUND
ENGINE GROUND
JOINT CONNECTOR A
FUSIBLE LINK
BATTERY
Harness layout
E.C.U. harness connector
Power transistor harness connector
Ignition coil harness connector
Resistor and condenser harness connector

ECCS SELF-DIAGNOSTIC CHART—1990 TRUCK/PATHFINDER WITH KA24E ENGINE

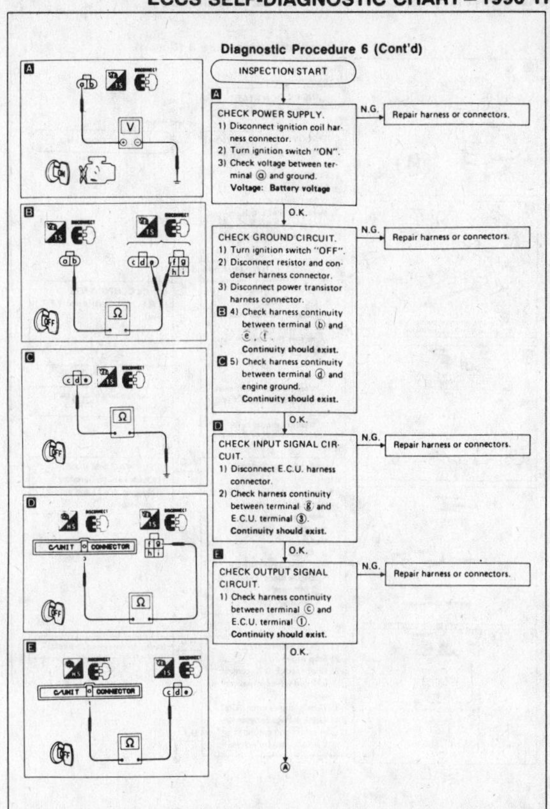
Diagnostic Procedure 6 (Cont'd)
INSPECTION START
A
CHECK POWER SUPPLY.
1) Disconnect ignition coil harness connector.
2) Turn ignition switch "ON".
3) Check voltage between terminal ⓐ and ground.
Voltage: Battery voltage
N.G.
Repair harness or connectors.
O.K.
B
CHECK GROUND CIRCUIT.
1) Turn ignition switch "OFF".
2) Disconnect resistor and condenser harness connector.
3) Disconnect power transistor harness connector.
B 4) Check harness continuity between terminal ⓑ and ⓔ, ⓕ.
Continuity should exist.
C 5) Check harness continuity between terminal ⓓ and engine ground.
Continuity should exist.
N.G.
Repair harness or connectors.
O.K.
D
CHECK INPUT SIGNAL CIRCUIT.
1) Disconnect E.C.U. harness connector.
2) Check harness continuity between terminal ⓖ and E.C.U. terminal ③.
Continuity should exist.
N.G.
Repair harness or connectors.
O.K.
E
CHECK OUTPUT SIGNAL CIRCUIT.
1) Check harness continuity between terminal ⓒ and E.C.U. terminal ①.
Continuity should exist.
N.G.
Repair harness or connectors.
O.K.
Ⓐ
C/UNIT
CONNECTOR

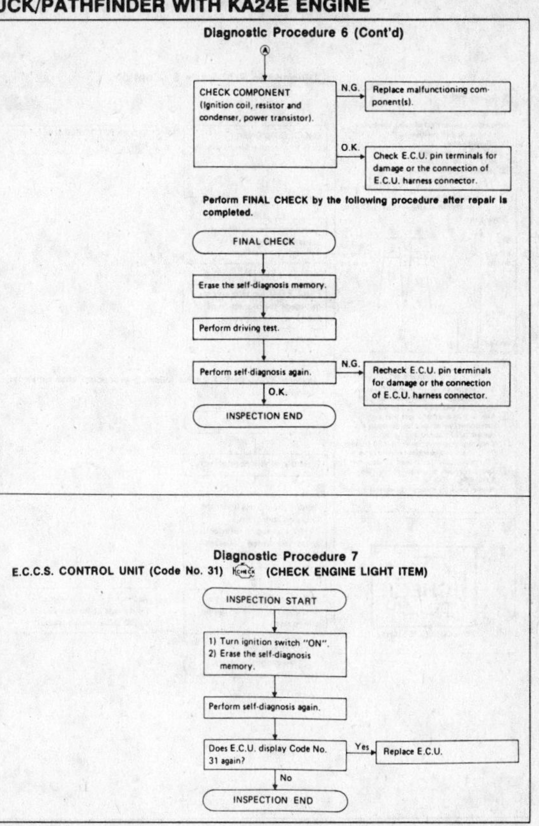
Diagnostic Procedure 6 (Cont'd)
Ⓐ
CHECK COMPONENT
(Ignition coil, resistor and condenser, power transistor).
N.G.
Replace malfunctioning component(s).
O.K.
Check E.C.U. pin terminals for damage or the connection of E.C.U. harness connector.
Perform FINAL CHECK by the following procedure after repair is completed.
FINAL CHECK
Erase the self-diagnosis memory.
Perform driving test.
Perform self-diagnosis again.
N.G.
Recheck E.C.U. pin terminals for damage or the connection of E.C.U. harness connector.
O.K.
INSPECTION END
Diagnostic Procedure 7
E.C.C.S. CONTROL UNIT (Code No. 31) (CHECK ENGINE LIGHT ITEM)
INSPECTION START
1) Turn ignition switch "ON".
2) Erase the self-diagnosis memory.
Perform self-diagnosis again.
Does E.C.U. display Code No. 31 again?
Yes
Replace E.C.U.
No
INSPECTION END

ECCS DIAGNOSTIC CHART—1990 TRUCK/PATHFINDER WITH KA24E ENGINE

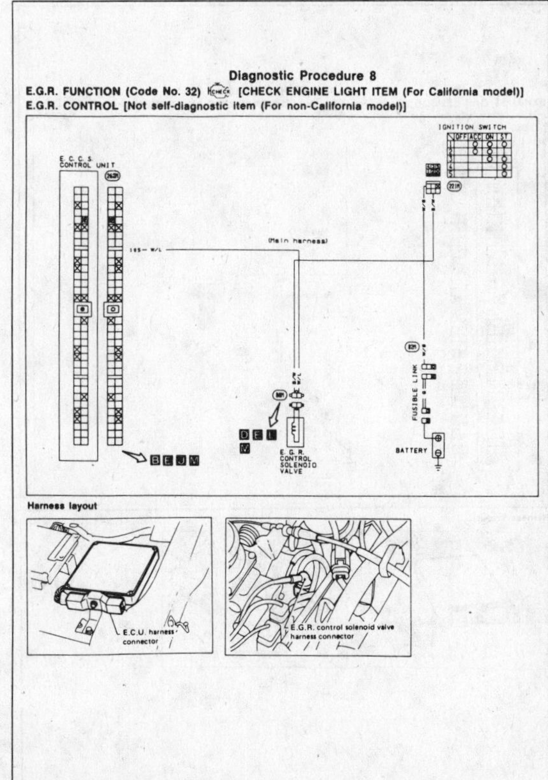
Diagnostic Procedure 8
E.G.R. FUNCTION (Code No. 32) [CHECK ENGINE LIGHT ITEM (For California model)]
E.G.R. CONTROL [Not self-diagnostic item (For non-California model)]
IGNITION SWITCH
E.C.C.S. CONTROL UNIT
Main harness
FUSIBLE LINK
BATTERY
E.G.R. CONTROL SOLENOID VALVE
Harness layout
E.C.U. harness connector
E.G.R. control solenoid valve harness connector

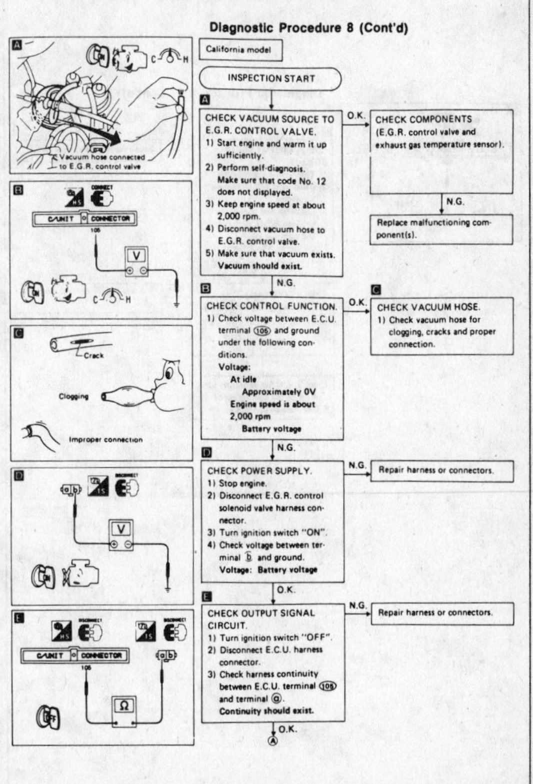
Diagnostic Procedure 8 (Cont'd)
California model
INSPECTION START
A
CHECK VACUUM SOURCE TO E.G.R. CONTROL VALVE.
1) Start engine and warm it up sufficiently.
2) Perform self-diagnosis. Make sure that code No. 12 does not displayed.
3) Keep engine speed at about 2,000 rpm.
4) Disconnect vacuum hose to E.G.R. control valve.
5) Make sure that vacuum exists.
Vacuum should exist.
O.K.
CHECK COMPONENTS
(E.G.R. control valve and exhaust gas temperature sensor).
N.G.
Replace malfunctioning component(s).
Vacuum hose connected to E.G.R. control valve
N.G.
B
CHECK CONTROL FUNCTION.
1) Check voltage between E.C.U. terminal ⑩⑤ and ground under the following conditions.
Voltage:
At idle
Approximately 0V
Engine speed is about 2,000 rpm
Battery voltage
O.K.
C
CHECK VACUUM HOSE.
1) Check vacuum hose for clogging, cracks and proper connection.
Crack
Clogging
Improper connection
N.G.
D
CHECK POWER SUPPLY.
1) Stop engine.
2) Disconnect E.G.R. control solenoid valve harness connector.
3) Turn ignition switch "ON".
4) Check voltage between terminal ⓑ and ground.
Voltage: Battery voltage
N.G.
Repair harness or connectors.
O.K.
E
CHECK OUTPUT SIGNAL CIRCUIT.
1) Turn ignition switch "OFF".
2) Disconnect E.C.U. harness connector.
3) Check harness continuity between E.C.U. terminal ⑩⑤ and terminal ⓐ.
Continuity should exist.
N.G.
Repair harness or connectors.
O.K.
Ⓐ
C/UNIT
CONNECTOR

ECCS DIAGNOSTIC CHART—1990 TRUCK/PATHFINDER WITH KA24E ENGINE

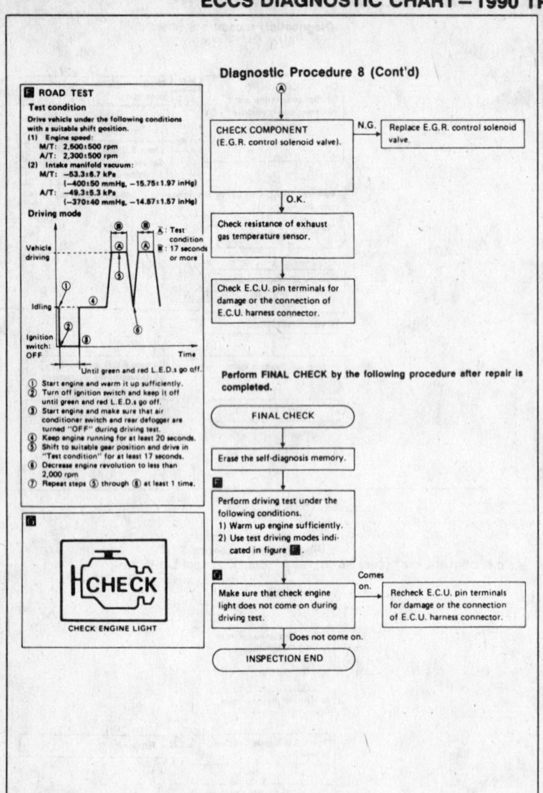
Diagnostic Procedure 8 (Cont'd)
A
CHECK COMPONENT (E.G.R. control solenoid valve).
N.G.
Replace E.G.R. control solenoid valve.
O.K.
Check resistance of exhaust gas temperature sensor.
Check E.C.U. pin terminals for damage or the connection of E.C.U. harness connector.
Perform FINAL CHECK by the following procedure after repair is completed.
FINAL CHECK
Erase the self-diagnosis memory.
Perform driving test under the following conditions.
1) Warm up engine sufficiently.
2) Use test driving modes indicated in figure E.
Make sure that check engine light does not come on during driving test.
Comes on.
Recheck E.C.U. pin terminals for damage or the connection of E.C.U. harness connector.
Does not come on.
INSPECTION END
E ROAD TEST
Test condition
Drive vehicle under the following conditions with a suitable shift position.
(1) Engine speed:
M/T: 2,500±500 rpm
A/T: 2,300±500 rpm
(2) Intake manifold vacuum:
M/T: −53.3±6.7 kPa (−400±50 mmHg, −15.75±1.97 inHg)
A/T: −49.3±5.3 kPa (−370±40 mmHg, −14.57±1.57 inHg)
Driving mode
A: Test condition
B: 17 seconds or more
Vehicle driving
Idling
Ignition switch: OFF
Time
Until green and red L.E.D.s go off.
① Start engine and warm it up sufficiently.
② Turn off ignition switch and keep it off until green and red L.E.D.s go off.
③ Start engine and make sure that air conditioner switch and rear defogger are turned "OFF" during driving test.
④ Keep engine running for at least 20 seconds.
⑤ Shift to suitable gear position and drive in "Test condition" for at least 17 seconds.
⑥ Decrease engine revolution to less than 2,000 rpm
⑦ Repeat steps ⑤ through ⑥ at least 1 time.
CHECK
CHECK ENGINE LIGHT

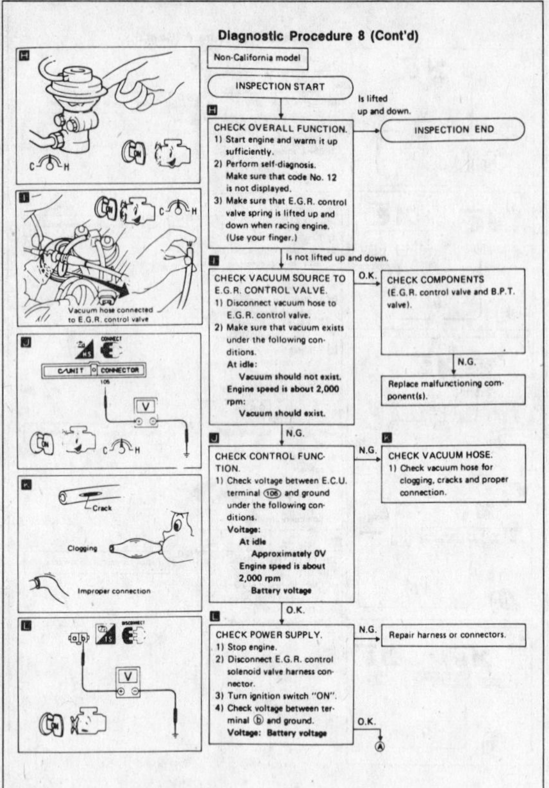
Diagnostic Procedure 8 (Cont'd)
Non-California model
INSPECTION START
H
CHECK OVERALL FUNCTION.
1) Start engine and warm it up sufficiently.
2) Perform self-diagnosis. Make sure that code No. 12 is not displayed.
3) Make sure that E.G.R. control valve spring is lifted up and down when racing engine. (Use your finger.)
Is lifted up and down.
INSPECTION END
Is not lifted up and down.
I
CHECK VACUUM SOURCE TO E.G.R. CONTROL VALVE.
1) Disconnect vacuum hose to E.G.R. control valve.
2) Make sure that vacuum exists under the following conditions.
At idle:
Vacuum should not exist.
Engine speed is about 2,000 rpm:
Vacuum should exist.
O.K.
CHECK COMPONENTS (E.G.R. control valve and B.P.T. valve).
N.G.
Replace malfunctioning component(s).
N.G.
J
CHECK CONTROL FUNCTION.
1) Check voltage between E.C.U. terminal 106 and ground under the following conditions.
Voltage:
At idle
Approximately 0V
Engine speed is about 2,000 rpm
Battery voltage
N.G.
K
CHECK VACUUM HOSE.
1) Check vacuum hose for clogging, cracks and proper connection.
O.K.
L
CHECK POWER SUPPLY.
1) Stop engine.
2) Disconnect E.G.R. control solenoid valve harness connector.
3) Turn ignition switch "ON".
4) Check voltage between terminal ⓑ and ground.
Voltage: Battery voltage
N.G.
Repair harness or connectors.
O.K.
A
Vacuum hose connected to E.G.R. control valve
C/UNIT
CONNECTOR
106
V
Crack
Clogging
Improper connection

ECCS DIAGNOSTIC CHART—1990 TRUCK/PATHFINDER WITH KA24E ENGINE

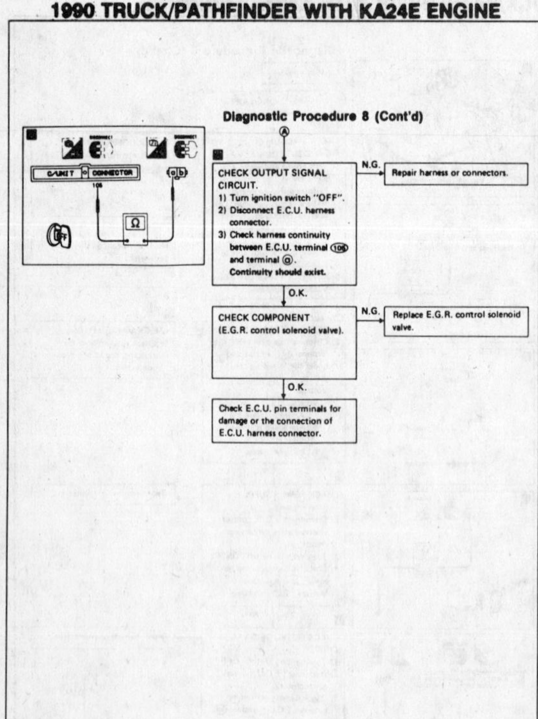
Diagnostic Procedure 8 (Cont'd)
A
M
CHECK OUTPUT SIGNAL CIRCUIT.
1) Turn ignition switch "OFF".
2) Disconnect E.C.U. harness connector.
3) Check harness continuity between E.C.U. terminal 106 and terminal ⓐ.
Continuity should exist.
N.G.
Repair harness or connectors.
O.K.
CHECK COMPONENT (E.G.R. control solenoid valve).
N.G.
Replace E.G.R. control solenoid valve.
O.K.
Check E.C.U. pin terminals for damage or the connection of E.C.U. harness connector.
C/UNIT
CONNECTOR
106
Ω

ECCS SELF-DIAGNOSTIC CHART—1990 TRUCK/PATHFINDER WITH KA24E ENGINE

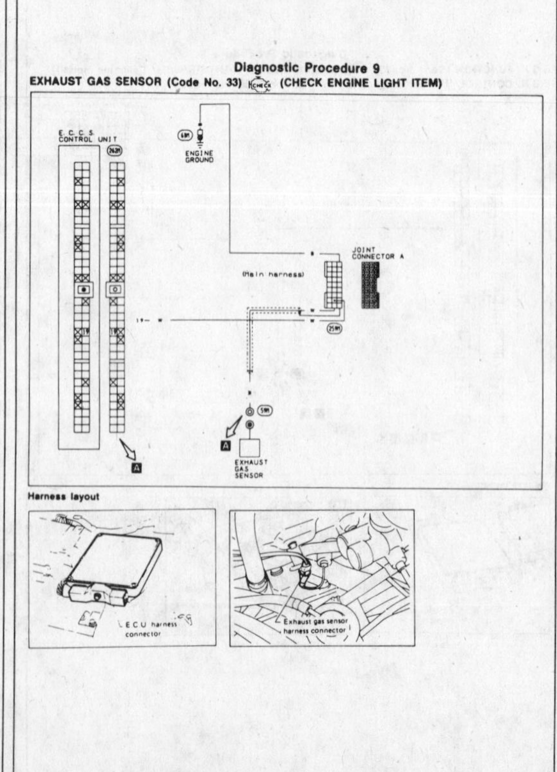
Diagnostic Procedure 9
EXHAUST GAS SENSOR (Code No. 33) (CHECK ENGINE LIGHT ITEM)
E.C.C.S. CONTROL UNIT
ENGINE GROUND
JOINT CONNECTOR A
(Main harness)
EXHAUST GAS SENSOR
Harness layout
E.C.U. harness connector
Exhaust gas sensor harness connector

ECCS SELF-DIAGNOSTIC CHART—1990 TRUCK/PATHFINDER WITH KA24E ENGINE

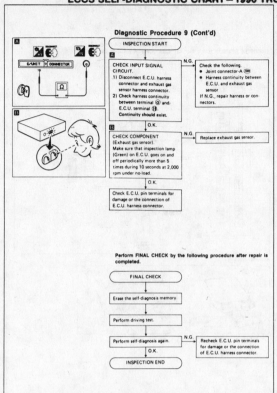

Diagnostic Procedure 9 (Cont'd)

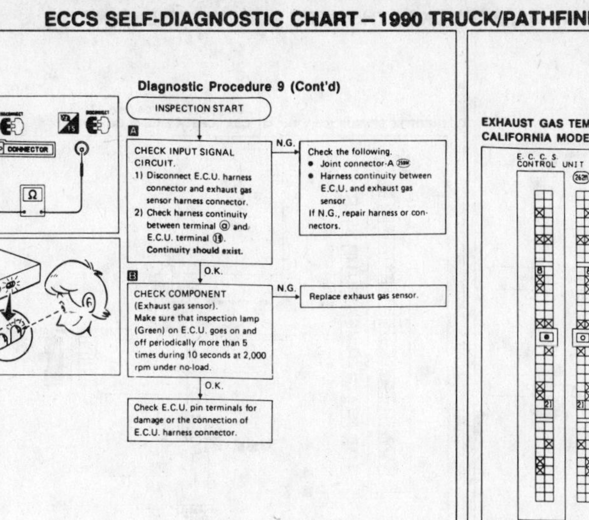
INSPECTION START
A
CHECK INPUT SIGNAL CIRCUIT.
1) Disconnect E.C.U. harness connector and exhaust gas sensor harness connector.
2) Check harness continuity between terminal ⓐ and E.C.U. terminal ⑲.
Continuity should exist.
N.G.
Check the following.
• Joint connector-A
• Harness continuity between E.C.U. and exhaust gas sensor
If N.G., repair harness or connectors.
O.K.
B
CHECK COMPONENT (Exhaust gas sensor).
Make sure that inspection lamp (Green) on E.C.U. goes on and off periodically more than 5 times during 10 seconds at 2,000 rpm under no-load.
N.G.
Replace exhaust gas sensor.
O.K.
Check E.C.U. pin terminals for damage or the connection of E.C.U. harness connector.
C/UNIT
CONNECTOR

Perform FINAL CHECK by the following procedure after repair is completed.

FINAL CHECK

Erase the self-diagnosis memory.

Perform driving test.

Perform self-diagnosis again. — N.G. → Recheck E.C.U. pin terminals for damage or the connection of E.C.U. harness connector.

O.K.

INSPECTION END

Diagnostic Procedure 10

EXHAUST GAS TEMPERATURE SENSOR (Code No. 35) (CHECK ENGINE LIGHT ITEM): CALIFORNIA MODEL ONLY

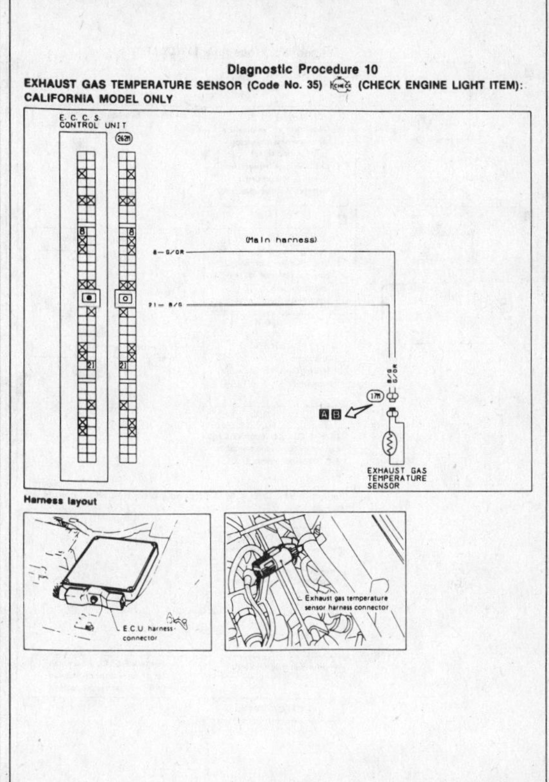
E.C.C.S. CONTROL UNIT
(Main harness)
8 — G/OR
21 — B/S
EXHAUST GAS TEMPERATURE SENSOR
Harness layout
E.C.U. harness connector
Exhaust gas temperature sensor harness connector

ECCS SELF-DIAGNOSTIC CHART—1990 TRUCK/PATHFINDER WITH KA24E ENGINE

Diagnostic Procedure 10 (Cont'd)

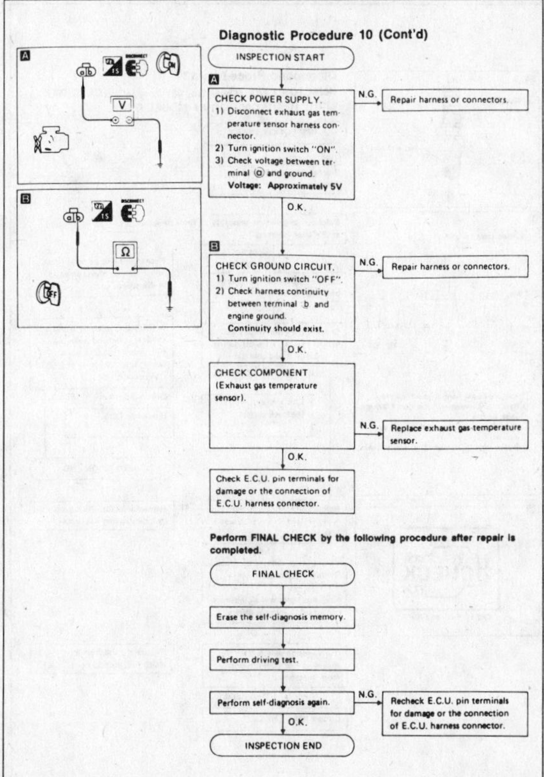
INSPECTION START
A
CHECK POWER SUPPLY.
1) Disconnect exhaust gas temperature sensor harness connector.
2) Turn ignition switch "ON".
3) Check voltage between terminal ⓐ and ground.
Voltage: Approximately 5V
N.G.
Repair harness or connectors.
O.K.
B
CHECK GROUND CIRCUIT.
1) Turn ignition switch "OFF".
2) Check harness continuity between terminal ⓑ and engine ground.
Continuity should exist.
N.G.
Repair harness or connectors.
O.K.
CHECK COMPONENT (Exhaust gas temperature sensor).
N.G.
Replace exhaust gas temperature sensor.
O.K.
Check E.C.U. pin terminals for damage or the connection of E.C.U. harness connector.
Perform FINAL CHECK by the following procedure after repair is completed.
FINAL CHECK
Erase the self-diagnosis memory.
Perform driving test.
Perform self-diagnosis again.
N.G.
Recheck E.C.U. pin terminals for damage or the connection of E.C.U. harness connector.
O.K.
INSPECTION END

Diagnostic Procedure 11

AIR TEMPERATURE SENSOR (Code No. 41)

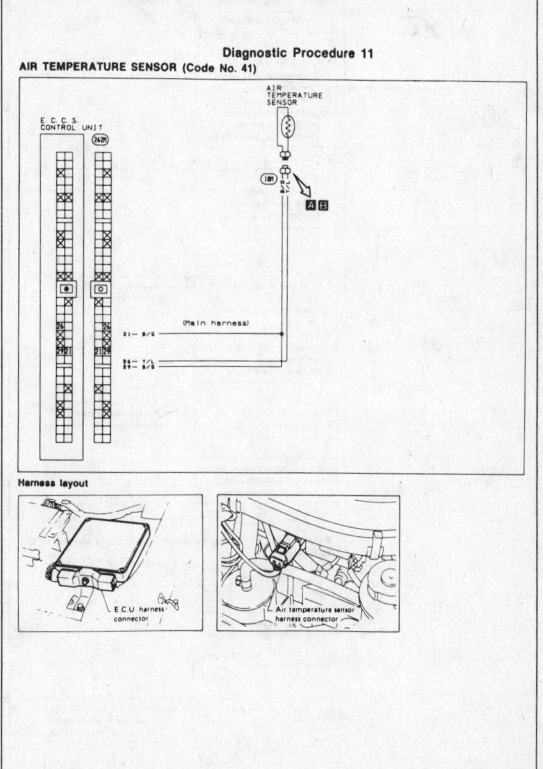
AIR TEMPERATURE SENSOR
E.C.C.S. CONTROL UNIT
(Main harness)
21 — B/S
Harness layout
E.C.U. harness connector
Air temperature sensor harness connector

ECCS SELF-DIAGNOSTIC CHART—1990 TRUCK/PATHFINDER WITH KA24E ENGINE

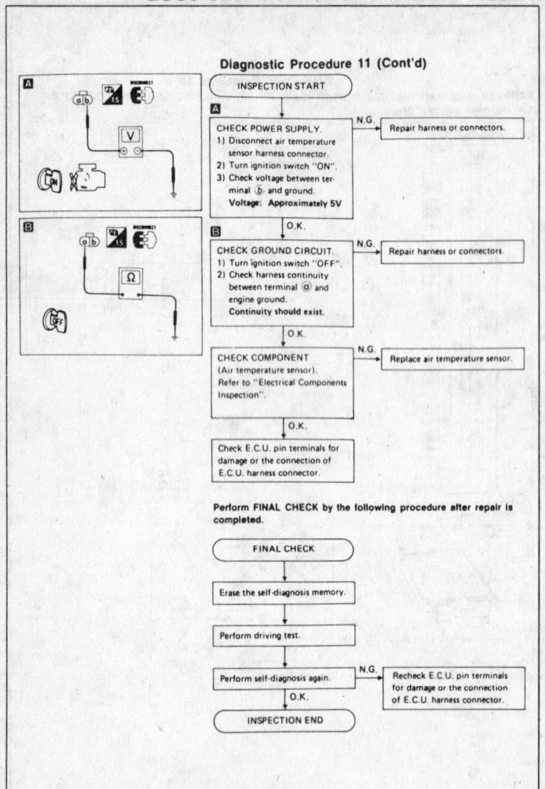
Diagnostic Procedure 11 (Cont'd)
INSPECTION START
CHECK POWER SUPPLY.
1) Disconnect air temperature sensor harness connector.
2) Turn ignition switch "ON".
3) Check voltage between terminal ⓑ and ground.
Voltage: Approximately 5V
N.G.
Repair harness or connectors.
O.K.
CHECK GROUND CIRCUIT.
1) Turn ignition switch "OFF".
2) Check harness continuity between terminal ⓐ and engine ground.
Continuity should exist.
N.G.
Repair harness or connectors.
O.K.
CHECK COMPONENT
(Air temperature sensor).
Refer to "Electrical Components Inspection".
N.G.
Replace air temperature sensor.
O.K.
Check E.C.U. pin terminals for damage or the connection of E.C.U. harness connector.
Perform FINAL CHECK by the following procedure after repair is completed.
FINAL CHECK
Erase the self-diagnosis memory.
Perform driving test.
Perform self-diagnosis again.
N.G.
Recheck E.C.U. pin terminals for damage or the connection of E.C.U. harness connector.
O.K.
INSPECTION END

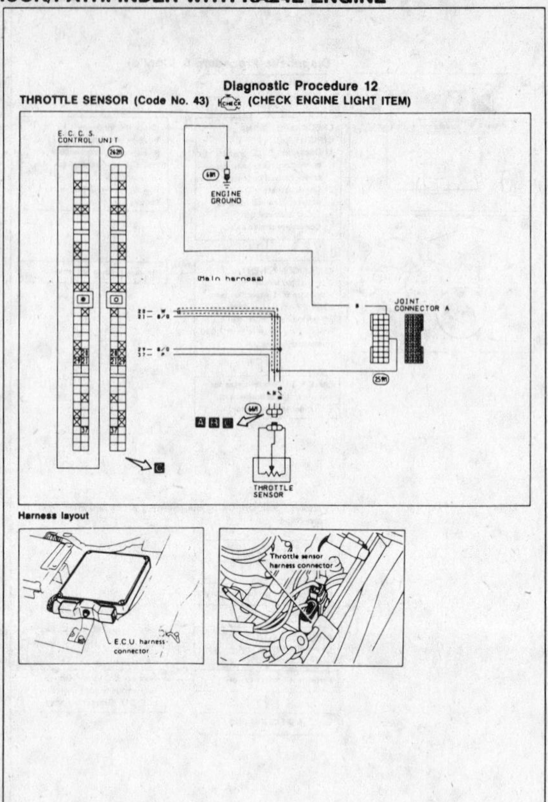
Diagnostic Procedure 12
THROTTLE SENSOR (Code No. 43) (CHECK ENGINE LIGHT ITEM)
E.C.C.S. CONTROL UNIT
ENGINE GROUND
Main harness
JOINT CONNECTOR A
THROTTLE SENSOR
Harness layout
E.C.U. harness connector
Throttle sensor harness connector

ECCS SELF-DIAGNOSTIC CHART—1990 TRUCK/PATHFINDER WITH KA24E ENGINE

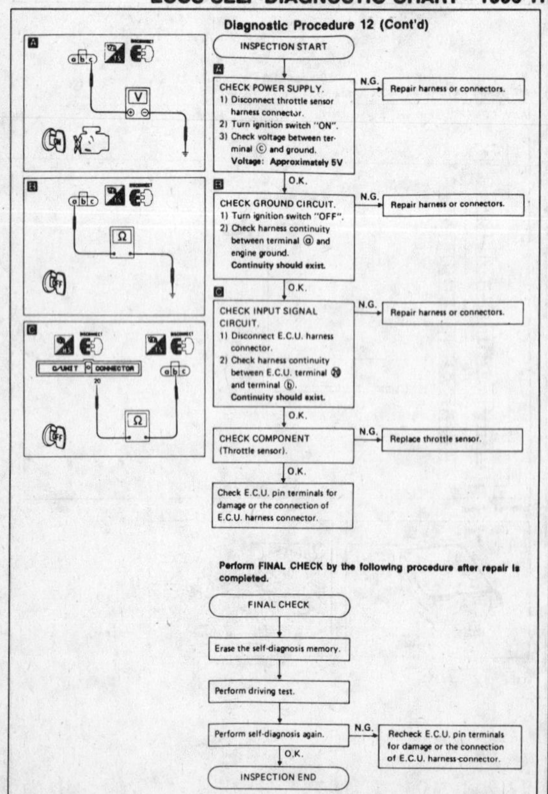
Diagnostic Procedure 12 (Cont'd)
INSPECTION START
CHECK POWER SUPPLY.
1) Disconnect throttle sensor harness connector.
2) Turn ignition switch "ON".
3) Check voltage between terminal ⓒ and ground.
Voltage: Approximately 5V
N.G.
Repair harness or connectors.
O.K.
CHECK GROUND CIRCUIT.
1) Turn ignition switch "OFF".
2) Check harness continuity between terminal ⓐ and engine ground.
Continuity should exist.
N.G.
Repair harness or connectors.
O.K.
CHECK INPUT SIGNAL CIRCUIT.
1) Disconnect E.C.U. harness connector.
2) Check harness continuity between E.C.U. terminal ⑳ and terminal ⓑ.
Continuity should exist.
N.G.
Repair harness or connectors.
O.K.
CHECK COMPONENT
(Throttle sensor).
N.G.
Replace throttle sensor.
O.K.
Check E.C.U. pin terminals for damage or the connection of E.C.U. harness connector.
Perform FINAL CHECK by the following procedure after repair is completed.
FINAL CHECK
Erase the self-diagnosis memory.
Perform driving test.
Perform self-diagnosis again.
N.G.
Recheck E.C.U. pin terminals for damage or the connection of E.C.U. harness connector.
O.K.
INSPECTION END

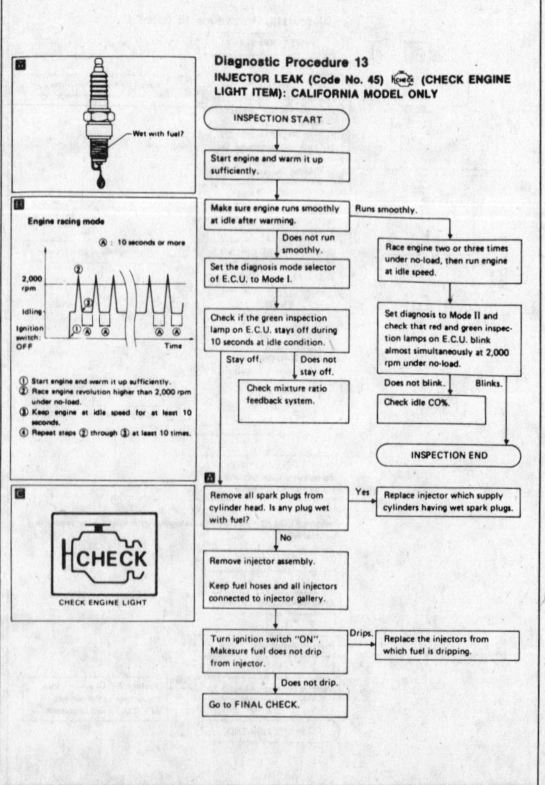
Diagnostic Procedure 13
INJECTOR LEAK (Code No. 45) (CHECK ENGINE LIGHT ITEM): CALIFORNIA MODEL ONLY
Wet with fuel?
Engine racing mode
Ⓐ : 10 seconds or more
2,000 rpm
Idling
Ignition switch: OFF
Time
① Start engine and warm it up sufficiently.
② Race engine revolution higher than 2,000 rpm under no-load.
③ Keep engine at idle speed for at least 10 seconds.
④ Repeat steps ② through ③ at least 10 times.
CHECK
CHECK ENGINE LIGHT
INSPECTION START
Start engine and warm it up sufficiently.
Make sure engine runs smoothly at idle after warming.
Runs smoothly.
Does not run smoothly.
Race engine two or three times under no-load, then run engine at idle speed.
Set the diagnosis mode selector of E.C.U. to Mode I.
Set diagnosis to Mode II and check that red and green inspection lamps on E.C.U. blink almost simultaneously at 2,000 rpm under no-load.
Check if the green inspection lamp on E.C.U. stays off during 10 seconds at idle condition.
Stay off.
Does not stay off.
Check mixture ratio feedback system.
Does not blink.
Blinks.
Check idle CO%.
INSPECTION END
Remove all spark plugs from cylinder head. Is any plug wet with fuel?
Yes
Replace injector which supply cylinders having wet spark plugs.
No
Remove injector assembly.
Keep fuel hoses and all injectors connected to injector gallery.
Turn ignition switch "ON". Makesure fuel does not drip from injector.
Drips.
Replace the injectors from which fuel is dripping.
Does not drip.
Go to FINAL CHECK.

ECCS SELF-DIAGNOSTIC CHART—1990 TRUCK/PATHFINDER WITH KA24E ENGINE

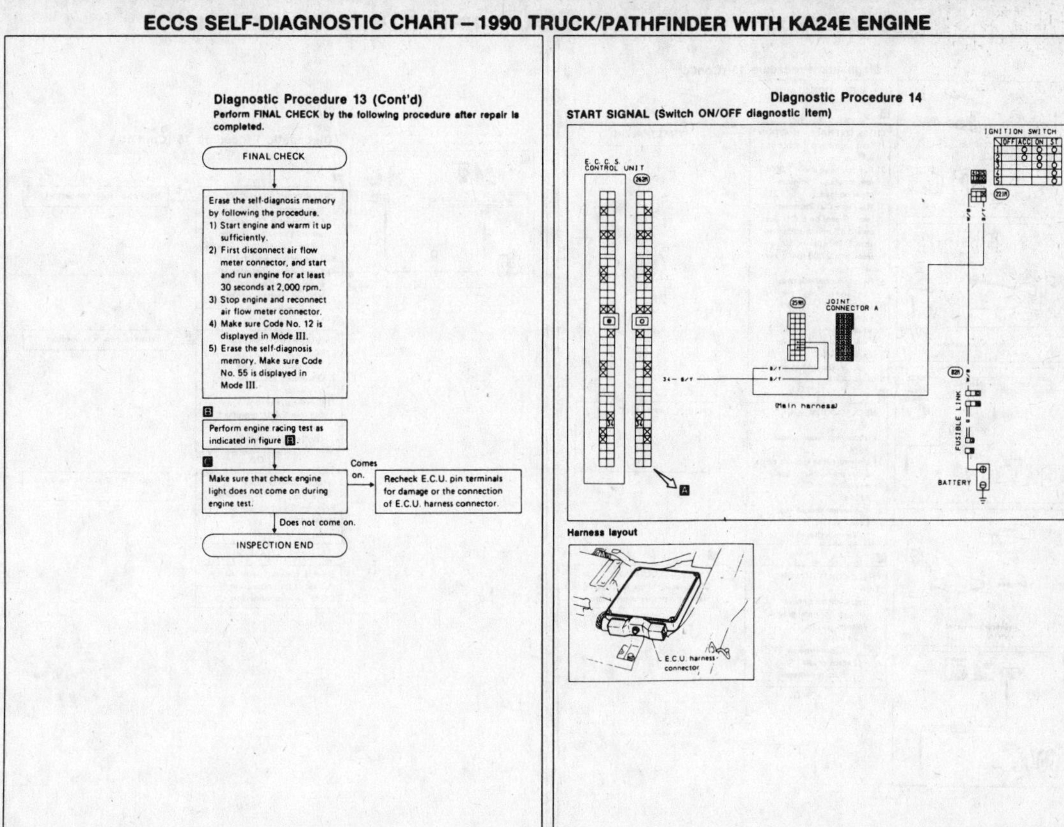
Diagnostic Procedure 13 (Cont'd)
Perform FINAL CHECK by the following procedure after repair is completed.
FINAL CHECK
Erase the self-diagnosis memory by following the procedure.
1) Start engine and warm it up sufficiently.
2) First disconnect air flow meter connector, and start and run engine for at least 30 seconds at 2,000 rpm.
3) Stop engine and reconnect air flow meter connector.
4) Make sure Code No. 12 is displayed in Mode III.
5) Erase the self-diagnosis memory. Make sure Code No. 55 is displayed in Mode III.
Perform engine racing test as indicated in figure
Make sure that check engine light does not come on during engine test.
Comes on.
Recheck E.C.U. pin terminals for damage or the connection of E.C.U. harness connector.
Does not come on.
INSPECTION END
Diagnostic Procedure 14
START SIGNAL (Switch ON/OFF diagnostic item)
E.C.C.S. CONTROL UNIT
IGNITION SWITCH
JOINT CONNECTOR A
Main harness
FUSIBLE LINK
BATTERY
Harness layout
E.C.U. harness connector

ECCS SELF-DIAGNOSTIC CHART—1990 TRUCK/PATHFINDER WITH KA24E ENGINE

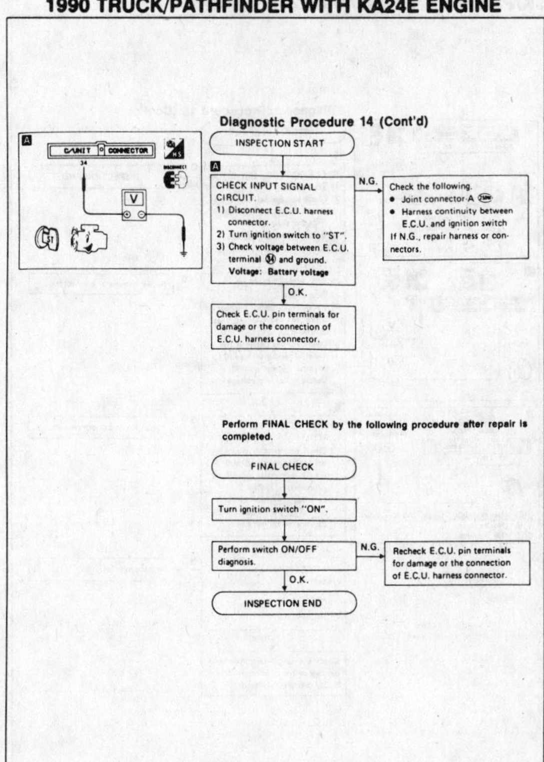
Diagnostic Procedure 14 (Cont'd)
C/UNIT CONNECTOR
INSPECTION START
CHECK INPUT SIGNAL CIRCUIT.
1) Disconnect E.C.U. harness connector.
2) Turn ignition switch to "ST".
3) Check voltage between E.C.U. terminal 34 and ground.
Voltage: Battery voltage
N.G.
Check the following.
• Joint connector-A
• Harness continuity between E.C.U. and ignition switch
If N.G., repair harness or connectors.
O.K.
Check E.C.U. pin terminals for damage or the connection of E.C.U. harness connector.
Perform FINAL CHECK by the following procedure after repair is completed.
FINAL CHECK
Turn ignition switch "ON".
Perform switch ON/OFF diagnosis.
N.G.
Recheck E.C.U. pin terminals for damage or the connection of E.C.U. harness connector.
O.K.
INSPECTION END

ECCS DIAGNOSTIC CHART—1990 TRUCK/PATHFINDER WITH KA24E ENGINE

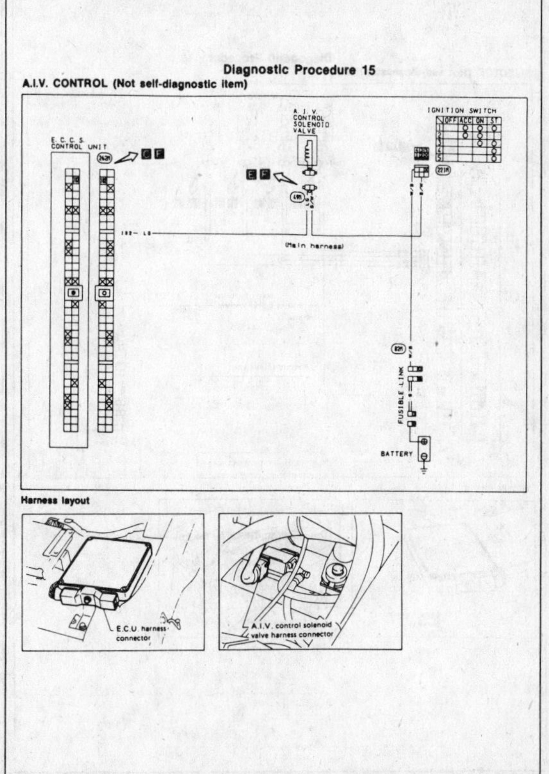
Diagnostic Procedure 15
A.I.V. CONTROL (Not self-diagnostic item)
E.C.C.S. CONTROL UNIT
A.I.V. CONTROL SOLENOID VALVE
IGNITION SWITCH
Main harness
FUSIBLE LINK
BATTERY
Harness layout
E.C.U. harness connector
A.I.V. control solenoid valve harness connector

ECCS DIAGNOSTIC CHART—1990 TRUCK/PATHFINDER WITH KA24E ENGINE

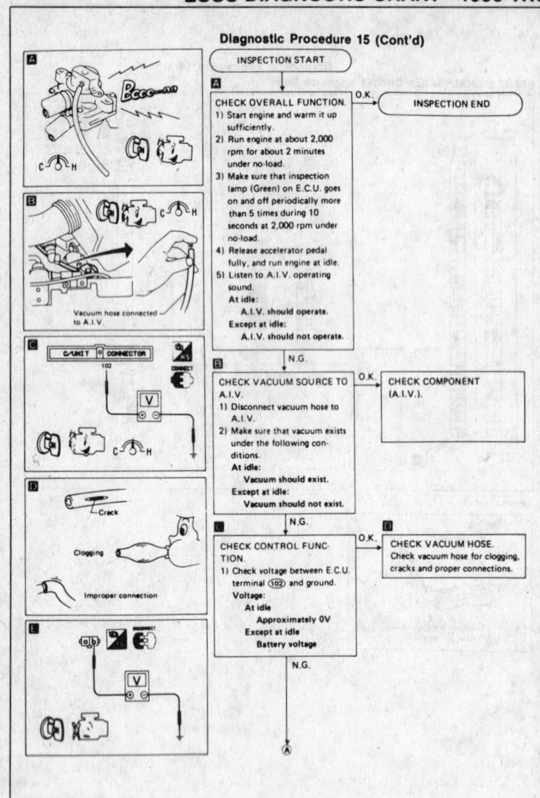
Diagnostic Procedure 15 (Cont'd)
INSPECTION START
A
CHECK OVERALL FUNCTION.
1) Start engine and warm it up sufficiently.
2) Run engine at about 2,000 rpm for about 2 minutes under no-load.
3) Make sure that inspection lamp (Green) on E.C.U. goes on and off periodically more than 5 times during 10 seconds at 2,000 rpm under no-load.
4) Release accelerator pedal fully, and run engine at idle.
5) Listen to A.I.V. operating sound.
At idle:
A.I.V. should operate.
Except at idle:
A.I.V. should not operate.
O.K.
INSPECTION END
N.G.
B
CHECK VACUUM SOURCE TO A.I.V.
1) Disconnect vacuum hose to A.I.V.
2) Make sure that vacuum exists under the following conditions.
At idle:
Vacuum should exist.
Except at idle:
Vacuum should not exist.
O.K.
CHECK COMPONENT (A.I.V.).
N.G.
C
CHECK CONTROL FUNCTION.
1) Check voltage between E.C.U. terminal 102 and ground.
Voltage:
At idle
Approximately 0V
Except at idle
Battery voltage
O.K.
D
CHECK VACUUM HOSE.
Check vacuum hose for clogging, cracks and proper connections.
N.G.
A
Vacuum hose connected to A.I.V.
C/UNIT CONNECTOR
102
Crack
Clogging
Improper connection

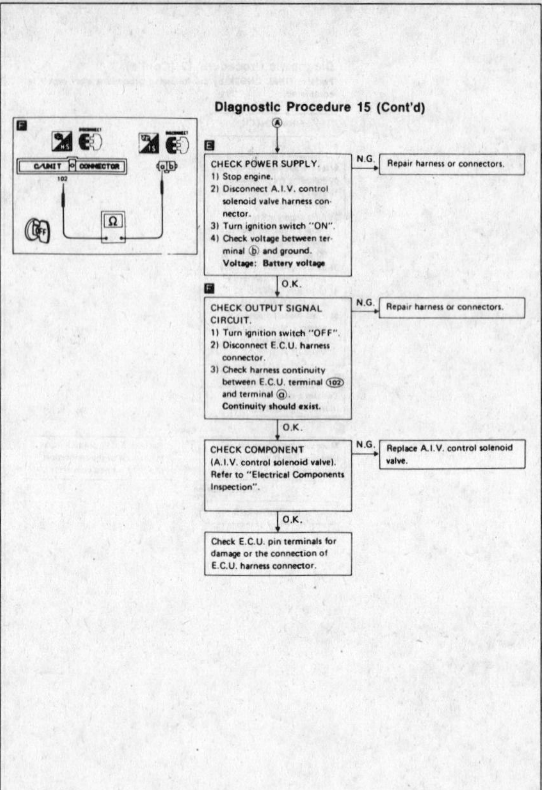
Diagnostic Procedure 15 (Cont'd)
A
E
CHECK POWER SUPPLY.
1) Stop engine.
2) Disconnect A.I.V. control solenoid valve harness connector.
3) Turn ignition switch "ON".
4) Check voltage between terminal ⓑ and ground.
Voltage: Battery voltage
N.G.
Repair harness or connectors.
O.K.
F
CHECK OUTPUT SIGNAL CIRCUIT.
1) Turn ignition switch "OFF".
2) Disconnect E.C.U. harness connector.
3) Check harness continuity between E.C.U. terminal 102 and terminal ⓐ.
Continuity should exist.
N.G.
Repair harness or connectors.
O.K.
CHECK COMPONENT
(A.I.V. control solenoid valve).
Refer to "Electrical Components Inspection".
N.G.
Replace A.I.V. control solenoid valve.
O.K.
Check E.C.U. pin terminals for damage or the connection of E.C.U. harness connector.
C/UNIT CONNECTOR
102

ECCS DIAGNOSTIC CHART—1990 TRUCK/PATHFINDER WITH KA24E ENGINE

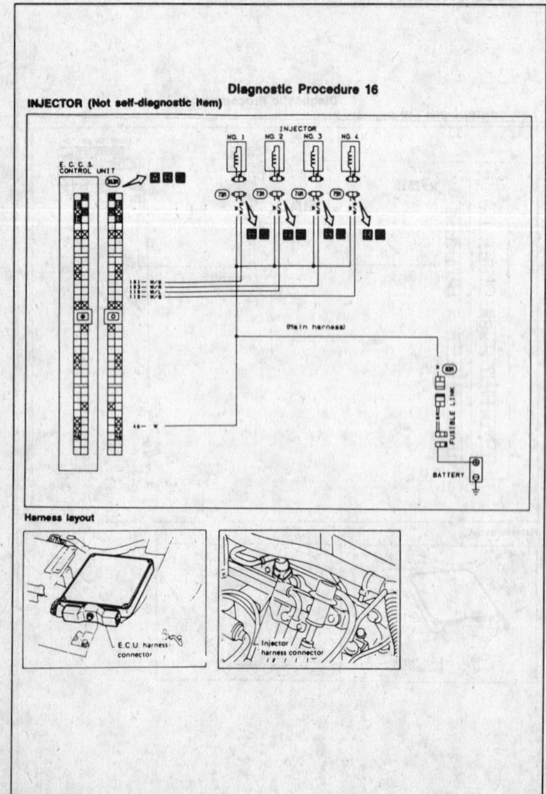
Diagnostic Procedure 16
INJECTOR (Not self-diagnostic item)
INJECTOR
NO. 1
NO. 2
NO. 3
NO. 4
ECCS CONTROL UNIT
Main harness
FUSIBLE LINK
BATTERY
Harness layout
E.C.U. harness connector
Injector harness connector

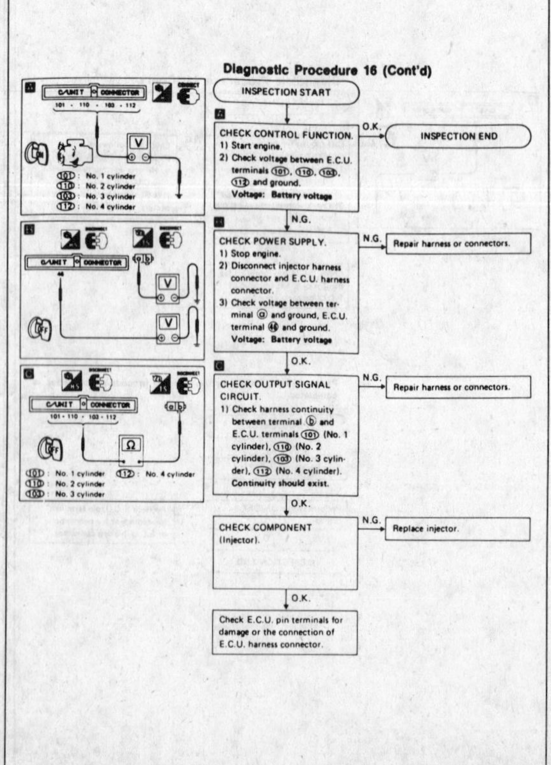
Diagnostic Procedure 16 (Cont'd)
INSPECTION START
A
CHECK CONTROL FUNCTION.
1) Start engine.
2) Check voltage between E.C.U. terminals 101, 110, 103, 112 and ground.
Voltage: Battery voltage
O.K.
INSPECTION END
N.G.
B
CHECK POWER SUPPLY.
1) Stop engine.
2) Disconnect injector harness connector and E.C.U. harness connector.
3) Check voltage between terminal ⓐ and ground, E.C.U. terminal 46 and ground.
Voltage: Battery voltage
N.G.
Repair harness or connectors.
O.K.
C
CHECK OUTPUT SIGNAL CIRCUIT.
1) Check harness continuity between terminal ⓑ and E.C.U. terminals 101 (No. 1 cylinder), 110 (No. 2 cylinder), 103 (No. 3 cylinder), 112 (No. 4 cylinder).
Continuity should exist.
N.G.
Repair harness or connectors.
O.K.
CHECK COMPONENT
(Injector).
N.G.
Replace injector.
O.K.
Check E.C.U. pin terminals for damage or the connection of E.C.U. harness connector.
C/UNIT CONNECTOR
101 · 110 · 103 · 112
101 : No. 1 cylinder
110 : No. 2 cylinder
103 : No. 3 cylinder
112 : No. 4 cylinder
46

ECCS DIAGNOSTIC CHART—1990 TRUCK/PATHFINDER WITH KA24E ENGINE

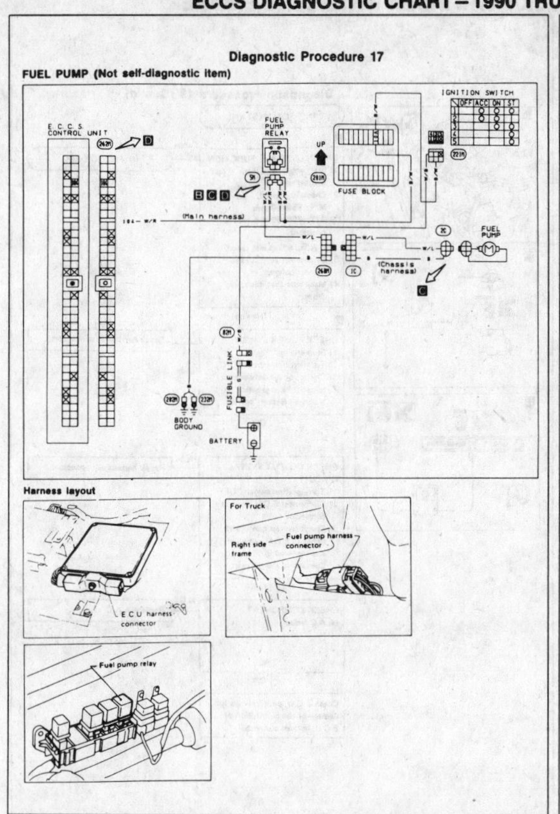
Diagnostic Procedure 17
FUEL PUMP (Not self-diagnostic item)
E.C.C.S. CONTROL UNIT
IGNITION SWITCH
FUEL PUMP RELAY
FUSE BLOCK
(Main harness)
(Chassis harness)
FUEL PUMP
BODY GROUND
FUSIBLE LINK
BATTERY
Harness layout
E.C.U. harness connector
For Truck
Right side frame
Fuel pump harness connector
Fuel pump relay

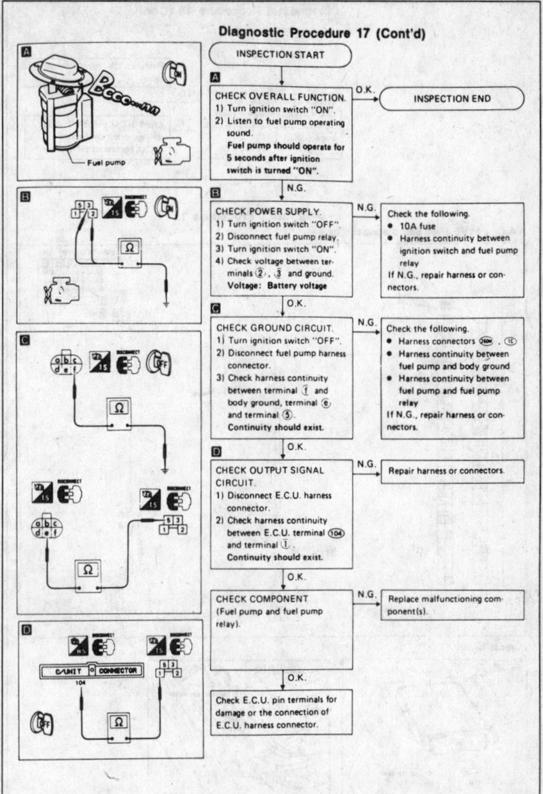
Diagnostic Procedure 17 (Cont'd)
INSPECTION START
A CHECK OVERALL FUNCTION.
1) Turn ignition switch "ON".
2) Listen to fuel pump operating sound.
Fuel pump should operate for 5 seconds after ignition switch is turned "ON".
O.K.
INSPECTION END
N.G.
B CHECK POWER SUPPLY.
1) Turn ignition switch "OFF".
2) Disconnect fuel pump relay.
3) Turn ignition switch "ON".
4) Check voltage between terminals ②, ③ and ground.
Voltage: Battery voltage
N.G.
Check the following.
• 10A fuse
• Harness continuity between ignition switch and fuel pump relay
If N.G., repair harness or connectors.
O.K.
C CHECK GROUND CIRCUIT.
1) Turn ignition switch "OFF".
2) Disconnect fuel pump harness connector.
3) Check harness continuity between terminal ① and body ground, terminal ⓔ and terminal ⑤.
Continuity should exist.
N.G.
Check the following.
• Harness connectors
• Harness continuity between fuel pump and body ground
• Harness continuity between fuel pump and fuel pump relay
If N.G., repair harness or connectors.
O.K.
D CHECK OUTPUT SIGNAL CIRCUIT.
1) Disconnect E.C.U. harness connector.
2) Check harness continuity between E.C.U. terminal ⑩④ and terminal ①.
Continuity should exist.
N.G.
Repair harness or connectors.
O.K.
CHECK COMPONENT
(Fuel pump and fuel pump relay).
N.G.
Replace malfunctioning component(s).
O.K.
Check E.C.U. pin terminals for damage or the connection of E.C.U. harness connector.
Fuel pump
DISCONNECT
C/UNIT CONNECTOR

ECCS DIAGNOSTIC CHART—1990 TRUCK/PATHFINDER WITH KA24E ENGINE

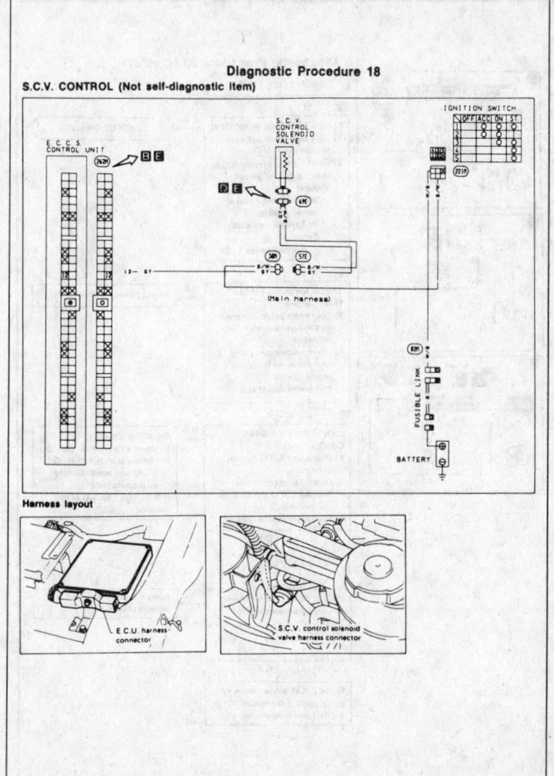
Diagnostic Procedure 18
S.C.V. CONTROL (Not self-diagnostic item)
E.C.C.S. CONTROL UNIT
IGNITION SWITCH
S.C.V. CONTROL SOLENOID VALVE
(Main harness)
FUSIBLE LINK
BATTERY
Harness layout
E.C.U. harness connector
S.C.V. control solenoid valve harness connector

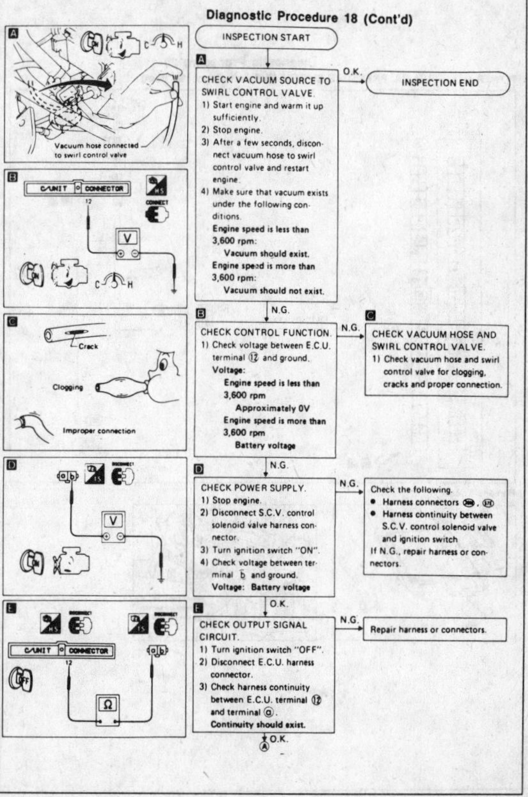
Diagnostic Procedure 18 (Cont'd)
INSPECTION START
A CHECK VACUUM SOURCE TO SWIRL CONTROL VALVE.
1) Start engine and warm it up sufficiently.
2) Stop engine.
3) After a few seconds, disconnect vacuum hose to swirl control valve and restart engine.
4) Make sure that vacuum exists under the following conditions.
Engine speed is less than 3,600 rpm:
Vacuum should exist.
Engine speed is more than 3,600 rpm:
Vacuum should not exist.
O.K.
INSPECTION END
N.G.
B CHECK CONTROL FUNCTION.
1) Check voltage between E.C.U. terminal ⑫ and ground.
Voltage:
Engine speed is less than 3,600 rpm
Approximately 0V
Engine speed is more than 3,600 rpm
Battery voltage
N.G.
C CHECK VACUUM HOSE AND SWIRL CONTROL VALVE.
1) Check vacuum hose and swirl control valve for clogging, cracks and proper connection.
N.G.
D CHECK POWER SUPPLY.
1) Stop engine.
2) Disconnect S.C.V. control solenoid valve harness connector.
3) Turn ignition switch "ON".
4) Check voltage between terminal ⓑ and ground.
Voltage: Battery voltage
N.G.
Check the following.
• Harness connectors
• Harness continuity between S.C.V. control solenoid valve and ignition switch
If N.G., repair harness or connectors.
O.K.
E CHECK OUTPUT SIGNAL CIRCUIT.
1) Turn ignition switch "OFF".
2) Disconnect E.C.U. harness connector.
3) Check harness continuity between E.C.U. terminal ⑫ and terminal ⓐ.
Continuity should exist.
N.G.
Repair harness or connectors.
O.K.
Ⓐ
Vacuum hose connected to swirl control valve
Crack
Clogging
Improper connection
C/UNIT CONNECTOR
DISCONNECT

ECCS DIAGNOSTIC CHART—1990 TRUCK/PATHFINDER WITH KA24E ENGINE

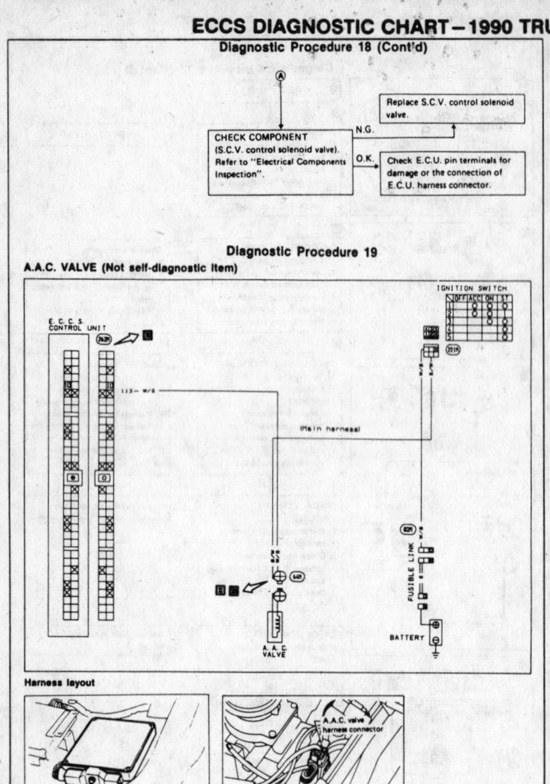
Diagnostic Procedure 18 (Cont'd)
CHECK COMPONENT (S.C.V. control solenoid valve). Refer to "Electrical Components Inspection".
N.G.
Replace S.C.V. control solenoid valve.
O.K.
Check E.C.U. pin terminals for damage or the connection of E.C.U. harness connector.
Diagnostic Procedure 19
A.A.C. VALVE (Not self-diagnostic item)
E.C.C.S. CONTROL UNIT
IGNITION SWITCH
(Main harness)
FUSIBLE LINK
BATTERY
A.A.C. VALVE
Harness layout
A.A.C. valve harness connector
E.C.U. harness connector

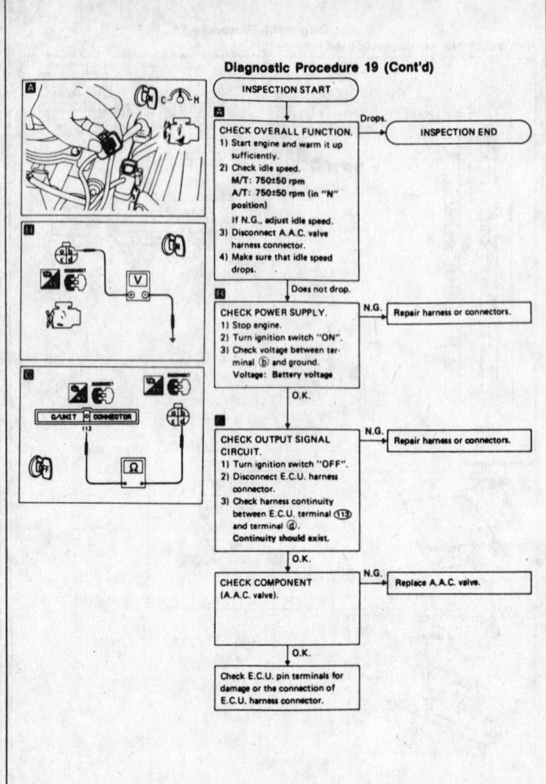
Diagnostic Procedure 19 (Cont'd)
INSPECTION START
CHECK OVERALL FUNCTION.
1) Start engine and warm it up sufficiently.
2) Check idle speed.
M/T: 750±50 rpm
A/T: 750±50 rpm (in "N" position)
If N.G., adjust idle speed.
3) Disconnect A.A.C. valve harness connector.
4) Make sure that idle speed drops.
Drops.
INSPECTION END
Does not drop.
CHECK POWER SUPPLY.
1) Stop engine.
2) Turn ignition switch "ON".
3) Check voltage between terminal ⓑ and ground.
Voltage: Battery voltage
N.G.
Repair harness or connectors.
O.K.
CHECK OUTPUT SIGNAL CIRCUIT.
1) Turn ignition switch "OFF".
2) Disconnect E.C.U. harness connector.
3) Check harness continuity between E.C.U. terminal 113 and terminal ⓓ.
Continuity should exist.
N.G.
Repair harness or connectors.
O.K.
CHECK COMPONENT (A.A.C. valve).
N.G.
Replace A.A.C. valve.
O.K.
Check E.C.U. pin terminals for damage or the connection of E.C.U. harness connector.

ECCS DIAGNOSTIC CHART—1990 TRUCK/PATHFINDER WITH KA24E ENGINE

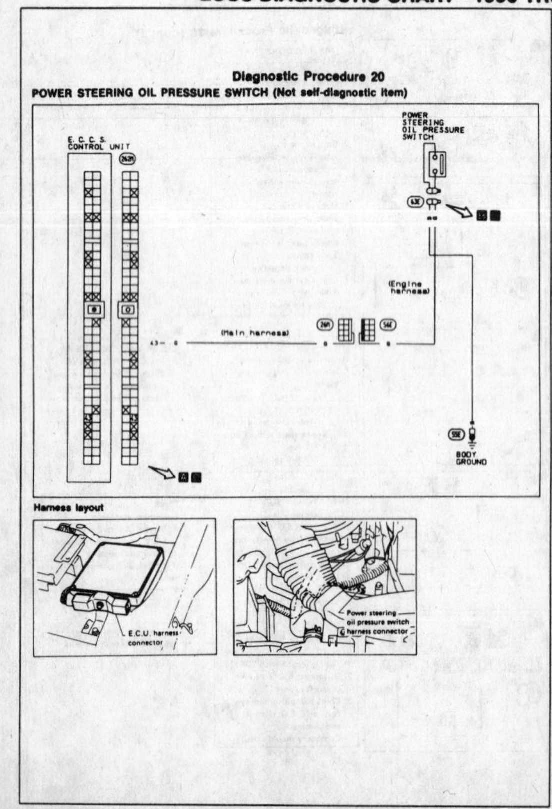
Diagnostic Procedure 20
POWER STEERING OIL PRESSURE SWITCH (Not self-diagnostic item)
E.C.C.S. CONTROL UNIT
POWER STEERING OIL PRESSURE SWITCH
(Engine harness)
(Main harness)
BODY GROUND
Harness layout
E.C.U. harness connector
Power steering oil pressure switch harness connector

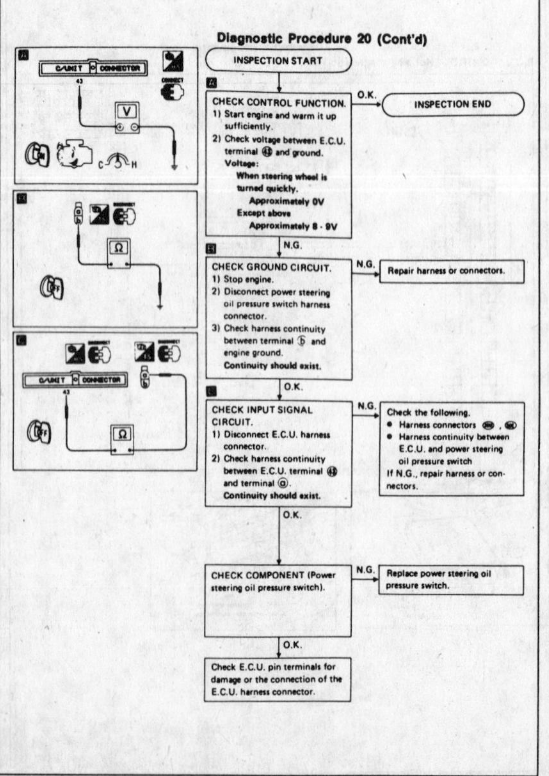
Diagnostic Procedure 20 (Cont'd)
INSPECTION START
CHECK CONTROL FUNCTION.
1) Start engine and warm it up sufficiently.
2) Check voltage between E.C.U. terminal 43 and ground.
Voltage:
When steering wheel is turned quickly.
Approximately 0V
Except above
Approximately 8 - 9V
O.K.
INSPECTION END
N.G.
CHECK GROUND CIRCUIT.
1) Stop engine.
2) Disconnect power steering oil pressure switch harness connector.
3) Check harness continuity between terminal ⓑ and engine ground.
Continuity should exist.
N.G.
Repair harness or connectors.
O.K.
CHECK INPUT SIGNAL CIRCUIT.
1) Disconnect E.C.U. harness connector.
2) Check harness continuity between E.C.U. terminal 43 and terminal ⓐ.
Continuity should exist.
N.G.
Check the following.
• Harness connectors
• Harness continuity between E.C.U. and power steering oil pressure switch
If N.G., repair harness or connectors.
O.K.
CHECK COMPONENT (Power steering oil pressure switch).
N.G.
Replace power steering oil pressure switch.
O.K.
Check E.C.U. pin terminals for damage or the connection of the E.C.U. harness connector.

ECCS DIAGNOSTIC CHART–1990 TRUCK/PATHFINDER WITH KA24E ENGINE

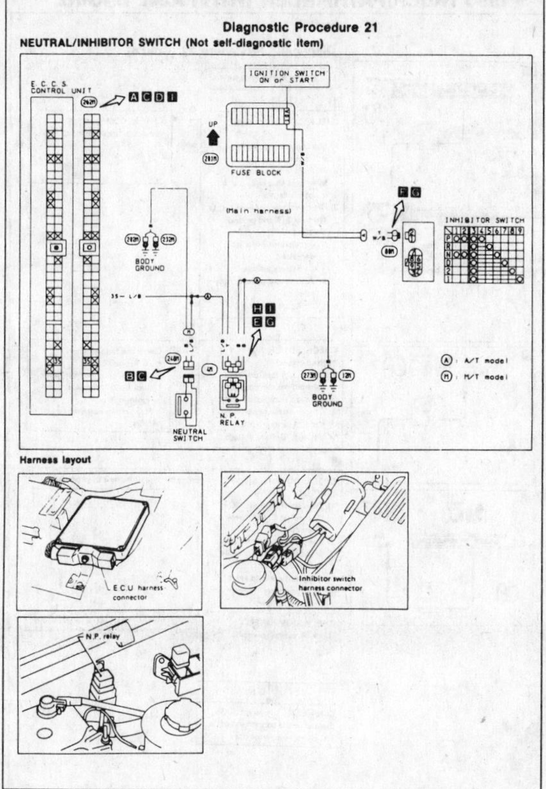
Diagnostic Procedure 21
NEUTRAL/INHIBITOR SWITCH (Not self-diagnostic item)
E.C.C.S. CONTROL UNIT
IGNITION SWITCH ON or START
UP
FUSE BLOCK
Main harness
BODY GROUND
INHIBITOR SWITCH
NEUTRAL SWITCH
N.P. RELAY
BODY GROUND
Ⓐ: A/T model
Ⓜ: M/T model
Harness layout
E.C.U. harness connector
Inhibitor switch harness connector
N.P. relay

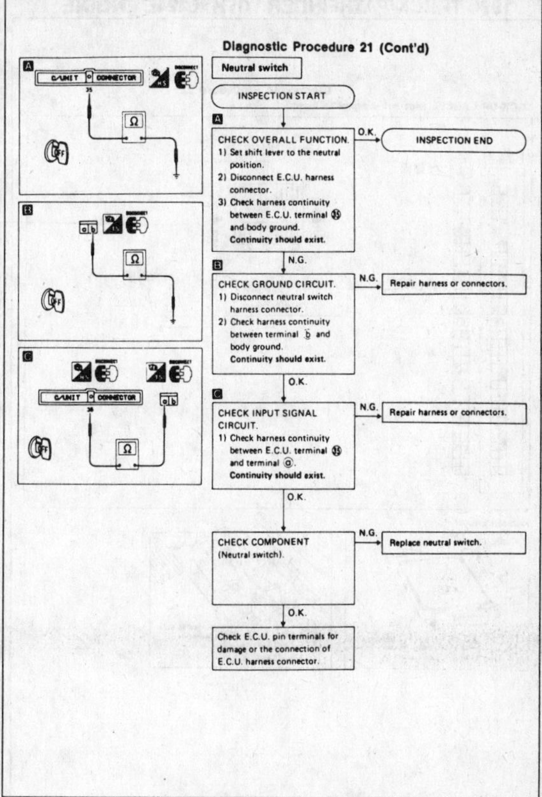
Diagnostic Procedure 21 (Cont'd)
C/UNIT CONNECTOR
Neutral switch
INSPECTION START
CHECK OVERALL FUNCTION.
1) Set shift lever to the neutral position.
2) Disconnect E.C.U. harness connector.
3) Check harness continuity between E.C.U. terminal ㉟ and body ground.
Continuity should exist.
O.K.
INSPECTION END
N.G.
CHECK GROUND CIRCUIT.
1) Disconnect neutral switch harness connector.
2) Check harness continuity between terminal ⓑ and body ground.
Continuity should exist.
N.G.
Repair harness or connectors.
O.K.
CHECK INPUT SIGNAL CIRCUIT.
1) Check harness continuity between E.C.U. terminal ㉟ and terminal ⓐ.
Continuity should exist.
N.G.
Repair harness or connectors.
O.K.
CHECK COMPONENT (Neutral switch).
N.G.
Replace neutral switch.
O.K.
Check E.C.U. pin terminals for damage or the connection of E.C.U. harness connector.

ECCS DIAGNOSTIC CHART–1990 TRUCK/PATHFINDER WITH KA24E ENGINE

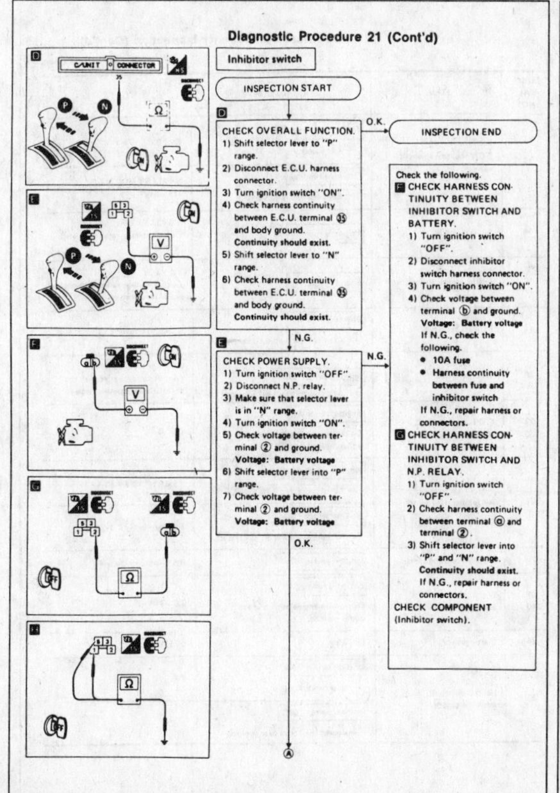
Diagnostic Procedure 21 (Cont'd)
C/UNIT CONNECTOR
Inhibitor switch
INSPECTION START
CHECK OVERALL FUNCTION.
1) Shift selector lever to "P" range.
2) Disconnect E.C.U. harness connector.
3) Turn ignition switch "ON".
4) Check harness continuity between E.C.U. terminal ㉟ and body ground.
Continuity should exist.
5) Shift selector lever to "N" range.
6) Check harness continuity between E.C.U. terminal ㉟ and body ground.
Continuity should exist.
O.K.
INSPECTION END
N.G.
CHECK POWER SUPPLY.
1) Turn ignition switch "OFF".
2) Disconnect N.P. relay.
3) Make sure that selector lever is in "N" range.
4) Turn ignition switch "ON".
5) Check voltage between terminal ② and ground.
Voltage: Battery voltage
6) Shift selector lever into "P" range.
7) Check voltage between terminal ② and ground.
Voltage: Battery voltage
N.G.
Check the following.
CHECK HARNESS CONTINUITY BETWEEN INHIBITOR SWITCH AND BATTERY.
1) Turn ignition switch "OFF".
2) Disconnect inhibitor switch harness connector.
3) Turn ignition switch "ON".
4) Check voltage between terminal ⓑ and ground.
Voltage: Battery voltage
If N.G., check the following.
• 10A fuse
• Harness continuity between fuse and inhibitor switch
If N.G., repair harness or connectors.
CHECK HARNESS CONTINUITY BETWEEN INHIBITOR SWITCH AND N.P. RELAY.
1) Turn ignition switch "OFF".
2) Check harness continuity between terminal ⓒ and terminal ②.
3) Shift selector lever into "P" and "N" range.
Continuity should exist.
If N.G., repair harness or connectors.
CHECK COMPONENT (Inhibitor switch).
O.K.
Ⓐ

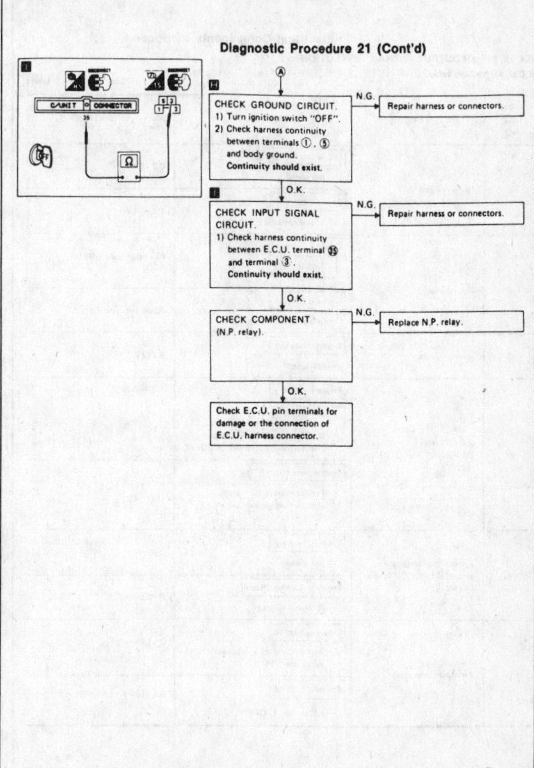
Diagnostic Procedure 21 (Cont'd)
C/UNIT CONNECTOR
Ⓐ
CHECK GROUND CIRCUIT.
1) Turn ignition switch "OFF".
2) Check harness continuity between terminals ①, ⑤ and body ground.
Continuity should exist.
N.G.
Repair harness or connectors.
O.K.
CHECK INPUT SIGNAL CIRCUIT.
1) Check harness continuity between E.C.U. terminal ㉟ and terminal ③.
Continuity should exist.
N.G.
Repair harness or connectors.
O.K.
CHECK COMPONENT (N.P. relay).
N.G.
Replace N.P. relay.
O.K.
Check E.C.U. pin terminals for damage or the connection of E.C.U. harness connector.

ECCS DIAGNOSTIC CHART—1990 TRUCK/PATHFINDER WITH KA24E ENGINE

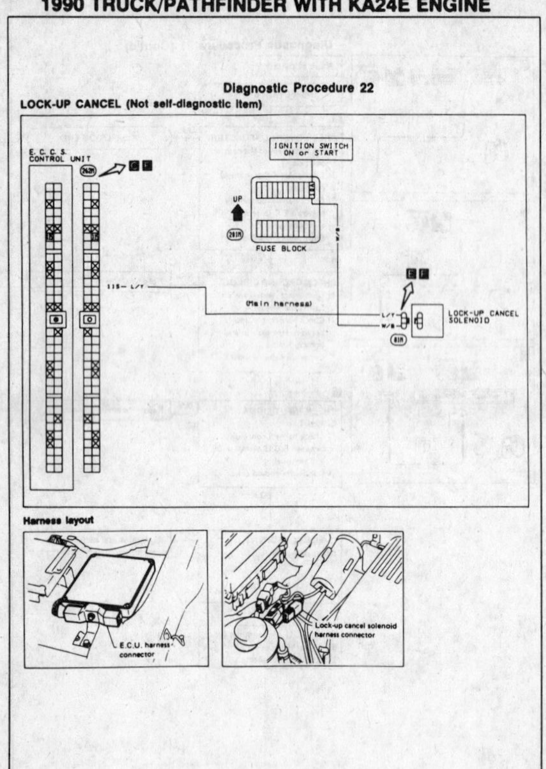
Diagnostic Procedure 22
LOCK-UP CANCEL (Not self-diagnostic item)
E.C.C.S. CONTROL UNIT
IGNITION SWITCH ON or START
UP
FUSE BLOCK
(Main harness)
LOCK-UP CANCEL SOLENOID
Harness layout
E.C.U. harness connector
Lock-up cancel solenoid harness connector

ECU SIGNAL INSPECTION—1990 TRUCK/PATHFINDER WITH KA24E ENGINE

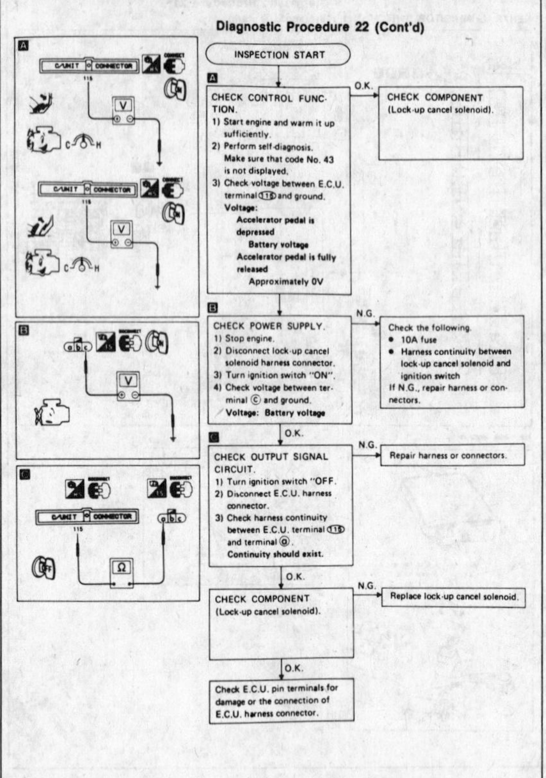
Diagnostic Procedure 22 (Cont'd)
INSPECTION START
A
CHECK CONTROL FUNCTION.
1) Start engine and warm it up sufficiently.
2) Perform self-diagnosis. Make sure that code No. 43 is not displayed.
3) Check voltage between E.C.U. terminal 115 and ground.
Voltage:
Accelerator pedal is depressed
Battery voltage
Accelerator pedal is fully released
Approximately 0V
O.K.
CHECK COMPONENT (Lock-up cancel solenoid).
B
CHECK POWER SUPPLY.
1) Stop engine.
2) Disconnect lock-up cancel solenoid harness connector.
3) Turn ignition switch "ON".
4) Check voltage between terminal © and ground.
Voltage: Battery voltage
N.G.
Check the following.
• 10A fuse
• Harness continuity between lock-up cancel solenoid and ignition switch
If N.G., repair harness or connectors.
O.K.
C
CHECK OUTPUT SIGNAL CIRCUIT.
1) Turn ignition switch "OFF.
2) Disconnect E.C.U. harness connector.
3) Check harness continuity between E.C.U. terminal 115 and terminal ⓐ.
Continuity should exist.
N.G.
Repair harness or connectors.
O.K.
CHECK COMPONENT (Lock-up cancel solenoid).
N.G.
Replace lock-up cancel solenoid.
O.K.
Check E.C.U. pin terminals for damage or the connection of E.C.U. harness connector.

ECU SIGNAL INSPECTION (CONT.)—1990 TRUCK/PATHFINDER WITH KA24E ENGINE

Electrical Components Inspection

E.C.U. INPUT/OUTPUT SIGNAL INSPECTION

E.C.U. Inspection table

*Data are reference values.

TERMINAL NO.	ITEM	CONDITION	*DATA
1	Ignition signal	Engine is running. — Idle speed	0.3 - 0.6V
		Engine is running. — Engine speed is 2,000 rpm.	1.2 - 1.5V
2	Tachometer	Engine is running. — Idle speed	Approximately 1.0V
		Engine is running. — Engine speed is 2,000 rpm.	Approximately 2.7V
3	Ignition check	Engine is running. — Idle speed	9 - 12V
4	E.C.C.S. relay (Main relay)	Engine is running. → Ignition switch "OFF" — Within approximately 1 second after turning ignition switch "OFF"	0 - 1 V
		Ignition switch "OFF" — For approximately 1 second after turning ignition switch "OFF"	BATTERY VOLTAGE (11 - 14V)
8	Exhaust gas temperature sensor (Only for California model)	Engine is running. — Idle speed	3.0 - 4.0V
		Engine is running. (Racing) — After warming up	0 - 3.0V
11	Air conditioner relay	Engine is running. — Both A/C switch and blower switch are "ON".	0 - 1.0V
		Engine is running. — A/C switch is "OFF".	BATTERY VOLTAGE (11 - 14V)

Electrical Components Inspection (Cont'd)

*Data are reference values.

TERMINAL NO.	ITEM	CONDITION	*DATA
12	S.C.V. control solenoid valve	Engine is running. — Idle speed	0 - 1.0V
		Engine is running. — Engine speed is 2,000 rpm.	BATTERY VOLTAGE (11 - 14V)
16	Air flow meter	Engine is running.	1.0 - 3.0V Output voltage varies with engine revolution.
18	Engine temperature sensor	Engine is running.	1.0 - 3.0V Output voltage varies with engine water temperature.
19	Exhaust gas sensor	Engine is running. — After warming up sufficiently.	0 - Approximately 1.0V
20	Throttle sensor	Ignition switch "ON" — After warming up sufficiently.	0.5 - Approximately 4V Output voltage varies with the throttle valve opening angle.
22 30	Crank angle sensor (Reference signal)	Engine is running. Do not run engine at high speed under no-load.	0.3 - 0.4V
26	Air temperature sensor	Ignition switch "ON" — Air temperature is 20°C (68°F).	Approximately 2.4V
		Ignition switch "ON" — Air temperature is 80°C (176°F).	Approximately 0.3V.
28	Throttle opening signal	Ignition switch "ON"	0.5 - Approximately 4V
31 40	Crank angle sensor (Position signal)	Engine is running. Do not run engine at high speed under no-load.	2.0 - 3.0V
34	Start signal	Cranking	8 - 12V
35	Neutral switch & Inhibitor switch	Ignition switch "ON" — Neutral/Parking	0V
		Ignition switch "ON" — Except the above gear position	6 - 7V

ECU SIGNAL INSPECTION (CONT.)–1990 TRUCK/PATHFINDER WITH KA24E ENGINE

Electrical Components Inspection (Cont'd)

*Data are reference values.

TERMINAL NO.	ITEM	CONDITION	*DATA
36	Ignition switch	Ignition switch "OFF"	0V
		Ignition switch "ON"	BATTERY VOLTAGE (11 - 14V)
37	Throttle sensor power supply	Ignition switch "ON"	Approximately 5V
38 47	Power supply for E.C.U.	Ignition switch "ON"	BATTERY VOLTAGE (11 - 14V)
41	Air conditioner switch	Engine is running. └ Both air conditioner switch and blower switch are "ON".	0V
		Engine is running. └ Air conditioner switch is "OFF".	BATTERY VOLTAGE (11 - 14V)
43	Power steering oil pressure switch	Engine is running. └ Steering wheel is being turned.	0.1 - 0.3V
		Engine is running. └ Steering wheel is not being turned.	Approximately 5V
46	Power supply (Back-up)	Ignition switch "OFF"	BATTERY VOLTAGE (11 - 14V)
101	Injector No. 1	Engine is running.	BATTERY VOLTAGE (11 - 14V)
103	Injector No. 3		
110	Injector No. 2		
112	Injector No. 4		
102	E.G.R. control solenoid valve	Engine is running. └ Engine is cold. [Water temperature is below 60°C (140°F).]	0.7 - 0.9V
		Engine is running. (Racing) └ After warming up [Water temperature is between 60°C (140°F) and 105°C (221°F).]	BATTERY VOLTAGE (11 - 14V)

Electrical Components Inspection (Cont'd)

*Data are reference values.

TERMINAL NO.	ITEM	CONDITION	*DATA
104	Fuel pump relay	Ignition switch "ON" └ For 5 seconds after turning ignition switch "ON" Engine is running.	0.7 - 0.9V
		Ignition switch "ON" └ Within 5 seconds after turning ignition switch "ON"	BATTERY VOLTAGE (11 - 14V)
113	A.A.C. valve	Engine is running. └ Idle speed	7 - 10V
		Engine is running. ├ Steering wheel is being turned. ├ Air conditioner is operating. ├ Rear defogger is "ON". └ Headlamps are in high position.	4 - 7V
115	Lock-up cancel solenoid	Engine is running └ Idle speed [Water temperature is below 40°C (104°F).]	Approximately 0V
		Engine is running ├ After warming up [Water temperature is above 40°C (104°F).] └ Engine speed is 2,000 rpm	BATTERY VOLTAGE (11 - 14V)

E.C.U. HARNESS CONNECTOR TERMINAL LAYOUT

101	102	103	104	105	106	107	108	1	2	3	4	5	6	7	15	16	17	18	19	20	21	22	31	32	33	34	35	36	37	38	39
109	110	111	112	113	114	115	116	8	9	10	11	12	13	14	23	24	25	26	27	28	29	30	40	41	42	43	44	45	46	47	48

H.S. CONNECT

ECCS SPECIFICATIONS—1990 TRUCK/PATHFINDER WITH KA24E ENGINE

Inspection and Adjustment

ENGINE TEMPERATURE SENSOR	20°C (68°F)	80°C (176°F)
Thermistor resistance kΩ	2.1 - 2.9	0.30 - 0.33
FUEL PRESSURE at idling (Measuring point: between fuel filter and fuel pipe)		
Vacuum hose is connected kPa (kg/cm², psi)	Approximately 226 (2.3, 33)	
Vacuum hose is disconnected kPa (kg/cm², psi)	Approximately 294 (3.0, 43)	
FUEL INJECTOR Coil resistance Ω	Approximately 10 - 15	
AIR REGULATOR Resistance Ω	Approximately 75	
EXHAUST GAS TEMPERATURE SENSOR	100°C (212°F)	
Thermistor resistance kΩ	85.3±8.53	

General Specifications

IGNITION TIMING	B.T.D.C.	15°±2°
IDLE SPEED	rpm	M/T 800±50 A/T 800±50 (in "N" position)

SUBARU MULTI-POINT FUEL INJECTION (MPFI) SYSTEM

NOTE: This section covers all multi-point fuel injected models except for the 1990 Legacy. The 1990 Legacy multi-point fuel injection system can be found in a following section. Please refer to the section index for the correct page.

General Information

The MPFI system supplies the optimum air/fuel mixture to the engine under all various operating conditions. System fuel, which is pressurized at a constant pressure, is injected into the intake air passage of the cylinder head. The amount of fuel injected is controlled by the intermittent injection system where the electro-magnetic injection valve (fuel injector) opens only for a short period of time, depending on the amount of air required for 1 cycle of operation. During system operation, the amount injection is determined by the duration of an electric pulse sent to the fuel injector, which permits precise metering of the fuel.

All the operating conditions of the engine are converted into electric signals, resulting in additional features of the system, such as improved adaptability and easier addition of compensating element. The MPFI system also incorporates the following features:

- Reduced emission of exhaust gases
- Reduction in fuel consumption
- Increased engine output
- Superior acceleration and deceleration
- Superior starting and warm-up performance in cold weather since compensation is made for coolant and intake air temperature
- Good performance with turbocharger, if equipped

SYSTEM COMPONENTS

Air Flow Meter

The MPFI system incorporates a hot wire type air flow meter. This meter converts the amount of air taken into the engine into an electric signal by using the heat transfer between the incoming air and a heating resistor (hot wire) located in the air intake. Features of the air flow meter are as follows:

- High altitude compensation is made automatically
- Ouick response
- There are no moving parts
- Its compact

Throttle Body

When depressing on the gas pedal, the throttle body opens/closes its valve to regulate the amount of air to be taken in to the combustion chamber. Negative pressure (positive pressure at supercharging) is generated according to the opening of the throttle valve, then is applied to the pressure ports for EGR control and canister purge. The pressure is used for controlling the EGR valve and canister purge.

On models equipped with 4 cylinder engines, during idling, the throttle valve is almost fully closed and the air flow through the throttle body is less than the air passing through the system. More than half of the air necessary for idling is supplied to the intake manifold through the idle bypass passage. Turning the idle adjust screw, on the idle bypass passage, can change the air flow to adjust the number of revolutions during idling. To prevent the number of revolutions from decreasing when the A/C is turned **ON**, the fast idle bypass passage is provided with a valve that is operated by the fast idle solenoid.

NOTE: Fast engine idle rpm can not be adjusted by turning the fast idle adjusting screw.

On models equipped with 6 cylinder engines, during idling, the throttle valve is almost fully closed and the air flow through the throttle body is less than the air passing through the system. More than half the air necessary for idling is supplied to the intake manifold through the bypass air control valve. The bypass air control valve properly controls the number of revolutions during idling, so no need for adjustment is necessary.

On all models, a throttle position sensor is incorporated with a potentiometer and an idle switch, interlocked with the throttle valve shaft. The throttle position sensor sends the MPFI control unit a potentiometer output signal corresponding to the opening of the throttle, and an idle switch signal that turns **ON** only when the throttle is opened the idle position. Using these signals, the MPFI control unit controls the air/fuel ratio during acceleration, deceleration and at idle.

By-Pass Air Control Valve

MODELS EQUIPPED WITH 6 CYLINDER ENGINE

The by-pass air control valve is controlled by a signal sent from the MPFI control unit. It controls the flow-rate of air passing through the by-pass passage, allowing the engine to operate at optimum speed under all various conditions. This results in stabilized exhaust gas emission, improved fuel economy and better performance. Features of the by-pass air control valve are as follows:

1. The by-pass passage opens partially when the engine is idling. When the engine is under a load (A/C turned **ON**), the by-pass passage opens wider to allow more air to pass through. This maintains the specified engine idling rpm.
2. When the engine is cold, the by-pass passage opens widely to allow more air to pass through. This speeds up engine warm-up.
3. Variations in engine idling rpm over time are compensated for.
4. When the throttle valve closes, a sudden drop in engine rpm is prevented to improve driving performance.

NOTE: When battery voltage drops momentarily, the by-pass valve sometimes activates to increase idling speed to 900 rpm. Before the engine reaches normal operating temperature, idling speed between N or P range and sometimes the D range sometimes differs. However, these are not problems.

Auxiliary Air Valve

MODELS EQUIPPED WITH 4 CYLINDER ENGINE

The auxiliary air valve is used to increase air flow when the engine is started at low temperature, and until the engine reaches normal operating temperature. It consists of a coiled bimetal, a bimetal-operated shutter valve and an electric heater element for bimetal. This passing air flow (during start-up) is increased as the temperature becomes lower. After the engine has been started, the heating is preformed by the heater element to which current is supplied from the fuel pump relay circuit. Thereby, the shutter valve turns gradually to decrease the air flow. After a certain elapsed time, the shutter valve is closed.

Ignition System

The ignition system consists of a battery, an ignition coil, dis-

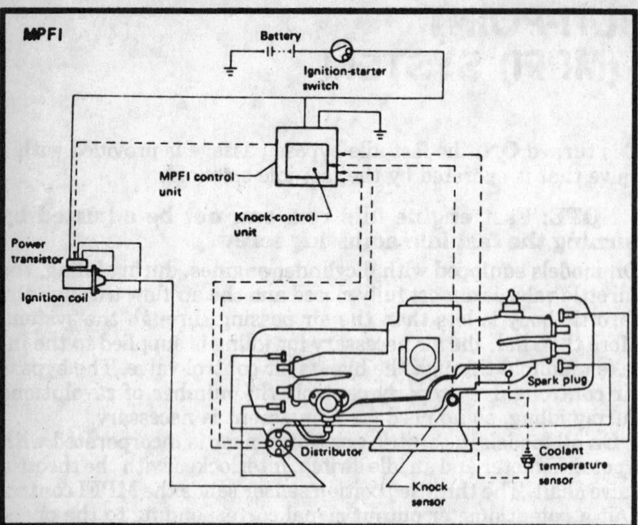
MPFI
Battery
Ignition-starter switch
MPFI control unit
Knock-control unit
Power transistor
Ignition coil
Spark plug
Distributor
Coolant temperature sensor
Knock sensor

MPFI ignition system schematic—Sedan and Wagon and 1990 Loyale

tributor, spark plugs, knock sensor, MPFI control unit and spark plug wires.

The crank angle sensor, built into the distributor, detects the reference crank angle and the positioned crank angle. An electronic signal of both angles is sent to the MPFI control unit which is used in conjuction with the fuel injection system. The MPFI control unit determines spark advance angle and spark timing. The signal from spark timing is determined by the control unit, and is sent to the power transistor where it makes the primary circuit to the ignition coil, where high voltage current is generated in the secondary circuit. The high voltage of the secondary circuit is sent to the spark plug of each cylinder.

Under normal operating conditions, the spark advance angle is calculated from the following 3 factors:

1. Engine speed compensation
2. Advanced when starting the engine
3. Advanced in all driving conditions, except when starting the engine, after engine rpm exceeds specifications

When knocking occurs, a signal is sent from the knock sensor to the MPFI control unit. The MPFI control unit then retards spark timing to prevent engine knocking.

A knock sensor is installed on the cylinder block, and senses knocking signals from each cylinder. The knock sensor is a piezo-electric type which converts knocking vibrations into electric signals. It consists of a piezo-electric element, a weight and a case. If knocking occurs in the engine, the weight in the case moves the piezo-electric element to generate voltage.

Air/Fuel Ratio Control System

This system stabilizes the quality of the hot-wire type air flow meter and the fuel injectors to maintain their orignal performance by correcting their variation and aging.

By learning the amount of feedback of the Oxygen (O_2) sensor, the system sends signals to the control unit to automatically set a coefficient of correction, thereby, the fuel injector always acheives fuel injection under all conditions.

OXYGEN (O_2) SENSOR

The O_2 sensor is installed in the center exhaust pipe and is used to sense oxygen concentration in the exhaust gas. If the fuel ratio is leaner than the optimum ratio (14.7:1), the exhaust gas contains more oxygen. If the fuel ratio is richer than the optimum ratio (14.7:1), the exhaust gas contains little oxygen. Therefore, the oxygen concentration in the exhaust gas makes it possible to determine whether the air/fuel ratio is richer or leaner than the optimum ratio (14.7:1).

The 0_2 sensor incorporates a zirconia tube (ceramic) which generates voltage, if there is a difference in oxygen concentration between the inside and outside of the tube. The inside and outside of the zirconia tube is coated with platinum for catalysis and electrode provision. The screw on the outside of the sensor is grounded to the exhaust pipe and its lead is connected to the MPFI control unit through the harness.

When a rich air/fuel mixture is burnt in the cylinder, the oxygen in the exhaust gases reacts almost completely through the catalytic action of the platinum coating on the surface of the zirconia tube. This results in a very large difference in the oxygen concentration between the inside and outside of the tube, and the electromotive force generated is large.

When a lean air/fuel mixture is burnt in the cylinder, oxygen remains in the exhaust gases even after the catalytic action, resulting in a small difference in the oxygen concentration. The electromotive force is very small.

The difference in oxygen concentration changes greatly in the area of the optimum air/fuel ratio, and changes in the electromotive force is also large. By inputting this information into the MPFI control unit, the air/fuel ratio can be determined easily. The O_2 sensor does not generate much electromotive force when the temperature is low. The electromotive force stabilizes at temperatures of approximately 572–752°F (300–400°C). California models use a ceramic heater to improve performance at low temperatures.

Fuel Injector

The fuel injector injects fuel according to the valve opening signal received from the MPFI control unit. The nozzle is attached on the top of the injector. The needle valve is lifted by the solenoid coil through the plunger when the valve opening signal is received.

Since the injection opening varies, the lifted level of the needle valve and the regulated controlled fuel pressure are kept constant, the amount of fuel to be injected can be controlled only by the valve opening signal from the MPFI control unit. At the fuel inlet of the injector, a filter is mounted to prevent dust from entering the system.

Coolant Thermosensor

The coolant thermosensor is installed on the waterpipe which is made of aluminum alloy. Its thermistor changes resistance with respect to temperature. The thermosensor sends the coolant temperature signal to the MPFI control unit, which determines the amount fuel volume to be injected.

Dropping Resistor

The dropping resistor used on 1988 XT models equipped with 4 cylinder engines serves as a voltage control to maintain optimum injector driving current.

Pressure Regulator

The pressure regulator is divided into 2 chambers: the fuel chamber and the spring chamber. Fuel is supplied to the fuel chamber, through the fuel inlet, connected to the injector. A difference in pressure between the fuel chamber and the spring chamber connected with the intake manifold causes the diaphragm to be pushed down, causing fuel to be fed back to the fuel tank through the return line. By the returning of the fuel, as to balance the above pressure difference and the spring force, fuel pressure is kept at a constant pressure of 36.3 psi against the intake manifold pressure.

Pressure Switch

1988 sedans and wagons equipped with 4 cylinder turbocharged engines are equipped with 2 positive pressure switches(which

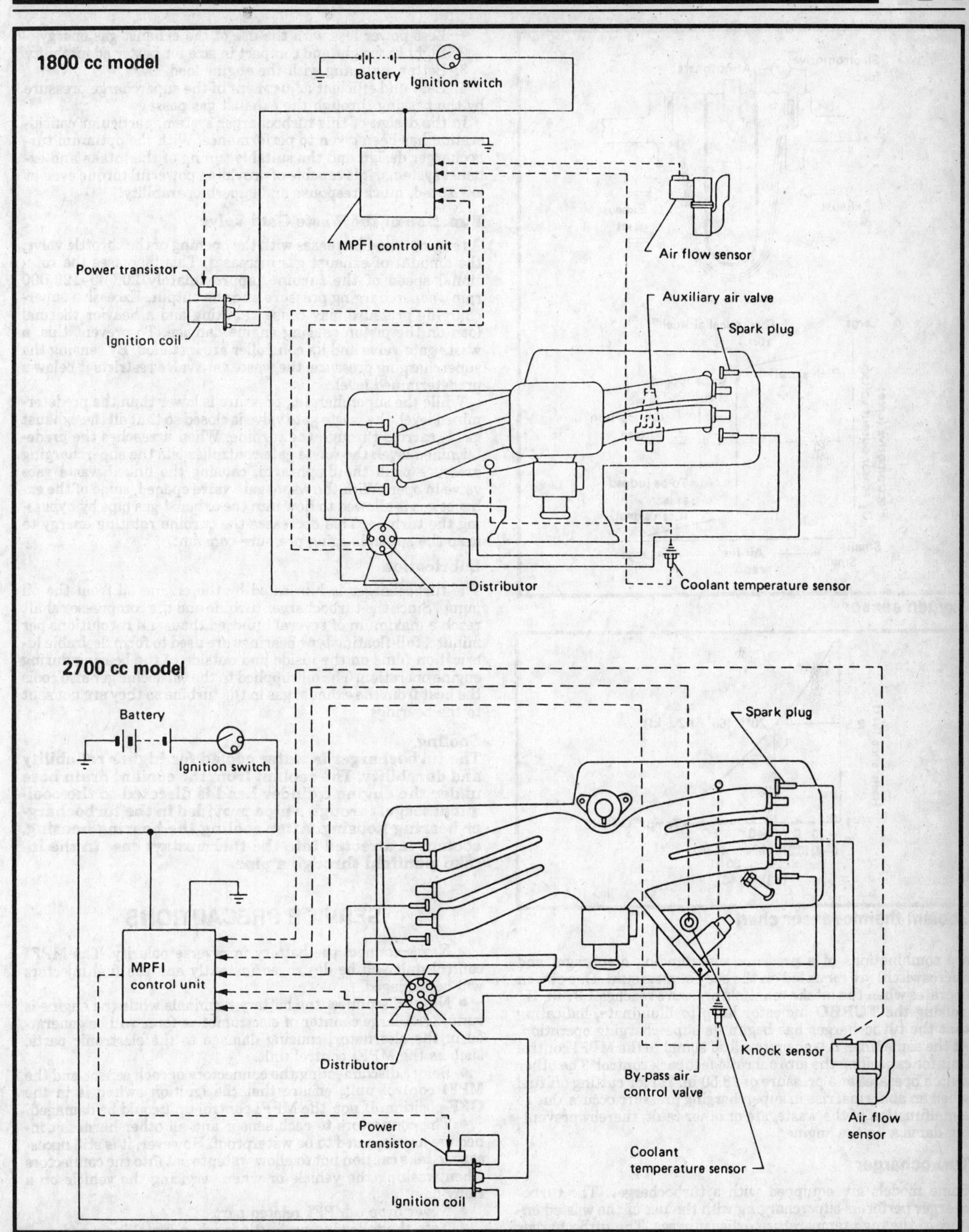
1800 cc model
Battery
Ignition switch
MPFI control unit
Power transistor
Ignition coil
Air flow sensor
Auxiliary air valve
Spark plug
Distributor
Coolant temperature sensor
2700 cc model
Battery
Ignition switch
Spark plug
MPFI control unit
Distributor
Knock sensor
By-pass air control valve
Air flow sensor
Power transistor
Ignition coil
Coolant temperature sensor

MPFI ignition system schematic—XT models

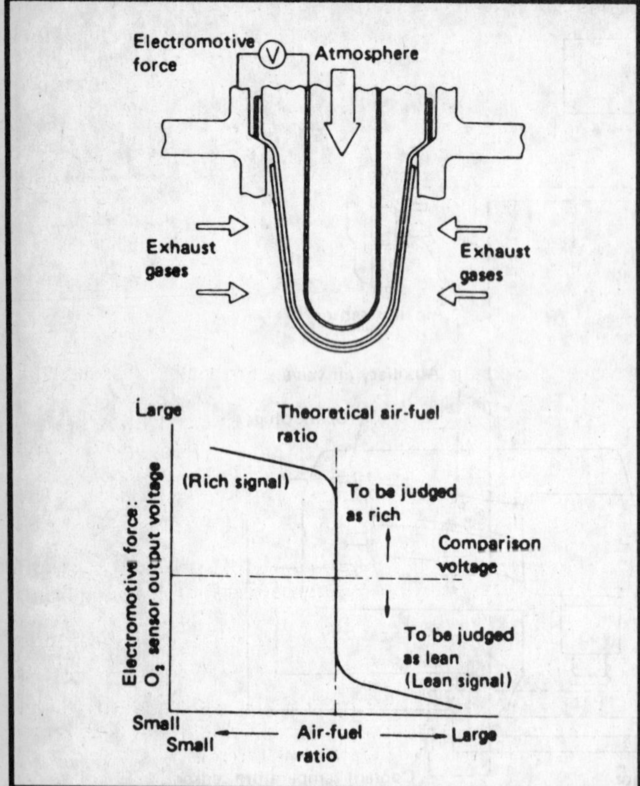
Electromotive force
Atmosphere
Exhaust gases
Exhaust gases
Large
Theoretical air-fuel ratio
(Rich signal)
To be judged as rich
Comparison voltage
To be judged as lean
(Lean signal)
Electromotive force: O_2 sensor output voltage
Small
Small
Air-fuel ratio
Large

Oxygen sensor

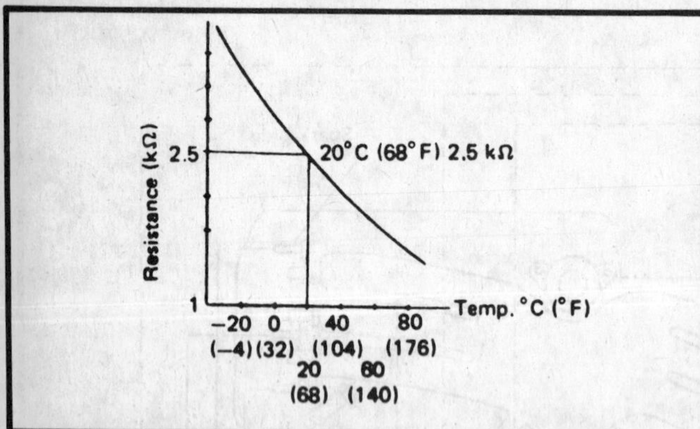
Resistance (kΩ)
2.5
20°C (68°F) 2.5 kΩ
1
Temp. °C (°F)
−20 0 40 80
(−4) (32) (104) (176)
20 60
(68) (140)

Coolant thermosensor chart

are combinations of a pressure withstanding diaphragm and microswitch) are mounted on the body strut mount. One switch operates when the intake manifold pressure reaches 1.97 in. Hg causing the **TURBO** indicator lamp to illuminate, indicating that the turbocharger has begun its supercharging operation. At the same time, it transmits a load signal to the MPFI control unit for cancelling the air/fuel ratio feedback control. The other switch operates at a pressure of 18.50 in. Hg for cutting off fuel when an abnormal rise in supercharging pressure occurs, due to a malfunction of the wastegate or other fault, thereby preventing damage to the engine.

Turbocharger

Some models are equipped with a turbocharger. The turbocharger performs supercharging with the use of the wasted energy in the high temperature exhaust gases. The turbocharger provides the following features:

1. Less power loss with the use of the exhaust gas energy
2. Light in weight and compact in size for better adaptability
3. Better matching with the engine load
4. Easy and efficient adjustment of the supercharge pressure by the passing through the exhaust gas passage

In the design of this turbocharger system, particular consideration has been given to performance. With the optimum turbocharger design and the suitable tuning of the intake and exhaust systems, it is capable of providing powerful torque even at low speed, quick response and superb operability.

Function of the Waste Gate Valve

As engine speed increases with the opening of the throttle valve, the amount of exhaust gas increases. This increases the rotational speed of the turbine (approximately 20,000–120,000 rpm), supercharging pressure and the output. Excessive supercharging pressure may cause knocking and a heavier thermal load on the piston causing engine damage. To prevent this, a waste gate valve and its controller are installed. By sensing the supercharging pressure, the waste gate valve restricts it below a predetermined level.

While the supercharging pressure is lower than the predetermined level, the waste gate valve is closed so that all the exhaust gas is carried through the turbine. When it reaches the predetermined level, the waste gate controller lets the supercharging pressure push the diaphragm, causing the linked waste gate valve to open. With the waste gate valve opened, some of the exhaust gas is allowed to flow into the exhaust gas pipe by bypassing the turbine. This decreases the turbine rotating energy to keep the supercharging pressure constant.

Lubrication

The turbocharger is lubricated by the engine oil from the oil pump. Since the turbocharger turbine and the compressor shaft reach a maximum of several hundred thousand revolutions per minute, full-floating type bearings are used to form desirable lubrication films on the inside and outside of the bearing during engine operation. The oil supplied to the turbocharger also cools the heat from the exhaust gas in the turbine so they are not sent to the bearings.

Cooling

The turbocharger is water cooled for higher reliability and durability. The coolant from the coolant drain hose under the engine cylinder head is directed to the coolant passage, through a pipe provided in the turbocharger bearing housing. After cooling the bearing housing, coolant is directed into the thermostart case in the intake manifold through a pipe.

SERVICE PRECAUTIONS

- Never connect the battery in reverse polarity. The MPFI control unit will be destroyed instantly and the fuel injectors will be damaged.
- Never disconnect the battery terminals while the engine is running. A large counter of electromotive force will be generated in the alternator, causing damage to the electronic parts, such as the MPFI control unit.
- Before disconnecting the connectors of each sensor and the MPFI control unit, ensure that the ignition switch is in the **OFF** position. If not, the MPFI control unit could be damaged.
- The connectors to each sensor and all other harness connectors are designed to be waterproof. However, it is still necessary to take caution not to allow water to get into the connectors when washing the vehicle or when servicing the vehicle on a rainy day.
- Never drop a MPFI related part.
- Note the following precautions when installing a radio:
  – Keep the antenna as far apart as possible from the con-

trol unit (the control unit is located under the steering column, inside the instrument panel lower trim panel).

–The antenna line must be positioned as far apart as possible from the MPFI control unit and MPFI harness.

–Carefully adjust the antenna for correct matching.

NOTE: Incorrect installing of the radio may affect the operation of the MPFI control unit.

• Before disconnecting the fuel line, disconnect the fuel pump connector, then crank the engine for approximately 5 seconds, to release the pressure in the fuel system. If the engine starts during this operation, let it run until it stops.

Diagnosis and Testing

SELF-DIAGNOSIS SYSTEM

The self-diagnosis system detects and indicates faults, in various inputs and outputs of the electronic control unit. The warning lamp (**CHECK ENGINE** light), located on the instrument panel indicates a fault or trouble, and the LED (light emitting diode) in the control unit indicates a trouble code. A fail-safe function is incorporated into the system to ensure minimal driveability if a failure of a sensor occurs.

Self-Diagnosis Function

The MPFI control unit carries out the computational processing of the input information received from various sensors and produces the output information for operating the fuel injectors, fuel pump, etc. This computational processing, reads out all the input/output information to examine matching with the predetermined levels (proper values or ranges). If a predetermined level is not satisfied, or a fault is found, the warning lamp signals the driver. When this occurs, the self-diagnosis function is performed.

Fail-Safe Function

The component which has been found faulty in the self-diagnosis function, the MPFI control unit receives the faulty signal and carries out the computational processing. When this occurs, the fail-safe function is performed.

Function Of Self-Diagnosis

The self-diagnosis function consists of 4 modes: U-check mode, Read memory mode, D-check mode and Clear code mode. Two connectors (Read memory and Test mode) and 2 lamps (**CHECK ENGINE** and O_2 monitor) are used. The connectors are used for mode selection and the lamps monitor the type of problem.

U-CHECK MODE

The U-Check mode diagnostics only the MPFI components necessary for start-up and driveabilty. When a fault occurs, the warning lamp (**CHECK ENGINE**) lights to indicate to the user that inspection is necessary. The diagnosis of other components, which do not effect start-up and driveabilty, are excluded from this mode.

READ MEMORY CODE

This mode is used to read past problems (even when the vehicle monitor lamps are off). It is most effective in detecting poor contacts or loose connections of the connectors, harness, etc.

D-CHECK MODE

This mode is used to check the entire MPFI system and to detect faulty components.

CLEAR MEMORY MODE

This mode is used to clear the trouble code from the memory after all faults have been corrected.

F-CHECK MODE (SELECT MONITOR)

This mode is used on 1990 Loyale models to measure the performance characteristics of system components when no trouble codes appear in the Read Memory, U-Check or D-Check mode (although problems have occurred or are occurring).

When working in this mode, use Select Monitor Cartridge No. 498347001 and Select Monitor Instruction Manual No.498327000, for details concerning select monitor and troubleshooting.

AIR FLOW METER

Inspection

1. Check for leaks or damage in the connection between the intake boot and the air flow meter. Repair if necessary.
2. Disconnect electrical connectors from the air flow meter, air intake boot and the air flow meter from the air cleaner case.
3. Check the exterior of the air flow meter for damage.
4. Check the interior for foreign particules, water or oil in the passages, especially in the by-pass. If any of the above are noted, replace the air flow sensor.
5. If none of the above defects were noted, proceed as follows:

a. Turn the ignition switch to the **OFF** position.

b. Install the air flow meter to the air cleaner.

c. Disconnect the air flow meter electrical connector, then slide back the rubber boot from the connector.

NOTE: Conduct the following test by connecting the tester pins to the connector terminals on the side from which the the rubber boot was removed.

d. Connect an ohmmeter between the body (B) and ground (BR) terminals on the connector. Ohmmeter should read 10 ohms. If reading is higher, check the harness and wiring to the control unit for discontinuity and the ground terminals on the intake manifold for poor contact.

e. Turn the ignition switch to the **ON** position, then connect the air flow meter connector.

f. On XT models, using a voltmeter, check the voltage across the power terminal (R) and ground. On all except XT models, using a voltmeter, check the voltage across the power terminal (SA) and ground. On all models, voltmeter should read 10 volts. If specifications are not as specified, check the condition of the components (battery, fuse, control unit harness, connector, etc.) in the power line.

g. On XT models, connect the positive lead of the voltmeter to signal terminal (W) and the negative lead to the ground terminal (BR). On all except XT models, connect the positive lead of the voltmeter to signal terminal (SA) and the negative lead to the ground terminal (BR). On all models, the voltmeter should read 1–2 volts. If specifications are not as specified, replace the air flow meter.

h. Remove the air flow meter from the air cleaner side, then connect a voltmeter between terminals (SA) and (B) on the connector. Blow air through the air flow meter and check that voltage is higher than specified in Step f. If not, replace the air flow meter.

i. Install the air flow meter on the air cleaner. Start and allow the engine to reach normal operating temperature.

j. Operate the vehicle at a speed greater than 15 mph for at least 1 minute (rev engine over 2000 rpm).

k. With engine idling, check voltage across terminal (LgR) on the air flow meter connector and ground. Voltmeter should read 0 volts.

l. Turn the ignition switch to the **OFF** position. Ensure that 12 volts are present across terminal (LgR) and ground as soon as the ignition switch is turned to the **OFF** position. If specifications are not as specified, check the harness to the control unit to the air folw meter for discontinuity.

THROTTLE SENSOR (IDLE CONTACT)

Inspection

4 CYLINDER ENGINE

1. Using an ohmmeter, check that continuity exists between terminals (A) and (C), (4 and 3 on XT models), when the throttle is fully closed.
2. Insert a feeler gauge (thickness gauge) of 0.0217 in (0.55mm) between the stopper screw on the throttle body and portion (G) (this corresponds to the throttle opening of 1.5 degrees). Ensure that continuity exists between (A) and (C), (4 and 3 on XT models).
3. Insert a feeler gauge (thickness gauge) of 0.0362 in (0.92mm) between the stopper screw on the throttle body and portion (G) (this corresponds to a throttle opening of 2.5 degrees). Ensure that continuity exists between terminals (A) and (C), (4 and 3 on XT models).
4. If above specifications are not as specified, loosen the throttle switch attaching screws, then turn throttle switch body until the correct adjustment is obtained.

THROTTLE SENSOR (THROTTLE OPENING SIGNAL)

Inspection

4 CYLINDER ENGINE

1. Using an ohmmeter, measure resistance between terminals (1) and (2), (3 and 2 on XT models). Ohmmeter should read 6–18 kohms. If specifications are not as specified, replace the sensor.
2. Using an ohmmeter, measure the resistance between terminals (1) and (3). Ohmmeter should read 5.8–17.8 kohms with the throttle closed, and 1.5–5.1 kohms with the throttle opened.

NOTE: Ensure that resistance changes smoothly between the fully-closed and fully-opened throttle positions.

3. If specifications are not as specified, replace the sensor.

THROTTLE SENSOR

Inspection

6 CYLINDER ENGINE

1. Disconnect the throttle sensor electrical connector.
2. Insert a feeler gauge (thickness gauge) of 0.0138 in (0.35mm) between the stopper screw and throttle lever (portion G). Using an ohmmter, measure the resistance between terminals (3) and (4) of the throttle sensor. Ohmmeter should read 5 kohms.
3. Insert a feeler gauge (thickness gauge) of 0.0295 in (0.75mm) between the stopper screw and throttle lever (portion G). Using an ohmmeter, measure the resistance between terminals (3) and (4) of the throttle sensor. Ohmmeter should read 1 Mohm.
4. Using an ohmmeter, measure resistance between terminals (1) and (4). Ohmmeter should read 3–7 kohms. Then, measure resistance between terminals (2) and (4). Ohmmeter should read 4.2–15 kohms with throttle valve fully closed and 0.1–11 kohms with throttle valve fully opened.

NOTE: Ensure that resistance changes smoothly between the fully closed and fully opened positions.

5. If specifications are not as specified, replace the sensor.

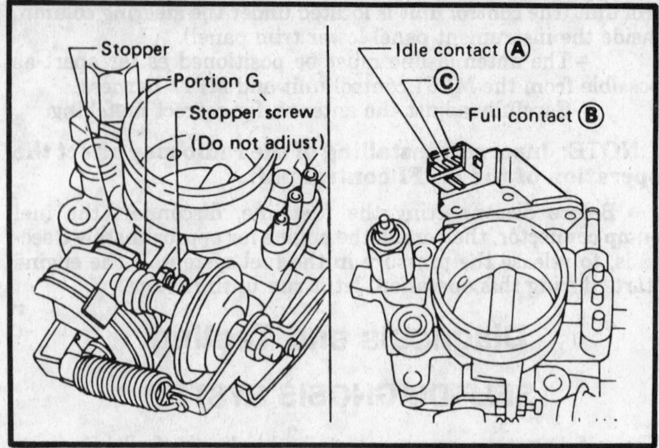
Stopper
Portion G
Stopper screw
(Do not adjust)
Idle contact Ⓐ
Ⓒ
Full contact Ⓑ

Throttle body testing and adjustment—Sedan and Wagon and 1990 Loyale

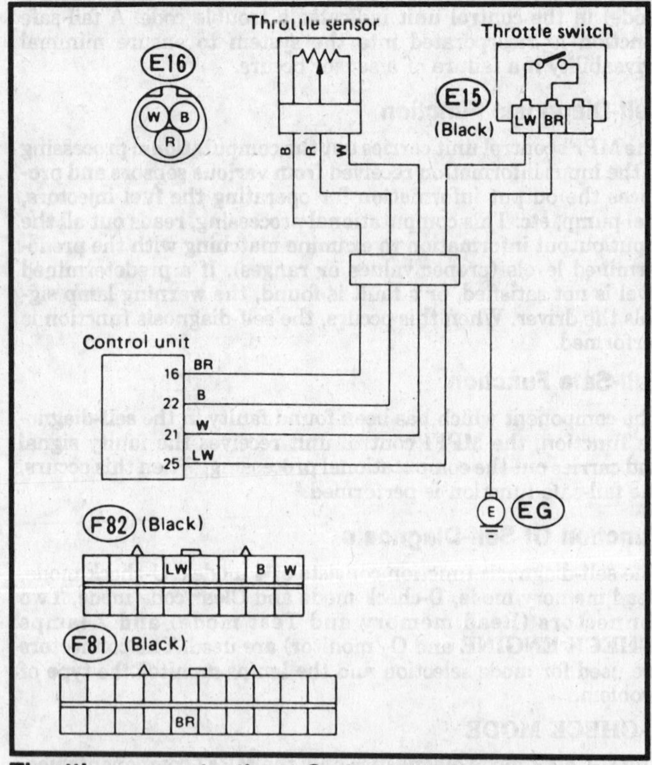
Throttle sensor
Throttle switch
E16
W
B
R
E15
(Black)
LW
BR
B
R
W
Control unit
16
BR
22
B
21
W
25
LW
E
EG
F82
(Black)
LW
B
W
F81
(Black)
BR

Throttle sensor testing—Sedan and Wagon and 1990 Loyale

DASH-POT

Inspection

1. Start, and bring engine up to normal operating temperature.
2. Under a non-loaded state, turn the throttle lever by hand to increase engine rpm until the end of the dash pot comes off the throttle cam.
3. Gradually return the throttle lever, then check engine rpm when the throttle cam contacts the end of the dash pot. Engine rpm should be 2800–3400.
4. If rpm is not as specified, loosen the dash pot lock nut, then turn the dash pot until engine rpm is within specification. Tighten locknut.

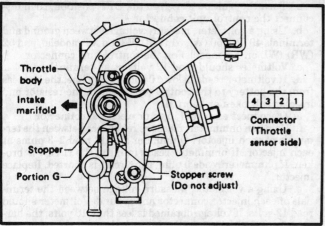

Throttle body testing and adjustment—XT models with 4 cylinder engine

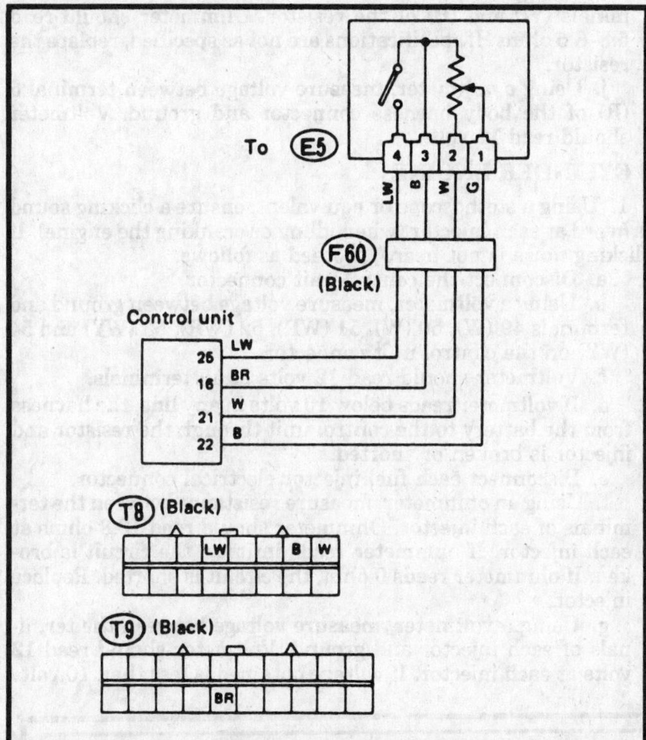
To E5
4 3 2 1
LW B W G
F60
(Black)
Control unit
25 LW
16 BR
21 W
22 B
T8 (Black)
LW B W
T9 (Black)
BR

Throttle sensor testing—XT models with 4 cylinder engine

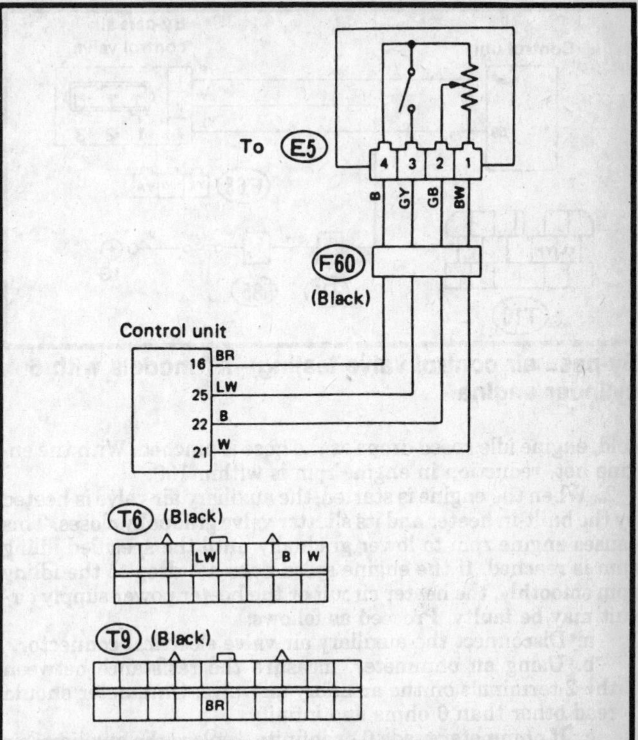
To E5
4 3 2 1
B GY GB BW
F60
(Black)
Control unit
16 BR
25 LW
22 B
21 W
T6 (Black)
LW B W
T9 (Black)
BR

Throttle sensor testing—XT models with 6 cylinder engine

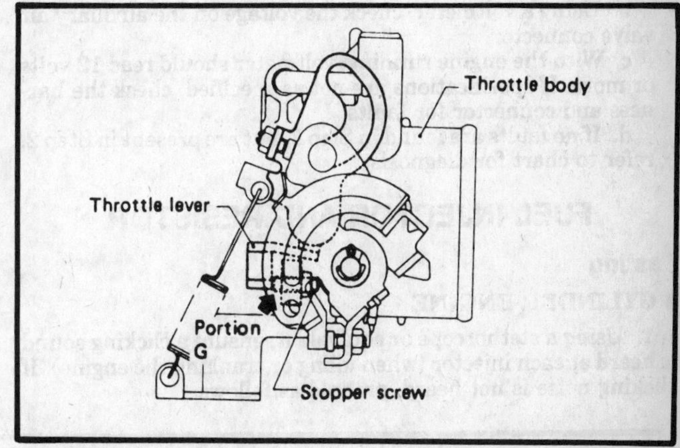
Throttle body
Throttle lever
Portion G
Stopper screw

Throttle body testing—XT models with 6 cylinder engine

5. After adjustment, rev the engine to ensure that the idle speed returns correctly as the throttle is released.

BY-PASS AIR CONTROL VALVE

Inspection

6 CYLINDER ENGINE

1. Disconnect the by-pass air control valve electrical connector.
2. Using an ohmmeter, measure the resistance between terminals (1) and (2) on the by-pass air control valve. Ohmmeter should read 9.5–11.5 ohms.
3. Using an ohmmeter, measure resistance between terminals (2) and (3) on the by-pass air control valve. Ohmmeter should read 8.5–10.5 ohms.
4. Using an ohmmeter, measure resistance between terminal (2) and the valve body, and terminal (3) and the valve body. Ohmmeter should read infinity in both tests.
5. If specifications are not as specified, replace the by-pass air control valve.

AUXILIARY AIR VALVE

Testing

4 CYLINDER ENGINE

1. Pinch the hose connecting the air intake duct and the auxiliary air valve, then note engine rpm change. With the engine

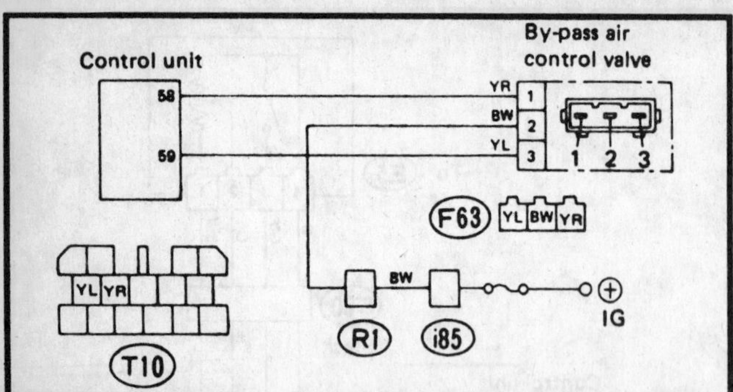
By-pass air control valve
Control unit
58
59
YR
BW
YL
1
2
3
1 2 3
F63
YL BW YR
YL YR
BW
IG
R1
i85
T10

By-pass air control valve testing – XT models with 6 cylinder engine

cold, engine idle speed drops as the hose is pinched. With the engine hot, reduction in engine rpm is within 100.

2. When the engine is started, the auxiliary air valve is heated by the built-in heater and its shutter valve gradually closes. This causes engine rpm to lower gradually until the specified idling rpm is reached. If the engine speed does not drop to the idling rpm smoothly, the heater circuit or the heater power supply circuit may be faulty. Proceed as follows:

a. Disconnect the auxiliary air valve electrical connector.

b. Using an ohmmeter, measure the resistance between the 2 terminals on the auxiliary air valve. Ohmmeter should read other than 0 ohms and infinity.

c. If ohmmeter reads 0 or infinity, replace the auxiliary air valve.

3. Check the voltage source as follows:

a. Disconnect the auxiliary air valve electrical connector.

b. Using a voltmeter, check the voltage on the auxiliary air valve connector.

c. With the engine running, voltmeter should read 12 volts or more. If specifications are not as specified, check the harness and connector for faults.

d. If no faults are found in Step 3, but are present in Step 2, refer to chart for diagnosis.

FUEL INJECTOR AND RESISTOR

Testing

4 CYLINDER ENGINE

1. Using a stethoscope or equivalent, ensure a clicking sound is heard at each injector (when idling or cranking the engine). If clicking noise is not heard, proceed as follows:

a. Turn the ignition switch to the **OFF** position, then disconnect the control unit connector.

b. Using a voltmeter, measure voltage between ground and terminals 49 (W), 50 (W), 51 (WR) (WL on XT models), and 52 (WR) (WL on XT models) on the control unit connector.

c. Voltmeter should read 12 volts at all terminals.

d. If voltmeter reads below 10 volts in any line, the harness from the battery to the control unit through the resistor and injector is broken or shorted.

e. Disconnect each fuel injector electrical connector.

f. Using an ohmmeter, measure resistance between the terminals of each injector. Ohmmeter should read 2–3 ohms at each injector. If ohmmeter reads infinity, the circuit is broken. If ohmmeter reads 0 ohm, the circuit is shorted. Replace injector.

g. Using a voltmeter, measure voltage between the terminals of each injector connector and ground. Voltmeter should read 12 volts. If voltage obtained is less than 10 volts, the harness from the battery to the injector through the resistor is disconnected or shorted.

h. Disconnect the electrical connector from the resistor.

i. Using an ohmmeter, measure the resistance between terminals (W) and (B) of the resistor. Ohmmeter should read 5.8–6.5 ohms. If specifications are not as specified, replace the resistor.

j. Using a voltmeter, measure voltage between terminal 5 (R) of the body harness connector and ground. Voltmeter should read 12 volts.

6 CYLINDER ENGINE

1. Using a stethoscope or equivalent, ensure a clicking sound is heard at each injector (when idling or cranking the engine). If clicking noise is not heard, proceed as follows:

a. Disconnect the control unit connector.

b. Using a voltmeter, measure voltage between ground and terminals 49 (W), 50 (W), 51 (WR), 52 (WR), 53 (WY) and 54 (WY) on the control unit connector.

c. Voltmeter should read 12 volts at all terminals.

d. If voltmeter reads below 10 volts in any line, the harness from the battery to the control unit through the resistor and injector is broken or shorted.

e. Disconnect each fuel injector electrical connector.

f. Using an ohmmeter, measure resistance between the terminals of each injector. Ohmmeter should read 13.8 ohms at each injector. If ohmmeter reads infinity, the circuit is broken. If ohmmeter reads 0 ohm, the circuit is shorted. Replace injector.

g. Using a voltmeter, measure voltage between the terminals of each injector and ground. Voltmeter should read 12 volts at each injector. If voltage obtained is less than 10 volts

	Cause of trouble	Symptom	Remedy
1	Sticking of shutter valve of auxiliary air valve. (Sticking in closed direction)	• Engine stalls easily when engine is cold.	Replace auxiliary air valve.
2	Sticking of shutter valve of auxiliary air valve. (Sticking in open direction)	• Engine rpm does not lower smoothly during warm-up operation. • Engine rpm remains high.	
3	Clogged air passage.	Same as ①	Check air passage, such as hose, etc. and clean.

Auxiliary air valve troubleshooting chart – models with 4 cylinder engine

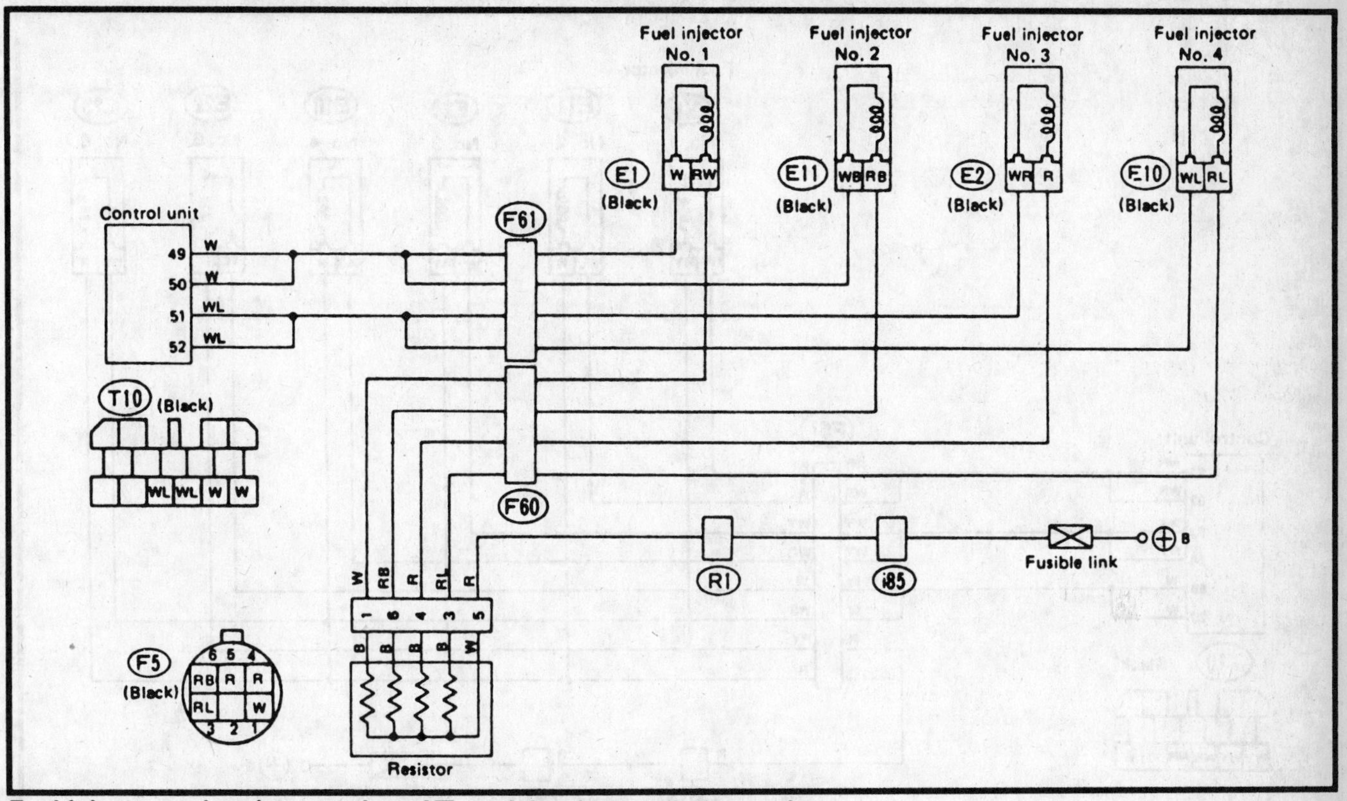

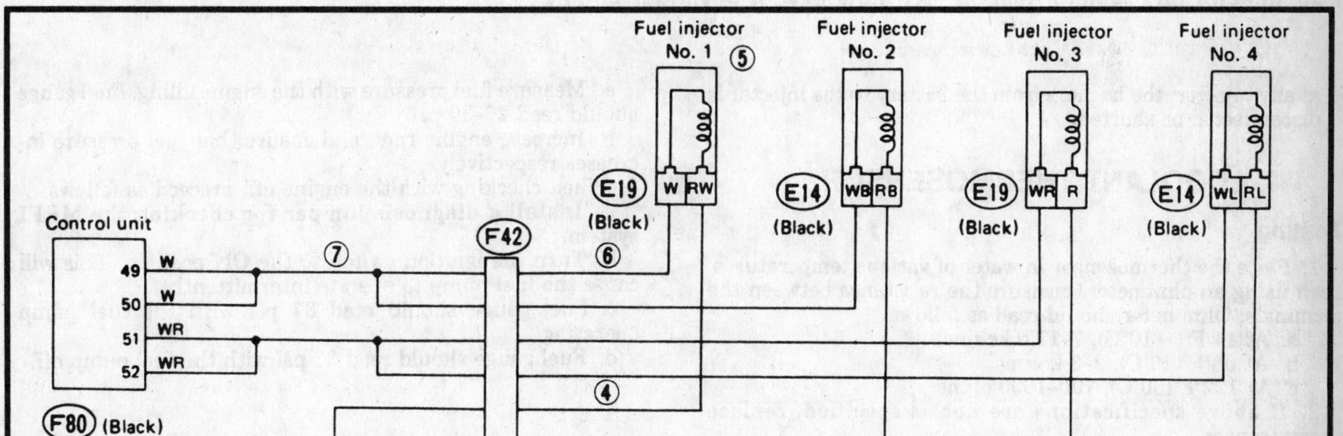

Fuel injector No. 1
Fuel injector No. 2
Fuel injector No. 3
Fuel injector No. 4
E1 (Black) W RW
E11 (Black) WB RB
E2 (Black) WR R
E10 (Black) WL RL
Control unit 49 W 50 W 51 WL 52 WL
F61
T10 (Black) WL WL W W
F60
R1
i85
Fusible link
⊕B
W RB R RL R
1 6 4 3 5
B B B B W
F5 (Black) 6 5 4 RB R R RL W 3 2 1
Resistor

Fuel injector and resistor testing – XT models with 4 cylinder engine

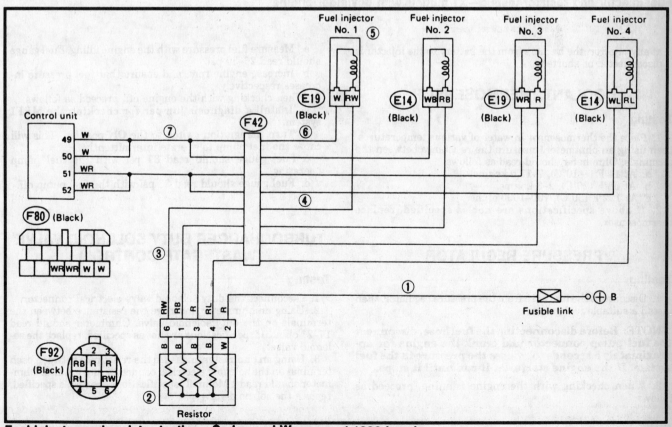

Fuel injector No. 1 ⑤
Fuel injector No. 2
Fuel injector No. 3
Fuel injector No. 4
E19 (Black) W RW
E14 (Black) WB RB
E19 (Black) WR R
E14 (Black) WL RL
Control unit 49 W 50 W 51 WR 52 WR
⑦
F42
⑥
④
F80 (Black) WR WR W W
③
①
Fusible link
⊕B
RW RB R RL R
6 5 3 4 2
B B B B W
F92 (Black) 1 2 3 RB R R RL RW 4 5 6
②
Resistor

Fuel injector and resistor testing – Sedan and Wagon and 1990 Loyale

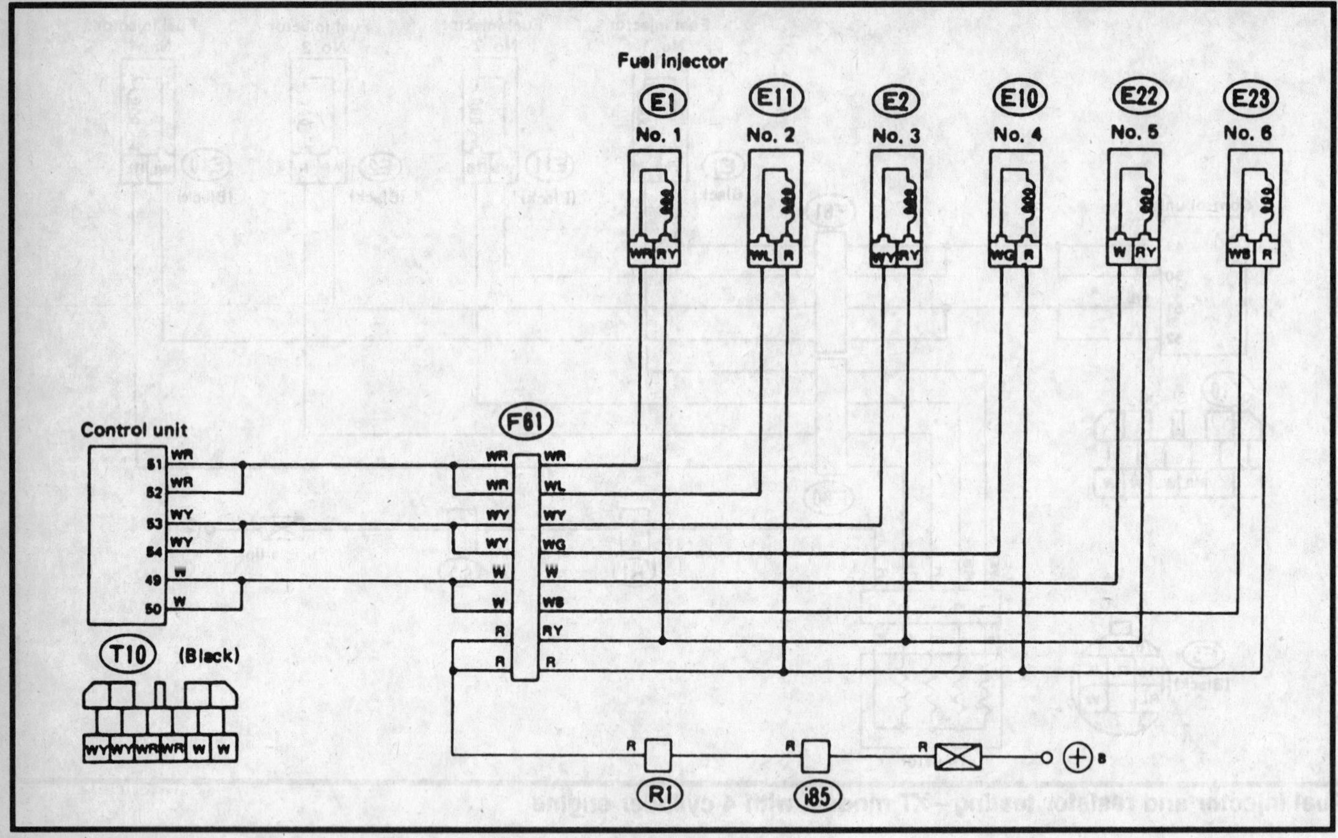
Fuel injector
E1
No. 1
E11
No. 2
E2
No. 3
E10
No. 4
E22
No. 5
E23
No. 6
Control unit
F61
T10
(Black)
R1
i85

Fuel injector and resistor testing—XT models with 6cylinder engine

at any injector, the harness from the battery to the injector is disconnected or shorted.

COOLANT THERMOSENSOR

Testing

1. Place the thermosensor in water of various temperatures, then using an ohmmeter, measure the resistance between the terminals. Ohmmeter should read as follows:
   a. At 14°F (−10°C), 7–11.5 kohms.
   b. At 68°F (20°C), 2–3 kohms.
   c. At 122°F (50°C), 700–1000 ohms.
2. If above specifications are not as specified, replace thermosensor.

PRESSURE REGULATOR

Testing

1. Disconnect the fuel hose from the pressure regulator, then install a suitable fuel gauge.

NOTE: Before disconnecting the fuel hose, disconnect the fuel pump connector and crank the engine for approximately 5 seconds to release the pressure in the fuel system. If the engine starts, let it run until it stops.

2. When checking with the engine running, proceed as follows:
   a. Measure fuel pressure with the engine idling. Fuel gauge should read 26–30 psi.
   b. Increase engine rpm, and ensure that fuel pressure increases respectively.
3. When checking with the engine off, proceed as follows:
   a. Install a diagnosis jumper for checking the MPFI system.
   b. Turn the ignition switch to the **ON** position. This will cause the fuel pump to operate intermittently.
   c. Fuel gauge should read 37 psi with the fuel pump operating.
   d. Fuel gauge should read 33 psi with the fuel pump off.

TURBOCHARGER DUTY SOLENOID VALVE (WASTEGATE CONTROL)

Testing

1. Disconnect the duty solenoid valve electrical connector.
2. Using an ohmmeter, measure the resistance between the terminals on the duty solenoid valve. Ohmmeter should read 17–21 ohms. If specifications are not as specified, replace the solenoid valve.
3. Using an ohmmeter, measure the resistance between each terminal on the solenoid electrical connector and ground. Ohmmeter should read 1 Mohm. If specifications are not as specified, replace the solenoid valve.

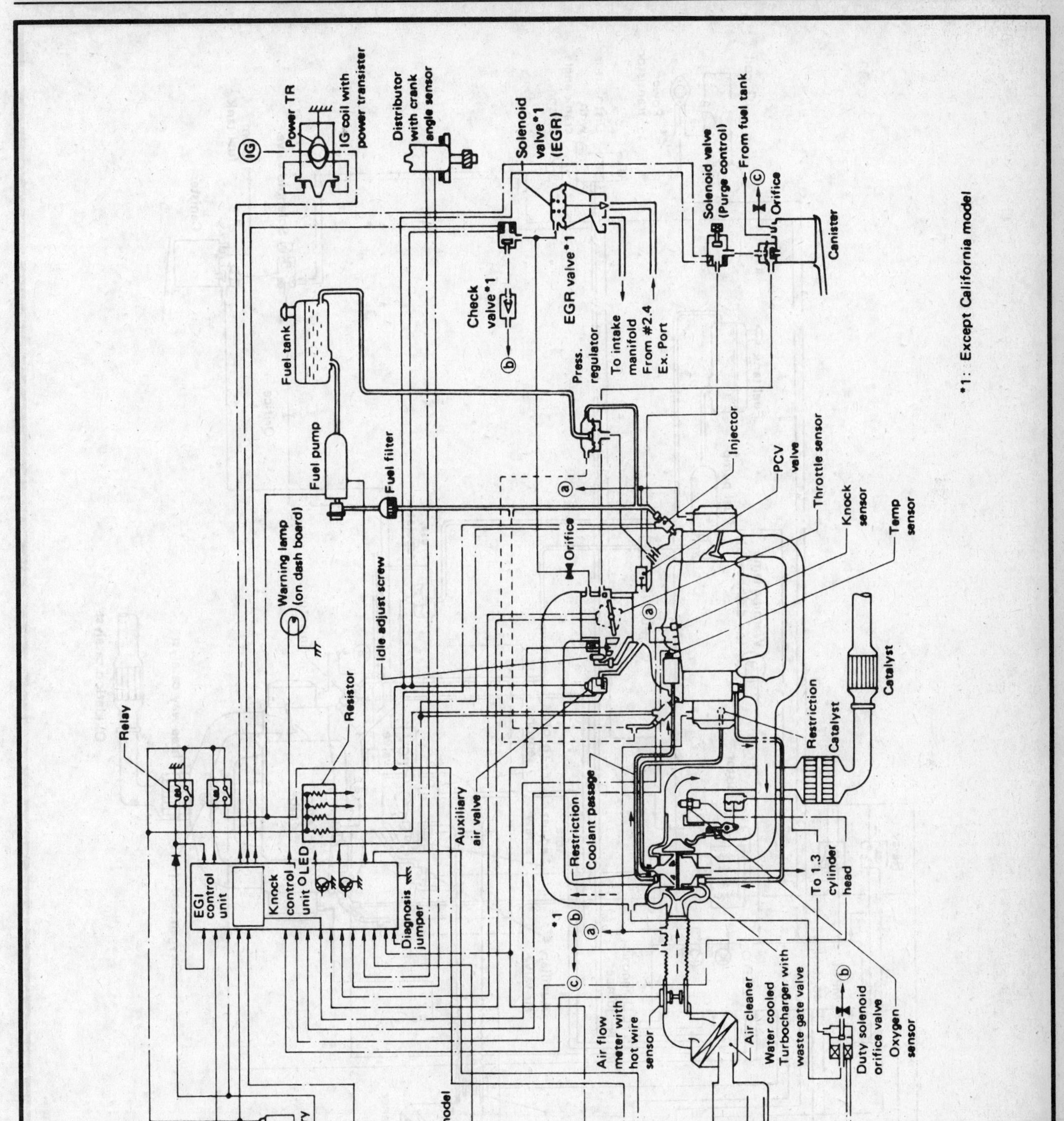
Speed sensor
IG. SW
Starter SW
Neutral SW
Air conditioner
Battery
MT model only
EGI control unit
Knock control unit
LED
Diagnosis jumper
Relay
Resistor
Warning lamp (on dash board)
Idle adjust screw
Auxiliary air valve
Restriction
Coolant passage
Fuel pump
Fuel filter
Fuel tank
IG
Power TR
IG coil with power transister
Distributor with crank angle sensor
Solenoid valve*1 (EGR)
Check valve*1
EGR valve*1
Press. regulator
To intake manifold
From #2.4 Ex. Port
Solenoid valve (Purge control)
From fuel tank
Orifice
Canister
Injector
PCV valve
Throttle sensor
Knock sensor
Temp sensor
Orifice
Catalyst
Restriction
Catalyst
To 1.3 cylinder head
Air flow meter with hot wire sensor
Air cleaner
Water cooled Turbocharger with waste gate valve
Duty solenoid orifice valve
Oxygen sensor
*1
*1: Except California model

MPFI system schematic diagram—Sedan and Wagon

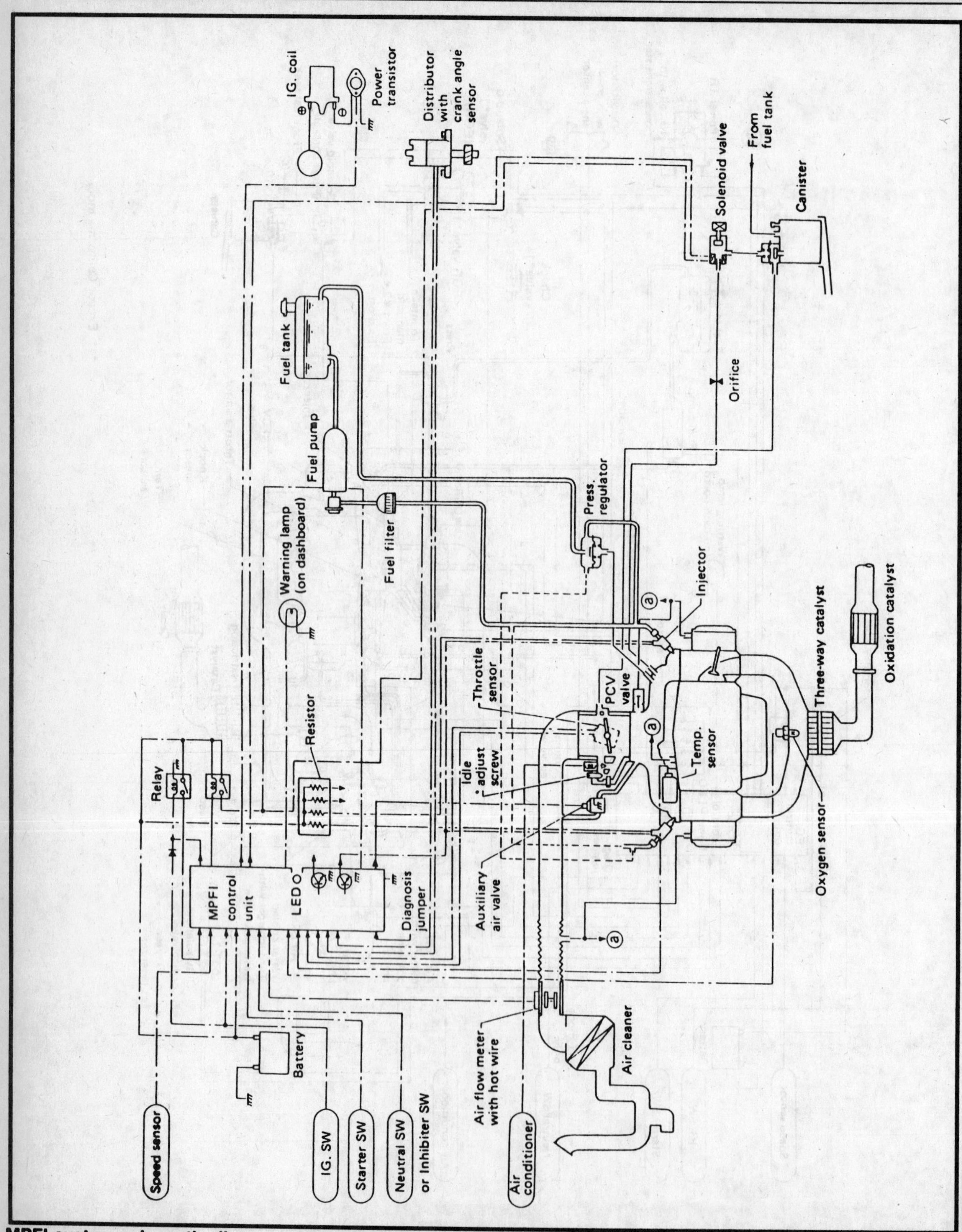
IG. coil
Power transistor
Distributor with crank angle sensor
Solenoid valve
From fuel tank
Canister
Fuel tank
Orifice
Fuel pump
Warning lamp (on dashboard)
Fuel filter
Press. regulator
Injector
Relay
Resistor
Throttle sensor
PCV valve
Idle adjust screw
Temp. sensor
Three-way catalyst
Oxidation catalyst
MPFI control unit
LED
Diagnosis jumper
Auxiliary air valve
Oxygen sensor
Battery
Air flow meter with hot wire
Air cleaner
Speed sensor
IG. SW
Starter SW
Neutral SW or Inhibiter SW
Air conditioner

MPFI system schematic diagram—XT models with 4 cylinder engine

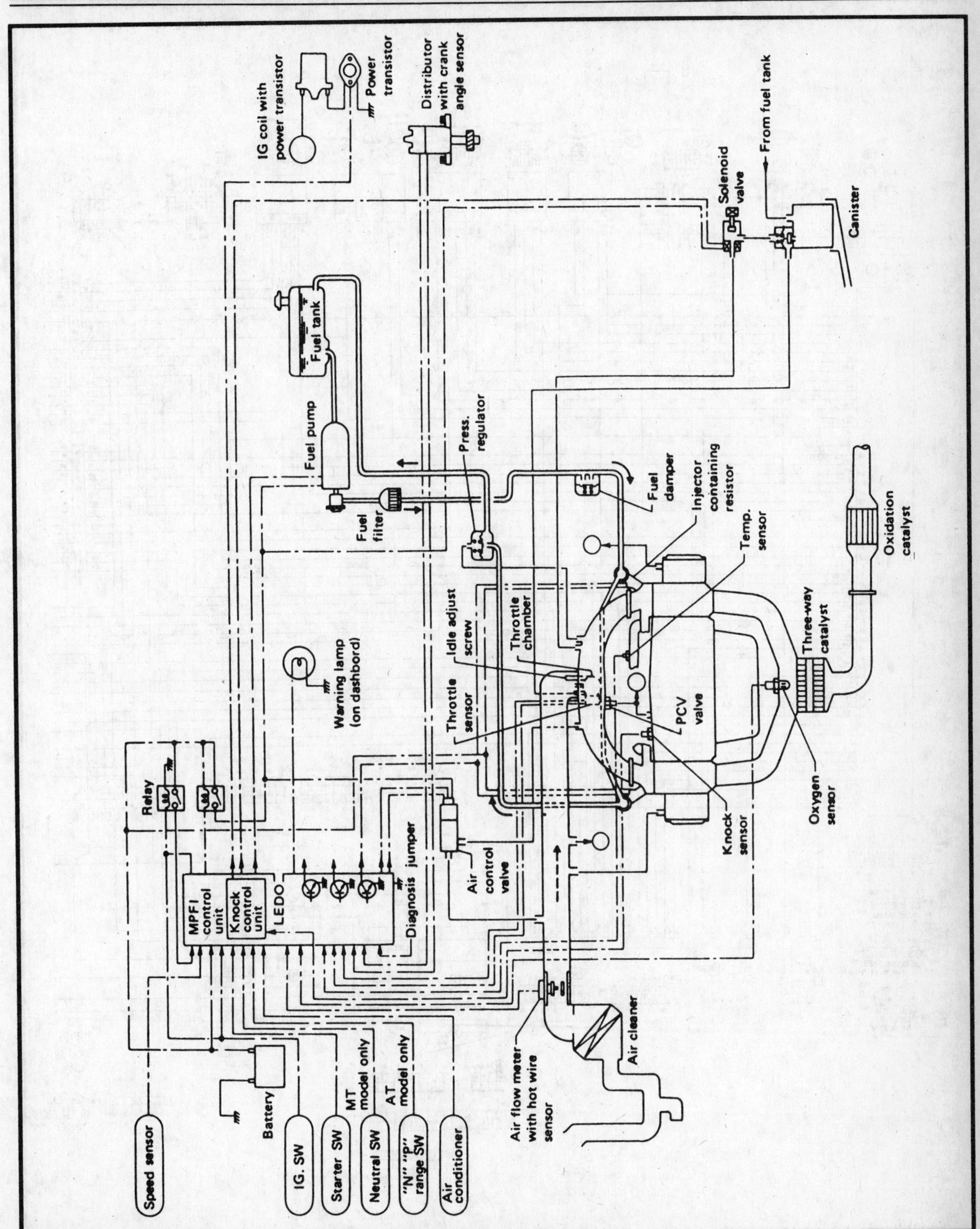
IG coil with power transistor
Power transistor
Distributor with crank angle sensor
From fuel tank
Solenoid valve
Canister
Fuel tank
Fuel pump
Press. regulator
Fuel filter
Fuel damper
Injector containing resistor
Temp. sensor
Oxidation catalyst
Relay
Warning lamp (on dashbord)
Idle adjust screw
Throttle chamber
Throttle sensor
PCV valve
Three-way catalyst
Oxygen sensor
Knock sensor
MPFI control unit
Knock control unit
LEDO
Diagnosis jumper
Air control valve
Speed sensor
Battery
IG. SW
Starter SW
MT model only
Neutral SW
AT model only
"N" "P" range SW
Air conditioner
Air flow meter with hot wire sensor
Air cleaner

MPFI system schematic diagram—XT models with 6 cylinder engine

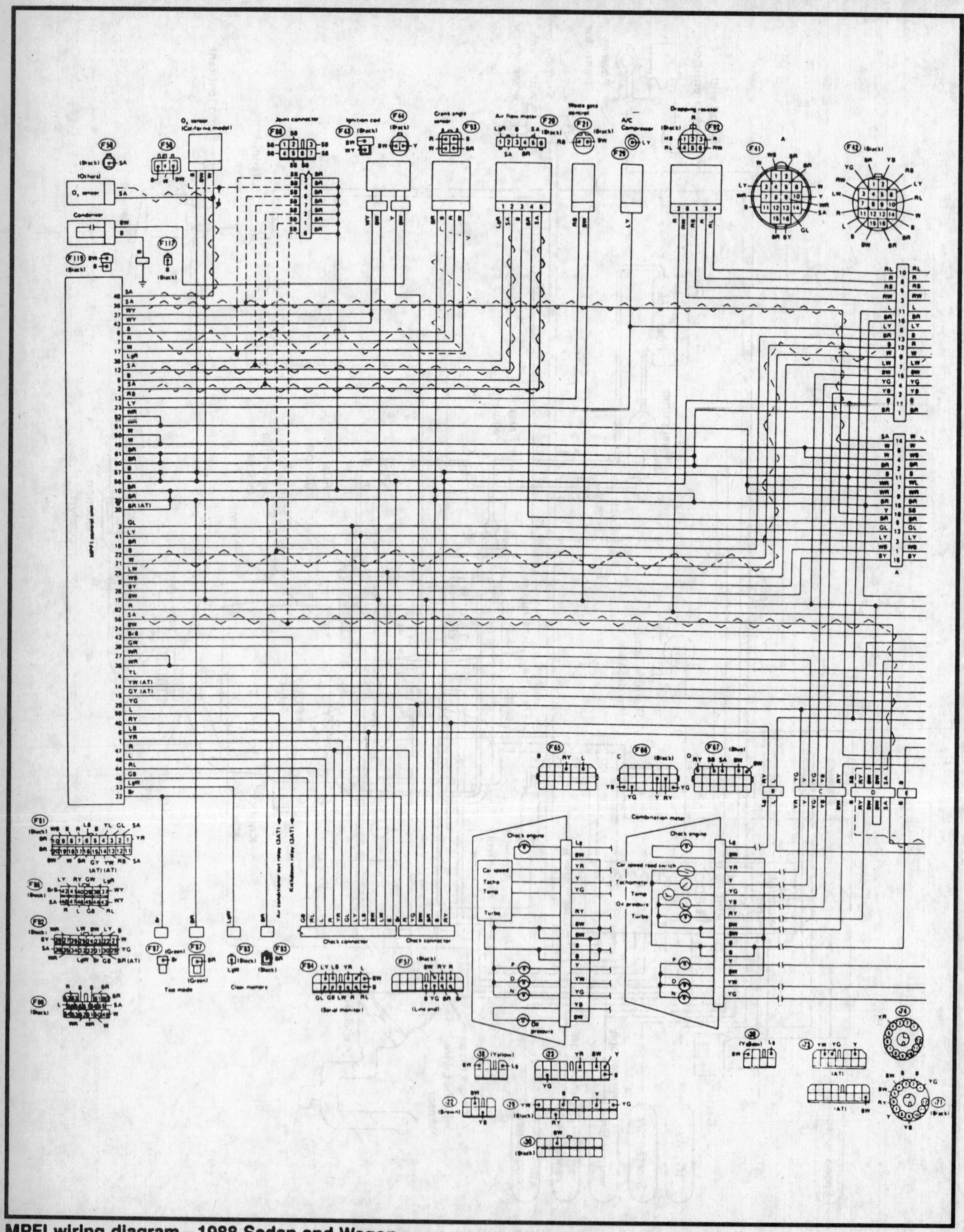

MPFI wiring diagram—1988 Sedan and Wagon

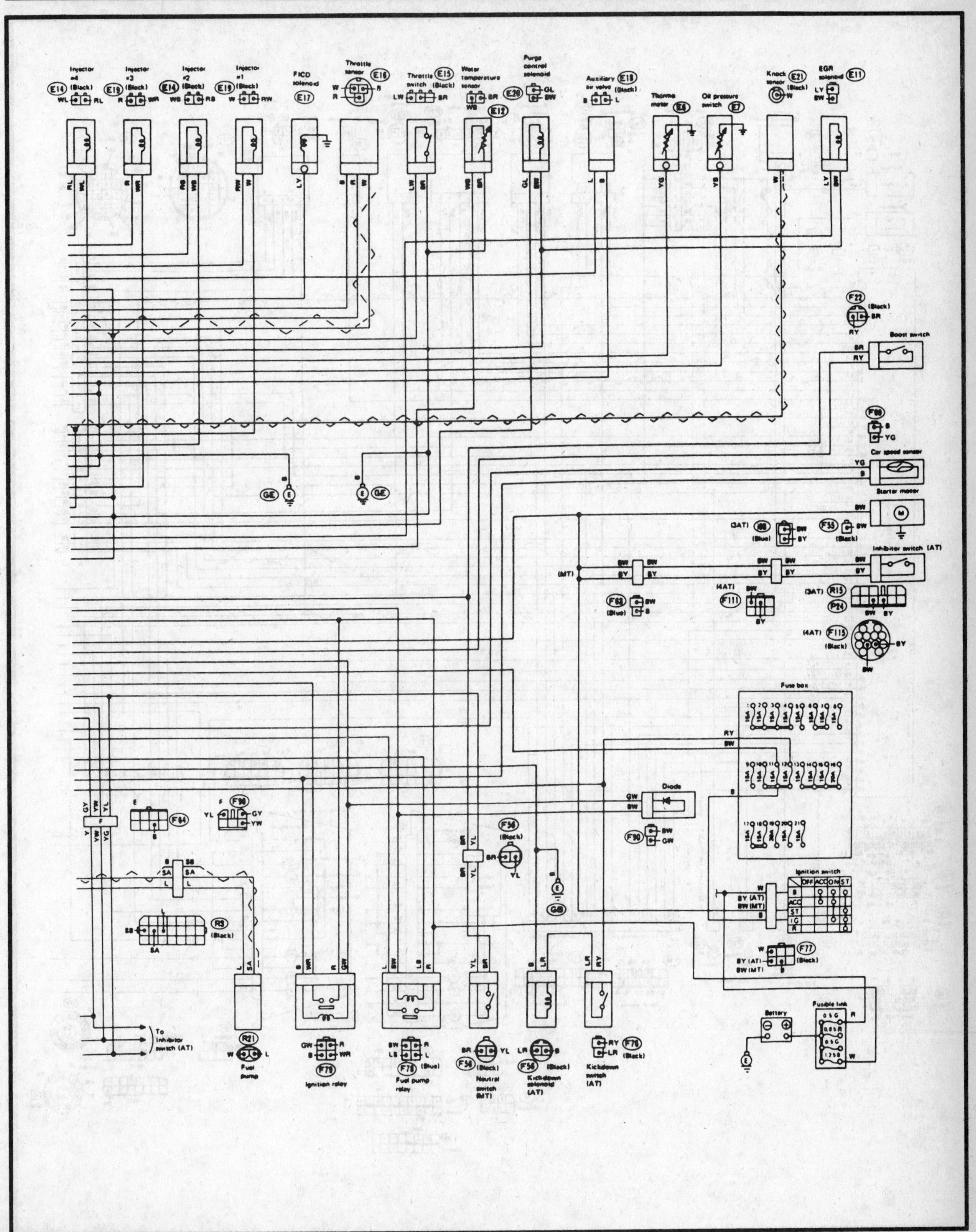

MPFI wiring diagram (cont.)—1988 Sedan and Wagon

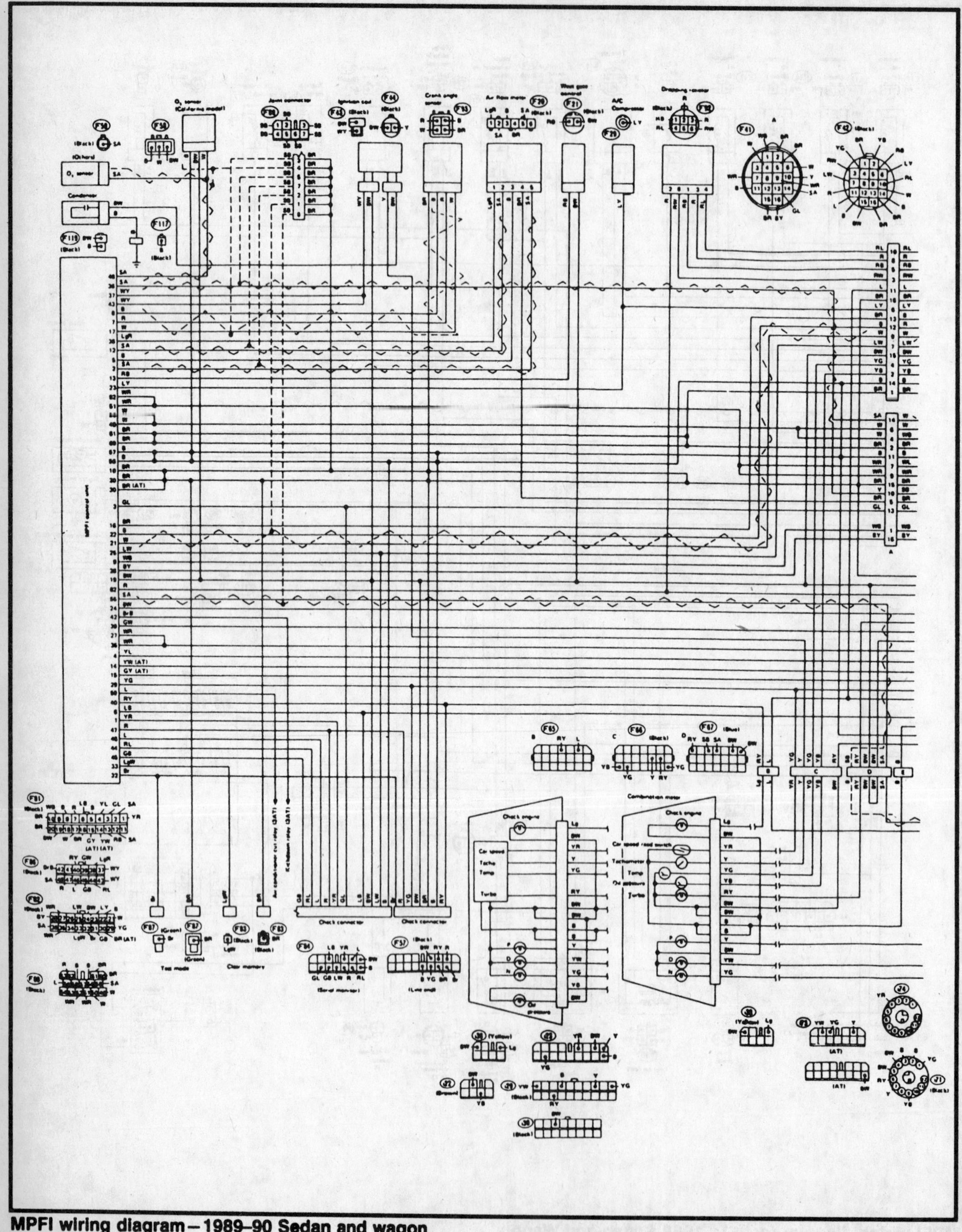

MPFI wiring diagram – 1989–90 Sedan and wagon

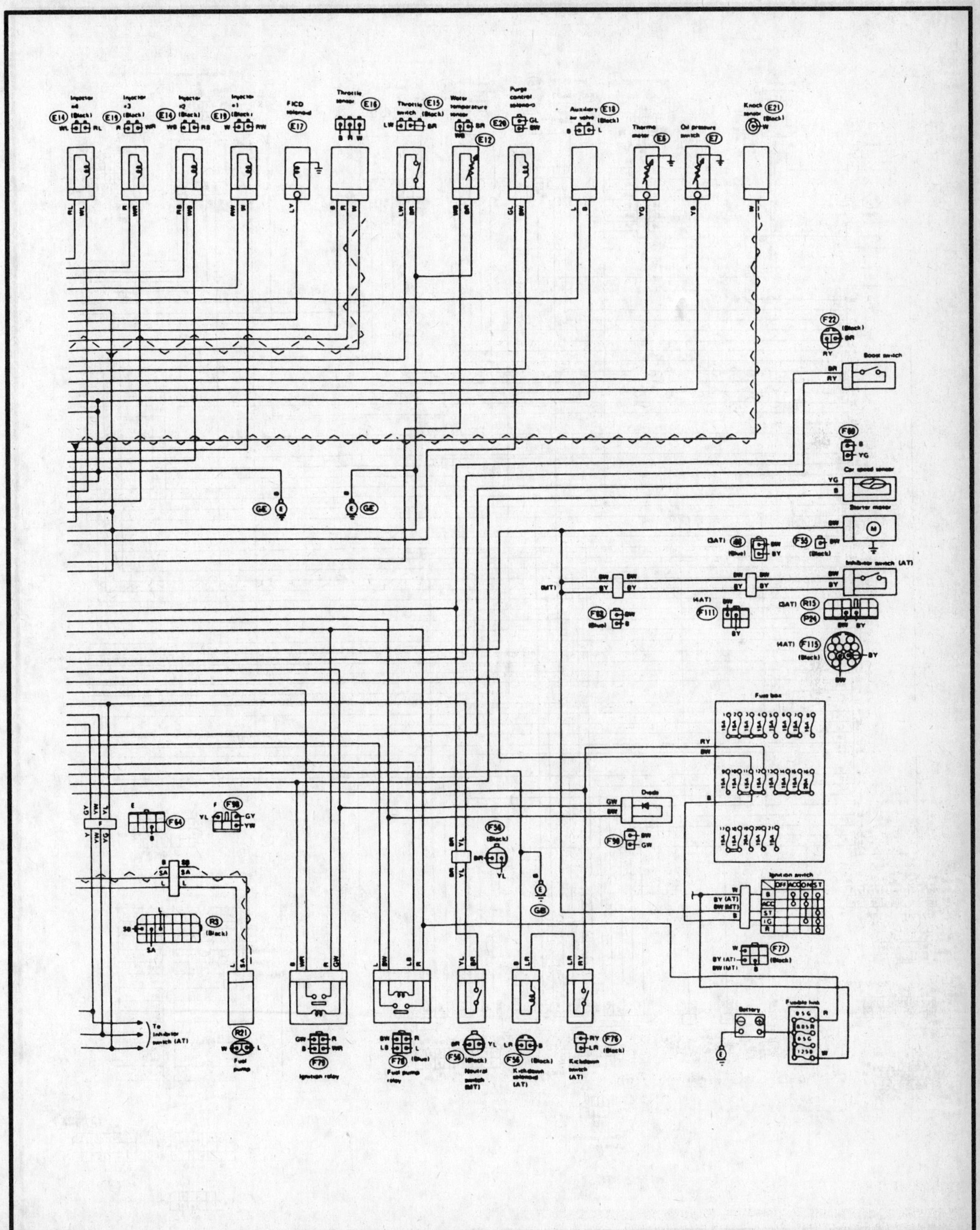

MPFI wiring diagram (cont.) – 1989–90 Sedan and Wagon

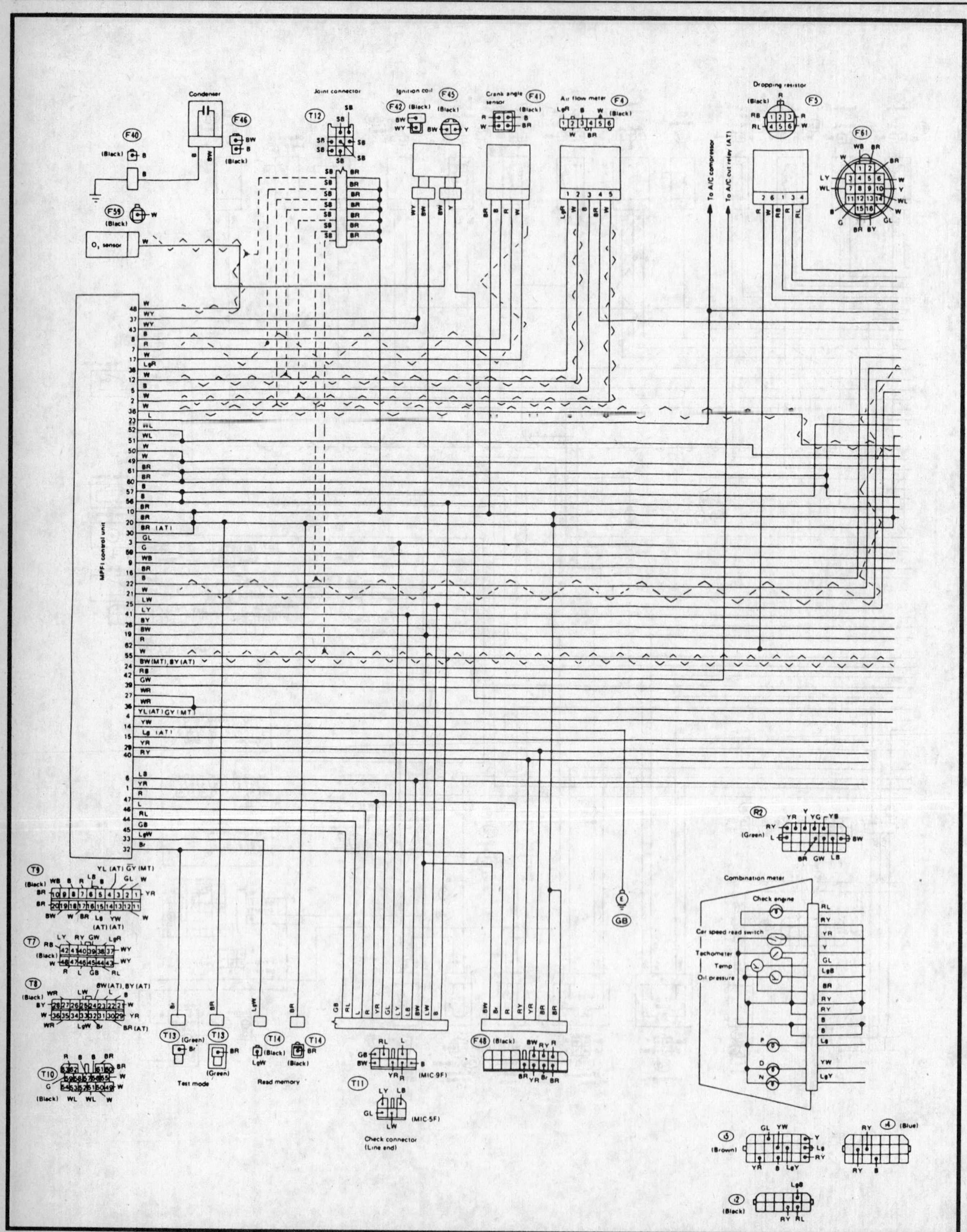
Condenser
Joint connector
Ignition coil
Crank angle sensor
Air flow meter
Dropping resistor
O₂ sensor
To A/C compressor
To A/C cut relay (AT)
MPFI control unit
Combination meter
Check engine
Car speed read switch
Tachometer
Temp
Oil pressure
Test mode
Read memory
Check connector (Line end)
(MIC 9F)
(MIC 5F)

MPFI wiring diagram—XT models with 4 cylinder engine

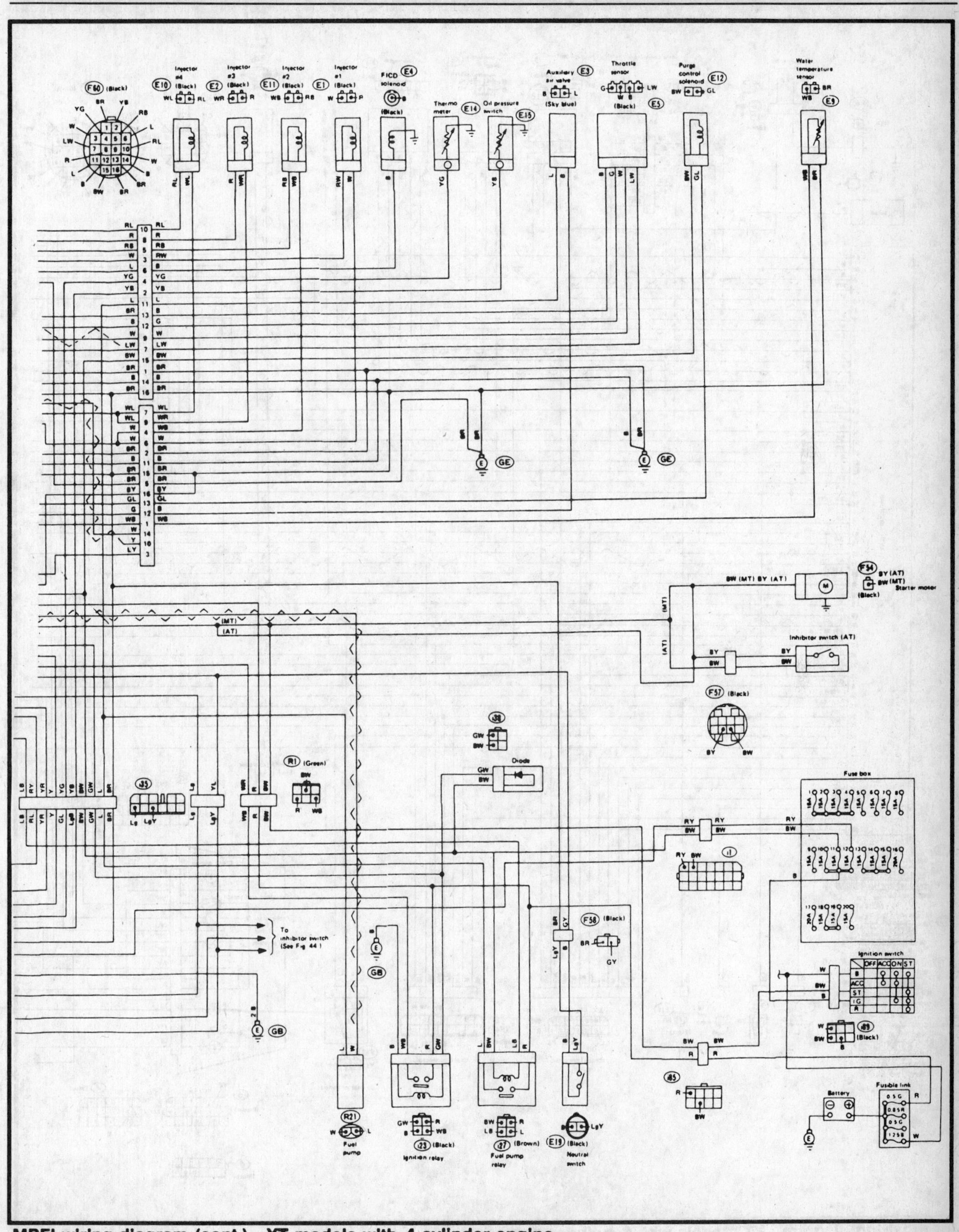
Injector #4
Injector #3
Injector #2
Injector #1
FICD solenoid
Thermo meter
Oil pressure switch
Auxiliary air valve
Throttle sensor
Purge control solenoid
Water temperature sensor
Starter motor
Inhibitor switch (AT)
Diode
Fuse box
Ignition switch
Battery
Fusible link
Fuel pump
Ignition relay
Fuel pump relay
Neutral switch
To inhibitor switch (See Fig 44)

MPFI wiring diagram (cont.)—XT models with 4 cylinder engine

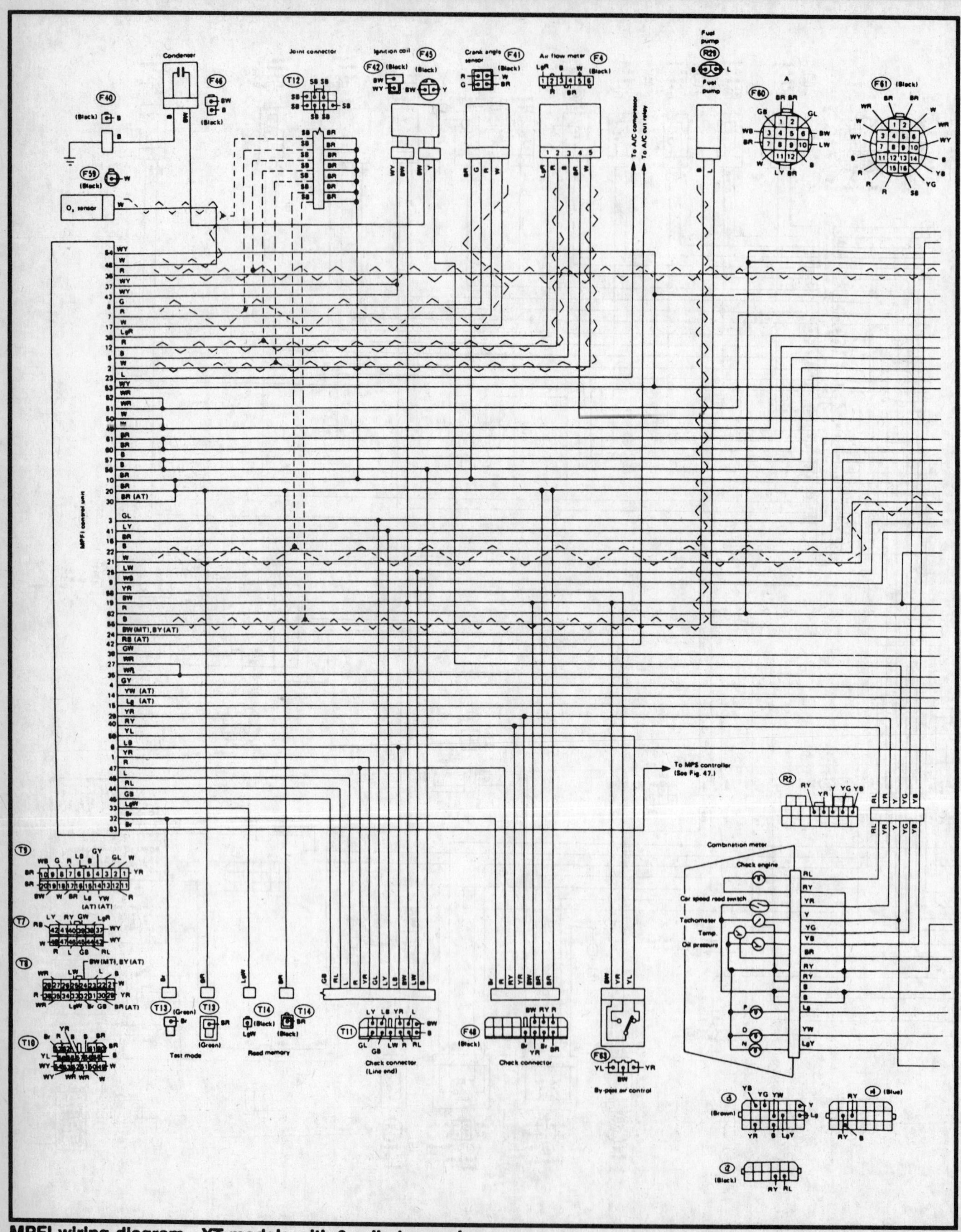

MPFI wiring diagram—XT models with 6 cylinder engine

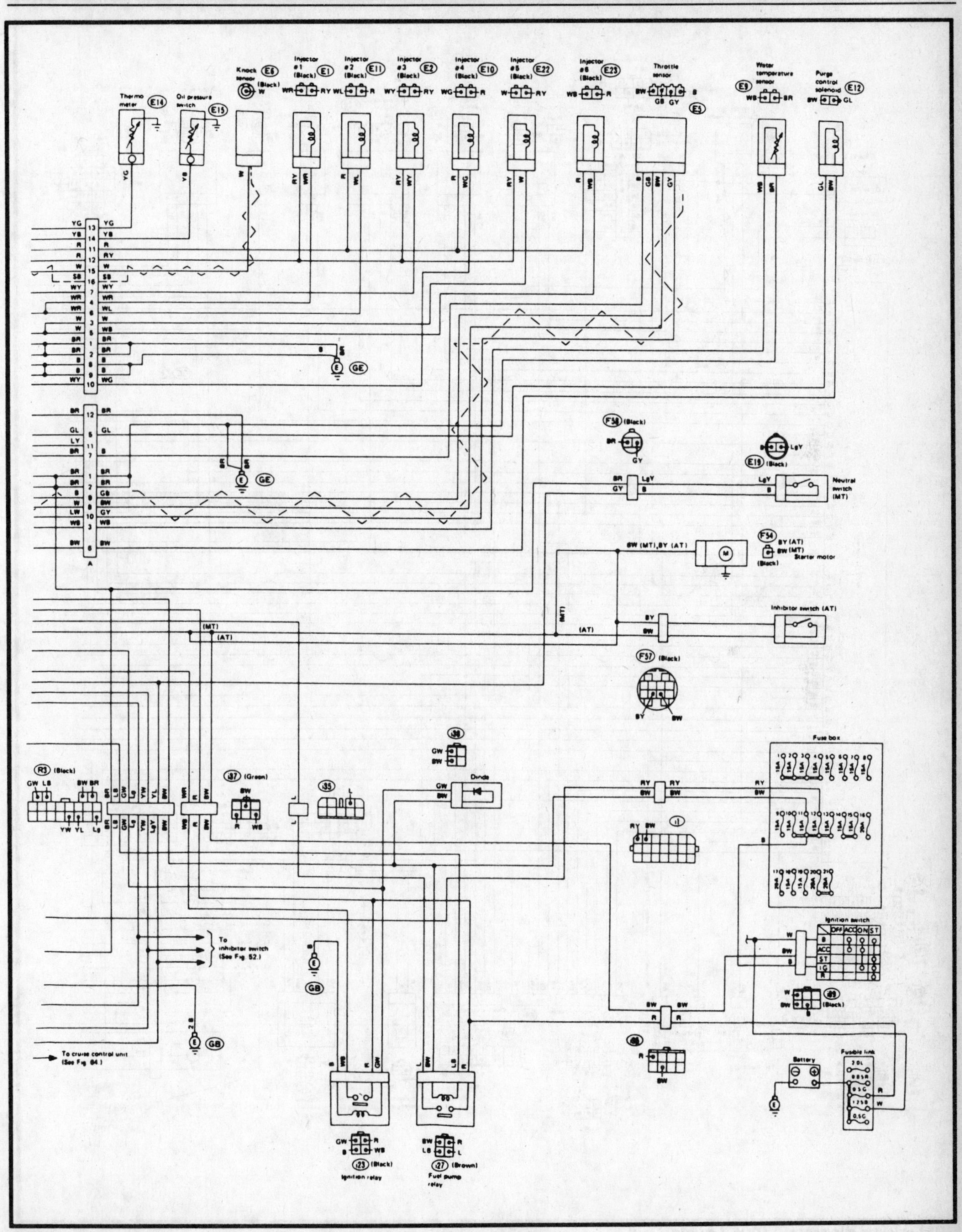

MPFI wiring diagram (cont.) – XT models with 6 cylinder engine

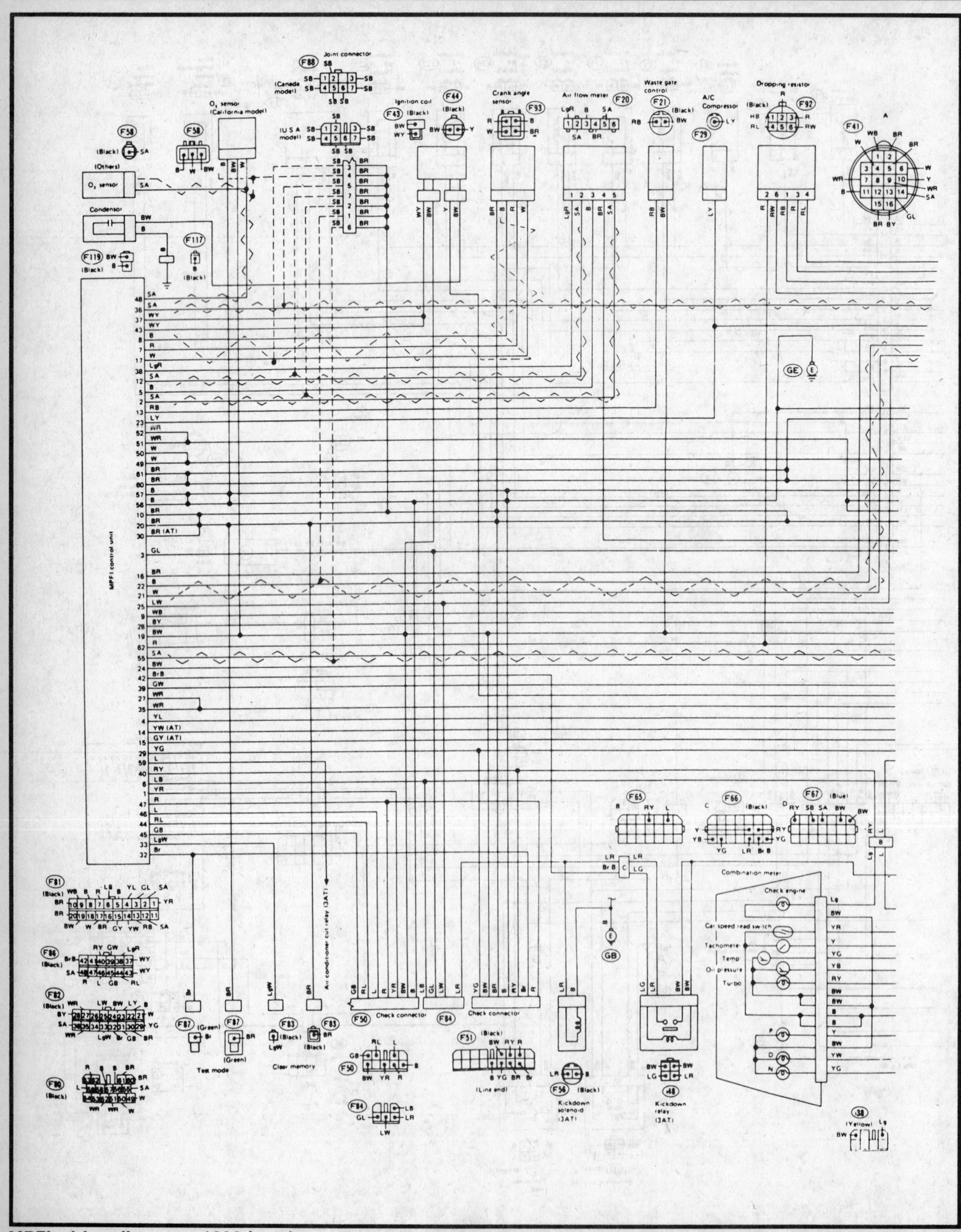
Joint connector
(Canada model)
(U.S.A. model)
O2 sensor (California model)
Ignition coil
Crank angle sensor
Air flow meter
Waste gate control
A/C Compressor
Dropping resistor
(Others)
O2 sensor
Condensor
MPFI control unit
Check connector
Test mode
Clear memory
Air conditioner cut relay (3AT)
Line end
Kickdown solenoid (3AT)
Kickdown relay (3AT)
Combination meter
Check engine
Car speed read switch
Tachometer
Temp
Oil pressure
Turbo

MPFI wiring diagram – 1990 Loyale

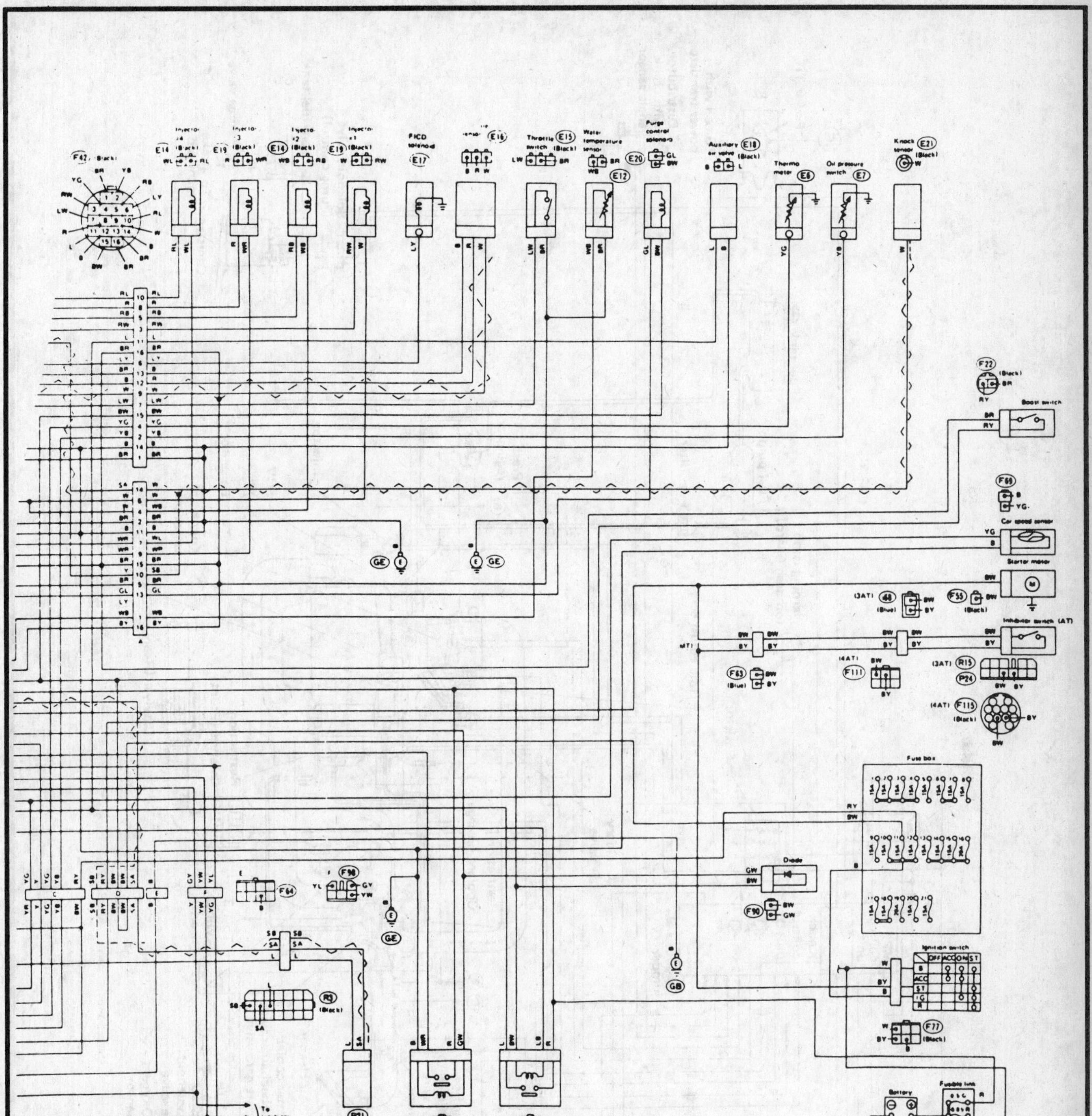

MPFI wiring diagram (cont.) – 1990 Loyale

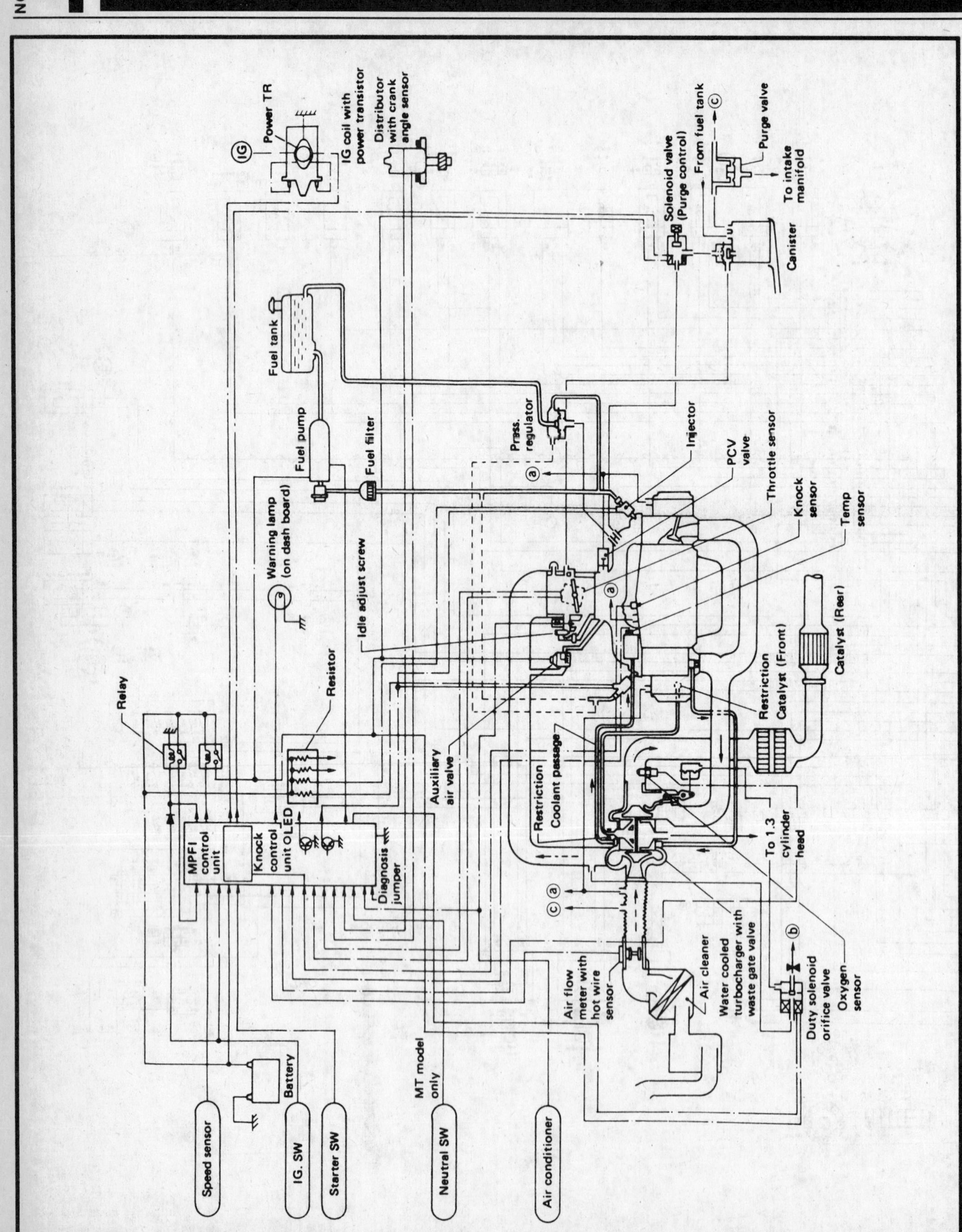
IG
Power TR
IG coil with power transistor
Distributor with crank angle sensor
Solenoid valve (Purge control)
From fuel tank
c
Purge valve
To intake manifold
Canister
Fuel tank
Fuel pump
Fuel filter
Press. regulator
a
Injector
PCV valve
Throttle sensor
Knock sensor
Temp sensor
Warning lamp (on dash board)
Idle adjust screw
Relay
Resistor
Auxiliary air valve
Restriction
Coolant passage
Restriction
Catalyst (Front)
Catalyst (Rear)
MPFI control unit
Knock control unit
LED
Diagnosis jumper
To 1.3 cylinder head
c a
b
Air flow meter with hot wire sensor
Air cleaner
Water cooled turbocharger with waste gate valve
Duty solenoid orifice valve
Oxygen sensor
Battery
MT model only
Speed sensor
IG. SW
Starter SW
Neutral SW
Air conditioner

MPFI system schematic diagram—Loyale

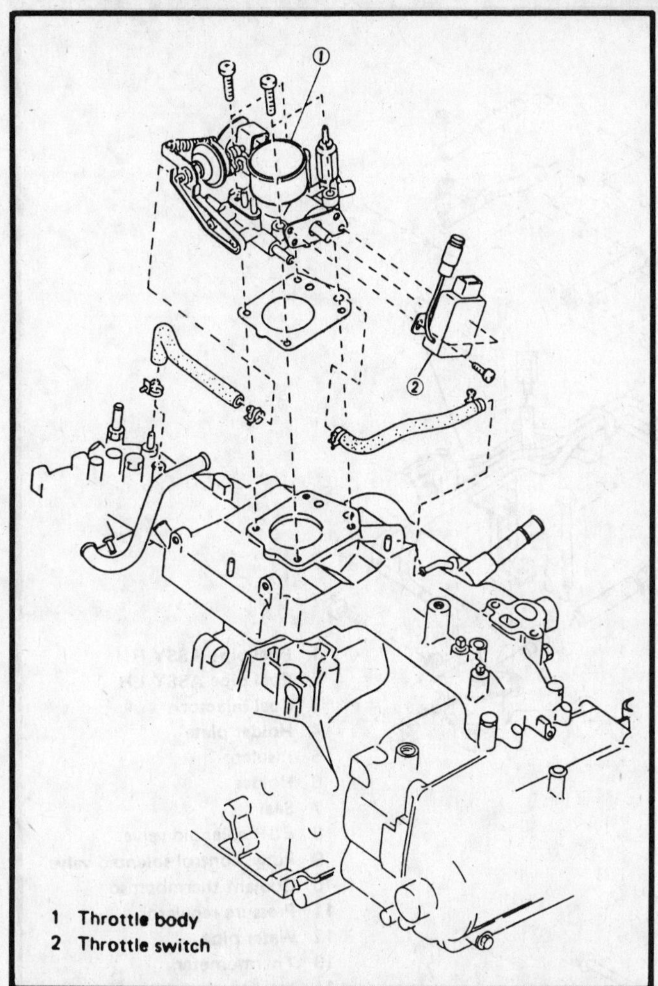
1 Throttle body
2 Throttle switch

Exploded view of throttle body—Sedan and Wagon and 1990 Loyale

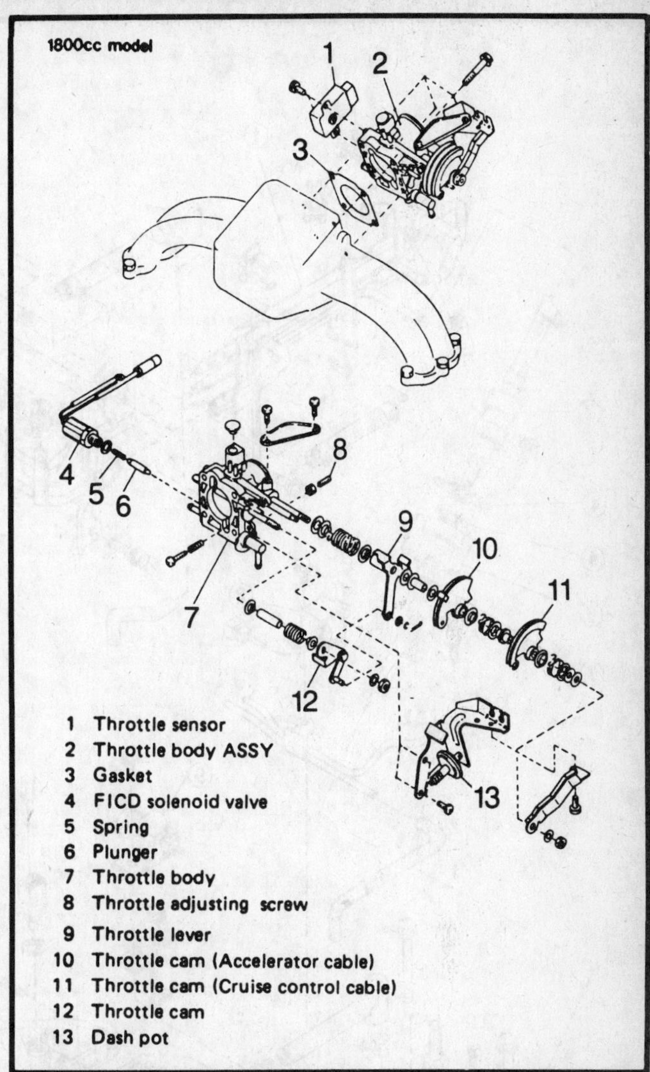
1800cc model
1 Throttle sensor
2 Throttle body ASSY
3 Gasket
4 FICD solenoid valve
5 Spring
6 Plunger
7 Throttle body
8 Throttle adjusting screw
9 Throttle lever
10 Throttle cam (Accelerator cable)
11 Throttle cam (Cruise control cable)
12 Throttle cam
13 Dash pot

Exploded view of throttle body—XT models with 4 cylinder engine

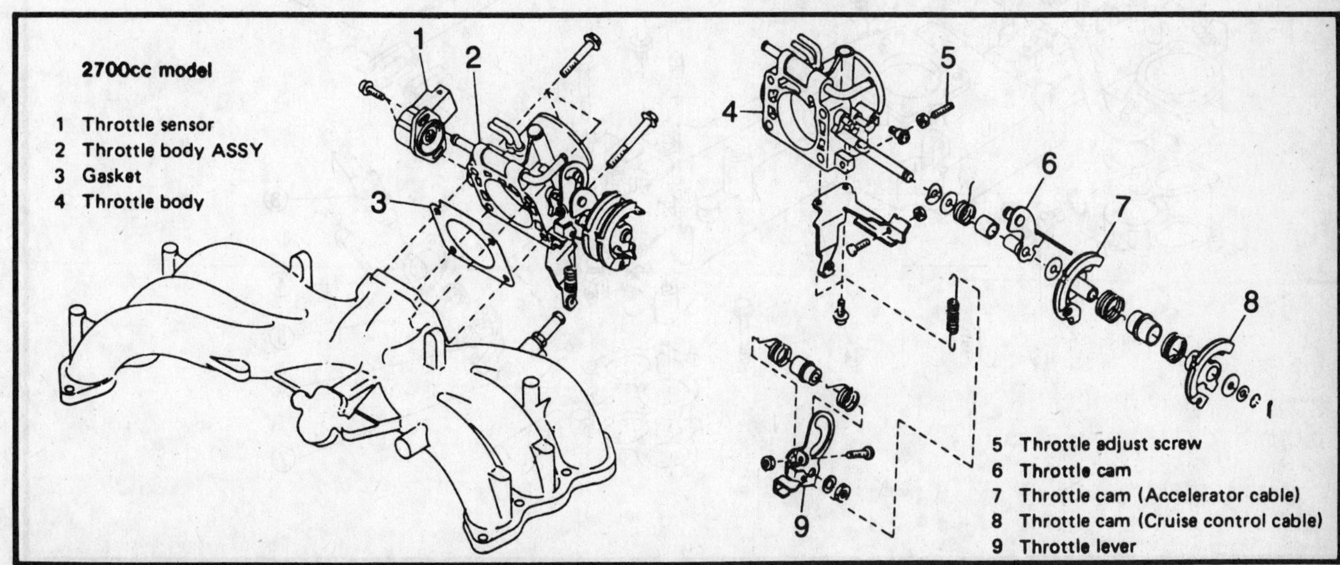
2700cc model
1 Throttle sensor
2 Throttle body ASSY
3 Gasket
4 Throttle body
5 Throttle adjust screw
6 Throttle cam
7 Throttle cam (Accelerator cable)
8 Throttle cam (Cruise control cable)
9 Throttle lever

Exploded view of throttle body—XT models with 6 cylinder engine

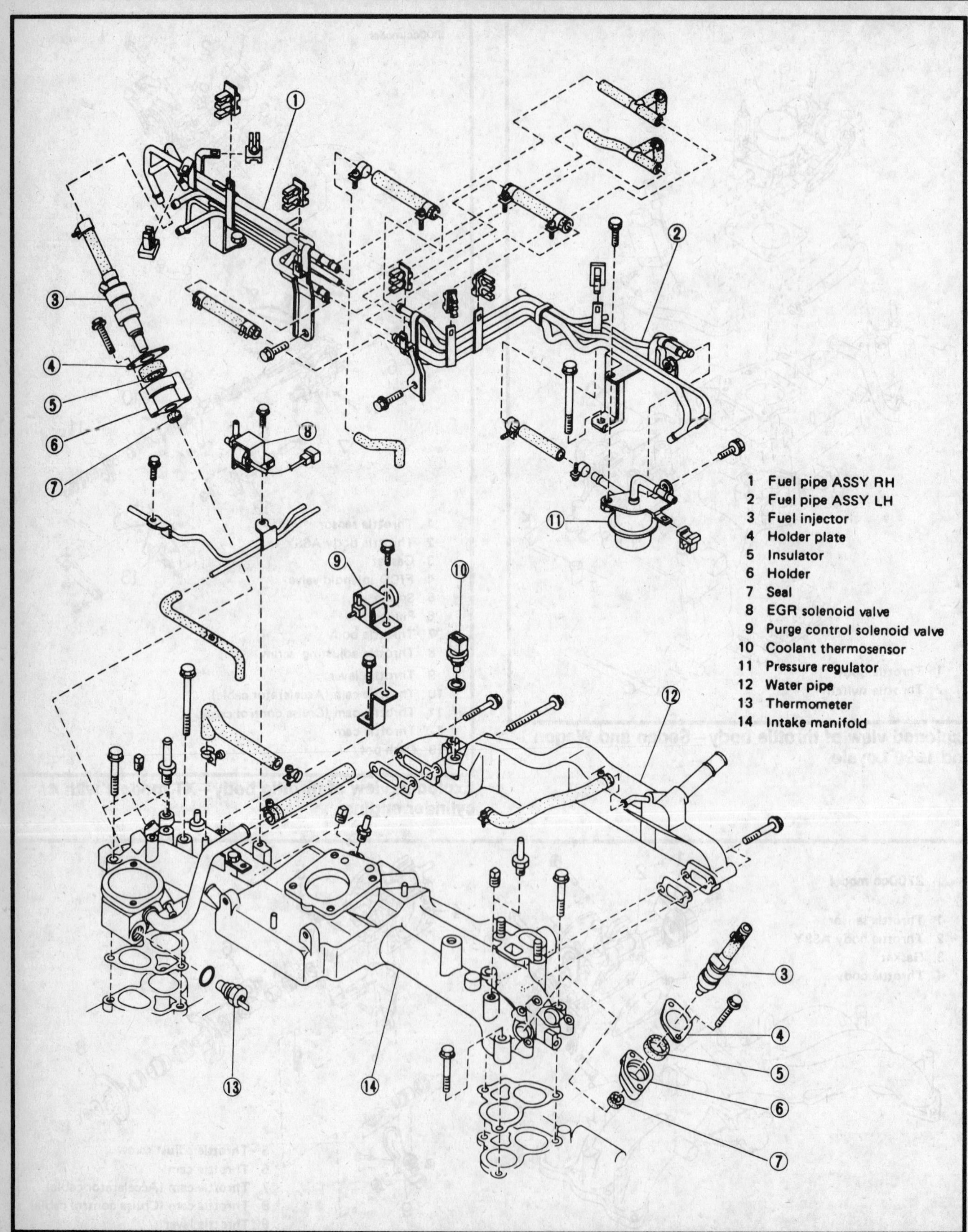
1 Fuel pipe ASSY RH
2 Fuel pipe ASSY LH
3 Fuel injector
4 Holder plate
5 Insulator
6 Holder
7 Seal
8 EGR solenoid valve
9 Purge control solenoid valve
10 Coolant thermosensor
11 Pressure regulator
12 Water pipe
13 Thermometer
14 Intake manifold

Exploded view of intake manifold—Sedan and Wagon

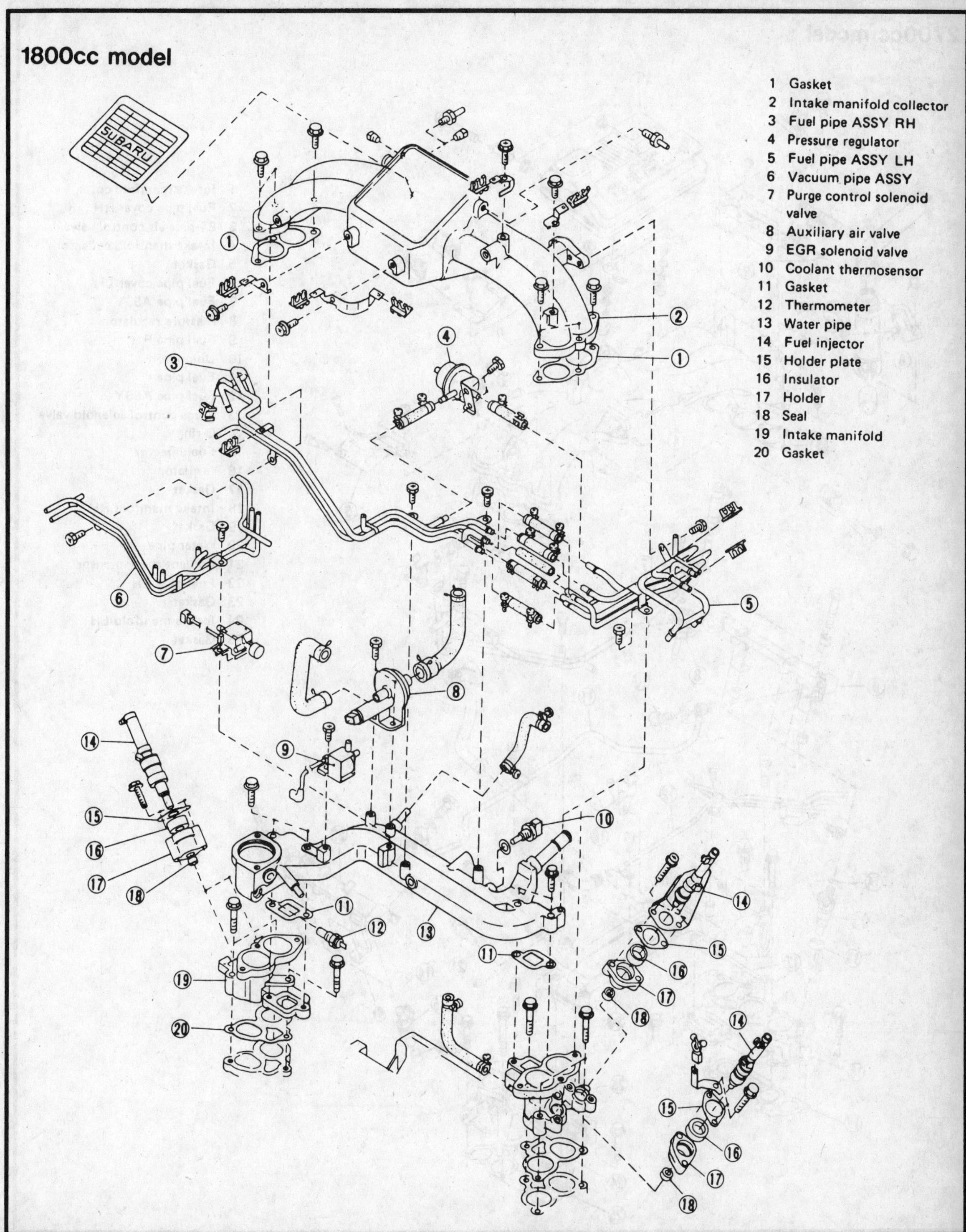
1800cc model
SUBARU
1 Gasket
2 Intake manifold collector
3 Fuel pipe ASSY RH
4 Pressure regulator
5 Fuel pipe ASSY LH
6 Vacuum pipe ASSY
7 Purge control solenoid valve
8 Auxiliary air valve
9 EGR solenoid valve
10 Coolant thermosensor
11 Gasket
12 Thermometer
13 Water pipe
14 Fuel injector
15 Holder plate
16 Insulator
17 Holder
18 Seal
19 Intake manifold
20 Gasket

Exploded view of intake manifold—XT models with 4 cylinder engine

2700cc model

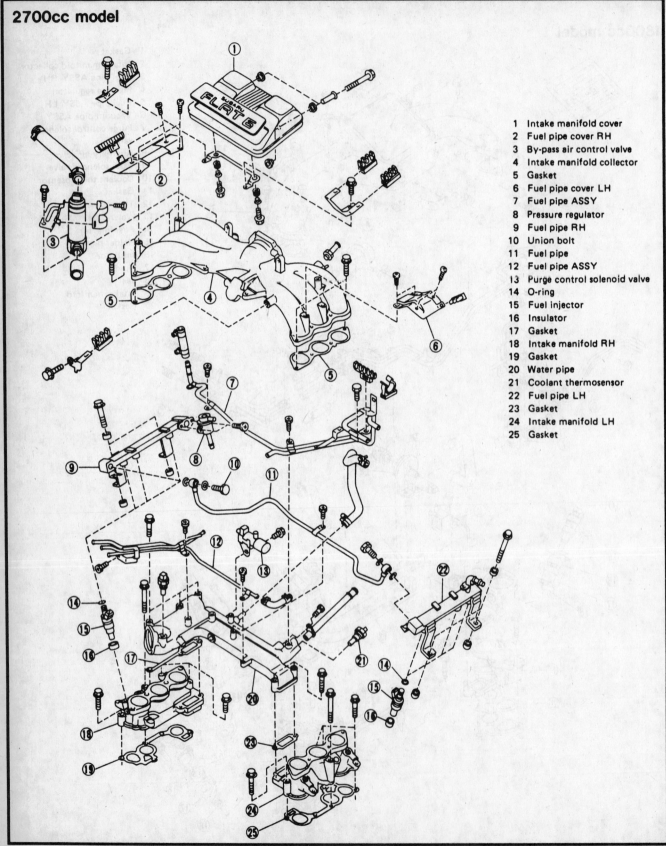

1 Intake manifold cover
2 Fuel pipe cover RH
3 By-pass air control valve
4 Intake manifold collector
5 Gasket
6 Fuel pipe cover LH
7 Fuel pipe ASSY
8 Pressure regulator
9 Fuel pipe RH
10 Union bolt
11 Fuel pipe
12 Fuel pipe ASSY
13 Purge control solenoid valve
14 O-ring
15 Fuel injector
16 Insulator
17 Gasket
18 Intake manifold RH
19 Gasket
20 Water pipe
21 Coolant thermosensor
22 Fuel pipe LH
23 Gasket
24 Intake manifold LH
25 Gasket

Exploded view of intake manifold – XT models with 6 cylinder engine

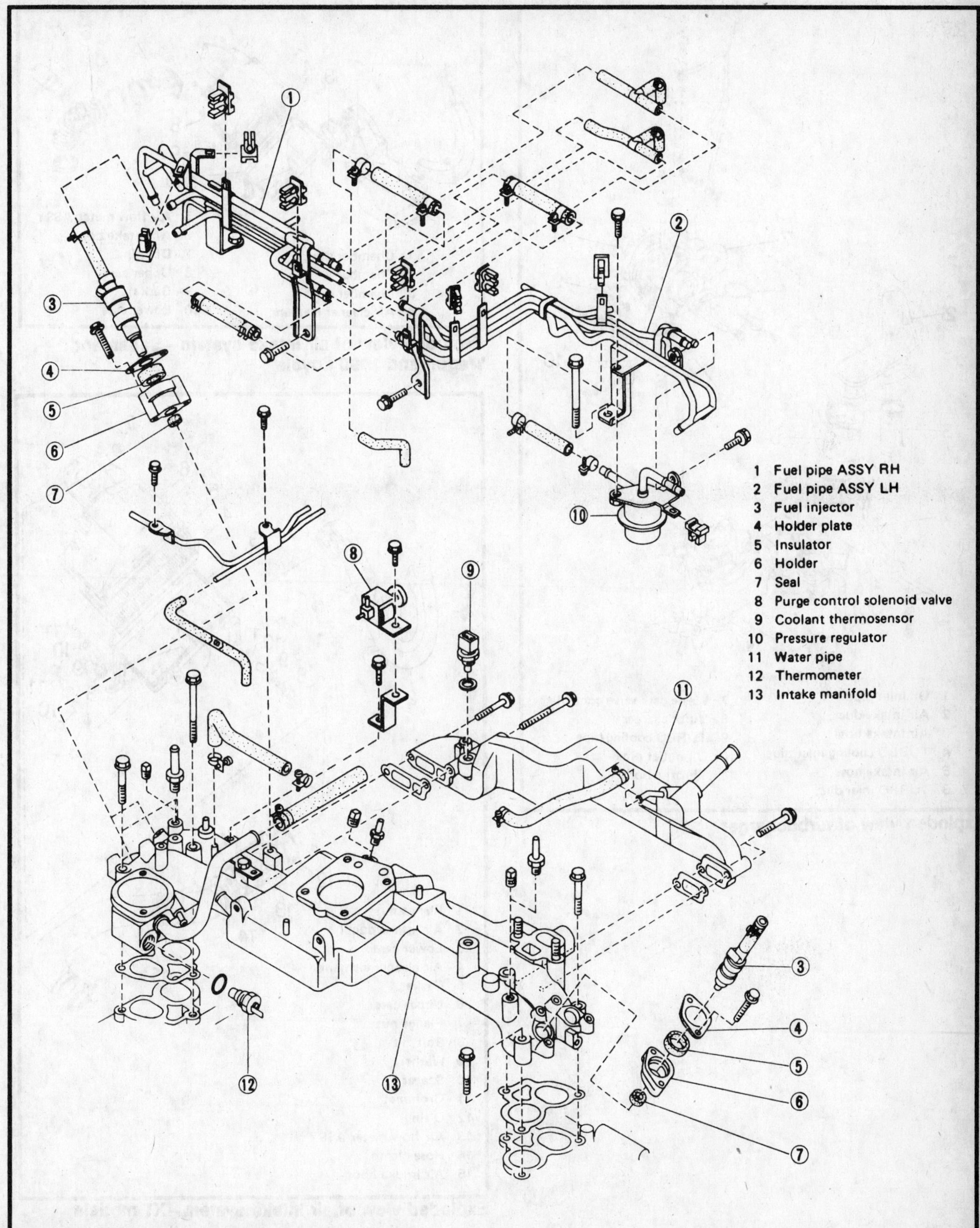
1 Fuel pipe ASSY RH
2 Fuel pipe ASSY LH
3 Fuel injector
4 Holder plate
5 Insulator
6 Holder
7 Seal
8 Purge control solenoid valve
9 Coolant thermosensor
10 Pressure regulator
11 Water pipe
12 Thermometer
13 Intake manifold

Exploded view of intake manifold – 1990 Loyale

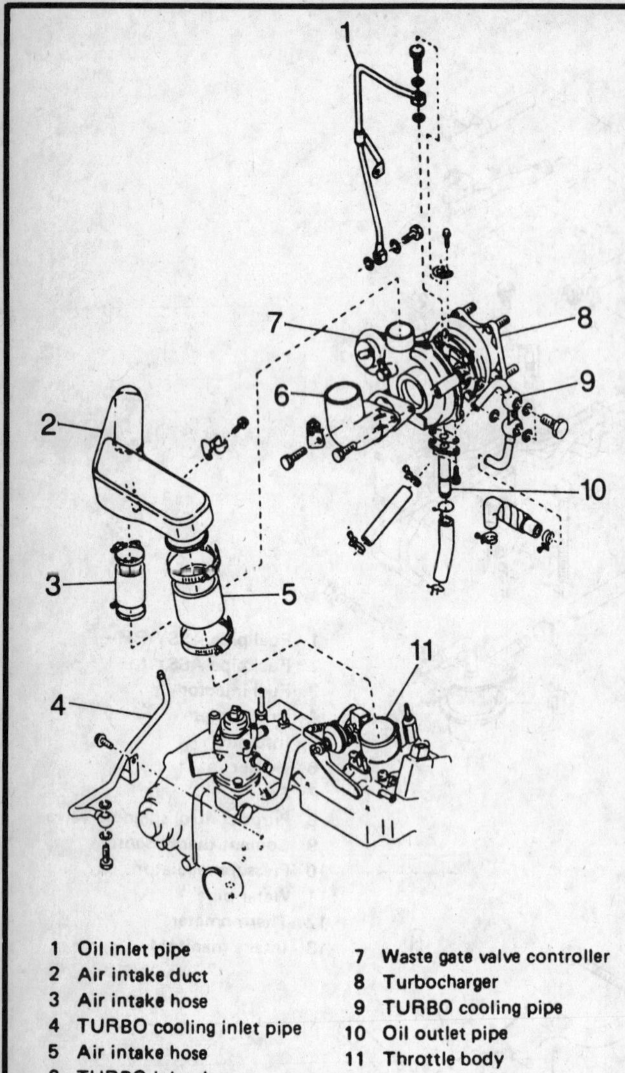
1 Oil inlet pipe
2 Air intake duct
3 Air intake hose
4 TURBO cooling inlet pipe
5 Air intake hose
6 TURBO inlet duct
7 Waste gate valve controller
8 Turbocharger
9 TURBO cooling pipe
10 Oil outlet pipe
11 Throttle body

Exploded view of turbocharger

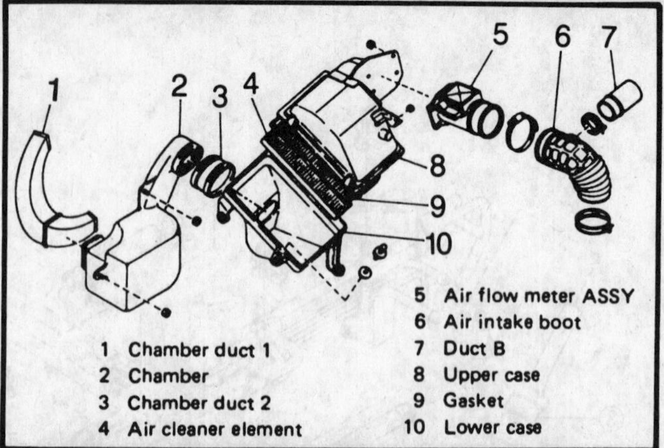
1 Chamber duct 1
2 Chamber
3 Chamber duct 2
4 Air cleaner element
5 Air flow meter ASSY
6 Air intake boot
7 Duct B
8 Upper case
9 Gasket
10 Lower case

Exploded view of air intake system – Sedan and Wagon and 1990 Loyale

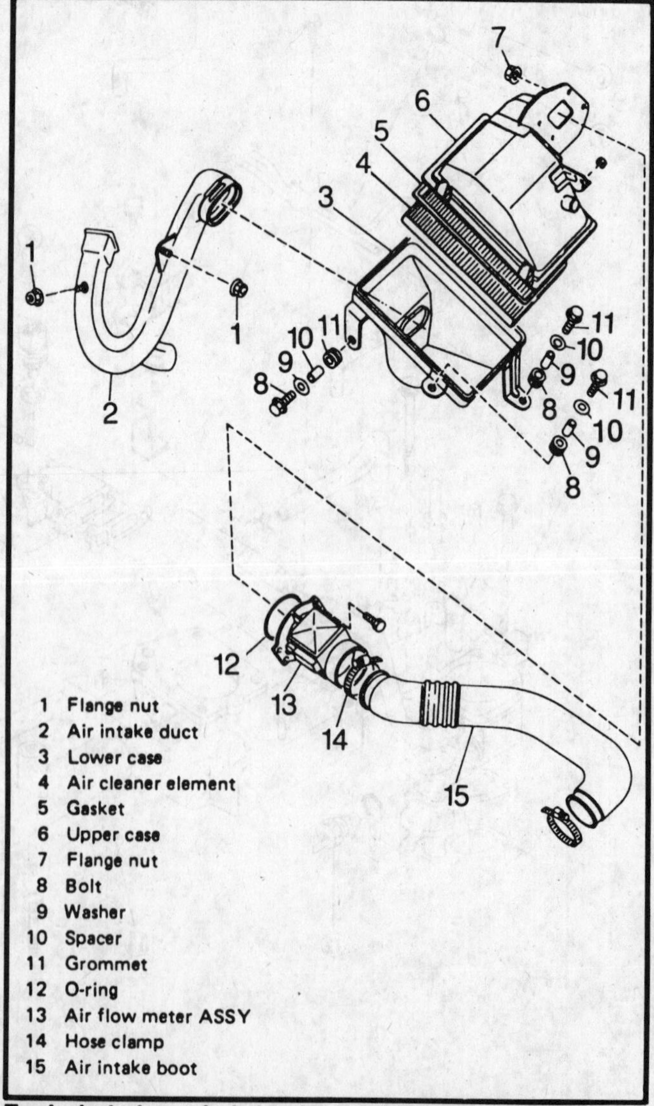
1 Flange nut
2 Air intake duct
3 Lower case
4 Air cleaner element
5 Gasket
6 Upper case
7 Flange nut
8 Bolt
9 Washer
10 Spacer
11 Grommet
12 O-ring
13 Air flow meter ASSY
14 Hose clamp
15 Air intake boot

Exploded view of air intake system – XT models

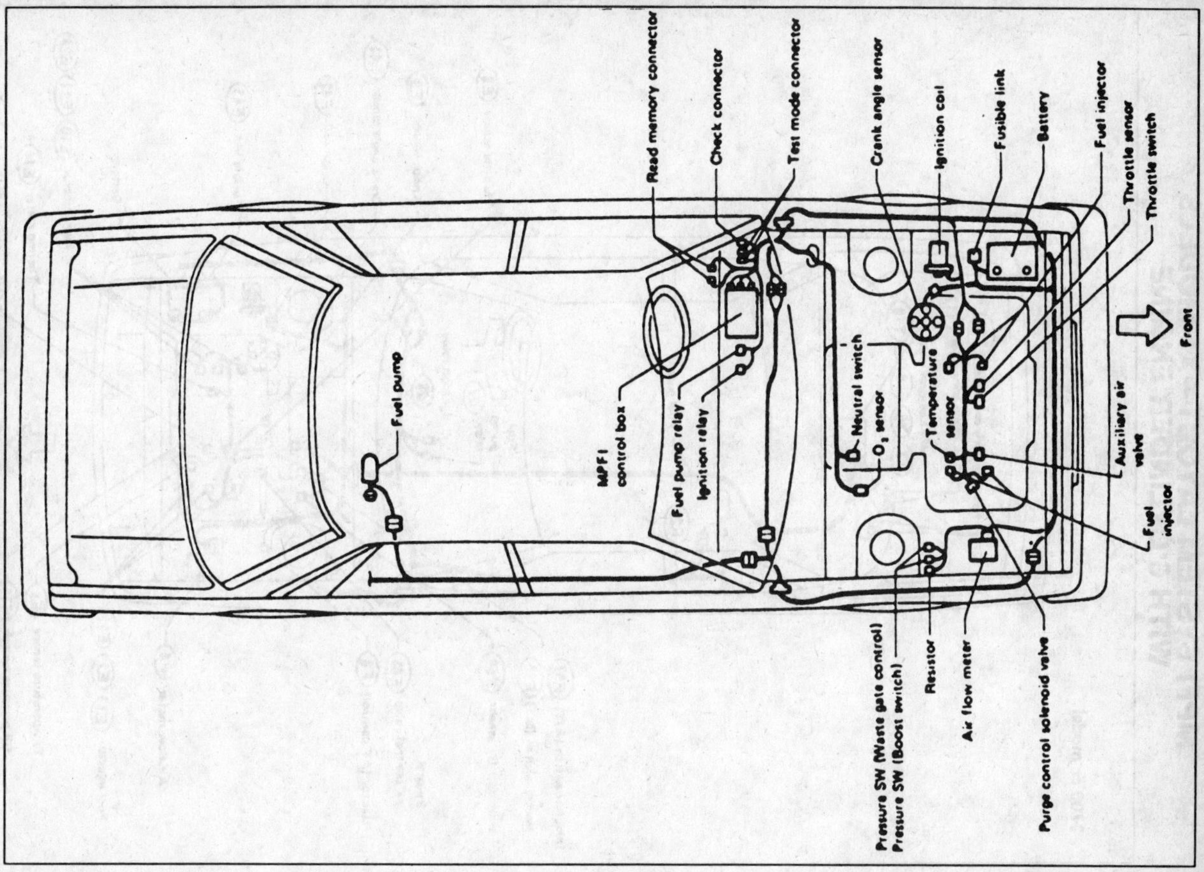

MPFI SYSTEM LAYOUT—1988 SEDAN/WAGON

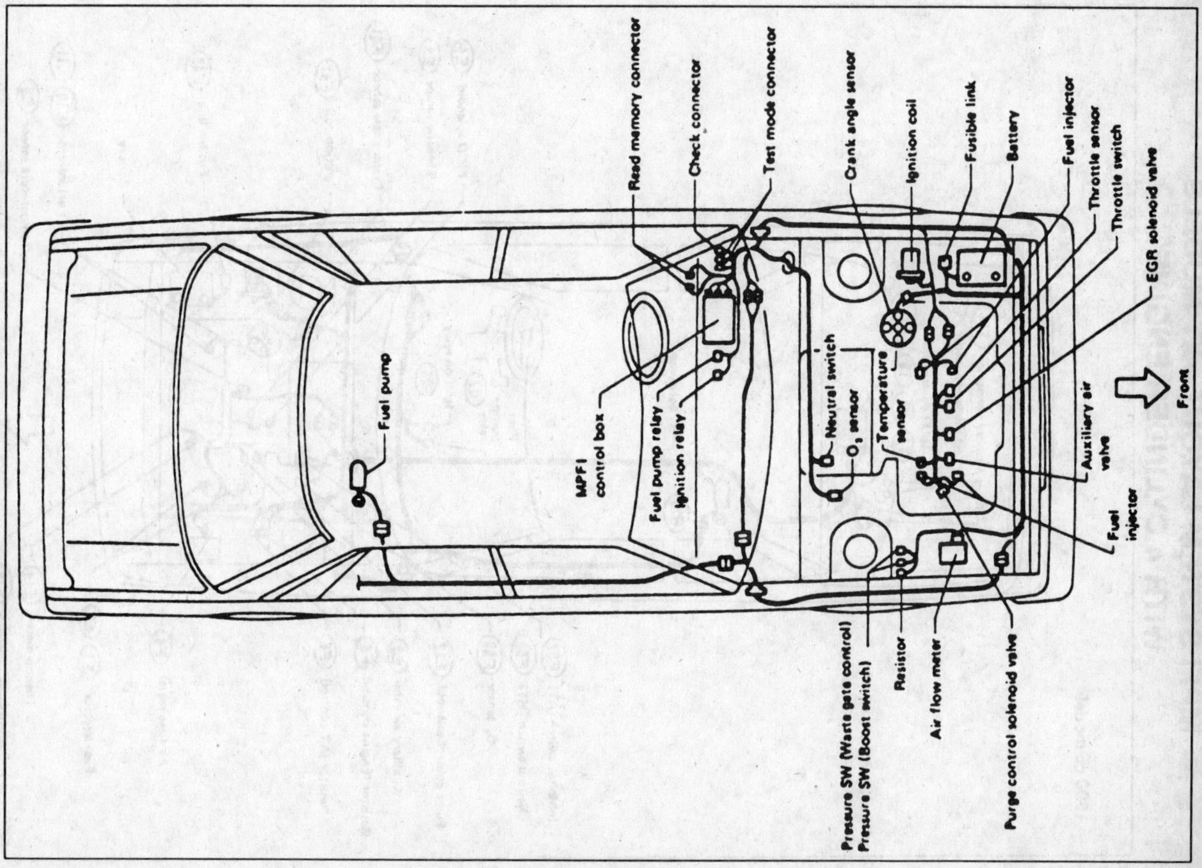

Read memory connector
Check connector
Test mode connector
Crank angle sensor
Ignition coil
Fusible link
Battery
Fuel injector
Throttle sensor
Throttle switch
EGR solenoid valve
Fuel pump
MPFI control box
Fuel pump relay
Ignition relay
Neutral switch
O_2 sensor
Temperature sensor
Auxiliary air valve
Fuel injector
Front
Pressure SW (Waste gate control)
Pressure SW (Boost switch)
Resistor
Air flow meter
Purge control solenoid valve

MPFI SYSTEM LAYOUT—1989–90 SEDAN/WAGON AND 1990 LOYALE

Read memory connector
Check connector
Test mode connector
Crank angle sensor
Ignition coil
Fusible link
Battery
Fuel injector
Throttle sensor
Throttle switch
Fuel pump
MPFI control box
Fuel pump relay
Ignition relay
Neutral switch
O_2 sensor
Temperature sensor
Auxiliary air valve
Fuel injector
Front
Pressure SW (Waste gate control)
Pressure SW (Boost switch)
Resistor
Air flow meter
Purge control solenoid valve

MPFI SYSTEM LAYOUT – XT MODELS WITH 4 CYLINDER ENGINE

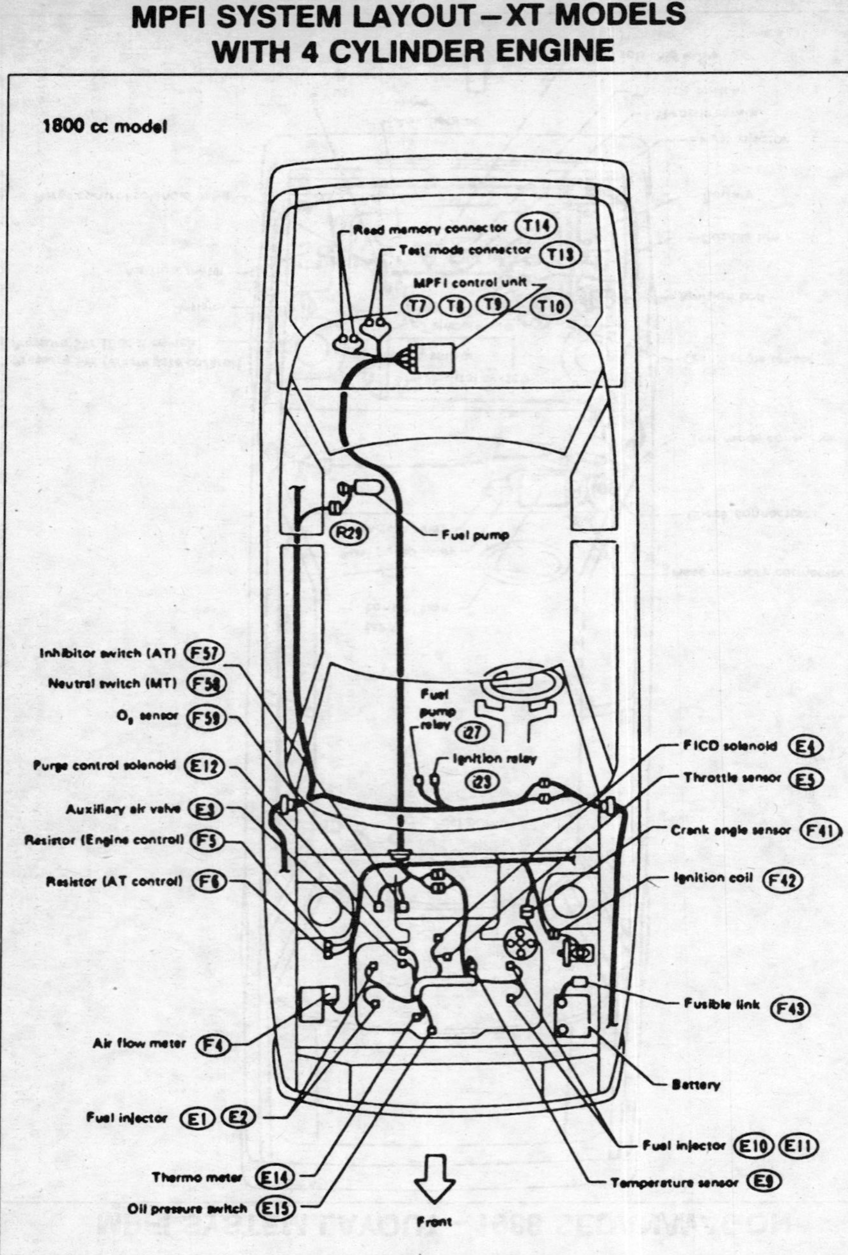
1800 cc model
Read memory connector T14
Test mode connector T13
MPFI control unit
T7 T8 T9 T10
R29
Fuel pump
Inhibitor switch (AT) F57
Neutral switch (MT) F58
O2 sensor F59
Purge control solenoid E12
Auxiliary air valve E3
Resistor (Engine control) F5
Resistor (AT control) F6
Fuel pump relay i27
Ignition relay i23
FICD solenoid E4
Throttle sensor E5
Crank angle sensor F41
Ignition coil F42
Fusible link F43
Air flow meter F4
Battery
Fuel injector E1 E2
Fuel injector E10 E11
Thermo meter E14
Temperature sensor E9
Oil pressure switch E15
Front

MPFI SYSTEM LAYOUT – XT MODELS WITH 6 CYLINDER ENGINE

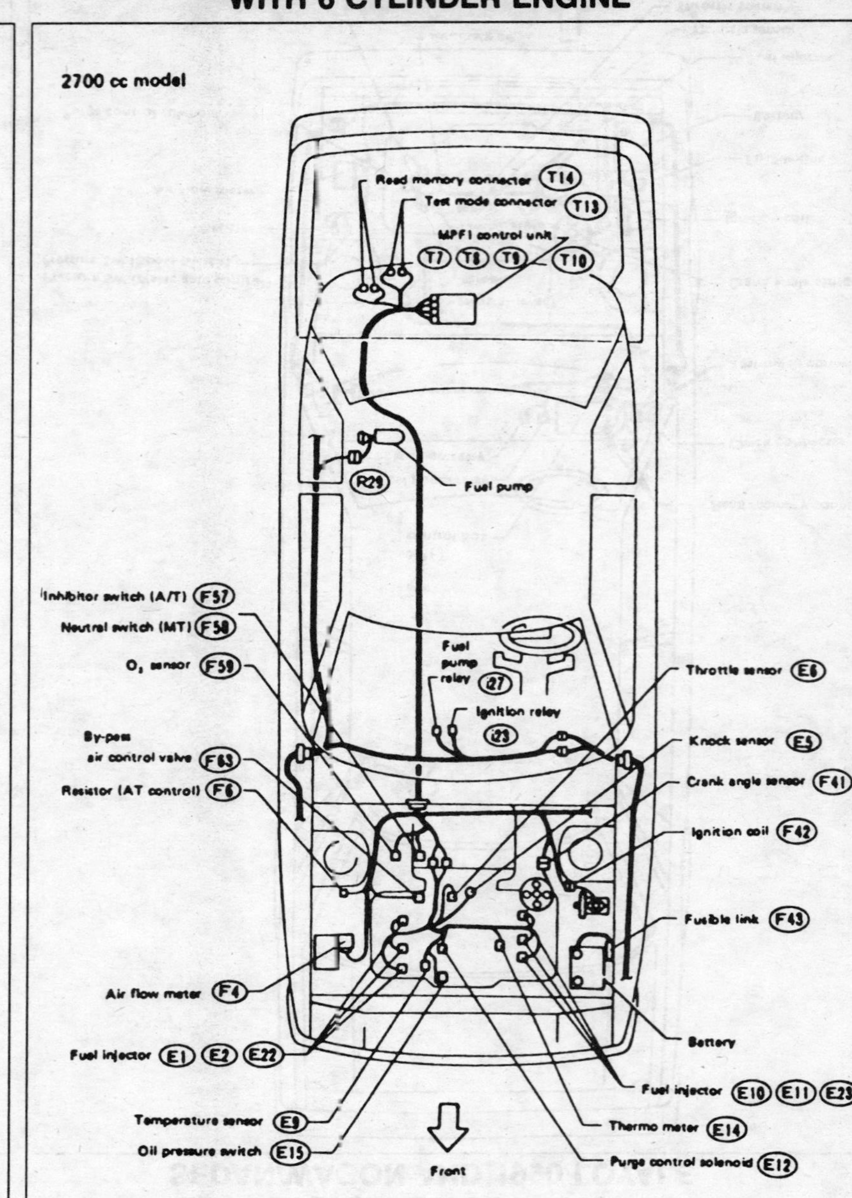
2700 cc model
Read memory connector T14
Test mode connector T13
MPFI control unit
T7 T8 T9 T10
R29
Fuel pump
Inhibitor switch (A/T) F57
Neutral switch (MT) F58
O2 sensor F59
By-pass air control valve F63
Resistor (AT control) F6
Fuel pump relay i27
Ignition relay i23
Throttle sensor E8
Knock sensor E5
Crank angle sensor F41
Ignition coil F42
Fusible link F43
Air flow meter F4
Battery
Fuel injector E1 E2 E22
Fuel injector E10 E11 E23
Temperature sensor E9
Thermo meter E14
Oil pressure switch E15
Purge control solenoid E12
Front

F81 (Black)

10	9	8	7	6	5	4	3	2	1
20	19	18	17	16	15	14	13	12	11

F86 (Black)

42	41	40	39	38	37
48	47	46	45	44	43

F82 (Black)

28	27	26	25	24	23	22	21
36	35	34	33	32	31	30	29

F80 (Black)

63	62				61	60
	59	58	57	56	55	
54	53	52	51	50	49	

1	YR	Check connector (used at line end only)	33	LgW	Read-memory connector
2	SA	Air flow meter (signal)	34	–	———
3	GL	Purge control solenoid	35	WR	Control unit power (input)
4	YL	Neutral switch (MT)	36	SA	Knock sensor
5	B	Air flow meter (ground)	37	WY	Ignition output (power transistor)
6	LB	Fuel pump relay	38	LgR	Air flow meter (burn-off output)
7	R	Crank angle sensor (power)	39	GW	Self-shutoff output
8	W	Crank angle sensor (reference)	40	RY	ECS lamp output
9	WB	Water temperature sensor	*41	LY	EGR solenoid
10	BR	Ground	42	BrB	Kick down relay (3AT)
11	–	———	43	WY	Ignition output (power transistor)
12	SA	Air flow meter (power)	44	RL	Check connector
13	RB	Duty solenoid (waste gate control)	45	GB	Check connector
14	YW	Inhibitor switch	46	L	Check connector
15	GY	Inhibitor switch	47	R	Check connector (used at line end only)
16	BR	Throttle sensor (ground)	48	SA	O_2 sensor
17	W	Crank angle sensor (position)	49	W	Fuel injector #1
18	–	———	50	W	Fuel injector #2
19	BW	Ignition switch signal	51	WR	Fuel injector #3
20	BR	Ground	52	WR	Fuel injector #4
21	W	Throttle sensor (power supply)	53	–	———
22	B	Throttle sensor (signal)	54	–	———
23	LY	Air conditioner signal	55	SA	Fuel pump control
24	BW	Starter signal	56	B	Ground
25	LW	Idle switch signal	57	B	Ground
26	–	———	58	–	———
27	WR	Control unit power (input)	59	L	A/C cut relay (3AT)
28	BY	Identification of specifications	60	BR	Ground
29	YG	Car-speed sensor	61	BR	Ground
30	BR	Identification of AT	62	R	Power (input)
31	–	———	63	–	———
32	Br	Test mode connector			

***: Except California model**

Control unit connector identification—Sedan and Wagon

T9 (Black)

10	9	8	7	6	5	4	3	2	1
20	19	18	17	16	15	14	13	12	11

T7 (Black)

42	41	40	39	38	37
48	47	46	45	44	43

T8 (Black)

28	27	26	25	24	23	22	21
36	35	34	33	32	31	30	29

T10 (Black)

63	62		61	60	
59	58	57	56	55	
54	53	52	51	50	49

1	YR	Check connector	32	Br	Test mode connector
2	W	Air flow meter (signal)	33	LgW	Read-memory connector
3	GL	Purge control solenoid	34	–	–
4	GY	Neutral switch (MT)	35	WR	Control unit power (input)
	YL	Inhibitor switch (AT)	36	–	–
5	B	Air flow meter (ground)	37	WY	Ignition output (power transistor)
6	LB	Fuel pump relay	38	LgR	Air flow meter (burn-off output)
7	R	Crank angle sensor (power)	39	GW	Self-shutoff output
8	B	Crank angle sensor (reference)	40	RY	CHECK ENGINE light
9	WB	Water temperature sensor	41	–	–
10	BR	Ground	42	RB	A/C cut control
11	–	–	43	WY	Ignition output (power transistor)
12	W	Air flow meter (power)	44	RL	Check connector
13	–	–	45	GB	Check connector
14	YW	Identification of FWD and 4WD	46	L	Check connector
15	Lg	Inhibitor switch (AT)	47	R	Trouble code output
16	BR	Throttle sensor (ground)	48	W	O_2 sensor
17	W	Crank angle sensor (position signal)	49	W	Fuel injector #1, #2
18	–	–	50	W	Fuel injector #1, #2
19	BW	Ignition switch signal	51	WL	Fuel injector #3, #4
20	BR	Ground	52	WL	Fuel injector #3, #4
21	W	Throttle sensor (power supply)	53	–	–
22	B	Throttle sensor (signal)	54	–	–
23	L	Air conditioner signal	55	W	Fuel pump control
24	BW (MT) BY (AT)	Starter signal	56	B	Ground (ignition)
25	LW	Idle switch signal	57	B	Ground (ignition)
26	–	–	58	–	–
27	WR	Control unit power (input)	59	G	Auxiliary air valve control
28	BY	Specification code	60	BR	Ground (fuel injector)
29	YR	Car-speed sensor	61	BR	Ground (fuel injector)
30	BR	Identification of AT and MT	62	R	Power (fuel injector)
31	–	–	63	–	–

Control unit connector identification – XT models with 4 cylinder engine

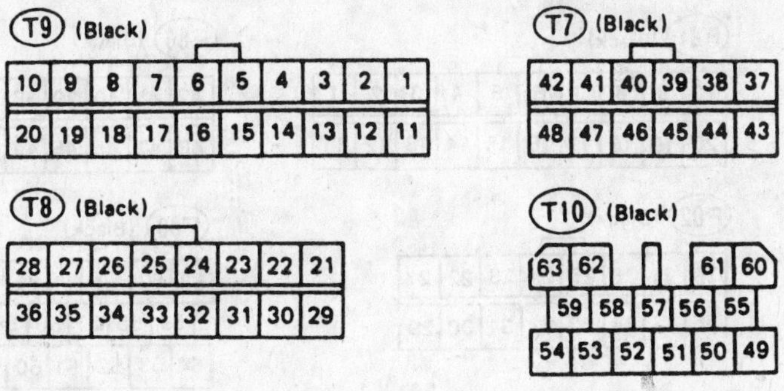

T9 (Black)

10	9	8	7	6	5	4	3	2	1
20	19	18	17	16	15	14	13	12	11

T7 (Black)

42	41	40	39	38	37
48	47	46	45	44	43

T8 (Black)

28	27	26	25	24	23	22	21
36	35	34	33	32	31	30	29

T10 (Black)

63	62		61	60	
59	58	57	56	55	
54	53	52	51	50	49

1	YR	Check connector	32	Br	Test mode connector
2	W	Air flow meter (signal)	33	LgW	Read-memory connector
3	GL	Purge control solenoid	34	—	—
4	GY	Neutral switch (MT)	35	WR	Control unit power (input)
	YL	Inhibitor switch (AT)	36	R	Knock sensor
5	B	Air flow meter (ground)	37	WY	Ignition output (power transistor)
6	LB	Fuel pump relay	38	LgR	Air flow meter (burn-off output)
7	R	Crank angle sensor (power)	39	GW	Self-shutoff output
8	G	Crank angle sensor (position signal)	40	RY	CHECK ENGINE light
9	WB	Water temperature sensor	41	—	—
10	BR	Ground	42	RB	A/C cut control
11	—	—	43	WY	Ignition output (power transistor)
12	R	Air flow meter (power)	44	RL	Check connector
13	—	—	45	GB	Check connector
14	YW	Inhibitor switch (AT)	46	L	Check connector
15	Lg	Inhibitor switch (AT)	47	R	Trouble code output
16	BR	Throttle sensor (ground)	48	W	O_2 sensor
17	W	Crank angle sensor (reference)	49	W	Fuel injector #5, #6
18	—	—	50	W	Fuel injector #5, #6
19	BW	Ignition switch signal	51	WR	Fuel injector #1, #2
20	BR	Ground	52	WR	Fuel injector #1, #2
21	W	Throttle sensor (power supply)	53	WY	Fuel injector #3, #4
22	B	Throttle sensor (signal)	54	WY	Fuel injector #3, #4
23	L	Air conditioner signal	55	B	Fuel pump control
24	BW (MT) BY (AT)	Starter signal	56	B	Ground (ignition)
25	LW	Idle switch signal	57	B	Ground (ignition)
26	—	—	58	YR	By-pass air control
27	WR	Control unit power (input)	59	YL	By-pass air control
28	—	—	60	BR	Ground (fuel injector)
29	YR	Car-speed sensor	61	BR	Ground (fuel injector)
30	BR	Identification of AT and MT	62	R	Power (fuel injector)
31	LB	Specification code	63	Br	Power steering control

Control unit connector identification—XT models with 6 cylinder engine

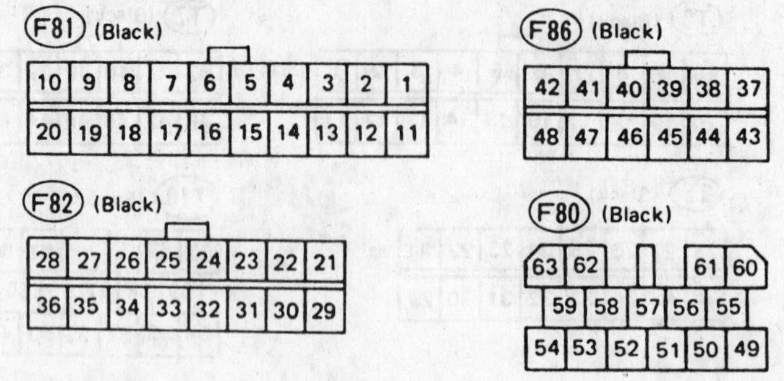

F81 (Black)
10 9 8 7 6 5 4 3 2 1
20 19 18 17 16 15 14 13 12 11
F86 (Black)
42 41 40 39 38 37
48 47 46 45 44 43
F82 (Black)
28 27 26 25 24 23 22 21
36 35 34 33 32 31 30 29
F80 (Black)
63 62 61 60
59 58 57 56 55
54 53 52 51 50 49

1	YR	Check connector (used at line end only)	33	LgW	Read-memory connector
2	SA	Air flow meter (signal)	34	—	———
3	GL	Purge control solenoid	35	WR	Control unit power (input)
4	—	———	36	SA	Knock sensor
5	B	Air flow meter (ground)	37	WY	Ignition output (power transistor)
6	LB	Fuel pump relay	38	LgR	Air flow meter (burn-off output)
7	R	Crank angle sensor (power)	39	GW	Self-shutoff output
8	W	Crank angle sensor (reference)	40	RY	ECS lamp output
9	WB	Water temperature sensor	41	—	———
10	BR	Ground	42	BrB	Kick down relay (3AT)
11	—	———	43	WY	Ignition output (power transistor)
12	SA	Air flow meter (power)	44	RL	Check connector
13	RB	Duty solenoid (waste gate control)	45	GB	Check connector
14	YW	Inhibitor switch	46	L	Check connector
15	GY	Inhibitor switch	47	R	Check connector (used at line end only)
16	BR	Throttle sensor (ground)	48	SA	O_2 sensor
17	W	Crank angle sensor (position)	49	W	Fuel injector #1
18	—	———	50	W	Fuel injector #2
19	BW	Ignition switch signal	51	WR	Fuel injector #3
20	BR	Ground	52	WR	Fuel injector #4
21	W	Throttle sensor (power supply)	53	—	———
22	B	Throttle sensor (signal)	54	—	———
23	LY	Air conditioner signal	55	SA	Fuel pump control
24	BW	Starter signal	56	B	Ground
25	LW	Idle switch signal	57	B	Ground
26	—	———	58	—	———
27	WR	Control unit power (input)	59	L	A/C cut relay (3AT)
28	BY	Identification of specifications	60	BR	Ground
29	YG	Car-speed sensor	61	BR	Ground
30	BR	Identification of AT	62	R	Power (input)
31	—	———	63	—	———
32	Br	Test mode connector			

Control unit connector identification—1990 Loyale

NO TROUBLE

O : CONNECT X : DISCONNECT

Engine	Read memory connector	Test mode connector	CHECK ENGINE light	O_2 monitor lamp	Remarks
ON	X	X	OFF	O_2 monitor	
ON	O	X	OFF	O_2 monitor	
*ON	X	O	** OFF → Blink	OFF	Vehicle specification code is outputted when CHECK ENGINE light is OFF.
*ON	O	O	OFF → Blink	OFF	All memory stored in control unit is cleared after CHECK ENGINE light blinks.
OFF (Ignition switch ON)	O	X	ON	Vehicle specification code	Before starting the engine, the self-diagnosis system assumes the engine to be in NO TROUBLE condition.
OFF (Ignition switch ON)	X	X	ON	Vehicle specification code	
OFF (Ignition switch ON)	X	O	ON	Vehicle specification code	
OFF (Ignition switch ON)	O	O	ON	Vehicle specification code	

TROUBLE

Engine	Read memory connector	Test mode connector	CHECK ENGINE light	O_2 monitor lamp	Remarks
ON	X	X	ON	Trouble code	
ON	O	X	ON	Trouble code (memory)	
*ON	X	O	** OFF → ON	Trouble code	Vehicle specification code is outputted when CHECK ENGINE light is OFF.
*ON	O	O	** OFF → ON	Trouble code	
OFF (Ignition switch ON)	O	X	ON	Trouble code (memory)	
STALL (Ignition switch ON)	X	X	ON	Trouble code	
STALL (Ignition switch ON)	X	O	ON	Trouble code	
STALL (Ignition switch ON)	O	O	ON	Trouble code	

*: Ignition timing is set to 20° BTDC (when the engine is on, test mode connector is connected, and idle switch is ON).

**: CHECK ENGINE light remains off until engine is operated at speed greater than 2,000 rpm for at least 40 seconds.

Operation of self-diagnosis system

GENERAL TROUBLESHOOTING TABLE—ALL MODELS

TROUBLESHOOTING

General Troubleshooting Table

*: The CHECK ENGINE light blinks.
*1: The CHECK ENGINE light blinks when contact is resumed during inspection (although poor contact is present in the D-check).
*2: The CHECK ENGINE light lights when the mixture is leaner than that specified and does not light (U-check) or blink (D-check) when the mixture is richer.
*3: The CHECK ENGINE light lights when abnormality is detected in the D-check mode if the idle switch persistently remains off with the acclerator pedal released.

Symbols shown in the table refer to the degree of possibility of the reason for the trouble ("Very often" to "Rarely").
◎ : Very often
○ : Sometimes
△ : Rarely
☆ : Occurs only in extremely low temperatures

	TROUBLE	
1	Engine will not start	No initial combustion
2		Initial combustion occurs.
3		Engine stalls after initial combustion.
4	Rough idle and engine stall	
5	Inability to drive at constant speed	
6	Inability to accelerate and decelerate	
7	Engine does not return to idle.	
8	Afterburning in exhaust system	
9	Knocking	
10	Excessive fuel consumption	
11		
U	CHECK ENGINE light operation	U-check mode & read memory mode
D		D-check mode

1	2	3	4	5	6	7	8	9	10		U (#)	D (#)	POSSIBLE CAUSE
													AIR FLOW METER
		☆	◎				△	△	○		ON	ON	• Connector not connected
		△	◎	◎	◎		◎	○	△		ON	*1	• Poor contact of terminal
		☆	◎				△	○	△		ON	ON	• Short circuit
		☆	◎				△	○	○		ON	ON	• Discontinuity of wiring harness
		○	○	○	○		△	◎	○		*2	*2	• Performance characteristics unusual
													COOLANT THERMOSENSOR
	☆	○	☆		○		○	○	○		ON	ON	• Connector not connected
	△	△	◎	○	○		◎	△	○		ON	*1	• Poor contact of terminal
	☆	○	☆		○		○	○	○		ON	ON	• Short circuit
	☆	○	☆		○		○	○	○		ON	ON	• Discontinuity of wiring harness
	☆	○	☆		○		◎	◎	◎		*2	*2	• Performance characteristics unusual
													IDLE SWITCH OF THROTTLE SENSOR
				○	◎		◎				OFF	ON	• Connector not connected
				○			○				ON	*1	• Poor contact of terminal
				○	△		○				ON	ON	• Short circuit
					△		○				OFF	ON	• Discontinuity of wiring harness
							○				OFF	*3	• Improper adjustment
1	2	3	4	5	6	7	8	9	10		U	D	

#: CHECK ENGINE light

GENERAL TROUBLESHOOTING TABLE—SEDAN AND WAGON AND 1990 LOYALE

1	2	3	4	5	6	7	8	9	10		U (CHECK ENGINE light)	D (CHECK ENGINE light)	POSSIBLE CAUSE
													THROTTLE SENSOR
					◎		◎				ON	ON	• Connector not connected
				◎	◎		◎				ON	*1	• Poor contact of terminal
△					◎		◎				ON	ON	• Short circuit
					◎		◎				ON	ON	• Discontinuity of wiring harness
	○	○	△	○	◎		◎				OFF	*	• Performance characteristics unusual
													PRESSURE REGULATOR
	○	○	◎	◎	◎	○		△			*2	*2	• Sensing hose not connected
	△				○		○		◎		OFF	*	• Fuel pressure too high
○	○	○	◎	◎	◎		◎				*2	*2	• Fuel pressure too low
													FUEL INJECTOR
	◎	◎	○	○	○		○	○			ON	*1	• Connector not connected
	○	○	○	◎	◎		○				ON	ON	• Poor contact of terminal
	◎	◎	○	○	○		○				ON	ON	• Short circuit
	◎	◎	○	○	○		○	○			ON	ON	• Discontinuity of wiring harness
	△	○	○	△	○		○	△	○		*2	*2	• Performance characteristics unusual
	△	○	△	△	○		○	△			*2	*2	• Clogged filter
	△	○	○	○	○		○	△			*2	*2	• Clogged nozzle
○											OFF	*	• Stuck open
			○				○		○		OFF	*	• Slight leakage from seat
													CRANK ANGLE SENSOR
◎											ON	ON	• Connector disconnected
	○	○	◎	◎	◎		○	○			ON	*1	• Poor contact of terminal
◎											ON	ON	• Short circuit
◎											ON	ON	• Discontinuity of wiring harness
													POWER TRANSISTOR OF IGNITION COIL
◎											OFF	*	• Connector not connected
	○	○	◎	◎	◎		○				OFF	*	• Poor contact of terminal
◎											OFF	*	• Short circuit
◎											OFF	*	• Discontinuity of wiring harness
													AIR REGULATOR
						◎					OFF	*	• Connector not connected
	○	◎	○								OFF	*	• Short circuit
						◎					OFF	*	• Discontinuity of wiring harness
													KNOCK SENSOR
								◎			ON	ON	• Connector not connected
				○	○						ON	ON	• Short circuit
								◎			ON	ON	• Discontinuity of wiring harness
													DUTY SOLENOID
					△						OFF	*	• Connector disconnected
											OFF	*	• Poor contact of terminal
								○			OFF	*	• Short circuit
					△						OFF	*	• Discontinuity of wiring harness
			○	○	○	○	○	○			OFF	*	• Disconnected or cracked hose
													ENGINE GROUNDING
○											ON		• Disconnected engine grounding terminal at intake manifold
◎	◎	○	◎	◎	◎						ON	*1	• Poor contact of engine grounding terminal
○											ON		• Discontinuity of wiring harness for engine grounding
1	2	3	4	5	6	7	8	9	10		U	D	

GENERAL TROUBLESHOOTING TABLE—XT MODELS

1	2	3	4	5	6	7	8	9	10		U (#)	D (#)	POSSIBLE CAUSE
													THROTTLE SENSOR
					◎		◎				ON	ON	• Connector not connected
				◎	◎		◎				ON	*1	• Poor contact of terminal
△					◎		◎				ON	ON	• Short circuit
					◎		◎				ON	ON	• Discontinuity of wiring harness
	○	○	△	○	◎		◎				OFF	*	• Performance characteristics unusual
													PRESSURE REGULATOR
	○	○	◎	◎	◎	○		△			*2	*2	• Sensing hose not connected
	△				○		○		◎		OFF	*	• Fuel pressure too high
○	○	○	◎	◎	◎		◎				*2	*2	• Fuel pressure too low
													FUEL INJECTOR
	○	○	○	○	○		○	○			ON	*1	• Connector not connected
	○	○	○	◎	◎		○				ON	ON	• Poor contact of terminal
	◎	◎	○	○	○		○				ON	ON	• Short circuit
	◎	◎	○	○	○		○	○			ON	ON	• Discontinuity of wiring harness
	△	○	○	△	○		○	△	○		*2	*2	• Performance characteristics unusual
	△	○	△	△	○		○	△			*2	*2	• Clogged filter
	△	○	○	○	○		○	△			*2	*2	• Clogged nozzle
○											OFF	*	• Stuck open
			○				○		○		OFF	*	• Slight leakage from seat
													CRANK ANGLE SENSOR
◎											ON	ON	• Connector disconnected
	○	○	◎	◎	◎		○	○			ON	*1	• Poor contact of terminal
◎											ON	ON	• Short circuit
◎											ON	ON	• Discontinuity of wiring harness
													POWER TRANSISTOR OF IGNITION COIL
◎											OFF	*	• Connector not connected
	○	○	◎	◎	◎		○				OFF	*	• Poor contact of terminal
◎											OFF	*	• Short circuit
◎											OFF	*	• Discontinuity of wiring harness
													AIR REGULATOR [1800 cc model only]
						◎					OFF	*	• Connector not connected
	○	◎	○								OFF	*	• Short circuit
						◎					OFF	*	• Discontinuity of wiring harness
													KNOCK SENSOR [2700 cc model only]
								◎			ON	ON	• Connector not connected
				○	○						ON	ON	• Short circuit
								◎			ON	ON	• Discontinuity of wiring harness
1	2	3	4	5	6	7	8	9	10		U	D	

#: CHECK ENGINE light

1	2	3	4	5	6	7	8	9	10	11	U (#)	D (#)	POSSIBLE CAUSE
													AIR CONTROL VALVE [2700 cc model only]
	○	△	◎								ON	ON	• Connector not connected
	△	○	◎								ON	*1	• Poor contact of terminal
			○				○				ON	ON	• Short circuit
	○	△	◎								ON	ON	• Discontinuity of wiring harness
			◎			◎					OFF	*	• IAS improperly adjusted
					◎	◎					ON	*	• Stuck open
	○	○	◎								OFF	*	• Stuck closed
													ENGINE GROUNDING
○											ON		• Disconnecting of engine grounding terminal at intake manifold
◎	◎	○	◎	◎	◎						ON	*1	• Poor contact of engine grounding terminal
○											ON		• Discontinuity of wiring harness for engine grounding
1	2	3	4	5	6	7	8	9	10	11	U	D	

#: CHECK ENGINE light

ENGINE TROUBLESHOOTING TABLE—ALL MODELS

TROUBLESHOOTING

Engine Trouble in General

Symbols shown in the chart refer to the possibility of reason for the trouble in order ("Very often" to "Rarely")
◎ – Very often
○ – Sometimes
△ – Rarely

No.	TROUBLE	
1	Engine will not start.	Starter does not turn.
2		Initial combustion does not occur.
3		Initial combustion occurs.
4		Engine stalls after initial combustion.
5	Rough idle and engine stall.	
6	Low output, hesitation and poor acceleration.	
7	Surging.	
8	Engine does not return to idle.	
9	Dieseling (Run-on).	
10	Afterburning in exhaust system.	
11	Knocking.	
12	Excessive engine oil consumption.	
13	Excessive fuel consumption.	

1	2	3	4	5	6	7	8	9	10	11	12	13	POSSIBLE CAUSE
													STARTER
○													• Defective battery-to-starter harness.
△													• Defective starter switch.
△													• Defective inhibitor switch.
○	△												• Defective starter.
													BATTERY
◎													• Poor terminal connection.
◎													• Run-down battery.
○													• Defective charging system.
	◎	◎	◎	◎	◎	◎	◎	◎	◎	◎		◎	MPFI SYSTEM
													IGNITION SYSTEM
	○	○	○	◎	◎	◎	◎	◎	◎	◎		◎	• Incorrect ignition timing.
	○	◎		○	○	○			△			△	• Disconnection of spark plug cord.
	○			△	◎	○	○		◎	○			• Defective distributor.
	○			△	○	○							• Defective ignition coil.
	○			△	△	△							• Defective cord or wiring.
	○	○		△	○	△			○				• Leakage of spark plug cord.
		○		○	○	○			◎				• Defective spark plug.
	○	○	○	◎	○	◎	△		○	○			• Incorrect cam timing.
1	2	3	4	5	6	7	8	9	10	11	12	13	

1	2	3	4	5	6	7	8	9	10	11	12	13	POSSIBLE CAUSE
													INTAKE SYSTEM
		○	○	◎	○	○	◎	○	○			◎	• Improper idle adjustment.
			○	◎	◎	◎			△	◎			• Loosened or cracked intake boot.
			○	◎	◎	◎			△	◎			• Loosened or cracked intake duct.
			△	◎	◎	◎			△	◎	◎		• Loosened or cracked blow-by hose.
			△	◎	○	◎	◎		○	◎			• Loosened or cracked vacuum hose.
			△	○	○	○				◎			• Defective air cleaner gasket.
		○	○	○	○	○				◎			• Defective intake manifold gasket.
		○	○	○	○	○				◎			• Defective throttle body gasket.
				△	○	○			○	○	○		• Defective PCV valve.
				○	○	○			△	○	△		• Loosened oil filler cap.
			△	△	◎	○				○		◎	• Dirty air cleaner element.
													FUEL LINE
	◎	△		△	○	○							• Defective fuel pump.
		△	△	△	○	○							• Clogged fuel line.
	○	○	○	○	△	△							• Lack of or insufficient fuel.
													BELT
	○	○	○										• Defective.
	○	○	○	△	○	○			○	○		○	• Defective timing.
													FRICTION
△													• Seizure of crankshaft and connecting-rod bearing.
△													• Seized camshaft.
△													• Seized or stuck piston and cylinder.
													COMPRESSION
	△	△	△	○	○	○			○	△		○	• Incorrect valve clearance.
	△	△	△	○	○	△			△			△	• Loosened spark plugs or defective gasket.
	△	△	△	○	○	△			△			△	• Loosened cylinder head nuts or defective gasket.
	△	△	△	○	○	△			○			○	• Improper valve seating.
	△	△	△	△	△	△			△		◎	△	• Defective valve stem.
	○	○	○	○	○	△			△			△	• Worn or broken valve spring.
	△	△	△	○	△	△			△		○	○	• Worn or stuck piston rings, cylinder and piston.
	○	○	○	◎	◎	◎			◎	○		○	• Incorrect valve timing.
	○	○	○	○	○	○							• Improper engine oil (low viscosity).
1	2	3	4	5	6	7	8	9	10	11	12	13	

ENGINE TROUBLESHOOTING TABLE—ALL MODELS

1	2	3	4	5	6	7	8	9	10	11	12	13	POSSIBLE CAUSE
													LUBRICATION SYSTEM
				○	○				△			△	• Incorrect oil pressure.
											○		• Loosened oil pump attaching bolts and defective gasket.
											○		• Defective oil filter seal.
											○		• Defective crankshaft oil seal.
				△							○		• Defective rocker cover gasket.
											○		• Loosened oil drain plug or defective gasket.
											○		• Loosened oil pan fitting bolts or defective oil pan.
													COOLING SYSTEM
				△	△	○		○		◎			• Overheating.
					△				△			△	• Over cooling.
													TURBOCHARGER
				△	◎	◎						○	• Malfunction of turbocharger.
					◎	◎				◎		○	• Malfunction of waste gate valve.
											○		• Defective oil pipe and hose.
													OTHERS
				○	◎	◎			○	◎		○	• Malfunction of EGR System.
				◎	◎	△			△				• Malfunction of Evaporative Emission Control System.
				○			◎						• Stuck or damaged throttle valve.
				△			○					○	• Dashpot out of adjustment.
				△			○	○				○	• Accelerator cable out of adjustment.
				○			○	○				○	• FICD out of adjustment.
				○			○						• Malfunction of FICD.
1	2	3	4	5	6	7	8	9	10	11	12	13	

TROUBLE CODE CHART—1988 SEDAN AND WAGON

List of Trouble Codes

Trouble code	Item
11	Crank angle sensor (No reference pulse)
12	Starter switch (Continuously in ON position or continuously in OFF position while cranking)
13	Crank angle sensor (No position pulse)
14	Fuel injectors #1 and #2 (Abnormal injector output)
15	Fuel injectors #3 and #4 (Abnormal injector output)
21	Water temperature sensor (Open or shorted circuit)
22	Knock sensor (Open or shorted circuit)
23	Air flow meter (Open or shorted circuit)
31	Throttle sensor (Open or shorted circuit)
32	O_2 sensor (Abnormal sensor signal)
33	Car-speed sensor (No signal is present during operation)
*34	EGR solenoid valve (Solenoid switch continuously in ON or OFF position)
35	Purge control solenoid valve (Solenoid switch continuously in ON or OFF position)
41	System too lean
42	Idle switch (Abnormal idle switch signal in relation to throttle sensor output)
51	Neutral switch (Continuously in ON position)

*: **Except California model**

List of Specification Codes

Specification codes	Specification
01	MT, 49-state and Canada
02	MT, California
03	AT, 49-state and Canada
04	AT, California

TROUBLE CODE CHART—1989–90 SEDAN AND WAGON

List of Trouble Codes

Trouble code	Item
11	Crank angle sensor (No reference pulse)
12	Starter switch (Continuously in ON position or continuously in OFF position while cranking)
13	Crank angle sensor (No position pulse)
14	Fuel injectors #1 and #2 (Abnormal injector output)
15	Fuel injectors #3 and #4 (Abnormal injector output)
21	Water temperature sensor (Open or shorted circuit)
22	Knock sensor (Open or shorted circuit)
23	Air flow meter (Open or shorted circuit)
31	Throttle sensor (Open or shorted circuit)
32	O_2 sensor (Abnormal sensor signal)
33	Car-speed sensor (No signal is present during operation)
*34	EGR solenoid valve (Solenoid switch continuously in ON or OFF position)
35	Purge control solenoid valve (Solenoid switch continuously in ON or OFF position)
41	System too lean
42	Idle switch (Abnormal idle switch signal in relation to throttle sensor output)
44	Duty solenoid valve (Waste gate control)
51	Neutral switch (Continuously in ON position)

*: Except California model

List of Specification Codes

Specification codes	Specification
01	MT, 49-state and Canada
02	MT, California
03	AT, 49-state and Canada
04	AT, California

TROUBLE CODE CHART—XT MODELS

List of Trouble Codes

Trouble code	Item	U-check	D-check
11	Crank angle sensor (No reference pulse)	O	O
12	Starter switch (Continuously in ON position or continuously in OFF position while cranking)	O	O
13	Crank angle sensor (No position pulse)	O	O
14	Fuel injectors *#1 and #2, **#5 and #6 (Abnormal injector output)	O	O
15	Fuel injectors *#3 and #4, **#1 and #2 (Abnormal injector output)	O	O
21	Water temperature sensor (Open or shorted circuit)	O	O
**22	Knock sensor (Open or shorted circuit)	O	O
23	Air flow meter (Open or shorted circuit)	O	O
**24	By-pass air control valve (Open or shorted circuit)	O	O
**25	Fuel injectors #3 and #4 (Abnormal injector output)	O	O
31	Throttle sensor (Open or shorted circuit)	O	O
32	O_2 sensor (Abnormal sensor signal)	O	O
33	Car-speed sensor (No signal is present during operation)	O	O
35	Purge control solenoid valve (Solenoid switch continuously in ON or OFF position)	–	O
41	System too lean	O	O
42	Idle switch (Abnormal idle switch signal in relation to throttle sensor output)	–	O
51	Neutral switch (No signal is present)	–	O

*: 1800 cc model **: 2700 cc model

List of Specification Codes

		1800 cc model		2700 cc model
		FWD	4WD	
MT	49-state and Canada	05		01
	California	06		02
AT	49-state and Canada	07	03	03
	California	08	04	04

TROUBLE CODE CHART—1990 LOYALE

List of Trouble Codes

Trouble code	Item
11	Crank angle sensor (No reference pulse)
12	Starter switch (Continuously in ON position or continuously in OFF position while cranking)
13	Crank angle sensor (No position pulse)
14	Fuel injectors #1 and #2 (Abnormal injector output)
15	Fuel injectors #3 and #4 (Abnormal injector output)
21	Water temperature sensor (Open or shorted circuit)
22	Knock sensor (Open or shorted circuit)
23	Air flow meter (Open or shorted circuit)
31	Throttle sensor (Open or shorted circuit)
32	O_2 sensor (Abnormal sensor signal)
33	Car-speed sensor (No signal is present during operation)
35	Purge control solenoid valve (Solenoid switch continuously in ON or OFF position)
41	System too lean
42	Idle switch (Abnormal idle switch signal in relation to throttle sensor output)
44	Duty solenoid valve (Waste gate control)

List of Specification Codes

Specification codes	Specification
01	MT, 49-state and Canada
02	MT, California
03	AT, 49-state and Canada
04	AT, California

HOW TO READ TROUBLE CODES

How to Read Trouble Codes (Flashing)

The O_2 monitor lamp flashes the code corresponding to the faulty part.
The long segment (1.2 sec on) indicates a "ten", and the short segment (0.2 sec on) signifies a "one".

Example:

When only one part has failed:
Flashing code 12
(unit: second)

0.2 0.2
1.2 1.8
0.3 0.3

When two or more parts have failed:
Flashing codes 12 and 21
(unit: second)

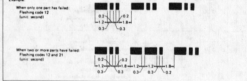

0.2 0.2
1.2 1.8 1.2 1.2 1.8
0.3 0.3 0.3 0.3 0.2

TROUBLESHOOTING CHART FOR SELF-DIAGNOSIS SYSTEM—ALL MODELS EXCEPT 1990 LOYALE

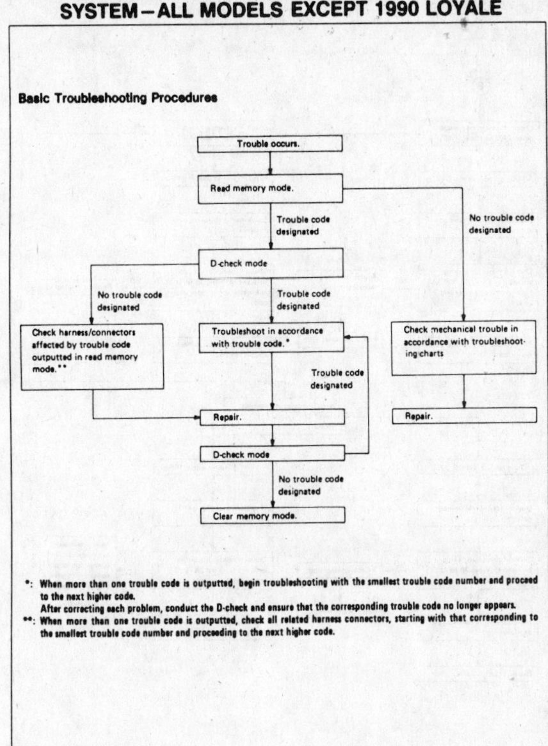
Basic Troubleshooting Procedures
Trouble occurs.
Read memory mode.
Trouble code designated
No trouble code designated
D-check mode
No trouble code designated
Trouble code designated
Check harness/connectors affected by trouble code outputted in read memory mode.**
Troubleshoot in accordance with trouble code.*
Check mechanical trouble in accordance with troubleshooting charts
Trouble code designated
Repair.
Repair.
D-check mode
No trouble code designated
Clear memory mode.
*: When more than one trouble code is outputted, begin troubleshooting with the smallest trouble code number and proceed to the next higher code.
After correcting each problem, conduct the D-check and ensure that the corresponding trouble code no longer appears.
**: When more than one trouble code is outputted, check all related harness connectors, starting with that corresponding to the smallest trouble code number and proceeding to the next higher code.

TROUBLESHOOTING CHART FOR SELF-DIAGNOSIS SYSTEM—1990 LOYALE

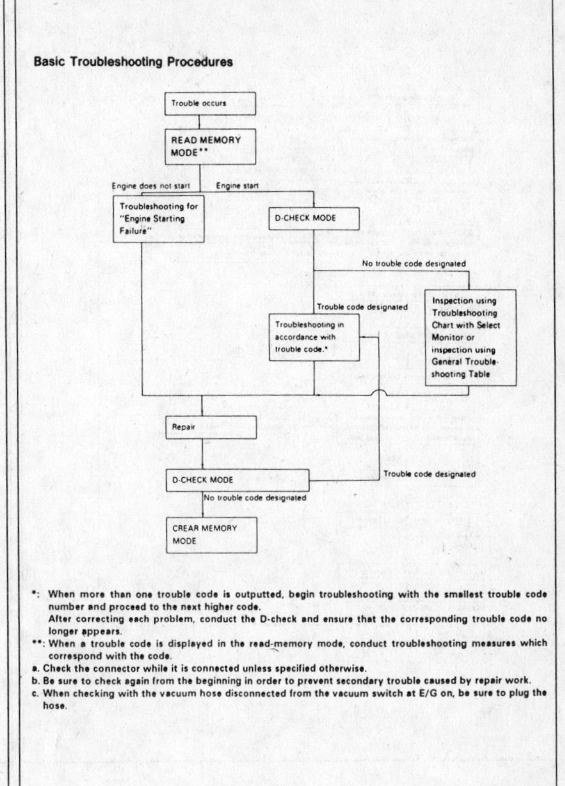
Basic Troubleshooting Procedures
Trouble occurs
READ MEMORY MODE**
Engine does not start
Engine start
Troubleshooting for "Engine Starting Failure"
D-CHECK MODE
No trouble code designated
Trouble code designated
Troubleshooting in accordance with trouble code.*
Inspection using Troubleshooting Chart with Select Monitor or inspection using General Trouble-shooting Table
Repair
D-CHECK MODE
Trouble code designated
No trouble code designated
CREAR MEMORY MODE
*: When more than one trouble code is outputted, begin troubleshooting with the smallest trouble code number and proceed to the next higher code.
After correcting each problem, conduct the D-check and ensure that the corresponding trouble code no longer appears.
**: When a trouble code is displayed in the read-memory mode, conduct troubleshooting measures which correspond with the code.
a. Check the connector while it is connected unless specified otherwise.
b. Be sure to check again from the beginning in order to prevent secondary trouble caused by repair work.
c. When checking with the vacuum hose disconnected from the vacuum switch at E/G on, be sure to plug the hose.

READ MEMORY MODE DIAGNOSIS CHART—SEDAN AND WAGON AND 1990 LOYALE

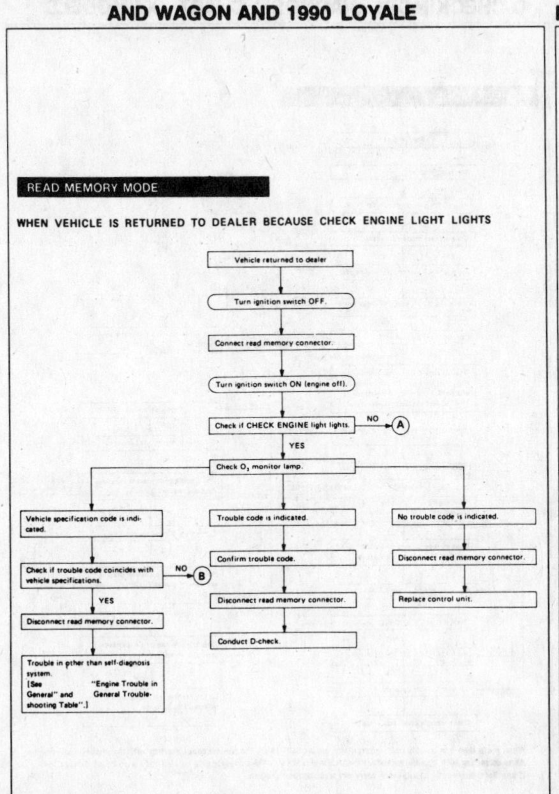
READ MEMORY MODE
WHEN VEHICLE IS RETURNED TO DEALER BECAUSE CHECK ENGINE LIGHT LIGHTS
Vehicle returned to dealer
Turn ignition switch OFF.
Connect read memory connector.
Turn ignition switch ON (engine off).
Check if CHECK ENGINE light lights.
NO
A
YES
Check O2 monitor lamp.
Vehicle specification code is indicated.
Trouble code is indicated.
No trouble code is indicated.
Confirm trouble code.
Disconnect read memory connector.
Check if trouble code coincides with vehicle specifications.
NO
B
Disconnect read memory connector.
Replace control unit.
YES
Disconnect read memory connector.
Conduct D-check.
Trouble in other than self-diagnosis system.
[See "Engine Trouble in General" and General Trouble-shooting Table".]

READ MEMORY MODE DIAGNOSIS CHART—XT MODELS

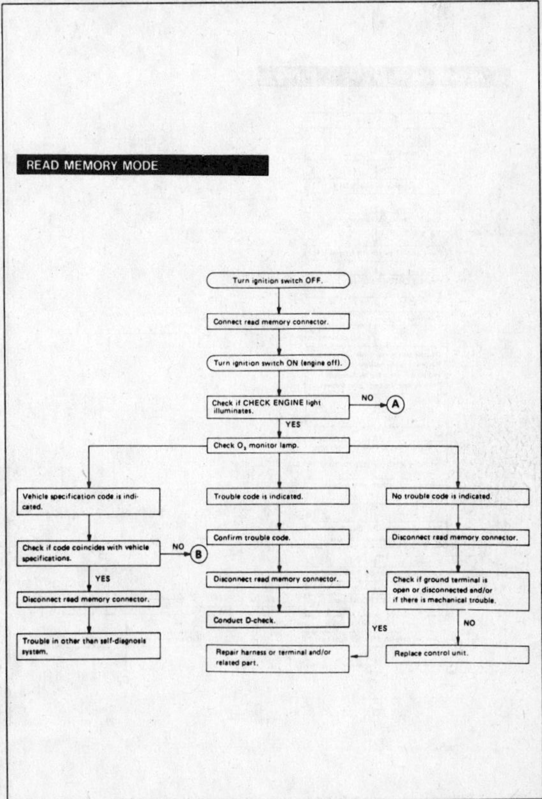
READ MEMORY MODE
Turn ignition switch OFF.
Connect read memory connector.
Turn ignition switch ON (engine off).
Check if CHECK ENGINE light illuminates.
NO
A
YES
Check O2 monitor lamp.
Vehicle specification code is indicated.
Trouble code is indicated.
No trouble code is indicated.
Confirm trouble code.
Disconnect read memory connector.
Check if code coincides with vehicle specifications.
NO
B
Disconnect read memory connector.
Check if ground terminal is open or disconnected and/or if there is mechanical trouble.
YES
Disconnect read memory connector.
Conduct D-check.
YES
NO
Trouble in other than self-diagnosis system.
Repair harness or terminal and/or related part.
Replace control unit.

READ MEMORY MODE DIAGNOSIS CHART—ALL MODELS

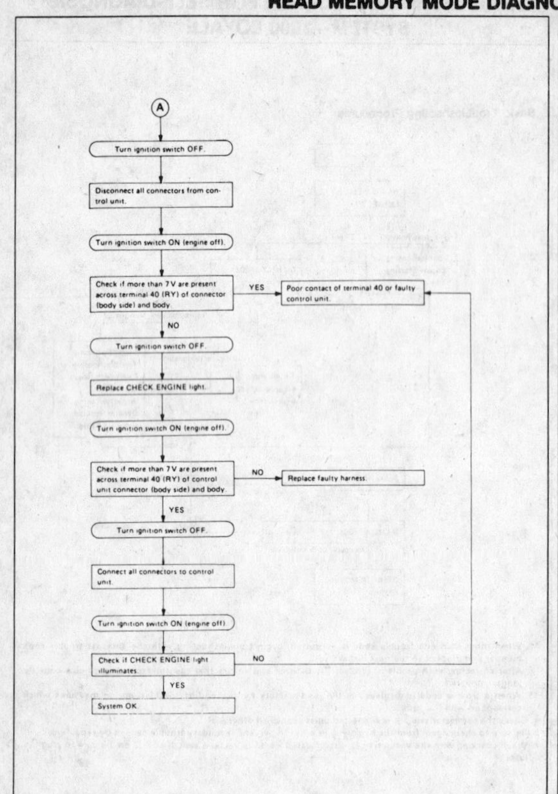
A
Turn ignition switch OFF.
Disconnect all connectors from control unit.
Turn ignition switch ON (engine off).
Check if more than 7V are present across terminal 40 (RY) of connector (body side) and body.
YES
Poor contact of terminal 40 or faulty control unit.
NO
Turn ignition switch OFF.
Replace CHECK ENGINE light.
Turn ignition switch ON (engine off).
Check if more than 7V are present across terminal 40 (RY) of control unit connector (body side) and body.
NO
Replace faulty harness.
YES
Turn ignition switch OFF.
Connect all connectors to control unit.
Turn ignition switch ON (engine off).
Check if CHECK ENGINE light illuminates.
NO
YES
System OK.

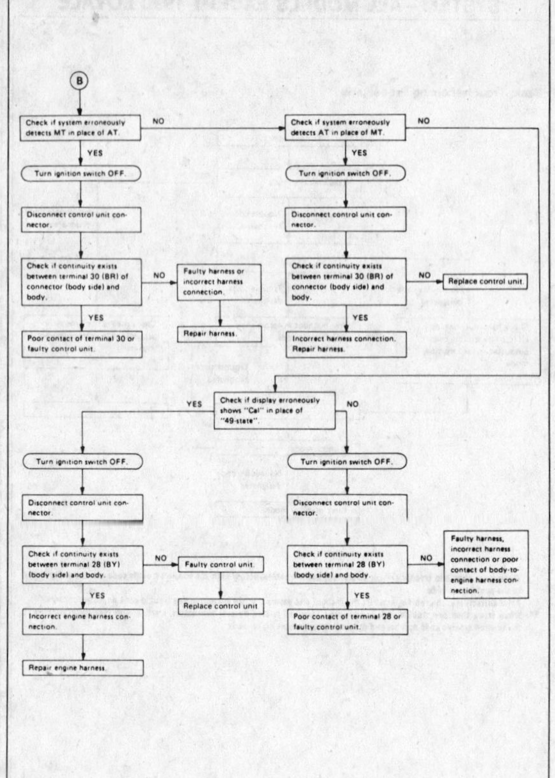
B
Check if system erroneously detects MT in place of AT.
NO
YES
Turn ignition switch OFF.
Disconnect control unit connector.
Check if continuity exists between terminal 30 (BR) of connector (body side) and body.
NO
Faulty harness or incorrect harness connection.
Repair harness.
YES
Poor contact of terminal 30 or faulty control unit.
Check if system erroneously detects AT in place of MT.
NO
YES
Turn ignition switch OFF.
Disconnect control unit connector.
Check if continuity exists between terminal 30 (BR) of connector (body side) and body.
NO
Replace control unit.
YES
Incorrect harness connection. Repair harness.
Check if display erroneously shows "Cal" in place of "49-state".
YES
NO
Turn ignition switch OFF.
Disconnect control unit connector.
Check if continuity exists between terminal 28 (BY) (body side) and body.
NO
Faulty control unit.
Replace control unit
YES
Incorrect engine harness connection.
Repair engine harness.
Turn ignition switch OFF.
Disconnect control unit connector.
Check if continuity exists between terminal 28 (BY) (body side) and body.
NO
Faulty harness, incorrect harness connection or poor contact of body-to-engine harness connection.
YES
Poor contact of terminal 28 or faulty control unit.

D-CHECK MODE DIAGNOSIS CHART—SEDAN AND WAGON AND 1990 LOYALE

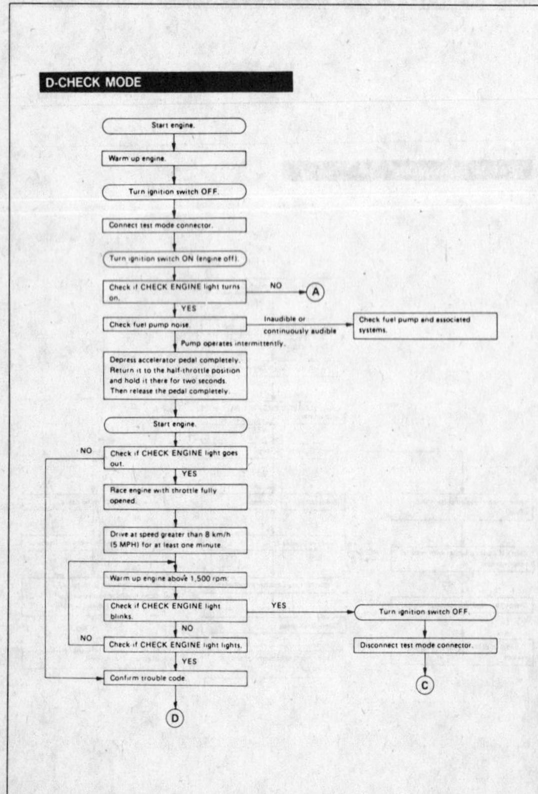
D-CHECK MODE
Start engine.
Warm up engine.
Turn ignition switch OFF.
Connect test mode connector.
Turn ignition switch ON (engine off).
Check if CHECK ENGINE light turns on.
NO
A
YES
Check fuel pump noise.
Inaudible or continuously audible
Check fuel pump and associated systems.
Pump operates intermittently.
Depress accelerator pedal completely. Return it to the half-throttle position and hold it there for two seconds. Then release the pedal completely.
Start engine.
Check if CHECK ENGINE light goes out.
NO
YES
Race engine with throttle fully opened.
Drive at speed greater than 8 km/h (5 MPH) for at least one minute.
Warm up engine above 1,500 rpm.
Check if CHECK ENGINE light blinks.
YES
Turn ignition switch OFF.
Disconnect test mode connector.
C
NO
Check if CHECK ENGINE light lights.
NO
YES
Confirm trouble code.
D

D-CHECK MODE DIAGNOSIS CHART—XT MODELS

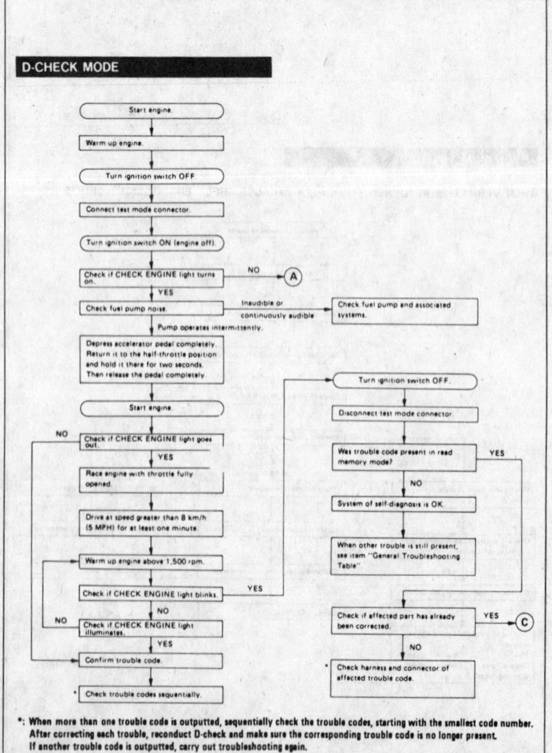
D-CHECK MODE
Start engine.
Warm up engine.
Turn ignition switch OFF.
Connect test mode connector.
Turn ignition switch ON (engine off).
Check if CHECK ENGINE light turns on.
NO
A
YES
Check fuel pump noise.
Inaudible or continuously audible
Check fuel pump and associated systems.
Pump operates intermittently.
Depress accelerator pedal completely. Return it to the half-throttle position and hold it there for two seconds. Then release the pedal completely.
Start engine.
Check if CHECK ENGINE light goes out.
NO
YES
Race engine with throttle fully opened.
Drive at speed greater than 8 km/h (5 MPH) for at least one minute.
Warm up engine above 1,500 rpm.
Check if CHECK ENGINE light blinks.
YES
NO
Check if CHECK ENGINE light illuminates.
NO
YES
Confirm trouble code.
* Check trouble codes sequentially.
Turn ignition switch OFF.
Disconnect test mode connector.
Was trouble code present in read memory mode?
YES
NO
System of self-diagnosis is OK.
When other trouble is still present, see item "General Troubleshooting Table".
Check if affected part has already been corrected.
YES
C
NO
* Check harness and connector of affected trouble code.

***: When more than one trouble code is outputted, sequentially check the trouble codes, starting with the smallest code number. After correcting each trouble, reconduct D-check and make sure the corresponding trouble code is no longer present. If another trouble code is outputted, carry out troubleshooting again.**

D-CHECK MODE DIAGNOSIS CHART—SEDAN AND WAGON AND 1990 LOYALE

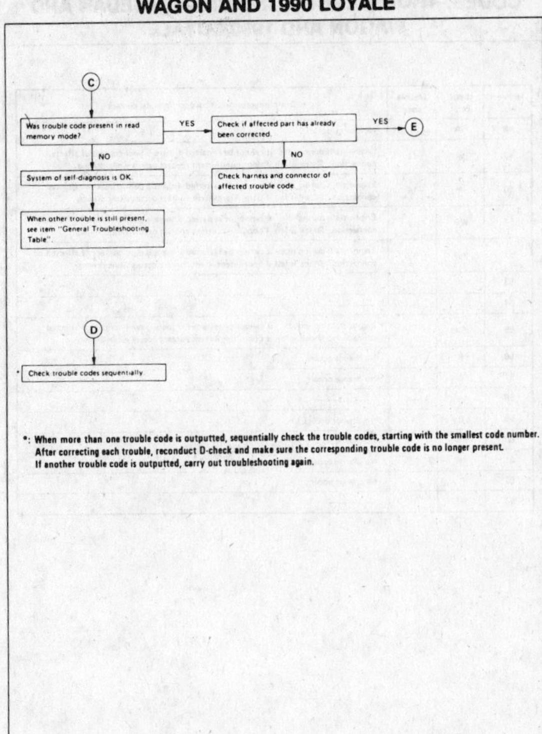
C
Was trouble code present in read memory mode?
YES
Check if affected part has already been corrected.
YES
E
NO
NO
System of self-diagnosis is OK.
Check harness and connector of affected trouble code
When other trouble is still present, see item "General Troubleshooting Table".
D
* Check trouble codes sequentially

*: When more than one trouble code is outputted, sequentially check the trouble codes, starting with the smallest code number. After correcting each trouble, reconduct D-check and make sure the corresponding trouble code is no longer present. If another trouble code is outputted, carry out troubleshooting again.

CLEAR MEMORY MODE DIAGNOSIS CHART—SEDAN AND WAGON AND 1990 LOYALE

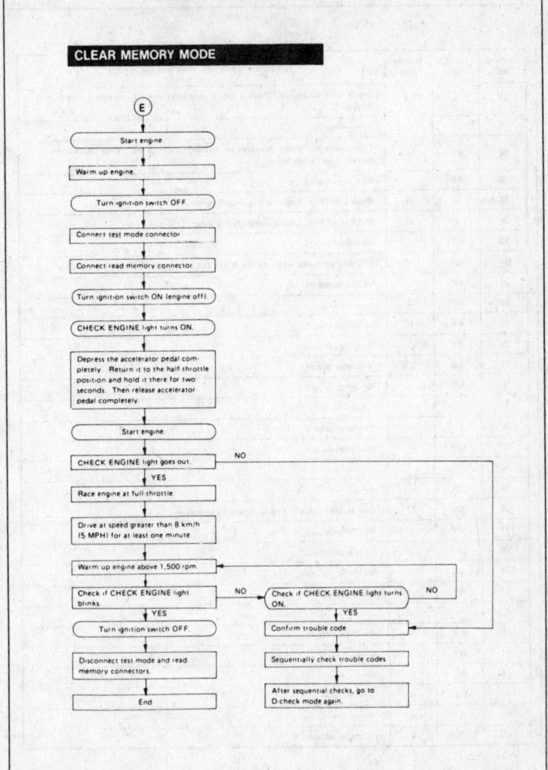
CLEAR MEMORY MODE
E
Start engine.
Warm up engine.
Turn ignition switch OFF.
Connect test mode connector
Connect read memory connector
Turn ignition switch ON (engine off).
CHECK ENGINE light turns ON.
Depress the accelerator pedal completely. Return it to the half-throttle position and hold it there for two seconds. Then release accelerator pedal completely.
Start engine.
CHECK ENGINE light goes out.
NO
YES
Race engine at full-throttle.
Drive at speed greater than 8 km/h (5 MPH) for at least one minute.
Warm up engine above 1,500 rpm.
Check if CHECK ENGINE light blinks.
NO
Check if CHECK ENGINE light turns ON.
NO
YES
YES
Turn ignition switch OFF.
Confirm trouble code.
Disconnect test mode and read memory connectors.
Sequentially check trouble codes.
End
After sequential checks, go to D-check mode again.

CLEAR MEMORY MODE DIAGNOSIS CHART XT MODELS

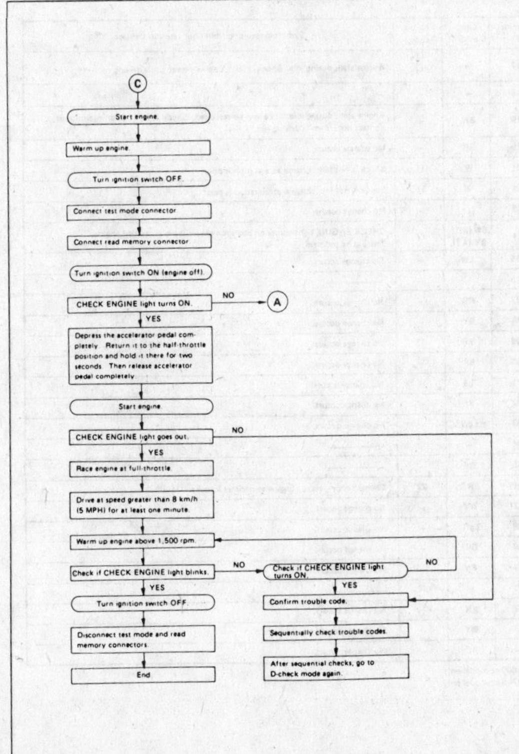
C
Start engine.
Warm up engine.
Turn ignition switch OFF.
Connect test mode connector
Connect read memory connector
Turn ignition switch ON (engine off).
CHECK ENGINE light turns ON.
NO
A
YES
Depress the accelerator pedal completely. Return it to the half-throttle position and hold it there for two seconds. Then release accelerator pedal completely.
Start engine.
CHECK ENGINE light goes out.
NO
YES
Race engine at full-throttle.
Drive at speed greater than 8 km/h (5 MPH) for at least one minute.
Warm up engine above 1,500 rpm.
Check if CHECK ENGINE light blinks.
NO
Check if CHECK ENGINE light turns ON.
NO
YES
YES
Turn ignition switch OFF.
Confirm trouble code.
Disconnect test mode and read memory connectors.
Sequentially check trouble codes.
End
After sequential checks, go to D-check mode again.

HARNESS AND CONNECTORS, RELATED TO TROUBLE CODES, TROUBLESHOOTING CHART—SEDAN AND WAGON AND 1990 LOYALE

Checking Harnesses and Connectors Related to Trouble Codes

When a trouble code is outputted in the read memory mode but not in the D-check mode, check the affected harness and connector terminal as described below.

CHECKING TERMINALS OF CONTROL UNIT CONNECTOR (BODY SIDE)

1) When terminals are not locked securely, insert into connectors until they lock.

2) When terminals are considered to be open:

(1) Method of determining "OK" or "Faulty":

a. Pull out the terminal from the connector (body side).

b. Insert this terminal (female) into the terminal (male) of the connector (control unit).

c. Check "pull" force required to disconnect the female terminal from the male terminal.

If the terminal is loose, it is "faulty".

(2) When terminals are faulty:

Pinch the terminal using a pair of nose pliers. If the terminal is still loose, replace it or the harness ASSY.

SYMPTOMS RESULTING FROM POOR CONTACT OF CONTROL UNIT CONNECTOR TERMINALS AND RELATED TROUBLE CODES

Terminal No.	Lead color	Trouble code	Symptom resulting from poor terminal contact
1	YR	–	Shock is not felt.
2	SA	23	Engine stalls during idle. It can be restarted. Problems do not occur while driving.
3	GL	35	Shock is not felt.
4	YL	–	Shock is not felt.
5	B	–	Engine runs at low speed and sometimes stalls during idle. It can be started but idle speed is low.
6	LB	–	Engine stalls during idle. It cannot be restarted if affected circuit remains opened. Engine lacks power while driving.
7	R	11	Engine stalls during idle. Shock is felt. Engine lacks power (poor acceleration) while driving.
8	W	11	Engine stalls during idle. Shock is felt. Engine speed decreases while driving.
9	WB	21	Slight shock is felt while engine is cold but is not felt after warm-up. Driving performance is affected while engine is cold but is not affected after warm-up.
10	BR	–	Shock is not felt.
11	–	–	————
12	SA	23	Engine stalls during idle. It can be restarted. Problems do not occur while driving.
13	RB	–	No change occurs.
14	YW	–	No change occurs
15	GY	–	No change occurs
16	BR	–	No change occurs.

HARNESS AND CONNECTORS, RELATED TO TROUBLE CODES, TROUBLESHOOTING CHART—SEDAN AND WAGON AND 1990 LOYALE

Terminal No.	Lead color	Trouble code	Symptom resulting from poor terminal contact
17	W	13	Engine stalls during idle. Shock is felt. Engine lacks power (poor acceleration) while driving.
18	–	–	———
19	BW	–	Engine stalls during idle. It cannot be restarted. Shock is felt if poor intermittent contact occurs while driving.
20	BR	–	No change occurs.
21	W	31	Shock is not felt. Engine acceleration is poor.
22	B	31	Shock is not felt. Engine acceleration is poor.
23	LY	–	No change occurs.
24	BW	12	ECS lamp comes on when starting engine and soon goes off. There is no problem.
25	LW	–	No change occurs.
26	–	–	———
27	WR	–	No change occurs.
28	BY	–	No change occurs.
29	YG	33	No change occurs.
30	BR	–	No change occurs.
31	–	–	———
32	Br	–	No change occurs.
33	LgW	–	No change occurs.
34	–	–	———
35	WR	–	———
36	SA	22	Change rarely occurs. Engine sometimes knocks.
37	WY	–	No change occurs.
38	LgR	–	No change occurs.
39	GW	–	No change occurs.
40	RY	–	No change occurs.
41	–	–	———
42	BrB	–	No change occurs.
43	WY	–	———
44	RL	–	———
45	GB	–	———
46	L	–	———
47	R	–	No change occurs.

HARNESS AND CONNECTORS, RELATED TO TROUBLE CODES, TROUBLESHOOTING CHART—SEDAN AND WAGON AND 1990 LOYALE

Terminal No.	Lead color	Trouble code	Symptom resulting from poor terminal contact
48	SA	–	No change occurs.
49	W	14	Engine stalls during idle. It cannot be restarted if there is poor contact of affected connection. Shock is felt if poor intermittent contact occurs while driving.
50	W	14	Engine stalls during idle. It cannot be restarted if there is poor contact of affected connection. Shock is felt if poor intermittent contact occurs while driving.
51	WR	15	Engine stalls during idle. It cannot be restarted if there is poor contact of affected connection. Shock is felt if poor intermittent contact occurs while driving.
52	WR	15	Engine stalls during idle. It cannot be restarted if there is poor contact of affected connection. Shock is felt if poor intermittent contact occurs while driving.
53	–	–	———
54	–	–	———
55	SA	–	Engine stalls during idle. It cannot be restarted if there is poor contact of affected connection. Shock is felt if poor intermittent contact occurs while driving.
56	B	–	No change occurs.
57	B	–	No change occurs.
58	–	–	———
59	L	–	No change occurs.
60	BR	–	No change occurs.
61	BR	–	No change occurs.
62	R	–	No change occurs.
63	–	–	———

HARNESS AND CONNECTORS, RELATED TO TROUBLE CODES, TROUBLESHOOTING CHART—XT MODELS

Checking Harnesses and Connectors Related to Trouble Codes

When a trouble code is outputted in the read memory mode but not in the D-check mode, check the affected harness and connector terminal as described below.

CHECKING TERMINALS OF CONTROL UNIT CONNECTOR (BODY SIDE)

1) When terminals are not locked securely, insert into connectors until they lock.

2) When terminals are considered to be open:

(1) Method of determining "OK" or "Faulty":

a. Pull out the terminal from the connector (body side).

b. Insert this terminal (female) into the terminal (male) of the connector (control unit).

c. Check "pull" force required to disconnect the female terminal from the male terminal.

If the terminal is loose, it is "faulty".

(2) When terminals are faulty:

Pinch the terminal using a pair of nose pliers. If the terminal is still loose, replace it or the harness ASSY.

SYMPTOMS RESULTING FROM POOR CONTACT OF CONTROL UNIT CONNECTOR TERMINALS AND RELATED TROUBLE CODES

Terminal No.	Lead color	Trouble code	Symptom resulting from poor terminal contact
1	YR	–	No change occurs.
2	W	23	Engine stalls during idle. It can be restarted. Problems do not occur while driving.
3	GL	35	No change occurs.
4	YL or GY	–	No change occurs.
5	B	–	Engine runs at low speed and sometimes stalls during idle. It can be started but idle speed is low.
6	LB	–	Engine stalls during idle. It cannot be restarted if affected circuit remains opened. Engine lacks power while driving.
7	R	11	Engine stalls during idle. Shock is felt. Engine lacks power (poor acceleration) while driving.
8	*B **G	*11 **13	Engine stalls during idle. Shock is felt. Engine lacks power while driving.
9	WB	21	Slight shock is felt while engine is cold but is not felt after warm-up. Driving performance is affected while engine is cold but is not affected after warm-up.
10	BR	–	No change occurs.
11	–	–	———
12	*W **R	23	Engine stalls during idle. It can be restarted. Problems do not occur while driving.
13	–	–	–
14	YW	–	No change occurs.
15	GY	–	No change occurs.
16	BR	–	No change occurs.

*: 1800 cc model only
**: 2700 cc model only

HARNESS AND CONNECTORS, RELATED TO TROUBLE CODES, TROUBLESHOOTING CHART—XT MODELS

Terminal No.	Lead color	Trouble code	Symptom resulting from poor terminal contact
17	W	*13 **11	Engine stalls during idle. Shock is felt. Engine power while driving.
18	–	–	–
19	BW	–	Engine stalls during idle. It cannot be restarted. Shock is felt if poor intermittent contact occurs while driving.
20	BR	–	No change occurs.
21	W	31	Shock is not felt. Engine acceleration is poor.
22	B	31	Shock is not felt. Engine acceleration is poor.
23	L	–	No change occurs.
24	BW (MT) BY (AT)	12	CHECK ENGINE light comes on and soon goes out when starting engine. There is no problem.
25	LW	–	No change occurs.
26	–	–	———
27	WR	–	No change occurs.
*28	BY	–	No change occurs.
29	YR	33	No change occurs.
30	BR	–	No change occurs.
**31	LB	–	No change occurs.
32	Br	–	No change occurs.
33	LgW	–	No change occurs.
34	–	–	———
35	WR	–	———
**36	R	22	Change rarely occurs. Engine sometimes knocks.
37	WY	–	No change occurs.
38	LgR	–	No change occurs.
39	GW	–	No change occurs.
40	RY	–	No change occurs.
41	–	–	———
42	RB	–	No change occurs.
43	WY	–	No change occurs.
44	RL	–	No change occurs.

*: 1800 cc model only
**: 2700 cc model only

HARNESS AND CONNECTORS, RELATED TO TROUBLE CODES, TROUBLESHOOTING CHART—XT MODELS

Terminal No.	Lead color	Trouble code	Symptom resulting from poor terminal contact
45	GB	–	No change occurs.
46	L	–	No change occurs.
47	R	–	No change occurs.
48	W	–	No change occurs.
49	W	*14 **25	Engine stalls during idle. It cannot be restarted if there is poor contact of affected connection. Shock is felt if poor intermittent contact occurs while driving.
50	W	*14 **25	Engine stalls during idle. It cannot be restarted if there is poor contact of affected connection. Shock is felt if poor intermittent contact occurs while driving.
51	*WL **WR	*15 **14	Engine stalls during idle. It cannot be restarted if there is poor contact of affected connection. Shock is felt if poor intermittent contact occurs while driving.
52	*WL **WR	*15 **14	Engine stalls during idle. It cannot be restarted if there is poor contact of affected connection. Shock is felt if poor intermittent contact occurs while driving.
**53	WY	15	Engine stalls during idle. It cannot be restarted if there is poor contact of affected connection. Shock is felt if poor intermittent contact occurs while driving.
**54	WY	15	Engine stalls during idle. If cannot be restarted if there is poor contact of affected connection. Shock is felt if poor intermittent contact occurs while driving.
55	W	–	Engine stalls during idle. It cannot be restarted if there is poor contact of affected connection. Shock is felt if poor intermittent contact occurs while driving.
56	B	–	No change occurs.
57	B	–	No change occurs.
**58	YR	24	No change occurs.
**59	YL	24	Engine stalls occasionally.
60	BR	–	No change occurs.
61	BR	–	No change occurs.
62	R	–	No change occurs.
**63	Br	–	Difficulty in engine starting while engine is cold.

*: 1800 cc model only
**: 2700 cc model only

GROUND AND CONTROL UNIT POWER SUPPLY DIAGNOSIS CHART—SEDAN AND WAGON AND 1990 LOYALE

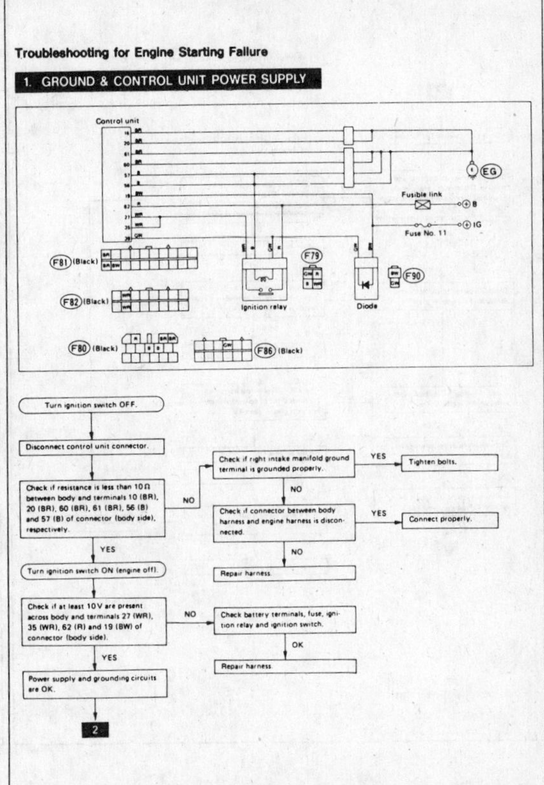
Troubleshooting for Engine Starting Failure
1. GROUND & CONTROL UNIT POWER SUPPLY
Control unit
Fusible link
Fuse No. 11
Ignition relay
Diode
Turn ignition switch OFF.
Disconnect control unit connector.
Check if resistance is less than 10 Ω between body and terminals 10 (BR), 20 (BR), 60 (BR), 61 (BR), 56 (B) and 57 (B) of connector (body side), respectively.
NO
Check if right intake manifold ground terminal is grounded properly.
YES
Tighten bolts.
NO
Check if connector between body harness and engine harness is disconnected.
YES
Connect properly.
NO
Repair harness.
YES
Turn ignition switch ON (engine off).
Check if at least 10 V are present across body and terminals 27 (WR), 35 (WR), 62 (R) and 19 (BW) of connector (body side).
NO
Check battery terminals, fuse, ignition relay and ignition switch.
OK
Repair harness.
YES
Power supply and grounding circuits are OK.
2

GROUND AND CONTROL UNIT POWER SUPPLY DIAGNOSIS CHART—XT MODELS

Troubleshooting for Engine Starting Failure

1. GROUND & CONTROL UNIT POWER SUPPLY

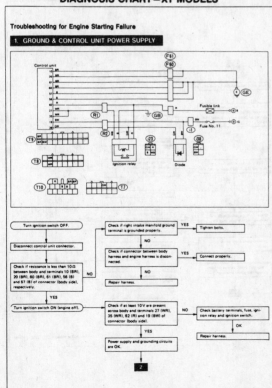

Control unit | Fusible link | Fuse No. 11 | Ignition relay | Diode

Turn ignition switch OFF.

Disconnect control unit connector.

Check if resistance is less than 10 Ω between body and terminals 10 (BR), 20 (BR), 60 (BR), 61 (BR), 56 (B) and 57 (B) of connector (body side), respectively.

- NO → Check if right intake manifold ground terminal is grounded properly.
  - YES → Tighten bolts.
  - NO → Check if connector between body harness and engine harness is disconnected.
    - YES → Connect properly.
    - NO → Repair harness.
- YES → Turn ignition switch ON (engine off).

Check if at least 10 V are present across body and terminals 27 (WR), 35 (WR), 62 (R) and 19 (BW) of connector (body side).

- NO → Check battery terminals, fuse, ignition relay and ignition switch.
  - OK → Repair harness.
- YES → Power supply and grounding circuits are OK. → 2

IGNITION CONTROL SYSTEM DIAGNOSIS CHART—SEDAN AND WAGON AND 1990 LOYALE

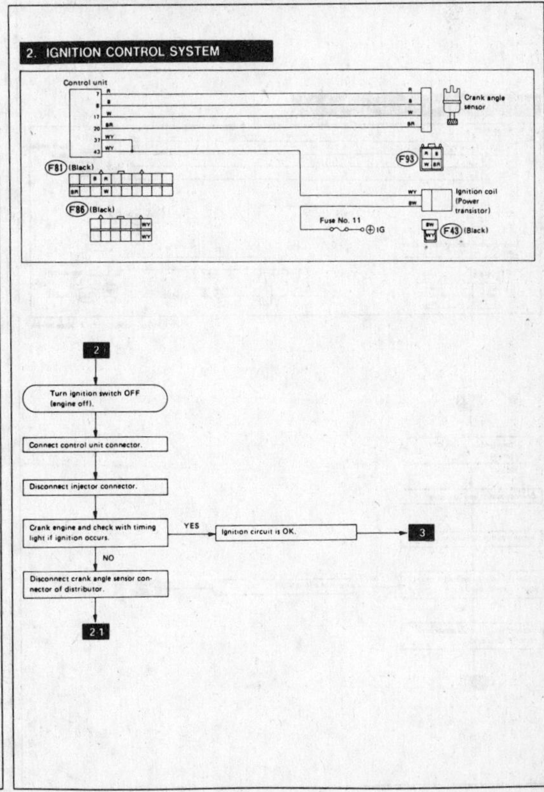
2. IGNITION CONTROL SYSTEM
Control unit
Crank angle sensor
Ignition coil (Power transistor)
Fuse No. 11
2
Turn ignition switch OFF (engine off).
Connect control unit connector.
Disconnect injector connector.
Crank engine and check with timing light if ignition occurs.
YES
Ignition circuit is OK.
3
NO
Disconnect crank angle sensor connector of distributor.
2-1

IGNITION CONTROL SYSTEM DIAGNOSIS CHART—SEDAN AND WAGON AND 1990 LOYALE

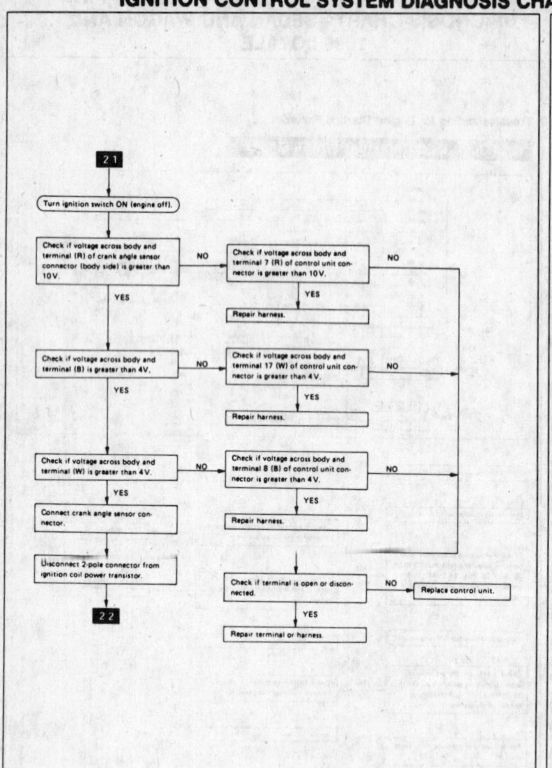
21
Turn ignition switch ON (engine off).
Check if voltage across body and terminal (R) of crank angle sensor connector (body side) is greater than 10V.
NO
Check if voltage across body and terminal 7 (R) of control unit connector is greater than 10V.
NO
YES
Repair harness.
YES
Check if voltage across body and terminal (B) is greater than 4V.
NO
Check if voltage across body and terminal 17 (W) of control unit connector is greater than 4V.
NO
YES
Repair harness.
YES
Check if voltage across body and terminal (W) is greater than 4V.
NO
Check if voltage across body and terminal 8 (B) of control unit connector is greater than 4V.
NO
YES
Repair harness.
YES
Connect crank angle sensor connector.
Disconnect 2-pole connector from ignition coil power transistor.
22
Check if terminal is open or disconnected.
NO
Replace control unit.
YES
Repair terminal or harness.

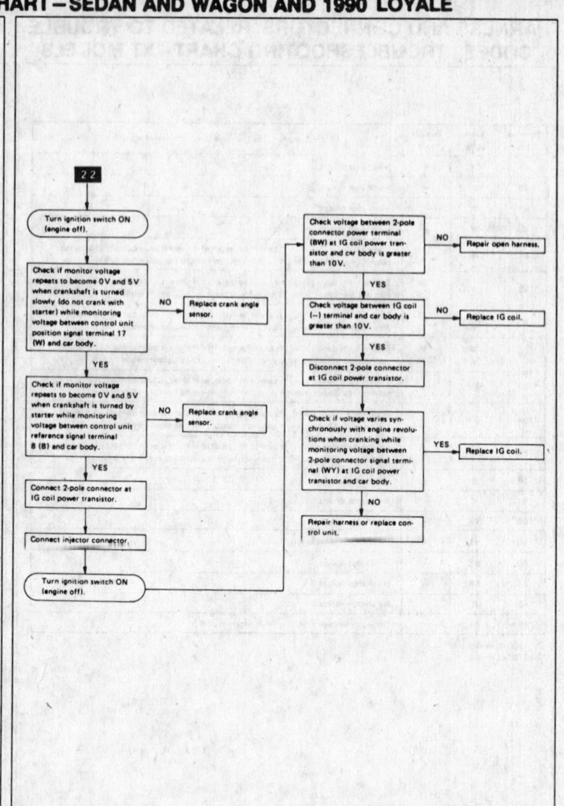
22
Turn ignition switch ON (engine off).
Check if monitor voltage repeats to become 0V and 5V when crankshaft is turned slowly (do not crank with starter) while monitoring voltage between control unit position signal terminal 17 (W) and car body.
NO
Replace crank angle sensor.
YES
Check if monitor voltage repeats to become 0V and 5V when crankshaft is turned by starter while monitoring voltage between control unit reference signal terminal 8 (B) and car body.
NO
Replace crank angle sensor.
YES
Connect 2-pole connector at IG coil power transistor.
Connect injector connector.
Turn ignition switch ON (engine off).
Check voltage between 2-pole connector power terminal (BW) at IG coil power transistor and car body is greater than 10V.
NO
Repair open harness.
YES
Check voltage between IG coil (−) terminal and car body is greater than 10V.
NO
Replace IG coil.
YES
Disconnect 2-pole connector at IG coil power transistor.
Check if voltage varies synchronously with engine revolutions when cranking while monitoring voltage between 2-pole connector signal terminal (WY) at IG coil power transistor and car body.
YES
Replace IG coil.
NO
Repair harness or replace control unit.

IGNITION CONTROL SYSTEM DIAGNOSIS CHART—XT MODELS

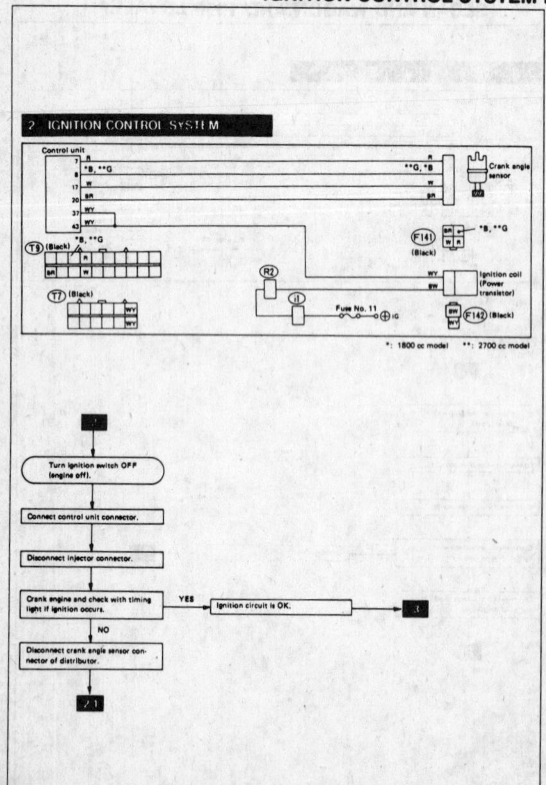
2 IGNITION CONTROL SYSTEM
Control unit
7
8
17
20
37
43
R
*B, **G
W
BR
WY
WY
**G, *B
Crank angle sensor
F141
(Black)
*B, **G
T9 (Black)
T7 (Black)
R2
i1
Fuse No. 11
IG
Ignition coil (Power transistor)
F142 (Black)
*: 1800 cc model
**: 2700 cc model
Turn ignition switch OFF (engine off).
Connect control unit connector.
Disconnect injector connector.
Crank engine and check with timing light if ignition occurs.
YES
Ignition circuit is OK.
3
NO
Disconnect crank angle sensor connector of distributor.
21

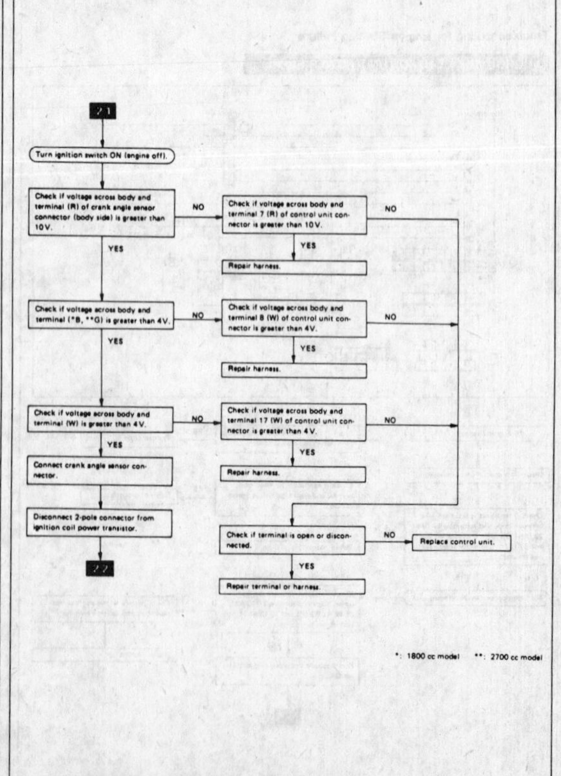
21
Turn ignition switch ON (engine off).
Check if voltage across body and terminal (R) of crank angle sensor connector (body side) is greater than 10V.
NO
Check if voltage across body and terminal 7 (R) of control unit connector is greater than 10V.
NO
YES
Repair harness.
YES
Check if voltage across body and terminal (*B, **G) is greater than 4V.
NO
Check if voltage across body and terminal 8 (W) of control unit connector is greater than 4V.
NO
YES
Repair harness.
YES
Check if voltage across body and terminal (W) is greater than 4V.
NO
Check if voltage across body and terminal 17 (W) of control unit connector is greater than 4V.
NO
YES
Repair harness.
YES
Connect crank angle sensor connector.
Disconnect 2-pole connector from ignition coil power transistor.
22
Check if terminal is open or disconnected.
NO
Replace control unit.
YES
Repair terminal or harness.
*: 1800 cc model
**: 2700 cc model

IGNITION CONTROL SYSTEM DIAGNOSIS CHART XT MODELS

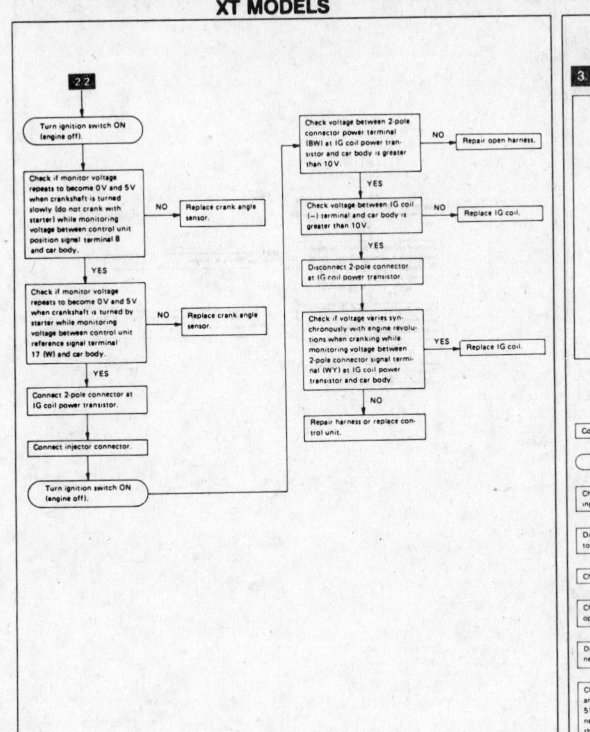

22
Turn ignition switch ON (engine off).
Check if monitor voltage repeats to become 0V and 5V when crankshaft is turned slowly (do not crank with starter) while monitoring voltage between control unit position signal terminal 8 and car body.
NO
Replace crank angle sensor.
YES
Check if monitor voltage repeats to become 0V and 5V when crankshaft is turned by starter while monitoring voltage between control unit reference signal terminal 17 (W) and car body.
NO
Replace crank angle sensor.
YES
Connect 2-pole connector at IG coil power transistor.
Connect injector connector.
Turn ignition switch ON (engine off).
Check voltage between 2-pole connector power terminal (BW) at IG coil power transistor and car body is greater than 10V.
NO
Repair open harness.
YES
Check voltage between IG coil (–) terminal and car body is greater than 10V.
NO
Replace IG coil.
YES
Disconnect 2-pole connector at IG coil power transistor.
Check if voltage varies synchronously with engine revolutions when cranking while monitoring voltage between 2-pole connector signal terminal (WY) at IG coil power transistor and car body.
YES
Replace IG coil.
NO
Repair harness or replace control unit.

FUEL PUMP CIRCUIT DIAGNOSIS CHART—SEDAN AND WAGON AND 1990 LOYALE

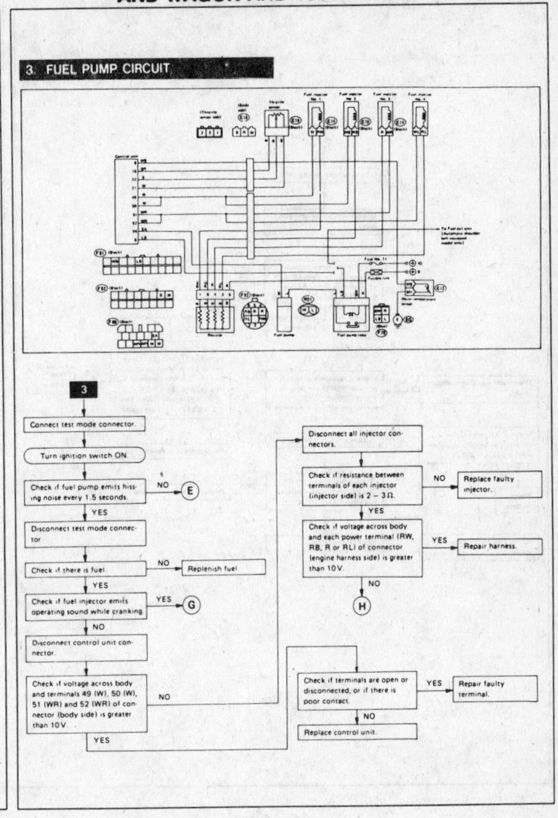

3. FUEL PUMP CIRCUIT
3
Connect test mode connector.
Turn ignition switch ON.
Check if fuel pump emits hissing noise every 1.5 seconds.
NO
E
YES
Disconnect test mode connector.
Check if there is fuel.
NO
Replenish fuel.
YES
Check if fuel injector emits operating sound while cranking.
YES
G
NO
Disconnect control unit connector.
Check if voltage across body and terminals 49 (W), 50 (W), 51 (WR) and 52 (WR) of connector (body side) is greater than 10V.
NO
YES
Disconnect all injector connectors.
Check if resistance between terminals of each injector (injector side) is 2 – 3Ω.
NO
Replace faulty injector.
YES
Check if voltage across body and each power terminal (RW, RB, R or RL) of connector (engine harness side) is greater than 10V.
YES
Repair harness.
NO
H
Check if terminals are open or disconnected, or if there is poor contact.
YES
Repair faulty terminal.
NO
Replace control unit.

FUEL PUMP CIRCUIT DIAGNOSIS CHART—SEDAN AND WAGON AND 1990 LOYALE

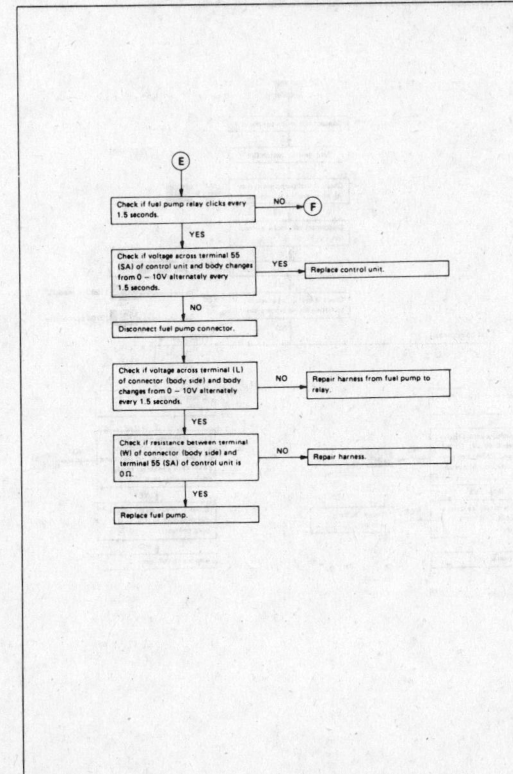

E
Check if fuel pump relay clicks every 1.5 seconds.
NO
F
YES
Check if voltage across terminal 55 (SA) of control unit and body changes from 0 – 10V alternately every 1.5 seconds.
YES
Replace control unit.
NO
Disconnect fuel pump connector.
Check if voltage across terminal (L) of connector (body side) and body changes from 0 – 10V alternately every 1.5 seconds.
NO
Repair harness from fuel pump to relay.
YES
Check if resistance between terminal (W) of connector (body side) and terminal 55 (SA) of control unit is 0Ω.
NO
Repair harness.
YES
Replace fuel pump.

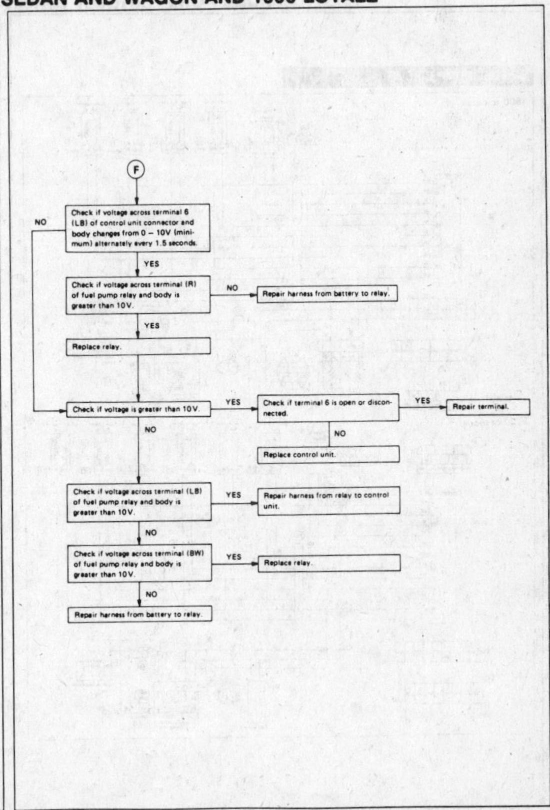

F
Check if voltage across terminal 6 (LB) of control unit connector and body changes from 0 – 10V (minimum) alternately every 1.5 seconds.
NO
YES
Check if voltage across terminal (R) of fuel pump relay and body is greater than 10V.
NO
Repair harness from battery to relay.
YES
Replace relay.
Check if voltage is greater than 10V.
YES
Check if terminal 6 is open or disconnected.
YES
Repair terminal.
NO
Replace control unit.
NO
Check if voltage across terminal (LB) of fuel pump relay and body is greater than 10V.
YES
Repair harness from relay to control unit.
NO
Check if voltage across terminal (BW) of fuel pump relay and body is greater than 10V.
YES
Replace relay.
NO
Repair harness from battery to relay.

FUEL PUMP CIRCUIT DIAGNOSIS CHART—SEDAN AND WAGON AND 1990 LOYALE

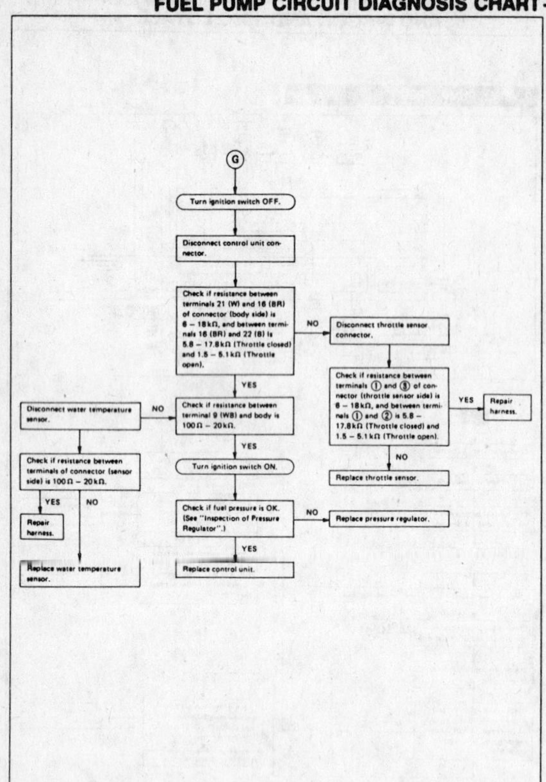
G
Turn ignition switch OFF.
Disconnect control unit connector.
Check if resistance between terminals 21 (W) and 16 (BR) of connector (body side) is 6 – 18kΩ, and between terminals 16 (BR) and 22 (B) is 5.8 – 17.8kΩ (Throttle closed) and 1.5 – 5.1kΩ (Throttle open).
NO
Disconnect throttle sensor connector.
Check if resistance between terminals ① and ③ of connector (throttle sensor side) is 6 – 18kΩ, and between terminals ① and ② is 5.8 – 17.8kΩ (Throttle closed) and 1.5 – 5.1kΩ (Throttle open).
YES
Repair harness.
NO
Replace throttle sensor.
YES
Disconnect water temperature sensor.
NO
Check if resistance between terminal 9 (WB) and body is 100Ω – 20kΩ.
Check if resistance between terminals of connector (sensor side) is 100Ω – 20kΩ.
YES
NO
Repair harness.
Replace water temperature sensor.
YES
Turn ignition switch ON.
Check if fuel pressure is OK. (See "Inspection of Pressure Regulator".)
NO
Replace pressure regulator.
YES
Replace control unit.

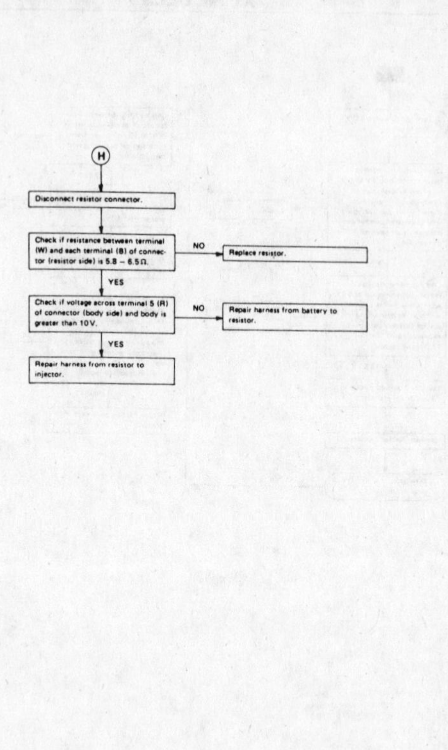
H
Disconnect resistor connector.
Check if resistance between terminal (W) and each terminal (B) of connector (resistor side) is 5.8 – 6.5Ω.
NO
Replace resistor.
YES
Check if voltage across terminal 5 (R) of connector (body side) and body is greater than 10V.
NO
Repair harness from battery to resistor.
YES
Repair harness from resistor to injector.

FUEL PUMP CIRCUIT DIAGNOSIS CHART—XT MODELS

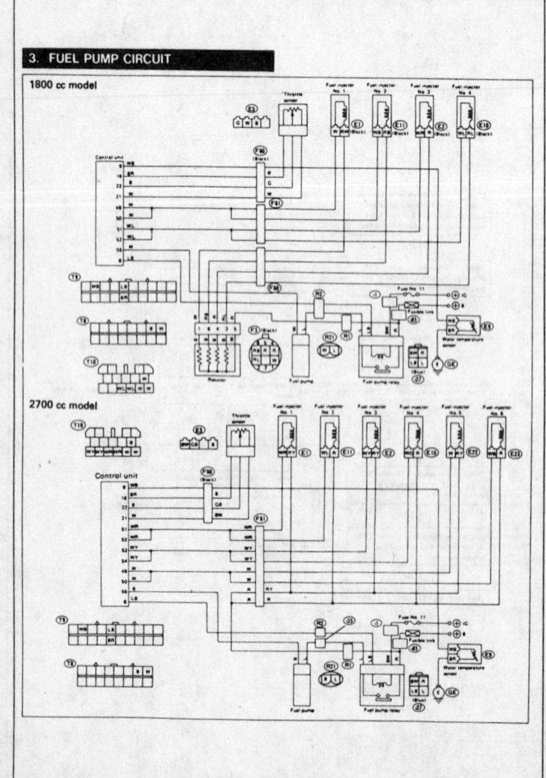
3. FUEL PUMP CIRCUIT
1800 cc model
2700 cc model
Control unit

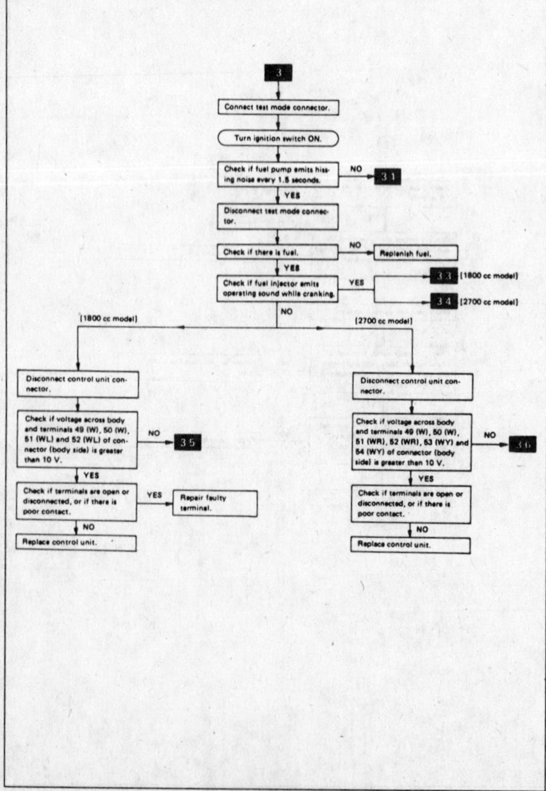
3
Connect test mode connector.
Turn ignition switch ON.
Check if fuel pump emits hissing noise every 1.5 seconds.
NO
31
YES
Disconnect test mode connector.
Check if there is fuel.
NO
Replenish fuel.
YES
Check if fuel injector emits operating sound while cranking.
YES
33 [1800 cc model]
34 [2700 cc model]
NO
[1800 cc model]
[2700 cc model]
Disconnect control unit connector.
Check if voltage across body and terminals 49 (W), 50 (W), 51 (WL) and 52 (WL) of connector (body side) is greater than 10 V.
NO
35
YES
Check if terminals are open or disconnected, or if there is poor contact.
YES
Repair faulty terminal.
NO
Replace control unit.
Disconnect control unit connector.
Check if voltage across body and terminals 49 (W), 50 (W), 51 (WR), 52 (WR), 53 (WY) and 54 (WY) of connector (body side) is greater than 10 V.
NO
36
YES
Check if terminals are open or disconnected, or if there is poor contact.
NO
Replace control unit.

FUEL PUMP CIRCUIT DIAGNOSIS CHART—XT MODELS

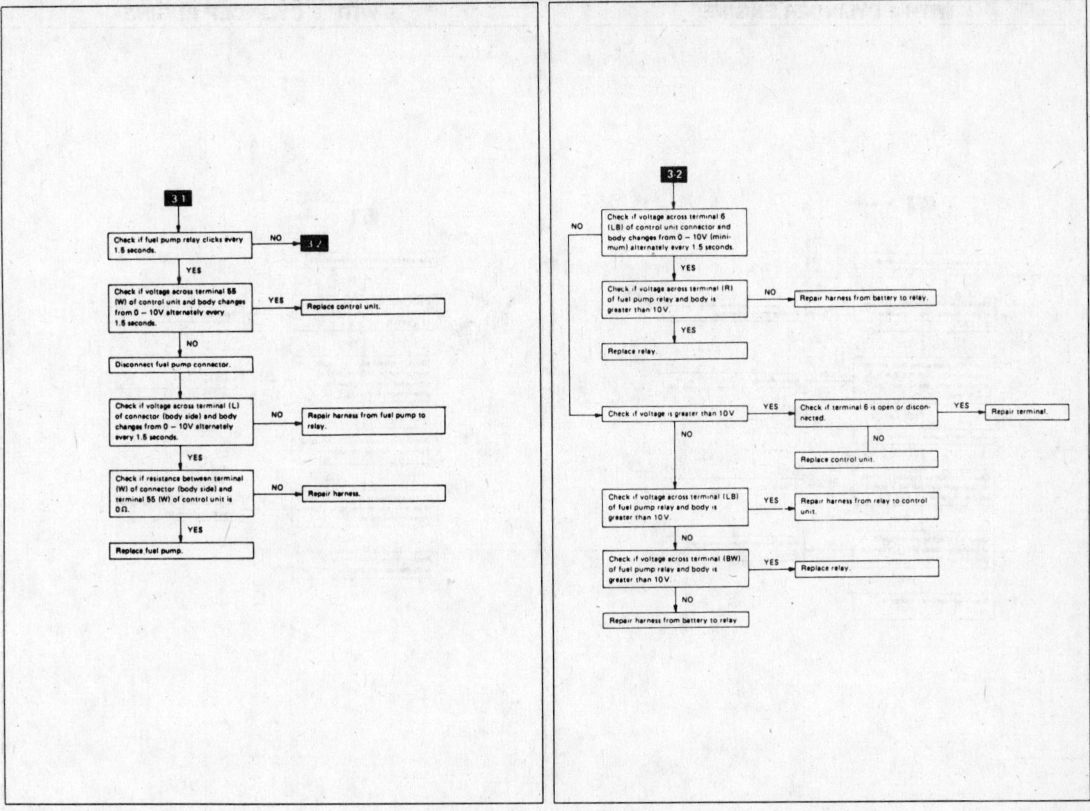
3-1
Check if fuel pump relay clicks every 1.5 seconds.
NO
3-2
YES
Check if voltage across terminal 55 (W) of control unit and body changes from 0 – 10V alternately every 1.5 seconds.
YES
Replace control unit.
NO
Disconnect fuel pump connector.
Check if voltage across terminal (L) of connector (body side) and body changes from 0 – 10V alternately every 1.5 seconds.
NO
Repair harness from fuel pump to relay.
YES
Check if resistance between terminal (W) of connector (body side) and terminal 55 (W) of control unit is 0Ω.
NO
Repair harness.
YES
Replace fuel pump.
3-2
Check if voltage across terminal 6 (LB) of control unit connector and body changes from 0 – 10V (minimum) alternately every 1.5 seconds.
NO
YES
Check if voltage across terminal (R) of fuel pump relay and body is greater than 10V.
NO
Repair harness from battery to relay.
YES
Replace relay.
Check if voltage is greater than 10V
YES
Check if terminal 6 is open or disconnected.
YES
Repair terminal.
NO
Replace control unit.
NO
Check if voltage across terminal (LB) of fuel pump relay and body is greater than 10V.
YES
Repair harness from relay to control unit.
NO
Check if voltage across terminal (BW) of fuel pump relay and body is greater than 10V.
YES
Replace relay.
NO
Repair harness from battery to relay

FUEL PUMP CIRCUIT DIAGNOSIS CHART—XT MODELS WITH 4 CYLINDER ENGINE

FUEL PUMP CIRCUIT DIAGNOSIS CHART—XT MODELS WITH 6 CYLINDER ENGINE

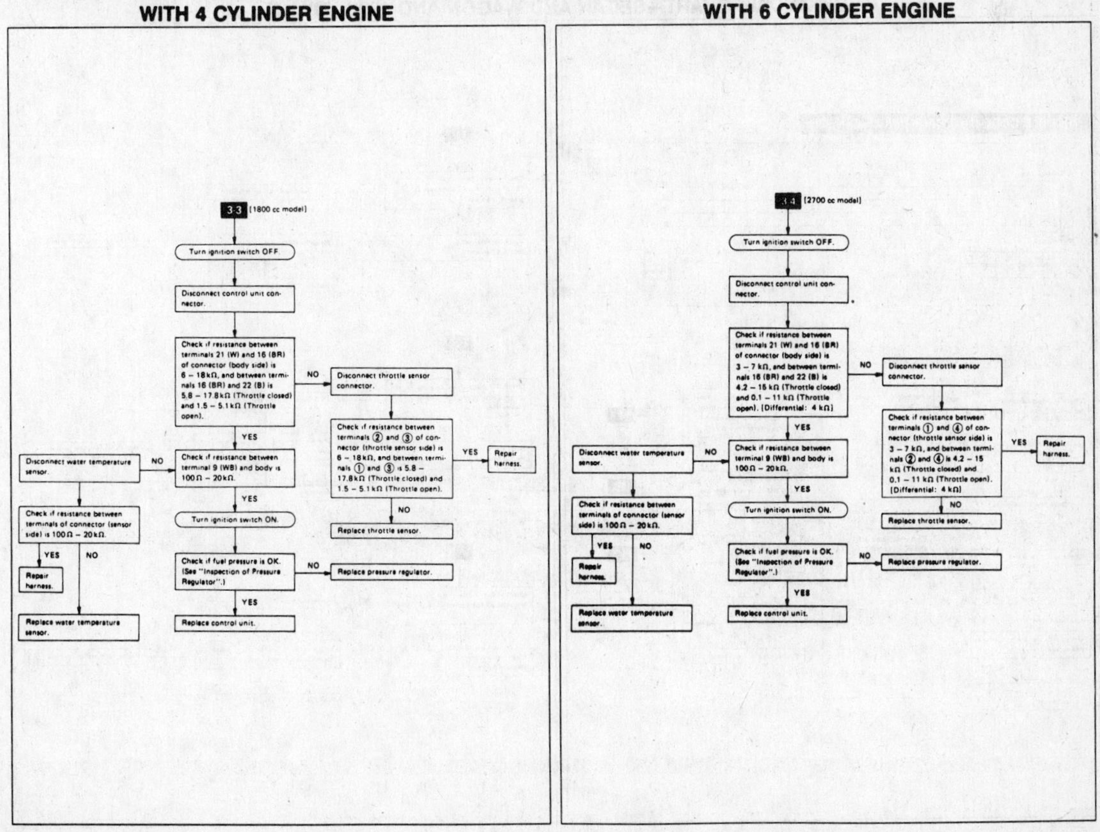
3-3 [1800 cc model]
Turn ignition switch OFF.
Disconnect control unit connector.
Check if resistance between terminals 21 (W) and 16 (BR) of connector (body side) is 6 – 18kΩ, and between terminals 16 (BR) and 22 (B) is 5.8 – 17.8kΩ (Throttle closed) and 1.5 – 5.1kΩ (Throttle open).
NO
Disconnect throttle sensor connector.
Check if resistance between terminals ② and ③ of connector (throttle sensor side) is 6 – 18kΩ, and between terminals ① and ③ is 5.8 – 17.8kΩ (Throttle closed) and 1.5 – 5.1kΩ (Throttle open).
YES
Repair harness.
NO
Replace throttle sensor.
YES
Check if resistance between terminal 9 (WB) and body is 100Ω – 20kΩ.
NO
Disconnect water temperature sensor.
Check if resistance between terminals of connector (sensor side) is 100Ω – 20kΩ.
YES
Repair harness.
NO
Replace water temperature sensor.
YES
Turn ignition switch ON.
Check if fuel pressure is OK. (See "Inspection of Pressure Regulator".)
NO
Replace pressure regulator.
YES
Replace control unit.
3-4 [2700 cc model]
Turn ignition switch OFF.
Disconnect control unit connector.
Check if resistance between terminals 21 (W) and 16 (BR) of connector (body side) is 3 – 7 kΩ, and between terminals 16 (BR) and 22 (B) is 4.2 – 15 kΩ (Throttle closed) and 0.1 – 11 kΩ (Throttle open). [Differential: 4 kΩ]
NO
Disconnect throttle sensor connector.
Check if resistance between terminals ① and ④ of connector (throttle sensor side) is 3 – 7 kΩ, and between terminals ② and ④ is 4.2 – 15 kΩ (Throttle closed) and 0.1 – 11 kΩ (Throttle open). [Differential: 4 kΩ]
YES
Repair harness.
NO
Replace throttle sensor.
YES
Check if resistance between terminal 9 (WB) and body is 100Ω – 20kΩ.
NO
Disconnect water temperature sensor.
Check if resistance between terminals of connector (sensor side) is 100Ω – 20kΩ.
YES
Repair harness.
NO
Replace water temperature sensor.
YES
Turn ignition switch ON.
Check if fuel pressure is OK. (See "Inspection of Pressure Regulator".)
NO
Replace pressure regulator.
YES
Replace control unit.

FUEL PUMP CIRCUIT DIAGNOSIS CHART—XT MODELS WITH 4 CYLINDER ENGINE

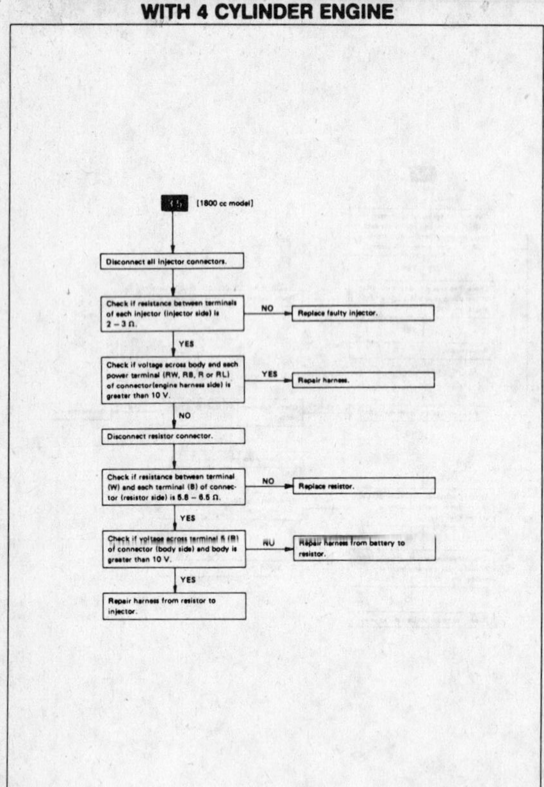
[1800 cc model]
Disconnect all injector connectors.
Check if resistance between terminals of each injector (injector side) is 2 – 3 Ω.
NO
Replace faulty injector.
YES
Check if voltage across body and each power terminal (RW, RB, R or RL) of connector(engine harness side) is greater than 10 V.
YES
Repair harness.
NO
Disconnect resistor connector.
Check if resistance between terminal (W) and each terminal (B) of connector (resistor side) is 5.8 – 6.5 Ω.
NO
Replace resistor.
YES
of connector (body side) and body is greater than 10 V.
Repair harness from battery to resistor.
YES
Repair harness from resistor to injector.

FUEL PUMP CIRCUIT DIAGNOSIS CHART—XT MODELS WITH 6 CYLINDER ENGINE

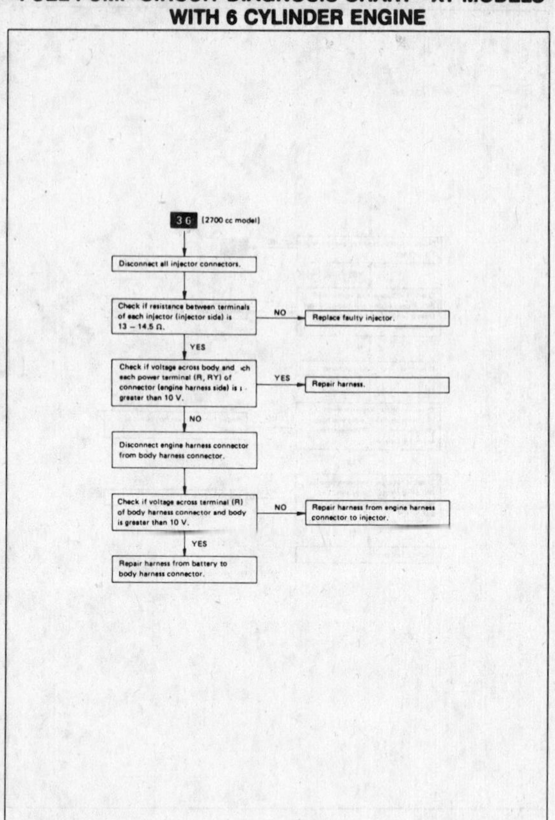
3 6
(2700 cc model)
Disconnect all injector connectors.
Check if resistance between terminals of each injector (injector side) is 13 – 14.5 Ω.
NO
Replace faulty injector.
YES
Check if voltage across body and each power terminal (R, RY) of connector (engine harness side) is greater than 10 V.
YES
Repair harness.
NO
Disconnect engine harness connector from body harness connector.
Check if voltage across terminal (R) of body harness connector and body is greater than 10 V.
NO
Repair harness from engine harness connector to injector.
YES
Repair harness from battery to body harness connector.

DIAGNOSIS CHART—SEDAN AND WAGON AND 1990 LOYALE

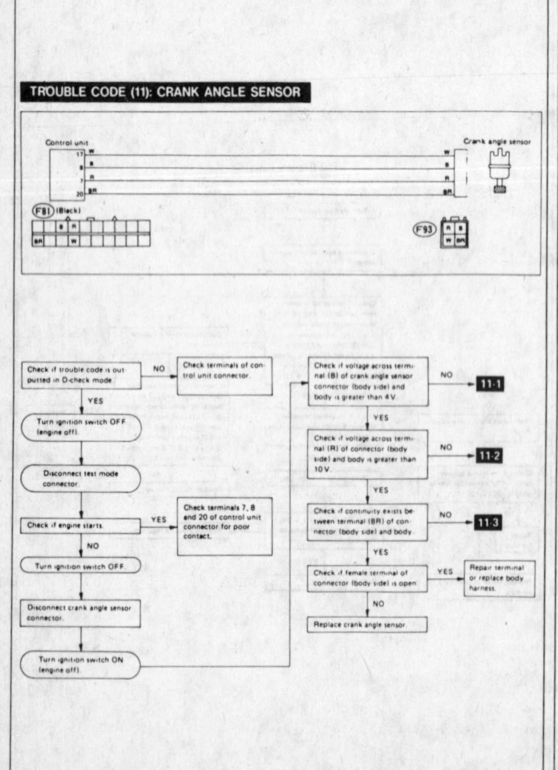
TROUBLE CODE (11): CRANK ANGLE SENSOR
Control unit
Crank angle sensor
17 W
8 B
7 R
20 BR
F81 (Black)
F93
Check if trouble code is outputted in D-check mode.
NO
Check terminals of control unit connector.
YES
Turn ignition switch OFF (engine off).
Disconnect test mode connector.
Check if engine starts.
YES
Check terminals 7, 8 and 20 of control unit connector for poor contact.
NO
Turn ignition switch OFF.
Disconnect crank angle sensor connector.
Turn ignition switch ON (engine off).
Check if voltage across terminal (B) of crank angle sensor connector (body side) and body is greater than 4 V.
NO
11·1
YES
Check if voltage across terminal (R) of connector (body side) and body is greater than 10 V.
NO
11·2
YES
Check if continuity exists between terminal (BR) of connector (body side) and body.
NO
11·3
YES
Check if female terminal of connector (body side) is open.
YES
Repair terminal or replace body harness.
NO
Replace crank angle sensor.

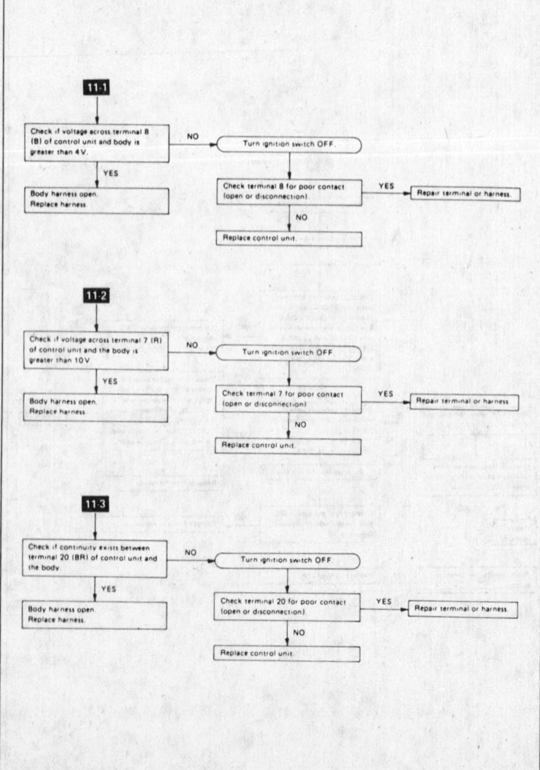
11·1
Check if voltage across terminal 8 (B) of control unit and body is greater than 4 V.
NO
Turn ignition switch OFF.
YES
Body harness open. Replace harness.
Check terminal 8 for poor contact (open or disconnection).
YES
Repair terminal or harness.
NO
Replace control unit.
11·2
Check if voltage across terminal 7 (R) of control unit and the body is greater than 10 V.
NO
Turn ignition switch OFF.
YES
Body harness open. Replace harness.
Check terminal 7 for poor contact (open or disconnection).
YES
Repair terminal or harness.
NO
Replace control unit.
11·3
Check if continuity exists between terminal 20 (BR) of control unit and the body.
NO
Turn ignition switch OFF.
YES
Body harness open. Replace harness.
Check terminal 20 for poor contact (open or disconnection).
YES
Repair terminal or harness.
NO
Replace control unit.

DIAGNOSIS CHART—XT MODELS

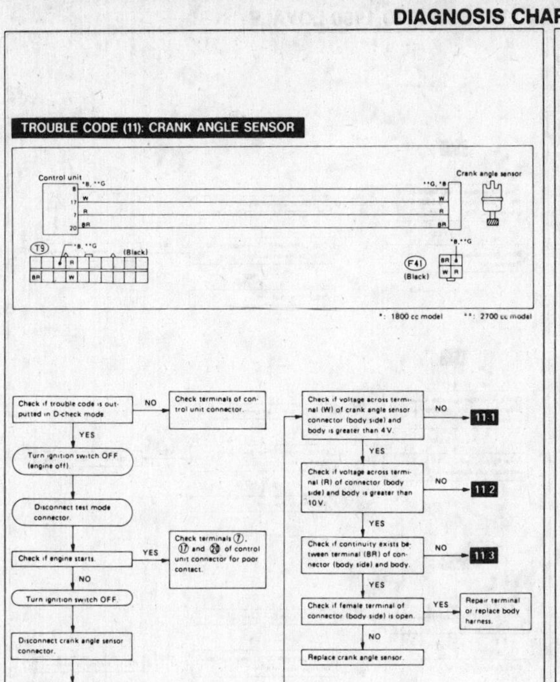
TROUBLE CODE (11): CRANK ANGLE SENSOR
Control unit
Crank angle sensor
(Black)
*: 1800 cc model
**: 2700 cc model
Check if trouble code is outputted in D-check mode.
NO
Check terminals of control unit connector.
YES
Turn ignition switch OFF (engine off).
Disconnect test mode connector.
Check if engine starts.
YES
Check terminals ⑦, ⑰ and ⑳ of control unit connector for poor contact.
NO
Turn ignition switch OFF.
Disconnect crank angle sensor connector.
Turn ignition switch ON (engine off).
Check if voltage across terminal (W) of crank angle sensor connector (body side) and body is greater than 4 V.
11 1
Check if voltage across terminal (R) of connector (body side) and body is greater than 10 V.
11 2
Check if continuity exists between terminal (BR) of connector (body side) and body.
11 3
Check if female terminal of connector (body side) is open.
Repair terminal or replace body harness.
Replace crank angle sensor.

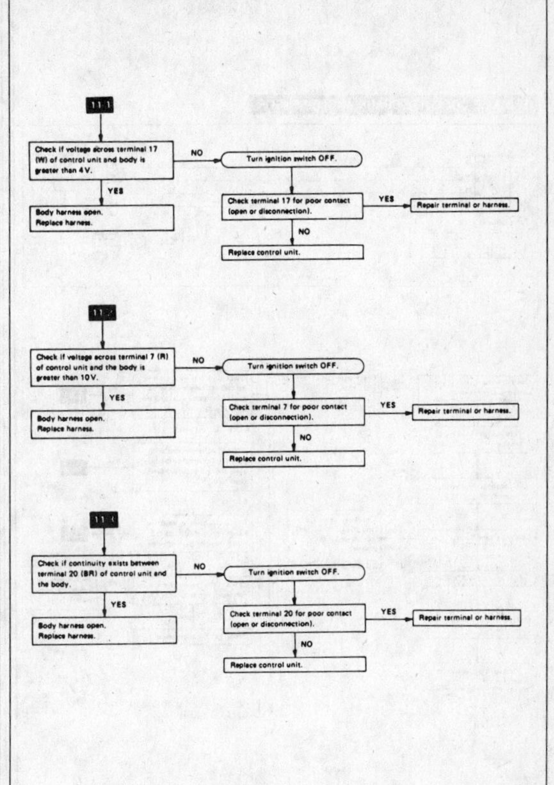
11 1
Check if voltage across terminal 17 (W) of control unit and body is greater than 4 V.
NO
Turn ignition switch OFF.
YES
Body harness open. Replace harness.
Check terminal 17 for poor contact (open or disconnection).
YES
Repair terminal or harness.
NO
Replace control unit.
11 2
Check if voltage across terminal 7 (R) of control unit and the body is greater than 10 V.
NO
Turn ignition switch OFF.
YES
Body harness open. Replace harness.
Check terminal 7 for poor contact (open or disconnection).
YES
Repair terminal or harness.
NO
Replace control unit.
11 3
Check if continuity exists between terminal 20 (BR) of control unit and the body.
NO
Turn ignition switch OFF.
YES
Body harness open. Replace harness.
Check terminal 20 for poor contact (open or disconnection).
YES
Repair terminal or harness.
NO
Replace control unit.

DIAGNOSIS CHART—SEDAN AND WAGON AND 1990 LOYALE

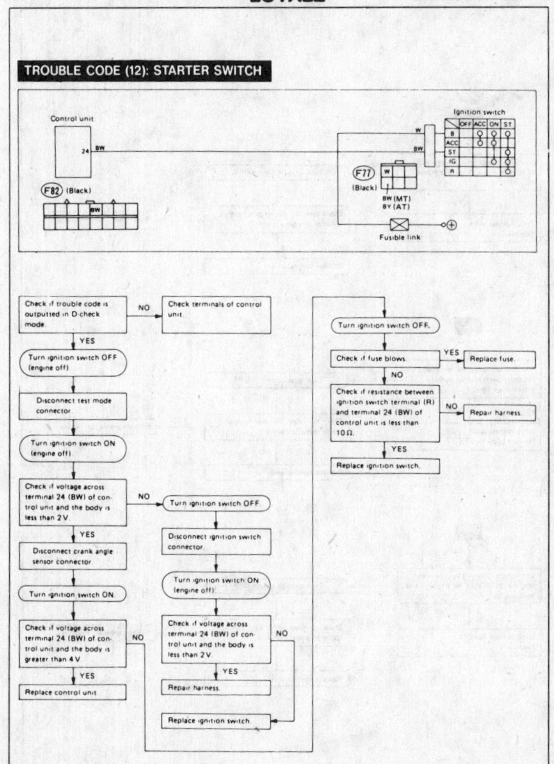
TROUBLE CODE (12): STARTER SWITCH
Control unit
Ignition switch
(Black)
Fusible link
Check if trouble code is outputted in D-check mode.
NO
Check terminals of control unit.
YES
Turn ignition switch OFF (engine off).
Disconnect test mode connector.
Turn ignition switch ON (engine off).
Check if voltage across terminal 24 (BW) of control unit and the body is less than 2 V.
NO
Turn ignition switch OFF.
Disconnect ignition switch connector.
Turn ignition switch ON (engine off).
Check if voltage across terminal 24 (BW) of control unit and the body is less than 2 V.
Repair harness.
Replace ignition switch.
YES
Disconnect crank angle sensor connector.
Turn ignition switch ON.
Check if voltage across terminal 24 (BW) of control unit and the body is greater than 4 V.
Replace control unit.
Turn ignition switch OFF.
Check if fuse blows.
Replace fuse.
Check if resistance between ignition switch terminal (R) and terminal 24 (BW) of control unit is less than 10 Ω.
Repair harness.
Replace ignition switch.

DIAGNOSIS CHART—XT MODELS

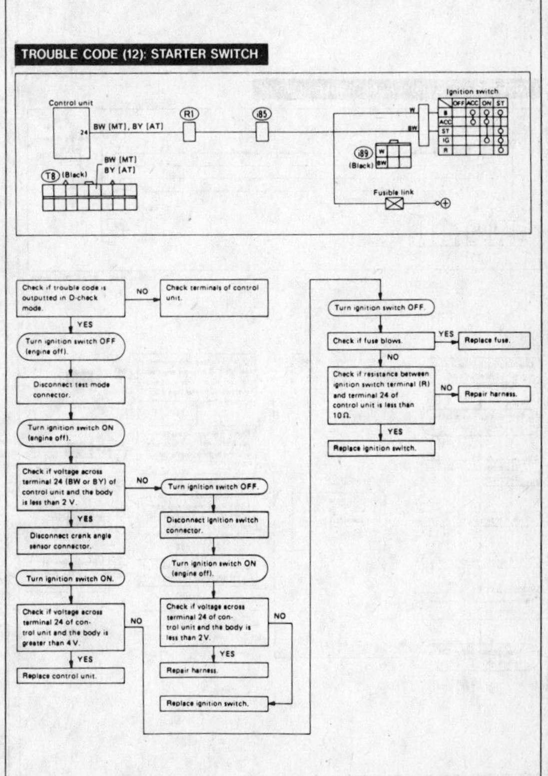
TROUBLE CODE (12): STARTER SWITCH
Control unit
Ignition switch
BW (MT), BY (AT)
(Black)
Fusible link
Check if trouble code is outputted in D-check mode.
NO
Check terminals of control unit.
YES
Turn ignition switch OFF (engine off).
Disconnect test mode connector.
Turn ignition switch ON (engine off).
Check if voltage across terminal 24 (BW or BY) of control unit and the body is less than 2 V.
NO
Turn ignition switch OFF.
Disconnect ignition switch connector.
Turn ignition switch ON (engine off).
Check if voltage across terminal 24 of control unit and the body is less than 2 V.
Repair harness.
Replace ignition switch.
YES
Disconnect crank angle sensor connector.
Turn ignition switch ON.
Check if voltage across terminal 24 of control unit and the body is greater than 4 V.
Replace control unit.
Turn ignition switch OFF.
Check if fuse blows.
Replace fuse.
Check if resistance between ignition switch terminal (R) and terminal 24 of control unit is less than 10 Ω.
Repair harness.
Replace ignition switch.

DIAGNOSIS CHART—SEDAN AND WAGON AND 1990 LOYALE

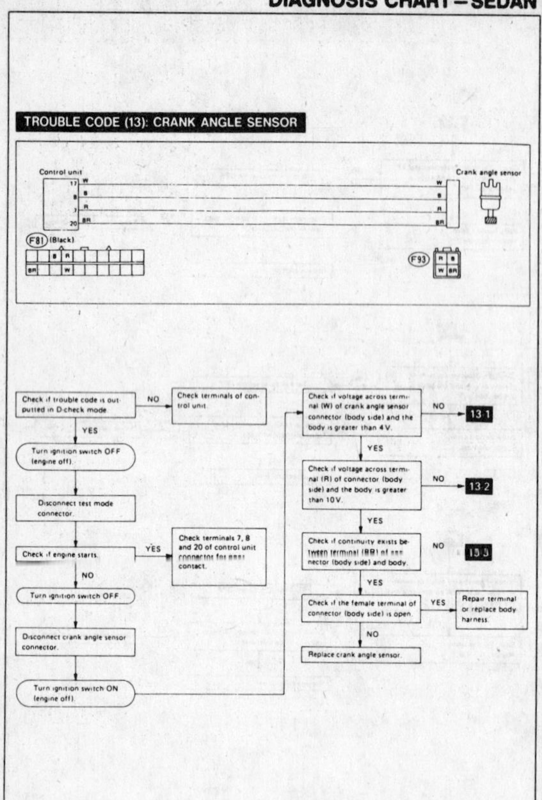
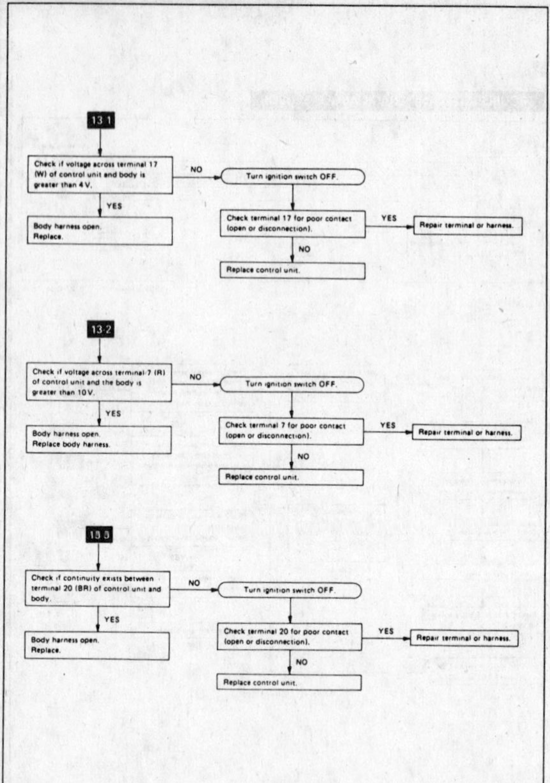

DIAGNOSIS CHART—XT MODELS

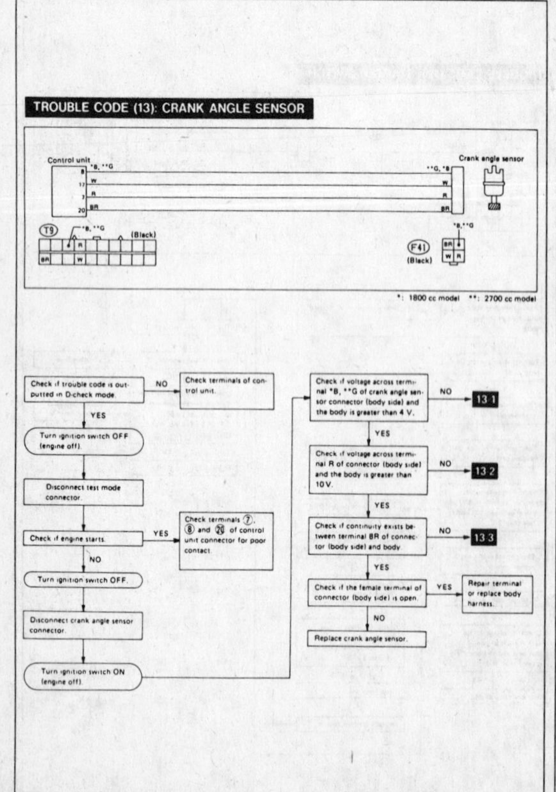
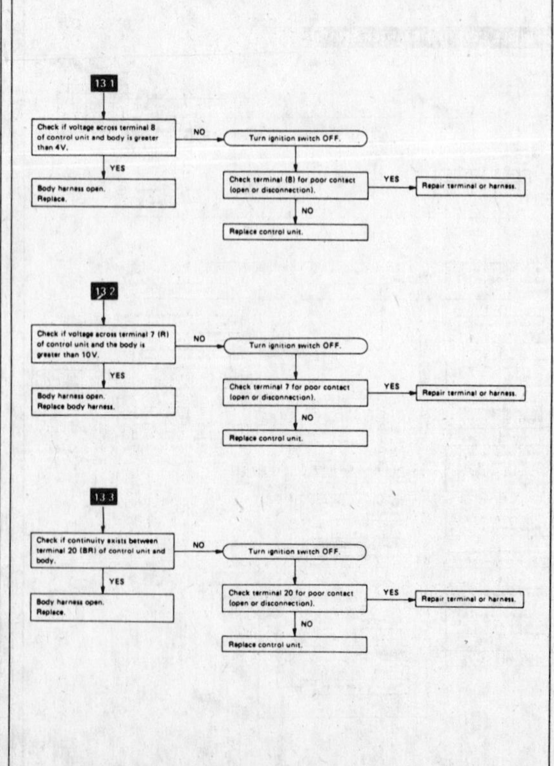

DIAGNOSIS CHART—SEDAN AND WAGON AND 1990 LOYALE

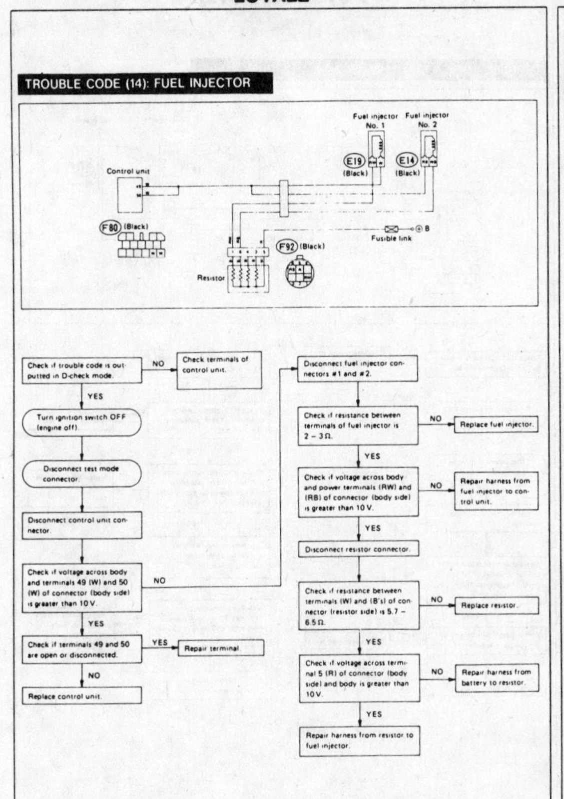
TROUBLE CODE (14): FUEL INJECTOR
Fuel injector No. 1
Fuel injector No. 2
Control unit
E19 (Black)
E14 (Black)
F80 (Black)
F92 (Black)
Fusible link
Resistor
Check if trouble code is outputted in D-check mode.
NO
Check terminals of control unit.
YES
Turn ignition switch OFF (engine off).
Disconnect test mode connector.
Disconnect control unit connector.
Check if voltage across body and terminals 49 (W) and 50 (W) of connector (body side) is greater than 10 V.
NO
YES
Check if terminals 49 and 50 are open or disconnected.
YES
Repair terminal.
NO
Replace control unit.
Disconnect fuel injector connectors #1 and #2.
Check if resistance between terminals of fuel injector is 2 – 3 Ω.
NO
Replace fuel injector.
YES
Check if voltage across body and power terminals (RW) and (RB) of connector (body side) is greater than 10 V.
NO
Repair harness from fuel injector to control unit.
YES
Disconnect resistor connector.
Check if resistance between terminals (W) and (B's) of connector (resistor side) is 5.7 – 6.5 Ω.
NO
Replace resistor.
YES
Check if voltage across terminal 5 (R) of connector (body side) and body is greater than 10 V.
NO
Repair harness from battery to resistor.
YES
Repair harness from resistor to fuel injector.

DIAGNOSIS CHART—XT MODELS WITH 4 CYLINDER ENGINE

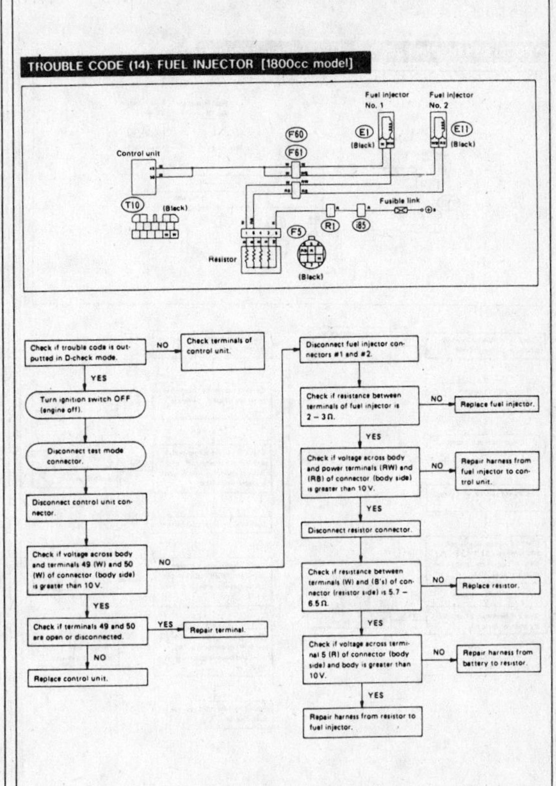
TROUBLE CODE (14): FUEL INJECTOR [1800cc model]
Fuel injector No. 1
Fuel injector No. 2
Control unit
F60
F61
E1 (Black)
E11 (Black)
T10 (Black)
Fusible link
R1
i85
F5
Resistor
(Black)
Check if trouble code is outputted in D-check mode.
NO
Check terminals of control unit.
YES
Turn ignition switch OFF (engine off).
Disconnect test mode connector.
Disconnect control unit connector.
Check if voltage across body and terminals 49 (W) and 50 (W) of connector (body side) is greater than 10 V.
NO
YES
Check if terminals 49 and 50 are open or disconnected.
YES
Repair terminal.
NO
Replace control unit.
Disconnect fuel injector connectors #1 and #2.
Check if resistance between terminals of fuel injector is 2 – 3 Ω.
NO
Replace fuel injector.
YES
Check if voltage across body and power terminals (RW) and (RB) of connector (body side) is greater than 10 V.
NO
Repair harness from fuel injector to control unit.
YES
Disconnect resistor connector.
Check if resistance between terminals (W) and (B's) of connector (resistor side) is 5.7 – 6.5 Ω.
NO
Replace resistor.
YES
Check if voltage across terminal 5 (R) of connector (body side) and body is greater than 10 V.
NO
Repair harness from battery to resistor.
YES
Repair harness from resistor to fuel injector.

DIAGNOSIS CHART—XT MODELS WITH 6 CYLINDER ENGINE

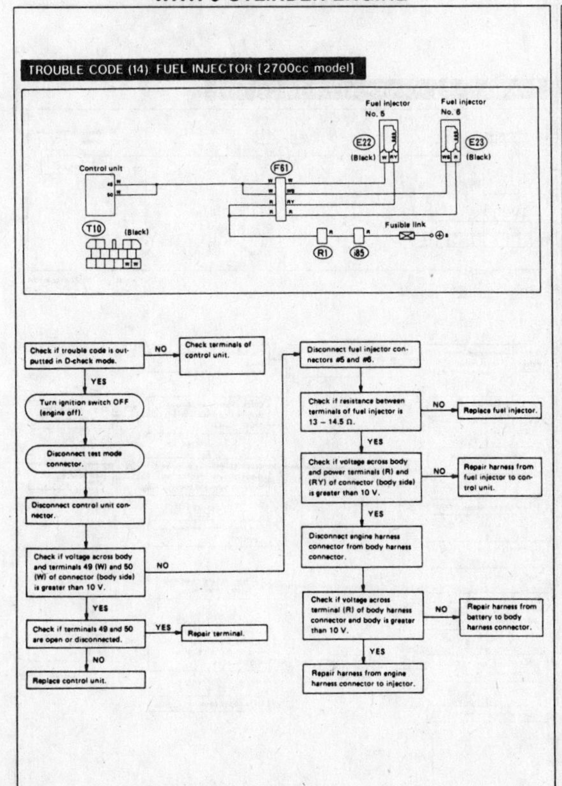
TROUBLE CODE (14): FUEL INJECTOR [2700cc model]
Fuel injector No. 5
Fuel injector No. 6
E22 (Black)
E23 (Black)
Control unit
F61
T10 (Black)
Fusible link
R1
i85
Check if trouble code is outputted in D-check mode.
NO
Check terminals of control unit.
YES
Turn ignition switch OFF (engine off).
Disconnect test mode connector.
Disconnect control unit connector.
Check if voltage across body and terminals 49 (W) and 50 (W) of connector (body side) is greater than 10 V.
NO
YES
Check if terminals 49 and 50 are open or disconnected.
YES
Repair terminal.
NO
Replace control unit.
Disconnect fuel injector connectors #5 and #6.
Check if resistance between terminals of fuel injector is 13 – 14.5 Ω.
NO
Replace fuel injector.
YES
Check if voltage across body and power terminals (R) and (RY) of connector (body side) is greater than 10 V.
NO
Repair harness from fuel injector to control unit.
YES
Disconnect engine harness connector from body harness connector.
Check if voltage across terminal (R) of body harness connector and body is greater than 10 V.
NO
Repair harness from battery to body harness connector.
YES
Repair harness from engine harness connector to injector.

DIAGNOSIS CHART—SEDAN AND WAGON AND 1990 LOYALE

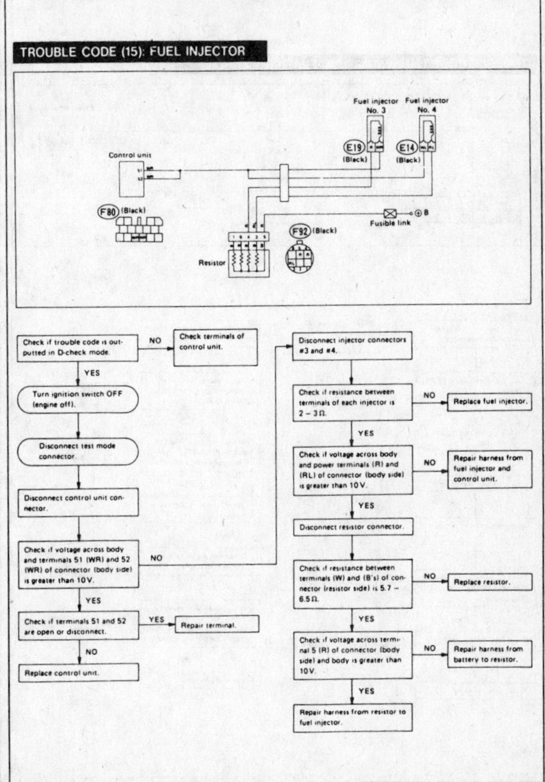
TROUBLE CODE (15): FUEL INJECTOR
Fuel injector No. 3
Fuel injector No. 4
Control unit
E19 (Black)
E14 (Black)
F80 (Black)
F92 (Black)
Fusible link
Resistor
Check if trouble code is outputted in D-check mode.
NO
Check terminals of control unit.
YES
Turn ignition switch OFF (engine off).
Disconnect test mode connector.
Disconnect control unit connector.
Check if voltage across body and terminals 51 (WR) and 52 (WR) of connector (body side) is greater than 10 V.
NO
YES
Check if terminals 51 and 52 are open or disconnect.
YES
Repair terminal.
NO
Replace control unit.
Disconnect injector connectors #3 and #4.
Check if resistance between terminals of each injector is 2 – 3 Ω.
NO
Replace fuel injector.
YES
Check if voltage across body and power terminals (R) and (RL) of connector (body side) is greater than 10 V.
NO
Repair harness from fuel injector and control unit.
YES
Disconnect resistor connector.
Check if resistance between terminals (W) and (B's) of connector (resistor side) is 5.7 – 6.5 Ω.
NO
Replace resistor.
YES
Check if voltage across terminal 5 (R) of connector (body side) and body is greater than 10 V.
NO
Repair harness from battery to resistor.
YES
Repair harness from resistor to fuel injector.

DIAGNOSIS CHART—XT MODELS WITH 4 CYLINDER ENGINE

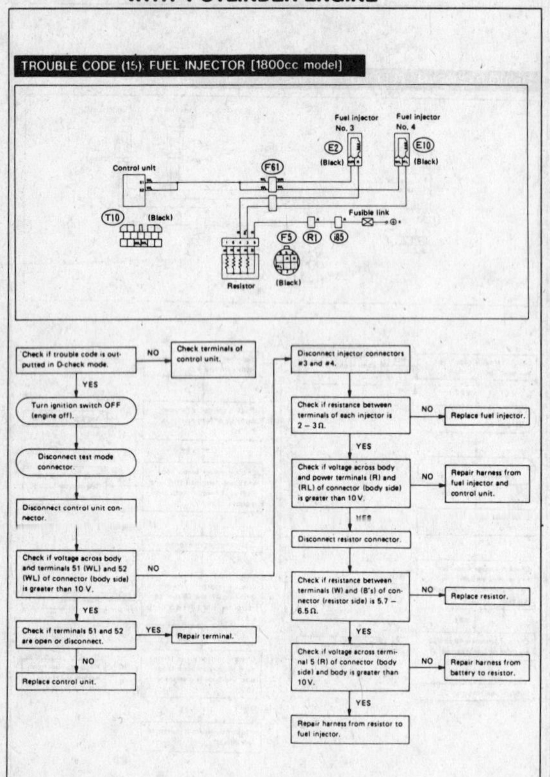
TROUBLE CODE (15): FUEL INJECTOR [1800cc model]
Fuel injector No. 3
Fuel injector No. 4
E2
(Black)
E10
(Black)
Control unit
F61
T10
(Black)
Fusible link
F5
R1
i85
Resistor
(Black)
Check if trouble code is outputted in D-check mode.
NO
Check terminals of control unit.
YES
Turn ignition switch OFF (engine off).
Disconnect test mode connector.
Disconnect control unit connector.
Check if voltage across body and terminals 51 (WL) and 52 (WL) of connector (body side) is greater than 10 V.
NO
YES
Check if terminals 51 and 52 are open or disconnect.
YES
Repair terminal.
NO
Replace control unit.
Disconnect injector connectors #3 and #4.
Check if resistance between terminals of each injector is 2 – 3 Ω.
NO
Replace fuel injector.
YES
Check if voltage across body and power terminals (R) and (RL) of connector (body side) is greater than 10 V.
NO
Repair harness from fuel injector and control unit.
Disconnect resistor connector.
Check if resistance between terminals (W) and (B's) of connector (resistor side) is 5.7 – 6.5 Ω.
NO
Replace resistor.
YES
Check if voltage across terminal 5 (R) of connector (body side) and body is greater than 10 V.
NO
Repair harness from battery to resistor.
YES
Repair harness from resistor to fuel injector.

DIAGNOSIS CHART—XT MODELS WITH 6 CYLINDER ENGINE

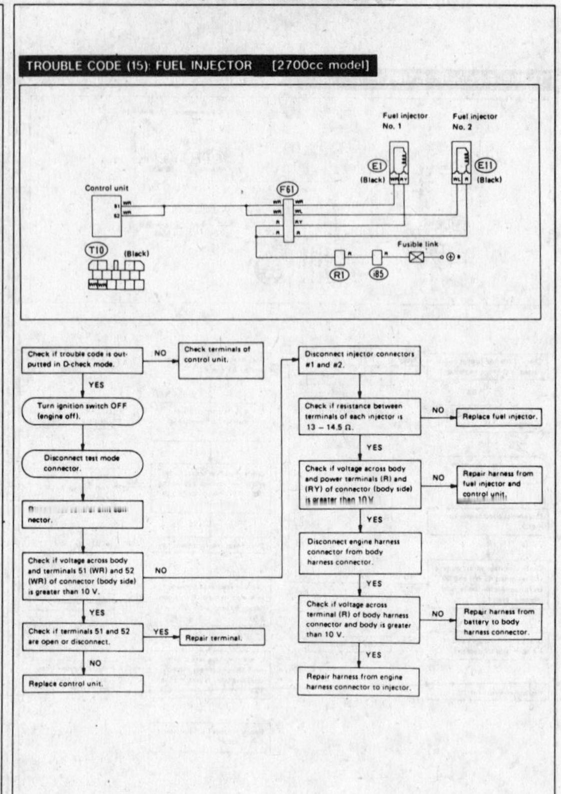
TROUBLE CODE (15): FUEL INJECTOR [2700cc model]
Fuel injector No. 1
Fuel injector No. 2
E1
(Black)
E11
(Black)
Control unit
F61
T10
(Black)
Fusible link
R1
i85
Check if trouble code is outputted in D-check mode.
NO
Check terminals of control unit.
YES
Turn ignition switch OFF (engine off).
Disconnect test mode connector.
nector.
Check if voltage across body and terminals 51 (WR) and 52 (WR) of connector (body side) is greater than 10 V.
NO
YES
Check if terminals 51 and 52 are open or disconnect.
YES
Repair terminal.
NO
Replace control unit.
Disconnect injector connectors #1 and #2.
Check if resistance between terminals of each injector is 13 – 14.5 Ω.
NO
Replace fuel injector.
YES
Check if voltage across body and power terminals (R) and (RY) of connector (body side) is greater than 10 V.
NO
Repair harness from fuel injector and control unit.
YES
Disconnect engine harness connector from body harness connector.
YES
Check if voltage across terminal (R) of body harness connector and body is greater than 10 V.
NO
Repair harness from battery to body harness connector.
YES
Repair harness from engine harness connector to injector.

DIAGNOSIS CHART—SEDAN AND WAGON AND 1990 LOYALE

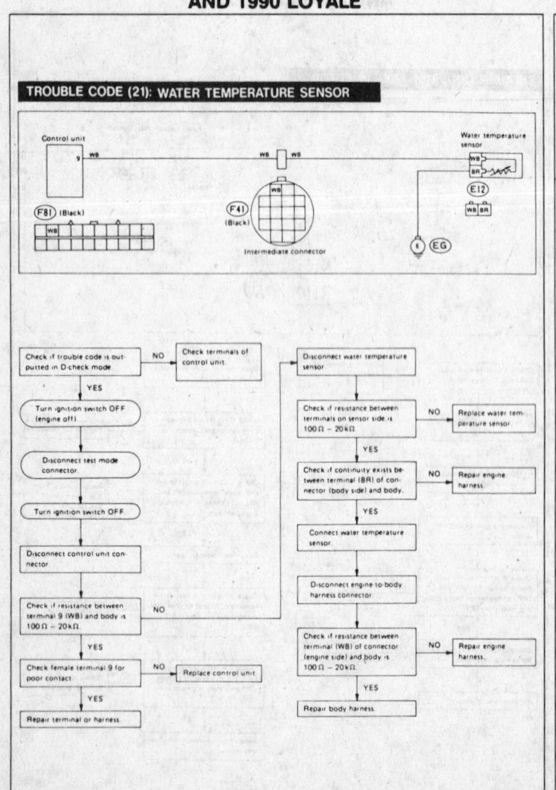
TROUBLE CODE (21): WATER TEMPERATURE SENSOR
Control unit
Water temperature sensor
E12
F81 (Black)
F41
(Black)
Intermediate connector
EG
Check if trouble code is outputted in D-check mode.
NO
Check terminals of control unit.
YES
Turn ignition switch OFF (engine off).
Disconnect test mode connector.
Turn ignition switch OFF.
Disconnect control unit connector.
Check if resistance between terminal 9 (WB) and body is 100 Ω – 20 kΩ.
NO
YES
Check female terminal 9 for poor contact.
NO
Replace control unit.
YES
Repair terminal or harness.
Disconnect water temperature sensor.
Check if resistance between terminals on sensor side is 100 Ω – 20 kΩ.
NO
Replace water temperature sensor.
YES
Check if continuity exists between terminal (BR) of connector (body side) and body.
NO
Repair engine harness.
YES
Connect water temperature sensor.
Disconnect engine-to-body harness connector.
Check if resistance between terminal (WB) of connector (engine side) and body is 100 Ω – 20 kΩ.
NO
Repair engine harness.
YES
Repair body harness.

DIAGNOSIS CHART—XT MODELS

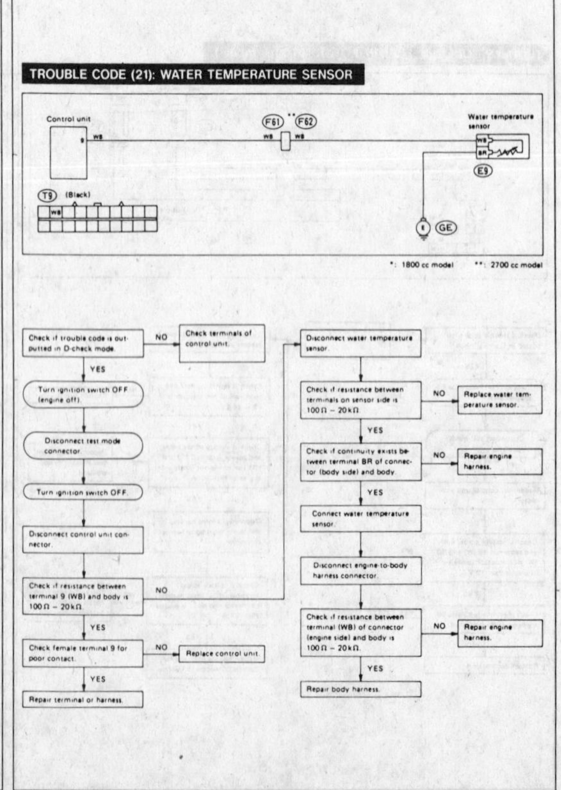
TROUBLE CODE (21): WATER TEMPERATURE SENSOR
Control unit
F61
F62
Water temperature sensor
E9
T9 (Black)
GE
*: 1800 cc model
**: 2700 cc model
Check if trouble code is outputted in D-check mode.
NO
Check terminals of control unit.
YES
Turn ignition switch OFF (engine off).
Disconnect test mode connector.
Turn ignition switch OFF.
Disconnect control unit connector.
Check if resistance between terminal 9 (WB) and body is 100 Ω – 20 kΩ.
NO
YES
Check female terminal 9 for poor contact.
NO
Replace control unit.
YES
Repair terminal or harness.
Disconnect water temperature sensor.
Check if resistance between terminals on sensor side is 100 Ω – 20 kΩ.
NO
Replace water temperature sensor.
YES
Check if continuity exists between terminal BR of connector (body side) and body.
NO
Repair engine harness.
YES
Connect water temperature sensor.
Disconnect engine-to-body harness connector.
Check if resistance between terminal (WB) of connector (engine side) and body is 100 Ω – 20 kΩ.
NO
Repair engine harness.
YES
Repair body harness.

DIAGNOSIS CHART—SEDAN AND WAGON AND 1990 LOYALE

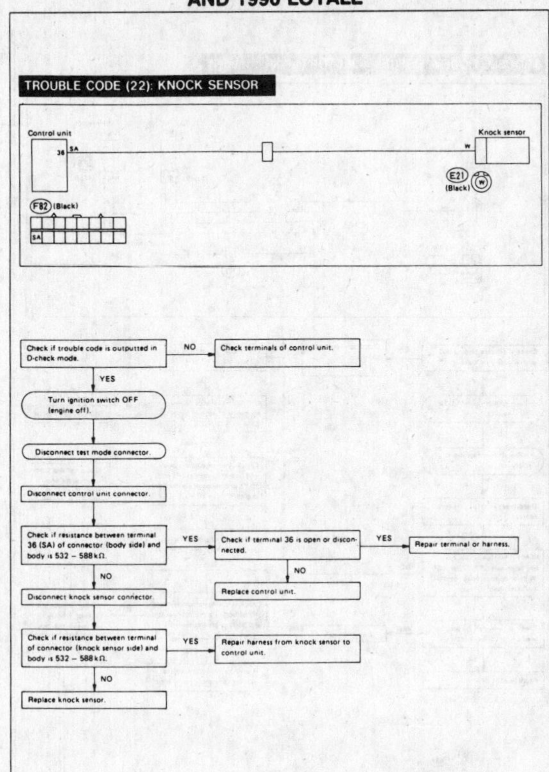
TROUBLE CODE (22): KNOCK SENSOR
Control unit
36
SA
W
Knock sensor
E21
(Black)
W
F82
(Black)
SA
Check if trouble code is outputted in D-check mode.
NO
Check terminals of control unit.
YES
Turn ignition switch OFF (engine off).
Disconnect test mode connector.
Disconnect control unit connector.
Check if resistance between terminal 36 (SA) of connector (body side) and body is 532 – 588kΩ.
YES
Check if terminal 36 is open or disconnected.
YES
Repair terminal or harness.
NO
Replace control unit.
NO
Disconnect knock sensor connector.
Check if resistance between terminal of connector (knock sensor side) and body is 532 – 588kΩ.
YES
Repair harness from knock sensor to control unit.
NO
Replace knock sensor.

DIAGNOSIS CHART—XT MODELS

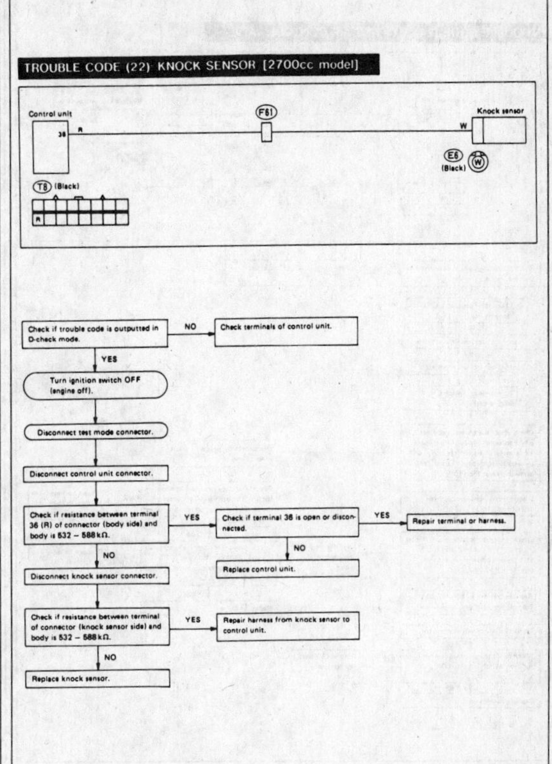
TROUBLE CODE (22) KNOCK SENSOR [2700cc model]
Control unit
36
R
F81
W
Knock sensor
E6
(Black)
W
T8
(Black)
R
Check if trouble code is outputted in D-check mode.
NO
Check terminals of control unit.
YES
Turn ignition switch OFF (engine off).
Disconnect test mode connector.
Disconnect control unit connector.
Check if resistance between terminal 36 (R) of connector (body side) and body is 532 – 588kΩ.
YES
Check if terminal 36 is open or disconnected.
YES
Repair terminal or harness.
NO
Replace control unit.
NO
Disconnect knock sensor connector.
Check if resistance between terminal of connector (knock sensor side) and body is 532 – 588kΩ.
YES
Repair harness from knock sensor to control unit.
NO
Replace knock sensor.

DIAGNOSIS CHART—SEDAN AND WAGON AND 1990 LOYALE

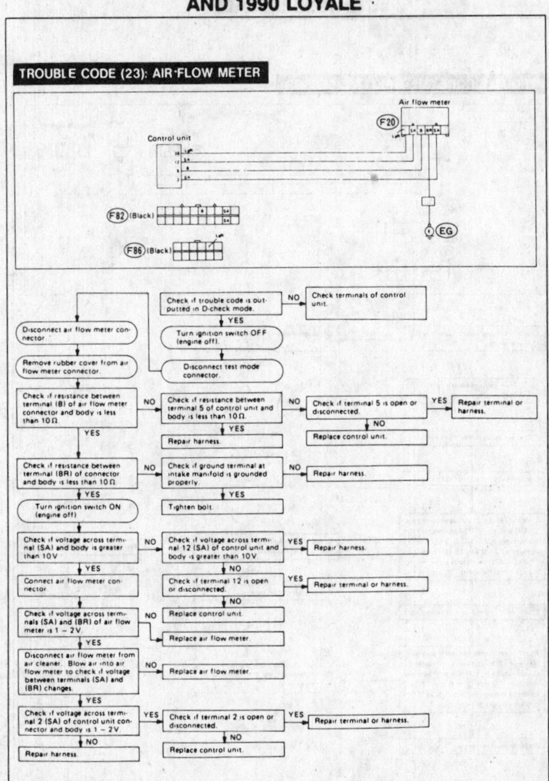
TROUBLE CODE (23): AIR FLOW METER
Air flow meter
F20
Control unit
F82 (Black)
F86 (Black)
EG
Check if trouble code is outputted in D-check mode.
NO
Check terminals of control unit.
YES
Turn ignition switch OFF (engine off).
Disconnect test mode connector.
Disconnect air flow meter connector.
Remove rubber cover from air flow meter connector.
Check if resistance between terminal (B) of air flow meter connector and body is less than 10Ω.
NO
Check if resistance between terminal 5 of control unit and body is less than 10Ω.
NO
Check if terminal 5 is open or disconnected.
YES
Repair terminal or harness.
NO
Replace control unit.
YES
Repair harness.
YES
Check if resistance between terminal (BR) of connector and body is less than 10Ω.
NO
Check if ground terminal at intake manifold is grounded properly.
NO
Repair harness.
YES
Tighten bolt.
YES
Turn ignition switch ON (engine off).
Check if voltage across terminal (SA) and body is greater than 10V.
NO
Check if voltage across terminal 12 (SA) of control unit and body is greater than 10V.
YES
Repair harness.
NO
Check if terminal 12 is open or disconnected.
YES
Repair terminal or harness.
NO
Replace control unit.
YES
Connect air flow meter connector.
Check if voltage across terminals (SA) and (BR) of air flow meter is 1 – 2V.
NO
Replace air flow meter.
YES
Disconnect air flow meter from air cleaner. Blow air into air flow meter to check if voltage between terminals (SA) and (BR) changes.
NO
Replace air flow meter.
YES
Check if voltage across terminal 2 (SA) of control unit connector and body is 1 – 2V.
YES
Check if terminal 2 is open or disconnected.
YES
Repair terminal or harness.
NO
Replace control unit.
NO
Repair harness.

DIAGNOSIS CHART—XT MODELS

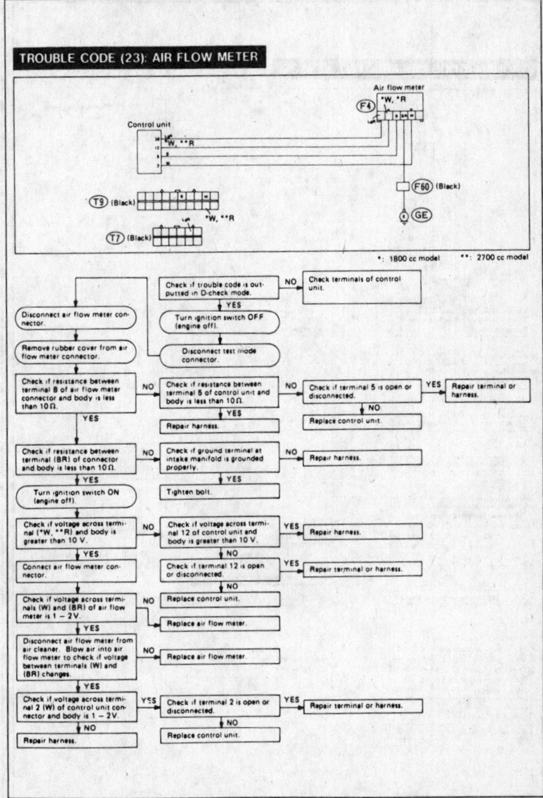
TROUBLE CODE (23): AIR FLOW METER
Air flow meter
*W, *R
F4
Control unit
*W, **R
F60 (Black)
T9 (Black)
*W, **R
T7 (Black)
GE
*: 1800 cc model
**: 2700 cc model
Check if trouble code is outputted in D-check mode.
NO
Check terminals of control unit.
YES
Turn ignition switch OFF (engine off).
Disconnect test mode connector.
Disconnect air flow meter connector.
Remove rubber cover from air flow meter connector.
Check if resistance between terminal B of air flow meter connector and body is less than 10Ω.
NO
Check if resistance between terminal 5 of control unit and body is less than 10Ω.
NO
Check if terminal 5 is open or disconnected.
YES
Repair terminal or harness.
NO
Replace control unit.
YES
Repair harness.
YES
Check if resistance between terminal (BR) of connector and body is less than 10Ω.
NO
Check if ground terminal at intake manifold is grounded properly.
NO
Repair harness.
YES
Tighten bolt.
YES
Turn ignition switch ON (engine off).
Check if voltage across terminal (*W, **R) and body is greater than 10 V.
NO
Check if voltage across terminal 12 of control unit and body is greater than 10 V.
YES
Repair harness.
NO
Check if terminal 12 is open or disconnected.
YES
Repair terminal or harness.
NO
Replace control unit.
YES
Connect air flow meter connector.
Check if voltage across terminals (W) and (BR) of air flow meter is 1 – 2V.
NO
Replace air flow meter.
YES
Disconnect air flow meter from air cleaner. Blow air into air flow meter to check if voltage between terminals (W) and (BR) changes.
NO
Replace air flow meter.
YES
Check if voltage across terminal 2 (W) of control unit connector and body is 1 – 2V.
YES
Check if terminal 2 is open or disconnected.
YES
Repair terminal or harness.
NO
Replace control unit.
NO
Repair harness.

DIAGNOSIS CHART—XT MODELS

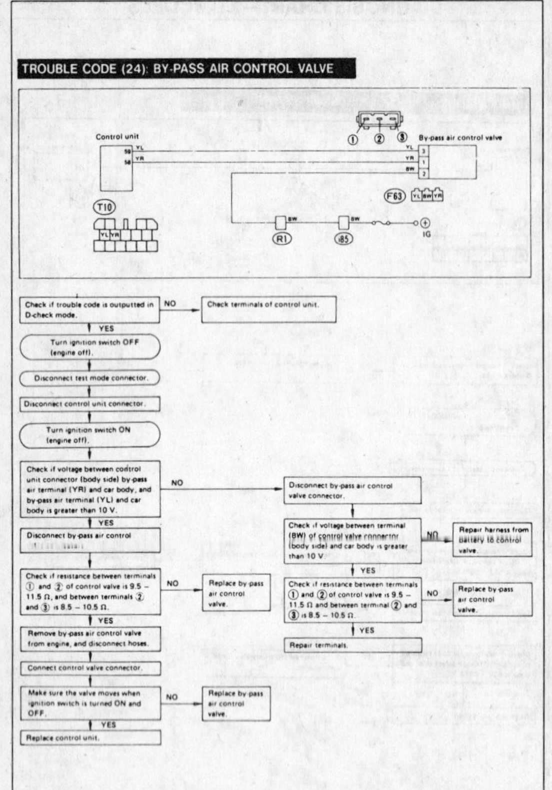
TROUBLE CODE (24): BY-PASS AIR CONTROL VALVE
Control unit
By-pass air control valve
T10
F63
R1
i85
IG
Check if trouble code is outputted in D-check mode.
NO
Check terminals of control unit.
YES
Turn ignition switch OFF (engine off).
Disconnect test mode connector.
Disconnect control unit connector.
Turn ignition switch ON (engine off).
Check if voltage between control unit connector (body side) by-pass air terminal (YR) and car body, and by-pass air terminal (YL) and car body is greater than 10 V.
NO
Disconnect by-pass air control valve connector.
Check if voltage between terminal (BW) of control valve connector (body side) and car body is greater than 10 V.
NO
Repair harness from battery to control valve.
YES
Check if resistance between terminals ① and ② of control valve is 9.5 – 11.5 Ω and between terminal ② and ③ is 8.5 – 10.5 Ω.
NO
Replace by-pass air control valve.
YES
Repair terminals.
YES
Disconnect by-pass air control
Check if resistance between terminals ① and ② of control valve is 9.5 – 11.5 Ω, and between terminals ② and ③ is 8.5 – 10.5 Ω.
NO
Replace by-pass air control valve.
YES
Remove by-pass air control valve from engine, and disconnect hoses.
Connect control valve connector.
Make sure the valve moves when ignition switch is turned ON and OFF.
NO
Replace by-pass air control valve.
YES
Replace control unit.

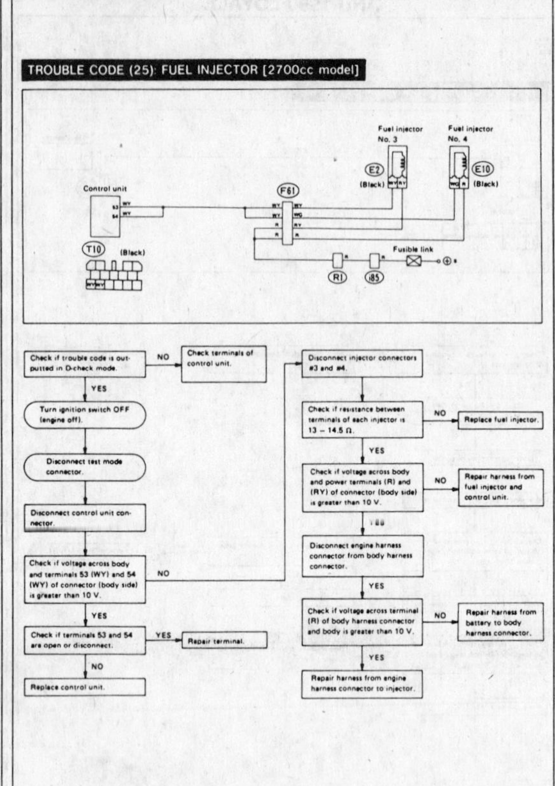
TROUBLE CODE (25): FUEL INJECTOR [2700cc model]
Fuel injector No. 3
Fuel injector No. 4
E2 (Black)
E10 (Black)
Control unit
F61
T10 (Black)
Fusible link
R1
i85
Check if trouble code is outputted in D-check mode.
NO
Check terminals of control unit.
YES
Turn ignition switch OFF (engine off).
Disconnect test mode connector.
Disconnect control unit connector.
Check if voltage across body and terminals 53 (WY) and 54 (WY) of connector (body side) is greater than 10 V.
NO
Disconnect injector connectors #3 and #4.
Check if resistance between terminals of each injector is 13 – 14.5 Ω.
NO
Replace fuel injector.
YES
Check if voltage across body and power terminals (R) and (RY) of connector (body side) is greater than 10 V.
NO
Repair harness from fuel injector and control unit.
YES
Disconnect engine harness connector from body harness connector.
YES
Check if voltage across terminal (R) of body harness connector and body is greater than 10 V.
NO
Repair harness from battery to body harness connector.
YES
Repair harness from engine harness connector to injector.
YES
Check if terminals 53 and 54 are open or disconnect.
YES
Repair terminal.
NO
Replace control unit.

DIAGNOSIS CHART—1988 SEDAN AND WAGON

TROUBLE CODE (31): THROTTLE SENSOR

Throttle sensor (Throttle sensor side)
(Body side)
E16
Control unit
F81 (Black)
F82 (Black)

Check if trouble code is outputted in D-check mode → NO → Check control unit terminals. (See page 116.)

YES ↓

Turn ignition switch OFF (engine off).

Disconnect test mode connector.

Disconnect control unit connector.

Check if resistance between terminals 21 (W) and 16 (BR) of connector (body side) is 6 – 18kΩ, and between terminals 22 (B) and 16 (BR) is 5.8 – 17.8kΩ (throttle closed) and 1.5 – 5.1kΩ (throttle open) → YES → Check if terminals 21, 16 and 22 are open or disconnected. → NO → Replace control unit.

Check if terminals 21, 16 and 22 are open or disconnected. → YES → Repair terminal or harness.

NO ↓

Disconnect throttle sensor connector.

Check if resistance between terminals 1 and 2 of connector (throttle sensor side) is 6 – 18kΩ, and between terminals 1 and 3 is 5.8 – 17.8kΩ (throttle closed) and 1.5 – 5.1kΩ (throttle open) → YES → Repair harness from throttle sensor to control unit.

NO ↓

Replace throttle sensor.

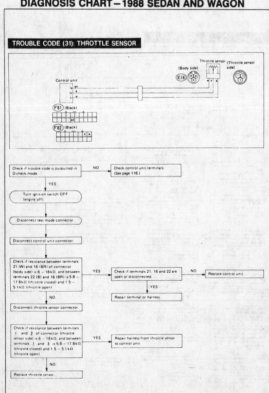

DIAGNOSIS CHART—1989–90 SEDAN AND WAGON AND 1990 LOYALE

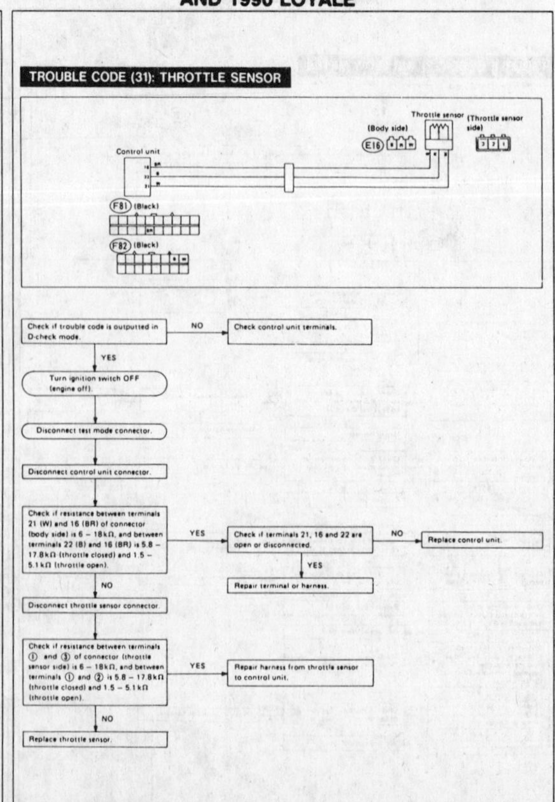
TROUBLE CODE (31): THROTTLE SENSOR
Throttle sensor
(Throttle sensor side)
(Body side)
E16
Control unit
F81 (Black)
F82 (Black)
Check if trouble code is outputted in D-check mode.
NO
Check control unit terminals.
YES
Turn ignition switch OFF (engine off).
Disconnect test mode connector.
Disconnect control unit connector.
Check if resistance between terminals 21 (W) and 16 (BR) of connector (body side) is 6 – 18kΩ, and between terminals 22 (B) and 16 (BR) is 5.8 – 17.8kΩ (throttle closed) and 1.5 – 5.1kΩ (throttle open).
YES
Check if terminals 21, 16 and 22 are open or disconnected.
NO
Replace control unit.
YES
Repair terminal or harness.
NO
Disconnect throttle sensor connector.
Check if resistance between terminals ① and ③ of connector (throttle sensor side) is 6 – 18kΩ, and between terminals ① and ② is 5.8 – 17.8kΩ (throttle closed) and 1.5 – 5.1kΩ (throttle open).
YES
Repair harness from throttle sensor to control unit.
NO
Replace throttle sensor.

DIAGNOSIS CHART—XT MODELS WITH 4 CYLINDER ENGINE

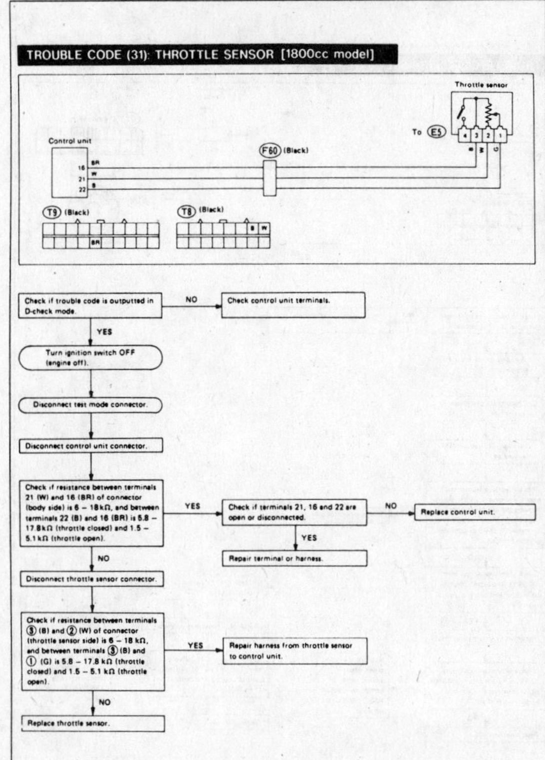
TROUBLE CODE (31): THROTTLE SENSOR [1800cc model]
Throttle sensor
To E5
Control unit
F60 (Black)
T9 (Black)
T8 (Black)
Check if trouble code is outputted in D-check mode.
NO
Check control unit terminals.
YES
Turn ignition switch OFF (engine off).
Disconnect test mode connector.
Disconnect control unit connector.
Check if resistance between terminals 21 (W) and 16 (BR) of connector (body side) is 6 – 18 kΩ, and between terminals 22 (B) and 16 (BR) is 5.8 – 17.8 kΩ (throttle closed) and 1.5 – 5.1 kΩ (throttle open).
YES
Check if terminals 21, 16 and 22 are open or disconnected.
NO
Replace control unit.
YES
Repair terminal or harness.
NO
Disconnect throttle sensor connector.
Check if resistance between terminals ③ (B) and ② (W) of connector (throttle sensor side) is 6 – 18 kΩ, and between terminals ③ (B) and ① (G) is 5.8 – 17.8 kΩ (throttle closed) and 1.5 – 5.1 kΩ (throttle open).
YES
Repair harness from throttle sensor to control unit.
NO
Replace throttle sensor.

DIAGNOSIS CHART—XT MODELS WITH 6 CYLINDER ENGINE

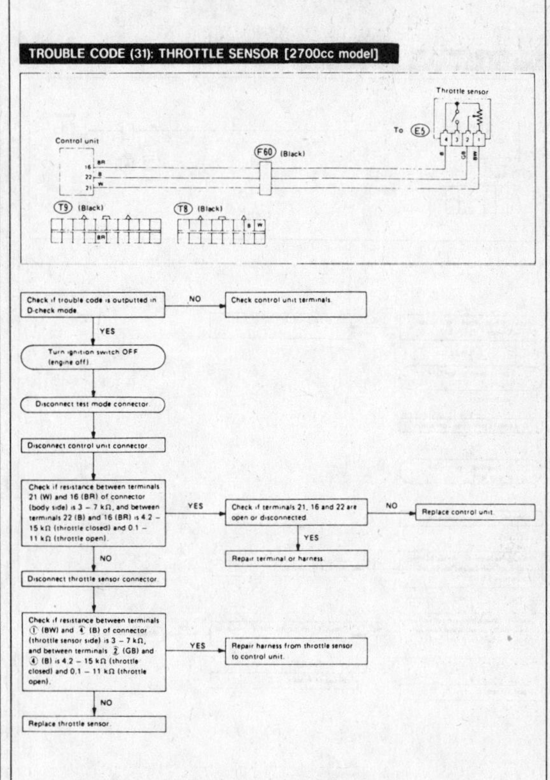
TROUBLE CODE (31): THROTTLE SENSOR [2700cc model]
Throttle sensor
To E5
Control unit
F60 (Black)
T9 (Black)
T8 (Black)
Check if trouble code is outputted in D-check mode.
NO
Check control unit terminals.
YES
Turn ignition switch OFF (engine off).
Disconnect test mode connector.
Disconnect control unit connector.
Check if resistance between terminals 21 (W) and 16 (BR) of connector (body side) is 3 – 7 kΩ, and between terminals 22 (B) and 16 (BR) is 4.2 – 15 kΩ (throttle closed) and 0.1 – 11 kΩ (throttle open).
YES
Check if terminals 21, 16 and 22 are open or disconnected.
NO
Replace control unit.
YES
Repair terminal or harness.
NO
Disconnect throttle sensor connector.
Check if resistance between terminals ① (BW) and ④ (B) of connector (throttle sensor side) is 3 – 7 kΩ, and between terminals ② (GB) and ④ (B) is 4.2 – 15 kΩ (throttle closed) and 0.1 – 11 kΩ (throttle open).
YES
Repair harness from throttle sensor to control unit.
NO
Replace throttle sensor.

DIAGNOSIS CHART—SEDAN AND WAGON AND 1990 LOYALE

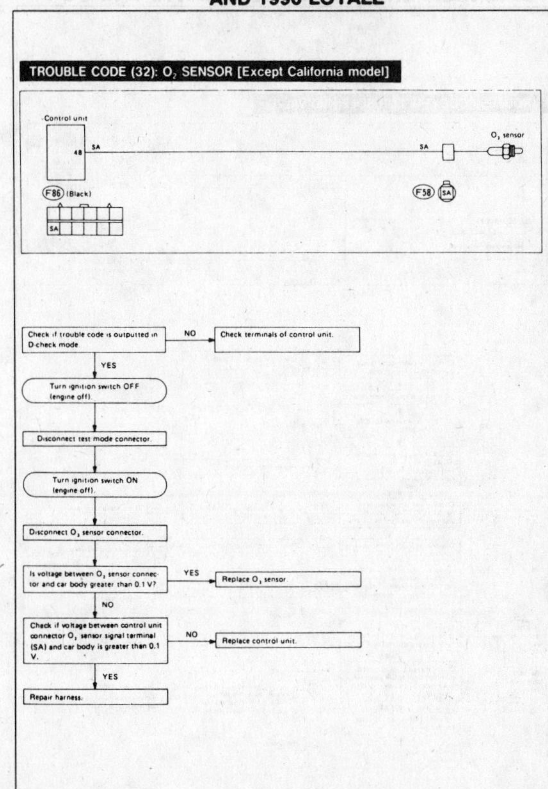
TROUBLE CODE (32): O2 SENSOR [Except California model]
Control unit
O2 sensor
F86 (Black)
F58
Check if trouble code is outputted in D-check mode.
NO
Check terminals of control unit.
YES
Turn ignition switch OFF (engine off).
Disconnect test mode connector.
Turn ignition switch ON (engine off).
Disconnect O2 sensor connector.
Is voltage between O2 sensor connector and car body greater than 0.1 V?
YES
Replace O2 sensor.
NO
Check if voltage between control unit connector O2 sensor signal terminal (SA) and car body is greater than 0.1 V.
NO
Replace control unit.
YES
Repair harness.

DIAGNOSIS CHART—SEDAN AND WAGON AND 1990 LOYALE

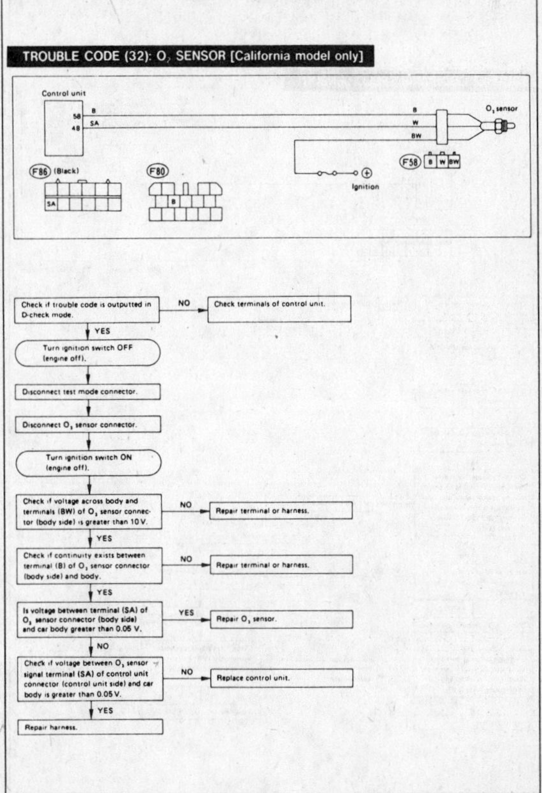
TROUBLE CODE (32): O2 SENSOR [California model only]
Control unit
O2 sensor
Ignition
F86 (Black)
F80
F58
Check if trouble code is outputted in D-check mode.
NO
Check terminals of control unit.
YES
Turn ignition switch OFF (engine off).
Disconnect test mode connector.
Disconnect O2 sensor connector.
Turn ignition switch ON (engine off).
Check if voltage across body and terminals (BW) of O2 sensor connector (body side) is greater than 10 V.
NO
Repair terminal or harness.
YES
Check if continuity exists between terminal (B) of O2 sensor connector (body side) and body.
NO
Repair terminal or harness.
YES
Is voltage between terminal (SA) of O2 sensor connector (body side) and car body greater than 0.05 V.
YES
Repair O2 sensor.
NO
Check if voltage between O2 sensor signal terminal (SA) of control unit connector (control unit side) and car body is greater than 0.05 V.
NO
Replace control unit.
YES
Repair harness.

DIAGNOSIS CHART—XT MODELS

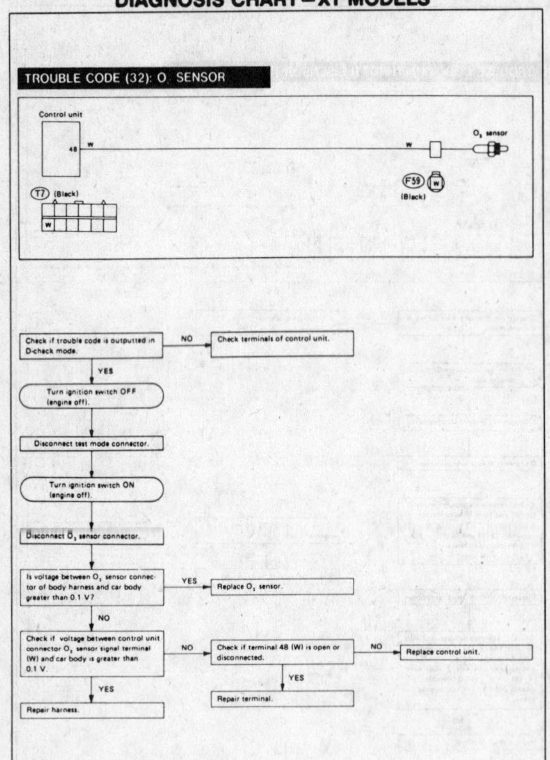
TROUBLE CODE (32): O₂ SENSOR
Control unit
48
W
W
O₂ sensor
T7 (Black)
F59
(Black)
Check if trouble code is outputted in D-check mode.
NO
Check terminals of control unit.
YES
Turn ignition switch OFF (engine off).
Disconnect test mode connector.
Turn ignition switch ON (engine off).
Disconnect O₂ sensor connector.
Is voltage between O₂ sensor connector of body harness and car body greater than 0.1 V?
YES
Replace O₂ sensor.
NO
Check if voltage between control unit connector O₂ sensor signal terminal (W) and car body is greater than 0.1 V.
NO
Check if terminal 48 (W) is open or disconnected.
NO
Replace control unit.
YES
Repair terminal.
YES
Repair harness.

DIAGNOSIS CHART—SEDAN AND WAGON AND 1990 LOYALE

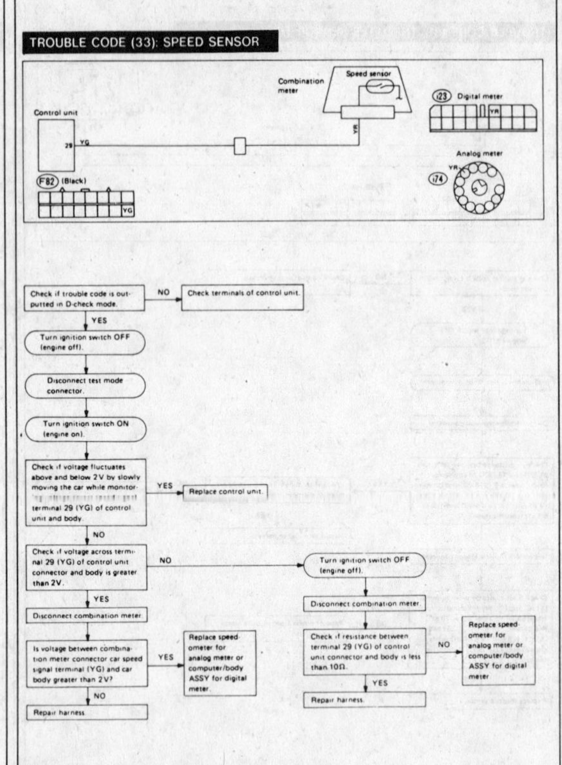
TROUBLE CODE (33): SPEED SENSOR
Combination meter
Speed sensor
i23 Digital meter
YR
Control unit
29
YG
Analog meter
i74
F82 (Black)
YG
Check if trouble code is outputted in D-check mode.
NO
Check terminals of control unit.
YES
Turn ignition switch OFF (engine off).
Disconnect test mode connector.
Turn ignition switch ON (engine on).
Check if voltage fluctuates above and below 2V by slowly moving the car while monitor- terminal 29 (YG) of control unit and body.
YES
Replace control unit.
NO
Check if voltage across terminal 29 (YG) of control unit connector and body is greater than 2V.
NO
Turn ignition switch OFF (engine off).
Disconnect combination meter.
Check if resistance between terminal 29 (YG) of control unit connector and body is less than 10Ω.
NO
Replace speedometer for analog meter or computer/body ASSY for digital meter.
YES
Repair harness.
YES
Disconnect combination meter.
Is voltage between combination meter connector car speed signal terminal (YG) and car body greater than 2V?
YES
Replace speedometer for analog meter or computer/body ASSY for digital meter.
NO
Repair harness.

DIAGNOSIS CHART—XT MODELS

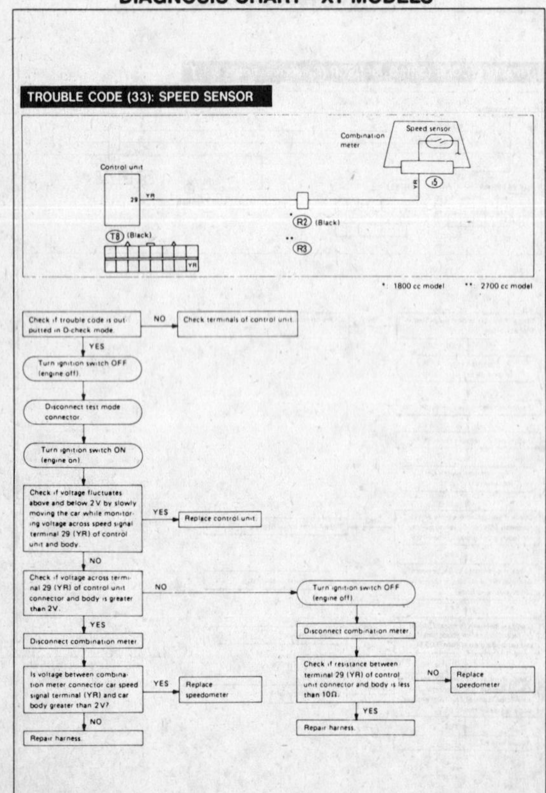
TROUBLE CODE (33): SPEED SENSOR
Combination meter
Speed sensor
Control unit
29
YR
i5
R2 (Black)
R3
T8 (Black)
*: 1800 cc model
**: 2700 cc model
Check if trouble code is outputted in D-check mode.
NO
Check terminals of control unit.
YES
Turn ignition switch OFF (engine off).
Disconnect test mode connector.
Turn ignition switch ON (engine on).
Check if voltage fluctuates above and below 2V by slowly moving the car while monitoring voltage across speed signal terminal 29 (YR) of control unit and body.
YES
Replace control unit.
NO
Check if voltage across terminal 29 (YR) of control unit connector and body is greater than 2V.
NO
Turn ignition switch OFF (engine off).
Disconnect combination meter.
Check if resistance between terminal 29 (YR) of control unit connector and body is less than 10Ω.
NO
Replace speedometer
YES
Repair harness.
YES
Disconnect combination meter
Is voltage between combination meter connector car speed signal terminal (YR) and car body greater than 2V?
YES
Replace speedometer
NO
Repair harness.

DIAGNOSIS CHART—1988–90 SEDAN AND WAGON

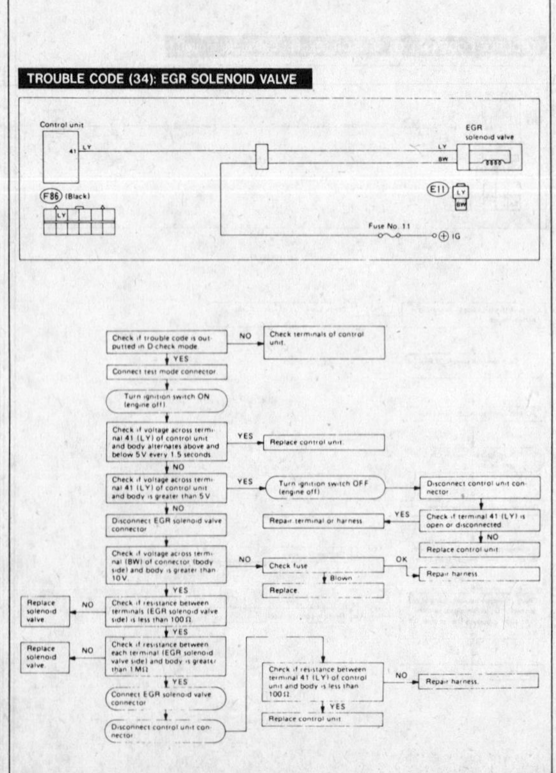
TROUBLE CODE (34): EGR SOLENOID VALVE
Control unit
41
LY
EGR solenoid valve
LY
BW
F86 (Black)
E11
Fuse No. 11
IG
Check if trouble code is outputted in D-check mode.
NO
Check terminals of control unit.
YES
Connect test mode connector.
Turn ignition switch ON (engine off).
Check if voltage across terminal 41 (LY) of control unit and body alternates above and below 5V every 1.5 seconds.
YES
Replace control unit.
NO
Check if voltage across terminal 41 (LY) of control unit and body is greater than 5V.
YES
Turn ignition switch OFF (engine off).
Disconnect control unit connector.
Check if terminal 41 (LY) is open or disconnected.
YES
Repair terminal or harness.
NO
Replace control unit.
NO
Disconnect EGR solenoid valve connector.
Check if voltage across terminal (BW) of connector (body side) and body is greater than 10V.
NO
Check fuse.
OK
Repair harness.
Blown
Replace.
YES
Check if resistance between terminals (EGR solenoid valve side) is less than 100Ω.
NO
Replace solenoid valve.
YES
Check if resistance between each terminal (EGR solenoid valve side) and body is greater than 1MΩ.
NO
Replace solenoid valve.
YES
Connect EGR solenoid valve connector.
Disconnect control unit connector.
Check if resistance between terminal 41 (LY) of control unit and body is less than 100Ω.
NO
Repair harness.
YES
Replace control unit.

DIAGNOSIS CHART – SEDAN AND WAGON AND 1990 LOYALE

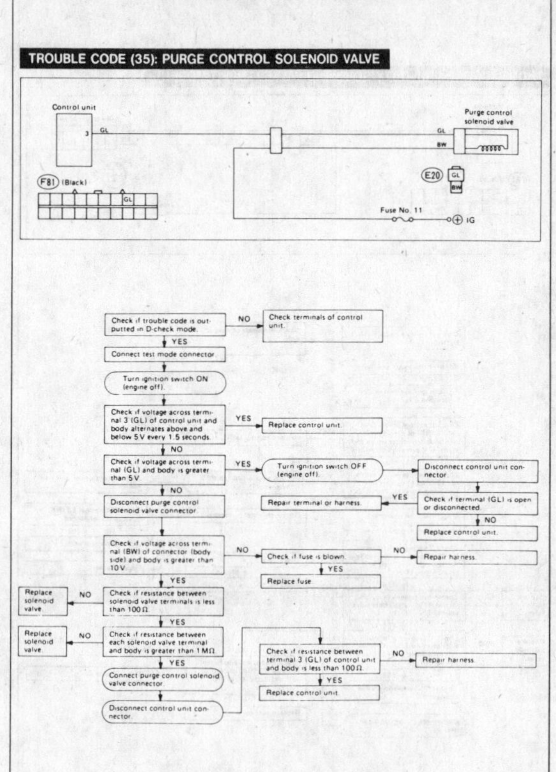
TROUBLE CODE (35): PURGE CONTROL SOLENOID VALVE
Control unit
Purge control solenoid valve
F81 (Black)
E20
Fuse No. 11
IG
Check if trouble code is outputted in D-check mode.
NO
Check terminals of control unit.
YES
Connect test mode connector.
Turn ignition switch ON (engine off).
Check if voltage across terminal 3 (GL) of control unit and body alternates above and below 5 V every 1.5 seconds.
YES
Replace control unit.
NO
Check if voltage across terminal (GL) and body is greater than 5 V.
YES
Turn ignition switch OFF (engine off).
Disconnect control unit connector.
Check if terminal (GL) is open or disconnected.
YES
Repair terminal or harness.
NO
Replace control unit.
NO
Disconnect purge control solenoid valve connector.
Check if voltage across terminal (BW) of connector (body side) and body is greater than 10 V.
NO
Check if fuse is blown.
NO
Repair harness.
YES
Replace fuse.
YES
Check if resistance between solenoid valve terminals is less than 100 Ω.
NO
Replace solenoid valve.
YES
Check if resistance between each solenoid valve terminal and body is greater than 1 MΩ.
NO
Replace solenoid valve.
YES
Connect purge control solenoid valve connector.
Disconnect control unit connector.
Check if resistance between terminal 3 (GL) of control unit and body is less than 100 Ω.
NO
Repair harness.
YES
Replace control unit.

DIAGNOSIS CHART – XT MODELS

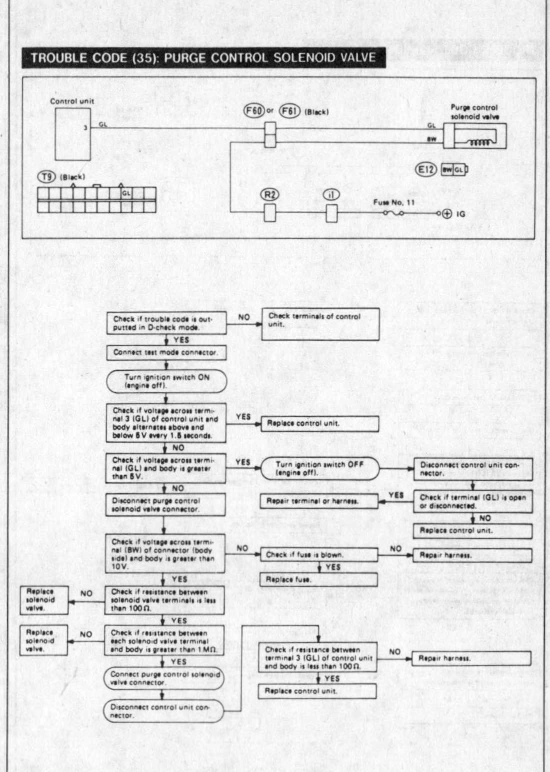
TROUBLE CODE (35): PURGE CONTROL SOLENOID VALVE
Control unit
F60 or F61 (Black)
Purge control solenoid valve
T9 (Black)
E12
R2
i1
Fuse No. 11
IG
Check if trouble code is outputted in D-check mode.
NO
Check terminals of control unit.
YES
Connect test mode connector.
Turn ignition switch ON (engine off).
Check if voltage across terminal 3 (GL) of control unit and body alternates above and below 5 V every 1.5 seconds.
YES
Replace control unit.
NO
Check if voltage across terminal (GL) and body is greater than 5 V.
YES
Turn ignition switch OFF (engine off).
Disconnect control unit connector.
Check if terminal (GL) is open or disconnected.
YES
Repair terminal or harness.
NO
Replace control unit.
NO
Disconnect purge control solenoid valve connector.
Check if voltage across terminal (BW) of connector (body side) and body is greater than 10 V.
NO
Check if fuse is blown.
NO
Repair harness.
YES
Replace fuse.
YES
Check if resistance between solenoid valve terminals is less than 100 Ω.
NO
Replace solenoid valve.
YES
Check if resistance between each solenoid valve terminal and body is greater than 1 MΩ.
NO
Replace solenoid valve.
YES
Connect purge control solenoid valve connector.
Disconnect control unit connector.
Check if resistance between terminal 3 (GL) of control unit and body is less than 100 Ω.
NO
Repair harness.
YES
Replace control unit.

DIAGNOSIS CHART – ALL MODELS

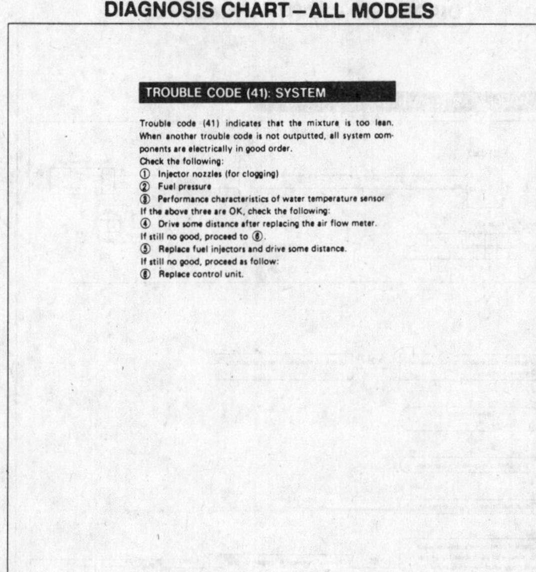

TROUBLE CODE (41): SYSTEM

Trouble code (41) indicates that the mixture is too lean. When another trouble code is not outputted, all system components are electrically in good order.

Check the following:

① Injector nozzles (for clogging)
② Fuel pressure
③ Performance characteristics of water temperature sensor

If the above three are OK, check the following:

④ Drive some distance after replacing the air flow meter.

If still no good, proceed to ⑥.

⑤ Replace fuel injectors and drive some distance.

If still no good, proceed as follow:

⑥ Replace control unit.

DIAGNOSIS CHART – SEDAN AND WAGON AND 1990 LOYALE

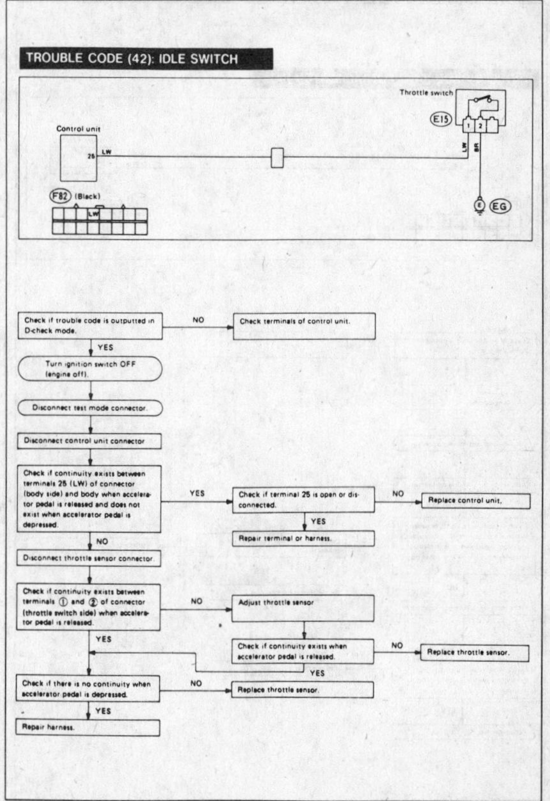
TROUBLE CODE (42): IDLE SWITCH
Throttle switch
E15
Control unit
F82 (Black)
EG
Check if trouble code is outputted in D-check mode.
NO
Check terminals of control unit.
YES
Turn ignition switch OFF (engine off).
Disconnect test mode connector.
Disconnect control unit connector
Check if continuity exists between terminals 25 (LW) of connector (body side) and body when accelerator pedal is released and does not exist when accelerator pedal is depressed.
YES
Check if terminal 25 is open or disconnected.
NO
Replace control unit.
YES
Repair terminal or harness.
NO
Disconnect throttle sensor connector.
Check if continuity exists between terminals ① and ② of connector (throttle switch side) when accelerator pedal is released.
NO
Adjust throttle sensor
Check if continuity exists when accelerator pedal is released.
NO
Replace throttle sensor.
YES
YES
Check if there is no continuity when accelerator pedal is depressed.
NO
Replace throttle sensor.
YES
Repair harness.

DIAGNOSIS CHART—XT MODELS

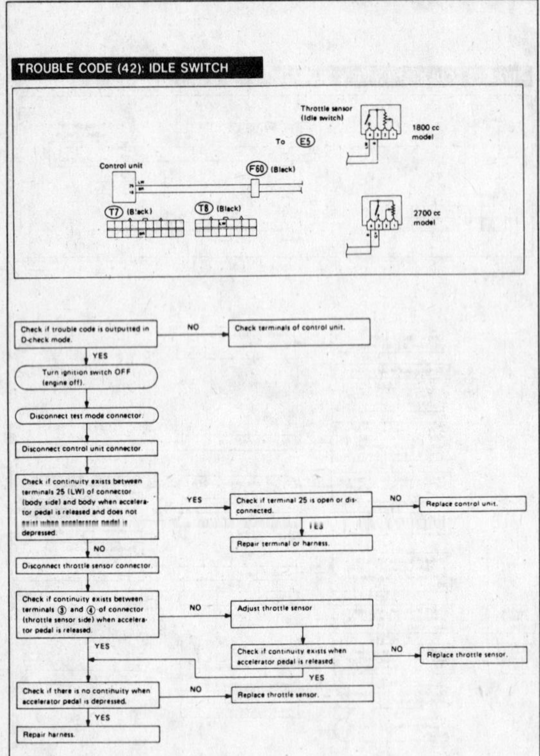
TROUBLE CODE (42): IDLE SWITCH
Throttle sensor (Idle switch)
1800 cc model
2700 cc model
Control unit
F60 (Black)
T7 (Black)
T8 (Black)
Check if trouble code is outputted in D-check mode.
NO
Check terminals of control unit.
YES
Turn ignition switch OFF (engine off).
Disconnect test mode connector.
Disconnect control unit connector.
Check if continuity exists between terminals 25 (LW) of connector (body side) and body when accelerator pedal is released and does not exist when accelerator pedal is depressed.
Check if terminal 25 is open or disconnected.
Replace control unit.
Repair terminal or harness.
Disconnect throttle sensor connector.
Check if continuity exists between terminals ③ and ④ of connector (throttle sensor side) when accelerator pedal is released.
Adjust throttle sensor.
Check if continuity exists when accelerator pedal is released.
Replace throttle sensor.
Check if there is no continuity when accelerator pedal is depressed.
Replace throttle sensor.
Repair harness.

DIAGNOSIS CHART—SEDAN AND WAGON AND 1990 LOYALE

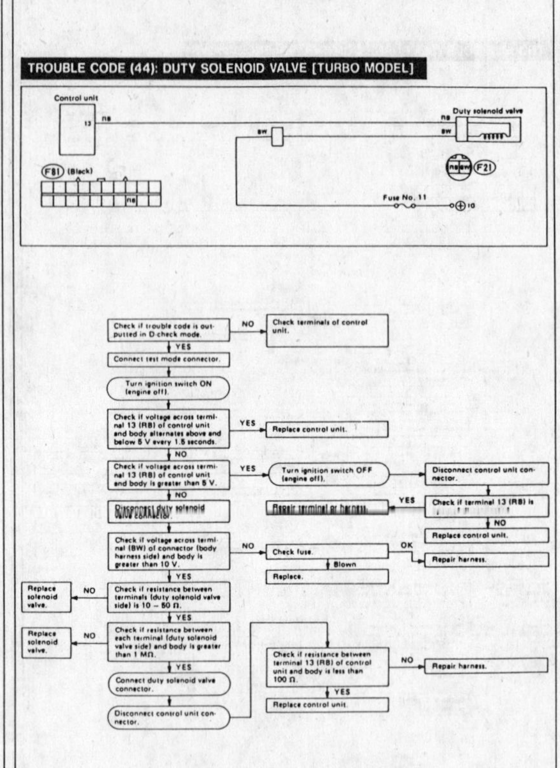
TROUBLE CODE (44): DUTY SOLENOID VALVE [TURBO MODEL]
Control unit
Duty solenoid valve
F81 (Black)
F21
Fuse No. 11
Check if trouble code is outputted in D-check mode.
NO
Check terminals of control unit.
YES
Connect test mode connector.
Turn ignition switch ON (engine off).
Check if voltage across terminal 13 (RB) of control unit and body alternates above and below 8 V every 1.5 seconds.
Replace control unit.
Check if voltage across terminal 13 (RB) of control unit and body is greater than 8 V.
Turn ignition switch OFF (engine off).
Disconnect control unit connector.
Disconnect duty solenoid valve connector.
Repair terminal or harness.
Check if terminal 13 (RB) is
Replace control unit.
Check if voltage across terminal (BW) of connector (body harness side) and body is greater than 10 V.
Check fuse.
OK
Repair harness.
Blown
Replace.
Replace solenoid valve.
Check if resistance between terminals (duty solenoid valve side) is 10 – 50 Ω.
Check if resistance between each terminal (duty solenoid valve side) and body is greater than 1 MΩ.
Connect duty solenoid valve connector.
Disconnect control unit connector.
Check if resistance between terminal 13 (RB) of control unit and body is less than 100 Ω.
Repair harness.
Replace control unit.

DIAGNOSIS CHART—1988–90 SEDAN AND WAGON

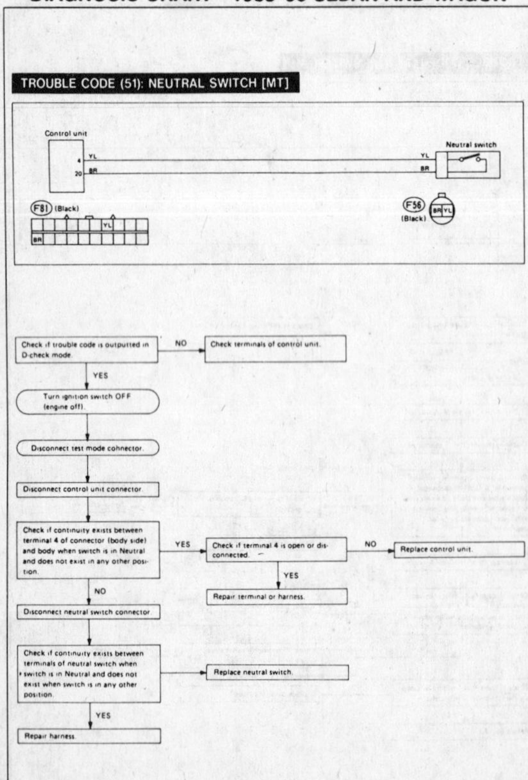
TROUBLE CODE (51): NEUTRAL SWITCH [MT]
Control unit
Neutral switch
F81 (Black)
F56 (Black)
Check if trouble code is outputted in D-check mode.
NO
Check terminals of control unit.
YES
Turn ignition switch OFF (engine off).
Disconnect test mode connector.
Disconnect control unit connector.
Check if continuity exists between terminal 4 of connector (body side) and body when switch is in Neutral and does not exist in any other position.
Check if terminal 4 is open or disconnected.
Replace control unit.
Repair terminal or harness.
Disconnect neutral switch connector.
Check if continuity exists between terminals of neutral switch when switch is in Neutral and does not exist when switch is in any other position.
Replace neutral switch.
Repair harness.

DIAGNOSIS CHART—XT MODELS

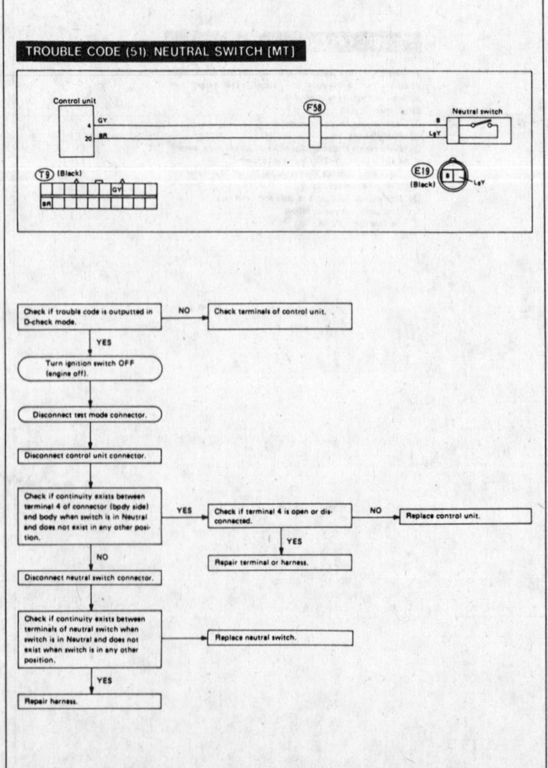
TROUBLE CODE (51) NEUTRAL SWITCH [MT]
Control unit
F58
Neutral switch
T9 (Black)
E19 (Black)
Check if trouble code is outputted in D-check mode.
NO
Check terminals of control unit.
YES
Turn ignition switch OFF (engine off).
Disconnect test mode connector.
Disconnect control unit connector.
Check if continuity exists between terminal 4 of connector (body side) and body when switch is in Neutral and does not exist in any other position.
Check if terminal 4 is open or disconnected.
Replace control unit.
Repair terminal or harness.
Disconnect neutral switch connector.
Check if continuity exists between terminals of neutral switch when switch is in Neutral and does not exist when switch is in any other position.
Replace neutral switch.
Repair harness.

TROUBLESHOOTING CHART WITH SELECT MONITOR 1990 LOYALE

Troubleshooting Chart with Select Monitor

BASIC TROUBLESHOOTING CHART

Problem occurs No trouble codes appear in Read Memory, U-Check or D-check mode.

↓

Measure each item in "F" mode (select-monitor function).

↓

Refer to "General Troubleshooting Table" for inspection order.

↓

Compare measured value with "Evaluation standard". Determine item which is outside "Evaluation standard".

↓

* Check sensor or actuator corresponding with item outside "Evaluation standard".

↓

* When item is not constantly outside "Evaluation standard", problem may be due to poor harness contact. Shake affected sensor/actuator connector or gently pull its harness to check if trouble code appears.

OUTPUT MODES OF SELECT MONITOR – 1990 LOYALE

OUTPUT MODES OF SELECT MONITOR

FUNCTION MODE

MODE	Contents	Abbr.	Unit	Contents of display
00	PROM ID Number	YEAR	–	Model year of vehicle to which select monitor is connected
02	Vehicle Speed Sensor	VSP	m/h	Vehicle speed inputted from vehicle speed sensor
03	Vehicle Speed Sensor	VSP	km/h	Vehicle speed inputted from vehicle speed sensor
05	Engine RPM	EREV	rpm	Engine revolution data based on "reference" signal sent from crank angle sensor
06	Water Temp Sensor	TW	deg F	Coolant temperature inputted from water temperature sensor
07	Water Temp Sensor	TW	deg C	Coolant temperature inputted from water temperature sensor
08	Ignition Timing	ADVS	deg	Ignition timing determined by control unit in relation to signals sent from various sensors
09	Air Flow Sensor	QA	V	Voltage inputted from air flow meter
10	Load Data	LDATA	–	Engine load value determined by related sensor signals
11	Throttle Sensor	THV	V	Voltage inputted from throttle position sensor
12	Injector Pulse Width	TIM	mS	Duration of pulse flowing through injectors
14	O_2 Sensor	O_2	V	Voltage outputted from O2 sensor
17	ALPHA	ALPHA	%	AF correction ratio determined in relation to signal outputted from O2 sensor
18	Knock Sensor	RTRD	deg	Ignition timing correction determined in relation to signal-inputted from knock sensor
19	Duty Sol. (Waste gate)	WGC	%	"Duty" ratio of waste gate solenoid valve
23	Fuel Pump Duty	FPDY	%	"Duty ratio of fuel pump control
A0 ~ A6	ON ↔ OFF signal			
B0	Trouble Code	DIAG	–	Trouble code in U- or D-check mode
B1	Trouble Code	DIAG	–	Trouble code memorized in U-check mode

ON/OFF SIGNAL LIST – 1990 LOYALE

ON ↔ OFF SIGNAL LIST

MODE	LED No.	Contents	Display	LED "ON" requirements
A0	3	Canister Purge Cont.	CN	Canister purge "ON"
	4	Fuel Pump Control	FP	Fuel pump relay in operation
	10	O_2 Monitor	O_2	O_2 monitor lamp "ON"
A1	1	Idle SW	ID	Idle switch "ON"
	3	A/C SW	AC	Air conditioner switch "ON"
	10	O_2 Monitor	O_2	O_2 monitor lamp "ON"
A2	2	Ignition SW	IG	Ignition switch "ON"
	10	O_2 Monitor	O_2	O_2 monitor lamp "ON"
A3	1	Neutral SW	NT	Neutral switch "ON"
	10	O_2 Monitor	O_2	O_2 monitor lamp "ON"
A4	2	EGR	ER	(Not applicable for 90MY)
	10	O_2 Monitor	O_2	O_2 monitor lamp "ON"
A6	1	Knock Signal	KS	Engine knocks occur
	10	O_2 Monitor	O_2	O_2 monitor lamp "ON"

LED No. 10 (O_2 monitor) operates in any mode.

EXPLANATION OF MODES – 1990 LOYALE

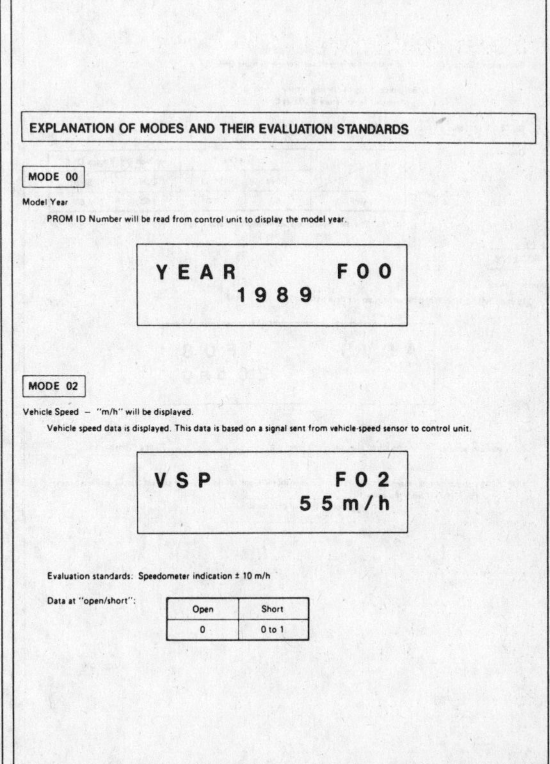

EXPLANATION OF MODES AND THEIR EVALUATION STANDARDS

MODE 00

Model Year

PROM ID Number will be read from control unit to display the model year.

YEAR F00
1989

MODE 02

Vehicle Speed – "m/h" will be displayed.

Vehicle speed data is displayed. This data is based on a signal sent from vehicle-speed sensor to control unit.

VSP F02
55 m/h

Evaluation standards: Speedometer indication ± 10 m/h

Data at "open/short":

Open	Short
0	0 to 1

EXPLANATION OF MODES—1990 LOYALE

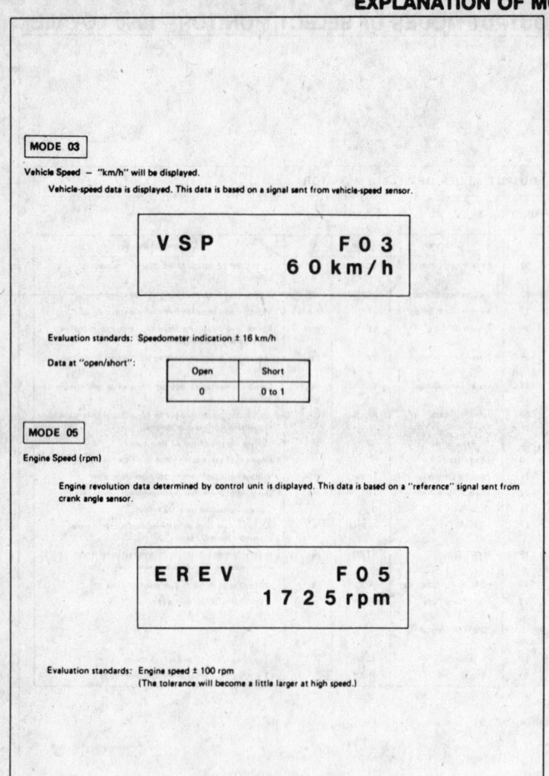

MODE 03

Vehicle Speed — "km/h" will be displayed.

Vehicle-speed data is displayed. This data is based on a signal sent from vehicle-speed sensor.

```
VSP           F03
            60km/h
```

Evaluation standards: Speedometer indication ± 16 km/h

Data at "open/short":

Open	Short
0	0 to 1

MODE 05

Engine Speed (rpm)

Engine revolution data determined by control unit is displayed. This data is based on a "reference" signal sent from crank angle sensor.

```
EREV          F05
          1725rpm
```

Evaluation standards: Engine speed ± 100 rpm
(The tolerance will become a little larger at high speed.)

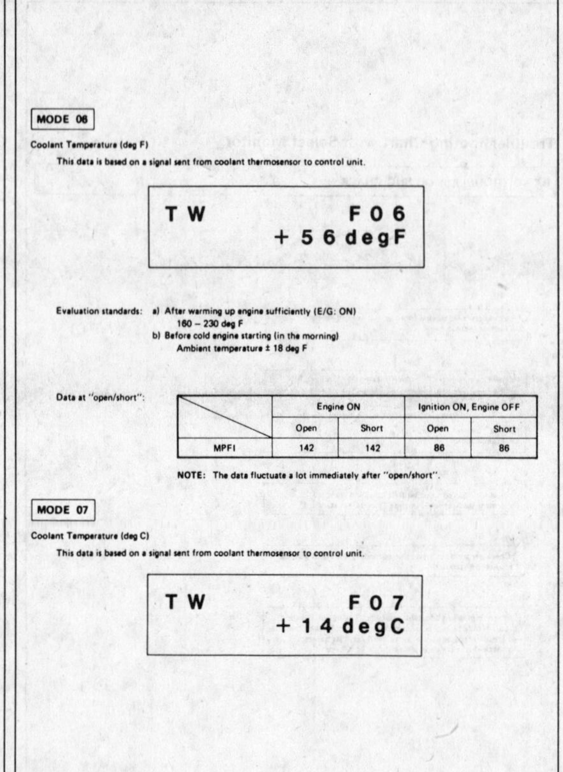

MODE 06

Coolant Temperature (deg F)

This data is based on a signal sent from coolant thermosensor to control unit.

```
TW            F06
         +56degF
```

Evaluation standards: a) After warming up engine sufficiently (E/G: ON)
160 – 230 deg F
b) Before cold engine starting (in the morning)
Ambient temperature ± 18 deg F

Data at "open/short":

	Engine ON		Ignition ON, Engine OFF	
	Open	Short	Open	Short
MPFI	142	142	86	86

NOTE: The data fluctuate a lot immediately after "open/short".

MODE 07

Coolant Temperature (deg C)

This data is based on a signal sent from coolant thermosensor to control unit.

```
TW            F07
         +14degC
```

EXPLANATION OF MODES—1990 LOYALE

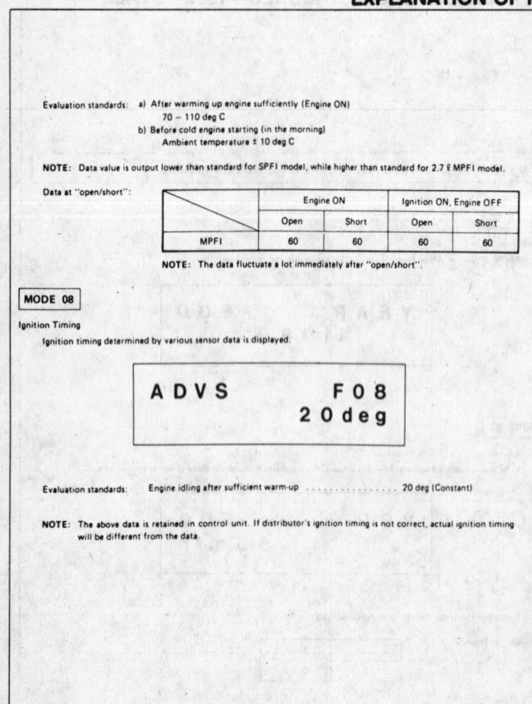

Evaluation standards: a) After warming up engine sufficiently (Engine ON)
70 – 110 deg C
b) Before cold engine starting (in the morning)
Ambient temperature ± 10 deg C

NOTE: Data value is output lower than standard for SPFI model, while higher than standard for 2.7 ℓ MPFI model.

Data at "open/short":

	Engine ON		Ignition ON, Engine OFF	
	Open	Short	Open	Short
MPFI	60	60	60	60

NOTE: The data fluctuate a lot immediately after "open/short".

MODE 08

Ignition Timing

Ignition timing determined by various sensor data is displayed.

```
ADVS          F08
            20deg
```

Evaluation standards: Engine idling after sufficient warm-up .... 20 deg (Constant)

NOTE: The above data is retained in control unit. If distributor's ignition timing is not correct, actual ignition timing will be different from the data.

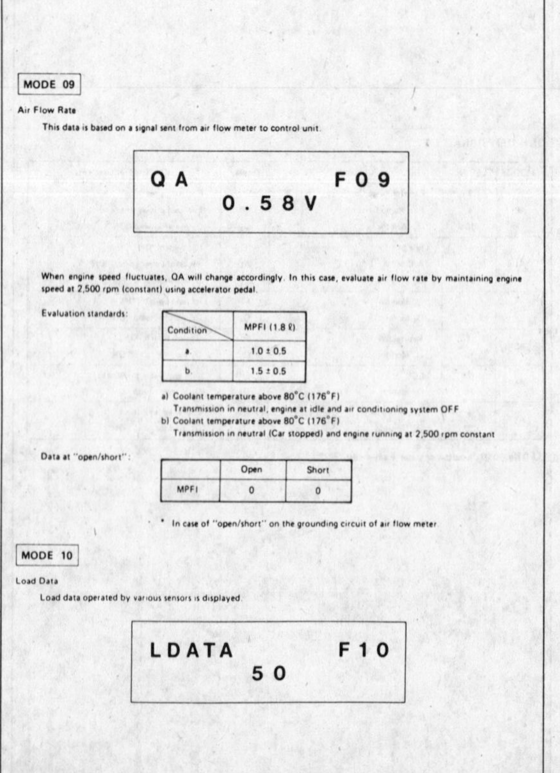

MODE 09

Air Flow Rate

This data is based on a signal sent from air flow meter to control unit.

```
QA            F09
       0.58V
```

When engine speed fluctuates, QA will change accordingly. In this case, evaluate air flow rate by maintaining engine speed at 2,500 rpm (constant) using accelerator pedal.

Evaluation standards:

Condition	MPFI (1.8 ℓ)
a.	1.0 ± 0.5
b.	1.5 ± 0.5

a) Coolant temperature above 80°C (176°F)
Transmission in neutral, engine at idle and air conditioning system OFF
b) Coolant temperature above 80°C (176°F)
Transmission in neutral (Car stopped) and engine running at 2,500 rpm constant

Data at "open/short":

	Open	Short
MPFI	0	0

* In case of "open/short" on the grounding circuit of air flow meter

MODE 10

Load Data

Load data operated by various sensors is displayed.

```
LDATA         F10
       50
```

EXPLANATION OF MODES—1990 LOYALE

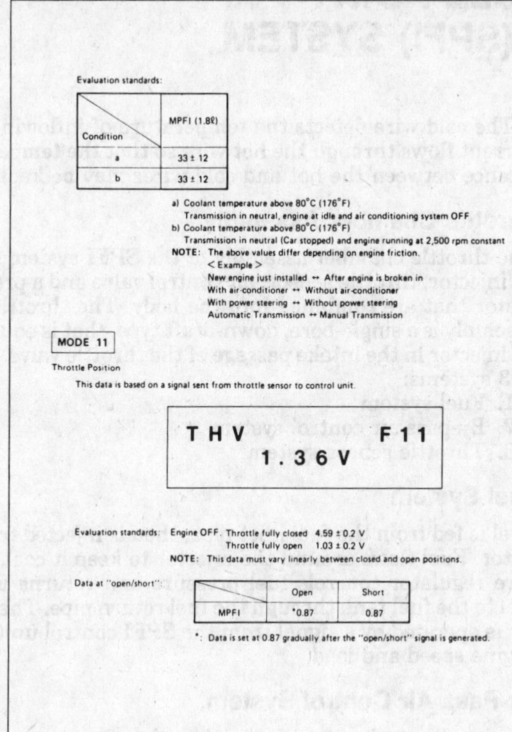

Evaluation standards:

Condition	MPFI (1.8ℓ)
a	33 ± 12
b	33 ± 12

a) Coolant temperature above 80°C (176°F)
Transmission in neutral, engine at idle and air conditioning system OFF
b) Coolant temperature above 80°C (176°F)
Transmission in neutral (Car stopped) and engine running at 2,500 rpm constant

NOTE: The above values differ depending on engine friction.
< Example >
New engine just installed ↔ After engine is broken in
With air-conditioner ↔ Without air-conditioner
With power steering ↔ Without power steering
Automatic Transmission ↔ Manual Transmission

MODE 11

Throttle Position

This data is based on a signal sent from throttle sensor to control unit.

THV F11
1.36V

Evaluation standards: Engine OFF. Throttle fully closed 4.59 ± 0.2 V
Throttle fully open 1.03 ± 0.2 V

NOTE: This data must vary linearly between closed and open positions.

Data at "open/short":

	Open	Short
MPFI	*0.87	0.87

*: Data is set at 0.87 gradually after the "open/short" signal is generated.

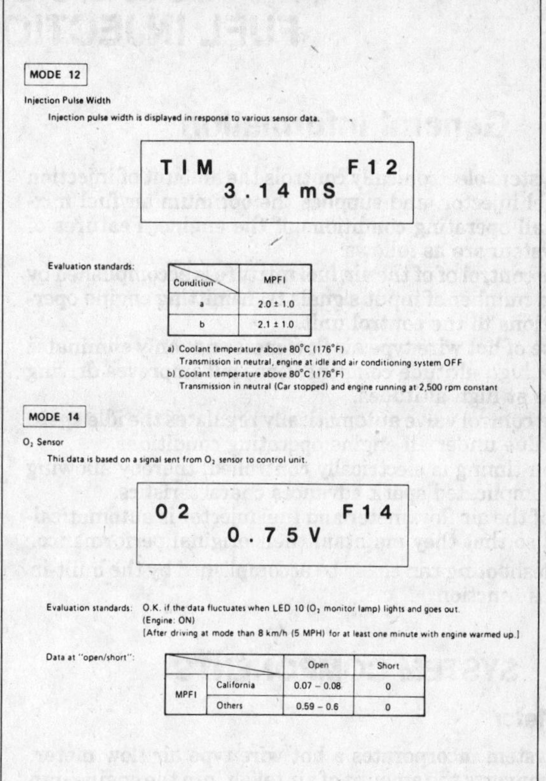

MODE 12

Injection Pulse Width

Injection pulse width is displayed in response to various sensor data.

TIM F12
3.14mS

Evaluation standards:

Condition	MPFI
a	2.0 ± 1.0
b	2.1 ± 1.0

a) Coolant temperature above 80°C (176°F)
Transmission in neutral, engine at idle and air conditioning system OFF
b) Coolant temperature above 80°C (176°F)
Transmission in neutral (Car stopped) and engine running at 2,500 rpm constant

MODE 14

O_2 Sensor

This data is based on a signal sent from O_2 sensor to control unit.

O2 F14
0.75V

Evaluation standards: O.K. if the data fluctuates when LED 10 (O_2 monitor lamp) lights and goes out.
(Engine: ON)
[After driving at mode than 8 km/h (5 MPH) for at least one minute with engine warmed up.]

Data at "open/short":

		Open	Short
MPFI	California	0.07 – 0.08	0
	Others	0.59 – 0.6	0

EXPLANATION OF MODES—1990 LOYALE

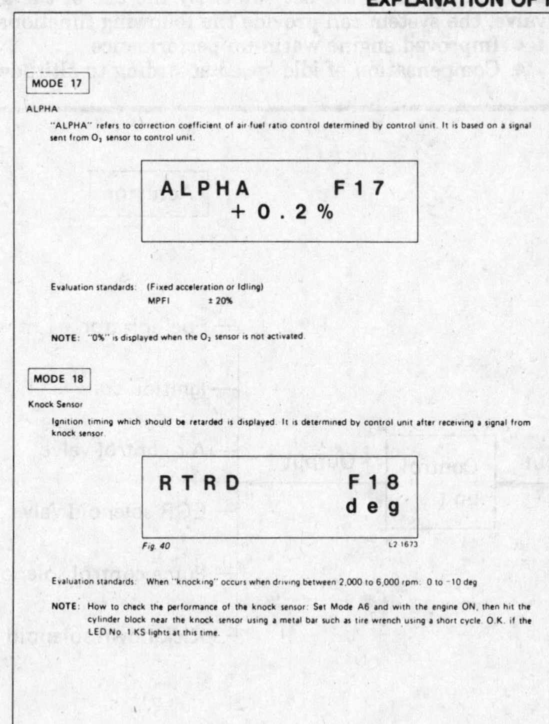

MODE 17

ALPHA

"ALPHA" refers to correction coefficient of air-fuel ratio control determined by control unit. It is based on a signal sent from O_2 sensor to control unit.

ALPHA F17
+0.2%

Evaluation standards: (Fixed acceleration or Idling)
MPFI ± 20%

NOTE: "0%" is displayed when the O_2 sensor is not activated.

MODE 18

Knock Sensor

Ignition timing which should be retarded is displayed. It is determined by control unit after receiving a signal from knock sensor.

RTRD F18
deg

Fig. 40 L2 1673

Evaluation standards: When "knocking" occurs when driving between 2,000 to 6,000 rpm. 0 to −10 deg

NOTE: How to check the performance of the knock sensor: Set Mode A6 and with the engine ON, then hit the cylinder block near the knock sensor using a metal bar such as tire wrench using a short cycle. O.K. if the LED No. 1 KS lights at this time.

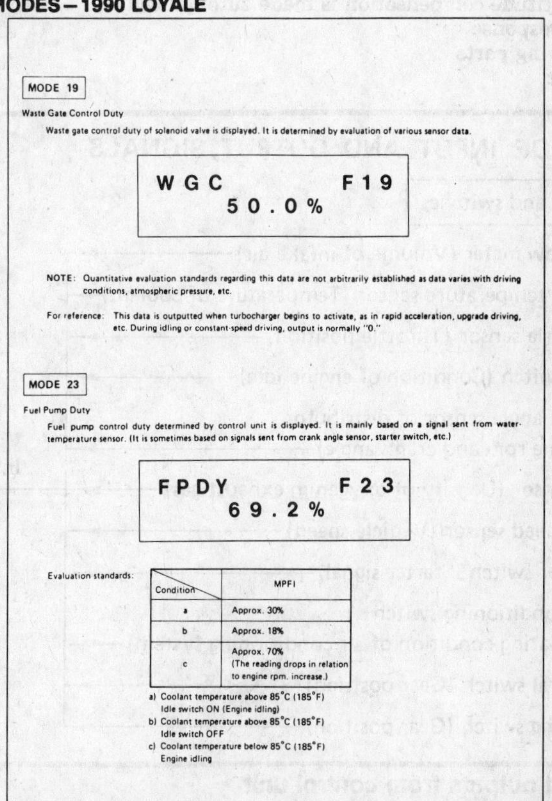

MODE 19

Waste Gate Control Duty

Waste gate control duty of solenoid valve is displayed. It is determined by evaluation of various sensor data.

WGC F19
50.0%

NOTE: Quantitative evaluation standards regarding this data are not arbitrarily established as data varies with driving conditions, atmospheric pressure, etc.

For reference: This data is outputted when turbocharger begins to activate, as in rapid acceleration, upgrade driving, etc. During idling or constant-speed driving, output is normally "0."

MODE 23

Fuel Pump Duty

Fuel pump control duty determined by control unit is displayed. It is mainly based on a signal sent from water-temperature sensor. (It is sometimes based on signals sent from crank angle sensor, starter switch, etc.)

FPDY F23
69.2%

Evaluation standards:

Condition	MPFI
a	Approx. 30%
b	Approx. 18%
c	Approx. 70% (The reading drops in relation to engine rpm. increase.)

a) Coolant temperature above 85°C (185°F)
Idle switch ON (Engine idling)
b) Coolant temperature above 85°C (185°F)
Idle switch OFF
c) Coolant temperature below 85°C (185°F)
Engine idling

SUBARU SINGLE POINT FUEL INJECTION (SPFI) SYSTEM

General Information

The SPFI system electronically controls the amount of injection from the fuel injector, and supplies the optimum air/fuel mixture under all operating conditions of the engine. Features of the SPFI system are as follows:

1. Precise control of of the air/fuel mixture is accomplished by an increased number of input signals transmitting engine operating conditions to the control unit.
2. The use of hot wire type air flow meter not only eliminates the need for high altitude compensation, but improves driving performance at high altitudes.
3. The air control valve automatically regulates the idle speed to the set value under all engine operating conditions.
4. Ignition timing is electrically controlled, thereby allowing the use of complicated spark advances characteristics.
5. Wear of the air flow meter and fuel injector is automatically corrected so that they maintain their original performance.
6. Troubleshooting can easily be accomplished by the built-in self-diagnosis function.

SYSTEM COMPONENTS

Air Flow Meter

The SPFI system incorporates a hot wire type air flow meter. This meter converts the amount of air taken into the engine into an electric signal by the use of the heat transfer between the incoming air and a heating resistor (hot wire) located in the air intake. Feature of the air flow meter are as follows:

- High altitude compensation is made automatically
- Quick response
- No moving parts
- Compact

The cold wire detects the temperature of inflowing air, then current flows through the hot wire so that the temperature difference between the hot and cold wires may be kept constant.

Throttle Chamber Assembly

The throttle chamber assembly of the SPFI system consists of an injector, throttle sensor, air control valve and a pressure regulator that are combined into the body. The throttle chamber assembly is a single-bore, down-draft type, that is equipped with an injector in the intake passage of the throttle valve. It consists of 3 systems:

1. Fuel system
2. By-pass air control system
3. Throttle sensor system

Fuel System

Fuel is fed from the fuel inlet pipe, then is injected from the injector. Fuel flows around the injector to keep it cool. The pressure regulator controls fuel pressure and returns un-injected fuel to the fuel tank through the fuel return pipe. The fuel injector is operated by a signal from the SPFI control unit, based on engine speed and load.

By-Pass Air Control System

An air passage, by-passing the throttle valve, is provided to direct the air into the lower portion of the throttle valve. The air control valve is located in the middle of this passage and controls the amount of air during engine starting, idle speed, etc.

The air control valve is controlled by signals from the SPFI control unit and regulates the opening of the by-pass to maintain idle speed at the set value. By the use of the air control valve, the system can provide the following functions:

- Improved engine warm-up performance
- Compensation of idle speed according to altitude

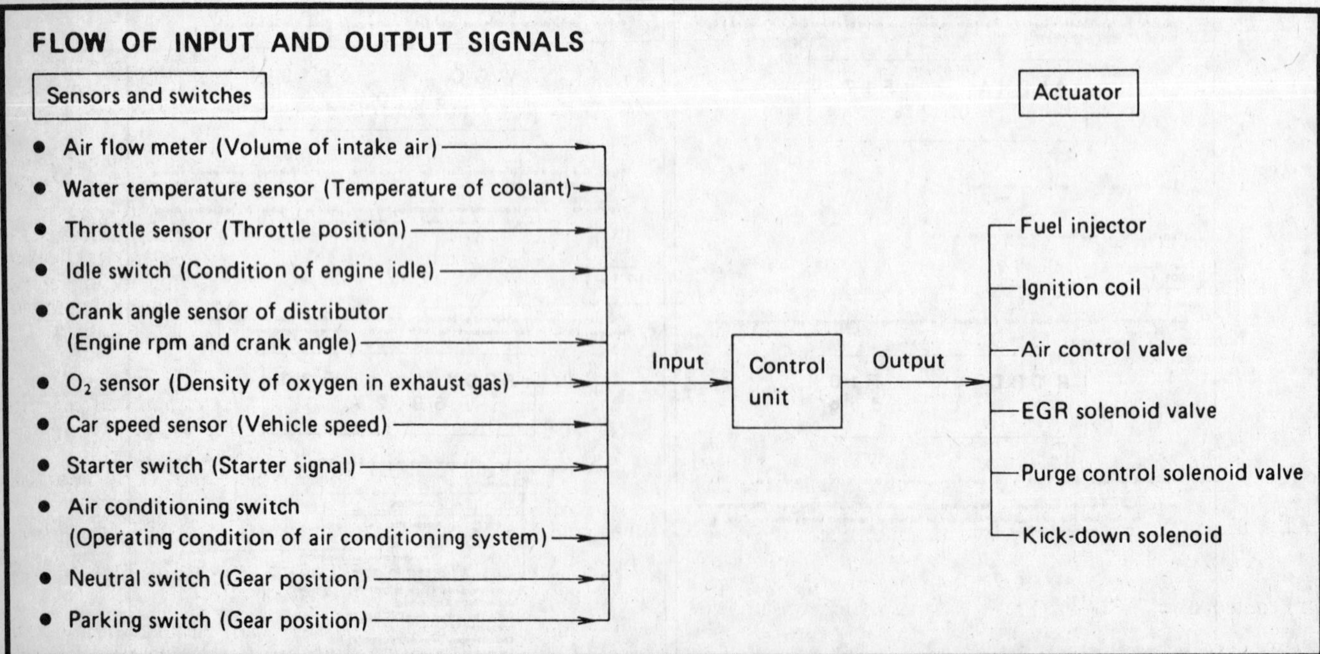
FLOW OF INPUT AND OUTPUT SIGNALS
Sensors and switches
Air flow meter (Volume of intake air)
Water temperature sensor (Temperature of coolant)
Throttle sensor (Throttle position)
Idle switch (Condition of engine idle)
Crank angle sensor of distributor
(Engine rpm and crank angle)
O2 sensor (Density of oxygen in exhaust gas)
Car speed sensor (Vehicle speed)
Starter switch (Starter signal)
Air conditioning switch
(Operating condition of air conditioning system)
Neutral switch (Gear position)
Parking switch (Gear position)
Input
Control unit
Output
Actuator
Fuel injector
Ignition coil
Air control valve
EGR solenoid valve
Purge control solenoid valve
Kick-down solenoid

Inputs and outputs from control unit

- Compensation of idle speed with the A/C system operating
- Compensation for idle speed fluctuation with wear

Throttle Sensor System

A throttle position sensor is incorporated with a potentiometer and an idle switch interlocked with the throttle valve shaft. This sensor sends the SPFI control unit a potentiometer output signal corresponding to the opening of the throttle valve and a idle switch signal that turns **ON** only when the throttle is opened to the idle position. Using these signals, the SPFI control unit precisely controls the air/fuel ratio during acceleration and deceleration as well as idling.

Ignition System

The ignition system consists of a distributor containing a photoelectric crank angle sensor, an ignition coil equipped with a power transistor and the SPFI control unit. The crank-angle signal and reference signal detected by the photoelectric crank-angle sensor are sent to the SPFI control unit. The SPFI control unit determines the optimum ignition timing from these signals and other engine operating paramaters, and transmits an ignition signal to the ignition coil igniter. The igniter amplifies this ignition signal and causes the primary current to flow intermittently in the ignition coil.

Air/Fuel Ratio Control System

This system has been designed to stabilize the quality of the hotwire type air flow meter and fuel injector, to maintain their original performance by correcting their qualitative variation and wear. By learning the feedback control amount of the O_2 sensor, the system controls the SPFI control unit to automatically set a coefficient of correction, thereby, the fuel injector always achieves fuel injection under all operating conditions.

Oxygen (O_2) Sensor

The O_2 sensor is mounted on the front exhaust pipe, and is used to sense oxygen concentration in the exhaust gas. If the fuel ratio is leaner than the optimum air/fuel mixture (14.7:1), the exhaust gas contains more oxygen. If the fuel ratio is richer than the optimum air/fuel ratio, the exhaust gas hardly contains oxygen. Therefore, examination of the oxygen concentration in the exhaust gas makes it possible to show whether the air/fuel ratio is leaner or richer than the optimum air/fuel mixture.

The O_2 sensor incorporates a zirconia tube (ceramic) which generates voltage if there is a difference in oxygen concentration between the inside and outside of the tube. The inside and outside of the zirconia tube is coated with platinum for the purpose of catalysis and electrode provisions. The hexagon screw on the outside is grounded to the exhaust pipe, and the inside is connected to the SPFI control unit through a harness.

When a rich air/fuel mixture is burnt in the cylinder in the oxygen in the exhaust gases reacts almost completely through the catalytic action of the platinum coating on the surface of the zirconia tube. This results in very large difference in the oxygen concentration between the inside and the outside, and the electromotive force generated is large.

When a lean air/fuel mixture is burnt in the cylinder, oxygen remains in the exhaust gases even after the catalytic action, and this results in a small difference in the oxygen concentration. The electromotive force is very small.

The difference in oxygen concentration changes greatly in the area of the optimum air/fuel ratio, and the change in the electromotive force is also large. By inputting this information to the SPFI control unit, the air/fuel ratio of the supplied mixture can be determined easily. The O_2 sensor does not generate much electromotive force when the temperature is low. The electromotive force stabilizes at temperatures of approximately 572–752°F (300–400°C).

Coolant Thermosensor

The coolant thermosensor is located on the thermocasing of the intake manifold. Its thermistor changes resistance with respect to temperature. A water temperature signal converted into resistance is transmitted to the control unit, to control the amount of fuel injection, ignition timing, purge control solenoid valve, etc.

EGR Gas Temperature Sensor

CALIFORNIA MODELS

The EGR gas temperature sensor is located in the EGR gas passage on the intake manifold. An EGR gas temperature signal converted into resistance is sent to the control unit for EGR system diagnosis.

Kick-Down Control System

On models equipped with automatic transmissions, a throttle sensor is used in place of the previous kick-down switch. It sends a signal to the control unit to set the throttle valve to a specified position. When the throttle valve is in that position, the kick-down control relay turns on.

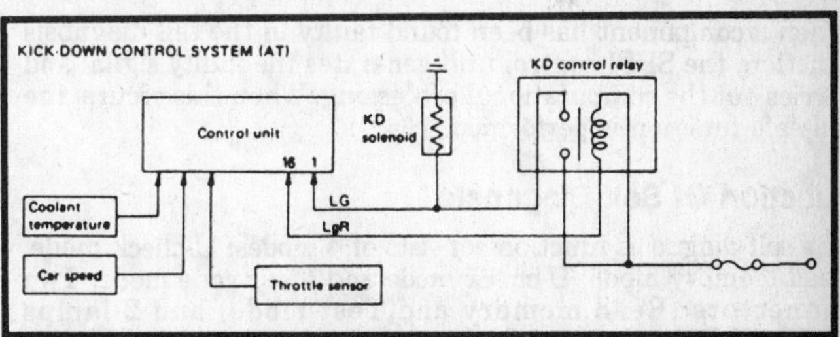
KICK-DOWN CONTROL SYSTEM (AT)
KD control relay
Control unit
KD solenoid
16
1
LG
LgR
Coolant temperature
Car speed
Throttle sensor

Kick-down control system

SERVICE PRECAUTIONS

- Never connect the battery in reverse polarity. The SPFI control unit will be destroyed instantly and the fuel injectors will be damaged.
- Never disconnect the battery terminals while the engine is running. A large counter of electromotive force will be generated in the alternator, causing damage to the electronic parts, such as the SPFI control unit.
- Before disconnecting the connectors of each sensor and the SPFI control unit, ensure that the ignition switch is in the **OFF** position. If not, the SPFI control unit could be damaged.
- The connectors to each sensor and all other harness connectors are designed to be waterproof. However, it is still necessary to take caution not to allow water to get into the connectors when washing the vehicle or when servicing the vehicle on a rainy day.
- Never drop a SPFI related part.
- Note the following precautions when installing a radio:
  - Keep the antenna as far apart as possible from the control unit (the control unit is located under the steering column, inside the instrument panel lower trim panel).
  - The antenna line must be positioned as far apart as possible from the SPFI control unit and SPFI harness.
  - Carefully adjust the antenna for correct matching.

NOTE: Incorrect installation of the radio may affect the operation of the SPFI control unit.

- Before disconnecting the fuel line, disconnect the fuel pump connector, then crank the engine for approximately 5 seconds the pressure in the fuel system. If the engine starts during this operation, let it run until it stops.

Diagnosis and Testing

SELF-DIAGNOSIS SYSTEM

The self-diagnosis system detects and indicates faults, in various inputs and outputs of the electronic control unit. The warning lamp (**CHECK ENGINE** light), located on the instrument panel indicates a fault or trouble, and the LED (light emitting diode) in the control unit indicates a trouble code. A fail-safe function is incorporated into the system to ensure minimal driveability if a failure of a sensor occurs.

Self-Diagnosis Function

The SPFI control unit carries out the computational processing of the input information received from various sensors and produces the output information for operating the fuel injectors, fuel pump, etc. This computational processing reads out all the input/output information to examine matching with the predetermined levels (proper values or ranges). If a predetermined level is not satisfied, or a fault is detected, the warning lamp signals the driver. When this occurs, the self-diagnosis function is performed.

Fail-Safe Function

When a component has been found faulty in the self-diagnosis function, the SPFI control unit generates the faulty signal and carries out the computational processing. When this occurs, the fail-safe function is performed.

Function Of Self-Diagnosis

The self-diagnosis function consists of 4 modes: U-check mode, Read memory mode, D-check mode and Clear code mode. Two connectors (Read memory and Test mode) and 2 lamps (**CHECK ENGINE** and O_2 monitor) are used. The connectors are used for mode selection and the lamps monitor the type of problem.

U-CHECK MODE

The U-Check mode diagnosis only the SPFI components necessary for start-up and driveabilty. When a fault occurs, the warning lamp (**CHECK ENGINE**) lights to indicate to the user that inspection is necessary. The diagnosis of other components which do not effect start-up and driveabilty are excluded from this mode.

READ MEMORY CODE

This mode is used to read past problems (even when the vehicle monitor lamps are off). It is most effective in detecting poor contacts or loose connections of the connectors, harness, etc.

D-CHECK MODE

This mode is used to check the entire SPFI system and to detect faulty components.

F-CHECK MODE (SELECT MONITOR)

This mode is used on 1990 Loyale models to measure the performance characteristics of system components when no trouble codes appear in the Read Memory, U-Check or D-Check mode (although problems have occurred or are occurring).

When working in this mode, use Select Monitor Cartridge No. 498347001 and Select Monitor Instruction Manual No.498327000, for details concerning select monitor and troubleshooting.

CLEAR MEMORY MODE

This mode is used to clear the trouble code from the memory after all faults have been corrected.

AIR FLOW METER

Testing

1. Check for leaks or damage in the connection between the air intake boot and the air flow meter. Repair if necessary.
2. Disconnect the electrical connector from the air flow meter, then remove the air intake boot, and the air flow meter from the air cleaner assembly.
3. Check the exterior of the air flow meter for damage.
4. Check the interior of the air flow meter for foreign particules, water, or oil in the passages, especially in the bypass. If any of the above faults are noted, replace the air flow meter.
5. If none of the above faults are noted, proceed as follows:
   a. Turn the ignition switch to the **OFF** position, then install the air flow meter onto air cleaner assembly.
   b. Disconnect the air flow meter electrical connector, then slide back the rubber boot.

NOTE: Conduct the following test by connecting the tester pins to the connector terminals on the side from which the rubber boot was removed.

   c. Using an ohmmeter, measure the resistance between terminal (B) on the connector and ground. Ohmmeter should read 10 ohms. If ohmmeter reading exceeds 10 ohms, check the harness and internal circuits of the control unit for discontinuity.
   d. Turn the ignition switch to the **ON** position.
   e. Using a voltmeter, measure voltage across power terminal (R) and ground. Voltmeter should read 10 volts. If specifications are not as specified, check the power line (battery, fuse, control unit, harness connector, etc).
   f. Connect the air flow meter electrical connector. Using a voltmeter, connect the positive lead of the tester to terminal (W) and the negative lead of the tester to terminal (B) and measure voltage across the 2 terminals. Voltmeter should read 0.1–0.5 volts. If specifications are not as specified, replace the air flow meter.
   g. Remove the upper section of the air cleaner assembly, then blow air in from the air cleaner side and check if voltage across terminals (W) and (B) is higher than specification obtained in Step f. If not, replace the air flow meter.

IGNITION TIMING

Inspection

1. Start, and bring engine up to normal operating temperature.
2. Turn the ignition switch to the **OFF** position.
3. Disconnect the throttle sensor electrical connector.
4. Using an ohmmeter, check that the resistance between terminal (A) and (B) is 0 ohms when the accelerator pedal is released. If ohmmeter reads infinity, proceed to "Throttle Sensor (Idle Contact)."
5. Connect the throttle sensor electrical connector, then a suitable test mode connector.
6. Start, and run engine at idle speed. Do not depress the accelerator pedal.
7. Check and adjust ignition timing, if necessary. Ignition timing should be 20 degrees BTDC.

NOTE: The specified ignition timing can be obtained regardless of engine speed when the engine is idling without depressing the accelerator pedal.

8. Stop engine, and disconnect the test mode connector.

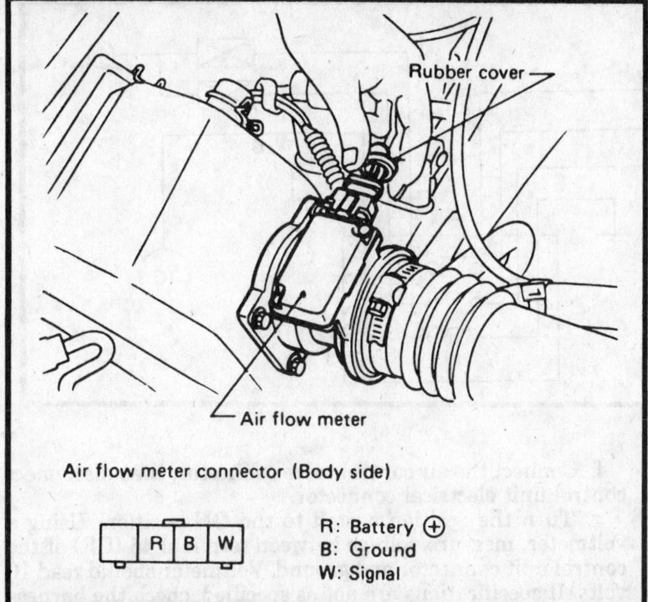
Rubber cover
Air flow meter
Air flow meter connector (Body side)
R B W
R: Battery ⊕
B: Ground
W: Signal
Checking air flow meter

IDLE SPEED

Inspection

1. Disconnect the air control valve electrical connector from the throttle chamber assembly.

2. Adjust idling speed to 550 ± 50 rpm by turning in or out the Idle Air Speed (IAS) screw.

3. Reconnect the air control valve electrical connector, and check that engine idles at 700 ± 100 rpm.

4. If engine rpm is less than 600 rpm, the connector is faulty or the harness is broken.

THROTTLE SENSOR (IDLE CONTACT)

Testing

1. Using an ohmmeter, check that continuity exists between terminals (A) and (B), when the throttle is fully closed, and that no continuity exists whren the throttle is fully opened.

2. Insert a feeler gauge (thickness gauge) of 0.0122 in (0.31mm) between the stopper screw on the throttle chamber and the stopper (this corresponds to the throttle opening of 1.0 degree). Ensure that continuity exists veen terminals (A) and (B).

3. Insert a feeler gauge (thickness uge) of 0.0311 in (0.79mm) between the stopper screw on throttle chamber and the stopper (this corresponds to a thro opening of 2.5 degrees). Ensure that continuity exists betwe terminals (A) and (B).

4. If above specifications are not as specii d, loosen throttle sensor attaching screws, then turn throttle ensor body until the correct adjustment is obtained.

THROTTLE SENSOR (THROTTLE OPENING SIGNAL)

Testing

1. Using an ohmmeter, measure resistance between terminals (B) and (D), then between (B) and (C) (changes with the opening of the throttle valve).

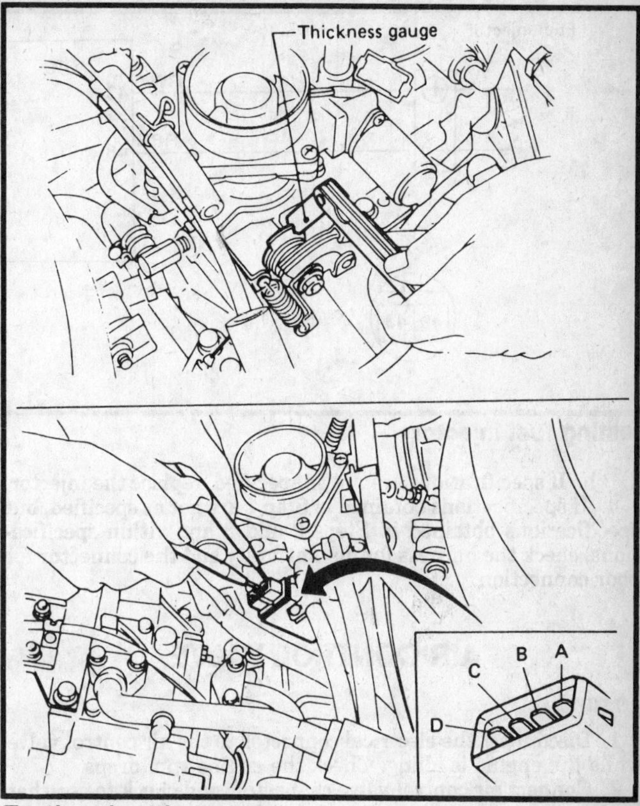
Thickness gauge
C
B
A
D
Testing throttle sensor

2. Ohmmeter should read 3.5–6.5 Kohm between terminals (B) and (D).

3. Check that ohmmeter reading between terminals (B) and (C) is less than 1 Kohm with the throttle valve fully closed, and 2.4 Kohm with the throttle valve fully opened.

4. When the throttle valve is moved from the fully closed to the fully opened position, check that resistance between terminals (B) and (C) increases continiously.

5. When the throttle valve is moved from the fully opened to the fully closed position, check that resistance between terminals (B) and (C) decreases continiously.

6. If any of the above faults are noted, replace the throttle sensor.

FUEL INJECTOR

Testing

1. Using a stethoscope or equivalent, ensure a clicking sound is heard at the injector (when idling or cranking the engine). If clicking noise is not heard, proceed as follows:

a. Turn the ignition switch to the **OFF** position, then disconnect the control unit connector.

b. Using an ohmmeter, measure resistance between terminal 43 (RW) and terminal 48 (RB) on the harness connector. Ohmmeter should read 0.5–2 ohms.

2. Check the injector for discontinuity as follows:

a. Disconnect the electrical connector from the injector.

b. Using an ohmmeter, measure the resistance between the terminals of the connector. Ohmmeter should read 0.5–2 ohms.

c. If specifications are not as specified, replace the injector.

3. Check the injector for insulation as follows:

a. Using an ohmmeter, measure resistance between each terminal of the connector on the injector side and ground. Ohmmeter should read 1 Mohm.

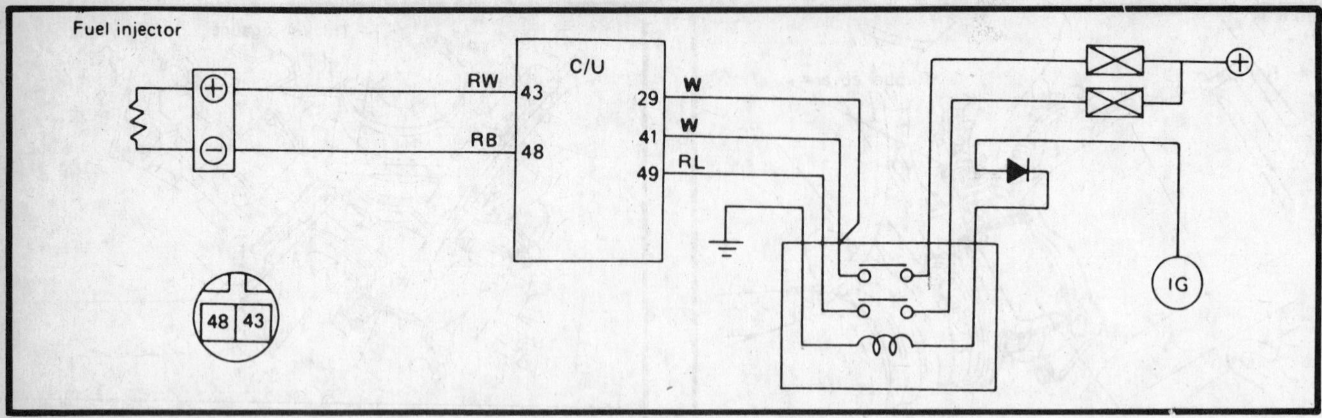
Fuel injector
C/U
RW
43
RB
48
29
W
41
W
49
RL
IG
48
43

Testing fuel injector

b. If specifications are not as specified, replace the injector.

4. If specifications obtained in Step 1 are not as specified, but specifications obtained in Steps 2 and 3 are within specifications, check the harness for discontinuity and the connector for poor connection.

AIR CONTROL VALVE

Testing

1. Disconnect the electrical connector to the air control valve while the engine is idling. Check the engine rpm drops.

2. Connect air control valve connector, and check to see that engine rpm resumes its original position.

NOTE: Disconnecting the connector causes a change in the engine rpm when the engine is cold. However, when the engine is warm, it causes a smaller change or almost no change at all.

3. If the engine shows no rpm change, proceed as follows:

a. Stop the engine, then disconnect the the electrical connector from the air control valve.

b. Turn the ignition switch to the **ON** position.

c. Using a voltmeter, measure the voltage across power terminal (BW) on the air control valve connector and ground. Voltmeter should read 10 volts. If voltage obtained is less than specified, check the harness.

d. Turn the ignition switch to the **OFF** position. Using an ohmmeter, measure the resistance between each terminal of the connector on the air control valve. Ohmmeter should read 7.3–13 ohms at −4–176°F (−20–80°C). If specifications are not as specified, replace the air control valve.

e. Using an ohmmeter, measure the insulation resistance between each terminal of the connector on the air control valve and ground. Ohmmeter should read 1 Mohm. If specifications are not as specified, replace the air control valve.

f. Connect the air control valve connector, then disconnect control unit electrical connector.

g. Turn the ignition switch to the **ON** position. Using a voltmeter, measure voltage between terminal 45 (GR) of the control unit connector and ground. Voltmeter should read 10 volts. If specifications are not as specified, check the harness between the air control valve and the control unit.

h. Turn the ignition switch to the **OFF** position, then connect the control unit connector.

i. Monitor the voltage across terminal 45 (GR) on the control unit connector and ground, when the ignition switch is turned to the **ON** position. Voltmeter should read 1 volt when for approximately 1 minute after the ignition switch is turned **ON**, and 10 volts after 1 minute. If specifications are not as specified, check for poor contact of the terminal or a faulty control unit.

j. Turn the ignition switch to the **OFF** position, then disconnect the air control valve hose.

k. Turn the ignition switch to the **ON** position. Look through the open end of the pipe (from which the air control valve was disconnected) and check that the valve moves from the fully closed position to the fully opened position, 1 minute after the ignition switch is turned to the **ON** position.

PRESSURE REGULATOR

Testing

1. Disconnect the fuel hose from the fuel delivery pipe of the throttle chamber, then install a suitable fuel gauge.

NOTE: Before disconnecting the fuel hose, disconnect the fuel pump connector and crank the engine for approximately 5 seconds to release the pressure in the fuel system. If the engine starts, let it run until it stops.

2. Measure fuel pressure with the engine idling. Fuel gauge should read 20–24 psi.

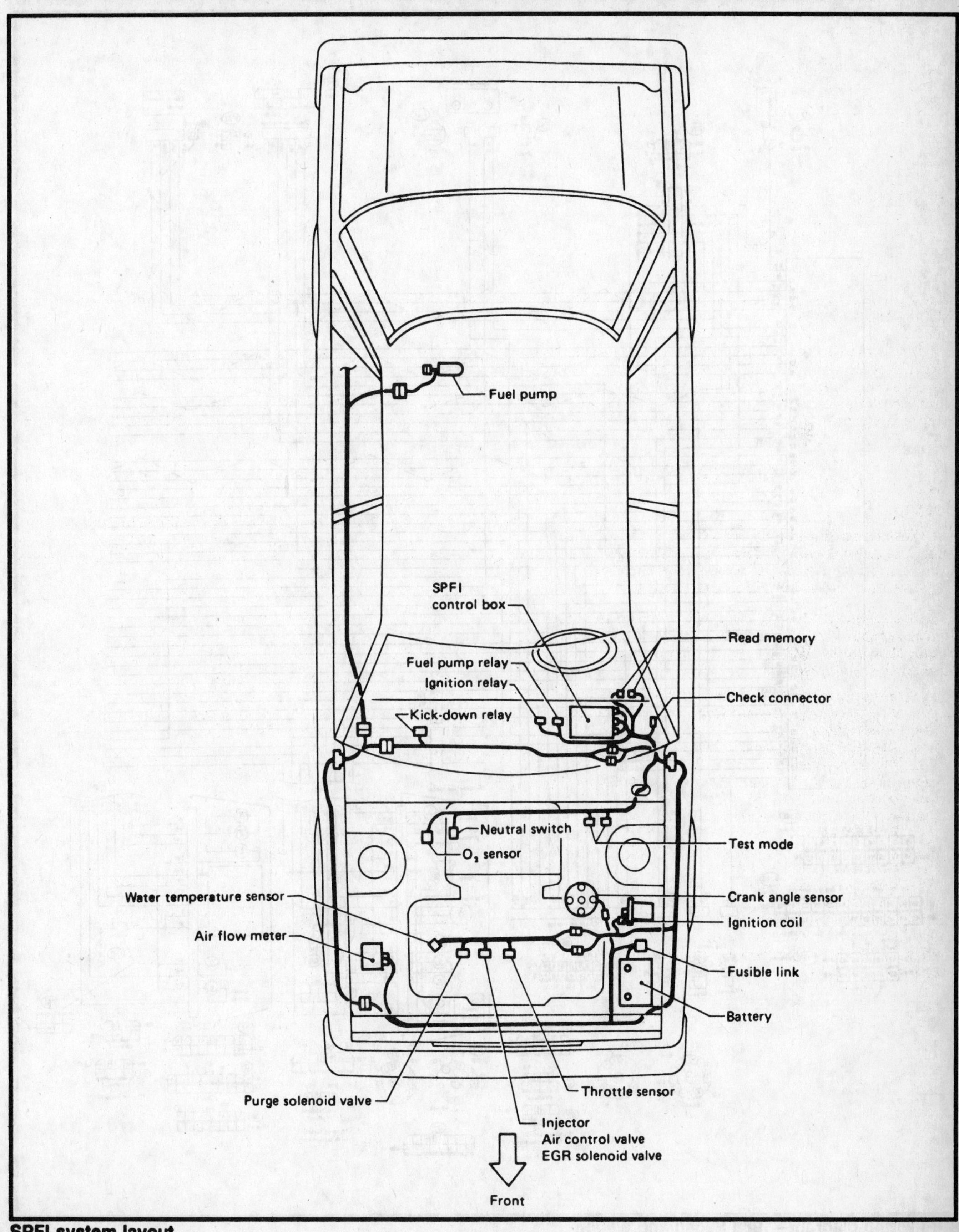
Fuel pump
SPFI
control box
Read memory
Fuel pump relay
Ignition relay
Check connector
Kick-down relay
Neutral switch
Test mode
O_2 sensor
Water temperature sensor
Crank angle sensor
Ignition coil
Air flow meter
Fusible link
Battery
Throttle sensor
Purge solenoid valve
Injector
Air control valve
EGR solenoid valve
Front

SPFI system layout

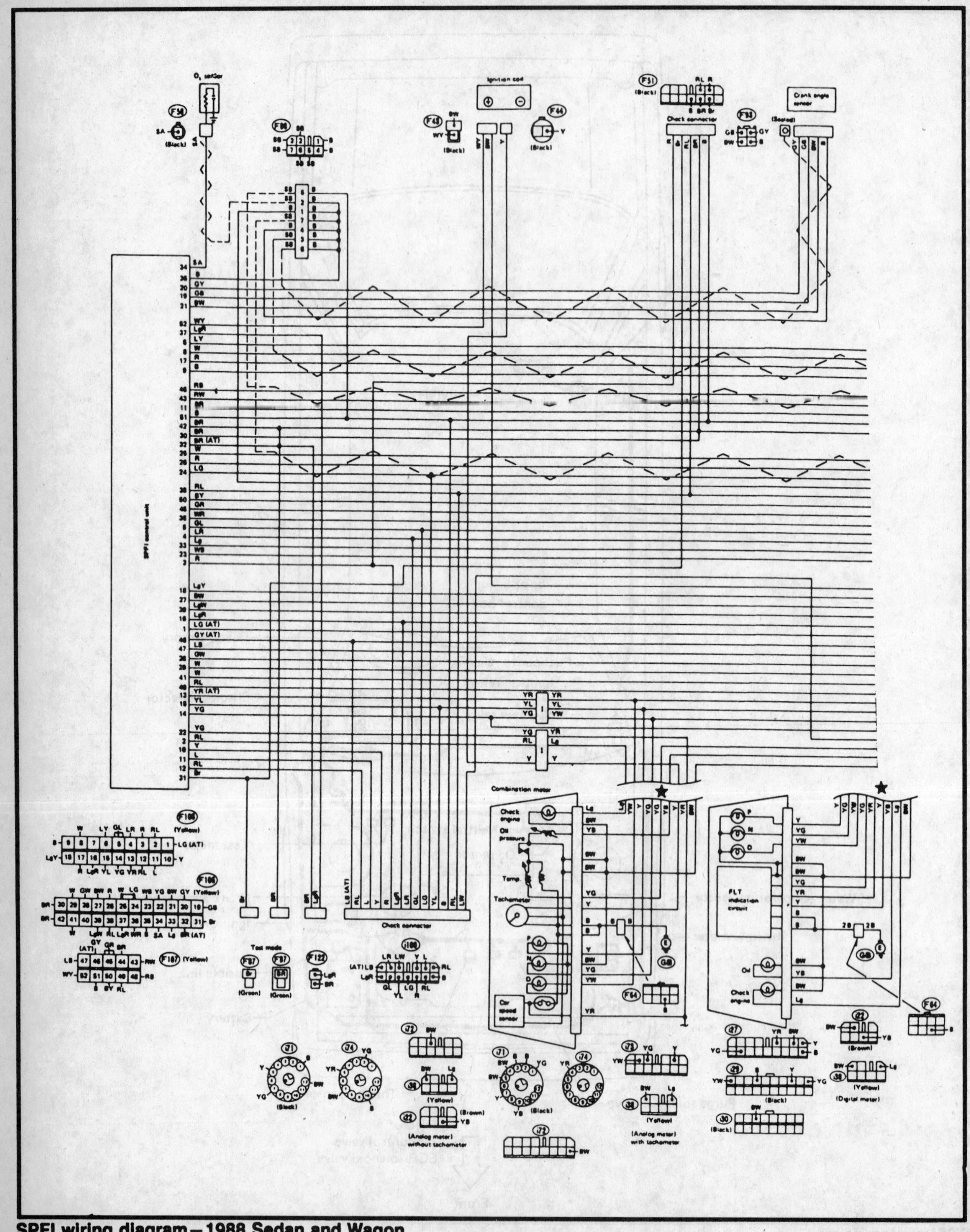
SPFI control unit
Ignition coil
Check connector
Crank angle sensor
Combination meter
Check engine
Oil pressure
Temp.
Tachometer
Car speed sensor
FLT indication circuit
Test mode
(Analog meter) without tachometer
(Analog meter) with tachometer
Digital meter

SPFI wiring diagram—1988 Sedan and Wagon

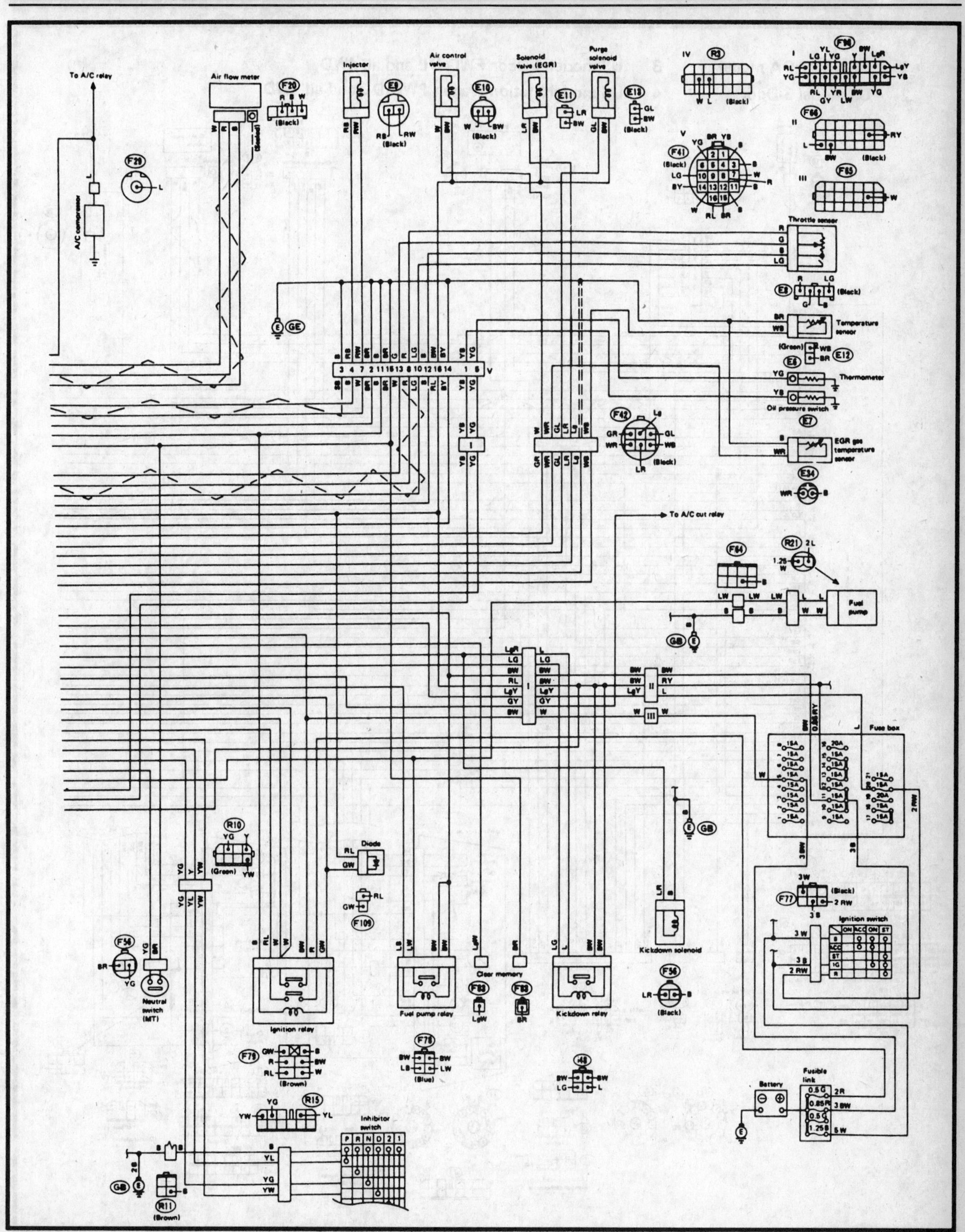

SPFI wiring diagram (cont.) – 1988 Sedan and Wagon

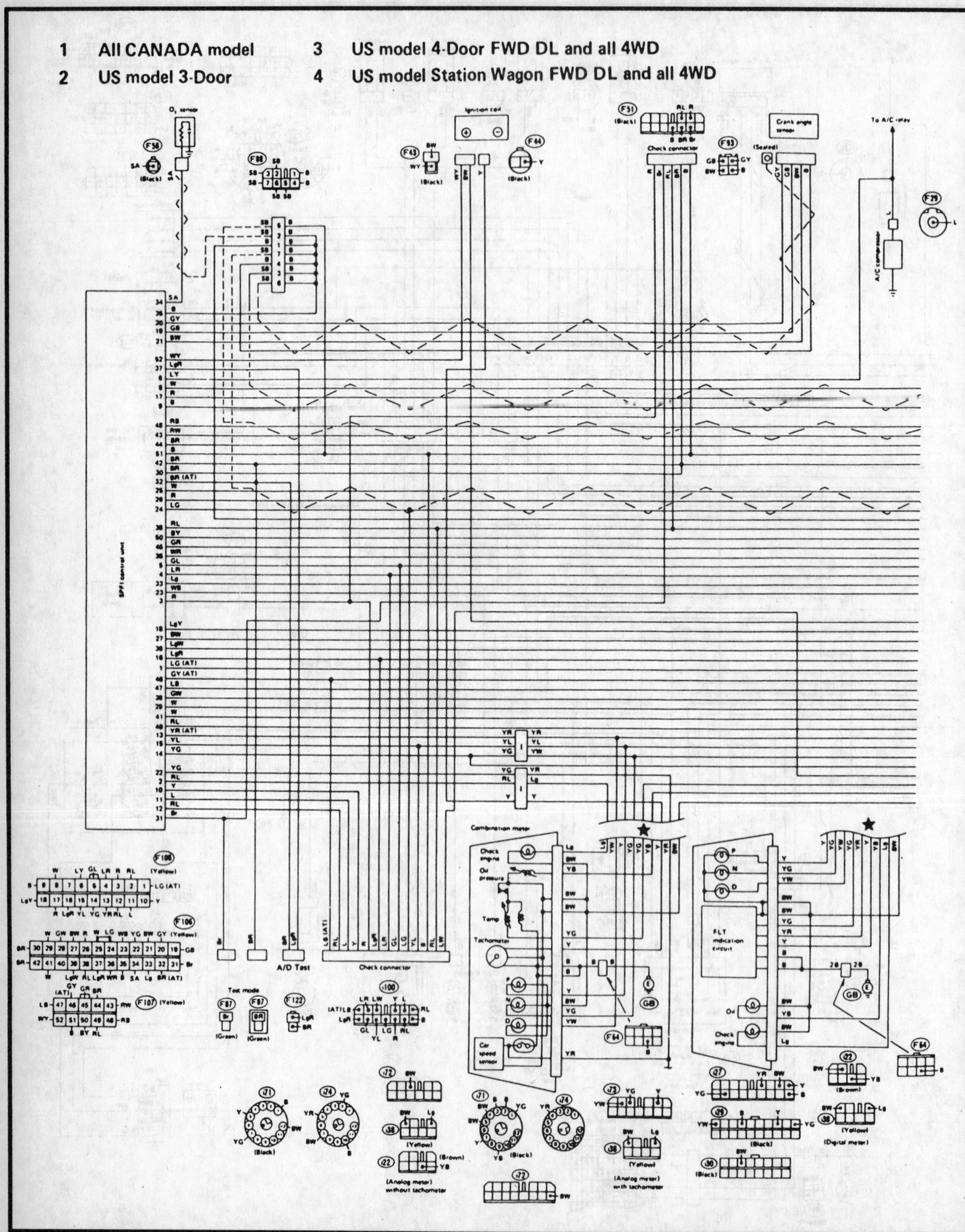
1 All CANADA model
2 US model 3-Door
3 US model 4-Door FWD DL and all 4WD
4 US model Station Wagon FWD DL and all 4WD
O2 sensor
Ignition coil
Check connector
Crank angle sensor
To A/C relay
A/C compressor
SPFI control unit
Combination meter
Check engine
Oil pressure
Temp
Tachometer
Car speed sensor
FLT indication circuit
A/D Test
Test mode
Check connector
(Analog meter) without tachometer
(Analog meter) with tachometer
(Digital meter)

SPFI wiring diagram—1989–90 Sedan and Wagon

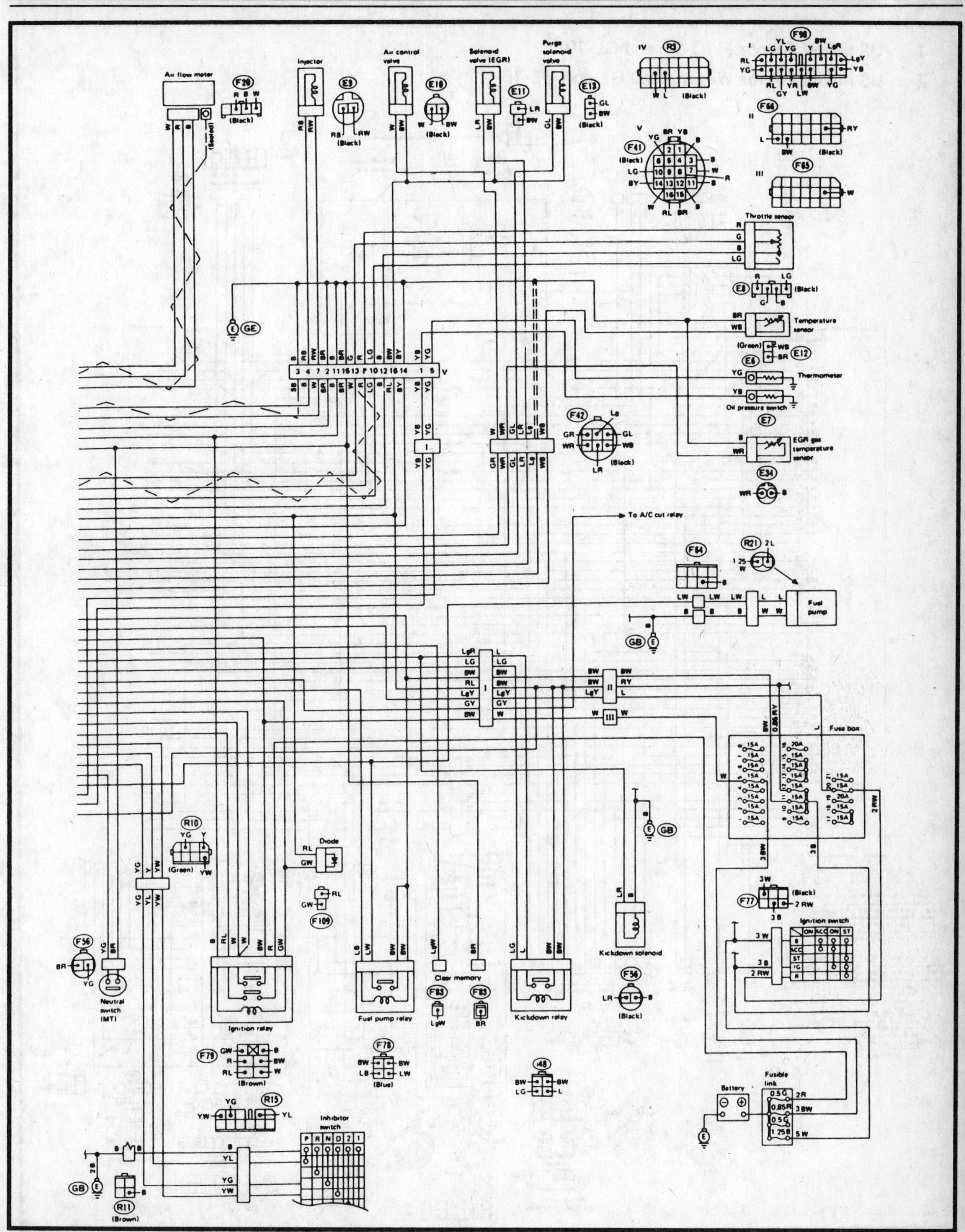
Air flow meter
Injector
Air control valve
Solenoid valve (EGR)
Purge solenoid valve
Throttle sensor
Temperature sensor
Thermometer
Oil pressure switch
EGR gas temperature sensor
To A/C cut relay
Fuel pump
Fuse box
Ignition switch
Diode
Clear memory
Ignition relay
Fuel pump relay
Kickdown relay
Kickdown solenoid
Neutral switch (MT)
Inhibitor switch
Battery
Fusible link

SPFI wiring diagram (cont.) – 1989–90 Sedan and Wagon

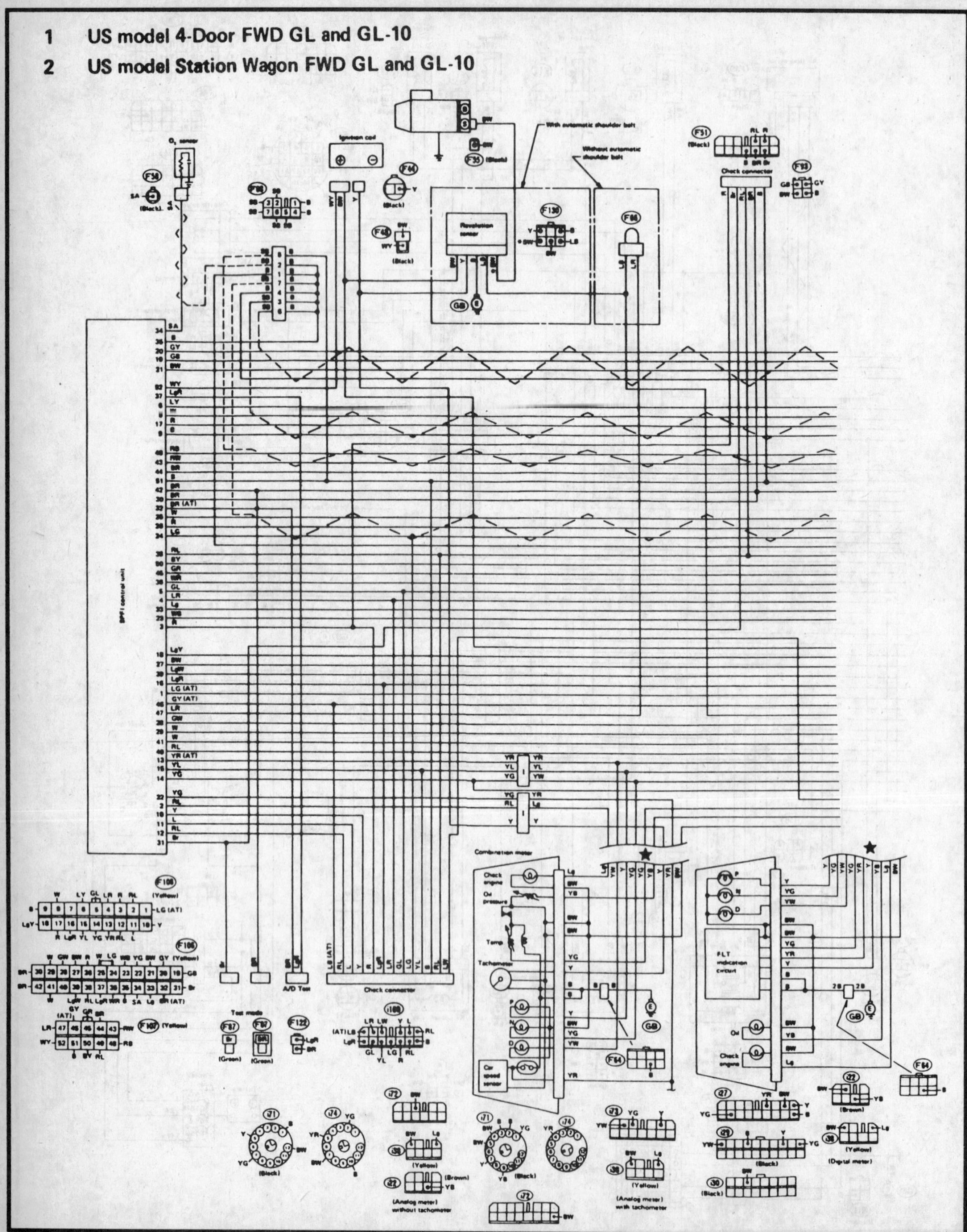
1 US model 4-Door FWD GL and GL-10
2 US model Station Wagon FWD GL and GL-10
Ignition coil
Revolution sensor
Check connector
SPFI control unit
Combination meter
Tachometer
A/D Test
Test mode
(Analog meter) without tachometer
(Analog meter) with tachometer
(Digital meter)

SPFI wiring diagram—1989–90 Sedan and Wagon

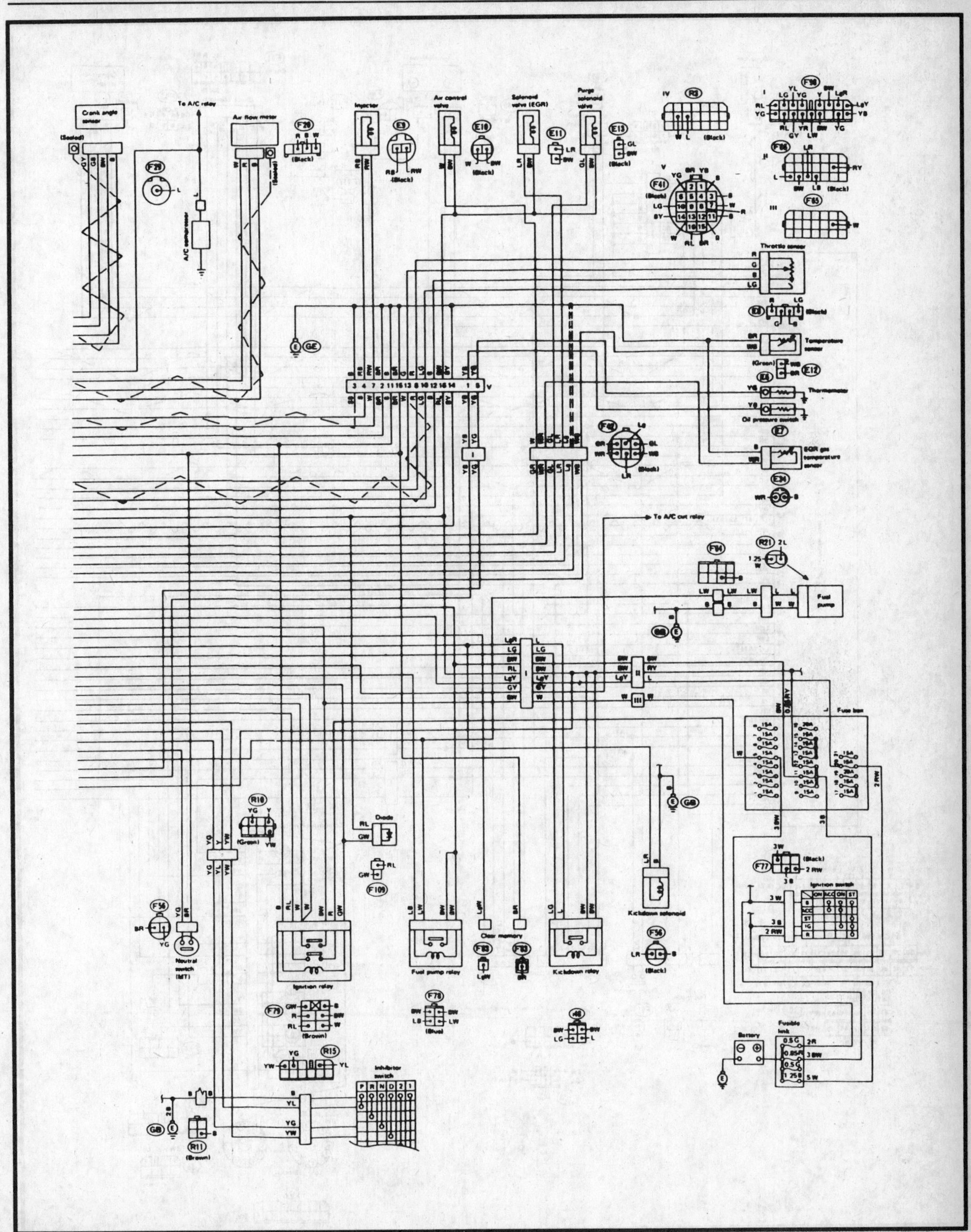

SPFI wiring diagram (cont.) – 1989–90 Sedan and Wagon

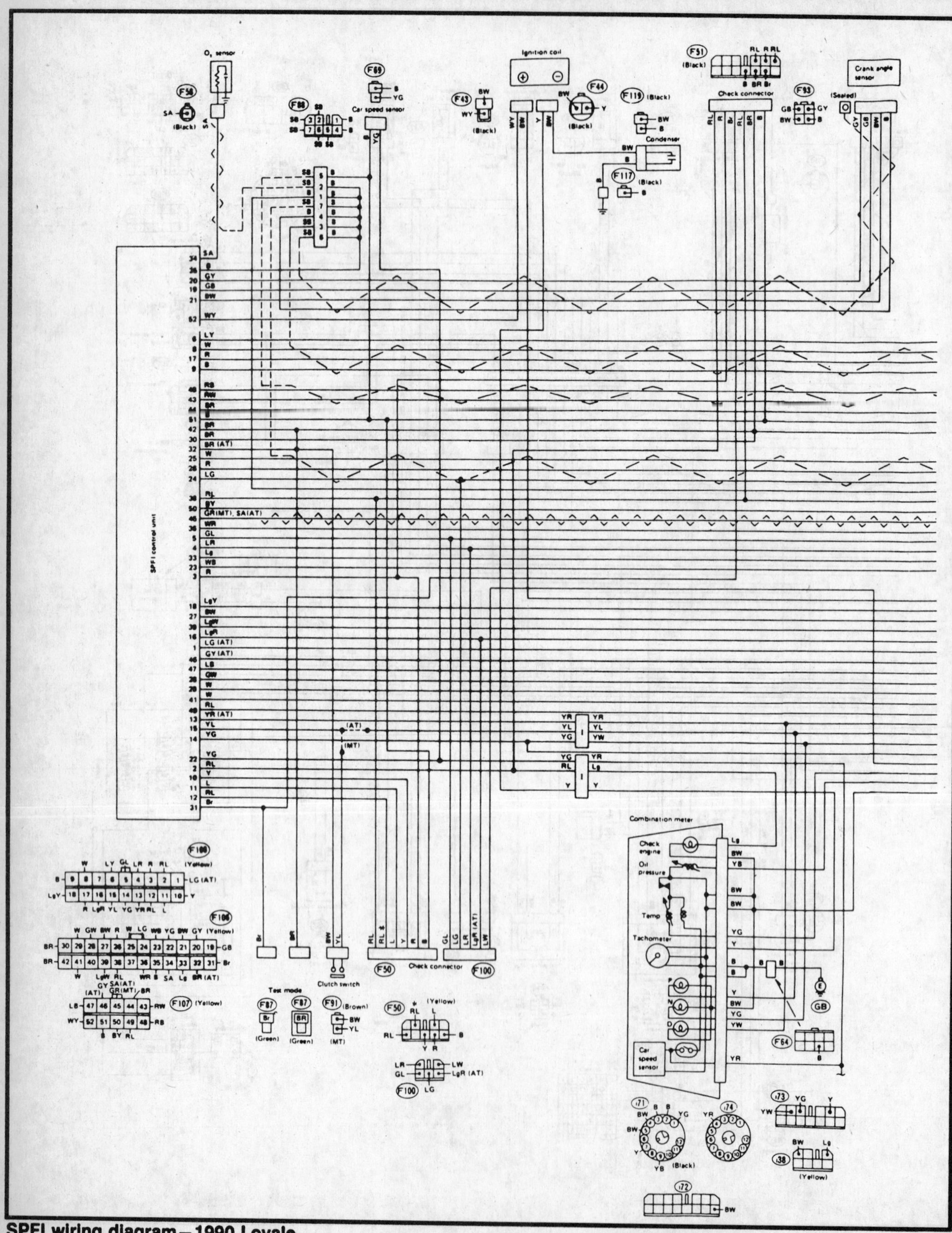
O2 sensor
Ignition coil
Car speed sensor
Condenser
Check connector
Crank angle sensor
SPFI control unit
Combination meter
Check engine
Oil pressure
Temp
Tachometer
Test mode
Clutch switch

SPFI wiring diagram – 1990 Loyale

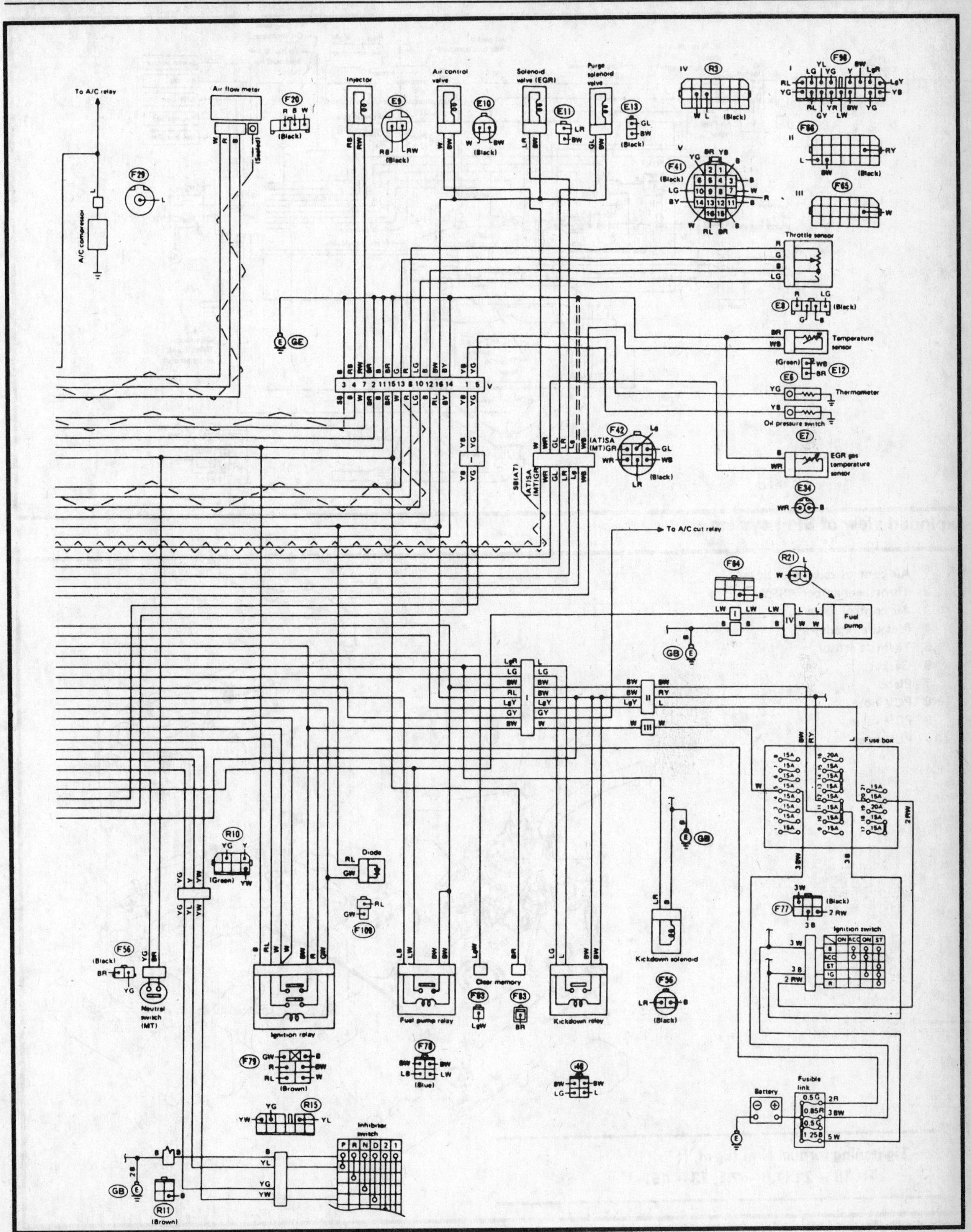
To A/C relay
Air flow meter
Injector
Air control valve
Solenoid valve (EGR)
Purge solenoid valve
Throttle sensor
Temperature sensor
Thermometer
Oil pressure switch
EGR gas temperature sensor
To A/C cut relay
Fuel pump
Fuse box
Diode
Clear memory
Kickdown solenoid
Ignition switch
Neutral switch (MT)
Ignition relay
Fuel pump relay
Kickdown relay
Battery
Fusible link
Inhibitor switch
A/C compressor

SPFI wiring diagram (cont.) – 1990 Loyale

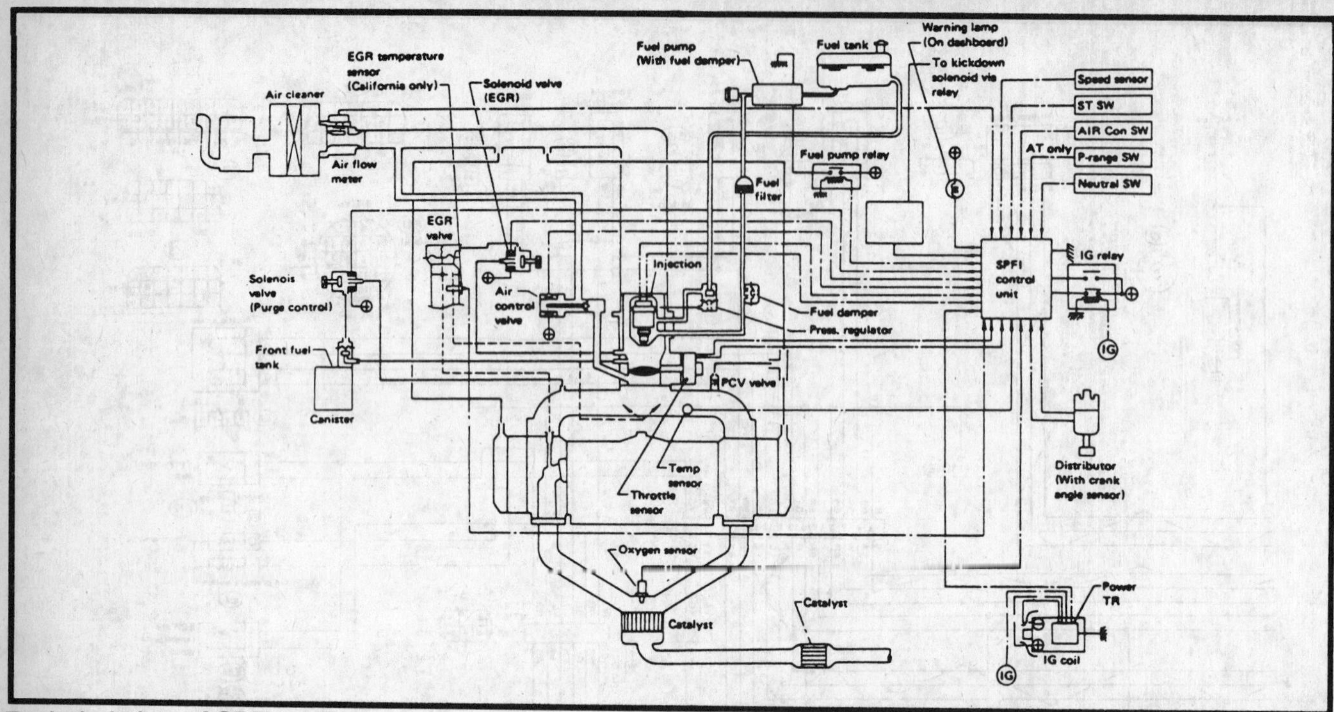
Air cleaner
EGR temperature sensor (California only)
Solenoid valve (EGR)
Fuel pump (With fuel damper)
Fuel tank
Warning lamp (On dashboard)
To kickdown solenoid via relay
Speed sensor
ST SW
AIR Con SW
AT only
P-range SW
Neutral SW
Air flow meter
Fuel pump relay
Fuel filter
EGR valve
Injection
SPFI control unit
IG relay
Solenois valve (Purge control)
Air control valve
Fuel damper
Press. regulator
Front fuel tank
Canister
PCV valve
IG
Temp sensor
Throttle sensor
Distributor (With crank angle sensor)
Oxygen sensor
Catalyst
Catalyst
Power TR
IG coil
IG

Exploded view of SPFI system

1 Air control valve inlet hose
2 Throttle chamber ASSY
3 Air control valve
4 Pressure regulator
5 Throttle sensor
6 Gasket
7 Plate
8 PCV hose
9 PCV valve
10 PCV hose

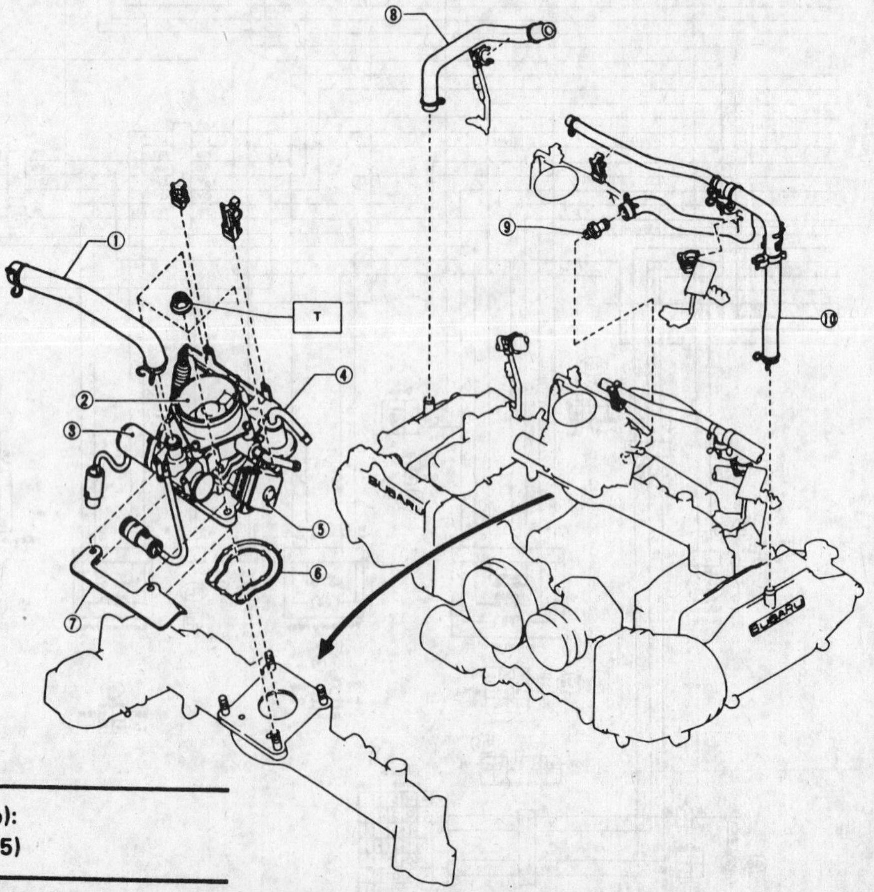

T

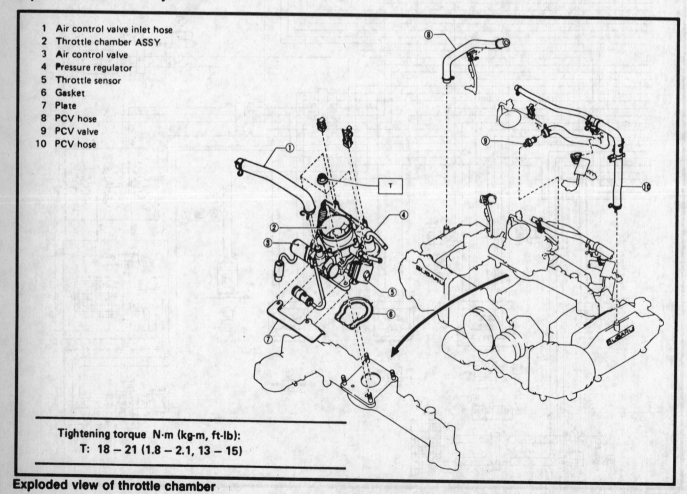

Tightening torque N·m (kg-m, ft-lb):
T: 18 – 21 (1.8 – 2.1, 13 – 15)

Exploded view of throttle chamber

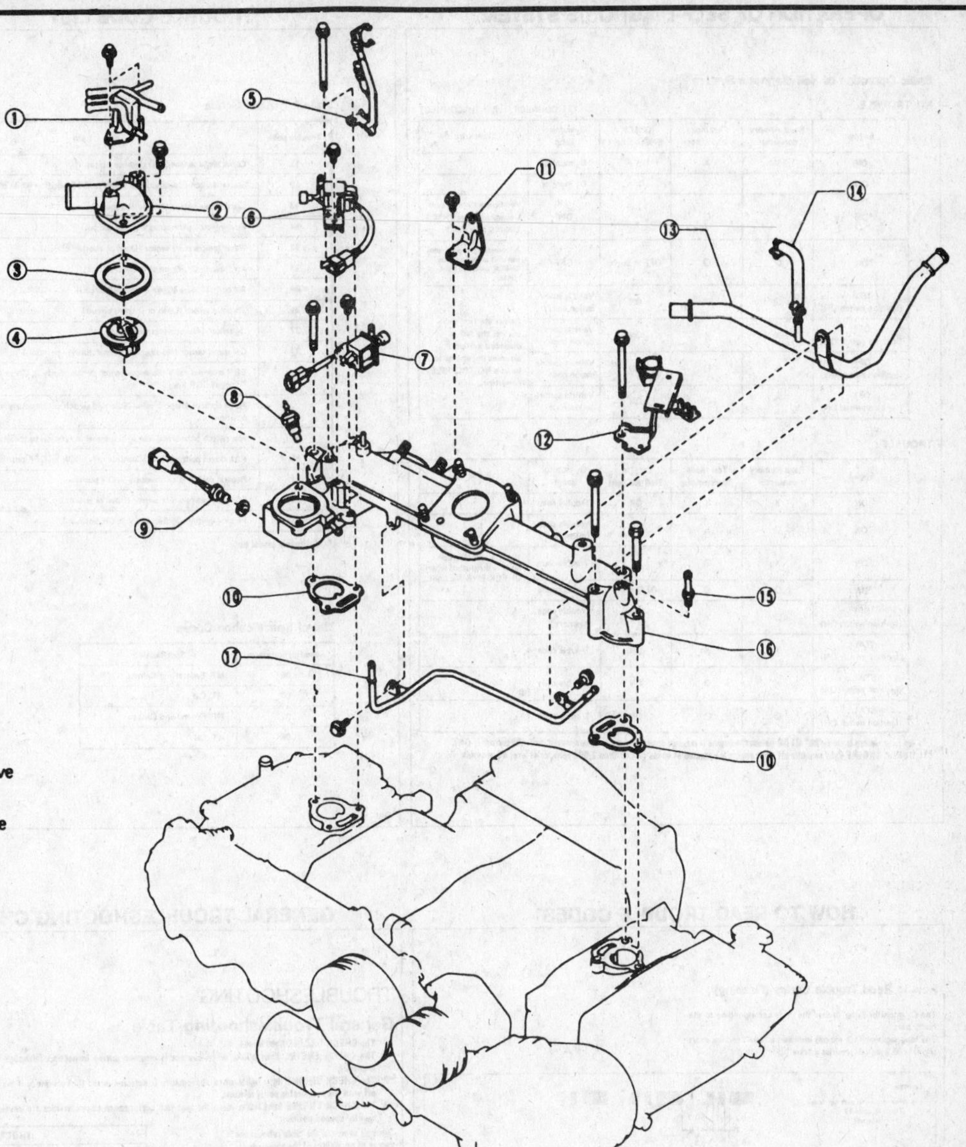

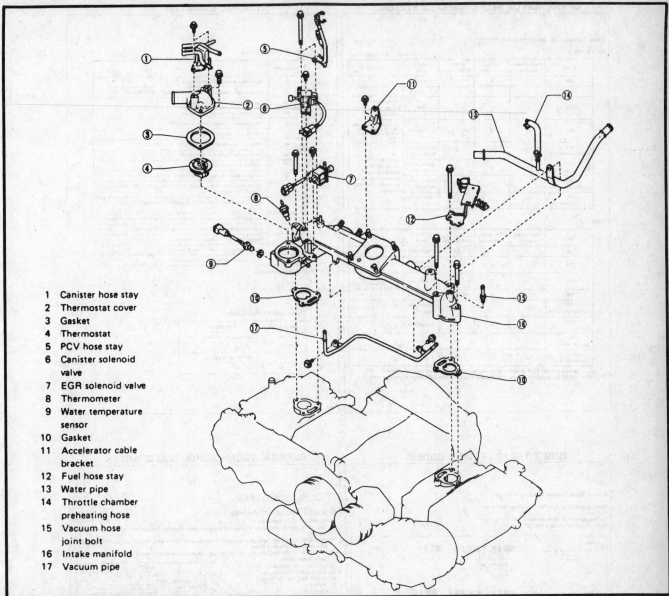

1 Canister hose stay
2 Thermostat cover
3 Gasket
4 Thermostat
5 PCV hose stay
6 Canister solenoid valve
7 EGR solenoid valve
8 Thermometer
9 Water temperature sensor
10 Gasket
11 Accelerator cable bracket
12 Fuel hose stay
13 Water pipe
14 Throttle chamber preheating hose
15 Vacuum hose joint bolt
16 Intake manifold
17 Vacuum pipe

Exploded view of intake manifold

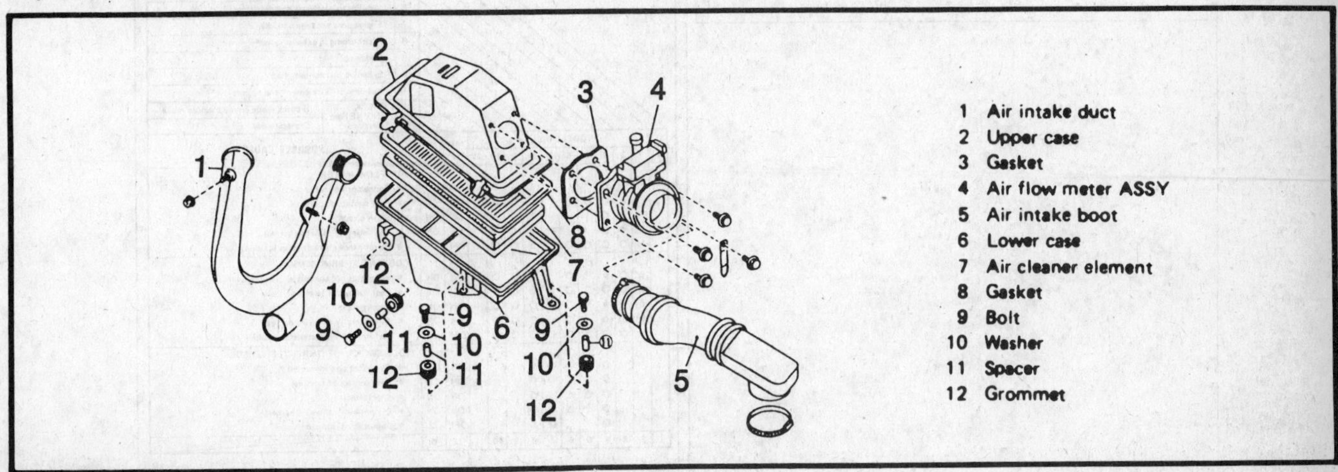
1 Air intake duct
2 Upper case
3 Gasket
4 Air flow meter ASSY
5 Air intake boot
6 Lower case
7 Air cleaner element
8 Gasket
9 Bolt
10 Washer
11 Spacer
12 Grommet

Exploded view of air intake system

OPERATION OF SELF-DIAGNOSIS SYSTEM

Basic Operation of Self-diagnosis System

NO TROUBLE

O : CONNECT X : DISCONNECT

Engine	Read memory connector	Test mode connector	CHECK ENGINE light	O_2 monitor lamp	Remarks
ON	X	X	OFF	O_2 monitor	
ON	O	X	OFF	O_2 monitor	
*ON	X	O	** OFF → Blink	OFF	Vehicle specification code is outputted when CHECK ENGINE light is OFF.
*ON	O	O	OFF → Blink	OFF	All memory stored in control unit is cleared after CHECK ENGINE light blinks.
OFF (Ignition switch ON)	O	X	ON	Vehicle specification code	Before starting the engine, the self-diagnosis system assumes the engine to be in a NO TROUBLE condition.
OFF (Ignition switch ON)	X	X	ON	Vehicle specification code	
OFF (Ignition switch ON)	X	O	ON	Vehicle specification code	
OFF (Ignition switch ON)	O	O	ON	Vehicle specification code	

TROUBLE

Engine	Read memory connector	Test mode connector	CHECK ENGINE light	O_2 monitor lamp	Remarks
ON	X	X	ON	Trouble code	
ON	O	X	ON	Trouble code (memory)	
*ON	X	O	** OFF → ON	Trouble code	Vehicle specification code is outputted when CHECK ENGINE light is OFF.
*ON	O	O	OFF → ON	Trouble code	
OFF (Ignition switch ON)	O	X	ON	Trouble code (memory)	
STALL (Ignition switch ON)	X	X	ON	Trouble code	
STALL (Ignition switch ON)	X	O	ON	Trouble code	
STALL (Ignition switch ON)	O	O	ON	Trouble code	

*: Ignition timing is set to 20° BTDC (when the engine is on, test mode connector is connected, and idle switch is ON).
**: CHECK ENGINE light remains off until engine is operated at speed greater than 2,000 rpm for at least 40 seconds.

TROUBLE CODE LIST

List of Trouble Codes

Trouble code	Item
11	Crank angle sensor (No reference pulse)
12	Starter switch (Continuously in ON or OFF position while cranking)
13	Crank angle sensor (No position pulse)
14	Fuel injector (Abnormal injector output)
21	Water temperature sensor (Open or shorted circuit)
23	Air flow meter (Open or shorted circuit)
24	Air control valve (Open or shorted circuit)
31	Throttle sensor (Open or shorted circuit)
32	O_2 sensor (Abnormal sensor signal)
33	Car-speed sensor (No signal is present during operation)
34	EGR solenoid valve (Solenoid switch continuously in ON or OFF position, or *clogged EGR line)
35	Purge control solenoid valve (Solenoid switch continuously in ON or OFF position)
42	Idle switch (Abnormal idle switch signal in relation to throttle sensor output)
45	Kick-down control relay (Continuously in ON or OFF position)
51	Neutral switch (Continuously in ON position)
*55	EGR gas temperature sensor (Open or short circuit)
61	Parking switch (Continuously in ON position)

*: California model only

List of Specification Codes

Specification codes	Specification
05	MT, Federal and Canada
06	MT, Cal
07	AT, Federal and Canada
08	AT, Cal

HOW TO READ TROUBLE CODES

How to Read Trouble Codes (Flashing)

The O_2 monitor lamp flashes the code corresponding to the faulty part.
The long segment (1.2 sec on) indicates a "ten", and the short segment (0.2 sec on) signifies a "one".

Example:

When only one part has failed:
Flashing code 12
(unit: second)

When two or more parts have failed:
Flashing codes 12 and 21
(unit: second)

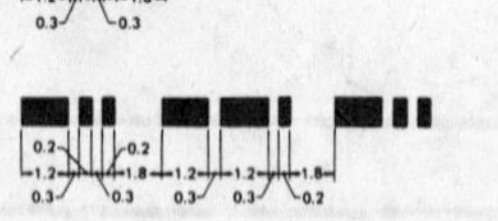
0.2
0.2
1.2
1.8
0.3
0.3
0.2
0.2
1.2
1.8
1.2
1.2
1.8
0.3
0.3
0.3
0.3
0.2

GENERAL TROUBLESHOOTING CHART

TROUBLESHOOTING

General Troubleshooting Table

*: The CHECK ENGINE light blinks.
*1: The CHECK ENGINE light blinks when contact is resumed during inspection (although poor contact is present in the D-check).
*2: The CHECK ENGINE light lights when abnormality is detected in the D-check mode if the idle switch persistently remains off with the accelerator pedal released.
*3: The CHECK ENGINE light lights when the specified performance characteristics are unusual with the throttle valve in the slightly-opened position.

Symbols shown in the table refer to the degree of possibility of the reason for the trouble ("Very often" to "Rarely").
◎ : Very often
○ : Sometimes
△ : Rarely
☆ : Occurs only in extremely low temperatures

TROUBLE		
1	Engine will not start.	No initial combustion
2		Initial combustion occur.
3		Engine stalls after initial combustion.
4	Rough idle and engine stall.	
5	Inability to drive at constant speed	
6	Inability to accelerate and decelerate	
7	Engine does not return to idle.	
8	Afterburning in exhaust system	
9	Knocking	
10	Excessive fuel consumption	
11	Inability to "kick-down" and upshift	
U	CHECK ENGINE light operation	U-check mode & read memory mode
D		D-check mode

TROUBLE No. 1	2	3	4	5	6	7	8	9	10	11	CHECK ENGINE light U	D	POSSIBLE CAUSE
													AIR FLOW METER
		☆	◎				△	△	○		ON	ON	• Connector not connected
		△	◎	◎	◎		◎	○	△		ON	*1	• Poor contact of terminal
		☆	◎				△	○	△		ON	ON	• Short circuit
		☆	◎				△	△	○		ON	ON	• Discontinuity of wiring harness
		○	○	○	○		△	◎	○		OFF	*	• Performance characteristics unusual
													COOLANT THERMOSENSOR
	☆	○	☆		○		○	○	○		ON	ON	• Connector not connected
	△	△	◎	○	○	△	◎	◎	○		ON	*1	• Poor contact of terminal
	☆	○	☆		○		○	○	○		ON	ON	• Short circuit
	☆	○	☆		○		○	◎	○		ON	ON	• Discontinuity of wiring harness
	☆	○	○	△	○	○	◎	◎	◎		OFF	*	• Performance characteristics unusual
													IDLE SWITCH OF THROTTLE SENSOR
			◎	○	◎	◎	◎				ON	ON	• Connector not connected
			◎	○		◎	◎				ON	*1	• Poor contact of terminal
			◎	◎	△		◎				ON	ON	• Short circuit
			◎		△	◎	◎				ON	ON	• Discontinuity of wiring harness
			◎			◎	◎				OFF	*2	• Improper adjustment
1	2	3	4	5	6	7	8	9	10	11	U	D	

GENERAL TROUBLESHOOTING CHART

1	2	3	4	5	6	7	8	9	10	11	CHECK ENGINE light U	D	POSSIBLE CAUSE
													THROTTLE SENSOR
				○	**○**		**○**			○	ON	*1	• Poor contact of terminal
△			**○**		**○**		**○**			**○**	ON	ON	• Short circuit
			△		**○**		**○**			**○**	ON	ON	• Discontinuity of wiring harness
	○	○	△	○	**○**		**○**			**○**	OFF	*3	• Performance characteristics unusual
													PRESSURE REGULATOR
								△			OFF	*	• Sensing hose cracked or disconnected
	△				○		○		**○**		OFF	*	• Fuel pressure too high
○	○	○	**○**	**○**	**○**		**○**				OFF	*	• Fuel pressure too low
													FUEL INJECTOR
○											ON	ON	• Connector not connected
	○	○	**○**	**○**	**○**		○				ON	*1	• Poor contact of terminal
○											ON	ON	• Short circuit
○											ON	ON	• Discontinuity of wiring harness
	○	○	**○**	**○**	**○**		**○**		**○**		OFF	*	• Performance characteristics unusual
	○	○	○	○	○						OFF	*	• Clogged filter
○	△										OFF	*	• Stuck open
			○				○		○		OFF	*	• Slight leakage from seat
													AIR CONTROL VALVE
	○	△	**○**								ON	ON	• Connector not connected
	△	○	**○**								ON	*1	• Poor contact of terminal
			○				○				ON	ON	• Short circuit
	○	△	**○**								ON	ON	• Discontinuity of wiring harness
			○			**○**					OFF	*	• IAS improperly adjusted
						○					OFF	*	• Stuck open
	○	○	**○**								OFF	*	• Stuck closed
													CRANK ANGLE SENSOR
○											ON	ON	• Connector not connected
	○	○	**○**	**○**	**○**		○	○			ON	*1	• Poor contact of terminal
○											ON	ON	• Short circuit
○											ON	ON	• Discontinuity of wiring harness
													POWER TRANSISTOR OF IGNITION COIL
○											OFF	*	• Connector not connected
	○	○	**○**	**○**	**○**		○	△			OFF	*	• Poor contact of terminal
○											OFF	*	• Short circuit
○											OFF	*	• Discontinuity of wiring harness
1	2	3	4	5	6	7	8	9	10	11	U	D	

ENGINE TROUBLESHOOTING CHART

TROUBLESHOOTING

Engine Trouble in General

Symbols shown in the chart refer to the possibility of reason for the trouble in order ("Very often" to "Rarely")

○ – Very often
○ – Sometimes
△ – Rarely

No.	TROUBLE	
1	Engine will not start.	Starter does not turn.
2		Initial combustion does not occur.
3		Initial combustion occurs.
4		Engine stalls after initial combustion.
5	Rough idle and engine stall.	
6	Low output, hesitation and poor acceleration.	
7	Surging.	
8	Engine does not return to idle.	
9	Dieseling (Run-on).	
10	Afterburning in exhaust system.	
11	Knocking.	
12	Excessive engine oil consumption.	
13	Excessive fuel consumption.	

1	2	3	4	5	6	7	8	9	10	11	12	13	POSSIBLE CAUSE
													STARTER
○													• Defective battery-to-starter harness.
△													• Defective starter switch.
△													• Defective inhibitor switch.
○	△												• Defective starter.
													BATTERY
○													• Poor terminal connection.
○													• Run-down battery.
○													• Defective charging system.
	○	**○**	**○**	**○**	**○**	**○**	**○**	**○**	**○**	**○**		**○**	SPFI SYSTEM (See Chap. 2-7.)
													IGNITION SYSTEM
	○	○	○	**○**	**○**	**○**	**○**	**○**	**○**	**○**		**○**	• Incorrect ignition timing.
	○	**○**		○	○	○			△			△	• Disconnection of spark plug cord.
	○			△	**○**	○	○		**○**	○			• Defective distributor.
	○			△	○	○							• Defective ignition coil.
	○			△	△	△							• Defective cord or wiring.
	○	○		△	○	△			○				• Leakage of spark plug cord.
		○		○	○	○			**○**				• Defective spark plug.
	○	○	○	**○**	○	**○**	△		○	○			• Incorrect cam timing.
1	2	3	4	5	6	7	8	9	10	11	12	13	

ENGINE TROUBLESHOOTING CHART

1	2	3	4	5	6	7	8	9	10	11	12	13	POSSIBLE CAUSE
													INTAKE SYSTEM
		○	○	**○**	○	○	**○**	○	○			**○**	• Improper idle adjustment.
			○	**○**	**○**	**○**			△	**○**			• Loosened or cracked intake boot.
			○	**○**	**○**	**○**			△	**○**			• Loosened or cracked intake duct.
			△	**○**	**○**	**○**			△	**○**	**○**		• Loosened or cracked blow-by hose.
			△	**○**	○	**○**	**○**		○	**○**			• Loosened or cracked vacuum hose.
			△	○	○	○				**○**			• Defective air cleaner gasket.
		○	○	○	○	○				**○**			• Defective intake manifold gasket.
		○	○	○	○	○				**○**			• Defective thrott'e body gasket.
				△	○	○			○	○	○		• Defective PCV valve.
				○	○	○			△	○	△		• Loosened oil filler cap.
			△	△	**○**	○				○		**○**	• Dirty air cleaner element.
													FUEL LINE
	○	△		△	○	○							• Defective fuel pump.
		△	△	△	○	○							• Clogged fuel line.
	○	○	○	○	△	△							• Lack of or insufficient fuel.
													BELT
	○	○	○										• Defective.
	○	○	○	△	○	○			○	○		○	• Defective timing.
													FRICTION
△													• Seizure of crankshaft and connecting-rod bearing.
△													• Seized camshaft.
△													• Seized or stuck piston and cylinder.
													COMPRESSION
	△	△	△	○	○	○			○	△		○	• Incorrect valve clearance.
	△	△	△	○	○	△			△			△	• Loosened spark plugs or defective gasket.
	△	△	△	○	○	△			△			△	• Loosened cylinder head nuts or defective gasket.
	△	△	△	○	○	△			○			○	• Improper valve seating.
	△	△	△	△	△	△			△		**○**	△	• Defective valve stem.
	○	○	○	○	○	△			△			△	• Worn or broken valve spring.
	△	△	△	○	△	△			△		**○**	○	• Worn or stuck piston rings, cylinder and piston.
	○	○	○	**○**	**○**	**○**			**○**	○		○	• Incorrect valve timing.
	○	○	○	○	○	○							• Improper engine oil (low viscosity).
1	2	3	4	5	6	7	8	9	10	11	12	13	

1	2	3	4	5	6	7	8	9	10	11	12	13	POSSIBLE CAUSE
													LUBRICATION SYSTEM
				○	○				△			△	• Incorrect oil pressure.
											○		• Loosened oil pump attaching bolts and defective gasket.
											○		• Defective oil filter seal.
											○		• Defective crankshaft oil seal.
				△							○		• Defective rocker cover gasket.
											○		• Loosened oil drain plug or defective gasket.
											○		• Loosened oil pan fitting bolts or defective oil pan.
													COOLING SYSTEM
				△	△	○		○		**○**			• Overheating.
					△				△			△	• Over cooling.
													TURBOCHARGER
				△	**○**	**○**						○	• Malfunction of turbocharger.
					○	**○**				**○**		○	• Malfunction of waste gate valve.
											○		• Defective oil pipe and hose.
													OTHERS
				○	**○**	**○**			○	**○**		○	• Malfunction of EGR System.
				○	**○**	△			△				• Malfunction of Evaporative Emission Control System.
				○			**○**						• Stuck or damaged throttle valve.
				△			○					○	• Dashpot out of adjustment.
				△			○	○				○	• Accelerator cable out of adjustment.
				○			○	○				○	• FICD out of adjustment.
				○			○						• Malfunction of FICD.
1	2	3	4	5	6	7	8	9	10	11	12	13	

THROTTLE CHAMBER TROUBLESHOOTING CHART

Troubleshooting Table for Throttle Chamber

1. Fuel leakage	
1) Damaged O-ring in fuel injector	Replace O-ring
2) Damaged O-ring in pressure regulator	Replace O-ring
3) Damaged gasket or loose screws	Replace gasket or tighten screws
4) Pressure regulator vacuum sensing hose out of place or cracked	Install or replace hose
2. No fuel is injected	
When fuel injector drive signal is normal and fuel pump and fuel line are in good condition	
1) Lead wire broken or improperly contacting in fuel injector	Replace fuel injector ASSY
2) Fuel injector seized	Repair cause of seizure and replace fuel injector ASSY
3) Pressure regulator regulating pressure too low	Replace pressure regulator ASSY
3. Excessive fuel consumption	
When fuel line (especially return circuit) is normal and both control system and ignition system are in good condition	
1) Pressure regulator regulating pressure too high	Replace pressure regulator ASSY
2) Fuel leaks	Refer to "Fuel leakage" in item 1 above
3) Throttle sensor improperly adjusted or abnormal output signal	Adjust or replace throttle sensor
4. Rough idle	
When fuel pump, fuel line, control system, ignition system and resistance are normal	
1) Damaged gasket or screws loose	Replace gasket or tighten screws
2) Pressure regulator vacuum sensing hose out of place or cracked	Install or replace hose
3) Improperly [illegible] malfunctioning idle switch	Adjust or replace throttle sensor
4) Lead wire broken or improperly contacting in air control valve	Replace air control valve ASSY
5) Foreign matter in air control valve metering unit or contamination	Remove foreign matter or replace air control valve ASSY
6) Fuel injector nozzle tip contaminated or deformed	Clean or replace injector ASSY
7) Fuel injector filter clogged	Replace injector
8) Pressure regulator regulating pressure too low or unstable	Replace pressure regulator ASSY
9) Throttle chamber throttle valve orifice clogged	Clean

5. Engine lacks power and/or high-speed performance	
When fuel pump, fuel line, control system, ignition system and intake system are in good condition.	
1) Pressure regulator regulating pressure too low	Replace pressure regulator ASSY
2) Fuel injector filter clogged	Replace fuel injector
3) Throttle sensor out of adjustment or abnormal output signal	Adjust or replace throttle sensor
4) Gasket damaged or screws loose	Replace gasket or tighten screws
5) Pressure regulator vacuum sensing hose out of place or cracked	Install or replace hose
6. Hesitation and/or insufficient acceleration performance	
When fuel pump, fuel line, control system, ignition system and intake system are in good condition	
1) Throttle sensor output signal abnormal	Replace throttle sensor ASSY
2) Pressure regulator regulating pressure too low	Replace pressure regulator ASSY
3) Fuel injector filter clogged	Replace fuel injector
4) Damaged gasket or screws loose	Replace gasket or tighten screws
5) Pressure regulator vacuum sensing hose out of place or cracked	Install or replace hose
7. Hard starting in cold weather	
When fuel pump, fuel line, control system and ignition system are in good condition	
1) Lead wire broken or improperly contacting in air control valve	Replace air control valve ASSY
2) Foreign matter caught in air control valve metering unit or contamination	Remove foreign matter or replace air control valve ASSY
3) Fuel injector filter clogged	Replace fuel injector
4) Pressure regulator regulating pressure too low	Replace pressure regulator ASSY
5) Throttle chamber throttle valve orifice clogged	Clean

CONTROL UNIT CONNECTOR IDENTIFICATION

Connector Terminal

CONTROL UNIT CONNECTOR

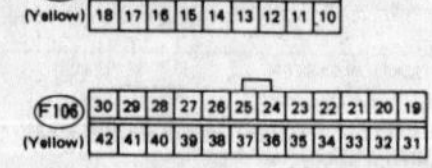
F108
(Yellow)
9 8 7 6 5 4 3 2 1
18 17 16 15 14 13 12 11 10
F106
(Yellow)
30 29 28 27 26 25 24 23 22 21 20 19
42 41 40 39 38 37 36 35 34 33 32 31

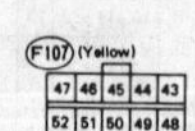
F107 (Yellow)
47 46 45 44 43
52 51 50 49 48

1	LG	Kick-down control	27	BW	Power (input)
2	RL	CHECK ENGINE light	28	GW	Self-shutoff signal
3	R	Test 4	29	W	Power (input)
4	LR	EGR solenoid (control)	30	BR	GND
5	GL	Purge control solenoid	31	Br	Test mode connector (used at line end only)
6	LY	Air conditioner signal	32	BR	Test mode connector (used at line end only)
7	–	———	33	Lg	49-state/Cal identification
8	W	Air flow meter (signal)	34	SA	O_2 sensor
9	B	Air flow meter (GND)	35	B	GND
10	Y	Line end cord output	36	WR	EGR monitor
11	L	Line end cord output	37	LgR	Test mode connector (used at line end only)
12	RL	Line end cord output	38	RL	Ignition switch
13	YR	Inhibitor switch (AT models only)	39	LgW	Clear memory
14	YG	Neutral switch	40	–	———
15	YL	Parking switch (AT models only)	41	W	Power (input)
16	LgR	Kick-down monitor	42	BR	GND
17	R	Air flow meter power (output)	43	RW	Injector ⊕
18	LgY	Starter switch	44	BR	GND
19	GB	Crank angle sensor power (output)	45	GR	Air control valve
20	GY	Crank angle sensor signal (reference)	46	GY	A/C control
21	BW	Crank angle sensor signal (position)	47	LB	Fuel pump
22	YG	Car-speed sensor	48	RB	Injector ⊖
23	WB	Water temperature sensor	49	RL	Power (input)
24	LG	Idle switch	50	BY	GND
25	W	Throttle sensor (signal)	51	B	GND
26	R	Throttle sensor power (output)	52	WY	Ignition control

CONNECTOR IDENTIFICATION—SEDAN AND WAGON

Intermediate connector I (body side) . . . F41

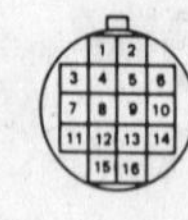

1	BR	Ground
2	YB	Oil pressure (to combination meter)
3	–	———
4	YG	Thermometer (to combination meter)
5	B	Injector ⊖
6	B	Shield
7	LG	Idle switch
8	–	———
9	R	Throttle sensor (power)
10	W	Injector ⊕
11	BY	Ground
12	W	Throttle sensor (signal)
13	B	Ground
14	B	Ground
15	RL	Power supply
16	BR	Ground

Intermediate connector II (body side) . . . F42 (Black)

1	GR	Air control valve (control)
2	Lg	Identification of specifications
3	GL	Purge solenoid (control)
4	WR	EGR gas temperature sensor
5	LR	EGR solenoid (control)
6	WB	Water temperature signal

CONNECTOR IDENTIFICATION – 1990 LOYALE

Intermediate connector I (body side) . . . (F41)

1	YB	Oil pressure (to combination meter)
2	BR	Ground
3	B	Shield
4	B	Injector ⊖
5	YG	Thermometer (to combination meter)
6	–	——
7	W	Injector ⊕
8	R	Throttle sensor (power)
9	–	——
10	LG	Idle switch
11	B	Ground
12	B	Ground
13	W	Throttle sensor (signal)
14	BY	Ground
15	BR	Ground
16	RL	Power supply

Intermediate connector II (body side) . . . (F42) (Black)

1	GR	Air control valve (control)
2	Lg	Identification of specifications
3	GL	Purge solenoid (control)
4	WR	EGR gas temperature sensor
5	LR	EGR solenoid (control)
6	WB	Water temperature signal

CONNECTOR IDENTIFICATION

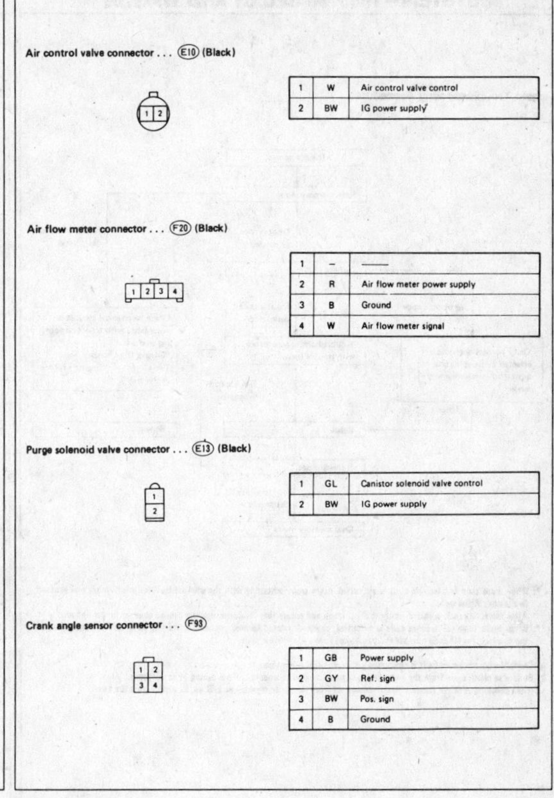

Air control valve connector . . . (E10) (Black)

1	W	Air control valve control
2	BW	IG power supply

Air flow meter connector . . . (F20) (Black)

1	–	——
2	R	Air flow meter power supply
3	B	Ground
4	W	Air flow meter signal

Purge solenoid valve connector . . . (E13) (Black)

1	GL	Canister solenoid valve control
2	BW	IG power supply

Crank angle sensor connector . . . (F93)

1	GB	Power supply
2	GY	Ref. sign
3	BW	Pos. sign
4	B	Ground

CONNECTOR IDENTIFICATION

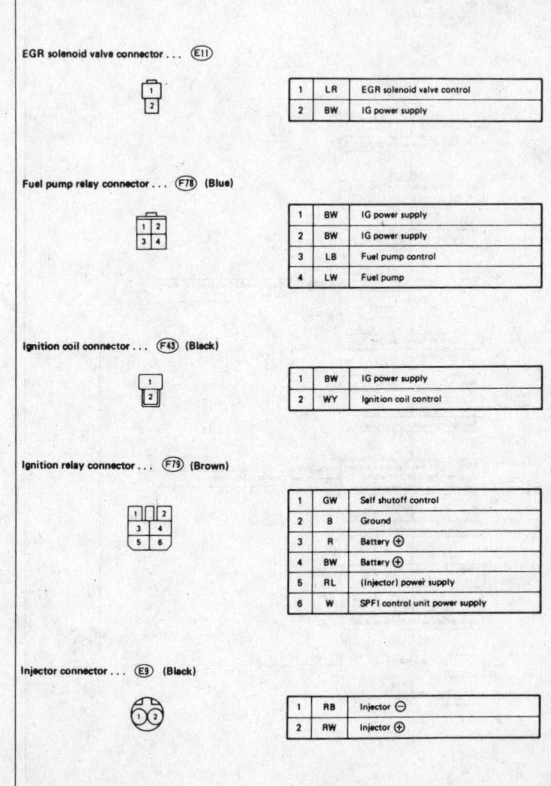

EGR solenoid valve connector . . . (E11)

1	LR	EGR solenoid valve control
2	BW	IG power supply

Fuel pump relay connector . . . (F78) (Blue)

1	BW	IG power supply
2	BW	IG power supply
3	LB	Fuel pump control
4	LW	Fuel pump

Ignition coil connector . . . (F43) (Black)

1	BW	IG power supply
2	WY	Ignition coil control

Ignition relay connector . . . (F79) (Brown)

1	GW	Self shutoff control
2	B	Ground
3	R	Battery ⊕
4	BW	Battery ⊕
5	RL	(Injector) power supply
6	W	SPFI control unit power supply

Injector connector . . . (E9) (Black)

1	RB	Injector ⊖
2	RW	Injector ⊕

KD relay connector . . . (i48)

1	BW	IG power supply
2	BW	IG power supply
3	LG	SPFI C/U (for KD control)
4	L	KD solenoid

Neutral switch connector (MT) . . . (F56)

1	BR	Ground
2	YG	Neutral signal

Throttle sensor connector . . . (E8) (Black)

1	R	Battery ⊕
2	G	Throttle position signal
3	B	Ground
4	LG	Idle switch signal

Water temperature sensor . . . (E12) (Black)

1	BR	Ground
2	WB	Water temperature signal

EGR gas temperature sensor . . . (F54)

1	WR	EGR gas temperature signal
2	B	Ground

TROUBLESHOOTING CHART FOR SELF-DIAGNOSIS SYSTEM – 1988–90 SEDAN AND WAGON

Basic Troubleshooting Procedures

Trouble occurs
Read memory mode
Trouble code designated
No trouble code designated
D-check mode
No trouble code designated
Trouble code designated
Check mechanical trouble in accordance with troubleshooting charts ("General Troubleshooting Table" "Engine Trouble in General").
Check harness/connectors affected by trouble code outputted in read memory mode.**
Troubleshoot in accordance with trouble code.*
Trouble code designated
Repair
Repair
D-check mode
No trouble code designated
Clear memory mode

*: When more than one trouble code is outputted, begin troubleshooting with the smallest trouble code number and proceed to the next higher code.
After correcting each problem, conduct the D-check and ensure that the corresponding trouble code no longer appears.

**: When more than one trouble code is outputted, check all related harness connectors, starting with that corresponding to the smallest trouble code number and proceeding to the next higher code.

a. Check the connector while it is connected unless specified otherwise.
b. Be sure to check again from the beginning in order to prevent secondary trouble caused by repair work.
c. When checking with the vacuum hose disconnected from the vacuum switch at E/G on, be sure to plug the hose.

TROUBLESHOOTING CHART FOR SELF-DIAGNOSIS SYSTEM – 1990 LOYALE

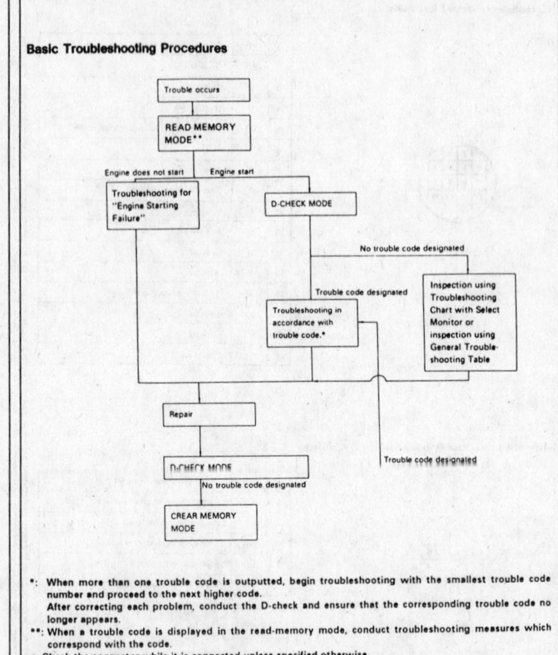
Basic Troubleshooting Procedures
Trouble occurs
READ MEMORY MODE**
Engine does not start
Engine start
Troubleshooting for "Engine Starting Failure"
D-CHECK MODE
No trouble code designated
Trouble code designated
Troubleshooting in accordance with trouble code.*
Inspection using Troubleshooting Chart with Select Monitor or inspection using General Trouble-shooting Table
Repair
Trouble code designated
D-CHECK MODE
No trouble code designated
CREAR MEMORY MODE

*: When more than one trouble code is outputted, begin troubleshooting with the smallest trouble code number and proceed to the next higher code.
After correcting each problem, conduct the D-check and ensure that the corresponding trouble code no longer appears.

**: When a trouble code is displayed in the read-memory mode, conduct troubleshooting measures which correspond with the code.

a. Check the connector while it is connected unless specified otherwise.
b. Be sure to check again from the beginning in order to prevent secondary trouble caused by repair work.
c. When checking with the vacuum hose disconnected from the vacuum switch at E/G on, be sure to plug the hose.

READ MEMORY MODE DIAGNOSIS CHART

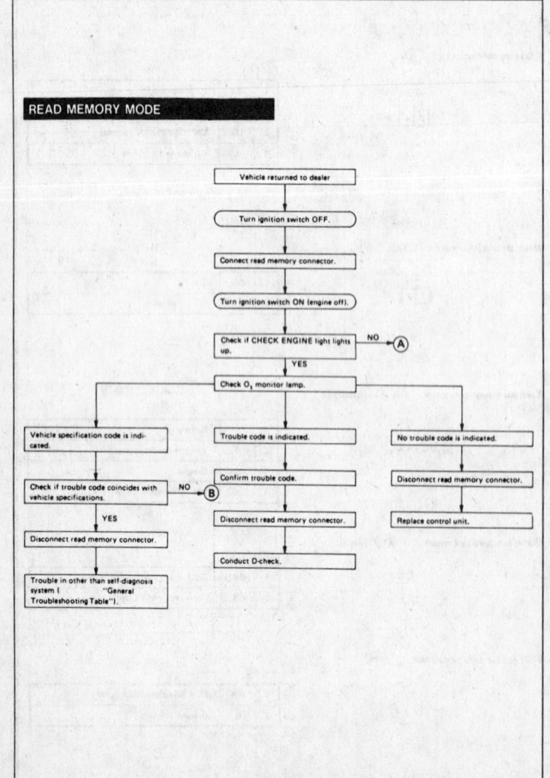
READ MEMORY MODE
Vehicle returned to dealer
Turn ignition switch OFF.
Connect read memory connector.
Turn ignition switch ON (engine off).
Check if CHECK ENGINE light lights up.
NO
A
YES
Check O2 monitor lamp.
Vehicle specification code is indicated.
Trouble code is indicated.
No trouble code is indicated.
Check if trouble code coincides with vehicle specifications.
NO
B
YES
Confirm trouble code.
Disconnect read memory connector.
Disconnect read memory connector.
Disconnect read memory connector.
Replace control unit.
Conduct D-check.
Trouble in other than self-diagnosis system ("General Troubleshooting Table").

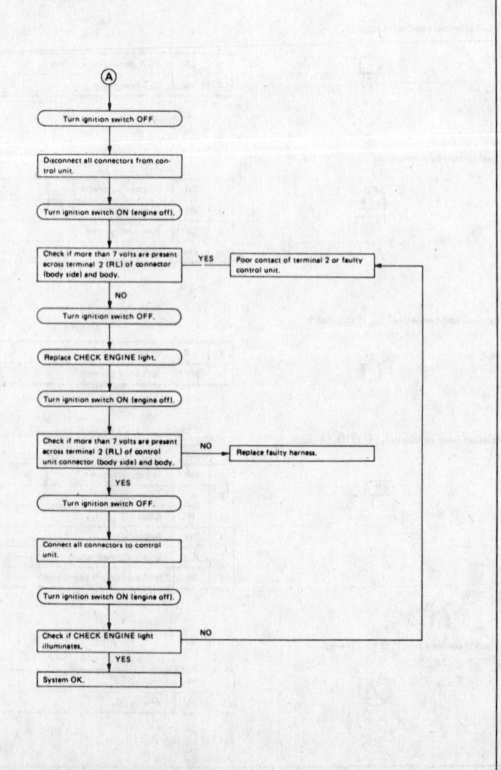
A
Turn ignition switch OFF.
Disconnect all connectors from control unit.
Turn ignition switch ON (engine off).
Check if more than 7 volts are present across terminal 2 (RL) of connector (body side) and body.
YES
Poor contact of terminal 2 or faulty control unit.
NO
Turn ignition switch OFF.
Replace CHECK ENGINE light.
Turn ignition switch ON (engine off).
Check if more than 7 volts are present across terminal 2 (RL) of control unit connector (body side) and body.
NO
Replace faulty harness.
YES
Turn ignition switch OFF.
Connect all connectors to control unit.
Turn ignition switch ON (engine off).
Check if CHECK ENGINE light illuminates.
NO
YES
System OK.

READ MEMORY MODE DIAGNOSIS CHART

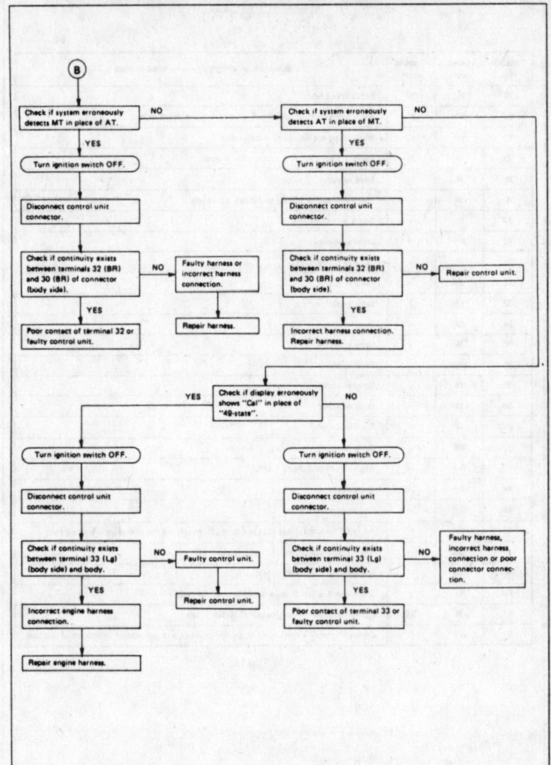
B
Check if system erroneously detects MT in place of AT.
NO
Check if system erroneously detects AT in place of MT.
NO
YES
Turn ignition switch OFF.
Disconnect control unit connector.
Check if continuity exists between terminals 32 (BR) and 30 (BR) of connector (body side).
NO
Faulty harness or incorrect harness connection.
Repair harness.
YES
Poor contact of terminal 32 or faulty control unit.
YES
Turn ignition switch OFF.
Disconnect control unit connector.
Check if continuity exists between terminals 32 (BR) and 30 (BR) of connector (body side).
NO
Repair control unit.
YES
Incorrect harness connection. Repair harness.
Check if display erroneously shows "Cal" in place of "49-state".
YES
NO
Turn ignition switch OFF.
Disconnect control unit connector.
Check if continuity exists between terminal 33 (Lg) (body side) and body.
NO
Faulty control unit.
Repair control unit.
YES
Incorrect engine harness connection.
Repair engine harness.
Turn ignition switch OFF.
Disconnect control unit connector.
Check if continuity exists between terminal 33 (Lg) (body side) and body.
NO
Faulty harness, incorrect harness connection or poor connector connection.
YES
Poor contact of terminal 33 or faulty control unit.

D-CHECK MODE DIAGNOSIS CHART

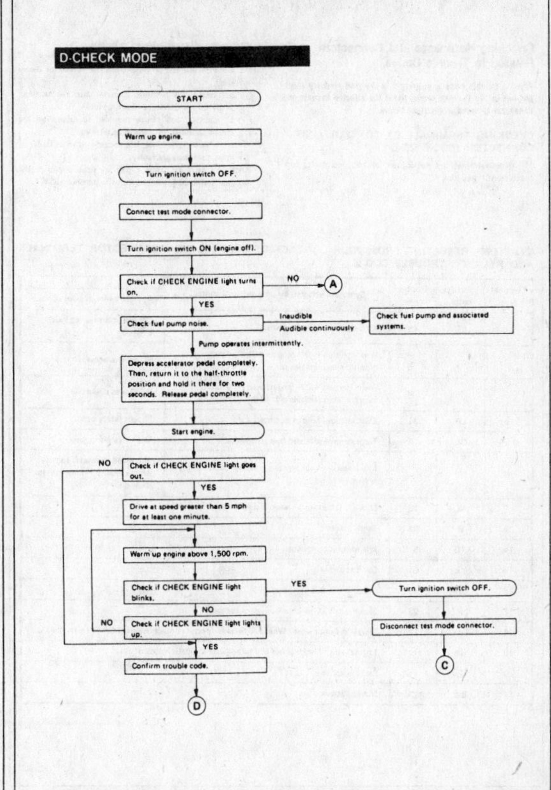
D-CHECK MODE
START
Warm up engine.
Turn ignition switch OFF.
Connect test mode connector.
Turn ignition switch ON (engine off).
Check if CHECK ENGINE light turns on.
NO
A
YES
Check fuel pump noise.
Inaudible
Audible continuously
Check fuel pump and associated systems.
Pump operates intermittently.
Depress accelerator pedal completely. Then, return it to the half-throttle position and hold it there for two seconds. Release pedal completely.
Start engine.
NO
Check if CHECK ENGINE light goes out.
YES
Drive at speed greater than 5 mph for at least one minute.
Warm up engine above 1,500 rpm.
Check if CHECK ENGINE light blinks.
YES
Turn ignition switch OFF.
Disconnect test mode connector.
C
NO
NO
Check if CHECK ENGINE light lights up.
YES
Confirm trouble code.
D

D-CHECK MODE DIAGNOSIS CHART

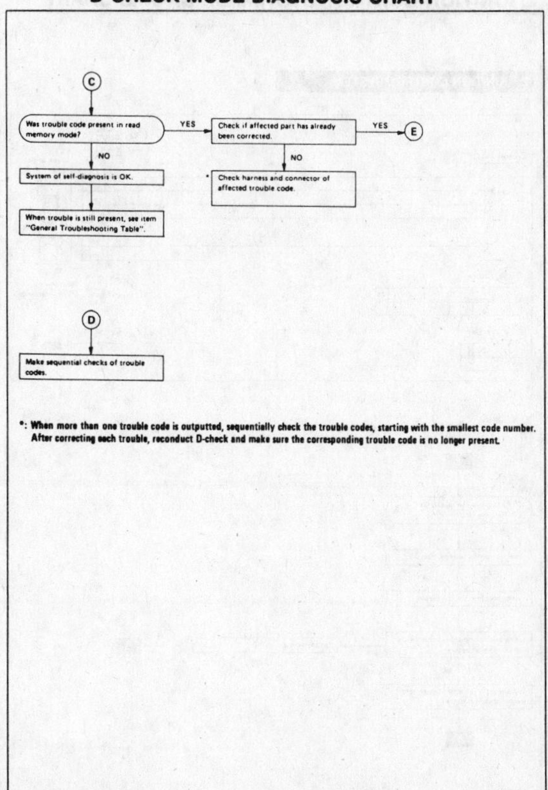
C
Was trouble code present in read memory mode?
YES
Check if affected part has already been corrected.
YES
E
NO
System of self-diagnosis is OK.
When trouble is still present, see item "General Troubleshooting Table".
NO
* Check harness and connector of affected trouble code.
D
Make sequential checks of trouble codes.

*: When more than one trouble code is outputted, sequentially check the trouble codes, starting with the smallest code number. After correcting each trouble, reconduct D-check and make sure the corresponding trouble code is no longer present.

CLEAR MEMORY MODE DIAGNOSIS CHART

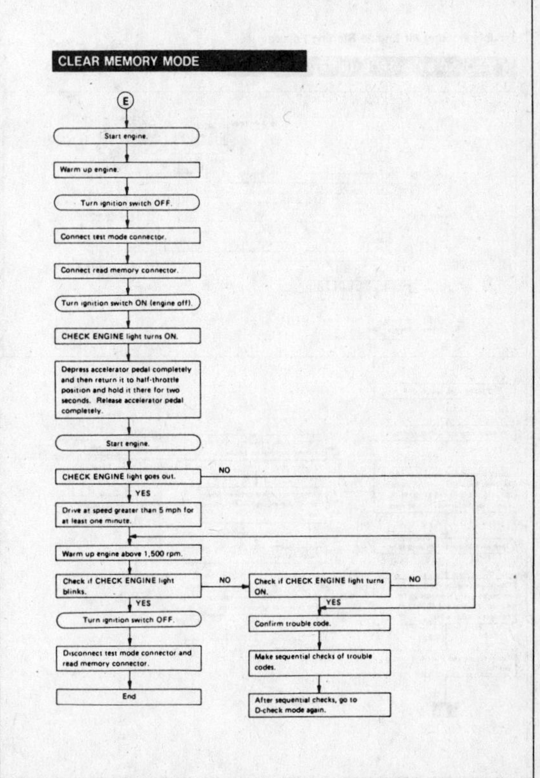
CLEAR MEMORY MODE
E
Start engine.
Warm up engine.
Turn ignition switch OFF.
Connect test mode connector.
Connect read memory connector.
Turn ignition switch ON (engine off).
CHECK ENGINE light turns ON.
Depress accelerator pedal completely and then return it to half-throttle position and hold it there for two seconds. Release accelerator pedal completely.
Start engine.
CHECK ENGINE light goes out.
NO
YES
Drive at speed greater than 5 mph for at least one minute.
Warm up engine above 1,500 rpm.
Check if CHECK ENGINE light blinks.
NO
Check if CHECK ENGINE light turns ON.
NO
YES
Turn ignition switch OFF.
Disconnect test mode connector and read memory connector.
End
YES
Confirm trouble code.
Make sequential checks of trouble codes.
After sequential checks, go to D-check mode again.

CHECKING HARNESSES AND CONNECTORS RELATED TO TROUBLE CODES

Checking Harnesses and Connectors Related to Trouble Codes

When a trouble code is outputted in the read memory mode but not in the D-check mode, check the affected harness and connector terminal as described below.

CHECKING TERMINALS OF CONTROL UNIT CONNECTOR (BODY SIDE)

1) When terminals are not locked securely, insert into connectors until they lock.

2) When terminals are considered to be open:

(1) Method of judging "OK" and "Faulty":

a. Pull out the terminal from the connector (body side).

b. Insert this terminal (female) into the terminal (male) of the connector (control unit).

c. Check "pull" force required to disconnect the female terminal from the male terminal.

If the terminal is loose, it is considered to be faulty.

(2) When terminals are faulty:

Pinch the terminal using a pair of nose pliers. If the terminal is still loose, replace it or the harness ASSY.

SYMPTOMS RESULTING FROM POOR CONTACT OF CONTROL UNIT CONNECTOR TERMINALS AND RELATED TROUBLE CODES

Terminal No.	Lead color	Trouble code	Symptom affected by poor terminal contact	At instantaneous poor contact
1	LG	45	Kick-down no longer occurs.	Shocks occur during kick-down.
2	RL	–	When ignition is ON (engine off), O_2 sensor monitor lamp remains off.	No shocks occur.
3	R	–	When ignition is ON (engine on), O_2 sensor monitor lamp remains off.	No shocks occur.
4	LR	34	EGR solenoid fails to operate.	Shocks rarely occur.
5	GL	35	Purge control solenoid fails to operate.	Shocks rarely occur.
6	LY	–	Idle speed does not increase when air conditioning system turns on.	Idle speed decreases slightly when air conditioning system turns on.
8	W	23	Shock is felt at instantaneous poor contact.	
9	B	22	Same as above.	
14	YG	51	Idle speed is erroneous.	
15	YL	61	Same as above.	
16	LgR	45	Shock is not felt.	
17	R	23	Shock is felt at instantaneous poor contact.	
18	LgY	12	Starter does not start. When instantaneous poor contact occurs shock is rarely felt.	
19	GB	11	Engine stalls. Shock is felt and tachometer indication goes down.	
20	GY	11	Same as above.	
21	BW	13	Same as above.	

Terminal No.	Lead color	Trouble code	Symptom affected by poor terminal contact
22	YG	33	Shock is not felt.
23	WB	21	While engine is cold, idle speed is erroneous and shock is felt.
24	LG	42	While idling engine, speed is erroneous.
25	W	31	Shock is rarely felt and acceleration is poor.
26	R	31	Same as above.
28	GW	–	Restarting ability is poor and shock is not felt.
29	W	–	Shock is not felt.
30	BR	–	Same as above.
31	Br	–	Same as above.
32	BR	–	Same as above.
33	Lg	–	Same as above.
34	SA	–	Same as above.
35	B	–	Same as above.
36	WR	34	Same as above.
38	RL	–	Same as above.
39	LgW	–	Same as above.
41	W	–	Same as above.
42	BR	–	Same as above.
43	RW	14	Engine stalls and shock is felt.
44	BR	14	Same as above.
45	GR	24	Engine speed decreases.
46	GY	–	Air conditioning system does not turn off though the throttle valve is opened fully.
47	LB	–	Engine lacks power, engine stalls, shock is felt.
48	RB	14	Engine stalls and shock is felt.
49	RL	–	Slight shock is felt at instantaneous poor contact.
51	B	34	Speed decreases and engine stalls. Shock is felt.
52	WY	–	Engine misfires. When engine stops, shock is felt and tachometer indication goes down.

GROUND AND CONTROL UNIT POWER SUPPLY DIAGNOSIS CHART

Troubleshooting for Engine Starting Failure

1. GROUND & CONTROL UNIT POWER SUPPLY

Turn ignition switch OFF.

Disconnect control unit connector.

Check if resistance is less than 10 Ω between body and terminals 30 (BR), 42 (BR), 44 (BR) and 51 (B) of connector (body side), respectively.
- NO → Check if right intake manifold ground terminal is grounded properly.
  - YES → Tighten bolts.
  - NO → Check if connector between body harness and engine harness is disconnected.
    - YES → Connect properly.
    - NO → Repair harness.
- YES → Turn ignition switch ON (engine off).

Check if at least 10 V DC is present across body and terminals 29 (W), 41 (W), 27 (BW) and 49 (RL) of connector (body side).
- NO → Check battery terminals, fuse, ignition relay and ignition switch.
  - OK → Repair harness.
- YES → Power supply and grounding circuits are OK. → 2

IGNITION CONTROL SYSTEM DIAGNOSIS CHART

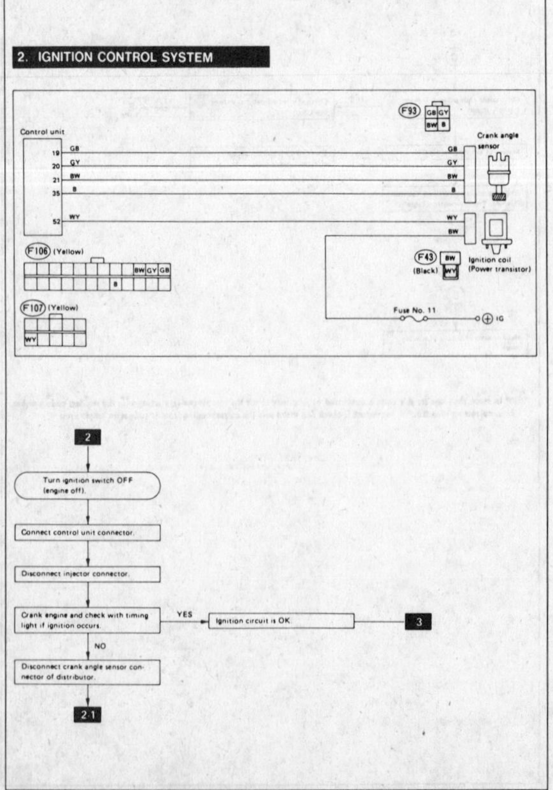
2. IGNITION CONTROL SYSTEM
F93
Control unit
Crank angle sensor
F106 (Yellow)
F107 (Yellow)
F43 (Black)
Ignition coil (Power transistor)
Fuse No. 11
IG
2
Turn ignition switch OFF (engine off).
Connect control unit connector.
Disconnect injector connector.
Crank engine and check with timing light if ignition occurs.
YES
Ignition circuit is OK.
3
NO
Disconnect crank angle sensor connector of distributor.
2-1

IGNITION CONTROL SYSTEM DIAGNOSIS CHART

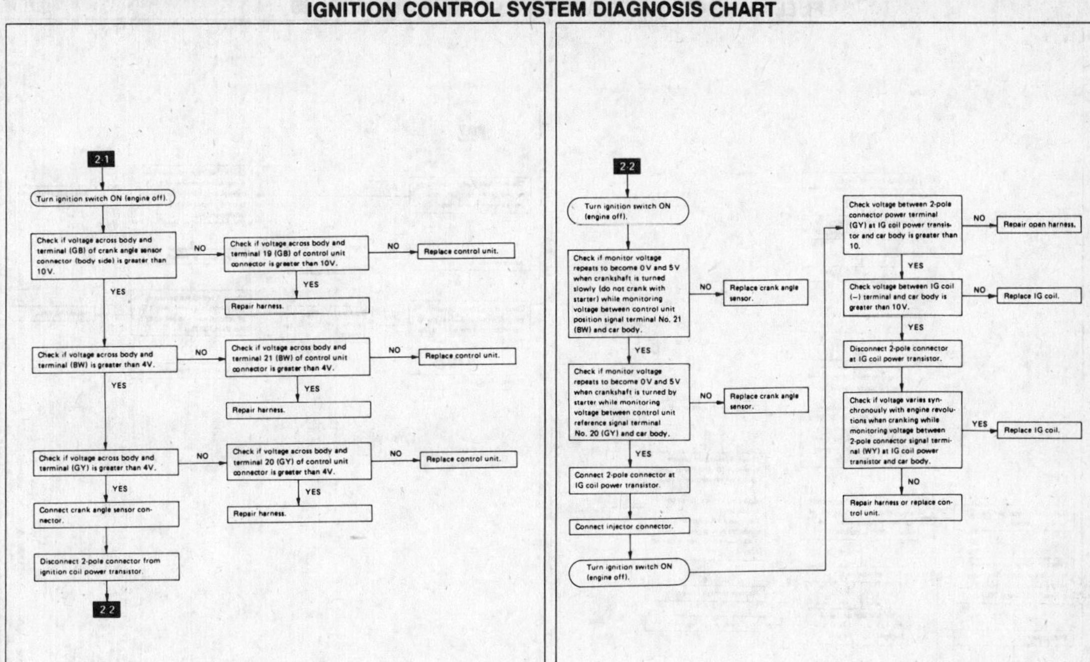
2-1
Turn ignition switch ON (engine off).
Check if voltage across body and terminal (GB) of crank angle sensor connector (body side) is greater than 10V.
NO
Check if voltage across body and terminal 19 (GB) of control unit connector is greater than 10V.
NO
Replace control unit.
YES
Repair harness.
YES
Check if voltage across body and terminal (BW) is greater than 4V.
NO
Check if voltage across body and terminal 21 (BW) of control unit connector is greater than 4V.
NO
Replace control unit.
YES
Repair harness.
YES
Check if voltage across body and terminal (GY) is greater than 4V.
NO
Check if voltage across body and terminal 20 (GY) of control unit connector is greater than 4V.
NO
Replace control unit.
YES
Repair harness.
YES
Connect crank angle sensor connector.
Disconnect 2-pole connector from ignition coil power transistor.
2-2
2-2
Turn ignition switch ON (engine off).
Check if monitor voltage repeats to become 0V and 5V when crankshaft is turned slowly (do not crank with starter) while monitoring voltage between control unit position signal terminal No. 21 (BW) and car body.
NO
Replace crank angle sensor.
YES
Check if monitor voltage repeats to become 0V and 5V when crankshaft is turned by starter while monitoring voltage between control unit reference signal terminal No. 20 (GY) and car body.
NO
Replace crank angle sensor.
YES
Connect 2-pole connector at IG coil power transistor.
Connect injector connector.
Turn ignition switch ON (engine off).
Check voltage between 2-pole connector power terminal (GY) at IG coil power transistor and car body is greater than 10.
NO
Repair open harness.
YES
Check voltage between IG coil (–) terminal and car body is greater than 10V.
NO
Replace IG coil.
YES
Disconnect 2-pole connector at IG coil power transistor.
Check if voltage varies synchronously with engine revolutions when cranking while monitoring voltage between 2-pole connector signal terminal (WY) at IG coil power transistor and car body.
YES
Replace IG coil.
NO
Repair harness or replace control unit.

FUEL PUMP CIRCUIT DIAGNOSIS CHART 1988 SEDAN AND WAGON

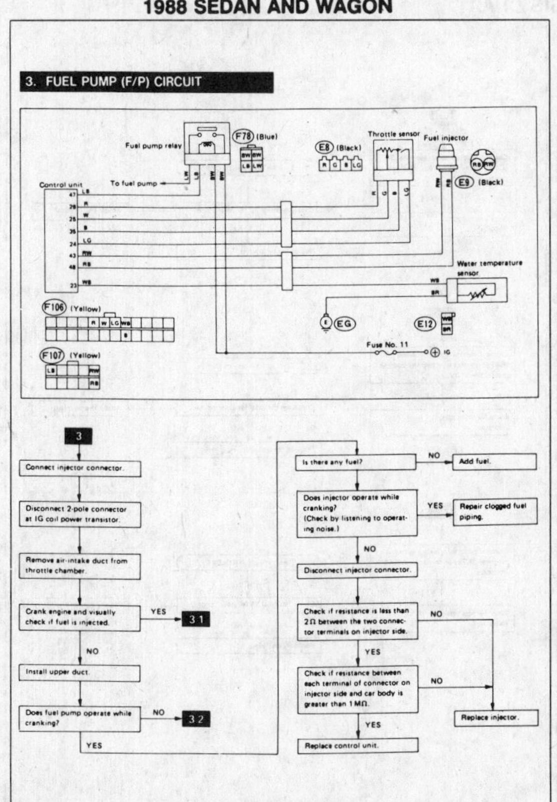
3. FUEL PUMP (F/P) CIRCUIT
Fuel pump relay
F78 (Blue)
E8 (Black)
Throttle sensor
Fuel injector
E9 (Black)
Control unit
To fuel pump
Water temperature sensor
F106 (Yellow)
F107 (Yellow)
EG
E12
Fuse No. 11
3
Connect injector connector.
Disconnect 2-pole connector at IG coil power transistor.
Remove air-intake duct from throttle chamber.
Crank engine and visually check if fuel is injected.
YES
3-1
NO
Install upper duct.
Does fuel pump operate while cranking?
NO
3-2
YES
Is there any fuel?
NO
Add fuel.
Does injector operate while cranking? (Check by listening to operating noise.)
YES
Repair clogged fuel piping.
NO
Disconnect injector connector.
Check if resistance is less than 2Ω between the two connector terminals on injector side.
NO
YES
Check if resistance between each terminal of connector on injector side and car body is greater than 1 MΩ.
NO
Replace injector.
YES
Replace control unit.

FUEL PUMP CIRCUIT DIAGNOSIS CHART 1989–90 SEDAN AND WAGON AND 1990 LOYALE

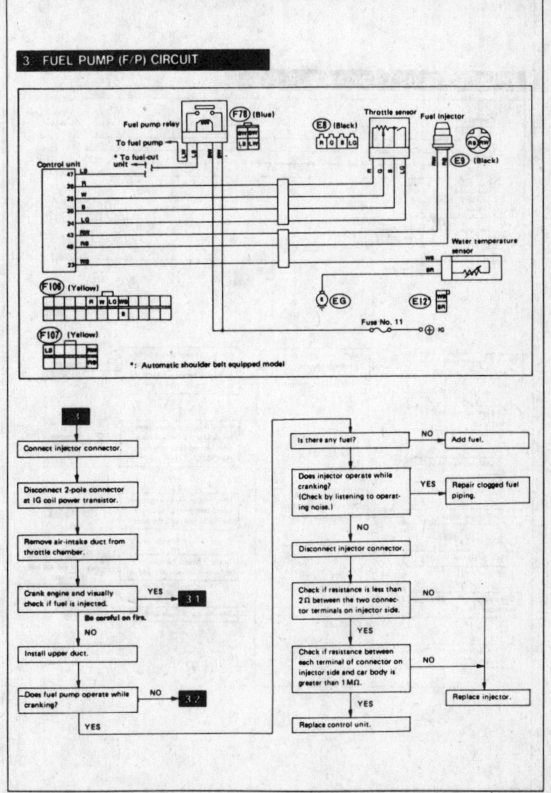
3. FUEL PUMP (F/P) CIRCUIT
Fuel pump relay
F78 (Blue)
To fuel pump
* To fuel-cut unit
E8 (Black)
Throttle sensor
Fuel injector
E9 (Black)
Control unit
Water temperature sensor
F106 (Yellow)
F107 (Yellow)
EG
E12
Fuse No. 11
*: Automatic shoulder belt equipped model
3
Connect injector connector.
Disconnect 2-pole connector at IG coil power transistor.
Remove air-intake duct from throttle chamber.
Crank engine and visually check if fuel is injected.
YES
3-1
Be careful on fire.
NO
Install upper duct.
Does fuel pump operate while cranking?
NO
3-2
YES
Is there any fuel?
NO
Add fuel.
Does injector operate while cranking? (Check by listening to operating noise.)
YES
Repair clogged fuel piping.
NO
Disconnect injector connector.
Check if resistance is less than 2Ω between the two connector terminals on injector side.
NO
YES
Check if resistance between each terminal of connector on injector side and car body is greater than 1 MΩ.
NO
Replace injector.
YES
Replace control unit.

FUEL PUMP CIRCUIT DIAGNOSIS CHART—ALL MODELS

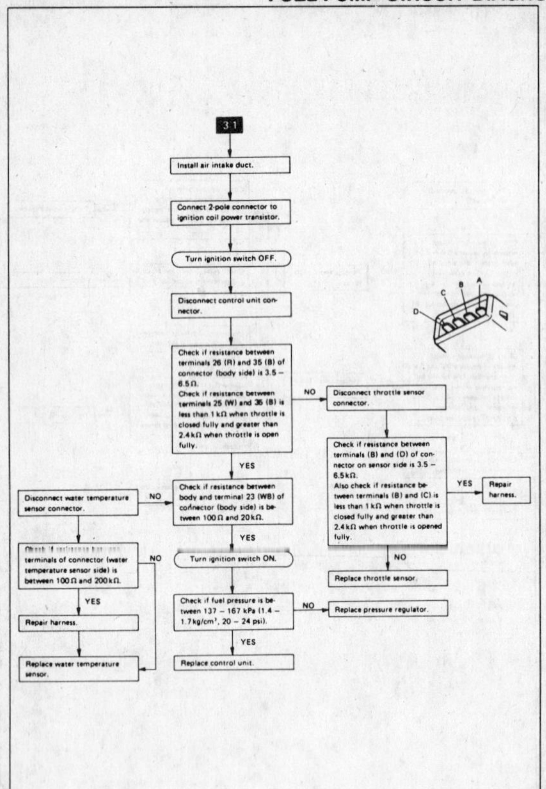
31
Install air intake duct.
Connect 2-pole connector to ignition coil power transistor.
Turn ignition switch OFF.
Disconnect control unit connector.
Check if resistance between terminals 26 (R) and 35 (B) of connector (body side) is 3.5 – 6.5Ω. Check if resistance between terminals 25 (W) and 36 (B) is less than 1 kΩ when throttle is closed fully and greater than 2.4kΩ when throttle is open fully.
NO
Disconnect throttle sensor connector.
Check if resistance between terminals (B) and (D) of connector on sensor side is 3.5 – 6.5kΩ. Also check if resistance between terminals (B) and (C) is less than 1 kΩ when throttle is closed fully and greater than 2.4kΩ when throttle is opened fully.
YES
Repair harness.
NO
Replace throttle sensor.
YES
Check if resistance between body and terminal 23 (WB) of connector (body side) is between 100Ω and 20kΩ.
NO
Disconnect water temperature sensor connector.
Check if resistance between terminals of connector (water temperature sensor side) is between 100Ω and 200kΩ.
NO
YES
Repair harness.
Replace water temperature sensor.
YES
Turn ignition switch ON.
Check if fuel pressure is between 137 – 167 kPa (1.4 – 1.7kg/cm², 20 – 24 psi).
NO
Replace pressure regulator.
YES
Replace control unit.
A
B
C
D

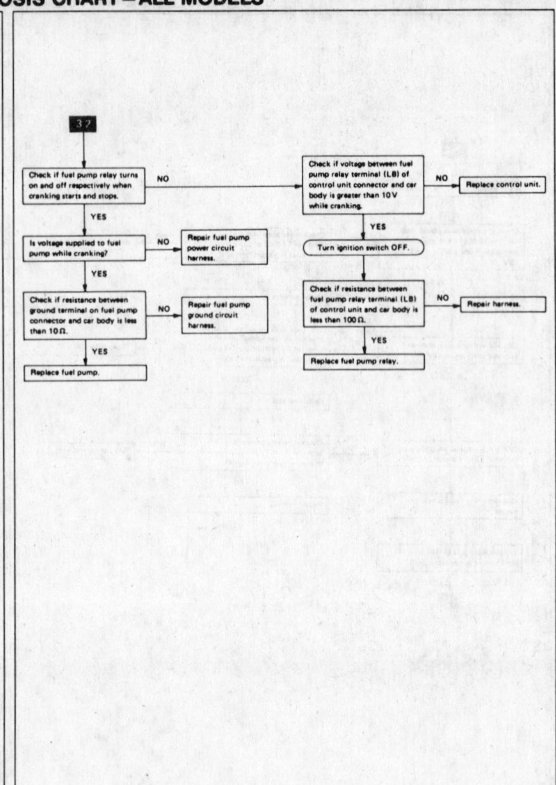
32
Check if fuel pump relay turns on and off respectively when cranking starts and stops.
NO
Check if voltage between fuel pump relay terminal (LB) of control unit connector and car body is greater than 10V while cranking.
NO
Replace control unit.
YES
Turn ignition switch OFF.
Check if resistance between fuel pump relay terminal (LB) of control unit and car body is less than 100Ω.
NO
Repair harness.
YES
Replace fuel pump relay.
YES
Is voltage supplied to fuel pump while cranking?
NO
Repair fuel pump power circuit harness.
YES
Check if resistance between ground terminal on fuel pump connector and car body is less than 10Ω.
NO
Repair fuel pump ground circuit harness.
YES
Replace fuel pump.

DIAGNOSIS CHART

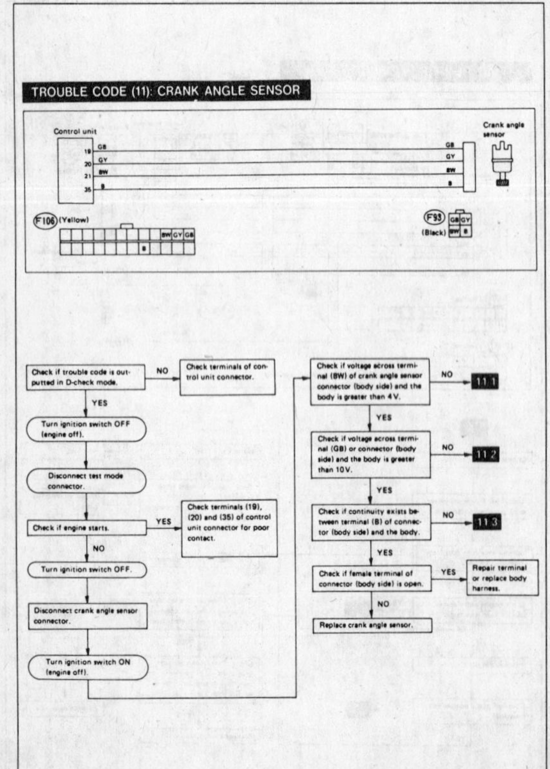
TROUBLE CODE (11): CRANK ANGLE SENSOR
Control unit
19
20
21
35
GB
GY
BW
B
Crank angle sensor
F106 (Yellow)
F93 (Black)
Check if trouble code is outputted in D-check mode.
NO
Check terminals of control unit connector.
YES
Turn ignition switch OFF (engine off).
Disconnect test mode connector.
Check if engine starts.
YES
Check terminals (19), (20) and (35) of control unit connector for poor contact.
NO
Turn ignition switch OFF.
Disconnect crank angle sensor connector.
Turn ignition switch ON (engine off).
Check if voltage across terminal (BW) of crank angle sensor connector (body side) and the body is greater than 4V.
NO
111
YES
Check if voltage across terminal (GB) or connector (body side) and the body is greater than 10V.
NO
112
YES
Check if continuity exists between terminal (B) of connector (body side) and the body.
NO
113
YES
Check if female terminal of connector (body side) is open.
YES
Repair terminal or replace body harness.
NO
Replace crank angle sensor.

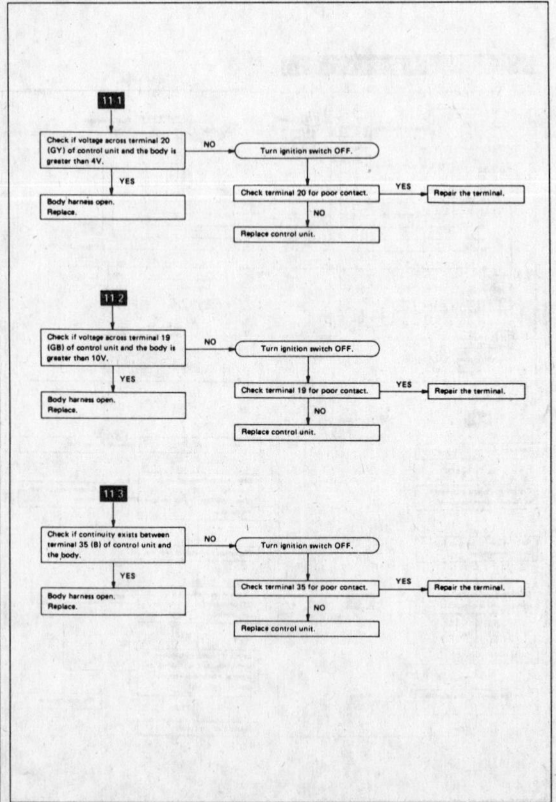
111
Check if voltage across terminal 20 (GY) of control unit and the body is greater than 4V.
NO
Turn ignition switch OFF.
Check terminal 20 for poor contact.
YES
Repair the terminal.
NO
Replace control unit.
YES
Body harness open. Replace.
112
Check if voltage across terminal 19 (GB) of control unit and the body is greater than 10V.
NO
Turn ignition switch OFF.
Check terminal 19 for poor contact.
YES
Repair the terminal.
NO
Replace control unit.
YES
Body harness open. Replace.
113
Check if continuity exists between terminal 35 (B) of control unit and the body.
NO
Turn ignition switch OFF.
Check terminal 35 for poor contact.
YES
Repair the terminal.
NO
Replace control unit.
YES
Body harness open. Replace.

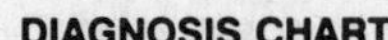

DIAGNOSIS CHART

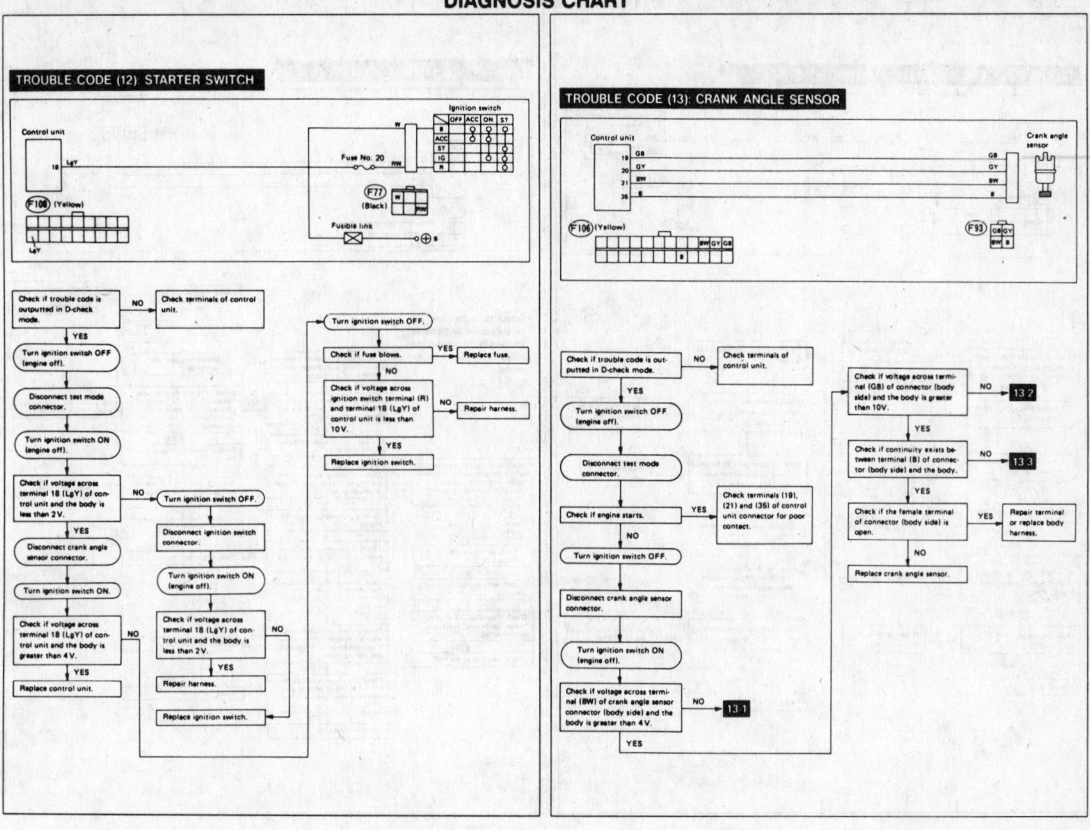

TROUBLE CODE (12): STARTER SWITCH
Control unit
Ignition switch
Fuse No. 20
(Yellow)
(Black)
Fusible link
Check if trouble code is outputted in D-check mode.
NO
Check terminals of control unit.
YES
Turn ignition switch OFF (engine off).
Disconnect test mode connector.
Turn ignition switch ON (engine off).
Check if voltage across terminal 18 (LgY) of control unit and the body is less than 2V.
Turn ignition switch OFF.
Disconnect ignition switch connector.
Turn ignition switch ON (engine off).
Check if voltage across terminal 18 (LgY) of control unit and the body is less than 2V.
Repair harness.
Replace ignition switch.
Disconnect crank angle sensor connector.
Turn ignition switch ON.
Check if voltage across terminal 18 (LgY) of control unit and the body is greater than 4V.
Replace control unit.
Turn ignition switch OFF.
Check if fuse blows.
Replace fuse.
Check if voltage across ignition switch terminal (R) and terminal 18 (LgY) of control unit is less than 10V.
Repair harness.
Replace ignition switch.
TROUBLE CODE (13): CRANK ANGLE SENSOR
Control unit
Crank angle sensor
(Yellow)
Check if trouble code is outputted in D-check mode.
Check terminals of control unit.
Turn ignition switch OFF (engine off).
Disconnect test mode connector.
Check if engine starts.
Check terminals (19), (21) and (35) of control unit connector for poor contact.
Turn ignition switch OFF.
Disconnect crank angle sensor connector.
Turn ignition switch ON (engine off).
Check if voltage across terminal (BW) of crank angle sensor connector (body side) and the body is greater than 4 V.
13.1
Check if voltage across terminal (GB) of connector (body side) and the body is greater than 10V.
13.2
Check if continuity exists between terminal (B) of connector (body side) and the body.
13.3
Check if the female terminal of connector (body side) is open.
Repair terminal or replace body harness.
Replace crank angle sensor.

DIAGNOSIS CHART

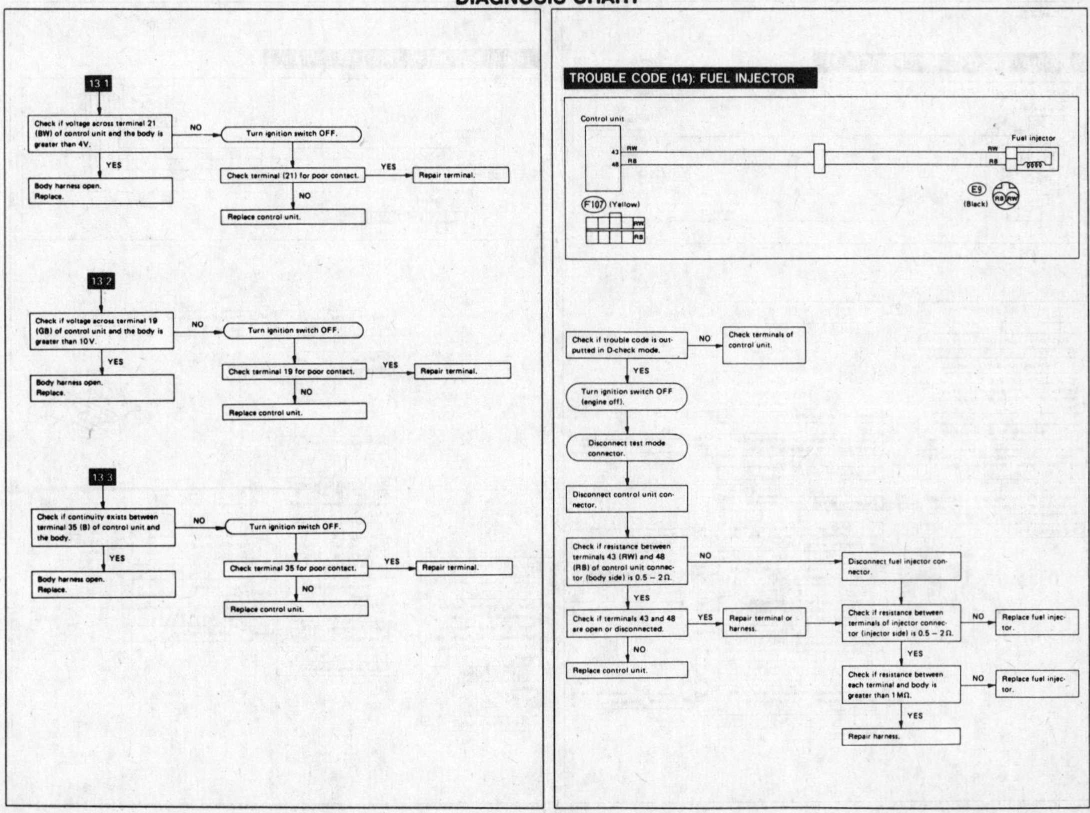

13.1
Check if voltage across terminal 21 (BW) of control unit and the body is greater than 4V.
Body harness open. Replace.
Turn ignition switch OFF.
Check terminal (21) for poor contact.
Repair terminal.
Replace control unit.
13.2
Check if voltage across terminal 19 (GB) of control unit and the body is greater than 10V.
Body harness open. Replace.
Turn ignition switch OFF.
Check terminal 19 for poor contact.
Repair terminal.
Replace control unit.
13.3
Check if continuity exists between terminal 35 (B) of control unit and the body.
Body harness open. Replace.
Turn ignition switch OFF.
Check terminal 35 for poor contact.
Repair terminal.
Replace control unit.
TROUBLE CODE (14): FUEL INJECTOR
Control unit
Fuel injector
(Yellow)
(Black)
Check if trouble code is outputted in D-check mode.
Check terminals of control unit.
Turn ignition switch OFF (engine off).
Disconnect test mode connector.
Disconnect control unit connector.
Check if resistance between terminals 43 (RW) and 48 (RB) of control unit connector (body side) is 0.5 – 2Ω.
Check if terminals 43 and 48 are open or disconnected.
Repair terminal or harness.
Replace control unit.
Disconnect fuel injector connector.
Check if resistance between terminals of injector connector (injector side) is 0.5 – 2Ω.
Replace fuel injector.
Check if resistance between each terminal and body is greater than 1MΩ.
Replace fuel injector.
Repair harness.

DIAGNOSIS CHART

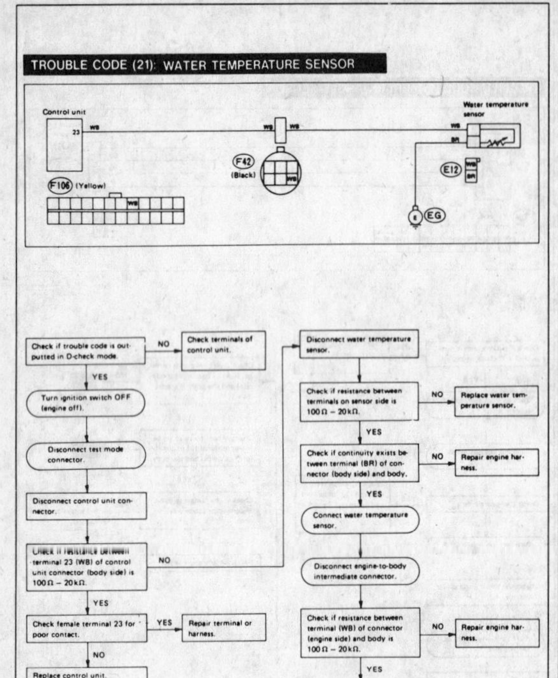
TROUBLE CODE (21): WATER TEMPERATURE SENSOR
Control unit
Water temperature sensor
F42 (Black)
E12
EG
F106 (Yellow)
Check if trouble code is outputted in D-check mode.
NO
Check terminals of control unit.
YES
Turn ignition switch OFF (engine off).
Disconnect test mode connector.
Disconnect control unit connector.
Check if resistance between terminal 23 (WB) of control unit connector (body side) is 100 Ω – 20 kΩ.
NO
YES
Check female terminal 23 for poor contact.
YES
Repair terminal or harness.
NO
Replace control unit.
Disconnect water temperature sensor.
Check if resistance between terminals on sensor side is 100 Ω – 20 kΩ.
NO
Replace water temperature sensor.
YES
Check if continuity exists between terminal (BR) of connector (body side) and body.
NO
Repair engine harness.
YES
Connect water temperature sensor.
Disconnect engine-to-body intermediate connector.
Check if resistance between terminal (WB) of connector (engine side) and body is 100 Ω – 20 kΩ.
NO
Repair engine harness.
YES
Repair body harness.

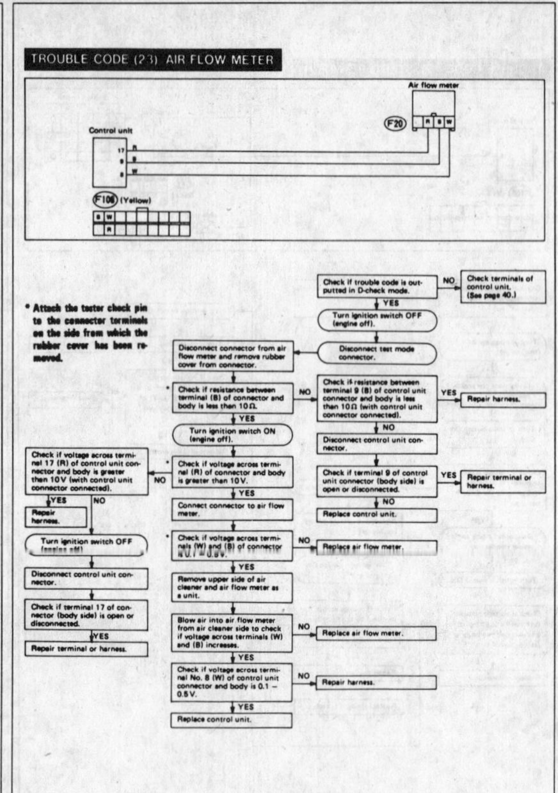
TROUBLE CODE (23): AIR FLOW METER
Air flow meter
F20
Control unit
F106 (Yellow)
* Attach the tester check pin to the connector terminals on the side from which the rubber cover has been removed.
Check if trouble code is outputted in D-check mode.
NO
Check terminals of control unit. (See page 40.)
YES
Turn ignition switch OFF (engine off).
Disconnect test mode connector.
Disconnect connector from air flow meter and remove rubber cover from connector.
* Check if resistance between terminal (B) of connector and body is less than 10 Ω.
NO
Check if resistance between terminal 9 (B) of control unit connector and body is less than 10 Ω (with control unit connector connected).
YES
Repair harness.
NO
Disconnect control unit connector.
Check if terminal 9 of control unit connector (body side) is open or disconnected.
YES
Repair terminal or harness.
NO
Replace control unit.
YES
Turn ignition switch ON (engine off).
* Check if voltage across terminal (R) of connector and body is greater than 10 V.
NO
Check if voltage across terminal 17 (R) of control unit connector and body is greater than 10 V (with control unit connector connected).
YES
Repair harness.
NO
Turn ignition switch OFF (engine off).
Disconnect control unit connector.
Check if terminal 17 of connector (body side) is open or disconnected.
YES
Repair terminal or harness.
YES
Connect connector to air flow meter.
* Check if voltage across terminals (W) and (B) of connector is 0.1 – 0.5 V.
NO
Replace air flow meter.
YES
Remove upper side of air cleaner and air flow meter as a unit.
Blow air into air flow meter from air cleaner side to check if voltage across terminals (W) and (B) increases.
NO
Replace air flow meter.
YES
Check if voltage across terminal No. 8 (W) of control unit connector and body is 0.1 – 0.5 V.
NO
Repair harness.
YES
Replace control unit.

DIAGNOSIS CHART

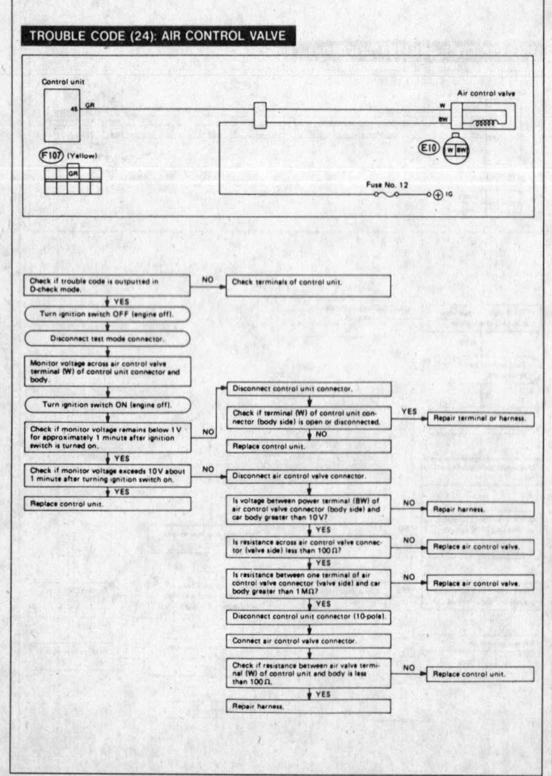
TROUBLE CODE (24): AIR CONTROL VALVE
Control unit
Air control valve
F107 (Yellow)
E10
Fuse No. 12
IG
Check if trouble code is outputted in D-check mode.
NO
Check terminals of control unit.
YES
Turn ignition switch OFF (engine off).
Disconnect test mode connector.
Monitor voltage across air control valve terminal (W) of control unit connector and body.
Turn ignition switch ON (engine off).
Check if monitor voltage remains below 1 V for approximately 1 minute after ignition switch is turned on.
NO
Disconnect control unit connector.
Check if terminal (W) of control unit connector (body side) is open or disconnected.
YES
Repair terminal or harness.
NO
Replace control unit.
YES
Check if monitor voltage exceeds 10V about 1 minute after turning ignition switch on.
YES
Replace control unit.
NO
Disconnect air control valve connector.
Is voltage between power terminal (BW) of air control valve connector (body side) and car body greater than 10V?
NO
Repair harness.
YES
Is resistance across air control valve connector (valve side) less than 100 Ω?
NO
Replace air control valve.
YES
Is resistance between one terminal of air control valve connector (valve side) and car body greater than 1 MΩ?
NO
Replace air control valve.
YES
Disconnect control unit connector (10-pole).
Connect air control valve connector.
Check if resistance between air valve terminal (W) of control unit and body is less than 100 Ω.
NO
Replace control unit.
YES
Repair harness.

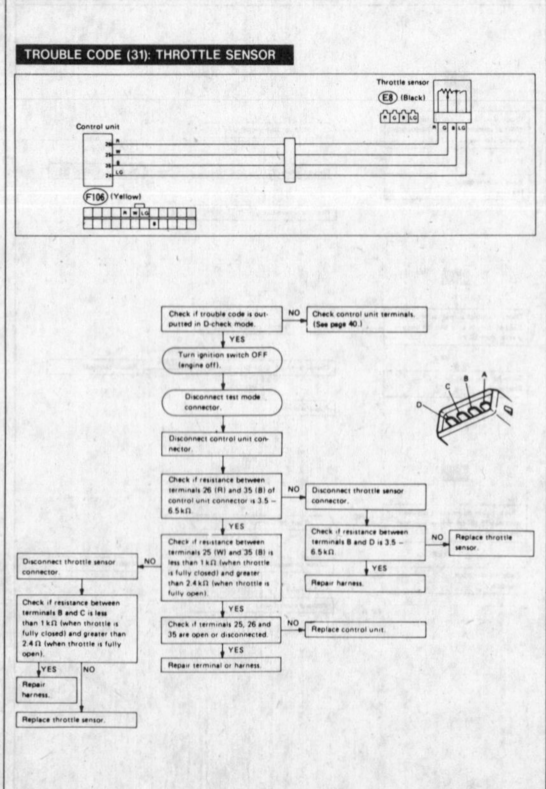
TROUBLE CODE (31): THROTTLE SENSOR
Throttle sensor
E8 (Black)
Control unit
F106 (Yellow)
Check if trouble code is outputted in D-check mode.
NO
Check control unit terminals. (See page 40.)
YES
Turn ignition switch OFF (engine off).
Disconnect test mode connector.
Disconnect control unit connector.
Check if resistance between terminals 26 (R) and 35 (B) of control unit connector is 3.5 – 6.5 kΩ.
NO
Disconnect throttle sensor connector.
Check if resistance between terminals B and D is 3.5 – 6.5 kΩ.
NO
Replace throttle sensor.
YES
Repair harness.
YES
Check if resistance between terminals 25 (W) and 35 (B) is less than 1 kΩ (when throttle is fully closed) and greater than 2.4 kΩ (when throttle is fully open).
NO
Disconnect throttle sensor connector.
Check if resistance between terminals B and C is less than 1 kΩ (when throttle is fully closed) and greater than 2.4 Ω (when throttle is fully open).
YES
Repair harness.
NO
Replace throttle sensor.
YES
Check if terminals 25, 26 and 35 are open or disconnected.
NO
Replace control unit.
YES
Repair terminal or harness.
A
B
C
D

DIAGNOSIS CHART

DIAGNOSIS CHART—1988 SEDAN AND WAGON

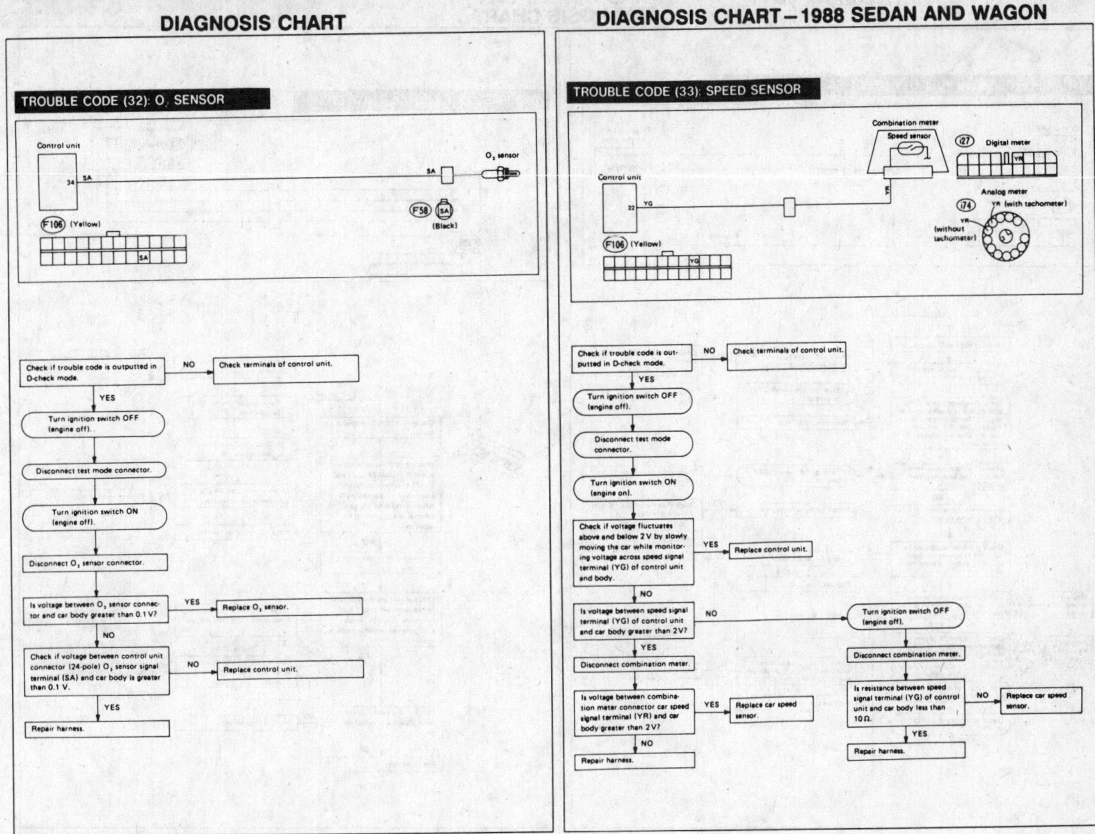
TROUBLE CODE (32): O_2 SENSOR
Control unit
O_2 sensor
F106 (Yellow)
F58 (Black)
Check if trouble code is outputted in D-check mode.
NO
Check terminals of control unit.
YES
Turn ignition switch OFF (engine off).
Disconnect test mode connector.
Turn ignition switch ON (engine off).
Disconnect O_2 sensor connector.
Is voltage between O_2 sensor connector and car body greater than 0.1 V?
YES
Replace O_2 sensor.
NO
Check if voltage between control unit connector (24 pole) O_2 sensor signal terminal (SA) and car body is greater than 0.1 V.
NO
Replace control unit.
YES
Repair harness.
TROUBLE CODE (33): SPEED SENSOR
Combination meter
Speed sensor
Digital meter
Control unit
Analog meter
YR (with tachometer)
YR (without tachometer)
F106 (Yellow)
Check if trouble code is outputted in D-check mode.
NO
Check terminals of control unit.
YES
Turn ignition switch OFF (engine off).
Disconnect test mode connector.
Turn ignition switch ON (engine on).
Check if voltage fluctuates above and below 2 V by slowly moving the car while monitoring voltage across speed signal terminal (YG) of control unit and body.
YES
Replace control unit.
NO
Is voltage between speed signal terminal (YG) of control unit and car body greater than 2V?
NO
Turn ignition switch OFF (engine off).
YES
Disconnect combination meter.
Disconnect combination meter.
Is voltage between combination meter connector car speed signal terminal (YR) and car body greater than 2V?
YES
Replace car speed sensor.
NO
Repair harness.
Is resistance between speed signal terminal (YG) of control unit and car body less than 10Ω.
NO
Replace car speed sensor.
YES
Repair harness.

DIAGNOSIS CHART—1989-90 SEDAN AND WAGON AND 1990 LOYALE

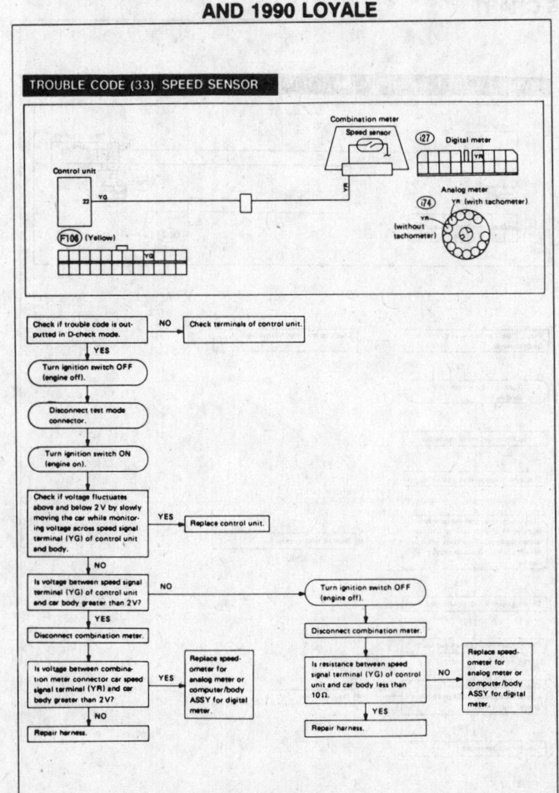
TROUBLE CODE (33): SPEED SENSOR
Combination meter
Speed sensor
Digital meter
Control unit
Analog meter
YR (with tachometer)
YR (without tachometer)
F106 (Yellow)
Check if trouble code is outputted in D-check mode.
NO
Check terminals of control unit.
YES
Turn ignition switch OFF (engine off).
Disconnect test mode connector.
Turn ignition switch ON (engine on).
Check if voltage fluctuates above and below 2 V by slowly moving the car while monitoring voltage across speed signal terminal (YG) of control unit and body.
YES
Replace control unit.
NO
Is voltage between speed signal terminal (YG) of control unit and car body greater than 2V?
NO
Turn ignition switch OFF (engine off).
YES
Disconnect combination meter.
Disconnect combination meter.
Is voltage between combination meter connector car speed signal terminal (YR) and car body greater than 2V?
YES
Replace speedometer for analog meter or computer/body ASSY for digital meter.
NO
Repair harness.
Is resistance between speed signal terminal (YG) of control unit and car body less than 10Ω.
NO
Replace speedometer for analog meter or computer/body ASSY for digital meter.
YES
Repair harness.

DIAGNOSIS CHART

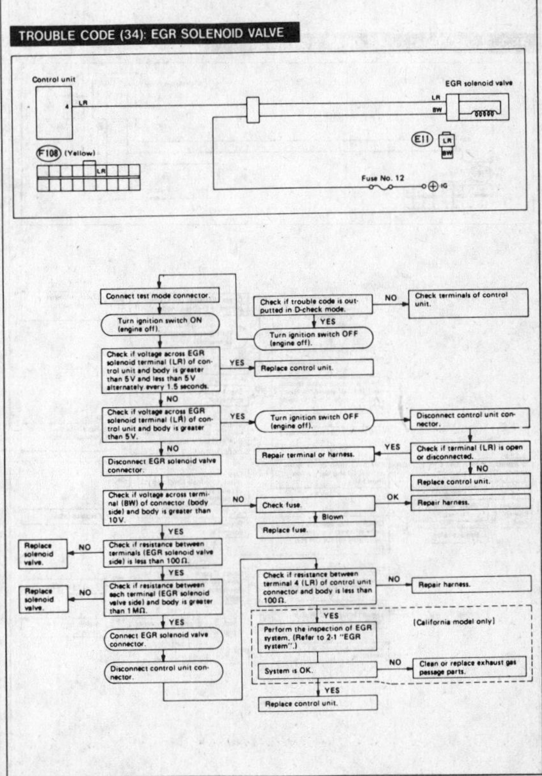
TROUBLE CODE (34): EGR SOLENOID VALVE
Control unit
EGR solenoid valve
F108 (Yellow)
Fuse No. 12
Connect test mode connector.
Turn ignition switch ON (engine off).
Check if voltage across EGR solenoid terminal (LR) of control unit and body is greater than 5V and less than 5V alternately every 1.5 seconds.
YES
Replace control unit.
NO
Check if voltage across EGR solenoid terminal (LR) of control unit and body is greater than 5V.
YES
Turn ignition switch OFF (engine off).
Disconnect control unit connector.
Check if terminal (LR) is open or disconnected.
YES
Repair terminal or harness.
NO
Replace control unit.
NO
Disconnect EGR solenoid valve connector.
Check if voltage across terminal (BW) of connector (body side) and body is greater than 10V.
NO
Check fuse.
OK
Repair harness.
Blown
Replace fuse.
YES
Check if resistance between terminals (EGR solenoid valve side) is less than 100 Ω.
NO
Replace solenoid valve.
YES
Check if resistance between each terminal (EGR solenoid valve side) and body is greater than 1 MΩ.
NO
Replace solenoid valve.
YES
Connect EGR solenoid valve connector.
Disconnect control unit connector.
Check if resistance between terminal 4 (LR) of control unit connector and body is less than 100 Ω.
NO
Repair harness.
YES
(California model only)
Perform the inspection of EGR system. (Refer to 2-1 "EGR system".)
System is OK.
NO
Clean or replace exhaust gas passage parts.
YES
Replace control unit.
Check if trouble code is outputted in D-check mode.
NO
Check terminals of control unit.
YES
Turn ignition switch OFF (engine off).

DIAGNOSIS CHART

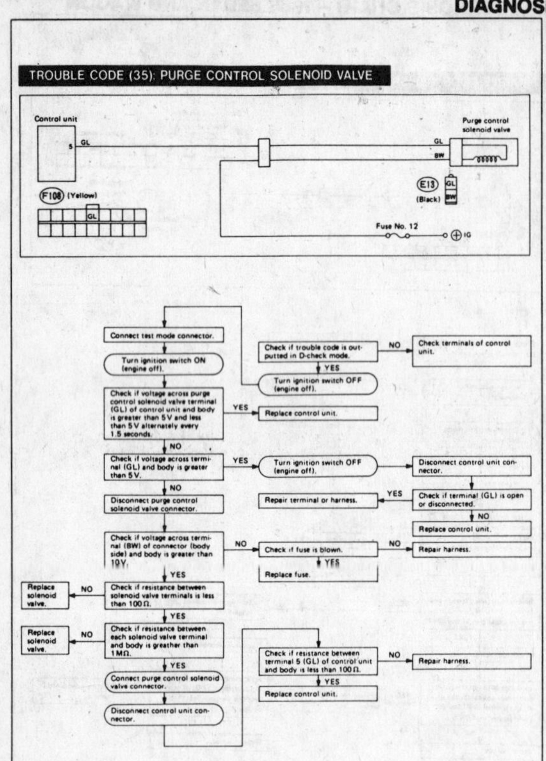
TROUBLE CODE (35): PURGE CONTROL SOLENOID VALVE
Control unit
Purge control solenoid valve
F108 (Yellow)
E13 (Black)
Fuse No. 12
Connect test mode connector.
Turn ignition switch ON (engine off).
Check if voltage across purge control solenoid valve terminal (GL) of control unit and body is greater than 5 V and less than 5 V alternately every 1.5 seconds.
Replace control unit.
Check if voltage across terminal (GL) and body is greater than 5 V.
Turn ignition switch OFF (engine off).
Disconnect control unit connector.
Check if terminal (GL) is open or disconnected.
Repair terminal or harness.
Replace control unit.
Disconnect purge control solenoid valve connector.
Check if voltage across terminal (BW) of connector (body side) and body is greater than 10 V.
Check if fuse is blown.
Repair harness.
Replace fuse.
Check if resistance between solenoid valve terminals is less than 100 Ω.
Replace solenoid valve.
Check if resistance between each solenoid valve terminal and body is greater than 1 MΩ.
Replace solenoid valve.
Connect purge control solenoid valve connector.
Disconnect control unit connector.
Check if resistance between terminal 5 (GL) of control unit and body is less than 100 Ω.
Repair harness.
Replace control unit.
Check if trouble code is outputted in D-check mode.
Check terminals of control unit.
YES
NO

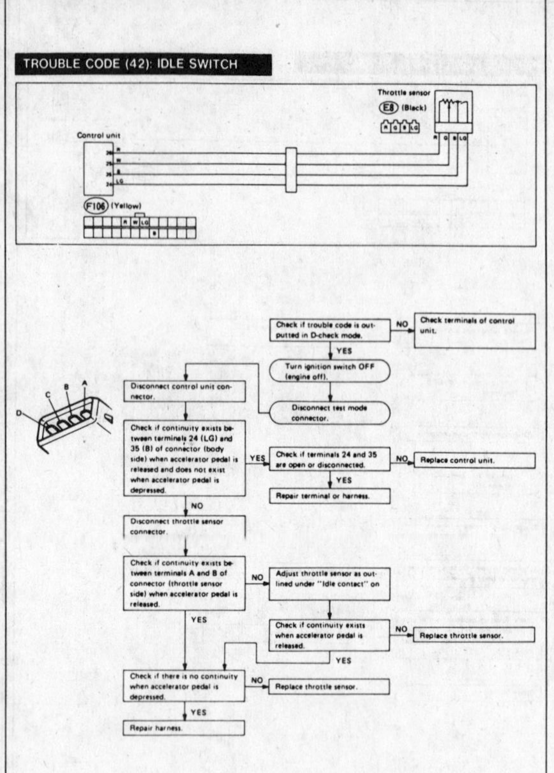
TROUBLE CODE (42): IDLE SWITCH
Throttle sensor
E8 (Black)
Control unit
F106 (Yellow)
Check if trouble code is outputted in D-check mode.
Check terminals of control unit.
Turn ignition switch OFF (engine off).
Disconnect control unit connector.
Disconnect test mode connector.
Check if continuity exists between terminals 24 (LG) and 35 (B) of connector (body side) when accelerator pedal is released and does not exist when accelerator pedal is depressed.
Check if terminals 24 and 35 are open or disconnected.
Replace control unit.
Repair terminal or harness.
Disconnect throttle sensor connector.
Check if continuity exists between terminals A and B of connector (throttle sensor side) when accelerator pedal is released.
Adjust throttle sensor as outlined under "Idle contact" on
Check if continuity exists when accelerator pedal is released.
Replace throttle sensor.
Check if there is no continuity when accelerator pedal is depressed.
Replace throttle sensor.
Repair harness.
YES
NO

DIAGNOSIS CHART

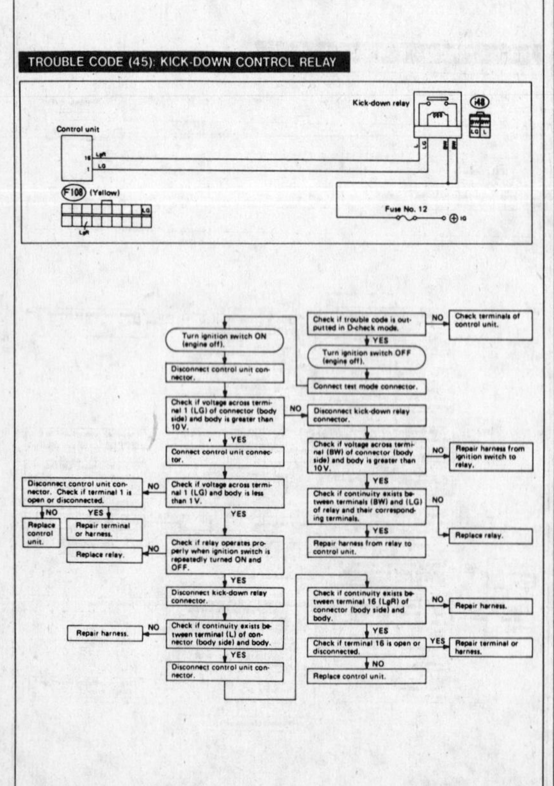
TROUBLE CODE (45): KICK-DOWN CONTROL RELAY
Kick-down relay
Control unit
F108 (Yellow)
Fuse No. 12
Check if trouble code is outputted in D-check mode.
Check terminals of control unit.
Turn ignition switch ON (engine off).
Turn ignition switch OFF (engine off).
Disconnect control unit connector.
Connect test mode connector.
Check if voltage across terminal 1 (LG) of connector (body side) and body is greater than 10 V.
Disconnect kick-down relay connector.
Check if voltage across terminal (BW) of connector (body side) and body is greater than 10 V.
Repair harness from ignition switch to relay.
Connect control unit connector.
Check if continuity exists between terminals (BW) and (LG) of relay and their corresponding terminals.
Replace relay.
Disconnect control unit connector. Check if terminal 1 is open or disconnected.
Check if voltage across terminal 1 (LG) and body is less than 1 V.
Replace control unit.
Repair terminal or harness.
Repair harness from relay to control unit.
Replace relay.
Check if relay operates properly when ignition switch is repeatedly turned ON and OFF.
Disconnect kick-down relay connector.
Check if continuity exists between terminal 16 (LgR) of connector (body side) and body.
Repair harness.
Repair harness.
Check if continuity exists between terminal (L) of connector (body side) and body.
Check if terminal 16 is open or disconnected.
Repair terminal or harness.
Disconnect control unit connector.
Replace control unit.
YES
NO

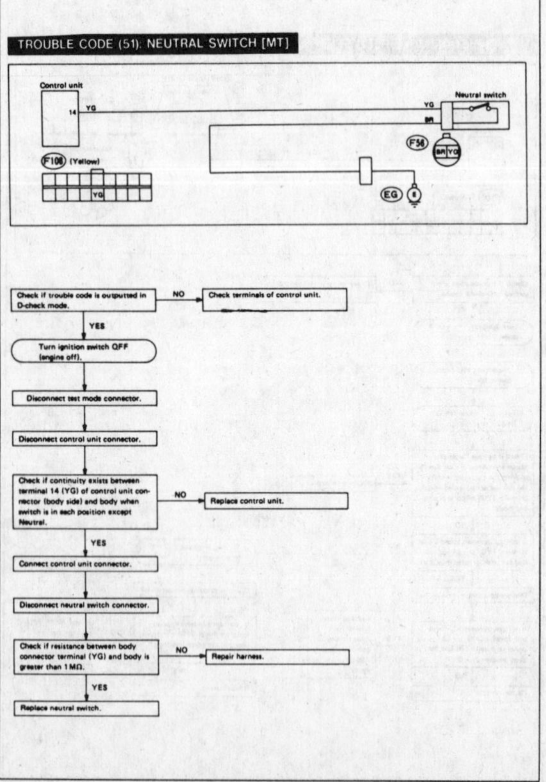
TROUBLE CODE (51): NEUTRAL SWITCH [MT]
Control unit
Neutral switch
F108 (Yellow)
F56
EG
Check if trouble code is outputted in D-check mode.
Check terminals of control unit.
Turn ignition switch OFF (engine off).
Disconnect test mode connector.
Disconnect control unit connector.
Check if continuity exists between terminal 14 (YG) of control unit connector (body side) and body when switch is in each position except Neutral.
Replace control unit.
Connect control unit connector.
Disconnect neutral switch connector.
Check if resistance between body connector terminal (YG) and body is greater than 1 MΩ.
Repair harness.
Replace neutral switch.
YES
NO

DIAGNOSIS CHART

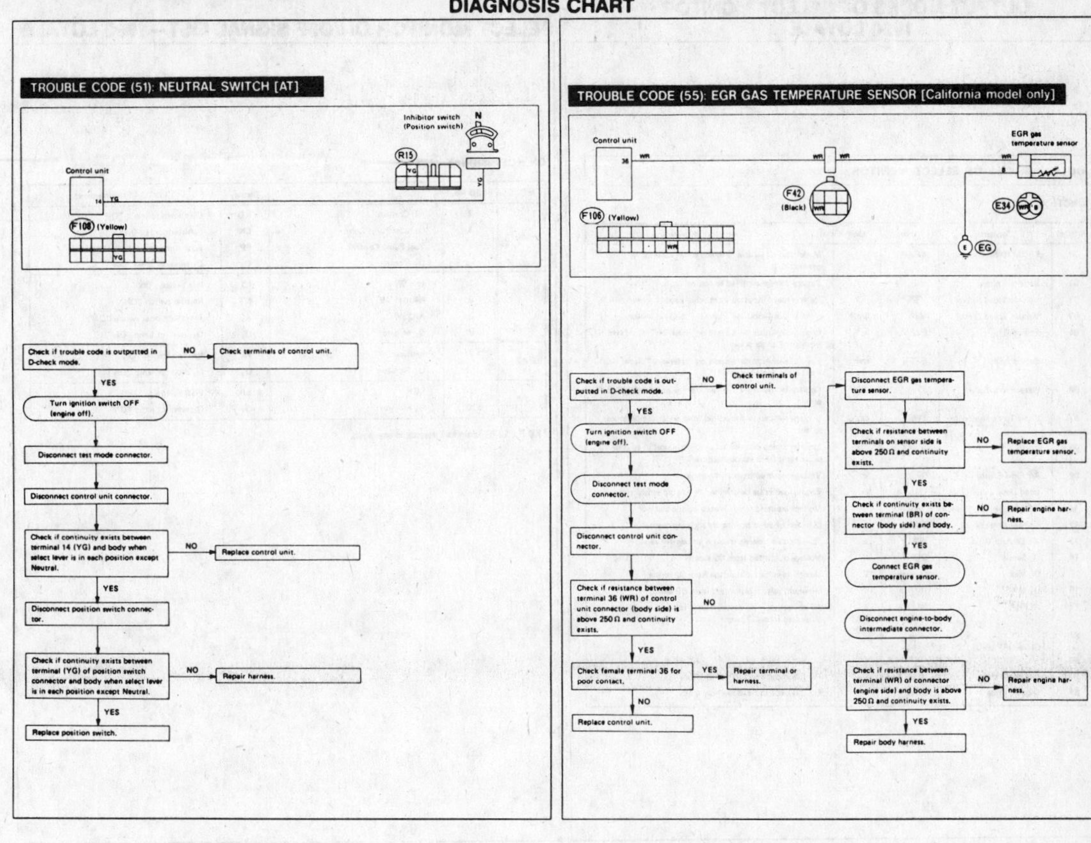
TROUBLE CODE (51): NEUTRAL SWITCH [AT]
Inhibitor switch (Position switch)
N
R15
Control unit
14
YG
F108 (Yellow)
Check if trouble code is outputted in D-check mode.
NO
Check terminals of control unit.
YES
Turn ignition switch OFF (engine off).
Disconnect test mode connector.
Disconnect control unit connector.
Check if continuity exists between terminal 14 (YG) and body when select lever is in each position except Neutral.
NO
Replace control unit.
YES
Disconnect position switch connector.
Check if continuity exists between terminal (YG) of position switch connector and body when select lever is in each position except Neutral.
NO
Repair harness.
YES
Replace position switch.
TROUBLE CODE (55): EGR GAS TEMPERATURE SENSOR [California model only]
Control unit
36
WR
EGR gas temperature sensor
F42 (Black)
F106 (Yellow)
E34
EG
Check if trouble code is outputted in D-check mode.
NO
Check terminals of control unit.
YES
Turn ignition switch OFF (engine off).
Disconnect test mode connector.
Disconnect control unit connector.
Check if resistance between terminal 36 (WR) of control unit connector (body side) is above 250 Ω and continuity exists.
NO
YES
Check female terminal 36 for poor contact.
YES
Repair terminal or harness.
NO
Replace control unit.
Disconnect EGR gas temperature sensor.
Check if resistance between terminals on sensor side is above 250 Ω and continuity exists.
NO
Replace EGR gas temperature sensor.
YES
Check if continuity exists between terminal (BR) of connector (body side) and body.
NO
Repair engine harness.
YES
Connect EGR gas temperature sensor.
Disconnect engine-to-body intermediate connector.
Check if resistance between terminal (WR) of connector (engine side) and body is above 250 Ω and continuity exists.
NO
Repair engine harness.
YES
Repair body harness.

DIAGNOSIS CHART

TROUBLE CODE (61): PARKING SWITCH [AT]

Inhibitor switch (Position switch) P

R15

Control unit 15 YL

F108 (Yellow)

- Check if trouble code is outputted in D-check mode. — NO → Check terminals of control unit.
- YES → Turn ignition switch OFF (engine off).
- Disconnect test mode connector.
- Disconnect control unit connector.
- Check if continuity exists between terminal 15 (YL) and body when select lever is in each position except "P". — NO → Replace control unit.
- YES → Disconnect position switch connector.
- Check if continuity exists between terminal (YL) of position switch connector and body when select lever is in each position except "P". — NO → Repair harness.
- YES → Replace position switch.

TROUBLESHOOTING CHART WITH SELECT MONITOR 1990 LOYALE

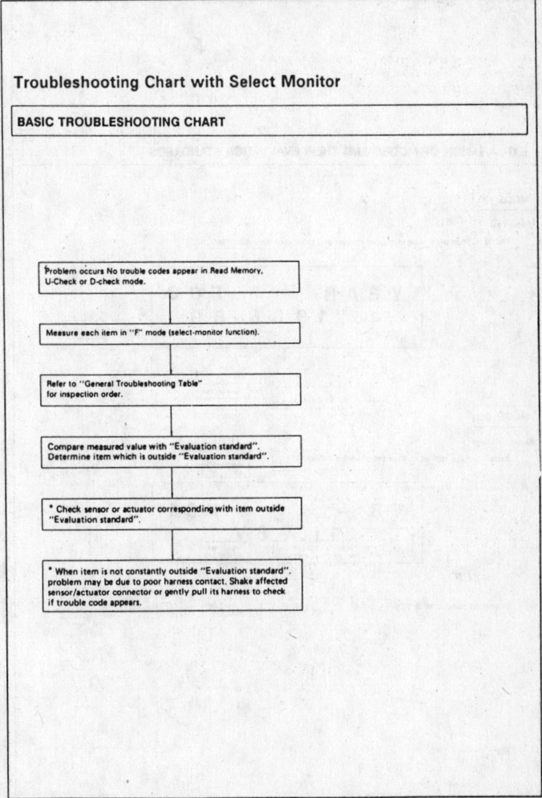
Troubleshooting Chart with Select Monitor
BASIC TROUBLESHOOTING CHART
Problem occurs No trouble codes appear in Read Memory, U-Check or D-check mode.
Measure each item in "F" mode (select-monitor function).
Refer to "General Troubleshooting Table" for inspection order.
Compare measured value with "Evaluation standard". Determine item which is outside "Evaluation standard".
* Check sensor or actuator corresponding with item outside "Evaluation standard".
* When item is not constantly outside "Evaluation standard", problem may be due to poor harness contact. Shake affected sensor/actuator connector or gently pull its harness to check if trouble code appears.

OUTPUT MODES OF SELECT MONITOR 1990 LOYALE

OUTPUT MODES OF SELECT MONITOR

FUNCTION MODE

MODE	Contents	Abbr.	Unit	Contents of display
00	PROM ID Number	YEAR	–	Model year of vehicle to which select monitor is connected
01	Battery Voltage	VB	V	Battery voltage supplied to control unit
02	Vehicle Speed Sensor	VSP	m/h	Vehicle speed inputted from vehicle speed sensor
03	Vehicle Speed Sensor	VSP	km/h	Vehicle speed inputted from vehicle speed sensor
04	Engine RPM	EREV	rpm	Engine revolution data based on "position" signal sent from crank angle sensor
05	Engine RPM	EREV	rpm	Engine revolution data based on "reference" signal sent from crank angle sensor
06	Water Temp Sensor	TW	deg F	Coolant temperature inputted from water temperature sensor
07	Water Temp Sensor	TW	deg C	Coolant temperature inputted from water temperature sensor
08	Ignition Timing	ADVS	deg	Ignition timing determined by control unit in relation to signals sent from various sensors
09	Air Flow Sensor	QA	V	Voltage inputted from air flow meter
10	Load Data	LDATA	–	Engine load value determined by related sensor signals
11	Throttle Sensor	THV	V	Voltage inputted from throttle position sensor
12	Injector Pulse Width	TIM	mS	Duration of pulse flowing through injectors
13	Air Control Valve	ISC	%	"Duty" ratio flowing through air control valve
14	O_2 Sensor	O_2	V	Voltage outputted from O2 sensor
15	O_2 Max	O_2 max	V	Maximum voltage outputted from O2 sensor
16	O_2 Min	O_2 min	V	Minimum voltage outputted from O2 sensor
17	ALPHA	ALPHA	%	AF correction ratio determined in relation to signal outputted from O2 sensor
A0 ~ A5	ON ↔ OFF signal			
B0	Trouble Code	DIAG	–	Trouble code in U- or D-check mode
B1	Trouble Code	DIAG	–	Trouble code in Read memoly mode

SELECT MONITOR ON/OFF SIGNAL LIST – 1990 LOYALE

ON ↔ OFF SIGNAL LIST

MODE	LED No.	Contents	Display	LED "ON" requirements
A0	2	KD Control	KD	Kickdown solenoid "ON"
	3	Canister Purge Cont.	CN	Canister purge "ON"
	4	Fuel Pump Control	FP	Fuel pump relay in operation
	10	O_2 Monitor	O_2	O_2 monitor lamp "ON"
A1	1	Idle SW	ID	Idle switch "ON"
	2	Neutral SW	NT	Neutral switch "ON"
	3	A/C SW	AC	Air conditioner switch "ON"
	10	O_2 Monitor	O_2	O_2 monitor lamp "ON"
A2	2	Ignition SW	IG	Ignition switch "ON"
	10	O_2 Monitor	O_2	O_2 monitor lamp "ON"
A4	2	EGR	ER	EGR solenoid "ON"
	10	O_2 Monitor	O_2	O_2 monitor lamp "ON"

LED No. 10 (O_2 monitor) operates in any mode.

EXPLANATION OF MODES – 1990 LOYALE

EXPLANATION OF MODES AND THEIR EVALUATION STANDARDS

MODE 00

Model Year

PROM ID Number will be read from control unit to display the model year.

```
YEAR        F00
    1988-89
```

MODE 01

Battery Voltage

Battery voltage applied to the control unit is displayed.

```
VB          F01
    11.76V
```

Evaluation standards: 8 – 16 V.

MODE 02

Vehicle Speed – "m/h" will be displayed.

Vehicle speed data is displayed. This data is based on a signal sent from vehicle-speed sensor to control unit.

```
VSP         F02
          55m/h
```

Evaluation standards: Speedometer indication ± 10 m/h

Data at "open/short":

Open	Short
0	0 to 1

MODE 03

Vehicle Speed – "km/h" will be displayed.

Vehicle-speed data is displayed. This data is based on a signal sent from vehicle-speed sensor.

```
VSP         F03
         60km/h
```

Evaluation standards: Speedometer indication ± 16 km/h

Data at "open/short":

Open	Short
0	0 to 1

EXPLANATION OF MODES—1990 LOYALE

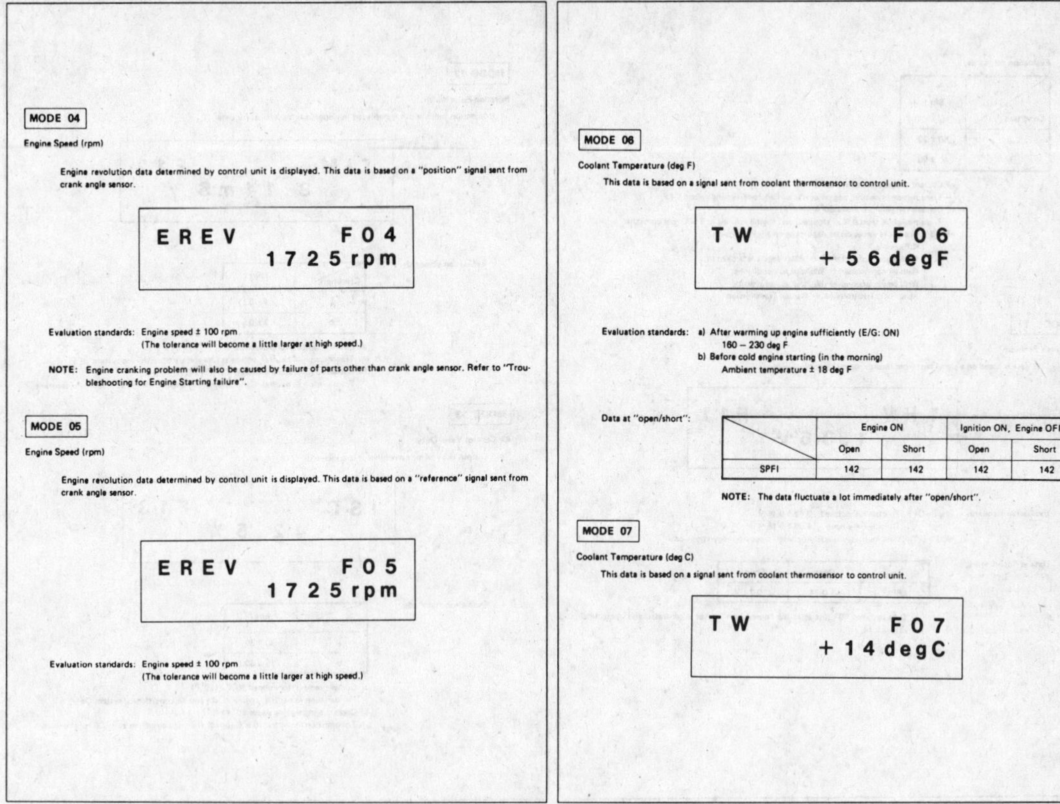

MODE 04

Engine Speed (rpm)

Engine revolution data determined by control unit is displayed. This data is based on a "position" signal sent from crank angle sensor.

```
EREV          F04
       1725rpm
```

Evaluation standards: Engine speed ± 100 rpm
(The tolerance will become a little larger at high speed.)

NOTE: Engine cranking problem will also be caused by failure of parts other than crank angle sensor. Refer to "Troubleshooting for Engine Starting failure".

MODE 05

Engine Speed (rpm)

Engine revolution data determined by control unit is displayed. This data is based on a "reference" signal sent from crank angle sensor.

```
EREV          F05
       1725rpm
```

Evaluation standards: Engine speed ± 100 rpm
(The tolerance will become a little larger at high speed.)

MODE 06

Coolant Temperature (deg F)

This data is based on a signal sent from coolant thermosensor to control unit.

```
TW            F06
        +56degF
```

Evaluation standards:
a) After warming up engine sufficiently (E/G: ON)
160 – 230 deg F
b) Before cold engine starting (in the morning)
Ambient temperature ± 18 deg F

Data at "open/short":

	Engine ON		Ignition ON, Engine OFF	
	Open	Short	Open	Short
SPFI	142	142	142	142

NOTE: The data fluctuate a lot immediately after "open/short".

MODE 07

Coolant Temperature (deg C)

This data is based on a signal sent from coolant thermosensor to control unit.

```
TW            F07
        +14degC
```

EXPLANATION OF MODES—1990 LOYALE

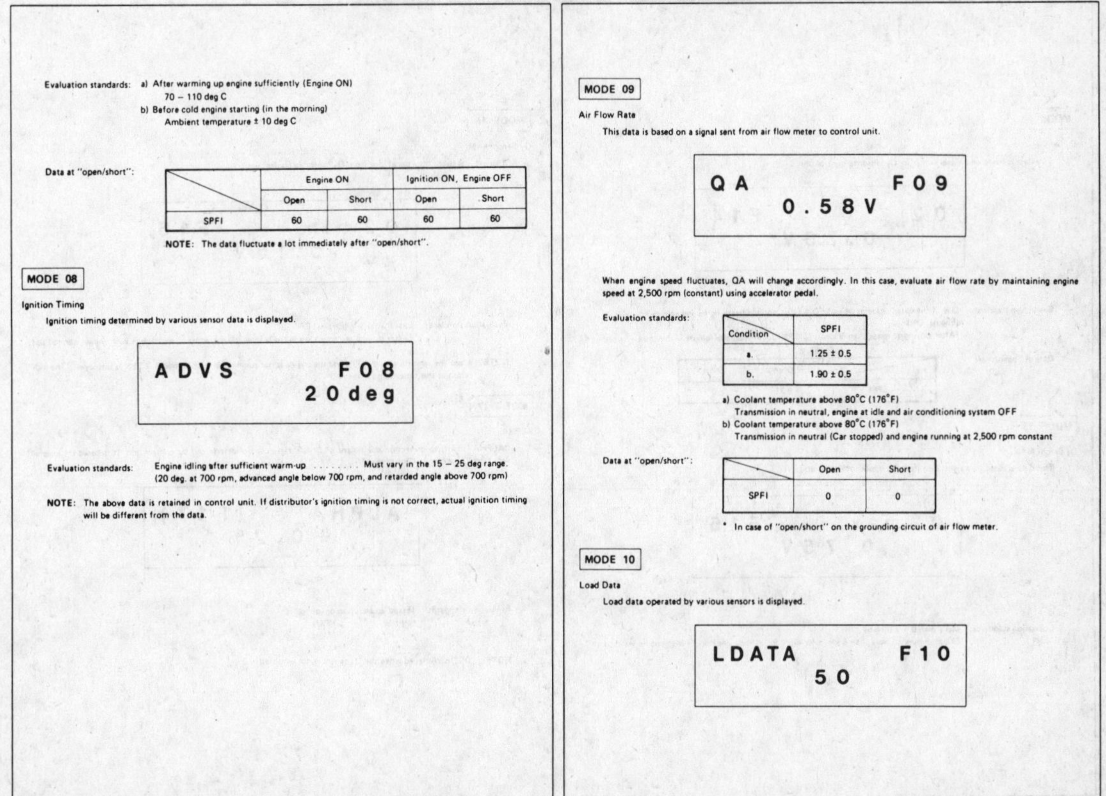

Evaluation standards:
a) After warming up engine sufficiently (Engine ON)
70 – 110 deg C
b) Before cold engine starting (in the morning)
Ambient temperature ± 10 deg C

Data at "open/short":

	Engine ON		Ignition ON, Engine OFF	
	Open	Short	Open	Short
SPFI	60	60	60	60

NOTE: The data fluctuate a lot immediately after "open/short".

MODE 08

Ignition Timing

Ignition timing determined by various sensor data is displayed.

```
ADVS          F08
          20deg
```

Evaluation standards: Engine idling after sufficient warm-up .... Must vary in the 15 – 25 deg range.
(20 deg. at 700 rpm, advanced angle below 700 rpm, and retarded angle above 700 rpm)

NOTE: The above data is retained in control unit. If distributor's ignition timing is not correct, actual ignition timing will be different from the data.

MODE 09

Air Flow Rate

This data is based on a signal sent from air flow meter to control unit.

```
QA            F09
      0.58V
```

When engine speed fluctuates, QA will change accordingly. In this case, evaluate air flow rate by maintaining engine speed at 2,500 rpm (constant) using accelerator pedal.

Evaluation standards:

Condition	SPFI
a.	1.25 ± 0.5
b.	1.90 ± 0.5

a) Coolant temperature above 80°C (176°F)
Transmission in neutral, engine at idle and air conditioning system OFF
b) Coolant temperature above 80°C (176°F)
Transmission in neutral (Car stopped) and engine running at 2,500 rpm constant

Data at "open/short":

	Open	Short
SPFI	0	0

* In case of "open/short" on the grounding circuit of air flow meter.

MODE 10

Load Data

Load data operated by various sensors is displayed.

```
LDATA         F10
        50
```

EXPLANATION OF MODES—1990 LOYALE

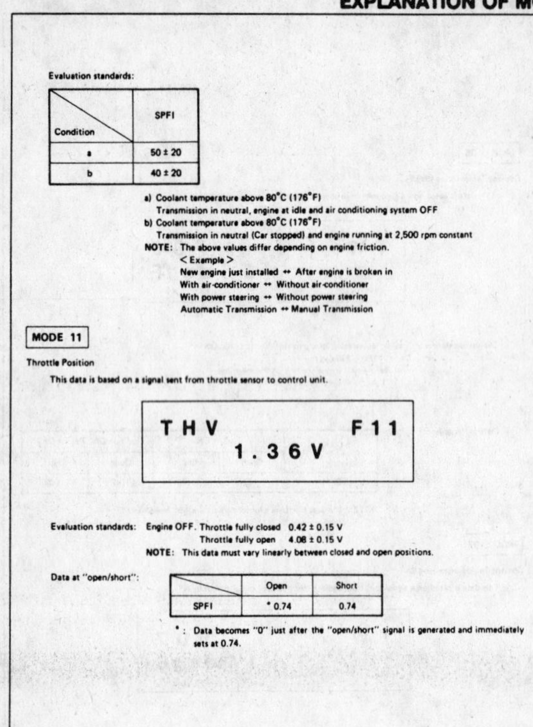

Evaluation standards:

Condition	SPFI
a	50 ± 20
b	40 ± 20

a) Coolant temperature above 80°C (176°F)
Transmission in neutral, engine at idle and air conditioning system OFF
b) Coolant temperature above 80°C (176°F)
Transmission in neutral (Car stopped) and engine running at 2,500 rpm constant

NOTE: The above values differ depending on engine friction.
< Example >
New engine just installed ↔ After engine is broken in
With air-conditioner ↔ Without air-conditioner
With power steering ↔ Without power steering
Automatic Transmission ↔ Manual Transmission

MODE 11

Throttle Position

This data is based on a signal sent from throttle sensor to control unit.

```
THV          F11
     1.36V
```

Evaluation standards: Engine OFF. Throttle fully closed 0.42 ± 0.15 V
Throttle fully open 4.08 ± 0.15 V

NOTE: This data must vary linearly between closed and open positions.

Data at "open/short":

	Open	Short
SPFI	* 0.74	0.74

* : Data becomes "0" just after the "open/short" signal is generated and immediately sets at 0.74.

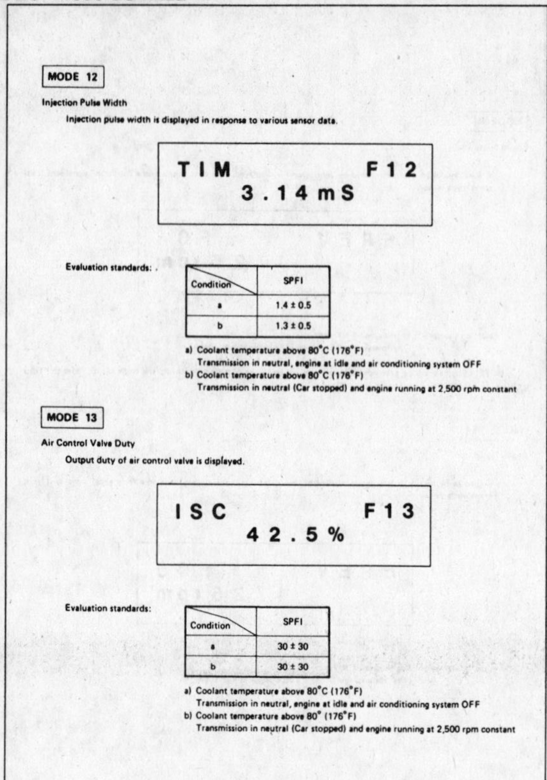

MODE 12

Injection Pulse Width

Injection pulse width is displayed in response to various sensor data.

```
TIM          F12
     3.14mS
```

Evaluation standards:

Condition	SPFI
a	1.4 ± 0.5
b	1.3 ± 0.5

a) Coolant temperature above 80°C (176°F)
Transmission in neutral, engine at idle and air conditioning system OFF
b) Coolant temperature above 80°C (176°F)
Transmission in neutral (Car stopped) and engine running at 2,500 rpm constant

MODE 13

Air Control Valve Duty

Output duty of air control valve is displayed.

```
ISC          F13
     42.5%
```

Evaluation standards:

Condition	SPFI
a	30 ± 30
b	30 ± 30

a) Coolant temperature above 80°C (176°F)
Transmission in neutral, engine at idle and air conditioning system OFF
b) Coolant temperature above 80° (176°F)
Transmission in neutral (Car stopped) and engine running at 2,500 rpm constant

EXPLANATION OF MODES—1990 LOYALE

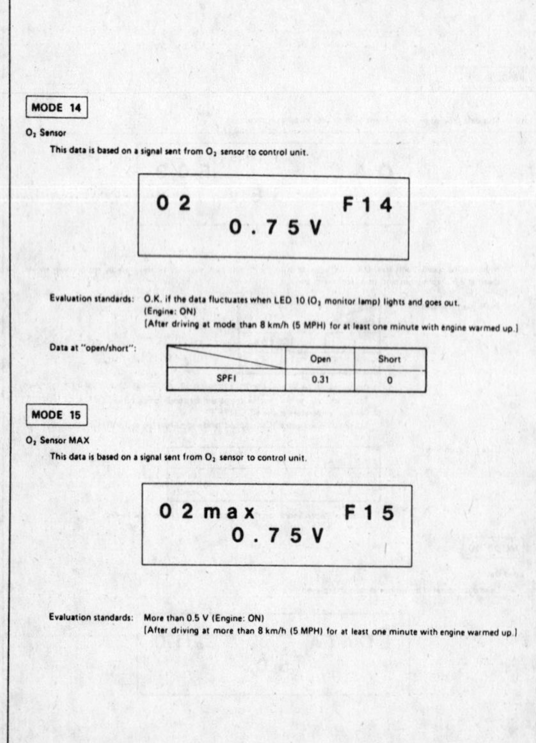

MODE 14

O_2 Sensor

This data is based on a signal sent from O_2 sensor to control unit.

```
O2           F14
     0.75V
```

Evaluation standards: O.K. if the data fluctuates when LED 10 (O_2 monitor lamp) lights and goes out.
(Engine: ON)
[After driving at mode than 8 km/h (5 MPH) for at least one minute with engine warmed up.]

Data at "open/short":

	Open	Short
SPFI	0.31	0

MODE 15

O_2 Sensor MAX

This data is based on a signal sent from O_2 sensor to control unit.

```
O2max        F15
     0.75V
```

Evaluation standards: More than 0.5 V (Engine: ON)
[After driving at more than 8 km/h (5 MPH) for at least one minute with engine warmed up.]

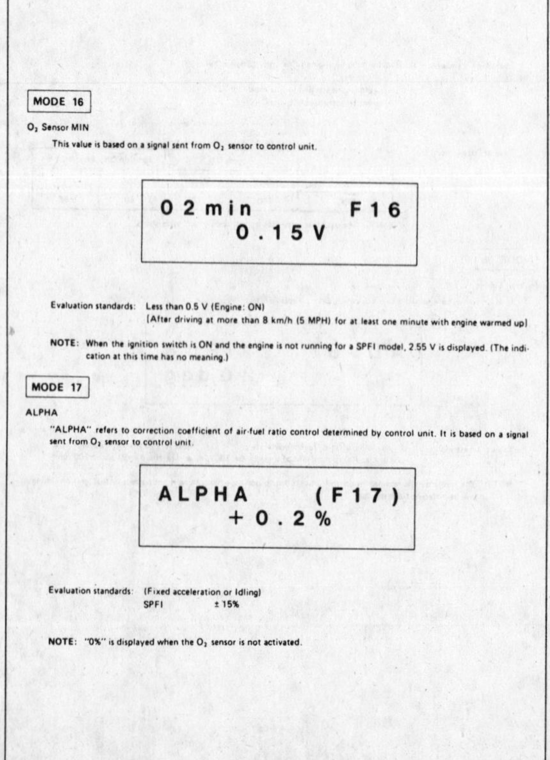

MODE 16

O_2 Sensor MIN

This value is based on a signal sent from O_2 sensor to control unit.

```
O2min        F16
     0.15V
```

Evaluation standards: Less than 0.5 V (Engine: ON)
[After driving at more than 8 km/h (5 MPH) for at least one minute with engine warmed up]

NOTE: When the ignition switch is ON and the engine is not running for a SPFI model, 2.55 V is displayed. (The indication at this time has no meaning.)

MODE 17

ALPHA

"ALPHA" refers to correction coefficient of air-fuel ratio control determined by control unit. It is based on a signal sent from O_2 sensor to control unit.

```
ALPHA       (F17)
     +0.2%
```

Evaluation standards: (Fixed acceleration or Idling)
SPFI ± 15%

NOTE: "0%" is displayed when the O_2 sensor is not activated.

1990 SUBARU LEGACY MULTI-POINT FUEL INJECTION (MPFI) SYSTEM

General Information

The MPFI system used on the 1990 Legacy supplies the optimum air/fuel mixture to the engine under all various operating conditions. System fuel, which is pressurized at a constant pressure, is injected into the intake air passage of the cylinder head. The amount of fuel is controlled by an intermittent injection system where the electro-magnetic injection valve (fuel injector) opens only for a short period of time, depending on the amount of air required for 1 cycle of operation. When operating, the amount of injection is determined by the duration of an electric pulse (signal) sent to the fuel injector, which permits precise metering of the fuel. All the operating conditions of the are converted into electric signals, which results in improved adaptability, easier addition of compensating element, etc. Characteristics of the MPFI system are as follows:

- Reduced emission of harmful exhaust gases
- Reduced in fuel consumption
- Increased engine output
- Superior acceleration and deceleration
- Superior startability and warm-up performance in cold weather since compensation is made for coolant and intake air temperature

SYSTEM COMPONENTS

Air Control System

Air, which is drawn in and filtered by the air cleaner, is metered and sent to the throttle body through the air intake boot. From the throttle body, the air is controlled by the open/close operation of the throttle valve and is then delivered to the intake manifold. From the intake manifold, it is distributed to the cylinders to mix with the fuel injected by the fuel injectors. Then, the air/fuel mixture is sent into the cylinders. Some of the air is sent to the by-pass air control valve which controls the engine idle speed.

AIR FLOW SENSOR

The MPFI system uses 2 different types of air flow sensors. On vehicles equipped with automatic transmission, a hot film type sensor is used. On vehicles equipped with manual transmission,

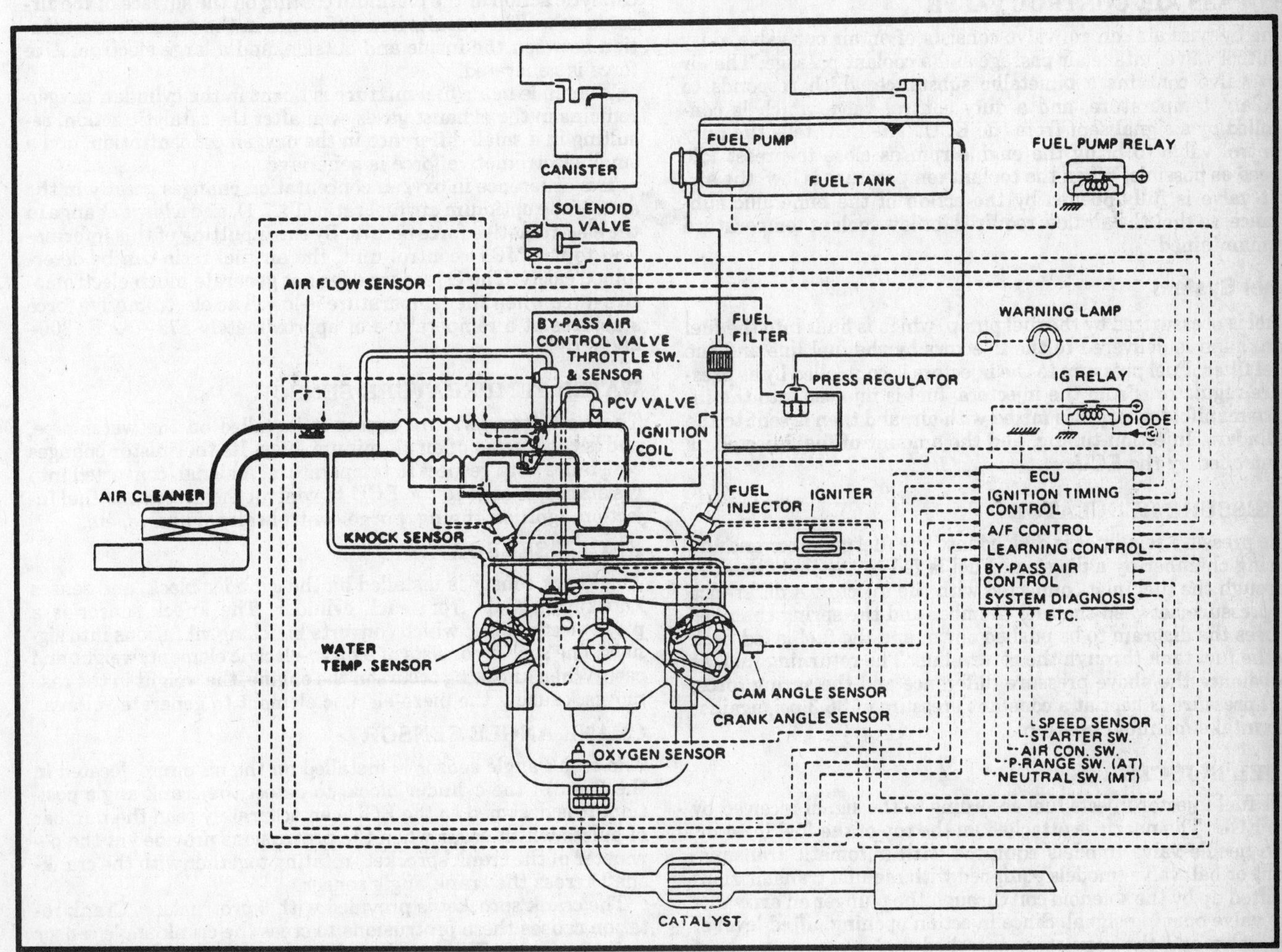
CANISTER
FUEL PUMP
FUEL TANK
FUEL PUMP RELAY
SOLENOID VALVE
AIR FLOW SENSOR
WARNING LAMP
BY-PASS AIR CONTROL VALVE
FUEL FILTER
THROTTLE SW. & SENSOR
PRESS REGULATOR
IG RELAY
PCV VALVE
IGNITION COIL
DIODE
ECU
IGNITION TIMING CONTROL
A/F CONTROL
LEARNING CONTROL
BY-PASS AIR CONTROL
SYSTEM DIAGNOSIS
ETC.
AIR CLEANER
FUEL INJECTOR
IGNITER
KNOCK SENSOR
WATER TEMP. SENSOR
CAM ANGLE SENSOR
CRANK ANGLE SENSOR
SPEED SENSOR
STARTER SW.
AIR CON. SW.
P-RANGE SW. (AT)
NEUTRAL SW. (MT)
OXYGEN SENSOR
CATALYST
Exploded view of MPFI system

a hot wire type air flow sensor is used. These air flow sensors convert the amount of air taken into the engine into an electric signal by using the heat transfer between the incoming air and a heating resistor (hot film or hot wire) located in the air intake. Characteristics of these flow sensors are as follows:

- High altitude compensation is made automatically
- Quick response
- There are no moving parts
- Compact

THROTTLE BODY

When depressing the accelerator pedal, the throttle body opens/closes its valve to control the amount of air to be taken into the combustion chamber. During idling, the throttle valve is almost fully closed. More than half the air necessary for idling is supplied to the intake manifold through the by-pass air control valve. The by-pass air control valve controls the number of revolutions during idling, so that no adjustment is necessary.

THROTTLE SENSOR

A throttle position sensor is incorporated with a potentiometer and an idle switch interlocked with the throttle valve shaft. It sends the MPFI control unit an potentiometer output signal, relating to the opening of the throttle valve, and an idle switch signal that turns **ON** only when the throttle is opened to the idle position. Using these signals, the MPFI control unit controls the air/fuel ratio during acceleration and deceleration as well as idling.

BY-PASS AIR CONTROL VALVE

The by-pass air control valve consists of an air cut valve, duty control valve, intake air passage and a coolant passage. The air cut valve contains a bimetallic substance which responds to coolant temperature, and a duty control valve which is controlled by a signal sent from the ECU. The ECU tells the duty control valve to bring the engine rpm as close to preset idle speed as possible. When the coolant temperature is low, the air-cut valve is fully opened by the action of the bimetallic substance, so that the air flow required for low coolant temperature is maintained.

Fuel System

Fuel is pressurized by the fuel pump, which is built into the fuel tank, and is delivered to the injectors by the fuel line and the fuel filter. Fuel pressure to the injectors is controlled by as pressure regulator. From the injectors, fuel is injected into the intake manifold where it is mixed with air, and then is sent to the cylinders. Injection timing, and the amount of fuel injected, is controlled by the ECU.

PRESSURE REGULATOR

The pressure regulator is divided into the fuel chamber and the spring chamber by a diagram. Fuel is fed to the fuel chamber through the fuel inlet connected with the injector. A difference in pressure between the fuel chamber and the spring chamber causes the diagram to be pushed down, and the fuel is fed back to the fuel tank through the return line. The returning fuel, as to balance the above pressure difference and the spring force, fuel pressure is kept at a constant pressure of 36.3 psi, against the intake manifold pressure.

FUEL INJECTOR

The fuel injector injects fuel according to the signal received by the ECU. The nozzle is attached on the top of the fuel injector. The needle valve (models equipped with automatic transmission) or ball valve (models equipped with manual transmission) is lifted up by the solenoid coil through the plunger on arrival of the valve opening signal. Since injection opening, lifted level of the valve and the regulator-controlled fuel pressure are kept constant, the amount of fuel to be injected can be controlled only by the valve opening signal from the ECU. Characteristics of the fuel injector are as follows:

- High heat resistance
- Low driving noise
- Easy to service
- Compact

Sensors and Switches

OXYGEN SENSOR

The Oxygen (O_2) sensor is used to determine the oxygen concentration in the exhaust gas. If the fuel ratio is leaner than optimum air/fuel ratio (14.7:1), the exhaust gas contains more oxygen. If the fuel ratio is richer than the optimum air/fuel ratio (14.7:1), the exhaust gas contains hardly any oxygen. Examination of the oxygen concentration in the exhaust gas makes it possible to determine wheter the air/fuel ratio is leaner or richer than the optimum air/fuel ratio (14.7:1).

The O_2 sensor incorporates a zirconia tube (cermamic) which generates voltage, if there is a difference in oxygen concentration between the inside and outside of the tube. The inside and outside zinconia tube is coated with platinum for the purpose of catalysis and electrode provision. The hexagon screw on the outside is grounded to the exhaust pipe, and the inside is connected to the ECU through the harness. A ceramic heater is installed to improve performance at low temperatures.

When a rich air/fuel mixture is burnt in the cylinder, the oxygen in the exhaust gases reacts almost completely through the catalytic action of the platinum coating on the surface of the zirconia tube. This results in a difference in the oxygen concentration between the inside and outside, and a large electromotive force is generated.

When a lean air/fuel mixture is burnt in the cylinder, oxygen remains in the exhaust gases even after the catalytic action, resulting in a small difference in the oxygen concentration, and a small electromotive force is generated.

The difference in oxygen concentation changes greatly in the area of the optimum air/fuel ratio (14.7:1), and a large change in the electromotive force occurs. By the inputting of this information to the MPFI control unit, the air/fuel ratio can be determined easily. The O_2 sensor does not generate much electromotive force when the temperature is low. The electromotive force stabilizes at a temperature of approximately 572–752°F (300–400°C).

WATER TEMPERATURE SENSOR

The water temperature sensor is installed on the water pipe, and is constucted of an aluminum alloy. Its thermistor changes resistance with respect to temperature. A signal, converted into resistance, is sent to the ECU to control the amount of fuel injection, ignition timing, purge control solenoid valve, etc.

KNOCK SENSOR

The knock sensor is installed on the cylinder block, and senses knocking signals from each cylinder. The knock sensor is a piezo-electric type, which converts knocking vibrations into signals. The sensor consists of a piezo-electric element, weight and case. When knocking occurs in the engine, the weight in the case moves, causing the piezo-electric element to generate voltage.

CRANK ANGLE SENSOR

The crank angle sensor is installed on the oil pump, located in the front of the cylinder block, to detect the crank angle position. It is designed so the ECU can accurately read the number of pulses which occur when the protrusions provided at the perimeter of the crank sprocket (rotating together with the crankshaft) cross the crank angle sensor.

The crank sprocket is provided with 6 protrusions. Crank rotation causes these protrusions to cross the crank angle sensor so the magnetic fluxes in the coil change with the change in the air gap between the sensor pickup and the sprocket. The change

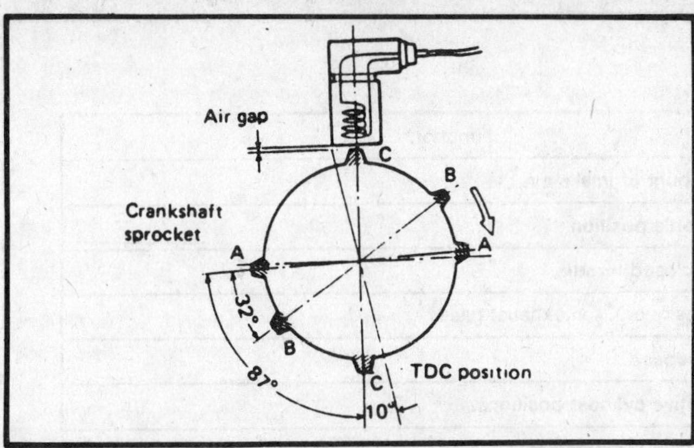
Air gap
C
B
Crankshaft sprocket
A
A
32°
B
87°
C
TDC position
10°
Crankshaft sprocket and crank angle sensor

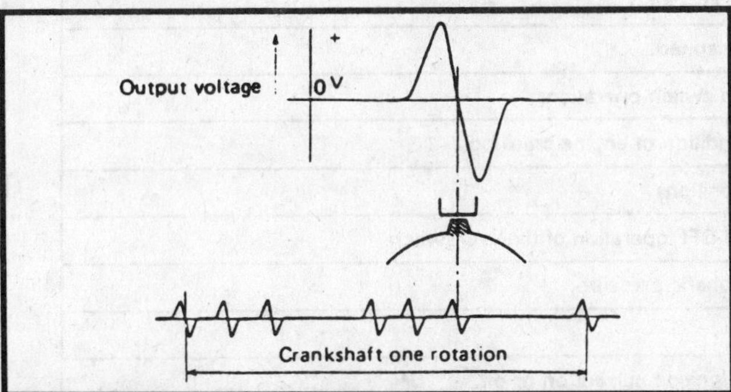
Output voltage
+
0 V
Crankshaft one rotation
Crank angle sensor operation

in the air gap, produces an electromotive force, which is then sent to the ECU.

CAM ANGLE SENSOR

The cam angle sensor is located on the left-hand camshaft support, to detect the combustion cylinder at any time. It is designed so the ECU can read the number of pulses which occur when the protrusions provided on the back of the LH camshaft drive sprocket cross the sensor.

Construction, and operating principle of the cam angle sensor are similar to those of the crank angle sensor. A total of 7 protrusions (1 each at 2 locations, 2 at 1 location and 3 at 1 location) are arranged in 4 equal parts on the sprocket.

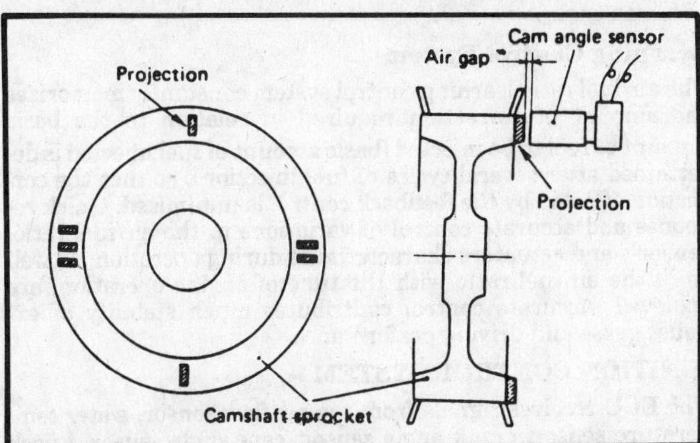
Cam angle sensor
Air gap
Projection
Projection
Camshaft sprocket
Cam angle sensor operation

SPEED SENSOR

The speed sensor consists of a magnetic rotor which is routed by a speedometer cable and a reed switch. It is built into the combination meter (speedometer). One rotation of the magnetic rotor turns the reed switch **ON/OFF** 4 times to produce a signal. The signal is used as a vehicle speed signal which is sent to the ECU.

ATMOSPHERIC PRESSURE SENSOR

The atmospheric pressure sensor is built into the ECU. It uses an "absolute" pressure sensor design. It purpose is to detect the atmospheric pressure used to compensate for pressure at high altitudes and to maintain driving stability.

The signal from the sensor is also used for "shift control," on vehicles equipped with automatic transmission at high altitudes.

AIR CONDITIONING SWITCH AND RELAY

The A/C switch turns the A/C system on/off. The on/offoperation of the switch is sent to the ECU. The A/C cut relay breaks the current flow to the compressor through the use of an output signal from the ECU, for a short period of time, when a full throttle signal (sent from the throttle sensor) enters the ECU while the compressor is operating. This prevents the degradation of acceleration performance and stabilizes driving performance. The A/C cut relay is installed in the fuse box, located at the left side of the engine compartment.

Control System

The ECU receives signals sent from various sensors and switches to determine engine operating conditions, and emits output signals to provide the optimum control and/or functioning of various systems. The ECU controls the following:

- Fuel injection control
- Ignition system cotrol
- By-pass air control (idle speed control)
- Canister purge control
- Radiator fan control
- Fuel pump control
- A/C cut control
- Self-diagnosis function
- Fail-safe function

FUEL INJECTION CONTROL

The ECU receives signals from various sensors to control the amount of fuel to be injected, and fuel injection timing. Sequential fuel injection control is used over the entire engine operating range, except when starting. The sequential fuel injection system is designed so that fuel is injected at a specific time to provide maximum air intake efficiency for each cylinder. Fuel injection is completed just before the intake valve begins to open.

The amount of fuel injected by the injector valve depends upon the length of time it remains open. Fuel injection timing is determined by a signal sent to the injector by the ECU according to varying engine operations. Feedback control is accomplished by the means of a learning control. As a result, the fuel injection control system is responsive and accurate in design and function.

Correction Coefficients

Correction coefficients are used to correct the basic duration of fuel injection so that the air/fuel ratio meets the requirements of varying engine operations. These correction coefficients are as follows:

1. Air/Fuel Ratio Coefficient: To provide the optimum air/fuel ratio in relation to engine speed and the basic amount of fuel injected.

2. Start Increment Coefficient: Increases the amount of fuel injected only when cranking the engine, which improves starting ability.

3. Water Temperature Coefficient: Used to increase the amount of fuel injected in relation to a signal from the water temperature sensor, for easier starting when the engine is cold. The lower the water temperature, the greater the increment rate.

INPUT AND OUTPUT SIGNALS

	Unit	Function
Input signal	Air flow sensor	Detects the amount of intake air.
	Throttle sensor	Detects the throttle position.
	Idle switch	Detects a fully-closed throttle.
	O_2 sensor	Detects the density of O_2 in exhaust gases.
	Crank angle sensor	Detects engine speed.
	Cam angle sensor	Detects the relative cylinder positions.
	Water temperature sensor	Detects the coolant temperature.
	Knock sensor	Detects engine knocking.
	Vehicle speed sensor	Detects vehicle speed.
	Ignition switch	Detects ignition switch operation.
	Starter switch	Detects the condition of engine cranking
	Inhibitor switch (A/T)	Detects shift positions.
	A/C switch	Detects the ON-OFF operation of the A/C switch.
	Atmospheric pressure sensor	Detects atmospheric pressure.
Output signal	Fuel injector	Inject fuel.
	Ignition signal	Turns primary ignition current on or off.
	Fuel pump relay	Turns the fuel pump relay on or off.
	A/C control relay	Turns A/C control relay on or off.
	Radiator fan control relay	Turns radiator fan control relay on or off.
	Air control valve	Adjusts the amount of bypass air flowing through the throttle valve.
	O_2 monitor light	Indicates rich/lean air-fuel ratio as well as trouble codes.
	Check engine light	Indicates trouble.
	Purge control solenoid valve	Controls the canister purge control solenoid valve.

ECU input and output signals

4. After-Start Increment Coefficient: Increases the amount of fuel injected for a certain period of time immediately after the engine starts, to stabilize engine operation.

5. Full Increment Coefficient: Increases the amount of fuel injected by a signal from the throttle sensor in relation to a signal from the air flow sensor.

6. Acceleration Increment Coefficient: Compensates for time lags of air flow measurement and/or fuel injection during acceleration, to provide quick response.

Air/Fuel Ratio Feedback Coefficient (Alpha)

This feedback coefficient uses the O_2 sensor's voltage (electromotive force) as a signal to be sent to the ECU. When low voltage is entered, the ECU determines it as a lean mixture, and when high voltage is entered, it determines it as a rich mixture. Therefore, when the air/fuel ratio is richer, the amount of fuel injected is decreased. When the air/fuel ratio leaner, the amount of fuel injected is increased. This means, the air/fuel ratio is compensated so that it comes as close to the optimum air/fuel ratio as possible (14.7:1), on which the 3-way catalyst operates most effectively.

Learning Control System

The air/fuel ratio learning control system constantly memorizes the amount of correction required in relation to the basic amount of fuel to be injected (basic amount of fuel injected is determined after several cycles of fuel injection), so that the correction affected by the feedback control is minimized. Quick response and accurate control of variations in the air/fuel ratio, sensor's and actuators characteristics during operation, as well as in the air/fuel ratio with the time of engine operation, are achieved. Accurate control contributes much stability of exhaust gases and driving performance.

IGNITION CONTROL SYSTEM

The ECU receives signals from the air flow sensor, water temperature sensor, crank angle sensor, cam angle sensor, knock sensor, etc., to determine the operating condition of the engine. It then selects the optimum ignition timing stored in the memory and immediately sends a signal to the igniter to control ignition timing.

When the ECU receives signals from the knock sensor, it ensures that advanced igniton timing is maintained immediately

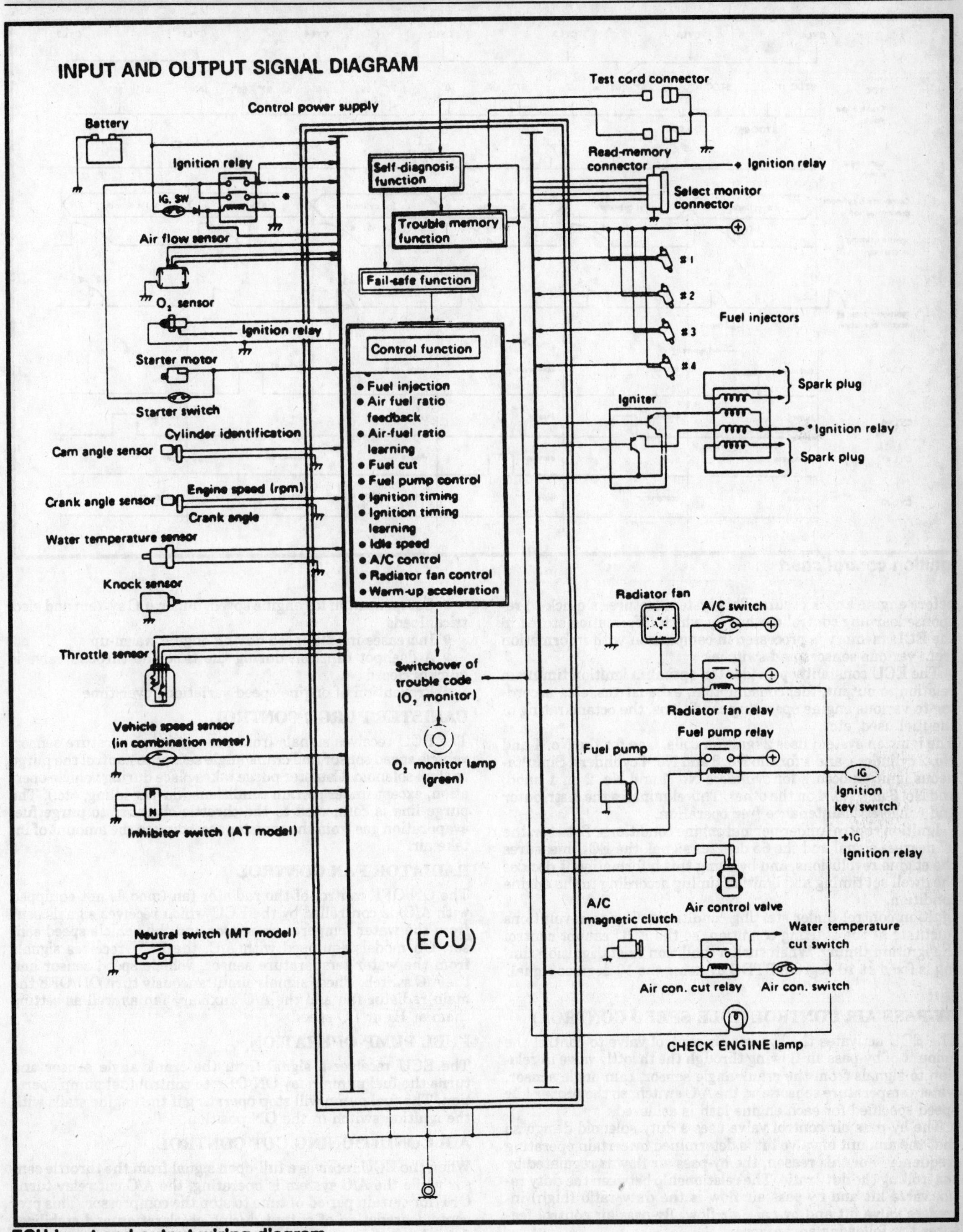
INPUT AND OUTPUT SIGNAL DIAGRAM
Control power supply
Battery
Ignition relay
IG. SW
Air flow sensor
O_2 sensor
Ignition relay
Starter motor
Starter switch
Cylinder identification
Cam angle sensor
Engine speed (rpm)
Crank angle sensor
Crank angle
Water temperature sensor
Knock sensor
Throttle sensor
Vehicle speed sensor
(in combination meter)
Inhibitor switch (AT model)
Neutral switch (MT model)
Self-diagnosis function
Trouble memory function
Fail-safe function
Control function
Fuel injection
Air fuel ratio feedback
Air-fuel ratio learning
Fuel cut
Fuel pump control
Ignition timing
Ignition timing learning
Idle speed
A/C control
Radiator fan control
Warm-up acceleration
Switchover of trouble code
(O_2 monitor)
O_2 monitor lamp
(green)
(ECU)
Test cord connector
Read-memory connector
Ignition relay
Select monitor connector
#1
#2
#3
#4
Fuel injectors
Igniter
Spark plug
Ignition relay
Spark plug
Radiator fan
A/C switch
Radiator fan relay
Fuel pump relay
Fuel pump
IG
Ignition key switch
Ignition relay
Air control valve
A/C magnetic clutch
Water temperature cut switch
Air con. cut relay
Air con. switch
CHECK ENGINE lamp

ECU input and output wiring diagram

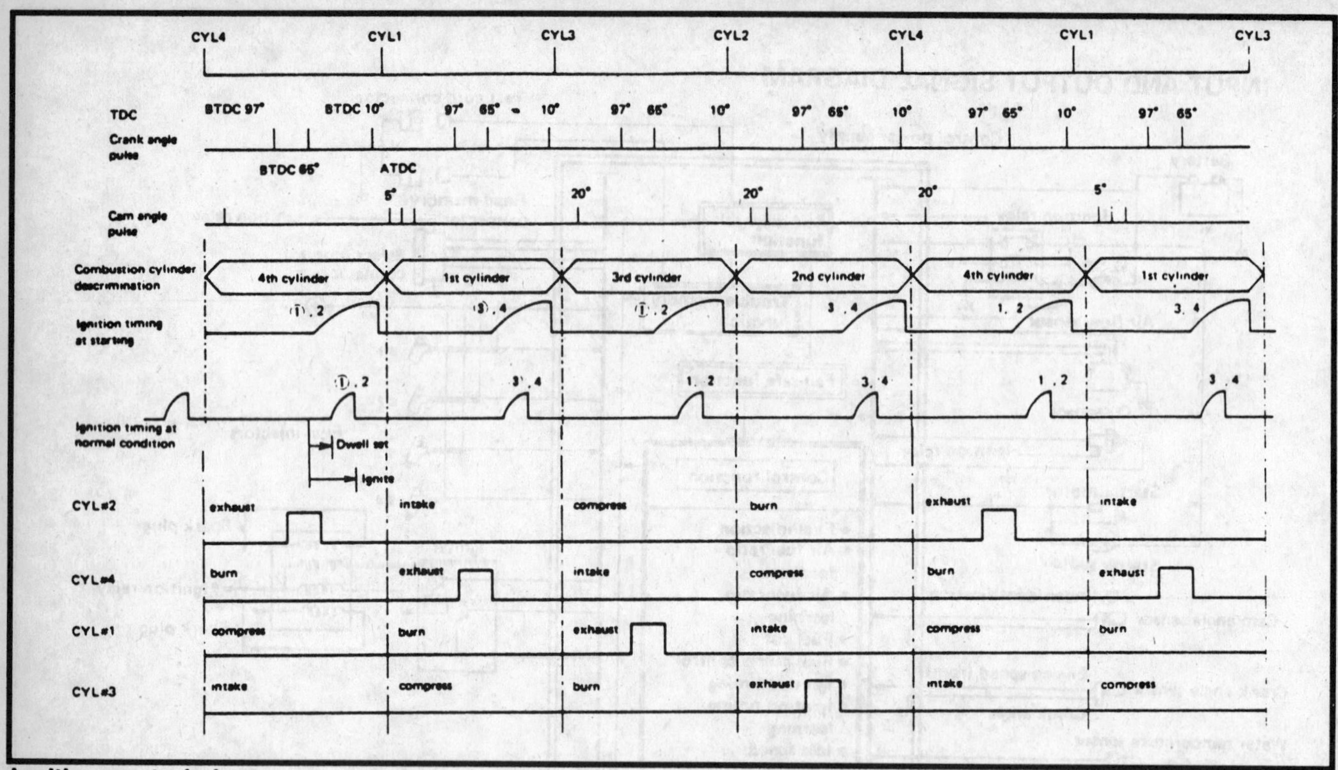
CYL4
CYL1
CYL3
CYL2
CYL4
CYL1
CYL3
TDC
BTDC 97°
BTDC 10°
97°
65°
10°
Crank angle pulse
BTDC 65°
ATDC
5°
20°
Cam angle pulse
Combustion cylinder descrimination
4th cylinder
1st cylinder
3rd cylinder
2nd cylinder
Ignition timing at starting
Ignition timing at normal condition
Dwell set
Ignite
CYL#2
CYL#4
CYL#1
CYL#3
exhaust
intake
compress
burn

Ignition control chart

before engine knock occurs. This system features a quick-to-response learning control method by which information stored in the ECU memory is processed in comparison with information from various sensors and switches.

The ECU constantly provides the optimum ignition timing in relation to output, fuel consumption, exhaust gas, etc., according to various engine operating conditions, the octane rating of the fuel used, etc.

The ignition system uses 2 ignition coils. One for the No. 1 and No. 2 cylinders, and 1 for the No. 3 and No. 4 cylinders. Simultaneous ignition occurs for cylinders No. 1 and No. 2 on 1 hand, and No. 3 and No. 4 on the other. This eliminates the distributor and achieves maintenance free operation.

Ignition control under normal engine conditions: Between the 97 degrees signal and the 65 degrees signal, the ECU measures the engine revolutions, and by using this information it decides the dwell set timing and ignition timing according to the engine condition.

Ignition control under starting conditions: Engine revolutions fluctuate at the starting condition, so the ECU cannot control the ignition timing. When such a condition exists, ignition timing is fixed at 10 degrees BTDC by using the 10 degrees signal.

BY-PASS AIR CONTROL (IDLE SPEED CONTROL)

The ECU activates the by-pass air control valve to control the amount of by-pass air flowing through the throttle valve in relation to signals from the crank angle sensor, cam angle sensor, water temperature sensor and the A/C switch, so the proper idle speed specified for each engine load is achieved.

The by-pass air control valve uses a duty solenoid design so that the amount of valve lift is determined by certain operating frequency. For this reason, the by-pass air flow is regulated by controlling the duty ratio. The relationship between the duty ratio, valve lift and by-pass air flow is the duty ratio (high) increases valve lift and by-pass air flow. By-pass air control features the following advantages:

- Compensation for engine speed, under A/C system and electrical loads
- Increases in idle speed during engine warm-up
- A dashpot function during the time the throttle valve is quickly closed
- Prevention of engine speed variations over time

CANISTER PURGE CONTROL

The ECU receives signals from the water temperature sensor, vehicle speed sensor and crank angle sensor to control the purge control solenoid. Canister purge takes place during vehicle operation, except under certain conditions (during idling, etc.). The purge line is connected to the throttle chamber to purge fuel evaporation gas from the canister according to the amount of intake air.

RADIATOR FAN CONTROL

The ON/OFF control of the radiator fan (models not equipped with A/C) is controlled by the ECU which receives signals sent from the water temperature sensor and the vehicle speed sensor. On models equipped with A/C, the ECU receives signals from the water temperature sensor, vehicle speed sensor and the A/C switch. These signals simultaneously turn ON/OFF the main radiator fan and the A/C auxiliary fan as well as setting them at HI or LO speed.

FUEL PUMP OPERATION

The ECU receives a signal from the crank angle sensor and turns the fuel pump relay ON/OFF to control fuel pump operation. The fuel pump will stop operating if the engine stalls with the ignition switch in the **ON** position.

AIR CONDITIONING CUT CONTROL

When the ECU receives a full-open signal from the throttle sensor while the A/C system is operating, the A/C cut relay turns OFF for certain period of time to stop the compressor. This prevents degradation of output during acceleration and stabilizes driveability.

POWER SUPPLY CONTROL

When the ECU receives an ON signal from the ignition switch, current flows through the ignition relay. This turns the ignition relay ON so that power is supplied to the injectors, air flow sensor, by-pass air control valve, etc. Power to the above components, except the fuel injectors is turned OFF for approximately 5 seconds after the ECU receives an OFF signal from the ignition switch. The fuel injectors stop fuel injection after the ignition switch is turned **OFF** because the injection signal is cut off.

SERVICE PRECAUTIONS

- Never connect the battery in reverse polarity. The MPFI control unit will be destroyed instantly and the fuel injectors will be damaged.
- Never disconnect the battery terminals while the engine is running. A large counter of electromotive force will be generated in the alternator, causing damage to the electronic parts, such as the ECU (MPFI control unit).
- Before disconnecting the connectors of each sensor and the ECU, ensure that the ignition switch is in the **OFF** position. If not, the ECU could be damaged.
- The connectors to each sensor and all other harness connectors are designed to be waterproof. However, it is still necessary to take caution not to allow water to get into the connectors when washing the vehicle or when serviching the vehicle on a rainy day.
- Never drop a MPFI related part.
- Note the following precautions when installing a radio:
  - Keep the antenna as far apart as possible from the control unit (the ECU is located under the steering column, inside the instrument panel lower trim panel).
  - The antenna line must be positioned as far apart as possible from the ECU and MPFI harness.
  - Carefully adjust the antenna for correct matching.

NOTE: Incorrect installing of the radio may affect the operation of the ECU.

- Before disconnecting the fuel line, disconnect the fuel pump connector, then crank the engine for approximately 5 seconds the pressure in the fuel system. If the engine starts during this operation, let it run until it stops.

Diagnosis and Testing

SELF-DIAGNOSIS SYSTEM

The self-diagnosis system detects and indicates faults, in various inputs and outputs of the electronic control unit. The warning lamp (**CHECK ENGINE** light), located on the instrument panel indicates a fault or trouble, and the LED (light emitting diode) in the ECU indicates a trouble code. A fail-safe function is incorporated into the system to ensure minimal driveability if a failure of a sensor occurs.

Function Of Self-Diagnosis

The self-diagnosis function consists of 4 modes: U-check mode, Read memory mode, D-check mode and Clear code mode. Two connectors (Read memory and Test mode) and 2 lamps (**CHECK ENGINE** and O_2 monitor) are used. The connectors are used for mode selection and the lamps monitor the type of problem.

Fail-Safe Function

When a component has been found faulty in the self-diagnosis function, the ECU generates the faulty signal and carries out the computational processing. When this occurs, the fail-safe function is performed.

U-CHECK MODE

The U-Check mode diagnosis only the MPFI components necessary for start-up and driveabilty. When a fault occurs, the warning lamp (**CHECK ENGINE**) lights to indicate to the user that inspection is necessary. The diagnosis of other components which do not effect start-up and driveabilty are excluded from this mode.

READ MEMORY CODE

This mode is used to read past problems (even when the vehicle monitor lamps are off). It is most effective in detecting poor contacts or loose connections of the connectors, harness, etc.

D-CHECK MODE

This mode is used to check the entire MPFI system and to detect faulty components.

F-CHECK MODE (SELECT MONITOR)

This mode is used to measure the performance characteristics of system components when no trouble codes appear in the Read Memory, U-Check or D-Check mode (although problems have occurred or are occurring).

When working in this mode, use Select Monitor Cartriege No. 498347001 and Select Monitor Instruction Manual No. 498327000, for details concerning select monitor and troubleshooting.

CLEAR MEMORY MODE

This mode is used to clear the trouble code from the memory after all faults have been corrected.

OPERATION CHART OF SELF-DIAGNOSIS SYSTEM

BASIC OPERATION OF SELF-DIAGNOSIS SYSTEM

• No TROUBLE

Mode	Read memory connector	Test mode connector	Condition	CHECK ENGINE light	O_2 monitor light (ECU)
U-check	×	×	Ignition switch ON (Engine OFF)	ON	Vehicle specification code
			Engine ON	OFF	O_2 monitor
Read memory	○	×	Ignition switch ON (Engine OFF)	Blink	Vehicle specification code
			Engine ON		O_2 monitor
D-check	×	○	Ignition switch ON (Engine OFF)	ON	Vehicle specification code
			Engine ON	Vehicle specification code → Blink*	
Clear memory	○	○	Ignition switch ON (Engine OFF)	ON	Vehicle specification code
			Engine ON	Vehicle specification code → Blink	

• TROUBLE

Mode	Read memory connector	Test mode connector	Condition	CHECK ENGINE light	O_2 monitor light (ECU)
U-check	×	×	Ignition switch ON	ON	Trouble code
Read memory	○	×	Ignition switch ON	Trouble code (memory)	Trouble code (memory)
D-check	×	○	Engine ON	Trouble code**	Trouble code**
Clear memory	○	○	Engine ON	Trouble code**	Trouble code**

* When the engine operates at a speed greater than 2,000 rpm for more than 40 seconds, the check engine light blinks. However, when all check items check out "O.K'," even before the 40 seconds is reached, the check engine light blinks.

** When the engine operates at a speed greater than 2,000 rpm for more than 40 seconds, a trouble code is emitted.

TROUBLE CODES AND FAIL-SAFE OPERATION CHART

TROUBLE CODES AND FAIL-SAFE OPERATION

Trouble code	Item	Contents of diagnosis	Fail-safe operation
11	Crank angle sensor	No signal entered from crank angle sensor, but signal (corresponding to at least one rotation of crank) entered from cam angle sensor.	–
12	Ignition switch	Abnormal signal emitted from ignition switch.	Turns ignition switch signal ON or OFF according to engine speed.
13	Cam angle sensor	No signal entered from cam angle sensor, but signal (corresponding to at least two rotations of cam) entered from crank angle sensor.	–
14	Injector #1	Fuel injector inoperative. (Abnormal signal emitted from monitor circuit.)	–
15	Injector #2		–
16	Injector #3		–
17	Injector #4		–
21	Water temperature sensor	Abnormal signal emitted from water temperature sensor.	Adjusts water to a specific temperature. Maintains radiator fan "ON" to prevent overheating.
22	Knock sensor	Abnormal voltage produced in knock sensor monitor circuit.	Sets in regular fuel map and retards ignition timing by 5°.
23	Air flow sensor	Abnormal voltage input entered from air flow sensor.	Controls the amount of fuel (injected) in relation to engine speed and throttle sensor position.
24	Air control valve	Air control valve inoperative. (Abnormal signal produced in monitor circuit.)	Prevents abnormal engine speed using "fuel cut" in relation to engine speed, vehicle speed and throttle sensor position.
31	Throttle sensor	Abnormal voltage input entered from throttle sensor.	Sets throttle sensor's voltage output to a fixed value.
32	O_2 sensor	O_2 sensor inoperative.	–
33	Vehicle speed sensor	Abnormal voltage input entered from vehicle speed sensor.	Sets vehicle speed signal to a fixed value.
35	Canister purge solenoid valve	Solenoid valve inoperative.	–
41	A/F learning control	Faulty learning control function.	–
42	Idle switch	Abnormal voltage input entered from idle switch.	Judges ON or OFF operation according to throttle sensor's signal.
45	Atmospheric pressure sensor	Faulty sensor.	Sets sensor to 760 mmHg.
49	Air flow sensor	Use of improper air flow sensor.	–
51	Neutral switch	Abnormal signal entered from neutral switch.	–
51	Inhibitor switch	Abnormal signal entered from inhibitor switch.	–
52	Parking switch	Abnormal signal entered from parking switch.	–

MPFI WIRING DIAGRAM

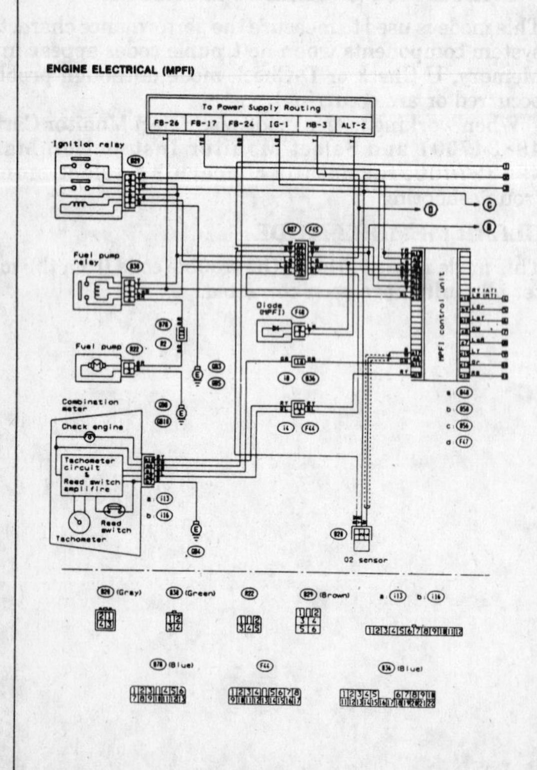
ENGINE ELECTRICAL (MPFI)
To Power Supply Routing
FB-26
FB-17
FB-24
IG-1
MB-3
ALT-2
Ignition relay
Fuel pump relay
Fuel pump
Diode (MPFI)
MPFI control unit
Combination meter
Check engine
Tachometer circuit
Reed switch amplifire
Reed switch
Tachometer
O2 sensor

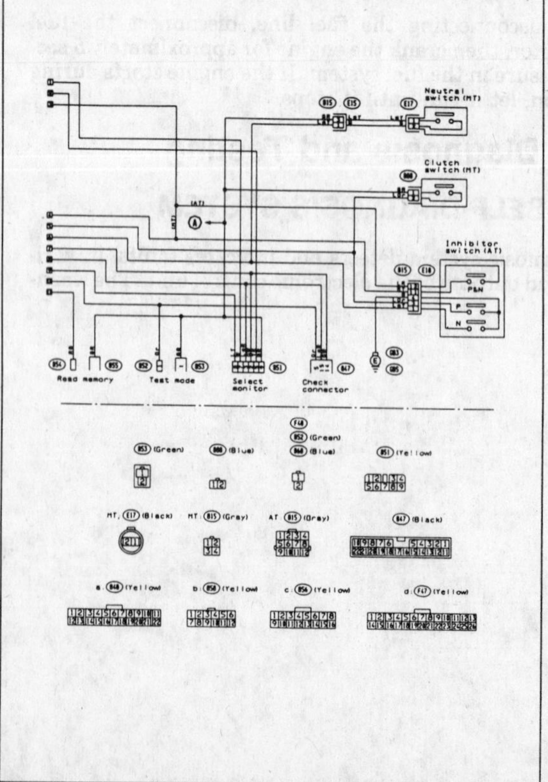
Neutral switch (MT)
Clutch switch (MT)
Inhibitor switch (AT)
Read memory
Test mode
Select monitor
Check connector

MPFI WIRING DIAGRAM

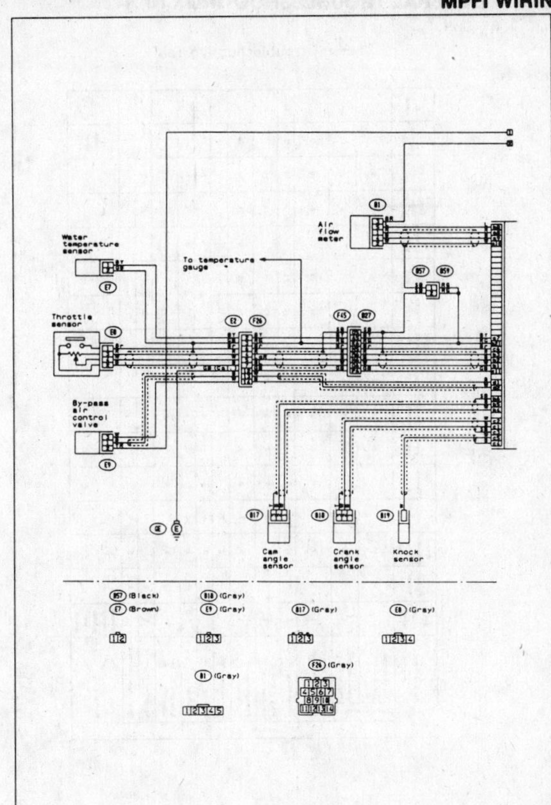
Water temperature sensor
To temperature gauge
Throttle sensor
By-pass air control valve
Air flow meter
Cam angle sensor
Crank angle sensor
Knock sensor
(B57) (Black)
(E7) (Brown)
(B18) (Gray)
(E9) (Gray)
(B17) (Gray)
(E8) (Gray)
(B1) (Gray)
(F26) (Gray)

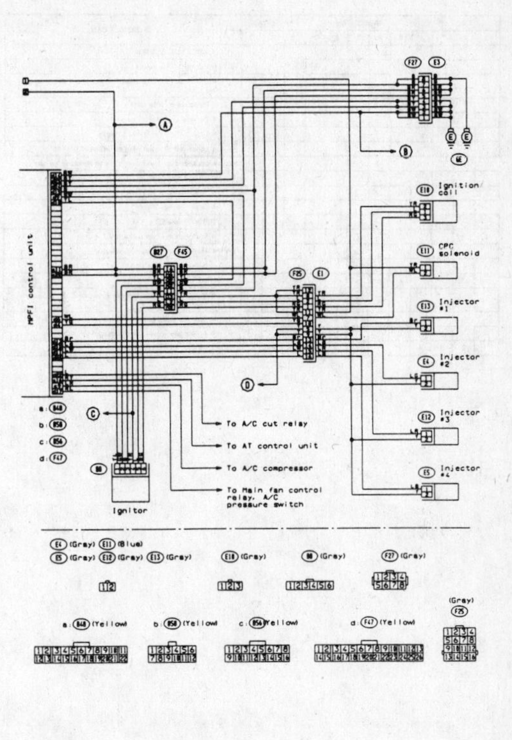
MPFI control unit
Ignition coil
CPC solenoid
Injector #1
Injector #2
Injector #3
Injector #4
To A/C cut relay
To AT control unit
To A/C compressor
To Main fan control relay, A/C pressure switch
Ignitor
(E4) (Gray)
(E11) (Blue)
(E5) (Gray)
(E12) (Gray)
(E13) (Gray)
(E18) (Gray)
(B8) (Gray)
(F27) (Gray)
(F25) (Gray)
a. (B48) (Yellow)
b. (B58) (Yellow)
c. (B56) (Yellow)
d. (F47) (Yellow)

ENGINE TROUBLESHOOTING CHART

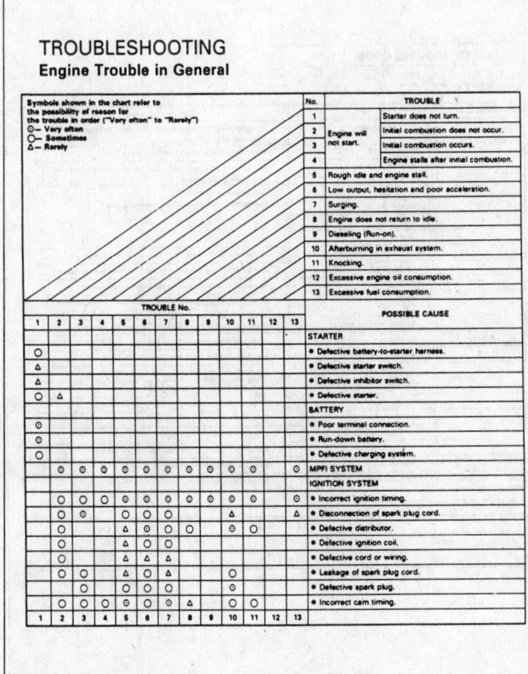

TROUBLESHOOTING

Engine Trouble in General

Symbols shown in the chart refer to the possibility of reason for the trouble in order ("Very often" to "Rarely")
⊙— Very often
○— Sometimes
Δ— Rarely

No.	TROUBLE	
1	Engine will not start.	Starter does not turn.
2		Initial combustion does not occur.
3		Initial combustion occurs.
4		Engine stalls after initial combustion.
5	Rough idle and engine stall.	
6	Low output, hesitation and poor acceleration.	
7	Surging.	
8	Engine does not return to idle.	
9	Dieseling (Run-on).	
10	Afterburning in exhaust system.	
11	Knocking.	
12	Excessive engine oil consumption.	
13	Excessive fuel consumption.	

TROUBLE No.													POSSIBLE CAUSE
1	2	3	4	5	6	7	8	9	10	11	12	13	
													STARTER
○													• Defective battery-to-starter harness.
Δ													• Defective starter switch.
Δ													• Defective inhibitor switch.
○	Δ												• Defective starter.
													BATTERY
⊙													• Poor terminal connection.
⊙													• Run-down battery.
○													• Defective charging system.
	⊙	⊙	⊙	⊙	⊙	⊙	⊙	⊙	⊙	⊙		⊙	MPFI SYSTEM
													IGNITION SYSTEM
	○	○	○	⊙	⊙	⊙	⊙	⊙	⊙	⊙		⊙	• Incorrect ignition timing.
	○	⊙		○	○	○			Δ			Δ	• Disconnection of spark plug cord.
	○			Δ	⊙	○	○		⊙	○			• Defective distributor.
	○			Δ	○	○							• Defective ignition coil.
	○			Δ	Δ	Δ							• Defective cord or wiring.
	○	○		Δ	○	Δ			○				• Leakage of spark plug cord.
		○		○	○	○			⊙				• Defective spark plug.
	○	○	○	⊙	○	⊙	Δ		○	○			• Incorrect cam timing.
1	2	3	4	5	6	7	8	9	10	11	12	13	

TROUBLE No.													POSSIBLE CAUSE
1	2	3	4	5	6	7	8	9	10	11	12	13	
													INTAKE SYSTEM
		○	○	⊙	○	○	⊙	○	○			⊙	• Improper idle adjustment.
			○	⊙	⊙	⊙			Δ	⊙			• Loosened or cracked intake boot.
			○	⊙	⊙	⊙			Δ	⊙			• Loosened or cracked intake duct.
			Δ	⊙	⊙	⊙			Δ	⊙	⊙		• Loosened or cracked blow-by hose.
			Δ	⊙	○	⊙	⊙		○	⊙			• Loosened or cracked vacuum hose.
			Δ	○	○	○				⊙			• Defective air cleaner gasket.
		○	○	○	○	○				⊙			• Defective intake manifold gasket.
		○	○	○	○	○				⊙			• Defective throttle body gasket.
				Δ	○	○			○	○	○		• Defective PCV valve.
				○	○	○			Δ	○	Δ		• Loosened oil filler cap.
			Δ	Δ	⊙	○				○		⊙	• Dirty air cleaner element.
													FUEL LINE
	⊙	Δ		Δ	○	○							• Defective fuel pump.
		Δ		Δ	○	○							• Clogged fuel line.
	○	○	○	○	Δ	Δ							• Lack of or insufficient fuel.
													BELT
	○	○	○										• Defective.
	○	○	○	Δ	○	○			○	○		○	• Defective timing.
													FRICTION
Δ													• Seizure of crankshaft and connecting-rod bearing.
Δ													• Seized camshaft.
Δ													• Seized or stuck piston and cylinder.
													COMPRESSION
	Δ	Δ	Δ	○	○	○			○	Δ		○	• Incorrect valve clearance.
	Δ	Δ	Δ	○	○	Δ			Δ			Δ	• Loosened spark plugs or defective gasket.
	Δ	Δ	Δ	○	○	Δ			Δ			Δ	• Loosened cylinder head nuts or defective gasket.
	Δ	Δ	Δ	○	○	Δ			○			○	• Improper valve seating.
	Δ	Δ	Δ	Δ	Δ	Δ			Δ		⊙	Δ	• Defective valve stem.
	○	○	○	○	○	Δ			Δ			Δ	• Worn or broken valve spring.
	Δ	Δ	Δ	○	Δ	Δ			Δ		⊙	○	• Worn or stuck piston rings, cylinder and piston.
	○	○	○	⊙	⊙	⊙			⊙	○		○	• Incorrect valve timing.
	○	○	○	○	○	○							• Improper engine oil (low viscosity).
1	2	3	4	5	6	7	8	9	10	11	12	13	

ENGINE TROUBLESHOOTING CHART

TROUBLE No. 1	2	3	4	5	6	7	8	9	10	11	12	13	POSSIBLE CAUSE
													LUBRICATION SYSTEM
				○	○				△			∠	• Incorrect oil pressure.
											○		• Loosened oil pump attaching bolts and defective gasket.
											○		• Defective oil filter seal.
											○		• Defective crankshaft oil seal.
											○		• Defective rocker cover gasket.
											○		• Loosened oil drain plug or defective gasket.
											○		• Loosened oil pan fitting bolts or defective oil pan.
													COOLING SYSTEM
				△	△	○		○		◎			• Overheating.
					∠				△			∠	• Over cooling.
													TURBOCHARGER
				∠	◎	◎						○	• Malfunction of turbocharger.
					◎	◎				◎		○	• Malfunction of waste gate valve.
											○		• Defective oil pipe and hose.
													OTHERS
				◎	◎	△			△				• Malfunction of Evaporative Emission Control System.
				○			◎						• Stuck or damaged throttle valve.
				△			○	○				○	• Accelerator cable out of adjustment.
1	2	3	4	5	6	7	8	9	10	11	12	13	

GENERAL TROUBLESHOOTING CHART

General Troubleshooting Table

Symptom \ Parts to check	ECU power supply	Air flow sensor	Water tempera-ture sensor	Idle switch	Throttle sensor	Fuel pump	Pressure regulator	Fuel injector	Igniter (power transis-tor)	Ignition coil	Spark plug	Knock sensor	Cam angle sensor	Crank angle sensor	Air control valve	O_2 sensor
Failure of engine to start: Initial combustion does not occur.	1	10	11			5	6	7	2	3	4		8	9		
Failure of engine to start: Initial combustion occurs.	1		10			2	3	4	5	6	7		8	9	11	
Failure of engine to start: Engine stalls after initial combustion.	1	2	7		8	4	5	6	11	12	13		9	10	3	
Rough idling	1	3	12	8	7	4	5	6	9	10	11		13	14	2	15
Hard to drive at constant speed	1	4	6	8	7	3	2	9	12	13	14		10	11		5
Poor acceleration/ deceleration	1	2	6	7	8	3	4	5	13	14	15	9	11	12	10	
Poor return to idle			3	2											1	
Backfire			3	4	5		6	7					2	1		
Knocking		1	2				4	6				3		7		
Excessive fuel consumption	-	3	4				1	2								
Shocks while driving	1	8						7	4	5	6		2	3		
Poor engine revving		2	3	4	5		1									
Remarks	Include engine grounding circuit.														Check hoses	

TROUBLESHOOTING CHART FOR SELF-DIAGNOSIS SYSTEM

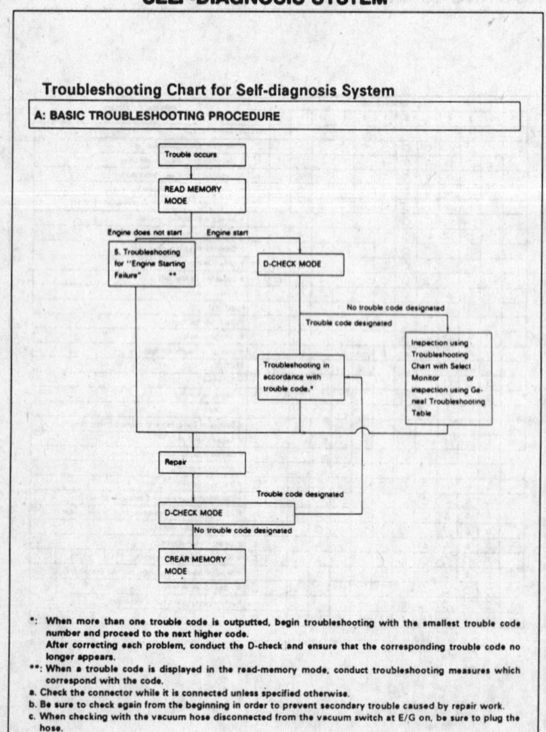
Troubleshooting Chart for Self-diagnosis System
A: BASIC TROUBLESHOOTING PROCEDURE
Trouble occurs
READ MEMORY MODE
Engine does not start
Engine start
5. Troubleshooting for "Engine Starting Failure" **
D-CHECK MODE
No trouble code designated
Trouble code designated
Inspection using Troubleshooting Chart with Select Monitor or inspection using General Troubleshooting Table
Troubleshooting in accordance with trouble code.*
Repair
Trouble code designated
D-CHECK MODE
No trouble code designated
CREAR MEMORY MODE

*: When more than one trouble code is outputted, begin troubleshooting with the smallest trouble code number and proceed to the next higher code.
After correcting each problem, conduct the D-check and ensure that the corresponding trouble code no longer appears.

**: When a trouble code is displayed in the read-memory mode, conduct troubleshooting measures which correspond with the code.

a. Check the connector while it is connected unless specified otherwise.
b. Be sure to check again from the beginning in order to prevent secondary trouble caused by repair work.
c. When checking with the vacuum hose disconnected from the vacuum switch at E/G on, be sure to plug the hose.

TROUBLE CODE LIST

B: LIST OF TROUBLE CODE

1. TROUBLE CODE

Trouble code	Item	Content of diagnosis
11.	Crank angle sensor	No signal entered from crank angle sensor, but signal entered from cam angle sensor.
12.	Starter switch	Abnormal signal emitted from ignition switch.
13.	Cam angle sensor	No signal entered from cam angle sensor, but signal entered from crank angle sensor.
14.	Injector #1	Fuel injector inoperative. (Abnormal signal emitted from monitor circuit.)
15.	Injector #2	
16.	Injector #3	
17.	Injector #4	
21.	Water temperature sensor	Abnormal signal emitted from water temperature sensor.
22.	Knock sensor	Abnormal voltage produced in knock sensor monitor circuit.
23.	Air flow sensor	Abnormal voltage input entered from air flow sensor.
24.	Air control valve	Air control valve inoperative. (Abnormal signal emitted from monitor circuit.)
31.	Throttle position sensor	Abnormal voltage input entered from throttle sensor.
32.	O_2 sensor	O_2 sensor inoperative.
33.	Vehicle speed sensor	Abnormal voltage input entered from speed sensor.
35.	Canister purge solenoid valve	Solenoid valve inoperative.
41.	AF (Air/fuel) learning control	Faulty learning control function.
42.	Idle switch	Abnormal voltage input entered from idle switch.
45.	Atmospheric sensor	Faulty sensor.
49.	Air flow sensor	Use of improper air flow sensor.
51.	Neutral switch (MT)	Abnormal signal entered from newtral switch.
51.	Inhibitor switch (AT)	Abnormal signal entered from inhibitor switch.
52.	Parking switch	Abnormal signal entered from parking switch.

2. SPECIFICATION CODE

	49-state and Canada	California
AT model	03	04
MT model	01	02

READ MEMORY MODE DIAGNOSIS CHART

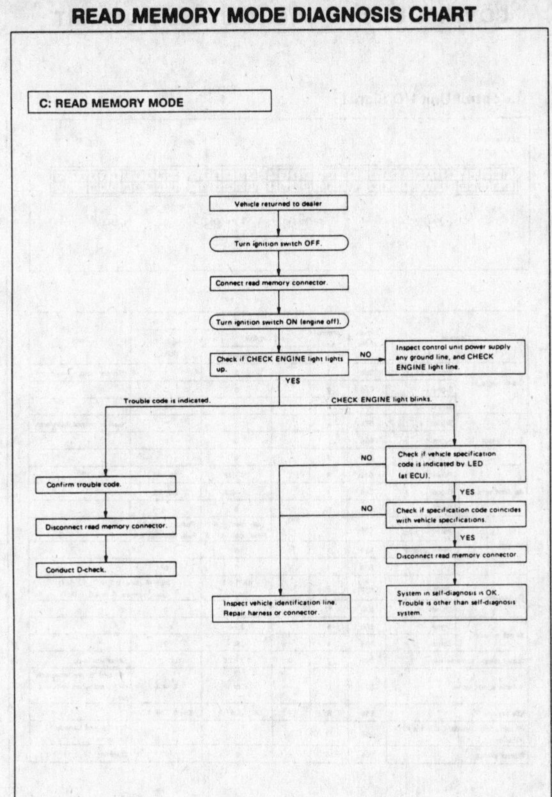
C: READ MEMORY MODE
Vehicle returned to dealer
Turn ignition switch OFF.
Connect read memory connector.
Turn ignition switch ON (engine off).
Check if CHECK ENGINE light lights up.
NO
Inspect control unit power supply any ground line, and CHECK ENGINE light line.
YES
Trouble code is indicated.
CHECK ENGINE light blinks.
Confirm trouble code.
Disconnect read memory connector.
Conduct D-check.
NO
Check if vehicle specification code is indicated by LED (at ECU).
YES
NO
Check if specification code coincides with vehicle specifications.
YES
Disconnect read memory connector.
System in self-diagnosis is OK. Trouble is other than self-diagnosis system.
Inspect vehicle identification line. Repair harness or connector.

D-CHECK MODE DIAGNOSIS CHART

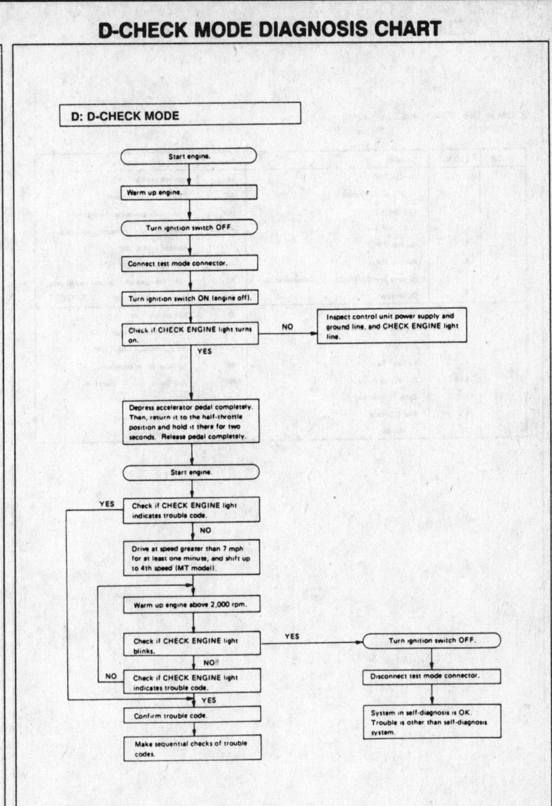
D: D-CHECK MODE
Start engine.
Warm up engine.
Turn ignition switch OFF.
Connect test mode connector.
Turn ignition switch ON (engine off).
Check if CHECK ENGINE light turns on.
NO
Inspect control unit power supply and ground line, and CHECK ENGINE light line.
YES
Depress accelerator pedal completely. Then, return it to the half-throttle position and hold it there for two seconds. Release pedal completely.
Start engine.
YES
Check if CHECK ENGINE light indicates trouble code.
NO
Drive at speed greater than 7 mph for at least one minute, and shift up to 4th speed (MT model).
Warm up engine above 2,000 rpm.
Check if CHECK ENGINE light blinks.
YES
Turn ignition switch OFF.
Disconnect test mode connector.
System in self-diagnosis is OK. Trouble is other than self-diagnosis system.
NO
NO
Check if CHECK ENGINE light indicates trouble code.
YES
Confirm trouble code.
Make sequential checks of trouble codes.

CLEAR MEMORY MODE DIAGNOSIS CHART

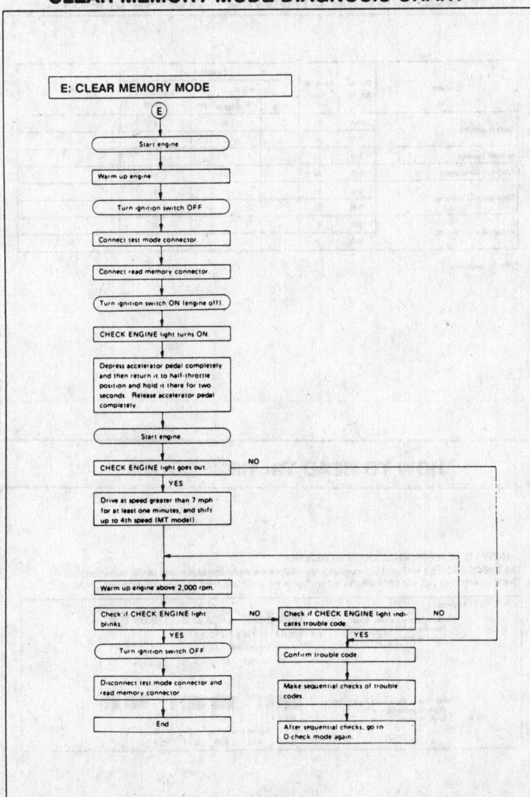
E: CLEAR MEMORY MODE
E
Start engine
Warm up engine
Turn ignition switch OFF
Connect test mode connector
Connect read memory connector
Turn ignition switch ON (engine off)
CHECK ENGINE light turns ON
Depress accelerator pedal completely and then return it to half-throttle position and hold it there for two seconds. Release accelerator pedal completely.
Start engine
CHECK ENGINE light goes out
NO
YES
Drive at speed greater than 7 mph for at least one minutes, and shift up to 4th speed (MT model).
Warm up engine above 2,000 rpm.
Check if CHECK ENGINE light blinks.
NO
Check if CHECK ENGINE light indicates trouble code
NO
YES
YES
Turn ignition switch OFF
Confirm trouble code
Disconnect test mode connector and read memory connector
Make sequential checks of trouble codes
End
After sequential checks, go to D-check mode again.

OUTPUT MODES OF SELECT MONITOR

Output Modes of Select Monitor

FUNCTION MODE

MODE	Contents	Abbr.	Unit	Contents of display
00	PROM ID Number	YEAR	–	Model year of vehicle to which select monitor is connected
01	Battery Voltage	VB	V	Battery voltage supplied to control unit
02	Vehicle Speed Sensor	VSP	m/h	Vehicle speed inputted from vehicle speed sensor
03	Vehicle Speed Sensor	VSP	km/h	Vehicle speed inputted from vehicle speed sensor
04	Engine RPM	EREV	rpm	Engine speed inputted from crank angle sensor
05	Water Temp Sensor	TW	deg F	Coolant temperature inputted from water temperature sensor
06	Water Temp Sensor	TW	deg C	Coolant temperature inputted from water temperature sensor
07	Ignition Timing	ADVS	deg	Ignition timing determined by ECV in relation to signals sent from various sensors
08	Air Flow Sensor	QA	V	Voltage inputted from air flow meter
09	Load Data	LDATA	–	Engine load value determined by related sensor signals
10	Throttle Sensor	THV	V	Voltage inputted from throttle position sensor
11	Injector Pulse Width	TIM	mS	Duration of pulse flowing through injectors
12	Air Control Valve	ISC	%	"Duty" ratio flowing through air control valve
13	O_2 Sensor	O_2	V	Voltage outputted from O2 sensor
14	O_2 Max	O_2 max	V	Maximum voltage outputted from O2 sensor
15	O_2 Min	O_2 min	V	Minimum voltage outputted from O2 sensor
16	ALPHA	ALPHA	%	AF correction ratio determined in relation to signal outputted from O2 sensor
17	Knock Sensor	RTRD	deg	Ignition timing correction determined in relation to signal inputted from knock sensor
19	Atmospheric Sensor	BARO.P	mm Hg	–
A0	ON ↔ OFF Signal	–	–	–
A1	ON ↔ OFF Signal	–	–	–
B0	Trouble Code	DIAG	–	Trouble code in U- or D-check mode
B1	Trouble Code	DIAG	–	Trouble code memorized in U-check mode
C0	Clear Memory	–	–	(Used to clear memory)

SELECT MONITOR ON/OFF SIGNAL LIST

2. ON ↔ OFF SIGNAL LIST

MODE	LED No.	Contents	Display	LED "ON" requirements
A0	1	Ignition SW	IG	Ignition switch "ON"
	2	AT/MT discrimination	AT	"ON" (AT models only)
	3	Test Mode	UD	Test mode connector connected
	4	Read Memory	RM	Read-memory connector connected
	7	Neutral SW	NT	Neutral switch "ON"
	8	Parking SW	PK	Parking switch "ON" [AT]
	9	49-state and Canada/California Discrimination	FC	"ON" (49-state and Canada model only)
	10	O_2 Monitor	O_2	O_2monitor lamp "ON"
A1	1	Idle SW	ID	Idle switch "ON"
	2	A/C SW	AC	Air conditioner switch "ON"
	3	A/C Relay	AR	Air conditioner relay "ON"
	4	Radiator Fan	RF	Radiator fan in operation
	6	Fuel Pump Relay	FP	Fuel pump relay in operation
	7	Canistor Solenoid	CN	Canister purge "ON"
	8	Knock Sensor	KS	Engine knocks occur
	10	O_2 Monitor	O_2	O_2 monitor lamp "ON"

CONTROL UNIT INPUT/OUTPUT SIGNAL CHART

Control Unit I/O Signal

ECU terminal

13	12	11	10	9	8	7	6	5	4	3	2	1
26	25	24	23	22	21	20	19	18	17	16	15	14

To F47

8	7	6	5	4	3	2	1
16	15	14	13	12	11	10	9

To B56

6	5	4	3	2	1
12	11	10	9	8	7

To B58

11	10	9	8	7	6	5	4	3	2	1
22	21	20	19	18	17	16	15	14	13	12

To B48

Content		Connector No.	Terminal No.	Signal (V) Ig SW OFF	Signal (V) Ig SW ON (Engine OFF)	Signal (V) Engine ON (Idling)	Note
Crank angle sensor	Signal (+)	B56	1	·	0	*	*Sensor output waveform
	Signal (-)	B56	2	·	0	0	·
	Shield	B56	3	·	0	0	·
Cam sensor	Signal (+)	B58	4	·	0	*	*Sensor output waveform
	Signal (-)	B58	5	·	0	0	·
	Shield	B58	6	·	0	0	·
Air flow sensor	Power supply	B48	8	·	10 - 13	13 - 14	·
	Signal	B48	9	·	0 - 0.3	0.8 -1.2	·
	GND	B48	10	·	0	0	·
Throttle sensor	Signal	B58	2	·	Fully closed: 4.7 Fully opened: 1.6	Fully closed: 4.7 Fully opened: 1.6	·
	Power supply	B58	3	·	5	5	·
	GND	B58	1	·	0	0	·
O_2 sensor	Signal	B48	6	·	(AT) 0.1 (MT) 0	Rich mixture: 0.8 (AT), 1.2 (MT) Lean mixture: 0.1 (AT), 0 (MT)	
	Shield	B48	17	·	0	0	·
Knock sensor	Signal	B56	5	·	3 - 4	3 - 4	·
	Shield	B56	4	·	0	0	·
Water temperature sensor		B48	7	0	0.7 - 1.0	0.7 - 1.0	*After warm-up
Vehicle speed sensor		B58	11	·	0 or 5	0 or 5	"5" and "0" are repeatedly displayed when vehicle is driven.
Idle switch		B56	6	·	ON:0, OFF:4.6	ON:0, OFF:4.6	·
49 and state Canada/ California identification		B56	11	·	·	·	49 state and Canada:12 California:0
Starter switch		B56	10	·	0	0	Cranking: 10 to 14

CONTROL UNIT INPUT/OUTPUT SIGNAL CHART

Content		Connector No.	Terminal No.	Signal (V) Ig SW OFF	Signal (V) Ig SW ON (Engine OFF)	Signal (V) Engine ON (Idling)	Note
Air conditioner switch		B56	9	·	ON:10 - 13, OFF:0	ON:13 - 14, OFF:0	·
Ignition switch		B58	12	0	10 -13	13 -14	·
Neutral switch		B58	10	·	[AT] N Range: 0 Other: 8, min. [MT] N Position: 8, min. Other: 0	[AT] N Range: 0 Other: 8, min. [MT] N Position: 8, min. Other: 0	
Parking switch [AT]		B58	9	·	P Range: 0 Other: 8, min.	P Range: 0 Other: 8, min.	·
Test mode connector		B56	13	·	[AT] 10 - 13 [MT] 5	[AT] 13 - 14 [MT] 5	[AT] When connected: 0
Read memory connector		B56	12	·	[AT] 10 - 13 [MT] 5	[AT] 13 - 14 [MT] 5	[AT] When connected: 0
AT/MT identification		B48	20	·	[AT] 0 [MT] 5	[AT] 0 [MT] 5	·
Back-up power supply		B48	15	10 - 13	10 - 13	13 - 14	
Control unit power supply		B48	2	0	10 - 13	13 - 14	·
		B48	13	0	10 -13	13 - 14	·
Ignition control	#1, #2	F47	10	·	[AT] 0.01 [MT] 0	·	·
	#3, #4	F47	9	·	[AT] 0.01 [MT] 0	·	·
Fuel injector	#1	F47	13	10 - 13	10 - 13	13 - 14	·
	#2	F47	12	10 - 13	10 - 13	13 - 14	·
	#3	F47	11	10 -13	10 - 13	13 - 14	·
	#4	F47	26	10 - 13	10 - 13	13 - 14-	
Air control valve	OPEN end	F47	2	·	[AT] 7 [MT] 12 → *0	·	*1 min. after ignition switch ON.
	CLOSE end	F47	1	·	[AT] 6 [MT] 0	·	·
Fuel pump relay control		F47	23	·	ON: 0.7 OFF: 10 - 13	0.7	·
Air conditioner cut relay control		F47	22	·	ON: 0 Off: 10 - 13	ON: 0 OFF: 13 - 14	·
Radiator fan control		F47	17	·	ON: 0 OFF: 10 - 13	ON: 0 OFF: 13 - 14	·
Self-shutoff control		F47	5	·	10 - 13	·	·
Trouble code output		B56	15	·	·	·	·
CHECK ENGINE light		F47	19	·	·	·	LED "ON": 1, max. LED "OFF": 10 - 14
Engine tachometer output		B56	16	·	·	·	·
TI monitor*		F47	18	·	·	·	·
Canister purge control		F47	6	·	ON: 0 OFF: 10 - 13	ON: 0 OF: 13 - 14	·
Altitude sensor		B48	16	·	·	·	
GND (sensors)		B48	21	·	0	0	·

CONTROL UNIT INPUT/OUTPUT SIGNAL CHART

Content	Connector No.	Terminal No.	Signal (V) Ig SW OFF	Signal (V) Ig SW ON (Engine OFF)	Signal (V) Engine ON (Idling)	Note
GND (injectors)	F47	24	·	0	0	·
	F47	25	·	0	0	·
Ignition system	F47	16	·	0	0	·
GND (power supply)	F47	14	·	0	0	·
GND (control systems)	B48	11	·	0	0	·
	B48	22	·	0	0	·
Select Monitor Signal	B56	8	·	·	·	·
	B56	7	·	·	·	·

*: For manufacture

HOW TO READ TROUBLE CODES

HOW TO READ TROUBLE CODE (FLASHING)

The CHECK ENGINE LAMP and O_2 monitor lamp flashes the code corresponding to the faulty part. The long segment (1.2 sec on) indicates a "ten", and the short segment (0.2 sec on) signifies "one".

Example:

When only one part has failed:
Flashing code 12
(unit: second)

0.2 0.2
1.2 1.8
0.3 0.3

When two or more parts have failed:
Flashing codes 12 and 21
(unit: second)

0.2 0.2
1.2 1.8 1.2 1.2 1.8
0.3 0.3 0.3 0.3 0.2

DIAGNOSIS CHART

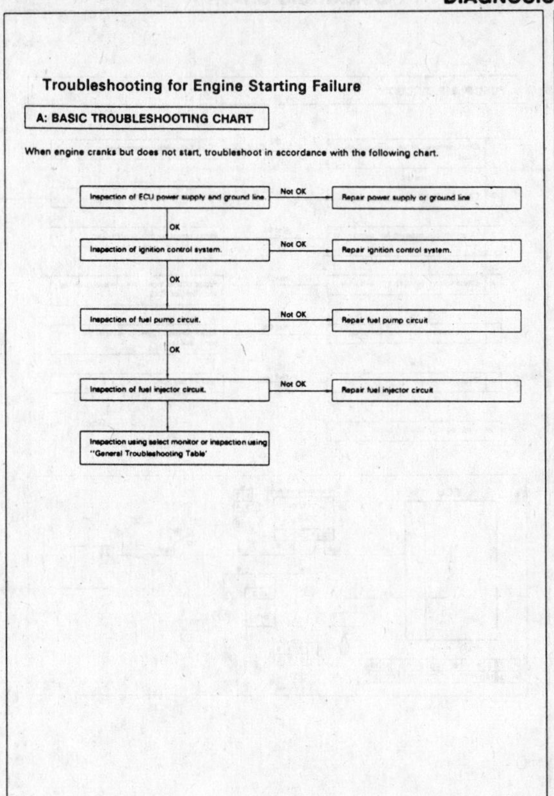

Troubleshooting for Engine Starting Failure

A: BASIC TROUBLESHOOTING CHART

When engine cranks but does not start, troubleshoot in accordance with the following chart.

Inspection of ECU power supply and ground line.
Not OK
Repair power supply or ground line
OK
Inspection of ignition control system.
Not OK
Repair ignition control system.
OK
Inspection of fuel pump circuit.
Not OK
Repair fuel pump circuit
OK
Inspection of fuel injector circuit.
Not OK
Repair fuel injector circuit
Inspection using select monitor or inspection using "General Troubleshooting Table'

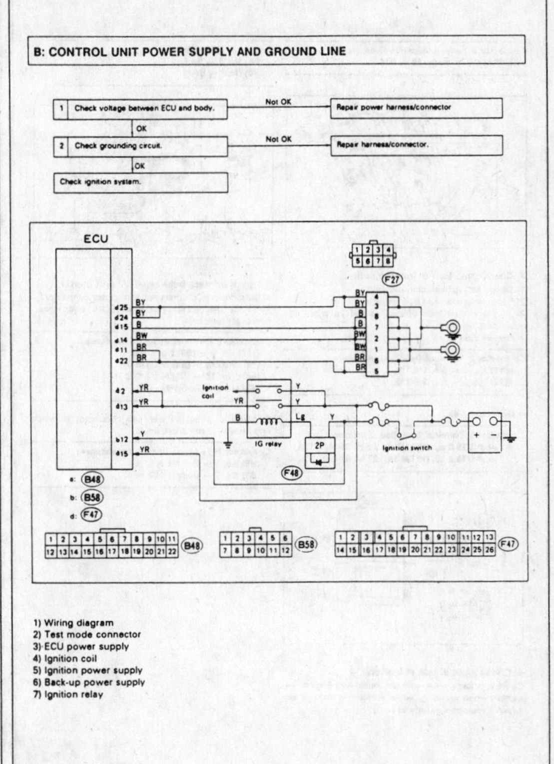

B: CONTROL UNIT POWER SUPPLY AND GROUND LINE

1 Check voltage between ECU and body.
Not OK
Repair power harness/connector
OK
2 Check grounding circuit.
Not OK
Repair harness/connector.
OK
Check ignition system.
ECU
F27
Ignition coil
IG relay
2P
F48
Ignition switch
a: B48
b: B58
d: F47
B48
B58
F47

1) Wiring diagram
2) Test mode connector
3) ECU power supply
4) Ignition coil
5) Ignition power supply
6) Back-up power supply
7) Ignition relay

DIAGNOSIS CHART TESTING

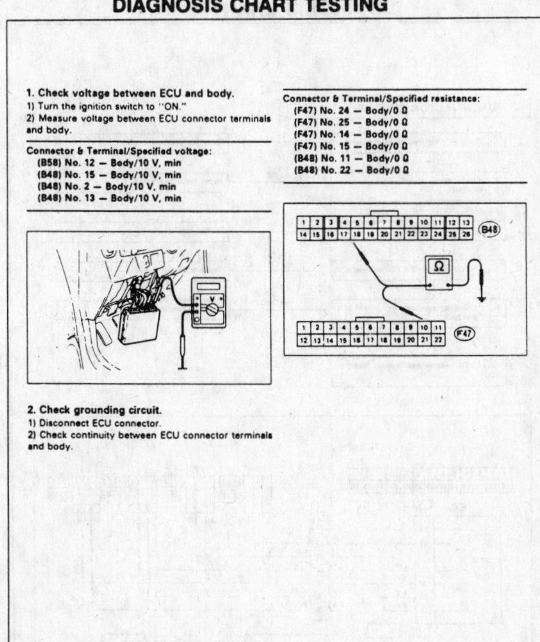

1. Check voltage between ECU and body.
1) Turn the ignition switch to "ON."
2) Measure voltage between ECU connector terminals and body.

Connector & Terminal/Specified voltage:
(B58) No. 12 — Body/10 V, min
(B48) No. 15 — Body/10 V, min
(B48) No. 2 — Body/10 V, min
(B48) No. 13 — Body/10 V, min

Connector & Terminal/Specified resistance:
(F47) No. 24 — Body/0 Ω
(F47) No. 25 — Body/0 Ω
(F47) No. 14 — Body/0 Ω
(F47) No. 15 — Body/0 Ω
(B48) No. 11 — Body/0 Ω
(B48) No. 22 — Body/0 Ω

B48
F47

2. Check grounding circuit.
1) Disconnect ECU connector.
2) Check continuity between ECU connector terminals and body.

DIAGNOSIS CHART

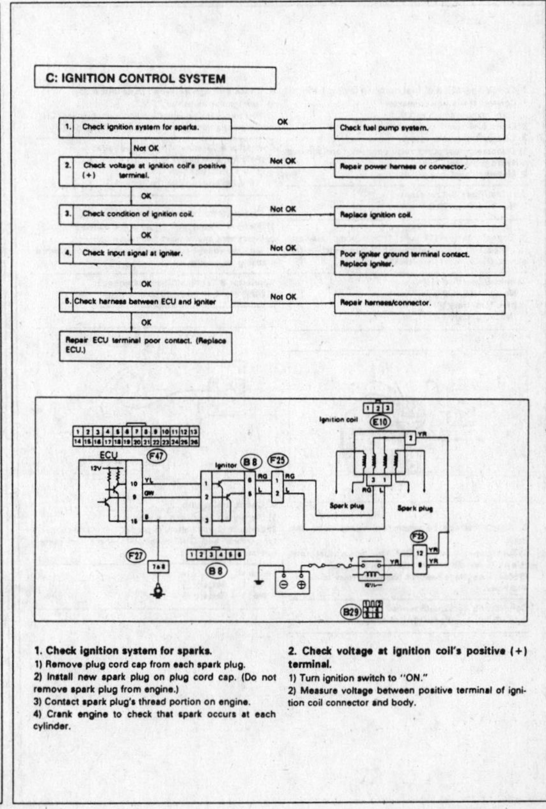

C: IGNITION CONTROL SYSTEM

1. Check ignition system for sparks.
OK
Check fuel pump system.
Not OK
2. Check voltage at ignition coil's positive (+) terminal.
Not OK
Repair power harness or connector.
OK
3. Check condition of ignition coil.
Not OK
Replace ignition coil.
OK
4. Check input signal at igniter.
Not OK
Poor igniter ground terminal contact. Replace igniter.
OK
5. Check harness between ECU and igniter
Not OK
Repair harness/connector.
OK
Repair ECU terminal poor contact. (Replace ECU.)
Ignition coil
E10
ECU
F47
Ignitor
B8
F25
Spark plug
Spark plug
F27
B8
F25
B29

1. Check ignition system for sparks.
1) Remove plug cord cap from each spark plug.
2) Install new spark plug on plug cord cap. (Do not remove spark plug from engine.)
3) Contact spark plug's thread portion on engine.
4) Crank engine to check that spark occurs at each cylinder.

2. Check voltage at ignition coil's positive (+) terminal.
1) Turn ignition switch to "ON."
2) Measure voltage between positive terminal of ignition coil connector and body.

DIAGNOSIS CHART TESTING

Connector & Terminal/Specified voltage:
(E10) No. 2 — Body/10 V, min.

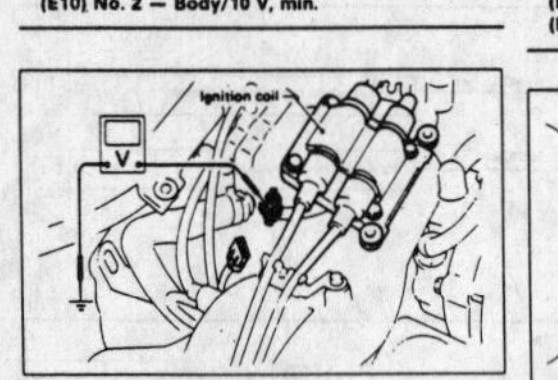
Ignition coil
V

3. Check condition of ignition coil.
1) Disconnect ignition coil connector.
2) Remove ignition coil from engine.
3) Measure resistance of ignition coil's primary and secondary windings.
• Primary side

Connector & Terminal/Specified resistance:
(E10) No. 2 — No. 1/0.7 Ω.
(E10) No. 2 — No. 3/0.7 Ω.

•Secondary side

Connector & Terminal/Specified resistance:
#1 — #2/13.8 kΩ. (HITACHI), 21 kΩ (DIAMOND)
#3 — #4/13.8 kΩ. (HITACHI), 21 kΩ (DIAMOND)

4. Check input signal at igniter.
Check if voltage varies synchronously with engine revolution when cranking, while monitoring voltage between igniter connector and body.

Connector & Terminal
(B8) No. 1 — Body
(B8) No.2 — Body

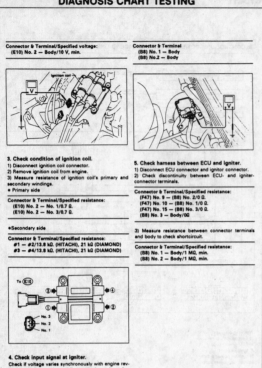

5. Check harness between ECU and igniter.
1) Disconnect ECU connector and ignitor connector.
2) Check discontinuity between ECU- and igniter-connector terminals.

Connector & Terminal/Specified resistance:
(F47) No. 9 — (B8) No. 2/0 Ω.
(F47) No. 10 — (B8) No. 1/0 Ω.
(F47) No. 15 — (B8) No. 3/0 Ω.
(B8) No. 3 — Body/0Ω

3) Measure resistance between connector terminals and body to check shortcircuit.

Connector & Terminal/Specified resistance:
(B8) No. 1 — Body/1 MΩ, min.
(B8) No. 2 — Body/1 MΩ, min.

DIAGNOSIS CHART

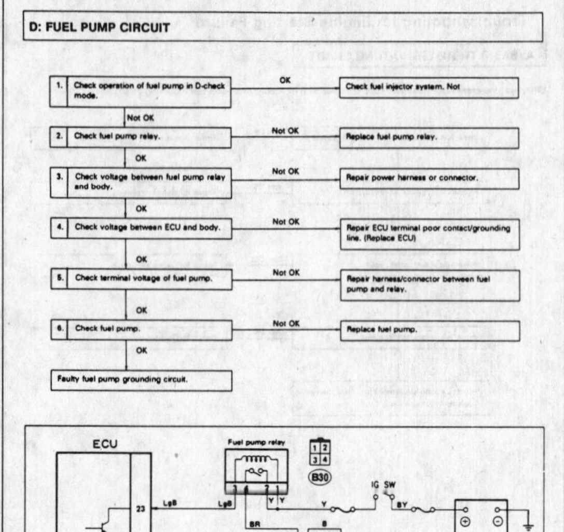
D: FUEL PUMP CIRCUIT
1. Check operation of fuel pump in D-check mode.
OK
Check fuel injector system. Not
Not OK
2. Check fuel pump relay.
Not OK
Replace fuel pump relay.
OK
3. Check voltage between fuel pump relay and body.
Not OK
Repair power harness or connector.
OK
4. Check voltage between ECU and body.
Not OK
Repair ECU terminal poor contact/grounding line. (Replace ECU)
OK
5. Check terminal voltage of fuel pump.
Not OK
Repair harness/connector between fuel pump and relay.
OK
6. Check fuel pump.
Not OK
Replace fuel pump.
OK
Faulty fuel pump grounding circuit.

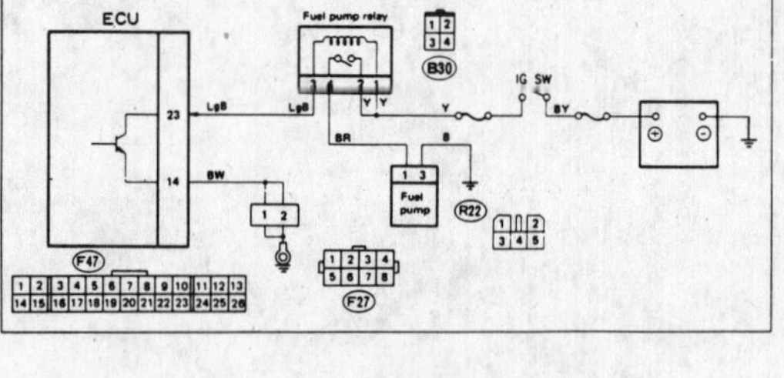
ECU
Fuel pump relay
B30
IG SW
Fuel pump
R22
F47
F27

DIAGNOSIS CHART TESTING

1. Check operation of fuel pump in D-check mode.
1) Connect test-mode connector.
2) Turn ignition switch to "ON."
3) Check fuel pump for proper operation.

2. Check fuel pump relay.
1) Disconnect fuel pump relay connector and remove relay from bracket.
2) Measure resistance of relay coil.

Terminal/Specified resistance:
No. 1 — No. 3/70 Ω.

3) Connect battery (12 volts) to fuel pump relay coil terminals and check continuity between switching terminals. (Relay must issue clicks).

Terminal/Specified resistance:
No. 2 — No. 4/0 Ω.
(No. 1: Battery ⊕)
(No. 3: Battery ⊖)

3. Check voltage between fuel pump relay and body.
1) Turn ignition switch to "OFF," and remove fuel pump relay. (Do not disconnect connector.)
2) Measure voltage between fuel pump relay connector and body.

Connector & Terminal/Specified voltage:
(B30) No. 1 — Body/10 V, min.

4. Check voltage between ECU and body.
1) Turn ignition switch to "ON."
2) Measure voltage when ignition switch is in "ON." Also measure voltage when cranking the engine.

Connector & Terminal/Specified voltage:
(F47) No. 9 — Body/
10 V, min. (Ignition ON)
0 V (when cranking the engine)

5. Check terminal voltage of fuel pump.
1) Remove access lid of fuel pump located in trunk compartment and remove fuel pump connector.
2) Measure voltage between connector and body while cranking the engine.

Connector & Terminal/Specified voltage:
(R22) No. 1 — Body/10 V, min.

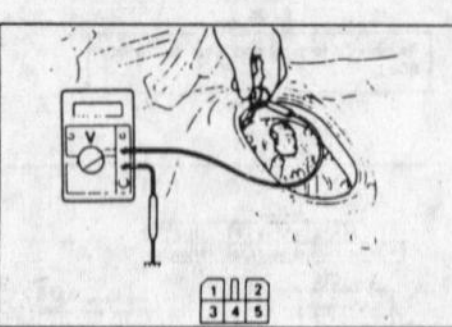

6. Check fuel pump.
1) Disconnect fuel pump connector.
2) Connect 12-volt battery to proper fuel pump connector terminal and GND terminal to check fuel pump operation.

Terminal:
No. 1 → Battery ⊕
No. 3 → Battery ⊖

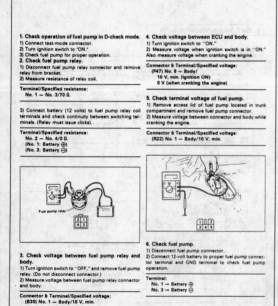

DIAGNOSIS CHART

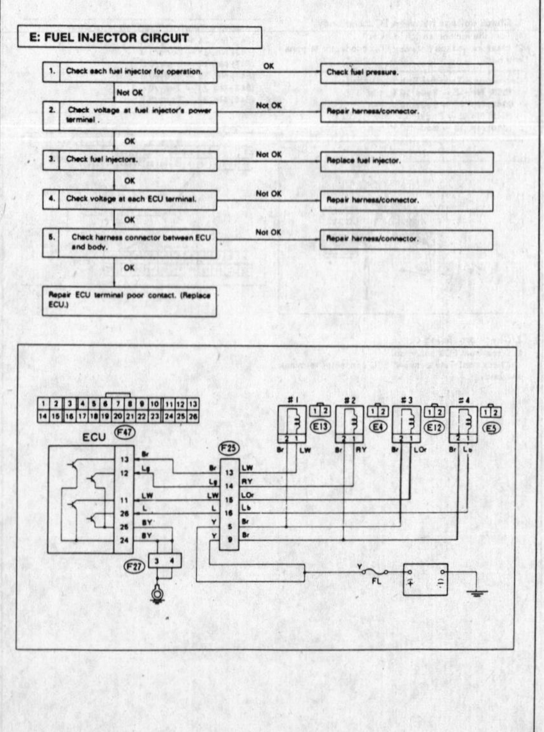
E: FUEL INJECTOR CIRCUIT
1. Check each fuel injector for operation.
OK
Check fuel pressure.
Not OK
2. Check voltage at fuel injector's power terminal .
Not OK
Repair harness/connector.
OK
3. Check fuel injectors.
Not OK
Replace fuel injector.
OK
4. Check voltage at each ECU terminal.
Not OK
Repair harness/connector.
OK
5. Check harness connector between ECU and body.
Not OK
Repair harness/connector.
OK
Repair ECU terminal poor contact. (Replace ECU.)
ECU
F47
F25
F27
E13
E4
E12
E5
FL

DIAGNOSIS CHART TESTING

1. Check each fuel injector for operation.
While cranking the engine, check that each fuel injector emits "operating" sound. Use a sound scope or attach a screwdriver to injector for this check.

2. Check voltage at fuel injector's power terminal.
1) Disconnect connector from injector.
2) Measure voltage between injector connector power terminal and body.

Connector & Terminal/Specified voltage:
(E12) No. 1 — Body/10 V, min.
(E4) No. 1 — Body/10 V, min.
(E13) No. 1 — Body/10 V, min.
(E5) No. 1 — Body/10 V, min.

3. Check fuel injectors.
1) Disconnect connector from injector.
2) Measure resistance between injector terminals.

Specified register:
11 ~ 12 Ω

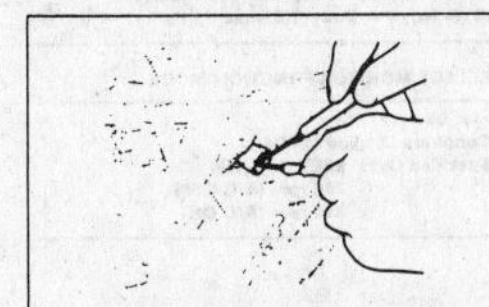

4. Check voltage at each ECU terminal.
Measure voltage between each fuel injector terminal of ECU connector and body.
(Fuel injector connector is connected.)

Connector & Terminal/Specified voltage:
(F47) No. 11 — Body/10 V, min.
(F47) No. 12 — Body/10 V, min.
(F47) No. 13 — Body/10 V, min.
(F47) No. 26 — Body/10 V, min.

5. Check harness connector between ECU and body.
1) Disconnect connector from ECU.
2) Measure resistance between ECU connector and body.

Connector & Terminal/Specified resistance:
(F47) No. 24/0 Ω
(F47) No. 25/0 Ω

DIAGNOSIS CHART

Troubleshooting Chart with Trouble Code

A: TROUBLE CODE (11) —CRANK ANGLE SENSOR—

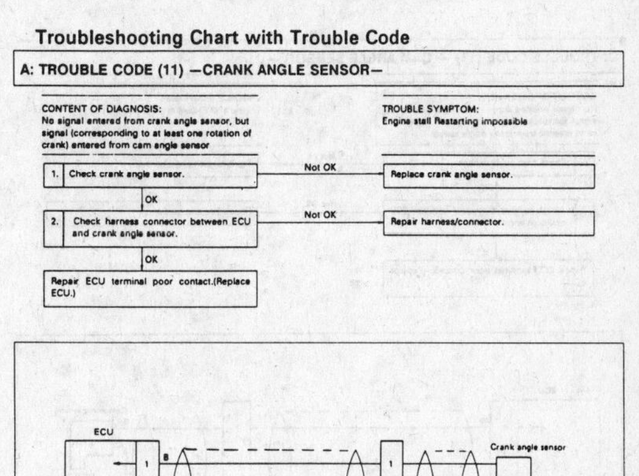
CONTENT OF DIAGNOSIS:
No signal entered from crank angle sensor, but signal (corresponding to at least one rotation of crank) entered from cam angle sensor
TROUBLE SYMPTOM:
Engine stall Restarting impossible
1. Check crank angle sensor.
Not OK
Replace crank angle sensor.
OK
2. Check harness connector between ECU and crank angle sensor.
Not OK
Repair harness/connector.
OK
Repair ECU terminal poor contact.(Replace ECU.)

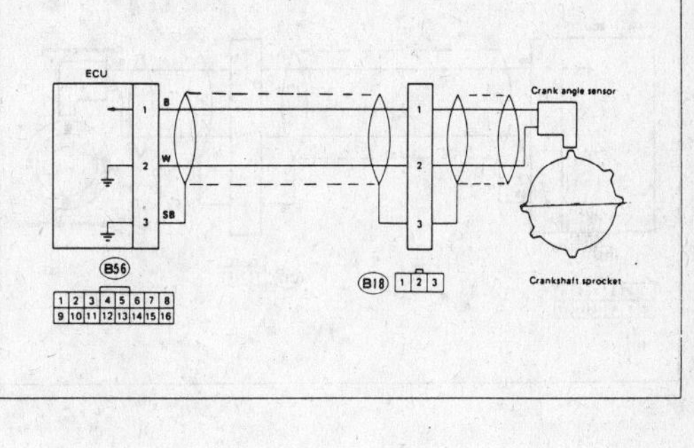
ECU
Crank angle sensor
B56
B18
Crankshaft sprocket

DIAGNOSIS CHART TESTING

1. Check crank angle sensor.
1) Disconnect crank angle sensor connector.
2) Check if voltage varies synchronously with engine revolutions when cranking, while monitoring voltage between crank angle sensor connector terminals (AC 0.1 V, min.).

Terminal:
No. 1 ~ No. 2

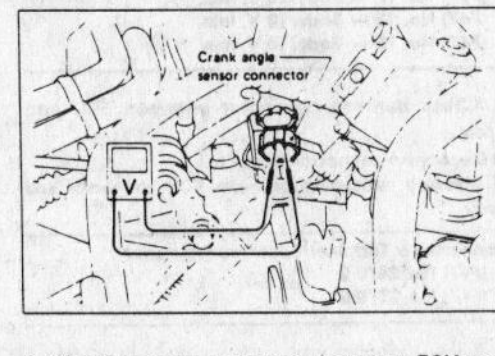
Crank angle sensor connector
V

2. Check harness connector between ECU and crank angle sensor.
1) Disconnect connectors from ECU and crank angle sensor.
2) Measure resistance between ECU connector and angle sensor connector.

Connector & Terminal/Specified resistance:
(B56) No. 1 — (B18) No. 1/0 Ω
(B56) No. 2 — (B18) No. 2/0 Ω
(B56) No. 3 — (B18) No. 3/1 Ω, max.

3) Measure resistance between crank angle sensor connector and body.

Connector & Terminal/Specified resistance:
(B18) No. 1 — Body/1 MΩ, min.
(B18) No. 2 — Body/1 MΩ, min.

4) Connect ECU connector and measure resistance between crank angle sensor sealed terminal and body.

Connector & Terminal/Specified resistance:
(B18) No. 3 — Body/1 Ω, max.

5) Disconnect cam angle sensor connector and measure resistance between sealed terminal and body.

Connector & Terminal/Specified resistance:
(B17) No. 3 — Body/1 Ω, max.

• SELECT MONITOR FUNCTION MODE

Mode: 04
Condition: Engine at idle
Specified Data: EREV F04
700 rpm (A/C OFF)
850 rpm (A/C ON)

DIAGNOSIS CHART

B: TROUBLE CODE (12) — STARTER SWITCH —

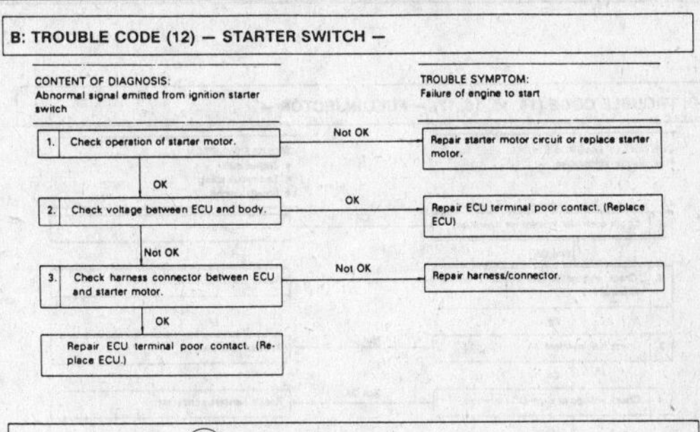
CONTENT OF DIAGNOSIS:
Abnormal signal emitted from ignition starter switch
TROUBLE SYMPTOM:
Failure of engine to start
1. Check operation of starter motor.
Not OK
Repair starter motor circuit or replace starter motor.
OK
2. Check voltage between ECU and body.
OK
Repair ECU terminal poor contact. (Replace ECU)
Not OK
3. Check harness connector between ECU and starter motor.
Not OK
Repair harness/connector.
OK
Repair ECU terminal poor contact. (Replace ECU.)

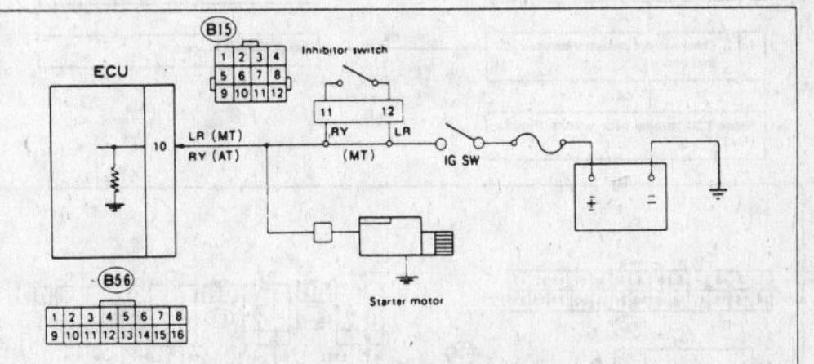
B15
Inhibitor switch
ECU
LR (MT)
RY (AT)
IG SW
Starter motor
B56

DIAGNOSIS CHART TESTING

1. Check operation of starter motor.
Turn ignition switch to "ST" to ensure that starter motor functions.

2. Measure voltage between ECU and body.
Measure voltage between ECU connector terminal and body while cranking the engine.

Connector & Terminal/Specified voltage:
(B56) No. 10 — Body/9 - 12 V.

3. Check harness connector between ECU and starter motor.
1) Disconnect connectors from ECU and starter motor.
2) Measure resistance between ECU connector and starter motor connector.

Connector & Terminal/Specified resistance:
(B56) No. 10 — (B14) No. 1/0 Ω.

3) Measure resistance between starter motor connector and body.

Connector & Terminal/Specified resistance:
(B14) No. 1 — Body/1 MΩ, min.

DIAGNOSIS CHART

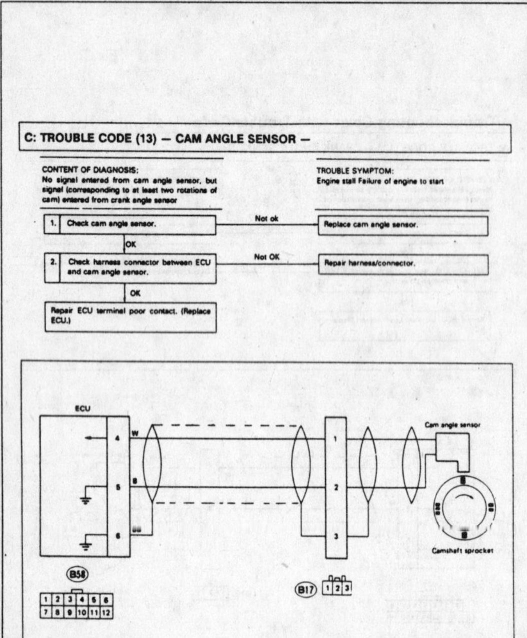
C: TROUBLE CODE (13) — CAM ANGLE SENSOR —
CONTENT OF DIAGNOSIS:
No signal entered from cam angle sensor, but signal (corresponding to at least two rotations of cam) entered from crank angle sensor
TROUBLE SYMPTOM:
Engine stall Failure of engine to start
1. Check cam angle sensor.
Not ok
Replace cam angle sensor.
OK
2. Check harness connector between ECU and cam angle sensor.
Not OK
Repair harness/connector.
OK
Repair ECU terminal poor contact. (Replace ECU.)
ECU
Cam angle sensor
Camshaft sprocket
B58
B17

DIAGNOSIS CHART TESTING

1. Check cam angle sensor.

1) Disconnect cam angle sensor connector.

2) Check if voltage varies synchronously with engine revolutions when cranking, while monitoring voltage between cam angle sensor connector terminals (AC 0.1 V, min.).

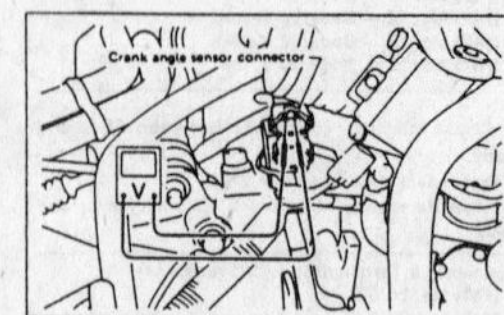
Crank angle sensor connector

2. Check harness connector between ECU and cam angle sensor.

1) Disconnect connectors from ECU and cam angle sensor.

2) Measure resistance between ECU connector and cam angle sensor connector.

Connector & Terminal/Specified resistance:
(B58) No. 4 — (B17) No. 1/0 Ω.
(B58) No. 5 — (B17) No. 2/0 Ω.
(B58) No. 6 — (B17) No. 3/1 Ω, max.

3) Measure resistance between cam angle sensor connector and body.

Connector & Terminal/Specified resistance:
(B17) No. 1 — Body/1 MΩ, min.
(B17) No. 2 — Body/1 MΩ, min.

4) Connect ECU connector and measure resistance between cam angle sensor sealed terminal and body.

Connector & Terminal/Specified resistance:
(B17) No. 3 — Body/1 Ω, max.

5) Disconnect crank angle sensor connector and measure resistance between sealed terminal and body.

Connector &Terminal/Specified resistance:
(B18) No. 3 — Body/1 Ω, max.

• SELECT MONITOR FUNCTION MODE

Mode: 04
Condition: Engine at idle
Specified Data: EREV F04
700 rpm (A/C OFF)
850 rpm (A/C ON)

DIAGNOSIS CHART

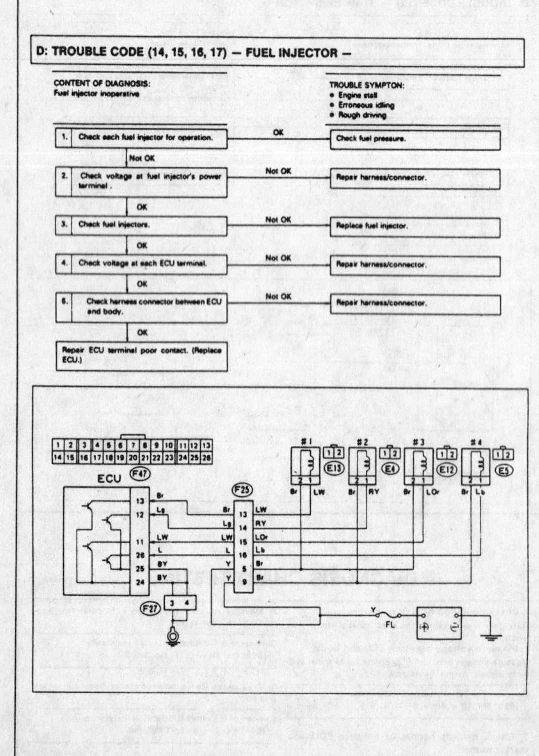
D: TROUBLE CODE (14, 15, 16, 17) — FUEL INJECTOR —
CONTENT OF DIAGNOSIS:
Fuel injector inoperative
TROUBLE SYMPTON:
• Engine stall
• Erroneous idling
• Rough driving
1. Check each fuel injector for operation.
OK
Check fuel pressure.
Not OK
2. Check voltage at fuel injector's power terminal.
Not OK
Repair harness/connector.
OK
3. Check fuel injectors.
Not OK
Replace fuel injector.
OK
4. Check voltage at each ECU terminal.
Not OK
Repair harness/connector.
OK
5. Check harness connector between ECU and body.
Not OK
Repair harness/connector.
OK
Repair ECU terminal poor contact. (Replace ECU.)
ECU
F47
F25
F27
E13
E4
E12
E5
FL

DIAGNOSIS CHART TESTING

1. Check each fuel injector for operation.
While cranking the engine, check that each fuel injector emits "operating" sound. Use a sound scope or attach a screwdriver to injector for this check.

2. Check voltage at fuel injector's power terminal.

1) Disconnect connector from injector.

2) Measure voltage between injector connector power terminal and body.

Connector & Terminal/Specified voltage:
(E12) No. 1 — Body/10 V, min.
(E4) No. 1 — Body/10 V, min.
(E13) No. 1 — Body/10 V, min.
(E5) No. 1 — Body/10 V, min.

3. Check fuel injectors.

1) Disconnect connector from injector.

2) Measure resistance between injector terminals.

Specified resistor:
11 – 12 Ω

4. Check voltage at each ECU terminal.
Measure voltage between each fuel injector terminal of ECU connector and body.
(Fuel injector connector is connected.)

Connector & Terminal/Specified voltage:
(F47) No. 11 — Body/10 V, min.
(F47) No. 12 — Body/10 V, min.
(F47) No. 13 — Body/10 V, min.
(F47) No. 26 — Body/10 V, min.

5. Check harness connector between ECU and body.

1) Disconnect connector from ECU.

2) Measure resistance between ECU connector and body.

Connector & Terminal/Specified resistance:
(F47) No. 24/0 Ω
(F47) No. 25/0 Ω

DIAGNOSIS CHART

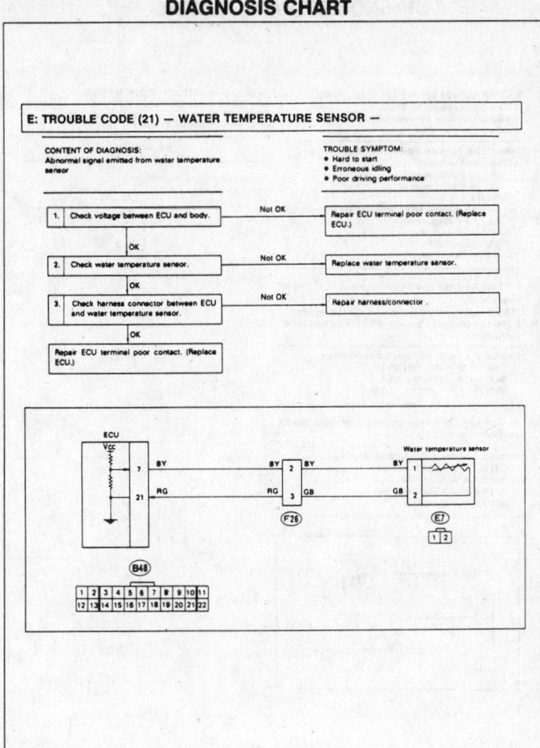

E: TROUBLE CODE (21) — WATER TEMPERATURE SENSOR —

CONTENT OF DIAGNOSIS:
Abnormal signal emitted from water temperature sensor

TROUBLE SYMPTOM:
- Hard to start
- Erroneous idling
- Poor driving performance

1. Check voltage between ECU and body. — Not OK → Repair ECU terminal poor contact. (Replace ECU.)
   - OK ↓
2. Check water temperature sensor. — Not OK → Replace water temperature sensor.
   - OK ↓
3. Check harness connector between ECU and water temperature sensor. — Not OK → Repair harness/connector .
   - OK ↓

Repair ECU terminal poor contact. (Replace ECU.)

ECU
Vcc
7
21
BY
RG
BY 2 BY
RG 3 GB
F26
Water temperature sensor
BY 1
GB 2
E7
1 2
B48
1 2 3 4 5 6 7 8 9 10 11
12 13 14 15 16 17 18 19 20 21 22

DIAGNOSIS CHART TESTING

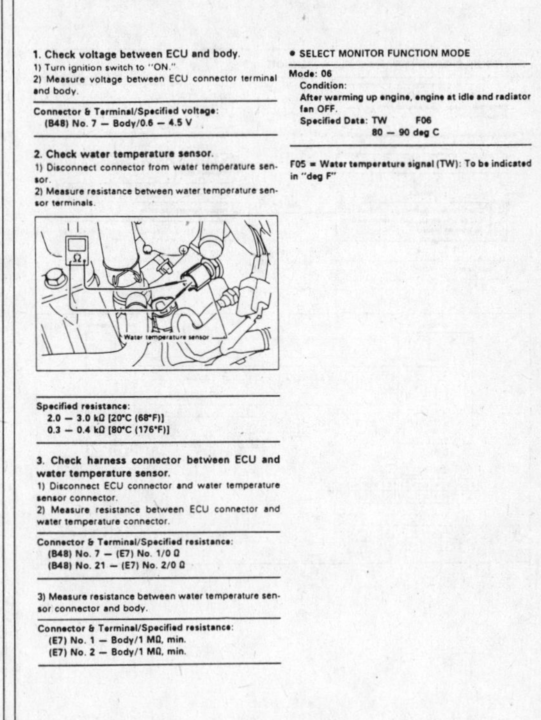

1. Check voltage between ECU and body.
1) Turn ignition switch to "ON."
2) Measure voltage between ECU connector terminal and body.

Connector & Terminal/Specified voltage:
(B48) No. 7 — Body/0.6 — 4.5 V

2. Check water temperature sensor.
1) Disconnect connector from water temperature sensor.
2) Measure resistance between water temperature sensor terminals.

Ω
Water temperature sensor

Specified resistance:
2.0 — 3.0 kΩ [20°C (68°F)]
0.3 — 0.4 kΩ [80°C (176°F)]

3. Check harness connector between ECU and water temperature sensor.
1) Disconnect ECU connector and water temperature sensor connector.
2) Measure resistance between ECU connector and water temperature connector.

Connector & Terminal/Specified resistance:
(B48) No. 7 — (E7) No. 1/0 Ω
(B48) No. 21 — (E7) No. 2/0 Ω

3) Measure resistance between water temperature sensor connector and body.

Connector & Terminal/Specified resistance:
(E7) No. 1 — Body/1 MΩ, min.
(E7) No. 2 — Body/1 MΩ, min.

• SELECT MONITOR FUNCTION MODE

Mode: 06
Condition:
After warming up engine, engine at idle and radiator fan OFF.
Specified Data: TW F06
80 — 90 deg C

F05 = Water temperature signal (TW): To be indicated in "deg F"

DIAGNOSIS CHART

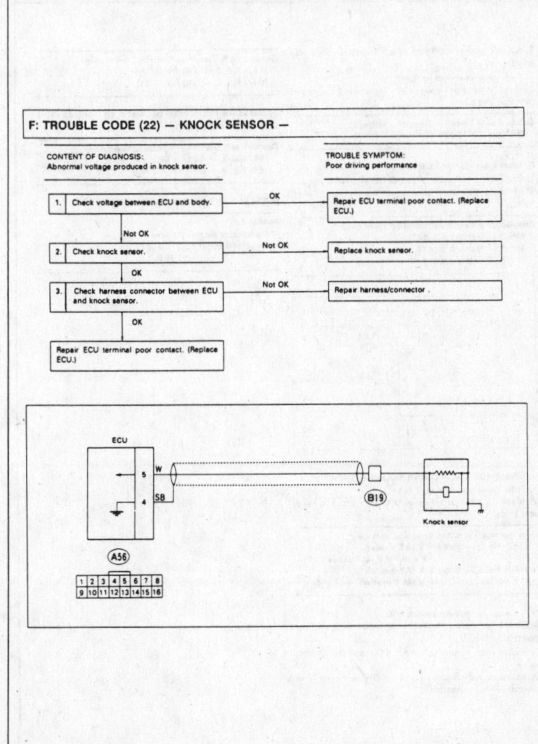

F: TROUBLE CODE (22) — KNOCK SENSOR —

CONTENT OF DIAGNOSIS:
Abnormal voltage produced in knock sensor.

TROUBLE SYMPTOM:
Poor driving performance

1. Check voltage between ECU and body. — OK → Repair ECU terminal poor contact. (Replace ECU.)
   - Not OK ↓
2. Check knock sensor. — Not OK → Replace knock sensor.
   - OK ↓
3. Check harness connector between ECU and knock sensor. — Not OK → Repair harness/connector .
   - OK ↓

Repair ECU terminal poor contact. (Replace ECU.)

ECU
5
4
W
SB
B19
Knock sensor
A56
1 2 3 4 5 6 7 8
9 10 11 12 13 14 15 16

DIAGNOSIS CHART TESTING

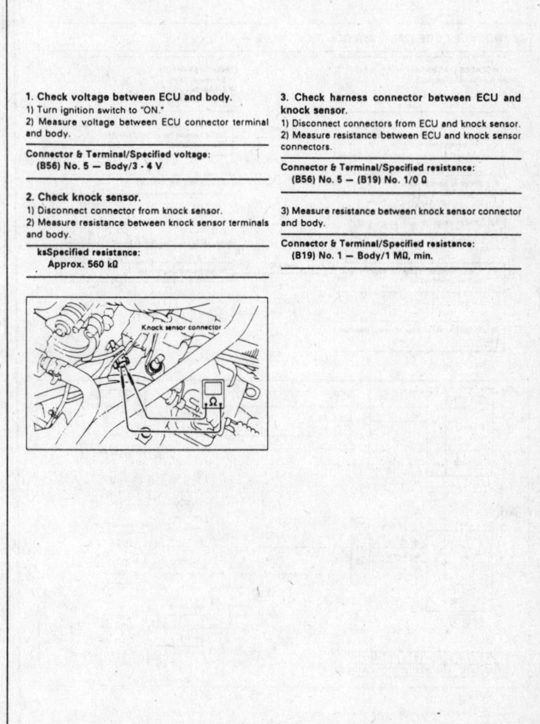

1. Check voltage between ECU and body.
1) Turn ignition switch to "ON."
2) Measure voltage between ECU connector terminal and body.

Connector & Terminal/Specified voltage:
(B56) No. 5 — Body/3 - 4 V

2. Check knock sensor.
1) Disconnect connector from knock sensor.
2) Measure resistance between knock sensor terminals and body.

ksSpecified resistance:
Approx. 560 kΩ

Knock sensor connector
Ω

3. Check harness connector between ECU and knock sensor.
1) Disconnect connectors from ECU and knock sensor.
2) Measure resistance between ECU and knock sensor connectors.

Connector & Terminal/Specified resistance:
(B56) No. 5 — (B19) No. 1/0 Ω

3) Measure resistance between knock sensor connector and body.

Connector & Terminal/Specified resistance:
(B19) No. 1 — Body/1 MΩ, min.

DIAGNOSIS CHART

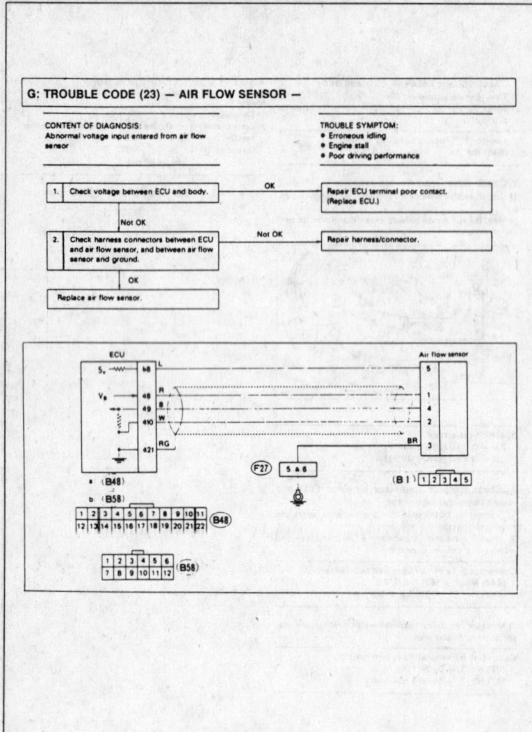
G: TROUBLE CODE (23) — AIR FLOW SENSOR —
CONTENT OF DIAGNOSIS:
Abnormal voltage input entered from air flow sensor
TROUBLE SYMPTOM:
• Erroneous idling
• Engine stall
• Poor driving performance
1. Check voltage between ECU and body.
OK
Repair ECU terminal poor contact. (Replace ECU.)
Not OK
2. Check harness connectors between ECU and air flow sensor, and between air flow sensor and ground.
Not OK
Repair harness/connector.
OK
Replace air flow sensor.
ECU
Air flow sensor
B48
B58

DIAGNOSIS CHART TESTING

1. Check voltage between ECU and body.

1) Turn ignition switch to "ON."
2) Measure voltage between ECU connector terminal and body.

Connector & Terminal/Specified voltage:
(B48) No. 8 — Body/
10 - 13 V (Engine OFF)
13 - 14 V (Engine at idle)
(B48) No. 9 — Body/
0 - 0.3 V (Engine OFF)
0.8 - 1.2 V (Engine at idle)
(B48) No. 10 — Body/
0 V (Engine OFF)
0 V (Engine at idle)

2. Check harness connector between ECU and air flow sensor.

1) Disconnect ECU and air flow sensor connectors.
2) Measure resistance between ECU and air flow sensor connectors.

Connector & Terminal/Specified resistance:
(B48) No. 8 — (B1) No. 1/0 Ω
(B48) No. 9 — (B1) No. 4/0 Ω
(B48) No. 10 — (B1) No. 2/0 Ω

3. Measure resistance between air flow sensor connector and body.

Connector & Terminal/Specified resistance:
(B1) No. 1 — Body/1 MΩ, min.
(B1) No. 4 — Body/1 MΩ, min.
(B1) No. 2 — Body/1 MΩ, min.
(B1) No. 3 — Body/0 Ω

• SELECT MONITOR FUNCTION MODE

Mode: 08
Condition: Engine at idle
Specified Data: QA F08
0.9 — 1.1 V

DIAGNOSIS CHART

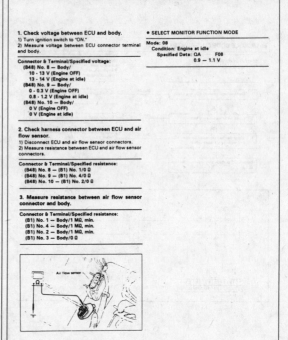

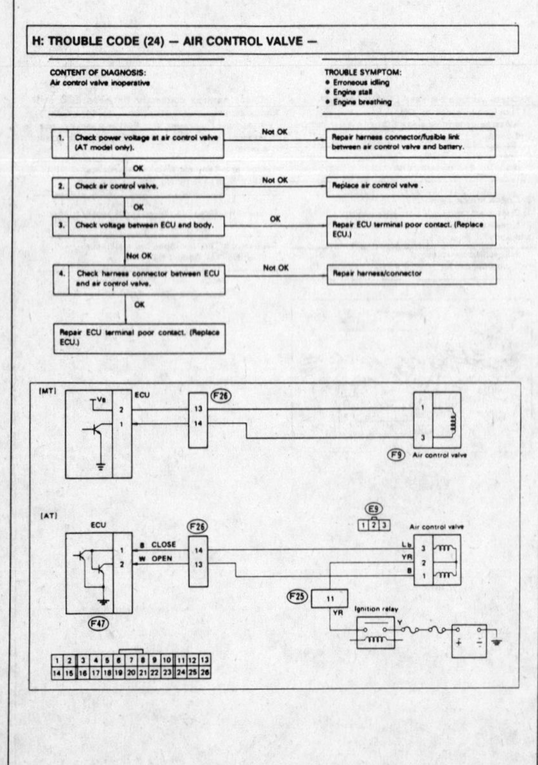
H: TROUBLE CODE (24) — AIR CONTROL VALVE —
CONTENT OF DIAGNOSIS:
Air control valve inoperative
TROUBLE SYMPTOM:
• Erroneous idling
• Engine stall
• Engine breathing
1. Check power voltage at air control valve (AT model only).
Not OK
Repair harness connector/fusible link between air control valve and battery.
OK
2. Check air control valve.
Not OK
Replace air control valve .
OK
3. Check voltage between ECU and body.
OK
Repair ECU terminal poor contact. (Replace ECU.)
Not OK
4. Check harness connector between ECU and air control valve.
Not OK
Repair harness/connector
OK
Repair ECU terminal poor contact. (Replace ECU.)
[MT]
ECU
F26
F9 Air control valve
[AT]
B CLOSE
W OPEN
E9
Air control valve
F25
Ignition relay
F47

DIAGNOSIS CHART TESTING

1. Check power voltage at air control valve (AT model only).

1) Turn ignition switch to "ON."
2) Measure voltage between air control valve connector terminal and body.

Connector & Terminal/Specified voltage:
(E9) No. 2 — Body/10 V, min.

2. Check air control valve.

1) Disconnect connector from air control valve.
2) Measure resistance between air control valve terminals.

Connector & Terminal/Specified resistance:
No. 1 — No. 3/9.6 Ω [MT]
No. 1 — No. 2/9 Ω [AT]
No. 2 — No. 3/9 Ω [AT]

3. Check voltage between ECU and body.

1) Turn ignition switch to "ON."
2) Measure voltage between ECU connector terminal and body.

Connector & Terminal/Specified voltage:
(F47) No. 2 — Body/7 V [AT model]
12 V → *0 V [MT model]
(F47) No. 1 — Body/6 V (AT model)
0 V (M T model)

*: 1 min after ignition switch ON.

4. Check harness connector between ECU and air control valve.

1) Disconnect connectors from ECU and air control valve.
2) Measure resistance between ECU connector and control valve connector.

Connector & Terminal/Specified resistance:
(F47) No. 2 — (E9) No. 1/0 Ω
(F47) No. 1 — (E9) No. 3/0 Ω

3) Measure resistance between air control valve connector and body.

Connector & Terminal/Specified resistance:
(E9) No. 1 — Body/1 MΩ, min.
(E9) No. 3 — Body/1 MΩ, min.

• SELECT MONITOR FUNCTION MODE

Mode: 12
Condition: Engine at idle
Specified Data: ISC F12
30 - 40%

DIAGNOSIS CHART

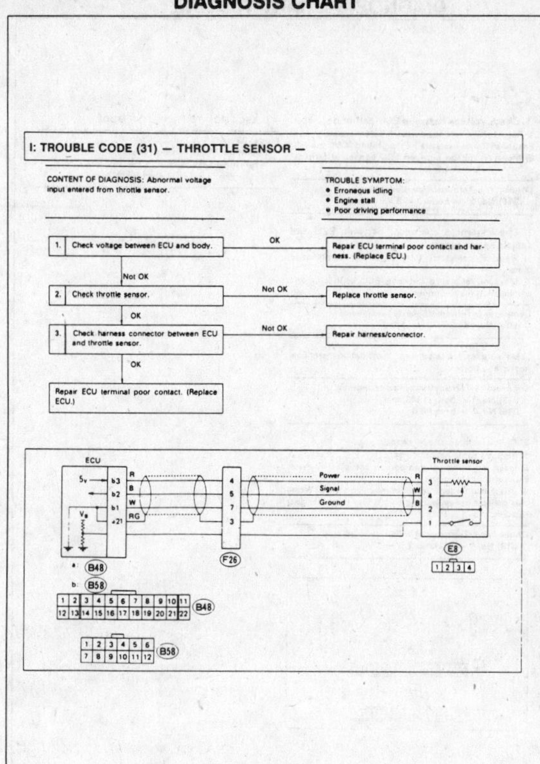
I: TROUBLE CODE (31) — THROTTLE SENSOR —
CONTENT OF DIAGNOSIS: Abnormal voltage input entered from throttle sensor.
TROUBLE SYMPTOM:
• Erroneous idling
• Engine stall
• Poor driving performance
1. Check voltage between ECU and body.
OK
Repair ECU terminal poor contact and harness. (Replace ECU.)
Not OK
2. Check throttle sensor.
Not OK
Replace throttle sensor.
OK
3. Check harness connector between ECU and throttle sensor.
Not OK
Repair harness/connector.
OK
Repair ECU terminal poor contact. (Replace ECU.)
ECU
Throttle sensor
Power
Signal
Ground
B48
B58
F26
E8

DIAGNOSIS CHART TESTING

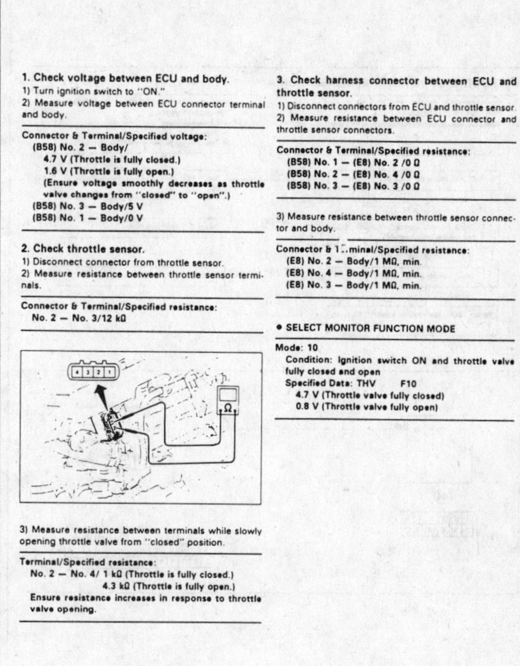

1. Check voltage between ECU and body.

1) Turn ignition switch to "ON."
2) Measure voltage between ECU connector terminal and body.

Connector & Terminal/Specified voltage:
(B58) No. 2 — Body/
4.7 V (Throttle is fully closed.)
1.6 V (Throttle is fully open.)
(Ensure voltage smoothly decreases as throttle valve changes from "closed" to "open".)
(B58) No. 3 — Body/5 V
(B58) No. 1 — Body/0 V

2. Check throttle sensor.

1) Disconnect connector from throttle sensor.
2) Measure resistance between throttle sensor terminals.

Connector & Terminal/Specified resistance:
No. 2 — No. 3/12 kΩ

3) Measure resistance between terminals while slowly opening throttle valve from "closed" position.

Terminal/Specified resistance:
No. 2 — No. 4/ 1 kΩ (Throttle is fully closed.)
4.3 kΩ (Throttle is fully open.)
Ensure resistance increases in response to throttle valve opening.

3. Check harness connector between ECU and throttle sensor.

1) Disconnect connectors from ECU and throttle sensor.
2) Measure resistance between ECU connector and throttle sensor connectors.

Connector & Terminal/Specified resistance:
(B58) No. 1 — (E8) No. 2 /0 Ω
(B58) No. 2 — (E8) No. 4 /0 Ω
(B58) No. 3 — (E8) No. 3 /0 Ω

3) Measure resistance between throttle sensor connector and body.

Connector & Terminal/Specified resistance:
(E8) No. 2 — Body/1 MΩ, min.
(E8) No. 4 — Body/1 MΩ, min.
(E8) No. 3 — Body/1 MΩ, min.

• SELECT MONITOR FUNCTION MODE

Mode: 10
Condition: Ignition switch ON and throttle valve fully closed and open
Specified Data: THV F10
4.7 V (Throttle valve fully closed)
0.8 V (Throttle valve fully open)

DIAGNOSIS CHART

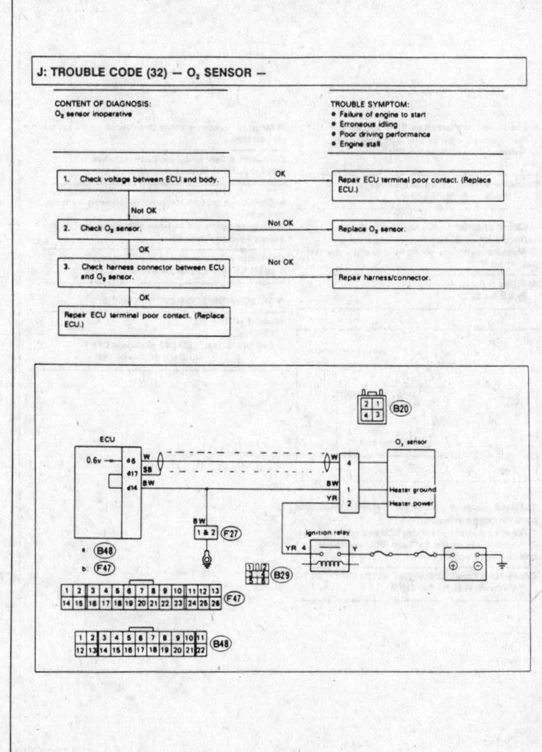
J: TROUBLE CODE (32) — O₂ SENSOR —
CONTENT OF DIAGNOSIS:
O₂ sensor inoperative
TROUBLE SYMPTOM:
• Failure of engine to start
• Erroneous idling
• Poor driving performance
• Engine stall
1. Check voltage between ECU and body.
OK
Repair ECU terminal poor contact. (Replace ECU.)
Not OK
2. Check O₂ sensor.
Not OK
Replace O₂ sensor.
OK
3. Check harness connector between ECU and O₂ sensor.
Not OK
Repair harness/connector.
OK
Repair ECU terminal poor contact. (Replace ECU.)
B20
ECU
O₂ sensor
Heater ground
Heater power
Ignition relay
F27
B29
B48
F47

DIAGNOSIS CHART TESTING

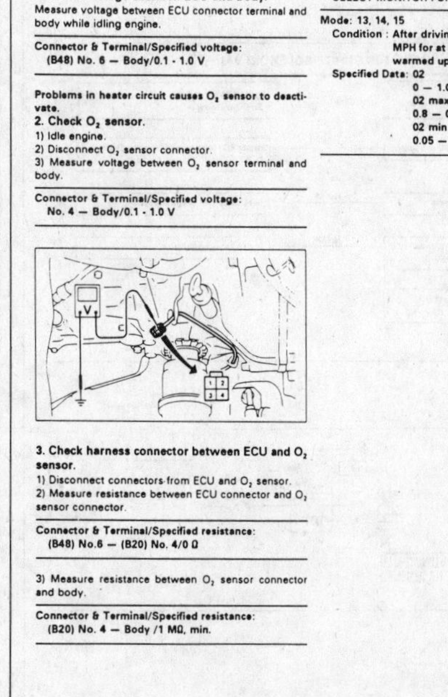

1. Check voltage between ECU and body.

Measure voltage between ECU connector terminal and body while idling engine.

Connector & Terminal/Specified voltage:
(B48) No. 6 — Body/0.1 - 1.0 V

Problems in heater circuit causes O_2 sensor to deactivate.

2. Check O_2 sensor.

1) Idle engine.
2) Disconnect O_2 sensor connector.
3) Measure voltage between O_2 sensor terminal and body.

Connector & Terminal/Specified voltage:
No. 4 — Body/0.1 - 1.0 V

3. Check harness connector between ECU and O_2 sensor.

1) Disconnect connectors from ECU and O_2 sensor.
2) Measure resistance between ECU connector and O_2 sensor connector.

Connector & Terminal/Specified resistance:
(B48) No.6 — (B20) No. 4/0 Ω

3) Measure resistance between O_2 sensor connector and body.

Connector & Terminal/Specified resistance:
(B20) No. 4 — Body /1 MΩ, min.

• SELECT MONITOR FUNCTION MODE

Mode: 13, 14, 15
Condition : After driving at more than 7 MPH for at least one minute with engine warmed up.
Specified Data: 02 F13
0 — 1.0 V
02 max F14
0.8 — 0.9 V
02 min F15
0.05 — 0.12V

DIAGNOSIS CHART

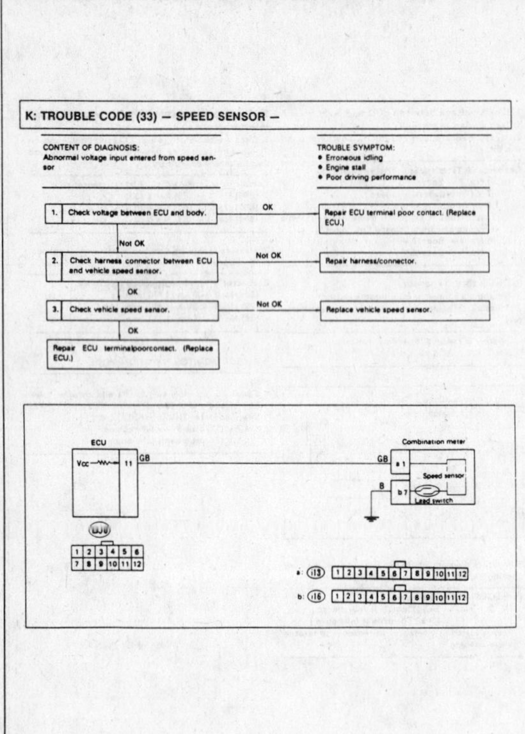
K: TROUBLE CODE (33) – SPEED SENSOR –
CONTENT OF DIAGNOSIS:
Abnormal voltage input entered from speed sensor
TROUBLE SYMPTOM:
• Erroneous idling
• Engine stall
• Poor driving performance
1. Check voltage between ECU and body.
OK
Repair ECU terminal poor contact. (Replace ECU.)
Not OK
2. Check harness connector between ECU and vehicle speed sensor.
Not OK
Repair harness/connector.
OK
3. Check vehicle speed sensor.
Not OK
Replace vehicle speed sensor.
OK
Repair ECU terminalpoorcontact. (Replace ECU.)
ECU
Vcc
11
GB
Combination meter
GB
a 1
Speed sensor
B
b 7
Lead switch
B58
1 2 3 4 5 6
7 8 9 10 11 12
a i13
b i16

DIAGNOSIS CHART TESTING

1. Check voltage between ECU and body.
1) Raise vehicle and support with safety stands. Ensure all four wheels are off the ground (4WD model).
2) Measure voltage between ECU connector terminal and body while slowly driving wheels.

Connector & Terminal/Specified voltage:
(B58) No. 11 — Body/0 ↔ 5 V

2. Check harness connector between ECU and vehicle speed sensor.
1) Remove connector from ECU and combination meter.
2) Measure resistance between ECU connector and combination meter connector.

Connector & Terminal/Specified resistance:
(B58) No. 11 — (i13) No. 1/0 Ω

3) Measure resistance between combination meter connector and body.

Connector & Terminal/Specified resistance:
(i13) No. 1 — Body/1 MΩ, min.
(i16) No. 7 — body/ 0 Ω

3. Check vehicle speed sensor.
1) Remove combination meter.
2) Insert a screwdriver into portion occupied by meter cable and rotate rotor.
3) Check that voltage across combination meter terminals deflects between 0 and 5 volts four times per gear rotation.

Connector & Terminal/Specified voltage:
(i13) No. 1 — (i16) No. 7/0 ↔ 5 V

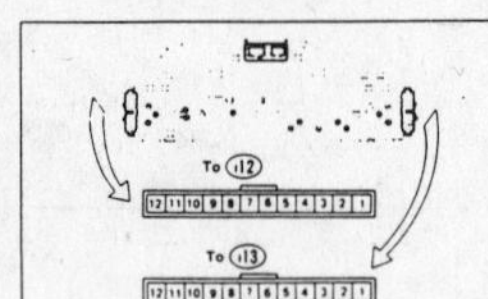
To i12
To i13

• SELECT MONITOR FUNCTION MODE

Mode: 03
Condition: While driving vehicle:
Specified data: VSP F02
(Car speed) km/h

DIAGNOSIS CHART

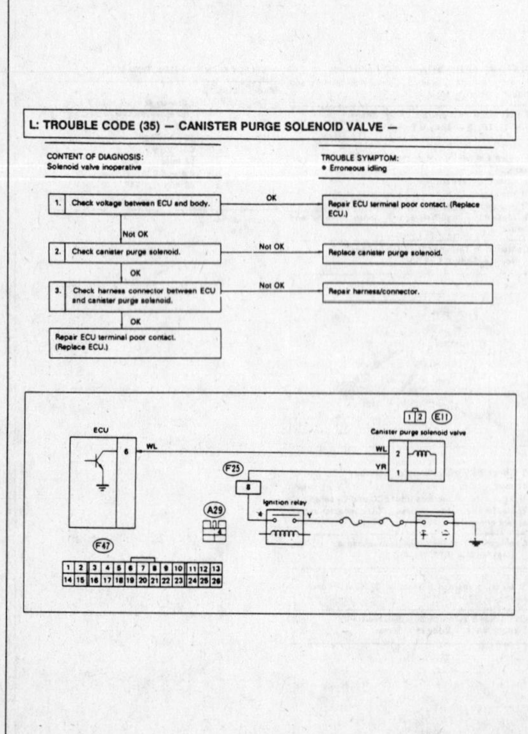
L: TROUBLE CODE (35) – CANISTER PURGE SOLENOID VALVE –
CONTENT OF DIAGNOSIS:
Solenoid valve inoperative
TROUBLE SYMPTOM:
• Erroneous idling
1. Check voltage between ECU and body.
OK
Repair ECU terminal poor contact. (Replace ECU.)
Not OK
2. Check canister purge solenoid.
Not OK
Replace canister purge solenoid.
OK
3. Check harness connector between ECU and canister purge solenoid.
Not OK
Repair harness/connector.
OK
Repair ECU terminal poor contact. (Replace ECU.)
ECU
6
WL
E11
Canister purge solenoid valve
WL
YR
F25
Ignition relay
A29
F47

DIAGNOSIS CHART TESTING

1. Check voltage between ECU and body.
1) Turn ignition switch to "ON" with engine OFF.
2) Measure voltage between ECU connecter terminal and body.

Connector & Terminal/Specified voltage:
F47 No. 6 — Body/10 - 13 V

2. Check canister purge solenoid valve.
1) Disconnect connector from solenoid valve.
2) Measure resistance between solenoid valve terminals.

Specified resistance:
35.5 Ω (at 20°C)

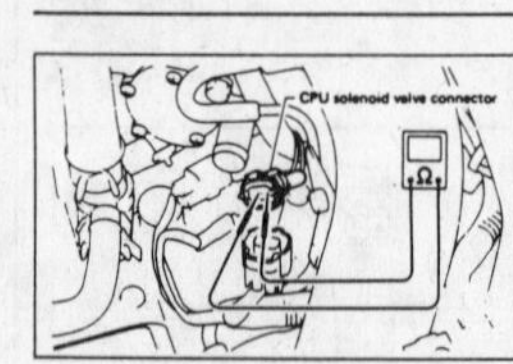
CPU solenoid valve connector

3. Check harness connector between ECU and canister purge solenoid.
1) Disconnect connectors from ECU and solenoid valve.
2) Measure resistance between ECU connector and solenoid valve bconnector.

Connector & Terminal/Specified resistance:
(F47) No. 6 — (E11) No. 2/0 Ω

3) Measure resistance between solenoid valve connector and body.

Connector & Terminal/Specified resistance:
(E11) No. 2 — Body/1 MΩ, min.

4) Disconnect ground and positive terminals from battery in that order.
5) Measure resistance between solenoid connector and battery's positive terminal.

Connector & Terminal/Specified resistance:
(E11) No. 1 — (+) terminal/0 Ω

• SELECT MONITOR FUNCTION MODE

Mode: A1
LED No.: 7
ON/OFF Signal: LED OFF (Solenoid OFF)
LED ON (Solenoid ON)

DIAGNOSIS CHART

M: TROUBLE CODE (41) — AIR-FUEL RATIO CONTROL SYSTEM —

CONTENT OF DIAGNOSIS:
Faulty learning control system

TROUBLE SYMPTOM:
- Erroneous idling
- Engine stall

	Not OK
1. Check operation of injectors.	Check harness. Replace injectors.
OK ↓	
2. Check air flow sensor.	Check harness. Replace air flow sensor.
OK ↓	
3. Check water temperature sensor.	Replace water temperature sensor.
OK ↓	
4. Check throttle sensor.	Replace throttle sensor.
OK ↓	
5. Check O_2 sensor.	Replace O_2 sensor.
OK ↓	
6. Check fuel pressure.	Replace pressure regulator and/or fuel pump.
OK ↓	
7. Check injectors.	Replace injector.
OK ↓	
Repair ECU terminal poor contact. (Replace ECU.)	

DIAGNOSIS CHART

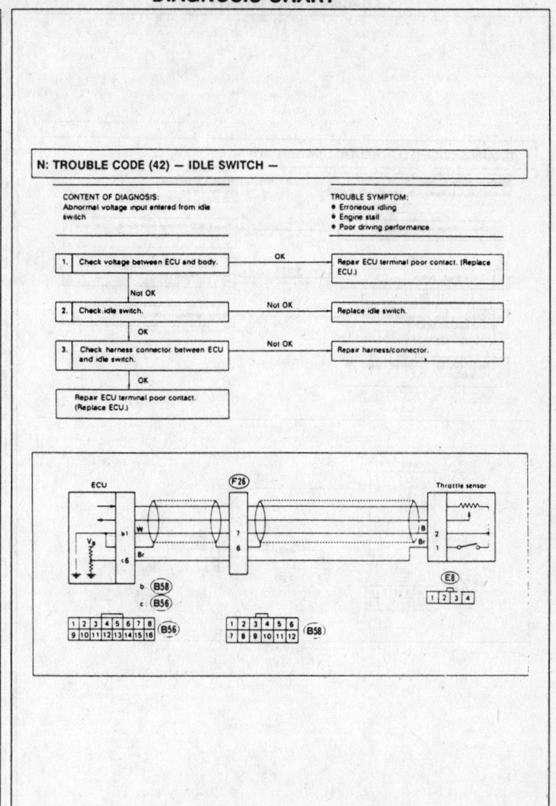
N: TROUBLE CODE (42) — IDLE SWITCH —
CONTENT OF DIAGNOSIS:
Abnormal voltage input entered from idle switch
TROUBLE SYMPTOM:
• Erroneous idling
• Engine stall
• Poor driving performance
1. Check voltage between ECU and body.
OK
Repair ECU terminal poor contact. (Replace ECU.)
Not OK
2. Check idle switch.
Not OK
Replace idle switch.
OK
3. Check harness connector between ECU and idle switch.
Not OK
Repair harness/connector.
OK
Repair ECU terminal poor contact. (Replace ECU.)
ECU
F26
Throttle sensor
b1
W
c6
Br
7
6
B
Br
2
1
b B58
c B56
E8
1 2 3 4
1 2 3 4 5 6 7 8
9 10 11 12 13 14 15 16
B56
1 2 3 4 5 6
7 8 9 10 11 12
B58

DIAGNOSIS CHART TESTING

1. Check voltage between ECU and body.
1) Turn ignition switch to "ON."
2) Measure voltage between ECU connector terminal and body.

Connector & Terminal/Specified voltage:
(B56) No. 6 — Body/ 0 V (Throttle is fully closed.)
Approx.4.6 V (Throttle is open.)

2. Check idle switch.
1) Disconnect connector from throttle sensor.
2) Check continuity between throttle sensor idle switch terminals.

Terminal/Specified resistance:
No. 1 — No. 2 /0 Ω (Throttle is fully closed.)
1 MΩ, min. (Throttle is fully open.)

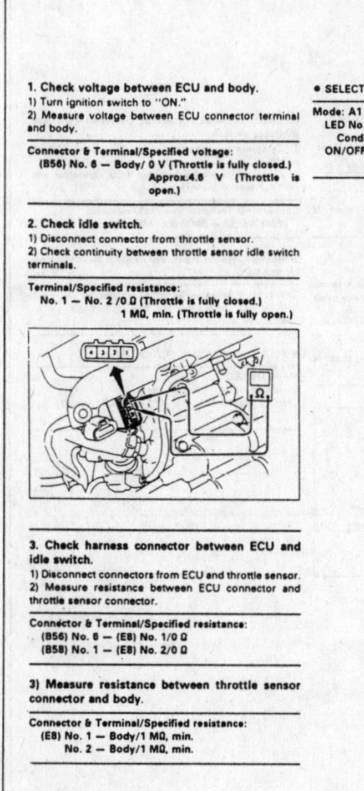
4 3 2 1
Ω

3. Check harness connector between ECU and idle switch.
1) Disconnect connectors from ECU and throttle sensor.
2) Measure resistance between ECU connector and throttle sensor connector.

Connector & Terminal/Specified resistance:
(B56) No. 6 — (E8) No. 1/0 Ω
(B58) No. 1 — (E8) No. 2/0 Ω

3) Measure resistance between throttle sensor connector and body.

Connector & Terminal/Specified resistance:
(E8) No. 1 — Body/1 MΩ, min.
No. 2 — Body/1 MΩ, min.

• SELECT MONITOR FUNCTION MODE

Mode: A1
LED No.: 1
Condition: Ignition switch ON
ON/OFF Signal: LED OFF (Idle switch OFF)
LED ON (Idle switch ON)

DIAGNOSIS CHART

O: TROUBLE CODE (45) — ATMOSPHERIC PRESSURE SENSOR —

CONTENT OF DIAGNOSIS:
Faulty atmospheric pressure sensor inside ECU

TROUBLE SYMPTOM:
- Erroneous idling
- Failure of engine to start

When trouble code 45 appears on display, replace ECU.

• SELECT MONITOR FUNCTION MODE

Mode: 19
Specified data: BAROP F19
(*Atomospheric pressure value) mmHg

*;760 mmHg-10 x [Alutitude (1/100 m)] ± 200 mmHg

P; TROUBLE CODE (49) — AIR FLOW SENSOR —

CONTENT OF DIAGNOSIS:
Use of improper air flow sensor

TROUBLE SYMPTOM:
- Erroneous idling
- Failure of engine to start

When trouble code 49 appears on display, check the specifications of air flow sensor and ECU. Replace air flow sensor (or ECU) with one of a proper type.
- AT model : Hot film type air flow sensor (JECS)
- MT model : Hot wire type air flow sensor (HITACHI)

DIAGNOSIS CHART

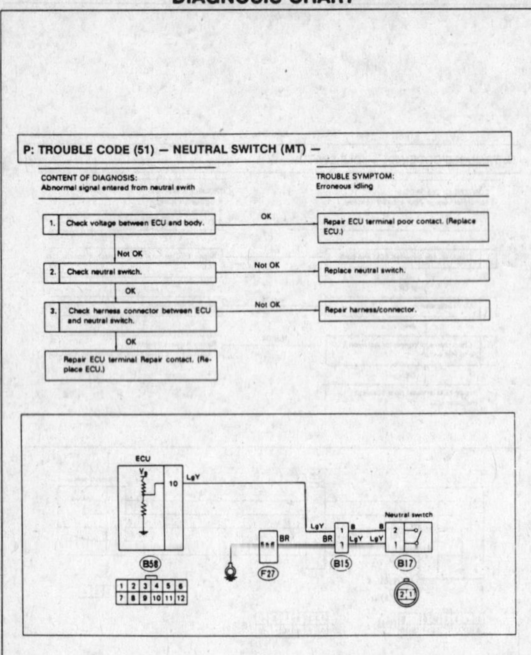
P: TROUBLE CODE (51) — NEUTRAL SWITCH (MT) —
CONTENT OF DIAGNOSIS:
Abnormal signal entered from neutral swith
TROUBLE SYMPTOM:
Erroneous idling
1. Check voltage between ECU and body.
OK
Repair ECU terminal poor contact. (Replace ECU.)
Not OK
2. Check neutral switch.
Not OK
Replace neutral switch.
OK
3. Check harness connector between ECU and neutral switch.
Not OK
Repair harness/connector.
OK
Repair ECU terminal Repair contact. (Replace ECU.)
ECU
10
LgY
BR
Neutral switch
B58
F27
B15
B17

DIAGNOSIS CHART TESTING

1. Check voltage between ECU and body.
1) Turn ignition switch to "ON."
2) Measure voltage between ECU connector terminal and body.

Connector & Terminal/Specified voltage:
(B58) No. 10 — Body/Approx. 8 V,min. (Neutral position)
0 V (Other than neutral position)

2. Check neutral switch.
1) Disconnect transmission connectors.
2) Measure resistance between neutral switch terminals while shifting shift lever from Neutral to any other position.

Connector & Terminal / Spcified resistance:
(E15) No. 1 — No. 3 / 1MΩ,min.(Neutral position)
0 Ω (Other than neutral position)

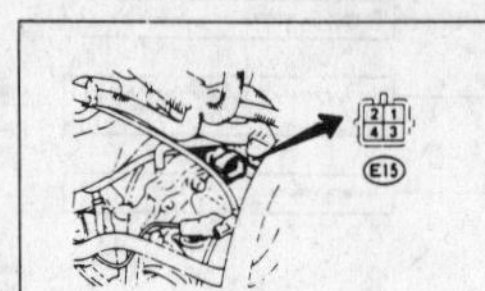
E15

3. Check harness connector between ECU and neutral switch.
1) Disconnect connectors from ECU and neutral switch.
2) Measure resistance between ECU connector and neutral switch connector.

Connector & Terminal/Specified resistance:
(B58) No. 10 — (E17) No. 1/0 Ω

3) Measure resistance between neutral switch connector and body.

Connector & Terminal/Specified resistance:
(E17) No. 1 — Body/1 MΩ, min.
(E17) No. 2 — Body/0 Ω

- **SELECT MONITOR FUNCTION MODE**

Mode: A0 LED No.: 7
Condition: Ignition switch ON
ON/OFF Signal:
LED OFF (Other than neutral position)
LED ON (Neutral position)

DIAGNOSIS CHART

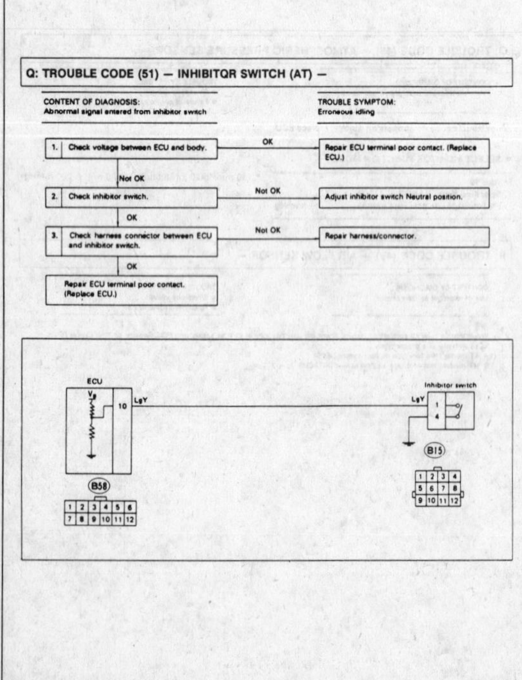
Q: TROUBLE CODE (51) — INHIBITQR SWITCH (AT) —
CONTENT OF DIAGNOSIS:
Abnormal signal entered from inhibitor switch
TROUBLE SYMPTOM:
Erroneous idling
1. Check voltage between ECU and body.
OK
Repair ECU terminal poor contact. (Replace ECU.)
Not OK
2. Check inhibitor switch.
Not OK
Adjust inhibitor switch Neutral position.
OK
3. Check harness connector between ECU and inhibitor switch.
Not OK
Repair harness/connector.
OK
Repair ECU terminal poor contact. (Replace ECU.)
ECU
10
LgY
Inhibitor switch
B58
B15

DIAGNOSIS CHART TESTING

1. Check voltage between ECU and body.
1) Turn ignition switch to "ON."
2) Measure voltage between ECU connector terminal and body.

Connector & Terminal/Specified voltage:
(B58) No. 10 — Body/0 V (N Range)
8 V, min. (Other than N Range)

2. Check inhibitor switch.
1) Disconnect transmission connectors.
2) Measure resistance between inhibitor switch terminals while shifting select lever from Neutral to any other position.

Connector & Terminal/Specified resistance:
(E15) No. 1 — No. 4/ 0 Ω (N Range)
1 MΩ, min. (Other than N Range)

E15

3. Check harness connector between ECU and inhibitor switch.
1) Disconnect connectors from ECU and inhibitor switch.
2) Measure resistance between ECU connector and inhibitor switch connector.

Connector & Terminal/Specified resistance:
(B58) No. 10 — (B15) No. 1/0 Ω

3) Measure resistance between inhibitor switch connector and body.

Connector & Terminal/Specified resistance:
(B15) No. 1 — Body/1 MΩ, min.
(B15) No. 4 — Body/0 Ω

- **SELECT MONITOR FUNCTION MODE**

Mode: A0 LED No.: 7
Condition: Ignition switch ON
ON/OFF Signal: LED OFF (Other than N Range)
LED ON (N Range)

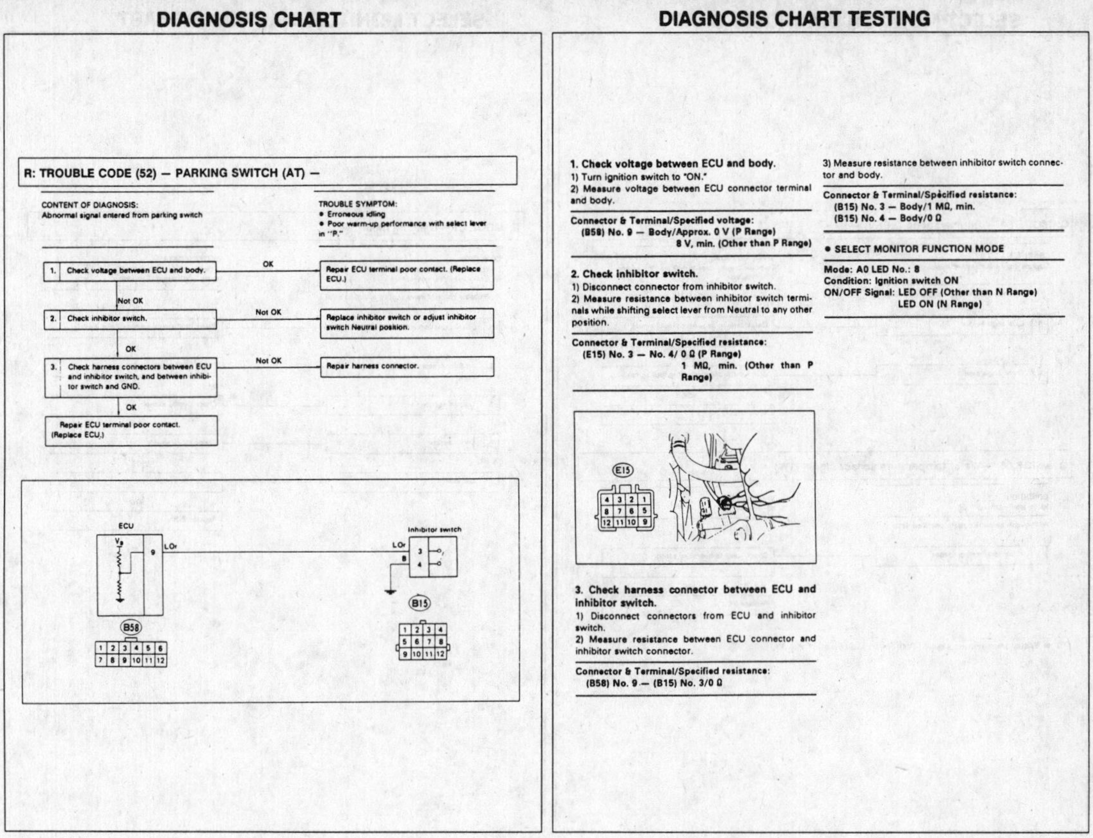

DIAGNOSIS CHART

R: TROUBLE CODE (52) — PARKING SWITCH (AT) —

CONTENT OF DIAGNOSIS:
Abnormal signal entered from parking switch

TROUBLE SYMPTOM:
- Erroneous idling
- Poor warm-up performance with select lever in "P."

1. Check voltage between ECU and body. — OK → Repair ECU terminal poor contact. (Replace ECU.)
   Not OK ↓
2. Check inhibitor switch. — Not OK → Replace inhibitor switch or adjust inhibitor switch Neutral position.
   OK ↓
3. Check harness connectors between ECU and inhibitor switch, and between inhibitor switch and GND. — Not OK → Repair harness connector.
   OK ↓
Repair ECU terminal poor contact. (Replace ECU.)

ECU
V
9
LOr
Inhibitor switch
LOr
3
4
B
B58
1 2 3 4 5 6
7 8 9 10 11 12
B15
1 2 3 4
5 6 7 8
9 10 11 12

DIAGNOSIS CHART TESTING

1. Check voltage between ECU and body.
1) Turn ignition switch to "ON."
2) Measure voltage between ECU connector terminal and body.

Connector & Terminal/Specified voltage:
(B58) No. 9 — Body/Approx. 0 V (P Range)
8 V, min. (Other than P Range)

2. Check inhibitor switch.
1) Disconnect connector from inhibitor switch.
2) Measure resistance between inhibitor switch terminals while shifting select lever from Neutral to any other position.

Connector & Terminal/Specified resistance:
(E15) No. 3 — No. 4/ 0 Ω (P Range)
1 MΩ, min. (Other than P Range)

E15
4 3 2 1
8 7 6 5
12 11 10 9

3. Check harness connector between ECU and inhibitor switch.
1) Disconnect connectors from ECU and inhibitor switch.
2) Measure resistance between ECU connector and inhibitor switch connector.

Connector & Terminal/Specified resistance:
(B58) No. 9 — (B15) No. 3/0 Ω

3) Measure resistance between inhibitor switch connector and body.

Connector & Terminal/Specified resistance:
(B15) No. 3 — Body/1 MΩ, min.
(B15) No. 4 — Body/0 Ω

• SELECT MONITOR FUNCTION MODE

Mode: A0 LED No.: 8
Condition: Ignition switch ON
ON/OFF Signal: LED OFF (Other than N Range)
LED ON (N Range)

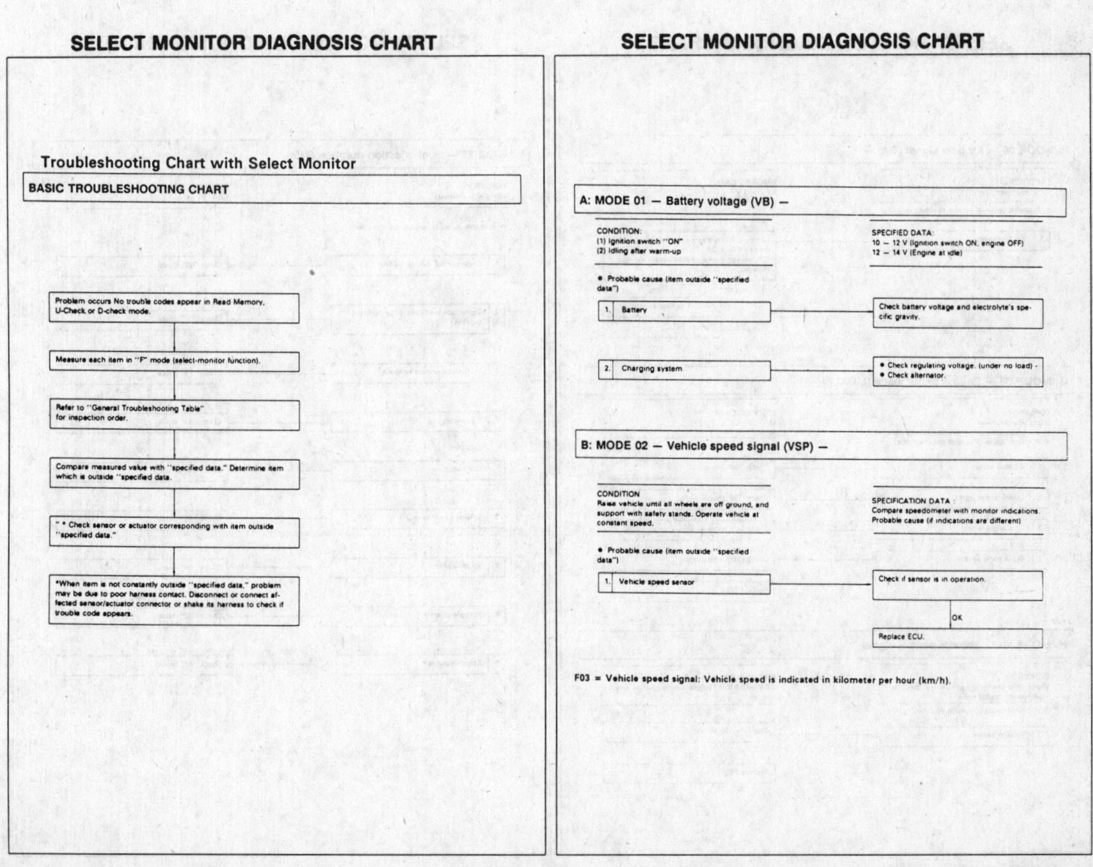

SELECT MONITOR DIAGNOSIS CHART

Troubleshooting Chart with Select Monitor

BASIC TROUBLESHOOTING CHART

Problem occurs No trouble codes appear in Read Memory, U-Check or D-check mode.
↓
Measure each item in "F" mode (select-monitor function).
↓
Refer to "General Troubleshooting Table" for inspection order.
↓
Compare measured value with "specified data." Determine item which is outside "specified data.
↓
" * Check sensor or actuator corresponding with item outside "specified data."
↓
*When item is not constantly outside "specified data," problem may be due to poor harness contact. Disconnect or connect affected sensor/actuator connector or shake its harness to check if trouble code appears.

SELECT MONITOR DIAGNOSIS CHART

A: MODE 01 — Battery voltage (VB) —

CONDITION:
(1) Ignition switch "ON"
(2) Idling after warm-up

SPECIFIED DATA:
10 — 12 V (Ignition switch ON, engine OFF)
12 — 14 V (Engine at idle)

- Probable cause (item outside "specified data")

1. Battery → Check battery voltage and electrolyte's specific gravity.
2. Charging system → • Check regulating voltage. (under no load) • Check alternator.

B: MODE 02 — Vehicle speed signal (VSP) —

CONDITION
Raise vehicle until all wheels are off ground, and support with safety stands. Operate vehicle at constant speed.

SPECIFICATION DATA :
Compare speedometer with monitor indications.
Probable cause (if indications are different)

- Probable cause (item outside "specified data")

1. Vehicle speed sensor → Check if sensor is in operation.
   OK ↓
   Replace ECU.

F03 = Vehicle speed signal: Vehicle speed is indicated in kilometer per hour (km/h).

SELECT MONITOR DIAGNOSIS CHART

C: MODE 04 — Engine speed (EREV) —

CONDITION:
Operate engine at constant speed.

SPECIFIED DATA:
Compare engine speeds indicated on engine tester monitor.

- Probable cause (if outside specified data)

1. Cam angle sensor → Check cam angle sensor output signal.
   OK
2. Crank angle sensor → Check crank angle sensor output signal.
   OK
   Replace ECU.

D: MODE 06 — Water temperature sensor signal (TW) —

CONDITION:
Idling after warm-up

SPECIFIED DATA:
80 — 90 deg C

- Probable cause (if outside specified data)

1. Water temperature sensor → Check water temperature sensor.
   OK
   Replace ECU.

F05 = Water temperature signal (TW): To be indicated in "deg F".

SELECT MONITOR DIAGNOSIS CHART

E: MODE 07 — Ignition timing —

CONDITION:
(1) While idling after warm-up
(2) Gears in neutral position

SPECIFIED DATA:
12 deg — 28 deg (AT model)
18 deg — 22 deg (MT model)

- Probable cause (if items outside specified data)

1. L Data (specified amount of fuel injection) → Check "09" mode (engine under loads).
   OK
2. Air flow sensor → Check "08" mode (air flow signal)
   OK
3. Throttle sensor → Check throttle sensor.
   OK
4. Knock sensor → Check knock sensor.
   OK
5. Idle switch → Check idle switch.
   OK
   Replace ECU.

SELECT MONITOR DIAGNOSIS CHART

F: MODE 08 — Air flow signal (QA) —

CONDITION:
Idling after warm-up

SPECIFIED DATA:
0.8 — 1.2 V

- Probable cause (if outside specified data)

1. Air flow sensor → Compare air flow sensor signal voltage with specified data QA. (QA voltage equals air flow sensor signal voltage.)
   OK → Replace air flow sensor.
   Not OK → Replace ECU.

G: MODE 09 — Engine under loads (L DATA) —

CONDITION:
Idling after warm-up

SPECIFIED DATA:
15 — 20 (AT)
40 — 50 (MT)

- Probable cause (if outside specified data)

1. Air flow sensor → Check "08" mode.
2. Engine speed (rpm) → Check "04" mode.
3. Cam angle sensor → Check cam angle sensor.
4. Crank angle sensor → Check crank angle sensor.
   Replace ECU.

H: MODE 10 — Throttle sensor signal

CONDITION:
Check while changing from "fully-closed" to "fully-open" throttle valve.

SPECIFIED DATA:
4.7 V → 1.6 V V *Engine throttle change must be smooth.

- Probable cause (if outside specified data)

1. Throttle sensor → Check throttle sensor.
   OK
   Replace ECU.

SELECT MONITOR DIAGNOSIS CHART

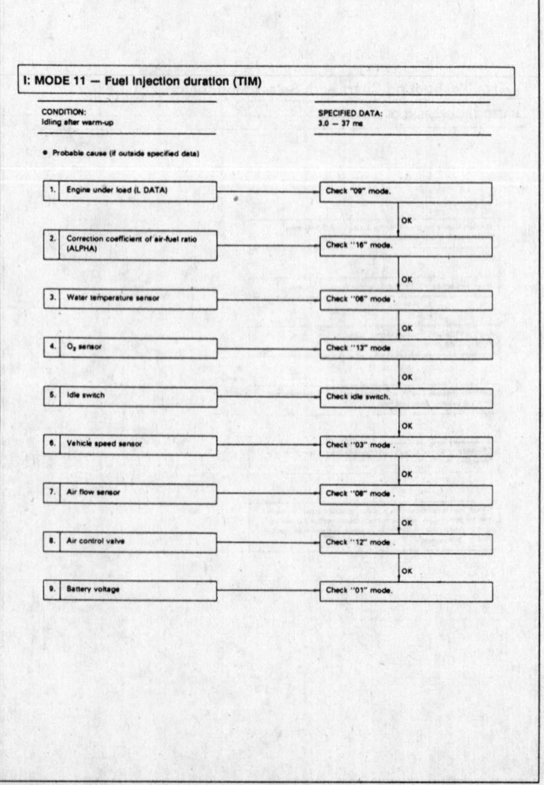
I: MODE 11 — Fuel injection duration (TIM)
CONDITION:
Idling after warm-up
SPECIFIED DATA:
3.0 — 37 ms
• Probable cause (if outside specified data)
1. Engine under load (L DATA)
Check "09" mode.
OK
2. Correction coefficient of air-fuel ratio (ALPHA)
Check "16" mode.
OK
3. Water temperature sensor
Check "08" mode .
OK
4. O2 sensor
Check "13" mode
OK
5. Idle switch
Check idle switch.
OK
6. Vehicle speed sensor
Check "03" mode .
OK
7. Air flow sensor
Check "08" mode .
OK
8. Air control valve
Check "12" mode .
OK
9. Battery voltage
Check "01" mode.

SELECT MONITOR DIAGNOSIS CHART

SELECT MONITOR DIAGNOSIS CHART

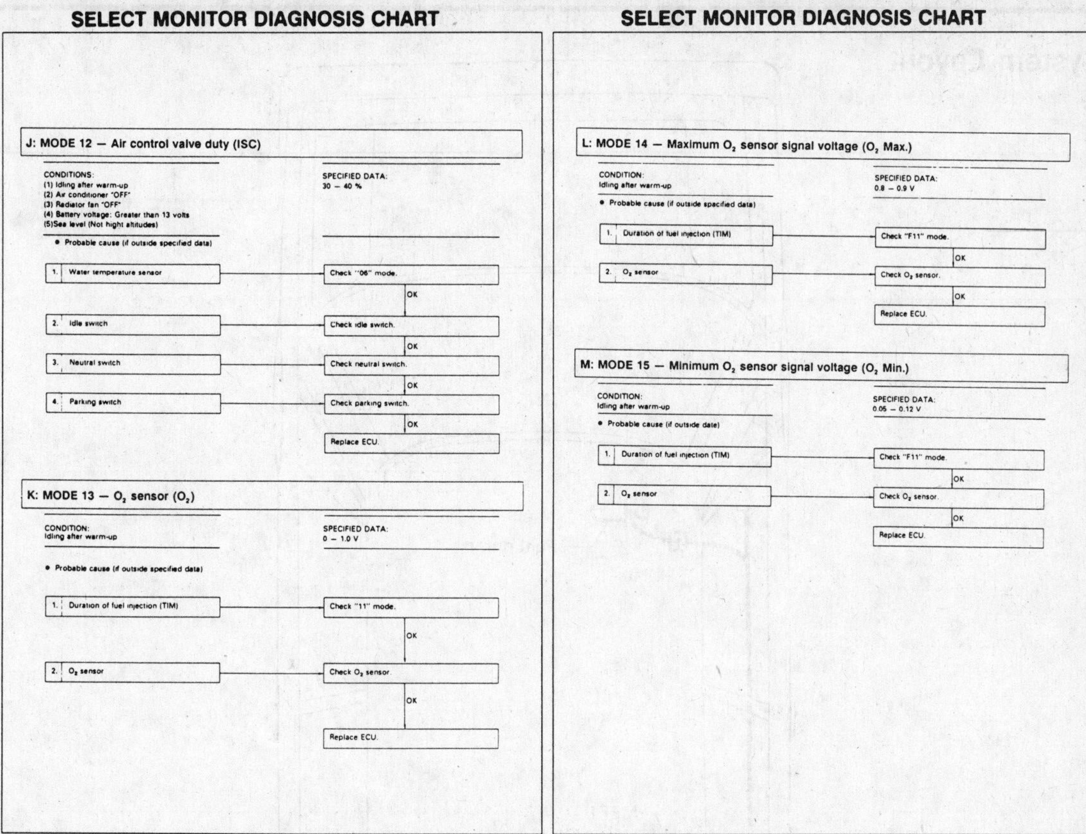
J: MODE 12 — Air control valve duty (ISC)
CONDITIONS:
(1) Idling after warm-up
(2) Air conditioner "OFF"
(3) Radiator fan "OFF"
(4) Battery voltage: Greater than 13 volts
(5)Sea level (Not hight altitudes)
SPECIFIED DATA:
30 – 40 %
• Probable cause (if outside specified data)
1. Water temperature sensor
Check "06" mode.
OK
2. Idle switch
Check idle switch.
OK
3. Neutral switch
Check neutral switch.
OK
4. Parking switch
Check parking switch.
OK
Replace ECU.
K: MODE 13 — O2 sensor (O2)
CONDITION:
Idling after warm-up
SPECIFIED DATA:
0 – 1.0 V
• Probable cause (if outside specified data)
1. Duration of fuel injection (TIM)
Check "11" mode.
OK
2. O2 sensor
Check O2 sensor.
OK
Replace ECU.
L: MODE 14 — Maximum O2 sensor signal voltage (O2 Max.)
CONDITION:
Idling after warm-up
SPECIFIED DATA:
0.8 – 0.9 V
• Probable cause (if outside specified data)
1. Duration of fuel injection (TIM)
Check "F11" mode.
OK
2. O2 sensor
Check O2 sensor.
OK
Replace ECU.
M: MODE 15 — Minimum O2 sensor signal voltage (O2 Min.)
CONDITION:
Idling after warm-up
SPECIFIED DATA:
0.05 – 0.12 V
• Probable cause (if outside date)
1. Duration of fuel injection (TIM)
Check "F11" mode.
OK
2. O2 sensor
Check O2 sensor.
OK
Replace ECU.

SELECT MONITOR DIAGNOSIS CHART

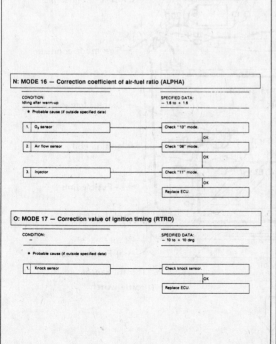

N: MODE 16 — Correction coefficient of air-fuel ratio (ALPHA)

CONDITION:
Idling after warm-up

SPECIFIED DATA:
– 1.6 to + 1.6

• Probable cause (if outside specified data)

1. O_2 sensor → Check "13" mode.
   OK
2. Air flow sensor → Check "08" mode.
   OK
3. Injector → Check "11" mode.
   OK
   Replace ECU.

O: MODE 17 — Correction value of ignition timing (RTRD)

CONDITION:
–

SPECIFIED DATA:
– 10 to + 10 deg

• Probable cause (if outside specified data)

1. Knock sensor → Check knock sensor.
   OK
   Replace ECU.

SELECT MONITOR DIAGNOSIS CHART

P: MODE 19 — Atmospheric pressure (BARO. P)

CONDITION:
Ground surface (not high altitudes)

SPECIFIED DATA:
*760 mmHg

• Probable cause (if outside specified data)

*"–9 to 10 mmHg" changes at an altitude of 100 meters.

1. Atmospheric sensor → Replace ECU.

MPFI System Layout

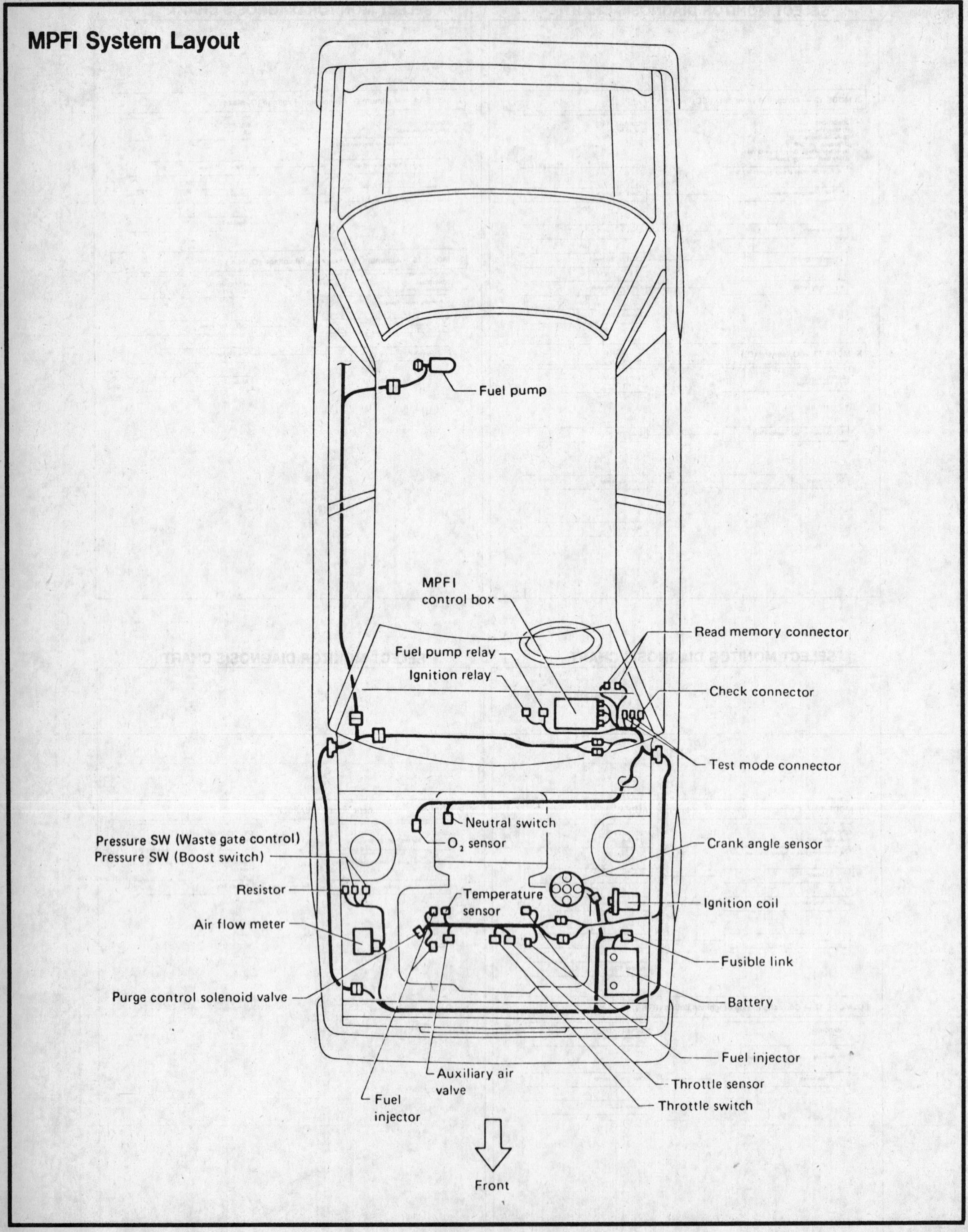
Fuel pump
MPFI
control box
Read memory connector
Fuel pump relay
Ignition relay
Check connector
Test mode connector
Neutral switch
O_2 sensor
Pressure SW (Waste gate control)
Pressure SW (Boost switch)
Crank angle sensor
Resistor
Temperature
sensor
Ignition coil
Air flow meter
Fusible link
Battery
Purge control solenoid valve
Fuel injector
Throttle sensor
Auxiliary air
valve
Throttle switch
Fuel
injector
Front

SUZUKI ELECTRONIC FUEL INJECTION (EFI) SYSTEMS

General Information

The Electronic Fuel Injection (EFI) system supplies the vehicle's combustion chambers with air/fuel mixture of optimized ratio under varying driving conditions.

THROTTLE BODY INJECTION (TBI) SYSTEM

The TBI system consists of a single injector which injects fuel into a throttle body bore. This system consists of 2 major sub-systems: An air/fuel delivery system and the electronic control system.

The main components of the air/fuel delivery system consists of the fuel tank, fuel pump, fuel filter, throttle body assembly, fuel feed and return lines, air cleaner and the Idle Speed Control (ISC) solenoid valve.

The electronic control system consists of the Electronic Control Module (ECM), which controls various devices according to signals received from sensors. Functionally, the air/fuel control system is divided into 5 sub-systems. These sub-systems are as follows:

- Fuel injection control system
- ISC solenoid valve control system
- Fuel pump control system
- EGR control system (California model only)
- Shift-up indicator light control system (if equipped)

Also, vehicles equipped with automatic transmission, the ECM sends a throttle valve opening signal to an automatic transmission control module to control the transmission.

MULTI-POINT FUEL INJECTION (MPFI) SYSTEM

The MPFI system consists of 1 injector per cylinder, which injects fuel into each intake port of the cylinder head. This system consists of 3 major sub-system: An air intake system, fuel delivery system and an electronic control system.

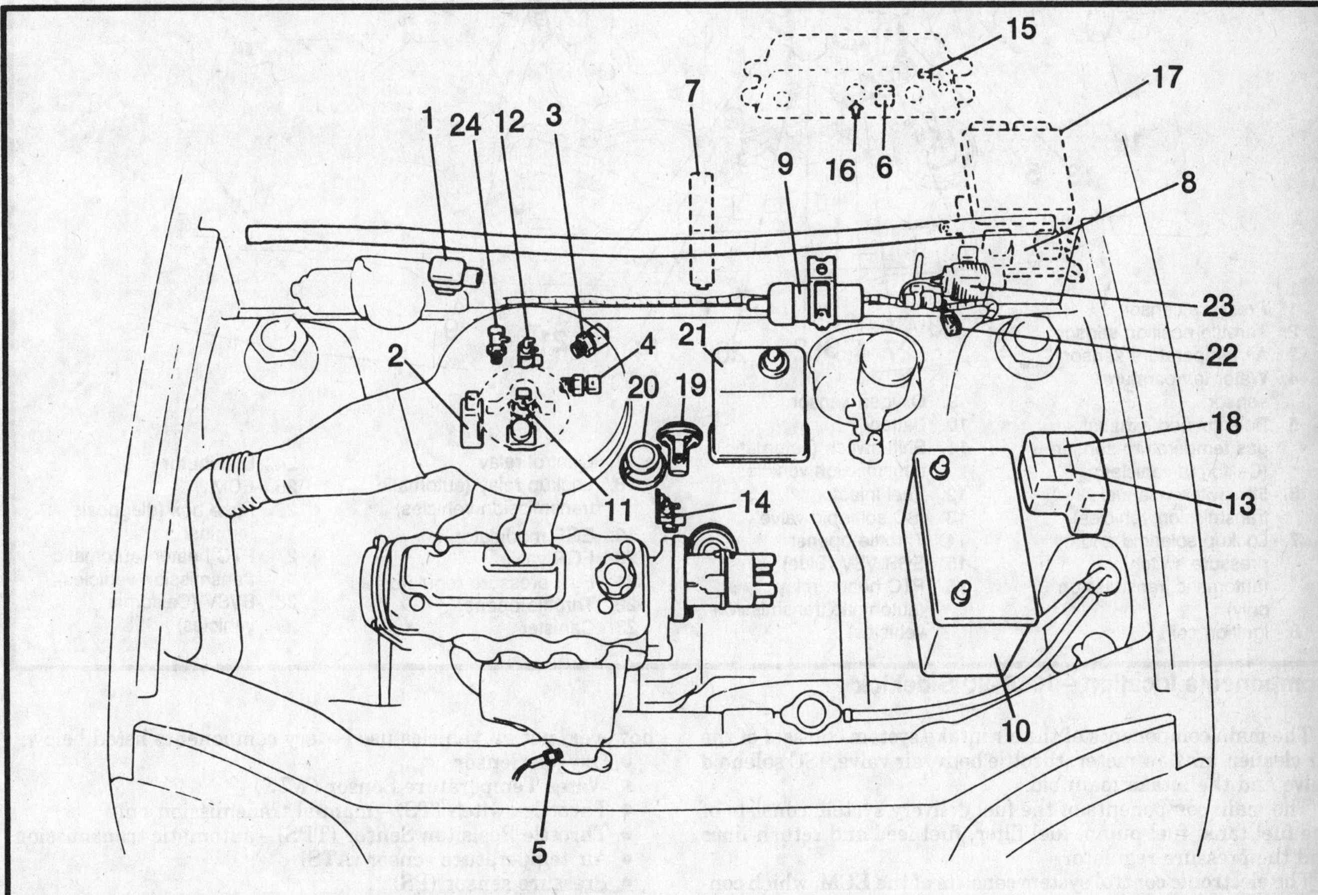
1
24
12
3
7
15
17
9
16
6
8
23
22
2
4
21
20
19
18
13
11
14
10
5
1. Pressure sensor
2. TS or TPS
3. ATS
4. WTS
5. Oxygen sensor
6. Speed sensor
7. Automatic transmission control module (automatic transmission only)
8. Junction/fuse box (diagnosis switch terminal)
9. Ignition coil
10. Battery
11. Fuel injector
12. ISC solenoid valve
13. Fuel pump relay
14. EGR VSV (California model only)
15. Check engine light
16. Shift-up indicator light (if equipped)
17. Electronic Control Module (ECM)
18. EFI main relay
19. EGR valve (California model only)
20. EGR modulator (California model only)
21. Canister
22. Monitor connector
23. Injector resistor
24. BVSV

Components location—1988–90 Swift

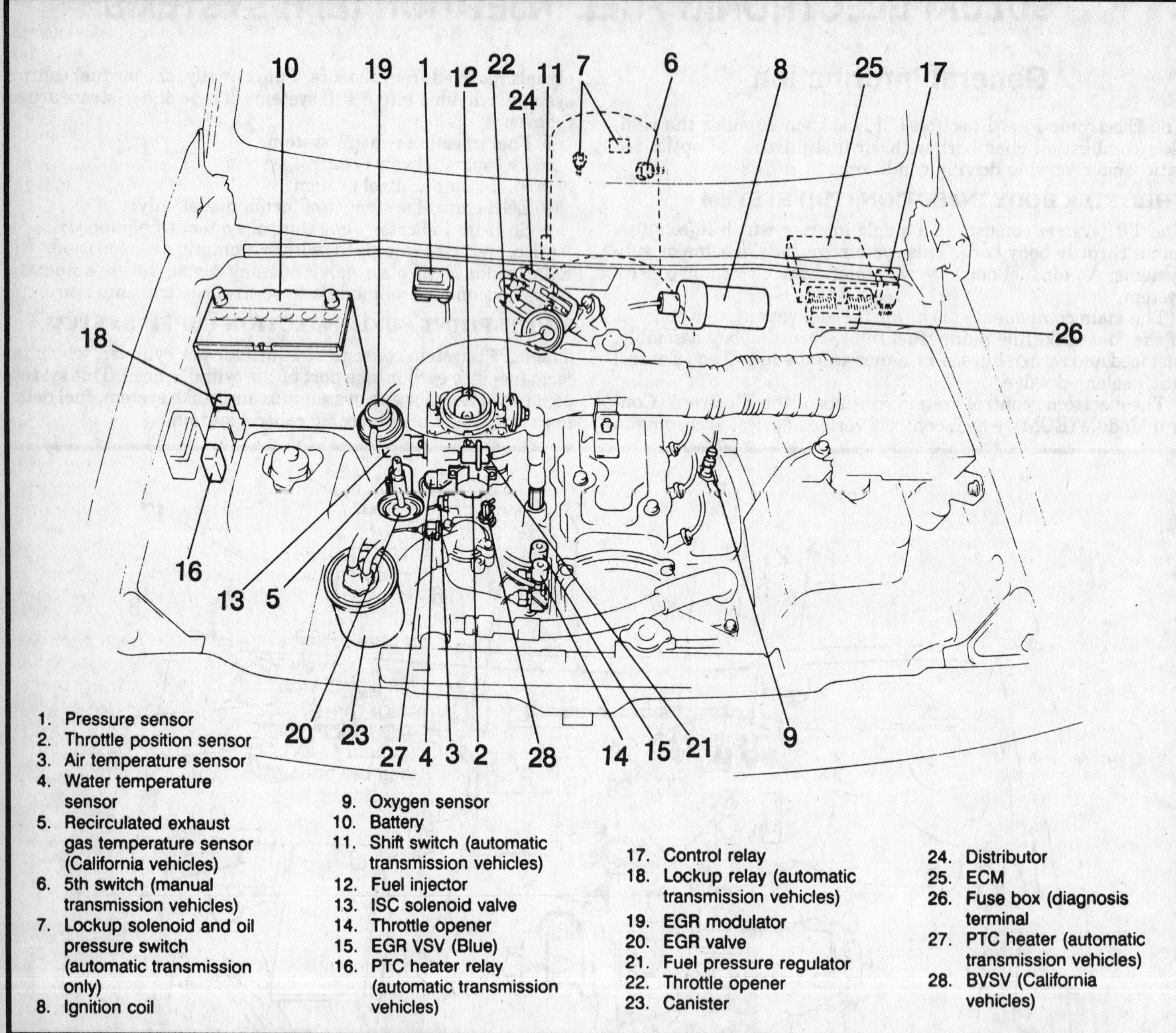
10
19
1
12
22
24
11
7
6
8
25
17
18
26
16
13
5
20
23
27
4
3
2
28
14
15
21
9
1. Pressure sensor
2. Throttle position sensor
3. Air temperature sensor
4. Water temperature sensor
5. Recirculated exhaust gas temperature sensor (California vehicles)
6. 5th switch (manual transmission vehicles)
7. Lockup solenoid and oil pressure switch (automatic transmission only)
8. Ignition coil
9. Oxygen sensor
10. Battery
11. Shift switch (automatic transmission vehicles)
12. Fuel injector
13. ISC solenoid valve
14. Throttle opener
15. EGR VSV (Blue)
16. PTC heater relay (automatic transmission vehicles)
17. Control relay
18. Lockup relay (automatic transmission vehicles)
19. EGR modulator
20. EGR valve
21. Fuel pressure regulator
22. Throttle opener
23. Canister
24. Distributor
25. ECM
26. Fuse box (diagnosis terminal
27. PTC heater (automatic transmission vehicles)
28. BVSV (California vehicles)

Components location – 1988–90 Sidekick

The main components of the air intake system consists of the air cleaner, air flow meter, throttle body, air valve, ISC solenoid valve and the intake manifold.

The main components of the fuel delivery system consists of the fuel tank, fuel pump, fuel filter, fuel feed and return lines and the pressure regulator.

The electronic control system consists of the ECM, which controls various devices according to signals received from sensors. Other controlling functions of the ECM are as follows:

- EGR control system, equipped in only California spec. vehicles
- Evaporative emission control system
- Throttle valve opening signal output for automatic transmission
- Electronic Spark Advance (ESA) system

Both the MPFI and TBI system includes a self-diagnosis function which is controlled by the ECM. If a fault is detected when the ignition switch is **ON** and the engine is running, the ECM will response by turning on or flashing the check engine light. The self-diagnosis system includes the following components; however, not all vehicles used every components listed below:

- Oxygen Sensor
- Water Temperature Sensor (WTS)
- Throttle switch (TS) – manual transmission only
- Throttle Posisiton Sensor (TPS) – automatic transmission
- Air temperature sensor (ATS)
- Pressure sensor (PS)
- Ignition signal
- Speed sensor
- EGR system
- Air Flow Meter (AFM)
- Crank Angle Sensor
- Idle Switch Circuit
- Lock-up Circuit – automatic transmission
- 5th Switch Circuit – manual transmission only
- Central Processing Unit (CPU) of the ECM

When the ignition switch is turned **ON** and the engine is stop, the check engine light will light. This is only to check the check engine light bulb and circuit. However, if the self-diagnosis system detects trouble in the EFI system, the ECM will turn **ON** the check engine light with the engine running to warn the driver of such trouble and at the same time it stroes the trouble area in the ECM backup memory. The check engine light will remain **ON** as long as the trouble exists but will turn **OFF** when the normal condition is restored.

The EFI system also includes a fail-safe function. Should a malfunction occur in the EFI system, the ECM will control such functions as the injector, ISC solenoid valve and others on the basic of a standard program pre-stored in the ECM. During a fail-safe condition, the ECM will ignore the failure signal and/or the CPU and thus provide the vehicle with a deminished level of engine performance.

SYSTEM OPERATION

THROTTLE BODY INJECTION (TBI) SYSTEM

When the ignition switch is turn **ON**, power is supply to the fuel pump via the fuel pump relay. The fuel pump is activated and the fuel system is pressurized. Simultaneously, an ignition signal is sent from the ignition coil primary circuit. An engine start signal is also sent to the ECM via the engine starter circuit. The ECM uses the engine start signal to determine whether the engine is cranking or not and thus control the fuel injector and fuel pump relay.

While the engine is cranking, the ECM keeps the Idle Speed Control (ISC) solenoid valve **ON**. This provides the engine with a better start. After the engine has started, the ECM gradually reduce the ISC solenoid valve **ON** time to maintain the specified idle speed.

When the injector (solenoid coil) is energized by the ECM, the needle valve which is incorporated with the plunger opens and the injector, which is under pressure, injects fuel in a conic dispersion into the throttle body bore. The injected fuel is mixed wit the air which has been filtered through the air cleaner in the throttle body. The air/fuel mixture is drawn through clearance between the throttle valve, throttle bore and an idle by-pass passage into the intake manifold. The intake manifold then distributes the air/fuel mixture to each combustion chamber. Should the fuel system pressure exceed a preset level, a valve, located in the fuel pressure regulator opens and excess fuel returns to the fuel tank via the return line.

An air valve, located in the throttle body supplies bypass air into the intake manifold without letting the air pass through the throttle valve when the engine is cold. This condition causes the engine speed to increase (fast idle state) and thus provides engine warm-up. As the engine is warmed up, a piston inside the air valve gradually blocks the amount of air passing through the air valve and simultaneously the engine speed is reduced. As the engine coolant temperature reaches approximately 176°F (80°C), the valve is fully open and the engine speed returns to normal idle speed.

The ECM also uses the following signals to compensates for engine speeds and/or fuel injection **ON** time:

- Air Conditioning Signal – The ECM uses this signal to determine whether the air conditioner is operating or not and uses it as 1 of the signals for controlling the ISC valve operation.
- Battery Voltage – The fuel injector is driven by its solenoid coil based upon the ECM output. However, there is some delay called "ineffective injection time", which doesn't provide fuel, between the ECM signal and the valve action. The ineffective injection time depends on the battery voltage signal. The ECM takes this information to compensate for fuel injection time.
- **R**, **D**, **2** or **L** Range Signal – When in these range, the automatic transmission (automatic transmission only) module sends a battery voltage signal to the ECM. The ECM uses this signal as 1 of the signals to control the fuel injector and the ISC solenoid valve.
- Illumination Light Signal – If equipped, this signal is sent from the illumination light circuit. It is used to reduce intensity of the shift-up indicator light when the illumination light is **ON**.

MULTI-POINT FUEL INJECTION (MPFI) SYSTEM

The Multi-Point Fuel Injection (MPFI) system operates basically the same as the Throttle Body Injection (TBI) system. When the ignition switch is turn **ON**, power is supply to the fuel pump via the fuel pump relay. The fuel pump is activated and the fuel system is pressurized. As the starter circuit is energized, an engine start signal is sent to the Electronic Control Module (ECM) via the engine starter circuit. The ECM uses the engine start signal to determine whether the engine is cranking or not and thus control the fuel injectors and fuel pump relay. Simultaneously, a pulse signal is sent to the ECM via a Crank Angel Sensor (CSA) located in the distributor.

While the engine is cranking, the ECM keeps the Idle Speed Control (ISC) solenoid valve **ON**. This provides the engine with a better start. After the engine has started, the ECM gradually reduce the ISC solenoid valve **ON** time to maintain the specified idle speed.

There are 4 injectors (1 for each cylinder), each of which is installed between the cylinder head and the delivery pipe. These electromagnetic type injection nozzle injects fuel into the intake port of the cylinder head according to the signal from the ECM. There are 2 type of injection timing. One is "synchronous injection" (during starting) in which injection is synchronous with the Crank Angle Sensor (CSA) and the other is "asynchronous injection" in which injection takes place independently of the CSA signal.

When the engine is cranking (synchronous injection), all 4 injectors start injecting the fuel simultaneously at every 6 degrees of CSA signal depending on the engine temperature.

If the accelerator pedal is depressed (asynchronous injection), when the idle switch turns **OFF** from **ON** and when the throttle valve opening increases suddenly) the 4 injectors inject fuel simultaneously once to a few times in addition to the synchronous injection and independently of the CAS signal.

The pressure regulator maintains the fuel pressure at 36.2 (255 kpa) higher than the pressure in the intake manifold. Excess fuel then returns to the fuel tank via the fuel return line.

SYSTEM COMPONENTS

Electronic Control Module (ECM)

The ECM, located at left side underside of the instrument panel (Swift), consists of a microcomputer, analog/digital converter and an Input/Output unit, etc. The ECM not only controls the such function as the fuel injectors, ISC solenoid valve, fuel pump relay, etc., but also the self-diagnosis function and the fail-safe function.

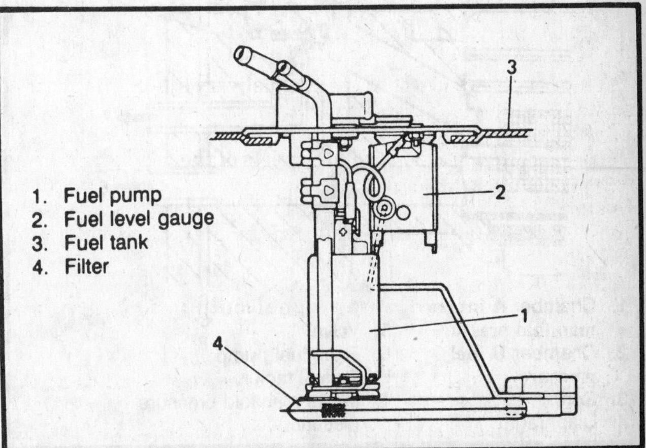
1. Fuel pump
2. Fuel level gauge
3. Fuel tank
4. Filter
1
2
3
4

Fuel pump assembly

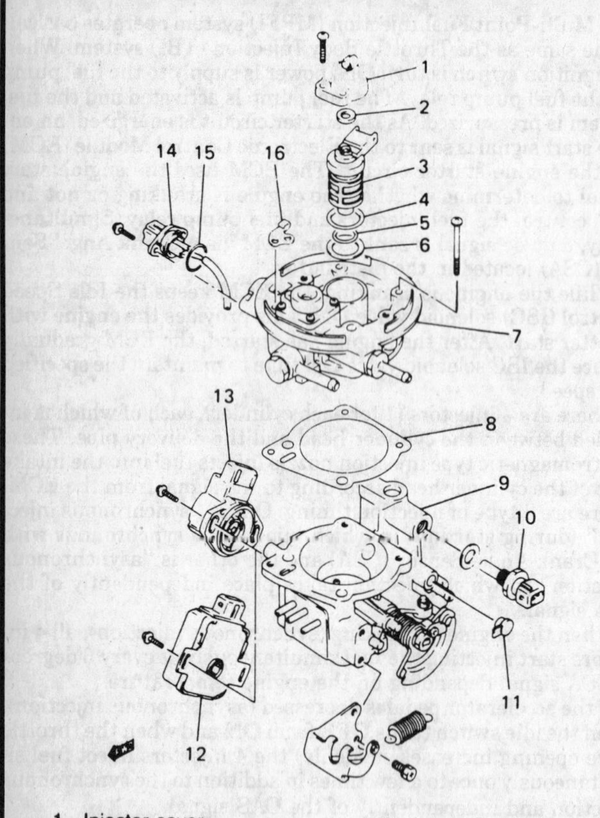
1
2
3
4
5
6
7
8
9
10
11
12
13
14
15
16
1. Injector cover
2. Upper insulator (small)
3. Fuel injector
4. Upper O-ring (large)
5. Fuel pressure regulator
6. Upper body
7. Lower insulator (large)
8. Lower O-ring (small)
9. Vacuum nozzle
10. Throttle valve
11. Lower body
12. Air valve
13. Fuel
14. Air

Throttle body cross-sectional view

Fuel Pump

The fuel pump, located in the fuel tank, consists of an armature, magnet, impeller, brush and a check valve. Fuel is drawn into the inlet port of the fuel pump and under pressure it is discharged through the outlet port. The fuel pump is also equipped with a check valve to keep the fuel system pressurized when the fuel pump is not operating. The ECM controls the **ON** and **OFF** operation of the fuel pump.

Throttle Body

The throttle body consists of the main bore, air and fuel passage, vacuum passage, air induction passage, fuel injector, fuel pressure regulator, throttle valve, air valve and a throttle position sensor (throttle switch, vehicles equipped with manual transmission).

Fuel Pressure Regulator

The fuel pressure regulator is a diaphragm operated relief valve consisting of a diaphragm, spring and valve. The fuel pressure regulator maintains the fuel pressure to the injector at 25.6 psi (180 kpa) higher than that in the intake manifold at all times. If the pressure increase more than 25.6 psi (180 kpa) higher that the intake manifold pressure, a valve in the regulator opens and excess fuel returns to the fuel tank via the return line.

Fuel Injector

The fuel injector is an electromagnetic type solenoid valve, which injects fuel in the throttle body bore according to a signal from the ECM. The amount of fuel injected at any given time is determined by the length of time (duration) the solenoid coil is energized.

Pressure Sensor (PS)

The pressure sensor, consisting of a semi-conductor element, senses pressure changes in the intake manifold and converts it into a signal voltage. As the manifold pressure changes, the elec-

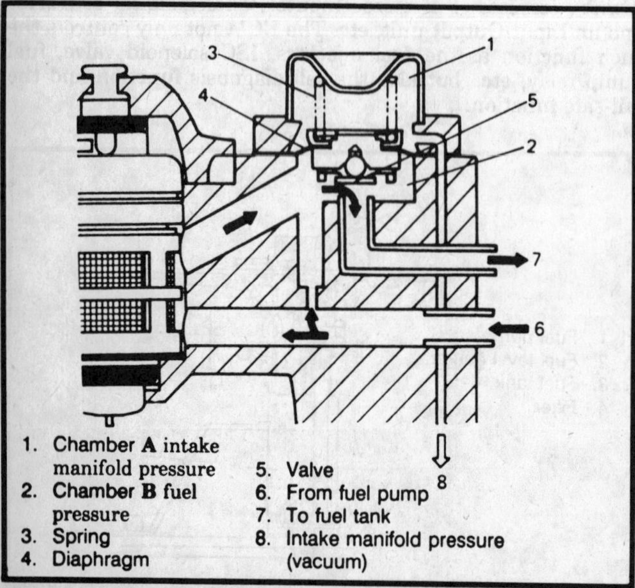
1
2
3
4
5
6
7
8
1. Chamber A intake manifold pressure
2. Chamber B fuel pressure
3. Spring
4. Diaphragm
5. Valve
6. From fuel pump
7. To fuel tank
8. Intake manifold pressure (vacuum)

Fuel pressure regulator cross-sectional view

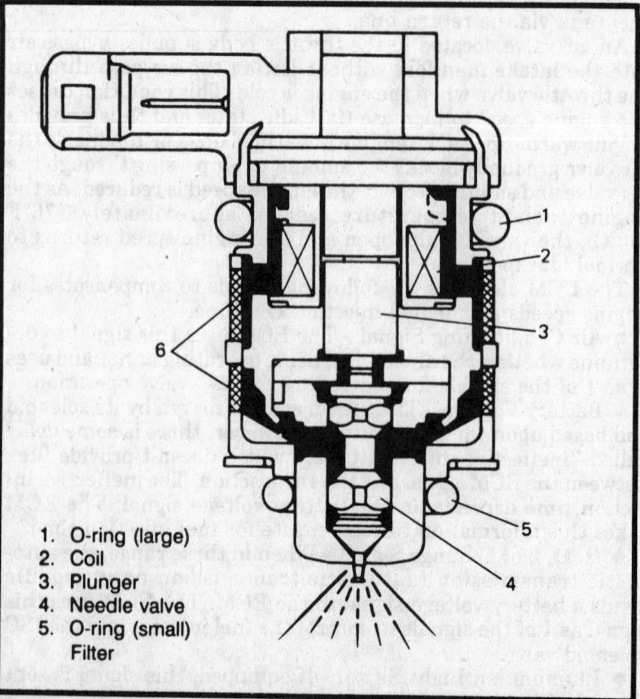
1
2
3
4
5
6
1. O-ring (large)
2. Coil
3. Plunger
4. Needle valve
5. O-ring (small)
Filter

Fuel injector cross-sectional view

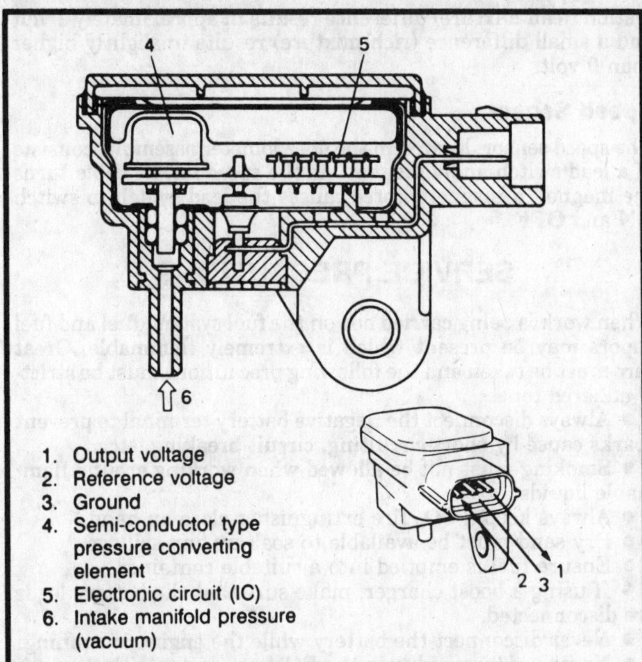
1. Output voltage
2. Reference voltage
3. Ground
4. Semi-conductor type pressure converting element
5. Electronic circuit (IC)
6. Intake manifold pressure (vacuum)

Pressure sensor

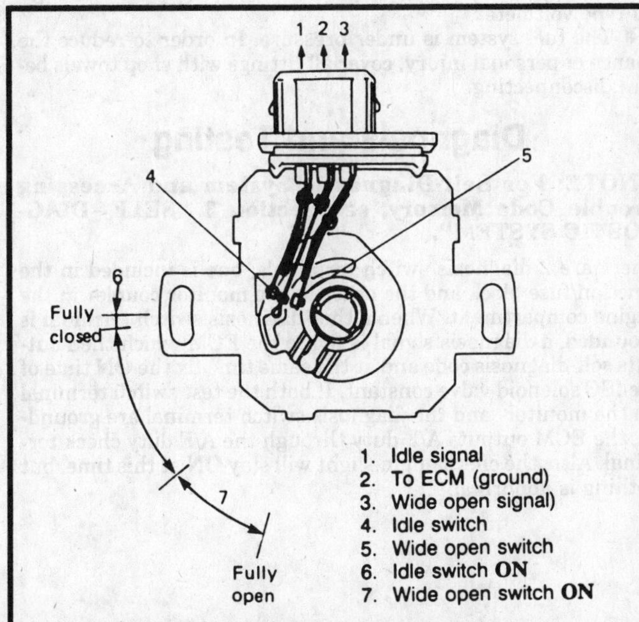
Fully closed
Fully open
1. Idle signal
2. To ECM (ground)
3. Wide open signal)
4. Idle switch
5. Wide open switch
6. Idle switch ON
7. Wide open switch ON

Throttle switch (manual transmission only)

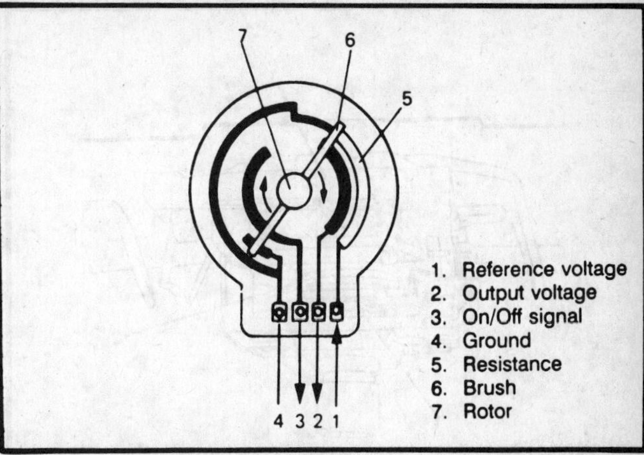
1. Reference voltage
2. Output voltage
3. On/Off signal
4. Ground
5. Resistance
6. Brush
7. Rotor

Throttle position sensor (TPS)

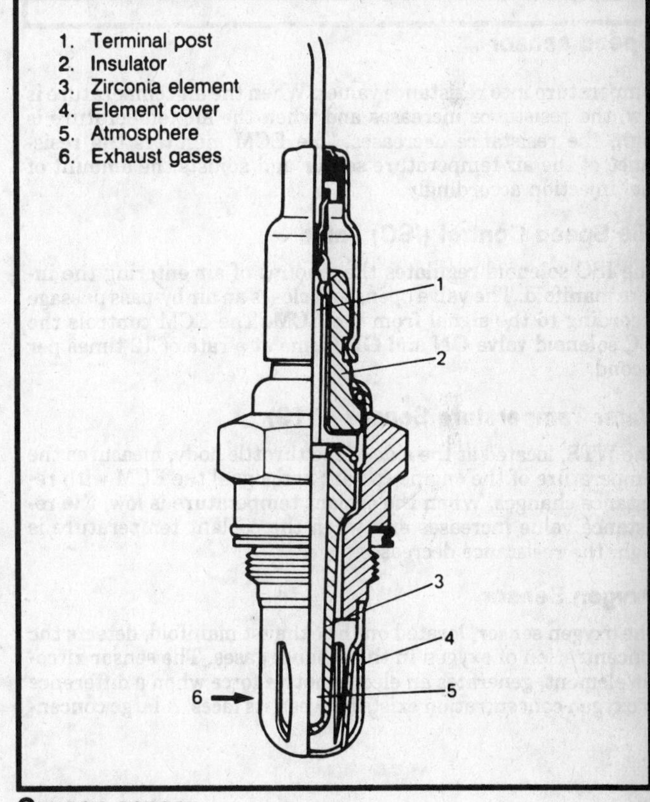
1. Terminal post
2. Insulator
3. Zirconia element
4. Element cover
5. Atmosphere
6. Exhaust gases

Oxygen sensor

trical resistance of the pressure sensor also changes. The ECM uses the voltage signal from the pressure sensor as 1 of the signals to control the fuel injectors.

Throttle Switch

MANUAL TRANSMISSION ONLY

The throttle switch, connected to the throttle valve shaft, consists of 2 contact points (idle switch and wide open switch) which detects throttle valve angle. The throttle position in the idle state is detected by the idle switch. When the throttle is in the wide open position, the ECM monitors the signal from the wide open switch.

Throttle Position Sensor (TPS)

AUTOMATIC TRANSMISSION ONLY

The TPS, connected to the throttle valve shaft, consists of a contact point (idle switch) and a potentiometer which detects throttle valve angle. The throttle position in the idle state is detected by the idle switch, but beyond idle position, it is detected by the output voltage across the the potentiometer. The ECM not only uses the TPS signal to control the injector, but also sends it to the automatic transmission control module, where it is used as 1 of the signals to control the automatic transmission.

Air Temperature Sensor (ATS)

The ATS, located on the side of the air cleaner assembly, measures the temperature of the incoming air and converts the air

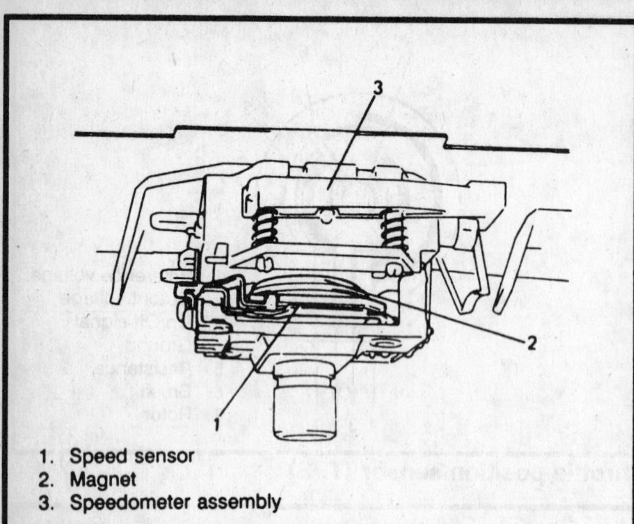
3
2
1
1. Speed sensor
2. Magnet
3. Speedometer assembly

Speed sensor

temperature into resistance values. When the air temperature is low, the resistance increases and when the air temperature is high, the resistance decreases. The ECM monitors the resistance of the air temperature sensor and adjusts the amount of fuel injection accordingly.

Idle Speed Control (ISC) Valve

The ISC solenoid regulates the amount of air entering the intake manifold. The valve opens and closes an air by-pass passage according to the signal from the ECM. The ECM controls the ISC solenoid valve **ON** and **OFF** time at a rate of 12 times per second.

Water Temperature Sensor (WTS)

The WTS, located at the side of the throttle body, measures the temperature of the engine coolant and signal the ECM with resistance changes. When the coolant temperature is low, the resistance value increases and when the coolant temperature is high, the resistance decreases.

Oxygen Sensor

The oxygen sensor, located on the exhaust manifold, detects the concentration of oxygen in the exhaust gases. The sensor zirconia element, generates an electromotive force when a difference in oxygen concentration exists between its faces. A large concentration (lean mixture) difference results in approximately 1 volt and a small difference (rich mixture) results in slightly higher than 0 volt.

Speed Sensor

The speed sensor, located in the speedometer assembly, consists of a lead switch and a magnet. As the speedometer cable turns the magnet, a magnetic force causes the lead switch to switch **ON** and **OFF**.

SERVICE PRECAUTIONS

When work is being carried out on the fuel system, fuel and fuel vapors may be present which is extremely flammable. Great care must be taken and the following precautions must be strictly adhered to:

- Always disconnect the negative battery terminal to prevent sparks cause by short-circuiting, circuit-breaking, etc.
- Smoking must not be allowed when working around flammable liquids.
- Always keep a CO_2 fire extinguisher close on hand.
- Dry sand must be available to soak up any spillage.
- Ensure fuel is emptied into a suitable container.
- If using a boost charger, make sure that the battery leads are disconnected.
- Never disconnect the battery while the engine is running.
- Never unplug or plug in the ECM connector with the ignition switched **ON**.
- Be sure to use a voltmeter with a high impedance or a digital type voltmeter.
- The fuel system is under pressure. In order to reduce the chance of personal injury, cover all fittings with shop towels before disconnecting.

Diagnosis and Testing

NOTE: For Self-Diagnostic System and Accessing Trouble Code Memory, see Section 3 "SELF–DIAGNOSTIC SYSTEM".

There are 2 diagnosis switch terminals; one is included in the junction/fuse block and the other in the monitor coupler in the engine compartment. When either diagnosis switch terminal is grounded, a diagnosis signal is fed to the ECM which then outputs self-diagnosis code and at the same time fix the **ON** time of the ISC solenoid valve constant. If both the test switch terminal (in the monitor) and the diagnosis switch terminal are grounded, the ECM outputs A/F duty through the A/F duty check terminal. Also, the check engine light will stay **ON** at this time, but nothing is abnormal.

DIAGNOSTIC FLOW CHART—1988–90 SWIFT

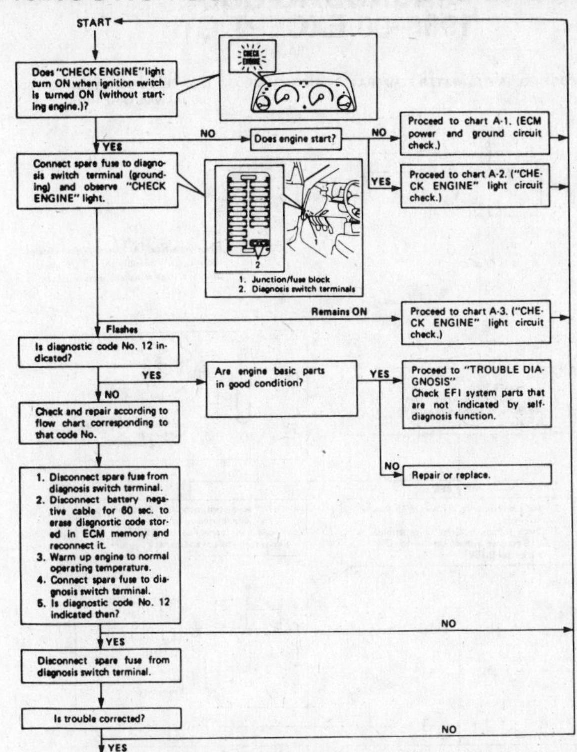
START
Does "CHECK ENGINE" light turn ON when ignition switch is turned ON (without starting engine.)?
NO
Does engine start?
NO
Proceed to chart A-1. (ECM power and ground circuit check.)
YES
Proceed to chart A-2. ("CHECK ENGINE" light circuit check.)
YES
Connect spare fuse to diagnosis switch terminal (grounding) and observe "CHECK ENGINE" light.
1. Junction/fuse block
2. Diagnosis switch terminals
Remains ON
Proceed to chart A-3. ("CHECK ENGINE" light circuit check.)
Flashes
Is diagnostic code No. 12 indicated?
YES
Are engine basic parts in good condition?
YES
Proceed to "TROUBLE DIAGNOSIS" Check EFI system parts that are not indicated by self-diagnosis function.
NO
Repair or replace.
NO
Check and repair according to flow chart corresponding to that code No.
1. Disconnect spare fuse from diagnosis switch terminal.
2. Disconnect battery negative cable for 60 sec. to erase diagnostic code stored in ECM memory and reconnect it.
3. Warm up engine to normal operating temperature.
4. Connect spare fuse to diagnosis switch terminal.
5. Is diagnostic code No. 12 indicated then?
NO
YES
Disconnect spare fuse from diagnosis switch terminal.
Is trouble corrected?
NO
YES
END

DIAGNOSTIC CODE TABLE 1988–90 EXCEPT GTI

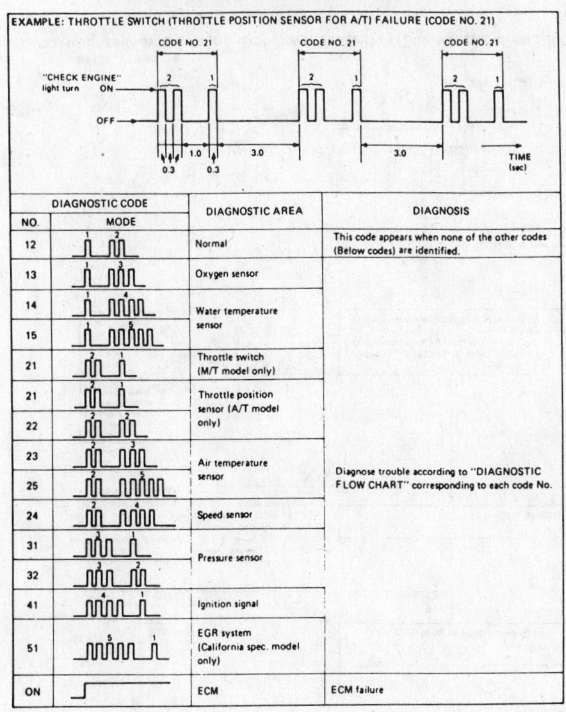
EXAMPLE: THROTTLE SWITCH (THROTTLE POSITION SENSOR FOR A/T) FAILURE (CODE NO. 21)

CODE NO. 21
"CHECK ENGINE" light turn ON
OFF
0.3
1.0
3.0
TIME (sec)

DIAGNOSTIC CODE NO.	MODE	DIAGNOSTIC AREA	DIAGNOSIS
12		Normal	This code appears when none of the other codes (Below codes) are identified.
13		Oxygen sensor	Diagnose trouble according to "DIAGNOSTIC FLOW CHART" corresponding to each code No.
14		Water temperature sensor	
15			
21		Throttle switch (M/T model only)	
21		Throttle position sensor (A/T model only)	
22			
23		Air temperature sensor	
25			
24		Speed sensor	
31		Pressure sensor	
32			
41		Ignition signal	
51		EGR system (California spec. model only)	
ON		ECM	ECM failure

DIAGNOSTIC CODE TABLE 1988–90 EXCEPT GTI

A-1 ECM POWER AND GROUND CIRCUIT CHECK
("CHECK ENGINE" LIGHT DOESN'T LIGHT AT IGNITION SWITCH ON AND ENGINE DOESN'T START THOUGH IT IS CRANKED UP.)

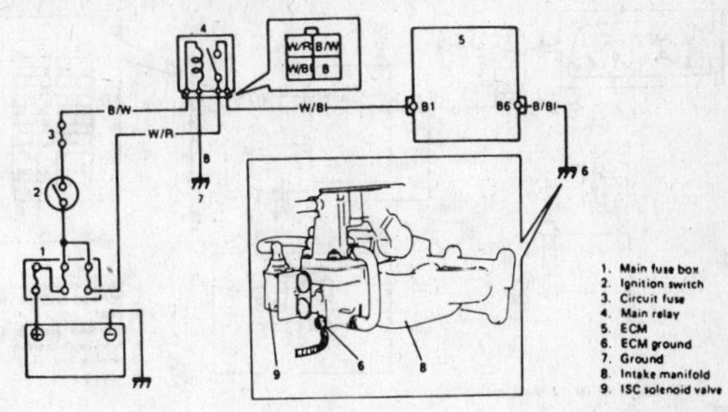
W/R
B/W
W/Bl
B
B/W
W/R
W/Bl
B1
B6
B/Bl
1. Main fuse box
2. Ignition switch
3. Circuit fuse
4. Main relay
5. ECM
6. ECM ground
7. Ground
8. Intake manifold
9. ISC solenoid valve

ECM Power and Ground Circuit

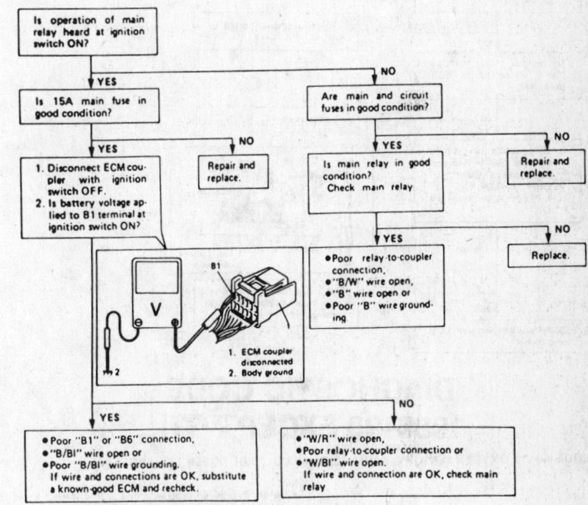
Is operation of main relay heard at ignition switch ON?
YES
NO
Is 15A main fuse in good condition?
Are main and circuit fuses in good condition?
YES
NO
Repair and replace.
YES
NO
Repair and replace.
1. Disconnect ECM coupler with ignition switch OFF.
2. Is battery voltage applied to B1 terminal at ignition switch ON?
Is main relay in good condition? Check main relay
YES
NO
Replace.
•Poor relay-to-coupler connection,
•"B/W" wire open,
•"B" wire open or
•Poor "B" wire grounding
B1
V
1. ECM coupler disconnected
2. Body ground
YES
NO
•Poor "B1" or "B6" connection,
•"B/Bl" wire open or
•Poor "B/Bl" wire grounding.
If wire and connections are OK, substitute a known-good ECM and recheck.
•"W/R" wire open,
•Poor relay-to-coupler connection or
•"W/Bl" wire open.
If wire and connection are OK, check main relay

Diagnostic Flow Chart A-1 for ECM Power and Ground Circuit

A-2 "CHECK ENGINE" LIGHT CIRCUIT CHECK
("CHECK ENGINE" LIGHT DOES NOT LIGHT BUT ENGINE STARTS.)

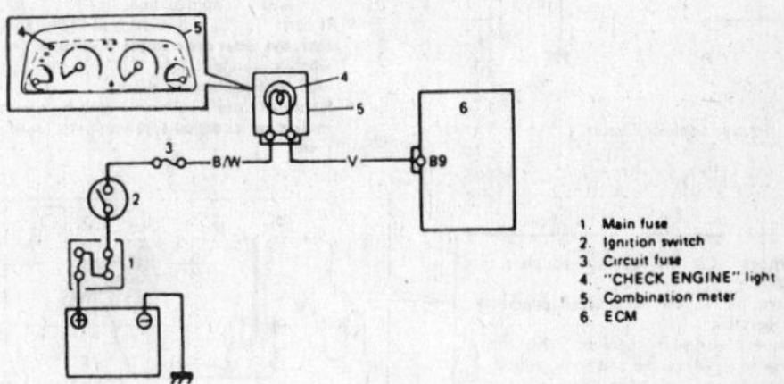
B/W
V
B9
1. Main fuse
2. Ignition switch
3. Circuit fuse
4. "CHECK ENGINE" light
5. Combination meter
6. ECM

Fig. 6E-46 "CHECK ENGINE" Light Circuit

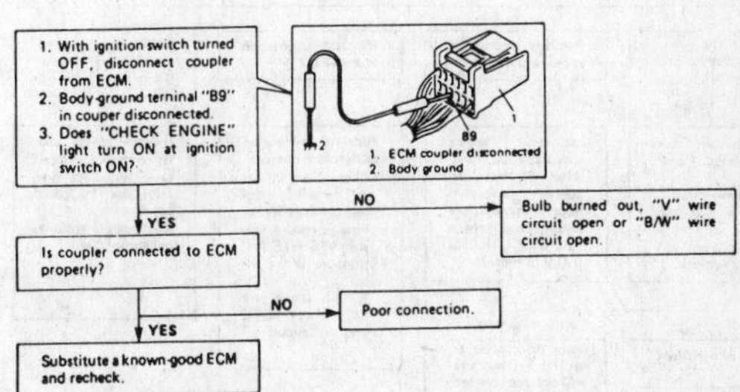
1. With ignition switch turned OFF, disconnect coupler from ECM.
2. Body-ground terminal "B9" in couper disconnected.
3. Does "CHECK ENGINE" light turn ON at ignition switch ON?
B9
1. ECM coupler disconnected
2. Body ground
NO
Bulb burned out, "V" wire circuit open or "B/W" wire circuit open.
YES
Is coupler connected to ECM properly?
NO
Poor connection.
YES
Substitute a known-good ECM and recheck.

DIAGNOSTIC CODE TABLE 1988–90 EXCEPT GTI

A-3 "CHECK ENGINE" LIGHT CIRCUIT CHECK
("CHECK ENGINE" LIGHT DOESN'T FLASH OR JUST REMAINS ON EVEN WITH SPARE FUSE CONNECTED TO DIAGNOSIS SWITCH TERMINAL.)

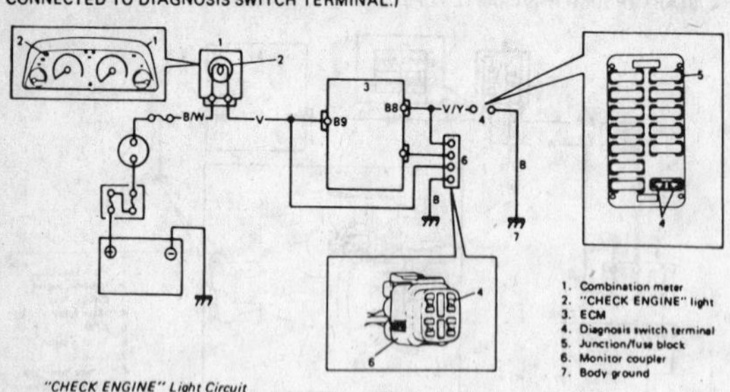
B/W
V
B9
B8
V/Y
B
1. Combination meter
2. "CHECK ENGINE" light
3. ECM
4. Diagnosis switch terminal
5. Junction/fuse block
6. Monitor coupler
7. Body ground

"CHECK ENGINE" Light Circuit

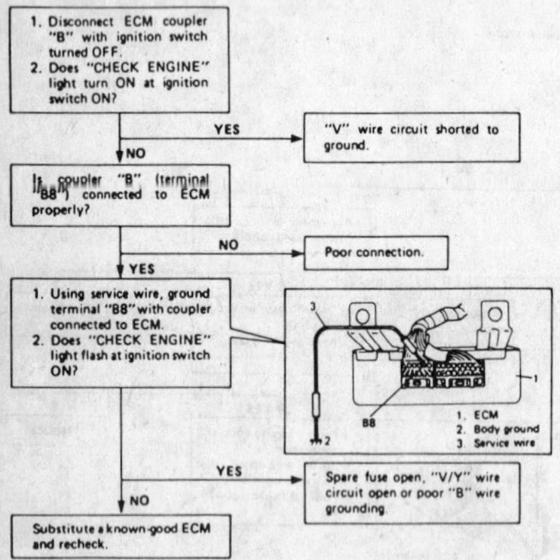
1. Disconnect ECM coupler "B" with ignition switch turned OFF.
2. Does "CHECK ENGINE" light turn ON at ignition switch ON?
YES
"V" wire circuit shorted to ground.
NO
Is coupler "B" (terminal "B8") connected to ECM properly?
NO
Poor connection.
YES
1. Using service wire, ground terminal "B8" with coupler connected to ECM.
2. Does "CHECK ENGINE" light flash at ignition switch ON?
B8
1. ECM
2. Body ground
3. Service wire
YES
Spare fuse open, "V/Y" wire circuit open or poor "B" wire grounding.
NO
Substitute a known-good ECM and recheck.

DIAGNOSTIC CODE 1988–90 EXCEPT GTI

CODE NO. 13 OXYGEN SENSOR CIRCUIT (SIGNAL VOLTAGE DOESN'T CHANGE)

NOTE:
- Before diagnosing trouble according to flow chart given below, check to make sure that following system and parts other than EFI system are in good condition.
  - Air cleaner (clogged)
  - Vacuum leaks (air inhaling)
  - Spark plugs (contamination, gap)
  - High-tension cords (crack, deterioration)
  - Distributor rotor or cap (wear, crack)
  - Ignition timing
  - Engine compression
  - Any other system and parts which might affect A/F mixture or combustion.
- If code No. 13 and another code No. are indicated together, the latter has priority. Therefore, check and correct what is represented by that code No. first and then proceed to the following check.
- Be sure to use a voltmeter with high impedance (MΩ/V minimum) or digital type voltmeter for accurate measurement.

1. ECM
2. ECM coupler
3. Oxygen sensor
4. Coupler

Oxygen Sensor Circuit

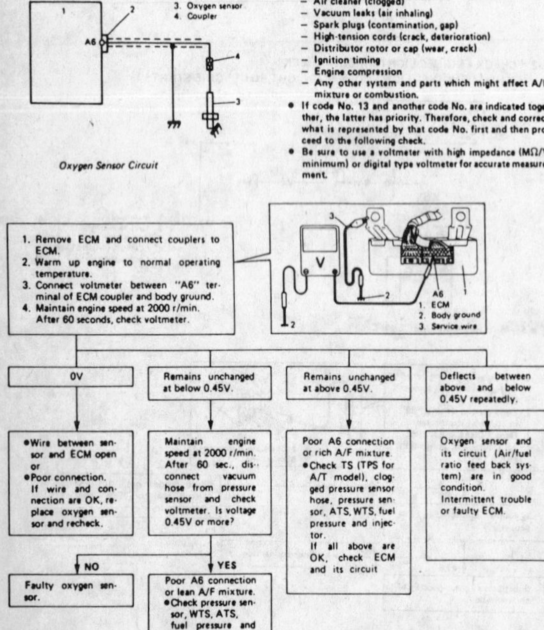
1. Remove ECM and connect couplers to ECM.
2. Warm up engine to normal operating temperature.
3. Connect voltmeter between "A6" terminal of ECM coupler and body ground.
4. Maintain engine speed at 2000 r/min. After 60 seconds, check voltmeter.
A6
1. ECM
2. Body ground
3. Service wire
0V
Remains unchanged at below 0.45V.
Remains unchanged at above 0.45V.
Deflects between above and below 0.45V repeatedly.
•Wire between sensor and ECM open or
•Poor connection. If wire and connection are OK, replace oxygen sensor and recheck.
Maintain engine speed at 2000 r/min. After 60 sec., disconnect vacuum hose from pressure sensor and check voltmeter. Is voltage 0.45V or more?
Poor A6 connection or rich A/F mixture.
•Check TS (TPS for A/T model), clogged pressure sensor hose, pressure sensor, ATS, WTS, fuel pressure and injector. If all above are OK, check ECM and its circuit
Oxygen sensor and its circuit (Air/fuel ratio feed back system) are in good condition. Intermittent trouble or faulty ECM.
NO
Faulty oxygen sensor.
YES
Poor A6 connection or lean A/F mixture.
•Check pressure sensor, WTS, ATS, fuel pressure and injector. If all above are OK, check ECM and its circuit

DIAGNOSTIC CODE 1988–90 EXCEPT GTI

CODE NO. 14 WTS (WATER TEMPERATURE SENSOR) CIRCUIT (LOW TEMPERATURE INDICATED, SIGNAL VOLTAGE HIGH)

1. ECM
2. WTS
3. WTS coupler
4. ECM coupler
5. To other sensors

WTS Circuit

NOTE:
When Code Nos. 14, 23 and 32 are indicated together, it is possible that "Lg/B" wire is open or A16 terminal connection is poor.

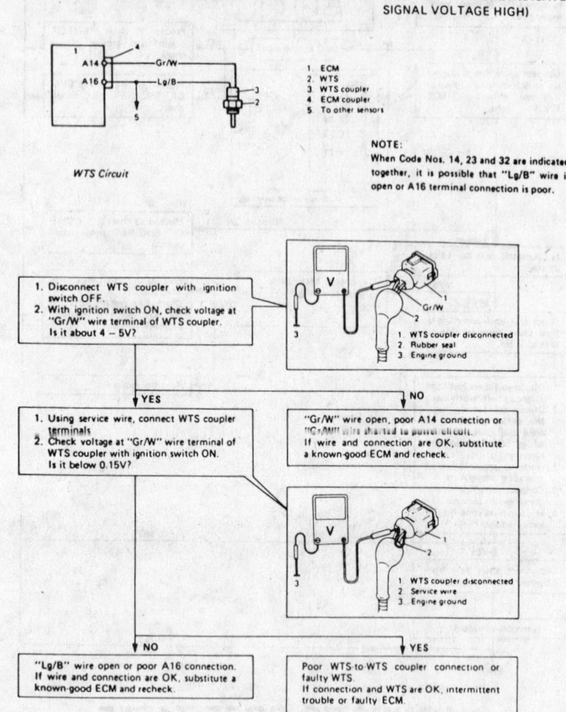
A14
A16
Gr/W
Lg/B
1. Disconnect WTS coupler with ignition switch OFF.
2. With ignition switch ON, check voltage at "Gr/W" wire terminal of WTS coupler. Is it about 4 – 5V?
1. WTS coupler disconnected
2. Rubber seal
3. Engine ground
YES
NO
1. Using service wire, connect WTS coupler terminals
2. Check voltage at "Gr/W" wire terminal of WTS coupler with ignition switch ON. Is it below 0.15V?
"Gr/W" wire open, poor A14 connection or "Gr/W" wire shorted to power circuit. If wire and connection are OK, substitute a known-good ECM and recheck.
1. WTS coupler disconnected
2. Service wire
3. Engine ground
NO
YES
"Lg/B" wire open or poor A16 connection. If wire and connection are OK, substitute a known-good ECM and recheck.
Poor WTS-to-WTS coupler connection or faulty WTS. If connection and WTS are OK, intermittent trouble or faulty ECM.

CODE NO. 15 WTS (WATER TEMPERATURE SENSOR) CIRCUIT (HIGH TEMPERATURE INDICATED, SIGNAL VOLTAGE LOW)

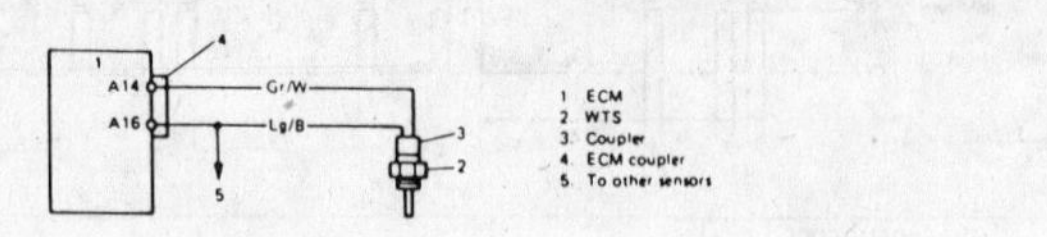
A14
A16
Gr/W
Lg/B
1. ECM
2. WTS
3. Coupler
4. ECM coupler
5. To other sensors

Fig. 6E-54 WTS Circuit

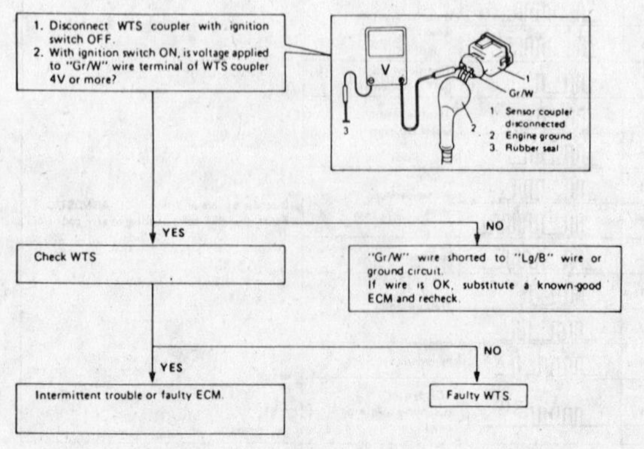
1. Disconnect WTS coupler with ignition switch OFF.
2. With ignition switch ON, is voltage applied to "Gr/W" wire terminal of WTS coupler 4V or more?
Gr/W
1. Sensor coupler disconnected
2. Engine ground
3. Rubber seal
YES
NO
Check WTS
"Gr/W" wire shorted to "Lg/B" wire or ground circuit. If wire is OK, substitute a known-good ECM and recheck.
YES
NO
Intermittent trouble or faulty ECM.
Faulty WTS

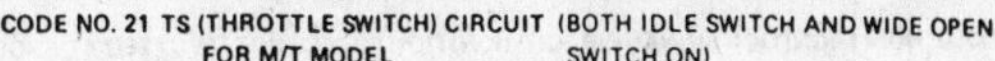

DIAGNOSTIC CODE – 1988–90 EXCEPT GTI

CODE NO. 21 TS (THROTTLE SWITCH) CIRCUIT (BOTH IDLE SWITCH AND WIDE OPEN SWITCH ON) FOR M/T MODEL

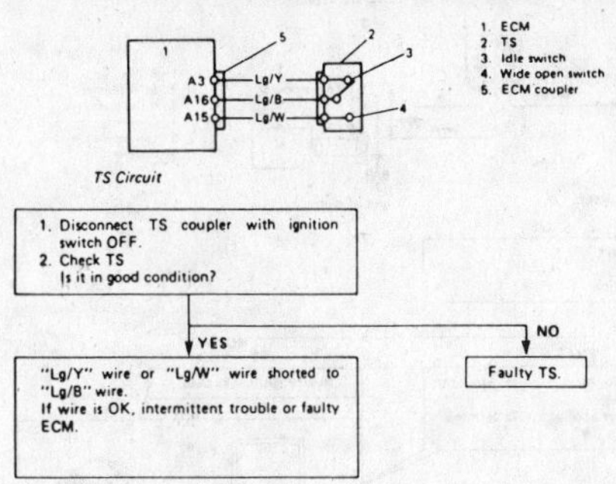
1. ECM
2. TS
3. Idle switch
4. Wide open switch
5. ECM coupler

TS Circuit

1. Disconnect TS coupler with ignition switch OFF.
2. Check TS
   Is it in good condition?

YES: "Lg/Y" wire or "Lg/W" wire shorted to "Lg/B" wire. If wire is OK, intermittent trouble or faulty ECM.

NO: Faulty TS.

CODE NO. 21 TPS (THROTTLE POSITION SENSOR) CIRCUIT (SIGNAL VOLTAGE HIGH) FOR A/T MODEL

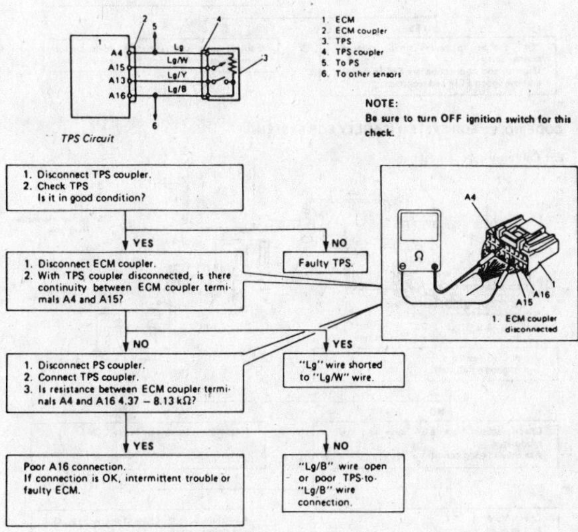
1. ECM
2. ECM coupler
3. TPS
4. TPS coupler
5. To PS
6. To other sensors

TPS Circuit

NOTE:
Be sure to turn OFF ignition switch for this check.

1. Disconnect TPS coupler.
2. Check TPS
   Is it in good condition?

NO: Faulty TPS.

YES:
1. Disconnect ECM coupler.
2. With TPS coupler disconnected, is there continuity between ECM coupler terminals A4 and A15?

1. ECM coupler disconnected

YES: "Lg" wire shorted to "Lg/W" wire.

NO:
1. Disconnect PS coupler.
2. Connect TPS coupler.
3. Is resistance between ECM coupler terminals A4 and A16 4.37 – 8.13 kΩ?

YES: Poor A16 connection. If connection is OK, intermittent trouble or faulty ECM.

NO: "Lg/B" wire open or poor TPS-to-"Lg/B" wire connection.

CODE NO. 22 TPS (THROTTLE POSITION SENSOR) CIRCUIT (SIGNAL VOLTAGE LOW) FOR A/T MODEL

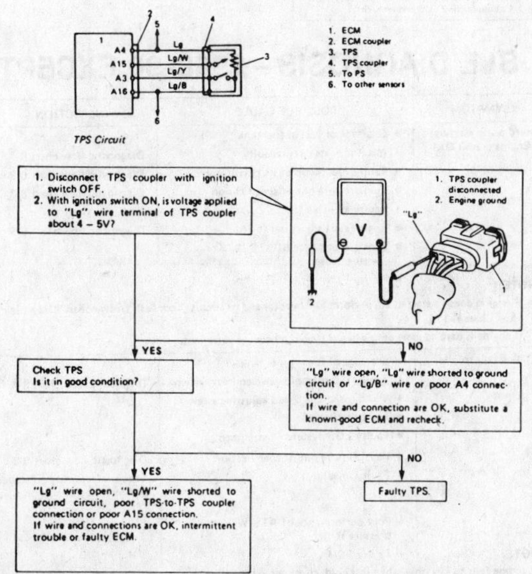
1. ECM
2. ECM coupler
3. TPS
4. TPS coupler
5. To PS
6. To other sensors

TPS Circuit

1. Disconnect TPS coupler with ignition switch OFF.
2. With ignition switch ON, is voltage applied to "Lg" wire terminal of TPS coupler about 4 – 5V?

1. TPS coupler disconnected
2. Engine ground

NO: "Lg" wire open, "Lg" wire shorted to ground circuit or "Lg/B" wire or poor A4 connection. If wire and connection are OK, substitute a known-good ECM and recheck.

YES: Check TPS
Is it in good condition?

NO: Faulty TPS.

YES: "Lg" wire open, "Lg/W" wire shorted to ground circuit, poor TPS-to-TPS coupler connection or poor A15 connection. If wire and connections are OK, intermittent trouble or faulty ECM.

CODE NO. 23 ATS (AIR TEMPERATURE SENSOR) CIRCUIT (LOW TEMPERATURE INDICATED, SIGNAL VOLTAGE HIGH)

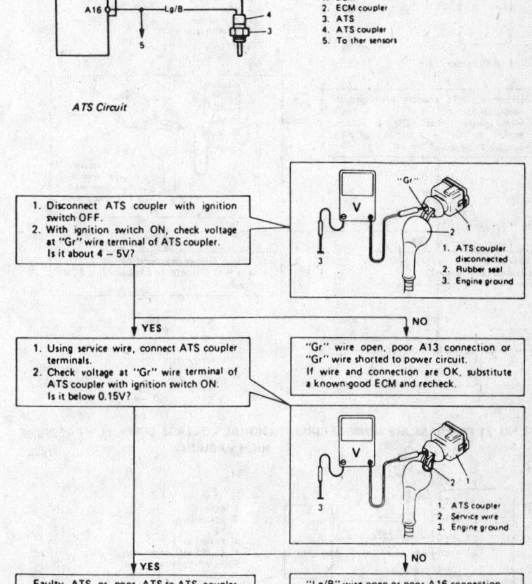
1. ECM
2. ECM coupler
3. ATS
4. ATS coupler
5. To ther sensors

ATS Circuit

1. Disconnect ATS coupler with ignition switch OFF.
2. With ignition switch ON, check voltage at "Gr" wire terminal of ATS coupler. Is it about 4 – 5V?

1. ATS coupler disconnected
2. Rubber seal
3. Engine ground

NO: "Gr" wire open, poor A13 connection or "Gr" wire shorted to power circuit. If wire and connection are OK, substitute a known-good ECM and recheck.

YES:
1. Using service wire, connect ATS coupler terminals.
2. Check voltage at "Gr" wire terminal of ATS coupler with ignition switch ON. Is it below 0.15V?

1. ATS coupler
2. Service wire
3. Engine ground

YES: Faulty ATS or poor ATS-to-ATS coupler connection. If ATS and connection are OK, intermittent trouble or faulty ECM.

NO: "Lg/B" wire open or poor A16 connection. If wire and connection are OK, substitute a known-good ECM and recheck.

CODE NO. 24 SPEED SENSOR CIRCUIT (SPEED SENSOR SIGNAL NOT INPUTTED ALTHOUGH FUEL IS KEPT CUT AT LOWER THAN 4000 r/min FOR LONGER THAN 4 SECONDS)

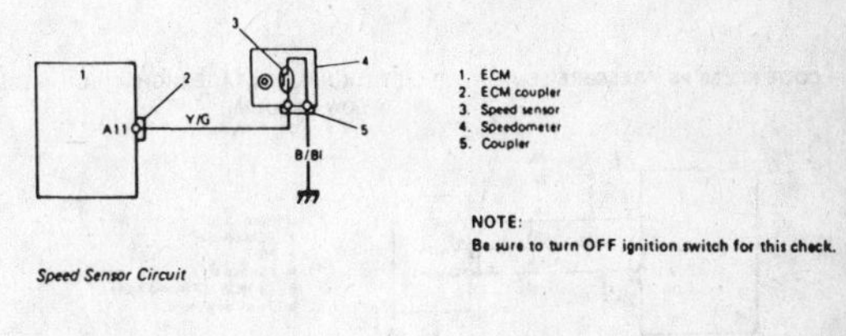
1. ECM
2. ECM coupler
3. Speed sensor
4. Speedometer
5. Coupler

Speed Sensor Circuit

NOTE:
Be sure to turn OFF ignition switch for this check.

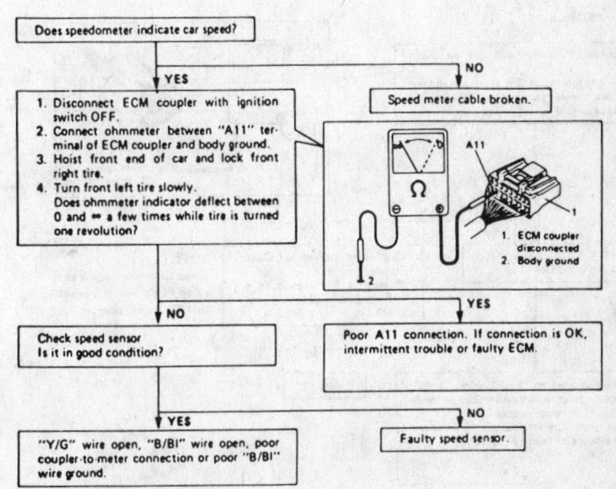

Does speedometer indicate car speed?

NO: Speed meter cable broken.

YES:
1. Disconnect ECM coupler with ignition switch OFF.
2. Connect ohmmeter between "A11" terminal of ECM coupler and body ground.
3. Hoist front end of car and lock front right tire.
4. Turn front left tire slowly. Does ohmmeter indicator deflect between 0 and ∞ a few times while tire is turned one revolution?

1. ECM coupler disconnected
2. Body ground

YES: Poor A11 connection. If connection is OK, intermittent trouble or faulty ECM.

NO: Check speed sensor
Is it in good condition?

NO: Faulty speed sensor.

YES: "Y/G" wire open, "B/Bl" wire open, poor coupler-to-meter connection or poor "B/Bl" wire ground.

DIAGNOSTIC CODE – 1988–90 EXCEPT GTI

CODE NO. 25 ATS (AIR TEMPERATURE SENSOR) CIRCUIT (HIGH TEMPERATURE INDICATED, SIGNAL VOLTAGE LOW)

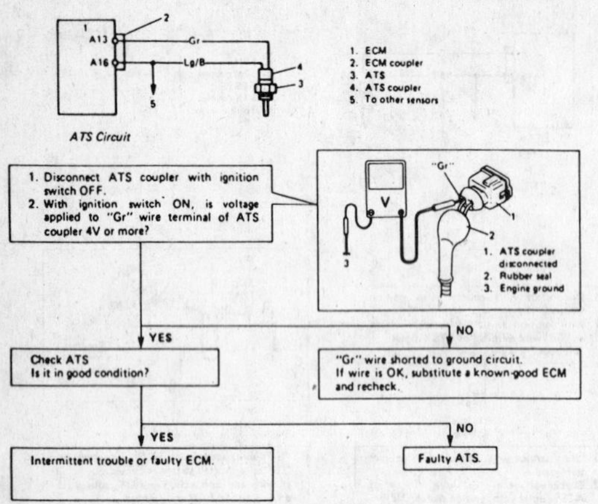
A13
A16
Gr
Lg/B
1. ECM
2. ECM coupler
3. ATS
4. ATS coupler
5. To other sensors
ATS Circuit
1. Disconnect ATS coupler with ignition switch OFF.
2. With ignition switch ON, is voltage applied to "Gr" wire terminal of ATS coupler 4V or more?
1. ATS coupler disconnected
2. Rubber seal
3. Engine ground
YES
NO
Check ATS
Is it in good condition?
"Gr" wire shorted to ground circuit.
If wire is OK, substitute a known-good ECM and recheck.
YES
NO
Intermittent trouble or faulty ECM.
Faulty ATS.

CODE NO. 31 PS (PRESSURE SENSOR) CIRCUIT (SIGNAL VOLTAGE LOW – LOW PRESSURE – HIGH VACUUM)

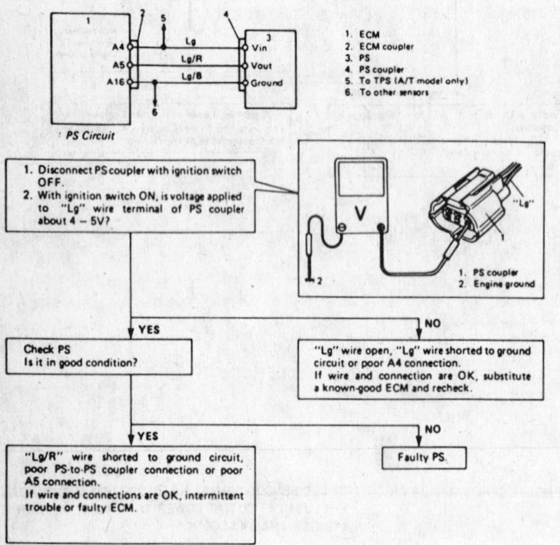
A4
A5
A16
Lg
Lg/R
Lg/B
Vin
Vout
Ground
1. ECM
2. ECM coupler
3. PS
4. PS coupler
5. To TPS (A/T model only)
6. To other sensors
PS Circuit
1. Disconnect PS coupler with ignition switch OFF.
2. With ignition switch ON, is voltage applied to "Lg" wire terminal of PS coupler about 4 – 5V?
1. PS coupler
2. Engine ground
YES
NO
Check PS
Is it in good condition?
"Lg" wire open, "Lg" wire shorted to ground circuit or poor A4 connection.
If wire and connection are OK, substitute a known-good ECM and recheck.
YES
NO
"Lg/R" wire shorted to ground circuit, poor PS-to-PS coupler connection or poor A5 connection.
If wire and connections are OK, intermittent trouble or faulty ECM.
Faulty PS.

CODE NO. 32 PS (PRESSURE SENSOR) CIRCUIT (SIGNAL VOLTAGE HIGH – HIGH PRESSURE – LOW VACUUM)

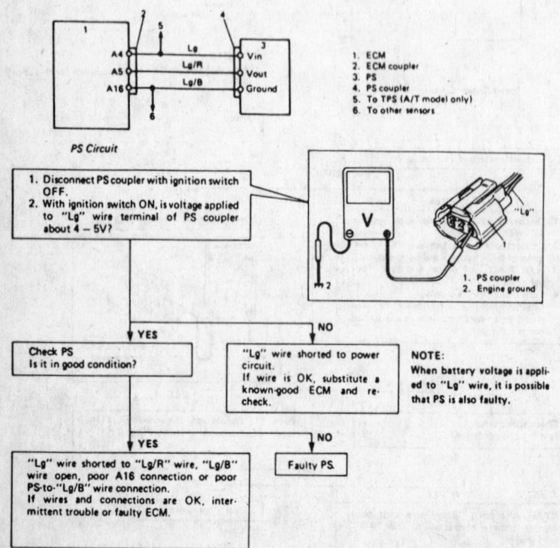
A4
A5
A16
Lg
Lg/R
Lg/B
Vin
Vout
Ground
1. ECM
2. ECM coupler
3. PS
4. PS coupler
5. To TPS (A/T model only)
6. To other sensors
PS Circuit
1. Disconnect PS coupler with ignition switch OFF.
2. With ignition switch ON, is voltage applied to "Lg" wire terminal of PS coupler about 4 – 5V?
1. PS coupler
2. Engine ground
YES
NO
Check PS
Is it in good condition?
"Lg" wire shorted to power circuit.
If wire is OK, substitute a known-good ECM and recheck.
NOTE:
When battery voltage is applied to "Lg" wire, it is possible that PS is also faulty.
YES
NO
"Lg" wire shorted to "Lg/R" wire, "Lg/B" wire open, poor A16 connection or poor PS-to-"Lg/B" wire connection.
If wires and connections are OK, intermittent trouble or faulty ECM.
Faulty PS.

CODE NO. 41 IGNITION SIGNAL CIRCUIT (IGNITION SIGNAL NOT INPUTTED AT ENGINE CRANKING)

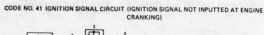

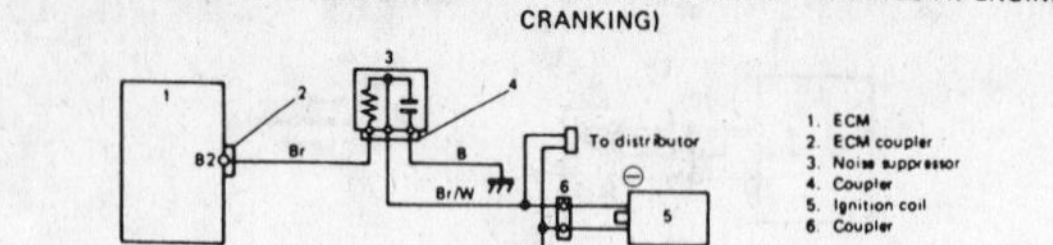
B2
Br
B
Br/W
B/W
To distributor
To ignition switch
1. ECM
2. ECM coupler
3. Noise suppressor
4. Coupler
5. Ignition coil
6. Coupler
Ignition Signal Circuit

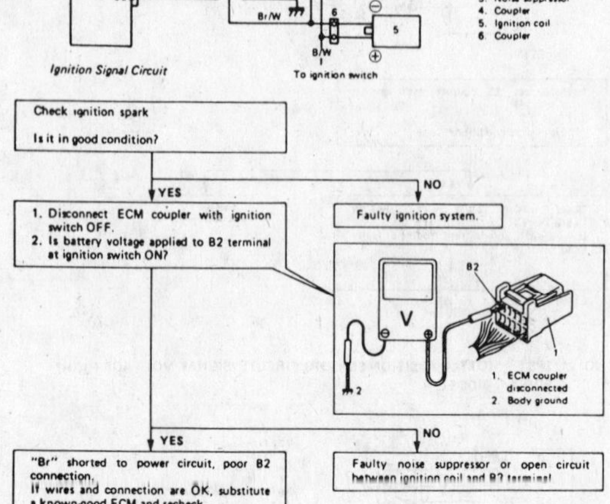
Check ignition spark
Is it in good condition?
YES
NO
1. Disconnect ECM coupler with ignition switch OFF.
2. Is battery voltage applied to B2 terminal at ignition switch ON?
Faulty ignition system.
B2
1. ECM coupler disconnected
2. Body ground
YES
NO
"Br" shorted to power circuit, poor B2 connection.
If wires and connection are OK, substitute a known-good ECM and recheck.
Faulty noise suppressor or open circuit between ignition coil and B2 terminal.

CODE NO. 51 EGR SYSTEM (FAULTY EGR SYSTEM)

For California spec. model only

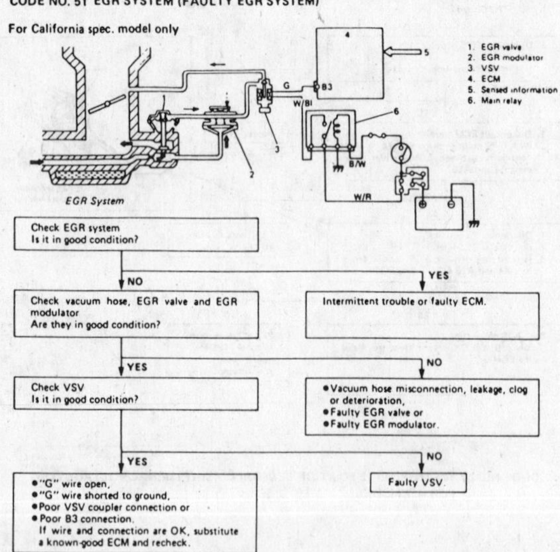
G
B3
W/Bl
B/W
W/R
1. EGR valve
2. EGR modulator
3. VSV
4. ECM
5. Sensed information
6. Main relay
EGR System
Check EGR system
Is it in good condition?
NO
YES
Check vacuum hose, EGR valve and EGR modulator
Are they in good condition?
Intermittent trouble or faulty ECM.
YES
NO
Check VSV
Is it in good condition?
• Vacuum hose misconnection, leakage, clog or deterioration.
• Faulty EGR valve or
• Faulty EGR modulator.
YES
NO
• "G" wire open,
• "G" wire shorted to ground,
• Poor VSV coupler connection or
• Poor B3 connection.
If wire and connection are OK, substitute a known-good ECM and recheck.
Faulty VSV.

TROUBLE DIAGNOSIS – 1988–90 EXCEPT GTI

SYMPTOM	POSSIBLE CAUSE	INSPECTION
Hard or no starting (Engine cranks OK)	• Shortage of fuel in fuel tank	
	• Injector or its circuit faulty	Diagnostic flow chart B-1
	• Faulty fuel pump or its circuit open	Diagnostic flow chart B-2
	• Fuel pressure out of specification	Diagnostic flow chart B-3
	• Faulty air valve	
	• Engine start signal not to fed	Diagnostic flow chart B-5
	• Poor performance of ATS, WTS or pressure sensor	
	• Faulty ECM	

NOTE:
- If engine doesn't start at all, perform fuel injector and its circuit check first. (Advance to "Diagnostic Flow Chart B-1".)
- If engine is hard to start only when it is cold, check air valve first.

SYMPTOM	POSSIBLE CAUSE	INSPECTION
Engine fails to idle	• Shortage of fuel in fuel tank	
	• Faulty ISC solenoid valve control system	Diagnostic flow chart B-4
	• Maladjusted idle speed adjusting screw	
	• Faulty air valve	
	• Faulty EGR system (if equipped.)	
	• Fuel pressure out of specification	Diagnostic flow chart B-3
	• Faulty injector	Check injector for resistance and injection condition
	• Poor performance of ATS, WTS or pressure sensor	
	• Faulty ECM	

NOTE:
If engine fails to idle only when it is cold, check air valve.

TROUBLE DIAGNOSIS CONTINUED—1988–90 EXCEPT GTI

SYMPTOM	POSSIBLE CAUSE	INSPECTION
Improper engine idle speed	• Maladjusted accelerator cable play • Clogged pressure sensor vacuum passage • Faulty ISC solenoid valve control system • Faulty air-conditioner VSV control system (if equipped) • Faulty idle switch (in TS or TPS) • Maladjusted idle speed adjusting screw • Faulty air valve • Fuel pressure out of specification • Poor performance of ATS, WTS or pressure sensor • Faulty ECM	Check vacuum hose and filter Diagnostic flow chart B-4 Diagnostic flow chart B-3

NOTE:
- With engine warmed up, if engine idle speed is high and ISC solenoid valve is not heard to operate, check accelerator cable play, idle switch and ISC solenoid valve control system in that order.
- If engine idle speed lowers below specification only when electric load is applied (e.g. headlight ON), check ISC solenoid valve control system first.
- With A/T model, if engine idle speed lowers below specification only when shifted to "R", "D", "2" or "L" range, check if "R", "D", "2" or "L" signal is inputted to ECM first.

SYMPTOM	POSSIBLE CAUSE	INSPECTION
Engine has no or poor power	• Maladjusted accelerator cable play • Faulty TS (wide open switch) for M/T model • Faulty EGR system (if equipped.) • Fuel pressure out of specification (Low fuel pressure) • Poor performance of TPS (For A/T model), ATS, WTS or pressure sensor • Faulty ECM	Diagnostic flow chart B-3
Engine hesitates when accelerating	• Clogged pressure sensor vacuum passage • Faulty EGR system (if equipped.) • Fuel pressure out of specification (Low fuel pressure) • Poor performance of ATS, WTS or pressure sensor • Faulty ECM	Check vacuum hose and filter Diagnostic flow chart B-3
Surges (Variation in car speed is felt although accelerator pedal is not operated)	• Variable fuel pressure (Clogged fuel filter, faulty fuel pressure regulator, etc.) • Poor performance of pressure sensor • Faulty ECM	Diagnostic flow chart B-3

SYMPTOM	POSSIBLE CAUSE	INSPECTION
Poor gasoline mileage	• High idle speed • Fuel pressure out of specification or fuel leakage • Faulty TS (M/T model) or TPS (A/T model) • Poor performance of ATS or WTS • Faulty ECM	Diagnostic flow chart B-3
Excessive hydrocarbon (HC) emission	• Engine not at normal operating temperature • Clogged air cleaner • Faulty ignition system • Vacuum leaks • Low compression • Lead contamination of catalytic converter • Fuel pressure out of specification • A/F feed back compensation fails – Faulty TS (M/T model) or TPS (A/T model) – Poor performance of WTS or pressure sensor • Poor performance of ATS • Faulty injector • Faulty ECM	Check for absence of filler neck restrictor Diagnostic flow chart B-3
Excessive carbon monoxide (CO)	• Engine not at normal operating temperature • Clogged air cleaner • Faulty ignition system • Low compression • Lead contamination of catalytic conveter • Fuel pressure out of specification • A/F feed back compensation fails – Faulty TS (M/T model) or TPS (A/T model) – Poor performance of WTS or pressure sensor • Poor performance of ATS • Faulty injector • Faulty ECM	Check for absence of filler neck restrictor Diagnostic flow chart B-3

SYMPTOM	POSSIBLE CAUSE	INSPECTION
Excessive nitrogen oxides (NOx) emission	• Improper ignition timing • Lead contamination of catalytic converter • Faulty EGR system (if equipped.) • Fuel pressure out of specification • A/F feed back compensation fails – Faulty TS (M/T model) or TPS (A/T model) – Poor performance of WTS or pressure sensor • Poor performance of ATS • Faulty injector • Faulty ECM	Check for adsence of filler neck restrictor Diagnostic flow chart B-3

B-1 FUEL INJECTOR AND ITS CIRCUIT CHECK (ENGINE NOT STARTING)

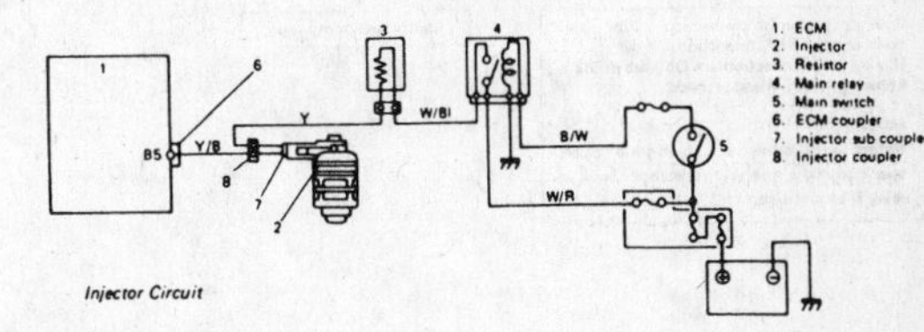
1. ECM
2. Injector
3. Resistor
4. Main relay
5. Main switch
6. ECM coupler
7. Injector sub coupler
8. Injector coupler
B5
Y/B
Y
W/Bl
B/W
W/R

Injector Circuit

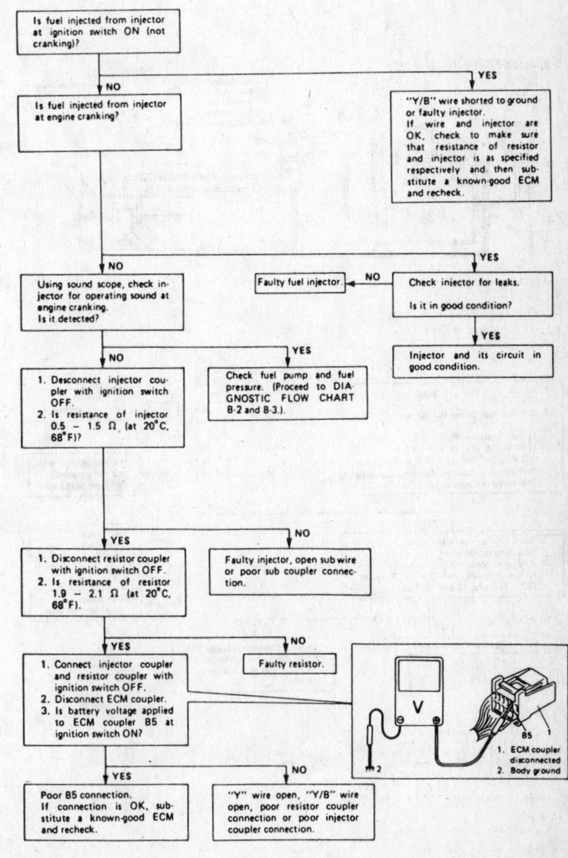
Is fuel injected from injector at ignition switch ON (not cranking)?
YES
"Y/B" wire shorted to ground or faulty injector. If wire and injector are OK, check to make sure that resistance of resistor and injector is as specified respectively and, then substitute a known-good ECM and recheck.
NO
Is fuel injected from injector at engine cranking?
YES
Check injector for leaks. Is it in good condition?
NO
Faulty fuel injector.
YES
Injector and its circuit in good condition.
NO
Using sound scope, check injector for operating sound at engine cranking. Is it detected?
YES
Check fuel pump and fuel pressure. (Proceed to DIAGNOSTIC FLOW CHART B-2 and B-3.).
NO
1. Desconnect injector coupler with ignition switch OFF.
2. Is resistance of injector 0.5 – 1.5 Ω (at 20°C, 68°F)?
NO
Faulty injector, open sub wire or poor sub coupler connection.
YES
1. Disconnect resistor coupler with ignition switch OFF.
2. Is resistance of resistor 1.9 – 2.1 Ω (at 20°C, 68°F).
NO
Faulty resistor.
YES
1. Connect injector coupler and resistor coupler with ignition switch OFF.
2. Disconnect ECM coupler.
3. Is battery voltage applied to ECM coupler B5 at ignition switch ON?
V
B5
1. ECM coupler disconnected
2. Body ground
YES
Poor B5 connection. If connection is OK, substitute a known-good ECM and recheck.
NO
"Y" wire open, "Y/B" wire open, poor resistor coupler connection or poor injector coupler connection.

TROUBLE DIAGNOSIS CONTINUED—1988–90 EXCEPT GTI

B-2 FUEL PUMP AND ITS CIRCUIT CHECK

1. ECM
2. ECM coupler
3. Fuel pump relay
4. Fuel pump
5. Main relay

Fuel Pump Circuit

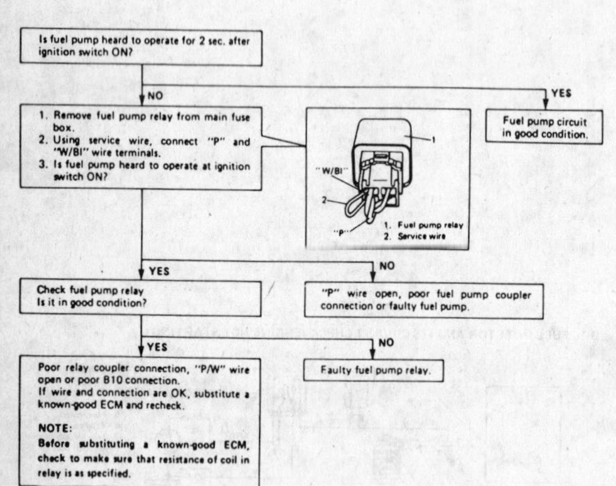
Is fuel pump heard to operate for 2 sec. after ignition switch ON?
NO
YES
Fuel pump circuit in good condition.
1. Remove fuel pump relay from main fuse box.
2. Using service wire, connect "P" and "W/Bl" wire terminals.
3. Is fuel pump heard to operate at ignition switch ON?
"W/Bl"
"P"
1. Fuel pump relay
2. Service wire
YES
NO
Check fuel pump relay
Is it in good condition?
"P" wire open, poor fuel pump coupler connection or faulty fuel pump.
YES
NO
Poor relay coupler connection, "P/W" wire open or poor B10 connection.
If wire and connection are OK, substitute a known-good ECM and recheck.
NOTE:
Before substituting a known-good ECM, check to make sure that resistance of coil in relay is as specified.
Faulty fuel pump relay.

B-3 FUEL PRESSURE CHECK

1. Fuel pump
2. Fuel filter
3. Throttle body
4. Fuel injector
5. Fuel pressure regulator
6. Special tool (Fuel pressure gauge & 3-way joint)
7. Fuel feed line
8. Fuel return line

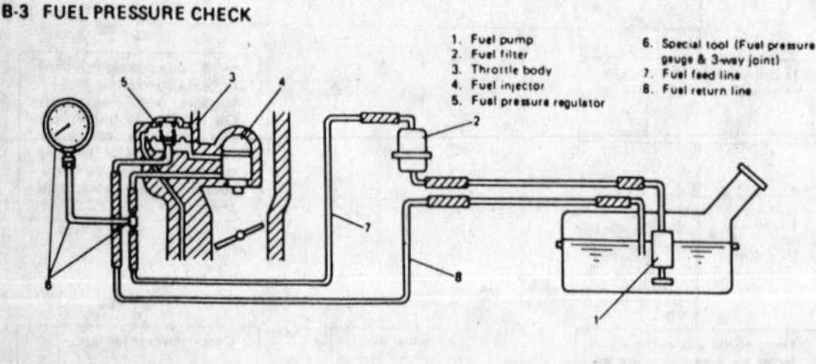

Fuel Pressure Check

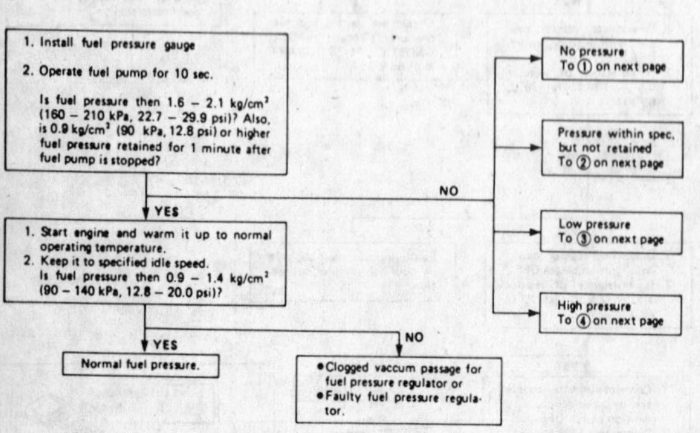
1. Install fuel pressure gauge
2. Operate fuel pump for 10 sec.
Is fuel pressure then 1.6 – 2.1 kg/cm² (160 – 210 kPa, 22.7 – 29.9 psi)? Also, is 0.9 kg/cm² (90 kPa, 12.8 psi) or higher fuel pressure retained for 1 minute after fuel pump is stopped?
NO
No pressure
To ① on next page
Pressure within spec. but not retained
To ② on next page
Low pressure
To ③ on next page
High pressure
To ④ on next page
YES
1. Start engine and warm it up to normal operating temperature.
2. Keep it to specified idle speed.
Is fuel pressure then 0.9 – 1.4 kg/cm² (90 – 140 kPa, 12.8 – 20.0 psi)?
YES
NO
Normal fuel pressure.
•Clogged vaccum passage for fuel pressure regulator or
•Faulty fuel pressure regulator.

B-3 FUEL PRESSURE CHECK (continued)

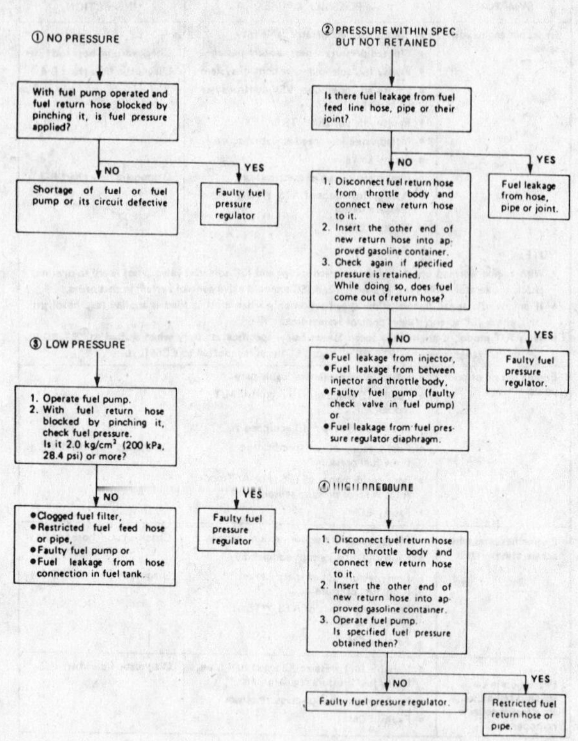
① NO PRESSURE
With fuel pump operated and fuel return hose blocked by pinching it, is fuel pressure applied?
NO
YES
Shortage of fuel or fuel pump or its circuit defective
Faulty fuel pressure regulator
③ LOW PRESSURE
1. Operate fuel pump.
2. With fuel return hose blocked by pinching it, check fuel pressure.
Is it 2.0 kg/cm² (200 kPa, 28.4 psi) or more?
NO
YES
•Clogged fuel filter,
•Restricted fuel feed hose or pipe,
•Faulty fuel pump or
•Fuel leakage from hose connection in fuel tank.
Faulty fuel pressure regulator
② PRESSURE WITHIN SPEC. BUT NOT RETAINED
Is there fuel leakage from fuel feed line hose, pipe or their joint?
NO
YES
Fuel leakage from hose, pipe or joint.
1. Disconnect fuel return hose from throttle body and connect new return hose to it.
2. Insert the other end of new return hose into approved gasoline container.
3. Check again if specified pressure is retained. While doing so, does fuel come out of return hose?
NO
YES
•Fuel leakage from injector,
•Fuel leakage from between injector and throttle body,
•Faulty fuel pump (faulty check valve in fuel pump) or
•Fuel leakage from fuel pressure regulator diaphragm.
Faulty fuel pressure regulator.
④ HIGH PRESSURE
1. Disconnect fuel return hose from throttle body and connect new return hose to it.
2. Insert the other end of new return hose into approved gasoline container.
3. Operate fuel pump.
Is specified fuel pressure obtained then?
NO
YES
Faulty fuel pressure regulator.
Restricted fuel return hose or pipe.

B-4 ISC SOLENOID VALVE CONTROL SYSTEM CHECK

1. ISC solenoid valve
2. ECM
3. Sensed information
4. Monitor coupler
5. Diagnosis switch terminal
6. Main relay
7. Silencer

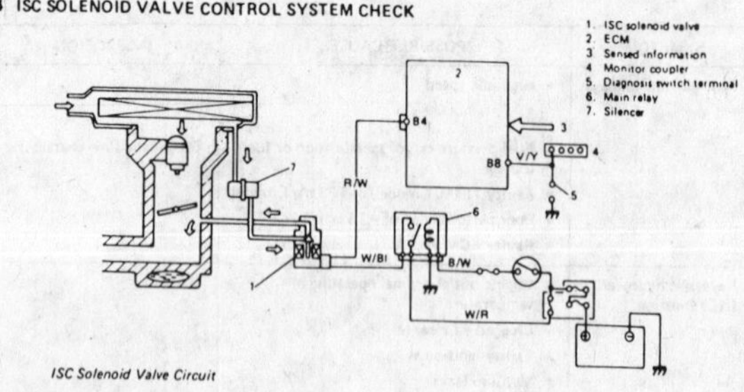

ISC Solenoid Valve Circuit

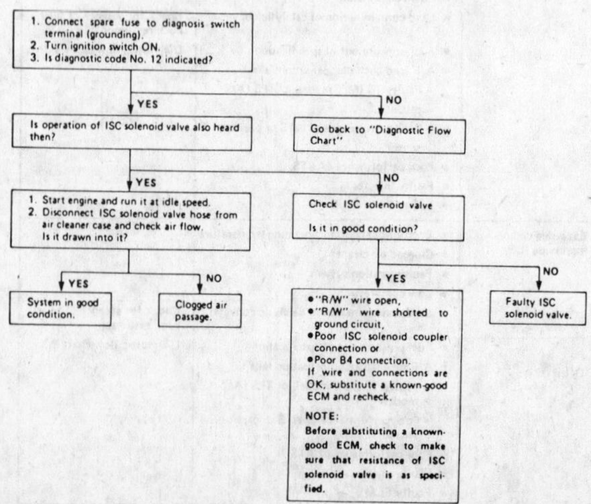
1. Connect spare fuse to diagnosis switch terminal (grounding).
2. Turn ignition switch ON.
3. Is diagnostic code No. 12 indicated?
YES
NO
Is operation of ISC solenoid valve also heard then?
Go back to "Diagnostic Flow Chart"
YES
NO
1. Start engine and run it at idle speed.
2. Disconnect ISC solenoid valve hose from air cleaner case and check air flow.
Is it drawn into it?
Check ISC solenoid valve
Is it in good condition?
YES
NO
System in good condition.
Clogged air passage.
YES
NO
•"R/W" wire open,
•"R/W" wire shorted to ground circuit,
•Poor ISC solenoid coupler connection or
•Poor B4 connection.
If wire and connections are OK, substitute a known-good ECM and recheck.
NOTE:
Before substituting a known-good ECM, check to make sure that resistance of ISC solenoid valve is as specified.
Faulty ISC solenoid valve.

TROUBLE DIAGNOSIS CONTINUED 1988–90 EXCEPT GTI

B-5 ENGINE START SIGNAL CHECK

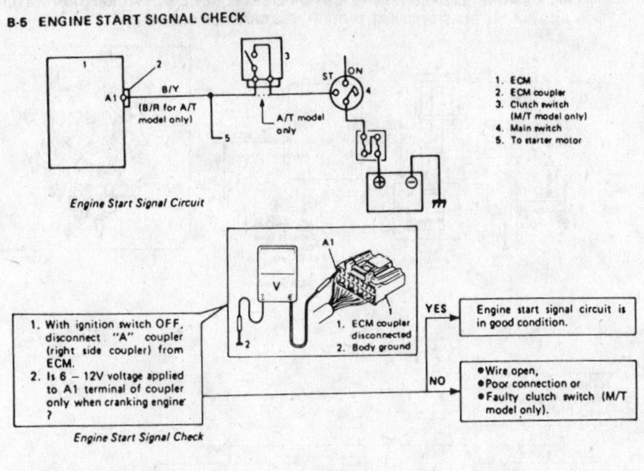
A1
B/Y
(B/R for A/T model only)
A/T model only
ST
ON
1. ECM
2. ECM coupler
3. Clutch switch (M/T model only)
4. Main switch
5. To starter motor
Engine Start Signal Circuit
1. With ignition switch OFF, disconnect "A" coupler (right side coupler) from ECM.
2. Is 6 – 12V voltage applied to A1 terminal of coupler only when cranking engine?
1. ECM coupler disconnected
2. Body ground
YES
Engine start signal circuit is in good condition.
NO
• Wire open,
• Poor connection or
• Faulty clutch switch (M/T model only).
Engine Start Signal Check

B-6 "R", "D", "2" OR "L" RANGE SIGNAL CHECK (A/T MODEL ONLY)

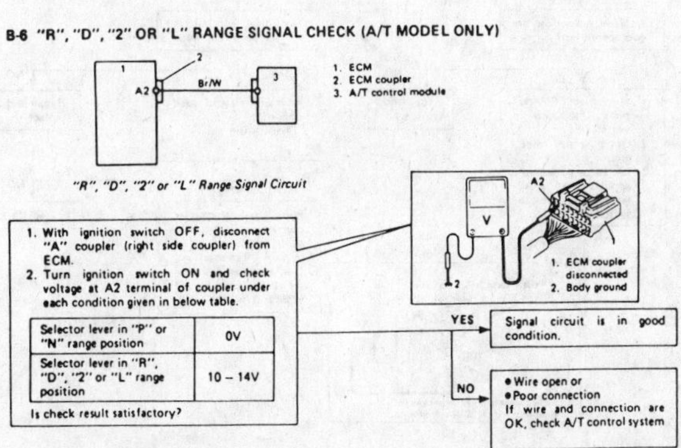
A2
B/W
1. ECM
2. ECM coupler
3. A/T control module
"R", "D", "2" or "L" Range Signal Circuit
1. With ignition switch OFF, disconnect "A" coupler (right side coupler) from ECM.
2. Turn ignition switch ON and check voltage at A2 terminal of coupler under each condition given in below table.
Selector lever in "P" or "N" range position | 0V
Selector lever in "R", "D", "2" or "L" range position | 10 – 14V
Is check result satisfactory?
1. ECM coupler disconnected
2. Body ground
YES
Signal circuit is in good condition.
NO
• Wire open or
• Poor connection
If wire and connection are OK, check A/T control system

ECM TERMINAL IDENTIFICATION 1988–90 EXCEPT GTI

TERMINAL	CIRCUIT	NORMAL VOLTAGE	CONDITION
A1	Engine start switch (Engine start signal)	6 – 12V	While engine cranking
		0V	Other than above
A2	Lighting switch (Illumination light signal, if equipped)	0V	Lighting switch OFF
		10 – 14V	Lighting switch ON
A2 (A/T model only)	A/T control module ("R", "D", "2" or "L" range signal)	0V	Ignition switch ON, Selector lever in "P" or "N" range position
		10 – 14V	Ignition switch ON, Selector lever in "R", "D", "2" or "L" range position
A3	Idle switch (in TS or TPS)	0V	Ignition switch ON Throttle valve at idle position
		4 – 5V	Ignition switch ON Throttle valve opens larger than idle position
A4	Power source of sensor (PS and TPS)	4.75–5.25V	Ignition switch ON
A5	Pressure sensor	3.5 – 4.1V	Ignition switch ON Barometric pressure: 760 mmHg
A6	Oxygen sensor	Indicator deflection repeated between over and under 0.45V	While engine running at 2,000 r/min after warmed up and kept running at 2,000 r/min for 1 minute
A7	Circuit ground	0V	Ignition switch ON
A8	Shift-up indicator light (if equipped)	1 – 2V	Ignition switch ON, Lighting switch OFF
		3 – 5V	Ignition switch ON, Lighting switch ON
		10 – 14V	While engine running at idle speed
A8	——	——	——
A9 (A/T model only)	A/T control module (Throttle valve opening signal)	10 – 14V	Ignition switch ON, Throttle valve at idle position
		10 – 14V ↕ 0.2 – 0.4V	Ignition switch ON, Voltage varies as specified at the left while throttle valve is opened gradually. (Refer to Fig. 6E-157 for relations between opening and voltage)
A10	Air conditioner circuit (If equipped)	10 – 14V	Ignition switch ON
		0 – 0.6V	While engnie running at idle speed Air-conditioner ON

ECM TERMINAL IDENTIFICATION CONTINUED – 1988–90 EXCEPT GTI

TERMINAL	CIRCUIT	NORMAL VOLTAGE	CONDITION
A11	Speed sensor	Indicator deflection repeated between 0V and 4 – 5V	Ignition switch ON, front left tire turned slowly with front right tire locked.
A12	Test switch terminal	4 – 5V	Ignition switch ON
		0V	Ignition switch ON, Test switch terminal grounded
A13	ATS	2.0 – 2.7V	Ignition switch ON Sensor ambient temp (Intake air temp.): 20°C (68°F)
A14	WTS	0.45–0.80V	Ignition switch ON Engine cooling water temp.: 80°C (176°F)
A15 (M/T model only)	Wide open switch (in TS)	4 – 5V	Ignition switch ON Throttle valve at idle position
		0V	Ignition switch ON Throttle valve at full open position
A15 (A/T model only)	TPS	0.25–0.85V	Ignition switch ON Throttle valve at idle position
		3.3 – 4.5V	Ignition switch ON Throttle valve at full open position
A16	Ground of sensors	0V	Ignition switch ON
A17 (A/T model only)	A/T control module (Throttle valve opening signal)	10 – 14V	Ignition switch ON, Throttle valve at idle position
		10 – 14V ↕ 0.2 – 0.4V	Ignition switch ON Voltage varies as specified at the left while throttle valve is opened gradually. (Refer to Fig. 6E-157 for relations between opening and voltage.)
A17	——	——	
A18 (A/T model only)	A/T control module (Throttle valve opening signal)	10 – 14V	Ignition switch ON, Throttle valve at idle position
		10 – 14V ↕ 0.2 – 0.4V	Ignition switch ON Voltage varies as specified at the left while throttle valve is opened gradually. (Refer to Fig. 6E-157 for relations between opening and voltage.)

TERMINAL	CIRCUIT	NORMAL VOLTAGE	CONDITION
B1	Power source	10 – 14V	Ignition switch ON
B2	Ignition coil (Ignition signal)	10 – 14V	Ignition switch ON
B3 (California spec. model only)	EGR VSV	10 – 14V	Ignition switch ON
B4	ISC solenoid valve	0.9 – 1.5V	Ignition switch ON Diagnosis switch terminal ungrounded
		10 – 13V	Ignition switch ON Diagnosis switch terminal grounded
B5	Injector	10 – 14V	Ignition switch ON
B6	Ground	0V	Ignition switch ON
B7	Power source for back-up circuit	10 – 14V	Ignition switch ON and OFF
B8	Diagnosis switch terminal	4 – 5V	Ignition switch ON
		0V	Ignition switch ON, Diagnosis switch terminal grounded
B9	"CHECK ENGINE" light	1.2 – 2.0 V	Ignition switch ON Diagnosis switch terminal ungrounded
		10 – 14V	When engine running Diagnosis switch terminal ungrounded
B10	Fuel pump relay	1.2 – 1.8V	For 2 seconds after ignition switch ON
		10 – 14V	When over 2 seconds after ignition switch ON
B11	——	——	——
B12	——	——	——

ECM TERMINAL IDENTIFICATION CONTINUED – 1988–90 EXCEPT GTI

TERMINALS	CIRCUIT	NORMAL RESISTANCE	CONDITION
A3 – A16	Idle switch (in TS or TPS)	0 (Zero)	Throttle valve is at idle position
		∞ (Infinity)	Throttle valve opens larger than idle position
A11 – Body ground	Speed sensor	Ohmmeter indicator deflects between 0 and ∞	While front left tire turned slowly with front right tire locked
A12 – Body ground	Test switch terminal	0	Test switch terminal ground
		∞	Test switch terminal ungrounded
A13 – A16	ATS	2.21 – 2.69 kΩ	Intake air temp. 20°C (68°F)
A14 – A16	WTS	290 – 354 Ω	Engine cooling water temp. 80°C (176°F)
A15 – A16 (M/T model only)	Wide open switch (in TS)	∞ (Infinity)	Throttle valve at idle position
		0 (Zero)	Throttle valve at full open position
A15 – A16 (A/T model only)	TPS	240 – 1140 Ω	Throttle valve at idle position
		3.17 – 6.60 kΩ	Throttle valve at full open position

TERMINALS	CIRCUIT	NORMAL RESISTANCE	CONDITION
B3 – B1 (California spec. model only)	EGR VSV	33 – 39 Ω	———
B4 – B1	ISC solenoid valve	30 – 33 Ω	———
B5 – B1	Injector and resistor	2.4 – 3.6 Ω	———
B6 – Body ground	ECM ground	0 (Zero)	———

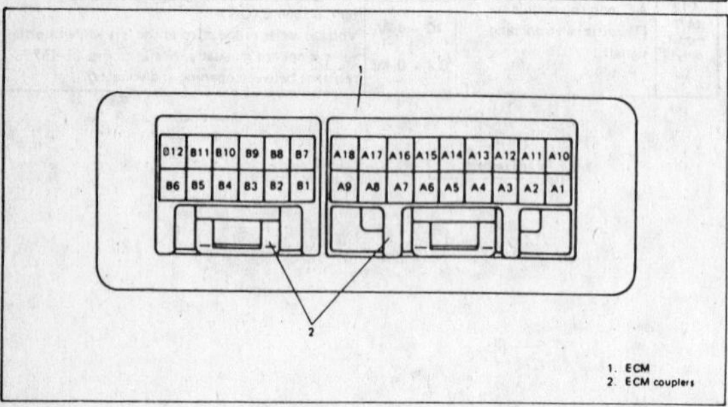
B12 B11 B10 B9 B8 B7
B6 B5 B4 B3 B2 B1
A18 A17 A16 A15 A14 A13 A12 A11 A10
A9 A8 A7 A6 A5 A4 A3 A2 A1
1
2
1. ECM
2. ECM couplers

DIAGNOSTIC CODE TABLE – 1988–90 GTI

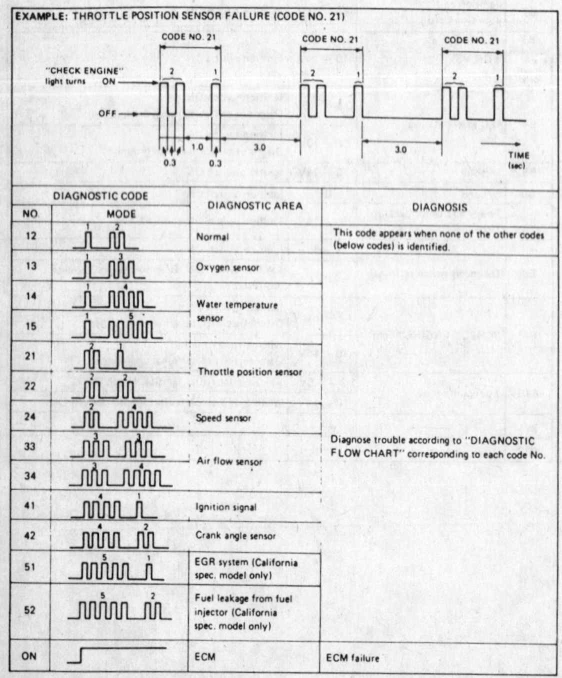

EXAMPLE: THROTTLE POSITION SENSOR FAILURE (CODE NO. 21)

"CHECK ENGINE" light turns ON / OFF — CODE NO. 21 — 0.3, 1.0, 0.3, 3.0, 3.0 — TIME (sec)

DIAGNOSTIC CODE NO.	MODE	DIAGNOSTIC AREA	DIAGNOSIS
12		Normal	This code appears when none of the other codes (below codes) is identified.
13		Oxygen sensor	Diagnose trouble according to "DIAGNOSTIC FLOW CHART" corresponding to each code No.
14		Water temperature sensor	
15			
21		Throttle position sensor	
22			
24		Speed sensor	
33		Air flow sensor	
34			
41		Ignition signal	
42		Crank angle sensor	
51		EGR system (California spec. model only)	
52		Fuel leakage from fuel injector (California spec. model only)	
ON		ECM	ECM failure

DIAGNOSTIC FLOW CHART – 1988–90 GTI

A-1 ECM POWER AND GROUND CIRCUIT CHECK
("CHECK ENGINE" LIGHT DOESN'T LIGHT OR LIGHTS ONLY DIMLY AT IGNITION SWITCH ON AND ENGINE DOESN'T START THOUGH IT IS CRANKED UP.)

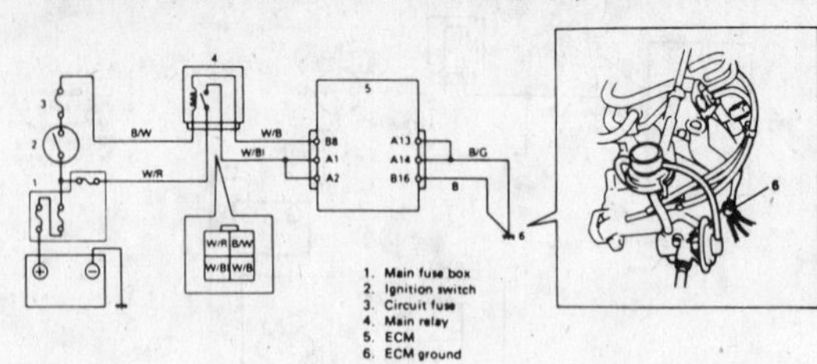
B/W
W/B
W/R
W/BI
W/R B/W
W/BI W/B
B8
A1
A2
A13
A14
B16
B/G
B
1. Main fuse box
2. Ignition switch
3. Circuit fuse
4. Main relay
5. ECM
6. ECM ground

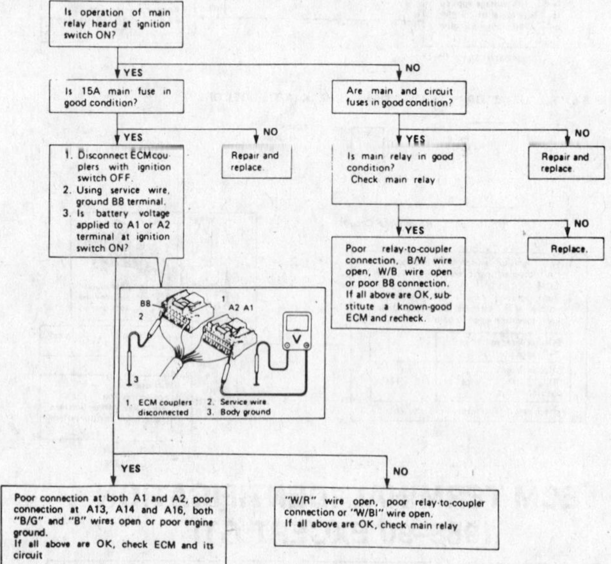
Is operation of main relay heard at ignition switch ON?
YES
NO
Is 15A main fuse in good condition?
Are main and circuit fuses in good condition?
YES
NO
1. Disconnect ECM couplers with ignition switch OFF.
2. Using service wire, ground B8 terminal.
3. Is battery voltage applied to A1 or A2 terminal at ignition switch ON?
Repair and replace.
Is main relay in good condition? Check main relay.
Repair and replace.
YES
NO
Poor relay-to-coupler connection, B/W wire open, W/B wire open or poor B8 connection. If all above are OK, substitute a known-good ECM and recheck.
Replace.
B8
A2 A1
1. ECM couplers disconnected
2. Service wire
3. Body ground
YES
NO
Poor connection at both A1 and A2, poor connection at A13, A14 and A16, both "B/G" and "B" wires open or poor engine ground. If all above are OK, check ECM and its circuit
"W/R" wire open, poor relay-to-coupler connection or "W/BI" wire open. If all above are OK, check main relay

DIAGNOSTIC FLOW CHART CONTINUED 1988–90 GTI

A-2 "CHECK ENGINE" LIGHT CIRCUIT CHECK
("CHECK ENGINE" LIGHT DOES NOT LIGHT BUT ENGINE STARTS.)

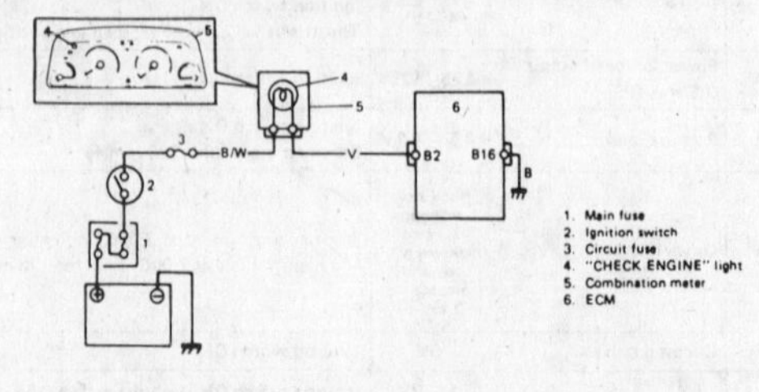
B/W
V
B2
B16
B
1. Main fuse
2. Ignition switch
3. Circuit fuse
4. "CHECK ENGINE" light
5. Combination meter
6. ECM

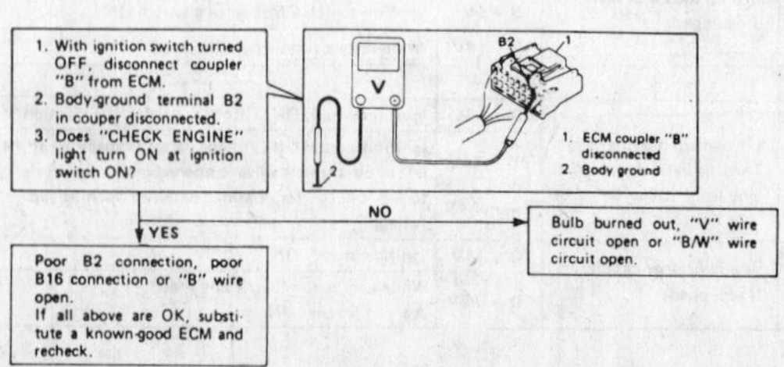
1. With ignition switch turned OFF, disconnect coupler "B" from ECM.
2. Body-ground terminal B2 in coupler disconnected.
3. Does "CHECK ENGINE" light turn ON at ignition switch ON?
B2
1. ECM coupler "B" disconnected
2. Body ground
YES
NO
Poor B2 connection, poor B16 connection or "B" wire open. If all above are OK, substitute a known-good ECM and recheck.
Bulb burned out, "V" wire circuit open or "B/W" wire circuit open.

DIAGNOSTIC FLOW CHART CONTINUED 1988–90 GTI

A-3 "CHECK ENGINE" LIGHT CIRCUIT CHECK
("CHECK ENGINE" LIGHT DOESN'T FLASH OR JUST REMAINS ON EVEN WITH SPARE FUSE CONNECTED TO DIAGNOSIS SWITCH TERMINAL.)

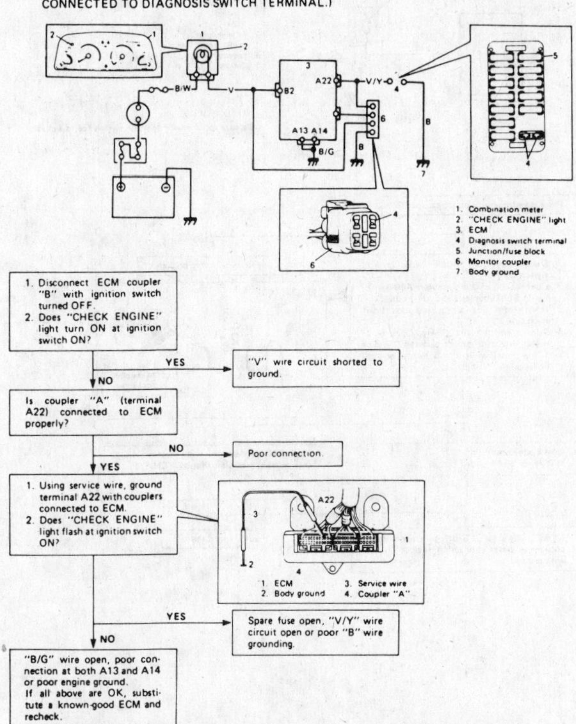
B/W
V
B2
A22
V/Y
A13 A14
B/G
B
1. Combination meter
2. "CHECK ENGINE" light
3. ECM
4. Diagnosis switch terminal
5. Junction/fuse block
6. Monitor coupler
7. Body ground
1. Disconnect ECM coupler "B" with ignition switch turned OFF.
2. Does "CHECK ENGINE" light turn ON at ignition switch ON?
YES
"V" wire circuit shorted to ground.
NO
Is coupler "A" (terminal A22) connected to ECM properly?
NO
Poor connection.
YES
1. Using service wire, ground terminal A22 with couplers connected to ECM.
2. Does "CHECK ENGINE" light flash at ignition switch ON?
1. ECM
2. Body ground
3. Service wire
4. Coupler "A"
YES
Spare fuse open, "V/Y" wire circuit open or poor "B" wire grounding.
NO
"B/G" wire open, poor connection at both A13 and A14 or poor engine ground.
If all above are OK, substitute a known-good ECM and recheck.

DIAGNOSTIC CODE – 1988–90 GTI

CODE NO. 13 OXYGEN SENSOR CIRCUIT (SIGNAL VOLTAGE DOESN'T CHANGE)

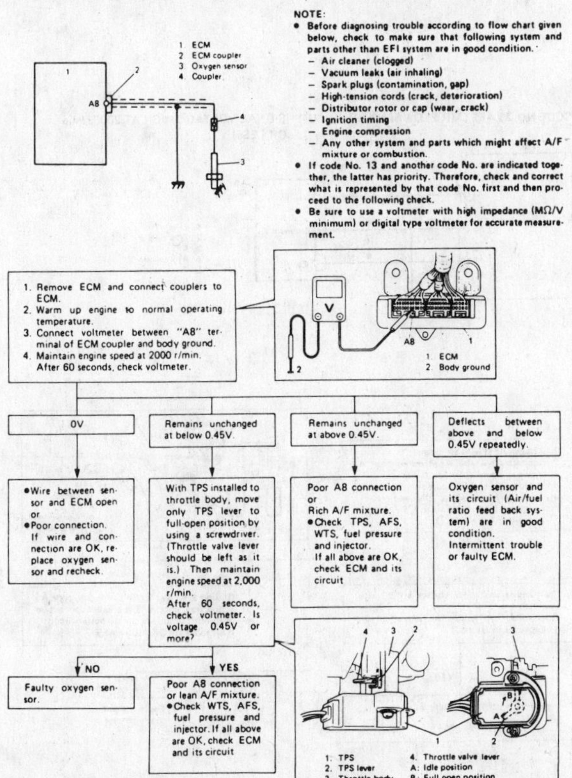
1. ECM
2. ECM coupler
3. Oxygen sensor
4. Coupler
A8
NOTE:
• Before diagnosing trouble according to flow chart given below, check to make sure that following system and parts other than EFI system are in good condition.
– Air cleaner (clogged)
– Vacuum leaks (air inhaling)
– Spark plugs (contamination, gap)
– High-tension cords (crack, deterioration)
– Distributor rotor or cap (wear, crack)
– Ignition timing
– Engine compression
– Any other system and parts which might affect A/F mixture or combustion.
• If code No. 13 and another code No. are indicated together, the latter has priority. Therefore, check and correct what is represented by that code No. first and then proceed to the following check.
• Be sure to use a voltmeter with high impedance (MΩ/V minimum) or digital type voltmeter for accurate measurement.
1. Remove ECM and connect couplers to ECM.
2. Warm up engine to normal operating temperature.
3. Connect voltmeter between "A8" terminal of ECM coupler and body ground.
4. Maintain engine speed at 2000 r/min. After 60 seconds, check voltmeter.
1. ECM
2. Body ground
0V
Remains unchanged at below 0.45V.
Remains unchanged at above 0.45V.
Deflects between above and below 0.45V repeatedly.
•Wire between sensor and ECM open or
•Poor connection.
If wire and connection are OK, replace oxygen sensor and recheck.
With TPS installed to throttle body, move only TPS lever to full-open position by using a screwdriver. (Throttle valve lever should be left as it is.) Then maintain engine speed at 2,000 r/min. After 60 seconds, check voltmeter. Is voltage 0.45V or more?
Poor A8 connection or Rich A/F mixture.
•Check TPS, AFS, WTS, fuel pressure and injector.
If all above are OK, check ECM and its circuit
Oxygen sensor and its circuit (Air/fuel ratio feed back system) are in good condition.
Intermittent trouble or faulty ECM.
NO
Faulty oxygen sensor.
YES
Poor A8 connection or lean A/F mixture.
•Check WTS, AFS, fuel pressure and injector. If all above are OK, check ECM and its circuit
1: TPS
2: TPS lever
3: Throttle body
4: Throttle valve lever
A: Idle position
B: Full open position

DIAGNOSTIC CODE – 1988–90 GTI

CODE NO. 14 WTS (WATER TEMPERATURE SENSOR) CIRCUIT (LOW TEMPERATURE INDICATED, SIGNAL VOLTAGE HIGH)

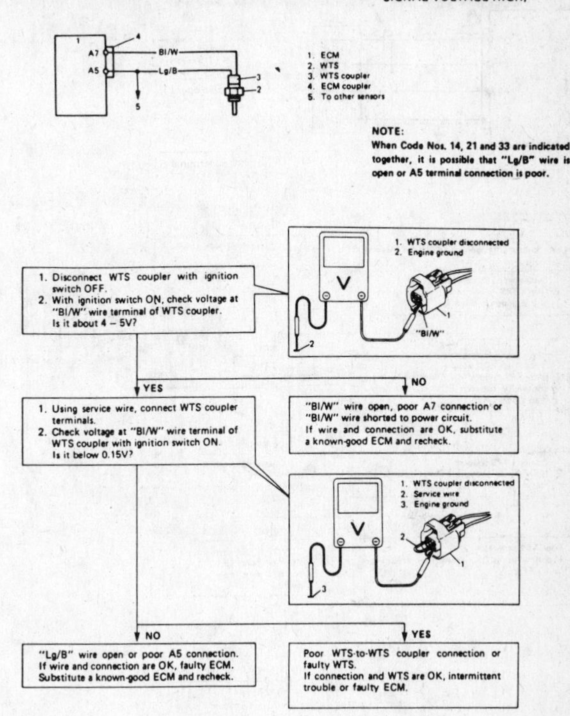
A7
A5
Bl/W
Lg/B
1. ECM
2. WTS
3. WTS coupler
4. ECM coupler
5. To other sensors
NOTE:
When Code Nos. 14, 21 and 33 are indicated together, it is possible that "Lg/B" wire is open or A5 terminal connection is poor.
1. Disconnect WTS coupler with ignition switch OFF.
2. With ignition switch ON, check voltage at "Bl/W" wire terminal of WTS coupler. Is it about 4 – 5V?
1. WTS coupler disconnected
2. Engine ground
YES
NO
1. Using service wire, connect WTS coupler terminals.
2. Check voltage at "Bl/W" wire terminal of WTS coupler with ignition switch ON. Is it below 0.15V?
"Bl/W" wire open, poor A7 connection or "Bl/W" wire shorted to power circuit.
If wire and connection are OK, substitute a known-good ECM and recheck.
1. WTS coupler disconnected
2. Service wire
3. Engine ground
NO
"Lg/B" wire open or poor A5 connection.
If wire and connection are OK, faulty ECM.
Substitute a known-good ECM and recheck.
YES
Poor WTS-to-WTS coupler connection or faulty WTS.
If connection and WTS are OK, intermittent trouble or faulty ECM.

CODE NO. 15 WTS (WATER TEMPERATURE SENSOR) CIRCUIT (HIGH TEMPERATURE INDICATED, SIGNAL VOLTAGE LOW)

1. ECM
2. WTS
3. Coupler
4. ECM coupler
5. To other sensors

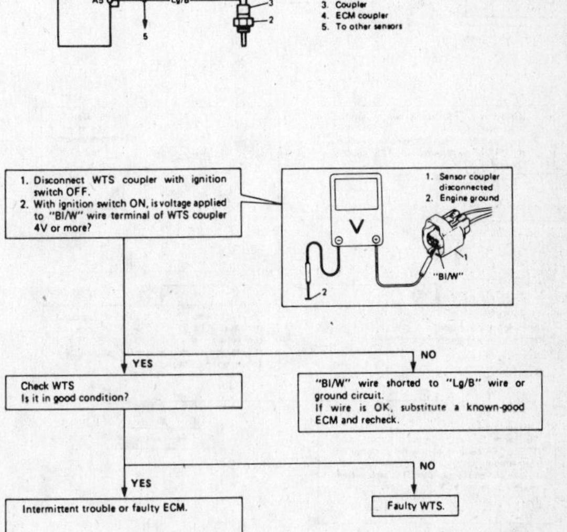
1. Disconnect WTS coupler with ignition switch OFF.
2. With ignition switch ON, is voltage applied to "Bl/W" wire terminal of WTS coupler 4V or more?
1. Sensor coupler disconnected
2. Engine ground
YES
NO
Check WTS
Is it in good condition?
"Bl/W" wire shorted to "Lg/B" wire or ground circuit.
If wire is OK, substitute a known-good ECM and recheck.
YES
Intermittent trouble or faulty ECM.
NO
Faulty WTS.

DIAGNOSTIC CODE – 1988–90 GTI

CODE NO. 21 TPS (THROTTLE POSITION SENSOR) CIRCUIT (SIGNAL VOLTAGE HIGH)

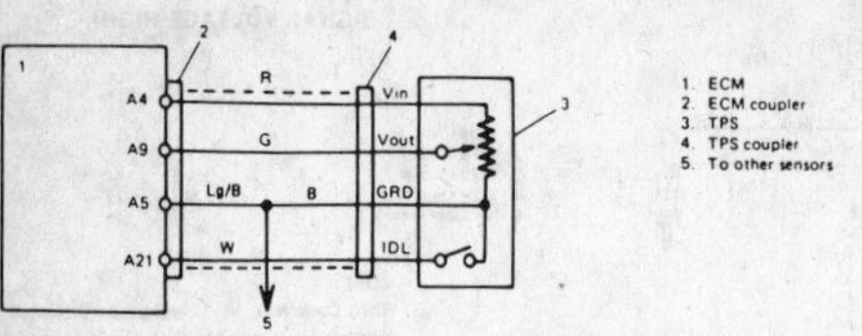
A4
A9
A5
A21
R
G
Lg/B
B
W
Vin
Vout
GRD
IDL

1. ECM
2. ECM coupler
3. TPS
4. TPS coupler
5. To other sensors

NOTE:
Be sure to turn OFF ignition switch for this check.

1. Disconnect TPS coupler.
2. Check TPS
   Is it in good condition?

NO → Faulty TPS.

YES ↓

1. Disconnect ECM coupler.
2. With TPS coupler disconnected, is there continuity between ECM coupler terminals A4 and A9?

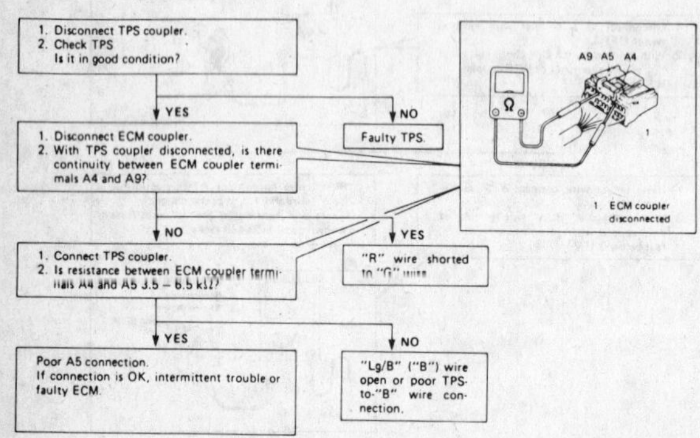
A9 A5 A4
1. ECM coupler disconnected

YES → "R" wire shorted to "G" wire

NO ↓

1. Connect TPS coupler.
2. Is resistance between ECM coupler terminals A4 and A5 3.5 – 6.5 kΩ?

YES ↓ Poor A5 connection. If connection is OK, intermittent trouble or faulty ECM.

NO → "Lg/B" ("B") wire open or poor TPS-to-"B" wire connection.

CODE NO. 24 SPEED SENSOR CIRCUIT (CAR SPEED LOWER THAN 1 km/h (0,6 mile/h) ALTHOUGH FUEL IS KEPT CUT AT LOWER THAN 4000 r/min FOR LONGER THAN 4 SECONDS)

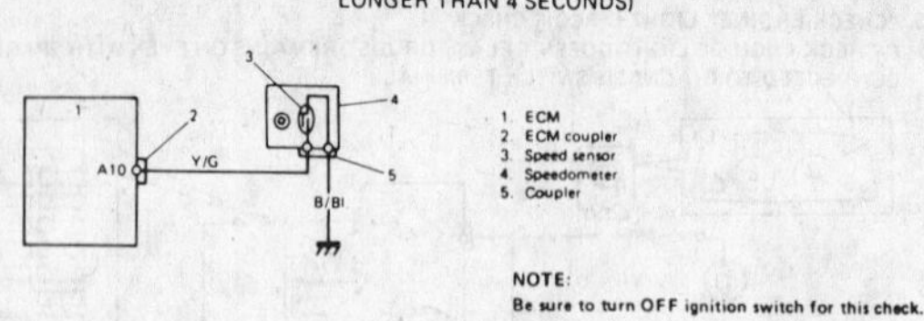
A10
Y/G
B/Bl

1. ECM
2. ECM coupler
3. Speed sensor
4. Speedometer
5. Coupler

NOTE:
Be sure to turn OFF ignition switch for this check.

Does speedometer indicate car speed?

NO → Speed meter cable broken.

YES ↓

1. Disconnect ECM coupler with ignition switch OFF.
2. Connect ohmmeter between A10 terminal of ECM coupler and body ground.
3. Hoist front end of car and lock front right tire.
4. Turn front left tire slowly. Does ohmmeter indicator deflect between 0 and ∞ a few times while tire is turned one revolution?

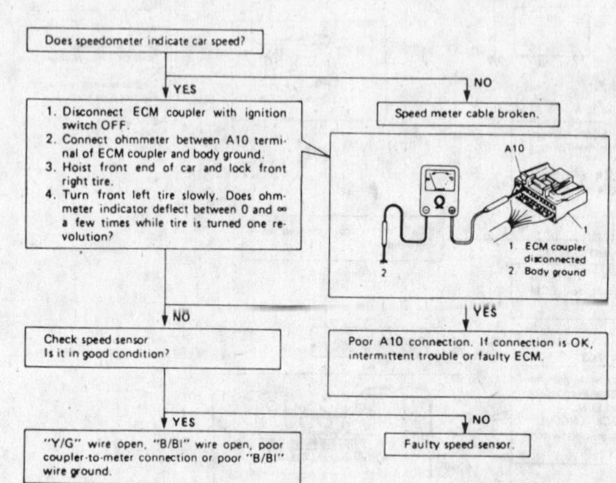
A10
1. ECM coupler disconnected
2. Body ground

YES → Poor A10 connection. If connection is OK, intermittent trouble or faulty ECM.

NO ↓

Check speed sensor
Is it in good condition?

YES ↓ "Y/G" wire open, "B/Bl" wire open, poor coupler-to-meter connection or poor "B/Bl" wire ground.

NO → Faulty speed sensor.

CODE NO. 22 TPS (THROTTLE POSITION SENSOR) CIRCUIT (SIGNAL VOLTAGE LOW)

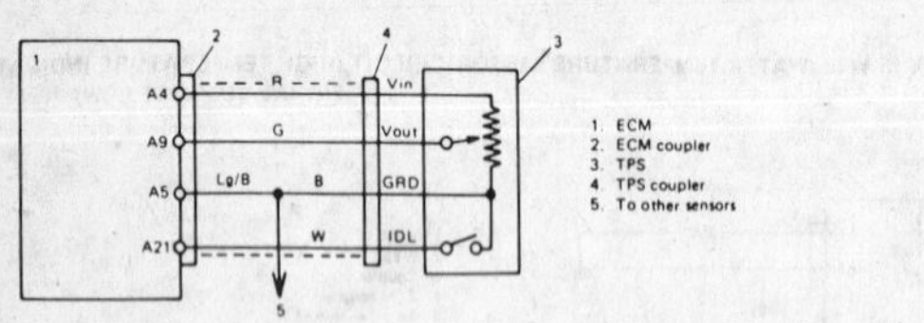
A4
A9
A5
A21
R
G
Lg/B
B
W
Vin
Vout
GRD
IDL

1. ECM
2. ECM coupler
3. TPS
4. TPS coupler
5. To other sensors

1. Disconnect TPS coupler with ignition switch OFF.
2. With ignition switch ON, is voltage applied to "R" wire terminal of TPS coupler about 4 – 5V?

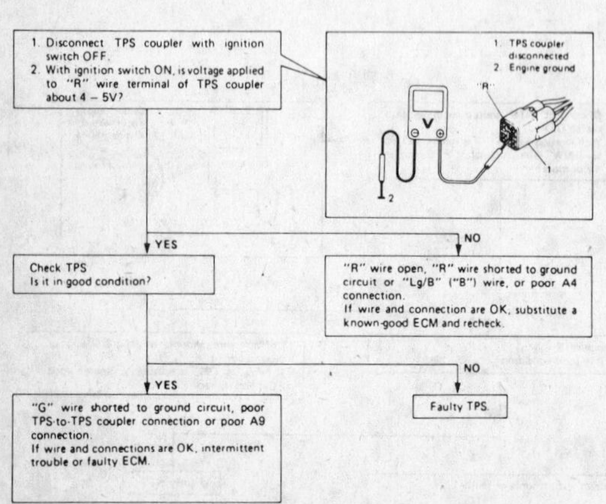
1. TPS coupler disconnected
2. Engine ground
"R"

NO → "R" wire open, "R" wire shorted to ground circuit or "Lg/B" ("B") wire, or poor A4 connection. If wire and connection are OK, substitute a known-good ECM and recheck.

YES ↓

Check TPS
Is it in good condition?

NO → Faulty TPS.

YES ↓ "G" wire shorted to ground circuit, poor TPS-to-TPS coupler connection or poor A9 connection. If wire and connections are OK, intermittent trouble or faulty ECM.

CODE NO. 33 AFS (AIR FLOW SENSOR) CIRCUIT (SIGNAL VOLTAGE HIGH AT 2000 r/min OR LESS)

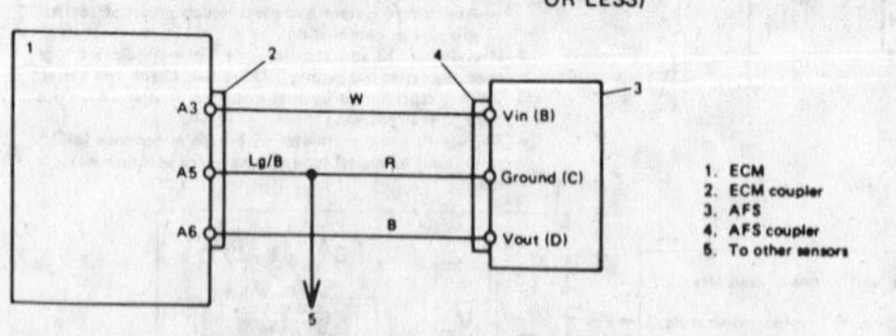
A3
A5
A6
W
Lg/B
R
B
Vin (B)
Ground (C)
Vout (D)

1. ECM
2. ECM coupler
3. AFS
4. AFS coupler
5. To other sensors

1. Remove AFM with air cleaner.
2. Connect AFM and AFM outlet hose as shown in figure.
3. Remove seal of AFS coupler.
4. Turn ignition switch ON and check volage at "B" wire terminal.
   Is it within 0.2 – 0.8V?

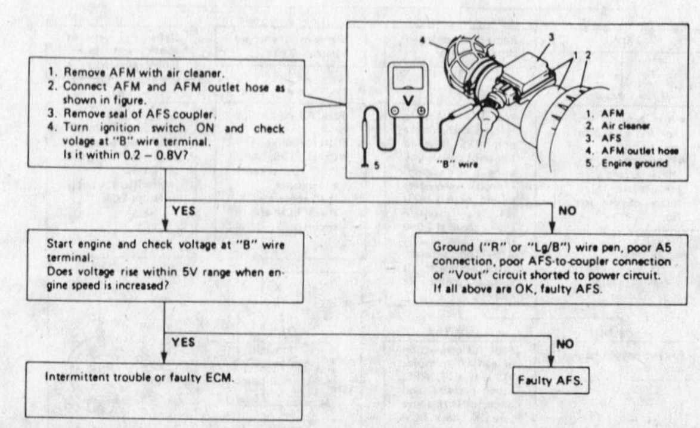
"B" wire
1. AFM
2. Air cleaner
3. AFS
4. AFM outlet hose
5. Engine ground

NO → Ground ("R" or "Lg/B") wire pen, poor A5 connection, poor AFS-to-coupler connection or "Vout" circuit shorted to power circuit. If all above are OK, faulty AFS.

YES ↓

Start engine and check voltage at "B" wire terminal.
Does voltage rise within 5V range when engine speed is increased?

YES ↓ Intermittent trouble or faulty ECM.

NO → Faulty AFS.

DIAGNOSTIC CODE – 1988–90 GTI

CODE NO. 34 AFS (AIR FLOW SENSOR) CIRCUIT (SIGNAL VOLTAGE LOW)

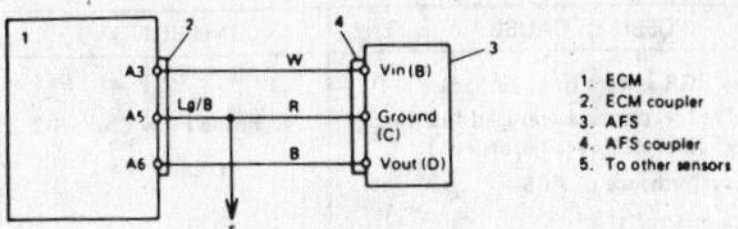
1
A3
A5
A6
W
Lg/B
R
B
Vin (B)
Ground (C)
Vout (D)
1. ECM
2. ECM coupler
3. AFS
4. AFS coupler
5. To other sensors

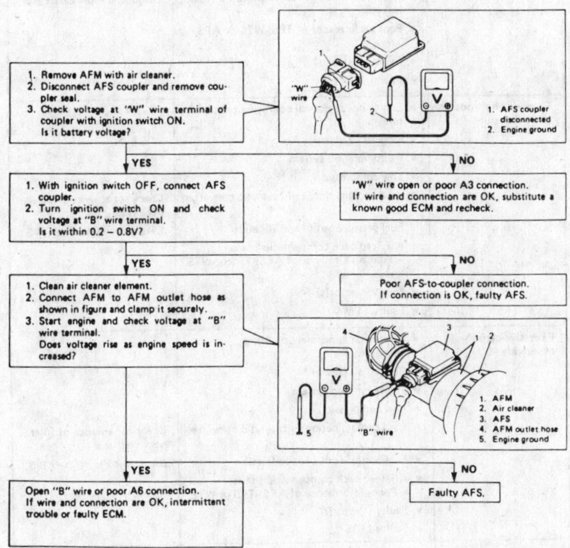
1. Remove AFM with air cleaner.
2. Disconnect AFS coupler and remove coupler seal.
3. Check voltage at "W" wire terminal of coupler with ignition switch ON.
Is it battery voltage?
"W" wire
1. AFS coupler disconnected
2. Engine ground
YES
NO
1. With ignition switch OFF, connect AFS coupler.
2. Turn ignition switch ON and check voltage at "B" wire terminal.
Is it within 0.2 – 0.8V?
"W" wire open or poor A3 connection.
If wire and connection are OK, substitute a known good ECM and recheck.
YES
NO
1. Clean air cleaner element.
2. Connect AFM to AFM outlet hose as shown in figure and clamp it securely.
3. Start engine and check voltage at "B" wire terminal.
Does voltage rise as engine speed is increased?
Poor AFS-to-coupler connection.
If connection is OK, faulty AFS.
"B" wire
1. AFM
2. Air cleaner
3. AFS
4. AFM outlet hose
5. Engine ground
YES
NO
Open "B" wire or poor A6 connection.
If wire and connection are OK, intermittent trouble or faulty ECM.
Faulty AFS.

CODE NO. 41 IGNITION SIGNAL CIRCUIT (IGNITION SIGNAL NOT INPUTTED)

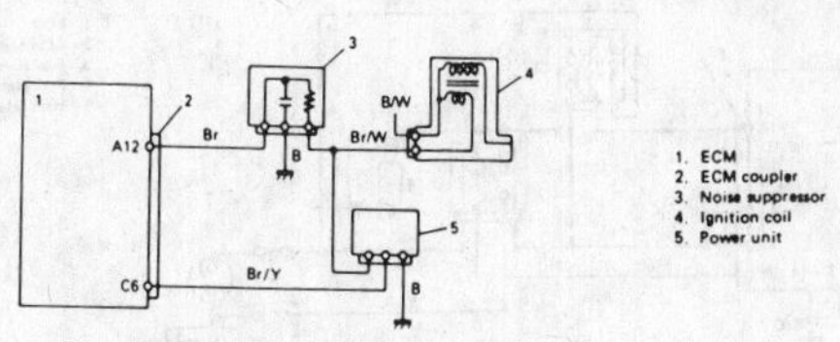
1
A12
C6
Br
B
B/W
Br/W
Br/Y
1. ECM
2. ECM coupler
3. Noise suppressor
4. Ignition coil
5. Power unit

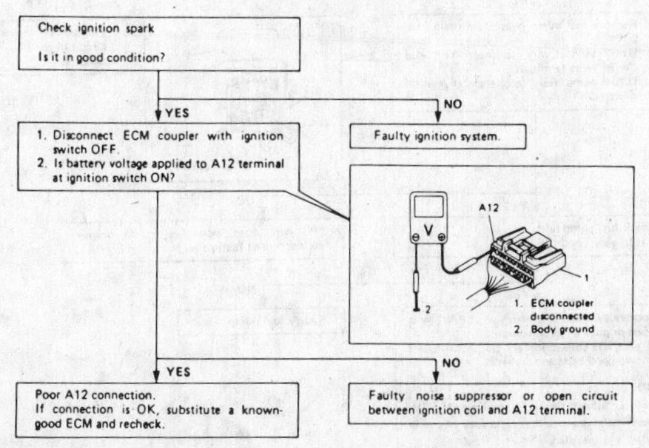
Check ignition spark
Is it in good condition?
YES
NO
1. Disconnect ECM coupler with ignition switch OFF.
2. Is battery voltage applied to A12 terminal at ignition switch ON?
Faulty ignition system.
A12
1. ECM coupler disconnected
2. Body ground
YES
NO
Poor A12 connection.
If connection is OK, substitute a known-good ECM and recheck.
Faulty noise suppressor or open circuit between ignition coil and A12 terminal.

CODE NO. 42 CRANK ANGLE SENSOR CIRCUIT (SENSOR SIGNAL NOT INPUTTED FOR 2 SECONDS AT ENGINE CRANKING)

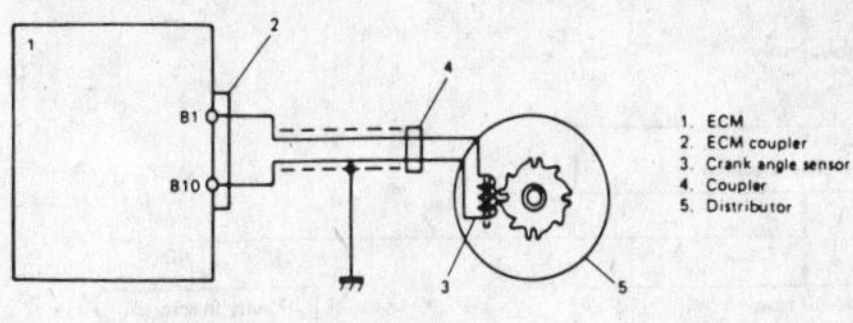
1
B1
B10
1. ECM
2. ECM coupler
3. Crank angle sensor
4. Coupler
5. Distributor

Crank Angle Sensor Circuit

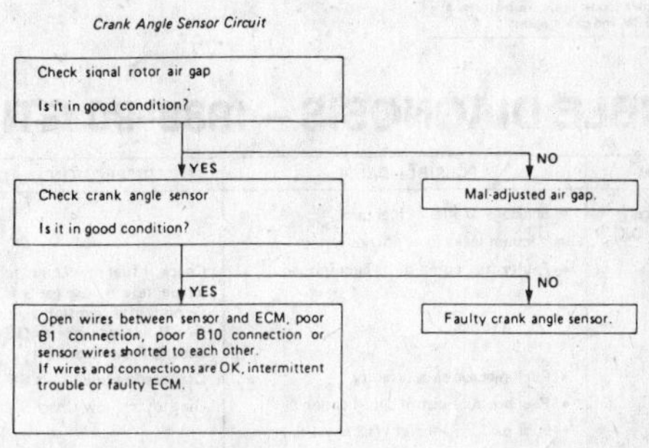
Check signal rotor air gap
Is it in good condition?
YES
NO
Mal-adjusted air gap.
Check crank angle sensor
Is it in good condition?
YES
NO
Faulty crank angle sensor.
Open wires between sensor and ECM, poor B1 connection, poor B10 connection or sensor wires shorted to each other.
If wires and connections are OK, intermittent trouble or faulty ECM.

CODE NO. 51 EGR SYSTEM AND REGTS (RECIRCULATED EXHAUST GAS TEMPERATURE SENSOR) CIRCUIT (RECIRCULATED EXHAUST GAS TEMP. DOESN'T CHANGE)

California spec. model only

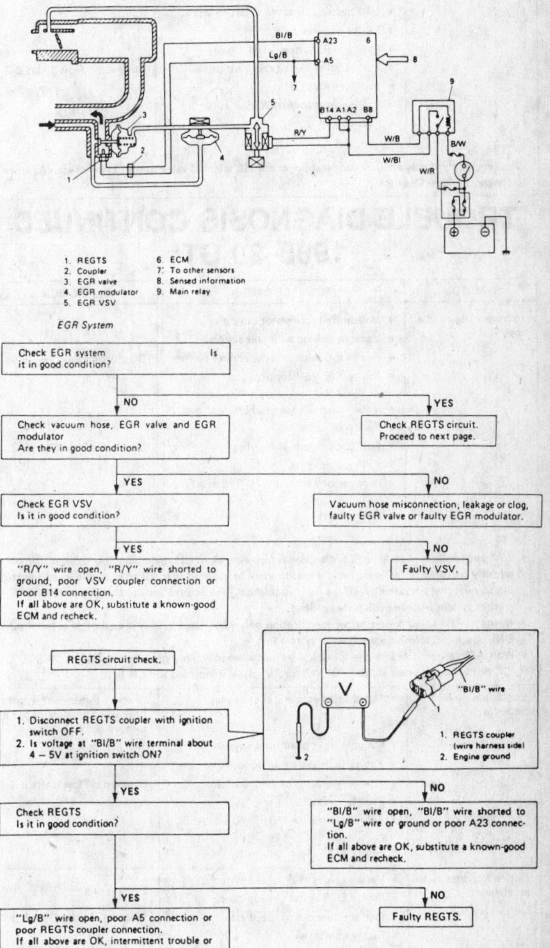
Bl/B
Lg/B
A23
A5
B14 A1 A2 B8
R/Y
W/B
W/Bl
B/W
W/R
1. REGTS
2. Coupler
3. EGR valve
4. EGR modulator
5. EGR VSV
6. ECM
7. To other sensors
8. Sensed information
9. Main relay
EGR System
Check EGR system
Is it in good condition?
NO
YES
Check vacuum hose, EGR valve and EGR modulator
Are they in good condition?
Check REGTS circuit.
Proceed to next page.
YES
NO
Check EGR VSV
Is it in good condition?
Vacuum hose misconnection, leakage or clog, faulty EGR valve or faulty EGR modulator.
YES
NO
"R/Y" wire open, "R/Y" wire shorted to ground, poor VSV coupler connection or poor B14 connection.
If all above are OK, substitute a known-good ECM and recheck.
Faulty VSV.
REGTS circuit check.
1. Disconnect REGTS coupler with ignition switch OFF.
2. Is voltage at "Bl/B" wire terminal about 4 – 5V at ignition switch ON?
"Bl/B" wire
1. REGTS coupler (wire harness side)
2. Engine ground
YES
NO
Check REGTS
Is it in good condition?
"Bl/B" wire open, "Bl/B" wire shorted to "Lg/B" wire or ground or poor A23 connection.
If all above are OK, substitute a known-good ECM and recheck.
YES
NO
"Lg/B" wire open, poor A5 connection or poor REGTS coupler connection.
If all above are OK, intermittent trouble or faulty ECM.
Faulty REGTS.

DIAGNOSTIC CODE 52 – 1988–90 GTI

CODE NO. 52 FUEL INJECTOR (FUEL LEAKAGE FROM FUEL INJECTOR, OXYGEN SENSOR OUTPUT VOLTAGE DOES NOT REDUCE TO LOWER THAN 0.35V WHEN FUEL IS CUT FOR LONGER THAN 1 SEC.)

California spec. model only

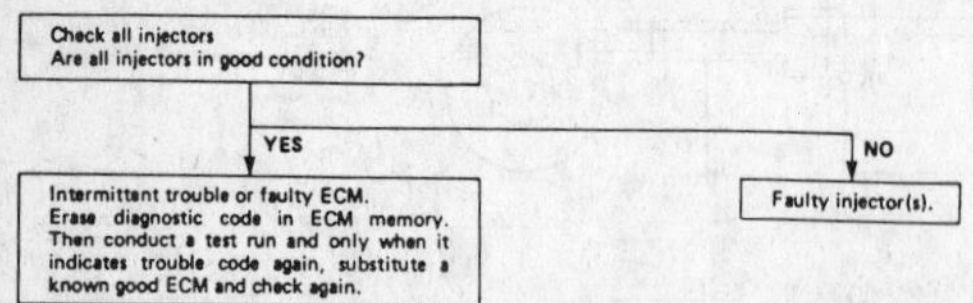
Check all injectors
Are all injectors in good condition?
YES
NO
Intermittant trouble or faulty ECM. Erase diagnostic code in ECM memory. Then conduct a test run and only when it indicates trouble code again, substitute a known good ECM and check again.
Faulty injector(s).

TROUBLE DIAGNOSIS – 1988–90 GTI

SYMPTOM	POSSIBLE CAUSE	INSPECTION
Hard or no starting (Engine cranks OK)	• Shortage of fuel in fuel tank • Vacuum leaks in air intake system • Faulty fuel pump or its circuit open • Fuel injector circuit faulty • Fuel pressure out of specification • Poor performance of WTS or AFS • Faulty ECM	Check if fuel pressure is felt at fuel return hose for 3 seconds after ignition switch ON. If not, advance to Diagnostic Flow Chart B-1. Diagnostic Flow Chart B-2 Diagnostic Flow Chart B-3
NOTE: If engine is warm and hard to start but starts easily with accelerator pedal depressed, check ISC solenoid valve control system first. (Advance to Diagnostic Flow Chart B-4.)		
Engine fails to idle	• Shortage of fuel in fuel tank • Vacuum leaks in air intake system • Maladjusted idle speed adjusting screw • Faulty air valve • Faulty EGR system (if equipped) • Fuel pressure out of specification • Faulty injector(s) • Poor performance of WTS or AFS • Faulty ECM	Diagnostic Flow Chart B-3
NOTE: When engine stops only at quick deceleration, check ISC solenoid valve control system first. (Advance to Diagnostic Flow Chart B-4.)		

TROUBLE DIAGNOSIS CONTINUED 1988–90 GTI

SYMPTOM	POSSIBLE CAUSE	INSPECTION
Improper engine idle speed	• Maladjusted accelerator cable play • Vacuum leaks in air intake system • Faulty ISC solenoid valve control system • Faulty A/C VSV control system (if equipped) • Maladjusted idle speed adjusting screw • Faulty air valve • Fuel pressure out of specification • Faulty injector(s) • Poor performance of WTS, TPS or AFS • Faulty ECM	Diagnostic Flow Chart B-4 Diagnostic Flow Chart B-3
NOTE: • With engine warmed up, if engine idle speed is high even when idle speed adjusting screw is tightened fully (in other words, if engine idle speed cannot be adjusted as specified with idle speed adjusting screw), check accelerator cable play, ISC solenoid valve control system, air valve and A/C VSV control system (if equipped) in that order. • If engine idle speed lowers below specification only when electric load is applied (e.g. headlight ON), check ISC solenoid valve control system first. • With A/T model, if engine idle speed lowers below specification only when shifted to "R", "D", "2" or "L" range, check if "R", "D", "2" or "L" signal is inputted to ECM first.		
Engine has no or poor power	• Maladjusted accelerator cable play • Vacuum leaks in air intake system • Faulty EGR system (if equipped) • Fuel pressure out of specification • Faulty injector(s) • Poor performance of TPS, WTS or AFS • Faulty ECM	Check if throttle valve opens fully when accelerator pedal is depressed fully. Diagnostic Flow Chart B-3
Engine hesitates when accelerating	• Vacuum leaks in air intake system • Faulty EGR system (if equipped) • Fuel pressure out of specification • Faulty injector(s) • Poor performance of TPS, WTS or AFS • Faulty ECM	Diagnostic Flow Chart B-3

TROUBLE DIAGNOSIS CONTINUED 1988–90 GTI

SYMPTOM	POSSIBLE CAUSE	INSPECTION
Surges (Variation in car speed is felt although accelerator pedal is not operated)	• Faulty EGR system (if equipped) • Variable fuel pressure (clogged fuel filter, faulty fuel pressure regulator, etc.) • Poor performance of AFS • Faulty injector(s) • Faulty ECM	Diagnostic Flow Chart B-3
Poor gasoline mileage	• High idle speed • Fuel pressure out of specification or fuel leakage • Poor performance of TPS, WTS or AFS • Faulty injector(s) • Faulty ECM	Diagnostic Flow Chart B-3
Excessive hydrocarbon (HC) emission	• Engine not at normal operating temperature • Clogged air cleaner • Faulty ignition system • Low compression • Lead contamination of catalytic converter • Fuel pressure out of specification • A/F feed back compensation fails – Poor performance of AFS, TPS or WTS • Faulty injector(s) • Faulty ECM	Check for absence of filler neck restrictor Diagnostic Flow Chart B-3
Excessive carbon monoxide (CO)	• Engine not at normal operating temperature • Clogged air cleaner • Faulty ignition system • Low compression • Lead contamination of catalytic converter • Fuel pressure out of specification • A/F feed back compensation fails – Poor performance of AFS, TPS or WTS • Faulty injector(s) • Faulty ECM	Check for absence of filler neck restrictor Diagnostic Flow Chart B-3
Excessive nitrogen oxides (NOx) emission	• Vacuum leaks in air intake system • Improper ignition timing • Lead contamination of catalytic converter • Faulty EGR system (if equipped) • Fuel pressure out of specification • A/F feed back compensation fails – Poor performance of AFS, TPS or WTS • Faulty injector(s) • Faulty ECM	Cehck for absence of filler neck restrictor Diagnostic Flow Chart B-3

FUEL SYSTEM DIAGNOSIS – 1988–90 GTI

B-1 FUEL PUMP CIRCUIT CHECK

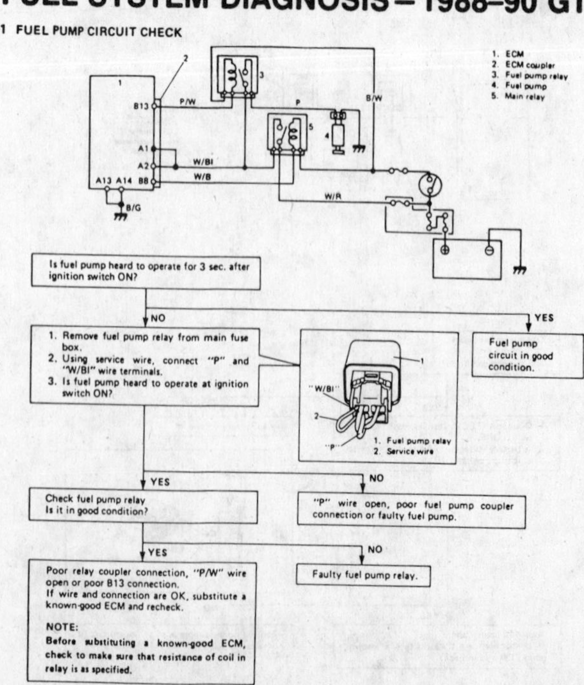
1. ECM
2. ECM coupler
3. Fuel pump relay
4. Fuel pump
5. Main relay
B13
P/W
P
B/W
A1
A2
W/Bl
W/B
A13 A14 B8
B/G
W/R
Is fuel pump heard to operate for 3 sec. after ignition switch ON?
NO
YES
1. Remove fuel pump relay from main fuse box.
2. Using service wire, connect "P" and "W/Bl" wire terminals.
3. Is fuel pump heard to operate at ignition switch ON?
Fuel pump circuit in good condition.
"W/Bl"
"P"
1. Fuel pump relay
2. Service wire
YES
NO
Check fuel pump relay
Is it in good condition?
"P" wire open, poor fuel pump coupler connection or faulty fuel pump.
YES
NO
Poor relay coupler connection, "P/W" wire open or poor B13 connection.
If wire and connection are OK, substitute a known-good ECM and recheck.
NOTE:
Before substituting a known-good ECM, check to make sure that resistance of coil in relay is as specified.
Faulty fuel pump relay.

FUEL SYSTEM DIAGNOSIS CONTINUED – 1988–90 GTi

B-2 FUEL INJECTOR CIRCUIT CHECK

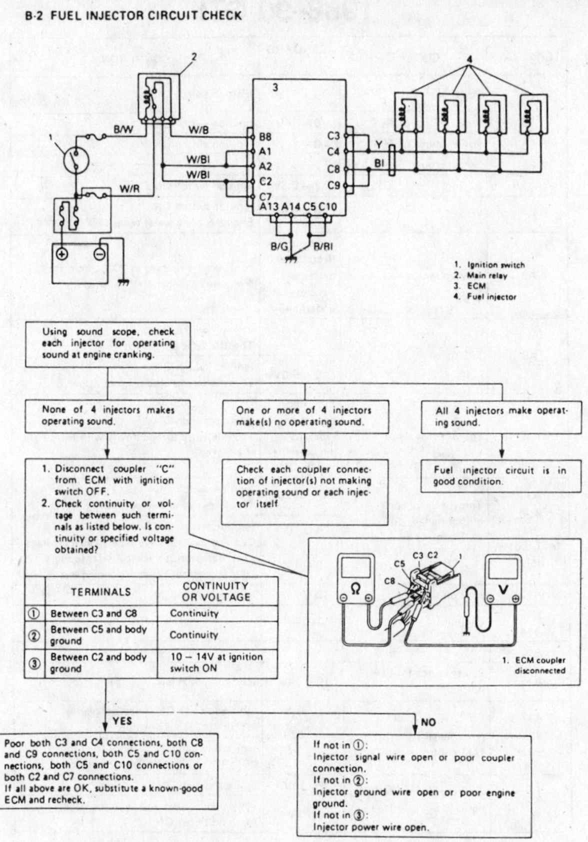
B/W
W/B
W/R
W/Bl
B8
A1
A2
C2
C7
A13 A14 C5 C10
C3
C4
C8
C9
Y
Bl
B/G
B/Bl
1. Ignition switch
2. Main relay
3. ECM
4. Fuel injector
Using sound scope, check each injector for operating sound at engine cranking.
None of 4 injectors makes operating sound.
One or more of 4 injectors make(s) no operating sound.
All 4 injectors make operating sound.
1. Disconnect coupler "C" from ECM with ignition switch OFF.
2. Check continuity or voltage between such terminals as listed below. Is continuity or specified voltage obtained?
Check each coupler connection of injector(s) not making operating sound or each injector itself
Fuel injector circuit is in good condition.
1. ECM coupler disconnected
YES
Poor both C3 and C4 connections, both C8 and C9 connections, both C5 and C10 connections, both C5 and C10 connections or both C2 and C7 connections. If all above are OK, substitute a known-good ECM and recheck.
NO
If not in ①: Injector signal wire open or poor coupler connection.
If not in ②: Injector ground wire open or poor engine ground.
If not in ③: Injector power wire open.

	TERMINALS	CONTINUITY OR VOLTAGE
①	Between C3 and C8	Continuity
②	Between C5 and body ground	Continuity
③	Between C2 and body ground	10 – 14V at ignition switch ON

B-3 FUEL PRESSURE CHECK (continued)

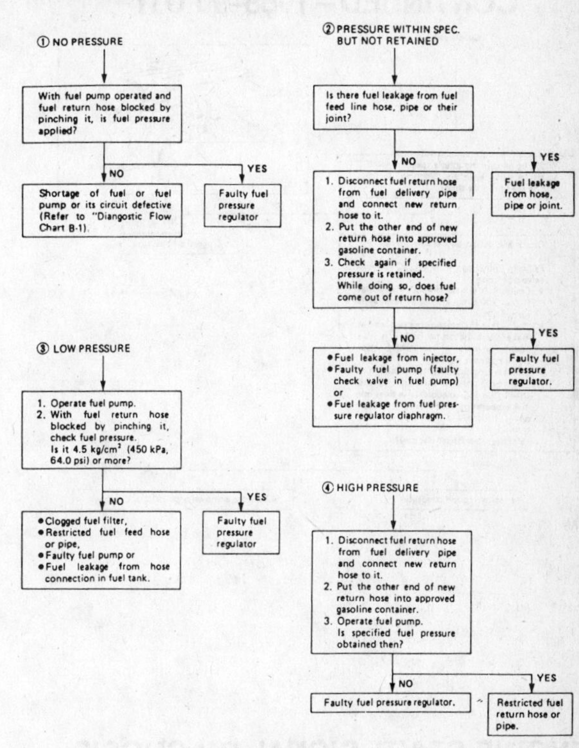
① NO PRESSURE
With fuel pump operated and fuel return hose blocked by pinching it, is fuel pressure applied?
NO
YES
Shortage of fuel or fuel pump or its circuit defective (Refer to "Diagnostic Flow Chart B-1).
Faulty fuel pressure regulator
② PRESSURE WITHIN SPEC. BUT NOT RETAINED
Is there fuel leakage from fuel feed line hose, pipe or their joint?
NO
YES
1. Disconnect fuel return hose from fuel delivery pipe and connect new return hose to it.
2. Put the other end of new return hose into approved gasoline container.
3. Check again if specified pressure is retained. While doing so, does fuel come out of return hose?
Fuel leakage from hose, pipe or joint.
NO
YES
• Fuel leakage from injector,
• Faulty fuel pump (faulty check valve in fuel pump) or
• Fuel leakage from fuel pressure regulator diaphragm.
Faulty fuel pressure regulator.
③ LOW PRESSURE
1. Operate fuel pump.
2. With fuel return hose blocked by pinching it, check fuel pressure. Is it 4.5 kg/cm² (450 kPa, 64.0 psi) or more?
NO
YES
• Clogged fuel filter,
• Restricted fuel feed hose or pipe,
• Faulty fuel pump or
• Fuel leakage from hose connection in fuel tank.
Faulty fuel pressure regulator
④ HIGH PRESSURE
1. Disconnect fuel return hose from fuel delivery pipe and connect new return hose to it.
2. Put the other end of new return hose into approved gasoline container.
3. Operate fuel pump. Is specified fuel pressure obtained then?
NO
YES
Faulty fuel pressure regulator.
Restricted fuel return hose or pipe.

ISC SOLENOID VALVE DIAGNOSIS 1988–90 GTi

B-4 ISC SOLENOID VALVE CONTROL SYSTEM CHECK

B-3 FUEL PRESSURE CHECK

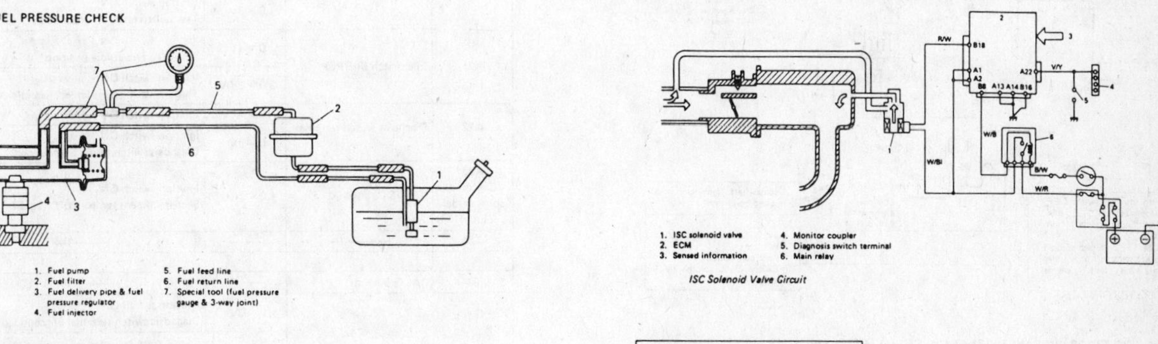

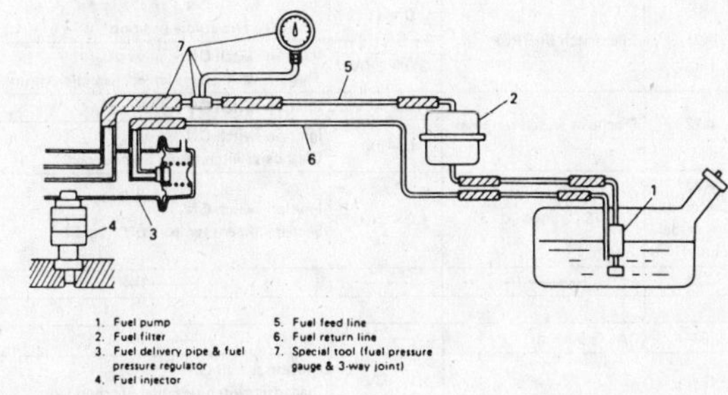
1. Fuel pump
2. Fuel filter
3. Fuel delivery pipe & fuel pressure regulator
4. Fuel injector
5. Fuel feed line
6. Fuel return line
7. Special tool (fuel pressure gauge & 3-way joint)

NOTE:
Before using following flow chart, check to make sure that battery voltage is higher than 11V. If battery voltage is low, pressure becomes lower than specification even if fuel pump and line are in good condition.

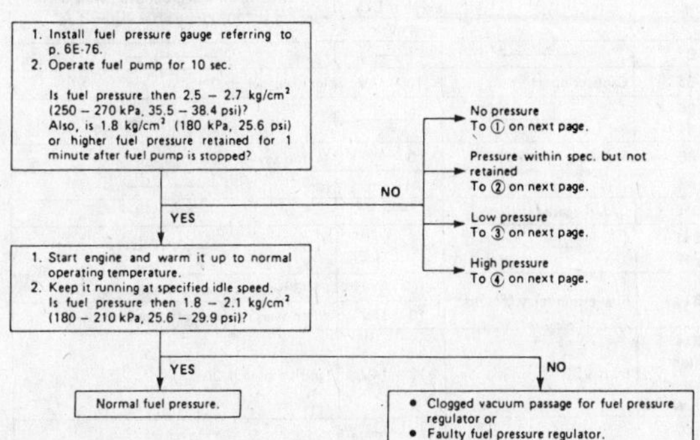
1. Install fuel pressure gauge referring to p. 6E-76.
2. Operate fuel pump for 10 sec.
Is fuel pressure then 2.5 – 2.7 kg/cm² (250 – 270 kPa, 35.5 – 38.4 psi)? Also, is 1.8 kg/cm² (180 kPa, 25.6 psi) or higher fuel pressure retained for 1 minute after fuel pump is stopped?
NO
No pressure To ① on next page.
Pressure within spec. but not retained To ② on next page.
Low pressure To ③ on next page.
High pressure To ④ on next page.
YES
1. Start engine and warm it up to normal operating temperature.
2. Keep it running at specified idle speed. Is fuel pressure then 1.8 – 2.1 kg/cm² (180 – 210 kPa, 25.6 – 29.9 psi)?
YES
NO
Normal fuel pressure.
• Clogged vacuum passage for fuel pressure regulator or
• Faulty fuel pressure regulator.

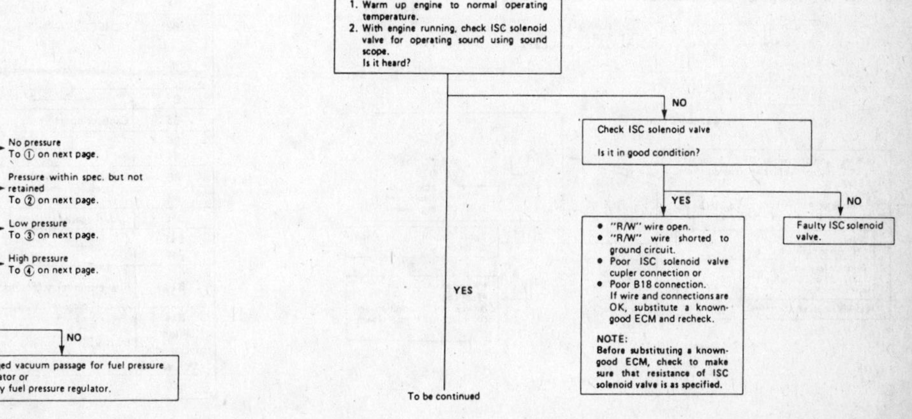
1. ISC solenoid valve
2. ECM
3. Sensed information
4. Monitor coupler
5. Diagnosis switch terminal
6. Main relay
ISC Solenoid Valve Circuit
1. Warm up engine to normal operating temperature.
2. With engine running, check ISC solenoid valve for operating sound using sound scope. Is it heard?
NO
Check ISC solenoid valve
Is it in good condition?
YES
NO
• "R/W" wire open.
• "R/W" wire shorted to ground circuit.
• Poor ISC solenoid valve cupler connection or
• Poor B18 connection. If wire and connections are OK, substitute a known-good ECM and recheck.
NOTE: Before substituting a known-good ECM, check to make sure that resistance of ISC solenoid valve is as specified.
Faulty ISC solenoid valve.
YES
To be continued

ISC SOLENOID VALVE DIAGNOSIS CONTINUED – 1988–90 GTI

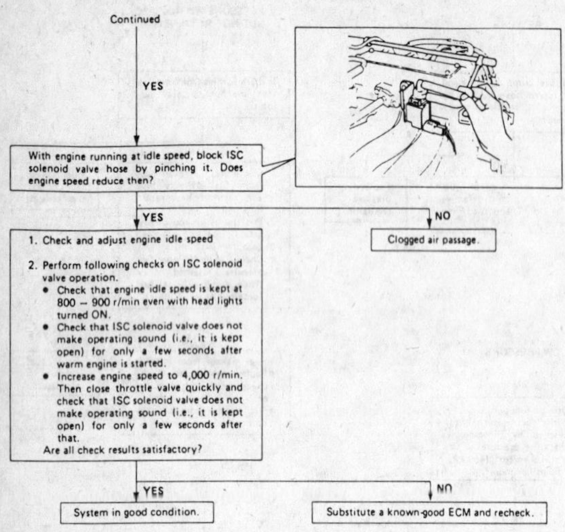
Continued
YES
With engine running at idle speed, block ISC solenoid valve hose by pinching it. Does engine speed reduce then?
YES
NO
Clogged air passage.
1. Check and adjust engine idle speed
2. Perform following checks on ISC solenoid valve operation.
• Check that engine idle speed is kept at 800 – 900 r/min even with head lights turned ON.
• Check that ISC solenoid valve does not make operating sound (i.e., it is kept open) for only a few seconds after warm engine is started.
• Increase engine speed to 4,000 r/min. Then close throttle valve quickly and check that ISC solenoid valve does not make operating sound (i.e., it is kept open) for only a few seconds after that.
Are all check results satisfactory?
YES
NO
System in good condition.
Substitute a known-good ECM and recheck.

ENGINE START SIGNAL DIAGNOSIS 1988–90 GTI

B-5 ENGINE START SIGNAL CHECK

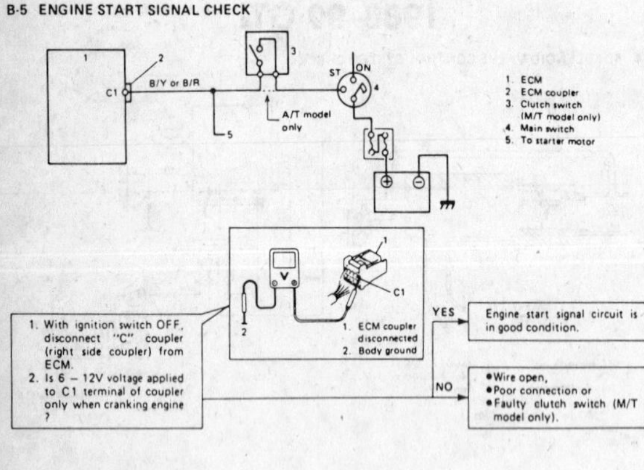
1
2
C1
B/Y or B/R
3
ST
ON
4
A/T model only
5
1. ECM
2 ECM coupler
3. Clutch switch (M/T model only)
4. Main switch
5. To starter motor
V
1
C1
2
1. ECM coupler disconnected
2. Body ground
1. With ignition switch OFF, disconnect "C" coupler (right side coupler) from ECM.
2. Is 6 – 12V voltage applied to C1 terminal of coupler only when cranking engine ?
YES
Engine start signal circuit is in good condition.
NO
•Wire open,
•Poor connection or
•Faulty clutch switch (M/T model only).

B-6 "R", "D", "2" OR "L" RANGE SIGNAL CHECK (A/T MODEL ONLY)

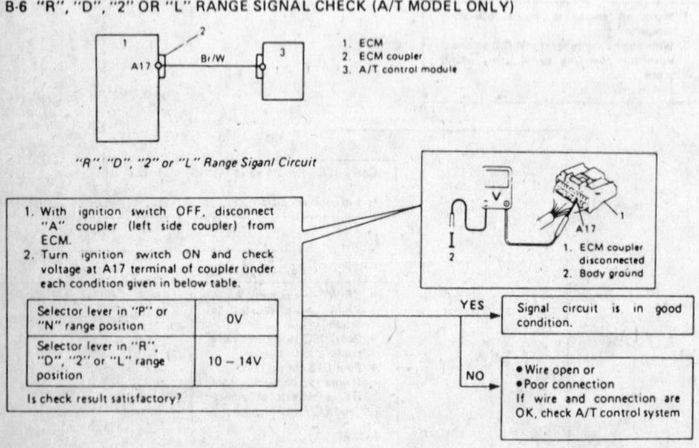
1
2
A17
Br/W
3
1. ECM
2. ECM coupler
3. A/T control module
"R", "D", "2" or "L" Range Siganl Circuit
V
A17
1
2
1. ECM coupler disconnected
2. Body ground
1. With ignition switch OFF, disconnect "A" coupler (left side coupler) from ECM.
2. Turn ignition switch ON and check voltage at A17 terminal of coupler under each condition given in below table.
Selector lever in "P" or "N" range position | 0V
Selector lever in "R", "D", "2" or "L" range position | 10 – 14V
Is check result satisfactory?
YES
Signal circuit is in good condition.
NO
•Wire open or
•Poor connection
If wire and connection are OK, check A/T control system

ECM TERMINAL IDENTIFICATION 1988–90 GTI

TERMINAL	CIRCUIT	STANDARD VOLTAGE	CONDITION
A1 A2	Power source	10 – 14V	Ignition switch ON
A3	Power source of AFS	10 – 14V	Ignition switch ON
A4	Power source of TPS	4.0 – 5.5V	Ignition switch ON
A5	Sensor ground	——	——
A6	AFS signal	0.2 – 0.8V	Ignition switch ON
A7	WTS	1.0 – 3.0V	Ignition switch ON Engine cooling water temp.: 80°C (176°F)
A8	Oxygen sensor	Indicator deflection repeated between over and under 0.45V	While engine running at 2000 r/min for 1 minute or longer after warmed up
A9	TPS signal	0 – 1V	Ignition switch ON Throttle valve at idle position
		3.0 – 5.0V	Ignition switch ON Throttle valve at full open position
A10	Speed sensor signal	Indicator deflection repeated between 0V and 3 – 5V	Ignition switch ON Front left tire turned slowly with front right tire locked
A11 A20 B4 (A/T model only)	Throttle valve opening output signal (A/T control module)	10 – 14V	Ignition switch ON Throttle valve at idle position
		0 – 1V ↕ 10 – 14V	Ignition switch ON Opening throttle valve slowly causes voltage to vary as given at the left. (Refer to Fig. 6E-160 for relations between opening and voltage)
A12	Ignition signal	10 – 14V	Ignition switch ON
A13 A14	Ground	——	——
A15	——	——	——
A16	Power source for back up circuit	10 – 14V	Ignition switch ON and OFF
A17 (A/T model only)	"R", "D", "2" or "L" range signal (A/T control module)	0 – 2V	Ignition switch ON, Selector lever in "P" or "N" range position
		10 – 14V	Ignition switch ON, Selector lever in "R", "D", "2" or "L" range position
A18	Air-conditioner ON/OFF signal (if equipped)	8 – 14V	Ignition switch ON
		0 – 2V	While engine running at idle speed, Air-conditioner ON

TERMINAL	CIRCUIT	STANDARD VOLTAGE	CONDITION
A19	Test switch terminal	10 – 14V	Ignition switch ON
		0 – 1V	Ignition switch ON Test switch terminal grounded
A21	Idle switch (in TPS)	0 – 1V	Ignition switch ON Throttle valve at idle position
		3.0 – 5.0V	Ignition switch ON Throttle valve opens larger than idle position
A22	Diagnosis switch terminal	10 – 14V	Ignition switch ON
		0 – 1V	Ignition switch ON Diagnosis switch terminal grounded
A23 (California spec. model only)	REGTS	4.0 – 5.0V	Ignition switch ON Sensor ambient temp.: 20°C (68°F)
A24	——	——	——
B1	CAS (positive)	——	——
B2	"CHECK ENGINE" light	0 – 3V	Ignition switch ON Diagnosis switch terminal ungrounded
		10 – 14V	Engine running Diagnosis switch terminal ungrounded
		Indicator deflection within 1.2V – 14V	Diagnosis switch terminal grounded Test switch terminal grounded, while engine running at 2000 r/min after warmed up
B3	——	——	——
B5	Canister purge VSV	10 – 14V	Ignition switch ON
B6 B7	——	——	——
B8	Main relay ground	0 – 2V	Ignition switch ON
B9	——	——	——
B10	CAS (negative)	——	——
B11 B12	——	——	——
B13	Fuel pump relay ground	0 – 4V	For 3 seconds after ignition switch ON
		10 – 14V	When over 3 seconds after ignition switch ON
B14 (California spec. modle only)	EGR VSV	10 – 14V	Ignition switch ON
B15	——	——	——

ECM TERMINAL IDENTIFICATION CONTINUED—1988-90 GTI

TERMINAL	CIRCUIT	STANDARD VOLTAGE	CONDITION
B16	Ground	—	—
B17		—	—
B18	ISC solenoid valve	10 – 14V	Ignition switch ON
C1	Engine start signal (Engine start switch)	6 – 12V	While engine cranking
		0 – 1V	Other than above
C2 C7	Power source for injector	10 – 14V	Ignition switch ON
C3 C4	Injector (positive)	—	—
C5 C10	Ground for injector	—	—
C6	Ignition output signal	0V	Ignition switch ON
		1 – 3V	While engine cranking
C8 C9	Injector (negative)	—	—

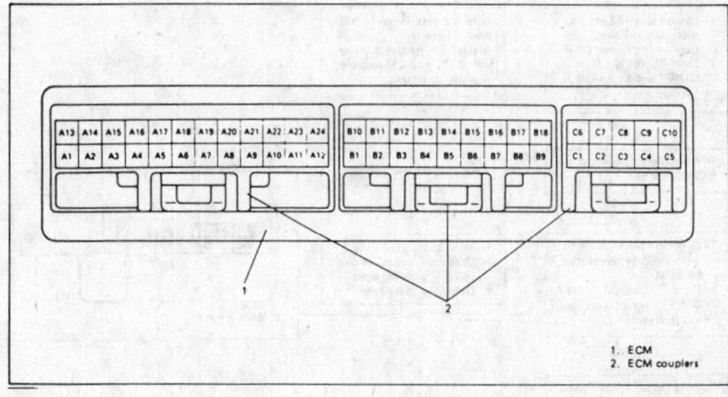
A13 A14 A15 A16 A17 A18 A19 A20 A21 A22 A23 A24
A1 A2 A3 A4 A5 A6 A7 A8 A9 A10 A11 A12
B10 B11 B12 B13 B14 B15 B16 B17 B18
B1 B2 B3 B4 B5 B6 B7 B8 B9
C6 C7 C8 C9 C10
C1 C2 C3 C4 C5
1
2
1. ECM
2. ECM couplers

TERMINALS	CIRCUIT	STANDARD RESISTANCE	CONDITION
A7 – A5	WTS	Approx. 320 Ω	Engine cooling water temp. 80°C (176°F)
A9 – A5	TPS	0 – 500 Ω	Throttle valve at idle position
		3.5 – 6.5 kΩ	Throttle valve at full open position
A10 – Body ground	Speed sensor	Ohmmeter indicator deflects between 0 and ∞	While front left tire turned slowly with front right tire loked
A13 – Body ground	Ground	0 (Zero)	—
A14 – Body ground	Ground	0 (Zero)	—
A19 – Body ground	Test switch terminal	∞ (Infinity)	Test switch terminal ungrounded
		0 (Zero)	Test switch terminal grounded
A21 – A5	Idle switch (in TPS)	0 (Zero)	Throttle valve is at idle position
		∞ (Infinity)	Throttle valve opens larger than idle position
A22 – Body ground	Diagnosis switch terminal	∞ (Infinity)	Diagnosis switch terminal ungrounded
		0 (Zero)	Diagnosis switch terminal grounded
B1 – B10	CAS	588 – 882 Ω	—
B5 – A1	Canister purge VSV	33 – 39 Ω	—
B14 – A1	EGR VSV	33 – 39 Ω	—
B16 – Body ground	Ground	0 (Zero)	—
B18 – A1	ISC solenoid valve	30 – 33 Ω	—
C5 – Body ground C10 – Body ground	Ground	0 (Zero)	—

DIAGNOSTIC FLOW CHART 1988-90 SIDEKICK

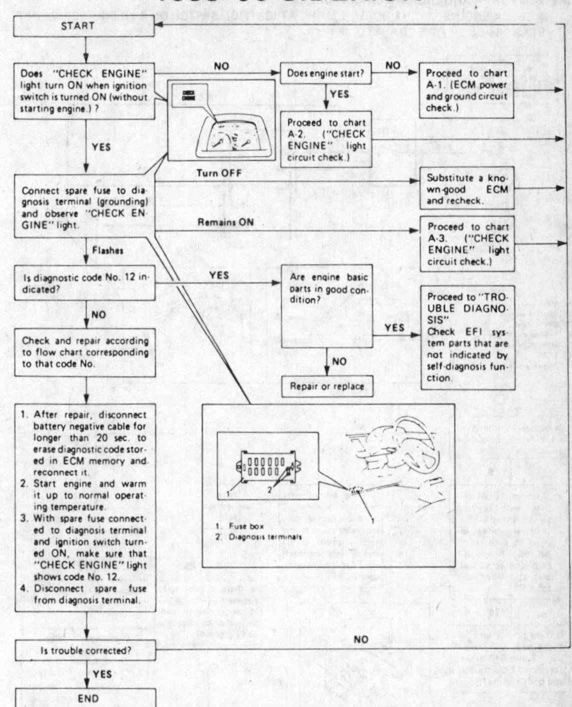
START
Does "CHECK ENGINE" light turn ON when ignition switch is turned ON (without starting engine.) ?
NO
Does engine start?
NO
Proceed to chart A-1. (ECM power and ground circuit check.)
YES
Proceed to chart A-2. ("CHECK ENGINE" light circuit check.)
YES
Turn OFF
Substitute a known-good ECM and recheck.
Connect spare fuse to diagnosis terminal (grounding) and observe "CHECK ENGINE" light.
Remains ON
Proceed to chart A-3. ("CHECK ENGINE" light circuit check.)
Flashes
Is diagnostic code No. 12 indicated?
YES
Are engine basic parts in good condition?
YES
Proceed to "TROUBLE DIAGNOSIS" Check EFI system parts that are not indicated by self-diagnosis function.
NO
NO
Repair or replace
Check and repair according to flow chart corresponding to that code No.
1. After repair, disconnect battery negative cable for longer than 20 sec. to erase diagnostic code stored in ECM memory and reconnect it.
2. Start engine and warm it up to normal operating temperature.
3. With spare fuse connected to diagnosis terminal and ignition switch turned ON, make sure that "CHECK ENGINE" light shows code No. 12.
4. Disconnect spare fuse from diagnosis terminal.
1. Fuse box
2. Diagnosis terminals
Is trouble corrected?
NO
YES
END

DIAGNOSIS CODE TABLE 1988-90 SIDEKICK

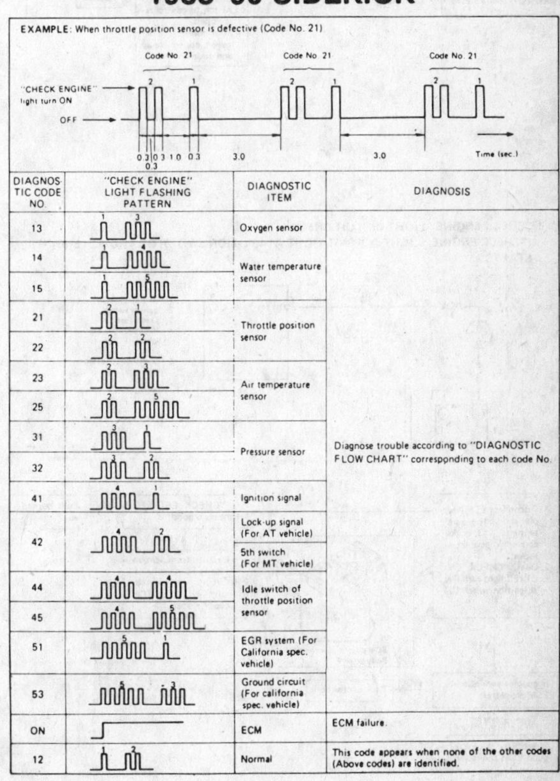
EXAMPLE: When throttle position sensor is defective (Code No. 21)
Code No. 21
"CHECK ENGINE" light turn ON
OFF
0.3 0.3 1.0 0.3 3.0 3.0
Time (sec.)

DIAGNOSTIC CODE NO.	"CHECK ENGINE" LIGHT FLASHING PATTERN	DIAGNOSTIC ITEM	DIAGNOSIS
13	1 3	Oxygen sensor	Diagnose trouble according to "DIAGNOSTIC FLOW CHART" corresponding to each code No.
14	1 4	Water temperature sensor	
15	1 5		
21	2 1	Throttle position sensor	
22	2 2		
23	2 3	Air temperature sensor	
25	2 5		
31	3 1	Pressure sensor	
32	3 2		
41	4 1	Ignition signal	
42	4 2	Lock-up signal (For AT vehicle) 5th switch (For MT vehicle)	
44	4 4	Idle switch of throttle position sensor	
45	4 5		
51	5 1	EGR system (For California spec. vehicle)	
53	5 3	Ground circuit (For california spec. vehicle)	
ON		ECM	ECM failure.
12	1 2	Normal	This code appears when none of the other codes (Above codes) are identified.

DIAGNOSTIC FLOW CHART—1988–90 SIDEKICK

A-1 ECM POWER AND GROUND CIRCUIT CHECK

("CHECK ENGINE" LIGHT DOESN'T LIGHT AT IGNITION SWITCH ON AND ENGINE DOESN'T START THOUGH IT IS CRANKED UP.)

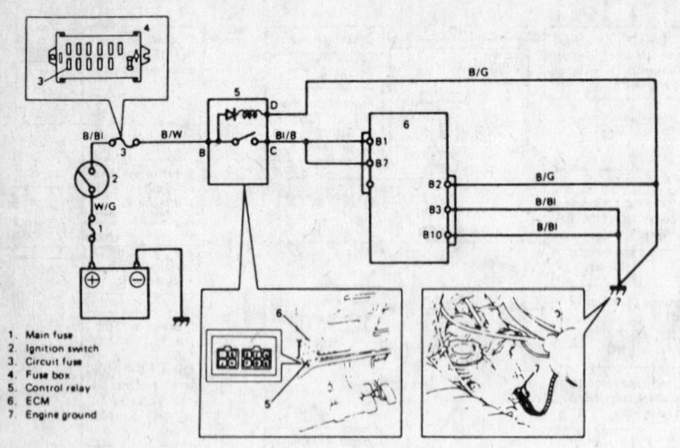
B/G
B/BI
B/W
W/G
B/B
B1
B7
B2
B3
B10
B/G
B/BI
B/BI
1. Main fuse
2. Ignition switch
3. Circuit fuse
4. Fuse box
5. Control relay
6. ECM
7. Engine ground

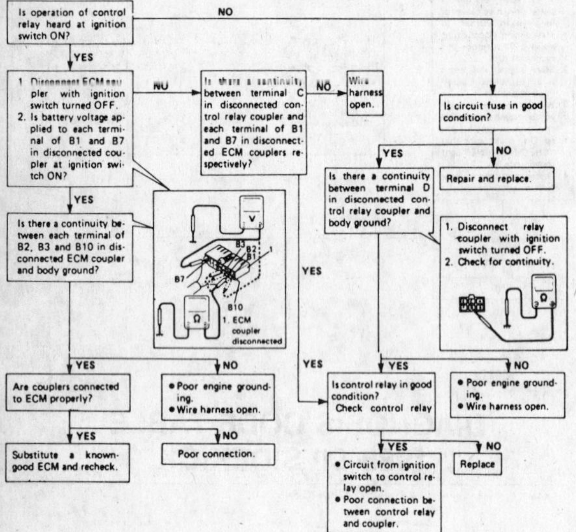
Is operation of control relay heard at ignition switch ON?
NO
YES
1. Disconnect ECM coupler with ignition switch turned OFF.
2. Is battery voltage applied to each terminal of B1 and B7 in disconnected coupler at ignition switch ON?
NO
Is there a continuity between terminal C in disconnected control relay coupler and each terminal of B1 and B7 in disconnected ECM couplers respectively?
NO
Wire harness open.
Is circuit fuse in good condition?
YES
NO
Is there a continuity between terminal D in disconnected control relay coupler and body ground?
Repair and replace.
1. Disconnect relay coupler with ignition switch turned OFF.
2. Check for continuity.
YES
Is there a continuity between each terminal of B2, B3 and B10 in disconnected ECM coupler and body ground?
B7
B10
1. ECM coupler disconnected
YES
YES
NO
YES
NO
Are couplers connected to ECM properly?
• Poor engine grounding.
• Wire harness open.
Is control relay in good condition? Check control relay
• Poor engine grounding.
• Wire harness open.
YES
NO
YES
NO
Substitute a known-good ECM and recheck.
Poor connection.
• Circuit from ignition switch to control relay open.
• Poor connection between control relay and coupler.
Replace

A-2 "CHECK ENGINE" LIGHT CIRCUIT CHECK

("CHECK ENGINE" LIGHT DOESN'T LIGHT AT IGNITION SWITCH ON THOUGH ENGINE STARTS.)

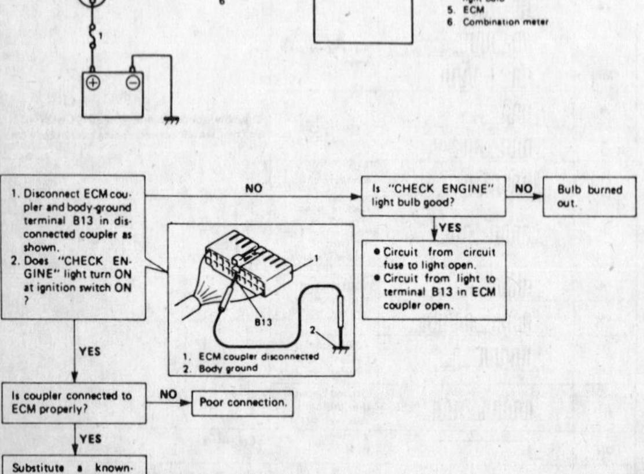
B/W
V/Y
B13
1. Main fuse
2. Ignition switch
3. Circuit fuse
4. "CHECK ENGINE" light bulb
5. ECM
6. Combination meter
1. Disconnect ECM coupler and body-ground terminal B13 in disconnected coupler as shown.
2. Does "CHECK ENGINE" light turn ON at ignition switch ON?
NO
Is "CHECK ENGINE" light bulb good?
NO
Bulb burned out.
YES
• Circuit from circuit fuse to light open.
• Circuit from light to terminal B13 in ECM coupler open.
B13
1. ECM coupler disconnected
2. Body ground
YES
Is coupler connected to ECM properly?
NO
Poor connection.
YES
Substitute a known-good ECM and recheck.

A-3 "CHECK ENGINE" LIGHT CIRCUIT CHECK

("CHECK ENGINE" LIGHT DOESN'T FLASH OR JUST REMAINS ON EVEN WITH SPARE FUSE CONNECTED TO DIAGNOSIS TERMINAL.)

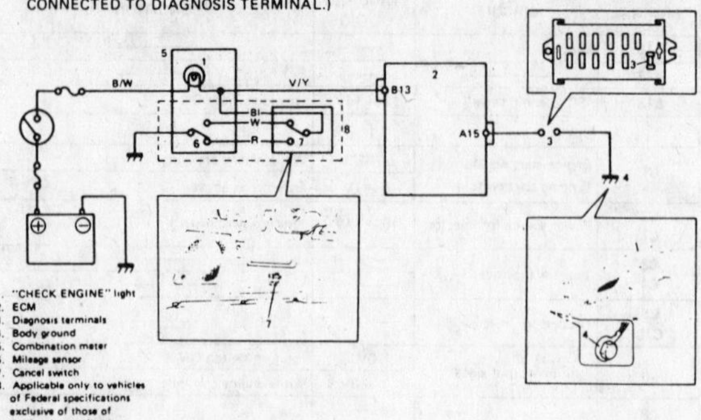
B/W
V/Y
B13
A15
1. "CHECK ENGINE" light
2. ECM
3. Diagnosis terminals
4. Body ground
5. Combination meter
6. Mileage sensor
7. Cancel switch
8. Applicable only to vehicles of Federal specifications exclusive of those of California specifications.

For California Specification

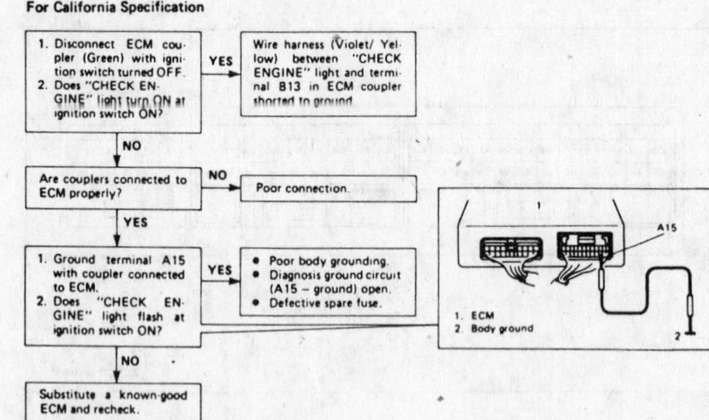
1. Disconnect ECM coupler (Green) with ignition switch turned OFF.
2. Does "CHECK ENGINE" light turn ON at ignition switch ON?
YES
Wire harness (Violet/Yellow) between "CHECK ENGINE" light and terminal B13 in ECM coupler shorted to ground.
NO
Are couplers connected to ECM properly?
NO
Poor connection.
YES
1. Ground terminal A15 with coupler connected to ECM.
2. Does "CHECK ENGINE" light flash at ignition switch ON?
YES
• Poor body grounding.
• Diagnosis ground circuit (A15 – ground) open.
• Defective spare fuse.
A15
1. ECM
2. Body ground
NO
Substitute a known-good ECM and recheck.

For Federal Specification Except California

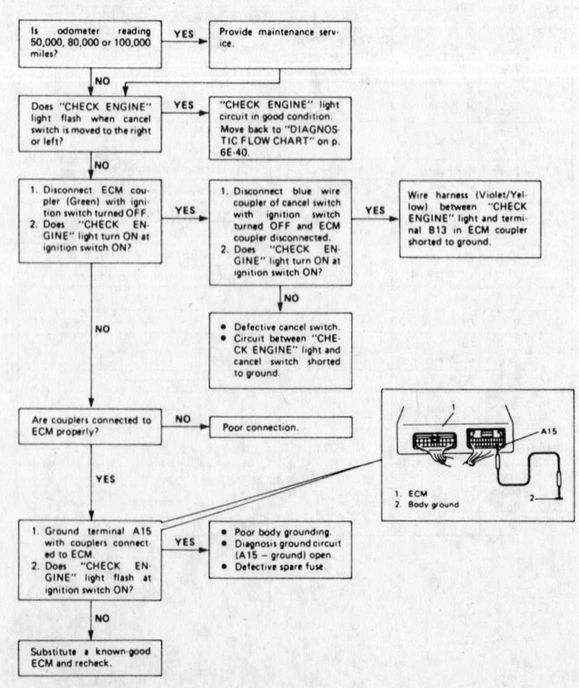
Is odometer reading 50,000, 80,000 or 100,000 miles?
YES
Provide maintenance service.
NO
Does "CHECK ENGINE" light flash when cancel switch is moved to the right or left?
YES
"CHECK ENGINE" light circuit in good condition. Move back to "DIAGNOSTIC FLOW CHART" on p. 6E-40.
NO
1. Disconnect ECM coupler (Green) with ignition switch turned OFF.
2. Does "CHECK ENGINE" light turn ON at ignition switch ON?
YES
1. Disconnect blue wire coupler of cancel switch with ignition switch turned OFF and ECM coupler disconnected.
2. Does "CHECK ENGINE" light turn ON at ignition switch ON?
YES
Wire harness (Violet/Yellow) between "CHECK ENGINE" light and terminal B13 in ECM coupler shorted to ground.
NO
• Defective cancel switch.
• Circuit between "CHECK ENGINE" light and cancel switch shorted to ground.
NO
Are couplers connected to ECM properly?
NO
Poor connection.
A15
1. ECM
2. Body ground
YES
1. Ground terminal A15 with couplers connected to ECM.
2. Does "CHECK ENGINE" light flash at ignition switch ON?
YES
• Poor body grounding.
• Diagnosis ground circuit (A15 – ground) open.
• Defective spare fuse.
NO
Substitute a known-good ECM and recheck.

DIAGNOSTIC CODE – 1988–90 SIDEKICK

CODE NO. 13 OXYGEN SENSOR CIRCUIT (SIGNAL VOLTAGE LOW AND DOESN'T CHANGE)

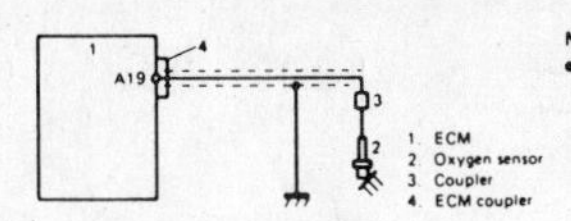
A19
1. ECM
2. Oxygen sensor
3. Coupler
4. ECM coupler

NOTE:
- Before diagnosing trouble according to flow chart given below, check to make sure that following system and parts other than EFI system are in good condition.
  - Air cleaner (clogged)
  - Vacuum leaks (air inhaling)
  - Spark plugs (contamination, gap)
  - High tension cords (crack, deterioration)
  - Distributor rotor or cap (wear, crack)
  - Ignition timing
  - Engine compression
  - Any other system and parts which might affect A/F mixture or combustion.
- If code No. 13 and another code No. are indicated together, the latter has priority. Therefore, check and correct what is represented by that code No. first and then proceed to the following check.

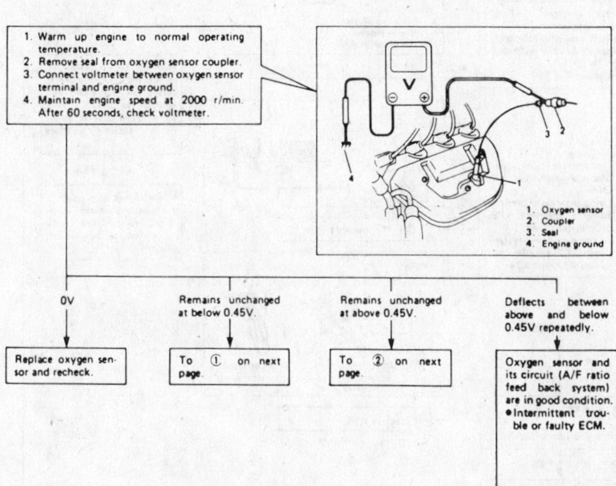
1. Warm up engine to normal operating temperature.
2. Remove seal from oxygen sensor coupler.
3. Connect voltmeter between oxygen sensor terminal and engine ground.
4. Maintain engine speed at 2000 r/min. After 60 seconds, check voltmeter.
1. Oxygen sensor
2. Coupler
3. Seal
4. Engine ground
0V
Remains unchanged at below 0.45V.
Remains unchanged at above 0.45V.
Deflects between above and below 0.45V repeatedly.
Replace oxygen sensor and recheck.
To ① on next page.
To ② on next page.
Oxygen sensor and its circuit (A/F ratio feed back system) are in good condition.
• Intermittent trouble or faulty ECM.

CODE NO. 14 WTS (WATER TEMPERATURE SENSOR) CIRCUIT (LOW TEMPERATURE INDICATED)

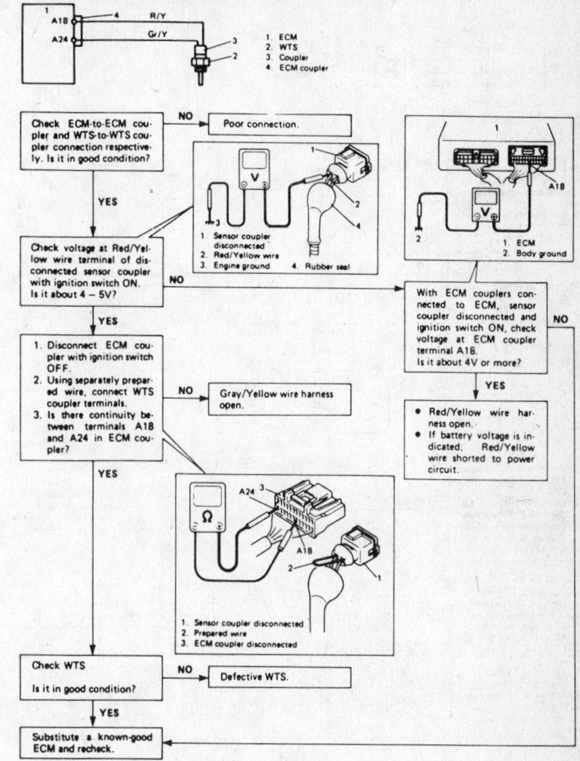
A18
A24
R/Y
Gr/Y
1. ECM
2. WTS
3. Coupler
4. ECM coupler
Check ECM-to-ECM coupler and WTS-to-WTS coupler connection respectively. Is it in good condition?
NO
Poor connection.
YES
1. Sensor coupler disconnected
2. Red/Yellow wire
3. Engine ground
4. Rubber seal
Check voltage at Red/Yellow wire terminal of disconnected sensor coupler with ignition switch ON. Is it about 4 – 5V?
NO
1. ECM
2. Body ground
With ECM couplers connected to ECM, sensor coupler disconnected and ignition switch ON, check voltage at ECM coupler terminal A18. Is it about 4V or more?
YES
• Red/Yellow wire harness open.
• If battery voltage is indicated, Red/Yellow wire shorted to power circuit.
YES
1. Disconnect ECM coupler with ignition switch OFF.
2. Using separately prepared wire, connect WTS coupler terminals.
3. Is there continuity between terminals A18 and A24 in ECM coupler?
NO
Gray/Yellow wire harness open.
YES
1. Sensor coupler disconnected
2. Prepared wire
3. ECM coupler disconnected
Check WTS
Is it in good condition?
NO
Defective WTS.
YES
Substitute a known-good ECM and recheck.

CODE NO. 13 OXYGEN SENSOR CIRCUIT (Continued)

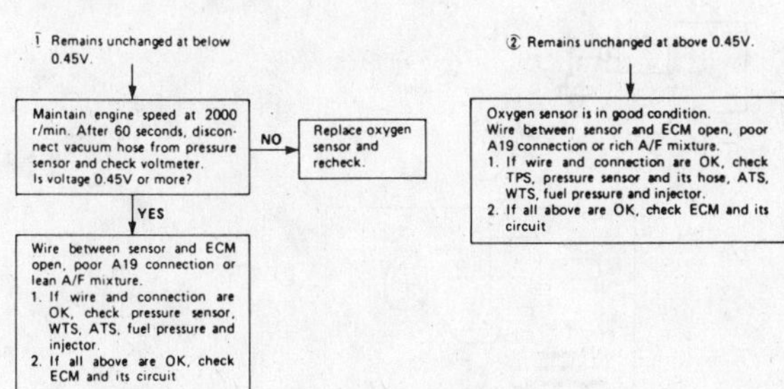
① Remains unchanged at below 0.45V.
Maintain engine speed at 2000 r/min. After 60 seconds, disconnect vacuum hose from pressure sensor and check voltmeter. Is voltage 0.45V or more?
NO
Replace oxygen sensor and recheck.
YES
Wire between sensor and ECM open, poor A19 connection or lean A/F mixture.
1. If wire and connection are OK, check pressure sensor, WTS, ATS, fuel pressure and injector.
2. If all above are OK, check ECM and its circuit
② Remains unchanged at above 0.45V.
Oxygen sensor is in good condition. Wire between sensor and ECM open, poor A19 connection or rich A/F mixture.
1. If wire and connection are OK, check TPS, pressure sensor and its hose, ATS, WTS, fuel pressure and injector.
2. If all above are OK, check ECM and its circuit

CODE NO. 15 WTS (WATER TEMPERATURE SENSOR) CIRCUIT (HIGH TEMPERATURE INDICATED)

A18
A24
R/Y
Gr/Y
1. ECM
2. WTS
3. Coupler
4. ECM coupler

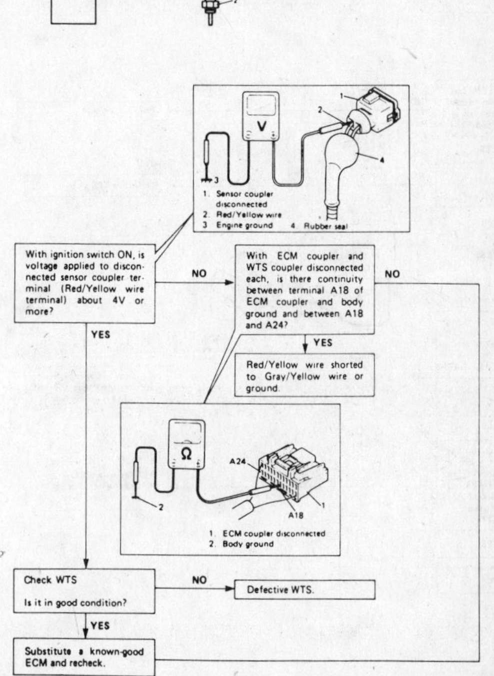
1. Sensor coupler disconnected
2. Red/Yellow wire
3. Engine ground
4. Rubber seal
With ignition switch ON, is voltage applied to disconnected sensor coupler terminal (Red/Yellow wire terminal) about 4V or more?
NO
With ECM coupler and WTS coupler disconnected each, is there continuity between terminal A18 of ECM coupler and body ground and between A18 and A24?
NO
YES
Red/Yellow wire shorted to Gray/Yellow wire or ground.
YES
A24
A18
1. ECM coupler disconnected
2. Body ground
Check WTS
Is it in good condition?
NO
Defective WTS.
YES
Substitute a known-good ECM and recheck.

DIAGNOSTIC CODE – 1988–90 SIDEKICK

CODE NO. 21 TPS (THROTTLE POSITION SENSOR) CIRCUIT (SIGNAL VOLTAGE HIGH)

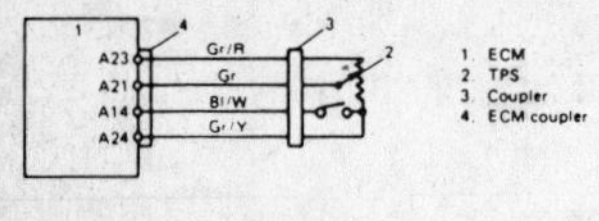
A23
A21
A14
A24
Gr/R
Gr
Bl/W
Gr/Y
1. ECM
2. TPS
3. Coupler
4. ECM coupler

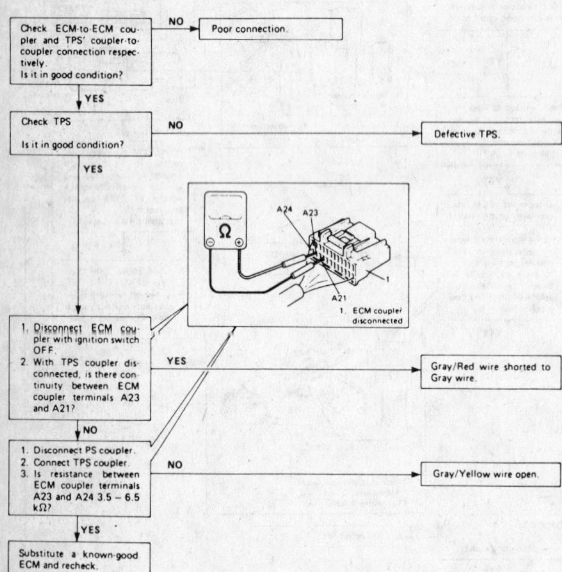
Check ECM-to-ECM coupler and TPS' coupler-to-coupler connection respectively. Is it in good condition?
NO
Poor connection.
YES
Check TPS
Is it in good condition?
NO
Defective TPS.
YES
A24
A23
A21
1. ECM coupler disconnected
1. Disconnect ECM coupler with ignition switch OFF.
2. With TPS coupler disconnected, is there continuity between ECM coupler terminals A23 and A21?
YES
Gray/Red wire shorted to Gray wire.
NO
1. Disconnect PS coupler.
2. Connect TPS coupler.
3. Is resistance between ECM coupler terminals A23 and A24 3.5 – 6.5 kΩ?
NO
Gray/Yellow wire open.
YES
Substitute a known-good ECM and recheck.

CODE NO. 23 ATS (AIR TEMPERATURE SENSOR) CIRCUIT (LOW TEMPERATURE INDICATED)

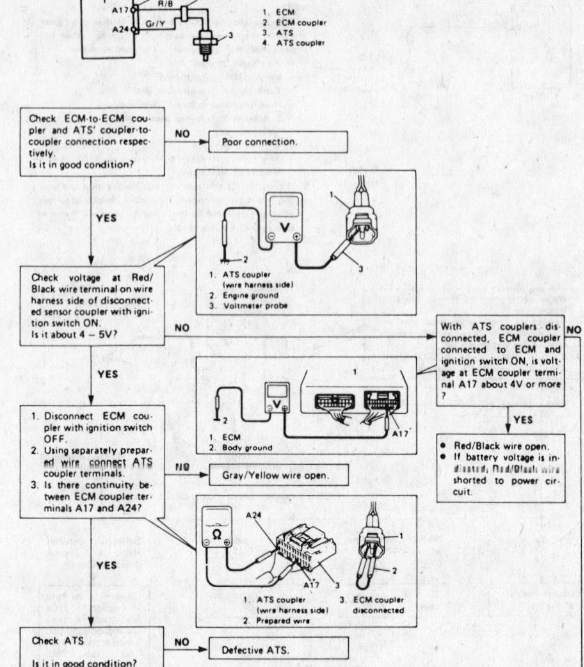
A17
A24
R/B
Gr/Y
1. ECM
2. ECM coupler
3. ATS
4. ATS coupler
Check ECM-to-ECM coupler and ATS' coupler-to-coupler connection respectively. Is it in good condition?
NO
Poor connection.
YES
1. ATS coupler (wire harness side)
2. Engine ground
3. Voltmeter probe
Check voltage at Red/Black wire terminal on wire harness side of disconnected sensor coupler with ignition switch ON. Is it about 4 – 5V?
NO
With ATS couplers disconnected, ECM coupler connected to ECM and ignition switch ON, is voltage at ECM coupler terminal A17 about 4V or more?
NO
YES
• Red/Black wire open.
• If battery voltage is indicated, Red/Black wire shorted to power circuit.
YES
1. ECM
2. Body ground
A17
1. Disconnect ECM coupler with ignition switch OFF.
2. Using separately prepared wire connect ATS coupler terminals.
3. Is there continuity between ECM coupler terminals A17 and A24?
NO
Gray/Yellow wire open.
YES
A24
A17
1. ATS coupler (wire harness side)
2. Prepared wire
3. ECM coupler disconnected
Check ATS
Is it in good condition?
NO
Defective ATS.
YES
Substitute a known-good ECM and recheck.

CODE NO. 22 TPS (THROTTLE POSITION SENSOR) CIRCUIT (SIGNAL VOLTAGE LOW)

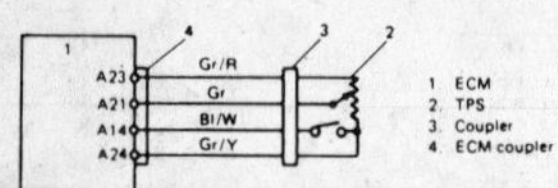
A23
A21
A14
A24
Gr/R
Gr
Bl/W
Gr/Y
1. ECM
2. TPS
3. Coupler
4. ECM coupler

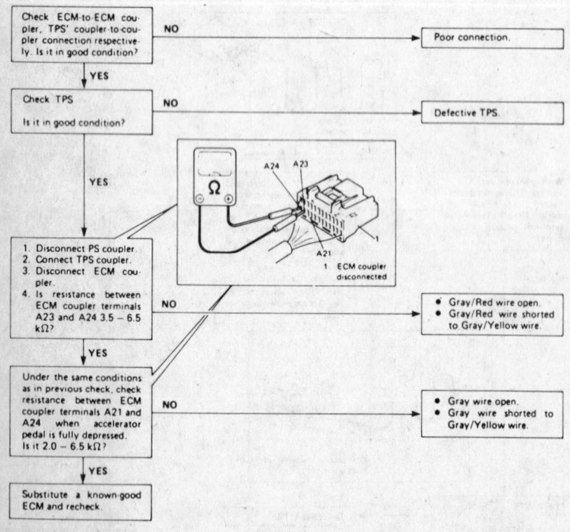
Check ECM-to-ECM coupler, TPS' coupler-to-coupler connection respectively. Is it in good condition?
NO
Poor connection.
YES
Check TPS
Is it in good condition?
NO
Defective TPS.
YES
A24
A23
A21
1. ECM coupler disconnected
1. Disconnect PS coupler.
2. Connect TPS coupler.
3. Disconnect ECM coupler.
4. Is resistance between ECM coupler terminals A23 and A24 3.5 – 6.5 kΩ?
NO
• Gray/Red wire open.
• Gray/Red wire shorted to Gray/Yellow wire.
YES
Under the same conditions as in previous check, check resistance between ECM coupler terminals A21 and A24 when accelerator pedal is fully depressed. Is it 2.0 – 6.5 kΩ?
NO
• Gray wire open.
• Gray wire shorted to Gray/Yellow wire.
YES
Substitute a known-good ECM and recheck.

CODE NO. 25 ATS (AIR TEMPERATURE SENSOR) CIRCUIT (HIGH TEMPERATURE INDICATED)

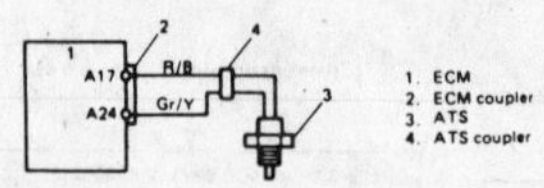
A17
A24
R/B
Gr/Y
1. ECM
2. ECM coupler
3. ATS
4. ATS coupler

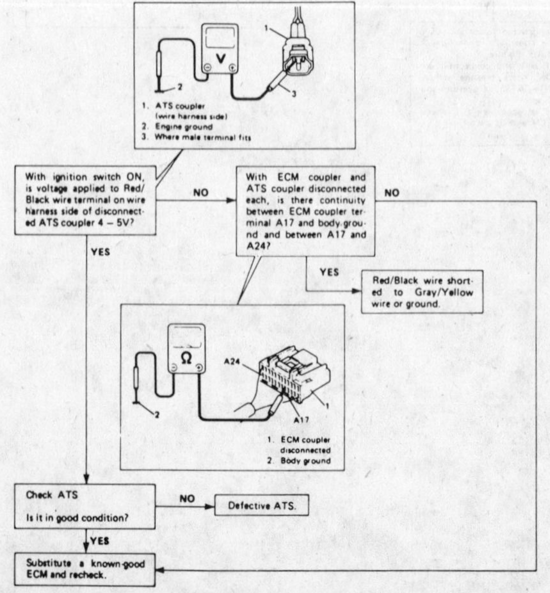
1. ATS coupler (wire harness side)
2. Engine ground
3. Where male terminal fits
With ignition switch ON, is voltage applied to Red/Black wire terminal on wire harness side of disconnected ATS coupler 4 – 5V?
NO
With ECM coupler and ATS coupler disconnected each, is there continuity between ECM coupler terminal A17 and body ground and between A17 and A24?
NO
YES
Red/Black wire shorted to Gray/Yellow wire or ground.
YES
A24
A17
1. ECM coupler disconnected
2. Body ground
Check ATS
Is it in good condition?
NO
Defective ATS.
YES
Substitute a known-good ECM and recheck.

DIAGNOSTIC CODE – 1988–90 SIDEKICK

CODE NO. 31 PS (PRESSURE SENSOR) CIRCUIT (SIGNAL VOLTAGE HIGH–LOW VACUUM)

Check output voltage of PS
Is proper voltage obtained? — YES →

NO ↓

Check ECM-to-ECM coupler and PS' coupler-to-coupler connection respectively.
Is it in good condition? — NO → Poor connection.

YES ↓

1. Disconnect ECM coupler with ignition switch OFF.
2. With PS coupler disconnected, is there continuity between ECM coupler terminals A22 and A23? — YES → Gray/Green wire shorted to Gray/Red wire.

NO ↓

1. Under the same conditions as in previous check, connect Gray/Green and Gray/Yellow wire terminals by using separately prepared wire.
2. Is there continuity between ECM coupler terminals A22 and A24? — YES → Gray/Yellow wire open.

NO ↓

Check PS individually
Is it in good condition? — NO → Defective PS.

YES ↓

Substitute a known-good ECM and recheck.

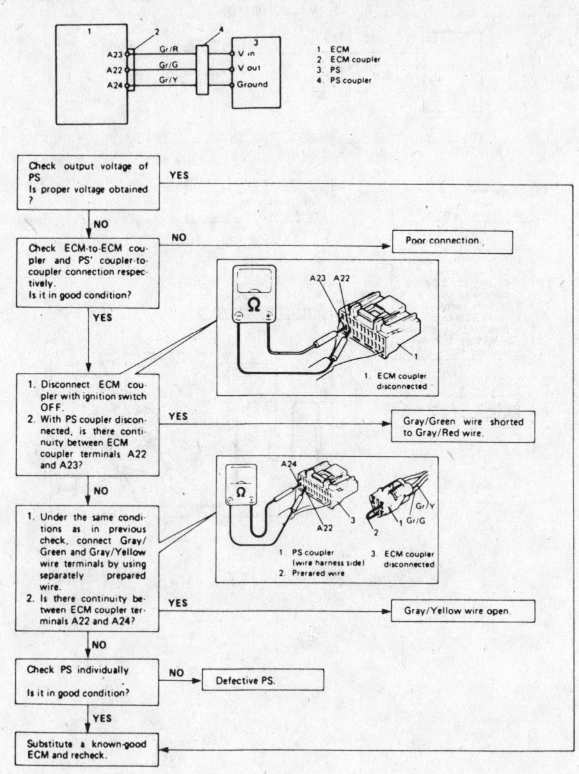

CODE NO. 41 IGNITION SIGNAL CIRCUIT (NO SIGNAL)
3. Noise suppressor
4. Suppressor coupler
5. Ignition coil
6. Ignition coil coupler

Check ignition system
Is it in good condition? — NO → Defective ignition system.

YES ↓

1. Disconnect ECM coupler with ignition switch OFF.
2. Is battery voltage applied to ECM coupler terminal A1 at ignition switch ON? — YES → 1. With ECM coupler disconnected, disconnect ignition coil coupler. 2. Is battery voltage applied to ECM coupler terminal A1 at ignition switch ON? — YES → Brown/White or Red/Blue wire shorted to power circuit.

NO ↓

1. Remove noise suppressor.
2. Is there continuity between suppressor terminals A and B? — NO → 1. Remove noise suppressor. 2. Is resistance between suppressor terminals B and C ∞ (infinity)? — YES → Is ECM coupler connected to ECM properly? — YES ... NO → Poor ECM-to-ECM coupler connection.

↓ (Defective noise suppressor.)

1. Connect Black/Green and Red/Blue terminals of suppressor coupler with separately prepared wire.
2. Is there continuity between ECM coupler terminal A1 and body ground? — NO → Blue or Brown wire open.

YES ↓

Is there continuity in Brown/White wire between suppressor coupler and ignition coil coupler? — NO → Brown/White wire open.

YES ↓

Is each coupler of ignition coil and suppressor connected properly? — NO → Poor connection.

YES ↓

Substitute a known-good ECM and recheck.

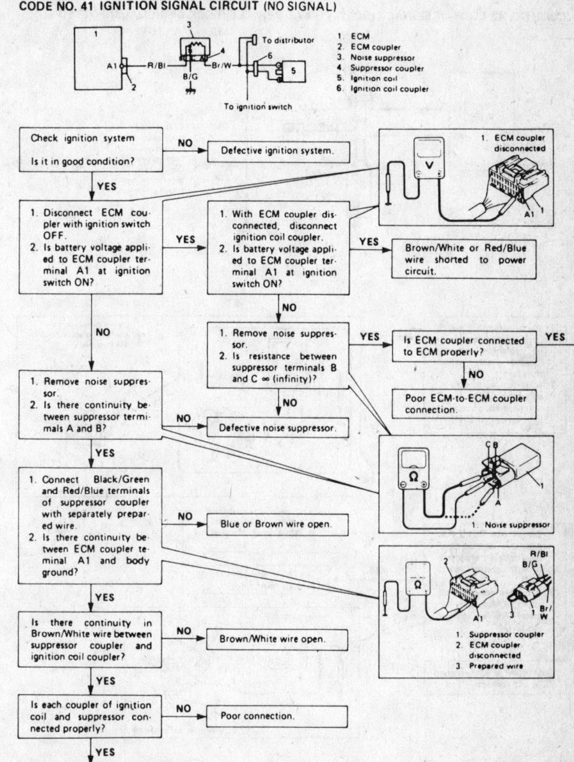

CODE NO. 32 PS (PRESSURE SENSOR) CIRCUIT (SIGNAL VOLTAGE LOW–HIGH VACCUM)

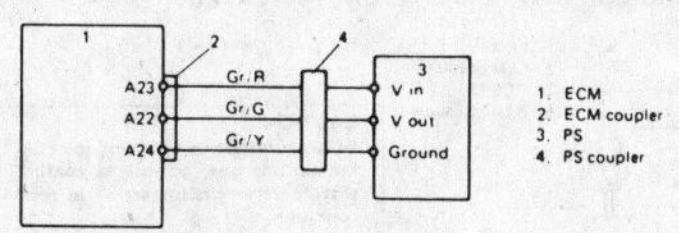
1. ECM
2. ECM coupler
3. PS
4. PS coupler

Check output voltage of PS
Is proper voltage obtained? — YES →

NO ↓

Check ECM-to-ECM coupler and PS' coupler-to-coupler connection respectively.
Is it in good condition? — NO → Poor connection.

YES ↓

1. Disconnect ECM coupler with ignition switch OFF.
2. With PS coupler disconnected, is there continuity between ECM coupler terminals A22 and A24 and between A23 and A24? — YES → Gray/Green or Gray/Red wire shorted to Gray/Yellow wire.

NO ↓

1. Under the same conditions as in previous check, connect Gray/Red and Gray/Green wire terminals on PS coupler (wire harness side) by using separately prepared wire.
2. Is there continuity between ECM coupler terminals A22 and A23? — YES → Gray/Green or Gray/Red wire open.

NO ↓

Check PS individually
Is it in good condition? — NO → Defective PS.

YES ↓

Substitute a known-good ECM and recheck.

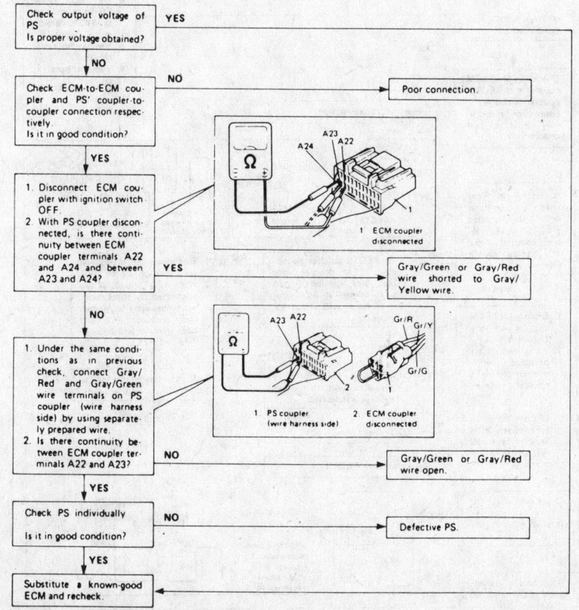

CODE NO. 42 5TH SWITCH CIRCUIT FOR MT VEHICLE (A5 TERMINAL GROUNDED CONSTANTLY)

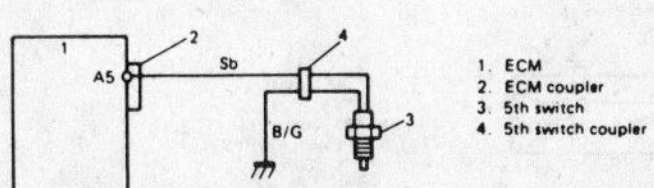
1. ECM
2. ECM coupler
3. 5th switch
4. 5th switch coupler

Check 5th switch
Is it in good condition? — NO → Defective 5th switch.

YES ↓

1. ECM coupler disconnected
2. Body ground

1. With ignition switch OFF, disconnect ECM coupler and 5th switch coupler respectively.
2. Is there continuity between ECM coupler terminal A5 and body ground? — YES → Skyblue wire shorted to ground.

NO ↓

Substitute a known-good ECM and recheck.

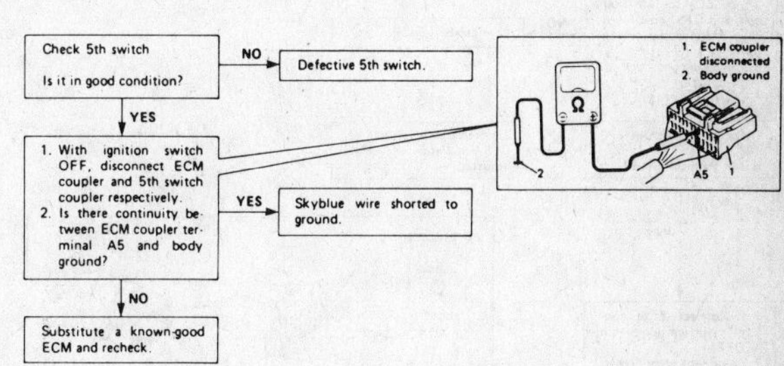

DIAGNOSTIC CODE – 1988–90 SIDEKICK

CODE NO. 42 LOCK-UP SIGNAL CIRCUIT FOR AT VEHICLE (LOCK-UP SIGNAL INPUTTED CONSTANTLY)

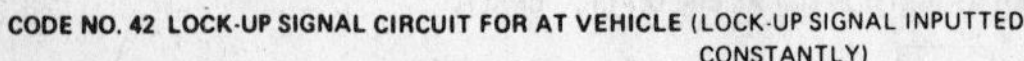

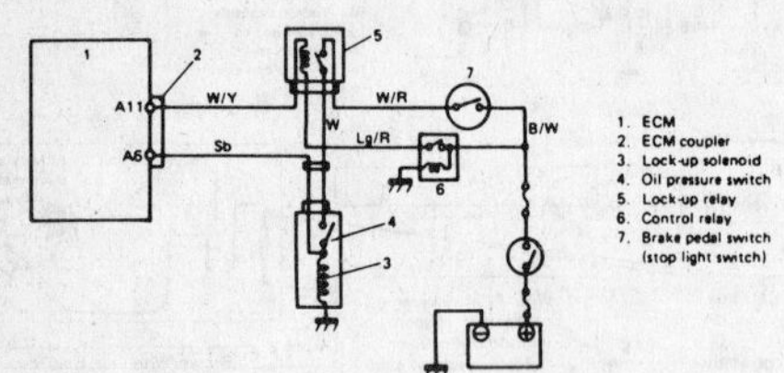
A11
A6
W/Y
Sb
W/R
W
Lg/R
B/W
1. ECM
2. ECM coupler
3. Lock-up solenoid
4. Oil pressure switch
5. Lock-up relay
6. Control relay
7. Brake pedal switch (stop light switch)

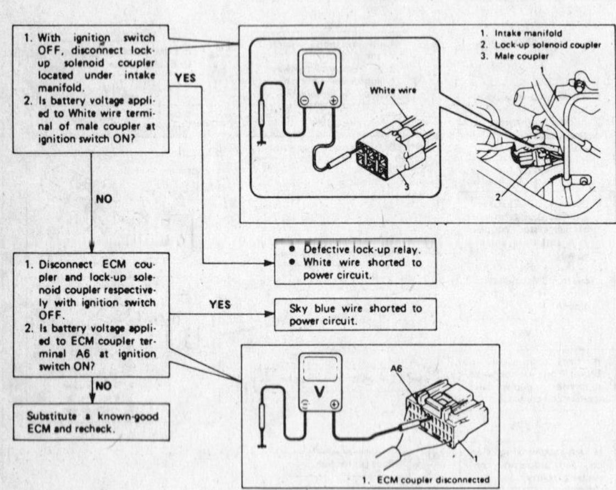
1. With ignition switch OFF, disconnect lock-up solenoid coupler located under intake manifold.
2. Is battery voltage applied to White wire terminal of male coupler at ignition switch ON?
YES
NO
1. Intake manifold
2. Lock-up solenoid coupler
3. Male coupler
White wire
• Defective lock-up relay.
• White wire shorted to power circuit.
1. Disconnect ECM coupler and lock-up solenoid coupler respectively with ignition switch OFF.
2. Is battery voltage applied to ECM coupler terminal A6 at ignition switch ON?
YES
Sky blue wire shorted to power circuit.
NO
Substitute a known-good ECM and recheck.
A6
1. ECM coupler disconnected

CODE NO. 45 IDLE SWITCH CIRCUIT (CIRCUIT SHORT OR TPS INSTALLATION ANGLE MALADJUSTED)

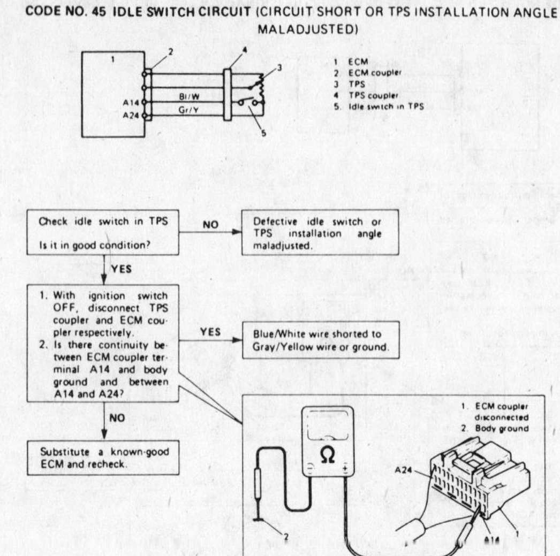
A14
A24
Bl/W
Gr/Y
1. ECM
2. ECM coupler
3. TPS
4. TPS coupler
5. Idle switch in TPS
Check idle switch in TPS
Is it in good condition?
NO
Defective idle switch or TPS installation angle maladjusted.
YES
1. With ignition switch OFF, disconnect TPS coupler and ECM coupler respectively.
2. Is there continuity between ECM coupler terminal A14 and body ground and between A14 and A24?
YES
Blue/White wire shorted to Gray/Yellow wire or ground.
NO
Substitute a known-good ECM and recheck.
1. ECM coupler disconnected
2. Body ground
A24
A14

CODE NO. 44 IDLE SWITCH CIRCUIT (CIRCUIT OPEN OR TPS INSTALLATION ANGLE MALADJUSTED)

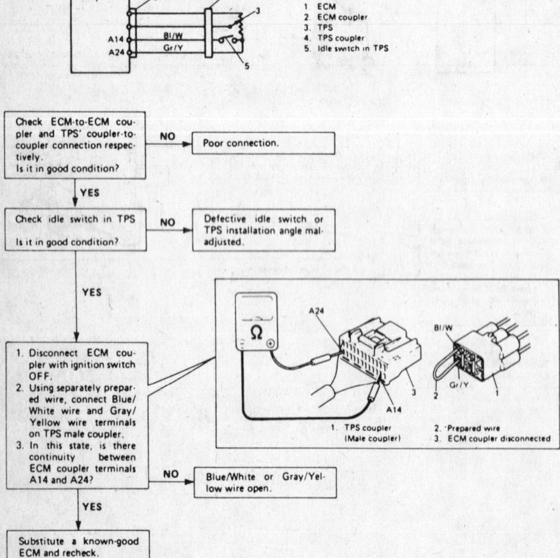
A14
A24
Bl/W
Gr/Y
1. ECM
2. ECM coupler
3. TPS
4. TPS coupler
5. Idle switch in TPS
Check ECM-to-ECM coupler and TPS' coupler-to-coupler connection respectively.
Is it in good condition?
NO
Poor connection.
YES
Check idle switch in TPS
Is it in good condition?
NO
Defective idle switch or TPS installation angle maladjusted.
YES
1. Disconnect ECM coupler with ignition switch OFF.
2. Using separately prepared wire, connect Blue/White wire and Gray/Yellow wire terminals on TPS male coupler.
3. In this state, is there continuity between ECM coupler terminals A14 and A24?
NO
Blue/White or Gray/Yellow wire open.
YES
Substitute a known-good ECM and recheck.
A24
A14
Bl/W
Gr/Y
1. TPS coupler (Male coupler)
2. Prepared wire
3. ECM coupler disconnected

CODE NO. 51 EGR SYSTEM AND REGTS (RECIRCULATED EXHAUST GAS TEMPERATURE SENSOR) CIRCUIT (HIGH OR LOW TEMPERATURE INDICATED)
(For California Specification Only)

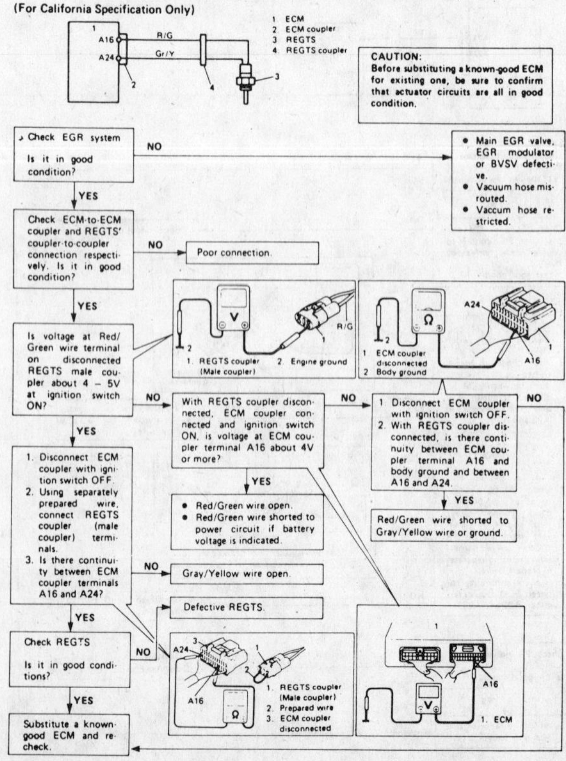
A16
A24
R/G
Gr/Y
1. ECM
2. ECM coupler
3. REGTS
4. REGTS coupler
CAUTION:
Before substituting a known-good ECM for existing one, be sure to confirm that actuator circuits are all in good condition.
Check EGR system
Is it in good condition?
NO
• Main EGR valve, EGR modulator or BVSV defective.
• Vacuum hose misrouted.
• Vaccum hose restricted.
YES
Check ECM-to-ECM coupler and REGTS' coupler-to-coupler connection respectively. Is it in good condition?
NO
Poor connection.
YES
R/G
1. REGTS coupler (Male coupler)
2. Engine ground
A24
A16
1. ECM coupler disconnected
2. Body ground
Is voltage at Red/Green wire terminal on disconnected REGTS male coupler about 4 – 5V at ignition switch ON?
NO
With REGTS coupler disconnected, ECM coupler connected and ignition switch ON, is voltage at ECM coupler terminal A16 about 4V or more?
NO
1. Disconnect ECM coupler with ignition switch OFF.
2. With REGTS coupler disconnected, is there continuity between ECM coupler terminal A16 and body ground and between A16 and A24.
NO
YES
Red/Green wire shorted to Gray/Yellow wire or ground.
YES
• Red/Green wire open.
• Red/Green wire shorted to power circuit if battery voltage is indicated.
YES
1. Disconnect ECM coupler with ignition switch OFF.
2. Using separately prepared wire, connect REGTS coupler (male coupler) terminals.
3. Is there continuity between ECM coupler terminals A16 and A24?
NO
Gray/Yellow wire open.
YES
Defective REGTS.
Check REGTS
Is it in good conditions?
NO
YES
Substitute a known-good ECM and recheck.
A24
A16
1. REGTS coupler (Male coupler)
2. Prepared wire
3. ECM coupler disconnected
A16
1. ECM

DIAGNOSTIC CODE – 1988–90 SIDEKICK

CODE NO. 53 GROUND CIRCUIT FOR CALIFORNIA SPEC. VEHICLE ONLY (CIRCUIT OPEN)

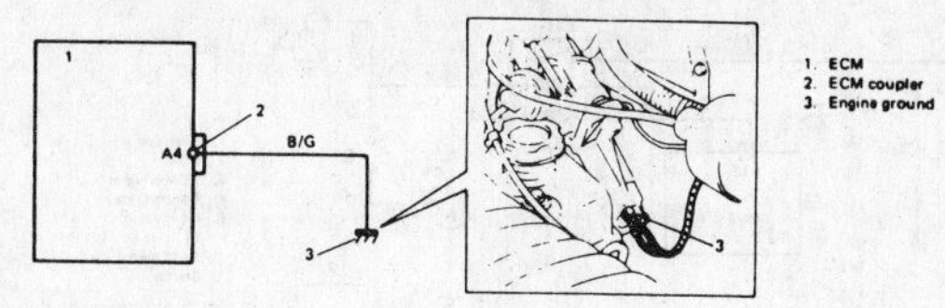
1
2
A4
B/G
3
1. ECM
2. ECM coupler
3. Engine ground

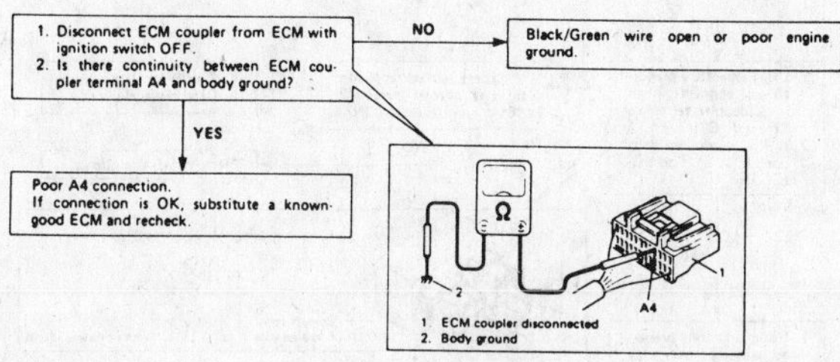
1. Disconnect ECM coupler from ECM with ignition switch OFF.
2. Is there continuity between ECM coupler terminal A4 and body ground?
NO
Black/Green wire open or poor engine ground.
YES
Poor A4 connection.
If connection is OK, substitute a known-good ECM and recheck.
Ω
A4
1. ECM coupler disconnected
2. Body ground

TROUBLE DIAGNOSIS – 1988–90 SIDEKICK

SYMPTOM	POSSIBLE CAUSE	INSPECTION
Hard or no starting (Engine cranks OK)	• Injector or its circuit defective	Diagnostic flow chart B-1
	• Defective fuel pump or its circuit open	Diagnostic flow chart B-2
	• Fuel pressure out of specification	Diagnostic flow chart B-3
	• Defective air valve	
	• Open starter signal circuit	Check voltage at ECM coupler terminal B11
	• Defective throttle opener system	Diagnostic flow chart B-4
	• Defective PTC heater system (For AT vehicle)	Diagnostic flow chart B-7
	• Poor performance of water temperature sensor, air temperature sensor or pressure sensor	
	• Faulty ECM	
NOTE: • If engine doesn't start at all, perform fuel injector and its circuit check first. (Advance to Diagnostic flow chart B-1.) • If engine is hard to start only when it is cold, check air valve first. • If engine starts easily with help of accelerator pedal operation, check throttle opener system first. (Advance to Diagnostic flow chart B-4.) • If engine in AT model doesn't start easily only in extremely cold environment, check its PTC heater system first. (Advance to Diagnositic flow chart B-7.)		
Improper engine idling or engine fails to idle	• Clogged pressure sensor vacuum passage	Check vacuum hose and gas filter
	• Defective throttle opener system	Diagnostic flow chart B-4
	• Maladjusted idle speed adjusting screw	
	• Defective air valve	
	• ISC solenoid valve or its circuit defective	Diagnostic flow chart B-5
	• Fuel pressure out of specification	Diagnostic flow chart B-3
	• Defective EGR system (For Federal specification except California)	Diagnostic flow chart B-6
	• Defective injector	Check injector for resistance, injection condition and fuel leakage
	• Poor performance of water temperature sensor, throttle position sensor, air temperature sensor or pressure sensor	
	• Faulty ECM	

TROUBLE DIAGNOSIS CONTINUED 1988–90 SIDEKICK

SYMPTOM	POSSIBLE CAUSE	INSPECTION
NOTE: • If engine fails to idle when it is cold, check air valve first. • If engine stops immediately after it started (i.e. when throttle opener moves throttle valve back to idle position) although it doesn't if accelerator pedal is used, check idle speed first and check ISC solenoid valve. (Advance to Diagnostic flow chart B-5.) • If idling speed is too high after engine is warmed up, first check if throttle opener has opened throttle valve when not supposed to. If it is still too high even after it was made sure that throttle opener system is in good condition and idle speed adjusting screw was loosened fully, check if air valve stays open or ISC solenoid valve fails to operate properly.		
Engine has no or poor power	• Clogged pressure sensor vacuum passage	Check vacuum hose and gas filter
	• Maladjusted accelerator cable play	
	• Maladjusted installation angle of throttle position sensor	
	• Fuel pressure out of specification (Low fuel pressure)	Diagnostic flow chart B-3
	• Defective EGR system (For Federal specification except California)	Diagnostic flow chart B-6
	• Defective injector	Check injector for resistance, injection condition and fuel leakage.
	• Poor performance of throttle position sensor, water temperature sensor, air temperature sensor or pressure sensor	
	• Faulty ECM	
Engine hesitates when acceleration	• Defective throttle valve operation	Check throttle valve for smooth operation
	• Poor performance throttle position sensor	
	• Fuel pressure out of specification (Low fuel pressure)	Diagnostic flow chart B-3
	• Defective EGR system	Diagnostic flow chart B-6 for Federal specification except California
	• Defective injector	Check injector for resistance, injection condition and fuel leakage
	• Poor performance of water temperature sensor or pressure sensor	
	• Faulty ECM	

SYMPTOM	POSSIBLE CAUSE	INSPECTION
Surges (Variation in vehicle speed is felt although accelerator pedal is not operated)	• Variable fuel pressure (Clogged fuel filter, defective fuel pressure regulator etc.)	Diagnostic flow chart B-3
	• Defective EGR system	Diagnostic flow chart B-6 for Federal specification except California
	• Defective injector	Check injector for resistance, injection condition and fuel leakage
	• Poor performance of throttle position sensor, water temperature sensor or pressure sensor	
	• Faulty ECM	
Excessive detonation (Engine makes sharp metallic knocks that change with throttle opening)	• Low fuel pressure	Diagnostic flow chart B-3
	• Defective EGR system (For Federal specification except California)	Diagnostic flow chart B-6
	• Defective injector	Check injector for resistance, injection condition and fuel leakage
	• Poor performance of throttle position sensor, water temperature sensor or pressure sensor	
	• Faulty ECM	
Poor gasoline mileage	• High idle speed (Malfunctioning air valve or abnormal ISC solenoid valve operation)	Diagnostic flow chart B-5 for ISC solenoid valve
	• High fuel pressure	Diagnostic flow chart B-3
	• Defective EGR system (For Federal specification except California)	Diagnostic flow chart B-6
	• Defective injector	Check injector for fuel leakage
	• Poor performance of throttle position sensor, water temperature sensor or pressure sensor	
	• Faulty ECM	
Excessive hydrocarbons (HC) emission (Rich or lean fuel mixture)	• Faulty basic engine parts (Clogged air cleaner, vacuum leaks, faulty ignition system, engine compression, etc..)	
	• Engine not at normal operating temperature	
	• Lead contamination of catalytic converter	Check for absence of filler neck restrictor
	• Fuel leakage from injector	
	• Fuel pressure out of specification	Diagnostic flow chart B-3
	• Poor performance of water temperature sensor, pressure sensor or air temperature sensor	
	• Faulty ECM	

TROUBLE DIAGNOSIS CONTINUED 1988–90 SIDEKICK

SYMPTOM	POSSIBLE CAUSE	INSPECTION
Excessive carbon monoxide (CO) emission (Righ fuel mixture)	• Faulty basic engine parts (Clogged air cleaner, vacuum leaks, faulty ignition system, engine compression, etc..)	
	• Engine not at normal operating temperature	
	• Lead contamination of catalytic converter	Check for absence of filler neck restrictor
	• Fuel leakage from injector	
	• Fuel pressure out of specification (High fuel pressure)	Diagnostic flow chart B-3
	• Poor performance of water temperature sensor, pressure sensor or air temperature sensor	
	• Faulty ECM	
Excessive nitrogen oxides (NOx) emission (Lean fuel mixture)	• Improper ignition timing	
	• Lead contamination of catalytic converter	Check for absence of filler neck restrictor
	• Misrouted vacuum hoses	
	• Defective EGR system (For Federal specification except California)	Diagnostic flow chart B-6
	• Fuel pressure out of specification (Low fuel pressure)	Diagnostic flow chart B-3
	• Poor performance of water temperature sensor, pressure sensor or air temperature sensor	
	• Faulty ECM	

FUEL PUMP AND CIRCUIT DIAGNOSIS 1988–90 SIDEKICK

B-2 FUEL PUMP AND ITS CIRCUIT CHECK

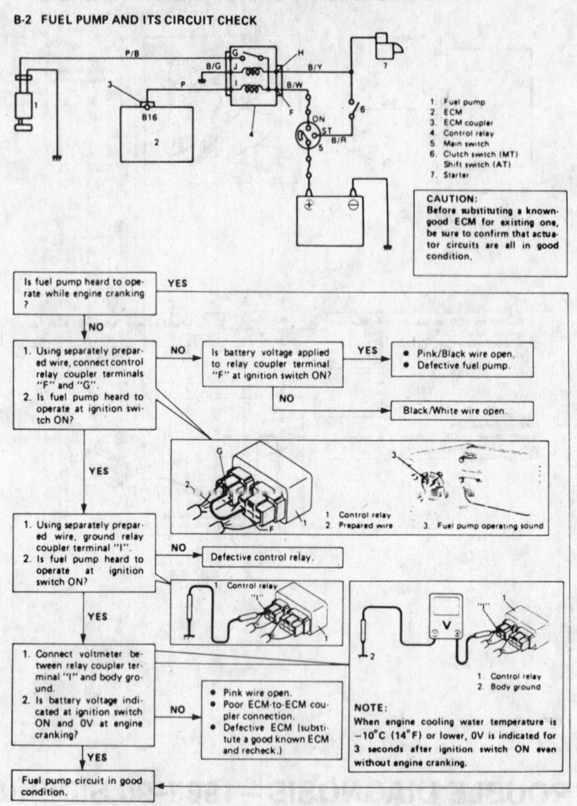
P/B
B/G
P
B/Y
B/W
B16
ON
ST
B/R
1. Fuel pump
2. ECM
3. ECM coupler
4. Control relay
5. Main switch
6. Clutch switch (MT) Shift switch (AT)
7. Starter
CAUTION:
Before substituting a known-good ECM for existing one, be sure to confirm that actuator circuits are all in good condition.
Is fuel pump heard to operate while engine cranking?
YES
NO
1. Using separately prepared wire, connect control relay coupler terminals "F" and "G".
2. Is fuel pump heard to operate at ignition switch ON?
NO
Is battery voltage applied to relay coupler terminal "F" at ignition switch ON?
YES
• Pink/Black wire open.
• Defective fuel pump.
NO
Black/White wire open.
YES
1. Control relay
2. Prepared wire
3. Fuel pump operating sound
1. Using separately prepared wire, ground relay coupler terminal "I".
2. Is fuel pump heard to operate at ignition switch ON?
NO
Defective control relay.
YES
1. Control relay
1. Connect voltmeter between relay coupler terminal "I" and body ground.
2. Is battery voltage indicated at ignition switch ON and 0V at engine cranking?
NO
• Pink wire open.
• Poor ECM-to-ECM coupler connection.
• Defective ECM (substitute a good known ECM and recheck.)
1. Control relay
2. Body ground
NOTE:
When engine cooling water temperature is −10°C (14°F) or lower, 0V is indicated for 3 seconds after ignition switch ON even without engine cranking.
YES
Fuel pump circuit in good condition.

FUEL INJECTOR AND CIRCUIT DIAGNOSIS 1988–90 SIDEKICK

B-1 FUEL INJECTOR AND ITS CIRCUIT CHECK (ENGINE NO STARTING)

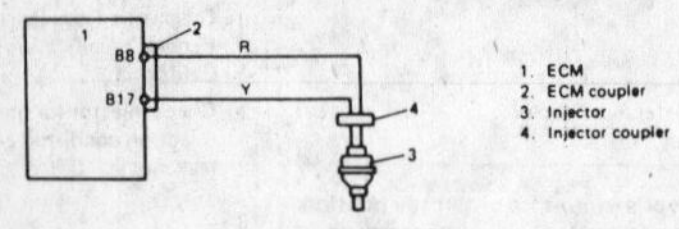
B8
B17
R
Y
1. ECM
2. ECM coupler
3. Injector
4. Injector coupler

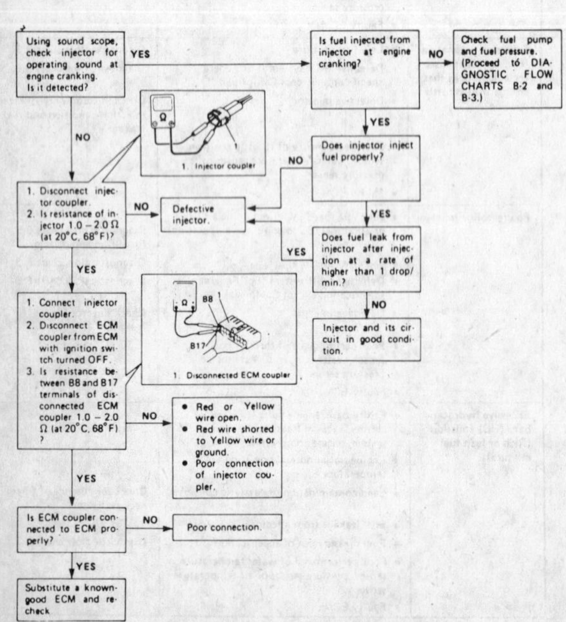
Using sound scope, check injector for operating sound at engine cranking. Is it detected?
YES
Is fuel injected from injector at engine cranking?
NO
Check fuel pump and fuel pressure. (Proceed to DIAGNOSTIC FLOW CHARTS B-2 and B-3.)
YES
Does injector inject fuel properly?
NO
1. Injector coupler
1. Disconnect injector coupler.
2. Is resistance of injector 1.0 – 2.0 Ω (at 20°C, 68°F)?
NO
Defective injector.
YES
YES
Does fuel leak from injector after injection at a rate of higher than 1 drop/min.?
NO
Injector and its circuit in good condition.
YES
1. Connect injector coupler.
2. Disconnect ECM coupler from ECM with ignition switch turned OFF.
3. Is resistance between B8 and B17 terminals of disconnected ECM coupler 1.0 – 2.0 Ω (at 20°C, 68°F)?
B8
B17
1. Disconnected ECM coupler
NO
• Red or Yellow wire open.
• Red wire shorted to Yellow wire or ground.
• Poor connection of injector coupler.
YES
Is ECM coupler connected to ECM properly?
NO
Poor connection.
YES
Substitute a known-good ECM and recheck.

FUEL PRESSURE DIAGNOSIS 1988–90 SIDEKICK

B-3 FUEL PRESSURE CHECK

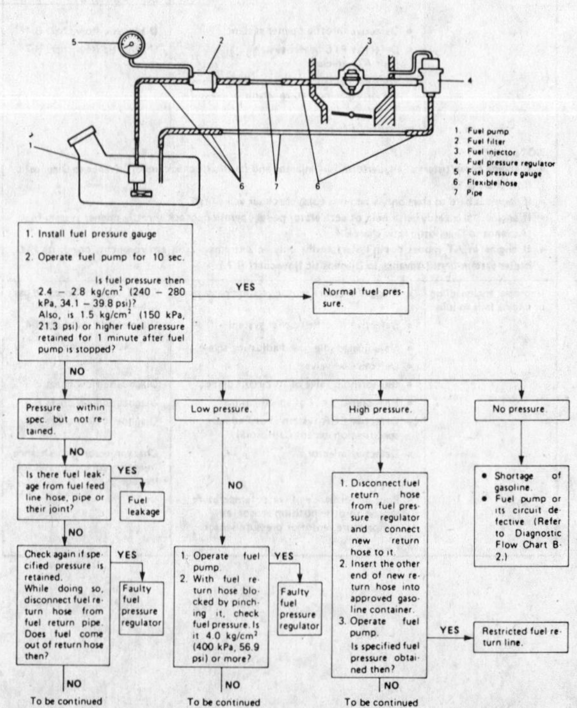
1. Fuel pump
2. Fuel filter
3. Fuel injector
4. Fuel pressure regulator
5. Fuel pressure gauge
6. Flexible hose
7. Pipe
1. Install fuel pressure gauge
2. Operate fuel pump for 10 sec.
Is fuel pressure then 2.4 – 2.8 kg/cm² (240 – 280 kPa, 34.1 – 39.8 psi)? Also, is 1.5 kg/cm² (150 kPa, 21.3 psi) or higher fuel pressure retained for 1 minute after fuel pump is stopped?
YES
Normal fuel pressure.
NO
Pressure within spec but not retained.
Low pressure.
High pressure.
No pressure.
NO
Is there fuel leakage from fuel feed line hose, pipe or their joint?
YES
Fuel leakage
NO
Check again if specified pressure is retained. While doing so, disconnect fuel return hose from fuel return pipe. Does fuel come out of return hose then?
YES
Faulty fuel pressure regulator
NO
To be continued
NO
1. Operate fuel pump.
2. With fuel return hose blocked by pinching it, check fuel pressure. Is it 4.0 kg/cm² (400 kPa, 56.9 psi) or more?
YES
Faulty fuel pressure regulator
NO
To be continued
1. Disconnect fuel return hose from fuel pressure regulator and connect new return hose to it.
2. Insert the other end of new return hose into approved gasoline container.
3. Operate fuel pump. Is specified fuel pressure obtained then?
YES
Restricted fuel return line.
NO
To be continued
• Shortage of gasoline.
• Fuel pump or its circuit defective (Refer to Diagnostic Flow Chart B-2.)

FUEL PRESSURE DIAGNOSIS CONTINUED 1988–90 SIDEKICK

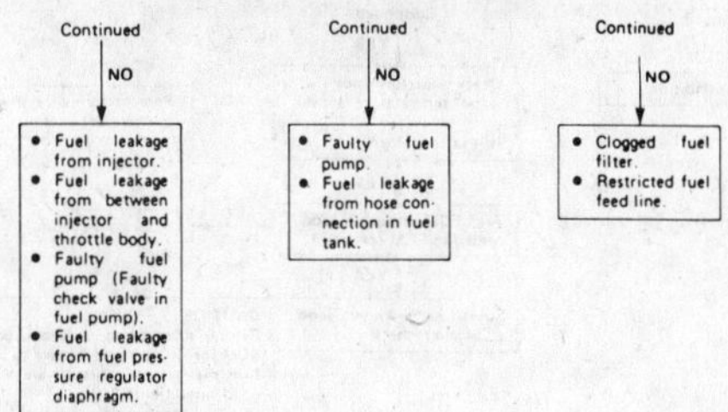
Continued
NO
• Fuel leakage from injector.
• Fuel leakage from between injector and throttle body.
• Faulty fuel pump (Faulty check valve in fuel pump).
• Fuel leakage from fuel pressure regulator diaphragm.
Continued
NO
• Faulty fuel pump.
• Fuel leakage from hose connection in fuel tank.
Continued
NO
• Clogged fuel filter.
• Restricted fuel feed line.

THROTTLE OPENER SYSTEM DIAGNOSIS 1988–90 SIDEKICK

B-4 THROTTLE OPENER SYSTEM CHECK

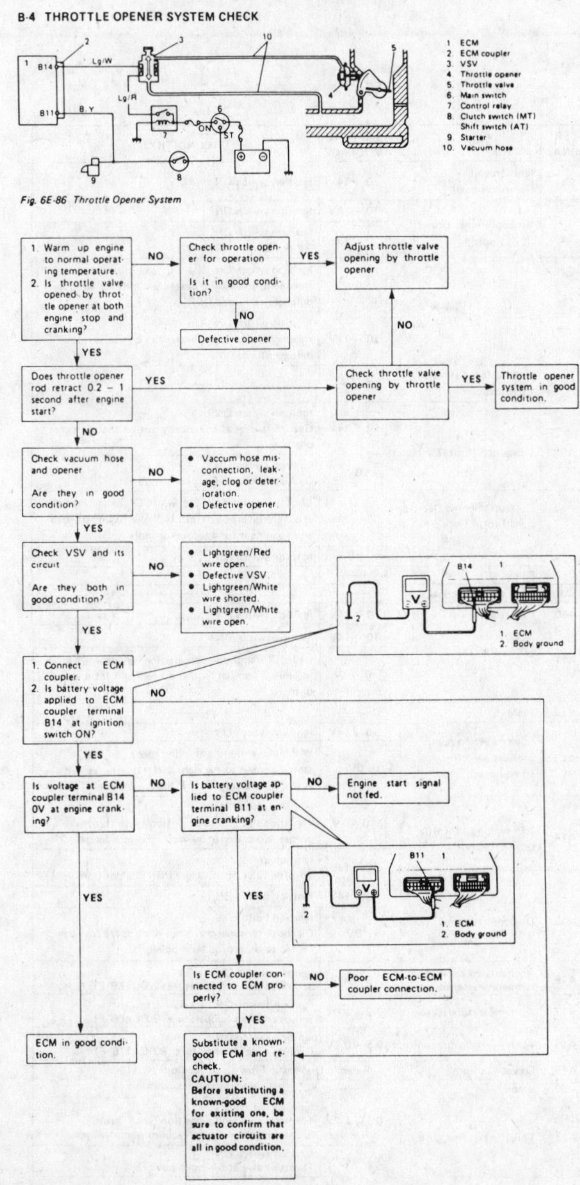
1. ECM
2. ECM coupler
3. VSV
4. Throttle opener
5. Throttle valve
6. Main switch
7. Control relay
8. Clutch switch (MT) Shift switch (AT)
9. Starter
10. Vacuum hose
B14
B11
Lg/W
Lg/R
B/Y
Fig. 6E-86 Throttle Opener System
1. Warm up engine to normal operating temperature.
2. Is throttle valve opened by throttle opener at both engine stop and cranking?
NO
Check throttle opener for operation
Is it in good condition?
YES
Adjust throttle valve opening by throttle opener
NO
Defective opener
YES
Does throttle opener rod retract 0.2 – 1 second after engine start?
YES
Check throttle valve opening by throttle opener
YES
Throttle opener system in good condition.
NO
Check vacuum hose and opener
Are they in good condition?
NO
• Vacuum hose misconnection, leakage, clog or deterioration.
• Defective opener
YES
Check VSV and its circuit
Are they both in good condition?
NO
• Lightgreen/Red wire open.
• Defective VSV.
• Lightgreen/White wire shorted.
• Lightgreen/White wire open.
YES
1. Connect ECM coupler.
2. Is battery voltage applied to ECM coupler terminal B14 at ignition switch ON?
NO
B14
1. ECM
2. Body ground
YES
Is voltage at ECM coupler terminal B14 0V at engine cranking?
NO
Is battery voltage applied to ECM coupler terminal B11 at engine cranking?
NO
Engine start signal not fed.
B11
1. ECM
2. Body ground
YES
YES
Is ECM coupler connected to ECM properly?
NO
Poor ECM-to-ECM coupler connection.
YES
ECM in good condition.
Substitute a known-good ECM and recheck.
CAUTION:
Before substituting a known-good ECM for existing one, be sure to confirm that actuator circuits are all in good condition.

IDLE SPEED SOLENOID VALVE DIAGNOSIS 1988–90 SIDEKICK

B-5 ISC SOLENOID VALVE CIRCUIT CHECK

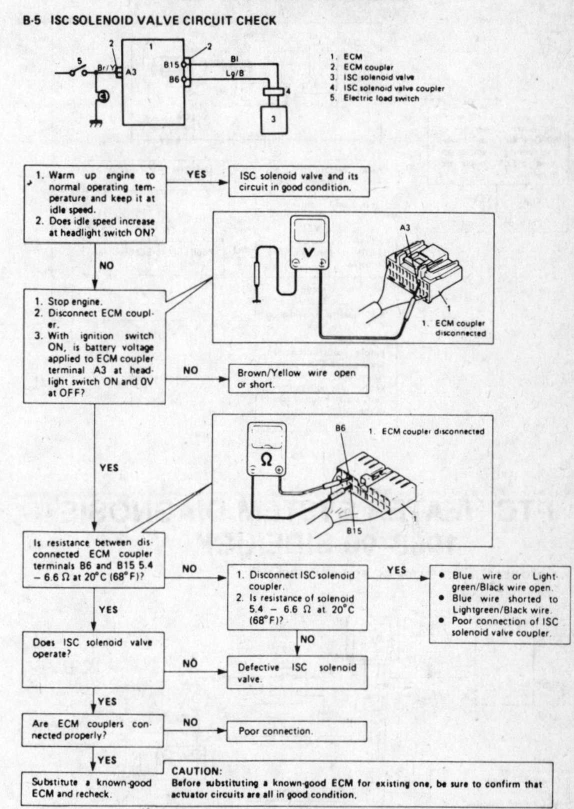
1. ECM
2. ECM coupler
3. ISC solenoid valve
4. ISC solenoid valve coupler
5. Electric load switch
Br/Y
A3
B15
B6
Bl
Lg/B
1. Warm up engine to normal operating temperature and keep it at idle speed.
2. Does idle speed increase at headlight switch ON?
YES
ISC solenoid valve and its circuit in good condition.
NO
1. Stop engine.
2. Disconnect ECM coupler.
3. With ignition switch ON, is battery voltage applied to ECM coupler terminal A3 at headlight switch ON and 0V at OFF?
1. ECM coupler disconnected
NO
Brown/Yellow wire open or short.
YES
1. ECM coupler disconnected
Is resistance between disconnected ECM coupler terminals B6 and B15 5.4 – 6.6 Ω at 20°C (68°F)?
NO
1. Disconnect ISC solenoid coupler.
2. Is resistance of solenoid 5.4 – 6.6 Ω at 20°C (68°F)?
YES
• Blue wire or Lightgreen/Black wire open.
• Blue wire shorted to Lightgreen/Black wire.
• Poor connection of ISC solenoid valve coupler.
NO
Defective ISC solenoid valve.
YES
Does ISC solenoid valve operate?
NO
YES
Are ECM couplers connected properly?
NO
Poor connection.
YES
Substitute a known-good ECM and recheck.
CAUTION:
Before substituting a known-good ECM for existing one, be sure to confirm that actuator circuits are all in good condition.

EGR SYSTEM DIAGNOSIS 1988–90 SIDEKICK

B-6 EGR SYSTEM CHECK
(For Federal Specification Except California)

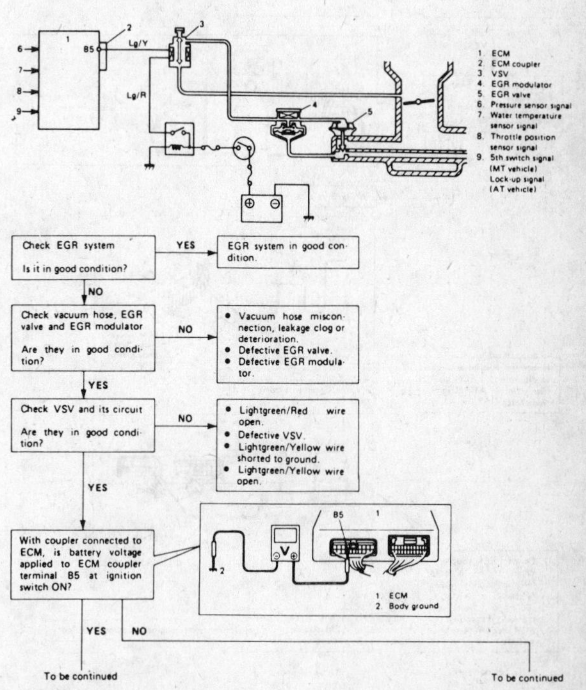
1. ECM
2. ECM coupler
3. VSV
4. EGR modulator
5. EGR valve
6. Pressure sensor signal
7. Water temperature sensor signal
8. Throttle position sensor signal
9. 5th switch signal (MT vehicle) Lock-up signal (AT vehicle)
B5
Lg/Y
Lg/R
Check EGR system
Is it in good condition?
YES
EGR system in good condition.
NO
Check vacuum hose, EGR valve and EGR modulator
Are they in good condition?
NO
• Vacuum hose misconnection, leakage clog or deterioration.
• Defective EGR valve.
• Defective EGR modulator.
YES
Check VSV and its circuit
Are they in good condition?
NO
• Lightgreen/Red wire open.
• Defective VSV.
• Lightgreen/Yellow wire shorted to ground.
• Lightgreen/Yellow wire open.
YES
With coupler connected to ECM, is battery voltage applied to ECM coupler terminal B5 at ignition switch ON?
B5
1. ECM
2. Body ground
YES
NO
To be continued
To be continued

EGR SYSTEM DIAGNOSIS CONTINUED 1988–90 SIDEKICK

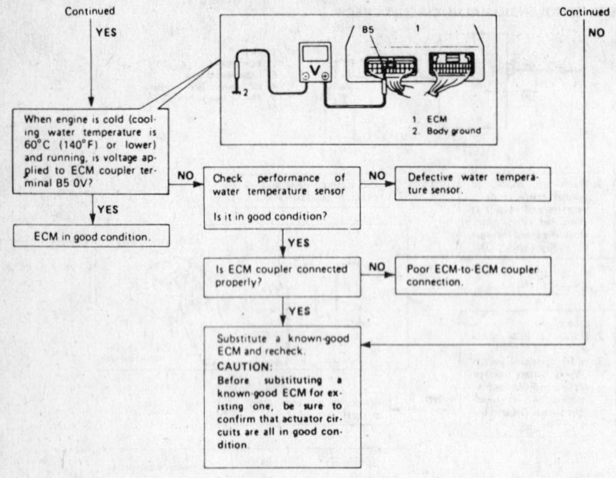
Continued
YES
Continued
NO
B5
1
2
1. ECM
2. Body ground
When engine is cold (cooling water temperature is 60°C (140°F) or lower) and running, is voltage applied to ECM coupler terminal B5 0V?
NO
Check performance of water temperature sensor
Is it in good condition?
NO
Defective water temperature sensor.
YES
ECM in good condition.
YES
Is ECM coupler connected properly?
NO
Poor ECM-to-ECM coupler connection.
YES
Substitute a known-good ECM and recheck.
CAUTION:
Before substituting a known-good ECM for existing one, be sure to confirm that actuator circuits are all in good condition.

PTC HEATER SYSTEM DIAGNOSIS 1988–90 SIDEKICK

B-7 PTC HEATER SYSTEM CHECK (FOR AT VEHICLE)

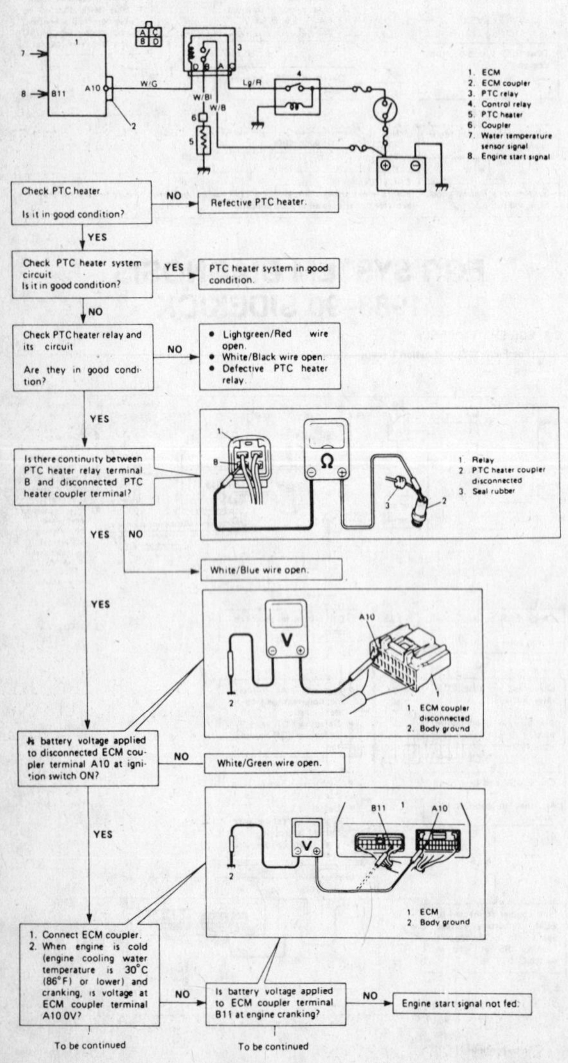
7
8
B11
A10
W/G
Lg/R
W/Bl
W/B
1. ECM
2. ECM coupler
3. PTC relay
4. Control relay
5. PTC heater
6. Coupler
7. Water temperature sensor signal
8. Engine start signal
Check PTC heater.
Is it in good condition?
NO
Refective PTC heater.
YES
Check PTC heater system circuit.
Is it in good condition?
YES
PTC heater system in good condition.
NO
Check PTC heater relay and its circuit
Are they in good condition?
NO
• Lightgreen/Red wire open.
• White/Black wire open.
• Defective PTC heater relay.
YES
Is there continuity between PTC heater relay terminal B and disconnected PTC heater coupler terminal?
1. Relay
2. PTC heater coupler disconnected
3. Seal rubber
YES
NO
White/Blue wire open.
YES
A10
1. ECM coupler disconnected
2. Body ground
Is battery voltage applied to disconnected ECM coupler terminal A10 at ignition switch ON?
NO
White/Green wire open.
YES
B11
A10
1. ECM
2. Body ground
1. Connect ECM coupler.
2. When engine is cold (engine cooling water temperature is 30°C (86°F) or lower) and cranking, is voltage at ECM coupler terminal A10 0V?
NO
Is battery voltage applied to ECM coupler terminal B11 at engine cranking?
NO
Engine start signal not fed.
To be continued
To be continued

ECM TERMINAL IDENTIFICATION 1988–90 SIDEKICK

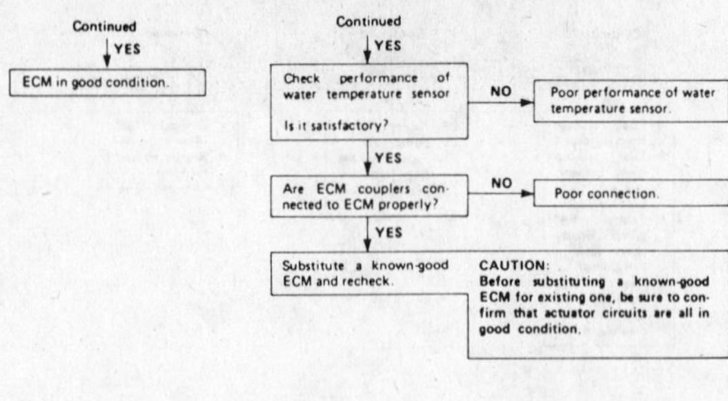
Continued
YES
ECM in good condition.
Continued
YES
Check performance of water temperature sensor
Is it satisfactory?
NO
Poor performance of water temperature sensor.
YES
Are ECM couplers connected to ECM properly?
NO
Poor connection.
YES
Substitute a known-good ECM and recheck.
CAUTION:
Before substituting a known-good ECM for existing one, be sure to confirm that actuator circuits are all in good condition.

TERMINAL	CIRCUIT	NORMAL VOLTAGE	CONDITION
A1	Ignition coil – (Ignition signal)	10 – 14V	Ignition switch ON
A2	Air-conditioner circuit (if equipped)	10 – 14V	Ignition switch ON
		0 – 1V	Ignition switch ON Air-conditioner switch ON
A3	Electric load switch	0V	Ignition switch ON Headlight, small light, heater fan and rear window defogger all turned OFF
		10 – 14V	Ignition switch ON Headlight, small light, heater fan or rear window defogger turned ON
A4	Ground (for California specification only)	——	——
A5	5th switch (for MT vehicle)	10 – 14V	Ignition switch ON Gear shift lever at any other position than 5th gear position
		0V	Ignition switch ON Gear shift lever at 5th gear position
A6	Lock-up solenoid (lock-up signal) for AT vehicle	10 – 14V	Lock up solenoid ON (i.e. with "D" range position, driving vehicle at 67 km/h (42 mile/h) on flat road and keeping it for 4 second or more.)
		0V	Ignition switch ON
A7, A8	Blank	——	——
A9	Duty check terminal	——	——
A10	PTC relay (for AT vehicle)	10 – 14V	Ignition switch ON
A11	Lock-up relay (for AT vehicle)	10 – 14V	Ignition switch ON
		0 – 1V	With "D" range position, driving vehicle at 67 km/h (42 mile/h) on flat road and keeping it for 4 second or more.
A12	Blank	——	——
A13	Power steering pump pressure switch (if equipped)	10 – 14V	Ignition switch ON
		0V	With engine running at idling speed, turning steering wheel to the right and left as far as it stops, repeating it a few times.
A14	Idle switch of throttle position sensor	0 – 1V	Ignition switch ON Throttle valve is at idle position (with throttle opener rod drawn in by vacuum pump gauge)
		10 – 14V	Ignition switch ON Throttle valve opens larger than idle position
A15	Diagnosis terminal	10 – 14V	Ignition switch ON
		0V	Ignition switch ON Diagnosis terminal grounded (with spare fuse connected to diagnostic terminals)
A16	Recirculated exhaust gas temperature sensor	3.8 – 4.5V	Ignition switch ON Sensor ambient temperature: 20°C (68°F)
A17	Air temperature sensor	2.2 – 3.0V	Ignition switch ON Sensor ambinet temperature: 20°C (68°F)
A18	Water temperature sensor	0.5 – 0.9V	Ignition switch ON Cooling water temperature: 80°C (176°F)
A19	Oxygen sensor	Refer to Diagnostic Flow Chart for Code No. 13	
A20	Blank		
A21	Throttle position sensor	0.5 – 1.2V	Ignition switch ON Throttle valve at idle position (with throttle opener rod drawn in by vacuum pump gauge)
		3.4 – 4.7V	Ignition switch ON Throttle valve at full open position

ECM TERMINAL IDENTIFICATION CONTINUED—1988–90 SIDEKICK

TERMINAL	CIRCUIT	NORMAL VOLTAGE	CONDITION
A22	Pressure sensor	Refer to p. 6E-94	
A23	Power source of sensors	4.75–5.25V	Ignition switch ON
A24	Ground of sensors	—	—
B1	Power source	10 – 14V	Ignition switch ON
B2	Ground	—	—
B3	Ground	—	—
B4	Blank	—	—
B5	EGR VSV	10 – 14V	Ignition switch ON
B6	ISC solenoid valve ⊖	—	—
B7	Power source	10 – 14V	Ignition switch ON
B8	Injector ⊕	—	—
B9	Power source for back-up circuit	10 – 14V	Ignition switch OFF and ON
B10	Ground	—	—
B11	Engine start switch (Engine start signal)	6 – 10V	While engine cranking
		0V	Other than above
B12	Shift switch (for AT vehicle)	0V	Ignition switch ON Selector lever in "P" or "N" range
		10 – 14V	Ignition switch ON Selector lever in any other range than "P" and "N"
B13	"CHECK ENGINE" light	0 – 1V	Ignition switch ON
		10 – 14V	When engine running
B14	Throttle opener VSV	10 – 14V	Ignition switch ON
B15	ISC solenoid valve ⊕	—	—
B16	Fuel pump relay in control relay	10 – 14V	Ignition switch ON (3 seconds after ignition switch ON when cooling water temp. is lower than −10°C or 14°F)
B17	Injector ⊖	—	—

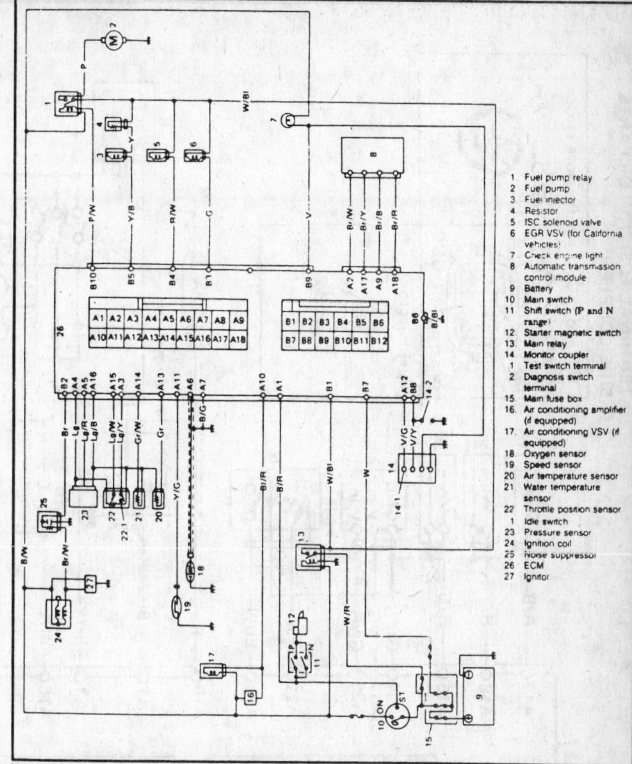
1 Fuel pump relay
2 Fuel pump
3 Fuel injector
4 Resistor
5 ISC solenoid valve
6 EGR VSV (for California vehicles)
7 Check engine light
8 Automatic transmission control module
9 Battery
10 Main switch
11 Shift switch (P and N range)
12 Starter magnetic switch
13 Main relay
14 Monitor coupler
1 Test switch terminal
2 Diagnosis switch terminal
15 Main fuse box
16 Air conditioning amplifier (if equipped)
17 Air conditioning VSV (if equipped)
18 Oxygen sensor
19 Speed sensor
20 Air temperature sensor
21 Water temperature sensor
22 Throttle position sensor
1 Idle switch
23 Pressure sensor
24 Ignition coil
25 Noise suppressor
26 ECM
27 Ignitor

EFI system wiring diagram—1988–90 Swift, except GTi with automatic transmission

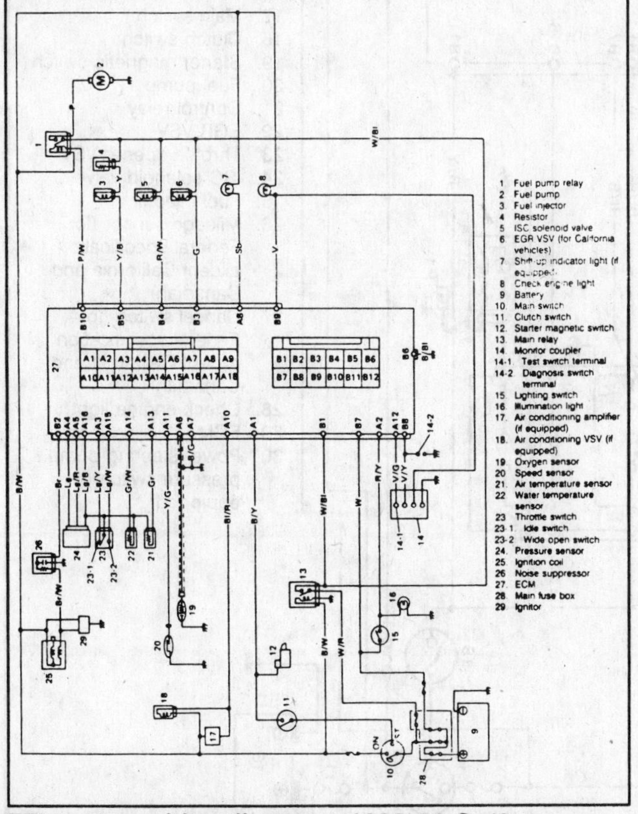
1 Fuel pump relay
2 Fuel pump
3 Fuel injector
4 Resistor
5 ISC solenoid valve
6 EGR VSV (for California vehicles)
7 Shift-up indicator light (if equipped)
8 Check engine light
9 Battery
10 Main switch
11 Clutch switch
12 Starter magnetic switch
13 Main relay
14 Monitor coupler
14-1 Test switch terminal
14-2 Diagnosis switch terminal
15 Lighting switch
16 Illumination light
17 Air conditioning amplifier (if equipped)
18 Air conditioning VSV (if equipped)
19 Oxygen sensor
20 Speed sensor
21 Air temperature sensor
22 Water temperature sensor
23 Throttle switch
23-1 Idle switch
23-2 Wide open switch
24 Pressure sensor
25 Ignition coil
26 Noise suppressor
27 ECM
28 Main fuse box
29 Ignitor

EFI system wiring diagram—1988–90 Swift, except GTi with manual transmission

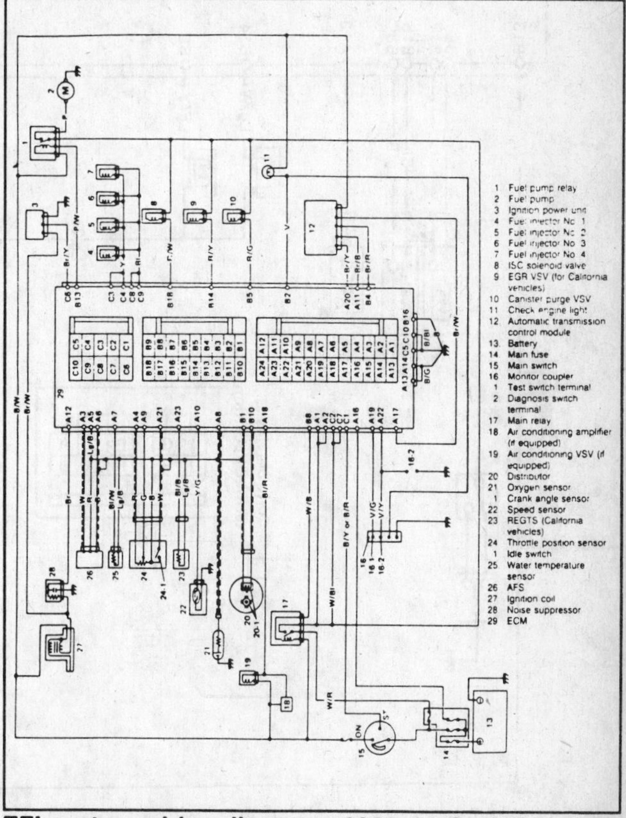
1 Fuel pump relay
2 Fuel pump
3 Ignition power unit
4 Fuel injector No 1
5 Fuel injector No 2
6 Fuel injector No 3
7 Fuel injector No 4
8 ISC solenoid valve
9 EGR VSV (for California vehicles)
10 Canister purge VSV
11 Check engine light
12 Automatic transmission control module
13 Battery
14 Main fuse
15 Main switch
16 Monitor coupler
1 Test switch terminal
2 Diagnosis switch terminal
17 Main relay
18 Air conditioning amplifier (if equipped)
19 Air conditioning VSV (if equipped)
20 Distributor
21 Oxygen sensor
1 Crank angle sensor
22 Speed sensor
23 REGTS (California vehicles)
24 Throttle position sensor
1 Idle switch
25 Water temperature sensor
26 AFS
27 Ignition coil
28 Noise suppressor
29 ECM

EFI system wiring diagram—1988–90 Swift GTi with automatic transmission

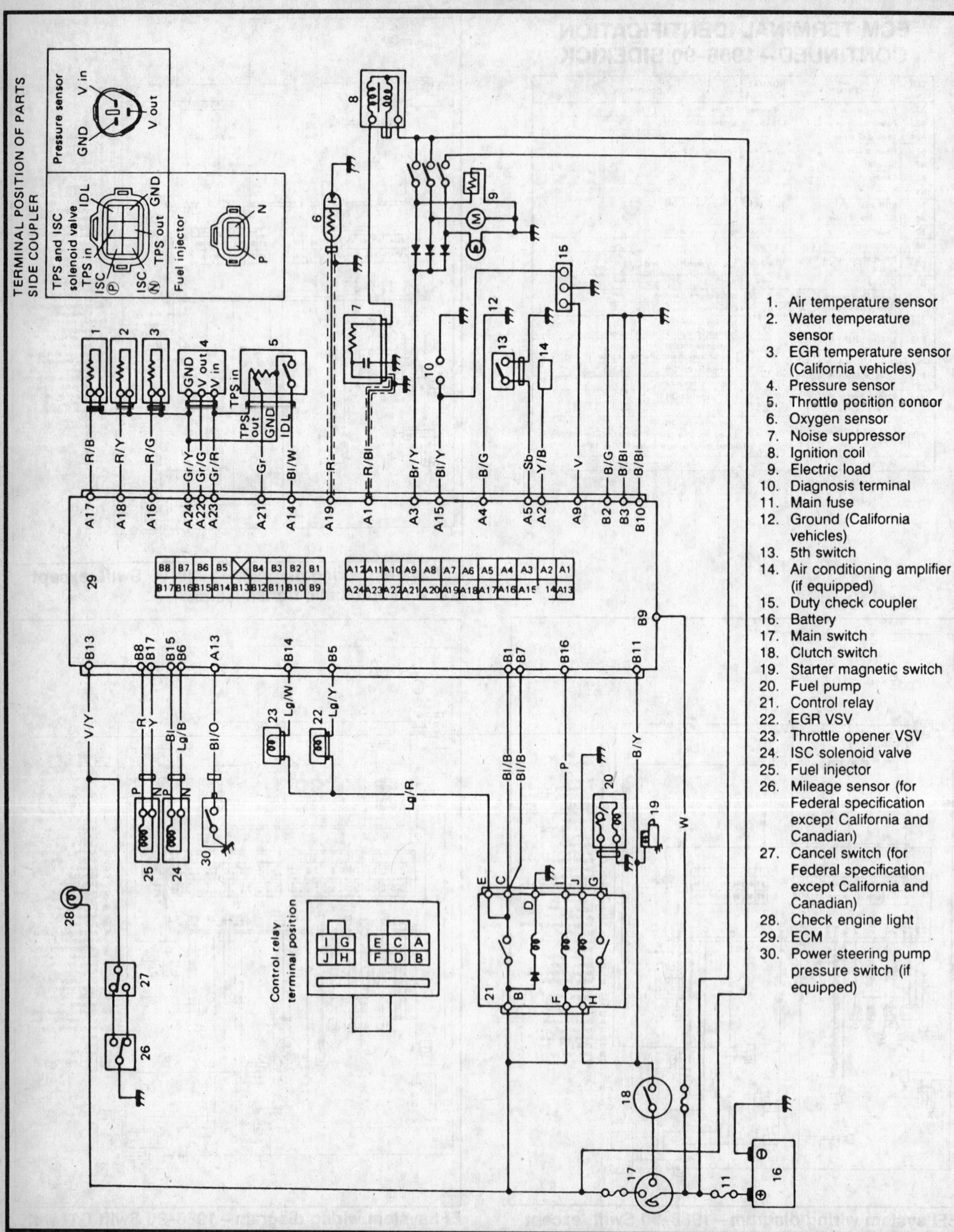
1. Air temperature sensor
2. Water temperature sensor
3. EGR temperature sensor (California vehicles)
4. Pressure sensor
5. Throttle position sensor
6. Oxygen sensor
7. Noise suppressor
8. Ignition coil
9. Electric load
10. Diagnosis terminal
11. Main fuse
12. Ground (California vehicles)
13. 5th switch
14. Air conditioning amplifier (if equipped)
15. Duty check coupler
16. Battery
17. Main switch
18. Clutch switch
19. Starter magnetic switch
20. Fuel pump
21. Control relay
22. EGR VSV
23. Throttle opener VSV
24. ISC solenoid valve
25. Fuel injector
26. Mileage sensor (for Federal specification except California and Canadian)
27. Cancel switch (for Federal specification except California and Canadian)
28. Check engine light
29. ECM
30. Power steering pump pressure switch (if equipped)

EFI system wiring diagram – 1988–90 Sidekick with manual transmission

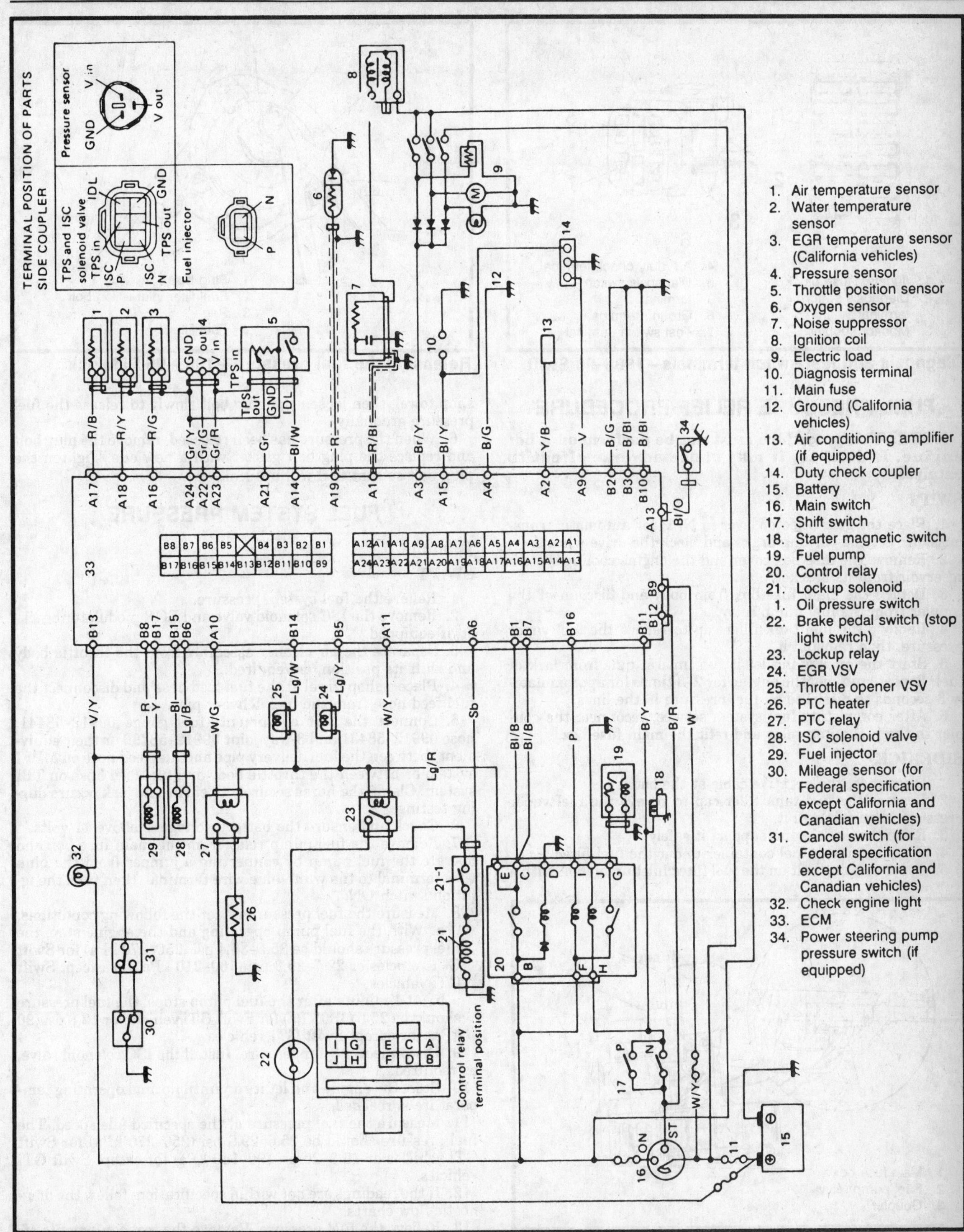
TERMINAL POSITION OF PARTS SIDE COUPLER
Pressure sensor
GND
V out
V in
TPS and ISC solenoid valve
TPS in
ISC
P
IDL
GND
ISC
N
TPS out
Fuel injector
N
P
1. Air temperature sensor
2. Water temperature sensor
3. EGR temperature sensor (California vehicles)
4. Pressure sensor
5. Throttle position sensor
6. Oxygen sensor
7. Noise suppressor
8. Ignition coil
9. Electric load
10. Diagnosis terminal
11. Main fuse
12. Ground (California vehicles)
13. Air conditioning amplifier (if equipped)
14. Duty check coupler
15. Battery
16. Main switch
17. Shift switch
18. Starter magnetic switch
19. Fuel pump
20. Control relay
21. Lockup solenoid
1. Oil pressure switch
22. Brake pedal switch (stop light switch)
23. Lockup relay
24. EGR VSV
25. Throttle opener VSV
26. PTC heater
27. PTC relay
28. ISC solenoid valve
29. Fuel injector
30. Mileage sensor (for Federal specification except California and Canadian vehicles)
31. Cancel switch (for Federal specification except California and Canadian vehicles)
32. Check engine light
33. ECM
34. Power steering pump pressure switch (if equipped)
Control relay terminal position
EFI system wiring diagram – 1988–90 Sidekick with automatic transmission

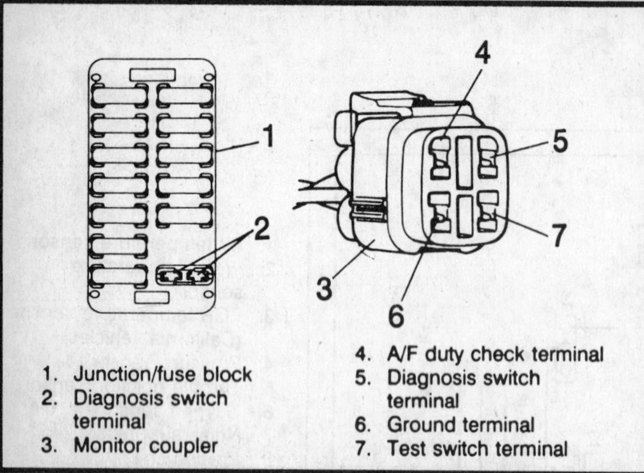
1
2
3
4
5
6
7
1. Junction/fuse block
2. Diagnosis switch terminal
3. Monitor coupler
4. A/F duty check terminal
5. Diagnosis switch terminal
6. Ground terminal
7. Test switch terminal

Diagnosis and test switch terminals – 1988–90 Swift

FUEL PRESSURE RELIEF PROCEDURE

NOTE: This procedure must not be perform on a hot engine. If done so, it may cause adverse effect to catalyst.

SWIFT

1. Place the gear selector lever in **N**, **P** for automatic transmission), set the parking brake and block the drive wheels.
2. Remove the fuse box cover and the engine cooling system reservoir from its bracket.
3. Detach the main fuse box from body and disconnect the coupler from the fuel pump relay.
4. Remove the fuel tank filler cap to relieve the fuel vapor pressure, then reinstall it.
5. Start the engine and let it run until it quit from lack of fuel. Repeat cranking the engine for 2–3 times for approximately 3 seconds to dissipate all fuel pressure in the lines.
6. After completing fuel system service, reconnect the coupler to the fuel pump relay and refit the main fuse box.

SIDEKICK

1. Disconnect the negative cable at the battery.
2. Remove the fuel tank filler cap to relieve the fuel vapor pressure, then reinstall it.
3. Raise the vehicle and support it safely.
4. Place a suitable fuel container under the fuel filter.
5. Cover the plug bolt on the fuel filter inlet union bolt with a shop towel, then loosen the plug bolt slowly to release the fuel pressure gradually.
6. When the pressure has been released, remove the plug bolt and replace the plug bolt gasket with a new one. Tighten the plug bolt so that fuel will not leak.

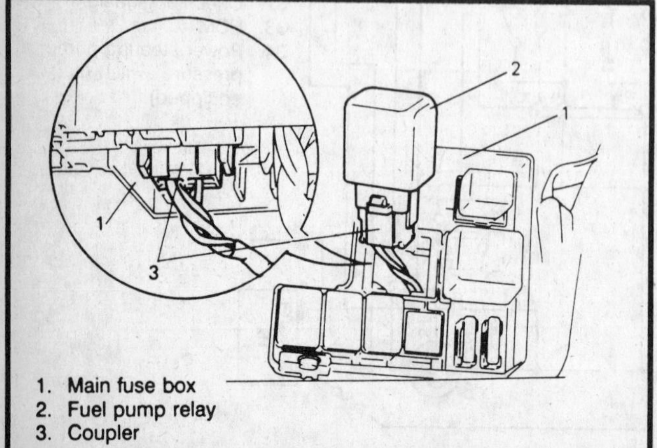
1
2
3
1. Main fuse box
2. Fuel pump relay
3. Coupler

Fuel pump relay coupler

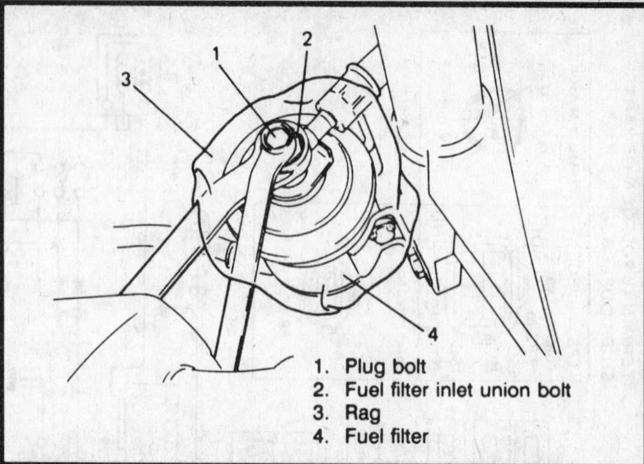
1
2
3
4
1. Plug bolt
2. Fuel filter inlet union bolt
3. Rag
4. Fuel filter

Releasing the fuel pressure – 1988–90 Sidekick

FUEL SYSTEM PRESSURE

Testing

SWIFT

1. Relieve the fuel system pressure.
2. Remove the ISC solenoid valve and EGR modulator bracket, if equipped.
3. Separate the air cleaner assembly from the throttle body and shift its position, as required.
4. Place a shop towel at the fuel feed hose and disconnect the fuel feed hose from the fuel delivery pipe.
5. Connect the test equipment (fuel gauge 09912-58441, hose 09912-58431 and 3-way joint 09912-58490 or their equivalent) between the fuel delivery pipe and fuel feed hose on MPFI system or between the throttle body and fuel feed hose on TBI system. Clamp the hoses securely to ensure no leak occurs during testing.
6. Check and ensure the battery voltage is above 11 volts.
7. Remove the fuel pump relay from the main fuse box and operate the fuel pump by connecting a jumper from the pink wire terminal to the white/blue wire terminal, then turn the ignition switch **ON**.
8. Measure the fuel pressure under the following conditions:
   a. With the fuel pump operating and the engine stop, the fuel pressure should be 35.5–38.4 psi (250–270 kPa) for Swift GTi vehicles or 22.7–29.9 psi (160–210 kPa) for except Swift GTi vehicles.
   b. At 1 minute after the fuel pump stops, the fuel pressure should be 25.6 (180 kPa) for Swift GTi vehicles or 12.8 psi (90 kPa) for except Swift GTi vehicles.
9. Disconnect the jumper wire. Install the ISC solenoid valve, as required.
10. Start the engine and let it run until normal operating temperature is reached.
11. Measure the fuel pressure at the specified idle speed. The fuel pressure should be 25.6–29.9 psi (250–270 kPa) for Swift GTi vehicles or 12.8–20 psi (90–140 kPa) for except Swift GTi vehicles.
12. If the readings are not within specification, follow the diagnostic flow charts.
13. Relieve the fuel pressure. Remove the test equipments after removing the ISC solenoid valve.

14. Connect the fuel feed hose to the fuel delivery pipe and clamp it securely.
15. Install the ISC solenoid valve and EGR modulator bracket, if equipped.
16. Install the air cleaner assembly, as required.
17. Start the engine and check for leaks.

SIDEKICK

1. Relieve the fuel system pressure.
2. Remove the plug bolt on the fuel filter union bolt and connect fuel pressure gauge set 09912-58412 or equivalent to the fuel filter inlet union bolt.
3. Start the engine and run until normal operating temperature is reached.
4. Measure the fuel pressure under each of the following conditions:
   a. At the recommended idle speed or with the engine stop and the fuel pump operating, the fuel pressure should be 39.8 psi (280 kPa).
   b. Within 1 minute after the engine has stop, the fuel pressure should be over 21.3 psi (150 kPa).
5. If the pressure readings are not within specification, follow the diagnosis flow chart B-3.
6. After checking the fuel pressure, release the fuel system pressure and remove the fuel pressure gauge.
7. Install the plug bolt to the fuel filter inlet union bolt using a new gasket.
8. Start the engine and check for leaks.

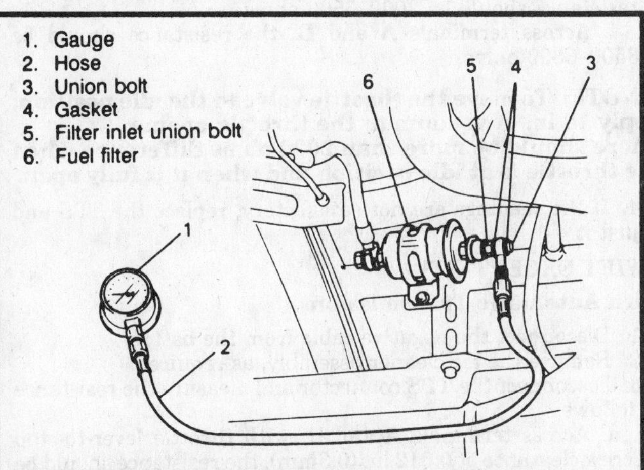
1. Gauge
2. Hose
3. Union bolt
4. Gasket
5. Filter inlet union bolt
6. Fuel filter
6
5
4
3
1
2

Checking the fuel system pressure – 1988–90 Sidekick

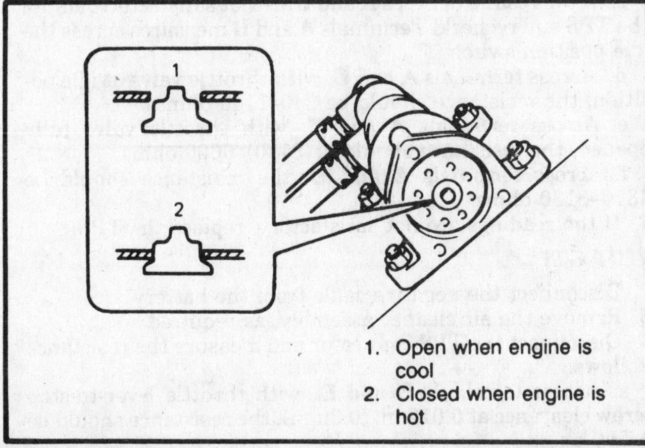
1
2
1. Open when engine is cool
2. Closed when engine is hot

Checking the air valve

AIR VALVE

Testing

1. With the engine stopped, remove the air valve cap.
2. Visually check that the air valve is open when the engine temperature is approximately 140°F (60°C) or lower and closed when the engine temperature is approximately 158°F (70°C) or higher.
3. Install the air valve cap using a new gasket.

FUEL INJECTOR

Testing

THROTTLE BODY INJECTION

1. Disconnect the negative battery cable from the battery.
2. Disconnect the injector electrical connector and measure the resistance accross the injector. Injector resistance should be as follows:
   Swift – 0.5–1.5 ohms at 68°F (20°C).
   Sidekick – 1.0–2.0 ohms at 68°F (20°C).
3. If the resistance reading is not within specification, replace the fuel injector.
4. Reconnect the injector leads and remove the air cleaner assembly without disconnecting the ATS connector.
5. Check that fuel is injected out in a cronical shape from the fuel injector when cranking or running the engine.
6. If no fuel is injected, check the wiring harness for continuity and the connector for proper connection.
7. If the fuel in not injected in a cronical shape, replace the fuel injector.

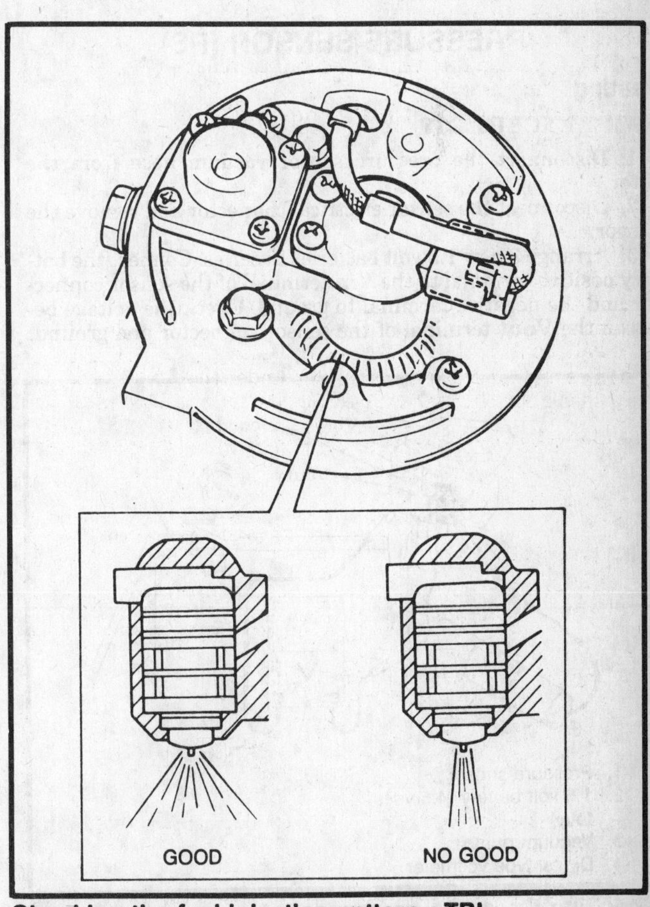
GOOD
NO GOOD

Checking the fuel injection pattern – TBI

8. Check that the fuel injector does not leaks after the engine is stopped. Replace if leakage is visible.
9. Reinstall the air cleaner assembly.

MULTI-POINT FUEL INJECTION

1. Check the operating sound of the injector when the engine is running or cranking, using a sound scope or equivalent. The cycle of the operating sound should vary according to the engine speed.
2. If no sound or an unusual sound is heard, check the injector circuit or injector.
3. Disconnect the injector connector and measure the resistance across the injector. The resistance of the injector should be 1.5–2.2 ohms.
4. If the resistance is not within specification, replace the injector.

IDLE SPEED CONTROL (ISC) SOLENOID VALVE

Testing

SIDEKICK

1. Disconnect the negative cable from the battery.
2. Disconnect the ISC solenoid valve connector and measure the resistance across the ISC solenoid valve. The resistance of the ISC solenoid valve should be 5.4–6.6 ohms at 68°F (20°C).
3. If the resistance reading is not within specification, replace the ISC solenoid valve.
4. Reconnect the ISC solenoid valve leads and the negative battery cable.

PRESSURE SENSOR (PS)

Testing

SWIFT EXCEPT GTi

1. Disconnect the pressure sensor vacuum hose from the filter.
2. Disconnect the sensor electrical connector and remove the sensor.
3. Arrange 3 new 1.5 volt batteries in series. Connect the battery positive terminal to the **Vin** terminal of the sensor connector and the negative terminal to ground. Check the voltage between the **Vout** terminal of the sensor connector and ground.

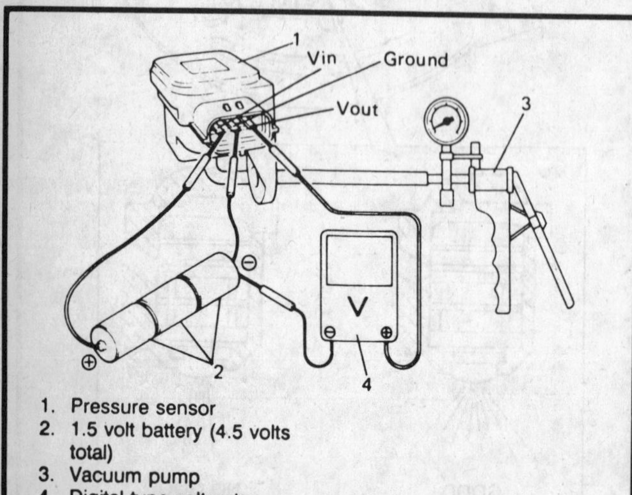
1
Vin
Ground
Vout
3
V
2
4

1. Pressure sensor
2. 1.5 volt battery (4.5 volts total)
3. Vacuum pump
4. Digital type voltmeter

Checking the pressure sensor

NOTE: Make absolutely sure all connections are as indicated, as connection to a wrong terminal will cause damage to the pressure sensor.

4. Apply 15 in. of vacuum and check that the voltage reduces.
5. If the result is not satisfactory, replace the pressure sensor.
6. Install the pressure sensor and connect the vacuum hose and electrical connector.

THROTTLE POSITION SENSOR (TPS)

Testing

SIDEKICK

1. Disconnect the negative cable from the battery.
2. Remove the air cleaner assembly, as required.
3. Disconnect the TPS connector and measure the resistance as follows:
   a. Across terminals **C** and **D**, with throttle at idle position, the resistance should be 0–500 ohms.
   b. Across terminals **C** and **D**, with throttle fully open, it should read infinity.
   c. If the results in Steps A and B are not satisfactory, adjust the TPS and recheck. Terminals **C** and **D** measures across the idle position switch.
   d. Across terminals **B** and **D**, with throttle at idle position, the resistance should be 0–2000 ohms.
   e. Across terminals **B** and **D**, with throttle fully open, the resistance should be 2000–6500 ohms.
   f. Across terminals **A** and **D**, the resistance should be 3500–6500 ohms.

NOTE: To move the throttle valve to the idle position, apply 15 in. of vacuum to the throttle opener. There should be more than 2000 ohms difference when the throttle is at idle position and when it is fully open.

4. If the readings are not satisfactory, replace the TPS and adjust it.

SWIFT EXCEPT GTi

With Automatic Transmission

1. Disconnect the negative cable from the battery.
2. Remove the air cleaner assembly, as required.
3. Disconnect the TPS connector and measure the resistance as follows:
   a. Across terminals **A** and **B**, with throttle lever-to-stop screw clearance at 0.012 in. (0.3mm), the resistance should be 0 ohm.
   b. Across terminals **A** and **B**, with throttle lever-to-stop screw clearance at 0.035 in. (0.9mm), it should read infinity.
   c. If the results in Steps A and B are not satisfactory, adjust the TPS and recheck. Terminals **A** and **B** measures across the idle position switch.
   d. Across terminals **A** and **C**, with throttle valve at idle position, the resistance should be 240–1140 ohms.
   e. Across terminals **A** and **C**, with throttle valve fully opened, the resistance should be 3170)–6600 ohms.
   f. Across terminals **A** and **D**, the resistance should be 4370–8130 ohms.
4. If the readings are not satisfactory, replace the TPS.

SWIFT GTi

1. Disconnect the negative cable from the battery.
2. Remove the air cleaner assembly, as required.
3. Disconnect the TPS connector and measure the resistance as follows:
   a. Across terminals **C** and **D**, with throttle lever-to-stop screw clearance at 0.012 in. (0.3mm), the resistance should be 0–500 ohms.
   b. Across terminals **C** and **D**, with throttle lever-to-stop

screw clearance at 0.035 in. (0.9mm), it should read infinity.

c. If the results in Steps A and B are not satisfactory, adjust the TPS and recheck. Terminals **C** and **D** measures across the idle switch.

d. Across terminals **B** and **C**, with throttle at idle position, the resistance should be 0–500 ohms.

e. Across terminals **B** and **C**, with throttle fully open, the resistance should be 3500–6500 ohms.

f. Across terminals **A** and **C**, the resistance should be 3500–6500 ohms.

4. If the readings are not satisfactory, replace the TPS and adjust it.

THROTTLE SWITCH (TS)

Testing

SWIFT EXCEPT GTi

With Manual Transmission

1. Disconnect the negative cable from the battery.
2. Remove the air cleaner assembly, as required.
3. Disconnect the TS connector and insert a feeler gauge between the throttle lever and the throttle stop screw. Using an ohmmeter, check the readings under the following conditions:

a. Across terminals **A** and **B**, with throttle lever-to-stop screw clearance at 0.012 in. (0.3mm), the should be continuity.

b. Across terminals **A** and **B**, with throttle lever-to-stop screw clearance at 0.035 in. (0.9mm), the should be no continuity.

c. Across terminals **A** and **B**, with throttle valve fully opened, the should be no continuity.

d. If the results in Steps A and B are not satisfactory, adjust the TS and recheck. Terminals **A** and **B** measures across the idle switch.

e. Across terminals **B** and **C**, with throttle lever-to-stop screw clearance at 0.012 in. (0.3mm), the should be no continuity.

f. Across terminals **B** and **C**, with throttle lever-to-stop screw clearance at 0.035 in. (0.9mm), the should be no continuity.

g. Across terminals **B** and **C**, with throttle valve fully opened, the should be continuity. Terminals **B** and **C** measures across the wide open switch.

h. Across terminals **A** and **C**, with throttle lever-to-stop screw clearance at 0.012 in. (0.3mm), the should be no continuity.

i. Across terminals **A** and **C**, with throttle lever-to-stop screw clearance at 0.035 in. (0.9mm), the should be no continuity.

j. Across terminals **A** and **C**, with throttle valve fully opened, the should be no continuity.

4. If the readings are not satisfactory, replace the TS.
5. Reconnect the TS electrical connector and install the air cleaner assembly.
6. Reconnect the negative battery cable.

Component Replacement

ELECTRONIC CONTROL MODULE (ECM)

Removal and Installation

1. Disconnect the negative cable from the battery. Locate the ECM at the underside of the LH instrument panel.
2. Remove the retaining bolts from the junction/fuse block. Lower it, as required.
3. Disconnect the connectors from the ECM while releasing the connector locks.
4. Remove the ECM from the vehicle.

To install:

5. Install the ECM to the vehicle and install the electrical connectors securely.
6. Raise the junction/fuse block, as required. Install the retaining bolts.
7. Connect the negative battery cable.

THROTTLE BODY

Removal and Installation

SWIFT WITH TBI

1. Relieve the fuel system pressure.
2. Disconnect the negative battery cable at the battery.
3. Drain the cooling system.
4. Remove the air cleaner assembly.
5. Disconnect the TS or TPS, fuel injector and WTS electrical leads.
6. Disconnect the return and feed hoses from the throttle body.
7. Disconnect and tag the vacuum hoses and remove the water hose from the throttle body.
8. Disconnect the accelerator cable from the throttle valve lever and cable bracket.
9. Remove the throttle body-to-intake manifold retaining bolts and nuts. Remove the throttle body from the intake manifold. Clean the mating surfaces.

To install:

10. Install a new gasket to the intake manifold. Position the throttle body on the intake manifold and install the throttle body-to-intake manifold retaining bolts and nuts. Tighten the bolts and nuts 13.0–20.0 ft. lbs. (18–28 Nm).
11. Install the accelerator cable to the throttle valve lever and cable bracket. Adjust the cable play to specification.
12. Connect the fuel feed and return hoses to the throttle body.
13. Connect the vacuum hoses, electrical connectors and water hose to the throttle body. Clamp the hose securely.
14. Install the air cleaner assembly, refill the cooling system, connect the negative battery cable, start the engine and check for leaks.

SWIFT WITH MPFI

1. Relieve the fuel system pressure.
2. Disconnect the negative battery cable at the battery.
3. Drain the cooling system.
4. Disconnect the accelerator cable from the throttle valve lever and cable bracket.
5. Remove the Air Flow Meter (AFM) outlet hose and disconnect the TPS connector.
6. Disconnect the vacuum hose and engine cooling water hose from the throttle body.
7. Remove the throttle body-to-intake manifold retaining bolts and nuts; then, remove the throttle body from the intake manifold. Remove the throttle body-to-intake manifold gasket.
8. Clean the gasket surfaces. Clean the idle bypass passage and vacuum passage using compressed air.

NOTE: Do not place the TPS, dash pot, throttle valve shaft seal and other components containing rubber in solvent or cleaner bath.

To install:

9. Install a new gasket to the intake manifold; then, position the throttle body on the intake manifold. Tighten the bolts and nut 13.5–20 ft. lbs. (18–28 Nm).
10. Connect the engine water hose, vacuum hose and TPS electrical connector.
11. Install the AFM outlet hose and clamp it securely.
12. Install the accelerator cable to the throttle valve lever and cable bracket. Adjust the cable paly.

13. Refill the cooling system, connect the negative battery cable, start the engine and check for leaks.

SIDEKICK

1. Disconnect the negative battery cable at the battery.
2. Drain the cooling system.
3. Remove the air intake case, accelerator cable, and the kickdown cable (automatic transmission only) from the throttle body.
4. Disconnect and tag all electrical harness from the throttle body.
5. Disconnect and tag all vacuum hoses from the throttle body.
6. Remove the water hose from the air valve.
7. Relieve the fuel system pressure.
8. Remove the fuel feed pipe from the throttle body and fuel return hoses from the fuel pressure regulator.
9. Remove the throttle body retaining bolts; then, remove the throttle body and gasket from the intake manifold. Clean the gasket surfaces.

To install:

10. Fit a new gasket on the intake manifold.
11. Position the throttle body on the intake manifold and EGR modulator bracket to the throttle body. Tighten the 4 throttle body retaining bolts 13.5–20 ft. lbs. (18–28 Nm).
12. Install the water hose to the air valve and the fuel return hose to the fuel pressure regulator. Lubricate a new O-ring and install it on the fuel feed pipe. Install the fuel feed pipe to the throttle body.
13. Reconnect electrical connectors to the injector, TPS and ISC solenoid valve.
14. Reconnect vacuum hoses to the throttle body and throttle opener.
15. Install the throttle cable to the throttle valve lever and adjust to specification. If the vehicle is equipped with an automatic transmission, install the kickdown cable.
16. Install the air intake assembly.
17. Refill the cooling system, connect the negative battery cable, start the engine and check for leaks.

Disassemble and Assemble

SWIFT WITH TBI

NOTE: Do not remove either the fuel pressure regulator or the air valve from the throttle body. They are factory adjusted precisely.

1. Remove the injector from the throttle body.
2. Remove the TS or TPS and the WTS.
3. Remove the lower-to-upper throttle body retaining screws and separate the upper and lower bodies.
4. Clean the passage and fuel injector chamber using compressed air.

NOTE: Do not place the TPS, throttle valve shaft seal and other components containing rubber in solvent or cleaner bath.

To assemble

5. Install a new gasket to the lower body.
6. Install the upper body on the gasket, using care not to allow the gasket to slip from its position.
7. Install the lower-to-upper throttle body retaining screws and torque them to 2.1–2.9 ft. lbs. (2.9–4.1 Nm).
8. Install the WTS, TS or TPS to the throttle body. Install the fuel injector.
9. Install the fuel injector subwire to the throttle body. Use a new O-ring.

SIDEKICK

1. Remove the injector from the throttle body.
2. Remove the TPS.
3. Remove the pressure regulator and the ISC solenoid valve from the throttle body.
4. Remove the lower-to-upper throttle body retaining screws and separate the upper and lower bodies.
5. Clean the idle bypass passage and vacuum passage using compressed air.

NOTE: Do not place the TPS, throttle valve shaft seal and other components containing rubber in solvent or cleaner bath.

To assemble

6. Install a new gasket to the lower body.
7. Install the upper body on the gasket, using care not to allow the gasket to slip from its position.
8. Install the lower-to-upper throttle body retaining screws and torque them to 2.5 ft. lbs. (3.5 Nm).
9. Install the fuel perssure regulator, ISC solenoid valve, the fuel injector and the TPS to the throttle body.

AIR VALVE

Removal and Installation

SWIFT

1. Disconnect the negative battery cable from the battery.
2. Drain the cooling system.
3. Disconnect the engine cooling water hoses from the air valve.
4. Disconnect the air hose and remove the air valve and gasket form the intake manifold.

To install:

5. Position a new gasket on the intake manifold and install the air valve.
6. Connect the air hose and engine water hoses. Clamp the hoses securely.
7. Refill the cooling system, connect the negative battery cable, start the engine and check for leaks.

FUEL INJECTOR

Removal and Installation

SWIFT WITH TBI

1. Relieve the fuel system pressure.
2. Disconnect the negative battery cable at the battery.
3. Remove the air cleaner assembly.
4. Remove the injector cover and upper insulator; then, open the claws of the injector after removing the connector cover and disconnect the connector from it. Remove the fuel injector.

NOTE: Use care not to break the claws by opening them too far outwards.

5. Check the fuel injector filter for evidence of dirt and contamination. If present, clean and check for present of dirt in the fuel lines and fuel tank.

To install:

6. Apply a thin coat of spindle oil or gasoline to the new upper and lower O-rings. Install the lower O-ring to the injector cavity and upper O-ring to the injector.
7. Install a new lower insulator to the injector cavity.
8. Install the injector by gently pushing it straight into the fuel injector cavity.
9. Install the new upper insulator and the new injector cover. Tighten the cover screw 2.1–2.9 ft. lbs. (2.9–4.1 Nm).
10. Connect the injector connector to the injector with its lug side upward. Install the connector with its cover and wire tube pushed against the injector connector and clamp the sub wire.

SWIFT WITH MPFI

1. Relieve the fuel system pressure.
2. Disconnect the negative cable at the battery.

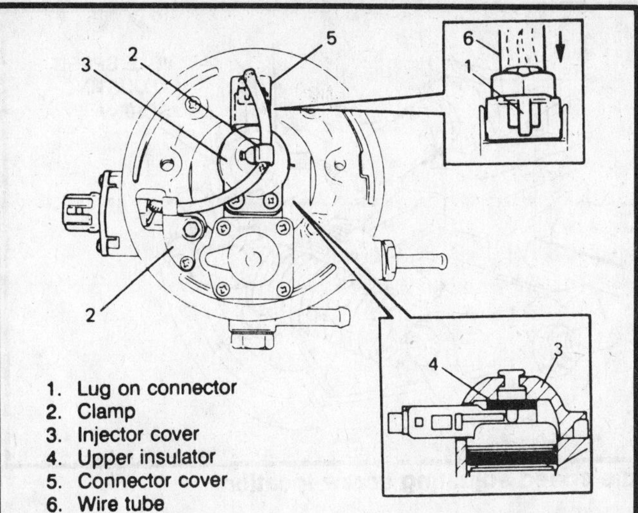
1. Lug on connector
2. Clamp
3. Injector cover
4. Upper insulator
5. Connector cover
6. Wire tube

Injector installed position

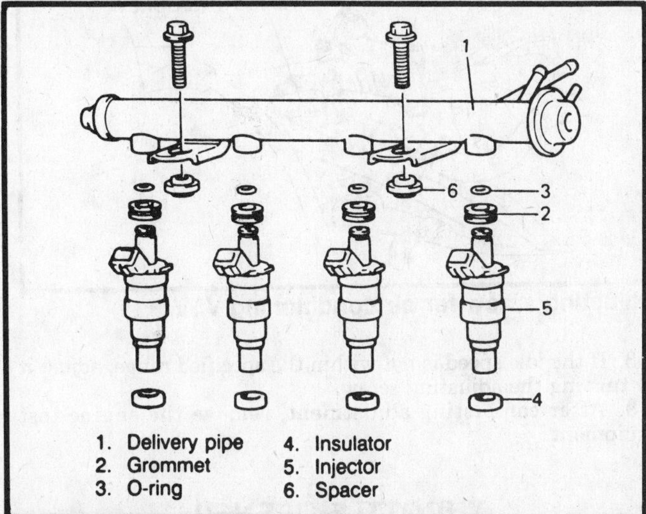
1. Delivery pipe
2. Grommet
3. O-ring
4. Insulator
5. Injector
6. Spacer

Fuel injectors and delivery pipe positioning

3. Remove the ISC solenoid valve and the EGR modulator bracket from the intake manifold.
4. Disconnect the fuel feed hose, return hose and vacuum hose from the delivery pipe.
5. Disconnect the electrical connector from each injector.
6. Remove the fuel delivery pipe with the fuel injectors. Do not drop the injectors.
7. Remove the delivery pipe form the injectors.

To install:

8. Install the grommet to the injector(s). Apply a thin coat of fuel to the O-rings and install on the injectors. Install the injectors into the delivery pipe. Make sure the injectors rotate smoothly. If not, the O-ring(s) may be incorrectly installed.
9. Inspect the insulators and spacers, if these parts are damaged or scored, replace them.
10. Install the insulators and spacers to the cylinder head.
11. Install the injectors with the delivery pipe and tighten the delivery pipe bolts 13.5–20.0 ft. lbs. (18–28 Nm).
12. Connect the fuel feed, return hose and vacuum hose to the delivery pipe.
13. Install the electrical connectors to the injectors.
14. Install the EGR modulator bracket and the ISC solenoid valve to the intake manifold.
15. Connect the negative battery cable.
16. With the engine **OFF** and the ignition **ON**, check for leaks.

SIDEKICK

1. Relieve the fuel system pressure.
2. Disconnect the negative cable at the battery.
3. Remove the air intake case.
4. Remove the fuel feed pipe clamp from the intake manifold and disconnect the fuel feed pipe from the throttle body.
5. Remove the injector cover and disconnect the injector connector.
6. Place a shop towel over the injector and position 1 hand on the top of it. Using a blow gun, apply approximately 72 psi (500 kPa) or less of compressed air into the fuel inlet port of the throttle body to remove the injector. Remove the injector.

NOTE: Do not exceed the recommended air pressure when removing the injector, as excessively high pressure may force the injector to jump out and cause damage to the injector or personal injury.

Do not immersed the injector in any type of liquid solvent or cleaner, as damage may occur.

7. Check the fuel injector filter for evidence of dirt and contamination. If present, clean and check for present of dirt in the fuel lines and fuel tank.

To install:

8. Apply a thin coat of spindle oil or gasoline to the new O-rings.
9. Install the injector to the throttle body. Fit the injector wire harness into the groove in the throttle body securely.
10. Install the injector cover. Apply thread locking cement 99000–32110 or equivalent to the threads of the cover screws and tighten the screws 1.4 ft. lbs. (2 Nm).
11. Apply a thin coat of engine oil to the O-ring on the fuel feed pipe and connect the fuel feed pipe to the throttle body.
12. Connect the injector connector and the negative battery cable, start the engine and check for leaks. If no leaks is found, install the air cleaner assembly.

FUEL PRESSURE REGULATOR

Removal and Installation

SIDEKICK

1. Disconnect the negative cable from the battery.
2. Relieve the fuel system pressure.
3. Disconnect the fuel return hose and vacuum hose from the pressure regulator.
4. Remove the fuel pressure regulator retaining screws and remove the regulator from the throttle body.

To install:

5. Apply a thin coat of spindle oil or gasoline to the new O-rings and install it on the fuel pressure regulator.
6. Install the fuel pressure regulator to the throttle body. Tighten the regulator screws to 2.5 ft. lbs. (3.5 Nm).
7. Install the fuel return hose and vacuum hose to the pressure regulator.
8. Reconnect the negative battery cable, start the engine and check for leaks.

IDLE SPEED CONTROL (ISC) SOLENOID VALVE

Removal and Installation

SIDEKICK

1. Disconnect the negative cable from the battery.
2. Disconnect the ISC solenoid valve connector. Pull out the wire harness terminals from the coupler after unlocking the terminal lock.

3. Remove the ISC solenoid valve and gasket form the throttle body.

NOTE: Do not immersed the ISC solenoid valve in any type of liquid solvent or cleaner, as damage may occur.

To install:

4. Install the ISC solenoid valve to the throttle body using a new gasket. Tighten the ISC solenoid valve retaining screws 2.5 ft. lbs. (3.5 Nm).
5. Connect the ISC solenoid valve connector and check to ensure it is locked securely.
6. Reconnect the negative battery cable.

BASIC IDLE SPEED

Adjustment

When adjustment of the idle speed or ignition timing is necessary, the vehicle emission label should be checked for up-to-date information.

NOTE: The EFI system automatically adjusts the engine idle speed to specification by means of the ECM. If the specified idle speed is not available, it is necessary to adjust the basic idle speed.

SWIFT

1. Connect a tachometer to the engine.
2. Turn all accessories, including the air conditioner, **OFF**.
3. Check and if necessary, adjust the accelerator cable.
4. Place the gear selector lever in **N** (**P** for automatic transmission), set the parking brake and block the drive wheels.
5. Start the engine and let it run until normal operating is reached.
6. Check and if necessary, adjust the ignition timing to specification.
7. Ground the diagnosis switch terminal by connecting a spare fuse across the terminal in the junction/fuse block. This will keep the amount of air supplied by the ISC solenoid valve at a constant.

NOTE: At this time, the check engine light should indicate diagnostic Code 12 and the ISC solenoid valve should be heard. If not, check and repair accordingly.

8. Remove the adjusting screw cap and using the idle speed adjusting screw, adjust the basic idle speed as follows:
   a. Manual transmission – 650 ± 50 rpm (air conditioning **OFF**) or 900 ± 50 rpm (air conditioning **ON**) and the gear selector in **N** range.
   b. Automatic transmission – 650 ± 50 rpm (air conditioning **OFF**) or 750 ± 50 rpm (air conditioning **ON**) and the gear selector in **R**, **D**, **2**, or **L** range.
   c. Automatic transmission – 750 ± 50 rpm (air conditioning **OFF**) or 850 ± 50 rpm (air conditioning **ON**) and the gear selector in **P** or **N** range.
9. After completing adjustment, reinstall the adjusting screw cap and disconnect the spare fuse from the diagnosis switch terminal. Remove the engine test equipment.

SIDEKICK

1. Connect a tachometer to the engine.
2. Turn all accessories, including the air conditioner, **OFF**.
3. Check and if necessary, adjust the accelerator cable.
4. Place the gear selector lever in **N** (**P** for automatic transmission), set the parking brake and block the drive wheels.
5. Start the engine and let it run until normal operating is reached.
6. Check and if necessary, adjust the ignition timing to specification.
7. Check to ensure that the idle speed is within specification. The engine isle speed should be 800 ± 50 rpm.

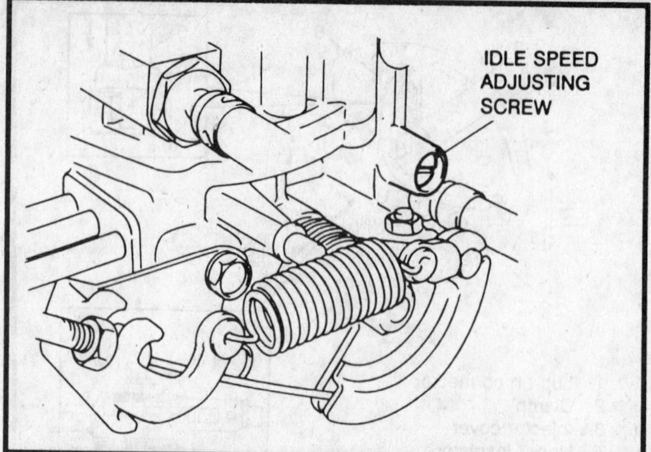
IDLE SPEED
ADJUSTING
SCREW

Idle speed adjusting screw location

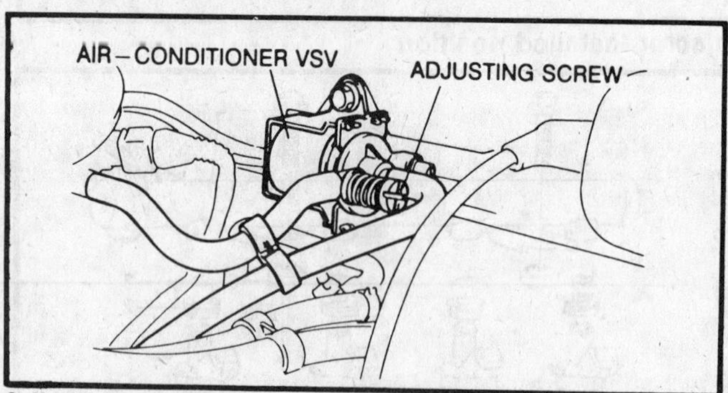
AIR – CONDITIONER VSV
ADJUSTING SCREW

Adjusting screw for air conditioning VSV

8. If the idle speed is not within the specified range, adjust it by turning the adjusting screw.
9. After completing adjustment, remove the engine test equipment.

THROTTLE OPENER

Adjustment

SIDEKICK

1. Connect a tachometer to the engine.
2. Place the gear selector lever in **N** (**P** for automatic transmission), set the parking brake and block the drive wheels.
3. Start the engine and let it run until normal operating is reached.
4. Check to ensure that no electrical load is **ON**.
5. Disconnect the vacuum hose from the throttle opener and plug the hose.
6. Check that the engine speed with the throttle opener operating is 1700–1800 rpm.
7. If the engine speed is not as specified, adjust it by turning the throttle opener adjusting screw.
8. After adjustment, connect the vacuum hose to the opener securely.

ACCELERATOR CABLE

Adjustment

1. Check the cable play with the engine **OFF** and the accelerator pedal released.
2. If the accelerator cable play is not within specification,

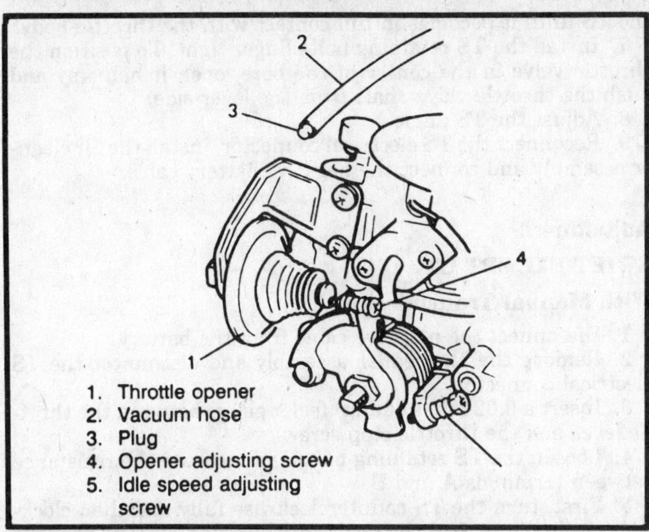
1
2
3
4
1. Throttle opener
2. Vacuum hose
3. Plug
4. Opener adjusting screw
5. Idle speed adjusting screw

Adjusting engine speed for throttle opener – 1988–90 Sidekick

loosening the locknut and turn the adjusting nut until the specified cable play is obtained. The accelerator cable play specification is as follows:

a. Swift – 0.12–0.20 in. (3–5mm)
b. Sidekick – 0.4–0.6 in. (10–15mm)

3. Tighten the locknut securely after adjustment.

THROTTLE POSITION SENSOR (TPS)

Removal and Installation

1. Disconnect the negative cable from the battery.
2. Remove the air cleaner assembly, as required.
3. Disconnect the TPS electrical connector. Pull out the wire harness terminals from the coupler after unlocking the terminal lock.
4. Remove the TPS from the throttle body.

To install:

5. Fit the TPS to the throttle body in such a way that the sensor adjusting holes are offset slightly counterclockwise from the retaining bolts holes. Then rotate the sensor clockwise so that the sensor adjusting holes align with the mounting bolts holes. Install the retaining bolts finger tight.
6. Connect the TPS connector and check to ensure it is locked securely.
7. Adjust the TPS.
8. Reconnect the negative battery cable.

Adjustment

SIDEKICK

1. Disconnect the negative cable from the battery.
2. Disconnect the throttle opener vacuum hose from the VSV and connect a vacuum pump to the hose.
3. Apply 15 in. of vacuum to the throttle opener to move the throttle valve to the idle position.
4. To close the throttle valve fully, loosen the idle speed adjusting screw, noting the number of turns, until there is clearance between the throttle valve lever and the idle speed adjusting screw. Then, tighten the screw until it just contacts the lever, again noting the number of turns. Subtract the number of turns counted while tightening the screw from that noted previously. The difference represents the number of turns by which the idle speed adjusting screw was actually loosened from the idle position. Use it as a guide when setting it back to the idle position after adjustment.
5. Insert a 0.086 in. (2.2mm) feeler gauge between the throttle valve lever and the idle speed adjusting screw (vehicles equipped with manual transmission). Vehicles equipped with automatic transmission, use a 0.094 in. (2.4mm) feeler gauge.
6. Loosen the TPS retaining bolts and measure the resistance between terminals **C** and **D**.
7. First, turn the TPS fully clockwise and then counterclockwise gradually to find the position where the reading changes from infinity to 0. Then, tighten the TPS retaining bolts to 2.5 ft. lbs. (3.5 Nm).
8. Reconnect the TPS electrical connector and the throttle opener vacuum hose to the VSV.
9. Tighten the idle speed adjusting screw the number of turns recorded in Step 4.
10. Reconnect the negative battery cable. Start the engine and adjust the idle speed.

SWIFT EXCEPT GTi

With Automatic Transmission

1. Disconnect the negative cable from the battery.
2. Remove the air cleaner assembly and disconnect the TPS connector.
3. Insert a 0.024 in. (0.6mm) feeler gauge between the throttle lever and the throttle stop screw.
4. Loosen the TPS retaining bolts and measure the resistance between terminals **A** and **B**.
5. First, turn the TPS counterclockwise fully and then clockwise gradually to find the position where the reading changes from 0 to infinity. Then, tighten the TPS retaining bolts 1.2–1.7 ft. lbs. (1.6–2.4 Nm).
6. Check that there is no continuity between terminals **A** and **B** when a 0.035 in. (0.9mm) feeler gauge is inserted.
7. Check that there is continuity between terminals **A** and **B** when a 0.012 in. (0.3mm) feeler gauge is inserted.
8. If the result is not satisfactory in Steps 6 and 7, it indicates that the installation angle of the TPS in not adjusted properly. Repeat the adjustment procedure.

NOTE: The throttle stop screw if factory adjusted precisely, do not remove or adjust it.

9. Reconnect the TPS electrical connector, install the air cleaner assembly and connect the negative battery cable.

SWIFT GTi

1. Disconnect the negative cable from the battery.
2. Remove the air cleaner assembly and disconnect the TPS connector.
3. Insert a 0.024 in. (0.6mm) feeler gauge between the throttle lever and the throttle stop screw.
4. Loosen the TPS retaining bolts and measure the resistance between terminals **C** and **D**.
5. First, turn the TPS clockwise fully and then counterclockwise gradually to find the position where the reading changes from 0 to no continuity. Then, tighten the TPS retaining bolts 2.5 ft. lbs. (3.5 Nm).
6. Check that there is no continuity between terminals **C** and **D** when a 0.035 in. (0.9mm) feeler gauge is inserted.
7. Check that there is continuity between terminals **C** and **D** when a 0.012 in. (0.3mm) feeler gauge is inserted.
8. If the result is not satisfactory in Steps 6 and 7, it indicates that the installation angle of the TPS in not adjusted properly. Repeat the adjustment procedure.

NOTE: The throttle stop screw if factory adjusted precisely, do not remove or adjust it.

9. Reconnect the TPS electrical connector, install the air cleaner assembly and connect the negative battery cable.

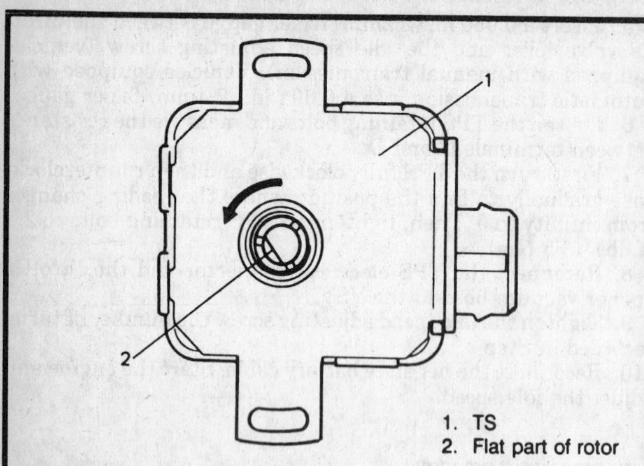
1
2
1. TS
2. Flat part of rotor

TS rotor position for installation

THROTTLE SWITCH (TS)

Removal and Installation

SWIFT EXCEPT GTi

1. Disconnect the negative cable from the battery.
2. Remove the air cleaner assembly and disconnect the TS electrical connector.
3. Remove the TPS from the throttle body.

To install:

4. Before installing the TS, check that the flat part of the TS rotor is positioned as indicated.
5. If not, turn the rotor counterclockwise until it stops.
6. Install the TS to the throttle valve shaft by aligning the flat part of the TS rotor with the cut part of the shaft and pushing the TS until it becomes in full contact with the throttle body.
7. Install the TS retaining bolts finger tight. To position the throttle valve in the center of the bore, open it half-way and push the throttle valve shaft from the lever side.
8. Adjust the TS angle.
9. Reconnect the TS electrical connector, install the air cleaner assembly and connect the negative battery cable.

Adjustment

SWIFT EXCEPT GTi

With Manual Transmission

1. Disconnect the negative cable from the battery.
2. Remove the air cleaner assembly and disconnect the TS electrical connector.
3. Insert a 0.024 in. (0.6mm) feeler gauge between the throttle lever and the throttle stop screw.
4. Loosen the TS retaining bolts and measure the resistance between terminals **A** and **B**.
5. First, turn the TS counterclockwise fully and then clockwise gradually to find the position where the reading changes from 0 to no continuity. Then, tighten the TS retaining bolts 1.2–1.7 ft. lbs. (1.6–2.4 Nm).
6. Check that there is no continuity between terminals **A** and **B** when a 0.035 in. (0.9mm) feeler gauge is inserted.
7. Check that there is continuity between terminals **A** and **B** when a 0.012 in. (0.3mm) feeler gauge is inserted.
8. If the result is not satisfactory in Steps 6 and 7, in indicates that the installation angle of the TS in not adjusted properly. Repeat the adjustment procedure.

NOTE: The throttle stop screw if factory adjusted precisely, do not remove or adjust it.

9. Reconnect the TS electrical connector, install the air cleaner assembly and connect the negative battery cable.

TOYOTA ELECTRONIC FUEL INJECTION (EFI) SYSTEM

ENGINE CONTROL SYSTEM APPLICATION CHART

Year	Model	Engine cc (liter)	Engine Code	Fuel System	Ignition System
1988	Camry	1998 (2.0)	3S-FE	EFI	ESA
	Celica	1998 (2.0)	3S-GE	EFI	ESA
		1998 (2.0)	3S-FE	EFI	ESA
	Corolla	1587 (1.6)	4A-GE	EFI	ESA
		1587 (1.6)	4A-C	Carb	11A
		1587 (1.6)	4A-F	Carb	11A
	Cressida	2759 (2.8)	5M-GE	EFI	ESA
	MR-2	1587 (1.6)	4A-GE	EFI	ESA
		1587 (1.6)	4A-GZE	EFI	ESA
	Supra	2954 (3.0)	7M-GE	EFI	ESA
		2954 (3.0)	7M-GTE	EFI	ESA
	Tercel	1452 (1.4)	3A-C	Carb	11A
		1456 (1.4)	3E	Carb	11A
	Van	2237 (2.2)	4Y-EC	EFI	ESA
	Truck & 4-Runner	2366 (2.4)	22R	Carb	ESA
		2366 (2.4)	22R-E	EFI	ESA
		2366 (2.4)	22R-TE	EFI	ESA
1989	Camry	1998 (2.0)	3S-FE	EFI	ESA
		2507 (2.5)	2VZ-FE	EFI	ESA
	Celica	1998 (2.0)	3S-FE	EFI	ESA
		1998 (2.0)	3S-GE	EFI	ESA
		1998 (2.0)	3S-GTE	EFI	ESA
	Corolla	1587 (1.6)	4A-F	Carb	11A
		1587 (1.6)	4A-FE	EFI	11A
		1587 (1.6)	4A-FE	EFI	ESA
		1587 (1.6)	4A-GE	EFI	ESA
	Cressida	2954 (3.0)	7M-GE	EFI	ESA
	MR-2	1587 (1.6)	4A-GE	EFI	ESA
		1587 (1.6)	4A-GZE	EFI	ESA
	Supra	2954 (3.0)	7M-GE	EFI	ESA
		2954 (3.0)	7M-GTE	EFI	ESA
	Tercel	1456 (1.4)	3E	Carb	11A
	Van	2237 (2.2)	4Y-EC	EFI	ESA
	Truck & 4-Runner	2366 (2.4)	22R	Carb	ESA
		2366 (2.4)	22R-E	EFI	ESA
		2958 (3.0)	3VZ-E	EFI	ESA

ENGINE CONTROL SYSTEM APPLICATION CHART

Year	Model	Engine cc (liter)	Engine Code	Fuel System	Ignition System
1990	Camry	1998 (2.0)	3S-FE	EFI	ESA
		2507 (2.5)	2VZ-FE	EFI	ESA
	Celica	1587 (1.6)	4A-FE	EFI	ESA
		2164 (2.2)	5S-FE	EFI	ESA
		1998 (2.0)	3S-GTE	EFI	ESA
	Corolla	1587 (1.6)	4A-FE	EFI	ESA
		1587 (1.6)	4A-GE	EFI	ESA
	Cressida	2954 (3.0)	7M-GE	EFI	ESA
	MR-2	2954 (3.0)	7M-GE	EFI	ESA
		2954 (3.0)	7M-GTE	EFI	ESA
	Supra	2954 (3.0)	7M-GE	EFI	ESA
		2954 (3.0)	7M-GTE	EFI	ESA
	Tercel	1456 (1.4)	3E	Carb	IIA
		1456 (1.4)	3E-E	EFI	ESA
	Truck & 4-Runner	2366 (2.4)	22R	Carb	ESA
		2366 (2.4)	22R-E	EFI	ESA
		2958 (3.0)	3VZ-E	EFI	ESA

General Information

This system is broken down into 3 major systems: the Fuel System, Air Induction System and the Electronic Control System.

FUEL SYSTEM

An electric fuel pump supplies sufficient fuel, under a constant pressure, to the EFI injectors. These injectors inject a metered quantity of fuel into the intake manifold in accordance with signals from the EFI computer. Each injector injects at the same time, ½ the fuel required for ideal combustion with each engine revolution.

AIR INDUCTION SYSTEM

The air induction system provides sufficient air for the engine operation. This includes the throttle body, air intake device and idle control components.

ELECTRONIC CONTROL SYSTEM

Most of the engines are equipped with a Toyota Computer Control System (TCCS) which centrally controls the electronic fuel injection, electronic spark advance and the exhaust gas recirculation valve. The systems can be diagnosed by means of an Electronic Control Unit (ECU) which employs a microcomputer. The ECU and the TCCS control the following functions:

Electronic Fuel Injection (EFI)

The ECU receives signals from the various sensors indicating changing engine operations conditions such as:

1. Intake air volume
2. Intake air temperature
3. Coolant temperature sensor
4. Engine rpm
5. Acceleration/deceleration
6. Exhaust oxygen content

These signals are utilized by the ECU to determine the injection duration necessary for an optimum air-fuel ratio.

The Electronic Spark Advance (ESA)

The ECU is programmed with data for optimum ignition timing during any and all operating conditions. Using the data provided by sensors which monitor various engine functions (rpm, intake air volume, coolant temperature, etc.), the microcomputer (ECU) triggers the spark at precisely the right moment.

Idle Speed Control (ISC)

The ECU is programmed with specific engine speed values to respond to different engine conditions (coolant temperature, air conditioner on/off, etc.). Sensors transmit signals to the ECU which controls the flow of air through the by-pass of the throttle valve and adjusts the idle speed to the specified value. Some vehicles use a ISC valve while others use an air valve to control throttle body by-pass air flow.

Exhaust Gas Recirculation (EGR)

The ECU detects the coolant temperature and controls the EGR operations accordingly.

Electronic Controlled Transmission (ECT)

AUTOMATIC TRANSMISSION ONLY

A serial signal is transmitted to the ECT computer to prevent up shifting to 3rd or overdrive during cold engine operation. Diagnostics, which are outlined below.

Fail-Safe Function

In the event of a computer malfunction, a backup circuit will take over to provide minimal driveability. Simultaneously, the "Check Engine" warning light is activated.

Turbo Indicator

The ECU detects turbocharger pressure, which is determined by the intake volume and the engine rpm, and lights a green colored turbocharger indicator light located in the combination meter. Moreover, if the turbocharger pressure increases abnormally, the ECU will light the "Check Engine" warning light on the instrument panel.

SERVICE PRECAUTIONS

- Do not operate the fuel pump when the fuel lines are empty.
- Do not reuse fuel hose clamps.
- Make sure all EFI harness connectors are fastened securely. A poor connection can cause an extremely high surge voltage in the coil and condenser and result in damage to integrated circuits.
- Keep the EFI harness at least 4 in. away from adjacent harnesses to prevent an EFI system malfunction due to external electronic "noise."
- Keep EFI all parts and harnesses dry during service.
- Before attempting to remove any parts, turn **OFF** the ignition switch and disconnect the battery ground cable.
- Always use a 12 volt battery as a power source.
- Do not attempt to disconnect the battery cables with the engine running.
- Do not depress the accelerator pedal when starting.
- Do not rev up the engine immediately after starting or just prior to shutdown.
- Do not attempt to disassemble the EFI control unit under any circumstances.
- If installing a 2-way or CB radio, keep the antenna as far as possible away from the electronic control unit. Keep the antenna feeder line at least 8 in. away from the EFI harness and do not let them run parallel for a long distance. Be sure to ground the radio to the vehicle body.
- Do not apply battery power directly to injectors.

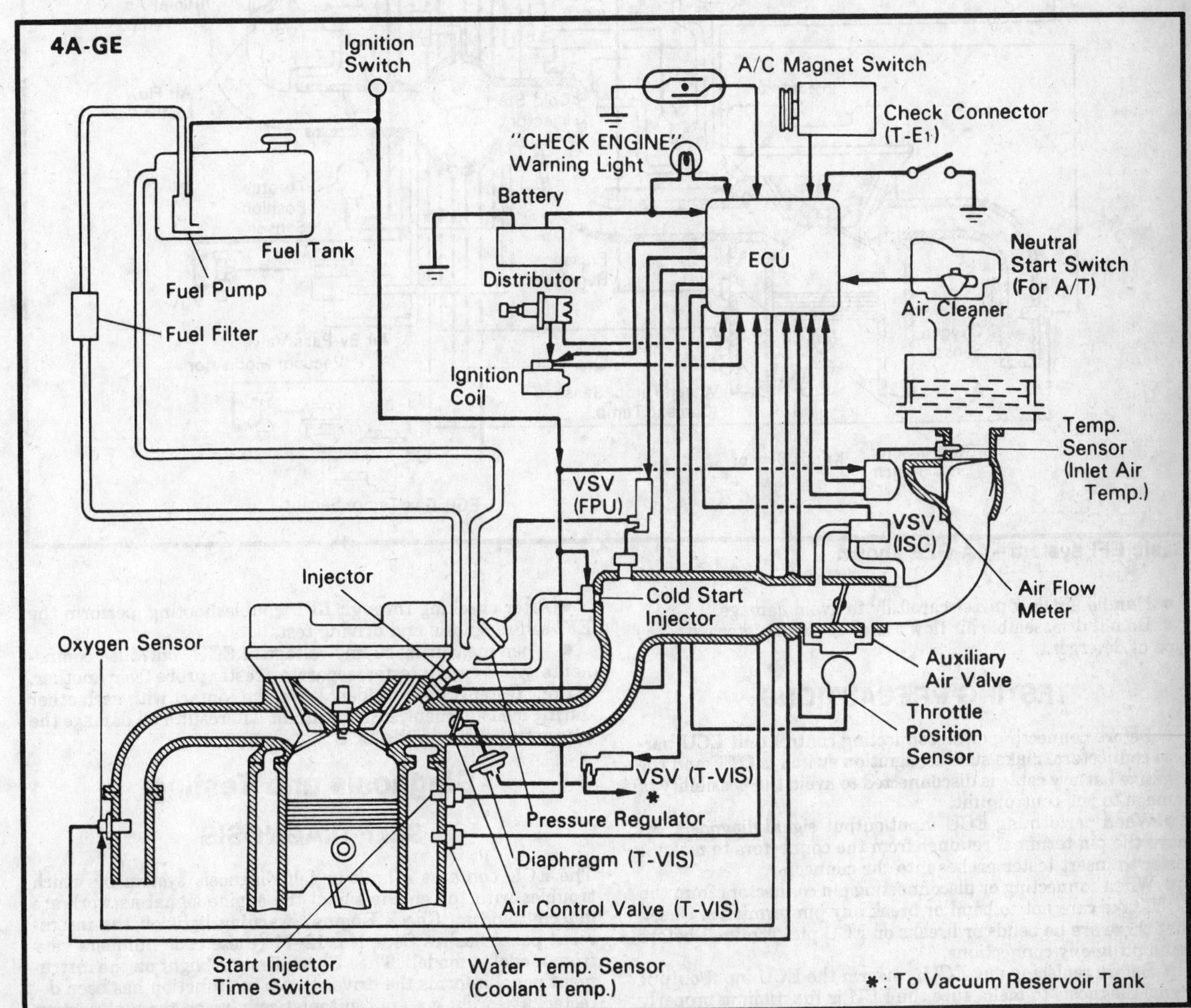
4A-GE
Ignition Switch
A/C Magnet Switch
Check Connector (T-E1)
"CHECK ENGINE" Warning Light
Battery
Fuel Tank
ECU
Neutral Start Switch (For A/T)
Fuel Pump
Distributor
Air Cleaner
Fuel Filter
Ignition Coil
Temp. Sensor (Inlet Air Temp.)
VSV (FPU)
VSV (ISC)
Injector
Cold Start Injector
Air Flow Meter
Oxygen Sensor
Auxiliary Air Valve
Throttle Position Sensor
VSV (T-VIS)
*
Pressure Regulator
Diaphragm (T-VIS)
Air Control Valve (T-VIS)
Start Injector Time Switch
Water Temp. Sensor (Coolant Temp.)
*: To Vacuum Reservoir Tank
Basic EFI system—4A-GE shown

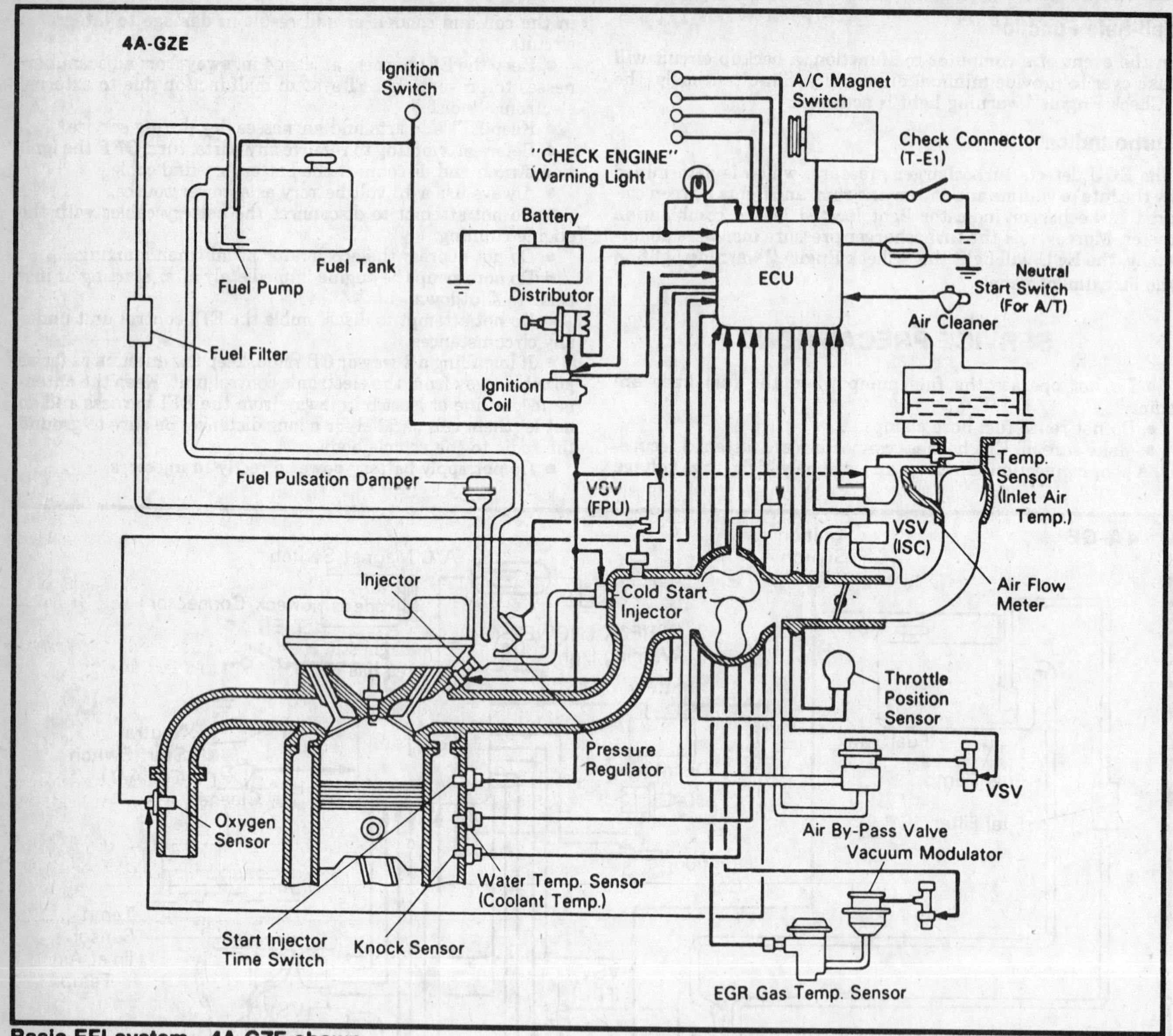
4A-GZE
Ignition Switch
A/C Magnet Switch
Check Connector (T-E1)
"CHECK ENGINE" Warning Light
Battery
Fuel Tank
Fuel Pump
Distributor
ECU
Neutral Start Switch (For A/T)
Air Cleaner
Fuel Filter
Ignition Coil
Temp. Sensor (Inlet Air Temp.)
Fuel Pulsation Damper
VSV (FPU)
VSV (ISC)
Injector
Cold Start Injector
Air Flow Meter
Throttle Position Sensor
Pressure Regulator
VSV
Oxygen Sensor
Air By-Pass Valve
Vacuum Modulator
Water Temp. Sensor (Coolant Temp.)
Start Injector Time Switch
Knock Sensor
EGR Gas Temp. Sensor

Basic EFI system—4A-GZE shown

- Handle air flow meter carefully to avoid damage.
- Do not disassemble air flow meter or clean meter with any type of detergent.

TESTING PRECAUTIONS

- Before connecting or disconnecting control unit ECU harness connectors, make sure the ignition switch is **OFF** and the negative battery cable is disconnected to avoid the possibility of damage to the control unit.
- When performing ECU input/output signal diagnosis, remove the pin terminal retainer from the connectors to make it easier to insert tester probes into the connector.
- When connecting or disconnecting pin connectors from the ECU, take care not to bend or break any pin terminals. Check that there are no bends or breaks on ECU pin terminals before attempting any connections.
- Before replacing any ECU, perform the ECU input/output signal diagnosis to make sure the ECU is functioning properly or not.
- After checking through EFI troubleshooting, perform the EFI self-diagnosis and driving test.
- When measuring supply voltage of ECU controlled components with a circuit tester, separate a tester probe from another. If the 2 tester probes accidentally make contact with each other during measurement, a short circuit will result and damage the power transistor in the ECU.

Diagnosis and Testing

SELF-DIAGNOSIS

The ECU contains a built-in self diagnosis system by which troubles with the engine signal the engine signal network are detected and a "Check Engine" warning light on the instrument panel flashes Code No. 12–71 (these code numbers vary from model to model). The "Check Engine" light on the instrument panel informs the driver that a malfunction has been detected. The light goes out automatically when the malfunction has been cleared.

The diagnostic code can be read by the number of blinks of the "Check Engine" warning light when the proper terminals of the check connector are short-circuited. If the vehicle is equipped with a super monitor display, the diagnostic code is indicated on the display screen.

Check Engine Warning Light

1. The "Check Engine" warning light will come on when the ignition switch is placed **ON** and the engine is not running.
2. When the engine is started, the "Check Engine" warning light should go out.
3. If the light remains on, the diagnosis system has detected a malfunction in the system.

NOTE: For Self-Diagnosis System and Accessing Trouble Code Memory, see Section 3 "Self-Diagnosis Systems".

Output of Diagnostic Codes

1. The battery voltage should be above 11 volts. Throttle valve fully closed (throttle position sensor IDL points closed).
2. Place the transmission in **P** or **N** range. Turn the A/C switch **OFF**. Start the engine and let it run to reach its normal operating temperature.

WITHOUT SUPER MONITOR DISPLAY

1. Turn the ignition switch to the **ON** position. Do not start the engine. Use a suitable jumper wire and short the terminals of the check connector.
2. Read the diagnostic code as indicated by the number of flashes of the "Check Engine" warning light.
3. If the system is operating normally (no malfunction), the light will blink once every 0.25 seconds.
4. In the event of a malfunction, the light will blink once every 0.5 seconds (some models it may 1 to 2 or 3 seconds). The 1st number of blinks will equal the 1st digit of a 2 digit diagnostic code. After a 1.5 second pause, the 2nd number of blinks will equal the 2nd number of a 2 digit diagnostic code. If there are 2 or more codes, there will be a 2.5 second pause between each.
5. After all the codes have been output, there will be a 4.5 second pause and they will be repeated as long as the terminals of the check connector are shorted.

NOTE: In event of multiple trouble codes, indication will begin from the smaller value and continue to the larger in order.

6. After the diagnosis check, remove the jumper wire from the check connector.

WITH SUPER MONITOR DISPLAY

1. Turn the ignition switch to the **ON** position. Do not start the engine.
2. Simultaneously push and hold in the SELECT and INPUT M keys for at least 3 seconds. The letters DIAG will appear on the screen.
3. After a short pause, hold the SET key in for at least 3 seconds. If the system is normal (no malfunctions), ENG-OK will appear on the screen.
4. If there is a malfunction, the code number for it will appear on the screen. In the event of 2 or more numbers, there will be a 3 second pause between each (Example ENG-42).
5. After confirmation of the diagnostic code, either turn **OFF** the ignition switch or push the super monitor display key **ON** so the time appears.

Canceling Out The Diagnostic Code

1. After repairing the trouble area, the diagnostic code that is retained in the ECU memory must be canceled out by removing the EFI (15A) fuse for 30 seconds or more, depending on the ambient temperature (the lower temperature, the longer the fuse must be left out with the ignition switch **OFF**).

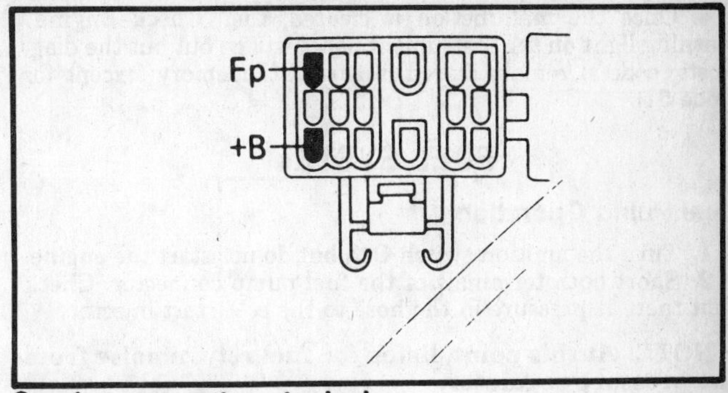
Fp
+B
Service connector—typical

NOTE: Cancellation can also be done by removing the battery negative terminal, but keep in mind, when removing the negative battery cable, the other memory systems (radio, ETR, clock, etc.) will also be canceled out.

If the diagnostic code is not canceled out, it will be retained by the ECU and appear along with a new code in event of future trouble. If it is necessary to work on engine components requiring removal of the battery terminal, a check must first be made to see if a diagnostic code is detected.

2. After cancellation, perform a road test, if necessary, confirm that a normal code is now read on the "Check Engine" warning light or super monitor display.
3. If the same diagnostic code is still indicated, it indicates that the trouble area has not been repaired thoroughly.

Diagnosis Indication

1. Including "Normal", the ECU is programmed several diagnostic codes.
2. When more than a single code is indicated, the lowest number code will appear first. However, no other code will appear along with Code 11.
3. All detected diagnostic codes, except Code 51 and 53, will be retained in memory by the ECU from the time of detection until canceled out.

Holding in the SELECT and INPUT buttons

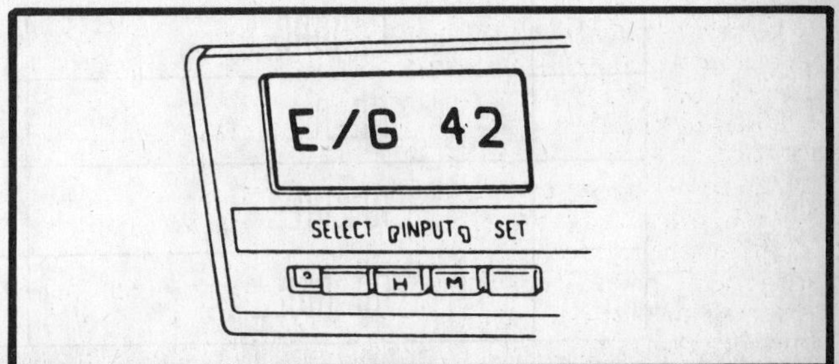
E/G 42
SELECT
INPUT
SET
H
M
Typical super monitor display of trouble codes

4. Once the malfunction is cleared, the "Check Engine" warning light on the instrument panel will go out but the diagnostic code(s) remain stored in the ECU memory (except for Code 51).

FUEL SYSTEM

Fuel Pump Operation

1. Turn the ignition switch **ON**, but do not start the engine.
2. Short both terminals of the fuel pump connector. Check that there is pressure in the hose to the cold start injector.

NOTE: At this point, listen for fuel return noise from the pressure regulator.

3. Remove the service wire and install the rubber cap on the fuel pump check connector. Turn **OFF** the ignition switch.
4. If there is no pressure, check the following components:
   a. Fusible link
   b. EFI and Ignition fuses
   c. EFI main relay
   d. Circuit opening relay
   e. Fuel pump
   f. Wiring connector

Fuel Pressure Check

1. Disconnect the negative battery cable. Disconnect the wiring connector from the cold start injector.

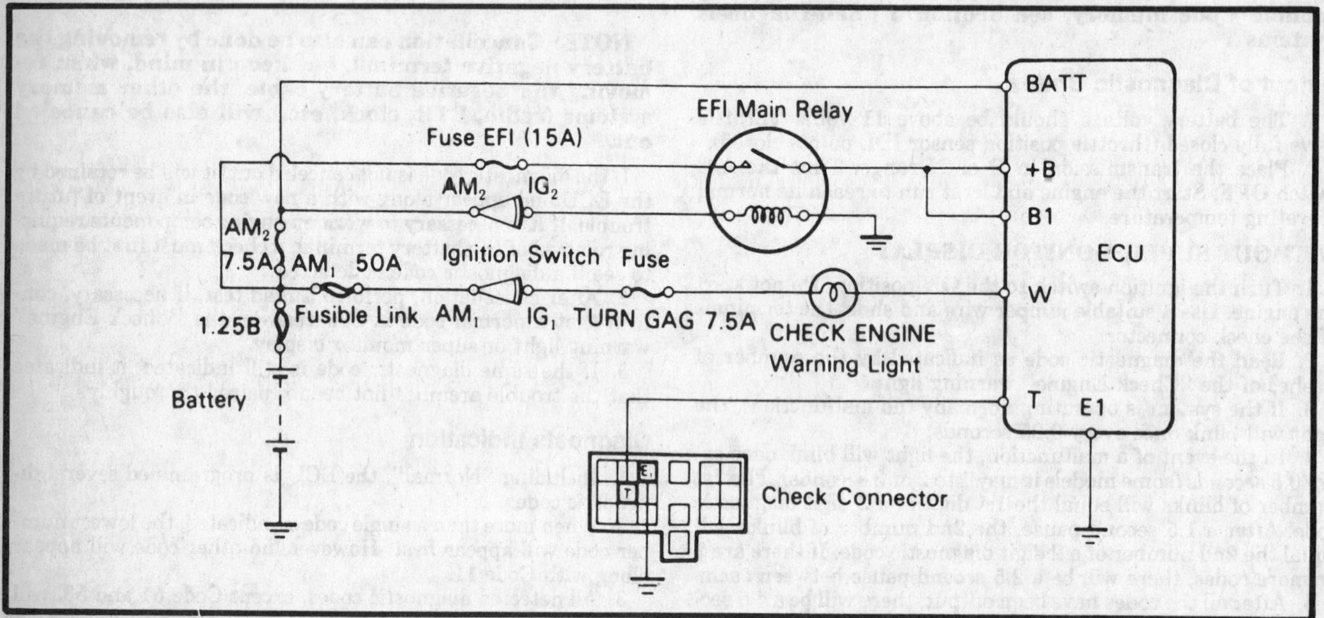
BATT
EFI Main Relay
Fuse EFI (15A)
+B
AM_2
IG_2
B1
AM_2
7.5A
AM_1
50A
Ignition Switch
Fuse
ECU
W
1.25B
Fusible Link
AM_1
IG_1
TURN GAG 7.5A
CHECK ENGINE
Warning Light
Battery
T
E1
E_1
T
Check Connector

Typical diagnostic circuit

Code No.	Light Pattern	Code No.	Light Pattern
–	ON / OFF	31	
11		32	
12		41	
13		42	
14		43	
21		51	
22		52	
23		53	

Typical trouble code (flashes) light pattern

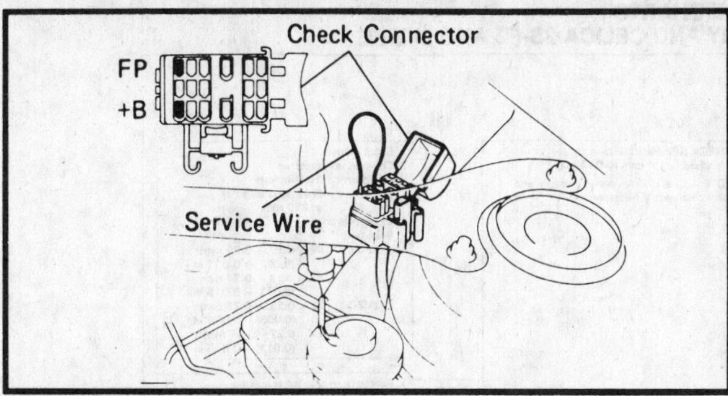
Check Connector
FP
+B
Service Wire

Jumping fuel pump terminals

2. Place a suitable container or shop towel under the rear end of the delivery pipe.

3. Slowly loosen the union bolt of the cold start injector (on some models it may be necessary to remove the cold start valve in order to connect the pressure gauge) hose and remove the bolt and 2 gaskets from the delivery pipe.

4. Drain the fuel from the delivery pipe. Install a gasket, a pressure gauge (tool SST-09268-45011 or equivalent), another gasket and the union bolt to the delivery pipe.

5. Wipe up any excess gasoline, reconnect the negative battery cable and start the engine.

6. Short the terminals on the fuel pump check connector (Fp and +B) and turn the ignition switch to the **ON** position, then take a fuel pressure reading.

7. The fuel pressure shoud read:
   Camry 3S-FE – 38–44 psi
   Celica 3S-FE – 38–44 psi
   Celica 3S-GE – 33–38 psi
   Celica 3S-GTE – 33–38 psi
   Corolla – 38–44 psi
   Cressida 5M-GE – 33–38 psi
   Cressida 7M-GE – 38–44 psi
   MR2 4A-GE – 38–44 psi
   MR2 4A-GZE – 33–40 psi
   Supra 7M-GE – 38–44 psi
   Supra 7M-GTE – 33–40 psi
   Van – 38–44 psi
   Truck and 4-Runner except 22R-TE – 38–44 psi
   Truck and 4-Runner 22R-TE – 33–38 psi

8. Remove the service wire from the check connector. Start the engine. Disconnect the vacuum sensing hose from the pressure regulator and pinch it off. Measure the fuel pressure at idle. The fuel pressure should be the same as the above readings.

9. If the fuel pressure is too high, replace the pressure regulator.

10. If the fuel pressure is low, check the following components:
   a. Fuel hoses and fuel connections
   b. Fuel pump
   c. Fuel filter
   d. Pressure regulator

11. Reconnect the vacuum sensing hose to the pressure regulator. Measure the fuel pressure at idling:
   Camry 3S-FE – 33–37 psi
   Celica 3S-FE – 33–37 psi
   Celica 3S-GE – 27–31 psi
   Celica 3S-GTE – 27–31 psi
   Corolla – 30–33 psi
   Cressida 5M-GE – 27–31 psi
   Cressida 7M-GE – 33–37 psi
   MR2 4A-GE – 30–33 psi
   MR2 4A-GZE – 20–27 psi
   Supra 7M-GE – 33–37 psi
   Supra 7M-GTE – 23–30 psi
   Van – 30–33 psi
   Truck and 4-Runner except 22R-TE – 33–37 psi
   Truck and 4-Runner 22R-TE – 27–31 psi

12. If there is no fuel pressure, check the vacuum sensing hose and pressure regulator. Stop the engine. Check that the fuel pressure remains above 21 psi for 5 minutes after the engine has been shut off.

13. If not within specifications, check the fuel pump, pressure regulator and or the injectors.

14. After checking the fuel pressure, disconnect the battery ground cable and carefully remove the pressure gauge to prevent gasoline from splashing.

15. Using new gaskets, reconnect the cold start injector hose to the delivery pipe (install the cold start injector valve if it was removed). Connect the wiring connector to the cold start injector. Check for fuel leakage.

DIAGNOSTIC TREE CHARTS
TROUBLESHOOTING PROCEDURES—CAMRY AND CELICA 3S-FE AND 2VZ-FE

Digital Type | Analog Type

Use a volt/ohmmeter with high impedance (10 kΩ/V minimum) for troubleshooting of the electrical circuit.

SYMPTOM — DIFFICULT TO START OR NO START (ENGINE WILL NOT CRANK OR CRANKS SLOWLY)

CHECK ELECTRIC SOURCE — BAD →
1. Battery
   (1) Connection
   (2) Gravity — Drive belt — Charging system
   (3) Voltage

↓ OK

CHECK STARTING SYSTEM — BAD →
1. Ignition switch
2. Neutral start switch (A/T)
3. Clutch start switch (M/T)
4. Starter relay (M/T)
5. Starter
6. Wiring/Connection

SYMPTOM — DIFFICULT TO START OR NO START (CRANKS OK)

CHECK DIAGNOSIS SYSTEM
Check for output of diagnosis code. — Malfunction code(s) → Diagnostic code(s)

↓ Normal code

DOES ENGINE START WITH ACCELERATOR PEDAL DEPRESSED? — OK →
ISC system
(1) ISC valve
(2) Wiring connection

↓ NO

CHECK FOR VACUUM LEAKS IN AIR INTAKE LINE — BAD →
1. Oil filler cap
2. Oil dipstick
3. Hose connections
4. PCV hoses
5. EGR system — EGR valve stays open

↓ OK

CHECK IGNITION SPARK — BAD →
1. High-tension cords
2. Distributor
3. Ignition coil
4. Igniter

↓ OK

CHECK SPARK PLUGS
Standard: 1.1 mm (0.043 in.)
NOTE: Check compression pressure and valve clearance if necessary.

NO →
1. Spark plugs
2. Compression pressure
   Minimum: 10.0 kg/cm² (142 psi, 981 kPa) at 250 rpm
3. Valve clearance
   Standard:
   3F-FE IN 0.19 — 0.29 mm (0.007 — 0.011 in.)
   EX 0.28 — 0.38 mm (0.011 — 0.015 in.)
   2VZ-FE IN 0.13 — 0.23 mm (0.005 — 0.009 in.)
   EX 0.27— 0.37 mm (0.011 — 0.015 in.)

BAD All Plugs WET →
1. Injector(s) — shorted or leaking
2. Injector wiring — short circuited
3. Cold start injector — leakage
4. Cold start injector time switch

↓ OK

CHECK FUEL SUPPLY TO INJECTOR
1. Fuel tank
2. Fuel pressure in fuel line
   (1) Connect terminals +B and FP of the check connector.
   (2) Fuel pressure at fuel hose of fuel filter can be felt.

BAD →
1. Fuel line — leakage — deformation
2. Fuse(s)
3. Fuel pump
4. Fuel filter
5. Fuel pressure regulator

↓ OK

CHECK FUEL PUMP SWITCH IN AIR FLOW METER
Check continuity between terminals FC and E1 while measuring plate of air flow meter is open. — BAD → Air flow meter

↓ OK

CHECK IGNITION TIMING
1. Connect terminals TE1 and E1 of the check connector.
2. Check ignition timing.
   Standard: 10° BTDC @ idle

BAD → Ignition timing — Adjust

↓ OK

TROUBLESHOOTING PROCEDURES—CAMRY AND CELICA 3S-FE AND 2VZ-FE

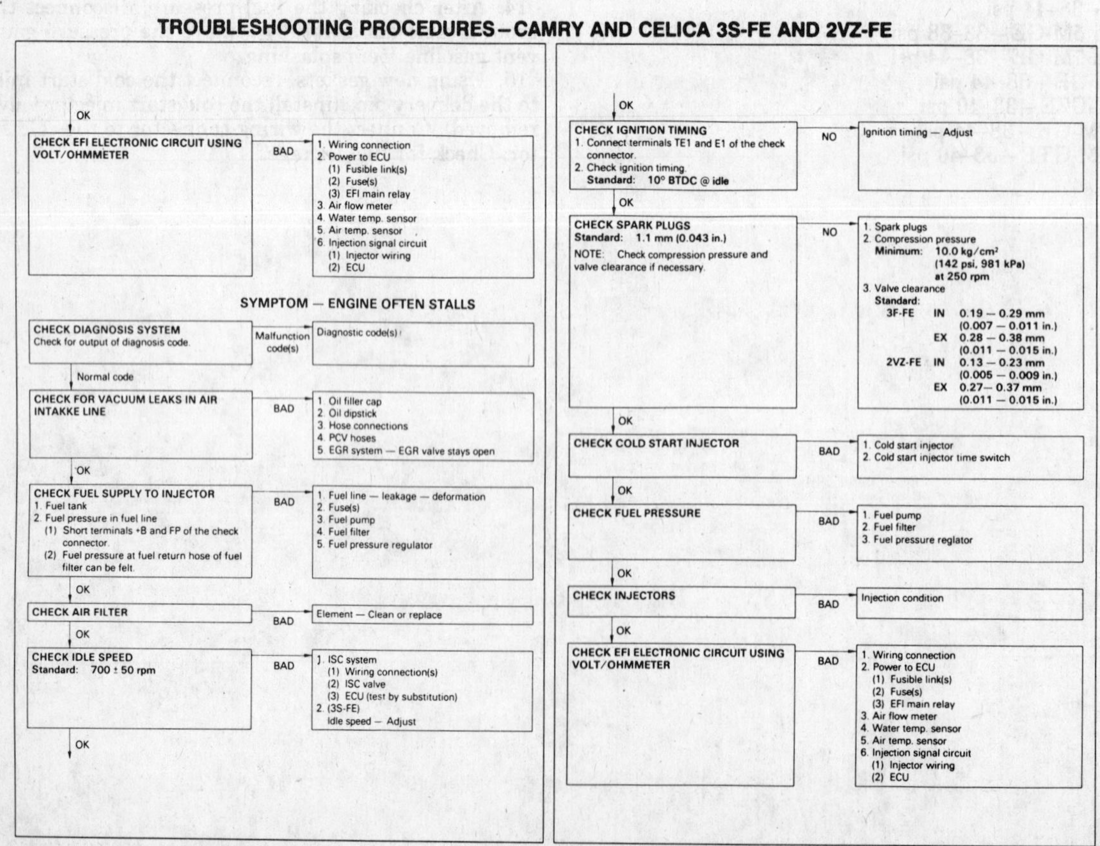
OK
CHECK EFI ELECTRONIC CIRCUIT USING VOLT/OHMMETER
BAD
1. Wiring connection
2. Power to ECU
(1) Fusible link(s)
(2) Fuse(s)
(3) EFI main relay
3. Air flow meter
4. Water temp. sensor
5. Air temp. sensor
6. Injection signal circuit
(1) Injector wiring
(2) ECU
SYMPTOM — ENGINE OFTEN STALLS
CHECK DIAGNOSIS SYSTEM
Check for output of diagnosis code.
Malfunction code(s)
Diagnostic code(s)
Normal code
CHECK FOR VACUUM LEAKS IN AIR INTAKKE LINE
BAD
1. Oil filler cap
2. Oil dipstick
3. Hose connections
4. PCV hoses
5. EGR system — EGR valve stays open
OK
CHECK FUEL SUPPLY TO INJECTOR
1. Fuel tank
2. Fuel pressure in fuel line
(1) Short terminals +B and FP of the check connector.
(2) Fuel pressure at fuel return hose of fuel filter can be felt.
BAD
1. Fuel line — leakage — deformation
2. Fuse(s)
3. Fuel pump
4. Fuel filter
5. Fuel pressure regulator
OK
CHECK AIR FILTER
BAD
Element — Clean or replace
OK
CHECK IDLE SPEED
Standard: 700 ± 50 rpm
BAD
1. ISC system
(1) Wiring connection(s)
(2) ISC valve
(3) ECU (test by substitution)
2. (3S-FE)
Idle speed — Adjust
OK
OK
CHECK IGNITION TIMING
1. Connect terminals TE1 and E1 of the check connector.
2. Check ignition timing.
Standard: 10° BTDC @ idle
NO
Ignition timing — Adjust
OK
CHECK SPARK PLUGS
Standard: 1.1 mm (0.043 in.)
NOTE: Check compression pressure and valve clearance if necessary.
NO
1. Spark plugs
2. Compression pressure
Minimum: 10.0 kg/cm²
(142 psi, 981 kPa)
at 250 rpm
3. Valve clearance
Standard:
3F-FE IN 0.19 — 0.29 mm
(0.007 — 0.011 in.)
EX 0.28 — 0.38 mm
(0.011 — 0.015 in.)
2VZ-FE IN 0.13 — 0.23 mm
(0.005 — 0.009 in.)
EX 0.27— 0.37 mm
(0.011 — 0.015 in.)
OK
CHECK COLD START INJECTOR
BAD
1. Cold start injector
2. Cold start injector time switch
OK
CHECK FUEL PRESSURE
BAD
1. Fuel pump
2. Fuel filter
3. Fuel pressure reglator
OK
CHECK INJECTORS
BAD
Injection condition
OK
CHECK EFI ELECTRONIC CIRCUIT USING VOLT/OHMMETER
BAD
1. Wiring connection
2. Power to ECU
(1) Fusible link(s)
(2) Fuse(s)
(3) EFI main relay
3. Air flow meter
4. Water temp. sensor
5. Air temp. sensor
6. Injection signal circuit
(1) Injector wiring
(2) ECU

TROUBLESHOOTING PROCEDURES—CAMRY AND CELICA 3S-FE AND 2VZ-FE

SYMPTOM — ENGINE SOMETIMES STALLS

CHECK DIAGNOSIS SYSTEM
Check for output of diagnosis code.
- Malfunction code(s) → Diagnostic code(s)
- Normal code ↓

CHECK AIR FLOW METER
- BAD → Air flow meter
- OK ↓

CHECK WIRING CONNECTORS AND RELAYS
Check for a signal change when the connector or relay is slightly tapped or wiggled.
- BAD → 1. Connector(s) 2. EFI main relay 3. Circuit opening relay

SYMPTOM — ROUGH IDLING AND/OR MISSING

CHECK DIAGNOSIS SYSTEM
Check for output of diagnosis code
- Malfunction code(s) → Diagnostic code(s)
- Normal code ↓

CHECK FOR VACUUM LEAKS IN AIR INTAKE LINE
- BAD → 1. Oil filler cap 2. Oil dipstick 3. Hose connections 4. PCV hoses 5. EGR system — EGR valve stays open
- OK ↓

CHECK AIR FILTER
- BAD → Element — Clean or replace
- OK ↓

CHECK IDLE SPEED
Standard: 700 ± 50 rpm
- BAD → 1. ISC system (1) Wiring connection(s) (2) ISC valve (3) ECU (test by substitution) 2. (3S-FE) Idle speed — Adjust
- OK ↓

CHECK IGNITION TIMING
1. Connect terminals TE1 and E1 of the check connector.
2. Check ignition timing
Standard: 10° BTDC @ idle
- NO → Ignition timing — Adjust
- OK ↓

CHECK SPARK PLUGS
Standard: 1.1 mm (0.043 in.)
NOTE: Check compression pressure and valve clearance if necessary.
- NO → 1. Spark plugs 2. Compression pressure Minumum: 10.0 kg/cm² (142 psi, 981 kPa) at 250 rpm 3. Valve clearance Standard: 3F-FE IN 0.19 – 0.29 mm (0.007 – 0.011 in.) EX 0.28 – 0.38 mm (0.011 – 0.016 in.) 2VZ-FE IN 0.13 – 0.23 mm (0.005 – 0.009 in.) EX 0.27– 0.37 mm (0.011 – 0.016 in.)
- OK ↓

CHECK COLD START INJECTOR
- BAD → 1. Cold start injector 2. Cold start injector time switch
- OK ↓

CHECK FUEL PRESSURE
- BAD → 1. Fuel pump 2. Fuel filter 3. Fuel pressure regulator
- OK ↓

CHECK INJECTORS
- BAD → Injection condition
- OK ↓

CHECK EFI ELECTRONIC CIRCUIT USING VOLT/OHMMETER
- BAD → 1. Wiring connection 2. Power to ECU (1) Fusible link(s) (2) Fuse(s) (3) EFI main relay 3. Air flow meter 4. Water temp. sensor 5. Air temp. sensor 6. Injection signal circuit (1) Injector wiring (2) ECU 7. Oxygen sensor(s)

TROUBLESHOOTING PROCEDURES—CAMRY AND CELICA 3S-FE AND 2VZ-FE

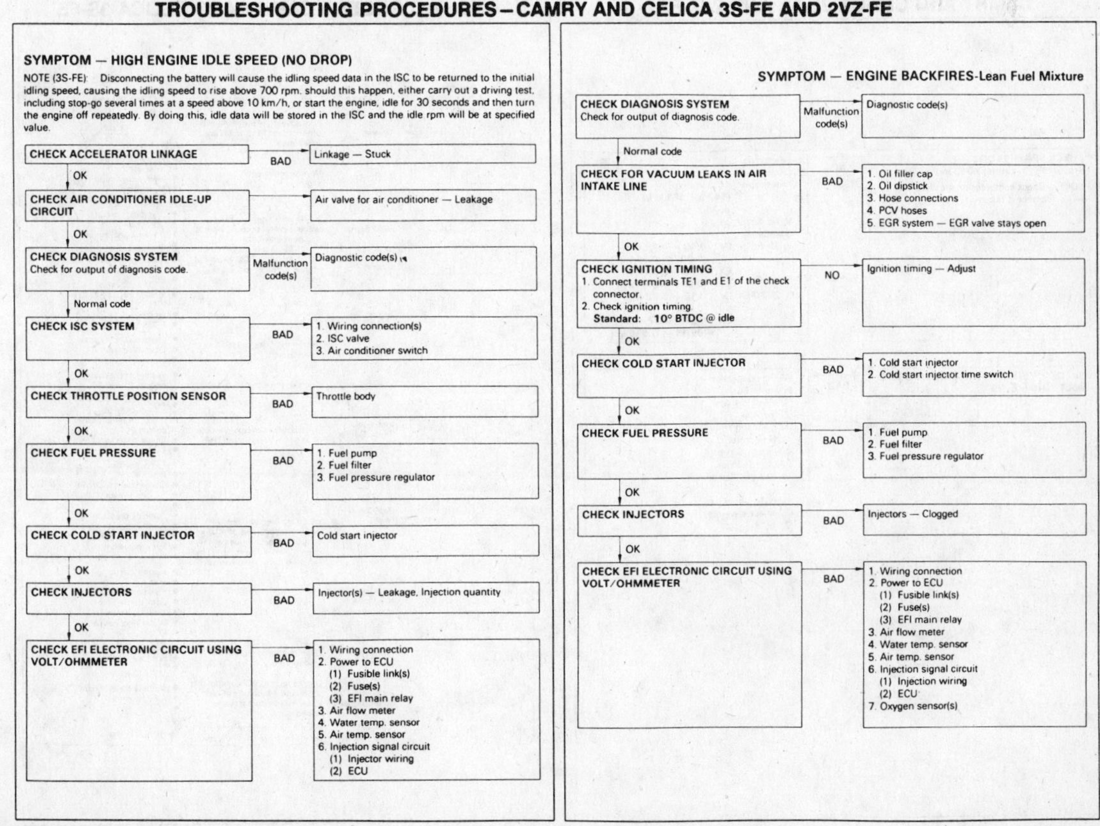
SYMPTOM — HIGH ENGINE IDLE SPEED (NO DROP)
NOTE (3S-FE): Disconnecting the battery will cause the idling speed data in the ISC to be returned to the initial idling speed, causing the idling speed to rise above 700 rpm. should this happen, either carry out a driving test, including stop-go several times at a speed above 10 km/h, or start the engine, idle for 30 seconds and then turn the engine off repeatedly. By doing this, idle data will be stored in the ISC and the idle rpm will be at specified value.
CHECK ACCELERATOR LINKAGE
BAD
Linkage — Stuck
OK
CHECK AIR CONDITIONER IDLE-UP CIRCUIT
Air valve for air conditioner — Leakage
OK
CHECK DIAGNOSIS SYSTEM
Check for output of diagnosis code.
Malfunction code(s)
Diagnostic code(s)
Normal code
CHECK ISC SYSTEM
BAD
1. Wiring connection(s)
2. ISC valve
3. Air conditioner switch
OK
CHECK THROTTLE POSITION SENSOR
BAD
Throttle body
OK
CHECK FUEL PRESSURE
BAD
1. Fuel pump
2. Fuel filter
3. Fuel pressure regulator
OK
CHECK COLD START INJECTOR
BAD
Cold start injector
OK
CHECK INJECTORS
BAD
Injector(s) — Leakage, Injection quantity
OK
CHECK EFI ELECTRONIC CIRCUIT USING VOLT/OHMMETER
BAD
1. Wiring connection
2. Power to ECU
(1) Fusible link(s)
(2) Fuse(s)
(3) EFI main relay
3. Air flow meter
4. Water temp. sensor
5. Air temp. sensor
6. Injection signal circuit
(1) Injector wiring
(2) ECU
SYMPTOM — ENGINE BACKFIRES-Lean Fuel Mixture
CHECK DIAGNOSIS SYSTEM
Check for output of diagnosis code.
Malfunction code(s)
Diagnostic code(s)
Normal code
CHECK FOR VACUUM LEAKS IN AIR INTAKE LINE
BAD
1. Oil filler cap
2. Oil dipstick
3. Hose connections
4. PCV hoses
5. EGR system — EGR valve stays open
OK
CHECK IGNITION TIMING
1. Connect terminals TE1 and E1 of the check connector.
2. Check ignition timing
Standard: 10° BTDC @ idle
NO
Ignition timing — Adjust
OK
CHECK COLD START INJECTOR
BAD
1. Cold start injector
2. Cold start injector time switch
OK
CHECK FUEL PRESSURE
BAD
1. Fuel pump
2. Fuel filter
3. Fuel pressure regulator
OK
CHECK INJECTORS
BAD
Injectors — Clogged
OK
CHECK EFI ELECTRONIC CIRCUIT USING VOLT/OHMMETER
BAD
1. Wiring connection
2. Power to ECU
(1) Fusible link(s)
(2) Fuse(s)
(3) EFI main relay
3. Air flow meter
4. Water temp. sensor
5. Air temp. sensor
6. Injection signal circuit
(1) Injection wiring
(2) ECU
7. Oxygen sensor(s)

TROUBLESHOOTING PROCEDURES—CAMRY AND CELICA 3S-FE AND 2VZ-FE

SYMPTOM — MUFFLER EXPLOSION (AFTER FIRE)-Rich Fuel Mixture-Misfire

CHECK DIAGNOSIS SYSTEM
Check for output of diagnosis code.
→ Malfunction code(s): Diagnostic code(s)

↓ Normal code

CHECK IGNITION TIMING
1. Connect terminals TE1 and E1 of the check connector.
2. Check ignition timing.
   Standard: 10° BTDC @ idle

→ NO: Ignition timing — Adjust

↓ OK

CHECK COLD START INJECTOR
→ BAD:
1. Cold start injector
2. Cold start injector time switch

↓ OK

CHECK FUEL PRESSURE
→ BAD: Fuel pressure regulator

↓ OK

CHECK INJECTORS
→ BAD: Injectors — Leakage

↓ OK

CHECK SPARK PLUGS
Standard: 1.1 mm (0.043 in.)
NOTE: Check compression pressure and valve clearance if necessary.

→ NO:
1. Spark plugs
2. Compression pressure
   Minimum: 10.0 kg/cm² (142 psi, 981 kPa) at 250 rpm
3. Valve clearance
   Standard:
   3F-FE IN 0.19 — 0.29 mm (0.007 — 0.011 in.)
   EX 0.28 — 0.38 mm (0.011 — 0.015 in.)
   2VZ-FE IN 0.13 — 0.23 mm (0.005 — 0.009 in.)
   EX 0.27 — 0.37 mm (0.011 — 0.015 in.)

↓ OK

CHECK EFI ELECTRONIC CIRCUIT USING VOLT/OHMMETER
→ BAD:
1. Throttle position sensor
2. Injection signal circuit
   (1) Injector wiring
   (2) Fuel cut RPM
   (3) ECU
3. Oxygen sensor(s)

SYMPTOM — ENGINE HESITATES AND/OR POOR ACCELERATION

CHECK CLUTCH AND BRAKES
→ BAD:
1. Clutch — Slips
2. Brakes — Drag

↓ OK

CHECK FOR VACUUM LEAKS IN AIR INTAKE LINE
→ BAD:
1. Oil filler cap
2. Oil dipstick
3. Hose connections
4. PCV hoses
5. EGR system — EGR valve stays open

↓ OK

CHECK AIR FILTER
→ BAD: Element — Clean or replace

↓ OK

CHECK DIAGNOSIS SYSTEM
Check for output of diagnosis code.
→ Malfunction code(s): Diagnostic code(s)

↓ Normal code

CHECK IGNITION SPARK
→ BAD:
1. High-tension cords
2. Distributor
3. Ignition coil
4. Igniter

↓ OK

CHECK IGNITION TIMING
1. Connect terminals TE1 and E1 of the check connector.
2. Check ignition timing.
   Standard: 10° BTDC @ idle

→ NO: Ignition timing — Adjust

↓ OK

CHECK FUEL PRESSURE
→ BAD:
1. Fuel pump
2. Fuel filter
3. Fuel pressure regulator

↓ OK

CHECK INJECTORS
→ BAD: Injector — Clogged

↓ OK

TROUBLESHOOTING PROCEDURES—CAMRY AND CELICA 3S-FE AND 2VZ-FE

↓ OK

CHECK SPARK PLUGS
Standard: 1.1 mm (0.043 in.)
NOTE: Check compression pressure and valve clearance if necessary.

→ NO:
1. Spark plug
2. Compression pressure
   Minimum: 10.0 kg/cm² (142 psi, 981 kPa) at 250 rpm
3. Valve clearance
   Standard:
   3F-FE IN 0.19 — 0.29 mm (0.007 — 0.011 in.)
   EX 0.28 — 0.38 mm (0.011 — 0.015 in.)
   2VZ-FE IN 0.13 — 0.23 mm (0.005 — 0.009 in.)
   EX 0.27 — 0.37 mm (0.011 — 0.015 in.)

↓ OK

CHECK EFI ELECTRONIC CIRCUIT USING VOLT/OHMMETER
→ BAD:
1. Wiring connection
2. Power to ECU
   (1) Fusible link(s)
   (2) Fuse(s)
   (3) EFI main relay
3. Air flow meter
4. Water temp. sensor
5. Air temp. sensor
6. Throttle position sensor
7. Injection signal circuit
   (1) Injector wiring
   (2) ECU

DIAGNOSTIC CODES—CAMRY AND CELICA 3S-FE

Code No.	Number of Check engine blinks	System	Diagnosis	Trouble area
—	ON/OFF	Normal	This appears when none of the other codes (11 thru 71) are indentified	—
11		ECU (+B)	Wire severance, however slight, in +B (ECU)	• IG switch circuit • IG switch • Main relay circuit • Main relay • ECU
12		RPM Signal	No NE or G signal to ECU within several seconds after engine is cranked	• Distributor circuit • Distributor • Starter signal circuit • ECU
13		RPM Signal	No NE signal to ECU when the engine speed is above 1,000 rpm	• Distributor circuit • Distributor • ECU
14		Ignition Signal	No IGF signal to ECU 4 — 5 times in succession	• Ignition circuit (+B, IGT, IGF) • Igniter • ECU
21		Oxygen Sensor Signal	Detection of oxygen sensor deterioration	• Oxygen sensor circuit • Oxygen sensor • ECU
22		Water temp Sensor Signal	Open or short circuit in water temp sensor signal (THW)	• Water temp sensor circuit • Water temp sensor • ECU
24		Intake air Temp Sensor Signal	Open or short circuit in intake air temp sensor signal (THA)	• Intake air temp sensor circuit • Intake air temp sensor • ECU
*25		Air-fuel Ratio Lean Malfunction	When air-fuel ratio feedback compensation valve or adative control value continues at the upper (lean) or lower (rich) limit renewed for a certain priod of time	• Injector circuit • Injector • Oxygen sensor circuit • Oxygen sensor • ECU • Fuel line pressure • Air flow meter • Water temp sensor • Ignition system • ECU
*26		Air-fuel Ratio Rich Malfunction		• Injector circuit • Injector • Fuel line pressure • Cold start injector • Air flow meter • Water temp ensor • ECU
*27		Sub-oxygen Sensor Signal	Open or short circuit in sub-oxygen sensor signal (OX2)	• Sub-oxygen sensor circuit • Sub-oxygen sensor • ECU
31		Air flow Meter Signal	Open circuir in VC signal or short circuit between VC and E2 when idle contacts are closed	• Air flow meter circuit • Air flow meter • ECU
32		Air flow Meter Signal	Open circuit in E2 or short circuit between VC and VS	Same as 31, above

*CALIF. only

DIAGNOSTIC CODES—CAMRY AND CELICA 3S-FE

Code No.	Number of Check engine blinks	System	Diagnosis	Trouble area
41		Throttle Position Sensor Signal	(w/o ECT) IDL and PSW signals being output simulaneously for several seconds. (w/ ECT) Open or short circuit in throttle position sensor signal (VTA).	• Throttle position sensor circuit • Throttle position sensor • ECU
42		Vehicle Speed Sensor Signal	No SPD signal for several seconds when engine speed is between 2,500 — 5,500 rpm and coolant temp. is below 80°C (176°F) except when racing the engine	• Vehicle speed sensor circuit • Vehicle speed sensor • ECU
43		Starter Signal	No STA signal to ECU when behicle stopped and engine running over 800 rpm.	• IG switch circuit • IG switch • ECU
*71		EGR Malfunction	EGR gas temp. below predetermined level for during EGR control.	• EGR system (EGR valve, EGR hose etc.) • EGR gas temp. sensor circuit • EGR gas temp. sensor • BVSV for EGR • BVSV for EGR circuit • ECU
51		Switch Signal	No IDL signal or A/C signal to ECU, with the check Terminals TE1 and E1 connected.	• A/C switch circuit • A/C amplifire • Throttle position sensor circuit • Throttle position sensor • Accelerator pedal and cable

*CALIF. only

DIAGNOSTIC CODES—CAMRY 2VZ-FE

Code No.	Number of Check engine blinks	System	Diagnosis	Trouble area
—		Normal	This appears when none of the other codes are identified	—
11		ECU (+B)	Momentary interruption in power supply to ECU	• Ignition switch circuit • Ignition switch • Main relay circuit • Main relay • ECU
12		RPM Signal	No NE or G signal to ECU within 2 seconds after engine has been cranked.	• Distributor circuit • Distributor • Starter signal circuit • ECU
13		RPM Signal	No NE signal to ECU when engine speed is above 1,000 rpm	• Distributor circuit • Distributor • ECU
14		Ignition Signal	No IGF signal to ECU 6 — 8 times in succession.	• Igniter and ignition coil circuit • Igniter and ignition coil • ECU
21		Oxygen Sensor Signal	Detection of oxygen sensor deterioration	• Oxygen sensor circuit • Oxygen sensor • ECU
		Oxygen Sensor Heater Signal	Open or short circuit in oxygen sensor heater signal (HT)	• Oxygen sensor heater circuit • Oxygen sensor heater • ECU
22		Water Temp Sensor Signal	Open or short circuit in water temp sensor signal (THW)	• Water temp sensor circuit • Water temp sensor • ECU
24		Intake Air Temp Sensor Signal	Open or short circuit in intake air temp sensor signal (THA)	• Intake air temp. sensor circuit • Intake air temp sensor • ECU
25		Air-fuel Ratio Lean Malfunction	(CALIF.) • When air-fuel ratio feed-back correction value or adaptive control value continues at the upper (lean) or lower (rich) limit for a certain period of time or adaptive control value is not renewed for a certain period of time	• Injector circuit • Injector • Fuel line pressure • Air flow meter • Air intake system • Oxygen sensor circuits • Oxygen sensors • Ignition system • ECU
26		Air-fuel Ratio Rich Malfunction	• When feedback frequency of air-fuel ratio feed-back correction or adaptive control is abnormally high during feedback condition (Others) • Oxygen sensor outputs a lean signal continuosly for several seconds during air-fuel ratio feedback correction. • Open circuit in oxygen sensor signal (OX)	• Injector circuit • Injector • Fuel line pressure • Air flow meter • Cold start injector • ECU
*27		Sub-oxygen Sensor Signal	Open or short circuit in sub-oxygen sensor signal (OX2)	• Sub-oxygen sensor circuit • Sub-oxygen sensor • ECU

* CALIF. only

DIAGNOSTIC CODES—CAMRY 2VZ-FE

Code No.	Number of Check engine blinks	System	Diagnosis	Trouble area
31		Air flow Meter Signal	Open circuit in VC signal or short circuit between VS and E2 when idle contacts are closed	• Air flow meter circuit • Air flow meter • ECU
32		Air Flow Meter Signal	Open circuit in E2 or short circuit between VC and VS	• Air flow meter circuit • Air flow meter • ECU
41		Throttle Position Sensor Signal	Open or short circuit in throttle position sensor signal (VTA)	• Throttle position sensor circuit • Throttle position sensor • ECU
42		Vehicle Speed Sensor Signal	No SP1 signal for 8 seconds when engine speed in between 2,500 rpm and 4,500 rpm and coolant temp is below 80°C (176°F) except when racing the engine	• No 1 vehicle speed sensor (Meter side) circuit • No 1 vehicle speed sensor (Meter side) • ECU
43		Starter Signal	No STA signal to ECU until engine speed reaches 800 rpm with vehicle not moving	• Ignition switch circuit • Ignition switch • ECU
*71		EGR System Malfunction	EGR gas temp below predetermined level during EGR operation	• EGR valve • EGR hose • EGR gas temp sensor circuit • EGR gas temp sensor • BVSV for EGR • BVSV circuit for EGR • ECU
51		Switch Signal	No IDL signal NSW signal or A/C signal to ECU, with the check terminals TE1 and E1 shorted	• A C switch circuit • A C switch • A C amplifire • Throttle position sensor circuit • Throttle position sensor • Neutral start switch circuit • Neutral start switch • Acceleration pedal and cable • ECU

* CALIF only

DIAGNOSTIC CIRCUITS—CAMRY

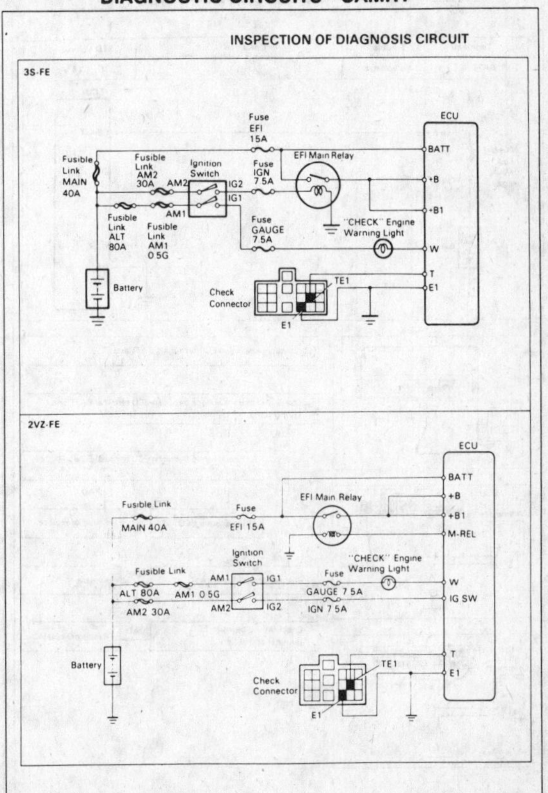
INSPECTION OF DIAGNOSIS CIRCUIT
3S-FE
ECU
Fuse EFI 15A
Fusible Link MAIN 40A
Fusible Link AM2 30A
Ignition Switch
EFI Main Relay
Fuse IGN 7.5A
BATT
+B
+B1
AM2
IG2
IG1
AM1
Fusible Link ALT 80A
Fusible Link AM1 0.5G
Fuse GAUGE 7.5A
"CHECK" Engine Warning Light
W
T
E1
Battery
Check Connector
TE1
E1
2VZ-FE
ECU
BATT
+B
+B1
M-REL
EFI Main Relay
Fusible Link MAIN 40A
Fuse EFI 15A
Ignition Switch
"CHECK" Engine Warning Light
Fusible Link
ALT 80A
AM1 0.5G
AM2 30A
AM1
AM2
IG1
IG2
Fuse GAUGE 7.5A
IGN 7.5A
W
IG SW
Battery
T
E1
TE1
Check Connector
E1

DIAGNOSTIC CIRCUITS TEST—CAMRY

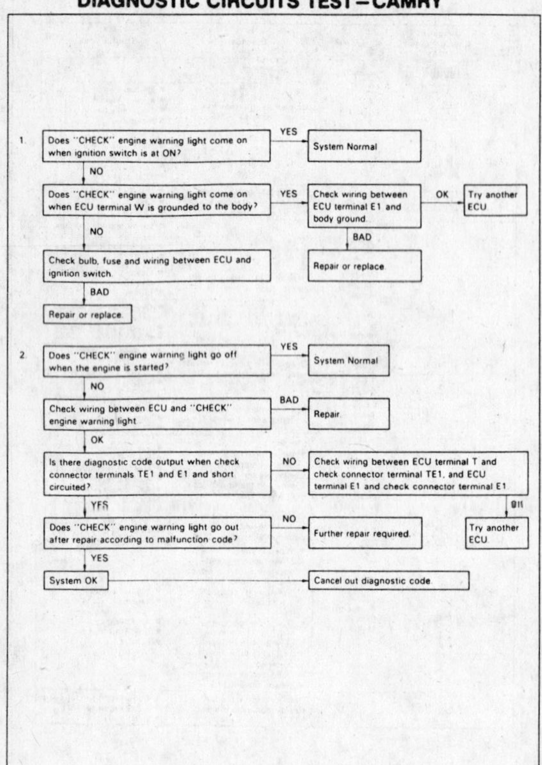
1. Does "CHECK" engine warning light come on when ignition switch is at ON?
YES
System Normal
NO
Does "CHECK" engine warning light come on when ECU terminal W is grounded to the body?
YES
Check wiring between ECU terminal E1 and body ground.
OK
Try another ECU
BAD
NO
Check bulb, fuse and wiring between ECU and ignition switch.
Repair or replace
BAD
Repair or replace.
2. Does "CHECK" engine warning light go off when the engine is started?
YES
System Normal
NO
Check wiring between ECU and "CHECK" engine warning light.
BAD
Repair.
OK
Is there diagnostic code output when check connector terminals TE1 and E1 and short circuited?
NO
Check wiring between ECU terminal T and check connector terminal TE1, and ECU terminal E1 and check connector terminal E1
YES
Does "CHECK" engine warning light go out after repair according to malfunction code?
NO
Further repair required.
Try another ECU
YES
System OK
Cancel out diagnostic code.

TROUBLESHOOTING CHART 1—CAMRY AND CELICA 3S-FE

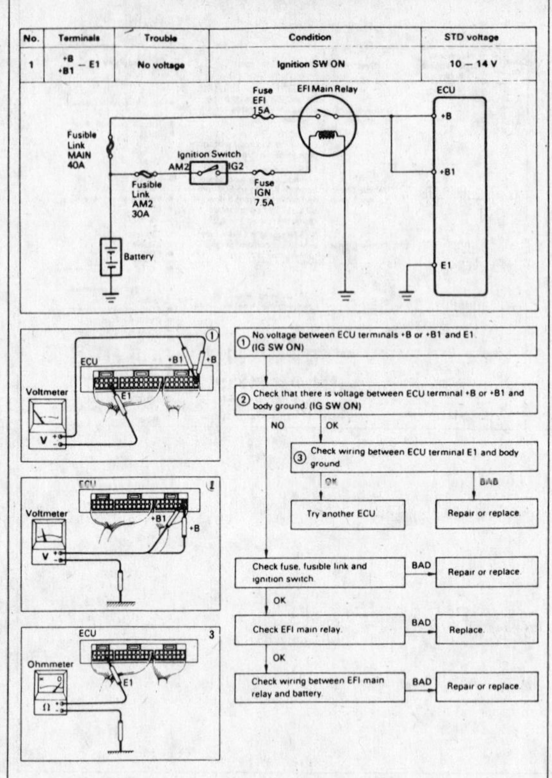
No. | Terminals | Trouble | Condition | STD voltage
1 | +B +B1 – E1 | No voltage | Ignition SW ON | 10 – 14 V
Fuse EFI 15A
EFI Main Relay
ECU
Fusible Link MAIN 40A
Ignition Switch
AM2
IG2
Fuse IGN 7.5A
Fusible Link AM2 30A
Battery
+B
+B1
E1
Voltmeter
1 No voltage between ECU terminals +B or +B1 and E1. (IG SW ON)
2 Check that there is voltage between ECU terminal +B or +B1 and body ground. (IG SW ON)
NO
OK
3 Check wiring between ECU terminal E1 and body ground
OK
BAD
Try another ECU.
Repair or replace.
Ohmmeter
Check fuse, fusible link and ignition switch.
BAD
Repair or replace.
OK
Check EFI main relay.
BAD
Replace.
OK
Check wiring between EFI main relay and battery.
BAD
Repair or replace.

TROUBLESHOOTING CHART 2—CAMRY AND CELICA 3S-FE

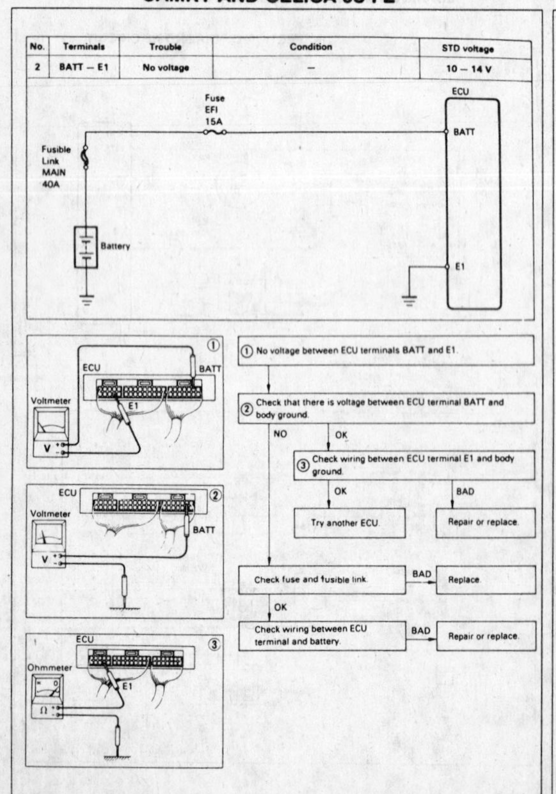
No. | Terminals | Trouble | Condition | STD voltage
2 | BATT – E1 | No voltage | – | 10 – 14 V
Fuse EFI 15A
ECU
BATT
Fusible Link MAIN 40A
Battery
E1
Voltmeter
Ohmmeter
1 No voltage between ECU terminals BATT and E1.
2 Check that there is voltage between ECU terminal BATT and body ground.
NO
OK
3 Check wiring between ECU terminal E1 and body ground.
OK
BAD
Try another ECU.
Repair or replace.
Check fuse and fusible link.
BAD
Replace.
OK
Check wiring between ECU terminal and battery.
BAD
Repair or replace.

TROUBLESHOOTING CHART 3—CAMRY AND CELICA 3S-FE

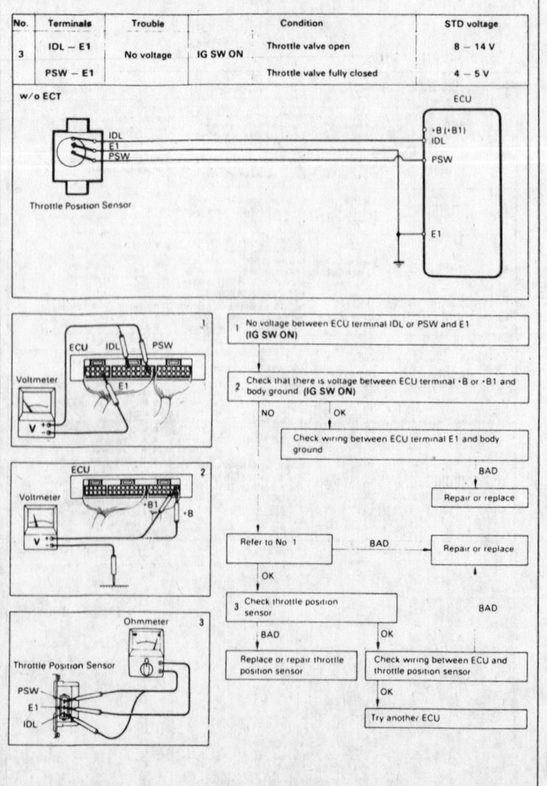
No. | Terminals | Trouble | Condition | STD voltage
3 | IDL – E1 | No voltage | IG SW ON | Throttle valve open | 8 – 14 V
PSW – E1 | Throttle valve fully closed | 4 – 5 V
w/o ECT
ECU
IDL
E1
PSW
+B (+B1)
Throttle Position Sensor
Voltmeter
Ohmmeter
1 No voltage between ECU terminal IDL or PSW and E1 (IG SW ON)
2 Check that there is voltage between ECU terminal +B or +B1 and body ground (IG SW ON)
NO
OK
Check wiring between ECU terminal E1 and body ground
BAD
Repair or replace
Refer to No. 1
BAD
Repair or replace
OK
3 Check throttle position sensor
BAD
OK
Replace or repair throttle position sensor
Check wiring between ECU and throttle position sensor
OK
Try another ECU

TROUBLESHOOTING CHART 4— CAMRY AND CELICA 3S-FE

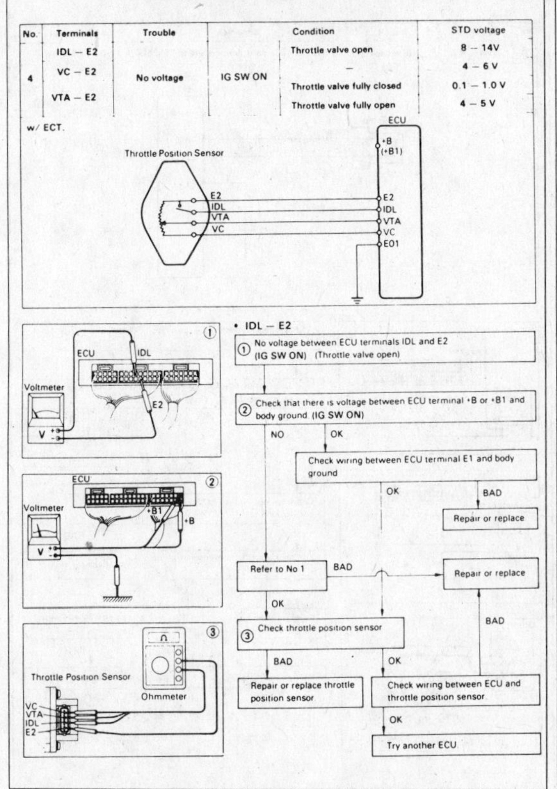

No.	Terminals	Trouble	Condition		STD voltage
4	IDL – E2	No voltage	IG SW ON	Throttle valve open	8 – 14V
	VC – E2			—	4 – 6 V
	VTA – E2			Throttle valve fully closed	0.1 – 1.0 V
				Throttle valve fully open	4 – 5 V

w/ ECT.

Throttle Position Sensor
E2
IDL
VTA
VC
ECU
+B
(+B1)
E2
IDL
VTA
VC
E01
ECU
IDL
Voltmeter
E2
ECU
Voltmeter
+B1
+B
Throttle Position Sensor
VC
VTA
IDL
E2
Ohmmeter
• IDL – E2
① No voltage between ECU terminals IDL and E2 (IG SW ON) (Throttle valve open)
② Check that there is voltage between ECU terminal +B or +B1 and body ground. (IG SW ON)
NO
OK
Check wiring between ECU terminal E1 and body ground
OK
BAD
Repair or replace
Refer to No 1
BAD
Repair or replace
OK
BAD
③ Check throttle position sensor
BAD
OK
Repair or replace throttle position sensor
Check wiring between ECU and throttle position sensor.
OK
Try another ECU

TROUBLESHOOTING CHART 4— CAMRY AND CELICA 3S-FE

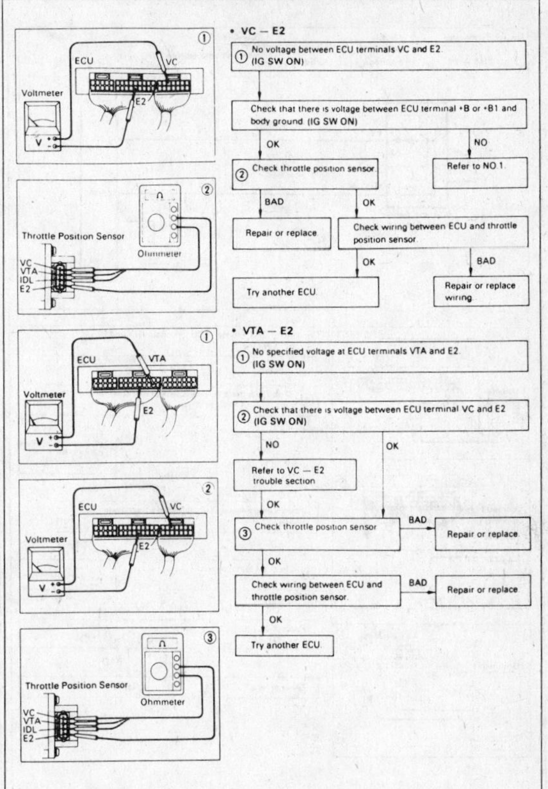

ECU
VC
Voltmeter
E2
Throttle Position Sensor
VC
VTA
IDL
E2
Ohmmeter
• VC – E2
① No voltage between ECU terminals VC and E2. (IG SW ON)
Check that there is voltage between ECU terminal +B or +B1 and body ground. (IG SW ON)
OK
NO
② Check throttle position sensor
Refer to NO.1.
BAD
OK
Repair or replace
Check wiring between ECU and throttle position sensor.
OK
BAD
Try another ECU.
Repair or replace wiring
ECU
VTA
Voltmeter
E2
ECU
VC
Voltmeter
E2
Throttle Position Sensor
Ohmmeter
VC
VTA
IDL
E2
• VTA – E2
① No specified voltage at ECU terminals VTA and E2. (IG SW ON)
② Check that there is voltage between ECU terminal VC and E2. (IG SW ON)
NO
OK
Refer to VC – E2 trouble section
OK
③ Check throttle position sensor
BAD
Repair or replace
OK
Check wiring between ECU and throttle position sensor.
BAD
Repair or replace
OK
Try another ECU.

TROUBLESHOOTING CHART 5— CAMRY AND CELICA 3S-FE

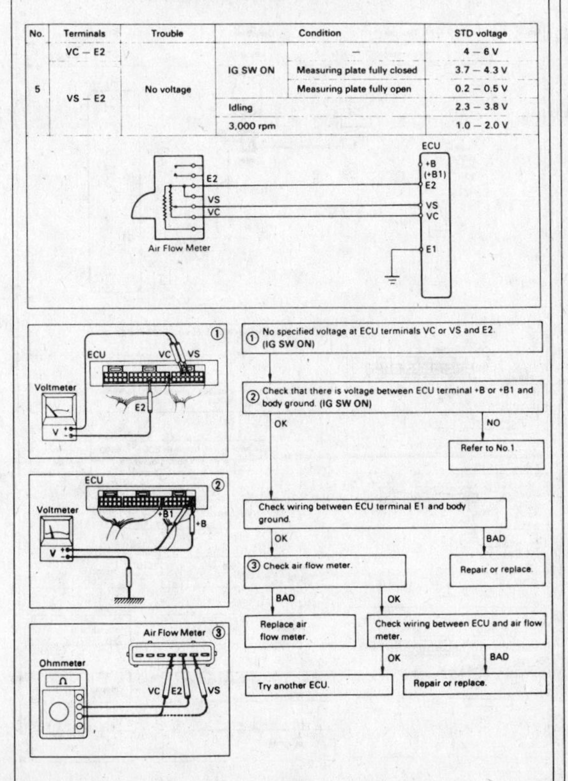

No.	Terminals	Trouble	Condition		STD voltage
5	VC – E2	No voltage		—	4 – 6 V
	VS – E2		IG SW ON	Measuring plate fully closed	3.7 – 4.3 V
				Measuring plate fully open	0.2 – 0.5 V
			Idling		2.3 – 3.8 V
			3,000 rpm		1.0 – 2.0 V

E2
VS
VC
Air Flow Meter
ECU
+B
(+B1)
E2
VS
VC
E1
ECU
VC
VS
Voltmeter
E2
ECU
Voltmeter
+B1
+B
Air Flow Meter
Ohmmeter
VC
E2
VS
① No specified voltage at ECU terminals VC or VS and E2. (IG SW ON)
② Check that there is voltage between ECU terminal +B or +B1 and body ground. (IG SW ON)
NO
OK
Refer to No.1.
Check wiring between ECU terminal E1 and body ground.
OK
BAD
③ Check air flow meter.
Repair or replace.
BAD
OK
Replace air flow meter.
Check wiring between ECU and air flow meter.
OK
BAD
Try another ECU.
Repair or replace.

TROUBLESHOOTING CHART 6— CAMRY AND CELICA 3S-FE

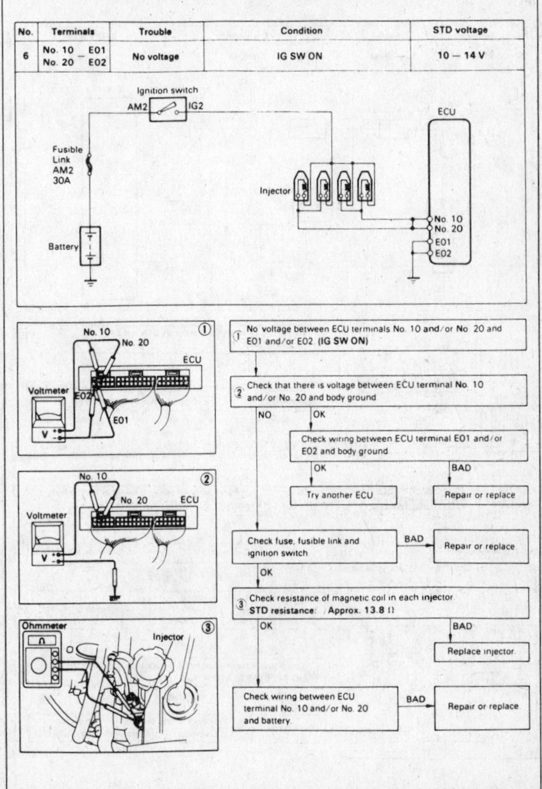

No.	Terminals	Trouble	Condition	STD voltage
6	No. 10 – E01 No. 20 – E02	No voltage	IG SW ON	10 – 14 V

Ignition switch
AM2
IG2
ECU
Fusible Link AM2 30A
Injector
Battery
No. 10
No. 20
E01
E02
No. 10
No. 20
ECU
Voltmeter
E02
E01
No. 10
No. 20
ECU
Voltmeter
Ohmmeter
Injector
① No voltage between ECU terminals No. 10 and/or No. 20 and E01 and/or E02. (IG SW ON)
② Check that there is voltage between ECU terminal No. 10 and/or No. 20 and body ground.
NO
OK
Check wiring between ECU terminal E01 and/or E02 and body ground.
OK
BAD
Try another ECU.
Repair or replace
Check fuse, fusible link and ignition switch
BAD
Repair or replace
OK
③ Check resistance of magnetic coil in each injector. STD resistance: Approx. 13.8 Ω
OK
BAD
Replace injector.
Check wiring between ECU terminal No. 10 and/or No. 20 and battery.
BAD
Repair or replace

TROUBLESHOOTING CHART 7— CAMRY AND CELICA 3S-FE

No.	Terminals	Trouble	Condition		STD voltage
7	THA – E2	No voltage	IG SW ON	Intake air temperature 20°C (68°F)	1 – 3 V

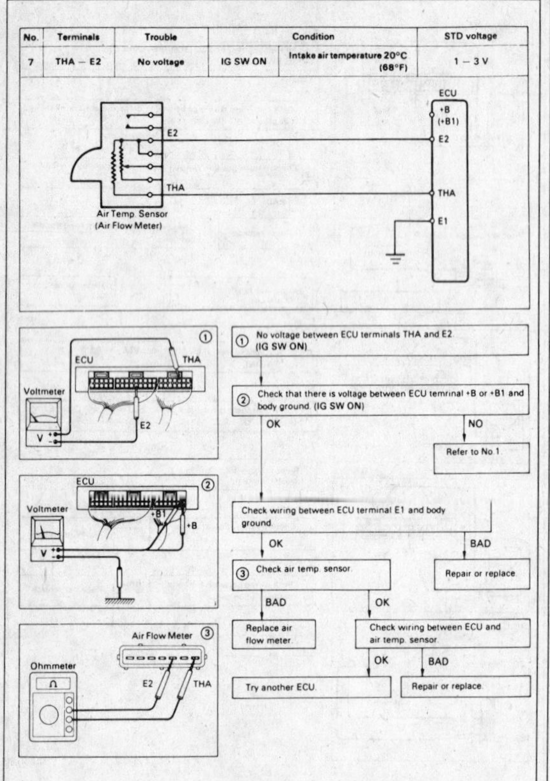
ECU
+B (+B1)
E2
THA
E1
Air Temp. Sensor (Air Flow Meter)
Voltmeter
Ohmmeter
Air Flow Meter
① No voltage between ECU terminals THA and E2 (IG SW ON)
② Check that there is voltage between ECU temrinal +B or +B1 and body ground. (IG SW ON)
OK
NO
Refer to No.1
Check wiring between ECU terminal E1 and body ground.
BAD
Repair or replace.
③ Check air temp. sensor.
Replace air flow meter.
Check wiring between ECU and air temp. sensor.
Try another ECU.
Repair or replace.

TROUBLESHOOTING CHART 8— CAMRY AND CELICA 3S-FE

No.	Terminals	Trouble	Condition		STD voltage
8	THW – E2	No voltage	IG SW ON	Coolant temperature 80° (176°F)	0.1 – 1.0 V

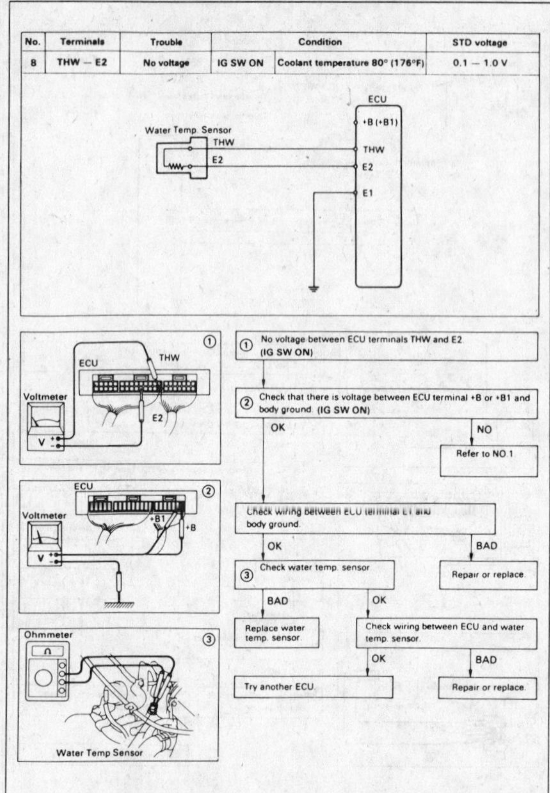
ECU
+B (+B1)
THW
E2
E1
Water Temp. Sensor
Voltmeter
Ohmmeter
Water Temp Sensor
① No voltage between ECU terminals THW and E2 (IG SW ON)
② Check that there is voltage between ECU terminal +B or +B1 and body ground. (IG SW ON)
OK
NO
Refer to NO.1
Check wiring between ECU terminal E1 and body ground.
BAD
Repair or replace.
③ Check water temp. sensor.
Replace water temp. sensor.
Check wiring between ECU and water temp. sensor.
Try another ECU.
Repair or replace.

TROUBLESHOOTING CHART 9— CAMRY AND CELICA 3S-FE

No.	Terminals	Trouble	Condition	STD voltage
9	STA – E1	No voltage	Cranking	6 – 14 V

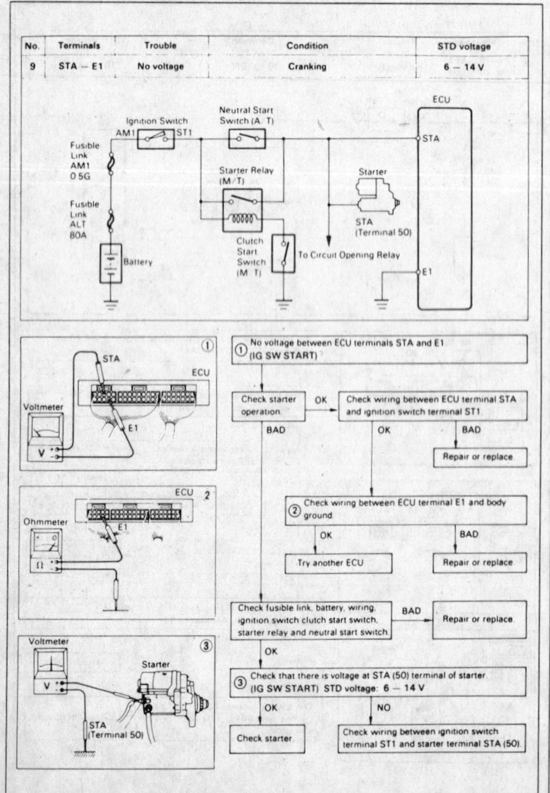
Ignition Switch
AM1
ST1
Neutral Start Switch (A/T)
Fusible Link AM1 0.5G
Fusible Link ALT 80A
Battery
Starter Relay (M/T)
Clutch Start Switch (M/T)
To Circuit Opening Relay
Starter
STA (Terminal 50)
ECU
STA
E1
Voltmeter
Ohmmeter
① No voltage between ECU terminals STA and E1 (IG SW START)
Check starter operation.
OK
Check wiring between ECU terminal STA and ignition switch terminal ST1
BAD
Repair or replace.
② Check wiring between ECU terminal E1 and body ground.
Try another ECU.
Repair or replace.
Check fusible link, battery, wiring, ignition switch clutch start switch, starter relay and neutral start switch.
Repair or replace.
③ Check that there is voltage at STA (50) terminal of starter. (IG SW START) STD voltage: 6 – 14 V
NO
Check starter.
Check wiring between ignition switch terminal ST1 and starter terminal STA (50).

TROUBLESHOOTING CHART 10— CAMRY AND CELICA 3S-FE

No.	Terminals	Trouble	Condition	STD voltage
10	IGT – E1	No voltage	Idling	0.7 – 1.0 V

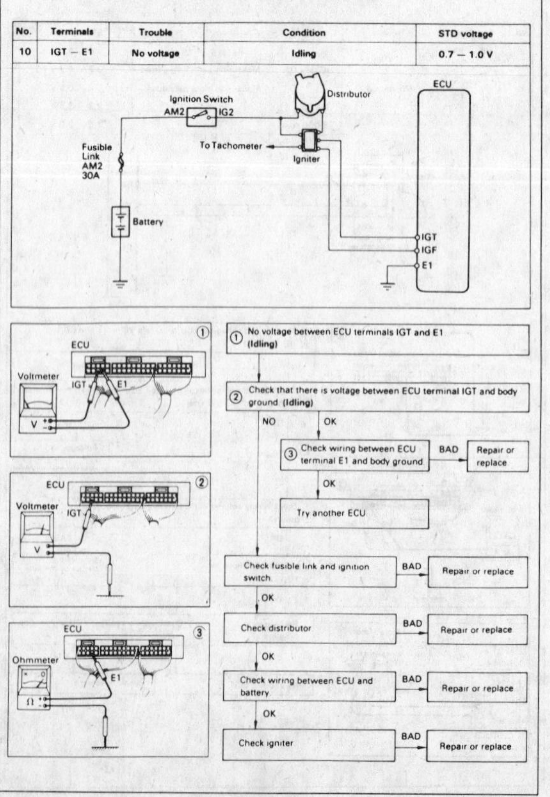
Ignition Switch
AM2
IG2
Distributor
To Tachometer
Igniter
Fusible Link AM2 30A
Battery
ECU
IGT
IGF
E1
Voltmeter
Ohmmeter
① No voltage between ECU terminals IGT and E1 (Idling)
② Check that there is voltage between ECU terminal IGT and body ground. (Idling)
NO
OK
③ Check wiring between ECU terminal E1 and body ground.
BAD
Repair or replace.
Try another ECU.
Check fusible link and ignition switch.
Repair or replace.
Check distributor.
Check wiring between ECU and battery.
Check igniter.

TROUBLESHOOTING CHART 11— CAMRY AND CELICA 3S-FE

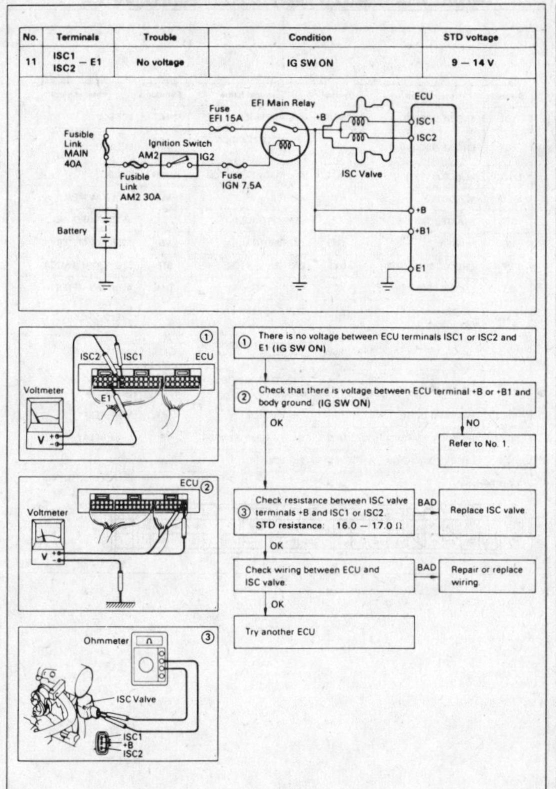

No.	Terminals	Trouble	Condition	STD voltage
11	ISC1 ISC2 — E1	No voltage	IG SW ON	9 — 14 V

ECU
Fuse EFI 15A
EFI Main Relay
+B
ISC1
ISC2
Fusible Link MAIN 40A
Ignition Switch
AM2
IG2
Fusible Link AM2 30A
Fuse IGN 7.5A
ISC Valve
Battery
+B
+B1
E1
ISC2 ISC1 ECU
Voltmeter
E1
V
Ohmmeter
ISC Valve
ISC1 +B ISC2
1 There is no voltage between ECU terminals ISC1 or ISC2 and E1 (IG SW ON)
2 Check that there is voltage between ECU terminal +B or +B1 and body ground. (IG SW ON)
OK
NO
Refer to No. 1.
3 Check resistance between ISC valve terminals +B and ISC1 or ISC2 STD resistance: 16.0 — 17.0 Ω
BAD
Replace ISC valve
OK
Check wiring between ECU and ISC valve.
BAD
Repair or replace wiring.
OK
Try another ECU

TROUBLESHOOTING CHART 12— CAMRY AND CELICA 3S-FE

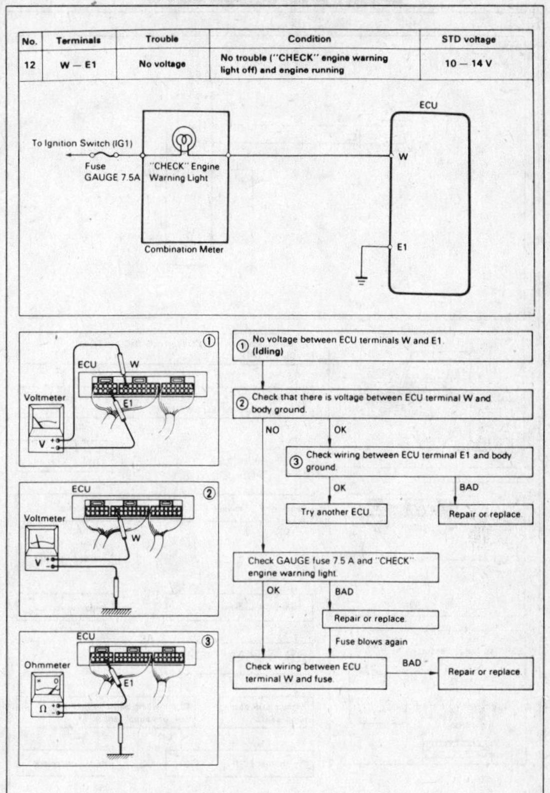

No.	Terminals	Trouble	Condition	STD voltage
12	W — E1	No voltage	No trouble ("CHECK" engine warning light off) and engine running	10 — 14 V

ECU
To Ignition Switch (IG1)
Fuse GAUGE 7.5A
"CHECK" Engine Warning Light
W
Combination Meter
E1
Voltmeter
Ohmmeter
1 No voltage between ECU terminals W and E1 (Idling)
2 Check that there is voltage between ECU terminal W and body ground.
NO
OK
3 Check wiring between ECU terminal E1 and body ground.
OK
BAD
Try another ECU.
Repair or replace.
Check GAUGE fuse 7.5 A and "CHECK" engine warning light.
OK
BAD
Repair or replace.
Fuse blows again
Check wiring between ECU terminal W and fuse.
BAD
Repair or replace.

TROUBLESHOOTING CHART 13— CAMRY AND CELICA 3S-FE

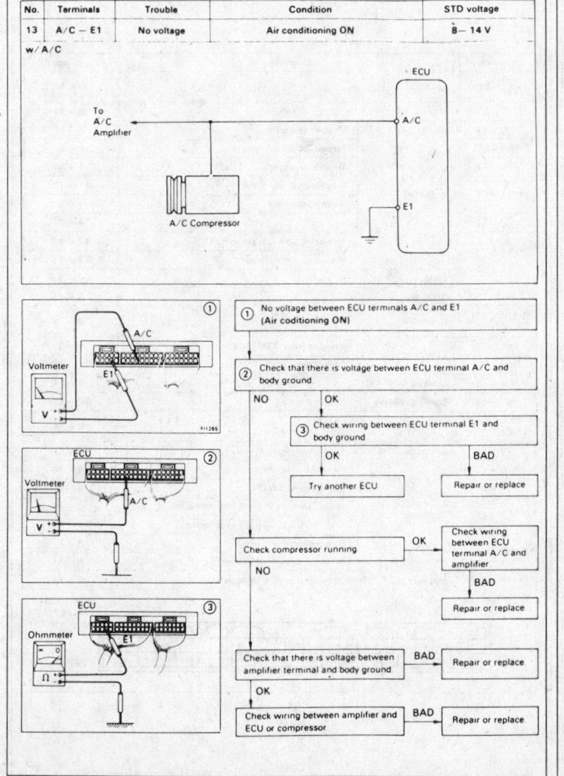

No.	Terminals	Trouble	Condition	STD voltage
13	A/C — E1	No voltage	Air conditioning ON	8— 14 V

w/A/C
ECU
To A/C Amplifier
A/C
E1
A/C Compressor
Voltmeter
Ohmmeter
1 No voltage between ECU terminals A/C and E1 (Air coditioning ON)
2 Check that there is voltage between ECU terminal A/C and body ground.
NO
OK
3 Check wiring between ECU terminal E1 and body ground.
OK
BAD
Try another ECU
Repair or replace
Check compressor running
OK
Check wiring between ECU terminal A/C and amplifier
BAD
Repair or replace
NO
Check that there is voltage between amplifier terminal and body ground.
BAD
Repair or replace
OK
Check wiring between amplifier and ECU or compressor
BAD
Repair or replace

TROUBLESHOOTING OXYGEN SENSOR— CAMRY AND CELICA 3S-FE

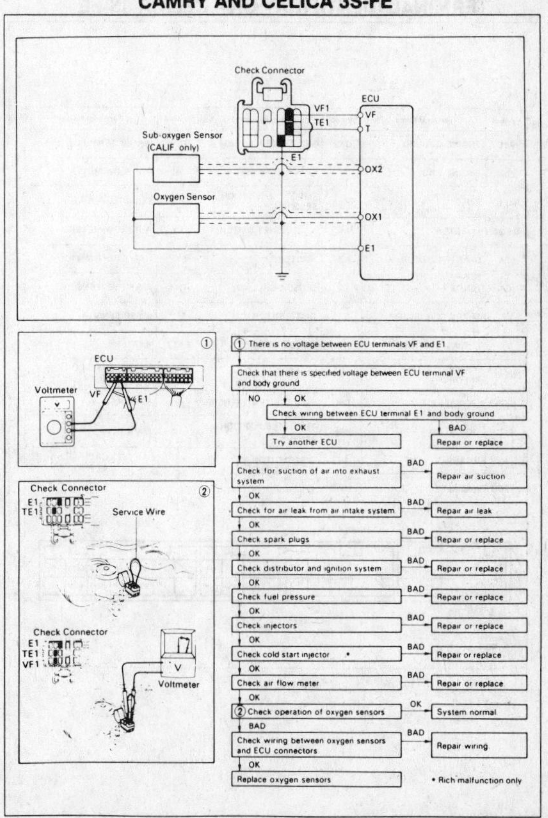
Check Connector
VF1
TE1
ECU
VF
T
Sub-oxygen Sensor (CALIF. only)
E1
OX2
Oxygen Sensor
OX1
E1
Voltmeter
Check Connector E1 TE1
Service Wire
Check Connector E1 TE1 VF1
V
1 There is no voltage between ECU terminals VF and E1
Check that there is specified voltage between ECU terminal VF and body ground
NO
OK
Check wiring between ECU terminal E1 and body ground
OK
BAD
Try another ECU
Repair or replace
Check for suction of air into exhaust system
BAD
Repair air suction
OK
Check for air leak from air intake system
BAD
Repair air leak
OK
Check spark plugs
BAD
Repair or replace
OK
Check distributor and ignition system
BAD
Repair or replace
OK
Check fuel pressure
BAD
Repair or replace
OK
Check injectors
BAD
Repair or replace
OK
Check cold start injector *
BAD
Repair or replace
OK
Check air flow meter
BAD
Repair or replace
OK
2 Check operation of oxygen sensors
OK
System normal
BAD
Check wiring between oxygen sensors and ECU connectors
BAD
Repair wiring
OK
Replace oxygen sensors
* Rich malfunction only

TROUBLESHOOTING EGR TEMPERATURE SENSOR—CAMRY AND CELICA 3S-FE

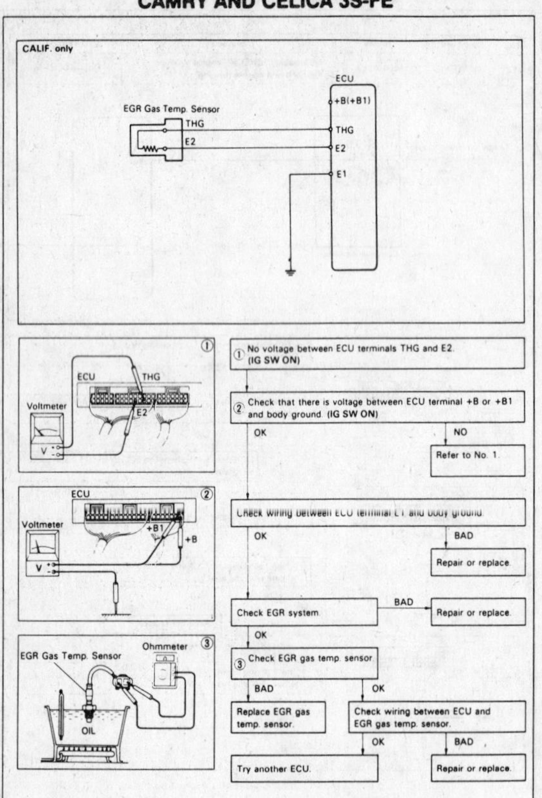
CALIF. only
ECU
+B(+B1)
EGR Gas Temp. Sensor
THG
E2
E1
ECU
THG
Voltmeter
E2
1
No voltage between ECU terminals THG and E2. (IG SW ON)
2
Check that there is voltage between ECU terminal +B or +B1 and body ground. (IG SW ON)
OK
NO
Refer to No. 1.
ECU
Voltmeter
+B1
+B
Check wiring between ECU terminal E1 and body ground.
OK
BAD
Repair or replace.
Check EGR system.
BAD
Repair or replace.
OK
3
EGR Gas Temp. Sensor
Ohmmeter
OIL
Check EGR gas temp. sensor.
BAD
OK
Replace EGR gas temp. sensor.
Check wiring between ECU and EGR gas temp. sensor.
OK
BAD
Try another ECU.
Repair or replace.

TERMINAL IDENTIFICATION—CAMRY 3S-FE

Symbol	Terminal name	Symbol	Terminal name	Symbol	Terminal name
E01	ENGINE GROUND	*4ACT	A/C AMPLIFIER	*2ECT	ECT ECU
E02	ENGINE GROUND	IDL	THROTTLE POSITION SENSOR	*2L1	ECT ECU
No. 10	INJECTOR	*4A/C	A/C MAGNET SWITCH	*2L2	ECT ECU
No. 20	INJECTOR	IGF	IGNITER	VC	AIR FLOW METER
STA	STARTER SWITCH	E2	SENSOR GROUND	E21	SENSOR GROUND
IGT	IGNITER	G⊖	DISTRIBUTOR	VS	AIR FLOW METER
VF	CHECK CONNECTOR	OX1	OXYGEN SENSOR	STP	STOP LIGHT SWITCH
E1	ENGINE GROUND	G	DISTRIBUTOR	THA	AIR FLOW METER
NSW	NEUTRAL START SWITCH	*3THG	EGR GAS TEMP. SENSOR	SPD	SPEED SENSOR
ISC1	ISC VALVE	*1PSW	THROTTLE POSITION SENSOR	BATT	BATTERY
ISC2	ISC VALVE	*2VTA	THROTTLE POSITION SENSOR	ELS	HEADLIGHT and DEFOGGER
W	WARNING LIGHT	NE	DISTRIBUTOR	+B1	MAIN RELAY
*3OX2	SUB-OXYGEN SENSOR	THW	WATER TEMP. SENSOR	+B	MAIN RELAY
T	CHECK CONNECTOR	*2L3	ECT ECU		

ECU Terminals

E01 | No 10 | STA | VF | NSW | ISC1 | W | T | IDL | IGF | G⊖ | G | | NE | L3 | L1 | VC | VS | THA | BATT | +B1
E02 | No 20 | IGT | E1 | | ISC2 | OX2 | ACT | A C | E2 | OX1 | THG | PSW & VTA | THW | ECT | L2 | E21 | STP | SPD | ELS | +B

*1 w/o ECT
*2 w/ ECT
*3 CALIF. only
*5 w/ A/C

TERMINAL IDENTIFICATION—CELICA 3S-FE

Symbol	Terminal name	Symbol	Terminal name	Symbol	Terminal name
E01	ENGINE GROUND	*1OX2	SUB-OXYGEN SENSOR	THW	WATER TEMP. SENSOR
E02	ENGINE GROUND	T	CHECK CONNECTOR	VC	AIR FLOW METER
No. 10	INJECTOR	IDL	THROTTLE POSITION SENSOR	E21	SENSOR GROUND
No. 20	INJECTOR	*2A/C	A/C MAGNET SWITCH	VS	AIR FLOW METER
STA	STARTER SWITCH	IGF	IGNITER	STP	STOP LIGHT SWITCH
IGT	IGNITER	E2	SENSOR GROUND	THA	AIR TEMP. SENSOR
VF	CHECK CONNECTOR	G⊖	DISTRIBUTOR	SPD	SPEED SENSOR
E1	ENGINE GROUND	OX1	OXYGEN SENSOR	BATT	BATTERY
NSW	NEUTRAL START SWITCH	G	DISTRIBUTOR	ELS	HEADLIGHT AND DEFOGGER
ISC1	ISC VALVE	*1THG	EGR GAS TEMP. SENSOR	+B1	MAIN RELAY
ISC2	ISC VALVE	PSW	THROTTLE POSITION SENSOR	+B	MAIN RELAY
W	WARNING LIGHT	NE	DISTRIBUTOR		

*1 CALIF. only
*2 w/ A/C

ECU Terminals

E01 | No 10 | STA | VF | NSW | ISC1 | W | T | IDL | IGF | G⊖ | G | NE | | | VC | VS | THA | BATT | +B1
E02 | No 20 | IGT | E1 | | ISC2 | OX2 | | A/C | E2 | OX1 | THG | PSW | THW | | | E21 | STP | SPD | ELS | +B

ECU TERMINAL VOLTAGES—CAMRY 3S-FE

Terminals	Condition		STD voltage (V)
+B, +B1 — E1	IG SW ON		10 — 14
BATT — E1	—		10 — 14
*1IDL — E1	IG SW ON	Throttle valve open	8 — 14
*1PSW — E1		Throttle valve fully closed	4 — 5
*2IDL — E2		Throttle valve open	8 — 14
*2VTA — E2		Throttle valve fully closed	0.1 — 1.0
		Throttle valve open	4 — 5
VC — E2		—	4 — 6
VS — E2		Measuring plate fully closed	3.7 — 4.3
		Measuring plate fully open	0.2 — 0.5
	Idling		2.3 — 3.8
	3,000 rpm		1.0 — 2.0
No.10 — E01, No.20 — E02	IG SW ON		10 — 14
THA — E2	IG SW ON	Intake air temp. 20°C (68°F)	1 — 3
THW — E2		Coolant temp. 80°C (176°F)	0.1 — 1.0
STA — E1	Cranking		6 — 14
IGT — E1	Cranking or idling		0.7 — 1.0
ISC1, ISC2 — E1	IG SW ON		9 — 14
W — E1	No trouble ("CHECK" engine warning light off) and engine running		10 — 14
*3A/C — E1	IG SW ON	Air conditioning ON	8 — 14
*3ACT — E1		Heater blower SW ON	4 — 6
T — E1		Check connector TE1 — E1 not connected	10 — 14
		Check connector TE1 — E1 connected	0.5 or less
NSW — E1		Shift position P or N range	0 — 2
		Ex. shift position P or N range	6 — 14
STP — E1	Stop light SW ON (Brake pedal depressed) or defogger SW ON		10 — 14

ECU Terminals

E01 | No 10 | STA | VF | NSW | ISC1 | W | T | IDL | IGF | G⊖ | G | | NE | L3 | L1 | VC | VS | THA | BATT | +B1
E02 | No 20 | IGT | E1 | | ISC2 | OX2 | ACT | A/C | E2 | OX | THG | PSW & VTA | THW | ECT | L2 | E21 | STP | SPD | ELS | +B

*1 w/o ECT
*2 w/ ECT
*3 w/ A/C

ECU TERMINAL VOLTAGES—CELICA 3S-FE

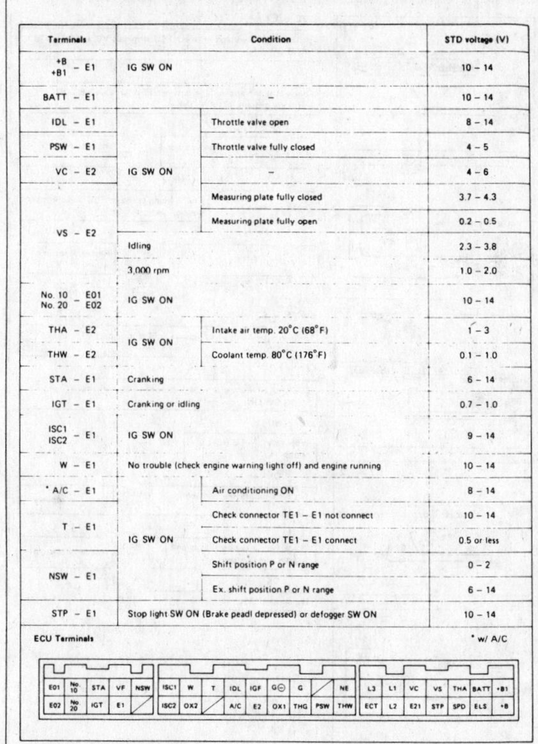

Terminals	Condition		STD voltage (V)
+B +B1 – E1	IG SW ON		10 – 14
BATT – E1	–		10 – 14
IDL – E1	IG SW ON	Throttle valve open	8 – 14
PSW – E1	IG SW ON	Throttle valve fully closed	4 – 5
VC – E2	IG SW ON	–	4 – 6
VS – E2	IG SW ON	Measuring plate fully closed	3.7 – 4.3
VS – E2	IG SW ON	Measuring plate fully open	0.2 – 0.5
VS – E2	Idling		2.3 – 3.8
VS – E2	3,000 rpm		1.0 – 2.0
No. 10 – E01, No. 20 – E02	IG SW ON		10 – 14
THA – E2	IG SW ON	Intake air temp. 20°C (68°F)	1 – 3
THW – E2	IG SW ON	Coolant temp. 80°C (176°F)	0.1 – 1.0
STA – E1	Cranking		6 – 14
IGT – E1	Cranking or idling		0.7 – 1.0
ISC1 ISC2 – E1	IG SW ON		9 – 14
W – E1	No trouble (check engine warning light off) and engine running		10 – 14
* A/C – E1	IG SW ON	Air conditioning ON	8 – 14
T – E1	IG SW ON	Check connector TE1 – E1 not connect	10 – 14
T – E1	IG SW ON	Check connector TE1 – E1 connect	0.5 or less
NSW – E1	IG SW ON	Shift position P or N range	0 – 2
NSW – E1	IG SW ON	Ex. shift position P or N range	6 – 14
STP – E1	Stop light SW ON (Brake peadl depressed) or defogger SW ON		10 – 14

ECU Terminals

* w/ A/C

E01
No. 10
STA
VF
NSW
ISC1
W
T
IDL
IGF
G⊖
G
NE
L3
L1
VC
VS
THA
BATT
+B1
E02
No. 20
IGT
E1
ISC2
OX2
A/C
E2
OX1
THG
PSW
THW
ECT
L2
E21
STP
SPD
ELS
+B

TROUBLESHOOTING CHART 1—CAMRY 2VZ-FE

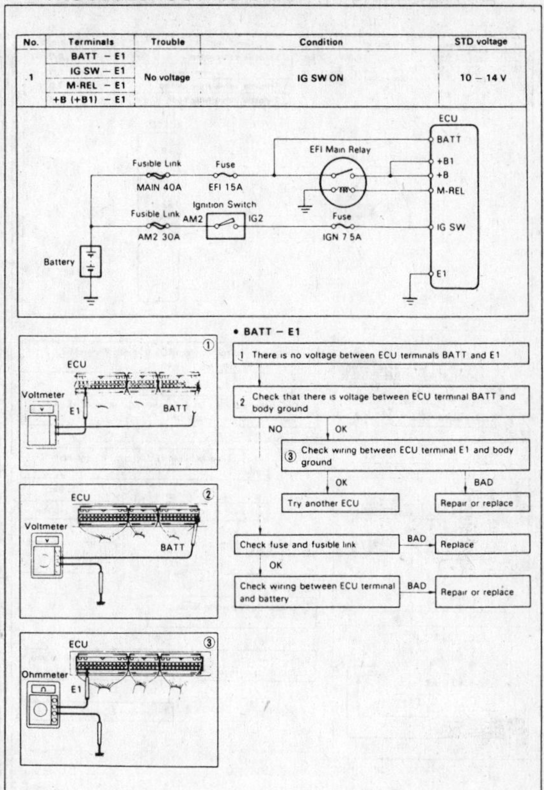

No.	Terminals	Trouble	Condition	STD voltage
1	BATT – E1 IG SW – E1 M-REL – E1 +B (+B1) – E1	No voltage	IG SW ON	10 – 14 V

ECU
BATT
EFI Main Relay
+B1
+B
M-REL
Fusible Link
MAIN 40A
Fuse
EFI 15A
Ignition Switch
Fusible Link
AM2
IG2
AM2 30A
Fuse
IGN 7.5A
IG SW
Battery
E1
① ECU Voltmeter E1 BATT
② ECU Voltmeter BATT
③ ECU Ohmmeter E1
• BATT – E1
1 There is no voltage between ECU terminals BATT and E1
2 Check that there is voltage between ECU terminal BATT and body ground
NO
OK
③ Check wiring between ECU terminal E1 and body ground
OK
BAD
Try another ECU
Repair or replace
Check fuse and fusible link
BAD
Replace
OK
Check wiring between ECU terminal and battery
BAD
Repair or replace

TROUBLESHOOTING CHART 1—CAMRY 2VZ-FE

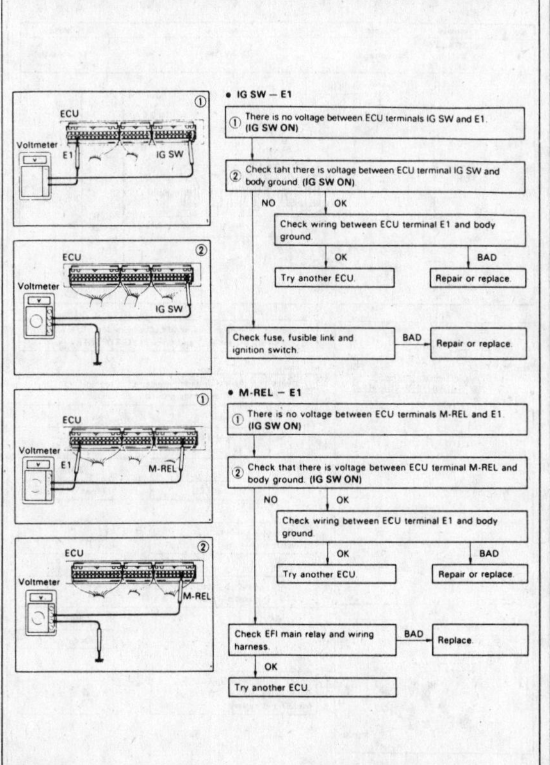

① ECU Voltmeter E1 IG SW
② ECU Voltmeter IG SW
• IG SW – E1
① There is no voltage between ECU terminals IG SW and E1. (IG SW ON)
② Check taht there is voltage between ECU terminal IG SW and body ground. (IG SW ON)
NO
OK
Check wiring between ECU terminal E1 and body ground.
OK
BAD
Try another ECU.
Repair or replace.
Check fuse, fusible link and ignition switch.
BAD
Repair or replace.
① ECU Voltmeter E1 M-REL
② ECU Voltmeter M-REL
• M-REL – E1
① There is no voltage between ECU terminals M-REL and E1. (IG SW ON)
② Check that there is voltage between ECU terminal M-REL and body ground. (IG SW ON)
NO
OK
Check wiring between ECU terminal E1 and body ground.
OK
BAD
Try another ECU.
Repair or replace.
Check EFI main relay and wiring harness.
BAD
Replace.
OK
Try another ECU.

TROUBLESHOOTING CHART 1—CAMRY 2VZ-FE

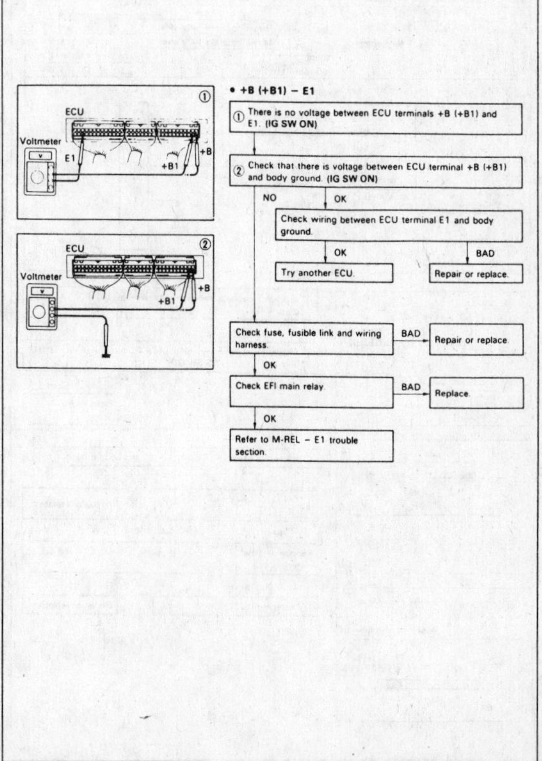

① ECU Voltmeter E1 +B1 +B
② ECU Voltmeter +B1 +B
• +B (+B1) – E1
① There is no voltage between ECU terminals +B (+B1) and E1. (IG SW ON)
② Check that there is voltage between ECU terminal +B (+B1) and body ground. (IG SW ON)
NO
OK
Check wiring between ECU terminal E1 and body ground.
OK
BAD
Try another ECU.
Repair or replace.
Check fuse, fusible link and wiring harness.
BAD
Repair or replace.
OK
Check EFI main relay.
BAD
Replace.
OK
Refer to M-REL – E1 trouble section.

TROUBLESHOOTING CHART 2—CAMRY 2VZ-FE

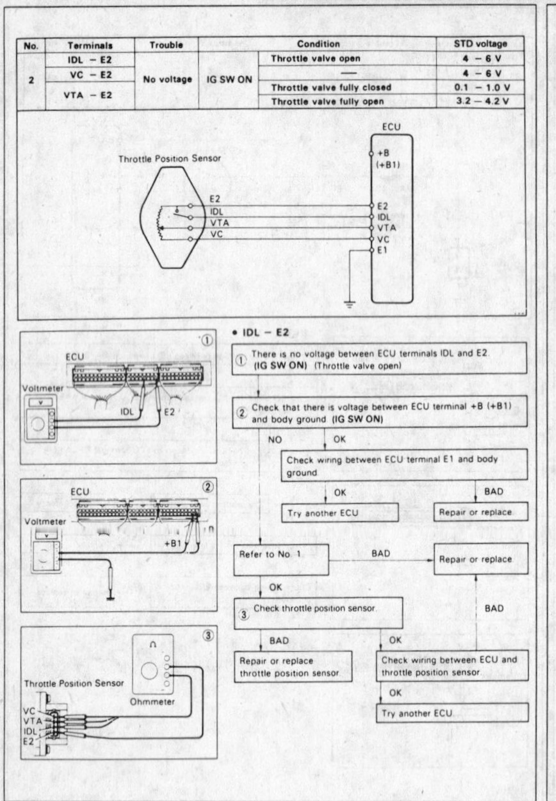

No.	Terminals	Trouble	Condition		STD voltage
2	IDL – E2	No voltage	IG SW ON	Throttle valve open	4 – 6 V
	VC – E2			—	4 – 6 V
	VTA – E2			Throttle valve fully closed	0.1 – 1.0 V
				Throttle valve fully open	3.2 – 4.2 V

ECU
+B
(+B1)
Throttle Position Sensor
E2
IDL
VTA
VC
E1
• IDL – E2
ECU
Voltmeter
IDL
E2
① There is no voltage between ECU terminals IDL and E2. (IG SW ON) (Throttle valve open)
② Check that there is voltage between ECU terminal +B (+B1) and body ground. (IG SW ON)
NO
OK
Check wiring between ECU terminal E1 and body ground
OK
BAD
Try another ECU
Repair or replace
ECU
Voltmeter
+B1
Refer to No. 1
BAD
Repair or replace
OK
③ Check throttle position sensor.
BAD
BAD
OK
Repair or replace throttle position sensor
Check wiring between ECU and throttle position sensor
OK
Try another ECU
Throttle Position Sensor
Ohmmeter
VC
VTA
IDL
E2

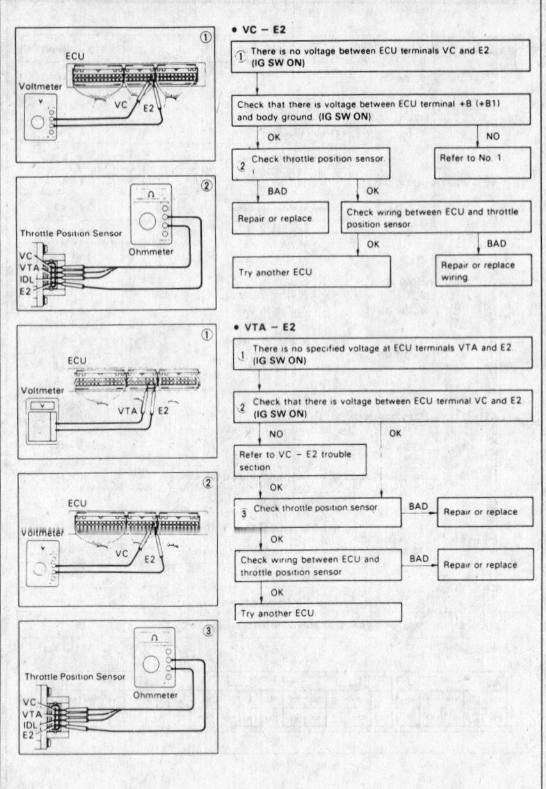

• VC – E2
ECU
Voltmeter
VC
E2
① There is no voltage between ECU terminals VC and E2. (IG SW ON)
Check that there is voltage between ECU terminal +B (+B1) and body ground. (IG SW ON)
OK
NO
② Check throttle position sensor.
Refer to No. 1
BAD
OK
Repair or replace
Check wiring between ECU and throttle position sensor
OK
BAD
Try another ECU
Repair or replace wiring
Throttle Position Sensor
Ohmmeter
VC
VTA
IDL
E2
• VTA – E2
ECU
Voltmeter
VTA
E2
① There is no specified voltage at ECU terminals VTA and E2. (IG SW ON)
② Check that there is voltage between ECU terminal VC and E2. (IG SW ON)
NO
OK
Refer to VC – E2 trouble section
OK
③ Check throttle position sensor.
BAD
Repair or replace
OK
Check wiring between ECU and throttle position sensor
BAD
Repair or replace
OK
Try another ECU
ECU
Voltmeter
VC
E2
Throttle Position Sensor
Ohmmeter
VC
VTA
IDL
E2

TROUBLESHOOTING CHART 3—CAMRY 2VZ-FE

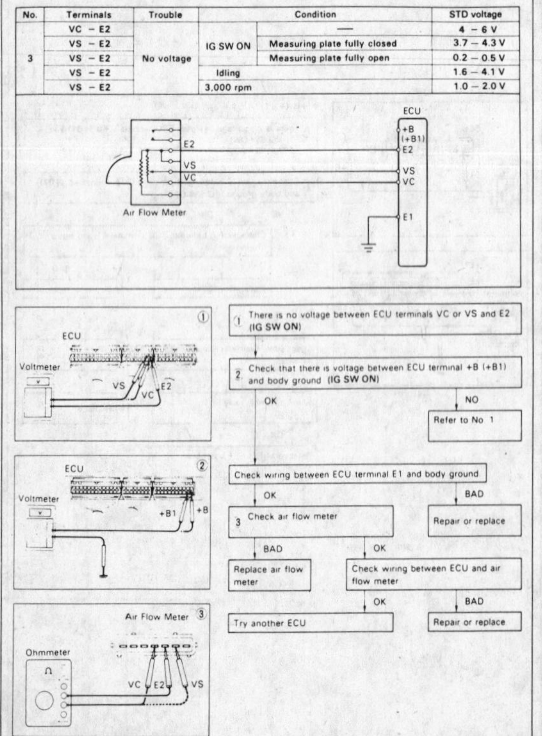

No.	Terminals	Trouble	Condition		STD voltage
3	VC – E2	No voltage	IG SW ON	—	4 – 6 V
	VS – E2			Measuring plate fully closed	3.7 – 4.3 V
	VS – E2			Measuring plate fully open	0.2 – 0.5 V
	VS – E2		Idling		1.6 – 4.1 V
	VS – E2		3,000 rpm		1.0 – 2.0 V

ECU
+B
(+B1)
E2
VS
VC
E1
E2
VS
VC
Air Flow Meter
ECU
Voltmeter
VS
VC
E2
① There is no voltage between ECU terminals VC or VS and E2. (IG SW ON)
② Check that there is voltage between ECU terminal +B (+B1) and body ground. (IG SW ON)
OK
NO
Refer to No. 1
ECU
Voltmeter
+B1
+B
Check wiring between ECU terminal E1 and body ground
OK
BAD
③ Check air flow meter
Repair or replace
BAD
OK
Replace air flow meter
Check wiring between ECU and air flow meter
OK
BAD
Try another ECU
Repair or replace
Air Flow Meter
Ohmmeter
VC
E2
VS

TROUBLESHOOTING CHART 4—CAMRY 2VZ-FE

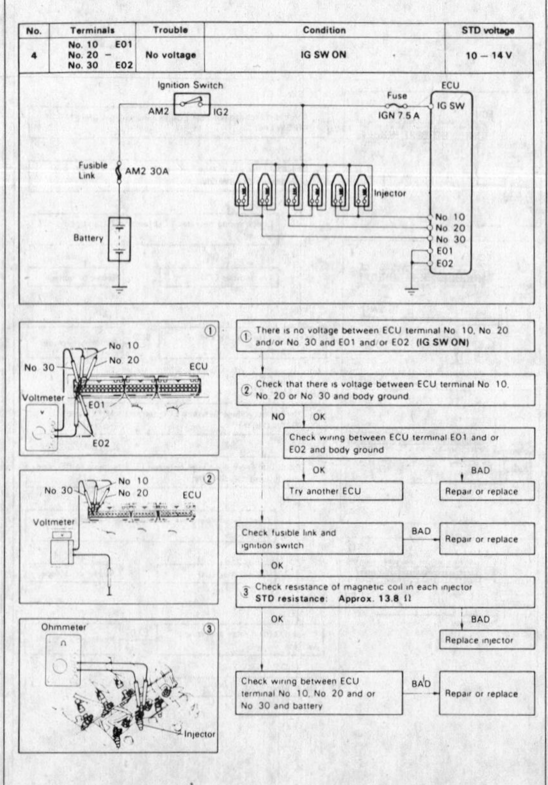

No.	Terminals	Trouble	Condition	STD voltage
4	No. 10 – E01 No. 20 – No. 30 – E02	No voltage	IG SW ON	10 – 14 V

Ignition Switch
AM2
IG2
Fuse
IGN 7.5 A
ECU
IG SW
Fusible Link
AM2 30A
Injector
Battery
No. 10
No. 20
No. 30
E01
E02
No. 10
No. 20
No. 30
ECU
Voltmeter
E01
E02
① There is no voltage between ECU terminal No. 10, No. 20 and/or No. 30 and E01 and/or E02. (IG SW ON)
② Check that there is voltage between ECU terminal No. 10, No. 20 or No. 30 and body ground.
NO
OK
Check wiring between ECU terminal E01 and or E02 and body ground
OK
BAD
Try another ECU
Repair or replace
No. 30
No. 10
No. 20
ECU
Voltmeter
Check fusible link and ignition switch
BAD
Repair or replace
OK
③ Check resistance of magnetic coil in each injector
STD resistance: Approx. 13.8 Ω
OK
BAD
Replace injector
Ohmmeter
Check wiring between ECU terminal No. 10, No. 20 and or No. 30 and battery
BAD
Repair or replace
Injector

TROUBLESHOOTING CHART 5—CAMRY 2VZ-FE

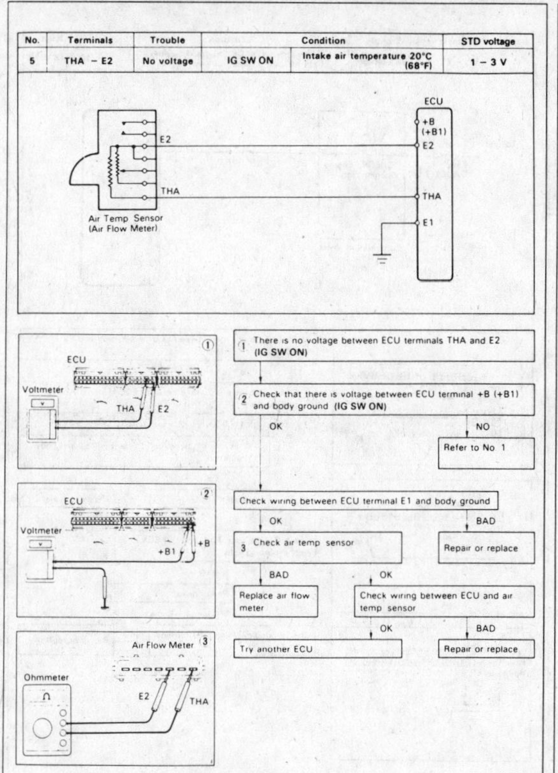

No.	Terminals	Trouble	Condition		STD voltage
5	THA – E2	No voltage	IG SW ON	Intake air temperature 20°C (68°F)	1 – 3 V

ECU
+B (+B1)
E2
THA
E1
Air Temp Sensor (Air Flow Meter)
Voltmeter
Ohmmeter
Air Flow Meter
① There is no voltage between ECU terminals THA and E2 (IG SW ON)
② Check that there is voltage between ECU terminal +B (+B1) and body ground (IG SW ON)
OK
NO
Refer to No. 1
Check wiring between ECU terminal E1 and body ground
BAD
Repair or replace
③ Check air temp. sensor
Replace air flow meter
Check wiring between ECU and air temp. sensor
Try another ECU
Repair or replace

TROUBLESHOOTING CHART 6—CAMRY 2VZ-FE

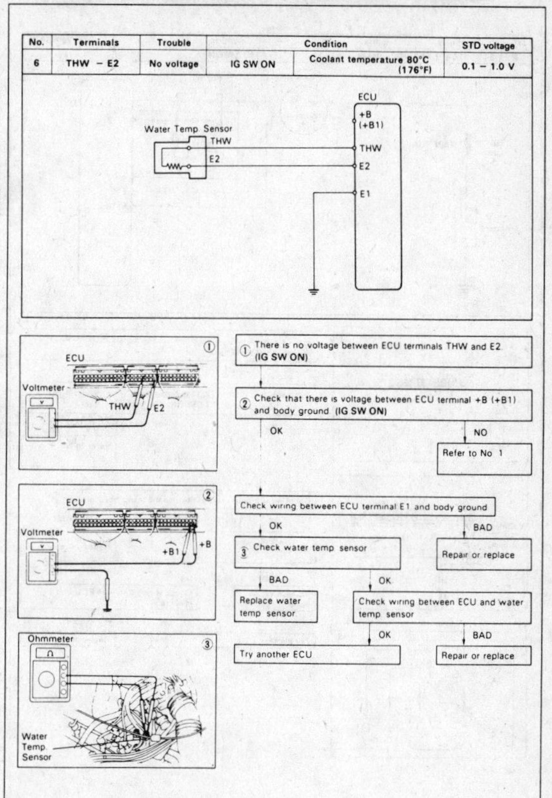

No.	Terminals	Trouble	Condition		STD voltage
6	THW – E2	No voltage	IG SW ON	Coolant temperature 80°C (176°F)	0.1 – 1.0 V

ECU
Water Temp Sensor
+B (+B1)
THW
E2
E1
Voltmeter
Ohmmeter
Water Temp. Sensor
① There is no voltage between ECU terminals THW and E2 (IG SW ON)
② Check that there is voltage between ECU terminal +B (+B1) and body ground (IG SW ON)
OK
NO
Refer to No. 1
Check wiring between ECU terminal E1 and body ground
BAD
Repair or replace
③ Check water temp. sensor
Replace water temp. sensor
Check wiring between ECU and water temp. sensor
Try another ECU
Repair or replace

TROUBLESHOOTING CHART 7—CAMRY 2VZ-FE

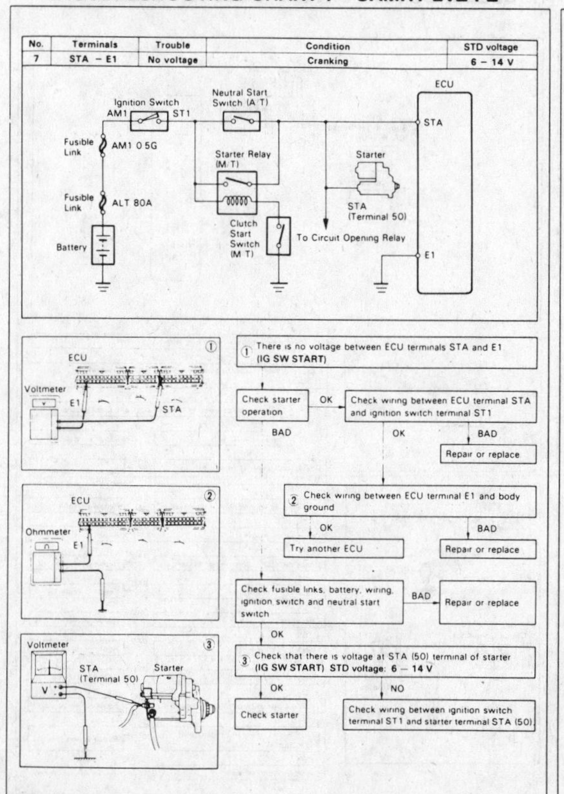

No.	Terminals	Trouble	Condition	STD voltage
7	STA – E1	No voltage	Cranking	6 – 14 V

ECU
Ignition Switch
AM1
ST1
Neutral Start Switch (A/T)
STA
Fusible Link
AM1 0.5G
ALT 80A
Starter Relay (M/T)
Starter
STA (Terminal 50)
Clutch Start Switch (M/T)
To Circuit Opening Relay
Battery
E1
Voltmeter
Ohmmeter
① There is no voltage between ECU terminals STA and E1 (IG SW START)
Check starter operation
OK
Check wiring between ECU terminal STA and ignition switch terminal ST1
BAD
Repair or replace
② Check wiring between ECU terminal E1 and body ground
Try another ECU
Repair or replace
Check fusible links, battery, wiring, ignition switch and neutral start switch
Repair or replace
③ Check that there is voltage at STA (50) terminal of starter (IG SW START) STD voltage: 6 – 14 V
NO
Check starter
Check wiring between ignition switch terminal ST1 and starter terminal STA (50)

TROUBLESHOOTING CHART 8—CAMRY 2VZ-FE

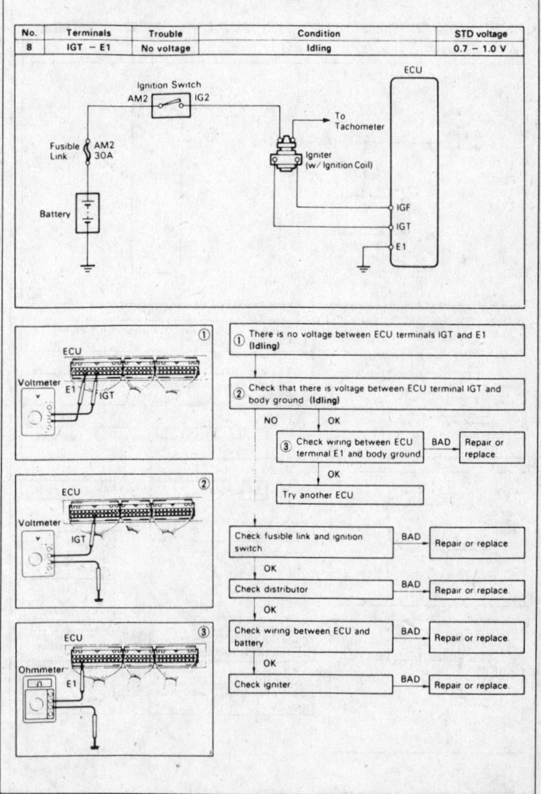

No.	Terminals	Trouble	Condition	STD voltage
8	IGT – E1	No voltage	Idling	0.7 – 1.0 V

ECU
Ignition Switch
AM2
IG2
To Tachometer
Fusible Link
AM2 30A
Igniter (w/ Ignition Coil)
Battery
IGF
IGT
E1
Voltmeter
Ohmmeter
① There is no voltage between ECU terminals IGT and E1 (Idling)
② Check that there is voltage between ECU terminal IGT and body ground (Idling)
NO
OK
③ Check wiring between ECU terminal E1 and body ground
BAD
Repair or replace
Try another ECU
Check fusible link and ignition switch
Check distributor
Check wiring between ECU and battery
Check igniter
Repair or replace

TROUBLESHOOTING CHART 9—CAMRY 2VZ-FE

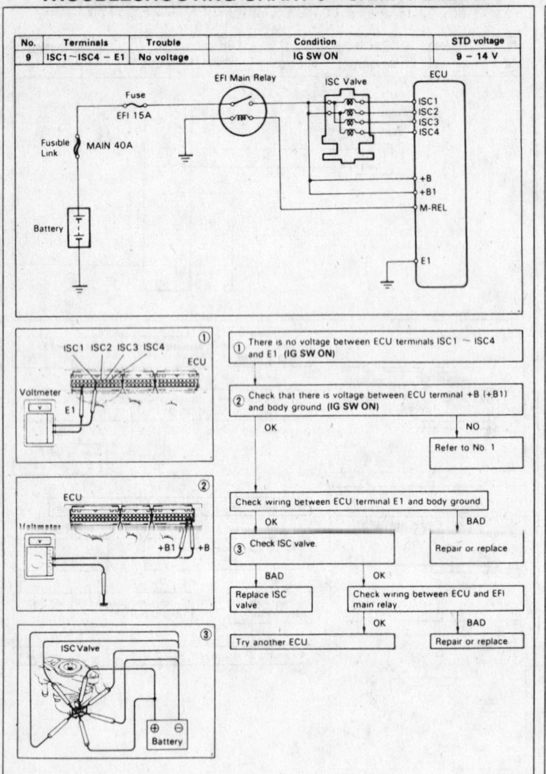

No.	Terminals	Trouble	Condition	STD voltage
9	ISC1~ISC4 – E1	No voltage	IG SW ON	9 – 14 V

Fuse
EFI 15A
EFI Main Relay
ISC Valve
ECU
ISC1
ISC2
ISC3
ISC4
Fusible Link
MAIN 40A
+B
+B1
M-REL
Battery
E1
ISC1 ISC2 ISC3 ISC4
Voltmeter
① There is no voltage between ECU terminals ISC1 ~ ISC4 and E1 (IG SW ON)
② Check that there is voltage between ECU terminal +B (+B1) and body ground (IG SW ON)
OK
NO
Refer to No. 1
Check wiring between ECU terminal E1 and body ground
OK
BAD
③ Check ISC valve
Repair or replace
BAD
OK
Replace ISC valve
Check wiring between ECU and EFI main relay
OK
BAD
Try another ECU
Repair or replace
ISC Valve
Battery

TROUBLESHOOTING CHART 10—CAMRY 2VZ-FE

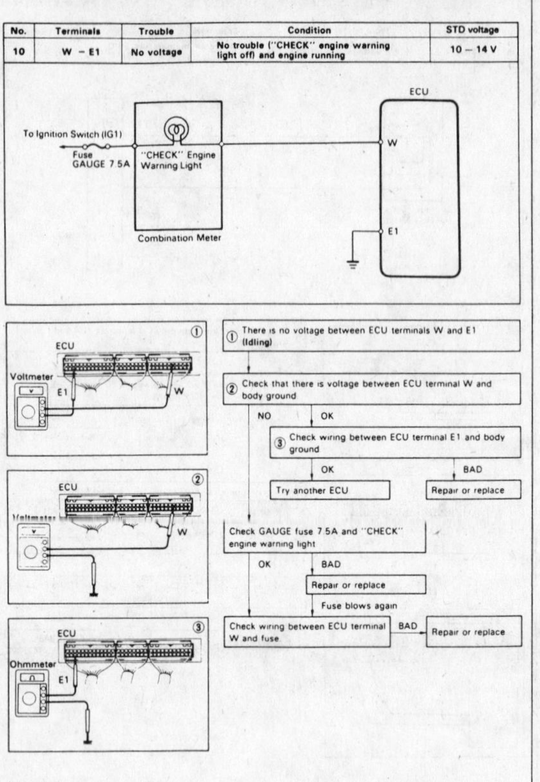

No.	Terminals	Trouble	Condition	STD voltage
10	W – E1	No voltage	No trouble ("CHECK" engine warning light off) and engine running	10 – 14 V

To Ignition Switch (IG1)
Fuse
GAUGE 7.5A
"CHECK" Engine Warning Light
Combination Meter
ECU
W
E1
Voltmeter
Ohmmeter
① There is no voltage between ECU terminals W and E1 (Idling)
② Check that there is voltage between ECU terminal W and body ground
NO
OK
③ Check wiring between ECU terminal E1 and body ground
OK
BAD
Try another ECU
Repair or replace
Check GAUGE fuse 7.5A and "CHECK" engine warning light
OK
BAD
Repair or replace
Fuse blows again
Check wiring between ECU terminal W and fuse
BAD
Repair or replace

TROUBLESHOOTING CHART 11—CAMRY 2VZ-FE

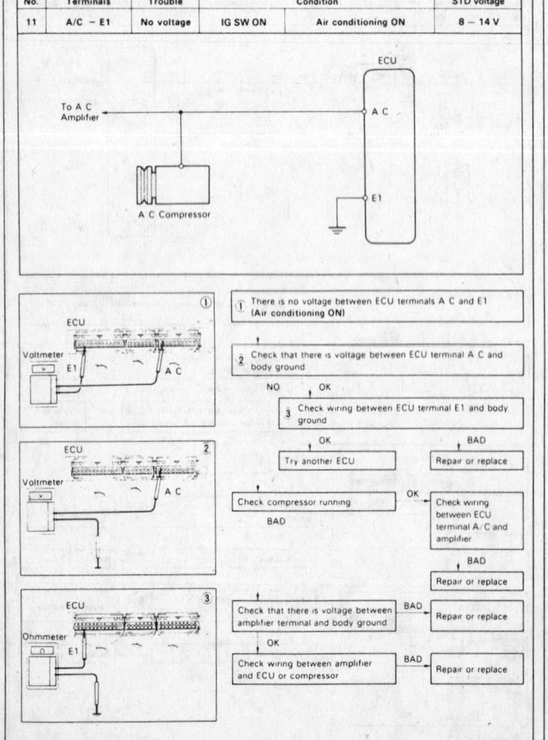

No.	Terminals	Trouble	Condition		STD voltage
11	A/C – E1	No voltage	IG SW ON	Air conditioning ON	8 – 14 V

ECU
To A C Amplifier
A C
A C Compressor
E1
Voltmeter
Ohmmeter
① There is no voltage between ECU terminals A C and E1 (Air conditioning ON)
2 Check that there is voltage between ECU terminal A C and body ground
NO
OK
3 Check wiring between ECU terminal E1 and body ground
OK
BAD
Try another ECU
Repair or replace
Check compressor running
OK
BAD
Check wiring between ECU terminal A/C and amplifier
BAD
Repair or replace
Check that there is voltage between amplifier terminal and body ground
BAD
Repair or replace
OK
Check wiring between amplifier and ECU or compressor
BAD
Repair or replace

TROUBLESHOOTING OXYGEN SENSOR—CAMRY 2VZ-FE

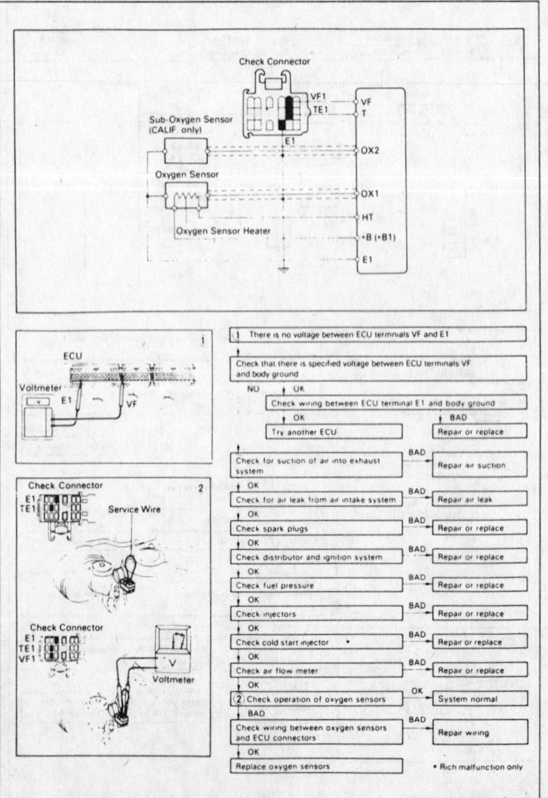

Check Connector
VF1
TE1
VF
T
E1
Sub-Oxygen Sensor (CALIF. only)
OX2
Oxygen Sensor
OX1
HT
Oxygen Sensor Heater
+B (+B1)
E1
ECU
Voltmeter
VF
Check Connector
Service Wire
E1
TE1
VF1
1 There is no voltage between ECU terminals VF and E1
Check that there is specified voltage between ECU terminals VF and body ground
NO
OK
Check wiring between ECU terminal E1 and body ground
OK
BAD
Try another ECU
Repair or replace
Check for suction of air into exhaust system
BAD
Repair air suction
OK
Check for air leak from air intake system
BAD
Repair air leak
OK
Check spark plugs
BAD
Repair or replace
OK
Check distributor and ignition system
BAD
Repair or replace
OK
Check fuel pressure
BAD
Repair or replace
OK
Check injectors
BAD
Repair or replace
OK
Check cold start injector •
BAD
Repair or replace
OK
Check air flow meter
BAD
Repair or replace
OK
② Check operation of oxygen sensors
OK
System normal
BAD
Check wiring between oxygen sensors and ECU connectors
BAD
Repair wiring
OK
Replace oxygen sensors
• Rich malfunction only

TROUBLESHOOTING EGR TEMPERATURE SENSOR—CAMRY 2VZ-FE

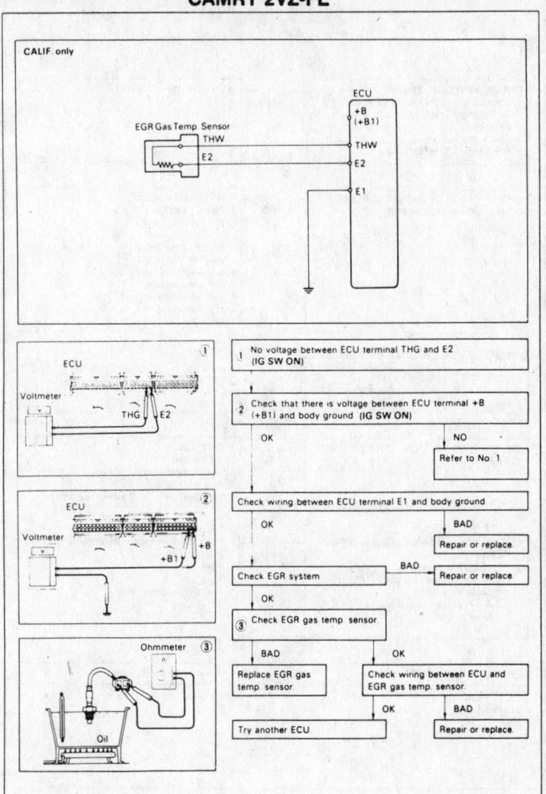
CALIF. only
ECU
EGR Gas Temp. Sensor
THW
E2
+B
(+B1)
THW
E2
E1
ECU
Voltmeter
THG
E2
1
ECU
Voltmeter
+B1
+B
2
Ohmmeter
3
Oil
1 No voltage between ECU terminal THG and E2 (IG SW ON)
2 Check that there is voltage between ECU terminal +B (+B1) and body ground (IG SW ON)
OK
NO
Refer to No. 1
Check wiring between ECU terminal E1 and body ground
OK
BAD
Repair or replace
Check EGR system
BAD
Repair or replace
OK
3 Check EGR gas temp. sensor
BAD
OK
Replace EGR gas temp. sensor
Check wiring between ECU and EGR gas temp. sensor
OK
BAD
Try another ECU
Repair or replace

ECU TERMINAL IDENTIFICATION—CAMRY 2VZ-FE

Symbol	Terminal Name	Symbol	Terminal Name	Symbol	Terminal Name
E01	POWER GROUND	G2	DISTRIBUTOR	A C	A C COMPRESSOR
E02	POWER GROUND	G1	DISTRIBUTOR	*OD1	CRUISE CONTROL COMPUTER
No 10	INJECTOR (No 1 and 6)	NE	DISTRIBUTOR	SP1	No 1 SPEED SENSOR (Meter side)
No 30	INJECTOR (No 4 and 5)	G−	DISTRIBUTOR	*OD2	OD MAIN SWITCH
No 20	INJECTOR (No 2 and 3)	VF	CHECK CONNECTOR	*SP2	No 2 SPEED SENSOR (A T side)
E1	COMPUTER GROUND		–	L1	TEMS ECU
STJ	COLD START INJECTOR	*PWR	PATTERN SELECT SWITCH	*DG	CHECK CONNECTOR
		T	CHECK CONNECTOR	L2	TEMS ECU
		OX1	OXYGEN SENSOR		
ACT	A C AMPLIFIER		–	L3	TEMS ECU
HT	OXYGEN SENSOR HEATER	*OX2	SUB OXYGEN SENSOR	W	WARNING LIGHT
		*BK	BRAKE SWITCH	*N	SIFT POSITION SWITCH
ISC1	ISC MOTOR NO 1 COIL	THW	WATER TEMP SENSOR	M-REL	EFI MAIN RELAY (COIL)
IGT	IGNITER	IDL	THROTTLE POSITION SENSOR	*2	SIFT POSITION SWITCH
ISC2	ISC MOTOR NO 2 COIL	THA	AIR TEMP SENSOR	*R	SIFT POSITION SWITCH
*S1	ECT SOLENOID	VTA	THROTTLE POSITION SENSOR	*L	SIFT POSITION SWITCH
ISC3	ISC MOTOR NO 3 COIL	VS	AIR FLOW METER	IG SW	IGNITION SWITCH
*S2	ECT SOLENOID	*THG	EGR GAS TEMP SENSOR	B1	EFI MAIN RELAY
ISC4	ISC MOTOR NO 4 COIL	VC	AIR FLOW METER THROTTLE POSITION SENSOR	BATT	BATTERY
*SL	ECT SOLENOID	E2	SENSOR GROUND	B	EFI MAIN RELAY
IGF	IGNITER	STA	STARTER SWITCH		
	–	*NSW	NEUTRAL START SWITCH		

*1 w/ ECT
*2 CALIF only

ECU Terminals

ECU TERMINAL VOLTAGES—CAMRY 2VZ-FE

Terminals	Condition		STD voltage (V)
BATT — E1	—		10 — 14
IG SW — E1	IG SW ON		
M-REL — E1			
+B / +B1 — E1			
IDL — E2	IG SW ON	Throttle valve open	4 — 6
VTA — E2		Throttle valve fully closed	0.1 — 1.0
		Throttle valve open	3.2 — 4.2
VC — E2		—	4 — 6
VS — E2		Measuring plate fully closed	3.7 — 4.3
		Measuring plate fully open	0.2 — 0.5
	Idling		1.6 — 4.1
	3,000 rpm		1.0 — 2.0
No.10 / No.20 / No.30 — E01 / E02	IG SW ON		10 — 14
THA — E2	IG SW ON	Intake air temp. 20°C (68°F)	1 — 3
THW — E2		Coolant temp. 80°C (176°F)	0.1 — 1.0
STA — E1	Cranking		6 — 14
IGT — E1	Cranking or idling		0.7 — 1.0
ISC1 / ISC2 / ISC3 / ISC4 — E1	IG SW ON		9 — 14
W — E1	No trouble ("CHECK" engine warning light off) and engine running		10 — 14
*1A/C — E1	IG SW ON	Air conditioning ON	8 — 14
*1ACT — E1		Heater blower SW ON	4 — 6
T — E1		Check connector TE1 — E1 not connected	4 — 6
		Check connector TE1 — E1 connected	0.5 or less
*2NSW — E1		Shift position P or N range	0 — 2
		Ex. shift position P or N range	10 — 14
*2BK — E1	Stop light SW ON (Brake pedal depressed)		10 — 14

ECU Terminals

*1 w/ A/C
*2 w/ ECT

TROUBLESHOOTING PROCEDURES—CELICA 3S-FE, 3S-GE AND 3S-GTE

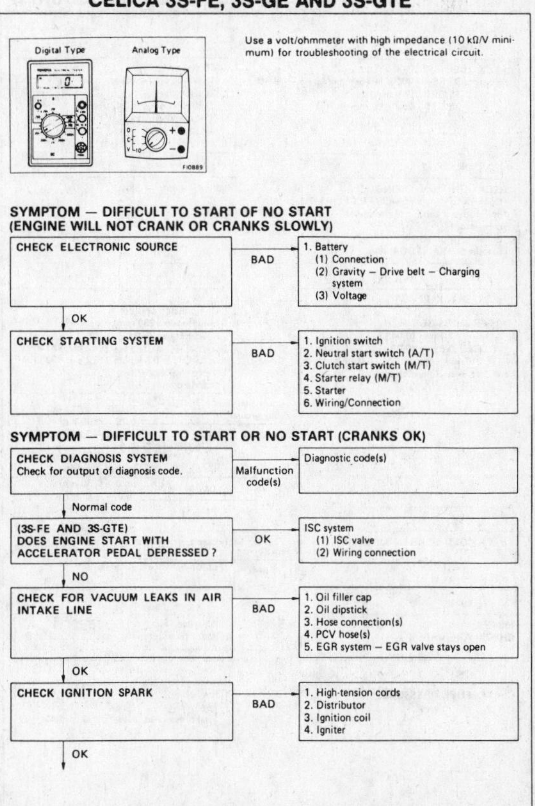
Digital Type
Analog Type
Use a volt/ohmmeter with high impedance (10 kΩ/V minimum) for troubleshooting of the electrical circuit.
SYMPTOM — DIFFICULT TO START OF NO START (ENGINE WILL NOT CRANK OR CRANKS SLOWLY)
CHECK ELECTRONIC SOURCE
BAD
1. Battery
(1) Connection
(2) Gravity – Drive belt – Charging system
(3) Voltage
OK
CHECK STARTING SYSTEM
BAD
1. Ignition switch
2. Neutral start switch (A/T)
3. Clutch start switch (M/T)
4. Starter relay (M/T)
5. Starter
6. Wiring/Connection
SYMPTOM — DIFFICULT TO START OR NO START (CRANKS OK)
CHECK DIAGNOSIS SYSTEM
Check for output of diagnosis code.
Malfunction code(s)
Diagnostic code(s)
Normal code
(3S-FE AND 3S-GTE)
DOES ENGINE START WITH ACCELERATOR PEDAL DEPRESSED?
OK
ISC system
(1) ISC valve
(2) Wiring connection
NO
CHECK FOR VACUUM LEAKS IN AIR INTAKE LINE
BAD
1. Oil filler cap
2. Oil dipstick
3. Hose connection(s)
4. PCV hose(s)
5. EGR system – EGR valve stays open
OK
CHECK IGNITION SPARK
BAD
1. High-tension cords
2. Distributor
3. Ignition coil
4. Igniter
OK

TROUBLESHOOTING PROCEDURES—CELICA 3S-FE, 3S-GE AND 3S-GTE

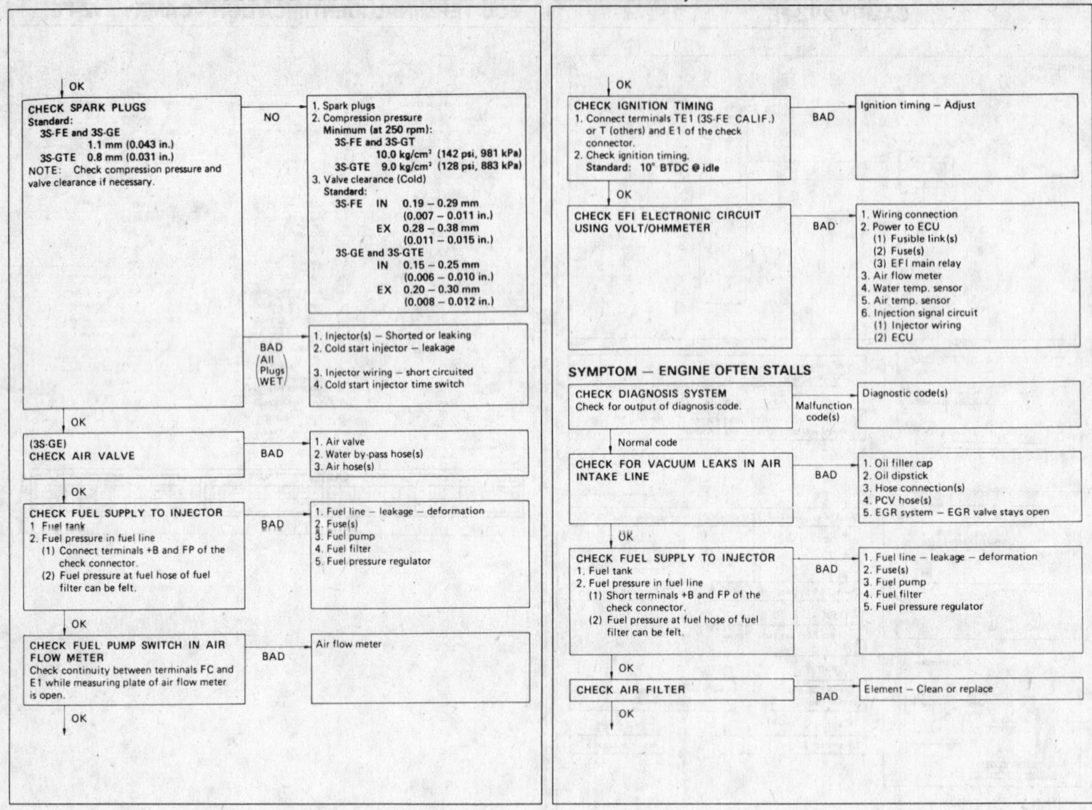
OK
CHECK SPARK PLUGS
Standard:
3S-FE and 3S-GE
1.1 mm (0.043 in.)
3S-GTE 0.8 mm (0.031 in.)
NOTE: Check compression pressure and valve clearance if necessary.
NO
1. Spark plugs
2. Compression pressure
Minimum (at 250 rpm):
3S-FE and 3S-GT
10.0 kg/cm² (142 psi, 981 kPa)
3S-GTE 9.0 kg/cm² (128 psi, 883 kPa)
3. Valve clearance (Cold)
Standard:
3S-FE IN 0.19 – 0.29 mm
(0.007 – 0.011 in.)
EX 0.28 – 0.38 mm
(0.011 – 0.015 in.)
3S-GE and 3S-GTE
IN 0.15 – 0.25 mm
(0.006 – 0.010 in.)
EX 0.20 – 0.30 mm
(0.008 – 0.012 in.)
BAD (All Plugs WET)
1. Injector(s) – Shorted or leaking
2. Cold start injector – leakage
3. Injector wiring – short circuited
4. Cold start injector time switch
OK
(3S-GE)
CHECK AIR VALVE
BAD
1. Air valve
2. Water by-pass hose(s)
3. Air hose(s)
OK
CHECK FUEL SUPPLY TO INJECTOR
1. Fuel tank
2. Fuel pressure in fuel line
(1) Connect terminals +B and FP of the check connector.
(2) Fuel pressure at fuel hose of fuel filter can be felt.
BAD
1. Fuel line – leakage – deformation
2. Fuse(s)
3. Fuel pump
4. Fuel filter
5. Fuel pressure regulator
OK
CHECK FUEL PUMP SWITCH IN AIR FLOW METER
Check continuity between terminals FC and E1 while measuring plate of air flow meter is open.
BAD
Air flow meter
OK
OK
CHECK IGNITION TIMING
1. Connect terminals TE1 (3S-FE CALIF.) or T (others) and E1 of the check connector.
2. Check ignition timing.
Standard: 10° BTDC @ idle
BAD
Ignition timing – Adjust
OK
CHECK EFI ELECTRONIC CIRCUIT USING VOLT/OHMMETER
BAD
1. Wiring connection
2. Power to ECU
(1) Fusible link(s)
(2) Fuse(s)
(3) EFI main relay
3. Air flow meter
4. Water temp. sensor
5. Air temp. sensor
6. Injection signal circuit
(1) Injector wiring
(2) ECU
SYMPTOM — ENGINE OFTEN STALLS
CHECK DIAGNOSIS SYSTEM
Check for output of diagnosis code.
Malfunction code(s)
Diagnostic code(s)
Normal code
CHECK FOR VACUUM LEAKS IN AIR INTAKE LINE
BAD
1. Oil filler cap
2. Oil dipstick
3. Hose connection(s)
4. PCV hose(s)
5. EGR system – EGR valve stays open
OK
CHECK FUEL SUPPLY TO INJECTOR
1. Fuel tank
2. Fuel pressure in fuel line
(1) Short terminals +B and FP of the check connector.
(2) Fuel pressure at fuel hose of fuel filter can be felt.
BAD
1. Fuel line – leakage – deformation
2. Fuse(s)
3. Fuel pump
4. Fuel filter
5. Fuel pressure regulator
OK
CHECK AIR FILTER
BAD
Element – Clean or replace
OK

TROUBLESHOOTING PROCEDURES—CELICA 3S-FE, 3S-GE AND 3S-GTE

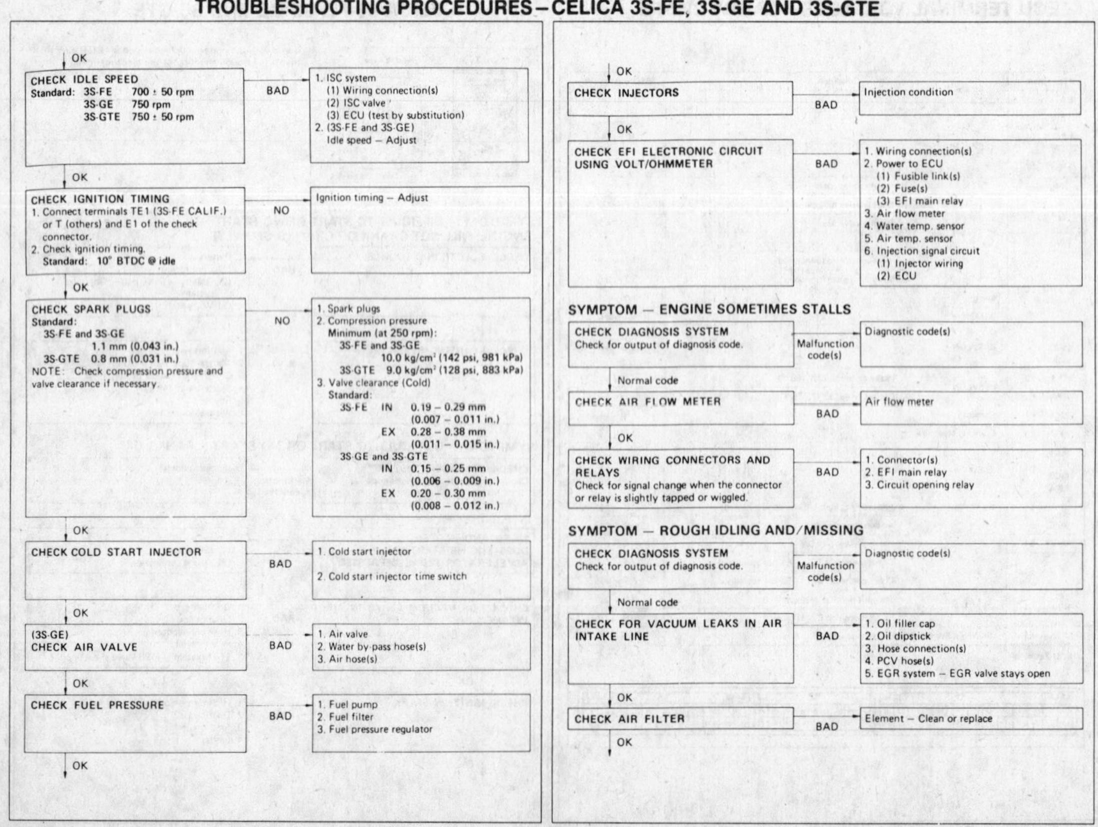
OK
CHECK IDLE SPEED
Standard: 3S-FE 700 ± 50 rpm
3S-GE 750 rpm
3S-GTE 750 ± 50 rpm
BAD
1. ISC system
(1) Wiring connection(s)
(2) ISC valve
(3) ECU (test by substitution)
2. (3S-FE and 3S-GE)
Idle speed – Adjust
OK
CHECK IGNITION TIMING
1. Connect terminals TE1 (3S-FE CALIF.) or T (others) and E1 of the check connector.
2. Check ignition timing.
Standard: 10° BTDC @ idle
NO
Ignition timing – Adjust
OK
CHECK SPARK PLUGS
Standard:
3S-FE and 3S-GE
1.1 mm (0.043 in.)
3S-GTE 0.8 mm (0.031 in.)
NOTE: Check compression pressure and valve clearance if necessary.
NO
1. Spark plugs
2. Compression pressure
Minimum (at 250 rpm):
3S-FE and 3S-GE
10.0 kg/cm² (142 psi, 981 kPa)
3S-GTE 9.0 kg/cm² (128 psi, 883 kPa)
3. Valve clearance (Cold)
Standard:
3S-FE IN 0.19 – 0.29 mm
(0.007 – 0.011 in.)
EX 0.28 – 0.38 mm
(0.011 – 0.015 in.)
3S-GE and 3S-GTE
IN 0.15 – 0.25 mm
(0.006 – 0.009 in.)
EX 0.20 – 0.30 mm
(0.008 – 0.012 in.)
OK
CHECK COLD START INJECTOR
BAD
1. Cold start injector
2. Cold start injector time switch
OK
(3S-GE)
CHECK AIR VALVE
BAD
1. Air valve
2. Water by-pass hose(s)
3. Air hose(s)
OK
CHECK FUEL PRESSURE
BAD
1. Fuel pump
2. Fuel filter
3. Fuel pressure regulator
OK
OK
CHECK INJECTORS
BAD
Injection condition
OK
CHECK EFI ELECTRONIC CIRCUIT USING VOLT/OHMMETER
BAD
1. Wiring connection(s)
2. Power to ECU
(1) Fusible link(s)
(2) Fuse(s)
(3) EFI main relay
3. Air flow meter
4. Water temp. sensor
5. Air temp. sensor
6. Injection signal circuit
(1) Injector wiring
(2) ECU
SYMPTOM — ENGINE SOMETIMES STALLS
CHECK DIAGNOSIS SYSTEM
Check for output of diagnosis code.
Malfunction code(s)
Diagnostic code(s)
Normal code
CHECK AIR FLOW METER
BAD
Air flow meter
OK
CHECK WIRING CONNECTORS AND RELAYS
Check for signal change when the connector or relay is slightly tapped or wiggled.
BAD
1. Connector(s)
2. EFI main relay
3. Circuit opening relay
SYMPTOM — ROUGH IDLING AND/MISSING
CHECK DIAGNOSIS SYSTEM
Check for output of diagnosis code.
Malfunction code(s)
Diagnostic code(s)
Normal code
CHECK FOR VACUUM LEAKS IN AIR INTAKE LINE
BAD
1. Oil filler cap
2. Oil dipstick
3. Hose connection(s)
4. PCV hose(s)
5. EGR system – EGR valve stays open
OK
CHECK AIR FILTER
BAD
Element – Clean or replace
OK

TROUBLESHOOTING PROCEDURES—CELICA 3S-FE, 3S-GE AND 3S-GTE

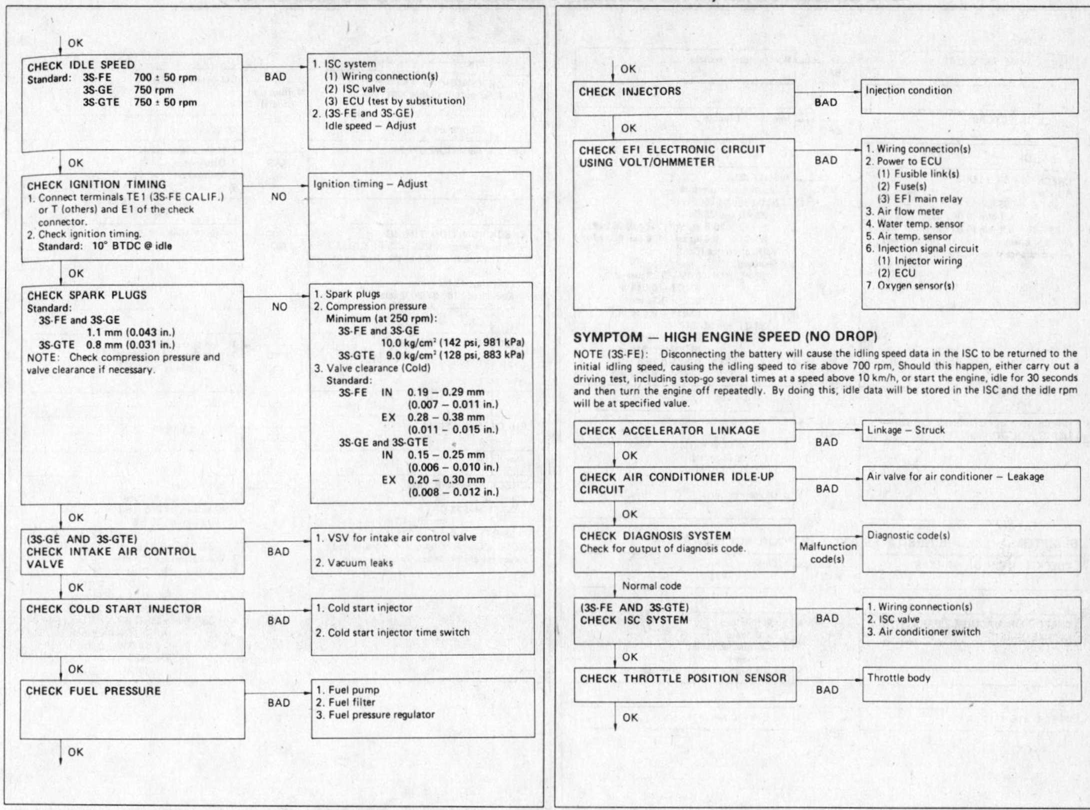
OK
CHECK IDLE SPEED
Standard: 3S-FE 700 ± 50 rpm
3S-GE 750 rpm
3S-GTE 750 ± 50 rpm
BAD
1. ISC system
(1) Wiring connection(s)
(2) ISC valve
(3) ECU (test by substitution)
2. (3S-FE and 3S-GE)
Idle speed – Adjust
OK
CHECK IGNITION TIMING
1. Connect terminals TE1 (3S-FE CALIF.) or T (others) and E1 of the check connector.
2. Check ignition timing.
Standard: 10° BTDC @ idle
NO
Ignition timing – Adjust
OK
CHECK SPARK PLUGS
Standard:
3S-FE and 3S-GE
1.1 mm (0.043 in.)
3S-GTE 0.8 mm (0.031 in.)
NOTE: Check compression pressure and valve clearance if necessary.
NO
1. Spark plugs
2. Compression pressure
Minimum (at 250 rpm):
3S-FE and 3S-GE
10.0 kg/cm² (142 psi, 981 kPa)
3S-GTE 9.0 kg/cm² (128 psi, 883 kPa)
3. Valve clearance (Cold)
Standard:
3S-FE IN 0.19 – 0.29 mm
(0.007 – 0.011 in.)
EX 0.28 – 0.38 mm
(0.011 – 0.015 in.)
3S-GE and 3S-GTE
IN 0.15 – 0.25 mm
(0.006 – 0.010 in.)
EX 0.20 – 0.30 mm
(0.008 – 0.012 in.)
OK
(3S-GE AND 3S-GTE)
CHECK INTAKE AIR CONTROL VALVE
BAD
1. VSV for intake air control valve
2. Vacuum leaks
OK
CHECK COLD START INJECTOR
BAD
1. Cold start injector
2. Cold start injector time switch
OK
CHECK FUEL PRESSURE
BAD
1. Fuel pump
2. Fuel filter
3. Fuel pressure regulator
OK
OK
CHECK INJECTORS
BAD
Injection condition
OK
CHECK EFI ELECTRONIC CIRCUIT USING VOLT/OHMMETER
BAD
1. Wiring connection(s)
2. Power to ECU
(1) Fusible link(s)
(2) Fuse(s)
(3) EFI main relay
3. Air flow meter
4. Water temp. sensor
5. Air temp. sensor
6. Injection signal circuit
(1) Injector wiring
(2) ECU
7. Oxygen sensor(s)
SYMPTOM — HIGH ENGINE SPEED (NO DROP)
NOTE (3S-FE): Disconnecting the battery will cause the idling speed data in the ISC to be returned to the initial idling speed, causing the idling speed to rise above 700 rpm, Should this happen, either carry out a driving test, including stop-go several times at a speed above 10 km/h, or start the engine, idle for 30 seconds and then turn the engine off repeatedly. By doing this, idle data will be stored in the ISC and the idle rpm will be at specified value.
CHECK ACCELERATOR LINKAGE
BAD
Linkage – Struck
OK
CHECK AIR CONDITIONER IDLE-UP CIRCUIT
BAD
Air valve for air conditioner – Leakage
OK
CHECK DIAGNOSIS SYSTEM
Check for output of diagnosis code.
Malfunction code(s)
Diagnostic code(s)
Normal code
(3S-FE AND 3S-GTE)
CHECK ISC SYSTEM
BAD
1. Wiring connection(s)
2. ISC valve
3. Air conditioner switch
OK
CHECK THROTTLE POSITION SENSOR
BAD
Throttle body
OK

TROUBLESHOOTING PROCEDURES—CELICA 3S-FE, 3S-GE AND 3S-GTE

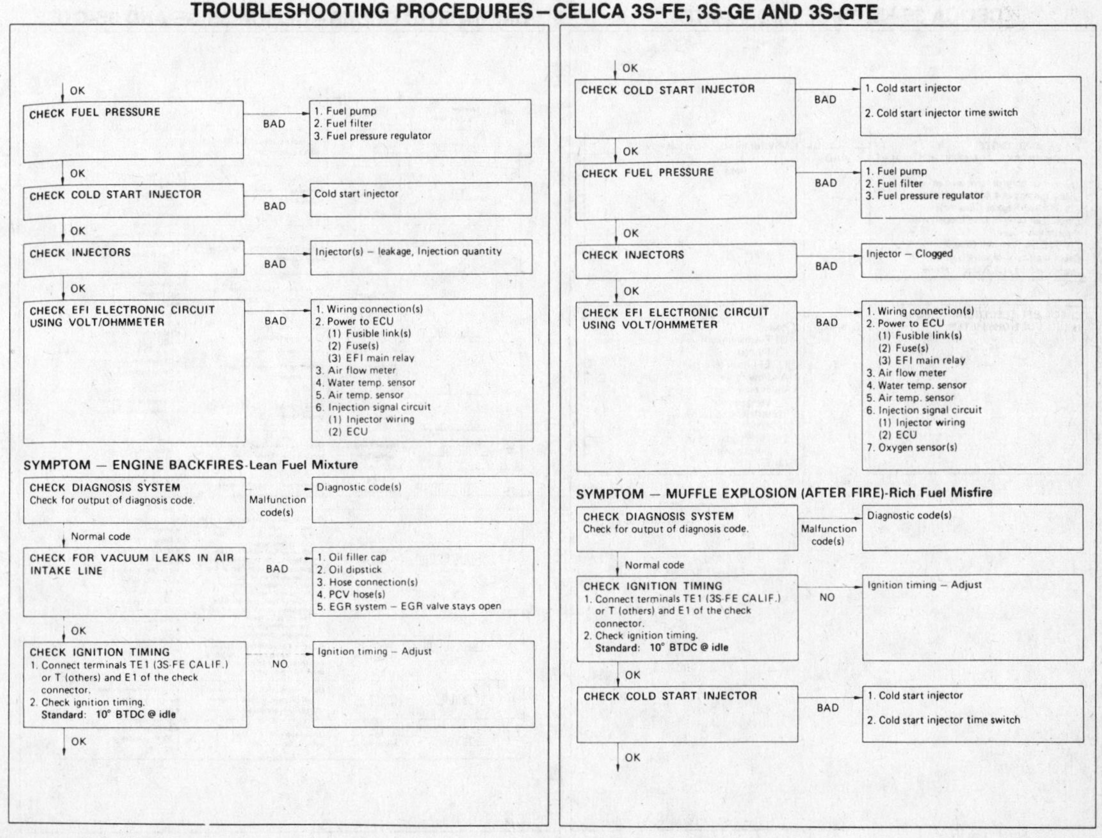
OK
CHECK FUEL PRESSURE
BAD
1. Fuel pump
2. Fuel filter
3. Fuel pressure regulator
OK
CHECK COLD START INJECTOR
BAD
Cold start injector
OK
CHECK INJECTORS
BAD
Injector(s) – leakage, Injection quantity
OK
CHECK EFI ELECTRONIC CIRCUIT USING VOLT/OHMMETER
BAD
1. Wiring connection(s)
2. Power to ECU
(1) Fusible link(s)
(2) Fuse(s)
(3) EFI main relay
3. Air flow meter
4. Water temp. sensor
5. Air temp. sensor
6. Injection signal circuit
(1) Injector wiring
(2) ECU
SYMPTOM — ENGINE BACKFIRES-Lean Fuel Mixture
CHECK DIAGNOSIS SYSTEM
Check for output of diagnosis code.
Malfunction code(s)
Diagnostic code(s)
Normal code
CHECK FOR VACUUM LEAKS IN AIR INTAKE LINE
BAD
1. Oil filler cap
2. Oil dipstick
3. Hose connection(s)
4. PCV hose(s)
5. EGR system – EGR valve stays open
OK
CHECK IGNITION TIMING
1. Connect terminals TE1 (3S-FE CALIF.) or T (others) and E1 of the check connector.
2. Check ignition timing.
Standard: 10° BTDC @ idle
NO
Ignition timing – Adjust
OK
OK
CHECK COLD START INJECTOR
BAD
1. Cold start injector
2. Cold start injector time switch
OK
CHECK FUEL PRESSURE
BAD
1. Fuel pump
2. Fuel filter
3. Fuel pressure regulator
OK
CHECK INJECTORS
BAD
Injector – Clogged
OK
CHECK EFI ELECTRONIC CIRCUIT USING VOLT/OHMMETER
BAD
1. Wiring connection(s)
2. Power to ECU
(1) Fusible link(s)
(2) Fuse(s)
(3) EFI main relay
3. Air flow meter
4. Water temp. sensor
5. Air temp. sensor
6. Injection signal circuit
(1) Injector wiring
(2) ECU
7. Oxygen sensor(s)
SYMPTOM — MUFFLE EXPLOSION (AFTER FIRE)-Rich Fuel Misfire
CHECK DIAGNOSIS SYSTEM
Check for output of diagnosis code.
Malfunction code(s)
Diagnostic code(s)
Normal code
CHECK IGNITION TIMING
1. Connect terminals TE1 (3S-FE CALIF.) or T (others) and E1 of the check connector.
2. Check ignition timing.
Standard: 10° BTDC @ idle
NO
Ignition timing – Adjust
OK
CHECK COLD START INJECTOR
BAD
1. Cold start injector
2. Cold start injector time switch
OK

TROUBLESHOOTING PROCEDURES—CELICA 3S-FE, 3S-GE AND 3S-GTE

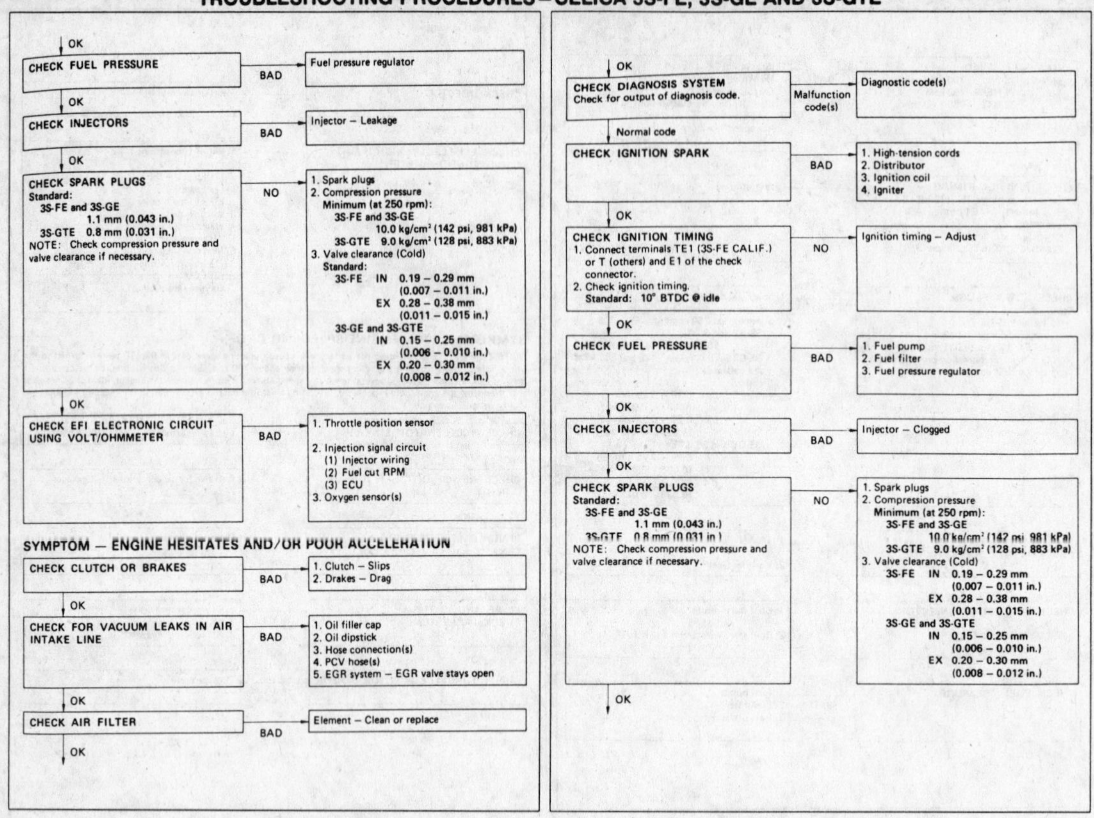
OK
CHECK FUEL PRESSURE
BAD
Fuel pressure regulator
OK
CHECK INJECTORS
BAD
Injector – Leakage
OK
CHECK SPARK PLUGS
Standard:
3S-FE and 3S-GE
1.1 mm (0.043 in.)
3S-GTE 0.8 mm (0.031 in.)
NOTE: Check compression pressure and valve clearance if necessary.
NO
1. Spark plugs
2. Compression pressure
Minimum (at 250 rpm):
3S-FE and 3S-GE
10.0 kg/cm² (142 psi, 981 kPa)
3S-GTE 9.0 kg/cm² (128 psi, 883 kPa)
3. Valve clearance (Cold)
Standard:
3S-FE IN 0.19 – 0.29 mm
(0.007 – 0.011 in.)
EX 0.28 – 0.38 mm
(0.011 – 0.015 in.)
3S-GE and 3S-GTE
IN 0.15 – 0.25 mm
(0.006 – 0.010 in.)
EX 0.20 – 0.30 mm
(0.008 – 0.012 in.)
OK
CHECK EFI ELECTRONIC CIRCUIT USING VOLT/OHMMETER
BAD
1. Throttle position sensor
2. Injection signal circuit
(1) Injector wiring
(2) Fuel cut RPM
(3) ECU
3. Oxygen sensor(s)
SYMPTOM — ENGINE HESITATES AND/OR POOR ACCELERATION
CHECK CLUTCH OR BRAKES
BAD
1. Clutch – Slips
2. Drakes – Drag
OK
CHECK FOR VACUUM LEAKS IN AIR INTAKE LINE
BAD
1. Oil filler cap
2. Oil dipstick
3. Hose connection(s)
4. PCV hose(s)
5. EGR system – EGR valve stays open
OK
CHECK AIR FILTER
BAD
Element – Clean or replace
OK
OK
CHECK DIAGNOSIS SYSTEM
Check for output of diagnosis code.
Malfunction code(s)
Diagnostic code(s)
Normal code
CHECK IGNITION SPARK
BAD
1. High-tension cords
2. Distributor
3. Ignition coil
4. Igniter
OK
CHECK IGNITION TIMING
1. Connect terminals TE1 (3S-FE CALIF.) or T (others) and E1 of the check connector.
2. Check ignition timing.
Standard: 10° BTDC @ idle
NO
Ignition timing – Adjust
OK
CHECK FUEL PRESSURE
BAD
1. Fuel pump
2. Fuel filter
3. Fuel pressure regulator
OK
CHECK INJECTORS
BAD
Injector – Clogged
OK
CHECK SPARK PLUGS
Standard:
3S-FE and 3S-GE
1.1 mm (0.043 in.)
3S-GTE 0.8 mm (0.031 in.)
NOTE: Check compression pressure and valve clearance if necessary.
NO
1. Spark plugs
2. Compression pressure
Minimum (at 250 rpm):
3S-FE and 3S-GE
10.0 kg/cm² (142 psi, 981 kPa)
3S-GTE 9.0 kg/cm² (128 psi, 883 kPa)
3. Valve clearance (Cold)
3S-FE IN 0.19 – 0.29 mm
(0.007 – 0.011 in.)
EX 0.28 – 0.38 mm
(0.011 – 0.015 in.)
3S-GE and 3S-GTE
IN 0.15 – 0.25 mm
(0.006 – 0.010 in.)
EX 0.20 – 0.30 mm
(0.008 – 0.012 in.)
OK

TROUBLESHOOTING PROCEDURES—CELICA 3S-FE, 3S-GE AND 3S-GTE

OK

(3S-GE AND 3S-GTE) CHECK INTAKE AIR CONTROL VALVE
(3S-GE)
Check if air control valve is open with engine running at 4,400 rpm above.
(3S-GTE (w/ Regular Gasoline))
Check if air control valve is open with throttle valve open.
(3S-GTE (w/ Premium Gasoline))
Check if air control valve is open with engine running at 4,200 rpm above.

BAD →
1. VSV for intake air control valve
2. Vacuum leaks

OK

CHECK EFI ELECTRONIC CIRCUIT USING VOLT/OHMMETER

BAD →
1. Wiring connection(s)
2. Power to ECU
   (1) Fusible link(s)
   (2) Fuse(s)
   (3) EFI main relay
3. Air flow meter
4. Water temp. sensor
5. Air temp. sensor
6. Throttle position sensor
7. Injection signal circuit
   (1) Injector wiring
   (2) ECU

DIAGNOSTIC CODES—CELICA 3S-GE AND 3S-GTE

Code No.	Number of check engine blinks	System	Diagnosis	Trouble area
–	ON / OFF	Normal	This appears when none of the other codes are identified.	–
11		ECU (+B)	Momentary interruption in power supply to ECU.	• IG switch circuit • IG switch • Main relay circuit • Main relay • ECU
12		RPM Signal	No NE or G signal to ECU within 2 seconds after engine has been cranked.	• Distributor circuit • Distributor • Starter signal circuit • ECU
13		RPM Signal	No NE signal to ECU when engine speed is above 1,000 rpm.	• Distributor circuit • Distributor • ECU
14		Ignition Signal	No IGF signal to ECU 8 – 11 times in succession.	• Igniter and ignition coil circuit • Igniter and ignition coil • ECU
21		Oxygen Sensor Signal	Detection of oxygen sensor deterioration.	• Oxygen sensor circuit • Oxygen sensor • ECU
		Oxygen Sensor Heater Circuit	Open or short circuit in oxygen sensor heater.	• Oxygen sensor heater circuit • Oxygen sensor heater • ECU
22		Water Temp. Sensor Signal	Open or short circuit in water temp. sensor signal (THW).	• Water temp. sensor circuit • Water temp. sensor • ECU
24		Intake Air Temp. Sensor Signal	Open or short circuit in intake air temp. sensor signal (THA)	• Intake air temp. sensor circuit • Intake air temp. sensor • ECU
25		Air-fuel Ratio Lean Malfunction	(1) When oxygen sensor signal at the upper (rich) or lower (lean) limit for a certain period of time during feedback condition. (2) When air-fuel ratio feedback compensation value or adaptive control value continues at the upper (rich) or lower (rich) limit renewed for a certain period of time. (3) When air-fuel ratio feedback compensation value or adaptive control value feedback frequency is abnormally high during feedback condition. NOTE: For conditions (3), since neither a lean (code No. 25) nor a rich (code No. 26) diagnosis displayed consecutively.	• Injector circuit • Injector • Oxygen sensor circuit • Oxygen sensor • ECU • Fuel line pressure • Air flow meter • Air intake system • Ignition system
26	F12563	Air-fuel Ratio Rich Malfunction		• Injector circuit • Injector • Fuel line pressure • Cold start injector • Air flow meter • ECU
31	F11394	Air-flow Meter Signal	Open circuit in VC signal or short circuit between VC and E2 when idle contacts are closed.	• Air flow meter circuit • Air flow meter • ECU
32	F11395	Air-flow Meter Signal	Open circuit in E2 or short circuit between VC and VS.	• Air flow meter circuit • Air flow meter • ECU

DIAGNOSTIC CODES—CELICA 3S-GE AND 3S-GTE

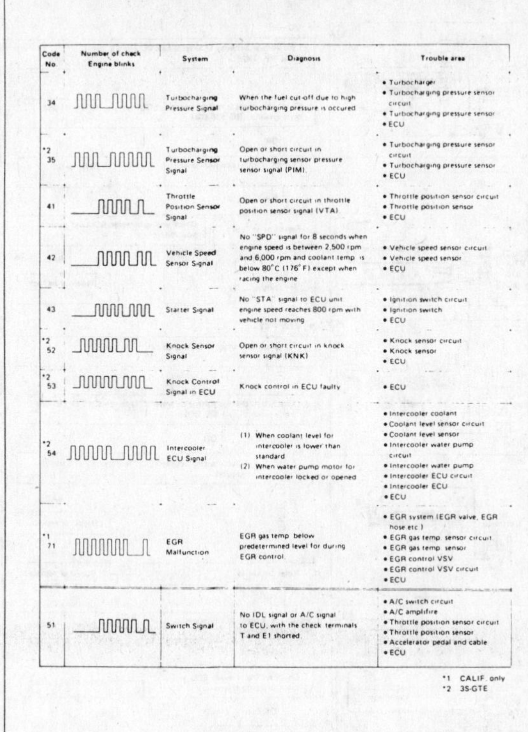

Code No.	Number of check Engine blinks	System	Diagnosis	Trouble area
34		Turbocharging Pressure Signal	When the fuel cut-off due to high turbocharging pressure is occured	• Turbocharger • Turbocharging pressure sensor circuit • Turbocharging pressure sensor • ECU
*2 35		Turbocharging Pressure Sensor Signal	Open or short circuit in turbocharging sensor pressure sensor signal (PIM).	• Turbocharging pressure sensor circuit • Turbocharging pressure sensor • ECU
41		Throttle Position Sensor Signal	Open or short circuit in throttle position sensor signal (VTA)	• Throttle position sensor circuit • Throttle position sensor • ECU
42		Vehicle Speed Sensor Signal	No "SPD" signal for 8 seconds when engine speed is between 2,500 rpm and 6,000 rpm and coolant temp. is below 80°C (176°F) except when racing the engine	• Vehicle speed sensor circuit • Vehicle speed sensor • ECU
43		Starter Signal	No "STA" signal to ECU unit engine speed reaches 800 rpm with vehicle not moving	• Ignition switch circuit • Ignition switch • ECU
*2 52		Knock Sensor Signal	Open or short circuit in knock sensor signal (KNK)	• Knock sensor circuit • Knock sensor • ECU
*2 53		Knock Control Signal in ECU	Knock control in ECU faulty	• ECU
*2 54		Intercooler ECU Signal	(1) When coolant level for intercooler is lower than standard (2) When water pump motor for intercooler locked or opened	• Intercooler coolant • Coolant level sensor circuit • Coolant level sensor • Intercooler water pump circuit • Intercooler water pump • Intercooler ECU circuit • Intercooler ECU • ECU
*1 71		EGR Malfunction	EGR gas temp. below predetermined level for during EGR control.	• EGR system (EGR valve, EGR hose etc.) • EGR gas temp. sensor circuit • EGR gas temp. sensor • EGR control VSV • EGR control VSV circuit • ECU
51		Switch Signal	No IDL signal or A/C signal to ECU, with the check terminals T and E1 shorted.	• A/C switch circuit • A/C amplifire • Throttle position sensor circuit • Throttle position sensor • Accelerator pedal and cable • ECU

*1 CALIF. only
*2 3S-GTE

DIAGNOSTIC CIRCUITS TEST—CELICA

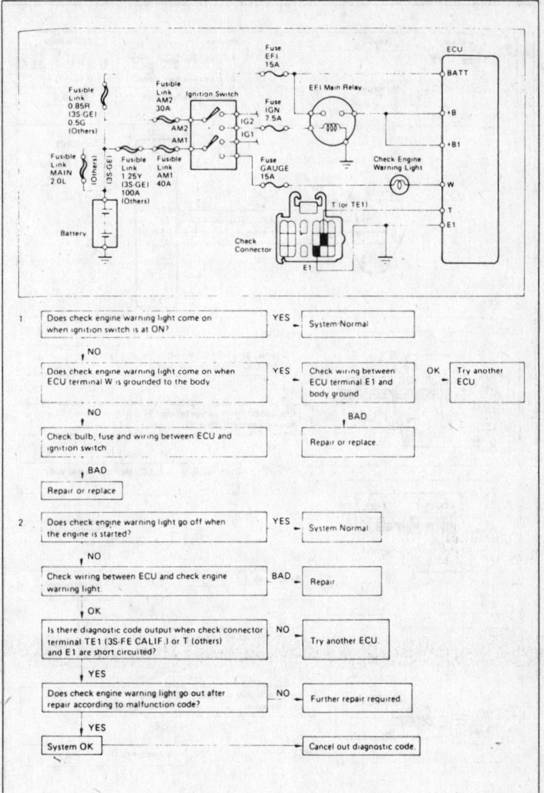
Fuse EFI 15A
ECU
BATT
Fusible Link 0.85R (3S-GE) 0.5G (Others)
Fusible Link AM2 30A
Ignition Switch
EFI Main Relay
Fuse IGN 7.5A
IG2
IG1
AM2
AM1
+B
+B1
Fusible Link MAIN 2.0L
(Others)
(3S-GE)
Fusible Link 1.25Y (3S-GE) 100A (Others)
Fusible Link AM1 40A
Fuse GAUGE 15A
Check Engine Warning Light
W
T (or TE1)
T
E1
Battery
Check Connector
E1
1 Does check engine warning light come on when ignition switch is at ON?
YES
System Normal
NO
Does check engine warning light come on when ECU terminal W is grounded to the body
YES
Check wiring between ECU terminal E1 and body ground
OK
Try another ECU
NO
BAD
Check bulb, fuse and wiring between ECU and ignition switch
Repair or replace.
BAD
Repair or replace
2 Does check engine warning light go off when the engine is started?
YES
System Normal
NO
Check wiring between ECU and check engine warning light.
BAD
Repair.
OK
Is there diagnostic code output when check connector terminal TE1 (3S-FE CALIF.) or T (others) and E1 are short circuited?
NO
Try another ECU.
YES
Does check engine warning light go out after repair according to malfunction code?
NO
Further repair required.
YES
System OK
Cancel out diagnostic code.

TROUBLESHOOTING CHART 1—CELICA 3S-GE AND 3S-GTE

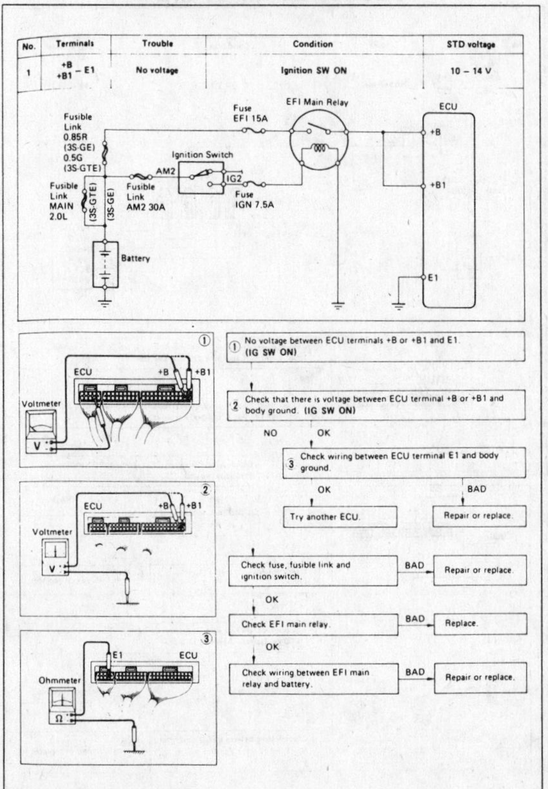

No.	Terminals	Trouble	Condition	STD voltage
1	+B +B1 – E1	No voltage	Ignition SW ON	10 – 14 V

Fusible Link 0.85R (3S-GE) 0.5G (3S-GTE)
Fuse EFI 15A
EFI Main Relay
ECU
+B
Ignition Switch
AM2
IG2
+B1
Fusible Link MAIN 2.0L
(3S-GTE)
(3S-GE)
Fusible Link AM2 30A
Fuse IGN 7.5A
Battery
E1
①
ECU
+B
+B1
Voltmeter
V
②
ECU
+B
+B1
Voltmeter
V
③
E1
ECU
Ohmmeter
Ω
① No voltage between ECU terminals +B or +B1 and E1. (IG SW ON)
2 Check that there is voltage between ECU terminal +B or +B1 and body ground. (IG SW ON)
NO
OK
3 Check wiring between ECU terminal E1 and body ground.
OK
BAD
Try another ECU.
Repair or replace.
Check fuse, fusible link and ignition switch.
BAD
Repair or replace.
OK
Check EFI main relay.
BAD
Replace.
OK
Check wiring between EFI main relay and battery.
BAD
Repair or replace.

TROUBLESHOOTING CHART 2—CELICA 3S-GE AND 3S-GTE

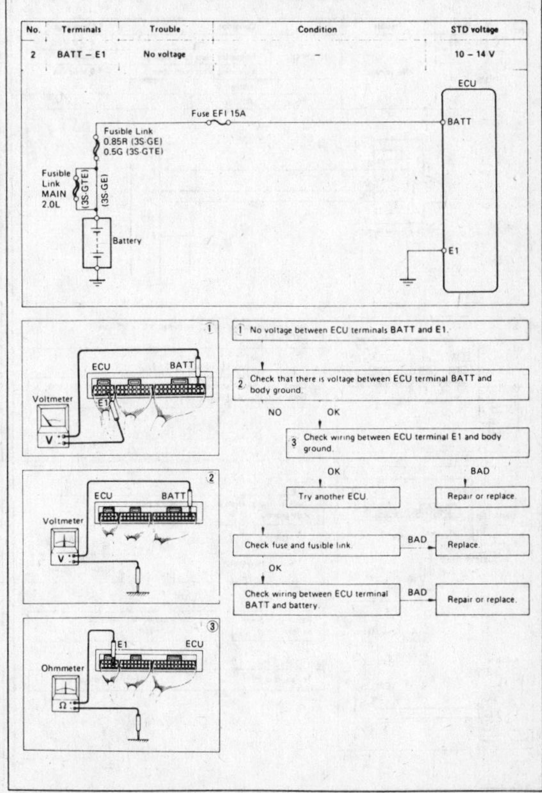

No.	Terminals	Trouble	Condition	STD voltage
2	BATT – E1	No voltage	–	10 – 14 V

ECU
Fuse EFI 15A
BATT
Fusible Link 0.85R (3S-GE) 0.5G (3S-GTE)
Fusible Link MAIN 2.0L
(3S-GTE)
(3S-GE)
Battery
E1
1
ECU
BATT
Voltmeter
E1
V
2
ECU
BATT
Voltmeter
V
3
E1
ECU
Ohmmeter
Ω
1 No voltage between ECU terminals BATT and E1.
2 Check that there is voltage between ECU terminal BATT and body ground.
NO
OK
3 Check wiring between ECU terminal E1 and body ground.
OK
BAD
Try another ECU.
Repair or replace.
Check fuse and fusible link.
BAD
Replace.
OK
Check wiring between ECU terminal BATT and battery.
BAD
Repair or replace.

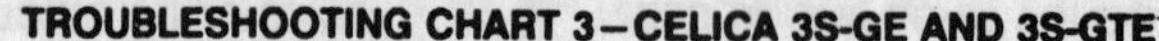

TROUBLESHOOTING CHART 3 – CELICA 3S-GE AND 3S-GTE

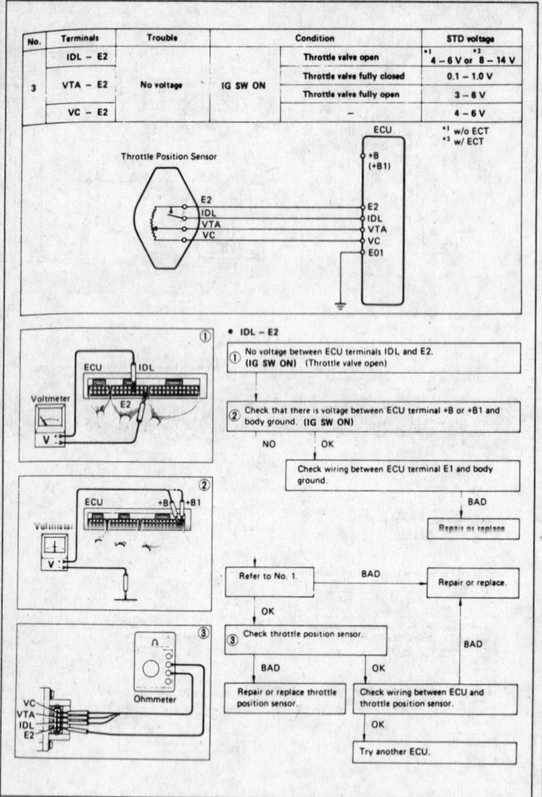

No.	Terminals	Trouble	Condition		STD voltage
3	IDL – E2	No voltage	IG SW ON	Throttle valve open	[*1] 4 – 6 V or [*2] 8 – 14 V
	VTA – E2			Throttle valve fully closed	0.1 – 1.0 V
				Throttle valve fully open	3 – 6 V
	VC – E2			–	4 – 6 V

*1 w/o ECT
*2 w/ ECT

ECU
Throttle Position Sensor
+B
(+B1)
E2
IDL
VTA
VC
E01
Voltmeter
ECU
IDL
E2
• IDL – E2
① No voltage between ECU terminals IDL and E2. (IG SW ON) (Throttle valve open)
② Check that there is voltage between ECU terminal +B or +B1 and body ground. (IG SW ON)
NO
OK
Check wiring between ECU terminal E1 and body ground.
BAD
Repair or replace
Refer to No. 1.
BAD
Repair or replace.
OK
③ Check throttle position sensor.
BAD
BAD
OK
Repair or replace throttle position sensor.
Check wiring between ECU and throttle position sensor.
OK
Try another ECU.
Ohmmeter
VC
VTA
IDL
E2

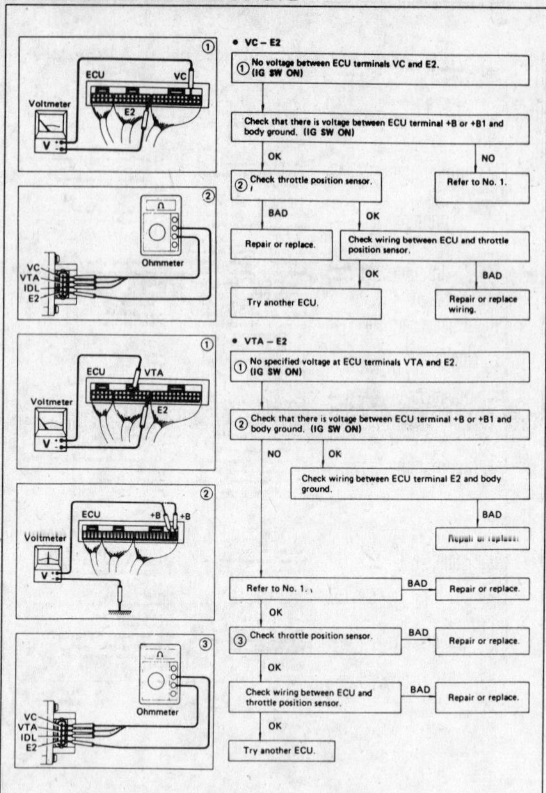

ECU
VC
E2
Voltmeter
• VC – E2
① No voltage between ECU terminals VC and E2. (IG SW ON)
Check that there is voltage between ECU terminal +B or +B1 and body ground. (IG SW ON)
OK
NO
② Check throttle position sensor.
Refer to No. 1.
BAD
OK
Repair or replace.
Check wiring between ECU and throttle position sensor.
OK
BAD
Try another ECU.
Repair or replace wiring.
Ohmmeter
VC
VTA
IDL
E2
• VTA – E2
① No specified voltage at ECU terminals VTA and E2. (IG SW ON)
② Check that there is voltage between ECU terminal +B or +B1 and body ground. (IG SW ON)
NO
OK
Check wiring between ECU terminal E2 and body ground.
BAD
Repair or replace
Refer to No. 1.
BAD
Repair or replace.
OK
③ Check throttle position sensor.
BAD
Repair or replace.
OK
Check wiring between ECU and throttle position sensor.
BAD
Repair or replace.
OK
Try another ECU.

TROUBLESHOOTING CHART 4 – CELICA 3S-GE AND 3S-GTE

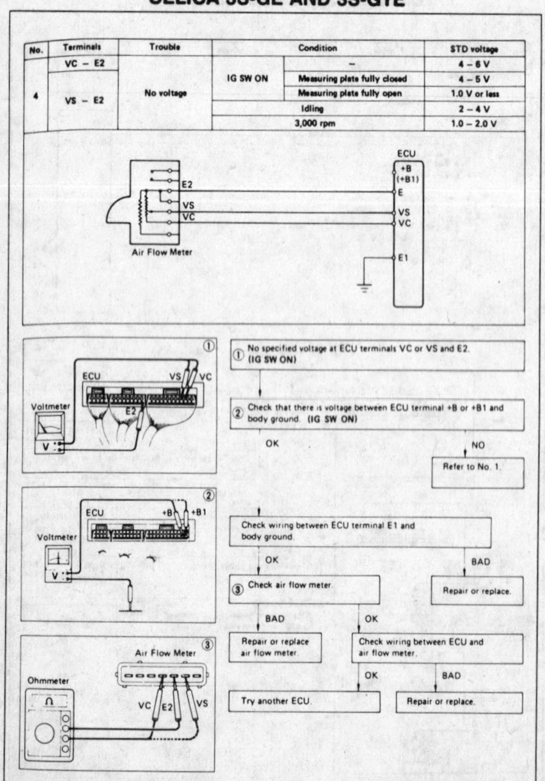

No.	Terminals	Trouble	Condition		STD voltage
4	VC – E2	No voltage	IG SW ON	–	4 – 6 V
	VS – E2			Measuring plate fully closed	4 – 5 V
				Measuring plate fully open	1.0 V or less
			Idling		2 – 4 V
			3,000 rpm		1.0 – 2.0 V

ECU
+B
(+B1)
E
VS
VC
E1
E2
VS
VC
Air Flow Meter
Voltmeter
ECU
VS
VC
E2
① No specified voltage at ECU terminals VC or VS and E2. (IG SW ON)
② Check that there is voltage between ECU terminal +B or +B1 and body ground. (IG SW ON)
OK
NO
Refer to No. 1.
Check wiring between ECU terminal E1 and body ground.
OK
BAD
③ Check air flow meter.
Repair or replace.
BAD
OK
Repair or replace air flow meter.
Check wiring between ECU and air flow meter.
OK
BAD
Try another ECU.
Repair or replace.
Air Flow Meter
Ohmmeter
VC
E2
VS

TROUBLESHOOTING CHART 5 – CELICA 3S-GE AND 3S-GTE

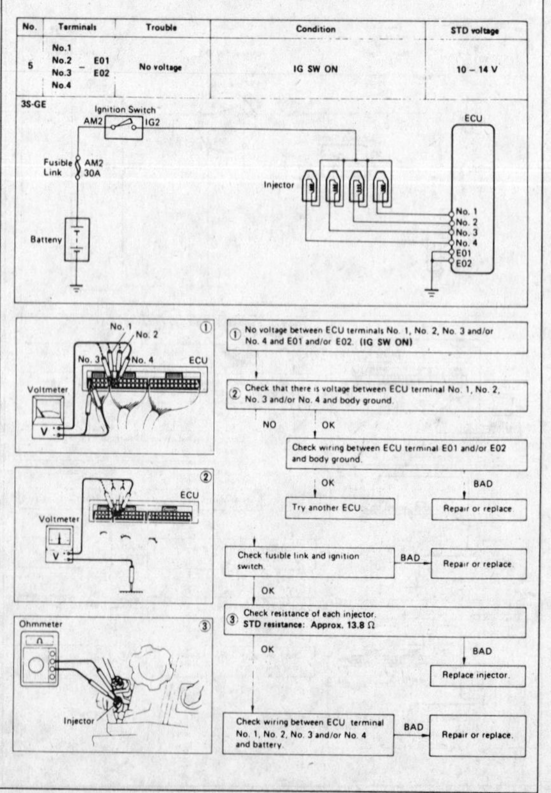

No.	Terminals	Trouble	Condition	STD voltage
5	No.1 No.2 – E01 No.3 – E02 No.4	No voltage	IG SW ON	10 – 14 V

3S-GE
Ignition Switch
AM2
IG2
Fusible Link
AM2
30A
Battery
Injector
ECU
No. 1
No. 2
No. 3
No. 4
E01
E02
Voltmeter
① No voltage between ECU terminals No. 1, No. 2, No. 3 and/or No. 4 and E01 and/or E02. (IG SW ON)
② Check that there is voltage between ECU terminal No. 1, No. 2, No. 3 and/or No. 4 and body ground.
NO
OK
Check wiring between ECU terminal E01 and/or E02 and body ground.
OK
BAD
Try another ECU.
Repair or replace
Check fusible link and ignition switch.
BAD
Repair or replace.
OK
③ Check resistance of each injector. STD resistance: Approx. 13.8 Ω
OK
BAD
Replace injector.
Check wiring between ECU terminal No. 1, No. 2, No. 3 and/or No. 4 and battery.
BAD
Repair or replace.
Ohmmeter
Injector

TROUBLESHOOTING CHART 6—CELICA 3S-GE AND 3S-GTE

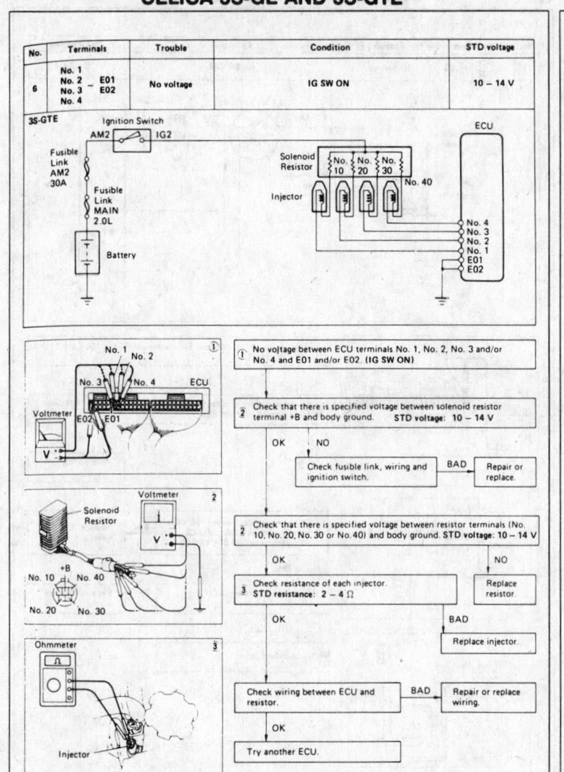

No.	Terminals	Trouble	Condition	STD voltage
6	No. 1 No. 2 — E01 No. 3 — E02 No. 4	No voltage	IG SW ON	10 – 14 V

3S-GTE
Ignition Switch
AM2
IG2
Fusible Link AM2 30A
Fusible Link MAIN 2.0L
Battery
Solenoid Resistor
No. 10
No. 20
No. 30
No. 40
Injector
ECU
No. 4
No. 3
No. 2
No. 1
E01
E02
No voltage between ECU terminals No. 1, No. 2, No. 3 and/or No. 4 and E01 and/or E02. (IG SW ON)
Check that there is specified voltage between solenoid resistor terminal +B and body ground. STD voltage: 10 – 14 V
OK
NO
Check fusible link, wiring and ignition switch.
BAD
Repair or replace.
Check that there is specified voltage between resistor terminals (No. 10, No. 20, No. 30 or No. 40) and body ground. STD voltage: 10 – 14 V
OK
NO
Replace resistor.
Check resistance of each injector. STD resistance: 2 – 4 Ω
OK
BAD
Replace injector.
Check wiring between ECU and resistor.
BAD
Repair or replace wiring.
OK
Try another ECU.
Voltmeter
Ohmmeter
Injector
+B

TROUBLESHOOTING CHART 7—CELICA 3S-GE AND 3S-GTE

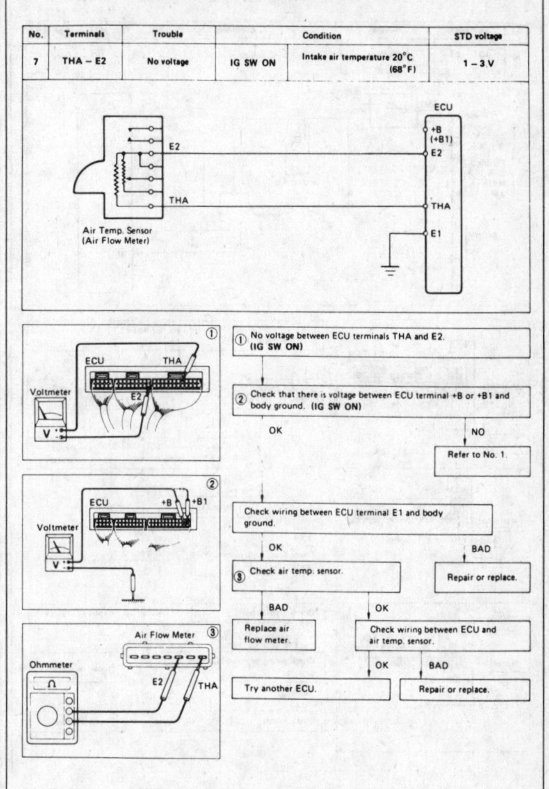

No.	Terminals	Trouble	Condition		STD voltage
7	THA – E2	No voltage	IG SW ON	Intake air temperature 20°C (68°F)	1 – 3 V

ECU
+B (+B1)
E2
THA
E1
Air Temp. Sensor (Air Flow Meter)
No voltage between ECU terminals THA and E2. (IG SW ON)
Check that there is voltage between ECU terminal +B or +B1 and body ground. (IG SW ON)
OK
NO
Refer to No. 1.
Check wiring between ECU terminal E1 and body ground.
OK
BAD
Check air temp. sensor.
Repair or replace.
BAD
OK
Replace air flow meter.
Check wiring between ECU and air temp. sensor.
OK
BAD
Try another ECU.
Repair or replace.
Voltmeter
Ohmmeter
Air Flow Meter

TROUBLESHOOTING CHART 8—CELICA 3S-GE AND 3S-GTE

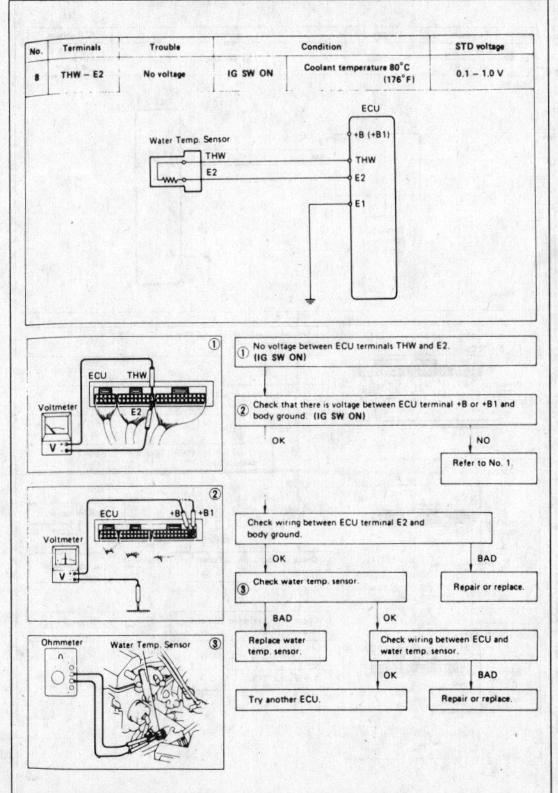

No.	Terminals	Trouble	Condition		STD voltage
8	THW – E2	No voltage	IG SW ON	Coolant temperature 80°C (176°F)	0.1 – 1.0 V

ECU
+B (+B1)
THW
E2
E1
Water Temp. Sensor
No voltage between ECU terminals THW and E2. (IG SW ON)
Check that there is voltage between ECU terminal +B or +B1 and body ground. (IG SW ON)
OK
NO
Refer to No. 1.
Check wiring between ECU terminal E2 and body ground.
OK
BAD
Check water temp. sensor.
Repair or replace.
BAD
OK
Replace water temp. sensor.
Check wiring between ECU and water temp. sensor.
OK
BAD
Try another ECU.
Repair or replace.
Voltmeter
Ohmmeter

TROUBLESHOOTING CHART 9—CELICA 3S-GE AND 3S-GTE

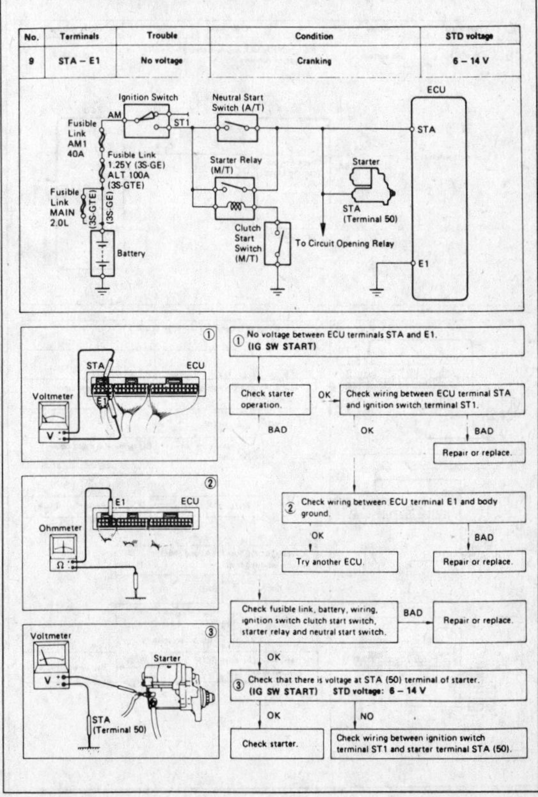

No.	Terminals	Trouble	Condition	STD voltage
9	STA – E1	No voltage	Cranking	6 – 14 V

Ignition Switch
AM
ST1
Neutral Start Switch (A/T)
ECU
STA
Fusible Link AM1 40A
Fusible Link 1.25Y (3S-GE) ALT 100A (3S-GTE)
Starter Relay (M/T)
Starter
Fusible Link MAIN 2.0L
STA (Terminal 50)
Clutch Start Switch (M/T)
To Circuit Opening Relay
Battery
E1
No voltage between ECU terminals STA and E1. (IG SW START)
Check starter operation.
OK
Check wiring between ECU terminal STA and ignition switch terminal ST1.
BAD
OK
BAD
Repair or replace.
Check wiring between ECU terminal E1 and body ground.
OK
BAD
Try another ECU.
Repair or replace.
Check fusible link, battery, wiring, ignition switch clutch start switch, starter relay and neutral start switch.
BAD
Repair or replace.
OK
Check that there is voltage at STA (50) terminal of starter. (IG SW START) STD voltage: 6 – 14 V
OK
NO
Check starter.
Check wiring between ignition switch terminal ST1 and starter terminal STA (50).
Voltmeter
Ohmmeter

TROUBLESHOOTING CHART 10— CELICA 3S-GE AND 3S-GTE

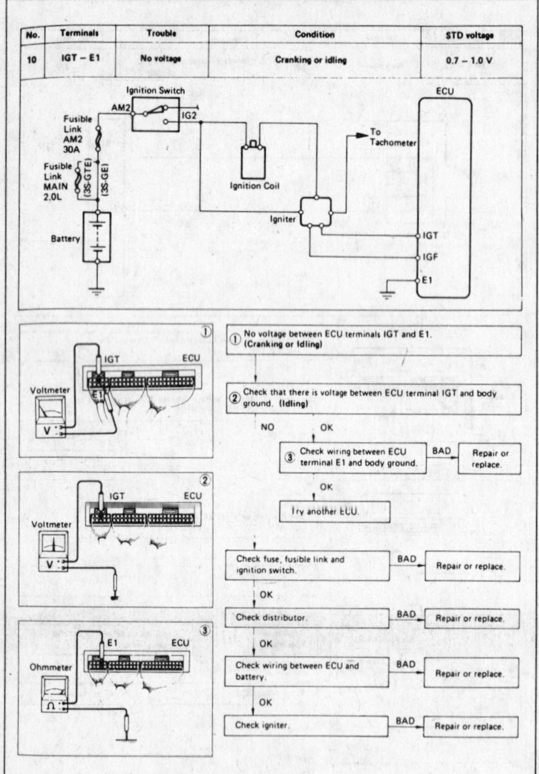
No.
Terminals
Trouble
Condition
STD voltage
10
IGT – E1
No voltage
Cranking or idling
0.7 – 1.0 V
Ignition Switch
AM2
IG2
Fusible Link AM2 30A
Fusible Link MAIN 2.0L
(3S-GTE)
(3S-GE)
Battery
Ignition Coil
To Tachometer
Igniter
ECU
IGT
IGF
E1
IGT
ECU
Voltmeter
E1
V
IGT
ECU
Voltmeter
V
E1
ECU
Ohmmeter
Ω
No voltage between ECU terminals IGT and E1. (Cranking or Idling)
Check that there is voltage between ECU terminal IGT and body ground. (Idling)
NO
OK
Check wiring between ECU terminal E1 and body ground.
BAD
Repair or replace.
OK
Try another ECU.
Check fuse, fusible link and ignition switch.
BAD
Repair or replace.
OK
Check distributor.
BAD
Repair or replace.
OK
Check wiring between ECU and battery.
BAD
Repair or replace.
OK
Check igniter.
BAD
Repair or replace.

TROUBLESHOOTING CHART 11— CELICA 3S-GE AND 3S-GTE

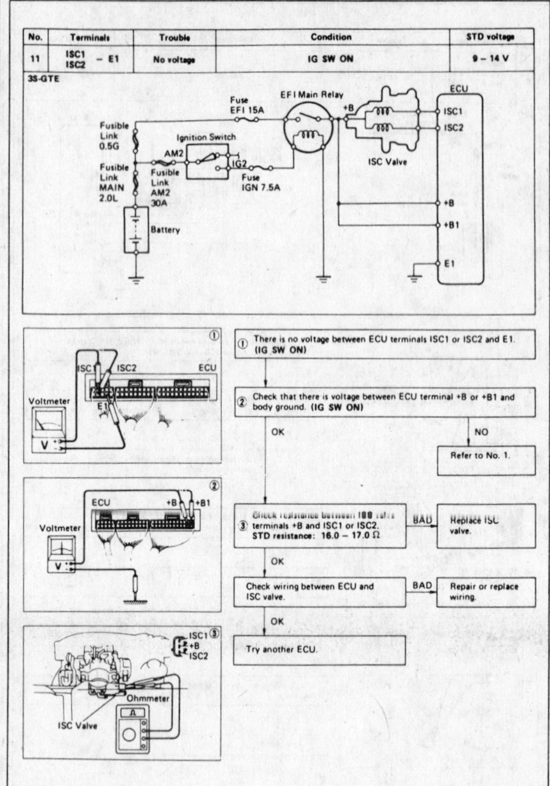
No.
Terminals
Trouble
Condition
STD voltage
11
ISC1 ISC2 – E1
No voltage
IG SW ON
9 – 14 V
3S-GTE
Fuse EFI 15A
EFI Main Relay
ECU
ISC1
ISC2
Fusible Link 0.5G
Ignition Switch
AM2
IG2
Fusible Link MAIN 2.0L
Fusible Link AM2 30A
Fuse IGN 7.5A
ISC Valve
+B
+B1
E1
Battery
ISC1
ISC2
ECU
Voltmeter
E1
V
ECU
+B
+B1
Voltmeter
V
ISC1
+B
ISC2
Ohmmeter
ISC Valve
There is no voltage between ECU terminals ISC1 or ISC2 and E1. (IG SW ON)
Check that there is voltage between ECU terminal +B or +B1 and body ground. (IG SW ON)
OK
NO
Refer to No. 1.
Check resistance between ISC valve terminals +B and ISC1 or ISC2. STD resistance: 16.0 – 17.0 Ω
BAD
Replace ISC valve.
OK
Check wiring between ECU and ISC valve.
BAD
Repair or replace wiring.
OK
Try another ECU.

TROUBLESHOOTING CHART 12— CELICA 3S-GE AND 3S-GTE

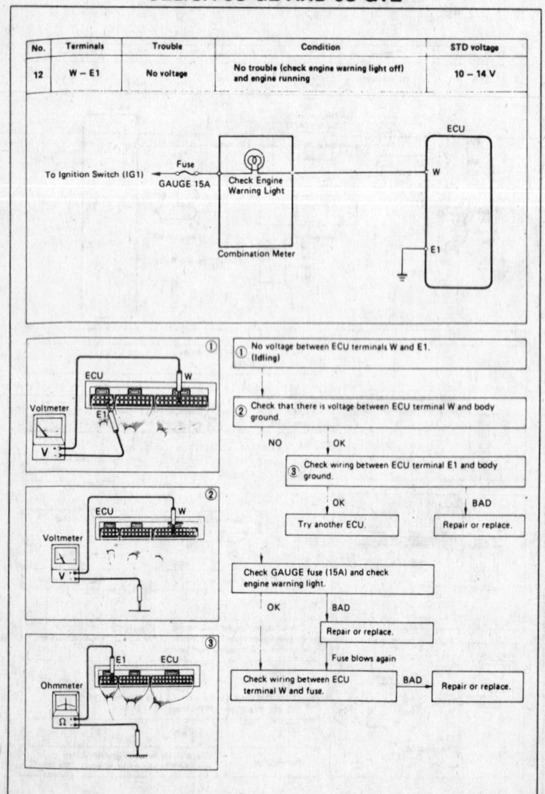
No.
Terminals
Trouble
Condition
STD voltage
12
W – E1
No voltage
No trouble (check engine warning light off) and engine running
10 – 14 V
To Ignition Switch (IG1)
Fuse GAUGE 15A
Check Engine Warning Light
Combination Meter
ECU
W
E1
ECU
W
Voltmeter
E1
V
ECU
W
Voltmeter
V
E1
ECU
Ohmmeter
Ω
No voltage between ECU terminals W and E1. (Idling)
Check that there is voltage between ECU terminal W and body ground.
NO
OK
Check wiring between ECU terminal E1 and body ground.
OK
BAD
Try another ECU.
Repair or replace.
Check GAUGE fuse (15A) and check engine warning light.
OK
BAD
Repair or replace.
Fuse blows again
Check wiring between ECU terminal W and fuse.
BAD
Repair or replace.

TROUBLESHOOTING CHART 13—CELICA 3S-GTE

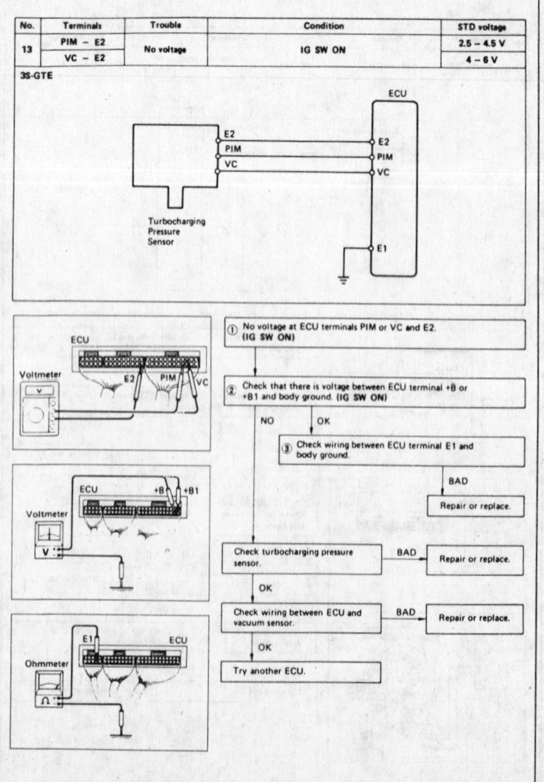
No.
Terminals
Trouble
Condition
STD voltage
13
PIM – E2
VC – E2
No voltage
IG SW ON
2.5 – 4.5 V
4 – 6 V
3S-GTE
ECU
E2
PIM
VC
Turbocharging Pressure Sensor
E1
ECU
Voltmeter
E2
PIM
VC
ECU
+B
+B1
Voltmeter
V
E1
ECU
Ohmmeter
No voltage at ECU terminals PIM or VC and E2. (IG SW ON)
Check that there is voltage between ECU terminal +B or +B1 and body ground. (IG SW ON)
NO
OK
Check wiring between ECU terminal E1 and body ground.
BAD
Repair or replace.
Check turbocharging pressure sensor.
BAD
Repair or replace.
OK
Check wiring between ECU and vacuum sensor.
BAD
Repair or replace.
OK
Try another ECU.

TROUBLESHOOTING CHART 14—CELICA 3S-GE AND 3S-GTE

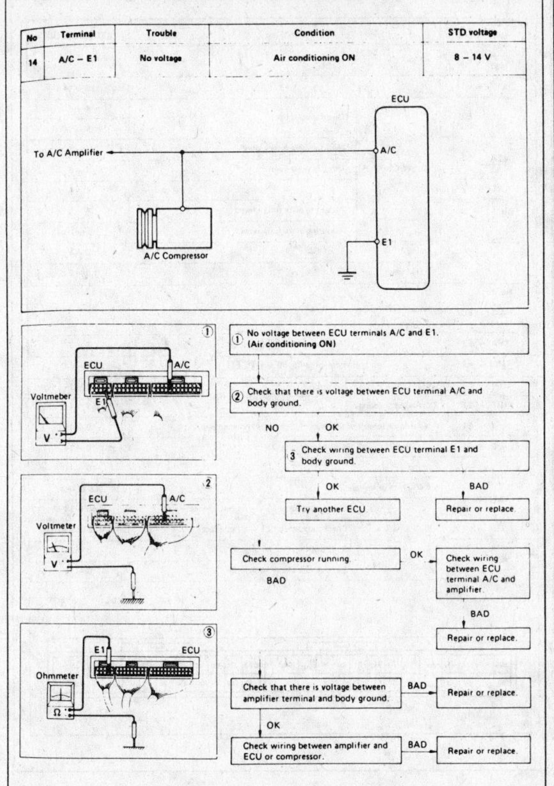
No | Terminal | Trouble | Condition | STD voltage
14 | A/C – E1 | No voltage | Air conditioning ON | 8 – 14 V
ECU
To A/C Amplifier
A/C
A/C Compressor
E1
1 No voltage between ECU terminals A/C and E1. (Air conditioning ON)
2 Check that there is voltage between ECU terminal A/C and body ground.
NO
OK
3 Check wiring between ECU terminal E1 and body ground.
OK
BAD
Try another ECU
Repair or replace.
Check compressor running.
OK
Check wiring between ECU terminal A/C and amplifier.
BAD
BAD
Repair or replace.
Check that there is voltage between amplifier terminal and body ground.
BAD
Repair or replace.
OK
Check wiring between amplifier and ECU or compressor.
BAD
Repair or replace.
Voltmeter
Ohmmeter

TROUBLESHOOTING OXYGEN SENSOR—CELICA 3S-GE AND 3S-GTE

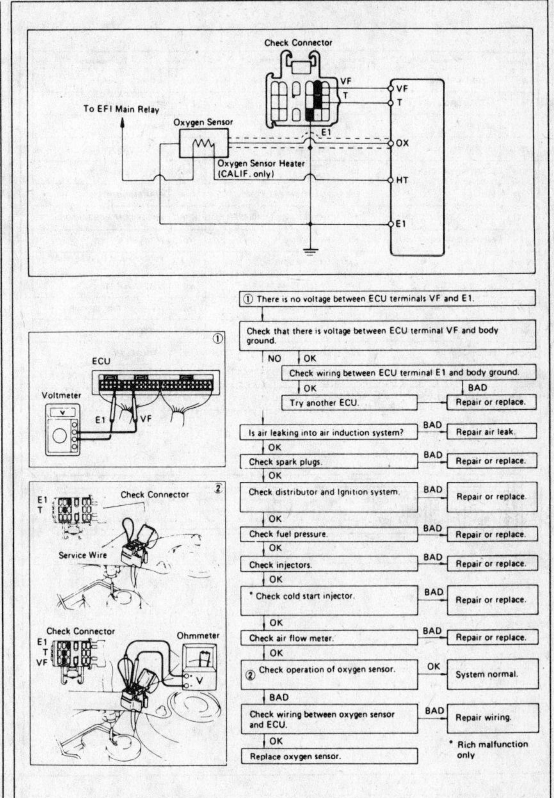
Check Connector
VF
T
To EFI Main Relay
Oxygen Sensor
E1
OX
Oxygen Sensor Heater (CALIF. only)
HT
E1
1 There is no voltage between ECU terminals VF and E1.
Check that there is voltage between ECU terminal VF and body ground.
NO
OK
Check wiring between ECU terminal E1 and body ground.
OK
BAD
Try another ECU.
Repair or replace.
Is air leaking into air induction system?
BAD
Repair air leak.
OK
Check spark plugs.
BAD
Repair or replace.
OK
Check distributor and Ignition system.
BAD
Repair or replace.
OK
Check fuel pressure.
BAD
Repair or replace.
OK
Check injectors.
BAD
Repair or replace.
OK
* Check cold start injector.
BAD
Repair or replace.
OK
Check air flow meter.
BAD
Repair or replace.
OK
2 Check operation of oxygen sensor.
OK
System normal.
BAD
Check wiring between oxygen sensor and ECU.
BAD
Repair wiring.
OK
Replace oxygen sensor.
* Rich malfunction only
ECU
Voltmeter
Service Wire
Ohmmeter

TROUBLESHOOTING EGR TEMPERATURE SENSOR—CELICA 3S-GE AND 3S-GTE

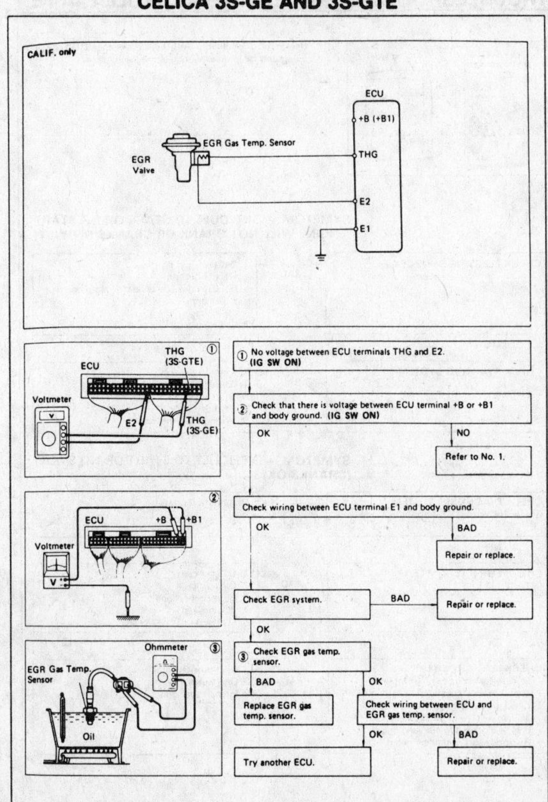
CALIF. only
ECU
+B (+B1)
EGR Gas Temp. Sensor
EGR Valve
THG
E2
E1
1 No voltage between ECU terminals THG and E2. (IG SW ON)
2 Check that there is voltage between ECU terminal +B or +B1 and body ground. (IG SW ON)
OK
NO
Refer to No. 1.
Check wiring between ECU terminal E1 and body ground.
OK
BAD
Repair or replace.
Check EGR system.
BAD
Repair or replace.
OK
3 Check EGR gas temp. sensor.
BAD
OK
Replace EGR gas temp. sensor.
Check wiring between ECU and EGR gas temp. sensor.
OK
BAD
Try another ECU.
Repair or replace.
THG (3S-GTE)
THG (3S-GE)
Voltmeter
Ohmmeter
EGR Gas Temp. Sensor
Oil

TERMINAL IDENTIFICATION—CELICA 3S-GE

Symbol	Terminal name	Symbol	Terminal name	Symbol	Terminal name
E01	ENGINE GROUND	G1	DISTRIBUTOR	L3	ECT ECU
E02	ENGINE GROUND	T	CHECK CONNECTOR	L2	ECT ECU
HT	OXYGEN SENSOR HEATER	G2	DISTRIBUTOR	OD1	ECT ECU
STA	STARTER SWITCH	VTA	THROTTLE POSITION SENSOR	*1 A/C	A/C MAGNET SWITCH
IGT	IGNITER	NE	DISTRIBUTOR	SPD	SPEED SENSOR
STJ	COLD START SWITCH	IDL	THROTTLE POSITION SENSOR	W	WARNING LIGHT
E1	ENGINE GROUND	V-ISC	IDLE-UP VSV	STP	STOP LIGHT SWITCH
NSW	NEUTRAL START SWITCH	IGF	IGNITER	THA	AIR TEMP. SWITCH
T-VIS	T-VIS VSV	OX	OXYGEN SENSOR	VS	AIR FLOW METER
No. 1	No. 1 INJECTOR	THW	WATER TEMP. SWITCH	*2 THG	EGR GAS TEMP. SENSOR
No. 2	No. 2 INJECTOR	E2	SENSOR GROUND	VC	AIR FLOW METER
No. 3	No. 3 INJECTOR	OX1	OXYGEN SENSOR	BATT	BATTERY
No. 4	No. 4 INJECTOR	E22	SENSOR GROUND	+B	MAIN RELAY
G⊖	DISTRIBUTOR	E11	ENGINE GROUND	+B1	MAIN RELAY
VF	CHECK CONNECTOR	L1	ECT ECU		

*1 w/ A/C
*2 CALIF. only

ECU Terminals

E01	HT	STA	STJ	NSW	No. 1	No. 2	G⊖	G1	G2	NE	V-ISC	IGF	THW	OX1	/	L1	L2	/	SPD	STP	THA	VS	VC	BATT	/
E02	/	IGT	E1	T-VIS	No. 3	No. 4	VF	T	VTA	IDL	/	OX	E2	E22	E11	L3	OD1	A/C	W	/	/	*THG	/	+B	+B1

TERMINAL IDENTIFICATION – CELICA 3S-GTE

Symbol	Terminal name	Symbol	Terminal name	Symbol	Terminal name
E01	ENGINE GROUND	T	CHECK CONNECTOR	*1 A/C	A/C MAGNET SWITCH
E02	ENGINE GROUND	G2	DISTRIBUTOR	SPD	SPEED SENSOR
ISC1	ISC VALVE	VTA	THROTTLE POSITION SENSOR	W	WARNING LIGHT
ISC2	ISC VALVE	NE	DISTRIBUTOR	STP	STOP LIGHT SWITCH
STA	STARTER SWITCH	IDL	THROTTLE POSITION SENSOR	FPU	FUEL PRESSURE VSV
IGT	IGNITER	FPR	FUEL PUMP RELAY	THA	AIR TEMP. SWITCH
STJ	COLD START INJECTOR	*2 HT	OXYGEN SENSOR HEATER	KNK	KNOCK CONTROL SENSOR
E1	ENGINE GROUND	IGF	IGNITER	VS	AIR FLOW METER
T-VIS	T-VIS VSV	OX	OXYGEN SENSOR	PIM	TURBOCHARGING PRESSURE SENSOR
No. 1	No. 1 INJECTOR	THW	WATER TEMP. SENSOR	VC	AIR FLOW METER
No. 2	No. 2 INJECTOR	E2	SENSOR GROUND	*2 THG	EGR GAS TEMP. SENSOR
No. 3	No. 3 INJECTOR	WIN	INTERCOOLER ECU	BATT	BATTERY
No. 4	No. 4 INJECTOR	E22	SENSOR GROUND	+B	MAIN RELAY
G⊝	DISTRIBUTOR	TPC	TURBOCHARGING PRESSURE VSV	ELS	HEADLIGHT AND DEFOGGER
VF	CHECK CONNECTOR	EGR	EGR CONTROL VSV	+B1	MAIN RELAY
G1	DISTRIBUTOR				

*1 w/ A/C
*2 CALIF. only

ECU Terminals

E02 ISC2 IGT E1 T-VIS | No. 3 No. 4 VF T VTA IDL HT OX E2 | E22 EGR A/C W FPU KNK PIM *2 THG +B +B1

ECU TERMINAL VOLTAGES – CELICA 3S-GE

Terminals	Condition		STD voltage (V)
+B, +B1 – E1	IG SW ON		10 – 14
BATT – E1	–		10 – 14
IDL – E2	IG SW ON	Throttle valve open	*2 4 – 6
			*3 8 – 14
VTA – E2	IG SW ON	Throttle valve fully closed	0.1 – 1.0
		Throttle valve open	3 – 6
VC – E2	IG SW ON	–	4 – 6
VS – E2	IG SW ON	Measuring plate fully closed	4 – 5
		Measuring plate fully open	1.0 or less
	Idling		2 – 4
	3,000 rpm		1.0 – 2.0
No. 1, No. 2, No. 3, No. 4 – E01, E02	IG SW ON		10 – 14
THA – E2	IG SW ON	Intake air temp. 20°C (68°F)	1 – 3
THW – E2	IG SW ON	Coolant temp. 80°C (176°F)	0.1 – 1.0
STA – E1	Cranking		6 – 14
IGT – E1	Cranking or idling		0.7 – 1.0
W – E1	No trouble (check engine warning light off) and engine running		10 – 14
*1 A/C – E1	IG SW ON	Air conditioning ON	8 – 14
T-VIS – E1	Idling		10 – 14
	4,400 rpm or more		2.0 or less
T – E1	IG SW ON	Check connector TE1 – E1 not connect	10 – 14
		Check connector TE1 – E1 connect	0.5 or less
NSW – E1	IG SW ON	Shift position P or N range	0 – 2
		Ex. shift position P or N range	6 – 14

ECU Terminals

*1 w/ A/C *2 w/o ECT *3 w/ ECT

E01 HT STA STJ NSW | No. 1 No. 2 G⊝ G1 G2 NE V-ISC IGF THW | OX1 L1 L2 SPD STP THA VS VC BATT
E0 IGT E2 T-VIS | No. 3 No. 4 VF T VTA IDL OX E2 | E22 E11 L3 OD1 A/C W *3 THG +B +B1

ECU TERMINAL VOLTAGES – CELICA 3S-GTE

Terminals	Condition		STD voltage (V)
+B, +B1 – E1	IG SW ON		10 – 14
BATT – E1	–		10 – 14
IDL – E2	IG SW ON	Throttle valve open	4 – 6
VTA – E2	IG SW ON	Throttle valve fully closed	0.1 – 1.0
		Throttle valve open	3 – 6
VC – E2	IG SW ON	–	4 – 6
VS – E2	IG SW ON	Measuring plate fully closed	4 – 5
		Measuring plate fully open	1.0 or less
	Idling		2 – 4
	3,000 rpm		1.0 – 2.0
No. 1, No. 2, No. 3, No. 4 – E01, E02	IG SW ON		10 – 14
THA – E2	IG SW ON	Intake air temp. 20°C (68°F)	1 – 3
THW – E2	IG SW ON	Coolant temp. 80°C (176°F)	0.1 – 1.0
STA – E1	Cranking		6 – 14
IGT – E1	Cranking or idling		0.7 – 1.0
ISC1, ISC2 – E1	IG SW ON		9 – 14
W – E1	No trouble (check engine warning light off) and engine running		10 – 14
PIM – E2	IG SW ON		2.5 – 4.5
*1 A/C – E1	IG SW ON	Air conditioning ON	8 – 14
*2 T-VIS – E1	IG SW ON	Throttle valve fully closed	2.0 or less
		Throttle valve open	10 – 14
*2 T-VIS – E1	Idling		2.0 or less
	4,200 rpm or more		10 – 14
T – E1	IG SW ON	Check connector TE1 – E1 not connect	10 – 14
		Check connector TE1 – E1 connect	0.5 or less

ECU Terminals

*1 w/ A/C *2 w/ Regular Gasoline *3 w/ Premium Gasoline

E01 ISC1 STA STJ | No. 1 No. 2 G⊝ G1 G2 NE FPR IGF THW | WIN TPC SPD STP THA VS VC BATT ELS
E02 ISC2 IGT E1 T-VIS | No. 3 No. 4 VF T VTA IDL HT OX E2 | E22 EGR A/C W FPU KNK PIM THG +B +B1

TROUBLESHOOTING PROCEDURES – COROLLA 4A-FE

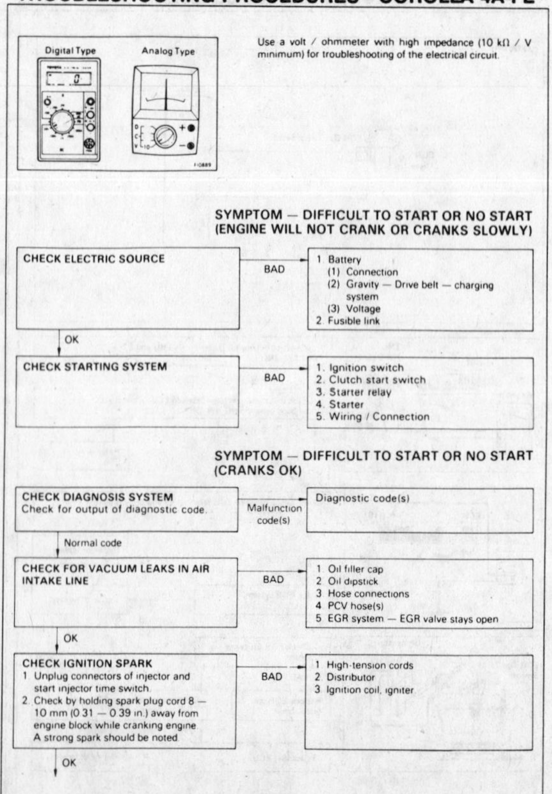
Digital Type
Analog Type
Use a volt / ohmmeter with high impedance (10 kΩ / V minimum) for troubleshooting of the electrical circuit.
SYMPTOM — DIFFICULT TO START OR NO START (ENGINE WILL NOT CRANK OR CRANKS SLOWLY)
CHECK ELECTRIC SOURCE
BAD
1 Battery
(1) Connection
(2) Gravity — Drive belt — charging system
(3) Voltage
2 Fusible link
OK
CHECK STARTING SYSTEM
BAD
1. Ignition switch
2. Clutch start switch
3. Starter relay
4. Starter
5. Wiring / Connection
SYMPTOM — DIFFICULT TO START OR NO START (CRANKS OK)
CHECK DIAGNOSIS SYSTEM
Check for output of diagnostic code.
Malfunction code(s)
Diagnostic code(s)
Normal code
CHECK FOR VACUUM LEAKS IN AIR INTAKE LINE
BAD
1 Oil filler cap
2 Oil dipstick
3 Hose connections
4 PCV hose(s)
5 EGR system — EGR valve stays open
OK
CHECK IGNITION SPARK
1 Unplug connectors of injector and start injector time switch.
2 Check by holding spark plug cord 8 — 10 mm (0.31 — 0.39 in.) away from engine block while cranking engine. A strong spark should be noted.
BAD
1 High-tension cords
2 Distributor
3 Ignition coil, igniter
OK

TROUBLESHOOTING PROCEDURES—COROLLA 4A-FE

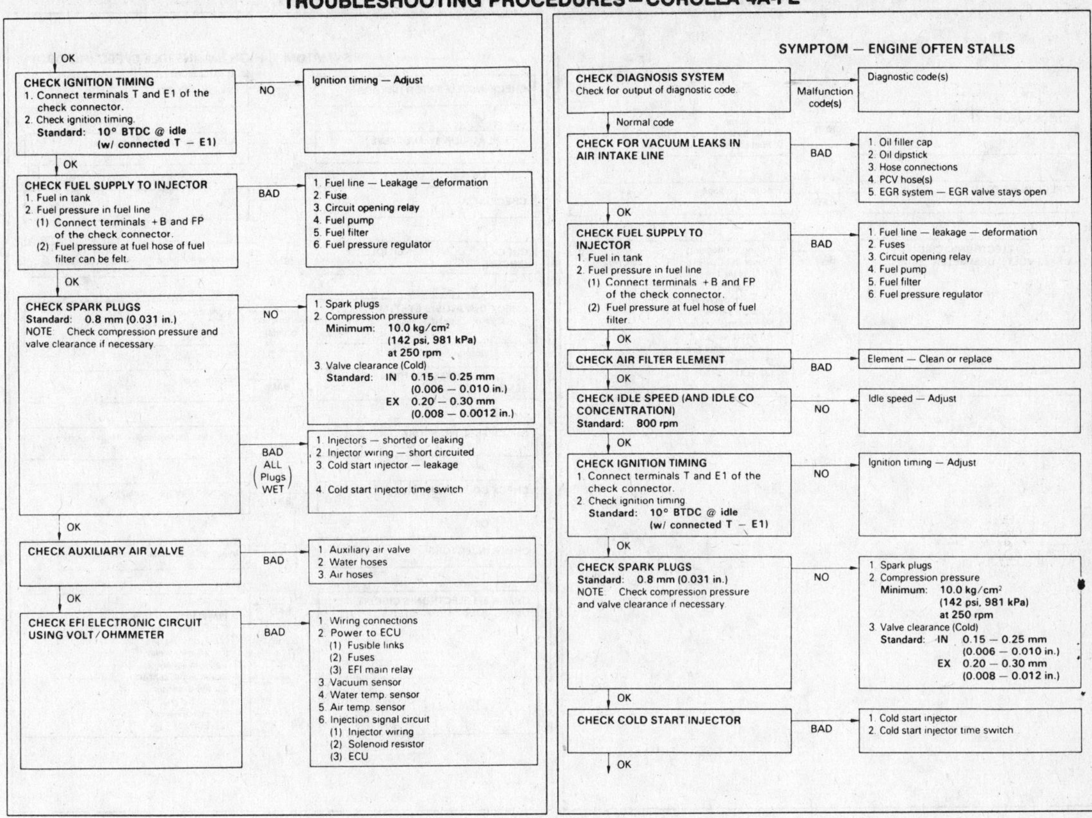
OK
CHECK IGNITION TIMING
1. Connect terminals T and E1 of the check connector.
2. Check ignition timing.
Standard: 10° BTDC @ idle (w/ connected T – E1)
NO
Ignition timing — Adjust
OK
CHECK FUEL SUPPLY TO INJECTOR
1. Fuel in tank
2. Fuel pressure in fuel line
(1) Connect terminals +B and FP of the check connector.
(2) Fuel pressure at fuel hose of fuel filter can be felt.
BAD
1. Fuel line — Leakage — deformation
2. Fuse
3. Circuit opening relay
4. Fuel pump
5. Fuel filter
6. Fuel pressure regulator
OK
CHECK SPARK PLUGS
Standard: 0.8 mm (0.031 in.)
NOTE Check compression pressure and valve clearance if necessary
NO
1. Spark plugs
2. Compression pressure
Minimum: 10.0 kg/cm² (142 psi, 981 kPa) at 250 rpm
3. Valve clearance (Cold)
Standard: IN 0.15 – 0.25 mm (0.006 – 0.010 in.)
EX 0.20 – 0.30 mm (0.008 – 0.0012 in.)
BAD ALL (Plugs WET)
1. Injectors — shorted or leaking
2. Injector wiring — short circuited
3. Cold start injector — leakage
4. Cold start injector time switch
OK
CHECK AUXILIARY AIR VALVE
BAD
1. Auxiliary air valve
2. Water hoses
3. Air hoses
OK
CHECK EFI ELECTRONIC CIRCUIT USING VOLT/OHMMETER
BAD
1. Wiring connections
2. Power to ECU
(1) Fusible links
(2) Fuses
(3) EFI main relay
3. Vacuum sensor
4. Water temp. sensor
5. Air temp. sensor
6. Injection signal circuit
(1) Injector wiring
(2) Solenoid resistor
(3) ECU
SYMPTOM — ENGINE OFTEN STALLS
CHECK DIAGNOSIS SYSTEM
Check for output of diagnostic code
Malfunction code(s)
Diagnostic code(s)
Normal code
CHECK FOR VACUUM LEAKS IN AIR INTAKE LINE
BAD
1. Oil filler cap
2. Oil dipstick
3. Hose connections
4. PCV hose(s)
5. EGR system — EGR valve stays open
OK
CHECK FUEL SUPPLY TO INJECTOR
1. Fuel in tank
2. Fuel pressure in fuel line
(1) Connect terminals +B and FP of the check connector.
(2) Fuel pressure at fuel hose of fuel filter
BAD
1. Fuel line — leakage — deformation
2. Fuses
3. Circuit opening relay
4. Fuel pump
5. Fuel filter
6. Fuel pressure regulator
OK
CHECK AIR FILTER ELEMENT
BAD
Element — Clean or replace
OK
CHECK IDLE SPEED (AND IDLE CO CONCENTRATION)
Standard: 800 rpm
NO
Idle speed — Adjust
OK
CHECK IGNITION TIMING
1. Connect terminals T and E1 of the check connector.
2. Check ignition timing
Standard: 10° BTDC @ idle (w/ connected T – E1)
NO
Ignition timing — Adjust
OK
CHECK SPARK PLUGS
Standard: 0.8 mm (0.031 in.)
NOTE Check compression pressure and valve clearance if necessary
NO
1. Spark plugs
2. Compression pressure
Minimum: 10.0 kg/cm² (142 psi, 981 kPa) at 250 rpm
3. Valve clearance (Cold)
Standard: IN 0.15 – 0.25 mm (0.006 – 0.010 in.)
EX 0.20 – 0.30 mm (0.008 – 0.012 in.)
OK
CHECK COLD START INJECTOR
BAD
1. Cold start injector
2. Cold start injector time switch
OK

TROUBLESHOOTING PROCEDURES—COROLLA 4A-FE

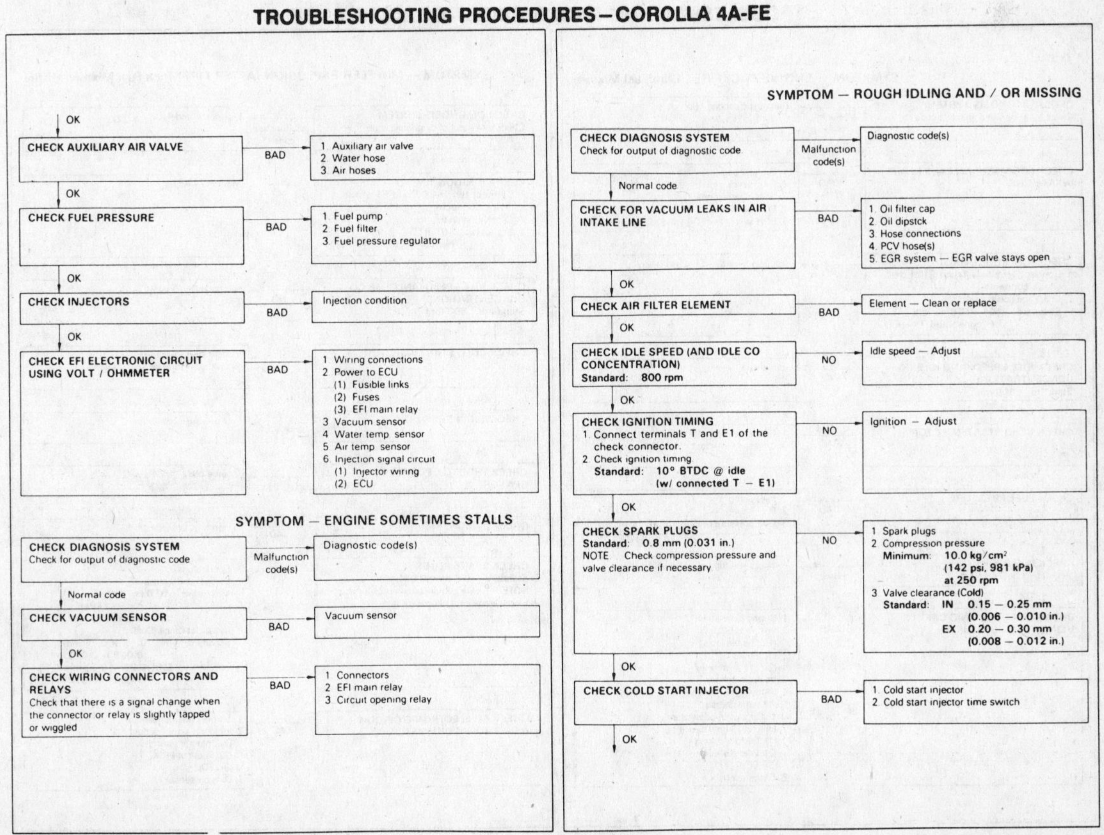
OK
CHECK AUXILIARY AIR VALVE
BAD
1. Auxiliary air valve
2. Water hose
3. Air hoses
OK
CHECK FUEL PRESSURE
BAD
1. Fuel pump
2. Fuel filter
3. Fuel pressure regulator
OK
CHECK INJECTORS
BAD
Injection condition
OK
CHECK EFI ELECTRONIC CIRCUIT USING VOLT / OHMMETER
BAD
1. Wiring connections
2. Power to ECU
(1) Fusible links
(2) Fuses
(3) EFI main relay
3. Vacuum sensor
4. Water temp. sensor
5. Air temp. sensor
6. Injection signal circuit
(1) Injector wiring
(2) ECU
SYMPTOM — ENGINE SOMETIMES STALLS
CHECK DIAGNOSIS SYSTEM
Check for output of diagnostic code
Malfunction code(s)
Diagnostic code(s)
Normal code
CHECK VACUUM SENSOR
BAD
Vacuum sensor
OK
CHECK WIRING CONNECTORS AND RELAYS
Check that there is a signal change when the connector or relay is slightly tapped or wiggled
BAD
1. Connectors
2. EFI main relay
3. Circuit opening relay
SYMPTOM — ROUGH IDLING AND / OR MISSING
CHECK DIAGNOSIS SYSTEM
Check for output of diagnostic code
Malfunction code(s)
Diagnostic code(s)
Normal code
CHECK FOR VACUUM LEAKS IN AIR INTAKE LINE
BAD
1. Oil filter cap
2. Oil dipstick
3. Hose connections
4. PCV hose(s)
5. EGR system — EGR valve stays open
OK
CHECK AIR FILTER ELEMENT
BAD
Element — Clean or replace
OK
CHECK IDLE SPEED (AND IDLE CO CONCENTRATION)
Standard: 800 rpm
NO
Idle speed — Adjust
OK
CHECK IGNITION TIMING
1. Connect terminals T and E1 of the check connector.
2. Check ignition timing
Standard: 10° BTDC @ idle (w/ connected T – E1)
NO
Ignition — Adjust
OK
CHECK SPARK PLUGS
Standard: 0.8 mm (0.031 in.)
NOTE Check compression pressure and valve clearance if necessary
NO
1. Spark plugs
2. Compression pressure
Minimum: 10.0 kg/cm² (142 psi, 981 kPa) at 250 rpm
3. Valve clearance (Cold)
Standard: IN 0.15 – 0.25 mm (0.006 – 0.010 in.)
EX 0.20 – 0.30 mm (0.008 – 0.012 in.)
OK
CHECK COLD START INJECTOR
BAD
1. Cold start injector
2. Cold start injector time switch
OK

TROUBLESHOOTING PROCEDURES—COROLLA 4A-FE

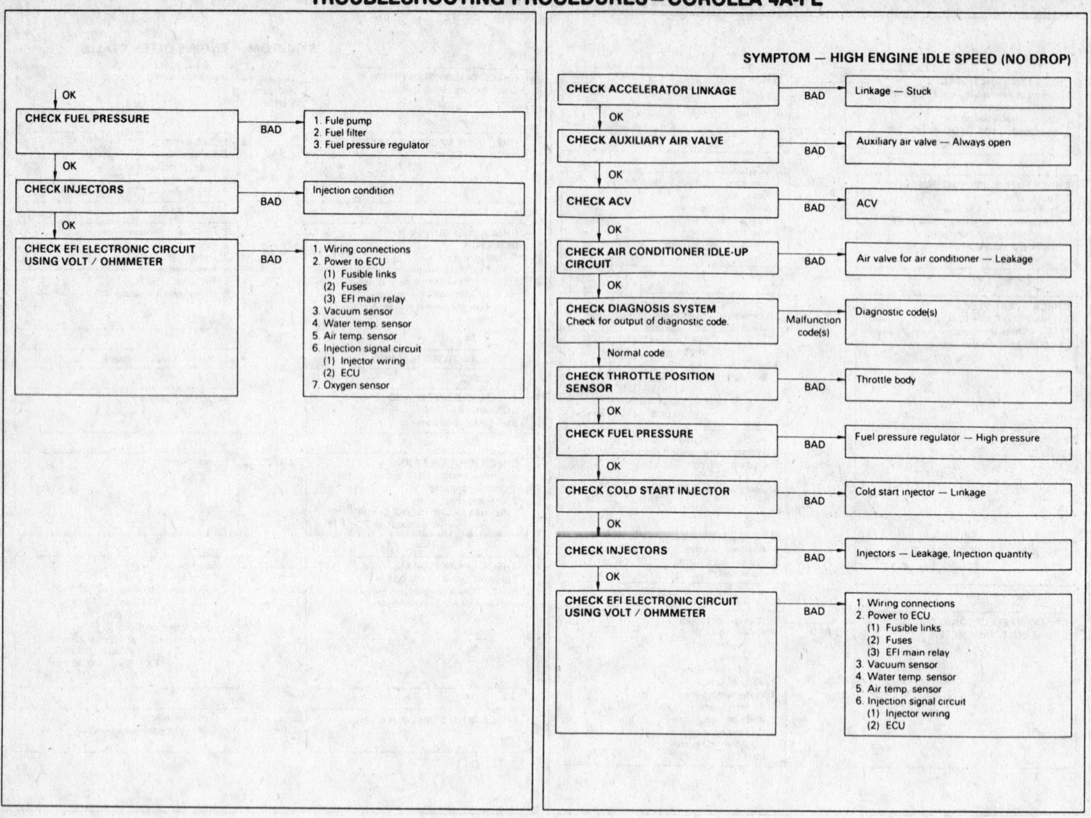
OK
CHECK FUEL PRESSURE
BAD
1. Fule pump
2. Fuel filter
3. Fuel pressure regulator
OK
CHECK INJECTORS
BAD
Injection condition
OK
CHECK EFI ELECTRONIC CIRCUIT USING VOLT / OHMMETER
BAD
1. Wiring connections
2. Power to ECU
(1) Fusible links
(2) Fuses
(3) EFI main relay
3. Vacuum sensor
4. Water temp. sensor
5. Air temp. sensor
6. Injection signal circuit
(1) Injector wiring
(2) ECU
7. Oxygen sensor
SYMPTOM — HIGH ENGINE IDLE SPEED (NO DROP)
CHECK ACCELERATOR LINKAGE
BAD
Linkage — Stuck
OK
CHECK AUXILIARY AIR VALVE
BAD
Auxiliary air valve — Always open
OK
CHECK ACV
BAD
ACV
OK
CHECK AIR CONDITIONER IDLE-UP CIRCUIT
BAD
Air valve for air conditioner — Leakage
OK
CHECK DIAGNOSIS SYSTEM
Check for output of diagnostic code.
Malfunction code(s)
Diagnostic code(s)
Normal code
CHECK THROTTLE POSITION SENSOR
BAD
Throttle body
OK
CHECK FUEL PRESSURE
BAD
Fuel pressure regulator — High pressure
OK
CHECK COLD START INJECTOR
BAD
Cold start injector — Linkage
OK
CHECK INJECTORS
BAD
Injectors — Leakage, Injection quantity
OK
CHECK EFI ELECTRONIC CIRCUIT USING VOLT / OHMMETER
BAD
1. Wiring connections
2. Power to ECU
(1) Fusible links
(2) Fuses
(3) EFI main relay
3. Vacuum sensor
4. Water temp. sensor
5. Air temp. sensor
6. Injection signal circuit
(1) Injector wiring
(2) ECU

TROUBLESHOOTING PROCEDURES—COROLLA 4A-FE

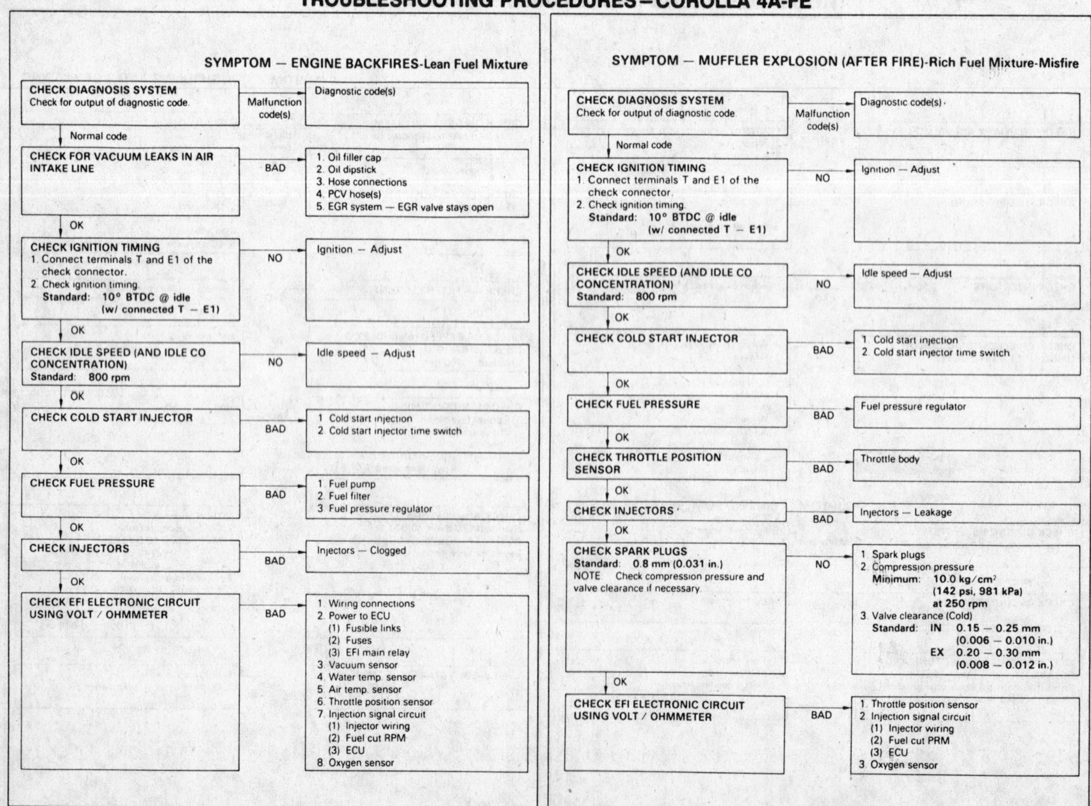
SYMPTOM — ENGINE BACKFIRES-Lean Fuel Mixture
CHECK DIAGNOSIS SYSTEM
Check for output of diagnostic code.
Malfunction code(s)
Diagnostic code(s)
Normal code
CHECK FOR VACUUM LEAKS IN AIR INTAKE LINE
BAD
1. Oil filler cap
2. Oil dipstick
3. Hose connections
4. PCV hose(s)
5. EGR system — EGR valve stays open
OK
CHECK IGNITION TIMING
1. Connect terminals T and E1 of the check connector.
2. Check ignition timing.
Standard: 10° BTDC @ idle (w/ connected T — E1)
NO
Ignition — Adjust
OK
CHECK IDLE SPEED (AND IDLE CO CONCENTRATION)
Standard: 800 rpm
NO
Idle speed — Adjust
OK
CHECK COLD START INJECTOR
BAD
1. Cold start injection
2. Cold start injector time switch
OK
CHECK FUEL PRESSURE
BAD
1. Fuel pump
2. Fuel filter
3. Fuel pressure regulator
OK
CHECK INJECTORS
BAD
Injectors — Clogged
OK
CHECK EFI ELECTRONIC CIRCUIT USING VOLT / OHMMETER
BAD
1. Wiring connections
2. Power to ECU
(1) Fusible links
(2) Fuses
(3) EFI main relay
3. Vacuum sensor
4. Water temp. sensor
5. Air temp. sensor
6. Throttle position sensor
7. Injection signal circuit
(1) Injector wiring
(2) Fuel cut RPM
(3) ECU
8. Oxygen sensor
SYMPTOM — MUFFLER EXPLOSION (AFTER FIRE)-Rich Fuel Mixture-Misfire
CHECK DIAGNOSIS SYSTEM
Check for output of diagnostic code.
Malfunction code(s)
Diagnostic code(s)
Normal code
CHECK IGNITION TIMING
1. Connect terminals T and E1 of the check connector.
2. Check ignition timing.
Standard: 10° BTDC @ idle (w/ connected T — E1)
NO
Ignition — Adjust
OK
CHECK IDLE SPEED (AND IDLE CO CONCENTRATION)
Standard: 800 rpm
NO
Idle speed — Adjust
OK
CHECK COLD START INJECTOR
BAD
1. Cold start injection
2. Cold start injector time switch
OK
CHECK FUEL PRESSURE
BAD
Fuel pressure regulator
OK
CHECK THROTTLE POSITION SENSOR
BAD
Throttle body
OK
CHECK INJECTORS
BAD
Injectors — Leakage
OK
CHECK SPARK PLUGS
Standard: 0.8 mm (0.031 in.)
NOTE Check compression pressure and valve clearance if necessary.
NO
1. Spark plugs
2. Compression pressure
Minimum: 10.0 kg/cm² (142 psi, 981 kPa) at 250 rpm
3. Valve clearance (Cold)
Standard: IN 0.15 — 0.25 mm (0.006 — 0.010 in.)
EX 0.20 — 0.30 mm (0.008 — 0.012 in.)
OK
CHECK EFI ELECTRONIC CIRCUIT USING VOLT / OHMMETER
BAD
1. Throttle position sensor
2. Injection signal circuit
(1) Injector wiring
(2) Fuel cut PRM
(3) ECU
3. Oxygen sensor

TROUBLESHOOTING PROCEDURES—COROLLA 4A-FE

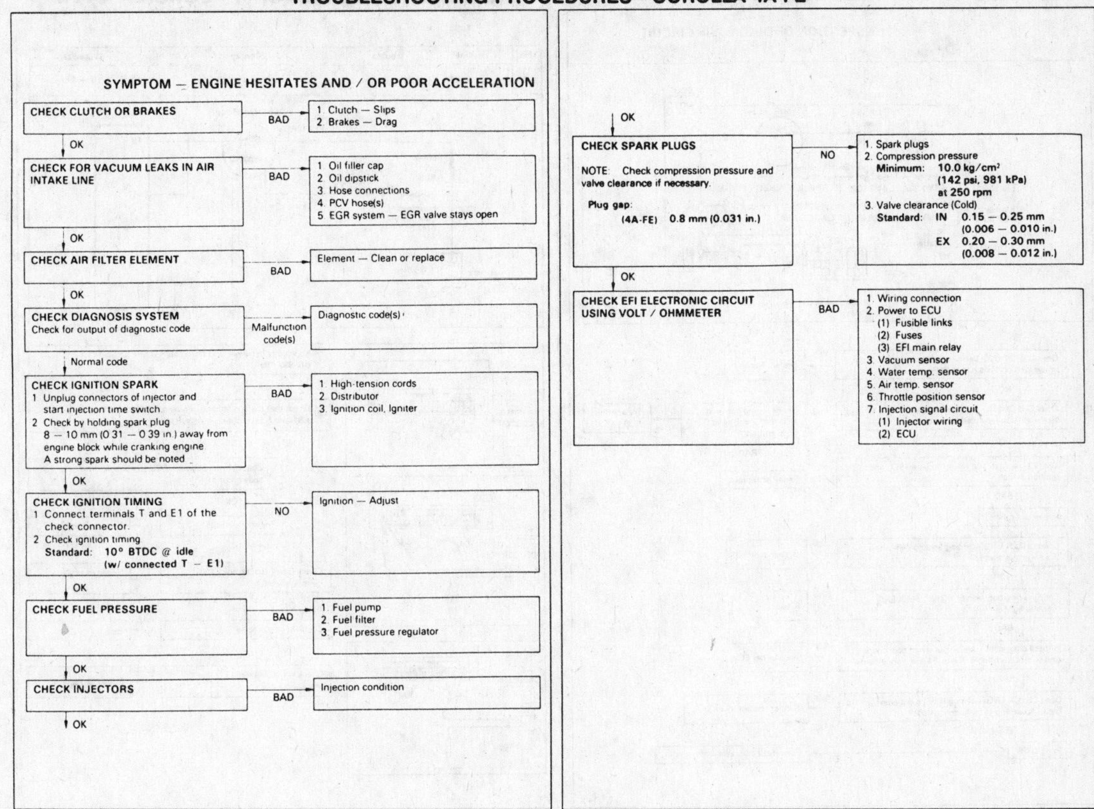
SYMPTOM — ENGINE HESITATES AND / OR POOR ACCELERATION
CHECK CLUTCH OR BRAKES
BAD
1 Clutch — Slips
2 Brakes — Drag
OK
CHECK FOR VACUUM LEAKS IN AIR INTAKE LINE
BAD
1 Oil filler cap
2 Oil dipstick
3 Hose connections
4 PCV hose(s)
5 EGR system — EGR valve stays open
OK
CHECK AIR FILTER ELEMENT
BAD
Element — Clean or replace
OK
CHECK DIAGNOSIS SYSTEM
Check for output of diagnostic code
Malfunction code(s)
Diagnostic code(s)
Normal code
CHECK IGNITION SPARK
1 Unplug connectors of injector and start injection time switch
2 Check by holding spark plug 8 – 10 mm (0.31 – 0.39 in.) away from engine block while cranking engine
A strong spark should be noted
BAD
1 High-tension cords
2 Distributor
3 Ignition coil, Igniter
OK
CHECK IGNITION TIMING
1 Connect terminals T and E1 of the check connector.
2 Check ignition timing
Standard: 10° BTDC @ idle (w/ connected T – E1)
NO
Ignition — Adjust
OK
CHECK FUEL PRESSURE
BAD
1 Fuel pump
2 Fuel filter
3 Fuel pressure regulator
OK
CHECK INJECTORS
BAD
Injection condition
OK
OK
CHECK SPARK PLUGS
NOTE: Check compression pressure and valve clearance if necessary.
Plug gap:
(4A-FE) 0.8 mm (0.031 in.)
NO
1. Spark plugs
2. Compression pressure
Minimum: 10.0 kg/cm² (142 psi, 981 kPa) at 250 rpm
3. Valve clearance (Cold)
Standard: IN 0.15 – 0.25 mm (0.006 – 0.010 in.)
EX 0.20 – 0.30 mm (0.008 – 0.012 in.)
OK
CHECK EFI ELECTRONIC CIRCUIT USING VOLT / OHMMETER
BAD
1 Wiring connection
2 Power to ECU
(1) Fusible links
(2) Fuses
(3) EFI main relay
3 Vacuum sensor
4 Water temp. sensor
5 Air temp. sensor
6 Throttle position sensor
7 Injection signal circuit
(1) Injector wiring
(2) ECU

DIAGNOSTIC CODES—COROLLA 4A-FE

Code No.	Number of blinks "CHECK ENGINE"	System	Diagnosis	Trouble area
—	ON / OFF	Normal	This apears when none of the other codes are identified.	—
11		ECU (+ B)	Momentary interruption in power supply to ECU.	• Ignition switch circuit • Ignition switch • Main relay circuit • Main relay • ECU
12		RPM Signal	No "Ne" or "G" signal to ECU within 2 seconds after engine has been cranked.	• Distributor circuit • Distributor • Starter signal circuit • Igniter circuit • Igniter • ECU
13		RPM Signal	No "Ne" signal to ECU when the engine speed is above 1,000 rpm.	• Distributor circuit • Distributor • ECU
14		Ignition Signal	No "IGf" signal to ECU 4 times in succession.	• Igniter circuit • Igniter • Igniter and ignition coil circuit • Igniter and ignition coil • ECU
21		Oxygen Sensor Signal	Detection of oxygen sensor detrioration	• Oxygen sensor circuit • Oxygen sensor • ECU
22		Water Temp. Sensor	Open or short circuit in water temp. sensor signal (THW)	• Water temp. sensor circuit • Water temp. sensor • ECU
24		Intake Air Temp. Sensor Signal	Open or short circuit in intake air temp. sensor signal (THA).	• Intake air temp. sensor circuit • Intake air temp. sensor • ECU

Code No.	Number of blinks "CHECK ENGINE"	System	Diagnosis	Trouble area
25		Air fuel Ratio Lean Malfunction	• Lean signal sent by oxygen sensor for several seconds during air-fuel ratio feed back correction • Open or short circuit oxygen sensor (Ox)	• Injection circuit • Injector • Oxygen sensor circuit • Oxygen sensor • ECU • Fuel line pressure • Air leak • Vacuum sensor • Ignition system
26		Air-fuel Ratio Rich Malfunction	Rich signal sent by oxygen sensor for several seconds during air-fuel ratio feedback correction.	• Injector circuit • Injector • Oxygen sensor circuit • Fuel line pressure • Vacuum sensor • Cold start injector • ECU
31		Vacuum Sensor Signal	Open or short circuit intake manifold pressure signal (PIM)	• Vacuum sensor circuit • Vacuum sensor • ECU
41		Throttle Position Sensor Signal	The "IDL" and "PSW" signals are output simultaneously for several seconds.	• Throttle position sensor circuit • Throttle position sensor • ECU
42		Vehicle Speed Sensor Signal	No "SPD" signal for 8 seconds when engine speed is between 2,300 rpm and 5,500 rpm and coolant temp. is below 80°C (176°F) except when racing the engine	• Vehicle speed sensor circuit • Vehicle speed sensor • ECU
43		Starter Signal	No "STA" signal to ECU until engine speed reaches 800 rpm with vehicle not moving	• IG switch circuit • IG switch • ECU
51		Switch Signal	No "IDL" signal, "NSW" signal or "A/C" signal to ECU, with the check terminals E1 and T connected.	• A/C switch circuit • A/C switch • A/C Amplifier • Throttle position sensor circuit • Throttle position sensor • ECU

DIAGNOSTIC CIRCUITS TEST—COROLLA 4A-FE

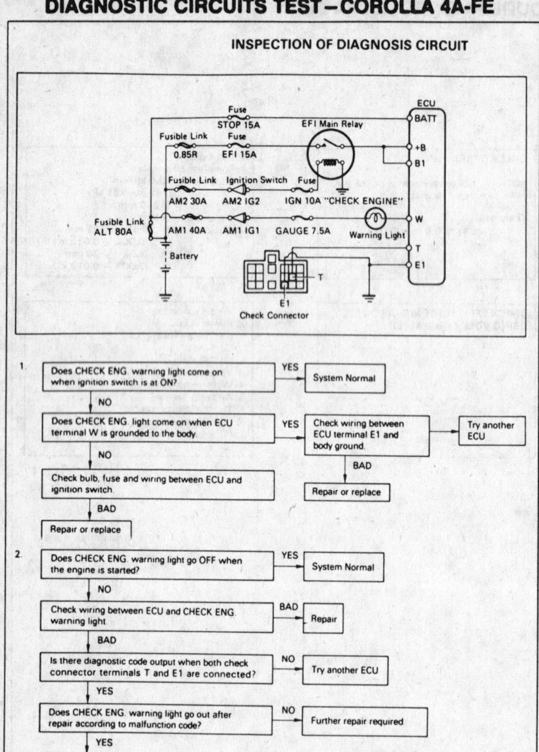
INSPECTION OF DIAGNOSIS CIRCUIT
ECU
BATT
Fuse
STOP 15A
EFI Main Relay
Fusible Link
0.85R
Fuse
EFI 15A
+B
B1
Fusible Link
AM2 30A
Ignition Switch
AM2 IG2
Fuse
IGN 10A "CHECK ENGINE"
Fusible Link
ALT 80A
AM1 40A
AM1 IG1
GAUGE 7.5A
W
Warning Light
Battery
T
E1
T
E1
Check Connector
1.
Does CHECK ENG. warning light come on when ignition switch is at ON?
YES
System Normal
NO
Does CHECK ENG. light come on when ECU terminal W is grounded to the body.
YES
Check wiring between ECU terminal E1 and body ground.
Try another ECU
BAD
Repair or replace
NO
Check bulb, fuse and wiring between ECU and ignition switch.
BAD
Repair or replace
2.
Does CHECK ENG. warning light go OFF when the engine is started?
YES
System Normal
NO
Check wiring between ECU and CHECK ENG. warning light.
BAD
Repair
BAD
Is there diagnostic code output when both check connector terminals T and E1 are connected?
NO
Try another ECU
YES
Does CHECK ENG. warning light go out after repair according to malfunction code?
NO
Further repair required
YES
System OK
Cancel out diagnosis code.

TROUBLESHOOTING CHART 1—COROLLA 4A-FE

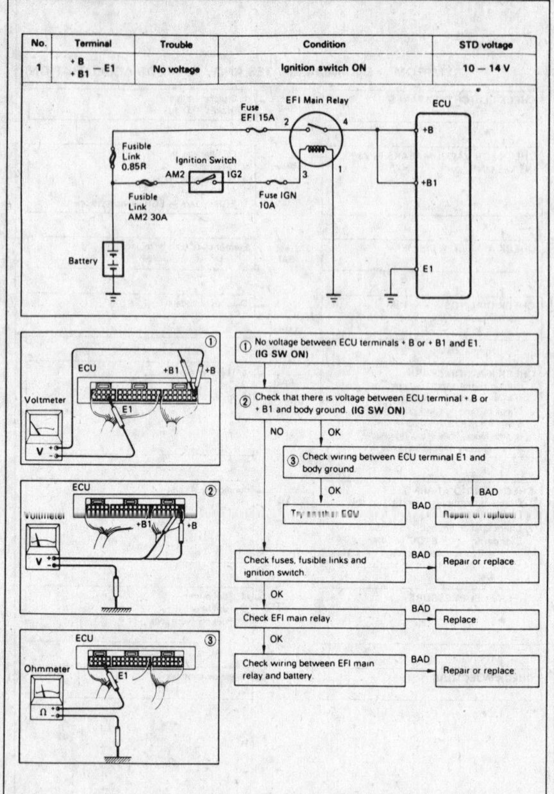

No.	Terminal	Trouble	Condition	STD voltage
1	+B +B1 — E1	No voltage	Ignition switch ON	10 – 14 V

Fuse
EFI 15A
EFI Main Relay
ECU
+B
Fusible Link
0.85R
Ignition Switch
AM2
IG2
+B1
Fusible Link
AM2 30A
Fuse IGN
10A
Battery
E1
1
ECU
+B1
+B
E1
Voltmeter
V
2
ECU
Voltmeter
+B1
+B
V
3
ECU
Ohmmeter
E1
Ω
1 No voltage between ECU terminals +B or +B1 and E1. (IG SW ON)
2 Check that there is voltage between ECU terminal +B or +B1 and body ground. (IG SW ON)
NO
OK
3 Check wiring between ECU terminal E1 and body ground.
OK
BAD
Try another ECU
BAD
Repair or replace
Check fuses, fusible links and ignition switch.
BAD
Repair or replace
OK
Check EFI main relay.
BAD
Replace
OK
Check wiring between EFI main relay and battery.
BAD
Repair or replace

TROUBLESHOOTING CHART 2—COROLLA 4A-FE

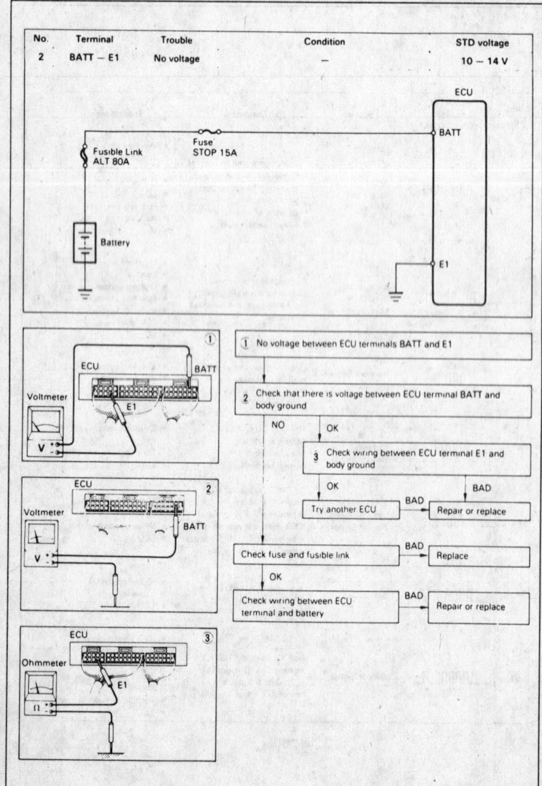

No.	Terminal	Trouble	Condition	STD voltage
2	BATT — E1	No voltage	—	10 – 14 V

ECU
BATT
Fuse
STOP 15A
Fusible Link
ALT 80A
Battery
E1
1
ECU
BATT
Voltmeter
E1
V
2
ECU
Voltmeter
BATT
V
3
ECU
Ohmmeter
E1
Ω
1 No voltage between ECU terminals BATT and E1
2 Check that there is voltage between ECU terminal BATT and body ground
NO
OK
3 Check wiring between ECU terminal E1 and body ground
OK
BAD
Try another ECU
BAD
Repair or replace
Check fuse and fusible link
BAD
Replace
OK
Check wiring between ECU terminal and battery
BAD
Repair or replace

TROUBLESHOOTING CHART 3—COROLLA 4A-FE

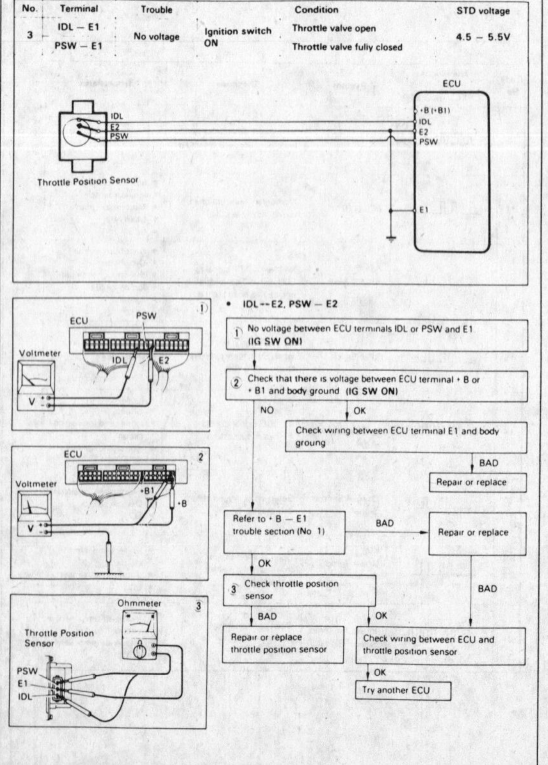

No.	Terminal	Trouble	Condition		STD voltage
3	IDL — E1	No voltage	Ignition switch ON	Throttle valve open	4.5 – 5.5V
	PSW — E1			Throttle valve fully closed	

ECU
+B (+B1)
IDL
E2
PSW
IDL
E2
PSW
E1
Throttle Position Sensor
1
ECU
PSW
Voltmeter
IDL
E2
V
2
ECU
Voltmeter
+B1
+B
V
3
Ohmmeter
Throttle Position Sensor
PSW
E1
IDL
• IDL — E2, PSW — E2
1 No voltage between ECU terminals IDL or PSW and E1. (IG SW ON)
2 Check that there is voltage between ECU terminal +B or +B1 and body ground. (IG SW ON)
NO
OK
Check wiring between ECU terminal E1 and body grounq
BAD
Repair or replace
Refer to +B — E1 trouble section (No. 1)
BAD
Repair or replace
OK
3 Check throttle position sensor
BAD
BAD
OK
Repair or replace throttle position sensor
Check wiring between ECU and throttle position sensor
OK
Try another ECU

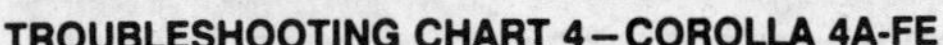

TROUBLESHOOTING CHART 4—COROLLA 4A-FE

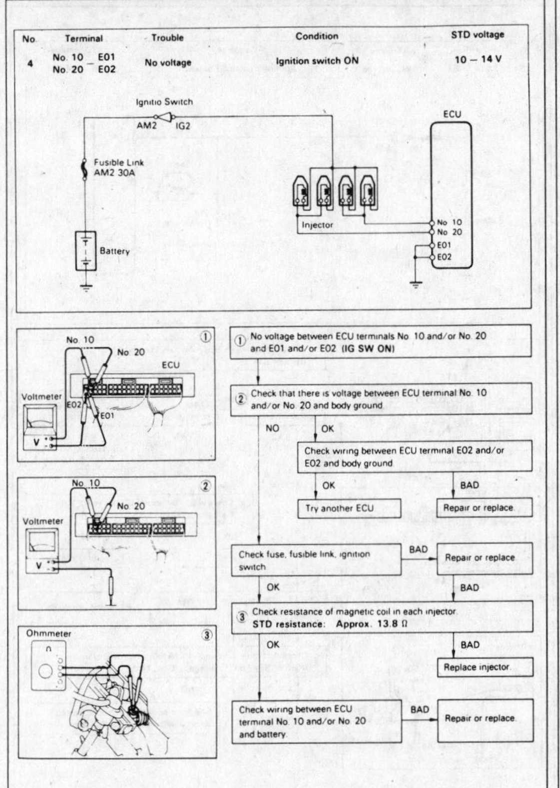
No. | Terminal | Trouble | Condition | STD voltage
4 | No. 10 — E01 No. 20 — E02 | No voltage | Ignition switch ON | 10 — 14 V
Ignition Switch
AM2
IG2
Fusible Link AM2 30A
Battery
Injector
ECU
No. 10
No. 20
E01
E02
No. 10
No. 20
ECU
Voltmeter
E02
E01
V
Ohmmeter
① No voltage between ECU terminals No. 10 and/or No. 20 and E01 and/or E02 (IG SW ON)
② Check that there is voltage between ECU terminal No. 10 and/or No. 20 and body ground
NO
OK
Check wiring between ECU terminal E02 and/or E02 and body ground
OK
BAD
Try another ECU
Repair or replace
Check fuse, fusible link, ignition switch
BAD
Repair or replace
OK
BAD
③ Check resistance of magnetic coil in each injector. STD resistance: Approx. 13.8 Ω
OK
BAD
Replace injector.
Check wiring between ECU terminal No. 10 and/or No. 20 and battery.
BAD
Repair or replace.

TROUBLESHOOTING CHART 5—COROLLA 4A-FE

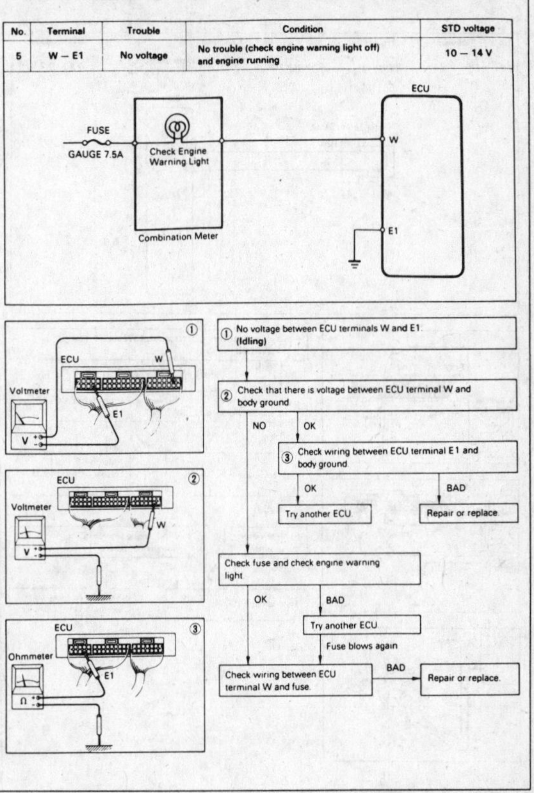
No. | Terminal | Trouble | Condition | STD voltage
5 | W — E1 | No voltage | No trouble (check engine warning light off) and engine running | 10 — 14 V
FUSE
GAUGE 7.5A
Check Engine Warning Light
Combination Meter
ECU
W
E1
Voltmeter
Ohmmeter
① No voltage between ECU terminals W and E1. (Idling)
② Check that there is voltage between ECU terminal W and body ground
NO
OK
③ Check wiring between ECU terminal E1 and body ground
OK
BAD
Try another ECU
Repair or replace.
Check fuse and check engine warning light
OK
BAD
Try another ECU
Fuse blows again
Check wiring between ECU terminal W and fuse.
BAD
Repair or replace.

TROUBLESHOOTING CHART 6—COROLLA 4A-FE

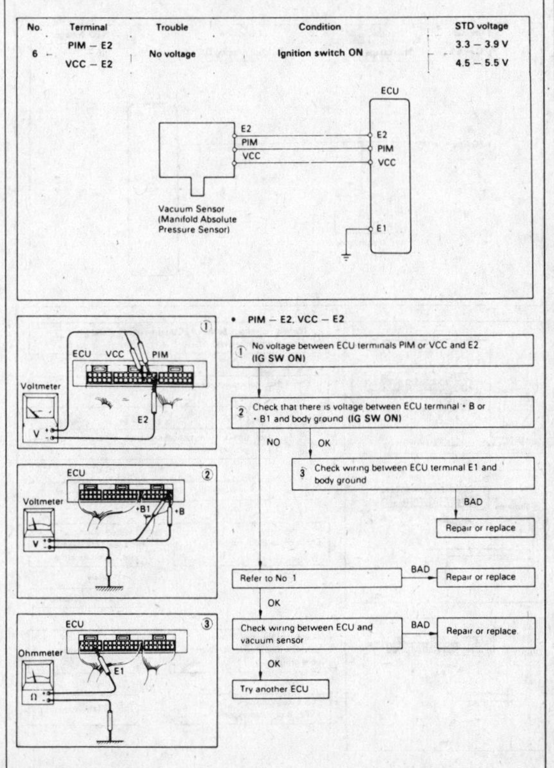
No. | Terminal | Trouble | Condition | STD voltage
6 | PIM — E2 / VCC — E2 | No voltage | Ignition switch ON | 3.3 — 3.9 V / 4.5 — 5.5 V
ECU
E2
PIM
VCC
E1
Vacuum Sensor (Manifold Absolute Pressure Sensor)
Voltmeter
Ohmmeter
+B1
+B
• PIM — E2, VCC — E2
① No voltage between ECU terminals PIM or VCC and E2 (IG SW ON)
② Check that there is voltage between ECU terminal + B or + B1 and body ground (IG SW ON)
NO
OK
③ Check wiring between ECU terminal E1 and body ground
BAD
Repair or replace
Refer to No. 1
BAD
Repair or replace
OK
Check wiring between ECU and vacuum sensor
BAD
Repair or replace
OK
Try another ECU

TRQUBLESHOOTING CHART 7—COROLLA 4A-FE

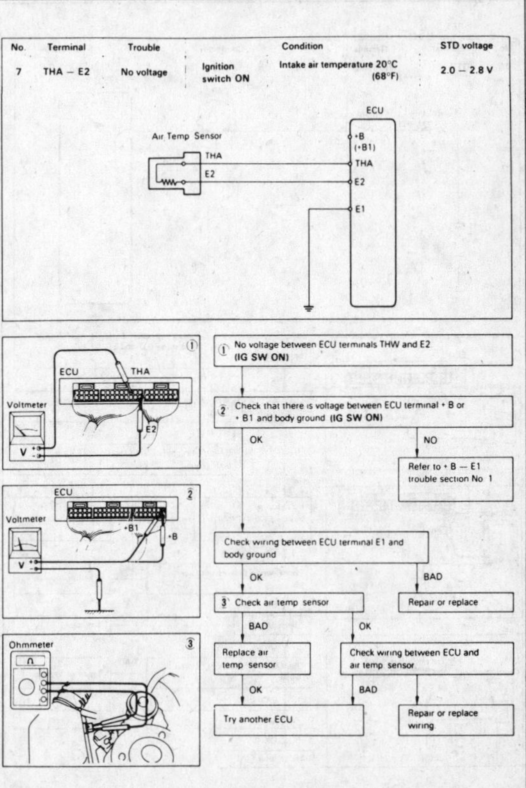
No. | Terminal | Trouble | Condition | STD voltage
7 | THA — E2 | No voltage | Ignition switch ON | Intake air temperature 20°C (68°F) | 2.0 — 2.8 V
Air Temp Sensor
THA
E2
ECU
+B (+B1)
THA
E2
E1
Voltmeter
Ohmmeter
① No voltage between ECU terminals THW and E2 (IG SW ON)
② Check that there is voltage between ECU terminal + B or + B1 and body ground (IG SW ON)
OK
NO
Refer to + B — E1 trouble section No. 1
Check wiring between ECU terminal E1 and body ground
OK
BAD
③ Check air temp sensor
Repair or replace
BAD
OK
Replace air temp sensor
Check wiring between ECU and air temp sensor
OK
BAD
Try another ECU
Repair or replace wiring

TROUBLESHOOTING CHART 8—COROLLA 4A-FE

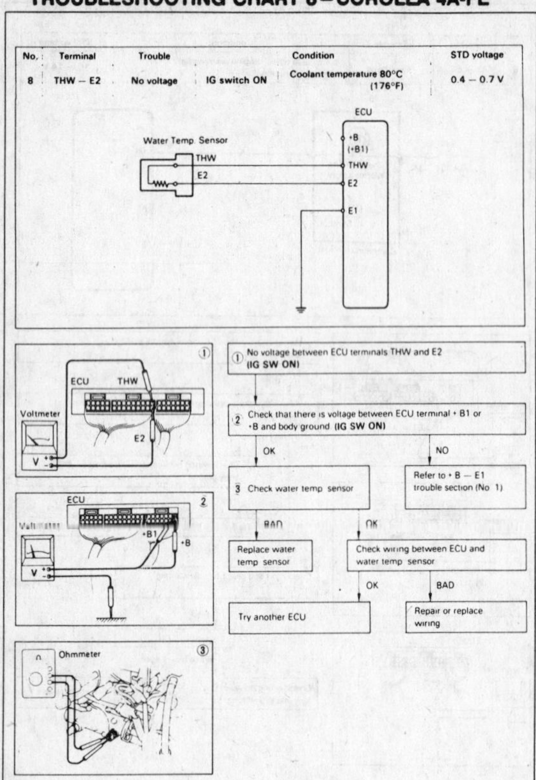
No. | Terminal | Trouble | Condition | STD voltage
8 | THW – E2 | No voltage | IG switch ON | Coolant temperature 80°C (176°F) | 0.4 – 0.7 V
ECU
Water Temp. Sensor
THW
E2
+B (+B1)
E1
Voltmeter
① No voltage between ECU terminals THW and E2 (IG SW ON)
② Check that there is voltage between ECU terminal +B1 or +B and body ground (IG SW ON)
OK
NO
③ Check water temp. sensor
Refer to +B – E1 trouble section (No. 1)
BAD
OK
Replace water temp. sensor
Check wiring between ECU and water temp. sensor
OK
BAD
Try another ECU
Repair or replace wiring
Ohmmeter

TROUBLESHOOTING CHART 9—COROLLA 4A-FE

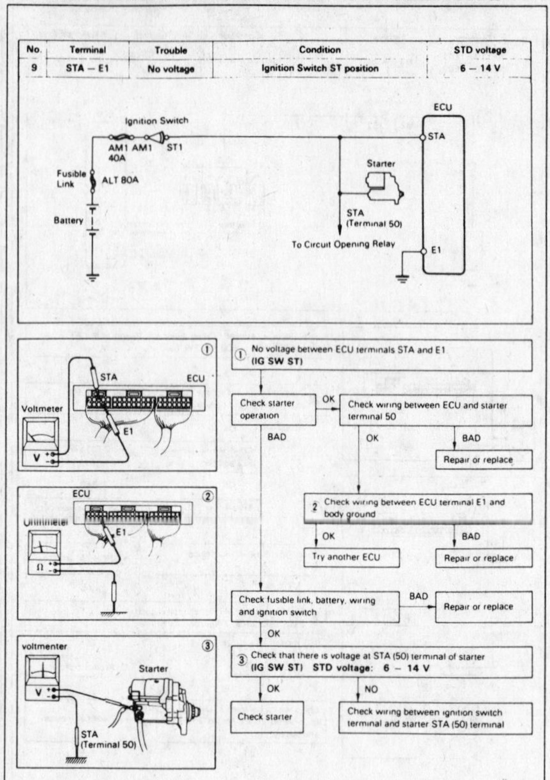
No. | Terminal | Trouble | Condition | STD voltage
9 | STA – E1 | No voltage | Ignition Switch ST position | 6 – 14 V
ECU
Ignition Switch
AM1 AM1 ST1
40A
Fusible Link
ALT 80A
Battery
STA
Starter
STA (Terminal 50)
To Circuit Opening Relay
E1
Voltmeter
Ohmmeter
① No voltage between ECU terminals STA and E1 (IG SW ST)
Check starter operation
OK
Check wiring between ECU and starter terminal 50
BAD
OK
BAD
Repair or replace
② Check wiring between ECU terminal E1 and body ground
OK
BAD
Try another ECU
Repair or replace
Check fusible link, battery, wiring and ignition switch
BAD
Repair or replace
OK
③ Check that there is voltage at STA (50) terminal of starter (IG SW ST) STD voltage: 6 – 14 V
OK
NO
Check starter
Check wiring between ignition switch terminal and starter STA (50) terminal

TROUBLESHOOTING CHART 10—COROLLA 4A-FE

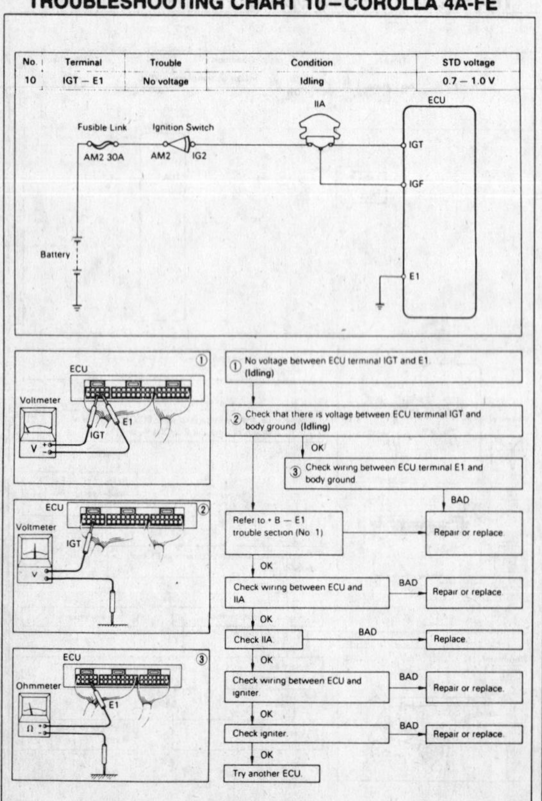
No. | Terminal | Trouble | Condition | STD voltage
10 | IGT – E1 | No voltage | Idling | 0.7 – 1.0 V
IIA
ECU
Fusible Link
Ignition Switch
AM2 30A
AM2
IG2
IGT
IGF
Battery
E1
Voltmeter
Ohmmeter
① No voltage between ECU terminal IGT and E1. (Idling)
② Check that there is voltage between ECU terminal IGT and body ground. (Idling)
OK
③ Check wiring between ECU terminal E1 and body ground.
BAD
Refer to +B – E1 trouble section (No. 1)
Repair or replace.
OK
Check wiring between ECU and IIA
BAD
Repair or replace.
OK
Check IIA.
BAD
Replace.
OK
Check wiring between ECU and igniter.
BAD
Repair or replace.
OK
Check igniter.
BAD
Repair or replace.
OK
Try another ECU.

TROUBLESHOOTING CHART 11—COROLLA 4A-FE

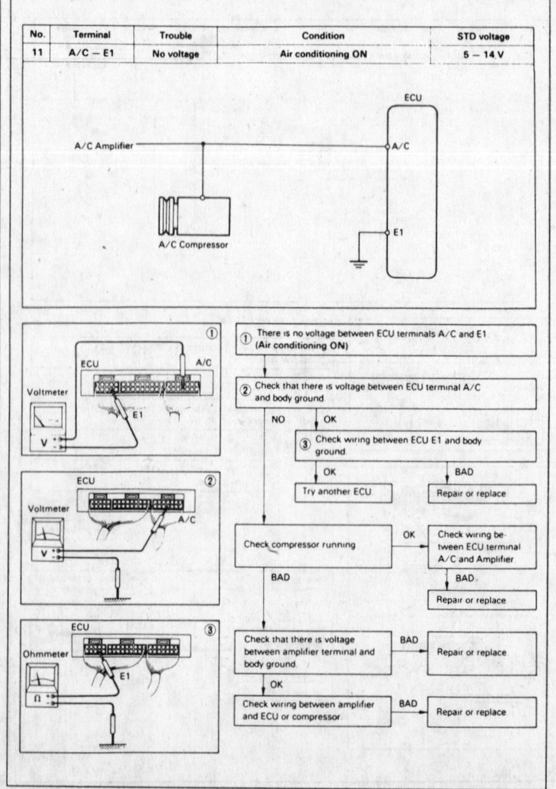
No. | Terminal | Trouble | Condition | STD voltage
11 | A/C – E1 | No voltage | Air conditioning ON | 5 – 14 V
ECU
A/C Amplifier
A/C
A/C Compressor
E1
Voltmeter
Ohmmeter
① There is no voltage between ECU terminals A/C and E1. (Air conditioning ON)
② Check that there is voltage between ECU terminal A/C and body ground.
NO
OK
③ Check wiring between ECU E1 and body ground.
OK
BAD
Try another ECU.
Repair or replace.
Check compressor running.
OK
Check wiring between ECU terminal A/C and Amplifier
BAD
BAD
Repair or replace.
Check that there is voltage between amplifier terminal and body ground.
BAD
Repair or replace.
OK
Check wiring between amplifier and ECU or compressor.
BAD
Repair or replace.

TROUBLESHOOTING OXYGEN SENSOR—COROLLA 4A-FE

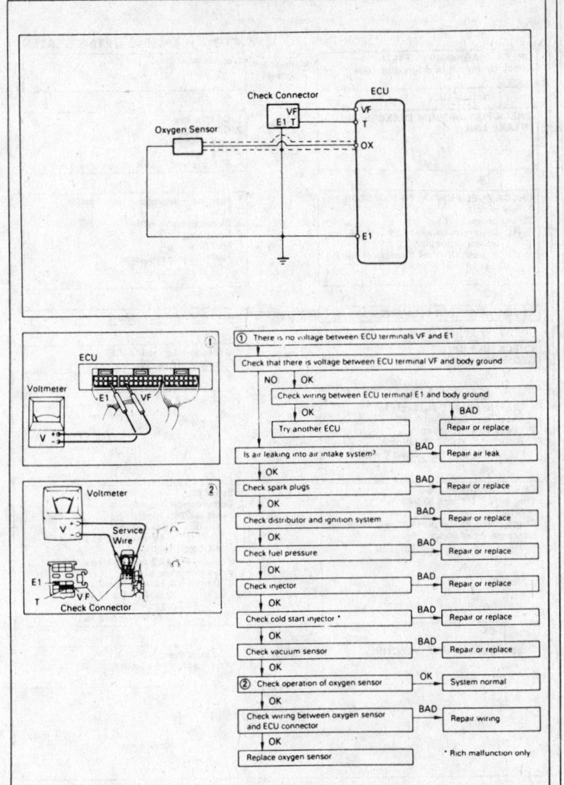
Check Connector
ECU
VF
E1 T
Oxygen Sensor
OX
E1
ECU
Voltmeter
V
1
Voltmeter
Service Wire
E1
T
VF
Check Connector
2
① There is no voltage between ECU terminals VF and E1
Check that there is voltage between ECU terminal VF and body ground
NO
OK
Check wiring between ECU terminal E1 and body ground
OK
BAD
Try another ECU
Repair or replace
Is air leaking into air intake system?
BAD
Repair air leak
OK
Check spark plugs
BAD
Repair or replace
OK
Check distributor and ignition system
BAD
Repair or replace
OK
Check fuel pressure
BAD
Repair or replace
OK
Check injector
BAD
Repair or replace
OK
Check cold start injector *
BAD
Repair or replace
OK
Check vacuum sensor
BAD
Repair or replace
OK
② Check operation of oxygen sensor
OK
System normal
OK
Check wiring between oxygen sensor and ECU connector
BAD
Repair wiring
OK
Replace oxygen sensor
* Rich malfunction only

ECU TERMINAL IDENTIFICATION—COROLLA 4A-FE

Connectors of ECU

Symbol	Terminal	Symbol	Terminal	Symbol	Terminal
E01	Engine ground (Power)	G⊖	Crank angle sensor	—	—
E02	Engine ground (Power)	E21	Sensor ground	—	—
No 10	No 1. 3 injector	G1	Crank angle sensor	—	—
No 20	No 2. 4 injector	NE	Engine revolution sensor	—	—
STA	Starter switch	IGF	Igniter	—	—
IGT	Igniter	—	—	SPD	Speedometer sensor
—	—	T	Check connector	FC	Circuit opening relay
E1	Engine ground	IDL	Throttle position sensor	A/C	A/C magnet clutch
*1 NSW	Neutral start switch	THA	Intake air temp. sensor	—	—
—	—	VCC	Vacuum sensor	—	—
V-ISC	VSV(ACV)	PIM	Vacuum sensor	BATT	Battery
—	—	PSW	Throttle position sensor	W	Warning light
OX	Oxygen sensor	THW	Water temp. sensor	+B1	EFI main relay
VF	Check connector	E2	Sensor ground	+B	EFI main relay

*1 For A/T

ECU connectors

E01	No 10	STA	—	*1 NSW	V-ISC	OX	G⊖	G1	IGF	T	THA	PIM	THW	—	—	—	FC	—	BATT	+B1
E02	No 20	IGT	E1	—	—	VF	E21	NE	—	IDL	VCC	PSW	E2	—	—	SPD	A/C	—	W	+B

ECU TERMINAL VOLTAGES—COROLLA 4A-FE

Voltage at ECU Wiring Connectors

Terminals	STD Voltage	Condition	
+B — E1	10 — 14	Ignition switch ON	
+B1 — E1			
BATT — E1		—	
IDL — E2	4.5 — 5.5	Ignition switch ON	Throttle valve open
PSW — E2			Throttle valve fully closed
No. 10 — E01 No. 20 — E02	10 — 14	Ignition switch ON	
W — E1	0	Ignition switch ON	
	10 — 14	Engine start	
PIM — E2	3.3 — 3.9	Ignition switch ON	
VCC — E2	4.5 — 5.5		
THA — E2	2.0 — 2.8	Ignition switch ON	Intake air temperature 20°C (68°F)
THW — E2	0.4 — 0.7		Coolant temperature 80°C (176°F)
STA — E1	6 — 14	IG switch ST position	
IGT — E1	0.7 — 1.0	Idling	
A/C — E1	5 — 14	IG switch ON	A/C switch ON
	0		A/C switch OFF
T — E1	10 — 14	IG switch ON	Check connector T — E1 not connect
	0		Check connector T — E1 connect

ECU Connectors

E01	No 10	STA	—	NSW	V-ISC	OX	G⊖	G1	IGF	T	THA	PIM	THW	—	—	—	FC	—	BATT	+B1
E02	No 20	IGT	E1	—	—	VF	E21	NE	—	IDL	VCC	PSW	E2	—	—	SPD	A/C	—	W	+B

TROUBLESHOOTING PROCEDURES—COROLLA AND MR-2 4A-GE AND 4A-GZE

Digital Type   Analog Type

Use a volt/ohmmeter with high impedance (10 kΩ/V minimum) for troubleshooting of the electrical circuit.

SYMPTOM – DIFFICULT TO START OR NO START (ENGINE WILL NOT CRANK OR CRANKS SLOWLY)

CHECK ELECTRIC SOURCE — BAD →
1. Battery
   (1) Connection
   (2) Gravity – drive belt–charging system
   (3) Voltage
2. Fusible link

↓ OK

CHECK STARTING SYSTEM — BAD →
1. Ignition switch
2. Clutch (M/T) or Neutral start switch (A/T)
3. Starter relay
4. Starter
5. Wiring/Connection

SYMPTOM – DIFFICULT TO START OR NO START (CRANKS OK)

CHECK DIAGNOSIS SYSTEM
Check for output of diagnosis code. — Malfunction code → Diagnosis code(s)

↓ Normal code

CHECK FOR VACUUM LEAKS IN AIR INDUCTION SYSTEM — BAD →
1. Oil filler cap
2. Oil level gauge
3. Hose connections
4. PCV hose
5. EGR system – EGR valve stays open

↓ OK

CHECK IGNITION SPARK
1. Unplug connectors of injector and start injector time switch.
2. Check by holding spark plug cord 8 – 10 mm (0.31 – 0.39 in.) away from engine block while cranking engine. A strong spark should be noted.

— BAD →
1. High-tension cords
2. Distributor
3. Ignition coil, igniter

↓ OK

CHECK IGNITION TIMING
1. Short terminals T and E_1 of the check connector.
2. Check ignition timing.
   STD: 10° BTDC @ Idling (w/short-circuited T and E_1)

— NO → Ignition timing – Adjust

↓ OK

TROUBLESHOOTING PROCEDURES—COROLLA AND MR-2 4A-GE AND 4A-GZE

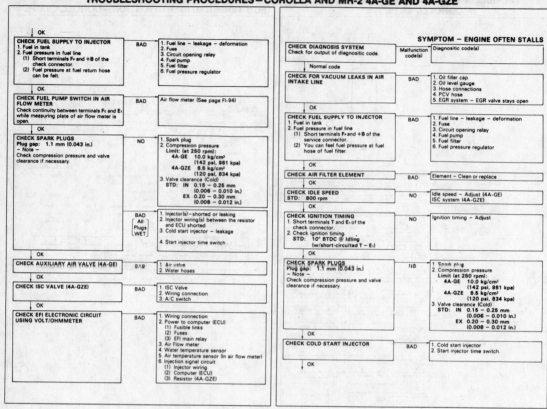

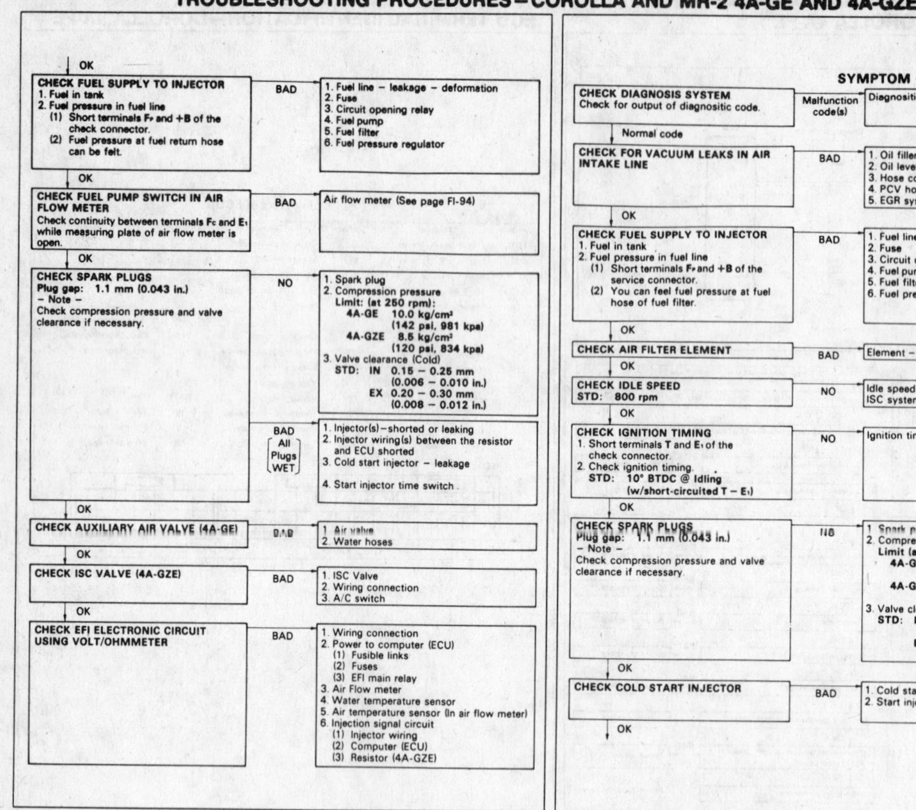
OK
CHECK FUEL SUPPLY TO INJECTOR
1. Fuel in tank
2. Fuel pressure in fuel line
(1) Short terminals Fp and +B of the check connector.
(2) Fuel pressure at fuel return hose can be felt.
BAD
1. Fuel line – leakage – deformation
2. Fuse
3. Circuit opening relay
4. Fuel pump
5. Fuel filter
6. Fuel pressure regulator
OK
CHECK FUEL PUMP SWITCH IN AIR FLOW METER
Check continuity between terminals Fc and E1 while measuring plate of air flow meter is open.
BAD
Air flow meter (See page FI-94)
OK
CHECK SPARK PLUGS
Plug gap: 1.1 mm (0.043 in.)
– Note –
Check compression pressure and valve clearance if necessary.
NO
1. Spark plug
2. Compression pressure
Limit: (at 250 rpm):
4A-GE 10.0 kg/cm² (142 psi, 981 kpa)
4A-GZE 8.5 kg/cm² (120 psi, 834 kpa)
3. Valve clearance (Cold)
STD: IN 0.15 – 0.25 mm (0.006 – 0.010 in.)
EX 0.20 – 0.30 mm (0.008 – 0.012 in.)
BAD All Plugs WET
1. Injector(s)—shorted or leaking
2. Injector wiring(s) between the resistor and ECU shorted
3. Cold start injector – leakage
4. Start injector time switch
OK
CHECK AUXILIARY AIR VALVE (4A-GE)
BAD
1. Air valve
2. Water hoses
OK
CHECK ISC VALVE (4A-GZE)
BAD
1. ISC Valve
2. Wiring connection
3. A/C switch
OK
CHECK EFI ELECTRONIC CIRCUIT USING VOLT/OHMMETER
BAD
1. Wiring connection
2. Power to computer (ECU)
(1) Fusible links
(2) Fuses
(3) EFI main relay
3. Air Flow meter
4. Water temperature sensor
5. Air temperature sensor (In air flow meter)
6. Injection signal circuit
(1) Injector wiring
(2) Computer (ECU)
(3) Resistor (4A-GZE)
SYMPTOM – ENGINE OFTEN STALLS
CHECK DIAGNOSIS SYSTEM
Check for output of diagnositic code.
Malfunction code(s)
Diagnositic code(s)
Normal code
CHECK FOR VACUUM LEAKS IN AIR INTAKE LINE
BAD
1. Oil filler cap
2. Oil level gauge
3. Hose connections
4. PCV hose
5. EGR system – EGR valve stays open
OK
CHECK FUEL SUPPLY TO INJECTOR
1. Fuel in tank
2. Fuel pressure in fuel line
(1) Short terminals Fp and +B of the service connector.
(2) You can feel fuel pressure at fuel hose of fuel filter.
BAD
1. Fuel line – leakage – deformation
2. Fuse
3. Circuit opening relay
4. Fuel pump
5. Fuel filter
6. Fuel pressure regulator
OK
CHECK AIR FILTER ELEMENT
BAD
Element – Clean or replace
OK
CHECK IDLE SPEED
STD: 800 rpm
NO
Idle speed – Adjust (4A-GE)
ISC system (4A-GZE)
OK
CHECK IGNITION TIMING
1. Short terminals T and E1 of the check connector.
2. Check ignition timing.
STD: 10° BTDC @ Idling (w/short-circuited T – E1)
NO
Ignition timing – Adjust
OK
CHECK SPARK PLUGS
Plug gap: 1.1 mm (0.043 in.)
– Note –
Check compression pressure and valve clearance if necessary.
NO
1. Spark plug
2. Compression pressure
Limit (at 250 rpm):
4A-GE 10.0 kg/cm² (142 psi, 981 kpa)
4A-GZE 8.5 kg/cm² (120 psi, 834 kpa)
3. Valve clearance (Cold)
STD: IN 0.15 – 0.25 mm (0.006 – 0.010 in.)
EX 0.20 – 0.30 mm (0.008 – 0.012 in.)
OK
CHECK COLD START INJECTOR
BAD
1. Cold start injector
2. Start injector time switch.
OK

TROUBLESHOOTING PROCEDURES—COROLLA AND MR-2 4A-GE AND 4A-GZE

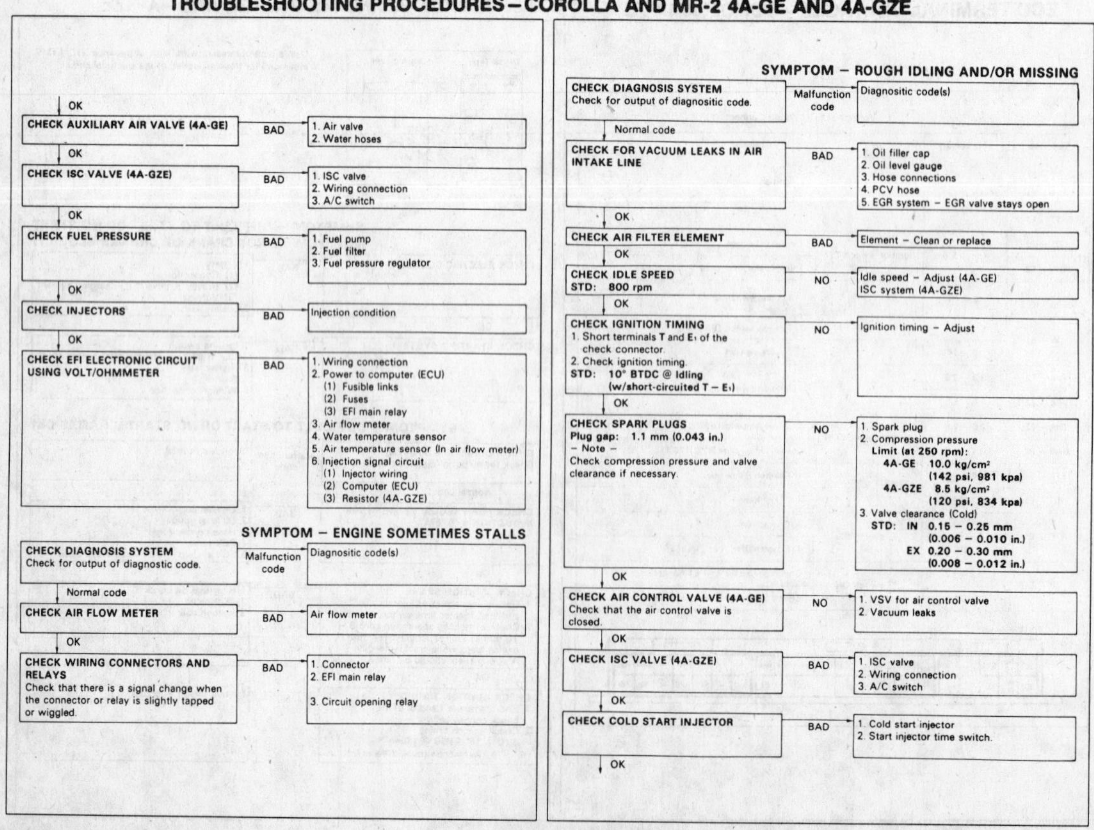
OK
CHECK AUXILIARY AIR VALVE (4A-GE)
BAD
1. Air valve
2. Water hoses
OK
CHECK ISC VALVE (4A-GZE)
BAD
1. ISC valve
2. Wiring connection
3. A/C switch
OK
CHECK FUEL PRESSURE
BAD
1. Fuel pump
2. Fuel filter
3. Fuel pressure regulator
OK
CHECK INJECTORS
BAD
Injection condition
OK
CHECK EFI ELECTRONIC CIRCUIT USING VOLT/OHMMETER
BAD
1. Wiring connection
2. Power to computer (ECU)
(1) Fusible links
(2) Fuses
(3) EFI main relay
3. Air flow meter
4. Water temperature sensor
5. Air temperature sensor (In air flow meter)
6. Injection signal circuit
(1) Injector wiring
(2) Computer (ECU)
(3) Resistor (4A-GZE)
SYMPTOM – ENGINE SOMETIMES STALLS
CHECK DIAGNOSIS SYSTEM
Check for output of diagnostic code.
Malfunction code
Diagnositic code(s)
Normal code
CHECK AIR FLOW METER
BAD
Air flow meter
OK
CHECK WIRING CONNECTORS AND RELAYS
Check that there is a signal change when the connector or relay is slightly tapped or wiggled.
BAD
1. Connector
2. EFI main relay
3. Circuit opening relay
SYMPTOM – ROUGH IDLING AND/OR MISSING
CHECK DIAGNOSIS SYSTEM
Check for output of diagnositic code.
Malfunction code
Diagnositic code(s)
Normal code
CHECK FOR VACUUM LEAKS IN AIR INTAKE LINE
BAD
1. Oil filler cap
2. Oil level gauge
3. Hose connections
4. PCV hose
5. EGR system – EGR valve stays open
OK
CHECK AIR FILTER ELEMENT
BAD
Element – Clean or replace
OK
CHECK IDLE SPEED
STD: 800 rpm
NO
Idle speed – Adjust (4A-GE)
ISC system (4A-GZE)
OK
CHECK IGNITION TIMING
1. Short terminals T and E1 of the check connector.
2. Check ignition timing.
STD: 10° BTDC @ Idling (w/short-circuited T – E1)
NO
Ignition timing – Adjust
OK
CHECK SPARK PLUGS
Plug gap: 1.1 mm (0.043 in.)
– Note –
Check compression pressure and valve clearance if necessary.
NO
1. Spark plug
2. Compression pressure
Limit (at 250 rpm):
4A-GE 10.0 kg/cm² (142 psi, 981 kpa)
4A-GZE 8.5 kg/cm² (120 psi, 834 kpa)
3. Valve clearance (Cold)
STD: IN 0.15 – 0.25 mm (0.006 – 0.010 in.)
EX 0.20 – 0.30 mm (0.008 – 0.012 in.)
OK
CHECK AIR CONTROL VALVE (4A-GE)
Check that the air control valve is closed.
NO
1. VSV for air control valve
2. Vacuum leaks
OK
CHECK ISC VALVE (4A-GZE)
BAD
1. ISC valve
2. Wiring connection
3. A/C switch
OK
CHECK COLD START INJECTOR
BAD
1. Cold start injector
2. Start injector time switch.
OK

TROUBLESHOOTING PROCEDURES—COROLLA AND MR-2 4A-GE AND 4A-GZE

OK

Check	Result	Possible cause
CHECK FUEL PRESSURE	BAD	1. Fuel pump 2. Fuel filter 3. Fuel pressure regulator
OK ↓		
CHECK INJECTORS	BAD	Injection condition
OK ↓		
CHECK EFI ELECTRONIC CIRCUIT USING VOLT/OHMMETER	BAD	1. Wiring connection 2. Power to computer (ECU) (1) Fusible link (2) Fuses (3) EFI main relay 3. Air flow meter 4. Water temperature sensor 5. Air temperature sensor (In air flow meter) 6. Injection signal circuit (1) Injector wirings (2) Computer (ECU) (3) Resistor (4A-GZE) 7. Oxygen sensor

SYMPTOM – HIGH ENGINE IDLE SPEED (NO DROP)

Check	Result	Possible cause
CHECK ACCELERATOR LINKAGE	BAD	Linkage – Stuck dash pot system
OK ↓		
CHECK AUXILIARY AIR VALVE (4A-GE)	BAD	1. Air valve 2. Water hoses
OK ↓		
CHECK ISC VALVE (4A-GZE)	BAD	1. ISC valve 2. Wiring connection 3. A/C switch
OK ↓		
CHECK AIR CONDITIONER IDLE-UP CIRCUIT	BAD	Air valve for air conditioner – Leakage VSV for air conditioner – Leakage
OK ↓		
CHECK DIAGNOSIS SYSTEM Check for output of diagnosis code.	Malfunction code	Diagnosis code(s)
Normal code ↓		
CHECK THROTTLE POSITION SENSOR	BAD	Throttle body
OK ↓		
CHECK FUEL PRESSURE	BAD	Fuel pressure regulator – High pressure
OK ↓		
CHECK COLD START INJECTOR	BAD	Cold start injector – Leakage
OK ↓		
CHECK INJECTORS	BAD	Injectors – Leakage, Injection quality
OK ↓		
CHECK EFI ELECTRONIC CIRCUIT USING VOLT/OHMMETER	BAD	1. Wiring connection 2. Power to computer (ECU) (1) Fusible links (2) Fuses (3) EFI main relay 3. Air flow meter 4. Water temperature sensor 5. Air temperature sensor (In air flow meter) 6. Injection signal circuit (1) Injector wiring (2) Computer (ECU) (3) Resistor (4A-GZE)

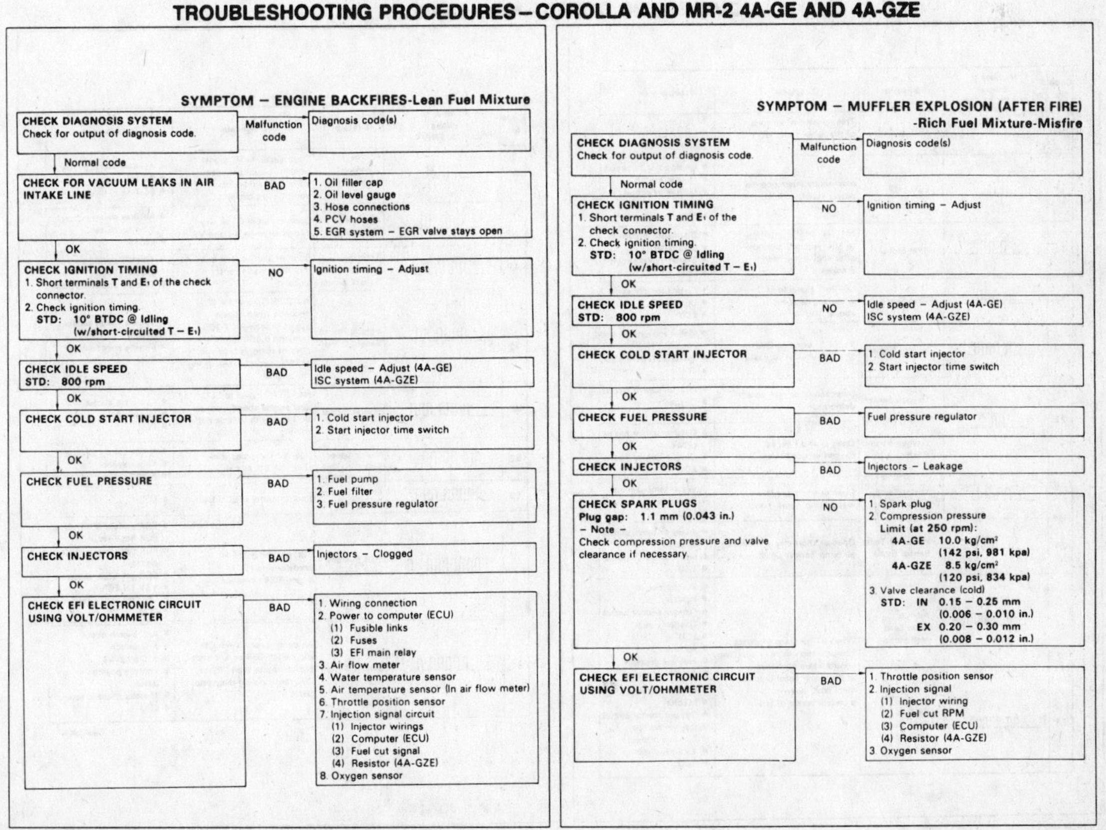

TROUBLESHOOTING PROCEDURES—COROLLA AND MR-2 4A-GE AND 4A-GZE
SYMPTOM – ENGINE BACKFIRES-Lean Fuel Mixture
CHECK DIAGNOSIS SYSTEM
Check for output of diagnosis code.
Malfunction code
Diagnosis code(s)
Normal code
CHECK FOR VACUUM LEAKS IN AIR INTAKE LINE
BAD
1. Oil filler cap
2. Oil level gauge
3. Hose connections
4. PCV hoses
5. EGR system – EGR valve stays open
OK
CHECK IGNITION TIMING
1. Short terminals T and E1 of the check connector.
2. Check ignition timing.
STD: 10° BTDC @ Idling
(w/short-circuited T – E1)
NO
Ignition timing – Adjust
OK
CHECK IDLE SPEED
STD: 800 rpm
BAD
Idle speed – Adjust (4A-GE)
ISC system (4A-GZE)
OK
CHECK COLD START INJECTOR
BAD
1. Cold start injector
2. Start injector time switch
OK
CHECK FUEL PRESSURE
BAD
1. Fuel pump
2. Fuel filter
3. Fuel pressure regulator
OK
CHECK INJECTORS
BAD
Injectors – Clogged
OK
CHECK EFI ELECTRONIC CIRCUIT USING VOLT/OHMMETER
BAD
1. Wiring connection
2. Power to computer (ECU)
(1) Fusible links
(2) Fuses
(3) EFI main relay
3. Air flow meter
4. Water temperature sensor
5. Air temperature sensor (In air flow meter)
6. Throttle position sensor
7. Injection signal circuit
(1) Injector wirings
(2) Computer (ECU)
(3) Fuel cut signal
(4) Resistor (4A-GZE)
8. Oxygen sensor
SYMPTOM – MUFFLER EXPLOSION (AFTER FIRE)
-Rich Fuel Mixture-Misfire
CHECK DIAGNOSIS SYSTEM
Check for output of diagnosis code.
Malfunction code
Diagnosis code(s)
Normal code
CHECK IGNITION TIMING
1. Short terminals T and E1 of the check connector.
2. Check ignition timing.
STD: 10° BTDC @ Idling
(w/short-circuited T – E1)
NO
Ignition timing – Adjust
OK
CHECK IDLE SPEED
STD: 800 rpm
NO
Idle speed – Adjust (4A-GE)
ISC system (4A-GZE)
OK
CHECK COLD START INJECTOR
BAD
1. Cold start injector
2. Start injector time switch
OK
CHECK FUEL PRESSURE
BAD
Fuel pressure regulator
OK
CHECK INJECTORS
BAD
Injectors – Leakage
OK
CHECK SPARK PLUGS
Plug gap: 1.1 mm (0.043 in.)
– Note –
Check compression pressure and valve clearance if necessary.
NO
1. Spark plug
2. Compression pressure
Limit (at 250 rpm):
4A-GE 10.0 kg/cm²
(142 psi, 981 kpa)
4A-GZE 8.5 kg/cm²
(120 psi, 834 kpa)
3. Valve clearance (cold)
STD: IN 0.15 – 0.25 mm
(0.006 – 0.010 in.)
EX 0.20 – 0.30 mm
(0.008 – 0.012 in.)
OK
CHECK EFI ELECTRONIC CIRCUIT USING VOLT/OHMMETER
BAD
1. Throttle position sensor
2. Injection signal
(1) Injector wiring
(2) Fuel cut RPM
(3) Computer (ECU)
(4) Resistor (4A-GZE)
3. Oxygen sensor

TROUBLESHOOTING PROCEDURES—COROLLA AND MR-2 4A-GE AND 4A-GZE

SYMPTOM – ENGINE HESITATES AND/OR POOR ACCELERATION

CHECK CLUTCH OR BRAKE — BAD → 1. Clutch – Slips 2. Brakes – Drag

↓ OK

CHECK FOR VACUUM LEAKS IN AIR INTAKE LINE — BAD → 1. Oil filler cap 2. Oil level gauge 3. Hose connections 4. PCV hose 5. EGR system – EGR valve stays open

↓ OK

CHECK AIR FILTER ELEMENT — BAD → Element – Clean or replace

↓ OK

CHECK DIAGNOSIS SYSTEM
Check for output of diagnosis code. — Malfunction code → Diagnosis code(s)

↓ Normal code

CHECK IGNITION SPARK
1. Unplug connectors of injector and start injection time switch.
2. Check by holding spark plug 8 – 10 mm (0.31 – 0.39 in.) away from engine block while cranking engine. A strong spark should be noted.

— BAD → 1. High-tension cords 2. Distributor 3. Ignition coil, Igniter

↓ OK

CHECK IGNITION TIMING
1. Short terminals T and E_1 of the check connector.
2. Check ignition timing.
STD: 10° BTDC @ Idling (w/short-circuited T – E_1)

— NO → Ignition timing – Adjust

↓ OK

CHECK FUEL PRESSURE — BAD → 1. Fuel pump 2. Fuel filter 3. Fuel pressure regulator

↓ OK

CHECK INJECTORS — BAD → Injection condition

↓ OK

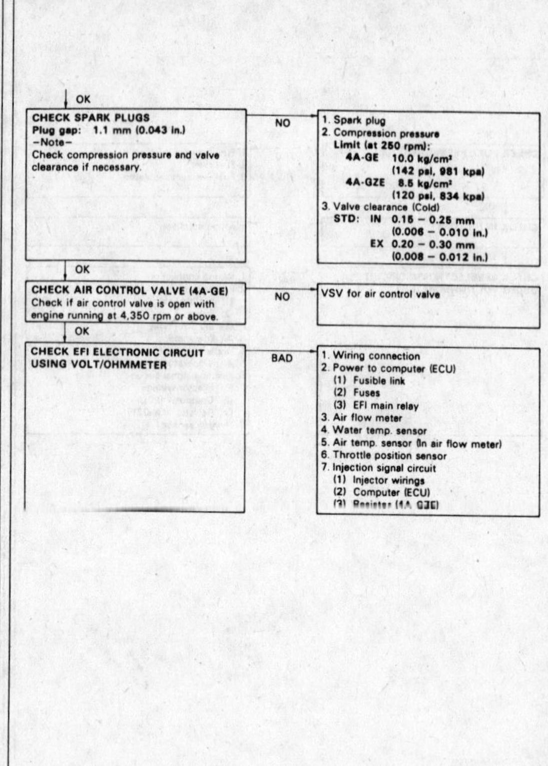
OK
CHECK SPARK PLUGS
Plug gap: 1.1 mm (0.043 in.)
–Note–
Check compression pressure and valve clearance if necessary.
NO
1. Spark plug
2. Compression pressure
Limit (at 250 rpm):
4A-GE 10.0 kg/cm² (142 psi, 981 kpa)
4A-GZE 8.5 kg/cm² (120 psi, 834 kpa)
3. Valve clearance (Cold)
STD: IN 0.15 – 0.25 mm (0.006 – 0.010 in.)
EX 0.20 – 0.30 mm (0.008 – 0.012 in.)
OK
CHECK AIR CONTROL VALVE (4A-GE)
Check if air control valve is open with engine running at 4,350 rpm or above.
NO
VSV for air control valve
OK
CHECK EFI ELECTRONIC CIRCUIT USING VOLT/OHMMETER
BAD
1. Wiring connection
2. Power to computer (ECU)
(1) Fusible link
(2) Fuses
(3) EFI main relay
3. Air flow meter
4. Water temp. sensor
5. Air temp. sensor (In air flow meter)
6. Throttle position sensor
7. Injection signal circuit
(1) Injector wirings
(2) Computer (ECU)
(3) Resistor (4A-GZE)

DIAGNOSTIC CODES—MR-2 4A-GZE

Code No.	Number of blinks "CHECK ENGINE"	System	Diagnosis	Trouble area
–	ON/OFF	Normal	This appears when none of the other codes are identified.	–
11		ECU (+B)	Momentary interruption in power supply to ECU.	• Ignition switch circuit • Ignition switch • Main relay circuit • Main relay • ECU
12		RPM Signal	No "Ne" or "G" signal to ECU within 2 seconds after engine has been cranked.	• Distributor circuit • Distributor • Starter signal circuit • ECU
13		RPM Signal	No "Ne" signal to ECU when engine speed is above 1,000 rpm.	• Distributor circuit • Distributor • ECU
14		Ignition Signal	No "IGf" signal to ECU 8 – 11 times in succession.	• Igniter and ignition coil circuit • Igniter and ignition coil • ECU
21		Oxygen Sensor	Detection of oxygen sensor detrioration.	• Oxygen sensor circuit • Oxygen sensor • ECU
		Oxygen Sensor Heater	Open or short circuit oxygen sensor heater.	• Oxygen sensor heater circuit • Oxygen sensor heater • ECU
22		Water Temp. Sensor Signal	Open or short circuit in water temp. screw signal.	• Water temp. sensor • Water temp. sensor • ECU
24		Intake Air Temp. Sensor Signal	Open or short circuit in intake air temp. sensor signal.	• Intake air temp. sensor Circuit • Intake air temp. sensor • ECU
25		Air-fuel Ratio Lean Malfunction	When air-fuel ratio feedback compensation valve or adaptive control value continues at the upper (lean) or lower (rich) limit renewed for a certain period of time.	• Injector circuit • Injector • Oxygen sensor circuit • Oxygen sensor • ECU • Air leak • Fuel line pressure • Air flow meter
26		Air-fuel Ratio Rich Malfunction		• Injector circuit • Injector • Oxygen sensor circuit • Oxygen sensor • Fuel line pressure • Air flow meter • ECU

Code No.	Number of blinks "CHECK ENGINE"	System	Diagnosis	Trouble area
31		Air-flow Meter Signal	Open circuit in VC signal or short circuit between VS and E2 when idle contacts are closed	• Air flow meter circuit • Air flow meter • ECU
32		Air-flow Meter Signal	Open circuit in E2 or short circuit between VC and VS	• Air flow meter circuit • Air-flow meter • ECU
41		Throttle Position Sensor Signal	Open or short circuit in throttle position sensor signal.	• Throttle position sensor circuit • Throttle position sensor
42		Vehicle Speed Sensor Signal	No "SPD" signal for 8 seconds when engine speed is between 2,300 rpm and 5,000 rpm and coolant temp. is below 80°C (176°F) except when racing the engine.	• Vehicle speed sensor circuit • Vehicle speed sensor • ECU
43		Starter Signal	No "STA" signal to ECU until engine speed reaches 800 rpm with Vehicle not moving	• IG switch circuit • IG switch • ECU
52		Knock Sensor Signal	Open or short circuit in knock sensor signal	• Knock sensor circuit • Knock sensor • ECU
53		Knock Control Signal in ECU	Knock control in ECU faulty	• ECU
*71		EGR System Malfunction	EGR gas temp. below predetermined level during EGR operation	• EGR system (EGR valve, EGR hose etc.,) • EGR gas temp. sensor circuit • EGR gas temp. sensor • VSV for EGR • VSV for EGR circuit • ECU
51		Switch Signal	Air conditioner switch ON, idle switch OFF or A/T shift position other than "P" or "N" range during diagnosis check.	• A/C switch circuit • A/C switch • A/C amplifire • Throttle position sensor circuit • Throttle position sensor • ECU

*: For California

DIAGNOSTIC CODES—COROLLA AND MR-2 4A-GE

Code No.	Number of check engine blinks	System	Diagnosis	Trouble area
–	ON/OFF	Normal	This appears when none of the other codes are identified.	–
12		RPM Signal	• No "Ne" signal to ECU within 2 seconds after the engine is cranked. • No "G" signal to ECU 2 times in succession when engine speed is between 500 rpm and 4000 rpm.	• Distributor circuit • Distributor • Starter signal circuit • Igniter circuit • Igniter • ECU
13		RPM Signal	No "Ne" signal to ECU when the engine speed is above 1500 rpm	• Distributor circuit • Distributor • ECU
14		Ignition Signal	No "IGF" signal to ECU 4 times in succession.	• Igniter circuit • Igniter • ECU
21		Oxygen Sensor Signal	Detection of oxygen sensor detrioration.	• Oxygen sensor circuit • Oxygen sensor • ECU
		Oxygen Sensor Heater Circuit	Open or short circuit oxygen sensor heater	• Oxygen sensor heater circuit • Oxygen sensor heater • ECU
22		Water Temp. Sensor Signal	Open or short circuit in water temp. sensor signal. (THW)	• Water temp. sensor circuit • Water temp. sensor • ECU
24		Intake air Temp. Sensor Signal	Open or short circuit in intake air temp. sensor signal. (THA)	• Intake air temp. sensor circuit • Intake air temp. sensor • ECU
25		Air-fuel Ratio Lean Malfunction	When air-fuel ratio feedback compensation value or adaptive control value continues at the upper (lean) or lower (rich) limit renewed for a certain period of time.	• Injector circuit • Injector • Oxygen sensor circuit • Oxygen sensor • ECU • Fuel line pressure • Air leak • Air-flow meter • Air intake system • Ignition system
26		Air-fuel Ratio Rich Malfunction		• Injector circuit • Injector • Oxygen sensor circuit • Oxygen sensor • Fuel line pressure • Air-flow meter • Cold start injector • ECU

DIAGNOSTIC CIRCUITS TEST—CAMRY

Dode No.	Number of check engine blinks	System	Diagnosis	Trouble area
*27		Sub-oxygen Sensor Signal	Open or short circuit in sub-oxygen sensor signal.	• Sub-oxygen sensor circuit • Sub-oxygen sensor • ECU
31		Air-flow Meter Signal	Short circuit between VC and VB, VC and E2, or VS and VC.	• Air flow meter circuit • Air flow meter • ECU
41		Throttle Position Sensor Signal	Open or short circuit in throttle position sensor signal. (VTA)	• Throttle position sensor circuit • Throttle position sensor • ECU
42		Vehicle Speed Sensor Signal	No "SPD" signal for 8 seconds when engine speed is above 2800 rpm.	• Vehicle speed sensor circuit • Vehicle speed sensor • ECU
43		Starter Signal	No "STA" signal to ECU until engine speed reaches 800 rpm with vehicle not moving.	• Starter signal circuit • Ignition switch, main relay circuit • ECU
*71		EGR Malfunction	EGR gas temp. below predeterminal level for during EGR control.	• EGR system (EGR valve, EGR hose etc.) • EGR gas temp. sensor circuit • EGR gas temp. sensor • VSV for EGR • VSV for EGR circuit • ECU
51		Switch Signal	Air conditioner switch ON, idle switch OFF during diagnosis check.	• A/C switch circuit • A/C switch • A/C Amplifire • Throttle position sensor circuit • Throttle position sensor • ECU

DIAGNOSTIC CIRCUITS TEST—COROLLA AND MR-2 4A-GE AND 4A-GZE

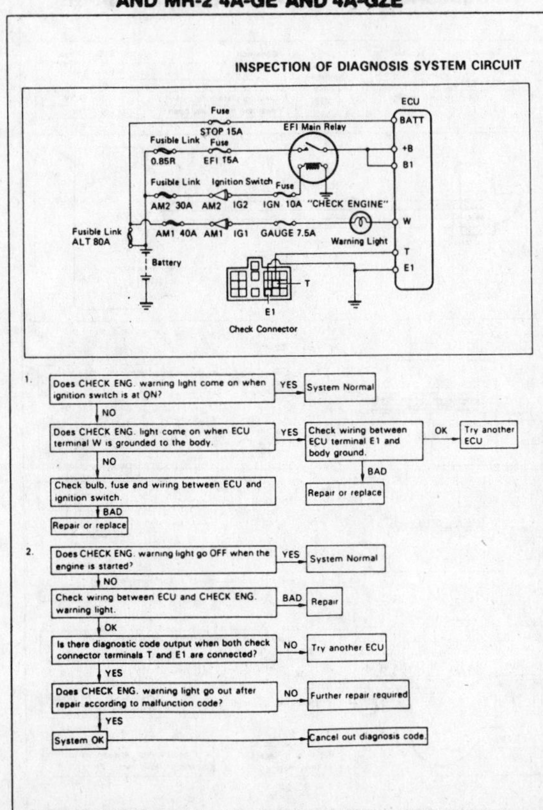
INSPECTION OF DIAGNOSIS SYSTEM CIRCUIT
ECU
Fuse
STOP 15A
EFI Main Relay
BATT
Fusible Link
Fuse
0.85R
EFI 15A
+B
B1
Fusible Link
Ignition Switch
Fuse
AM2 30A
AM2
IG2
IGN 10A
"CHECK ENGINE"
Fusible Link ALT 80A
AM1 40A
AM1
IG1
GAUGE 7.5A
W
Warning Light
T
E1
Battery
T
E1
Check Connector
1.
Does CHECK ENG. warning light come on when ignition switch is at ON?
YES
System Normal
NO
Does CHECK ENG. light come on when ECU terminal W is grounded to the body.
YES
Check wiring between ECU terminal E1 and body ground.
OK
Try another ECU
BAD
NO
Check bulb, fuse and wiring between ECU and ignition switch.
Repair or replace
BAD
Repair or replace
2.
Does CHECK ENG. warning light go OFF when the engine is started?
YES
System Normal
NO
Check wiring between ECU and CHECK ENG. warning light.
BAD
Repair
OK
Is there diagnostic code output when both check connector terminals T and E1 are connected?
NO
Try another ECU
YES
Does CHECK ENG. warning light go out after repair according to malfunction code?
NO
Further repair required
YES
System OK
Cancel out diagnosis code.

TROUBLESHOOTING CHART 1—COROLLA AND MR-2 4A-GE

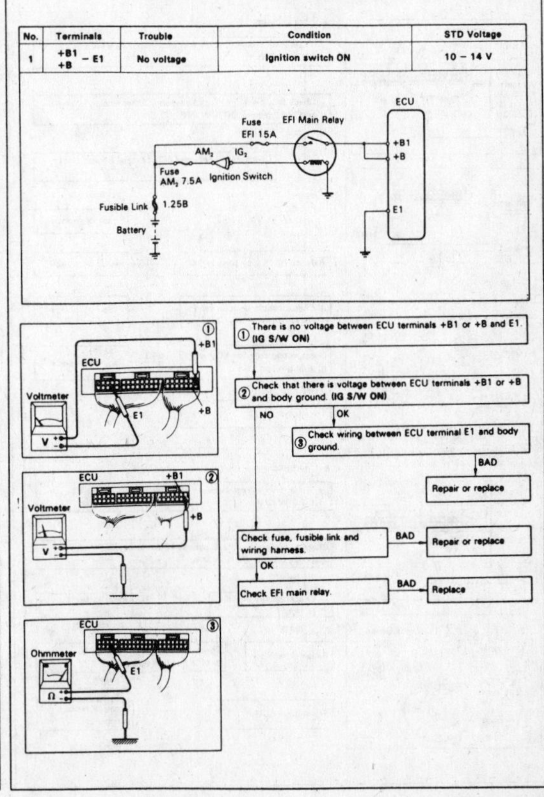

No.	Terminals	Trouble	Condition	STD Voltage
1	+B1 +B – E1	No voltage	Ignition switch ON	10 – 14 V

ECU
Fuse
EFI 15A
EFI Main Relay
AM2
IG2
+B1
+B
Fuse
AM2 7.5A
Ignition Switch
Fusible Link
1.25B
Battery
E1
ECU
Voltmeter
+B1
E1
+B
ECU
+B1
Voltmeter
+B
ECU
Ohmmeter
E1
① There is no voltage between ECU terminals +B1 or +B and E1. (IG S/W ON)
② Check that there is voltage between ECU terminals +B1 or +B and body ground. (IG S/W ON)
NO
OK
③ Check wiring between ECU terminal E1 and body ground.
BAD
Repair or replace
Check fuse, fusible link and wiring harness.
BAD
Repair or replace
OK
Check EFI main relay.
BAD
Replace

TROUBLESHOOTING CHART 2— COROLLA AND MR-2 4A-GE

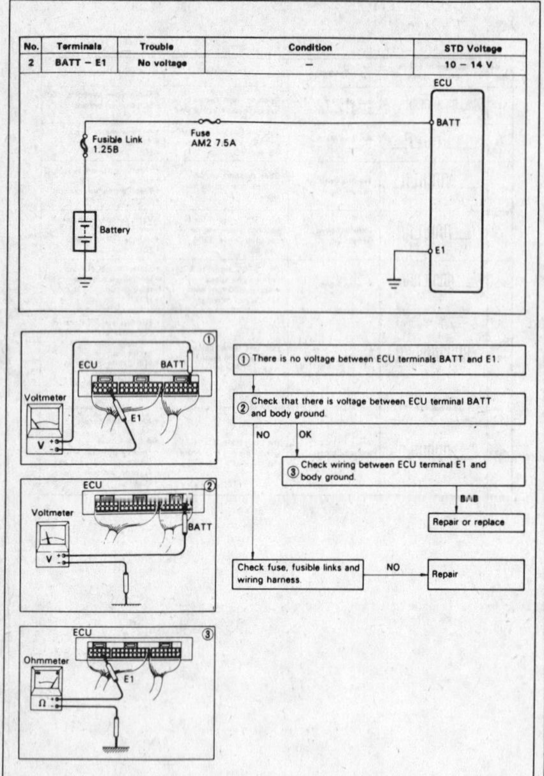

No.	Terminals	Trouble	Condition	STD Voltage
2	BATT – E1	No voltage	–	10 – 14 V

ECU
BATT
Fusible Link 1.25B
Fuse AM2 7.5A
Battery
E1
Voltmeter
ECU
BATT
E1
① There is no voltage between ECU terminals BATT and E1.
② Check that there is voltage between ECU terminal BATT and body ground.
NO
OK
③ Check wiring between ECU terminal E1 and body ground.
BAD
Repair or replace
Check fuse, fusible links and wiring harness.
NO
Repair
Ohmmeter

TROUBLESHOOTING CHART 3— COROLLA AND MR-2 4A-GE

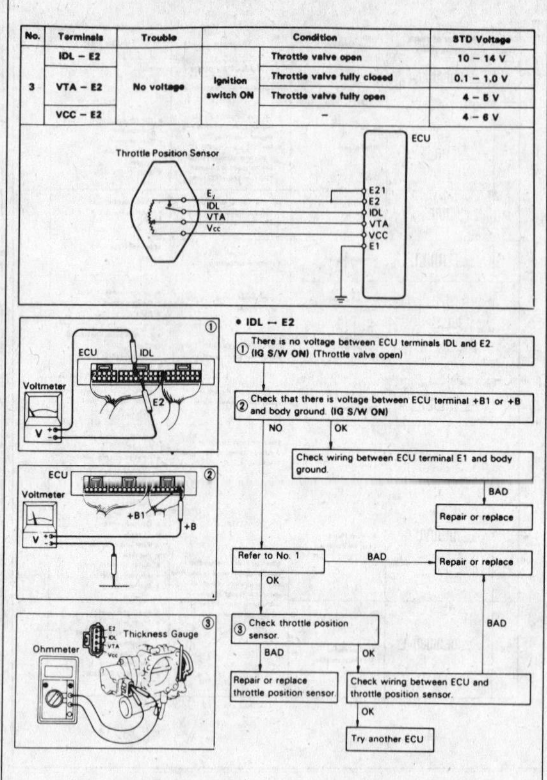

No.	Terminals	Trouble	Condition		STD Voltage
3	IDL – E2	No voltage	Ignition switch ON	Throttle valve open	10 – 14 V
	VTA – E2			Throttle valve fully closed	0.1 – 1.0 V
				Throttle valve fully open	4 – 5 V
	VCC – E2			–	4 – 6 V

Throttle Position Sensor
ECU
E2
IDL
VTA
Vcc
E21
E2
IDL
VTA
VCC
E1
• IDL ↔ E2
① There is no voltage between ECU terminals IDL and E2. (IG S/W ON) (Throttle valve open)
② Check that there is voltage between ECU terminal +B1 or +B and body ground. (IG S/W ON)
NO
OK
Check wiring between ECU terminal E1 and body ground.
BAD
Repair or replace
Refer to No. 1
BAD
Repair or replace
OK
③ Check throttle position sensor.
BAD
OK
BAD
Repair or replace throttle position sensor.
Check wiring between ECU and throttle position sensor.
OK
Try another ECU
Voltmeter
Ohmmeter
Thickness Gauge
+B1
+B

TROUBLESHOOTING CHART 3— COROLLA AND MR-2 4A-GE

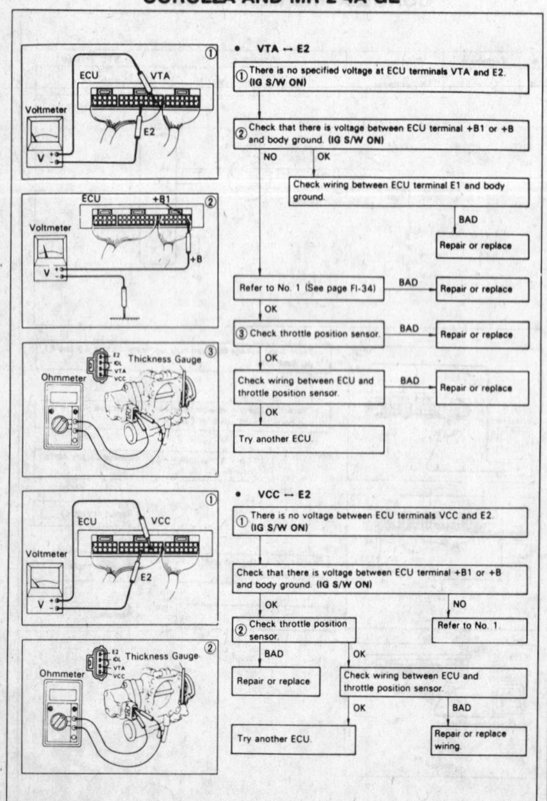

• VTA ↔ E2
① There is no specified voltage at ECU terminals VTA and E2. (IG S/W ON)
② Check that there is voltage between ECU terminal +B1 or +B and body ground. (IG S/W ON)
NO
OK
Check wiring between ECU terminal E1 and body ground.
BAD
Repair or replace
Refer to No. 1 (See page FI-34)
BAD
Repair or replace
OK
③ Check throttle position sensor.
BAD
Repair or replace
OK
Check wiring between ECU and throttle position sensor.
BAD
Repair or replace
OK
Try another ECU.
• VCC ↔ E2
① There is no voltage between ECU terminals VCC and E2. (IG S/W ON)
Check that there is voltage between ECU terminal +B1 or +B and body ground. (IG S/W ON)
OK
NO
② Check throttle position sensor.
Refer to No. 1.
BAD
OK
Repair or replace
Check wiring between ECU and throttle position sensor.
OK
BAD
Try another ECU.
Repair or replace wiring.
ECU
VTA
VCC
E2
+B1
+B
Voltmeter
Ohmmeter
Thickness Gauge

TROUBLESHOOTING CHART 4—COROLLA 4A-GE

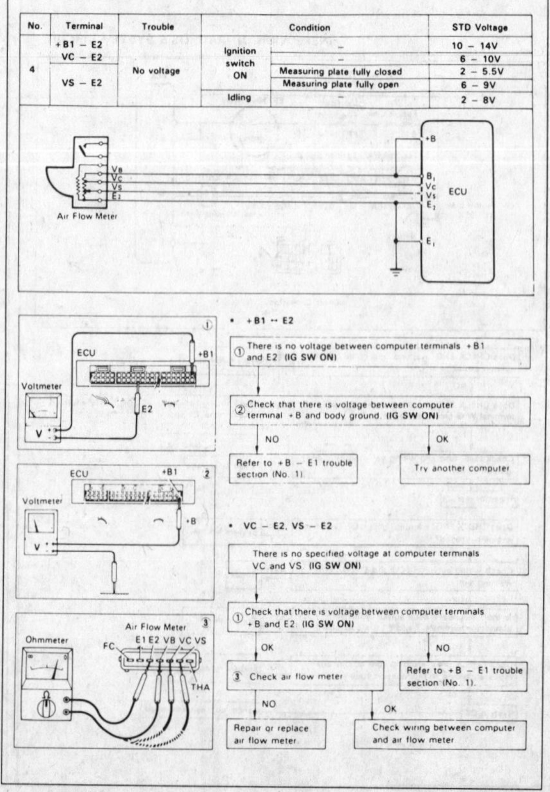

No.	Terminal	Trouble	Condition		STD Voltage
4	+B1 – E2	No voltage	Ignition switch ON	–	10 – 14V
	VC – E2			–	6 – 10V
	VS – E2			Measuring plate fully closed	2 – 5.5V
				Measuring plate fully open	6 – 9V
			Idling	–	2 – 8V

Air Flow Meter
VB
VC
VS
E2
+B
B1
Vc
Vs
E2
E1
ECU
• +B1 ↔ E2
① There is no voltage between computer terminals +B1 and E2. (IG SW ON)
② Check that there is voltage between computer terminal +B and body ground. (IG SW ON)
NO
OK
Refer to +B – E1 trouble section (No. 1).
Try another computer.
• VC – E2, VS – E2
There is no specified voltage at computer terminals VC and VS. (IG SW ON)
① Check that there is voltage between computer terminals +B and E2. (IG SW ON)
OK
NO
③ Check air flow meter
Refer to +B – E1 trouble section (No. 1).
NO
OK
Repair or replace air flow meter.
Check wiring between computer and air flow meter.
Voltmeter
Ohmmeter
Air Flow Meter
FC
E1 E2 VB VC VS
THA

TROUBLESHOOTING CHART 4—MR-2 4A-GE

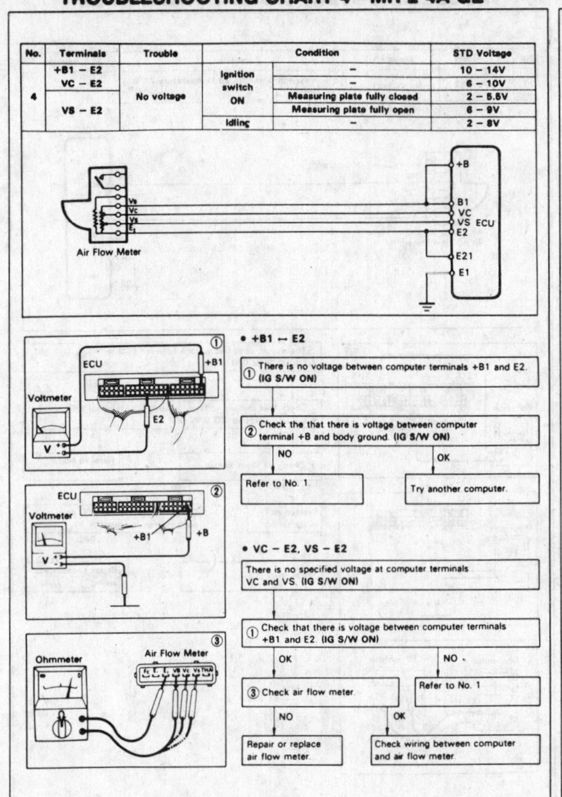

No.	Terminals	Trouble	Condition		STD Voltage
4	+B1 — E2	No voltage	Ignition switch ON	—	10 — 14V
	VC — E2			—	6 — 10V
	VS — E2			Measuring plate fully closed	2 — 5.5V
				Measuring plate fully open	6 — 9V
			Idling	—	2 — 8V

Air Flow Meter
+B
B1
VC
VS
E2
ECU
E21
E1
ECU
Voltmeter
+B1
E2
V
Ohmmeter
Air Flow Meter
• +B1 — E2
① There is no voltage between computer terminals +B1 and E2. (IG S/W ON)
② Check the that there is voltage between computer terminal +B and body ground. (IG S/W ON)
NO
OK
Refer to No. 1
Try another computer.
• VC — E2, VS — E2
There is no specified voltage at computer terminals VC and VS. (IG S/W ON)
① Check that there is voltage between computer terminals +B1 and E2. (IG S/W ON)
OK
NO
③ Check air flow meter.
Refer to No. 1
NO
OK
Repair or replace air flow meter.
Check wiring between computer and air flow meter.

TROUBLESHOOTING CHART 5—COROLLA 4A-GE

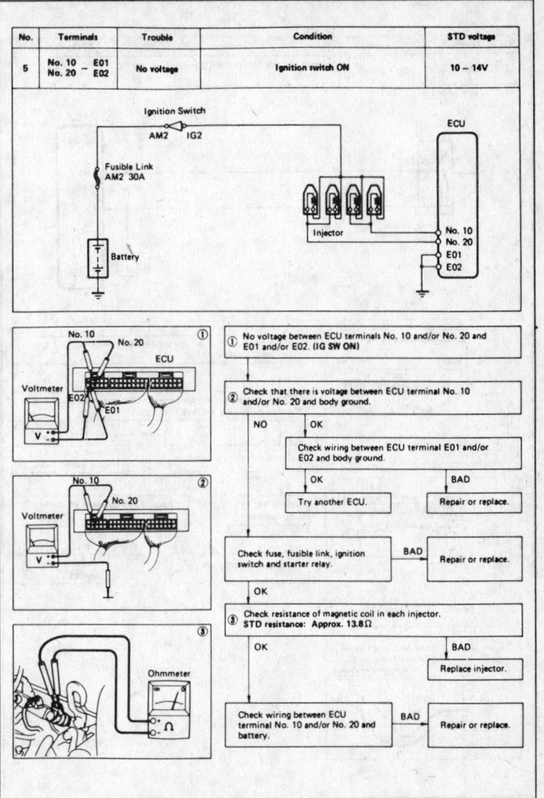

No.	Terminals	Trouble	Condition	STD voltage
5	No. 10 — E01 No. 20 — E02	No voltage	Ignition switch ON	10 – 14V

Ignition Switch
AM2
IG2
ECU
Fusible Link
AM2 30A
Injector
Battery
No. 10
No. 20
E01
E02
ECU
Voltmeter
Ohmmeter
① No voltage between ECU terminals No. 10 and/or No. 20 and E01 and/or E02. (IG SW ON)
② Check that there is voltage between ECU terminal No. 10 and/or No. 20 and body ground.
NO
OK
Check wiring between ECU terminal E01 and/or E02 and body ground.
OK
BAD
Try another ECU.
Repair or replace.
Check fuse, fusible link, ignition switch and starter relay.
BAD
Repair or replace.
OK
③ Check resistance of magnetic coil in each injector. STD resistance: Approx. 13.8Ω
BAD
OK
Replace injector.
Check wiring between ECU terminal No. 10 and/or No. 20 and battery.
BAD
Repair or replace.

TROUBLESHOOTING CHART 5—MR-2 4A-GE

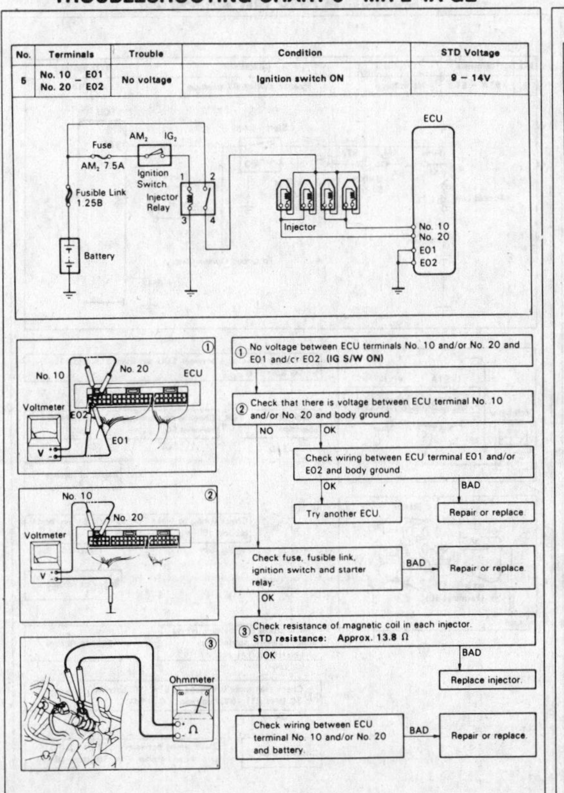

No.	Terminals	Trouble	Condition	STD Voltage
5	No. 10 — E01 No. 20 — E02	No voltage	Ignition switch ON	9 – 14V

ECU
Fuse
AM₂ 7.5A
AM₂
IG₂
Ignition Switch
Fusible Link 1.25B
Injector Relay
Battery
Injector
No. 10
No. 20
E01
E02
ECU
Voltmeter
Ohmmeter
① No voltage between ECU terminals No. 10 and/or No. 20 and E01 and/or E02. (IG S/W ON)
② Check that there is voltage between ECU terminal No. 10 and/or No. 20 and body ground.
NO
OK
Check wiring between ECU terminal E01 and/or E02 and body ground.
OK
BAD
Try another ECU.
Repair or replace.
Check fuse, fusible link, ignition switch and starter relay
BAD
Repair or replace.
OK
③ Check resistance of magnetic coil in each injector. STD resistance: Approx. 13.8 Ω
OK
BAD
Replace injector.
Check wiring between ECU terminal No. 10 and/or No. 20 and battery.
BAD
Repair or replace.

TROUBLESHOOTING CHART 6—COROLLA AND MR-2 4A-GE

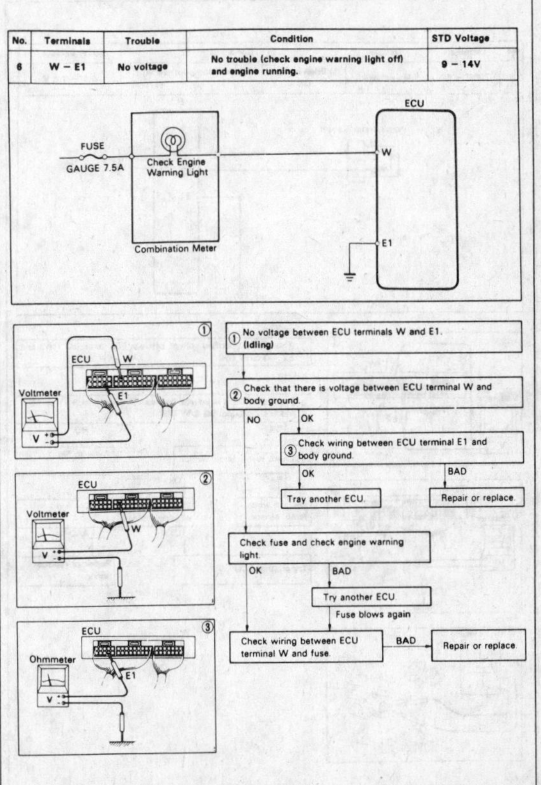

No.	Terminals	Trouble	Condition	STD Voltage
6	W — E1	No voltage	No trouble (check engine warning light off) and engine running.	9 – 14V

ECU
FUSE
GAUGE 7.5A
Check Engine Warning Light
Combination Meter
W
E1
ECU
Voltmeter
Ohmmeter
① No voltage between ECU terminals W and E1. (Idling)
② Check that there is voltage between ECU terminal W and body ground.
NO
OK
③ Check wiring between ECU terminal E1 and body ground.
OK
BAD
Tray another ECU.
Repair or replace.
Check fuse and check engine warning light.
OK
BAD
Try another ECU
Fuse blows again
Check wiring between ECU terminal W and fuse.
BAD
Repair or replace.

TROUBLESHOOTING CHART 7—COROLLA 4A-GE

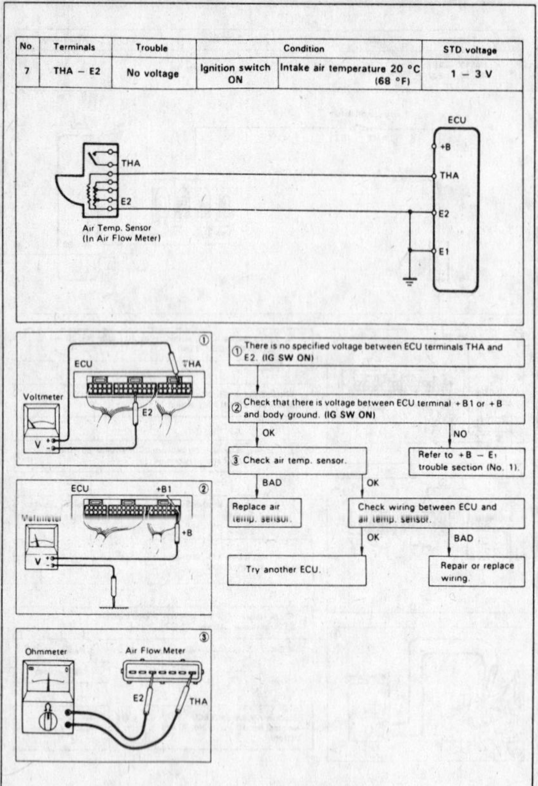

No.	Terminals	Trouble	Condition		STD voltage
7	THA – E2	No voltage	Ignition switch ON	Intake air temperature 20 °C (68 °F)	1 – 3 V

ECU
+B
THA
E2
E1
Air Temp. Sensor
(In Air Flow Meter)
Voltmeter
+B1
Ohmmeter
Air Flow Meter
① There is no specified voltage between ECU terminals THA and E2. (IG SW ON)
② Check that there is voltage between ECU terminal +B1 or +B and body ground. (IG SW ON)
OK
NO
③ Check air temp. sensor.
Refer to +B – E1 trouble section (No. 1).
BAD
OK
Replace air temp. sensor.
Check wiring between ECU and air temp. sensor.
OK
BAD
Try another ECU.
Repair or replace wiring.

TROUBLESHOOTING CHART 7—MR-2 4A-GE

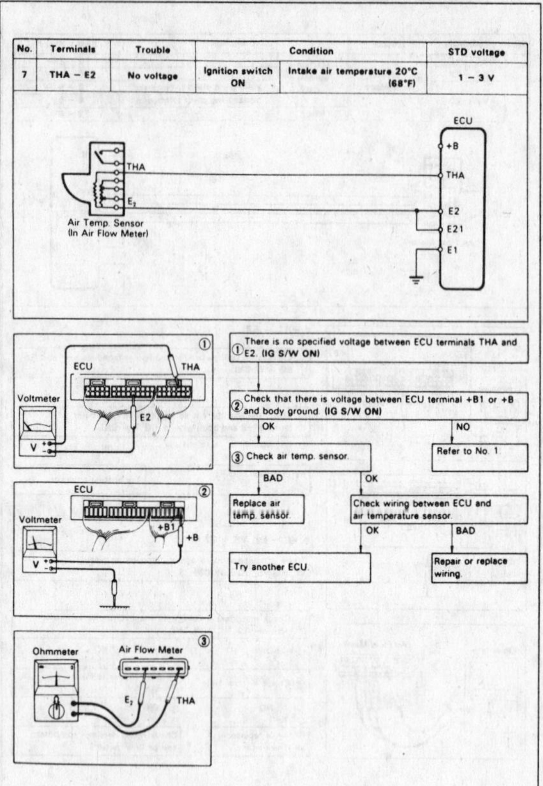

No.	Terminals	Trouble	Condition		STD voltage
7	THA – E2	No voltage	Ignition switch ON	Intake air temperature 20°C (68°F)	1 – 3 V

ECU
+B
THA
E2
E21
E1
Air Temp. Sensor
(In Air Flow Meter)
Voltmeter
+B1
Ohmmeter
Air Flow Meter
① There is no specified voltage between ECU terminals THA and E2. (IG S/W ON)
② Check that there is voltage between ECU terminal +B1 or +B and body ground. (IG S/W ON)
OK
NO
③ Check air temp. sensor.
Refer to No. 1
BAD
OK
Replace air temp. sensor.
Check wiring between ECU and air temperature sensor.
OK
BAD
Try another ECU.
Repair or replace wiring.

TROUBLESHOOTING CHART 8—COROLLA AND MR-2 4A-GE

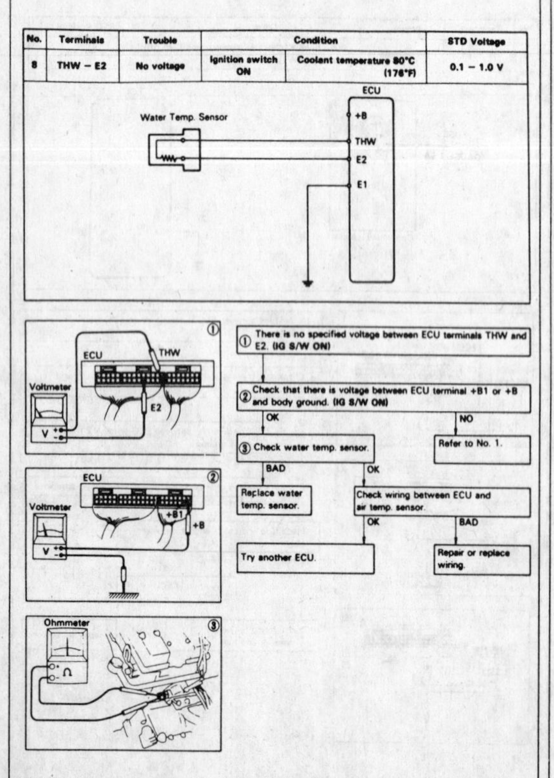

No.	Terminals	Trouble	Condition		STD Voltage
8	THW – E2	No voltage	Ignition switch ON	Coolant temperature 80°C (176°F)	0.1 – 1.0 V

ECU
Water Temp. Sensor
+B
THW
E2
E1
Voltmeter
+B1
Ohmmeter
① There is no specified voltage between ECU terminals THW and E2. (IG S/W ON)
② Check that there is voltage between ECU terminal +B1 or +B and body ground. (IG S/W ON)
OK
NO
③ Check water temp. sensor.
Refer to No. 1.
BAD
OK
Replace water temp. sensor.
Check wiring between ECU and air temp. sensor.
OK
BAD
Try another ECU.
Repair or replace wiring.

TROUBLESHOOTING CHART 9—COROLLA AND MR-2 4A-GE

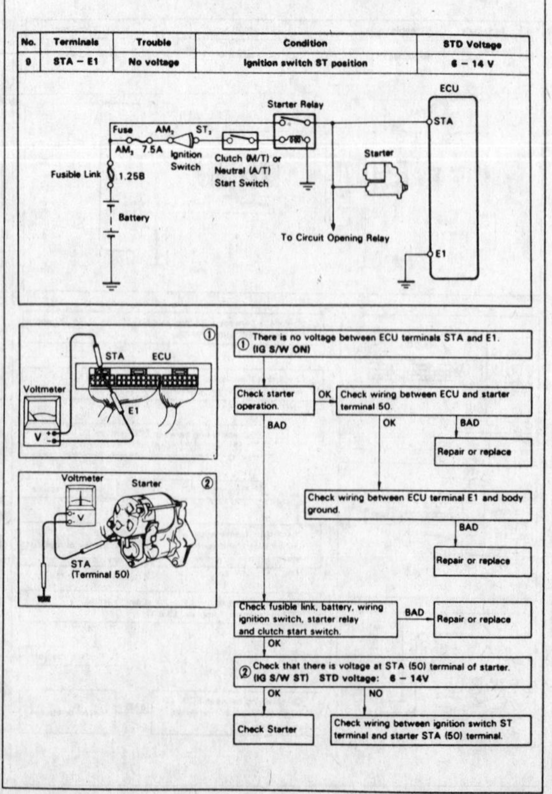

No.	Terminals	Trouble	Condition	STD Voltage
9	STA – E1	No voltage	Ignition switch ST position	6 – 14 V

ECU
Starter Relay
STA
Fuse
AM2
7.5A
ST2
Ignition Switch
Clutch (M/T) or Neutral (A/T) Start Switch
Starter
Fusible Link
1.25B
Battery
To Circuit Opening Relay
E1
Voltmeter
STA (Terminal 50)
① There is no voltage between ECU terminals STA and E1. (IG S/W ON)
Check starter operation.
OK
Check wiring between ECU and starter terminal 50.
BAD
OK
BAD
Repair or replace
Check wiring between ECU terminal E1 and body ground.
BAD
Repair or replace
Check fusible link, battery, wiring ignition switch, starter relay and clutch start switch.
BAD
Repair or replace
OK
② Check that there is voltage at STA (50) terminal of starter. (IG S/W ST) STD voltage: 6 – 14V
OK
NO
Check Starter
Check wiring between ignition switch ST terminal and starter STA (50) terminal.

TROUBLESHOOTING CHART 10—COROLLA 4A-GE

TROUBLESHOOTING CHART 10—MR-2 4A-GE

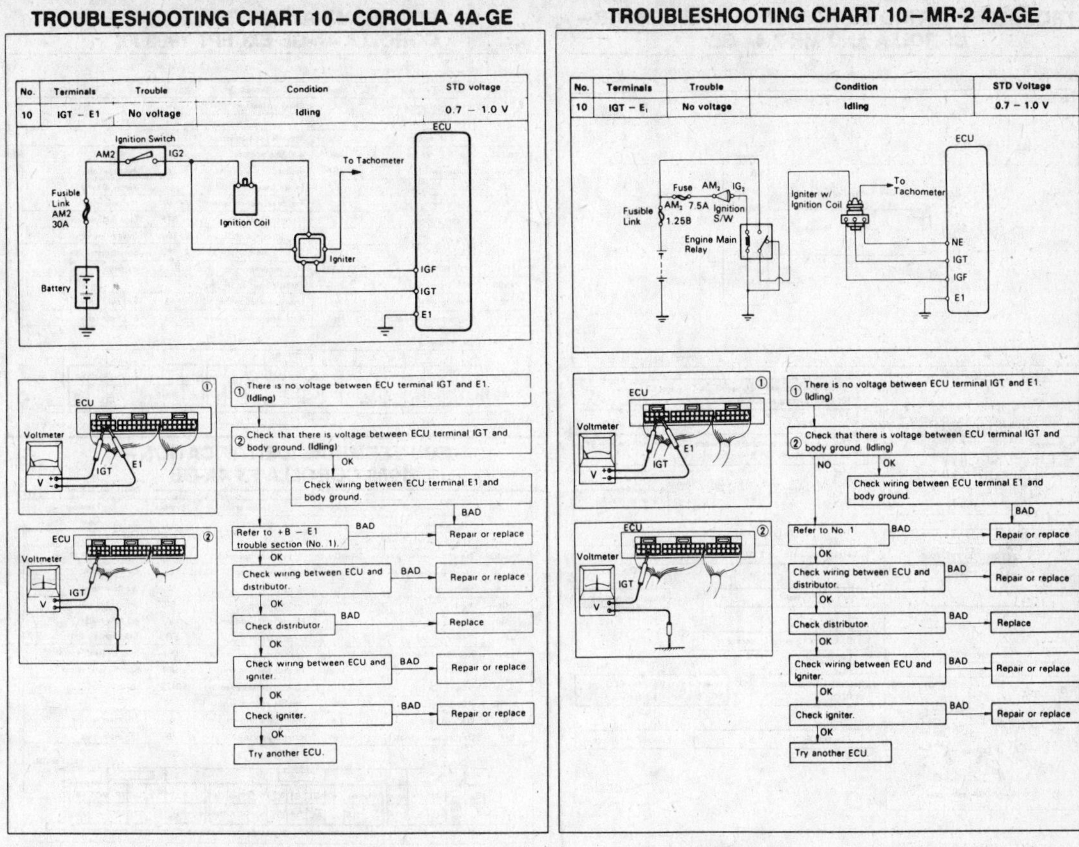
No. | Terminals | Trouble | Condition | STD voltage
10 | IGT – E1 | No voltage | Idling | 0.7 – 1.0 V
Ignition Switch
AM2
IG2
To Tachometer
ECU
Fusible Link AM2 30A
Ignition Coil
Igniter
Battery
IGF
IGT
E1
① There is no voltage between ECU terminal IGT and E1. (Idling)
② Check that there is voltage between ECU terminal IGT and body ground. (Idling)
OK
Check wiring between ECU terminal E1 and body ground.
BAD
Refer to +B – E1 trouble section (No. 1).
Repair or replace
Check wiring between ECU and distributor.
Repair or replace
Check distributor.
Replace
Check wiring between ECU and igniter.
Repair or replace
Check igniter.
Repair or replace
Try another ECU.
Voltmeter
No. | Terminals | Trouble | Condition | STD Voltage
10 | IGT – E1 | No voltage | Idling | 0.7 – 1.0 V
Fuse AM2 7.5A
IG2
Ignition S/W
Fusible Link AM2 1.25B
Engine Main Relay
Igniter w/ Ignition Coil
To Tachometer
NE
IGT
IGF
E1
① There is no voltage between ECU terminal IGT and E1. (Idling)
② Check that there is voltage between ECU terminal IGT and body ground. (Idling)
NO
OK
Check wiring between ECU terminal E1 and body ground.
BAD
Refer to No. 1
Repair or replace
Check wiring between ECU and distributor.
Repair or replace
Check distributor.
Replace
Check wiring between ECU and Igniter.
Repair or replace
Check igniter.
Repair or replace
Try another ECU.

TROUBLESHOOTING CHART 11—COROLLA AND MR-2 4A-GE

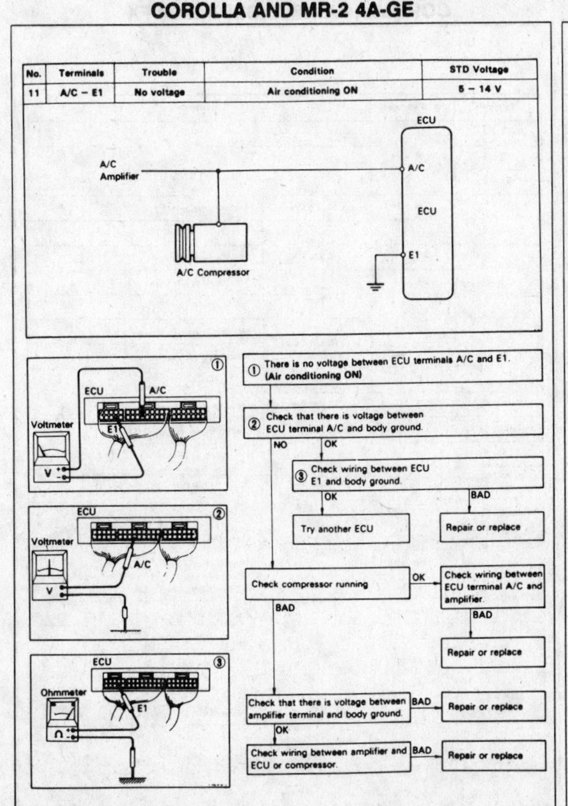
No. | Terminals | Trouble | Condition | STD Voltage
11 | A/C – E1 | No voltage | Air conditioning ON | 5 – 14 V
ECU
A/C Amplifier
A/C
E1
A/C Compressor
Voltmeter
Ohmmeter
① There is no voltage between ECU terminals A/C and E1. (Air conditioning ON)
② Check that there is voltage between ECU terminal A/C and body ground.
NO
OK
③ Check wiring between ECU E1 and body ground.
OK
BAD
Try another ECU
Repair or replace
Check compressor running
OK
Check wiring between ECU terminal A/C and amplifier.
BAD
Repair or replace
Check that there is voltage between amplifier terminal and body ground.
Repair or replace
Check wiring between amplifier and ECU or compressor.
Repair or replace

TROUBLESHOOTING OXYGEN SENSOR—COROLLA AND MR-2 4A-GE

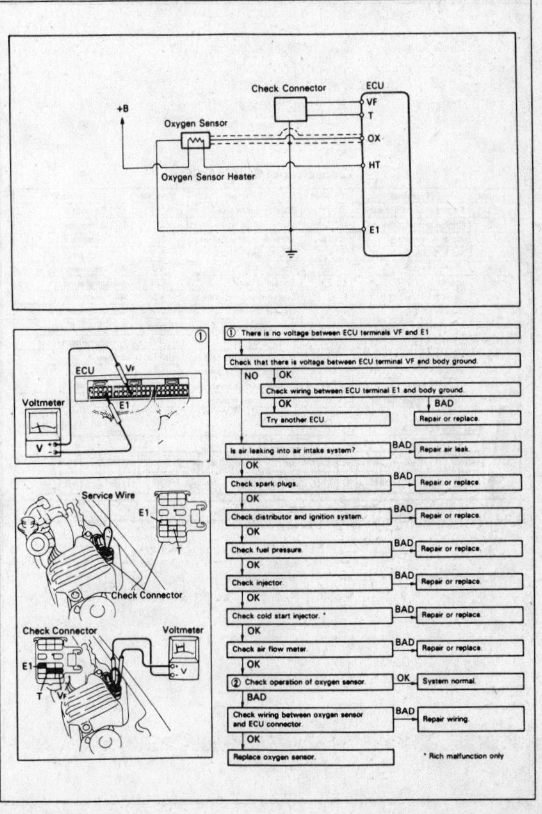
Check Connector
ECU
+B
VF
T
Oxygen Sensor
OX
HT
Oxygen Sensor Heater
E1
Voltmeter
Service Wire
Check Connector
① There is no voltage between ECU terminals VF and E1
Check that there is voltage between ECU terminal VF and body ground.
NO
OK
Check wiring between ECU terminal E1 and body ground.
OK
BAD
Try another ECU.
Repair or replace.
Is air leaking into air intake system?
Repair air leak.
Check spark plugs.
Repair or replace.
Check distributor and ignition system.
Repair or replace.
Check fuel pressure.
Repair or replace.
Check injector.
Repair or replace.
Check cold start injector. *
Repair or replace.
Check air flow meter.
Repair or replace.
② Check operation of oxygen sensor.
System normal.
Check wiring between oxygen sensor and ECU connector.
Repair wiring.
Replace oxygen sensor.
* Rich malfunction only

TROUBLESHOOTING EGR TEMPERATURE SENSOR – COROLLA AND MR-2 4A-GE

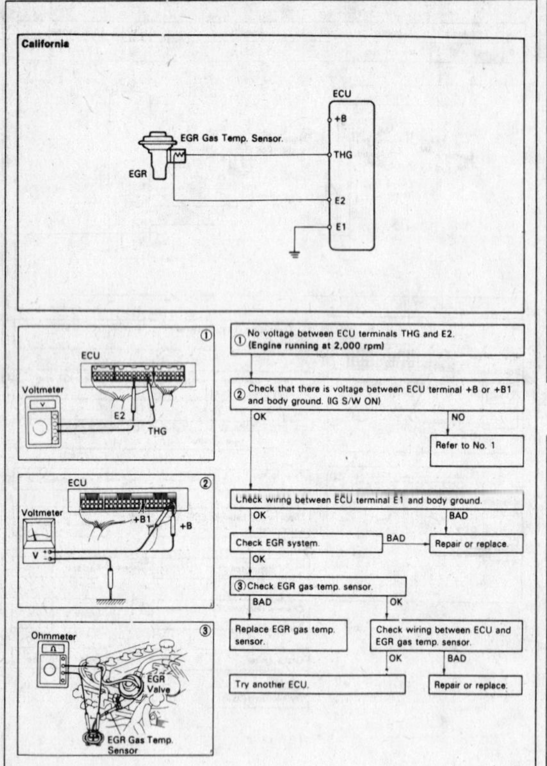
California
ECU
+B
EGR Gas Temp. Sensor
THG
EGR
E2
E1
① ECU
Voltmeter
E2
THG
② ECU
Voltmeter
+B1
+B
V
③ Ohmmeter
EGR Valve
EGR Gas Temp. Sensor
① No voltage between ECU terminals THG and E2. (Engine running at 2,000 rpm)
② Check that there is voltage between ECU terminal +B or +B1 and body ground. (IG S/W ON)
OK
NO
Refer to No. 1
Check wiring between ECU terminal E1 and body ground.
OK
BAD
Check EGR system.
BAD
Repair or replace.
OK
③ Check EGR gas temp. sensor.
BAD
OK
Replace EGR gas temp. sensor.
Check wiring between ECU and EGR gas temp. sensor.
OK
BAD
Try another ECU.
Repair or replace.

ECU TERMINAL IDENTIFICATION – COROLLA 4A-GE EXCEPT 1988 FX

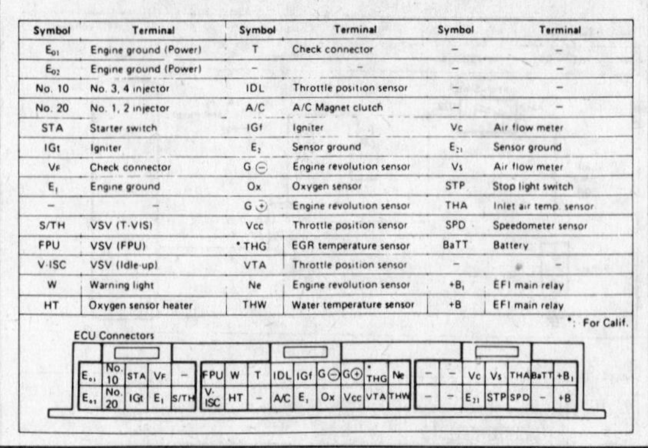

Symbol	Terminal	Symbol	Terminal	Symbol	Terminal
E_{01}	Engine ground (Power)	T	Check connector	–	–
E_{02}	Engine ground (Power)	–	–	–	–
No. 10	No. 3, 4 injector	IDL	Throttle position sensor	–	–
No. 20	No. 1, 2 injector	A/C	A/C Magnet clutch	–	–
STA	Starter switch	IGf	Igniter	Vc	Air flow meter
IGt	Igniter	E_2	Sensor ground	E_{21}	Sensor ground
V_F	Check connector	G ⊖	Engine revolution sensor	Vs	Air flow meter
E_1	Engine ground	Ox	Oxygen sensor	STP	Stop light switch
–	–	G ⊕	Engine revolution sensor	THA	Inlet air temp. sensor
S/TH	VSV (T-VIS)	Vcc	Throttle position sensor	SPD	Speedometer sensor
FPU	VSV (FPU)	*THG	EGR temperature sensor	BaTT	Battery
V-ISC	VSV (Idle-up)	VTA	Throttle position sensor	–	–
W	Warning light	Ne	Engine revolution sensor	$+B_1$	EFI main relay
HT	Oxygen sensor heater	THW	Water temperature sensor	+B	EFI main relay

*: For Calif.

ECU Connectors
E_{01} No. 10 STA V_F – FPU W T IDL IGf G⊖ G⊕ *THG Ne – – Vc Vs THA BaTT $+B_1$
E_{02} No. 20 IGt E_1 S/TH V-ISC HT – A/C E_1 Ox Vcc VTA THW – – E_{21} STP SPD – +B

ECU TERMINAL IDENTIFICATION – 1988 COROLLA FX 4A-GE

Symbol	Terminal	Symbol	Terminal	Symbol	Terminal
E_{01}	Engine ground (Power)	T	Check connector	*1 L_3	ECT computer
E_{02}	Engine ground (Power)	–	–	*1 ECT	ECT computer
No. 10	No. 3, 4 injector	IDL	Throttle position sensor	*1 L_1	ECT computer
No. 20	No. 1, 2 injector	A/C	A/C Magnet clutch	*1 L_2	ECT computer
STA	Starter switch	IGf	Igniter	Vc	Air flow meter
IGt	Igniter	E_2	Sensor ground	E_{21}	Sensor ground
V_F	Check connector	G ⊖	Engine revolution sensor	Vs	Air flow meter
E_1	Engine ground	Ox	Oxygen sensor	STP	Stop light switch
*1 NSW	Neutral start switch	G ⊕	Engine revolution sensor	THA	Inlet air temp. sensor
S/TH	VSV (T-VIS)	Vcc	Throttle position sensor	SPD	Speedometer sensor
FPU	VSV (FPU)	*2 THG	EGR temperature sensor	BaTT	Battery
V-ISC	VSV (Idle-up)	VTA	Throttle position sensor	–	–
W	Warning light	Ne	Engine revolution sensor	$+B_1$	EFI main relay
HT	Oxygen sensor heater	THW	Water temperature sensor	+B	EFI main relay

*1: For A/T
*2: For Calif.

ECU Connectors
E_{01} No. 10 STA V_F *1 NSW FPU W T IDL IGf G⊖ G⊕ *2 THG Ne *1 L_3 *1 L_1 Vc Vs THA BaTT $+B_1$
E_{02} No. 20 IGt E_1 S/TH V-ISC HT – A/C E_1 Ox Vcc VTA THW *1 ECT *1 L_2 E_{21} STP SPD – +B

ECU TERMINAL IDENTIFICATION – MR-2 4A-GE

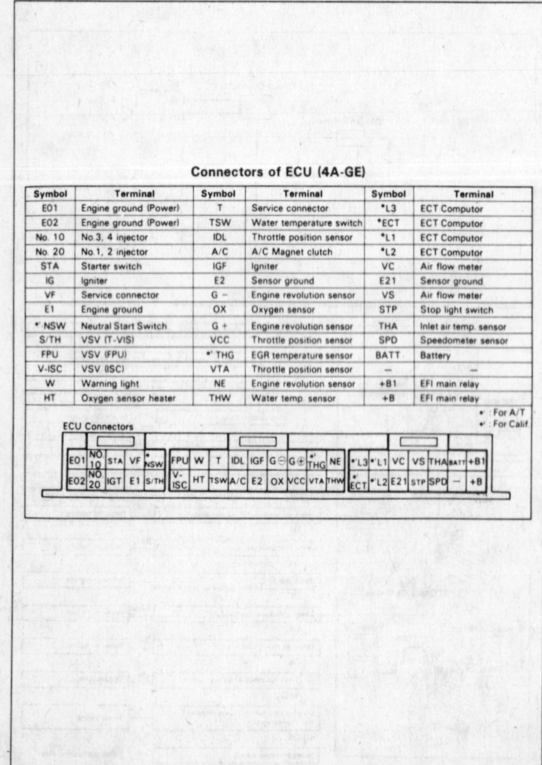

Connectors of ECU (4A-GE)

Symbol	Terminal	Symbol	Terminal	Symbol	Terminal
E01	Engine ground (Power)	T	Service connector	*L3	ECT Computor
E02	Engine ground (Power)	TSW	Water temperature switch	*ECT	ECT Computor
No. 10	No.3, 4 injector	IDL	Throttle position sensor	*L1	ECT Computor
No. 20	No.1, 2 injector	A/C	A/C Magnet clutch	*L2	ECT Computor
STA	Starter switch	IGF	Igniter	VC	Air flow meter
IG	Igniter	E2	Sensor ground	E21	Sensor ground
VF	Service connector	G –	Engine revolution sensor	VS	Air flow meter
E1	Engine ground	OX	Oxygen sensor	STP	Stop light switch
*' NSW	Neutral Start Switch	G +	Engine revolution sensor	THA	Inlet air temp. sensor
S/TH	VSV (T-VIS)	VCC	Throttle position sensor	SPD	Speedometer sensor
FPU	VSV (FPU)	*' THG	EGR temperature sensor	BATT	Battery
V-ISC	VSV (ISC)	VTA	Throttle position sensor	–	–
W	Warning light	NE	Engine revolution sensor	+B1	EFI main relay
HT	Oxygen sensor heater	THW	Water temp. sensor	+B	EFI main relay

*' : For A/T
*' : For Calif.

ECU Connectors
E01 NO. 10 STA VF *NSW FPU W T IDL IGF G⊖ G⊕ *'THG NE *'L3 *'L1 VC VS THA BATT +B1
E02 NO. 20 IGT E1 S/TH V-ISC HT TSW A/C E2 OX VCC VTA THW *'ECT *'L2 E21 STP SPD – +B

ECU TERMINAL VOLTAGES – COROLLA 4A-GE EXCEPT 1988 FX

Terminals	STD Voltage	Condition	
BaTT – E_1	10 – 14	–	
+B – E_1		IG S/W ON	
$+B_1$ – E_1			
IDL – E_2	10 – 14	IG S/W ON	Throttle valve open
Vc – E_2	6 – 10	IG S/W ON	–
VTA – E_2	0.1 – 1.0		Throttle valve fully closed
	4 – 5		Throttle valve fully open
Vcc – E_2	4 – 6	IG S/W ON	
Vs – E_2	2 – 5.5	IG S/W ON	Measuring plate fully closed
	6 – 9		Measuring plate fully open
	2 – 8	Idling	
THA – E_2	1 – 3	IG S/W ON	Intake air temperature 20°C (68° F)
THW – E_2	0.1 – 1.0	IG S/W ON	Coolant temperature 80°C (176°F)
STA – E_1	6 – 14	IG S/W ST position	
No.10 – E_{01} No.20 – E_{02}	9 – 14	IG S/W ON	
IGt – E_1	0.7 – 1.0	Idling	
T – E_1	10 – 14	IG S/W ON	Check connector T ↔ E_1 not short
	0.5 or less		Check connector T ↔ E_1 short
A/C – E_1	5 – 14	IG S/W ON	A/C switch ON
	0.5 or less		A/C switch OFF
W – E_1	0.5 or less	IG S/W ON	
	9 – 14	Engine start	
S/TH – E_1	0 – 2	Idling	
	10 – 14	More than 4,350 rpm	

ECU TERMINAL VOLTAGES—1988 COROLLA FX 4A-GE

Terminals	STD Voltage	Condition	
BaTT – E_1	10 – 14	–	
+B – E_1	10 – 14	IG S/W ON	
$+B_1$ – E_1	10 – 14	IG S/W ON	
IDL – E_2	10 – 14	IG S/W ON	Throttle valve open
Vc – E_2	6 – 10		–
VTA – E_2	0.1 – 1.0	IG S/W ON	Throttle valve fully closed
	4 – 5		Throttle valve fully open
Vcc – E_2	4 – 6	IG S/W ON	
Vs – E_2	2 – 5.5	IG S/W ON	Measuring plate fully closed
	6 – 9		Measuring plate fully open
	2 – 8	Idling	
THA – E_2	1 – 3	IG S/W ON	Intake air temperature 20°C (68°F)
THW – E_2	0.1 – 1.0	IG S/W ON	Coolant temperature 80°C (176°F)
STA – E_1	6 – 14	IG S/W ST position and press on the cluch pedal (M/T)	
No.10 – E_{01} No.20 – E_{02}	9 – 14	IG S/W ON	
IGt – E_1	0.7 – 1.0	Idling	
T – E_1	10 – 14	IG S/W ON	Check connector T ↔ E_1 not short
	0		Check connector T ↔ E_1 short
A/C – E_1	5 – 14	IG S/W ON	A/C switch ON
	0.5 or less		A/C switch OFF
W – E_1	0.5 or less	IG S/W ON	
	9 – 14	Engine start	
S/TH – E_1	0 – 2	Idling	
	10 – 14	More than 4,350 rpm	
* NSW – E_1	0	IG S/W ON	Shift position P or N range
	10 – 14		Ex. P or N range
	6 – 11	Cranking	

*: For A/T

ECU TERMINAL VOLTAGES—MR-2 4A-GE

Terminals	STD voltage	Condition	
BATT – E1	10 – 14	–	
+B – E1	10 – 14	Ignition S/W ON	
+B1 – E1	10 – 14	Ignition S/W ON	
IDL – E1	10 – 14	Ignition S/W ON	Throttle valve open
VTA – E2	0.1 – 1.0		Throttle valve fully closed
	4 – 5		Throttle valve fully open
VCC – E2	4 – 6	Ignition S/W ON	
VC – E2	6 – 10	Ignition S/W ON	–
VS – E2	2 – 5.5		Measuring plate fully closed
	6 – 9		Measuring plate fully open
	2 – 8	Idling	–
THA – E2	1 – 3	Ignition S/W ON	Intake air temperature 20°C (68°F)
THW – E2	0.1 – 1.0	Ignition S/W ON	Coolant temperature 80°C (176°F)
STA – E1	6 – 14	Ignition S/W ST position and press on the clutch pedal (M/T)	
No. 10 – E01 No. 20 – E02	9 – 14	Ignition S/W ON	
IGT – E1	0.7 – 1.0	Idling	
T – E1	10 – 14	Ignition S/W ON	Service connector T ↔ E_1 not short
	0.5 or less		Service connector T ↔ E_1 short
A/C – E1	5 – 14	Ignition S/W ON	A/C switch ON
	0.5 or less		A/C switch OFF
W – E1	0.5 or less	Ignition S/W ON	
	9 – 14	Engine start	
S/TH – E1	0 – 2	Idling	
	10 – 14	More than 4,350 rpm	
* NSW – E1	0	Ignition S/W ON	Shift position P or N range
	10 – 14		Ex. P or N range
	6 – 11	Cranking	

*: For A/T

TROUBLESHOOTING CHART 1—MR-2 4A-GZE

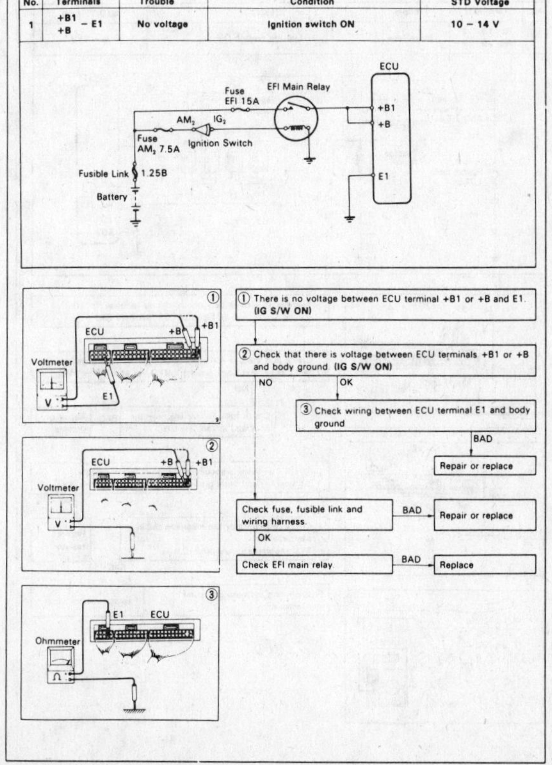
No.
Terminals
Trouble
Condition
STD Voltage
1
+B1 – E1
+B
No voltage
Ignition switch ON
10 – 14 V
ECU
Fuse EFI 15A
EFI Main Relay
+B1
+B
AM2
IG2
Fuse AM2 7.5A
Ignition Switch
Fusible Link 1.25B
Battery
E1
① There is no voltage between ECU terminal +B1 or +B and E1. (IG S/W ON)
② Check that there is voltage between ECU terminals +B1 or +B and body ground. (IG S/W ON)
NO
OK
③ Check wiring between ECU terminal E1 and body ground.
BAD
Repair or replace
Check fuse, fusible link and wiring harness.
BAD
Repair or replace
OK
Check EFI main relay.
BAD
Replace
Voltmeter
Ohmmeter

TROUBLESHOOTING CHART 2—MR-2 4A-GZE

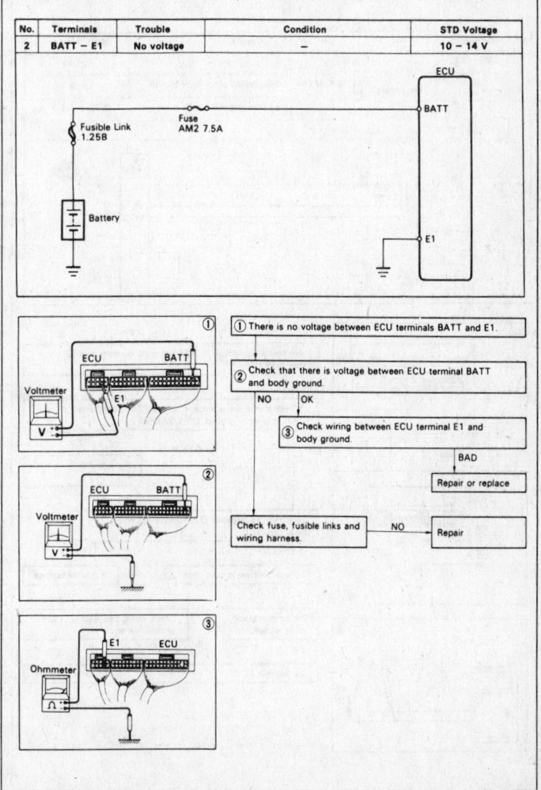
No.
Terminals
Trouble
Condition
STD Voltage
2
BATT – E1
No voltage
–
10 – 14 V
ECU
BATT
Fuse AM2 7.5A
Fusible Link 1.25B
Battery
E1
① There is no voltage between ECU terminals BATT and E1.
② Check that there is voltage between ECU terminal BATT and body ground.
NO
OK
③ Check wiring between ECU terminal E1 and body ground.
BAD
Repair or replace
Check fuse, fusible links and wiring harness.
NO
Repair
Voltmeter
Ohmmeter

TROUBLESHOOTING CHART 3 – MR-2 4A-GZE

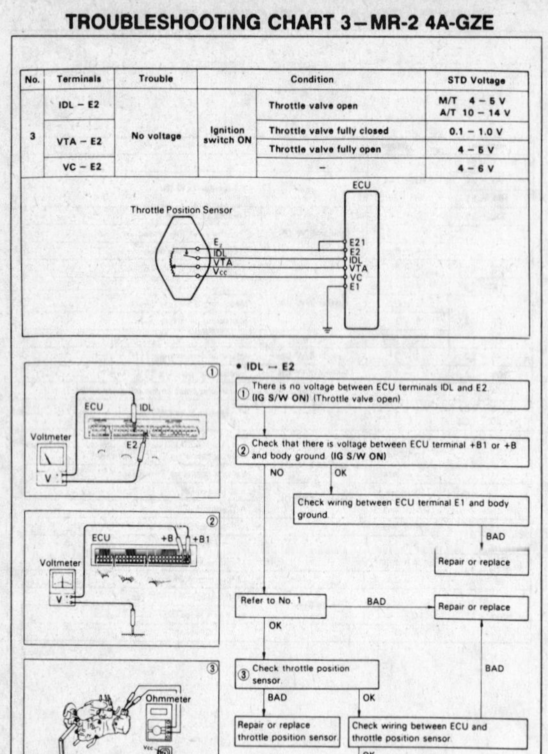

No.	Terminals	Trouble	Condition		STD Voltage
3	IDL – E2	No voltage	Ignition switch ON	Throttle valve open	M/T 4 – 5 V A/T 10 – 14 V
	VTA – E2			Throttle valve fully closed	0.1 – 1.0 V
				Throttle valve fully open	4 – 5 V
	VC – E2			–	4 – 6 V

ECU
Throttle Position Sensor
E1
IDL
VTA
Vcc
E21
E2
IDL
VTA
VC
E1
• IDL – E2
① There is no voltage between ECU terminals IDL and E2 (IG S/W ON) (Throttle valve open)
② Check that there is voltage between ECU terminal +B1 or +B and body ground. (IG S/W ON)
NO
OK
Check wiring between ECU terminal E1 and body ground.
BAD
Repair or replace
Refer to No. 1
BAD
Repair or replace
OK
③ Check throttle position sensor.
BAD
OK
BAD
Repair or replace throttle position sensor
Check wiring between ECU and throttle position sensor.
OK
Try another ECU
ECU
IDL
Voltmeter
E2
+B
+B1
Ohmmeter
Thickness Gauge

TROUBLESHOOTING CHART 3 – MR-2 4A-GZE

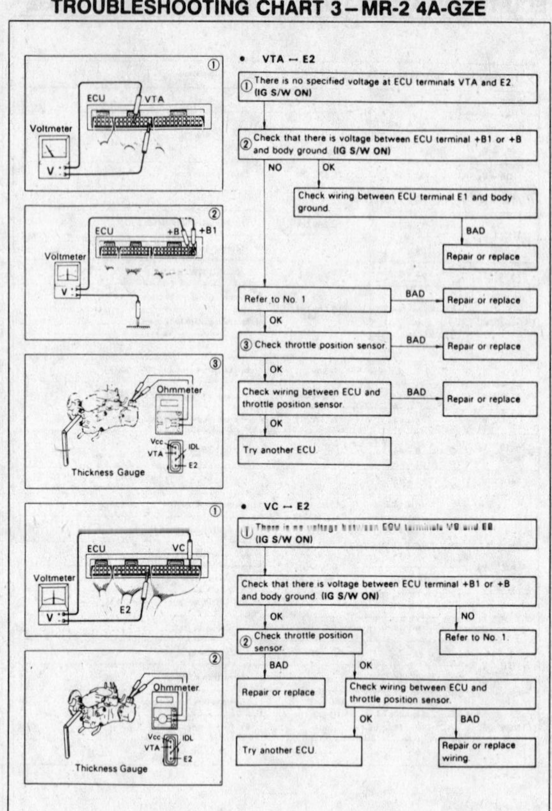

• VTA – E2
① There is no specified voltage at ECU terminals VTA and E2. (IG S/W ON)
② Check that there is voltage between ECU terminal +B1 or +B and body ground. (IG S/W ON)
NO
OK
Check wiring between ECU terminal E1 and body ground.
BAD
Repair or replace
Refer to No. 1
BAD
Repair or replace
OK
③ Check throttle position sensor.
BAD
Repair or replace
OK
Check wiring between ECU and throttle position sensor.
BAD
Repair or replace
OK
Try another ECU
• VC – E2
① There is no voltage between ECU terminals VC and E2 (IG S/W ON)
Check that there is voltage between ECU terminal +B1 or +B and body ground. (IG S/W ON)
OK
NO
② Check throttle position sensor.
Refer to No. 1.
BAD
OK
Repair or replace
Check wiring between ECU and throttle position sensor.
OK
BAD
Try another ECU.
Repair or replace wiring.
ECU
VTA
Voltmeter
+B
+B1
Ohmmeter
Vcc
IDL
VTA
E2
Thickness Gauge
VC

TROUBLESHOOTING CHART 4 – MR-2 4A-GZE

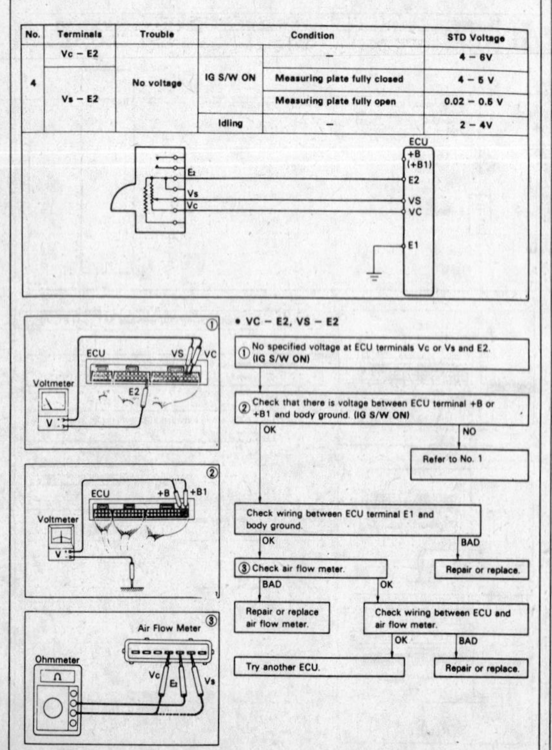

No.	Terminals	Trouble	Condition		STD Voltage
4	Vc – E2	No voltage	IG S/W ON	–	4 – 6V
	Vs – E2			Measuring plate fully closed	4 – 5 V
				Measuring plate fully open	0.02 – 0.5 V
			Idling	–	2 – 4V

ECU
+B (+B1)
E2
VS
VC
E1
E2
Vs
Vc
• VC – E2, VS – E2
① No specified voltage at ECU terminals Vc or Vs and E2. (IG S/W ON)
② Check that there is voltage between ECU terminal +B or +B1 and body ground. (IG S/W ON)
OK
NO
Refer to No. 1
Check wiring between ECU terminal E1 and body ground.
OK
BAD
③ Check air flow meter.
Repair or replace.
BAD
OK
Repair or replace air flow meter.
Check wiring between ECU and air flow meter.
OK
BAD
Try another ECU.
Repair or replace.
ECU
VS
VC
Voltmeter
E2
+B
+B1
Air Flow Meter
Ohmmeter
Vc
E1
Vs

TROUBLESHOOTING CHART 5 – MR-2 4A-GZE

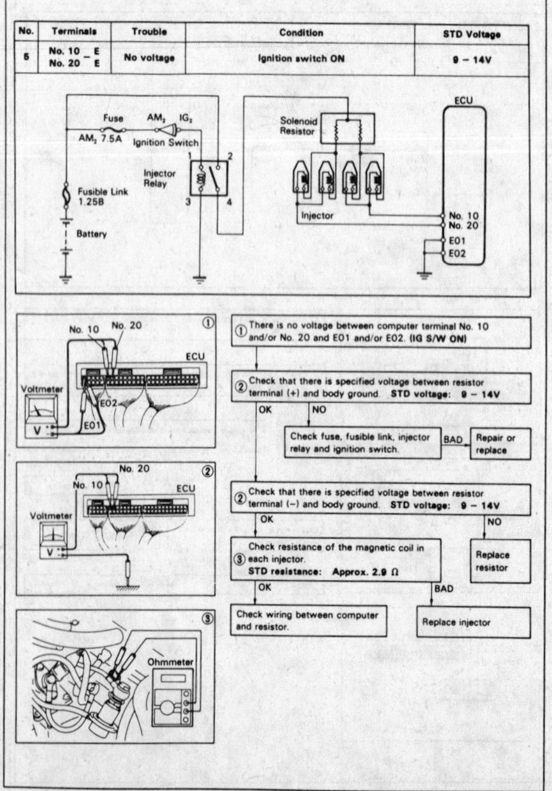

No.	Terminals	Trouble	Condition	STD Voltage
5	No. 10 – E No. 20 – E	No voltage	Ignition switch ON	9 – 14V

ECU
Fuse
AM2 7.5A
AM2
IG2
Ignition Switch
Solenoid Resistor
Injector Relay
Fusible Link 1.25B
Battery
Injector
No. 10
No. 20
E01
E02
① There is no voltage between computer terminal No. 10 and/or No. 20 and E01 and/or E02. (IG S/W ON)
② Check that there is specified voltage between resistor terminal (+) and body ground. STD voltage: 9 – 14V
OK
NO
Check fuse, fusible link, injector relay and ignition switch.
BAD
Repair or replace
② Check that there is specified voltage between resistor terminal (–) and body ground. STD voltage: 9 – 14V
OK
NO
③ Check resistance of the magnetic coil in each injector. STD resistance: Approx. 2.9 Ω
Replace resistor
OK
BAD
Check wiring between computer and resistor.
Replace injector
Voltmeter
Ohmmeter

TROUBLESHOOTING CHART 6–MR-2 4A-GZE

No.	Terminals	Trouble	Condition	STD Voltage
6	W – E1	No voltage	No trouble ("CHECK ENGINE" warning light off) and engine running.	9 – 14V

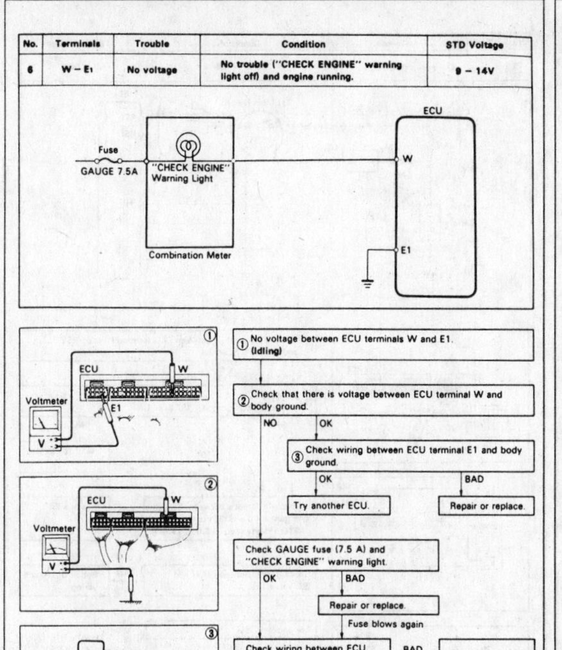
ECU
Fuse
GAUGE 7.5A
"CHECK ENGINE" Warning Light
Combination Meter
W
E1
① No voltage between ECU terminals W and E1. (Idling)
② Check that there is voltage between ECU terminal W and body ground.
NO
OK
③ Check wiring between ECU terminal E1 and body ground.
OK
BAD
Try another ECU.
Repair or replace.
Check GAUGE fuse (7.5 A) and "CHECK ENGINE" warning light.
OK
BAD
Repair or replace.
Fuse blows again
Check wiring between ECU terminal W and fuse.
BAD
Repair or replace.
Voltmeter
Ohmmeter

TROUBLESHOOTING CHART 7–MR-2 4A-GZE

No.	Terminals	Trouble	Condition		STD voltage
7	THA – E2	No voltage	Ignition switch ON	Intake air temperature 20°C (68°F)	1 – 3 V

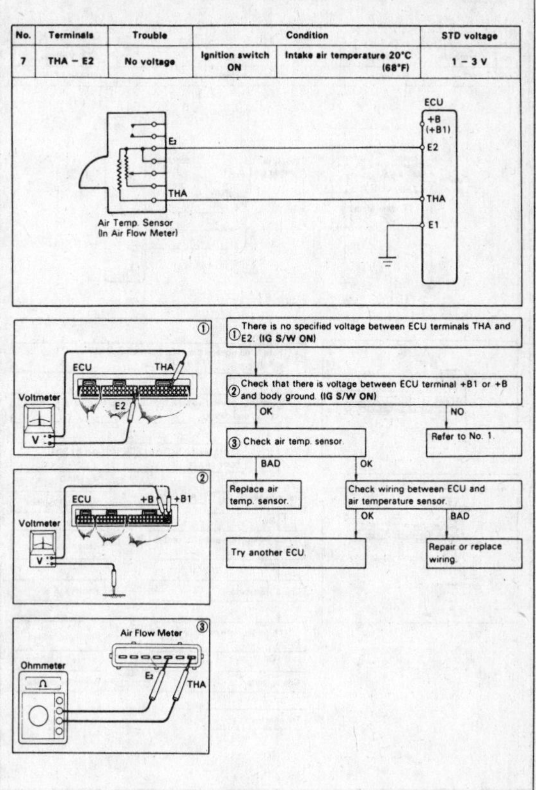
ECU
+B (+B1)
E2
THA
E1
Air Temp. Sensor (In Air Flow Meter)
① There is no specified voltage between ECU terminals THA and E2. (IG S/W ON)
② Check that there is voltage between ECU terminal +B1 or +B and body ground. (IG S/W ON)
OK
NO
Refer to No. 1.
③ Check air temp. sensor.
BAD
OK
Replace air temp. sensor.
Check wiring between ECU and air temperature sensor.
OK
BAD
Try another ECU.
Repair or replace wiring.
Voltmeter
Ohmmeter
Air Flow Meter

TROUBLESHOOTING CHART 8–MR-2 4A-GZE

No.	Terminals	Trouble	Condition		STD Voltage
8	THW – E2	No voltage	Ignition switch ON	Coolant temperature 80°C (176°F)	0.1 – 1.0 V

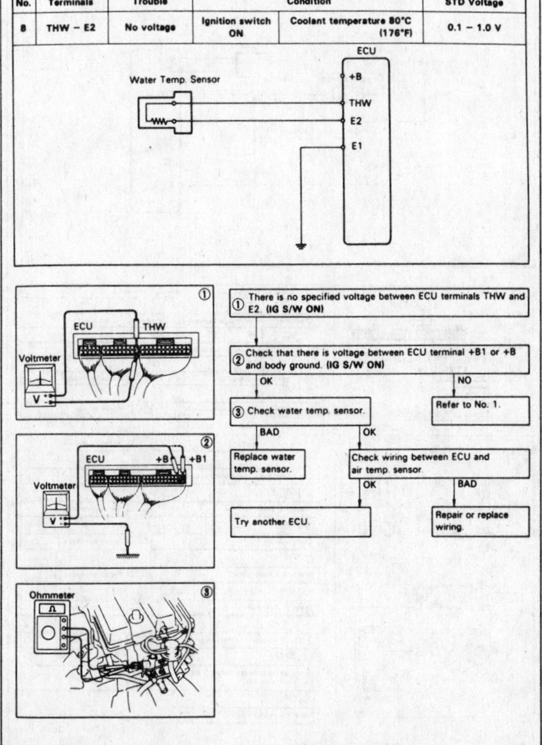
ECU
Water Temp. Sensor
+B
THW
E2
E1
① There is no specified voltage between ECU terminals THW and E2. (IG S/W ON)
② Check that there is voltage between ECU terminal +B1 or +B and body ground. (IG S/W ON)
OK
NO
Refer to No. 1.
③ Check water temp. sensor.
BAD
OK
Replace water temp. sensor.
Check wiring between ECU and air temp. sensor.
OK
BAD
Try another ECU.
Repair or replace wiring.
Voltmeter
Ohmmeter

TROUBLESHOOTING CHART 9–MR-2 4A-GZE

No.	Terminals	Trouble	Condition	STD Voltage
9	STA – E1	No voltage	Ignition switch ST position	6 – 14 V

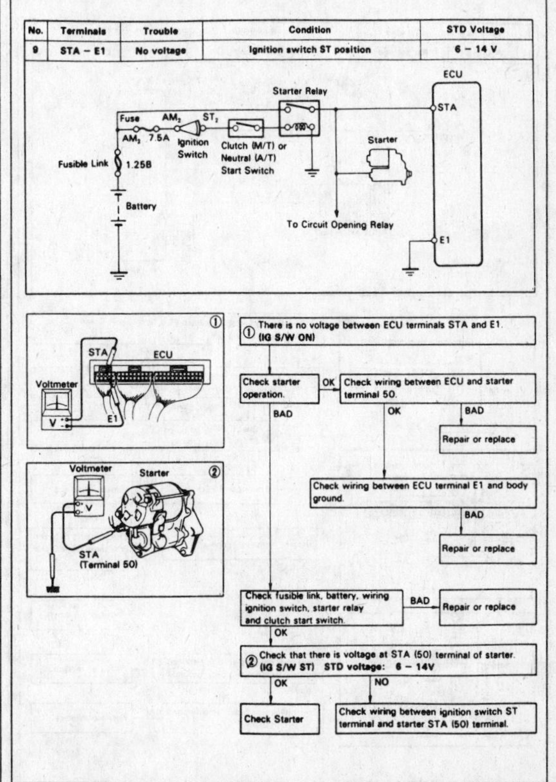
ECU
Starter Relay
STA
Fuse
AM2
ST2
7.5A
Ignition Switch
Clutch (M/T) or Neutral (A/T) Start Switch
Starter
Fusible Link
1.25B
Battery
To Circuit Opening Relay
E1
① There is no voltage between ECU terminals STA and E1. (IG S/W ON)
Check starter operation.
OK
Check wiring between ECU and starter terminal 50.
BAD
OK
BAD
Repair or replace
Check wiring between ECU terminal E1 and body ground.
BAD
Repair or replace
Check fusible link, battery, wiring ignition switch, starter relay and clutch start switch.
BAD
Repair or replace
OK
② Check that there is voltage at STA (50) terminal of starter. (IG S/W ST) STD voltage: 6 – 14V
OK
NO
Check Starter
Check wiring between ignition switch ST terminal and starter STA (50) terminal.
Voltmeter
STA (Terminal 50)

TROUBLESHOOTING CHART 10—MR-2 4A-GZE

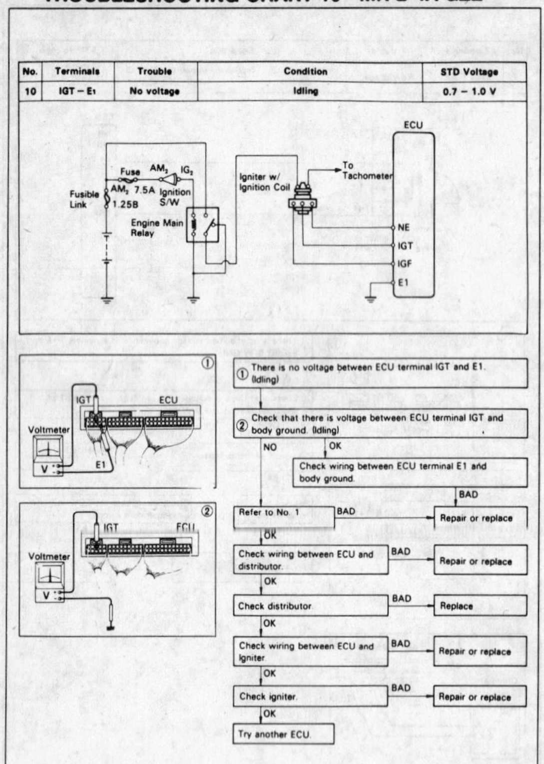

No.	Terminals	Trouble	Condition	STD Voltage
10	IGT – E1	No voltage	Idling	0.7 – 1.0 V

ECU
Fuse
AM2
IG2
Fusible Link
AM2 7.5A
Ignition S/W
1.25B
Engine Main Relay
Igniter w/ Ignition Coil
To Tachometer
NE
IGT
IGF
E1
IGT
ECU
Voltmeter
E1
IGT
ECU
Voltmeter
① There is no voltage between ECU terminal IGT and E1. (Idling)
② Check that there is voltage between ECU terminal IGT and body ground. (Idling)
NO
OK
Check wiring between ECU terminal E1 and body ground.
BAD
Refer to No. 1
BAD
Repair or replace
OK
Check wiring between ECU and distributor.
BAD
Repair or replace
OK
Check distributor
BAD
Replace
OK
Check wiring between ECU and Igniter.
BAD
Repair or replace
OK
Check igniter.
BAD
Repair or replace
OK
Try another ECU.

TROUBLESHOOTING CHART 11—MR-2 4A-GZE

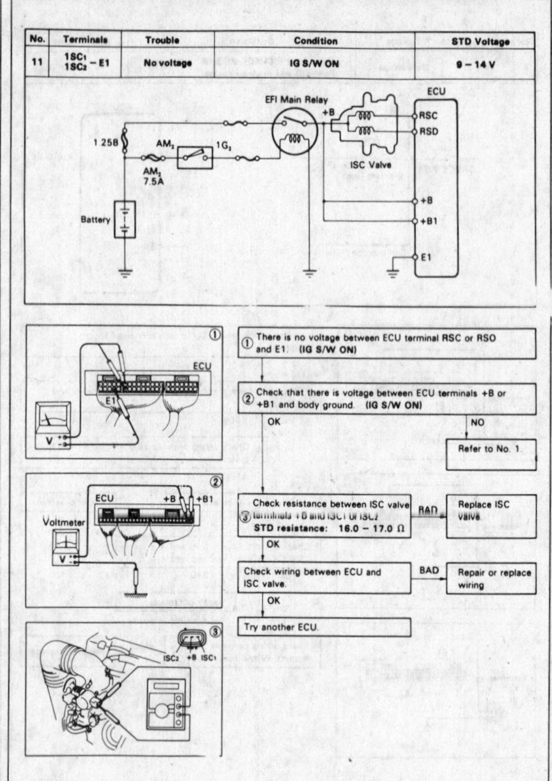

No.	Terminals	Trouble	Condition	STD Voltage
11	ISC1 ISC2 – E1	No voltage	IG S/W ON	9 – 14 V

ECU
EFI Main Relay
+B
RSC
RSD
1.25B
AM2
IG2
AM2 7.5A
ISC Valve
Battery
+B
+B1
E1
ECU
+B
+B1
ECU
Voltmeter
ISC2
+B
ISC1
① There is no voltage between ECU terminal RSC or RSO and E1. (IG S/W ON)
② Check that there is voltage between ECU terminals +B or +B1 and body ground. (IG S/W ON)
OK
NO
Refer to No. 1.
③ Check resistance between ISC valve terminals +B and ISC1 or ISC2
STD resistance: 16.0 – 17.0 Ω
BAD
Replace ISC valve
OK
Check wiring between ECU and ISC valve.
BAD
Repair or replace wiring
OK
Try another ECU.

TROUBLESHOOTING CHART 12—MR-2 4A-GZE

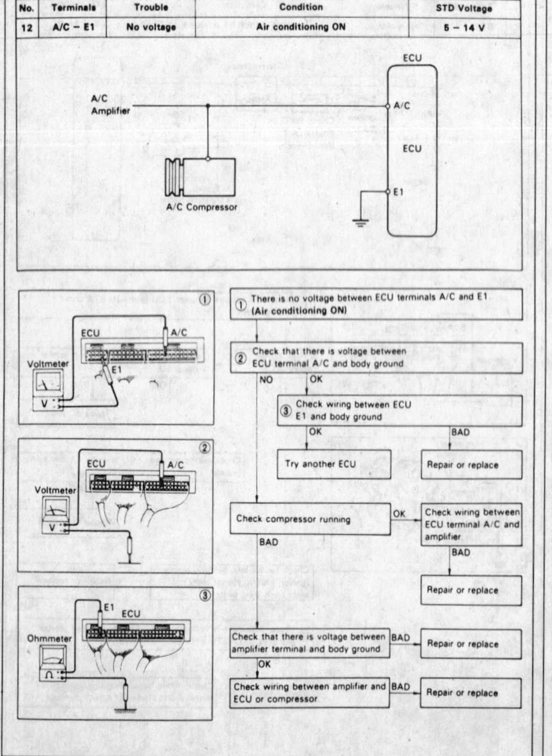

No.	Terminals	Trouble	Condition	STD Voltage
12	A/C – E1	No voltage	Air conditioning ON	5 – 14 V

ECU
A/C Amplifier
A/C
ECU
E1
A/C Compressor
ECU
A/C
Voltmeter
E1
ECU
A/C
Voltmeter
E1
ECU
Ohmmeter
① There is no voltage between ECU terminals A/C and E1. (Air conditioning ON)
② Check that there is voltage between ECU terminal A/C and body ground
NO
OK
③ Check wiring between ECU E1 and body ground
OK
BAD
Try another ECU
Repair or replace
Check compressor running
OK
Check wiring between ECU terminal A/C and amplifier
BAD
BAD
Repair or replace
Check that there is voltage between amplifier terminal and body ground
BAD
Repair or replace
OK
Check wiring between amplifier and ECU or compressor
BAD
Repair or replace

TROUBLESHOOTING OXYGEN SENSOR—MR-2 4A-GZE

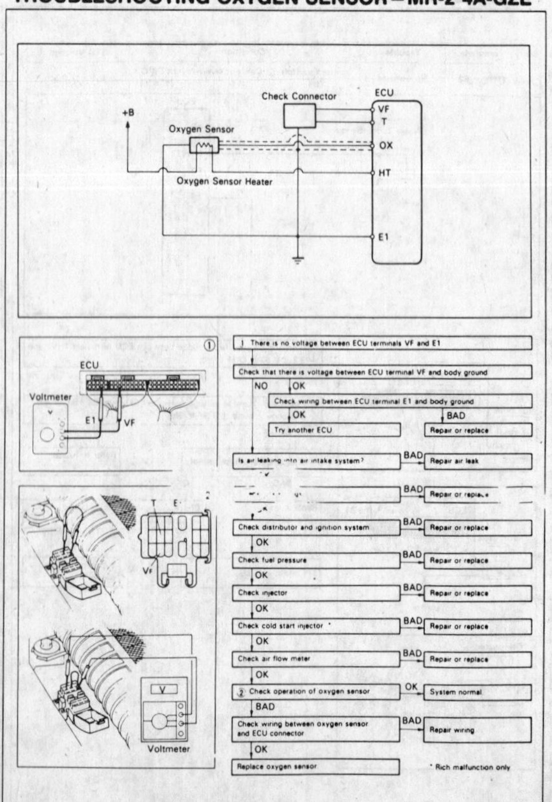

Check Connector
ECU
+B
VF
T
Oxygen Sensor
OX
HT
Oxygen Sensor Heater
E1
ECU
Voltmeter
E1
VF
T
E1
VF
Voltmeter
① There is no voltage between ECU terminals VF and E1
Check that there is voltage between ECU terminal VF and body ground
NO
OK
Check wiring between ECU terminal E1 and body ground
OK
BAD
Try another ECU
Repair or replace
Is air leaking into air intake system?
BAD
Repair air leak
BAD
Check distributor and ignition system
BAD
Repair or replace
OK
Check fuel pressure
BAD
Repair or replace
OK
Check injector
BAD
Repair or replace
OK
Check cold start injector *
BAD
Repair or replace
OK
Check air flow meter
BAD
Repair or replace
OK
② Check operation of oxygen sensor
OK
System normal
BAD
Check wiring between oxygen sensor and ECU connector
BAD
Repair wiring
OK
Replace oxygen sensor
* Rich malfunction only

TROUBLESHOOTING EGR TEMPERATURE SENSOR—MR-2 4A-GZE

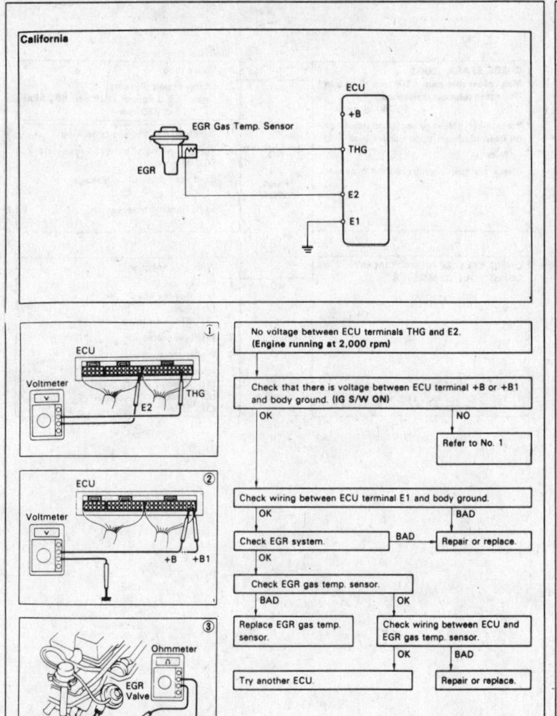
California
ECU
+B
EGR Gas Temp. Sensor
THG
EGR
E2
E1
①
ECU
Voltmeter
THG
E2
②
ECU
Voltmeter
+B
+B1
③
Ohmmeter
EGR Valve
EGR Gas Temp. Sensor
No voltage between ECU terminals THG and E2. (Engine running at 2,000 rpm)
Check that there is voltage between ECU terminal +B or +B1 and body ground. (IG S/W ON)
OK
NO
Refer to No. 1
Check wiring between ECU terminal E1 and body ground.
OK
BAD
Check EGR system.
BAD
Repair or replace.
OK
Check EGR gas temp. sensor.
BAD
OK
Replace EGR gas temp. sensor.
Check wiring between ECU and EGR gas temp. sensor.
OK
BAD
Try another ECU.
Repair or replace.

ECU TERMINAL IDENTIFICATION—MR-2 4A-GZE

Symbol	Terminal	Symbol	Terminal	Symbol	Terminal
E01	Engine ground (Power)	G2	Engine revolution sensor	TIL	Super charger indicator lamp.
E02	Engine ground (Power)	VTA	Throttle position sensor	A/C	A/C Magnet clutch
RSC	ISC valve	NE	Engine revolution sensor	SPD	Speedometer sensor
RSO	ISC valve	IDL	Throttle position sensor	W	Warning light
STA	Starter switch	VSV3	VSV (Air bypass)	ELS1	Stop light
IGT	Igniter	HT	Oxygen sensor heater	FPU	VSV (FPU)
EGR	EGR valve	IGF	Igniter	THA	Inlet air temp. sensor
E1	Engine ground	OX	Oxygen sensor	KNK	Knock sensor
•' NSW	Neutral start switch	THW	Water temperature sensor	VS	Air flow meter
SMC	Super charger relay	E2	Sensor ground	–	–
No. 10	No. 3, 4 injector	R/P	Fuel control switch	VC	Air flow meter
–	–	E22	Sensor ground	•' THG	EGR temperature sensor
No. 20	No. 1, 2 injector	VSV2	VSV (Air bleed)	BATT	Battery
–	–	–	–	+B	EFI main relay
G ⊖	Engine revolution sensor	•' L1	ECT computer	ELS2	Accessory switch
VF	Service connector	•' L3	ECT computer	+B1	EFI main relay
G1	Engine revolution sensor	•' L2	ECT computer	–	–
T	Service connector	ECT1	ECT computer	–	–

•: For A/T
•': For Calif.

E01	RSC	STA	EGR	NSW	No 10	No. 20	G⊖	G1	G2	NE	VSV 3	IGF	THW	R/P	VSV 2	L1	L2	TIL	SPD	ELS 1	THA	VS	VC	BATT	ELS 2
E02	RSO	IGT	E1	SMC	–	–	VF	T	VTA	IDL	HT	OX	E2	E22	–	L3	ECT 1	A/C	W	FPU	KNK	–	THG	+ B	+B1

ECU TERMINAL VOLTAGES—MR-2 4A-GZE

Terminals	STD voltage	Condition	
BATT – E1	10 – 14	–	
+B1, +B – E1	10 – 14	Ignition S/W ON	
IDL – E2	M/T 4 – 5 A/T 10 – 14	Ignition S/W ON	Throttle valve open
VTA – E2	0.1 – 1.0	Ignition S/W ON	Throttle valve fully closed
	4 – 5		Throttle valve fully open
VC – E2	4 – 6		–
VS – E2	4 – 5	Ignition S/W ON	Measuring plate fully closed
	0.02 – 0.5		Measuring plate fully open
	2 – 4	Idling	
THA – E2	1 – 3	Ignition S/W ON	Intake air temperature 20°C (68°F)
THW – E2	0.1 – 1.0	Ignition S/W ON	Coolant temperature 80°C (176°F)
STA – E1	6 – 14	Ignition S/W ST position and press on the clutch pedal (M/T)	
No 10 – E01 No 20 – E02	9 – 14	Ignition S/W ON	
IGT – E1	0.7 – 1.0	Idling	
T – E1	10 – 14	Ignition S/W ON	Service connector T — E_1 not short
	0.5 or less		Service connector T — E_1 short
A/C – E1	5 – 14	Ignition S/W ON	A/C switch ON
	0.5 or less		A/C switch OFF
W – E1	0.5 or less	Ignition S/W ON	
	9 – 14	Engine start	
RSC, RSO – E1	9 – 14	Ignition S/W ON	
*NSW – E1	0	Ignition S/W ON	Shift position P or N range
	10 – 14		Ex. P or N range
	6 – 11	Cranking	

* For A/T

TROUBLESHOOTING PROCEDURES—CRESSIDA 5M-GE

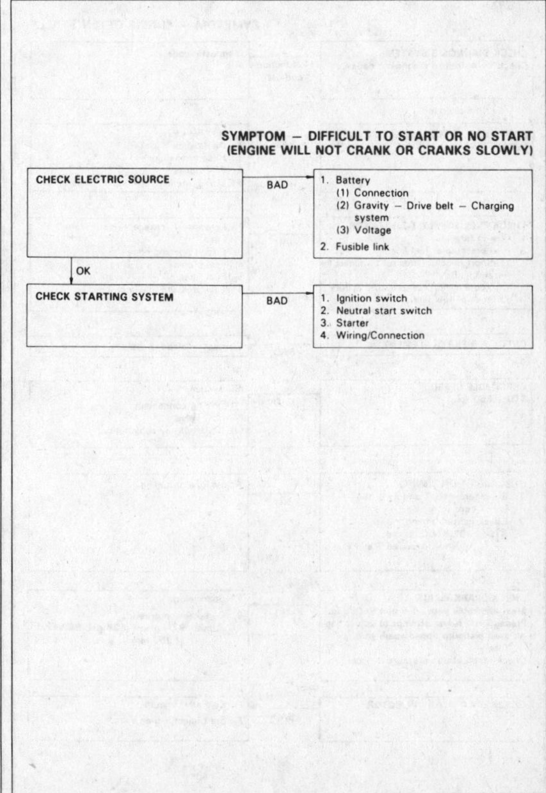
SYMPTOM – DIFFICULT TO START OR NO START (ENGINE WILL NOT CRANK OR CRANKS SLOWLY)
CHECK ELECTRIC SOURCE
BAD
1. Battery
(1) Connection
(2) Gravity – Drive belt – Charging system
(3) Voltage
2. Fusible link
OK
CHECK STARTING SYSTEM
BAD
1. Ignition switch
2. Neutral start switch
3. Starter
4. Wiring/Connection

TROUBLESHOOTING PROCEDURES—CRESSIDA 5M-GE

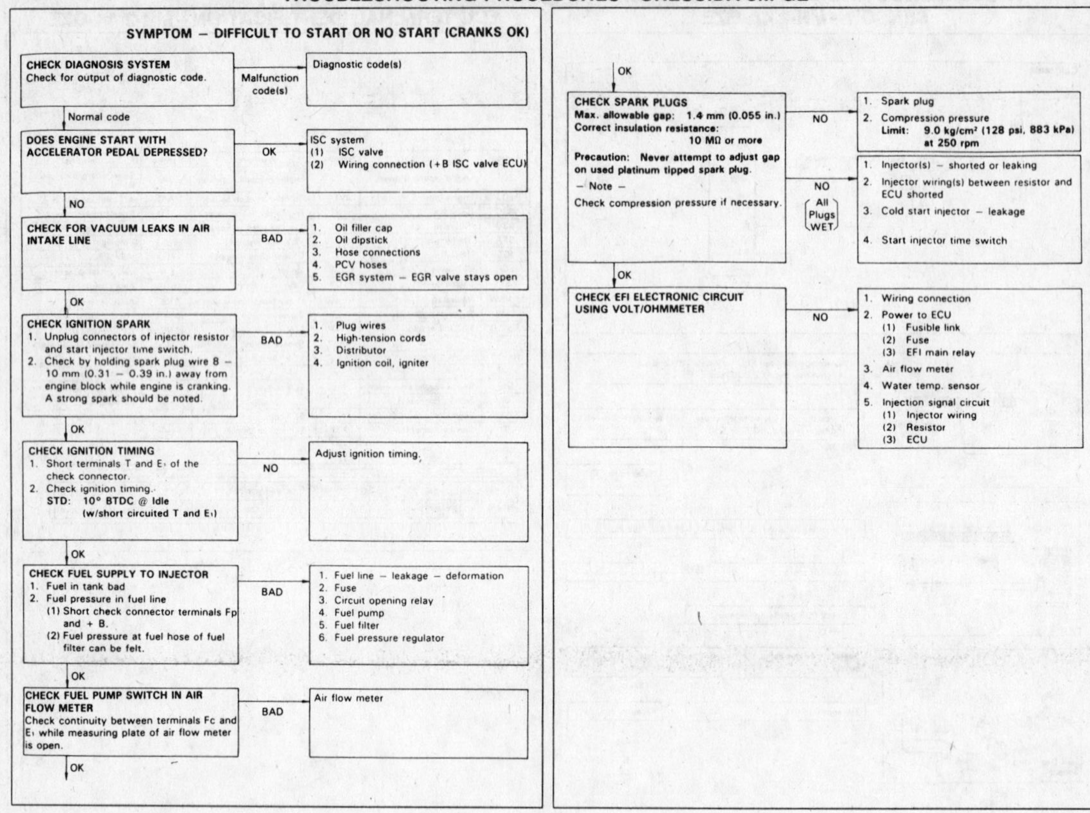
SYMPTOM – DIFFICULT TO START OR NO START (CRANKS OK)
CHECK DIAGNOSIS SYSTEM
Check for output of diagnostic code.
Malfunction code(s)
Diagnostic code(s)
Normal code
DOES ENGINE START WITH ACCELERATOR PEDAL DEPRESSED?
OK
ISC system
(1) ISC valve
(2) Wiring connection (+B ISC valve ECU)
NO
CHECK FOR VACUUM LEAKS IN AIR INTAKE LINE
BAD
1. Oil filler cap
2. Oil dipstick
3. Hose connections
4. PCV hoses
5. EGR system – EGR valve stays open
OK
CHECK IGNITION SPARK
1. Unplug connectors of injector resistor and start injector time switch.
2. Check by holding spark plug wire 8 – 10 mm (0.31 – 0.39 in.) away from engine block while engine is cranking. A strong spark should be noted.
BAD
1. Plug wires
2. High-tension cords
3. Distributor
4. Ignition coil, igniter
OK
CHECK IGNITION TIMING
1. Short terminals T and E₁ of the check connector.
2. Check ignition timing.
STD: 10° BTDC @ Idle (w/short circuited T and E₁)
NO
Adjust ignition timing.
OK
CHECK FUEL SUPPLY TO INJECTOR
1. Fuel in tank bad
2. Fuel pressure in fuel line
(1) Short check connector terminals Fp and + B.
(2) Fuel pressure at fuel hose of fuel filter can be felt.
BAD
1. Fuel line – leakage – deformation
2. Fuse
3. Circuit opening relay
4. Fuel pump
5. Fuel filter
6. Fuel pressure regulator
OK
CHECK FUEL PUMP SWITCH IN AIR FLOW METER
Check continuity between terminals Fc and E₁ while measuring plate of air flow meter is open.
BAD
Air flow meter
OK
OK
CHECK SPARK PLUGS
Max. allowable gap: 1.4 mm (0.055 in.)
Correct insulation resistance: 10 MΩ or more
Precaution: Never attempt to adjust gap on used platinum tipped spark plug.
– Note –
Check compression pressure if necessary.
NO
1. Spark plug
2. Compression pressure
Limit: 9.0 kg/cm² (128 psi, 883 kPa) at 250 rpm
NO
(All Plugs WET)
1. Injector(s) – shorted or leaking
2. Injector wiring(s) between resistor and ECU shorted
3. Cold start injector – leakage
4. Start injector time switch
OK
CHECK EFI ELECTRONIC CIRCUIT USING VOLT/OHMMETER
NO
1. Wiring connection
2. Power to ECU
(1) Fusible link
(2) Fuse
(3) EFI main relay
3. Air flow meter
4. Water temp. sensor
5. Injection signal circuit
(1) Injector wiring
(2) Resistor
(3) ECU

TROUBLESHOOTING PROCEDURES—CRESSIDA 5M-GE

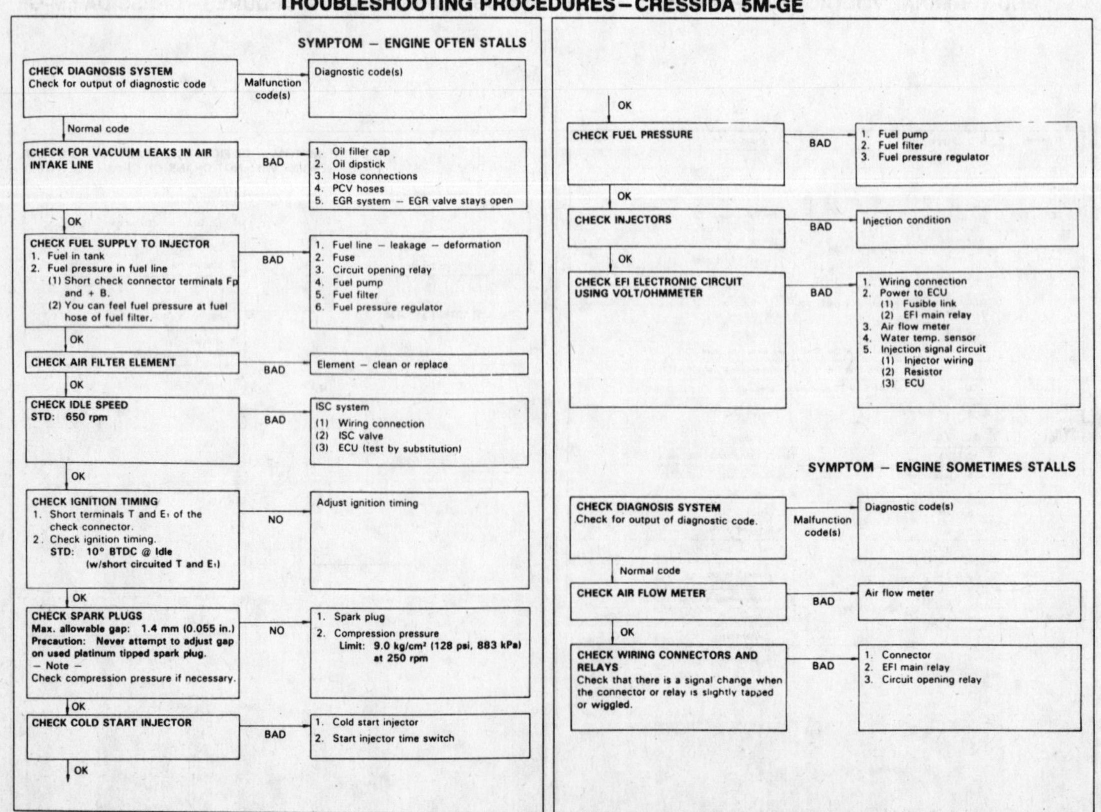
SYMPTOM – ENGINE OFTEN STALLS
CHECK DIAGNOSIS SYSTEM
Check for output of diagnostic code
Malfunction code(s)
Diagnostic code(s)
Normal code
CHECK FOR VACUUM LEAKS IN AIR INTAKE LINE
BAD
1. Oil filler cap
2. Oil dipstick
3. Hose connections
4. PCV hoses
5. EGR system – EGR valve stays open
OK
CHECK FUEL SUPPLY TO INJECTOR
1. Fuel in tank
2. Fuel pressure in fuel line
(1) Short check connector terminals Fp and + B.
(2) You can feel fuel pressure at fuel hose of fuel filter.
BAD
1. Fuel line – leakage – deformation
2. Fuse
3. Circuit opening relay
4. Fuel pump
5. Fuel filter
6. Fuel pressure regulator
OK
CHECK AIR FILTER ELEMENT
BAD
Element – clean or replace
OK
CHECK IDLE SPEED
STD: 650 rpm
BAD
ISC system
(1) Wiring connection
(2) ISC valve
(3) ECU (test by substitution)
OK
CHECK IGNITION TIMING
1. Short terminals T and E₁ of the check connector.
2. Check ignition timing.
STD: 10° BTDC @ Idle (w/short circuited T and E₁)
NO
Adjust ignition timing
OK
CHECK SPARK PLUGS
Max. allowable gap: 1.4 mm (0.055 in.)
Precaution: Never attempt to adjust gap on used platinum tipped spark plug.
– Note –
Check compression pressure if necessary.
NO
1. Spark plug
2. Compression pressure
Limit: 9.0 kg/cm² (128 psi, 883 kPa) at 250 rpm
OK
CHECK COLD START INJECTOR
BAD
1. Cold start injector
2. Start injector time switch
OK
OK
CHECK FUEL PRESSURE
BAD
1. Fuel pump
2. Fuel filter
3. Fuel pressure regulator
OK
CHECK INJECTORS
BAD
Injection condition
OK
CHECK EFI ELECTRONIC CIRCUIT USING VOLT/OHMMETER
BAD
1. Wiring connection
2. Power to ECU
(1) Fusible link
(2) EFI main relay
3. Air flow meter
4. Water temp. sensor
5. Injection signal circuit
(1) Injector wiring
(2) Resistor
(3) ECU
SYMPTOM – ENGINE SOMETIMES STALLS
CHECK DIAGNOSIS SYSTEM
Check for output of diagnostic code.
Malfunction code(s)
Diagnostic code(s)
Normal code
CHECK AIR FLOW METER
BAD
Air flow meter
OK
CHECK WIRING CONNECTORS AND RELAYS
Check that there is a signal change when the connector or relay is slightly tapped or wiggled.
BAD
1. Connector
2. EFI main relay
3. Circuit opening relay

TROUBLESHOOTING PROCEDURES—CRESSIDA 5M-GE

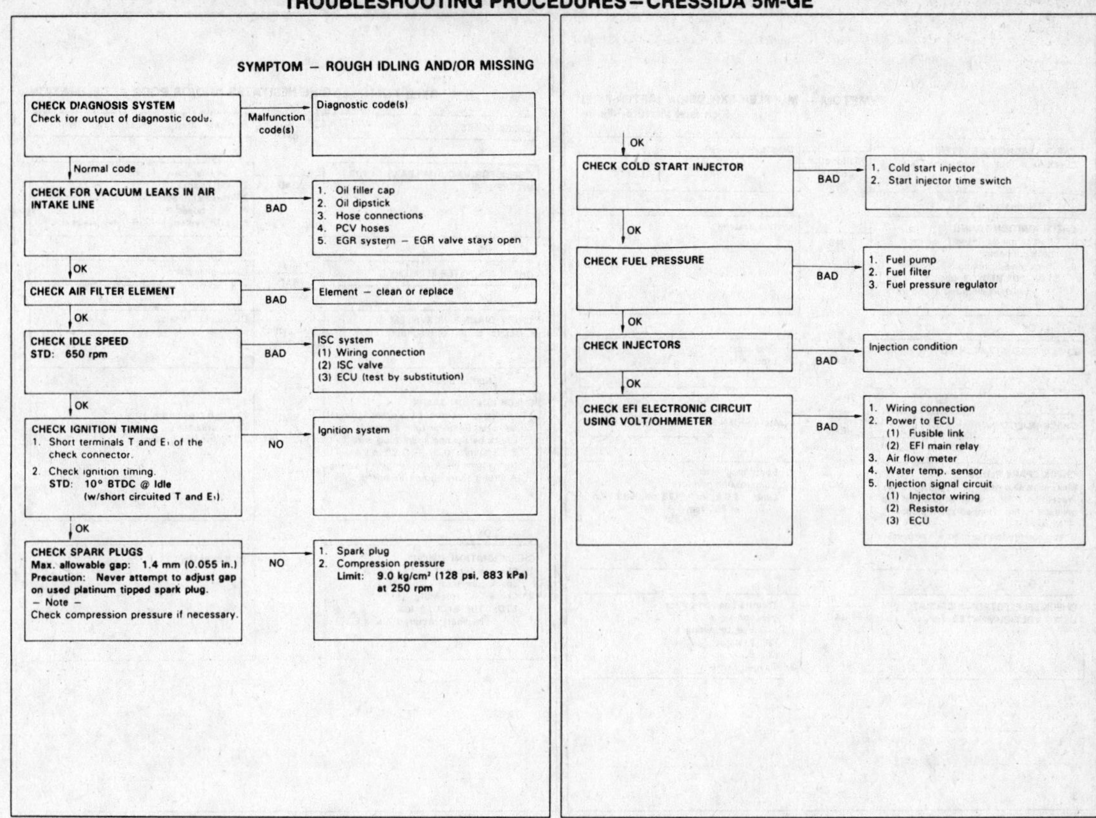
SYMPTOM – ROUGH IDLING AND/OR MISSING
CHECK DIAGNOSIS SYSTEM
Check for output of diagnostic code.
Malfunction code(s)
Diagnostic code(s)
Normal code
CHECK FOR VACUUM LEAKS IN AIR INTAKE LINE
BAD
1. Oil filler cap
2. Oil dipstick
3. Hose connections
4. PCV hoses
5. EGR system – EGR valve stays open
OK
CHECK AIR FILTER ELEMENT
BAD
Element – clean or replace
OK
CHECK IDLE SPEED
STD: 650 rpm
BAD
ISC system
(1) Wiring connection
(2) ISC valve
(3) ECU (test by substitution)
OK
CHECK IGNITION TIMING
1. Short terminals T and E1 of the check connector.
2. Check ignition timing.
STD: 10° BTDC @ Idle
(w/short circuited T and E1)
NO
Ignition system
OK
CHECK SPARK PLUGS
Max. allowable gap: 1.4 mm (0.055 in.)
Precaution: Never attempt to adjust gap on used platinum tipped spark plug.
– Note –
Check compression pressure if necessary.
NO
1. Spark plug
2. Compression pressure
Limit: 9.0 kg/cm² (128 psi, 883 kPa) at 250 rpm
OK
CHECK COLD START INJECTOR
BAD
1. Cold start injector
2. Start injector time switch
OK
CHECK FUEL PRESSURE
BAD
1. Fuel pump
2. Fuel filter
3. Fuel pressure regulator
OK
CHECK INJECTORS
BAD
Injection condition
OK
CHECK EFI ELECTRONIC CIRCUIT USING VOLT/OHMMETER
BAD
1. Wiring connection
2. Power to ECU
(1) Fusible link
(2) EFI main relay
3. Air flow meter
4. Water temp. sensor
5. Injection signal circuit
(1) Injector wiring
(2) Resistor
(3) ECU

TROUBLESHOOTING PROCEDURES—CRESSIDA 5M-GE

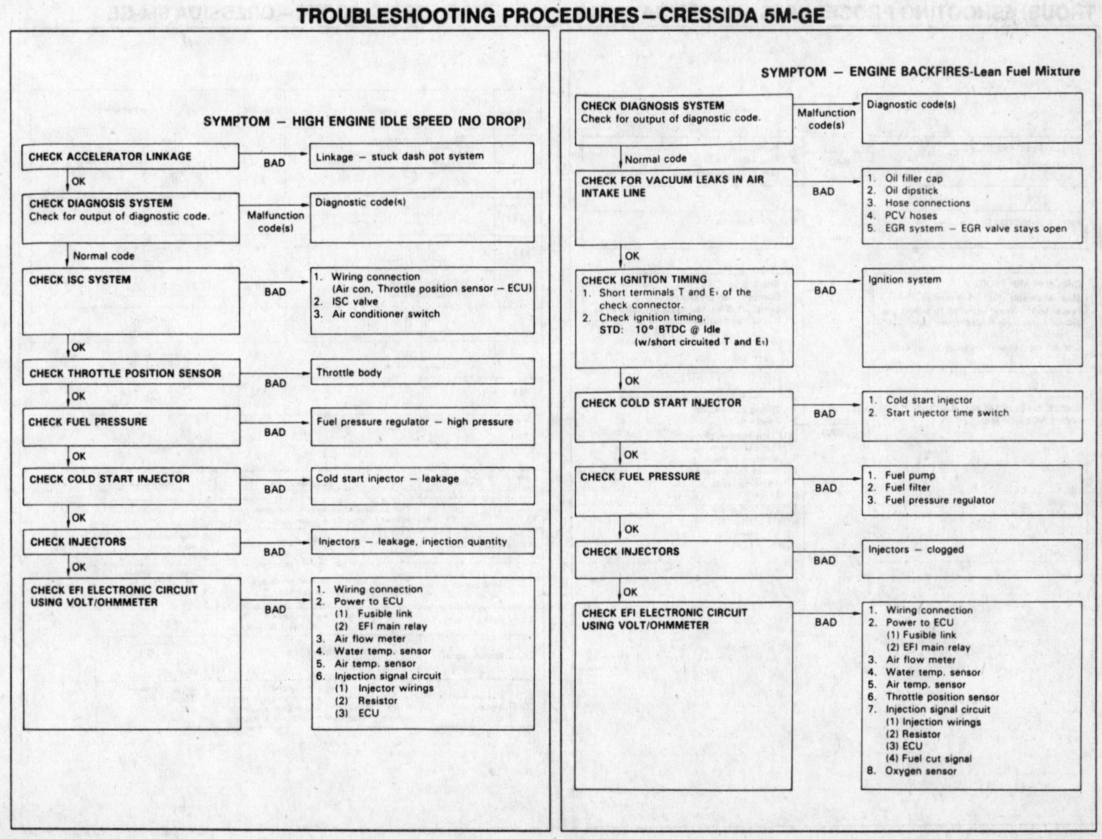
SYMPTOM – HIGH ENGINE IDLE SPEED (NO DROP)
CHECK ACCELERATOR LINKAGE
BAD
Linkage – stuck dash pot system
OK
CHECK DIAGNOSIS SYSTEM
Check for output of diagnostic code.
Malfunction code(s)
Diagnostic code(s)
Normal code
CHECK ISC SYSTEM
BAD
1. Wiring connection
(Air con, Throttle position sensor – ECU)
2. ISC valve
3. Air conditioner switch
OK
CHECK THROTTLE POSITION SENSOR
BAD
Throttle body
OK
CHECK FUEL PRESSURE
BAD
Fuel pressure regulator – high pressure
OK
CHECK COLD START INJECTOR
BAD
Cold start injector – leakage
OK
CHECK INJECTORS
BAD
Injectors – leakage, injection quantity
OK
CHECK EFI ELECTRONIC CIRCUIT USING VOLT/OHMMETER
BAD
1. Wiring connection
2. Power to ECU
(1) Fusible link
(2) EFI main relay
3. Air flow meter
4. Water temp. sensor
5. Air temp. sensor
6. Injection signal circuit
(1) Injector wirings
(2) Resistor
(3) ECU
SYMPTOM – ENGINE BACKFIRES-Lean Fuel Mixture
CHECK DIAGNOSIS SYSTEM
Check for output of diagnostic code.
Malfunction code(s)
Diagnostic code(s)
Normal code
CHECK FOR VACUUM LEAKS IN AIR INTAKE LINE
BAD
1. Oil filler cap
2. Oil dipstick
3. Hose connections
4. PCV hoses
5. EGR system – EGR valve stays open
OK
CHECK IGNITION TIMING
1. Short terminals T and E1 of the check connector.
2. Check ignition timing.
STD: 10° BTDC @ Idle
(w/short circuited T and E1)
BAD
Ignition system
OK
CHECK COLD START INJECTOR
BAD
1. Cold start injector
2. Start injector time switch
OK
CHECK FUEL PRESSURE
BAD
1. Fuel pump
2. Fuel filter
3. Fuel pressure regulator
OK
CHECK INJECTORS
BAD
Injectors – clogged
OK
CHECK EFI ELECTRONIC CIRCUIT USING VOLT/OHMMETER
BAD
1. Wiring connection
2. Power to ECU
(1) Fusible link
(2) EFI main relay
3. Air flow meter
4. Water temp. sensor
5. Air temp. sensor
6. Throttle position sensor
7. Injection signal circuit
(1) Injection wirings
(2) Resistor
(3) ECU
(4) Fuel cut signal
8. Oxygen sensor

TROUBLESHOOTING PROCEDURES—CRESSIDA 5M-GE

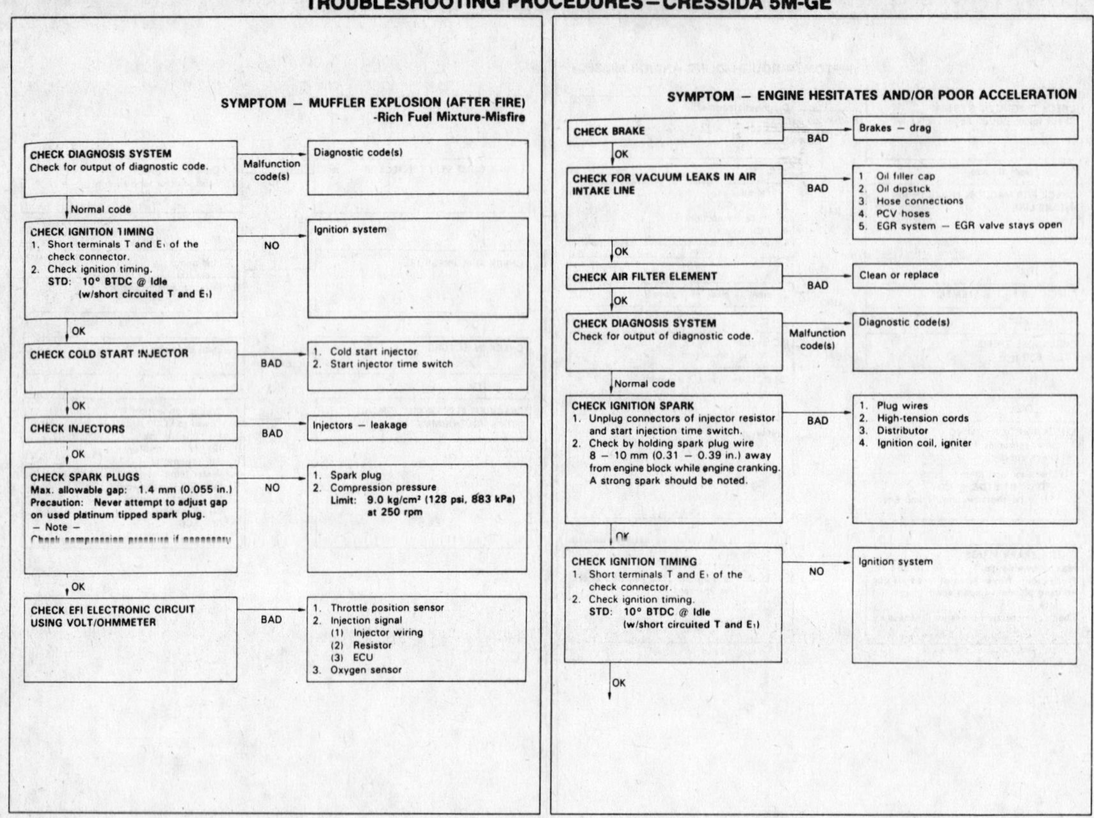
SYMPTOM — MUFFLER EXPLOSION (AFTER FIRE)
-Rich Fuel Mixture-Misfire
CHECK DIAGNOSIS SYSTEM
Check for output of diagnostic code.
Malfunction code(s)
Diagnostic code(s)
Normal code
CHECK IGNITION TIMING
1. Short terminals T and E1 of the check connector.
2. Check ignition timing.
STD: 10° BTDC @ Idle
(w/short circuited T and E1)
NO
Ignition system
OK
CHECK COLD START INJECTOR
BAD
1. Cold start injector
2. Start injector time switch
OK
CHECK INJECTORS
BAD
Injectors — leakage
OK
CHECK SPARK PLUGS
Max. allowable gap: 1.4 mm (0.055 in.)
Precaution: Never attempt to adjust gap on used platinum tipped spark plug.
— Note —
Check compression pressure if necessary
NO
1. Spark plug
2. Compression pressure
Limit: 9.0 kg/cm² (128 psi, 883 kPa) at 250 rpm
OK
CHECK EFI ELECTRONIC CIRCUIT USING VOLT/OHMMETER
BAD
1. Throttle position sensor
2. Injection signal
(1) Injector wiring
(2) Resistor
(3) ECU
3. Oxygen sensor
SYMPTOM — ENGINE HESITATES AND/OR POOR ACCELERATION
CHECK BRAKE
BAD
Brakes — drag
OK
CHECK FOR VACUUM LEAKS IN AIR INTAKE LINE
BAD
1. Oil filler cap
2. Oil dipstick
3. Hose connections
4. PCV hoses
5. EGR system — EGR valve stays open
OK
CHECK AIR FILTER ELEMENT
BAD
Clean or replace
OK
CHECK DIAGNOSIS SYSTEM
Check for output of diagnostic code.
Malfunction code(s)
Diagnostic code(s)
Normal code
CHECK IGNITION SPARK
1. Unplug connectors of injector resistor and start injection time switch.
2. Check by holding spark plug wire 8 — 10 mm (0.31 — 0.39 in.) away from engine block while engine cranking. A strong spark should be noted.
BAD
1. Plug wires
2. High-tension cords
3. Distributor
4. Ignition coil, igniter
OK
CHECK IGNITION TIMING
1. Short terminals T and E1 of the check connector.
2. Check ignition timing.
STD: 10° BTDC @ Idle
(w/short circuited T and E1)
NO
Ignition system
OK

TROUBLESHOOTING PROCEDURES—CRESSIDA 5M-GE

OK

Check	Result	Trouble area
CHECK FUEL PRESSURE	BAD	1. Fuel pump 2. Fuel filter 3. Fuel pressure regulator
OK		
CHECK INJECTORS	BAD	Injection condition
OK		
CHECK SPARK PLUGS Max. allowable gap: 1.4 mm (0.055 in.) Precaution: Never attempt to adjust gap on used platinum tipped spark plug. — Note — Check compression pressure if necessary.	NO	1. Spark plug 2. Compression pressure Limit: 9.0 kg/cm² (128 psi, 883 kPa) at 250 rpm
OK		
CHECK EFI ELECTRONIC CIRCUIT USING VOLT/OHMMETER	BAD	1. Wiring connection 2. Power to ECU (1) Fusible link (2) EFI main relay 3. Air flow meter 4. Water temp. sensor 5. Air temp. sensor 6. Throttle position sensor 7. Injection signal circuit (1) Injector wirings (2) Resistor (3) ECU

DIAGNOSTIC CODES—CRESSIDA 5M-GE

Code No.	Light Pattern	Code No.	Light Pattern
—	ON / OFF	31	
11		32	
12		41	
13		42	
14		43	
21		51	
22		52	
23		53	

Code No.	System	Diagnosis	Trouble Area
	Normal	This appears when none of the other codes (11 thru 53) are identified	
11	ECU (+ B)	Wire severence, however slight, in + B (ECU).	1. Main relay circuit 2. Main relay 3. ECU
12	RPM Signal	No Ne, G signal to ECU within several seconds after engine is cranked.	1. Distributor circuit 2. Distributor 3. Starter signal circuit 4. ECU
13	RPM Signal	No Ne signal to ECU within several seconds after engine reaches 1,000 rpm.	Same as 12, above.
14	Ignition Signal	No signal from igniter six times in succession.	1. Igniter circuit (+B, IGt, IGf) 2. Igniter 3. ECU
21	Oxygen Sensor Signal	Oxygen sensor gives a lean signal for several seconds even when coolant temperature is above 50°C (122°F) and engine is running under high load conditions above 1,500 rpm.	1. Oxygen sensor circuit 2. Oxygen sensor 3. ECU

DIAGNOSTIC CODES—CRESSIDA 5M-GE

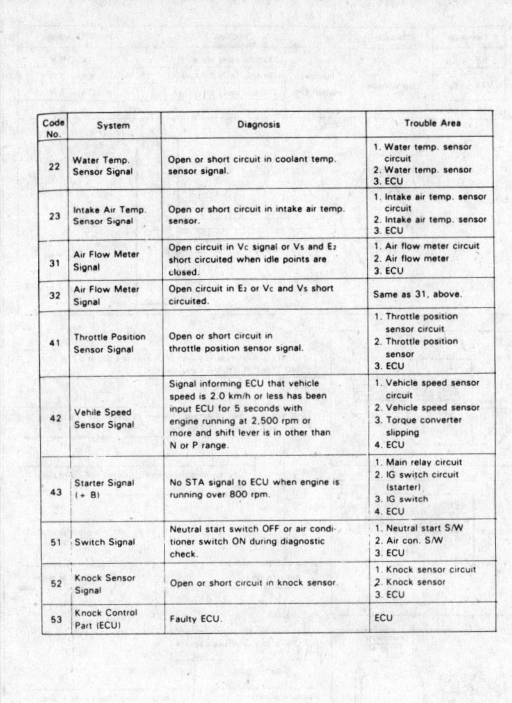

Code No.	System	Diagnosis	Trouble Area
22	Water Temp. Sensor Signal	Open or short circuit in coolant temp. sensor signal.	1. Water temp. sensor circuit 2. Water temp. sensor 3. ECU
23	Intake Air Temp. Sensor Signal	Open or short circuit in intake air temp. sensor.	1. Intake air temp. sensor circuit 2. Intake air temp. sensor 3. ECU
31	Air Flow Meter Signal	Open circuit in Vc signal or Vs and E_2 short circuited when idle points are closed.	1. Air flow meter circuit 2. Air flow meter 3. ECU
32	Air Flow Meter Signal	Open circuit in E_2 or Vc and Vs short circuited.	Same as 31, above.
41	Throttle Position Sensor Signal	Open or short circuit in throttle position sensor signal.	1. Throttle position sensor circuit 2. Throttle position sensor 3. ECU
42	Vehile Speed Sensor Signal	Signal informing ECU that vehicle speed is 2.0 km/h or less has been input ECU for 5 seconds with engine running at 2,500 rpm or more and shift lever is in other than N or P range.	1. Vehicle speed sensor circuit 2. Vehicle speed sensor 3. Torque converter slipping 4. ECU
43	Starter Signal (+ B)	No STA signal to ECU when engine is running over 800 rpm.	1. Main relay circuit 2. IG switch circuit (starter) 3. IG switch 4. ECU
51	Switch Signal	Neutral start switch OFF or air conditioner switch ON during diagnostic check.	1. Neutral start S/W 2. Air con. S/W 3. ECU
52	Knock Sensor Signal	Open or short circuit in knock sensor.	1. Knock sensor circuit 2. Knock sensor 3. ECU
53	Knock Control Part (ECU)	Faulty ECU.	ECU

DIAGNOSTIC CIRCUITS TEST—CRESSIDA 5M-GE

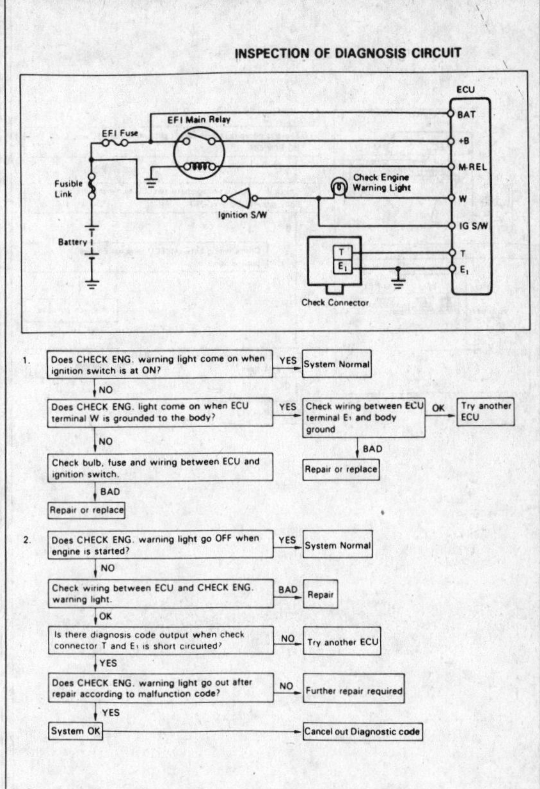

INSPECTION OF DIAGNOSIS CIRCUIT
ECU
BAT
+B
M-REL
W
IG S/W
T
E₁
EFI Main Relay
EFI Fuse
Fusible Link
Check Engine Warning Light
Ignition S/W
Battery
Check Connector
1.
Does CHECK ENG. warning light come on when ignition switch is at ON?
YES
System Normal
NO
Does CHECK ENG. light come on when ECU terminal W is grounded to the body?
YES
Check wiring between ECU terminal E₁ and body ground
OK
Try another ECU
BAD
Repair or replace
NO
Check bulb, fuse and wiring between ECU and ignition switch.
BAD
Repair or replace
2.
Does CHECK ENG. warning light go OFF when engine is started?
YES
System Normal
NO
Check wiring between ECU and CHECK ENG. warning light.
BAD
Repair
OK
Is there diagnosis code output when check connector T and E₁ is short circuited?
NO
Try another ECU
YES
Does CHECK ENG. warning light go out after repair according to malfunction code?
NO
Further repair required
YES
System OK
Cancel out Diagnostic code

TROUBLESHOOTING CHART 1—CRESSIDA 5M-GE

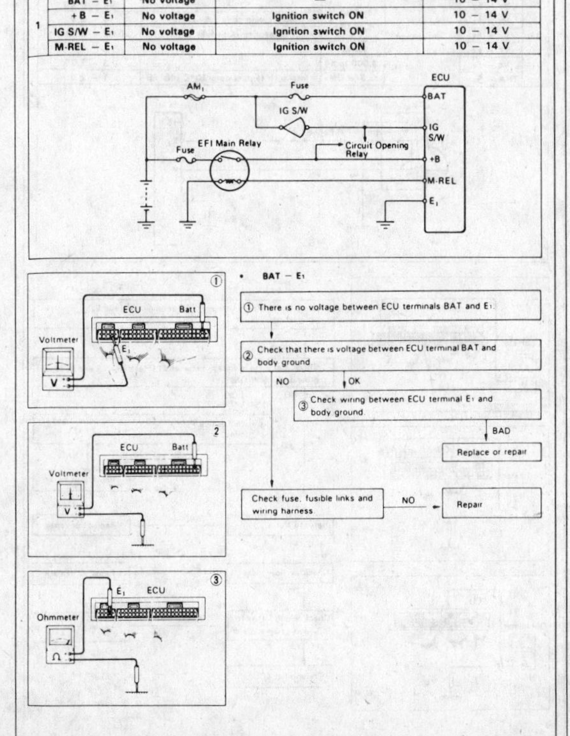

No.	Terminals	Trouble	Condition	STD Voltage
1	BAT – E_1	No voltage	—	10 – 14 V
	+B – E_1	No voltage	Ignition switch ON	10 – 14 V
	IG S/W – E_1	No voltage	Ignition switch ON	10 – 14 V
	M-REL – E_1	No voltage	Ignition switch ON	10 – 14 V

ECU
AM₁
Fuse
BAT
IG S/W
IG S/W
Circuit Opening Relay
Fuse
EFI Main Relay
+B
M-REL
E₁
• BAT – E₁
① There is no voltage between ECU terminals BAT and E₁.
② Check that there is voltage between ECU terminal BAT and body ground.
NO
OK
③ Check wiring between ECU terminal E₁ and body ground.
BAD
Replace or repair
Check fuse, fusible links and wiring harness.
NO
Repair
Voltmeter
ECU
Batt
E₁
Ohmmeter

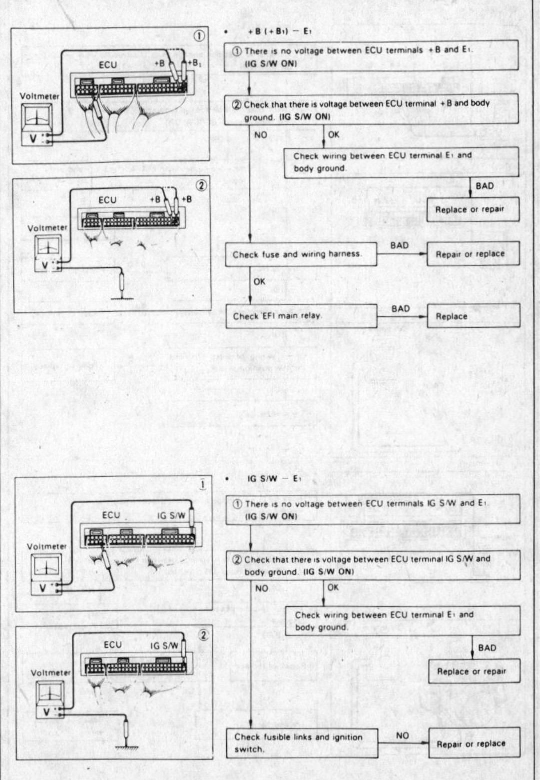

• +B (+B₁) – E₁
① There is no voltage between ECU terminals +B and E₁. (IG S/W ON)
② Check that there is voltage between ECU terminal +B and body ground. (IG S/W ON)
NO
OK
Check wiring between ECU terminal E₁ and body ground.
BAD
Replace or repair
Check fuse and wiring harness.
BAD
Repair or replace
OK
Check EFI main relay.
BAD
Replace
• IG S/W – E₁
① There is no voltage between ECU terminals IG S/W and E₁. (IG S/W ON)
② Check that there is voltage between ECU terminal IG S/W and body ground. (IG S/W ON)
NO
OK
Check wiring between ECU terminal E₁ and body ground.
BAD
Replace or repair
Check fusible links and ignition switch.
NO
Repair or replace
Voltmeter
ECU
+B
+B₁
IG S/W

TROUBLESHOOTING CHART 1—CRESSIDA 5M-GE

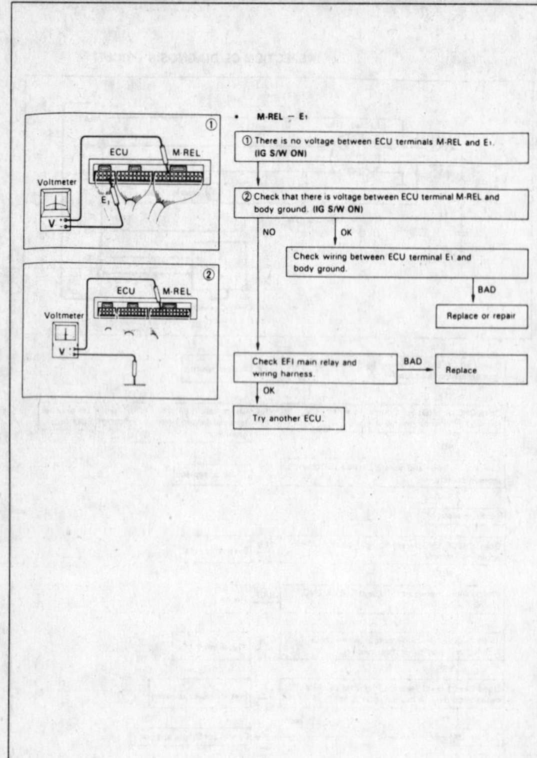
• M-REL – E1
① There is no voltage between ECU terminals M-REL and E1. (IG S/W ON)
② Check that there is voltage between ECU terminal M-REL and body ground. (IG S/W ON)
NO
OK
Check wiring between ECU terminal E1 and body ground.
BAD
Replace or repair
Check EFI main relay and wiring harness.
BAD
Replace
OK
Try another ECU.
ECU
M-REL
Voltmeter
E1
V

TROUBLESHOOTING CHART 2—CRESSIDA 5M-GE

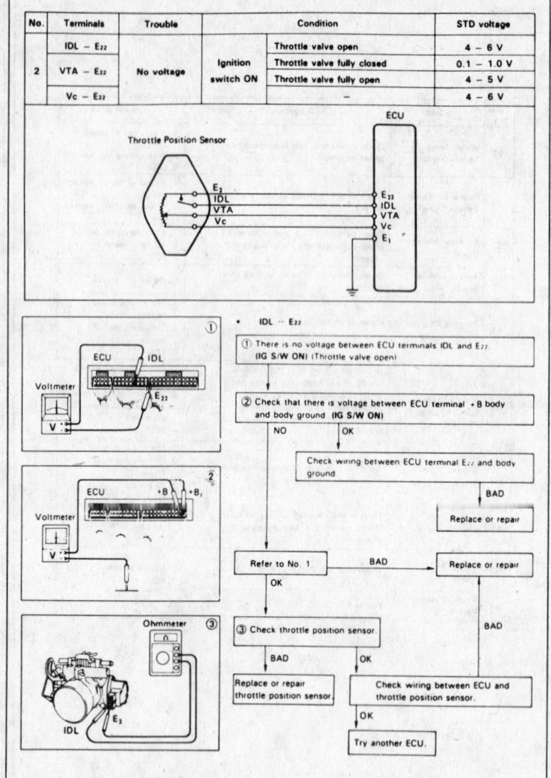

No.	Terminals	Trouble	Condition		STD voltage
2	IDL – E_{22}	No voltage	Ignition switch ON	Throttle valve open	4 – 6 V
	VTA – E_{22}			Throttle valve fully closed	0.1 – 1.0 V
				Throttle valve fully open	4 – 5 V
	Vc – E_{22}			—	4 – 6 V

Throttle Position Sensor
ECU
E2
IDL
VTA
Vc
E22
E1
• IDL – E22
① There is no voltage between ECU terminals IDL and E22. (IG S/W ON) (Throttle valve open)
② Check that there is voltage between ECU terminal +B body and body ground. (IG S/W ON)
NO
OK
Check wiring between ECU terminal E22 and body ground
BAD
Replace or repair
Refer to No. 1.
BAD
Replace or repair
OK
BAD
③ Check throttle position sensor.
BAD
OK
Replace or repair throttle position sensor.
Check wiring between ECU and throttle position sensor.
OK
Try another ECU.
Voltmeter
Ohmmeter
+B
IDL
E2

TROUBLESHOOTING CHART 2—CRESSIDA 5M-GE

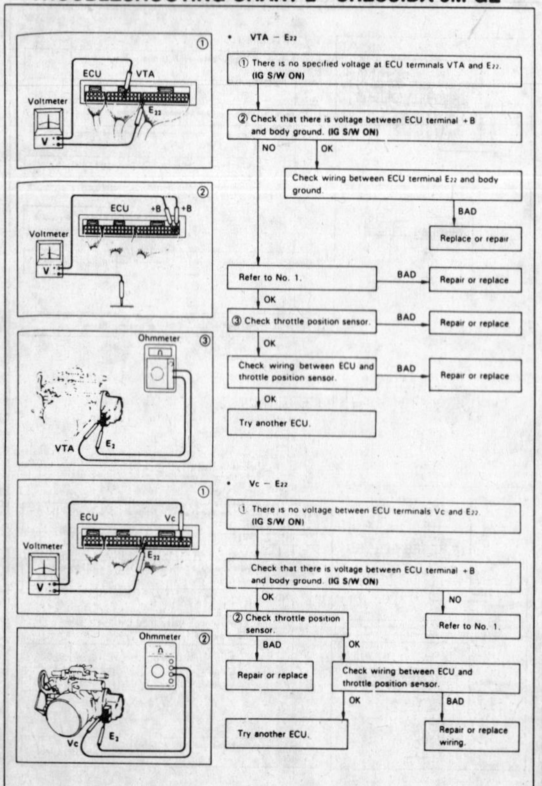
• VTA – E22
① There is no specified voltage at ECU terminals VTA and E22. (IG S/W ON)
② Check that there is voltage between ECU terminal +B and body ground. (IG S/W ON)
NO
OK
Check wiring between ECU terminal E22 and body ground.
BAD
Replace or repair
Refer to No. 1.
BAD
Repair or replace
OK
③ Check throttle position sensor.
BAD
Repair or replace
OK
Check wiring between ECU and throttle position sensor.
BAD
Repair or replace
OK
Try another ECU.
• Vc – E22
① There is no voltage between ECU terminals Vc and E22. (IG S/W ON)
Check that there is voltage between ECU terminal +B and body ground. (IG S/W ON)
OK
NO
② Check throttle position sensor.
Refer to No. 1.
BAD
OK
Repair or replace
Check wiring between ECU and throttle position sensor.
OK
BAD
Try another ECU.
Repair or replace wiring.
ECU
VTA
Vc
+B
E22
E2
Voltmeter
Ohmmeter

TROUBLESHOOTING CHART 3—CRESSIDA 5M-GE

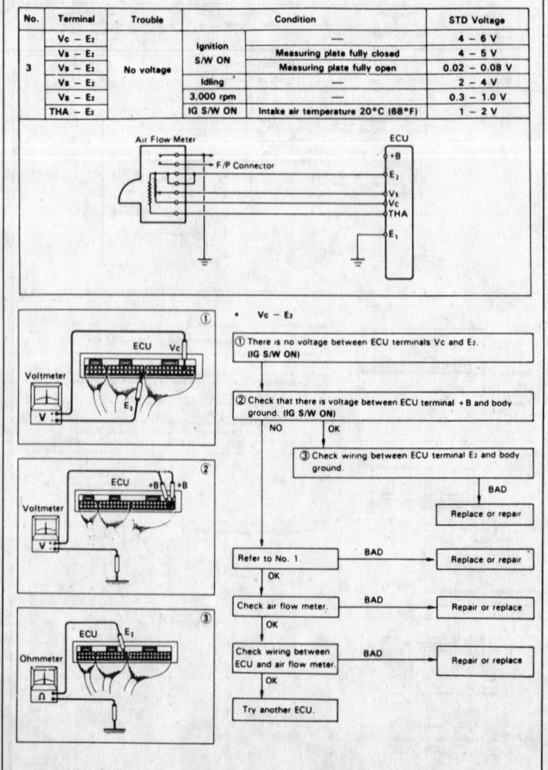

No.	Terminal	Trouble	Condition		STD Voltage
3	Vc – E_2	No voltage	Ignition S/W ON	—	4 – 6 V
	Vs – E_2			Measuring plate fully closed	4 – 5 V
	Vs – E_2			Measuring plate fully open	0.02 – 0.08 V
	Vs – E_2		Idling	—	2 – 4 V
	Vs – E_2		3,000 rpm	—	0.3 – 1.0 V
	THA – E_2		IG S/W ON	Intake air temperature 20°C (68°F)	1 – 2 V

Air Flow Meter
ECU
F/P Connector
+B
E2
Vs
Vc
THA
E1
• Vc – E2
① There is no voltage between ECU terminals Vc and E2. (IG S/W ON)
② Check that there is voltage between ECU terminal +B and body ground. (IG S/W ON)
NO
OK
③ Check wiring between ECU terminal E2 and body ground.
BAD
Replace or repair
Refer to No. 1.
BAD
Replace or repair
OK
Check air flow meter.
BAD
Repair or replace
OK
Check wiring between ECU and air flow meter.
BAD
Repair or replace
OK
Try another ECU.
Voltmeter
Ohmmeter

TROUBLESHOOTING CHART 3—CRESSIDA 5M-GE

TROUBLESHOOTING CHART 4—CRESSIDA 5M-GE

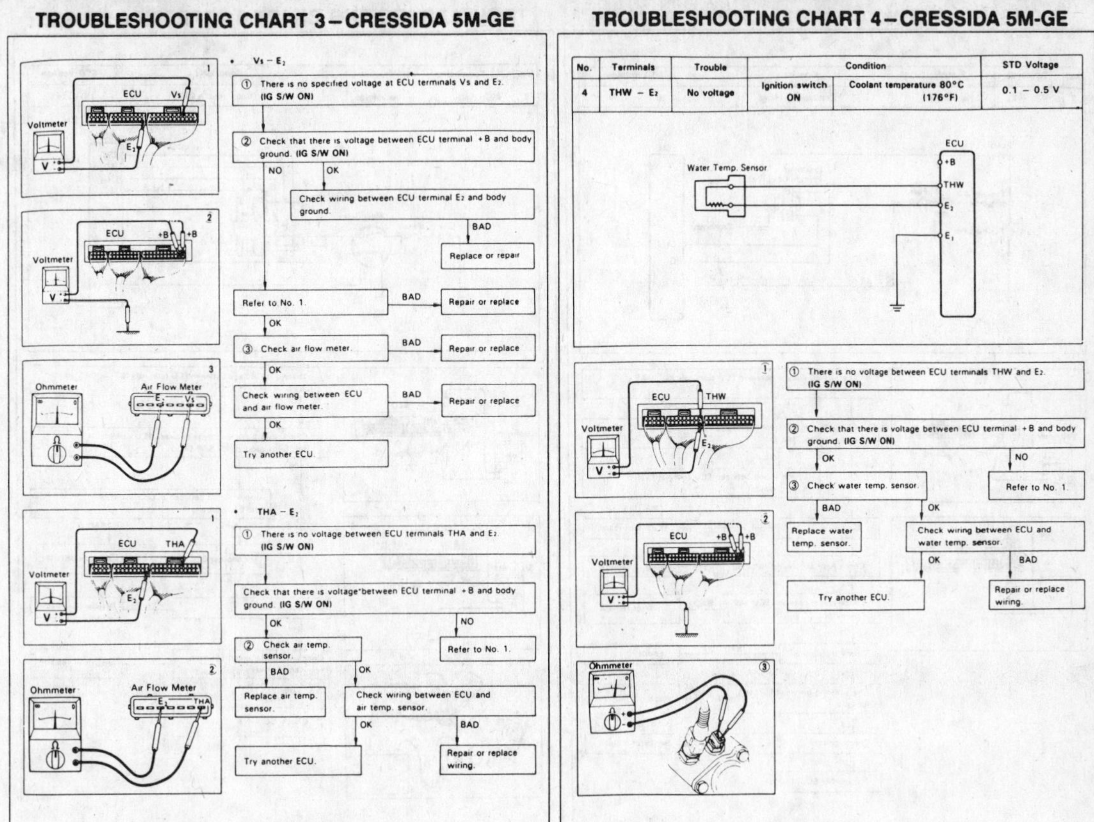
• Vs – E₂
① There is no specified voltage at ECU terminals Vs and E₂. (IG S/W ON)
② Check that there is voltage between ECU terminal +B and body ground. (IG S/W ON)
NO
OK
Check wiring between ECU terminal E₂ and body ground.
BAD
Replace or repair
Refer to No. 1.
BAD
Repair or replace
OK
③ Check air flow meter.
BAD
Repair or replace
OK
Check wiring between ECU and air flow meter.
BAD
Repair or replace
OK
Try another ECU.
ECU
Vs
Voltmeter
E₂
+B
Ohmmeter
Air Flow Meter
• THA – E₂
① There is no voltage between ECU terminals THA and E₂. (IG S/W ON)
Check that there is voltage between ECU terminal +B and body ground. (IG S/W ON)
OK
NO
② Check air temp. sensor.
Refer to No. 1.
BAD
OK
Replace air temp. sensor.
Check wiring between ECU and air temp. sensor.
OK
BAD
Try another ECU.
Repair or replace wiring.
THA
Water Temp. Sensor
THW
E₁
① There is no voltage between ECU terminals THW and E₂. (IG S/W ON)
② Check that there is voltage between ECU terminal +B and body ground. (IG S/W ON)
OK
NO
③ Check water temp. sensor.
Refer to No. 1.
BAD
OK
Replace water temp. sensor.
Check wiring between ECU and water temp. sensor.
OK
BAD
Try another ECU.
Repair or replace wiring.

No.	Terminals	Trouble	Condition		STD Voltage
4	THW – E_2	No voltage	Ignition switch ON	Coolant temperature 80°C (176°F)	0.1 – 0.5 V

TROUBLESHOOTING CHART 5—CRESSIDA 5M-GE

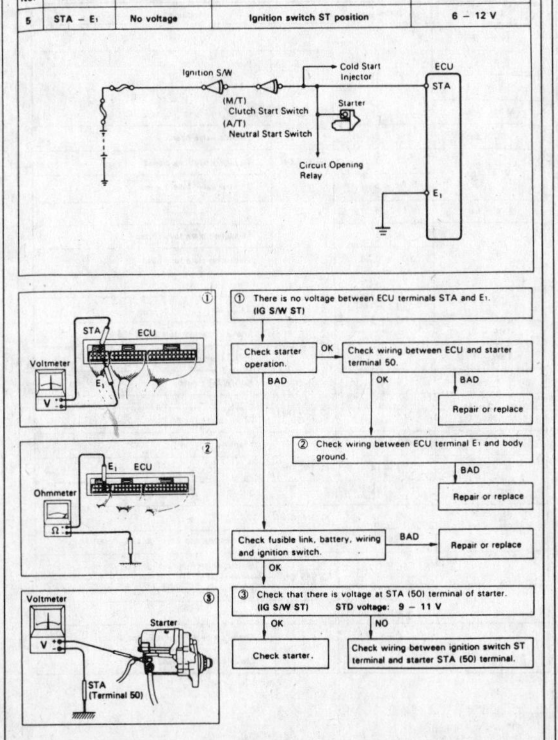

No.	Terminals	Trouble	Condition	STD Voltage
5	STA – E_1	No voltage	Ignition switch ST position	6 – 12 V

Ignition S/W
Cold Start Injector
ECU
STA
(M/T) Clutch Start Switch
(A/T) Neutral Start Switch
Starter
Circuit Opening Relay
E₁
① There is no voltage between ECU terminals STA and E₁. (IG S/W ST)
Check starter operation.
OK
Check wiring between ECU and starter terminal 50.
BAD
OK
BAD
Repair or replace
② Check wiring between ECU terminal E₁ and body ground.
BAD
Repair or replace
Check fusible link, battery, wiring and ignition switch.
BAD
Repair or replace
OK
③ Check that there is voltage at STA (50) terminal of starter. (IG S/W ST) STD voltage: 9 – 11 V
OK
NO
Check starter.
Check wiring between ignition switch ST terminal and starter STA (50) terminal.
Voltmeter
Ohmmeter
Starter
STA (Terminal 50)

TROUBLESHOOTING CHART 6—CRESSIDA 5M-GE

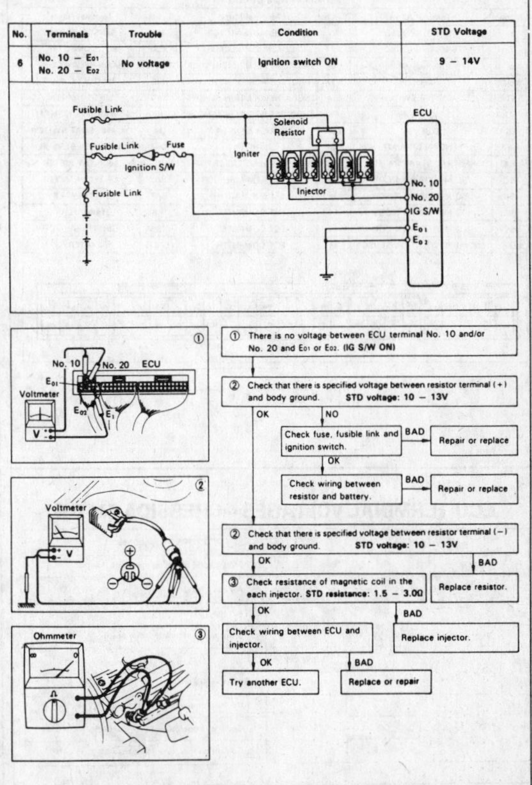

No.	Terminals	Trouble	Condition	STD Voltage
6	No. 10 – E_{01} No. 20 – E_{02}	No voltage	Ignition switch ON	9 – 14V

Fusible Link
Solenoid Resistor
ECU
Fuse
Igniter
Ignition S/W
Injector
No. 10
No. 20
IG S/W
E₀₁
E₀₂
① There is no voltage between ECU terminal No. 10 and/or No. 20 and E₀₁ or E₀₂. (IG S/W ON)
② Check that there is specified voltage between resistor terminal (+) and body ground. STD voltage: 10 – 13V
OK
NO
Check fuse, fusible link and ignition switch.
BAD
Repair or replace
OK
Check wiring between resistor and battery.
BAD
Repair or replace
② Check that there is specified voltage between resistor terminal (–) and body ground. STD voltage: 10 – 13V
OK
BAD
③ Check resistance of magnetic coil in the each injector. STD resistance: 1.5 – 3.0Ω
Replace resistor.
OK
BAD
Check wiring between ECU and injector.
Replace injector.
OK
BAD
Try another ECU.
Replace or repair
Voltmeter
Ohmmeter

TROUBLESHOOTING CHART 7—CRESSIDA 5M-GE

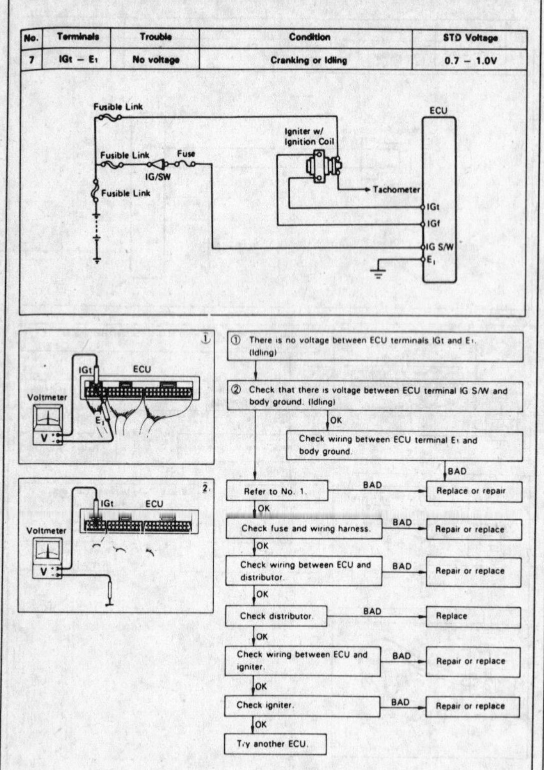
No. | Terminals | Trouble | Condition | STD Voltage
7 | IGt – E1 | No voltage | Cranking or Idling | 0.7 – 1.0V
Fusible Link
Fusible Link
Fuse
IG/SW
Fusible Link
Igniter w/ Ignition Coil
ECU
Tachometer
IGt
IGf
IG S/W
E1
IGt
ECU
Voltmeter
V
① There is no voltage between ECU terminals IGt and E1. (Idling)
② Check that there is voltage between ECU terminal IG S/W and body ground. (Idling)
OK
Check wiring between ECU terminal E1 and body ground.
BAD
Refer to No. 1.
Replace or repair
Check fuse and wiring harness.
Repair or replace
Check wiring between ECU and distributor.
Repair or replace
Check distributor.
Replace
Check wiring between ECU and igniter.
Repair or replace
Check igniter.
Repair or replace
Try another ECU.

TROUBLESHOOTING CHART 8—CRESSIDA 5M-GE

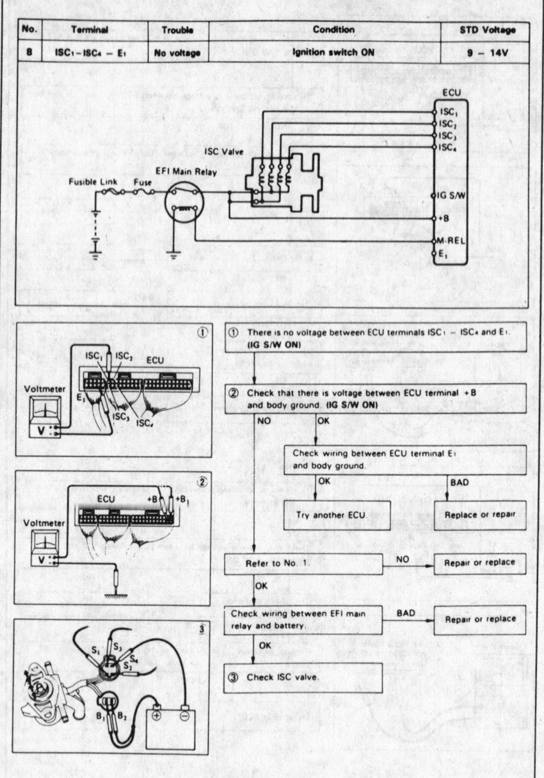
No. | Terminal | Trouble | Condition | STD Voltage
8 | ISC1 – ISC4 – E1 | No voltage | Ignition switch ON | 9 – 14V
ECU
ISC1
ISC2
ISC3
ISC4
ISC Valve
EFI Main Relay
Fusible Link
Fuse
IG S/W
+B
M-REL
E1
Voltmeter
V
① There is no voltage between ECU terminals ISC1 – ISC4 and E1. (IG S/W ON)
② Check that there is voltage between ECU terminal +B and body ground. (IG S/W ON)
NO
OK
Check wiring between ECU terminal E1 and body ground.
BAD
Try another ECU.
Replace or repair
Refer to No. 1.
Repair or replace
Check wiring between EFI main relay and battery.
Repair or replace
③ Check ISC valve.

ECU TERMINAL IDENTIFICATION—CRESSIDA 5M-GE

Connectors of ECU

Symbol	Terminal Name	Symbol	Terminal Name	Symbol	Terminal Name
E_{01}	ENGINE GROUND	V_F	CHECK CONNECTOR	E_{22}	SENSOR EARTH
E_{02}	ENGINE GROUND	T	CHECK CONNECTOR	TCD	ECT COMPUTER
No. 10	INJECTOR	G	ENGINE REVOLUTION SENSOR	M-REL	MAIN RELAY COIL
No. 20	INJECTOR	VTA	THROTTLE POSITION SENSOR	A/C	A/C MAGNETIC SWITCH
STA	STARTER SWITCH	Ne	ENGINE REVOLUTION SENSOR	SPD	SPEEDOMETER
IGt	IGNITER	IDL	THROTTLE POSITION SENSOR	W	WARNING LIGHT
EGR	EGR VSV	KNK	KNOCK SENSOR	OIL	OIL PRESSURE SWITCH
E_1	ENGINE GROUND	IGf	IGNITER	THA	AIR TEMP. SENSOR
N/C	NEUTRAL START SWITCH	Ox	OXYGEN SENSOR	V_s	AIR FLOW METER
ISC_1	ISC MOTOR NO. 1 COIL	THW	WATER TEMP. SENSOR	V_c	AIR FLOW METER
ISC_2	ISC MOTOR NO. 2 COIL	E_2	SENSOR EARTH	BAT	BATTERY +B
ISC_3	ISC MOTOR NO. 3 COIL	L_1	ECT COMPUTER	+B	MAIN RELAY
ISC_4	ISC MOTOR NO. 4 COIL	L_2	ECT COMPUTER	IG S/W	IGNITION SWITCH
G⊖	ENGINE REVOLUTION SENSOR	L_3	ECT COMPUTER	$+B_1$	BATTERY

E_{01}	No 10	STA	EGR	N/C	ISC 1	ISC 2	G⊖	/	G	Ne	/	IGf	THW	L_1	L_2	L_3	M REL	/	SPD	/	THA	V_s	V_c	BAT	IG S/W
E_{02}	No 20	IGt	E_1	/	ISC 3	ISC 4	V_F	T	VTA	IDL	KNK	Ox	E_2	E_{22}	/	TCD	/	A/C	W	OIL	/	/	/	+B	$+B_1$

ECU TERMINAL VOLTAGES—CRESSIDA 5M-GE

Terminals	STD Voltage	Condition	
OIL – E_1	4 – 6		IG S/W ON (Warning light on)
	0		Start engine (Warning light out)
A/C – E_1	10 – 13	IG S/W ON	Air con S/W ON
	0		Air con S/W OFF
V_F ↔ E_1	0 – 5		Start engine (Throttle valve open)
W – E_1	0	IG S/W ON	—
	10 – 13		Start engine
TCD – E_1	2 – 3	IG S/W ON	Coolant temperature Less than 35°C (95°F)
	0		Coolant temperature 35 – 60°C (95 – 140°F)
	4 – 6		Coolant temperature More than 60°C (140°F)

ECU TERMINAL VOLTAGES—CRESSIDA 5M-GE

Terminals	STD Voltage	Condition	
BAT – E_1	10 – 14		—
+B – E_1		IG S/W ON	—
IG S/W – E_1			
M-REL – E_1			
IDL – E_{22}	4 – 6		Throttle valve open
VTA – E_{22}	0.1 – 1.0		Throttle valve fully closed
	4 – 5		Throttle valve fully opened
V_c – E_{22}	4 – 6		—
V_s – E_2	4 – 5		Measuring plate fully closed
	0.02 – 0.08		Measuring plate fully open
	2 – 4		Idling
	0.3 – 1.0		3,000 rpm
THA – E_2	1 – 2	IG S/W ON	Intake air temperature 20°C (68°F)
THW – E_1	0.1 – 0.5		Coolant temperature 80°C (176°F)
STA – E_1	6 – 12		IG S/W ST position
No. 10 – E_{01} No. 20 – E_{02}	9 – 14	IG S/W ON	—
IGt – E_1	0.7 – 1.0		Cranking or Idling
ISC_1 ~ ISC_4 – E_1	9 – 14	IG S/W ON	—
	9 – 14		2 – 3 secs. after engine off
+B – EGR	10 – 13	IG S/W ON	—
	0		Start engine and warm up oxygen sensor
N/C – E_1	0	IG S/W ON	Shift position P or N range
	10 – 14		Ex. P or N range
	9 – 11		Cranking
T – E_1	4 – 6	IG S/W ON	Check connector T and E_1 not short
	0		Check connector T and E_1 short

TROUBLESHOOTING PROCEDURES—CRESSIDA 7M-GE

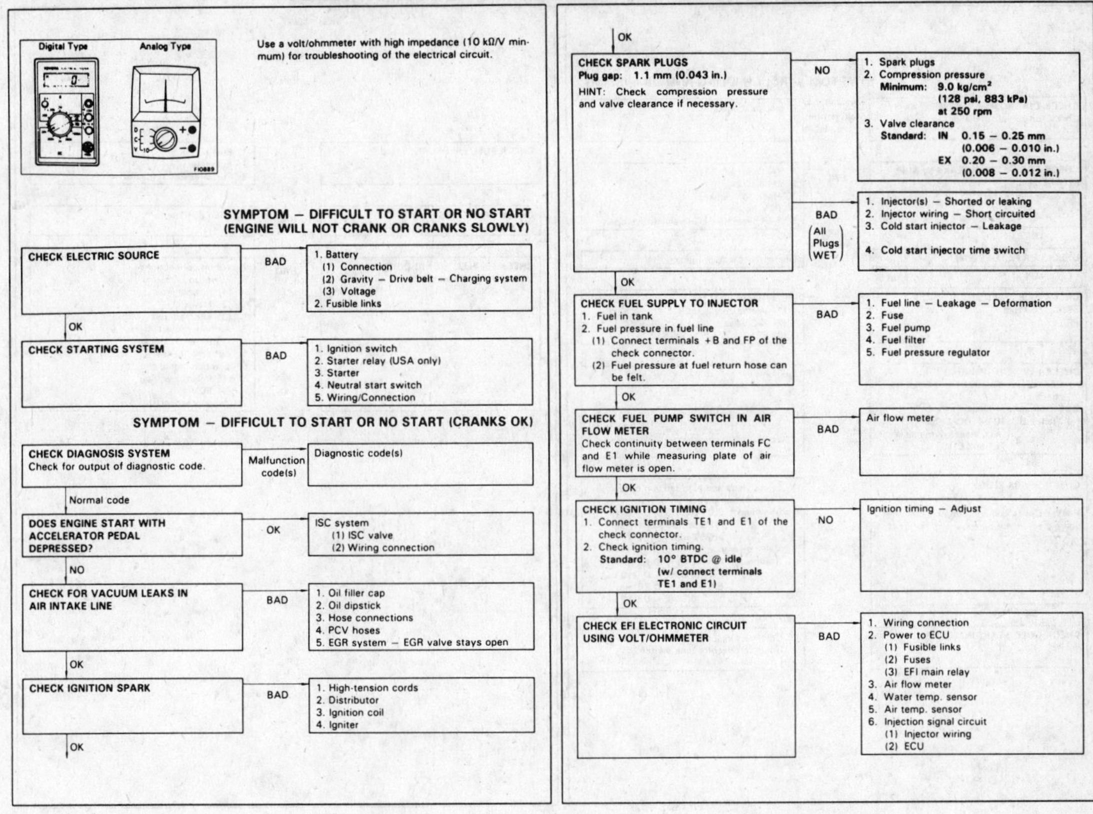
Digital Type
Analog Type
Use a volt/ohmmeter with high impedance (10 kΩ/V minimum) for troubleshooting of the electrical circuit.
SYMPTOM — DIFFICULT TO START OR NO START (ENGINE WILL NOT CRANK OR CRANKS SLOWLY)
CHECK ELECTRIC SOURCE
BAD
1. Battery
(1) Connection
(2) Gravity — Drive belt — Charging system
(3) Voltage
2. Fusible links
OK
CHECK STARTING SYSTEM
BAD
1. Ignition switch
2. Starter relay (USA only)
3. Starter
4. Neutral start switch
5. Wiring/Connection
SYMPTOM — DIFFICULT TO START OR NO START (CRANKS OK)
CHECK DIAGNOSIS SYSTEM
Check for output of diagnostic code.
Malfunction code(s)
Diagnostic code(s)
Normal code
DOES ENGINE START WITH ACCELERATOR PEDAL DEPRESSED?
OK
ISC system
(1) ISC valve
(2) Wiring connection
NO
CHECK FOR VACUUM LEAKS IN AIR INTAKE LINE
BAD
1. Oil filler cap
2. Oil dipstick
3. Hose connections
4. PCV hoses
5. EGR system — EGR valve stays open
OK
CHECK IGNITION SPARK
BAD
1. High-tension cords
2. Distributor
3. Ignition coil
4. Igniter
OK
OK
CHECK SPARK PLUGS
Plug gap: 1.1 mm (0.043 in.)
HINT: Check compression pressure and valve clearance if necessary.
NO
1. Spark plugs
2. Compression pressure
Minimum: 9.0 kg/cm²
(128 psi, 883 kPa)
at 250 rpm
3. Valve clearance
Standard: IN 0.15 — 0.25 mm
(0.006 — 0.010 in.)
EX 0.20 — 0.30 mm
(0.008 — 0.012 in.)
BAD (All Plugs WET)
1. Injector(s) — Shorted or leaking
2. Injector wiring — Short circuited
3. Cold start injector — Leakage
4. Cold start injector time switch
OK
CHECK FUEL SUPPLY TO INJECTOR
1. Fuel in tank
2. Fuel pressure in fuel line
(1) Connect terminals +B and FP of the check connector.
(2) Fuel pressure at fuel return hose can be felt.
BAD
1. Fuel line — Leakage — Deformation
2. Fuse
3. Fuel pump
4. Fuel filter
5. Fuel pressure regulator
OK
CHECK FUEL PUMP SWITCH IN AIR FLOW METER
Check continuity between terminals FC and E1 while measuring plate of air flow meter is open.
BAD
Air flow meter
OK
CHECK IGNITION TIMING
1. Connect terminals TE1 and E1 of the check connector.
2. Check ignition timing.
Standard: 10° BTDC @ idle
(w/ connect terminals TE1 and E1)
NO
Ignition timing — Adjust
OK
CHECK EFI ELECTRONIC CIRCUIT USING VOLT/OHMMETER
BAD
1. Wiring connection
2. Power to ECU
(1) Fusible links
(2) Fuses
(3) EFI main relay
3. Air flow meter
4. Water temp. sensor
5. Air temp. sensor
6. Injection signal circuit
(1) Injector wiring
(2) ECU

TROUBLESHOOTING PROCEDURES—CRESSIDA 7M-GE

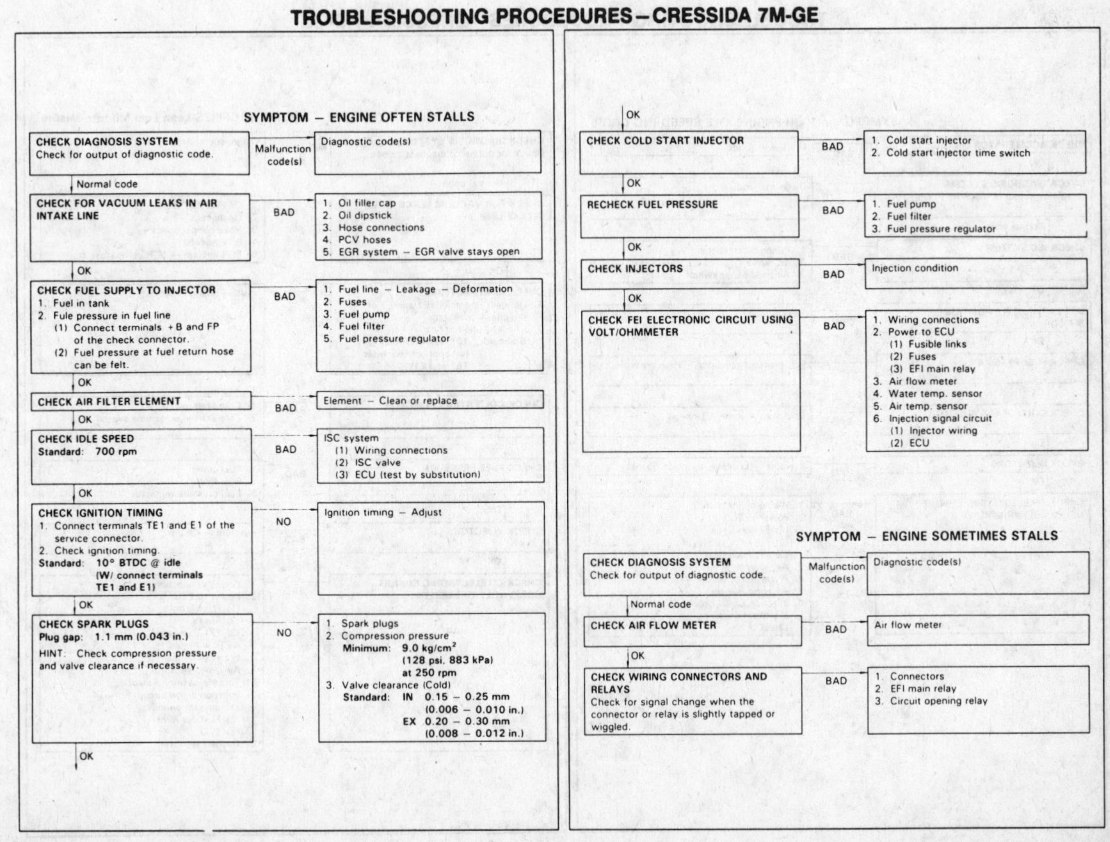
SYMPTOM — ENGINE OFTEN STALLS
CHECK DIAGNOSIS SYSTEM
Check for output of diagnostic code.
Malfunction code(s)
Diagnostic code(s)
Normal code
CHECK FOR VACUUM LEAKS IN AIR INTAKE LINE
BAD
1. Oil filler cap
2. Oil dipstick
3. Hose connections
4. PCV hoses
5. EGR system — EGR valve stays open
OK
CHECK FUEL SUPPLY TO INJECTOR
1. Fuel in tank
2. Fule pressure in fuel line
(1) Connect terminals +B and FP of the check connector.
(2) Fuel pressure at fuel return hose can be felt.
BAD
1. Fuel line — Leakage — Deformation
2. Fuses
3. Fuel pump
4. Fuel filter
5. Fuel pressure regulator
OK
CHECK AIR FILTER ELEMENT
BAD
Element — Clean or replace
OK
CHECK IDLE SPEED
Standard: 700 rpm
BAD
ISC system
(1) Wiring connections
(2) ISC valve
(3) ECU (test by substitution)
OK
CHECK IGNITION TIMING
1. Connect terminals TE1 and E1 of the service connector.
2. Check ignition timing.
Standard: 10° BTDC @ idle
(W/ connect terminals TE1 and E1)
NO
Ignition timing — Adjust
OK
CHECK SPARK PLUGS
Plug gap: 1.1 mm (0.043 in.)
HINT: Check compression pressure and valve clearance if necessary.
NO
1. Spark plugs
2. Compression pressure
Minimum: 9.0 kg/cm²
(128 psi, 883 kPa)
at 250 rpm
3. Valve clearance (Cold)
Standard: IN 0.15 — 0.25 mm
(0.006 — 0.010 in.)
EX 0.20 — 0.30 mm
(0.008 — 0.012 in.)
OK
OK
CHECK COLD START INJECTOR
BAD
1. Cold start injector
2. Cold start injector time switch
OK
RECHECK FUEL PRESSURE
BAD
1. Fuel pump
2. Fuel filter
3. Fuel pressure regulator
OK
CHECK INJECTORS
BAD
Injection condition
OK
CHECK FEI ELECTRONIC CIRCUIT USING VOLT/OHMMETER
BAD
1. Wiring connections
2. Power to ECU
(1) Fusible links
(2) Fuses
(3) EFI main relay
3. Air flow meter
4. Water temp. sensor
5. Air temp. sensor
6. Injection signal circuit
(1) Injector wiring
(2) ECU
SYMPTOM — ENGINE SOMETIMES STALLS
CHECK DIAGNOSIS SYSTEM
Check for output of diagnostic code.
Malfunction code(s)
Diagnostic code(s)
Normal code
CHECK AIR FLOW METER
BAD
Air flow meter
OK
CHECK WIRING CONNECTORS AND RELAYS
Check for signal change when the connector or relay is slightly tapped or wiggled.
BAD
1. Connectors
2. EFI main relay
3. Circuit opening relay

TROUBLESHOOTING PROCEDURES—CRESSIDA 7M-GE

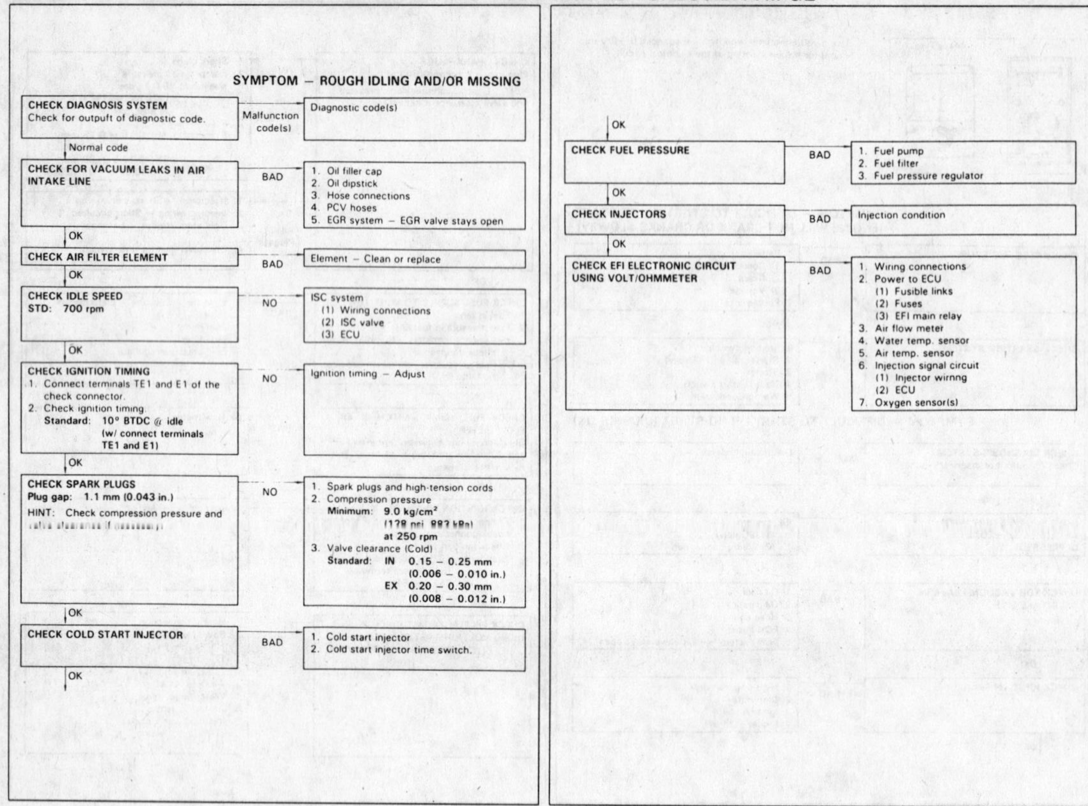
SYMPTOM – ROUGH IDLING AND/OR MISSING
CHECK DIAGNOSIS SYSTEM
Check for outpuft of diagnostic code.
Malfunction code(s)
Diagnostic code(s)
Normal code
CHECK FOR VACUUM LEAKS IN AIR INTAKE LINE
BAD
1. Oil filler cap
2. Oil dipstick
3. Hose connections
4. PCV hoses
5. EGR system – EGR valve stays open
OK
CHECK AIR FILTER ELEMENT
BAD
Element – Clean or replace
OK
CHECK IDLE SPEED
STD: 700 rpm
NO
ISC system
(1) Wiring connections
(2) ISC valve
(3) ECU
OK
CHECK IGNITION TIMING
1. Connect terminals TE1 and E1 of the check connector.
2. Check ignition timing.
Standard: 10° BTDC @ idle (w/ connect terminals TE1 and E1)
NO
Ignition timing – Adjust
OK
CHECK SPARK PLUGS
Plug gap: 1.1 mm (0.043 in.)
HINT: Check compression pressure and
NO
1. Spark plugs and high-tension cords
2. Compression pressure
Minimum: 9.0 kg/cm²
at 250 rpm
3. Valve clearance (Cold)
Standard: IN 0.15 – 0.25 mm (0.006 – 0.010 in.)
EX 0.20 – 0.30 mm (0.008 – 0.012 in.)
OK
CHECK COLD START INJECTOR
BAD
1. Cold start injector
2. Cold start injector time switch.
OK
OK
CHECK FUEL PRESSURE
BAD
1. Fuel pump
2. Fuel filter
3. Fuel pressure regulator
OK
CHECK INJECTORS
BAD
Injection condition
OK
CHECK EFI ELECTRONIC CIRCUIT USING VOLT/OHMMETER
BAD
1. Wiring connections
2. Power to ECU
(1) Fusible links
(2) Fuses
(3) EFI main relay
3. Air flow meter
4. Water temp. sensor
5. Air temp. sensor
6. Injection signal circuit
(1) Injector wirnng
(2) ECU
7. Oxygen sensor(s)

TROUBLESHOOTING PROCEDURES—CRESSIDA 7M-GE

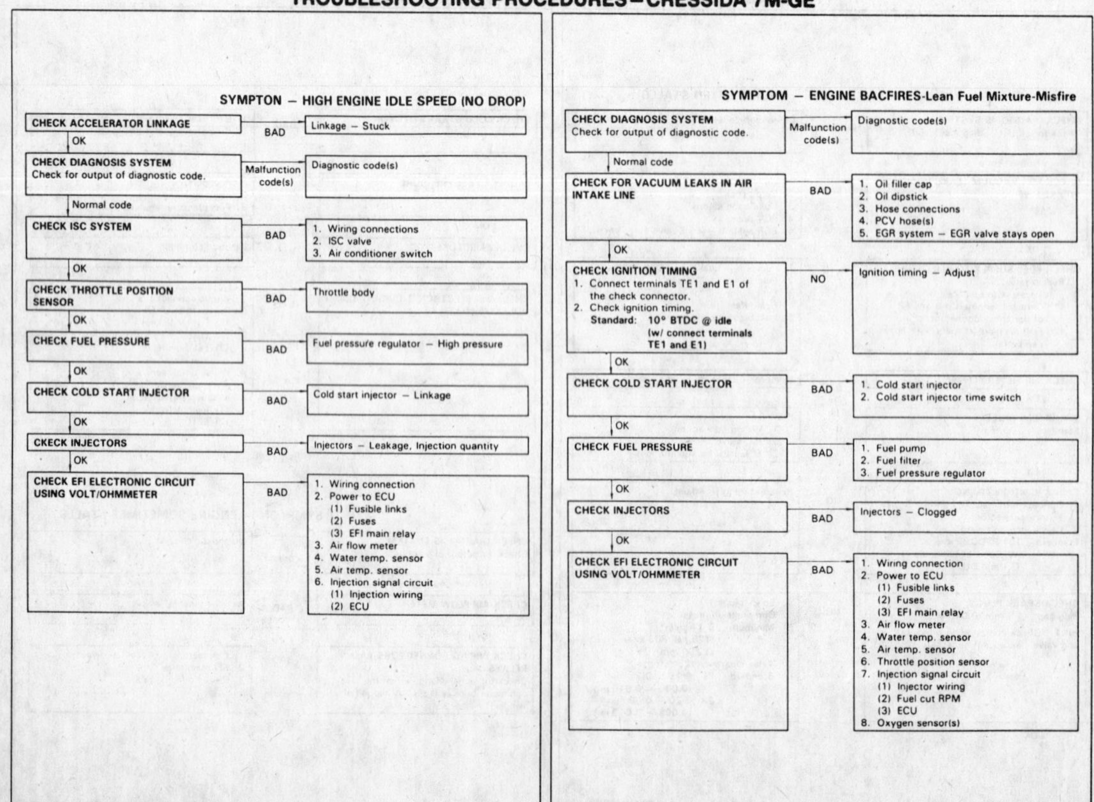
SYMPTON – HIGH ENGINE IDLE SPEED (NO DROP)
CHECK ACCELERATOR LINKAGE
BAD
Linkage – Stuck
OK
CHECK DIAGNOSIS SYSTEM
Check for output of diagnostic code.
Malfunction code(s)
Diagnostic code(s)
Normal code
CHECK ISC SYSTEM
BAD
1. Wiring connections
2. ISC valve
3. Air conditioner switch
OK
CHECK THROTTLE POSITION SENSOR
BAD
Throttle body
OK
CHECK FUEL PRESSURE
BAD
Fuel pressure regulator – High pressure
OK
CHECK COLD START INJECTOR
BAD
Cold start injector – Linkage
OK
CKECK INJECTORS
BAD
Injectors – Leakage, Injection quantity
OK
CHECK EFI ELECTRONIC CIRCUIT USING VOLT/OHMMETER
BAD
1. Wiring connection
2. Power to ECU
(1) Fusible links
(2) Fuses
(3) EFI main relay
3. Air flow meter
4. Water temp. sensor
5. Air temp. sensor
6. Injection signal circuit
(1) Injection wiring
(2) ECU
SYMPTOM – ENGINE BACFIRES-Lean Fuel Mixture-Misfire
CHECK DIAGNOSIS SYSTEM
Check for output of diagnostic code.
Malfunction code(s)
Diagnostic code(s)
Normal code
CHECK FOR VACUUM LEAKS IN AIR INTAKE LINE
BAD
1. Oil filler cap
2. Oil dipstick
3. Hose connections
4. PCV hose(s)
5. EGR system – EGR valve stays open
OK
CHECK IGNITION TIMING
1. Connect terminals TE1 and E1 of the check connector.
2. Check ignition timing.
Standard: 10° BTDC @ idle (w/ connect terminals TE1 and E1)
NO
Ignition timing – Adjust
OK
CHECK COLD START INJECTOR
BAD
1. Cold start injector
2. Cold start injector time switch
OK
CHECK FUEL PRESSURE
BAD
1. Fuel pump
2. Fuel filter
3. Fuel pressure regulator
OK
CHECK INJECTORS
BAD
Injectors – Clogged
OK
CHECK EFI ELECTRONIC CIRCUIT USING VOLT/OHMMETER
BAD
1. Wiring connection
2. Power to ECU
(1) Fusible links
(2) Fuses
(3) EFI main relay
3. Air flow meter
4. Water temp. sensor
5. Air temp. sensor
6. Throttle position sensor
7. Injection signal circuit
(1) Injector wiring
(2) Fuel cut RPM
(3) ECU
8. Oxygen sensor(s)

TROUBLESHOOTING PROCEDURES—CRESSIDA 7M-GE

SYMPTOM — MUFFLER EXPLOSION (AFTER FIRE)-Rich Fuel Mixture-Misfire

Check	Result	Action
CHECK DIAGNOSIS SYSTEM Check for output of diagnostic code.	Malfunction code(s)	Diagnostic code(s)
↓ Normal code		
CHECK IGNITION TIMING 1. Connect terminals TE1 and E1 of the check connector. 2. Check ignition timing. **Standard: 10° BTDC @ idle (w/ connect terminals TE1 and E1)**	NO	Ignition timing — Adjust
↓ OK		
CHECK COLD START INJECTOR	BAD	1. Cold start injector 2. Cold start injector time switch
↓ OK		
CHECK FUEL PRESSURE	BAD	Fuel pressure regulator
↓ OK		
CHECK INJECTORS	BAD	Injectors — Leakage
↓ OK		
CHECK SPARK PLUGS **Plug gap: 1.1 mm (0.043 in.)** HINT: Check compression pressure and valve clearance if necessary.	NO	1. Spark plugs 2. Compression pressure **Minimum: 9.0 kg/cm² (128 psi, 883 kPa) at 250 rpm** 3. Valve clearance (cold) **Standard: IN 0.15 — 0.25 mm (0.006 — 0.010 in.)** **EX 0.20 — 0.30 mm (0.008 — 0.012 in.)**
↓ OK		
CHECK EFI ELECTRONIC CIRCUIT USING VOLT/OHMMETER	BAD	1. Throttle positioin sensor 2. Injection signal circuit (1) Injector wiring (2) Fuel cut RPM (3) ECU 3. Oxygen sensor(s)

SYMPTOM — ENGINE HESITATES AND/OR POOR ACCELERATION

Check	Result	Action
CHECK CLUTCH OR BRAKES	BAD	1. Clutch — Slips 2. Brakes — Drag
↓ OK		
CHECK FOR VACUUM LEAKS IN AIR INTAKE LINE	BAD	1. Oil filler cap 2. Oil dipstick 3. Hose connections 4. PCV hose(s) 5. EGR system — EGR valve stays open
↓ OK		
CHECK AIR FILTER ELEMENT	BAD	Element — Clean or replace
↓ OK		
CHECK DIAGNOSIS SYSTEM Check for output of diagnostic code.	Malfunction code(s)	Diagnostic code(s)
↓ Normal code		
CHECK IGNITION SPARK	BAD	1. High-tension cords 2. Distributor 3. Ignition coil 4. Igniter
↓ OK		
CHECK IGNITION TIMING 1. Connect terminals TE1 and E1 of the check connector. 2. Check igntion timing. **Standard: 10° BTDC @ idle (w/ connect terminals TE1 and E1)**	NO	Ignition timing — Adjust
↓ OK		
CHECK FUEL PRESSURE	BAD	1. Fuel pump 2. Fuel filter 3. Fuel pressure regulator
↓ OK		
CHECK INJECTORS	BAD	Injection condition
↓ OK		
CHECK SPARK PLUGS **Plug gap: 1.1 mm (0.043 in.)** HINT: Check compression pressure and valve clearance if necessary.	NO	1. Spark plugs 2. Compression pressure **Minimum: 9.0 kg/cm² (128 psi, 883 kPa) at 250 rpm** 3. Valve clearance (Cold) **Standard: IN 0.15 — 0.25 mm (0.006 — 0.010 in.)** **EX 0.20 — 0.30 mm (0.008 — 0.012 in.)**
↓ OK		
CHECK EFI ELECTRONIC CIRCUIT USING VOLT/OHMMETER	BAD	1. Wiring connections 2. Power to ECU (1) Fusible links (2) Fuses (3) EFI main relay 3. Air flow meter 4. Water temp. sensor 5. Air temp. sensor 6. Throttle position sensor 7. Injection signal circuit (1) Injector wiring (2) ECU

DIAGNOSTIC CODES—CRESSIDA 7M-GE

Code No.	Number of Check engine blinks	System	Diagnosis	Trouble area
–		Normal	This appears when none of the other codes are identified.	–
11		ECU (+B)	Momentary interruption in power supply to ECU.	• Ignition switch circuit • Ignition switch • Main relay circuit • Main relay • ECU
12		RPM Singal	No "NE" or "G" signal to ECU within 2 seconds after engine has been cranked.	• Distributor circuit • Distributor • Starter signal circuit • ECU
13		RPM Signal	No "NE" signal to ECU when engine speed is above 1,000 rpm.	• Distibutor circuit • Distributor • ECU
14		Ignition Singal	No "IGF" signal to ECU 6 – 8 times in succession.	• Igniter and ignition coil circuit • Igniter and ignition coil • ECU
16		ECT Control Signal	ECT control program faulty.	• ECU
21		Oxygen Sensor Singal	Deterioration of the oxygen sensor.	• Oxygen sensor circuit • Oxygen sensor • ECU
22		Water Temp. Sensor Signal	Open or short circuit in water temp. sensor signal (THW).	• Water temp. sensor circut • Water temp. sensor • ECU
24		Intake Air Temp. Sensor Singal	Open or short circuit in intake air temp. sensor signal (THA).	• Intake air temp. sensor circuit • Intake air temp. sensor • ECU
25		Air-Fuel Ratio Lean Malfunction	• When air-fuel ratio feedback correction value is not renewed for a certain period of time. • When air-fuel ratio feedback compensation value or adaptive control value feedback frequency is abnormally high during idle switch on and feedback condition. • When marked variation is detected in engine revolutions for each cylinder during idle switch on and feedback condition.	• Injector circuit • Injector • Fuel line pressure • Air flow meter • Air intake system • Oxygen sensor circuit • Oxygen sensor • Ignition system • Water temp. sensor • ECU
26		Air-Fuel Ratio Rich Malfunction		• Oxygen sensor circuit • Oxygen sensor • Injector circuit • Injector • Fuel line pressure • Air flow meter • Cold start injector • Water temp. sensor • ECU
*27		Sub-Oxygen Sensor Signal	Open or short circuit in sub-oxygen sensor signal (OX2).	• Sub-oxygen sensor circuit • Sub-oxygen sensor • ECU
			Open or short circuit in sub-oxygen sensor heater signal (HT).	• Sub-oxygen sensor heater circuit • Sub-oxygen sensor heater • ECU
31		Air Flow Meter Signal	Open circuit in VC signal or short circuit between VS and E2 when idle contacts are closed.	• Air flow meter circuit • Air flow meter • ECU
32		Air Flow Meter Signal	Open circuit in E2 or short circuit between VC and VS.	• Air flow meter circuit • Air flow meter • ECU
41		Throttle Position Sensor Signal	Open or short circuit in throttle position sensor signal (VTA).	• Throttle position sensor circuit • Throttle position sensor • ECU
42		Vehicle Speed Sensor Singnal	No "SP1" signal to ECU for 8 seconds when engine speed is above 2,500 rpm and neutral start switch is off.	• No.1 vehicle speed sensor (Meter side) circuit • No.1 vehicle speed sensor (Meter side) • ECU
43		Starter Signal	No "STA" signal to ECU until engine speed reaches 400 rpm with vehicle not moving.	• Ignition switch circuit • Ignition switch • ECU
52		Knock Sensor Singal	Open or short circuit in knock sensor singal (KNK).	• Knock sensor circuit • Knock sensor • ECU
53		Knock Control Singal	Knock control program faulty.	• ECU
*71		EGR System Malfuncition	• EGR gas temp. below predetermined level during EGR operation. • Open circuit in EGR gas temp. sensor signal (THG).	• EGR valve • EGR hose • EGR gas temp. sensor circuit • EGR gas temp. sensor • VSV for EGR • VSV circuit for EGR • ECU
51		Switch Signal	No "IDL" signal, "NSW" signal or "A/C" signal to ECU, during diagnosis check for test mode.	• A/C switch circuit • A/C switch • A/C amplifier • Throttle position sensor circuit • Throttle position sensor • Neutral start switch circuit • Neutral start switch • Accelerater pedal and cable • ECU

* California specification vehicles only

DIAGNOSTIC CODES—CRESSIDA 7M-GE

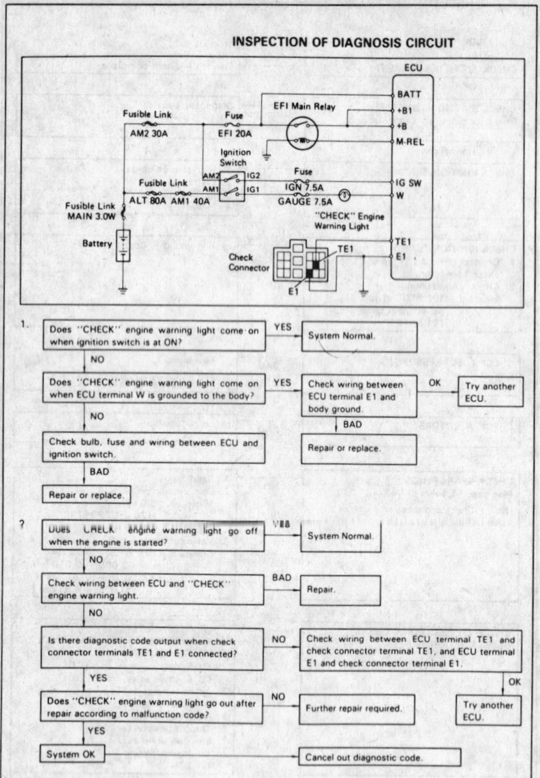
INSPECTION OF DIAGNOSIS CIRCUIT
ECU
BATT
+B1
+B
M-REL
IG SW
W
TE1
E1
EFI Main Relay
Fusible Link
AM2 30A
Fuse
EFI 20A
Ignition Switch
AM2
IG2
AM1
IG1
Fusible Link
ALT 80A
AM1 40A
Fusible Link
MAIN 3.0W
Fuse
IGN 7.5A
GAUGE 7.5A
"CHECK" Engine Warning Light
Battery
Check Connector
TE1
E1
1.
Does "CHECK" engine warning light come on when ignition switch is at ON?
YES
System Normal.
NO
Does "CHECK" engine warning light come on when ECU terminal W is grounded to the body?
YES
Check wiring between ECU terminal E1 and body ground.
OK
Try another ECU.
BAD
Repair or replace.
NO
Check bulb, fuse and wiring between ECU and ignition switch.
BAD
Repair or replace.
2.
Does "CHECK" engine warning light go off when the engine is started?
YES
System Normal.
NO
Check wiring between ECU and "CHECK" engine warning light.
BAD
Repair.
NO
Is there diagnostic code output when check connector terminals TE1 and E1 connected?
NO
Check wiring between ECU terminal TE1 and check connector terminal TE1, and ECU terminal E1 and check connector terminal E1.
OK
YES
Does "CHECK" engine warning light go out after repair according to malfunction code?
NO
Further repair required.
Try another ECU.
YES
System OK
Cancel out diagnostic code.

DIAGNOSTIC CIRCUITS TEST—CRESSSIDA 7M-GE

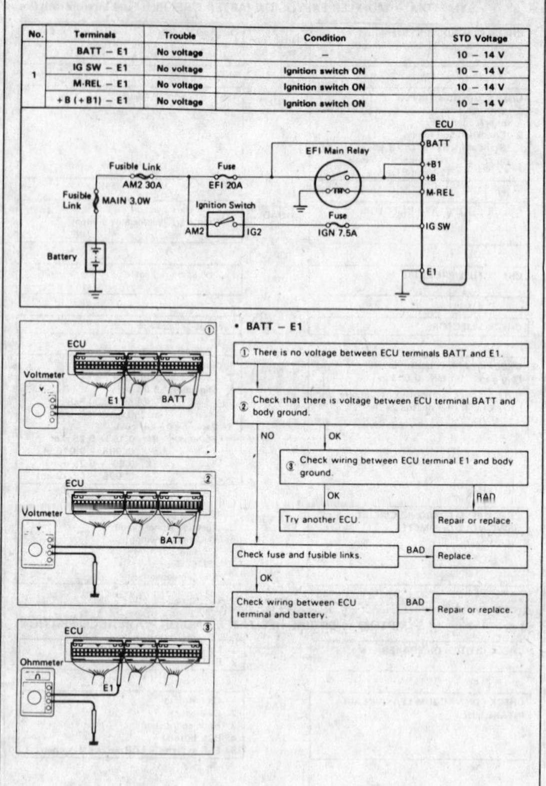

No.	Terminals	Trouble	Condition	STD Voltage
1	BATT – E1	No voltage	–	10 – 14 V
	IG SW – E1	No voltage	Ignition switch ON	10 – 14 V
	M-REL – E1	No voltage	Ignition switch ON	10 – 14 V
	+B (+B1) – E1	No voltage	Ignition switch ON	10 – 14 V

ECU
BATT
+B1
+B
M-REL
IG SW
E1
EFI Main Relay
Fusible Link
AM2 30A
Fuse
EFI 20A
Fusible Link
MAIN 3.0W
Ignition Switch
AM2
IG2
Fuse
IGN 7.5A
Battery
• BATT – E1
ECU
Voltmeter
E1
BATT
① There is no voltage between ECU terminals BATT and E1.
② Check that there is voltage between ECU terminal BATT and body ground.
NO
OK
③ Check wiring between ECU terminal E1 and body ground.
OK
BAD
Try another ECU.
Repair or replace.
Check fuse and fusible links.
BAD
Replace.
OK
Check wiring between ECU terminal and battery.
BAD
Repair or replace.
Ohmmeter

TROUBLESHOOTING CHART 1—CRESSIDA 7M-GE

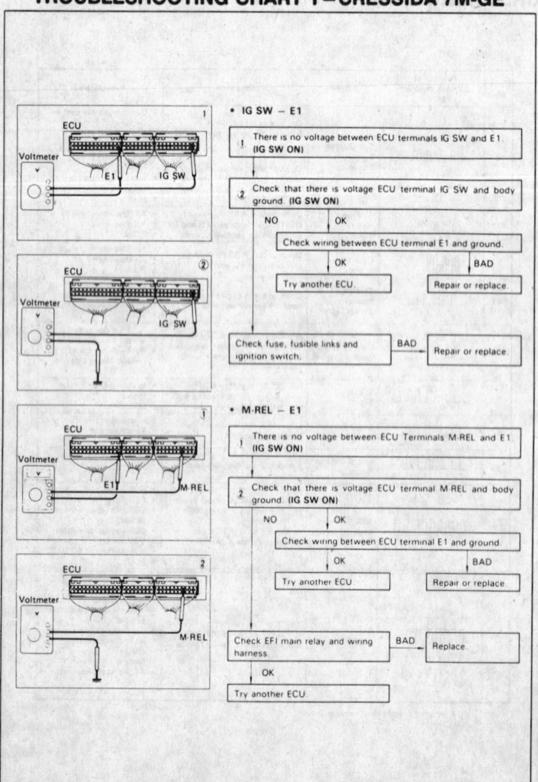
• IG SW – E1
ECU
Voltmeter
E1
IG SW
1 There is no voltage between ECU terminals IG SW and E1. (IG SW ON)
2 Check that there is voltage ECU terminal IG SW and body ground. (IG SW ON)
NO
OK
Check wiring between ECU terminal E1 and ground.
OK
BAD
Try another ECU.
Repair or replace.
Check fuse, fusible links and ignition switch.
BAD
Repair or replace.
• M-REL – E1
M-REL
1 There is no voltage between ECU Terminals M-REL and E1. (IG SW ON)
2 Check that there is voltage ECU terminal M-REL and body ground. (IG SW ON)
NO
OK
Check wiring between ECU terminal E1 and ground.
OK
BAD
Try another ECU.
Repair or replace.
Check EFI main relay and wiring harness.
BAD
Replace.
OK
Try another ECU.

TROUBLESHOOTING CHART 1—CRESSIDA 7M-GE

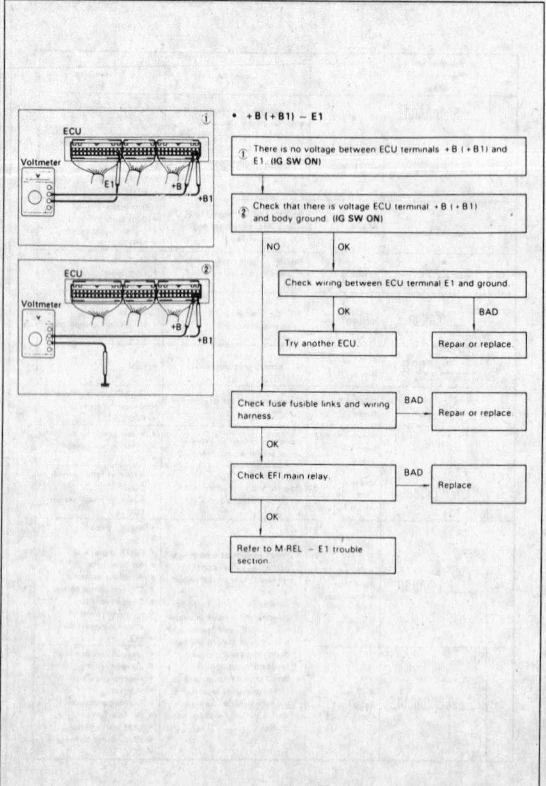
• +B (+B1) – E1
ECU
Voltmeter
E1
+B
+B1
① There is no voltage between ECU terminals +B (+B1) and E1. (IG SW ON)
② Check that there is voltage ECU terminal +B (+B1) and body ground. (IG SW ON)
NO
OK
Check wiring between ECU terminal E1 and ground.
OK
BAD
Try another ECU.
Repair or replace.
Check fuse fusible links and wiring harness.
BAD
Repair or replace.
OK
Check EFI main relay.
BAD
Replace.
OK
Refer to M-REL – E1 trouble section.

TROUBLESHOOTING CHART 2—CRESSIDA 7M-GE

No.	Terminals	Trouble	Condition		STD Voltage
2	IDL – E2	No voltage	Ignition SW ON	Throttle valve open	4 – 6 V
	VC – E2			—	4 – 6 V
	VTA – E2			Throttle valve fully closed	0.1 – 1.0 V
				Throttle valve fully open	3.2 – 4.2 V

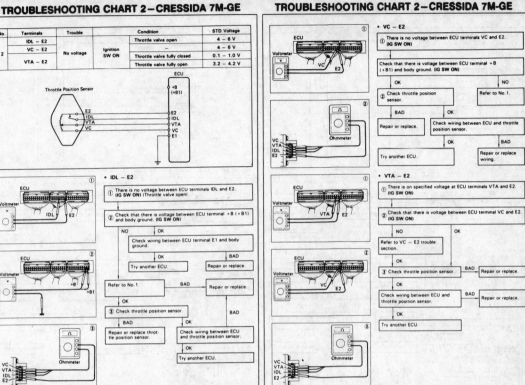

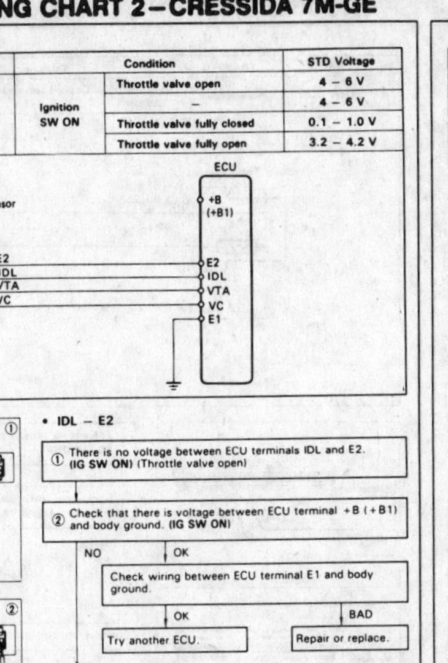
ECU
Throttle Position Sensor
+B
(+B1)
E2
IDL
VTA
VC
E1
Voltmeter
Ohmmeter

• IDL – E2

① There is no voltage between ECU terminals IDL and E2. (IG SW ON) (Throttle valve open)

② Check that there is voltage between ECU terminal +B (+B1) and body ground. (IG SW ON)
- OK: Check wiring between ECU terminal E1 and body ground.
  - OK: Try another ECU.
  - BAD: Repair or replace.
- NO: Refer to No. 1.
  - BAD: Repair or replace.
  - OK: ③ Check throttle position sensor.
    - BAD: Repair or replace throttle position sensor.
    - OK: Check wiring between ECU and throttle position sensor.
      - BAD: Repair or replace.
      - OK: Try another ECU.

TROUBLESHOOTING CHART 2—CRESSIDA 7M-GE

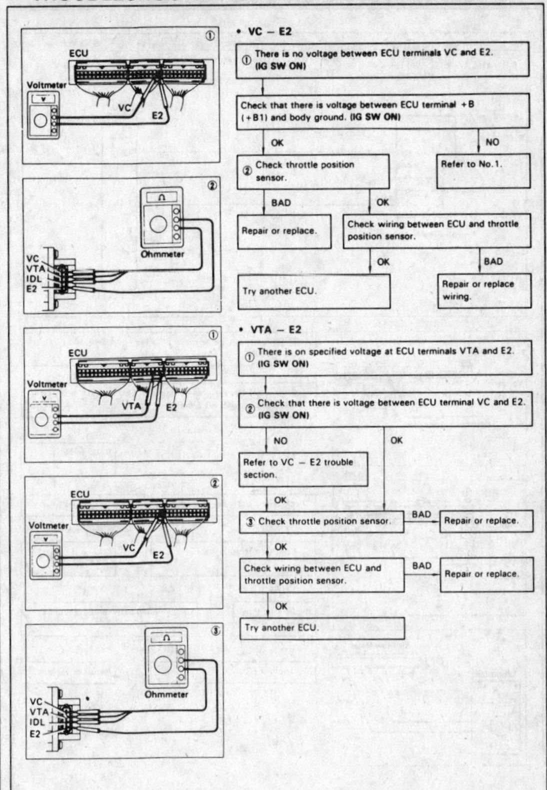
ECU
Voltmeter
VC
E2
VTA
IDL
Ohmmeter

• VC – E2

① There is no voltage between ECU terminals VC and E2. (IG SW ON)

Check that there is voltage between ECU terminal +B (+B1) and body ground. (IG SW ON)
- NO: Refer to No. 1.
- OK: ② Check throttle position sensor.
  - BAD: Repair or replace.
  - OK: Check wiring between ECU and throttle position sensor.
    - OK: Try another ECU.
    - BAD: Repair or replace wiring.

• VTA – E2

① There is on specified voltage at ECU terminals VTA and E2. (IG SW ON)

② Check that there is voltage between ECU terminal VC and E2. (IG SW ON)
- NO: Refer to VC – E2 trouble section.
- OK: ③ Check throttle position sensor.
  - BAD: Repair or replace.
  - OK: Check wiring between ECU and throttle position sensor.
    - BAD: Repair or replace.
    - OK: Try another ECU.

TROUBLESHOOTING CHART 3—CRESSIDA 7M-GE

No.	Terminals	Trouble	Condition		STD Voltage
3	VC – E2	No voltage	Ignition SW ON	—	4 – 6 V
	VS – E2			Measuring plate fully closed	3.7 – 4.3 V
				Measuring plate fully open	0.2 – 0.5 V
			Idling	—	2.3 – 2.8 V
			3,000 rpm	—	1.0 – 2.0 V

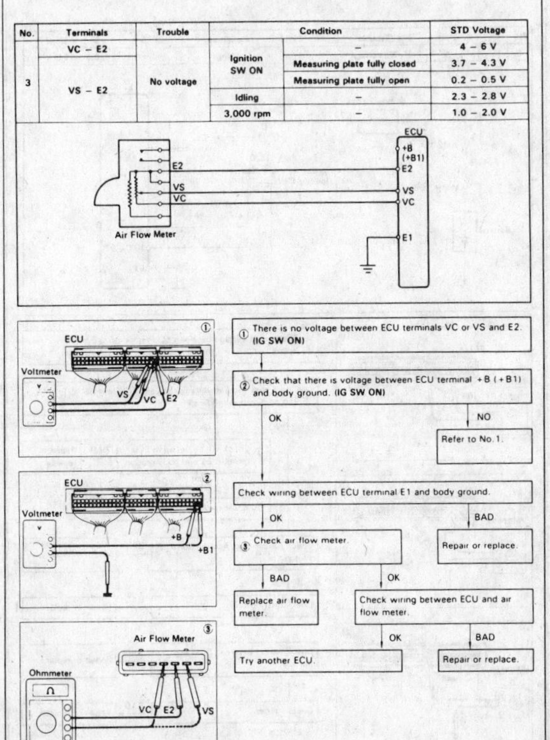
ECU
+B
(+B1)
E2
VS
VC
E1
Air Flow Meter
Voltmeter
Ohmmeter

① There is no voltage between ECU terminals VC or VS and E2. (IG SW ON)

② Check that there is voltage between ECU terminal +B (+B1) and body ground. (IG SW ON)
- NO: Refer to No. 1.
- OK: Check wiring between ECU terminal E1 and body ground.
  - BAD: Repair or replace.
  - OK: ③ Check air flow meter.
    - BAD: Replace air flow meter.
    - OK: Check wiring between ECU and air flow meter.
      - OK: Try another ECU.
      - BAD: Repair or replace.

TROUBLESHOOTING CHART 4—CRESSIDA 7M-GE

No.	Terminals	Trouble	Condition	STD Voltage
4	No.10 No.20 – E01 No.30 E02	No voltage	Ignition SW ON	10 – 14 V

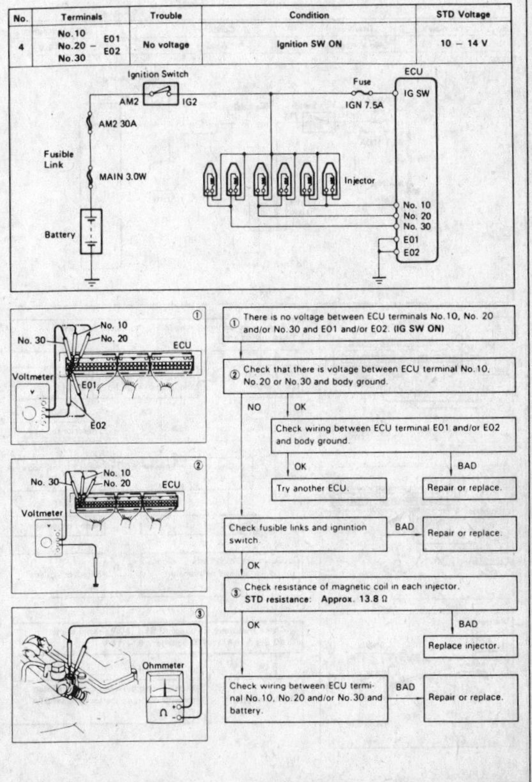
Ignition Switch
AM2
IG2
AM2 30A
Fusible Link
MAIN 3.0W
Battery
Fuse
IGN 7.5A
ECU
IG SW
Injector
No. 10
No. 20
No. 30
E01
E02
Voltmeter
Ohmmeter

① There is no voltage between ECU terminals No. 10, No. 20 and/or No. 30 and E01 and/or E02. (IG SW ON)

② Check that there is voltage between ECU terminal No. 10, No. 20 or No. 30 and body ground.
- OK: Check wiring between ECU terminal E01 and/or E02 and body ground.
  - OK: Try another ECU.
  - BAD: Repair or replace.
- NO: Check fusible links and ignition switch.
  - BAD: Repair or replace.
  - OK: ③ Check resistance of magnetic coil in each injector. STD resistance: Approx. 13.8 Ω
    - BAD: Replace injector.
    - OK: Check wiring between ECU terminal No. 10, No. 20 and/or No. 30 and battery.
      - BAD: Repair or replace.

TROUBLESHOOTING CHART 5—CRESSIDA 7M-GE

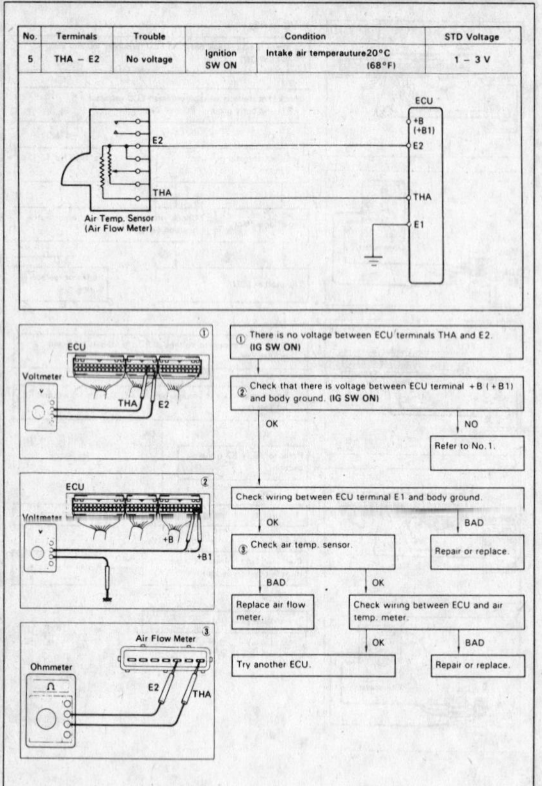
No.
Terminals
Trouble
Condition
STD Voltage
5
THA – E2
No voltage
Ignition SW ON
Intake air temperauture20°C (68°F)
1 – 3 V
ECU
+B (+B1)
E2
THA
E1
Air Temp. Sensor (Air Flow Meter)
Voltmeter
Ohmmeter
Air Flow Meter
① There is no voltage between ECU terminals THA and E2. (IG SW ON)
② Check that there is voltage between ECU terminal +B (+B1) and body ground. (IG SW ON)
OK
NO
Refer to No.1.
Check wiring between ECU terminal E1 and body ground.
BAD
Repair or replace.
③ Check air temp. sensor.
Replace air flow meter.
Check wiring between ECU and air temp. meter.
Try another ECU.
Repair or replace.

TROUBLESHOOTING CHART 6—CRESSIDA 7M-GE

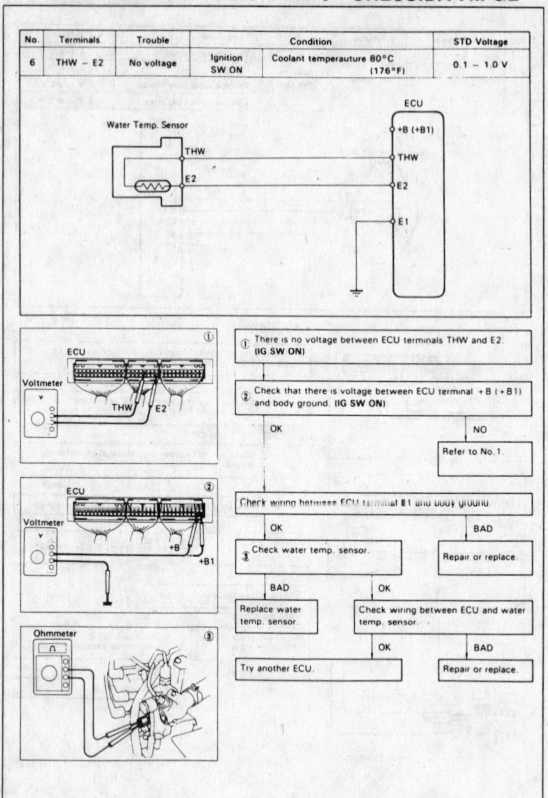
No.
Terminals
Trouble
Condition
STD Voltage
6
THW – E2
No voltage
Ignition SW ON
Coolant temperauture 80°C (176°F)
0.1 – 1.0 V
ECU
Water Temp. Sensor
+B (+B1)
THW
E2
E1
Voltmeter
Ohmmeter
① There is no voltage between ECU terminals THW and E2. (IG SW ON)
② Check that there is voltage between ECU terminal +B (+B1) and body ground. (IG SW ON)
OK
NO
Refer to No.1.
Check wiring between ECU terminal E1 and body ground.
BAD
Repair or replace.
③ Check water temp. sensor.
Replace water temp. sensor.
Check wiring between ECU and water temp. sensor.
Try another ECU.
Repair or replace.

TROUBLESHOOTING CHART 7—CRESSIDA 7M-GE

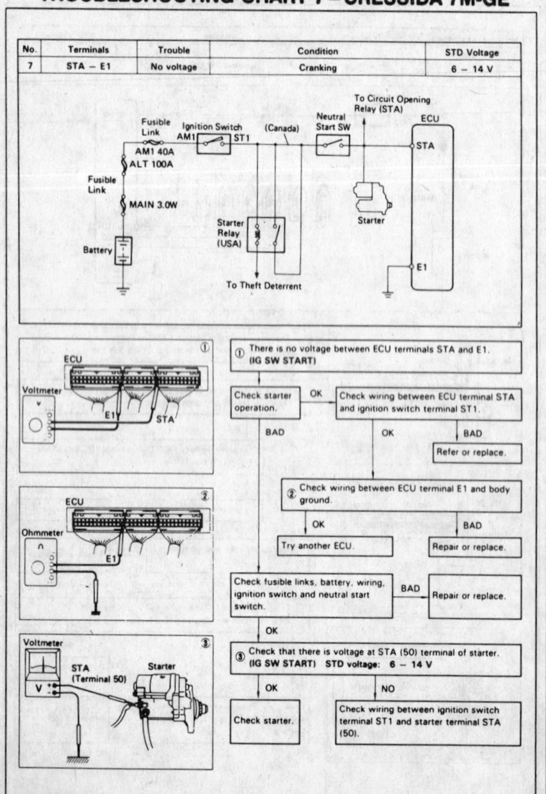
No.
Terminals
Trouble
Condition
STD Voltage
7
STA – E1
No voltage
Cranking
6 – 14 V
Fusible Link
AM1 40A
ALT 100A
MAIN 3.0W
Battery
Ignition Switch
AM1
ST1
(Canada)
Neutral Start SW
To Circuit Opening Relay (STA)
ECU
STA
E1
Starter
Starter Relay (USA)
To Theft Deterrent
Voltmeter
Ohmmeter
STA (Terminal 50)
① There is no voltage between ECU terminals STA and E1. (IG SW START)
Check starter operation.
OK
Check wiring between ECU terminal STA and ignition switch terminal ST1.
BAD
Refer or replace.
② Check wiring between ECU terminal E1 and body ground.
Try another ECU.
Repair or replace.
Check fusible links, battery, wiring, ignition switch and neutral start switch.
③ Check that there is voltage at STA (50) terminal of starter. (IG SW START) STD voltage: 6 – 14 V
NO
Check starter.
Check wiring between ignition switch terminal ST1 and starter terminal STA (50).

TROUBLESHOOTING CHART 8—CRESSIDA 7M-GE

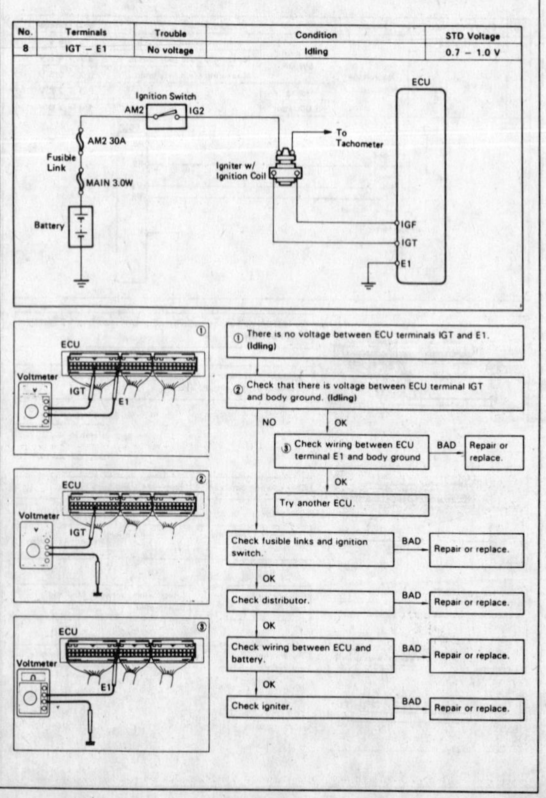
No.
Terminals
Trouble
Condition
STD Voltage
8
IGT – E1
No voltage
Idling
0.7 – 1.0 V
Ignition Switch
AM2
IG2
AM2 30A
Fusible Link
MAIN 3.0W
Battery
To Tachometer
Igniter w/ Ignition Coil
ECU
IGF
IGT
E1
Voltmeter
① There is no voltage between ECU terminals IGT and E1. (Idling)
② Check that there is voltage between ECU terminal IGT and body ground. (Idling)
NO
OK
③ Check wiring between ECU terminal E1 and body ground
BAD
Repair or replace.
Try another ECU.
Check fusible links and ignition switch.
Check distributor.
Check wiring between ECU and battery.
Check igniter.

TROUBLESHOOTING CHART 9–CRESSIDA 7M-GE

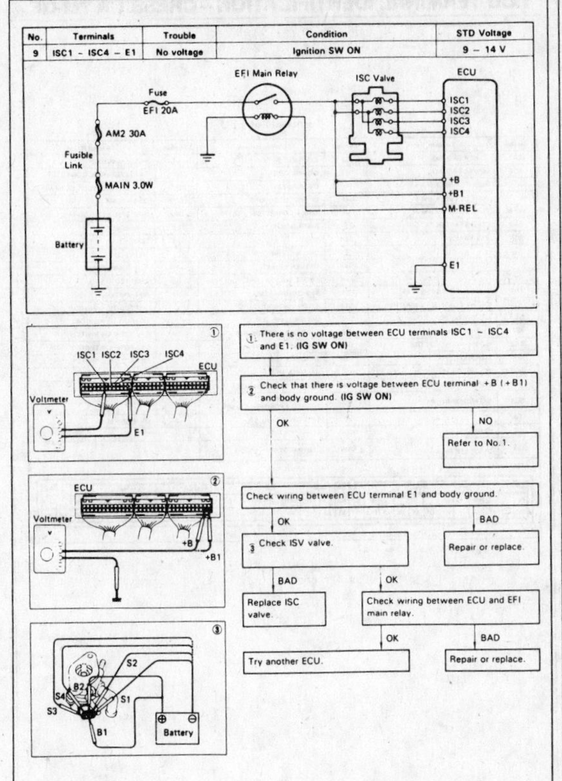
No. | Terminals | Trouble | Condition | STD Voltage
9 | ISC1 – ISC4 – E1 | No voltage | Ignition SW ON | 9 – 14 V
EFI Main Relay
ISC Valve
ECU
Fuse
EFI 20A
AM2 30A
Fusible Link
MAIN 3.0W
Battery
ISC1
ISC2
ISC3
ISC4
+B
+B1
M-REL
E1
Voltmeter
① There is no voltage between ECU terminals ISC1 ~ ISC4 and E1. (IG SW ON)
② Check that there is voltage between ECU terminal +B (+B1) and body ground. (IG SW ON)
OK
NO
Refer to No.1.
Check wiring between ECU terminal E1 and body ground.
OK
BAD
③ Check ISV valve.
Repair or replace.
BAD
OK
Replace ISC valve.
Check wiring between ECU and EFI main relay.
OK
BAD
Try another ECU.
Repair or replace.
S2
S4
B2
S1
S3
B1
Battery

TROUBLESHOOTING CHART 10–CRESSIDA 7M-GE

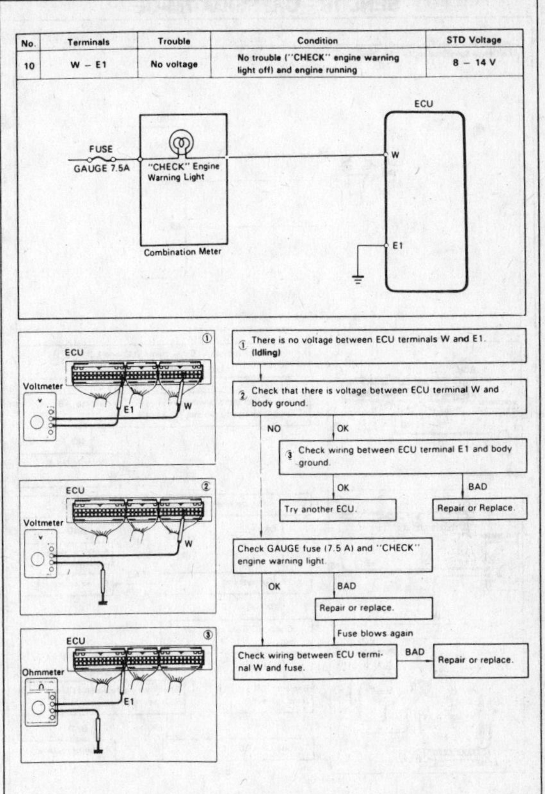
No. | Terminals | Trouble | Condition | STD Voltage
10 | W – E1 | No voltage | No trouble ("CHECK" engine warning light off) and engine running | 8 – 14 V
ECU
FUSE
GAUGE 7.5A
"CHECK" Engine Warning Light
Combination Meter
W
E1
Voltmeter
Ohmmeter
① There is no voltage between ECU terminals W and E1. (Idling)
② Check that there is voltage between ECU terminal W and body ground.
NO
OK
③ Check wiring between ECU terminal E1 and body ground.
OK
BAD
Try another ECU.
Repair or Replace.
Check GAUGE fuse (7.5 A) and "CHECK" engine warning light.
OK
BAD
Repair or replace.
Fuse blows again
Check wiring between ECU terminal W and fuse.
BAD
Repair or replace.

TROUBLESHOOTING CHART 11–CRESSIDA 7M-GE

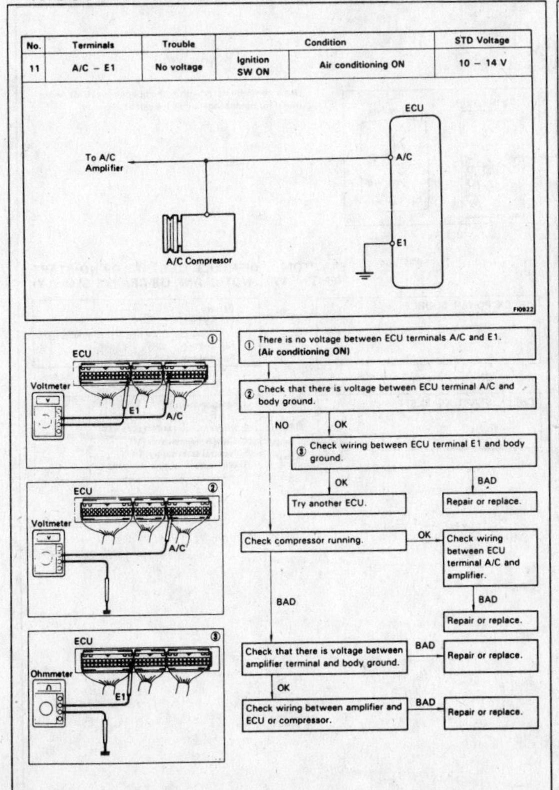
No. | Terminals | Trouble | Condition | STD Voltage
11 | A/C – E1 | No voltage | Ignition SW ON | Air conditioning ON | 10 – 14 V
ECU
To A/C Amplifier
A/C
E1
A/C Compressor
Voltmeter
Ohmmeter
① There is no voltage between ECU terminals A/C and E1. (Air conditioning ON)
② Check that there is voltage between ECU terminal A/C and body ground.
NO
OK
③ Check wiring between ECU terminal E1 and body ground.
OK
BAD
Try another ECU.
Repair or replace.
Check compressor running.
OK
Check wiring between ECU terminal A/C and amplifier.
BAD
BAD
Repair or replace.
Check that there is voltage between amplifier terminal and body ground.
BAD
Repair or replace.
OK
Check wiring between amplifier and ECU or compressor.
BAD
Repair or replace.

TROUBLESHOOTING OXYGEN SENSOR–CRESSIDA 7M-GE

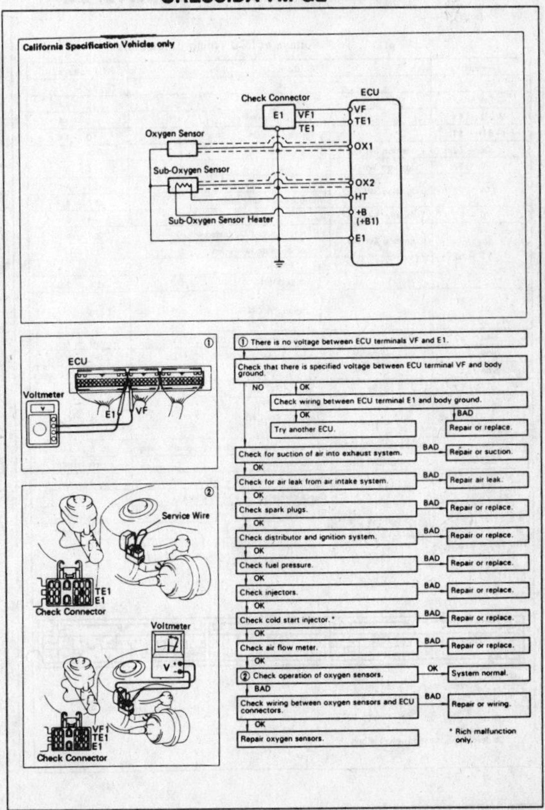
California Specification Vehicles only
Check Connector
ECU
E1
VF1
TE1
VF
OX1
Oxygen Sensor
Sub-Oxygen Sensor
OX2
HT
+B (+B1)
Sub-Oxygen Sensor Heater
E1
Voltmeter
Service Wire
Check Connector
① There is no voltage between ECU terminals VF and E1.
Check that there is specified voltage between ECU terminal VF and body ground.
NO
OK
Check wiring between ECU terminal E1 and body ground.
OK
BAD
Try another ECU.
Repair or replace.
Check for suction of air into exhaust system.
BAD
Repair or suction.
Check for air leak from air intake system.
Repair air leak.
Check spark plugs.
Repair or replace.
Check distributor and ignition system.
Check fuel pressure.
Check injectors.
Check cold start injector. *
Check air flow meter.
② Check operation of oxygen sensors.
System normal.
Check wiring between oxygen sensors and ECU connectors.
Repair or wiring.
Repair oxygen sensors.
* Rich malfunction only.

TROUBLESHOOTING EGR TEMPERATURE SENSOR—CRESSIDA 7M-GE

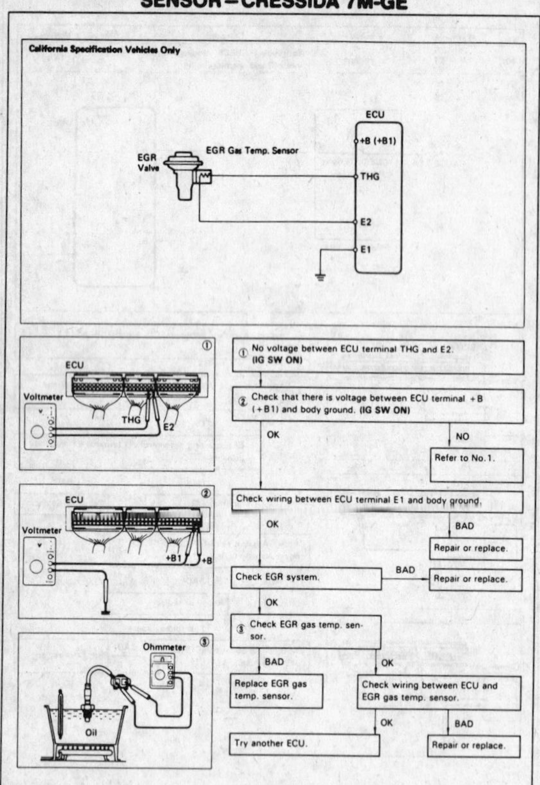
California Specification Vehicles Only
ECU
+B (+B1)
EGR Gas Temp. Sensor
EGR Valve
THG
E2
E1
ECU
Voltmeter
THG
E2
ECU
Voltmeter
+B1
+B
Ohmmeter
Oil
① No voltage between ECU terminal THG and E2. (IG SW ON)
② Check that there is voltage between ECU terminal +B (+B1) and body ground. (IG SW ON)
OK
NO
Refer to No.1.
Check wiring between ECU terminal E1 and body ground.
OK
BAD
Repair or replace.
Check EGR system.
BAD
Repair or replace.
OK
③ Check EGR gas temp. sensor.
BAD
OK
Replace EGR gas temp. sensor.
Check wiring between ECU and EGR gas temp. sensor.
OK
BAD
Try another ECU.
Repair or replace.

ECU TERMINAL IDENTIFICATION—CRESSIDA 7M-GE

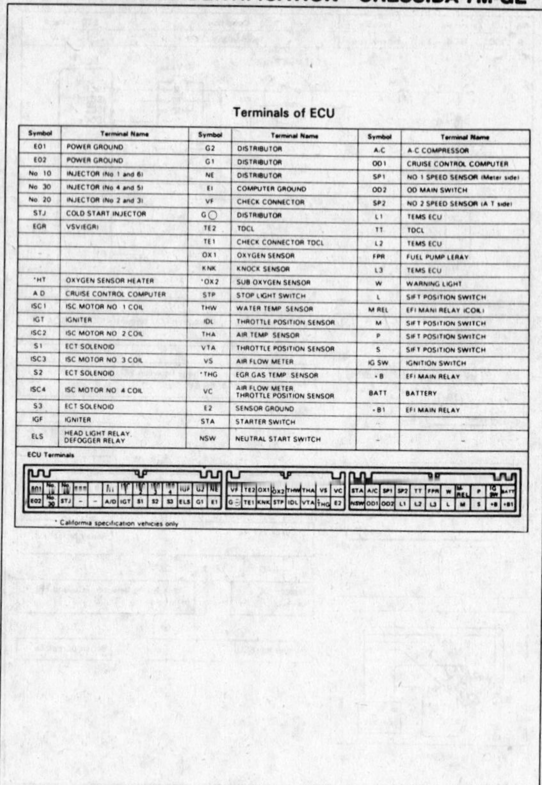

Terminals of ECU

Symbol	Terminal Name	Symbol	Terminal Name	Symbol	Terminal Name
E01	POWER GROUND	G2	DISTRIBUTOR	A.C	A.C COMPRESSOR
E02	POWER GROUND	G1	DISTRIBUTOR	OD1	CRUISE CONTROL COMPUTER
No. 10	INJECTOR (No 1 and 6)	NE	DISTRIBUTOR	SP1	NO 1 SPEED SENSOR (Meter side)
No. 30	INJECTOR (No 4 and 5)	EI	COMPUTER GROUND	OD2	OD MAIN SWITCH
No. 20	INJECTOR (No 2 and 3)	VF	CHECK CONNECTOR	SP2	NO 2 SPEED SENSOR (A T side)
STJ	COLD START INJECTOR	G⊖	DISTRIBUTOR	L1	TEMS ECU
EGR	VSV(EGR)	TE2	TDCL	TT	TDCL
		TE1	CHECK CONNECTOR TDCL	L2	TEMS ECU
–		OX1	OXYGEN SENSOR	FPR	FUEL PUMP LERAY
	–	KNK	KNOCK SENSOR	L3	TEMS ECU
*HT	OXYGEN SENSOR HEATER	*OX2	SUB OXYGEN SENSOR	W	WARNING LIGHT
A D	CRUISE CONTROL COMPUTER	STP	STOP LIGHT SWITCH	L	SIFT POSITION SWITCH
ISC1	ISC MOTOR NO 1 COIL	THW	WATER TEMP SENSOR	M REL	EFI MANI RELAY (COIL)
IGT	IGNITER	IDL	THROTTLE POSITION SENSOR	M	SIFT POSITION SWITCH
ISC2	ISC MOTOR NO 2 COIL	THA	AIR TEMP SENSOR	P	SIFT POSITION SWITCH
S1	ECT SOLENOID	VTA	THROTTLE POSITION SENSOR	S	SIFT POSITION SWITCH
ISC3	ISC MOTOR NO 3 COIL	VS	AIR FLOW METER	IG SW	IGNITION SWITCH
S2	ECT SOLENOID	*THG	EGR GAS TEMP SENSOR	+B	EFI MAIN RELAY
ISC4	ISC MOTOR NO 4 COIL	VC	AIR FLOW METER, THROTTLE POSITION SENSOR	BATT	BATTERY
S3	ECT SOLENOID	E2	SENSOR GROUND	+B1	EFI MAIN RELAY
IGF	IGNITER	STA	STARTER SWITCH	–	
ELS	HEAD LIGHT RELAY, DEFOGGER RELAY	NSW	NEUTRAL START SWITCH	–	–

ECU Terminals

E01 No. 10 No. 20 *HT ISC1 ISC2 ISC3 ISC4 IGF G2 NE
E02 No. 30 STJ – – A/D IGT S1 S2 S3 ELS G1 E1
VF TE2 OX1 *OX2 THW THA VS VC
G⊖ TE1 KNK STP IDL VTA *THG E2
STA A/C SP1 SP2 TT FPR W M-REL P IG SW BATT
NSW OD1 OD2 L1 L2 L3 L M S +B +B1

* California specification vehicles only

ECU TERMINAL VOLTAGES—CRESSIDA 7M-GE

Voltage at ECU Wiring Connectors

Terminals	Condition		STD Voltage (V)
BATT – E1	–		10 – 14
IG SW – E1	Ignition SW ON		10 – 14
M-REL – E1	Ignition SW ON		10 – 14
+B (+B1) – E1	Ignition SW ON		10 – 14
IDL – E2	Ignition SW ON	Throttle valve open	4 – 6
VC – E2	Ignition SW ON	–	4 – 6
VTA – E2	Ignition SW ON	Throttle valve fully closed	0.1 – 1.0
		Throttle valve fully open	3.2 – 4.2
VC – E2	Ignition SW ON	–	4 – 6
VS – E2	Ignition SW ON	Measuring plate fully closed	3.7 – 4.3
		Measuring plate fully open	0.2 – 0.5
	Idling		2.3 – 2.8
	3,000 rpm		1.0 – 2.0
No. 10, No. 20, No. 30 – E01, E02	Ignition SW ON		10 – 14
THA – E2	Ignition SW ON	Intake air temperature 20°C (68°F)	1 – 3
THW – E2	Ignition SW ON	Coolant temperature 80°C (176°F)	0.1 – 1.0
STA – E1	Cranking		6 – 14
IGT – E1	Ignition SW ON		0.7 – 1.0
ISC1 ∫ ISC4 – E1	Ignition SW ON		9 – 14
W – E1	No trouble ("CHECK" engine warning light off) and engine running		8 – 14
A/C – E1	Ignition SW ON	Air conditioning ON	10 – 14
TE1 – E1	Ignition SW ON	Check connector TE1 – E1 not connect	4 – 6
		Check connector TE1 – E1 connect	0
NSW – E1	Ignition SW ON	Shift position P or N range	0
		Ex. P or N range	10 – 14

ECU Terminals

E01 No. 10 No. 20 EGR – *HT ISC1 ISC2 ISC3 ISC4 IGF G2 NE
E02 No. 30 STJ – – ND IGT S1 S2 S3 ELS G1 E1
VF TE2 OX1 *OX2 THW THA VS VC
G⊖ TE1 KNK STP IDL VTA *THG E2
STA A/C SP1 SP2 TT FPR W M-REL P IG SW BATT
NSW OD1 OD2 L1 L2 L3 L M S +B +B1

*California specification vehicles only

TROUBLESHOOTING PROCEDURES—SUPRA 7M-GE AND 7M-GTE

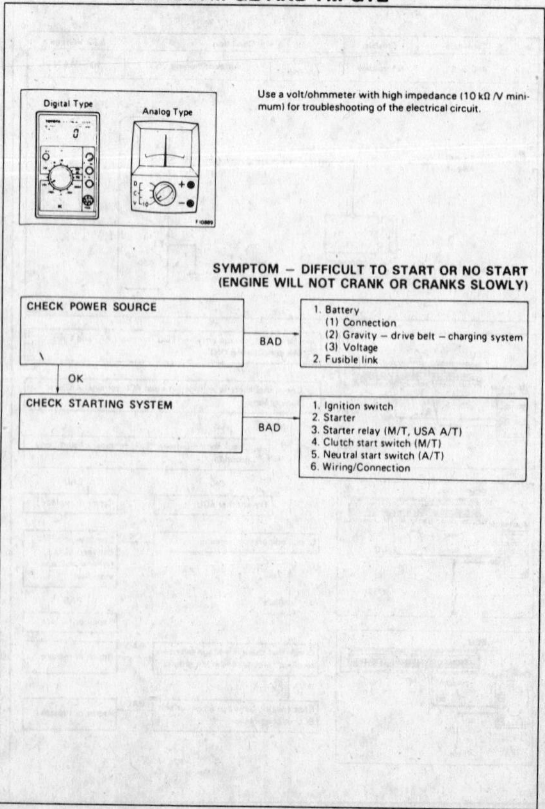
Digital Type
Analog Type
Use a volt/ohmmeter with high impedance (10 kΩ/V minimum) for troubleshooting of the electrical circuit.
SYMPTOM – DIFFICULT TO START OR NO START (ENGINE WILL NOT CRANK OR CRANKS SLOWLY)
CHECK POWER SOURCE
BAD
1. Battery
(1) Connection
(2) Gravity – drive belt – charging system
(3) Voltage
2. Fusible link
OK
CHECK STARTING SYSTEM
BAD
1. Ignition switch
2. Starter
3. Starter relay (M/T, USA A/T)
4. Clutch start switch (M/T)
5. Neutral start switch (A/T)
6. Wiring/Connection

TROUBLESHOOTING PROCEDURES—SUPRA 7M-GE AND 7M-GTE

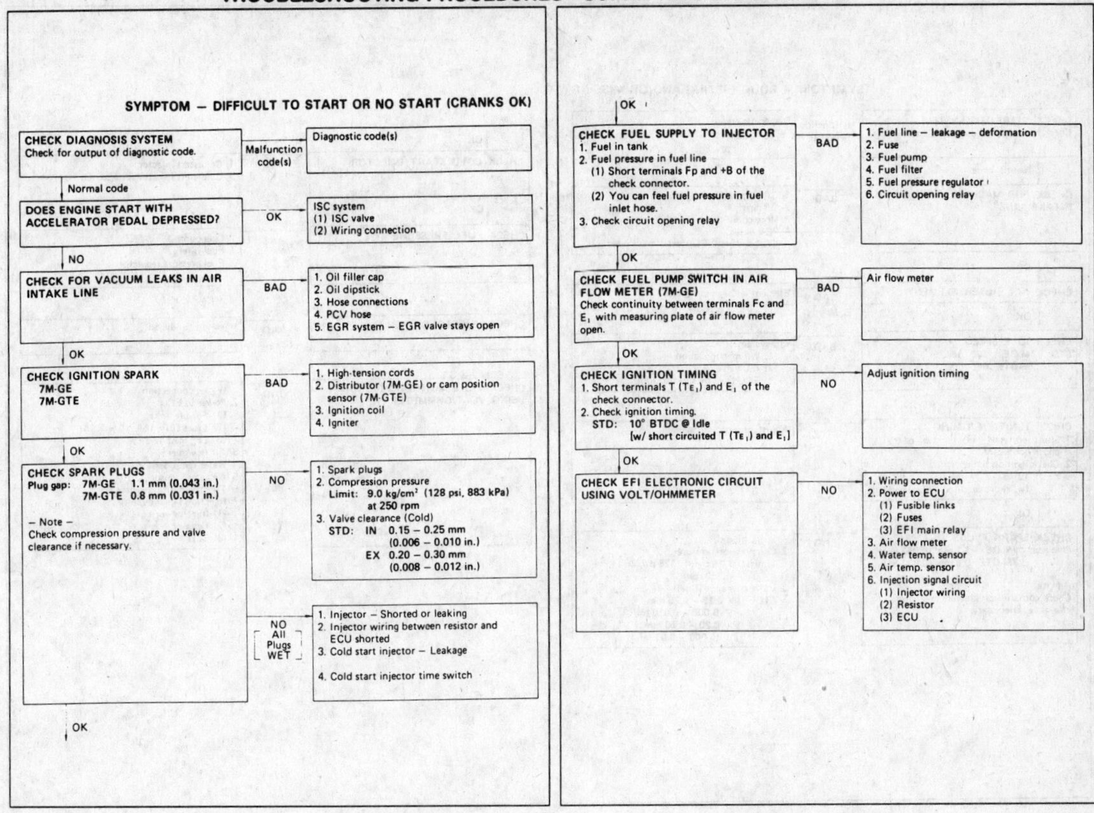
SYMPTOM – DIFFICULT TO START OR NO START (CRANKS OK)
CHECK DIAGNOSIS SYSTEM
Check for output of diagnostic code.
Malfunction code(s)
Diagnostic code(s)
Normal code
DOES ENGINE START WITH ACCELERATOR PEDAL DEPRESSED?
OK
ISC system
(1) ISC valve
(2) Wiring connection
NO
CHECK FOR VACUUM LEAKS IN AIR INTAKE LINE
BAD
1. Oil filler cap
2. Oil dipstick
3. Hose connections
4. PCV hose
5. EGR system – EGR valve stays open
OK
CHECK IGNITION SPARK
7M-GE
7M-GTE
BAD
1. High-tension cords
2. Distributor (7M-GE) or cam position sensor (7M-GTE)
3. Ignition coil
4. Igniter
OK
CHECK SPARK PLUGS
Plug gap: 7M-GE 1.1 mm (0.043 in.)
7M-GTE 0.8 mm (0.031 in.)
– Note –
Check compression pressure and valve clearance if necessary.
NO
1. Spark plugs
2. Compression pressure
Limit: 9.0 kg/cm² (128 psi, 883 kPa) at 250 rpm
3. Valve clearance (Cold)
STD: IN 0.15 – 0.25 mm (0.006 – 0.010 in.)
EX 0.20 – 0.30 mm (0.008 – 0.012 in.)
NO
All Plugs WET
1. Injector – Shorted or leaking
2. Injector wiring between resistor and ECU shorted
3. Cold start injector – Leakage
4. Cold start injector time switch
OK
OK
CHECK FUEL SUPPLY TO INJECTOR
1. Fuel in tank
2. Fuel pressure in fuel line
(1) Short terminals Fp and +B of the check connector.
(2) You can feel fuel pressure in fuel inlet hose.
3. Check circuit opening relay
BAD
1. Fuel line – leakage – deformation
2. Fuse
3. Fuel pump
4. Fuel filter
5. Fuel pressure regulator
6. Circuit opening relay
OK
CHECK FUEL PUMP SWITCH IN AIR FLOW METER (7M-GE)
Check continuity between terminals Fc and E1 with measuring plate of air flow meter open.
BAD
Air flow meter
OK
CHECK IGNITION TIMING
1. Short terminals T (TE1) and E1 of the check connector.
2. Check ignition timing.
STD: 10° BTDC @ Idle
[w/ short circuited T (TE1) and E1]
NO
Adjust ignition timing
OK
CHECK EFI ELECTRONIC CIRCUIT USING VOLT/OHMMETER
NO
1. Wiring connection
2. Power to ECU
(1) Fusible links
(2) Fuses
(3) EFI main relay
3. Air flow meter
4. Water temp. sensor
5. Air temp. sensor
6. Injection signal circuit
(1) Injector wiring
(2) Resistor
(3) ECU

TROUBLESHOOTING PROCEDURES—SUPRA 7M-GE AND 7M-GTE

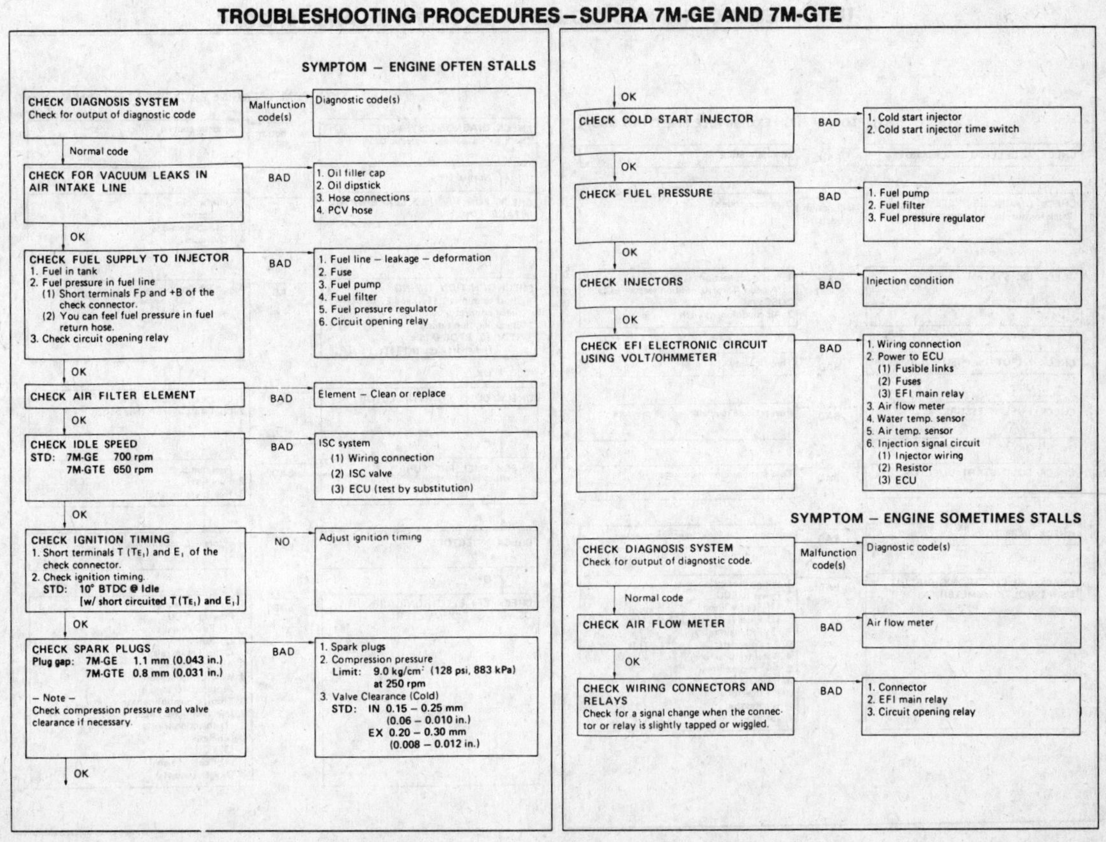
SYMPTOM – ENGINE OFTEN STALLS
CHECK DIAGNOSIS SYSTEM
Check for output of diagnostic code
Malfunction code(s)
Diagnostic code(s)
Normal code
CHECK FOR VACUUM LEAKS IN AIR INTAKE LINE
BAD
1. Oil filler cap
2. Oil dipstick
3. Hose connections
4. PCV hose
OK
CHECK FUEL SUPPLY TO INJECTOR
1. Fuel in tank
2. Fuel pressure in fuel line
(1) Short terminals Fp and +B of the check connector.
(2) You can feel fuel pressure in fuel return hose.
3. Check circuit opening relay
BAD
1. Fuel line – leakage – deformation
2. Fuse
3. Fuel pump
4. Fuel filter
5. Fuel pressure regulator
6. Circuit opening relay
OK
CHECK AIR FILTER ELEMENT
BAD
Element – Clean or replace
OK
CHECK IDLE SPEED
STD: 7M-GE 700 rpm
7M-GTE 650 rpm
BAD
ISC system
(1) Wiring connection
(2) ISC valve
(3) ECU (test by substitution)
OK
CHECK IGNITION TIMING
1. Short terminals T (TE1) and E1 of the check connector.
2. Check ignition timing.
STD: 10° BTDC @ Idle
[w/ short circuited T (TE1) and E1]
NO
Adjust ignition timing
OK
CHECK SPARK PLUGS
Plug gap: 7M-GE 1.1 mm (0.043 in.)
7M-GTE 0.8 mm (0.031 in.)
– Note –
Check compression pressure and valve clearance if necessary.
BAD
1. Spark plugs
2. Compression pressure
Limit: 9.0 kg/cm² (128 psi, 883 kPa) at 250 rpm
3. Valve Clearance (Cold)
STD: IN 0.15 – 0.25 mm (0.06 – 0.010 in.)
EX 0.20 – 0.30 mm (0.008 – 0.012 in.)
OK
OK
CHECK COLD START INJECTOR
BAD
1. Cold start injector
2. Cold start injector time switch
OK
CHECK FUEL PRESSURE
BAD
1. Fuel pump
2. Fuel filter
3. Fuel pressure regulator
OK
CHECK INJECTORS
BAD
Injection condition
OK
CHECK EFI ELECTRONIC CIRCUIT USING VOLT/OHMMETER
BAD
1. Wiring connection
2. Power to ECU
(1) Fusible links
(2) Fuses
(3) EFI main relay
3. Air flow meter
4. Water temp. sensor
5. Air temp. sensor
6. Injection signal circuit
(1) Injector wiring
(2) Resistor
(3) ECU
SYMPTOM – ENGINE SOMETIMES STALLS
CHECK DIAGNOSIS SYSTEM
Check for output of diagnostic code.
Malfunction code(s)
Diagnostic code(s)
Normal code
CHECK AIR FLOW METER
BAD
Air flow meter
OK
CHECK WIRING CONNECTORS AND RELAYS
Check for a signal change when the connector or relay is slightly tapped or wiggled.
BAD
1. Connector
2. EFI main relay
3. Circuit opening relay

TROUBLESHOOTING PROCEDURES—SUPRA 7M-GE AND 7M-GTE

SYMPTOM – ROUGH IDLING AND/OR MISSING

CHECK DIAGNOSIS SYSTEM
Check for output of diagnostic code.
- Malfunction code(s) → Diagnostic code(s)
- Normal code ↓

CHECK FOR VACUUM LEAKS IN AIR INTAKE LINE
- BAD →
  1. Oil filler cap
  2. Oil dipstick
  3. Hose connections
  4. PCV hoses
  5. EGR system – EGR valve stays open
- OK ↓

CHECK AIR FILTER ELEMENT
- BAD → Element – Clean or replace
- OK ↓

CHECK IDLE SPEED
STD: 7M-GE 700 rpm
7M-GTE 650 rpm
- BAD → ISC system
  (1) Wiring connection
  (2) ISC valve
  (3) ECU (test by substitution)
- OK ↓

CHECK IGNITION TIMING
1. Short terminals T (T_{E_1}) and E_1 of the check connector.
2. Check ignition timing.
STD: 10° BTDC @ Idle
[w/ short circuited T(T_{E_1}) and E_1]
- NO → Adjust ignition timing
- OK ↓

CHECK SPARK PLUGS
Plug gap: 7M-GE 1.1 mm (0.043 in.)
7M-GTE 0.8 mm (0.031 in.)
– Note –
Check compression pressure and valve clearance if necessary.
- NO →
  1. Spark plug
  2. Compression pressure
     Limit: 9.0 kg/cm² (128 psi, 883 kPa) at 250 rpm
  3. Valve clearance (Cold)
     STD: IN 0.15 – 0.25 mm (0.006 – 0.010 in.)
     EX 0.20 – 0.30 mm (0.008 – 0.012 in.)
- OK ↓

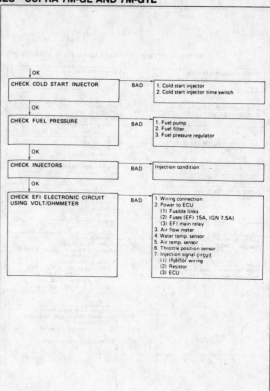

CHECK COLD START INJECTOR
- BAD →
  1. Cold start injector
  2. Cold start injector time switch
- OK ↓

CHECK FUEL PRESSURE
- BAD →
  1. Fuel pump
  2. Fuel filter
  3. Fuel pressure regulator
- OK ↓

CHECK INJECTORS
- BAD → Injection condition
- OK ↓

CHECK EFI ELECTRONIC CIRCUIT USING VOLT/OHMMETER
- BAD →
  1. Wiring connection
  2. Power to ECU
     (1) Fusible links
     (2) Fuses (EFI 15A, IGN 7.5A)
     (3) EFI main relay
  3. Air flow meter
  4. Water temp. sensor
  5. Air temp. sensor
  6. Throttle position sensor
  7. Injection signal circuit
     (1) Injector wiring
     (2) Resistor
     (3) ECU

TROUBLESHOOTING PROCEDURES—SUPRA 7M-GE AND 7M-GTE

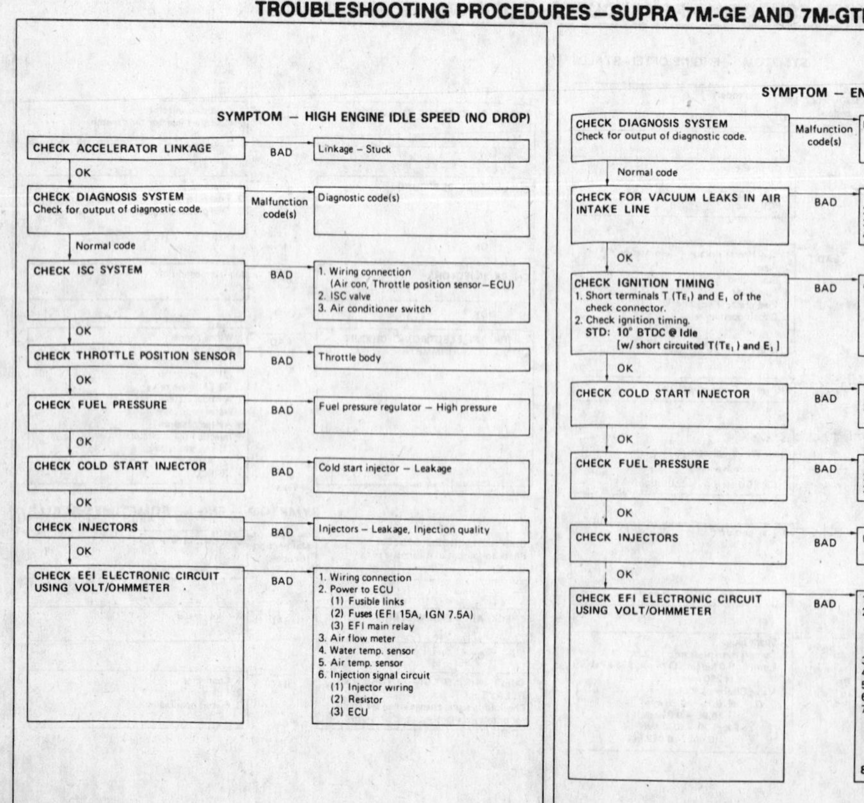
SYMPTOM – HIGH ENGINE IDLE SPEED (NO DROP)
CHECK ACCELERATOR LINKAGE
BAD
Linkage – Stuck
OK
CHECK DIAGNOSIS SYSTEM
Check for output of diagnostic code.
Malfunction code(s)
Diagnostic code(s)
Normal code
CHECK ISC SYSTEM
BAD
1. Wiring connection
(Air con, Throttle position sensor–ECU)
2. ISC valve
3. Air conditioner switch
OK
CHECK THROTTLE POSITION SENSOR
BAD
Throttle body
OK
CHECK FUEL PRESSURE
BAD
Fuel pressure regulator – High pressure
OK
CHECK COLD START INJECTOR
BAD
Cold start injector – Leakage
OK
CHECK INJECTORS
BAD
Injectors – Leakage, Injection quality
OK
CHECK EEI ELECTRONIC CIRCUIT USING VOLT/OHMMETER
BAD
1. Wiring connection
2. Power to ECU
(1) Fusible links
(2) Fuses (EFI 15A, IGN 7.5A)
(3) EFI main relay
3. Air flow meter
4. Water temp. sensor
5. Air temp. sensor
6. Injection signal circuit
(1) Injector wiring
(2) Resistor
(3) ECU

SYMPTOM – ENGINE BACKFIRES-Lean Fuel Mixture

CHECK DIAGNOSIS SYSTEM
Check for output of diagnostic code.
- Malfunction code(s) → Diagnostic code(s)
- Normal code ↓

CHECK FOR VACUUM LEAKS IN AIR INTAKE LINE
- BAD →
  1. Oil filler cap
  2. Oil dipstick
  3. Hose connections
  4. PCV hoses
- OK ↓

CHECK IGNITION TIMING
1. Short terminals T (T_{E_1}) and E_1 of the check connector.
2. Check ignition timing.
STD: 10° BTDC @ Idle
[w/ short circuited T(T_{E_1}) and E_1]
- BAD → Adjust ignition timing
- OK ↓

CHECK COLD START INJECTOR
- BAD →
  1. Cold start injector
  2. Cold start injector time switch
- OK ↓

CHECK FUEL PRESSURE
- BAD →
  1. Fuel pump
  2. Fuel filter
  3. Fuel pressure regulator
- OK ↓

CHECK INJECTORS
- BAD → Injectors – Clogged
- OK ↓

CHECK EFI ELECTRONIC CIRCUIT USING VOLT/OHMMETER
- BAD →
  1. Wiring connection
  2. Power to ECU
     (1) Fusible links
     (2) Fuses
     (3) EFI main relay
  3. Air flow meter
  4. Water temp. sensor
  5. Air temp. sensor
  6. Throttle position sensor
  7. Injection signal circuit
     (1) Injection wiring
     (2) Resistor
     (3) ECU
     (4) Fuel cut signal
  8. Oxygen sensor(s)

TROUBLESHOOTING PROCEDURES—SUPRA 7M-GE AND 7M-GTE

SYMPTOM – MUFFLER EXPLOSION (AFTER FIRE) -Rich Fuel Mixture-Misfire

CHECK DIAGNOSIS SYSTEM
Check for output of diagnostic code.
- Malfunction code(s) → Diagnostic code(s)
- Normal code ↓

CHECK IGNITION TIMING
1. Short terminals T (T_{E_1}) and E_1 of the check connector.
2. Check ignition timing.
   STD: 10° BTDC @ Idle
   [w/ short circuited T (T_{E_1}) and E_1]
- NO → Adjust ignition timing
- OK ↓

CHECK COLD START INJECTOR
- BAD → 1. Cold start injector 2. Cold start injector time switch
- OK ↓

CHECK INJECTORS
- BAD → Injectors – Leakage
- OK ↓

CHECK SPARK PLUGS
Plug gap: 7M-GE 1.1 mm (0.043 in.)
7M-GTE 0.8 mm (0.031 in.)
– Note –
Check compression pressure and valve clearance if necessary.
- NO → 1. Spark plugs
  2. Compression pressure
     Limit: 9.0 kg/cm² (128 psi, 883 kPa) at 250 rpm
  3. Valve clearance (Cold)
     STD: IN 0.15 – 0.25 mm (0.006 – 0.010 in.)
     EX 0.20 – 0.30 mm (0.008 – 0.012 in.)
- OK ↓

CHECK EFI ELECTRONIC CIRCUIT USING VOLT/OHMMETER
- BAD → 1. Throttle position sensor
  2. Injection signal circuit
     (1) Injector wiring
     (2) Resistor
     (3) ECU
  3. Oxygen sensor(s)

SYMPTOM – ENGINE HESITATES AND/OR POOR ACCELERATION

CHECK CLUTCH AND BRAKE
- BAD → 1. Clutch – Slips 2. Brakes – Drag
- OK ↓

CHECK FOR VACUUM LEAKS IN AIR INTAKE LINE
- BAD → 1. Oil filler cap 2. Oil dipstick 3. Hose connections 4. PCV hoses 5. EGR system – EGR valve system open
- OK ↓

CHECK AIR FILTER ELEMENT
- BAD → Element – Clean or replace
- OK ↓

CHECK DIAGNOSIS SYSTEM
Check for output of diagnostic code.
- Malfunction code(s) → Diagnostic code(s)
- Normal code ↓

CHECK IGNITION SPARK
7M-GE
7M-GTE
- BAD → 1. High-tension cords 2. Distributor (7M-GE) or cam position sensor (7M-GTE) 3. Ignition coil 4. Igniter
- OK ↓

CHECK IGNITION TIMING
1. Short terminals T (T_{E_1}) and E_1 of the check connector.
2. Check ignition timing.
   STD: 10° BTDC @ Idle
   [w/ short circuited T (T_{E_1}) and E_1]
- NO → Adjust ignition timing
- OK ↓

CHECK FUEL PRESSURE
- BAD → 1. Fuel pump 2. Fuel filter 3. Fuel pressure regulator
- OK ↓

CHECK INJECTORS
- BAD → Injection condition
- OK ↓

TROUBLESHOOTING PROCEDURES—SUPRA 7M-GE AND 7M-GTE

OK ↓

CHECK SPARK PLUGS
Plug gap: 7M-GE 1.1 mm (0.043 in.)
7M-GTE 0.8 mm (0.031 in.)
– Note –
Check compression pressure and valve clearance if necessary.
- NO → 1. Spark plug
  2. Compression pressure
     Limit: 9.0 kg/cm² (128 psi, 883 kPa) at 250 rpm
  3. Valve clearance (cold)
     STD: IN 0.15 – 0.25 mm (0.006 – 0.010 in.)
     EX 0.20 – 0.30 mm (0.008 – 0.012 in.)
- OK ↓

CHECK EFI ELECTRONIC CIRCUT USING VOLT/OHMMETER
- BAD → 1. Wiring connection
  2. Power to ECU
     (1) Fusible links
     (2) Fuses
     (3) EFI main relay
  3. Air flow meter
  4. Water temp. sensor
  5. Air temp. sensor
  6. Throttle position sensor
  7. Injection signal circuit
     (1) Injector wiring
     (2) Resistor
     (3) ECU

DIAGNOSTIC CODES—SUPRA 7M-GE AND 7M-GTE

Code No.	Number of check engine blinks	System	Diagnosis	Trouble area
–		Normal	This appears when none of the other codes are identified.	–
11		ECU (+B)	Momentary interruption in power supply to ECU.	• Ignition switch circuit • Ignition switch • Main relay circuit • Main relay • ECU
12		RPM Signal	No "Ne" or "G" signal to ECU within 2 seconds after engine has been cranked.	• Distributor circuit • Distributor • Starter signal circuit • ECU
13		RPM Signal	No "Ne" signal to ECU when engine speed is above 1,000 rpm.	• Distributor circuit • Distributor • ECU
14		Ignition Signal	No "IGf" signal to ECU 6 – 8 times in succession.	• Igniter and ignition coil circuit • Igniter and ignition coil • ECU
21		Oxygen Sensor Signal	Detection of oxygen sensor detrioration.	• Oxygen sensor circuit • Oxygen sensor • ECU
22		Water Temp. Sensor Signal	Open or short circuit in water temp. sensor signal.	• Water temp. sensor circuit • Water temp. sensor • ECU
24		Intake air Temp. Sensor Signal	Open or short circuit in intake air temp. sensor signal.	• Intake air temp. sensor circuit • Intake air temp. sensor • ECU
*1 25		Air-fuel Ratio Lean Malfunction	(1) When air-fuel ratio feedback compensation value or adaptive control value continues at the upper (lean) or lower (rich) limit renewed for a certain period of time. (2) When air-fuel ratio feedback compensation value or adaptive control value feedback frequency is abnormally high during feedback condition.	• Injector circuit • Injector • Fuel line pressure • Air flow meter • Air intake system • Oxygen sensor circuit • Oxygen sensor • Ignition system • ECU
*1 26		Air-fuel Ratio Rich Malfunction	(3) When marked variation is detected in engine revolutions for each cylinder during idle switch on and feedback condition. NOTE: For conditions (2) and (3), since neither a lean (Code No. 25) nor a rich (Code No. 26) diagnosis displayed consecutively.	• Injector circuit • Injector • Fuel line pressure • Air flow meter • Cold start injector • ECU

DIAGNOSTIC CODES—SUPRA 7M-GE AND 7M-GTE

Code No.	Number of check engine blinks	System	Diagnosis	Trouble area
*1 27		Sub-oxygen Sensor Signal	Open or short circuit in sub-oxygen sensor signal.	• Sub-oxygen sensor circuit • Sub-oxygen sensor • ECU
		Sub-oxygen Sensor Heater	Open or short circuit in sub-oxygen sensor heater.	• Sub-oxygen sensor heater circuit • Sub-oxygen sensor heater circuit • ECU
31		Air Flow Meter Signal	Open circuit in Vc signal or short circuit between Vs and E_2 when idle contacts are closed.	• Air flow meter circuit • Air flow meter • ECU
32		(7M-GE) Air Flow Meter Signal	Open circuit in E_2 or short circuit between Vc and Vs.	• Air flow meter circuit • Air flow meter • ECU
		(7M-GTE) HAC Sensor Signal	Open or short circuit in HAC sensor signal.	• HAC sensor circuit • HAC sensor • ECU
*2 34		Turbocharger Pressure	*3 The turbocharger pressure is abnormal.	• Turbocharger • Air flow meter • ECU • Intercooler system
41		Throttle Position Sensor Signal	Open or short circuit in throttle position sensor signal.	• Throttle position sensor circuit • Throttle position sensor • ECU
42		Vehicle Speed Sensor Signal	No "SPD" signal for 5 seconds when engine speed is between 2,500 rpm and 4,500 rpm and coolant temp. is below 80°C (176°F) except when racing the engine.	• Vehicle speed sensor circuit • Vehicle speed sensor • ECU
43		Starter Signal	No "STA" signal to ECU until engine speed reaches 800 rpm with vehicle not moving.	• Ignition switch circuit • Ignition switch • ECU
52		Knock Sensor Signal	Open or short circuit in knock sensor signal.	• Knock sensor circuit • Knock sensor • ECU
53		Knock Control Signal in ECU	Knock control in ECU faulty.	• ECU

DIAGNOSTIC CODES (Cont'd)

Code No.	Number of check engine blinks	System	Diagnosis	Trouble area
*1 71		EGR System Malfunction	EGR gas temp. below predetermined level during EGR operation.	• EGR system (EGR valve, EGR hose etc.) • EGR gas temp. sensor circuit • EGR gas temp. sensor • VSV for EGR • VSV for EGR circuit • ECU
51		Switch Signal	No "IDL" signal, "NSW" signal or "A/C" signal to ECU, with the check terminals E_1 and T (TE_1) shorted.	• A/C switch circuit • A/C switch • A/C Amplifire • Throttle position sensor circuit • Throttle position sensor • ECU

*1 7M-GE California vehicles only
*2 7M-GTE only
*3 Abnormalities in the air flow meter may also be detected.

DIAGNOSTIC CIRCUITS TEST—SUPRA 7M-GE AND 7M-GTE

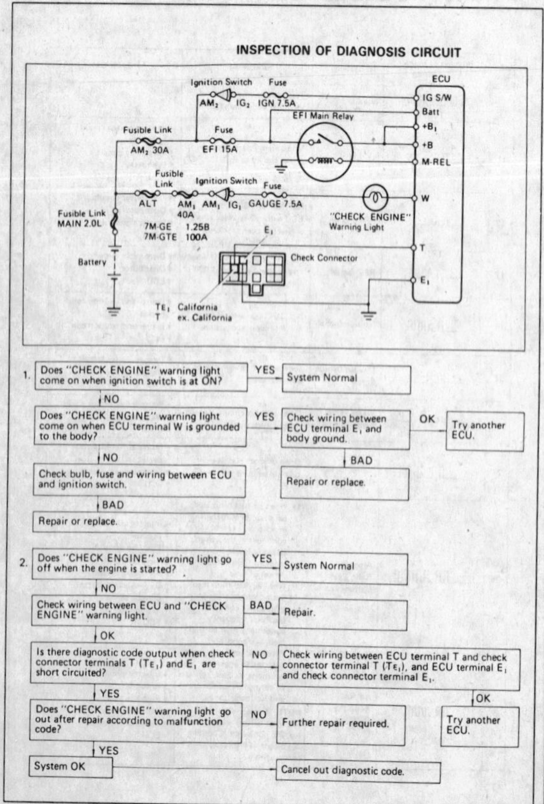
INSPECTION OF DIAGNOSIS CIRCUIT
Ignition Switch
Fuse
ECU
AM₂ IG₂ IGN 7.5A
IG S/W
Batt
EFI Main Relay
+B₁
Fusible Link
Fuse
AM₂ 30A
EFI 15A
+B
M-REL
Fusible Link
Ignition Switch
Fuse
ALT
AM₁ 40A
AM₁ IG₁
GAUGE 7.5A
W
Fusible Link MAIN 2.0L
7M-GE 1.25B
7M-GTE 100A
"CHECK ENGINE" Warning Light
E₁
T
Battery
Check Connector
TE₁ California
T ex. California
1. Does "CHECK ENGINE" warning light come on when ignition switch is at ON?
YES
System Normal
NO
Does "CHECK ENGINE" warning light come on when ECU terminal W is grounded to the body?
YES
Check wiring between ECU terminal E₁ and body ground.
OK
Try another ECU.
BAD
Repair or replace.
NO
Check bulb, fuse and wiring between ECU and ignition switch.
BAD
Repair or replace.
2. Does "CHECK ENGINE" warning light go off when the engine is started?
YES
System Normal
NO
Check wiring between ECU and "CHECK ENGINE" warning light.
BAD
Repair.
OK
Is there diagnostic code output when check connector terminals T (TE₁) and E₁ are short circuited?
NO
Check wiring between ECU terminal T and check connector terminal T (TE₁), and ECU terminal E₁ and check connector terminal E₁.
OK
Try another ECU.
YES
Does "CHECK ENGINE" warning light go out after repair according to malfunction code?
NO
Further repair required.
YES
System OK
Cancel out diagnostic code.

TROUBLESHOOTING CHART 1—1988 SUPRA 7M-GE AND 7M-GTE

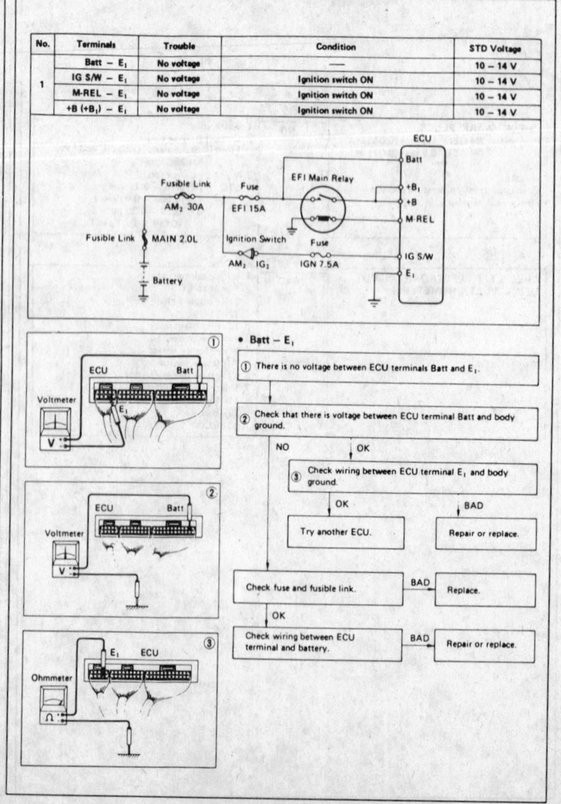

No.	Terminals	Trouble	Condition	STD Voltage
1	Batt – E_1	No voltage	—	10 – 14 V
	IG S/W – E_1	No voltage	Ignition switch ON	10 – 14 V
	M-REL – E_1	No voltage	Ignition switch ON	10 – 14 V
	+B (+B_1) – E_1	No voltage	Ignition switch ON	10 – 14 V

ECU
Batt
EFI Main Relay
Fusible Link
Fuse
+B₁
AM₂ 30A
EFI 15A
+B
M-REL
Fusible Link
MAIN 2.0L
Ignition Switch
Fuse
AM₂ IG₂
IGN 7.5A
IG S/W
E₁
Battery
• Batt – E₁
① There is no voltage between ECU terminals Batt and E₁.
② Check that there is voltage between ECU terminal Batt and body ground.
NO
OK
③ Check wiring between ECU terminal E₁ and body ground.
OK
BAD
Try another ECU.
Repair or replace.
Check fuse and fusible link.
BAD
Replace.
OK
Check wiring between ECU terminal and battery.
BAD
Repair or replace.
①
ECU
Batt
Voltmeter
E₁
②
ECU
Batt
Voltmeter
③
E₁
ECU
Ohmmeter

TROUBLESHOOTING CHART 1—1988 SUPRA 7M—GE AND 7M-GTE

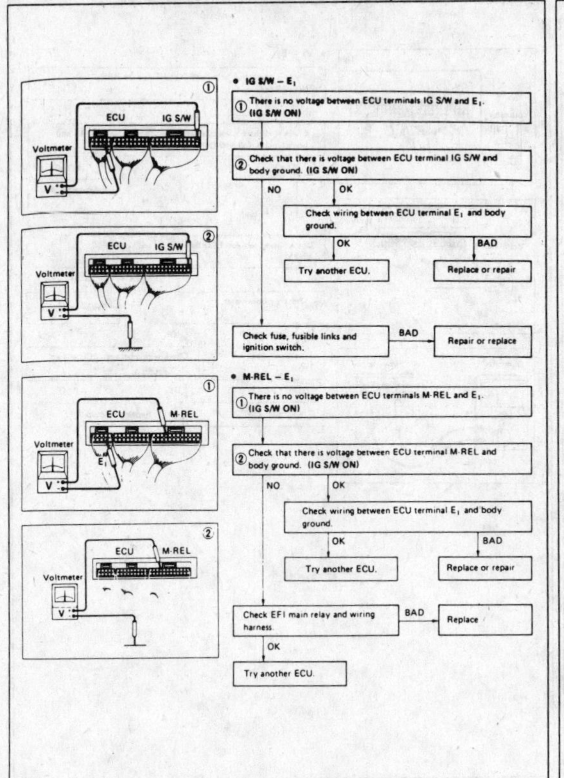
• IG S/W – E_1
① There is no voltage between ECU terminals IG S/W and E_1. (IG S/W ON)
② Check that there is voltage between ECU terminal IG S/W and body ground. (IG S/W ON)
NO
OK
Check wiring between ECU terminal E_1 and body ground.
OK
BAD
Try another ECU.
Replace or repair
Check fuse, fusible links and ignition switch.
BAD
Repair or replace
• M-REL – E_1
① There is no voltage between ECU terminals M-REL and E_1. (IG S/W ON)
② Check that there is voltage between ECU terminal M-REL and body ground. (IG S/W ON)
NO
OK
Check wiring between ECU terminal E_1 and body ground.
OK
BAD
Try another ECU.
Replace or repair
Check EFI main relay and wiring harness.
BAD
Replace
OK
Try another ECU.
ECU
IG S/W
M-REL
Voltmeter
V
E_1

TROUBLESHOOTING CHART 1—1988 SUPRA 7M-GE AND 7M-GTE

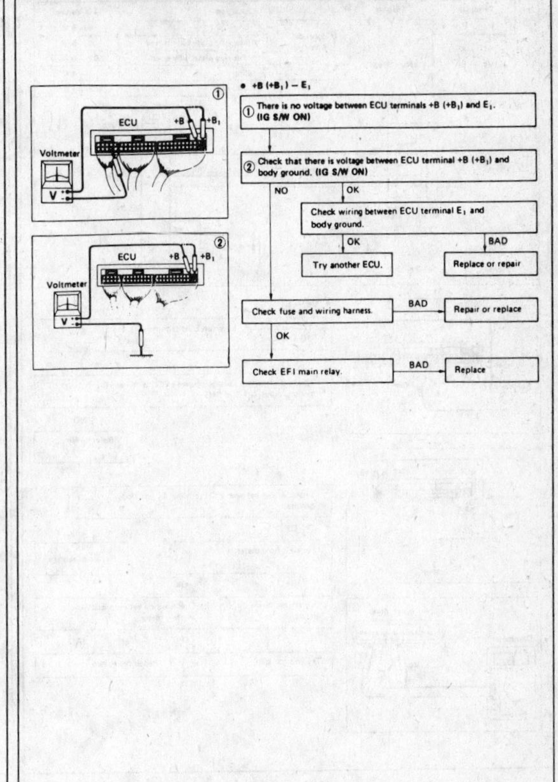
• +B (+B_1) – E_1
① There is no voltage between ECU terminals +B (+B_1) and E_1. (IG S/W ON)
② Check that there is voltage between ECU terminal +B (+B_1) and body ground. (IG S/W ON)
NO
OK
Check wiring between ECU terminal E_1 and body ground.
OK
BAD
Try another ECU.
Replace or repair
Check fuse and wiring harness.
BAD
Repair or replace
OK
Check EFI main relay.
BAD
Replace
ECU
+B
+B_1
Voltmeter
V

TROUBLESHOOTING CHART 2—1988 SUPRA 7M-GE AND 7M-GTE

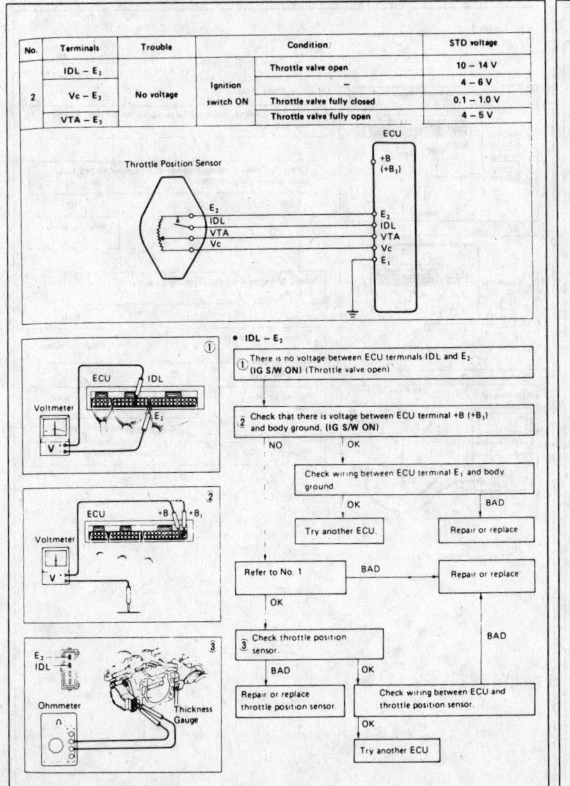

No.	Terminals	Trouble	Condition		STD voltage
2	IDL – E_2	No voltage	Ignition switch ON	Throttle valve open	10 – 14 V
	Vc – E_2			–	4 – 6 V
	VTA – E_2			Throttle valve fully closed	0.1 – 1.0 V
				Throttle valve fully open	4 – 5 V

Throttle Position Sensor
ECU
+B
(+B_1)
E_2
IDL
VTA
Vc
E_1
• IDL – E_2
① There is no voltage between ECU terminals IDL and E_2. (IG S/W ON) (Throttle valve open)
2 Check that there is voltage between ECU terminal +B (+B_1) and body ground. (IG S/W ON)
NO
OK
Check wiring between ECU terminal E_1 and body ground.
OK
BAD
Try another ECU.
Repair or replace
Refer to No. 1
BAD
Repair or replace
OK
3 Check throttle position sensor.
BAD
OK
Repair or replace throttle position sensor.
Check wiring between ECU and throttle position sensor.
BAD
OK
Try another ECU
Voltmeter
V
Ohmmeter
Thickness Gauge

TROUBLESHOOTING CHART 2—1988 SUPRA 7M-GE AND 7M-GTE

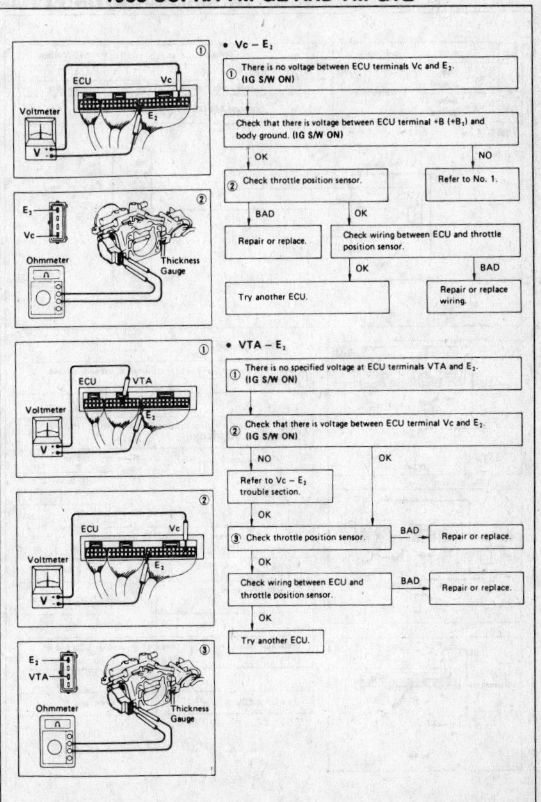
• Vc – E_2
① There is no voltage between ECU terminals Vc and E_2. (IG S/W ON)
Check that there is voltage between ECU terminal +B (+B_1) and body ground. (IG S/W ON)
OK
NO
② Check throttle position sensor.
Refer to No. 1.
BAD
OK
Repair or replace.
Check wiring between ECU and throttle position sensor.
OK
BAD
Try another ECU.
Repair or replace wiring.
• VTA – E_2
① There is no specified voltage at ECU terminals VTA and E_2. (IG S/W ON)
② Check that there is voltage between ECU terminal Vc and E_2. (IG S/W ON)
NO
OK
Refer to Vc – E_2 trouble section.
OK
③ Check throttle position sensor.
BAD
Repair or replace.
OK
Check wiring between ECU and throttle position sensor.
BAD
Repair or replace.
OK
Try another ECU.
ECU
Vc
VTA
Voltmeter
V
Ohmmeter
Thickness Gauge

TROUBLESHOOTING CHART 3—1988 SUPRA 7M-GE

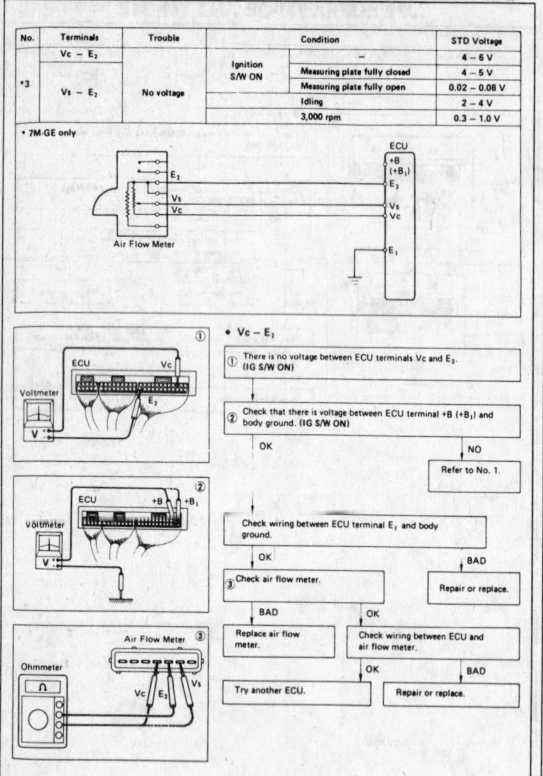

No.	Terminals	Trouble	Condition		STD Voltage
*3	Vc – E_2	No voltage	Ignition S/W ON	–	4 – 6 V
	Vs – E_2			Measuring plate fully closed	4 – 5 V
				Measuring plate fully open	0.02 – 0.08 V
			Idling		2 – 4 V
			3,000 rpm		0.3 – 1.0 V

* 7M-GE only

ECU
+B
(+B_1)
E_2
Vs
Vc
E_1
Air Flow Meter
• Vc – E_2
Voltmeter
ECU
① There is no voltage between ECU terminals Vc and E_2. (IG S/W ON)
② Check that there is voltage between ECU terminal +B (+B_1) and body ground. (IG S/W ON)
OK
NO
Refer to No. 1.
Check wiring between ECU terminal E_1 and body ground.
BAD
Repair or replace.
③ Check air flow meter.
Replace air flow meter.
Check wiring between ECU and air flow meter.
Try another ECU.
Repair or replace.
Air Flow Meter
Ohmmeter

TROUBLESHOOTING CHART 3—1988 SUPRA 7M-GE

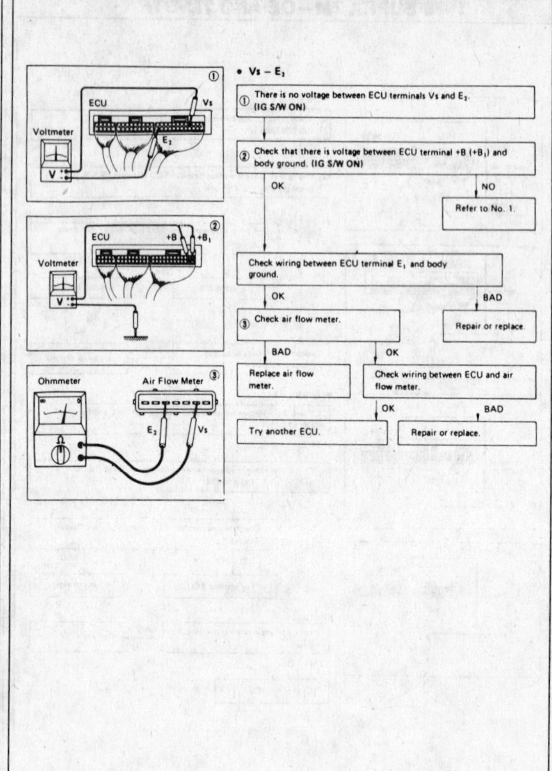

• Vs – E_2
ECU
Voltmeter
① There is no voltage between ECU terminals Vs and E_2. (IG S/W ON)
② Check that there is voltage between ECU terminal +B (+B_1) and body ground. (IG S/W ON)
OK
NO
Refer to No. 1.
Check wiring between ECU terminal E_1 and body ground.
BAD
Repair or replace.
③ Check air flow meter.
Replace air flow meter.
Check wiring between ECU and air flow meter.
Try another ECU.
Repair or replace.
Ohmmeter
Air Flow Meter

TROUBLESHOOTING CHART 4—1988 SUPRA 7M-GTE

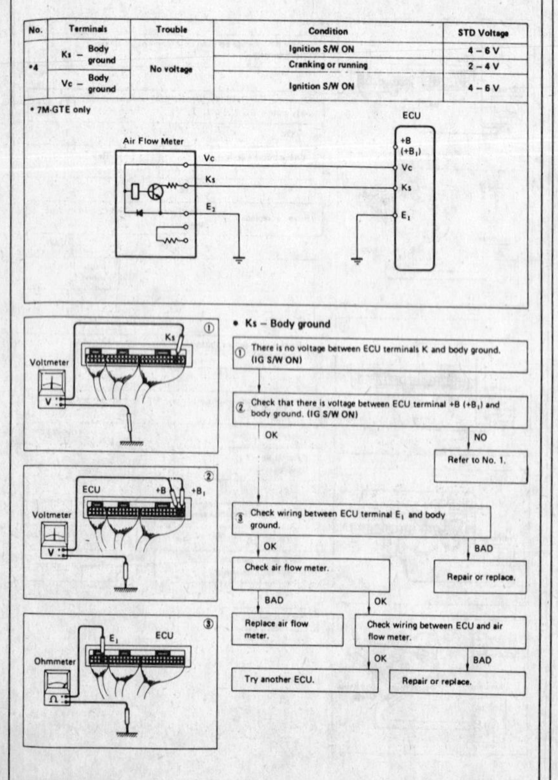

No.	Terminals	Trouble	Condition	STD Voltage
*4	Ks – Body ground	No voltage	Ignition S/W ON	4 – 6 V
			Cranking or running	2 – 4 V
	Vc – Body ground		Ignition S/W ON	4 – 6 V

* 7M-GTE only

Air Flow Meter
ECU
+B
(+B_1)
Vc
Ks
E_2
E_1
• Ks – Body ground
Voltmeter
① There is no voltage between ECU terminals K and body ground. (IG S/W ON)
② Check that there is voltage between ECU terminal +B (+B_1) and body ground. (IG S/W ON)
OK
NO
Refer to No. 1.
③ Check wiring between ECU terminal E_1 and body ground.
BAD
Check air flow meter.
Repair or replace.
Replace air flow meter.
Check wiring between ECU and air flow meter.
Try another ECU.
Repair or replace.
Ohmmeter

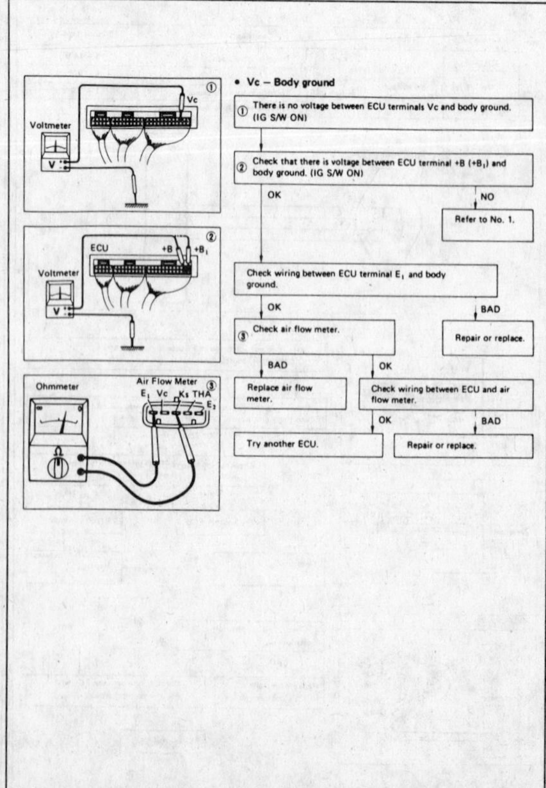

• Vc – Body ground
Voltmeter
ECU
① There is no voltage between ECU terminals Vc and body ground. (IG S/W ON)
② Check that there is voltage between ECU terminal +B (+B_1) and body ground. (IG S/W ON)
OK
NO
Refer to No. 1.
Check wiring between ECU terminal E_1 and body ground.
BAD
③ Check air flow meter.
Repair or replace.
Replace air flow meter.
Check wiring between ECU and air flow meter.
Try another ECU.
Repair or replace.
Ohmmeter
Air Flow Meter
E_1 Vc Ks THA

TROUBLESHOOTING CHART 5—1988 SUPRA 7M-GE AND 7M-GTE

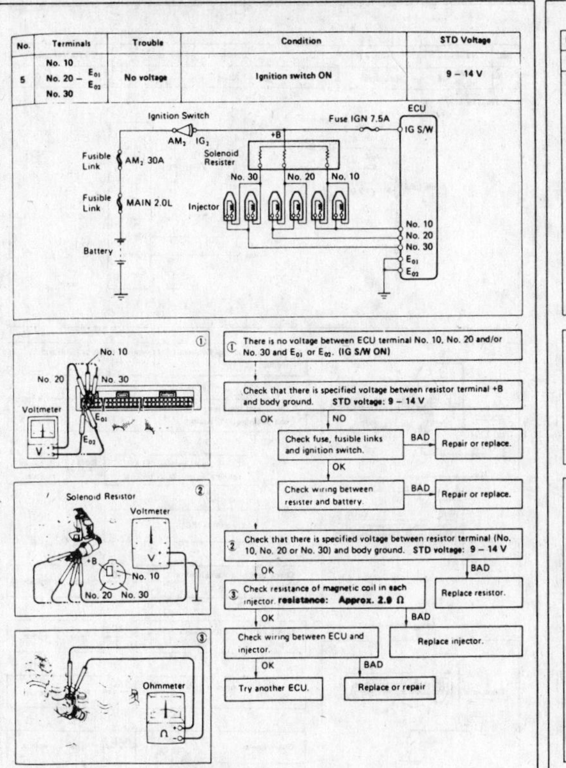

No.	Terminals	Trouble	Condition	STD Voltage
5	No. 10 No. 20 – E_{01} / E_{02} No. 30	No voltage	Ignition switch ON	9 – 14 V

Ignition Switch
AM₂ IG₂
Fuse IGN 7.5A
ECU
IG S/W
+B
Fusible Link
AM₂ 30A
Solenoid Resister
No. 30 No. 20 No. 10
Fusible Link
MAIN 2.0L
Injector
No. 10
No. 20
No. 30
E01
E02
Battery
① There is no voltage between ECU terminal No. 10, No. 20 and/or No. 30 and E01 or E02. (IG S/W ON)
Check that there is specified voltage between resistor terminal +B and body ground. STD voltage: 9 – 14 V
OK
NO
Check fuse, fusible links and ignition switch.
BAD
Repair or replace.
OK
Check wiring between resister and battery.
BAD
Repair or replace.
② Check that there is specified voltage between resistor terminal (No. 10, No. 20 or No. 30) and body ground. STD voltage: 9 – 14 V
OK
BAD
③ Check resistance of magnetic coil in each injector. resistance: Approx. 2.9 Ω
Replace resistor.
OK
BAD
Check wiring between ECU and injector.
Replace injector.
OK
BAD
Try another ECU.
Replace or repair
No. 10
No. 20
No. 30
Voltmeter
Solenoid Resistor
Voltmeter
+B
No. 10
No. 20 No. 30
Ohmmeter

TROUBLESHOOTING CHART 6—1988 SUPRA 7M-GE AND 7M-GTE

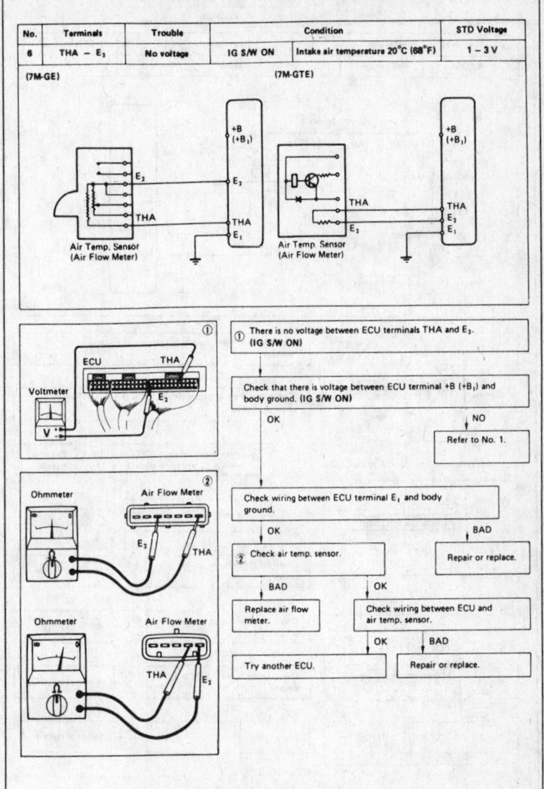

No.	Terminals	Trouble	Condition		STD Voltage
6	THA – E_2	No voltage	IG S/W ON	Intake air temperature 20°C (68°F)	1 – 3 V

(7M-GE)
(7M-GTE)
+B (+B₁)
E₂
THA
E₁
Air Temp. Sensor (Air Flow Meter)
Air Temp. Sensor (Air Flow Meter)
① There is no voltage between ECU terminals THA and E₂. (IG S/W ON)
ECU
THA
Voltmeter
E₂
Check that there is voltage between ECU terminal +B (+B₁) and body ground. (IG S/W ON)
OK
NO
Refer to No. 1.
Check wiring between ECU terminal E₁ and body ground.
OK
BAD
② Check air temp. sensor.
Repair or replace.
BAD
OK
Replace air flow meter.
Check wiring between ECU and air temp. sensor.
OK
BAD
Try another ECU.
Repair or replace.
Ohmmeter
Air Flow Meter
E₂
THA
Ohmmeter
Air Flow Meter
THA
E₂

TROUBLESHOOTING CHART 7—1988 SUPRA 7M-GE AND 7M-GTE

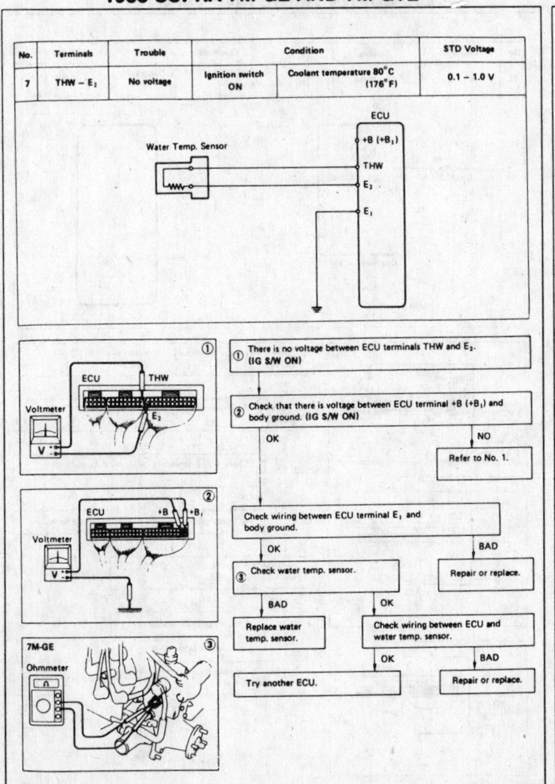

No.	Terminals	Trouble	Condition		STD Voltage
7	THW – E_2	No voltage	Ignition switch ON	Coolant temperature 80°C (176°F)	0.1 – 1.0 V

ECU
Water Temp. Sensor
+B (+B₁)
THW
E₂
E₁
① There is no voltage between ECU terminals THW and E₂. (IG S/W ON)
ECU
THW
Voltmeter
E₂
② Check that there is voltage between ECU terminal +B (+B₁) and body ground. (IG S/W ON)
OK
NO
Refer to No. 1.
ECU
+B
+B₁
Voltmeter
Check wiring between ECU terminal E₁ and body ground.
OK
BAD
③ Check water temp. sensor.
Repair or replace.
BAD
OK
Replace water temp. sensor.
Check wiring between ECU and water temp. sensor.
OK
BAD
Try another ECU.
Repair or replace.
7M-GE
Ohmmeter

TROUBLESHOOTING CHART 8—1988 SUPRA 7M-GE AND 7M-GTE

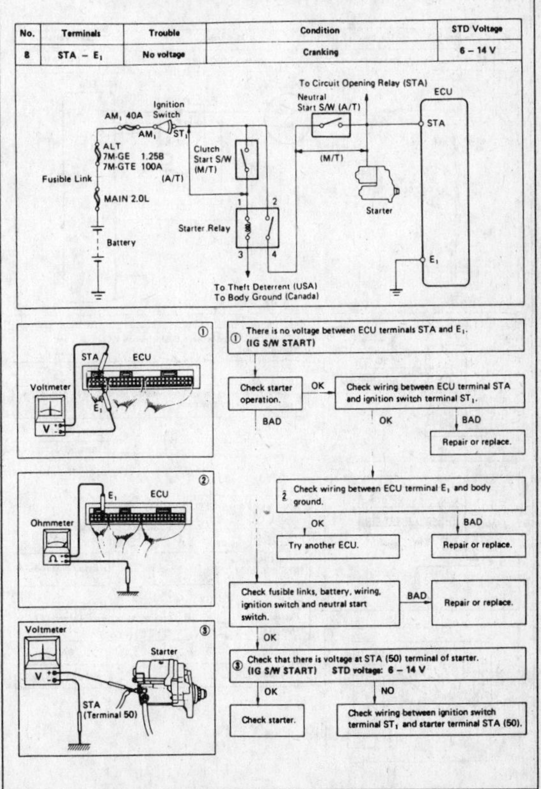

No.	Terminals	Trouble	Condition	STD Voltage
8	STA – E_1	No voltage	Cranking	6 – 14 V

To Circuit Opening Relay (STA)
Neutral Start S/W (A/T)
ECU
STA
AM₁ 40A
Ignition Switch
AM₁
ST₁
ALT 7M-GE 1.25B 7M-GTE 100A
Fusible Link
Clutch Start S/W (M/T)
(A/T)
(M/T)
Starter
MAIN 2.0L
Starter Relay
Battery
E₁
To Theft Deterrent (USA)
To Body Ground (Canada)
① There is no voltage between ECU terminals STA and E₁. (IG S/W START)
STA
ECU
Voltmeter
E₁
Check starter operation.
OK
Check wiring between ECU terminal STA and ignition switch terminal ST₁.
BAD
OK
BAD
Repair or replace.
② Check wiring between ECU terminal E₁ and body ground.
OK
BAD
Try another ECU.
Repair or replace.
E₁
ECU
Ohmmeter
Check fusible links, battery, wiring, ignition switch and neutral start switch.
BAD
Repair or replace.
OK
③ Check that there is voltage at STA (50) terminal of starter. (IG S/W START) STD voltage: 6 – 14 V
OK
NO
Check starter.
Check wiring between ignition switch terminal ST₁ and starter terminal STA (50).
Voltmeter
Starter
STA (Terminal 50)

TROUBLESHOOTING CHART 9— 1988 SUPRA 7M-GE AND 7M-GTE

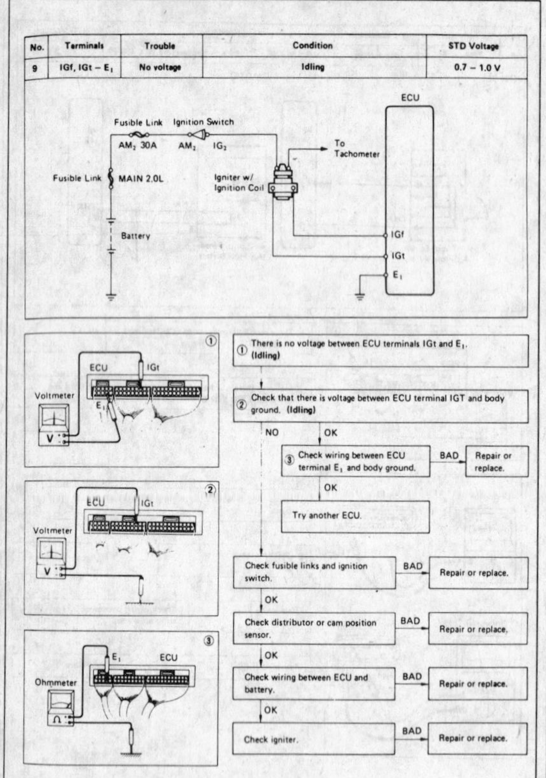
No. | Terminals | Trouble | Condition | STD Voltage
9 | IGf, IGt – E1 | No voltage | Idling | 0.7 – 1.0 V
Fusible Link
Ignition Switch
AM2 30A
AM2
IG2
Fusible Link
MAIN 2.0L
Battery
Igniter w/ Ignition Coil
To Tachometer
ECU
IGf
IGt
E1
Voltmeter
Ohmmeter
1 There is no voltage between ECU terminals IGt and E1. (Idling)
2 Check that there is voltage between ECU terminal IGT and body ground. (Idling)
NO
OK
3 Check wiring between ECU terminal E1 and body ground.
BAD
Repair or replace.
Try another ECU.
Check fusible links and ignition switch.
Check distributor or cam position sensor.
Check wiring between ECU and battery.
Check igniter.
Repair or replace.

TROUBLESHOOTING CHART 10— 1988 SUPRA 7M-GTE

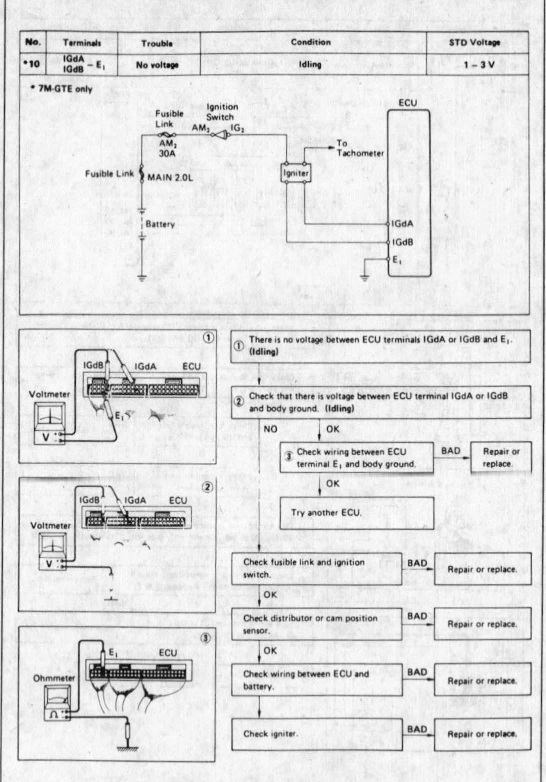
No. | Terminals | Trouble | Condition | STD Voltage
*10 | IGdA IGdB – E1 | No voltage | Idling | 1 – 3 V
* 7M-GTE only
Fusible Link
Ignition Switch
AM2 30A
IG2
Fusible Link
MAIN 2.0L
Battery
Igniter
To Tachometer
ECU
IGdA
IGdB
E1
Voltmeter
Ohmmeter
1 There is no voltage between ECU terminals IGdA or IGdB and E1. (Idling)
2 Check that there is voltage between ECU terminal IGdA or IGdB and body ground. (Idling)
NO
OK
3 Check wiring between ECU terminal E1 and body ground.
BAD
Repair or replace.
Try another ECU.
Check fusible link and ignition switch.
Check distributor or cam position sensor.
Check wiring between ECU and battery.
Check igniter.
Repair or replace.

TROUBLESHOOTING CHART 11— 1988 SUPRA 7M-GE AND 7M-GTE

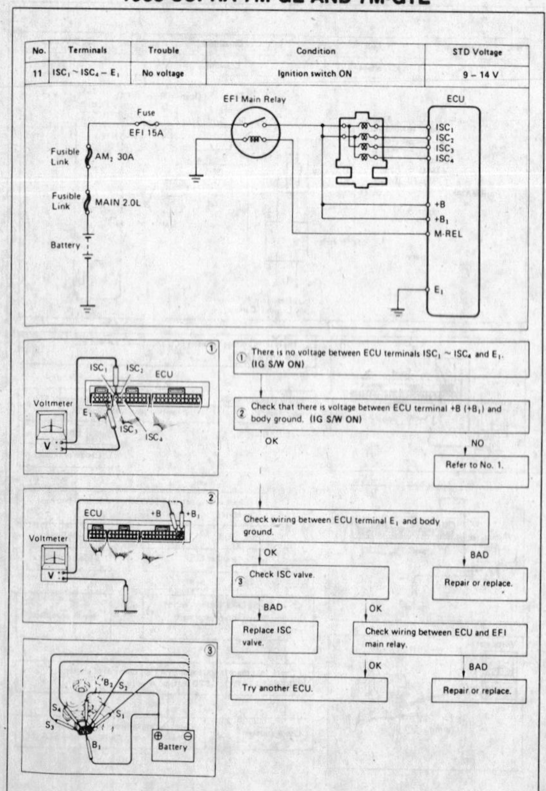
No. | Terminals | Trouble | Condition | STD Voltage
11 | ISC1 ~ ISC4 – E1 | No voltage | Ignition switch ON | 9 – 14 V
EFI Main Relay
Fuse
EFI 15A
Fusible Link
AM2 30A
Fusible Link
MAIN 2.0L
Battery
ECU
ISC1
ISC2
ISC3
ISC4
+B
+B1
M-REL
E1
Voltmeter
1 There is no voltage between ECU terminals ISC1 ~ ISC4 and E1. (IG S/W ON)
2 Check that there is voltage between ECU terminal +B (+B1) and body ground. (IG S/W ON)
OK
NO
Refer to No. 1.
Check wiring between ECU terminal E1 and body ground.
BAD
3 Check ISC valve.
Repair or replace.
Replace ISC valve.
Check wiring between ECU and EFI main relay.
Try another ECU.
Repair or replace.
Battery

TROUBLESHOOTING CHART 12— 1988 SUPRA 7M-GE AND 7M-GTE

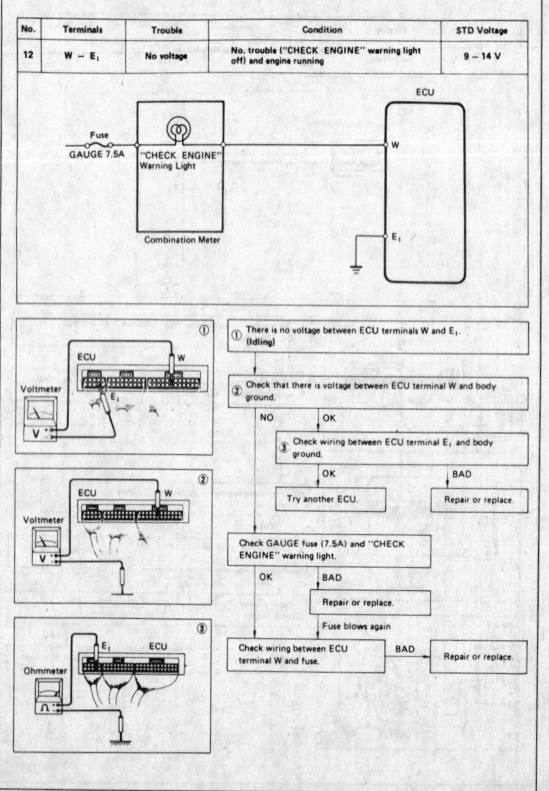
No. | Terminals | Trouble | Condition | STD Voltage
12 | W – E1 | No voltage | No trouble ("CHECK ENGINE" warning light off) and engine running | 9 – 14 V
Fuse
GAUGE 7.5A
"CHECK ENGINE" Warning Light
Combination Meter
ECU
W
E1
Voltmeter
Ohmmeter
1 There is no voltage between ECU terminals W and E1. (Idling)
2 Check that there is voltage between ECU terminal W and body ground.
NO
OK
3 Check wiring between ECU terminal E1 and body ground.
BAD
Try another ECU.
Repair or replace.
Check GAUGE fuse (7.5A) and "CHECK ENGINE" warning light.
Fuse blows again
Check wiring between ECU terminal W and fuse.

TROUBLESHOOTING CHART 13—1988 SUPRA 7M-GE AND 7M-GTE

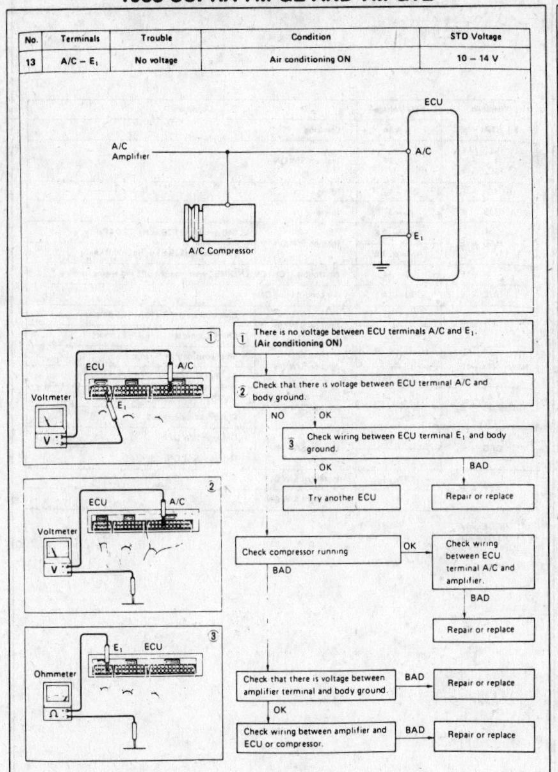

TROUBLESHOOTING CHART 14—1988 SUPRA 7M-GTE

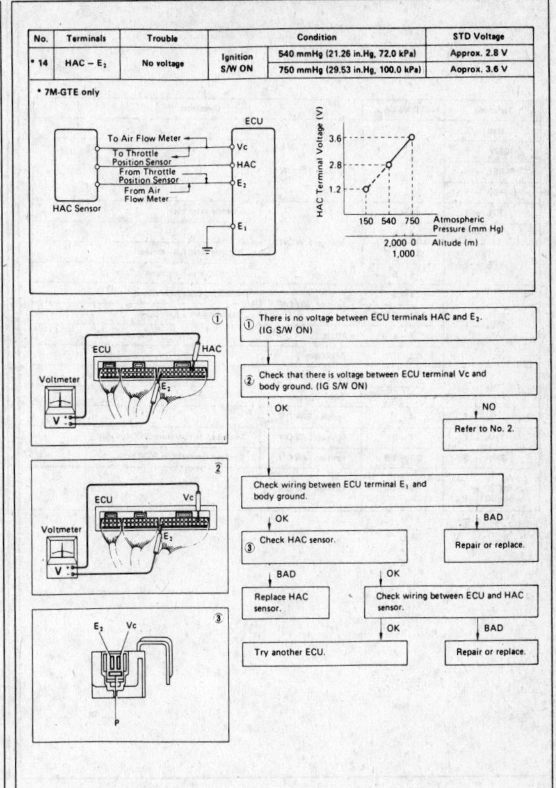

TROUBLESHOOTING OXYGEN SENSOR—1988 SUPRA 7M-GE

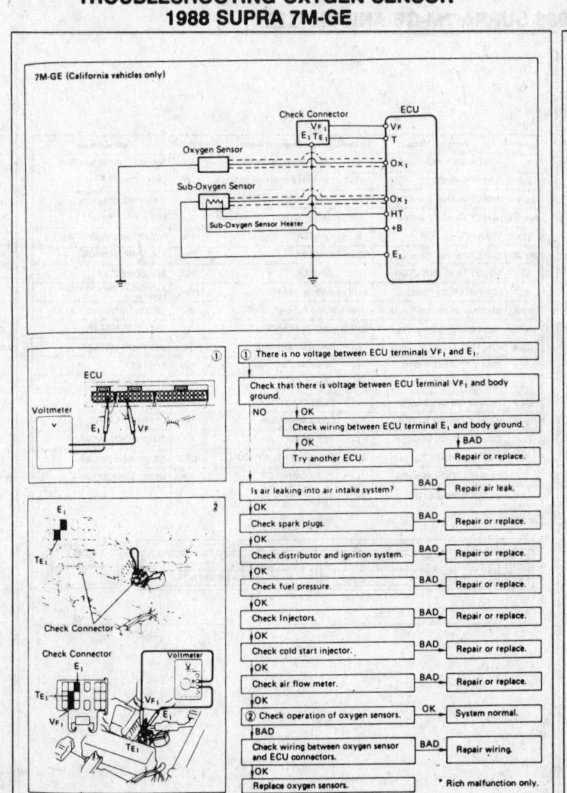

TROUBLESHOOTING EGR TEMPERATURE SENSOR—1988 SUPRA 7M-GE

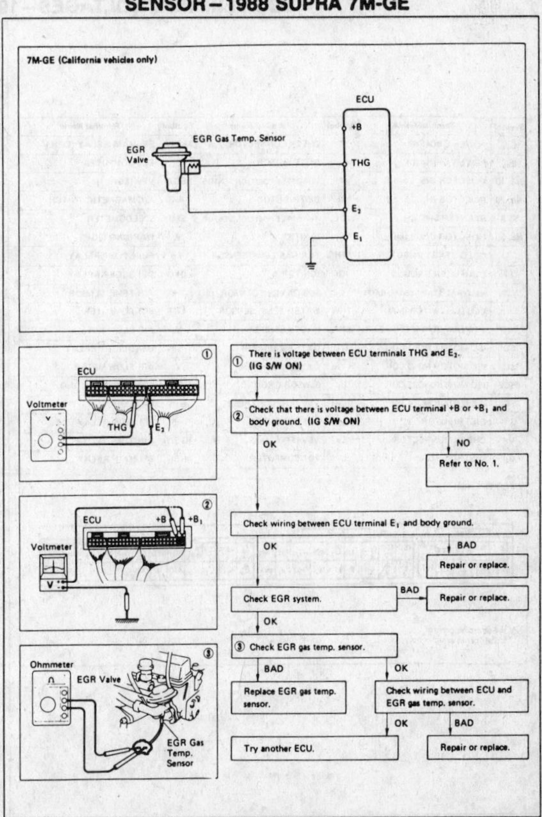

ECU TERMINAL IDENTIFICATION—1988 SUPRA 7M-GTE

Terminals	STD Voltage	Condition	
Batt – E_1	10 – 14	–	
IGS/W – E_1		Ignition S/W ON	
M-REL – E_1			
+B (+B_1) – E_1			
IDL – E_2	4 – 6	Ignition S/W ON	Throttle valve open
Vc – E_2	4 – 6		–
VTA – E_2	0.1 – 1.0		Throttle valve fully closed
	4 – 5		Throttle valve fully open
*1 Vs – E_2	4 – 5	Ignition S/W ON	Measuring plate fully closed
	0.02 – 0.08		Measuring plate fully open
	2 – 4	Idling	
	0.3 – 1.0	3,000 rpm	
*2 Ks – E_2	4 – 6	Ignition S/W ON	
	2 – 4	Cranking or running	
Vc – E_2	4 – 6	Ignition S/W ON	
THA – E_2	1 – 3	Ignition S/W ON	Intake air temperature 20°C (68°F)
THW – E_2	0.1 – 1.0	Ignition S/W ON	Coolant temperature 80°C (176°F)
No. 10 / No. 20 / No 30 – E_{01} / E_{02}	9 – 14	Ignition S/W ON	

Terminals	STD Voltage	Condition	
STA – E_1	6 – 14	Cranking	
ISC_1 ~ ISC_4 – E_{01} / E_{02}	9 – 14	Ignition S/W ON	
IGf, IGt – E_1	0.7 – 1.0	Idling	
*2 IGdA, IGdB – E_1	1 – 3	Idling	
*2 HAC – E_2	Approx. 2.8	Ignition S/W ON	540 mm Hg (21.26 in.Hg, 72.0 kPa)
	Approx. 3.6		750 mm Hg (129.53 in.Hg, 100.0 kPa)
W – E_1	9 – 14	No trouble ("CHECK ENGINE" warning light off) and engine running	
A/C – E_1	10 – 14	Air conditioning ON	
T – E_1	4 – 6	Ignition S/W ON	Check connector T (TE_1) – E_1 not short
	0		Check connector T (TE_1) – E_1 short
NSW (A/T) – E_1	0	Ignition S/W ON	Shift position P or N range
	10 – 14		Ex. P or N range
N/C (M/T) – E_1	0	Ignition S/W ON	Clutch pedal not depressed
	10 – 14		Clutch pedal depressed
DFG – E_1	10 – 14	Ignition S/W ON	Defogger S/W OFF
	0		Defogger S/W ON
LP – E_1	10 – 14	Headlight S/W OFF	
	U	Headlight S/W ON	

*1 7M-GE only
*2 7M-GTE only

ECU TERMINAL VOLTAGES—1988 SUPRA 7M-GE AND 7M-GTE

Symbol	Terminal Name	Symbol	Terminal Name	Symbol	Terminal Name
E_{01}	POWER GROUND	T	CHECK CONNECTOR	M-REL	EFI MAIN RELAY (COIL)
E_{02}	POWER GROUND	G_2	DISTRIBUTOR	L_3	ECT COMPUTER
No. 10	INJECTOR (No. 1 and 4)	VTA	THROTTLE POSITION SENSOR	EGT	VSV (EGR)
No. 20	INJECTOR (No. 2 and 6)	Ne	DISTRIBUTOR	A/C	A/C MAGNETIC SWITCH
STA	STARTER SWITCH	IDL	THROTTLE POSITION SENSOR	SPD	SPEEDOMETER
No. 30	INJECTOR (No. 3 and 5)	IGt	IGNITER	W	WARNING LIGHT
STJ	COLD START INJECTOR	*2 THG	EGR GAS TEMP. SENSOR	Fp	FUEL PUMP RELAY
E_1	COMPUTER GROUND	IGf	IGNITER	DFG	DEFOGGER RELAY
NSW	NEUTRAL START SWITCH(A/T)	*2 Ox_2	SUB-OXYGEN SENSOR	THA	AIR TEMP. SENSOR
*1 N/C	CLUTCH SWITCH (M/T)	THW	WATER TEMP. SENSOR	ECT	ECT COMPUTER
*2 HT	OXYGEN SENSOR HEATER	KNK	KNOCK SENSOR	Vs	AIR FLOW METER
ISC1	ISC MOTOR NO. 1 COIL	Ox	OXYGEN SENSOR	LP	HEADLIGHT RELAY
ISC3	ISC MOTOR NO. 3 COIL	*2 Ox_1	OXYGEN SENSOR	Vc	AIR FLOW METER
ISC2	ISC MOTOR NO. 2 COIL	E_2	SENSOR GROUND	*2 E_{11}	COMPUTER GROUND
ISC4	ISC MOTOR NO. 4 COIL	VSV1	VSV (AIR CONTROL)	Batt	BATTERY
G ⊖	DISTRIBUTOR	L_1	ECT COMPUTER	+B	EFI MAIN RELAY
VF	CHECK CONNECTOR	VSV2	VSV (FPU)	IG S/W	IGNITION SWITCH
G_1	DISTRIBUTOR	L_2	ECT COMPUTER	+B_1	EFI MAIN RELAY

ECU Terminals

E_{01}	No. 10	STA	STJ	NSW or N/C	ISC 1	ISC 2	G⊖	G_1	G_2	Ne	IGt	IGf	THW	Ox or Ox_1	VSV 1	VSV 2	M-REL	EGR	SPD	Fp	THA	Vs	Vc	Batt	IG S/W
E_{02}	No. 20	No. 30	E_1	HT	ISC 3	ISC 4	VF	T	VTA	IDL	THG	Ox_2	KNK	E_2	L_1	L_2	L_3	A/C	W	DGF	ECT	LP	E_{11}	+B	+B_1

*1 For cruise control
*2 California vehicles only

Symbol	Terminal Name	Symbol	Terminal Name	Symbol	Terminal Name
E_{01}	POWER GROUND	G_2	CAM POSITION SENSOR	A/C	A/C MAGNETIC SWITCH
E_{02}	POWER GROUND	VTA	THROTTLE POSITION SENSOR	SPD	SPEEDOMETER
No. 10	INJECTOR (No. 1 and 4)	Ne	CAM POSITION SENSOR	W	WARNING LIGHT
No. 20	INJECTOR (No. 2 and 6)	IDL	THROTTLE POSITION SENSOR	Fp	FUEL PUMP RELAY
STA	STARTER SWITCH	IGt	IGNITER	OIL	OIL PRESSURE SWITCH
No. 30	INJECTOR (No. 3 and 5)	IGdA	IGNITER	THA	AIR TEMP. SENSOR
STJ	COLD START INJECTOR	IGf	IGNITER	ECT	ECT COMPUTER
E_1	COMPUTER GROUND	KNK1	KNOCK SENSOR	HAC	ALTITUDE COMPENSATION SENSOR
NSW	NEUTRAL START SWITCH (A/T)	THW	WATER TEMP. SENSOR	Fc	CIRCUIT OPENING RELAY
* N/C	CLUTCH SWITCH (M/T)	KNK2	KNOCK SENSOR	Vc	AIR FLOW METER
IGdB	IGNITER	Ox	OXYGEN SENSOR	Ks	AIR FLOW METER
ISC1	ISC MOTOR NO. 1 COIL	E_2	SENSOR GROUND	Batt	BATTERY
ISC3	ISC MOTOR NO. 3 COIL	VSV	VSV (FPU)	+B	EFI MAIN RELAY
ISC2	ISC MOTOR NO. 2 COIL	L_1	ECT COMPUTER	IG S/W	IGNITION SWITCH
ISC4	ISC MOTOR NO. 4 COIL	HT	OXGEN SENSOR	+B_1	EFI MAIN RELAY
G⊖	CAM POSITION SENSOR	L_2	ECT COMPUTER	TIL	TURBO INDICATOR
VF	CHECK CONNECTOR	M-REL	EFI MAIN RELAY (COIL)	DFG	DEFOGGER RELAY
G_1	CAM POSITION SENSOR	L_3	ECT COMPUTER	LP	HEADLIGHT RELAY
T	CHECK CONNECTOR	EGR	VSV (EGR)		

ECU Terminals

E_{01}	No 10	STA	STJ	NSW / N/C	ISC 1	ISC 2	G⊖	G_1	G_2	Ne	IGt	IGf	THW	Ox	VSV 2	HT	M-REL	EGR	SPD	Fp	THA	HAC	Vc	Batt	IG S/W
E_{02}	No 20	No. 30	E_1	IGdB	ISC 3	ISC 4	VF	T	VTA	IDL	IGdA	KNK1	KNK2	E_2	L_1	L_2	L_3	A/C	W	OIL	ECT	Fc	KS	+B	+B_1

		DFG
	TIL	LP

* For cruise control

TROUBLESHOOTING CHART 1—1989–90 SUPRA 7M-GE

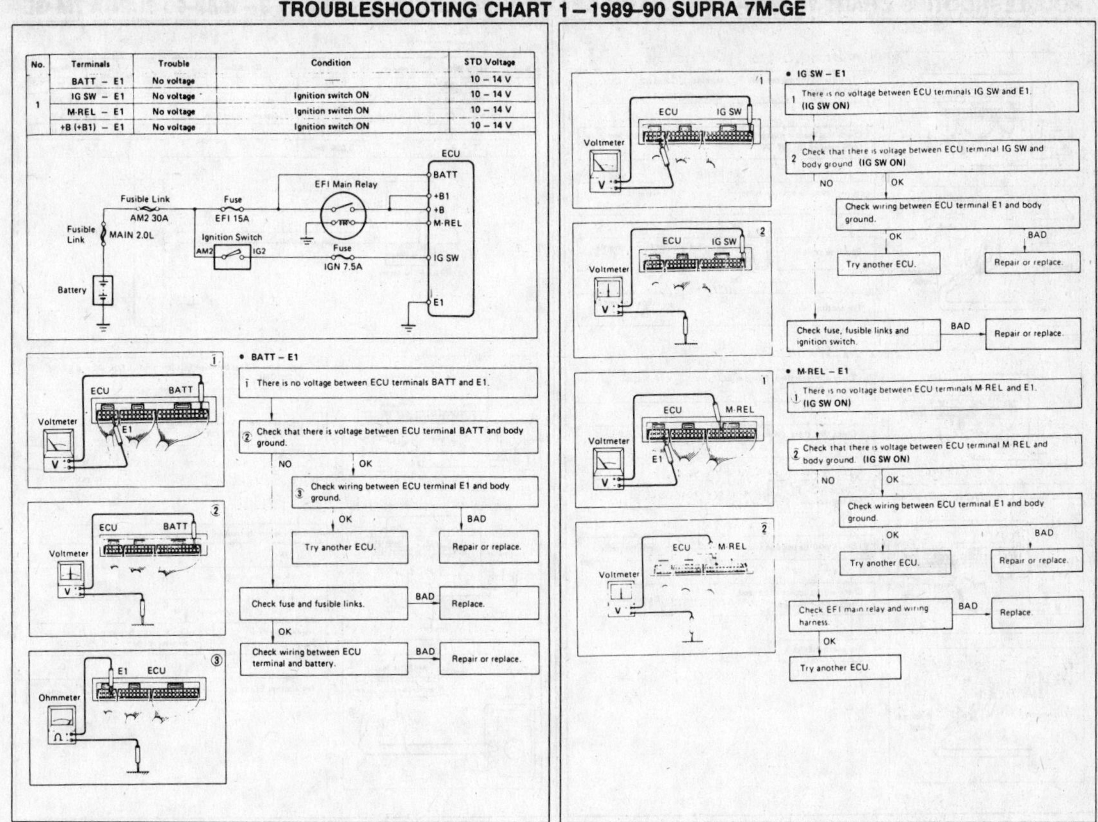

No.	Terminals	Trouble	Condition	STD Voltage
1	BATT – E1	No voltage	—	10 – 14 V
	IG SW – E1	No voltage	Ignition switch ON	10 – 14 V
	M-REL – E1	No voltage	Ignition switch ON	10 – 14 V
	+B (+B1) – E1	No voltage	Ignition switch ON	10 – 14 V

ECU
BATT
EFI Main Relay
+B1
+B
M-REL
IG SW
E1
Fusible Link
AM2 30A
Fuse
EFI 15A
Fusible Link
MAIN 2.0L
Ignition Switch
AM2
IG2
Fuse
IGN 7.5A
Battery
• BATT – E1
1 There is no voltage between ECU terminals BATT and E1.
2 Check that there is voltage between ECU terminal BATT and body ground.
NO
OK
3 Check wiring between ECU terminal E1 and body ground.
OK
BAD
Try another ECU.
Repair or replace.
Check fuse and fusible links.
BAD
Replace.
OK
Check wiring between ECU terminal and battery.
BAD
Repair or replace.
Voltmeter
Ohmmeter
• IG SW – E1
1 There is no voltage between ECU terminals IG SW and E1. (IG SW ON)
2 Check that there is voltage between ECU terminal IG SW and body ground. (IG SW ON)
NO
OK
Check wiring between ECU terminal E1 and body ground.
OK
BAD
Try another ECU.
Repair or replace.
Check fuse, fusible links and ignition switch.
BAD
Repair or replace.
• M-REL – E1
1 There is no voltage between ECU terminals M-REL and E1. (IG SW ON)
2 Check that there is voltage between ECU terminal M-REL and body ground. (IG SW ON)
NO
OK
Check wiring between ECU terminal E1 and body ground.
OK
BAD
Try another ECU.
Repair or replace.
Check EFI main relay and wiring harness.
BAD
Replace.
OK
Try another ECU.

TROUBLESHOOTING CHART 1—1989–90 SUPRA 7M-GE

TROUBLESHOOTING CHART 2—1989–90 SUPRA 7M-GE

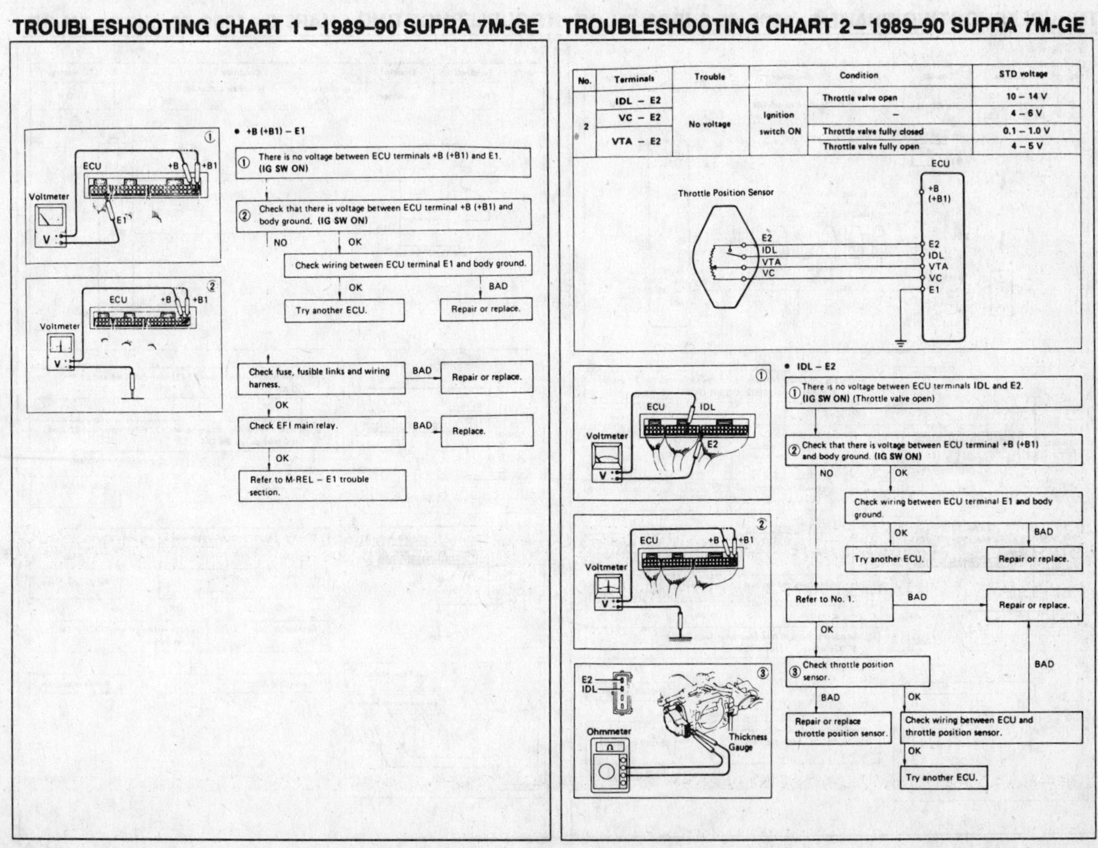

No.	Terminals	Trouble	Condition		STD voltage
2	IDL – E2	No voltage	Ignition switch ON	Throttle valve open	10 – 14 V
	VC – E2			–	4 – 6 V
	VTA – E2			Throttle valve fully closed	0.1 – 1.0 V
				Throttle valve fully open	4 – 5 V

• +B (+B1) – E1
1 There is no voltage between ECU terminals +B (+B1) and E1. (IG SW ON)
2 Check that there is voltage between ECU terminal +B (+B1) and body ground. (IG SW ON)
NO
OK
Check wiring between ECU terminal E1 and body ground.
OK
BAD
Try another ECU.
Repair or replace.
Check fuse, fusible links and wiring harness.
BAD
Repair or replace.
OK
Check EFI main relay.
BAD
Replace.
OK
Refer to M-REL – E1 trouble section.
ECU
+B
+B1
E1
Voltmeter
Throttle Position Sensor
E2
IDL
VTA
VC
+B (+B1)
• IDL – E2
1 There is no voltage between ECU terminals IDL and E2. (IG SW ON) (Throttle valve open)
2 Check that there is voltage between ECU terminal +B (+B1) and body ground. (IG SW ON)
NO
OK
Check wiring between ECU terminal E1 and body ground.
OK
BAD
Try another ECU.
Repair or replace.
Refer to No. 1.
BAD
Repair or replace.
OK
3 Check throttle position sensor.
BAD
OK
BAD
Repair or replace throttle position sensor.
Check wiring between ECU and throttle position sensor.
OK
Try another ECU.
Ohmmeter
Thickness Gauge

TROUBLESHOOTING CHART 2—1989–90 SUPRA 7M-GE

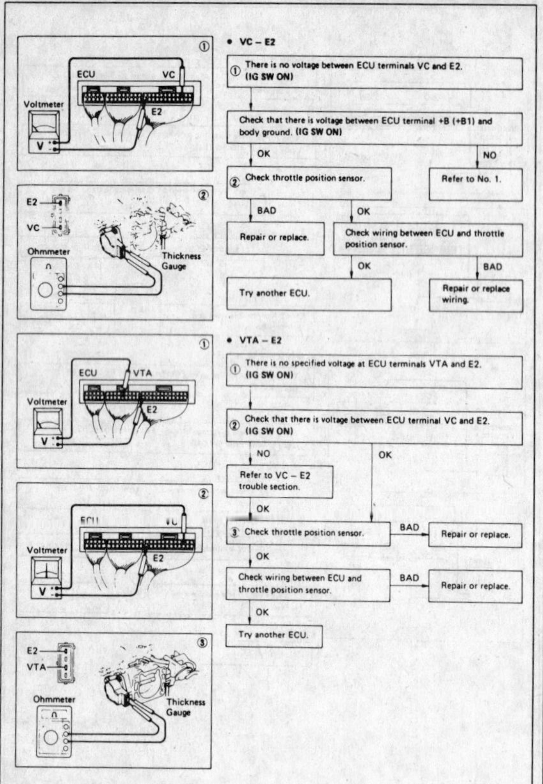
ECU
VC
E2
Voltmeter
V
Ohmmeter
Thickness Gauge
VTA
• VC – E2
① There is no voltage between ECU terminals VC and E2. (IG SW ON)
Check that there is voltage between ECU terminal +B (+B1) and body ground. (IG SW ON)
OK
NO
② Check throttle position sensor.
Refer to No. 1.
BAD
OK
Repair or replace.
Check wiring between ECU and throttle position sensor.
OK
BAD
Try another ECU.
Repair or replace wiring.
• VTA – E2
① There is no specified voltage at ECU terminals VTA and E2. (IG SW ON)
② Check that there is voltage between ECU terminal VC and E2. (IG SW ON)
NO
OK
Refer to VC – E2 trouble section.
OK
③ Check throttle position sensor.
BAD
Repair or replace.
OK
Check wiring between ECU and throttle position sensor.
BAD
Repair or replace.
OK
Try another ECU.

TROUBLESHOOTING CHART 3—1989–90 SUPRA 7M-GE

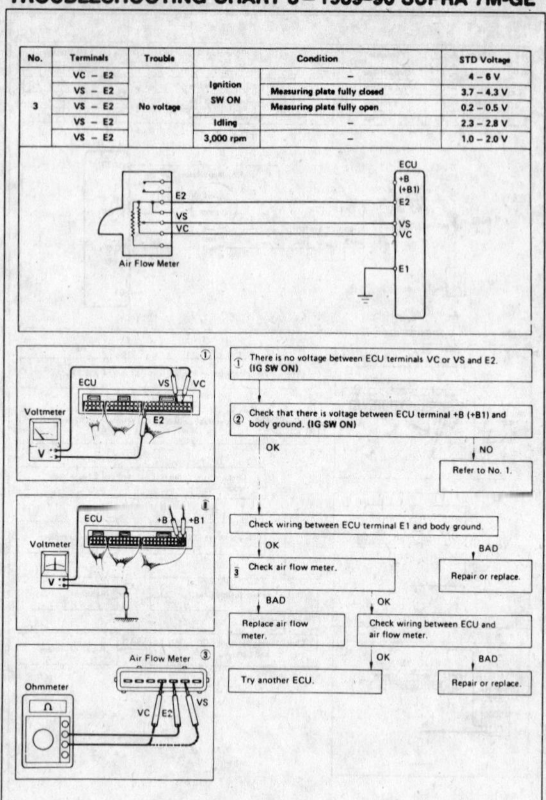

No.	Terminals	Trouble	Condition		STD Voltage
3	VC – E2	No voltage	Ignition SW ON	–	4 – 6 V
	VS – E2			Measuring plate fully closed	3.7 – 4.3 V
	VS – E2			Measuring plate fully open	0.2 – 0.5 V
	VS – E2		Idling	–	2.3 – 2.8 V
	VS – E2		3,000 rpm	–	1.0 – 2.0 V

ECU
+B (+B1)
E2
VS
VC
E1
Air Flow Meter
Voltmeter
Ohmmeter
① There is no voltage between ECU terminals VC or VS and E2. (IG SW ON)
② Check that there is voltage between ECU terminal +B (+B1) and body ground. (IG SW ON)
OK
NO
Refer to No. 1.
Check wiring between ECU terminal E1 and body ground.
OK
BAD
③ Check air flow meter.
Repair or replace.
BAD
OK
Replace air flow meter.
Check wiring between ECU and air flow meter.
OK
BAD
Try another ECU.
Repair or replace.

TROUBLESHOOTING CHART 4—1989–90 SUPRA 7M-GE

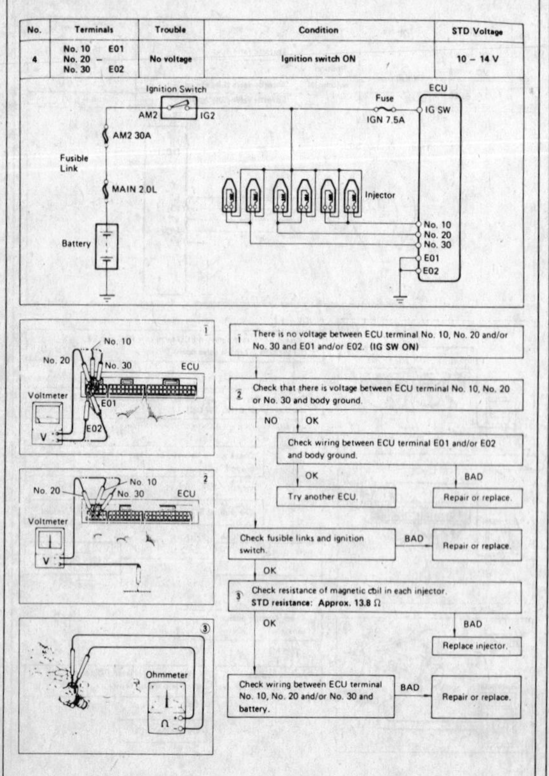

No.	Terminals	Trouble	Condition	STD Voltage
4	No. 10 E01 No. 20 – No. 30 E02	No voltage	Ignition switch ON	10 – 14 V

Ignition Switch
AM2
IG2
AM2 30A
Fusible Link
MAIN 2.0L
Battery
Fuse
IGN 7.5A
ECU
IG SW
Injector
No. 10
No. 20
No. 30
E01
E02
Voltmeter
Ohmmeter
1 There is no voltage between ECU terminal No. 10, No. 20 and/or No. 30 and E01 and/or E02. (IG SW ON)
2 Check that there is voltage between ECU terminal No. 10, No. 20 or No. 30 and body ground.
NO
OK
Check wiring between ECU terminal E01 and/or E02 and body ground.
OK
BAD
Try another ECU.
Repair or replace.
Check fusible links and ignition switch.
BAD
Repair or replace.
OK
3 Check resistance of magnetic coil in each injector. STD resistance: Approx. 13.8 Ω
OK
BAD
Replace injector.
Check wiring between ECU terminal No. 10, No. 20 and/or No. 30 and battery.
BAD
Repair or replace.

TROUBLESHOOTING CHART 5—1989–90 SUPRA 7M-GE

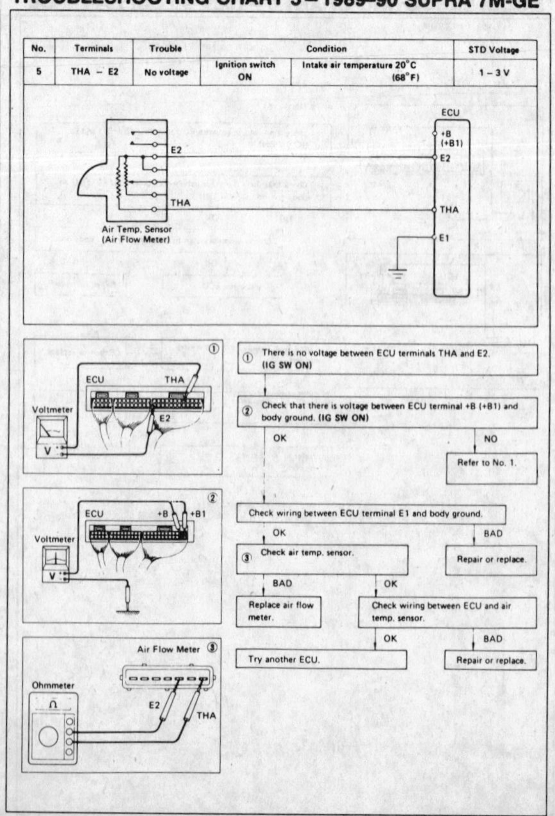

No.	Terminals	Trouble	Condition		STD Voltage
5	THA – E2	No voltage	Ignition switch ON	Intake air temperature 20°C (68°F)	1 – 3 V

ECU
+B (+B1)
E2
THA
E1
Air Temp. Sensor (Air Flow Meter)
Voltmeter
Ohmmeter
Air Flow Meter
① There is no voltage between ECU terminals THA and E2. (IG SW ON)
② Check that there is voltage between ECU terminal +B (+B1) and body ground. (IG SW ON)
OK
NO
Refer to No. 1.
Check wiring between ECU terminal E1 and body ground.
OK
BAD
③ Check air temp. sensor.
Repair or replace.
BAD
OK
Replace air flow meter.
Check wiring between ECU and air temp. sensor.
OK
BAD
Try another ECU.
Repair or replace.

4
SECTION

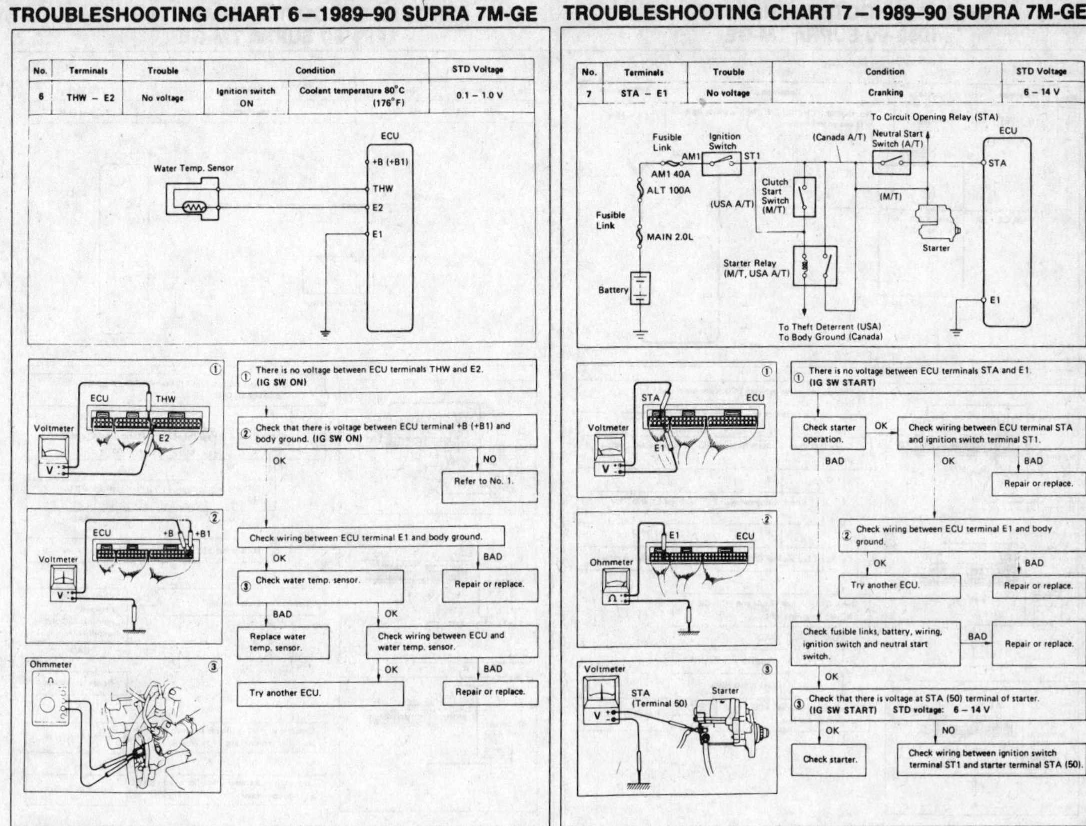

TROUBLESHOOTING CHART 6 – 1989–90 SUPRA 7M-GE

No.	Terminals	Trouble	Condition		STD Voltage
6	THW – E2	No voltage	Ignition switch ON	Coolant temperature 80°C (176°F)	0.1 – 1.0 V

ECU
+B (+B1)
THW
E2
E1
Water Temp. Sensor
ECU
THW
Voltmeter
E2
ECU
+B
+B1
Voltmeter
Ohmmeter
① There is no voltage between ECU terminals THW and E2. (IG SW ON)
② Check that there is voltage between ECU terminal +B (+B1) and body ground. (IG SW ON)
OK
NO
Refer to No. 1.
Check wiring between ECU terminal E1 and body ground.
OK
BAD
③ Check water temp. sensor.
Repair or replace.
BAD
OK
Replace water temp. sensor.
Check wiring between ECU and water temp. sensor.
OK
BAD
Try another ECU.
Repair or replace.

TROUBLESHOOTING CHART 7 – 1989–90 SUPRA 7M-GE

No.	Terminals	Trouble	Condition	STD Voltage
7	STA – E1	No voltage	Cranking	6 – 14 V

To Circuit Opening Relay (STA)
ECU
Fusible Link
Ignition Switch
AM1
ST1
(Canada A/T)
Neutral Start Switch (A/T)
STA
AM1 40A
ALT 100A
Clutch Start Switch (M/T)
(USA A/T)
(M/T)
Fusible Link
MAIN 2.0L
Starter
Starter Relay (M/T, USA A/T)
Battery
E1
To Theft Deterrent (USA)
To Body Ground (Canada)
STA
ECU
Voltmeter
E1
E1
ECU
Ohmmeter
Voltmeter
STA (Terminal 50)
Starter
① There is no voltage between ECU terminals STA and E1. (IG SW START)
Check starter operation.
OK
Check wiring between ECU terminal STA and ignition switch terminal ST1.
BAD
OK
BAD
Repair or replace.
② Check wiring between ECU terminal E1 and body ground.
OK
BAD
Try another ECU.
Repair or replace.
Check fusible links, battery, wiring, ignition switch and neutral start switch.
BAD
Repair or replace.
OK
③ Check that there is voltage at STA (50) terminal of starter. (IG SW START) STD voltage: 6 – 14 V
OK
NO
Check starter.
Check wiring between ignition switch terminal ST1 and starter terminal STA (50).

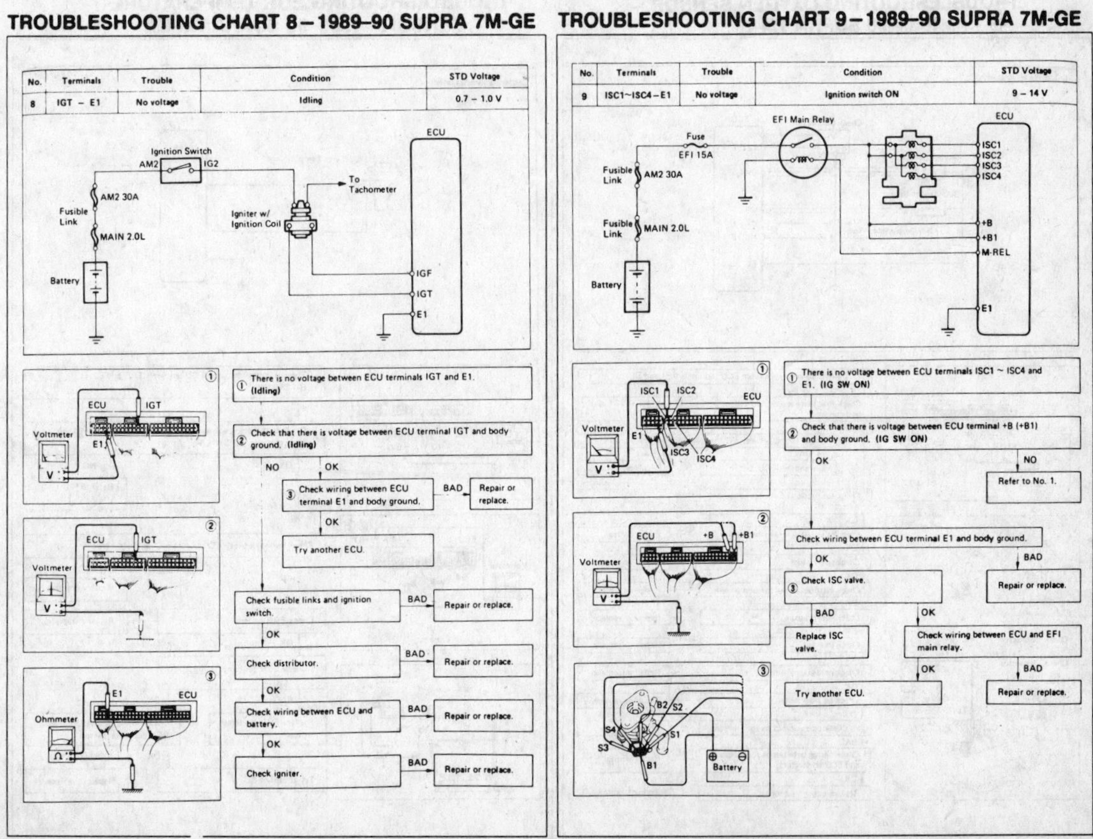

TROUBLESHOOTING CHART 8 – 1989–90 SUPRA 7M-GE

No.	Terminals	Trouble	Condition	STD Voltage
8	IGT – E1	No voltage	Idling	0.7 – 1.0 V

ECU
Ignition Switch
AM2
IG2
To Tachometer
AM2 30A
Fusible Link
Igniter w/ Ignition Coil
MAIN 2.0L
Battery
IGF
IGT
E1
ECU
IGT
Voltmeter
E1
ECU
IGT
Voltmeter
E1
ECU
Ohmmeter
① There is no voltage between ECU terminals IGT and E1. (Idling)
② Check that there is voltage between ECU terminal IGT and body ground. (Idling)
NO
OK
③ Check wiring between ECU terminal E1 and body ground.
BAD
Repair or replace.
OK
Try another ECU.
Check fusible links and ignition switch.
BAD
Repair or replace.
OK
Check distributor.
BAD
Repair or replace.
OK
Check wiring between ECU and battery.
BAD
Repair or replace.
OK
Check igniter.
BAD
Repair or replace.

TROUBLESHOOTING CHART 9 – 1989–90 SUPRA 7M-GE

No.	Terminals	Trouble	Condition	STD Voltage
9	ISC1~ISC4 – E1	No voltage	Ignition switch ON	9 – 14 V

ECU
EFI Main Relay
Fuse
EFI 15A
Fusible Link
AM2 30A
ISC1
ISC2
ISC3
ISC4
Fusible Link
MAIN 2.0L
+B
+B1
M-REL
Battery
E1
ISC1
ISC2
ECU
Voltmeter
E1
ISC3
ISC4
ECU
+B
+B1
Voltmeter
B2
S2
S4
S1
S3
B1
Battery
① There is no voltage between ECU terminals ISC1 ~ ISC4 and E1. (IG SW ON)
② Check that there is voltage between ECU terminal +B (+B1) and body ground. (IG SW ON)
OK
NO
Refer to No. 1.
Check wiring between ECU terminal E1 and body ground.
OK
BAD
③ Check ISC valve.
Repair or replace.
BAD
OK
Replace ISC valve.
Check wiring between ECU and EFI main relay.
OK
BAD
Try another ECU.
Repair or replace.

TROUBLESHOOTING CHART 10– 1989–90 SUPRA 7M-GE

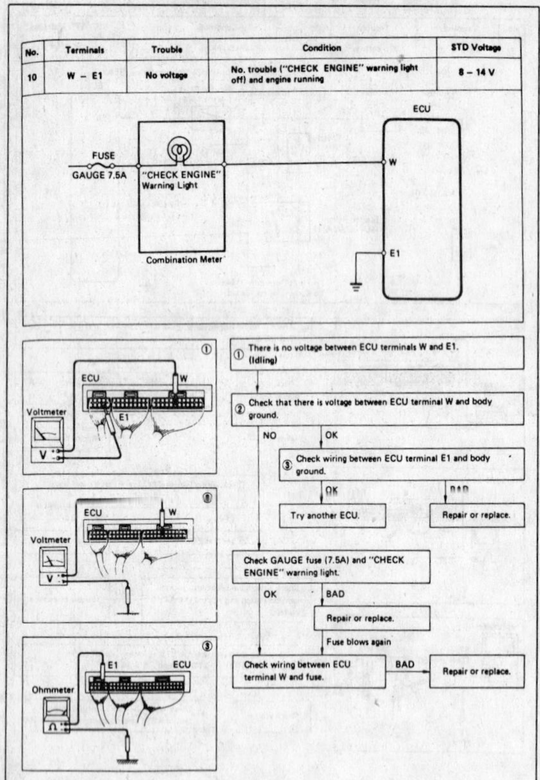
No. | Terminals | Trouble | Condition | STD Voltage
10 | W – E1 | No voltage | No. trouble ("CHECK ENGINE" warning light off) and engine running | 8 – 14 V
ECU
FUSE
GAUGE 7.5A
"CHECK ENGINE" Warning Light
Combination Meter
W
E1
Voltmeter
Ohmmeter
① There is no voltage between ECU terminals W and E1. (Idling)
② Check that there is voltage between ECU terminal W and body ground.
NO
OK
③ Check wiring between ECU terminal E1 and body ground.
OK
BAD
Try another ECU.
Repair or replace.
Check GAUGE fuse (7.5A) and "CHECK ENGINE" warning light.
OK
BAD
Repair or replace.
Fuse blows again
Check wiring between ECU terminal W and fuse.
BAD
Repair or replace.

TROUBLESHOOTING CHART 11– 1989–90 SUPRA 7M-GE

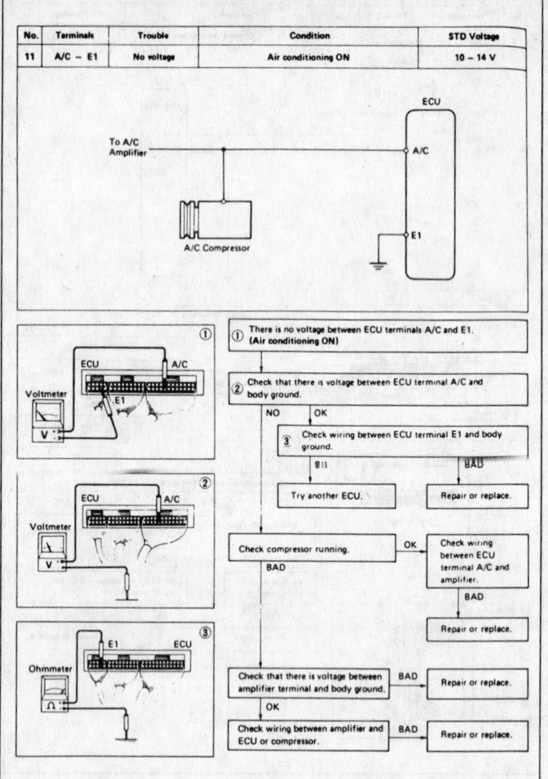
No. | Terminals | Trouble | Condition | STD Voltage
11 | A/C – E1 | No voltage | Air conditioning ON | 10 – 14 V
ECU
To A/C Amplifier
A/C
E1
A/C Compressor
Voltmeter
Ohmmeter
① There is no voltage between ECU terminals A/C and E1. (Air conditioning ON)
② Check that there is voltage between ECU terminal A/C and body ground.
NO
OK
③ Check wiring between ECU terminal E1 and body ground.
BAD
Try another ECU.
Repair or replace.
Check compressor running.
OK
Check wiring between ECU terminal A/C and amplifier.
BAD
Repair or replace.
Check that there is voltage between amplifier terminal and body ground.
BAD
Repair or replace.
OK
Check wiring between amplifier and ECU or compressor.
BAD
Repair or replace.

TROUBLESHOOTING OXYGEN SENSOR– 1989–90 SUPRA 7M-GE

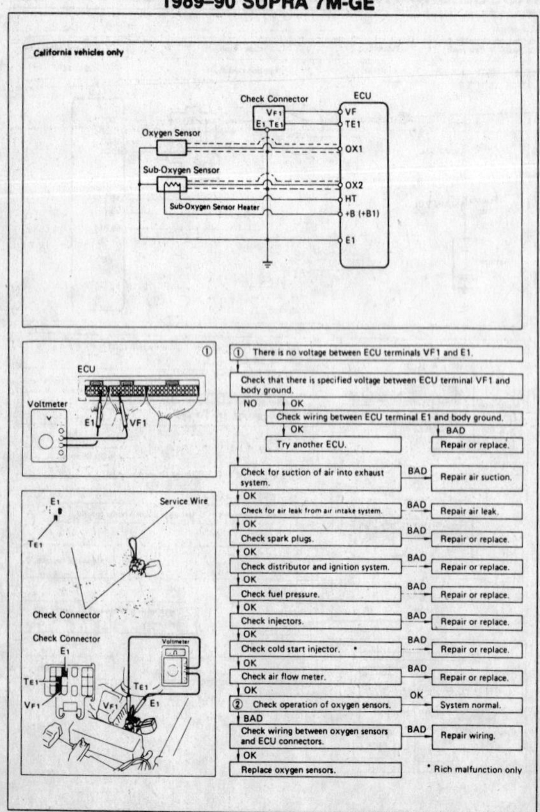
California vehicles only
Check Connector
ECU
VF
TE1
Oxygen Sensor
OX1
Sub-Oxygen Sensor
OX2
HT
Sub-Oxygen Sensor Heater
+B (+B1)
E1
Voltmeter
VF1
① There is no voltage between ECU terminals VF1 and E1.
Check that there is specified voltage between ECU terminal VF1 and body ground.
NO
OK
Check wiring between ECU terminal E1 and body ground.
OK
BAD
Try another ECU.
Repair or replace.
Check for suction of air into exhaust system.
BAD
Repair air suction.
Check for air leak from air intake system.
Repair air leak.
Check spark plugs.
Repair or replace.
Check distributor and ignition system.
Repair or replace.
Check fuel pressure.
Repair or replace.
Check injectors.
Repair or replace.
Check cold start injector. *
Repair or replace.
Check air flow meter.
Repair or replace.
② Check operation of oxygen sensors.
System normal.
Check wiring between oxygen sensors and ECU connectors.
Repair wiring.
Replace oxygen sensors.
* Rich malfunction only
Service Wire
Check Connector

TROUBLESHOOTING EGR TEMPERATURE SENSOR–1989–90 SUPRA 7M-GE

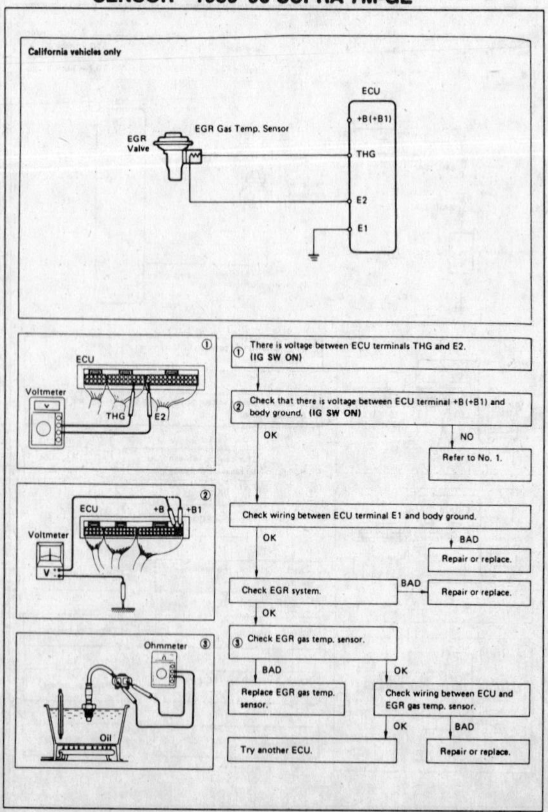
California vehicles only
ECU
EGR Gas Temp. Sensor
EGR Valve
+B (+B1)
THG
E2
E1
Voltmeter
Ohmmeter
Oil
① There is voltage between ECU terminals THG and E2. (IG SW ON)
② Check that there is voltage between ECU terminal +B (+B1) and body ground. (IG SW ON)
OK
NO
Refer to No. 1.
Check wiring between ECU terminal E1 and body ground.
OK
BAD
Repair or replace.
Check EGR system.
BAD
Repair or replace.
OK
③ Check EGR gas temp. sensor.
BAD
OK
Replace EGR gas temp. sensor.
Check wiring between ECU and EGR gas temp. sensor.
OK
BAD
Try another ECU.
Repair or replace.

TROUBLESHOOTING CHART 1—1989–90 SUPRA 7M-GTE

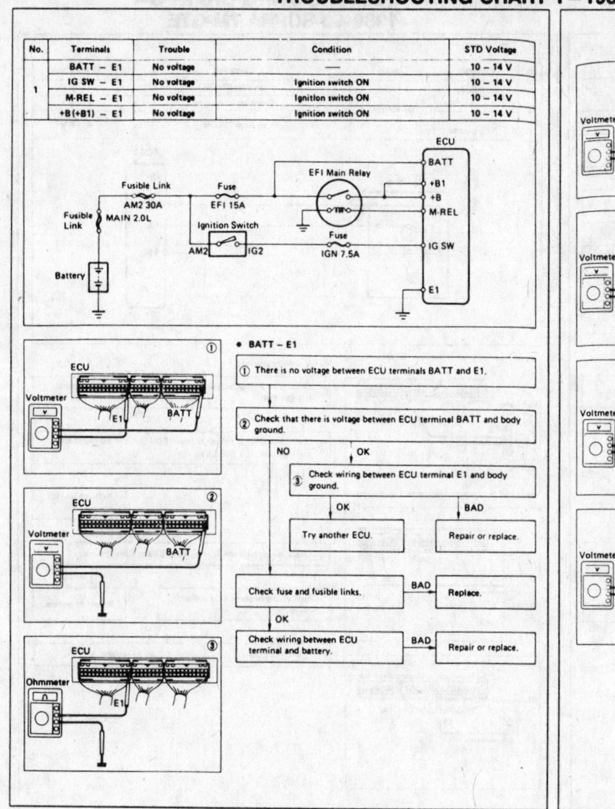

No.	Terminals	Trouble	Condition	STD Voltage
1	BATT – E1	No voltage	—	10 – 14 V
	IG SW – E1	No voltage	Ignition switch ON	10 – 14 V
	M-REL – E1	No voltage	Ignition switch ON	10 – 14 V
	+B(+B1) – E1	No voltage	Ignition switch ON	10 – 14 V

ECU
BATT
EFI Main Relay
+B1
+B
M-REL
Fusible Link
AM2 30A
Fuse
EFI 15A
Fusible Link
MAIN 2.0L
Ignition Switch
AM2
IG2
Fuse
IGN 7.5A
IG SW
Battery
E1
Voltmeter
Ohmmeter
• BATT – E1
① There is no voltage between ECU terminals BATT and E1.
② Check that there is voltage between ECU terminal BATT and body ground.
NO
OK
③ Check wiring between ECU terminal E1 and body ground.
OK
BAD
Try another ECU.
Repair or replace.
Check fuse and fusible links.
BAD
Replace.
OK
Check wiring between ECU terminal and battery.
BAD
Repair or replace.

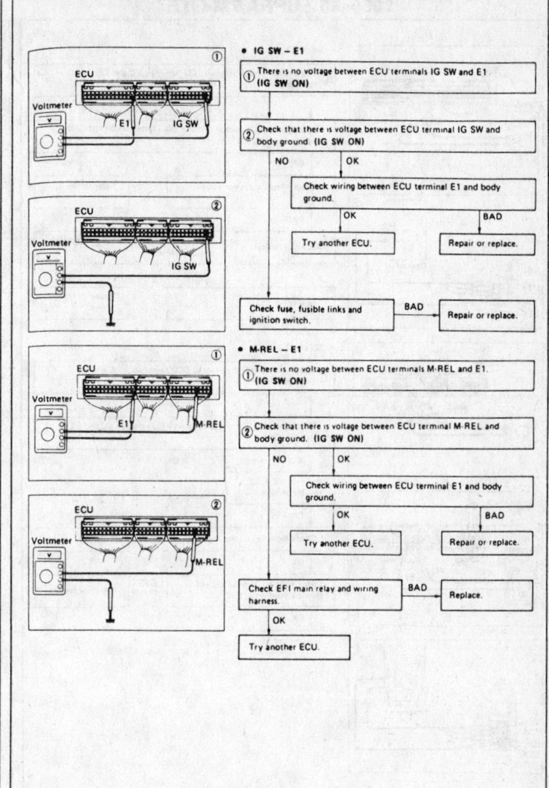

ECU
Voltmeter
E1
IG SW
M-REL
• IG SW – E1
① There is no voltage between ECU terminals IG SW and E1 (IG SW ON)
② Check that there is voltage between ECU terminal IG SW and body ground. (IG SW ON)
NO
OK
Check wiring between ECU terminal E1 and body ground.
OK
BAD
Try another ECU.
Repair or replace.
Check fuse, fusible links and ignition switch.
BAD
Repair or replace.
• M-REL – E1
① There is no voltage between ECU terminals M-REL and E1. (IG SW ON)
② Check that there is voltage between ECU terminal M-REL and body ground. (IG SW ON)
NO
OK
Check wiring between ECU terminal E1 and body ground.
OK
BAD
Try another ECU.
Repair or replace.
Check EFI main relay and wiring harness.
BAD
Replace.
OK
Try another ECU.

TROUBLESHOOTING CHART 1—1989–90 SUPRA 7M-GTE

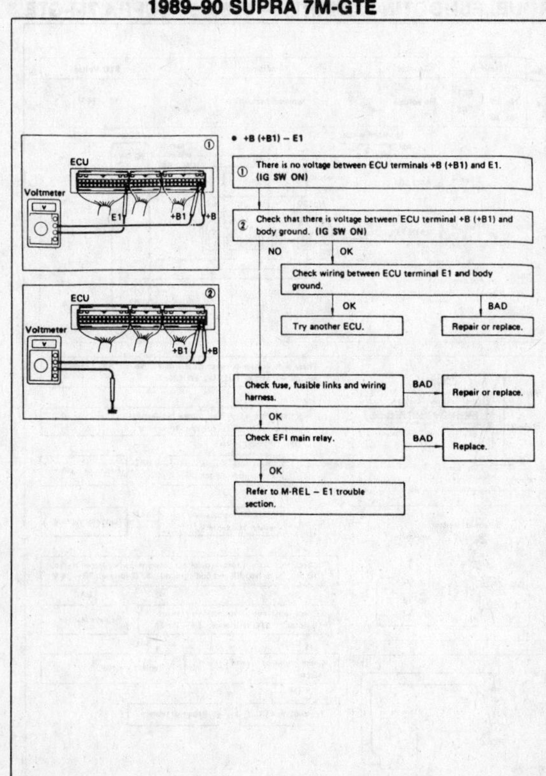

ECU
Voltmeter
E1
+B1
+B
• +B (+B1) – E1
① There is no voltage between ECU terminals +B (+B1) and E1. (IG SW ON)
② Check that there is voltage between ECU terminal +B (+B1) and body ground. (IG SW ON)
NO
OK
Check wiring between ECU terminal E1 and body ground.
OK
BAD
Try another ECU.
Repair or replace.
Check fuse, fusible links and wiring harness.
BAD
Repair or replace.
OK
Check EFI main relay.
BAD
Replace.
OK
Refer to M-REL – E1 trouble section.

TROUBLESHOOTING CHART 2—1989–90 SUPRA 7M-GTE

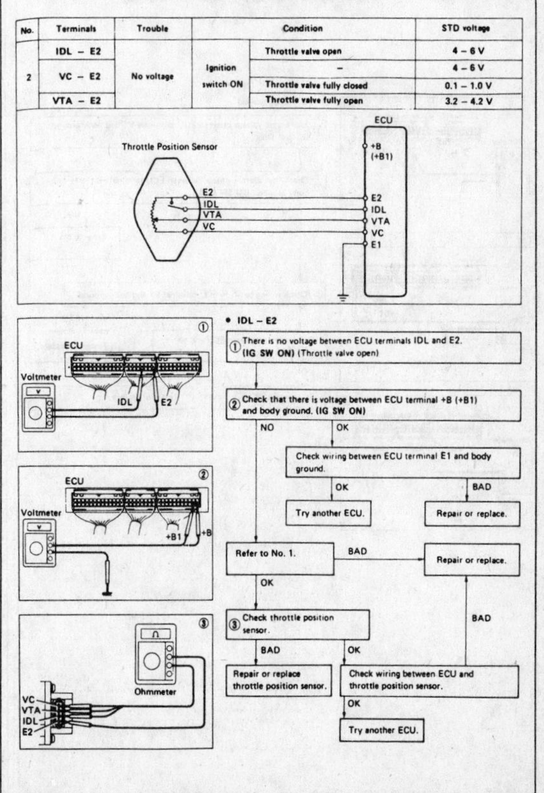

No.	Terminals	Trouble	Condition		STD voltage
2	IDL – E2	No voltage	Ignition switch ON	Throttle valve open	4 – 6 V
	VC – E2			–	4 – 6 V
	VTA – E2			Throttle valve fully closed	0.1 – 1.0 V
				Throttle valve fully open	3.2 – 4.2 V

ECU
Throttle Position Sensor
+B (+B1)
E2
IDL
VTA
VC
E1
Voltmeter
Ohmmeter
• IDL – E2
① There is no voltage between ECU terminals IDL and E2. (IG SW ON) (Throttle valve open)
② Check that there is voltage between ECU terminal +B (+B1) and body ground. (IG SW ON)
NO
OK
Check wiring between ECU terminal E1 and body ground.
OK
BAD
Try another ECU.
Repair or replace.
Refer to No. 1.
BAD
Repair or replace.
OK
③ Check throttle position sensor.
BAD
OK
BAD
Repair or replace throttle position sensor.
Check wiring between ECU and throttle position sensor.
OK
Try another ECU.

TROUBLESHOOTING CHART 2—1989–90 SUPRA 7M-GTE

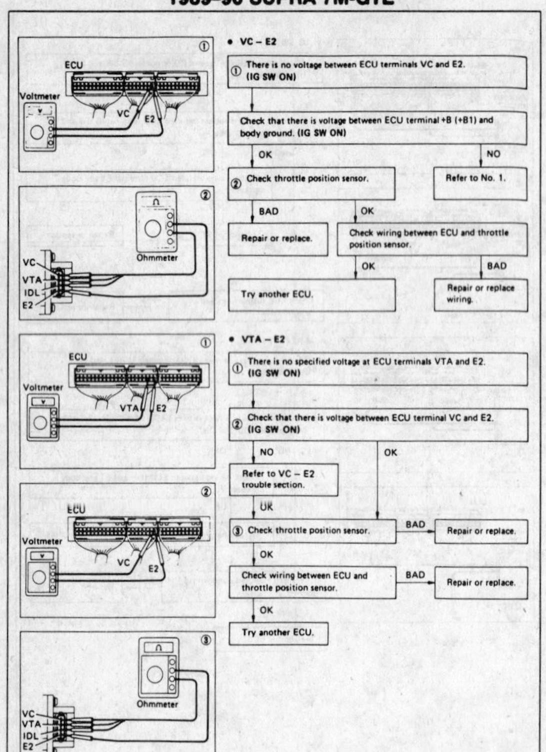
• VC – E2
① There is no voltage between ECU terminals VC and E2. (IG SW ON)
Check that there is voltage between ECU terminal +B (+B1) and body ground. (IG SW ON)
OK
NO
② Check throttle position sensor.
Refer to No. 1.
BAD
OK
Repair or replace.
Check wiring between ECU and throttle position sensor.
OK
BAD
Try another ECU.
Repair or replace wiring.
• VTA – E2
① There is no specified voltage at ECU terminals VTA and E2. (IG SW ON)
② Check that there is voltage between ECU terminal VC and E2. (IG SW ON)
NO
OK
Refer to VC – E2 trouble section.
OK
③ Check throttle position sensor.
BAD
Repair or replace.
OK
Check wiring between ECU and throttle position sensor.
BAD
Repair or replace.
OK
Try another ECU.
ECU
Voltmeter
VC
E2
Ohmmeter
VTA
IDL

TROUBLESHOOTING CHART 3—1989–90 SUPRA 7M-GTE

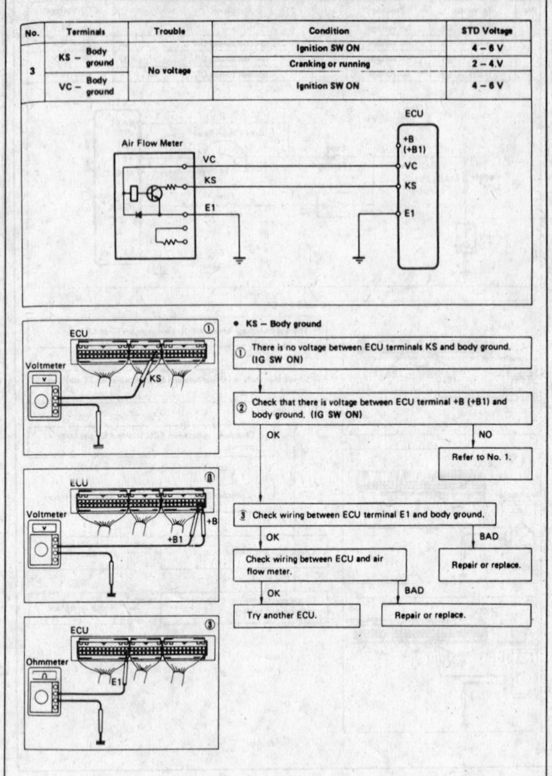

No.	Terminals		Trouble	Condition	STD Voltage
3	KS –	Body ground	No voltage	Ignition SW ON	4 – 6 V
				Cranking or running	2 – 4 V
	VC –	Body ground		Ignition SW ON	4 – 6 V

Air Flow Meter
ECU
+B (+B1)
VC
KS
E1
• KS – Body ground
① There is no voltage between ECU terminals KS and body ground. (IG SW ON)
② Check that there is voltage between ECU terminal +B (+B1) and body ground. (IG SW ON)
OK
NO
Refer to No. 1.
③ Check wiring between ECU terminal E1 and body ground.
OK
BAD
Check wiring between ECU and air flow meter.
Repair or replace.
OK
BAD
Try another ECU.
Repair or replace.
Voltmeter
Ohmmeter

TROUBLESHOOTING CHART 3—1989–90 SUPRA 7M-GTE

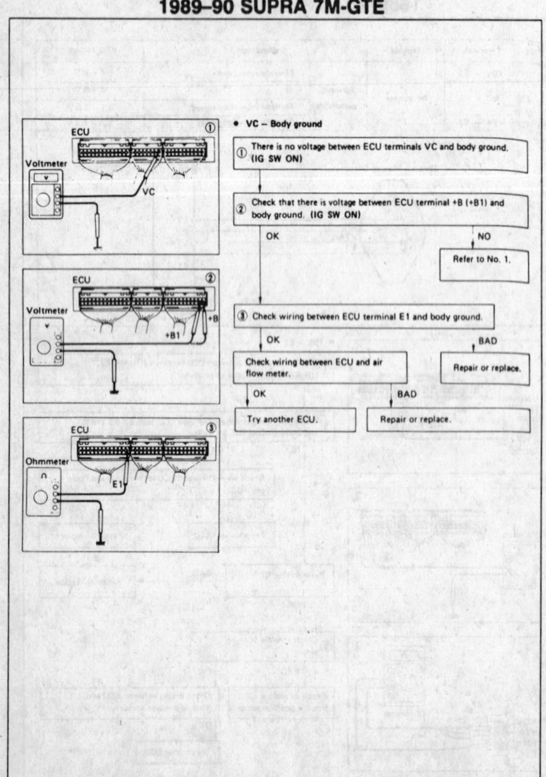
• VC – Body ground
① There is no voltage between ECU terminals VC and body ground. (IG SW ON)
② Check that there is voltage between ECU terminal +B (+B1) and body ground. (IG SW ON)
OK
NO
Refer to No. 1.
③ Check wiring between ECU terminal E1 and body ground.
OK
BAD
Check wiring between ECU and air flow meter.
Repair or replace.
OK
BAD
Try another ECU.
Repair or replace.
ECU
Voltmeter
VC
+B
+B1
Ohmmeter
E1

TROUBLESHOOTING CHART 4—1989–90 SUPRA 7M-GTE

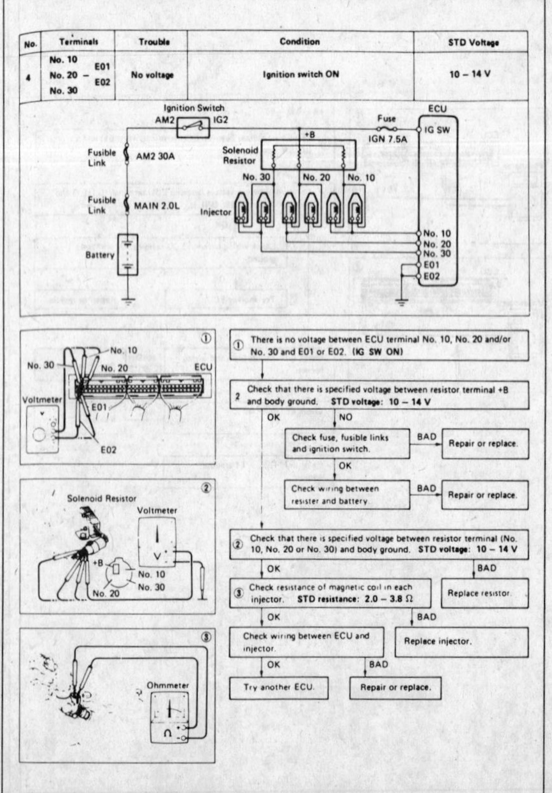

No.	Terminals		Trouble	Condition	STD Voltage
4	No. 10 No. 20 – No. 30	E01 E02	No voltage	Ignition switch ON	10 – 14 V

Ignition Switch
AM2
IG2
Fuse
IGN 7.5A
ECU
IG SW
Fusible Link
AM2 30A
+B
Solenoid Resistor
No. 30
No. 20
No. 10
Fusible Link
MAIN 2.0L
Injector
Battery
No. 10
No. 20
No. 30
E01
E02
① There is no voltage between ECU terminal No. 10, No. 20 and/or No. 30 and E01 or E02. (IG SW ON)
2 Check that there is specified voltage between resistor terminal +B and body ground. STD voltage: 10 – 14 V
OK
NO
Check fuse, fusible links and ignition switch.
BAD
Repair or replace.
OK
Check wiring between resister and battery.
BAD
Repair or replace.
② Check that there is specified voltage between resistor terminal (No. 10, No. 20 or No. 30) and body ground. STD voltage: 10 – 14 V
OK
BAD
③ Check resistance of magnetic coil in each injector. STD resistance: 2.0 – 3.8 Ω
Replace resistor.
OK
BAD
Check wiring between ECU and injector.
Replace injector.
OK
BAD
Try another ECU.
Repair or replace.
Voltmeter
Solenoid Resistor
Ohmmeter

TROUBLESHOOTING CHART 5— 1989–90 SUPRA 7M-GTE

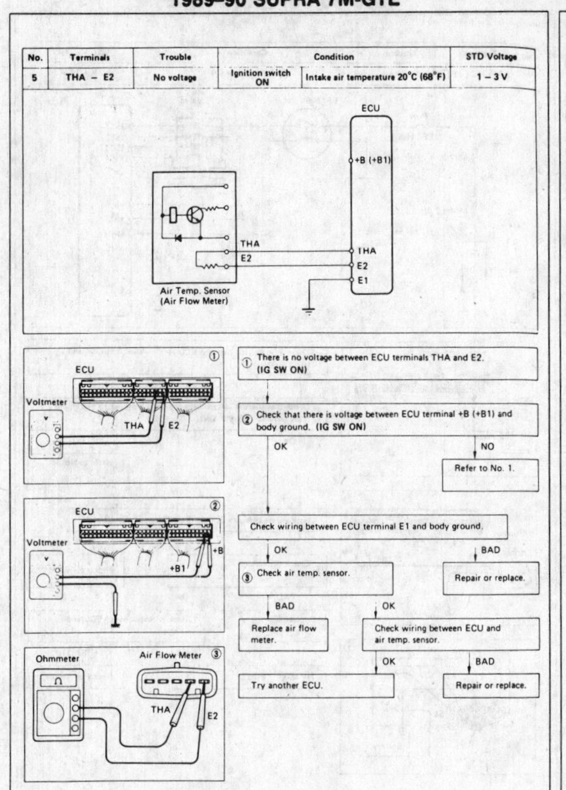
No. | Terminals | Trouble | Condition | STD Voltage
5 | THA – E2 | No voltage | Ignition switch ON | Intake air temperature 20°C (68°F) | 1 – 3 V
ECU
+B (+B1)
THA
E2
E1
Air Temp. Sensor (Air Flow Meter)
There is no voltage between ECU terminals THA and E2. (IG SW ON)
Check that there is voltage between ECU terminal +B (+B1) and body ground. (IG SW ON)
OK
NO
Refer to No. 1.
Check wiring between ECU terminal E1 and body ground.
BAD
Check air temp. sensor.
Repair or replace.
Replace air flow meter.
Check wiring between ECU and air temp. sensor.
Try another ECU.
Repair or replace.
Voltmeter
Ohmmeter
Air Flow Meter

TROUBLESHOOTING CHART 6— 1989–90 SUPRA 7M-GTE

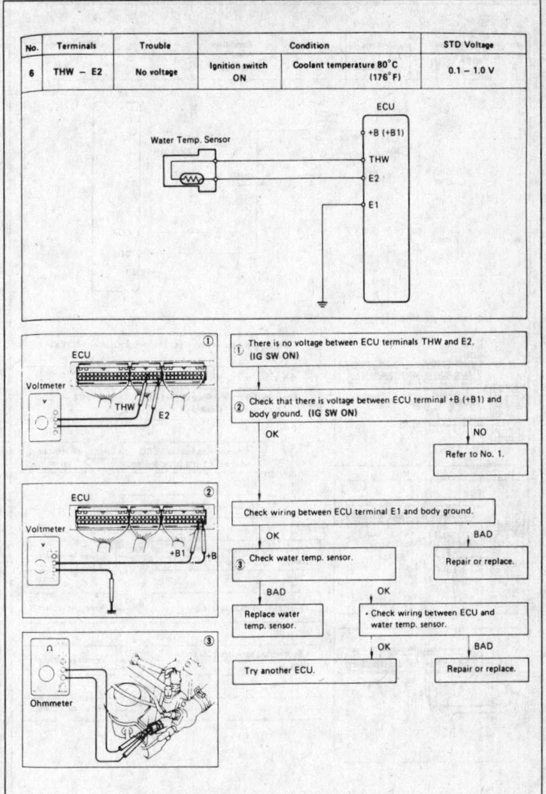
No. | Terminals | Trouble | Condition | STD Voltage
6 | THW – E2 | No voltage | Ignition switch ON | Coolant temperature 80°C (176°F) | 0.1 – 1.0 V
ECU
+B (+B1)
THW
E2
E1
Water Temp. Sensor
There is no voltage between ECU terminals THW and E2. (IG SW ON)
Check that there is voltage between ECU terminal +B (+B1) and body ground. (IG SW ON)
OK
NO
Refer to No. 1.
Check wiring between ECU terminal E1 and body ground.
BAD
Check water temp. sensor.
Repair or replace.
Replace water temp. sensor.
Check wiring between ECU and water temp. sensor.
Try another ECU.
Repair or replace.
Voltmeter
Ohmmeter

TROUBLESHOOTING CHART 7— 1989–90 SUPRA 7M-GTE

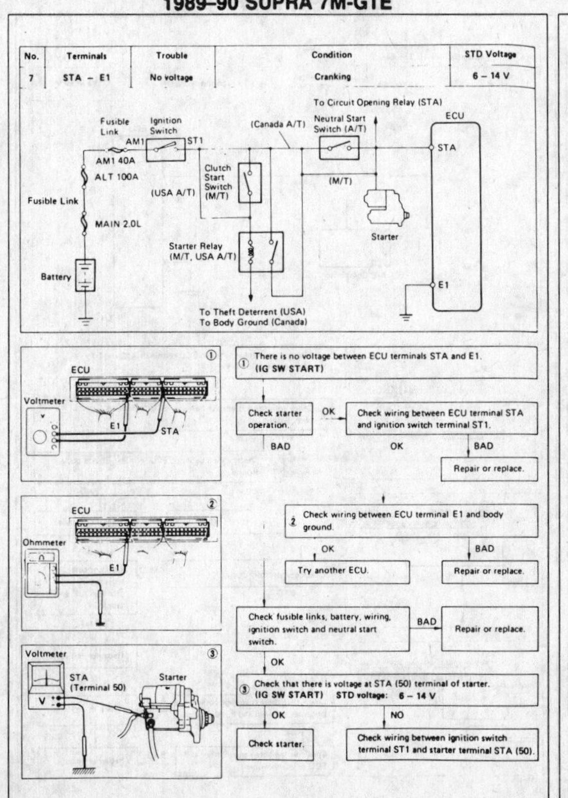
No. | Terminals | Trouble | Condition | STD Voltage
7 | STA – E1 | No voltage | Cranking | 6 – 14 V
To Circuit Opening Relay (STA)
Fusible Link
Ignition Switch
AM1
ST1
AM1 40A
ALT 100A
MAIN 2.0L
Battery
(Canada A/T)
Neutral Start Switch (A/T)
(USA A/T)
Clutch Start Switch (M/T)
(M/T)
Starter Relay (M/T, USA A/T)
Starter
ECU
STA
E1
To Theft Deterrent (USA)
To Body Ground (Canada)
There is no voltage between ECU terminals STA and E1. (IG SW START)
Check starter operation.
OK
Check wiring between ECU terminal STA and ignition switch terminal ST1.
BAD
Repair or replace.
Check wiring between ECU terminal E1 and body ground.
Try another ECU.
Repair or replace.
Check fusible links, battery, wiring, ignition switch and neutral start switch.
Repair or replace.
Check that there is voltage at STA (50) terminal of starter. (IG SW START) STD voltage: 6 – 14 V
NO
Check starter.
Check wiring between ignition switch terminal ST1 and starter terminal STA (50).
Voltmeter
Ohmmeter
STA (Terminal 50)

TROUBLESHOOTING CHART 8— 1989–90 SUPRA 7M-GTE

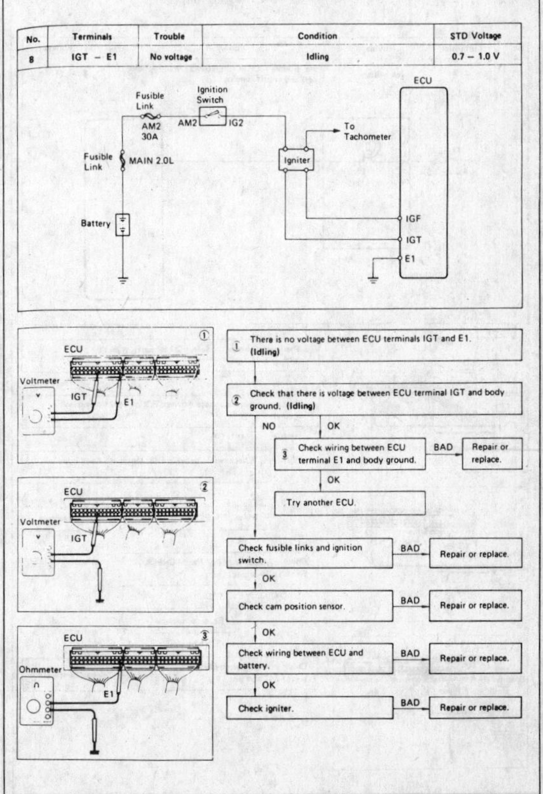
No. | Terminals | Trouble | Condition | STD Voltage
8 | IGT – E1 | No voltage | Idling | 0.7 – 1.0 V
Fusible Link
Ignition Switch
AM2 30A
AM2
IG2
MAIN 2.0L
Battery
Igniter
To Tachometer
ECU
IGF
IGT
E1
There is no voltage between ECU terminals IGT and E1. (Idling)
Check that there is voltage between ECU terminal IGT and body ground. (Idling)
NO
OK
Check wiring between ECU terminal E1 and body ground.
BAD
Repair or replace.
Try another ECU.
Check fusible links and ignition switch.
Check cam position sensor.
Check wiring between ECU and battery.
Check igniter.
Voltmeter
Ohmmeter

TROUBLESHOOTING CHART 9— 1989-90 SUPRA 7M-GTE

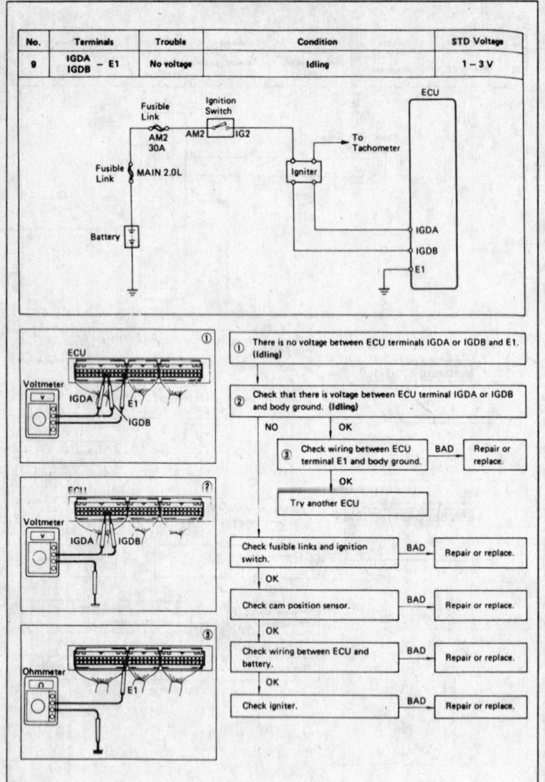

No.	Terminals	Trouble	Condition	STD Voltage
9	IGDA IGDB – E1	No voltage	Idling	1 – 3 V

ECU
Fusible Link
AM2 30A
Ignition Switch
AM2
IG2
To Tachometer
Fusible Link
MAIN 2.0L
Igniter
Battery
IGDA
IGDB
E1
Voltmeter
Ohmmeter
① There is no voltage between ECU terminals IGDA or IGDB and E1. (Idling)
② Check that there is voltage between ECU terminal IGDA or IGDB and body ground. (Idling)
NO
OK
③ Check wiring between ECU terminal E1 and body ground.
BAD
Repair or replace.
Try another ECU
Check fusible links and ignition switch.
Repair or replace.
Check cam position sensor.
Check wiring between ECU and battery.
Check igniter.

TROUBLESHOOTING CHART 10— 1989-90 SUPRA 7M-GTE

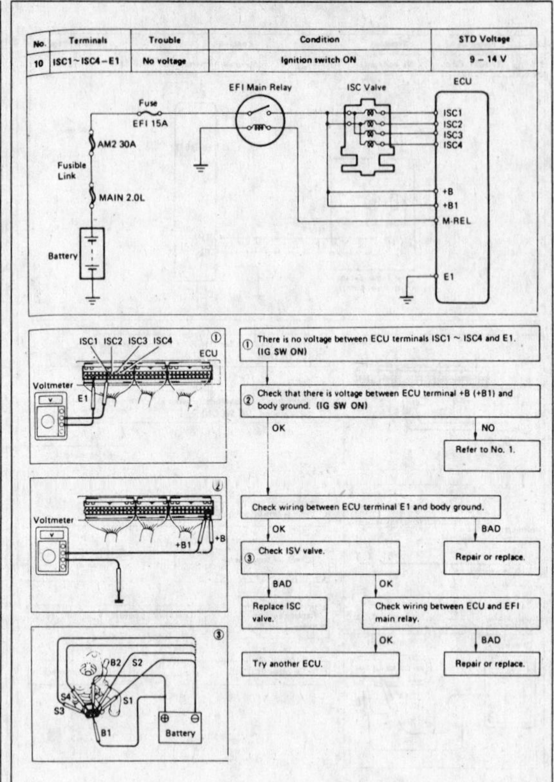

No.	Terminals	Trouble	Condition	STD Voltage
10	ISC1 ~ ISC4 – E1	No voltage	Ignition switch ON	9 – 14 V

ECU
EFI Main Relay
ISC Valve
Fuse
EFI 15A
AM2 30A
Fusible Link
MAIN 2.0L
Battery
ISC1
ISC2
ISC3
ISC4
+B
+B1
M-REL
E1
Voltmeter
① There is no voltage between ECU terminals ISC1 ~ ISC4 and E1. (IG SW ON)
② Check that there is voltage between ECU terminal +B (+B1) and body ground. (IG SW ON)
OK
NO
Refer to No. 1.
Check wiring between ECU terminal E1 and body ground.
BAD
Repair or replace.
③ Check ISV valve.
Replace ISC valve.
Check wiring between ECU and EFI main relay.
Try another ECU.
B2
S2
S4
S3
S1
B1
Battery

TROUBLESHOOTING CHART 11— 1989-90 SUPRA 7M-GTE

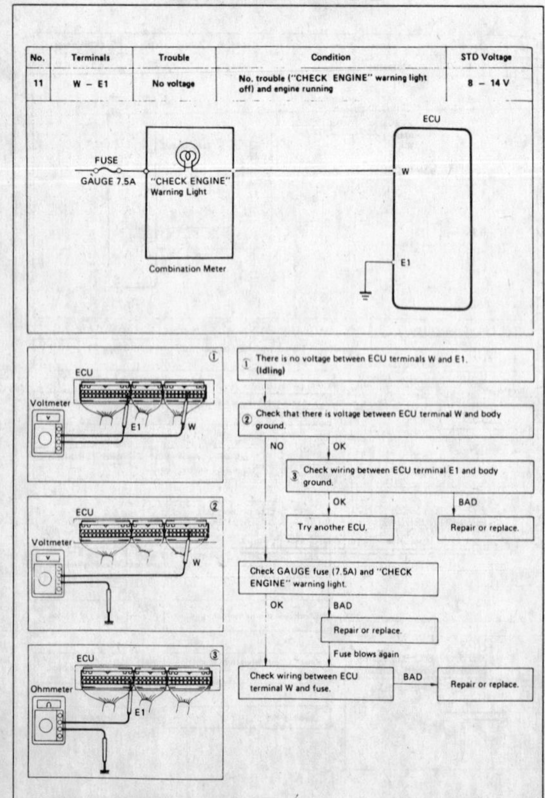

No.	Terminals	Trouble	Condition	STD Voltage
11	W – E1	No voltage	No. trouble ("CHECK ENGINE" warning light off) and engine running	8 – 14 V

ECU
FUSE
GAUGE 7.5A
"CHECK ENGINE" Warning Light
Combination Meter
W
E1
Voltmeter
Ohmmeter
① There is no voltage between ECU terminals W and E1. (Idling)
② Check that there is voltage between ECU terminal W and body ground.
NO
OK
③ Check wiring between ECU terminal E1 and body ground.
BAD
Try another ECU.
Repair or replace.
Check GAUGE fuse (7.5A) and "CHECK ENGINE" warning light.
Fuse blows again
Check wiring between ECU terminal W and fuse.

TROUBLESHOOTING CHART 12— 1989-90 SUPRA 7M-GTE

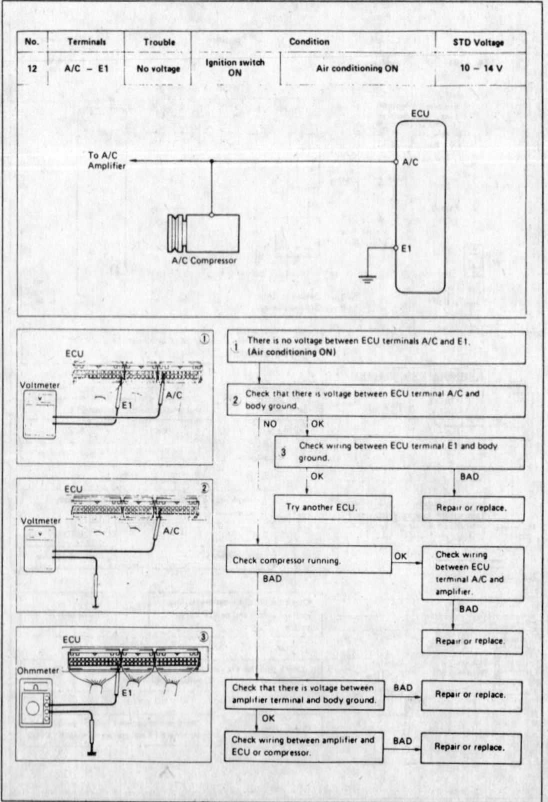

No.	Terminals	Trouble	Condition		STD Voltage
12	A/C – E1	No voltage	Ignition switch ON	Air conditioning ON	10 – 14 V

ECU
To A/C Amplifier
A/C
A/C Compressor
E1
Voltmeter
Ohmmeter
1 There is no voltage between ECU terminals A/C and E1. (Air conditioning ON)
2 Check that there is voltage between ECU terminal A/C and body ground.
NO
OK
3 Check wiring between ECU terminal E1 and body ground.
BAD
Try another ECU.
Repair or replace.
Check compressor running.
Check wiring between ECU terminal A/C and amplifier.
Check that there is voltage between amplifier terminal and body ground.
Check wiring between amplifier and ECU or compressor.

TROUBLESHOOTING OXYGEN SENSOR – 1989–90 SUPRA 7M-GTE

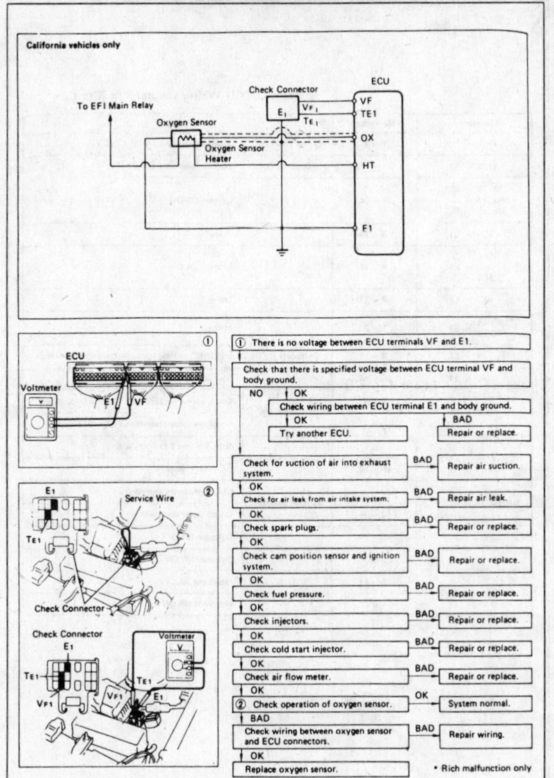
California vehicles only
To EFI Main Relay
Check Connector
ECU
Oxygen Sensor
Oxygen Sensor Heater
VF
TE1
OX
HT
E1
Voltmeter
Service Wire
Check Connector
① There is no voltage between ECU terminals VF and E1.
Check that there is specified voltage between ECU terminal VF and body ground.
NO
OK
Check wiring between ECU terminal E1 and body ground.
OK
BAD
Try another ECU.
Repair or replace.
Check for suction of air into exhaust system.
BAD
Repair air suction.
OK
Check for air leak from air intake system.
BAD
Repair air leak.
OK
Check spark plugs.
BAD
Repair or replace.
OK
Check cam position sensor and ignition system.
BAD
Repair or replace.
OK
Check fuel pressure.
BAD
Repair or replace.
OK
Check injectors.
BAD
Repair or replace.
OK
Check cold start injector.
BAD
Repair or replace.
OK
Check air flow meter.
BAD
Repair or replace.
OK
② Check operation of oxygen sensor.
OK
System normal.
BAD
Check wiring between oxygen sensor and ECU connectors.
BAD
Repair wiring.
OK
Replace oxygen sensor.
• Rich malfunction only

TROUBLESHOOTING EGR TEMPERATURE SENSOR – 1989–90 SUPRA 7M-GTE

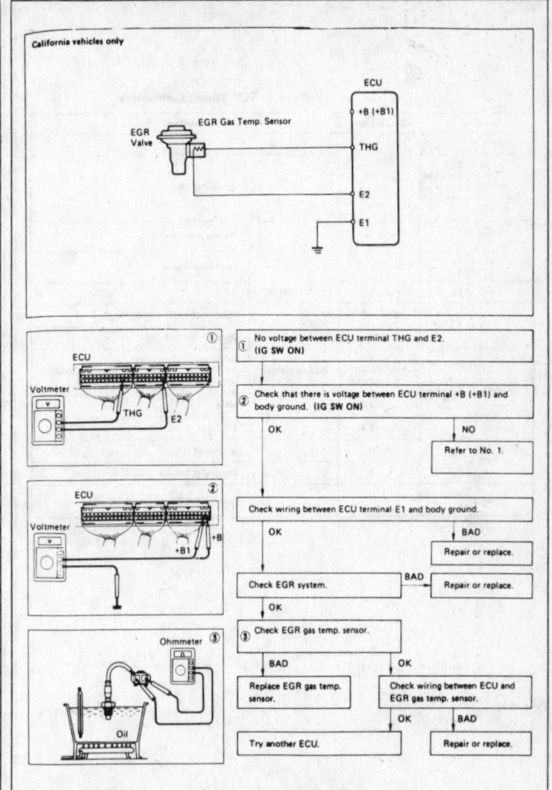
California vehicles only
ECU
EGR Gas Temp. Sensor
EGR Valve
+B (+B1)
THG
E2
E1
Voltmeter
THG
E2
+B1
+B
Ohmmeter
Oil
① No voltage between ECU terminal THG and E2. (IG SW ON)
② Check that there is voltage between ECU terminal +B (+B1) and body ground. (IG SW ON)
OK
NO
Refer to No. 1.
Check wiring between ECU terminal E1 and body ground.
OK
BAD
Repair or replace.
Check EGR system.
BAD
Repair or replace.
OK
③ Check EGR gas temp. sensor.
BAD
OK
Replace EGR gas temp. sensor.
Check wiring between ECU and EGR gas temp. sensor.
OK
BAD
Try another ECU.
Repair or replace.

ECU TERMINAL IDENTIFICATION – 1989–90 SUPRA 7M-GE

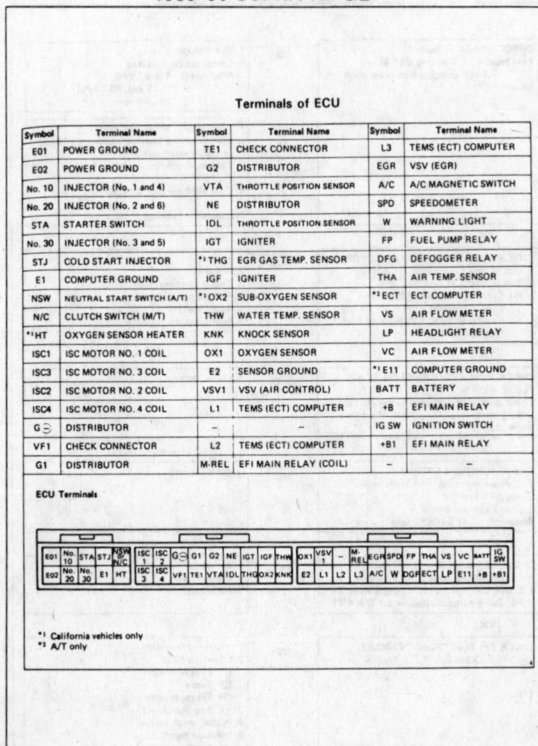

Terminals of ECU

Symbol	Terminal Name	Symbol	Terminal Name	Symbol	Terminal Name
E01	POWER GROUND	TE1	CHECK CONNECTOR	L3	TEMS (ECT) COMPUTER
E02	POWER GROUND	G2	DISTRIBUTOR	EGR	VSV (EGR)
No. 10	INJECTOR (No. 1 and 4)	VTA	THROTTLE POSITION SENSOR	A/C	A/C MAGNETIC SWITCH
No. 20	INJECTOR (No. 2 and 6)	NE	DISTRIBUTOR	SPD	SPEEDOMETER
STA	STARTER SWITCH	IDL	THROTTLE POSITION SENSOR	W	WARNING LIGHT
No. 30	INJECTOR (No. 3 and 5)	IGT	IGNITER	FP	FUEL PUMP RELAY
STJ	COLD START INJECTOR	*1 THG	EGR GAS TEMP. SENSOR	DFG	DEFOGGER RELAY
E1	COMPUTER GROUND	IGF	IGNITER	THA	AIR TEMP. SENSOR
NSW	NEUTRAL START SWITCH (A/T)	*1 OX2	SUB-OXYGEN SENSOR	*2 ECT	ECT COMPUTER
N/C	CLUTCH SWITCH (M/T)	THW	WATER TEMP. SENSOR	VS	AIR FLOW METER
*1 HT	OXYGEN SENSOR HEATER	KNK	KNOCK SENSOR	LP	HEADLIGHT RELAY
ISC1	ISC MOTOR NO. 1 COIL	OX1	OXYGEN SENSOR	VC	AIR FLOW METER
ISC3	ISC MOTOR NO. 3 COIL	E2	SENSOR GROUND	*1 E11	COMPUTER GROUND
ISC2	ISC MOTOR NO. 2 COIL	VSV1	VSV (AIR CONTROL)	BATT	BATTERY
ISC4	ISC MOTOR NO. 4 COIL	L1	TEMS (ECT) COMPUTER	+B	EFI MAIN RELAY
G⊖	DISTRIBUTOR	–	–	IG SW	IGNITION SWITCH
VF1	CHECK CONNECTOR	L2	TEMS (ECT) COMPUTER	+B1	EFI MAIN RELAY
G1	DISTRIBUTOR	M-REL	EFI MAIN RELAY (COIL)	–	–

ECU Terminals

E01 No. 10 STA STJ NSW or N/C ISC 1 ISC 2 G⊖ G1 G2 NE IGT IGF THW OX1 VSV 1 – M-REL EGR SPD FP THA VS VC BATT IG SW
E02 No. 20 No. 30 E1 HT ISC 3 ISC 4 VF1 TE1 VTA IDL THG OX2 KNK E2 L1 L2 L3 A/C W DFG ECT LP E11 +B +B1

*1 California vehicles only
*2 A/T only

ECU TERMINAL IDENTIFICATION – 1989–90 SUPRA 7M-GTE

Terminals of ECU

Symbol	Terminal Name	Symbol	Terminal Name	Symbol	Terminal Name
E01	POWER GROUND	G2	CAM POSITION SENSOR	A/C	A/C COMPRESSOR
E02	POWER GROUND	G1	CAM POSITION SENSOR	*1 LP	HEADLIGHT RELAY
No. 10	INJECTOR (No. 1 and 6)	NE	CAM POSITION SENSOR	SPD	SPEED SENSOR
No. 30	INJECTOR (No. 4 and 5)	E1	COMPUTER GROUND	*3 ECT	ECT COMPUTER
No. 20	INJECTOR (No. 2 and 3)	VF	CHECK CONNECTOR	*1 DFG	DEFOGGER RELAY
STJ	COLD START INJECTOR	G⊖	CAM POSITION SENSOR	L1	TEMS COMPUTER
HT	OXYGEN SENSOR HEATER	OIL	OIL PRESSURE SWITCH	FC	CIRCUIT OPENING RELAY
–	–	TE1	CHECK CONNECTOR	L2	TEMS ECU COMPUTER
VSV2	VSV (FPU)	OX	OXYGEN SENSOR	FP	FUEL PUMP RELAY
–	–	–	–	L3	TEMS COMPUTER
EGR	VSV (EGR)	KNK1	KNOCK SENSOR	W	WARNING LIGHT
–	–	KNK2	KNOCK SENSOR	TIL	TURBO INDICATOR
ISC1	ISC MOTOR NO. 1 COIL	THW	WATER TEMP. SENSOR	M-REL	EFI MAIN RELAY (COIL)
IGT	IGNITER	IDL	THROTTLE POSITION SENSOR	–	–
ISC2	ISC MOTOR NO. 2 COIL	THA	AIR TEMP. SENSOR	–	–
IGDA	IGNITER	VTA	THROTTLE POSITION SENSOR	–	–
ISC3	ISC MOTOR NO. 3 COIL	KS	AIR FLOW METER	IG SW	IGNITION SWITCH
IGDB	IGNITER	–	–	+B1	EFI MAIN RELAY
ISC4	ISC MOTOR NO. 4 COIL	VC	AIR FLOW METER THROTTLE POSITION SENSOR	BATT	BATTERY
–	–	E2	SENSOR GROUND	+B	EFI MAIN RELAY
IGF	IGNITER	STA	STARTER SWITCH		
*1 THG	EGR GAS TEMP. SENSOR	N/C (NSW)	CLUTCH SWITCH (M/T) NEUTRAL START SWITCH (A/T)		

*1 California vehicles only
*2 M/T only
*3 A/T only

ECU Terminals

E01	No. 10	No. 20	HT	VSV 2	EGR	ISC 1	ISC 2	ISC 3	ISC 4	IGF	G2	NE	VF	OIL	OX	KNK 1	THW	THA	KS	VC	STA	A/C	SPD	DFG	FC	FP	W	M-REL	–	IG SW	BATT
E02	No. 30	STJ	–	–	–	IGT	IGDA	IGDB	–	THG	G1	E1	G⊖	TE1	–	KNK 2	IDL	VTA	–	E2	N/C (NSW)	LP	ECT	L1	L2	L3	TIL	–	–	+B1	+B

ECU TERMINAL VOLTAGES— 1989–90 SUPRA 7M-GE AND 7M-GTE

Voltage at ECU Wiring Connectors

Terminals	STD Voltage	Condition	
BATT – E1	10 – 14	–	
IG SW – E1		Ignition SW ON	
M-REL – E1			
+B (+B1) – E1			
IDL – E2 (7M-GE)	10 – 14	Ignition SW ON	Throttle valve open
IDL – E2 (7M-GTE)	4 – 6		Throttle valve open
VC – E2	4 – 6		–
VTA – E2	0.1 – 1.0		Throttle valve fully closed
	4 – 5		Throttle valve fully open
*1 VS – E2	3.7 – 4.3	Ignition SW ON	Measuring plate fully closed
	0.2 – 0.5		Measuring plate fully open
	2.3 – 2.8	Idling	
	1.0 – 2.0	3,000 rpm	
*2 KS – Body ground	4 – 6	Ignition SW ON	
	2 – 4	Cranking or running	
*1 VL – Body ground	4 – 6	Ignition SW ON	

ECU TERMINAL VOLTAGES— 1989–90 SUPRA 7M-GE AND 7M-GTE

Voltage at ECU Wiring Connectors (Cont'd)

Terminals	STD Voltage	Condition	
THA – E2	1 – 3	Ignition SW ON	Intake air temperature 20°C (68°F)
THW – E2	0.1 – 1.0	Ignition SW ON	Coolant temperature 80°C (176°F)
No. 10 – E01 No. 20 – No. 30 – E01	10 – 14	Ignition SW ON	
STA – E1	6 – 14	Cranking	
ISC1 ~ ISC4 – E1	9 – 14	Ignition SW ON	
IGT – E1	0.7 – 1.0	Idling	
*2 IGDA, IGDB – E1	1 – 3	Idling	
W – E1	8 – 14	No trouble ("CHECK ENGINE" warning light off) and engine running.	
A/C – E1	10 – 14	Air conditioning ON	
TE1 – E1	4 – 6	Ignition SW ON	Check connector terminals TE1 – E1 not connect
	0		Check connector terminals TE1 – E1 connect
NSW (A/T) – E1	0	Ignition SW ON	Shift position P or N range
	10 – 14		Ex. P or N range
N/C (M/T) – E1	0	Ignition SW ON	Clutch pedal not depressed
	10 – 14		Clutch pedal depressed
*3 DFG – E1	10 – 14	Ignition SW ON	Defogger SW OFF
	0		Defogger SW ON
*3 LP – E1	10 – 14		Headlight SW OFF
	0		Headlight SW ON

*1 7M-GE only
*2 7M-GTE only
*3 7M-GTE (M/T) only

TROUBLESHOOTING PROCEDURES—VAN 4Y-EC

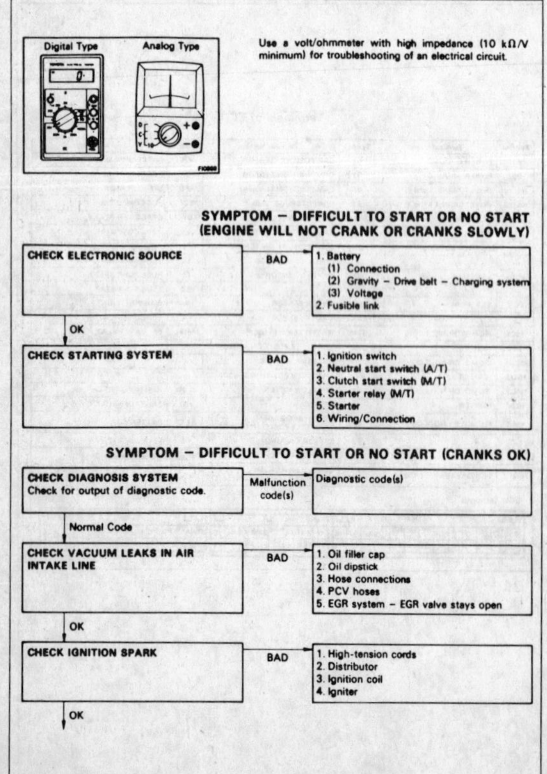
Digital Type
Analog Type
Use a volt/ohmmeter with high impedance (10 kΩ/V minimum) for troubleshooting of an electrical circuit.
SYMPTOM – DIFFICULT TO START OR NO START (ENGINE WILL NOT CRANK OR CRANKS SLOWLY)
CHECK ELECTRONIC SOURCE
BAD
1. Battery
(1) Connection
(2) Gravity – Drive belt – Charging system
(3) Voltage
2. Fusible link
OK
CHECK STARTING SYSTEM
BAD
1. Ignition switch
2. Neutral start switch (A/T)
3. Clutch start switch (M/T)
4. Starter relay (M/T)
5. Starter
6. Wiring/Connection
SYMPTOM – DIFFICULT TO START OR NO START (CRANKS OK)
CHECK DIAGNOSIS SYSTEM
Check for output of diagnostic code.
Malfunction code(s)
Diagnostic code(s)
Normal Code
CHECK VACUUM LEAKS IN AIR INTAKE LINE
BAD
1. Oil filler cap
2. Oil dipstick
3. Hose connections
4. PCV hoses
5. EGR system – EGR valve stays open
OK
CHECK IGNITION SPARK
BAD
1. High-tension cords
2. Distributor
3. Ignition coil
4. Igniter
OK

OK

CHECK SPARK PLUGS
Maximum: 1.3 mm (0.051 in.)
NOTE: Check compression pressure if necessary.

NO →
1. Spark plugs
2. Compression pressure
Minimum: 9.0 kg/cm² (128 psi, 883 kPa) at 250 rpm

NO (All Plugs WET) →
1. Injector(s) – Short circuited
2. Injector wiring between solenoid resistor and ECU short circuited
3. Cold start injector – Leakage
4. Cold start injector time switch
5. Injector – Leakage

OK

CHECK FUEL SUPPLY TO INJECTOR
1. Fuel in tank
2. Fuel pressure in fuel line
(1) Connect terminals of fuel pump check connector.
(2) Fuel pressure at fuel hose of fuel filter can be felt.

BAD →
1. Fuel line – Leakage – Deformation
2. Fuses
3. Circuit opening relay
4. Fuel pump
5. Fuel filter
6. Fuel pressure regulator

OK

CHECK FUEL PUMP SWITCH IN AIR FLOW METER
Check continuity between terminals FC and E1 while measuring plate of air flow meter is open.

BAD → Air flow meter

OK

CHECK IGNITION TIMING
1. Connect terminals (T—E1) of check engine connector.
2. Check ignition timing.
Standard: 12° BTDC @ idle

NO → Ignition system

OK

CHECK AIR VALVE
1. Disconnect air hose from air valve.
2. Check that the air valve opens when cold.
Air valve temp.: Below 20°C (68°F)

NO →
1. Air valve
2. Water hoses
3. Air hoses

OK

CHECK EFI ELECTRONIC CIRCUIT
Use VOLT/OHMMETER

NO →
1. Wiring connection
2. Power to ECU
(1) Fusible links
(2) Fuses
(3) EFI main relay
3. Air flow meter
4. Water temp. sensor
5. Injection signal circuit
(1) Injector wiring
(2) Solenoid resistor
(3) ECU

TROUBLESHOOTING PROCEDURES—VAN 4Y-EC

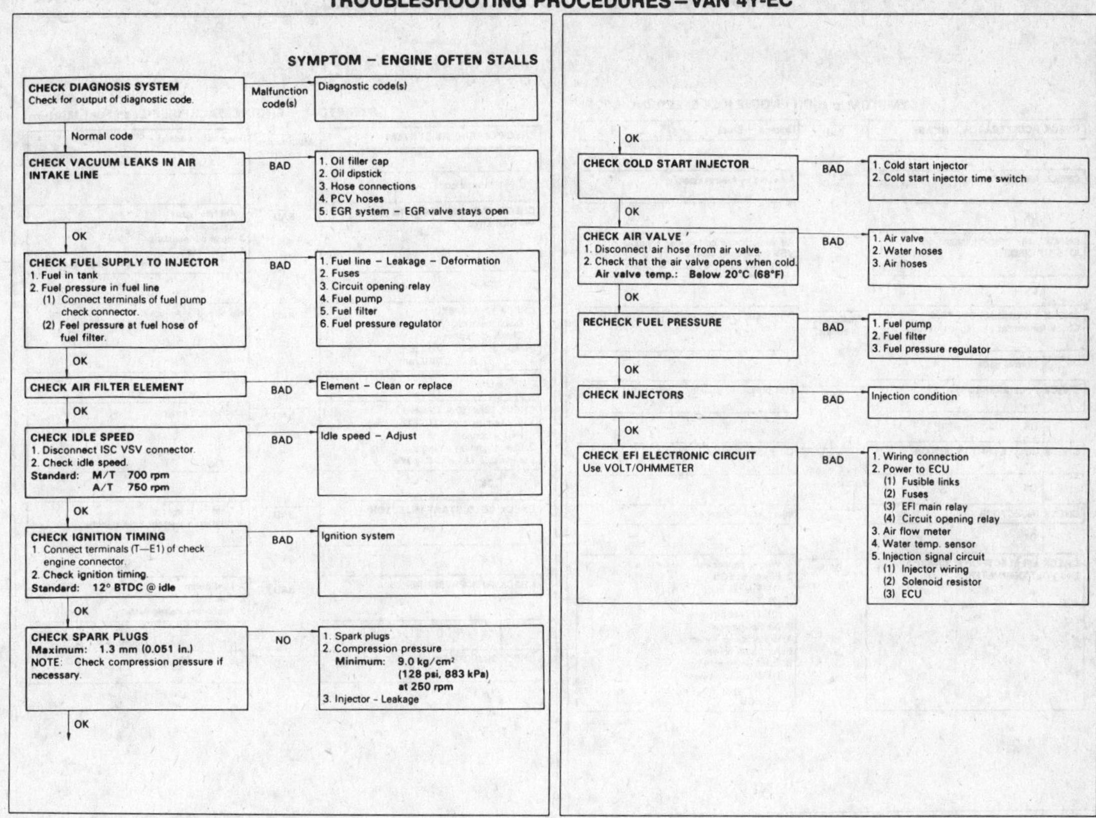
SYMPTOM – ENGINE OFTEN STALLS
CHECK DIAGNOSIS SYSTEM
Check for output of diagnostic code.
Malfunction code(s)
Diagnostic code(s)
Normal code
CHECK VACUUM LEAKS IN AIR INTAKE LINE
BAD
1. Oil filler cap
2. Oil dipstick
3. Hose connections
4. PCV hoses
5. EGR system – EGR valve stays open
OK
CHECK FUEL SUPPLY TO INJECTOR
1. Fuel in tank
2. Fuel pressure in fuel line
(1) Connect terminals of fuel pump check connector.
(2) Feel pressure at fuel hose of fuel filter.
BAD
1. Fuel line – Leakage – Deformation
2. Fuses
3. Circuit opening relay
4. Fuel pump
5. Fuel filter
6. Fuel pressure regulator
OK
CHECK AIR FILTER ELEMENT
BAD
Element – Clean or replace
OK
CHECK IDLE SPEED
1. Disconnect ISC VSV connector.
2. Check idle speed.
Standard: M/T 700 rpm
A/T 750 rpm
BAD
Idle speed – Adjust
OK
CHECK IGNITION TIMING
1. Connect terminals (T—E1) of check engine connector.
2. Check ignition timing.
Standard: 12° BTDC @ idle
BAD
Ignition system
OK
CHECK SPARK PLUGS
Maximum: 1.3 mm (0.051 in.)
NOTE: Check compression pressure if necessary.
NO
1. Spark plugs
2. Compression pressure
Minimum: 9.0 kg/cm²
(128 psi, 883 kPa)
at 250 rpm
3. Injector – Leakage
OK
OK
CHECK COLD START INJECTOR
BAD
1. Cold start injector
2. Cold start injector time switch
OK
CHECK AIR VALVE
1. Disconnect air hose from air valve.
2. Check that the air valve opens when cold.
Air valve temp.: Below 20°C (68°F)
BAD
1. Air valve
2. Water hoses
3. Air hoses
OK
RECHECK FUEL PRESSURE
BAD
1. Fuel pump
2. Fuel filter
3. Fuel pressure regulator
OK
CHECK INJECTORS
BAD
Injection condition
OK
CHECK EFI ELECTRONIC CIRCUIT
Use VOLT/OHMMETER
BAD
1. Wiring connection
2. Power to ECU
(1) Fusible links
(2) Fuses
(3) EFI main relay
(4) Circuit opening relay
3. Air flow meter
4. Water temp. sensor
5. Injection signal circuit
(1) Injector wiring
(2) Solenoid resistor
(3) ECU

TROUBLESHOOTING PROCEDURES—VAN 4Y-EC

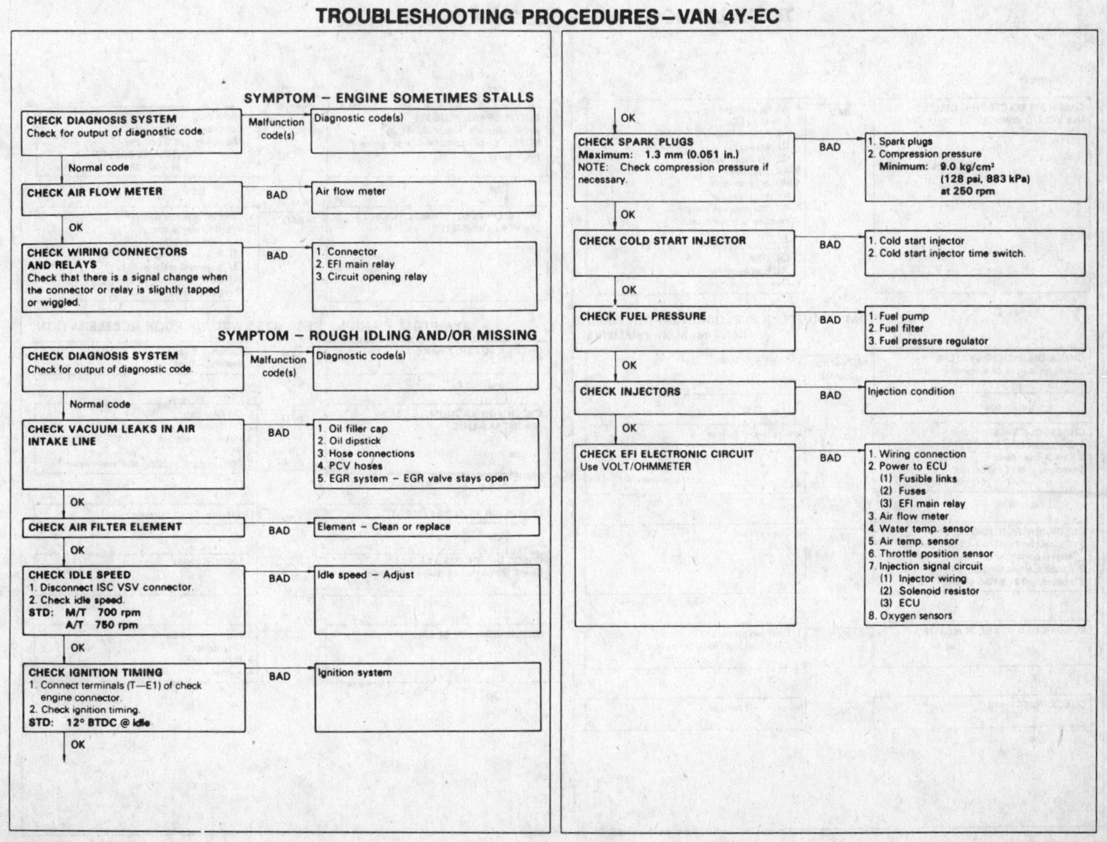
SYMPTOM – ENGINE SOMETIMES STALLS
CHECK DIAGNOSIS SYSTEM
Check for output of diagnostic code.
Malfunction code(s)
Diagnostic code(s)
Normal code
CHECK AIR FLOW METER
BAD
Air flow meter
OK
CHECK WIRING CONNECTORS AND RELAYS
Check that there is a signal change when the connector or relay is slightly tapped or wiggled.
BAD
1. Connector
2. EFI main relay
3. Circuit opening relay
SYMPTOM – ROUGH IDLING AND/OR MISSING
CHECK DIAGNOSIS SYSTEM
Check for output of diagnostic code.
Malfunction code(s)
Diagnostic code(s)
Normal code
CHECK VACUUM LEAKS IN AIR INTAKE LINE
BAD
1. Oil filler cap
2. Oil dipstick
3. Hose connections
4. PCV hoses
5. EGR system – EGR valve stays open
OK
CHECK AIR FILTER ELEMENT
BAD
Element – Clean or replace
OK
CHECK IDLE SPEED
1. Disconnect ISC VSV connector.
2. Check idle speed.
STD: M/T 700 rpm
A/T 750 rpm
BAD
Idle speed – Adjust
OK
CHECK IGNITION TIMING
1. Connect terminals (T—E1) of check engine connector.
2. Check ignition timing.
STD: 12° BTDC @ idle
BAD
Ignition system
OK
OK
CHECK SPARK PLUGS
Maximum: 1.3 mm (0.051 in.)
NOTE: Check compression pressure if necessary.
BAD
1. Spark plugs
2. Compression pressure
Minimum: 9.0 kg/cm²
(128 psi, 883 kPa)
at 250 rpm
OK
CHECK COLD START INJECTOR
BAD
1. Cold start injector
2. Cold start injector time switch.
OK
CHECK FUEL PRESSURE
BAD
1. Fuel pump
2. Fuel filter
3. Fuel pressure regulator
OK
CHECK INJECTORS
BAD
Injection condition
OK
CHECK EFI ELECTRONIC CIRCUIT
Use VOLT/OHMMETER
BAD
1. Wiring connection
2. Power to ECU
(1) Fusible links
(2) Fuses
(3) EFI main relay
3. Air flow meter
4. Water temp. sensor
5. Air temp. sensor
6. Throttle position sensor
7. Injection signal circuit
(1) Injector wiring
(2) Solenoid resistor
(3) ECU
8. Oxygen sensors

TROUBLESHOOTING PROCEDURES—VAN 4Y-EC

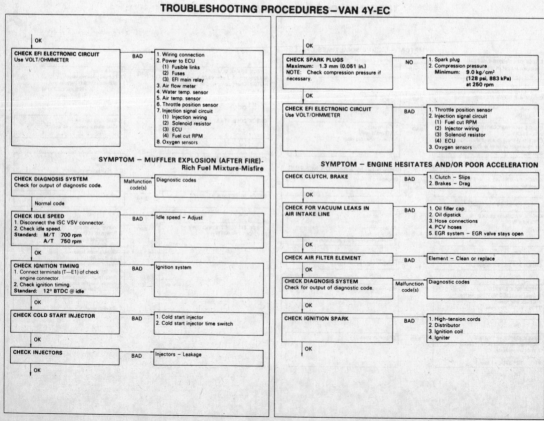

SYMPTOM – HIGH ENGINE IDLE SPEED (NO DROP)

- **CHECK ACCELERATOR LINKAGE** — BAD → Linkage – Stuck
  - OK
- **CHECK AIR VALVE** — BAD → Air valve – Always open
  - OK
- **CHECK AIR CONDITIONER IDLE-UP CIRCUIT** — BAD → Air valve for air conditioner – Leakage; VSV for A/C – Leakage; Water temperature switch – Short
  - OK
- **CHECK DIAGNOSIS SYSTEM** Check for output of diagnostic code. — Malfunction code(s) → Diagnostic code(s)
  - Normal code
- **CHECK FUEL PRESSURE** — BAD → Fuel pressure regulator – High pressure
  - OK
- **CHECK COLD START INJECTOR** — BAD → Cold start injector – Leakage
  - OK
- **CHECK INJECTORS** — BAD → Injectors – Leakage
  - OK
- **CHECK EFI ELECTRONIC CIRCUIT** Use VOLT/OHMMETER — BAD →
  1. Wiring connection
  2. Power to ECU
     (1) Fusible links
     (2) Fuses
     (3) EFI main relay
  3. Air flow meter
  4. Water temp. sensor
  5. Air temp. sensor
  6. Injection signal circuit
     (1) Injector wiring
     (2) Solenoid resistor
     (3) ECU

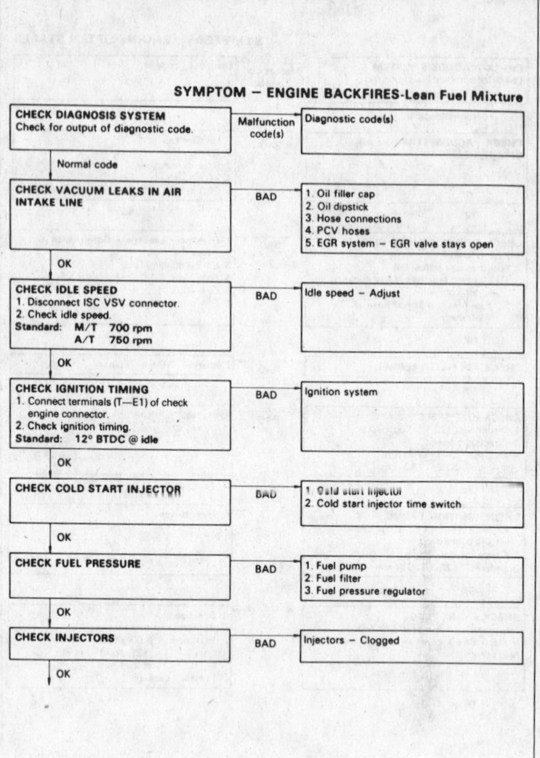
SYMPTOM – ENGINE BACKFIRES-Lean Fuel Mixture
CHECK DIAGNOSIS SYSTEM
Check for output of diagnostic code.
Malfunction code(s)
Diagnostic code(s)
Normal code
CHECK VACUUM LEAKS IN AIR INTAKE LINE
BAD
1. Oil filler cap
2. Oil dipstick
3. Hose connections
4. PCV hoses
5. EGR system – EGR valve stays open
OK
CHECK IDLE SPEED
1. Disconnect ISC VSV connector.
2. Check idle speed.
Standard: M/T 700 rpm
A/T 750 rpm
BAD
Idle speed – Adjust
OK
CHECK IGNITION TIMING
1. Connect terminals (T—E1) of check engine connector.
2. Check ignition timing.
Standard: 12° BTDC @ idle
BAD
Ignition system
OK
CHECK COLD START INJECTOR
BAD
1. Cold start injector
2. Cold start injector time switch
OK
CHECK FUEL PRESSURE
BAD
1. Fuel pump
2. Fuel filter
3. Fuel pressure regulator
OK
CHECK INJECTORS
BAD
Injectors – Clogged
OK

TROUBLESHOOTING PROCEDURES—VAN 4Y-EC

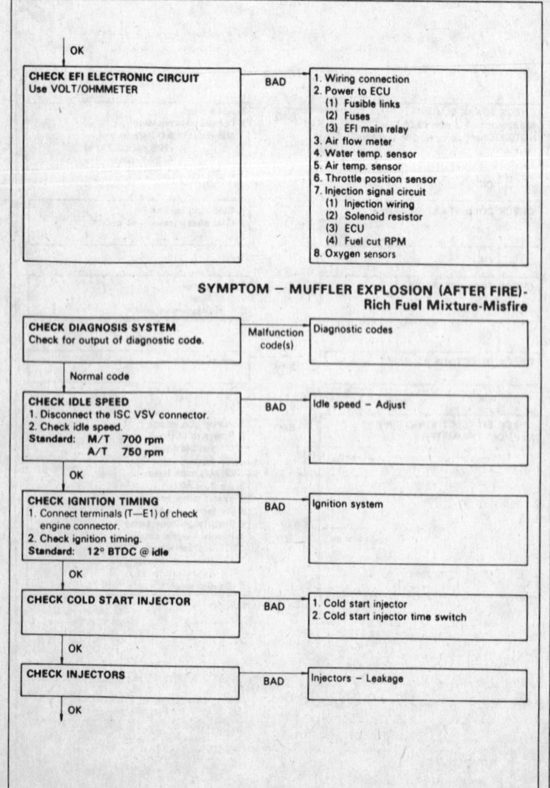
OK
CHECK EFI ELECTRONIC CIRCUIT
Use VOLT/OHMMETER
BAD
1. Wiring connection
2. Power to ECU
(1) Fusible links
(2) Fuses
(3) EFI main relay
3. Air flow meter
4. Water temp. sensor
5. Air temp. sensor
6. Throttle position sensor
7. Injection signal circuit
(1) Injection wiring
(2) Solenoid resistor
(3) ECU
(4) Fuel cut RPM
8. Oxygen sensors
SYMPTOM – MUFFLER EXPLOSION (AFTER FIRE)-Rich Fuel Mixture-Misfire
CHECK DIAGNOSIS SYSTEM
Check for output of diagnostic code.
Malfunction code(s)
Diagnostic codes
Normal code
CHECK IDLE SPEED
1. Disconnect the ISC VSV connector.
2. Check idle speed.
Standard: M/T 700 rpm
A/T 750 rpm
BAD
Idle speed – Adjust
OK
CHECK IGNITION TIMING
1. Connect terminals (T—E1) of check engine connector.
2. Check ignition timing.
Standard: 12° BTDC @ idle
BAD
Ignition system
OK
CHECK COLD START INJECTOR
BAD
1. Cold start injector
2. Cold start injector time switch
OK
CHECK INJECTORS
BAD
Injectors – Leakage
OK

- OK
- **CHECK SPARK PLUGS** Maximum: 1.3 mm (0.051 in.) NOTE: Check compression pressure if necessary. — NO →
  1. Spark plug
  2. Compression pressure Minimum: 9.0 kg/cm² (128 psi, 883 kPa) at 250 rpm
  - OK
- **CHECK EFI ELECTRONIC CIRCUIT** Use VOLT/OHMMETER — BAD →
  1. Throttle position sensor
  2. Injection signal circuit
     (1) Fuel cut RPM
     (2) Injector wiring
     (3) Solenoid resistor
     (4) ECU
  3. Oxygen sensors

SYMPTOM – ENGINE HESITATES AND/OR POOR ACCELERATION

- **CHECK CLUTCH, BRAKE** — BAD → 1. Clutch – Slips; 2. Brakes – Drag
  - OK
- **CHECK FOR VACUUM LEAKS IN AIR INTAKE LINE** — BAD →
  1. Oil filler cap
  2. Oil dipstick
  3. Hose connections
  4. PCV hoses
  5. EGR system – EGR valve stays open
  - OK
- **CHECK AIR FILTER ELEMENT** — BAD → Element – Clean or replace
  - OK
- **CHECK DIAGNOSIS SYSTEM** Check for output of diagnostic code. — Malfunction code(s) → Diagnostic codes
  - OK
- **CHECK IGNITION SPARK** — BAD →
  1. High-tension cords
  2. Distributor
  3. Ignition coil
  4. Igniter
  - OK

TROUBLESHOOTING PROCEDURES—VAN 4Y-EC

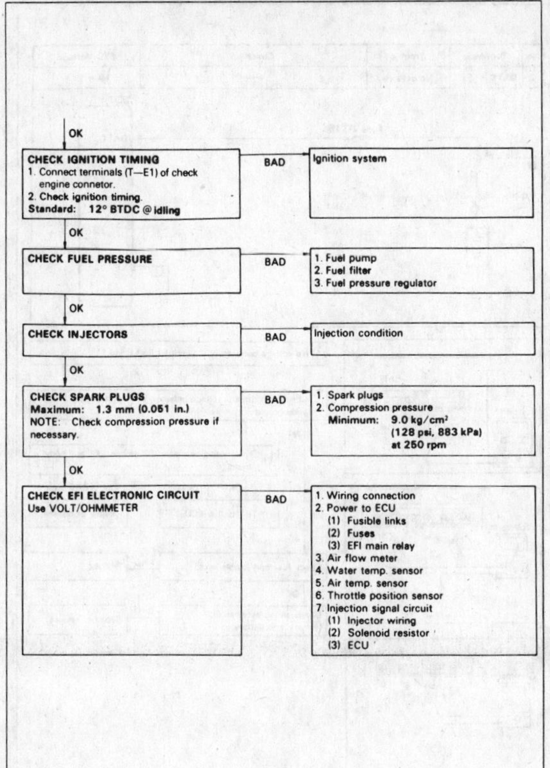
OK
CHECK IGNITION TIMING
1. Connect terminals (T—E1) of check engine connector.
2. Check ignition timing.
Standard: 12° BTDC @ idling
BAD
Ignition system
OK
CHECK FUEL PRESSURE
BAD
1. Fuel pump
2. Fuel filter
3. Fuel pressure regulator
OK
CHECK INJECTORS
BAD
Injection condition
OK
CHECK SPARK PLUGS
Maximum: 1.3 mm (0.051 in.)
NOTE: Check compression pressure if necessary.
BAD
1. Spark plugs
2. Compression pressure
Minimum: 9.0 kg/cm²
(128 psi, 883 kPa)
at 250 rpm
OK
CHECK EFI ELECTRONIC CIRCUIT
Use VOLT/OHMMETER
BAD
1. Wiring connection
2. Power to ECU
(1) Fusible links
(2) Fuses
(3) EFI main relay
3. Air flow meter
4. Water temp. sensor
5. Air temp. sensor
6. Throttle position sensor
7. Injection signal circuit
(1) Injector wiring
(2) Solenoid resistor
(3) ECU

DIAGNOSTIC CODES—VAN 4Y-EC

Code No.	Number of Check engine blinks	System	Diagnosis	Trouble area
–		Normal	This appears when none of the other codes are identified.	–
12		RPM Signal	No "NE" signal to ECU within 2 seconds after engine has been cranked.	• Distributor circuit • Distributor • Stater signal circuit • Igniter circuit • Igniter • ECU
13		RPM Signal	No "NE", signal to ECU when engine speed is above 1,500 rpm	• Distributor circuit • Distributor • Igniter circuit • Igniter • ECU
14		Ignition Signal	No "IGF" signal to ECU 4-5 times in succession.	• Igniter and ignition coil circuit • Igniter and ignition coil • ECU
21		Oxygen Sensor Signal	Detection of oxygen sensor detrioration.	• Oxygen sensor circuit • Oxygen sensor • ECU
		Oxygen Sensor Heater Signal	Open or short circuit in oxygen sensor heater signal	• Oxygen sensor heater circuit • Oxygen sensor heater • ECU
22		Water Temp. Sensor Signal	Open or short circuit in water temp. sensor signal.	• Water temp. sensor circuit • Water temp. sensor • ECU
24		Intake Air Temp. Sensor Signal	Open or short circuit in intake air temp. sensor signal.	• Intake air temp sensor circuit • Intake air temp. sensor • ECU
25		Air-fuel Ratio Lean Malfunction	(1) When air-fuel ratio feedback compensation value or adaptive control value continues at the upper (lean) or lower (rich) limit renewed for a certain period of time. (2) When air-fuel ratio feedback compensation value or addptive control value feedback frequency is abnromally high during feedback condition.	• Injector circuit • Injector • Fuel line pressure • Air flow meter • Air intake system • Oxygen sensor circuit • Oxygen sensor • Ignition system • ECU
26		Air-fuel Ratio Rich Malfunction	NOTE: For condition (2), since neither a lean (Code No. 25) nor a rich (Code No. 26) diagnosis displayed consecutively.	• Injector circuit • Injector • Fuel line pressure • Air flow meter • Cold start injector • ECU

DIAGNOSTIC CODES—VAN 4Y-EC

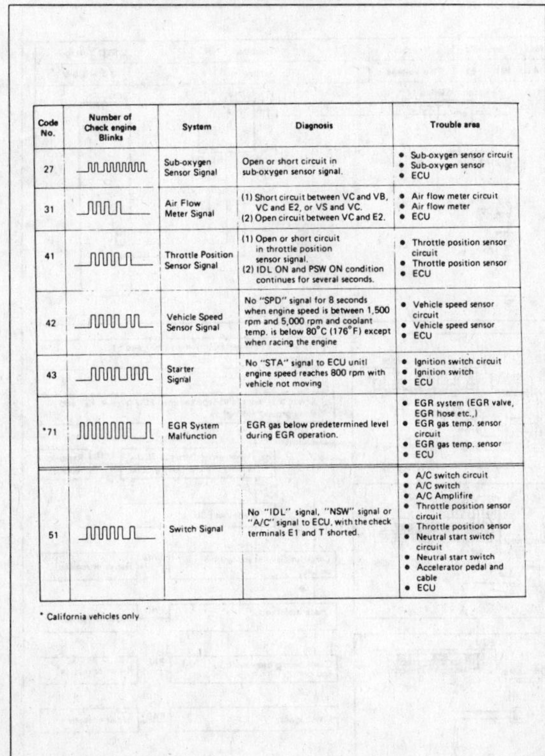

Code No.	Number of Check engine Blinks	System	Diagnosis	Trouble area
27		Sub-oxygen Sensor Signal	Open or short circuit in sub-oxygen sensor signal.	• Sub-oxygen sensor circuit • Sub-oxygen sensor • ECU
31		Air Flow Meter Signal	(1) Short circuit between VC and VB, VC and E2, or VS and VC. (2) Open circuit between VC and E2.	• Air flow meter circuit • Air flow meter • ECU
41		Throttle Position Sensor Signal	(1) Open or short circuit in throttle position sensor signal. (2) IDL ON and PSW ON condition continues for several seconds.	• Throttle position sensor circuit • Throttle position sensor • ECU
42		Vehicle Speed Sensor Signal	No "SPD" signal for 8 seconds when engine speed is between 1,500 rpm and 5,000 rpm and coolant temp. is below 80°C (176°F) except when racing the engine	• Vehicle speed sensor circuit • Vehicle speed sensor • ECU
43		Starter Signal	No "STA" signal to ECU unitl engine speed reaches 800 rpm with vehicle not moving	• Ignition switch circuit • Ignition switch • ECU
*71		EGR System Malfunction	EGR gas below predetermined level during EGR operation.	• EGR system (EGR valve, EGR hose etc.,) • EGR gas temp. sensor circuit • EGR gas temp. sensor • ECU
51		Switch Signal	No "IDL" signal, "NSW" signal or "A/C" signal to ECU, with the check terminals E1 and T shorted.	• A/C switch circuit • A/C switch • A/C Amplifire • Throttle position sensor circuit • Throttle position sensor • Neutral start switch circuit • Neutral start switch • Accelerator pedal and cable • ECU

* California vehicles only

DIAGNOSTIC CIRCUITS TEST—VAN 4Y-EC

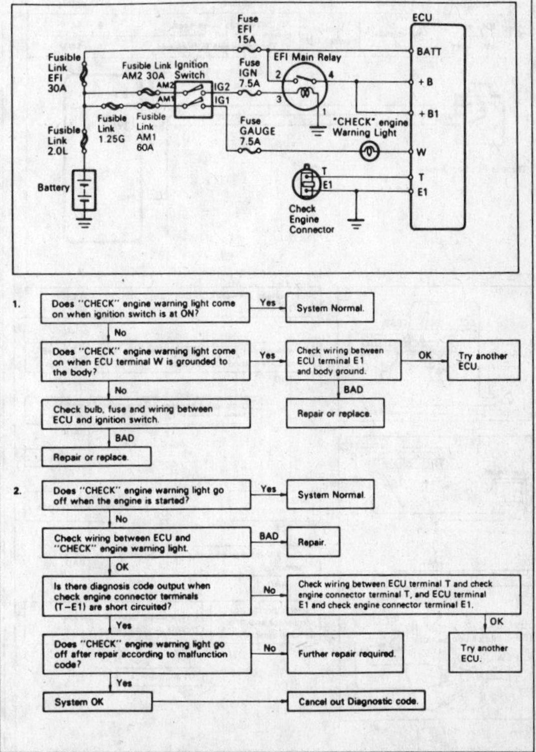
INSPECTION OF DIAGNOSIS CIRCUIT
Fuse EFI 15A
ECU
Fusible Link EFI 30A
Fusible Link AM2 30A
Ignition Switch
AM2
AM1
IG2
IG1
Fuse IGN 7.5A
EFI Main Relay
2
4
3
1
BATT
+B
+B1
Fusible Link 2.0L
Fusible Link 1.25G
Fusible Link AM1 60A
Fuse GAUGE 7.5A
"CHECK" engine Warning Light
W
T
E1
Battery
Check Engine Connector
1.
Does "CHECK" engine warning light come on when ignition switch is at ON?
Yes
System Normal.
No
Does "CHECK" engine warning light come on when ECU terminal W is grounded to the body?
Yes
Check wiring between ECU terminal E1 and body ground.
OK
Try another ECU.
BAD
Repair or replace.
No
Check bulb, fuse and wiring between ECU and ignition switch.
BAD
Repair or replace.
2.
Does "CHECK" engine warning light go off when the engine is started?
Yes
System Normal.
No
Check wiring between ECU and "CHECK" engine warning light.
BAD
Repair.
OK
Is there diagnosis code output when check engine connector terminals (T–E1) are short circuited?
No
Check wiring between ECU terminal T and check engine connector terminal T, and ECU terminal E1 and check engine connector terminal E1.
OK
Try another ECU.
Yes
Does "CHECK" engine warning light go off after repair according to malfunction code?
No
Further repair required.
Yes
System OK
Cancel out Diagnostic code.

TROUBLESHOOTING CHART 1—VAN 4Y-EC

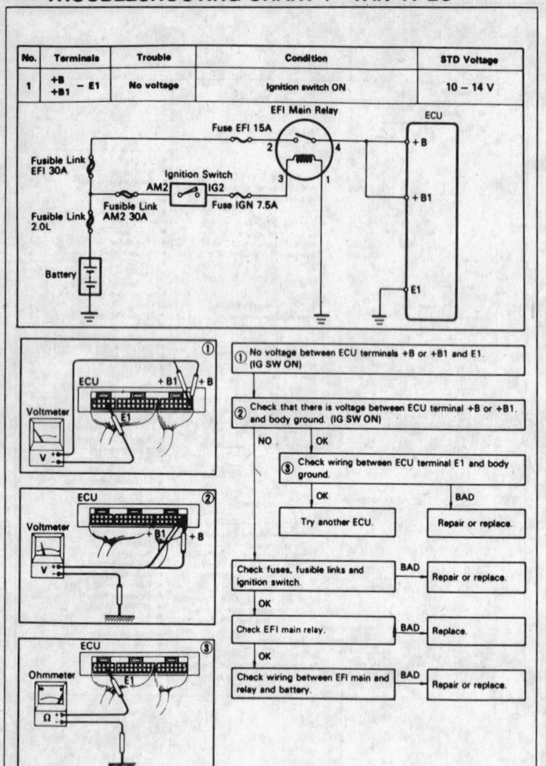

No.	Terminals	Trouble	Condition	STD Voltage
1	+B +B1 – E1	No voltage	Ignition switch ON	10 – 14 V

EFI Main Relay
Fuse EFI 15A
ECU
+B
+B1
E1
Fusible Link EFI 30A
Ignition Switch
AM2
IG2
Fuse IGN 7.5A
Fusible Link AM2 30A
Fusible Link 2.0L
Battery
Voltmeter
Ohmmeter
① No voltage between ECU terminals +B or +B1 and E1. (IG SW ON)
② Check that there is voltage between ECU terminal +B or +B1 and body ground. (IG SW ON)
NO
OK
③ Check wiring between ECU terminal E1 and body ground.
OK
BAD
Try another ECU.
Repair or replace.
Check fuses, fusible links and ignition switch.
BAD
Repair or replace.
OK
Check EFI main relay.
BAD
Replace.
OK
Check wiring between EFI main and relay and battery.
BAD
Repair or replace.

TROUBLESHOOTING CHART 2—VAN 4Y-EC

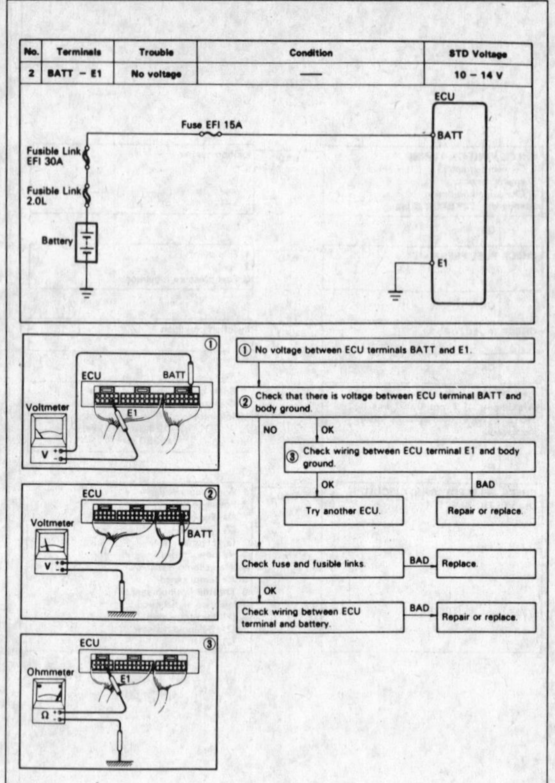

No.	Terminals	Trouble	Condition	STD Voltage
2	BATT – E1	No voltage	—	10 – 14 V

ECU
Fuse EFI 15A
BATT
Fusible Link EFI 30A
Fusible Link 2.0L
Battery
E1
Voltmeter
Ohmmeter
① No voltage between ECU terminals BATT and E1.
② Check that there is voltage between ECU terminal BATT and body ground.
NO
OK
③ Check wiring between ECU terminal E1 and body ground.
OK
BAD
Try another ECU.
Repair or replace.
Check fuse and fusible links.
BAD
Replace.
OK
Check wiring between ECU terminal and battery.
BAD
Repair or replace.

TROUBLESHOOTING CHART 3—VAN 4Y-EC

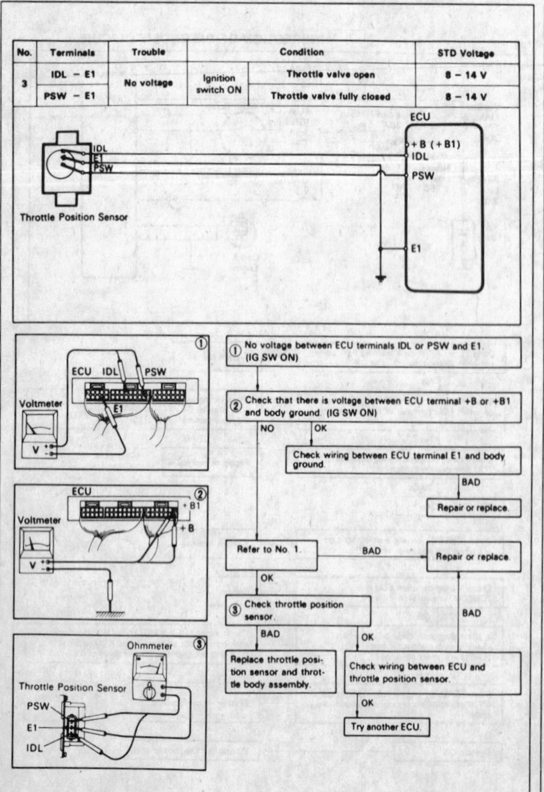

No.	Terminals	Trouble	Condition		STD Voltage
3	IDL – E1	No voltage	Ignition switch ON	Throttle valve open	8 – 14 V
	PSW – E1			Throttle valve fully closed	8 – 14 V

ECU
+B (+B1)
IDL
PSW
E1
Throttle Position Sensor
Voltmeter
Ohmmeter
① No voltage between ECU terminals IDL or PSW and E1. (IG SW ON)
② Check that there is voltage between ECU terminal +B or +B1 and body ground. (IG SW ON)
NO
OK
Check wiring between ECU terminal E1 and body ground.
BAD
Repair or replace.
Refer to No. 1.
BAD
Repair or replace.
OK
③ Check throttle position sensor.
BAD
BAD
OK
Replace throttle position sensor and throttle body assembly.
Check wiring between ECU and throttle position sensor.
OK
Try another ECU.

TROUBLESHOOTING CHART 4—VAN 4Y-EC

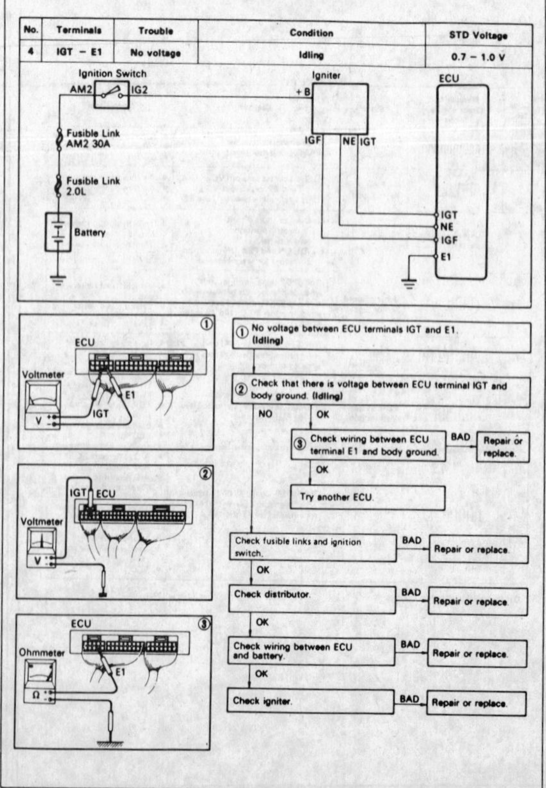

No.	Terminals	Trouble	Condition	STD Voltage
4	IGT – E1	No voltage	Idling	0.7 – 1.0 V

Ignition Switch
AM2
IG2
Igniter
ECU
+B
IGF
NE
IGT
Fusible Link AM2 30A
Fusible Link 2.0L
Battery
E1
Voltmeter
Ohmmeter
① No voltage between ECU terminals IGT and E1. (Idling)
② Check that there is voltage between ECU terminal IGT and body ground. (Idling)
NO
OK
③ Check wiring between ECU terminal E1 and body ground.
BAD
Repair or replace.
OK
Try another ECU.
Check fusible links and ignition switch.
BAD
Repair or replace.
OK
Check distributor.
BAD
Repair or replace.
OK
Check wiring between ECU and battery.
BAD
Repair or replace.
OK
Check igniter.
BAD
Repair or replace.

TROUBLESHOOTING CHART 5—VAN 4Y-EC

TROUBLESHOOTING CHART 6—VAN 4Y-EC

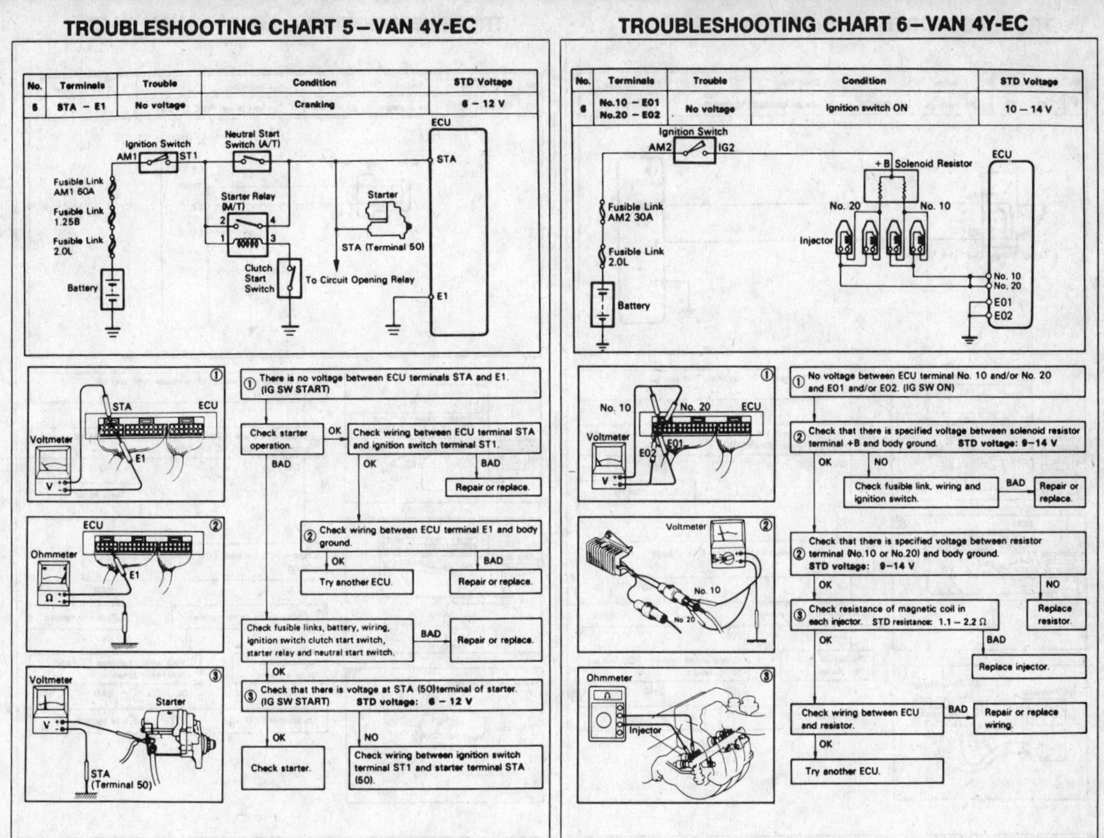

No.	Terminals	Trouble	Condition	STD Voltage
5	STA — E1	No voltage	Cranking	6 — 12 V

ECU
Ignition Switch
AM1
ST1
Neutral Start Switch (A/T)
STA
Fusible Link AM1 60A
Fusible Link 1.25B
Fusible Link 2.0L
Battery
Starter Relay (M/T)
Starter
STA (Terminal 50)
Clutch Start Switch
To Circuit Opening Relay
E1
Voltmeter
Ohmmeter
① There is no voltage between ECU terminals STA and E1. (IG SW START)
Check starter operation.
OK
Check wiring between ECU terminal STA and ignition switch terminal ST1.
BAD
OK
BAD
Repair or replace.
② Check wiring between ECU terminal E1 and body ground.
OK
BAD
Try another ECU.
Repair or replace.
Check fusible links, battery, wiring, ignition switch clutch start switch, starter relay and neutral start switch.
BAD
Repair or replace.
OK
③ Check that there is voltage at STA (50)terminal of starter. (IG SW START) STD voltage: 6 — 12 V
OK
NO
Check starter.
Check wiring between ignition switch terminal ST1 and starter terminal STA (50).
STA (Terminal 50)

No.	Terminals	Trouble	Condition	STD Voltage
6	No.10 — E01 No.20 — E02	No voltage	Ignition switch ON	10 — 14 V

Ignition Switch
AM2
IG2
+B Solenoid Resistor
ECU
Fusible Link AM2 30A
Fusible Link 2.0L
Battery
No. 20
No. 10
Injector
No. 10
No. 20
E01
E02
Voltmeter
Ohmmeter
① No voltage between ECU terminal No. 10 and/or No. 20 and E01 and/or E02. (IG SW ON)
② Check that there is specified voltage between solenoid resistor terminal +B and body ground. STD voltage: 9—14 V
OK
NO
Check fusible link, wiring and ignition switch.
BAD
Repair or replace.
② Check that there is specified voltage between resistor terminal (No.10 or No.20) and body ground. STD voltage: 9—14 V
OK
NO
Replace resistor.
③ Check resistance of magnetic coil in each injector. STD resistance: 1.1 — 2.2 Ω
OK
BAD
Replace injector.
Check wiring between ECU and resistor.
BAD
Repair or replace wiring.
OK
Try another ECU.

TROUBLESHOOTING CHART 7—VAN 4Y-EC

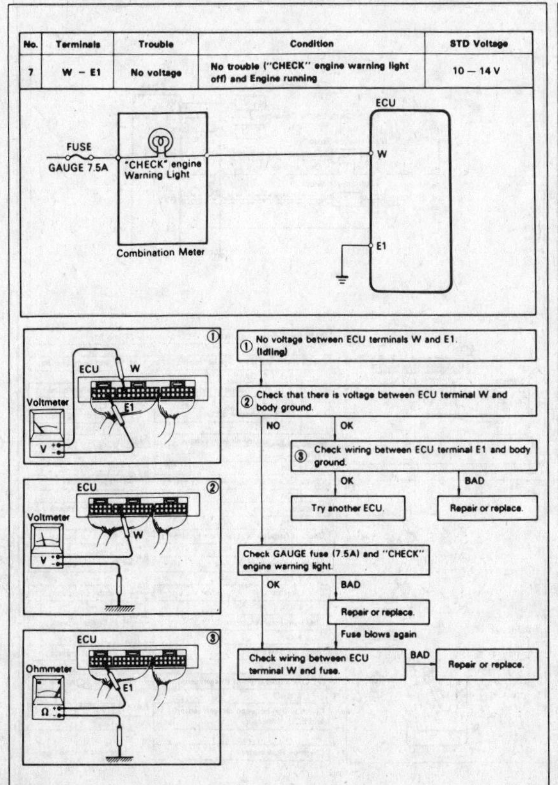

No.	Terminals	Trouble	Condition	STD Voltage
7	W — E1	No voltage	No trouble ("CHECK" engine warning light off) and Engine running	10 — 14 V

ECU
FUSE
GAUGE 7.5A
"CHECK" engine Warning Light
W
Combination Meter
E1
Voltmeter
Ohmmeter
① No voltage between ECU terminals W and E1. (Idling)
② Check that there is voltage between ECU terminal W and body ground.
NO
OK
③ Check wiring between ECU terminal E1 and body ground.
OK
BAD
Try another ECU.
Repair or replace.
Check GAUGE fuse (7.5A) and "CHECK" engine warning light.
OK
BAD
Repair or replace.
Fuse blows again
Check wiring between ECU terminal W and fuse.
BAD
Repair or replace.

TROUBLESHOOTING CHART 8—VAN 4Y-EC

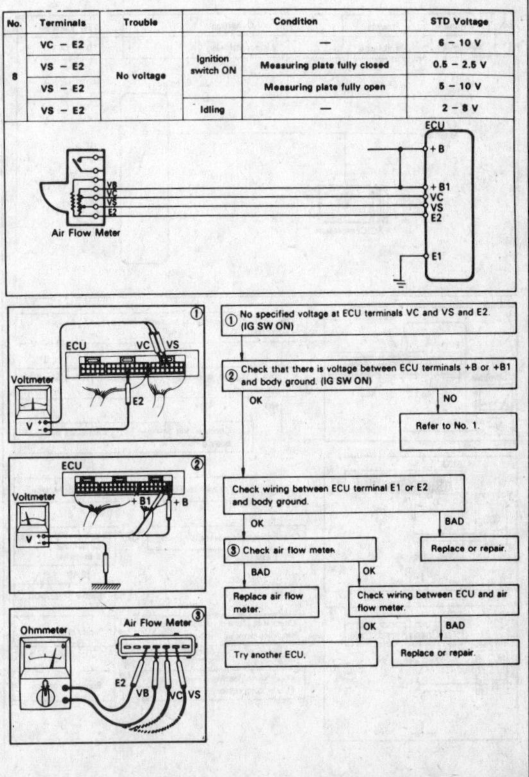

No.	Terminals	Trouble	Condition		STD Voltage
8	VC — E2	No voltage	Ignition switch ON	—	6 — 10 V
	VS — E2			Measuring plate fully closed	0.5 — 2.5 V
	VS — E2			Measuring plate fully open	5 — 10 V
	VS — E2		Idling	—	2 — 8 V

ECU
+B
+B1
VB
VC
VS
E2
E1
Air Flow Meter
Voltmeter
Ohmmeter
① No specified voltage at ECU terminals VC and VS and E2. (IG SW ON)
② Check that there is voltage between ECU terminals +B or +B1 and body ground. (IG SW ON)
OK
NO
Refer to No. 1.
Check wiring between ECU terminal E1 or E2 and body ground.
OK
BAD
③ Check air flow meter.
Replace or repair.
BAD
OK
Replace air flow meter.
Check wiring between ECU and air flow meter.
OK
BAD
Try another ECU.
Replace or repair.

TROUBLESHOOTING CHART 9—VAN 4Y-EC

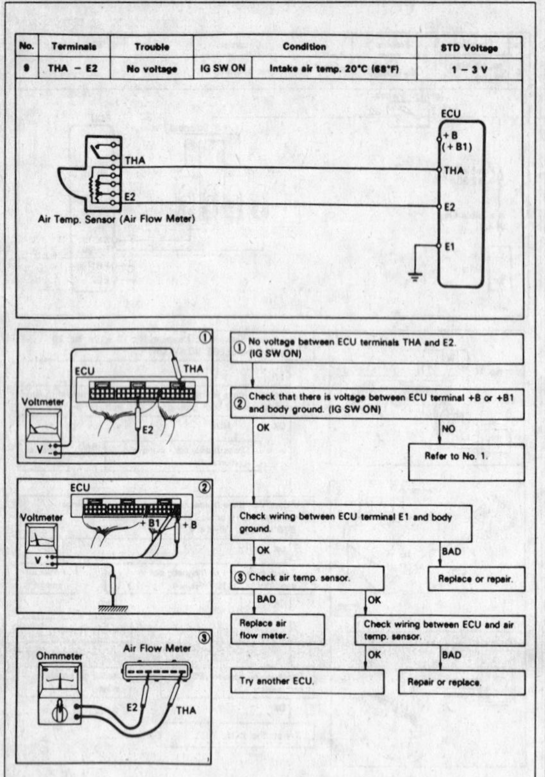

No.	Terminals	Trouble	Condition		STD Voltage
9	THA – E2	No voltage	IG SW ON	Intake air temp. 20°C (68°F)	1 – 3 V

ECU
+B (+B1)
THA
E2
E1
Air Temp. Sensor (Air Flow Meter)
Voltmeter
V
Ohmmeter
Air Flow Meter
① No voltage between ECU terminals THA and E2. (IG SW ON)
② Check that there is voltage between ECU terminal +B or +B1 and body ground. (IG SW ON)
OK
NO
Refer to No. 1.
Check wiring between ECU terminal E1 and body ground.
OK
BAD
③ Check air temp. sensor.
Replace or repair.
BAD
OK
Replace air flow meter.
Check wiring between ECU and air temp. sensor.
OK
BAD
Try another ECU.
Repair or replace.

TROUBLESHOOTING CHART 10—VAN 4Y-EC

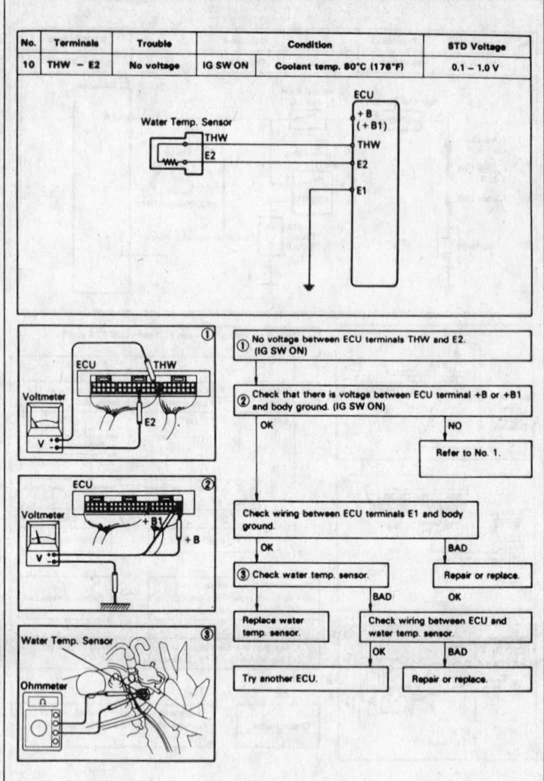

No.	Terminals	Trouble	Condition		STD Voltage
10	THW – E2	No voltage	IG SW ON	Coolant temp. 80°C (176°F)	0.1 – 1.0 V

ECU
Water Temp. Sensor
+B (+B1)
THW
E2
E1
Voltmeter
V
Ohmmeter
① No voltage between ECU terminals THW and E2. (IG SW ON)
② Check that there is voltage between ECU terminal +B or +B1 and body ground. (IG SW ON)
OK
NO
Refer to No. 1.
Check wiring between ECU terminals E1 and body ground.
OK
BAD
③ Check water temp. sensor.
Repair or replace.
BAD
OK
Replace water temp. sensor.
Check wiring between ECU and water temp. sensor.
OK
BAD
Try another ECU.
Repair or replace.

TROUBLESHOOTING CHART 11—VAN 4Y-EC

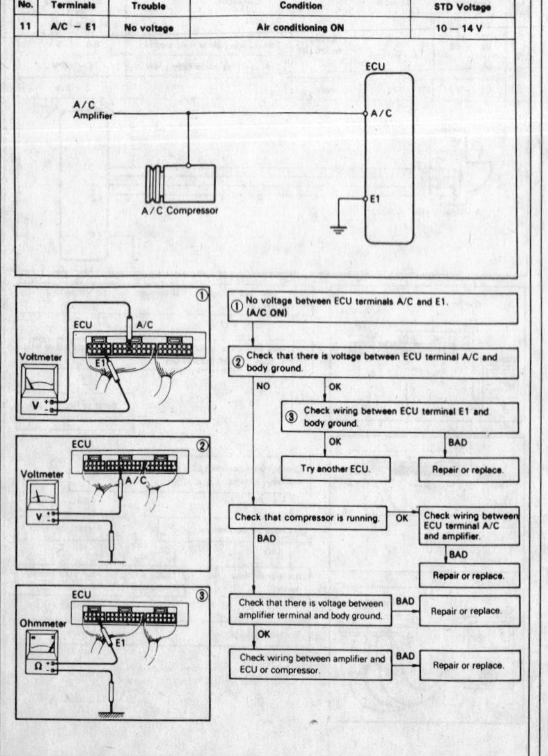

No.	Terminals	Trouble	Condition	STD Voltage
11	A/C – E1	No voltage	Air conditioning ON	10 – 14 V

ECU
A/C Amplifier
A/C
E1
A/C Compressor
Voltmeter
V
Ohmmeter
① No voltage between ECU terminals A/C and E1. (A/C ON)
② Check that there is voltage between ECU terminal A/C and body ground.
NO
OK
③ Check wiring between ECU terminal E1 and body ground.
OK
BAD
Try another ECU.
Repair or replace.
Check that compressor is running.
OK
Check wiring between ECU terminal A/C and amplifier.
BAD
BAD
Repair or replace.
Check that there is voltage between amplifier terminal and body ground.
BAD
Repair or replace.
OK
Check wiring between amplifier and ECU or compressor.
BAD
Repair or replace.

TROUBLESHOOTING OXYGEN SENSOR—VAN 4Y-EC

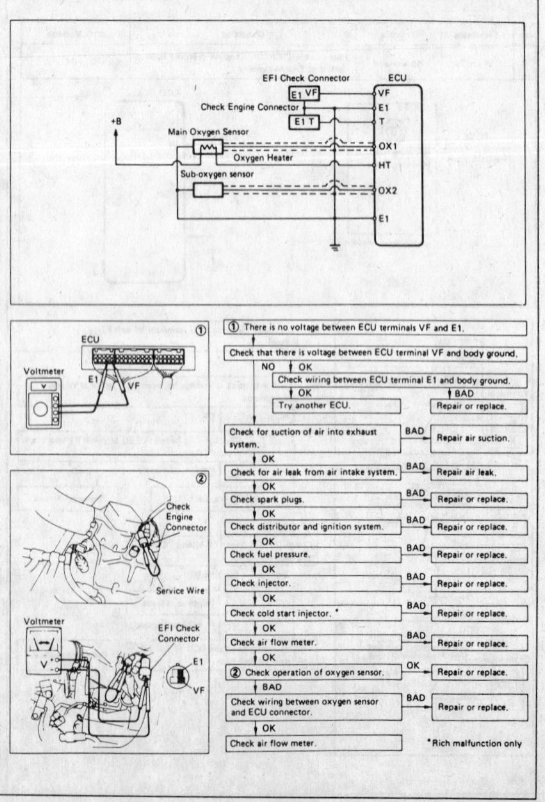

EFI Check Connector
ECU
E1 VF
Check Engine Connector
E1 T
+B
Main Oxygen Sensor
Oxygen Heater
Sub-oxygen sensor
VF
E1
T
OX1
HT
OX2
E1
Voltmeter
Check Engine Connector
Service Wire
EFI Check Connector
① There is no voltage between ECU terminals VF and E1.
Check that there is voltage between ECU terminal VF and body ground.
NO
OK
Check wiring between ECU terminal E1 and body ground.
OK
BAD
Try another ECU.
Repair or replace.
Check for suction of air into exhaust system.
BAD
Repair air suction.
OK
Check for air leak from air intake system.
BAD
Repair air leak.
OK
Check spark plugs.
BAD
Repair or replace.
OK
Check distributor and ignition system.
BAD
Repair or replace.
OK
Check fuel pressure.
BAD
Repair or replace.
OK
Check injector.
BAD
Repair or replace.
OK
Check cold start injector. *
BAD
Repair or replace.
OK
Check air flow meter.
BAD
Repair or replace.
OK
② Check operation of oxygen sensor.
OK
Repair or replace.
BAD
Check wiring between oxygen sensor and ECU connector.
BAD
Repair or replace.
OK
Check air flow meter.
*Rich malfunction only

TROUBLESHOOTING EGR TEMPERATURE SENSOR—VAN 4Y-EC

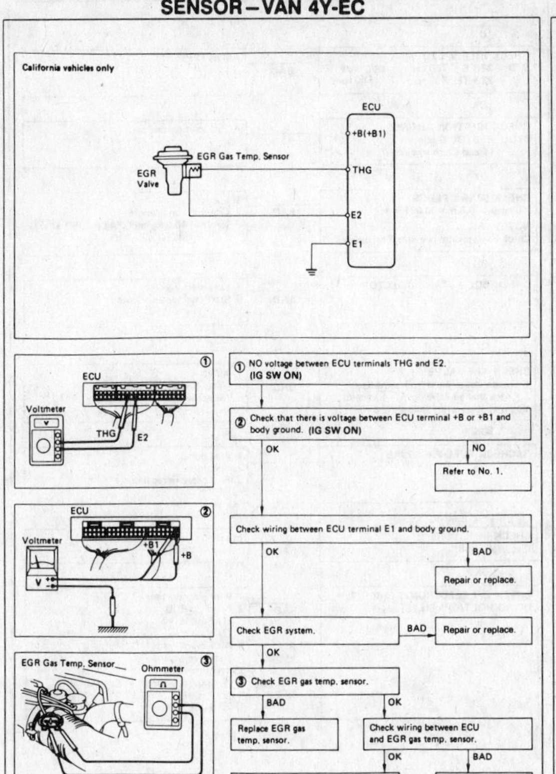
California vehicles only
ECU
+B(+B1)
EGR Gas Temp. Sensor
EGR Valve
THG
E2
E1
ECU
Voltmeter
THG
E2
① NO voltage between ECU terminals THG and E2. (IG SW ON)
② Check that there is voltage between ECU terminal +B or +B1 and body ground. (IG SW ON)
OK
NO
Refer to No. 1.
ECU
Voltmeter
+B1
+B
V
Check wiring between ECU terminal E1 and body ground.
OK
BAD
Repair or replace.
Check EGR system.
BAD
Repair or replace.
OK
EGR Gas Temp. Sensor
Ohmmeter
③ Check EGR gas temp. sensor.
BAD
OK
Replace EGR gas temp. sensor.
Check wiring between ECU and EGR gas temp. sensor.
OK
BAD
Try another ECU.
Repair or replace.

ECU TERMINAL IDENTIFICATION—VAN 4Y-EC ECU TERMINAL VOLTAGES

Symbol	Terminal Name	Symbol	Terminal Name
E01	ENGINE GROUND	IGF	IGNITER
E02	ENGINE GROUND	E2	SENSOR GROUND
No. 10	INJECTOR	OX2	SUB-OXYGEN SENSOR
No. 20	INJECTOR	OX1	MAIN OXYGEN SENSOR
STA	STARTER SWITCH	FPU	VSV (FUEL PRESSURE UP)
IGT	IGNITER	HT	OXYGEN SENSOR HEATER
STJ	START INJECTOR	PSW	THROTTLE POSITION SENSOR
E1	ENGINE GROUND	NE	ENGINE REVOLUTION SENSOR
NSW	NEUTRAL START SWITCH	THW	WATER TEMP. SENSOR
VF	CHECK CONNECTOR	VC	AIR FLOW METER
V-ISC	ISC VSV	E21	SENSOR GROUND
ACV	A/C VSV	VS	AIR FLOW METER
W	CHECK ENGINE WARNING LIGHT	THA	AIR TEMP. SENSOR
TSW	WATER TEMP. SWITCH	SPD	SPEED SENSOR
T	CHECK ENGINE CONNECTOR	BATT	BATTERY
*THG	EGR GAS TEMP. SENSOR	+B1	MAIN RELAY
IDL	THROTTLE POSITION SENSOR	+B	MAIN RELAY
A/C	A/C MAGNET CLUTCH	—	—

Terminals	Condition		Voltage (V)
+B +B1 – E1	Ignition switch ON		10 – 14
BATT – E1	——		10 – 14
IDL – E1	Ignition switch ON	Throttle valve open	8 – 14
PSW – E1	Ignition switch ON	Throttle valve closed	8 – 14
IGT – E1	Idling		0.7 – 1.0
STA – E1	Cranking		6 – 12
No. 10 No. 20 – E1	Ignition switch ON		10 – 14
W – E1	No trouble ("CHECK" engine warning light off) and engine running		10 – 14
VC – E2	Ignition switch ON		6 – 10
VS – E2 E21	Ignition switch ON	Measuring plate fully closed	0.5 – 2.5
		Measuring plate fully open	5 – 10
	Idling		2 – 8
THA – E2	Ignition switch ON	Intake temperature 20°C (68°F)	1 – 3
THW – E2	Ignition switch ON	Coolant temperature 80°C (176°F)	0.1 – 1.0
A/C – E1	Ignition switch ON	Air conditioning ON	10 – 14
T – E1	Ignition switch ON	Check engine connector (T—E1) not connect	10 – 14
		Check connector (T—E1) connect	0

ECU Terminals

E01	No 10	STA	STJ	NSW	V-ISC	W	T	IDL	IGF	OX2	—	HT	NE	—	—	VC	VS	THA	BATT	+B1
E02	No 20	IGT	E1	VF	ACV	TSW	*THG	A/C	E2	OX1	FPU	PSW	THW	—	—	E21	—	SPD	—	+B

* California vehicles only

TROUBLESHOOTING PROCEDURES—PICK-UP AND 4-RUNNER 22R-E AND 22R-TE

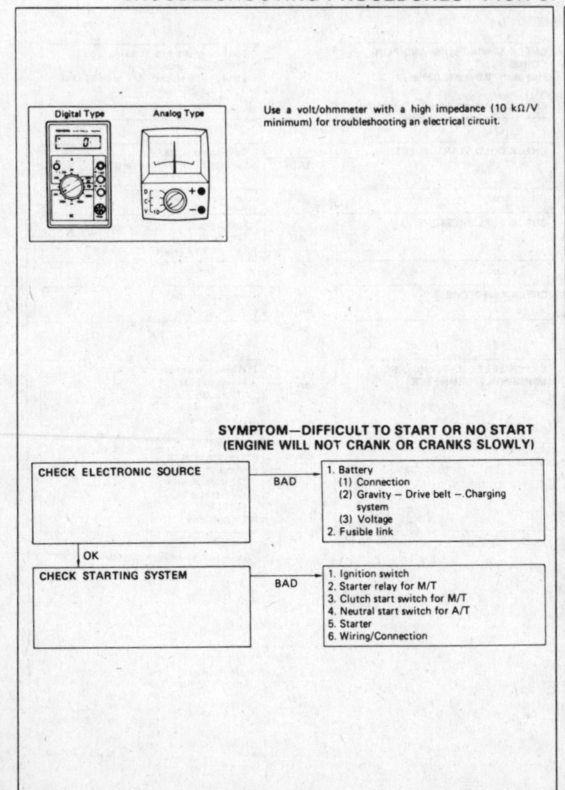
Digital Type
Analog Type
Use a volt/ohmmeter with a high impedance (10 kΩ/V minimum) for troubleshooting an electrical circuit.
SYMPTOM—DIFFICULT TO START OR NO START (ENGINE WILL NOT CRANK OR CRANKS SLOWLY)
CHECK ELECTRONIC SOURCE
BAD
1. Battery
(1) Connection
(2) Gravity – Drive belt – Charging system
(3) Voltage
2. Fusible link
OK
CHECK STARTING SYSTEM
BAD
1. Ignition switch
2. Starter relay for M/T
3. Clutch start switch for M/T
4. Neutral start switch for A/T
5. Starter
6. Wiring/Connection

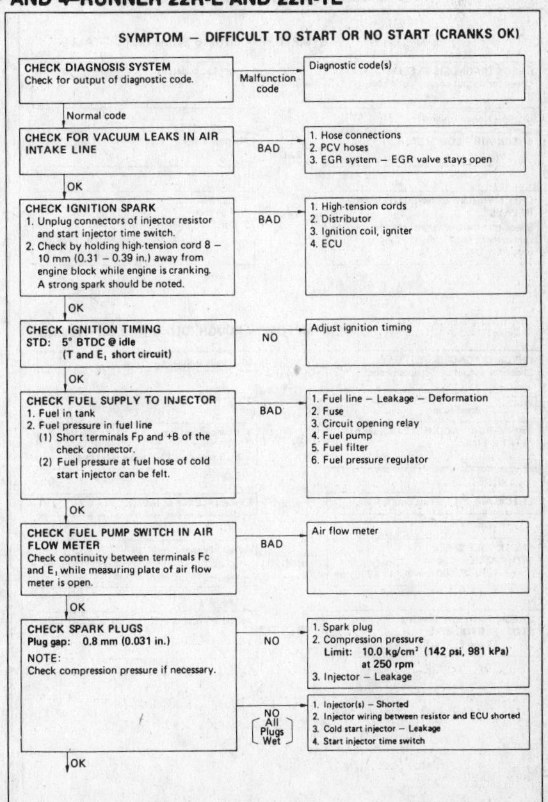
SYMPTOM – DIFFICULT TO START OR NO START (CRANKS OK)
CHECK DIAGNOSIS SYSTEM
Check for output of diagnostic code.
Malfunction code
Diagnostic code(s)
Normal code
CHECK FOR VACUUM LEAKS IN AIR INTAKE LINE
BAD
1. Hose connections
2. PCV hoses
3. EGR system – EGR valve stays open
OK
CHECK IGNITION SPARK
1. Unplug connectors of injector resistor and start injector time switch.
2. Check by holding high-tension cord 8 – 10 mm (0.31 – 0.39 in.) away from engine block while engine is cranking. A strong spark should be noted.
BAD
1. High-tension cords
2. Distributor
3. Ignition coil, igniter
4. ECU
OK
CHECK IGNITION TIMING
STD: 5° BTDC @ idle
(T and E1 short circuit)
NO
Adjust ignition timing
OK
CHECK FUEL SUPPLY TO INJECTOR
1. Fuel in tank
2. Fuel pressure in fuel line
(1) Short terminals Fp and +B of the check connector.
(2) Fuel pressure at fuel hose of cold start injector can be felt.
BAD
1. Fuel line – Leakage – Deformation
2. Fuse
3. Circuit opening relay
4. Fuel pump
5. Fuel filter
6. Fuel pressure regulator
OK
CHECK FUEL PUMP SWITCH IN AIR FLOW METER
Check continuity between terminals Fc and E1 while measuring plate of air flow meter is open.
BAD
Air flow meter
OK
CHECK SPARK PLUGS
Plug gap: 0.8 mm (0.031 in.)
NOTE:
Check compression pressure if necessary.
NO
1. Spark plug
2. Compression pressure
Limit: 10.0 kg/cm² (142 psi, 981 kPa) at 250 rpm
3. Injector – Leakage
NO All Plugs Wet
1. Injector(s) – Shorted
2. Injector wiring between resistor and ECU shorted
3. Cold start injector – Leakage
4. Start injector time switch
OK

TROUBLESHOOTING PROCEDURES—PICK-UP AND 4-RUNNER 22R-E AND 22R-TE

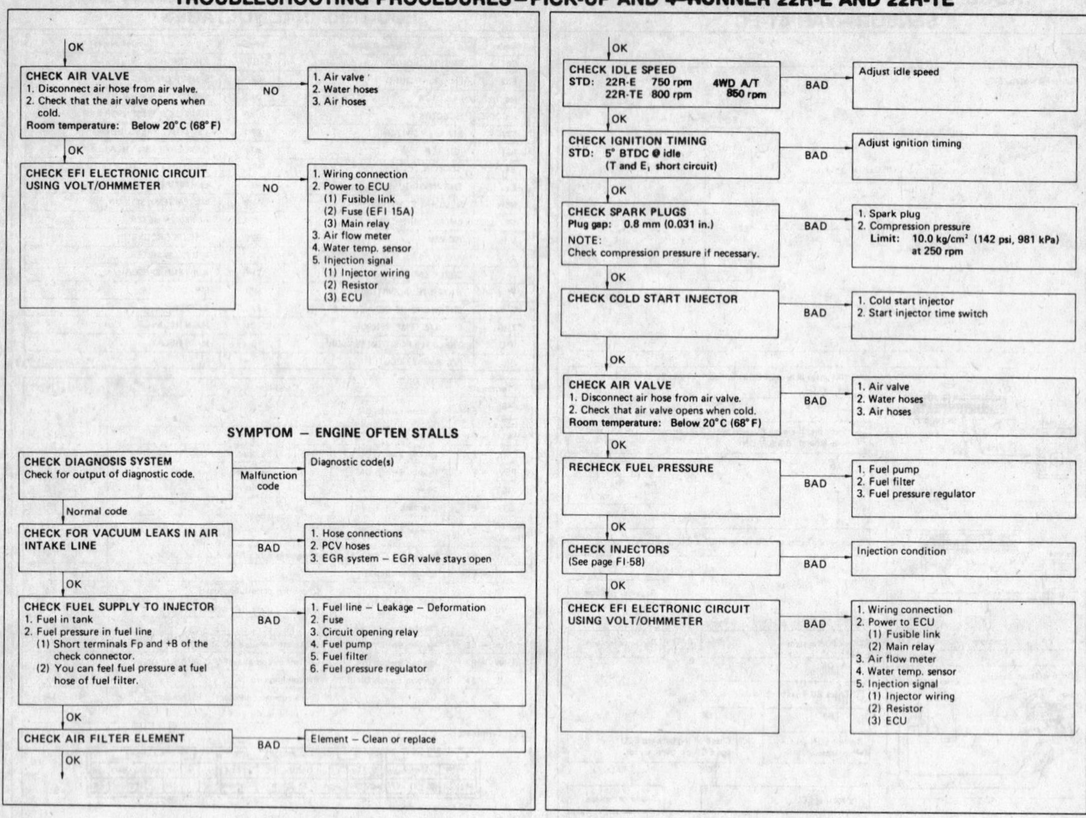
OK
CHECK AIR VALVE
1. Disconnect air hose from air valve.
2. Check that the air valve opens when cold.
Room temperature: Below 20°C (68°F)
NO
1. Air valve
2. Water hoses
3. Air hoses
OK
CHECK EFI ELECTRONIC CIRCUIT USING VOLT/OHMMETER
NO
1. Wiring connection
2. Power to ECU
(1) Fusible link
(2) Fuse (EFI 15A)
(3) Main relay
3. Air flow meter
4. Water temp. sensor
5. Injection signal
(1) Injector wiring
(2) Resistor
(3) ECU
SYMPTOM – ENGINE OFTEN STALLS
CHECK DIAGNOSIS SYSTEM
Check for output of diagnostic code.
Malfunction code
Diagnostic code(s)
Normal code
CHECK FOR VACUUM LEAKS IN AIR INTAKE LINE
BAD
1. Hose connections
2. PCV hoses
3. EGR system – EGR valve stays open
OK
CHECK FUEL SUPPLY TO INJECTOR
1. Fuel in tank
2. Fuel pressure in fuel line
(1) Short terminals Fp and +B of the check connector.
(2) You can feel fuel pressure at fuel hose of fuel filter.
BAD
1. Fuel line – Leakage – Deformation
2. Fuse
3. Circuit opening relay
4. Fuel pump
5. Fuel filter
6. Fuel pressure regulator
OK
CHECK AIR FILTER ELEMENT
BAD
Element – Clean or replace
OK
OK
CHECK IDLE SPEED
STD: 22R-E 750 rpm 4WD A/T
22R-TE 800 rpm 850 rpm
BAD
Adjust idle speed
OK
CHECK IGNITION TIMING
STD: 5° BTDC @ idle
(T and E1 short circuit)
BAD
Adjust ignition timing
OK
CHECK SPARK PLUGS
Plug gap: 0.8 mm (0.031 in.)
NOTE:
Check compression pressure if necessary.
BAD
1. Spark plug
2. Compression pressure
Limit: 10.0 kg/cm² (142 psi, 981 kPa) at 250 rpm
OK
CHECK COLD START INJECTOR
BAD
1. Cold start injector
2. Start injector time switch
OK
CHECK AIR VALVE
1. Disconnect air hose from air valve.
2. Check that air valve opens when cold.
Room temperature: Below 20°C (68°F)
BAD
1. Air valve
2. Water hoses
3. Air hoses
OK
RECHECK FUEL PRESSURE
BAD
1. Fuel pump
2. Fuel filter
3. Fuel pressure regulator
OK
CHECK INJECTORS
(See page FI-58)
BAD
Injection condition
OK
CHECK EFI ELECTRONIC CIRCUIT USING VOLT/OHMMETER
BAD
1. Wiring connection
2. Power to ECU
(1) Fusible link
(2) Main relay
3. Air flow meter
4. Water temp. sensor
5. Injection signal
(1) Injector wiring
(2) Resistor
(3) ECU

TROUBLESHOOTING PROCEDURES—PICK-UP AND 4-RUNNER 22R-E AND 22R-TE

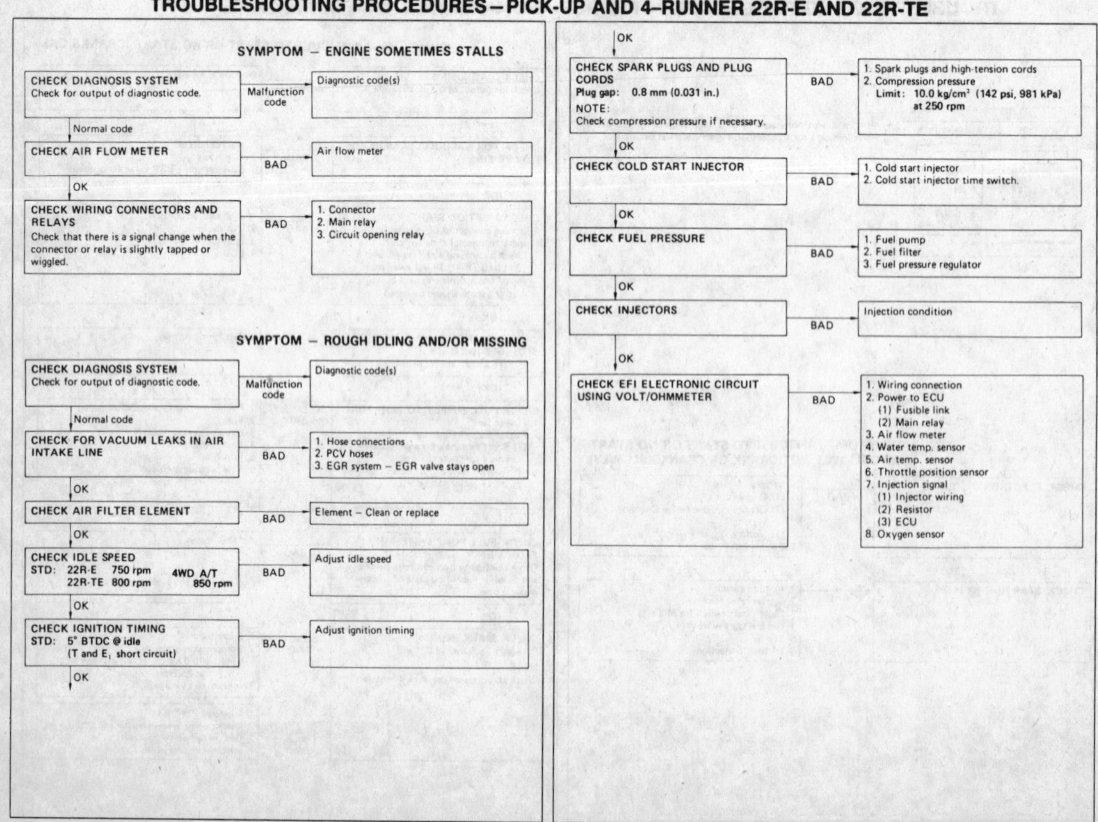
SYMPTOM – ENGINE SOMETIMES STALLS
CHECK DIAGNOSIS SYSTEM
Check for output of diagnostic code.
Malfunction code
Diagnostic code(s)
Normal code
CHECK AIR FLOW METER
BAD
Air flow meter
OK
CHECK WIRING CONNECTORS AND RELAYS
Check that there is a signal change when the connector or relay is slightly tapped or wiggled.
BAD
1. Connector
2. Main relay
3. Circuit opening relay
SYMPTOM – ROUGH IDLING AND/OR MISSING
CHECK DIAGNOSIS SYSTEM
Check for output of diagnostic code.
Malfunction code
Diagnostic code(s)
Normal code
CHECK FOR VACUUM LEAKS IN AIR INTAKE LINE
BAD
1. Hose connections
2. PCV hoses
3. EGR system – EGR valve stays open
OK
CHECK AIR FILTER ELEMENT
BAD
Element – Clean or replace
OK
CHECK IDLE SPEED
STD: 22R-E 750 rpm 4WD A/T
22R-TE 800 rpm 850 rpm
BAD
Adjust idle speed
OK
CHECK IGNITION TIMING
STD: 5° BTDC @ idle
(T and E1 short circuit)
BAD
Adjust ignition timing
OK
OK
CHECK SPARK PLUGS AND PLUG CORDS
Plug gap: 0.8 mm (0.031 in.)
NOTE:
Check compression pressure if necessary.
BAD
1. Spark plugs and high-tension cords
2. Compression pressure
Limit: 10.0 kg/cm² (142 psi, 981 kPa) at 250 rpm
OK
CHECK COLD START INJECTOR
BAD
1. Cold start injector
2. Cold start injector time switch.
OK
CHECK FUEL PRESSURE
BAD
1. Fuel pump
2. Fuel filter
3. Fuel pressure regulator
OK
CHECK INJECTORS
BAD
Injection condition
OK
CHECK EFI ELECTRONIC CIRCUIT USING VOLT/OHMMETER
BAD
1. Wiring connection
2. Power to ECU
(1) Fusible link
(2) Main relay
3. Air flow meter
4. Water temp. sensor
5. Air temp. sensor
6. Throttle position sensor
7. Injection signal
(1) Injector wiring
(2) Resistor
(3) ECU
8. Oxygen sensor

TROUBLESHOOTING PROCEDURES—PICK-UP AND 4-RUNNER 22R-E AND 22R-TE

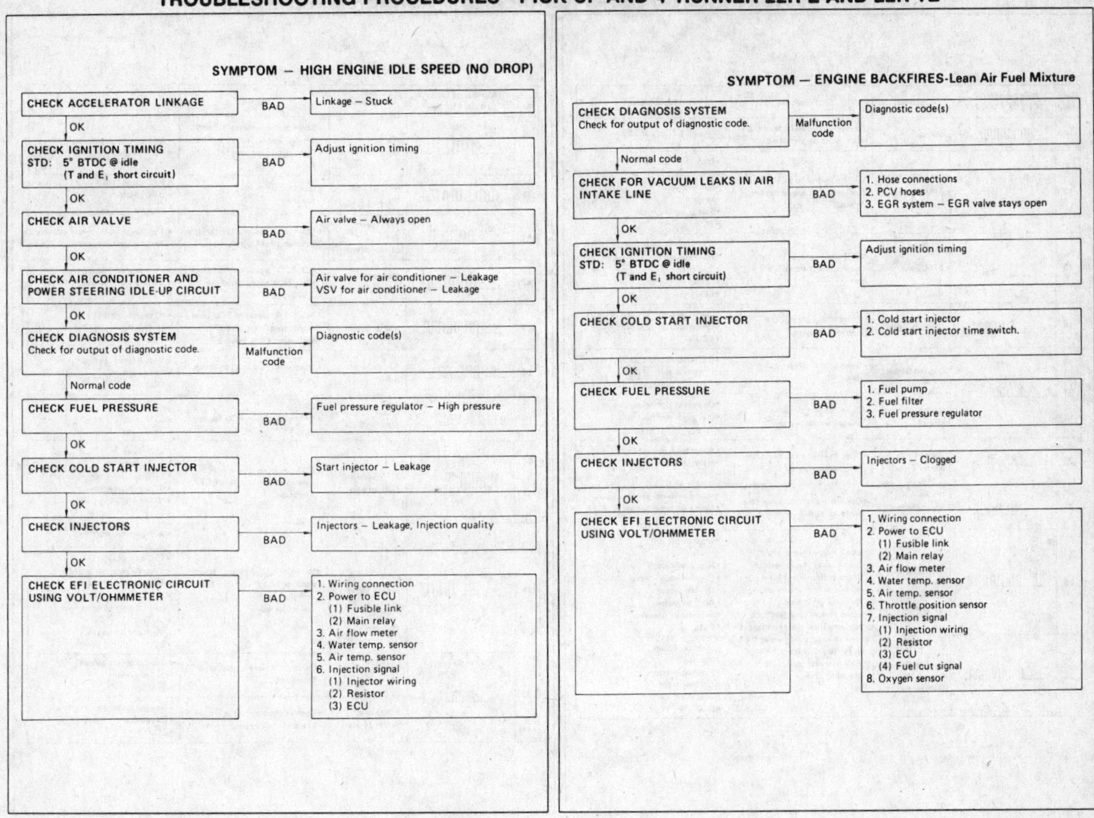
SYMPTOM — HIGH ENGINE IDLE SPEED (NO DROP)
CHECK ACCELERATOR LINKAGE
BAD
Linkage — Stuck
OK
CHECK IGNITION TIMING
STD: 5° BTDC @ idle
(T and E1 short circuit)
BAD
Adjust ignition timing
OK
CHECK AIR VALVE
BAD
Air valve — Always open
OK
CHECK AIR CONDITIONER AND POWER STEERING IDLE-UP CIRCUIT
BAD
Air valve for air conditioner — Leakage
VSV for air conditioner — Leakage
OK
CHECK DIAGNOSIS SYSTEM
Check for output of diagnostic code.
Malfunction code
Diagnostic code(s)
Normal code
CHECK FUEL PRESSURE
BAD
Fuel pressure regulator — High pressure
OK
CHECK COLD START INJECTOR
BAD
Start injector — Leakage
OK
CHECK INJECTORS
BAD
Injectors — Leakage, Injection quality
OK
CHECK EFI ELECTRONIC CIRCUIT USING VOLT/OHMMETER
BAD
1. Wiring connection
2. Power to ECU
(1) Fusible link
(2) Main relay
3. Air flow meter
4. Water temp. sensor
5. Air temp. sensor
6. Injection signal
(1) Injector wiring
(2) Resistor
(3) ECU
SYMPTOM — ENGINE BACKFIRES-Lean Air Fuel Mixture
CHECK DIAGNOSIS SYSTEM
Check for output of diagnostic code.
Malfunction code
Diagnostic code(s)
Normal code
CHECK FOR VACUUM LEAKS IN AIR INTAKE LINE
BAD
1. Hose connections
2. PCV hoses
3. EGR system — EGR valve stays open
OK
CHECK IGNITION TIMING
STD: 5° BTDC @ idle
(T and E1 short circuit)
BAD
Adjust ignition timing
OK
CHECK COLD START INJECTOR
BAD
1. Cold start injector
2. Cold start injector time switch.
OK
CHECK FUEL PRESSURE
BAD
1. Fuel pump
2. Fuel filter
3. Fuel pressure regulator
OK
CHECK INJECTORS
BAD
Injectors — Clogged
OK
CHECK EFI ELECTRONIC CIRCUIT USING VOLT/OHMMETER
BAD
1. Wiring connection
2. Power to ECU
(1) Fusible link
(2) Main relay
3. Air flow meter
4. Water temp. sensor
5. Air temp. sensor
6. Throttle position sensor
7. Injection signal
(1) Injection wiring
(2) Resistor
(3) ECU
(4) Fuel cut signal
8. Oxygen sensor

TROUBLESHOOTING PROCEDURES—PICK-UP AND 4-RUNNER 22R-E AND 22R-TE

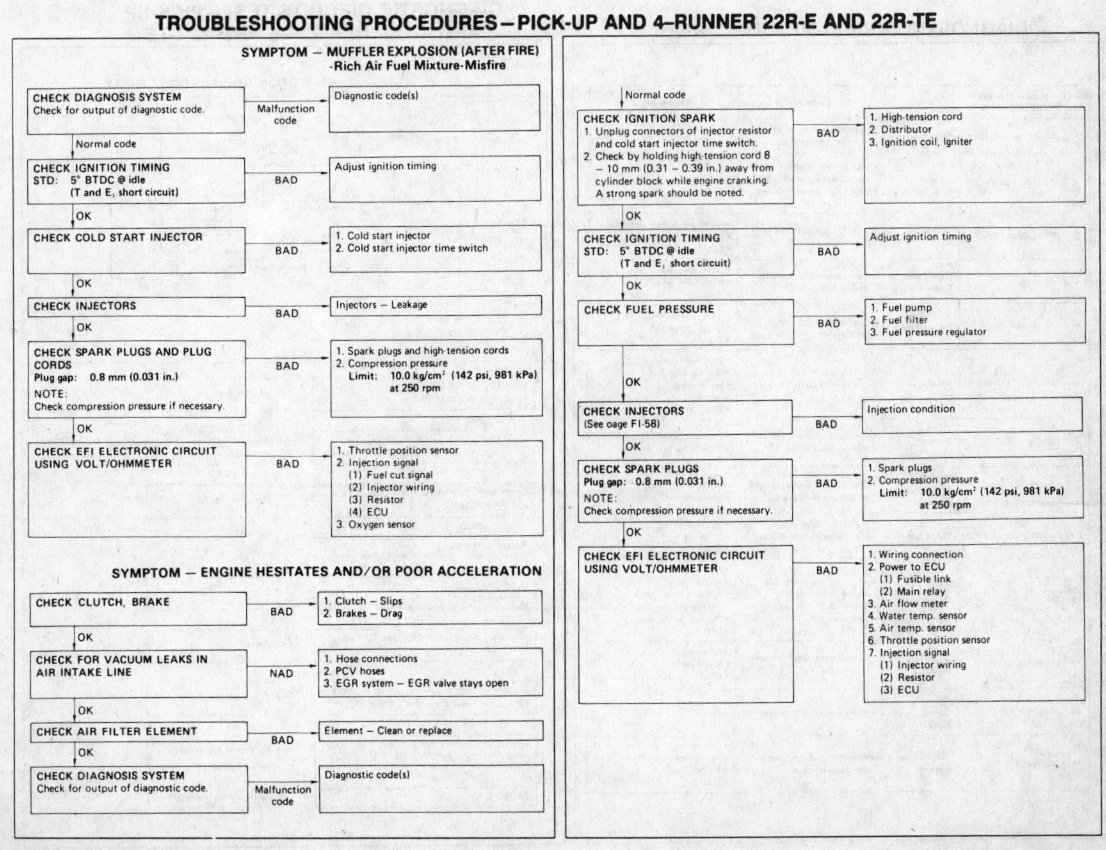
SYMPTOM — MUFFLER EXPLOSION (AFTER FIRE)
-Rich Air Fuel Mixture-Misfire
CHECK DIAGNOSIS SYSTEM
Check for output of diagnostic code.
Malfunction code
Diagnostic code(s)
Normal code
CHECK IGNITION TIMING
STD: 5° BTDC @ idle
(T and E1 short circuit)
BAD
Adjust ignition timing
OK
CHECK COLD START INJECTOR
BAD
1. Cold start injector
2. Cold start injector time switch
OK
CHECK INJECTORS
BAD
Injectors — Leakage
OK
CHECK SPARK PLUGS AND PLUG CORDS
Plug gap: 0.8 mm (0.031 in.)
NOTE:
Check compression pressure if necessary.
BAD
1. Spark plugs and high-tension cords
2. Compression pressure
Limit: 10.0 kg/cm² (142 psi, 981 kPa) at 250 rpm
OK
CHECK EFI ELECTRONIC CIRCUIT USING VOLT/OHMMETER
BAD
1. Throttle position sensor
2. Injection signal
(1) Fuel cut signal
(2) Injector wiring
(3) Resistor
(4) ECU
3. Oxygen sensor
SYMPTOM — ENGINE HESITATES AND/OR POOR ACCELERATION
CHECK CLUTCH, BRAKE
BAD
1. Clutch — Slips
2. Brakes — Drag
OK
CHECK FOR VACUUM LEAKS IN AIR INTAKE LINE
NAD
1. Hose connections
2. PCV hoses
3. EGR system — EGR valve stays open
OK
CHECK AIR FILTER ELEMENT
BAD
Element — Clean or replace
OK
CHECK DIAGNOSIS SYSTEM
Check for output of diagnostic code.
Malfunction code
Diagnostic code(s)
Normal code
CHECK IGNITION SPARK
1. Unplug connectors of injector resistor and cold start injector time switch.
2. Check by holding high-tension cord 8 – 10 mm (0.31 – 0.39 in.) away from cylinder block while engine cranking. A strong spark should be noted.
BAD
1. High-tension cord
2. Distributor
3. Ignition coil, Igniter
OK
CHECK IGNITION TIMING
STD: 5° BTDC @ idle
(T and E1 short circuit)
BAD
Adjust ignition timing
OK
CHECK FUEL PRESSURE
BAD
1. Fuel pump
2. Fuel filter
3. Fuel pressure regulator
OK
CHECK INJECTORS
(See oage FI-58)
BAD
Injection condition
OK
CHECK SPARK PLUGS
Plug gap: 0.8 mm (0.031 in.)
NOTE:
Check compression pressure if necessary.
BAD
1. Spark plugs
2. Compression pressure
Limit: 10.0 kg/cm² (142 psi, 981 kPa) at 250 rpm
OK
CHECK EFI ELECTRONIC CIRCUIT USING VOLT/OHMMETER
BAD
1. Wiring connection
2. Power to ECU
(1) Fusible link
(2) Main relay
3. Air flow meter
4. Water temp. sensor
5. Air temp. sensor
6. Throttle position sensor
7. Injection signal
(1) Injector wiring
(2) Resistor
(3) ECU

DIAGNOSTIC CODES—PICK-UP AND 4–RUNNER 22R-E

Code No.	Number of blinks "CHECK ENGINE"	System	Diagnosis	Trouble area
–	ON OFF	Normal	This appears when none of the other codes are identified.	———
12		RPM Signal	NO "Ne" signal to ECV within 2 seconds after engine has been cranked.	• Distributor circuit • Distributor • Igniter circuit • Igniter • Starter signal circuit • ECV
13		RPM Signal	NO "Ne signal to ECU when engine speed is above 1,500 rpm.	• Distributor circuit • Distributor • Igniter circuit • Igniter • ECU
14		Ignition Signal	NO "IGf" signal to ECU 4 – 5 times in succession.	• Igniter and ignition coil circuit • Igniter and ignition coil • ECU
21		Oxygen Sensor	• Detection of oxygen sensor detrioration.	• Oxygen sensor circuit • Oxygen sensor • ECU
		Oxygen Sensor Heater	• Open or short circuit in oxygen sensor heater.	• Oxygen sensor heater circuit • Oxygen sensor heater • ECU
22		Water Temp. Sensor Signal	• Open or short circuit in water temp. sensor signal.	• Water temp. sensor circuit • Water temp. sensor • ECU
24		Intake Air Temp. Sensor Signal	• Open or short circuit in intake air temp. sensor signal.	• Intake air temp. sensor circuit • Intake air temp. sensor • ECU
25		Air-fuel Ratio Lean Malfunction	• When oxygen sensor signal continues at the upper (rich) or lower (lean) limit for a certain period of time during feedback condition. • When air-fuel ratio feedback compensation value or adaptive control value continues at the upper (lean) or lower (rich) limit renewed for a certain period of time.	• Injector circuit • Injector • Fuel line pressure • Oxygen sensor circuit • Oxygen sensor • Air flow meter • Air intake system • ECU
26		Air-fuel Ratio Rich Malfunction		• Injector circuit • Injector • Fuel line pressure • Air-flow meter • Cold start injector • ECU

Code No.	Number of blinks "CHECK ENGINE"	System	Diagnosis	Trouble area
31		Air flow. Meter Signal	• Short circuit between VC and VB, VC and E_2, or VS and VC. • Open circuit between VC and E_2.	• Air flow meter circuit • Air flow meter
35		HAC Sensor Signal	• Open circuit in HAC sensor signal	• HAC sensor circuit • HAC sensor • ECU
41		Throttle Position Sensor Signal	• Open or short circuit in throttle position sensor signal.	• Throttle position sensor circuit • Throttle position sensor
42		Vehicle Speed Sensor Signal	• NO "SPD" signal for 5 seconds when engine speed is above 2,500 rpm.	• Vehicle speed sensor circuit • Vehicle speed sensor • ECU
43		Starter Signal	• NO "STA" signal to ECU until engine speed reaches 800 rpm with vehicle not moving.	• IG switch circuit • IG switch • ECU
52		Knock Sensor Signal	• Open or short circuit in knock sensor signal.	• Knock sensor circuit • Knock sensor • ECU
53		Knock Control Signal in ECU	• Knock control in ECU faulty.	• ECU
71		EGR System Malfunction	• EGR gas temp. below pre-determined level during EGR operation.	• EGR valve • EGR hose • EGR gas temp. sensor circuit • EGR gas temp. sensor • VSV for EGR circuit • ECU
51		Switch Signal	• No "IDL" signal or No "NSW" signal or "A/C" signal to ECU, with the check terminals E_1 and T shorted.	• A/C switch circuit • A/C switch • A/C Amplifire • Throttle position sensor circuit • Throttle position sensor • (A/T) Neutral start switch circuit • (A/T) Neutral start switch • ECU
27	(California only)	Sub Oxygen Sensor	• Detection of oxygen sensor detrioration.	• Oxygen sensor circuit • Oxygen sensor • ECU
		Sub Oxygen Sensor Heater	• Open or short circuit in oxygen sensor heater.	• Oxygen sensor heater circuit • Oxygen sensor heater • ECU

DIAGNOSTIC CODES—PICK-UP 22R-TE

Code No.	Number of blinks "CHECK ENGINE"	System	Diagnosis	Trouble area
1	ON OFF	Normal	This appears when none of the other codes are identified.	———
2		Air flow. Meter Signal	• Short circuit between VC and VB, VC and E_2, or VS and VC. • Open circuit between VC and E_2.	• Air flow meter circuit • Air flow meter
3		Ignition Signal	NO "IGf" signal to ECU 4–5 times in succession.	• Igniter and ignition coil circuit • Igniter and ignition coil • ECU
4		Water Temp. Sensor Signal	• Open or short circuit in water temp. sensor signal.	• Water temp. sensor circuit • Water temp. sensor • ECU
5		Oxygen Sensor	• Open or short circuit in oxygen sensor	• Oxygen sensor circuit • Oxygen sensor • ECU
6		RPM Signal	NO "Ne signal to ECU when engine speed is above 1,500 rpm.	• Distributor circuit • Distributor • Igniter circuit • Igniter • ECU
7		Throttle Position Sensor Signal	• Open or short circuit in throttle position sensor signal.	• Throttle position sensor circuit • Throttle position sensor
8		Intake Air Temp. Sensor Signal	• Open or short circuit in intake air temp. sensor signal.	• Intake air temp. sensor circuit • Intake air temp. sensor ECU
10		Starter Signal	• NO "STA" signal to ECU until engine speed reaches 800 rpm with vehicle not moving.	• IG switch circuit • IG switch • ECU
11		Switch Signal	• Air conditioner switch ON, neutral start switch OFF, idle switch OFF during diagnosis check.	• A/C switch circuit • A/C switch • A/C Amplifire • Throttle position sensor circuit • Throttle position sensor • ECU
12		Knock Sensor Signal	• Open or short circuit in knock sensor signal.	• Knock sensor circuit • Knock sensor • ECU
13		Knock Control Signal	• Knock control ECU faulty.	• ECU
14		Turbocharger Pressure	• When the fuel cut-off due to high boost is occured.	• Turbocharger • Air flow meter circuit • Air flow meter • ECU

DIAGNOSTIC CIRCUITS TEST—PICK-UP AND 4–RUNNER 22R-E AND 22R-TE

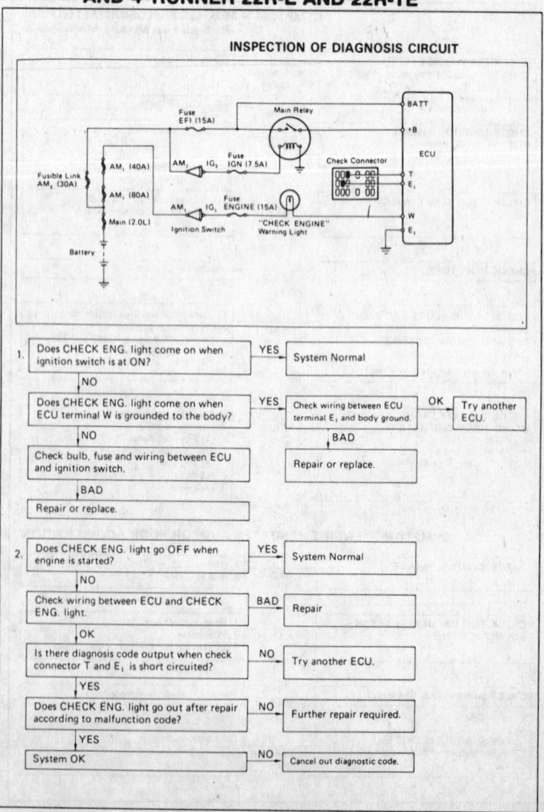
INSPECTION OF DIAGNOSIS CIRCUIT
Fuse EFI (15A)
Main Relay
BATT
+B
ECU
Fusible Link AM_2 (30A)
AM_1 (40A)
AM_1 (80A)
Main (2.0L)
Battery
AM_2
IG_2
Fuse IGN (7.5A)
Check Connector
T
E_1
AM_1
IG_1
Fuse ENGINE (15A)
Ignition Switch
"CHECK ENGINE" Warning Light
W
E_1
1. Does CHECK ENG. light come on when ignition switch is at ON?
YES
System Normal
NO
Does CHECK ENG. light come on when ECU terminal W is grounded to the body?
YES
Check wiring between ECU terminal E_1 and body ground.
OK
Try another ECU.
BAD
Repair or replace.
NO
Check bulb, fuse and wiring between ECU and ignition switch.
BAD
Repair or replace.
2. Does CHECK ENG. light go OFF when engine is started?
YES
System Normal
NO
Check wiring between ECU and CHECK ENG. light.
BAD
Repair
OK
Is there diagnosis code output when check connector T and E_1 is short circuited?
NO
Try another ECU.
YES
Does CHECK ENG. light go out after repair according to malfunction code?
NO
Further repair required.
YES
System OK
NO
Cancel out diagnostic code.

TROUBLESHOOTING CHART 1—1988 PICK-UP AND 4-RUNNER 22R-E AND 22R-TE

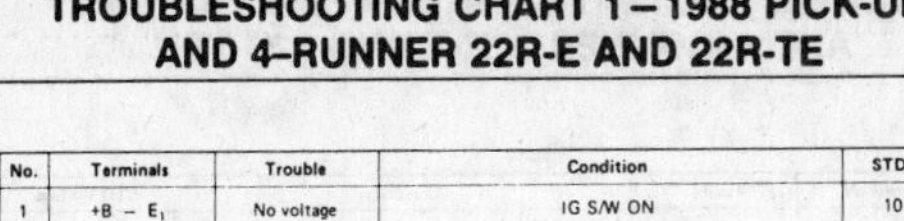

No.	Terminals	Trouble	Condition	STD Voltage
1	+B – E_1	No voltage	IG S/W ON	10 – 14 V

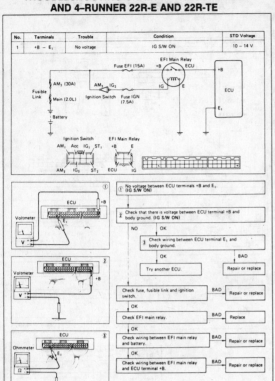

Fuse EFI (15A), EFI Main Relay, +B, ECU, AM_2 (30A), Fusible Link, Main (2.0L), Battery, AM_2, IG_2, Ignition Switch, Fuse IGN (7.5A), IG, E, E_1

Ignition Switch: AM_1, Acc, IG_1, ST_1, AM_2, IG_2, ST_2 — EFI Main Relay: +B, E, ECU, IG

1. No voltage between ECU terminals +B and E_1. (IG S/W ON)
2. Check that there is voltage between ECU terminal +B and body ground. (IG S/W ON)
   - OK → 3. Check wiring between ECU terminal E_1 and body ground.
     - OK → Try another ECU.
     - BAD → Repair or replace
   - NO → Check fuse, fusible link and ignition switch.
     - BAD → Repair or replace
     - OK → Check EFI main relay.
       - BAD → Replace
       - OK → Check wiring between EFI main relay and battery.
         - BAD → Repair or replace
         - OK → Check wiring between EFI main relay and ECU terminal +B.
           - BAD → Repair or replace

(Test illustrations: ① Voltmeter, ECU, +B, E_1; ② Voltmeter, ECU, +B; ③ Ohmmeter, ECU, E_1)

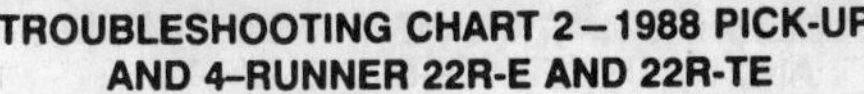

TROUBLESHOOTING CHART 2—1988 PICK-UP AND 4-RUNNER 22R-E AND 22R-TE

No.	Terminals	Trouble	Condition	STD Voltage
2	BATT – E_1	No voltage	—	10 – 14 V

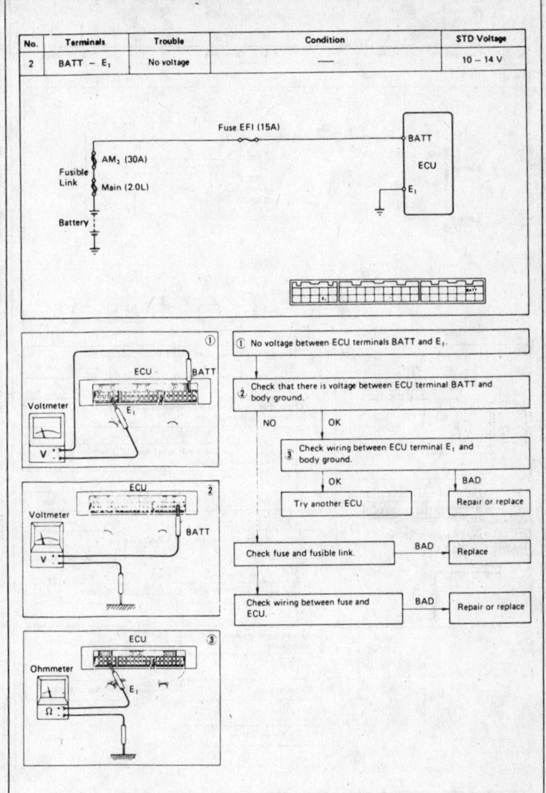

Fuse EFI (15A)
BATT
ECU
E_1
AM_2 (30A)
Fusible Link
Main (2.0L)
Battery
① No voltage between ECU terminals BATT and E_1.
② Check that there is voltage between ECU terminal BATT and body ground.
NO
OK
③ Check wiring between ECU terminal E_1 and body ground.
OK
BAD
Try another ECU.
Repair or replace
Check fuse and fusible link.
BAD
Replace
Check wiring between fuse and ECU.
BAD
Repair or replace
Voltmeter
Ohmmeter

TROUBLESHOOTING CHART 3—1988 PICK-UP AND 4-RUNNER 22R-E AND 22R-TE

No.	Terminals	Trouble	Condition		STD Voltage
3	IDL – E_2	No voltage	Ignition switch ON	Throttle valve open	8 – 14 V
	Vcc – E_2			–	4 – 6 V
	VTA – E_2			Throttle valve fully closed	0.1 – 1.0 V
				Throttle valve fully open	4 – 5 V

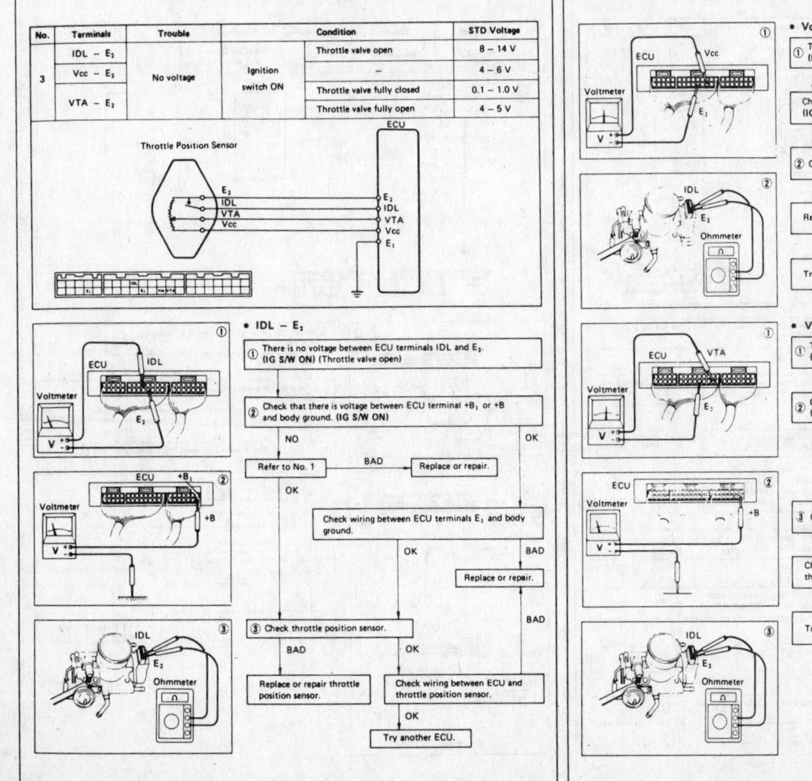

Throttle Position Sensor
ECU
E_2
IDL
VTA
Vcc
E_1
• IDL – E_2
① There is no voltage between ECU terminals IDL and E_2. (IG S/W ON) (Throttle valve open)
② Check that there is voltage between ECU terminal $+B_1$ or +B and body ground. (IG S/W ON)
NO
OK
Refer to No. 1
BAD
Replace or repair.
OK
Check wiring between ECU terminals E_1 and body ground.
OK
BAD
Replace or repair.
BAD
③ Check throttle position sensor.
BAD
OK
Replace or repair throttle position sensor.
Check wiring between ECU and throttle position sensor.
OK
Try another ECU.
Voltmeter
Ohmmeter

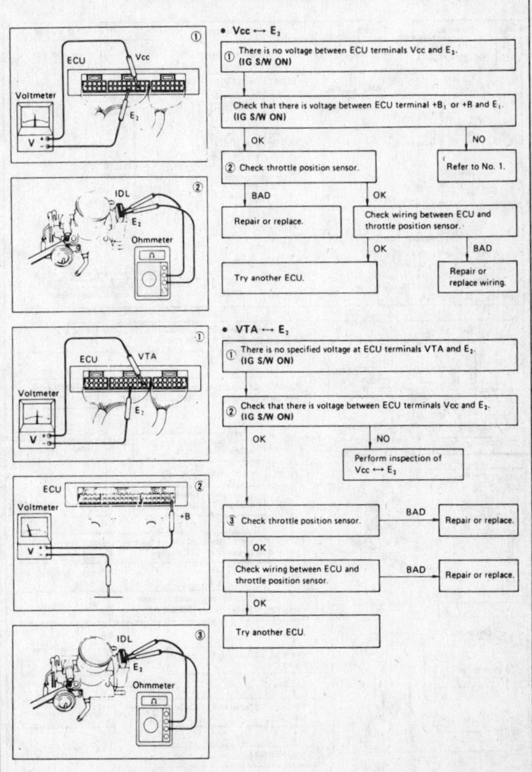

• Vcc ↔ E_2
① There is no voltage between ECU terminals Vcc and E_2. (IG S/W ON)
Check that there is voltage between ECU terminal $+B_1$ or +B and E_1. (IG S/W ON)
OK
NO
② Check throttle position sensor.
Refer to No. 1.
BAD
OK
Repair or replace.
Check wiring between ECU and throttle position sensor.
OK
BAD
Try another ECU.
Repair or replace wiring.
• VTA ↔ E_2
① There is no specified voltage at ECU terminals VTA and E_2. (IG S/W ON)
② Check that there is voltage between ECU terminals Vcc and E_2. (IG S/W ON)
OK
NO
Perform inspection of Vcc ↔ E_2
③ Check throttle position sensor.
BAD
Repair or replace.
OK
Check wiring between ECU and throttle position sensor.
BAD
Repair or replace.
OK
Try another ECU.
Voltmeter
Ohmmeter

TROUBLESHOOTING CHART 4–1988 PICK-UP AND 4–RUNNER 22R-E AND 22R-TE

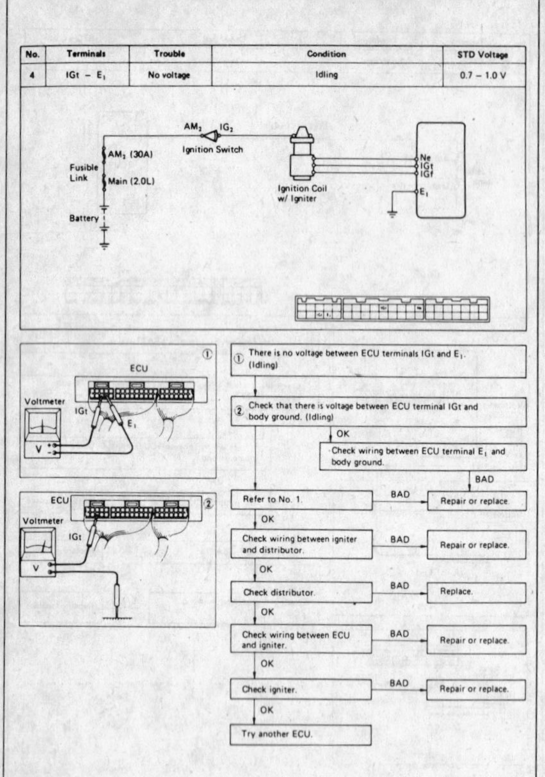

No.	Terminals	Trouble	Condition	STD Voltage
4	IGt – E_1	No voltage	Idling	0.7 – 1.0 V

AM₂ IG₂
Ignition Switch
AM₂ (30A)
Fusible Link
Main (2.0L)
Battery
Ignition Coil w/ Igniter
Ne
IGt
IGf
E₁
ECU
Voltmeter
IGt
E₁
① There is no voltage between ECU terminals IGt and E₁. (Idling)
② Check that there is voltage between ECU terminal IGt and body ground. (Idling)
OK
Check wiring between ECU terminal E₁ and body ground.
BAD
Refer to No. 1.
BAD
Repair or replace.
OK
Check wiring between igniter and distributor.
BAD
Repair or replace.
OK
Check distributor.
BAD
Replace.
OK
Check wiring between ECU and igniter.
BAD
Repair or replace.
OK
Check igniter.
BAD
Repair or replace.
OK
Try another ECU.

TROUBLESHOOTING CHART 5–1988 PICK-UP AND 4–RUNNER 22R-E AND 22R-TE

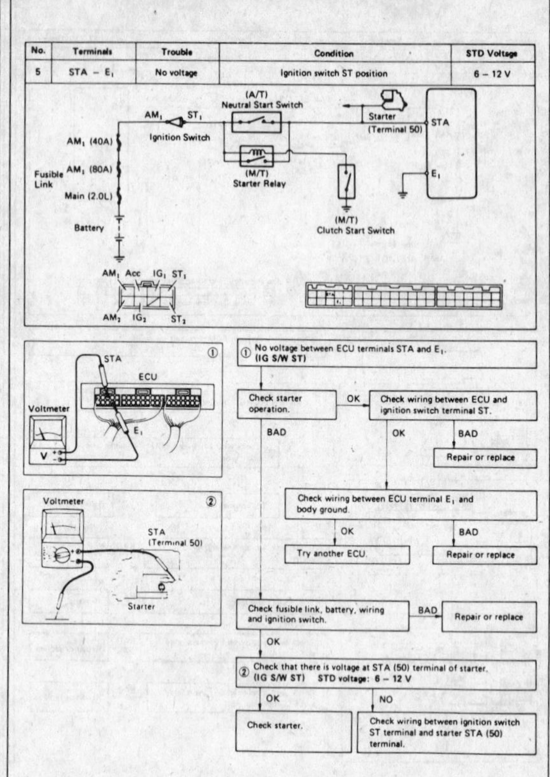

No.	Terminals	Trouble	Condition	STD Voltage
5	STA – E_1	No voltage	Ignition switch ST position	6 – 12 V

(A/T)
Neutral Start Switch
AM₁ ST₁
Ignition Switch
AM₁ (40A)
AM₁ (80A)
Fusible Link
Main (2.0L)
Battery
(M/T)
Starter Relay
Starter (Terminal 50)
STA
E₁
(M/T)
Clutch Start Switch
AM₁ Acc IG₁ ST₁
AM₂ IG₂ ST₂
STA
ECU
Voltmeter
E₁
STA (Terminal 50)
Starter
① No voltage between ECU terminals STA and E₁. (IG S/W ST)
Check starter operation.
OK
Check wiring between ECU and ignition switch terminal ST.
BAD
OK
BAD
Repair or replace
Check wiring between ECU terminal E₁ and body ground.
OK
BAD
Try another ECU.
Repair or replace
Check fusible link, battery, wiring and ignition switch.
BAD
Repair or replace
OK
② Check that there is voltage at STA (50) terminal of starter. (IG S/W ST) STD voltage: 6 – 12 V
OK
NO
Check starter.
Check wiring between ignition switch ST terminal and starter STA (50) terminal.

TROUBLESHOOTING CHART 6–1988 PICK-UP AND 4–RUNNER 22R-E AND 22R-TE

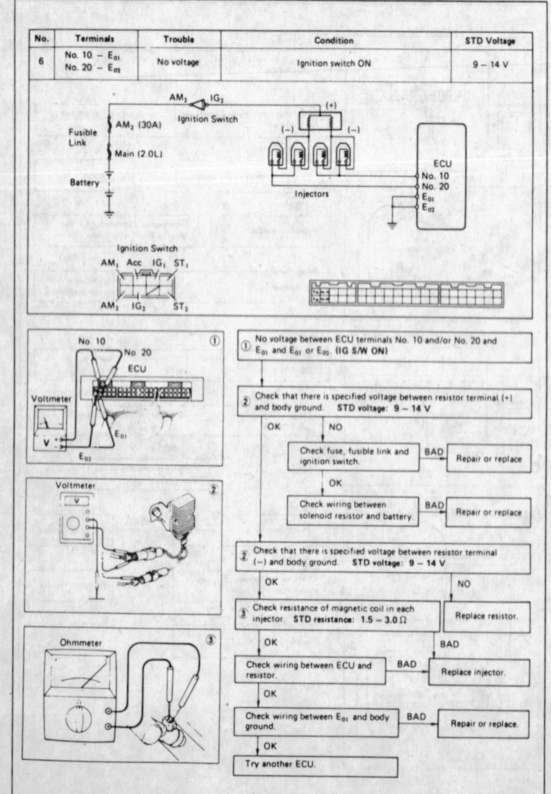

No.	Terminals	Trouble	Condition	STD Voltage
6	No. 10 – E_{01} No. 20 – E_{02}	No voltage	Ignition switch ON	9 – 14 V

AM₂ IG₂
Ignition Switch
AM₂ (30A)
Fusible Link
Main (2.0L)
Battery
(+)
(–)
(–)
Injectors
ECU
No. 10
No. 20
E₀₁
E₀₂
Ignition Switch
AM₁ Acc IG₁ ST₁
AM₂ IG₂ ST₂
No. 10
No. 20
ECU
Voltmeter
E₀₁
E₀₂
Voltmeter
Ohmmeter
① No voltage between ECU terminals No. 10 and/or No. 20 and E₀₁ and E₀₁ or E₀₂. (IG S/W ON)
② Check that there is specified voltage between resistor terminal (+) and body ground. STD voltage: 9 – 14 V
OK
NO
Check fuse, fusible link and ignition switch.
BAD
Repair or replace
OK
Check wiring between solenoid resistor and battery.
BAD
Repair or replace
② Check that there is specified voltage between resistor terminal (–) and body ground. STD voltage: 9 – 14 V
OK
NO
③ Check resistance of magnetic coil in each injector. STD resistance: 1.5 – 3.0 Ω
Replace resistor.
OK
BAD
Check wiring between ECU and resistor.
BAD
Replace injector.
OK
Check wiring between E₀₁ and body ground.
BAD
Repair or replace.
OK
Try another ECU.

TROUBLESHOOTING CHART 7–1988 PICK-UP AND 4–RUNNER 22R-E AND 22R-TE

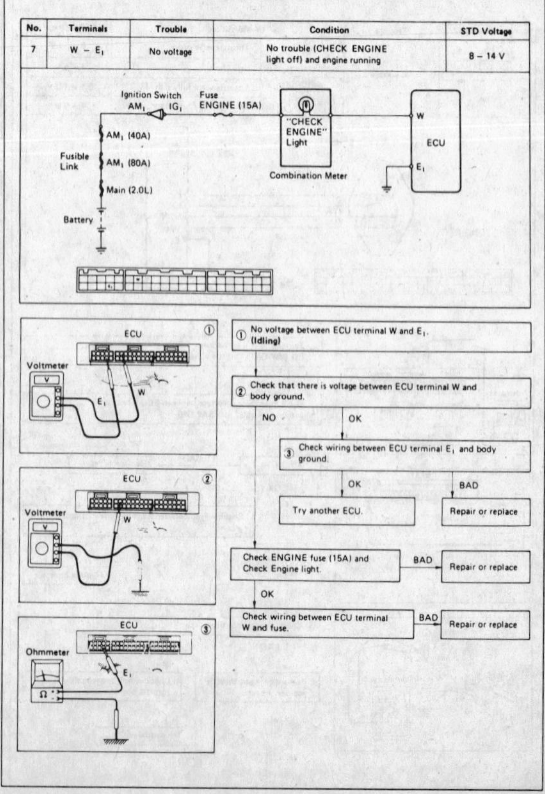

No.	Terminals	Trouble	Condition	STD Voltage
7	W – E_1	No voltage	No trouble (CHECK ENGINE light off) and engine running	8 – 14 V

Ignition Switch
AM₁ IG₁
Fuse ENGINE (15A)
"CHECK ENGINE" Light
Combination Meter
W
ECU
E₁
AM₁ (40A)
Fusible Link
AM₁ (80A)
Main (2.0L)
Battery
ECU
Voltmeter
E₁
W
Ohmmeter
① No voltage between ECU terminal W and E₁. (Idling)
② Check that there is voltage between ECU terminal W and body ground.
NO
OK
③ Check wiring between ECU terminal E₁ and body ground.
OK
BAD
Try another ECU.
Repair or replace
Check ENGINE fuse (15A) and Check Engine light.
BAD
Repair or replace
OK
Check wiring between ECU terminal W and fuse.
BAD
Repair or replace

TROUBLESHOOTING CHART 8—1988 PICK-UP AND 4-RUNNER 22R-E AND 22R-TE

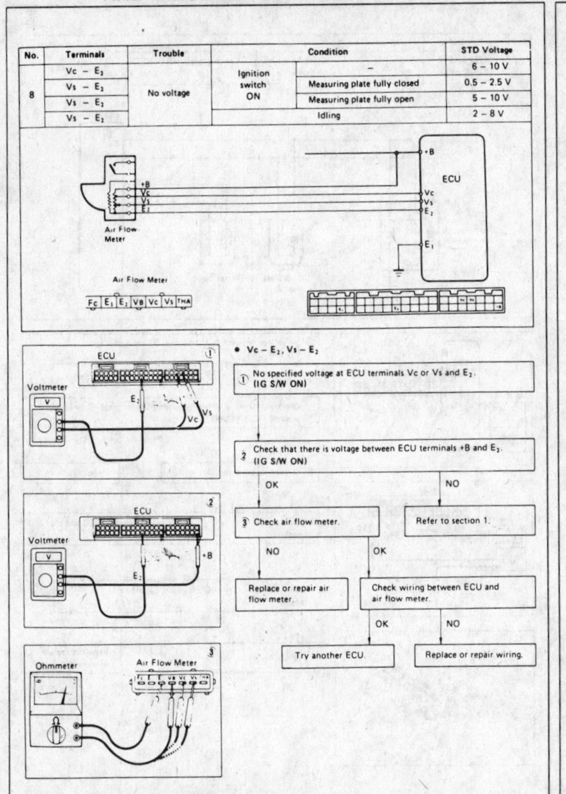

No.	Terminals	Trouble	Condition		STD Voltage
8	Vc – E₂	No voltage	Ignition switch ON	–	6 – 10 V
	Vs – E₂			Measuring plate fully closed	0.5 – 2.5 V
	Vs – E₂			Measuring plate fully open	5 – 10 V
	Vs – E₂			Idling	2 – 8 V

+B
ECU
Vc
Vs
E₂
E₁
Air Flow Meter
Air Flow Meter
Fc E₁ E₂ VB Vc Vs THA
ECU
Voltmeter
E₂
Vs
Vc
+B
Ohmmeter
• Vc – E₂, Vs – E₂
① No specified voltage at ECU terminals Vc or Vs and E₂. (IG S/W ON)
2 Check that there is voltage between ECU terminals +B and E₂. (IG S/W ON)
OK
NO
③ Check air flow meter.
Refer to section 1.
NO
OK
Replace or repair air flow meter.
Check wiring between ECU and air flow meter.
OK
NO
Try another ECU.
Replace or repair wiring.

TROUBLESHOOTING CHART 9—1988 PICK-UP AND 4-RUNNER 22R-E AND 22R-TE

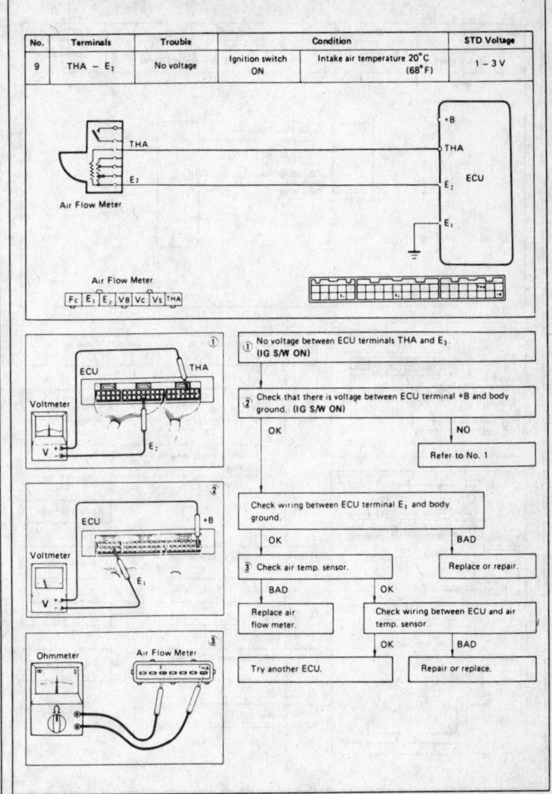

No.	Terminals	Trouble	Condition		STD Voltage
9	THA – E₂	No voltage	Ignition switch ON	Intake air temperature 20°C (68°F)	1 – 3 V

+B
THA
ECU
E₂
E₁
Air Flow Meter
Air Flow Meter
Fc E₁ E₂ VB Vc Vs THA
ECU
THA
Voltmeter
E₂
+B
E₁
Ohmmeter
① No voltage between ECU terminals THA and E₂. (IG S/W ON)
② Check that there is voltage between ECU terminal +B and body ground. (IG S/W ON)
OK
NO
Refer to No. 1
Check wiring between ECU terminal E₁ and body ground.
OK
BAD
③ Check air temp. sensor.
Replace or repair.
BAD
OK
Replace air flow meter.
Check wiring between ECU and air temp. sensor.
OK
BAD
Try another ECU.
Repair or replace.

TROUBLESHOOTING CHART 10—1988 PICK-UP AND 4-RUNNER 22R-E AND 22R-TE

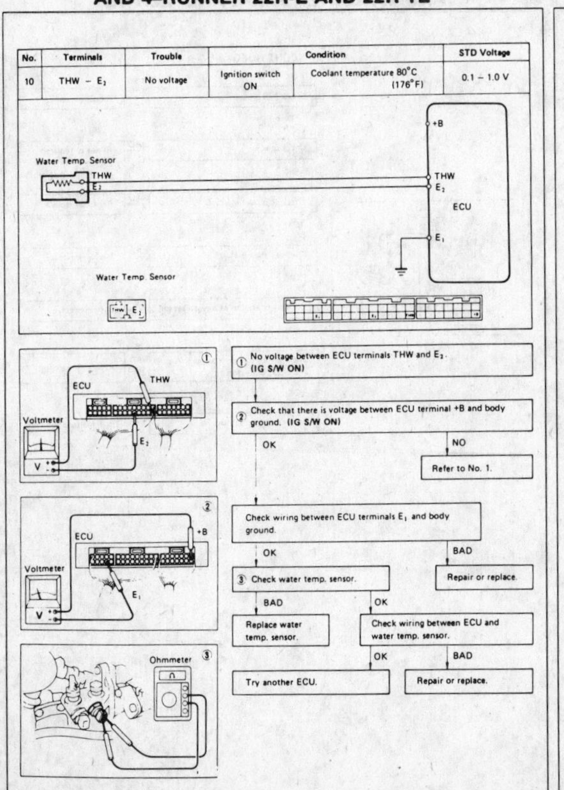

No.	Terminals	Trouble	Condition		STD Voltage
10	THW – E₂	No voltage	Ignition switch ON	Coolant temperature 80°C (176°F)	0.1 – 1.0 V

+B
Water Temp. Sensor
THW
E₂
ECU
E₁
Water Temp. Sensor
ECU
THW
Voltmeter
E₂
+B
E₁
Ohmmeter
① No voltage between ECU terminals THW and E₂. (IG S/W ON)
② Check that there is voltage between ECU terminal +B and body ground. (IG S/W ON)
OK
NO
Refer to No. 1.
Check wiring between ECU terminals E₁ and body ground.
OK
BAD
③ Check water temp. sensor.
Repair or replace.
BAD
OK
Replace water temp. sensor.
Check wiring between ECU and water temp. sensor.
OK
BAD
Try another ECU.
Repair or replace.

TROUBLESHOOTING CHART 11—1988 PICK-UP AND 4-RUNNER 22R-E AND 22R-TE

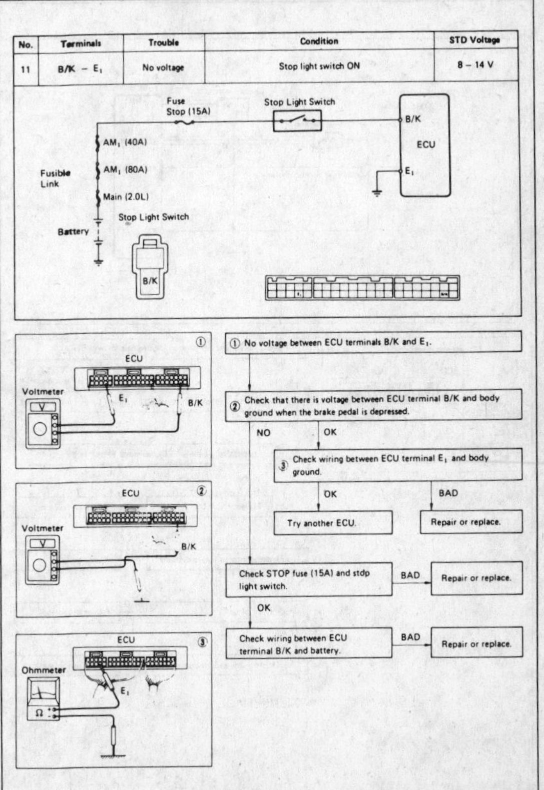

No.	Terminals	Trouble	Condition	STD Voltage
11	B/K – E₁	No voltage	Stop light switch ON	8 – 14 V

Fuse Stop (15A)
Stop Light Switch
B/K
ECU
E₁
AM₁ (40A)
AM₂ (80A)
Fusible Link
Main (2.0L)
Stop Light Switch
Battery
B/K
ECU
Voltmeter
E₁
B/K
Ohmmeter
① No voltage between ECU terminals B/K and E₁.
② Check that there is voltage between ECU terminal B/K and body ground when the brake pedal is depressed.
NO
OK
③ Check wiring between ECU terminal E₁ and body ground.
OK
BAD
Try another ECU.
Repair or replace.
Check STOP fuse (15A) and stop light switch.
BAD
Repair or replace.
OK
Check wiring between ECU terminal B/K and battery.
BAD
Repair or replace.

TROUBLESHOOTING CHART 12–1988 PICK-UP AND 4-RUNNER 22R-E AND 22R-TE

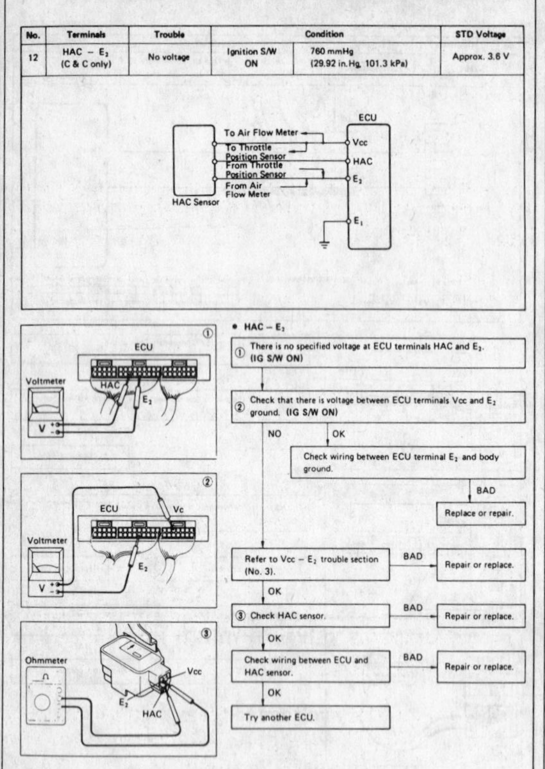

No.	Terminals	Trouble	Condition		STD Voltage
12	HAC – E_2 (C & C only)	No voltage	Ignition S/W ON	760 mmHg (29.92 in.Hg, 101.3 kPa)	Approx. 3.6 V

ECU
To Air Flow Meter
To Throttle Position Sensor
From Throttle Position Sensor
From Air Flow Meter
HAC Sensor
Vcc
HAC
E_2
E_1
Voltmeter
Ohmmeter
• HAC – E_2
① There is no specified voltage at ECU terminals HAC and E_2. (IG S/W ON)
② Check that there is voltage between ECU terminals Vcc and E_2 ground. (IG S/W ON)
NO
OK
Check wiring between ECU terminal E_2 and body ground.
BAD
Replace or repair.
Refer to Vcc – E_2 trouble section (No. 3).
BAD
Repair or replace.
OK
③ Check HAC sensor.
BAD
Repair or replace.
OK
Check wiring between ECU and HAC sensor.
BAD
Repair or replace.
OK
Try another ECU.

TROUBLESHOOTING CHART 13–1988 PICK-UP AND 4-RUNNER 22R-E

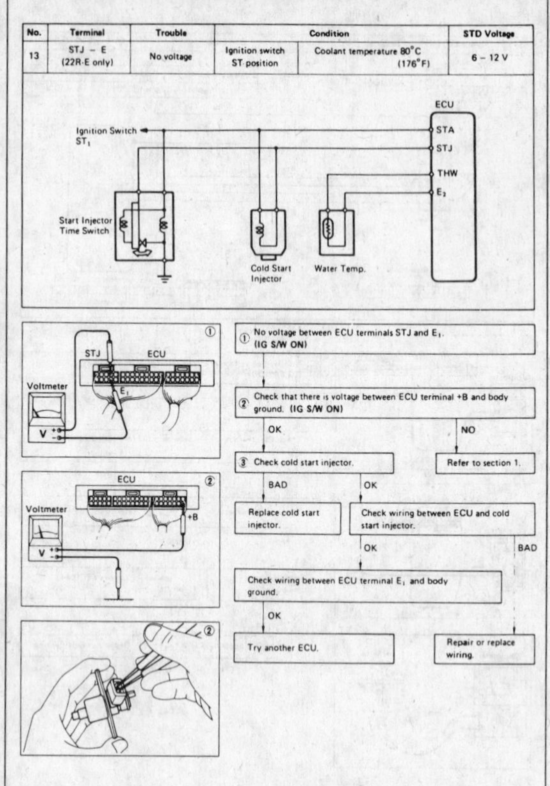

No.	Terminal	Trouble	Condition		STD Voltage
13	STJ – E (22R-E only)	No voltage	Ignition switch ST position	Coolant temperature 80°C (176°F)	6 – 12 V

ECU
Ignition Switch ST_1
STA
STJ
THW
E_2
Start Injector Time Switch
Cold Start Injector
Water Temp.
Voltmeter
STJ
E_1
+B
① No voltage between ECU terminals STJ and E_1. (IG S/W ON)
② Check that there is voltage between ECU terminal +B and body ground. (IG S/W ON)
OK
NO
③ Check cold start injector.
Refer to section 1.
BAD
OK
Replace cold start injector.
Check wiring between ECU and cold start injector.
OK
BAD
Check wiring between ECU terminal E_1 and body ground.
OK
Try another ECU.
Repair or replace wiring.

TROUBLESHOOTING OXYGEN SENSOR–1988 PICK-UP AND 4-RUNNER 22R-E AND 22R-TE

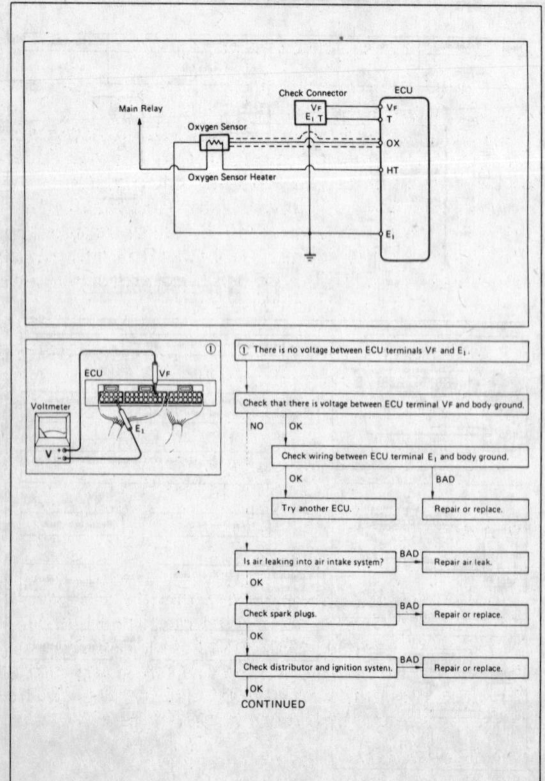
Main Relay
Check Connector
ECU
Oxygen Sensor
Oxygen Sensor Heater
V_F
E_1
T
OX
HT
Voltmeter
① There is no voltage between ECU terminals V_F and E_1.
Check that there is voltage between ECU terminal V_F and body ground.
NO
OK
Check wiring between ECU terminal E_1 and body ground.
OK
BAD
Try another ECU.
Repair or replace.
Is air leaking into air intake system?
BAD
Repair air leak.
OK
Check spark plugs.
BAD
Repair or replace.
OK
Check distributor and ignition system.
BAD
Repair or replace.
OK
CONTINUED

TROUBLESHOOTING OXYGEN SENSOR–1988 PICK-UP AND 4-RUNNER 22R-E AND 22R-TE

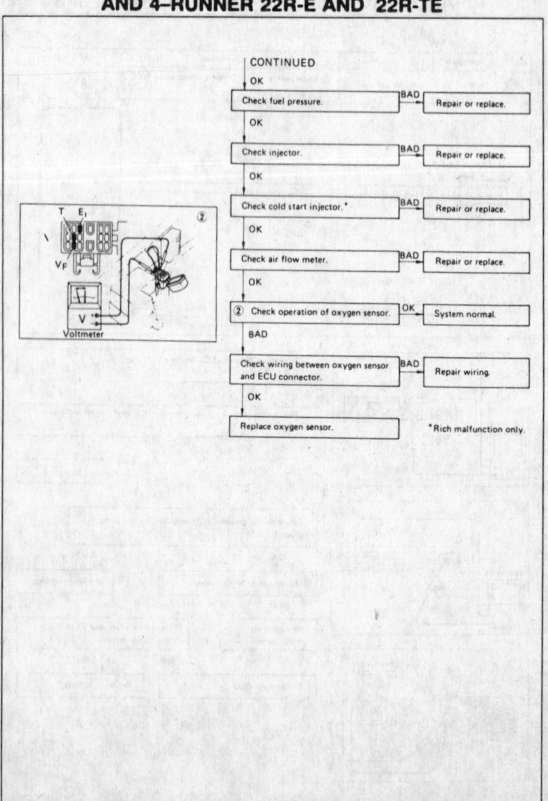
CONTINUED
OK
Check fuel pressure.
BAD
Repair or replace.
OK
Check injector.
BAD
Repair or replace.
OK
Check cold start injector.*
BAD
Repair or replace.
OK
Check air flow meter.
BAD
Repair or replace.
OK
② Check operation of oxygen sensor.
OK
System normal.
BAD
Check wiring between oxygen sensor and ECU connector.
BAD
Repair wiring.
OK
Replace oxygen sensor.
*Rich malfunction only.
T
E_1
V_F
V
Voltmeter

IDLE SPEED CONTROL (ISC) VALVE

Testing

2VZ-FE ENGINE

1. Check for operating sound from the ISC valve.
2. Disconnect the ISC valve connector.
3. Using an ohmmeter, measure the resistance between the B1 terminal (upper middle terminal) and the S1 or S3 (upper outer terminals), and between B2 terminal (lower middle terminal) and the S2 or S4 (lower outer terminals).
4. If the resistance is not 10–30 ohms, replace the ISC valve.
5. Apply battery voltage to terminals B1 and B2 while repeatedly grounding S1, S2, S3, S4, S1 in sequence. Check that valve moves toward the closed position.
6. Apply battery voltage to terminals B1 and B2 while repeatedly grounding S4, S3, S2, S1, S4 in sequence. Check that valve moves toward the opened position.
7. If valve does not function as specified, replace the ISC valve.

3S-FE ENGINE

1. Bring engine to normal operating temperature with transmission in **N**.
2. Using a jumper wire, connect the terminals TE1 or T and E1 of the check connector.
3. Raise engine rpm to 1000–1300 rpm for 5 seconds. If rpm returns as specified, the valve is functioning properly.
4. If rpm did not return, stop engine, disconnect the jumper wire.
5. Disconnect the ISC valve connector.
6. Using an ohmmeter, measure the resistance between the +B terminal (middle terminal) and the ISC1 and ISC2 (outer terminals).
7. If the resistance is not 16–17 ohms, replace the ISC valve.

3S-GTE ENGINE

1. Bring engine to normal operating temperature with transmission in **N**.
2. Disconnect the ISC valve connector.
3. Raise engine rpm to 1000–1300 rpm.
4. Reconnect the ISC valve. If rpm returns to 700–800 rpm the valve is functioning properly.
5. If rpm did not return, stop engine and disconnect the ISC valve connector.
6. Using an ohmmeter, measure the resistance between the +B terminal (middle terminal) and the ISC1 and ISC2 (outer terminals).
7. If the resistance is not 16–17 ohms, replace the ISC valve.

4A-GZE ENGINE

1. Disconnect the ISC valve connector.
2. Using an ohmmeter, measure the resistance between the +B terminal (middle terminal) and the ISC1 and ISC2 (outer terminals).
3. If the resistance is not 16–17 ohms, replace the ISC valve.
4. Apply battery voltage to the +B terminal and ground the ISC1 terminal; the valve should move toward the closed position.
5. Apply battery voltage to the +B terminal and ground the ISC2 terminal; the valve should move toward the opened position.

5M-GE

1. Confirm that there is a clicking sound from valve immediately after stopping the engine.
2. Disconnect the ISC valve connector.
3. Using an ohmmeter, measure the resistance between the B1 terminal and S1 or S3, and between B2 terminal and the S2 or S4 terminals.
4. If the resistance is not 10–30 ohms, replace the ISC valve.

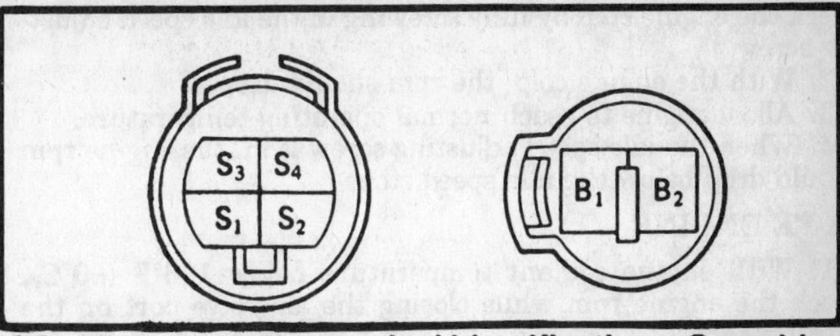
S3
S4
S1
S2
B1
B2

ISC valve connector terminal identification – Cressida and 1987-88 Supra

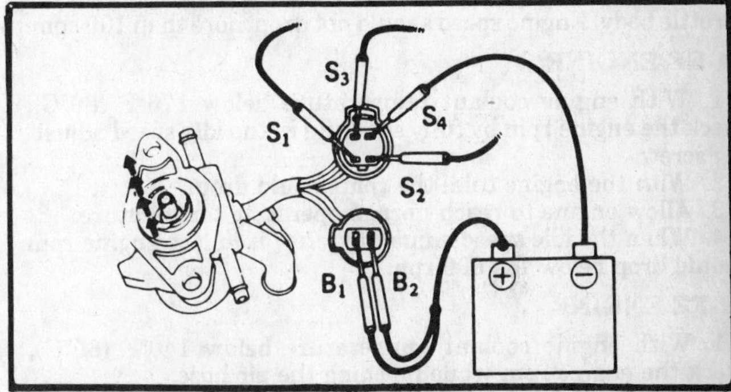
S3
S4
S1
S2
B1
B2

ISC inspection valve terminal identification – Cressida and 1987-88 Supra

5. Apply battery voltage to terminals B1 and B2 while repeatedly grounding S1, S2, S3, S4, S1 in sequence. Check that valve moves toward the closed position.
6. Apply battery voltage to terminals B1 and B2 while repeatedly grounding S4, S3, S2, S1, S4 in sequence. Check that valve moves toward the opened position.
7. If valve does not function as specified, replace the ISC valve.

7M-GE AND 7M-GTE ENGINES

1. Check for operating sound from the ISC valve.
2. Disconnect the ISC valve connector.
3. Using an ohmmeter, measure the resistance between the B1 terminal (upper middle terminal) and the S1 or S3 (upper outer terminals), and between B2 terminal (lower middle terminal) and the S2 or S4 (lower outer terminals).
4. If the resistance is not 10–30 ohms, replace the ISC valve.
5. Apply battery voltage to terminals B1 and B2 while repeatedly grounding S1, S2, S3, S4, S1 in sequence. Check that valve moves toward the closed position.
6. Apply battery voltage to terminals B1 and B2 while repeatedly grounding S4, S3, S2, S1, S4 in sequence. Check that valve moves toward the opened position.
7. If valve does not function as specified, replace the ISC valve.

AIR VALVE

Testing

3S-GE ENGINE

1. With engine coolant temperature below 176°F (80°C), check the engine rpm while pinching the air hose.
2. With the engine cold, the rpm should drop.
3. Allow engine to reach normal operating temperature.
4. Check engine rpm while pinching the air hose. Engine speed should not drop more than 100 rpm.

3VZ-E ENGINE

1. With engine coolant temperature below 176°F (80°C),

check the engine rpm by fully screwing in the idle speed adjusting screw.

2. With the engine cold, the rpm should drop.
3. Allow engine to reach normal operating temperature.
4. When the idle speed adjusting screw is in, the engine rpm should drop below the idle speed stop.

4A-FE ENGINE

1. With engine coolant temperature below 176°F (80°C), check the engine rpm while closing the air valve port on the throttle body.
2. With the engine cold, the rpm should drop.
3. Allow engine to reach normal operating temperature.
4. Check engine rpm while closing the air valve port on the throttle body. Engine speed should not drop more than 100 rpm.

4A-GE ENGINE

1. With engine coolant temperature below 176°F (80°C), check the engine rpm by fully screwing in the idle speed adjusting screw.
2. With the engine cold, the rpm should drop.
3. Allow engine to reach normal operating temperature.
4. When the idle speed adjusting screw is in, the engine rpm should drop below the 600 rpm.

4Y-FE ENGINE

1. With engine coolant temperature below 140°F (60°C), check the engine rpm while pinching the air hose.
2. With the engine cold the rpm should drop.
3. Allow engine to reach normal operating temperature.
4. Check engine rpm while pinching the air hose. Engine speed should not drop more than 50 rpm.
5. Using an ohmmeter check the resistance of air valve heating coil.
6. The resistance should be 40–60 ohms with coolant temperature above 176°F (80°C) and the valve closed.
7. If the air valve fails any test, replace the valve.

1989–90 22R-E ENGINE

1. With engine coolant temperature below 176°F (80°C), check the engine rpm by fully screwing in the idle speed adjusting screw.
2. With the engine cold, the rpm should drop.
3. Allow engine to reach normal operating temperature.
4. When the idle speed adjusting screw is in, the engine rpm should drop below the idle speed stop.

1988 22R-E AND 22R-TE ENGINE

1. With engine coolant temperature below 140°F (60°C), check the engine rpm while pinching the air hose.
2. With the engine cold, the rpm should drop.
3. Allow engine to reach normal operating temperature.
4. Check engine rpm while pinching the air hose. Engine speed should not drop more than 50 rpm.
5. Using an ohmmeter check the resistance of air valve heating coil.
6. The resistance should be 40–60 ohms, with coolant at normal operating temperature.
7. If the air valve fails any test, replace the valve.

AUXILIARY AIR VALVE

Some models are equipped with an auxiliary air valve. This air valve operates in the same manner as the main air valve. To remove the auxiliary air valve, the throttle body must first be removed and then the auxiliary air valve can be removed. This valve can be tested in the same manner as the main air valve.

THROTTLE POSITION SENSOR

Testing

1. Disconnect the sensor connector.
2. Insert a feeler gauge between the throttle stop screw and the stop lever.
3. Using a ohmmeter measure the resistance between the terminals.

AIR FLOW METER

The air flow meter and temperature sensor are tested and serviced as a single unit. If only testing the air temperature sensor, perform the resistance tests on terminals THA and E2.

Testing

1. Disconnect the air flow meter connector.
2. Using a ohmmeter measure the resistance between the terminals.
3. If the resistance is not as specified, replace the air flow meter. The air temperature sensor and air flow meter are replaced as a single unit.

MANIFOLD ABSOLUTE PRESSURE (MAP) SENSOR

Testing

1. Disconnect the vacuum sensor connector.
2. Turn the ignition switch **ON**.
3. Measure the voltage between terminals VCC and E2 (outer terminals) of the vacuum sensor connector.

THROTTLE POSITION SENSOR TESTING—3 WIRE

Engine	Clearance between lever and stop screw	Continuity between terminals		
		IDL-E1	PSW-E1	IDL-PSW
4Y-EC	0.57 mm (0.0224 in.)	Continuity	No continuity	No continuity
	0.85 mm (0.0335 in.)	No continuity	No continuity	No continuity
	Throttle valve fully opened position	No continuity	Continuity	No continuity
4A-FE	0.60 mm (0.0236 in.)	Continuity	No continuity	No continuity
	0.80 mm (0.0315 in.)	No continuity	No continuity	No continuity
	Throttle valve fully opened position	No continuity	Continuity	No continuity
3S-FE	0.50 mm (0.020 in.)	Continuity	No continuity	—
	0.90 mm (0.035 in.)	No continuity	No continuity	—
	Throttle valve fully opened	No continuity	Continuity	—

THROTTLE POSITION SENSOR TESTING—4 WIRE

Engine	Clearance between lever and stop screw	Between terminals	Resistance kilo ohms
2VZ-FE	(0 in.) 0 mm	VTA-E2	0.3–6.3
	(0.012 in.) 0.30 mm	IDL-E2	2.3 or less
	(0.028 in.) 0.70 mm	IDL-E2	Infinity
	Throttle valve fully opened	VTA-E2	3.5–10.3
	—	VC-E2	4.25–8.25
3S-FE, 3S-GE	(0 in.) 0 mm	VTA-E2	0.2–0.8
	(0.020 in.) 0.50 mm	IDL-E2	2.3 or less
	(0.028 in.) 0.70 mm	IDL-E2	Infinity
	Throttle valve fully opened	VTA-E2	3.3–10
	—	VC-E2	3–7
3S-GTE	(0 in.) 0 mm	VTA-E2	0.2–0.8
	(0.020 in.) 0.50 mm	IDL-E2	2.3 or less
	(0.028 in.) 0.70 mm	IDL-E2	Infinity
	Throttle valve fully opened	VTA-E2	3.3–10.3
	—	VC-E2	3–8.3
3VZ-E	(0 in.) 0 mm	VTA-E_2	0.2–0.8
	(0.0197 in.) 0.50 mm	IDL-E_2	Less than 2.3
	(0.0303 in.) 0.77 mm	IDL-E_2	Infinity
	Throttle valve fully opened position	VTA-E_2	3.3–10
	—	Vcc-E_2	4–9
4A-GE	(0 in.) 0 mm	VTA-E_2	0.2–0.8
	(0.0138 in.) 0.35 mm	IDL-E_2	2.3 or less
	(0.0232 in.) 0.59 mm	IDL-E_2	Infinity
	Throttle valve fully opened position	VTA-E_2	3.3–10
		Vcc-E_2	3–7
4A-GZE	(0 in.) 0 mm	VTA-E_2	0.2–0.8
	(0.0157 in.) 0.40 mm	IDL-E_2	2.3 or less
	(0.0256 in.) 0.65 mm	IDL-E_2	Infinity
	Throttle valve fully opened position	VTA-E_2	3.3–10
		Vcc-E_2	3–7
5M-GE	(0 in.) 0 mm	VTA-E_2	0.2–0.8
	(0.0197 in.) 0.50 mm	IDL-E_2	0
	(0.0354 in.) 0.90 mm	IDL-E_2	Infinity
	Throttle valve fully opened position	VTA-E_2	3.3–10
	—	Vc-E_2	3–7
7M-GE Cressida 7M-GTE Supra	(0 in.) 0 mm	VTA-E2	① 0.3–6.3
	(0.0197 in.) 0.50 mm	IDL-E2	2.3 or less
	(0.0354 in.) 0.90 mm	IDL-E2	Infinity
	Throttle valve fully opened	VTA-E2	3.5–10.3
	—	VC-E2	4.25–8.25
7M-GE Supra	(0 in.) 0 mm	VTA-E_2	① 0.3–6.3
	(0.0157 in.) 0.40 mm	IDL-E_2	Less than 2.3
	(0.0295 in.) 0.75 mm	IDL-E_2	Infinity
	Throttle valve fully opened position	VTA-E_2	3.5–10.3
	—	Vc-E_2	4.25–8.25

THROTTLE POSITION SENSOR TESTING—4 WIRE

Engine	Clearance between lever and stop screw	Between terminals	Resistance kilo ohms
22R-E, 22R-TE	(0 in.) 0 mm	VTA-E_2	0.2–0.8
	(0.0224 in.) 0.57 mm	IDL-E_2	Less than 2.3
	(0.0335 in.) 0.85 mm	IDL-E_2	Infinity
	Throttle valve fully opened position	VTA-E_2	3.3–10
	—	Vcc-E_2	② 4–9

NOTE: Resistance between terminals E_2 and VS will change in a wave pattern as the measuring plate slowly opens.

① 1989–90 Supra 0.2–1.2 kilo ohms.

② 1988 22R-E and 22R-TE 3–7 kilo ohms.

AIR FLOW METER TESTING

Engine	Terminals	Resistance (ohms)	Measuring plate opening
2YZ-FE, 3S-FE 3S-GE, 3S-GTE 3YZ-E, 5M-GE 7M-GE, 4A-GZE	FC–E1	Infinity	Fully closed
		Zero	Other than closed position
	VS–E2	200–600 ①	Fully closed
		20–1,200 ②	Fully open
	E2–VC	200–400	—
4Y-EC, 4A-GE 22R-E, 22R-TE	E1–FC	Infinity	Fully closed
		Zero	Other than closed position
	E2–VS	20–400	Fully closed
		20–1,000 ③	Fully open
	E2–VC	100–300	—
	E2–VB	200–400	—
7M-GTE	Ks–E_1	Infinity	—
	E_1–Ks	5,000–10,000	—
	Vc–E_1	10,000–15,000	—
	E_1–Vc	5,000–10,000	—

① 1988 Cressida 20–400 ohms

② Camry 20–1000 ohms

③ 4A-GE 20–300 ohms

AIR TEMPERATURE SENSOR TESTING

Between terminals	Resistance (ohms)	Temp. °F (°C)
E2–THA	10,000–20,000	−4 (−20)
	4,000–7,000	32 (0)
	2,000–3,000	68 (20)
	900–1,300	104 (40)
	400–700	140 (60)

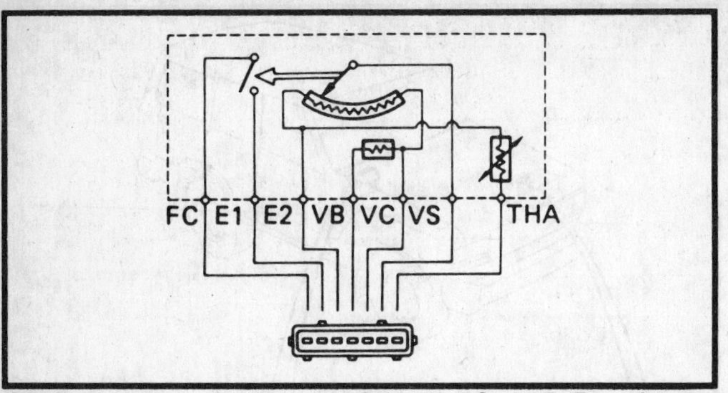
FC
E1
E2
VB
VC
VS
THA
Air flow meter circuit – 4A-GE, 4Y-EC, 22R-E and 22R-TE

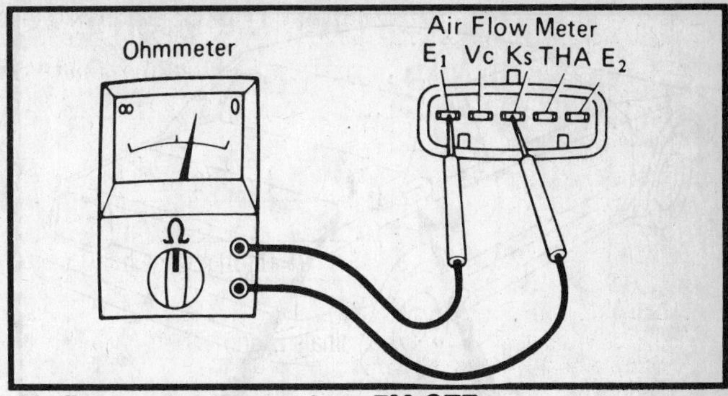
Ohmmeter
Air Flow Meter
E_1 Vc Ks THA E_2
Air flow meter connector – 7M-GTE

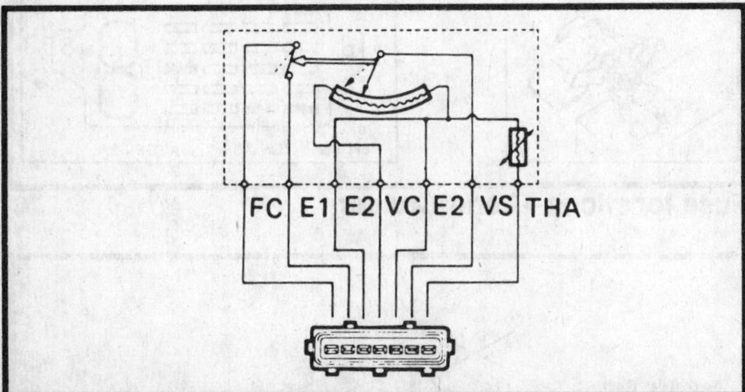
FC
E1
E2
VC
E2
VS
THA
Air flow meter circuit – 2VZ-FE, 3F-E, 3S-FE, 3S-GE, 3S-GTE, 3VZ-E, 4A-GZE and 5M-GE

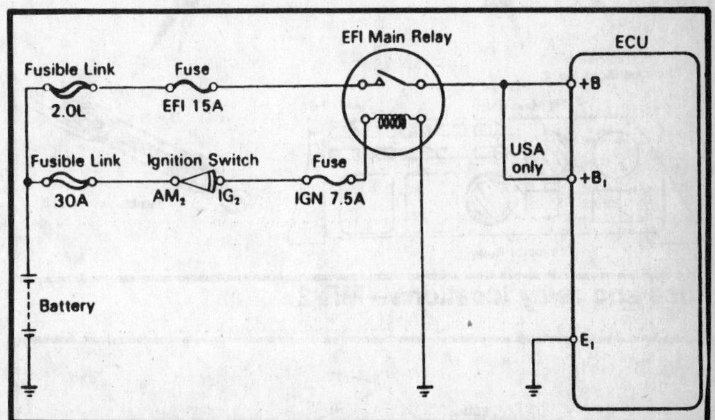
EFI Main Relay
ECU
Fusible Link
2.0L
Fuse
EFI 15A
+B
Fusible Link
30A
Ignition Switch
AM_2
IG_2
Fuse
IGN 7.5A
USA only
$+B_1$
Battery
E_1
Typical EFI main relay wiring schematic

4. The voltage should read 4–6 volts.
5. Turn ignition switch **OFF** and reconnect the sensor.
6. Disconnect the vacuum hose of the intake chamber side.
7. Turn the ignition switch **ON**.
8. Connect a voltmeter to terminals PIM and E2 of the ECU. Apply vacuum to the sensor and record the voltage drop.
9. The voltmeter should read:
   0.3–0.5 volts – 3.94 in. Hg
   0.7–0.9 volts – 7.87 in. Hg
   1.1–1.3 volts – 11.8 in. Hg
   1.5–1.7 volts – 15.6 in. Hg
   1.9–2.1 volts – 19.7 in. Hg

MAIN RELAY

Testing

EXCEPT 22R AND 5M-GE ENGINE

1. When the ignition switch is turn to the **ON** position, there should be a noise heard from the EFI main relay.
2. Locate the EFI main relay and remove it from the relay block. Using a suitable ohmmeter, check that there is continuity between terminals 1 and 3.
3. Check that there is no continuity between terminals 2 and 4.
4. If the continuity is not a specified, replace the relay.
5. Apply battery voltage across terminals 1 and 3. Check that there is continuity between terminals 2 and 4.
6. If the continuity is not a specified, replace the relay.

22R AND 5M-GE ENGINES

1. When the ignition switch is turn to the **ON** position, there should be a noise heard from the EFI main relay.
2. Locate the EFI main relay and remove it from the relay block. Using a suitable ohmmeter, check that the resistance between 1 and 2 is 40–60 ohms.

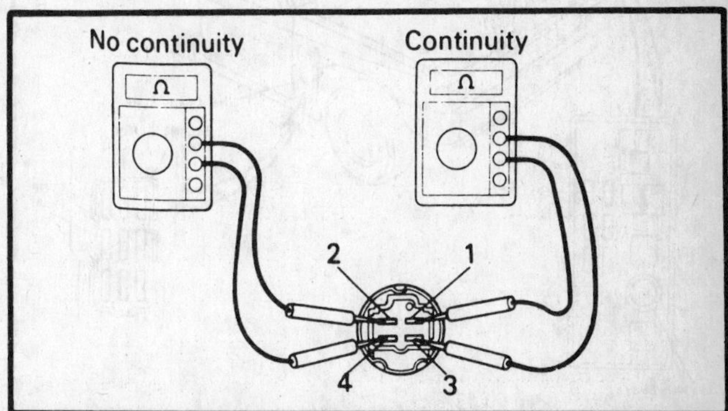
No continuity
Continuity
2
1
4
3
Testing the EFI main relay

3. Check that there is no continuity between terminals 3 and 4.
4. If the continuity is not a specified, replace the relay.

CIRCUIT OPENING RELAY

Testing

1. Start the engine.
2. Check for battery voltage at the Fp connector of the relay.
3. Turn ignition switch **OFF**.
4. Disconnect the relay connector.
5. Measure the resistance between terminals. Resistance should be:
   STA – E1 – 17–25 ohms
   +B – Fc – 88–112 ohms
   +B – Fp – infinity

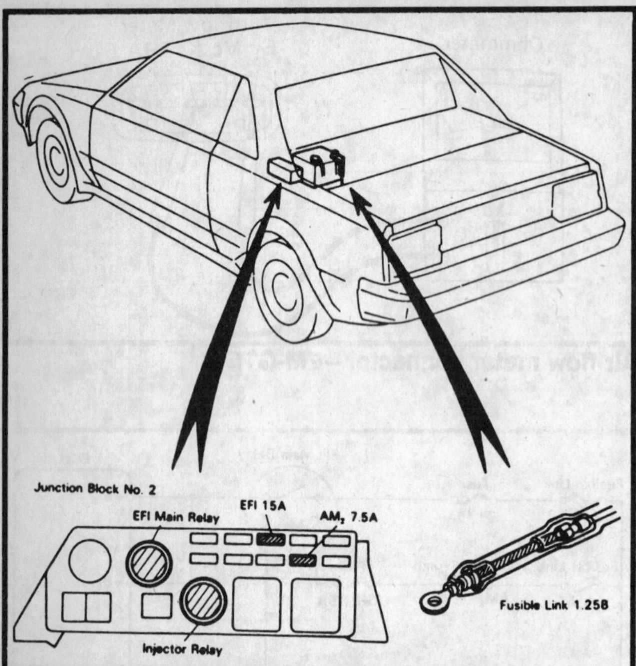
Junction Block No. 2
EFI Main Relay
EFI 15A
AM, 7.5A
Fusible Link 1.25B
Injector Relay
Fuse and relay locations – MR-2

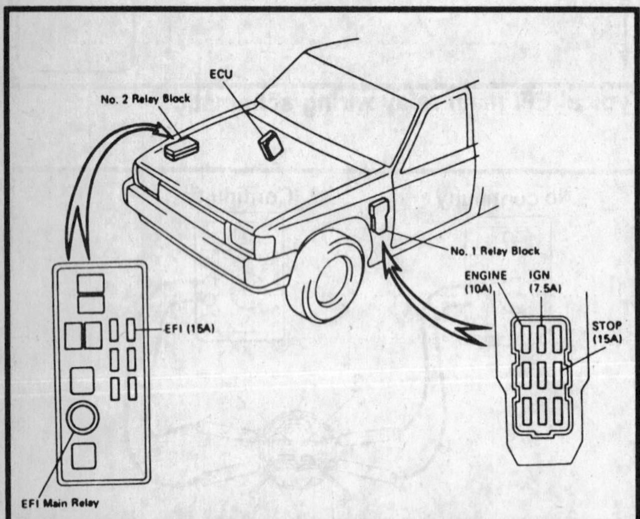
ECU
No. 2 Relay Block
No. 1 Relay Block
ENGINE (10A)
IGN (7.5A)
STOP (15A)
EFI (15A)
EFI Main Relay
Fuse and relay locations – 1989–90 Pick-Up

6. If the continuity is not a specified, replace the relay.

START INJECTOR TIME SWITCH

Testing

1. Disconnect the start injector time switch electrical connector.
2. Using a suitable ohmmeter, measure the resistance between each terminal.
3. If the resistance is not correct, replace the timer.

WATER THERMO SENSOR

Testing

1. Disconnect the electrical connector to the water thermo sensor.

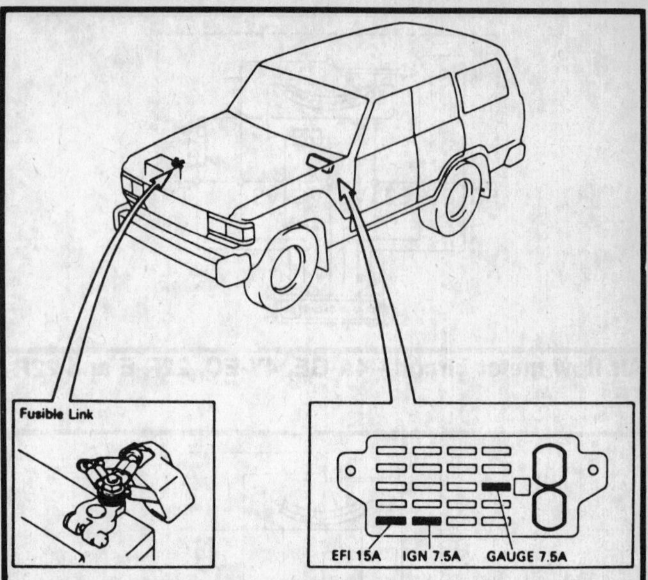
Fusible Link
EFI 15A
IGN 7.5A
GAUGE 7.5A
Fuse locations – Land Cruiser

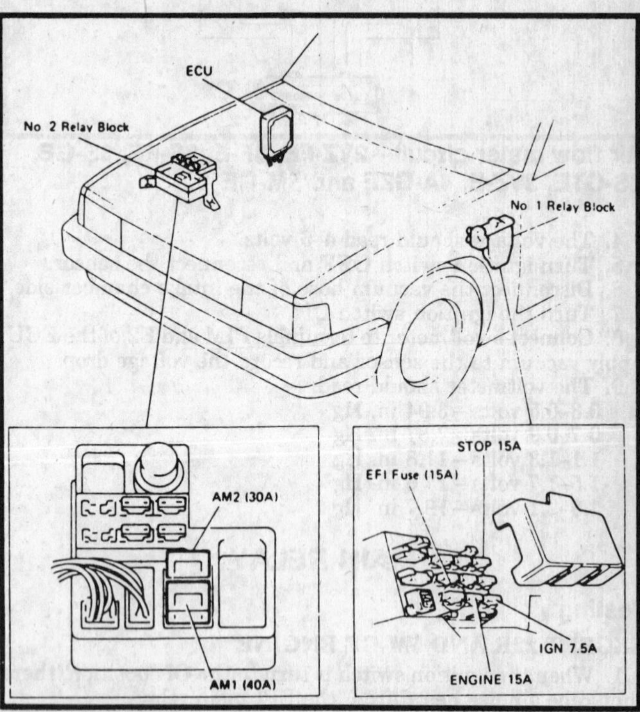
ECU
No. 2 Relay Block
No. 1 Relay Block
AM2 (30A)
AM1 (40A)
STOP 15A
EFI Fuse (15A)
IGN 7.5A
ENGINE 15A
Fuse and relay locations – 1988 Pick-Up and 4-Runner

2. Using a suitable ohmmeter, measure the resistance between both charts. Refer to the water thermo sensor resistance chart.

FUEL CUT RPM

Testing

1. Start the engine and let it run until it reaches normal operating temperature.
2. Disconnect the throttle position sensor connector from the throttle position sensor.

COLD START INJECTOR TIME SWITCH RESISTANCE TESTING

Engine	Between Terminals	Resistance (ohms)	Temperature °F (°C)
3S-GE, 3S-GTE	STA-STJ	30–50	50 (10)
3V2-E, 22RE	STA-STJ	70–90	77 (25)
3F-E	STA-Ground	30–90	—
3S-FE, 4A-GE	STA-STJ	20–40	86 (30)
4A-GZE, 5M-GE	STA-STJ	40–60	104 (40)
22R-TE	STA-Ground	20–80	—
2VZ-FE, 4Y-EC	STA-STJ	25–45	59 (15)
7M-GE, 7M-GTE	STA-STJ	65–85	86 (30)
	STA-Ground	25–85	—

FUEL CUT RPM

Engine	Fuel Cut (rpm)	Fuel Return (rpm)
2VZ-FE	1800	1200
3S-FE	1700	1300
3S-GE, 3S-GTE	2000	1600
3F-E, 3VZ-E	1300	1000
4A-FE, 4A-GE, 4A-GZE	1600 A/C Off	1200 A/C Off
	1900 A/C Off	1500 A/C On
4Y-EC	2200	1800
5M-GE, 7M-GE	1800	1200
7M-GTE	1600	1200
22R-E 2WD A.T. Stoplight On	1300	1000
22R-E exc. 2WD A.T. w/Stoplight On	1900	1600
22R-TE	1800	1600

3. Short circuit terminals E_1 (or E_2) and IDL on the wire connector side. Gradually raise the engine rpm and check that there is fluctuation between the fuel cut and fuel return points. The fuel cut rpm should be as specified.

NOTE: The vehicle should be stopped.

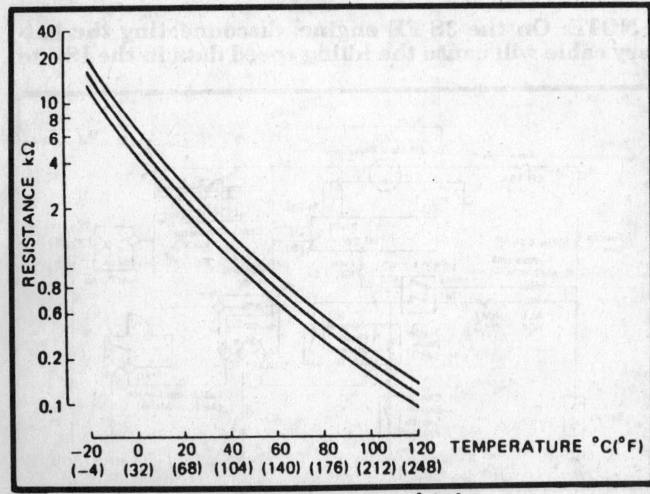
RESISTANCE kΩ
40
20
10
8
6
4
2
1
0.8
0.6
0.2
0.1
−20 (−4)
0 (32)
20 (68)
40 (104)
60 (140)
80 (176)
100 (212)
120 (248)
TEMPERATURE °C(°F)

Water thermo sensor resistance chart

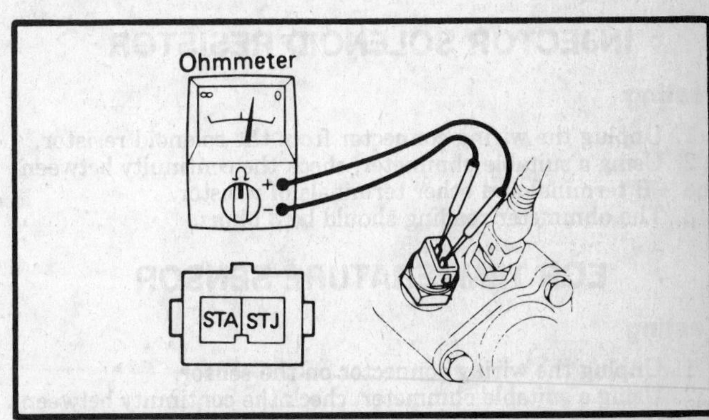
Ohmmeter
STA
STJ

Making a resistance test on the start injection time switch

COLD START INJECTOR

Testing

1. Unplug the wiring connector on the cold start injector.
2. Using a suitable ohmmeter, check the continuity of both terminals.
3. The ohmmeter reading should be as follows:
   2VZ-FE engine – 2–4 ohms

3S-FE engine – 2–4 ohms
3F-E engine – 2–4 ohms
3S-GE engine – 3–5 ohms
3S-GTE engine – 2–4 ohms
3VZ-E engine – 2–4 ohms
4A-FE engine – 3–5 ohms
4A-GE engine – 3–5 ohms
4A-GZE engine – 2–4 ohms
4Y-EC engine – 3–5 ohms
5M-GE engine – 3–5 ohms
7M-GE engine – 2–4 ohms
7M-GTE engine – 2–4 ohms
22R-E engine – 2–4 ohms
22R-TE engine – 2–4 ohms

4. If the resistance is not as specified, replace the injector.

INJECTOR

Resistance Test

1. Unplug the wiring connector on the injector.
2. Using a suitable ohmmeter, check the continuity of both terminals.
3. The ohmmeter reading should be as follows:

2VZ-FE engine – 13.8 ohms
3S-FE engine – 13.8 ohms
3F-E engine – 13.8 ohms
3S-GE engine – 13.8 ohms
3S-GTE engine – 2–4 ohms
3VZ-E engine – 13.8 ohms
4A-FE engine – 13.8 ohms
4A-GE engine – 13.8 ohms
4A-GZE engine – 2.9 ohms
4Y-EC engine – 1.1–2.2 ohms
5M-GE engine – 1.5–3.0 ohms
7M-GE engine – 1.8–3.4 ohms 1988
7M-GE engine – 13.8 ohms except 1988
7M-GTE engine – 2.0–4.0 ohms
22R-E engine – 1.75 ohms 1988
22R-E engine – 13.8 ohms except 1988
22R-TE engine – 1.6 ohms

4. If the resistance is not as specified, replace the injector.

INJECTOR SOLENOID RESISTOR

Testing

1. Unplug the wiring connector from the solenoid resistor.
2. Using a suitable ohmmeter, check the continuity between the +B terminal and other terminals of resistor.
3. The ohmmeter reading should be 3 ohms.

EGR TEMPERATURE SENSOR

Testing

1. Unplug the wiring connector on the sensor.
2. Using a suitable ohmmeter, check the continuity between the terminals.
3. The ohmmeter reading should be as follows:

69.40–88.50 ohms – 112°F (50°C)
11.89–14.37 ohms – 212°F (100°C)
2.79–3.59 ohms – 302°F (150°C)

4. If the resistance is not as specified, replace the sensor.

VARIABLE INDUCTION SYSTEM (VIS) VSV

Testing

3S-GE

1. Disconnect the VIS Vacuum Solenoid Valve (VSV).
2. Check the resistance between the terminal. The resistance should be 33–39 ohms cold.
3. Check that there is no continuity between terminals and valve housing.
4. Check that vacuum flows through the hose connections without battery voltage.
5. Check that vacuum is bled off through the filter when battery voltage is applied to terminals.

3S-GTE

1. Disconnect the VIS Vacuum Solenoid Valve (VSV).
2. Check the resistance between the terminal. The resistance should be 33–39 ohms cold.
3. Check that there is no continuity between terminals and valve housing.
4. Check that vacuum is bled off through the filter when no battery voltage is applied.
5. Check that flows through the hoses and is not bled off when battery voltage is applied to terminals.

OXYGEN HEATER SENSOR

Testing

1. Unplug the wiring connector on the sensor.
2. Using a suitable ohmmeter, check the continuity between the +B and HT terminals.
3. The ohmmeter reading should be 5.1–6.3 ohms at 68°F (20°C).
4. If the resistance is not as specified, replace the sensor.

SUB-OXYGEN SENSOR

Testing

NOTE: The sub-oxygen sensor is only used on California vehicles. Only follow these procedures if a Code 27 is displayed.

1. Clear all codes.
2. Bring engine up to normal operating temperature.
3. Drive the vehicle at least 5 minutes at less than 50 mph in **D** range.
4. Fully depress the accelerator for 2 seconds or more.
5. Stop the vehicle and turn the ignition switch **OFF**.
6. Repeat Steps 2 through 5 to test acceleration action.
7. Check ECU codes. If Code 27 appears, check the sensor circuit. If circuit is normal, replace the sub-oxygen sensor.

Component Replacement

NOTE: On the 3S-FE engine, disconnecting the battery cable will cause the idling speed data in the ISC to

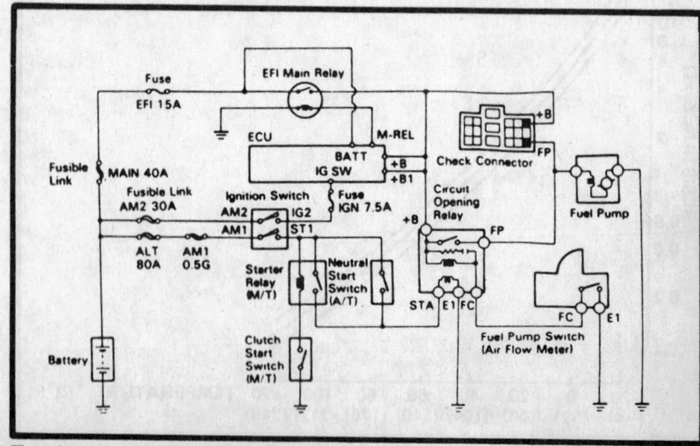
Fuse
EFI 15A
EFI Main Relay
ECU
M-REL
BATT
IG SW
+B
+B1
Check Connector
FP
Fusible Link
MAIN 40A
Fusible Link
AM2 30A
Ignition Switch
AM2
IG2
AM1
ST1
Fuse
IGN 7.5A
Circuit Opening Relay
Fuel Pump
ALT
80A
AM1
0.5G
Starter Relay (M/T)
Neutral Start Switch (A/T)
STA
E1
FC
Fuel Pump Switch (Air Flow Meter)
Clutch Start Switch (M/T)
Battery

Fuel pump circuit – 2VZ-FE

OXYGEN SENSOR TESTING—TYPICAL

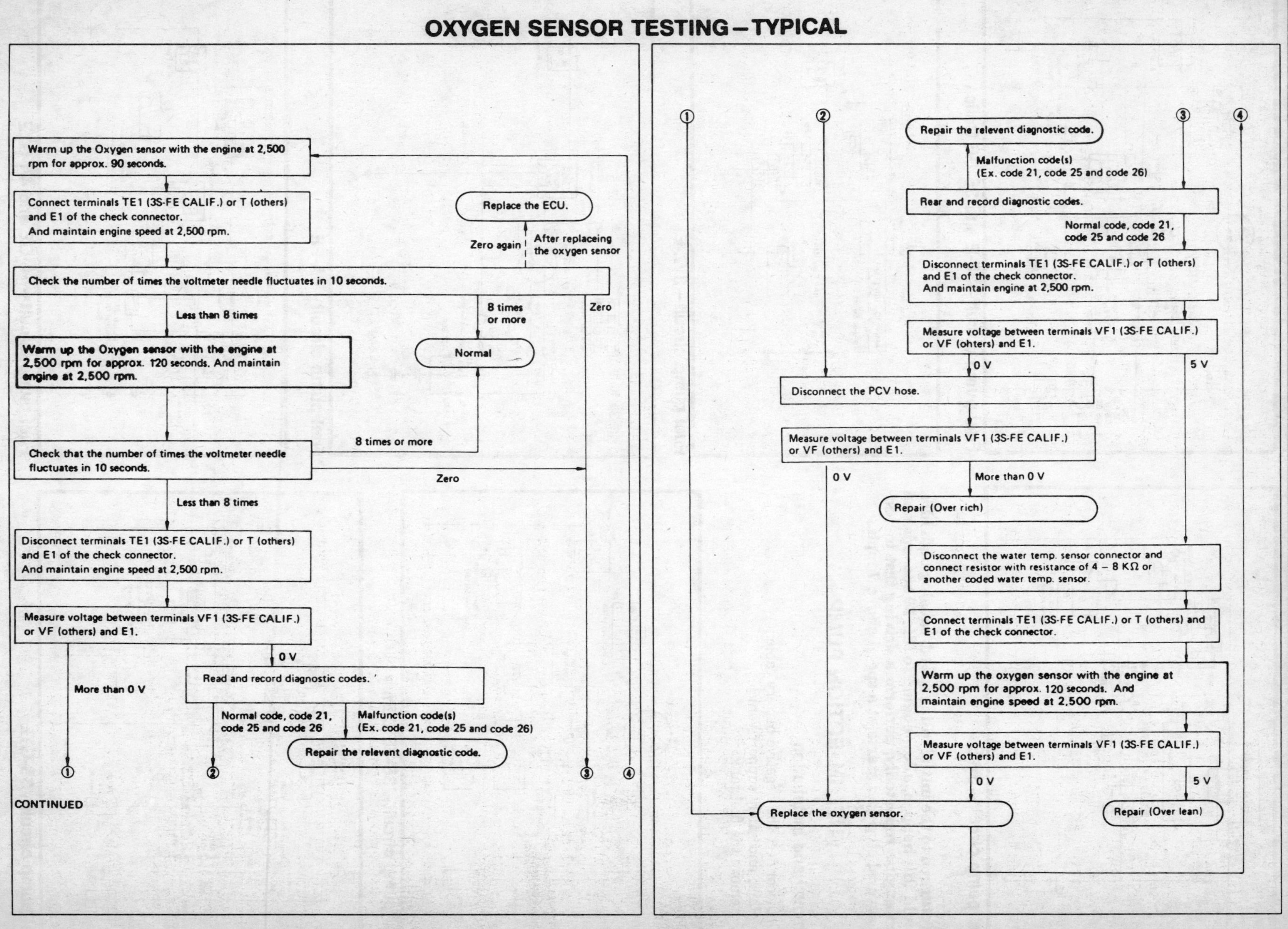
Warm up the Oxygen sensor with the engine at 2,500 rpm for approx. 90 seconds.
Connect terminals TE1 (3S-FE CALIF.) or T (others) and E1 of the check connector. And maintain engine speed at 2,500 rpm.
Check the number of times the voltmeter needle fluctuates in 10 seconds.
Less than 8 times
8 times or more
Zero
Normal
Replace the ECU.
Zero again
After replaceing the oxygen sensor
Warm up the Oxygen sensor with the engine at 2,500 rpm for approx. 120 seconds. And maintain engine at 2,500 rpm.
Check that the number of times the voltmeter needle fluctuates in 10 seconds.
8 times or more
Zero
Less than 8 times
Disconnect terminals TE1 (3S-FE CALIF.) or T (others) and E1 of the check connector. And maintain engine speed at 2,500 rpm.
Measure voltage between terminals VF1 (3S-FE CALIF.) or VF (others) and E1.
0 V
More than 0 V
Read and record diagnostic codes.
Normal code, code 21, code 25 and code 26
Malfunction code(s) (Ex. code 21, code 25 and code 26)
Repair the relevent diagnostic code.
1
2
3
4
CONTINUED
1
2
3
4
Repair the relevent diagnostic code.
Malfunction code(s) (Ex. code 21, code 25 and code 26)
Rear and record diagnostic codes.
Normal code, code 21, code 25 and code 26
Disconnect terminals TE1 (3S-FE CALIF.) or T (others) and E1 of the check connector. And maintain engine at 2,500 rpm.
Measure voltage between terminals VF1 (3S-FE CALIF.) or VF (ohters) and E1.
0 V
5 V
Disconnect the PCV hose.
Measure voltage between terminals VF1 (3S-FE CALIF.) or VF (others) and E1.
0 V
More than 0 V
Repair (Over rich)
Disconnect the water temp. sensor connector and connect resistor with resistance of 4 – 8 KΩ or another coded water temp. sensor.
Connect terminals TE1 (3S-FE CALIF.) or T (others) and E1 of the check connector.
Warm up the oxygen sensor with the engine at 2,500 rpm for approx. 120 seconds. And maintain engine speed at 2,500 rpm.
Measure voltage between terminals VF1 (3S-FE CALIF.) or VF (others) and E1.
0 V
5 V
Replace the oxygen sensor.
Repair (Over lean)

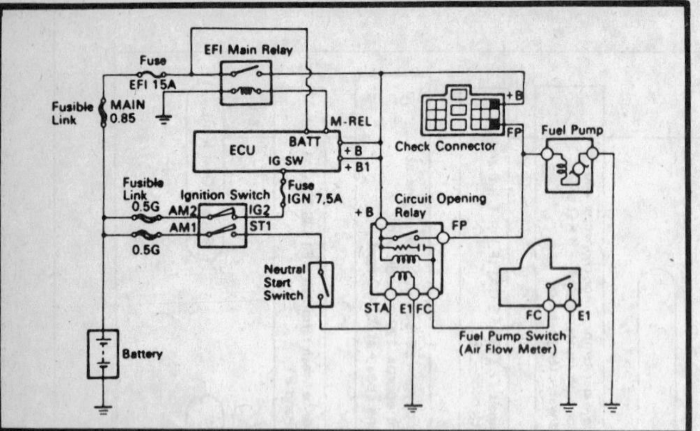
Fuel pump circuit—3F-E

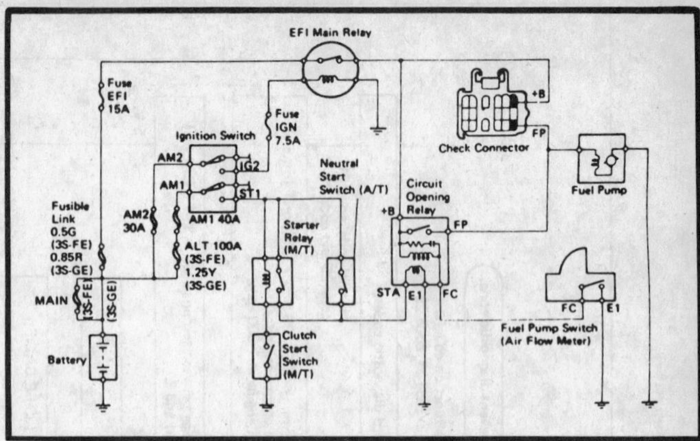
Fuel pump circuit—3F-E and 3S-GE Celica

be returned to the initial idling speed, causing high idle. Should this happen, allow engine to idle 30 seconds and start engine repeatedly, perform a driving test including several stop and starts at a speed above 7 mph.

FUEL INJECTION PUMP

Removal and Installation

1. Disconnect the negative battery cable.
2. Raise and safely support vehicle.
3. Remove the fuel tank.

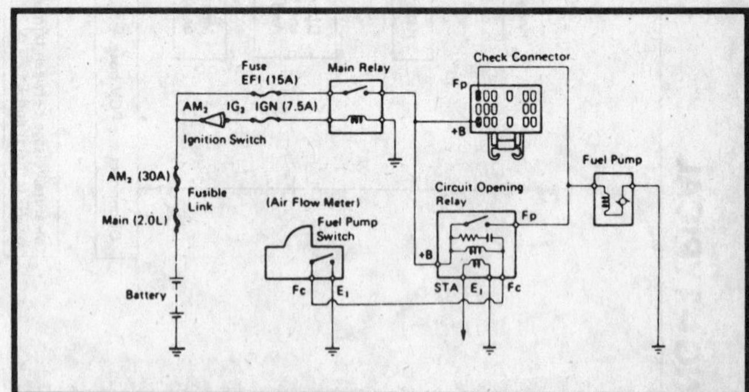
Fuel pump circuit—3VZ-E

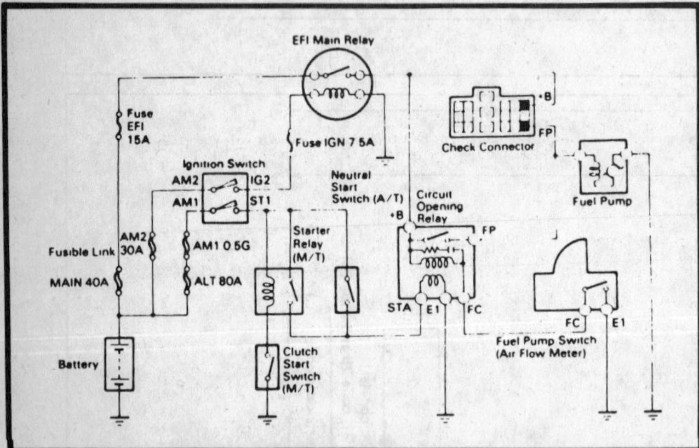
Fuel pump circuit—3S-FE Camry

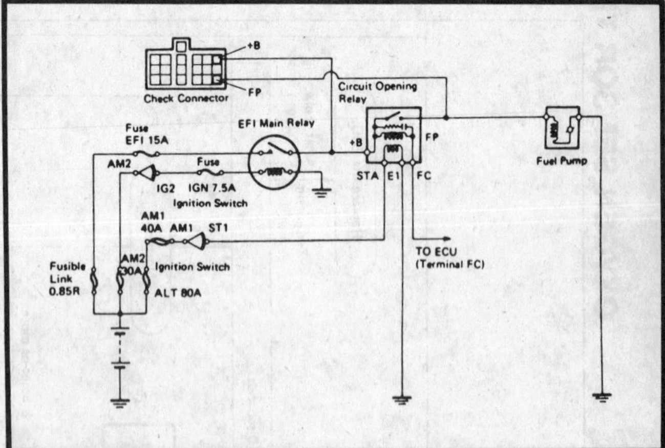
Fuel pump circuit—4A-FE

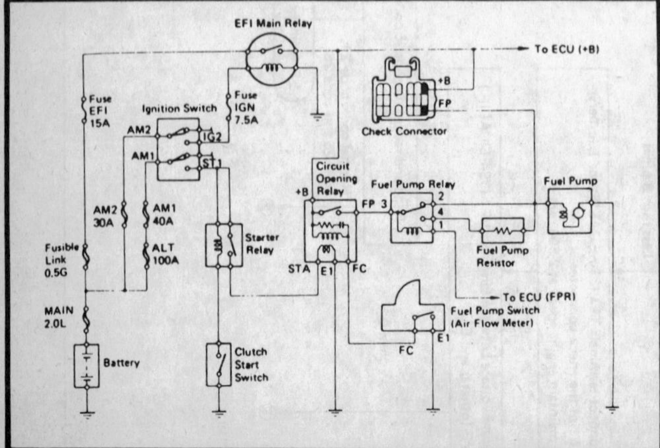
Fuel pump circuit—3S-GTE

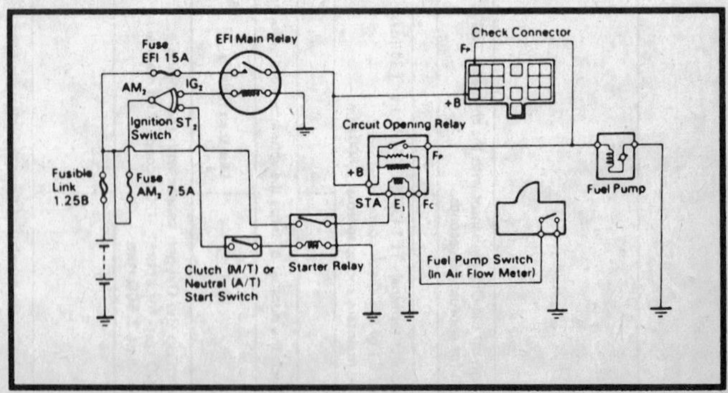
Fuel pump circuit—4A-GE and 4A-GZE

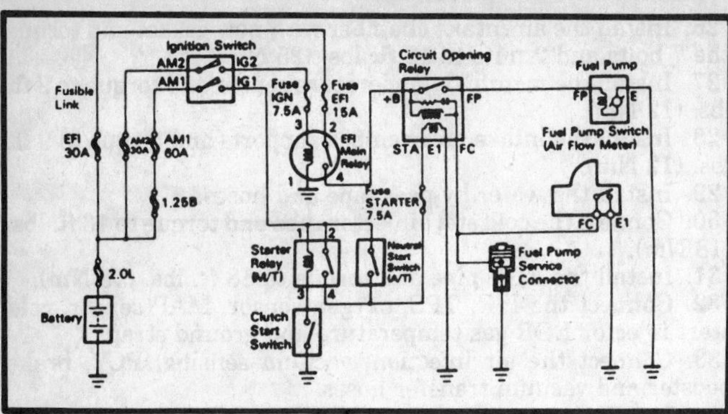
Ignition Switch
AM2
IG2
AM1
IG1
Fusible Link
EFI 30A
AM2 30A
AM1 60A
Fuse IGN 7.5A
Fuse EFI 15A
EFI Main Relay
Circuit Opening Relay
Fuel Pump
+B
FP
STA
E1
FC
Fuel Pump Switch (Air Flow Meter)
Fuse STARTER 7.5A
1.25B
2.0L
Battery
Starter Relay (M/T)
Neutral Start Switch (A/T)
Clutch Start Switch
Fuel Pump Service Connector

Fuel pump circuit – 4Y-CE

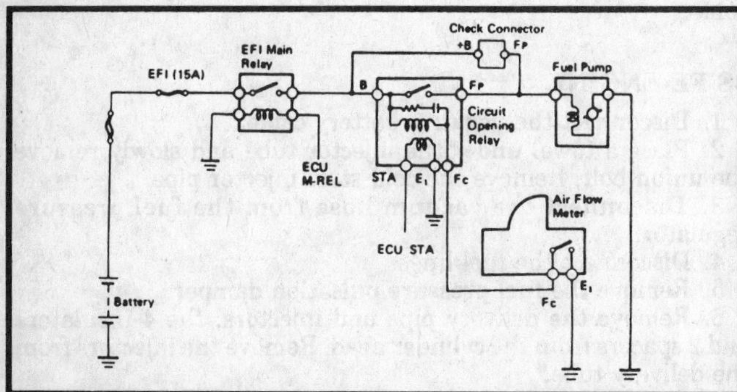
Check Connector
EFI Main Relay
EFI (15A)
Fuel Pump
Circuit Opening Relay
ECU M-REL
STA
Air Flow Meter
ECU STA
Battery

Fuel pump circuit – 5M-GE

4. Remove the retaining screws which secure the pump access plate to the fuel tank. Withdraw the plate, gasket and pump assembly.
5. Disconnect the leads and hoses from the fuel pump.
6. Install pump assembly with new gasket.
7. Install the fuel tank.

FUEL INJECTOR

Removal and Installation

NOTE: Before removal, tag all the hoses and electrical connections for reference during reassembly.

2VZ-FE ENGINE

1. Disconnect the negative battery cable.
2. Drain the engine coolant.
3. Disconnect the accelerator cable and throttle cable from the throttle body, if equipped.
4. Remove the air cleaner, air flow meter and air cleaner hose.
5. Disconnect PCV, vacuum sensing, emission control and water by-pass hoses.
6. Disconnect ISC, TPS and EGR temperature sensor connectors.
7. Remove the No. 1 right engine mount.
8. Disconnect the cold start injector.
9. Place a towel under the injector tube, slowly remove the union bolt. Remove the injector tube.
10. Disconnect the brake booster hose, PS vacuum and air hoses and cruise control hose.
11. Disconnect the ground strap and remove the nut to disconnect the wire clamp.
12. Disconnect the EGR pipe.
13. Remove the 2 bolts and the No. 1 engine hanger.
14. Remove the bolt, and disconnect the air intake chamber support from the air intake chamber.

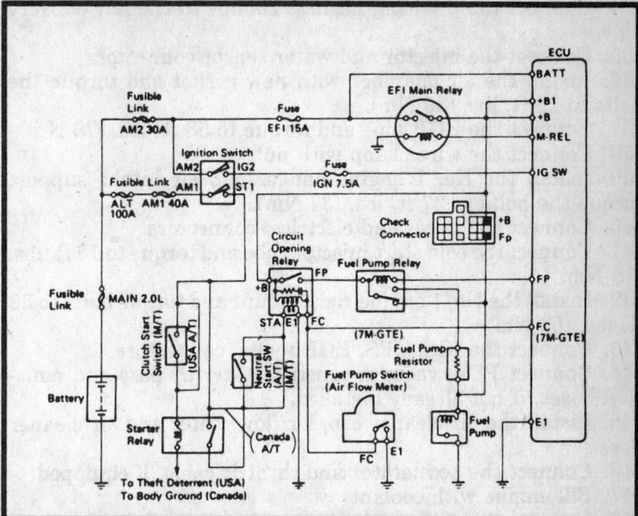
ECU
BATT
EFI Main Relay
+B1
+B
M-REL
IG SW
FP
FC
(7M-GTE)
E1
Fusible Link
AM2 30A
Fuse EFI 15A
Ignition Switch
AM2
IG2
AM1
ST1
Fuse IGN 7.5A
Fusible Link ALT 100A
AM1 40A
Check Connector
Circuit Opening Relay
Fuel Pump Relay
MAIN 2.0L
STA
Clutch Start Switch (M/T)
(USA A/T)
Neutral Start SW (A/T)
Fuel Pump Resistor
Fuel Pump Switch (Air Flow Meter)
Battery
Starter Relay
(Canada) A/T
Fuel Pump
To Theft Deterrent (USA)
To Body Ground (Canada)

Fuel pump circuit – 7M-GE and 7M-GTE

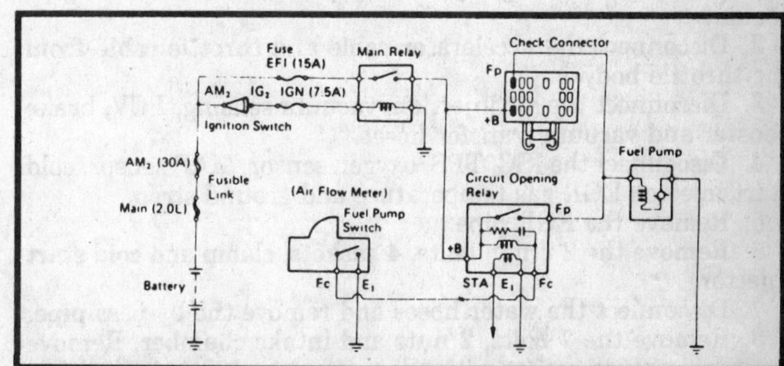
Fuse EFI (15A)
Main Relay
Check Connector
IGN (7.5A)
Ignition Switch
AM2 (30A)
Fusible Link
Main (2.0L)
(Air Flow Meter)
Fuel Pump Switch
Circuit Opening Relay
Fuel Pump
Battery
STA

Fuel pump circuit – 22R-E and 22R-TE

15. Remove the air intake chamber.
16. Disconnect the injector connectors and the water temperature sensor connector.
17. Disconnect the fuel lines.
18. Remove the fuel pipe.
19. Remove the 2 bolts and the left delivery pipe together with the injectors.
20. Remove the 2 bolts and the right delivery pipe together with the injectors.
21. Remove the injectors from the delivery pipe.
22. Remove the 6 insulators and 4 spacers from the intake manifold.

To install:

23. Install a new grommet to the injector.
24. Apply a light coat of gasoline to the new O-ring and install it to the injector.
25. While twisting the injector, install it into the delivery pipe.
26. Place the 6 insulators and 4 spacers in position on the intake manifold.
27. Install the right side injectors with the delivery pipe on the intake manifold.
28. Install the left side injectors with the delivery pipe on the intake manifold.
29. Check that the injectors rotate smoothly; if not, replace O-rings.
30. Position the injector connector upward.
31. Install the 5 bolts and torque to 9 ft. lbs. (13 Nm).
32. Install the No. 2 fuel pipe with 4 new gaskets and torque bolts to 21 ft. lbs. (29 Nm).
33. Install the fuel lines.

34. Connect the 3 wiring harness clamps to the left delivery pipe.
35. Connect the injector and water sensor connectors.
36. Install the air chamber with new gasket and torque the bolts to 29 ft. lbs. (39 Nm).
37. Connect the EGR pipe and torque to 58 ft. lbs. (78 Nm).
38. Connect the wire clamp with nut.
39. Install the No. 1 engine hanger and air intake support torque the bolts to 27 ft. lbs. (37 Nm).
40. Connect the hoses and electrical connectors.
41. Connect the cold start injector tube and torque to 13 ft. lbs. (18 Nm).
42. Install the No. 1 engine right mount and torque bolts to 38 ft. lbs. (52 Nm).
43. Connect the ISC, TPS, EGR temp., connectors.
44. Connect PCV, vacuum sensing, water by-pass and emission hoses, if not already installed.
45. Install the air cleaner cap, air flow meter and air cleaner hose.
46. Connect the accelerator and throttle cable, if equipped.
47. Fill engine with coolant.
48. Connect the negative battery cable.

3F-E ENGINE

1. Disconnect the negative battery cable and drain the engine coolant.
2. Disconnect the accelerator cable and throttle cable from the throttle body.
3. Disconnect the air injection, vacuum sensing, PCV, brake booster and vacuum transfer hoses.
4. Disconnect the ISC, TPS, oxygen sensor, MAP sensor, cold start injector EGR gas temperature and ground strap.
5. Remove the EGR pipe.
6. Remove the 2 union bolts, 4 gaskets, clamp and cold start injector.
7. Disconnect the water hoses and remove the by-pass pipe.
8. Remove the 7 bolts, 2 nuts and intake chamber. Remove the temperature sensor with the support and the air chamber gasket.
9. Disconnect the cold start injector time switch, water temperature sensor, water temperature sender, water temperature switch, and the injector connectors.
10. Tag and disconnect the fuel and vacuum hoses.
11. Disconnect the fuel lines, pulsation damper and pressure regulator.
12. Remove the 3 nuts, plate washers, spacers and delivery pipe with the injectors.
13. Remove the 6 insulators, 6 spacers and 3 collars from the intake manifold.

To install:

14. Install a new grommet to the injector.
15. Apply a light coat of gasoline to the new O-ring and install it to the injector.
16. While twisting the injector, install it into the delivery pipe.
17. Place the 6 insulators and 6 spacers and 3 collars in position on the intake manifold.
18. Install the injectors with the delivery pipe on the intake manifold.
19. Check that the injectors rotate smoothly; if not, replace O-rings.
20. Position the injector connector upward.
21. Install the 3 spacers, plate washer and nuts. Torque to 9 ft. lbs. (12 Nm).
22. Install the fuel line to the pressure regulator and torque to 14 ft. lbs. (19 Nm).
23. Install the fuel line to the pressure damper and torque to 22 ft. lbs. (29 Nm).
24. Connect the vacuum and fuel hoses.
25. Connect the cold start injector, water temperature sensor, water temperature sender, water temperature switch and injector connectors.
26. Install the air intake chamber with new gasket and torque the 7 bolts and 2 nuts to 18 ft. lbs. (25 Nm0.
27. Install the manifold temperature sensor and torque to 9 ft. lbs. (12 Nm).
28. Install the intake air chamber supports and torque to 9 ft. lbs. (12 Nm).
29. Install the water by-pass pipe and hoses.
30. Connect the cold start injector tube and torque to 13 ft. lbs. (18 Nm).
31. Install the EGR pipe and torque to 58 ft. lbs. (78 Nm).
32. Connect the ISC, TPS, oxygen sensor, MAP sensor, cold start injector EGR gas temperature and ground strap.
33. Connect the air injection, vacuum sensing, PCV, brake booster and vacuum transfer hoses.
34. Install the air intake hose.
35. Connect the accelerator and throttle cables.
36. Fill the cooling system and connect the negative battery cable.

3S-FE ENGINE

1. Disconnect the negative battery cable.
2. Place a towel under the injector tube and slowly remove the union bolt. Remove the cold start injector pipe.
3. Disconnect the vacuum hose from the fuel pressure regulator.
4. Disconnect the fuel lines.
5. Remove the fuel pressure pulsation damper.
6. Remove the delivery pipe and injectors, the 4 insulators and 2 spacers from the cylinder head. Remove the injectors from the delivery tube.

To install:

7. Install a new grommet to the injector.
8. Apply a light coat of gasoline to the new O-ring and install it to the injector.
9. While twisting the injector, install it into the delivery pipe.
10. Place the 6 insulators and 4 spacers in position on the intake manifold.
11. Install the injectors with the delivery pipe onto the cylinder head.
12. Check that the injectors rotate smoothly; if not, replace O-rings.
13. Position the injector connector upward.
14. Install the 2 bolts and torque to 9 ft. lbs. (13 Nm).
15. Connect the fuel lines.
16. Connect the vacuum hoses.
17. Connect the cold start injector tube and torque to 13 ft. lbs. (18 Nm).
18. Connect the negative battery cable.

3S-GE ENGINE

1. Disconnect the negative battery cable.
2. Drain the engine coolant.
3. Disconnect the accelerator cable and throttle cable from the throttle body, if equipped.
4. Disconnect the coil connector and coil wire and remove the suspension upper brace.
5. Remove the air cleaner hose.
6. Remove the igniter.
7. Remove the suspension lower crossmember.
8. Disconnect the throttle position sensor.
9. Tag and disconnect the PCV, water by-pass, air tube and emission control hoses.
10. Remove the throttle body.
11. Disconnect the EGR pipe.
12. Remove the intake manifold by the following procedure:
   a. Remove the No. 2 engine hanger and No. 2 intake manifold support.
   b. Disconnect the cold start injector connector.
   c. Place a towel under the injector tube, slowly remove the union bolt. Remove the injector tube.

d. Disconnect the brake booster hose, PS vacuum and air hoses and cruise control hose.

e. Disconnect the VSV hose from the V-ISC and the vacuum sensing hose.

f. Disconnect the injector connectors.

g. Disconnect the fuel lines.

h. Remove the No. 1 and No. 3 intake manifold supports.

i. Disconnect the ground strap and remove the nut to disconnect the wire clamp.

j. Disconnect the VSV connector and power steering hoses.

k. Remove the intake manifold and air control valve.

13. Remove the fuel delivery pipe with the injectors.

14. Remove the injectors from the delivery pipe.

15. Remove the 4 insulators and 3 spacers from the cylinder head.

To install:

16. Install a new grommet to the injector.

17. Apply a light coat of gasoline to the new O-ring and install it to the injector.

18. While twisting the injector, install it into the delivery pipe.

19. Place the 4 insulators and 3 spacers in position on the cylinder head.

20. Check that the injectors rotate smoothly; if not, replace O-rings.

21. Position the injector connector upward.

22. Install the 3 bolts and torque to 14 ft. lbs. (19 Nm).

23. Install the intake manifold by the following procedure:

a. Install the intake manifold and air control valve. Torque the bolts to 14 ft. lbs. (19 Nm).

b. Connect the VSV connector and power steering hoses.

c. Connect the ground strap and nut to the wire clamp.

d. Install the No. 1 and No. 3 intake manifold supports, torque intake manifold side to 14 ft. lbs. (19 Nm) or torque cylinder block side to 19 ft. lbs. (25 Nm).

e. Connect the fuel lines.

f. Connect the injector connectors.

g. Connect the VSV hose to the V-ISC and the vacuum sensing hose.

h. Connect the brake booster hose, PS vacuum and air hoses and cruise control hose.

i. Install the cold start injector and torque to 48 inch lbs. (5.4 Nm).

j. Install the cold start injector pipe and torque the bolts to 13 ft. lbs. (18 Nm).

k. Connect the cold start injector connector.

l. Install the the No. 2 engine hanger and No. 2 intake manifold support. Torque the 12mm bolts to 14 ft. lbs. (19 Nm) and the 14mm bolts to 29 ft. lbs. (39 Nm).

24. Connect the EGR pipe.

25. Install the throttle body and throttle position sensor.

26. Connect vacuum, emission and water hoses.

27. Install the suspension lower crossmember. Torque the 5 crossmember bolts to 154 ft. lbs. (208 Nm) and the center mounting bolt to 29 ft. lbs. (39 Nm).

28. Install the igniter.

29. Connect the coil connector and coil wire and suspension upper brace.

30. Install the air cleaner hose.

31. Install the accelerator cable and throttle cable from the throttle body, if equipped.

32. Fill coolant system and connect the negative battery cable.

3S-GTE ENGINE

1. Disconnect the negative battery cable.
2. Drain coolant.
3. Remove the throttle body.
4. Remove the fuel pressure regulator.
5. Disconnect the injector connectors.
6. Disconnect the fuel lines.
7. Remove the delivery pipe and injectors, the 4 insulators and 3 spacers from the cylinder head. Remove the injectors from the delivery tube.

To install:

8. Install a new grommet to the injector.
9. Apply a light coat of gasoline to the new O-ring and install it to the injector.
10. While twisting the injector, install it into the delivery pipe.
11. Place the 4 insulators and 3 spacers in position on the intake manifold.
12. Install the injectors with the delivery pipe onto the cylinder head.
13. Check that the injectors rotate smoothly; if not, replace O-rings.
14. Position the injector connector upward.
15. Install the 3 bolts and torque to 14 ft. lbs. (19 Nm).
16. Connect the fuel lines. Torque the pulsation damper fuel bolt to 22 ft. lbs. (29 Nm).
17. Connect the vacuum hoses.
18. Connect the injector connectors, EGR vacuum modulator, fuel pressure regulator and throttle body.
19. Fill engine with coolant and connect the negative battery cable.

3VZ-E ENGINE

1. Disconnect the negative battery cable.
2. Drain coolant.
3. Tag and disconnect the vacuum hoses to throttle body, pressure regulator, PCV, and from fuel filter.
4. Disconnect the water by-pass hoses.
5. Disconnect the cold start injector.
6. Disconnect the fuel lines.
7. Place a towel under the injector tube, slowly remove the union bolt. Remove the injector tube.
8. Disconnect the EGR hose, modulator, air pipe and temperature sensor, if equipped.
9. Remove the support nut and bolt.
10. Disconnect the EGR water hoses, if equipped.
11. Remove the 5 nuts and EGR valve with the pipes and 2 gaskets.
12. Disconnect the air hose from the reed valve.
13. Remove the 6 bolts, 2 nuts, intake chamber and gasket.
14. Disconnect the knock sensor, water temperature switch, cold start injector timer, water temperature sensor, water temperature sender, ground strap and injector connectors.
15. Remove the 2 bolts and wiring harness.
16. Disconnect the vacuum hose from the BVSV, remove the 4 union bolts, fuel pipes and 8 gaskets.
17. Remove the injectors from the delivery pipe.

To install:

18. Connect the vacuum hose from the BVSV, remove the 4 union bolts, fuel pipes and 8 gaskets. Install the wiring harness.
19. Connect the knock sensor, water temperature switch, cold start injector timer, water temperature sensor, water temperature sender, ground strap and injector connectors.
20. Install the 6 bolts, 2 nuts, intake chamber and gasket.
21. Connect the air hose from the reed valve.
22. Connect the 5 nuts and EGR valve with the pipes and 2 gaskets.
23. Connect the EGR water hoses, if equipped.
24. Connect the EGR hose, modulator, air pipe and temperature sensor, if equipped and the support.
25. Connect the fuel lines and the cold start injector.
26. Connect the vacuum hoses to throttle body, pressure regulator, PCV, and fuel filter. Connect the water by-pass hoses.
27. Fill the cooling system adn connect the negative battery cable.

4A-FE AND 4A-GE ENGINE

The following procedures are for the 4A-GE engine; the 4A-FE is very similar.

1. Disconnect the negative battery cable.

2. Place a towel under the injector tube and slowly remove the union bolt. Remove the cold start injector pipe.
3. Disconnect the vacuum hose from the fuel pressure regulator and the PCV hose.
4. Remove the fuel pressure regulator.
5. Disconnect the fuel lines.
6. Remove the delivery pipe and injectors, the 4 insulators and 3 collars from the cylinder head. Remove the injectors from the delivery tube.

To install:

7. Install a new grommet to the injector.
8. Apply a light coat of gasoline to the new O-ring and install it to the injector.
9. While twisting the injector, install it into the delivery pipe.
10. Place the 4 insulators and 3 collars in position on the cylinder head.
11. Install the injectors with the delivery pipe onto the cylinder head.
12. Check that the injectors rotate smoothly; if not, replace O-rings.
13. Position the injector connector upward.
14. Install the 3 thinner spacers and bolts and torque to 13 ft. lbs. (17 Nm).
15. Connect the fuel lines and torque union to 22 ft. lbs. (29 Nm).
16. Install the pressure regulator and connect all the vacuum hoses.
17. Connect the cold start injector tube and torque to 13 ft. lbs. (18 Nm).
18. Fill the cooling system and connect the negative battery cable.

4A-GZE ENGINE

1. Disconnect the negative battery cable.
2. Drain the engine coolant.
3. Disconnect the accelerator cable and throttle cable from the throttle body, if equipped.
4. Remove the throttle body.
5. Disconnect the cold start injector connector.
6. Place a towel under the injector tube and slowly remove the union bolt. Remove the injector tube.
7. Loosen the air outlet duct and remove the accelerator cable bracket.
8. Disconnect the injector connectors.
9. Disconnect the fuel lines and vacuum lines from pressure regulator and EGR modulator.
10. Remove the vacuum pipe.
11. Remove the fuel pressure regulator.
12. Remove the fuel delivery pipe, 4 insulators and 3 collars.
13. Remove the injectors from the delivery pipe.

To install:

14. Install a new grommet to the injector.
15. Apply a light coat of gasoline to the new O-ring and install it to the injector.
16. While twisting the injector, install it into the delivery pipe.
17. Place the 4 insulators and 3 collars in position on the cylinder head.
18. Check that the injectors rotate smoothly; if not, replace O-rings.
19. Position the injector connector upward.
20. Install the 3 thinner spacers and bolts and torque to 13 ft. lbs. (17 Nm).
21. Install the fuel pipe with 2 new gaskets on the delivery pipe and torque to 22 ft. lbs. (29 Nm). Install the clamp.
22. Install the pressure regulator.
23. Install the vacuum pipe.
24. Connect the vacuum hoses and fuel lines.
25. Connect the injector connectors.
26. Install the air outlet duct and torque the 2 bolts and nuts to 7 ft. lbs. (10 Nm).
27. Connect the cold start injector, install the union and torque to 11 ft. lbs. (15 Nm).
28. Install the throttle body.
29. Refill the cooling system and connect the negative battery cable.

4Y-EC ENGINE

1. Disconnect the negative battery cable.
2. Drain the engine coolant.
3. Disconnect the accelerator cable and throttle cable from the throttle body, if equipped.
4. Remove the air cleaner hose.
5. Disconnect the cold start injector, air valve, PCV, brake booster and canister and emission hoses.
6. Remove the EGR pipe.
7. Remove the 2 union bolts, 4 gaskets, clamp and cold start injector.
8. Disconnect the water by-pass hoses and remove the pressure regulator hose.
9. Remove the air intake chamber with the air valve.
10. Disconnect the fuel line from the pulsation damper and disconnect the fuel outlet pipe.
11. Remove the delivery pipe with the injectors.
12. Remove the injectors from the delivery pipe.

To install:

13. Install a new grommet to the injector.
14. Apply a light coat of gasoline to the new O-ring and install it to the injector.
15. While twisting the injector, install it into the delivery pipe.
16. Place the 4 insulators and 2 spacers in position on the intake manifold.
17. Install the injectors with the delivery pipe on the intake manifold.
18. Check that the injectors rotate smoothly; if not, replace O-rings.
19. Position the injector connector upward.
20. Install the fuel line to the pulsation damper and torque to 22 ft. lbs. (29 Nm).
21. Connect the outlet pipe with the union bolt and new gaskets and toruque the union bolt to 14 ft. lbs. (20 Nm).
22. Install the air intake chamber with air valve using a new gasket.
23. Install the bolt with the bond cable and nuts and torque to 9 ft. lbs. (12 Nm).
24. Connect the cold start injector, water by-pass and pressure regulator hoses.
25. Install the EGR valve and pipes.
26. Connect the brake booster, canister, emission control, PCV, water by-pass, cold start injector and air valve hoses.
27. Install the air cleaner hose.
28. Connect the accelerator cable and throttle cable.
29. Fill the cooling system and connect the negative battery cable.

5M-GE ENGINE

1. Disconnect the negative battery cable and drain the engine coolant.
2. Disconnect the accelerator cable and throttle cable from the throttle body.
3. Tag and disconnect the ISC valve, actuator vacuum, PCV, brake booster and emission hoses.
4. Disconnect the ISC, TPS, cold start injector EGR gas temperature and VSV connectors.
5. Remove the air intake chamber support.
6. Remove the vacuum pipe subassembly.
7. Disconnect the EGR pipe.
8. Disconnect the cold start fuel hose from the delivery pipe.
9. Remove the distributor.
10. Remove the fuel pipe.
11. Disconnect and remove the wiring harness.
12. Remove the delivery pipe with the injectors.

13. Remove the 6 insulators from the intake manifold.

To install:

14. Install a new grommet to the injector.
15. Apply a light coat of gasoline to the new O-ring and install it to the injector.
16. While twisting the injector, install it into the delivery pipe.
17. Place the 6 insulators in position on the intake manifold.
18. Install the injectors with the delivery pipe on the intake manifold.
19. Check that the injectors rotate smoothly; if not, replace O-rings.
20. Position the injector connector upward.
21. Install the 4 bolts and torque to 10 ft. lbs. (14 Nm).
22. Connect and install the wiring harness.
23. Install the fuel pipe.
24. Install the distributor.
25. Install the air intake chamber.
26. Connect the cold start fuel hose.
27. Connect the EGR pipe.
28. Install the vacuum subassembly.
29. Install the air intake chamber support.
30. Connect the VSV, ISV, TPS and cold start injector connectors.
31. Connect the accelerator linkage and throttle cable.
32. Connect the emission control, actuator, brake booster, PCV, ISC and ISC water by-pass hoses.
33. Install the air intake connector.
34. Filling the cooling system and connect the negative battery cable.

7M-GE AND 7M-GTE ENGINES

1. Disconnect the negative battery cable and drain the engine coolant.
2. Disconnect the by-pass water hoses or throttle body water hoses, as needed.
3. Disconnect the ISC, pressure regulator and throttle body vacuum hoses.
4. Disconnect the ISC, TPS and cold start injector.
5. Disconnect the accelerator connecting rods.
6. Remove the air intake connector. On the 7M-GTE engine, it is necessary to remove the accelerator and throttle cables, ISC pipe and PCV hoses.
7. On the 7M-GTE engine, remove the throttle body.
8. Remove the ISC valve and disconnect the injector connectors.
9. Place a towel under the injector tube and slowly remove the union bolt. Remove the injector tube.
10. Remove the fuel pipes.
11. Remove the pressure regulator.
12. Remove the fuel delivery pipe.
13. Remove the 6 insulators and 3 spacers from the cylinder head.

To install:

14. Install a new grommet to the injector.
15. Apply a light coat of gasoline to the new O-ring and install it to the injector.
16. While twisting the injector, install it into the delivery pipe.
17. Place the 6 insulators into the hole of the cylinder head.
18. Install the black rings on the upper portion of each of the 3 spaces and install the spacers on the delivery pipe mounting hole of the cylinder head.
19. Check that the injectors rotate smoothly; if not, replace O-rings.
20. Position the injector connector upward.
21. Install the 3 spacers and bolts and torque to 13 ft. lbs. (18 Nm).
22. Fully loosen the locknut of the pressure regulator. Apply a light coat of gasoline to the new O-ring. Thrust the pressure regulator completely into the delivery pipe. Turn the regulator counterclockwise until the outlet faces perpendicular to the delivery pipe. Torque the locknut to 18 ft. lbs. (25 Nm).
23. Install the fuel lines and torque to 20 ft. lbs. (28 Nm). Torque the fuel line to the 7M-GTE pulsation damper to 29 ft. lbs. (39 Nm).
24. Connect the injectors.
25. Connect the cold start injector tube and torque to 13 ft. lbs. (18 Nm) cold injector side 22 ft. lbs. (29 Nm) delivery side.
26. Install the ISC valve and torque the bolts to 9 ft. lbs. (13 Nm). Connect the by-pass hoses.
27. On the 7M-GTE, install the throttle body and connect the water hoses.
28. Install the air intake connector and torque the bolts to 13 ft. lbs. (18 Nm).
29. Connect the accelerator connecting rods.
30. Connect the TPS, ISC and cold start injector.
31. Connect the throttle body vacuum hoses, ISC hoses and pressure regulator hoses.
32. Fill the cooling system and connect the negative battery cable.

22R-E AND 22R-TE ENGINES

1. Disconnect the negative battery cable.
2. Drain the engine coolant.
3. Disconnect the accelerator cable and throttle cable from the throttle body, if equipped.
4. Disconnect the PCV, brake booster, air control valve, VSV, EVAP, EGR, fuel pressure, reed valve, fuel, water by-pass and throttle body hoses.
5. Remove the EGR vacuum modulator.
6. Disconnect the cold start, TPS and EGR temperature sensor, if equipped.
7. Remove the air chamber with the throttle body as an assembly.
8. Disconnect the knock sensor, oil pressure sender, starter, transmission, A/C compressor, water temperature sender, igniter, VSV, cold start timer and water temperature sensor wires.
9. Disconnect the fuel hose from the delivery pipe.

To install:

10. Install a new grommet to the injector.
11. Apply a light coat of gasoline to the new O-ring and install it to the injector.
12. While twisting the injector, install it into the delivery pipe.
13. Check that the injectors rotate smoothly; if not, replace O-rings.
14. Turn the injector so the injector guide is aligned with the positioning rib of the delivery pipe.
15. Install and torque bolts to 14 ft. lbs. (19 Nm).
16. Connect the knock sensor, oil pressure sender, starter, transmission, A/C compressor, water temperature sender, igniter, VSV, cold start timer and water temperature sensor wires.
17. Install the air chamber with the throttle body as an assembly.
18. Install the EGR vacuum modulator and connect the cold start, TPS and EGR temperature sensor, if equipped.
19. Connect the PCV, brake booster, air control valve, VSV, EVAP, EGR, fuel pressure, reed valve, fuel, water by-pass and throttle body hoses.
20. Connect the accelerator cable and throttle cable from the throttle body, if equipped.
21. Fill the cooling system and connect the negative battery cable.

COLD START INJECTOR

Removal and Installation

NOTE: Before removal, tag all the hoses and electrical connections for reference during reassembly.

EXCEPT 5M-GE ENGINE

1. Disconnect the negative battery cable.
2. Disconnect the cold start injector.

3. Place a towel under the injector tube and slowly remove the union bolt. Remove the injector tube.
4. On the 3S-GTE, remove the throttle body.
5. Remove the 2 bolts, cold start injector and gasket.

To install:

6. Install a new gasket, the injector and bolts. Torque the injector bolts to:
   2VZ-E – 48 inch lbs. (5.4 Nm)
   3F-E – 43 inch lbs. (4.9 Nm)
   3S-FE – 82 inch lbs. (9.3 Nm)
   3S-GE – 48 inch lbs. (5.4 Nm)
   3S-GTE – 52 inch lbs. (5.9 Nm)
   3VZ-E – 69 inch lbs. (7.8 Nm)
   4A-GE, 4A-GZE – 65 inch lbs. (7.4 Nm)
   4Y-EC – 52 inch lbs. (5.9 Nm)
   7M-GE, 7M-GTE – 48 inch lbs. (5.4 Nm)
   22R-E, 22R-TE – 43 inch lbs. (4.9 Nm)
7. Connect the injector tube to clamp.
8. Connect the injector tube with the 2 new gaskets and the union bolt and torque to 13 ft. lbs. (18 Nm).
9. Connect the cold start injector connector.
10. Connect the negative battery cable.

5M-GE ENGINE

1. Disconnect the negative battery cable.
2. Drain the engine coolant.
3. Remove the air intake connector.
4. Disconnect the water by-pass hoses from the throttle body and ISC.
5. Disconnect the hose from the ISC valve body.
6. Tag and disconnect the PCV, brake booster actuator and emission hoses.
7. Disconnect the accelerator linkage and the cable from the throttle body.
8. Disconnect the cold start injector, TPS, ISC and VSV connectors.
9. Remove the air intake chamber support.
10. Disconnect the EGR pipe.
11. Disconnect the cold start fuel hose from the delivery pipe.
12. Remove the air intake chamber.
13. Remove the cold start injector from the air intake manifold.

To install:

14. Place the new gasket and install the cold start injector and bolts.
15. Install the air chamber.
16. Connect the cold start fuel hose to the delivery pipe.
17. Tighten the EGR connecting nut.
18. Install the vacuum subassembly.
19. Install the air intake chamber support.
20. Connect the VSV, ISV, TPS and cold start injector connectors.
21. Connect the accelerator linkage and throttle cable.
22. Connect the emission control, actuator, brake booster, PCV, ISC and ISC water by-pass hoses.
23. Install the air intake connector.
24. Filling the cooling system and connect the negative battery cable.

THROTTLE BODY

Removal and Installation

ALL MODELS

Procedures may vary slightly dependent on models and options.

1. Disconnect the negative battery cable.
2. Drain the engine coolant.
3. Disconnect the accelerator cable and throttle cable from the throttle body, if equipped.

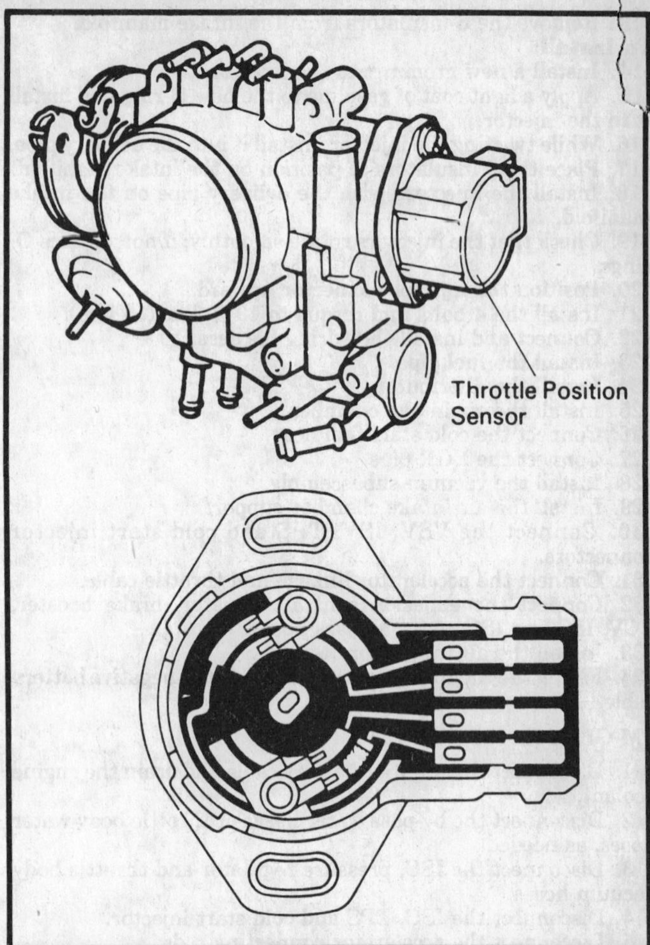
Throttle Position
Sensor

Typical throttle body assembly

4. Remove the air cleaner, air flow meter and air cleaner hose.
5. Disconnect PCV, vacuum sensing, emission control and water by-pass hoses.
6. Disconnect ISC and TPS.
7. Remove the throttle cable bracket, if equipped.
8. Remove the 4 – manual transmission or 3 – automatic transmission throttle body retaining bolts.
9. Remove the throttle body.

To install:

10. Install the throttle body and new gasket.
11. Install the the throttle cable bracket, if equipped.
12. Connect the ISC and TPS.
13. Connect the PCV, vacuum sensing, emission control and water by-pass hoses.
14. Install the air cleaner, air flow meter and air cleaner hose.
15. Connect the accelerator cable and throttle cable.
16. Fill the cooling system and connect the negative battery cable.

AIR FLOW METER

Removal and Installation

1. Disconnect the negative battery cable and the air flow meter connector.
2. Disconnect the vacuum hoses and air cleaner hose. On some vehicles, remove the air cleaner cap.
3. Disconnect the air flow meter wiring.

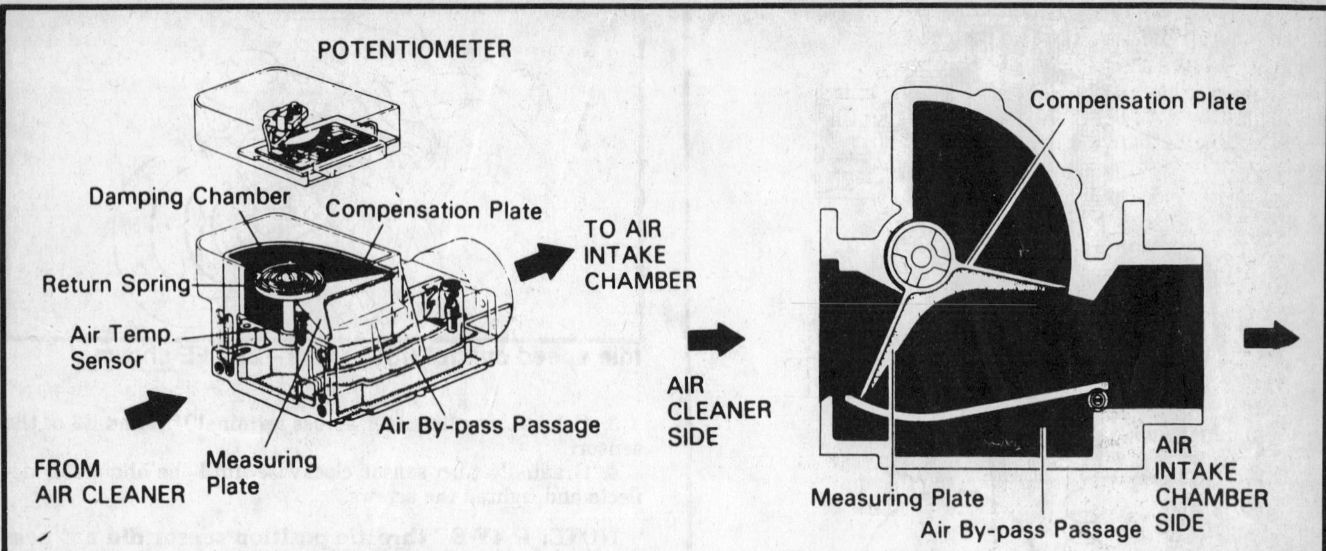
POTENTIOMETER
Damping Chamber
Compensation Plate
Return Spring
Air Temp. Sensor
TO AIR INTAKE CHAMBER
FROM AIR CLEANER
Measuring Plate
Air By-pass Passage
AIR CLEANER SIDE
Compensation Plate
Measuring Plate
Air By-pass Passage
AIR INTAKE CHAMBER SIDE

Exploded view of a typical air flow meter

4. Pry off the lock plate in order to gain access to the retaining bolts. Remove the air flow meter retaining bolts and nuts and remove the air flow meter.
5. On the 7M-GTE engine, remove the air flow meter with the air cleaner case and remove the air flow meter from the air cleaner assembly.

To install:

6. Install air flow meter with gasket, lock plate, plate washers and nuts.
7. Pry lock plate on nut.
8. Install the air cleaner cap and air flow meter assembly.
9. Connect the air flow meter wiring.
10. Connect the negative battery cable.

PRESSURE REGULATOR

Removal and Installation

2VZ-FE, 3VZ-E, 5M-GE, 7M-GE AND 7M-GTE

1. Disconnect the vacuum hose from the regulator.
2. Put a contain or towel under pressure regulator and slowly disconnect the fuel hose.
3. Loosen the locknut and remove the pressure regulator.

To install:

4. Fully loosen the locknut of the pressure regulator.
5. Apply a light coat of gasoline to the regulator O-ring and install it to the regulator.
6. Install the pressure regulator completely into the fuel delivery pipe.
7. Turn the pressure regulator counterclockwise, were applicable, unit the vacuum fitting is vertical.
8. Tighten the locknut to:
   2VZ-FE – 18 ft. lbs. (25 Nm)
   3VZ-E – 22 ft. lbs. (29 Nm)
   22R-E, 22R-TE – 22 ft. lbs. (29 Nm)
   5M-GE – 22 ft. lbs. (29 Nm)
   7M-GE, 7M-GTE – 18 ft. lbs. (25 Nm)
9. Connect fuel hose and vacuum line.

3S-GE, 3S-GTE

1. Remove the throttle body.
2. Disconnect the cold start injector.
3. Disconnect the vacuum hose from the regulator.
4. Put a contain or towel under pressure regulator and slowly disconnect the fuel hose.
5. Loosen the locknut and remove the pressure regulator.

To install:

6. Fully loosen the locknut of the pressure regulator.
7. Apply a light coat of gasoline to the regulator O-ring and install it to the regulator.
8. Thrust the pressure regulator completely into the fuel delivery pipe.
9. Turn the pressure regulator counterclockwise unit the vacuum fitting is vertical.
10. Tighten the locknut to 22 ft. lbs. (29 Nm)
11. Connect the cold start injector pipe.
12. Install the throttle body.

3F-E, 3S-FE

1. On Van, raise vehicle.
2. Disconnect the vacuum hose from the regulator.
3. Put a contain or towel under pressure regulator and slowly disconnect the fuel line.
4. Remove the 2 bolts and pull the regulator from the delivery pipe.

To install:

5. Apply a light coat of gasoline to the regulator O-ring and install it to the regulator.
6. Install the regulator and torque the bolts to:
   3F-E – 43 inch lbs. (4.9 Nm)
   3S-FE – 48 inch lbs. (5.4 Nm)
   4A-GE – 65 inch lbs. (7.4 Nm)
   4Y-EC – 52 inch lbs. (5.9 Nm)
7. Torque fuel line to:
   4A-GE – 22 ft. lbs. (29 Nm)
   4A-GZE – 11 ft. lbs. (15 Nm)
   4Y-EC – 14 ft. lbs. (20 Nm)
8. Connect vacuum hose to regulator.
9. Lower vehicle.

THROTTLE POSITION SENSOR

Adjustment

ALL MODELS

1. Loosen the sensor screws.
2. Insert the appropriate feeler gauge between the throttle stop screw and the stop lever. Correct size feeler gauge is:
   2VZ-FE – 0.020 in. (0.5mm)
   3F-E – 0.037 in. (0.93mm)

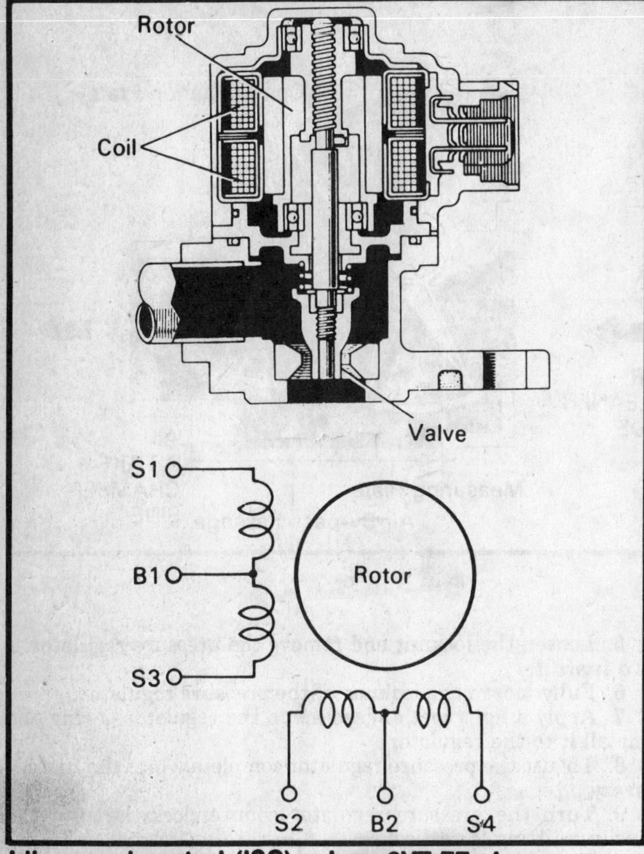
Rotor
Coil
Valve
S1
B1
Rotor
S3
S2
B2
S4
Idle speed control (ISC) valve – 2VZ-FE shown

3S-FE w/ECT – 0.024 in. (0.6mm)
3S-FE wo/ECT – 0.028 in. (0.7mm)
3S-GE, 3S-GTE – 0.024 in. (0.6mm)
3VZ-E – 0.024 in. (0.6mm)
4A-FE – 0.028 in. (0.7mm)
4A-GE – 0.019 in. (0.5mm)
4A-GZE – 0.021 in. (0.53mm)
5M-GE – 0.020 in. (0.50mm)
7M-GE Supra – 0.023 in. (0.58mm)
7M-GE Cressida – 0.028 in. (0.7mm)
7M-GTE – 0.028 in. (0.7mm)
22R-E, 22R-TE – 0.028 in. (0.7mm)

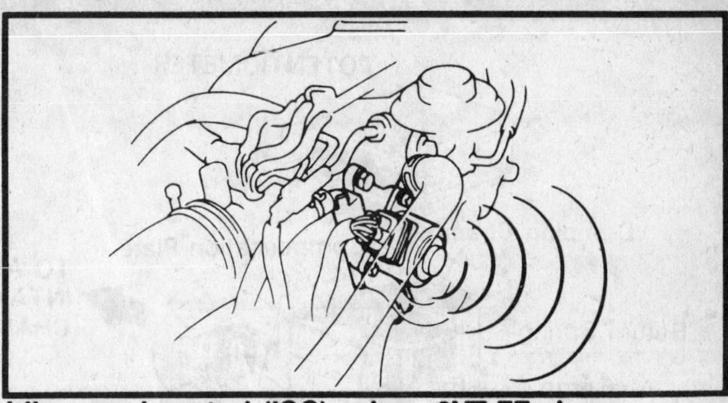
Idle speed control (ISC) valve – 2VZ-FE shown

3. Connect an ohmmeter across terminal IDL and E2 of the sensor.
4. Gradually turn sensor clockwise until the ohmmeter deflects and tighten the screws.

NOTE: If 4Y-EC throttle position sensor did not pass testing, replace throttle body and sensor as an assembly.

AIR OR IDLE SPEED CONTROL VALVE

Removal and Installation

2F-E, 2VZ-FE, 4A-GZE, 7M-GE AND 7M-GTE ENGINES

1. Drain engine coolant.
2. Disconnect ISC valve connector. On the 7M-GTE engine, remove the seal washer and check valve from the intake chamber.
3. Disconnect air hose and water by-pass hoses.
4. Remove the ISC valve and gasket.

To install:

5. Install the ISC valve and gasket.
6. Connect the air hose and water by-pass hoses.
7. Connect the ISC valve connector.
8. Fill the cooling system.

3S-FE, 3S-GE, 3S-GTE, 4A-FE, 4A-GE AND 22R-E ENGINES

1. Remove the throttle body.
2. Remove the ISC valve screws and remove the ISC valve and gasket.

To install:

3. Install a gasket and the ISC valve to the throttle body.
4. Install the throttle body.